VACCINES FOR BIODEFENSE AND EMERGING AND NEGLECTED DISEASES

VACCINES FOR BIODEFENSE AND EMERGING AND NEGLECTED DISEASES

Edited by

ALAN D.T. BARRETT

University of Texas Medical Branch
Sealy Center for Vaccine Development
Galveston, Texas

LAWRENCE R. STANBERRY

Columbia University
College of Physicians and Surgeons
New York, New York

ELSEVIER

AMSTERDAM • BOSTON • HEIDELBERG • LONDON
NEW YORK • OXFORD • PARIS • SAN DIEGO
SAN FRANCISCO • SINGAPORE • SYDNEY • TOKYO

Academic Press is an imprint of Elsevier
84 Theobald's Road, London WC1X 8RR, UK
Radarweg 29, PO Box 211, 1000 AE Amsterdam, The Netherlands
30 Corporate Drive, Suite 400, Burlington, MA 01803, USA
525 B Street, Suite 1900, San Diego, CA 92101-4495, USA

First edition 2009

The editors thank Dr Andrew McNees (UTMB) for his initial artwork in the design
of the syringe needle in the globe

British Library Cataloguing in Publication Data
A catalogue record for this book is available from the British Library

Library of Congress Cataloguing in Publication Data
A catalogue record for this book is available from Library of Congress

ISBN 978-0-12-369408-9

Typeset by Charon Tec Ltd., A Macmillan Company.
(www.macmillansolutions.com)

09 10 11 12 10 9 8 7 6 5 4 3 2 1

Printed and bound in Italy

Contents

5. Immunology of Vaccines: Consideration of the Neonate as a Target for Vaccination

MONICA R. FOOTE AND D. MARK ESTES

6. Molecular Approaches to Bacterial Vaccines

VLADIMIR L. MOTIN AND ALFREDO G. TORRES

7. Viral Vectors

DAVID H. HOLMAN, DANHER WANG, JAN WORARATANADHARM AND JOHN Y. DONG

8. DNA Vaccines for Biodefense and Emerging and Neglected Infectious Diseases

SHAN LU, JILL M. GRIMES-SERRANO AND SHIXIA WANG

9. Vaccine Adjuvants

CARL R. ALVING

10. Preclinical Vaccine Development: Implementing the Food and Drug Administration's Good Laboratory Practices in a Biocontainment Environment—A University BSL3/BSL4 Laboratory Perspective

MELISSA M. EITZEN, ANDREW G. McNEES AND DAVID W.C. BEASLEY

11. Manufacturing Vaccines for an Emerging Viral Infection–Specific Issues Associated with the Development of a Prototype SARS Vaccine

INCA C. KUSTERS, JAMES MATTHEWS AND JEAN FRANÇOIS SALUZZO

12. Clinical Trials of Vaccines for Biodefense and Emerging and Neglected Diseases

WENDY A. KEITEL

13. The US Food and Drug Administration Pre-IND and IND Process for Vaccines

JEANNE M. NOVAK, JUDY RUCKMAN AND DENNIS W. TRENT

14. U.S. Regulatory Issues Related to Vaccines against Emerging Infectious Diseases

PHILIP R. KRAUSE

15. Vaccine Safety

JOHN D. GRABENSTEIN

16. Special Issues around Veterinary Vaccines

L. GARRY ADAMS, LORNE BABIUK, DAVID McGAVIN AND ROBERT NORDGREN

17. Misinformation about Vaccines

MARTIN G. MYERS AND DIEGO PINEDA

18. Vaccines for Military Application

ANDREW W. ARTENSTEIN

III
VIRAL VACCINES

19. Dengue

DAVID W. VAUGHN, STEPHEN S. WHITEHEAD AND ANNA P. DURBIN

20. Ebola and Marburg

GARY P. KOBINGER, MARIA CROYLE AND HEINZ FELDMANN

21. Alphaviral Encephalitides

SCOTT C. WEAVER AND SLOBODAN PAESSLER

22. Foot-and-Mouth Disease

PETER W. MASON AND MARVIN J. GRUBMAN

23. Hantavirus

CHRISTOPHER D. HAMMERBECK, VICTORIA WAHL-JENSEN AND JAY W. HOOPER

24. Hepatitis C

MEITAL GAL-TANAMY, CHRISTOPHER WALKER, STEVEN FOUNG AND STANLEY M. LEMON

25. Human Immunodeficiency Virus

FARAH CASSIS-GHAVAMI, MARCEL CURLIN, ROBERT GEISE AND ANN DUERR

26. Human Papillomavirus

WILLIAM BONNEZ

27. Influenza

JOAN E. NICHOLS AND JAMES W. LeDUC

28. Japanese Encephalitis

ICHIRO KURANE

29. Junin (Argentine Hemorrhagic Fever)

KELLY T. McKEE, DELIA A. ENRIA AND JULIO G. BARRERA ORO

30. Measles

WILLIAM J. MOSS, DIANE E. GRIFFIN AND W. HARRY FEINSTONE

31. Hendra and Nipah

VINCENT DEUBEL AND T. FABIAN WILD

32. O'nyong-nyong and Chikungunya

ANN M. POWERS AND AARON C. BRAULT

33. Rabies

ANTHONY R. FOOKS, NICHOLAS JOHNSON AND CHARLES E. RUPPRECHT

34. Ross River Virus: Epidemic Polyarthritis

JOHN AASKOV

35. Rotavirus

MONICA MALONE McNEAL AND DAVID I. BERNSTEIN

36. SARS

BART L. HAAGMANS AND ALBERT D.M.E. OSTERHAUS

37. Smallpox

RICK KENNEDY AND GREGORY A. POLAND

43. Burkholderia *mallei* and Burkholderia *pseudomallei*

MITALI SARKAR-TYSON AND RICHARD W. TITBALL

44. Chlamydia

ROGER G. RANK

45. Cholera

CHANDRABALI GHOSE, STEPHEN B. CALDERWOOD AND EDWARD T. RYAN

46. Botulinum Toxin

LANCE L. SIMPSON

47. *Ehrlichia*

JERE W. McBRIDE

48. Extraintestinal Pathogenic *Escherichia coli*

THOMAS A. RUSSO AND JAMES R. JOHNSON

49. Gonorrhea

MICHAEL W. RUSSELL AND EDWARD W. HOOK III

50. *Helicobacter pylori*

VICTOR E. REYES, GIOVANNI SUÁREZ, JOHANNA C. SIERRA AND ELLEN J. BESWICK

51. Intestinal Pathogenic *Escherichia coli*

ALFREDO G. TORRES

52. Lyme Disease

CHRISTOPHER G. EARNHART AND RICHARD T. MARCONI

53. *Neisseria meningitidis*

NADINE G. ROUPHAEL, SHANTA M. ZIMMER AND DAVID S. STEPHENS

54. Plague

E. DIANE WILLIAMSON

55. Ricin: A Type II Ribosome-Inactivating Protein

ROBERT N. BREY AND NICHOLAS J. MANTIS

56. *Rickettsia*

XUEJIE YU, NAHED ISMAIL AND DAVID H. WALKER

57. Typhoid Fever

ROBERT A. KINGSLEY AND GORDON DOUGAN

58. Shigellosis

MALABI VENKATESAN, ROBERT W. KAMINSKI AND RYAN T. RANALLO

59. Syphilis

LORENZO GIACANI, SHEILA LUKEHART AND ARTURO CENTURION-LARA

60. *Mycobacterium tuberculosis*

LEONARD MOISE, JULIE McMURRY AND ANNE S. DE GROOT

61. *Francisella tularensis*

RICHARD W. TITBALL AND PETRA C.F. OYSTON

V
PARASITIC VACCINES

62. African Trypanosomiasis
ANDREW J. NOK

63. Leishmaniasis
LYNN SOONG

64. Leptospirosis
NOBUO KOIZUMI AND HARUO WATANABE

65. Malaria
THOMAS L. RICHIE AND FALGUNEE K. PAREKH

66. Hookworm Infection
PETER J. HOTEZ AND DAVID DIEMERT

67. Onchocerciasis
SARA LUSTIGMAN AND DAVID ABRAHAM

68. Schistosomiasis

CHRISTOPHER L. KING

69. American Trypanosomiasis

VANDANAJAY BHATIA, JIAN-JUN WEN, MICHELE A. ZACKS AND NISHA JAIN GARG

Preface

"Quiet the little feet that trod
So merrily the floor
The little hands that clasped my neck
Will clasp my neck no more.
Ah! Children mine and yet not mine
For a few years were given
And then recalled to draw my heart,
Nearer to God and Heaven."

Henry Whinery, who drove the stage coach between San Jose and Santa Cruz, California saw two of his children Harry F. Whinery, (age 5 yrs, 9 mo) and Martha Whinery (age 7 yrs, 5 mo) die of Diphtheria on December 6, 1876.

Vaccines have had a greater impact on human health than any other biomedical development. Over the past three centuries vaccines have facilitated the eradication of one infectious disease, smallpox, and have limited the scourge caused by more than 25 enteric, respiratory, genitourinary and zoonotic and vector-borne pathogens including measles, polio, hepatitis B, yellow fever, and *Haemophilus influenza* type B. Despite the successes there remains enormous work to be done. Infectious diseases continue to be the number one cause of death globally claiming 20 million lives yearly. Old threats such as malaria, dengue, and syphilis have yet to be tamed. Over the past 25 years there has been a dramatic increase in newly recognized or emerging pathogens including human immunodeficiency virus, hepatitis C virus and *Helicobacter pylori*; however the development of successful vaccines has not kept pace with the discovery of these new perils. Increases in international travel and migration have made re-emerging diseases such as tuberculosis and influenza global menaces. As the world strives to achieve the ambitious Millennium Development Goals, resource poor countries struggle to overcome the burden caused by neglected diseases like trypanosomiasis and hookworm. In addition, the risk of microbes as biothreat agents, such as smallpox, anthrax and plague, continues to present a danger to mankind in the 21st century.

This comprehensive new textbook authored by leading authorities is intended to inform researchers, clinicians, public health specialists and policy makers regarding the current state of development of vaccines for emerging and neglected diseases and biothreat agents. Detailed information regarding epidemiology, clinical disease, management, immunology, pathogenesis, as well as vaccinology is presented for each pathogen. The book also provides in-depth information regarding the processes that are critical to the generation of these vaccines to the marketplace. This includes chapters on vaccine platforms, preclinical development, regulatory approval, clinical trials, manufacture and post licensure issues.

The editors would like to thank our many colleagues whose outstanding contributions made this book a reality. We would also like express our appreciation to Lisa Tickner and Carrie Bolger of Elsevier for their able and professional assistance.

Alan D.T. Barrett, Ph.D.
Lawrence R. Stanberry, M.D., Ph.D.

Foreword

Rino Rappuoli
Global Head of Vaccines Research, Novartis
Vaccines & Diagnostics, Sienna, Italy

Vaccination is the medical practice that together with clean water had the greatest impact on human health during the 20th century. Thanks to vaccines, devastating diseases that had been responsible for much of the morbidity and mortality recorded were eliminated. These include smallpox, poliomyelitis, tetanus, diphtheria, pertussis, hepatitis B, *Haemophilus influenzae*, measles, mumps, and rubella, which killed or disabled millions of people. Meningococcal meningitis, perhaps the last disease that can attack in a few hours and kill healthy children and young people, is also on its way to being conquered by vaccination. Now that all these diseases have been controlled, is there a role for vaccination in the 21st century?

Clearly the 70 chapters of this book show that we still have a long way to go before we can conquer all the diseases that can be addressed by vaccination. There is increased interest in emerging and neglected diseases, which have not been a priority for vaccine developers, thanks to the technological revolution of the last decades that increased the feasibility of developing new vaccines and the renewed business interest on vaccines that emerged during the last 5 years.

Let us look at the increased feasibility of vaccines first. A recent review of the history of vaccination concluded that the probability of success in vaccine development is highest when protection is mediated by antibodies and antigens that have no or limited antigenic variability (Rappuoli, 2007). In a graph representing the type of immunity and the antigenic variability on the vertical and horizontal axes, vaccines that have the highest probability of success are in the upper right quadrant and the risk in vaccine development increases moving toward the intersection of the two (Fig. 1). Accordingly, vaccines for which T-cell immunity is critical for protection or protective antigens are highly variable have an increased degree of failure during development. In Fig. 1 I have tried to rank the vaccines discussed in this book based on the above-mentioned criteria. From the figure it is clear that the majority of the vaccines reside in the upper right quadrant, which belongs to the pathogens that can be addressed by antigens that meet the validated criteria for vaccine success, that is, they induce an antibody-mediated protection and they are not highly variable. Therefore, in most cases, vaccines for neglected and emerging diseases can be developed using well-established technologies.

Thanks to technologies such as recombinant DNA, conjugation, genomics, and the increased understanding of the immune system, the feasibility of vaccine development has increased a great deal. Good examples are the several recombinant and conjugate vaccines already licensed, and several vaccines developed using genomics (reverse vaccinology) that are in development. Reverse vaccinology is perhaps the most powerful tool for vaccine development that has become available lately. In fact, the availability of genomic sequences allowed us to use computers to search the entire genetic repertoire for protective antigens, thus increasing by several orders of magnitude the numbers of antigens available for vaccine development. With many more antigens available for each pathogen, it is now possible to select those antigens that respond to validated principles, such as limited or absent antigenic variability and antibody-mediated protection.

An important minority of the vaccines described in this book resides in the middle or left quadrant of Fig. 1. These are the vaccines that are not yet within

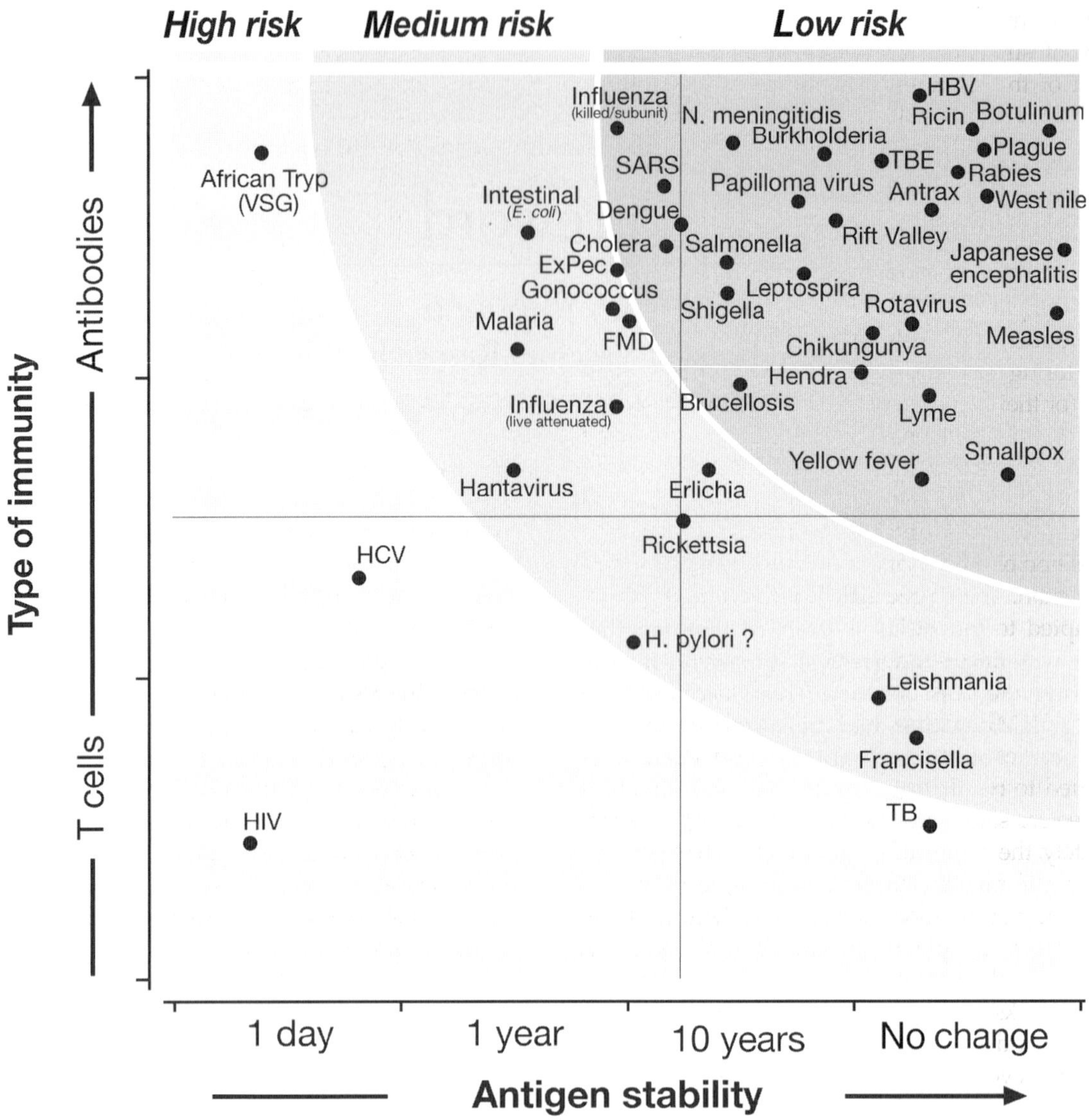

FIGURE 1 Graph representing the type of immunity and the antigenic variability.

the reach of today's technologies and developing these vaccines requires bridging science gaps such as learning to develop vaccines based on T-cell-mediated protection. Today these vaccines are mostly addressed by using innovative immunostimulatory molecules and adjuvants, replicating or nonreplicating viral vectors, prime-boost regimens, and so forth. An alternative approach can be to bring them to the comfort zone of upper right quadrant by learning how to engineer immunosilent conserved epitopes to become immunodominant. A good example of this is HIV wherein we know that antibodies such as b12 that recognize the conserved CD4 binding site would be able to protect from infection. However, we are unable to make this epitope immunodominant and therefore we cannot yet make this vaccine. We believe that a systematic approach to the structural properties of immunodominant and immunosilent epitopes can provide the scientific rationale that in the future may allow us to engineer immunodominant epitopes. A rational approach to the 3D structure of antigens (structural vaccinology) is one of the basic aspects of vaccine research that should be a priority.

In conclusion, with a few exceptions, the majority of the vaccines addressed in this book are within the reach of today's technologies; the question is therefore whether or when they will be developed. Unfortunately, technical feasibility is only one of the hurdles in vaccine development. Even more important is often whether there is a market that can justify the huge investment that is necessary to bring vaccines to licensure. Vaccines today are developed by few global vaccine manufacturers that can only afford to invest in those vaccines that have a high probability of success in the market. With few exceptions, most of the vaccines addressed in this book are "neglected" because they do not have a market that justifies their development; as a consequence, the probability that they will be developed is low. In fact, for these vaccines the basic research and initial clinical trials will be carried out in academia and small biotechs funded by the public sector. However, the gestation of vaccines, the phase during which a discovery is transformed into a potential product, is unlikely to happen. As discussed in this book, during this phase of the vaccine development, pathogens (or their genomes) are systematically screened to identify the best antigens, the antigens are prioritized using in vitro assays and animal models that are relevant to making decisions, the selected antigens are engineered and expressed in hosts suitable for industrial scale, and the composition of the vaccine is defined using adjuvants suitable for man. Then, production of the candidate vaccine is adapted to industrial needs; it is scaled up and produced in cell lines suitable for GMP manufacturing. A robust formulation is developed; the vaccine is produced under GMP conditions and tested for stability. Toxicology studies are performed, regulatory documents are submitted to regulatory agencies, and the vaccine is finally tested for safety and for proof of concept in man. Unfortunately, the results of this phase, which is perhaps the most critical for vaccine development, are not suitable for publication in good journals and therefore there is no incentive for people in academia to embark in this area. In addition, the knowledge to do the work required in this phase lacks in academia; it is not even present in most biotechs and vaccine manufacturers in developing countries. However, five or six large vaccine manufacturers have the traditional knowledge and the necessary investment to carry out this obscure but essential part of the work. In fact many of the failures in vaccine development are due to the poor understanding and underestimation of this phase. In order to fill this gap in vaccine development, recently a new institute has been established. This is the Novartis Vaccines Institute for Global Health (NVGH), an institute with a nonprofit mission dedicated to the development of effective and affordable vaccines for neglected infectious diseases in developing countries. By having access to all the know-how and technologies available with a large vaccine manufacturer, and having neglected diseases as a mission, this institute has prospects of being successful in this area.

Despite difficulties in filling technical and structural gaps the field has never been as promising as today. Thanks to the vast progress in technology and the renewed recognition of the enormous value that vaccines bring to the society, the field of vaccine development is moving beyond its 20th century renaissance. Vaccines today are usually given to prevent diseases that parents and pediatricians have never seen; in their minds vaccines are no longer immediate lifesavers but tools that improve the quality and duration of life. In agreement with this new vision, thanks to the scientific progress, new vaccines target infections such as HBV and HPV that can result in cancer decades after infection; they hold great promise for preventing devastating outbreaks such as pandemic influenza and are being explored for the control of chronic, metabolic, and neurological diseases. Vaccines targeting hypertension, drug addictions, and cigarette smoking are examples of cases that are not traditionally tackled by vaccination. Although some difficult targets such as HIV presently remain beyond the reach of technical feasibility, a new vision is emerging that views vaccines as friendly and safe tools for a global health for all ages and all populations around the world. In this vision, one of the most important missions is to be ready and promptly tackle those diseases that are traditional targets for vaccination, that is diseases caused by emerging infectious agents, traditional infectious agents causing diseases in developing countries that cannot afford the cost of vaccine development, and those infectious agents that have the potential to be improperly used as biological weapons. Although investing in the development of these vaccines is going to be a cost to our society, this can be one of the best investments for improving the quality of life of people globally and for minimizing the risk of biological agents as weapons of terror.

The developmental risk of the vaccines discussed in this book is based on the predicted type of immunity and antigen stability. The graph is adapted from Rappuoli (2007).

References

Abbott, A. Neglected diseases get vaccine research boost. Nature 2008; 451:1037.

Rappuoli, R. Bridging the knowledge gaps in vaccine design. Nat. Biotechnol. 2007; 25:1361–1366.

Contributors

John Aaskov Institute of Health and Biomedical Innovation; World Health Organisation, Collaborating Centre for Arbovirus Reference and Research, Queensland University of Technology, Brisbane, Australia

David Abraham Department of Microbiology and Immunology, Thomas Jefferson University, Philadelphia, PA

L. Garry Adams College of Veterinary Medicine, Texas A&M University, College Station, TX

Carl R. Alving Department of Adjuvant and Antigen Research, Division of Retrovirology, Walter Reed Army Institute of Research, U.S. Military HIV Research Program, Rockville, MD

Kevin Anderson National Biodefense Analysis and Countermeasures Center, Science and Technology Directorate, U.S. Department of Homeland Security, Frederick, MD

Andrew W. Artenstein Department of Medicine and Center for Biodefense and Emerging Pathogens, Memorial Hospital of Rhode Island, Pawtucket; The Warren Alpert Medical School of Brown University, Providence, RI

Lorne Babiuk University of Alberta, Edmonton, Alberta, Canada

Julio G. Barrera Oro GoldbeltRaven LLC/USAMRIID, Fort Detrick, MD

Alan D.T. Barrett Department of Pathology, Sealy Center for Vaccine Development, Center for Biodefense and Emerging Infectious Diseases, and Institute for Human Infections and Immunity, University of Texas Medical Branch, Galveston, TX

David W.C. Beasley Institute for Human and Infections and Immunity; Galveston National Laboratory Regulatory Services Core; Department of Microbiology and Immunology, Sealy Center for Vaccine Development, University of Texas Medical Branch, Galveston, TX

David I. Bernstein Division of Infectious Diseases, Cincinnati Children's Hospital Medical Center, University of Cincinnati, Cincinnati, OH

Ellen J. Beswick Department of Pediatrics, University of Texas Medical Branch, Galveston, TX

Vandanajay Bhatia Department of Microbiology and Immunology, University of Texas Medical Branch, Galveston, TX

William Bonnez School of Medicine and Dentistry, University of Rochester, Rochester, NY

Aaron C. Brault Departments of Pathology and Microbiology and Immunology, Center for Vector-Borne Diseases, School of Veterinary Medicine, University of California, Davis, CA

Robert N. Brey DOR BioPharma, Inc., Ewing, NJ

Stephen B. Calderwood Division of Infectious Diseases, Marsachusetts General Hospital, Harvard Medical School, Boston, MA, USA

Farah Cassis-Ghavami HIV Vaccine Trials Network, Seattle, WA

Arturo Centurion-Lara Department of Medicine, School of Medicine, University of Washington, Seattle, WA

Jason E. Comer Department of Microbiology and Immunology, Sealy Center for Vaccine Development, University of Texas Medical Branch, Galveston, TX

Maria Croyle Division of Pharmaceutics, College of Pharmacy, The University of Texas at Austin, Austin, TX

Marcel Curlin HIV Vaccine Trials Network, Seattle, WA

Anne S. De Groot Collaborative for Vaccine Research and Development, Institute for Immunology and Informatics, University of Rhode Island; EpiVax, Inc., Providence, RI

Vincent Deubel Institut Pasteur of Shanghai, Chinese Academy of Sciences, Shanghai, China

David Diemert Department of Microbiology, Immunology and Tropical Medicine, The George Washington University; Albert B. Sabin Vaccine Institute, Washington, DC

John Y. Dong Division of Biodefense Vaccines, GenPhar, Inc., Mt. Pleasant, SC; Department of Microbiology and Immunology, Medical University of South Carolina, Charleston, SC

Gordon Dougan The Wellcome Trust Sanger Institute, The Wellcome Trust Genome Campus, Hinxton, Cambridge, UK

Ann Duerr HIV Vaccine Trials Network, Seattle, WA

Anna P. Durbin Centre for Immunization Research, Department of International Health, Johns Hopkins Bloomberg School of Public Health, Baltimore, MD

Christopher G. Earnhart Department of Microbiology and Immunology, Medical College of

Virginia at Virginia Commonwealth University, Richmond, VA

Melissa M. Eitzen GLP Program Office, Department of Research Services, University of Texas Medical Branch, Galveston, TX

Delia A. Enria Instituto Nacional de Enfermedades Virales Humanas (INEVH) "Dr. Julio I. Maiztegui", Pergamino, Buenos Aires Province, Argentina

D. Mark Estes Department of Pediatrics and Microbiology and Immunology, Sealy Center for Vaccine Development, University of Texas Medical Branch, Galveston, TX

Heinz Feldmann Special Pathogens Program, National Microbiology Laboratory, Public Health Agency of Canada; Department of Medical Microbiology, University of Manitoba, Winnipeg, MB

Thomas A. Ficht College of Veterinary Medicine, Texas A&M University, College Station, TX

Anthony R. Fooks Rabies and Wildlife Zoonoses Group, Veterinary Laboratories Agency (Weybridge), Addlestone, Surrey, UK

Monica R. Foote Department of Animal Science, Cornell University, Ithaca, NY

Steven Foung Department of Pathology, Stanford University School of Medicine, Palo Alto, CA

Sareen E. Galbraith Brain Infections Group, Divisions of Neurological Science and Medical Microbiology, University of Liverpool, Liverpool, UK; Department of Pathology, Sealy Center for Vaccine Development, Center for Biodefense and Emerging Infectious Diseases, and Institute for Human Infections and Immunity, University of Texas Medical Branch, Galveston, TX

Meital Gal-Tanamy Center for Hepatitis Research, Institute for Human Infections and Immunity, and the Department of Microbiology and Immunology, University of Texas Medical Branch, Galveston, TX

Nisha Jain Garg Departments of Microbiology and Immunology and Pathology, Center for Biodefense and Emerging Infectious Diseases, Sealy Center for Vaccine Development, University of Texas Medical Branch, Galveston, TX

Robert Geise HIV Vaccine Trials Network, Seattle, WA

Chandrabali Ghose Division of Infectious Diseases, Massachusetts General Hospital, Boston, MA; Aaron Diamond AIDS Research Center, New York, NY

Lorenzo Giacani Department of Medicine, School of Medicine, University of Washington, Seattle, WA

John D. Grabenstein Merck Vaccine Division, West Point, PA

Diane E. Griffin Department of Molecular Microbiology and Immunology, Johns Hopkins Bloomberg School of Public Health, Baltimore, MD

Jill M. Grimes-Serrano Laboratory of Nucleic Acid Vaccines, Department of Medicine, University of Massachusetts Medical School, Worcester, MA

Marvin J. Grubman U.S. Department of Agriculture, Agricultural Research Service, Plum Island Animal Disease Center, Greenport, NY

Bart L. Haagmans Department of Virology, Erasmus MC, Rotterdam, The Netherlands

Christopher D. Hammerbeck Virology Division, U.S. Army Medical Research Institute of Infectious Diseases, Fort Detrick, MD

W. Harry Feinstone Department of Molecular Microbiology and Immunology, John Hopkins Bloomberg School of Public Health, Baltimore, MD

Michael R. Holbrook Department of Pathology, Center for Biodefense and Emerging Diseases, Sealy Center for Vaccine Development, and Institute for Human Infections and Immunity, University of Texas Medical Branch, Galveston, TX

David H. Holman Division of Biodefense Vaccines, GenPhar, Inc., Mt. Pleasant, SC

Edward W. Hook Department of Medicine, Division of Infectious Diseases, University of Alabama at Birmingham, Birmingham, AL

Jay W. Hooper Virology Division, U.S. Army Medical Research Institute of Infectious Diseases, Fort Detrick, MD

Peter J. Hotez Department of Microbiology, Immunology and Tropical Medicine, The George Washington University; Albert B. Sabin Vaccine Institute, Washington, DC

Nahed Ismail Department of Pathology, Sealy Center for Vaccine Development, Center for Biodefense and Emerging Infectious Diseases, University of Texas Medical Branch, Galveston, TX

Nicholas Johnson Rabies and Wildlife Zoonoses Group, Veterinary Laboratories Agency (Weybridge), Addlestone, Surrey, UK

James R. Johnson Medical Service, VA Medical Center, and Department of Medicine, University of Minnesota, Minneapolis, MN

Robert W. Kaminski Department of Pediatrics, Division of Bacterial and Rickettsial Diseases, Walter Reed Army Institute of Research, Silver Spring, MD

Wendy A. Keitel Baylor College of Medicine, Houston, TX

Rick Kennedy Mayo Vaccine Research Group, Mayo Clinic College of Medicine, Rochester, MN

Christopher L. King Center for Global Health and Diseases, Case Western Reserve University School of Medicine, Cleveland, OH

Robert A. Kingsley The Wellcome Trust Sanger Institute, The Wellcome Trust Genome Campus, Hinxton, Cambridge, UK

Gary P. Kobinger Special Pathogens Program, National Microbiology Laboratory, Public Health Agency of Canada; Department of Medical Microbiology, University of Manitoba, Winnipeg, MB

Nobuo Koizumi Department of Bacteriology, National Institute of Infectious Diseases, Toyama Shinjuku-ku, Toyama, Tokyo, Japan

Philip R. Krause Division of Viral Products, Office of Vaccines Research and Review, Center for Biologics Evaluation and Research, Food and Drug Administration, Bethesda, MD

Ichiro Kurane Department of Virology, National Institute of Infectious Diseases, Tokyo, Japan

Inca C. Kusters Sanofi Pasteur, Marcy l'Etoile, France

James W. LeDuc Institute for Human Infections and Immunity and Department of Microbiology and Immunology, University of Texas Medical Branch, Galveston, TX

Axel T. Lehrer PanThera Biopharma, Aiea, HI

Stanley M. Lemon Center for Hepatitis Research, Institute for Human Infections and Immunity, and the Department of Microbiology and Immunology, University of Texas Medical Branch, Galveston, TX

Shan Lu Laboratory of Nucleic Acid Vaccines, Department of Medicine, University of Massachusetts Medical School, Worcester, MA

Sheila Lukehart Department of Medicine, School of Medicine, University of Washington, Seattle, WA

Sara Lustigman Laboratory of Molecular Parasitology, Lindsley F. Kimball Research Institute, New York Blood Center, New York, NY

Monica Malone McNeal Division of Infectious Diseases, Cincinnati Children's Hospital Medical Center, University of Cincinnati, Cincinnati, OH

Nicholas J. Mantis Division of Infectious Disease, Wadsworth Center, New York State Department of Health, Albany, NY

Richard T. Marconi Department of Microbiology and Immunology; Center for the Study of Biological Complexity, Medical College of Virginia at Virginia Commonwealth University, Richmond, VA

Peter W. Mason Department of Pathology; Sealy Center for Vaccine Development, University of Texas Medical Branch, Galveston, TX

James Matthews Sanofi Pasteur, Swiftwater, PA

Jere W. McBride Departments of Pathology and Microbiology and Immunology, Institute for Human Infections and Immunity, Center for Biodefense and Emerging Infectious Diseases, Sealy Center for Vaccine Development, University of Texas Medical Branch, Galveston, TX

David McGavin Pfizer Animal Health, Kalamazoo, MI

Kelly T. McKee Quintiles Public Health and Government Services, Quintiles Transnational Corporation, Durhan, NC

Julie McMurry EpiVax, Inc., Providence, RI

Andrew G. McNees Institute for Human Infections and Immunity; Sealy Center for Vaccine Development, Galveston National Laboratory Regulatory Services Core, University of Texas Medical Branch, Galveston, TX

Leonard Moise Collaborative for Vaccine Research and Development, Institute for Immunology and Informatics, University of Rhode Island; EpiVax, Inc., Providence, RI

William J. Moss Department of Epidemiology, International Health and Molecular Microbiology and Immunology, Johns Hopkins Bloomberg School of Public Health, Baltimore, MD

Vladimir L. Motin Departments of Pathology and Microbiology and Immunology, Sealy Center for Vaccine Development, University of Texas Medical Branch, Galveston, TX

Martin G. Myers Sealy Center for Vaccine Development, University of Texas Medical Branch; National Network for Immunization Information, Galveston, TX

Joan E. Nichols Department of Internal Medicine, Division of Infectious Diseases, Department of Microbiology and Immunology, and Sealy Center for Vaccine Development, University of Texas Medical Branch, Galveston, TX

Andrew J. Nok Department of Biochemistry, Ahmadu Bello University, Zaria, Nigeria

Robert Nordgren Technology Acquisition and North American Biological Development, Merial Limited, Duluth, GA

Jeanne M. Novak CBR International Corporation, Boulder, CO

Juan P. Olano Departments of Pathology and Microbiology and Immunology, Center for Biodefense and Emerging infectious Diseases, Sealy Center for Vaccine Development, University of Texas Medical Branch, Galveston, TX, USA

Albert D.M.E. Osterhaus Department of Virology, Erasmus MC, Rotterdam, The Netherlands

Petra C.F. Oyston Defence Science and Technology Laboratory, Salisbury, Wiltshire, UK

Slobodan Paessler Department of Pathology, Centers for Tropical Diseases, and Biodefense and Emerging Infectious Diseases, and Sealy Center for Vaccine

Development, University of Texas Medical Branch, Galveston, TX

Falgunee K. Parekh U.S. Military Malaria Vaccine Program, Naval Medical Research Center/Walter Reed Army Institute of Research, Silver Spring, MD

Johnny W. Peterson Department of Microbiology and Immunology, Sealy Center for Vaccine Development, University of Texas Medical Branch, Galveston, TX

Diego Pineda National Network for Immunization Information, Galveston, TX

Gregory A. Poland Mayo Vaccine Research Group; Program in Translational Immunovirology and Biodefense, Mayo Clinic College of Medicine, Rochester, MN

Ann M. Powers Division of Vector-Borne Infectious Diseases, Centers for Disease Control and Prevention, Rampart Road, Fort Collins, CO

Ryan T. Ranallo Department of Pediatrics, Division of Bacterial and Rickettsial Diseases, Walter Reed Army Institute of Research, Silver Spring, MD

Roger G. Rank Department of Microbiology and Immunology, University of Arkansas for Medical Sciences and Arkansas Children's Hospital Research Institute, Little Rock, AR

Victor E. Reyes Department of Pediatrics; Departments of Microbiology and Immunology, Sealy Center for Vaccine Development, University of Texas Medical Branch, Galveston, TX

Thomas L. Richie U.S. Military Malaria Vaccine Program, Naval Medical Research Center/Walter Reed Army Institute of Research, Silver Spring, MD

John T. Roehrig Arboviral Diseases Branch, Division of Vector-Borne Infectious Diseases, Centers for Diseases Control and Prevention, U.S. Public Health Service, Department of Health and Human Services, Fort Collins, CO

Nadine G. Rouphael Division of Infectious Diseases, Department of Medicine, Emory University School of Medicine, Atlanta, GA

Judy Ruckman CBR International Corporation, Boulder, CO

Charles E. Rupprecht Centers for Disease Control and Prevention, Atlanta, GA

Michael W. Russell Department of Microbiology and Immunology, Witebsky Center for Microbial Pathogenesis and Immunology, University at Buffalo, Buffalo, NY

Thomas A. Russo VA Medical Center, Department of Medicine, Department of Microbiology, The Center for Microbial Pathogenesis, and The Center for Excellence in Bioinformatics and Life Sciences, University at Buffalo, Buffalo, NY

Edward T. Ryan Division of Infectious Diseases; Tropical and Geographic Medicine Center, Massachusetts General Hospital; Harvard Medical School; Department of Immunology and Infectious Diseases, Harvard School of Public Health, Boston, MA

Jean François Saluzzo Sanofi Pasteur, Marcy l'Etoile, France

Mitali Sarkar-Tyson Defence Science and Technology Laboratory, Porton Down, Salisbury, Wiltshire, UK

Johanna C. Sierra Department of Pediatrics, University of Texas Medical Branch, Galveston, TX

Lance L. Simpson Center for Research on Bioterrorism and Biodefense, Environmental Health Sciences, Jefferson Medical College, Philadelphia, PA

Lynn Soong Departments of Microbiology and Immunology and Pathology, Center for Biodefense and Emerging Infections Diseases, Sealy Center for Vaccine Development, University of Texas Medical Branch, Galveston, TX

Lawrence R. Stanberry Department of Pediatrics, College of Physicians and Surgeons, Columbia University, New York, NY

David S. Stephens Division of Infectious Diseases, Department of Medicine, Emory University School of Medicine, Atlanta; Laboratories of Microbial Pathogenesis, Atlanta VA Medical Center, Decatur, GA

Giovanni Suárez Departments of Microbiology and Immunology, Sealy Center for Vaccine Development, University of Texas Medical Branch, Galveston, TX

Richard W. Titball School of Biosciences, Geoffrey Pope Building, University of Exeter, Exeter, Devon, UK

Alfredo G. Torres Departments of Pathology and Microbiology and Immunology, Sealy Center for Vaccine Development, University of Texas Medical Branch, Galveston, TX

Dennis W. Trent D. Trent and Associates, American Fork, UT

David W. Vaughn Military Infectious Diseases Research Program, US Army Medical Research and Development Command, Fort Detrick, MD

Malabi Venkatesan Department of Pediatrics, Division of Bacterial and Rickettsial Diseases, Walter Reed Army Institute of Research, Silver Spring, MD

Victoria Wahl-Jensen Virology Division, U.S. Army Medical Research Institute of Infectious Diseases, Fort Detrick, MD

Christopher Walker The Research Institute at Nationwide Children's Hospital, Columbus, OH

David H. Walker Departments of Pathology and Microbiology and Immunology, Center for Biodefense and Emerging Infectious Diseases, Sealy Center for Vaccine Development, University of Texas Medical Branch, Galveston, TX

Danher Wang Division of Biodefense Vaccines, GenPhar, Inc., Mt. Pleasant, SC

Shixia Wang Laboratory of Nucleic Acid Vaccines, Department of Medicine, University of Massachusetts Medical School, Worcester, MA

Haruo Watanabe Deputy of Director General, National Institute of Infectious Diseases, Toyama, Shinjuku-ku, Tokyo, Japan

Scott C. Weaver Department of Pathology, Centers for Tropical Diseases, and Biodefense and Emerging Infectious Diseases, and Sealy Center for Vaccine Development, University of Texas Medical Branch, Galveston, TX

Jian-Jun Wen Department of Microbiology and Immunology, University of Texas Medical Branch, Galveston, TX

Stephen S. Whitehead Laboratory of Infectious Diseases, National Institute of Allergy and Infectious Diseases, Bethesda, MD

Fabian T. Wild Inserm, U758, IFR128 BioSciences Lyon-Gerland Lyon, France

E. Diane Williamson DSTL Porton Down, Salisbury, Wiltshire, UK

Jan Woraratanadharm Division of Biodefense Vaccines, GenPhar, Inc., Mt. Pleasant, SC

Xuejie Yu Department of Pathology, Sealy Center for Vaccine Development, Center for Biodefense and Emerging Infectious Diseases, University of Texas Medical Branch, Galveston, TX

Michele A. Zacks Department of Pathology, University of Texas Medical Branch, Galveston, TX

Shanta M. Zimmer Division of Infectious Diseases, Department of Medicine, Emory University School of Medicine, Atlanta, GA

SECTION I

BIOTHREATS AND EMERGING INFECTIOUS DISEASES

CHAPTER

1

Agents of Emerging Infectious Diseases

Juan P. Olano and David H. Walker
Departments of Pathology and Microbiology and Immunology, Center for Biodefense and Emerging Infectious Diseases, Sealy Center for Vaccine Development, University of Texas Medical Branch, Galveston, TX, USA

OUTLINE

ABSTRACT

Dramatic improvements in the control of infectious diseases in developed countries owing to socioeconomic changes, vaccines, and antibiotics during the first seven decades of the 20th century led to the mistakened concept that infectious diseases would no longer be a concern. Since the declaration of victory in the war against infectious diseases in 1967, approximately 50 new disease agents have been identified. Nearly every type of etiologic agent and clinical manifestation have been involved including acute respiratory infections (e.g., H5N1 influenza A, SARS, hantaviral cardiopulmonary syndrome, and Legionnaires' disease), central nervous system involvement (e.g., West Nile encephalitis, Nipah virus encephalitis, and prion diseases), enteric infections (e.g., *Helicobacter pylori* gastric and duodenal diseases, cryptosporidiosis, microsporidioses, and Shiga toxin diseases), systemic bacterial diseases (e.g., Lyme disease, six new rickettsioses, three new human ehrlichioses, bartonelloses, and staphylococcal and streptococcal toxic shock syndrome), viral hemorrhagic fevers (e.g., Marburg, Ebola, Lassa, Bolivian, Argentine, and Venezuelan hemorrhagic fevers), human retroviral infections (e.g., HIV1 and 2 and HTLV-I and II), new human herpesviruses (HHV6, HHV7, and HHV8), and the viral agents of hepatitis A, B, C, D, and E.

There are the reciprocal threats that a bioterror agent (e.g., smallpox virus) could cause a newly emerging infectious disease (EID) and that an agent of emerging infections (e.g., SARS-coronavirus or Rift Valley fever virus) could be disseminated by terrorists.

Vaccines offer a critically important potential countermeasure against the effects of these and future EIDs. An aggressive approach to developing prototype vaccines against each new class of etiologic agent must be driven by public health initiatives because commercial interests will not undertake these projects. The microbe must be completely characterized biologically, molecularly, and genetically. An accurate animal model of the human infectious disease should be developed. The mechanisms of vaccine-induced protective immunity must be elucidated and the antigens that stimulate these mechanisms of protective immune memory identified. Preclinical testing of vaccine candidates should then be completed in the animal models.

It would be most effective if subunit vaccine platforms were developed in which new antigen cassettes could be inserted and FDA approval obtained using one or more prototypes. Experience in manufacturing and a track record of effectiveness and safety for vaccines against numerous emerging infectious agents could be achieved for veterinary diseases caused by organisms that also cause emerging human infections (e.g., West Nile virus and ehrlichioses).

In the United States, these approaches are driven currently by individual investigator initiative in pursuing the scientific questions through grants from the National Institute of Allergy and Infectious Diseases. Progress occurs, but not at the desired level. An emerging infection with high transmissibility (e.g., $R_o = 10$) and a case-fatality rate of 15% would cause global devastating effects at a level on the order of magnitude of a nuclear war. Our efforts to prepare for EIDs fall far short of nuclear attack preparedness during the Cold War.

INTRODUCTION

A rather optimistic statement in 1967 from the US Surgeon General proclaimed stated that "the war against infectious diseases has been won." The emergence and re-emergence of infectious diseases during the past four decades has been astounding and has obviously proved this highly publicized statement wrong. A former director of the National Institute of Allergy and Infectious Diseases stated in 1981 that microbial diversity and evolutionary vigor were still dynamic forces threatening mankind. With the advent of revolutionary research tools, our view of the world of microbial diversity continues to expand astronomically. As an example, a random sample of seawater near Bermuda yielded DNA sequences never known before. The most accepted definition of an emerging and re-emerging infectious disease is "a disease that has newly appeared in the population or has existed previously, but is rapidly increasing in incidence and geographic range." Another widely cited definition is a "new emerging or drug resistant infection whose incidence in humans has increased within the last two decades, or whose incidence threatens to increase in the near future." Strictly speaking all infectious diseases have "emerged" at one time or another. Thus, the definitions are rather arbitrary. The emergence of *Mycobacterium tuberculosis* and *Plasmodium falciparum* as human pathogens probably occurred within the last few tens of thousands of years. Other pathogens required rather large human populations interacting closely so that human–human transmission would occur such as with measles and smallpox viruses. HIV probably made the jump to humans 60–75 years ago. Emerging infections are further classified by others as newly emerging, re-emerging/resurging, and deliberately emerging. The factors involved in the process of emergence and re-emergence are extremely complex and have to do with the microbes themselves, the environment, and the hosts, a triad well known to any scientist who works in the field of infectious agents. The Convergence Model tries to explain all factors that affect the host–microbe interaction. This interaction is affected by interlocking domains, namely genetic and biologic factors, ecologic factors, physical environmental factors, and social, political and economic factors.

EMERGING INFECTIOUS DISEASES SINCE 1967

We have attempted to create a comprehensive list of infectious diseases that fall into this category (Table 1.1). A discussion of the clinico-epidemiological characteristics of the most representative emerging infectious diseases (EIDs) follows. We have classified EIDs based on clinical presentation (syndromes such as pneumonia, encephalitis, systemic bacterial infection, enteric infection, and viral hemorrhagic fever).

EMERGING INFECTIONS CAUSING ACUTE RESPIRATORY INFECTIONS

Acute respiratory infections are the leading cause of mortality from infectious diseases around the world. In 2003, 37 million deaths were due to acute lower respiratory tract infections. Upper respiratory tract infections, although important from the point of view of morbidity, are rarely fatal. In developed countries the main culprits are influenza and bacterial pneumonias. In underdeveloped countries, the highest mortality rates are seen in children less than 5 years of age with viral pneumonia caused by respiratory syncytial virus, parainfluenza viruses, influenza viruses, and adenoviruses. Bacterial respiratory agents include *Streptococcus pneumoniae*, *Hemophilus influenzae*, and *Staphylococcus aureus*.

VIRAL PULMONARY SYNDROMES

Influenza Viruses

These viruses belong to the family Orthomyxoviridae and are enveloped, negative sense, single-stranded, segmented, RNA viruses responsible for recurrent epidemics of febrile respiratory disease every 1–2 years. Influenza viruses are further classified as A, B, and C based on antigenic characteristics. Influenza B and C viruses only undergo antigenic drift as compared with influenza A viruses that can undergo antigenic shift in addition to antigenic drift. Antigenic shifts are responsible for the major pandemics seen with influenza A viruses, three of which occurred in the 20th century, including the highly lethal 1918/1919 pandemic that killed tens of millions of people around the world. Pandemics do not occur with influenza B and C viruses. In fact influenza C causes a mild disease without seasonality, and influenza B causes severe disease confined to the elderly and other persons at high risk. Influenza A viruses are divided into subtypes based on the presence of different hemagglutinin (H) and neuraminidase (N) molecules on their surface. Sixteen distinct hemagglutinins and nine different neuraminidases have been described to date. Influenza viruses are among the most contagious human viruses with attack rates of 10–40% during epidemics. Epidemics are seasonal (winter) in temperate regions and can occur year round in tropical areas. Since the last influenza pandemic in 1968, the predominant virus circulating around the world is the H3N2 subtype that undergoes minor mutations in the H and N proteins and is responsible for yearly outbreaks. In 1977, the H1N1 virus (the same subtype that was responsible for the 1918 pandemic) was reintroduced and is currently co-circulating with H3N2 virus. However, isolated cases of human influenza due to avian viruses such as H5N1 are cause for concern among health professionals around the world because of the high lethality and the possibility that development of efficient

TABLE 1.1 Infectious diseases with agents identified since "The end of the war against infectious diseases" 1967–2004

Year	Agent	Disease
1967	Marburg virus	Marburg hemorrhagic fever
1969	Lassa virus	Lassa fever
1971	JC virus	Progressive multifocal leukoencephalopathy
1972	Norovirus	Norwalk diarrheal illness
1973	Rotavirus	Major cause of infantile diarrhea worldwide
1975	Parvovirus B19	Fifth disease; aplastic crisis in chronic hemolytic anemia; hydrops fetalis; chronic anemia of immunosuppressed patient
1976	*Vibrio vulnificus*	Sepsis and necrotizing fasciitis
1976	*Cryptosporidium parvum*	Human cryptosporidiosis
1977	Ebola virus	Ebola hemorrhagic fever
1977	*Clostridium difficile*	Pseudomembranous colitis
1977	*Legionella pneumophila*	Legionnaires' disease, Pontiac fever
1977	Hantaan virus	Hemorrhagic fever with renal syndrome
1977	Delta viral hepatitis	Hepatitis B virus-associated hepatitis
1977	*Campylobacter* sp.	Enteric pathogens distributed globally
1979	*Cyclospora cayetanensis*	Diarrheal illness
1980	HTLV-I	T-cell lymphoma-leukemia; tropical spastic paresis
1981	*Staphylococcus aureus* toxin	Toxic shock syndrome
1982	*Borrelia burgdorferi*	Lyme disease
1982	*Escherichia coli* 0157:H7	Hemorrhagic diarrhea, hemolytic uremic syndrome
1982	HTLV-II	Associated with neurologic syndromes
1983	HIV-1	AIDS
1983	*Helicobacter pylori*	Gastric and duodenal ulcers
1984	*Hemophilus influenzae aegyptius*	Brazilian purpuric fever
1985	*Enterocytozoon bieneusi*	Microsporidiosis
1986	*Chlamydia pneumoniae*	A major cause of pneumonia
1988	Human herpesvirus 6	Exanthem subitum (roseola infantum)
1989	*Rickettsia japonica*	Japanese spotted fever
1989	Hepatitis C virus	Parenterally transmitted nonA–nonB hepatitis
1990	Hepatitis E virus	Enteric nonA, nonB hepatitis
1990	*Balamuthia mandrillaris*	Leptomyxid amebic meningoencephalitis
1990	Human herpesvirus 7	Another cause of exanthem subitum
1991	Guanarito virus	Venezuelan arenaviral hemorrhagic fever
1991	*Encephalitozoon hellem*	Microsporidiosis
1991	*Ehrlichia chaffeensis*	Human monocytotropic ehrlichiosis
1992	Barmah Forest virus	Febrile polyarthralgia
1992	*Vibrio cholerae* 0139	New strain associated with epidemic cholera
1992	*Bartonella henselae*	Cat scratch disease; bacillary angiomatosis; endocarditis
1992	*Rickettsia honei*	Flinders Island spotted fever
1992	*Tropheryma whippelii*	Whipple's disease
1993	Sin Nombre hantavirus	Hantavirus cardiopulmonary syndrome
1994	*Anaplasma phagocytophilum*	Human granulocytic anaplasmosis
1994	Hendra virus	Acute respiratory syndrome, meningitis
1996	Human herpesvirus 8	Kaposi's sarcoma, Castleman's disease, primary effusion based B-cell lyphoma

(Continued)

TABLE 1.1 (Continued)

Year	Agent	Disease
1997	*Rickettsia slovaca*	Tick-borne lymphadenopathy
1999	*Ehrlichia ewingii*	Ehrlichiosis ewingii
1999	Nipah virus	Encephalitis
2001	Human metapneumovirus	Upper and lower respiratory infections
2003	SARS-CoV	Severe acute respiratory syndrome (SARS)
2004	Monkeypox virus	Human monkeypox (USA outbreak)

person-to-person transmission could trigger another pandemic. The first case of human influenza due to H5N1 was described in 1997, and the infection re-emerged in Southeast Asia in 2004. Death rates in domestic birds in Southeast Asia are staggering. Other avian influenza viruses have also emerged in other parts of the world such as H7N7 in the Netherlands causing an outbreak of hemorrhagic conjunctivitis in humans. Influenza A viruses are capable of infecting a wide variety of animals including birds, horses, marine mammals, swine, and humans. The most important reservoir is wild aquatic birds (water fowl, ducks, and shorebirds). These animals can carry different subtypes in their gut, and some viruses are capable of infecting domestic poultry. The ability of avian influenza viruses to infect humans is rather restricted as is the ability for transmission from person-to-person. Because of their segmented genomes, influenza A viruses have the potential to swap or reassort genes when present in the same host or "mixing vessel" such as pigs (these animals have cell surface receptors that allow cellular entry of both bird and human influenza viruses). If the favorable environmental conditions are present, human influenza viruses can coexist with avian viruses in pigs, and new reassortants may be produced with potentially enhanced ability to infect humans.

Clinically, the human disease manifests as a severe febrile illness with abrupt onset, myalgias, severe malaise, dry cough, arthralgias and nasal discharge. Most patients recover from the acute phase of the disease in 3–5 days. Complications include primary influenza viral pneumonitis, which was first well documented in the 1957–1958 pandemic. However, it is widely accepted that this and bacterial superinfections were the causes of death of millions of persons in the 1918 pandemic. These patients progress to acute respiratory failure rapidly, and the main pathologic finding is that of diffuse alveolar damage with a hemorrhagic component. Other complications include secondary bacterial pneumonias.

Severe Acute Respiratory Syndrome (SARS)

An explosive outbreak of severe adult respiratory distress syndrome (ARDS) occurred in residents of Hong Kong and persons who had visited Hong Kong early in 2003. A cluster of cases that had started in November 2002 also occurred in the Guangdong Province of China. After a few months of excellent "detective" work by health agencies and research laboratories worldwide, a new coronavirus was identified as the culprit. Between November 2002 and July 2003, 8096 cases were reported from nearly 30 countries, but the vast majority of cases occurred in mainland China and Hong Kong. The case-fatality rate was between 7 and 17% and approached 50% in the elderly. High mortality rates were also observed in patients with preexisting conditions such as diabetes and cardiopulmonary diseases.

An animal reservoir for SARS-CoV appears likely to be bats, but the virus also has been isolated from civet cats and other wild animals in markets in China where cross-species infections may have occurred. Other coronaviruses in humans have cyclical patterns of circulation. It is suspected that SARS-CoV could reappear in humans. In severe cases, the main finding is diffuse alveolar damage leading to acute respiratory insufficiency. In milder cases, fever, malaise and mylagia are the main manifestations.

Human Metapneumovirus

This viral pathogen was first identified in 2001 in an outbreak of upper and lower respiratory illness in the Netherlands. The agent is a paramyxovirus, subfamily Pneumovirinae in which two genera have been established: Pneumovirus (respiratory syncytial virus, RSV) and Metapneumovirus. The disease spectrum is not completely known but can manifest as severe upper and lower respiratory infections in children and adults especially in patients with previous cardiopulmonary conditions. The disease is similar to RSV infection and can account for up to 10% of acute upper respiratory tract infections in which another etiologic agent has not been identified. Asthmatic exacerbations have also been related to metapneumovirus infections. Retrospective serological studies have established that this virus has been circulating in humans for at least 50 years.

Hantavirus Cardiopulmonary Syndrome (HCPS)

Sin Nombre virus (SNV) was the first hantavirus in the Americas associated with HCPS. The disease was first described in an outbreak in the Four Corners area of the southwestern US in 1993. SNV belongs to the genus Hantavirus within the family Bunyaviridae. The vast majority of viruses within this family are arthropod-borne zoonoses, with the exception of hantaviruses, which are not vector-borne. Hantaviruses are found in wild rodents, which excrete the virus in urine and saliva for months. SNV is the most important pathogenic hantavirus in North America and causes chronic infections in deer mice (*Peromyscus maniculatus*). In North America, other viruses within the same genus associated with HCPS have been described such as New York virus found in the white footed deer mouse (*Peromyscus leucopus*). In South America, the main representative of the genus is Andes virus, which is responsible for HCPS in Chile and Argentina, and is the only hantavirus for which human-to-human transmission has been documented. In 1978, another hantavirus associated with hemorrhagic fever and renal syndrome (HFRS) in the Korean peninsula was isolated. This infection is endemic across the Asian continent.

The main mode of transmission is through aerosol spread of infected excreta (urine and possibly feces) and less frequently by bites of infected rodents. HCPS is an explosive febrile illness accompanied by myalgias and sometimes abdominal pain. In 4–5 days, respiratory symptoms appear and rapidly progress to severe noncardiogenic pulmonary edema with subsequent hypoxia and shock within hours. Cardiac dysfunction also occurs.

BACTERIAL PULMONARY SYNDROMES

Legionnaires' Disease

An outbreak of pneumonia in 1976 in Philadelphia, Pennsylvania during the state American Legion convention affected 221 people, 34 of whom died from the infection. In 1977, Dr. McDade and Dr. Shepard at the CDC isolated the etiologic agent, a fastidious gram negative organism later named *Legionella pneumophila*. This bacterium has been responsible for subsequent epidemics and sporadic cases, and retrospective studies determined that it had been responsible for outbreaks of pneumonia in the 1950s and 1960s. *Legionella* species are naturally occurring aquatic bacteria that grow in warm water, especially in cooling towers, water heaters and plumbing, hence the propensity to cause nosocomial and community outbreaks (hotels and other facilities). Free-living amebas also support the intracellular growth of *Legionella spp*. More than 20 *Legionella* species have been described affecting humans, thus the general name of legionellosis.

The disease occurs both sporadically (65–75% of cases) and in outbreaks or epidemics. Recent epidemics have occurred in Spain (2001, 700 cases), England (2002, 130 cases), and the Netherlands (1999, 188 cases).

Transmission is by aerosolization of contaminated water sources. Once in the alveoli of the lung, the bacteria are phagocytosed by alveolar macrophages through coiling phagocytosis, and multiplication occurs. The bacteria are then released into the alveolar space by dying macrophages where they can invade other alveolar macrophages. A type IV secretion system is important in promoting intracellular infection including inhibition of phagolysosomal fusion. The lungs reveal patchy-to-confluent bronchopneumonia that may be complicated by pleural effusions or cavitation in a minority of cases. Extrapulmonary infection is rare.

Tuberculosis

Tuberculosis (TB) is as old as civilization itself as demonstrated by evidence of spinal tuberculosis in Egyptian mummies and Neolithic and pre-Columbian bones. However, tuberculosis did not become a major public health problem until the 17th and 18th centuries during the Industrial Revolution. Tuberculosis also ravaged (and still does) the native American populations after Columbus' voyages to the Americas. In the United States, tuberculosis saw a steady decline in the middle part of the 20th century until 1985 when the incidence climbed again principally due to the appearance of HIV. Other factors included deterioration of living conditions, intravenous drug abuse, and underfunding of tuberculosis control programs. However, since the mid-1990s rates have declined and in 2002 reached the lowest incidence in history from the time statistics became available. This control is the result of better anti-HIV therapies, intensified diagnosis, aggressive and monitored anti-TB treatment and prevention efforts. However, not all is good news in the TB world. The appearance of multidrug resistant strains of *M. tuberculosis* is a challenge recalling the preantibiotic era. In addition, *M. tuberculosis* is the number one killer worldwide (2 million deaths and 8 million new cases diagnosed each year), and approximately 2 billion people are infected.

More than 95% of infections caused by *M. tuberculosis* are acquired by inhalation of aerosols generated from an infectious patient. The initial focus of infection (Gohn's lesion) later develops into Gohn's complex (accompanying infected draining hilar lymph node lesions), and in most cases the infection is contained by the immune system. In the most severe cases, the infection disseminates hematogenously in the lungs and/or systemically leading to miliary tuberculosis. Individuals who control the primary infection may undergo reactivation of the latent infection due to multiple factors including AIDS, malnourishment, alcoholism, and cancer. In these cases, reactivation can involve the lungs (cavitary, endobronchial, pneumonic, and/or bronchopneumonic) and other organs including the spleen, liver, bone marrow, kidneys, CNS, and bones. Primary and secondary mycobacterial resistance to antituberculosis medication in certain subpopulations in New York City and California makes tuberculosis a public health priority. Recent trends have shown that the percentage of isolates with resistance to antituberculosis drugs is decreasing due to vigorous public health efforts.

Viral Hemorrhagic Fevers (VHF)

It is estimated that 75% of EIDs in humans originate in animals, and VHFs are remarkable examples. The etiologic agents of this syndrome are a heterogenous group of RNA viruses belonging to three families, namely filoviruses (Ebola and Marburg viruses), arenaviruses (Lassa, Junin, Machupo, Guanarito, and Sabia viruses), and bunyaviruses (Crimean-Congo hemorrhagic fever and Rift Valley fever (RVF) viruses).

All these viruses have limited geographic ranges due to their specific natural reservoirs and vectors. In all cases, humans are accidental hosts. Pathogenetically, all VHFs lead to dramatically increased vascular permeability systemically with a hemorrhagic diathesis and edema in multiple organ systems, including the lungs and brain. These diseases all occur in localized outbreaks usually with very high case-fatality rates. Clinically, subtle differences exist between the syndromes such as more prominent hemorrhagic manifestations and terminal disseminated intravascular coagulation (DIC) in infections caused by filoviruses and CCHF, and prominent CNS and hemorrhagic manifestations in infections by the New World arenaviruses (Junin, Machupo, Sabia, and Guanarito), and prominent liver disease in RVF.

Rift Valley Fever

This is a mosquito-borne viral disease that affects newborn ruminants, especially sheep. Other affected animals include lambs, calves, goats, kittens, mice, and hamsters. The virus was first isolated in 1930 from sheep in Kenya, and the known distribution was limited to the African continent for decades. However, in 2000, a large epidemic in the Arabian peninsula led to high case-fatality ratios in humans, which had usually been around 5% in previous epidemics. Complications include hepatorenal failure, encephalitis and DIC leading to shock and multiorgan failure. The vector range for the virus is impressive and includes at least 30 species of mosquitoes in eight genera. Transovarial transmission occurs in the mosquitoes. Most outbreaks are related to climatic events that favor floods leading to increased vector populations. Outbreaks in Africa, besides the original one in Kenya, have occurred in Egypt (Aswan Dam construction), Mauritania in 1987 (Diama Dam construction), Kenya and Somalia in 1997–1998 (increased rainfall due to El Niño oscillation), and Kenya in 2007.

Ebola Hemorrhagic Fever (EHF)

Close to 20 outbreaks have occurred in Africa since Ebola virus was identified in 1976 in central Africa. The reservoir has been elusive although human outbreaks are usually preceded by severe primate die-offs. Recent studies of animals in the areas of primate mortality have identified bats as a likely reservoir host. Primary mechanisms of transmission from nonhuman primates to humans include contact with dead carcasses or handling or consumption of bushmeat. The virus has also been linked to lethal outbreaks in duikers, a wild ruminant. Other factors associated with EHF outbreaks include increases in rainfall. The viruses are rather stable genetically, since isolates obtained 20 years apart have limited genetic variation. The disease is rapidly progressive and leads to hemorrhagic manifestations very quickly.

South American Hemorrhagic Fevers

South American hemorrhagic fevers (HF) are caused by arenaviruses whose geographic distribution is limited. Most arenaviruses in Africa and the Americas do not cause human disease. Rodents develop a chronic infection that is most times asymptomatic but with persistent viremia that can be lifelong. Vertical transmission occurs in mice. Both New World and Old World arenaviruses display a high specificity for their particular host. In the New World, reservoir rodents belong to the family Muridae, subfamily Sigmodontinae (rats and mice). Human infections occur via inhalation of aerosol particles containing infected urine from rodents. Ecologic factors associated with the emergence of these viruses as human pathogens include factors leading to increased rodent populations and increased contact between humans and rodents, e.g., deforestation and encroachment of farming into these areas. Human-to-human transmission is rare but can occur through contact with infected body fluids.

The New World viruses belong to the Tacaribe complex and include Junin (Argentine HF), Machupo (Bolivian HF), and Guanarito (Venezuelan HF) viruses. Few cases of HF due to Sabia virus (including two laboratory infections) have been described in Brazil, but the disease spectrum is largely unknown. Severe systemic disease with hemorrhage and prominent neurologic manifestations are the rule.

Dengue Fever

Dengue virus has been responsible for outbreaks of acute febrile illness (break-bone fever) in the tropics, and for the last 25 years, geographic expansion has occurred due to multiple environmental factors. In the United States, two large outbreaks occurred in Florida (1934) and New Orleans (1945). Four distinct dengue virus species are currently recognized, and the main vector is *Aedes aegypti*. The current geographic distribution ranges between 35° of latitude north and south. A prominent resurgence of dengue virus infections has occurred in the Caribbean basin, where the vector had been eradicated because of extensive campaigns against yellow fever. However, *A. aegypti* was reintroduced to the area, and large outbreaks of dengue fever (DF) and dengue hemorrhagic fever (DHF) have occurred in Cuba, Venezuela, Colombia, Central American countries, Mexico, and the Caribbean islands. A portion of dengue viral infections develop a life-threatening hemorrhagic fever.

EMERGING ENCEPHALITIC SYNDROMES

Nipah Virus Infection

This agent is a recently discovered paramyxovirus that, along with Hendra virus, comprises a new genus (*Henipavirus*) in the family Paramyxoviridae. In 1999, an outbreak of an acute febrile encephalitic syndrome in Malaysia was traced to Nipah virus. The epidemic was preceded by an epizootic of severe respiratory disease in pigs that was terminated after epidemiological control measures were instituted, including the culling of millions of pigs. A total of 283 cases of human encephalitis were diagnosed with a case-fatality rate close to 40%. A relapsing/remitting neurologic syndrome has also been associated with Nipah virus. A subsequent outbreak occurred in Bangladesh, but swine were not associated with this event, suggesting different ecological factors. The natural reservoir is thought to be fruit bats of the *Pteropus* genus. Other animals such as dogs, cats, and horses could also serve as hosts.

Hendra virus was first described in 1994 in Australia during a highly lethal epidemic of acute respiratory disease in horses. The only two human cases that have been associated with Hendra viruses manifested as acute respiratory syndrome and leptomeningitis, respectively. Both cases were fatal. The natural reservoir for Hendra virus is the fruit bat.

West Nile Virus (WNV) Infection

WNV was first isolated from humans in 1939 in Uganda and remained limited to the Middle East, Eastern Europe, and Africa until 1999, when an outbreak occurred in New York. By 2002, WNV spread across North America to the west coast, and annual outbreaks are now the rule. In 2002 and 2003, approximately 4000 and 9000 (case definition was modified in 2003) cases occurred, respectively, leading to CNS disease in 3000 and 2700 cases each year, respectively. Previous outbreaks of West Nile fever have occurred in Israel (1957), South Africa (1974), Algeria (1994), Tunisia (1997), and Congo (1998). WNV circulates enzootically between birds and mosquitoes. At least 300 species of birds and 62 species of mosquitoes can be infected by WNV. Most susceptible birds are the *Corvidae* (blue jays and crows), and the most important mosquito in the enzootic cycle is *Culex pipiens*. Humans and horses are dead-end hosts, both developing febrile and/or encephalitic syndromes. Other forms of transmission are through transplanted organs, transfusion, and breast milk. In fact, blood testing by PCR is now a component of the screening process for donated blood.

Prion Diseases (Transmissible Neurodegenerative Diseases)

Initially known as "atypically slow infections" (Sigurdsson, 1954), the concept of prion (proteinaceous infectious agents) was first introduced by Prussiner in 1982. Currently a prion is known as "a small infectious pathogen that contains protein and is resistant to procedures that modify or hydrolyze nucleic acids." The pathogenesis of these diseases involves the generation and accumulation of abnormal prion protein from a normal isoform. The common denominator in these diseases in the presence of progressive neuronal degeneration (incubation periods measured in years or decades) accompanied by reactive astrocytosis and absence of inflammation. In most cases, prominent vacuolation of cells or neuropil is observed microscopically. Accumulation of abnormal prions in CNS tissue is present in all cases.

In humans, prions are responsible for several diseases of which Creutzfeldt–Jakob disease (CJD) and its new variant (linked to "mad cow disease" or bovine spongiform encephalopathy) are the most common. Other diseases include kuru, familial fatal insomnia (FFI), and Gerstmann–Straussler–Scheinker syndrome (GSS). CJD has several forms including sporadic (source of infection unknown), familial (mutations in the PrP gene), and iatrogenic (cadaveric grafts from patients with CJD). Kuru was related to anthropophagic practices in isolated tribes in Papua New Guinea, and new variant CJD has been linked to BSE. GSS and FFI are due to genetic mutations in the PrP gene in most cases.

ARTHROPOD TRANSMITTED BACTERIAL DISEASES

Lyme Borreliosis (Lyme Disease)

Lyme borreliosis is a zoonosis transmitted by hard ticks of the genus *Ixodes*. The vector in the eastern United States is *Ixodes scapularis* and in the western United States is *Ixodes pacificus. Ixodes ricinus* is the vector in Europe whereas *Ixodes persulcatus* is the vector in Russia and northern Asia. *B. burgdorferi* sensu lato includes *Borrelia burgdorferi* sensu stricto, *B. afzelii*, and *B. garinii*. The latter two are responsible for most cases of Lyme borreliosis in Europe, Russia, and northern Asia. Lyme disease has not been documented in tropical areas. The main hosts of *B. burgdorferi* in nature are rodents. Larval forms of ticks acquire the infection from small mammals, and the spirochetes are transmitted transtadially to nymphal and adult ticks. Human infections usually occur secondarily to nymph bites that go unnoticed very easily due to their small size.

Lyme borreliosis is a disease with an acute phase characterized by erythema migrans and nonspecific symptoms such as fever, headaches, myalgias, and arthralgias. The chronic phase (weeks to years) is characterized by oligoarthritis, central and peripheral nervous system sequelae, myocarditis, and other manifestations.

Rickettsioses

Rickettsia are obligately intracellular bacteria with a gram negative cell wall and a characteristic lipopolysaccharide. Their main target is the microvascular endothelium and different cells in their arthropod host/vector (fleas, ticks, lice, mites). Spotted fever group rickettsiae (with the exception of *Rickettsia felis* [fleas] and *Rickettsia akari* [mites]) are transmitted transovarially and transstadially in their tick vector and circulate in nature via small mammals. Fleas and lice host typhus group rickettsiae although they do not transmit the infection vertically to their offspring.

Newly described rickettsioses include *Rickettsia africae* (African tick bite fever), a closely related agent named *Rickettsia parkeri* (clusters of a relatively mild febrile disease in North and South America), *Rickettsia slovaca* (tick-borne lymphadenopathy or DEBONEL [*Dermacentor*-borne necrosis, eschar, lymphadenopathy]), *R. felis* (flea-borne spotted fever), *Rickettsia honei* (Flinders Island spotted fever), and *Rickettsia japonica* (Japanese spotted fever). The geographic distribution of these agents continues to expand. In fact, the distribution of organisms that appear to be variants of *R. japonica* now includes China and Korea. *R. honei* has been described in Thailand, and *R. felis* is probably distributed worldwide.

Ehrlichioses and Anaplasmosis

Obligately intracellular bacteria in the family Anaplasmataceae, which are related to the genus *Rickettsia*, include four human pathogens, *Ehrlichia chaffeensis* (human monocytotropic ehrlichiosis, HME), *Anaplasma phagocytophilum* (human granulocytotropic anaplasmosis, HGA), *Ehrlichia ewingii* (ehrlichiosis ewingii), and *Neorickettsia sennetsu* (mononucleosis-like illness in Japan). *N. sennetsu* is limited geographically to Japan, and its trematode host cycle differs greatly from the other pathogens in this family. The other three agents are tick-borne and have been considered until recently as veterinary pathogens. HME was first described in 1987, followed by HGA in 1994 and ehrlichiosis ewingii in 1999.

HME is predominantly transmitted by *Amblyomma americanum* ticks, and its main mammal reservoir in nature is the white-tailed deer (*Odocoileus virginianus*). Patients develop a febrile illness after a nymphal or adult tick bite accompanied by headache, myalgias, malaise, and other nonspecific symptoms. A maculopapular skin rash is present in only 30–40% of cases. Common laboratory findings are the presence of leukopenia and/or thrombocytopenia during the acute phase of the disease. Complications include meningoencephalitis, hepatitis, and diffuse alveolar damage.

HGA is transmitted by *I. scapularis* ticks in the eastern United States and *I. ricinus* in Europe. The clinical picture is similar to HME although the case-fatality ratio and complications are less severe. The target cell of *A. phagocytophilum* is the polymorphonuclear neutrophil as compared with monocytes and macrophages for *E. chaffeensis*. The main reservoir in the eastern United States is the white footed mouse.

Ehrlichiosis ewingii caused by *E. ewingii* is the least known of the human ehrlichioses. It is transmitted by the same vector tick as *E. chaffeensis* and has a milder course. The target cell is the neutrophil. The majority of cases have been diagnosed in immunocompromised patients.

Scrub Typhus

A disease that ravaged American troops in the Pacific and Southeastern Asia during World War II is re-emerging in southern India, Sri Lanka, the Maldives, and Micronesia. The agent is *Orientia tsutsugamushi*, an obligately intracellular bacterium formerly known as *R.ickettsia tsutsugamushi*. The disease is transmitted by chiggers in their larval stage and occurs mostly in tropical Asia, the western Pacific islands, northern Australia, and temperate zones in Kashmir, Korea, Japan, and the lower Himalayas. In nature, wild rats maintain the chigger population (*Leptotrombidium* spp.), and *Orientia* is transmitted transovarially and transstadially. However, rats are not reservoirs for *O. tsutsugamushi*.

The disease is characterized by fever, headache, malaise, and a variable incidence of eschar formation at the bite site, lymphadenopathy, and a transient maculopapular rash. Severe manifestations include diffuse alveolar damage and meningoencephalitis. Poor responses to conventional treatments (tetracyclines and chloramphenicol) have been described in Thailand.

Bartonelloses

The genus *Bartonella* contains five recognized human pathogens: *B. bacilliformis* (Oroya fever and verruga peruana transmitted by sandflies of the genus *Lutzomyia*), *B. quintana* (trench fever, bacillary angiomatosis, and endocarditis transmitted by the human body louse, *Pediculus humanus corporis*), *B. henselae* (cat scratch disease, bacillary angiomatosis, and endocarditis transmitted by cat fleas and cats), and *B. elizabethae* and *B. vinsonii* (endocarditis, transmitted possibly by fleas and ticks).

The reservoir host for *B. bacilliformis* is humans who reside in endemic areas in Peru, Ecuador, and Colombia. Clinically the disease has an acute phase in which the main manifestation is severe hemolytic anemia followed by a chronic phase in which lesions composed of prominent capillary proliferations are present on the skin.

Trench fever affected millions of soldiers during World Wars I and II, and infections continue to be diagnosed worldwide especially in patients with HIV or malnourished, homeless alcoholics.

The feline ectoparasite flea *Ctenocephalides felis* is the vector of *B. henselae,* and humans acquire infection from contact with flea-infested cats. Endocarditis is a serious form of infection that can occur with *B. quintana*, *B. henselae*, *B. elizabethae*, and *B. vinsonii*. The epidemiology of the latter two infections is largely unknown.

EMERGING ENTERIC PATHOGENS

Cholera

Vibrio spp. are gram negative, oxidase positive, free-living bacteria found in warm, salty waters around the world. In these places, *Vibrio* is usually isolated from shellfish, and their concentration in the tissue is far greater than in the surrounding waters. *Vibrio cholerae*, *Vibrio vulnificus*, and *Vibrio parahemolyticus* are all associated with acute enteric infections in humans although the only one causing epidemics or pandemics is *V. cholerae*. The latter two are also associated with wound infections in warm salty waters or ingestion of raw shellfish with involvement of subcutaneous tissue and skeletal muscle and septicemia. Infections caused by *V. parahemolyticus* and *V. vulnificus* are more severe in immunocompromised patients and in persons with underlying diseases such as cirrhosis, hemochromatosis, and diabetes.

V. cholerae is classified by O antigens (>150), biotypes (classical and El Tor), and serotypes (Ogawa, Inaba). Six of the seven pandemics since the 19th century have originated in the Bengal basin, and the seventh pandemic began in 1961 in Indonesia. In 1991 this pandemic extended to Peru and other countries on the South American Pacific coast. In the US, the rare cholera cases have been linked to consumption of raw oysters or undercooked crab. All cases of *V. cholerae* infection were caused by O1 strains, but since

1992, 0139 strain has been isolated from clinical cases of cholera in the Indian subcontinent. Other non-O1 serogroups are sporadically associated with acute gastroenteritis around the world.

The majority of people infected with *V. cholerae* are asymptomatic or mildly ill (75% for the classical biotype and 93% for El Tor biotype). Presentation of the classic illness is explosive after a short incubation period (12h–5 days) and consists of abundant watery diarrhea that leads to dehydration and circulatory collapse if not treated promptly.

Nontyphoidal Salmonelloses

In the United States, infections caused by nontyphoidal *Salmonella* spp. affect 1.4 million/year and kill approximately 600/year. Most patients are mildly symptomatic or asymptomatic. *Salmonella* are gram negative bacilli, nonlactose fermentors. The classification and nomenclature are extremely complex, but phylogenetic studies based on DNA sequencing reveal that *Salmonella* spp. associated with human illness are considered *Salmonella choleraesuis*, which has approximately 2500 serotypes including Typhimurium and Typhi. The name *Salmonella enterica* has been proposed to replace *S. choleraesuis*.

Virtually all cases of *Salmonella* infection are foodborne and are second only to enteric infections caused by *Campylobacter* spp. in the US. Foods associated with outbreaks in the US include undercooked ground beef, eggs, cheese, ice cream, fresh sprouts, juice, and other vegetables.

Clinical syndromes in nontyphoidal salmonellosis include gastroenteritis, bacteremia/septicemia with or without distant focal infections (infectious endarteritis in arteries with large atherosclerotic plaques, septic arthritis, and endocarditis) and asymptomatic carriage. Patients with immunosuppression of any etiology are at greater risk for systemic illness than immunocompetent patients.

Antimicrobial resistance (ampicillin, tetracyclines, chloramphenicol, streptomycin, and sulfonamides) in nontyphoidal salmonellosis is spreading throughout the world due to a specific phage infecting *Salmonella* type Typhimurium (DT104) that first appeared in 1990. Even more recently, resistance to fluoroquinolones and third and fourth generation cephalosporins has been reported.

Shiga-toxin Producing *Escherichia coli* Infection

The first human cases due to this organism were described in 1982 during two outbreaks of hemorrhagic colitis in the US due to *E. coli* serotype O157:H7. The association of hemorrhagic gastroenteritis due to O157:H7 and hemolytic uremic syndrome (microangiopathic hemolytic anemia, acute renal failure, and thrombocytopenia) was established soon thereafter. More than 30 serotypes of *E. coli* can produce *Shigella*-like toxins, but the vast majority of cases in the US are due to serotype O157:H7. The main vehicles of transmission are contaminated food (hamburgers, uncooked vegetables, and others), water, including swimming pools, person-to-person, and animal contacts.

Clinical presentation is rapid and includes vomiting, diarrhea, and cramping. The diarrhea becomes blood-streaked after a few days. One of the main complications is hemolytic uremic syndrome, mostly in patients under 5 years of age.

Helicobacter pylori Infection

The isolation of *H. pylori* (initially classified as *Campylobacter pylori* or *pyloridis*) from a human with active gastritis represents a turning point in how the medical field views peptic ulcer disease, acute and chronic gastritis, and gastric carcinoma (especially the intestinal variant). *H. pylori* is a motile, microaerophilic, gram negative bacillus capable of surviving in the harsh environment of the human stomach due to its microaerophilicity, capacity to penetrate the mucus layer overlying the gastric mucosa, and production of ammonia from urea via urease to neutralize the low pH in the environment. *H. pylori* infections seem to be limited to humans, but other *Helicobacter* spp. have been isolated from almost every mammal studied. Rates of infection in human populations are very high (in developing countries up to 70% by age 10 and nearly 100% by age 20), but clinical disease is not always apparent. Therefore, microbial and host factors are important in pathogenicity. The so-called *cag* pathogenicity island, the *vacA* gene, and their polymorphisms seem to play important roles in pathogenesis.

Human infections have been present for thousands of years based on studies of *H. pylori* alleles and human migrations. Once infection is acquired it persists for years or decades. Clinical syndromes associated with *H. pylori* infection include chronic diffuse superficial gastritis, intestinal metaplasia and atrophic gastritis, peptic, especially duodenal, ulcers (cagA-containing organisms), noncardiac gastric adenocarcinoma (cagA-containing bacteria), and gastric MALT-type B-cell lymphomas.

Cryptosporidiosis

Crysptosporidium parvum is a protozoan in the phylum Apicomplexa, class Sporozoae, subclass Coccidia.

Recent molecular studies have shown that *C. parvum* (the species thought to cause most human infections) contains various genotypes. In humans two genotypes cause infection: bovine and human genotypes. The name *Crysptosporidium hominis* has been proposed for the human genotype. The first human case was described in 1976, and the first outbreak of cryptosporidiosis in the US was documented in 1987 in Carroll County, GA where an estimated 13,000 cases of gastroenteritis were diagnosed. The largest reported outbreak occurred in Milwaukee, WI where there were an estimated 403,000 cases in 1993. *Cryptosporidium* has been described throughout the world. Infection occurs via the fecal–oral route mostly by ingestion of contaminated water. The infectious thick walled oocysts are resistant to chlorination and survive in moist environments for long periods of time. The cycle is completed in the same host, and multiplication is both sexual and asexual. Once inside the host, *Cryptosporidium* organisms invade the microvillous surface of the epithelial cells in the terminal ileum and proximal colon leading to self-limited watery diarrhea in immunocompetent patients and chronic/persistent diarrhea in immunocompromised patients (especially with AIDS).

Microsporidioses

Phylum Microsporidia are spore-forming, obligate intracellular protozoans that reside in the intestine, liver, kidneys, brain, and other tissues of wild and domesticated mammals and several other animal species. Eight genera out of more than 144 (containing more than 1000 species) have been documented as human pathogens which include *Encephalitozoon*, *Enterocytozoon*, *Pleistophora*, *Brachiola*, *Nosema*, *Trachipleistophora*, *Vittaforma*, and *Microsporidium*. The two most common species involved in humans are *Enterocytozoon bieneusi* and *Encephalitozoon intestinalis*. Microsporidioses, in general, are rare diseases in humans that have received attention due to the increased incidence of infections present in patients with AIDS.

Clinical syndromes in immunocompetent patients consist of self-limited diarrhea, ocular infections (keratitis), and meningoencephalitis. In immunocompromised patients (especially with AIDS), persistent watery diarrhea is the most common presentation. Severe CNS infections and disseminated infections have also been described including hepatitis, peritonitis, nephritis, and pneumonitis.

OTHER EMERGING BACTERIAL PATHOGENS

Diphtheria

Disease caused by infection with *Corynebacterium diphtheriae*, a gram positive, pleomorphic bacterium that was kept under good control in most parts of the world until 1990, re-emerged in the form of large epidemics in the former Soviet Union, extending into parts of Eastern Europe and Asia. The epidemic peaked in 1995 when 50,000 cases were reported in the Russian Federation. Vaccination lapses and disarray of the public health infrastructure are to blame for its reappearance. Humans are the only known reservoir of *C. diphtheriae*, and the primary route of spread is via respiratory droplets. Skin infection is less common. Asymptomatic pharyngeal carriers play an important role in maintaining the organisms in communities. The disease is mediated by a toxin encoded by a lysogenic phage. The toxin inhibits elongation factor 2 (EF2) in the ribosomes, thereby impairing protein synthesis. The main target organs are the nerves, heart, and kidneys. Before diphtheria was controlled in the mid-20th century, mainly infants were affected. However, in recent outbreaks young adults are more frequently affected, suggesting that infants in such populations are still somewhat protected.

Bordetella Infections

The main human pathogens in this group are *B. pertussis* (agent of whooping cough) and *B. parapertussis* (the agent of a milder form of whooping cough). Pertussis continues to be a public health problem worldwide with millions of cases reported every year. In the United States approximately 8000 cases occurred in 2002, affecting mostly adolescents and young adults. The disease seems to be largely underdiagnosed in these population groups, who are more susceptible than children because of lack of booster immunization for pertussis beyond childhood. Pertussis starts as an upper respiratory tract infection that progresses in a few days to paroxysmal cough that lasts for weeks or months. The bacterium is not invasive and remains in the respiratory tract. Secondary infections in the form of bacterial pneumonia can occur. Other complications include pneumothoraces, hernias, seizures, and otitis media.

Staphylococcus aureus Infections

This gram positive, aerobic coccus is one of the most ubiquitous microorganisms in human populations and is responsible for a wide spectrum of human diseases that range from simple skin "boils" to life-threatening generalized infections. In the last few decades, new clinical entities associated with *S. aureus* have emerged such as staphylococcal toxic shock syndrome (STSS), which is a potentially lethal disease mediated by a "superantigen" toxin (TSST-1) and is characterized by fever, hypotension, multisystem organ failure, and an erythematous rash that desquamates during recovery. TSST-1 is a nonspecific T-cell mitogen that leads to a dramatic increase of T-cells and subsequent increase in circulating cytokines. This syndrome was initially described in the setting of vaginal colonization by toxin-producing strains and use of superabsorbent tampons in 1978. This type of tampon was withdrawn from the market, and the incidence declined. However, nonmenstrual forms do exist and are associated with nasal packing, surgical wounds, and other focal *S. aureus* infections. A related condition present in neonates has been described recently in Japan and is known as neonatal toxic shock syndrome-like exanthematous disease.

Antibiotic resistance is a major problem in nosocomial infections due to *S. aureus*. Both methicillin-resistant and glycopeptide-resistant strains are a major threat in the hospital setting. *S. aureus* may be part of the normal flora, especially of the nares and skin. Other sites include the genitourinary tract, mucous membranes, GI tract, and upper respiratory tract. In fact, most infections are considered to be endogenous in origin.

Coagulase negative *Staphylococci* have also emerged as human pathogens in the setting of indwelling catheters and bioprosthetic devices. These bacteria are capable of forming extensive biofilms on inert surfaces leading to local or systemic infections.

GROUP A β-HEMOLYTIC STREPTOCOCCAL (GABHS) INFECTIONS

Streptococcus pyogenes Infections

This complex group of bacterial pathogens is responsible for a wide spectrum of infections in humans ranging from focal infections such as acute tonsillitis to the highly lethal streptococcal necrotizing fasciitis (better known as flesh-eating disease). The classification of *S. pyogenes* is based on antigenic properties and presence or absence of hemolysis on blood agar plates.

Classical diseases caused by this group of streptococci include its nonsuppurative complications (acute rheumatic fever and glomerulonephritis), tonsillitis, and scarlet fever. During the last two decades, GABHS have been associated with severe necrotizing fasciitis or streptococcal gangrene and streptococcal toxic shock syndrome. These conditions are associated with highly virulent streptococci capable of producing toxin superantigens (streptococcal pyrogenic exotoxins, SPE-A, SPE-B, and SPE-C) and other virulence factors capable of destroying host tissues. Fortunately, antibiotic resistance is not a major concern in infections caused by GABHS since the strains are still susceptible to most penicillin-related antibiotics. However, resistance to clindamycin and macrolides has been documented. Similar to *Staphylococci*, these organisms are part of the normal flora of multiple body sites including the skin, vagina, and pharynx.

Streptococcus agalactiae Infections

Group B streptococcal (*S. agalactiae*) infections have also emerged as a major cause of neonatal sepsis related to carriage of the bacterium in the vagina of pregnant women. Other conditions include focal and generalized infections in nonpregnant adults. Underlying conditions usually present in these patients are diabetes mellitus, neoplastic diseases, immobility, and chronic liver disease.

EMERGING CHRONIC VIRAL DISEASES

Acquired Immunodeficiency Syndrome

A pathogenic retrovirus infecting one of the most important cells orchestrating the acquired immune response in humans was something no scientist had in mind at the beginning of the 1980s. We now know that human immunodeficiency virus (HIV) infections were probably occurring sporadically for decades before the initial cases of HIV infection/AIDS were documented in the early 1980s. Unfortunately, HIV infection turns out to be one of the worst epidemics in modern history with more than 42 million people currently infected, most of whom live in sub-Saharan Africa (cumulative number of cases since the epidemic began is approximately 70 million). AIDS has already claimed at least 25 million deaths around the world, and although major therapeutic advances have been made, the possibility of a vaccine or a cure remains elusive.

An evolving problem within the HIV emergence as a pathogen is viral resistance to antiretroviral drug therapy. Due to the high rates of recombination and mutation in HIV genes, drug resistance is very common in patients receiving only one anti-HIV medication. Therefore, treatment regimens usually employ multiple medications. Even so, development of resistance still occurs. Rates of resistant infections vary from 3 to 4% in Australia to 23% in some areas of the United States.

HIV infection is transmitted sexually, through exposure to infected blood (intravenous drug use and accidental exposures) or vertically from mother to fetus. Virus gains access to target cells via the interaction of viral glycoprotein gp120 and the cellular receptor CD4 and chemokine receptor CCR5, both present in macrophages and CD4+ T-cells. During the acute infection some patients develop a flu-like illness that is in some cases accompanied by a maculopapular rash. Viremia is then reduced dramatically after a few weeks by both neutralizing antibody and cytotoxic T-cells. A variable latency period then ensues and ranges from years to decades during which there is progressive deterioration of the immune system due to declining levels of CD4+ T-cells. As the population of CD4+ T-cells declines, the occurrence of opportunistic infections and certain neoplasm increases, marking the appearance of AIDS per se.

HIV-2

This virus entered human populations more recently than HIV-1 and came directly most likely from sooty mangabeys. The virus is most prevalent in west Africa. Transmission from such animals probably occurs through bites and scratches or contact with infected tissues by hunters. Viremic levels in patients with HIV-2 infections are lower than HIV-1 patients, and the proportion of nonprogressers is higher.

HTLV-I and II

HTLV-I was isolated from a patient with a T-cell leukemia/lymphoma and was the first human retrovirus that was associated with human neoplasia. A few years later, HTLV-II was isolated from a patient with hairy cell leukemia, an association that has not been observed subsequently. HTLV-I is endemic in Japan (Okinawa), East Asia, Papua New Guinea, and the Caribbean basin. Routes of infection include sexual intercourse, vertical transmission from mothers to offspring, blood transfusions, and use of contaminated needles. Typically, patients with adult T-cell lymphoma/leukemia have hypercalcemia. Tropical spastic paraparesis (Caribbean) and HTLV-associated myelopathy (Japan) are associated with both HTLV-I and HTLV-II.

Human Herpesvirus (HHV) Infections

These agents are large, enveloped, double-stranded DNA viruses that infect a wide variety of species. So far, eight herpesviruses have been associated with human illness, namely herpes simplex viruses I and II (HHV1 and 2, respectively), varicella-zoster virus (HHV3), Epstein–Barr virus (EBV or HHV4), cytomegalovirus (CMV or HHV5), human herpesvirus 6 (HHV6), human herpesvirus 7 (HHV7), and Kaposi sarcoma virus (HHV8). The common denominator for all human herpesvirus infections is lifelong latent infection. Clinical syndromes are varied. The main pathogenic mechanisms include direct cytotoxicity, immunopathologic responses, and neoplastic transformation of host cells (HHV4 and HHV8).

Hepatotropic Viruses

Well-defined viral pathogens causing acute and chronic hepatitis belong to different families including hepatitis A (Picornaviridae), hepatitis B (Hepadnaviridae), hepatitis C (Flaviviridae), and hepatitis E (Hepeviridae).

Hepatitis A is transmitted by the fecal–oral route and is a common cause of acute hepatitis in developing countries. Chronicity does not occur, and most patients recover from acute infections. In fact, the majority of patients are anicteric during the acute infection. An effective parenteral vaccine is currently available.

Hepatitis B virus currently infects more than 500 million people worldwide with the highest incidence in the Far East. Routes of transmission are varied and include vertical (prenatal and perinatal) and horizontal (unprotected sex, intravenous drug use, exposure to contaminated fluids and blood products). Infection in humans ranges from asymptomatic carriers to late complications including cirrhosis and hepatocellular carcinoma. In between, patients can develop acute and chronic hepatitis. Some patients develop extrahepatic manifestations, which are immune-mediated including glomerulonephritic syndromes, vasculitides, and polyarthritis. An effective vaccine is available and has controlled the incidence and prevalence of hepatitis B infections in the industrialized world.

Hepatitis delta virus (HDV) is a satellite virus known as delta (δ) hepatitis virus that requires hepatitis B virus to complete its life cycle in the eukaryotic host cell. HDV replication needs HBsAg for packaging of the HDV genome. The exact classification of the virus remains elusive, but it is related to viroids, plant satellite viruses, and simple infectious RNA molecules. The current classification is that of Deltaviridae family. Infection with HDV occurs either as a co-infection with HBV or a superinfection in a patient already infected with HBV. When both viruses are present, the risk of severe disease is fourfold when compared to patients infected with HBV alone. The spectrum of disease ranges from asymptomatic to end-stage liver disease (cirrhosis). In addition, superinfected patients are at higher risk for developing fulminant acute hepatitis.

Hepatitis C virus is an enveloped, positive-stranded RNA virus that has become one of the leading causes of chronic hepatitis, cirrhosis and hepatocellular carcinoma in the world. The virus has recently been cultivated in cell lines in vitro, and animal models of infection are available. A sensitive diagnostic test based on antibody detection in serum and RT-PCR detection of the viral RNA revolutionized diagnosis of hepatitis C infection during the mid-to late-1990s. Most cases develop chronicity leading in many patients to subsequent fibrosis, cirrhosis, and an increased risk of hepatocellular carcinoma. However, even among patients with chronic hepatitis C viral infections, only 5–20% develop cirrhosis. Clearly, several host and viral factors are in play. Consumption of alcohol is a synergistic factor in developing complications such as cirrhosis. Other factors include viral genotype, HIV co-infection and HLA phenotypes. Transmission is through percutaneous exposure and transfusion of contaminated blood products, although the latter is rare in countries where blood products are tightly regulated. Sexual and vertical transmission is rare. Current estimates suggest that the virus has infected 170 million people worldwide and close to 4 million persons in the US.

Hepatitis E virus is responsible for enteric non-A, non-B hepatitis epidemics in developing countries (especially in Southeast and Central Asia, Middle East, and North Africa). Outbreaks have also been described in Mexico. However, infections are mostly endemic. The disease is transmitted through the fecal–oral route, and the disease is a self-limited acute hepatitis with no known chronicity. In pregnant women, the incidence of fulminant hepatitis is high. HEV is a non-enveloped, single-stranded, positive-sense RNA virus, formerly known as enterically transmitted non-A, non-B hepatitis virus. The viral genome was cloned in 1990. Morphologically, the closest related viruses are those of the family Caliciviridae. However, the genomic organization is closer to the rubella virus which is in the family Togaviridae genus *Alphavirus*.

BIOTERRORISM AS A MECHANISM OF EMERGENCE OF INFECTIOUS DISEASE

For most potential bioterror agents such as *Bacillus anthracis*, humans are a dead-end host. However, some viruses and bacteria that are considered in the high priority categories for use in bioterrorism have the capability for establishment in the human population via person-to-person transmission or in zoonotic cycles. These diseases could spread widely geographically and affect many persons, creating public terror as the perpetrators intended. Although smallpox was eradicated nearly 30 years ago, Soviet scientists in the mammoth Biopreparat program weaponized variola virus and produced tons of the agent. If even a single person were infected with smallpox by a terrorist, there would be the potential for widespread transmission around the world where vaccination ceased decades ago. The outcome of re-establishment of a viral agent that is transmitted from person-to-person by aerosol and small droplet spread is evident in the reintroduction of influenzavirus A H1N1, the origin of which some sources attribute to release from a Russian laboratory. This virus that had ceased to circulate after the H2N2 influenzavirus A pandemic in the late 1950s is now co-circulating with the 1968 pandemic H3N2 influenzavirus A. Intentional release of agents capable of person-to-person transmission could result in their emergence or re-emergence.

The potential for establishment of an agent introduced into widely geographically dispersed zoonotic cycles from which transmission to humans occurs is exemplified by the events after the appearance of WNV in New York City in 1999. Although there is no evidence that WNV was introduced into the US by a terrorist, it could have been done, and there are other potential bioterror agents that could be brought to the US and become established in expanding zoonotic cycles. These include other mosquito-borne viruses such as RVF virus and Venezuelan equine encephalitis virus. These would threaten both domestic animals and humans. Other agents of bioterror such as foot and mouth disease virus or *Ehrlichia ruminantium* could target our food animals and ruin our livestock industry. Food and mouth disease and heartwater do not occur in the US, and their establishment would be catastrophic, particularly if wild ruminants became reservoirs of infection.

The recent occurrence of global SARS and domestic monkeypox epidemics demonstrate the capacity of EIDs to be used in the future for bioterrorism. These viruses exist in natural cycles in Southeast Asia and Central and West Africa, respectively. SARS-coronavirus is a particularly serious threat because of person-to-person spread and significant mortality. Strains of monkeypox virus from Central Africa also cause significant mortality, and the danger of its intentional establishment in a wildlife reservoir such as prairie dogs would create prolonged terror. Thus, EID agents could be used for bioterrorism, and agents of bioterror could become emerging infections.

RATIONALE FOR VACCINES AGAINST EMERGING INFECTIOUS DISEASES

The rationale for not developing vaccines against EIDs and not developing platforms that can be rapidly applied to newly emerging infections is that apparently small problems will not grow and that what we do not know will not hurt us. The fallacies of these concepts are quite evident in the ongoing tragedy of HIV/AIDS.

Vaccines are the province of industry. The sole rationale for commercial vaccine development is profit calculated as return on investment to stockholders. A particular vaccine would have to be more profitable than a particular drug in order to have resources for development and manufacture devoted to it. Successful vaccines from the perspective of preventive medicine are given only once and induce lifelong protection, translating as a single sale per customer. Drugs for prevalent chronic diseases require enormous numbers of doses for decades. The money and effort invested in development of a vaccine that could have produced a blockbuster drug represent lost opportunity cost to the pharmaceutical enterprise. The effects of not developing more effective and safer vaccines against whooping cough and tuberculosis are measured in days of active life lost, medical costs, and deaths that are borne by society, most often in developing countries. The same metrics apply to DF, DHF, Lassa fever, louse-borne typhus, scrub typhus, RVF, Venezuelan equine encephalitis, Ebola HF, Marburg HF, malaria, plague, and leptospirosis.

The development of vaccines against newly emerging infectious agents demands initiation of the project at the time of the first appearance of the emerging infection, well before the problem has reached significant public health attention. It would be logical to establish research programs to elucidate the principles of stimulating protective immunity against each family of pathogens. The goal would be to develop a prototype vaccine before the disease became a public health catastrophe. Ideally platforms for vaccines would be developed that stimulated memory cells to produce each of the components of immunity that are critical to protection against the particular agent, e.g., Th_1 antibodies or CD8 cytotoxic T lymphocytes. The new infectious agent could be characterized rapidly and the components predicted to be protective used with the appropriate vaccine platform. A disease with high lethality, rapid efficient transmission, and no treatment might justify manufacture and deployment of the vaccine with field use evaluation of safety and efficacy. Of course, preclinical animal testing for safety and efficacy and accelerated phase 1 and 2 trials would be preferred.

The principles that would determine the priority of development of vaccines would more logically be driven by public health concerns than merely financial return on investment. At the top of the list of criteria would be the severity of illness and incidence or threatened increase in incidence of the disease. Consideration of the severity of illness would include the ratio of symptomatic to asymptomatic infections, case-fatality ratio, incidence and nature of permanent sequelae, duration and medical costs of the illness in survivors, and other societal and economic impacts of the disease. Some diseases that would merit consideration for vaccine development on the criteria of disease severity and incidence include infections that currently occur nearly exclusively in developing countries such as Lassa fever, RVF, Ebola HF, louse-borne typhus, scrub typhus, leptospirosis, and Venezuelan equine encephalitis.

Other pathogens for which effective vaccine development effort should be undertaken are agents for which antibiotic resistance is a significant threat, potentially water-borne agents that are resistant to control by chlorination, and agents of veterinary diseases that would be potential agents of devastating agroterrorism. The rising problem of bacterial antimicrobial resistance is evident to all, particularly for nosocomial infection. What is generally ignored is the creation of weapons for biowarfare or bioterrorism that are resistant to antimicrobial treatment. For example, the confidence that rickettsial diseases respond well to early treatment with doxycycline should be reconsidered in light of the development of tetracycline-resistant *Rickettsia prowazekii* in a Russian laboratory. Similarly contamination of municipal water supplies with chlorine-resistant organisms such as *Cryptosporidium parvum* would cause severe havoc. After several widespread outbreaks, the public would be receptive to prevention by an effective vaccine.

Development of veterinary vaccines is an important goal on its own merit. The lessons that would be learned by the development of vaccines against foot and mouth disease virus, *E. ruminantium*, and *Burkholderia mallei* would have additional impact on human diseases caused by related agents. Other veterinary pathogens such as Venezuelan equine encephalitis virus, *Brucella melitensis*, *Brucella abortus*, *Brucella suis*, and RVF virus are also human disease agents.

Respiratory diseases deserve special consideration for preparation for prevention by development of vaccines. The potential reappearance of SARS-coronavirus should energize efforts to produce an effective vaccine. The threat of pandemic influenza A combines nearly every reason that could be marshaled for all agents together to create a system for rapid vaccine formulation, testing, and manufacture.

THE CHALLENGE OF DEVELOPING VACCINES AGAINST EMERGING AND RE-EMERGING INFECTIOUS DISEASES

The development of a new vaccine is not easy by any means. Even when the pharmaceutical industry recognizes the market potential of a vaccine against an emerging pathogen, such as HIV or hepatitis C virus, there are many difficult steps between the desire to develop the vaccine and its availability for purchase. There are more steps from the discovery of a newly EID agent to the availability of a vaccine against it than for a pathogen that has been studied for decades or more than a century. First, the pathogen must be characterized at the levels of biology, molecular composition, and its genome. Even well into the contemporary molecular, proteomic, and genomic era, knowledge of the proteome and genetic sequence provides no more complete understanding of the agent than possession of an encyclopedia, which written in a language that has been translated only partly, would impart omniscience. Second, appropriate animal models that adequately mimic the pathology of the human disease, anatomic distribution of the agent, and the human immune response to the infection must be developed. There are numerous agents that are so highly adapted to humans that no adequate animal model exists. Other models utilize animal species that do not permit effective analysis of the protective immune response owing to lack of reagents to evaluate the immune cells and proteins and absence of animals that are genetically inbred or have knockout of specific genes. Third, studies of animal models and of humans and their cells, cytokines, and antibodies are employed to elucidate the mechanisms of protective immunity. Development of a vaccine that stimulates the presence of memory cells that generate the immune components required to kill the invading pathogen rapidly requires determining which elements (e.g., antibodies, CD4, and/or CD8 T lymphocytes) are critical. Fourth, the antigens of the particular infectious agent that stimulate the protective memory immune components must be identified. For subunit vaccines, some antigens stimulate protective responses, and others suppress protective immunity. Fifth, the animal models are utilized to develop and test vaccine candidates that stimulate protective memory immune responses without harmful side effects. At the point of the primary identification of the etiology of the emerging infection, it would be unlikely that any of these five steps would have been addressed.

Because many emerging infections are zoonotic diseases that have naturally occurring animal counterparts, a strategy of developing related veterinary vaccines on the pathway toward future development of a vaccine for humans may be feasible owing to financial and regulatory considerations. Imagine how much better off we would be today if an effective vaccine against a simian immunodeficiency virus had been produced in the 1970s or a changeable molecular cassette subunit vaccine had been approved for avian influenza viruses a decade ago. The crucial message regarding vaccines for emerging infections diseases is that it is never too early to support research to achieve the five steps outlined above.

Suggested Reading

1. Burroughs, T., Knobler, S. and Lederberg, J. The Emergence of Zoonotic Diseases: Understanding the Impact on Animal and Human Health. The National Academies Press, Washington, DC, 2002.
2. Conway, D.J. and Roper, C. Microevolution and emergence of pathogens. Int. J. Parasitol. 2000; 30:1423–1430.
3. Knobler, S., Mahmoud, A., Lemon, S. and Pray, L. The impact of globalization on infectious disease emergence and control: exploring the consequences and opportunities. The National Academies Press, Washington, DC, 2006.
4. Lederberg, J. Infectious disease as an evolutionary paradigm. Emerg. Infect. Dis. 1997; 3:417–423.
5. Lederberg, J. and Davis, J.R. Emerging infectious diseases from the global to the local perspective. The National Academies Press, Washington, DC, 2001.
6. Lederberg, J., Shope, R.E. and Oaks, S.C. Emerging Infections: Microbial Threats to Health in the United States. The National Academies Press, Washington, DC, 1992.

7. Morens, D.M., Folkers, G.K. and Fauci, A.S. The challenge of emerging and re-emerging infectious diseases. Nature 2004; 430:242–249.
8. Scheld, W.M., Armstrong, D. and Hughes, J.M. Emerging Infections 1. Washington, DC: ASM Press, 1997.
9. Scheld, W.M., Craig, W.A. and Hughes, J.M. Emerging Infections 2. Washington, DC: ASM Press, 1998.
10. Scheld, W.M., Craig, W.A. and Hughes, J.M. Emerging Infections 3. Washington, DC: ASM Press, 1999.
11. Scheld, W.M., Craig, W.A. and Hughes, J.M. Emerging Infections 4. Washington, DC: ASM Press, 2000.
12. Scheld, W.M., Craig, W.A. and Hughes, J.M. Emerging Infections 5. Washington, DC: ASM Press, 2001.
13. Scheld, W.M., Hughes, J.M. and Murray, B.E. Emerging Infections 6. Washington, DC: ASM Press, 2004.
14. Scheld, W.M., Hooper, D.C. and Hughes, J.M. Emerging Infections 7. Washington, DC: ASM Press, 2006.
15. Steele, J.C.H. Emerging infections and their causative agents. Clin. Lab. Med. 2004; 24:xi–xiii.
16. Sigurdsson, B. and Rida, A. Chronic Encephalitis of Sheep: With General Remarks on Infections which Develop Slowly and Some of Their Speicial Characteristics. British Veterinary Journal. 1954; 110:341–354.

CHAPTER

2

Bacterial Biothreat Agents

Richard W. Titball

School of Biosciences, Geoffrey Pope Building, University of Exeter, Exeter, Devon, UK

OUTLINE

ABSTRACT

A range of bacterial pathogens might be used illegitimately to cause disease in humans or animals. The properties of these pathogens are dissimilar. Much of our information on the properties of these pathogens comes from past programs to develop biological weapons. In principle, disease caused by bacterial biothreat agents could be prevented or treated using antibiotics. However, there are limitations associated with the use of antibiotics in this way. Against this background vaccines will play a key role in protecting vulnerable populations from bacterial biothreats.

INTRODUCTION

One simple definition of a bacterial biothreat agent is a bacterium that is used to deliberately cause disease in human, animal, or plant populations to the economic or social detriment of the host nation. Based on this type of definition there are a number of published lists of bacterial biothreat agents, ranging from those that inform export controls, such as the Australia group list, to those compiled to protect the U.S. Agricultural industry from disruption (Table 2.1). The differences in threat agents of concern reflect the difficulties in devising a single definitive list of biothreat agents. For example, agents that are likely to cause disruption of the agricultural industry are not necessarily those which would be used against a human population. Similarly, agents which would be used against civilian populations would not necessarily be used to attack the armed forces.

Despite these differences, there are some agents that consistently appear on all of the lists. The observation

TABLE 2.1 Bacterial pathogens listed as possible biowarfare or bioterrorism agents

List	*Bacillus anthracis*	*Bartonella quintana*	*Brucella abortus*	*Brucella melitensis*	*Brucella suis*	*Burkholderia mallei*	*Burkholderia pseudomallei*	*Chlamydia psittaci*	*Clostridium botulinum*	*Clostridium perfringens*[a]	*Clostridium tetani*	*Coxiella burnetii*
Australia group list[c]	Yes	Yes	Yes	Yes	Yes	Yes	Yes	Yes	Yes	Yes	Yes[d]	Yes
HHS[f]	Yes	No	Yes	Yes	Yes	Yes	Yes	No	Yes[e]	No	No	Yes
USDA[g]	Yes	No	Yes	Yes	Yes	Yes	Yes	No	Yes[e]	No	No	Yes
WHO	Yes	No	Yes	Yes	Yes	Yes	Yes	No	No	No	No	Yes
NATO	Yes	No	Yes	Yes	Yes	No	Yes	Yes	No	No	No	Yes

List	EHEC[b]	*Francisella tularensis*	*Legionella pneumophila*	*Orienta tsutsugamushi*	*Rickettsia prowazekii*	*Rickettsia rickettsii*	*Salmonella enterica* serovar Typhi	*Shigella dysenteriae*	*Vibrio cholerae*	*Yersinia pestis*	*Yersinia pseudotuberculosis*
Australia group list[c]	Yes	Yes	Yes[d]	No	Yes	Yes	Yes	Yes	Yes	Yes	Yes[d]
HHS[f]	No	Yes	No	No	Yes	Yes	No	No	No	Yes	No
USDA[g]	No	Yes	No	No	No	No	No	No	No	No	No
WHO	No	Yes	No	No	Yes	Yes	Yes	No	Yes	Yes	No
NATO	No	Yes	No	Yes	Yes	Yes	Yes	Yes	Yes	Yes	No

Note: EHEC, enterohemorrhagic *Escherichia coli*. *Source*: Anon (1970, 1996).
[a] Epsilon-toxin producing types.
[b] Enterohemorrhagic *Escherichia coli*, serotype O157 and other verotoxin-producing serotypes.
[c] List of agents subject to export control, 2005; http://www.australiagroup.net/en/control_list/bio_agents.htm.
[d] Included on the warning list.
[e] Neurotoxin-producing strains.
[f] Department of Health and Human Sciences, National Institute for Health, USA, 2005.
[g] U.S. Department of Agriculture, 2005.

that these agents are of concern to such a wide range of authorities clearly suggests that these pathogens would be preferred for vaccine development and licensing.

This chapter highlights the ways in which bacteria might be used to deliberately cause disease and reviews the utility of vaccines for the prevention of disease. Finally, the prospects for improved vaccines against bacterial biothreat agents are considered.

PAST USES OF BACTERIAL PATHOGENS TO DELIBERATELY CAUSE DISEASE

There are a number of reported examples of the use of bacteria to deliberately cause disease in humans or animals, although most of these involve small-scale and relatively localized incidents. A review of 270 cases of alleged bioterrorism until 2001 has been compiled by Carus (2001). In addition to these incidents, there are a number of proven or suspected uses of biological agents during warfare, either against civilian or military targets.

Arguably, the program that resulted in the greatest number of human casualties involved the Japanese Army (Unit 731) during the 1930s and 1940s and was the creation of the bacteriologist General Ishii Shiro (Mangold and Goldberg, 1999; Guillemin, 2004). It is reported that human trials were carried out with a wide range of potential biothreat agents (Table 2.2). Many of the casualties occurred after prisoners were deliberately exposed to pathogens during trials of biological weapons. For example, one trial used 10 Chinese prisoners who were exposed to a shrapnel bomb containing *C. perfringens* (Mangold and Goldberg, 1999). Reportedly, all of the victims died of gas gangrene within a week (Mangold and Goldberg, 1999). Other trials involved fleas infected with *Y. pestis* which were dropped from aircraft onto Chinese villages. One attack, on the hamlet of Congshan, reportedly resulted in an outbreak of plague that killed one quarter of the population (Mangold and Goldberg, 1999). In spite of this significant military program, which continued for a decade, several reports indicate that the Japanese program failed to devise a range of effective biological weapons which could be used for large-scale attacks on military or civilian populations (Mangold and Goldberg, 1999; Guillemin, 2004). In part, this reflects the difficulty of moving from the observation that pathogens cause disease to the large-scale production and successful weaponization of these bacteria.

B. anthracis is one of the bacterial pathogens that is frequently cited as a likely biological warfare or bioterrorism agent, and there is a significant body of information supporting this suggestion. In the past both the United Kingdom and the United States developed weapons containing this bacterium. In trials on Gruinard Island off the west coast of Scotland during the 1940s *B. anthracis* spores were tested as means of inducing pulmonary anthrax in sheep (Manchee et al., 1981, 1982). These trials demonstrated the feasibility of such a weapon but left the island contaminated. Subsequently, a major exercise during the 1980s resulted in the decontamination of the island (Manchee et al., 1983) which was then returned to grazing use (Aldhous, 1990). The programs to develop biological weapons in the United Kingdom and the United States ceased in the mid-1950s and late 1970s, respectively. However, there is evidence that other nations continued to develop anthrax as a biological weapon (Inglesby et al., 1999). An outbreak of human anthrax in Sverdlosk in the former Soviet Union was originally attributed to the ingestion of anthrax-contaminated meat (Sternbach, 2003; Meselson et al., 1994). However, subsequent investigations by a team of U.S. scientists revealed the cause to be the accidental airborne release of anthrax spores from a military facility (Meselson et al., 1994). This resulted in the largest outbreak of human inhalation anthrax documented with at least 79 cases of anthrax and 68 deaths (Inglesby et al.,

TABLE 2.2 Examples of the alleged use of biological agents to deliberately cause disease

Agent	Year	Location	References
B. anthracis	2001	USA	Jernigan et al. (2002)
B. anthracis	1990–1991	Tokyo, Japan	Kaplan and Marshall (1996)
S. dysenteriae	1996	Texas, USA	Kolavic et al. (1997)
S. dysenteriae	1964	Chiba, Japan	Carus (2001)
Salmonella enteritidis	1984	Oregon, USA	Carus (2001); Torok et al. (1997)
B. anthracis, Y. pestis, S. enterica serovar Typhi, *Salmonella paratyphi, R. prowazeki, F. tularensis, C. perfringens, C. tetani, V. cholerae, S. dysenteriae*	1931–1945	Manchuria (occupied by Japan)	Mangold and Goldberg (1999), Guillemin (2004)

1999). Additional cases of anthrax occurred in livestock outside of the city limits (Meselson et al., 1994).

B. anthracis has also attracted the attention of terrorist groups. During the early 1990s the Aum Shinrikyo group attempted to release significant quantities of *B. anthracis* spores over Tokyo. The failure of these attacks to cause disease has been attributed both to failures of the dissemination system and the mistaken use of the attenuated vaccine strain of *B. anthracis* (Kaplan and Marshall, 1996; Carus, 2001). However, more recently, *B. anthracis* has been used in the United States to cause a number of cases of disease. During 2001, *B. anthracis* spores were released into the U.S. Mail system resulting in 22 cases of anthrax (Jernigan et al., 2002). Of the 11 cases of inhalation, 5 were fatal. The high fatality rate for these inhalational cases of disease reflects the difficulties of treating symptomatic inhalation anthrax (Inglesby et al., 1999).

Other reported examples of the use of bacteria to deliberately cause disease involve minor events. For example, 12 cases of shigellosis occurred in laboratory staff working at St Paul's Medical Center Hospital in Dallas, Texas (Kolavic et al., 1997). Reportedly these were the result of the deliberate contamination of pastries with *S. dysenteriae* type 2 by one of the laboratory staff. These pastries were offered to other laboratory workers and 11 consumed them between 07.15 a.m. and 13.30 p.m. One victim became ill after consuming a contaminated pastry that was taken home by one of the laboratory staff. All of the individuals developed shigellosis within three days of consuming the pastries. Subsequent investigations showed that the organism was stored in the laboratory on coated beads within a vial, and that some of the beads were missing. The perpetrator of this crime was identified as one of the laboratory staff and subsequently sentenced to 20 years in prison. This crime is very similar to the so-called "sponge cake incident" which was reportedly carried out in 1964 in Japan, when a physician contaminated a cake with *S. dysenteriae* and offered it to four colleagues who became ill and required hospitalization (Carus, 2001). The perpetrator of this crime was eventually charged with 13 incidents of deliberate infection of a total of 66 individuals with *S. dysenteriae* or *S. enterica* serovar Typhi.

The examples cited above illustrate the ways in which bacterial pathogens might be used to cause disease in humans. However, the possibility that pathogens might be used to cause disease in animals should not be overlooked. For example, during World War II the United Kingdom produced cattle cakes containing *B. anthracis*, as a potential weapon against German domesticated livestock. This weapon was never used; however, there is evidence that during World War I *B. mallei* (formerly *Pseudomonas mallei*) was used to cause disease in horses that were essential to the Allied supply chains (Wheelis, 1999; Geissler et al., 1999). Although it is documented that outbreaks of disease did occur in horses, the precise incidence of disease caused by the deliberate release of agent is not known (Wheelis, 1999).

POTENTIAL FOR AIRBORNE BACTERIAL PATHOGENS TO CAUSE DISEASE

Most of the examples of the illegitimate use of bacteria to cause disease involve the relatively small-scale release or use of bacterial pathogens to cause disease. However, the scenario that is of greatest concern involves the airborne release of a bacterial agent over an area of high population density.

There are little data on the likely incidence of disease following the airborne dissemination of a pathogen. However, a report from the WHO during 1970 (Anon, 1970) made an attempt at predicting the numbers of casualties caused by the release of 50 kg of dried biowarfare agent released as a line source upwind of the population. The study considered six different targets. Three typified economically developed cities with populations of 500,000; 1,000,000; and 5,000,000. Three typified cities with similar populations but in economically developing areas of the world.

The estimation by the WHO made several assumptions. With the exception of the spores of *B. anthracis*, bacterial biological agents are generally unstable in the airborne state and especially when exposed to UV irradiation. The attacks were therefore assumed to be carried out at night with dried bacteria mixed with stabilizing compounds to limit the extent of airborne decay. It was assumed that the agent was dispersed from an aircraft at a height of 10 m above ground and over a distance of 2 km upwind of the target site. The wind speed was assumed to be 2.5 m/s. Under these conditions, the number of casualties was related to the infectious dose of the pathogen and the stability of the bacterium in the airborne state. Therefore, agents that are aerosol stable and have a low infectious dose caused the greatest numbers of casualties (Fig. 2.1). However, caution should be exercised when viewing these estimates of casualties—changes to any of the variables outlined above, such as height of release, time of release, wind speed, or the properties of any stabilizing agent could impact significantly on the number of predicted casualties.

Notwithstanding these concerns, and the issues highlighted above concerning the technical challenges

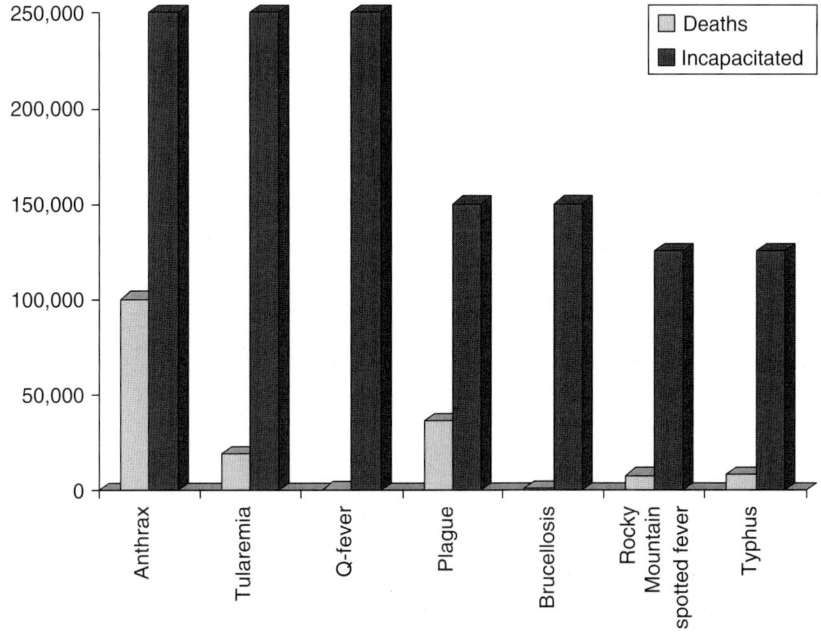

FIGURE 2.1 Predicted numbers of casualties following the airborne release of selected bacterial pathogens over an area of high population density (5 million) in an economically developed country.

associated with mounting such a large-scale attack, the exercise carried out by the WHO does illustrate the different properties of different bacterial agents. For example, not all of these pathogens would cause significant numbers of deaths. Some like *Brucella* sp. causes severely debilitating diseases, which would likely place significant requirements on the healthcare systems. Some agents cause diseases that are transmissible from person to person (for example, plague), meaning that there would be a need to restrict the movement of individuals within the area covered by the agent. Others like *B. anthracis* spores would leave the environment contaminated for long periods of time, requiring costly and technically challenging decontamination processes.

VACCINES AND ANTIBIOTICS FOR THE PREVENTION AND TREATMENT OF DISEASE

Both antibiotics and vaccines have been proposed to have a role to play in the prevention or treatment of disease caused by bacterial biothreat agents. In principle, antibiotics might be used either prophylactically or therapeutically but surprisingly their effectiveness in either scenario is not certain. The use of antibiotics for the treatment of disease is especially problematic because of the acute nature of many of the infections which occur after the inhalation of bacteria (Inglesby et al., 1999, 2000; Dennis et al., 2001). By the time most individuals are symptomatic the effective treatment of disease is difficult, and for diseases with a high mortality rate, many of these symptomatic individuals will die even with aggressive antibiotic therapy and other interventions. The problems of treating inhalational anthrax with antibiotics were clearly highlighted during the 2001 mail-associated outbreak of disease. Of the 11 cases of inhalational disease, 5 died. In contrast, there were no deaths in the 11 individuals who contracted cutaneous anthrax (Jernigan et al., 2002).

The prophylactic use of antibiotics also presents some significant problems, which were again highlighted during the mail-associated outbreak of anthrax in the United States. It is estimated that 32,000 individuals initiated antibiotic prophylaxis (Jernigan et al., 2002). However, the numbers of individuals who completed the recommended 60-day course was relatively low. In some cases termination of the course was the appropriate course of action when the full extent of the risk to the individual was quantified (Jernigan et al., 2002). An alarming number of individuals failed to complete the course, even though they were considered to be at a significant risk of contracting the disease (Stein et al., 2004). In a more detailed study, Shepard et al., (2002) found that the proportion of individuals who completed a 60-day course of antimicrobial prophylaxis ranges from 21 to 64% depending on the site (Fig. 2.2). Adverse events were the most frequently

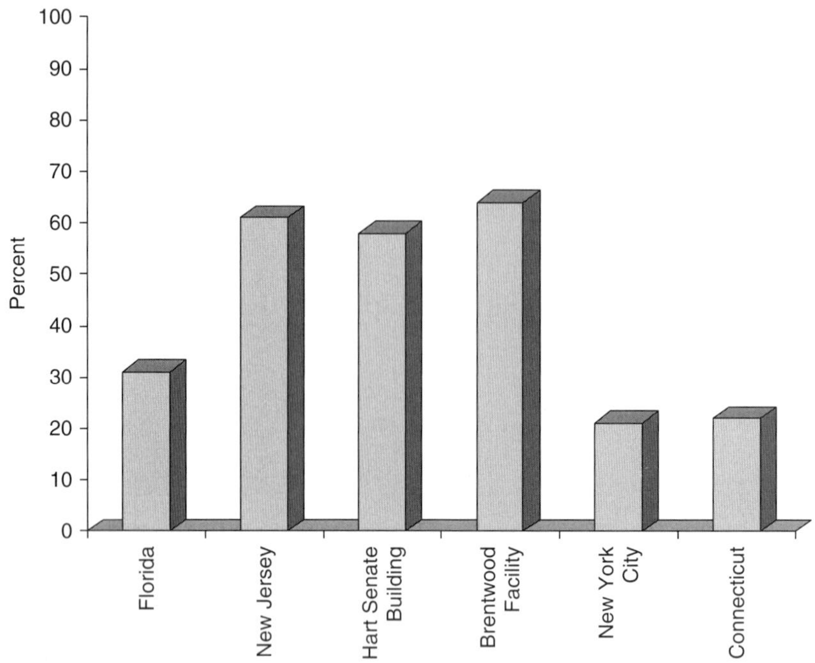

FIGURE 2.2 Percentage of individuals completing at least 60 days of antimicrobial prophylaxis following possible exposure to *B. anthracis* spores in the United States during 2001. Total numbers of individuals at each site were as follows: Florida, 1082; New Jersey, 1402; Hart Senate Building, Washington, D.C., 600; Brentwood facility, Washington, D.C. 2743; New York City, 2259; Connecticut, 1217.

cited reason for failing to complete the course of antibiotics. These ranged from gastrointestinal disturbances to seizures (Shepard et al., 2002)—although the proportion of these adverse events which are definitely attributable to antimicrobial prophylaxis is not known. Overall, it is clear that the prophylactic use of antibiotics may not be an effective way of protecting target populations from disease caused by the deliberate release of a bacterial pathogen.

Against this background, the use of vaccines to protect target populations appears to be an attractive proposition. However, the concept of use for these vaccines is not yet established. It may be possible to include vaccines against biothreat agents in childhood immunization programs. However, the cost of such an immunization program may be prohibitive. Alternatively, vaccines against bacterial biothreat agents may be given only when the risk of an attack is judged to be high. Vaccines used in this way might need to rapidly induce protective immunity.

AVAILABLE VACCINES AGAINST BACTERIAL BIOTHREAT AGENTS

Bearing in mind the ways in which bacterial biothreat agents might be used to cause disease, the suitability of existing vaccines against bacterial agents can be assessed. Clearly, for the majority of agents these vaccines need to provide good protection against an airborne challenge. For pathogens which might be used to contaminate food or water supplies vaccines must protect against this route of infection. Ideally, vaccines should result in the induction of protective immunity after a single dose. In some scenarios it may be desirable to immunize large numbers of individuals within a short period of time. Finally, these vaccines should have been tested according to national or international guidelines for pharmaceutical products. With reference to the possible biothreat agents as shown in Table 2.1, it is clear that relatively few commercially available or fully licensed vaccines exist (Titball and Williamson, 2003) (Table 2.3). In addition, some of the available vaccines do not meet the criteria outlined above (Titball and Williamson, 2003). The performance of some of these vaccines are detailed in subsequent chapters in this book.

The limited availability of vaccines can be ascribed largely to the minimal investment in this area of research during the past 50 years. The costs associated with working with dangerous pathogens under containment level 3 or 4 conditions are significant. Not only are the costs of construction of suitable laboratory facilities high (typically \$50 million–\$100 million) but the costs of operating these facilities is likely to be several million dollars per year. Much of this cost is

TABLE 2.3 Licensed/commercially available vaccines available in Europe and N. America against bacterial biothreat agents

Pathogen	Composition
B. anthracis	Live attenuated Sterne strain (for veterinary use only) or live attenuated STI-1 strain (used only in humans in the former Soviet Union (Brachman et al., 2004)
B. anthracis	Anthrax vaccine adsorbed (Brachman et al., 2004)
B. abortus	Live attenuated strain 19 (veterinary use only) (Schurig et al., 2002)
B. melitensis	Live attenuated strain Rev. 1 (veterinary use only) (Schurig et al., 2002; Blasco, 1997)
B. melitensis	Live attenuated strain RB51 (veterinary use only) (Schurig et al., 1995; Schurig et al., 2002)
S. enterica serovar Typhi	Killed whole cell (Levine, 2004)
S. enterica serovar Typhi	Live attenuated (TY21a) (Levine, 2004)
S. enterica serovar Typhi	Capsular polysaccharide (Vi) (Levine, 2004)
V. cholerae	Killed whole cell (injectable) (Sack and Lang, 2004)
V. cholerae	Killed whole cell+CTB (oral) (Sack and Lang, 2004)
V. cholerae	Live attenuated (oral) (Sack and Lang, 2004)
Y. pestis	Killed whole cell (Titball et al., 2004)

related to the regular servicing of laboratory air and water handling systems and the cost of heating air for laboratory ventilation systems which is eventually lost to the atmosphere.

These problems are compounded by the relatively low level of financial return on the sale of vaccines. Bacterial biothreat agents do not normally cause significant disease in the developed world and there is a small market for these vaccines outside of the biodefense arena. In addition, the commitment of nation-states in the developed world to purchase stocks of vaccines is not yet determined. Against this background we cannot assume that the commercial sector will bear most of the risks and costs associated with the development of biothreat agent vaccines.

NEW VACCINES AND NEW VACCINE TECHNOLOGIES

During the past decade there has been a marked increase in the level of government-funded research into improved vaccines against bacterial biothreat agents. This work can broadly be divided into four areas: studies on mechanisms of virulence of pathogens, studies on the host response to bacterial pathogens, work to develop and test vaccine candidates, and work to devise and evaluate new vaccine delivery platforms. There is every reason to believe that this initiative will be successful and that a suite of vaccines will become available over the next two decades. However, the development of vaccines against biothreat agents does present some significant technical challenges (Titball and Williamson, 2003).

Firstly, there is a paucity of data on mechanisms of virulence or biochemical makeup of many of these pathogens. For other pathogens this basic information has often been the starting point for the development of vaccines. Although some of the research currently underway will address these questions, it is unlikely to be mature enough in the near future to be directly exploitable. One alternative strategy might be to exploit genome sequence information to kick-start vaccine development programs (Titball and Williamson, 2003). This information might be used to support whole genome mutagenesis approaches to identify virulence determinants, predict metabolic pathways, identify targets for the construction of rationally attenuated mutants, and identify candidate protein antigens using "reverse vaccinology" (Titball and Williamson, 2003; Capecchi et al., 2004; Masignani and Rappuoli, 2004; Plotkin, 2005; Rappuoli and Covacci, 2003; Hensel et al., 1995; Hou et al., 2003). Fortunately, genome sequences are now available for almost all of the bacterial agents of concern (Table 2.4).

The technology of reverse vaccinology merits particular comment, since this approach has been successfully used to identify candidate vaccine antigens from a range of pathogens including *Helicobacter pylori* (Lissolo and Quentin-Millet, 2000), *Neisseria meningitidis* (Pizza et al., 2000), *Porphyromonas gingivalis* (Ross et al., 2001), *Streptococcus pneumoniae* (Wizemann et al., 2001), and more recently *B. anthracis* (Ariel et al., 2002, 2003). The approach relies on the assumption that vaccine antigens are located on the surface of the bacterial cell. Therefore, the predicted proteome of the pathogen of interest is screened to identify proteins that possess motifs indicating that they are destined for a cell surface location. The most frequently used algorithm is PSORT (http://psort.nibb.ac.jp/) which provides an integrated software package for this purpose. This conceptually simple approach does have some limitations. Although programs such as PSORT are excellent for the prediction of proteins which are exported via the type II secretion pathway, they are much less effective at predicting proteins exported via other pathways, and potential vaccine antigen might not be identified using this approach. In addition, the

TABLE 2.4 Reported genome sequences of possible biothreat agents

Pathogen	Strain(s)	References	Pathogen	Strain(s)	References
B. anthracis	Ames	Read et al. (2003b)	*F. tularensis*	Subspecies *tularensis* SchuS4	Larsson et al. (2005)
	Sterne	Read et al. (2002)		Subspecies *tularensis* FSC198	Chaudhuri et al. (2007)
	Ames ancestor	Genbank AE017334		Subspecies *tularensis* WY96-3418	Genbank CP000608
B. quintana	Toulouse	Alsmark et al. (2004)		Subspecies *holarctica* OSU18	Petrosino et al. (2006)
B. abortus	9-941	Halling et al. (2005)		Subspecies *holarctica* LVS	AM233362
B. melitensis	16M	DelVecchio et al. (2002)		Subspecies *novicida* U112	Genbank CP000439
	Biovar Abortus 2308	Chain et al. (2005)	*L. pneumophila*	Lens	Cazalet et al. (2004)
B. suis	1330	Paulsen et al. (2002)		Paris	Cazalet et al. (2004)
B. mallei	ATCC 23344	Nierman et al. (2004)		Philadelphia 1	Chien et al. (2004)
	NCTC10229	Genbank CP000545.1		Corby	Genbank CP000675
	NCTC10247	Genbank CP000547.1	*O. tsutsugamushi*	Boryong	Cho et al. (2007)
	SAVP1	Genbank CP000525.1	*R. prowazekii*	Madrid E	Andersson et al. (1998)
B. pseudomallei	K96243	Holden et al. (2004)	*R. rickettsii*	Wilmington	McLeod et al. (2004)
	1106a	Genbank CP000572	*S. enterica* serovar Typhi	CT18	Parkhill et al. (2001a)
	1710b	Genbank CP000124		TY2	Deng et al. (2003)
	668	Genbank CP000570.1	*S. dysenteriae*	Sd197	Yang et al. (2005)
C. psittaci	GPIC	Read et al. (2003a)	*V. cholerae*	El Tor N16961	Heidelberg et al. (2000)
C. botulinum	Hall A[a]	Minton et al. (2002)		O395	Genbank CP000626
C. perfringens	Strain 13	Shimizu et al. (2002)	*Y. pestis*	CO92	Parkhill et al. (2001b)
	SM101	Myers et al. (2006)		91001	Song et al. (2004)
	NCTC8237	Myers et al. (2006)		KIM	Deng et al. (2002)
C. tetani	E88	Bruggemann et al. (2003)		Antiqua	Chain et al. (2006)
C. burnettii	None	None		Nepal 516	Chain et al. (2006)
EHEC	0157:H7 EDL933 0157:H7 Sakai	Perna et al. (2001) Hayashi et al. (2001)		Pestoides F	Genbank CP000668

[a] Reference to ongoing sequencing project.

output from PSORT is a list of proteins with the potential to be located on the cell surface. Each of these proteins must be produced and tested.

Even when vaccine candidates are identified, produced under manufacturing conditions, and tested in animal models of disease, the licensing of vaccines for use in humans may be a difficult process. It is certainly possible to gain data on the safety of these vaccines via human clinical trials. However, with the exceptions of some pathogens that cause relatively mild diseases, efficacy studies in humans are not feasible. In these cases it will be necessary to gain significant information on mechanisms of protection in different animal models of disease, and to then show that similar immune responses are induced in immunized human volunteers.

CONCLUSIONS

It is clear that vaccines are likely to play a major role in plans to limit the impact of biothreat agents on human populations, and they are especially valuable when airborne releases of significant quantities of the pathogen are anticipated. However, currently there are few commercially available vaccines that are suitable for the immunization of at-risk populations. This situation is likely to change over the next decade—there are already improved vaccines against some biothreat pathogens which are in the advanced stages of development. The greatest challenge today is the development of appropriate plans for the effective use of these vaccines. Should they become part of the normal childhood vaccination program or should they be used only when the likelihood of an attack is considered to be high? Should a prime-boost strategy be adopted where individuals are primed to allow the rapid development of protective immunity on boosting? These are important questions because they might drive not only the selection of vaccine antigens but also the selection of appropriate vaccine delivery systems.

Finally, we should not lose sight of the fact that many of the diseases caused by bacterial biothreat agents are a significant cause of morbidity and mortality in economically deprived areas of the world. The vaccines that are developed to protect against biothreat agents are likely to find additional utility in the protection of these disadvantaged populations of humans.

References

Aldhous, P. Biological warfare. Gruinard Island handed back. Nature 1990; 344:801.

Alsmark, C.M., Frank, A.C., Karlberg, E.O., Legault, B.A., Ardell, D.H., Canback, B., Eriksson, A.S., Naslund, A.K., Handley, S.A., Huvet, M., La Scola, B., Holmberg, M. and Andersson, S.G.E. The louse-borne human pathogen *Bartonella quintana* is a genomic derivative of the zoonotic agent *Bartonella henselae*. Proc. Natl. Acad. Sci. U.S.A. 2004; 101:9716–9721.

Andersson, S.G.E., Zomorodipour, A., Andersson, J.O., Sicheritz-Ponten, T., Alsmark, U.C.M., Podowski, R.M., Naslund, A.K., Eriksson, A.S., Winkler, H.H. and Kurland, C.G. The genome sequence of *Rickettsia prowazekii* and the origin of mitochondria. Nature 1998; 396:133–140.

Anon Health Aspects of Chemical and Biological Weapons. Geneva: World Health Organization, 1970.

Anon. NATO handbook on the medical aspects of NBC defensive operations. *AmedP-6(B),* Part II- Biological, 1996.

Ariel, N., Zvi, A., Grosfeld, H., Gat, O., Inbar, Y., Velan, B., Cohen, S. and Shafferman, A. Search for potential vaccine candidate open reading frames in the *Bacillus anthracis* virulence plasmid pXO1: In silico and in vitro screening. Infect. Immun. 2002; 70:6817–6827.

Ariel, N., Zvi, A., Makarova, K.S., Chitlaru, T., Elhanany, E., Velan, B., Cohen, S., Friedlander, A.M. and Shafferman, A. Genome-based bioinformatic selection of chromosomal *Bacillus anthracis* putative vaccine candidates coupled with proteomic identification of surface-associated antigens. Infect. Immun. 2003; 71:4563–4579.

Blasco, J.M. A review of the use of *B. melitensis* Rev 1 vaccine in adult sheep and goats. Prev. Vet. Med. 1997; 31:275–283.

Brachman, P.S., Friedlander, A.M. and Grabenstein, J.D. Anthrax vaccine. In: Vaccines (S.A. Plotkin and W.A. Orenstein, Eds.). Philadelphia: Saunders, 2004.

Bruggemann, H., Baumer, S., Fricke, W.F., Wiezer, A., Liesegang, H., Decker, I., Herzberg, C., Martinez-Arias, R., Merkl, R., Henne, A. and Gottschalk, G. The genome sequence of *Clostridium tetani*, the causative agent of tetanus disease. Proc. Natl. Acad. Sci. U.S.A. 2003; 100:1316–1321.

Capecchi, B., Serruto, D., Bobie, J.A., Rappuoli, R. and Pizza, M. The genome revolution in vaccine research. Curr. Issues Mol. Biol. 2004; 6:17–27.

Carus, W.S. Bioterrorism and Biocrimes. The Illicit Use of Biological Agents Since 1900. Washington: Center for Counterproliferation Research, National Defense University, 2001.

Cazalet, C., Rusniok, C., Bruggemann, H., Zidane, N., Magnier, A., Ma, L., Tichit, M., Jarraud, S., Bouchier, C., Vandenesch, F., Kunst, F., Etienne, J., Glaser, P. and Buchrieser, C. Evidence in the *Legionella pneumophila* genome for exploitation of host cell functions and high genome plasticity. Nat. Genet. 2004; 36:1165–1173.

Chain, P.S., Comerci, D.J., Tolmasky, M.E., Larimer, F.W., Malfatti, S.A., Vergez, L.M., Aguero, F., Land, M.L., Ugalde, R.A. and Garcia, E. Whole-genome analyses of speciation events in pathogenic *Brucellae*. Infect. Immun. 2005; 73:8353–8361.

Chain, P.S., Hu, P., Malfatti, S.A., Radnedge, L., Larimer, F., Vergez, L.M., Worsham, P., Chu, M.C. and Andersen, G.L. Complete genome sequence of *Yersinia pestis* strains Antiqua and Nepal516: evidence of gene reduction in an emerging pathogen. J. Bacteriol. 2006; 188:4453–4463.

Chaudhuri, R.R., Ren, C.P., Desmond, L., Vincent, G.A., Silman, N.J., Brehm, J., Elmore, M.J., Hudson, M.J., Forsman, M., Isherwood, K.E., Gurycova, D., Minton, N.P., Titball, R.W., Pallen, M.J. and Vipond, R. Genome sequencing shows that European isolates of *Francisella tularensis* subspecies *tularensis* are almost identical to US laboratory strain Schu S4. PLoS ONE. 2007; 2:e352.

Chien, M.C., Morozova, I., Shi, S.D., Sheng, H.T., Chen, J., Gomez, S.M., Asamani, G., Hill, K., Nuara, J., Feder, M., Rineer, J., Greenberg, J.J., Steshenko, V., Park, S.H., Zhao, B.H., Teplitskaya, E., Edwards, J.R., Pampou, S., Georghiou, A., Chou, I.C., Iannuccilli, W., Ulz, M.E., Kim, D.H., Geringer-Sameth, A., Goldsberry, C., Morozov, P., Fischer, S.G., Segal, G., Qu, X.Y., Rzhetsky, A., Zhang, P.S., Cayanis, E., De Jong, P.J., Ju, J.Y., Kalachikov, S., Shuman, H.A. and Russo, J.J. The genomic sequence of the accidental pathogen *Legionella pneumophila*. Science 2004; 305:1966–1968.

Cho, N.H., Kim, H.R., Lee, J.H., Kim, S.Y., Kim, J., Cha, S., Kim, S.Y., Darby, A.C., Fuxelius, H.H., Yin, J., Kim, J.H., Kim, J., Lee, S.J., Koh, Y.S., Jang, W.J., Park, K.H., Andersson, S.G., Choi, M.S. and Kim, I.S. The *Orientia tsutsugamushi* genome reveals massive proliferation of conjugative type IV secretion system and host-cell interaction genes. Proc. Natl. Acad. Sci. U.S.A. 2007; 104:7981–7986.

Delvecchio, V.G., Kapatral, V., Redkar, R.J., Patra, G., Mujer, C., Los, T., Ivanova, N., Anderson, I., Bhattacharyya, A., Lykidis, A., Reznik, G., Jablonski, L., Larsen, N., D'souza, M., Bernal, A., Mazur, M., Goltsman, E., Selkov, E., Elzer, P.H., Hagius, S., O'callaghan, D., Letesson, J.J., Haselkorn, R., Kyrpides, N. and Overbeek, R. The genome sequence of the facultative intracellular pathogen *Brucella melitensis*. Proc. Natl. Acad. Sci. U.S.A. 2002; 99:443–448.

Deng, W., Burland, V., Plunkett, G., Boutin, A., Mayhew, G.F., Liss, P., Perna, N.T., Rose, D.J., Mau, B., Zhou, S.G., Schwartz, D.C., Fetherston, J.D., Lindler, L.E., Brubaker, R.R., Plano, G.V., Straley, S.C., Mcdonough, K.A., Nilles, M.L., Matson, J.S., Blattner, F.R. and Perry, R.D. Genome sequence of *Yersinia pestis* KIM. J. Bacteriol. 2002; 184:4601–4611.

Deng, W., Liou, S.R., Plunkett, G., Mayhew, G.F., Rose, D.J., Burland, V., Kodoyianni, V., Schwartz, D.C. and Blattner, F.R. Comparative genomics of *Salmonella enterica* serovar *typhi* strains Ty2 and CT18. J. Bacteriol. 2003; 185:2330–2337.

Dennis, D.T., Inglesby, T.V., Henderson, D.A., Bartlett, J.G., Ascher, M.S., Eitzen, E., Fine, A.D., Friedlander, A.M., Hauer, J., Layton, M., Lillibridge, S.R., Mcdade, J.E., Osterholm, M.T., O'toole, T., Parker, G., Perl, T.M., Russell, P.K. and Tonat, K. Tularemia as a biological weapon: medical and public health management. JAMA 2001; 285:2763–2773.

Geissler, E., Moon, J.E.V.C. and Pearson, G.S. Lessons from the history of biological and toxin warfare. In: Biological and Toxin Weapons: Research, Development and Use from the Middle Ages to 1945. SIPRI Chemical & Biological Warfare Studies No. 18 (E. Geissler and J.E. Van Courtland Moon, Eds.). Oxford: Oxford University Press, 1999.

Guillemin, J. Biological Weapons. New York: Columbia University Press, 2004.

Halling, S.M., Peterson-Burch, B.D., Bricker, B.J., Zuerner, R.L., Qing, Z., Li, L.L., Kapur, V., Alt, D.P. and Olsen, S.C. Completion of the genome sequence of *Brucella abortus* and comparison to the highly similar genomes of *Brucella melitensis* and *Brucella suis*. J. Bacteriol. 2005; 187:2715–2726.

Hayashi, T., Makino, K., Ohnishi, M., Kurokawa, K., Ishii, K., Yokoyama, K., Han, C.G., Ohtsubo, E., Nakayama, K., Murata, T., Tanaka, M., Tobe, T., Iida, T., Takami, H., Honda, T., Sasakawa, C., Ogasawara, N., Yasunaga, T., Kuhara, S., Shiba, T., Hattori, M. and Shinagawa, H. Complete genome sequence of enterohemorrhagic *Escherichia coli* O157: H7 and genomic comparison with a laboratory strain K-12. DNA Res. 2001; 8:11–22.

Heidelberg, J.F., Eisen, J.A., Nelson, W.C., Clayton, R.A., Gwinn, M.L., Dodson, R.J., Haft, D.H., Hickey, E.K., Peterson, J.D., Umayam, L., Gill, S.R., Nelson, K.E., Read, T.D., Tettelin, H., Richardson, D., Ermolaeva, M.D., Vamathevan, J., Bass, S., Qin, H.Y., Dragoi, I., Sellers, P., Mcdonald, L., Utterback, T., Fleishmann, R.D., Nierman, W.C., White, O., Salzberg, S.L., Smith, H.O., Colwell, R.R., Mekalanos, J.J., Venter, J.C. and Fraser, C.M. DNA sequence of both chromosomes of the cholera pathogen *Vibrio cholerae*. Nature 2000; 406:477–483.

Hensel, M., Shea, J.E., Gleeson, C., Jones, M.D., Dalton, E. and Holden, D.W. Simultaneous identification of bacterial virulence genes by negative selection. Science 1995; 269:400–403.

Holden, M.T.G., Titball, R.W., Peacock, S.J., Cerdeno-Tarraga, A.M., Atkins, T., Crossman, L.C., Pitt, T., Churcher, C., Mungall, K., Bentley, S.D., Sebaihia, M., Thomson, N.R., Bason, N., Beacham, I.R., Brooks, K., Brown, K.A., Brown, N.F., Challis, G.L., Cherevach, I., Chillingworth, T., Cronin, A., Crossett, B., Davis, P., Deshazer, D., Feltwell, T., Fraser, A., Hance, Z., Hauser, H., Holroyd, S., Jagels, K., Keith, K.E., Maddison, M., Moule, S., Price, C., Quail, M.A., Rabbinowitsch, E., Rutherford, K., Sanders, M., Simmonds, M., Songsivilai, S., Stevens, K., Tumapa, S., Vesaratchavest, M., Whitehead, S., Yeats, C., Barrell, B.G., Oyston, P.C.F. and Parkhill, J. Genomic plasticity of the causative agent of melioidosis, *Burkholderia pseudomallei*. Proc. Natl. Acad. Sci. U.S.A. 2004; 101:14240–14245.

Hou, B.K., Wackett, L.P. and Ellis, L.B.M. Microbial pathway prediction: a functional group approach. J. Chem. Inf. Comput. Sci. 2003; 43:1051–1057.

Inglesby, T.V., Dennis, D.T., Henderson, D.A., Bartlett, J.G., Ascher, M.S., Eitzen, E., Fine, A.D., Friedlander, A.M., Hauer, J., Koerner, J.F., Layton, M., Mcdade, J., Osterholm, M.T., O'toole, T., Parker, G., Perl, T.M., Russell, P.K., Schoch-Spana, M. and Tonat, K. Plague as a biological weapon: medical and public health management. JAMA 2000; 283:2281–2290.

Inglesby, T.V., Henderson, D.A., Bartlett, J.G., Ascher, M.S., Eitzen, E., Friedlander, A.M., Hauer, J., Mcdade, J., Osterholm, M.T., O'toole, T., Parker, G., Perl, T.M., Russell, P.K. and Tonat, K. Anthrax as a biological weapon: medical and public health management. JAMA 1999; 281:1735–1745.

Jernigan, D.B., Raghunathan, P.L., Bell, B.P., Brechner, R., Bresnitz, E.A., Butler, J.C., Cetron, M., Cohen, M., Doyle, T., Fischer, M., Greene, C., Griffith, K.S., Guarner, J., Hadler, J.L., Hayslett, J.A., Meyer, R., Petersen, L.R., Phillips, M., Pinner, R., Popovic, T., Quinn, C.P., Reefhuis, J., Reissman, D., Rosenstein, N., Schuchat, A., Shieh, W.J., Siegal, L., Swerdlow, D.L., Tenover, F.C., Traeger, M., Ward, J.W., Weisfuse, I., Wiersma, S., Yeskey, K., Zaki, S., Ashford, D.A., Perkins, B.A., Ostroff, S., Hughes, J., Fleming, D., Koplan, J.P., Gerberding, J.L. and Team, N.A.E.I. Investigation of bioterrorism-related anthrax, United States, 2001: epidemiologic findings. Emerg. Infect. Dis. 2002; 8:1019–1028.

Kaplan, D.E. and Marshall, A. The Cult at the End of the World. New York: Crown Publishers, 1996.

Kolavic, S.A., Kimura, A., Simons, S.L., Slutsker, L., Barth, S. and Haley, C.E. An outbreak of *Shigella dysenteriae* type 2 among laboratory workers due to intentional food contamination. JAMA 1997; 278:396–398.

Larsson, P., Oyston, P.C.F., Chain, P., Chu, M.C., Duffield, M., Fuxelius, H.H., Garcia, E., Halltorp, G., Johansson, D., Isherwood, K.E., Karp, P.D., Larsson, E., Liu, Y., Michell, S., Prior, J., Prior, R., Malfatti, S., Sjostedt, A., Svensson, K., Thompson, N., Vergez, L., Wagg, J.K., Wren, B.W., Lindler, L.E., Andersson, S.G.E., Forsman, M. and Titball, R.W. The complete genome sequence of *Francisella tularensis*, the causative agent of tularemia. Nat. Genet. 2005; 37:153–159.

Levine, M.M. Typhoid fever vaccines. In: Vaccines (S.A. Plotkin and W.A. Orenstein, Eds.). Philadelphia: Saunders, 2004.

Lissolo, L. and Quentin-Millet, M.J. New strategies for identifying and developing novel vaccines: genome based discoveries. In: Fighting Infection in the 21st Century (P.W. Andrew, P.C.F. Oyston, G.L. Smith and D.E. Stewart-Tull, Eds.). Oxford: Blackwell, Science, 2000.

Manchee, R.J., Broster, M.G., Anderson, I.S., Henstridge, R.M. and Melling, J. Decontamination of *Bacillus anthracis* on Gruinard Island? Nature 1983; 303:239–240.

Manchee, R.J., Broster, M.G., Henstridge, R.M., Stagg, A.J. and Melling, J. Anthrax island. Nature 1982; 296:598.

Manchee, R.J., Broster, M.G., Melling, J., Henstridge, R.M. and Stagg, A.J. *Bacillus anthracis* on Gruinard Island. Nature 1981; 294:254–255.

Mangold, T. and Goldberg, J. Plague Wars: A True Story of Biological Warfare. London: Macmillan, 1999.

Masignani, V. and Rappuoli, R. Reverse vaccinology: using genome information to develop new vaccines. Tissue Antigens 2004; 64:357.

Mcleod, M.P., Qin, X., Karpathy, S.E., Gioia, J., Highlander, S.K., Fox, G.E., Mcneill, T.Z., Jiang, H.Y., Muzny, D., Jacob, L.S., Hawes, A.C., Sodergren, E., Gill, R., Hume, J., Morgan, M., Fan, G.W., Amin, A.G., Gibbs, R.A., Hong, C., Yu, X.J., Walker, D.H. and Weinstock, G.M. Complete genome sequence of *Rickettsia typhi* and comparison with sequences of other rickettsiae. J. Bacteriol. 2004; 186:5842–5855.

Meselson, M., Guillemin, J., Hugh-Jones, M., Langmuir, A., Popova, I., Shelokov, A. and Yampolskaya, O. The Sverdlovsk anthrax outbreak of 1979. Science 1994; 266:1202–1208.

Minton, N., Elmore, M., Holden, M., Barrell, B., Mauchline, M., Patel, S., Peck, M., Hutson, R. and Parkhill, J. Genome sequencing of *Clostridium botulinum* Hall A strain (ATCC 3502), an update. Naunyn Schmiedebergs Arch. Pharmacol. 2002; 365:R29.

Myers, G.S., Rasko, D.A., Cheung, J.K., Ravel, J., Seshadri, R., Deboy, R.T., Ren, Q., Varga, J., Awad, M.M., Brinkac, L.M., Daugherty, S.C., Haft, D.H., Dodson, R.J., Madupu, R., Nelson, W.C., Rosovitz, M.J., Sullivan, S.A., Khouri, H., Dimitrov, G.I., Watkins, K.L., Mulligan, S., Benton, J., Radune, D., Fisher, D.J., Atkins, H.S., Hiscox, T., Jost, B.H., Billington, S.J.,

Songer, J.G., Mcclane, B.A., Titball, R.W., Rood, J.I., Melville, S.B. and Paulsen, I.T. Skewed genomic variability in strains of the toxigenic bacterial pathogen, *Clostridium perfringens*. Genome Res. 2006; 16:1031–1040.

Nierman, W.C., Deshazer, D., Kim, H.S., Tettelin, H., Nelson, K.E., Feldblyum, T., Ulrich, R.L., Ronning, C.M., Brinkac, L.M., Daugherty, S.C., Davidsen, T.D., Deboy, R.T., Dimitrov, G., Dodson, R.J., Durkin, A.S., Gwinn, M.L., Haft, D.H., Khouri, H., Kolonay, J.F., Madupu, R., Mohammoud, Y., Nelson, W.C., Radune, D., Romero, C.M., Sarria, S., Selengut, J., Shamblin, C., Sullivan, S.A., White, O., Yu, Y., Zafar, N., Zhou, L.W. and Fraser, C.M. Structural flexibility in the *Burkholderia mallei* genome. Proc. Natl. Acad. Sci. U.S.A. 2004; 101:14246–14251.

Parkhill, J., Dougan, G., James, K.D., Thomson, N.R., Pickard, D., Wain, J., Churcher, C., Mungall, K.L., Bentley, S.D., Holden, M.T.G., Sebaihia, M., Baker, S., Basham, D., Brooks, K., Chillingworth, T., Connerton, P., Cronin, A., Davis, P., Davies, R.M., Dowd, L., White, N., Farrar, J., Feltwell, T., Hamlin, N., Haque, A., Hien, T.T., Holroyd, S., Jagels, K., Krogh, A., Larsen, T.S., Leather, S., Moule, S., O'Gaora, P., Parry, C., Quail, M., Rutherford, K., Simmonds, M., Skelton, J., Stevens, K., Whitehead, S. and Barrell, B.G. Complete genome sequence of a multiple drug resistant *Salmonella enterica* serovar Typhi CT18. Nature 2001a; 413:848–852.

Parkhill, J., Wren, B.W., Thomson, N.R., Titball, R.W., Holden, M.T.G., Prentice, M.B., Sebaihia, M., James, K.D., Churcher, C., Mungall, K.L., Baker, S., Basham, D., Bentley, S.D., Brooks, K., Cerdeno-Tarraga, A.M., Chillingworth, T., Cronin, A., Davies, R.M., Davis, P., Dougan, G., Feltwell, T., Hamlin, N., Holroyd, S., Jagels, K., Karlyshev, A.V., Leather, S., Moule, S., Oyston, P.C.F., Quail, M., Rutherford, K., Simmonds, M., Skelton, J., Stevens, K., Whitehead, S. and Barrell, B.G. Genome sequence of *Yersinia pestis*, the causative agent of plague. Nature 2001b; 413: 523–527.

Paulsen, I.T., Seshadri, R., Nelson, K.E., Eisen, J.A., Heidelberg, J.F., Read, T.D., Dodson, R.J., Umayam, L., Brinkac, L.M., Beanan, M.J., Daugherty, S.C., Deboy, R.T., Durkin, A.S., Kolonay, J.F., Madupu, R., Nelson, W.C., Ayodeji, B., Kraul, M., Shetty, J., Malek, J., Van Aken, S.E., Riedmuller, S., Tettelin, H., Gill, S.R., White, O., Salzberg, S.L., Hoover, D.L., Lindler, L.E., Halling, S.M., Boyle, S.M. and Fraser, C.M. The *Brucella suis* genome reveals fundamental similarities between animal and plant pathogens and symbionts. Proc. Natl. Acad. Sci. U.S.A. 2002; 99:13148–13153.

Perna, N.T., Plunkett, G., Burland, V., Mau, B., Glasner, J.D., Rose, D.J., Mayhew, G.F., Evans, P.S., Gregor, J., Kirkpatrick, H.A., Posfai, G., Hackett, J., Klink, S., Boutin, A., Shao, Y., Miller, L., Grotbeck, E.J., Davis, N.W., Limk, A., Dimalanta, E.T., Potamousis, K.D., Apodaca, J., Anantharaman, T.S., Lin, J.Y., Yen, G., Schwartz, D.C., Welch, R.A. and Blattner, F.R. Genome sequence of enterohaemorrhagic *Escherichia coli* O157: H7. Nature 2001; 409:529–533.

Petrosino, J.F., Xiang, Q., Karpathy, S.E., Jiang, H., Yerrapragada, S., Liu, Y., Gioia, J., Hemphill, L., Gonzalez, A., Raghavan, T.M., Uzman, A., Fox, G.E., Highlander, S., Reichard, M., Morton, R.J., Clinkenbeard, K.D. and Weinstock, G.M. Chromosome rearrangement and diversification of *Francisella tularensis* revealed by the type B (OSU18) genome sequence. J. Bacteriol. 2006; 188:6977–6985.

Pizza, M., Scarlato, V., Masignani, V., Giuliani, M.M., Arico, B., Comanducci, M., Jennings, G.T., Baldi, L., Bartolini, E., Capecchi, B., Galeotti, C.L., Luzzi, E., Manetti, R., Marchetti, E., Mora, M., Nuti, S., Ratti, G., Santini, L., Savino, S., Scarselli, M., Storni, E., Zuo, P.J., Broeker, M., Hundt, E., Knapp, B., Blair, E., Mason, T., Tettelin, H., Hood, D.W., Jeffries, A.C., Saunders, N.J., Granoff, D.M., Venter, J.C., Moxon, E.R., Grandi, G. and Rappuoli, R. Identification of vaccine candidates against serogroup B meningococcus by whole-genome sequencing. Science 2000; 287:1816–1820.

Plotkin, S.A. Six revolutions in vaccinology. Pediatr. Infect. Dis. J. 2005; 24:1–9.

Rappuoli, R. and Covacci, A. Reverse vaccinology and genomics. Science 2003; 302:602.

Read, T.D., Myers, G.S.A., Brunham, R.C., Nelson, W.C., Paulsen, I.T., Heidelberg, J., Holtzapple, E., Khouri, H., Federova, N.B., Carty, H.A., Umayam, L.A., Haft, D.H., Peterson, J., Beanan, M.J., White, O., Salzberg, S.L., Hsia, R.C., Mcclarty, G., Rank, R.G., Bavoil, P.M. and Fraser, C.M. Genome sequence of *Chlamydophila caviae* (*Chlamydia psittaci* GPIC): examining the role of niche-specific genes in the evolution of the *Chlamydiaceae*. Nucleic Acids Res. 2003a; 31:2134–2147.

Read, T.D., Peterson, S.N., Tourasse, N., Baillie, L.W., Paulsen, I.T., Nelson, K.E., Tettelin, H., Fouts, D.E., Eisen, J.A., Gill, S.R., Holtzapple, E.K., Okstad, O.A., Helgason, E., Rilstone, J., Wu, M., Kolonay, J.F., Beanan, M.J., Dodson, R.J., Brinkac, L.M., Gwinn, M., Deboy, R.T., Madpu, R., Daugherty, S.C., Durkin, A.S., Haft, D.H., Nelson, W.C., Peterson, J.D., Pop, M., Khouri, H.M., Radune, D., Benton, J.L., Mahamoud, Y., Jiang, L.X., Hance, I.R., Weidman, J.F., Berry, K.J., Plaut, R.D., Wolf, A.M., Watkins, K.L., Nierman, W.C., Hazen, A., Cline, R., Redmond, C., Thwaite, J.E., White, O., Salzberg, S.L., Thomason, B., Friedlander, A.M., Koehler, T.M., Hanna, P.C., Kolsto, A.B. and Fraser, C.M. The genome sequence of *Bacillus anthracis* Ames and comparison to closely related bacteria. Nature 2003b; 423:81–86.

Read, T.D., Salzberg, S.L., Pop, M., Shumway, M., Umayam, L., Jiang, L.X., Holtzapple, E., Busch, J.D., Smith, K.L., Schupp, J.M., Solomon, D., Keim, P. and Fraser, C.M. Comparative genome sequencing for discovery of novel polymorphisms in *Bacillus anthracis*. Science 2002; 296:2028–2033.

Ross, B.C., Czajkowski, L., Hocking, D., Margetts, M., Webb, E., Rothel, L., Patterson, M., Agius, C., Camuglia, S., Reynolds, E., Littlejohn, T., Gaeta, B., Ng, A., Kuczek, E.S., Mattick, J.S., Gearing, D. and Barr, I.G. Identification of vaccine candidate antigens from a genomic analysis of *Porphyromonas gingivalis*. Vaccine 2001; 19:4135–4142.

Sack, D.A. and Lang, D.R. Cholera vaccine. In: Vaccines (S.A. Plotkin and W.A. Orenstein, Eds.). Philadelphia: Saunders, 2004.

Schurig, G., Boyle, S. and Sriranganathan, N. Brucella abortus vaccine strain RB51: a brief review. Arch. Med. Vet. 1995; 27: 19–22.

Schurig, G.G., Sriranganathan, N. and Corbel, M.J. Brucellosis vaccines: past, present and future. Vet. Microbiol. 2002; 90:479–496.

Shepard, C.W., Soriano-Gabarro, M., Zell, E.R., Hayslett, J., Lukacs, S., Goldstein, S., Factor, S., Jones, J., Ridzon, R., Williams, I., Rosenstein, N. and Group., C.A.E.W. Antimicrobial postexposure prophylaxis for anthrax: adverse events and adherence. Emerg. Infect. Dis. 2002; 8:1124–1132.

Shimizu, T., Ohtani, K., Hirakawa, H., Ohshima, K., Yamashita, A., Shiba, T., Ogasawara, N., Hattori, M., Kuhara, S. and Hayashi, H. Complete genome sequence of *Clostridium perfringens*, an anaerobic flesh-eater. Proc. Natl. Acad. Sci. U.S.A. 2002; 99:996–1001.

Song, Y.J., Tong, Z.Z., Wang, J., Wang, L., Guo, Z.B., Han, Y.P., Zhang, J.G., Pei, D.C., Zhou, D.S., Qin, H., Pang, X., Han, Y.J., Zhai, J.H., Li, M., Cui, B.Z., Qi, Z.Z., Jin, L.X., Dai, R.X., Chen, F., Li, S.T., Ye, C., Du, Z.M., Lin, W., Yu, J., Yang, H.M., Huang, P.T. and Yang, R.F. Complete genome sequence of *Yersinia pestis* strain 91001, an isolate avirulent to humans. DNA Res. 2004; 11:179–197.

Stein, B.D., Tanielian, T.L., Ryan, G.W., Rhodes, H.J., Young, S.D. and Blanchard, J.C. A bitter pill to swallow: nonadherence with prophylactic antibiotics during the anthrax attacks and the role of private physicians. Biosecur. Bioterror. 2004; 2:175–185.

Sternbach, G. The history of anthrax. J. Emerg. Med. 2003; 24:463–467.

Titball, R.W. and Williamson, E.D. Vaccine development for potential bioterrorism agents. Curr. Drug Targets Infect. Disord. 2003; 3:255–262.

Titball, R.W., Williamson, E.D. and Dennis, D.T. Plague. In: Vaccines (S.A. Plotkin and W.A. Orenstein, Eds.). Philadelphia: Saunders, 2004.

Torok, T.J., Tauxe, R.V., Wise, R.P., Livengood, J.R., Sokolow, R., Mauvais, S., Birkness, K.A., Skeels, M.R., Horan, J.M. and Foster, L.R. A large community outbreak of salmonellosis caused by intentional contamination of restaurant salad bars. JAMA 1997; 278:389–395.

Wheelis, M. Biological sabotage in World War I. In: Biological and Toxin Weapons: Research, Development and Use from the Middle Ages to 1945. SIPRI Chemical & Biological Warfare Studies No. 18 (E. Geissler and J.E.V.C. Moon, Eds.). Oxford: Oxford University Press, 1999.

Wizemann, T.M., Heinrichs, J.H., Adamou, J.E., Erwin, A.L., Kunsch, C., Choi, G.H., Barash, S.C., Rosen, C.A., Masure, H.R., Tuomanen, E., Gayle, A., Brewah, Y.A., Walsh, W., Barren, P., Lathigra, R., Hanson, M., Langermann, S., Johnson, S. and Koenig, S. Use of a whole genome approach to identify vaccine molecules affording protection against *Streptococcus pneumoniae* infection. Infect. Immun. 2001; 69:1593–1598.

Yang, F., Yang, J., Zhang, X., Chen, L., Jiang, Y., Yan, Y., Tang, X., Wang, J., Xiong, Z., Dong, J., Xue, Y., Zhu, Y., Xu, X., Sun, L., Chen, S., Nie, H., Peng, J., Xu, J., Wang, Y., Yuan, Z., Wen, Y., Yao, Z., Shen, Y., Qiang, B., Hou, Y., Yu, J. and Jin, Q. Genome dynamics and diversity of *Shigella* species, the etiologic agents of bacillary dysentery. Nucl. Acids Res. 2005; 33:6445–6458.

CHAPTER

3

Viral Biothreat Agents

Kevin Anderson

National Biodefense Analysis and Countermeasures Center, Science and Technology Directorate, U.S. Department of Homeland Security, Frederick, MD, USA

OUTLINE

ABSTRACT

The risk posed by viruses as biological threat agents is discussed primarily from a public health perspective, with the potential occurrence of significant morbidity and mortality as a result of infection via natural or intentional exposure. Parameters of risk associated with the spectrum of viruses considered as biological threat agents are discussed, to include examples of intentional use. In consideration of the threat posed by viruses, mitigating illness and preventing death are the principal goals of medical countermeasure development efforts. The existence of safe and efficacious vaccines is critical to establishing a robust posture of preparedness to address the spectrum of viral threat agents.

INTRODUCTION

Given the primary focus of this text is on vaccines, the term biological threat will be defined primarily from a public health perspective. Therefore, viruses as biological threat agents are defined as those that cause harm to public health as manifested by significant morbidity and/or mortality, which can occur via natural or intentional exposure.

Natural exposure to viruses that cause human disease can occur via direct or indirect contact. Direct contact would include person-to-person spread, such as with community-acquired viral infections; animal-to-person spread, as with viral zoonoses; and spread to an unborn child from its mother, also known as vertical transmission. Also, viruses can be transmitted directly by vectors, such as mosquitoes and ticks. Indirect contact would include environmentally acquired viral infections (via food, water, air, or fomites) that are not spread by close contact. Infectious viral diseases can be spread through the air either by direct or indirect contact via droplet or particle transmission, respectively.

Intentional exposure to viruses resulting in human disease would include biowarfare, bioterrorism, and biocrimes (Swearengen, 2006). Biowarfare is defined as the military use of biological agents, where targets of agents are predominately military personnel, government officials, or resources that might hinder a nation's ability to attack and/or defend itself. Bioterrorism is defined as the threat or use of biological agents by individuals or groups motivated by political, religious, ecological, or other ideological objective. Viruses of concern from a biowarfare or bioterrorism perspective represent a major theme of this chapter. A biocrime is defined as the threat or use of biological agents for individual objectives such as revenge or financial gain. Use of viruses in biocrimes against people have been documented, with sources limited primarily to clinical samples containing viruses such as HIV-1 and hepatitis A (Carus, 2001). Viruses have also been used to reduce animal populations unlawfully and by definition, such action would be categorized as a biocrime (Carus, 2001).

The extent to which particular viruses are considered as biological threats is largely dependent on the risk they pose. The risk posed by a virus is defined as the probability of it causing infections of significant consequence, with high morbidity and mortality, as a result of either naturally occurring epidemic or pandemic disease, or malevolent intent, in the case of biocrime, bioterrorism, or biowarfare.

Viruses as credible, high-impact threats are defined by the properties or characteristics of the virus, the vulnerability of those subject to exposure and also, in the case of malevolent use, the capability and intent of perpetrators. Credible, high-impact threats include viruses that could inflict harm by either natural exposure, in which parameters of risk include agent characteristics and vulnerabilities, or intentional exposure, in which adversary capability and intent also become factors to consider (Fig. 3.1). Parameters of risk associated with properties or characteristics of viruses would include high infectivity, high morbidity and mortality, environmental stability, airborne transmissibility, and high contagiousness. Parameters associated with vulnerabilities include limited or unavailable medical countermeasures, such as vaccines and therapeutics, with cost and increasing public resistance to vaccination as contributing factors, limited or unavailable means to detect and identify viruses, and the ability of a virus to cause disease that is difficult to treat. Parameters of risk associated with adversary capability and intent include ease of acquisition, production, and dissemination of viral agents; and evidence of prior research and development as a biological weapon (Center for Biosecurity, UPMC, 2008).

In human history, the consequences of natural exposure to viruses have been much greater with respect to morbidity and mortality than those realized by intentional release, as depicted in Fig. 3.1. With the exception of smallpox, which some experts speculate is responsible for more deaths throughout modern history than all other infectious diseases combined, the greatest human and economic consequences derived from virus infections via natural exposure have been realized from emerging diseases, all of which have been derived from infected animals via zoonoses. For example, influenza pandemics that occurred in 1918 (H1N1), 1957 (H2N2), and 1968 (H3N2) resulted in approximately 40–50, 2, and 1 million deaths worldwide, respectively (World Health Organization, 2008). It is thought that the 1918 H1N1 virus was an avian virus that was introduced into pigs, and subsequently transmitted to human carriers, which spread the virus along trade routes and shipping lines resulting in a global pandemic (Taubenberger and Morens, 2006). In the past 25 years, approximately 60 million people have been infected by HIV-1, resulting in greater than 25 million deaths (Avert, 2008). HIV-1 most likely originated in wild chimpanzees living in

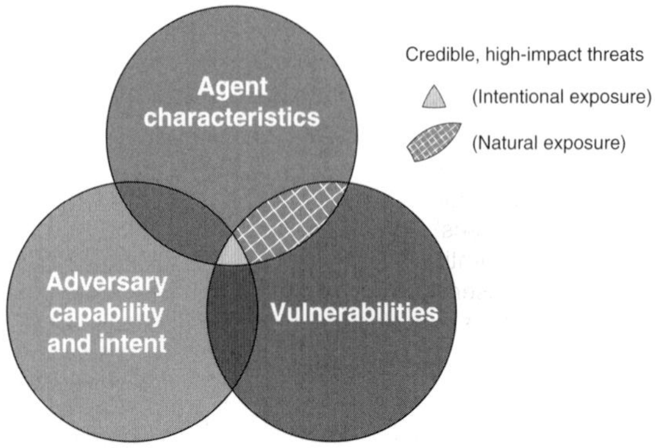

FIGURE 3.1 Parameters of risk.

tropical rain forests of equatorial Africa, with infection of humans likely occurring during the French colonial period from 1919 to 1960 (Keele et al., 2006).

THE SPECTRUM OF BIOLOGICAL THREATS

Although the societal effects of emerging infectious diseases continue to shape the landscape of medical preparedness, the proliferation of biological materials, technologies, and expertise has increased the potential for adversaries to use viruses for biological attacks. Not only does society have a need for developing vaccines, therapeutics, and diagnostics to counter naturally occurring and emerging infectious diseases, but countermeasures must keep pace with potentially devastating adversarial efforts to subvert developing or existing biomedical technologies. Moreover, caution must be exercised to prevent biomedical advances that benefit mankind from being applied to create enhanced or advanced viral threat agents generated through the use of these technologies (Petro and Carus, 2005).

The spectrum of viruses considered as biological threat agents resulting from intentional exposure encompasses four categories as shown in Fig. 3.2, each of which presents unique challenges and significant opportunities for development of medical countermeasures (Bush, 2007). Traditional agents would include viruses that could be disseminated to cause mass casualties in susceptible populations. Although risk posed by traditional viral agents is significant, a finite number of viruses fit in this category. Of particular concern are variola virus (smallpox) and viruses that cause hemorrhagic fever such as filoviruses (Lake Victoria marburgvirus and ebola viruses), arenaviruses (Lassa, Junin, Machupo, Guanarito, and Sabia viruses), bunyaviruses (hantaviruses, Rift Valley fever virus, and Crimean-Congo hemorrhagic fever virus), and flaviviruses (yellow fever virus, tick-borne encephalitis virus, and dengue viruses) (NIAID, 2008). These viruses can be easily disseminated or transmitted from person-to-person, resulting in high case-fatality rates and have the potential for major public health impact (Peters, 2002); might cause public panic and social disruption; and require special action for public health preparedness (Centers for Disease Control and Prevention, 2008; WHO, 2005). Therefore, the development or existence of targeted medical countermeasures against these agents will reduce the current risk posed or the likelihood of mass casualties resulting from intentional use.

Emerging agents would include those viruses which undergo rapid co-evolution with their hosts and the environment. Emerging viruses are for the most part, zoonotic in nature and spillover via natural exposure to humans as incidental hosts: examples of which are SARS coronavirus and avian influenza virus. Improved disease surveillance, to include early detection, is essential for containing the spread of an outbreak and is dependent on strong, flexible public health systems at local, state, and federal levels (Centers for Disease Control and Prevention, 2008).

Enhanced agents would include selection of naturally occurring viruses to enhance their virulence or host range or to facilitate their delivery beyond parameters that define natural exposure. For example, viruses used in an aerosol attack may exhibit altered patterns of infectivity, resulting in a different outcome than that observed in naturally occurring outbreaks. Enhanced viral agents may also be capable of producing disease with a more rapid onset of symptoms, making timely diagnosis, treatment, and containment more difficult (Schmaljohn and Hevey, 2005). Increasingly, medical countermeasure development

Traditional	***Enhanced***	***Emerging***	***Advanced***
Naturally occurring microorganisms (e.g., viruses) or toxins with the potential to be disseminated to cause mass casualties.	Traditional agents that have been modified or selected for enhanced ability to harm human populations or circumvent current countermeasures.	Previously unrecognized pathogens that might be naturally occurring and present a serious risk to human populations.	Novel pathogens or other materials of biological nature that have been artificially engineered to bypass traditional countermeasures or produce a more severe or enhanced spectrum of disease.

FIGURE 3.2 The spectrum of biological threats.

strategies should include protection against agent enhancements, such as aerosol exposure as mentioned above.

Advanced agents would include viruses that have been artificially engineered in the laboratory to circumvent countermeasures or produce a more severe or enhanced spectrum of disease. Of particular concern have been examples of viruses engineered to contain host biological response modulators, which in animal models have demonstrated an ability to circumvent levels of protective immunity obtained from prior vaccination, a primary example being ectromelia virus engineered to express IL-4 (Jackson et al., 2001).

Given the challenges of developing medical countermeasures against enhanced, emerging, and advanced agents, which are confounded to a great extent by unknowns, a posture of preparedness is largely based on the ability to rapidly identify and characterize the agent, understand the extent of exposure, and to evoke appropriate levels of patient care and social distancing to limit the spread of infection (World Health Organization and Robinson, 2004).

VIRUSES OF CONCERN

Several lists of threat agents, which include viruses, have been generated by groups or agencies within governments for a variety of purposes, including relative risk, biosecurity, and export control (select agents), as shown in Table 3.1. Although all of the agents listed represent traditional or emerging agents, the likelihood of their use for intentional exposure is a key consideration for prioritizing development of biodefense medical countermeasures, including vaccines.

Smallpox is a bioterrorism threat due to its potential to cause severe morbidity in a nonimmune population and because it can be transmitted via inhalation, ingestion, abrasion, or injection (Darling et al., 2005). A single confirmed case of smallpox identified anywhere in the world would be considered a public health emergency (Saks and Karras, 2006).

The first documented attempt to use variola virus as a biological weapon occurred during the French and Indian Wars (1754–1767) in North America. Blankets from smallpox victims were given as gifts to susceptible North American Indians, resulting in epidemics among native populations leading to case-fatality rates approaching 50% in some areas. During World War II, the Japanese Military conducted experiments with variola virus for use as a biological weapon. The former Soviet Union conducted biological weapons programs to include large-scale industrial production of variola virus that was stable in an aerosol and was claimed to be deliverable in warheads (Alibek and Handelman, 1999).

Subsequent to the World Health Organization (WHO) declaration of the eradication of smallpox in 1980, member nations agreed to destroy all laboratory stocks of the virus and/or provide them to one of the two officially sanctioned WHO reference laboratories: The Centers for Disease Control and Prevention in Atlanta, GA, or the Institute of Virus Preparations in Moscow, Russia. Although remaining stocks at these laboratories were scheduled to be destroyed in 1996, the World Health Assembly subsequently accepted a recommendation to retain the remaining stocks and to review a decision to destroy them on a periodic basis. Should the existing repositories be destroyed, other potential sources of smallpox may exist. Conceivably, sources of the variola virus may include unknown stores, cadavers in permafrost, and recreation of the viral genome or strains in laboratories utilizing genetic engineering, recombination, and data from published genomic sequences as a basis for design (Darling et al., 2005).

In addition to variola virus, a number of other viruses have been weaponized or studied for possible use as biological weapons against humans by state programs, including Lake Victoria marburgvirus, Venezuelan equine encephalitis virus, ebola viruses, Junin virus (Argentinian hemorrhagic fever), Machupo virus (Bolivian hemorrhagic fever), yellow fever virus, Lassa virus, Japanese encephalitis virus, and tick-borne encephalitis virus (Alibek and Handelman, 1999). Most of these viruses can be readily acquired from naturally occurring outbreaks and can be grown in the laboratory to high titers. The hemorrhagic fever viruses associated with this group are notorious for causing aerosol-mediated laboratory infections with case-fatality rates as great as 30% (Peters, 2002).

A number of viruses of concern are viewed primarily as threats to agriculture and their introduction to domestic livestock would have devastating effects such as reduced food production, economic loss in food and farm sectors, export embargoes, destabilization of markets, and loss of public confidence in the safety of the food supply (CIDRP, 2008). Such viruses would include foot-and-mouth disease, Rift Valley fever, Venezuelan equine encephalitis, and other viruses that may employ invertebrate hosts as vectors, from which virus can be subsequently amplified in animal hosts and also can be transmitted to humans. In addition, Nipah and Hendra viruses have been transmitted from domestic animals or livestock to

TABLE 3.1 List of threat agents (viruses) generated by groups or agencies within governments for relative risk, biosecurity, and export control

List	African horse sickness virus	African swine fever virus	Akabane virus	Alcelaphine herpesvirus type 1	Andes virus	Bluetongue virus	Caliciviruses	California encephalitis virus	Camel pox virus
Australia Group	Yes	Yes	No	No	No	Yes	No	No	No
HHS	No	No	No	No	No	No	No	No	No
USDA	Yes	Yes	Yes	Yes	Yes	Yes	No	No	Yes
WHO	No	No	No	No	No	No	No	No	No
NATO	No	No	No	No	No	No	No	No	No
NIAID	No	No	No	No	Yes (C)	No	Yes (B)	Yes (B)	No

List	Cercopithecine herpesvirus	Chikungunya virus	Classical swine fever virus	Crimean-Congo hemorrhagic fever virus	Dengue viruses	Dobrava virus	Eastern equine encephalomyelitis virus	Ebola viruses	Flexal virus
Australia Group	No	Yes	No	Yes	Yes	Yes	Yes	Yes	Yes
HHS	Yes	No	No	Yes	No	No	Yes	Yes	Yes
USDA	No	No	Yes	No	No	No	Yes	No	No
WHO	No	Yes	No	Yes	Yes	No	Yes	No	No
NATO	No	Yes	No	Yes	Yes	No	Yes	Yes	No
NIAID	No	Yes (C)	No	Yes (C)	Yes (A)	Yes (C)	Yes (B)	Yes (A)	No

List	Foot and mouth disease virus	Goat pox virus	Guanarito virus	Hantaan virus	Hendra virus	Hepatitis A virus	Herpes virus (Aujeszky's disease)	Hog cholera virus	Influenza viruses
Australia Group	Yes	Yes	Yes	Yes	No	No	Yes	Yes	No
HHS	No	No	Yes	No	Yes	No	No	No	Yes
USDA	Yes	Yes	No	No	Yes	No	No	No	Yes
WHO	No	No	No	Yes	No	No	No	No	Yes
NATO	No	No	No	Yes	No	No	No	No	Yes
NIAID	No	No	Yes (A)	Yes (A)	Yes (C)	Yes (B)	No	No	Yes (C)

(Continued)

TABLE 3.1 (Continued)

List	Japanese encephalitis virus	Junin virus	Kyasanur Forest disease virus	LaCrosse virus	Lake Victoria marburgvirus	Lassa virus	Louping ill virus	Lumpy skin disease virus	Lymphocytic choriomeningitis virus
Australia Group	Yes	Yes	Yes	No	Yes	Yes	Yes	Yes	Yes
HHS	No	Yes	Yes	No	Yes	Yes	No	No	No
USDA	Yes	No	No	No	No	No	No	Yes	No
WHO	Yes	No	No	No	Yes	No	No	No	No
NATO	No	Yes	No	No	No	Yes	No	No	No
NIAID	Yes (B)	Yes (A)	Yes (B)	Yes (B)	Yes (A)	Yes (A)	No	No	Yes (A)

List	Lyssavirus	Machupo virus	Menangle virus	Monkey pox virus	Murray Valley encephalitis virus	Newcastle disease virus	Nipah virus	O'nyong-nyong virus	Omsk hemorrhagic fever virus
Australia Group	Yes	Yes	No	Yes	Yes	Yes	Yes	No	Yes
HHS	No	Yes	No	Yes	No	No	Yes	No	Yes
USDA	No	No	Yes	No	No	Yes	Yes	No	No
WHO	No	No	No	No	No	No	No	Yes	No
NATO	No	Yes	No	No	No	No	No	No	Yes
NIAID	No	Yes (A)	No	Yes (A)	No	No	Yes (C)	No	Yes (C)

List	Oropouche virus	Peste des petits ruminants virus	Porcine enterovirus virus type 9	Powassan virus	Puumala virus	Rabies virus	Rift Valley fever virus	Rinderpest virus	Rocio virus
Australia Group	Yes	Yes	Yes	Yes	Yes	No	Yes	Yes	Yes
HHS	No	No	No	No	No	No	Yes	No	No
USDA	No	Yes	No	No	Yes	No	Yes	Yes	No
WHO	No	No	No	No	No	No	Yes	No	No
NATO	No	No	No	No	No	No	Yes	No	No
NIAID	No	No	No	No	Yes (C)	Yes (C)	Yes (A)	No	No

List	Sabia virus	SARS corona virus	Seoul virus	Sheep pox virus	Sin Nombre virus	St. Louis encephalitis virus	Swine vesicular disease virus	Teschen disease virus	Tick-borne encephalitis virus
Australia Group	Yes	No	Yes	No	No	Yes	No	Yes	Yes
HHS	Yes	No	No	No	No	No	No	No	Yes
USDA	No	No	Yes	Yes	No	No	Yes	No	No
WHO	No	No	Yes	No	No	No	No	No	Yes
NATO	No	No	Yes	No	No	No	No	No	Yes
NIAID	Yes (A)	Yes (C)	Yes (A)	No	Yes (C)	No	No	No	Yes (C)

List	Variola virus	Venezuelan equine encephalomyelitis virus	Vesicular stomatitis virus	West Nile virus	Westewn equine encephalomyelitis virus	White pox agent	Yellow fever virus
Australia Group	Yes	Yes	Yes	No	Yes	Yes	Yes
HHS	Yes	Yes	No	No	No	No	No
USDA	No	Yes	Yes	No	No	No	No
WHO	Yes	Yes	No	No	Yes	No	Yes
NATO	Yes	Yes	No	No	No	No	Yes
NIAID	Yes (A)	Yes (B)	No	Yes (B)	Yes (B)	No	Yes (C)

Notes: (A) NIAID Category A Priority Pathogen; (B) NIAID Category B Priority Pathogen; and (C) NIAID Category C Priority Pathogen.

humans leading to high case-fatality rates and therefore, are of concern from both an agricultural and a public health perspective.

PREPAREDNESS

Biodefense vaccine development represents, in part, a defensive strategy to protect the public from biological terrorism through the development of safe and effective countermeasures to mitigate illness and death, protect critical infrastructure, and minimize economic consequences (Gay et al., 2007). The need to accelerate and expand development of current and new vaccines to prepare for intentional introduction or natural occurrence of catastrophic human and zoonotic diseases is reflected by the list of vaccines shown in Table 3.2. In summary, licensed vaccines exist for only a few viruses of concern and are of limited supply in case of catastrophic need (Schmaljohn and Hevey, 2005). For other viruses, vaccines are available under investigational new drug (IND) status, which have not been licensed for a variety of reasons, primarily the lack of market incentive to develop through to licensure (Schmaljohn and Hevey, 2005). In light of the paucity of medical countermeasures effective against viruses of concern, subsequent chapters of this text are devoted to review of vaccine development efforts against biodefense, emerging and neglected diseases.

Adoption of a national strategy to rapidly develop effective biodefense countermeasures, which includes safe and efficacious vaccines, requires periodic threat and risk assessments to understand the evolving biological threat (Bush, 2002). Such assessment must be performed to define gaps in knowledge or vulnerabilities in our preparedness posture, thus guiding prioritization of investments in biodefense-related research, development, planning, and overall preparedness. Vaccines represent a necessary component of biodefense countermeasure development to prepare for and to minimize the consequences of a biological attack. Risk assessments have provided a means of identifying and prioritizing vaccine research and development efforts for biological agents capable of being used in biological attacks. In support of vaccine research and development, current biodefense strategies include the development of an enduring infrastructure to test and evaluate existing, proposed, or promising countermeasures, assess their safety and effectiveness, expedite their development, and ensure rapid licensure (Gronvall et al., 2007c; NIAID, 2008; U.S. Department of Health & Human Services, 2008).

Response to bioterror events involves rapid agent detection and diagnosis, implementation of exposure control measures, and use of vaccines, when appropriate, to minimize consequences to human and animal health (Gronvall et al., 2007b). Current approaches to biodefense vaccine development include use of existing applications for development and manufacturing, and development of new technologies to rapidly design, test and evaluate, and manufacture desired products. Given the urgency with which vaccines designed to minimize consequences of bioterror events are needed, new technologies and infrastructure are mandated to shorten timelines for their development. New technologies applicable to the design of vaccines would include rapid methods for the identification and selection of protective immunogens, and the development of vaccines that allow for differentiation of vaccinated hosts from those infected by exposure to virus, with options to employ improved adjuvants that stimulate innate immune responses and/or promote adaptive immune responses (Capua, et al., 2004). New infrastructure would include biocontainment facilities to increase current capacities for vaccine testing and evaluation, to optimize vaccine dosage and delivery in conjunction with routes of exposure, and enhance manufacturing capacities to ensure availability of a minimum number of doses for intervention and reduce average cycle times to manufacture vaccine lots. Rapid deployment of biodefense vaccines will require both infrastructures, to include facilities for storage of vaccine stockpiles, and technologies that ensure validated storage conditions for such stockpiles (Gronvall et al., 2007a).

CONCLUSIONS

Programs to develop better medical countermeasures against viruses must increasingly reflect the potential use of enhanced, emerging, or genetically engineered viruses as biological weapons and scenarios that require countermeasures to provide broad-spectrum coverage to prevent illness postexposure. Support for research and development of biodefense countermeasures, including vaccines, should result in improved methods for vaccine discovery and development, and enhance preparedness against viruses, as biological threats, to the benefit of public and animal health.

ACKNOWLEDGMENTS

Many thanks to David Hodge, Sara Klucking, and Jens Kuhn for critical review and to Ed Marino and Cynthia Sheffield for exceptional technical assistance in preparation of this chapter.

TABLE 3.2 Available countermeasures against viruses considered as biological threat agents

Virus	Countermeasure(s)
Alphavirus	
Venezuelan equine encephalitis virus	Live attenuated and killed vaccines (IND)
Eastern equine encephalitis virus	Killed vaccine (IND)
Western equine encephalitis virus	Killed vaccine (IND)
Chikungunya virus	Live attenuated vaccine (IND)
Arenavirus	
Lassa virus	No vaccine, Ribavirin
Junin virus	Live attenuated vaccine (IND), Ribavirin
Machupo virus	No vaccine, Ribavirin
Bunyaviridae	
Crimean-Congo hemorrhagic fever virus	No vaccine, Ribavirin
Hantaan, Seoul, Puumala, Dobrava viruses	No vaccine, Ribavirin
Sin Nombre, Andes viruses	No vaccine, Ribavirin
Rift Valley fever virus	Killed vaccine (IND)
Flavivirus	
Tick-borne encephalitis virus	Killed vaccine (licensed Europe)
Yellow fever virus	Live attenuated vaccine (licensed)
Japanese encephalitis virus	Killed vaccine (licensed)
Dengue viruses	Live attenuated vaccine (IND),
Filoviridae	
Ebola viruses	None
Lake Victoria marburgvirus	None
Henipavirus	
Nipah and Hendra viruses	None
Orthopoxvirus	
Variola virus	Live attenuated vaccine (licensed), Cidofovir
Monkeypox virus	Live attenuated vaccine (licensed), Cidofovir

References

Alibek, K., and Handelman, S. Biohazard: the Chilling True Story of the Largest Covert Biological Weapons Program in the World, Told From the Inside by the Man Who Ran It. 1st ed. xi, 319 p, [8] p of plates vols. New York: Random House, 1999.

Avert. Worldwide AIDS & HIV Statistics Including Deaths. Apr. 13, 2008. http://www.avert.org/worldstats.htm.

Bush, G.W. Biodefense for the 21st Century: Homeland Security Presidential Directive/HSPD-10. Apr. 13, 2008. 2002. http://www.whitehouse.gov/news/releases/2004/04/20040428-6.html.

Bush, G.W. Medical Countermeasures against Weapons of Mass Destruction: Homeland Security Presidential Directive/HSPD-18. Apr. 13, 2008. 2007. http://www.whitehouse.gov/news/releases/2007/02/20070207-2.html.

Capua, I., Cattoli, G. and Marangon, S. DIVA: a vaccination strategy enabling the detection of field exposure to avian influenza. Dev. Biol. (Basel) 2004; 119:229–233.

Carus ,W.S.. Bioterrorism and Biocrimes. The Illicit Use of Biological Agents Since 1900. Apr. 13, 2008. 2001. National Defense University http://stinet.dtic.mil/oai/oai?verb=getRecord&metadataPrefix=html&identifier=ADA402108.

Center for Biosecurity. Agents and Diseases Background Information: Categorization and Ongoing Assessment of Biological Agents. Apr. 13, 2008. http://www.upmc-biosecurity.org/website/focus/agents_diseases/background.html.

Centers for Disease Control and Prevention. Public Health Emergency Response: The CDC Role. Apr. 13, 2008. http://www.bt.cdc.gov/.

CIDRP. Overview of Agricultural BioSecurity. Apr. 13, 2008. http://www.cidrap.umn.edu/cidrap/content/biosecurity/ag-biosec/biofacts/agbiooview.html.

Darling, R.G., Burgess, T.H., Lawler, J.V. and Endy, T.P. Smallpox as a biological weapon. Biological weapons defense infectious diseases and counterbioterrorism. In: Infectious Disease (L.E. Lindler, F.J. Lebeda, and G. Korch, Eds.), xix, 597 p vols, pp. 99–120. Totowa, NJ: Humana Press, 2005.

Gay, C.G., Richie, T.L., Pastoret, P.P., Minguez-Tudela, I., de Baetselier, P., Gobel, T., Goddeeris, B., Kaiser, P., Morrison, I., Sanchez-Vizcaino, J.M., Anderson, K., Baillie, L.W., Brown, W.C., Estes, D.M., Herrera, E., Nara, P.L., Ockenhouse, C.F., Roth, J.A. and Sztein, M.B. Advances in immunology and vaccine discovery report of the United States–European Commission workshop. Vaccine 2007; 25(41): 7007–7011.

Gronvall, G.K., Fitzgerald, J., Chamberlain, A., Inglesby, T.V. and O'Toole, T. High-containment biodefense research laboratories: meeting report and center recommendations. Biosecur. Bioterror. 2007a; 5(1):75–85.

Gronvall, G.K., Matheny, J., Smith, B.T., Mair, M., Chamberlain, A., Deitch, S., Borio, L., Inglesby, T.V. and O'Toole, T. Flexible defenses roundtable meeting: promoting the strategic innovation of medical countermeasures. Biosecur. Bioterror. 2007b; 5(3):271–277.

Gronvall, G.K., Smith, B.T., Matheny, J., Mair, M., Chamberlain, A., Deitch, S., Borio, L., Inglesby, T.V. and O'Toole, T. Biomedical Advanced Research and Development Authority (BARDA) roundtable. Biosecur. Bioterror. 2007c; 5(2):174–179.

Jackson, R.J., Ramsay, A.J., Christensen, C.D., Beaton, S., Hall, D.F. and Ramshaw, I.A. Expression of mouse interleukin-4 by a recombinant ectromelia virus suppresses cytolytic lymphocyte responses and overcomes genetic resistance to mousepox. J. Virol. 2001; 75(3):1205–1210.

Keele, B.F., Van Heuverswyn, F., Li, Y., Bailes, E., Takehisa, J., Santiago, M.L., Bibollet-Ruche, F., Chen, Y., Wain, L.V., Liegeois, F., Loul, S., Ngole, E.M., Bienvenue, Y., Delaporte, E., Brookfield, J.F., Sharp, P.M., Shaw, G.M., Peeters, M. and Hahn, B.H. Chimpanzee reservoirs of pandemic and nonpandemic HIV-1. Science 2006; 313(5786):523–526.

NIAID. NIAID Category A, B, and C Priority Pathogens. Apr. 13, 2008. http://www3.niaid.nih.gov/topics/Biodefense Related/Biodefense/research/CatA.htm.

Peters, C.J. Many Viruses Are Potential Agents of Bioterrorism. ASM News, Apr. 2002: 168–173.

Petro, J.B. and Carus, W.S. Biological threat characterization research: a critical component of national biodefense. Biosecur. Bioterror. 2005; 3(4):295–308.

Saks, M.A. and Karras, D. Emergency medicine and the public's health: emerging infectious diseases. Emerg. Med. Clin. North Am. 2006; 24(4):1019–1033.

Schmaljohn, A. and Hevey, M. Medical countermeasures for filoviruses and other viral agents. Biological weapons defense infectious diseases and counterbioterrorism. In: Infectious Disease (L.E. Lindler, F.J. Lebeda, and G. Korch, Eds.), xix, 597 p vols. pp. 239–253. Totowa, NJ: Humana Press, 2005.

Swearengen, J.R. Biodefense Research Methodology and Animal Models. 331 p vols. Boca Raton, FL: Taylor & Francis, 2006.

Taubenberger, J.K. and Morens, D.M. 1918 Influenza: the mother of all pandemics. Emerg. Infect. Dis. 2006; 12(1):15–22.

U.S. Department of Health & Human Services. Biomedical Advanced Research and Development Authority. Apr. 13, 2008. http://www.hhs.gov/aspr/barda/index.html.

WHO. WHO Checklist for Influenza Pandemic Preparedness Planning. Apr. 13, 2008. 2005. Geneva: World Health Organization. http://www.who.int/csr/resources/en/.

World Health Organization. Ten Things you Need to Know About Pandemic Influenza. Apr. 13, 2008. Geneva: World Health Organization. http://www.who.int/csr/disease/influenza/pandemic10things/en/index.html.

World Health Organization and Julian Perry Robinson. Public Health Response to Biological and Chemical Weapons WHO Guidance. 2nd ed xix, 340 p vols. Geneva: World Health Organization, 2004.

SECTION II

FUNDAMENTAL ASPECTS OF VACCINOLOGY

CHAPTER

4

Vaccine Development Pathway

Lawrence R. Stanberry[1] *and Alan D.T. Barrett*[2]

[1]Department of Pediatrics, College of Physicians and Surgeons, Columbia University, New York, NY, USA

[2]Department of Pathology, Sealy Center for Vaccine Development, Center for Biodefense and Emerging Infectious Diseases, and Institute for Human Infections and Immunity, University of Texas Medical Branch, Galveston, TX, USA

OUTLINE

ABSTRACT

The process to develop a vaccine becomes longer and more complex each year due to the many steps required to prepare and get a product licensed. Here, we overview the complex pathway that is needed to develop a vaccine. The process starts with basic science/discovery, proceeds through preclinical development, to clinical trials, manufacturing, registration/licensure, implementation, and finally post-marketing surveillance.

INTRODUCTION

Development of a vaccine is a very complex process involving multiple steps and a long period of time. It has been estimated that it takes 15 years for a vaccine to move from the initial discovery through to licensure. Why does it take so long? This can be explained by the need to have a vaccine that is both safe and efficacious. In the 20th century the focus was on preventing infectious diseases while in the 21st century,

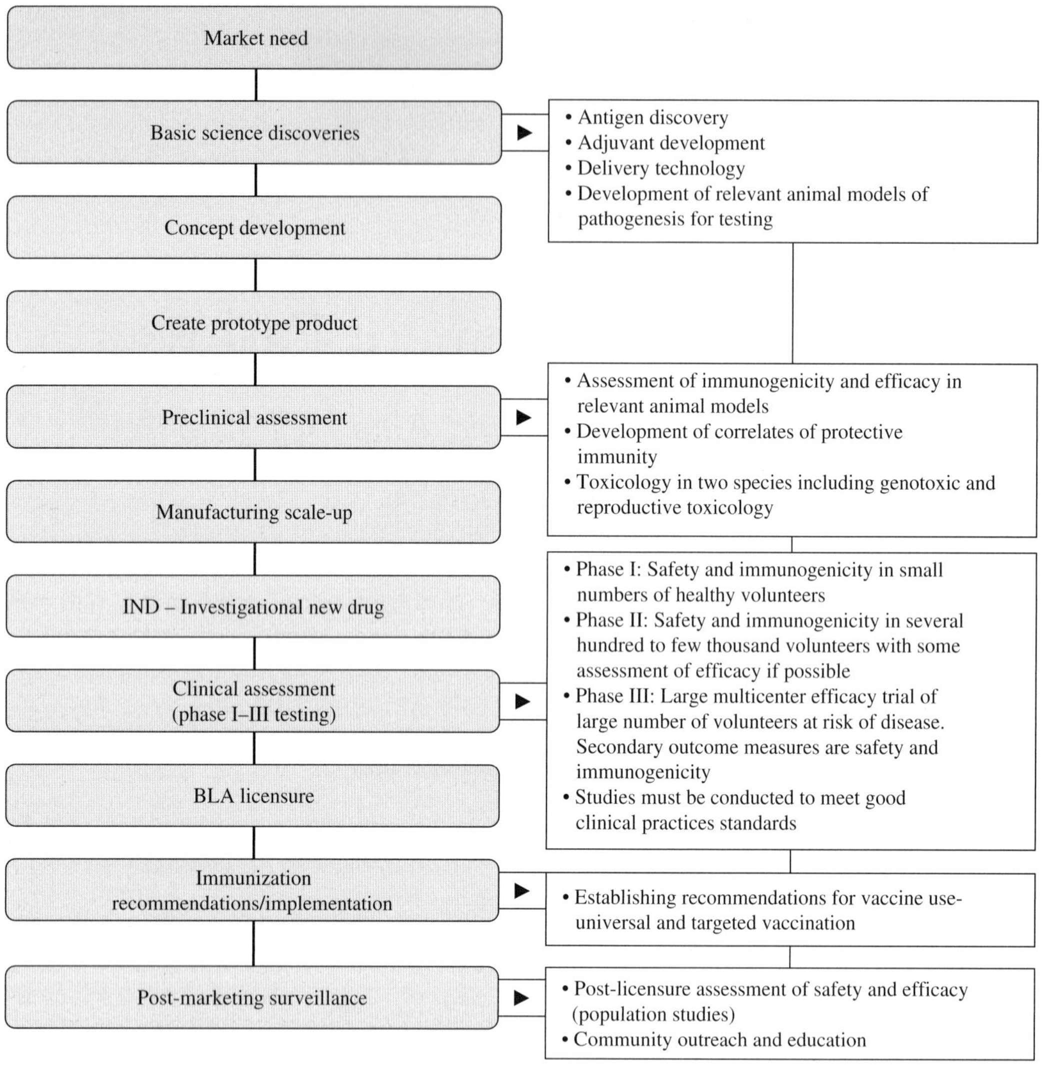

FIGURE 4.1 The vaccine development pathway.

society has changed its views on vaccines and now focuses on the need to have a vaccine that is 100% safe. In part, this is due to the successes of science and public health in the 20th century where many of the major infectious diseases have greatly reduced in incidence and rarely seen in developed countries such that the vaccine is now considered to be a potential cause of disease. Thus, in general, the steps to develop a vaccine fall into three categories: financial, scientific, and regulatory. The pathway is illustrated in Fig. 4.1.

MARKET NEED

The first step in vaccine development is determination of the need for the proposed vaccine (Batson, 2002). Need is not simply based on the number of people who might benefit from the product but rather it is based on the number of people who likely could pay for the vaccine (or have it purchased on their behalf). Consequently, estimates of market need are then influenced by the anticipated vaccine purchase price. Although there may be more individuals in a developing country that would benefit from a vaccine there may be fewer financial resources available to buy the product and hence the market need may be perceived as insufficient to support the product development, i.e., the vaccine producer has to recover the costs of research and development plus the costs involved in manufacturing and marketing, and also make a profit for shareholders. While scientific feasibility, and political and marketplace issues, heavily influence the decision to develop a vaccine, it is economics that ultimately drives the decision. The investment costs required for development, manufacturing, and product distribution are weighed against anticipated revenue to determine the potential financial return on investment. The development of a vaccine may take more than 10 years and cost between $200–500 million (Struck, 1996; Andre, 2002). The decision to proceed with development is influenced by the projected return on investment and by anticipated risks that could impact successful implementation of the vaccine to the general population. Governments that fund the development of vaccines may accept a lower return on investment, or the return may be calculated to include the saving associated with monies required to treat and prevent the disease by other modalities. For example, a mosquito-borne disease such as dengue which has no vaccine currently requires monies to support control of the mosquito vector plus medical support for individuals who have the disease. For vaccines developed in the private sector, the so-called "push and pull" mechanisms can impact the economic formula (Pauly, 2005; Batson, 2005; Grabowski, 2005). Push mechanisms are those that subsidize or reduce research and development costs or risks; e.g., tax credits for working on government targeted diseases. Pull mechanisms pledge purchase of the vaccine once developed and hence reduce uncertainty regarding potential revenue.

TABLE 4.1 Examples of university and hospital contributions to vaccine development

Contribution	Investigators	Institution
Growth of viruses in cell cultures	Enders, Weller, and Robbins	Harvard
Inactivated polio vaccine	Salk	University of Pittsburgh
Inactivated influenza vaccine	Frances	University of Michigan
Oral polio vaccine	Sabin	Cincinnati Children's Hospital
Haemophilus influenza type B vaccine	Smith and Anderson	University of Rochester
Measles vaccine	Enders and Katz	Harvard University
Varicella vaccine	Takahashi	Niigata University
Yellow fever vaccine	Theiler	Harvard/Rockefeller Foundation
Rubella vaccine	Plotkin	University of Pennsylvania
Rabies vaccine	Koprowski	Wistar Institute/University of Pennsylvania
Rotavirus vaccine	Bernstein and Ward	Cincinnati Children's Hospital
Papillomavirus vaccine	Fraser	University of Queensland
Rotavirus vaccine	Clark, Offit, and Plotkin	Children's Hospital of Philadelphia
Papillomavirus vaccine	Reichman, Bonnez, and Rose	University of Rochester

BASIC SCIENCE DISCOVERIES

Basic science discoveries are key to the development of new vaccines (Poland et al., 2002). Breakthroughs or insights resulting from fundamental research can inform use-inspired science (The Biodesign Institute, http://www.biodesign.asu.edu/news/New-ASU-Interdisciplinary-PhD-Graduate-Program-in-Biological-Design-Set-for-Launch). Industry-based vaccine scientists constantly monitor the ongoing research at universities and in government, institute and research-oriented hospital laboratories seeking to identify advances that can be applied to vaccine development. Historically, university- and hospital-based researchers have been instrumental in the development of vaccines (Table 4.1). Increasingly, academic and government researchers are focusing more effort in product-oriented research with greater emphasis on protecting the intellectual property developed in their laboratories.

Not all discoveries have an obvious application at the time they are made. Discoveries in molecular genetics that established the field of genetic engineering (Winther and Dougan, 1984) led to the development of recombinant hepatitis B virus (HBV) vaccines (Zahradnik et al., 1987; Kuwert et al., 1985), which replaced a HBV vaccine derived from purification of particles from individuals infected with the virus. The HBV vaccine consists of the viral S protein produced by yeast genetically engineered to express the viral gene encoding the S protein. Examples of other licensed or experimental recombinant vaccines include the human papillomavirus vaccines (Koutsky et al., 2002; Harper et al., 2004), herpes simplex virus (HSV) vaccines (Corey et al., 1999; Stanberry et al., 2002), a malaria vaccine (Polhemus et al., 2007), and a cytomegalovirus vaccine (Bernstein et al., 2002). Advances in genetic engineering, including reverse genetics (see Chapters 6 and 7), have permitted the development of genetically attenuated experimental viral vaccines including vaccines against HSV (de Bruyn et al., 2006), influenza (Leroux-Roels et al., 2007), dengue (Edelman et al., 2003), and chimeric vaccines such as ChimeriVax-Den where dengue genes are incorporated into the yellow fever 17D vaccine virus backbone (Guirakhoo et al., 2006). Genetic engineering also permitted the development of vector systems, including vaccinia virus (Smith et al., 1984), canarypox (Taylor et al., 1994), adenovirus (Graham, 1990), yellow fever virus (Galler et al., 1997), and salmonella (Schödel and Curtiss, 1995; Kwon et al., 2007). These vector platforms are being used in a variety of new experimental vaccines against influenza, human immunodeficiency virus (HIV), Japanese encephalitis, and tuberculosis (Russell et al., 2007; Goonetilleke et al., 2006; Van Kampen et al., 2005; Monath et al., 2003; Xing and Lichty, 2006).

The discovery that the intramuscular injection of a DNA plasmid coding for an immunogenic protein could elicit an immune response to the encoded protein (the so-called genetic immunization) led to the development of DNA vaccines (Tang et al., 1992; Ulmer et al., 1993). The application of the DNA vaccine platform to the development of vaccines for biodefense and emerging and neglected diseases is reviewed by Lu and colleagues in Chapter 8. The extension of the discovery of genetic immunization has permitted antigen discovery through immunization with expression libraries of pathogen DNA (Barry et al., 1995).

An early recognition (Avery and Goebel, 1929) that linking a poorly immunogenic carbohydrate antigen to a more potent protein antigen increased the immunogenicity of the carbohydrate antigen formed the basis of conjugate vaccines (Rappuoli, 2001). Application of this discovery has led to the development of vaccines that protect against infections caused by *Haemophilus influenzae* type B (Heath, 1998) and selected serotypes of *Streptococcus pneumonia* (Oosterhuis-Kafeja et al., 2007) and *Neisseria meningitides* (Keyserling et al., 2006).

Recent discoveries in the area of basic immunology concerning toll-like receptors and control of innate immunity (Kopp and Medzhitov, 1999) is providing a framework for understanding adjuvant mechanisms of action (Kwissa et al., 2007) and rational adjuvant discovery programs (Hauguel and Hackett, 2008), which will ultimately lead to vaccines with improved immunogenicity.

CONCEPT DEVELOPMENT

Concept development is a complex and iterative process that examines the technical and marketing issues relevant to a proposed new product. In general, the process should produce a product profile, identify essential technology with assessment of intellectual property issues and evidence supporting feasibility, address manufacturing issues, and examine market need. In general, organizations involved in product development have somewhat different approaches to the process of concept development. This phase of the vaccine development pathway has been described by some as the "Fuzzy Front End," the stage of the development process between when an idea is first considered and when it is determined to be ready for development (Koen et al., 2001).

CREATE PROTOTYPE PRODUCT

The creation of the prototype vaccine is a critical step in the development pathway. This stage not only demonstrates the technical feasibility of producing the product but should also lay the groundwork for meeting the regulatory requirements that will later be required for a U.S. investigational new drug (IND) or European clinical trial application (CTA), and ultimately licensure. Elements considered in prototype development include production of vaccine components and formulation with an eye toward eventual scale-up and manufacturing, and use of components that will meet regulatory requirements such as those outlined in U.S. FDA guidance documents such as, "Characterization and Qualification of Cell Substrates and Other Biological Starting Materials Used in the Production of Viral Vaccines for the Prevention and Treatment of Infectious Diseases" or "Considerations for Plasmid DNA Vaccines for Infectious Disease Indications" (see Chapters 13 and 14). Industry is well attuned to addressing early the regulatory requirements that will delay or derail the downstream product development; academic scientists, however, are generally less aware of the requirements to get a vaccine to licensure and prototype vaccines developed in a research laboratory may lack detailed information on the derivation and passage history of cell cultures, media used, use of materials from bovine sources, testing for adventitious agents, and origin and history of viral or bacterial agents used in the development of the prototype vaccine. Lack of attention to meeting downstream regulatory requirements can be costly both in terms of time and money. Sufficient quantities of the prototype vaccine are prepared, ideally under conditions that meet good laboratory practices (GLP) (US Food and Drug Administration, http://www.fda.gov/ora/compliance_ref/bimo/glp/default.htm) (see Chapter 10), and the prototype vaccine then moves to the preclinical assessment phase.

PRECLINICAL ASSESSMENT

Before entering clinical trials, prototype vaccines are tested to evaluate safety (toxicology), immunogenicity, and, when possible, efficacy. Toxicology testing must be conducted in a GLP-compliant laboratory and ideally the immunogenicity and animal efficacy studies also conducted under GLP. Nonclinical safety assessment includes tests to determine the direct toxicity of the product, the lack of contaminants, impurities, and adventitious agents, and the potential of the vaccine to trigger preexisting conditions (Verdier, 2002). In general, vaccine toxicology is assessed in single- and repeat-dose studies in two mammalian species, typically one rodent and one non-rodent (often a nonhuman primate). The vaccine is usually given by the route of administration planned for clinical use and tested at the maximum proposed human dose. The repeat dose studies typically include one dose more than anticipated for use in humans, e.g., if a three-dose regimen is planned for human studies then animal studies would examine the safety of a four-dose regimen. The repeat-dose studies are aimed at determining a no-observed-adverse-effect-level (NOAEL). The toxicology assessments generally include mortality, clinical signs, food consumption, body weight, clinical chemistry, hematology, gross pathology, and histopathology. Vaccines intended for pregnant women require genetic and reproductive toxicology studies (Verdier et al., 2003). The other product-specific tests may be required such as potential of a live viral vaccine to establish a persistent infection or of a DNA vaccine to integrate into the host genome. With the development of new vaccine platforms, novel adjuvant systems, and unique delivery systems it is likely that expanded safety testing will be required (Brennan and Dougan, 2005). As one might expect, the more detailed the studies, the longer it takes for a vaccine to reach licensure.

Preclinical immunogenicity testing determines the ability of the experimental vaccine to induce humoral and/or cellular immune responses, and this will be developed into correlate(s) of protection. In the case of most vaccines developed to date, this has revolved around neutralizing antibodies. The types of assays to be used may include well-established tests previously used in the development of licensed vaccines or new, yet-to-be validated assays. Ultimately, this will become a surrogate of protection. In general, the animal immunogenicity testing should use the same assays that will be used in the clinical trials.

Demonstration of efficacy in relevant animal models lends strong support for advancing a vaccine to evaluation in humans. Ideally the animal model utilizes the human pathogen and appropriate routes of challenge, e.g., intravaginal or ocular inoculation with HSV rather than injection of the virus into the footpad. It is also preferable that the pathogenesis and outcome measures should reflect what is observed in humans (Stanberry, 1991). Under some specific circumstances, animal studies may substitute for human studies in demonstrating vaccine effectiveness. The FDA Animal Efficacy Rule (the so-called Animal Rule), finalized in May 2002, allows for the development and licensure of products where human efficacy trials are not feasible or ethical. Such circumstances exist for many biothreat agents and emerging and neglected infectious diseases where there are insufficient numbers of cases of the disease to undertake efficacy studies. The rule permits the FDA to rely on evidence from animal studies to provide substantial evidence of vaccine effectiveness when the effect is demonstrated in either more than one animal species expected to react with a response predictive for humans, or one well-characterized animal species model (adequately evaluated for its responsiveness in humans) for predicting the response in humans, and when the animal study endpoint is clearly related to the desired benefit in humans (Food and Drug Administration, 2005). To date, no vaccine has been approved under the Animal Rule but there is optimism that this pathway will facilitate the development of products needed for biodefense and for the control of emerging and neglected infectious diseases (Rosenthal and Clifford, 2002; Gronvall et al., 2007).

MANUFACTURING SCALE-UP AND CURRENT GOOD MANUFACTURING PRACTICES (cGMP)

The small quantities of vaccine material required for preclinical studies are generally produced under GLP conditions in research laboratories or pilot plants. Such materials are currently acceptable for clinical trials, although it is probable that in the future vaccine materials intended for use in clinical studies will be produced according to current good manufacturing practices (cGMP). Transitioning from the production of small lots in the laboratory or pilot facilities to large-scale production in manufacturing plants requires extensive planning and specialized expertise, and is resource intense (see Chapter 11). Also, the large-scale commercial production of vaccines must comply with increasingly complex cGMP regulations (Milstien, 2004). In the United States the Food and Drug Administration has the responsibility for developing and promulgating the regulations, currently detailed in Sections 210 and 211 and applicable parts of Sections 600–680 of the U.S. Code of Federal Regulations (CFR); in Australia, regulations are developed by the Therapeutic Goods Administration; in Canada, by Health Canada; and in Europe, by the European Medicines Evaluation Agency (EMEA). In general, cGMP regulations require the manufacturer to define and follow the conditions and controls required throughout the production process to consistently produce a safe, pure, and potent product that is free of any environmental contamination. Each step of the production process must be documented and validated. Areas included in cGMP regulations include personnel education and training; facility design and maintenance; equipment design, calibration, maintenance, and operation; raw material sources and specifications; aseptic processing; segregation of pre- and post-inactivation steps; safety/purity assessment; potency assessment; analytical laboratory procedures including instrument calibration and maintenance; packaging procedures; and storage and shipping procedures. Because of the global nature of the vaccine industry, the multiple regulatory bodies and rules present increasing challenges to the development and use of new vaccines. To address this growing problem regulatory agencies and manufacturers in the United States, Europe, and Japan established the International Conference on Harmonization (ICH) (Milstien et al., 2006).

INVESTIGATIONAL NEW DRUG (IND) APPLICATION

In the United States in order to initiate clinical studies of a new experimental vaccine the FDA's Center for Biologics Evaluation and Research (CBER) must review and approve an IND application (see Chapter 14 and 13) (Code of Federal Regulations, 2005; Stehlin, 1996; National Institute of Allergy and Infectious Diseases, 2003; Davenport, 1995). The application must contain detailed information regarding the vaccine including the rationale for the vaccine, its production, characterization, purity, safety, animal immunogenicity and efficacy data and any human experience, clinical trial protocols, an investigators' brochure, and any other data relevant to the review (Table 4.2). Upon receipt, the FDA has 30 days to review the IND application. Under the IND the clinical trial protocols must undergo an ethics review and receive approval from an Institutional Review Board (IRB). In the clinical development pathway the candidate vaccine advances through increasingly complex phase 1, 2, and 3 trials.

TABLE 4.2 Elements of an investigational new drug application

The background and discovery of the vaccine
The chemistry, manufacturing, and controls of the vaccine and information on the final product formulation
Documentation of controlled and validated assay results supporting the product characterization including safety, consistency, stability, potency, and lot release criteria for the vaccine
The results of the preclinical (animal) toxicology and immunogenicity studies to support safety and animal efficacy data if available
Any human experience with either the experimental vaccine or similar products
Detailed clinical trial protocols
An investigators' brochure
Any additional information necessary to facilitate review and evaluation

Source: Adapted from http://www3.niaid.nih.gov/daids/vaccine/Science/VRTT/08_Pre-IND.htm.

As part of the IND process all clinical trials are monitored by the FDA and IRB to ensure safety. If safety concerns develop during a study the FDA can put the trial on hold until the concerns are thoroughly investigated and the issues resolved (Miller and Ross, 2005).

CLINICAL ASSESSMENT (PHASE 1, 2, AND 3 TESTING)

The pre-licensure evaluation of vaccines in humans proceeds in a stepwise manner through an increasingly complex series of clinical trials (see Chapters 12 and 14). The first studies, referred to as phase 1 trials, are safety and immunogenicity studies performed in a small number of closely monitored healthy volunteers. These are generally short-term studies that may examine a range of vaccine dosages. The study design may be uncontrolled open label or more often are randomized, double blind, and placebo controlled. Inclusion of a placebo group is important in assessing adverse events that may occur during the trial. Phase 2 studies involve larger numbers of healthy volunteers, possibly volunteers at risk for the target disease. The design may examine optimum dosage and dosing regimen, and focus on safety and immunogenicity, although depending on the prevalence of the target disease some efficacy data may be collected. These larger sample sizes provide greater power to detect less common adverse events. Phase 2 trials are randomized, blinded, and placebo controlled. Phase 3 trials may enroll hundreds to thousands of individuals depending on the prevalence of the target disease. Generally, thousands of volunteers are included in phase 3 trials in order to assess for rare adverse events. The phase 3 trials must be randomized, double blinded, and placebo controlled. Two large studies, or under some circumstances a single, large, multicenter phase 3 field trial designed to demonstrate the efficacy of the vaccine against the target disease is pivotal for licensure. As was discussed earlier, for target diseases that are rare, or where clinical trials cannot be ethically conducted, the FDA has promulgated an alternative pathway that substitutes efficacy data from relevant animal models but still requires extensive human safety data (Food and Drug Administration, 2005).

Throughout the world, regulatory agencies require that clinical trials be conducted to meet good clinical practice (GCP) standards promulgated by the ICH. GCP is an ethical and scientific quality standard for designing, conducting, recording, and reporting trials that involve the participation of human subjects. The standard was developed to provide the public assurance that the rights, safety, and well-being of subjects participating in clinical trials are protected and that the clinical trial data are credible (FDA (CBER), 1996).

BLA-LICENSURE

Following successful completion of the clinical trial program, the next step in the vaccine development pathway is submission of a Biologics License Application (BLA) (FDA (CBER), 2007). The applications are expected to contain all the information necessary for the FDA's multidisciplinary review team to evaluate the product safety and efficacy so as to make determination of the vaccine's risks and benefits. According to the U.S. federal regulations licensure requires that vaccines demonstrate safety, purity, and potency (see Chapter 14). During the BLA review process, the FDA conducts a pre-approval inspection of the planned manufacturing facility with detailed examination of the manufacturing processes and GLP compliance. In considering the application, the FDA may also conduct audits of the IND clinical studies, including site visits, to evaluate the conduct of studies and to ensure satisfactory record keeping. During the licensure process advice regarding the safety and efficacy of the vaccine for the proposed indication may be sought from the FDA's Vaccines and Related Biological Products Advisory Committee (VRBPAC), an independent panel of 12 experts. Adequate product labeling is a requirement for vaccine licensure. The labeling should permit health care providers to understand the vaccine's proper use, its potential risks and benefits, information to be conveyed to patients and parents, and how to safely deliver the vaccine.

Because the safety of a vaccine may not be apparent until it is used widely in the population, FDA approval of the new product may be contingent upon the conduct of phase 4 studies, also referred to as post-marketing studies.

IMMUNIZATION RECOMMENDATIONS/ IMPLEMENTATION

In the United States, after the FDA has licensed a new vaccine several committees of experts make recommendations to health care providers and public health programs for use of the vaccine. The committees of experts include the Advisory Committee on Immunization Practices (ACIP) of the Centers for Disease Control and Prevention, the Committee on Infectious Diseases (COID) of the American Academy of Pediatrics (AAP), and the American Academy of Family Physicians (AAFP) Commission on Clinical Policies and Research (CCPR). These committees independently assess the evidence to recommend what populations should receive the vaccine, at what ages, the number of doses and dosing interval, and precautions and contraindications (National Network for Immunization Information, www.immunizationinfo.org/immunization_policy_detail.cfv?id = 41#3).

The ACIP consists of 15 experts in immunization-related fields. Members are appointed by the Secretary of the U.S. Department of Health and Human Services. Their role is to provide advice and guidance to the Secretary, the Assistant Secretary for Health, and the Centers for Disease Control and Prevention (CDC) on the control of vaccine-preventable diseases. The ACIP is the only entity in the federal government that makes immunization recommendations (Centers for Disease Control and Prevention, www.cdc.gov/vaccines/recs/ACIP/default.htm). The Director of CDC (and/or the Secretary of Health and Human Services) makes immunization recommendations after consideration of the advice provided by the ACIP. These are then published in the CDC publication MMWR. Individual states consider the ACIP recommendations when setting their vaccine policies.

The AAP's COID (also called the "Redbook Committee") develops and reviews policy recommendations on the use of vaccines. Members of the committee are pediatricians selected by the AAP leadership for their knowledge of infectious diseases and expertise in vaccines. The recommendations by the COID are approved by the AAP leadership before publication in the Report of the COID (also known as "The Red Book" of the AAP).

The AAFP CCPR is a commission whose members are selected by the AAFP leadership for their scientific knowledge of conditions seen in family medicine. The CCPR meeting includes a liaison from AAP's COID. Recommendations by the CCPR are approved by the AAFP leadership before publication and promotion to AAFP's members.

Similar groups of experts advise and assist in the development of recommendations in other counties and for the World Health Organization.

POST-MARKETING SURVEILLANCE

Until a vaccine has been widely used in the general population rare or uncommon adverse events may be unrecognized. In many countries surveillance programs have been developed in order to identify problems that may develop following licensure and marketing. In the United States, the FDA may mandate specific post-marketing studies as part of the licensure agreement. In addition, the U.S. government has developed the vaccine adverse event reporting system (VAERS) and the vaccine safety datalink to identify problems after marketing has begun (Jacobson et al., 2001; National Network for Immunization Information, www.immunizationinfo.org/vaccine_safety_detail.cfv?id = 106; National Network for Immunization Information, www.immunizationinfo.org/vaccine_safety_detail.cfv?id = 35). The effectiveness of the VAERS system is illustrated by its success in detecting unexpected cases of intussusception associated with rhesus rotavirus vaccine, Rotashield®, which were eventually determined to have occurred at a rate of less than 1 in 10,000, and which in turn quickly led to the withdrawal of the vaccine (Centers for Disease Control and Prevention (CDC), 2004).

References

Andre, F.E. How the research-based industry approaches vaccine development and establishes priorities. Dev. Biol. 2002; 110: 25–29.

Avery, O.T. and Goebel, W.F. Chemo-immunological studies on conjugated carbohydrate-proteins. II. Immunological specificity of synthetic sugar-protein antigen. J. Exp. Med. 1929; 50:533–550.

Barry, M.A., Lai, W.C. and Johnston, S.A. Protection against mycoplasma infection using expression-library immunization. Nature 1995; 377(6550):632–635.

Batson, A. The costs and economics of modern vaccine development. Dev. Biol. 2002; 110:15–24.

Batson, A. The problems and promise of vaccine markets in developing countries. Health Aff. 2005; 24:690–693.

Bernstein, D.I., Schleiss, M.R., Berencsi, K., Gonczol, E., Dickey, M., Khoury, P., Cadoz, M., Meric, C., Zahradnik, J., Duliege, A.M. and Plotkin, S. Effect of previous or simultaneous immunization with canarypox expressing cytomegalovirus (CMV) glycoprotein B (gB) on response to subunit gB vaccine plus MF59 in healthy CMV-seronegative adults. J. Infect. Dis. 2002; 185:686–690.

Brennan, F.R. and Dougan, G. Non-clinical safety evaluation of novel vaccines and adjuvants: new products, new strategies. Vaccine 2005; 23:3210–3222.

Centers for Disease Control and Prevention (CDC) Suspension of rotavirus vaccine after reports of intussusception--United States, 1999. MMWR 2004; 53:786–789.

Centers for Disease Control and Prevention. Advisory Committee on Immunization Practices (ACIP). www.cdc.gov/vaccines/recs/ACIP/default.htm.

Code of Federal Regulations, in 21 CFR 312. 2005.

Corey, L., Langenberg, A.G., Ashley, R., Sekulovich, R.E., Izu, A.E., Douglas, J.M., Jr., Handsfield, H.H., Warren, T., Marr, L., Tyring, S., DiCarlo, R., Adimora, A.A., Leone, P., Dekker, C.L., Burke, R.L., Leong, W.P. and Straus, S.E. Recombinant glycoprotein vaccine for the prevention of genital HSV-2 infection: two randomized controlled trials. Chiron HSV Vaccine Study Group. JAMA 1999; 282:331–340.

Davenport, L.W. Regulatory considerations in vaccine design. Pharm. Biotechnol. 1995; 6:81–96.

de Bruyn, G., Vargas-Cortez, M., Warren, T., Tyring, S.K., Fife, K.H., Lalezari, J., Brady, R.C., Shahmanesh, M., Kinghorn, G., Beutner, K.R., Patel, R., Drehobl, M.A., Horner, P., Kurtz, T.O., McDermott, S., Wald, A. and Corey, L. A randomized controlled trial of a replication defective (gH deletion) herpes simplex virus vaccine for the treatment of recurrent genital herpes among immunocompetent subjects. Vaccine 2006; 24:914–920.

Edelman, R., Wasserman, S.S., Bodison, S.A., Putnak, R.J., Eckels, K.H., Tang, D., Kanesa-Thasan, N., Vaughn, D.W., Innis, B.L. and Sun, W. Phase I trial of 16 formulations of a tetravalent live-attenuated dengue vaccine. Am. J. Trop. Med. Hyg. 2003; 69(Suppl.):48–60.

FDA (CBER). Guidance for Industry: E6 Good Clinical Practice: Consolidated Guidance. US Department of Health and Human Services, Food and Drug Administration, Center for Drug Evaluation and Research (CDER), Center for Biologics Evaluation and Research (CBER), April 1996.

FDA (CBER). Manual of Standard Operating Procedures and Policies Regulatory—License Applications; Administrative Processing of Biologics License Application (BLA), SOPP 8401 Version 6 Date: May 1, 2007. www.fda.gov/Cber/regsopp/8401.htm.

Food and Drug Administration, HHS Registration of food facilities under the Public Health Security and Bioterrorism Preparedness and Response Act of 2002. Final rule. Fed. Regist. 2005; 70:57505–57509.

Galler, R., Freire, M.S., Jabor, A.V. and Mann, G.F. The yellow fever 17D vaccine virus: molecular basis of viral attenuation and its use as an expression vector. Braz. J. Med. Biol. Res. 1997; 30:157–168.

Grabowski, H. Encouraging the development of new vaccines. Health Aff. 2005; 24:697–700.

Graham, F.L. Adenoviruses as expression vectors and recombinant vaccines. Trends Biotechnol. 1990; 8:85–87.

Goonetilleke, N., Moore, S., Dally, L., Winstone, N., Cebere, I., Mahmoud, A., Pinheiro, S., Gillespie, G., Brown, D., Loach, V., Roberts, J., Guimaraes-Walker, A., Hayes, P., Loughran, K., Smith, C., De Bont, J., Verlinde, C., Vooijs, D., Schmidt, C., Boaz, M., Gilmour, J., Fast, P., Dorrell, L., Hanke, T. and McMichael, A.J. Induction of multifunctional human immunodeficiency virus type 1 (HIV-1)-specific T cells capable of proliferation in healthy subjects by using a prime-boost regimen of DNA- and modified vaccinia virus Ankara-vectored vaccines expressing HIV-1 Gag coupled to CD8+ T-cell epitopes. J. Virol. 2006; 80:4717–4728.

Gronvall, G.K., Trent, D., Borio, L., Brey, R. and Nagao, L.Alliance for Biosecurity The FDA animal efficacy rule and biodefense. Nat. Biotechnol. 2007; 25:1084–1087.

Guirakhoo, F., Kitchener, S., Morrison, D., Forrat, R., McCarthy, K., Nichols, R., Yoksan, S., Duan, X., Ermak, T.H., Kanesa-Thasan, N., Bedford, P., Lang, J., Quentin-Millet, M.J. and Monath, T.P. Live attenuated chimeric yellow fever dengue type 2 (ChimeriVax-DEN2) vaccine: Phase I clinical trial for safety and immunogenicity: effect of yellow fever pre-immunity in induction of cross neutralizing antibody responses to all 4 dengue serotypes. Hum. Vaccine 2006; 2:60–67.

Harper, D.M., Franco, E.L., Wheeler, C., Ferris, D.G., Jenkins, D., Schuind, A., Zahaf, T., Innis, B., Naud, P., De Carvalho, N. S., Roteli-Martins, C.M., Teixeira, J., Blatter, M.M., Korn, A.P., Quint, W. and Dubin, G.GlaxoSmithKline HPV Vaccine Study Group Efficacy of a bivalent L1 virus-like particle vaccine in prevention of infection with human papillomavirus types 16 and 18 in young women: a randomised controlled trial. Lancet 2004; 364(9447):1757–1765.

Hauguel, T.M. and Hackett, C.J. Rationally-designed vaccine adjuvants: separating efficacy from toxicity. Front. Biosci. 2008; 13:2806–2813.

Heath, P.T. Haemophilus influenzae type b conjugate vaccines: a review of efficacy data. Pediatr. Infect. Dis. J. 1998; 17(Suppl.):S117–S122.

Jacobson, R.M., Adegbenro, A., Pankratz, V.S. and Poland, G.A. Adverse events and vaccination-the lack of power and predictability of infrequent events in pre-licensure study. Vaccine 2001; 19:2428–2433.

Keyserling, H.L., Pollard, A.J., DeTora, L.M. and Gilmet, G.P. Experience with MCV-4, a meningococcal, diphtheria toxoid conjugate vaccine against serogroups A, C, Y and W-135. Expert Rev. Vaccines 2006; 5:445–459.

Koen, P., Ajamian, G., Burkart, R., Clamen, A., Davidson, J., D'Amore, R., Elkins, C., Herald, K., Incorvia, M., Johnson, A., Karol, R., Seibert, R., Slavejkov, A. and Wagner, K. Providing clarity and a common language to the "Fuzzy Front End". Res. Technol. Manage. 2001; 44:46–55.

Kopp, E.B. and Medzhitov, R. The Toll-receptor family and control of innate immunity. Curr. Opin. Immunol. 1999; 11:13–18.

Koutsky, L.A., Ault, K.A., Wheeler, C.M., Brown, D.R., Barr, E., Alvarez, F.B., Chiacchierini, L.M. and Jansen, K.U.Proof of Principle Study Investigators A controlled trial of a human papillomavirus type 16 vaccine. N. Engl. J. Med. 2002; 347:1645–1651.

Kuwert, E., Scheiermann, N., Gesemann, M., Paar, D., Safary, A., Simoen, E., Hauser, P. and Andre, F. Dose range study in healthy volunteers of a hepatitis B vaccine produced in yeast. Antiviral Res. 1985; 4(Suppl. 1):281–288.

Kwissa, M., Kasturi, S.P. and Pulendran, B. The science of adjuvants. Expert Rev. Vaccines 2007; 6(5):673–684.

Kwon, Y.M., Cox, M.M. and Calhoun, L.N. Salmonella-based vaccines for infectious diseases. Expert Rev. Vaccines 2007; 6:147–152.

Leroux-Roels, I., Borkowski, A., Vanwolleghem, T., Dramé, M., Clement, F., Hons, E., Devaster, J.M. and Leroux-Roels, G. Antigen sparing and cross-reactive immunity with an adjuvanted rH5N1 prototype pandemic influenza vaccine: a randomised controlled trial. Lancet 2007; 370(9587):580–589.

Miller, D.L. and Ross, J.J. Vaccine INDs: review of clinical holds. Vaccine 2005; 23:1099–1101.

Milstien, J.B. Regulation of vaccines: strengthening the science base. J. Public Health Policy 2004; 25:173–189.

Milstien, J.B., Kaddar, M. and Kieny, M.P. The impact of globalization on vaccine development and availability. Health Aff. 2006; 25:1061–1069.

Monath, T.P., Guirakhoo, F., Nichols, R., Yoksan, S., Schrader, R., Murphy, C., Blum, P., Woodward, S., McCarthy, K., Mathis, D., Johnson, C. and Bedford, P. Chimeric live, attenuated vaccine against Japanese encephalitis (ChimeriVax-JE): phase 2 clinical trials for safety and immunogenicity, effect of vaccine dose and schedule, and memory response to challenge with inactivated Japanese encephalitis antigen. J. Infect. Dis. 2003; 188:1213–1230.

National Institute of Allergy and Infectious Diseases. Vaccines: What They are, How They Work. US Department of Health and Human Services, National Institutes of Health, NIH Publication No. 03-4219, July 2003.

National Network for Immunization Information. Immunization Policy: Indications, Recommendations and Immunization Mandates. www.immunizationinfo.org/immunization_policy_detail.cfv?id = 41#3.

National Network for Immunization Information. Vaccine Safety: Vaccine Safety Datalink. www.immunizationinfo.org/vaccine_safety_detail.cfv?id = 106.

National Network for Immunization Information. Vaccine Safety: Vaccine Safety Datalink. Vaccine Safety: Vaccine Adverse Event Reporting System. www.immunizationinfo.org/vaccine_safety_detail.cfv?id = 35.

Oosterhuis-Kafeja, F., Beutels, P. and Van Damme, P. Immunogenicity, efficacy, safety and effectiveness of pneumococcal conjugate vaccines (1998–2006). Vaccine 2007; 25(12):2194–2212.

Pauly, M.V. Improving vaccine supply and development: who needs what? Health Aff. 2005; 24:680–689.

Poland, G.A., Murray, D. and Bonilla-Guerrero, R. New vaccine development. BMJ 2002; 324:1315–1319.

Polhemus, M.E., Magill, A.J., Cummings, J.F., Kester, K.E., Ockenhouse, C.F., Lanar, D.E., Dutta, S., Barbosa, A., Soisson, L., Diggs, C.L., Robinson, S.A., Haynes, J.D., Stewart, V.A., Ware, L.A., Brando, C., Krzych, U., Bowden, R.A., Cohen, J.D., Dubois, M.C., Ofori-Anyinam, O., De-Kock, E., Ballou, W.R. and Heppner, D.G., Jr. Phase I dose escalation safety and immunogenicity trial of Plasmodium falciparum apical membrane protein (AMA-1) FMP2.1, adjuvanted with AS02A, in malaria-naïve adults at the Walter Reed Army Institute of Research. Vaccine 2007; 25:4203–4212.

Rappuoli, R. Conjugates and reverse vaccinology to eliminate bacterial meningitis. Vaccine 2001; 19:2319–2322.

Rosenthal, S.R. and Clifford, J.C. Development of vaccines for biowarfare agents. Dev. Biol. (Basel), 2002; 110:99–105.

Russell, N.D., Graham, B.S., Keefer, M.C., McElrath, M.J., Self, S.G., Weinhold, K.J., Montefiori, D.C., Ferrari, G., Horton, H., Tomaras, G.D., Gurunathan, S., Baglyos, L., Frey, S.E., Mulligan, M.J., Harro, C.D., Buchbinder, S.P., Baden, L.R., Blattner, W.A., Koblin, B.A. and Corey, L. National Institute of Allergy and Infectious Diseases HIV Vaccine Trials Network Phase 2 study of an HIV-1 canarypox vaccine (vCP1452) alone and in combination with rgp120: negative results fail to trigger a phase 3 correlates trial. J. Acquir. Immune Defic. Syndr. 2007; 44:203–212.

Schödel, F. and Curtiss, R., 3rd Salmonellae as oral vaccine carriers. Dev. Biol. Stand. 1995; 84:245–253.

Smith, G.L., Mackett, M. and Moss, B. Recombinant vaccinia viruses as new live vaccines. Biotechnol. Genet. Eng. Rev. 1984; 2:383–407.

Stanberry, L.R. Herpes simplex virus vaccine evaluation in animals: the guinea pig model. Rev. Infect. Dis. 1991; 13(Suppl. 11):S920–S923.

Stanberry, L.R., Spruance, S.L., Cunningham, A.L., Bernstein, D.I., Mindel, A., Sacks, S., Tyring, S., Aoki, F.Y., Slaoui, M., Denis, M., Vandepapeliere, P. and Dubin, G.GlaxoSmithKline Herpes Vaccine Efficacy Study Group Glycoprotein-D-adjuvant vaccine to prevent genital herpes. N. Engl. J. Med. 2002; 347(21):1652–1661.

Stehlin, I.B. How FDA works to ensure vaccine safety. FDA Consumer. Magazine, Vol 29 (December) 1995. http://magazine-directory.com/FDA-Consumer.htm

Struck, M.M. Vaccine research and product development success rates and development times. Nat. Biotechnol. 1996; 14:591–593.

Tang, D.C., DeVit, M. and Johnston, S.A. Genetic immunization is a simple method for eliciting an immune response. Nature 1992; 356(6365):152–154.

Taylor, J., Tartaglia, J., Rivière, M., Duret, C., Languet, B., Chappuis, G. and Paoletti, E. Applications of canarypox (ALVAC) vectors in human and veterinary vaccination. Dev. Biol. Stand. 1994; 82:131–135.

The Biodesign Institute. http://www.biodesign.asu.edu/news/New-ASU-Interdisciplinary-PhD-Graduate-Program-in-Biological-Design-Set-for-Launch.

Ulmer, J.B., Donnelly, J.J., Parker, S.E., Rhodes, G.H., Felgner, P.L., Dwarki, V.J., Gromkowski, S.H., Deck, R.R., DeWitt, C.M., Friedman, A. et al. Heterologous protection against influenza by injection of DNA encoding a viral protein. Science 1993; 259(5102):1745–1749.

US Food and Drug Administration. http://www.fda.gov/ora/compliance_ref/bimo/glp/default.htm.

Van Kampen, K.R., Shi, Z., Gao, P., Zhang, J., Foster, K.W., Chen, D.T., Marks, D., Elmets, C.A. and Tang, D.C. Safety and immunogenicity of adenovirus-vectored nasal and epicutaneous influenza vaccines in humans. Vaccine 2005; 23:1029–1036.

Verdier, F. Non-clinical vaccine safety assessment. Toxicology 2002; 174:37–43.

Verdier, F., Barrow, P.C. and Burge, J. Reproductive toxicity testing of vaccines. Toxicology 2003; 185:213–219.

Winther, M.D. and Dougan, G. The impact of new technologies on vaccine development. Biotechnol. Genet. Eng. Rev. 1984; 2:1–39.

Xing, Z. and Lichty, B.D. Use of recombinant virus-vectored tuberculosis vaccines for respiratory mucosal immunization. Tuberculosis (Edinb) 2006; 86:211–217.

Zahradnik, J.M., Couch, R.B. and Gerin, J.L. Safety and immunogenicity of a purified hepatitis B virus vaccine prepared by using recombinant DNA technology. J. Infect. Dis. 1987; 155:903–908.

CHAPTER

5

Immunology of Vaccines: Consideration of the Neonate as a Target for Vaccination

Monica R. Foote[1] *and D. Mark Estes*[2]

[1]Department of Animal Science, Cornell University, Ithaca, NY, USA

[2]Department of Pediatrics and Microbiology and Immunology, Sealy Center for Vaccine Development, University of Texas Medical Branch, Galveston, TX, USA

OUTLINE

ABSTRACT

The use of vaccination as a strategy to reduce the impact of disease on society has a long-standing history of success, particularly for viral and bacterial-mediated diseases. The demand for safe and effective vaccine products is increasing due to the potential use of biothreat agents, and the emergence and reemergence of various pathogens. The implementation and use of vaccines in immunologically naive recipients in outbreak settings places extraordinary demands on the timing of the onset of immunity in a relatively broad population base, including those populations which may not always be ideal for delivery of vaccines (e.g., neonates, pregnant women, and the elderly). Each of these populations has its own unique demands for the induction of protective immunity and safety. For example, changes in immune function are present as the immune system matures or declines in neonates and the elderly, respectively. In addition, alterations in hormonal status and to functional changes necessary to accommodate the developing fetus are factors in the scheme used to promote immediate development of effector cells plus the capacity to induce long-term memory pools in pregnant women.

The development of immunologic memory is the standard for disease prophylaxis and the basis for most protective vaccines with the induction of both T-cell responses (CD4 and/or CD8 responses) and the production of neutralizing antibody. The kinetics of induction of primary adaptive immune responses in immunologically naive subjects is dictated by many factors including cell cycle transit time to increase the frequency of antigen-reactive cells and the relative rate at which effector cells acquire partial or full functional activities. Memory responses in B- and T-cell compartments are influenced by the phenotype, frequency, and persistence of cells over time (Robinson and Amara, 2005). Generation of effector and memory T cells, which contribute to both humoral and cell-mediated immune responses, requires antigen processing and presentation by cells of the innate immune system on class I or class II major histocompatibility molecules (MHC). In addition, the production of essential cytokines, crucial for early steps in the differentiation of lymphocytes is necessary for the generation of effector and memory T cells. An understanding of how the immune response status is altered from the "normal state" during different physiological states (e.g., pregnancy, lactation, growth), stages of maturity, and nutritional levels is necessary in order to develop effective and safe memory responses.

INTRODUCTION

Vaccination is a low-cost and effective strategy for the prevention and therapeutic reduction of infectious diseases. Protective host responses encompass the spectrum from adaptive humoral and cell-mediated immune responses in concert with a robust innate immune response (Table 5.1). The generation of secondary antibody responses as a result of natural exposure or vaccination can be attributed primarily to distinct populations of cells, short- and long-lived antibody-secreting plasma cells, and memory B cells (McHeyzer-Williams and McHeyzer-Williams, 2005; McHeyzer-Williams and Ahmed, 1999). Both T-dependent (T-D) and T-independent (T-I) pathways may be operational depending on the vaccine type (Table 5.2). T-I responses are generated via B-1, marginal zone cells, and dendritic cells (DC) (Vos et al., 2000; Mond et al., 1995a, b). B-1 responses generate rapid antibody responses and are partially dependent on innate-derived factors such as cytokines and pathogen-associated molecular pattern receptors (PAMPs) (Baumgarth et al., 2005). B-1 cells develop relatively low affinity responses (Herzenberg et al., 2000). Recent studies in mice suggest that B-1a cells account for much of the natural antibody produced as a component of innate defense, and B-1b cells are capable of generating anamnestic responses to vaccination with *Streptococcus pneumonia* (Haas et al., 2005). A linkage between DC and T-I B cell responses reinforces the need to enhance DC targeting for both T-I and T-D antibody responses (Austin et al., 2003; He et al., 2007; Litinskiy et al., 2002). T-I responses in neonates have also been shown to be less robust relative to adults (Adkins et al., 2004). Developments in our understanding of T-I immunity and the potential to generate vaccines that can generate rapid serum neutralizing antibody responses in addition to anticarbohydrate responses would represent major advances in the development of vaccines for biodefense applications and for several pediatric diseases. T-D responses rely on the coordinated interaction of T helper cells and B-2 cells at specific sites within lymphoid tissues (Randolph et al., 1999; Vinuesa et al., 2005; Cyster, 1999, 2003). Human neonates exhibit delayed T-D responses that are both qualitatively (isotype) and quantitatively different relative to adults (Adkins et al., 2004; Elliott et al., 2000; MacLennan et al., 1997; Gray et al., 1994; Inaba et al., 1983).

Studies in the T-cell compartment have shown that the memory population of $CD4^+$ and $CD8^+$ T cells can be subdivided into two populations: (1) effector memory T cells (T_{EM}) and (2) central memory T cells (T_{CM}) (Northrop and Shen, 2004). $CD4^+$ T cells can be subsetted based on surface phenotype, anatomical distribution,

TABLE 5.1 Examples of protective host responses to vaccination

Humoral immune responses
Neutralizing antibody to viruses, bacterial toxins
Nonneutralizing antibodies which may interfere with uptake through steric hindrance and other mechanisms
Complement-dependent and independent mechanism
Cell-mediated responses
T-dependent help provided via cognate and noncognate interactions with helper T cells
Lytic and soluble factor activities derived from $CD8^+$ and other T-cell populations including $\gamma\delta$ TCR-bearing cells

TABLE 5.2 Examples of different types of potential immune responses as a result of vaccination with various vaccine platforms

	TI Ab response[a]	TD Ab response[b]	CMI[c]
Live attenuated viral vaccines	+	+	+
Inactivated whole virus vaccines	+	+	±
Protein subunit antigens with adjuvant	+	+	±
Bacterial capsular polysaccharides	+		
Live intracellular bacterial vaccines		+	+
DNA vaccines	+	+	+

[a] T-independent antibody response.
[b] T-dependent antibody response.
[c] cell-mediated immune response conferred by cytolytic T cells.

and immune function. T_{EM} express low levels of CD62L, CCR7, and CD27, and are primarily found in the peripheral tissues such as the lung and liver. T_{CM} express high levels of CD62L, CCR7, and CD27, and are found in secondary lymphatic organs. T_{EM} are capable of producing effector cytokines immediately upon antigen recognition, whereas T_{CM} have greater proliferative potential upon restimulation. It has been hypothesized that T_{EM} confer immediate protection, whereas T_{CM} are a reservoir for generating a large pool of secondary effectors upon a second encounter with the pathogen. As recently reviewed by Reiner et al. (2007), the outcome of T-cell differentiation is dictated by many factors including costimulation, cytokines, antigen concentration, the relative frequency of antigen-specific T cells, and antigen presenting cells (DCs). T effector cells needed for an immediate response post vaccination can also be heterogeneous in terms of their activities, including cytokine production (e.g., Th1, Th2, Th17, or T regulatory cells). Additional cell populations have been defined, including follicular Th cells that produce IL-10 and IL-21 (Vinuesa et al., 2005), and IFN-γ and IL-10 (O'Garra and Vieira, 2007) or IFN-γ and IL-17 (Ivanov et al., 2006).

NEONATAL RESPONSES TO VACCINATION: LESSONS FROM ANIMAL MODELS

It is generally assumed that the immaturity of the neonatal immune system contributes to an increased susceptibility of the young host to infectious disease, limiting its capacity to develop a protective response to vaccination. Several differences between the neonatal and adult immune response have been identified. The mechanisms governing these differences, however, are not yet fully understood. Certain differences that have thus far been characterized include the observation that the neonatal T-cell population is abundant with a greater proportion of naïve T cells that can suppress immunoglobulin (Ig) production (Clement et al., 1990). Neonates also have higher proportions of antigen-presenting cells with defective costimulatory activity (Ridge et al., 1996). Importantly, the neonatal immune compartment is lymphopenic at birth. Lymphocyte development is a relatively late event in murine fetal development. At birth, the murine neonate has substantially fewer T cells in peripheral lymphoid organs including the lymph node and the spleen, and adult percentages of $CD4^+$ T cells are not reached in these organs until day 7 and 15 of age, respectively (Garcia et al., 2000). It has been shown recently that the neonatal environment supports lymphopenia-driven proliferation of $CD4^+$ T cells, resulting in an expansion of $CD4^+$ T cells with memory phenotypic and functional characteristics by the 17th day of life (Min et al., 2003). This lymphopenia-driven proliferation may equip the neonate with a large repertoire of memory cells (Min et al., 2003).

Although maternally derived immune factors, such as immunoglobulin, provide early protection to the neonate, they may interfere with postnatal activation of the neonate's immune system and its capacity to mount a protective response to vaccination or infection (Barrington and Parish, 2001; Endsley et al., 2003; Glezen, 2003; Siegrist, 2003). In humans, inhibition of the infant's responses to vaccination by maternal antibody is B-cell-specific, depends on antibody titer and dose of vaccine antigen, and does not appear to influence T-cell responses (reviewed by Siegrist, 2003). For example, neonatal calves fail to develop Ig responses to *Brucella abortus* if antigen-specific maternal Ig is present at the time of vaccination (Husband and Lascelles, 1975). Colostrum-deprived calves, however, produce relatively robust antibody responses to *B. abortus*. In addition, neonatal calves sensitized to ovalbumin (OVA) (an antigen to which there was no maternal Ig) at birth mount humoral responses similar to older cattle, regardless of colostral status (Husband and Lascelles, 1975). In addition, calves given a bovine viral diarrhea virus vaccine fail to develop a primary humoral response, but do generate memory B cells, in the presence of maternal Ig (Endsley et al., 2003). Likewise, neonatal calves vaccinated with *M. bovis* BCG fail to mount antigen-specific humoral responses, but calves do generate certain in vitro B cell responses to BCG following vaccination (Foote et al., 2007). These results suggest that although maternal Ig may block endogenous Ig production, B cells retain other functional capacities (i.e., differentiation into memory B cells.).

The altered immune reactivity of the neonate may reflect the demands of in utero conditions and the need to avoid unwanted reactivity in this environment (Levy, 2005; Levy et al., 2004). This idea is supported by observations indicating that neonatal immune cells frequently have a reduced ability to produce proinflammatory cytokines. For example, neonatal immune cells produce less TNF-α and IFN-γ in response to lipopolysaccharide (LPS) relative to adult PBMC (Levy et al., 2004; Yan et al., 2004).

Protective adaptive immunity requires an appropriate balance between Th1 and Th2 responses. Th1 cells secrete inflammatory cytokines (e.g., IFN-γ, lymphotoxin, and TNF-α) that activate components of the innate and adaptive immune system to defend against intracellular pathogens. In contrast, Th2 cells secrete cytokines (e.g., IL-4, IL-5, IL-13) that are generally antiinflammatory in nature and are important for evoking potent humoral (i.e., antibody) immune responses. Neonates are capable of developing mature inflammatory responses characterized by adult-level Th1 and CTL functions following antigenic exposure under potent Th1-driving conditions, such as complete Freund's adjuvant (reviewed in Adkins et al., 2004). More often, however, neonates develop strong Th2 responses to antigenic stimulation (reviewed in Min et al., 2003; Siegrist, 2000; Adkins, 2000). For example, neonatal mice immunized with the protein antigen KLH develop responses characterized with enhanced in vitro production of IL-4 and in vivo production of IgG1 (Adkins et al., 2003b; Adkins and Du, 1998). This phenomenon has been well established for recall responses. Mice initially immunized as neonates mount Th2-dominant memory recall responses as adults. Early development of Th2-dominant primary responses by $CD4^+$ cells likely contributes to observed Th2-dominant memory responses (Adkins et al., 2003a). The mechanisms governing the neonatal Th2 bias are now coming to light. Studies suggest that IL-4 production by neonatal T cells drives apoptosis of Th1 cells, thus skewing the neonatal recall or secondary response to Th2 (Li et al., 2004). In addition, new data suggest that developmental-specific epigenetic phenomena occurring at the Th2 locus contribute to the Th2 bias (Rose et al., 2007). The Th2 regulatory region CNS-1 preexists in a hypomethylated state in neonatal naïve $CD4^+$ T cells and $CD4^+$ thymocytes, suggesting that neonatal T cells are poised for rapid Th2 effector function (Rose et al., 2007).

Data in support of a Th2 bias in human neonates, indicate that human-cord-blood-derived DC, essential for priming and differentiation of naïve T cells into Th1 cells, have a profound defect in the expression of the Th1 cytokine, IL-12 (Goriely et al., 2001). Responses of human infants to viral vaccines are also characterized by reduced IFN-γ secretion and elevated serum antibody titers (Ota et al., 2004; Vekemans et al., 2002). Consequences of a Th2-biased response may include a reduced capacity of the neonate to respond effectively to vaccines that rely on a Th1 response for their efficacy. In contrast, vaccination of human newborns with *M. bovis* BCG, a potent Th1-driving stimulant, induces a potent Th1 response characterized by adult-like IFN-γ responses and reduced secretion of IL-4 and IL-5 by $CD4^+$ lymphocytes (Marchant et al., 1999; Vekemans et al., 2001). Similarly, studies indicate that the dairy calf, like the human infant, can generate an adult-like Th1 response when vaccinated during the first week of life with *M. bovis* BCG (Buddle et al., 2003; Nonnecke et al., 2005). In addition, vaccination of the neonatal calf with *M. bovis* BCG provides protection upon subsequent challenge (Buddle et al., 2003). *M. bovis* BCG is routinely administered to newborns for the prevention of tuberculosis, and human trials will ultimately be conducted in this population. It is therefore essential to determine safety and efficacy of candidate vaccines in this target population. Importantly, the neonatal bovine may provide a useful model for the development of efficacious tuberculosis vaccines.

Studies carried out in newborn mice demonstrate that DNA vaccines induce adult-like, protective $CD8^+$ T-cell responses to LCMV (Zhang et al., 2002). Likewise, DNA vaccination of nonhuman primates induces significant humoral and cellular immunity to influenza (Bot et al., 2001). Thus, genetic immunization is an appealing strategy to both prime and boost immune responses against pathogens in the neonate. A more developed understanding of the immunomodulatory properties of genetic immunization are needed to enhance the prospects of effective vaccines based on this platform in nonrodent species.

Protein-energy malnutrition (PEM) is the major cause of immunodeficiency worldwide (Delafuente, 1991), and is still a major concern in developing countries, particularly in young children. It has negative effects on both innate and adaptive immunity (reviewed by Woodward et al., 1999). Substantial evidence supports the negative effects of PEM on adaptive immune responses. The lymphoid involution (decrease in size and cellularity of secondary lymphoid organs) that occurs in wasting PEM is considered contributory to PEM-associated immunodepression (Woodward et al., 1999; Lee and Woodward, 1996). Lymphoid involution associated with PEM is characterized by drastic decreases in mononuclear cell numbers in blood, spleen, and mesenteric lymph nodes (Lee and Woodward, 1996). A low $CD4^+/CD8^+$ T-cell ratio in the blood has been commonly associated with PEM in humans and weanling mice (Chandra, 2001, 2002). However, imbalances within, rather than between, the two main T-cell subsets may be of greater importance. In mice, wasting PEM is associated with an overabundance of $CD4^+CD45RA^+$ T cells ($CD4^+$ naïve phenotype) and $CD8^+CD45RA^+CD62L^+$ T cells ($CD8^+$ naïve phenotype) that are quiescent compared with $CD45^-$ effector and memory phenotypes (Woodward et al., 1999; ten Bruggencate et al., 2001). In addition, wasting because of severe protein deficiency decreases proliferation and expression of IFN-γ, IL-2, and IL-2 receptor mRNA by splenic mononuclear cells (Mengheri et al., 1992). PEM also affects humoral responses negatively. Mucosal secretory IgA antibody responses are more sensitive to wasting than systemic humoral immunity, leading to the hypotheses that

PEM affects epithelial IgA transport (Chandra, 1991, 1993; Chandra and Sarchielli, 1993).

Obesity also has been associated with decreased immune function. In developed countries, the United States in particular, obesity has become an epidemic, affecting even children. Thirty-eight percent of obese children and adolescents show impairment of cell-mediated immune (CMI) responses including decreased DTH, abnormal lymphoproliferative responses to mitogens, and decreases in bacterial killing by neutrophils (Chandra and Scrimshaw, 1980). Obesity also has been related to elevated leukocyte and lymphocyte subsets (excluding NK and cytotoxic/suppressor T cells), lower T- and B-cell mitogen-induced lymphocyte proliferation, but increased phagocytosis and oxidative burst activity by monocytes and neutrophils (Nieman et al., 1999). An additive negative effect of obesity and old-age on immune function has also been reported (Moriguchi, 1995, p. 169).

Immune responses to vaccination with *M. bovis* BCG have been characterized well in the bovine (Nonnecke et al., 2005; Maue et al., 2004). Sensitization to *M. bovis* BCG and elicitation of recall responses (i.e., IFN-γ and NO) by mycobacterial antigen (i.e., PPD) have been used to model CMI responses in cattle (Fiske and Adams, 1985; Nickerson and Nonnecke, 1986; Nonnecke et al., 1986). Recent studies have investigated the effects of protein and energy status on adaptive immune responses in neonates using this BCG/PPD model in cattle (Foote et al., 2005, 2007).

Neonatal life is a unique period with regard to immune function and development. Additional studies are needed to elucidate what specific immune-related deficiencies exist as the neonate develops, and when each of these components is fully functional. The neonatal bovine may provide a useful model, particularly with regard to development of efficacious tuberculosis vaccines. Vaccination strategies that are designed to boost immunity of neonates must consider that the immune system is continuously changing during development. Strategies to enhance the immune response to vaccination must not only be properly targeted, but also be appropriate for the stage of immune development in order to avoid unwanted effects, such as the generation of an ineffective or negative response (i.e., the potential for development of tolerance or induction of self-reactivity).

References

Adkins, B. Development of neonatal Th1/Th2 function. Int. Rev. Immunol. 2000; 19(2–3):157–171.

Adkins, B., Bu, Y., Vincek, V. and Guevara, P. The primary responses of murine neonatal lymph node $CD4^+$ cells are Th2-skewed and are sufficient for the development of Th2-biased memory. Clin. Dev. Immunol. 2003a; 10(1):43–51.

Adkins, B. and Du, R.Q. Newborn mice develop balanced Th1/Th2 primary effector responses in vivo but are biased to Th2 secondary responses. J. Immunol. 1998; 160(9):4217–4224.

Adkins, B., Leclerc, C. and Marshall-Clarke, S. Neonatal adaptive immunity comes of age. Nat. Rev. Immunol. 2004; 4(7):553–564.

Adkins, B., Williamson, T., Guevara, P. and Bu, Y. Murine neonatal lymphocytes show rapid early cell cycle entry and cell division. J. Immunol. 2003b; 170(9):4548–4556.

Austin, A.S., Haas, K.M., Naugler, S.M., Bajer, A.A., Garcia-Tapia, D. and Estes, D.M. Identification and characterization of a novel regulatory factor: IgA-inducing protein. J. Immunol. 2003; 171(3):1336–1342.

Barrington, G.M. and Parish, S.M. Bovine neonatal immunology. Vet. Clin. North Am. Food Anim. Pract. 2001; 17(3):463–476.

Baumgarth, N., Tung, J.W. and Herzenberg, L.A. Inherent specificities in natural antibodies: a key to immune defense against pathogen invasion. Springer Semin. Immunopathol. 2005; 26(4):347–362.

Bot, A., Shearer, M., Bot, S., Avriette, M., Garcia-Sastre, A., White, G. et al. Induction of immunological memory in baboons primed with DNA vaccine as neonates. Vaccine 2001; 19(15–16):1960–1967.

Buddle, B.M., Wedlock, D.N., Parlane, N.A., Corner, L.A.L., De Lisle, G.W. and Skinner, M.A. Revaccination of neonatal calves with *Mycobacterium bovis* BCG reduces the level of protection against bovine tuberculosis induced by a single vaccination. Infect. Immun. 2003; 71(11):6411–6419.

Chandra, R.K. Interactions between early nutrition and the immune system. Ciba Found. Symp. 1991; 156:77–89, discussion 89–92.

Chandra, R.K. Nutrition and the immune system. Proc. Nutr. Soc. 1993; 52(1):77–84.

Chandra, R.K. Effect of vitamin and trace-element supplementation on cognitive function in elderly subjects. Nutrition 2001; 17(9): 709–712. [retraction in Meguid M.M. Nutrition 2005; 21(2): 286, PMID: 15793927]

Chandra, R.K. Nutrition and the immune system from birth to old age. Eur. J. Clin. Nutr. 2002; 56(Suppl. 3):S73–S76.

Chandra, R.K. and Sarchielli, P. Nutritional status and immune responses. Clin. Lab. Med. 1993; 13(2):455–461.

Chandra, R.K. and Scrimshaw, N.S. Immunocompetence in nutritional assessment. Am. J. Clin. Nutr. 1980; 33(12):2694–2697.

Clement, L.T., Vink, P.E. and Bradley, G.E. Novel immunoregulatory functions of phenotypically distinct subpopulations of $CD4^+$ cells in the human neonate. J. Immunol. 1990; 145(1):102–108.

Cyster, J.G. Chemokines and cell migration in secondary lymphoid organs. Science 1999; 286(5447):2098–2102.

Cyster, J.G. Homing of antibody secreting cells. Immunol. Rev. 2003; 194:48–60.

Delafuente, J.C. Nutrients and immune responses. Rheum. Dis. Clin. North Am. 1991; 17(2):203–212.

Elliott, S.R., Roberton, D.M., Zola, H. and Macardle, P.J. Expression of the costimulator molecules, CD40 and CD154, on lymphocytes from neonates and young children. Hum. Immunol. 2000; 61(4):378–388.

Endsley, J.J., Roth, J.A., Ridpath, J. and Neill, J. Maternal antibody blocks humoral but not T cell responses to BVDV. Biologicals 2003; 31(2):123–125.

Fiske, R.A. and Adams, L.G. Immune responsiveness and lymphoreticular morphology in cattle fed hypo- and hyperalimentative diets. Vet. Immunol. Immunopathol. 1985; 8(3):225–244.

Foote, M.R., Nonnecke, B.J., Beitz, D.C. and Waters, W.R. Antigen-specific B-cell responses by neonatal calves after early vaccination. J. Dairy Sci. 2007; 90:5208–5217.

Foote, M.R., Nonnecke, B.J., Waters, W.R., Palmer, M.V., Beitz, D.C., Fowler, M.A. et al. Effects of increased dietary protein and energy on composition and functional capacities of blood mononuclear cells from vaccinated, neonatal calves. Int. J. Vitam. Nutr. Res. 2005; 75(5):357–368.

Garcia, A.M., Fadel, S.A., Cao, S. and Sarzotti, M. T cell immunity in neonates. Immunol. Res. 2000; 22(2–3):177–190.

Glezen, W.P. Effect of maternal antibodies on the infant immune response. Vaccine 2003; 21(24):3389–3392.

Goriely, S., Vincart, B., Stordeur, P., Vekemans, J., Willems, F., Goldman, M. et al. Deficient IL-12(p35) gene expression by dendritic cells derived from neonatal monocytes. J. Immunol. 2001; 166(3):2141–2146.

Gray, D., Siepmann, K. and Wohlleben, G. CD40 ligation in B cell activation, isotype switching and memory development. Semin. Immunol. 1994; 6(5):303–310.

Haas, K.M., Poe, J.C., Steeber, D.A. and Tedder, T.F. B-1a and B-1b cells exhibit distinct developmental requirements and have unique functional roles in innate and adaptive immunity to S. pneumoniae. Immunity 2005; 23(1):7–18.

He, B., Xu, W., Santini, P.A., Polydorides, A.D., Chiu, A., Estrella, J. et al. Intestinal bacteria trigger T cell-independent immunoglobulin A(2) class switching by inducing epithelial-cell secretion of the cytokine APRIL. Immunity 2007; 26(6):812–826.

Herzenberg, L.A., Baumgarth, N. and Wilshire, J.A. B-1 cell origins and VH repertoire determination. Curr. Top. Microbiol. Immunol. 2000; 252:3–13.

Husband, A.J. and Lascelles, A.K. Antibody responses to neonatal immunisation in calves. Res. Vet. Sci. 1975; 18(2):201–207.

Inaba, K., Steinman, R.M., Van Voorhis, W.C. and Muramatsu, S. Dendritic cells are critical accessory cells for thymus-dependent antibody responses in mouse and in man. Proc. Natl. Acad. Sci. USA 1983; 80(19):6041–6045.

Ivanov, I.I., McKenzie, B.S., Zhou, L., Tadokoro, C.E., Lepelley, A., Lafaille, J.J. et al. The orphan nuclear receptor RORgammat directs the differentiation program of proinflammatory IL-17$^+$ T helper cells. Cell 2006; 126(6):1121–1133.

Lee, W.H. and Woodward, B.D. The CD4/CD8 ratio in the blood does not reflect the response of this index in secondary lymphoid organs of weanling mice in models of protein-energy malnutrition known to depress thymus-dependent immunity. J. Nutr. 1996; 126(4):849–859.

Levy, O. Innate immunity of the human newborn: distinct cytokine responses to LPS and other Toll-like receptor agonists. J. Endotoxin Res. 2005; 11(2):113–116.

Levy, O., Zarember, K.A., Roy, R.M., Cywes, C., Godowski, P.J. and Wessels, M.R. Selective impairment of TLR-mediated innate immunity in human newborns: neonatal blood plasma reduces monocyte TNF-alpha induction by bacterial lipopeptides, lipopolysaccharide, and imiquimod, but preserves the response to R-848. J. Immunol. 2004; 173(7):4627–4634.

Li, L., Lee, H.-H., Bell, J.J., Gregg, R.K., Ellis, J.S., Gessner, A. et al. IL-4 utilizes an alternative receptor to drive apoptosis of Th1 cells and skews neonatal immunity toward Th2. Immunity 2004; 20(4):429–440.

Litinskiy, M.B., Nardelli, B., Hilbert, D.M., He, B., Schaffer, A., Casali, P. et al. DCs induce CD40-independent immunoglobulin class switching through BLyS and APRIL. Nat. Immunol. 2002; 3(9):822–829.

MacLennan, I.C., Gulbranson-Judge, A., Toellner, K.M., Casamayor-Palleja, M., Chan, E., Sze, D.M. et al. The changing preference of T and B cells for partners as T-dependent antibody responses develop. Immunol. Rev. 1997; 156:53–66.

Marchant, A., Goetghebuer, T., Ota, M.O., Wolfe, I., Ceesay, S.J., De Groote, D. et al. Newborns develop a Th1-type immune response to *Mycobacterium bovis* bacillus Calmette-Guerin vaccination. J. Immunol. 1999; 163(4):2249–2255.

Maue, A.C., Waters, W.R., Palmer, M.V., Whipple, D.L., Minion, F.C., Brown, W.C. et al. CD80 and CD86, but not CD154, augment DNA vaccine-induced protection in experimental bovine tuberculosis. Vaccine 2004; 23(6):769–779.

McHeyzer-Williams, L.J. and McHeyzer-Williams, M.G. Antigen-specific memory B cell development. Annu. Rev. Immunol. 2005; 23:487–513.

McHeyzer-Williams, M.G. and Ahmed, R. B cell memory and the long-lived plasma cell. Curr. Opin. Immunol. 1999; 11(2):172–179.

Mengheri, E., Nobili, F., Crocchioni, G. and Lewis, J.A. Protein starvation impairs the ability of activated lymphocytes to produce interferon-gamma. J. Interferon Res. 1992; 12(1):17–21.

Min, B., McHugh, R., Sempowski, G.D., Mackall, C., Foucras, G. and Paul, W.E. Neonates support lymphopenia-induced proliferation. Immunity 2003; 18(1):131–140.

Mond, J.J., Lees, A. and Snapper, C.M. T cell-independent antigens type 2. Annu. Rev. Immunol. 1995a; 13:655–692.

Moriguchi, S., Oonishi, K., Kato, M. and Kishino, Y. Obesity is a risk factor for deteriorating cellular immune functions with aging. Nutr. Res. 1995; 15:151–160.

Mond, J.J., Vos, Q., Lees, A. and Snapper, C.M. T cell independent antigens. Curr. Opin. Immunol. 1995b; 7(3):349–354.

Nickerson, S.C. and Nonnecke, B.J. Tuberculin elicited cellular immune response in the lactating bovine mammary gland vaccinated intramammarily with *Mycobacterium bovis*. Vet. Immunol. Immunopathol. 1986; 13(1–2):39–50.

Nieman, D.C., Henson, D.A., Nehlsen-Cannarella, S.L., Ekkens, M., Utter, A.C., Butterworth, D.E. et al. Influence of obesity on immune function. J. Am. Diet. Assoc. 1999; 99(3):294–299.

Nonnecke, B.J., Elsken, L.A. and Kehrli, M.E., Jr. Local and systemic immune response in the cow after intramammary vaccination during lactation. Vet. Immunol. Immunopathol. 1986; 11(1):31–44.

Nonnecke, B.J., Waters, W.R., Foote, M.R., Palmer, M.V., Miller, B.L., Johnson, T.E. et al. Development of an adult-like cell-mediated immune response in calves after early vaccination with *Mycobacterium bovis* bacillus Calmette-Guerin. J. Dairy Sci. 2005; 88(1):195–210.

Northrop, J.K. and Shen, H. CD8$^+$ T-cell memory: only the good ones last. Curr. Opin. Immunol. 2004; 16(4):451–455.

O'Garra, A. and Vieira, P. T(H)1 cells control themselves by producing interleukin-10. Nat. Rev. Immunol. 2007; 7(6):425–428.

Ota, M.O.C., Vekemans, J., Schlegel-Haueter, S.E., Fielding, K., Whittle, H., Lambert, P.H. et al. Hepatitis B immunisation induces higher antibody and memory Th2 responses in newborns than in adults. Vaccine 2004; 22(3–4):511–519.

Randolph, D.A., Huang, G., Carruthers, C.J., Bromley, L.E. and Chaplin, D.D. The role of CCR7 in TH1 and TH2 cell localization and delivery of B cell help in vivo. Science 1999; 286(5447):2159–2162.

Reiner, S.L., Sallusto, F. and Lanzavecchia, A. Division of labor with a workforce of one: challenges in specifying effector and memory T cell fate. Science 2007; 317(5838):622–625.

Ridge, J.P., Fuchs, E.J. and Matzinger, P. Neonatal tolerance revisited: turning on newborn T cells with dendritic cells. Science 1996; 271(5256):1723–1726.

Robinson, H.L. and Amara, R.R. T cell vaccines for microbial infections. Nat. Med. 2005; 11(Suppl. 4):S25–S32.

Rose, S., Lichtenheld, M., Foote, M.R. and Adkins, B. Murine neonatal CD4$^+$ cells are poised for rapid Th2 effector-like function. J. Immunol. 2007; 178(5):2667–2678.

Siegrist, C.A. Vaccination in the neonatal period and early infancy. Int. Rev. Immunol. 2000; 19(2–3):195–219.

Siegrist, C.A. Mechanisms by which maternal antibodies influence infant vaccine responses: review of hypotheses and definition of main determinants. Vaccine 2003; 21(24):3406–3412.

ten Bruggencate, S.J., Hillyer, L.M. and Woodward, B.D. The proportion of CD45RA(+)CD62L(+) (quiescent-phenotype) T cells within the CD8(+) subset increases in advanced weight loss in the protein- or energy-deficient weanling mouse. J. Nutr. 2001; 131(12):3266–3269.

Vekemans, J., Amedei, A., Ota, M.O., D'Elios, M.M., Goetghebuer, T., Ismaili, J. et al. Neonatal bacillus Calmette-Guerin vaccination induces adult-like IFN-gamma production by $CD4^+$ T lymphocytes. Eur. J. Immunol. 2001; 31(5):1531–1535.

Vekemans, J., Ota, M.O.C., Wang, E.C.Y., Kidd, M., Borysiewicz, L.K., Whittle, H. et al. T cell responses to vaccines in infants: defective IFN gamma production after oral polio vaccination. Clin. Exp. Immunol. 2002; 127(3):495–498.

Vinuesa, C.G., Tangye, S.G., Moser, B. and Mackay, C.R. Follicular B helper T cells in antibody responses and autoimmunity. Nat. Rev. Immunol. 2005; 5(11):853–865.

Vos, Q., Lees, A., Wu, Z.Q., Snapper, C.M. and Mond, J.J. B-cell activation by T-cell-independent type 2 antigens as an integral part of the humoral immune response to pathogenic microorganisms. Immunol. Rev. 2000; 176:154–170.

Woodward, B., Hillyer, L. and Hunt, K. T cells with a quiescent phenotype ($CD45RA^+$) are overabundant in the blood and involuted lymphoid tissues in wasting protein and energy deficiencies. Immunology 1999; 96(2):246–253.

Yan, S.R., Qing, G., Byers, D.M., Stadnyk, A.W., Al-Hertani, W. and Bortolussi, R. Role of MyD88 in diminished tumor necrosis factor alpha production by newborn mononuclear cells in response to lipopolysaccharide. Infect. Immun. 2004; 72(3):1223–1229.

Zhang, J., Silvestri, N., Whitton, J.L. and Hassett, D.E. Neonates mount robust and protective adult-like CD8(+)-T-cell responses to DNA vaccines. J. Virol. 2002; 76(23):11911–11919.

CHAPTER

6

Molecular Approaches to Bacterial Vaccines

Vladimir L. Motin and Alfredo G. Torres
Departments of Pathology and Microbiology and Immunology,
Sealy Center for Vaccine Development, University of Texas Medical Branch,
Galveston, TX, USA

OUTLINE

ABSTRACT

Recent advances in molecular techniques provided a whole-genome perspective in designing novel and improving existing vaccines against bacterial infections. However, a tremendous amount of information derived from genome sequencing, comparative genomics, microarrays, proteomics, and other large-scale technologies is still not completely utilized to provide a systematic mining of new virulence determinants and prospective vaccine candidates. Therefore, the major task in coming years will be to develop cross-interacting databases which can quickly translate the functional genomic information into potential vaccines. Yet, in many cases, these approaches have already allowed identification of more promising targets than conventional approaches did in the past decades. The bottleneck for moving forward with vaccine technologies resides in the development of high-throughput systems for testing the properties of selected candidate vaccines. However, the more highly developed systems such as bacterial cell surface display are continuing to be improved. Together with the traditionally used outer membrane protein, fimbriae and flagella displays, novel platforms using autotransporters and lipoproteins have brought additional capabilities for the carriers to expose foreign molecules on the cell surface. Particularly important has been the creation of "second generation" vaccine delivery vehicles for mucosal immunization by the addition of codisplayed adhesins for targeting cells to specific immunoreactive sites. Bacterial peptide display has successfully competed with other epitope mapping techniques and, in combination with protein arrays, proven useful in identification of the entire set of antigens of the pathogen targeted by the immune system. In addition, the development of novel delivery systems has significantly advanced by using intracellular pathogens for genetic immunization. Delivery of DNA vaccine plasmid and even translation-competent mRNA directly to the cytosol of the antigen-presenting cells can be efficiently achieved by self-destructing live bacterial carriers. A bacterial ghost delivery system has become popular for heterologous antigen presentation, and also has been tested for DNA vaccination. Yet another technique, the use of protein secretion systems such as types I, III, and IV, had opened new horizons of using "inverted pathogenicity" for vaccine purposes. Further, genetic approaches have improved the area of identification of essential genes, characterization of functions required for growth under certain conditions (e.g., in vivo–induced), and importantly, in screening methods of direct selection of protective candidates. Finally, several novel stabilization systems for the foreign genes expressed in bacterial carriers have been generated and tested with different vaccine carriers.

REVERSE VACCINOLOGY

The term "reverse vaccinology" was proposed by Rappuoli (2000) and represents a genome-based approach to vaccine development. In comparison with the conventional approach, which requires a laborious process of selection of individual components important for immunity, reverse vaccinology offers the possibility of using genomic information derived from in silico analysis of sequenced organisms. This approach can significantly reduce the time necessary for the identification of candidate vaccines, and enables systematic identification of all potential antigens of pathogens, including those which are difficult or currently impossible to culture. Of course, in limiting analysis to predicted open reading frames (ORFs) of the microorganism, one has to assume that the pathogen can express any of them at a given time. In reality, the knowledge of whether the antigen is expressed in vivo during infection, its abundance and immunogenicity is crucial for determining the feasibility of using this antigen as a vaccine candidate. Moreover, this approach is limited to the identification of protein antigens, omitting such important vaccine components as polysaccharides and glycolipids. Nevertheless, reverse vaccinology can enable us to systemically classify the potential protective antigens, thereby helping to improve existing vaccines and develop efficient preparations virtually against any pathogen that has had its genome sequence determined. With the number of sequenced genomes progressively increasing every year, this approach, combined with recent advances in bioinformatics, comparative and functional genomics and proteomics, will be the method of choice for vaccinology studies at the beginning of the 21st century. The main limitation of this approach is the lack of a high-throughput system to estimate protective immunity of selected candidates.

According to Rappuoli (2000), reverse vaccinology was first successfully applied in the identification of vaccine candidates against serogroup B meningococcus (MenB). For this study, the entire genome sequence of *Neisseria meningitidis* was available, having been computer-analyzed, and in silico-selected vaccine candidates were expressed in *E. coli* followed by immunization of mice and testing of sera for bactericidal activity in vitro (complement-mediated in vitro killing of bacteria). A total of 600 novel genes were predicted to code for surface-exposed or exported proteins, and 350 of them were successfully expressed, purified, and used for immunization. Finally, 25 novel, surface-exposed proteins were shown to induce bactericidal antibodies, and most of these vaccine candidates had not been discovered by all previously used techniques. The foregoing approach has identified more potential vaccine candidates, including lipoproteins or other types of surface-associated proteins without membrane-spanning domains, than conventional approaches did for the past 40 years. A recent review by a group of scientists from Chiron Vaccine, Italy (Scarselli et al., 2005) contains additional examples of successfully applied reverse vaccinology techniques for *Bacillus anthracis*, *Streptococcus pneumonia*, *Staphylococcus aureus*, *Chlamydia pneumoniae*, *Porphyromonas gingivalis*, *Edwardsiella tarda*, and *Mycobacterium tuberculosis*. The most recent publication on this topic described identification of a universal Group B *Streptococcus* (GBS) vaccine by multiple genome screens. Eight sequenced genomes of GBS strains belonging to different serotypes were analyzed to find the genes shared by all strains ("core" genome), as well as genes that were not present in all strains ("variable" genome). Then the genes encoding the putative surface-associated and secreted proteins were uncovered within two subgenomes (589 total genes). Of these identified genes, 312 were successfully expressed in *E. coli*, and the antigens were purified in a soluble form and used to immunize animals. Four proteins elicited protection in mice, and their combination was highly protective against 12 virulent strains, which represented the major pathogenic GBS serotypes (Maione et al., 2005). In addition to using such newly discovered antigens for vaccine purpose, it is important to characterize these proteins in terms of their biological function. Therefore, the reverse vaccinology approach provides a powerful tool to discover previously unknown proteins that have a significant role in the biology of the microorganism.

The reverse vaccinology approach is hard to imagine without comparative genomics and structural biology of bacterial pathogens. By comparing whole genomes of pathogens with their less-virulent variants, non-pathogenic relatives, unrelated pathogens, and even commensals, it is possible to identify either unique, disease-related genes or conserved vaccine targets. Comparative genomics also provides crucial insight into the evolution of bacterial pathogenesis, and therefore, the vaccine candidates that are selected could be properly evaluated with respect to their antigenic appearance, e.g., in terms of criteria such as phase variation, gene duplication, and tendency to loss rudimentary functions (Scarselli et al., 2005). Comparison of the three-dimensional structures of the proteins implicated in bacterial pathogenesis can provide additional information on the motifs involved in adhesion and host recognition, as well as the mode of action of enzymatic virulence factors, including toxins (Waksman et al., 2005).

FUNCTIONAL GENOMICS AND PROTEOMICS

Recently developed DNA microarrays (or microchips) provided a powerful tool for understanding the complex regulation of the virulence network and selection of vaccine candidates. Ideally, transcriptome (the complete series of mRNA transcripts of an organism) of both pathogen and host should be determined during the course of natural infection. In reality, researchers use a set of in vitro growth conditions that mimic, to some extent, the environment pathogens encounter in the host, the appropriate in vivo models (cell cultures and/or animals), or expression analysis in clinical samples. Meinke et al., recently provided several examples describing the successful identification of new vaccine targets as a result of the use of transcriptional profiling. Among them were previously unconsidered components required for the hyperinfectious state and dissemination of *Vibrio cholerae*, genes of *Streptococcus pyogenes* encoding proteins that mediate invasion of innate defense mechanisms, and genes of *M. tuberculosis* that were differently expressed under acidic conditions resembling a phagosome environment (Meinke et al., 2004). Another review described the use of gene array technology to search for vaccine candidates and new virulence genes after interaction of *N. meningitidis* to human epithelial cells. A wide range of surface proteins were induced under the experimental conditions, and this approach, in combination with reverse vaccinology, could provide novel candidates for a meningococcal vaccine. This review also contains many other examples of the application of functional genomics to bacterial pathogens, including the interactions between intestinal epithelial cells and *Salmonella*, human promyelocytic cells and *Listeria monocytogenes*, bronchial epithelial cells and *Bordetella pertussis*, epithelial cells and *Pseudomonas aeruginosa*, mouse gastric epithelial cells and *Helicobacter pylori*, HeLa cells infected with *Chlamydia trachomatis*, human macrophage response to infection with *M. tuberculosis*, and the response of human intestinal epithelial cells to *Shigella flexneri* invasion (Serruto et al., 2004).

Recently, DNA gene arrays were used to characterize temporal global regulatory changes occurring in the *Yersinia pestis* genome after a temperature shift from 26 to 37°C. Plague bacilli exist in nature at ambient temperature within the flea vector or in the mammalian host at 37°C, and thermoregulated changes likely reflect introduction into these niches. In addition to numerous global regulators, transport systems, and metabolic genes, modulators of innate immunity (e.g., invasins, adhesins, cytotoxins, and inhibitors of proinflammatory cytokines) were found to be thermoregulated. Therefore, differential transcription during temperature shift identified a useful list of putative virulence-associated genes to examine as novel candidates for future research on the control of this pathogen (Motin et al., 2004). Besides identification of novel vaccine targets, microarray technology could be instrumental in understanding which pattern of host response is correlated with protective immunity. Using gene expression profiling in combination with immunohistochemical analysis, the mechanism of protection of a whole-cell sonicate vaccine of *Helicobacter felis* was unraveled (Mueller et al., 2003).

Although transcriptional profiling can shed light on host-pathogen interaction which occurs during the infection state, there is no direct correlation between the level of the transcript and corresponding amount of synthesized protein. Therefore, a gene array expression study should be, if possible, accompanied by analysis at the protein level. Recent achievements in combining two-dimensional gel electrophoresis and mass spectrometry made feasible the separation and identification of the proteome, a full set of proteins expressed by a genome (Nilsson, 2002). In addition to the use of proteomics in complementation with gene expression profiling, the combination of proteome analysis with serology have already provided the first vaccine candidate antigens. The significant limitation of such proteome analysis is the production of the complete proteome itself, since not all proteins are expressed at given experimental conditions. This bottleneck might be possible to overcome by using recently developed proteome microarray technology which relies on high-throughput amplification of each predicted ORF by using gene-specific primers, expression in an *E. coli*–based, cell-free in vitro transcription/translation system, followed by direct printing of the crude reactions on the microarray slide. Then the protein microarrays could be probed with the sera derived from vaccinated/infected humans and animals to determine a complete antigen-specific profile (Davies et al., 2005). A novel approach for the identification of bacterial vaccine candidate antigens called antigenome technology was recently applied to several bacterial pathogens of the genera *Staphylococcus* and *Streptococcus*. In order to select immunoreactive proteins, peptides of different sizes derived from the bacterial genome were displayed on the surface of *E. coli* via outer membrane proteins and were followed by screening with serum samples against the target pathogen (Meinke et al., 2005). The typical size of antigenome described in these experiments consisted of approximately 100 antigens, most of them surface and secreted proteins (Meinke et al., 2004). Similarly, another in vitro protein selection method, ribosome display, was employed to comprehensively identify and map the immunologically relevant proteins of *S. aureus* (Weichhart et al., 2003). Thus, together with protein arrays and protein display techniques, serological proteomics will allow identification of the entire set of antigens and epitopes (i.e., the immunome) targeted by the immune system and sufficient for protection (Klade, 2002; De Groot, 2004).

BACTERIAL CELL SURFACE DISPLAY SYSTEMS

Surface display of protein structures has two main applications in vaccine development. First, the use of cell surface–exposed heterologous antigens has been always considered beneficial for the eliciting antigen-specific immune response. Although this may not be an absolute truth, live bacterial delivery systems are traditionally used to fulfill this goal. Live vaccines could be loaded simultaneously with different antigens or their epitopes, receptor-specific ligands to change vaccine-targeting sites, and codisplayed adhesins that may increase the efficiency of the immune response. In general, either attenuated pathogens or nonpathogenic commensal or food-grade bacteria are used as live vaccine carriers. The display of different colonization factors on the surface of nonpathogens could be particularly advantageous for mucosal immunization, since these vaccine vehicles do not invade the host. Therefore, the codisplay of adhesins can increase colonization of mucosal epithelium, leading to better stimulation of the local immune response. Second, bacterial surface display expressing various polypeptide libraries has been utilized for mapping of epitope sequences for the antibodies elicited to the particular antigen(s). The use of these systems represents the frequent alternative/complement approach to phage display of random peptide libraries and synthetic peptide-based techniques traditionally employed for epitope mapping. The most common systems for bacterial cell surface display described in two recent reviews on the topic will be listed below (Samuelson et al., 2002; Wernerus and Stahl, 2004).

Gram-Negative Display Systems

Both Gram-negative and Gram-positive bacteria are used to present foreign molecules on their surface. The systems based on former microorganisms appeared earlier and employed outer membrane proteins as carriers for the heterologous peptides or proteins to be displayed. Since integral outer membrane proteins besides membrane-spanning regions contain surface-exposed loops, the task was to identify the "permissive" sites within these loops that can tolerate insertions without the loss of all biological functions. Therefore, there are limitations both with respect to the size and nature of the foreign part (passenger domain) used to modify the carrier protein. The examples of carrier outer membrane proteins which were used to display the passenger amino acid (aa) sequence of different sizes are OmpA (15–514 aa), OprF (17–43 aa), LamB (11–232 aa), OmpS (38–115 aa), and PhoE (8–32 aa). Most of these carrier proteins were expressed in enteric bacteria, and successfully used to elicit a strong immune response to the foreign part, followed by delivery as a live vaccine.

In addition to the outer membrane protein-based constructs tolerating insertions within an internal permissive site of the surface loop, a more advanced system termed Lpp'OmpA has been developed. This system utilized the N-terminal part of the major *E. coli* lipoprotein (Lpp) and transmembrane region of OmpA to allow delivery of C-terminal fusions on the surface of microbial cell. The Lpp'OmpA construct could display relatively large inserts (up to 40 kDa), while remaining sensitive to secondary and tertiary structures of the passenger. Another example of outer membrane proteins employed for the bacterial display is invasin of *Yersinia pseudotuberculosis*. A decamer random peptide library expressed as a fusion hybrid with the N-terminal part of invasin residing in the outer membrane was screened against human cultured cells. Several clones were identified for their human cell binding ability to human cells, which was mediated by the surface-displayed peptides. This approach could be applied to select antiadhesion agents to use in antiadhesion therapy of bacterial diseases (Ofek et al., 2003). In addition to outer membrane proteins, surface-exposed lipoproteins were tested as bacterial display. Among them were plasmid-encoded lipoprotein TraT involved in conjugation, the peptidoglycan-associated lipoprotein (PAL) from *E. coli*, and ice-nucleation protein Inp of *Pseudomonas syringae* which is attached to the membrane via the glycosylphosphatidylinositol anchor sequence. The mentioned systems have some advantages such as a very high number of copies (>10,000 for TraT), or ability to display enzymes, single-chain antibodies, and intact antigens (Inp system).

Recently, autotransporters have been employed for surface display. Since the autotransporter is an autonomous structure promoting its own transport by assembling a channel in the outer membrane, the system is capable of translocating relatively large segments. The examples of autotransporters that were used for display of heterologous antigens are the IgA1 protease from *Neisseria gonorrhoeae*, VirG of *Shigella*, and AIDA-I adhesin of *E. coli*. Moreover, the foreign protein expressed on the surface by an autotransporter could be released by autoproteolytic cleavage, if necessary. In addition, other systems have been developed which allow cell surface–bound antigens to be subsequently released extracellularly. One example of such a system is pullulanase from *Klebsiella pneumoniae*, which is released to the medium at the stationary growth phase. Finally, filamentous surface structures, such as flagella (major flagellin subunit, FliC) and fimbriae (P fimbrillin of uropathogenic *E. coli* and subunits of Type 1 fimbriae of *E. coli*), have been used for surface display. The common feature of these systems is that the insertion of foreign antigens should not disrupt the biogenesis of the filamentous structures that put significant limitations on the size and conformation of the passenger domains.

Gram-Positive Display Systems

Gram-positive microorganisms have several advantages for surface display applications, since their cells have only a single membrane, there is a common mechanism for surface anchoring, and bacteria can withstand rigorous manipulation conditions due to the thicker cell wall. The display systems of Gram-positive bacteria may be classified into three groups: cell wall–bound, cell membrane–anchored, and cell surface–associated proteins. The examples of the first group (most commonly used) are protein A of *S. aureus*, which has provided the surface exposure of heterologous proteins of various lengths (15–397 aa) in *Staphylococcus* and *Lactococcus lactis*; M6 protein of *S. pyogenes* successfully used for display of 15–438 aa epitopes in *Streptococcus gordonii* and *Lactobacillus*; the surface protein antigen P1 from *Streptococcus mutans* that expresses surface-located 179 aa antigen in *S. gordonii*; and extracellular cell wall–bound proteinase PrtP of *L. lactis* which tolerates the display of 52 kDa tetanus toxin fragment. The representatives of the second group are lipoprotein DppE of *Bacillus subtilis*, which can express a 192 kDa antigen which, however, was not surface-accessible unless the bacterial cell wall was stripped out; autolysin modifier protein CwbA of *B. subtilis*, exposing large 192–497 aa antigen; and *M. tuberculosis* lipoprotein Mtb19 fused with ~31 kDa outer surface protein A (OspA) of *Borrelia burgdorferi* and expressed on the surface of the *Mycobacterium bovis* strain bacilli Calmette-Guerin (BCG). Examples of the third group are the beta-D-fructosyltransferase Ffts of *Streptococcus salivarius*, which has been expressed as a carrier of heterologous proteins in *S. gordonii* and subsequently released from the surface when desired; and S-layer proteins EA1 and Sap of *B. anthracis* containing S-layer homology motifs (SLH) and stably expressing SLH-fused ~50 kDa antigen on the surface of *B. anthracis*.

NOVEL DELIVERY SYSTEMS

Listeriolysin O (LLO)

A novel approach for eliciting a cytotoxic T-cell response has been developed using *L. monocytogenes* (LLO). LLO is a pore-forming hemolysin which allows *L. monocytogenes* to escape from the phagolysosomal environment into the host cell cytosol by lysis of the phagosomal membrane. Therefore, the incorporation of LLO into vaccine formulations provides access to the cytosol of professional antigen-presenting cells (APC) by vaccine components, an activity that is crucial for MHC class I presentation of antigen. LLO was tested in combination with purified protein antigens, attenuated bacterial vaccine strains, DNA vaccines, and the contents of liposomes. Coadministration of LLO with the model antigen chicken ovalbumin resulted in efficient presentation of epitopes of this antigen, together with MHC class I molecules. Similarly, LLO introduced the nucleoprotein of influenza virus and beta-galactosidase directly into the host cell cytosol. Introduction of secreted LLO in attenuated live vaccine carriers such as *Salmonella*, mycobacteria (BCG), vaccinal strain of *B. anthracis*, and *E. coli* expressing recombinant antigen stimulated a class I–restricted $CD8^+$ T-cell response (Dietrich et al., 2001).

Genetic Vaccination

Genetic vaccination is a promising new technology that has emerged in recent years. In general, DNA vaccines are composed of bacterial plasmids encoding antigens under the control of strong eukaryotic promoters. Typically DNA vaccines are injected intramuscularly or intradermally in their naked form and induce antigen-specific humoral and cellular immune responses. However, delivery of DNA vaccine plasmids into mammalian cells by intracellular bacteria (bactofection) might allow more specific targeting of a variety of host cells, including professional APCs. In most cases, bacteria used as DNA vaccine carriers are attenuated pathogens which can be classified into three categories: extracellular pathogens (some strains of *E. coli*, *Yersinia* spp.); intraphagosomal pathogens (*Salmonella* spp.); and intracytosolic pathogens (*Shigella* spp., *L. monocytogenes*). *Y. enterocolitica* is an extracellular enteropathogen multiplying in abdominal lymphoid tissue. Normally, the pathogen is resistant to phagocytosis due to the presence of a virulence plasmid-encoding type III secretion system. Nevertheless, a virulence plasmid-cured variant could be phagocytosed by APC, thus providing a basis for the DNA vaccine delivery. *Salmonella* cells can promote their own uptake into phagocytes and avoid the killing mechanism of phagocytes following internalization. Virulent strains will multiply in the phagosomal compartment. However, attenuated live vaccine carriers of *Salmonella* die there, resulting in release of the DNA vaccine vectors into the phagosome and consecutively from the phagosome via phagosomal leakage into the cytosol. Upon invasion of APC, *Shigella* spp., unlike *Salmonella* spp., can rapidly escape from endocytic vacuoles to the cytosol of the mammalian cell. Therefore, attenuated vaccine carriers, which are lysed in the cytosol, can deliver the plasmids directly to this intracellular compartment. *L. monocytogenes*, the only Gram-positive carrier for bacterial DNA vaccine, can multiply within the cytosol and disseminate through tissue by intercellular spread. Like the *Shigella*-based vaccine delivery platform, *Listeria* carriers deliver DNA vaccine to the cytosol of a broad spectrum of host cell types. The nonpathogenic bacterium *Bifidobacterium longum* (found in the lower small intestine of humans) and *Agrobacterium tumefaciens* (soil plant pathogen) are also tested as DNA vaccine delivery systems. It is important to mention that both effective invasion of host cells and escape of the bacterial carrier from the phagosome are crucial for efficient DNA vaccination. When a noninvasive laboratory strain of *E. coli* used for DNA vaccination was modified to express the *inv* gene (invasin) of *Y. pseudotuberculosis*, DNA transfer occurred more efficiently. Likewise, inactivation of *L. monocytogenes* genes involved in invasion led to a reduced ability of the *L. monocytogenes*–based carrier to deliver DNA vaccine. To promote phagosomal escape, *Salmonella* carrier strains were created, either to secrete LLO or to produce a pore-forming hemolysin from *E. coli*, and this alteration improved gene transfer efficacy. Moreover, most of the bacterial carriers for DNA vaccination were designed to lyse when bacteria entered the host cell. The impaired cell wall synthesis (*Shigella*, *Salmonella*, invasive *E. coli* carriers) and production of a phage lysin (*Listeria*) are the common ways to achieve specific autolysis of the bacteria into the cytosol compartment (Loessner and Weiss, 2004; Schoen et al., 2004).

One of the major limiting factors in using bacterial carriers for DNA vaccination is the import of plasmid DNA from the cytosol into the nucleus where the expression takes place. The nuclear membrane represents a significant barrier for efficient gene transfer. Recently, a self-destructing *L. monocytogenes* carrier was used to release translation-competent messenger RNA (mRNA) directly into the cytosol of epithelial cells, macrophages, and human dendritic cells. The transfer of functional mRNA into mammalian cells using a bacterial carrier represents a totally novel delivery system for vaccine development (Schoen et al., 2005).

Bacterial Ghost System

The bacterial ghost (BG) system is another example of a recently developed antigen delivery vehicle. To produce BG of Gram-negative bacteria, the controlled nondenaturing method of lysis by the PhiX174 bacteriophage protein E was used. Protein E can form a specific lysis tunnel spanning the whole cell wall complex through which the cytoplasmic content of the cell is expelled. The resulting BG is generally free from nucleic acids, ribosomes, and other intracellular components; however, outer and inner membranes largely remain intact. In contrast to heat, irradiation, or chemical inactivation of pathogens for vaccine purposes, BG preserves its native antigenic structure, and LPS and peptidoglycan retain their immunostimulatory activity, while production and storage of BG can be achieved at low cost. This attractive platform has been adapted for antigen delivery by insertion of the antigens into the BG envelope. The BG system has been applied for envelope and/or heterologous antigen presentation of many medically important bacterial agents such as *V. cholerae*, pathogenic *E. coli*, *Salmonella* spp., *Actinobacillus pleuropneumoniae*, *Francisella tularensis*, *Brucella melitensis*, *Bordetella bronchiseptica*, etc. (Mayr et al., 2005). BG has also been successfully tested as a platform technology for DNA delivery following the loading of lyophilized BG in the DNA-containing buffer or by developing a procedure for the targeted immobilization of plasmid DNA in the cytoplasmic membrane of the carrier (Mayrhofer et al., 2005; Paukner et al., 2005).

Inverted Pathogenicity

The "inverted pathogenicity" concept means the use of virulence mechanisms for prevention or therapy of disease. With respect to the development of new bacterial delivery systems, the classical example of inverted pathogenicity is the employment of a type III secretion system (T3SS) for vaccination. T3SS has been identified in a number of Gram-negative mammalian and plant pathogens; its structure resembles the basal body of flagella and forms a "needle complex." This system not only provides efficient secretion of virulence factors by the pathogen but also delivers them directly into the cytosol of eukaryotic cells by the contact-mediated translocation (Hueck, 1998; Cornelis, 2002). Therefore, the ability to inject virulence factors (effectors) could be utilized for the cytosolic delivery of heterologous antigens and their immunorelevant fragments. For this purpose, TTSS of *Salmonella* pathogenicity island 1 (SPI-1) was employed that promotes uptake of the pathogen and formation of a membrane-bound compartment (macropinosome) in which *Salmonella* cells further multiply. Importantly, *Salmonella* continues to translocate type III effectors into cytosol of the host, as it resides in the macropinosome. Insertion of immunodominant heterologous CD8 epitopes into one of the effectors indeed led to the strong induction of specific MHC class I–restricted immune response, followed by immunization with attenuated *Salmonella* live carrier. The T3SS-based delivery system was further improved by using the *Yersinia* effector's YopE secretion and translocation N-terminal sequence, which is functional in *Salmonella* as well, thus allowing the cytosolic injection of significantly larger antigens. Recently, attenuated *Yersinia* and its T3SS were also used for live vaccine delivery which led to efficient antigen-specific CD4 and CD8 T-cell responses (Russmann, 2004). Similarly, the type IV bacterial secretion system of *A. tumefaciens*, which is capable of mediating DNA transfer into human cells, could be used for genetic vaccination (Loessner and Weiss, 2004). Finally, a type I secretion system can be an effective delivery system in virulence-attenuated, Gram-negative bacteria. For example, the use of the alpha-hemolysin secretion pathway provided an export of heterologous antigens from the cytoplasm of bacterial carrier into either the phagosome or the cytosol of infected cells (Gentschev et al., 2002).

GENETIC APPROACHES

The availability of genome-scale sequence data, the development of new tools for high-throughput gene expression monitoring, and improvements in the ability to analyze large data sets have revolutionized the field of vaccine development. In this section, we discuss several high-throughput genomics approaches that have been developed to discover genes whose expression takes place specifically when the bacteria infect their host and which encoded those antigens with potential applications as vaccine candidates.

Identification of Essential Genes: Nontransposon- and Transposon-Based Methods

Nontransposon Methods

A simple way to define an essential gene is derived from the functional analysis of the protein product. While a mutation (transposon-based or not) within a nonessential gene would not be expected to affect the growth rate of the cell, the mutation within an essential gene would probably result in a measurably diminished growth rate or fitness of the mutant compared to the wild-type bacterium (Lehoux et al., 2001). There are several methods to identify essential genes that do not rely on transposons, including comparative genomics and the directed knockout of genes. More traditional analysis uses the generation of conditional mutations that affect growth, such as temperature-sensitive (TS) mutations (Judson and Mekalanos, 2000). TS mutations are generated by chemical mutagenesis and result in unmarked point mutations that allow the identification of essential genes by growth at a permissive temperature, but not at a nonpermissive temperature. The location of the mutation is determined by a mapping and/or complementation cloning strategy. Another approach that builds on comparative genomics involves the targeted knockout of specific genes after bioinformatics analysis. This approach uses the premise that genes of unknown function that are conserved between organisms are more likely to be essential than nonconserved genes. Gene-by-gene PCR-product design, cloning, and attempted knockout of the gene in question are used to determine if the gene is essential (Judson and Mekalanos, 2000).

Transposon-Based Methods

There are a number of different transposon-based approaches available for defining essential genes (Lehoux et al., 2001). The utility of each approach varies, depending on the type of information one is interested in obtaining and the genetic systems available in the organism of interest. There are two ways to identify essential genes or regions of the bacterial chromosome which can be based on the location of the transposon insertion: the "negative" and the "positive" approaches (Judson and Mekalanos, 2000). A problem in trying to identify essential genes is that a knockout of an essential gene is lethal. Therefore, the use of the negative approach will help with the identification of many regions that are not essential and enable one to say with some certainty that regions in which transposon insertions are not observed are likely to be essential. These types of approaches that allow saturating in vitro transposon mutagenesis of PCR-amplified chromosomal segments of naturally competent organisms, uptake of these mutagenized fragments, and homologous recombination into the chromosome have being used to identify essential genes in *Haemophilus influenzae*, *Mycoplasma*, and *S. pneumoniae* (Judson and Mekalanos, 2000).

The positive approach defines genes that are essential instead of defining genes or regions of the chromosome that are nonessential by generating a conditional mutation and showing that it has a lethal phenotype. This approach defines essential genes by replacing the gene's natural promoter with an inducible one. Transposition with a transposon that has an outward-facing inducible promoter at one end of the transposon into the promoter region of a gene creates a gene in which the function of the natural promoter is replaced by an inducible promoter. If the gene is essential, the bacterial strain is now dependent on the inducer for growth or survival. This approach was first used with Tn*5tac1*, a derivative of transposon Tn*5* that generates conditional mutations, and modified methods have subsequently been used to identify a number of genes with conditional growth phenotypes (Judson and Mekalanos, 2000; Lehoux et al., 2001).

Functional Inhibition of Genes: STM, GAMBIT, TraSH

Signature-Tagged Mutagenesis (STM)

STM is a functional genomics technique that identifies microbial genes required for infection within an animal host, or within host cells, using extensive collections of mutants generated by insertional mutagenesis. DNA tags are usually incorporated into the mutagenesis vector (transposon or insertional mutagenesis plasmid) to label each mutant with a unique identifying DNA "signature-tag." Pools of mutants are then screened through an animal model or cell culture to identify clones in which a mutation has impaired multiplication. Mutants that fail to be recovered after the screen are likely to be attenuated and therefore altered within virulence genes. The presence or absence of individual mutations within a pool of mutants can then be detected by PCR amplification and detection of the signature-tags, and their identification enables negative selection screens to be performed. STM was originally used in the enteric pathogen *Salmonella enterica* serovar Typhimurium, in which the investigators identified known virulence determinants, thus demonstrating the utility of the technique, as well as the identification of a pathogenicity island (*Salmonella* pathogenicity island 2, SPI-2) containing a type III secretion system (Hensel et al., 1995). Since then, STM has been applied so far to a variety of pathogens, including *M. tuberculosis*, *S. aureus*, *V. cholerae*, *Y. enterocolitica*, *S. pneumoniae*, and *N. meningitidis* (Mecsas, 2002).

Genomic Analysis and Mapping by In Vitro Transposition (GAMBIT)

This method is a variation of the transposon-based approaches which use the transposon TnAraOut to perform high-density mutagenesis of restricted regions of the genome (Akerley and Lampe, 2002).

Using this technique, the complete set of genes required by *H. influenzae* for growth and viability in vitro has been identified. Further, genes essential for growth in *Mycoplasma* were defined using the same approach (Hutchison et al., 1999).

Transposon Site Hybridization (TraSH)

The combined use of high-density transposon mutagenesis and microarray hybridization mapping of pools of mutants resulted in the development of TraSH, a method suitable for the rapid functional characterization of bacterial genes required for growth under different conditions (Sassetti et al., 2001). In TraSH, each mutant contains a single transposon insertion, and the bacterial library contains mutations in each gene in the genome. After a growth phase, mutants harboring insertions in genes that are required for survival are lost from the library. The TraSH probe is generated from the selected library, and this "insertion probe" consists only of sequences complementary to genes that contain insertions in the selected library. Randomly labeled chromosomal DNA (genomic probe) will hybridize to every gene represented on the array. Spots that hybridize to the genomic probe, but not the insertion probe represent genes that are required for growth. TraSH has been applied to identify conditionally essential genes in *M. bovis* BCG which represent promising targets for rational attenuation (Sassetti et al., 2003).

Gene Expression In Vivo: IVET, RIVET, DFI

In Vitro Expression Technology (IVET)

IVET is designed to identify bacterial genes that are induced when a pathogen infects its host. A subset of these induced genes should include those that encode virulence factors. The system is based on complementation of a mutation in a biosynthetic gene (auxotrophic mutation) by gene fusion, and is designed to be used in a wide variety of pathogenic organisms (Slauch et al., 1994). The positively selected fusions are then sequenced to identify in vivo–induced genes. The original IVET made use of a *S. typhimurium* library consisting of a promoterless, two-reporter operon fused randomly in the bacterial chromosome (Mahan et al., 1993). One of the two reporters encoded an enzyme whose expression was necessary for survival in mice. In this study, a mouse was infected with the library, and only bacteria containing in vivo–induced fusions survived, since only they expressed the reporter enzyme. Bacteria were then harvested from the animal and grown on laboratory media. At this step, the second reporter gene, e.g., *lacZ*, became important. When the *lacZ* activity of bacteria grown on indicator media is determined, only those fusions with low *lacZ* activity (those with promoters active in the mouse but inactive in vitro) are kept for further study. As indicated, this method requires the existence of an attenuating and complementable auxotrophy, which might not be available in all microbial systems. Variations to the original IVET method have been developed using different reporting systems (e.g., gene fusions to green fluorescent protein and the use of flow cytometry to screen for the promoters; Valdivia and Ramakrishnan, 2000).

Recombination-Based In Vivo Expression Technology (RIVET)

This is a new version of the IVET method, which enables the detection of genes that are transiently turned on during adaptation to a new environment. The RIVET system is based on recombinase gene fusions, which, on induction during infection, mediate a site-specific recombination, the product of which can be screened for after recovery of bacteria from host tissues. RIVET is very sensitive to low or transient expression of in vivo–induced (*ivi*) genes during infection and is therefore capable of identifying members of this potentially interesting class of genes. RIVET is also uniquely designed for the postidentification analysis of *ivi* genes; e.g., it has been used to analyze the temporal and spatial patterns of virulence gene induction during infection and to dissect the regulatory requirements of in vivo induction with respect to both bacterial regulatory factors and host-inducing environments (Slauch and Camilli, 2000). RIVET was first successfully used with *V. cholerae* to identify genes that are induced transcriptionally during infection of the gastrointestinal tract of infant mice. A recent improvement in this technology has allowed the same investigators to increase the efficiency of the screening method and to reduce the frequency of false positives (Osorio et al., 2005).

Differential Fluorescence Induction (DFI)

DFI utilized a fluorescence-enhanced green fluorescent protein (GFP) and fluorescence activated cell sorter (FACS) to separate bacteria or infected cells on the basis of GFP fluorescence (Valdivia and Ramakrishnan, 2000). In the original application, the technique was used to screen a *S. typhimurium* library for promoters that are upregulated at pH 4.5 (Valdivia et al., 1996). Macrophage-like cell lines were also infected with the same library of *S. typhimurium*-bearing random gene fusions. Using this approach, cells that become fluorescent, due to their association with a *gfp*-expressing *S. typhimurium*, are collected by FACS. Then, bacteria cells are recovered, grown again in the absence of cells, and sampled by FACS. Bacteria that are no longer fluorescent in the extracellular environment are sorted and used for a second round of macrophage infection. *S. typhimurium* present within these fluorescent macrophages contained *gfp* fusions that were upregulated in the host's cell intracellular environment (Valdivia et al., 1996; Valdivia and Falkow, 1997). One of the advantages of DFI is that there is available automation for initial screening compared to manual screening. Another advantage is that DFI allows the study of upregulation, as opposed to on-off types of gene induction.

Capturing Transcribed Bacterial Genes During In Vitro Infection: SCOTS

Selective Capture of Transcribed Sequences (SCOTS)

The differential gene expression in bacteria within infected host cells or tissues has been limited by the low number of bacteria in these systems and the instability of bacterial mRNA. There are also difficulties involved in separating bacterial mRNA from ribosomal RNA and host RNA. Development of SCOTS has allowed identification of bacterial genes that are expressed within macrophages (Graham and Clark-Curtiss, 1999; Morrow et al., 1999). SCOTS allows the selective capture of bacterial cDNAs from total cDNA, prepared from infected cells or tissues, using hybridization to biotinylated, bacterial, genomic DNA. The cDNA mixtures obtained are then enriched for sequences that are transcribed preferentially during growth in the host, using additional hybridizations to bacterial genomic DNA in the presence of cDNA similarly prepared from bacteria grown in vitro.

Antigenomics: Antigenome Technology, IVIAT, ELI

Evaluation of the immune response against any candidate antigen is a crucial validation task and cannot be circumvented. Therefore, techniques using human immunogenicity as their primary screening and selecting parameter on a genome-wide basis seems to be particularly valuable in vaccine development. Moreover, the antibodies that are induced in the host are molecular proofs of the in vivo expression of the corresponding antigenic proteins.

Antigenome Technology

Combining the advantages of full genome coverage and serological antigen identification, antigenome technology is an approach for identification of antigenic protein epitopes from bacterial pathogens. This approach defines the most-relevant antigenic proteins that are targeted by the human immune system, including their antibody-binding sites (Etz et al., 2002), through the use of comprehensive small-fragment genomic-surface display libraries and antibodies from human serum. Antigenome technology offers an integrated approach for antigen validation to select the most-promising candidates for the development of subunit vaccines against targeted bacterial diseases, as selected clones can be used directly for the generation of specific immune sera for surface localization and in vitro functional assays, without the demanding task of high-throughput recombinant protein production.

In Vivo–Induced Antigen Technology (IVIAT)

This antibody-based selection method has been developed to identify those gene products targeted by the host immune system (Handfield et al., 2000). Sera from patients are first pre-adsorbed with cells of the pathogen grown in vitro, to select antibodies recognizing antigens specifically expressed in vitro, then used to probe expression libraries of the DNA of the pathogen to identify antigens expressed specifically during infection, but not during growth in standard laboratory media. Reactive clones are purified and the cloned DNA sequenced to identify the corresponding gene products. This technique does not require direct genetic manipulation of the pathogen or an animal model, but it is necessary that a gene product is immunogenic to be identified. IVIAT has been applied, e.g., for the identification of *Vibrio*, *E. coli* O157:H7, and *Mycobacterium* antigens that are uniquely expressed during human infection (Rollins et al., 2005).

Expression Library Immunization Technology (ELI)

Another major challenge in the development of efficient screening methods is direct selection for protective candidates. The ELI technology is designed to discover vaccine candidate genes, by using the immune system to identify protective antigens from complex organisms such as fungi, protozoa, and ticks (Barry et al., 2004). However, this technology has also been applied, in combination with in vitro assays, to select immunogenic and protective antigens from bacterial pathogens, such as *M. tuberculosis* (Delcayre et al., 2003), *Mycoplasma* (Barry et al., 1995), and recently,

Chlamydia abortus (Stemke-Hale et al., 2005). ELI is based on the idea that pools of genes from a given pathogen could be introduced into the test animal by genetic immunization and then screened for whether they contained one or more protective antigens by directly challenging the host with the pathogen. Although the method represents the most straightforward approach to select protective candidates, it is limited to the study of genes that can be expressed in eukaryotic cells, and necessitates animal models that are suitable for screening purposes.

Stability of Protective Antigen Expression for Recombinant Bacterial Vaccines

Eliciting the desired immune response against a heterologous antigen is influenced by the appropriate levels of antigen expression; insufficient expression levels will abrogate the immune response, but expression levels that are too high might result in increased metabolic load and toxicity problems. One method of avoiding inappropriate high expression levels is chromosomal integration of heterologous antigens (Galen and Levine, 2001). Although this approach has proven successful for some antigens, in other cases it failed to provide the expected protective immune responses. Since in vivo stability of protective antigen expression is important for recombinant bacterial vaccine efficacy, the optimization of carrier strains by using stabilized and antibiotic-resistant free-expression plasmids, capable of expressing high levels of heterologous antigen, is one of most common approaches when developing new bacterial live vaccines (Spreng and Viret, 2005). Further, the use of in vivo–inducible promoters to limit heterologous gene expression has been shown to enhance the stable expression and immunogenicity of foreign antigens, while reducing metabolic load during in vivo propagation (Galen and Levine, 2001). In addition, the plasmid-based expression of heterologous genes provides the flexibility of introducing heterologous expression cassettes into a variety of existing live vectors efficiently, without the need for cumbersome chromosomal integrations for each live vector. In the following sections, we will discuss some of the most popular plasmid maintenance systems used for the stabilization of plasmids in vivo.

Plasmid Maintenance Systems Based on Auxotrophies in DNA and Cell Wall Synthesis

In plasmid-based balanced lethal systems, plasmids express an essential protein required for the bacteria to grow and replicate. The loss of such plasmids destroys the ability of the bacterium to express this protein and results in cell death. This process of segregational plasmid loss is known as "postsegregational killing" (Spreng and Viret, 2005).

One of the first balanced lethal plasmid systems which rely on the ability to complement a mutation in an essential gene (auxotrophic mutation) was developed by Curtiss et al. (1990) and successfully employed in *S. typhimurium*. This system is based on expression of the *asd* gene encoding aspartate beta-semialdehyde dehydrogenase (Asd), which is involved in the biosynthesis of diaminopimelic acid (DAP) from aspartate. Asd is a critical enzyme involved in the synthesis of structural components essential for the formation of the cell wall in Gram-negative bacteria, so that *asd* mutants lyse in a growth medium deprived of DAP. However, cells deleted for the chromosomal *asd* gene, but harboring a complementing Asd^+ plasmid are restored to normal growth, and loss of the plasmid during segregation leads to cell lysis. Since DAP is not prevalent in the animal host, essentially 100% of the surviving avirulent *Salmonella* recovered from an immunized animal host still contains the recombinant plasmid and expresses the foreign antigen (Curtiss et al., 1990). Asd-stabilized plasmid systems encoding heterologous antigens have been widely evaluated in *S. typhimurium* and *S. typhi* vaccine strains (Spreng and Viret, 2005). One drawback from this system is that toxicity has been observed due to the expression of some foreign proteins in live bacterial vaccine carriers, which generally leads to the overgrowth of a plasmid-free bacterial population. This toxicity is believed to be caused by a vastly increased metabolic burden on the cell, and can be exacerbated by the use of plasmids that have high copy numbers. Therefore, attempts to reduce the level of Asd protein and obtain a more balanced expression have been made, e.g., depleting the *asd* -35 and -10 promoter region on the multicopy Asd^+ plasmids still harboring the Shine–Dalgarno sequence for ribosome recognition, which results in a significant reduction in expression of Asd protein and enhance immunogenicity of the foreign antigen (Spreng and Viret, 2005).

Another plasmid system uses the thymidylate synthase (*thyA*), which is an essential gene involved in the de novo synthesis of the DNA precursor thymidine monophosphate from uridine monophosphate. Isogenic *thyA* auxotrophic mutants are incapable of intracellular growth and survival in vitro and undergo so-called thymidineless death. The *thyA* lethal balanced system is a very efficient method to achieve stable plasmid inheritance both in vitro and in vivo, and this plasmid system can easily be adapted to any host–vector combinations, since *thyA* chromosomal mutations can readily be selected by trimethoprim selection (Morona et al., 1991). Such a stabilization system has been used, e.g., to generate a candidate carrier strain against cholera (Spreng and Viret, 2005).

A third balanced lethal plasmid system, based on the *purB* gene, has been developed for *Salmonella* spp. for the heterologous expression of *H. pylori* urease. The *purB* coding for adenylosuccinate lyase catalyzes an essential step in the de novo synthesis of adenosine monophosphate involved in the purine biosynthetic pathway. Using this system, it was found that despite an almost equal protein expression, the in vitro stability of the urease-encoding plasmid differed markedly between carrier strains, and the level of the immune response induced against the heterologous antigen seems to be conditioned by a stable antigen expression in vivo (Spreng and Viret, 2005).

Plasmid Maintenance Systems Based on Auxotrophies in Biosynthetic Pathways

Genes involved in amino acid and protein synthesis steps have also been employed for the design of balanced lethal plasmid stabilization systems. Glutamine synthetase, encoded by *glnA*, is required for glutamine synthesis, which serves together with glutamate as a primary nitrogen source in bacterial metabolism. For example, a *V. cholerae glnA* mutant was complemented in trans with a plasmid coding for the heterologous *S. typhimurium glnA* gene, and then the strain was used for oral inoculation of germ-free mice. Bacteria were recoverable for 10–12 days from the stools. However, a problem with this approach was that plasmid-free carrier bacteria could be recovered for 8 days after primary inoculation (Ryan et al., 2000).

Other example is the suicide *L. monocytogenes* carrier system developed for bactofection DNA delivery into mammalian cells and which takes advantage of a phage lysin which is specifically expressed once the bacteria enter the host cell cytosol. This results in the complete lysis of the bacteria and in the consequential release of plasmids encoding for heterologous antigens under the control of a eukaryotic promoter (Dietrich et al., 1998). This balanced lethal plasmid system was shown to efficiently stabilize the recombinant plasmid both in vitro and in vivo.

A third plasmid stabilization system has been described for *E. coli* based on the *infA* gene coding for translation initiation factor 1 (IF1), a small intracellular protein essential for cell viability. By deleting the chromosomal *infA* gene and placing it instead on a plasmid, the latter could be kept totally stable for at least 120 generations of growth in the absence of antibiotics. Although developed for large-scale cultures that are usually applied in bioreactors for microbial production of bio-molecules, this system does not seem to be limited to *E. coli*, and further evaluation in attenuated live bacterial carriers may clarify its use for vaccine development (Hägg et al., 2004).

Plasmid Maintenance Systems Based on Toxin–Antitoxin Modules

Toxin–antitoxin systems also known as "postsegregational cell killing" and "addiction" systems ensure plasmid maintenance during replication by conferring an advantage on plasmid-retaining cells due to reduced competitiveness of their plasmid-free counterparts. The best-characterized toxin–antitoxin module is the *hok-sok* locus of *E. coli* plasmid R1. *hok-sok* is a two-component system in which *hok* encodes a lethal pore-forming Hok protein. Hok translation is blocked by binding of the *sok* transcript that is complementary to the 5′ end of the *hok* mRNA. However, *sok* mRNA is highly susceptible to degradation by nucleases, and its protective intracellular concentration must be maintained at a critical level by constitutive transcription from a resident plasmid carrying the *hok-sok* locus. Therefore, the absence of a renewable source of the *sok* antitoxin RNA in bacteria that spontaneously lose such plasmids results in the accumulation of the more stable *hok* mRNA and thereby to the sustained synthesis of the Hok protein and subsequent cell death by cell membrane depolarization (Gerdes et al., 1997). The noncatalytic, postsegregation killing function of the *hok-sok* module has been incorporated in attenuated strains of *S. typhi* and *S. flexneri*, causing an enhanced inheritance of plasmids, but was not sufficient to maintain plasmids coding for toxic proteins (Galen and Levine, 2001). Therefore, plasmid maintenance systems combining a postsegregational system and partition functions significantly enhance plasmid segregational stability and minimize plasmid loss in a population of actively growing bacteria, and are becoming the next generation of plasmid maintenance systems used in bacterial-based vaccine development.

CONCLUSIONS

From the foundation for modern vaccinology established by Edward Jenner in the late 1700s, the field of bacterial vaccines has seen dramatic changes, particularly in recent years, due to tremendous advances in molecular biology, immunology, cell biology, biochemistry, and the use of innovative scientific approaches. The emergence of multidrug resistance of many clinically relevant bacterial pathogens has renewed an interest in the development of efficient, safe, and affordable vaccines. As presented in this chapter, to date, weakened microbes, killed microbes, inactivated toxins, and purified proteins or polysaccharides derived from microbes, represent some of the most common components used in vaccine development strategies. However, advances in biotechnology linked to the progress in genome sequencing of pathogens and high-throughput screening technologies are aiding the development of a new generation of vaccines, capable of activating the immune system, including eliciting both humoral and cell-mediated immunity against a variety of disease-producing organisms. Despite the progress, bacterial infections are still killing millions of people worldwide. Therefore, there is a need for the development of new and innovative vaccines to prevent the most challenging of diseases and to respond to the threat of bioterrorism.

References

Akerley, B.J. and Lampe, D.J. Analysis of gene function in bacterial pathogens by GAMBIT. Methods Enzymol. 2002; 358:100–108.

Barry, M.A., Howell, D.P., Andersson, H.A., Chen, J.L. and Singh, R.A. Expression library immunization to discover and improve vaccine antigens. Immunol. Rev. 2004; 199:68–83.

Barry, M.A., Lai, W.C. and Johnston, S.A. Protection against *Mycoplasma* infection using expression-library immunization. Nature 1995; 377:632–635.

Cornelis, G.R. The Yersinia Ysc-Yop 'type III' weaponry. Nat. Rev. Mol. Cell Biol. 2002; 3:742–752.

Curtiss, R., III, Galan, J.E., Nakayama, K. and Kelly, S.M. Stabilization of recombinant avirulent vaccine strains in vivo. Res. Microbiol. 1990; 141:797–805.

Davies, D.H., Liang, X., Hernandez, J.E., Randall, A., Hirst, S., Mu, Y., Romero, K.M., Nguyen, T.T., Kalantari-Dehaghi, M., Crotty, S., Baldi, P., Villarreal, L.P. and Felgner, P.L. Profiling the humoral immune response to infection by using proteome microarrays: high-throughput vaccine and diagnostic antigen discovery. Proc. Natl. Acad. Sci. U. S. A. 2005; 102:547–552.

De Groot, A.S. Immunome-derived vaccines. Expert. Opin. Biol. Ther. 2004; 4:767–772.

Delcayre, A., Peake, J.S., White, D.J., Yuan, S., McDonald, M.K., Liang, A., Tan, P.L. and Watson, J.D. A genome-based functional screening approach to vaccine development that combines *in vitro* assays and DNA immunization. Vaccine 2003; 21:3259–3264.

Dietrich, G., Bubert, A., Gentschev, I., Sokolovic, Z., Simm, A., Catic, A., Kaufmann, S.H., Hess, J., Szalay, A.A. and Goebel, W. Delivery of antigen-encoding plasmid DNA into the cytosol of macrophages by attenuated suicide *Listeria monocytogenes*. Nat. Biotechnol. 1998; 16:181–185.

Dietrich, G., Hess, J., Gentschev, I., Knapp, B., Kaufmann, S.H. and Goebel, W. From evil to good: a cytolysin in vaccine development. Trends Microbiol. 2001; 9:23–28.

Etz, H., Minh, D.B., Henics, T., Dryla, A., Winkler, B., Triska, C., Boyd, A.P., Sollner, J., Schmidt, W., von Ahsen, U., Buschle, M., Gill, S.R., Kolonay, J., Khalak, H., Fraser, C.M., von Gabain, A., Nagy, E. and Meinke, A. Identification of in vivo expressed vaccine candidate antigens from *Staphylococcus aureus*. Proc. Natl. Acad. Sci. U. S. A. 2002; 99:6573–6578.

Galen, J.E. and Levine, M.M. Can a 'flawless' live vector vaccine strain be engineered? Trends Microbiol. 2001; 9:372–376.

Gentschev, I., Dietrich, G. and Goebel, W. The *E. coli* alpha-hemolysin secretion system and its use in vaccine development. Trends Microbiol. 2002; 10:39–45.

Gerdes, K., Gultyaev, A.P., Franch, T., Pedersen, K. and Mikkelsen, N.D. Antisense RNA-regulated programmed cell death. Annu. Rev. Genet. 1997; 31:1–31.

Graham, J.E. and Clark-Curtiss, J.E. Identification of *Mycobacterium tuberculosis* RNAs synthesized in response to phagocytosis by human macrophages by selective capture of transcribed sequences (SCOTS). Proc. Natl. Acad. Sci. U. S. A. 1999; 96:11554–11559.

Hägg, P., de Pohl, J.W., Abdulkarim, F. and Isaksson, L.A. A host/plasmid system that is not dependent on antibiotics and antibiotic resistance genes for stable plasmid maintenance in *Escherichia coli*. J. Biotechnol. 2004; 111:17–30.

Handfield, M., Brady, L.J., Progulske-Fox, A. and Hillman, J.D. IVIAT: a novel method to identify microbial genes expressed specifically during human infections. Trends Microbiol. 2000; 8:336–339.

Hensel, M., Shea, J.E., Gleeson, C., Jones, M.D., Dalton, E. and Holden, D.W. Simultaneous identification of bacterial virulence genes by negative selection. Science 1995; 269:400–403.

Hueck, C.J. Type III protein secretion systems in bacterial pathogens of animals and plants. Microbiol. Mol. Biol. Rev. 1998; 62:379–433.

Hutchison, C.A., Peterson, S.N., Gill, S.R., Cline, R.T., White, O., Fraser, C.M., Smith, H.O. and Venter, J.C. Global transposon mutagenesis and a minimal *Mycoplasma* genome. Science 1999; 286:2165–2169.

Judson, N. and Mekalanos, J.J. Transposon-based approaches to identify essential bacterial genes. Trends Microbiol. 2000; 8:521–526.

Klade, C.S. Proteomics approaches towards antigen discovery and vaccine development. Curr. Opin. Mol. Ther. 2002; 4:216–223.

Lehoux, D.E., Sanschagrin, F. and Levesque, R.C. Discovering essential and infection-related genes. Curr. Opin. Microbiol. 2001; 4:515–519.

Loessner, H. and Weiss, S. Bacteria-mediated DNA transfer in gene therapy and vaccination. Expert. Opin. Biol. Ther. 2004; 4:157–168.

Mahan, M.J., Slauch, J.M. and Mekalanos, J.J. Selection of bacterial virulence genes that are specifically induced in host tissues. Science 1993; 259:686–688.

Maione, D., Margarit, I., Rinaudo, C.D., Masignani, V., Mora, M., Scarselli, M., Tettelin, H., Brettoni, C., Iacobini, E.T., Rosini, R., D'Agostino, N., Miorin, L., Buccato, S., Mariani, M., Galli, G., Nogarotto, R., Nardi Dei, V., Vegni, F., Fraser, C., Mancuso, G., Teti, G., Madoff, L.C., Paoletti, L.C., Rappuoli, R., Kasper, D.L., Telford, J.L. and Grandi, G. Identification of a universal Group B *Streptococcus* vaccine by multiple genome screen. Science 2005; 309:148–150.

Mayr, U.B., Walcher, P., Azimpour, C., Riedmann, E., Haller, C. and Lubitz, W. Bacterial ghosts as antigen delivery vehicles. Adv. Drug Deliv. Rev. 2005; 57:1381–1391.

Mayrhofer, P., Tabrizi, C.A., Walcher, P., Haidinger, W., Jechlinger, W. and Lubitz, W. Immobilization of plasmid DNA in bacterial ghosts. J. Control Release 2005; 102:725–735.

Mecsas, J. Use of signature-tagged mutagenesis in pathogenesis studies. Curr. Opin. Microbiol. 2002; 5:33–37.

Meinke, A., Henics, T., Hanner, M., Minh, D.B. and Nagy, E. Antigenome technology: a novel approach for the selection of bacterial vaccine candidate antigens. Vaccine 2005; 23:2035–2041.

Meinke, A., Henics, T. and Nagy, E. Bacterial genomes pave the way to novel vaccines. Curr. Opin. Microbiol. 2004; 7:314–320.

Morona, R., Yeadon, J., Considine, A., Morona, J.K. and Manning, P.A. Construction of plasmid vectors with a non-antibiotic selection system based on the *Escherichia coli thyA*+ gene: application to cholera vaccine development. Gene 1991; 107:139–144.

Morrow, B.J., Graham, J.E. and Curtiss, R.R. Genomic subtractive hybridization and selective capture of transcribed sequences identify a novel *Salmonella typhimurium* fimbrial operon and putative transcriptional regulator that are absent from the *Salmonella typhi* genome. Infect. Immun. 1999; 67:5106–5116.

Motin, V.L., Georgescu, A.M., Fitch, J.P., Gu, P.P., Nelson, D.O., Mabery, S.L., Garnham, J.B., Sokhansanj, B.A., Ott, L.L., Coleman, M.A., Elliott, J.M., Kegelmeyer, L.M., Wyrobek, A.J., Slezak, T.R., Brubaker, R.R. and Garcia, E. Temporal global changes in gene expression during temperature transition in *Yersinia pestis*. J. Bacteriol. 2004; 186:6298–6305.

Mueller, A., O'Rourke, J., Chu, P., Kim, C.C., Sutton, P., Lee, A. and Falkow, S. Protective immunity against *Helicobacter* is characterized by a unique transcriptional signature. Proc. Natl. Acad. Sci. U. S. A. 2003; 100:12289–12294.

Nilsson, C.L. Bacterial proteomics and vaccine development. Am. J. Pharmacogenomics 2002; 2:59–65.

Ofek, I., Hasty, D.L. and Sharon, N. Anti-adhesion therapy of bacterial diseases: prospects and problems. FEMS Immunol. Med. Microbiol. 2003; 38:181–191.

Osorio, C.G., Crawford, J.A., Michalski, J., Martinez-Wilson, H., Kaper, J.B. and Camilli, A. Second-generation recombination-based in vivo expression technology for large-scale screening for *Vibrio cholerae* genes induced during infection of the mouse small intestine. Infect Immun. 2005; 73:972–980.

Paukner, S., Kudela, P., Kohl, G., Schlapp, T., Friedrichs, S. and Lubitz, W. DNA-loaded bacterial ghosts efficiently mediate reporter gene transfer and expression in macrophages. Mol. Ther. 2005; 11:215–223.

Rappuoli, R. Reverse vaccinology. Curr. Opin. Microbiol. 2000; 3:445–450.

Rollins, S.M., Peppercorn, A., Hang, L., Hillman, J.D., Calderwood, S.B., Handfield, M. and Ryan, E.T. In vivo induced antigen technology (IVIAT). Cell Microbiol. 2005; 7:1–9.

Russmann, H. Inverted pathogenicity: the use of pathogen-specific molecular mechanisms for prevention or therapy of disease. Int. J. Med. Microbiol. 2004; 293:565–569.

Ryan, E.T., Crean, T.I., Kochi, S.K., John, M., Luciano, A.A., Killeen, K.P., Klose, K.E. and Calderwood, S.B. Development of a Delta*glnA* balanced lethal plasmid system for expression of heterologous antigens by attenuated vaccine vector strains of *Vibrio cholerae*. Infect Immun. 2000; 68:221–226.

Samuelson, P., Gunneriusson, E., Nygren, P.A. and Stahl, S. Display of proteins on bacteria. J. Biotechnol. 2002; 96:129–154.

Sassetti, C.M., Boyd, D.H. and Rubin, E.J. Comprehensive identification of conditionally essential genes in mycobacteria. Proc. Natl. Acad. Sci. U. S. A. 2001; 98:12712–12717.

Sassetti, C.M., Boyd, D.H. and Rubin, E.J. Genes required for mycobacterial growth defined by high density mutagenesis. Mol. Microbiol. 2003; 48:77–84.

Scarselli, M., Giuliani, M.M., Adu-Bobie, J., Pizza, M. and Rappuoli, R. The impact of genomics on vaccine design. Trends Biotechnol. 2005; 23:84–91.

Schoen, C., Kolb-Maurer, A., Geginat, G., Loffler, D., Bergmann, B., Stritzker, J., Szalay, A.A., Pilgrim, S. and Goebel, W. Bacterial delivery of functional messenger RNA to mammalian cells. Cell Microbiol. 2005; 7:709–724.

Schoen, C., Stritzker, J., Goebel, W. and Pilgrim, S. Bacteria as DNA vaccine carriers for genetic immunization. Int. J. Med. Microbiol. 2004; 294:319–335.

Serruto, D., Adu-Bobie, J., Capecchi, B., Rappuoli, R., Pizza, M. and Masignani, V. Biotechnology and vaccines: application of functional genomics to *Neisseria meningitidis* and other bacterial pathogens. J. Biotechnol. 2004; 113:15–32.

Slauch, J.M. and Camilli, A. IVET and RIVET: use of gene fusions to identify bacterial virulence factors specifically induced in host tissues. Methods Enzymol. 2000; 326:73–96.

Slauch, J.M., Mahan, M.J. and Mekalanos, J.J. In vivo expression technology for selection of bacterial genes specifically induced in host tissues. Methods Enzymol. 1994; 235:481–492.

Spreng, S. and Viret, J.F. Plasmid maintenance systems suitable for GMO-based bacterial vaccines. Vaccine 2005; 23:2060–2065.

Stemke-Hale, K., Kaltenboeck, B., DeGraves, F.J., Sykes, K.F., Huang, J., Bu, C.H. and Johnston, S.A. Screening the whole genome of a pathogen in vivo for individual protective antigens. Vaccine 2005; 23:3016–3025.

Valdivia, R.H. and Falkow, S. Probing bacterial gene expression within host cells. Trends Microbiol. 1997; 5:360–363.

Valdivia, R.H., Hromockyj, A.E., Monack, D., Ramakrishnan, L. and Falkow, S. Application of green fluorescent protein (GFP) in the study of host-pathogen interactions. Gene 1996; 173:47–52.

Valdivia, R.H. and Ramakrishnan, L. Applications of gene fusions to green fluorescent protein and flow cytometry to the study of bacterial gene expression in host cells. Methods Enzymol. 2000; 326:47–73.

Waksman, G., Caparon, M. and Hultgren, S. Structural Biology of Bacterial Pathogenesis. Washington, DC: ASM Press, 2005.

Weichhart, T., Horky, M., Sollner, J., Gangl, S., Henics, T., Nagy, E., Meinke, A., von Gabain, A., Fraser, C.M., Gill, S.R., Hafner, M. and von Ahsen, U. Functional selection of vaccine candidate peptides from *Staphylococcus aureus* whole-genome expression libraries in vitro. Infect Immun. 2003; 71:4633–4641.

Wernerus, H. and Stahl, S. Biotechnological applications for surface-engineered bacteria. Biotechnol. Appl. Biochem. 2004; 40:209–228.

CHAPTER

7

Viral Vectors

David H. Holman[1], *Danher Wang*[1], *Jan Woraratanadharm*[1] *and John Y. Dong*[1,2]

[1]Division of Biodefense Vaccines, GenPhar, Inc., Mt. Pleasant, SC, USA

[2]Department of Microbiology and Immunology, Medical University of South Carolina, Charleston, SC, USA

OUTLINE

ABSTRACT

Traditional vaccine development platforms such as live-attenuated virus, killed virus, or recombinant subunit-based vaccines are often effective in eliciting long-term immunity to a number of infectious human pathogens. However, for many human pathogens, vaccine platforms such as these are unsuitable for human use due to safety concerns, poor efficacy, or simple impracticality. As a result, much work has focused on the use of recombinant virus vectors as a means for vaccination against human pathogens. Viral vectors can express foreign proteins at high levels in host cells, resulting in strong, long-lasting immune responses against the target proteins. This chapter describes the use of virus vectors in the context of vaccination against human pathogens. Various vector platforms are discussed, compared, and contrasted.

INTRODUCTION

Traditional vaccination platforms such as live-attenuated or killed virus vaccines have been used successfully for decades. This approach is applied to many different human diseases including smallpox, polio, measles, mumps, yellow fever, rubella, influenza, varicella, and hepatitis A. This form of vaccine development can be very effective, with particular support demonstrated by the eradication of smallpox and near eradication of polio through global vaccination efforts.

Compared to inactivated/killed virus vaccines, live-attenuated vaccines are generally thought to provide longer-lasting immunity without the need for repeated booster immunizations, and are an economical and simple means for high efficacy vaccine production. However, development of live-attenuated virus vaccines for some human pathogens can encounter safety concerns due to under-attenuation of the virus or even a reversion to its pathogenic state. For example, in efforts to develop HIV vaccines, researchers have investigated live-attenuated simian immunodeficiency virus (SIV) vaccines as proof of principle and found strong levels of protection in nonhuman primates (Johnson and Desrosiers, 1998). Unfortunately, live-attenuated SIV vaccines can also cause AIDS in vaccinated monkeys due to reversion to pathogenic form (Baba et al., 1999; Berkhout et al., 1999), which is a frequent source of controversy for those in the field (Murphey-Corb, 1997). Similar findings are noted for other experimental live-attenuated vaccines. For example, live-attenuated dengue virus vaccines also show protective immunity in animal models (Blaney et al., 2005). However, a Phase I clinical trial involving a live-attenuated dengue virus vaccine formulation was halted due to side effects caused by under-attenuation of the vaccine virus strain (Kitchener et al., 2006). Obviously, there is a fine balance between developing a safe, attenuated virus strain and developing one that maintains a high level of immunogenicity. Unfortunately, the immunogenicity of live-attenuated vaccines is often gained at the cost of clinical safety, and vice versa.

Even licensed live-attenuated vaccines that have been approved and used for decades are not without a risk of vaccine-mediated disease. A classical example of this is smallpox vaccination. In the 19th–20th century, variola vaccinations were traditionally performed using the related but less pathogenic poxvirus, vaccinia virus. However, the early vaccine strains of vaccinia are well known for inducing significant side effects in vaccine recipients, some of which can even be life-threatening (Parrino and Graham, 2006). With the increased threat of bioterrorism, many recent research efforts have focused on developing new vaccine strains of vaccinia with more favorable safety profiles or even alternative vaccine platforms for smallpox vaccination (see Chapter 22). Global efforts for polio vaccinations, either as a live-attenuated oral vaccine (oral polio vaccine, OPV) or an inactivated parenteral vaccine (inactivated polio vaccine, IPV), have nearly eradicated this disease from the planet. Yet, despite the high levels of efficacy, the live-attenuated OPV formulation has been shown to cause paralytic poliomyelitis in some cases (Blume and Geesink, 2000; Henderson et al., 1964).

Some human pathogens, such as the Ebola and Marburg viruses of the Filoviridae family, are so deadly that live-attenuated vaccines are not even considered a possibility, since they would pose too great a risk if the vaccine strain of virus were under-attenuated or reverted to pathogenic state. As a result of constant pressure to find newer, more effective, and safer vaccine platforms, investigators have sought alternative approaches to vaccine development, such as recombinant protein subunit vaccines, DNA vaccines (see Chapter 8), and viral vectored vaccines.

Protein subunit vaccines can be an effective means for generating strong immune responses, such as in the recent anthrax vaccines (Miller et al., 1998) or the traditional hepatitis B vaccines produced by GlaxoSmithKline (Engerix-B®) or Merck (Recombivax HB®). However, subunit vaccines often require inclusion of an adjuvant to increase the immunogenicity (Berthold et al., 2005; Jendrek et al., 2003; Putnak et al., 2005; Qin et al., 2007). Often, the recombinant proteins are synthesized in insect or bacteria cells prior to purification, which can result in different levels of glycosylation, protein folding, and other tertiary modifications that would occur during protein expression in a natural infection. These differences can alter the antigenicity of the recombinant protein so that it differs from the natural wild-type protein. Even recombinant proteins synthesized in mammalian cells, which may be processed correctly, can be damaged during the subsequent purification processes. Thus, when delivered as a vaccine, the immune responses generated against the recombinant proteins can be quite different from those that would be generated against the natural protein of the target pathogen. Finally, the recombinant proteins can be degraded by the host immune system soon after injection into the vaccine recipient, which results in only a brief period of antigen presentation to the immune system. Therefore, while protein subunit vaccines are promising in certain areas of vaccine development, they may not be ideal for all vaccine applications.

DNA vaccines entail delivery of naked plasmid DNA that contains selected gene(s) of the target pathogen.

Upon vaccination, the plasmid DNA is taken up by local cells at the injection site, and the gene(s) are expressed to produce the pathogen's proteins. This form of de novo protein expression eliminates the need for purification of large amounts of the protein, since the actual protein is expressed from the cells in vivo. The potential efficacy of this vaccine platform has been successfully demonstrated for numerous pathogens, including Ebola virus (Vanderzanden et al., 1998; Xu et al., 1998), avian influenza (Kodihalli et al., 1999; Tompkins et al., 2007), dengue virus (Apt et al., 2006), West Nile virus (Bunning et al., 2007), and HIV (Cherpelis et al., 2001). A major limitation of DNA vaccines is the level of expression of the gene(s) of interest. Compared to other means of gene expression, such as that from viral vectors, expression from DNA vaccines is notoriously weak in vivo, resulting in lower immunogenicity of the vaccine. Thus, many DNA vaccine studies in animal models employ a system of DNA vaccine priming followed by a viral vector or protein subunit booster immunization (Kong et al., 2005; Shu et al., 2007; Sullivan et al., 2000; Wu et al., 2005; Zanin et al., 2007).

One of the most promising recombinant vaccine technology platforms is the viral vector. The premise behind using viral vectors as a means for vaccination is to present the naturally occurring form of the target pathogen antigens to the immune system in the absence of the infectious pathogen itself. This process induces an immune response against the target pathogen's antigens, similar to a natural infection, however in the absence of disease. One of the earliest demonstrations of this was in 1984, when scientists at the Wistar Institute engineered a vaccinia virus vector to express the rabies virus glycoprotein, and successfully protected mice and rabbits from lethal rabies virus challenge (Wiktor et al., 1984). This vaccine has since been licensed and marketed as a veterinary vaccine by Merial (Raboral VR-G®), which is commonly used for mass vaccination of certain species of wildlife in order to curtail the spread of rabies in the wild (Mackowiak et al., 1999) (see Chapter 33). As technology advances, so does the list of viral vectors used in experimental vaccines. Vectors based on adenovirus, adeno-associated virus (AAV), vesicular stomatitis virus (VSV), and alphavirus have all been studied on numerous fronts as vaccine tools (Brave et al., 2007). These and others will be discussed in this chapter.

ADENOVIRUS

Perhaps the most widely studied viral vector for vaccines and gene therapy applications is the adenovirus (Goncalves and de Vries, 2006). Adenoviruses (Ad) belong to the family, Adenoviridae, and are double-stranded DNA viruses with a genome size of approximately 36 kbp. Ad infect a wide range of species, but maintain a high level of species specificity. Of the human Ad, there are at least 51 different serotypes identified (Shenk, 2001), of which, the most studied in terms of viral vectors are serotypes 2 and 5 (Ad2 and Ad5, respectively). Ad are common, have well-characterized genomes that are easy to manipulate, can be grown and purified to high titers in cell culture, are able to infect a wide variety of dividing and non-dividing cell types, express high levels of recombinant transgene in infected cells, and have a favorable safety profile. These properties have all culminated in the use of Ad vectors in hundreds of human clinical trials around the world for a variety of medical applications, including cancer gene therapy and vaccine vectors.

Early work in Ad molecular biology found that the viruses were able to package genomes with sizes up to 105% of the wild-type genome length into infectious viral particles. This equated to approximately 1.8 kb of exogenous DNA (transgene) that could be inserted into a wild-type Ad genome (Ghosh-Choudhury et al., 1987; Graham et al., 1977). However, increasing the genome length beyond 105% of the wild-type genome length can result in genetic instability and loss of the transgene (Bett et al., 1993). Researchers quickly discovered that more genome space could be created for larger transgenes by specific deletions of the Ad genome. The first of these deletions were directed toward the Ad early genes E1a and E1b, which also created a replication-incompetent vector as loss of E1 subsequently inhibits downstream transcription of other Ad genes necessary for replication (Jones and Shenk, 1979). E1-deleted replication-incompetent vectors only replicate in permissive cell lines, such as the human embryonic kidney cell line HEK293 (Graham et al., 1977), which provides the missing E1 gene function for the Ad vector in trans. By additionally deleting the E3 region of the Ad genome, ample space exists for inserting foreign DNA sequences. E1/E3 deleted Ad vectors are referred to as first-generation vectors, and usually have 4–5 kb of genome space available for a transgene insert (Ghosh-Choudhury et al., 1987; Graham, 1984). First-generation Ad vectors are effective tools for gene delivery, but the duration of transgene expression is fairly limited due to host immune response against the Ad proteins expressed by the vector (such as E2 and E4 proteins). Therefore, efforts were made to reduce the amount of Ad protein expressed from the vector, while maintaining high levels of transgene expression. This was achieved through partial deletions in the E2 region (Hodges et al., 2000; Lusky et al., 1998; Moorhead et al., 1999) or the E4 region (Dedieu et al., 1997; Gao et al., 1996) of the Ad genome, which reduced or eliminated the expression of E2 or E4 proteins. These types of Ad vectors are referred to as second-generation vectors, and have an increased transgene insert capacity of approximately 6–7 kb. The transgene capacity can be increased even further in advanced Ad vectors with deletions in E1, E3, and most of E4 (excluding open reading frame 6) (Wang et al., 2006b). Ad vectors with the highest capacity for exogenous DNA are called "gutless" vectors, and consist solely of the exogenous transgene DNA flanked by the Ad inverted terminal repeat (ITR) sequences and the Ad packaging signal. These vectors can accommodate 30–35 kb of foreign DNA (Table 7.1), although they must be propagated using a helper virus to provide the missing Ad genes necessary for replication in the packaging cell line (Kochanek et al., 2001; Kumar-Singh and Chamberlain, 1996; Parks et al., 1996).

Ad vectors have been studied on numerous fronts, such as avian influenza (Gao et al., 2006), Ebola and Marburg viruses (Sullivan et al., 2000, 2003; Wang et al., 2006a, 2006b), West Nile virus (Schepp-Berglind et al., 2007), dengue virus (Holman et al., 2007; Jaiswal et al., 2003; Raja et al., 2007), SARS-CoV (Ma et al., 2006), HIV (Barouch and Nabel, 2005; Catanzaro et al., 2006), and anthrax (McConnell et al., 2007). Additionally, Ad vectors have been studied as potential gene therapy vectors for many types of cancer and other diseases in hundreds of human clinical trials (www.clinicaltrials.gov).

One of the primary criticisms of Ad vector use in humans is the issue of pre-existing immunity. Ad infections are quite common in humans, resulting in 35–55% of the population having neutralizing antibodies, in particular against the common serotype used in experimental Ad vectors, Ad5 (Chirmule et al., 1999; Nwanegbo et al., 2004). It is thought that these circulating Ad-neutralizing antibodies might limit any Ad-based vaccine vector's efficacy by neutralization of the vector prior to efficient transgene expression. Experimentally, this has been shown in various animal models of Ad5 immunity. Most of these studies establish an Ad5-immune animal by inoculating the animals multiple times with high doses of wild-type Ad5 or an unrelated Ad5-based vector. This is followed by vaccination with the experimental Ad5-based vector, which invariably fails due to pre-existing anti-Ad5 immunity (Capone et al., 2006; Hashimoto et al., 2005; Kobinger et al., 2006; Xiang et al., 2003). Most of these studies demonstrate the feasibility of overcoming Ad5 pre-existing immunity by using Ad vectors based on alternate serotypes that are antigenically distinct and thus are not neutralized by the anti-Ad5 immune response. However, conflicting data have been generated from human clinical trials. Phase I/II clinical trials involving Merck's Ad-based HIV vaccine found that while pre-existing Ad5 immunity did limit the vaccine's efficacy at lower vaccine doses, this limitation could be completely overcome by increasing the dose of vaccine (Cohen, 2006). Further support was produced from a clinical trial studying an Ad-based influenza vaccine, which showed no correlation between the levels of Ad5-neutralizing antibodies and the immunogenicity of the vaccine (Van Kampen et al., 2005). Finally, it is suggested that vaccination by alternate routes of administration (such as oral or intranasal) rather than injection can overcome pre-existing vector immunity (Appaiahgari et al., 2006; Xiang et al., 2003), which is supported by data from a human clinical trial (Van Kampen et al., 2005).

TABLE 7.1 Common viral vectors in recombinant vaccine development

Virus[a]	Family	Species	Genome (kb)	Transgene capacity (kb)[b]	PEI[a,b]
Ad	Adenoviridae	Human, chimp	36	7–35	+++
AAV	Parvoviridae	Human	5	5	+++
Alphavirus	Togaviridae	Zoonotic/mammals	11.8	5	+
NDV	Paramyxoviridae	Zoonotic/birds	15	3.2–4.5	–
Vaccinia	Poxviridae	Zoonotic…bovine?	192	25	+++
Avipox	Poxviridae	Zoonotic/birds	260	25	–
VSV	Rhabdoviridae	Zoonotic/mammals	11.1	4.5	+

[a] Abbreviations: Ad, adenovirus; AAV, adeno-associated virus; NDV, Newcastle disease virus; VSV, vesicular stomatitis virus; PEI, pre-existing immunity.
[b] Based on references cited in this chapter.

Pre-existing Ad vector immunity is a source of frequent debate, and is a significant factor for Ad-based vaccine vectors under certain conditions. However, experimental data exist for support on both sides of the argument, and it remains to be seen if the difficulties demonstrated by pre-existing immunity in animal models will translate to actual scenarios of human vaccination. As more data are produced from clinical trials, we will have a better understanding of the true significance of pre-existing immunity for Ad-based vaccine vectors.

ADENO-ASSOCIATED VIRUS

AAV belong to the family, Parvoviridae, and are small, single-stranded DNA viruses with a genome size of approximately 5 kb. There are eight known AAV serotypes, with AAV-2 being the most commonly studied (Chiorini et al., 1999, 1997; Gao et al., 2002; Muramatsu et al., 1996; Rutledge et al., 1998; Xiao et al., 1999). AAV are quite unique in that in order to replicate, they require the co-infection of a helper virus, such as Ad (Atchison et al., 1965) or herpes virus (Buller et al., 1981). In the absence of helper virus, AAV infection becomes latent and does not produce progeny virus. However, a productive infection can be achieved if latently infected cells are subsequently infected with Ad or herpes helper viruses (Berns and Linden, 1995). Wild-type AAV do not cause human disease, giving them an excellent safety profile as a potential therapeutic or vaccine vector. AAV have a broad host range, persistent transgene expression in host cells, and generate very weak antivector immune responses. These qualities have led to the intense study of AAV as tools for human gene therapy and recombinant vaccines vectors. Recombinant AAV vectors have been studied as vaccine vectors against herpes simplex virus type 2 (Manning et al., 1997), human papilloma virus (Liu et al., 2000), HIV (Xin et al., 2001), and cytomegalovirus (Gallez-Hawkins et al., 2004). Each of these studies found the recombinant AAV (rAAV) platform to be an effective means for inducing potent immune responses against the target pathogen.

Due to the small size of the AAV genome, rAAV vectors can only accommodate approximately 5 kb of exogenous DNA. This is usually done by removal of all viral genes between the 5′ and 3′ terminal repeat (TR) sequences and replacing them with the desired transgene(s) (Dong et al., 1996). Since the only viral DNA sequences the rAAV vectors retain are the TRs, the vectors must be propagated in a special packaging cell lines that express the AAV Rep and Cap proteins, in addition to the helper virus co-infection (Vincent et al., 1997). This has both advantages and disadvantages. An advantage of a recombinant viral vector completely void of parent virus genes is the lack of AAV protein expression upon transduction of host cells. This results in low levels of antivector immunity generated after vaccination. However, the disadvantages are the small genome size and requirement for helper virus during propagation. Additionally, AAV infections occurring in the absence of helper virus become latent by integrating the viral genome into host DNA, on specific sites of human chromosome 19 (Cheung et al., 1980; Samulski et al., 1991). Any virus that integrates its DNA into the host genome raises safety concerns as to the genetic consequences of the integration. Integration of viral DNA can be beneficial if the desired transgene is persistently expressed (the desired outcome for gene therapy studies); however, it may also be detrimental to the host if the integration results in mutations or deletions to host DNA (McCarty et al., 2004).

Similar to Ad vectors, a potential pitfall for rAAV vectors is the issue of pre-existing vector immunity. It was initially thought that AAV was nonpathogenic and only weakly immunogenic, thus creating the perfect gene transfer vehicle that is immunologically invisible. These initial suggestions spearheaded the movement for using rAAV vectors in gene therapy and vaccine vector development. However, it is now known that over 90% of humans do indeed have circulating antibodies that cross-react with AAV, and over 30% are seropositive for AAV neutralizing antibodies (Chirmule et al., 1999; Kotin, 1994). Furthermore, the highest level of human AAV antibodies are directed toward the capsid protein of the AAV-2 serotype, which is the most common serotype used as a backbone for vaccine vectors. However, as there are eight different AAV serotypes, using an alternate serotype as the vector backbone may be a means of overcoming the pre-existing immunity hurdle (Davidoff et al., 2005). More human clinical trials data are needed in order to better gauge the significance of pre-existing vector immunity in using rAAV as vaccine vectors.

It should be noted that on July 26, 2007, the U.S. Food and Drug Administration announced the death of a clinical trial participant involving an AAV-based rheumatoid arthritis treatment (FDA, 2007). As of the date of this chapter review, the details of this tragedy have yet to be released by the FDA or the clinical trial sponsor.

ALPHAVIRUSES

The alphaviruses belong to the family, Togaviridae, and are small, enveloped viruses with a single-stranded positive RNA genome of approximately 11.8 kb. Alphaviruses, which are arthropod-borne viruses (arboviruses), are grouped into six clades based on antigenic homology of the E1 glycoprotein: Barmah Forest (BF), Ndumu (NDU), Semliki Forest (SF), western equine encephalitis (WEE), eastern equine encephalitis (EEE), and Venezuelan equine encephalitis (VEE) (Griffin, 2001). The three primary alphaviruses used as backbones for recombinant vaccine vectors are SF, Sindbis virus (SIN), and VEE (Liljestrom and Garoff, 1991; Pushko et al., 1997; Xiong et al., 1989).

Alphaviruses are known to have a broad host range, and can infect a variety of cell types, including dendritic cells, which are major antigen-presenting cells (Gardner et al., 2000; MacDonald and Johnston, 2000). This is thought to be an important advantage of using alphaviruses as vaccine vectors in that they can directly target dendritic cells in vivo and induce rapid and strong immune responses. The primary method of using alphaviruses as vaccine vectors is to create what are known as replicons. This entails deleting the alphavirus structural genes from the virus genome and replacing them with the transgene(s) of interest. The recombinant RNA molecule must then be co-expressed in packaging cell lines with helper RNA containing the missing structural genes in order to effectively package the recombinant RNA into infectious viral particles (Bredenbeek et al., 1993; Liljestrom and Garoff, 1991). Thus, the replicons are enveloped viral particles containing the recombinant genome that, in transduced cells, expresses the transgene at high levels. Since the recombinant genome lacks the alphavirus structural genes, no progeny virus is produced from the replicon transduction. One pitfall noticed early on in alphavirus replicon development was recombination between the helper RNA (containing the structural genes) and the recombinant RNA (lacking the structural genes but containing the packaging signal) in the packaging cell line, which produced replication-competent alphavirus particles (Berglund et al., 1993). The generation of contaminating replication-competent virus is a serious safety concern for this vaccine vector platform. However, endeavors have been made in designing newer and more advanced helper RNA molecules and packaging cell lines to reduce the possibility of producing replication-competent alphavirus particle contaminants (Polo et al., 1999; Smerdou and Liljestrom, 1999). An additional shortcoming of the alphavirus vaccine vector is the genome capacity for a transgene insert. Despite complete removal of the alphavirus structural genes from the vector genome, a recombinant replicon can only accommodate transgene inserts of up to approximately 5 kb (Smerdou and Liljestrom, 1999).

Alphavirus vaccine vectors have been studied as vaccines for avian influenza (Schultz-Cherry et al., 2000), Ebola virus (Olinger et al., 2005; Pushko et al., 2000), Marburg virus (Hevey et al., 1998; Lee et al., 2006), HIV (Megede et al., 2006), cytomegalovirus (Reap et al., 2007), SARS-CoV (Deming et al., 2006), anthrax, and botulinum toxin (Lee et al., 2006). Additionally, VEE replicons containing HIV genes were tested in Phase I clinical trials and were well tolerated in vaccine recipients, although the immunogenicity was only moderate (Chulay et al., 2006). Other Phase I studies of alphavirus replicons for HIV and cytomegalovirus vaccinations are currently recruiting subjects (www.clinicaltrials.gov).

Alphavirus vaccine vectors must also face the issue of pre-existing immunity, although it is not likely to be as significant a hurdle as for other vectors, such as adenovirus described above. This is largely because the alphaviruses are zoonotic mosquito-borne viruses that are endemic only in certain geographical regions of the world. Furthermore, human alphavirus epidemics occur very infrequently. This suggests that the general population will have little pre-existing vector immunity against an alphavirus-based vaccine vector. However, it was suggested early on that in horses, pre-existing antibodies against one strain of alphavirus can interfere with infection from other alphavirus strains (Calisher et al., 1973). This could pose problems for repeated vaccination in humans using alphavirus replicons. For example, antivector immunity generated after vaccination with a VEE replicon could subsequently impede booster vaccination with the same or even a different alphavirus replicon. Indeed, there is evidence in humans that the immune response generated from vaccination with one alphavirus can interfere with the immune response generated from subsequent vaccinations with different alphaviruses (McClain et al., 1998).

NEWCASTLE DISEASE VIRUS

Newcastle disease virus (NDV) belongs to the family, Paramyxoviridae, and is a zoonotic virus that naturally infects all species of birds. The NDV genome is a nonsegmented single strand of negative-sense RNA of approximately 15 kb in length. NDV is antigenically distinct from any of the human paramyxoviruses, such as human parainfluenza viruses, which has led to interest in its use as a vaccine vector in humans. NDV are categorized into three groups based on their levels of pathogenicity in chickens: the avirulent lentogenic strains, the moderately pathogenic mesogenic strains, and the highly pathogenic velogenic strains. Lentogenic strains are widely used for NDV live-attenuated virus vaccines in the poultry industry. An outbreak of Newcastle disease can be quite severe in poultry, and is a constant threat to the industry worldwide. However, NDV are nonpathogenic in primates (Bukreyev et al., 2005), which has led to their study as vaccine vectors.

Construction of recombinant NDV vaccine vectors generally entails first modifying packaging cells in culture, either through stable expression techniques (Romer-Oberdorfer et al., 1999) or infection with a helper virus (Huang et al., 2003; Nakaya et al., 2001), to express T7 DNA-dependent-RNA polymerase. These cells are then transfected with a full-length NDV reverse sense cDNA that contains an exogenous transgene, as well as several DNA support plasmids carrying the genes of the NDV transcriptase complex: nucleocapsid protein (NP), phosphoprotein (P), and RNA-dependent-RNA polymerase protein (L) (Huang et al., 2003, 2001; Nakaya et al., 2001). After incubation, the produced recombinant virus is "rescued" and propagated in embryonated chicken eggs. NDV have been studied as recombinant vaccine vectors against SARS-CoV (DiNapoli et al., 2007), respiratory syncytial virus (Martinez-Sobrido et al., 2006), SIV (Nakaya et al., 2004), and influenza virus (Nakaya et al., 2001). Despite the dangers of NDV in poultry, these viruses have a favorable safety profile for use in humans. This is supported by the high levels of attenuation seen in nonhuman primate animal models of vaccination (Bukreyev et al., 2005), as well as the safe administration of oncolytic NDV (PV701) in human Phase I clinical trials (Lorence et al., 2003, 2007). Other paramyxoviruses have also been studied as experimental vaccine vectors, such as Sendai virus (Griesenbach et al., 2005; Takimoto et al., 2005), and simian virus 5 (Parks and Alexander-Miller, 2002).

Since NDV is an avian paramyxovirus, the issue of antivector pre-existing immunity is not considered a significant factor for this vaccine platform, which is a major advantage. Additionally, several NDV strains are licensed for use as veterinary vaccines, making them well characterized and readily available. Most recombinant NDV vectors can be propagated to high titers in chicken eggs and even in some cell lines, although inclusion of even small exogenous transgenes can significantly lower the yield of recombinant virus (Krishnamurthy et al., 2000). Also, the "rule of six," according to Sendai virus studies, indicates that the total viral genome length (transgene included) must be divisible by six in order for proper NP function in the transcriptase complex (Calain and Roux, 1993; Nakaya et al., 2001). This may complicate strategies for vaccine vector construction. Furthermore, NDV vectors may be somewhat limited in their capacity for large transgene inserts. Early studies of Sendai virus indicated a maximum insert size of 3.2 kb without significant genetic instability or reduction of viral yield (Sakai et al., 1999), although more recently Sendai virus vectors have been shown to accommodate 4.5 kb of exogenous DNA (Ferrari et al., 2007).

POXVIRUSES

Poxviruses are large, double-stranded DNA viruses of the family, Poxviridae. Compared to the other viruses mentioned in this chapter, the poxviruses are by far the largest in viron size (350 × 270 nm) as well as genome length, which can reach up to 300 kb in some types of poxviruses. The most notorious (and most virulent) of all poxviruses is variola virus, of the *Orthopoxvirus* genus, which is an obligate human pathogen that causes smallpox. The three other notable orthopoxviruses, vaccinia, monkeypox, and cowpox viruses, are zoonotic viruses and cause less severe disease in humans compared to variola. Vaccinia virus is often considered the prototype poxvirus, as it is heavily investigated in a number of research endeavors, and has thus been well characterized. Much of vaccinia's fame stems from its use as a live-attenuated smallpox vaccine, which, through global vaccination efforts, ultimately led to the eradication of smallpox in 1977.

Poxviruses of the genus, *Avipoxvirus*, receive a great deal of research attention as potential vaccine vectors. These viruses are zoonotic arboviruses that naturally infect birds, and are nonpathogenic in humans. This quality, in addition to their large 260 kb genomes, make avipox viruses attractive research tools. For example, fowlpox and canarypox viruses have been tested in animal models as a vaccine vectors for rabies (Taylor et al., 1991, 1988), H5N1 avian influenza (Steensels et al., 2007), nipah virus (Weingartl et al., 2006), and HIV (Gilbert et al., 2003; Radaelli et al., 2007; Zhang et al., 2007). The safety of avipox virus vectors in humans has been demonstrated in clinical trials as vaccine vectors for HIV (Emery et al., 2005; Russell et al., 2007) and malaria (Walther et al., 2006), as well as numerous cancer gene therapy clinical trials (www.clinicaltrials.gov).

Vaccinia virus has been used for decades as a smallpox vaccine in humans, and while earlier strains of vaccinia were known for significant side effects in vaccine recipients, newer strains have been developed or modified from previously existing vaccine strains with more favorable safety profiles for human use (Parrino and Graham, 2006) (see Chapter 22). The improved safety profiles of vaccinia virus strains has led to study of their use as vaccine vectors for other diseases. For example, the replication defective vaccinia strain, modified vaccinia Ankara (MVA), has been genetically modified to express genes from HIV (Zhang et al., 2007), herpes simplex virus type 2 (Meseda et al., 2006), rabies virus (Weyer et al., 2007), and cytomegalovirus (Wang et al., 2007). A significant difference between vaccinia and the avipox viruses is the possibility for pre-existing vector immunity. Since the avipox viruses are zoonotic and only naturally infect birds, the general population has little to no pre-existing immunity to these viruses. Like the avipox viruses, vaccinia is also considered a zoonotic virus, possibly distantly related to cowpox or buffalopox viruses (the exact origin is unknown). In contrast to the avipox viruses, however, much of the adult population today is seropositive for vaccinia due to childhood smallpox vaccinations. Data suggest that a vaccinia-based vaccine vector would be ineffective in these individuals due to antivector immunity (Belyakov et al., 1999). It is suggested that vaccination through mucosal routes of administration may be an effective means to overcome vaccinia-directed vector immunity (Belyakov et al., 1999; Naito et al., 2007).

The large genome size of the poxviruses is somewhat of a double-edged sword for vector development. On one hand, there is ample capacity for large transgene inserts, with sizes reaching to at least 25 kb for recombinant vaccinia virus (Smith and Moss, 1983). On the other hand, these vectors express hundreds of native virus proteins in addition to the transgene(s) of interest. This induces a very strong immune response against the vector proteins, which subsequently reduces the capacity for a strong immune response against the transgene product(s) due to an "immune dilution" effect. This could be an explanation for the poor performance of poxvirus vectors in human clinical trials. For example, fowlpox, canarypox, and vaccinia virus vectors have all been tested in humans and found to be safe. However, fowlpox and canarypox vectors containing HIV genes have induced little to no anti-HIV cellular immune responses in humans (Emery et al., 2005; Russell et al., 2007). Similar results were reported for a fowlpox-based malaria vaccine that induced moderate immune responses but showed extremely poor efficacy (Walther et al., 2006). Vaccinia virus vectors have also been documented as inducing weak cellular immune responses. In a direct comparison, a recombinant Ad vector expressing the nucleoprotein gene from Sin Nombre virus (a member of the *Hantavirus* genus) induced much stronger cellular immune responses in mice than a vaccinia virus vector expressing the same transgene (Maeda et al., 2005).

VESICULAR STOMATITIS VIRUS

VSV is a zoonotic arbovirus that belongs to the family, Rhabdoviridae, the same family as the rabies viruses. VSV has an 11 kb genome that consists of a single strand of negative-sense RNA. VSV transmission in animals occurs through insect bites, and can cause severe disease in cattle, horses, and swine with symptoms similar to foot and mouth disease (Rodriguez, 2002). Human VSV infections do occur, although much less frequently than in animals and with much less severe disease symptoms (usually a mild flu-like illness). Many human infections even go unnoticed, as they are completely asymptomatic; however, in rare cases, severe disease has been reported (Quiroz et al., 1988). Recombinant VSV can accommodate a 40% increase in genome size (approximately a 4.5 kb transgene insert) with only a slight reduction in infectivity titer (Haglund et al., 2000). In addition to ease of genome modifications, VSV vectors have an added advantage in that the virus can efficiently incorporate and express foreign transmembrane proteins on the surface of recombinant viral particles (Schubert et al., 1992). This, coupled with its relatively safe background and rare seroprevalance in humans, makes VSV an attractive platform for vaccine vector development.

Concerns for VSV vector safety are related to the possibility of severe human disease (Quiroz et al., 1988) as well as the neurovirulence and 50% mortality rate from experimental intranasal mouse infections (Reiss et al., 1998). Additionally, asymptomatic brain infections have been noted in cotton rats after intransal delivery of a recombinant VSV vector (Schlereth et al., 2003). However, more recent studies have shown that intranasal delivery of VSV did not spread to CNS tissues in nonhuman primates, indicating that these vectors may be safer than initially suggested (Johnson et al., 2007). Thus, recombinant VSV vectors have been studied as vaccines against H5N1 avian influenza virus (Schwartz et al., 2007), Ebola and Marburg viruses (Daddario-DiCaprio et al., 2006; Feldmann et al., 2007; Jones et al., 2005), plague (Palin et al., 2007), hepatitis C virus (Ezelle et al., 2002), and HIV (Haglund et al., 2000, 2002). While VSV vaccine vectors have yet to be tested in humans, results from these and other animal models of vaccination are encouraging.

Pre-existing vector immunity is not thought to be a significant factor for use of VSV vaccine vectors in humans. Since VSV is a zoonotic virus, human infections are generally limited to isolated geographical regions of the world or to individuals having close contact with infected animals. Thus, the global seroprevalance for VSV is considered to be low. However, animal models show that vaccination with a recombinant VSV vector induces neutralizing antibodies against the VSV envelope protein (Roberts et al., 1999). Therefore, in all likelihood, humans receiving a VSV-based vaccine vector will develop neutralizing immune responses against the viral envelope protein, which would reduce the efficacy of subsequent VSV-based booster vaccinations. Similar to the Ad and AAV vector strategies described above, a serotype rotation of different VSV glycoproteins is reported to bypass any level of antivector immunity induced after the primary vaccination (Rose et al., 2000).

OTHER VIRUSES

In addition to the viral vectors described above, there are several others being investigated for utility in gene therapy and vaccination applications. Herpes simplex viruses (HSV) are double-stranded DNA viruses of the family, Herpesviridae. HSV subtypes HSV-1 and HSV-2 are well known for causing human herpes infections, but have also been studied as vaccine vectors for SIV (Kaur et al., 2007), HIV (Santos et al., 2007), and intracellular bacterial pathogens (Lauterbach et al., 2004).

Measles virus (MV), of the Paramyxoviridae family, is a well-known childhood pathogen that, despite availability of effective vaccines, still causes significant mortality in children of developing countries. By using rescue techniques similar to those described above for construction of NDV vectors, MV vectors have been developed as a number of vaccine candidates. An advantage to MV vectors is thought to be the possibility of multivalence. Vaccination of children with a MV vector expressing genes from other pathogens, such as hepatitis B virus (Reyes Del Valle et al., 2007), West Nile virus (Despres et al., 2005), or HIV (Lorin et al., 2004) could serve as a bivalent vaccine against both MV and the other pathogen. However, much like vaccinia vectors, a major portion of the adult population will be immune to a recombinant measles vector due to childhood measles infections or measles vaccinations.

Polioviruses are members of the Picornaviridae family, the same family containing foot and mouth disease and hepatitis A viruses. Poliovirus is well known for causing paralytic poliomyelitis, which has nearly been eradicated due to development of successful and safe inactivated and live-attenuated poliovirus vaccines. Since the attenuated virus strains are safe for use in humans, researchers have studied possibly using poliovirus as a vector for other pathogens (Andino et al., 1994), such as hepatitis B virus (Yim et al., 1996) or SIV (Tang et al., 1997). However, the vast majority of the world's population has immunity to poliovirus from childhood vaccination, making a poliovirus-vectored vaccine somewhat impractical for adults.

CONCLUSIONS

In conclusion, viral vectors can be very useful for the presentation of naturally formed viral antigens to the immune system. Generally speaking, viral vectors have a more favorable safety profile than many live-attenuated virus vaccines and are more immunogenic than inactivated/killed virus vaccines. Furthermore, viral vectored vaccines present the desired antigens in the natural, correct conformation to the immune system, a process very poorly achieved by recombinant protein subunit vaccines. Finally, viral vectored vaccines express higher levels of foreign genes in vivo and for a longer duration when compared to DNA vaccines.

Three major considerations exist for the use of viral vectored vaccines in humans, the first of which is safety. Some platforms, such as Ad vectors, have been tested extensively in human clinical trials for many years and are accepted as being safe for human use. Other platforms, such as VSV, are still in their infancy and human safety remains an unknown factor. As more clinical trials take place, we will gain a better understanding of these vectors and their performance in humans. The second consideration is pre-existing vector immunity, which varies in significance depending on the vector of choice. For Ad or vaccinia vectors, this may have a serious impact on vaccine vector efficacy in humans; for the zoonotic virus vectors, such as avipox viruses or NDV, pre-existing immunity is not likely to play as significant a role in developmental and clinical applications. Finally, the third consideration for vaccine vector development is the vector's genomic capacity for a transgene insert. Depending on the vector, these exogenous DNA sequences can range in length from less than 1 kb to 35 kb. However, with increased transgene size often comes genetic instability and decreased virus yield in production. These must all be factored together to determine the ideal virus vaccine vector that is suitable for use in humans, but also maintaining a high level of efficacy.

The viral vectors mentioned in this chapter represent a portion of the ever-growing field of viral vector research. Some of these vectors are well characterized from the basic virology of the vector to the complexity of the immune responses elicited after vaccination. Others are still incompletely understood. The study and use of viral vectors has advanced the fields of human gene therapy for genetic diseases, cancer gene therapy, and vaccine development, and will continue to expand the scientific horizon.

References

Andino, R., Silvera, D., Suggett, S.D., Achacoso, P.L., Miller, C.J., Baltimore, D. and Feinberg, M.B. Engineering poliovirus as a vaccine vector for the expression of diverse antigens. Science 1994; 265(5177):1448–1451.

Appaiahgari, M.B., Saini, M., Rauthan, M., Jyoti, and Vrati, S. Immunization with recombinant adenovirus synthesizing the secretory form of Japanese encephalitis virus envelope protein protects adenovirus-exposed mice against lethal encephalitis. Microbes Infect. 2006; 8(1):92–104.

Apt, D., Raviprakash, K., Brinkman, A., Semyonov, A., Yang, S., Skinner, C., Diehl, L., Lyons, R., Porter, K. and Punnonen, J. Tetravalent neutralizing antibody response against four dengue serotypes by a single chimeric dengue envelope antigen. Vaccine 2006; 24(3):335–344.

Atchison, R.W., Casto, B.C. and Hammon, W.M. Adenovirus-associated defective virus particles. Science 1965; 149:754–756.

Baba, T.W., Liska, V., Khimani, A.H., Ray, N.B., Dailey, P.J., Penninck, D., Bronson, R., Greene, M.F., McClure, H.M., Martin, L.N. and Ruprecht, R.M. Live attenuated, multiply deleted simian immunodeficiency virus causes AIDS in infant and adult macaques. Nat. Med. 1999; 5(2):194–203.

Barouch, D.H. and Nabel, G.J. Adenovirus vector-based vaccines for human immunodeficiency virus type 1. Hum. Gene Ther. 2005; 16(2):149–156.

Belyakov, I.M., Moss, B., Strober, W. and Berzofsky, J.A. Mucosal vaccination overcomes the barrier to recombinant vaccinia immunization caused by preexisting poxvirus immunity. Proc. Natl. Acad. Sci. U.S.A. 1999; 96(8):4512–4517.

Berglund, P., Sjoberg, M., Garoff, H., Atkins, G.J., Sheahan, B.J. and Liljestrom, P. Semliki Forest virus expression system: production of conditionally infectious recombinant particles. Biotechnology (N.Y.) 1993; 11(8):916–920.

Berkhout, B., Verhoef, K., van Wamel, J.L. and Back, N.K. Genetic instability of live, attenuated human immunodeficiency virus type 1 vaccine strains. J. Virol. 1999; 73(2):1138–1145.

Berns, K.I. and Linden, R.M. The cryptic life style of adeno-associated virus. Bioessays 1995; 17(3):237–245.

Berthold, I., Pombo, M.L., Wagner, L. and Arciniega, J.L. Immunogenicity in mice of anthrax recombinant protective antigen in the presence of aluminum adjuvants. Vaccine 2005; 23(16):1993–1999.

Bett, A.J., Prevec, L. and Graham, F.L. Packaging capacity and stability of human adenovirus type 5 vectors. J. Virol. 1993; 67(10):5911–5921.

Blaney, J.E., Jr., Matro, J.M., Murphy, B.R. and Whitehead, S.S. Recombinant, live-attenuated tetravalent dengue virus vaccine formulations induce a balanced, broad, and protective neutralizing antibody response against each of the four serotypes in rhesus monkeys. J. Virol. 2005; 79(9):5516–5528.

Blume, S. and Geesink, I. A brief history of polio vaccines. Science 2000; 288(5471):1593–1594.

Brave, A., Ljungberg, K., Wahren, B. and Liu, M.A. Vaccine delivery methods using viral vectors. Mol. Pharm. 2007; 4(1):18–32.

Bredenbeek, P.J., Frolov, I., Rice, C.M. and Schlesinger, S. Sindbis virus expression vectors: packaging of RNA replicons by using defective helper RNAs. J. Virol. 1993; 67(11):6439–6446.

Bukreyev, A., Huang, Z., Yang, L., Elankumaran, S., St Claire, M., Murphy, B.R., Samal, S.K. and Collins, P.L. Recombinant Newcastle disease virus expressing a foreign viral antigen is attenuated and highly immunogenic in primates. J. Virol. 2005; 79(21):13275–13284.

Buller, R.M., Janik, J.E., Sebring, E.D. and Rose, J.A. Herpes simplex virus types 1 and 2 completely help adenovirus-associated virus replication. J. Virol. 1981; 40(1):241–247.

Bunning, M.L., Fox, P.E., Bowen, R.A., Komar, N., Chang, G.J., Speaker, T.J., Stephens, M.R., Nemeth, N., Panella, N.A., Langevin, S.A., Gordy, P., Teehee, M., Bright, P.R. and Turell, M.J. DNA vaccination of the American crow (*Corvus brachyrhynchos*) provides partial protection against lethal challenge with West Nile virus. Avian Dis. 2007; 51(2):573–577.

Calain, P. and Roux, L. The rule of six, a basic feature for efficient replication of Sendai virus defective interfering RNA. J. Virol. 1993; 67(8):4822–4830.

Calisher, C.H., Sasso, D.R. and Sather, G.E. Possible evidence for interference with Venezuelan equine encephalitis virus vaccination of equines by pre-existing antibody to Eastern or Western Equine encephalitis virus, or both. Appl. Microbiol. 1973; 26(4):485–488.

Capone, S., Meola, A., Ercole, B.B., Vitelli, A., Pezzanera, M., Ruggeri, L., Davies, M.E., Tafi, R., Santini, C., Luzzago, A., Fu, T.M., Bett, A., Colloca, S., Cortese, R., Nicosia, A. and Folgori, A. A novel adenovirus type 6 (Ad6)-based hepatitis C virus vector that overcomes preexisting anti-ad5 immunity and induces potent and broad cellular immune responses in rhesus macaques. J. Virol. 2006; 80(4):1688–1699.

Catanzaro, A.T., Koup, R.A., Roederer, M., Bailer, R.T., Enama, M.E., Moodie, Z., Gu, L., Martin, J.E., Novik, L., Chakrabarti, B.K., Butman, B.T., Gall, J.G., King, C.R., Andrews, C.A., Sheets, R., Gomez, P.L., Mascola, J.R., Nabel, G.J. and Graham, B.S. Phase 1 safety and immunogenicity evaluation of a multiclade HIV-1 candidate vaccine delivered by a replication-defective recombinant adenovirus vector. J. Infect. Dis. 2006; 194(12):1638–1649.

Cherpelis, S., Jin, X., Gettie, A., Ho, D.D., Barnett, S.W., Shrivastava, I. and Stamatatos, L. DNA-immunization with a V2 deleted HIV-1 envelope elicits protective antibodies in macaques. Immunol. Lett. 2001; 79(1–2):47–55.

Cheung, A.K., Hoggan, M.D., Hauswirth, W.W. and Berns, K.I. Integration of the adeno-associated virus genome into cellular DNA in latently infected human Detroit 6 cells. J. Virol. 1980; 33(2):739–748.

Chiorini, J.A., Kim, F., Yang, L. and Kotin, R.M. Cloning and characterization of adeno-associated virus type 5. J. Virol. 1999; 73(2):1309–1319.

Chiorini, J.A., Yang, L., Liu, Y., Safer, B. and Kotin, R.M. Cloning of adeno-associated virus type 4 (AAV4) and generation of recombinant AAV4 particles. J. Virol. 1997; 71(9):6823–6833.

Chirmule, N., Propert, K., Magosin, S., Qian, Y., Qian, R. and Wilson, J. Immune responses to adenovirus and adeno-associated virus in humans. Gene Ther. 1999; 6(9):1574–1583.

Chulay, J., Burke, D., Karim, S.S.A., Russel, N., Wecker, M., Allen, M., Ferrari, G. and Gilbert, P. Safety and immunogenicity of an alphavirus replicon HIV gag vaccine (AVX101) in healthy HIV-uninfected adults. Antiviral Ther. 2006; 11(Suppl. 2).

Cohen, P. Immunity's yin and yang. A successful vaccine must first avoid being eliminated by pre-existing immunity before it can promote a protective immune response. International AIDS Vaccine Initiative Report 2006; 10(1):1–5.

Daddario-DiCaprio, K.M., Geisbert, T.W., Geisbert, J.B., Stroher, U., Hensley, L.E., Grolla, A., Fritz, E.A., Feldmann, F., Feldmann, H. and Jones, S.M. Cross-protection against Marburg virus strains by using a live, attenuated recombinant vaccine. J. Virol. 2006; 80(19):9659–9666.

Davidoff, A.M., Gray, J.T., Ng, C.Y., Zhang, Y., Zhou, J., Spence, Y., Bakar, Y. and Nathwani, A.C. Comparison of the ability of adeno-associated viral vectors pseudotyped with serotype 2, 5, and 8 capsid proteins to mediate efficient transduction of the liver in murine and nonhuman primate models. Mol. Ther. 2005; 11(6):875–888.

Dedieu, J.F., Vigne, E., Torrent, C., Jullien, C., Mahfouz, I., Caillaud, J.M., Aubailly, N., Orsini, C., Guillaume, J.M., Opolon, P., Delaere, P., Perricaudet, M. and Yeh, P. Long-term gene delivery into the livers of immunocompetent mice with E1/E4-defective adenoviruses. J. Virol. 1997; 71(6):4626–4637.

Deming, D., Sheahan, T., Heise, M., Yount, B., Davis, N., Sims, A., Suthar, M., Harkema, J., Whitmore, A., Pickles, R., West, A., Donaldson, E., Curtis, K., Johnston, R. and Baric, R. Vaccine efficacy in senescent mice challenged with recombinant SARS-CoV bearing epidemic and zoonotic spike variants. Public Libr. Sci. Med. 2006; 3(12):e525.

Despres, P., Combredet, C., Frenkiel, M.P., Lorin, C., Brahic, M. and Tangy, F. Live measles vaccine expressing the secreted form of the West Nile virus envelope glycoprotein protects against West Nile virus encephalitis. J. Infect. Dis. 2005; 191(2):207–214.

DiNapoli, J.M., Kotelkin, A., Yang, L., Elankumaran, S., Murphy, B.R., Samal, S.K., Collins, P.L. and Bukreyev, A. Newcastle disease virus, a host range-restricted virus, as a vaccine vector for intranasal immunization against emerging pathogens. Proc. Natl. Acad. Sci. U.S.A. 2007; 104(23):9788–9793.

Dong, J.Y., Fan, P.D. and Frizzell, R.A. Quantitative analysis of the packaging capacity of recombinant adeno-associated virus. Hum. Gene Ther. 1996; 7(17):2101–2112.

Emery, S., Workman, C., Puls, R.L., Bloch, M., Baker, D., Bodsworth, N., Anderson, J., Crowe, S.M., French, M.A., Hoy, J., Aichelburg, A., Ward, L.D., Boyle, D.B., Law, M.G., Kelleher, A.D. and Cooper, D.A. Randomized, placebo-controlled, phase I/IIa evaluation of the safety and immunogenicity of fowlpox virus expressing HIV gag-pol and interferon-gamma in HIV-1 infected subjects. Hum. Vaccines 2005; 1(6):232–238.

Ezelle, H.J., Markovic, D. and Barber, G.N. Generation of hepatitis C virus-like particles by use of a recombinant vesicular stomatitis virus vector. J. Virol. 2002; 76(23):12325–12334.

FDA. FDA Statement on Gene Therapy Clinical Trial (July 26, 2007). U.S. Food and Drug Administration. 2007.

Feldmann, H., Jones, S.M., Daddario-DiCaprio, K.M., Geisbert, J.B., Stroher, U., Grolla, A., Bray, M., Fritz, E.A., Fernando, L., Feldmann, F., Hensley, L.E. and Geisbert, T.W. Effective post-exposure treatment of Ebola infection. Public Libr. Sci. Pathog. 2007; 3(1):e2.

Ferrari, S., Griesenbach, U., Iida, A., Farley, R., Wright, A.M., Zhu, J., Munkonge, F.M., Smith, S.N., You, J., Ban, H., Inoue, M., Chan, M., Singh, C., Verdon, B., Argent, B.E., Wainwright, B., Jeffery, P.K., Geddes, D.M., Porteous, D.J., Hyde, S.C., Gray, M.A., Hasegawa, M. and Alton, E.W. Sendai virus-mediated CFTR gene transfer to the airway epithelium. Gene Ther. 2007; 14(19):1371–1379.

Gallez-Hawkins, G., Li, X., Franck, A.E., Thao, L., Lacey, S.F., Diamond, D.J. and Zaia, J.A. DNA and low titer, helper-free, recombinant AAV prime-boost vaccination for cytomegalovirus induces an immune response to CMV-pp65 and CMV-IE1 in transgenic HLA A*0201 mice. Vaccine 2004; 23(6):819–826.

Gao, G.P., Alvira, M.R., Wang, L., Calcedo, R., Johnston, J. and Wilson, J.M. Novel adeno-associated viruses from rhesus monkeys as vectors for human gene therapy. Proc. Natl. Acad. Sci. U.S.A. 2002; 99(18):11854–11859.

Gao, G.P., Yang, Y. and Wilson, J.M. Biology of adenovirus vectors with E1 and E4 deletions for liver-directed gene therapy. J. Virol. 1996; 70(12):8934–8943.

Gao, W., Soloff, A.C., Lu, X., Montecalvo, A., Nguyen, D.C., Matsuoka, Y., Robbins, P.D., Swayne, D.E., Donis, R.O., Katz, J.M., Barratt-Boyes, S.M. and Gambotto, A. Protection of mice and poultry from lethal H5N1 avian influenza virus through adenovirus-based immunization. J. Virol. 2006; 80(4):1959–1964.

Gardner, J.P., Frolov, I., Perri, S., Ji, Y., MacKichan, M.L., zur Megede, J., Chen, M., Belli, B.A., Driver, D.A., Sherrill, S., Greer, C.E., Otten, G.R., Barnett, S.W., Liu, M.A., Dubensky, T.W. and Polo, J.M. Infection of human dendritic cells by a sindbis virus replicon vector is determined by a single amino acid substitution in the E2 glycoprotein. J. Virol. 2000; 74(24): 11849–11857.

Ghosh-Choudhury, G., Haj-Ahmad, Y. and Graham, F.L. Protein IX, a minor component of the human adenovirus capsid, is essential for the packaging of full length genomes. EMBO J. 1987; 6(6):1733–1739.

Gilbert, P.B., Chiu, Y.L., Allen, M., Lawrence, D.N., Chapdu, C., Israel, H., Holman, D., Keefer, M.C., Wolff, M. and Frey, S.E. Long-term safety analysis of preventive HIV-1 vaccines evaluated in AIDS vaccine evaluation group NIAID-sponsored Phase I and II clinical trials. Vaccine 2003; 21(21–22):2933–2947.

Goncalves, M.A. and de Vries, A.A. Adenovirus: from foe to friend. Rev. Med. Virol. 2006; 16(3):167–186.

Graham, F.L. Covalently closed circles of human adenovirus DNA are infectious. EMBO J. 1984; 3(12):2917–2922.

Graham, F.L., Smiley, J., Russell, W.C. and Nairn, R. Characteristics of a human cell line transformed by DNA from human adenovirus type 5. J. Gen. Virol. 1977; 36(1):59–74.

Griesenbach, U., Inoue, M., Hasegawa, M. and Alton, E.W. Sendai virus for gene therapy and vaccination. Curr. Opin. Mol. Ther. 2005; 7(4):346–352.

Griffin, D.E. Alphaviruses. In: Fields Virology (D.M. Knipe and P.M. Howley, Eds.), Vol. 1, pp. 917–962. Philadelphia: Lippincott Williams & Wilkins, 2001.

Haglund, K., Forman, J., Krausslich, H.G. and Rose, J.K. Expression of human immunodeficiency virus type 1 Gag protein precursor and envelope proteins from a vesicular stomatitis virus recombinant: high-level production of virus-like particles containing HIV envelope. Virology 2000; 268(1):112–121.

Haglund, K., Leiner, I., Kerksiek, K., Buonocore, L., Pamer, E. and Rose, J.K. High-level primary CD8(+) T-cell response to human immunodeficiency virus type 1 gag and env generated by vaccination with recombinant vesicular stomatitis viruses. J. Virol. 2002; 76(6):2730–2738.

Hashimoto, M., Boyer, J.L., Hackett, N.R., Wilson, J.M. and Crystal, R.G. Induction of protective immunity to anthrax lethal toxin with a nonhuman primate adenovirus-based vaccine in the presence of preexisting anti-human adenovirus immunity. Infect. Immun. 2005; 73(10):6885–6891.

Henderson, D.A., Witte, J.J., Morris, L. and Langmuir, A.D. Paralytic disease associated with oral polio vaccines. J. Am. Med. Assoc. 1964; 190:41–48.

Hevey, M., Negley, D., Pushko, P., Smith, J. and Schmaljohn, A. Marburg virus vaccines based upon alphavirus replicons protect guinea pigs and nonhuman primates. Virology 1998; 251(1):28–37.

Hodges, B.L., Serra, D., Hu, H., Begy, C.A., Chamberlain, J.S. and Amalfitano, A. Multiply deleted [E1, polymerase-, and pTP-] adenovirus vector persists despite deletion of the preterminal protein. J. Gene Med. 2000; 2(4):250–259.

Holman, D.H., Wang, D., Raviprakash, K., Raja, N.U., Luo, M., Zhang, J., Porter, K.R. and Dong, J.Y. Two complex, adenovirus-based vaccines that together induce immune responses to all four dengue virus serotypes. Clin. Vaccine Immunol. 2007; 14(2).

Huang, Z., Elankumaran, S., Panda, A. and Samal, S.K. Recombinant Newcastle disease virus as a vaccine vector. Poult. Sci. 2003; 82(6):899–906.

Huang, Z., Krishnamurthy, S., Panda, A. and Samal, S.K. High-level expression of a foreign gene from the most 3′-proximal locus of a recombinant Newcastle disease virus. J. Gen. Virol. 2001; 82 (Pt. 7):1729–1736.

Jaiswal, S., Khanna, N. and Swaminathan, S. Replication-defective adenoviral vaccine vector for the induction of immune responses to dengue virus type 2. J. Virol. 2003; 77(23):12907–12913.

Jendrek, S., Little, S.F., Hem, S., Mitra, G. and Giardina, S. Evaluation of the compatibility of a second generation recombinant anthrax vaccine with aluminum-containing adjuvants. Vaccine 2003; 21(21–22):3011–3018.

Johnson, J.E., Nasar, F., Coleman, J.W., Price, R.E., Javadian, A., Draper, K., Lee, M., Reilly, P.A., Clarke, D.K., Hendry, R.M. and Udem, S.A. Neurovirulence properties of recombinant vesicular stomatitis virus vectors in non-human primates. Virology 2007; 360(1):36–49.

Johnson, R.P. and Desrosiers, R.C. Protective immunity induced by live attenuated simian immunodeficiency virus. Curr. Opin. Immunol. 1998; 10(4):436–443.

Jones, N. and Shenk, T. An adenovirus type 5 early gene function regulates expression of other early viral genes. Proc. Natl. Acad. Sci. U.S.A. 1979; 76(8):3665–3669.

Jones, S.M., Feldmann, H., Stroher, U., Geisbert, J.B., Fernando, L., Grolla, A., Klenk, H.D., Sullivan, N.J., Volchkov, V.E., Fritz, E.A., Daddario, K.M., Hensley, L.E., Jahrling, P.B. and Geisbert, T. W. Live attenuated recombinant vaccine protects nonhuman primates against Ebola and Marburg viruses. Nat. Med. 2005; 11(7):786–790.

Kaur, A., Sanford, H.B., Garry, D., Lang, S., Klumpp, S.A., Watanabe, D., Bronson, R.T., Lifson, J.D., Rosati, M., Pavlakis, G.N., Felber, B.K., Knipe, D.M. and Desrosiers, R.C. Ability of herpes simplex virus vectors to boost immune responses to DNA vectors and to protect against challenge by simian immunodeficiency virus. Virology 2007; 357(2):199–214.

Kitchener, S., Nissen, M., Nasveld, P., Forrat, R., Yoksan, S., Lang, J. and Saluzzo, J.F. Immunogenicity and safety of two live-attenuated tetravalent dengue vaccine formulations in healthy Australian adults. Vaccine 2006; 24(9):1238–1241.

Kobinger, G.P., Feldmann, H., Zhi, Y., Schumer, G., Gao, G., Feldmann, F., Jones, S. and Wilson, J.M. Chimpanzee adenovirus vaccine protects against Zaire Ebola virus. Virology 2006; 346(2):394–401.

Kochanek, S., Schiedner, G. and Volpers, C. High-capacity 'gutless' adenoviral vectors. Curr. Opin. Mol. Ther. 2001; 3(5):454–463.

Kodihalli, S., Goto, H., Kobasa, D.L., Krauss, S., Kawaoka, Y. and Webster, R.G. DNA vaccine encoding hemagglutinin provides protective immunity against H5N1 influenza virus infection in mice. J. Virol. 1999; 73(3):2094–2098.

Kong, W.P., Xu, L., Stadler, K., Ulmer, J.B., Abrignani, S., Rappuoli, R. and Nabel, G.J. Modulation of the immune response to the severe acute respiratory syndrome spike glycoprotein by gene-based and inactivated virus immunization. J. Virol. 2005; 79(22):13915–13923.

Kotin, R.M. Prospects for the use of adeno-associated virus as a vector for human gene therapy. Hum. Gene Ther. 1994; 5(7):793–801.

Krishnamurthy, S., Huang, Z. and Samal, S.K. Recovery of a virulent strain of Newcastle disease virus from cloned cDNA: expression of a foreign gene results in growth retardation and attenuation. Virology 2000; 278(1):168–182.

Kumar-Singh, R. and Chamberlain, J.S. Encapsidated adenovirus minichromosomes allow delivery and expression of a 14 kb dystrophin cDNA to muscle cells. Hum. Mol. Genet. 1996; 5(7):913–921.

Lauterbach, H., Kerksiek, K.M., Busch, D.H., Berto, E., Bozac, A., Mavromara, P., Manservigi, R., Epstein, A.L., Marconi, P. and Brocker, T. Protection from bacterial infection by a single vaccination with replication-deficient mutant herpes simplex virus type 1. J. Virol. 2004; 78(8):4020–4028.

Lee, J.S., Groebner, J.L., Hadjipanayis, A.G., Negley, D.L., Schmaljohn, A.L., Welkos, S.L., Smith, L.A. and Smith, J.F. Multiagent vaccines vectored by Venezuelan equine encephalitis virus replicon elicits immune responses to Marburg virus and protection against anthrax and botulinum neurotoxin in mice. Vaccine 2006; 24(47–48):6886–6892.

Liljestrom, P. and Garoff, H. A new generation of animal cell expression vectors based on the Semliki Forest virus replicon. Biotechnology (N.Y.) 1991; 9(12):1356–1361.

Liu, D.W., Tsao, Y.P., Kung, J.T., Ding, Y.A., Sytwu, H.K., Xiao, X. and Chen, S.L. Recombinant adeno-associated virus expressing human papillomavirus type 16 E7 peptide DNA fused with heat shock protein DNA as a potential vaccine for cervical cancer. J. Virol. 2000; 74(6):2888–2894.

Lorence, R.M., Pecora, A.L., Major, P.P., Hotte, S.J., Laurie, S.A., Roberts, M.S., Groene, W.S. and Bamat, M.K. Overview of phase I studies of intravenous administration of PV701, an oncolytic virus. Curr. Opin. Mol. Ther. 2003; 5(6):618–624.

Lorence, R.M., Roberts, M.S., O'Neil, J.D., Groene, W.S., Miller, J.A., Mueller, S.N. and Bamat, M.K. Phase 1 clinical experience using intravenous administration of PV701, an oncolytic Newcastle disease virus. Curr. Cancer Drug Targets 2007; 7(2):157–167.

Lorin, C., Mollet, L., Delebecque, F., Combredet, C., Hurtrel, B., Charneau, P., Brahic, M. and Tangy, F. A single injection of recombinant measles virus vaccines expressing human immunodeficiency virus (HIV) type 1 clade B envelope glycoproteins induces neutralizing antibodies and cellular immune responses to HIV. J. Virol. 2004; 78(1):146–157.

Lusky, M., Christ, M., Rittner, K., Dieterle, A., Dreyer, D., Mourot, B., Schultz, H., Stoeckel, F., Pavirani, A. and Mehtali, M. In vitro and in vivo biology of recombinant adenovirus vectors with E1, E1/E2A, or E1/E4 deleted. J. Virol. 1998; 72(3):2022–2032.

Ma, C., Yao, K., Zhou, F. and Zhu, M. Comparative immunization in BALB/c mice with recombinant replication-defective adenovirus vector and DNA plasmid expressing a SARS-CoV nucleocapsid protein gene. Cell. Mol. Immunol. 2006; 3(6):459–465.

MacDonald, G.H. and Johnston, R.E. Role of dendritic cell targeting in Venezuelan equine encephalitis virus pathogenesis. J. Virol. 2000; 74(2):914–922.

Mackowiak, M., Maki, J., Motes-Kreimeyer, L., Harbin, T. and Van Kampen, K. Vaccination of wildlife against rabies: successful use of a vectored vaccine obtained by recombinant technology. Adv. Vet. Med. 1999; 41:571–583.

Maeda, K., West, K., Hayasaka, D., Ennis, F.A. and Terajima, M. Recombinant adenovirus vector vaccine induces stronger cytotoxic T-cell responses than recombinant vaccinia virus vector, plasmid DNA, or a combination of these. Viral Immunol. 2005; 18(4):657–667.

Manning, W.C., Paliard, X., Zhou, S., Pat Bland, M., Lee, A.Y., Hong, K., Walker, C.M., Escobedo, J.A. and Dwarki, V. Genetic immunization with adeno-associated virus vectors expressing herpes simplex virus type 2 glycoproteins B and D. J. Virol. 1997; 71(10):7960–7962.

Martinez-Sobrido, L., Gitiban, N., Fernandez-Sesma, A., Cros, J., Mertz, S.E., Jewell, N.A., Hammond, S., Flano, E., Durbin, R.K., Garcia-Sastre, A. and Durbin, J.E. Protection against respiratory syncytial virus by a recombinant Newcastle disease virus vector. J. Virol. 2006; 80(3):1130–1139.

McCarty, D.M., Young, S.M., Jr. and Samulski, R.J. Integration of adeno-associated virus (AAV) and recombinant AAV vectors. Annu. Rev. Genet. 2004; 38:819–845.

McClain, D.J., Pittman, P.R., Ramsburg, H.H., Nelson, G.O., Rossi, C.A., Mangiafico, J.A., Schmaljohn, A.L. and Malinoski, F.J. Immunologic interference from sequential administration of live attenuated alphavirus vaccines. J. Infect. Dis. 1998; 177(3):634–641.

McConnell, M.J., Hanna, P.C. and Imperiale, M.J. Adenovirus-based prime-boost immunization for rapid vaccination against anthrax. Mol. Ther. 2007; 15(1):203–210.

Megede, J.Z., Otten, G.R., Doe, B., Liu, H., Srivastava, R., Greer, C., Legg, H., Tang, T., Polo, J.M., Donnelly, J.J., Ulmer, J.B. and Barnett, S.W. Evaluation of human immunodeficiency virus type 1 subtype C gag, pol, and gagpol DNA and alphavirus replicon vaccines. Vaccine 2006; 24(15):2755–2763.

Meseda, C.A., Stout, R.R. and Weir, J.P. Evaluation of a needle-free delivery platform for prime-boost immunization with DNA and modified vaccinia virus Ankara vectors expressing herpes simplex virus 2 glycoprotein D. Viral Immunol. 2006; 19(2):250–259.

Miller, J., McBride, B.W., Manchee, R.J., Moore, P. and Baillie, L.W. Production and purification of recombinant protective antigen and protective efficacy against *Bacillus anthracis*. Lett. Appl. Microbiol. 1998; 26(1):56–60.

Moorhead, J.W., Clayton, G.H., Smith, R.L. and Schaack, J. A replication-incompetent adenovirus vector with the preterminal protein gene deleted efficiently transduces mouse ears. J. Virol. 1999; 73(2):1046–1053.

Muramatsu, S., Mizukami, H., Young, N.S. and Brown, K.E. Nucleotide sequencing and generation of an infectious clone of adeno-associated virus 3. Virology 1996; 221(1):208–217.

Murphey-Corb, M. Live-attenuated HIV vaccines: how safe is safe enough? Nat. Med. 1997; 3(1):17–18.

Naito, T., Kaneko, Y. and Kozbor, D. Oral vaccination with modified vaccinia virus Ankara attached covalently to TMPEG-modified cationic liposomes overcomes pre-existing poxvirus immunity from recombinant vaccinia immunization. J. Gen. Virol. 2007; 88(Pt. 1):61–70.

Nakaya, T., Cros, J., Park, M.S., Nakaya, Y., Zheng, H., Sagrera, A., Villar, E., Garcia-Sastre, A. and Palese, P. Recombinant Newcastle disease virus as a vaccine vector. J. Virol. 2001; 75(23):11868–11873.

Nakaya, Y., Nakaya, T., Park, M.S., Cros, J., Imanishi, J., Palese, P. and Garcia-Sastre, A. Induction of cellular immune responses to simian immunodeficiency virus gag by two recombinant negative-strand RNA virus vectors. J. Virol. 2004; 78(17):9366–9375.

Nwanegbo, E., Vardas, E., Gao, W., Whittle, H., Sun, H., Rowe, D., Robbins, P.D. and Gambotto, A. Prevalence of neutralizing antibodies to adenoviral serotypes 5 and 35 in the adult populations of The Gambia, South Africa, and the United States. Clin. Diagn. Lab. Immunol. 2004; 11(2):351–357.

Olinger, G.G., Bailey, M.A., Dye, J.M., Bakken, R., Kuehne, A., Kondig, J., Wilson, J., Hogan, R.J. and Hart, M.K. Protective cytotoxic T-cell responses induced by Venezuelan equine encephalitis virus replicons expressing Ebola virus proteins. J. Virol. 2005; 79(22):14189–14196.

Palin, A., Chattopadhyay, A., Park, S., Delmas, G., Suresh, R., Senina, S., Perlin, D.S. and Rose, J.K. An optimized vaccine vector based on recombinant vesicular stomatitis virus gives high-level, long-term protection against *Yersinia pestis* challenge. Vaccine 2007; 25(4):741–750.

Parks, G.D. and Alexander-Miller, M.A. High avidity cytotoxic T lymphocytes to a foreign antigen are efficiently activated following immunization with a recombinant paramyxovirus, simian virus 5. J. Gen. Virol. 2002; 83(Pt. 5):1167–1172.

Parks, R.J., Chen, L., Anton, M., Sankar, U., Rudnicki, M.A. and Graham, F.L. A helper-dependent adenovirus vector system: removal of helper virus by Cre-mediated excision of the viral packaging signal. Proc. Natl. Acad. Sci. U.S.A. 1996; 93(24):13565–13570.

Parrino, J. and Graham, B.S. Smallpox vaccines: past, present, and future. J. Allergy Clin. Immunol. 2006; 118(6):1320–1326.

Polo, J.M., Belli, B.A., Driver, D.A., Frolov, I., Sherrill, S., Hariharan, M.J., Townsend, K., Perri, S., Mento, S.J., Jolly, D.J., Chang, S.M., Schlesinger, S. and Dubensky, T.W., Jr. Stable alphavirus packaging cell lines for Sindbis virus and Semliki Forest virus-derived vectors. Proc. Natl. Acad. Sci. U.S.A. 1999; 96(8):4598–4603.

Pushko, P., Bray, M., Ludwig, G.V., Parker, M., Schmaljohn, A., Sanchez, A., Jahrling, P.B. and Smith, J.F. Recombinant RNA replicons derived from attenuated Venezuelan equine encephalitis virus protect guinea pigs and mice from Ebola hemorrhagic fever virus. Vaccine 2000; 19(1):142–153.

Pushko, P., Parker, M., Ludwig, G.V., Davis, N.L., Johnston, R.E. and Smith, J.F. Replicon-helper systems from attenuated Venezuelan equine encephalitis virus: expression of heterologous genes in vitro and immunization against heterologous pathogens in vivo. Virology 1997; 239(2):389–401.

Putnak, R., Coller, B.A., Voss, G., Vaughn, D.W., Clements, D., Peters, I., Bignami, G., Houng, H.S., Chen, R.C., Barvir, D.A., Seriwatana, J., Cayphas, S., Garcon, N., Gheysen, D., Kanesa-Thasan, N., McDonell, M., Humphreys, T., Eckels, K.H., Prieels, J.P. and Innis, B.L. An evaluation of dengue type-2 inactivated, recombinant subunit, and live-attenuated vaccine candidates in the rhesus macaque model. Vaccine 2005; 23(35):4442–4452.

Qin, L., Wang, X., Wu, S., Yuan, S., Huang, D., Lei, M., Pan, H. and Lin, Q. The immunity induced by recombinant spike proteins of SARS coronavirus in Balb/c mice. J. Huazhong Uni. Sci. Technol. 2007; 27(3):221–224.

Quiroz, E., Moreno, N., Peralta, P.H. and Tesh, R.B. A human case of encephalitis associated with vesicular stomatitis virus (Indiana serotype) infection. Am. J. Trop. Med. Hyg. 1988; 39(3): 312–314.

Radaelli, A., Bonduelle, O., Beggio, P., Mahe, B., Pozzi, E., Elli, V., Paganini, M., Zanotto, C., De Giuli Morghen, C. and Combadiere, B. Prime-boost immunization with DNA, recombinant fowlpox virus and VLP(SHIV) elicit both neutralizing antibodies and IFNgamma-producing T cells against the HIV-envelope protein in mice that control env-bearing tumour cells. Vaccine 2007; 25(11):2128–2138.

Raja, N.U., Holman, D.H., Wang, D., Raviprakash, K., Juompan, L.Y., Deitz, S.B., Luo, M., Zhang, J., Porter, K.R. and Dong, J.Y. Induction of bivalent immune responses by expression of dengue virus type 1 and type 2 antigens from a single complex adenoviral vector. Am. J. Trop. Med. Hyg. 2007; 76(4):743–751.

Reap, E.A., Dryga, S.A., Morris, J., Rivers, B., Norberg, P.K., Olmsted, R.A. and Chulay, J.D. Cellular and humoral immune responses to alphavirus replicon vaccines expressing cytomegalovirus pp65, IE1, and gB proteins. Clin. Vaccine Immunol. 2007; 14(6):748–755.

Reiss, C.S., Plakhov, I.V. and Komatsu, T. Viral replication in olfactory receptor neurons and entry into the olfactory bulb and brain. Ann. N.Y. Acad. Sci. 1998; 855:751–761.

Reyes Del Valle, J., Devaux, P., Hodge, G., Wegner, N.J., McChesney, M.B. and Cattaneo, R. A vectored measles virus induces hepatitis B surface antigen antibodies while protecting macaques against measles challenge. J. Virol. 2007; Oct; 10597–10605. .

Roberts, A., Buonocore, L., Price, R., Forman, J. and Rose, J.K. Attenuated vesicular stomatitis viruses as vaccine vectors. J. Virol. 1999; 73(5):3723–3732.

Rodriguez, L.L. Emergence and re-emergence of vesicular stomatitis in the United States. Virus Res. 2002; 85(2):211–219.

Romer-Oberdorfer, A., Mundt, E., Mebatsion, T., Buchholz, U.J. and Mettenleiter, T.C. Generation of recombinant lentogenic Newcastle disease virus from cDNA. J. Gen. Virol. 1999; 80(Pt. 11):2987–2995.

Rose, N.F., Roberts, A., Buonocore, L. and Rose, J.K. Glycoprotein exchange vectors based on vesicular stomatitis virus allow effective boosting and generation of neutralizing antibodies to a primary isolate of human immunodeficiency virus type 1. J. Virol. 2000; 74(23):10903–10910.

Russell, N.D., Graham, B.S., Keefer, M.C., McElrath, M.J., Self, S.G., Weinhold, K.J., Montefiori, D.C., Ferrari, G., Horton, H., Tomaras, G.D., Gurunathan, S., Baglyos, L., Frey, S.E., Mulligan, M.J., Harro, C.D., Buchbinder, S.P., Baden, L.R., Blattner, W.A., Koblin, B.A. and Corey, L. Phase 2 study of an HIV-1 canarypox vaccine (vCP1452) alone and in combination with rgp120: negative results fail to trigger a phase 3 correlates trial. J. Acquir. Immune Defic. Syndr. 2007; 44(2):203–212.

Rutledge, E.A., Halbert, C.L. and Russell, D.W. Infectious clones and vectors derived from adeno-associated virus (AAV) serotypes other than AAV type 2. J. Virol. 1998; 72(1):309–319.

Sakai, Y., Kiyotani, K., Fukumura, M., Asakawa, M., Kato, A., Shioda, T., Yoshida, T., Tanaka, A., Hasegawa, M. and Nagai, Y. Accommodation of foreign genes into the Sendai virus genome: sizes of inserted genes and viral replication. FEBS Lett. 1999; 456(2):221–226.

Samulski, R.J., Zhu, X., Xiao, X., Brook, J.D., Housman, D.E., Epstein, N. and Hunter, L.A. Targeted integration of adeno-associated virus (AAV) into human chromosome 19. EMBO J. 1991; 10(12):3941–3950.

Santos, K., Duke, C.M., Rodriguez-Colon, S.M., Dakwar, A., Fan, S., Keefer, M.C., Federoff, H.J., Frelinger, J.G., Bowers, W.J. and Dewhurst, S. Effect of promoter strength on protein expression and immunogenicity of an HSV-1 amplicon vector encoding HIV-1 Gag. Vaccine 2007; 25(9):1634–1646.

Schepp-Berglind, J., Luo, M., Wang, D., Wicker, J.A., Raja, N., Hoel, B., Holman, D.H., Barrett, A.D. and Dong, J.Y. Complex adenovirus-mediated expression of C, PreM, E, and NS1 proteins of West Nile Virus induces both humoral and cellular immune responses. Clin. Vaccine Immunol. 2007; Sep; 14(9):1117–1126.

Schlereth, B., Buonocore, L., Tietz, A., Meulen Vt, V., Rose, J.K. and Niewiesk, S. Successful mucosal immunization of cotton rats in the presence of measles virus-specific antibodies depends on degree of attenuation of vaccine vector and virus dose. J. Gen. Virol. 2003; 84(Pt. 8):2145–2151.

Schubert, M., Joshi, B., Blondel, D. and Harmison, G.G. Insertion of the human immunodeficiency virus CD4 receptor into the envelope of vesicular stomatitis virus particles. J. Virol. 1992; 66(3):1579–1589.

Schultz-Cherry, S., Dybing, J.K., Davis, N.L., Williamson, C., Suarez, D.L., Johnston, R. and Perdue, M.L. Influenza virus (A/HK/156/97) hemagglutinin expressed by an alphavirus replicon system protects chickens against lethal infection with Hong Kong-origin H5N1 viruses. Virology 2000; 278(1):55–59.

Schwartz, J.A., Buonocore, L., Roberts, A., Suguitan, A., Jr., Kobasa, D., Kobinger, G., Feldmann, H., Subbarao, K. and Rose, J.K. Vesicular stomatitis virus vectors expressing avian influenza H5 HA induce cross-neutralizing antibodies and long-term protection. Virology 2007; Sep 15; 366(1):166–173.

Shenk, T.E. Adenoviridae: the viruses and their replication. In: Fields Virology (D.M. Knipe and P.M. Howley, Eds.), Vol. 2, pp. 2265–2300. Philadelphia: Lippincott Williams & Wilkins, 2001.

Shu, Y., Winfrey, S., Yang, Z.Y., Xu, L., Rao, S.S., Srivastava, I., Barnett, S.W., Nabel, G.J. and Mascola, J.R. Efficient protein boosting after plasmid DNA or recombinant adenovirus immunization with HIV-1 vaccine constructs. Vaccine 2007; 25(8):1398–1408.

Smerdou, C. and Liljestrom, P. Two-helper RNA system for production of recombinant Semliki forest virus particles. J. Virol. 1999; 73(2):1092–1098.

Smith, G.L. and Moss, B. Infectious poxvirus vectors have capacity for at least 25,000 base pairs of foreign DNA. Gene 1983; 25(1):21–28.

Steensels, M., Van Borm, S., Lambrecht, B., De Vriese, J., Le Gros, F.X., Bublot, M. and van den Berg, T. Efficacy of an inactivated and a fowlpox-vectored vaccine in Muscovy ducks against an Asian H5N1 highly pathogenic avian influenza viral challenge. Avian Dis. 2007; 51(Suppl. 1):325–331.

Sullivan, N.J., Geisbert, T.W., Geisbert, J.B., Xu, L., Yang, Z.Y., Roederer, M., Koup, R.A., Jahrling, P.B. and Nabel, G.J. Accelerated vaccination for Ebola virus haemorrhagic fever in non-human primates. Nature 2003; 424(6949):681–684.

Sullivan, N.J., Sanchez, A., Rollin, P.E., Yang, Z.Y. and Nabel, G.J. Development of a preventive vaccine for Ebola virus infection in primates. Nature 2000; 408(6812):605–609.

Takimoto, T., Hurwitz, J.L., Zhan, X., Krishnamurthy, S., Prouser, C., Brown, B., Coleclough, C., Boyd, K., Scroggs, R.A., Portner, A. and Slobod, K.S. Recombinant Sendai virus as a novel vaccine candidate for respiratory syncytial virus. Viral Immunol. 2005; 18(2):255–266.

Tang, S., van Rij, R., Silvera, D. and Andino, R. Toward a poliovirus-based simian immunodeficiency virus vaccine: correlation between genetic stability and immunogenicity. J. Virol. 1997; 71(10):7841–7850.

Taylor, J., Trimarchi, C., Weinberg, R., Languet, B., Guillemin, F., Desmettre, P. and Paoletti, E. Efficacy studies on a canarypox-rabies recombinant virus. Vaccine 1991; 9(3):190–193.

Taylor, J., Weinberg, R., Languet, B., Desmettre, P. and Paoletti, E. Recombinant fowlpox virus inducing protective immunity in non-avian species. Vaccine 1988; 6(6):497–503.

Tompkins, S.M., Zhao, Z.S., Lo, C.Y., Misplon, J.A., Liu, T., Ye, Z., Hogan, R.J., Wu, Z., Benton, K.A., Tumpey, T.M. and Epstein, S.L. Matrix protein 2 vaccination and protection against influenza viruses, including subtype H5N1. Emerg. Infect. Dis. 2007; 13(3):426–435.

Vanderzanden, L., Bray, M., Fuller, D., Roberts, T., Custer, D., Spik, K., Jahrling, P., Huggins, J., Schmaljohn, A. and Schmaljohn, C. DNA vaccines expressing either the GP or NP genes of Ebola virus protect mice from lethal challenge. Virology 1998; 246(1):134–144.

Van Kampen, K.R., Shi, Z., Gao, P., Zhang, J., Foster, K.W., Chen, D.T., Marks, D., Elmets, C.A. and Tang, D.C. Safety and immunogenicity of adenovirus-vectored nasal and epicutaneous influenza vaccines in humans. Vaccine 2005; 23(8):1029–1036.

Vincent, K.A., Piraino, S.T. and Wadsworth, S.C. Analysis of recombinant adeno-associated virus packaging and requirements for rep and cap gene products. J. Virol. 1997; 71(3):1897–1905.

Walther, M., Thompson, F.M., Dunachie, S., Keating, S., Todryk, S., Berthoud, T., Andrews, L., Andersen, R.F., Moore, A.,

Gilbert, S.C., Poulton, I., Dubovsky, F., Tierney, E., Correa, S., Huntcooke, A., Butcher, G., Williams, J., Sinden, R.E. and Hill, A.V. Safety, immunogenicity, and efficacy of prime-boost immunization with recombinant poxvirus FP9 and modified vaccinia virus Ankara encoding the full-length *Plasmodium falciparum* circumsporozoite protein. Infect. Immun. 2006; 74(5):2706–2716.

Wang, D., Hevey, M., Juompan, L.Y., Trubey, C.M., Raja, N.U., Deitz, S.B., Woraratanadharm, J., Luo, M., Yu, H., Swain, B.M., Moore, K.M. and Dong, J.Y. Complex adenovirus-vectored vaccine protects guinea pigs from three strains of Marburg virus challenges. Virology 2006a; 353(2):324–332.

Wang, D., Raja, N.U., Trubey, C.M., Juompan, L.Y., Luo, M., Woraratanadharm, J., Deitz, S.B., Yu, H., Swain, B.M., Moore, K.M., Pratt, W.D., Hart, M.K. and Dong, J.Y. Development of a cAdVax-based bivalent Ebola vaccine that induces immune responses against both the Sudan and Zaire species of Ebola virus. J. Virol. 2006b; 80(6):2738–2746.

Wang, Z., Rosa, C.L., Li, Z., Ly, H., Krishnan, A., Martinez, J., Britt, W.J. and Diamond, D.J. Vaccine properties of a novel marker gene-free recombinant modified vaccinia Ankara expressing immunodominant CMV antigens pp65 and IE1. Vaccine 2007; 25(6):1132–1141.

Weingartl, H.M., Berhane, Y., Caswell, J.L., Loosmore, S., Audonnet, J.C., Roth, J.A. and Czub, M. Recombinant nipah virus vaccines protect pigs against challenge. J. Virol. 2006; 80(16):7929–7938.

Weyer, J., Rupprecht, C.E., Mans, J., Viljoen, G.J. and Nel, L.H. Generation and evaluation of a recombinant modified vaccinia virus Ankara vaccine for rabies. Vaccine 2007; 25(21):4213–4222.

Wiktor, T.J., Macfarlan, R.I., Reagan, K.J., Dietzschold, B., Curtis, P. J., Wunner, W.H., Kieny, M.P., Lathe, R., Lecocq, J.P., Mackett, M. et al. Protection from rabies by a vaccinia virus recombinant containing the rabies virus glycoprotein gene. Proc. Natl. Acad. Sci. U.S.A. 1984; 81(22):7194–7198.

Wu, L., Kong, W.P. and Nabel, G.J. Enhanced breadth of CD4 T-cell immunity by DNA prime and adenovirus boost immunization to human immunodeficiency virus Env and Gag immunogens. J. Virol. 2005; 79(13):8024–8031.

Xiang, Z.Q., Gao, G.P., Reyes-Sandoval, A., Li, Y., Wilson, J.M. and Ertl, H.C. Oral vaccination of mice with adenoviral vectors is not impaired by preexisting immunity to the vaccine carrier. J. Virol. 2003; 77(20):10780–10789.

Xiao, W., Chirmule, N., Berta, S.C., McCullough, B., Gao, G. and Wilson, J.M. Gene therapy vectors based on adeno-associated virus type 1. J. Virol. 1999; 73(5):3994–4003.

Xin, K.Q., Urabe, M., Yang, J., Nomiyama, K., Mizukami, H., Hamajima, K., Nomiyama, H., Saito, T., Imai, M., Monahan, J., Okuda, K., Ozawa, K. and Okuda, K. A novel recombinant adeno-associated virus vaccine induces a long-term humoral immune response to human immunodeficiency virus. Hum. Gene Ther. 2001; 12(9):1047–1061.

Xiong, C., Levis, R., Shen, P., Schlesinger, S., Rice, C.M. and Huang, H.V. Sindbis virus: an efficient, broad host range vector for gene expression in animal cells. Science 1989; 243(4895):1188–1191.

Xu, L., Sanchez, A., Yang, Z., Zaki, S.R., Nabel, E.G., Nichol, S.T. and Nabel, G.J. Immunization for Ebola virus infection. Nat. Med. 1998; 4(1):37–42.

Yim, T.J., Tang, S. and Andino, R. Poliovirus recombinants expressing hepatitis B virus antigens elicited a humoral immune response in susceptible mice. Virology 1996; 218(1):61–70.

Zanin, M.P., Webster, D.E. and Wesselingh, S.L. A DNA prime, orally delivered protein boost vaccination strategy against viral encephalitis. J. Neurovirol. 2007; 13(3):284–289.

Zhang, X., Cassis-Ghavami, F., Eller, M., Currier, J., Slike, B.M., Chen, X., Tartaglia, J., Marovich, M. and Spearman, P. Direct comparison of antigen production and induction of apoptosis by canarypox virus- and modified vaccinia virus Ankara-human immunodeficiency virus vaccine vectors. J. Virol. 2007; 81(13):7022–7033.

CHAPTER

8

DNA Vaccines for Biodefense and Emerging and Neglected Infectious Diseases

Shan Lu, Jill M. Grimes-Serrano and Shixia Wang
Laboratory of Nucleic Acid Vaccines, Department of Medicine, University of Massachusetts Medical School, Worcester, MA, USA

OUTLINE

ABSTRACT

Continued improvements in vaccination technologies have led to remarkable progress in the control of human infectious diseases. Our fundamental view of the nature of a vaccine was changed with the discovery of DNA immunization in the early 1990s when it was determined that the genetic material that encodes for antigens, rather than the actual antigens themselves, can be effective in eliciting an immune response. Given the ever increasing threat of emerging and reemerging infectious diseases and a renewed concern regarding the use of biological agents for bioterrorism purposes, the opportunities that DNA vaccine technology provides could not have come at a more critical time in history. Since its inception, DNA vaccination technology has undergone significant advancements and many candidate human vaccine formulations have already been developed. Improved modes of administration, the use of codon and antigen gene optimization, and the implementation of vaccination DNA prime/boost regimens have led to the quick progression of DNA vaccines from research laboratory benches to human clinical trials. Significant progress has been made in developing DNA vaccines against various biodefense and emerging infectious disease targets, such as HIV-1, influenza, severe acute respiratory syndrome associated coronavirus (SARS-CoV), Ebola, the viral encephalitides, anthrax, plague, and botulism, among others with some already moving into early phase clinical trials with promising results. Having the ability to respond to a potential bioterrorism threat or to some other emerging infectious disease outbreak is crucial and the advancements associated with DNA vaccination technology will allow us to do so in a prompt and rational matter.

INTRODUCTION

In the early 1990s, several research groups discovered, independently, that the direct inoculation of a DNA plasmid coding for a specific protein antigen could elicit an immune response against that antigen (Tang et al., 1992; Robinson et al., 1993; Ulmer et al., 1993; Wang et al., 1993; Lu et al., 1995). DNA or nucleic acid vaccination is now considered to be the fourth generation of vaccines after live attenuated vaccines, inactivated vaccines, and recombinant protein-based vaccines. This new vaccine technology could not have come at a more important time in history given the increasing threat of emerging and reemerging infectious diseases and a renewed concern regarding the use of biological agents for bioterrorism purposes. The need for immunological protection for large populations would require the production or stockpiling of large quantities of vaccines. Future vaccines need to be effective while maintaining a high safety profile. The designs of vaccines have to be flexible to counter antigen alterations and to protect people from more than one type of pathogen. DNA vaccines are an ideal vaccine system to employ in the face of such challenges.

DNA vaccines utilize a eukaryotic expression vector to express a gene product or multiple gene products in vivo. It is this expression system that serves as a delivery system to produce either secreted or cell-associated antigens with the goal of eliciting optimized antigen-specific humoral and cell-mediated immune responses in the host. Over the past 15 years, DNA vaccine technology has gone through significant improvements. The focus of research has gradually shifted from proof-of-concept studies in animal models to the demonstration of immunogenicity in humans. DNA vaccine technology provided researchers with a unique platform that is effective in achieving three important objectives in the development of the next generation of vaccines against emerging infectious diseases and pathogens that have the potential to be used as bioterrorism agents. First, DNA vaccines are a simple alternative option against those pathogens to which there were effective vaccines in the past but the original manufacturing process is no longer considered safe. Second, DNA vaccines are able to elicit improved cell-mediated immune responses in addition to protective antibody responses, which dominate the design of previous generations of vaccines. Finally, DNA vaccines are an ideal tool to screen for new or optimal protective antigens (PA) to formulate new or improved vaccines against emerging and neglected infectious diseases or bioterrorism.

ADVANCEMENTS IN DNA VACCINE TECHNOLOGY

Since the discovery of DNA vaccines in the early 1990s, it is well recognized that this novel immunization technology possesses the following key strengths: (1) the ease of construction and production of the constructs; (2) the native conformation of in vivo expressed protein antigens including any associated post-translational processing of newly synthesized proteins, such as glycosylation; (3) the flexibility of modifying antigen sequences to match with mutated pathogen genes or to produce optimized "designer's antigens"; (4) the ability to elicit both cell-mediated and humoral immune responses; (5) overall good safety profile; and (6) the potential to deliver multiple antigen genes in one construct or as a mixture of multiple plasmids.

Over the past 15 years, the immunogenicity of DNA vaccines has been further improved through the utilization of several key technological advancements including antigen gene codon and coding sequence optimizations, various DNA delivery methods, and the adoption of "prime-boost" regimens.

Codon and Antigen Gene Optimizations

The critical step for DNA vaccination is to establish a high level of gene expression in a mammalian host. One way to improve the immunogenicity of DNA vaccines is to alter the nucleic acid sequences of the viral DNA so that they better complement the genetic makeup of the mammalian host while keeping the same amino acid—a process called codon optimization. The basis for using codon optimization in DNA vaccine design stems from the idea that many amino acids can be coded by more than one codon and that the frequency by which a mammalian host may use a particular codon may differ from usage by a virus. Various studies have used this knowledge in order to improve the ability of mammalian cells to express viral proteins (Haas et al., 1996; Andre et al., 1998; Kotsopoulou et al., 2000; Liu et al., 2004; Apt et al., 2006; Wang et al., 2006a, 2006c). It is this method of altering the genetic sequence that has proven to be effective in increasing the immunogenicity of various DNA vaccines against human immunodeficiency virus (HIV) (Haas et al., 1996; Andre et al., 1998; Kotsopoulou et al., 2000; Liu et al., 2004; Wang et al., 2006a), influenza virus (Wang et al., 2006c), and dengue virus (Apt et al., 2006) due to increased overall antigen production as a result of enhanced mRNA stability (Wang et al., 2006a) and better utilization of host-cell tRNAs (Haas et al., 1996). Due to the adaptation of codon optimization, an otherwise low efficient DNA expression vector can achieve high-level antigen expression (Wang et al., 2006a).

The efficacy of DNA vaccines can be further optimized if the antigen genes are modified. The full-length gene sequences of some candidate antigens cannot be directly used as the antigen gene inserts of the DNA vaccine and may require modification in order to achieve a high-level expression of functional antigenic proteins. For example, an HIV DNA vaccine is not efficient in expressing the full-length glycoenvelope protein (Env) of HIV-1, a gp160 kDa protein; however, two truncated versions of Env, the gp120 and gp140 proteins, that only express the extracellular portion of Env are capable of producing large amounts of Env antigens and eliciting high Env-specific antibody responses (Lu et al., 1998). Sometimes a signal peptide sequence different from the natural leader sequence may also improve antigen expression and immunogenicity of DNA vaccines (Wang et al., 2006a). This process of modifying the original antigen coding sequence is called "antigen engineering" (Lu et al., 1999).

Mode of Administration

The route of DNA vaccine inoculation can greatly alter the immunogenicity of DNA vaccines. Various methods to deliver DNA into the hosts have been developed, using various administration methods, with or without adjuvants. The delivery methodology for DNA vaccines started with the intramuscular injection of so-called "naked DNA" into the host allowing for the uncoated DNA to be taken up into the cells, transcribed, the antigen produced and finally, an immune response elicited. Further development of DNA vaccine technology include facilitated DNA vaccines that employ the use of carriers made of various chemical materials (e.g., lipid polymers and cationic poly(lactide-coglycolide) [PLG] microparticles) (Felgner and Ringold 1989; Wheeler et al., 1996; Stopeck et al., 1998; Herrmann et al., 1999; Klavinskis et al., 1999; O'Hagan et al., 2001; Otten et al., 2005). DNA vaccines, naked or formulated with the above carriers, can be delivered intradermally, intramuscularly, and subcutaneously by a conventional needle injection. Mucosal delivery of DNA vaccines has also been tested including intranasal route; however, in some studies, an emulsion-mediated technique was required in order to induce immunity (Kim et al., 2006).

Other physical delivery methods (such as "gene gun" and electroporation) can improve the efficiency of DNA vaccine uptake by cells and/or increase activation of the immune system. While significant immune responses have been observed in small mammals using the more traditional routes of administration (i.e., intramuscular or intradermal needle injections), these inoculation methods have proven less successful in larger animals and in human studies. On the other hand, the physical delivery approach using particle mediated epidermal delivery (PMED) device has proven to be effective in eliciting immune responses in humans against hepatitis B virus (Tacket et al., 1999; Roy et al., 2000; Rottinghaus et al., 2003), influenza (Drape et al., 2006), and malaria (McConkey et al., 2003).

Prime and Boost Approach

Since the most successful DNA vaccine inoculation method (PMED) for eliciting immune responses in nonhuman primates and in humans is not commercially available for clinical use, it has become important to utilize other techniques and methods to increase the immunogenicity of DNA vaccines that are currently in development. One such method is the prime-boost approach. The rationale underlying this strategy is that gene-based vaccines, presented by a recombinant viral vector or as DNA plasmids, elicit immune responses by producing antigens in vivo; however, these immune responses have not been strong, particularly in humans. In order to increase the levels of immune responses, in particular the cell-mediated immune responses, strategies have been developed to prime the immune system with the DNA vaccine and boost with a recombinant viral vector (Pancholi et al., 2000; Ramshaw and Ramsay, 2000; McShane et al., 2001; Schneider et al., 2001; Gonzalo et al., 2002; Mellquist-Riemenschneider et al., 2003; Woodland, 2004; Seaman et al., 2005; Wu et al., 2005; Perkins et al., 2006), or an attenuated virus (Yuan et al., 2005). This DNA–viral vector combination has proven effective in eliciting high-level cell-mediated immune responses in nonhuman primate studies (Letvin et al., 2004; Boyer et al., 2005) and even in humans (Mwau et al., 2004; Vuola et al., 2005; Goonetilleke et al., 2006; Hanke et al., 2007).

Alternatively, recombinant protein can also be used to boost the immune system that has been primed with DNA vaccines (Barnett et al., 1997; Letvin et al., 1997; Richmond et al., 1998; Pal et al., 2005, 2006; Wang et al., 2005b, 2006b; Lu, 2006; Law et al., 2007). The DNA prime-protein boost combination has been proven effective in eliciting balanced humoral and cell-mediated immune responses even in human studies (Wang et al., 2008a). The DNA prime or protein boost components when used alone have been shown to be less effective in inducing immune responses when compared to prime plus boost approach (Wang et al., 2005b, 2006b).

REPRESENTATIVE DNA VACCINES FOR BIODEFENSE AND EMERGING INFECTIOUS DISEASES

A large number of studies, using various animal models, have been conducted over the past 15 years to examine DNA vaccine immunogenicity against a wide range of pathogens including viruses, bacteria, and parasites. Contrary to the popular belief, DNA vaccines are more complicated than simply inserting a randomly selected pathogen gene into a DNA vaccine vector. The process of developing a successful DNA vaccine is no different from any scientific discovery: trial and error. In the following sections, results from selected DNA vaccine studies are presented to illustrate various aspects of the development of DNA vaccines against many important biodefense and emerging infectious disease targets over the past decade.

DNA Vaccines against Viral Agents

Human Immunodeficiency Virus Type 1 (HIV-1)

Acquired immunodeficiency syndrome (AIDS) is the number one emerging infectious disease in the 20th century. It continues to be a worldwide pandemic in the new millennium. The development of an effective AIDS vaccine has been proven more challenging than initially anticipated. There does not appear to be natural adaptive immune protection against HIV-1 infection; hence, the correlate of protection to be induced by a vaccine is uncertain. As a retrovirus, HIV-1 integrates into the genome of host dividing cells, which in turn becomes the source of continuing infection. Its proteins go through constant mutations due to the lack of proofreading function of retroviral polymerases. The structure of envelope glycoprotein (Env), the immunodominant antigen of HIV-1, is very complicated and highly variable to escape from antibody responses.

The recombinant protein HIV-1 vaccines in the form of gp120 antigens have failed to protect a high-risk population in two phase III efficacy trials (Flynn et al., 2005; Gilbert et al., 2005; Pitisuttithum et al., 2006). While the live attenuated vaccination approach was successful in simian immunodeficiency virus (SIV) studies in nonhuman primates (Desrosiers et al., 1989; Daniel et al., 1990), due to safety concerns this approach has little hope of being tested in humans. HIV was among the first pathogens for which the DNA vaccine approach was explored (Wang et al., 1993; Lu et al., 1995) and the HIV vaccine field quickly became the major testing ground for the optimization and application of DNA vaccines. DNA vaccines have been used to express many key HIV-1 antigens including Env, Gag, Pol, Nef, Rev, and Tat for the induction of both antibody- and cell-mediated immune responses. Currently, DNA vaccines are one of the key components in almost every major HIV vaccine development program. DNA vaccines are used mainly as priming immunizations and are typically combined with either a viral vector boost (Mwau et al., 2004; Graham et al., 2006; Tavel et al., 2007) or a protein boost (Wang et al., 2008a). More human studies have been planned or are ongoing. The results from these studies will provide important guidance for the future development of DNA vaccines for other emerging infectious disease and biodefense applications.

Influenza Viruses

Influenza virus infection remains a major health threat to both humans and animals. Influenza A viruses infects both humans and other animal species, including swine, horses, and avian. The predominant human influenza virus vaccine is the trivalent inactivated influenza vaccine (TIV), which incorporates the predominant viruses that are predicted to circulate during the next influenza season. The vaccine typically includes an H1 subtype and an H3 subtype from influenza A plus an influenza B virus. The immunogenicity of these vaccines is relatively low and requires annual booster immunizations to maintain the levels of protective immune responses and to develop new protective antibodies against mutated viral antigens that occasionally emerge. A newly developed cold-adapted live influenza virus vaccine has much improved immunogenicity but it cannot be used for young children and older populations due to a lack of safety data in these groups. This restriction greatly reduces the utility of this new vaccine because young children and older people are particularly vulnerable to influenza infection.

The emerging threat of avian influenza as a potential cause of a future pandemic has raised issues of vaccine efficacy, production, and stockpiling. The old influenza vaccine technologies may not be sufficient in facing new challenges.

Influenza DNA vaccine development was another early key area for the field. Studies demonstrated the immunogenicity of DNA vaccines using either nucleocapsid (NP) or hemagglutinin (HA) antigens (Ulmer et al., 1993; Robinson et al., 1993; Fynan et al., 1993). A landmark study in the early 1990s showed that a DNA vaccine expressing NP was effective in eliciting NP-specific CTL and protection in mice from a subsequent challenge with a heterologous strain of influenza A virus, as measured by decreased viral lung titers, inhibition of mass loss, and increased survival (Ulmer et al., 1993). Subsequent studies from this and other groups confirmed the protective immune responses induced by NP DNA vaccines (Ulmer et al., 1998; Cox et al., 2002) or by the combination of DNA prime-adenoviral vector boost vaccines (Epstein et al., 2005), but there were also reports showing that NP DNA vaccine alone was not effective in protecting pigs against challenges with homologous virus (Macklin et al., 1998). However, later studies in ferrets and non-human primates demonstrated the efficacy of a DNA vaccine encoding three proteins, NP, HA, and M1 antigens (Donnelly et al., 1995, 1997). DNA vaccine expressing other influenza viral antigens have also been reported including the co-expression of both M1 and M2 antigens (Okuda et al., 2001), the co-delivery of NP and M2 DNA vaccines plus a low dose of HA-expressing DNA vaccines (Jimenez et al., 2007), and the use of NA-expressing DNA vaccines (Chen et al., 1998, 2000). Various levels of protection with these influenza DNA vaccines were achieved in mouse models. Protection against influenza B virus infection was also achieved with DNA vaccines expressing HA or NA antigens but not with NB or NP DNA vaccines (Chen et al., 2001).

The HA-expressing DNA vaccines appear to be the most protective influenza DNA vaccine and protection elicited by HA DNA vaccines correlates with titers of neutralizing antibody (Robinson et al., 1997). However, recent studies suggested that the HA antigens from H1 and H3 serotypes of influenza A viruses require different antigen designs for the induction of optimal protective antibody responses (Wang et al., 2006c). This may further affect the selection of HA antigens from avian influenza viruses. Until now, HA DNA vaccines have successfully elicited antibody responses and protection against several avian influenza viruses including H5N1 (Kodihalli et al., 1999), H5N2 (Kodihalli et al., 1997; Jiang et al., 2007), and H9N2 (Qiu et al., 2006). The H5 HA DNA vaccine has been shown to induce immune protection against not only a homologous, lethal H5 virus but also two antigenic variants in chickens (Kodihalli et al., 1997). The level of protection may vary in different animal models. Despite the success of inducing protective immunity against heterologous viral challenge in chickens, an HA DNA vaccine encoding for heterologous viruses did not prevent infection in mice. However, this HA DNA vaccine did prevent death associated with infection (Kodihalli et al., 1999).

New DNA vaccine technologies have also been incorporated into the design of improved influenza DNA vaccine designs. Codon optimization was shown to improve the immunogenicity and protective antibody responses (both HI and microneutralization antibodies) of HA DNA vaccines (Wang et al., 2006c; Jiang et al., 2007). Consensus-based DNA vaccines against H5N1 avian influenza viruses have been produced with initial results showing the possibility of inducing cross-reactive cellular immunity (Laddy et al., 2007). However, whether such consensus antigens can induce protective immune responses against different H5N1 viruses remains unconfirmed.

One of the key contributions of influenza DNA vaccine research was the demonstration that an HA DNA vaccine could induce HA-specific antibody responses induced in humans. Using a particle-mediated epidermal delivery (PMED) device, 3 groups of 12 healthy adult subjects received a single dose of 1, 2, or 4 μg of an HA DNA vaccine. The PMED influenza DNA vaccine elicited serum hemagglutination-inhibition antibody responses at all three dose levels with the highest and most consistent responses in subjects vaccinated with the highest dose. On day 56, sera HA-specific antibody responses in both 2 and 4 μg groups reached the levels required for influenza vaccine approval established by the Committee for Proprietary Medical Products (CPMP) in the European Union (Drape et al., 2006). This result was important because it demonstrated that a DNA vaccine was capable of eliciting protective immune response levels. One notable factor in this study was the use of the "gene gun" delivery method, which has proven to be more effective than needle injection of DNA vaccines in inducing immune responses in humans. The availability of the gene gun or other epidermal delivery methods will be important in the successful development of DNA vaccines for biodefense and emerging and neglected diseases.

Smallpox

The original smallpox vaccine, based on the vaccinia virus (VACV), facilitated the worldwide eradication of smallpox with the last case of natural smallpox infection reported in Somalia in 1977 (WHO, 1980). Despite the elimination of naturally occurring smallpox infection, the threat of its use as a biological weapon is compelling enough to develop a vaccine strategy to protect the large percentage of the population that no longer has immunity against this disease. Poxviruses are large viruses with a genome that encodes about 200 proteins. The size of the virus has complicated efforts to delineate the major PAs. Furthermore, the relative importance of humoral and cellular immunity in vaccine-induced protective immunity is uncertain.

In well-designed animal studies a DNA vaccination approach has recently identified several protective pox antigens. There are two forms of infectious poxvirus: the intracellular mature virus (IMV) and the extracellular enveloped virus (EEV). Studies have shown that vaccinia virus IMV-specific antigens, A27, L1, and D8, and EEV-specific antigens, A33 and B5, are immunogenic and protective, albeit variably, against VACV infection in animal models (Galmiche et al., 1999; Hooper et al., 2000, 2004; Fogg et al., 2004; Pulford et al., 2004; Sakhatskyy et al., 2006, 2008; Xiao et al., 2007). These antigens induced high-titer serum antibodies and/or neutralizing antibodies against the vaccinia virus (Galmiche et al., 1999; Hooper et al., 2000, 2003, 2004, 2007; Heraud et al., 2006; Sakhatskyy et al., 2006, 2008). A four-gene combination DNA vaccine that encodes two IMV (A27 and L1) and two EEV (A33 and B5) antigens was protective against VACV challenge in mice and against monkeypox virus challenge in nonhuman primates (Hooper et al., 2003). In a separate study, IMV antigen D8 was shown to be a highly immunogenic antigen that induced high-level protection, alone or in combination with the other pox antigens (Sakhatskyy et al., 2006, 2008). However, these pox DNA vaccines, as well as the recombinant protein-based pox subunit vaccines, were not as protective as the live-attenuated vaccinia virus vaccine unless polyvalent formulations and/or multiple immunizations are used (Hooper et al., 2000, 2003; Fogg et al., 2004; Pulford et al., 2004; Sakhatskyy et al., 2006, 2008).

Only limited studies have measured cell-mediated immune responses following immunization with various DNA vaccines against VACV challenge (Otero et al., 2006). A recent study demonstrated that mice immunized with the DNA plasmid encoding for the B5R antigen of VACV elicited strong EEV-induced IFN-γ responses (Pulford et al., 2004). Interestingly, contrary to previous studies, this study did not observe increased antibody responses in vaccinated mice but rather found that cell-mediated responses correlated with protection. In a DNA prime-protein boost study, rhesus macaques were immunized with a four-antigen formulation (L1R, A27L, A33R, and B5R) in the form of either DNA vaccines alone, *Escherichia coli* produced recombinant protein vaccines alone, or a combination of both vaccines. Animals that received the prime-boost immunization achieved the best protection (Heraud et al., 2006).

The DNA vaccination approach also provides a unique technical advantage in developing vaccines that are more closely related to smallpox antigens. It can permit the direct cloning and expression of antigen from variola major (Sakhatskyy et al., 2008; Aldaz-Carroll et al., 2007) since the protective antigen genes can be chemically synthesized, even codon optimized, without directly working with the highly pathogenic variola major viruses, which are no longer available for research applications. In summary, DNA vaccination is particularly useful for the next generation of smallpox vaccines based on subunit antigens. It can deliver multiple protective antigens in one mixed formulation, elicit both antibody- and cell-mediated immune responses, and optimize the antigen sequences and designs for the protective antigens. Optimized genes can also be used to produce recombinant protein-based subunit vaccines.

Severe Acute Respiratory Syndrome Associated Coronavirus (SARS-CoV)

Severe acute respiratory syndrome (SARS), an acute respiratory disease cause by a newly recognized human coronavirus (SARS-CoV), was first reported in Asia in 2003. During the 2003 outbreak, approximately 8000 people worldwide became ill with 774 fatalities before the outbreak was finally contained (CDC, 2005). SARS is the first major emerging infectious disease in the 21st century. Although no final human vaccine has been developed, several candidate antigens have been discovered and some have been tested as DNA vaccines against SARS infection. Both the nucleocapsid protein (N) and the envelope spike protein (S) have been examined for their ability to induce protective immunity against SARS-CoV in mice.

DNA plasmids, encoding the N protein, have been shown effective in eliciting both humoral and cellular immunity in mice after two or three immunizations (Zhu et al., 2004; Zhao et al., 2005). Since the N antigen is not known to be a target for antibody-mediated protection, the majority of SARS-CoV DNA vaccines have incorporated the S protein as the key protective antigen. No matter whether the entire S protein or fragments of the S protein are incorporated into the DNA vector for immunization, significant levels of S-specific immune responses were elicited including high-titer binding antibody (Yang et al., 2004; Kong et al., 2005; Zakhartchouk et al., 2005; Huang et al., 2006; Zhao et al., 2006; Hu et al., 2007; Wang et al., 2005b) or neutralizing antibody responses (Yang et al., 2004; Woo et al., 2005; Zakhartchouk et al., 2005; Wang et al., 2005b). In addition, S-specific cellular immune responses including both CD4+ and CD8+ T cell responses were detected with S DNA vaccines (Yang et al., 2004; Huang et al., 2006; Zhao et al., 2006; Hu et al., 2007). Even greater levels of immunity were observed when these S DNA vaccines were included as part of a prime-boost regimen (Kong et al., 2005; Woo et al., 2005; Zakhartchouk et al., 2005) or when a cytokine (IL-2) was included as the molecular adjuvant (Hu et al., 2007).

Although it has not been possible to determine the efficacy of most SARS DNA vaccines due to the limited availability of animal testing facilities capable of working with the SARS-CoV, it has been shown that some S DNA vaccines (i.e., deleted transmembrane domain or a truncated cytoplasmic domain) were effective in significantly reducing the viral load in the lungs of infected mice (Yang et al., 2004). However, a limitation of the SARS-CoV mouse model is its lack of clinical endpoints (e.g., disease or death) and the field is further limited by the lack of a generally accepted nonhuman primate model.

Results have been promising in the development of a DNA vaccine against SARS-CoV. However, more studies are needed, using better animal models that can assess the ability of the experimental vaccines to protect against infection and disease. It is important to note that use of the DNA vaccine platform proved to be nimble in facilitating the rapid development of highly immunogenic candidate vaccines against the sudden emergence of a new human infectious disease.

The Filoviruses: Ebola and Marburg

Infection with either Ebola or Marburg typically results in severe hemorrhagic fever in both humans and nonhuman primates. Four species of Ebola have been identified (i.e., Zaire, Sudan, Reston, and Ivory Coast) with the first known case occurring in Zaire in 1976. Marburg virus was identified in 1967 and named after the site where the infection occurred, Marburg, Germany. There is no effective prophylaxis or treatment for infection with either virus. While filoviruses are uncommon causes of human infection, the development of filovirus vaccines are warranted based on their remarkable lethality with mortality rates ranging from 23% (Marburg) to 90% (Ebola). There has been limited work on the development of live attenuated Ebola virus vaccines, but safety issues make this approach somewhat problematic.

Significant progress has been made in the development of DNA vaccines for both Ebola and Marburg viruses including the completion of a recent phase I human trial for an Ebola DNA vaccine. These vaccine strategies have targeted various viral proteins including both the secreted (sGP) and transmembrane (GP) forms of the glycoprotein and the nucleoprotein (NP). These DNA vaccines were tested in various animal models including mice, guinea pigs, and nonhuman primates, and in humans with promising immunogenicity results. In an early study, all three viral proteins were tested to determine their immunogenicity in mice and their protective efficacy in guinea pigs (Xu et al., 1998). Three $50\mu g$ injections of the NP and sGP DNA plasmids were effective, although to varying levels, in eliciting an antibody response while the GP DNA did not elicit a neutralizing response. However, the NP construct induced only minimal CTL responses, whereas the sGP and GP plasmids induced strong CTL responses. Furthermore, guinea pigs challenged approximately 2 months after the first immunization showed nearly complete protection: GP (6/6), sGP (5/6), and NP (4/4). Similar results were observed in mouse protection studies with the GP protein although as many as four immunizations were required (Vanderzanden et al., 1998; Riemenschneider et al., 2003). Likewise a GP DNA prime-recombinant baculovirus-derived GP protein boost regimen afforded guinea pigs good protection. A separate study showed that DNA immunization and subsequent boost with adenovirus vectors that encoded either GP, NP, or a combination of the two conferred 100% protection in guinea pigs (Sullivan et al., 2000). Furthermore, antibody titers were increased in mice that received the GP DNA prime plus adenovirus vector boost when compared to DNA alone. Since the Ebola virus has to be modified in order to produce disease in rodents, primate models of infection would more closely emulate infection in humans. After three injections of naked GP DNA vaccines, several months of rest, and recombinant adenovirus based vaccine boost, antibody titers were increased 10- to 20-fold when compared to DNA immunization alone and there was 9- to 20-fold increase in cellular immune responses that appear to be driven by CD4+ T cells. This vaccination regimen was able to confer protection in four out of four monkeys and sterilizing immunity was observed in three out of four.

Recently, a phase I clinical trial was conducted to determine the safety and immunogenicity of a DNA vaccine for the Ebola virus in healthy volunteers

(Martin et al., 2006). The DNA vaccine contained three plasmids in equal concentrations for the NP (Zaire strain), and two GP (one from the Zaire strain and the other from the Sudan/Gulu strain). This vaccine formulation was safe and well tolerated. DNA vaccine formulation elicited antigen-specific antibody responses to at least one of the three antigens in 100% of vaccinees and 19 out of 20 vaccines showed an antibody response to either GP antigen at one or more time points during the vaccination schedule. Similar to what was observed for monkey (Sullivan et al., 2000), CD4+ T cell responses were more robust than CD8+ responses with the GP antigens producing the greatest response when compared to NP.

The DNA vaccine platform has also been used to develop a vaccine against the Marburg virus although few studies have been conducted to date. Marburg virus GP-expressing DNA vaccine was administered to guinea pigs and serum antibodies were detected after three immunizations. This vaccine prevented death in 8 out of 10 animals that exhibited a 10-fold or higher increase in anti-Marburg virus antibodies 1 month after infection. However, a combination DNA prime plus baculovirus-expressed recombinant GP boost in adjuvant (one DNA prime and two protein boosts) conferred complete protection whereas neither vaccine modality did so when administered alone (Hevey et al., 2001). A second study observed complete protection in guinea pigs and protection in two of three monkeys vaccinated with a Marburg virus GP DNA plasmid (Riemenschneider et al., 2003). The antibody responses observed in the two monkeys that survived challenge were not significantly different from those in the monkey that did not survive, indicating that antibody levels alone do not necessarily predict protection against Marburg infection.

Results from the filovirus vaccine studies are promising in that they all indicate that vaccination with related DNA vaccines which encode for various Ebola or Marburg antigens can induce both cellular and humoral immune responses and that these vaccines can confer some level of protection from infection. More human trials will determine the efficacy of these vaccines.

Flaviviruses: West Nile Virus, Japanese and Tick-Borne Encephalitis Viruses, Dengue Virus

West Nile Virus At the end of the summer of 1999, there was an outbreak of viral encephalitis in the New York City and surrounding areas, which was later determined to be caused by West Nile virus (WNV). This outbreak continued in the subsequent years in the United States in humans and in horses. No human or veterinary vaccines existed for WNV at the time of the initial outbreak. Different versions of DNA vaccines were developed and they all showed good immunogenicity. One version expresses the WNV capsid protein, and in vivo study in mice showed Th1 type immune responses in immunized animals (Yang et al., 2001). Another used a novel approach to express a full-length infectious Kunjin virus RNA. Kunjin virus is a genetically stable Australian flavivirus originally shown to be genetically and antigenically very closely related to several WNV strains, which later was reclassified as a WNV (Heinz et al., 2000). Sequence analysis revealed 98–99% amino acid homology between Kunjin and NY99, the WNV causing the initial 1999 New York outbreak. By using a further modified molecular clone with mutated *NS1* gene, an attenuated Kunjin strain was produced and used as a vaccine. The use of DNA vaccine directing in vivo transcription of the full-length attenuated but infectious KUN viral RNA in mice led to full protection against the WNV NY strain (Hall et al., 2003).

Scientists at US CDC and Fort Dodge Laboratories, Inc., a division of Wyeth, developed the pCBWN DNA vaccine that expresses the WNV prM and E proteins (Davis et al., 2001). A single intramuscular injection of pCBWN DNA vaccine induced protective immunity, preventing WNV infection in both mice and horses (Davis et al., 2001). This vaccine was licensed by the US Department of Agriculture on July 18, 2005, becoming the first DNA vaccine approved by a regulatory agency. In granting full licensure, USDA's Center for Veterinary Biologics determined that the vaccine's safety and efficacy have been satisfactorily demonstrated.

In April of 2006, a DNA vaccine targeting WNV, developed by scientists at the NIH's Vaccine Research Center (VRC) in collaboration with Vical, Inc., entered into a phase I clinical trial with a targeted completion date of October of 2007 (NIH, 2007). This experimental vaccine encodes two key WNV surface proteins, the precursor transmembrane (prM) protein and the envelope (E) protein. The vaccine was administered via a Biojector 2000 device. Participants received three injections (i.m., 1 ml of vaccine per immunization for a total of 3 ml) at study days 0, 28, and 56. The primary study endpoint is vaccine safety, and secondary immunogenicity endpoints include ELISA and neutralizing antibody responses and ELISPOT and ICS staining assays for WNV-specific T cell responses. Results of the phase I clinical trial are not yet available (Martin et al., in press).

Japanese and Tick-Borne Encephalitis The viral encephalitides, in particular, Japanese encephalitis (JE) and Tick-borne encephalitis (TBE), are transmitted to humans and animals by infected mosquitoes or ticks. Infection with these viruses results in various symptoms ranging from fever to encephalitis. While there are inactivated vaccines against these infections, the immunization regimens require multiple doses without producing long-term immunity. Repeated immunizations increase the cost of vaccination campaigns in the developing countries. Some of these inactivated vaccines are not licensed in the United States. Efforts have been made to utilize DNA vaccine technology in order to produce a vaccine against these encephalitides with improved efficacy, greater ease, and lower cost of production and administration.

JE, a virus widely distributed throughout Asia, causes disease in the human central nervous system. Three viral proteins have been reported to provide protective immunity: prM, E, and NS1. DNA vaccines encoding for these viral proteins have been successful in inducing antibodies and protection against JE in mouse models. In one report an NS1 DNA vaccine was shown to afford greater protection from JE challenge and to induce a higher level of antibodies compared to a DNA vaccine encoding the E and prM proteins (Lin et al., 1998). Another study found that a construct encoding prM and E produced neutralizing antibodies and protected mice against JE challenge while constructs encoding capsid (C) or non-structural proteins, NS1, NS2A, NS2B, NS3, or NS5 induced cytotoxic T lymphocyte responses but afforded only partial protection against JE challenge (Konishi et al., 2003a). These results were interpreted to indicate the relative importance of antibody vis-à-vis T cells in vaccine-induced protection. Other investigators reported that a DNA vaccine encoding both the E and prM proteins was as effective or elicited better responses when compared to the commercially available JE vaccine, JEVAX (Chang et al., 2000). Additional studies have shown that needle-free injector administration of DNA vaccines encoding E and prM proteins elicited increased neutralizing antibodies earlier when compared to intramuscular administration in monkey (Tanabayashi et al., 2003) and mouse (Konishi et al., 2003b, 2003c).

Prime-boost regimens have proven to be superior to other methods of maximizing vaccine immunogenicity. Two studies showed that boosting an E/prM DNA vaccine formulation with either sub-viral extracellular particles (Imoto and Konishi, 2005) or inactivated commercially available JEVAX (Konishi et al., 2003b, 2003c) was superior to either vaccine modality alone in inducing neutralizing antibody. The DNA prime-JEVAX boost regimen induced long-lasting immunity, as titers were still 1:80 to 1:160, as of 21 weeks following the last immunization (Imoto and Konishi, 2005). Furthermore, DNA vaccines encoding for prM/E viral proteins afforded greater cross protection against heterologous JE viral strains when compared to inactivated virus vaccines (Wu et al., 2003a).

TBE is most frequently caused by Russian spring summer encephalitis (RSSE) and Central European encephalitis (CEE). DNA vaccines developed against TBE have been successful in animal models in eliciting protective immune responses against viral challenge. As with the other flaviviruses, the most effective vaccines have encoded the prM and E genes. DNA vaccines expressing the prM and E genes of RSSE and CEE, delivered by a gene gun, have been shown to protect both mice (Schmaljohn et al., 1997) and monkeys (Schmaljohn et al., 1999) against homologous and heterologous viral challenge while inducing long-lasting neutralizing antibody responses. Protection appears to be mediated by neutralizing antibody based on passive transfer studies in mice using high neutralizing titer antibodies generated in monkeys by RSSE or CEE DNA immunization (Schmaljohn et al., 1999).

Dengue Viruses Currently, there is no approved vaccine against dengue viruses (DEN). These viruses, spread by the *Aedes aegypti* mosquito, are endemic throughout tropical and sub-tropical regions of the world. There are approximately 100 million infections per year, resulting in both dengue fever and the more serious, dengue hemorrhagic fever (DHF). Although various vaccine approaches including live attenuated, inactivated whole virion and recombinant protein-based subunit vaccines have been tested, the presence of four serotypes of DEN and complicated immunopathologic mechanisms involved in DHF have greatly reduced the speed of developing an effective DEN vaccine. Several DNA vaccine formulations have been developed against all four serotypes (DEN-1, DEN-2, DEN-3, and DEN-4), and in some cases, a tetravalent formulation has induced memory immune responses against both homologous and heterologous viral challenges (Konishi et al., 2006). Typically, the DNA vaccines that most effectively elicit both humoral and cell-mediated immune responses contain the DEN premembrane/membrane (prM) and envelope (E) genes. Immunization with plasmids encoding these genes have been shown to engender antigen-specific antibody and neutralizing antibody responses and protection against DEN-1 in mice (Raviprakash et al., 2000a; Konishi et al., 2003b, 2003c) and nonhuman primates (Kochel et al., 2000; Raviprakash et al., 2000b) and against DEN-2 in mice (Konishi et al., 2000, 2003b, 2003c; Putnak et al., 2003) and nonhuman primates (Putnak et al., 2003). Additional studies have examined the ability of a tetravalent DNA vaccine formulation, encoding the prM and E genes of all four dengue serotypes, to induce immune responses and found that antibodies specific for each serotype could be induced following immunization (Mota et al., 2005; Apt et al., 2006; Konishi et al., 2006; Raviprakash et al., 2006). Furthermore, immunization with such a tetravalent DNA vaccine was able to elicit neutralizing antibody responses 4 days following the viral challenge against both the homologous virus used for the challenge assay and against the heterologous viruses (i.e., not used for challenge) (Konishi et al., 2006).

The ability of the dengue virus nonstructural protein 1 (NS1) to elicit immune response has also been examined with contradictory results. In one study, a DNA vaccine encoding for the NS1 failed to induce antigen-specific antibodies after one to three immunizations; however, greater responses were observed following challenge with DEN-2 in vaccinated mice when compared to the unvaccinated group (Wu et al., 2003b). In contrast, mice vaccinated with a DNA vaccine based on the NS1 protein from DEN-2 displayed high levels of antibodies (Costa et al., 2006). While it appears as though these two DNA vaccines encoded the same antigen from the DEN-2 virus, slight differences in immunization schedule and/or dosing may have led to the different results. Despite the fact that antibodies were not detected after one to three immunizations with the NS-1 from DEN-2 (Wu et al., 2003b), the NS-1 DNA vaccine conferred protection against lethal challenge in both studies. Furthermore, newborn mice born to dams that were either infected with the DEN-2 virus (100% of litters) or immunized with the NS-1 DNA vaccine (75% of litters) were also protected against the dengue virus following challenge (Wu et al., 2003b).

Alphaviruses: The Equine Encephalitides

The equine encephalitides, including Venezuelan equine encephalitis (VEE) and western equine encephalitis (WEE), are mosquito-borne viruses that can be transmitted to birds, horses, and humans. The clinical findings of these infections range from mild flu-like symptoms to encephalitis, coma, and death. While equine vaccines against some of these viruses have been developed, no human vaccines are currently available. DNA vaccine technology has been used to develop vaccines against VEE and WEE, which is of particular importance as these viruses have the potential to be used as biological warfare agents. In a first study, DNA immunization with a plasmid encoding the major structural proteins of WEE (capsid, E3, E2, 6K, and E1) afforded protection to 50–100% of mice challenged with a lethal dose of one of three strains of WEE. Furthermore, immunized mice showed a strong T-cell proliferation response to purified E2 protein and a partial response to E1 when compared to controls (Nagata et al., 2005). A DNA vaccine has also been developed against VEE. In this study, a replication-defective adenovirus (Ad5) was used to boost the immune responses following the delivery of the DNA vaccine that expressed a region encoding for the E3, E2, and 6K structural proteins (Perkins et al., 2006). Following DNA immunization, no difference was observed in VEE virus-specific IgG response; however, following one intranasal boost with a homologous Ad-based vaccine a significant increase in VEE virus-specific IgG was observed. The prime-boost strategy also significantly enhanced protection against aerosol VEE challenge.

Rabies

Infection with the rabies virus causes acute encephalitis with an extremely high fatality rate. Both inactivated and live-attenuated human rabies vaccines exist but each has its own limitations. DNA vaccination has proven successful against rabies infection in various animal models. In a first series of studies, mice immunized with the rabies glycoprotein displayed rabies virus glycoprotein-specific cell-mediated and humoral immune responses, in addition to complete protection against viral challenge (Xiang et al., 1994, 1995). Subsequent studies showed that a DNA vaccine encoding for a rabies glycoprotein induced a 176-fold increase in geometric mean neutralizing antibody titers following a boost 190 days after the primary immunization in nonhuman primates (Lodmell et al., 1998a, 1998b). Additional studies by this group showed that gene gun administration afforded mice protection against rabies virus challenge and that protective levels of antibodies persisted for over 300 days (Lodmell et al., 1998a, 1998b). They also showed that different prime-boost regimens elicited varying levels of neutralizing antibody responses: priming with a DNA vaccine, a recombinant vaccinia virus vaccine (RVV), or the commercially available inactivated rabies vaccine followed by boosting with either the DNA or inactivated vaccines resulted in rapid and long-lasting antibody responses (Lodmell and Ewalt, 2000). This group also demonstrated that DNA vaccination could be used as an effective post-exposure vaccine (Lodmell and Ewalt, 2001; Lodmell et al., 2002).

Arenaviruses

The arenaviruses are a family of viruses that are usually transmitted from rodents to humans and include viruses such as lymphocytic choriomeningitis (LCM), Lassa virus, Junin virus, and Machupo virus. Infection with these viruses presents as a wide range of symptoms including fever and flu-like symptoms, meningitis, encephalitis and severe, fatal hemorrhagic fever. Few studies have been conducted in the development of DNA vaccines against these viral agents. In one study, a single immunization with a DNA plasmid encoding the full-length Lassa nucleoprotein was able to induce CD8+ T cell responses in mice and afforded protection against two arenaviruses, LCMV and the Pichinde virus (Rodriguez-Carreno et al., 2005). Unfortunately, DNA immunization against LCMV in another study did not show favorable results following intracranial viral challenge. Not only were some vaccinated mice not afforded protection against viral challenge but they also displayed greater immunopathology instead as evidenced by clinical symptoms of encephalitis and eventual death (Zarozinski et al., 1995). This study represented a rare example of where a DNA vaccine may worsen the outcome of viral infection. Although it is believed that such unexpected results are part of the unique immunopathogenesis of LCMV infection, careful selection of PAs may also be important for the development of DNA vaccines.

DNA Vaccines against Bacterial Agents

Anthrax

Anthrax is one of the early model pathogens used to assess the feasibility of applying DNA vaccine technology to the development of a vaccine against bacterial infection. During the early days of DNA vaccine research, many considered DNA vaccine as a useful tool for the induction of cell-mediated immune responses but did not appreciate the potential of DNA vaccine to induce high-quality protective antibody responses. Because antibody plays a central role in the control of most of the bacterial infections, initially there were only limited DNA vaccine studies for bacterial agents, and the detailed analysis of DNA vaccine against anthrax provided an important milestone in this area.

Anthrax infection, caused by *Baccillus anthracis*, is a disease with high mortality. It attracted great attention in the United States following intentional release of the agent in 2001. Although a licensed vaccine against anthrax has been developed (Bioport Corporation, Lansing, MI), multiple immunizations are required in order to maintain protective levels of immunity, and safety of previous manufacturing process has been an issue. In a first study to establish the possibility of using a DNA vaccine to induce a protective immune response against anthrax, Gu et al. (1999) developed a DNA vaccine that contained a portion of the protective antigen (PA), the protein responsible for initial binding of the bacteria to the target cell. Antibodies against the PA have been confirmed to protect guinea pigs and rabbits against anthrax challenge (Little et al., 1997; Kobiler et al., 2002; Riemenschneider et al., 2003; Mohamed et al., 2005). After multiple inoculations with the PA-expressing DNA plasmid, anti-PA IgG antibodies were observed in all immunized mice, levels of which increased with subsequent boosting. Using an in vitro neutralization assay, this study showed that pooled serum from twice-boosted immunized mice was 100% protective against anthrax at a 1:20 dilution while 50% protection was observed at a 1:100 dilution. In an in vivo protection assay, the authors found that mice immunized with the plasmid PA DNA were completely protected against challenge while no control mice survived. Subsequent studies showed that plasmid DNA prime encoding for the PA antigen followed by a recombinant protein (rPA) boost significantly increased antibody levels and protective efficacy against anthrax challenge in both mice (Williamson et al., 2002) and rabbits (Galloway et al., 2004). However, increases in antibody titers and protective efficacy were also observed in rabbits without the recombinant protein boost (DNA alone) and these effects were long-lasting (Hermanson et al., 2004). In an examination of passive immunotherapy of anthrax infection, rabbits were immunized with a codon-optimized plasmid DNA vaccine encoding PA, with the most effective immunization strategy being the twice DNA prime + one time protein boost regimen. Immune rabbit sera, shown to contain ELISA and neutralization antibodies, were transferred to mice either 1 h prior to or 1 h after lethal anthrax challenge. All mice that received the hyperimmune rabbit sera 1 h prior to anthrax challenge survived 14 days after challenge (end of observation period), while in mice that received the sera 1 h after challenge, only one out of 5 mice died (2 days after challenge), indicating that protection against anthrax could be achieved through passive administration of antibodies that resulted from DNA immunization (Herrmann et al., 2006). Together, these studies indicate that DNA immunization was effective in producing high-titer antibody responses, which conferred protection against aerosolized anthrax challenge in both mice and rabbits.

Plague

While *Yersinia pestis* (*Y. pestis*), the causative agent of plague, is endemic in many rodent populations throughout the world, it is of great concern due to its potential as a biological weapon. Plague can be transmitted by flea bite causing bubonic plague or through aerosolization to cause pneumonic plague. It is pneumonic plague that is the greatest concern for biodefense as this form of the disease progresses very rapidly, is highly transmissible, and requires therapy shortly after infection. Without early intervention, pneumonic plague infection is most certainly fatal. These features of pneumonic plague along with its possible use as a biological weapon present a strong argument for the development of an effective vaccine.

Currently, there is no known effective clinical vaccine against pneumonic plague. While a variety of whole-cell, killed plague vaccines were developed, their efficacy in protection against pneumonic plague is limited in small animal models (Williamson et al., 1997, 2000). DNA vaccine technology offers advantages in establishing a safe and effective platform by which protection against a variety of plague antigens can be achieved. A first study compared two DNA vaccine delivery methods (i.e., gene gun delivery of 0.6 or 4μg plasmid DNA versus intramuscular or intradermal delivery of 50μg plasmid DNA) for the administration of a plasmid DNA encoding for the V antigen of *Y. pestis*. Antibody responses, obtained at 21 days post vaccination, were observed in all vaccinated groups. However, the highest geometric mean titers were observed in mice that were vaccinated with 4μg DNA via gene gun, with the least effective being intradermal administration of 50μg DNA (Bennett et al., 1999). Very little protection was afforded in mice inoculated (i.m. or gene gun) with this "optimized" V antigen-expressing DNA vaccine (addition of a CMV promoter) as only two out of six mice (gene gun) survived a subcutaneous challenge with *Y. pestis* (Garmory et al., 2004).

Greater success was achieved when mice were administered a DNA vaccine with a modified V antigen (LcrV) using a human tissue plasminogen activator (tPA) signal sequence (Wang et al., 2004b). In this study, not only were strong V antigen-specific antibodies elicited but mice were also protected against an intranasal challenge with *Y. pestis*. This is of particular importance as this model reflects infection with the more deadly mucosal challenge of plague. The tPA-V DNA vaccine was able to induce excellent secreted V antigens in oligomeric form, which has been hypothesized to be important in eliciting protective antibody responses.

In addition to the V antigen, immunity to the F1 capsular protein has been shown to afford protection against plague and F1-expressing DNA vaccines have been reported effective in eliciting protection. Grosfeld et al. (2003) compared three F1 DNA vaccines that carried different signals for cellular localization for their ability to induce antibodies and protection against *Y. pestis*. Results showed that although the full-length form of the F1 antigen was a poor immunogen, one altered plasmid in particular (deF1, containing the coding sequence of the F1 protein but a 21-amino acid long sequence that codes for a putative bacterial signal peptide is missing) induced substantial antibody titers and protective immunity (38 out of 39 mice) against a subcutaneous *Y. pestis* challenge. It was determined that mice that received a gene gun inoculation of deF1 were protected against 4000 LD_{50} *Y. pestis* challenge and 10 out of 10 animals survived this challenge.

By using the DNA vaccine technology, a group of "2nd line" antigens from *Y. pestis* have been studied for their ability in eliciting protective antibody responses (Wang et al., 2008b). *Y. pestis* outer proteins (YopB, YopD, and YopO), plasminogen activator protease (Pla), and Yop secretion protein F (YscF) are necessary for the full virulence of *Y. pestis*. They have been proposed as potential PAs for vaccines. DNA immunization was used as a tool to study the relative protective immunity of these individual antigens with a standardized intranasal challenge system in BALB/c mice. The natural full-length gene sequences for most of these *Y. pestis* proteins did not display a good level of protein expression in vitro when delivered by a DNA vaccine vector into mammalian cells. As a result, the overall immunogenicity of these wild-type gene DNA vaccines was low in eliciting antigen-specific antibody responses. However, gene sequence modifications, including the removal of hydrophobic regions and the addition of signal peptide sequences with high secretion potential, significantly improved the level of protein expression, as well as the immunogenicity of these antigens in eliciting high-level antigen-specific antibody responses in mice. Modified YopD, YopO, and YscF antigens were able to protect immunized mice, to varying degrees, against the lethal challenge of *Y. pestis* Kim strain (5000 cfu) by intranasal inoculation while no protection was observed with either the YopB or Pla antigens (Wang et al., 2008b). Results of this study demonstrate that DNA immunization is useful in screening, optimizing, and comparing the antigen design and immunogenicity of candidate antigens for the development of a subunit-based plague vaccine. This approach will be useful to expand the search for additional protective antigens against highly virulent *Y. pestis* infection.

Results from these studies indicate that protection against both subcutaneous and intranasal plague, indicative of bubonic and pneumonic plague, respectively, can be conferred through administration of DNA vaccines encoding different *Y. pestis* antigens. The development of broadly active vaccines against plague is important to defend against any ill use of this pathogen as a biological weapon, with or without any intentional modification of the antigens to escape from the protection elicited by the original vaccination.

Botulism

Botulism is a very rare but serious paralytic disease caused by the bacterium, *Clostridium botulinum*. Historically, three natural forms of botulism existed: food-borne, wound, and infant. However, there also exits a man-made, inhalation form of botulism that is lethal in the microgram dose range and hence represents a highly dangerous potential biothreat. Seven types of non-cross-reactive botulinum neurotoxin exist (A through G) and a licensed trivalent antitoxin, available from the CDC, contains neutralizing antibodies against types A, B, and E, for treatment against botulism toxin poisoning (CDC, 2007). An additional investigational heptavalent antitoxin has been shown to be effective against all known strains of botulism. It is this heptavalent formulation that has been contracted into the Strategic National Stockpile beginning in 2007 (HHS, 2006). However, it has been shown that some strains of botulism are capable of producing mixtures of two toxin types (Cordoba et al., 1995). It is also now known that there are subtypes under each of the serotypes, indicating that the understanding of cross protection and the development of appropriate vaccine strategies will remain a great challenge.

Few studies have examined the feasibility of using DNA vaccines to protect against botulism. In an early study, mice were immunized with a DNA vaccine expressing the carboxyl terminal 50 kDa fragment of the type A toxin, which was placed in two designs: pJT-1, a construct that leads to MHC I presentation, or pJT-2 that leads to MHC II presentation (Clayton and Middlebrook, 2000). After challenge with botulinum serotype A (or in a cross-protection study serotypes B or E), it was observed that the pJT-1 construct was ineffective in inducing protection against challenge with botulism (serotype A 25 LD_{50}). However, mice that received the pJT-2 construct (at least 10 μg DNA) were afforded some level of protection when challenged at week 11. While 50 μg of DNA increased the rate of survival (9 out of 10 challenged with serotype A 25 LD_{50}), a minimum of 11 weeks had to pass prior to challenge as no protection was afforded if challenge was presented at 4 or 7 weeks after immunization. Furthermore, these results showed that no cross protection was afforded by pJT-2 DNA immunization against serotypes B or E, as none of the 10 animals survived the challenge.

Additional studies showed that a DNA vaccine against botulinium subtype F, which is a very rare form of botulism, was found to protect mice against lethal challenge with the same serotype (Bennett et al., 2003; Jathoul et al., 2004). In the first study by this group, a minimum of three i.m. injections of the F subtype-specific DNA vaccine, pABFHc2, which contained a signal sequence after the Hc domain, given over a 4-week period, was sufficient to protect 100% of mice (10 out of 10) against 10^4 MLD of botulinium toxin subtype F, and a single dose protected 60% of mice when they were challenged no less than 28 days after vaccination (Bennett et al., 2003). Additional results from this study indicated that a relationship may exist between FHc-specific antibody levels and survival in the challenge assay, and that a DNA prime-protein boost regimen induced greater levels of serum antibodies when compared to both DNA alone and protein alone. Based on these findings, DNA vaccine appears useful for the design and optimization of a future botulinum vaccine that can protect from multiple serotype toxins.

Helicobacter pylori

Gastritis, peptic ulcers and gastric cancer have all been linked to infection with the *Helicobacter pylori* bacterium. While only two studies have examined the possibility of using DNA vaccine technology to induce antibodies and protection against *H. pylori* infection, the studies were successful (Todoroki et al., 2000; Miyashita et al., 2002). The difference between studies was the DNA vaccine encoding antigen, the enzyme catalase, or the heat shock protein A or B (HspA and HspB, respectively). Both DNA vaccine formulations induced high-titer serum antibodies when compared to the empty DNA vector control. For the DNA vaccine encoding catalase, serum antibody levels were detected beginning 4 weeks after the final immunization with peak levels by 3 months post immunization. Additional analyses included measurement of bacterial colonization in stomach after challenge with *H. pylori*, and all three DNA vaccine formulations (catalase, HspA, and HspB) afforded protection against *H. pylori* replication. The results support the potential use of a DNA vaccine to protect against *H. pylori* infection. As more antibiotic-resistant strains of this bacterium develop, there will be a greater need for the development of an *H. pylori* vaccine.

Tuberculosis

The *Mycobacterium tuberculosis* (TB) vaccine Bacille Calmette-Guerin (BCG) is widely used outside the United States in settings of high TB prevalence. Unfortunately, BCG is not completely effective in preventing TB infection, it can complicate interpretation of the tuberculin skin test, and it has not prevented the emergence of extensively drug-resistant TB strains. Despite its shortcomings, BCG will likely continue to be used in afflicted countries as it does reduce the mortality rates against the more serious forms of childhood TB (e.g., disseminated and meningeal TB) (ACET, 1998). It is clear that more effective TB vaccines are needed and numerous studies have explored the potential of using DNA vaccination technology to facilitate the development of new and more effective vaccines.

The majority of TB DNA vaccine studies have focused on a family of three protein complexes of the Ag complex: Ag85A, Ag85B, and Ag85C. These antigens are prime targets because they are both found on the cell surface and are also secreted from the bacterium (Wiker and Harboe, 1992). An initial study examined the immunogenicity of a DNA vaccine encoding all three forms of the Ag85 complex (Ag85A, B, and C) and observed high-titer antibody responses in immunized mice (Ulmer et al., 1997). A subsequent study showed that the secreted form of Ag85A induced higher levels of serum antibodies at lower DNA doses and conferred protection against a highly virulent strain of *M. tuberculosis* when compared to the mature form of the protein (Baldwin et al., 1999). DNA vaccines encoding the Ag85A and Ag85B proteins but not Ag85C induced strong Th1-like responses (e.g., increased levels of IL-2, IFN-γ, and TNF-α) towards native Ag85 protein (Lozes et al., 1997), a broader T cell response than observed with natural *M. tuberculosis* infection and increased CD8 + -mediated cytotoxicity (Denis et al., 1998). In addition to these cellular immune responses, protection against challenge was conferred by DNA vaccines encoding the Ag85A protein in mice (Lozes et al., 1997; Denis et al., 1998) and guinea pigs by intramuscular injection (Baldwin et al., 1998) and gene gun inoculation (Sugawara et al., 2003) and for the Ag85B protein in mice (Lozes et al., 1997).

Additional studies have utilized various "prime-boost" strategies to increase both the immunogenicity and protective efficacy of Ag85-encoded DNA vaccines. One such study primed with a DNA vaccine encoding the Ag85 protein and subsequently boosted with purified Ag85 protein. This "prime-boost" strategy increased IL-2 and IFN-γ responses, increased IFN-γ producing CD4+ cells, increased IgG2a isotype antibody responses, and increased the efficacy of the DNA vaccine against *M. tuberculosis* challenge (Tanghe et al., 2001). A second study examined the effects of DNA prime and subsequent boost with the current TB vaccine, BCG, and found that the DNA + BCG regimen was more effective than BCG alone in protecting against TB challenge (Feng et al., 2001; Ferraz et al., 2004) and that CD8+ T cell depletion attenuated this protection (Feng et al., 2001).

DNA vaccines have also been developed to target other TB antigens including hsp65 (Tascon et al., 1996; Ferraz et al., 2004), hsp70 (Lowrie et al., 1999; Ferraz et al., 2004), the alanine–proline-rich antigen (Apa) (Ferraz et al., 2004), and MPT-63, MPT-64, MPT-83, ESAT-6, KatG, and other antigens, either alone or in various combinations (Morris et al., 2000; Repique et al., 2002; Cai et al., 2005a, 2005b). High levels of specific IgG antibodies were observed against Ag85A, MTP-64, and MPT-83 and mice immunized with individual DNA vaccines (Cai et al., 2005b) or a construct encoding all the antigens (Cai et al., 2005a) were protected against TB challenge. While individual DNA vaccines encoding the MPT-63 and MPT-83 antigens provided only partial protection against TB challenge, the multivalent MPT-63, MPT-83, ESAT-6, and KatG DNA vaccine produced antigen-specific cell-mediated and humoral immune responses and a stronger protective response against TB challenge when compared to the current BCG vaccine (Morris et al., 2000). Studies examining the efficacy of DNA vaccines that encode for the heat shock proteins of *M. tuberculosis* found that these proteins were able to decrease the numbers of live bacteria in both the spleen and the lungs of infected mice after only one dose of the hsp65 DNA vaccine, with lower effects observed with the hsp70 and ESAT6 DNA vaccine formulations (Lowrie et al., 1999).

DNA Vaccines against Parasites

Malaria

Malaria is transmitted through mosquitoes and remains a major global public health problem. Each year 350–500 million new cases of malaria occur throughout the world with approximately 1 million deaths occurring mostly in children in sub-Saharan Africa. Although antimalarial drugs are effective in the prevention and treatment of malaria infections, the development of vaccines continues to be the ultimate goal for malaria control.

Significant developments have been made through clinical trials that have tested various designs of DNA vaccines against malaria. In the first set of human DNA vaccine studies, a plasmid DNA vaccine encoding the *Plasmodium falciparum* circumsporozoite protein (PfCSP) produced low CMI responses in human volunteers even with the use of a PMED (Wang et al., 1998, 2001). High DNA dose (2500 μg per immunization) was effective in increasing positive responses, and antigen-specific CTL responses were observed (Wang et al., 1998). Measurable antibody responses were not detected in any of the volunteers who had received this safe and well-tolerated malaria DNA vaccine. In later studies, a DNA prime-recombinant vaccinia virus Ankara (MVA) boost approach was developed to further enhance the immunogenicity of malaria DNA vaccines [for recent review, see (Gilbert et al., 2006)]. A unique chimeric antigen, ME-TRAP, was produced to express the thrombospondin related adhesion protein (TRAP) coupled with a string of multiple epitopes (ME) (18 T-cell epitopes and 2 B-cell epitopes from various pre-erythrocytic *P. falciparum* antigens) (Moorthy et al., 2003). DNA immunization elicited low-level T-cell responses that were amplified by a boost with MVA ME-TRAP. More significantly, intramuscular DNA priming was more effective than gene gun-mediated priming when combined with a MVA boost. The combination immunization regimen produced partial protection, which manifested as delayed parasitemia after sporozoite challenge with a different strain of *P. falciparum*. Recent studies have further demonstrated that the DNA prime-MVA boost vaccine encoding TRAP produced stronger ex vivo IFN-γ ELISPOT responses than did a similar DNA prime-MVA boost vaccine encoding the circumsporozoite (CS) antigen. Importantly, the latter vaccine could not protect healthy malaria-naïve adults against *P. falciparum* sporozoite challenge as did the TRAP-expressing DNA prime-MVA boost vaccine (Dunachie et al., 2006).

A different type of malaria vaccine prime-boost approach was tested in humans. The combination of a PfCSP DNA vaccine prime with the boost of a known protective recombinant protein RTS,S, which is a fusion protein of the *P. falciparum* PfCSP protein and the hepatitis B virus surface antigen (HBsAg). This construct encodes several CD4+ and CD8+ T-cell epitopes. It was administered to volunteers who had received the DNA vaccine (prime) in an earlier trial (Le et al., 2000) or naïve volunteers. Volunteers who were primed with PfCSP DNA vaccine had protective CD4+ T-cell and antibody responses (effects of protein boost) in addition to the previously observed CD8+ CTL and IFN-γ responses (effects of DNA prime). The naïve volunteer group exhibited only antibody and CD4+ T-cell responses, and not CD8+ T-cell responses (Epstein et al., 2004; Wang et al., 2004a). This finding suggested that the PfCSP DNA vaccine was not effective for the induction of antibody responses, either alone or as a priming vaccine.

In an attempt to expand the breadth of protective immune responses, volunteers were immunized with a mixture of five malaria DNA vaccines expressing different *P. falciparum* pre-erythrocyte stage antigens in conjunction with an "adjuvant" plasmid encoding human granulocyte–macrophage colony-stimulating factor (hGM-GSF) (Wang et al., 2005a). IFN-γ responses were detected to multiple class I and/or class II restricted T-cell epitopes from all five antigens (CSP, EXP-1, LSA-1, SSP2, and LSA-3) with no statistical difference between the groups. Additional analyses revealed that in volunteers who received DNA alone (i.e. without hGM-CSF) IFN-γ responses to both class I and class II peptides were boosted after challenge with *P. falciparum* parasites while only responses to class II peptides were boosted in volunteers who were vaccinated with DNA and hGM-GSF. Volunteers who received hGM-CSF actually had a reduced frequency of IFN-γ responses to class I peptides. More importantly, none were protected against malaria challenge, further raising the issue of antigen selection, since previous studies have demonstrated that only a DNA prime-MVA boost vaccine expressing ME-TRAP antigen was able to induce partial protection (Gilbert et al., 2006).

CONCLUSIONS

Tremendous strides have been made in the last two centuries in the control of human infectious diseases through the development of continuously improved vaccination technologies. The discovery of DNA immunization in 1992–1993 fundamentally changed our view of the nature of a vaccine: genetic material that encodes for antigens, rather than the actual antigens themselves, as the effective components of DNA vaccines. This new vaccine technology could not come at a more important time in history given the increasing threat of emerging and reemerging infectious diseases and a renewed concern regarding the use of biological agents for purposes of bioterrorism. Since its inception, DNA vaccination technology has undergone significant advancements and many candidate human vaccine formulations have already been developed. Improved modes of administration, the use of codon and antigen gene optimization, and the implementation of vaccination DNA prime/boost regimens have led to the quick progression of DNA vaccines from research laboratory benches to human clinical trials. Biodefense and emerging infectious disease targets have made a significant contribution to this important process.

References

ACET, A. Development of new vaccines for tuberculosis recommendations of the advisory council for the elimination of tuberculosis (ACET). MMWR 1998; 47:1–6.

Aldaz-Carroll, L., Xiao, Y., Whitbeck, J.C., Ponce de Leon, M., Lou, H., Kim, M., Yu, J., Reinherz, E.L., Isaacs, S.N., Eisenberg, R.J. and Cohen, G.H. Major neutralizing sites on vaccinia virus glycoprotein B5 are exposed differently on variola virus ortholog B6. J. Virol. 2007; 81(15):8131–8139.

Andre, S., Seed, B., Eberle, J., Schraut, W., Balltmann, A. and Haas, J. Increased immune response elicited by DNA vaccination with a synthetic gp120 sequence with optimized codon usage. J. Virol. 1998; 72(2):1497–1503.

Apt, D., Raviprakash, K., Brinkman, A., Semyonov, A., Yang, S., Skinner, C., Diehl, L., Lyons, R., Porter, K. and Punnonen, J. Tetravalent neutralizing antibody response against four dengue serotypes by a single chimeric dengue envelope antigen. Vaccine 2006; 24(3):335–344.

Baldwin, S.L., D'Souza, C., Roberts, A.D., Kelly, B.P., Frank, A.A., Liu, M.A., Ulmer, J.B., Huygen, K., McMurray, D.M. and Orme, I.M. Evaluation of new vaccines in the mouse and guinea pig model of tuberculosis. Infect. Immun. 1998; 66(6):2951–2959.

Baldwin, S.L., D'Souza, C.D., Orme, I.M., Liu, M.A., Huygen, K., Denis, O., Tang, A., Zhu, L., Montgomery, D. and Ulmer, J.B. Immunogenicity and protective efficacy of DNA vaccines encoding secreted and non-secreted forms of Mycobacterium tuberculosis Ag85A. Tuber. Lung Dis. 1999; 79(4):251–259.

Barnett, S.W., Rajasekar, S., Legg, H., Doe, B., Fuller, D.H., Haynes, J.R., Walker, C.M. and Steimer, K.S. Vaccination with HIV-1 gp120 DNA induces immune responses that are boosted by a recombinant gp120 protein subunit. Vaccine 1997; 15(8):869–873.

Bennett, A.M., Perkins, S.D. and Holley, J.L. DNA vaccination protects against botulinum neurotoxin type F. Vaccine 2003; 21(23):3110–3117.

Bennett, A.M., Phillpotts, R.J., Perkins, S.D., Jacobs, S.C. and Williamson, E.D. Gene gun mediated vaccination is superior to manual delivery for immunisation with DNA vaccines expressing protective antigens from Yersinia pestis or Venezuelan Equine Encephalitis virus. Vaccine 2000; 18(7-8):588–596.

Boyer, J.D., Robinson, T.M., Maciag, P.C., Peng, X., Johnson, R.S., Pavlakis, G., Lewis, M.G., Shen, A., Siliciano, R., Brown, C.R., Weiner, D.B. and Paterson, Y. DNA prime Listeria boost induces a cellular immune response to SIV antigens in the rhesus macaque model that is capable of limited suppression of SIV239 viral replication. Virology 2005; 333(1):88–101.

Cai, H., Hu, X.D., Yu, D.H., Li, S.X., Tian, X. and Zhu, Y.X. Combined DNA vaccine encapsulated in microspheres enhanced protection efficacy against Mycobacterium tuberculosis infection of mice. Vaccine 2005; 23(32):4167–4174.

Cai, H., Tian, X., Hu, X.D., Li, S.X., Yu, D.H. and Zhu, Y.X. Combined DNA vaccines formulated either in DDA or in saline protect cattle from Mycobacterium bovis infection. Vaccine 2005; 23(30):3887–3895.

CDC. Basic Information About SARS. 2005, May 3, 2005. Retrieved July 30, 2007, from http://www.cdc.gov/ncidod/sars/factsheet.htm.

CDC. Products Distributed by The Center for Diseases Control and Prevention. 2007, 11/05/2007. Retrieved 11/05/2007, 2007, from http://www.cdc.gov/ncidod/srp/drugs/formulary.html.

Chang, G.J., Hunt, A.R. and Davis, B. A single intramuscular injection of recombinant plasmid DNA induces protective immunity and prevents Japanese encephalitis in mice. J. Virol. 2000; 74(9):4244–4252.

Chen, Z., Kadowaki, S., Hagiwara, Y., Yoshikawa, T., Matsuo, K., Kurata, T. and Tamura, S. Cross-protection against a lethal influenza virus infection by DNA vaccine to neuraminidase. Vaccine 2000; 18(28):3214–3222.

Chen, Z., Kadowaki, S., Hagiwara, Y., Yoshikawa, T., Sata, T., Kurata, T. and Tamura, S. Protection against influenza B virus infection by immunization with DNA vaccines. Vaccine 2001; 19(11-12):1446–1455.

Chen, Z., Sahashi, Y., Matsuo, K., Asanuma, H., Takahashi, H., Iwasaki, T., Suzuki, Y., Aizawa, C., Kurata, T. and Tamura, S. Comparison of the ability of viral protein-expressing plasmid DNAs to protect against influenza. Vaccine 1998; 16(16):1544–1549.

Clayton, J. and Middlebrook, J.L. Vaccination of mice with DNA encoding a large fragment of botulinum neurotoxin serotype A. Vaccine 2000; 18(17):1855–1862.

Cordoba, J.J., Collins, M.D. and East, A.K. Studies on the gene encoding botulinum neurotoxin type A of Clostridium botulinum from a variety of sources. Systematic and Applied Microbiology 1995; 18:12–22.

Costa, S.M., Paes, M.V., Barreto, D.F., Pinhão, A.T., Barth, O.M., Queiroz, J.L., Armôa, G.R., Freire, M.S. and Alves, A.M. Protection against dengue type 2 virus induced in mice immunized with a DNA plasmid encoding the non-structural 1 (NS1) gene fused to the tissue plasminogen activator signal sequence. Vaccine 2006; 24(2):195–205.

Cox, R.J., Mykkeltvedt, E., Robertson, J. and Haaheim, L.R. Non-lethal viral challenge of influenza haemagglutinin and nucleoprotein DNA vaccinated mice results in reduced viral replication. Scand. J. Immunol. 2002; 55(1):14–23.

Daniel, M., Sehgal, P., Kodama, T., Wyand, M., Ringler, D., King, N., Schmidt, D., Troup, C. and Desrosiers, R. Use of simian immunodeficiency virus for vaccine research. J. Med. Primatol. 1990; 19(3-4):395–399.

Davis, B.S., Chang, G.J., Cropp, B., Roehrig, J.T., Martin, D.A., Mitchell, C.J., Bowen, R. and Bunning, M.L. West Nile virus recombinant DNA vaccine protects mouse and horse from virus challenge and expresses in vitro a noninfectious recombinant antigen that can be used in enzyme-linked immunosorbent assays. J. Virol. 2001; 75(9):4040–4047.

Denis, O., Tanghe, A., Palfliet, K., Jurion, F., van den Berg, T.P., Vanonckelen, A., Ooms, J., Saman, E., Ulmer, J.B., Content, J. and Huygen, K. Vaccination with plasmid DNA encoding mycobacterial antigen 85A stimulates a CD4+ and CD8+ T-cell epitopic repertoire broader than that stimulated by Mycobacterium tuberculosis H37Rv infection. Infect. Immun. 1998; 66(4):1527–1533.

Desrosiers, R.C., Wyand, M.S., Kodama, T., Ringler, D.J., Arthur, L.O., Sehgal, P.K., Letvin, N.L., King, N.W. and Daniel, M.D. Vaccine Protection against Simian Immunodeficiency Virus Infection. Proc. Natl. Acad. Sci. USA 1989; 86(16):6353–6357.

Donnelly, J.J., Friedman, A., Martinez, D., Montgomery, D.L., Shiver, J.W., Motzel, S.L., Ulmer, J.B. and Liu, M.A. Preclinical efficacy of a prototype DNA vaccine: enhanced protection against antigenic drift in influenza virus. Nat. Med. 1995; 1(6):583–587.

Donnelly, J.J., Friedman, A., Ulmer, J.B. and Liu, M.A. Further protection against antigenic drift of influenza virus in a ferret model by DNA vaccination. Vaccine 1997; 15(8):865–868.

Drape, R.J., Macklin, M.D., Barr, L.J., Jones, S., Haynes, J.R. and Dean, H.J. Epidermal DNA vaccine for influenza is immunogenic in humans. Vaccine 2006; 24(21):4475–4481.

Dunachie, S.J., Walther, M., Epstein, J.E., Keating, S., Berthoud, T., Andrews, L., Andersen, R.F., Bejon, P., Goonetilleke, N., Poulton, I., Webster, D.P., Butcher, G., Watkins, K., Sinden, R.E., Levine, G.L., Richie, T.L., Schneider, J., Kaslow, D., Gilbert, S.C., Carucci, D.J. and Hill, A.V. A DNA prime-modified vaccinia virus ankara boost vaccine encoding thrombospondin-related adhesion protein but not circumsporozoite protein partially protects healthy malaria-naive adults against Plasmodium falciparum sporozoite challenge. Infect. Immun. 2006; 74(10):5933–5942.

Epstein, J.E., Charoenvit, Y., Kester, K.E., Wang, R., Newcomer, R., Fitzpatrick, S., Richie, T.L., Tornieporth, N., Heppner, D.G., Ockenhouse, C., Majam, V., Holland, C., Abot, E., Ganeshan, H., Berzins, M., Jones, T., Freydberg, C.N., Ng, J., Norman, J., Carucci, D.J., Cohen, J. and Hoffman, S.L. Safety, tolerability, and antibody responses in humans after sequential immunization with a PfCSP DNA vaccine followed by the recombinant protein vaccine RTS,S/AS02A. Vaccine 2004; 22(13-14):1592–1603.

Epstein, S., Kong, W.-P., Misplon, J., Lo, C.-Y., Tumpey, T., Xu, L. and Nabel, G. Protection against multiple influenza A subtypes by vaccination with highly conserved nucleoprotein. Vaccine 2005; 23(46-47):5404–5410.

Felgner, P.L. and Ringold, G.M. Cationic liposome-mediated transfection. Nature 1989; 337(6205):387–388.

Feng, C.G., Palendira, U., Demangel, C., Spratt, J.M., Malin, A.S. and Britton, W.J. Priming by DNA immunization augments protective efficacy of Mycobacterium bovis Bacille Calmette-Guerin against tuberculosis. Infect. Immun. 2001; 69(6):4174–4176.

Ferraz, J.C., Stavropoulos, E., Yang, M., Coade, S., Espitia, C., Lowrie, D.B., Colston, M.J. and Tascon, R.E. A heterologous DNA priming-Mycobacterium bovis BCG boosting immunization strategy using mycobacterial Hsp70, Hsp65, and Apa antigens improves protection against tuberculosis in mice. Infect. Immun. 2004; 72(12):6945–6950.

Flynn, N.M., Forthal, D.N., Harro, C.D., Judson, F.N., Mayer, K.H. and Para, M.F. Placebo-controlled phase 3 trial of a recombinant glycoprotein 120 vaccine to prevent HIV-1 infection. J. Infect. Dis. 2005; 191(5):654–665.

Fogg, C., Lustig, S., Whitbeck, J.C., Eisenberg, R.J., Cohen, G.H. and Moss, B. Protective immunity to vaccinia virus induced by vaccination with multiple recombinant outer membrane proteins of intracellular and extracellular virions. J. Virol. 2004; 78(19):10230–10237.

Fynan, E.F., Webster, R.G., Fuller, D.H., Haynes, J.R., Santoro, J.C. and Robinson, H.L. DNA vaccines: protective immunizations by parenteral, mucosal, and gene-gun inoculations. Proc. Natl. Acad. Sci. USA 1993; 90(24):11478–11482.

Galloway, D., Liner, A., Legutki, J., Mateczun, A., Barnewall, R. and Estep, J. Genetic immunization against anthrax. Vaccine 2004; 22(13-14):1604–1608.

Galmiche, M.C., Goenaga, J., Wittek, R. and Rindisbacher, L. Neutralizing and protective antibodies directed against vaccinia virus envelope antigens. Virology 1999; 254(1):71–80.

Garmory, H., Freeman, D., Brown, K. and Titball, R. Protection against plague afforded by immunisation with DNA vaccines optimised for expression of the Yersinia pestis V antigen. Vaccine 2004; 22(8):947–957.

Gilbert, P.B., Peterson, M.L., Follmann, D., Hudgens, M.G., Francis, D.P., Gurwith, M., Heyward, W.L., Jobes, D.V., Popovic, V., Self, S.G., Sinangil, F., Burke, D. and Berman, P.W. Correlation between immunologic responses to a recombinant glycoprotein 120 vaccine and incidence of HIV-1 infection in a phase 3 HIV-1 preventive vaccine trial. J. Infect. Dis. 2005; 191(5):666–677.

Gilbert, S.C., Moorthy, V.S., Andrews, L., Pathan, A.A., McConkey, S.J., Vuola, J.M., Keating, S.M., Berthoud, T., Webster, D., McShane, H. and Hill, A.V. Synergistic DNA-MVA prime-boost vaccination regimes for malaria and tuberculosis. Vaccine 2006; 24(21):4554–4561.

Gonzalo, R.M., del Real, G., Rodriguez, J.R., Rodriguez, D., Heljasvaara, R., Lucas, P., Larraga, V. and Esteban, M. A heterologous prime-boost regime using DNA and recombinant vaccinia virus expressing the Leishmania infantum P36/LACK antigen protects BALB/c mice from cutaneous leishmaniasis. Vaccine 2002; 20(7-8):1226–1231.

Goonetilleke, N., Moore, S., Dally, L., Winstone, N., Cebere, I., Mahmoud, A., Pinheiro, S., Gillespie, G., Brown, D., Loach, V., Roberts, J., Guimaraes-Walker, A., Hayes, P., Loughran, K., Smith, C., De Bont, J., Verlinde, C., Vooijs, D., Schmidt, C., Boaz, M., Gilmour, J., Fast, P., Dorrell, L., Hanke, T. and McMichael, A.J. Induction of multifunctional human immunodeficiency virus type 1 (HIV-1)-specific T cells capable of proliferation in healthy subjects by using a prime-boost regimen of DNA- and modified vaccinia virus Ankara-vectored vaccines expressing HIV-1 Gag coupled to CD8+ T-cell epitopes. J. Virol. 2006; 80(10):4717–4728.

Graham, B.S., Koup, R.A., Roederer, M., Bailer, R.T., Enama, M.E., Moodie, Z., Martin, J.E., McCluskey, M.M., Chakrabarti, B.K., Lamoreaux, L., Andrews, C.A., Gomez, P.L., Mascola, J.R. and Nabel, G.J. Phase 1 safety and immunogenicity evaluation of a multiclade HIV-1 DNA candidate vaccine. J. Infect. Dis. 2006; 194(12):1650–1660.

Grosfeld, H., Cohen, S., Bino, T., Flashner, Y., Ber, R., Mamroud, E., Kronman, C., Shafferman, A. and Velan, B. Effective protective immunity to Yersinia pestis infection conferred by DNA vaccine coding for derivatives of the F1 capsular antigen. Infection and immunity 2003; 71(1):374–383.

Gu, M.L., Leppla, S.H. and Klinman, D.M. Protection against anthrax toxin by vaccination with a DNA plasmid encoding anthrax protective antigen. Vaccine 1999; 17(4):340–344.

Haas, J., Park, E.C. and Seed, B. Codon usage limitation in the expression of HIV-1 envelope glycoprotein. Curr. Bio. 1996; 6(3):315–324.

Hall, R., Nisbet, D., Pham, K., Pyke, A., Smith, G. and Khromykh, A. DNA vaccine coding for the full-length infectious Kunjin virus RNA protects mice against the New York strain of West Nile virus. Proc. Natl. Acad. Sci. USA 2003; 100(18):10460–10464.

Hanke, T., McMichael, A.J. and Dorrell, L. Clinical experience with plasmid DNA- and modified vaccinia virus Ankara-vectored human immunodeficiency virus type 1 clade A vaccine focusing on T-cell induction. J. Gen. Virol. 2007; 88(Pt 1):1–12.

Heinz, F.X., Collett, M.S., Purcell, R.H., Gould, E.A., Howard, C.R., Houghton, M., Moormann, R.J.M., Rice, C.M. and Thiel, H.-J., Eds. 7th Report of the International Committee on Taxonomy of Viruses. San Diego, Academic, 2000.

Heraud, J.M., Edghill-Smith, Y., Ayala, V., Kalisz, I., Parrino, J., Kalyanaraman, V.S., Manischewitz, J., King, L.R., Hryniewicz, A., Trindade, C.J., Hassett, M., Tsai, W.P., Venzon, D., Nalca, A., Vaccari, M., Silvera, P., Bray, M., Graham, B.S., Golding, H., Hooper, J.W. and Franchini, G. Subunit recombinant vaccine protects against monkeypox. J. Immunol. 2006; 177(4):2552–2564.

Hermanson, G., Whitlow, V., Parker, S., Tonsky, K., Rusalov, D., Ferrari, M., Lalor, P., Komai, M., Mere, R., Bell, M., Brenneman, K., Mateczun, A., Evans, T., Kaslow, D., Galloway, D. and Hobart, P. A cationic lipid-formulated plasmid DNA vaccine confers sustained antibody-mediated protection against aerosolized anthrax spores. Proc. Natl. Acad. Sci. USA 2004; 101(37):13601–13606.

Herrmann, J.E., Chen, S.C., Jones, D.H., Tinsley-Bown, A., Fynan, E.F., Greenberg, H.B. and Farrar, G.H. Immune responses and protection obtained by oral immunization with rotavirus VP4 and VP7 DNA vaccines encapsulated in microparticles. Virology 1999; 259(1):148–153.

Herrmann, J.E., Wang, S., Zhang, C., Panchal, R.G., Bavari, S., Lyons, C.R., Lovchik, J.A., Golding, B., Shiloach, J. and Lu, S. Passive immunotherapy of Bacillus anthracis pulmonary infection in mice with antisera produced by DNA immunization. Vaccine 2006; 24(31-32):5872–5880.

Hevey, M., Negley, D., VanderZanden, L., Tammariello, R.F., Geisbert, J., Schmaljohn, C., Smith, J.F., Jahrling, P.B. and Schmaljohn, A.L. Marburg virus vaccines: comparing classical and new approaches. Vaccine 2001; 20(3-4):586–593.

HHS. HHS Awards BioShield Contract For Botulism Antitoxin. 2006, 06/01/2006. Retrieved 11/05/2007, 2007, from http://www.dhhs.gov/news/press/2006pres/20060601.html.

Hooper, J.W., Custer, D.M., Schmaljohn, C.S. and Schmaljohn, A. L. DNA vaccination with vaccinia virus L1R and A33R genes protects mice against a lethal poxvirus challenge. Virology 2000; 266(2):329–339.

Hooper, J.W., Custer, D.M. and Thompson, E. Four-gene-combination DNA vaccine protects mice against a lethal vaccinia virus challenge and elicits appropriate antibody responses in nonhuman primates. Virology 2003; 306(1):181–195.

Hooper, J.W., Golden, J., Ferro, A. and King, A. Smallpox DNA vaccine delivered by novel skin electroporation device protects mice against intranasal poxvirus challenge. Vaccine 2007; 25(10):1814–1823.

Hooper, J.W., Thompson, E., Wilhelmsen, C., Zimmerman, M., Ichou, M., Steffen, S.E., Schmaljohn, C.S., Schmaljohn, A.L. and Jahrling, P.B. Smallpox DNA vaccine protects nonhuman primates against lethal monkeypox. J. Virol. 2004; 78(9): 4433–4443.

Hu, H., Lu, X., Tao, L., Bai, B., Zhang, Z., Chen, Y., Zheng, F., Chen, J., Chen, Z. and Wang, H. Induction of specific immune responses by severe acute respiratory syndrome coronavirus spike DNA vaccine with or without interleukin-2 immunization using different vaccination routes in mice. Clin. Vaccine Immunol. 2007; 14(7):894–901.

Huang, J., Ma, R. and Wu, C.-Y. Immunization with SARS-CoV S DNA vaccine generates memory CD4+ and CD8+ T cell immune responses. Vaccine 2006; 24(23):4905–4913.

Imoto, J.-I. and Konishi, E. Needle-free jet injection of a mixture of Japanese encephalitis DNA and protein vaccines: a strategy to effectively enhance immunogenicity of the DNA vaccine in a murine model. Viral Immunol. 2005; 18(1):205–212.

Jathoul, A.P., Holley, J.L. and Garmory, H.S. Efficacy of DNA vaccines expressing the type F botulinum toxin Hc fragment using different promoters. Vaccine 2004; 22(29-30):3942–3946.

Jiang, Y., Yu, K., Zhang, H., Zhang, P., Li, C., Tian, G., Li, Y., Wang, X., Ge, J., Bu, Z. and Chen, H. Enhanced protective efficacy of H5 subtype avian influenza DNA vaccine with codon optimized HA gene in a pCAGGS plasmid vector. Antivir. Res. 2007; 75(3):234–241.

Jimenez, G.S., Planchon, R., Wei, Q., Rusalov, D., Geall, A., Enas, J., Lalor, P., Leamy, V., Vahle, R., Luke, C.J., Rolland, A., Kaslow, D.C. and Smith, L.R. Vaxfectin-formulated influenza DNA vaccines encoding NP and M2 viral proteins protect mice against lethal viral challenge. Hum. Vaccin. 2007; 3(5):157–164.

Kim, T.W., Chung, H., Kwon, I.C., Sung, H.C., Kang, T.H., Han, H.D. and Jeong, S.Y. Induction of immunity against hepatitis B virus surface antigen by intranasal DNA vaccination using a cationic emulsion as a mucosal gene carrier. Mol. Cells 2006; 22(2):175–181.

Klavinskis, L.S., Barnfield, C., Gao, L. and Parker, S. Intranasal immunization with plasmid DNA-lipid complexes elicits mucosal immunity in the female genital and rectal tracts. J. Immunol. 1999; 162(1):254–262.

Kobiler, D., Gozes, Y., Rosenberg, H., Marcus, D., Reuveny, S. and Altboum, Z. Efficiency of protection of guinea pigs against infection with Bacillus anthracis spores by passive immunization. Infect. Immun. 2002; 70(2):544–560.

Kochel, T.J., Raviprakash, K., Hayes, C.G., Watts, D.M., Russell, K.L., Gozalo, A.S., Phillips, I.A., Ewing, D.F., Murphy, G.S. and Porter, K.R. A dengue virus serotype-1 DNA vaccine induces virus neutralizing antibodies and provides protection from viral challenge in Aotus monkeys. Vaccine 2000; 18(27):3166–3173.

Kodihalli, S., Goto, H., Kobasa, D.L., Krauss, S., Kawaoka, Y. and Webster, R.G. DNA vaccine encoding hemagglutinin provides protective immunity against H5N1 influenza virus infection in mice. J. Virol. 1999; 73(3):2094–2098.

Kodihalli, S., Haynes, J.R., Robinson, H.L. and Webster, R.G. Cross-protection among lethal H5N2 influenza viruses induced by DNA vaccine to the hemagglutinin. J. Virol. 1997; 71(5):3391–3396.

Kong, W.-P., Xu, L., Stadler, K., Ulmer, J., Abrignani, S., Rappuoli, R. and Nabel, G. Modulation of the immune response to the severe acute respiratory syndrome spike glycoprotein by gene-based and inactivated virus immunization. J. Virol. 2005; 79(22):13915–13923.

Konishi, E., Ajiro, N., Nukuzuma, C., Mason, P. and Kurane, I. Comparison of protective efficacies of plasmid DNAs encoding Japanese encephalitis virus proteins that induce neutralizing antibody or cytotoxic T lymphocytes in mice. Vaccine 2003a; 21(25-26):3675–3683.

Konishi, E., Kosugi, S. and Imoto, J.-I. Dengue tetravalent DNA vaccine inducing neutralizing antibody and anamnestic responses to four serotypes in mice. Vaccine 2006; 24(12):2200–2207.

Konishi, E., Terazawa, A. and Fujii, A. Evidence for antigen production in muscles by dengue and Japanese encephalitis DNA vaccines and a relation to their immunogenicity in mice. Vaccine 2003b; 21(25-26):3713–3720.

Konishi, E., Terazawa, A. and Imoto, J.-I. Simultaneous immunization with DNA and protein vaccines against Japanese encephalitis or dengue synergistically increases their own abilities to induce neutralizing antibody in mice. Vaccine 2003c; 21(17-18):1826–1832.

Konishi, E., Yamaoka, M., Kurane, I. and Mason, P.W. A DNA vaccine expressing dengue type 2 virus premembrane and envelope genes induces neutralizing antibody and memory B cells in mice. Vaccine 2000; 18(11-12):1133–1139.

Kotsopoulou, E., Kim, V.N., Kingsman, A.J., Kingsman, S.M. and Mitrophanous, K.A. A Rev-independent human immunodeficiency virus type 1 (HIV-1)-based vector that exploits a codon-optimized HIV-1 gag-pol gene. J. Virol. 2000; 74(10):4839–4852.

Laddy, D., Yan, J., Corbitt, N., Kobasa, D., Kobinger, G. and Weiner, D. Immunogenicity of novel consensus-based DNA vaccines against avian influenza. Vaccine 2007; 25(16):2984–2989.

Law, M., Cardoso, R.M., Wilson, I.A. and Burton, D.R. Antigenic and Immunogenic Study of Membrane-Proximal External Region-Grafted gp120 Antigens by a DNA Prime-Protein Boost Immunization Strategy. J. Virol. 2007; 81(8):4272–4285.

Le, T.P., Coonan, K.M., Hedstrom, R.C., Charoenvit, Y., Sedegah, M., Epstein, J.E., Kumar, S., Wang, R., Doolan, D.L., Maguire, J. D., Parker, S.E., Hobart, P., Norman, J. and Hoffman, S.L. Safety, tolerability and humoral immune responses after intramuscular administration of a malaria DNA vaccine to healthy adult volunteers. Vaccine 2000; 18(18):1893–1901.

Letvin, N.L., Huang, Y., Chakrabarti, B.K., Xu, L., Seaman, M.S., Beaudry, K., Korioth-Schmitz, B., Yu, F., Rohne, D., Martin, K.L., Miura, A., Kong, W.P., Yang, Z.Y., Gelman, R.S., Golubeva, O.G., Montefiori, D.C., Mascola, J.R. and Nabel, G.J. Heterologous envelope immunogens contribute to AIDS vaccine protection in rhesus monkeys. J. Virol. 2004; 78(14):7490–7497.

Letvin, N.L., Montefiori, D.C., Yasutomi, Y., Perry, H.C., Davies, M.E., Lekutis, C., Alroy, M., Freed, D.C., Lord, C.I., Handt, L.K., Liu, M.A. and Shiver, J.W. Potent, protective anti-HIV immune responses generated by bimodal HIV envelope DNA plus protein vaccination. Proc. Natl. Acad. Sci. USA 1997; 94(17):9378–9383.

Lin, Y.L., Chen, L.K., Liao, C.L., Yeh, C.T., Ma, S.H., Chen, J.L., Huang, Y.L., Chen, S.S. and Chiang, H.Y. DNA immunization with Japanese encephalitis virus nonstructural protein NS1 elicits protective immunity in mice. J. Virol. 1998; 72(1):191–200.

Little, S.F., Ivins, B.E., Fellows, P.F. and Friedlander, A.M. Passive protection by polyclonal antibodies against Bacillus anthracis infection in guinea pigs. Infect. Immun. 1997; 65(12):5171–5175.

Liu, F., Mboudjeka, I., Shen, S., Chou, T.H., Wang, S., Ross, T.M. and Lu, S. Independent but not synergistic enhancement to the immunogenicity of DNA vaccine expressing HIV-1 gp120 glycoprotein by codon optimization and C3d fusion in a mouse model. Vaccine 2004; 22(13-14):1764–1772.

Lodmell, D.L. and Ewalt, L.C. Rabies vaccination: comparison of neutralizing antibody responses after priming and boosting with different combinations of DNA, inactivated virus, or recombinant vaccinia virus vaccines. Vaccine 2000; 18(22):2394–2398.

Lodmell, D.L. and Ewalt, L.C. Post-exposure DNA vaccination protects mice against rabies virus. Vaccine 2001; 19(17-19):2468–2473.

Lodmell, D., Parnell, M., Bailey, J., Ewalt, L. and Hanlon, C. Rabies DNA vaccination of non-human primates: post-exposure studies using gene gun methodology that accelerates induction of neutralizing antibody and enhances neutralizing antibody titers. Vaccine 2002; 20(17-18):2221–2228.

Lodmell, D.L., Ray, N.B. and Ewalt, L.C. Gene gun particle-mediated vaccination with plasmid DNA confers protective immunity against rabies virus infection. Vaccine 1998a; 16(2-3):115–118.

Lodmell, D.L., Ray, N.B., Parnell, M.J., Ewalt, L.C., Hanlon, C.A., Shaddock, J.H., Sanderlin, D.S. and Rupprecht, C.E. DNA immunization protects nonhuman primates against rabies virus. Nat. Med. 1998b; 4(8):949–952.

Lowrie, D.B., Tascon, R.E., Bonato, V.L., Lima, V.M., Faccioli, L.H., Stavropoulos, E., Colston, M.J., Hewinson, R.G., Moelling, K. and Silva, C.L. Therapy of tuberculosis in mice by DNA vaccination. Nature 1999; 400(6741):269–271.

Lozes, E., Huygen, K., Content, J., Denis, O., Montgomery, D.L., Yawman, A.M., Vandenbussche, P., Van Vooren, J.P., Drowart, A., Ulmer, J.B. and Liu, M.A. Immunogenicity and efficacy of a tuberculosis DNA vaccine encoding the components of the secreted antigen 85 complex. Vaccine 1997; 15(8):830–833.

Lu, S. Combination DNA plus protein HIV vaccines. Springer Semin. Immunopathol. 2006; 28(3):255–265.

Lu, S., Manning, S. and Arthos, J. Antigen engineering in DNA immunization. In: Methods in Molecular Medicine DNA Vaccines (D.B. Lowrie and R. Whalen, Eds.), pp. 355–374. Secaucus, NJ: Humana Press, 1999.

Lu, S., Santoro, J.C., Fuller, D.H., Haynes, J.R. and Robinson, H.L. Use of DNAs expressing HIV-1 Env and noninfectious HIV-1 particles to raise antibody responses in mice. Virology 1995; 209(1):147–154.

Lu, S., Wyatt, R., Richmond, J., Mustafa, F., Wang, S., Weng, J., Montefiori, D.C., Sodroski, J. and Robinson, H.L. Immunogenicity of DNA vaccines expressing human immunodeficiency virus type 1 envelope glycoprotein with and without deletions in the V1/2 and V3 regions. AIDS Res. Hum. Retroviruses 1998; 14(2):151–155.

Macklin, M.D., McCabe, D., McGregor, M.W., Neumann, V., Meyer, T., Callan, R., Hinshaw, V.S. and Swain, W.F. Immunization of pigs with a particle-mediated DNA vaccine to influenza A virus protects against challenge with homologous virus. J. Virol. 1998; 72(2):1491–1496.

Martin, J., Sullivan, N., Enama, M., Gordon, I., Roederer, M., Koup, R., Bailer, R., Chakrabarti, B., Bailey, M., Gomez, P., Andrews, C., Moodie, Z., Gu, L., Stein, J., Nabel, G. and Graham, B. A DNA vaccine for Ebola virus is safe and immunogenic in a phase I clinical trial. Clin. Vacc. Immunol. 2006; 13(11):1267–1277.

Martin, J.E., Pierson, T.C., Hubka, S., Rucker, S., Gordon, I.J., Enama, M.E., Andrews, C.A., Xu, Q., Davis, B.S., Nason, M., Fay, M., Koup, R.A., Rodederer, M., Bailer, R.T., Gomez, P.L., Mascola, J. R., Chang, G.J., Nabel, G.J. and Graham, B.S. A West Nile Virus DNA Vaccine Induces Neutralizing Antibody in Healthy Adults in a Phase I Clinical Trial. J. Infect. Dis. 2007; 196(2):1732–1740.

McConkey, S.J., Reece, W.H., Moorthy, V.S., Webster, D., Dunachie, S., Butcher, G., Vuola, J.M., Blanchard, T.J., Gothard, P., Watkins, K., Hannan, C.M., Everaere, S., Brown, K., Kester, K.E., Cummings, J., Williams, J., Heppner, D.G., Pathan, A., Flanagan, K., Arulanantham, N., Roberts, M.T., Roy, M., Smith, G.L., Schneider, J., Peto, T., Sinden, R.E., Gilbert, S.C. and Hill, A. V. Enhanced T-cell immunogenicity of plasmid DNA vaccines boosted by recombinant modified vaccinia virus Ankara in humans. Nat. Med. 2003; 9(6):729–735.

McShane, H., Brookes, R., Gilbert, S.C. and Hill, A.V. Enhanced immunogenicity of CD4(+) t-cell responses and protective efficacy of a DNA-modified vaccinia virus Ankara prime-boost vaccination regimen for murine tuberculosis. Infect. Immun. 2001; 69(2):681–686.

Mellquist-Riemenschneider, J.L., Garrison, A.R., Geisbert, J.B., Saikh, K.U., Heidebrink, K.D., Jahrling, P.B., Ulrich, R.G. and Schmaljohn, C.S. Comparison of the protective efficacy of DNA and baculovirus-derived protein vaccines for EBOLA virus in guinea pigs. Virus Res 2003; 92(2):187–193.

Miyashita, M., Joh, T., Watanabe, K., Todoroki, I., Seno, K., Ohara, H., Nomura, T., Miyata, M., Kasugai, K., Tochikubo, K., Itoh, M. and Nitta, M. Immune responses in mice to intranasal and intracutaneous administration of a DNA vaccine encoding Helicobacter pylori-catalase. Vaccine 2002; 20(17-18):2336–2342.

Mohamed, N., Clagett, M., Li, J., Jones, S., Pincus, S., D'Alia, G., Nardone, L., Babin, M., Spitalny, G. and Casey, L. A high-affinity monoclonal antibody to anthrax protective antigen passively protects rabbits before and after aerosolized Bacillus anthracis spore challenge. Infect. Immun. 2005; 73(2):795–802.

Moorthy, V.S., McConkey, S., Roberts, M., Gothard, P., Arulanantham, N., Degano, P., Schneider, J., Hannan, C., Roy, M., Gilbert, S.C., Peto, T.E. and Hill, A.V. Safety of DNA and modified vaccinia virus Ankara vaccines against liver-stage P. falciparum malaria in non-immune volunteers. Vaccine 2003; 21(17-18):1995–2002.

Morris, S., Kelley, C., Howard, A., Li, Z. and Collins, F. The immunogenicity of single and combination DNA vaccines against tuberculosis. Vaccine 2000; 18(20):2155–2163.

Mota, J., Acosta, M., Argotte, R., Figueroa, R., MÃ©ndez, A. and Ramos, C. Induction of protective antibodies against dengue virus by tetravalent DNA immunization of mice with domain III of the envelope protein. Vaccine 2005; 23(26):3469–3476.

Mwau, M., Cebere, I., Sutton, J., Chikoti, P., Winstone, N., Wee, E. G., Beattie, T., Chen, Y.H., Dorrell, L., McShane, H., Schmidt, C., Brooks, M., Patel, S., Roberts, J., Conlon, C., Rowland-Jones, S.L., Bwayo, J.J., McMichael, A.J. and Hanke, T. A human immunodeficiency virus 1 (HIV-1) clade A vaccine in clinical trials: stimulation of HIV-specific T-cell responses by DNA and recombinant modified vaccinia virus Ankara (MVA) vaccines in humans. J. Gen. Virol. 2004; 85(Pt 4):911–919.

Nagata, L., Hu, W.-G., Masri, S., Rayner, G., Schmaltz, F., Das, D., Wu, J., Long, M., Chan, C., Proll, D., Jager, S., Jebailey, L., Suresh, M. and Wong, J. Efficacy of DNA vaccination against western equine encephalitis virus infection. Vaccine 2005; 23(17-18):2280–2283.

NIH. Vaccine to Prevent West Nile Virus Disease. 2005, 02/24/2007. Retrieved 11/5/2007, 2007, from http://clinicaltrials.gov/show/NCT00106769.

NIH. West Nile Virus Vaccine Development. 2007, 07/31/2007. Retrieved 11/5/2007, 2007, from http://www.vrc.nih.gov/scientificupdates_westnile.htm.

O'Hagan, D., Singh, M., Ugozzoli, M., Wild, C., Barnett, S., Chen, M., Schaefer, M., Doe, B., Otten, G.R. and Ulmer, J.B. Induction of potent immune responses by cationic microparticles with adsorbed human immunodeficiency virus DNA vaccines. J. Virol. 2001; 75(19):9037–9043.

Okuda, K., Ihata, A., Watabe, S., Okada, E., Yamakawa, T., Hamajima, K., Yang, J., Ishii, N., Nakazawa, M., Okuda, K., Ohnari, K., Nakajima, K. and Xin, K.Q. Protective immunity against influenza A virus induced by immunization with DNA plasmid containing influenza M gene. Vaccine 2001; 19(27):3681–3691.

Otero, M., Calarota, S., Dai, A., De Groot, A., Boyer, J. and Weiner, D. Efficacy of novel plasmid DNA encoding vaccinia antigens in improving current smallpox vaccination strategy. Vaccine 2006; 24(21):4461–4470.

Otten, G.R., Schaefer, M., Doe, B., Liu, H., Srivastava, I., Megede, J., Kazzaz, J., Lian, Y., Singh, M., Ugozzoli, M., Montefiori, D., Lewis, M., Driver, D.A., Dubensky, T., Polo, J.M., Donnelly, J., O'Hagan, D.T., Barnett, S. and Ulmer, J.B. Enhanced potency of plasmid DNA microparticle human immunodeficiency virus vaccines in rhesus macaques by using a priming-boosting regimen with recombinant proteins. J. Virol. 2005; 79(13):8189–8200.

Pal, R., Wang, S., Kalyanaraman, V.S., Nair, B.C., Whitney, S., Keen, T., Hocker, L., Hudacik, L., Rose, N., Cristillo, A., Mboudjeka, I., Shen, S., Wu-Chou, T.H., Montefiori, D., Mascola, J., Lu, S. and Markham, P. Polyvalent DNA prime and envelope protein boost HIV-1 vaccine elicits humoral and cellular responses and controls plasma viremia in rhesus macaques following rectal challenge with an R5 SHIV isolate. J. Med. Primatol. 2005; 34(5-6):226–236.

Pal, R., Wang, S., Kalyanaraman, V.S., Nair, B.C., Whitney, S., Keen, T., Hocker, L., Hudacik, L., Rose, N., Mboudjeka, I., Shen, S., Wu-Chou, T.H., Montefiori, D., Mascola, J., Markham, P. and Lu, S. Immunization of rhesus macaques with a polyvalent DNA prime/protein boost human immunodeficiency virus type 1 vaccine elicits protective antibody response against simian human immunodeficiency virus of R5 phenotype. Virology 2006; 348(2):341–353.

Pancholi, P., Liu, Q., Tricoche, N., Zhang, P., Perkus, M.E. and Prince, A.M. DNA prime-canarypox boost with polycistronic hepatitis C virus (HCV) genes generates potent immune responses to HCV structural and nonstructural proteins. J. Infect. Dis. 2000; 182(1):18–27.

Perkins, S., O'Brien, L. and Phillpotts, R. Boosting with an adenovirus-based vaccine improves protective efficacy against Venezuelan equine encephalitis virus following DNA vaccination. Vaccine 2006; 24(17):3440–3445.

Pitisuttithum, P., Gilbert, P., Gurwith, M., Heyward, W., Martin, M., van Griensven, F., Hu, D., Tappero, J.W. and Choopanya, K. Randomized, double-blind, placebo-controlled efficacy trial of a bivalent recombinant glycoprotein 120 HIV-1 vaccine among injection drug users in Bangkok, Thailand. J. Infect. Dis. 2006; 194(12):1661–1671.

Pulford, D.J., Gates, A., Bridge, S.H., Robinson, J.H. and Ulaeto, D. Differential efficacy of vaccinia virus envelope proteins administered by DNA immunisation in protection of BALB/c mice from a lethal intranasal poxvirus challenge. Vaccine 2004; 22(25-26):3358–3366.

Putnak, R., Fuller, J., Vanderzanden, L., Innis, B. and Vaughn, D. Vaccination of rhesus macaques against dengue-2 virus with a plasmid DNA vaccine encoding the viral pre-membrane and envelope genes. Am. J. Trop. Med. Hyg. 2003; 68(4):469–476.

Qiu, M., Fang, F., Chen, Y., Wang, H., Chen, Q., Chang, H., Wang, F., Wang, H., Zhang, R. and Chen, Z. Protection against avian influenza H9N2 virus challenge by immunization with hemagglutinin- or neuraminidase-expressing DNA in BALB/c mice. Biochem. Biophys. Res. Commun. 2006; 343(4):1124–1131.

Ramshaw, I.A. and Ramsay, A.J. The prime-boost strategy: exciting prospects for improved vaccination. Immun. Today 2000; 21(4):163–165.

Raviprakash, K., Apt, D., Brinkman, A., Skinner, C., Yang, S., Dawes, G., Ewing, D., Wu, S.-J., Bass, S., Punnonen, J. and Porter, K. A chimeric tetravalent dengue DNA vaccine elicits neutralizing antibody to all four virus serotypes in rhesus macaques. Virology 2006; 353(1):166–173.

Raviprakash, K., Kochel, T.J., Ewing, D., Simmons, M., Phillips, I., Hayes, C.G. and Porter, K.R. Immunogenicity of dengue virus type 1 DNA vaccines expressing truncated and full length envelope protein. Vaccine 2000a; 18(22):2426–2434.

Raviprakash, K., Porter, K.R., Kochel, T.J., Ewing, D., Simmons, M., Phillips, I., Murphy, G.S., Weiss, W.R. and Hayes, C.G. Dengue virus type 1 DNA vaccine induces protective immune responses in rhesus macaques. J. Gen. Virol. 2000b; 81(Pt 7):1659–1667.

Repique, C.J., Li, A., Collins, F.M. and Morris, S.L. DNA immunization in a mouse model of latent tuberculosis: effect of DNA vaccination on reactivation of disease and on reinfection with a secondary challenge. Infect. Immun. 2002; 70(7):3318–3323.

Richmond, J.F., Lu, S., Santoro, J.C., Weng, J., Hu, S.L., Montefiori, D.C. and Robinson, H.L. Studies of the neutralizing activity and avidity of anti-human immunodeficiency virus type 1 Env antibody elicited by DNA priming and protein boosting. J. Virol. 1998; 72(11):9092–9100.

Riemenschneider, J., Garrison, A., Geisbert, J., Jahrling, P., Hevey, M., Negley, D., Schmaljohn, A., Lee, J., Hart, M., Vanderzanden, L., Custer, D., Bray, M., Ruff, A., Ivins, B., Bassett, A., Rossi, C. and Schmaljohn, C. Comparison of individual and combination DNA vaccines for B. anthracis, Ebola virus, Marburg virus and Venezuelan equine encephalitis virus. Vaccine 2003; 21(25-26):4071–4080.

Robinson, H.L., Boyle, C.A., Feltquate, D.M., Morin, M.J., Santoro, J.C. and Webster, R.G. DNA immunization for influenza virus: studies using hemagglutinin- and nucleoprotein-expressing DNAs. J. Infect. Dis. 1997; 176(Suppl 1):50–55.

Robinson, H.L., Hunt, L.A. and Webster, R.G. Protection against a lethal influenza virus challenge by immunization with a haemagglutinin-expressing plasmid DNA. Vaccine 1993; 11(9):957–960.

Rodriguez-Carreno, M., Nelson, M., Botten, J., Smith-Nixon, K., Buchmeier, M. and Whitton, J. Evaluating the immunogenicity and protective efficacy of a DNA vaccine encoding Lassa virus nucleoprotein. Virology 2005; 335(1):87–98.

Rottinghaus, S.T., Poland, G.A., Jacobson, R.M., Barr, L.J. and Roy, M.J. Hepatitis B DNA vaccine induces protective antibody responses in human non-responders to conventional vaccination. Vaccine 2003; 21(31):4604–4608.

Roy, M.J., Wu, M.S., Barr, L.J., Fuller, J.T., Tussey, L.G., Speller, S., Culp, J., Burkholder, J.K., Swain, W.F., Dixon, R.M., Widera, G., Vessey, R., King, A., Ogg, G., Gallimore, A., Haynes, J.R. and Fuller, D. Induction of antigen-specific CD8+ T cells, T helper cells, and protective levels of antibody in humans by particle-mediated administration of a hepatitis B virus DNA vaccine. Vaccine 2000; 19(7-8):764–778.

Sakhatskyy, P., Wang, S., Chou, T.H. and Lu, S. Immunogenicity and protection efficacy of monovalent and polyvalent poxvirus vaccines that include the D8 antigen. Virology 2006; 355(2):164–174.

Sakhatskyy, P., Wang, S., Chou, T.H. and Lu, S. Immunogenicity and protection efficacy of subunit-based smallpox vaccines using Variola Major antigens. Virology 2008; 371(1):98–107.

Schmaljohn, C., Custer, D., Vanderzanden, L., Spik, K., Rossi, C. and Bray, M. Evaluation of tick-borne encephalitis DNA vaccines in monkeys. Virology 1999; 263(1):166–174.

Schmaljohn, C., Vanderzanden, L., Bray, M., Custer, D., Meyer, B., Li, D., Rossi, C., Fuller, D., Fuller, J., Haynes, J. and Huggins, J. Naked DNA vaccines expressing the prM and E genes of Russian spring summer encephalitis virus and Central European encephalitis virus protect mice from homologous and heterologous challenge. J. Virol. 1997; 71(12):9563–9569.

Schneider, J., Langermans, J.A., Gilbert, S.C., Blanchard, T.J., Twigg, S., Naitza, S., Hannan, C.M., Aidoo, M., Crisanti, A., Robson, K.J., Smith, G.L., Hill, A.V. and Thomas, A.W. A prime-boost immunisation regimen using DNA followed by recombinant modified vaccinia virus Ankara induces strong cellular immune responses against the Plasmodium falciparum TRAP antigen in chimpanzees. Vaccine 2001; 19(32):4595–4602.

Seaman, M.S., Xu, L., Beaudry, K., Martin, K.L., Beddall, M.H., Miura, A., Sambor, A., Chakrabarti, B.K., Huang, Y., Bailer, R., Koup, R.A., Mascola, J.R., Nabel, G.J. and Letvin, N.L. Multiclade human immunodeficiency virus type 1 envelope immunogens elicit broad cellular and humoral immunity in rhesus monkeys. J. Virol. 2005; 79(5):2956–2963.

Stopeck, A.T., Hersh, E.M., Brailey, J.L., Clark, P.R., Norman, J. and Parker, S.E. Transfection of primary tumor cells and tumor cell lines with plasmid DNA/lipid complexes. Cancer Gene Ther. 1998; 5(2):119–126.

Sugawara, I., Yamada, H., Udagawa, T. and Huygen, K. Vaccination of guinea pigs with DNA encoding Ag85A by gene gun bombardment. Tuberculosis (Edinb.) 2003; 83(6):331–337.

Sullivan, N.J., Sanchez, A., Rollin, P.E., Yang, Z.Y. and Nabel, G.J. Development of a preventive vaccine for Ebola virus infection in primates. Nature 2000; 408(6812):605–609.

Tacket, C.O., Roy, M.J., Widera, G., Swain, W.F., Broome, S. and Edelman, R. Phase 1 safety and immune response studies of a DNA vaccine encoding hepatitis B surface antigen delivered by a gene delivery device. Vaccine 1999; 17(22):2826–2829.

Tanabayashi, K., Mukai, R., Yamada, A., Takasaki, T., Kurane, I., Yamaoka, M., Terazawa, A. and Konishi, E. Immunogenicity of a Japanese encephalitis DNA vaccine candidate in cynomolgus monkeys. Vaccine 2003; 21(19–20):2338–2345.

Tang, D.C., Devit, M. and Johnston, S.A. Genetic immunization is a simple method for eliciting an immune response. Nature 1992; 356(6365):152–154.

Tanghe, A., D'Souza, S., Rosseels, V., Denis, O., Ottenhoff, T.H., Dalemans, W., Wheeler, C. and Huygen, K. Improved immunogenicity and protective efficacy of a tuberculosis DNA vaccine encoding Ag85 by protein boosting. Infect. Immun. 2001; 69(5):3041–3047.

Tascon, R.E., Colston, M.J., Ragno, S., Stavropoulos, E., Gregory, D. and Lowrie, D.B. Vaccination against tuberculosis by DNA injection. Nat. Med. 1996; 2(8):888–892.

Tavel, J.A., Martin, J.E., Kelly, G.G., Enama, M.E., Shen, J.M., Gomez, P.L., Andrews, C.A., Koup, R.A., Bailer, R.T., Stein, J.A., Roederer, M., Nabel, G.J. and Graham, B.S. Safety and Immunogenicity of a Gag-Pol Candidate HIV-1 DNA Vaccine Administered by a Needle-Free Device in HIV-1-Seronegative Subjects. J. Acquir. Immune Defic. Syndr. 2007; 44(5):601–605.

Todoroki, I., Joh, T., Watanabe, K., Miyashita, M., Seno, K., Nomura, T., Ohara, H., Yokoyama, Y., Tochikubo, K. and Itoh, M. Suppressive effects of DNA vaccines encoding heat shock protein on Helicobacter pylori-induced gastritis in mice. Biochem. Biophys. Res. Commun. 2000; 277(1):159–163.

Ulmer, J.B., Donnelly, J.J., Parker, S.E., Rhodes, G.H., Felgner, P.L., Dwarki, V.J., Gromkowski, S.H., Deck, R.R., Dewitt, C.M. and Friedman, A. Heterologous protection against influenza by injection of DNA encoding a viral protein. Science 1993; 259(5102):1745–1749.

Ulmer, J.B., Liu, M.A., Montgomery, D.L., Yawman, A.M., Deck, R.R., DeWitt, C.M., Content, J. and Huygen, K. Expression and immunogenicity of Mycobacterium tuberculosis antigen 85 by DNA vaccination. Vaccine 1997; 15(8):792–794.

Ulmer, J.B., Fu, T.M., Deck, R.R., Friedman, A., Guan, L., DeWitt, C., Liu, X., Wang, S., Liu, M.A., Donnelly, J.J. and Caulfield, M.J. Protective CD4+ and CD8+ T cells against influenza virus induced by vaccination with nucleoprotein DNA. J. Virol. 1998; 72(7):5648–5653.

Vanderzanden, L., Bray, M., Fuller, D., Roberts, T., Custer, D., Spik, K., Jahrling, P., Huggins, J., Schmaljohn, A. and Schmaljohn, C. DNA vaccines expressing either the GP or NP genes of Ebola virus protect mice from lethal challenge. Virology 1998; 246(1):134–144.

Vuola, J.M., Keating, S., Webster, D.P., Berthoud, T., Dunachie, S., Gilbert, S.C. and Hill, A.V. Differential immunogenicity of various heterologous prime-boost vaccine regimens using DNA and viral vectors in healthy volunteers. J. Immunol. 2005; 174(1):449–455.

Wang, B., Ugen, K.E., Srikantan, V., Agadjanyan, M.G., Dang, K., Refaeli, Y., Sato, A.I., Boyer, J., Williams, W.V. and Weiner, D.B. Gene inoculation generates immune responses against human immunodeficiency virus type 1. Proc. Natl. Acad. Sci. USA 1993; 90(9):4156–4160.

Wang, R., Doolan, D.L., Le, T.P., Hedstrom, R.C., Coonan, K.M., Charoenvit, Y., Jones, T.R., Hobart, P., Margalith, M., Ng, J., Weiss, W.R., Sedegah, M., De Taisne, C., Norman, J.A. and Hoffman, S.L. Induction of antigen-specific cytotoxic T lymphocytes in humans by a malaria DNA vaccine. Science 1998; 282(5388):476–480.

Wang, R., Epstein, J., Baraceros, F.M., Gorak, E.J., Charoenvit, Y., Carucci, D.J., Hedstrom, R.C., Rahardjo, N., Gay, T., Hobart, P., Stout, R., Jones, T.R., Richie, T.L., Parker, S.E., Doolan, D.L., Norman, J. and Hoffman, S.L. Induction of CD4(+) T cell-dependent CD8(+) type 1 responses in humans by a malaria DNA vaccine. Proc. Natl. Acad. Sci. USA 2001; 98(19):10817–10822.

Wang, R., Epstein, J., Charoenvit, Y., Baraceros, F.M., Rahardjo, N., Gay, T., Banania, J.G., Chattopadhyay, R., de la Vega, P., Richie, T.L., Tornieporth, N., Doolan, D.L., Kester, K.E., Heppner, D.G., Norman, J., Carucci, D.J., Cohen, J.D. and Hoffman, S.L. Induction in humans of CD8+ and CD4+ T cell and antibody responses by sequential immunization with malaria DNA and recombinant protein. J. Immunol. 2004a; 172(9):5561–5569.

Wang, S., Arthos, J., Lawrence, J.M., Van Ryk, D., Mboudjeka, I., Shen, S., Chou, T.H., Montefiori, D.C. and Lu, S. Enhanced immunogenicity of gp120 protein when combined with recombinant DNA priming to generate antibodies that neutralize the JR-FL primary isolate of human immunodeficiency virus type 1. J. Virol. 2005a; 79(12):7933–7937.

Wang, S., Chou, T.H., Sakhatskyy, P., Huang, S., Lawrence, J., Cao, H., Huang, X. and Lu, S. Identification of two neutralizing regions on SARS-CoV spike glycoprotein produced from mammalian expression system. J. Virol. 2005b; 79(3):1906–1910.

Wang, S., Farfan-Arribas, D.J., Shen, S., Chou, T.H., Hirsch, A., He, F. and Lu, S. Relative contributions of codon usage, promoter efficiency and leader sequence to the antigen expression and immunogenicity of HIV-1 Env DNA vaccine. Vaccine 2006a; 24(21):4531–4540.

Wang, S., Heilman, D., Liu, F., Giehl, T., Joshi, S., Huang, X., Chou, T.H., Goguen, J. and Lu, S. A DNA vaccine producing LcrV antigen in oligomers is effective in protecting mice from lethal mucosal challenge of plague. Vaccine 2004b; 22(25–26):3348–3357.

Wang, S., Kennedy, J., West, K., Montefiori, D.C., Bansal, A., Goepfert, P., Coley, S., Lawrence, J., Shen, S., Green, S., Rothman, A., Ennis, F., Pal, R., Markham, P. and Lu, S. Balanced cellular and antibody responses induced by the polyvalent DNA prime-protein boost HIV-1 vaccine formulation DP6-001 in healthy human volunteers. Vaccine 2008a; 26(31):3947–3957.

Wang, S., Joshi, S., Mboudjeka, I., Liu, F., Ling, T., Goguen, J.D. and Lu, S. Relative immunogenicity and protection potential of candidate Yersinia Pestis antigens against lethal mucosal plague challenge in Balb/C mice. Vaccine 2008b; 26(13):1664–1674.

Wang, S., Pal, R., Mascola, J.R., Chou, T.H., Mboudjeka, I., Shen, S., Liu, Q., Whitney, S., Keen, T., Nair, B.C., Kalyanaraman, V.S., Markham, P. and Lu, S. Polyvalent HIV-1 Env vaccine formulations delivered by the DNA priming plus protein boosting approach are effective in generating neutralizing antibodies against primary human immunodeficiency virus type 1 isolates from subtypes A, B, C, D and E. Virology 2006b; 350(1):34–47.

Wang, S., Taaffe, J., Parker, C., Solórzano, A., Cao, H., García-Sastre, A. and Lu, S. Hemagglutinin (HA) proteins from H1 and H3 serotypes of influenza A viruses require different antigen designs for the induction of optimal protective antibody responses as studied by codon-optimized HA DNA vaccines. J. Virol. 2006c; 80(23):11628–11637.

Wheeler, C.J., Felgner, P.L., Tsai, Y.J., Marshall, J., Sukhu, L., Doh, S.G., Hartikka, J., Nietupski, J., Manthorpe, M., Nichols, M., Plewe, M., Liang, X., Norman, J., Smith, A. and Cheng, S.H. A novel cationic lipid greatly enhances plasmid DNA delivery and expression in mouse lung. Proc. Natl. Acad. Sci. USA 1996; 93(21):11454–11459.

WHO. The global eradication of smallpox. Final report of the global commission for the certification of smallpox eradication. In: History of International Public Health. Geneva, World Health Organization. 1980; **4**.

Wiker, H.G. and Harboe, M. The antigen 85 complex: a major secretion product of Mycobacterium tuberculosis. Microbiol. Rev. 1992; 56(4):648–661.

Williamson, E.D., Bennett, A.M., Perkins, S.D., Beedham, R.J., Miller, J. and Baillie, L.W. Co-immunisation with a plasmid DNA cocktail primes mice against anthrax and plague. Vaccine 2002; 20(23–24):2933–2941.

Williamson, E.D., Eley, S.M., Stagg, A.J., Green, M., Russell, P. and Titball, R.W. A sub-unit vaccine elicits IgG in serum, spleen cell cultures and bronchial washings and protects immunized animals against pneumonic plague. Vaccine 1997; 15(10):1079–1084.

Williamson, E.D., Eley, S.M., Stagg, A.J., Green, M., Russell, P. and Titball, R.W. A single dose sub-unit vaccine protects against pneumonic plague. Vaccine 2000; 19(4–5):566–571.

Woo, P.C.Y., Lau, S.K.P., Tsoi, H.-W., Chen, Z.-W., Wong, B.H.L., Zhang, L., Chan, J.K.H., Wong, L.-P., He, W., Ma, C., Chan, K.-H., Ho, D.D. and Yuen, K.-Y. SARS coronavirus spike polypeptide DNA vaccine priming with recombinant spike polypeptide from Escherichia coli as booster induces high titer of neutralizing antibody against SARS coronavirus. Vaccine 2005; 23(42):4959–4968.

Woodland, D.L. Jump-starting the immune system: prime-boosting comes of age. Trends Immunol. 2004; 25(2):98–104.

Wu, C.-J., Huang, H.-W. and Tao, M.-H. Induction of cross-protection against two wild-type Taiwanese isolates of Japanese encephalitis virus using Beijing-1 strain DNA vaccine. Vaccine 2003; 21(25–26):3938–3945.

Wu, L., Kong, W.P. and Nabel, G.J. Enhanced breadth of CD4 T-cell immunity by DNA prime and adenovirus boost immunization to human immunodeficiency virus Env and Gag immunogens. J. Virol. 2005; 79(13):8024–8031.

Wu, S.F., Liao, C.L., Lin, Y.L., Yeh, C.T., Chen, L.K., Huang, Y.F., Chou, H.Y., Huang, J.L., Shaio, M.F. and Sytwu, H.K. Evaluation of protective efficacy and immune mechanisms of using a non-structural protein NS1 in DNA vaccine against dengue 2 virus in mice. Vaccine 2003; 21(25–26):3919–3929.

Xiang, Z.Q., Spitalnik, S., Tran, M., Wunner, W.H., Cheng, J. and Ertl, H.C.J. Vaccination with a Plasmid Vector Carrying the Rabies Virus Glycoprotein Gene Induces Protective Immunity against Rabies Virus. Virology 1994; 199(1):132–140.

Xiang, Z.Q., Spitalnik, S.L., Cheng, J., Erikson, J., Wojczyk, B. and Ertl, H.C.J. Immune Responses to Nucleic Acid Vaccines to Rabies Virus. Virology 1995; 209(2):569–579.

Xiao, Y., Aldaz-Carroll, L., Ortiz, A.M., Whitbeck, J.C., Alexander, E., Lou, H., Davis, H.L., Braciale, T.J., Eisenberg, R.J., Cohen, G.H. and Isaacs, S.N. A protein-based smallpox vaccine protects mice from vaccinia and ectromelia virus challenges when given as a prime and single boost. Vaccine 2007; 25(7):1214–1224.

Xu, L., Sanchez, A., Yang, Z., Zaki, S.R., Nabel, E.G., Nichol, S.T. and Nabel, G.J. Immunization for Ebola virus infection. Nat. Med. 1998; 4(1):37–42.

Yang, J.S., Kim, J.J., Hwang, D., Choo, A.Y., Dang, K., Maguire, H., Kudchodkar, S., Ramanathan, M.P. and Weiner, D.B. Induction of potent Th1-type immune responses from a novel DNA vaccine for West Nile virus New York isolate (WNV-NY1999). J. Infect. Dis. 2001; 184(7):809–816.

Yang, Z.Y., Kong, W.P., Huang, Y., Roberts, A., Murphy, B.R., Subbarao, K. and Nabel, G.J. A DNA vaccine induces SARS coronavirus neutralization and protective immunity in mice. Nature 2004; 428(6982):561–564.

Yuan, L., Azevedo, M.S.P., Gonzalez, A.M., Jeong, K.-i., Van Nguyen, T., Lewis, P., Iosef, C., Herrmann, J.E. and Saif, L.J. Mucosal and systemic antibody responses and protection induced by a prime/boost rotavirus-DNA vaccine in a gnotobiotic pig model. Vaccine 2005; 23(30):3925–3936.

Zakhartchouk, A., Liu, Q., Petric, M. and Babiuk, L. Augmentation of immune responses to SARS coronavirus by a combination of DNA and whole killed virus vaccines. Vaccine 2005; 23(35):4385–4391.

Zarozinski, C.C., Fynan, E.F., Selin, L.K., Robinson, H.L. and Welsh, R.M. Protective CTL-dependent immunity and enhanced immunopathology in mice immunized by particle bombardment with DNA encoding an internal virion protein. J. Immunol. 1995; 154(8):4010–4017.

Zhao, B., Jin, N.-Y., Wang, R.-L., Zhang, L.-S. and Zhang, Y.-J. Immunization of mice with a DNA vaccine based on severe acute respiratory syndrome coronavirus spike protein fragment 1. Viral Immunol. 2006; 19(3):518–524.

Zhao, P., Cao, J., Zhao, L.-J., Qin, Z.-L., Ke, J.-S., Pan, W., Ren, H., Yu, J.-G. and Qi, Z.-T. Immune responses against SARS-coronavirus nucleocapsid protein induced by DNA vaccine. Virology 2005; 331(1):128–135.

Zhu, M.-S., Pan, Y., Chen, H.-Q., Shen, Y., Wang, X.-C., Sun, Y.-J. and Tao, K.-H. Induction of SARS-nucleoprotein-specific immune response by use of DNA vaccine. Immunology letters 2004; 92(3):237–243.

CHAPTER

9

Vaccine Adjuvants*

Carl R. Alving
Department of Adjuvant and Antigen Research,
Division of Retrovirology, Walter Reed Army Institute of Research,
U.S. Military HIV Research Program,
Rockville, MD, USA

OUTLINE

ABSTRACT

It is widely believed that adjuvants are important, and in some cases critical, for the success of most modern vaccines, particularly for new types of vaccines that have highly purified or synthetic antigens. Although aluminum salts are the most commonly used type of adjuvant for human vaccines, they are weak adjuvants that have complex mechanisms that favor induction of antibodies rather than cellular immunity. Aluminum salts do have a long record of relative safety, but they are also often responsible for local reactions at the site of injection, particularly for reactions that are associated with subcutaneous administration. Adjuvant selection for human vaccines still relies strongly on direct empirical testing of candidate adjuvants for safety and efficacy in humans. However, principles of innate immunity have been developed that provide some guidance for rational selection of adjuvants. New forms of vaccine adjuvants that have been proposed for various vaccines feature oil-based emulsions; bacterial products, such as lipid A, heat-labile *E. coli* enterotoxin, or CpG nucleotides; viral

*The views expressed are those of the author and should not be construed to represent the positions of the Department of the Army or Department of Defense.

products, such as virus-like particles; plant products, such as saponin derivatives; biodegradable particles, such as liposomes; molecular adjuvants; and synthetic adjuvants. Adjuvant mechanisms include depot effects, recruitment of innate immunity mechanisms, specific targeting mechanisms, and carrier functions that hold the antigen in an appropriate conformation. The safety of proposed adjuvants is a primary consideration, and for this purpose it is often necessary to devise methods to reduce or eliminate the reactogenic effects of an adjuvant while preserving the efficacy. The most effective use of adjuvants for certain types of vaccines, particularly for stimulating mucosal immunity, may be to combine the adjuvant with a particular mode of delivery, such as oral, intranasal, or transcutaneous immunization. Carriers that carry and combine both the adjuvant and the antigen in a single formulation can serve as the basis for creation of important formulations for improved vaccines.

THE REQUIREMENT FOR VACCINE ADJUVANTS

A common observation in the process of recruiting the immune system to produce specific or nonspecific immunity is that the antigen by itself may not be adequate as a stimulating agent. Many potential antigens may even have no apparent immunizing activity at all when tested alone. This is particularly true for most small peptides that serve as antigens (Alving et al., 1995), but it can also be true for virtually any antigen. Because of this, the immunologist is often forced to reach into a bag of tricks to find an adjuvant, or an adjuvant strategy, to optimize the immune response. In the context of this chapter, a vaccine adjuvant is defined as any substance or strategy that improves the adaptive immune response or that stimulates the innate immune system to induce desired effectors or mediators.

Because of the need for potent adjuvants that are required for vaccines against highly purified antigens and for vaccines to chronic diseases that are difficult to control, such as HIV, parasitic diseases, and cancer, there has been considerable recent research directed toward the discovery and development of new adjuvants that are safer and stronger. Much of this recent scientific discovery activity has been stimulated both by new understanding of molecular mechanisms underlying certain types of innate immunity (Pulendran, 2004), and by the introduction of DNA that can code both for the adjuvant (sometimes referred to as a "molecular adjuvant"), and for the antigen, as components of DNA-based vaccines (Calarota and Weiner, 2004).

The definition of the term "adjuvant" that is used in the context of this chapter is quite broad and encompasses numerous mechanisms. The heterogeneity of potential adjuvants and adjuvant mechanisms has stimulated many attempts to organize and classify groups of adjuvants, often according to perceived mechanism of action, so that rational selection of adjuvants can be made from adjuvant groups. This is a useful exercise that serves to identify huge numbers of candidate adjuvants (Edelman and Tacket, 1990; Cox and Coulter, 1997; Schijns, 2000; Kenney et al., 2002; Vogel and Alving, 2002). One example of such a listing is given in Table 9.1. However, such classifications are inherently limited because individual adjuvants often have more than one mechanism or more than one dominant characteristic, and therefore, may not lend themselves easily to overarching principles of selection. Commercial considerations, such as ease of manufacture, cost, or toxic side effects are often the deciding factors for an individual adjuvant in a given vaccine. From a practical standpoint, for vaccine development there would seem to be an obvious benefit to utilizing multiple adjuvant mechanisms to either cause additive effects, or reduce the amount of expensive antigens required, or lower the side effects and enhance the safety of individual adjuvants.

In past years, and even to the present, an adjuvant function has been accomplished with human vaccines mainly through adsorption of the antigen to an aluminum salt (Glenny et al, 1926; Aprile and Wardlaw, 1966; Gupta, 1998; Lindblad, 2004). However, it should be noted that aluminum salts generally perform only weakly as adjuvants, particularly for booster immunizations, even though they do have a relatively good record of safety (Aprile and Wardlaw, 1966; Gupta et al., 1993; Baylor et al., 2002; Lindblad, 2004). Studies in which aluminum salts have been directly compared with modern experimental adjuvants in humans have shown that the aluminum adjuvants by themselves usually have lower levels of potency than many newer adjuvants (Gordon, 1993; McElrath, 1995). It may be important for certain vaccines, particularly those that address difficult diseases, such as malaria or HIV infection, to include modern adjuvants that are much stronger than the weak adjuvant activity obtained from simple adsorption of antigen to aluminum salts.

It should not be surprising that adjuvants may be required for optimization of vaccines when one considers that most elements of the immune system rely on various amplification systems as key strategies, and a highly purified antigen may lack the

TABLE 9.1 Types of immunological adjuvants (modified from Vogel and Alving, 2002)

Type of adjuvant	General type	Specific examples
Gel-type	Aluminum salts ("alum adjuvants") Calcium salts	Aluminum hydroxide, aluminum phosphate Calcium phosphate
Microbial	Muramyl dipeptide (MDP) derivatives Bacterial endotoxin Bacterial DNA Bacterial exotoxins	Murabutide, threonyl-MDP Monophosphoryl lipid A CpG Cholera toxin (CT), *Escherichia coli* heat-labile enterotoxin (LT)
Particulate	Biodegradable polymer microspheres, immuno stimulatory complexes (ISCOMs), liposomes	Virosomes
Oil emulsions and surfactants	Freund's incomplete adjuvant Microfluidized emulsions Saponins	Montanide ISA 720 MF59, AS02A QS-21
Synthetic	Nonionic block copolymers, polyphosphazene (PCPP), synthetic polynucleotides	Poly A:U, Poly A:C
Cytokines		Interleukin (IL)-2; IL-12; granulocyte-macrophage colony-stimulating factor (GM-CSF); interferon gamma (IFN-γ)
Genetic	Cytokine genes or genes encoding costimulatory molecules delivered as plasmid DNA	IL-12, IL-2, IFN-γ

ability to activate some or all of the important or most useful amplification systems. However, the older belief that adjuvants are materials that simply increase the potency of, and serve to enhance, the specific immune characteristics embodied in the vaccine antigen itself is often probably wrong. When antigens are taken out of the complex environment in which they exist in nature, whether they are normally present on bacteria, viruses, or cancer cells, the purified antigens often have a reduced ability to stimulate the immune system in the desired manner. Foreign invaders, such as bacteria, viruses, and parasites not only have antigens against which the host immune system reacts, but they also contain built-in adjuvants that carry and present the antigen in an appropriate manner and channel the immune response in an effective direction. When purified antigens are used it is often necessary to utilize adjuvants or adjuvant strategies to achieve the most efficacious presentation and amplification mechanisms.

Examples of current vaccines containing crude antigen preparations that have built-in adjuvant properties include parenterally injected killed inactivated whole cell bacterial vaccines, such as inactivated whole cell *Bordetella pertussis*, *Salmonella typhi*, or *Vibrio cholera* vaccines, which may be contaminated with bacterial endotoxin (Gupta et al., 1993). Endotoxin is a compound that can have strong natural adjuvant activity, and all of the adjuvant activity of endotoxin is due to the lipid A moiety of the bacterial surface lipopolysaccharide (LPS) (Alving and Rao, 2008).

There are, however, some unusual examples in which purified antigens themselves also have adjuvant activity. Three examples, all involving bacterial toxins, are as follows. Cholera toxin (CT) from *V. cholera* and heat-labile enterotoxin (LT) from *Escherichia coli* are examples of antigens that can exhibit both antigenic and adjuvant activity (Elson and Ealding, 1984; Snider, 1995; Glenn et al., 1998, 1999; Lavelle et al., 2004a, 2004b), and CT directly activates dendritic cells (Kawamura et al., 2003). Recombinant protective antigen, a component of the toxin of *Bacillus anthracis*, exhibited apparent adjuvant activity for inducing antibodies to itself by transcutaneous immunization, a process that normally requires an adjuvant in order to be effective (Matyas et al., 2004).

But if amplification is required beyond that supplied by the antigen itself, then the direction of amplification may also be critical. For creation of vaccines, amplification of T regulatory lymphocytes that suppress the desired immune response presumably would be an unwanted outcome (Pulendran, 2004; Pennington et al., 2005). Certain adjuvants may have complex immunomodulatory activities (Lavelle et al., 2004a, 2004b). Likewise, enhancement of IgE production by adjuvants such as by aluminum salts (Baylor et al., 2002), or by ADP-ribosylating toxins (Freytag and Clements, 2005), could be viewed as unfavorable adjuvant effects. Furthermore, in some instances the turning on of the amplification mechanisms in the host can result in unintended attacks on the host itself, and the side effects and toxicities of adjuvants are important factors in balancing the usefulness of an adjuvant. The engine of amplification not only has to be turned on, but it also has to be harnessed and tuned to guide it in the appropriate direction.

It is also important to remember that the initiation, quality, and magnitude of the immune response to a vaccine antigen can be influenced by numerous nonspecific factors. These factors include the type and dose of the antigen, the timing of immunization and boosting, the route of immunization, and even the age and general health of the vaccinee. In some instances, the utilization of these other factors may be considered to be types of adjuvant strategies. For example, delivery of antigens by needle-free methods, such as transcutaneous immunization (Glenn et al., 2000), or oral delivery (e.g., plant-derived vaccines), or vaccines delivered by intranasal routes (McGhee et al., 1992; González et al., 2004; Freytag and Clements, 2005; Rimoldi and Rescigno, 2005) can result in the induction of mucosal immunity, an effect that directs the course of the immune response simply by the route of vaccine delivery. However, additional adjuvants may also be required for induction of mucosal immunity (Eriksson and Holmgren, 2002).

Two Examples of How Adjuvants can Affect the Qualitative Type of Immune Response

Example 1

Differences in protection against *Mycobacterium tuberculosis* challenge in mice were induced by different adjuvant formulations (Lima et al., 2004). In this study, it was found that injection of recombinant mycobacterial 65 kDa heat shock protein (hsp65) as an antigen by itself was unable to protect mice, even with Freund's incomplete adjuvant. However, when the antigen was administered as a DNA vaccine or as a recombinant protein entrapped in cationic liposomes, protection was observed. In contrast, when hsp65 was encapsulated in polymeric microspheres for sustained release of antigen, high levels of specific antibody were observed but the mice were not protected after challenge. As the authors stated, the vaccine adjuvant "makes the difference."

Example 2

Cytokines are often cited as important adjuvants, but it has been found that mixtures of cytokines can direct the immune responses in various, perhaps even unanticipated, directions. Interleukin-12 (IL-12) together with granulocyte-macrophage colony-stimulating factor (GM-CSF) or tumor necrosis factor-alpha emulsified in Freund's adjuvant was used to steer immune responses to induction of cytotoxic T lymphocytes in BALB/c mice (i.e., toward a Th1 response) (Ahlers et al., 1997). However, GM-CSF alone gave mixed Th1 and Th2 responses, while GM-CSF plus IL-7 gave predominantly a Th2 response.

FREUND'S ADJUVANT AND ALUMINUM SALTS

Freund's Adjuvant

Starting in the 1930s, Jules Freund provided evidence that antigens emulsified in water-in-oil emulsions often had a marked enhancing effect on the immune response (Freund, 1956; Hilleman, 1966). Because of Freund's interest in tuberculosis, his vaccines are referred to as either complete (containing whole-killed mycobacteria) or incomplete (lacking mycobacteria). Although complete Freund's adjuvant is universally regarded as too toxic for use in humans, incomplete Freund's has been used as a potent adjuvant formulation for a diversity of vaccines in more than a million people (Davenport, 1968; Stuart-Harris, 1969; Alving, 2002). Research and commercial applications continue to fuel an interest in Freund's adjuvant, and it is currently being developed for certain human and veterinary vaccines (Chang et al., 1998; Miles et al., 2005; Saul et al., 2005). Although incomplete Freund's adjuvant has a widespread reputation for toxicity in humans (Davenport, 1968; Chang et al., 1998), this is probably largely undeserved because it is based mainly on faulty animal models that do not reflect or translate to expression of human toxicity (Stuart-Harris, 1969; Alving, 2002)

Aluminum Salts

The early introduction of aluminum salts as adjuvants (Glenny et al., 1926) and their endurance as components of numerous types of current human vaccines (Baylor et al., 2002) has led to a huge literature addressing the chemical, physical, and immunostimulatory characteristics and behavior of this class of compounds (Aprile and Wardlaw, 1966; Gupta et al., 1995; Gupta, 1998; Lindblad, 2004). The adjuvant activity of an aluminum salt is often relatively weak, so much so that even with certain established vaccines the presence of enhanced immunological potency provided by aluminum has sometimes been inconsistent (Aprile and Wardlaw, 1966). However, because of long experience with hundreds of millions of doses in human vaccines, and the lack of parenteral side effects, and relatively benign, albeit irritating, local side effects at the site of injection, observed with aluminum salts, this class of adjuvants has occupied a favored place as a frequent first choice for many vaccines.

The commonly used aluminum salt adjuvants include various physical forms of aluminum phosphate ($AlPO_4$), aluminum hydroxide [($Al(OH)_3$], and aluminum-precipitates of protein (Shirodkar et al., 1990). These compounds are currently so widely referred to informally as a group by the term "alum" that this term has become an accepted generic shorthand in the vaccine field for aluminum salt adjuvants even though the name "alum" was originally given to potassium aluminum sulfate [$KAl(SO_4)_2$], a chemical entity that apparently might be used in the formulation of certain vaccines in the United States (Baylor et al., 2002), but is not a favored entity (Gupta, 1998).

The dose of aluminum adjuvant is an important consideration. It is important to understand that there may be an optimal level of aluminum for a given vaccine antigen, with decreased immune response at higher levels of aluminum (Gupta, 1998). Although it has been stated that aluminum remains at the site of injection for a long period (Gupta, 1998), small particles of aluminum salts can be internalized by dendritic cells (Morefield et al., 2005), and are known to be removed from the body with time (Hem, 2002). The acceptable dose of aluminum in adjuvants ranges up to 1.25 mg both by US and by WHO regulations, but currently licensed US vaccines have a range of only 0.125–0.85 mg (Baylor et al., 2002). The latter reference can also be consulted for a relatively current list of all licensed US vaccines containing aluminum and the aluminum content of each.

The most common current method of using aluminum is by simple adsorption, and this is accomplished by mixing of antigen with preformed aluminum gel. As deceptively simple as this may seem, considerable difficulty in adjuvant reproducibility may be encountered because of the huge complexity of aluminum–protein interactions. For example, it has been noted that different lots or different brands of the same type of aluminum salt can be inconsistent in adsorptive capacity (Wassef et al., 1994). The physiochemical complexity of interactions of aluminum salts with antigens extends not only to the adsorption of antigen during vaccine manufacture, but also to potential exchange of protein antigen with interstitial proteins and other constituents of interstitial fluids after injection (Seeber et al., 1991b; Heimlich et al., 1999; Iyer et al., 2003).

One of the major benefits of adsorption of antigen to aluminum salts is that combination vaccines can be formulated on a particulate adjuvant. Such vaccines include diphtheria–pertussis–tetanus (DPT), DPT–hemophilis influenzae, or hepatitis B–hemophilus influenzae vaccines (Baylor et al., 2002). It has been shown that cycles of deaggregation and reaggregation of aluminum hydroxide adjuvant lead to homogeneous mixing of multiple antigens on aluminum salt particles in vaccines containing more that one antigen (Morefield et al., 2004). Adsorption to alum salt might also have the interesting benefit of stabilizing and retarding the degradation of an antigen that degrades with a pH-dependent mechanism (Wittayanukulluk et al., 2004).

The adsorptive binding of proteins to aluminum is often due to electrostatic interactions, and adsorption can be strongly affected by the age of the gel (Burrell, et al., 2000), the charge of the gel, the charge of the adsorbed protein, the physical structure of the aluminum compound, and the pH and other characteristics of the buffer that is used (Wittayanukulluk et al., 2004; Seeber et al., 1991a; Callahan et al., 1991; Feldkamp et al., 1981; reviewed by Gupta, 1998). It is possible to manipulate the electrostatic adsorption in certain cases. Aluminum hydroxide gel is positively charged at pH 7.4 and it adsorbs negatively charged proteins. However, pretreatment of aluminum hydroxide with buffer containing phosphate can lower, or even reverse, the positive charge, even allowing the electrostatic adsorption of cationic protein (Rinella et al., 1996).

Many other mechanisms also play a role in the adsorptive process, including hydrophobic interactions, van der Waals forces, and hydrogen bonding (reviewed by Gupta, 1998). Recently it was suggested that ligand exchange between the phosphate groups of phospholipids in the hepatitis B surface antigen (HBsAg) with the hydroxyl groups of aluminum hydroxide provides a mechanism of high affinity binding (Iyer et al., 2004) and that ligand exchange is a stronger force than electrostatic attraction for adsorption of antigens to aluminum hydroxide (Jiang et al., 2006). Presumably this mechanism could also account for the binding of phospholipid liposomes containing monophosphoryl lipid A to aluminum hydroxide adjuvant (Wassef et al., 1994; Alving et al., 1993).

ADJUVANT MECHANISMS

Depot Effects

This strategy encompasses numerous potential formulations including aluminum salts (see above), complex emulsions (Muderhwa et al., 1999; Podda, 2001; Matyas et al., 2003; Richards et al., 2004), various microparticles (O'Hagan et al., 2004; Peyre et al., 2004; Rydell and Sjöholm, 2005; Mossman, et al.,

2005; Westwood et al., 2005; Cai et al., 2005), and other similar materials. A strong argument has been made that one of the major mechanisms of adjuvant activity of oil-based emulsions (although by no means the only mechanism) is through the formation of a depot at the site of subcutaneous (SC) injection that gradually releases small amounts of antigen over a very long time. In one study, after injection of ovalbumin emulsified with IFA, ovalbumin was detectable in the SC IFA emulsion recovered from injected mice for up to 544 days after inoculation and the half-life of ovalbumin within the emulsion was found to be about 90 days (Herbert, 1968). Neither the oil, nor the emulsifier, nor the emulsion itself when injected at a separate site from the antigen, served as an adjuvant for inducing high antibody titers.

Emulsion science, as it relates to oil-based adjuvants, relies heavily on theoretical mechanisms that control the effects of altering the relative amounts of oil and emulsifier compounds in the formulation (Hunter et al., 1981; Woodard, 1990). Variation in the ratio of the oil and emulsifier provide a mechanism for allowing different amounts of stability of the emulsion prior to slow breakup of the emulsion and gradual release of antigen. A common experimental mistake in the formulation of emulsions, such as IFA or other types of oil-based emulsions, is to ignore the emulsifying properties of individual protein antigens that are present in the emulsion. Protein constituents that are present at certain concentrations in emulsions can sometimes cause breakdown of the emulsion. The stability of an emulsion is related to relative chemical properties of emulsifying agents that are known as the "hydrophile-lipophile balance" (HLB), and the adjuvant effects of emulsions were proposed to be related to the HLB (Hunter et al., 1981; Woodard, 1990). Figure 9.1 illustrates a complex emulsion in which different amounts of liposomes containing lipid A were used as an emulsifier for creating oil-in-water emulsions having different degrees of stability. The adjuvant goals in the depot approach can be several. There might be a need to protect the antigen from degradation until it gains access to an antigen-presenting cell (APC). Particles might be able to enter an APC and direct the antigen into an appropriate antigen presentation pathway (Rock, 1996). The material might also act either as an extracellular depot that attracts APCs to the site of the depot, or as an intracellular depot that protects the antigen until it can reach the intracellular location required to generate an immune response. In the case of aluminum salts, which can be taken up by dendritic cells (Morefield et al., 2005), but do not deliver antigen to an intracellular MHC class I pathway for promotion of Th1-type immunity, it would seem more likely that the antigen is either being held as a depot for ingestion by APCs to induce Th2-type immunity, or is being released for uptake by APCs (Jiang et al., 2006), and is promoting the proliferation of B-cell clones for secretion of antibodies.

Recent work has suggested that immunostimulatory effects of emulsions can be partly or wholly independent of both the HLB and the relative stability of the emulsion (Yang et al., 2005). There are controversies in the adjuvant field as to whether one or another

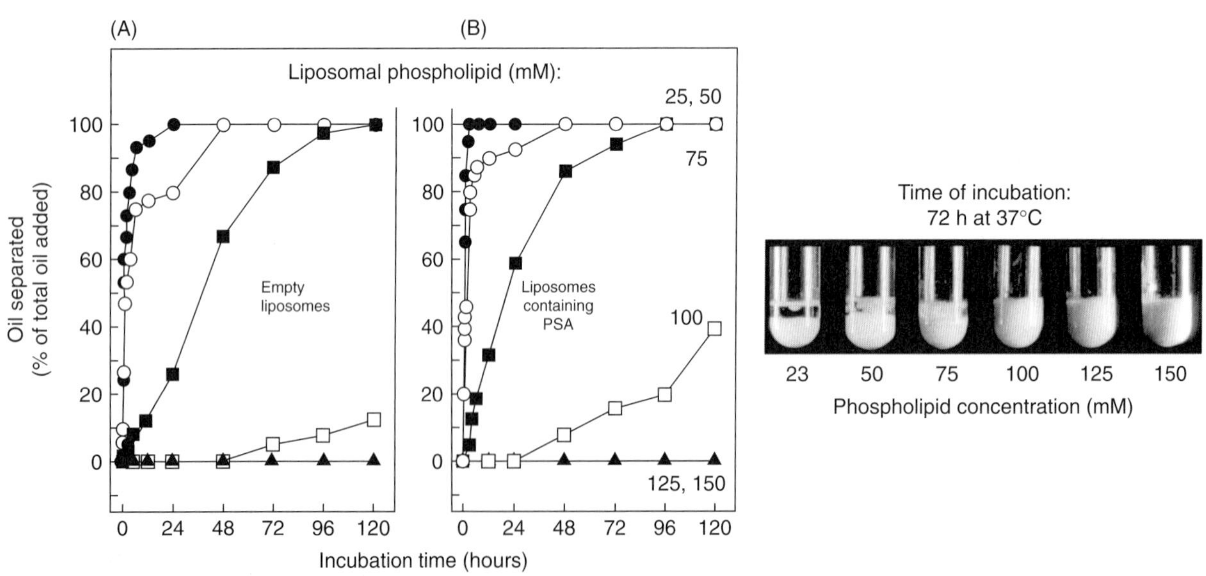

FIGURE 9.1 Time course of the degree of stability of emulsions formulated with 42.5% (v/v) of mineral oil and liposomes containing different concentrations of phospholipid and either (A) lacking or (B) containing prostate-specific antigen. Modified from Muderhwa et al. (1999) (with permission).

function of a given adjuvant represents the dominant mechanism, but it does seem unlikely that an emulsion, or a deposited particle, could simply sit as an inert material at the site of injection and still exhibit substantial adjuvant effects. As with many adjuvants, including aluminum salts (Carroll, 1998; Ulanova et al., 2001; HogenEsch, 2002), the depot may also stimulate complex innate immunity functions, such as the complement cascade, cytokines, and other factors that would recruit diverse cellular elements of the immune system to the site of antigen deposit. A depot microparticle can even exhibit complex direct stimulatory effects on dendritic cells (Westwood et al., 2005).

Innate Immunity

A major factor in the mechanism that is commonly used as the basis in the selection or creation of many different types of adjuvants involves the initiation, activation, or enhancement of innate immunity. This would include engagement of Toll-like receptors (TLRs) and other pathogen-recognition receptors (PRRs) and the co-receptors and intracellular adaptor signaling proteins with which they are associated (Pulendran, 2004; Beutler, 2004; Miller et al., 2005). The existence of such innate stimulatory materials from pathogens in the environment is consistent with the "danger theory" of immune activation (Matzinger, 1994). Adjuvant developers often seek to utilize these materials to harness the power of innate immunity to channel the immune response in a desired direction. This is a growing field that will undoubtedly lead to interesting adjuvant combinations. Scores of adjuvants utilize initiation mechanisms that involve complex interactions of intracellular signaling pathways that are the cause of, or the consequences of, initiation of the innate immune system. TLR recognition causes activation of transcription factors that drive cytokine expression, proliferation, survival, and inflammatory mediator expression. A recent summary of TLRs and PRRs and associated intracellular signaling mechanisms is illustrated in Fig. 9.2.

Gram-negative bacterial LPSs, of which the lipid A moiety is the region primarily recognized by the innate immune system, reacts mainly with TLR4 (Miller et al., 2005). Unmethylated CpG DNA dinucleotide sequences derived from bacteria reacts with TLR9 (Klinman et al., 2004). Thus, lipid A from LPS and CpG are two examples of materials from pathogens that interact with PRRs, and both of these substances are being developed

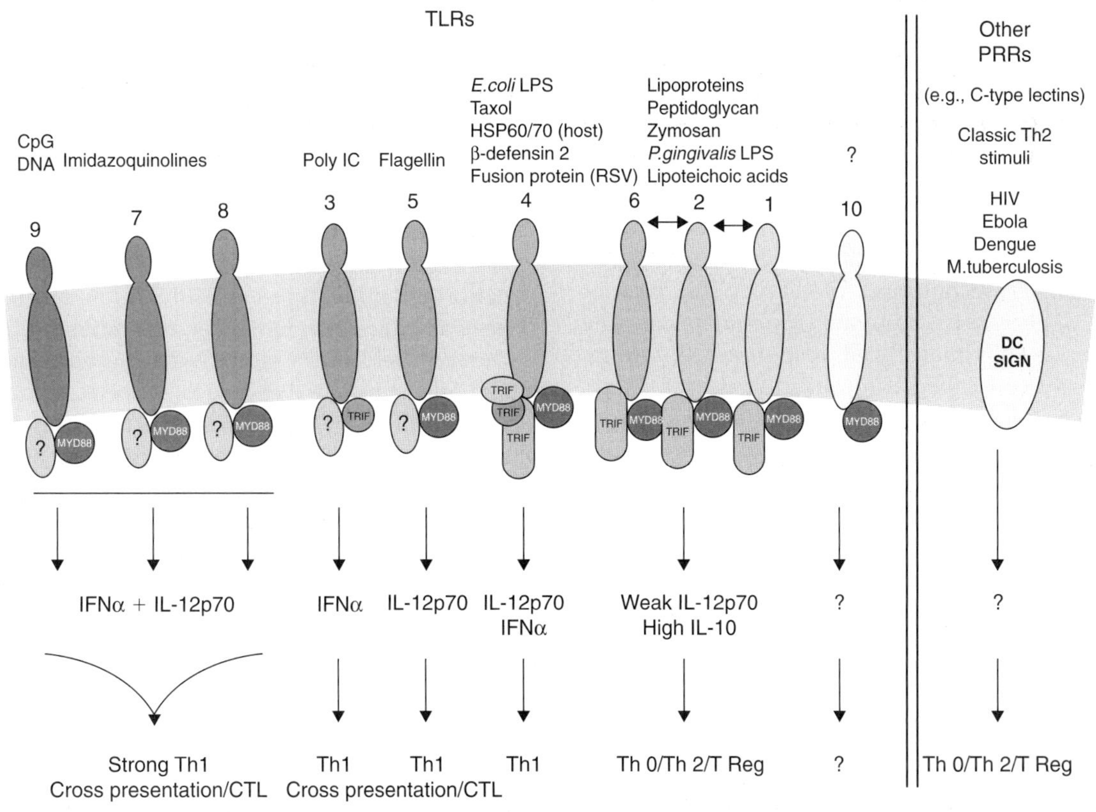

FIGURE 9.2 Regulation of adaptive immune responses by Toll-like receptors (TLRs) and adapter proteins. Reproduced from Pulendran (2004) (with permission).

as potential vaccine adjuvants. Lipid A and LPS can also cause strong complement activation (Carroll, 1998). Activation of complement, which can occur by numerous mechanisms, is a major mediator of B-cell enhancement, and activation of complement represents an important factor in the efficacy of numerous adjuvants (Carroll, 1998).

The emergence of detailed information regarding the existence and mechanisms of TLRs and PRRs has led to a new interest in the possibility of devising a rational means for selection of combinations of vaccine adjuvants that might have complementary, additive, or synergistic immunostimulatory effects (Pulendran, 2004). This is potentially a powerful and useful direction for expanding the future of adjuvant discovery, but it must also be placed in the context that there are many adjuvant mechanisms that are not necessarily related only to signaling pathways.

Secretion of cytokines is often a prominent consequence of initiation or activation of the innate immune system, and cytokines, or DNA that codes for cytokines, is also a focus of adjuvant development (Chen et al., 2001; Sasaki et al., 2003; Hoffman et al., 2003; Calarota and Weiner, 2004; Chattergoon et al., 2004; Ansari et al., 2004; Toka et al., 2004; Melchionda et al., 2005). When utilizing cytokines for enhancement of vaccine efficacy, it is important to understand which immunological parameters are important since the effects of cytokines on certain parts of the immune system can be highly focused and may not necessarily result in the desired effective immune response (Koopman et al., 2005).

Despite the possible importance of TLRs for various adjuvants, considerable caution must also be maintained in believing that adjuvant effects are driven only by recognition of TLRs. The role of TLRs as a necessary adjuvant mechanism in the generation of maximal B-cell responses to foreign proteins has been called into question (Gavin et al., 2006). Mice genetically deficient in both MYD88 and TRIF ($Myd88^{-/-}$; $Trif^{Lps2/Lps2}$ mice) have a complete lack of known TLR signaling, and because of this they were used to evaluate the role of TLRs in the mechanisms of several well-known adjuvants for enhancing the generation of induced antibodies. When T-cell dependent antigen was tested with four adjuvants: aluminum salt, Freund's complete adjuvant, Freund's incomplete adjuvant, and the "Ribi Adjuvant System" (monophosphoryl lipid A/trehalose dicorynomycolate adjuvant), robust antibody responses were observed (Gavin et al, 2006). It was thus suggested that the TLR system is not required for adjuvant effects with many classical adjuvants in the generation of adaptive immunity. If this is true, it raises the possibility that the cytokine production that is associated with TLRs may be more responsible for side effects than for adjuvant activity. The role of TLRs in the generation of adaptive immunity is an important question that has not yet been completely resolved. It should be emphasized that the exact role of TLRs as part of adjuvant mechanisms for B-cell immunology is the subject of investigations in which controversial areas still exist (Wickelgren, 2006; Nemazee et al., 2006; Pasare and Medzhitov, 2006).

An example of the complexity of adjuvant effects related to innate immunity is given by the role of apoptotic cell death induced by certain adjuvants (Dupuis et al., 2001; Yang et al., 2004). It appears that certain surfactant adjuvants, such as MF59 (an oil-in-water emulsion), or other types of emulsions, can exert their adjuvant effects at least partly by inducing apoptosis and necrosis of macrophages, dendritic cells, and perhaps other cells. It has been suggested that an emulsion can lead to apoptosis, and that phagocytosis of apoptotic bodies (e.g., from macrophages serving as first responders) can induce dendritic cells to mature and stimulate immunity (Dupuis et al., 2001). It has also been suggested that adjuvant-induced apoptosis can occur via a "nonclassical" caspase-independent pathway (Yang et al., 2004).

ADDITIONAL ADJUVANT CHARACTERISTICS AND MECHANISMS

Separation of Toxicity and Adjuvanticity

Substances that have the capacity to attach to membranes to induce nonspecific irritation, inflammation, or other types of stimulation can sometimes serve as adjuvants. However, as understanding in the adjuvant field of the underlying mechanisms of adjuvant effects has evolved, such as the realization of the potential roles of TLRs and PRRs, and dendritic cells, the importance of nonspecific irritation as a mechanism has given way to more sophisticated concepts, two examples of which are described below.

Lipid A

Native lipid A (diphosphoryl), and even monophosphoryl lipid A (Fig. 9.3), have varying degrees of toxic effect, a characteristic that is summarized by their well-known association with pyrogenicity. However, it was shown that although liposomes containing native lipid A or monophosphoryl lipid A completely lacked toxicity, both in animals (Richards et al., 1989; Alving, 1993) and humans (Fries et al., 1992), the lipid

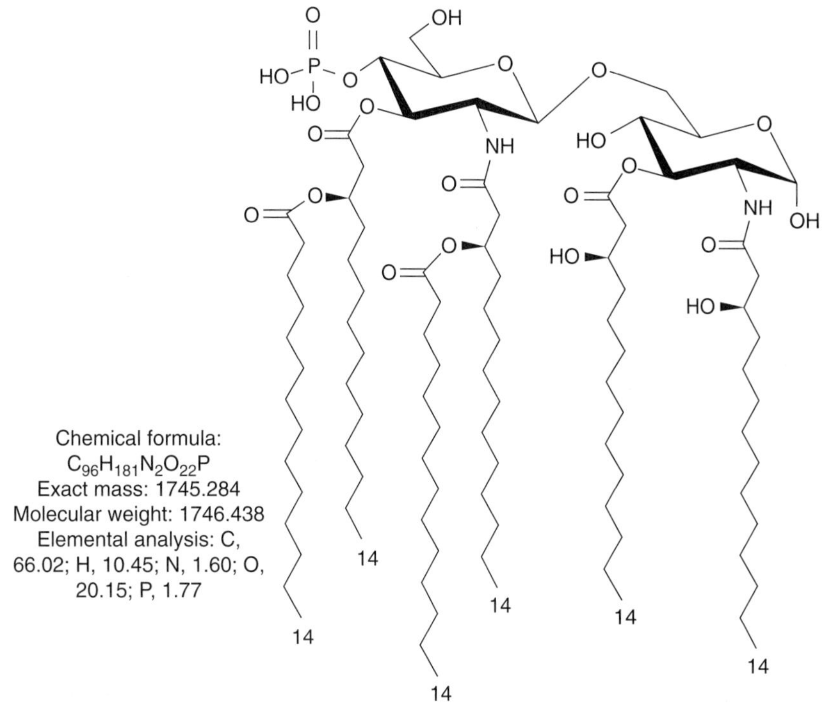

FIGURE 9.3 Synthetic monophosphoryl lipid A. Structure and chemical analysis kindly supplied by Dr. Walter Shaw, Avanti Polar Lipids, Inc., Birmingham, AL.

A in the liposomes still expressed potent adjuvant activity. Most of the toxicity of lipid A thus resides in the fatty acid portions of the molecule, and the fatty acyl groups are hidden in the hydrophobic region of the lipid bilayers of liposomes. This observation, therefore, led to the realization that toxicity could be separated from adjuvanticity. As noted earlier, it is now known that lipid A is a potent activator of TLR-4 and most, or perhaps all of the toxic effects of lipid A and monophosphoryl lipid are due to activation of TLR-4, and it is possible that the adjuvant effects has other mechanisms, such as increase of antigen presentation from inside APCs (Verma et al., 1992).

Monophosphoryl lipid A at nontoxic doses adsorbed to aluminum salt served as an effective adjuvant for inducing Th1-type immunity that led to apparent effective effectiveness of a Herpes simplex vaccine in a phase III clinical trial (Stanberry et al., 2002). It is interesting that the adjuvant appeared to be the pivotal factor in the latter study in that a previous study with a similar antigen, but with MF59 emulsion as an adjuvant, failed to exhibit significant efficacy in a phase III trial (Corey et al., 1999).

Saponins

Saponins are complex glycosides, containing a ring structure and carbohydrates, derived from tree bark and other plants that bind specifically to cholesterol and result in membrane pore formation (Bomford et al., 1992). Although the mechanisms of adjuvant activities compared to toxic effects has not been well-worked out, saponin-based materials presumably would have to balance potential toxicity (such as hemolysis of erythrocytes caused by saponin) with beneficial adjuvant effects. One material derived from saponin, termed Quil-A, is the basis for creation of so-called immunostimulating complexes (ISCOMs) (Morein, 1988; Sjölander et al., 1998). The adjuvant mechanism of Quil-A has been postulated to represent an interaction of an amphipathic antigen with the Quil-A/cholesterol complex (Sjölander et al., 1998), and it is effective for stimulating induction of cytotoxic T lymphocytes (Takahashi et al., 1990). Another prominent adjuvant in this category is QS-21, a defined structure derived from crude triterpene glycoside saponin that exhibits greatly reduced toxicity (Kensil et al., 1991). QS-21 has been very successfully used as an adjuvant in humans, particularly when given in combination with monophosphoryl lipid A in an oil-in-water emulsion known as AS02A (Stoute et al., 1997; Bojang et al., 2005). Recent research aimed at dissociating toxic effects and adjuvant effects of saponin has resulted in a new semisynthetic saponin that has reduced toxicity and enhanced adjuvant activity compared to QS-21 in humans (Slovin et al., 2005).

Specific Targeting Mechanisms

Antibodies, lectins, or toxins that carry antigens to APCs can promote adjuvant activities according to the definition of an adjuvant as used in this chapter. CT or LT from *E. coli*, and nontoxic analogues that bind to ganglioside GM1, which serve as adjuvants both for mucosal vaccines and also for transcutaneous immunization through binding to Langerhans cells in the skin (Glenn et al., 2000) would be prime examples adjuvant targeting, as would a tetanus toxin epitope that binds to ganglioside and also provides T-cell help (Boucher et al., 1994). Transcutaneous immunization using LT as an adjuvant has proven to be a promising preclinical modality in the development of new generations of vaccines to anthrax (Matyas et al., 2004; Peachman et al., 2006). An interesting twist is to attach a targeting molecule, such as a lectin, on the surface of the particle to direct an oral mucosal vaccine to intestinal M cells in the intestine (Roth-Walter et al., 2005).

Carriers that Present the Antigen in a Form that Emulates the Natural State of the Antigen

Virus-like particles (VLPs) are self-assembling particles derived from a major viral protein. They are DNA-free particles that display the antigen on their surfaces in an immunogenic form and, in keeping with the concepts in this chapter, they are therefore a type of antigen/adjuvant combination. Development of a pan-filovirus vaccine based on VLPs has been proposed (Swenson et al., 2005). A recent example of the potential of VLPs as a vaccine formulation in humans is given by the success of VLPs from human papilloma virus in a vaccine (known as Gardasil™, Merck) designed to target cervical cancer (Stanley, 2003; Stern, 2005; Villa et al, 2005). The papilloma virus VLPs, while effective by themselves, were greatly improved by adding aluminum salt as an additional adjuvant.

Certain other types of particles can be used to reconstitute an antigen in a form that exhibits the conformation that would be expected to be found in nature. Examples of this would include cells that express a surface antigen (Harshyne et al., 2001), or liposomes (such as virosomes) (Glück and Metcalfe, 2003; de Bruijn, et al., 2005) or proteosome particles (Jones et al., 2003; Treanor et al., 2006) that have antigen embedded in their surfaces. The HBsAg is an antigenic enveloped virus structure that is liposome-like in that it carries antigen in its lipid bilayer (Gavilanes et al., 1982). Recombinant HBsAg expressed in yeast has been used as the basis for creating a unique carrier in an emulsion containing two adjuvants (ASA02A) for proper conformation of unrelated recombinant antigens (Bojang et al., 2005). Although the HBsAg carrier does serve as a useful carrier/adjuvant, the adjuvant activity was greatly increased by utilizing a larger adjuvant formulation that contained MPL and QS-21 formulated in a squalene oil-in-water emulsion (Stoute et al., 1997; Bojang et al., 2005). It should be remembered that combinations of adjuvants and adjuvant strategies can often be the most effective, and the safest, method to achieve practical immunostimulation.

Phage Display as a Nanocarrier Strategy for Antigen Delivery and Presentation

Preclinical studies have suggested that T4 bacteriophage can be a useful nanocarrier for delivery and presentation of certain types antigens (Sathaliyawala et al., 2006). These particles are particularly useful for antigens that can be expressed in *E. coli*, and effectiveness has been reported in preclinical studies on anthrax and HIV vaccines (Shivachandra et al., 2006, 2007; Sathaliyawala et al., 2006). One potential benefit of a nanocarrier is that it might be used in conjunction with molecular adjuvants that are co-displayed on the T4, or with transcutaneous immunization using LT as an adjuvant.

SIDE EFFECTS OF WIDELY USED HUMAN ADJUVANTS

It is well-known that immunostimulators in general, and adjuvants in particular, can sometimes be double-edged swords. Side effects and toxicities of adjuvants can be widespread and commonplace and the toxicity and reactogenicity of a given adjuvant is often cited as having equal or greater importance than the potency as a criterion to be considered in the selection of an adjuvant (Gupta et al., 1993; Gupta and Siber, 1995). It is the fear of introduction of side effects or toxicities in otherwise healthy persons that is the most troubling aspect of potential vaccine side effects, and this is particularly true for pediatric vaccines. However, the risk of not using a vaccine, including one that contains an adjuvant, is always much higher than any possible vaccine or adjuvant-related side effect.

Some local effects, such as local ulceration induced by live Vaccinia organisms that are used for smallpox or the Bacille Calmette-Guérin (BCG) organism that is widely used around the world as a tuberculosis vaccine, and the familiar local redness and swelling at the site of injection induced by aluminum salts

(Jefferson et al., 2004), are usually tolerated as levels of toxicity that are generally not so stressful as to preclude the use of the offending materials. It is generally believed that the local inflammation associated with intramuscular injection of aluminum salts is due to the escape of adjuvant along the injection track into the SC space (Edelman, 1980). Pain after injection and the occurrence of painless nodules after multiple injections of aluminum-adjuvanted vaccines are associated with the SC route of injection (Frost et al., 1985; Pittman, 2002).

Systemic effects of various adjuvant materials sometimes can be more serious, and these can include: nausea and vomiting that are associated with a small fraction of vaccinated cohorts; pyrogenic effects due to endotoxin that are almost universally found with crude vaccines containing endotoxin; or even uveitis, a rare occurrence that may be associated with certain adjuvants (Goldenthal et al., 1993).

Sometimes the adjuvant can produce a set of symptoms that partly mimics the disease against which the vaccine is directed. Such is the case of orally administered CT or LT, each of which can induce diarrhea similar to that found in cholera or in *E. coli* diarrhea. However, a much more dramatic and dangerous example of toxicity of LT was recently discovered in a licensed Swiss vaccine that was administered intranasally (Mutsch et al., 2004; Couch, 2004). When LT was included in a liposomal (virosome) vaccine containing hemagglutinin and neuraminidase surface antigens of influenza virus, intranasal administration of the vaccine resulted in a strong association with an increased risk of facial nerve paralysis (Bell's palsy). The increased risk was highest at 31–60 days after vaccination and the toxic effect resulted in withdrawal of the vaccine from the market.

It is known that ganglioside GM1 serves as a high affinity receptor for LT and CT and, although the natural targets for CT and LT are the cells of the intestine, GM1 is widely distributed in many other tissues, including neural tissues (Eidels et al., 1983). Thus, although the mechanism of the induction of Bell's palsy was not yet completely clear, it would seem likely that the interaction of LT with GM1 on the olfactory nerve played an important role (Couch, 2004). This unfortunate neurological toxicity involving the facial nerve was probably uniquely related to the intranasal administration of enterotoxin. Fear of facial nerve toxicity should not have a negative impact on the use of native or modified enterotoxin vaccine adjuvants when given by other routes for immunomodulation and adjuvant effects. Furthermore, there is no reason to suspect that the virosome itself exhibited any toxicity independent of its role as a carrier of LT.

Although the safety of vaccine adjuvants is a primary concern because adjuvants are often applied to healthy individuals, it is important to understand that toxicity is not required in order to achieve beneficial adjuvant effects with many formulations. As noted earlier, the mediators generated by binding to TLRs do not contribute to the adjuvant effects of many commonly used adjuvants (Gavin et al., 2006). In addition, pronounced adjuvant effects have also been described in humans, without toxicity, by using liposomes containing lipid A as an adjuvant (Fries et al., 1992), or by transcutaneous immunization using LT adjuvant (Glenn et al., 2000).

References

Ahlers, J.D., Dunlop, N., Alling, D.W., Nara, P.L. and Berzofsky, J.A. Cytokine-in-adjuvant steering of the immune response phenotype to HIV-1 vaccine constructs. Granulocyte-macrophage colony-stimulating factor and TNF-alpha synergize with IL-12 to enhance induction of cytotoxic t lymphocytes. J. Immunol. 1997; 158:3947–3958.

Alving, C.R. Lipopolysaccharide, lipid A, and liposomes containing lipid A as immunologic adjuvants. Immunobiology 1993; 187:430–446.

Alving, C.R. Design and selection of vaccine adjuvants: animal models and human trials. Vaccine 2002; 20:S56–S64.

Alving, C.R and Rao, M. Lipid A and liposomes containing lipid A as antigens and adjuvants. Vaccine 2008; 26:3036–3045.

Alving, C.R., Koulchin, V., Glenn, G.M. and Rao, M. Liposomes as carriers of peptide antigens: induction of antibodies and cytotoxic T lymphocytes to conjugated and unconjugated peptides. Immunol. Rev. 1995; 145:5–31.

Alving, C.R., Shichijo, S., Mattsby-Baltzer, I., Richards, R.L. and Wassef, N.M. Preparation and use of liposomes in immunological studies. In: Liposome Technology, Interactions of Liposomes with the Biological Milieu (G. Gregoriadis, Ed.), 2nd ed., Vol. III, pp. 317–343. Boca Raton, FL: CRC Press, 1993.

Ansari, A.A., Mayne, A.E., Ohlamoon, N., Pattanapanyasat, K., Mori, K. and Villinger, F. Use of recombinant cytokines for optimized induction of antiviral immunity against SIV in the nonhuman primate model of human AIDS. Immunol. Res. 2004; 29:1–17.

Aprile, M.A. and Wardlaw, A.C. Aluminium compounds as adjuvants for vaccines and toxoids in man: a review. Can. J. Public Health 1966; 57:343–354.

Baylor, N.W., Egan, W. and Richman, P. Aluminum salts in vaccines—US perspective. Vaccine 2002; 20:S18–S23.

Beutler, B. Innate immunity: an overview. Mol. Immunol. 2004; 40:845–859.

Bojang, K.A., Olodude, F., Pinder, M., Ofori-Anyinam, O., Vigneron, L., Fitzpatrick, S., Njie, F., Kassanga, A., Leach, A., Milman, J., Rabinovich, R., McAdam, K.P.W.J., Kester, K.E., Heppner, D.G., Cohen, J.D., Tornieporth, N. and Milligan, P.J.M. Safety and immunogenicity of RTS,S/AS02A candidate malaria vaccine in Gambian children. Vaccine 2005; 23:4148–4157.

Bomford, R., Stapleton, M., Winsor, S., Beesley, J.E., Jessup, E.A., Price, K.R. and Fenwick, G.R. Adjuvanticity and ISCOM formation by structurally diverse saponins. Vaccine 1992; 10:572–577.

Boucher, P., Sato, H., Sato, Y. and Locht, C. Neutralizing antibodies and immunoprotection against pertussis and tetanus obtained by use of a recombinant pertussis toxin-tetanus toxin fusion protein. Infect. Immun. 1994; 62:449–456.

Burrell, L.S., White, J.L. and Hem, S.L. Stability of aluminium-containing adjuvants during aging at room temperature. Vaccine 2000; 18:2188–2192.

Cai, H., Hu, X.D., Yu, D.H., Li, S.X., Tian., X. and Zhu, Y.X. Combined DNA vaccine encapsulated in microspheres enhanced protection efficacy against *Mycobacterium tuberculosis* infection of mice. Vaccine 2005; 23:4167–4174.

Calarota, S.A. and Weiner, D.B. Enhancement of human immunodeficiency virus type 1-DNA vaccine potency through incorporation of T-helper 1 molecular adjuvants. Immunol. Rev. 2004; 199:84–99.

Callahan, P.M., Shorter, A.L. and Hem, S.L. The importance of surface charge in the optimization of antigen-adjuvant interactions. J. Pharm. Res. 1991; 8:851–858.

Carroll, M.C. The role of complement and complement receptors in induction and regulation of immunity. Ann. Rev. Immunol. 1998; 16:545–568.

Chang, J.C.C., Diveley, J.P., Savary, J.R. and Jensen, F.C. Adjuvant activity of incomplete Freund's adjuvant. Adv. Drug Deliv. Rev. 1998; 32:173–186.

Chattergoon, M.A., Saulino, V., Shames, J.P., Stein, J., Montaner, L.J. and Weiner, D.B. Co-immunization with plasmid IL-12 generates a strong T-cell memory response in mice. Vaccine 2004; 22:1744–1750.

Chen, H.-W., Pan, C.-H., Huan, H.-W., Liau, M.-Y., Chiang, J.-R. and Tao, M.-H. Suppression of immune response and protective immunity to a Japanese encephalitis virus DNA vaccine by coadministration of an IL-12-expressing plasmid. J. Immunol. 2001; 166:7419–7426.

Corey, L., Langenberg, A.G.M., Ashley, R., Sekulovich, R.E., Izu, A.E., Douglas, J.M., Handsfield, H.H., Warren, T., Marr, L., Tyring, S., DiCarlo, R., Adimora, A.A., Leone, P., Dekker, C.L., Burke, R.L., Leong, W.P. and Straus, S.E. Recombinant glycoprotein vaccine for the prevention of genital HSV-2 infection. Two randomized controlled trials. J. Am. Med. Assoc. 1999; 382:331–340.

Couch, R.B. Nasal vaccination, *Escherichia coli* enterotoxin, and Bell's palsy. N. Engl. J. Med. 2004; 350:860–861.

Cox, J.C. and Coulter, A.R. Adjuvants—a classification and review of their modes of action. Vaccine 1997; 15:248–256.

Davenport, F.M. Seventeen years' experience with mineral oil adjuvant influenza virus vaccines. Ann. Allergy 1968; 26:288–292.

de Bruijn, I.A., Nauta, J., Cramer, W.C.M., Gerez, L. and Palache, A.M. Clinical experience with inactivated, virosomal influenza vaccine. Vaccine 2005; 23(Suppl. 1):S39–S49.

Dupuis, M., Denis-Mize, K., LaBarbara, A., Peters, W., Charo, I.F., McDonald, D.M. and Ott, G. Immunization with the adjuvant MF59 induces macrophage trafficking and apoptosis. Eur. J. Immunol. 2001; 31:2910–2918.

Edelman, R. Vaccine adjuvants. Rev. Infect. Dis. 1980; 2:370–383.

Edelman, R. and Tacket, C.O. Adjuvants. Int. Rev. Immunol. 1990; 7:51–66.

Eidels, L., Proia, R.L. and Hart, D.A. Membrane receptors for bacterial toxins. Microbiol. Rev. 1983; 47:596–620.

Elson, C.O. and Ealding, W. Cholera toxin feeding did not induce oral tolerance in mice and abrogated oral tolerance to an unrelated protein antigen. J. Immunol. 1984; 133:2892–2897.

Eriksson, K. and Holmgren, J. Recent advances in mucosal vaccines and adjuvants. Curr. Opin. Immunol. 2002; 14:666–672.

Feldkamp, J.R., Shah, D.N., Meyer, S.L., White, J.L. and Hem, S.L. Effect of adsorbed carbonate on surface charge characteristics and physical properties of aluminum hydroxide gel. J. Pharm. Sci. 1981; 70:638–640.

Freund, J. The mode of action of immunologic adjuvants. Adv. Tuberc. Res. 1956; 1:130–148.

Freytag, L.C. and Clements, J.D. Mucosal adjuvants. Vaccine 2005; 23:1804–1813.

Fries, L.F., Gordon, D.M., Richards, R.L., Egan, J.E., Hollingdale, M.R., Gross, M., Silverman, C. and Alving, C.R. Liposomal malaria vaccine in humans: a safe and potent adjuvant strategy. Proc. Natl. Acad. Sci. 1992; 89:358–362.

Frost, L., Johansen, P., Pedersen, S., Veien, N., Østergaard, P.A. and Nielsen, M.H. Persistent subcutaneous nodules in children hyposensitized with aluminium-containing allergen extracts. Allergy 1985; 40:368–372.

Gavilanes, F., Gonzalez-Ros, J.M. and Peterson, D.L. Structure of hepatitis B surface antigen. Characterization of the lipid components and their association with the viral proteins. J. Biol. Chem. 1982; 257:7770–7777.

Gavin, A.L., Hoebe, K., Duong, B., Ota, T., Martin, C., Beutler, B. and Nemazee, D. Adjuvant-enhanced antibody responses in the absence of toll-like receptor signaling. Science 2006; 314:1936–1938.

Glenn, G.M., Rao, M., Matyas, G.R. and Alving, C.R. Skin immunization made possible by cholera toxin. Nature 1998; 391:851.

Glenn, G.M., Scharton-Kersten, T., Vassell, R., Matyas, G.R. and Alving, C.R. Transcutaneous immunization with bacterial ADP-ribosylating exotoxins as antigens and adjuvants. Infect. Immun. 1999; 67:1100–1106.

Glenn, G.M., Taylor, D.N., Li, X., Frankel, S., Montemarano, A. and Alving, C.R. Transcutaneous immunization: a human vaccine delivery strategy using a patch. Nat. Med. 2000; 6:1403–1406.

Glenny, A.T., Pope, C.G., Waddington, H. and Wallace, U. Immunological notes. XVII. The antigenic value of the toxin-antitoxin precipitate of Raman. XXIII. The antigenic value of toxoid precipitated by potassium alum. XXIV. The antigenic value of toxoids that will not flocculate in the presence of antitoxin. J. Pathol. Bacteriol. 1926; 29:31–40.

Glück, R. and Metcalfe, I.C. Novel approaches in the development of immunopotentiating reconstituted influenza virosomes as efficient antigen carrier systems. Vaccine 2003; 21:611–615.

Goldenthal, K.L., Cavagnaro, J.A., Alving, C.R. and Vogel, F.R. Safety evaluation of vaccine adjuvants. AIDS Res. Hum. Retroviruses 1993; 9(Suppl. 1):S47–S51.

González, A.M., Nguyen, T.V., Azevedo, M.S.P., Jeong, K., Agarib, F., Iosef, C., Chang, K., Lovgren-Bengtsson, K., Morein, B. and Saif, L.J. Antibody responses to human rotavirus (HRV) in gnotobiotic pigs following a new prime/boost vaccine strategy using oral attenuated HRV priming and intranasal VP2/6 rotavirus-like particle (VLP) boosting with ISCOM. Clin. Exp. Immunol. 2004; 135:361–372.

Gordon, D.M. Use of novel adjuvants and delivery systems to improve the humoral and cellular immune response to malaria vaccine candidate antigens. Vaccine 1993; 11:591–593.

Gupta, R.K. Aluminum compounds as vaccine adjuvants. Adv. Drug Deliv. Rev. 1998; 32:155–172.

Gupta, R.K., Relyveld, E.H., Lindblad, E.B., Bizzini, B., Ben-Efraim, S. and Gupta, C.K. Adjuvants—a balance between toxicity and adjuvanticity. Vaccine 1993; 11:293–306.

Gupta, R.K., Rost, B.E., Relyveld, E. and Siber, G.R. Adjuvant properties of aluminum and calcium compounds. In: Vaccine design, the subunit and adjuvant approach (M.F. Powell and M.J. Newman, Eds.), pp. 229–247. New York: Plenum Press, 1995.

Gupta, R.K. and Siber, G.R. Adjuvants for human vaccines—current status, problems and future prospects. Vaccine 1995; 13:1263–1276.

Harshyne, L.A., Watkins, S.C., Gambotto, A. and Barratt-Boyes, S.M. Dendritic cells acquire antigens from live cells for cross-presentation to CTL. J. Immunol. 2001; 166:3717–3723.

Heimlich, J.M., Regnier, F.E., White, J.L. and Hem, S.L. The in vitro displacement of adsorbed model antigens from aluminium-containing adjuvants by interstitial proteins. Vaccine 1999; 17:2873–2881.

Hem, S.L. Elimination of aluminum adjuvants. Vaccine 2002; 20(Suppl. 3):S40–S43.

Herbert, W.J. The mode of action of mineral oil emulsion adjuvants on antibody production in mice. Immunology 1968; 14:301–318.

Hilleman, M.R. Critical appraisal of emulsified oil adjuvants applied to viral vaccines. Prog. Med. Virol. 1966; 8:131–182.

Hoffman, S.J., Polack, F.P., Hauer, D.A., Singh, M., Billeter, M.A., Adams, R.J. and Griffin, D.E. Vaccination of rhesus macaques with a recombinant measles virus expressing interleukin-12 alters humoral and cellular immune responses. J. Infect. Dis. 2003; 188:1553–1561.

HogenEsch, H. Mechanisms of stimulation of the immune response by aluminum adjuvants. Vaccine 2002; 20:S34–S39.

Hunter, R., Strickland, F. and Kézdy, F. The adjuvant activity of nonionic block polymer surfactants. I. The role of hydrophile-lipophile balance. J. Immunol. 1981; 127:1244–1250.

Iyer, S., HogenEsch, H. and Hem, S.L. Relationship between the degree of antigen adsorption to aluminum hydroxide adjuvant in interstitial fluid and antibody production. Vaccine 2003; 21:1219–1223.

Iyer, S., Robinett, R.S.R., HogenEsch, H. and Hem, S.L. Mechanism of adsorption of hepatitis B surface antigen by aluminum hydroxide adjuvant. Vaccine 2004; 22:1475–1479.

Jefferson, T., Rudin, M. and Di Pietrantonj, C. Adverse events after immunisation with aluminium-containing DTP vaccines: systematic review of the evidence. Lancet Infect. Dis. 2004; 4:84–90.

Jiang, D., Morefield, G.L., HogenEsch, H. and Hem, S.L. Relationship of adsorption mechanism of antigens by aluminum-containing adjuvants to in vitro elution in interstitial fluid. Vaccine 2006; 24:1665–1669.

Jones, T., Allard, F., Cyr, S.L., Tran, S.P., Plante, M., Gauthier, J., Bellerose, N., Lowell, G.H. and Burt, D.S. A nasal Proteosome™ influenza vaccine containing baculovirus-derived hemagglutinin induces protective mucosal and systemic immunity. Vaccine 2003; 21:3706–3712.

Kawamura, Y.I.., Kawashima, R., Shirai, Y., Kato, R., Hambata, T., Yamamoto, M., Furukawa, K., Fujihashi, K., McGhee, J.R., Hayashi, H. and Dohi, T. Cholera toxin activates dendritic cells through dependence on GM1-ganglioside which is mediated by NF-kappaB translocation. Eur. J. Immunol. 2003; 33:3205–3212.

Kenney, R.T., Rabinovich, N.R., Pichyangkul, S., Price, V.L. and Engers, H.D. Meeting report. 2nd meeting on novel adjuvants currently in/close to human clinical testing, World Health Organization, Fondation Mérieux, Annecy, France, 5–7 June 2000. Vaccine 2002; 20:2155–2163.

Kensil, C.R., Patel, U., Lennick, M. and Marciani, D. Separation and characterization of saponins with adjuvant activity from *Quillaja saponaria* Molina cortex. J. Immunol. 1991; 146:431–437.

Klinman, D.M., Currie, D., Gursel, I. and Verthelyi, D. Use of CpG oligodeoxynucleotides as immune adjuvants. Immunol. Rev. 2004; 199:201–216.

Koopman, G., Mortier, D., Niphuis, H., Farese, A.M., Kahn, L.E., Mann, D., Wagner, R., MacVittie, T.J., Woulfe, S.L. and Heeney, J.L. Systemic mobilization of antigen presenting cells, with a chimeric Flt-3 and G-CSF receptor agonist, during immunization of *Macaca mulatta* with HIV-1 antigens is insufficient to modulate immune responses or vaccine efficacy. Vaccine 2005; 23:4195–4202.

Lavelle, E.C., Jarnicki, A., McNeela, E., Armstrong, M.E., Higgins, S.C., Leavy, O. and Mills, K.H. Effects of cholera toxin on innate and adaptive immunity and its application as an immunomodulatory agent. J. Leukoc. Biol. 2004a; 75:756–763.

Lavelle, E.C., McGuirk, P. and Mills, K.H.G. Molecules of infectious agents as immunomodulatory drugs. Curr. Top. Med. Chem. 2004b; 4:499–508.

Lindblad, E.B. Aluminium adjuvants—in retrospect and prospect. Vaccine 2004; 22:3658–3668.

Lima, K.M., dos Santos, S.A., Rodrigues, J.M., Jr. and Silva, C. L. Vaccine adjuvant: it makes the difference. Vaccine 2004; 22:2374–2379.

Matyas, G.R., Friedlander, A.M., Glenn, G.M., Little, S., Yu, J. and Alving, C.R. A needle-free skin patch vaccination method for anthrax. Infect. Immun. 2004; 72:1181–1183.

Matyas, G.R., Muderhwa, J.M. and Alving, C.R. Oil-in water liposomal emulsions for vaccine delivery. Methods Enzymol. 2003; 373:34–50.

Matzinger, P. Tolerance, danger, and the extended family. Ann. Rev. Immunol. 1994; 12:991–1045.

McElrath, M.J. Selection of potent immunological adjuvants for vaccine construction. Sem. Cancer Biol. 1995; 6:375–385.

McGhee, J.R., Mestecky, J., Dertzbaugh, M.T., Eldridge, J.H., Hirasawa, M. and Kiyono, H. The mucosal immune system: from fundamental concepts to vaccine development. Vaccine 1992; 10:75–88.

Melchionda, F., Fry, T.J., Milliron, M.J., McKirdy, M.A., Tagaya, Y. and Mackall, C.L. Adjuvant IL-7 or IL-15 overcomes immunodominance and improves survival of the $CD8^+$ memory cell pool. J. Clin. Invest. 2005; 115:1177–1187.

Miles, A.P., McClellan, H.A., Rausch, K.M., Zhu, D., Whitmore, M.D., Singh, S., Martin, L.B., Wu, Y., Giersing, B.K., Stowers, A.W., Long, C.A. and Saul, A. Montanide® ISA 720 vaccines: quality control of emulsions, stability of formulated antigens, and comparative immunogenicity of vaccine formulations. Vaccine 2005; 23:2530–2539.

Miller, S.I., Ernst, R.K. and Bader, M.W. LPS, TLR4 and infectious diversity. Nat. Rev. Microbiol. 2005; 3:36–46.

Morefield, G.L., HogenEsch, H., Robinson, J.P. and Hem, S.L. Distribution of adsorbed antigen in mono-valent and combination vaccines. Vaccine 2004; 22:1973–1984.

Morefield, G.L., Sokolovska, A., Jiang, D., HogenEsch, H., Robinson, J.P. and Hem, S.L. Vaccine 2005; 23:1588–1595.

Morein, B. The iscom antigen-presenting system. Nature 1988; 332:287–288.

Mossman, S.P., Evans, L.S., Fang, H., Staas, J., Tice, T., Raychaudhuri, S., Grabstein, K.H., Cheever, M.A. and Johnson, M.E. Development of a CTL vaccine for HER-2/neu using peptide-microspheres and adjuvants. Vaccine 2005; 23:3545–3554.

Muderhwa, J.M., Matyas, G.R., Spitler, L.E. and Alving, C.R. Oil-in-water liposomal emulsions: characterization and potential use in vaccine delivery. J. Pharm. Sci. 1999; 88:1332–1339.

Mutsch, M., Zhou, W., Rhodes, P., Bopp, M., Chen, R.T., Linder, T., Spyr, C. and Steffen, R. Use of the inactivated intranasal influenza vaccine and the risk of Bell's palsy in Switzerland. N. Engl. J. Med. 2004; 350:896–903.

Nemazee, D., Gavin, A., Hoebe, K. and Beutler, B. Immunology: Toll-like receptors and antibody responses. Nature 2006; 441:E4, discussion E4.

O'Hagan, D.T., Singh, M. and Ulmer, J.B. Microparticles for the delivery of DNA vaccines. Immunol. Rev. 2004; 199:191–2004.

Pasare, C. and Medzhitov, R. Control of B-cell responses by Toll-like receptors. Nature 2005; 438:364–368.

Pasare, C. and Medzhitov, R. Toll-like receptors and antibody responses (Reply). Nature 2006; 441:E4, discussion E4.

Peachman, K.K., Rao, M., Alving, C.R., Burge, R., Leppla, S.H., Rao, V.B. and Matyas, G.R. Correlation between lethal toxin neutralizing antibody titers and protection from intranasal challenge with *Bacillus anthracis* Ames strain spores in mice after transcutaneous immunization with recombinant anthrax protective antigen. Infect. Immun. 2006; 74:794–797.

Pennington, D.J., Vermijlen, D., Wise, E.L., Clarke, S.L., Tigelaar, R.E. and Hayday, A.C. The integration of conventional and unconventional T cells that characterizes cell-mediated responses. Adv. Immunol. 2005; 87:27–59.

Peyre, M., Fleck, R., Hockley, D., Gander, B. and Sesardic, D. In vivo uptake of an experimental microencapsulated diphtheria vaccine following sub-cutaneous immunisation. Vaccine 2004; 22:2430–2437.

Pittman, P.R. Aluminum-containing vaccine associated adverse events: role of route of administration and gender. Vaccine 2002; 20:S48–S50.

Podda, A. The adjuvanted influenza vaccines with novel adjuvants: experience with the MF59-adjuvanted vaccine. Vaccine 2001; 19:2673–2680.

Pulendran, B. Modulating vaccine responses with dendritic cells and Toll-like receptors. Immunol. Rev. 2004; 199:227–250.

Richards, R.L., Rao, M., VanCott, T.C., Matyas, G.R., Birx, D.L. and Alving, C.R. Liposome-stabilized oil-in-water emulsions as adjuvants: increased emulsion stability promotes induction of cytotoxic T lymphocytes against an HIV envelope antigen. Immunol. Cell Biol. 2004; 82:531–538.

Richards, R.L., Swartz, G.M., Jr., Schultz, C., Hayre, M.D., Ward, G.S., Ballou, W.R., Chulay, J.D., Hockmeyer, W.T., Berman, S.L. and Alving, C.R. Immunogenicity of liposomal malaria sporozoite antigen in monkeys: adjuvant effects of aluminum hydroxide and nonpyrogenic liposomal lipid A. Vaccine 1989; 7:506–512.

Rimoldi, M. and Rescigno, M. Uptake and presentation of orally administered antigens. Vaccine 2005; 23:1793–1796.

Rinella, J.V., Jr., White, J.L. and Hem, S.L. Treatment of aluminium hydroxide adjuvant to optimize the adsorption of basic proteins. Vaccine 1996; 14:298–300.

Rock, K.L. A new foreign policy: MHC class I molecules monitor the outside world. Immunol. Today 1996; 17:131–137.

Roth-Walter, F., Schöll, I., Untersmayr, E., Ellinger, A., Boltz-Nitulescu, G., Scheiner, O., Gabor, F. and Jensen-Jarolim, E. Mucosal targeting of allergen-loaded microspheres by *Aleuria aurantia* lectin. Vaccine 2005; 23:2703–2710.

Rydell, N. and Sjöholm, I. Mucosal vaccination against diphtheria using starch microparticles as adjuvant for cross-reacting material (CRM197) of diphtheria toxin. Vaccine 2005; 23:2775–2783.

Sasaki, M.D.G., Foccacia, R. and de Messias-Reason, I.J. Efficacy of granulocyte-macrophage colony-stimulating factor (GM-CSF) as a vaccine adjuvant for hepatitis B virus in patients with HIV infection. Vaccine 2003; 21:4545–4549.

Sathaliyawala, T., Rao, M., Maclean, D.M., Birx, D.L., Alving, C.R. and Rao, V.B. Assembly of HIV antigens on bacteriophage T4: a novel in vitro approach to construct multicomponent HIV vaccines. J. Virol. 2006; 80:7688–7698.

Saul, A., Lawrence, G., Allworth, A., Elliott, S., Anderson, K., Rzepczyk, C., Martin, L.B., Taylor, D., Eisen, D.P., Irving, D.O., Pye, D., Crewther, P.E., Hodder, A.N., Murphy, V.J. and Anders, R.F. A human phase 1 vaccine clinical trial of the *Plasmodium falciparum* malaria vaccine candidate apical membrane antigen 1 in Montanide ISA720 adjuvant. Vaccine 2005; 23:3076–3083.

Schijns, V.E.J.C. Immunological concepts of vaccine activity. Curr. Opin. Immunol. 2000; 12:456–463.

Seeber, S.J., White, J.L. and Hem, S.L. Predicting the adsorption of proteins by aluminium-containing adjuvants. Vaccine 1991a; 9:201–203.

Seeber, S.J., White, J.L. and Hem, S.L. Solubilization of aluminum-containing adjuvants by constituents of interstitial fluid. J. Parenter. Sci. Technol. 1991b; 45:156–159.

Shirodkar, S., Hutchinson, R.L., Perry, D.L., White, J.L. and Hem, S.L. Aluminum compounds used as adjuvants in vaccines. Pharm. Res. 1990; 7:1282–1288.

Shivachandra, S.B., Li, Q., Peachman, K.K., Matyas, G.R., Leppla, S.H., Alving, C.R., Rao, M. and Rao, V.B. Multicomponent anthrax toxin display and delivery using bacteriophage T4. Vaccine 2007; 25:1225–1235.

Shivachandra, S.B., Rao, M., Janosi, L., Sathaliyawala, T., Matyas, G.R., Alving, C.R. and Leppla, S.H. In vitro binding of anthrax protective antigen on bacteriophage T4 capsid surface through Hoc-capsid interactions: a strategy for efficient display of large full-length proteins. Virology 2006; 345: 190–198.

Sjölander, A., Cox, J.C. and Barr, I.G. ISCOMs: an adjuvant with multiple functions. J. Leukoc. Biol. 1998; 64:713–723.

Slovin, S.F., Ragupathi, G., Fernandez, C., Jefferson, M.P., Diani, M., Wilton, A.S., Powell, S., Spassova, M., Reis, C., Clausen, H., Danishefsky, S., Livingston, P. and Scher, H.I. A bivalent conjugate vaccine in the treatment of biochemically relapsed prostate cancer: a study of glycosylated MUC-2-KLH and Globo H-KLH conjugate vaccines given with the new semi-synthetic saponin immunological adjuvant GPI-0100 or QS-21. Vaccine 2005; 23:3114–3122.

Snider, D.P. The mucosal adjuvant activities of ADP-ribosylating bacterial enterotoxins. Crit. Rev. Immunol. 1995; 15:317–348.

Stanberry, L.R., Spruance, S.L., Cunningham, A.L., Bernstein, D.I., Mindel, A., Sacks, S., Tyring, S., Aoki, F.Y., Slaoui, M., Denis, M., Vandepapeliere, P. and Dubin, G. Glycoprotein-D-adjuvant vaccine to prevent genital herpes. N. Engl. J. Med. 2002; 347:1652–1661.

Stanley, M.A. Progress in prophylactic and therapeutic vaccines for human papillomavirus infection. Expert Rev. Vaccines 2003; 2:381–389.

Stern, P.L. Immune control of human papillomavirus (HPV) associated anogenital disease and potential for vaccination. J. Clin. Virol. 2005; 32S:S72–S81.

Stoute, J.A., Slaoui, M., Heppner, D.G., Momin, P., Kester, K.E., Desmons, P., Wellde, B.T., Garcon, N., Krzych, U. and Marchand, M. A preliminary evaluation of a recombinant circumsporozoite protein vaccine against *Plasmodium falciparum* malaria. RTS,S Malaria Vaccine Evaluation Group. N. Engl. J. Med. 1997; 336:86–91.

Stuart-Harris, C.H. Adjuvant influenza vaccines. Bull. World Health Organ. 1969; 41:617–621.

Swenson, D.L., Warfield, K.L., Negley, D.L., Schmaljohn, A., Aman, M.J. and Bavari, S. Virus-like particles exhibit potential as a pan-filovirus vaccine for both Ebola and Marburg viral infections. Vaccine 2005; 23:3033–3042.

Takahashi, H., Takeshita, T., Morein, B., Putney, S., Germain, R.N. and Berzofsky, J.A. Induction of $CD8^+$ cytotoxic T cells by immunization with purified HIV-1 envelope protein in ISCOMs. Nature 1990; 344:873–875.

Toka, F.N., Pack, C.D. and Rouse, B.T. Molecular adjuvants for mucosal immunity. Immunol. Rev. 2004; 199:100–112.

Treanor, J., Nolan, C., O'Brien, D., Burt, D., Burt, D., Lowell, G., Linden, J. and Fries, L. Intranasal administration of a proteosome-influenza vaccine is well-tolerated and induces serum and nasal secretion influenza antibodies in healthy human subjects. Vaccine 2006; 24:254–262.

Ulanova, M., Tarkowski, A., Hahn-Zoric, M. and Hanson, L.Å. The common vaccine adjuvant aluminum hydroxide up-regulates accessory properties of human monocytes via an interleukin-4-dependent mechanism. Infect. Immun. 2001; 69:1151–1159.

Verma, J.N., Rao, M., Amselem, S., Krzych, U., Alving, C.R., Green, S.J. and Wassef, N. Adjuvant effects of liposomes containing lipid A: enhancement of liposomal antigen presentation and recruitment of macrophages. Infect. Immun. 1992; 60:2438–2444.

Villa, L.L., Costa, R.L.R., Petta, C.A., Andrade, R., Ault, K.A., Giuliano, A.R., Wheeler, C.M., Koutsky, L.A., Malm, C., Lehtinen, M., Skjeldestad, F.E., Olsson, S.E., Steinwall, M., Brown, D.R., Kurman, R.J., Ronnett, B.M., Stoler, M.H., Ferenczy, A., Harper, D.M., Tamms, G.M., Yu, J., Lupinacci, L., Railkar, R., Taddeo, F.J., Jansen, K.U., Esser, M.T., Sings, H.L., Saah, A.J. and Barr, E. Prophylactic quadrivalent human papillomavirus (types 6, 11, 16, and 18) L1 virus-like particle vaccine in young women: a randomised double-blind placebo-controlled multicentre phase II efficacy trial. Lancet Oncol. 2005; 6:271–278.

Vogel, F.R. and Alving, C.R. Progress in immunologic adjuvant development: 1982–2002. In: The Jordan Report, 20th Anniversary. Accelerated Development of Vaccines 2002. (C. Heilman, P. McInnis and S. Landry, Eds.), pp. 47–51. Bethesda: US Department of Health and Human Services, National Institutes of Health, National Institute of Allergy and Infectious Diseases, 2002. As of August 2005, this document was available on the web at: www.niaid.nih.gov/dmid/vaccines/jordan20/

Wassef, N.M., Alving, C.R. and Richards, R.L. Liposomes as carriers for vaccines. Immunomethods 1994; 4:217–222.

Westwood, A., Healey, G.D., Williamson, E.D. and Eyles, J.E. Activation of dendritic cells by microparticles containing *Bacillus anthracis* protective antigen. Vaccine 2005; 23:3857–3863.

Wickelgren, I. Mouse studies question importance of Toll-like receptors to vaccines. Science 2006; 314:1859–1860.

Wittayanukulluk, A., Jiang, D., Regnier, F.E. and Hem, S.L. Effect of microenvironment pH of aluminum hydroxide adjuvant on the chemical stability of adsorbed antigen. Vaccine 2004; 22:1172–1176.

Woodard, L.F. Surface chemistry and classification of vaccine adjuvants and vehicles. In: Bacterial Vaccines (A. Mizrahi, Ed.), pp. 281–306. New York: Alan R. Liss, 1990, [Adv. Biotechnol. Process. 13, 281–306.].

Yang, Y.-W., Wei, A.-C. and Shen, S.-S. The immunogenicity-enhancing effect of emulsion vaccine adjuvants is independent of the dispersion type and antigen release rate—a revisit of the role of the hydrophile-lipophile balance (HLB) value. Vaccine 2005; 23:2665–2675.

Yang, Y.-W., Wu, C.-A. and Morrow, W.J.W. Cell death induced by vaccine adjuvants containing surfactants. Vaccine 2004; 22:1524–1536.

CHAPTER

10

Preclinical Vaccine Development: Implementing the Food and Drug Administration's Good Laboratory Practices in a Biocontainment Environment—A University BSL3/BSL4 Laboratory Perspective

Melissa M. Eitzen[1], *Andrew G. McNees*[2,3] *and David W.C. Beasley*[2–4]

[1]GLP Program Office, Department of Research Services, University of Texas Medical Branch, Galveston, TX, USA

[2]Institute for Human and Infections and Immunity, Sealy Center for Vaccine Development, University of Texas Medical Branch, Galveston, TX, USA

[3]Galveston National Laboratory Regulatory Services Core, Sealy Center for Vaccine Development, University of Texas Medical Branch, Galveston, TX, USA

[4]Department of Microbiology and Immunology, Sealy Center for Vaccine Development, University of Texas Medical Branch, Galveston, TX, USA

OUTLINE

ABSTRACT

The recent surge in awareness of biodefense-related and emerging infectious disease agents has created a demand for development and subsequent evaluation and testing of vaccines as well as therapeutic and diagnostic products for pathogens that require high (biosafety level 3) and maximum (biosafety level 4) biocontainment. For many of these products, human clinical trials may never be practical, so significant emphasis has also been placed on the potential for their approval by the US Food and Drug Administration using new mechanisms such as the "Animal Rule" and Emergency Use Authorization. As a result, preclinical in vitro and animal testing performed in compliance with the Good Laboratory Practice (GLP) regulations may be the sole source of data to support licensure of these products. This chapter aims to provide a background to the scope, intentions, and requirements of the US GLP regulations, as defined by 21 CFR Part 58, and provide some practical considerations for establishing the necessary policies and procedures to conduct GLP-compliant studies under high and maximum biocontainment.

INTRODUCTION

The past decade has seen a considerable surge in public and political concern regarding the potential for rapid emergence or deliberate release of infectious disease agents. This has led to significant investment by governments in research focused on "biodefense and emerging infectious diseases." In the United States, the National Institute of Allergy and Infectious Diseases (NIAID) of the National Institutes of Health (NIH) developed a Priority Pathogen list as a focus for these research activities, and both the U.S. Department of Health and Human Services and the U.S. Department of Agriculture designated many of the same agents and others as "Select Agents," with significant restrictions on their use and distribution.

While the advanced stages of product development have historically been conducted by pharmaceutical companies and contract research organizations, the eminent need for the development of vaccines, therapeutics, and diagnostics to counter diseases due to emerging and potentially weaponized pathogens has necessitated that the United States Government increasingly look at the academic sector for solutions. In 2002, Dr. Elias A. Zerhouni, Director, NIH, announced the NIH Roadmap, which seeks to transform discoveries arising from basic research to clinical application in a more streamlined and timely approach. As NIH set out to "reengineer" the clinical research enterprise, it sought to leverage its prior success of funding medical research and focus on translational research as the "most challenging, but critically important" theme of the roadmap. Dr. Zerhouni noted that at the center of this vision is "the need to develop new partnerships of research with organized patient communities, community-based health care providers, and academic researchers" and "to build better integrated networks of academic centers linked to a qualified body of community-based health care providers who care for sufficiently large groups of patients interested in working with researchers to quickly develop, test, and deliver new interventions" (US DHHS NIH, 2006).

Additionally in 2002, the US Food and Drug Administration (FDA) amended the new drug and biological product regulations to allow, in certain cases, appropriate animal studies to provide substantial evidence of the effectiveness of new drug and biological products used to reduce or prevent the toxicity of chemical, biological, radiological, or nuclear substances (FDA, DHHS 21 CFR 314 and 601: New Drug and Biological Drug Products, Evidence Needed to Demonstrate Effectiveness of New Drugs When Human Efficacy Studies are not Ethical or Feasible. Federal Register 67: 37995–37998, May 31, 2002). Whereas historically the FDA required products for use in humans to have been proven both safe and effective in clinical trials involving humans as a condition for approval by the FDA, under this amended "animal rule," the FDA can rely on data from animal studies to provide substantial evidence of product efficacy.

The US FDA also has authority to issue an Emergency Use Authorization (EUA) for the use of unapproved medical products, or for unapproved uses of approved products, in treatment of life-threatening disease or conditions caused by chemical, biological, radiological, or nuclear agents. Issuance of an EUA requires that an emergency has been declared by the Secretary of the Department of Health and Human Services, according to specific criteria. Candidate products may then be evaluated, and data supporting an EUA application should indicate that the product may be effective in diagnosing, treating, or preventing the life-threatening condition or disease. Recommended safety and efficacy data for an EUA submission includes preclinical testing data for *in vitro* and animal toxicity, *in vitro* and animal data from at least one species for effectiveness in treating the subject disease, data to support the proposed dosage, along with any clinical data demonstrating safety and efficacy in humans. Any submission for consideration for an EUA should indicate whether nonclinical laboratory studies were conducted in compliance with the US FDA Good Laboratory Practice (GLP) regulations (21 CFR Part 58 Good Laboratory Practice for Nonclinical Laboratory Studies).

These new pathways for product development are well suited for products intended for the prevention or treatment of many of the emerging infectious disease and biological threat agents, including many of the NIAID Priority Pathogens. The GLP regulations for preclinical research studies will most probably be the quality system required for much of the development and critical testing work that is to be performed in support of these products. Results from developmental and preclinical studies may be the only data available to evaluate a new product under the Animal Rule or EUA. As work with many of these Priority Pathogens and Select Agents requires high or maximum biocontainment (biosafety level [BSL] 3 or 4), critical development, and testing work of new vaccines and therapeutics must occur in those facilities and must be done with appropriate quality systems in place to ensure the quality and integrity of the data that is generated.

This chapter provides an overview of the requirements of GLP, with emphasis on the US FDA regulations, and identifies some specific considerations related to studies requiring high biocontainment (BSL3/BSL4). The information presented here is the authors' opinion and reflects both the specific requirements set out in the GLP regulations and our experiences during the planning and implementation of a GLP program to support biodefense and emerging infectious disease research activities at the University of Texas Medical Branch. At the time of writing, this program remains a work in progress.

HISTORY, INTENT, AND PENALTIES—*WHY REGULATE?*

In the mid-1970s, the FDA began investigating the conduct of preclinical research in response to allegations of impropriety. Since product licensure is granted based on the review of data submitted to the Agency, the investigation focused on the integrity of the data. FDA Officials investigated studies performed at G.D. Searle & Co. and Hazleton Laboratories; both produced data in support of Searle products. The initial investigation included review of over 20 completed and ongoing studies supporting new submissions, marketed drug products, an approved food additive, a pending new drug application, and a veterinary drug. Agency in-house reviews preceded the lengthy on-site investigation, which started in early October of 1975 and concluded in mid-December of 1975.

An employee of G.D. Searle & Co. wrote of his experience: "At times, there were more than 20 FDA employees present at Searle during this massive and unprecedented investigation. The FDA investigators were divided into five teams, with each team looking at the studies supporting different compounds. Searle, on its part, had as many as 300 individuals collecting and organizing documents on all its drug products in preparation for the FDA investigation…Over 200 interviews were conducted of Searle employees…" (Thies, 1978).

U.S. Senate subcommittees were formed to hear testimony from the FDA task force members who investigated Searle. The hearings raised serious questions about the testing of new drugs on animals.

At the center was the concern over the quality and the integrity of the data used to support applications. Based on the findings from the investigation, FDA Commissioner Alexander M. Schmidt prepared statements that included evidence of data omission, discrepancies between the protocol and reports submitted to the FDA, dosing errors, solubility problems, incomplete records, lack of animal identification, lack of established written procedures, high animal mortality rates, lack of trained personnel, improper data management, lack of management oversight, and significant quality control problems. His report included the Agency's action plan for additional inspections, the promulgation of standards for GLPs, and a method to monitor preclinical studies.

Additional investigations in the late 1970s involving the US Environmental Protection Agency (EPA) and the National Cancer Institute focused on Industrial Bio-Test Laboratories (IBT), an independent contract research organization. At the time approximately 22,000 toxicology studies in support of product licensure had been performed by this company, including studies for the Pentagon (Marshall, 1983). This number represented approximately 40% of all such studies that had been carried out (Novak, 2001). Results of the investigations revealed numerous findings of fraud and poor animal housing facilities including an improperly working automatic watering system that drenched some animals while allowing others to die of thirst; the animal facility was in such poor condition that it was referred to as "the swamp." After the investigation at IBT, millions of dollars were spent on the retesting of thousands of industrial chemicals and three IBT company officials were found guilty of mail fraud and making false statements to the government and were sentenced to jail time (Baldeshwiler, 2003).

The result of these findings was that the U.S. Congress proposed and then enacted the GLP Regulations for FDA as part of the Federal Food, Drug, and Cosmetic Act. The proposed regulations for GLP were published in the Code of Federal Regulations (21 CFR Part 58) on November 19, 1976 and the Final Rule was published on December 22, 1978. The latter includes preambles documenting the Agency's intent of the GLP regulations. Only two minor revisions have been made since the original enactment.

Subsequently, Congress proposed and eventually enacted (in 1983) the GLP Regulations for EPA. Two sets of EPA GLP regulations exist: 40 CFR Part 160 Federal Insecticide, Fungicide, and Rodenticide Act (FIFRA) and 40 CFR Part 792 Toxic Substances Control Act (TSCA). FIFRA governs pesticides, whereas TSCA governs most other toxic chemicals. Based on the U.S. FDA GLP regulations, the Organization for Economic Co-operation and Development (OECD), an international organization, published a policy series on the Principles of GLP and Compliance Monitoring. To date, there are 15 documents in the OECD series on GLP. The directives apply to safety testing of chemicals and allow for the mutual acceptance of data by member countries.

The FDA GLP regulations define a quality system that aims to ensure the quality and integrity of study data. They set the minimum standards for study personnel, facility management, animal and laboratory facilities, equipment, animal care, administration of the test agent, recording and handling of data, record retention, standard operating procedures (SOPs), reagents, quality assurance (QA), inspections, and provide a mechanism for facility disqualification. The standards outlined in the regulations are process oriented, as opposed to product oriented, and provide a mechanism of documentation that allows for reconstruction of the study. Each of these specific areas will be addressed in the remainder of this chapter.

APPLICATION AND SCOPE—*WHEN AND WHY SHOULD COMPLIANCE WITH THE GLP REGULATIONS BE IMPLEMENTED?*

There are common misconceptions that "GLP" refers to a simple principle of scientific performance and integrity that is applicable to all researchers, or perhaps represents a basic checklist of things to do above and beyond the activities of a "normal" research laboratory to improve the quality of the data generated: prepare some SOPs, buy the right GLP-compliant equipment, make sure you use carbon notebooks and ink pens, and start work.

In reality, the FDA GLP regulations define a complex quality system, with specific requirements for performance, documentation, oversight and review of preclinical research, and include penalties for noncompliance. All of the GLP regulations apply to data submitted to an official governing body that will determine if the candidate product is safe for use. The US FDA GLP regulations are also the quality system of choice for developmental and preclinical testing data that may be submitted in support of animal rule and EUA applications. While the intent of the respective regulations/principles is to assure the reliability of the data, the defined scope for the U.S. regulations mandates the type of data that must be submitted to the Agency in compliance with the law. The defined scope of the FDA, EPA (TSCA) regulations, and the OECD Principles are listed below.

U.S. FDA GLP (safety data for nonclinical laboratory studies in support of product licensure):

(a) This part prescribes GLPs for conducting nonclinical laboratory studies that support or are intended to support applications for research or marketing permits for products regulated by the FDA, including food and color additives, animal food additives, human and animal drugs, medical devices for human use, biological products, and electronic products. Compliance with this part is intended to assure the quality and integrity of the safety data filed pursuant to *sections 406, 408, 409, 502, 503, 505, 506, 507, 510, 512–516, 518–520, 706, and 801* of the Federal Food, Drug, and Cosmetic Act and *sections 351 and 354–360F* of the Public Health Service Act (US DHHS FDA, 1987).

U.S. EPA-TSCA (toxic chemicals):

(a) This part prescribes GLPs for conducting studies relating to health effects, environmental effects, and chemical fate testing. This part is intended to ensure the quality and integrity of data submitted pursuant to testing consent agreements and test rules issued under *section 4* of the TSCA.
(b) This part applies to any study described by paragraph "a" of this section which any person conducts, initiates, or supports on or after September 18, 1989.
(c) It is EPA's policy that all data developed under section 5 of TSCA is in accordance with provisions of this part. If data are not developed in accordance with provisions of this part, EPA will consider such data insufficient to evaluate the health and environmental effects of the chemical substances unless the submitter provides additional information demonstrating that the data are reliable and adequate (US EPA, 1989).

OECD (safety testing of chemicals):

These Principles of GLP should be applied to the nonclinical safety testing of test items contained in pharmaceutical products, pesticide products, cosmetic products, veterinary drugs as well as food additives, feed additives, and industrial chemicals. These test items are frequently synthetic chemicals, but may be of natural or biological origin and, in some circumstances, may be living organisms. The purpose of testing these test items is to obtain data on their properties and/or their safety with respect to human health and/or the environment.

Nonclinical health and environmental safety studies covered by the Principles of GLP include work conducted in the laboratory, in greenhouses, and in the field.

Unless specifically exempted by national legislation, these Principles of GLP apply to all nonclinical health and environmental safety studies required by regulations for the purpose of registering or licensing pharmaceuticals, pesticides, food and feed additives, cosmetic products, veterinary drug products and similar products, and for the regulation of industrial chemicals (OECD, 1997).

However, many areas that are identified in the regulations as requiring compliance do not include specific benchmarks but rather employ terms such as "appropriate," "adequate," or "suitable." This lack of specific definition means that there is no one-size-fits-all answer to GLP compliance. The specific policies and procedures must be defined by facility management and study staff to ensure they are appropriate to and compatible with the requirements of the study and the limitations of the facility. This becomes particularly apparent when considering performance of GLP-compliant research under high biocontainment conditions. Furthermore, unlike some other quality systems, such as certification of clinical laboratories or ISO certification of manufacturing and testing facilities, there is no GLP certification process. In many cases, the first test of compliance will be an FDA audit following submission of a completed study. At the simplest level, the GLP regulations represent a standard for performance and documentation of research studies to ensure the quality and integrity of the data generated, and to capture all information and key reagents that would be necessary to allow the study to be reproduced. The mantra of many GLP consultants is, "If it wasn't documented, it didn't happen."

Clearly there are a number of additional activities required to conduct GLP-compliant studies in a biocontainment environment and these should be given careful consideration from the moment a study is conceptualized. From a biological safety perspective, one must take into account the increased time and effort necessary to obtain security clearances, health status verifications, and vaccinations as well as training that may be necessary to allow individuals to work in BSL3/BSL4 laboratories. From a regulatory perspective, standardizing all routine procedures, validating and maintaining equipment, training study personnel, maintaining proper documentation, participating in and responding to QA inspections/audits, and compiling reports also add substantial time to the length of the study. Study schedules should, therefore, be scrutinized to assure that sufficient time is allowed for completing these tasks. One should also consider the possible need for additional personnel to assist with the increased workload associated with performing regulated studies.

While the U.S. GLP regulations mandate that the study sponsor must notify contracted laboratories of their intent to submit data to the Agency, the contract laboratory must not be negligent of the scope of the regulations/principles. If a facility is performing any type of testing that generates data for submission to the FDA in support of product licensure, the work must be done in compliance with the applicable regulations. In addition to regulatory mandates, there may be contractual obligations that apply the GLP regulations to what would otherwise be nonregulated studies. In order to promote the quality and integrity of data derived from studies that would not normally require compliance with the GLP regulations, a sponsor may contractually require a testing facility to comply with the regulations. The sponsor may argue that this compliance will expedite market approval reviews by the governing agency and decrease the possibility of repeat studies.

Inspections by Regulatory Authorities

Testing facilities that perform studies meeting the scope of the GLP regulations/principles are subject to inspection by the Agency or the appropriate country's compliance monitoring program. Inspection of a testing facility is performed by FDA, EPA, or by National Monitoring Program officials, each of which has published procedures for conducting inspections. The OECD published procedures are mutually accepted by its membership countries in order to assure harmonization and promote the mutual acceptance of data. Testing facilities that perform regulated work must accommodate inspections by Agency officials. Failure to comply with an Agency inspection will disallow the review of an application for a research or marketing permit that relies on data generated as part of the regulated work.

During inspections, Agency officials can request or make copies of documents; therefore, the testing facility should have a standardized operating procedure for the copy and release of restricted or regulated documents associated with select agents and select agent facilities to ensure appropriate compliance with Select Agent regulations. Furthermore, access to high containment areas by Agency officials may be impossible due to security restrictions, facility training requirements, vaccination requirements, etc. This potential inability to access an area by an Agency inspector may result in a request to inspect similar space at a lower containment level. In response to the limitation of inspecting the facility directly, the inspector may scrutinize the study protocol, procedures, and data in greater detail in order to reconstruct the processes that occur in these areas.

Definitions and Terms—*The Same Words Don't Always Mean the Same to Different Regulatory Authorities and Study Sponsors*

In the study planning phases, it is essential to align the country of the intended marketing permit submission with the appropriate set of GLP regulations/principles. The subsequent development of study documents should include common language and terminology. Before creating any documents, the definition of terms that are included in the applicable set of the GLP regulations/principles must be reviewed. Each document begins with these definitions, and they are crucial for effective communication between the governing agencies, the sponsor, and the testing sites or contracted laboratories. The terms should be used appropriately and consistently in all written documentation such as the protocol/study plan, SOPs, and the final report.

Some definitions are unique to a particular set of the GLP regulations/principles. For example, one difference is in the terminology used to define the compound or product being tested during the preclinical phase. This test compound is referred to as the test article (FDA GLP), test substance (EPA GLP), or the test item (OECD). Another definition difference is that the OECD Principles of GLP include a definition of principal investigator (PI)—an individual who acts on behalf of the study director in a multisite study. Neither the FDA nor EPA regulations include a definition that equates to a PI. Another variance applies to the definition of study start dates. While all the GLP regulations/principles define the study initiation date as the date the Study Director signs the Protocol/Study Plan, the EPA GLP regulations, and the OECD Principles of GLP include definitions for the experimental start and stop dates. The EPA GLP regulations define the experimental start date as the first date that the test substance is applied to the test system, whereas the OECD Principles of GLP define the experimental start date as the date on which the first study-specific data are collected.

Standardized terminology will facilitate communication between the Agency, the sponsor, and the testing facility. Utilize the respective definitions sections when negotiating contracts, during study planning, and in all study documentation for the duration of the study.

ORGANIZATION AND PERSONNEL—*WHO DOES WHAT AND HOW DO WE KNOW THEY CAN DO THAT JOB?*

Reducing variables that may affect study data is critical and consideration must be given to the human element in this regard. Proper training, education, and/or experience of study personnel combined with SOPs can reduce risks while streamlining efficiency and improving staff proficiency. Standardization and identification of personnel roles and responsibilities, along with proper personnel organization and appropriate training of study personnel, are key components of the GLP regulations/principles.

Defined roles and responsibilities for study personnel must be recorded and the completion of training necessary to perform these roles and responsibilities must be documented. This information should be organized in such a manner that an auditor can match training records to the assigned tasks for any given GLP study. Suggestions for organizing this documentation include the establishment of standardized training files for all GLP personnel and the creation of an organization chart that identifies the roles and illustrates the interactive relationships. GLP study personnel include any individuals who have impact on the study data such as GLP management, study directors, PIs, laboratory managers, research technicians, graduate students, veterinarians, animal handlers, statisticians (data handlers), and QA auditors. Training of study personnel should include any necessary technical training, biocontainment training, SOP training, and GLP training relevant to the job responsibilities.

For studies in high biocontainment facilities, the issues of appropriate training are significant. Not only must all laboratory and biocontainment engineering staff be appropriately trained to work safely in that environment, they must also be trained to meet the requirements of the GLP regulations. Likewise, GLP management and QA staff may require specialized training for access to those laboratory facilities to ensure appropriate oversight and review of the research activities. Consideration must also be given to the requirements placed on all individuals by other relevant regulations, such as those governing the use of Select Agents. Training of new staff for a GLP study in a high biocontainment laboratory is a process that requires significant investment of time, effort, and resources.

For any GLP study, there must be a defined testing facility management, a study director, and an independent quality assurance unit (QAU)/program. Further, the GLP regulations/principles specifically define the minimum responsibilities for study personnel, including GLP management, the study director, the QAU/program, and the PI (OECD only). SOPs should define how each role functions within the overall business structure of the testing facility along with the mechanisms followed to perform their responsibilities. These procedures should also address sanitation and health precautions that personnel must follow in order to decrease the potential of contaminating the test and control articles and test systems, including the reporting of any illness that may adversely affect the data. For many biocontainment laboratories, these issues overlap with precautions and reporting required from an employee health and safety perspective. Procedures governing these issues may also be addressed in facility policies that are not specific to the GLP program, as employee health records are subject to other regulatory requirements that maintain the privacy of medical records.

GLP Management

It is crucial for a facility to define their management structure prior to performance of GLP-compliant studies. Testing facilities that function primarily as contract research organizations may have a business hierarchy that directly aligns with the GLP management structure, whereas academic and other large institutions may need to have a separately defined infrastructure in place to support GLP studies. Effective GLP management structures in large institutions should possess the capacity to bridge multiple departments as studies may require cooperation from animal resource facilities, environmental monitoring and facilities maintenance departments, information technology groups, statisticians, and veterinary pathology laboratories. While appropriate allocation of management roles is a business decision, it is imperative to identify GLP management and equally important for GLP management to understand their roles and responsibilities. Failure to identify a functioning GLP management structure and failure of proper training of GLP management is a violation of the GLP regulations.

GLP management is responsible for providing the appropriate resources (facilities, equipment, procedures, and personnel) necessary to perform a regulated study and must be capable of appointing and/or replacing the study director. The latter may not necessarily transpire from poor performance, but may be necessary if the study director has insufficient time to devote to a GLP study, or may be unavailable for a significant portion of the study. Management must also assure that a sufficient number of personnel are available to conduct the study in accordance with the protocol. Furthermore, management is tasked with assuring that the QAU/program is completely independent of the study conduct and that QA inspection/audit findings are communicated to the study director.

The Study Director

The study director serves as the central point of contact for a GLP study and is responsible for the technical conduct of the study, including assuring that the GLP regulations are followed. Since the study director serves as the central point of contact for a GLP study, only one study director is allowed for each study; however, the OECD Principles of GLP allow for PIs to serve, under the direction of the study director, as points of contact for multisite studies. It is important to note that the OECD defined role and responsibility of the PI does not necessarily align with the U.S. terminology used to annotate the PI for grants and contracts. Care should be taken to identify which GLP regulations/principles are applicable to the study when defining the responsibilities of an individual serving in the role as a PI.

Since study directors serve as the central point of contact for the study, they must be knowledgeable in the GLP regulations, familiar with the science of the project, and possess skills in project management. However, it is acceptable to delegate tasks during a study and often the appointment of a laboratory manager or supervisor can assist in study organization. Care should be taken to clearly delineate these task assignments and, while delegating tasks can streamline efficiency, it must be remembered that the ultimate responsibility of the study conduct lies with the individual designated as the study director.

The study director is also responsible for all experimental data including any deviations from procedures or unforeseen circumstances that occur during the study. As documentation should allow for the reconstruction of the study, it is imperative that the study director assess all study circumstances for possible effects on the data. Facility or study-specific SOPs should describe how the study director reviews study documentation and how deviations from SOPs and/or the study protocol are managed. Furthermore, the study director determines the study initiation and closing dates by signing and dating the protocol and final report respectively, and is responsible for assuring that all records and specimens are transferred to archives during or at the close of the study.

Quality Assurance

The GLP regulations/principles also delineate a QA requirement. While individual studies may plan for appropriate quality control of procedures and data management, the QAU assesses the overall conduct of a GLP study and adherence to the regulations, study protocol/plan, and approved SOPs. This function involves facility and study-specific inspections and/or audits performed by an independent unit or program. Members of the QAU or QA program are referred to as QA auditors and, in order to remain unbiased, the QA auditor(s) for a specific study may not be involved in the conduct of that study. The frequency and type of inspections depend on the length of the study and the critical phases of the study; communication between the study director and the QA auditor facilitates the process. Inspections and data audit findings are compiled in a QA report that is distributed to the GLP testing facility management and the study director. The content of these reports is not subject to inspection by a government agency representative during a routine Agency audit; however, QA must keep a log of these inspections/audits as documentation of their occurrence. In order to properly manage the performance of inspections and audits, the QAU maintains copies of study protocols and a copy of the master schedule. The latter is a log of all GLP studies performed at the facility and is used by the Agency representatives to organize their facility inspections and to assess study director workload.

QA of GLP studies differs from the QA requirements included in the Good Manufacturing Practice (GMP) regulations. Significant findings during a GMP QA inspection/audit can halt production. The intent of the GLP regulations is to have an internal QA process that allows for correction during the study, thereby decreasing the possibility of repeat testing or validation, or study rejection. The QA auditor reports any gaps and/or deviations from the regulations, protocol, and approved SOPs to the study director and management. It is then the responsibility of management and the study director to correct any applicable issues as stated in the QA findings. In a high biocontainment laboratory environment, where physical access to study areas for audits by sponsors or regulatory agencies is likely to be limited, it is reasonable to expect that QAU personnel should be trained for access to the laboratories to ensure that they can adequately fulfill their oversight responsibilities.

Archivist

Another important role in a GLP program is that of the archivist since study records will be transferred from the supervision of the study director to an archive facility. The latest document from OECD actually includes a definition of an archivist, stating that an archivist is "an individual designated by test facility or test site management to be responsible for the management of the archive, i.e., for the operations and procedures for archiving" (OECD, 2007). Although the archive facility is discussed in the "records and reports" section of this chapter, study planning should include the identification of the individual who will be responsible for the archive.

FACILITIES—*WHAT MAKES A LABORATORY SUITABLE FOR GLP?*

Unlike other regulations governing laboratory or manufacturing facilities, the GLP regulations place relatively few specific requirements on the physical infrastructure of the laboratory. Testing facilities should be of "suitable size and construction" and designed to prevent any "adverse effect" on the study (21 CFR Subpart 58.41). In general, the intent is to separate various components of the study to reduce or eliminate risks of contamination.

Facilities for animal care, laboratory procedures, supply storage, and specimen and data storage must operate under standardized procedures and must be properly managed and maintained in order to decrease the risk of contamination of the study data. In high biocontainment laboratories, where available space and laboratory design may not permit physical separation of study components of different research activities, study design should consider methods to decrease contamination risks when establishing the workflow. Methods can include the use of separate rooms, biosafety cabinets, ventilated cage rack systems, procedures for decontamination, and appropriate personal protective equipment for study staff. Additionally, supplies such as feed, bedding, and animal enrichments should be stored separately from the animal housing rooms. Workflow should address all aspects of the study including animal quarantine prior to study placement, requirements and location of animal isolation, disposal of waste, and operation of shared areas in order to prevent study mix-ups and cross contamination. A facility floor plan can be used as a diagram to illustrate the workflow; however, floor plans for areas used for Select Agent studies should have SOPs established for their secured distribution since they may be considered sensitive documents under Select Agent regulations.

Facility design should also address isolation of areas used to prepare test/control articles in order to prevent mistakes and cross contamination. Care should be taken to separate areas for the receipt, handling, mixing, and storage of the test article. The storage areas must be separate from the test system and must meet the environmental specifications needed in order to maintain the integrity of the compound.

Biocontainment laboratories are complex facilities, often having sophisticated building automation systems (BAS) for control and monitoring of the biocontainment environment and facility security. Careful consideration must be given to the operation of these systems and their potential use in primary monitoring of laboratory spaces used for GLP-compliant studies. This will be discussed further in the following section.

In addition to laboratory and animal facilities necessary to perform GLP studies, space needs to be identified for specimen and data retention. Specimen and data retention during a study can be in proximity to the study location or can be in a central location; however, an archiving facility (or facilities) is required for long-term retention of study data and specimens. Archive facilities can be located at the sponsor facility, at an external facility, or internally at the testing site, but must be under the control of an archivist in order to expedite retrieval of the records. During the study, the data and specimen integrity is the responsibility of the study director; however, once transferred to the archive, the archivist is tasked with maintaining the integrity by assuring that the data and/or specimens are secure and stored in the appropriate environmental conditions necessary to prevent degradation. The indexing and operation of the archive should be defined in SOPs and retrieval of items should occur in a timely manner during the course of an inspection. If an off-site facility is utilized as a GLP archive, it is recommended that a QA inspection be performed in order for management and the study director to determine if the archive facility is adequate.

EQUIPMENT

Despite the common perception that performance of GLP-compliant studies requires certain specialized equipment, the US FDA regulations specify only that equipment used in the "generation, measurement, or assessment of data" and equipment for facility environmental control be of "appropriate design and adequate capacity to function according to the protocol" (21 CFR Part 58). As with planning for many other areas of GLP compliance, selection, and operation of equipment should be driven by the requirements of the study, within the limitations of the particular laboratory environment.

In order to increase data integrity, equipment should be designed to meet the needs of the study, and should be calibrated and maintained throughout the study. Equipment documentation is generally maintained from the point of acquisition until the point of retirement—i.e., throughout the equipment life cycle. Acquisition can be procurement of a new piece of equipment or the assessment of an existing piece of equipment. The process of documenting that a piece of equipment meets the needs of the study, and is operating and performing correctly, is generally referred to as equipment validation or equipment qualification. For purposes of discussion, this chapter will discuss the validation of equipment using a four-step process utilizing the terms design, installation, operational, and performance qualification (PQ).

Generally, a facility should develop high level policies governing the approach for equipment acquisition, validation, and operation for GLP-compliant studies. Before purchasing or using a piece of equipment during a GLP study, the system should be assessed to see if it meets the needs of the study. Documentation should align the manufacturer's specifications with the study needs. This process is referred to as design qualification (DQ). Next, the equipment must be installed and shown to be correctly functioning in its environment; this is called installation qualification (IQ). Once working, the equipment should be tested to see if it is operating according to the manufacturer's specifications (operational qualification, or OQ). If there are any special performance requirements specific to the needs of the study, these must also be tested and documented (performance qualification, or PQ). For simple pieces of equipment that do not collect data, a simplified validation/qualification process that combines these processes may be employed. This is sometimes referred to as equipment verification. As mentioned above, the requirements of and processes to be used for equipment validation or verification should ideally be defined at a facility or institutional level.

Planning for the equipment validation process should also include the preparation of specific SOPs for calibrating and maintaining equipment in a validated state. The type and frequency of the calibration, preventive maintenance, and general cleaning will be determined based on the use of the equipment, the requirements of the study, and the specific limitations of the operating environment. In a high biocontainment environment, equipment maintenance, and calibration schedules and procedures may be largely dictated by broader facility operational requirements and the limited access to laboratory spaces. The specific maintenance and calibration that is performed on the same piece of equipment will likely differ between BSL2, BSL3, and BSL4 laboratories within the same facility due to the unique nature of each laboratory environment. In any case, documentation of the performance of these functions should be retained with the study data. Finally, procedures should be in place addressing the retirement of equipment. Before retiring, assess the impact of the retirement on current studies, and determine if electronic data was generated. Plans for data storage and future data migration should be developed prior to the retirement of computerized equipment.

Validation of complicated systems can be a lengthy process. When procuring new pieces of equipment, requests can be made for validation documents and support from the manufacturer. Often the IQ and OQ portions may be completed by a vendor representative, but the supporting documentation may need to be evaluated to assure that sufficient documentation is obtained according to the facility's validation policies. Complicated systems that are critical to the study data collection may need to be validated under the direction of an experienced validation team or outside consultant. The length of time necessary to validate equipment should be considered in study planning and contract negotiations with the sponsor. Furthermore, retrospective evaluation of existing equipment for DQ prior to the start of the study may indicate that the equipment is not suitable for the needs of the study and a new piece of equipment may need to be purchased and validated before the study can begin.

Selection and validation of systems or equipment used for environmental monitoring will also depend upon the specific requirements of the facility and the studies that require GLP compliance. In a large biocontainment facility where only a subset of laboratories or rooms may be set aside for GLP-compliant research, validation of the BAS used for control and monitoring of the entire building may be impractical or deemed to be cost ineffective. In those cases, local monitoring using stand-alone devices or networked supervisory control and data acquisition (SCADA) systems may be a more practical solution for appropriate monitoring of laboratories used in performance of GLP-compliant studies.

If an area also contains equipment not used for GLP studies, which is likely to be the case in many BSL3 and BSL4 laboratories, it is recommended that some type of labeling process be established and documented to differentiate the equipment. This will clearly define to an Agency inspection official what supportive documentation should (and should not) be audited. In areas of high containment, it is recommended that annual or semiannual schedules for equipment calibration and preventive maintenance are coordinated with area/unit decontamination and shutdown times. This will increase efficiency of space utilization and allow for more accurate study timelines.

Another consideration for equipment in high biocontainment laboratories is the collection and storage of raw data. In many cases, a solely paper-based data collection system will be impractical due to the difficulty in transferring materials out of the laboratory. Capturing raw data in an electronic format may simplify the transfer and storage of that data, but requires that all components of the network and electronic systems be appropriately validated and function in compliance with 21 CFR Part 11 Electronic Records; Electronic Signatures. Suitable systems and processes for storage and backup of raw data must also be established to eliminate or otherwise minimize the risk of data loss.

TESTING FACILITIES OPERATION—*HOW DO WE DO WHAT WE DO?*

All routine methods and processes for GLP studies and critical supporting activities must follow standardized procedures that are approved by GLP management. Decisions on the extent of documentation required for supporting activities should be based on risk assessments that determine which functions may directly impact the performance and outcome of a study. SOPs govern day-to-day practices, whereas the protocol/study plan (discussed later) serves as the overall blueprint for the study. Each set of GLP regulations/principles list the minimum required SOPs to support a GLP study, but the list is not all-inclusive. Standard business policies should exist for facility operation and are subject to inspection; however, they should be reviewed prior to the start of a GLP study to assure that they are compliant with the regulations and appropriate to the requirements of the study. The QAU/program audit schedule should include a general facility inspection in addition to one or several study-specific inspections and/or audits. SOPs must ultimately be approved by management and implementation of subsequent changes made to an SOP should follow a standardized procedure that also includes management approval. Historical files of all SOPs and their respective versions should be retained and are subject to Agency inspections. Business definitions should identify if there is a difference between an SOP approval date and the actual effective date of the SOP. The latter date should be established after consideration of the length of time necessary to train appropriate personnel on the SOP or the new version of the SOP.

Current versions of the SOPs must be available in the area where the procedure is performed. While it is appropriate for a study director to maintain copies of all approved SOPs for his/her study, copies of applicable SOPs need only be present where the procedure occurs. Care should be taken in high containment areas and SOPs should be in place to handle contaminated paper documentation, including raw data or SOPs. An SOP may be available in a printed form with an expiration date or an SOP may be viewable from a computer screen. The handling and distribution of the SOPs should be addressed in an approved administrative or document-handling SOP. No expired or outdated versions of SOPs should be located in areas where procedures are performed. A paper-based system for preparation and distribution of SOPs may be impractical for high biocontainment laboratories, and even for GLP programs that support multiple concurrent studies. Suitable electronic document management systems are available that can be configured and validated for compliance with 21 CFR Part 11 and that will coordinate and control the processes for preparation, approval, distribution, and archiving of SOPs, as well as documenting SOP training.

Facility SOPs should include the operation of the animal care facility as well as the specific animal husbandry and handling that occurs within the biocontainment laboratory environment. It should be clear how a facility handles cage cleaning and change-outs, animal feeding and handling, animal isolation, animal treatment, finding of moribund animals, storage of feed and bedding, and use of any pest controls. Documentation of activities is necessary for study reconstruction and calibration, maintenance and quality control procedures need to be established not only for equipment that will generate study data, but also for other key equipment that supports an animal study (e.g., cage washing and sterilizing equipment). For instance, scales that are used to determine animal weights should be validated, cleaned, maintained, and calibrated if the data they generate will be used to calculate dosing concentrations.

Study planning should also include the type of feed, water, and bedding to be used. The study director, along with the sponsor, should evaluate these items for the presence of any interfering substances that should not be present in the feed, water, or bedding and should maintain documentation that demonstrates the absence of these interfering substances. Vendors may be able to provide documentation of testing for certified feed and bedding. Animal facility SOPs should state the frequency and types of testing performed on potable water. If additional testing is required to support a regulated study, communication with animal care facility management is critical and these issues should also be addressed in the budgeting and planning phases of project development.

Reagents used during a GLP study must be labeled with the identity, titer or concentration, storage requirements, and expiration date. Use of expired reagents is a violation of the regulations/principles; however, SOPs can be in place that allow for extension of the expiration date based on scientific evaluation or analysis. The OECD Principles of GLP also require that the source, preparation date, and stability information be accessible. Reagent preparations that include dilutions or mixing with other reagents must also be appropriately labeled. Generally, the expiration date of mixed reagents is the earliest expiration date of any one of the combined chemicals. Documentation of reagent lot numbers, reagent volumes, and identification of personnel preparing reagents may be critical pieces of information depending on the nature of the study and use of the reagent.

TEST AND CONTROL ARTICLES—*WHAT ARE WE TESTING AND HOW DO WE HANDLE IT?*

Since data obtained during the study will be used to support the product licensure of the test article, characterization and handling of the compound is critical. SOPs should describe the labeling and handling of the test and control articles from receipt to final discard or retention. Documentation should include the condition upon receipt as well as the storage location, and the storage location should align with the storage requirements and provide appropriate environmental conditions in order to retain the compound integrity. Storage of the test and control articles must be in a secured location and procedures should be established that allow complete accountability of handling and volume depletions and/or distributions.

Analysis of the test and control articles prior to study initiation includes identification of the compound along with its strength, purity, and composition. This analysis must be conducted for each batch that is manufactured. Sample derivation is also required; such documentation can be provided by the sponsor or, in the case of a marketed compound, be identified by the label (FDA only). The above characteristics must be identified prior to the initiation of the study; however, compound stability can be determined either prior to or concomitantly with the study. The responsibility for characterization of test and control articles (i.e., provided by the sponsor, or carried out by the testing laboratory) should be clearly established during contract negotiations and study planning. SOPs should exist to describe how compounds are mixed with carriers, and the mixtures must be tested in order to determine the uniformity and subsequent concentration. Care should be taken when handling or mixing the compounds with carriers to ensure that no contamination is introduced during the process. Preparation of test and control articles should occur in separate areas as appropriate to decrease the introduction of contaminants during the process, and to prevent any inadvertent mixing or substitution of test and control articles. Standard decontamination practices performed prior to compound preparation should be documented in order to demonstrate that the SOPs are being followed. For high biocontainment facilities, storage and even preparation of test and control articles should generally occur at BSL2. However, the specific requirements of the study and the characteristics of the test and control articles may require that final reagent preparation occur in the BSL3/BSL4 laboratory.

In addition to proper receipt, handling, and mixing of test and control articles, the storage containers must be labeled by name, chemical abstract number or code number, batch number, expiration date (if known), and storage conditions. Test and control article storage containers should be kept for the duration of the study (FDA, EPA) and an aliquot of the test/control compounds must be retained for long-term studies (defined as longer than 4 weeks by FDA, EPA). The Agency has the authority to transport a test compound and the SOPs in order to reproduce the study in their laboratories if that is deemed necessary and if appropriate facilities are available.

STUDY CONDUCT—*THE EXPERIMENTAL PLAN*

The document that states the overall plan for the study is called the protocol (FDA, EPA) or the study plan (OECD). All of the GLP regulations/principles state the minimum requirements for the information that must be included in the protocol/study plan. In general, the protocol/study plan should include the identification of the sponsor and the testing facility, key personnel involved in the study, the test system, the test and control articles, the study methods, the data collection methods, and the statistical manipulations. This document often serves as a contract between the sponsor and the testing facility and should serve as a communication tool stating the agreed parameters of the study to the sponsor, study director, testing facility management, QAU, and study personnel. Study director signature (FDA, EPA, and OECD) is required on the document as is the signature of the sponsor (FDA, EPA). For OECD studies, the signature of the sponsor and testing facility management is a requirement that varies by national regulation or legislation. Failure to perform the study in accordance with the protocol is a violation of the GLP regulations/principles.

As outlined earlier in this Chapter, under the FDA GLP regulations the date that the study director signs the protocol/study plan is the study initiation date. Any changes made to the protocol/study plan after this time must be made in the form of an amendment that also requires the dated signature of the study director. In general, the protocol/study plan along with any amendments outline the study objectives, the SOPs describe in greater detail the methods used to accomplish the study objectives, and the final report summarizes the study data and findings. During the study, any contradicting statements between the protocol/study plan and an SOP are decided by the protocol/study plan as the statements contained within a protocol/study plan supersede the SOPs. Writing effective protocols/study plans, SOPs, and final reports can lengthen the time needed to complete contract negotiations and study timelines. Training on SOP writing can be beneficial for individuals who possess minimal experience writing these types of documents.

Failure to perform the study in compliance with the protocol is a regulatory violation. While deviations from protocols/study plans or SOPs can happen, these deviations must be properly documented. SOP deviations must also be documented and the study director should assess whether or not the deviation impacted the study. The study director should also review the deviations for trends. If the same deviation is occurring frequently, it is possible that an SOP needs to be revised or the personnel are not properly trained to adequately perform their job duties. If SOP deviations are properly handled and the documentation is included in the study records, then they do not need to be listed individually in the final report; however, all protocol deviations must be identified in the final report. If SOP deviations are not appropriately handled or if the deviation had a direct impact on the quality and integrity of the data, then the circumstances must be disclosed in the final study report.

Recording data properly during the study is a critical aspect since the data generated will be used to reconstruct the study and subsequently used to support the application for product licensure. Unless the initial data is captured via electronic systems, collection of all study data must be documented promptly and in ink. All data must be accompanied by the signature or initials of the person entering the data and the date the data was recorded. The first capture of information is considered the raw data so care must be taken when observations are recorded. Trained study personnel should be aware of the appropriate places to record study data. The creation of standardized forms can assist in the capturing of raw data promptly. All original data must be retained. If changes need to be made to recorded data, the original data must not be obscured or deleted. The change must also be accompanied by a reason for the change, the signature/initials of the person making the change, and the date the change was made. These rules are applicable to manually recorded data as well as electronically captured data. As discussed earlier in this Chapter, paper-based data collection may not be practical for some studies, particularly in high biocontainment laboratories. However, the use of electronic data collection and handling raises compliance issues related to 21 CFR Part 11, and may also be subject to restrictions related to select agent or other regulations.

Proper study conduct also includes the correct labeling of specimens collected during the study. Specimens collected for studies following the FDA or EPA GLP regulations must be identified by test system, study nature, and date of collection. This information can be recorded on the specimen container directly or identified and linked in a manner that allows for traceability to the information. Specimens collected for studies following the OECD Principles on GLP must be labeled with a unique study identifier as well as the specimen origin.

RECORDS AND REPORTS—*WHAT DID WE DO AND WHAT DID WE FIND?*

Once the experimental data has been collected, a final report is written to summarize the study findings. While the minimum requirements for the report content are listed in the regulations/principles, any contributing scientific reports should be included in their entirety and not changed by the study director or sponsor. After the final report is written, the regulations require a review of the final report and experimental data by the QAU/program in order to determine if the data presented in the final report are an accurate reflection of the raw data collected during the study. The QAU will also provide a statement that specifies the dates study inspections were made and the dates the findings were reported to management and to the study director.

All study records must be retained and archived prior to or at the close of the study. The study director is responsible for identifying all data, specimens, and study records that must be transferred to the archives.

Once the records and specimens are moved to the archives it is no longer the study director's responsibility, as the archivist is tasked with the organization, storage location, and security of the archives. All study records must be easily retrievable during an audit. Several archive locations can be identified, but the organization of materials for expedited retrieval must be under the direction of one archivist. Off-site archives are appropriate for use, but due diligence by management must occur. It is recommended that off-site archive facilities be inspected by the QAU/program prior to initial record transfers and periodically, as decided by GLP management. Required duration periods are specified in each regulation/principle; however, business procedures should be developed for facility record retention policies. It is also appropriate for some or all of the study records or specimens to be transferred to the sponsor. In this regard, the final disposition of archival materials should ideally be established during initial contract negotiations with the sponsor, particularly in situations where on-site storage is limited or the timetable for product submission (and hence the timeframe for data retention) is not clear. Documentation of this transfer agreement should be retained with the study data as location of all records and specimens must be included.

DISQUALIFICATION OF TESTING FACILITIES—*WHAT HAPPENS IF WE GET IT WRONG?*

During data review, product marketing applications can be rejected or approving Agencies can request further validation of studies if there are concerns about the performance of the study or the quality and integrity of the data. However, disqualification of a testing facility is a unique part of the FDA GLP regulations. When a facility is disqualified, the FDA can refuse to accept any data submitted by the facility in support of an application for a research or marketing permit. In addition, any study data historically submitted to the Agency by the disqualified testing facility can be rereviewed. Disqualification proceedings occur only after a facility has failed to comply with the regulations, the noncompliance directly affected study data, and regulatory warnings have been inadequately addressed or ignored. The FDA GLP regulations also address the issue of the sponsor initiating the termination of studies at a testing facility. If a sponsor terminates or suspends a testing facility during a GLP study, the sponsor must notify the FDA in writing within 15 working days of the termination.

As an example, in 1999, FDA performed an inspection at the Coulston Foundation, a testing facility in the state of New Mexico, U.S. The inspection revealed significant findings that were detailed in a subsequent Warning Letter to the company Chief Executive Officer. The Coulston Foundation did not provide sufficient proof to the FDA that the findings had been addressed and a subsequent site visit in 2000 was conducted. During the review process, FDA instructed Coulston not to initiate any new studies and required Coulston to request reinspection after the deficiencies had been corrected. Coulston did not request such a reinspection and received another Warning Letter in 2001 stating that if they did not address the violations and correct the deficiencies, action may be taken to initiate disqualification. By 2003, FDA proceeded with disqualification and sent notice to Coulston of their opportunity to request a regulatory hearing; but by late 2002, the New Mexico Business Weekly was already reporting that the Coulston Foundation was shutting down and the animals were being relocated (Anonymous, 2002).

Once disqualified, the regulations provide a mechanism for facility reinstatement. The request for reinstatement must be made in writing to the FDA Commissioner and must contain proof that the deficiencies have been addressed in order to prevent violation of the regulations.

CONCLUSIONS

The intent of the GLP regulations/principles is to assure the quality and integrity of the data. They are standards that describe the minimum requirements necessary to assure integrity of the data and allow for the reconstruction of a study. The current emphasis of governments on developing products to protect the public from emerging infectious disease pathogens or deliberate release of a biological agent has led to an increased role in this process for the academic sector and those institutions with the appropriate facilities to house these agents.

The implementation of compliant studies in a university setting, where the organization of personnel may not immediately align with the requirements of the regulations, may require definition of new organizational structures, and the prospect of carrying out GLP-compliant studies in BSL3 and/or BSL4 laboratories adds additional hurdles. Among these are the limited amounts of space available in these laboratories for records retention, the difficulty or impossibility of transferring certain types of data and specimens out of the laboratories, the increased level of scrutiny on personnel who may access these facilities, and other regulations governing the conduct of work in these laboratories (e.g., Select Agent regulations).

All of these factors should be carefully considered from the moment a GLP study is contemplated and negotiations with potential study sponsors should take into account the increased time and resources required to complete regulated studies in these environments. In addition, it is imperative that investigators who consider studies that require compliance with the GLP regulations fully understand the additional effort that will be required of them and of their staff.

While the intent of this chapter is to introduce the reader to the GLP regulations and guidelines and the potential impact that regulated study planning will have on vaccine development, particularly in studies requiring high biocontainment, it is also important to understand that the field of compliance in the 21st century is changing. Currently, the US FDA is using enforcement discretion with regards to 21 CFR Part 11, the OECD continues to publish additional directorates, modernization of the US FDA regulations is under discussion, and global harmonization is increasing communication between regulating bodies. In the meantime, research and testing organizations should make compliance policy decisions utilizing a risk-based approach, keeping in mind that ignorance of the regulations is not accepted by the US FDA or other regulating agencies as an excuse for noncompliance.

ACKNOWLEDGMENTS

The authors thank Dr. Wayne Patterson (Senior Assistant Vice President for Research, Research Services, UTMB) for helpful review and critique of the chapter.

References

Anonymous. Coulston Lab Shuttered; Monkeys get New Caretakers. New Mexico Business Week, 18 September, 2002.

Baldeshwiler, A. History of FDA good laboratory practices. Qual. Assur. J. 2003; 7:157–161.

Marshall, E. The murky world of toxicity testing. Science 1983; 220(4602):1130–1132.

Novak, R. The long arm of the lab laws. Today Chemist 2001; 10(11):45–46.

Organisation for Economic Co-operation and Development. OECD Principles on Good Laboratory Practice. Rev. 1997.

Organisation for Economic Co-operation and Development. Advisory Document of the Working Group on Good Laboratory Practice, Establishment and Control of Archives that Operate in Compliance with the Principles of GLP. June, 2007.

Thies, R.C. Good laboratory practices: a legal perspective. Clin. Toxicol. 1978; 12(2):207–212.

U.S. Department of Health and Human Services Food and Drug Administration. Good Laboratory Practice Regulations; Final Rule (Docket No. 83N-0142). Federal Register 21 CFR Part 58. 4 September 1987.

U.S. Department of Health and Human Services National Institutes of Health. Overview of the NIH Roadmap. 2006. Retrieved 9/24/07 from http://nihroadmap.nih.gov/overview.asp.

U.S. Environmental Protection Agency Toxic Substance Control Act (TSCA); Good Laboratory Practice Standards. Federal Register 40 CFR Part 792. 17 August 1989.

Abad, X., Bosch, A. and Navarro, C. Implementation of good laboratory practice in a university research unit. Qual. Assur. J. 2005; 9:304–311.

Anonymous. The Scandal in Chemical Testing. The New York Times. 16 May 1983.

Broad, W. Who Tests the Product-Testing Labs? The New York Times. 22 May 1983.

Kennedy Hearings: The Searle Story. Provided by the Society of Quality Assurance. Charlottesville, VA, 1976.

Nightingale, S., Prasher, J. and Simonson, S. Emergency use authorization (EUA) to enable use of needed products in civilian and military emergencies, United States. Emerg. Infect. Dis. 2007; 13(7):1046–1051.

U.S. Department of Health and Human Services Food and Drug Administration. Good Laboratory Practice Regulations (Docket No. 76N-0400). Federal Register 21 CFR Part 58, 22 December 1978.

U.S. Department of Health and Human Services Food and Drug Administration. Guidance for Industry: Good Laboratory Practice Regulations Management Briefings Post Conference Report. August, 1979.

U.S. Department of Health and Human Services Food and Drug Administration. Guidance for Industry: Good Laboratory Practices Questions and Answers. June, 1981.

U.S. Department of Health and Human Services Food and Drug Administration. Good Laboratory Practice Regulations; Minor Amendment (Docket No. 89N-0054). Federal Register 21 CFR Part 58. 20 April 1989.

U.S. Department of Health and Human Services Food and Drug Administration. Fraud, Untrue Statements of Material Facts, Bribery, and Illegal Gratuities: Final Policy (Docket 90N-0332). Federal Register 56 FR 46191. 10 September 1991.

U.S. Department of Health and Human Services Food and Drug Administration. (1997). Electronic Records, Electronic Signatures; Final Rule (Docket 92N-0251). Federal Register 21 CFR Part 11. 20 March 1997.

U.S. Department of Health and Human Services Food and Drug Administration. Technical Implementation of Electronic Records and Electronic Signatures; Public Meeting and Request for Presentation Abstracts. (Docket No. 00N-0358). Federal Register 21 CFR Part 11. 22 February 2000.

U.S. Department of Health and Human Services Food and Drug Administration. Warning Letter, Coulston Foundation. 11 October 2001.

U.S. Department of Health and Human Services Food and Drug Administration. New Drug and Biological Drug Products; Evidence Needed to Demonstrate Effectiveness of New Drugs When Human Efficacy Studies are not Ethical or Feasible. Federal Register 21 CFR Parts 314 and 601. 31 May 2002.

U.S. Department of Health and Human Services Food and Drug Administration. Warning Letter, Coulston Foundation. 18 March 2003.

U.S. Department of Health and Human Services Food and Drug Administration. Comparison Chart of FDA and EPA Good

Laboratory Practice (GLP) Regulations and the OECD Principles of GLP. June, 2004.

U.S. Department of Health and Human Services Food and Drug Administration. (2005). Authorization of Emergency Use of Anthrax Vaccine Absorbed for Prevention of Inhalation Anthrax by Individuals at Heightened Risk of Exposure due to Attack with Anthrax; Availability. 3 August 2005.

U.S. Department of Health and Human Services Food and Drug Administration. Current Good Manufacturing Practice for Finished Pharmaceuticals. Federal Register 21 CFR Part 211, 1 April 2006.

U.S. Department of Health and Human Services Food and Drug Administration. Building a stronger defense against bioterrorism, FDA Consumer Health Information, 29 August 2007. 2007a.

U.S. Department of Health and Human Services Food and Drug Administration. Guidance for Emergency Use Authorization of Medical Products. July, 2007. 2007b.

U.S. Department of Health and Human Services Food and Drug Administration CDER CBER. Guidance for Industry: Fast Track Development Programs: Designation, Development, and Application Review. July, 2004.

U.S. Department of Health and Human Services Food and Drug Administration, CDER, CBER, CDRH, CFSAN, CVN, ORA. Guidance for Industry Part 11, Electronic Records: Electronic Signatures—Scope and Application. August, 2003.

U.S. Department of Health and Human Services Food and Drug Administration Office of Regulatory Affairs. Bioresearch Monitoring Good Laboratory Practice Compliance Program 7348.808 (Nonclinical Laboratories). 1 October 2000.

U.S. Department of Health and Human Services National Institutes of Health. NIAID Strategic Plan for Biodefense Research 2007 Update. September, 2007.

U.S. Environmental Protection Agency. Federal Insecticide, Fungicide and Rodenticide Act (FIFRA); Good Laboratory Practice Standards. Federal Register 40 CFR Part 160, 17 August 1989.

Van Houveling, C.D., Norcross, M.A. and Lepore, P.D. An overview of good laboratory practices. Clin. Toxicol. 1979; 15(5):515–526.

Useful Websites

British Association of Research Quality Assurance (BARQA). http://www.barqa.com.

FDA/ORA. Bioresearch Monitoring Information Page. http://www.fda.gov/ora/compliance_ref/bimo/.

Organisation for Economic Co-operation and Development. OECD Series on Principles of Good Laboratory Practice and Compliance Monitoring. http://www.oecd.org/documentprint/0,2744,en_2649_34381_2346175_1_1_1_1,00.html.

The Society of Quality Assurance (SQA). http://www.sqa.org.

CHAPTER

11

Manufacturing Vaccines for an Emerging Viral Infection–Specific Issues Associated with the Development of a Prototype SARS Vaccine

Inca C. Kusters[1], *James Matthews*[2] *and Jean François Saluzzo*[1]
[1]Sanofi Pasteur, Marcy l'Etoile, France
[2]Sanofi Pasteur, Swiftwater, PA, USA

OUTLINE

ABSTRACT

The world was struck by surprise when a Severe Acute Respiratory Syndrome (SARS) epidemic started in 2003 in China. This disease had never been observed in man before; the SARS-Coronavirus causing the disease was unknown. With the uncertainty about the future impact of this epidemic, an important international collaboration started spontaneously sharing scientific knowledge and reagents. Resources became quickly available, and public and private efforts were undertaken to develop rapidly a vaccine. We will discuss here the importance of the international collaboration and the availability of funding. Moreover, we will review the most important and challenging steps during the industrial development of the SARS vaccine highlighting the difficulties in terms of safety working with such a highly pathogenic, unknown virus. We will emphasize the industrial perspectives on inactivation and decontamination experiments, the selection of the most promising vaccine candidate, the production process and the choice and use of animal models in such a pressing and difficult situation. Finally, we will briefly review the unique regulatory environment created during this period for the development of a SARS vaccine.

INTRODUCTION

In late 2002, several hundred cases of a severe atypical pneumonia were reported in the Guangdong Province of the People's Republic of China. By the first quarter of 2003, similar cases were reported in Hong Kong and sporadically throughout South-East Asia and Canada. This severe disease, with a mortality of 5–10%, spread rapidly around the world and in April 2003, 25 countries on 5 different continents had reported cases [World Health Organization (WHO), accumulative SARS cases]. As a result, the WHO issued on March 2003 a global alert for the illness that would be known as "Severe Acute Respiratory Syndrome" (SARS) (WHO SARS alert). Within the same time frame, secondary cases of SARS were being identified in health care workers and family members who had close contact with patients suffering from this severe respiratory illness.

By the second week of June, using the WHO case definition (WHO case description), approximately 8000 SARS cases and 774 SARS-related deaths had been reported to the WHO. While the first wave of the SARS epidemic seemed to have reached its conclusion, it was completely unclear how the spread of the virus would evolve. There was no clear understanding of the animal reservoir and the impact of virus mutation and unapparent infections. Different situations could be envisaged; one scenario was that virulent SARS-coronavirus (SARS-CoV) would persist and become endemic. Another possibility was that other epidemic waves would occur or, finally, that the virus would disappear. Taking into account all the uncertainties and anticipating the worst-case scenario, many laboratories and vaccine manufacturers started working on a vaccine approach against SARS infection, largely based on what was known from animal CoVs.

In this chapter, we will discuss the necessity for international cooperation and the importance of discretionary funding for rapidly developing a prototype vaccine candidate. We will review the decision-making process, the strategic choices made in terms of vaccine candidate, adjuvants, working conditions, and the safety precautions implemented at the beginning and throughout the entire production process of the SARS vaccine. In addition, we will discuss the unique challenges associated with moving a vaccine such as SARS through the regulatory process.

ROLE OF INTERNATIONAL COLLABORATION AND U.S. NATIONAL INSTITUTE OF ALLERGY AND INFECTIOUS DISEASES (NIAID)—LESSONS LEARNED

With the uncertainty about the possible extent of the spread of the SARS outbreak in early 2003, there was substantial international collaboration first in the isolation and identification of the CoV, sharing of sequence data, and eventually the sharing of seed virus that would be suitable for vaccine development. In the case of the sanofi pasteur vaccine, the prototype virus was supplied by the Centers for Disease Control and Prevention (CDC) (Utah Virus P2/Vero cell P143). Furthermore, in this spirit of collaboration, sanofi pasteur provided stocks of the Vero cells for the virus isolation to better ensure that any vaccine made from these cells would be acceptable from a regulatory perspective. These were the same Vero cells as are routinely used in the commercial production of inactivated polio vaccine. In addition, there was rapid sharing of immunological reagents to ensure that the plaque-purified prototype seed virus accurately reflected the circulating epidemic SARS-CoV.

One particular difficulty with rapidly emerging infections such as SARS, of course, is that they may occur anytime of the year and are unlikely to be aligned with the annual process for assigning financial and human resources and project prioritization. In order to circumvent these problems, once it became clear that the SARS-CoV could grow in Vero cells, the National Institute for Allergy and Infectious Diseases (NIAID) in the United States focused their request on several companies that had experience in large-scale Vero cell culture and had the experience to work with viruses at biosafety level 3 (BSL3) containment. Our first contact with NIAID on the possibility of developing an inactivated vaccine occurred in the first quarter of 2003. Sanofi pasteur was one of only two companies that satisfied the NIAID's eligibility criteria. The solicitation and funding mechanism that NIAID employed was a directed Request for Proposals (RFP). This is sometimes called a Letter Contract by virtue of the fact that NIAID initiated contact with the companies first by letter outlining the project objectives, infrastructure requirements, and deliverables. The advantage of this mechanism, for diseases such as SARS, is that the absolute eligibility criteria, vaccine characteristics, and project objectives are clearly defined; the proposal review cycle is very short; and the necessary financial resources are immediately available. In this particular case, the RFP was received by us on May 23, and we submitted our project plan and budget by June 16. Preliminary work began in July and August, and by the third week of September 2003, we had a fully executed contract. By October 2004, our inactivated prototype SARS vaccine was filled and available for clinical assessment. Another positive aspect of this mechanism is that involvement of such a large research organization provides the potential to access the expertise of many different investigators, as technical problems may arise. Unlike traditional research grants, these types of contracts have very short duration and, consistent with the sense of urgency, progress against objectives is monitored on a weekly basis, perhaps, contributing to the fact that our project finished almost a month ahead of schedule. Importantly, this type of mechanism does not compete with academic researchers for funding but allows these researchers to develop more basic science proposals of longer duration that can be reviewed and funded by other established mechanisms.

Perhaps, among others, there are three important lessons about responding to emerging threats that can be learned from the SARS experience. First, information must be shared quickly and transparently. The somewhat surprising observation that the SARS-CoV grew well in Vero cells was important in prioritizing the development of an inactivated, Vero cell–derived virus. Second, there should be an ongoing dialogue between the national research organizations and their potential industrial partners to understand what capacity and experience could be urgently brought to bear in a crisis situation. Although there are potential issues with competitive intelligence, it remains important that information about industrial capability should be shared. Third, funding organizations should have in place a mechanism for rapid proposal solicitation, review and monitoring, and also adequate discretionary funding that could support new vaccine projects in an urgent manner.

PRIOR TO THE START OF THE LABORATORY WORK

As early as April 25, 2003, the WHO and the CDC in the United States (WHO guidelines, CDC) recommended laboratories and vaccine manufacturers to handle SARS-CoV specimens using Standard BSL3 work practices. It is not exceptional for vaccine manufacturers to work in BSL3 facilities for the production of certain vaccines, e.g., rabies, Japanese encephalitits, or polio viruses. Some of these commercial vaccines have been made for decades. For these vaccines, robust production processes and standard operating procedures have been put in place. The experience gained on the decontamination and inactivation of viral vaccines during the production process and the quality control (QC) are all important. Moreover, the personnel working at the different stages of such a vaccine production are typically vaccinated, and revaccination procedures are in place in case of major accidents (i.e., boost after potential rabies contamination).

In case of an emerging virus such as SARS, the situation is completely different. Although the BSL3 experience will be fully exploited, all processes and procedures need to be discussed, evaluated, validated, and implemented. For each emerging virus, this exercise needs to be repeated and specific conditions must to be adopted according to the unique characteristics of the new virus.

Consistent with the WHO and CDC recommendations, at sanofi pasteur live SARS-CoV was handled in a BSL3 facility. Since no treatment or vaccine was available against SARS, we decided to work in "BSL3 plus" laboratory conditions. Essentially our basic level of containment was BSL3 incorporating several BSL4 working practices: clothing change before entering the facilities, shower on exit, and all material decontaminated before exit from the facilities. Also, all steps, where the product was handled in open phase, were performed within Class III Biological Safety Cabinets (CSB). To prevent accidental contamination, the laboratory workers wore a positive pressure personal mask. These precautions were considered necessary taking into consideration the large amount of live virus handled (20–50 l) and finally turned out to be valid as the risk for contamination existed as was shown in the laboratory accidents in Singapore, Taiwan, and Beijing in China (Senior, 2003; Normile, 2004; Orellana, 2004). Furthermore, as described below, the decontamination experiments demonstrated that the SARS-CoV is an extremely resistant virus.

For the medical follow-up of the BSL3-trained personal involved in the SARS vaccine development, it was of critical importance to be able to distinguish the symptoms of respiratory distress caused by SARS or other respiratory agents. It was therefore decided that the BSL3 laboratory workers should be selected in accordance with their immune status and would be immunized with *Streptococcus pneumoniae* and influenza vaccines, if appropriate. In case a worker would present symptoms, a procedure had been put in place to isolate the worker using a high-efficiency particulate air (HEPA) filter mask before being transported to a nearby hospital where a special, dedicated negative pressure hospital room had been prepared.

STARTING THE LABORATORY WORK: FIRST STEPS

Definition of the Vaccine Profile

The very first decision concerned the choice of vaccine immunization strategy. In the intense early days of the epidemic a variety of approaches were considered. These included inactivated vaccines, subunit products, DNA (either alone or in combination as part of a prime-boost strategy), vectored vaccines, and live attenuated candidates. Similarly, alternative routes of vaccine administration (inoculation, aerosol, etc.) and formulation (adjuvants, etc.) were considered. In fact, several live attenuated and killed vaccines for veterinary CoVs are already marketed: e.g., a live attenuated and killed vaccine against the chicken infectious bronchitis virus (Ladman et al., 2002), a live modified virus against canine CoV (Pratelli et al., 2004). Ultimately, however, the crucial factor in the decision-making process was likely development timelines. At the time of this decision, the epidemic was prevalent and there was a sense of urgency that a prototype vaccine had to be developed as soon as possible. Embedded in this decision-making process was the realization that we were seeking a vaccine candidate that largely used existing, conventional technology and that could be made in significant quantities, if necessary.

In the context of the likelihood for rapid development and the possibility for large-scale production, the decision was taken to produce a monovalent, whole, inactivated, aluminum-adjuvanted SARS vaccine for intramuscular injection. From our perspective, a whole, inactivated vaccine was a logical first choice taking into consideration our extensive experience with inactivated vaccines such as inactivated poliovirus, rabies, and hepatitis A vaccines. We were also encouraged by the fact that SARS-CoV grew very well in Vero cells. Sanofi pasteur has a long industrial experience with these particular cells and developed cell banks that are validated and registered around the world. Even for such a direct experience-based approach, we realized that it would take 12 months to make a good manufacturing practice (GMP) clinical lot for Phase 1 and 2 clinical studies.

Killed vaccines need, in many cases, adjuvantation. In our case, the adjuvant of choice was aluminum hydroxide. Again, in the situation where a new virus is emerging, there is no time to evaluate new adjuvants. Our preference was to use aluminum hydroxide since it is well known, used in many vaccines, and well accepted by regulatory authorities.

SARS Inactivation and Testing Reagents

When the laboratory work on the SARS-CoV vaccine development started, no data were available on the inactivation characteristics of the virus. One of the priorities was to identify the conditions to fully inactivate the virus for vaccine development, but also for decontamination of equipment, facilities, and waste decontamination. The results, as described in more detail below, were unexpected. The SARS-CoV is an extremely resistant virus and several of the routine decontamination working practices cannot be applied for this virus. These results demonstrate how important it is to immediately perform decontamination testing and adapt decontamination practices and strategies for each specific (emerging) virus.

At the time the development of a vaccine against an emerging virus is initiated, it is very unlikely that routine tests and reagents are available. This was indeed the case for SARS. Therefore, our first experiments were dedicated to develop routine tests that are required for vaccine development and the analysis of the host response after immunization. These include neutralization tests, enzyme-linked immunosorbent assays (ELISA), polymerase chain reactions (PCRs), and others. Well-validated reagents are needed as reference standards for essential laboratory tests as polyclonal and monoclonal antibodies, recombinant proteins, and PCR primer pairs. One of the most sensitive issues was how to select an appropriate animal model to evaluate the candidate vaccines. For example, we observed that NMRI mice gave a very heterogeneous response whereas Balb-C or C57BL/6J mice responded uniformly to immunization with the vaccine candidate. Also, guinea pigs appeared to be a good model for the evaluation of immune responses. Beyond immunogenicity, it is important to work with an animal model that is appropriate for challenge studies, as an assessment of vaccine efficacy. This is especially important for emerging infections since efficacy studies in humans may not be possible. In case of the SARS, the *Macaca fascicularis* was identified very early on (Fouchier et al., 2003) as a likely predictive, challenge model.

All of the work was done in constant communication with the regulatory authorities. Our experience with SARS reinforces the idea that manufacturers should be encouraged to open and maintain an active dialogue with regulatory officials, very early in the development process.

Results of the Decontamination Experiments

Viruses can be inactivated by several methods, based on either physical or chemical mechanisms. We investigated five decontamination methods that are currently used for equipment, facilities, and waste decontamination: heat, alkaline treatment, sodium hypochlorite treatment, gaseous formol fumigation, and drying.

It should be stressed that the data presented here have been obtained under our specific experimental conditions. Indeed, the virus sensitivity to inactivation depends on the virus environment and concentration. Thus, the methods presented here were validated with our specific suspensions and experimental conditions, and appropriate precautions should be taken when manipulating SARS-CoV under other laboratory conditions.

Inactivation experiments using different pH values gave rise to unexpected results. When the pH was adjusted to pH13 or pH13.5, a strong decrease in the viral infectivity titer can be observed. However, the virus is not totally inactivated and even after 6h of alkaline treatment, infectious particles can still be observed. Total inactivation is observed only after 24h of treatment. These data are surprising as enveloped viruses are usually sensitive to such drastic alkaline conditions.

The results from the experiments performed to evaluate the viral loss of the SARS-CoV due to drying on glass surface were also surprising: 35–42 days were necessary to inactivate the virus to the detection limit of the technique. Other viruses, such as polio or rabies, can be inactivated by drying durations of approximately 72 and 144h, respectively. In our experience, the SARS-CoV is the most resistant virus ever described in an industrial setting.

The gaseous formaldehyde fumigation is a viral decontamination technique widely used throughout the world. We found that formol fumigation is totally inefficient on dried virus. However, virus in solution is efficiently inactivated. The two decontamination techniques tested here, drying and formaldehyde fumigation, reinforce the necessity to decontaminate the working areas in the laboratory, as well as the equipment, very frequently in order to avoid any drying of viral suspension onto glass or any other surface.

Finally, sodium hypochlorite (2°Cl) and heat treatment were evaluated. The effect of sodium hypochlorite on dried virus is very rapid and efficient, as no infectious viral particles were recovered after washing the surface with sodium hypochlorite. Thermal decontamination was shown to be efficient at both 58 and 68°C. To achieve total inactivation of the SARS-CoV, 1h heating at 68°C and 2h heating at 58°C are necessary.

Considering the decontamination data, the following strategy was put in place. All solid waste, as well as the equipment that could resist such a drastic treatment, was autoclaved. For liquid waste, the solutions were subjected to alkaline pH treatment for at least 2h and then transferred into a tank for thermal decontamination at 105°C for 30min. The laboratory facilities, and the equipment that could not be autoclaved, were decontaminated with 2°Cl sodium hypochlorite solutions, before being fumigated with gaseous formol.

The data on decontamination of the SARS-CoV presented here show that existing decontamination strategies cannot be directly extrapolated to emerging viruses and that these inactivation conditions should be determined empirically for each virus.

THE PRODUCTION PROCESS

Viral Source and Raw Material

The SARS-CoV seed virus isolate was provided on August 25, 2003, by the CDC. This isolate (the so-called UTAH strain) was made from the sputum from an acutely ill U.S. traveler who had apparently been exposed in Hong Kong. This isolate was fully sequenced by the CDC and shown to be virtually identical to the Urbani strain of SARS-CoV.

To obtain an original seed virus, in full accordance with Food and Drug Administration (FDA) requirements, sanofi pasteur provided certified Vero cells to the CDC, who performed the isolation of the virus and made two passages before sending the virus to sanofi pasteur. Upon receipt, the virus underwent two additional passages, and was plaque purified. This plaque purification step is of importance to limit the risk of adventitious agents during the subsequent expansion of the virus. In contrast, there is also a risk that in selecting a plaque, it may differ significantly from the uncloned vaccine. To evaluate the latter, it was decided to compare cloned *vs.* uncloned virus in terms of virus sequence and immunogenicity in guinea pigs. It was demonstrated that the two candidate vaccines were totally similar.

Following this verification, the selected clone (VVNFL11) was passaged eight additional times for adaptation. From our experience with viruses to be used to prepare an inactivated viral vaccine at industrial scale, there is a need to reach a titer of virus >7.0 log/ml. Indeed, with the SARS-CoV, we obtained a consistent titer of around 7.3 $\log_{10}$ $TCID_{50}$/ml. In Vero cells, distinct cytopathic effect (CPE) is always observed at day 2 or 3 post-infection. These first experiments encouraged us that the prototype vaccine could be produced in Vero cells using a single harvest totally compatible with our experience with inactivated poliovirus.

Raw materials used to develop the candidate vaccine (serum and trypsin) were selected in accordance with current regulation. Calf serum was imported from Australia, and gamma irradiated prior to use. The trypsin was from porcine origin and also gamma irradiated. Extensive evaluation for adventitious agents was performed on the raw material, which included the search for cytopathogenic agents, hemadsorbing agents, and specific viral contaminants as bluetongue, reovirus, rabies, parainfluenza type 3, specific bovine viruses [adenovirus, parvovirus, respiratory syncytial virus (RSV)], bovine viral diarrhea, rhinotracheitis virus), and porcine viruses (parvovirus, adenovirus; transmissible gastroenteritis virus; hepatitis E virus, rabies virus, and porcine pestis virus).

Manufacturing of the Viral Seed Lots

The viral seed lots were produced in Vero cells. QC testing is a major step in the qualification of such a seed. Of all the different tests performed (identification of the Vero cells, sterility, mycoplasma, titer, and contaminating viruses), the research of contaminating virus(es), also called adventitious agents, represents the most crucial step. To detect adventitious agents, sensitive cell culture monolayers of CV-1 cells, human diploid MRC-5 cells, and chick embryo fibroblasts (CEFs) were inoculated with the crude viral suspension and are observed for induced CPE and/or hemadsorption. CV-1 cells are used in these tests, as they are of the same species and the same origin as the cells used for vaccine manufacturing. MRC-5 cells are human diploid cells and can potentially reveal other viruses able to infect human cells. And finally, taking into consideration the origin of the specimen (pulmonary syndrome) we added CEF, which are known to be sensitive to the infection of several respiratory viruses such as influenza virus and RSV. At the same time, adventitious agent testing was performed on control cells (search for hemagglutining or hemadsorbant viruses) and on the supernatant of the control cells (search for adventitious agents) by inoculation of the three cell lines: Vero, MRC-5, and CEF. Adventitious agent testing was also done in vivo by inoculation in suckling mice, mice, and guinea pigs, as well as the allantoïc cavity and yolk sac of embryonated chicken eggs.

Complementary to the conventional methods to qualify viral seed, as described above, extensive characterization was performed using PCR testing. The PCR tests are listed in Table 11.1.

TABLE 11.1 List of viruses for which the crude SARS-CoV viral suspension was monitored by PCR

Detection of viruses by PCR
Adenovirus
Bovine and porcine circovirus
Bovine herpesvirus—I, IV
Bovine polyoma virus
Cytomegalovirus
Epstein-Barr virus
Hepatitis A virus
Hepatitis B virus
Hepatitis C virus
Human herpes virus (HHV)-6 and HHV-7
HHV-8
Human immunodeficiency virus (HIV)-1 and HIV-2
Human respiratory syncytial virus
Herpes simplex virus
Human T-cell leukemia virus type-1 (HTLV-1)
Human T-cell leukemia virus type-2 (HTLV-2)
Human papilloma virus
Human parainfluenza virus—1, 2, 3
Human polyoma virus
Metapneumovirus
Parvovirus B19
Simian spumavirus
Simian immunodeficiency virus
Simian retrovirus—1, 2, 3/SMRV
Simian T-cell lymphotropic virus
Simian virus 40

All tests were performed in accordance with international requirements, showing that no adventitious agents were detected.

Results of the Inactivation Experiments

The difficulty with inactivating viruses remains the balance between fully validated inactivation and preservation of immunogenicity or epitopes associated with protection. It is well known that reagents used to inactivate viruses [betapropiolactone (BPL), formaldehyde] can change the outer membrane antigens with the risk of a reduced immunogenicity of the vaccine. For the inactivation of SARS-CoV, we chose to test BPL. Assays were performed to determine the best BPL concentration for inactivation while maintaining a good immune response in mice. Virus inactivation was performed using three different BPL concentrations: 1/2000, 1/3000, and 1/4000 (v/v). In our experience, a 1/4000 dilution of BPL was the optimal concentration for the inactivation of SARS-CoV based on a balance between total inactivation and maintenance of antigenic properties. The 1/4000 BPL dilution is similar to what is used for rabies vaccine inactivation.

The usual way to measure viral inactivation is by kinetic studies, i.e., reduction in virus infectivity. This technique has a detection limit of 1.5 $\log_{10}$ $TCID_{50}$/ml. The kinetics of inactivation was performed at 0, 5, 30 min, and 1, 2, 3, 4, 6, 8, and 24 h, and it was shown that inactivation below the limit of detection was obtained after 6 h. In order to increase the detection sensitivity, an amplification test was also performed. For this amplification, Vero cells are incubated with a portion of the viral solution following inactivation for different periods of time. After 7 days of incubation, the cells are trypsinized and cultivated for additional 14 days. At the different stages of amplification, the cells were microscopically observed for CPE and at the end of the incubation (day 21) an immunocolorimetric assay test was performed. Our data demonstrated that the virus is fully inactivated by 12 h, not 6 h of BPL treatment, demonstrating that the amplification test is much more sensitive.

To complete the validation of the amplification test, the minimum limit of detectable infectious viral particles was determined. This was done by spiking the inactivated vaccine with different concentrations of live virus and incubating with Vero cells. Using this approach, it was possible to establish the minimum virus detection as 1 pfu.

Based on these data, it was concluded that inactivation with 1/4000 BPL dilution for 12 h fully inactivates the SARS-CoV batches. To ensure a very large safety margin, we adopted an inactivation period of 24 h for the SARS-CoV vaccine production process.

ANIMAL MODELS

Since a direct efficacy trial in humans will be impossible, because of a lack of naturally circulating SARS-CoV, the licensure of a SARS-CoV vaccine will depend on surrogate markers. Recently (as described below) the FDA adopted the Animal Efficacy Rule that envisions that under such circumstances, demonstration of efficacy can be performed in two animal models. For SARS-CoV vaccine development, monkeys and ferrets can be used to evaluate candidate vaccine. Both animal models show pathology in the lungs upon autopsy.

The immunogenicity of the SARS vaccine was evaluated in nonhuman primates, *M. fascicularis,* and ferrets. Both animal models are susceptible to infection, do show some signs of disease (lethargy), and show signs of pulmonary lesions upon histological examination (Fouchier et al., 2003; Martina et al., 2003; ter Meulen et al., 2004; Rowe et al., 2004; McAuliffe et al., 2004). Different doses of the SARS vaccine (6 or 7 $\log_{10}$ $TCID_{50}$/ml) were injected in the presence or absence of aluminum hydroxide. Two intramuscular injections were performed at a one month interval.

Regarding the humoral response, sustained levels of ELISA and serum-neutralizing virus-specific antibodies were elicited in vaccinated monkeys and ferrets. A significant dose–effect relationship could be demonstrated. Moreover, a strong adjuvant effect of aluminum hydroxide was evidenced for each vaccine dose and proved in most cases to be highly significant.

In order to evaluate the efficacy of the SARS vaccine, immunized monkeys and ferrets were challenged intratracheally with a heterologous Hong Kong SARS-CoV strain (Coronovative, Rotterdam, The Netherlands). Monkeys immunized with 6 or 7 $\log_{10}$ of inactivated virus were protected as measured by RT-PCR and viral titration on lung samples five days post-challenge. The ferrets were protected at the lower immunization dose of 5 or 6 $\log_{10}$.

Based on our experience to date, the inactivated, adjuvanted SARS-CoV prototype vaccine seems to be a good candidate for further evaluation in Phase 1 studies.

REGULATORY APPROVAL PROCESS—SOME UNIQUE ASPECTS OF SARS AND OTHER EMERGING DISEASES

As with other vaccines, vaccines for SARS and other emerging threats need to follow a structured pattern of regulatory development. The initial stages would be very similar to those followed for vaccines under development for conventional infectious diseases. In the United States, the earliest stages would include the development of sufficient preclinical information about the vaccine to allow the preparation of an investigational new drug (IND) application for submission to the FDA (see Chapter 13). The IND may have information unique to the vaccine candidate but should include information about the rationale for the vaccine design, the source of the virus and other components, the manufacture of the active vaccine component, formulation, preliminary characterization of the vaccine including purity and potential contaminants, immunological testing, and animal testing, including toxicology. Even at this early stage, the vaccine should be made under GMP conditions and other laboratory work conducted under good laboratory practice (GLP) conditions, as appropriate. The IND application should also include important information about the Phase 1 clinical design, focusing on how the safety will be monitored and a discussion of any potential adverse reactions based on the experience with vaccines that have similar components or methods of preparation. There should be an opportunity to outline the vaccine concept and Phase 1 clinical study at a pre-IND meeting that often provides the opportunity to receive the input and concerns of the regulatory agency.

Following approval of the IND, vaccines such as SARS can progress to a conventionally designed Phase 1 study. Typically, this Phase 1 clinical study is descriptive and would include a small number of healthy young adults with the emphasis on monitoring the safety of the vaccine (local and systemic reactions). Often, the first immunologic assessment is part of this study. Following successful completion of the Phase 1, as with other vaccines, a SARS vaccine candidate could move forward to Phase 2. During this phase, in addition to safety monitoring, dose-ranging studies are conducted in much larger groups of individuals and the vaccine should meet predefined primary and secondary endpoints. If the Phase 2 is successful, following a pre-Phase 3 meeting, clinical studies are conducted in a greater number of subjects during which less frequent reactions can be detected and the efficacy or effectiveness of the vaccine determined. As part of this Phase 3 evaluation, the consistency of sequential lots of the vaccine are typically compared in order to ensure that the vaccine can be reproducibly manufactured.

Obviously, as the vaccine progresses clinically from Phase 1 to Phase 3, the size of the lots of vaccine often increases and the manufacturing and in-process and release testing specifications become increasingly well defined, so as to guarantee that the vaccine can consistently be made at a commercially useful scale. If all three phases of clinical development are successful, the manufacturer may then submit a Biologics License Application (BLA), which is a very extensive compilation of all of the information relating to the development and manufacture of the vaccine.

As suggested in the Animal Models section above, the unique challenge for SARS and other emerging threats, whether anthrax or Ebola viruses, is that it may not be possible to conduct Phase 3 clinical studies to determine the effectiveness of the vaccine. For these agents, it would be too dangerous to conduct challenge studies in humans and the prevalence of the disease is either nonexistent, sporadic, or too small to allow the development of a reasonable clinical protocol. In anticipation of this problem, largely in the face of potential bioterrorist agents, in 2002, the FDA adapted the so-called Animal Rule [see Federal Register, May 31, 2002 (Volume 67, Number 105)]. Under this guidance, new drugs or biological products that are intended to prevent serious or life-threatening conditions may be approved on evidence of effectiveness derived from appropriate studies in animals and any additional supporting data, if controlled clinical studies cannot be conducted in human volunteers and field trials are not possible. In order to satisfy this alternative mechanism, however, several criteria must be met. First, there is a reasonably well-understood pathophysiological mechanism that can be ameliorated or prevented by the product. Second, the effect is demonstrated in more than one animal species, unless it is demonstrated in a single species that represents a sufficiently well-characterized animal model. Third, the animal study endpoint is clearly related to the desired benefit in humans. And finally, the data are sufficiently well understood to allow selection of an effective dose in humans. It is therefore reasonable to expect that the effectiveness of the product in animal model(s) is a reliable indicator of its effectiveness in humans.

Obviously, it is too early to know whether the SARS vaccine candidate as described in this chapter will move forward and be able to meet all of the criteria of the Animal Rule. In particular, SARS vaccine development is hindered by relatively little information about human CoVs in general. Until the rapid emergence of SARS, most of the basic research was focused on animal CoVs and our inactivated SARS vaccine candidate described in this chapter is exclusively based on experience with vaccines to animal CoVs. Certainly, it is too early to conclude whether the ferret and/or *M. fascicularis* is/are the most appropriate model(s) for human SARS infections. As a result, except for clinical cases documented during the outbreak, there is relatively little information about SARS pathogenesis and correlates of immunity. Another difficult aspect is that a feline infectious peritonitis (FIP) vaccine was actually harmful to the health of the immunized cats upon challenge with wild-type FIP virus (Weiss and Scott, 1981). Before moving forward with approval, therefore, it will be very important to determine whether these adverse outcomes can be prompted or mimicked by any of the SARS vaccine candidates.

CONCLUSIONS

What Were the Parameters for the Rapid Development of the SARS Vaccine?

The production of a GMP clinical lot of a monovalent, whole, inactivated, aluminum hydroxide-adjuvanted SARS-CoV vaccine took 12 months. In terms of vaccine development, this is extremely rapid. Several factors contributed to these short timelines.

The grants made available by the NIAID for the development of a SARS vaccine completely changed the classical environment, allowing vaccine industries to start almost immediately the development of a new vaccine. Indeed, the development of a new vaccine can only be done to the detriment of other vaccine developments, mobilizing teams and facilities.

From a technical point of view, the choice of a classical vaccine development strategy using conventional procedures, such as Vero cell culture for viral propagation and BPL inactivation, was a decisive factor to success. Importantly, we were able to quickly recruit a volunteer workforce that was both familiar with the technology and trained to work in a BSL3-plus environment.

A close collaboration with the reference laboratory, the CDC's Influenza Branch, where the SARS-CoV was isolated, was essential. We provided certified Vero cells to the CDC, which allowed us, upon receipt of the purified strain from the CDC, to re-isolate the SARS-CoV under conditions making the prompt start of a vaccine development possible.

When initiating vaccine development against a new emerging infectious agent, the problem of availability of reagents and routine tests to perform biological and molecular studies must be addressed. It is obvious, that at the beginning of such development, there are no such reagents or commercial kits available. As a consequence, the first step in the SARS-CoV project was to prepare the different reagents (antisera and monoclonal antibodies) and the appropriate tests (viral titration, PCR, ELISA, immunofluorescence assay, etc.). Finally, a series of preliminary experiments on monkeys (three months after the start of the project) had given guidance whether it was appropriate to use an inactivated vaccine, as well as to the choice of the adjuvant.

Constant communication with regulatory authorities has allowed the validation of this strategy from the beginning of the project. This communication was also very important for the qualification of the viral seed lots. The qualification of the Vero cells was not an issue as several vaccines are already produced in Vero cells, but this was obviously not the case for the viral seeds. Two major obstacles had to be overcome: (1) the realization that the animal testing had to be done in BSL3 facilities by BSL3-trained personnel, and (2) the search for adventitious agents using general classical tests (search for adventitious viruses on cells and in animals) and specific tests (PCR). For the latter, there was no list available and the final testing to be performed was under the responsibility of health authorities. This resulted in a rather exhaustive list of PCR testing.

Can We Shorten the Timelines Even More When Facing an Emerging Pathogen?

It is likely that epidemics will emerge in the future from unrecognized sources and some of these will be highly pathogenic for humans. These pathogens will be categorized as BSL3 or BSL4 pathogens needing high security level laboratories as well as specialized personnel. How to manipulate these pathogens that are highly pathogenic, in large quantities? To face the emergence of new pathogens, dedicated structures are needed with the right equipment and trained personnel. From an industrial perspective, this seems not compatible with the need and use of trained personnel and facilities that do have a constant activity to assure the production of existing vaccines and the development of new vaccines. Such emergency structures could be set up and maintained by national reference centers respecting the BSL requirements as well as the GMP conditions. It would be very beneficial for industries to collaborate with such reference centers that provide purified pathogens and reagents allowing a prompt start of a vaccine development.

References

Fouchier, R.A., Kuiken, T., Schutten, M. et al. Aetiology: Koch's postulates fulfilled for SARS virus. Nature 2003; 423(6937):240.

Ladman, B.S., Pope, C.R., Ziegler, A.F. et al. Protection of chickens after live and inactivated virus vaccination against challenge with nephropathogenic infectious bronchitis virus PA/Wolgemuth/98. Avian Dis. 2002; 46(4):938–944.

Martina, B.E., Haagmans, B.L., Kuiken, T. et al. Virology: SARS virus infection of cats and ferrets. Nature 2003; 425(6961):915.

McAuliffe, J., Vogel, L., Roberts, A. et al. Replication of SARS coronavirus administered into the respiratory tract of African Green, rhesus and cynomolgus monkeys. Virology 2004; 330(1):8–15.

Normile, D. Infectious diseases. Second lab accident fuels fears about SARS. Science 2004; 303(5654):26.

Orellana, C. Laboratory-acquired SARS raises worries on biosafety. Lancet Infect. Dis. 2004; 4(2):64.

Pratelli, A., Tinelli, A., Decaro, N. et al. Safety and efficacy of a modified-live canine coronavirus vaccine in dogs. Vet. Microbiol. 2004; 99(1):43–49.

Rowe, T., Gao, G., Hogan, R.J. et al. Macaque model for severe acute respiratory syndrome. J. Virol. 2004; 78(20):11401–11404.

Senior, K. Recent Singapore SARS case a laboratory accident. Lancet Infect. Dis. 2003; 3(11):679.

ter Meulen, J., Bakker, A.B., van den Brink, E.N. et al. Human monoclonal antibody as prophylaxis for SARS coronavirus infection in ferrets. Lancet 2004; 363(9427):2139–2141.

Weiss, R.C. and Scott, F.W. Antibody-mediated enhancement of disease in feline infectious peritonitis: comparisons with dengue hemorrhagic fever. Comp. Immunol. Microbiol. Infect. Dis. 1981; 4(2):175–189.

CHAPTER

12

Clinical Trials of Vaccines for Biodefense and Emerging and Neglected Diseases

Wendy A. Keitel
Baylor College of Medicine, Houston, TX, USA

OUTLINE

ABSTRACT

The development of safe and effective vaccines for the prevention and control of emerging and neglected infectious diseases is an international priority, as is the development of similar control measures for bioterrorist threats. International standards have been established to ensure the protection of human subjects and the scientific integrity of clinical trial study design and data collection. A general overview of the history of the regulatory guidelines and ethical principles underpinning the vaccine development process is provided, with an emphasis on clinical trials conducted in the United States.

The contemporary vaccine development process involves sequential assessment of safety, immunogenicity, and efficacy in phase1, 2, and 3 clinical trials. The process is not always linear or straightforward, and adaptive flexible clinical trial designs can increase the likelihood of a successful outcome. Because the efficacy of vaccines for a number of the diseases discussed in this volume cannot be tested directly in humans, alternative approaches have been developed to address this challenge. The major elements of typical clinical trial protocol are discussed, and a checklist of essential documents supporting the trial is provided. As novel vaccine approaches and technologies emerge, the regulatory and ethical considerations will need to be revisited and adapted to respond to the ever-changing landscape.

INTRODUCTION

Vaccines are considered among the most valuable and cost-effective tools for the control of infectious diseases; indeed, universal immunization of infants and children against a variety of pathogens was considered one of the top ten achievements of the 20th century (CDC, 1999). Major scientific, technical, ethical, and regulatory principles and/or processes underpinning the vaccine development and evaluation process have been defined; however, challenges posed by the threats of bioterrorism and emerging infectious diseases, as well as technological advances in the composition, formulation, and delivery of vaccines, expose gaps in policy and practice that require ongoing consideration and refinement of approaches to the evaluation of new vaccines. The purpose of this chapter is to provide a brief overview of the approach to the evaluation of candidate vaccines in humans, with a particular attention to issues related to vaccines for biodefense and emerging and neglected diseases. The major focus will be on clinical trials and protocol development in the US, although many of the principles, regulations, and practices have been harmonized across the globe. A brief overview of related regulatory issues and preclinical vaccine development is provided; however, detailed discussions of these topics can be found elsewhere in this volume.

HISTORICAL CONSIDERATIONS

In the US, vaccines are regulated as biologicals, although vaccines are legally defined as drugs under the Food, Drug and Cosmetic Act (FDCA). The ethical principles, and the regulatory agencies and requirements that guide product development and clinical research in the 21st century evolved progressively over the 20th century. In many cases, the development of codified guidelines and the establishment of a regulatory authority were spurred by a tragic accident or a clear ethical breech (for reviews of Historical Considerations, see Baylor and Midthun, 2004; Borchers et al., 2007).

Regulatory Issues

In 1901, a number of children became ill and died after treatment with diphtheria antitoxin contaminated with tetanus toxin. This episode resulted in the first legislation designed to regulate the purity and potency of biologicals—the Biologics Control Act (BCA) of 1902. In 1938, over 100 people died after ingesting an elixir of sulfanilamide containing diethylene glycol (antifreeze), leading to the enactment of the FDCA (Ballentine, 1981). Provisions of the FDCA required manufacturers to submit safety data to the Food and Drug Administration (FDA) prior to registration. The 1962 Kefauver–Harris Amendments to the FDCA required that efficacy data also be submitted. The1944 Public Health Service Act incorporated the BCA into section 351 of the US Code of Federal Regulations (CFR), which gave the federal government the authority to license biologicals and manufacturing facilities. In 1955, incomplete inactivation of a poliovirus vaccine resulted in the development of polio in a number of children (the "Cutter Incident"); this led to the establishment of the Division of Biologics Standards (DBS) within the NIH (Offit, 2005). The authority vested in the DBS was transferred to the FDA in 1972. The organization that is currently responsible for regulating biologics is the Center for Biologics Evaluation and Research (CBER) at the FDA. Title 21 of the CFR contains regulations pertaining to biologicals, including product standards, manufacturing, labeling, licensing, advertising, investigational use, and protection of human subjects (informed consent, nonclinical laboratory studies, and Institutional Review Boards, or IRBs). Part 312 of Title 21 CFR contains regulations related to the Investigational New Drug Application (IND). The CFR is updated each year to reflect changes in policies and procedures. Additional guidance documents related to vaccines are published by the FDA, as appropriate.

During the 1990s, an international group comprised of scientists, regulators, ethicists, and pharmaceutical representatives convened to discuss standards for designing, implementing, documenting and reporting clinical trials. The International Conference on

Harmonisation of Technical Requirements for Registration of Pharmaceuticals for Human Use (ICH) established guidelines for the conduct of clinical research (Good Clinical Practice, or GCP), many of which have been adopted by the US FDA (FDA, 1996). The ICH GCP provides a unified standard to ensure the quality of clinical trial data and the protection of human subjects for the European Union, Japan, and the US for clinical trials that will support licensure of new vaccines and drugs. For research related to the evaluation of vaccines for emerging and neglected diseases, implementation of GCP in developing countries may pose unique challenges. Acosta et al. (2007) conducted a multinational clinical trial of Vi polysaccharide typhoid fever vaccine in Asia among 200,000 individuals, during which implementation of GCP required adaptations in order to comply with the goals of the guidelines.

Ethical Issues

The revelation that Nazi physicians tortured and experimented on prisoners during World War II culminated in the publication of the Nuremberg Code in 1947 (Macrae, 2007). The Code is recognized as the first set of ethical guidelines for human research to be recognized by the international community. In 1965, the Declaration of Helsinki articulated additional responsibilities of investigators to research subjects, and paved the way for the development of IRBs.

Continued violations of informed consent were summarized in a *New England Journal of Medicine* article written by Dr. Henry Beecher (1966). This article and the subsequent expose of the ethical violations perpetrated during the Tuskeege Syphilis Study led to the 1974 National Research Act, which established IRBs and the National Commission for the Protection of Human Subjects of Biomedical and Behavioral Research. After considering the principles and practices of human research in the US, the commission published the Belmont Report in 1979. This report defined the three major ethical principles (and their related clinical trial applications) pertaining to human research: respect for persons (informed consent); beneficence (risk-benefit assessment); and justice (equitable subject selection). Recommendations of the National Bioethics Commission from 1981 related to research involving human subjects were adopted into federal law in 1991. The so-called Common Rule (45 CFR 46 Subpart A) articulated the federal policy for protection of human subjects who are participating in clinical trials. Additional protections for pregnant women and fetuses, prisoners, and children are outlined in Subparts B, C, and D of 45 CFR 46. The Council for International Organizations and Medical Sciences, formed in 1949, recently published updated international ethical guidelines for research involving human subjects (CIOMS, 2002).

It is abundantly clear that the ethical considerations and regulatory guidelines for the conduct of clinical research are inextricably intertwined: ethical considerations drive the need to regulate the conduct of clinical trials to ensure the protection of human subjects and the quality and integrity of the data generated. Seven ethical requirements that should be fulfilled in the conduct of clinical research have been proposed by Emanuel et al. (2000): the research should provide information that will advance science and knowledge (value); scientific validity; fair subject selection; favorable risk-benefit ratio; independent review; informed consent; and respect for enrolled subjects. The increasing complexity and redundancy of the oversight of contemporary clinical trials reflects the commitment to accomplish these goals. A simplified prototypical organizational structure for a vaccine research clinical trial is shown in Fig. 12.1. Further details regarding the roles and responsibilities of the participants in the clinical trials process are provided below.

STAGES OF VACCINE DEVELOPMENT

Overview

The major stages of vaccine development are outlined in Fig. 12.2. The initial stage is referred to as the pre-IND (investigational new drug) stage. Activities conducted in this stage culminate in the production of a vaccine that can be evaluated in humans—the IND stage. Vaccines that are shown to be safe, pure, and effective in humans may then be licensed for use (licensing and postmarketing stage).

Pre-IND Stage

Once a public health need for control of a disease by means of vaccination has been identified, detailed studies designed to understand the pathogenesis of infection and to identify immune responses associated with protection are conducted. Studies may involve assessment of infection and immunity in humans if the disease occurs naturally at a frequency high enough to permit their evaluation (interpandemic influenza, malaria, tuberculosis, and others); however, detailed evaluations of other rare and lethal diseases—particularly those included in the Centers for Disease Control and Prevention (CDC) Category A biothreat list—must

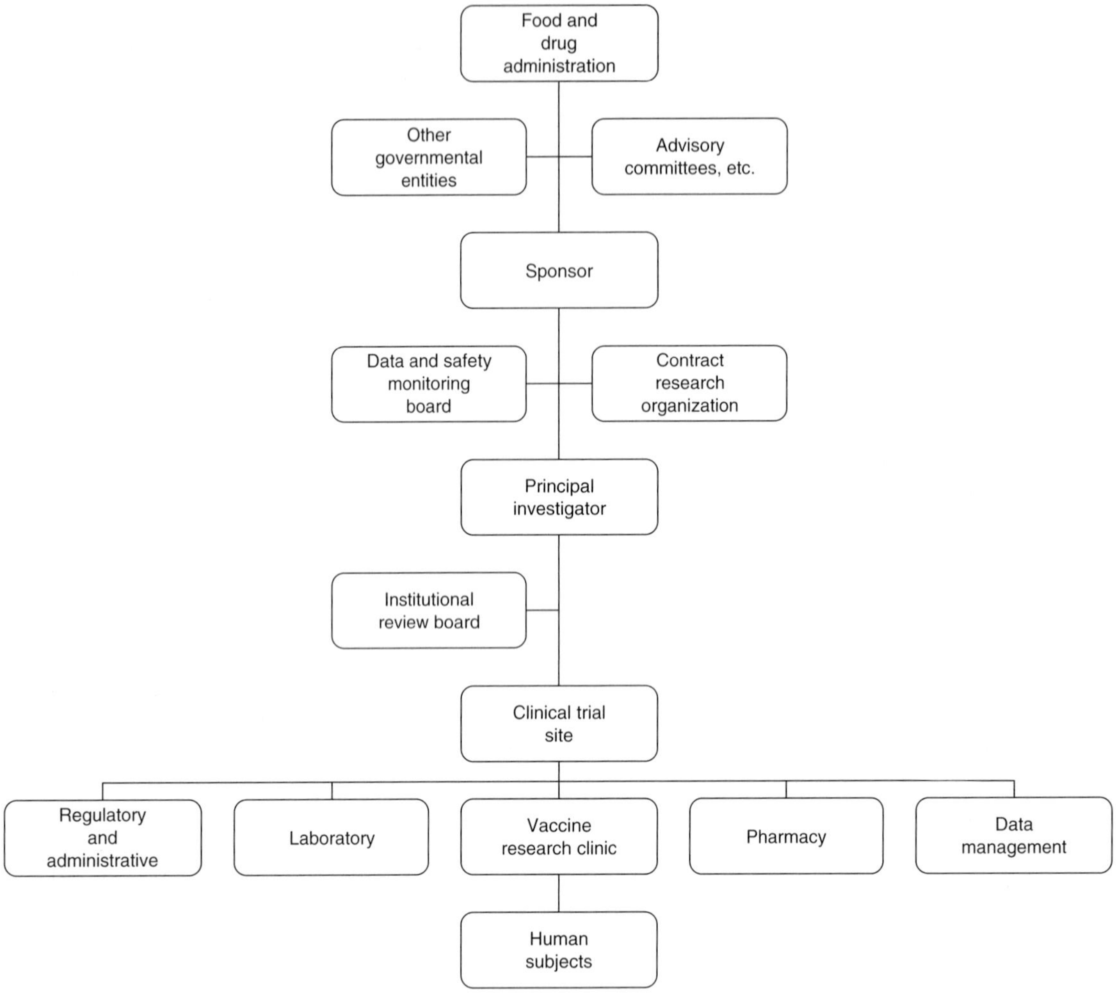

FIGURE 12.1 Organizational structure of a typical single-center vaccine clinical trial.

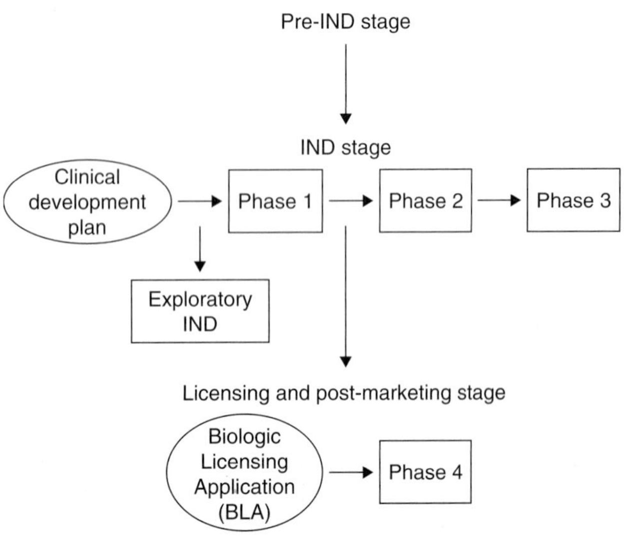

FIGURE 12.2 Overview of the major stages of vaccine development; IND, investigational new drug.

be studied in animal models that might predict the pathology and immune responses that occur in humans (smallpox, tularemia, inhalational anthrax, hemorrhagic fever viruses, etc.). Candidate vaccine formulations containing or expressing epitopes that can elicit immune responses associated with protection are then designed and evaluated in vitro and in animals, where their safety and immunogenicity profiles are established. A critical part of the pre-IND stage is the development and validation of a manufacturing process.

IND Stage

Candidate vaccines typically undergo a sequential series of evaluations in humans. Three major phases are identified, as summarized in Table 12.1. In practice, the scheme is not so straightforward: phase 1 and phase 2 clinical trials may overlap or be combined,

TABLE 12.1 Phases of clinical vaccine development

Clinical trial phase	Major objectives	Typical number of subjects	Type of subjects	Comments
I	Safety and tolerability; preliminary immunogenicity	20–80	Healthy adults	First introduction of a candidate vaccine into humans
II	Immunogenicity; safety	Hundreds	Healthy adults, followed by target populations	Immunization regimen defined (dosage levels, number of doses)
III	Efficacy; safety	Hundreds to thousands	At-risk population(s)	May be limited to immune response determinations[a]
IV	Safety; duration of protection; efficacy in other groups or for other indications	Thousands to millions	Vaccine recipients	Postmarketing studies

[a] If efficacy cannot be assessed in humans because disease does not occur naturally or at a high enough frequency (smallpox, inhalational anthrax, etc.), then efficacy may be established in animal models (see text) and correlates of protection identified in humans.

and multiple phase 1 and/or phase 2 evaluations may be necessary for adjustment of dosage, change in regimen, and evaluation in other age and/or risk groups, even after pivotal phase 3 studies have demonstrated efficacy. Close and ongoing communication between the sponsor, regulatory agencies, investigators, and IRBs is critical throughout the IND stage, including pre-IND meetings, end-of-phase 2 meeting, pre-Biologic Licensing Application (BLA), and BLA meetings to facilitate the timely progression through the clinical trials process.

Phase 1 Clinical Trials

The first evaluation of a candidate vaccine in humans is referred to as a phase 1 clinical trial. Typically less than 100 healthy young adult subjects are enrolled in phase 1 trials. In fact, the number may be considerably smaller for novel products, in which case investigators may elect to inoculate only a handful of subjects to exclude the possibility of unexpected reactogenicity or toxicity (Keitel et al., 1993a). The major goal of the phase 1 trial is to assess safety and tolerability; however, preliminary immunogenicity assessments and dose-ranging information can provide valuable information regarding dosage selection for subsequent evaluations. Although the design of phase 1 clinical trials of drugs often is open-label, many contemporary phase 1 clinical trials of candidate vaccines are randomized, double-blind, placebo-controlled trials. The phase 1 trial may be performed in stages, where a small cohort of subjects is vaccinated and observed for a period of 1–4 or more weeks before the remainder of the study cohort is vaccinated in stage 2. This approach reduces the exposure of subjects to the occurrence of unexpected toxicity. For vaccines that are produced in the US whose ultimate use will be in other countries (such as malaria vaccines), the initial evaluation occurs in the US (phase 1a), followed by phase 1b testing in a healthy population in the target country. Inclusion of placebos in this and other phases of development reduces bias in the assessment of adverse events (AEs) and serious AEs (SAEs), and provides internal controls for laboratory assessments of immune responses following immunization.

Because the number of subjects enrolled in a phase 1 trial is small, these studies lack statistical power to detect AEs that occur at a low rate. Nevertheless, small, carefully monitored phase 1 trials can identify unexpected and/or unacceptable toxicities that require reformulation or reevaluation of a potential candidate (Keitel et al., 1999; Edelman et al., 2002). In some circumstances, specific monitoring of subjects in phase 1 or phase 2 trials for evidence of infection after immunization is relevant, particularly if there are concerns that the vaccine might elicit immunopathologic responses when the vaccinated subject is naturally infected. Immunopatholgy resulting in severe atypical infection and/or death occurred among persons immunized with an inactivated measles vaccine and in infants given an inactivated vaccine to prevent respiratory syncytial virus (RSV) vaccine (Polack, 2007). Similar concerns exist regarding the potential for dengue virus and SARS coronavirus vaccines to elicit immunopathologic responses (Edelman et al., 2003; Deming et al., 2006).

Although not specifically relevant for this chapter, an exploratory IND study may precede a typical phase 1 evaluation for the assessment of drugs and therapeutic biologicals (FDA, 2006). The exploratory IND study option is characterized as a small early phase 1 trial that has no therapeutic or diagnostic intent; rather, the goal is to assess feasibility for further development of a drug or biological. Goals of this type of study may include determination of the mechanism of action of a drug in humans; evaluation of the pharmacokinetic profile of the agent; selection of one from a group of promising candidates; and exploring biodistribution characteristics using imaging techniques. Because these studies pose lower potential risk to subjects, exploratory IND studies require less or different preclinical support than typical phase 1 studies.

Phase 2 Clinical Trials

Vaccine candidates that have favorable safety and immunogenicity profiles in phase 1 trials may progress to expanded phase 2 trials. Several hundred healthy subjects may be enrolled into phase 2 trials. The major goals of phase 2 trials are to assess safety and to develop optimal regimens for immunization. Typically several dosage levels are evaluated; different immunization regimens (number of doses and interval between doses) also may be explored. Most phase 2 trials are randomized, double-blind, placebo-controlled trials. Although larger numbers of subjects are evaluated, phase 2 trials still lack statistical power to detect events that occur in a low proportion of subjects. Initial phase 2 trials are usually conducted among persons who are at low risk for acquiring the target disease; therefore, additional phase 1 and phase 2 trials may be necessary to assess safety and immunogenicity among populations who are at high risk for the disease, in which pivotal efficacy trials will be conducted. For several emerging and neglected diseases, the preliminary efficacy of a vaccine can be assessed in a human challenge model, including malaria, influenza, and cholera (Couch et al., 1971; Epstein et al., 2007; Tacket et al., 1999). These carefully controlled experimental inoculations of subjects with the target pathogen can provide proof of concept that immunization with a vaccine candidate confers protection prior to large phase 3 clinical trials. In the pediatric population, rechallenge of vaccinated subjects with homologous or related live attenuated influenza virus vaccines (LAIV) provided supportive evidence that intranasal immunization with LAIV would confer protection against subsequent infection with wild-type influenza viruses (Belshe et al., 2000). Phase 2 trials may also incorporate plans to assess the concomitant administration of other vaccines, biologicals, or medications. For example, administration of a new vaccine to infants in the context of the increasingly complicated childhood immunization schedule poses complex problems; evidence that simultaneous administration of licensed and experimental vaccines does not interfere with protective responses to components in either product is necessary (Rennels et al., 2000).

Phase 3 Clinical Trials

Once a vaccine candidate has been shown to be safe and to possess acceptable reactogenicity and immunogenicity profiles, assessment of the efficacy can be undertaken in phase 3 clinical trials. The type and size of the population to be studied will depend on epidemiology of the target disease (population at risk for disease and the disease attack rate) and the level of protection conferred via immunization. A phase 3 clinical trial for prevention of disease that occurs at a low incidence may require hundreds of thousands of subjects, such as the US field trial of inactivated poliovirus vaccine in the 1950s (Francis, 1955); whereas phase 3 trials of vaccines for prevention of high-incidence infections (such as RSV in infants) theoretically would require no more than hundreds of subjects, particularly if the vaccine was expected to be highly efficacious.

The assessment of efficacy of vaccines against agents of bioterrorism, as well as vaccines for candidate pandemic influenza, may be problematic. For example, smallpox, inhalational anthrax, hemorrhagic fever viruses, tularemia, plague, and influenza A/H5N1 do not occur naturally at a frequency high enough to permit controlled evaluation of clinical vaccine efficacy prior to licensure of vaccines. The FDA final rule entitled "New Drug and Biological Drug Products: Evidence Needed to Demonstrate Effectiveness of New Drugs When Human Efficacy Studies Are Not Ethical or Feasible" (otherwise referred to as the "Animal Rule") was published to address this circumstance (FDA, 2002). The rule permits use of animal efficacy data when collection of human efficacy data is not feasible. Safety and immunogenicity data in humans and animal efficacy data can be used to support licensure when several conditions are met, as follows: (1) the pathophysiology of the infection is reasonably well understood; (2) the pathogenesis and immune responses in one or two animal species are expected to predict these responses in humans; (3) the immunogenicity endpoint(s) correlated with protection in the animal(s) are related to human immune responses; and (4) immunogenicity endpoints in animals and humans are sufficiently well understood to permit selection of a regimen that would be expected to predict protection in humans. For other diseases where correlates of protection are reasonably well defined (for example, serum HAI antibody level against influenza viruses), surrogate markers can be used to support licensure (FDA, 2007a). Finally, in order to facilitate the approval of vaccines for such severe, life-threatening illnesses, additional mechanisms for expedited review and accelerated approval have been developed by the FDA.

Phase 4 Clinical Trials

Regulatory authorities are increasingly requiring additional studies to be conducted after market approval. Phase 4 studies may also be initiated by the sponsor for a variety of reasons. While a vaccine appears safe after it has been studied in thousands of individuals, rare adverse events may only be observed after hundreds of thousands or even millions of people have been vaccinated, as occurred with the first-generation live attenuated rotavirus vaccine (Murphy et al., 2001). The long-term safety or continued efficacy of a vaccine may be unknown at the time of licensure, and the benefit of immunization of special populations that may have been under-represented or not studied in the IND stage may be of interest. Unfortunately, phase 4 clinical trials typically are not randomized, and they may be uncontrolled. Nevertheless, several safety surveillance systems have been established to facilitate the early detection of potential rare, serious reactions to vaccines, including the Vaccine Adverse Event Reporting System (VAERS), the Vaccine Safety Datalink (VSD), and the American Academy of Pediatrics Practice Research Office Settings (PROS) (Ellenberg et al., 2005). Design options for phase 4 studies include case-control studies and cohort studies.

CLINICAL TRIAL PROTOCOL DEVELOPMENT

The design of protocols for the evaluation of candidate vaccines requires consideration of the pathogen, the disease pathogenesis and immune response profile, and the characteristics of the vaccine candidate itself. The success of a clinical trial will depend on implementation of a well-designed protocol, followed by careful monitoring and retention of subjects, close adherence to the protocol, and accurate reporting and documentation of observations made during the trial (for Comprehensive discussions, see Chow and Liu, 2004; Meinert, 1986; Wang and Bakhai, 2006).

A brief outline of essential elements for inclusion in clinical trial protocols is shown in Table 12.2. A detailed template for protocol development has been prepared by the National Institute of Allergy and Infectious Diseases that can be used to guide protocol development (NIAID, 2006). Novel constructs such as live, attenuated vaccines may require additional discussion of risks and special facilities for isolation, such as the evaluation of live attenuated vaccines based on potential pandemic influenza viruses. Note that the basic protocol elements are mirrored in the Informed Consent document.

Background and Rationale

The background and rationale should provide information regarding the current understanding of the disease epidemiology, pathogenesis, and immune responses relevant for protection against infection; and need for development of control measures. The

TABLE 12.2 Checklist of essential clinical trial protocol elements

Title page
Statement of compliance
Signature page
Protocol summary
Background and rationale
Purpose and objectives
Study design and endpoints
Study population: Description, inclusion/exclusion criteria; recruitment and retention
Study agent/interventions
Study procedures/evaluations
Study schedule
Assessment of safety: Safety parameters, reporting requirements, halting rules
Clinical monitoring structure: Site/safety monitoring plan, and safety reviews
Statistical consideration: Sample size; data analyses, etc.
Quality control and quality assurance
Ethics/protection of human subjects: IRB, informed consent; confidentiality, etc.
Data management
Appendices: Personnel roster; table of procedures; lab processing flow sheet, etc.

Source: Adapted from NIAID protocol template guidance (NIAID, 2006).

Courtesy: NIAID.

scientific rationale for selection of the vaccine candidate should be discussed. A concise description of the study agent should be provided, including summaries of preclinical studies and relevant clinical studies. Finally, potential risks and benefits of immunization with the investigational agent should be delineated.

Objectives and Purpose

The objectives and purpose of the trial should be clearly and explicitly stated. For phase 1 trials, the primary objectives will be to assess the safety, tolerability, and reactogenicity of a vaccine, whereas assessment of immunogenicity is a secondary objective. For combined phase 1/2 and phase 2 clinical trials, safety and immunogenicity may be coprimary endpoints. Exploratory endpoints may also be included, such as the effect of age, race, or gender on immune responses (Keitel et al., 2006). Combined phase 1/2 trials may be proposed when the vaccine candidate represents a variant of a previously licensed construct, such as a subvirion influenza vaccine for prevention of avian influenza (Treanor et al., 2006). Efficacy and safety typically are the primary endpoints for phase 3 trial.

Study Design

The study design then should be described. For clinical trials of candidate vaccines, the study design typically is a randomized, double-blind, controlled clinical trial—one example of a parallel group design. Phase 1 safety and tolerability studies often utilize a titration design, where ascending dosages of the experimental agent are sequentially administered to new cohorts of subjects (Gorse et al., 2006). Cluster-randomized clinical trial design is occasionally employed to assess vaccine efficacy. In this circumstance, larger groups of individuals (such as nursing homes or schools) are randomized to an intervention, rather than individual subjects, and the clinical endpoints are ascertained for vaccinated subjects (Rodrigues et al., 2005), or for a subset of the cluster, such as the contacts of healthcare workers in a closed setting (Hayward et al., 2006). Phase 1 clinical trials historically were open-label; however, in recent years most phase 1 clinical trials of vaccines have been randomized and blinded. The value of a placebo control in clinical trials has been described; however, a licensed control vaccine may be used rather than a placebo, particularly for phase 3 clinical trials in children. In a recent phase 3 clinical trial of pneumococcal conjugate vaccine in infants, a meningococcal type C vaccine served as the control vaccine (Black et al., 2000). In this case, the meningococcal control vaccine provided potential benefits to the study participants. Flexible adaptive design methods frequently are employed in the development of vaccines; these incorporate plans for modifications of the clinical trial design that are made before or during the conduct of the research (Chow and Chang, 2007). Adaptations to ongoing trials may include prospective adaptations, such as interim analysis, stopping rules for early termination due to futility/safety concerns/efficacy, or sample size re-estimation; ad hoc adaptations such as changes in inclusion and exclusion criteria, dosage or regimen alteration, and trial duration; or retrospective adaptations at the end of the study but before unblinding, including changing the study endpoint or altering the statistical hypothesis (superiority to noninferiority). The goal of this approach is to permit modification based on accumulated evidence to alter trial design to increase the probability of success without undermining the validity of the trial (Gallo et al., 2006). Such adaptations may require modifications of the study hypotheses, protocol amendments, and sample size recalculations.

Clinical trials are also classified as single-center or multicenter studies. Phase 1 studies often are single-center studies; however, multicenter trial design may be used for any phase of clinical vaccine development. Multicenter study design provides several advantages: enrollment of subjects is expedited, and the results of the trial are likely to be more generalizable.

Study endpoints need to be clearly identified. For phase 1 trials, safety, tolerability, and reactogenicity primary endpoints may include the frequencies and severities of injection site reactions (pain, tenderness, redness, and swelling) and systemic reactions (fever, chills, headache, myalgia, arthralgia, etc.), as well a laboratory evidence of adverse reactions (hematologic, biochemical, and other). For phase 2 clinical trials, specific immune responses at defined time points after immunization typically characterize the primary endpoints; safety and reactogenicity may be primary or secondary endpoints. For phase 3 clinical trials, protection against laboratory-confirmed infection and/or disease is the primary endpoint, and the major safety assessment may be the frequency of SAEs associated with administration of the investigational agent.

Study Population

A detailed description of the proposed study population and the number of subjects to be studied must be provided; specifically, characteristics (age range, health status, ability to provide informed consent, etc.) of potentially eligible persons (Inclusion Criteria) and factors that would render an individual ineligible (Exclusion Criteria) should be explicitly enumerated. For some phase 1 trials, screening for eligibility may include medical history, physical examination, and laboratory screening for evidence of good health (normal hematologic and biochemical parameters, and no evidence of active hepatitis B, hepatitis C, or HIV infection). Information regarding serosusceptibility to the candidate pathogen may be necessary. For example, a phase 1 clinical trial of a dengue virus vaccine may require evidence of no prior infections caused by these viruses (Edelman et al., 2003), and a phase 1 or 2 clinical trial assessing the immunogenicity of LAIV may focus on persons with low or absent levels of preexisting immunity to the candidate vaccine (Keitel et al., 1993b). For phase 3 clinical trials, it is necessary to identify a population in which the infection or disease occurs at a high enough frequency to assess the ability of a vaccine to protect. For example, pivotal phase 3 trials of an inactivated hepatitis A vaccine were conducted in specific US communities where the rate of hepatitis A infections in children was high (Werzberger et al., 1992).

Human subjects considerations may include a description of certain behaviors and/or concomitant medications that would exclude a subject. For most clinical trials of vaccines, women who are capable of bearing children must consent to certain birth control measures. The US Department of Health and Human Services (DHHS) has published a guidance regarding research in pregnant women: for research conducted in this population, there must be direct benefit to the woman or her fetus or there must be only minimal risk to the fetus, and information cannot be obtained any other way. For many phase 1 clinical trials, use of prescription medications is not permitted. Clinical trials of vaccines that potentially could be transmitted to others in the community raise special concerns. Recent reevaluations of smallpox vaccines posed concerns with regard to transmission of the vaccine virus from subjects to their contacts (Frey et al., 2002). In this case, persons who had household or other significant contacts with young infants, people with eczema, pregnant women, and immunocompromised individuals were excluded from participation.

The methods for test article allocation should also be described. For most phase 1 and phase 2 clinical trials of vaccines, the subjects are randomized to receive one of several dosage levels of vaccine or placebo. Ideally, randomization should not occur until the subjects have been qualified for participation. Typically the randomization occurs in blocks of a prespecified number that represents a multiple of the number of test articles. For example, if there were four dosage levels of vaccine and a placebo, then the block size might be 5, 10, or 15. If the block size chosen were 5, then the subjects would be randomized 1:1:1:1:1. Block randomization can reduce the risk of unequal group sizes. In some circumstances the probability of receiving one product differs from the probability of receiving another. For example, in an efficacy study to be conducted in children, an investigator may wish to reduce the number of subjects randomized to receive the placebo, and the randomization scheme selected may be 2:1 (vaccine:placebo). The vaccine group assignments for subjects should be concealed from the subjects and from investigators to reduce bias in the assessments performed after vaccination (so-called double blinding). Additional measures can be taken to reduce the potential for imbalances in baseline characteristics of enrolled subjects, such as stratification of subjects according to age, prior receipt of a related vaccine, etc., prior to randomization.

Study Agent/Interventions

The clinical trial protocol should contain basic information regarding the characterization of the vaccine formulations—including dosage(s), packaging, labeling, storage; preparation, administration, dosing, and accountability methods for each study product, including placebo and/or control preparations. More complete descriptions of study vaccines, including manufacturing information, preclinical and clinical safety, immunogenicity, and efficacy should be provided separately in the Investigators' Brochure (IB). Information regarding the use of concomitant medications, including prohibited medications, should be detailed. For example, during phase 1 clinical trials concomitant use of prescription medications may be prohibited. During phase 2 clinical trials, concomitant use of certain medications may be allowed, such as antihypertensive medications or antidepressants. In general, concomitant use of immunosuppressive, immunomodulatory, or cytotoxic drugs would be prohibited in any clinical trials of live attenuated vaccines.

Study Procedures and Evaluations

A description of the proposed clinical evaluations then follows. In phase 1 or 2 clinical trials, detailed and frequent physical assessments of the injection site and systemic responses may be indicated, as well as review of subject records of clinical responses following immunization. The intensity of study assessments will vary according to the nature of the study product: more frequent and detailed assessments would be indicated for novel products whose safety profile is undefined. Periodic collection of blood, nasal, fecal, or other samples to assess for the occurrence of toxicity, or to determine the frequency, magnitude, and/or duration of shedding of a live vaccine candidate may also be indicated (Piedra et al., 1993; Taylor et al., 1997). These laboratory assessments should be tailored to the particular needs of the protocol, and should be based on the pathogenesis of the disease, the vaccine under evaluation, and information collected in the pre-IND stage. Brief descriptions of the type(s) of specimens to be collected, methods for specimen collection, preparation, handling, storing, and shipping (if applicable) should be outlined; detailed procedures should be provided in a separate Manual of Procedures (MOP) for each study. For phase 3 trials, clinical follow-up is specifically targeted at ascertaining whether the vaccine prevents infection or disease and capturing the occurrence of SAEs; however, limited prospective safety assessments may be included (perhaps only in a subset of subjects) to expand the safety database (Oxman et al., 2005).

Study Schedule

Once the specific clinical and laboratory procedures for assessing safety, immunogenicity, and/or efficacy have been described, a detailed study schedule indicating the timing of various assessments/procedures/interventions should be outlined, including those that will be conducted at screening, enrollment, and each follow-up visit. Clinical assessments and laboratory procedures that should be completed in the event the subject's participation in the trial is terminated early should be described, as should those for women who become pregnant during the study. Finally, a description of clinical and laboratory assessments to be performed in the event the subject returns for an unscheduled visit (for example, for a severe vaccine reaction) may be appropriate. A summary table or figure outlining study procedures is particularly useful.

Assessment of Safety

As described above, the clinical trial protocol should detail the specific safety parameters that will be monitored during the course of the trial, along with the methods and timing for their assessment and reporting. Standardized definitions for AEs and SAEs have been formulated (FDA, 1996): an AE is defined as any untoward medical event that occurs in a subject who has received a study agent. SAEs include death, life-threatening events, hospitalization or prolongation of hospitalization, congenital anomalies, and events that result in permanent disability. Each AE and SAE that is recorded should be assessed for its severity and relationship to vaccination. The grading of AEs is usually based on the interference with activity, where 1 = mild, does not interfere with usual activities; 2 = moderate, interferes somewhat with usual activities; and 3 = severe, incapacitating. The severity grading system of laboratory abnormalities should be established before the initiation of the trial. The FDA has suggested guidelines for toxicity grading scale for healthy adults and adolescents enrolled in vaccine clinical trials (FDA, 2007b).

Contemporary standards for assessing the relationship of the AE to immunization have narrowed the choices to "associated" and "not associated." All AEs that are temporally related to administration of the investigational agent and have no alternative etiologies to explain the event are considered associated. SAEs should be reported promptly to the sponsor; in turn, the Sponsor must notify the FDA in a timely fashion.

Prospective guidelines for discontinuation of individuals from further vaccinations, as well as halting rules for the clinical trial should be outlined. Typical circumstances for discontinuation of an individual include severe and/or hypersensitivity reactions following immunization, development of an Exclusion Criterion during the trial, and failure or inability of the subject to comply with study procedures. Note that every attempt should be made to continue to follow these subjects for safety assessments. Rules for halting a phase 1 clinical trial may include the occurrence of one or two hypersensitivity reactions or SAEs associated with the investigational agent or the occurrence of moderate or severe reactogenicity among a predefined proportion of subjects.

Finally, the safety oversight plan should be described. For single-center trials or other small clinical trials, a Safety Monitoring Committee (SMC) consisting an Independent Safety Monitor (ISM) from each site and an independent member with expertize relevant to the protocol has the responsibility to review trial results periodically and on an ad hoc basis, and to make recommendations to the Sponsor regarding the conduct of the clinical trial. For larger, multicenter trials, a Data and Safety Monitoring Board (DSMB) consisting of site ISMs and individuals with clinical and statistical expertize relevant to the protocol is constituted to review study progress and advise the Sponsor. The SMC or DSMB may recommend terminating a trial because of unexpected or severe toxicity; continuation of a trial after review of AEs that triggered halting rules; alteration of sample size based on interim analyses; or any other protocol modifications that are deemed necessary to complete the trial successfully.

Clinical Monitoring

A clear plan for monitoring the conduct of the clinical trial site(s) should be outlined in the study protocol. Specific objectives of site-monitoring visits are to review all study documentation to ensure protection of human subjects; compliance with GCP, clinical and laboratory procedures, test article administration and accountability guidelines; and accurate and complete data collection and documentation. The Sponsor may conduct monitoring visits, and may also designate an independent Contract Research Organization (CRO) to conduct monitoring visits on a regular basis throughout the trial. Early monitoring visits can be particularly valuable in order to identify systematic, unintentional deviations from the protocol.

Statistical Considerations

A detailed Statistical Analysis Plan (SAP) should be prepared that restates the study hypotheses, objective and endpoints, describes the statistical basis for the sample size selected, and outlines the methods that will be used to analyze safety and efficacy. If interim analyses are planned, then the statistical issues related to this should be discussed.

Quality Management

The clinical trial site is responsible for protocol compliance and accurate and complete data collection and recording. In a separate document, site-specific Standard Operating Procedures (SOPs) should outline the methods that will be used to ensure that these activities are accomplished, as well as methods to ensure appropriate training of the study staff. The overall quality management plan should be described in the protocol.

Ethics/Protection of Human Subjects

A description of the ethical standards that will be followed to ensure protection of human subjects should be described: in the US, compliance with 45 CFR Part 46 and ICH E6 GCP is expected. The protocol should indicate that no research (including screening) will begin until the protocol and the consent form have been IRB-approved. The consent process should be described, as should the provisions for subject confidentiality. The Health Insurance Portability and Accountability Act of 1996 (HIPAA) was enacted to improve portability and continuity of health insurance coverage; nevertheless, it contains regulations that have direct relevance for clinical research (DHHS, 2007). For example, informed consent documents are required to include extensive information on how the participant's protected health information (PHI) will be kept private. Administrative, physical, and technical safeguards must be adopted to ensure the security of electronic PHI. Finally, special issues related to exclusion of special populations (women, minorities, and children) should be addressed.

Data Management and Record Keeping

The goals of data management are to ensure the accuracy, completeness, and timeliness of clinical trial data collection. While the primary responsibilities for data collection rest with the clinical trial site, additional oversight and data management responsibilities (quality review, analysis, and reporting) may be shared with the sponsor and a data coordinating center (DCC). The organizational structure of data management should be described in this part of the protocol. Data capture methods and internal quality checks, types of data, timing of reports, study records retention, and identification of protocol deviations and corrective actions should be addressed here, as well as in greater detail in the MOP.

Other Considerations

In addition to the major protocol elements described above, a full protocol should include a title page, a statement of compliance with GCP, and other regulatory guidelines, a signature page, a table of contents, a list of abbreviations; a protocol summary; a list of key personnel and their roles, a description of unique facilities, if applicable, a list of references, publication policy; and appropriate appendices. A list of essential documents that should be on file before the trial starts is shown in Table 12.3; additional documents should be added to the trial documentation as new information becomes available (protocol amendments, updates, IRB approvals, training certificates, CVs, screening and enrollment logs, test article accountability and shipment logs, monitoring reports, consent forms, source documents, CRFs, communications with the sponsor, etc.).

TABLE 12.3 Checklist of essential documents on file at the clinical trial site before the clinical trial begins

Signed full clinical trial protocol, and amendments, if applicable
Sample case report forms (CRFs)
IRB-approved informed consent document
Investigators' brochure
Manual of procedures (MOP)
Information that will be given to subjects
Recruiting materials (text of advertisements, flyers, etc.)
IRB approval letter
Copy of IRB/IEC Federal Wide Assurance
Composition of IRB
FDA Form 1572 (Principal Investigator Responsibilities)
Curriculum vitae of participating investigators
Financial disclosures; other clinical trial agreements
Copy of the principal investigator's medical license
Laboratory credentials/certifications
Laboratory reference ranges
Sample labels for investigational product
Instructions for handling investigational product and other trial materials
Shipping records for trial-related materials
Clinical trial site initiation monitoring report

Source: Adapted from ICH E6 GCP guidance (FDA, 1996).
Courtesy: U.S. FDA.

CONCLUSIONS

The success of any clinical trial hinges on the development of a carefully designed protocol. Although discussion of clinical trial implementation is beyond the scope of this chapter, it is clear that the protection of human subjects and scientific integrity of the trial design and documentation are the overarching goals of any clinical research protocol. The clinical trial protocol must document the processes that will be used to ensure that these goals are attained. Conscientious supervision of clinical trial activities is a shared responsibility of the investigators, the IRB, the sponsor, the CRO, the DCC, the safety oversight committee, and all other partners who are participating in the trial. While meticulous attention to detail and strict protocol compliance are essential, the entire study team must be flexible and prepared to respond in a timely fashion to unexpected findings. Novel approaches to the development of vaccines for biodefense and emerging and neglected diseases will continue to evolve and will require ongoing reconsideration of the ethical and regulatory principles and practices that guide the conduct of clinical trials.

ACKNOWLEDGMENTS

I gratefully acknowledge Dr. Robert Atmar and Connie Rangel, R.N., for their thoughtful reading of the manuscript, and Yvette Rugeley and Michelle Thomas-Sturm for assistance with manuscript preparation.

References

Acosta, C.J., Galindo, C.M., Ochiai, R.L., Danovaro-Holliday, M.C., Laure-Page, A., Thiem, V.D., Jin, Y., Khan, M.I., Sahito, S.M., Hamza, H.B. et al. Implementation of good clinical practice guidelines in vaccine trials in developing countries. Vaccine 2007; 25:2852–2857.

Ballentine, C. Taste of Raspberries, Taste of Death. The 1937 Elixir Sulfanilamide Incident. US Food and Drug Administration–FDA Consumer Magazine, June 1981:Issue 1–4. 1981. http://www.fda.gov/oc/history/elixir.html.

Baylor, N. and Midthun, K. Regulation and testing of vaccines. In: Vaccines (S. Plotkin et al., Eds.), pp. 1539–1556. Philadelphia: Elsevier Inc., 2004.

Beecher, H. Ethics and clinical research. N. Engl. J. Med. 1966; 274:1354–1360.

Belshe, R.B., Gruber, W.C., Mendelman, P.M., Mehta, H.B., Mahmood, K., Reisinger, K., Treanor, J., Zangwill, K., Hayden, F.G., Bernstein, D.I., Kotloff, K., King, J., Piedra, P.A., Block, S.L., Yan, L. and Wolff, M. Correlates of immune protection induced by live, attenuated, cold-adapted, trivalent, intranasal influenza virus vaccine. J. Infect. Dis. 2000; 181:1133–1137.

Black, S., Shinefield, H., Fireman, B., Lewis, E., Ray, P., Hansen, J.R., Elvin, L., Ensor, K.M., Hackell, J., Siber, G., Malinoski, F., Madore, D., Chang, I., Kohberger, R., Watson, W., Austrian, R. and Edwards, K. Efficacy, safety and immunogenicity of heptavalent pneumococcal conjugate vaccine in children. Northern California Kaiser permanente vaccine study center group. Pediatr. Infect. Dis. J. 2000; 19:187–195.

Borchers, A.T., Hagie, F., Keen, C.L. and Gershwin, M.E. The history and contemporary challenges of the US Food and Drug Administration. Clin. Ther. 2007; 29:1–16.

CDC. Ten great public health achievements – United States, 1900–1999. MMWR Morb. Mortal. Wkly. Rep. 1999; 48:241–243.

Chow, S.-C. and Chang, M. Adaptive Design Methods in Clinical Trials. Boca Raton: Chapman and Hall/CRC; Taylor and Francis Group, 2007.

Chow, S.-C. and Liu, J.-P. Design and Analysis of Clinical Trials. Concepts and Methodologies. Hoboken: Wiley, 2004.

CIOMS. International Ethical Guidelines for Biomedical Research Involving Human Subjects. Council for International Organizations of Medical Sciences. 2002: 1–59. http://www.cioms.ch/frame_guidelines_nov_2002.htm.

Couch, R.B., Douglas, R.G., Jr., Fedson, D.S. and Kasel, J.A. Correlated studies of a recombinant influenza-virus vaccine. 3. Protection against experimental influenza in man. J. Infect. Dis. 1971; 124:473–480.

Deming, D., Sheahan, T., Heise, M., Yount, B., Davis, N., Sims, A., Suthar, M., Harkema, J., Whitmore, A., Pickles, R., West, A., Donaldson, E., Curtis, K., Johnston, R. and Baric, R. Vaccine efficacy in senescent mice challenged with recombinant SARS-CoV bearing epidemic and zoonotic spike variants. PLoS Med. 2006; 3:e525.

Department of Health and Human Services. Office for Civil Rights-HIPAA Medical Privacy-National Standards to Protect the Privacy of Personal Health Information. United States Department of Health & Human Services. 2007. http://www.hhs.gov/ocr/hipaa/

Edelman, R., Wasserman, S.S., Bodison, S.A., Putnak, R.J., Eckels, K.H., Tang, D., Kanesa-Thasan, N., Vaughn, D.W., Innis, B.L. and Sun, W. Phase I trial of 16 formulations of a tetravalent live-attenuated dengue vaccine. Am. J. Trop. Med. Hyg. 2003; 69:48–60.

Edelman, R., Wasserman, S.S., Kublin, J.G., Bodison, S.A., Nardin, E.H., Oliveira, G.A., Ansari, S., Diggs, C.L., Kashala, O.L., Schmeckpeper, B.J. and Hamilton, R.G. Immediate-type hypersensitivity and other clinical reactions in volunteers immunized with a synthetic multi-antigen peptide vaccine (PfCS-MAP1NYU) against *Plasmodium falciparum* sporozoites. Vaccine 2002; 21:269–280.

Ellenberg, S.S., Foulkes, M.A., Midthun, K. and Goldenthal, K.L. Evaluating the safety of new vaccines: summary of a workshop. Am. J. Public Health 2005; 95:800–807.

Emanuel, E.J., Wendler, D. and Grady, C. What makes clinical research ethical? J. Am. Med. Assoc. 2000; 283:2701–2711.

Epstein, J.E., Rao, S., Williams, F., Freilich, D., Luke, T., Sedegah, M., de la Vega, P., Sacci, J., Richie, T.L. and Hoffman, S.L. Safety and clinical outcome of experimental challenge of human volunteers with *Plasmodium falciparum*-infected mosquitoes: an update. J. Infect. Dis. 2007; 196:145–154.

FDA. E6 Good Clinical Practice: Consolidated Guidance. Guidance for Industry- ICH1-58. US Department of Health and Human Services; Food and Drug Administration; Center for Drug Evaluation and Research (CDER); Center for Biologics Evaluation and Research (CBER). 1996. http://www.fda.gov/cder/guidelines.htm.

FDA. New Drug and Biological Drug Products; Evidence Needed to Demonstrate Effectiveness of New Drugs When Human Efficacy Studies Are Not Ethical or Feasible. HHS37989-37998. 2002. http://www.fda.gov/cber/rules/humeffic.htm.

FDA. Guidance for Industry, Investigators, and Reviewers; Exploratory IND Studies. US Department of Health and Human Services Food and Drug Administration Center for Drug Evaluation and Research1-13. 2006. http://www.fda.gov/cder/guidance/index.htm.

FDA. Clinical Data Needed to Support the Licensure of Pandemic Influenza Vaccines. Guidance for Industry, 1-18. 2007. US Department of Health and Human Services; Food and Drug Administration; Center for Biologics Evaluation and Research. 2007a. http://www.fda.gov/cber.guidelines.htm.

FDA. Toxicity Grading Scale for Healthy Adult and Adolescent Volunteers Enrolled in Preventive Vaccine Clinical Trials. US Department of Health and Human Services. Food Drug Administration. Center for Biologics Evaluation and Research 1–8. 2007 Guidance for Industry. 2007b. http://www.fda.gov/cber/guidelines.htm.

Francis, T., Jr. Evaluation of the 1954 poliomyelitis vaccine field trial; further studies of results determining the effectiveness of poliomyelitis vaccine (Salk) in preventing paralytic poliomyelitis. J. Am. Med. Assoc. 1955; 158:1266–1270.

Frey, S.E., Couch, R.B., Tacket, C.O., Treanor, J.J., Wolff, M., Newman, F.K., Atmar, R.L., Edelman, R., Nolan, C.M. and Belshe, R.B. Clinical responses to undiluted and diluted smallpox vaccine. N. Engl. J. Med. 2002; 346:1265–1274.

Gallo, P., Chuang-Stein, C., Dragalin, V., Gaydos, B., Krams, M. and Pinheiro, J. Adaptive designs in clinical drug development—an executive summary of the PhRMA Working Group. J. Biopharm. Stat. 2006; 16:275–283.

Gorse, G.J., Keitel, W., Keyserling, H., Taylor, D.N., Lock, M., Alves, K., Kenner, J., Deans, L. and Gurwith, M. Immunogenicity and tolerance of ascending doses of a recombinant protective antigen (rPA102) anthrax vaccine: a randomized, double-blinded, controlled, multicenter trial. Vaccine 2006; 24:5950–5959.

Hayward, A.C., Harling, R., Wetten, S., Johnson, A.M., Munro, S., Smedley, J., Murad, S. and Watson, J.M. Effectiveness of an influenza vaccine programme for care home staff to prevent death, morbidity, and health service use among residents: cluster randomised controlled trial. Br. Med. J. 2006; 333:1241.

Keitel, W., Couch, R., Bond, N., Adair, S., Van Nest, G. and Dekker, C. Pilot evaluation of influenza virus vaccine (IVV) combined with adjuvant. Vaccine 1993a; 11:909–913.

Keitel, W.A., Atmar, R.L., Cate, T.R., Petersen, N.J., Greenberg, S.B., Ruben, F. and Couch, R.B. Safety of high doses of influenza vaccine and effect on antibody responses in elderly persons. Arch. Intern. Med. 2006; 166:1121–1127.

Keitel, W.A., Couch, R.B., Quarles, J.M., Cate, T.R., Baxter, B. and Maassab, H.F. Trivalent attenuated cold-adapted influenza virus vaccine: reduced viral shedding and serum antibody responses in susceptible adults. J. Infect. Dis. 1993b; 167:305–311.

Keitel, W.A., Kester, K.E., Atmar, R.L., White, A.C., Bond, N.H., Holland, C.A., Krzych, U., Palmer, D.R., Egan, A., Diggs, C., Ballou, W.R., Hall, B.F. and Kaslow, D. Phase I trial of two recombinant vaccines containing the 19 kd carboxy terminal fragment of *Plasmodium falciparum* merozoite surface protein 1 (msp-1(19)) and T helper epitopes of tetanus toxoid. Vaccine 1999; 18:531–539.

Macrae, D.J. The Council for International Organizations and Medical Sciences (CIOMS) guidelines on ethics of clinical trials. Proc. Am. Thorac. Soc. 2007; 4:176–178.

Meinert, C. Clinical Trials Design, Conduct, and Analysis. New York: Oxford University Press, 1986.

Murphy, T.V., Gargiullo, P.M., Massoudi, M.S., Nelson, D.B., Jumaan, A.O., Okoro, C.A., Zanardi, L.R., Setia, S., Fair, E., LeBaron, C.W., Wharton, M. and Livengood, J.R. Intussusception among infants given an oral rotavirus vaccine. N. Engl. J. Med. 2001; 344:564–572.

NIAID. NIAID Protocol Template Extramural Guidance. 2006. http://www3.niaid.nih.gov/research/resources/toolkit/attachments/ProtocolTemplateEXTRAMURAL.pdf.

Offit, P.A. The Cutter incident, 50 years later. N. Engl. J. Med. 2005; 352:1411–1412.

Oxman, M.N., Levin, M.J., Johnson, G.R., Schmader, K.E., Straus, S.E., Gelb, L.D., Arbeit, R.D., Simberkoff, M.S., Gershon, A.A., Davis, L.E. et al. A vaccine to prevent herpes zoster and postherpetic neuralgia in older adults. N. Engl. J. Med. 2005; 352:2271–2284.

Piedra, P.A., Glezen, W.P., Mbawuike, I., Gruber, W.C., Baxter, B.D., Boland, F.J., Byrd, R.W., Fan, L.L., Lewis, J.K., Rhodes, L.J. et al. Studies on reactogenicity and immunogenicity of attenuated bivalent cold recombinant influenza type A (CRA) and inactivated trivalent influenza virus (TI) vaccines in infants and young children. Vaccine 1993; 11:718–724.

Polack, F.P. A typical measles and enhanced respiratory syncytial virus disease (ERD) made simple. Pediatr. Res. 2007; 62:111–115.

Rennels, M.B., Englund, J.A., Bernstein, D.I., Losonsky, G.A., Anderson, E.L., Pichichero, M.E., Munoz, F.M. and Wolff, M.C. Diminution of the anti-polyribosylribitol phosphate response to a combined diphtheria-tetanus-acellular pertussis/Haemophilus influenzae type b vaccine by concurrent inactivated poliovirus vaccination. Pediatr. Infect. Dis. J. 2000; 19:417–423.

Rodrigues, L.C., Pereira, S.M., Cunha, S.S., Genser, B., Ichihara, M.Y., de Brito, S.C., Hijjar, M.A., Dourado, I., Cruz, A.A., Sant'Anna, C.

et al. Effect of BCG revaccination on incidence of tuberculosis in school-aged children in Brazil: the BCG-REVAC cluster-randomised trial. Lancet 2005; 366:1290–1295.

Tacket, C.O., Cohen, M.B., Wasserman, S.S., Losonsky, G., Livio, S., Kotloff, K., Edelman, R., Kaper, J.B., Cryz, S.J., Giannella, R.A. et al. Randomized, double-blind, placebo-controlled, multicentered trial of the efficacy of a single dose of live oral cholera vaccine CVD 103-HgR in preventing cholera following challenge with *Vibrio cholerae* O1 El tor inaba three months after vaccination. Infect. Immun. 1999; 67:6341–6345.

Taylor, D.N., Tackett, C.O., Losonsky, G., Castro, O., Gutierrez, J., Meza, R., Nataro, J.P., Kaper, J.B., Wasserman, S.S., Elderman, R., Levine, M.M. and Cryz, S.J. Evaluation of a Bivalen (CVD 103-HgR/CVD 111) live oral cholera vaccine in adult volunteers from the United States and Peru. Infect. Immun. 1997; 65: 3852–3856.

Treanor, J.J., Campbell, J.D., Zangwill, K.M., Rowe, T. and Wolff, M. Safety and immunogenicity of an inactivated subvirion influenza A (H5N1) vaccine. N. Engl. J. Med. 2006; 354:1343–1351.

Wang, D. and Bakhai, A. Clinical Trials: A Practical Guide to Design Analysis and Reporting. London: Remedica, 2006.

Werzberger, A., Mensch, B., Kuter, B., Brown, L., Lewis, J., Sitrin, R., Miller, W., Shouval, D., Wiens, B., Calandra, G. et al. A controlled trial of a formalin-inactivated hepatitis A vaccine in healthy children. N. Engl. J. Med. 1992; 327:453–457.

CHAPTER

13

The US Food and Drug Administration Pre-IND and IND Process for Vaccines

Jeanne M. Novak[1], *Judy Ruckman*[1] *and Dennis W. Trent*[2]

[1]CBR International Corporation, Boulder, CO, USA

[2]D. Trent and Associates, American Fork, UT, USA

OUTLINE

ABSTRACT

Vaccines provide the most effective means of preventing potentially serious infectious diseases and play a critical role in public health management. Vaccines are regulated in the United States by the Office of Vaccines Research and Review in the Center for Biologics Evaluation and Research of the Food and Drug Administration (FDA). To conduct a human clinical trial of a new vaccine, the sponsor of the trial must submit an original Investigational New Drug Application (IND) to the FDA. The required content of an original IND is described in the Code of Federal Regulations. The IND provides the Agency with detailed information pertaining to the manufacture and testing of the vaccine to demonstrate its identity, safety, purity, potency and other quality attributes, the preclinical studies conducted in animals to establish the safety of the product, and the protocol for the proposed clinical study. The FDA will assemble a team of experts to review the information in the IND and determine whether the new vaccine is sufficiently safe for administration to human subjects. Sponsors can increase the likelihood of receiving a positive review decision from the Agency by ensuring that the original IND is detailed and complete, and addresses any potential safety concerns associated with the investigational product. Many resources are available to assist sponsors in the preparation of a successful IND. Numerous guidance documents have been issued by the FDA and other regulatory authorities that summarize FDA and worldwide expectations for the manufacture, testing, and clinical evaluation of specific product types. Most importantly, sponsors have the opportunity to meet with members of the FDA review staff via a formal "pre-IND" telephone conference (meeting) to seek specific guidance from the Agency regarding their product development plans. Optimally, the pre-IND meeting establishes a working relationship between the sponsor and the FDA that facilitates ongoing dialogue leading up to the IND submission and throughout the product development life cycle. After the submission of the original IND and initiation of clinical studies, communication with the FDA is primarily maintained through the submission of IND amendments, such as new clinical protocols, safety reports, or updated manufacturing information. Regulations also provide for additional meetings between the sponsor and FDA at prescribed points in product development, and at other times if issues arise that require detailed discussion. Ongoing communication with the FDA is an essential element of product and clinical development, and provides the most efficient mechanism to meet all regulatory requirements for product approval in the shortest possible time frame.

INTRODUCTION

The submission of an original Investigational New Drug Application (IND) to the US Food and Drug Administration (FDA) represents the first step under the US regulatory process toward evaluation of a new drug, biologic product, or vaccine in a human clinical trial. The original IND application provides all of the data and other information necessary for the FDA to assess the safety of an investigational product for administration to human volunteers. As product development progresses to more advanced clinical studies, the FDA will maintain regulatory oversight "to assure the safety and rights of [clinical trial] subjects" and "to help assure that the quality of the scientific evaluation of drugs is adequate to permit an evaluation of the drug's effectiveness and safety" (Code of Federal Regulations, Title 21, Part 312.22).

Deliberate and clear communication with FDA will result in the most streamlined and cost-effective development program with the greatest assurance of a successful outcome. The purpose of this chapter is to provide an overview of the IND process for sponsors of investigational vaccines, as well as to provide helpful hints to facilitate effective communication with the FDA.

REGULATORY OVERVIEW

FDA Regulations

An IND application must be submitted to the FDA for review and approval before the initiation of the first clinical study of a novel investigational drug or vaccine. The content of an original IND application and the sponsor's obligations throughout the IND process are described in the US Code of Federal Regulations, Title 21, Part 312 (21 CFR 312). In legal terms, the IND is a request for an exemption from the federal statute that prohibits an unapproved drug product from being shipped in interstate commerce. Practically, the IND provides the Agency with a complete description of the investigational product, including the results of any proof of concept studies justifying its evaluation in human subjects, the methods for product manufacture and analytical testing, the results of nonclinical safety studies in animals, and proposals for the clinical investigations planned in the first year of product development.

The majority of investigational products are developed through the submission of a routine or "commercial" IND. Typically, the entities that submit, or "sponsor," a commercial IND are pharmaceutical or biotechnology companies, or individuals or organizations with the intent that the investigational product will be commercialized. In addition to pharmaceutical or biotechnology firms, individual clinical investigators can sponsor an IND. The sponsor-investigator IND is subject to the same regulations as a commercially sponsored IND, but differs in that the investigator assumes the responsibilities of both the sponsor and the investigator, as defined in 21 CFR 312.3.

The sponsor of an IND need not have the intention to commercialize the investigational product in the United States. In fact, FDA has recently issued guidance to clarify its position that, for vaccines intended to treat global infectious diseases, submission of an IND application is encouraged regardless of the intended target population for the licensed vaccine. Specifically, "FDA encourages submission of an IND so that it can provide input on manufacturing, quality testing, assay validation, non-clinical and clinical trial design issues, statistical analysis plans, endpoints, and other important aspects of vaccine development. The IND process will allow sponsors to obtain important scientific and regulatory advice on products that are critical to the advancement of world health" (FDA Guidance for Industry, September 2008).

One exception to the regulatory IND requirements outlined above is provided by the Project Bioshield Act of 2004. The Act amended the Food Drug and Cosmetic Act to permit the introduction into interstate commerce of an unapproved drug, device, or biological product for use in an actual or potential emergency (Emergency Use Authorization, or EUA). In the event that there is a government determination of an emergency, or the potential for an emergency, due to exposure to a specific biologic, chemical, radiological, or nuclear agent with the potential to cause a serious or life-threatening disease or condition, the FDA may issue an EUA for a specific product or products. A product may be eligible for EUA if it is reasonable to believe, based on the scientific evidence available, that the product may be effective in treating the disease or condition, the known and potential benefits outweigh the known and potential risks, and there is no adequate, approved, and available alternative to the product for treating the disease or condition. FDA's guidance on "Emergency Use Authorization of Medical Products" was issued in September 2007.

FDA regulations include multiple elements to facilitate the development and licensure of high priority products intended to treat or prevent serious or life-threatening diseases or conditions. For example, a sponsor may seek, and FDA may grant, "fast track" designation for a specific product development program (FDA, 2006). Fast track designation entitles the sponsor to early and frequent communication with the FDA throughout product development, in order to optimize the design of Phase 2 and Phase 3 clinical studies, and expedite the resolution of any product development issues that may arise. Such products may receive "priority review" status at the time of Biologics License Application (BLA) submission which provides for a shorter, 6-month review of the license application, rather than the standard 10-month review clock. In addition, the license application may utilize a "rolling submission" process, in which sections of the application may be submitted sequentially rather than all at once.

The regulations governing licensure of biologic products (21 CFR 601) provide for approval of certain products intended to treat serious or life-threatening illnesses for which clinical efficacy data cannot readily be obtained (e.g., a vaccine for a potentially pandemic influenza virus strain). At its discretion, FDA may approve such a product based upon a surrogate endpoint (such as immunogenicity) that is reasonably likely, based on existing evidence, to predict clinical benefit. Accelerated approval may include a stipulation that the applicant "study the biological product further [in post-approval studies] to verify and describe its clinical benefit" (21 CFR 601.41). In addition, certain products intended to treat or prevent "serious or life-threatening conditions caused by exposure to lethal or permanently disabling toxic biological, chemical, radiological, or nuclear substances" (21 CFR 601.90) for which human clinical efficacy studies are not ethical or feasible, may be approved under the "animal rule." In this case, reasonable evidence of clinical benefit may be provided by the results from animal studies, where the animal species can be justified to provide an adequate model of the human response to both the toxic substance and the investigational biologic product. FDA has recently issued a concept paper to identify the critical characteristics of an animal model necessary for efficacy evaluation under the Animal Rule (FDA Concept Paper, September 2008).

FDA Organization and Review Responsibility

The FDA includes two centers of expertise with responsibility for review of IND applications. The

Center for Biologics Evaluation and Research (CBER) reviews applications pertaining to therapeutic and prophylactic vaccines, blood and blood products, and tissue, cellular and gene therapy products, while the Center for Drug Evaluation and Research (CDER) has primary responsibility for review of therapeutic drug and biologics applications.

CBER is organized into the following offices: the Office of Vaccines Research and Review (OVRR), the Office of Blood Research and Review, and the Office of Cellular, Tissue and Gene Therapies. Within the Office of Vaccines Research and Review, there is a discreet applications division, the Division of Vaccines and Related Products Applications (DVRPA) that is responsible for receiving IND submissions, routing the submissions for review and corresponding with the IND sponsor. An original IND submission will be routed to a specific project manager within DVRPA who will serve as the primary point of contact for the sponsor during the IND review and throughout the subsequent period of product development.

Guidance Documents

A variety of guidance materials are available to assist sponsors with the submission of an original IND. CBER makes available an "IND packet" that provides applicable regulations, forms, and articles helpful in the preparation of original INDs. These and other guidance materials may be obtained from CBER's Office of Communication, Training and Manufacturers Assistance (OCTMA), which is an office dedicated to providing general guidance and information to individuals and organizations interested in the regulatory process. FDA guidance is also available on the internet and may be accessed by visiting the FDA (http://www.fda.gov/) or CBER (http://www.fda.gov/cber/index.html) home pages and following the links for guidance documents. IND guidance from the OCTMA is available at http://www.fda.gov/cber/ind/ind.htm.

In addition to guidance materials for the IND submission process, the FDA has issued a wide variety of guidance documents pertaining to product manufacture, analytical testing, preclinical evaluation, and clinical trial design for vaccine products. The International Conference on Harmonization or ICH (an organization dedicated to the harmonization of regulatory requirements between the United States, Europe, and Japan) has also developed guidance documents in these areas, many of which have been adopted as official guidance of the US FDA. Links to FDA guidance documents pertaining to vaccine development can be found on the CBER website (http://www.fda.gov/cber/vaccine/vacpubs.htm), while current ICH guidelines may be downloaded from the ICH website (http://www.ich.org/cache/compo/276-254-1.html).

COMMUNICATIONS WITH THE FDA

Meetings and Teleconferences

To facilitate effective and timely communications regarding the development of investigational products, FDA will participate in certain meetings and telephone calls with IND sponsors. There are three basic types of meetings that can be requested by an IND sponsor. The Type A meeting is requested when a sponsor feels the development of their product or clinical program has been "stalled," and they require urgent input from the Agency in order to proceed. The Type A meeting must be scheduled by FDA within 30 days after the receipt of the formal written request for such meeting. Type B meetings are routine meetings that may be requested at critical milestones during the clinical development of an investigational product (21 CFR 312.47). The pre-IND, the End of Phase 2, and the pre-BLA meetings are all classified as Type B meetings. Type B meetings must be scheduled within 60 days after FDA receives the request for a meeting. The Type C meeting is reserved for less time-critical issues and must be scheduled within 75 days after FDA receives the request for a meeting. The FDA Guidance for Industry document entitled, "Formal Meetings with Sponsors and Applicants for PDUFA Products" (February 2000) provides detailed instructions for requesting, scheduling, conducting, and documenting these meetings.

Pre-IND Meeting or Teleconference

To facilitate submission of a successful IND, the FDA will conduct formal meetings with sponsors referred to as "pre-IND meetings." As mentioned previously, the pre-IND meeting is a Type B meeting and is discussed in more detail below. It should be noted that in recent years the OVRR will grant pre-IND telephone conferences, but will rarely grant face-to-face meetings due to limited time and resources. However, sponsors are encouraged to request a face-to-face meeting with the Agency if they feel the scientific, clinical, or regulatory issues surrounding their new vaccine warrant this type of interaction.

The pre-IND teleconference should be used by sponsors to discuss any product, nonclinical, clinical, or regulatory issues that the investigational vaccine may present. If a sponsor would like to request a pre-IND teleconference, they should follow the FDA meeting request guidance document mentioned above (February 2000) and refer to the SOPP 8101.1 issued by CBER. These documents describe exactly what is required for a written request for a meeting or teleconference with the Agency and outline the format for the request. Note that the Agency can deny a request for a meeting if the sponsor's written request is incomplete.

Most importantly, the meeting or teleconference request should include a statement of purpose of the meeting and a draft of a list of questions that the sponsor wishes to discuss with the Agency. This means that, before making a pre-IND meeting request, the sponsor should be able to identify the key issues for which guidance or concurrence from the Agency are necessary before submission of the IND.

The sponsor should prepare and submit a briefing document at least 4 weeks in advance of the scheduled teleconference date. Depending on the issues to be discussed, the briefing document may include background information on the product, its method of manufacture and analytical testing, preclinical testing completed and/or planned, and a brief outline of the proposed clinical trial (discussed below). A list of questions critical to further product development and the initial clinical trial(s) should be included in the briefing document for Agency review and consideration. The allotted time for the meeting or teleconference may be from 1 to 2h. Due to time constraints, sponsors are not usually permitted to make formal presentations at a meeting or have lengthy introductory remarks on a telephone conference. Therefore, emphasis should be placed on ensuring that the briefing document is clear, complete, succinct, and easy to navigate.

The FDA review team will make every effort at the pre-IND teleconference to provide responses to the questions submitted by the sponsor and will frequently advise the sponsor of information needed in the original IND submission. Typically, FDA representatives will include scientists with product, pharmacology/toxicology, and clinical expertise. Most importantly, the emphasis of the telephone conference should be on the questions and issues that the sponsor would like to address rather than on information already provided in the meeting materials.

It is now the norm that written responses to the questions posed in the briefing document will be supplied by FDA one or more days in advance of the scheduled telephone conference. It is recommended that the sponsor communicate to the FDA project manager their interest in receiving such written responses before the teleconference, if possible, as this will enable both the sponsor and the Agency to focus the meeting on those issues requiring further discussion. Upon forwarding their written responses to questions, the FDA will offer to cancel the teleconference if the responses are deemed satisfactory by the sponsor; however, it is recommended that the sponsor proceed with the call in order to discuss any issues requiring clarification. In addition, the pre-IND call provides an opportunity for the sponsor to "meet" the FDA reviewers and to potentially obtain additional useful advice. Sponsors are encouraged to submit a meeting summary describing their understanding of the FDA's comments and recommendations to the pre-IND file. The FDA will issue an official record of the meeting in the form of meeting minutes within 30 days after the meeting date.

End of Phase 2 Meeting and Special Protocol Assessment

For investigational products with the potential to treat or prevent a serious or life-threatening disease, the sponsor may request an "End of Phase 1" meeting (see section Other Meetings with FDA1 below). More typically, however, the next Type B meeting occurs after the completion of Phase 2 studies when the pivotal Phase 3 studies are being planned. This "End of Phase 2" meeting is held to: (1) determine the safety of proceeding to Phase 3 efficacy studies; (2) evaluate the Phase 3 plan and protocol(s) (a well-written draft protocol is sufficient); (3) identify any additional information necessary to support a license application, such as additional nonclinical safety studies; and (4) discuss relevant product issues, such as plans for manufacturing scale-up to produce product for commercial sale and the need for comparability testing relative to the product used in earlier Phase 1 and Phase 2 clinical studies. It may be necessary to schedule a separate meeting to address all of the product-related issues for Phase 3 and future commercial manufacture (FDA Guidance for Industry, May 2001).

Another important issue to be discussed at the End of Phase 2 meeting is whether a Special Protocol Assessment (SPA) agreement is appropriate for the Phase 3 pivotal study program. SPA, implemented in 1997, allows for a "binding agreement" between the sponsor and FDA that the design, endpoints, conduct, statistical analysis plan, and other details of the Phase 3 protocol(s) will be adequate to support efficacy claims for the vaccine being tested. The formal process for requesting and obtaining a SPA is described in the FDA Guidance for Industry entitled, "Special Protocol Assessment," issued in May 2002.

Pre-BLA Meeting

The pre-BLA meeting is held after the completion of Phase 3 efficacy studies and prior to the submission of a license application. This is a very important meeting for sponsors as it affords the opportunity to present the FDA with a complete overview of the proposed contents of the planned license application and to obtain guidance from the Agency regarding the information that will be sufficient to support filing of the BLA and Agency review of the license application. The meeting should be used to discuss all safety and efficacy data from clinical trials, potential post-approval obligations of the sponsor, product release specifications, stability data to support expiration dating, and the claims and indication statement in the vaccine product label.

Although the required contents of a BLA are stipulated by regulation in 21 CFR 601.2, the organization and format of a BLA most often follows the outline of the "Common Technical Document," or "CTD." The CTD was developed by the ICH to provide a common format for license applications to the United States, Europe, and Japan. Its use is not yet mandatory for license applications submitted to CBER but is strongly preferred by the Agency. The sponsor's plans for organizing the content of the application (traditional BLA format or CTD format) and the submission type (paper vs. electronic) should be discussed at the pre-BLA meeting.

The results of pivotal efficacy and safety studies will be presented at the pre-BLA meeting and the adequacy of these data to support the intended label claims will be discussed. The sponsor's plans for analysis and formatting of the integrated clinical data for submission in the license application should also be presented for Agency review and comment. As a result of these discussions, additional analyses or specific data tabulations that are critical to review of the application may be requested by the Agency.

The Pediatric Research Equity Act of 2003, as amended in 2007, requires that license applicants submit with their application either an assessment of the safety and efficacy of the investigational product in the pediatric population, an agreement from FDA for deferral of such studies until a specified date after license approval, or a full or partial waiver of the pediatric research requirement due to the specific nature of the product and/or indication. If the sponsor wishes to defer pediatric studies to the post-licensure period (or obtain a full or partial waiver), this plan should be discussed with the FDA at the pre-BLA meeting. The Agency's concurrence with the pediatric research plan as captured in the formal minutes from the meeting should be submitted as evidence of the agreement at the time of BLA submission.

The pre-BLA meeting is also an appropriate time to discuss plans for any post-approval clinical study or studies that will be performed as an element of a pharmacovigilence plan for the new vaccine. Frequently, observational studies are conducted post-licensure to obtain confirmatory evidence of efficacy under conditions of actual use, and an enhanced understanding of safety, including rare adverse events. For certain products with known or potential risks based on nonclinical or clinical studies or safety information for related products, appropriate strategies for risk minimization in the post-licensure period should be discussed (see FDA Guidance for Industry, Development and Use of Risk Minimization Action Plans [RiskMAP], March 2005).

The pre-BLA meeting is a critical meeting for discussion of any issues related to the transition from clinical to commercial manufacture. Any changes in the manufacturer, facility, or equipment that accompany the scale-up from the Phase 3 process to the commercial process should be presented, and the appropriate approach to establishing the comparability of the Phase 3 and commercial products should be discussed. Other topics may include the final specification for release of commercial product, the procedure for lot release by CBER, and the timing of the Pre-Approval Inspection of the manufacturing facility or facilities.

Finally, the pre-BLA meeting may be used by the sponsor to obtain clarification as needed of the regulatory procedures for submission, filing, and review of license applications in BLA or CTD format, including information related to electronic applications (eBLA or eCTD) if the sponsor plans to submit electronically.

Other Meetings with FDA

Other meetings (Type A and Type C) may be requested during the IND process. However, due to time and resource limitations, CBER will determine whether or not a face-to-face meeting is required to address the issue(s) under consideration. Examples of such issues include major changes in the manufacture or testing of product that require CBER input, major changes in clinical investigational plans, or major changes in protocol design or statistical analysis plans. Often, telephone conversations are adequate to address these issues, and can circumvent the need to expend resources on a face-to-face meeting.

Certain products with the potential to treat or prevent serious or life-threatening disease, or products identified for priority development as medical countermeasures to a bioterrorist attack, are eligible for additional meetings with FDA in order to accelerate the path to licensure. As mentioned previously, such products may be granted "fast track" designation which provides for regular and frequent communication between the sponsor and FDA throughout product development. An End of Phase 1 meeting is recommended for such products in order to discuss the design and scope of Phase 2 studies. Potentially, for high priority products, a well-designed and well-controlled Phase 2 study may provide sufficient data for licensure, with commitments to collect confirmatory information regarding the safety and efficacy of the product in post-approval studies.

Other Types of FDA Communication

The most important methods of communication used by CBER during the IND process are the clinical hold letter and the "non-hold" comment letter. A clinical hold letter is issued when FDA determines that insufficient information has been provided or a clinical study has been inadequately designed to ensure the safety of subjects participating in the trial (21 CFR 312.42). Even when no clinical hold is imposed for a protocol, comment letters may be provided and may be comprehensive. CBER reviewers will attempt to identify clinical and product manufacturing issues early enough in the IND process so that the sponsor can begin to make any changes or other plans necessary to support a potential license application. Extensive CBER comments should not necessarily be viewed in a negative light by the sponsor. If at any time during the IND process the sponsor does not understand a reviewer's comment or some aspect of the IND process, the sponsor should contact the IND project manager in order to discuss the most appropriate way to get advice or clarification.

THE CONTENT OF THE IND

The initial IND submission proposing the first Phase 1 study to be conducted with a new investigational product is referred to as an "original" submission. The FDA has 30 days in which to review the original IND and determine whether the investigational product is sufficiently safe for human clinical trials. Although during the life of an IND, FDA reviews all protocols submitted to determine their adequacy to demonstrate safety and effectiveness of an investigational product, the 30-day review of the original IND submission is the only waiting period required of sponsors. If FDA finds that the information provided in the original IND is not sufficient to ensure the safety of the clinical trial subjects, the IND may be placed on clinical hold, as discussed above. FDA will inform the sponsor of this decision, usually by telephone, within the initial 30-day review period, and will provide a written description of the issues that require resolution within 30 days following the imposition of the clinical hold. If, following the 30-day IND review period, the sponsor has not been notified of a clinical hold, the study may be initiated. Thus, for time-efficient and cost-effective product development, it is critical that the original IND contain all of the information necessary to establish the safety of the investigational product, the adequacy of the proposed clinical trial design, and the measures in place to protect the rights and safety of human subjects.

21 CFR 312.23 describes the elements specified by regulation to be included in an original IND. Each of these elements is listed in Table 13.1 and is discussed in detail in the paragraphs below. Although the regulations at 21 CFR 312.23 also prescribe the order in which each element should appear in the application, INDs are now also accepted in CTD format. The order of the IND elements in CTD format will differ modestly from that specified by the IND regulations as shown in Table 13.1. The CTD is discussed in more detail later in this chapter.

Form FDA 1571

Form FDA 1571 (1571) must accompany the original IND application and all subsequent submissions to the IND. This form serves as the cover sheet for the submission, and specifies the nature and contents of

TABLE 13.1 The elements of an Investigational New Drug Application

IND Section	Content
1	Form FDA 1571 and Form FDA 3674
2	Table of Contents
3	Introductory Statement
4	General Investigational Plan
5	Investigator's Brochure
6	Clinical Protocol
7	Chemistry, Manufacturing, and Controls Information
8	Pharmacology and Toxicology Information
9	Previous Human Experience
10	Additional Information

the submission. The 1571 is also a legal document that must include the signature of the sponsor's authorized representative. For an IND that is sponsored by an organization or individual outside of the United States, the 1571 must be countersigned by "an attorney, agent, or other authorized official who resides or maintains a place of business within the United States" [21 CFR 312.23(a)(1)(ix)]. Either the sponsor (with US representative countersignature) or the US representative can sign the 1571. An electronic version of Form FDA 1571 and instructions for its completion are available on the FDA website at <http://www.fda.gov/opacom/morechoices/fdaforms/CBER.html>.

Form FDA 3674

With the passage of the FDA Amendments Act of 2007, a new requirement was placed on sponsors of Phase 2 or Phase 3 clinical trials to submit a summary of information about the trial to the public clinical trials registry previously established by the US National Institutes of Health (http://www.clinicaltrials.gov). Sponsors are further required to certify their compliance with the registration requirement by completing and submitting Form FDA 3674 with certain regulatory submissions. This form should be included in any original IND submission. Form FDA 3674 may be downloaded from http://www.fda.gov/opacom/morechoices/fdaforms/FDA-3674_508.pdf.

Table of Contents

An original IND application may consist of several volumes and hundreds of pages of information. Thus, a comprehensive Table of Contents is essential to aid navigation through the submission. Typically, the submission will include "free-standing" documents, such as study protocols or test reports, that are independently paginated. Serial pagination, with numbering from the first through the last page of the entire submission, provides a mechanism for both the sponsor and the FDA to unequivocally reference specific pages or documents within the IND. The comprehensive or serial page numbering should be added to each page, in addition to any pre-existing page numbers, and should be cited in the comprehensive Table of Contents.

Introductory Statement

The technical sections of the IND (e.g., the clinical protocol or the nonclinical study reports) will be reviewed and evaluated primarily by FDA experts within the relevant disciplines. The introductory statement is therefore important to provide the entire FDA review team with a brief orientation to the investigational product, including its composition and proposed use, and the broad objectives of the clinical studies that are proposed. According to 21 CFR 312.23(a)(3), the introductory statement should include the name and biochemical description of the investigational vaccine, the formulation and planned route of administration, a description of any previous human experience with the product, including investigational or marketing experience in other countries, and a brief statement of the overall plan for clinical studies during the first year.

General Investigational Plan

The general investigational plan should provide a brief description of the clinical studies planned for the investigational product through the first year. The rationale for the proposed studies and their objectives, as well as a brief description of the study design (e.g., blinded vs. unblinded, inactive or active controls to be used, number of centers and approximate number of subjects) should be provided. Any known or potential risks of particular severity or seriousness should be described. It would also be appropriate to include a description of the longer-term clinical development plan, beyond the first year of study, to the extent that this can be anticipated. The information provided in this section is particularly useful to help the reviewer anticipate the need for resolution of difficult issues well in advance of upcoming protocols, for example, if the product must be approved under the "animal rule" (21 CFR 601.90-95; Federal Register Notice, 2002) because clinical efficacy studies are either unethical or impractical. If a sponsor is considering applying for fast track designation, it is appropriate to indicate that intention in this section of the IND.

Investigator's Brochure

The Investigator's Brochure is an informational document that must be provided to each clinical investigator participating in a clinical trial of the investigational vaccine. The Investigator's Brochure provides all available information on the investigational product that will assist the investigator in the conduct of the study and in the assessment of expected versus unexpected adverse events. Information to be included in the Investigator's Brochure is described in the IND regulations [21 CFR 312.23(a)(5)]. The document should include a brief description of the product, a summary of the pharmacologic, toxicologic, and pharmacokinetic behavior of the product in animals (and in humans, if known), a summary of information relating to the safety and effectiveness of the product in humans obtained from prior clinical studies, and a description of possible risks and side effects of the product. The ICH E6(R1) document, "Guideline for Good Clinical Practice" (June 1996), provides details on the appropriate content and structure of the Investigator's Brochure.

Although an Investigator's Brochure is provided to each clinical investigator before a clinical study is initiated, new observations regarding the product, particularly in regard to adverse effects, must be communicated to all participating investigators. Thus, regular revisions to the Investigator's Brochure are expected, and should be prepared and issued as needed, and at least on an annual basis if clinical trials are ongoing. The Investigator's Brochure should be updated before submitting and initiating additional clinical studies under the IND. The updated Investigator's Brochure should also be submitted to the FDA as an amendment to the IND.

Clinical Protocol

A complete clinical protocol must be included in the original IND submission. In most cases, the original submission contains a Phase 1 protocol in which the sponsor proposes to evaluate the safety of the investigational new product in a small number of healthy, young adult subjects (e.g., 18–40 years of age) and, for a vaccine, to obtain an initial indication of its immunogenicity in humans (Goldenthal et al., 2004). Dose escalation or dose-ranging studies may also be performed during Phase 1, if appropriate design elements, such as staggered enrollment and interim safety data review, are incorporated to ensure the safety of study volunteers.

Phase 2 and Phase 3 studies are designed to assess immunogenicity and efficacy (protection from disease), in addition to safety (Goldenthal et al., 2004). Phase 2 studies may be used to further assess the immune response to a vaccine, potentially evaluating alternative immunological endpoints, dose, schedule, or route of administration. Phase 2 studies may also provide information regarding the most common side effects (adverse events) associated with product use.

Phase 3 studies are pivotal safety and/or efficacy studies that will serve as the basis for licensure. In certain instances, an original IND submission may include a Phase 2 or Phase 3 protocol, for example, if earlier stage clinical trials were conducted outside of the United States under the oversight of a foreign regulatory agency. If the original submission contains a pivotal Phase 3 study protocol, it is strongly recommended that the sponsor contact CBER to arrange a meeting prior to submission of the IND (see discussion on meetings above). In addition, details of all completed clinical trials should be included in the IND under Section 9, Previous Human Experience (see Table 13.1).

It should be noted that the prospective design of clinical studies and the strategy for overall clinical development will differ for each vaccine. Thus, it is essential to formulate a product-specific development program that maximizes the data from each study and reduces the overall timeline to BLA submission.

A clinical protocol must be submitted to the FDA for each study to be conducted under the IND, and should describe the procedures by which the trial will be conducted. The elements of a clinical protocol can be found in 21 CFR 312.23(a)(6)(iii). At a minimum, a protocol should include a statement of the objectives and purpose of the study, the criteria for patient inclusion and exclusion, the approximate number of subjects to be enrolled, a description of the design of the study, including the use of control groups or blinding procedures, the plan for dose administration and the duration of individual patient exposure to the investigational product, a description of the parameters to be monitored during the study, and the plan for evaluating the data to asses the safety and activity of the investigational product in human subjects. Detailed guidance regarding the content and organization of clinical protocols is provided in the ICH E6(R1) document, "Guideline for Good Clinical Practice" (June 1996). A well-organized and complete protocol in the original IND submission will facilitate the FDA's review and will provide the best assurance that FDA will permit the clinical study to proceed following its initial 30-day review.

It is recommended that a sponsor prepare a clinical trial outline, or CTO, internally as an initial step in the preparation of a protocol. The CTO may be used to facilitate internal discussion regarding optimal protocol design and overall clinical development program planning. A brief description of the elements of a CTO is provided in Table 13.2.

Each clinical protocol submitted to an IND file is typically accompanied by one or more completed copies of Form FDA 1572, "Statement of Investigator." A 1572 form must be completed for each clinical investigator participating in the study. The purpose of the form, in part, is to provide the sponsor with verification of the investigator's qualifications and to obtain the investigator's commitment to adhere to the protocol and to pertinent FDA regulations governing clinical trials. Although submission of the 1572 form to the IND is not formally required, IND regulations under 21 CFR 312.23(a)(6)(iii)(b), require that clinical protocols include the information collected on the 1572 (including the name and address of the

TABLE 13.2 Clinical trial outline sections and contents

Section	Contents
Protocol Title	Title and protocol number of the study. The protocol number is helpful for communication purposes.
Investigator	Name of Principal Investigator for the overall study or Principal Investigator at each participating site.
Sites	List of all sites participating in the study.
Objectives	Primary and secondary objectives. Is the study intended to assess safety, immunogenicity, or efficacy? Is the study focused on a specific sub-population?
Rationale	Description of the phase of development (Phase 1, 2, 3, or 4) and how this study fits into the clinical development plan for the product. If the study is being conducted solely for scientific purposes, and will not be used to support licensure, the scientific rationale should be supplied.
Preclinical Data	Summary of all relevant animal studies, including a brief description of each protocol and the results. GLP, non-GLP, and proof of concept studies should be described.
Human Data	Summary of any previous human experience with the product from studies conducted under the same or different INDs, from studies not conducted under IND, or from post-marketing experience in other countries where the product is already licensed. Relevant clinical experience with related products may also be included in this section.
Study Design	Brief description of the study design, including treatment groups (active and control groups), randomization (including the method), blinding (open-label, single or double-blind), and phase of study (Phase 1, 2, 3, or 4).
Use of Investigational Product	Dose, schedule, and route/method of administration.
Controls	Description of control group, if any.
Subjects	Number, age, gender, and health status.
Entry Criteria	Inclusion and exclusion criteria.
Monitored Parameters	Description of all clinical, laboratory, and other parameters used to determine how safety, activity, and/or efficacy will be monitored. This should include those parameters determined at the screening/baseline visit as well as any follow-up visits. It is helpful to include a table that summarizes all monitored parameters for each clinic visit over the course of the study.
Endpoints	Description of the primary and secondary clinical endpoints that will be determined to address the objectives of the study (e.g., geometric mean titer of vaccine-specific antibodies, or incidence of disease). If data are to be collected at multiple times during the course of the study, the timepoint(s) at which data will be evaluated for the primary endpoint should be prospectively identified. Information pertaining to the analytical method for endpoint analysis, including validation of the method, should be included, especially if the analytical method is unique.
Sample Size	Clinical and statistical justification for the sample size. Note that Phase 1 and Phase 2 studies are not usually "powered" for statistical significance.
Data Analysis	Plans for analysis of each primary and secondary endpoint. It is particularly important that Phase 3 studies have all analysis plans prospectively defined in a Statistical Analysis Plan (SAP). This section should identify the study populations to be included in each analysis (e.g., intent-to-treat, per protocol, or other subgroups) and the statistical tests to be used.
Interim Analyses	Plans for interim analyses of safety or efficacy data, if any, and how multiple "looks" at the data will be addressed in the statistical analysis.
Data Monitoring	If a data monitoring committee (DMC)[a] will review the ongoing conduct of the study (safety and/or efficacy), the responsibilities of this committee and any plans for communication between the committee and the sponsor and/or investigators should be described.
Other	Dose escalation, subject withdrawal, management of adverse events and any other prospectively defined procedures should be described.

[a]See FDA Guidance for Clinical Trial Sponsors: Establishment and Operation of Clinical Trial Data Monitoring Committees, March 2006.

investigator and a statement of their qualifications, the names of any sub-investigators working under the supervision of the investigator, the name and address of the research facilities where the study will be conducted, and the name and address of the reviewing Institutional Review Board, IRB). Submission of the 1572 form with the clinical protocol provides a convenient means to comply with this element of the IND regulations. An electronic copy of Form FDA 1572 and instructions for its completion can be found at <http://www.fda.gov/opacom/morechoices/fdaforms/CBER.html>. A draft FDA guidance pertaining to the use of Form FDA 1572 has recently been issued (July 2008).

Sponsors should be familiar with all of the FDA regulations pertaining to the conduct of clinical trials in human subjects. In addition to the IND regulations describing the content of clinical protocols, 21 CFR 312 Subpart D delineates the responsibilities of both sponsors and investigators in conducting clinical trials with investigational products. In addition, the FDA has issued regulations governing protection of human subjects under 21 CFR Part 50 (including the requirement for informed consent), the disclosure of financial interests by clinical investigators (21 CFR Part 54), and the function, organization, and operation of IRBs (21 CFR Part 56). Note that IRB approval is not a prerequisite for submitting an IND to the FDA, but is required prior to the initiation of any clinical study.

Finally, sponsors should be aware of FDA guidance pertaining to the design of clinical studies for investigational vaccines. Recent guidance (September 2007) has been issued pertaining to appropriate toxicity grading scales to be used for the healthy populations typically enrolled in preventive vaccine clinical trials. Other guidance documents for the clinical development of specific types of vaccines have also been issued (Table 13.3). Sponsors should check the CBER website for the most current vaccine clinical trails-related guidance (http://www.fda.gov/cber/vaccine/vacpubs.htm).

TABLE 13.3 Selected guidance documents pertaining to vaccine clinical development

Source	Title
FDA	Providing Clinical Evidence of Effectiveness for Human Drug and Biological Products (May 1998)
FDA	Guidance for Industry: Toxicity Grading Scale for Healthy Adult and Adolescent Volunteers Enrolled in Preventive Vaccine Clinical Trials (September 2007)
FDA	Guidance for Industry: Clinical Data Needed to Support the Licensure of Seasonal Inactivated Influenza Vaccines (May 2007)
FDA	Guidance for Industry: Clinical Data Needed to Support the Licensure of Pandemic Influenza Vaccines (May 2007)
FDA	Guidance for Industry: Development of Preventive HIV Vaccines for Use in Pediatric Populations (May 2006)

Chemistry, Manufacturing, and Controls Data

The Chemistry, Manufacturing, and Controls (CMC) section describes the methods used to manufacture the investigational product and the analytical methods and test results used to control the manufacturing process and the quality of the final product. The information to be included in this section is specified in 21 CFR 312.23(a)(7), and is summarized below:

(1) Drug substance. This section should include detailed information pertaining to the physical, chemical, or biological characterization of the drug substance; the name and address of the manufacturer and the general method for preparation of the drug substance; the analytical methods used to assure the identity, strength, quality, and purity and acceptable limits for these tests; and information to support stability of the drug substance during the toxicological studies and the planned clinical studies.

(2) Drug product. This section should include information regarding all components used in the manufacture of the investigational drug product and its quantitative composition; the name and address of the drug product manufacturer and a brief general description of the manufacturing and packaging procedure; the analytical methods used to assure the identity, strength, quality, and purity of the drug product and acceptable limits for these tests; and information sufficient to assure the product's stability during the planned clinical studies.

(3) Placebo. A brief general description of the composition, manufacture, and control of any placebo to be used in a controlled clinical trial.

(4) Labeling. A copy of all labels and labeling to be provided to each investigator.

(5) Environmental analysis requirements. A claim for categorical exclusion under 21 CFR 25.31 or an environmental assessment under 21 CFR 25.40.

The FDA product reviewers will thoroughly evaluate all information related to manufacture and testing of the investigational product, including:

- the origin and development of recombinant biological products;
- the manufacture and testing of master and working cells banks or virus banks;
- the source and quality attributes of raw materials;
- the certification of any animal-derived materials for freedom from adventitious agents;
- the detailed manufacturing process, including in-process testing to verify process performance;
- the analytical testing and acceptance criteria for release of bulk drug substance and the finished and filled, final container drug product;
- qualification (or validation) of analytical methods.

All drug or biological products, including investigational products, should be manufactured according to current Good Manufacturing Practice (cGMP) guidelines. However, the implementation of cGMP is expected to occur in a phased manner as product development advances from early Phase 1 studies to pivotal studies enrolling hundreds or thousands of subjects. FDA has issued guidance defining the elements of cGMP that are applicable to the manufacture of product lots for Phase 1 clinical studies (FDA Guidance for Industry, July 2008). General cGMP guidelines applying to later stage clinical trial material and licensed drug or biological products are defined under 21 CFR 211.

The requirements for testing of licensed biological products prior to their release for distribution are described in 21 CFR 610. These regulations will apply to most vaccine products and should be consulted early in development when devising lot release specifications for clinical trial material. Release testing of biological products should include, at a minimum, tests for identity, potency, purity, sterility, and general safety. Table 13.4 provides some examples of analytical tests that might be applied to different types of vaccine products to provide assurance in the IND of their identity, potency, purity, safety, and other quality attributes.

Some additional considerations for analytical testing information to be supplied in the CMC section of an IND are discussed below.

i. In all cases, in-process and release testing must be conducted on the bulk vaccine drug substance, as well as the final, formulated and filled drug product. Tests conducted on the bulk drug substance need not be repeated on the drug product if the manufacturing steps for formulation and fill of the drug product would not affect the tested parameter (e.g., the relative content of an impurity).

ii. All vaccines require a test for potency. Potency is defined in the regulations (21 CFR 600.3(s)) as "the specific ability or capacity of the product, as indicated by appropriate laboratory tests or by adequately controlled clinical data obtained through the administration of the product in the manner intended, to effect a given result." ICH

TABLE 13.4 Examples of analytical tests to be included in the Chemistry, Manufacturing, and Controls section of an IND

Analytical tests	Recombinant or native purified protein antigen or inactivated viral or bacterial infectious agent	Live, attenuated infectious agent	Live, replication-defective, virally-vectored vaccine antigen	DNA plasmid-vectored vaccine antigen
Identity	Western blot	Nucleotide sequence	Nucleotide sequence of gene expression cassette	Restriction digest
	Amino terminal sequence	Specific neutralization	Specific antigen expression	Nucleotide sequence
	Mass spectrometry			
	Antibody binding assays (inactivated vaccines)			
Potency	In vivo immunogenicity	In vivo immunogenicity	In vivo immunogenicity	In vivo immunogenicity
	Antigen content	Infectious titer	Antigen expression in vitro	In vitro transfection efficiency
		Particle count		
Purity	SDS-PAGE	SDS-PAGE	SDS-PAGE	Agarose gel electrophoresis
	Immunoblot			
	HPLC			
Safety	General Safety	General Safety	General Safety	General Safety
	Sterility	Sterility	Sterility	Sterility
	Endotoxin	Endotoxin	Endotoxin	Endotoxin
	Mycoplasma	Mycoplasma	Mycoplasma	Host cell DNA
	Adventitious agents	Adventitious agents	Adventitious agents	
	Host cell DNA	Host cell DNA	Host cell DNA	
	Residual active infectious agent	Attenuation phenotype	Recombinant wild-type (replication competent) vector	
Quality	Residual process impurities (e.g., residual inactivating agent)	Residual host cell protein	Percentage of empty capsids Genetic stability	Percentage supercoiled

guidance Q6B further defines potency as "the quantitative measure of biological activity based on the attribute of the product which is linked to the relevant biological properties." For a vaccine, the potency test result would optimally be shown to correlate with the protective efficacy of the product in humans. Potency assays may be conducted in vivo (challenge-protection assays or serological immunogenicity assays) or in vitro (e.g., infectivity assays for live viral or bacterial vaccines, antigen content for inactivated viral or bacterial vaccines, or induction of antigen expression for vectored vaccines). In vivo assays require large numbers of animals and the need to include a reference standard in every assay run in order to calculate the relative potency of the test sample. In vivo assays are also inherently more variable (less precise) than alternative in vitro methods (Gisonni-Lex, 2005; Petricciani et al., 2007). Some sponsors are therefore seeking to develop in vitro potency test methods early in development and to validate these methods based on clinical efficacy data or by comparison to previously validated potency assays (Schalk et al., 2005; Capen et al., 2007).

iii. The General Safety test is required for all biological products, including vaccines (21 CFR 610.11). The test is conducted on the vaccine final product to detect extraneous toxic substances that might not otherwise be detected by the other release tests conducted. Vaccine manufacturers may request an exemption from the general safety test requirement under 21 CFR 610.11(g)(2) at the time of BLA submission if it can be established that "a test of general safety is unnecessary to assure the safety, purity, and potency of the product or cannot be performed."

iv. Stability testing of the vaccine must be conducted to provide evidence that the vaccine remains stable throughout the duration of nonclinical studies conducted in support of human clinical testing, as well as the full duration of the planned clinical trial. The stability of both the bulk drug substance and the final container vaccine must be evaluated [21 CFR 312.23(a)(7)(iv)(a)]. Since extensive stability data for the clinical lot of vaccine are frequently not available at the time of an original IND submission, the FDA will often accept stability data from a representative lot of vaccine as initial evidence of stability with a commitment to conduct stability evaluation of the clinical lot of vaccine concurrently with the Phase 1study.

CBER has prepared numerous guidelines and "Points to Consider" documents to provide regulatory and technical guidance for sponsors (Table 13.5). Pertinent ICH guidelines include the Q5 quality guidelines for biotechnological products and the Q6B guideline for specifications for biotechnological/biological products. The United States Pharmacopeia also includes general chapters pertaining to the testing of biological products (Table 13.5).

Pharmacology and Toxicology Data

The purpose of this IND section is to present the results of preclinical studies performed in animal models or *in vitro*, which have led the sponsor to conclude that the investigational product is sufficiently safe for the initiation of clinical studies. The scope and design of the preclinical studies necessary to support the first Phase 1study of an investigational product will vary according to the nature of the product and the duration and design of the proposed clinical trials. Typically, pre-IND studies of a vaccine will include immunogenicity studies to establish the proposed dose or dose range to be evaluated in the clinic, evaluation of alternative routes of administration, assessment of any immune enhancement provided by the use of an adjuvant, and toxicological evaluation of the vaccine in at least one animal species at or above the planned maximum human clinical dose. For vaccine products, toxicology studies are typically conducted by administration of a multiple of the human clinical dose in absolute terms; in other words, the dose is not scaled to body weight or body surface area. For products formulated with a novel adjuvant, toxicity studies may need to include safety assessment of the adjuvant alone, as well as the antigen/adjuvant combination. While studies to evaluate the immunogenicity of the vaccine product do not necessarily need to be conducted according to GLP guidelines (21 CFR Part 58), any toxicology study intended to establish the safety of the product for human clinical use must comply with GLP regulations.

In some cases, "proof of concept" studies can be conducted in animal models to demonstrate that the investigational product displays the biological activity or efficacy intended in humans. However, the demonstration of activity in animals does not have to be, and often cannot be, absolutely analogous to the clinical activity that is ultimately desired. For example, if an investigational vaccine is shown to be immunogenic in an animal model, this may serve as sufficient evidence for conducting Phase 1 studies. Demonstration of protection against challenge following vaccination is usually not required prior to Phase 1.

The necessary package of preclinical studies required to support entry into human clinical trials

TABLE 13.5 Selected guidance documents pertaining to IND chemistry, manufacturing, and controls information for vaccine products

Source	Title
FDA	Guidance for industry: Content and Format of Investigational New Drug Applications (INDs) for Phase 1 studies of drugs, including well-characterized, therapeutic, biotechnology-derived products (November 1995)
	Guidance for industry for the submission of chemistry, manufacturing, and controls information for a therapeutic Recombinant DNA-derived product or a monoclonal antibody product for in vivo use (August 1996)
	Guidance for industry: Content and format of chemistry, manufacturing and controls information and establishment description information for a vaccine or related product (January 1999)
	Guidance for industry: Considerations for plasmid DNA vaccines for infectious disease indications (November 2007)
	Guidance for industry: Manufacturing biological intermediates and biological drug substances using spore-forming microorganisms (September 2007)
	Guidance for industry: cGMP for phase 1 investigational drugs (July 2008)
	Draft Guidance for Industry: Characterization and qualification of cell substrates and other biological starting materials used in the production of viral vaccines for the prevention and treatment of infectious diseases (September 2006)
ICH	Q5A(R1): Viral safety evaluation of biotechnology products derived from cell lines of human or animal origin (September 1999)
	Q5B: Quality of biotechnological products: Analysis of the expression construct in cells used for production of r-DNA derived protein products (November 1995)
	Q5C: Quality of biotechnological products: Stability testing of biotechnological/biological products (November 1995)
	Q5D: Derivation and characterisation of cell substrates used for production of biotechnological/biological products (July 1997)
	Q5E: Comparability of biotechnological/biological products subject to changes in their manufacturing process (November 2004)
	Q6B: Specifications: Test procedures and acceptance criteria for biotechnological/biological products (March 1999)
USP	< 1041 > Biologics < 1045 > Biotechnology derived articles

is an appropriate topic for discussion at the pre-IND meeting with FDA. Specific questions pertaining to the design of individual studies may also be discussed. Sponsors are encouraged to seek agreement from FDA to review draft preclinical protocols in advance of their execution, especially if the study is to be conducted in an unusual model system or with an unconventional study design.

Finally, the sponsor should consult 21 CFR Part 58—Good Laboratory Practice for Nonclinical Laboratory Studies, ICH guidelines regarding safety, as appropriate, and the World Health Organization Guidance on Nonclinical Evaluation of Vaccines (http://www.who.int/biologicals/publications/nonclinical_evaluation_vaccines_nov_2003.pdf) when conducting preclinical studies in support of an original IND submission (Table 13.6).

Although initial preclinical testing of a vaccine product may support entry into human clinical trials, additional nonclinical studies may be requested by the FDA as product development continues under the IND. The nonclinical studies required to support later stages of clinical development and the final license application should be discussed in detail with the Agency after the Phase 1 study is complete and further clinical evaluation of the vaccine is warranted. For prophylactic vaccines to be used in women of childbearing potential, an assessment of the potential reproductive toxicity of the product is often required. Note that the FDA has issued a specific guidance pertaining to developmental toxicity studies for preventive and therapeutic vaccines (see Table 13.6).

TABLE 13.6 Selected guidance pertaining to nonclinical toxicology and pharmacology testing of vaccines

Source	Title
FDA	Guidance for industry: considerations for developmental toxicity studies for preventive and therapeutic vaccines for infectious disease indications (February 2006)
FDA	Guidance for industry: considerations for plasmid DNA vaccines for infectious disease indications (November 2007)
WHO	World Health Organization guidance on nonclinical evaluation of vaccines (November 2003)
ICH	S6: Preclinical safety evaluation of biotechnology-derived pharmaceuticals (July 1997)

Genotoxicity and carcinogenicity studies are normally not needed for vaccines, but might be required if the vaccine vector has oncogenic potential or if a novel adjuvant is included in the vaccine formulation. Safety pharmacology studies are also generally not required unless data from other studies suggest that the vaccine may have an effect on a physiological function other than the immune system (World Health Organization Guidance, 2003). Thus, the requirements for nonclinical evaluation of an investigational vaccine may evolve as new nonclinical and human clinical data are collected and evaluated.

The Pharmacology and Toxicology section of the IND should provide FDA reviewers with the relevant data pertaining to the immunological and toxicological effects of the vaccine observed in animal models. This section should include a detailed description of the design of the studies conducted (including any deviations from the design during study conduct), and a summary presentation of the results obtained. In addition, according to 21 CFR 312.23(a)(8)(ii), the toxicology portion of the IND should include "for each toxicology study that is intended primarily to support the safety of the proposed clinical investigation, a full tabulation of data suitable for detailed review." This should consist of complete line listings of the individual data points for each animal in the study, along with summary tabulations of the data. Usually, sponsors will comply with this requirement by appending the study report to the IND submission. Note that unaudited draft reports are acceptable for submission in the original IND since FDA has acknowledged that there are rarely any significant changes between the unaudited and final, audited study results (FDA Guidance for Industry, November 1995). However, note that if an unaudited draft report is supplied with the original IND, the final audited report must be submitted to the IND file within 120 days of the initiation of clinical studies. The updated information should highlight any differences between the unaudited report submitted for review and the final audited study report. Individual data listings are usually not required for studies focused on the immunological response to vaccination. The appropriate contents of the Pharmacology and Toxicology section of the original IND should be discussed with CBER prior to IND submission.

Previous Human Experience

This section of the IND should provide summaries of previous clinical trials of the investigational vaccine, whether conducted in the United States or overseas, or relevant information from trials of related products or components. For example, this section might include information on foreign clinical studies not conducted under an IND, or safety data from studies using the same vaccine vector to be used in the proposed Phase 1 study, but with modest changes to the encoded antigen. For products administered with a novel adjuvant, any prior clinical trials experience with the adjuvant should be described. The summary information should include a detailed outline of the study design as well as a thorough summary of the results, including statistical analysis, where applicable. It is understood that for many Phase 1 products, there will be no prior human experience to be reported in the original IND.

Additional Information

This section of the IND provides sponsors with an avenue to submit any other relevant information in support of the proposed clinical study, including publications and meeting abstracts. Prior IND correspondence may also be included in this section, for example, to provide documentation of specific agreements reached during the pre-IND meeting.

ELECTRONIC INDs

CBER has developed programs and policies to accommodate electronic submission of INDs (eINDs). In March 2002, the FDA published its final guidance on eINDs entitled, "Providing Regulatory Submissions to CBER in Electronic Format—Investigational New Drug Applications (INDs)," which provides instructions for the organization and assembly of eINDs. eINDs can be submitted via the FDA's electronic submissions gateway (ESG) or via CD-ROM in portable document format (PDF).

The guidance recommends that companies interested in submitting an eIND should contact CBER's Electronic Submissions Coordinator(s) at least 3 months prior to the planned eIND submission date in order to arrange for an informal demonstration or "test" submission of the eIND. The coordinators have been willing to arrange demonstration submissions with less advance notice, but CBER will require approximately 45–60 days to review the electronic files in the test submission and determine whether the files have been generated in accordance with the FDA eIND guidance. The content of the eIND is not considered during this review. If the review indicates that the eIND has been adequately generated, then the sponsor will be notified and allowed to proceed with the formal submission of the eIND. It should be noted that, once a sponsor has initiated submission of an IND in electronic format, all subsequent amendments to the IND file must also be made electronically.

INDs IN COMMON TECHNICAL DOCUMENT FORMAT

The ICH has developed a CTD that may be used for submission of licensing applications for drug products in the United States, the European Union, Japan, and other nations adhering to the ICH framework. The CTD is composed of five "modules" that include administrative information (Module 1), summary information (Module 2), detailed product quality information (Module 3), pharmacology and toxicology information (Module 4), and clinical information (Module 5). The organization of the CTD document is described in the 2004 ICH guideline M4(R3): Organization of the Common Technical Document for the Registration of Pharmaceuticals for Human Use. Although the CTD was designed for submission of licensing applications at the completion of clinical development, the FDA will accept original IND submissions that are organized according to the CTD outline. The advantage of this approach is that, as information amendments (described below) are submitted to the IND over the course of clinical development, the CTD exists as a "living document" that becomes increasingly detailed and populated as the product approaches licensure. Available guidance for mapping the required elements of the IND to the CTD Table of Contents is provided by the "FDA eCTD Table of Contents Headings and Hierarchy" (Updated July 2005), available on the FDA website at http://www.fda.gov/Cder/regulatory/ersr/ectd.htm. INDs in CTD format may also be submitted electronically as an "eCTD." (See FDA's June 2008 guidance entitled, "Providing Regulatory Submissions in Electronic Format—Human Pharmaceutical Applications and Related Submissions Using the eCTD Specifications.") Sponsors should discuss plans for submission of an original IND in CTD format with the FDA review team at the pre-IND meeting.

MAINTENANCE OF THE IND

An IND becomes "in effect" or "active" within 30 days of its receipt by FDA unless the FDA notifies the sponsor that the IND has been placed on clinical hold. The IND will remain in effect throughout product development, and will be updated or "amended" by the sponsor with new information as clinical development advances to Phase 2 and Phase 3 studies. The IND file should be a central repository for all information relevant to the safety and efficacy of the investigational product, including information regarding product manufacture and testing, toxicological evaluation in animal models, and testing for safety and efficacy in human clinical trials. From the FDA's perspective, maintenance of the IND file is essential for the performance of its regulatory mission to ensure the safety and adequacy of human clinical trials. For the sponsor, maintaining an up-to-date body of data and information in the IND will enable FDA's review staff to provide meaningful advice on product development issues as they arise, or to anticipate and avert potential problems before they occur.

Certain types of information, by regulation, must be submitted as amendments to the IND. All new clinical trial protocols must be submitted for FDA review as "protocol amendments" (21 CFR 312.30). For all protocols submitted subsequent to the original IND, the sponsor is not obligated to wait for the Agency's comments or approval prior to study initiation; however, it is strongly recommended that the sponsor allow time for Agency review in order to avoid the potential for a clinical hold after the study has begun enrolling subjects.

Significant changes to an existing protocol or the addition of a new investigator to an ongoing study are also protocol amendments that must be submitted to the IND.

"IND safety reports" [21 CFR 312.32(c)] are used to inform the Agency of any unexpected and serious adverse reaction associated with the use of the investigational product or any new finding from nonclinical studies that suggests a significant risk for human subjects. This category of information must be reported to the Agency as an "initial written report" as soon as possible (and no later than 15 calendar days after the sponsor's receipt of the information). Note that unexpected, fatal, or life-threatening adverse experiences must be communicated to the Agency by telephone or facsimile within 7 days of receipt of the information. Serious and unexpected adverse events should be evaluated for their significance in light of other safety information and followed-up to determine their outcome. The results of subsequent investigations should also be reported promptly to the Agency in the form of a "follow-up to a written report." The FDA will review the information received and determine whether additional data are required or there should be changes in the design or conduct of the ongoing clinical study.

A third category of IND amendment is the "annual report." No later than 60 days after the anniversary date on which the IND went into effect, the sponsor should submit an annual report for the FDA review staff that summarizes the status of ongoing clinical investigations, updates the IND with a current summary of CMC information, safety data from completed nonclinical studies, and safety and efficacy information from human clinical trials. The annual report should also preview the nonclinical and clinical studies planned for the coming year. The annual report serves as an administrative summary of IND activity over the previous year and should not be used to convey new information about product or clinical development. The appropriate content of an IND annual report is outlined in 21 CFR 312.33.

Finally, "information amendments" (21 CFR 312.31) are amendments that provide other types of essential information not included in a protocol amendment, IND safety report or annual report. Such amendments may contain CMC information such as significant changes to the manufacturing process or to the specifications for product release; nonclinical information, such as a toxicology study report; or clinical information, including clinical study reports, or updates to the Investigator's Brochure. Most product development data will be submitted to the IND in an information amendment.

Other types of IND amendments include "response to clinical hold," "response to FDA request for information," and "general correspondence," such as a change to the sponsor's authorized representative or the submission of a meeting request. All IND amendments must be accompanied by a completed Form FDA 1571, which specifies the type of information included in the submission. It is also advisable to include a cover letter with the amendment that clarifies the context and purpose of the submission.

PREPARING FOR THE BIOLOGICS LICENSE APPLICATION

In the later stages of product development under the IND process, sponsors must begin preparations for submission of a BLA/CTD and for commercial manufacture of the vaccine upon approval. As mentioned previously, the pre-BLA meeting with the FDA should be scheduled well in advance of the intended BLA filing date, and may be used to present preliminary data from the pivotal Phase 3 clinical trial(s) establishing the safety and efficacy of the product, and to discuss the format and content of the application itself. It may be necessary to schedule a separate meeting to discuss CMC issues, including (for example) final specifications for commercial product and any necessary changes in the commercial manufacturing scale or process. Questions specifically related to commercial facility design may be addressed by requesting a Type C meeting with the Division of Manufacturing and Product Quality at CBER. In the months leading up to submission of the licensing application, sponsors must plan for completion of analytical method validation, manufacturing process validation, and the manufacture of consistency lots according to the planned commercial process. The manufacturing facility should begin preparations for a pre-approval inspection (PAI) by FDA. In addition, Clinical Study Reports for pivotal clinical trials must be prepared for submission to the IND file, and/or with the original license application, and clinical trial records should be organized for inspection by FDA's Bioresearch Monitoring group.

ICH has issued guidance documents to assist sponsors in the preparation of a license application in CTD format (Table 13.7) many of which have been adopted by FDA; however, IND sponsors should be aware of regional requirements specific to the United States. These include specific administrative information to be included in Module 1, region-specific quality documentation to be provided in Module 3, and an integrated summary of clinical effectiveness and safety to be included in Module 5 (Table 13.7). Additional guidance has been provided by FDA for applications in electronic format (available on the FDA website at http://www.fda.gov/cder/Regulatory/ersr/ectd.htm).

Submission of an original BLA opens a new file and a new chapter in regulatory interactions with the Agency, and marks the transition from the IND process described in this chapter to the independent process for marketing application review and approval, and product commercialization.

CONCLUSIONS

The submission of an original IND is a significant step in the development of a new vaccine for the treatment or prevention of infectious disease. In this chapter, we have emphasized the importance of regular and ongoing communication throughout the IND process between sponsors and the CBER review staff through meetings, teleconferences and written submissions to the IND file. Sponsors should also review and stay up to date with FDA and ICH guidance documents that provide essential product and clinical development information. By maintaining positive and productive interactions with FDA,

TABLE 13.7 Guidance documents pertaining to license applications in CTD format

CTD topic	Document source	Title
CTD organization	ICH	M4(R3) Annex: Organisation of the Common Technical Document for Registration of Pharmaceuticals for Human Use (January 2004)
	ICH	M4 Implementation Working Group Questions and Answers (R3) (June 2004)
Administrative information	FDA	Draft Guidance for Industry: Submitting Marketing Applications According to the ICH-CTD Format—General Considerations (August 2001)[a]
Quality	ICH	M4Q(R1): Quality Overall Summary of Module 2; Module 3: Quality (September 2002)
	ICH	M4Q Implementation Working Group Questions and Answers (R1) (July 2003)
Nonclinical	ICH	M4S(R2): Nonclinical Overview and Nonclinical Summaries of Module 2; Organisation of Module 4 (December 2002)
	ICH	M4S Implementation Working Group Questions and Answers (R4) (November 2003)
Clinical	ICH	M4E(R1): Clinical Overview and Clinical Summary of Module 2; Module 5: Clinical Study Reports (September 2002)
	ICH	M4E Implementation Working Group Questions and Answers (R4) (June 2004)
	FDA	Draft Guidance for Industry: Integrated Summaries of Effectiveness and Safety: Location Within the Common Technical Document (June 2007)

[a]See also: FDA eCTD Table of Contents Headings and Hierarchy (updated July 2005) at http://www.fda.gov/cder/Regulatory/ersr/5640CTOC-v1.2.pdf.

sponsors can capitalize upon the broad experience of FDA staff while ensuring that their product development programs meets the regulatory and scientific expectations of the Agency in the most cost- and time-efficient manner. A well-planned and well-executed vaccine development program will expedite the preparation of a complete BLA/CTD that is the final step to marketing of a new vaccine.

ACKNOWLEDGMENTS

We would like to thank Lee Buttrill, Ernest Schipman, and Alison Pienciak for their technical and administrative assistance during the preparation of this chapter and Dr. David Hines and Dr. Michael Strauss for critical review of the manuscript.

References

Capen, R., Shank-Retzlaff, M.L., Sings, H.L., Esser, M., Sattler, C., Washabaugh, M. and Sitrin, R. Establishing potency specifications for antigen vaccines. Bioprocess Int. 2007; 5:30–43.

CBER SOPP 8101.1. Scheduling and Conduct of Regulatory Review Meetings with Sponsors and Applicants, May 2007.

FDA Guidance for Industry. Content and Format of Investigational New Drug Applications (INDs) for Phase 1 Studies of Drugs, Including Well-Characterized, Therapeutic, Biotechnology-Derived Products, November 1995.

FDA Guidance for Industry for the Submission of Chemistry, Manufacturing, and Controls Information for a Therapeutic Recombinant DNA-Derived Product or a Monoclonal Antibody Product for In Vivo Use, August 1996.

FDA Guidance for Industry. Providing Clinical Evidence of Effectiveness for Human Drug and Biological Products, May 1998.

FDA Guidance for Industry. Content and Format of Chemistry, Manufacturing and Controls Information and Establishment Description Information for a Vaccine or Related Product, January 1999.

FDA Guidance for Industry. Formal Meetings with Sponsors and Applicants for PDUFA Products, February 2000.

FDA Guidance for Industry. IND Meetings for Human Drugs and Biologics. Chemistry, Manufacturing and Controls Information, May 2001.

FDA Draft Guidance for Industry. Submitting Marketing Applications According to the ICH-CTD Format—General Considerations, August 2001.

FDA Guidance for Industry. Providing Regulatory Submissions to CBER in Electronic Format—Investigational New Drug Applications (INDs), March 2002.

FDA Guidance for Industry. Fast Track Drug Development Programs—Designation, Development, and Application Review, January 2006.

FDA Guidance for Industry. Considerations for Developmental Toxicity Studies for Preventive and Therapeutic Vaccines for Infectious Disease Indications, February 2006.

FDA Guidance for Industry. Development of Preventive HIV Vaccines for Use in Pediatric Populations, May 2006.

FDA Draft Guidance for Industry. Characterization and Qualification of Cell Substrates and Other Biological Starting Materials Used in the Production of Viral Vaccines for the Prevention and Treatment of Infectious Diseases, September 2006.

FDA Guidance for Industry. Clinical Data Needed to Support the Licensure of Seasonal Inactivated Influenza Vaccines, May 2007.

FDA Guidance for Industry. Clinical Data Needed to Support the Licensure of Pandemic Influenza Vaccines, May 2007.

FDA Draft Guidance for Industry. Integrated Summaries of Effectiveness and Safety: Location Within the Common Technical Document, June 2007.

FDA Guidance for Industry. Toxicity Grading Scale for Healthy Adult and Adolescent Volunteers Enrolled in Preventive Vaccine Clinical Trials, September 2007.

FDA Guidance for Industry. Considerations for Plasmid DNA Vaccines for Infectious Disease Indications, November 2007.

FDA Guidance for Industry. Providing Regulatory Submissions in Electronic Format-Human Pharmaceutical Product Applications and Related Submissions Using the eCTD Specifications, Revision 2, June 2008.

FDA Guidance for Industry. cGMP for Phase I Investigational Drugs, July 2008.

FDA Guidance for Industry. General Principles for the Development of Vaccines to Protect Against Global Infectious Diseases, September 2008.

FDA eCTD Table of Contents Headings and Hierarchy. Updated July 2005. Accessed December 2007 at http://www.fda.gov/Cder/regulatory/ersr/ectd.htm.

Federal Register Notice. New Drug and Biological Drug Products; Evidence Needed to Demonstrate Effectiveness of New Drugs When Human Efficacy Studies are Not Ethical or Feasible; Final Rule. May 2002; FR 67:37988-37998.

Gisonni-Lex, L. Meeting Report. Cancer Vaccine Consortium—Potency Testing Workshop, San Francisco, November 8, 2004. Part I: Historical perspective of vaccine potency tests. Biologicals 2005; 33:123–128.

Goldenthal, K.L., McVittie, L.D., Kleppinger, C. and Geber, A. The clinical evaluation of preventive vaccines for infectious disease indications. In: Biologics Development: A Regulatory Overview (M. Mathieu, Ed.), 3rd ed., pp. 99–114. Waltham, MA: Parexel International Corporation, 2004.

ICH Harmonised Tripartite Guideline. Q5C. Quality of Biotechnological Products: Stability Testing of Biotechnological/Biological Products, November 1995.

ICH Harmonised Tripartite Guideline. Q5B. Quality of Biotechnological Products: Analysis of the Expression Construct in Cells Used for Production of r-DNA Derived Protein Products, November 1995.

ICH Harmonised Tripartite Guideline. E6(R1). Guideline for Good Clinical Practice, June 1996.

ICH Harmonised Tripartite Guideline. Q5D. Derivation and Characterisation of Cell Substrates Used for Production of Biotechnological/Biological Products, July 1997.

ICH Harmonised Tripartite Guideline. S6. Preclinical Safety Evaluation of Biotechnology-Derived Pharmaceuticals, July 1997.

ICH Harmonised Tripartite Guideline. Q6B. Specifications: Test Procedures and Acceptance Criteria for Biotechnological/Biological Products, March 1999.

ICH Harmonised Tripartite Guideline. Q5A(R1). Viral Safety Evaluation of Biotechnology Products Derived from Cell Lines of Human or Animal Origin, September 1999.

ICH Harmonised Tripartite Guideline. M4E(R1). Clinical Overview and Clinical Summary of Module 2; Module 5: Clinical Study Reports, September 2002.

ICH Harmonised Tripartite Guideline. M4Q(R1). Quality Overall Summary of Module 2; Module 3: Quality, September 2002.

ICH Harmonised Tripartite Guideline. M4S(R2). Nonclinical Overview and Nonclinical Summaries of Module 2; Organisation of Module 4, December 2002.

ICH Harmonised Tripartite Guideline. M4Q. Implementation Working Group Questions and Answers (R1), July 2003.

ICH Harmonised Tripartite Guideline. M4S Implementation Working Group Questions and Answers (R4), November 2003.

ICH Harmonised Tripartite Guideline. M4(R3). Organisation of the Common Technical Document for the Registration of Pharmaceuticals for Human Use, January 2004.

ICH Harmonised Tripartite Guideline. M4. Implementation Working Group Questions and Answers (R3), June 2004.

ICH Harmonised Tripartite Guideline. M4E. Implementation Working Group Questions and Answers (R4), June 2004.

ICH Harmonised Tripartite Guideline. Q5E. Comparability of Biotechnological/Biological Products Subject to Changes in their Manufacturing Process, November 2004.

Petricciani, J., Egan, W., Vicari, G., Furesz, J. and Schild, G. Potency assays for therapeutic live whole cell cancer vaccines. Biologicals 2007; 35:107–113.

Schalk, J.A.C., deVries, C.G.J.C.A. and Jongen, P.M.J.M. Potency estimation of measles, mumps and rubella trivalent vaccines with quantitative PCR infectivity assay. Biologicals 2005; 33:71–79.

United States Pharmacopeia, General Chapter <1041> Biologics.

United States Pharmacopeia, General Chapter <1045> Biotechnology Derived Articles.

World Health Organization Guidance. WHO Guidance on Nonclinical Evaluation of Vaccines, November 2003.

CHAPTER

14

U.S. Regulatory Issues Related to Vaccines against Emerging Infectious Diseases

Philip R. Krause

Division of Viral Products, Office of Vaccines Research and Review, Center for Biologics Evaluation and Research, Food and Drug Administration, Bethesda, MD, USA

OUTLINE

ABSTRACT

The primary goal of the U.S. vaccine regulatory process is to provide assurance that vaccines are consistently safe and effective. This chapter summarizes the issues and regulatory process for development and licensure of vaccines in the United States, with particular emphasis on emerging infectious diseases. As is the case for all vaccines, vaccines to prevent emerging infectious diseases must fulfill the science-based regulatory requirements for safety and efficacy, but the special nature of these infections can make vaccine development more challenging, both from scientific and regulatory standpoints. Regulatory requirements are generally organized by phase of clinical development, with emphasis on safety throughout the development process, and on efficacy and immunogenicity later in development. Specific issues associated with assays and cell substrates must be considered throughout development. Frequent reference to the underlying regulations and sources of guidance is useful to assure that important issues are addressed. Regulatory processes that may have particular relevance to emerging infectious diseases include the Orphan Drug Program, Accelerated Approval, the Animal Rule, the Fast Track Program, Priority Review, Treatment IND, and Emergency Use Authorizations.

INTRODUCTION

In the United States, vaccines are regulated as biological products by the Center for Biologics Evaluation and Research (CBER) of the Food and Drug Administration (FDA). The primary responsibility for review of vaccine applications lies within CBER's Office of Vaccines Research and Review.

The laws related to vaccine development are implemented by Title 21 of the Code of Federal Regulations, the section that includes regulations for foods and drugs. Subchapter A covers general requirements for product development, including environmental impact considerations (Part 25), human subject protection (Parts 50 and 56), and regulations governing nonclinical laboratory studies (Part 58). Subchapter C includes general requirements for all drug products, including their labeling (Part 201), advertising (Part 202), and current good manufacturing practices (cGMP, Parts 210–211). Subchapter D covers requirements for drugs for human use, including regulations for investigational new drugs (INDs, Part 312), adequate and well-controlled clinical trials (Part 314), and Orphan Drugs (Part 316). Subchapter F (Sections 600–680) covers biological products, including establishment standards, inspection, and adverse experience reporting (Part 600), licensing (Part 601), and general biological product standards (Part 610).

In addition, CBER publishes "Guidance Documents" (previously called "Points to Consider" documents) that represent current thinking on various issues related to the review of biological products (including vaccines). These include nonbinding recommendations for fulfilling regulatory requirements, and are meant to help sponsors of biological products understand CBER's interpretation of the regulations. Guidance documents with special relevance to vaccine manufacture are summarized in Table 14.1.

The International Conference on Harmonization (ICH) (International Conference on Harmonization, 2008, http://www.ich.org) also provides general guidance on manufacture and testing of pharmaceuticals, including vaccines. The ICH documents represent an international consensus regarding acceptable approaches to product manufacture, including that of vaccines. While ICH documents are published in series depending on whether they address quality, safety, efficacy, or multidisciplinary issues, the documents related to quality of biotechnology products (prefixed with a "Q5") are quite relevant to vaccine manufacture and testing. These include Q5A on viral safety evaluation, Q5B on analysis of expression constructs in cells, Q5C on stability testing, Q5D on cell substrates, and Q5E on comparability of products subject to changes in manufacturing processes.

In addition, the World Health Organization (WHO) has an "Expert Committee on Biological Standardization," which publishes international guidelines for vaccines (World Health Organization, 2008). These guidelines do not have the force of law in the United States, but do represent an international scientific and regulatory consensus regarding good practices in manufacture and testing of biological products, including vaccines.

This chapter summarizes regulatory issues impacting the development of vaccines against emerging infectious diseases, in the context of a summary of some nonregulatory vaccine development issues.

NONREGULATORY ISSUES IMPACTING U.S. DEVELOPMENT OF VACCINES AGAINST EMERGING INFECTIOUS DISEASES

Emerging infectious diseases present special problems with regard to vaccine development. Although regulatory issues are sometimes cited as obstacles to rapid product development (Stern and Markel, 2005), and it does take some time to fulfill regulatory requirements to demonstrate the safety and efficacy of a candidate vaccine, the major obstacles to development of vaccines against emerging infectious diseases have generally not been regulatory in nature. There are scientific challenges associated with vaccine development, and these are often greater for vaccines against emerging infectious diseases. These scientific challenges may translate into economic challenges.

TABLE 14.1 CBER guidance documents with special relevance to vaccine manufacture

Guidances related to clinical studies
Guidance for Industry: Toxicity Grading Scale for Healthy Adult and Adolescent Volunteers Enrolled in Preventive Vaccine Clinical Trials
Guidance for Industry: Clinical Data Needed to Support the Licensure of Pandemic Influenza Vaccines
Guidance for Industry: Clinical Data Needed to Support the Licensure of Seasonal Inactivated Influenza Vaccines
Guidance for Industry: Development of Preventive HIV Vaccines for Use in Pediatric Populations
Draft Guidance for Industry: Postmarketing Safety Reporting for Human Drug and Biological Products Including Vaccines
Guidance for Industry: How to Complete the Vaccine Adverse Reporting System Form (VAERS-1)
Guidance for Industry: Establishing Pregnancy Exposure Registries
Guidance for Industry: Financial Disclosure by Clinical Investigators
Guidances related to product testing and manufacture
Guidance for Industry: Considerations for Plasmid DNA Vaccines for Infectious Disease Indications
Revision of the Requirements for Live Vaccine Processing: Direct Final Rule
Revision of the Requirements for Live Vaccine Processing: Companion to Direct Final Rule: Proposed Rule
Draft Guidance for Industry: Characterization and Qualification of Cell Substrates and Other Biological Starting Materials Used in the Production of Viral Vaccines for the Prevention and Treatment of Infectious Diseases
Guidance for Industry: Considerations for Developmental Toxicity Studies for Preventive and Therapeutic Vaccines for Infectious Disease Indications
Draft Guidance for Industry: Considerations for Reproductive Toxicity Studies for Preventive Vaccines for Infectious Disease Indications
Draft Guidance for Industry: INDs—Approaches to Complying with CGMP During Phase 1
Guidance for Industry for the Evaluation of Combination Vaccines for Preventable Diseases: Production, Testing and Clinical Studies
Draft Points to Consider in the Characterization of Cell Lines Used to Produce Biologicals
Guidance for Industry: Nonclinical Studies for the Safety Evaluation of Pharmaceutical Excipients
Draft Guidance for Industry: Manufacturing Biological Drug Substances, Intermediates, or Products Using Spore-Forming Microorganisms
Guidances related to product labeling
Draft Guidance for Industry: Patient-Reported Outcome Measures: Use in Medical Product Development to Support Labeling Claims
Guidance for Industry: Adverse Reactions Section of Labeling for Human Prescription Drug and Biological Products—Content and Format
Guidance for Industry: Clinical Studies Section of Labeling for Human Prescription Drug and Biological Products—Content and Format
Draft Guidance for Industry: Dosage and Administration Section of Labeling for Human Prescription Drug and Biological Products—Content and Format
Draft Guidance for Industry: Warnings and Precautions, Contraindications, and Boxed Warning Sections of Labeling for Human Prescription Drug and Biological Products—Content and Format
Draft Guidance for Industry: Labeling for Human Prescription Drug and Biological Products—Implementing the New Content and Format Requirements
Draft Guidance for Industry: Warnings and Precautions, Contraindications, and Boxed Warning Sections of Labeling for Human Prescription Drug and Biological Products—Content and Format
Draft Guidance for Industry: Labeling for Human Prescription Drug and Biological Products—Implementing the New Content and Format Requirements
Guidance for Industry: FDA Review of Vaccine Labeling Requirements for Warnings, Use Instructions, and Precautionary Information
Guidance regarding special regulatory situations
Guidance for Industry: Fast Track Drug Development Programs—Designation, Development, and Application Review
Guidance for Industry: Emergency Use Authorization of Medical Products
Guidance for Industry: Special Protocol Assessment

Note: All documents are available online at http://www.fda.gov/cber/guidelines.htm/.

Economic factors influence the development of all vaccines, and may be more likely to influence vaccines against emerging infectious diseases than other products. In recognition of the special considerations associated with emerging infectious diseases, programs and pathways have been put in place to address these obstacles.

Novel vaccines, especially those against emerging infectious diseases, are more likely to use newer technologies. Developing assurance that these new technologies lead to safe and effective vaccines requires a deep level of scientific understanding. This review covers regulatory assessment of some of these scientific issues—including the need to address issues associated with using new types of cells to produce viral vaccines (see "Other Nonclinical Issues That Affect All Stages of Development"), the need to develop potency assays that predict vaccine safety and efficacy (see "Other Nonclinical Issues That Affect All Stages of Development"), and the need to develop "surrogate markers" in humans or animals that predict vaccine efficacy (see "Phase III"). However, in many cases, the development of the underlying science is more a research than a regulatory question. Other scientific obstacles may affect vaccine development at an even earlier level—for example, antigens in combination with vaccines may interfere with one another, related vaccines could theoretically cross-react, and new adjuvants may be necessary to develop vaccines that effectively prevent some diseases (but stronger adjuvants may have increased risk of reactogenicity). In addition, it is always a greater challenge to develop vaccines in the United States against diseases that do not affect people in the United States, whether they are global diseases or emerging infectious diseases.

In many cases, the scope of the threat from an emerging infectious disease is unknown at the time when vaccine development would optimally occur. Some emerging infectious diseases may produce worldwide pandemics, as occurred with HIV, but the scope of this threat would have been difficult to predict in the early 1980s, as AIDS was being defined as a clinical syndrome. Nonetheless, it is widely accepted that there would be a robust U.S. and global market even for a partially effective HIV vaccine. However, SARS initially evoked worldwide concern, leading to rapid deployment of public health countermeasures including quarantine of infected and exposed individuals, and was relatively rapidly contained with the last reported human case in 2004. At the present time, the level of market demand for a potential SARS vaccine is unclear. Given this market uncertainty, it is not difficult to understand the reluctance of some pharmaceutical companies to invest in the development of vaccines for emerging infectious diseases.

Reports from the Institute of Medicine (IOM) repeatedly highlighted the importance of market issues in addressing vaccine availability. A 1992 IOM report highlighted the unfavorable ratio between profits and risk of failure, and the risks of potential liability claims for injuries or deaths attributed to vaccine administration (Lederberg et al., 1992). This report suggested a need for additional economic incentives to develop vaccines, stockpiling of selected vaccines, and the development of "surge" vaccine production capabilities.

Liability concerns are also a potential obstacle to development of vaccines against emerging infectious diseases. While these liability concerns may in part be addressed by the National Vaccine Injury Compensation Program [reviewed in Evans (2006)], which provides compensation to individuals injured by covered vaccines and provides liability protection for industry, it may not initially be clear whether a vaccine under development will ultimately qualify for this program, especially if it is directed at an emerging infectious disease.

Although it is nearly universally accepted that vaccines against potential pandemic influenza strains would be widely used once strains with the ability to be efficiently transmitted from human to human are isolated in the field, it is unclear what the full antigenic composition of those strains will be. Any vaccine development that occurs too far in advance of the emergence of the actual strains that the vaccine is intended to immunize against runs the risk of suboptimal immunogenicity (Smith, 2006). Moreover, the lead times necessary to develop such vaccines may permit some antigenic drift, leading to lower efficacy than desired, even in the context of routine influenza immunization. The uncertainty of the timing of the event that would trigger the market for pandemic influenza vaccines makes it difficult to decide on the ideal timing for vaccine production. This uncertainty, in turn, makes it unlikely that private market forces alone would lead to the widespread availability of pandemic influenza vaccines (Stohr and Esveld, 2004). Recent legislation has made government funds available to support development of the next generation of influenza vaccines, with an emphasis on pandemic influenza issues (Department of Health and Human Services, 2005, http://www.hhs.gov/pandemicflu/plan/). Heightened concern about pandemic influenza has led to substantial government activity in this area, including the development of a pandemic influenza plan.

In the United States, the federal government, mostly through the National Institutes of Health (NIH) (Fauci, 2005), the FDA (Zoon, 1999), the Centers for Disease Control and Prevention (CDC), and the National Vaccine Program Office (NVPO) has played a pivotal role in preparing the nation's public health infrastructure for the threat of emerging infectious diseases. By funding research in these areas, the government stimulates the development of scientifically valid approaches to developing vaccines in this critical public health arena. Should a larger need (and an accompanying larger market) arise, a basis for more rapid development and manufacture of these vaccines already exists.

Recent concerns about bioterrorism have increased the likelihood that vaccines against some emerging infections will find a market in the United States. When new infectious diseases emerge in the United States, it may not always be immediately clear whether or not their emergence resulted from the deliberate introduction of a pathogen. The Project Bioshield legislation of 2004 established a secure funding source for purchase and stockpiling of vaccines related to national security, increased the authority and flexibility of NIH to expedite national security-related biological research, and established an Emergency Use Authorization (described in more detail in "Project Bioshield and

Emergency Use Authorizations") for critical national security-related biological countermeasures. While the emphasis of Project Bioshield is on national security and bioterrorism, vaccines against etiologic agents of emerging infectious diseases that overlap with those targeted by Project Bioshield may find markets in the United States as a result of its government sponsorship of vaccine purchases.

Nonetheless, the additional obstacles to development of vaccines against emerging infectious diseases are superimposed on those associated with more routine vaccine development. Even routine vaccines may not be as profitable as other pharmaceutical products. The incentives and disincentives associated with the development of vaccines against emerging infectious diseases will play an important role in their ultimate availability in the United States.

OVERVIEW OF THE U.S. VACCINE REGULATORY REVIEW PROCESS

The primary purpose of the regulatory review process is to assure that marketed products are safe and effective. The regulatory process also plays an important role in the protection of human subjects, helping to assure the relative safety of investigational products and the appropriateness of clinical studies. To accomplish this, the manufacturer must demonstrate safety, purity, potency, efficacy, manufacturing reproducibility, and compliance with cGMP. Legally binding regulations relevant to vaccines are published in the Federal Register, and are summarized each year in the Code of Federal Regulations. The U.S. regulatory review process is described in detail in the underlying regulations and in other publications (Baylor and Midthun, 2004).

In the United States, vaccines are evaluated for possible licensure by the FDA. The Advisory Committee on Immunization Practices provides official guidelines for the use of licensed vaccines (Centers for Disease Control and Prevention, 2008, http://www.cdc.gov/vaccines/recs/nip/default.htm). The Centers for Disease Control and Prevention (CDC) also plays an important role in studying the epidemiology of vaccine-preventable diseases, and in coordinating the responses of state and local public health authorities.

Vaccines differ substantially from other tools in the public health armamentarium. Vaccines are among the most effective public health measures ever devised, having successfully been used to eradicate natural smallpox from the globe and natural poliovirus infections from the Western hemisphere. The morbidity in the United States associated with every vaccine-preventable disease has decreased dramatically with the introduction of vaccines (even against those agents that have not been eradicated), including measles, mumps, rubella, pertussis, Haemophilus influenzae type B meningitis, meningococcus, pneumococcal infections, varicella, and influenza. Some of the very high effectiveness of vaccines is attributable to herd immunity, when vaccination rates are sufficiently high, interruption of transmission among the vaccinated prevents infection even in the unvaccinated, who are less likely to be exposed to the pathogen (Anderson and May, 1991).

However, vaccines designed to prevent infectious diseases are generally administered to healthy individuals, in many cases children—thus, in spite of their success, there is low public tolerance for safety concerns with vaccines (Plotkin, 1991). Moreover, vaccines are unusual among FDA-regulated products because many states mandate universal vaccination of children prior to attendance at day care or school—thus, except when declined for religious reasons, vaccination is not a matter of choice for most U.S. citizens.

These circumstances lead to a need for vaccines to be very safe products (Chen et al., 2001). Maintenance of herd immunity depends on high vaccination rates, which in turn depend on high levels of public confidence in the safety of vaccines. Thus, the incredible success of vaccines depend largely on the public's confidence that approved products have been thoroughly reviewed by the FDA, and that the best, most current scientific principles have been applied to that review.

Over time, as concerns regarding rare or even theoretical safety issues with vaccines have evolved, manufacturers and the U.S. government have acted to assure that vaccines are as safe as possible. For example, although very effective, oral polio vaccine is no longer used in the United States, due mostly to an approximately one in a million chance of vaccine strain reversion that can cause vaccine-associated paralytic polio in vaccine recipients. When polio was widespread, this risk was considered a reasonable trade-off for the benefits of the vaccine, but this clearly is no longer the case. The U.S. government authorities moved quickly to reduce the number of thimerosal-containing vaccines, when concern was raised regarding the possibility that ethyl mercury in thimerosal could cause adverse consequences in vaccine recipients, even in the absence of clear scientific evidence of harm (Center for Biologics Evaluation and Research, 2008, http://www.fda.gov/Cber/vaccine/thimerosal.htm).

The thoroughness of the FDA's vaccine regulatory review process thus provides the public with a high degree of confidence that vaccines are safe. While regulatory requirements may occasionally be seen by some as burdensome, all regulatory decisions regarding vaccines take into account the best available scientific practices. The direct engagement of active laboratory-based researchers at CBER in the vaccine review process helps to assure that all decisions have a scientific foundation. When CBER scientists identify areas in which more data are needed in order to support regulatory decision making or to simplify review, intramural funding of CBER research allows these data to be objectively gathered (Korn, 1997; FDA Science Board Subcommittee on Science and Technology, 2007, http://www.fda.gov/ohrms/dockets/ac/07/briefing/2007-4329b_02_01_FDA%20Report%20on%20Science%20and%20Technology.pdf). By providing its objective pre- and post-marketing review of vaccines, the FDA thus plays an important role in assuring not only that vaccines are safe, but also that the public's confidence in their safety is justified.

In the following discussion of the U.S. regulatory process, reference is made to regulations and to guidance documents. As scientific advances are made, and as new needs are identified, these references may be updated, new laws may be passed, and new guidance documents may be issued. Updates on available CBER-issued guidance documents are available on the Internet (Center for Biologics Evaluation and Research, 2008, http://www.fda.gov/Cber/vaccine/thimerosal.htm). Table 14.1 lists current (as of January 2008) CBER guidance relevant to vaccines.

Pre-IND Vaccine Development

Prior to filing of an IND application, sufficient data to support a first use of a product in humans must be gathered. While FDA review focuses heavily on safety at this stage of development, the expense of thoroughly evaluating the safety of a potential vaccine may not be considered justified by vaccine sponsors unless preclinical data also support the likely efficacy of the product in humans. Prior to filing an IND, it is also necessary to develop a plan and the necessary facilities to manufacture a vaccine that meets standards for use in human clinical trials. General principles on nonclinical evaluation of vaccines have been promulgated by the WHO (World Health Organization, 2005).

It is a general principle of drug development that new drugs should be shown to be safe in appropriate animal models before they are first tested in humans. Appropriate toxicology tests, including animal autopsies and microscopic examination of organs, can provide assurance that vaccine products do not contain toxic proteins or other components. The general goals of toxicology tests are to determine whether it is reasonable to test the product in humans via a given route and schedule of administration, to determine appropriate starting doses, and to provide guidance regarding clinical parameters that should be more closely monitored during the human safety studies. Although FDA has not issued official guidance on toxicology studies for vaccines, general recommendations coauthored by FDA toxicology reviewers are available (Chang et al., 2003). These recommendations suggest administration of the product under good laboratory practices (GLP) conditions to a minimum of 20 animals of a relevant species (10 males, 10 females) via the intended route and schedule (plus one additional dose) at a minimum of the (non-mass adjusted) human dose, with intervening physical examination, clinical observations, and laboratory tests, and with animal sacrifices and thorough histopathology examinations of half the animals at 1–3 days after the last inoculation and half at 21 days after the final inoculation. These recommendations are consistent with those issued by the WHO on preclinical testing of vaccines (World Health Organization, 2005).

Potential virulence of the vaccine strain is a major safety issue for live attenuated vaccines. To assure that a vaccine will be safe in humans, there should be a strong scientific basis for believing that any infection with the vaccine strain will be mild, generally milder than that caused by the wild-type pathogen. This issue may be addressed in appropriate animal models, when such models are available and can be scientifically justified. For attenuated viruses, the genetic stability of the organism, including demonstration that the virus does not revert to a more virulent phenotype, may be particularly important.

For viruses that have the potential to be neurotropic (usually based on an understanding of natural disease with wild-type virus, and/or on preclinical toxicology studies), neurovirulence is another potential safety issue. Neurovirulence testing originally was developed by Dr. Albert Sabin to permit evaluation of live-attenuated polio vaccine. In the context of this vaccine, Dr. Sabin demonstrated that the inoculation of monkey brains made it possible to identify virus strains that had reverted to a phenotype making them more likely to cause vaccine-associated paralytic polio in humans (Sabin, 1985). Although neurovirulence testing was performed on a lot-by-lot basis for oral poliovirus vaccine, this issue has been addressed for most live neurotropic virus-based vaccines by showing that vaccine strains or master seed viruses have been sufficiently attenuated in appropriate animal models.

As new vaccine technologies are developed, FDA may provide guidance documents regarding the appropriate evaluation of these technologies. For example, a guidance document on DNA vaccines provides specific information regarding appropriate in vitro and in vivo studies to support the use of these vaccines in clinical trials (Center for Biologics Evaluation and Research, Guidance for Industry: Considerations for Plasmid DNA Vaccines for Infectious Disease Indications, 2007).

Although pregnant women are not the primary target for most vaccines, the potential for vaccination of pregnant women must be considered. FDA has issued specific guidance on the topic of reproductive toxicology studies for vaccines indicated for women of reproductive age (Center for Biologics Evaluation and Research, 2000, http://www.fda.gov/cber/gdlns/reprotox.pdf). Unless the vaccine is specifically intended for use in pregnant women, reproductive toxicology studies may be performed in parallel with clinical development, as long as the clinical studies include appropriate exclusion criteria. By the time clinical studies that could include pregnant women are initiated, these studies should be completed. In some cases, the creation of a pregnancy registry allows the compilation of additional data post-licensure (Food and Drug Administration, 2002, http://www.fda.gov/cber/gdlns/pregexp.pdf).

Pre-Phase I testing should also address the safety of any unusual components or novel adjuvants that a vaccine contains. When there is prior human experience with the same substance in the same or a different context, this information can be helpful.

Cell substrate issues and the setting of specifications are relevant to all stages of product development, and are discussed in "Other Nonclinical Issues that Affect All Stages of Development."

The Phases of an Investigational New Drug (IND) Investigation for Vaccines

The conduct of studies under IND is governed by regulations (21 CFR 312 Investigational New Drug Application, 2007). Prior to licensure, clinical studies are designated as Phase I, Phase II, or Phase III. Phase I studies represent the first human uses of the product, and focus on sufficient demonstration of safety to support expanded Phase II studies. In Phase II, additional data (focusing both on safety and immunogenicity) generally are obtained for proof of concept to support the performance of a Phase III efficacy study. The bulk of data to support licensure, though, come from Phase III studies.

Phase I

The primary focus of Phase I clinical studies is evaluation of safety. Phase I clinical studies generally represent the first human use of a candidate vaccine. Thus, these studies often are performed with dose escalation, starting with lower and increasing toward higher doses. Because of the focus on safety, there is generally no need for Phase I studies to be placebo controlled.

Although for many drugs, one purpose of Phase I studies is to obtain information regarding the maximum tolerated dose, vaccine candidates are ideally sufficiently safe so that a maximum tolerated dose is not reached in a Phase I study. Thus, maximum doses in vaccine Phase I studies are often chosen based on practical considerations, such as manufacturing capabilities, as well as the need to assure that the safety of doses required for optimal immunogenicity has been adequately studied. Doses used in Phase I studies must be justified by appropriate preclinical studies.

An original submission IND is required to adhere to the content and format specified in the relevant regulations (21 CFR 312.23 IND content and format, 2007). The IND should include all the results from the pre-IND phase of testing, together with all the information needed to evaluate the proposed Phase I clinical trial. This includes information about the specific lots of vaccine to be used and their testing, including proper identification, quality, purity, strength, and source materials used. The description of the clinical trial should include information about the study design, clearly defined eligibility (and exclusion) criteria, and information regarding the safety follow-up of vaccinees. Information regarding the qualifications of the investigators and adherence to the regulations pertaining to Institutional Review Boards (IRBs) is also required. Unless the sponsor is also the investigator, a copy of the investigator brochure also must be included (21 CFR 312.55 Informing Investigators, 2007).

Because the major focus of Phase I studies is on safety, the safety criteria utilized either for dose escalation or for suspension of the study receive particular scrutiny. These criteria will, of course, depend on the inclusion criteria for the study. However, it is expected that most vaccines should be sufficiently safe that they do not induce major laboratory abnormalities, and if such abnormalities occur during Phase I studies, there should be a plan to investigate them. Safety of individuals not in the study also may be an issue. For example, it may be necessary to defer vaccine recipients from blood donation or to restrict the receipt of certain live vaccines to individuals who are unlikely to be exposed to immunocompromised individuals.

There are regulatory requirements (21 CFR 56.108 IRB functions and operations, 2007; 21 CFR 312.32 IND safety reports, 2007) to report certain serious adverse events in a clinical trial to the IRB and FDA, within specified periods of time. Toxicity grading scales used in Phase I studies should reflect the population being immunized and expected adverse reactions. Adverse events are often graded on a 4-point scale, ranging from grade I (mild) to grade IV (severe). In general, any abnormal laboratory value is reported as at least a grade I (mild) toxicity. Prospective definition of the potential toxicities is important, because continuation of the clinical trial generally depends on the absence of significant toxicities. In some cases, specific toxicities may influence dose escalation, while in others, they may be tied to "stopping rules," that lead to an evaluation of whether the study should be continued or terminated. A CBER guidance document on toxicity scales for normal subjects in vaccine trials was recently issued (Center for Biologics Evaluation and Research, 2007, http://www.fda.gov/cber/gdlns/toxvac.pdf). In blinded studies (generally after Phase I), an independent data safety monitoring board may examine adverse reactions to determine whether a study should continue (International Conference on Harmonization, 1996, http://www.ich.org/LOB/media/MEDIA482.pdf).

Once an original submission IND, including a Phase I study, is received by the FDA, CBER will make a determination within 30 days regarding whether this study may proceed, or will be placed on "clinical hold" (21 CFR 312.42 Clinical Holds and Requests for Modification, 2007). The placement of a Phase I study on "clinical hold" generally indicates that important safety issues, best addressed during the pre-IND phase of development, have not yet been resolved sufficiently to assure that the product is safe enough to use in a clinical study. Other reasons for a Phase I clinical hold include insufficient qualifications of clinical investigators, inadequacy of the investigator's brochure, insufficient information to assess risk, or exclusion of individuals who may benefit on the grounds of potential reproductive or developmental toxicity.

Phase II

According to the IND regulations (21 CFR 312.21 Phases of an investigation, 2007), Phase II studies include controlled clinical studies conducted to evaluate the effectiveness of the drug for a particular indication in patients with the disease under study, and to determine common short-term side effects. These studies are typically randomized, well controlled, closely monitored, and include no more than several hundred subjects. Phase II studies must be supported by safety data from appropriately designed Phase I studies, and the general considerations for Phase I studies generally also apply.

Vaccine Phase II studies thus tend to focus on immunogenicity and expansion of the safety database. These studies are generally designed to provide the information necessary to support a Phase III efficacy study, often including sufficient proof of concept for the manufacturer to justify progression to Phase III. This may include final dose selection. Phase II studies also support assay development and the development of surrogate markers that may be helpful in evaluating the results of clinical trials. Phase II study populations are generally chosen to predict responses in larger Phase III studies intended to lead to licensure. Phase II or Phase III studies may be placed on clinical hold for "design issues" if the protocol design is considered inadequate to meet the objectives of the study. These studies may also be placed on clinical hold for the same reasons as Phase I studies.

Phase III

Phase III studies are performed to determine efficacy of the product. Phase III also generally contributes the bulk of the safety data in support of licensure. Manufacturing reproducibility is normally also addressed during Phase III, through the production and testing of three consistency lots.

Although two clinical trials are considered the standard for licensure, one well-designed study can be adequate if the result is compelling, (e.g., based on robust data from a multicenter trial) (Food and Drug Administration, 1998, http://www.fda.gov/cber/gdlns/clineff.pdf).

The choice of endpoint of Phase III studies deserves special consideration in the context of emerging infectious diseases. The most commonly used endpoint for Phase III studies is efficacy in a randomized, double-blind, placebo-controlled field efficacy study. However, in some cases, the prevalence of the disease is too low to easily perform field efficacy studies. In other cases, the time required for severe complications of disease to develop may be long, thus making true efficacy studies more difficult to perform.

Where the disease is not life threatening and alternative treatments are available, it may sometimes be possible to perform challenge studies to demonstrate vaccine efficacy. For example, a challenge study was used to evaluate efficacy of live attenuated influenza vaccine (Treanor et al., 1999). In such challenge studies, the ability of the vaccine product to prevent disease may be evaluated against disease rates in an appropriate comparison group. However, challenge studies (whether placebo-controlled or not) can sometimes be unethical, especially when the disease is severe, the spectrum of disease is not well understood, or treatments are not available.

In some cases, scientifically justified surrogate markers that predict efficacy may be used as endpoints in Phase III studies. For example, immunological surrogates were used to license vaccines against hepatitis B, Hemophilus influenzae type B, and meningococcus. Every year, vaccines against new strains of influenza virus are licensed based primarily on immunological data showing antigenic relatedness between the vaccine and circulating influenza strains. Recently performed studies of human papillomavirus vaccine used prevention of infection with oncogenic serotypes of virus, together with prevention of pre-cancerous lesions, as a predictor of the vaccine's ability to prevent cancer (Koutsky et al., 2002). Considerable discussion has occurred regarding the development of appropriate surrogate markers for HIV vaccines, especially therapeutic vaccines, for which parameters associated with immune system function may well provide reasonable surrogate endpoints (Nabel, 2002). When surrogate markers are not well established, licensure may be pursued under the "Accelerated Approval" rules (see "Accelerated Approval").

Even if surrogate markers are not a part of the initial licensure, development of surrogate markers for protection against disease is critical both in vaccine development and as endpoints in subsequent comparability studies, as may be done to support post-licensure vaccine manufacturing changes or concomitant administration studies. Surrogate markers, which usually are validated in Phase III, provide clinical trial endpoints that can be used to predict outcomes without requiring actual efficacy trials to be done. Without such surrogate markers of efficacy it becomes very difficult to maintain an understanding of the product over time.

There is no requirement that efficacy studies be performed in the United States. Thus, it may be possible to design foreign vaccine efficacy studies for diseases that are not sufficiently prevalent in the United States to study. For example, substantial efficacy data for hepatitis A, Japanese encephalitis, and pertussis vaccines were obtained overseas. When the bulk of vaccine development has occurred overseas, the FDA carefully examines the data to assure that sufficient information exists to assure vaccine safety and efficacy in domestic U.S. populations, including explicit consideration of ethnic factors. Such information may be obtained by performing bridging studies in U.S. populations.

Guidance documents addressing the evaluation of vaccines in special populations or for specific indications are also sometimes issued. Guidances on clinical studies were recently made available by CBER for evaluating new influenza vaccines (Center for Biologics Evaluation and Research, 2006, http://www.fda.gov/Cber/gdlns/trifluvac.htm; Center for Biologics Evaluation and Research, 2007, http://www.fda.gov/Cber/gdlns/trifluvac.pdf). An additional guidance on HIV vaccine trials for pediatric populations was also recently issued (Center for Biologics Evaluation and Research, 2006, http://www.fda.gov/Cber/gdlns/pedhiv.pdf). A recently passed law requires the evaluation of products in pediatric populations, when there is a likelihood that they may benefit children (FDA Amendments Act of 2007, 2007). Later-phase vaccine studies may need to be reported to the Clinical Trials Databank (Food and Drug Administration, 2004; FDA Amendments Act of 2007, 2007).

It is not always possible to design Phase III studies that meet these criteria. Under those circumstances, the Animal Rule may be used to show efficacy (see "Animal Rule").

Meetings with FDA during Product Development

FDA generally encourages and values effective communication between sponsors and reviewers. FDA does not generally recommend piecemeal discussion of issues associated with a product, but generally prefers to consider as much information as possible, in order to evaluate all aspects of the development plan in context. However, meetings may separately focus on chemistry, manufacturing and control issues (Food and Drug Administration, 2002, http://www.fda.gov/cber/gdlns/pregexp.pdf), or on clinical issues. Even when meetings do not take place, there may be significant back-and-forth discussion between a sponsor and FDA, in the form of written communication in the context of the IND.

Meetings between the FDA and sponsors often occur at the pre-IND stage, the end of Phase I (for products that have been designated as "fast track"), end of Phase II, and pre-Biologics License Application (BLA) filing. Specific regulations govern the end of Phase II and pre-BLA meetings (21 CFR 312.47 Meetings, 2007). Additional guidance exists on meetings for products covered under the Prescription Drug User Fee Act (Food and Drug Administration, 2000, http://www.fda.gov/cber/gdlns/mtpdufa.pdf).

Biologics License Application (BLA)

Biological products, including vaccines for human use, are licensed under the authority of Section 351 of the Public Health Service Act (42 USC. 262). The

Code of Federal Regulations specifies general requirements for biological products (21 CFR 600 Biological Products, 2007), and also summarizes licensing regulations (21 CFR 601 Licensing, 2007). Under these provisions, licensure requires the demonstration of safety, purity, and potency. Safety is defined as "the relative freedom from harmful effect to persons affected, directly or indirectly, by a product, when prudently administered, taking into consideration the character of the product in relation to the condition of the recipient at the time." Purity is defined as "relative freedom from extraneous matter in the finished product, whether or not harmful to the recipient or deleterious to the product. Purity includes but is not limited to relative freedom from residual moisture or other volatile substances and pyrogenic substances." Additional requirements regarding purity are specified by regulation (21 CFR 610.13, Purity, in Code of Federal Regulations, 2007). Potency "is interpreted to mean the specific ability or capacity of the product, as indicated by appropriate laboratory tests or by adequately controlled clinical data obtained through the administration of the product in the manner intended, to effect a given result." It is necessary for the product to have sufficient stability to maintain these attributes throughout the dating period. Vaccines must also adhere to the "General Biological Products Standards" (21 CFR 610 General Biological Products Standards, 2007).

BLAs are accepted in electronic Common Technical Document form, which allows manufacturers to file harmonized applications in multiple countries (Food and Drug Administration, 2005, http://www.fda.gov/cber/gdlns/esubapp.htm). License applications generally include all information required to evaluate the product, and thus stand alone, without needing to rely on reference to previously submitted materials.

During the licensure evaluation process, the conduct of the IND clinical studies may be closely examined in a process called "bioresearch monitoring," which involves inspection of IND sites to assure appropriate conduct of studies and record keeping. In addition, particular scrutiny is devoted to the label claims (including indication) made by the sponsor (and the regulations regarding labeling were revised in 2006) (21 CFR 610, subpart G: Labeling Standards, 2007; 21 CFR 201, Labeling, 2007). The manufacturing establishment (21 CFR 600, Subpart B Establishment Standards, 2007) and ability to adhere to current good manufacturing processes (21 CFR 210 and 211 Current Good Manufacturing Practice in Manufacturing, Packing, or Holding of Drugs, 2007) are also closely examined, including a pre-licensure inspection of the facilities.

In evaluation of a BLA, FDA will often consult with the Vaccines and Related Biological Products Advisory Committee (VRBPAC) [see "Vaccines and Related Biological Products Advisory Committee (VRBPAC)"].

Post-Marketing Phase

Under conditions of real-world use, the manufacturer is required to obtain and report information regarding the safety of licensed products (21 CFR 600, Subpart D Reporting of Adverse Experiences, 2007; Food and Drug Administration, 2001, http://www.fda.gov/cber/gdlns/safety031201.pdf).

Additional post-marketing safety surveillance is performed through the Vaccine Adverse Events Reporting System (VAERS), which is maintained by the CDC and the FDA to collect patient- and physician-initiated reports related to vaccine safety (Iskander et al., 2004). Data from VAERs are periodically analyzed to identify potential vaccine safety signals that merit further follow-up.

Recent legislation increased the ability of the FDA to require post-marketing studies (FDA Amendments Act of 2007, 2007). Even prior to this legislation, at the time of licensure, many vaccine manufacturers committed to perform additional post-licensure studies. Some of these studies may be in the context of Accelerated Approvals or Animal Rule approvals, while others may be related to obtaining more detailed information about the safety or duration of efficacy of a vaccine. For example, the manufacturer of a rotavirus vaccine agreed to perform additional studies of intussusceptions at the time of licensure, and the manufacturer of varicella vaccine agreed to perform long-term studies to better understand the duration of vaccine efficacy (Krause and Klinman, 1995), to also allow a better understanding of product performance once wild-type boosting of vaccine responses became less prevalent. Annual reports of post-marketing studies are required (21 CFR 601.70 Annual Progress Reports of Post-marketing Studies, 2007).

Once a product is licensed, changes to the indication or method of manufacture may be submitted as supplements to a BLA (21 CFR 601.12 Changes to an approved application, 2007), which (depending on the scope of the changes) often require FDA approval prior to implementaion. The FDA also performs inspections of the facilities used for the manufacture of licensed products (21 CFR 600, Subpart C Establishment Inspection, 2007). The FDA also examines test results on vaccine lots prior to marketing, and may also perform its own testing, providing additional assurance of product performance.

Other Nonclinical Issues that Affect All Stages of Development

Chemistry, manufacturing, and control issues that need to be addressed in order to obtain licensure are summarized in a guidance document (Center for Biologics Evaluation and Research, 1999, http://www.fda.gov/Cber/gdlns/cmcvacc.pdf). In this section, a summary of cell substrate issues and setting of specifications is provided.

Cell Substrate Issues

Because emerging infectious diseases may be associated with new viruses, the possibility exists that the vaccine may require novel cell substrates for optimal growth and manufacturing efficiency. This may be true for vectored vaccines, for which a replication-defective or replication-deficient virus will grow only in specific cell types that provide a missing growth factor, and may also be true for live-attenuated or inactivated vaccines.

Guidance documents particularly relevant to cell substrate issues include the Draft Guidance for Industry: Characterization and Qualification of Cell Substrates and Other Biological Starting Materials Used in the Production of Viral Vaccines for the Prevention and Treatment of Infectious Diseases (Center for Biologics Evaluation and Research, 2006, http://www.fda.gov/cber/gdlns/vaccsubstrates.htm) and ICH document Q5A (International Conference on Harmonization, 1999, http://www.ich.org/LOB/media/MEDIA425.pdf).

Introduction of new cell substrates for vaccine manufacture, especially for live-attenuated vaccines, is a complex matter. The earliest viral vaccines were produced in primary cells obtained directly from animals, such as calf lymph used to produce smallpox vaccines, or monkey kidney cells used to produce polio vaccines. In the 1950s, vaccine experts believed that production of vaccines in nonhuman, normal cells would be the best way to assure vaccine safety and freedom from the theoretical risk of oncogenicity associated with cell substrate components that would inevitably be a part of the vaccine. When SV40 was discovered as a contaminant both of Rhesus monkey kidney cells and of early poliovirus and adenovirus vaccines, the wisdom of this approach was questioned. Although poliovirus vaccine strains were treated with antibody to remove SV40, and production was changed to African Green monkey kidney cells (in which SV40 contamination, if it occurs, is more readily detected), reports that SV40 DNA could be detected by polymerase chain reaction in samples from individuals with various human cancers led to renewed concern over this historical contamination event. Although an IOM report concluded that there was insufficient evidence to accept or reject a hypothesis of causality between SV40-containing polio vaccines and cancer (Stratton et al., 2002), the fact that these concerns developed shows the high degree of importance of assuring that vaccine cell substrates are free of adventitious agents and are safe to use for vaccine production.

Over time, new vaccines, such as the rubella vaccine and the varicella vaccine, were produced using normal human diploid cells. The ability to readily bank these cells allows the performance of one set of exhaustive tests on a master cell bank to provide a high degree of assurance that adventitious infectious agents are not present. Other vaccines continue to be produced in primary cells, including eggs and other avian-derived cells. A recently approved rotavirus vaccine is produced in continuous Vero cells. Some investigational vaccines are produced in weakly tumorigenic cells, under defined circumstances that quantify the theoretical risk of oncogenic events, with limitations on the total quantity of cell substrate DNA and with thorough testing to assure that there is no reasonable chance that residual components of the cell substrate could be oncogenic.

The need to pay attention to the cell substrate used for vaccine production extends to a need to pay attention to the cells and the conditions used for development of a vaccine seed. If a vaccine seed has been passaged in poorly characterized cells, or may have been exposed to untested biological reagents (such as serum or trypsin), the potential for contamination with viruses that might infect the cells or biological reagents exists. Thus, even very early in vaccine development, it is critical to document sources and lot numbers of such reagents, and if complete testing of cell substrates is not feasible, retention of cell substrate samples to allow later testing could be very useful. Using carefully devised protocols, vaccine seeds that have been exposed to unknown cells or reagents may sometimes be purified (e.g., by serial plaque purification). However, the recommended approach is to consider and address cell substrate issues as early in vaccine development as possible.

To assure that vaccines are free of agents of transmissible spongiform encephalopathy (TSE), it is important to consider the sourcing of serum and other animal or human-derived vaccine materials used in vaccine manufacture. CBER's current thinking on this topic is made available on the internet (Center for Biologics Evaluation and Research, 2007, http://www.fda.gov/cber/bse/bse.htm)

Summaries of recent international meetings further address issues associated with introduction of new cell substrates (Sheets and Petricciani, 2006).

Specifications, Potency, and Stability

Specifications for biological products are defined (21 CFR 600.10 Personnel, 2007) as "the quality standards (i.e., tests, analytical procedures, and acceptance criteria) provided in an approved application to confirm the quality of products, intermediates, raw process materials, container closure systems, and other materials used in the production of a product." Adherence to specifications at the time of lot release helps to assure that a product will be safe and effective throughout its dating period. Thus, development of appropriate specifications for a product is a critical component of product development.

According to regulation, potency refers to the specific ability or capacity of the product, as indicated by appropriate laboratory tests or by adequately controlled clinical data obtained through the administration of the product in the manner intended, to effect a given result. Potency assays are laboratory tests that may be applied to individual lots of product, in order to predict that they will achieve the desired effect. Normally, for a vaccine or other biological product, the potency reflects the quantity of active ingredient. Because of the complexity of vaccines, though, it often is important for the potency assay to provide not only an indication of quantity, but also the total activity of that active ingredient. For many products, it is important to distinguish potency from strength; strength refers to the quantity of the main ingredient in the product, while the term "potency" implies that information is available to predict the capability of that ingredient to achieve clinical benefit.

Potency assays play an important role throughout the product development life cycle. While fully validated assays may not be required until licensure, it is important to have a good understanding of vaccine potency starting even before Phase I. Each progressive phase of development is based on the assumption that potency was correctly determined in previous studies. Thus, doses tested in Phase I studies rely on the assumption that potency was accurately determined in the supporting animal studies. In turn, Phase II and Phase III studies rely on the assumption that potency was accurately determined in supporting safety and immunogenicity studies. Once the product is licensed, potency assays are used to assure that marketed product contains a quantity of active ingredient similar to that which was shown to be safe and effective in the clinical trials. When there are gaps in this understanding of potency, interpretation of the data, and possibly the progress of the product through the regulatory process, are in jeopardy.

As clinical development progresses, measurement of potency may change. At the time a product is first developed, it may not be known what in vitro characteristics are likely to correlate with clinical benefit. This may be particularly unpredictable before clinical trials have tested product with a range of potencies, sometimes measured in different ways. The need to understand the performance of assays over time and through clinical development suggests the value of maintaining stable reference preparations against which potency results may be standardized, and of retaining (under conditions of high stability) samples of material used in clinical development, in the event retesting of potency in updated assays should become necessary.

In addition to prediction of clinical benefit, potency assays ideally have other attributes. The need to ultimately validate these assays implies that they should have characteristics that are amenable to validation [for a more detailed discussion of assay validation, see ICH document Q2(R1)] (International Conference on Harmonization, 2005, http://www.ich.org/LOB/media/MEDIA417.pdf). Larger amounts of potency ideally should correlate with greater activity or effect (preferably, with linear or first-order kinetics). Potency assays should also be useful in monitoring product stability—degradation of product over time should be detectable and measurable using the selected potency assays. In addition, potency assays should be selected such that they can be reproducibly performed with reasonable precision—if assay variability is too high, it may become too difficult to manufacture product within the potency range necessary to assure both adequate efficacy at end-expiry and safety at release.

For vaccines, potency assays often detect the ability of the product to achieve a desired effect in a cell culture system or other in vitro assay. For example, for live virus vaccines, potency assays may be related to infectivity in cell culture assays. For inactivated virus vaccines, potency assays may be related to the total amount of viral antigen present in the product, sometimes using conformation-dependent immunoassays that are capable of detecting product degradation. Potency may also be assayed in animal models. However, animal models are more difficult to use, are time consuming, and frequently have greater variability than in vitro assays, suggesting the value of attempting to identify in vitro predictors of efficacy and safety early in product development.

To assure that vaccines remain potent and safe through the dating period, it is important to understand the stability of the vaccine. Stability studies normally focus on potency and safety measures and on factors that are likely to affect potency or safety, but product stability also can be determined based on the presence of a potentially deleterious degradation product. A recent WHO guidance further addresses vaccine stability (World Health Organization, 2006, http://www.who.int/biologicals/publications/trs/areas/vaccines/stability/en/index.html).

For some relatively labile vaccines, it may be important to use information about product stability to understand the doses of vaccine that are actually given during a clinical trial. End-expiry potency specifications are generally set based on a vaccine dose that is effective in a clinical trial—the specification implies that doses below this level may not be effective. Vaccine release potency specifications are generally set based on a statistical analysis of end-expiry potency, the dating period of the vaccine, the stability of the vaccine through that dating period, and the potential errors in measurement of vaccine potency and stability. Specifications on other vaccine parameters, such as moisture or pH, may be based on the conditions under which vaccine stability was studied. Still other specifications may represent product quality parameters, such as residual components of manufacture (such as albumin) or safety parameters, such as sterility and adventitious agent tests. Generally, specifications describe the range of quality parameters under which the product is shown to be safe and effective.

Additional information regarding product quality may be obtained by monitoring the manufacturing process, independently of final specification results. It may be desirable to use test results to monitor the production process, to provide early warnings that some aspect of production may be drifting out of control, even if the final product is within the range of the specifications. Attributes appropriate for monitoring the manufacturing process are generally selected based on a good understanding of the scientific principles underlying product manufacture, and may be subject to validation, as appropriate (Lubiniecki et al., 2003; Ganzer et al., 2005).

If a licensed product fails to meet a specification, it cannot be released to the public without specific FDA approval, generally obtained via a BLA supplement. Considerations in setting lot release specifications for live virus vaccines are discussed in recent articles (Gombold et al., 2006a, 2006b).

Vaccines and Related Biological Products Advisory Committee (VRBPAC)

The Vaccines and Related Biological Products Advisory Committee (VRBPAC) serves as the principal external advisory board to the FDA in considering issues associated with the study and licensure of vaccines. In addition, the VRBPAC provides external oversight of OVRR's laboratory research programs, to help assure their excellence and continued contributions to public health.

Questions for the VRBPAC typically relate to design of studies under IND (including the question of whether safety issues have been adequately addressed), to fundamental issues underlying licensure of new vaccines (such as safety, efficacy, testing, or manufacture), or to the expansion of indications for existing vaccines into new populations.

By charter, the VRBPAC includes 12 voting members including a Chair. These members typically include one technically qualified person who is associated with consumer interests. A non-voting industry representative, as well as additional outside experts relevant to the topics under discussion may also supplement the Committee.

The VRBPAC schedule, as well as minutes and materials associated with past meetings, are available at the FDA Web site (Center for Biologics Evaluation and Research, 2006, http://www.fda.gov/Cber/advisory/vrbp/vrbpmain.htm).

SPECIAL REGULATORY PATHWAYS THAT CAN ADDRESS EMERGING INFECTIOUS DISEASES VACCINE ISSUES

Certain special regulatory pathways are potentially of relevance to the development of vaccines against emerging infectious diseases. These include the Orphan Drug Program, Accelerated Approval, the Animal Rule, Fast Track Programs, Priority Reviews, Treatment INDs, and Emergency Use Authorizations.

Orphan Drug Program

The Orphan Drug Act (Orphan Drug Act, 1983) specifies several advantages for developers of products designed to treat diseases or conditions that are rare in the United States. A rare disease or condition is defined as one that either affects fewer than 200,000 people in the United States, or if it affects more, there is no reasonable expectation that development costs could be recovered from U.S. sales alone.

The Act provides that the secretary of HHS will provide written recommendations regarding clinical and nonclinical development programs for products designated as Orphan Drugs. Moreover, the Act provides that if an Orphan Drug is licensed, no competing product will be licensed for seven years, unless it is shown to be clinically superior to the original

Orphan Drug. Sponsors of Orphan Drugs are encouraged to make these products available under Treatment INDs, even prior to licensure, to individuals who have the rare disease, and for whom an alternative treatment is not available. In addition, the Act provides for government funding to support the development of Orphan Drugs.

The implementation of the Act is described in the Code of Federal Regulations (21 CFR Section 316 Orphan Drugs, 2007). Major Orphan Product activities are coordinated by the FDA's Office of Orphan Products Development, including the administration of a grants program that funds clinical studies in support of the development of Orphan Products. This mechanism has been used to fund vaccine studies, including clinical studies for vaccines against Q fever, and of a live-attenuated cytomegalovirus vaccine.

Vaccines against infectious diseases may be designated as orphan products, although this designation has historically been rare, and obstacles to development of orphan vaccines are perceived to exist (Lang and Wood, 1999). Because most vaccines are preventive, rather than therapeutic, FDA interprets the 200,000 person cutoffs in the Act to refer to the number of individuals for whom the vaccine would be indicated as of the date of designation of the product as an Orphan Drug. Examples of such products (as of January 2008) are shown in Table 14.2. These include an investigational live-attenuated vaccine directed against Japanese encephalitis virus, manufactured by Boran Pharmaceuticals in Seoul, Korea, which was designated as an Orphan Product in 1999, and a baculovirus-produced protein subunit vaccine against parvovirus B19, manufactured by Medimmune, Inc., indicated to prevent aplastic crises in individuals with sickle cell anemia. More recently, an Epstein-Barr virus gp350 vaccine, designed to prevent post-transplantation lymphoproliferative disease, was designated as an Orphan Product. An example of a licensed vaccine-related Orphan Product is Vaccinia Immune Globulin, used to prevent complications of vaccinia immunization, manufactured by Dynport Vaccine Company, in Frederick, MD. This product was designated as an Orphan Drug in 2004, and received licensure in 2005.

Accelerated Approval

Accelerated Approval rules permit licensure of a product based on surrogate endpoints, where it is not practical to rapidly study endpoints that directly indicate clinical benefit. Rules governing Accelerated Approval of biologics, including vaccines, are promulgated in the regulations (21 CFR 601, subpart E Accelerated Approval of Biological Products for Serious or Life-threatening Illnesses, 2007). These rules apply to biological products that have been studied for their safety and effectiveness in treating serious or life-threatening illnesses and that provide meaningful therapeutic benefit to patients over existing treatments. The Accelerated Approval regulations authorize FDA to license a biological product on the basis of adequate and well-controlled clinical trials showing an effect on a surrogate endpoint that is reasonably likely, based on epidemiologic, therapeutic, pathophysiologic, or other evidence to predict clinical benefit, or an effect on a clinical endpoint other than survival or irreversible morbidity. Where there is uncertainty as to the relationship between the study endpoints and actual clinical benefit, licenses granted under the Accelerated Approval mechanism must also

TABLE 14.2 Vaccine-related products that have been designated as orphan products

Product	Indication	Sponsor
Cytomegalovirus DNA vaccine with plasmids expressing pp65 and gB genes	Prevention of cytomegalovirus (CMV) viremia, and asociated clinically significant CMV disease complications in at-risk	Vical Incorporated
Japanese encephalitis vaccine (live, attenuated)	Prevention of Japanese encephalitis	Boran Pharmaceuticals
Vaccinia Immune Globulin (Human) Intravenous	Treatment of severe complications from the smallpox vaccine	DynPort Vaccine Company LLC
Recombinant Epstein-Barr virus gp350 glycoprotein vaccine: Brachet, 12, B-6041-Gosselies	Prevention of post-transplantation lymphoproliferative disorders in pediatric recipients of solid-organ transplantation	Henogen S.A.
Parvovirus B19 (recombinant VP1 and VP2; S.frugiperda cells) vaccine	Prevention of transient aplastic crisis in patients with sickle cell anemia	MedImmune, Inc.

be supported with post-marketing studies to verify and describe the actual clinical benefit of the product.

In cases where a well-established surrogate marker exists and is known to predict clinical benefit, evaluation under the Accelerated Approval mechanism is unnecessary, and does not lead to a requirement for post-marketing studies. However, for most emerging infectious diseases, the establishment of surrogate markers for efficacy may be very difficult early in the course of product development to allow their use without verification by post-marketing studies, as specified under the Accelerated Approval rules.

In addition, the FDA has the authority, for safety reasons, to restrict the distribution and use of products approved using the Accelerated Approval rule, prior to verification of clinical benefit in the post-marketing studies. This may include limitation of product use to physicians who possess certain training or experience, to certain types of facilities, or to certain types of medical procedures. However, any such limitations must be commensurate with the actual safety concerns presented by the product.

Licenses granted under Accelerated Approval may be withdrawn if the post-marketing studies fail to show the anticipated clinical benefit, if the postmarketing studies are not performed with due diligence, if post-marketing restrictions cannot assure the safe use of the product, if the sponsor fails to adhere to any post-marketing restrictions, if the promotional materials are false or misleading, or if other evidence shows the product not to be safe and effective under its conditions of use.

Animal Rule

Vaccine efficacy may generally be shown either in clinical trials under field conditions, by demonstration of efficacy in challenge studies (where ethical), or by demonstration that the vaccine induces an immunological correlate of human protection (subject to verification in post-marketing studies if the correlate is not well established and licensure is obtained under Accelerated Approval). In some cases, it may be necessary to develop a vaccine under conditions where none of these three pathways are feasible. For example, a disease may not be prevalent enough to allow scientifically valid vaccine efficacy trials in humans to be performed, performance of challenge studies may be unethical, and human correlates of protection may not be well understood. The Animal Rule was promulgated (21 CFR 601.90 subpart H Approval of Biological Products when Human Efficacy Studies Are Not Ethical or Feasible, 2007; 21 CFR 314.610 Approval based on evidence of effectiveness in studies in animals, 2007) in order to provide a pathway to licensure for such products. Vaccines designed to prevent many emerging infections may fall into this category.

According to the Animal Rule, the FDA may grant a license based on animal efficacy data when definitive human efficacy studies cannot be performed due either to ethical or practical reasons. The Animal Rule also may not be used if any other regulatory pathway to approval exists, such as Accelerated Approval based on surrogate markers. Moreover, the Animal Rule is not necessarily a faster pathway to approval than other regulatory pathways. The animal efficacy data must establish that there is a reasonable likelihood that the product would produce clinical benefit in humans. The use of the Animal Rule requires the following: (1) reasonably well-understood pathophysiological mechanisms by which the disease agent causes toxicity (i.e., disease) and by which the product would prevent or substantially reduce that toxicity, (2) demonstration of efficacy in more than one animal model reasonably expected to predict effect in humans, or in a single well-characterized animal model that has been shown to predict human response, (3) a clear relationship between the endpoint of the animal study and the desired outcome in humans (i.e., prevention of death or major morbidity), and (4) sufficient pharmacokinetic, pharmacodynamic, or other data exist to support selection of an appropriate human dose.

Animal Rule studies supporting an efficacy claim must be performed in accordance with preexisting requirements for GLP (21 CFR 58 Good Laboratory Practices for nonclinical laboratory studies, 2007) and the Animal Welfare Act (7 USC. 2131).

The Animal Rule does not eliminate any of the other requirements for U.S. licensure, but is merely intended to allow the substitution of animal for human efficacy data under circumstances where there is no alternative. Thus, a sufficient human safety database must be obtained. Moreover, the use of the Animal Rule is subject to three additional conditions: (1) submission of a plan for performance of post-marketing studies that verify and describe the biological product's clinical benefit, and a requirement to perform such studies with due diligence when they become feasible and ethical, (2) the potential for FDA to restrict the distribution and use of the product, to assure its safety, and (3) the requirement to provide specific patient labeling describing the basis for approval, risks, and other relevant information to each patient who receives a product licensed by the Animal Rule.

In designing Animal Rule efficacy studies, it is thus particularly important to consider the desired label indication and its relationship to the study endpoints, the appropriate vaccine dose for the animal studies, and the appropriate challenge dose and its route of exposure. If the development plan includes the use of laboratory assays to bridge animal and human data (e.g., immunogenicity assays), these assays should be validated for use both in human and animal samples. Statistical considerations, including sample size and power calculations, also influence the design of Animal Rule efficacy studies.

FDA generally recommends extensive discussion with the Agency if the use of the Animal Rule to support product efficacy is anticipated.

Fast Track Program

FDA also has mechanisms to facilitate the review of products that demonstrate the potential to address unmet medical needs and are intended to treat life-threatening or serious conditions. These programs are called "fast track" programs, and were originally codified by law in 21 USC 356, Section 506.

A guidance document (Food and Drug Administration, 2006) provides further details regarding the definition of a serious condition (at a minimum, the condition should be associated with morbidity that has a substantial impact on day-to-day functioning). A drug is deemed to be intended to treat a serious condition if it is being evaluated for its effect on a serious manifestation or symptom of the condition or its ability to prevent the condition, when it is scientifically reasonable to assume that preventing the condition will prevent the serious consequences. A medical need is considered to be unmet when there is no previously existing available therapy for the condition, when a benefit to using the new product instead of or in combination with previously existing therapies is shown, or when all previously existing therapies were approved under the Accelerated Approval regulations. FDA also will evaluate the potential of the product to address these unmet medical needs. A product may be designated for Fast Track at any time prior to licensure in the drug development process.

Designation as a Fast Track product provides several advantages to a sponsor. Meetings and frequent communication between the sponsor and FDA are strongly encouraged, including an end-of-Phase-I meeting. Because Fast Track products are intended for rapid evaluation, this end-of-Phase-I meeting is particularly important for mapping out the development and regulatory plan.

The major advantage of Fast Track designation occurs during the application review process. Because the criteria for entry into Fast Track and for obtaining Priority Reviews are similar, Fast Track products are often eligible for Priority Review. In addition, FDA may consider permitting partial submission of Fast Track license applications (i.e., early submission of some sections, such as the CMC or Toxicology Sections) when the clinical studies are finished or near completion but are not yet ready to be submitted to the license application, and when FDA agrees that preliminary evaluation of the clinical data supports a determination that the product may be effective. While early submission of some portions of the application does not change FDA's review timelines, it often permits early communication regarding important review issues, and thus increases the likelihood that the review can be completed in a single cycle.

Fast Track products also are eligible for approval under the Accelerated Approval regulation (see "Special Regulatory Pathways that can Address Emerging Infectious Diseases Vaccine Issues: Accelerated Approval"), when they meet those criteria.

Priority Review

The CBER, which is responsible for regulation of vaccines, designates a BLA as a "priority" if the product, if approved, would be a significant improvement in the safety or effectiveness of the treatment, diagnosis, or prevention of a serious or life-threatening disease. Products selected for Priority Review receive regulatory Action Letters (either Complete Response letters or Approval Letters) on a more rapid timetable than those designated as standard applications, in conformance with the Prescription Drug User Fee Act mandated timelines. In 2005, a Priority Review was assigned a timeframe of 6 months, while a standard review was assigned 10 months to the time of regulatory action.

The similarity in the language that describes criteria for Fast Track and Priority Review suggests that Fast Track applications would normally be considered for Priority Review. However, the language of the Priority Review criteria, requiring that the product be a significant advantage in safety, effectiveness, diagnosis, or prevention would appear to set a higher standard than that for Fast Track, in which an unmet medical need is defined on the basis of demonstrable benefit relative to previously existing treatments, rather than on the basis of "significant improvement."

Nonetheless, many vaccine products designed to address emerging infectious disease issues could be interpreted as meeting the basic criteria for Priority Review.

Treatment IND

Investigational drugs may sometimes be used for unapproved indications, not as part of a clinical trial, under a Treatment IND (21 CFR 312.34 Treatment use of an investigational new drug, 2007). The purpose of the Treatment IND regulation is to facilitate the availability of promising new drugs to seriously ill patients as early in the drug development as possible, and to obtain additional data on the drug's safety and effectiveness. Normally, drugs may be made available under Treatment IND during or after Phase 3 investigations, although this may sometimes occur during Phase 2, in the case of a life-threatening illness. To qualify, the drug must be intended to treat a serious or immediately life-threatening disease, there must be no available comparable or satisfactory alternative to treat that stage of the disease in the intended patient population, the drug must be under investigation in clinical trials (or all clinical trials must be completed), and the sponsor must be pursuing marketing approval with due diligence. Use of a drug under Treatment IND requires compliance with IND process safeguards, including those governing informed consent and institutional review boards. The regulations also provide conditions under which the FDA may deny Treatment INDs.

Early in the 2004 influenza vaccination season, vaccine shortages caused by the shutdown of one manufacturer's production led to markedly increased demand relative to supply. To meet this demand, some influenza vaccine manufacturers with vaccines licensed in other countries offered the use of their vaccines to U.S. populations under IND (HHS Buys 1.2 Million Doses of Flu Vaccine, 2005). Thus, Treatment IND was shown to be a potential option for distribution of investigational vaccines to at-risk U.S. citizens.

Project Bioshield and Emergency Use Authorizations

In the case of public health emergencies, use of unapproved products (or use of approved products for unapproved indications), including a waiver of GMP requirements, may be authorized under the Project Bioshield Act of 2004, codified as FD&C Act Section 564, or 21 USC. 360bbb-3. Such an Emergency Use Authorization requires declaration of an actual or potential emergency by the Secretary of HHS. This declaration may follow the determination by the Secretary of Defense that a military emergency or the potential for a military emergency exists. The declaration lasts until the emergency is considered to no longer exist, or for one year. The Secretary of HHS may renew the declaration in one-year increments. Each such declaration must be published in the Federal Register. Emergency Use Authorizations under this law require the Secretary of HHS to determine, after consultation with the NIH and the CDC Directors, that the specified agent can cause a life-threatening or serious disease or condition, it is reasonable to believe that the product may be effective, and that the known and potential benefits of the product outweigh the potential risks, and that there is no reasonable alternative. Such products may only be used if the health care providers and recipients are adequately informed of the circumstances and the known risks and benefits of the product, if appropriate means of monitoring for adverse events are in place, and if appropriate manufacturer record keeping takes place. The legislation does not confer on the government the authority to require either manufacturers to produce vaccines or individuals to take them. A recently issued guidance document provides further information on Emergency Use Authorizations (Food and Drug Administration, 2007, http://www.fda.gov/oc/guidance/emergencyuse.html).

In January 2005, the Secretary of HHS published an Emergency Use Authorization for the use by the military of anthrax vaccine for the indication of preventing inhalation anthrax. In this case, a court injunction had previously prevented the military from continuing its anthrax vaccination program, due to legal concerns about the original basis for licensure of the vaccine. The Emergency Use Authorization thus allowed the military to resume its anthrax vaccination program (on a voluntary basis).

Emergency Use Authorization under the Project Bioshield Act thus provides a mechanism by which products intended to address rapidly evolving public health emergencies may be quickly made available to the public. This does not, however, address the means to obtain the manufacturing capacity that would be required for widespread vaccination, even if authorized under the authority granted by this law.

CONCLUSIONS

The U.S. regulatory process balances the need to make new and innovative vaccines, that positively influence public health, with the need to assure that the vaccines and the procedures used to test them are safe. While emerging infectious diseases present special challenges to the regulatory process, new approaches provide pathways by which these vaccines may be licensed.

As challenges like those posed by emerging infectious diseases are further discussed, it is likely that the regulatory process will evolve to further clarify and perhaps even provide new pathways to licensure for vaccines to prevent them. It will remain important for those working in rapidly occurring disease areas to keep abreast of these regulatory innovations.

ACKNOWLEDGMENTS

I thank Jerry Weir, Robin Levis, Erik Henchal, Anissa Cheung, and Michael Brennan for their helpful suggestions and their critical reading of this manuscript.

References

Anderson, R. and May, R. Infectious Diseases of Humans: Dynamics and Control. Oxford: Oxford University Press, 1991.

Baylor, N. and Midthun, K. Regulation and testing of vaccines. Vaccines, Ed. Plotkin, S.A. and Mortimer, E.A. 2004.

Center for Biologics Evaluation and Research, Guidance for Industry: Content and Format of Chemistry, Manufacturing, and Controls Information and Establishment Description Information for a Vaccine or Related Product, 1999, http://www.fda.gov/Cber/gdlns/cmcvacc.pdf, Accessed: 2008.

Center for Biologics Evaluation and Research, Draft Guidance for Industry: Considerations for Reproductive Toxicity Studies for Preventive Vaccines for Infectious Disease Indications, 2000, http://www.fda.gov/cber/gdlns/reprotox.pdf, Accessed: 2008.

Center for Biologics Evaluation and Research, Draft Guidance for Industry: Clinical Data Needed to Support the Licensure of Pandemic Influenza Vaccines, 2006, http://www.fda.gov/Cber/gdlns/trifluvac.htm, Accessed: 2008.

Center for Biologics Evaluation and Research, Guidance for Industry: Development of Preventive HIV Vaccines for Use in Pediatric Populations, 2006, http://www.fda.gov/Cber/gdlns/pedhiv.pdf, Accessed: 2008.

Center for Biologics Evaluation and Research, Draft Guidance for Industry: Characterization and Qualification of Cell Substrates and Other Biological Starting Materials Used in the Production of Viral Vaccines for the Prevention and Treatment of Infectious Diseases,2006,http://www.fda.gov/cber/gdlns/vaccsubstrates.htm, Accessed: 2008.

Center for Biologics Evaluation and Research, Vaccines and Related Biological Products Advisory Committee, 2006, http://www.fda.gov/Cber/advisory/vrbp/vrbpmain.htm, Accessed: 2006.

Center for Biologics Evaluation and Research, Guidance for Industry: Considerations for Plasmid DNA Vaccines for Infectious Disease Indications, 2007.

Center for Biologics Evaluation and Research, Guidance for Industry: Toxicity Grading Scale for Healthy Adult and Adolescent Volunteers Enrolled in Preventive Vaccine Clinical Trials, 2007, http://www.fda.gov/cber/gdlns/toxvac.pdf, Accessed: 2008.

Center for Biologics Evaluation and Research, Guidance for Industry: Clinical Data Needed to Support the Licensure of Seasonal Inactivated Influenza Vaccines, 2007, http://www.fda.gov/Cber/gdlns/trifluvac.pdf, Accessed: 2008.

Center for Biologics Evaluation and Research, Bovine Spongiform Encephalopathy (BSE), 2007, http://www.fda.gov/cber/bse/bse.htm, Accessed: 2008.

Center for Biologics Evaluation and Research, Thimerosal in Vaccines, 2008, http://www.fda.gov/Cber/vaccine/thimerosal.htm, Accessed: 2008.

Centers for Disease Control, Advisory Committee on Immunization Practices, 2008, http://www.cdc.gov/vaccines/recs/nip/default.htm, Accessed: 2008

21 CFR 56.108 IRB functions and operations, in Code of Federal Regulations. Washington, DC: US Government Printing Office, 2007.

21 CFR 58 Good Laboratory Practices for nonclinical laboratory studies, in Code of Federal Regulations. Washington, DC: US Government Printing Office, 2007.

21 CFR 201 Labeling, in Code of Federal Regulations. Washington, DC: US Government Printing Office, 2007.

21 CFR 210 and 211 Current Good Manufacturing Practice in Manufacturing, Packing, or Holding of Drugs; General and Current Good Manufacturing Practice for Finished Pharmaceuticals, in Code of Federal Regulations. Washington, DC: US Government Printing Office, 2007.

21 CFR 312 Investigational New Drug Application, in Code of Federal Regulations. Washington, DC: US Government Printing Office, 2007.

21 CFR 312.21 Phases of an investigation, in Code of Federal Regulations. Washington, DC: US Government Printing Office, 2007.

21 CFR 312.23 IND content and format, in Code of Federal Regulations. Washington, DC: US Government Printing Office, 2007.

21 CFR 312.32 IND safety reports, in Code of Federal Regulations. Washington, DC: US Government Printing Office, 2007.

21 CFR 312.34 Treatment use of an investigational new drug, in Code of Federal Regulations. Washington, DC: US Government Printing Office, 2007.

21 CFR 312.42 Clinical Holds and Requests for Modification, in Code of Federal Regulations. Washington, DC: US Government Printing Office, 2007.

21 CFR 312.47 Meetings, in Code of Federal Regulations. Washington, DC: US Government Printing Office, 2007.

21 CFR 312.55 Informing Investigators, in Code of Federal Regulations. Washington, DC: US Government Printing Office, 2007.

21 CFR 314.610 Approval based on evidence of effectiveness in studies in animals, in Code of Federal Regulations. Washington, DC: US Government Printing Office, 2007.

21 CFR Section 316 Orphan Drugs, in Code of Federal Regulations. Washington, DC: US Government Printing Office, 2007.

21 CFR 600 Biological Products, General, in Code of Federal Regulations. Washington, DC: US Government Printing Office, 2007.

21 CFR 600, Subpart B Establishment Standards, in Code of Federal Regulations. Washington, DC: US Government Printing Office, 2007.

21 CFR 600, Subpart C Establishment Inspection, in Code of Federal Regulations. Washington, DC: US Government Printing Office, 2007.

21 CFR 600, Subpart D Reporting of Adverse Experiences, in Code of Federal Regulations. Washington, DC: US Government Printing Office, 2007.

21 CFR 600.10 Personnel, in Code of Federal Regulations. Washington, DC: US Government Printing Office, 2007.

21 CFR 601 Licensing, in Code of Federal Regulations. Washington, DC: US Government Printing Office, 2007.

21 CFR 601.12 Changes to an approved application, in Code of Federal Regulations. Washington, DC: US Government Printing Office, 2007.
21 CFR 601, subpart E Accelerated Approval of Biological Products for Serious or Life-threatening Illnesses, in Code of Federal Regulations. Washington, DC: US Government Printing Office, 2007.
21 CFR 601.70 Annual Progress Reports of Post-marketing Studies, in Code of Federal Regulations. Washington, DC: US Government Printing Office, 2007.
21 CFR 601.90 subpart H Approval of Biological Products when Human Efficacy Studies Are Not Ethical or Feasible, in Code of Federal Regulations. Washington, DC: US Government Printing Office, 2007.
21 CFR 610 General Biological Products Standards, in Code of Federal Regulations. Washington, DC: US Government Printing Office, 2007.
21 CFR 610, subpart G: Labeling Standards, in Code of Federal Regulations. Washington, DC: US Government Printing Office, 2007.
21 CFR 610.13 Purity, in Code of Federal Regulations. Washington, DC: US Government Printing Office, 2007.
Chang, P.Y., Sheets, R., Shapiro, S., Hargus, S. and Gruber, M. Vaccine pre-clinical toxicology testing, 2003, http://www3.niaid.nih.gov/daids/vaccine/Science/VRTT/00_Main.htm, Accessed: 2008.
Chen, R.T., Destefano, F., Pless, R., Mootrey, G., Kramarz, P. and Hibbs, B. Challenges and controversies in immunization safety. Infect. Dis. Clin. North Am. 2001; 15:21–39.
Department of Health and Human Services, HHS Pandemic Influenza Plan, 2005, http://www.hhs.gov/pandemicflu/plan/, Accessed: 2006.
Evans, G. Update on vaccine liability in the United States: presentation at the National Vaccine Program Office Workshop on Strengthening the Supply of Routinely Recommended Vaccines in the United States, 12 February 2002. Clin. Infect. Dis. 2006; 42:S130–S137.
Fauci, A. The global challenge of infectious diseases: the evolving role of the National Institutes of Health in basic and clinical research. Nat. Immunol. 2005; 6:743–747.
FDA Amendments Act of 2007, in HR3580. 2007.
FDA Science Board Subcommittee on Science and Technology, FDA Science and Mission at Risk, 2007, http://www.fda.gov/ohrms/dockets/ac/07/briefing/2007-4329b_02_01_FDA%20Report%20on%20Science%20and%20Technology.pdf, Accessed: 2008.
Food and Drug Administration, Guidance for Industry: Providing Clinical Evidence of Effectiveness for Human Drugs and Biological Products, 1998, http://www.fda.gov/cber/gdlns/clineff.pdf, Accessed: 2008.
Food and Drug Administration, Guidance for Industry: Formal Meetings with Sponsors and Applicants for PDUFA Products, 2000, http://www.fda.gov/cber/gdlns/mtpdufa.pdf, Accessed:
Food and Drug Administration, Draft Guidance for Industry: Postmarketing Safety Reporting for Human Drug and Biological Products Including Vaccines, 2001, http://www.fda.gov/cber/gdlns/safety031201.pdf, Accessed: 2008.
Food and Drug Administration, Guidance for Industry: Establishing Pregnancy Exposure Registries, 2002, http://www.fda.gov/cber/gdlns/pregexp.pdf, Accessed: 2008.
Food and Drug Administration, Draft Guidance for Industry: Information Program on Clinical Trials for Serious or Life-Threatening Diseases and Conditions (Revision 1), 2004, http://www.fda.gov/cber/gdlns/clintrial.htm, Accessed: 2008.
Food and Drug Administration, Providing Regulatory Submissions in Electronic Format-Human Pharmaceutical Product Applications and Related Submissions Using the eCTD Specifications, 2005, http://www.fda.gov/cber/gdlns/esubapp.htm, Accessed: 2008.
Food and Drug Administration, Fast Track Drug Development Programs—Designation, Development, and Application Review, 2006.
Food and Drug Administration, Guidance: Emergency Use Authorization of Medical Products, 2007, http://www.fda.gov/oc/guidance/emergencyuse.html, Accessed: 2008
Ganzer, W.P., Materna, J.A., Mitchell, M.B. and Wall, L.K. Current thoughts on critical process parameters and API synthesis. Pharm. Technol. 2005; 29:46–66.
Gombold, J., Peden, K., Gavin, D., Wei, Z., Baradaran, K., Mire-Sluis, A. and Schenerman, M. Lot release and characterization testing of live-virus-based vaccines and gene therapy products, part 1: factors influencing assay choices. Bioprocess Int. 2006a; 4(4):46–56.
Gombold, J., Peden, K. Gavin, D., Wei, Z., Baradaran, K., Mire-Sluis, A. and Schenerman, M. Lot release and characterization testing of live-virus-based vaccines and gene therapy products, part 2: case studies and discussion. Bioprocess Int. 2006b; 4(5):56–65.
HHS Buys 1.2 Million Doses of Flu Vaccine. FDA Consumer 2005; 39(1).
International Conference on Harmonization, ICH harmonised tripartite guideline: Guideline for good clinical practice E6(R1), 1996, http://www.ich.org/LOB/media/MEDIA482.pdf, Accessed: 2006.
International Conference on Harmonization, Viral Safety Evaluation of Biotechnology Products Derived from Cell Lines of Human or Animal Origin Q5A(R1), 1999, http://www.ich.org/LOB/media/MEDIA425.pdf, Accessed: 2006.
International Conference on Harmonization, Validation of Analytical Procedures: Text and Methodology Q2(R1), 2005, http://www.ich.org/LOB/media/MEDIA417.pdf, Accessed: 2006.
International Conference on Harmonization, International Conference on Harmonization Web Site, 2008, http://www.ich.org, Accessed: 2008.
Iskander, J., Miller, E. and Chen, R. The role of the vaccine adverse event reporting system (VAERS) in monitoring vaccine safety. Pediatr. Ann. 2004; 33:599–606.
Korn, D. FDA under siege: the public at risk. Science 1997; 276:1627.
Koutsky, L., Ault, K., Wheeler, C., Brown, D., Barr, E., Alvarez, F., Chiacchierine, L. and Proof of Principle Study Investigators. A controlled trial of a human papillomavirus type 16 vaccine. N. Engl. J. Med. 2002; 347:1645–1651.
Krause, P. and Klinman, D. Efficacy, immunogenicity, safety, and use of live attenuated chickenpox vaccine. J. Pediatr. 1995; 127:518–525.
Lang, J. and Wood, S. Development of orphan vaccines: an industry perspective. Emerg. Infect. Dis. 1999; 5:749–756.
Lederberg, J., Shope, R. and Oaks, S., Jr. Emerging Infections: Microbial Threats to Health in the United States. Washington DC: National Academy Press, 1992, p.294.
Lubiniecki, A.S., Gardner, A.R., Smith, T.M., Wang, W.K., McAllister, P.R. and Federici, M.M. Validation of fermentation processes. Dev. Biol. (Basel) 2003; 113:37–44.
Nabel, G. HIV vaccine strategies. Vaccine 2002; 20:1945–1947.
Orphan Drug Act, in P.L. 97-414, as amended, 1983.
Plotkin, S. Lessons learned concerning vaccine safety. Vaccine 1991; 20:S16–S19.
Sabin, A. Oral poliovirus vaccine: history of its development and use and current challenge to eliminate poliomyelitis from the world. J. Infect. Dis. 1985; 151:420–436.
Sheets, R. and Petricciani, J. Vaccine cell substrates 2004. Dev. Biol. (Basel) 2006; 123:1–363.

Smith, D.J. Predictability and preparedness in influenza control. Science 2006; 312:392–394.
Stern, A. and Markel, H. The history of vaccines and immunization: familiar patterns, new challenges. Health Aff. 2005; 24:611–621.
Stohr, K. and Esveld, M. Will influenza vaccines be available for the next influenza pandemic? Science 2004; 306:2195–2196.
Stratton, K., Almario, D. and McCormick, M. (Eds.). Immunization Safety Review: SV40 Contamination of Polio Vaccine and Cancer. Immunization Safety Review Committee, Washington, DC: National Academies Press, 102, 2002.
Treanor, J.J., Kotloff, K., Betts, R.F., Belshe, R., Newman, F., Iacuzio, D., Wittes, J. and Bryant, M. Evaluation of trivalent, live, cold-adapted (CAIV-T) and inactivated (TIV) influenza vaccines in prevention of virus infection and illness following challenge of adults with wild-type influenza A (H1N1), A (H3N2), and B viruses. Vaccine 1999; 18:899–906.
World Health Organization, WHO guidelines on nonclinical evaluation of vaccines. WHO Technical Report Series, 2005; 927, Annex 1.
World Health Organization: Expert Committee on Biological Standardisation, Guidelines on Stability Evaluation of Vaccines, 2006, http://www.who.int/biologicals/publications/trs/areas/vaccines/stability/en/index.html, Accessed: 2008.
World Health Organization, WHO: Biologicals, 2008, www.who.int/biologicals/en, Accessed: 2008.
Zoon, K. Vaccines, pharmaceutical products, and bioterrorism: challenges for the US food and drug administration. Emerg. Infect. Dis. 1999; 5:534–536.

CHAPTER

15

Vaccine Safety

John D. Grabenstein

Merck Vaccine Division, West Point, PA, USA

OUTLINE

ABSTRACT

Vaccines are typically given to healthy people to keep them healthy. As a result, vaccines are the class of medications least tolerant of adverse reactions. If all medications were arrayed along a spectrum portraying degrees of safety, today's licensed vaccines would appear toward the end labeled "safest." To provide vaccines with the highest standards of safety, vaccine developers, regulators, and immunization program administrators work to minimize the types and frequency of adverse reactions that occur after vaccination. The key to assessing vaccine safety effectively is to be objective and to assess carefully all available data. Figuratively, one's eyes and ears must always be open to new findings. In assessing vaccine safety, it is important to differentiate cause-and-effect relationships from mere temporal relationships. This helps avoid the post hoc, ergo propter hoc fallacy. The incidence and severity of adverse reactions to immunization can be measured with greatest objectivity in randomized, double-blinded, placebo-controlled trials. But the acceptability of a vaccine's adverse-reaction profile is not fixed in an absolute sense across all time and space. Rather, a series of judgments are made, especially in relation to differing degrees of disease burden (and hence, vaccine value) over time or in different settings. This chapter describes multiple historical examples of vaccine-safety issues, including issues that evolved over the course of years or decades. These examples are instructive to scientists planning vaccine-safety evaluations for future vaccines or vaccination programs. Vaccines have succeeded so well that many of these vaccine-preventable diseases are virtually unheard of by clinicians or the public. Separating true causes from false ones is vitally important for both society and individuals. Using a disciplined, rational, methodical approach to attributing blame where appropriate, or clearing a vaccine's reputation where there is no basis for a link, continues to be the proper approach. Using the tools of science, scientists are most likely to find the truth and least likely to fool themselves about vaccine safety. The key to assessing vaccine safety is to be objective, assess carefully all available data, and be open to new findings.

Vaccines are typically given to healthy people to keep them healthy. Because of this, vaccines are the class of medications least tolerant of adverse reactions. To create vaccines with the highest standards of safety, vaccine developers, regulators, and immunization program administrators work to minimize the types and frequency of adverse reactions that occur after vaccination. These actions and judgments are undertaken in light of a vaccine's benefit in preventing an infectious disease (Nordenberg, 1999; Meadows, 2002; Grabenstein, 1999).

In pharmaceutical development, "efficacy" and "safety" are terms of art frequently used in tandem. Just as efficacy is an intricate construct, qualitatively and quantitatively, so is safety. The key to assessing vaccine safety effectively is to be objective and to assess carefully all available data. Figuratively, one's eyes and ears must always be open to new findings.

If all medications were arrayed along a spectrum portraying degrees of safety, today's licensed vaccines would appear toward the end labeled "safest." In contrast, cancer chemotherapeutic agents, renowned for their adverse-reaction profiles, would appear toward the end labeled "least safe." Yet the actions of pharmaceutical regulators [e.g., the U.S. Food and Drug Administration (FDA)] to license or approve medications involve a determination of "safety" for both ends of this spectrum. Regulators define safety as relative freedom from harmful effects when prudently administered, taking into account the characteristics of the medication relative to the condition of the patient (Nordenberg, 1999; Meadows, 2002; Title 21 U.S. Code of Federal Regulations). In other words, vaccines and chemotherapeutics are held to different standards, and yet each is acceptably safe under likely circumstances of clinical practice.

Within the category of vaccines, different vaccines have various degrees of safety (Grabenstein, 2006; Joellenbeck et al., 2002). Modern, highly purified versions of influenza vaccines would be among those considered "safest." Smallpox vaccine would be at the other end of the safety spectrum for vaccines, given the rare and unusual adverse reactions vaccinia viruses can produce (Advisory Committee on Immunization Practices, 2001, 2003; Centers for Disease Control and Prevention, 2003a, 2003b). And yet smallpox vaccine can still be given to many people with an acceptable level of safety, given sufficient screening, exemptions, and education (Grabenstein and Winkenwerder, 2003; Poland et al., 2005).

Because vaccines contain exogenous proteins or other large molecules, they rarely share a propensity to cause anaphylaxis (Grabenstein, 1997). All parenteral vaccines cause injection-site swelling, erythema, and other localized reactions at rates that vary by route of administration, degree of antigenic purity, adjuvant content, and other factors. All vaccines are typically associated with systemic symptoms, such as headache, myalgia, and fatigue (Grabenstein, 2006; Joellenbeck et al., 2002). The risk and consequence of these adverse reactions is considered acceptable in relation to preventing the morbidity and mortality of the infections the vaccines prevent (Grabenstein, 2006).

It is interesting to consider the rates of injection-site and systemic symptoms that occur among the placebo control arms of vaccine clinical trials. For example, a clinical trial of a vaccine to prevent Lyme disease (Lymerix, GlaxoSmithKline) included a double-blinded study arm of volunteers who received a placebo containing aluminum hydroxide (Steere et al., 1998). In this study, injection-site soreness occurred in 7.6% of aluminum hydroxide recipients along with arthralgia in 3.5%. Diary cards recorded symptoms related or possibly related to injection during the first 3 days in 53% of the "placebo" recipients. Similar rates for adverse reactions from aluminum-containing placebo injections were seen in trials of a quadrivalent human papillomavirus vaccine (Gardasil, Merck) (Joura et al., 2007; Garland et al., 2007; FUTURE II Study Group, 2007).

Such findings (as well as the more common use of isotonic sodium chloride placeboes) illustrate several points. Merely inserting a needle beneath the skin (perhaps into a muscle compartment) and injecting a liquid result in biological and mechanical effects, even in the absence of antigen. When people are asked to record symptoms and sensations over an interval of time, they may note things that otherwise would pass unnoticed during their daily activities. These include the background rates of headache, myalgia, fatigue, and other transient symptoms that occur in the absence of antigenic stimulation.

RATIONAL SAFETY

The incidence and severity of adverse reactions to immunization can be measured with objectivity in randomized, double-blinded, placebo-controlled trials. But the acceptability of a vaccine's adverse-reaction profile relies on human judgment. This judgment for any given vaccine is not fixed in an absolute sense across all time and space. Rather, a series of judgments can be made, especially in relation to differing degrees of disease burden (and hence, vaccine value) over time or in different settings.

When smallpox (variola infection) was commonplace, the rare adverse reactions resulting from smallpox vaccination were widely considered acceptable (Fenner et al., 1988). Once smallpox was eradicated as an endemic disease, national authorities judged that it was appropriate to limit smallpox vaccinations to laboratorians, military personnel, and others who might confront maliciously released variola virus (Advisory Committee on Immunization Practices, 2001, 2003; Centers for Disease Control and Prevention, 2003; Grabenstein and Winkenwerder, 2003).

Similarly, when poliomyelitis was a substantial domestic risk, national authorities judged the use of the attenuated poliovirus vaccine to be appropriate, even though the viruses in the vaccine could rarely revert to virulence and cause paralytic poliomyelitis (about one per 2.4 million vaccine doses distributed) (Advisory Committee on Immunization Practices, 1997). Once poliovirus infections were contained globally, poliovirus immunization policy in the United States shifted first to a combined schedule of inactivated and oral vaccines and then to an inactivated poliovirus vaccine exclusively (Advisory Committee on Immunization Practices, 1997, 1999b, 2000). Yet the ability of the attenuated poliovirus vaccine to be delivered orally continues to be advantageous in endemic areas. This illustrates how two different policies can be rational in the same place at two points in time (e.g., the United States in the 1980s versus 2000 and later) or in two places at the same point in time (e.g., the United States and Nigeria in 2007).

Rotavirus gastroenteritis hospitalizes over 1% of American children and kills one per 200,000 children in the United States. An oral rotavirus vaccine was licensed in August 1998 (RotaShield, Wyeth) (Advisory Committee on Immunization Practices, 1999). Clinical trials detected intussusception (a telescoping of the intestines) in both the vaccine and placebo groups, enough to warrant mention of the condition in the vaccine's labeling (i.e., package insert), but regulators did not consider this risk sufficient to withhold licensing. A few months after distribution of the vaccine expanded widely in clinical use (i.e., transitioning from tens of thousands to millions of vaccinees), routine safety surveillance systems within the United States identified an elevated but still rare risk of intussusception (Centers for Disease Control and Prevention, 1999a, 2004a; Advisory Committee on Immunization Practices, 1999). Those controlled studies revealed an attributable risk of one case of intussusception per 5,000–12,000 vaccine recipients. Attributable risk refers to the additional risk contributed by the vaccine, above the baseline risk of life without the vaccine. The manufacturer responded by stopping distribution of RotaShield in the U.S. commercial market in July 1999. A second rotavirus vaccine (RotaTeq, Merck) was licensed only after evaluation of three trials involving more than 70,000 volunteers to assess the attributable risk of intussusception (Advisory Committee on Immunization Practices, 2006).

Given that the morbidity and mortality burden of rotavirus gastroenteritis in developing countries with inadequate sanitation far exceeds the level within the United States, a situational dilemma develops. The U.S. domestic situation apparently left Wyeth unwilling or unable to distribute RotaShield in developing countries, as the vaccine was considered unacceptably risky in a premier market. But that "riskiness" was judged against the background of the United States, where the burden of rotavirus gastroenteritis is substantially less than in the developing world (i.e., a lesser achievable benefit to contribute to a vaccine's benefit-to-risk ratio).

ASSESSING SAFETY

Development of new vaccines progresses through animal and laboratory studies (i.e., preclinical studies) and then a series of clinical trials of escalating size and complexity.

Although no absolute lines delineate the stages of clinical investigation, the progress of drug research is usually conceived in a three-phased manner (see Chapter 12) (Grabenstein, 2006; Beatrice, 1991; Food and Drug Administration, 2007a). Phase I typically involves safety testing and pharmacologic profiling of the safe dosage range in 10–100 healthy human volunteers. Phase II provides the first controlled studies, the first tests of effectiveness against a disease in several dozen to several hundred volunteers. Phase II studies begin defining proper doses and regimens, assessing efficacy and safety. Phase III involves expanded clinical trials in several hundred to several thousand patient volunteers. These studies provide evidence of safety and efficacy in adequate and well-controlled clinical trials, sometimes called pivotal clinical trials. When a manufacturer concludes that it has scientifically established the safety and efficacy of a vaccine, it submits a Biological License Application (BLA) to the FDA or analogous documents to regulators in other countries (see Chapter 13).

Safety assessment continues after a vaccine is licensed, in what are called phase IV studies or post-marketing surveillance. "Post-marketing" could more properly be termed "post-licensing" because these data accumulate during (not after) widespread marketing of the vaccine for routine clinical use. After licensing, a manufacturer may undertake additional clinical studies to obtain information about longer follow-up intervals, other populations, and other uses or to assess benefits or adverse effects in greater detail. Some of these studies are tracked at an FDA Web site (Food and Drug Administration, 2007).

Another component of post-marketing or post-licensing surveillance takes the form of retrospective pharmacoepidemiologic research into automated clinical databases. The Vaccine Safety Datalink (VSD), coordinated by the Centers for Disease Control and Prevention (CDC), is a well-known example of following the healthcare encounters of vaccine recipients and similar nonrecipients in the databases of a consortium of health maintenance organizations (DeStefano, 2001; Mullooly et al., 2004; Davis et al., 2005). A similar effort of the U.S. Department of Defense (DoD) to assess the health of U.S. service members is called the Defense Medical Surveillance System (Joellenbeck et al., 2002; Lange et al., 2003, Arness et al., 2004; Payne et al., 2007). Relevant to immunization programs for emerging biothreats, the VSD is developing the capacity for earlier detection of vaccine-safety signals, known as rapid-cycle analysis (Davis et al., 2005).

Earlier in this chapter, we considered the allusion of keeping one's eyes and ears always open to new findings. The always-open avenue for submitting adverse events after immunization in the United States (even if no clinical trial is underway) involves the Vaccine Adverse Event Reporting System (VAERS) coordinated by the CDC and the FDA (Singleton et al., 1999; Centers for Disease Control and Prevention, 2003; Varricchio et al., 2004). VAERS collects reports of unexpected adverse events from clinicians and the public. Similar programs are conducted in Canada (the Vaccine-Associated Adverse Events Surveillance System, VAAESS), the United Kingdom (the "yellow card" system), and other countries, often under the rubric of pharmacovigilance systems.

Reports to VAERS and its counterparts result from spontaneous decisions of a clinician or a patient to inform the CDC and the FDA of an adverse event. This requirement for action by the reporter is a significant limitation due to human factors related to reporting behaviors and information bias. Reports of medical conditions to systems like VAERS are certainly related to immunization in time, but they do not establish that a cause-and-effect relationship exists. Reporting rates cannot be assumed to be true incidence rates (Singleton et al., 1999; Centers for Disease Control and Prevention, 2003; Varricchio et al., 2004). Even so, VAERS and its counterparts are an important component of a vaccine-safety surveillance system. These systems are valuable for their role in collecting an open-ended array of information from anywhere at any time and ability to identify signals worthy of further evaluation using less biased study designs. Spontaneous reporting systems like VAERS are important for identifying rare, unexpected events.

Summing up all the people enrolled in phase I through III studies, few vaccines have been assessed in more than 50,000 people at the point of licensing. Still, this number is an order of magnitude larger than the few thousand people in the average clinical trial dataset for other pharmaceutical development programs. Larger study sizes in vaccine trials help rule out adverse events rarer than with cardiac, respiratory, or other categories of medications.

If a vaccine causes an adverse reaction, it is important for society to recognize it as soon as possible and respond appropriately. But jumping to the conclusion that two events are causally related in the absence of objective evidence is scientifically improper. Jumping to a conclusion can delay finding the true cause for an illness. For example, assuming that egg allergy was linked to anaphylaxis after measles–mumps–rubella (MMR II, Merck) immunization may have delayed recognition of the role of gelatin allergy (Khakoo and Lack, 2000; Pickering, 2006).

SAFETY THROUGH DIALOGUE AND EDUCATION

One of the principal ways to prevent adverse reactions is to withhold a vaccine from people who should not receive it (Grabenstein, 2006). Who is susceptible to anaphylaxis usually cannot be predicted. But the replicating characteristics of live vaccines provide a rational basis for assuming that such vaccines are unsuitable for people whose immune systems are compromised and who might not be able to staunch viremia or bacteremia after immunization. Thus, yellow fever, smallpox, and measles–mumps–rubella vaccines may be withheld from people with suppressed immune systems.

People who provide immunization services are not always the primary-care providers, who know the medical histories of vaccine-eligible people. The principal way to determine the underlying health conditions that might contraindicate (bar) specific immunizations is by interviewing ("screening") the vaccine candidate and/or by reading the candidate's earlier medical records. The most common temporary or permanent contraindications to immunization include the following:

- Acute illness on the day of evaluation for immunization (typically of a magnitude to warrant additional medical care): Defer immunization until the candidate feels better.

- Allergies (i.e., immediate hypersensitivity) to vaccine components (e.g., anaphylaxis, laryngeal swelling): If this is a life-threatening allergy, the contraindication may be permanent.
- Prior reactions to immunization (e.g., substantial swelling at immunization site): Depending on the specifics of the patient's case, the contraindication may be temporary or permanent.
- Pregnancy, most often in the case of live vaccines: Certain subunit vaccines may be actively recommended for pregnant women (e.g., influenza, hepatitis B, meningococcal, tetanus–diphtheria), based on her personal risk of exposure to a preventable infection (Grabenstein, 2006).
- Immune deficiency: With regard to live viral or bacterial vaccines, related to serious disease (e.g., acquired immune deficiency syndrome, certain cancers), immune-suppressing medications (e.g., alkylating agents, antimetabolites, prolonged high-dose corticosteroids), or radiation therapy: For subunit vaccines, immune deficiency may impair efficacy, without safety implications per se. The degree and nature of the suppression will bear on whether the contraindication is temporary or permanent. In all cases, individual risk–benefit assessments are warranted (Grabenstein, 2006).
- Recent receipt of a blood product or immune globulin, applicable to certain live vaccines (e.g., MMR, varicella): Immunization should be deferred several months to allow time for catabolism of neutralizing antibodies (Grabenstein, 2006).
- Underlying disease conditions: Some underlying conditions will identify indications for immunization (e.g., diabetes and pneumococcal and influenza vaccines) and other conditions may identify contraindications to immunization (e.g., neurologic or rheumatologic conditions still evolving at the time of screening for immunization) (Grabenstein, 2006; Fenichel, 1999).

Another way to avert or mitigate adverse reactions after immunization is through education of the vaccinee. For example, vaccinees should be given clear expectations of what adverse reactions to expect after immunization (e.g., sore arms, fever, headache). They should be instructed to seek medical assistance if symptoms after immunization persist more than a few days or if they exceed a threshold that seems reasonable. In the case of smallpox (vaccinia) vaccination, education is important for preventing transfer of live vaccinia viruses from the vaccination site to some other site on the body or a close contact of the vaccinee (Grabenstein, 1997; Centers for Disease Control and Prevention, 1999). In this example, education averts adverse reactions.

Appropriate counseling should lead to understanding among vaccinees that no vaccine should be expected to be perfectly protective. Breakthrough (relatively milder) cases of chickenpox may occur in varicella-vaccinated people (Advisory Committee on Immunization Practices, 2007b). Breakthrough influenza infection may occur in influenza-vaccinated people (Advisory Committee on Immunization Practices, 2007), as can adenoviral, rhinoviral, and other viral infections that may be misattributed to influenza virus.

PROBLEMS INHERENT TO A PRODUCT

History records several problems in vaccine manufacturing that manifested early in the experience with certain vaccines. In early 1942, some of the first lots of yellow fever vaccine were manufactured with human serum albumin that inadvertently contained hepatitis B virus (Sawyer et al., 1944a, 1944b; Parish, 1965). Among ~2.5 million service members immunized from suspect lots, ~330,000 were infected with hepatitis B virus and at least 84 died (Parish, 1965; Seeff et al., 1987; Norman et al., 1993).

In April 1954, some early lots of the "Salk" vaccine, inactivated poliovirus vaccine, were not processed appropriately, allowing live polioviruses to contaminate the vaccine (the so-called Cutter incident). Over 200 people were paralyzed by this industrial accident (Parish, 1965; Nathanson and Langmuir, 1963; Smith, 1990; Offit, 2007).

Also in the 1950s, poliovirus and adenovirus vaccines were recognized to be contaminated with the simian virus 40 (SV40). Multidecade follow-up studies have not found evidence of any delayed adverse effect of the SV40 exposure (Brown and Lewis, 1998; Strickler et al., 1998; Institute of Medicine, 2002). Similarly, evaluation of avian leucosis virus and endogenous avian retrovirus contamination of yellow fever and measles–mumps–rubella vaccines has not identified any related harm in vaccine recipients (Waters et al., 1972; Hussain et al., 2001, 2003).

Although no other examples similar to the hepatitis B and poliovirus contamination incidents have occurred in modern times, it is wise to keep one's eyes and ears open whenever a new vaccine is introduced. The FDA requires much more extensive clinical studies and testing than was applied in the 1940s and 1950s. Even so, the rotavirus vaccine example illustrates how the number of people given a vaccine after licensure permits assessment of rare events that may not have been fully understood during clinical trials.

Safety is not the only thing that needs to be monitored with the initial widespread use of a new vaccine. Influenza vaccines were first widely used in the mid-1940s. Vaccine experts were surprised in the late 1940s to realize that influenza immunization was not as effective in preventing infection as it had been a few years earlier (Parish, 1965). From this surprising finding came the realization that influenza viruses change their surface proteins, with a corresponding requirement for a new influenza vaccine formulation each year to keep pace.

EXPANDED USE

During clinical trials, before licensing, the average vaccine is given to thousands of people. Adverse reactions occurring with a frequency of one per 50,000 vaccinees or more rarely would not be expected to be recognized during clinical trials of that magnitude. Large-scale immunization programs may provide additional safety data to help identify previously unrecognized adverse reactions to immunizations.

When the smallpox vaccination program resumed in the United States in December 2002, the U.S. Department of Health & Human Services (DHHS) and the DoD collaborated on parallel but mutually reinforcing surveillance projects among the civilian and military cohorts to be vaccinated (Grabenstein and Winkenwerder, 2003; Poland et al., 2005; Baciu et al., 2005; Centers for Disease Control and Prevention, 2004). Each department implemented a vigorous program to identify adverse events among widely dispersed vaccinees across the country for the DHHS and civilian vaccinees and on four continents and dozens of ships at sea for the DoD and military vaccinees. The DHHS and the DoD pooled the adverse-event cases they identified and provided the full case series to an independent panel of civilian physicians who comprised a joint work group drawn from the members of the CDC's Advisory Committee on Immunization Practices and the DoD's Armed Forces Epidemiological Board. Although the civilian and military vaccinees differed in demographic characteristics, the collaborative approach subjected all the adverse-event data to a common analytic framework.

Within this massive vaccinia safety surveillance project, multiple subanalyses were conducted. For example, the DoD commissioned an electronic symptom diary that accepted telephonic and web-based entries, which provided the first analysis of the progression of the lesion at the vaccination site in a kinetic fashion (Olmstead et al., 2005, 2006). Both the DoD and the DHHS conducted traditional symptom surveys that added to the knowledge base for primary smallpox vaccination of adults. In several ways, the 2002–2007 experience has provided a greater evidence base to the accumulated knowledge through the 1970s (Fenner et al., 1988).

Although sporadic cases of myocarditis had been reported after smallpox vaccination in Europe and Australia (notably in Finnish military trainees), it was not recognized with the widespread use of the New York City Board of Health (NYCBOH) strain of vaccinia virus used in the Dryvax brand (Wyeth) (Fenner et al., 1988; Lane et al., 1971; Lane and Millar, 1971). The conventional wisdom was that the NYCBOH strain was a safer vaccinial strain in this regard, with a lesser risk of encephalitis, compared to vaccinial strains used in Europe (Fenner et al., 1988). Nonetheless, the DoD reported its first case of myocarditis after smallpox vaccination in February 2003, the 10th case in March, and then published a case series of 18 patients a few months later (Halsell et al., 2003). The cases typically presented with severe chest pain in the second week after smallpox vaccination disproportionately among adult male Caucasian primary vaccinees (Arness et al., 2004). Recovery has generally been prompt, judged by recovery by electrocardiogram, echocardiogram, cardiac enzymes, and graded exercise stress test, although a subset reported continued fatigue and chest discomfort (Eckart et al., 2004). Additional study of these patients is underway.

Myopericarditis may have resulted from NYCBOH vaccination in the 1960s but been underrecognized because cardiac enzymes and imaging technologies were immature. In the 1960s, the first time most people received smallpox vaccination was as infants, unlike the recent experience with primary vaccination of young adults. This serves as yet another reminder to keep one's eyes and ears open, even if a substantial body of historical data is available.

CAUSE AND EFFECT?

In assessing vaccine safety, it is important to differentiate cause-and-effect relationships from mere temporal relationships (Grabenstein and Wilson, 1999; Grabenstein, 1999, 2000, 2001a). This helps avoid the post hoc, ergo propter hoc fallacy.

What should be expected in the health of someone who just received an immunization? What is that person's likelihood of developing asthma, Guillain-Barré syndrome (GBS), a heart attack, cancer, infertility, epilepsy, lupus, or any other condition? As a starting point, the vaccinee should have the same expectation of good health as a comparable unvaccinated person. Conversely, the vaccinee has the same expectation of a future medical problem as the unvaccinated person as well. It is unreasonable to expect zero cases of asthma, GBS, heart attack, or other conditions in a group of 1000 or 10,000 or 100,000 vaccinees. Nonetheless, surveillance data collected from vaccination programs may be helpful in assessing whether medical problems in vaccinees exceed the baseline expectation.

If there is no increased rate of illness among vaccinated people, relative to a proper comparison with similar unvaccinated people, there is evidence of no cause-and-effect relationship, at least according to the statistical power of the cohort evaluated.

A cause-and-effect relationship cannot and should not be ruled out entirely, however. The studies may not have been large enough to rule out rare effects or effects of small magnitude. Scientists must keep an open mind, of course, striving to avoid the words "never" or "always."

Although GBS has a reputation as being caused by immunization, its most common cause is an infection, especially with *Campylobacter* bacteria (Pickering, 2006; Grabenstein, 2001b; Kuwabara, 2004). GBS is a demyelinating disease that damages nerves, causing temporary weakness. From 80% to 85% patients recover fully from GBS. The general population has a risk of about one case of GBS per 60,000 people per year. Therefore, our baseline expectation is that GBS will occur among a large cohort of vaccinees, although an appropriate time interval (or "time window") must be applied to assure a fair comparison. The risk of GBS has been slightly elevated among influenza vaccine recipients in some years and some cohorts (perhaps 1–10 cases per million vaccinees), but not others (Grabenstein, 2001b; Lasky et al., 1998; Haber et al., 2004).

As this chapter is being written, the CDC and the FDA are investigating a cluster of GBS cases in recipients of a meningococcal protein-conjugated vaccine (Menactra, Sanofi Pasteur), attempting to measure the attributable risk in an unbiased manner and in perspective to the risk of a meningococcal disease in the population for whom immunization is recommended (Centers for Disease Control and Prevention, 2006a, 2006b). The analysis is complicated by seasonal variation in GBS incidence and Menactra administration, which both appear to increase independently in summer months.

Hepatitis B immunization can prevent liver cancer (Grabenstein, 2006), and human papillomavirus immunization can prevent cervical cancer (Advisory Committee on Immunization Practices, 2007). Can immunization cause cancer? Soon after SV40 was discovered in 1960, SV40 was found to be a contaminant of monkey kidney cells that is used to manufacture inactivated (and to a lesser extent oral) poliovirus vaccines. In 1961, the government required all poliovirus vaccines to be free of SV40, but more than a million people had already received vaccine doses containing this virus (Brown and Lewis, 1998; Strickler et al., 1998; Institute of Medicine, 2002). SV40 can cause some cancers in rodents. Recently, researchers found SV40 in people with rare cancers (e.g., ependymomas, osteosarcomas, mesotheliomas), but many of these people were too young to have received poliovirus vaccines containing SV40. So if unvaccinated people have SV40 virus in their tumors, one needs to ask whether vaccinated people with these rare cancers are more likely to involve SV40 than unvaccinated people. Evidence shows that the risk is the same, not elevated. There is no indication to date that an increased risk due to SV40 exists. (Brown and Lewis, 1998; Strickler et al., 1998; Institute of Medicine, 2002).

The same logic applies to other adverse events. A fraction of unvaccinated pregnant women will spontaneously miscarry their fetuses. With this rate as a starting point, we then expect a comparable fraction of vaccinated pregnant women to miscarry. Similarly, we need to know how often a heart attack, leukemia, thyroid disease, diabetes, or any other health problem happens among unvaccinated people before we can begin to assess whether vaccination increases the baseline (or background) rate.

The ideal situation is to know how often the adverse event happens among people who share every other risk factor except immunization. In such a setting, the comparison group would have the same age distribution, gender mix, underlying health status, and other personal characteristics as those of people who get the vaccine. Returning to the GBS example, it might help to know the specific incidence rates of GBS among adolescents, 20- to 45-year-old adults, as well as people 65 years or older, depending on the specific comparison being made. This helps compare apples to apples, and preferably red delicious apples to red delicious apples. Only from the baseline can elevations above baseline be recognized.

Whenever the symptom or condition is known to occur among people who have not been vaccinated, one needs to know whether the condition happens more often in vaccinated people than in a comparable group of unvaccinated people. A stepwise process has been proposed for this kind of analysis (Grabenstein, 1999, 2000, 2001a).

Any vaccine program administrator should plan for the occurrence of heart attacks among adult vaccinees. Shortly after the 1976 swine influenza immunization program began, several heart attacks on the same day in the same city among influenza vaccinees triggered a series of media reports that shook public confidence in the vaccine (Rubin and Hendy, 1977). Objective analyses showed that the deaths corresponded to the expected background death rate among unvaccinated people, but delayed reporting of this perspective did little to restore confidence in 1976.

Similarly, several smallpox vaccinees developed angina, heart attacks, or other ischemic events during 2003. Data among U.S. military personnel indicated that smallpox-vaccinated and -unvaccinated personnel developed ischemic events at comparable rates (Centers for Disease Control and Prevention, 2004). The DoD smallpox vaccinee population is younger on average than the DHHS population, but the DoD population included more than 80,000 vaccinees 40 years or older. An epidemiologic analysis of observed ischemic cardiac events occurring within 3 weeks of immunization among civilian vaccinees between January 24, 2003, and August 22, 2003, compared to best estimates of expected numbers of incident cardiac events in a similar population, did not support an increase in ischemic cardiac events within this population (Baciu et al., 2005; Centers for Disease Control and Prevention, 2003c). The myocardial infarction (MI) cases and MI-associated deaths observed among DoD and DHHS vaccinees within 3 weeks of vaccination at that point exceeded the point estimates of expected MIs and deaths, but both observations remained within the 95% predictive interval for these events. Still, smallpox vaccination program administrators kept in place the cardiac exemption criteria hastily imposed when the ischemic cases were first recognized. The conservative approach was adopted in a setting in which the incidence rate of variola infection was still zero, the backdrop against which to compare the vaccine's benefit-to-risk ratio.

During the comparisons and contrasts of military and civilian smallpox vaccinees, the issue of generalizability came up several times. Fitness requirements for military service exceed those of some civilian occupations, but not all. People fit for military service are generalizable to civilian first responders (e.g., police, fire and ambulance personnel) and other employment sectors with expectations for cardiovascular stamina. What group is analogous to civilian healthcare workers? Physicians and nurses' health studies offer possibilities, as would employed beneficiaries of health maintenance organizations and other occupational groups.

Anthrax vaccine adsorbed (BioThrax, Emergent BioSolutions) has been the target of prolonged skepticism, with assertions that the vaccine is associated with an extraordinary variety of adverse health effects. To provide an objective basis for assessing the vaccine's safety profile, the DoD conducted an extraordinary array of post-marketing safety studies. These studies involved cohort studies of acute symptoms (Pittman et al., 2001a, 2001b, 2002a, 2002b; Wasserman et al., 2003; Hoffman et al., 2003; Rehme et al., 2002; McNeil et al., 2007), hospitalizations (Joellenbeck et al., 2002; Lange et al., 2003), disability evaluations (Sulsky et al., 2004), and reproductive outcomes (Wiesen and Littell, 2002; Catherino et al., 2005) as well as secondary review of the spontaneous reports to VAERS (Sever et al., 2002, 2004). Several of the cohort studies span multiple years after immunization (Joellenbeck et al., 2002; Lange et al., 2003; Sulsky et al., 2004; Catherino et al., 2005; Pittman et al., 2004). Multiple studies employed active surveillance, defined as data collected at fixed time points, without relying on a recipient to take a special effort to report a symptom or condition (Pittman et al., 2002a, 2002b, 2004; Wasserman et al., 2003). Other studies featured systematic surveillance, that is, data collected automatically and electronically, without any reporting action required by a clinician or vaccinee (Lange et al., 2003; Rehme et al., 2002; Sulsky et al., 2004; Catherino et al., 2005).

After the National Academy of Sciences heard from vaccinees and comprehensively reviewed the accumulated scientific data, the academy concluded that anthrax vaccine has an adverse-reaction profile similar to other adult vaccines (Joellenbeck et al., 2002). The anthrax immunization program also pointed out needed improvements in the way the DoD exchanges information with military personnel and provides clinical immunization services in general. Skepticism about the value of anthrax immunization among military personnel was largely resolved after the anthrax bio-attacks of fall 2001 showed the public the lethality of anthrax spores.

PLANNING APPLIED TO PANDEMIC INFLUENZA

To assess the general principles discussed above, let us now consider a pandemic influenza scenario. Similar scenarios involving other public health emergencies could be envisioned. This scenario assumes use of two hypothetical vaccines to prevent A/H9N9 influenza infection. One vaccine is a subunit (split-virion) influenza vaccine administered intramuscularly, and the other is a live, attenuated influenza virus vaccine administered as an oral capsule. Neither of the products actually exists.

In a hypothetical pandemic of A/H9N9 influenza, the news media would likely herald a rising number of human cases and deaths due to the pandemic, as the virus spread across the globe. But the time needed to manufacture and distribute hundreds of millions of vaccine doses could lag behind pandemic spread of a lethal virus. This would likely create a scenario of "vaccine euphoria" and (perhaps irrational) vaccine demand, which are further discussed below.

With more than 50 years of experience with injectable whole- and subunit-virion A/H1N1, A/H2N2, A/H3N2, and other influenza virus vaccines, program administrators would have a substantial body of knowledge from which to base policy decisions. And yet the influenza vaccine formula has differed almost every year since 1945 (Grabenstein, 2006), so in some sense, we have 1–2 years of experience at the viral strain level with ~40 different (albeit similar) influenza vaccines.

Even with the immense experience with smallpox vaccine in the mid-20th century, new findings related to myopericarditis came to light in 2003 (Grabenstein and Winkenwerder, 2003; Poland et al., 2005; Arness et al., 2004; Halsell et al., 2003; Eckart et al., 2004). The intermittently elevated risk of GBS after influenza immunization, including the 1976 swine influenza vaccine incident, means that program administrators must focus on a potential association with A/H9N9 (Grabenstein, 2001b; Lasky et al., 1998; Haber et al., 2004). Similarly, heart attacks and ischemic disease must be monitored closely, for reasons cited above.

The United States has an increasing amount of experience with live, attenuated intranasal influenza vaccines, first licensed in 2003 (FluMist, MedImmune) (Grabenstein, 2006). The hypothetical A/H9N9 oral vaccine capsules in this scenario may be difficult for small children to swallow. The relative safety of a live, attenuated vaccine for pregnant women and immune-compromised people will likely be unknown. While the subunit A/H9N9 vaccine could be preferred for these subgroups, it is possible that local shortages of the subunit vaccine might force people to face uncertain risks in the setting of local incidence of the pandemic infection.

While many people would be glad to have access to either of the hypothetical A/H9N9 vaccines, some would shun them out of distrust of the government or for other reasons. Behavioral scientists and risk communicators will be important members of the vaccine program team at all levels: national, regional, state, and local.

What are the common features of infectious crises? Decades ago, when hundreds of thousands of Americans contracted diphtheria, poliomyelitis, or measles each year, few people stopped to ask about vaccine side effects. Vaccines have succeeded so well that many of these vaccine-preventable diseases are virtually unheard of by clinicians or the public. This phenomenon has caused some people to scrutinize side effects of vaccines, forgetting about the disease that the vaccines have successfully vanquished. Some aspects of this situation warrant the label "vaccine phobia."

On the other hand, what might be termed "vaccine euphoria" has occurred with meningococcal polysaccharide vaccine in the midst of meningococcal outbreaks (Grabenstein, 2001a; Gold, 1992; Hume, 1992; Centers for Disease Control and Prevention, 1999). Demand for meningococcal vaccine in community outbreaks arises periodically, even if the desire for the vaccine in such settings may exceed an objective assessment of the vaccine's utility.

For example, public clamoring for meningococcal polysaccharide vaccine during an early 1998 outbreak in Rhode Island exceeded the health department's plans (Centers for Disease Control and Prevention, 1999). The health department planned to offer meningococcal vaccine to all state residents between 2 and 22 years of age over the course of 6–12 months. Long lines and media attention caused that plan to be condensed into 6–8 weeks.

If called upon to oversee the safety surveillance system for a new vaccine or a vaccine suddenly expanded to a larger population, it may be helpful for a vaccine program administrator to consider a multifaceted approach. There are several useful data collection methods that include active (or systematic) and spontaneous surveillance covering all key demographic categories of interest (e.g., adults, children, elderly, pregnant women) over a variety of time windows (e.g., 72h, 2 weeks, longer). Obviously, a thorough literature search of accumulated or "established" knowledge to inform subsequent study designs is also important.

To collect common and acute symptom data, active surveillance systems could be implemented in selected clinics, perhaps using electronic symptom diaries (e.g., entries via telephone or Internet). To collect data about rare adverse events, large linked databases can be used to survey large populations. Ideally, these databases will permit comparison and contrast of similar vaccinated and unvaccinated cohorts. To complement the two approaches described in the paragraph above, a spontaneous data collection system (e.g., VAERS) will allow unanticipated adverse events to come to light for definitive evaluation by other means.

RISK COMMUNICATION

There is a divergence, a tension, between the way scientists approach vaccine safety and the way many members of the public do so. Good scientists gather evidence in a manner that minimizes bias. They take care to rule out competing explanations for an observation before they would assert that "A" caused "B" (Grabenstein and Wilson, 1999; Grabenstein, 1999, 2000, 2001a; Brewer et al., 2007; Glik, 2007; Dittmann, 2001).

Such self-discipline is not required, of course, among the general public. People believe what they wish. And in our society, they also are free to voice, and even shout, their opinions whether or not those opinions have an objective basis in fact.

When it comes to hearing about new situations, some people are entranced, and even mesmerized, by anecdotes, stories of individuals like themselves. Human nature tends to identify with individuals. As a result, documentaries and news magazines often begin by introducing the audience to a man, woman, or child affected by the subject of the story. Depicting events in the lives of these people personalizes the story, helping the author or scriptwriter connect with the audience.

Media stories exploring relationships between vaccines and their adverse events often focus on a family. Some member of such a family may have autism, multiple sclerosis, or some other serious health condition. This family will be searching for an explanation for the tragedy. The family may state, or the narrator may imply, that immunization could or should be considered the reason for the tragedy.

However, how can we tell whether a vaccine caused the problem? Is some other process really to be blamed? Separating true causes from false ones is vitally important for both society and individuals. Above, we have considered cause-and-effect in general and several health problems (e.g., GBS) in particular. Using a disciplined, rational, methodical approach to attributing blame where appropriate, or clearing a vaccine's reputation where there is no basis for a link, continues to be the proper approach. Using the tools of science, we are most likely to find the truth and least likely to fool ourselves about vaccine safety.

The key to assessing vaccine safety is to be objective, assess carefully all available data, and be open to new findings.

References

Advisory Committee on Immunization Practices Poliomyelitis prevention in the United States: introduction of a sequential vaccination schedule of inactivated poliovirus vaccine followed by oral poliovirus vaccine. MMWR 1997; 46(RR-3):1–25, errata 183. www.cdc.gov/mmwr/pdf/rr/rr4603.pdf.

Advisory Committee on Immunization Practices Rotavirus vaccine for the prevention of rotavirus gastroenteritis among children. MMWR 1999a; 48(RR-2):1–23, ftp://ftp.cdc.gov/pub/Publications/mmwr/rr/rr4802.pdf.

Advisory Committee on Immunization Practices Revised recommendations for routine poliomyelitis vaccination. MMWR 1999b; 48:590, www.cdc.gov/epo/mmwr/preview/mmwrhtml/mm4827a4.htm.

Advisory Committee on Immunization Practices Withdrawal of rotavirus vaccine recommendation. MMWR 1999c; 48:1007, ftp://ftp.cdc.gov/pub/Publications/mmwr/wk/mm4843.pdf.

Advisory Committee on Immunization Practices Poliomyelitis prevention in the United States: updated recommendations. MMWR 2000; 49(RR-5):1–22, ftp://ftp.cdc.gov/pub/Publications/mmwr/rr/rr4905.pdf.

Advisory Committee on Immunization Practices Vaccinia (smallpox) vaccine. MMWR 2001; 50(RR-10):1–25, www.cdc.gov/mmwr/PDF/rr/rr5010.pdf.

Advisory Committee on Immunization Practices Prevention of rotavirus gastroenteritis among infants and children: recommendations of the Advisory Committee on Immunization Practices (ACIP). MMWR 2006; 55(RR-12):1–13.

Advisory Committee on Immunization Practices Quadrivalent human papillomavirus vaccine: recommendations of the Advisory Committee on Immunization Practices. MMWR 2007a; 56(RR-2):1–24.

Advisory Committee on Immunization Practices Prevention of varicella: recommendations of the Advisory Committee on Immunization Practices. MMWR 2007b; 56(RR-4):1–40, http://www.cdc.gov/mmwr/PDF/rr/rr5604.pdf.

Advisory Committee on Immunization Practices Prevention and control of influenza: recommendations of the Advisory Committee on Immunization Practices (ACIP). MMWR 2007c; 56(RR-5):1–54, http://www.cdc.gov/mmwr/PDF/rr/rr5606.pdf.

Advisory Committee on Immunization Practices, Healthcare Infection Control Practices Advisory Committee Recommendations for using smallpox vaccine in pre-event vaccination program: supplemental recommendations. MMWR 2003; 52(RR-7):1–16, www.cdc.gov/mmwr/PDF/rr/rr5207.pdf.

Arness, M.K., Eckart, R.E., Love, S.S., Atwood, J.E., Wells, T.S., Engler, R.J., Collins, L.C., Ludwig, S.L., Riddle, J.R., Grabenstein, J.D. and Tornberg, D.N. Myopericarditis following smallpox vaccination. Am. J. Epidemiol. 2004; 160:642–651.

Baciu, A., Anason, A.P., Stratton, K. and Strom, B.Committee on Smallpox Vaccination Program Implementation The Smallpox Vaccination Program: Public Health in an Age of Terrorism. Washington, D.C.: Institute of Medicine, 2005.

Beatrice, M.G. Regulation, licensing and inspection of biological products. Pharm. Eng. 1991; 10:29–35.

Brewer, N.T., Chapman, G.B., Gibbons, F.X., Gerrard, M., McCaul, K.D. and Weinstein, N.D. Meta-analysis of the relationship between risk perception and health behavior: the example of vaccination. Health Psychol. 2007; 26:136–145.

Brown, F. and Lewis, A.M. (Eds.). Simian virus 40 (SV40): a possible human polyomavirus, symposium proceedings. Dev. Biol. Stand. 1998; 94:1–406.

Catherino, W.H., Levi, A., Kao, T.-C., Leondires, M.P., McKeeby, J. and Segars, J.H. The anthrax vaccine does not affect semen parameters, embryo quality, or pregnancy outcome in couples with a vaccinated male military service member. Fertil. Steril. 2005; 83:480–483.

Centers for Disease Control and Prevention Intussusception among recipients of rotavirus vaccine—United States, 1998–1999. MMWR 1999a; 48:577–581, www.cdc.gov/epo/mmwr/preview/mmwrhtml/mm4827a1.htm.

Centers for Disease Control and Prevention Meningococcal disease—New England, 1993–1998. MMWR 1999b; 48:629–633.

Centers for Disease Control and Prevention Smallpox vaccination and adverse events: guidance for clinicians. MMWR 2003a; 52(RR-4):1–30, www.cdc.gov/mmwr/PDF/rr/rr5204.pdf.

Centers for Disease Control and Prevention Surveillance for safety after immunization: Vaccine Adverse Event Reporting System (VAERS)—United States, 1991–2001. MMWR 2003b; 52(SS-01):1–24, Errata 52:113. www.cdc.gov/mmwr/PDF/ss/ss5201.pdf.

Centers for Disease Control and Prevention Update: cardiac and other adverse events following civilian smallpox vaccination—United States, 2003. MMWR 2003c; 52:639–642.

Centers for Disease Control and Prevention Suspension of rotavirus vaccine after reports of intussusception—United States, 1999. MMWR 2004a; 52:786–789, www.cdc.gov/mmwr/PDF/wk/mm5334.pdf.

Centers for Disease Control and Prevention Record of the Meeting of the Advisory Committee on Immunization Practices, October 27–28, 2004. Atlanta: CDC, 2004b, www.cdc.gov/nip/acip/minutes/acip-min-oct04.pdf. Accessed July 15, 2007.

Centers for Disease Control and Prevention Guillain-Barré syndrome among recipients of Menactra meningococcal conjugate vaccine—United States, October 2005–February 2006. MMWR 2006a; 55:364–367.

Centers for Disease Control and Prevention Update: Guillain-Barré syndrome among recipients of Menactra® meningococcal conjugate vaccine—United States, June 2005–September 2006. MMWR 2006b; 55:1120–1124.

Davis, R.L., Kolczak, M., Lewis, E., Nordin, J., Goodman, M., Shay, D.K., Platt, R., Black, S., Shinefield, H. and Chen, R.T. Active surveillance of vaccine safety: a system to detect early signs of adverse events. Epidemiology 2005; 16:336–341.

DeStefano, F. Vaccine safety datalink research group. The vaccine safety datalink project. Pharmacoepidemiol. Drug Saf. 2001; 10:403–406.

Dittmann, S. Vaccine safety: risk communication—a global perspective. Vaccine 2001; 19:2446–2456.

Eckart, R.E., Love, S.S., Atwood, J.E., Arness, M.K., Cassimatis, D.C., Campbell, C.L., Boyd, S.Y., Murphy, J.G., Swerdlow, D.L., Collins, L.C., Riddle, J.R., Tornberg, D.N., Grabenstein, J.D. and Engler, R.J.M. DoD smallpox vaccination clinical evaluation team. Incidence and follow-up of inflammatory cardiac complications following smallpox vaccination. J. Am. Coll. Cardiol. 2004; 44:201–205.

Fenichel, G.M. Neurologic risk of immunization: report of the therapeutics and technology assessment subcommittee of the American Academy of Neurology. Neurology 1999; 52:1546–1552.

Fenner, F., Henderson, D.A., Arita, I., Jezek, Z. and Ladnyi, I.D. Smallpox and Its Eradication. Geneva: World Health Organization, 1988, www.who.int/emc/diseases/smallpox/Smallpoxeradication.html.

Food and Drug Administration. Vaccine Product Approval Process. www.fda.gov/cber/vaccine/vacappr.htm. Accessed July 15, 2007a.

Food and Drug Administration. Postmarketing Study Commitments. www.fda.gov/cder/pmc/. Accessed August 22, 2007b.

FUTURE II Study Group Quadrivalent vaccine against human papillomavirus to prevent high-grade cervical lesions. N. Engl. J. Med. May 10, 2007; 356(19):1915–1927.

Garland, S.M., Hernandez-Avila, M., Wheeler, C.M., Perez, G., Harper, D.M., Leodolter, S., Tang, G.W., Ferris, D.G., Steben, M., Bryan, J., Taddeo, F.J., Railkar, R., Esser, M.T., Sings, H.L., Nelson, M., Boslego, J., Sattler, C., Barr, E. and Koutsky, L.A. Females united to unilaterally reduce endo/ectocervical disease (FUTURE) I investigators. Quadrivalent vaccine against human papillomavirus to prevent anogenital diseases. N. Engl. J. Med. May 10, 2007; 356(19):1928–1943.

Glik, D.C. Risk communication for public health emergencies. Annu. Rev. Public Health 2007; 28:33–54.

Gold, R. Meningococcal disease in Canada: 1991–1992. Can. J. Public Health 1992; 83:5–6.

Grabenstein, J.D. Anaphylaxis: epinephrine and emergency responses. Hosp. Pharm. 1997; 32:1377–1378, 1382–1384, 1387–1389.

Grabenstein, J.D. Vaccine side effects: separating mirage from reality. J. Am. Pharm. Assoc. 1999; 39:417–419.

Grabenstein, J.D. Side effects of vaccination: a consumer's guide to cause and effect. Hosp. Pharm. 2000; 35:1308–1318.

Grabenstein, J.D. The individual perspective: did this vaccine harm this person? Hosp. Pharm. 2001a; 36:413–418, 420, 464.

Grabenstein, J.D. Guillain-Barré syndrome and vaccination: usually unrelated. Hosp. Pharm. 2001b; 36:199–200, 202–204, 206–207.

Grabenstein, J.D. ImmunoFacts: Vaccines and Immunologic Drugs. Saint Louis: Wolters Kluwer Health, 2006.

Grabenstein, J.D. and Wilson, J.P. Are vaccines safe? Risk communication applied to vaccines. Hosp. Pharm. 1999; 34:713–714, 717–718, 721–723, 727–729.

Grabenstein, J.D. and Winkenwerder, W., Jr. U.S. military smallpox vaccination program experience. JAMA 2003; 289:3278–3282, jama.ama-assn.org/cgi/reprint/289/24/3278.pdf.

Haber, P., DeStefano, F., Angulo, F.J., Iskander, J., Shadomy, S.V., Weintraub, E. and Chen, R.T. Guillain-Barré syndrome following influenza vaccination. JAMA 2004; 292:2478–2481.

Halsell, J.S., Riddle, J.R., Atwood, J.E., Gardner, P., Shope, R., Poland, G.A., Gray, G.C., Ostroff, S., Eckart, R.E., Gibson, R.L., Grabenstein, J.D., Arness, M.K. and Tornberg, D.N. Department of Defense smallpox vaccination clinical evaluation team. Myopericarditis following smallpox vaccination among vaccinia-naive U.S. military personnel. JAMA 2003; 289:3283–3289.

Hoffman, K., Costello, C., Menich, M., Grabenstein, J.D. and Engler, R.J.M. Using a structured medical note for determining the safety profile of anthrax vaccine for U.S. soldiers in Korea. Vaccine 2003; 21:4399–4409, www.vaccines.mil/documents/library/Hoffman2003Vaccine4399.pdf.

Hume, S.E. Mass voluntary immunization campaigns for meningococcal disease in Canada: media hysteria. JAMA 1992; 267:1833–1834, 1837–1838.

Hussain, A.I., Johnson, J.A., Da Silva Freire, M. and Heneine, W. Identification and characterization of avian retroviruses in chick-embryo-derived yellow fever vaccines: investigation of transmission to vaccine recipients. J. Virol. 2003; 77: 1105–1111.

Hussain, A.I., Shanmugam, V., Switzer, W.M., Tsang, S.X., Fadly, A., Thea, D., Helfand, R., Bellini, W.J., Folks, T.M. and Heneine, W. Lack of evidence of endogenous avian leucosis virus and endogenous avian retrovirus transmission to measles mumps rubella vaccine recipients. Emerg. Infect. Dis. 2001; 7:66–72.

Institute of Medicine Immunization Safety Review: Simian Virus-40 Contamination of Polio Vaccine and Cancer. Washington, D.C.: National Academy of Sciences, October 2002. www.nap.edu/catalog/10534.html.

Joellenbeck, L.M., Zwanziger, L., Durch, J.S. and Strom, B.L. (Eds.) The Anthrax Vaccine: Is It Safe? Does It Work? Washington, D.C.: National Academy Press, April 2002.

Joura, E.A., Leodolter, S., Hernandez-Avila, M., Wheeler, C.M., Perez, G., Koutsky, L.A., Garland, S.M., Harper, D.M., Tang, G.W., Ferris, D.G., Steben, M., Jones, R.W., Bryan, J., Taddeo, F.J., Bautista, O.M., Esser, M.T., Sings, H.L., Nelson, M., Boslego, J.W., Sattler, C., Barr, E. and Paavonen, J. Efficacy of a quadrivalent prophylactic human papillomavirus (types 6, 11, 16, and 18) L1 virus-like-particle vaccine against high-grade vulval and vaginal lesions: a combined analysis of three randomised clinical trials. Lancet 2007; 369:1693–1702.

Khakoo, G.A. and Lack, G. Recommendations for using MMR vaccine in children allergic to eggs. BMJ 2000; 320:929–932.

Kuwabara, S. Guillain-Barré syndrome: epidemiology, pathophysiology, and management. Drugs 2004; 64:597–610.

Lane, J.M. and Millar, J.D. Risks of smallpox vaccination complications in the United States. Am. J. Epidemiol. 1971; 93:238–240.

Lane, J.M., Millar, J.D. and Neff, J.M. Smallpox and smallpox vaccination policy. Annu. Rev. Med. 1971; 22:251–272.

Lange, J.L., Lesikar, S.E., Brundage, J.F. and Rubertone, M.V. Comprehensive systematic surveillance for adverse effects of anthrax vaccine adsorbed, U.S. Armed Forces, 1998–2000. Vaccine 2003; 21:1620–1628, www.anthrax.mil/documents/library/science.pdf.

Lasky, T., Terracciano, G.J., Magder, L., Koski, C.L., Ballesteros, M., Nash, D., Clark, S., Haber, P., Stolley, P.D., Schonberger, L.B. and Chen, R.T. The Guillain-Barré syndrome and the 1992–1993 and 1993–1994 influenza vaccines. N. Engl. J. Med. 1998; 339:1797–1802.

McNeil, M.M., Chiang, I.S., Wheeling, J.T. and Zhang, Y. Short-term reactogenicity and gender effect of anthrax vaccine: analysis of a 1967–1972 study and review of the 1955–2005 medical literature. Pharmacoepidemiol. Drug Saf. 2007; 16:259–274.

Meadows, M. Why drugs get pulled off the market. FDA Consumer 2002; 36(Jan–Feb):4, www.fda.gov/fdac/features/2002/102_drug.html.

Mullooly, J., Drew, L., DeStefano, F., Maher, J., Bohlke, K., Immanuel, V., Black, S., Lewis, E., Ray, P., Vadheim, C., Lugg, M. and Chen, R. Quality assessments of HMO diagnosis databases used to monitor childhood vaccine safety. Methods Inf. Med. 2004; 43:163–170.

Nathanson, N. and Langmuir, A.D. The Cutter incident: poliomyelitis following formaldehyde-inactivated poliovirus vaccination in the United States during the spring of 1955. Am. J. Hyg. 1963; 78:16–81.

Nordenberg, T. When is a medical product too risky? FDA Consumer 1999; 33(Sep–Oct):1, www.fda.gov/fdac/features/1999/599_med.html.

Norman, J.E., Beebe, G.W., Hoofnagle, J.H. and Seeff, L.B. Mortality follow-up of the 1942 epidemic of hepatitis B in the U.S. Army. Hepatology 1993; 18:790–797.

Offit, P.A. The Cutter Incident: How America's First Polio Vaccine Led to the Growing Vaccine Crisis. New Haven: Yale University Press, 2007.

Olmstead, S.S., Grabenstein, J.D., Jain, A.K., Comerford, W., Giambo, P., Johnson, P., Mopsik, J., Zimmerman, B. and Lurie, N. Use of an electronic monitoring system for self-reporting smallpox vaccine reactions. Biosecur. Bioterror. 2005; 3:198–206.

Olmstead, S.S., Grabenstein, J.D., Jain, A.K. and Lurie, N. Patient experience with an electronic monitoring system to assess vaccination responses. Health Expect. 2006; 9:110–117.

Parish, H.J. A history of immunizations. Edinburgh: E&S Livingstone Ltd, 1965.

Payne, D.C., Franzke, L.H., Stehr-Green, P.A., Schwartz, B. and McNeil, M.M. Development of the Vaccine Analytic Unit's research agenda for investigating potential adverse events associated with anthrax vaccine adsorbed. Pharmacoepidemiol. Drug Saf. 2007; 16:46–54.

Pickering, L.K. (Ed.) Red Book: Report of the Committee on Infectious Diseases, 27th ed. Elk Grove Village, IL: American Academy of Pediatrics, 2006.

Poland, G.A., Grabenstein, J.D. and Neff, J.M. The U.S. smallpox vaccination program: a review of a large modern era smallpox vaccination implementation program. Vaccine 2005; 23(Mar 18):2078–2081.

Pittman, P.R., Coonan, K.M., Gibbs, P.H., Scott, H.M., Cannon, T.L. and McKee, K.T., Jr. Long-term health effects of repeated exposure to multiple vaccines. Vaccine 2004; 23:525–536, www.vaccines.mil/documents/library/Longtermhealtheffects.pdf.

Pittman, P.R., Gibbs, P.H., Cannon, T.L. and Friedlander, A.M. Anthrax vaccine: short-term safety experience in humans. Vaccine 2001a; 20:972–978, www.vaccines.mil/documents/library/Pittman2002Vaccine972.pdf.

Pittman, P.R., Hack, D., Mangiafico, J., Gibbs, P., McKee, K.T., Jr., Eitzen, E.M., Friedlander, A.M. and Sjogren, M.H. Antibody response to a delayed booster dose of anthrax vaccine and botulinum toxoid. Vaccine 2002a; 20(May 15):2107–2115, www.vaccines.mil/documents/library/Antibody_resp.pdf.

Pittman, P.R., Kim-Ahn, G., Pifat, D.Y., Coonan, K., Gibbs, P., Little, S., Pace-Templeton, J.G., Myers, R., Parker, G.W. and Friedlander, A.M. Anthrax vaccine: safety and immunogenicity of a dose-reduction, route comparison study in humans. Vaccine 2002b; 20:1412–1420, www.vaccines.mil/documents/library/Immunogenicity.pdf.

Pittman, P.R., Mangiafico, J.A., Rossi, C.A., Cannon, T.L., Gibbs, P.H., Parker, G.W. and Friedlander, A.M. Anthrax vaccine: increasing intervals between the first two doses enhances antibody response in humans. Vaccine 2001b; 18:213–216, www.vaccines.mil/documents/library/Vaccine19(2-3)213-6.pdf.

Rehme, P.A., Williams, R. and Grabenstein, J.D. Ambulatory medical visits among anthrax vaccinated and unvaccinated personnel after return from southwest Asia. Mil. Med. 2002; 167:205–210, www.anthrax.mil/documents/library/SWasia.pdf.

Rubin, D.R. and Hendy, V. Swine influenza and the news media. Ann. Intern. Med. 1977; 87:769–774.

Sawyer, W.A., Meyer, K.F., Eaton, M.D., Bauer, J.H., Putnam, P. and Schwentker, F.F. Jaundice in Army personnel in the western region of the United States and its relation to vaccination against yellow fever: Part I. Am. J. Hyg. 1944a; 39:337–430.

Sawyer, W.A., Meyer, K.F., Eaton, M.D., Bauer, J.H., Putnam, P. and Schwentker, F.F. Jaundice in Army personnel in the western region of the United States and its relation to vaccination against yellow fever: Parts II, III and IV. Am. J. Hyg. 1944b; 40:35–107.

Seeff, L.B., Beebe, G.W., Hoofnagle, J.H., Norman, J.E., Buskell-Bales, Z., Waggoner, J.G., Kaplowitz, N., Koff, R.S., Petrini, J.L., Jr., Schiff, E.R. et al. A serologic follow-up of the 1942 epidemic of post-vaccination hepatitis in the United States Army. N. Engl. J. Med. 1987; 316:965–970.

Sever, J.L., Brenner, A.I., Gale, A.D., Lyle, J.M., Moulton, L.H., Ward, B.J. and West, D.J. Safety of anthrax vaccine: an expanded review and evaluation of adverse events reported to the Vaccine Adverse Event Reporting System (VAERS). Pharmacoepidemiol. Drug Saf. 2004; 13:825–840, www.anthrax.mil/media/pdf/SeverArticle.pdf.

Sever, J.L., Brenner, A.I., Gale, A.D., Lyle, J.M., Moulton, L.H. and West, D.J. Safety of anthrax vaccine: a review by the Anthrax Vaccine Expert Committee (AVEC) of adverse events reported to the Vaccine Adverse Event Reporting System (VAERS). Pharmacoepidemiol. Drug Saf. 2002; 11:189–202, www.anthrax.mil/media/pdf/AVEC_ms.pdf.

Singleton, J.A., Lloyd, J.C., Mootrey, G.T., Salive, M.E. and Chen, R.T. VAERS Working Group. An overview of the vaccine adverse event reported system (VAERS) as a surveillance system. Vaccine 1999; 17:2908–2917.

Smith, J.S. Patenting the Sun: Polio & the Salk Vaccine. New York: William Morrow & Co., 1990.

Steere, A.C., Sidand, V.K., Meurice, F., Parenti, D.L., Fikrig, E., Schoen, R.T., Nowakowski, J., Schmid, C.H., Laukamp, S., Buscarino, C. and Krause, D.S. Lyme disease vaccine study group. N. Engl. J. Med. 1998; 339:209–215.

Strickler, H.D., Rosenberg, P.S., Devesa, S.S., Hertel, J., Fraumeni, J.F. and Goedert, J.J. Contamination of poliovirus vaccines with simian virus 40 (1955–1963) and subsequent cancer rates. JAMA 1998; 279:292–295.

Sulsky, S.I., Grabenstein, J.D. and Delbos, R.G. Disability among U.S. Army personnel vaccinated against anthrax. J. Occup. Environ. Med. 2004; 46:1065–1075, www.anthrax.mil/documents/library/Anthrax2004.pdf.

Title 21 U.S. Code of Federal Regulations, Section 600.3.

Varricchio, F., Iskander, J., DeStefano, F., Ball, R., Pless, R., Braun, M.M. and Chen, R.T. Understanding vaccine safety information from the Vaccine Adverse Events Reporting System. Pediatr. Infect. Dis. J. 2004; 23:287–294.

Wasserman, G.M., Grabenstein, J.D., Pittman, P.R., Rubertone, M.V., Gibbs, P.P., Wang, L.Z. and Golder, L.G. Analysis of adverse events after anthrax vaccination in U.S. Army medical personnel. J. Occup. Environ. Med. 2003; 45:222–233, www.anthrax.mil/documents/library/AnthraxVaccineEvaluation.pdf.

Waters, T.D., Anderson, P.S., Jr., Beebe, G.W. and Miller, R.W. Yellow fever vaccination, avian leucosis virus, and cancer risk in man. Science 1972; 177:76–77.

Wiesen, A.R. and Littell, C.T. Relationship between prepregnancy anthrax vaccination and pregnancy and birth outcomes among U.S. Army women. JAMA 2002; 287:1556–1560, jama.ama-assn.org/cgi/reprint/287/12/1556.pdf.

CHAPTER

16

Special Issues around Veterinary Vaccines

L. Garry Adams[1], *Lorne Babiuk*[2], *David McGavin*[3] *and Robert Nordgren*[4]

[1]College of Veterinary Medicine, Texas A&M University, College Station, TX, USA

[2]University of Alberta, Edmonton, Alberta, Canada

[3]Pfizer Animal Health, Kalamazoo, MI, USA

[4]Technology Acquisition and North American Biological Development, Merial Limited, Duluth, GA, USA

OUTLINE

ABSTRACT

The majority of vaccines licensed for controlling infectious disease of veterinary species today are based on technology that was introduced by Jenner using live vaccines and Pasteur using killed whole organism vaccines 200 and 100 years ago, respectively, yet this former technology has not stopped several successful vaccination programs from being developed. Much of veterinary vaccinology is driven by the realities that exist in raising production animals or working in veterinary practice, where making a living depends on keeping the animals healthy, because it is an industry where vaccines are like insurance policies—protection from events that one hopes never happen. For example, the USDA recognizes these varying levels of protection in the way that they allow label claims: (1) "aids in disease control," (2) "for the prevention of disease," and (3) "for the prevention of infection." Additionally there may be indirect protection, or herd immunity, that results from vaccination of sufficient numbers of animals in a given population resulting in the reduction of the ability of a disease to transmit through the vaccinated individuals. The perception that vaccines provide sterilizing immunity, where the disease agent does not establish an infection, while widely held, is generally unfounded and largely unrealistic. Recent advances, especially in the last 15 years in genomics, proteomics, biotechnology, immunology, pathogenesis, and vaccine formulation and delivery have dramatically changed our approach to vaccine development. When used optimally, vaccines have been shown to prevent disease, reduce the need for pharmaceutical intervention, and improve the health and welfare of animals, and indirectly people as well. The challenge in developing an optimal vaccination program is in dealing with the great diversity that exists within the animal world, and as such there probably is no single optimal program for all occasions. While there is no magic solution to optimizing vaccination programs for animals, nonetheless, a solid understanding of the animal's innate and environmental risk factors as well as the variables such as stress will enable the development of tailored vaccination schedules that best meets the needs of the animal. The use of vaccines in animal health is not restricted to the protection of morbidity and mortality of the animal hosts themselves, but they are regularly employed as key elements in public health programs. When appropriate biopreparedness, management modeling strategies, and contingency plans of the future are linked with (1) protective DIVA vaccines against zoonoses, (2) effective predictive modeling, and (3) deployable implementation policies, control, and prevention of serious zoonotic diseases of man and animals will become more achievable at local, state, and national levels.

MAXIMIZING STRENGTH AND DURATION OF PROTECTIVE IMMUNITY

Novel Vaccine and Delivery Systems

Novel Vaccines

The majority of vaccines licensed for controlling infectious disease of veterinary species today are based on technology that was introduced by Jenner using live vaccines and Pasteur using whole killed vaccines 200 and 100 years ago, respectively. Although these vaccines have been very useful in reducing economic losses to the industry, they were primarily produced by empirical processes with little knowledge of animal physiology, host responses to vaccines including the importance or relevance of Th1 versus Th2-like immune responses in controlling infection or the importance of specific virulence factors of the pathogen. However, recent advances, especially in the last 15 years in genomics, proteomics, biotechnology, immunology, pathogenesis, and vaccine formulation and delivery have dramatically changed our approach to vaccine development. As a result of these advances, we believe that the next generation of vaccines should be safer and more cost-effective.

Live Attenuated Vaccines

Historically, live attenuated vaccines were generated by repeated passage of the pathogen in vitro under different conditions that randomly generate mutants that have reduced virulence. Indeed, the majority of the live attenuated vaccines on the market today have been generated by this approach (van Oirschot, 1997). The selection of mutants can be further enhanced by exposing the pathogen to mutagenic agents or growing the pathogen at low temperatures. Under any of these conditions there is no control of which specific genes are mutated, nor is there control over the extent of attenuation. Thus, it requires testing each of these individual passaged agents to finally arrive at a level of attenuation that is acceptable.

Finally, since many mutations can be point mutations, there is a high probability of back mutation and reversion to virulence once introduced into the animals. Possibly the best example of this ability to revert to virulence is with polio virus where the oral polio virus given to humans contains a single-point mutation which can rapidly revert to a virulent form upon replication in vivo (Greensfelder, 2000). However, even with these potential problems, many very effective live vaccines are licensed and have had a major role in disease reduction even though the safety of these vaccines is of concern.

Fortunately, with more knowledge of each pathogen and the genes involved in inducing protection and virulence, it is possible to target specific genes involved in attenuation. By deleting an entire gene from the pathogen to attenuate the organism using genetic engineering techniques, the safety profile of the vaccines can be dramatically improved. This approach is not only safer, because it is more difficult to acquire a whole gene which would lead to reversion than it is to modify a point mutation, but it is also possible to modulate the degree of attenuation. This is important to ensure that adequate replication of the attenuated vaccine occurs, to induce optimal levels of protective immunity but not cause any pathology. A final advantage of using live vaccines is that they can be delivered by the natural route of infection to provide protection at the natural site of entry (i.e., mucosal protection). However, even with all these advantages, these approaches are under much greater scrutiny by regulatory agencies than the less safe conventional methodology of attenuation, thereby, adding significantly more expense to the licensing of many potentially valuable vaccines needed by the industry.

Currently, the more popular method of attenuation is to delete or incapacitate an entire gene that is not required for replication. However, it is also possible to introduce temperature-sensitive (TS) mutations which would reduce the pathogens ability to replicate at temperatures present in the lower respiratory tract. Thus, the vaccine could infect the upper respiratory tract and induce immunity, but have limited replication in the lower respiratory tract where damage could lead to complications such as pneumonia and possibly death. In addition to introducing this single TS mutation, one could introduce multiple TS mutations, thereby providing a much safer and stable attenuation than is currently possible with conventional TS mutants. This approach is very attractive for many respiratory infections, but especially for influenza viruses where TS mutants are currently in use for control of equine influenza (Townsend et al., 2001).

By deleting nonessential genes, a vaccine can be designed to reduce the virulence of a pathogen. However, it is also possible to make vaccines by deleting essential genes, which will completely prevent the replication of the vaccine in vivo. This type of mutation results in replication defective organisms, making the vaccine totally safe. Such an approach has been used for a variety of different prototype vaccines. The principle behind this approach is that it is possible to culture the vaccine in vitro by supplying the elements required for replication. Once the agent is introduced in vivo, where the element is absent, the agent undergoes an abortive infection. However, during this abortive infection, sufficient antigens are produced to induce immunity. These types of defective vaccines can be produced against a number of different organisms. A variation on the replication defective vaccines approach is one where a specific gene is introduced requiring a specific metabolite or compound to replicate (Donnenberg and Kaper, 1991). This allows controlled replication in vitro as well as in vivo as long as the required compounds are introduced with the vaccine or individuals are fed the specific compound for a period of time to allow the vaccine to replicate in vivo. On removal of the required compound, the agent dies. These types of approaches are easy to introduce into bacteria for making bacterial vaccines. These types of vaccines should remove the concern about shedding of organisms in environments as they would rapidly die in the environment because of lack of complementary growth components.

Possibly the greatest advantage of live genetically engineered vaccines is the fact that these vaccines can be used as vectors to carry protective proteins from other pathogens. Thus, it is possible to immunize an individual against not only the organism that has been attenuated, but also against antigens which are being carried by the vector. Thus, a single immunization can protect individuals from a variety of diseases. This approach is gaining considerable interest from the veterinary biopharmaceutical industry since it would make vaccine production and delivery much more economical with a single vector carrying multiple vaccines manufactured in a single step. Furthermore, there should be no interference between the antigens as is often seen with multiple agents at the time of vaccination. Currently, over a dozen viral vaccines based on a pox virus vector are licensed in veterinary medicine (Gerdts et al., 2006). Similarly, a number of companies are developing adenovirus-based vectors (Reddy et al., 1999a, 1999b). In the case of poxvirus-based vectors, one can either use mammalian or avian poxviruses as vectors. Since avian poxviruses undergo abortive replication in mammalian cells, they are gaining significant market share where they have been licensed. Thus, the safety concerns outweigh any genetic engineering concerns. The most recent report of developing solid immunity of monkeys to Ebola/Marburg and lassa virus following a single immunization with a VSV recombinant has created significant excitement using VSV as a potential vector (Jones et al., 2005). However, in all these instances of vectored vaccines, one is always concerned about the impact of preexisting immunity to the vector on the potential efficacy of the vaccine. The possible reason that the monkeys were totally protected from Ebola following VSV-vectored immunization versus adenovirus-vectored immunization is the fact that the monkeys were naïve to VSV.

Currently, it is possible to develop live vectored vaccines using bacteria or viruses as carriers, which then can be delivered by natural routes of infection, and by mass administration by aerosol or in feed or water, which is critical for the poultry industry. This approach overcomes a need to handle individual animals and also overcomes the disadvantage of needle delivery. In addition, these vectors can be used to induce immunity in ovo (Rautenschlein et al., 1999). Currently, in ovo vaccination has gained significant popularity due to the ease of handling and providing immunity very early in life of poultry. In addition to carrying genes from other pathogens, these vectors can also carry immunomodulatory genes to act as adjuvants to enhance the efficacy of the vaccine (Raggo et al., 1996; Giavedoni et al., 1997). Although these vectored vaccines have many advantages, consideration must be given to the antigen being delivered. For example, one would not choose a viral vector when immune responses to carbohydrate or lipopolysaccharide (LPS) antigens are needed. In contrast, a bacterial vector would be ideal for such a vaccine. Similarly, one would not choose a bacterial vector to deliver a viral protein whose confirmation is very sensitive to posttranslational modification. Thus, the appropriate vector is as important as the antigens being delivered by that vector.

Killed and Subunit Vaccines

Conventional killed vaccines are produced by inactivating the infectious agents so it cannot replicate in the host, but without altering immunogenicity of the protective proteins. Thus, all the proteins of the infectious agent are present in the vaccine, but the nucleic acid has been damaged to the extent that the agent cannot replicate. Unfortunately, many inactivating agents do have an impact on immunogenicity by damaging or altering specific proteins, therefore, it is important to achieve a balance between full inactivation and reduction of immunogenicity (Ferguson et al., 1993). Unfortunately, with some infectious agents, especially those that may aggregate, it is not possible to have inactivating agents penetrate the aggregate and, as a result, some outbreaks of disease have occurred due to improper inactivation (Brown, 1993). Thus, there continues to be a search for new inactivating agents. For all inactivated vaccines it is critical that the protective antigens are present at the concentration sufficient to induce an immune response. In bacteria, many virulence factors may be extracellular secreted proteins, therefore, they are not present in sufficient quantities in whole inactivated organisms to induce protective immunity. Possibly, the best examples are the RTX toxins in the *Mannheimia*, *Actinobacillus*, and *Pasteurellae* species (Shewen and Wilkie, 1982). These antigens are critical for effective vaccination, but are secreted and, therefore, are not present in whole, killed inactivated vaccines. To overcome this problem, companies have "spiked" the whole killed inactivated vaccines with culture supernatant containing the RTX toxins. Second, some virulence factors are only expressed under specific physiological conditions, i.e., iron-regulated proteins, thus, care must be taken to ensure that bacteria are cultured under the appropriate conditions to express the putative protective antigens (Deneer and Potter, 1989b, 1989a).

Since most inactivated vaccines are poorly immunogenic, they often require adjuvants to enhance their immunity. Many of these adjuvants and formulations cause tissue reactions (see below). Furthermore, they generally induce a good humoral immune response with minimal cell-mediated or mucosal immunity. In many cases, these later responses are critical for optimal protection. In addition, many of these vaccines are delivered by needles that may break and remain in the animal for long periods of time. This is both painful for the animal and is of concern to the consumer who might encounter such a foreign object in the meat. However, many very successful vaccines have used these technologies and continue to be used by the industry.

A variation on the theme of killed vaccines is that of subunit vaccines. Instead of developing vaccines containing all the antigens of the pathogen, it is possible to identify the few critical proteins that are involved in inducing protection and use them as a vaccine. Theoretically, it is possible to purify the proteins from any conventionally produced vaccine. However, this is very expensive, therefore, it is much more economical to clone the gene and produce it as a subunit vaccine using biotechnology approaches. Before embarking on a subunit vaccine program, it is critical to identify the critical antigens for incorporation into a vaccine and then produce these antigens in a commercial setting. Fortunately, our knowledge regarding the critical proteins and ability to isolate the genes coding for these proteins has become relatively routine in the last decade. Thus, the protective antigens for most important infectious disease agents in veterinary medicine are now known. For those agents where we still need to identify the putative protective antigen, we can use comparative biology and genomics to identify them relatively quickly. However, identification is only the first step in a vaccine production program. Once the antigen is identified, it must be produced in a commercially viable system.

For bacterial antigens, production in bacterial systems is relatively straightforward. These proteins can be produced either as secreted proteins or as inclusion bodies. However, for viral or parasitic antigens, the extent of posttranslational modification required to retain the three-dimensional structures dictates that they can be produced in eukaryotic systems to ensure their immunogenicity. Although these are often more expensive and difficult to scale-up, they are required especially for viral glycoproteins. Since viral glycoproteins are generally anchored in cell membranes, this requires destruction of cells to purify the glycoprotein. To overcome this impediment, it is possible to remove the transmembrane anchor and allow secretion of the protein (Kowalski et al., 1993). This not only increases the yield, but also reduces the purification cost, thereby making the vaccines affordable for livestock application. Significant advances in production of subunit vaccines are being made by expressing antigens in plants, which confer most of the posttranslational events present in mammalian cells. Further advantages are that it is possible to target the vaccine antigen to oil bodies in the plant (van Rooijen and Moloney, 1995). This makes the vaccine very thermostable, allowing ease of purification and, more importantly, the vaccine is already incorporated in the metabolizable oil adjuvant. Although there are currently no licensed vaccines produced in plants, it is envisaged that this will occur shortly as a number of companies are currently testing a number of such vaccines.

DNA Vaccines

The newest addition to our armamentarium in vaccination is genetic vaccination or polynucleotide vaccines, commonly referred to as DNA vaccines. The concept of DNA vaccination is extremely simple in that a plasmid expressing a gene encoding the antigen of interest is injected into an animal, which then expresses the protein (vaccine) of interest. Thus, the animal acts as the bioreactor to produce the vaccine and, consequently, the animal is immunized. This simple process does not require any downstream protein purification, although it does require plasmid purification and abolishes the need for adjuvants. Another advantage, depending on the route of administration, is that these vaccines generally induce both humoral and cellular immunity. This is primarily because the antigen is produced endogenously and all posttranslational modifications are similar to those seen following regular viral infections. Furthermore, improvements in both enhancing immunity as well as driving the cell-mediated arm of the immune system has been achieved by coadministration of these DNA-based vaccines with plasmids encoding IL-12 and IL-15 (Hanlon et al., 2001).

Although the concept is simple, the actual events that must occur to make DNA vaccination possible are many. First, efficiency of transfection is very low. Thus, few plasmids actually get into the nucleus and express antigen. To overcome this problem various methods have been used to increase transfection including: (1) electroporation, (2) formulation in different polymers or lipid-based vesicles, and (3) mechanical delivery using air or gene guns (reviewed by van Drunen Littel-van den Hurk et al. (2004)). Second, in the case of RNA viruses or bacterial genes, it is often necessary to introduce introns and remove cryptic splice sites to ensure expression occurs even after entry of the plasmid into the nucleus. Altering codon bias has also proven in some cases to enhance expression. The effects made to enhance the efficacy of DNA vaccines in livestock have been reviewed on a number of occasions (Babiuk et al., 2000, 2002). Possibly the most efficient method of delivery of genetic vaccines is the use of alphavirus replicons (Schultz-Cherry et al., 2000). Since alphaviruses naturally target cells for infection, they are ideal delivery vehicles. Using the alphavirus replicons system, the envelope can be engineered in such a way that it more efficiently targets the antigen-presenting cells (dendritic cells) for active targeting of vaccines. Since the replicons are not able to reproduce they are deemed to be safe.

Although there are still many hurdles to overcome to make DNA vaccination routine in many species, the first DNA-based vaccine has already been licensed to control West Nile virus in horses. With the regulatory path having been set, it is highly likely others will be licensed shortly.

Differentiating Infected from Vaccinated Animals (DIVA) Vaccines

Previously, agricultural disease surveillance and eradication programs were largely based on the serological confirmation of infection and the destruction or quarantine of herds to limit the spread of disease. Due to the reliance on serological confirmation, most countries banned vaccination against exotic diseases. Possibly, the best example of such an approach is with foot-and-mouth disease. The introduction of foot-and-mouth diseases into countries that are normally free of this disease results in quarantine and slaughter to eradicate disease rather than to conduct ring vaccination. Since ring vaccination would result in serological positive animals the country would continue to be considered "infected" as long as there were serologically positive animals in the country. However, the application of a serological test to differentiate between vaccinated and infected animals has a potential to change this practice. As countries move to eradication of endemic diseases or controlling novel introduced diseases through vaccination programs it is critical to develop robust detection systems which differentiate vaccinated (immune) animals from naturally infected animals which might be carriers of the disease. This aspiration has led to the development of DIVA vaccines where vaccines lack one or more highly immunogenic proteins, but still contain the critical proteins needed for inducing protection. Since the vaccine lacks a specific protein, a companion diagnostic test is developed which can serologically differentiate infected from vaccinated animals (DIVA). Although DIVA vaccines/diagnostic tests can be used with conventional killed vaccines, for example, the killed FMD vaccines do not induce antibodies to the viral nonstructural proteins (NSPs) (Clavijo et al., 2004), the trend is to use gene deleted or subunit DIVA vaccines.

The first gene deleted vaccine was used to eradicate pseudorabies virus. This vaccine, was the result of a natural deletion of the glycoprotein (gE) gene from the virus (van Oirschot et al., 1986). This vaccine was used for decades to control pseudorabies virus in pigs and was known as the Barta strain. Coincidentally, it also has a gE deletion, which then could be used as a marker for differentiating vaccinated animals from latent carriers of the disease. As a result of the first DIVA vaccine, many areas are now free of pseudorabies virus. Although the pseudorabies virus was a naturally occurring virus, the availability of recombinant DNA technology, combined with our knowledge of gene sequences, and the ability to identify nonessential genes in many pathogens and delete them provides us with a great opportunity to develop DIVA vaccines for almost any infectious disease. As a result of these initial successes, several marker vaccines are already commercially available and their role and contribution to disease eradication appears promising. However, for DIVA vaccines to be embraced by the industry there will need for support by various levels of governments including the livestock industry to ensure that the biopharmacy industry develops these vaccines. For example, there are opportunities to develop DIVA vaccines for foot-and-mouth disease, but the international community does not yet appear to be willing to embrace vaccination against foot-and-mouth disease in countries which are normally foot-and-mouth disease free (Clavijo et al., 2004). In the case of foot-and-mouth disease, a DIVA test is available based on the detection of antibodies to NSPs that is the preferred diagnostic method to distinguish viral infected carrier animals from vaccinated animals, but is not yet embraced by the international trade community. Thus, even though the better technology is available, this will not guarantee that the technology will be embraced by the industry or the global community. Thus, some of these vaccines and their acceptance will not only be driven by economics to the individual producer, but it is also often politically motivated by trade interests and macroeconomic issues. Thus, once a country is declared disease-free that country can use its disease-free status as an effective trade barrier to keep out livestock from countries that are not disease-free. As more confidence is developed in DIVA vaccines and the diagnostic tests are simpler to perform, it is hoped that the entire globe would embrace these types of vaccines and reduce the accidental introduction of exotic diseases into countries or if introduced, will reduce the number of animals that are needlessly slaughtered to return the country to disease-free status.

Formulation and Delivery

Regardless of the type of vaccine developed, if it is not formulated or delivered properly, the full benefit of the vaccine will not be achieved. This is important because how it is formulated and delivered will dramatically influence the quality as well as the magnitude of the immune response generated. For example, a parenterally administered nonreplicating vaccine will induce only a systemic immune response, but have very little ability to induce mucosal immunity. Currently, there is extensive evidence to suggest that delivering the vaccine to mucosal sites is the most effective way to induce mucosal immunity. Another advantage of mucosal delivery is that the animal also develops systemic immunity, in addition to mucosal immunity. In addition to broadening the immune response, mucosal immunity has other advantages such as avoiding the pain and injection site reactions, and broken needles, etc. associated with parenteral injection. This route of administration can also be less expensive, but depending on management systems, may be more difficult to perform. For example, wild range cattle are not very receptive to intranasal delivery of vaccines; therefore, oral delivery may be more attractive in these circumstances. Unfortunately, despite many experimental reports few oral vaccines are licensed for livestock. The reason for this is the lack of effective delivery systems that target the gut-associated lymphoid tissue (GALT). The difficulty of delivering vaccines to the GALT is a result of a combination of many factors including degradation of the antigen and various physiological barriers. For example, in the ruminant, traversing the rumen to the intestine where the GALT is provides a major challenge. However, despite these challenges a number of promising developments are being pursued.

Since many viruses and bacteria can survive the gastrointestinal tract they are being used as vectors for vaccine development for providing oral or mucosal delivery. Possibly, the best example of such a vaccine is the vaccinia-vectored rabies vaccine. In this case, the rabies glycoprotein is inserted into a vaccinia virus which is then incorporated into bait for the control of rabies virus in wildlife (Brochier et al., 1991, 1995). The advantage of such an approach is that if one chooses a thermostable vector, the vaccine remains viable in the environment for significant lengths of time. Second, one does not need to trap the individual animals since they routinely encounter the bait during their normal activities, as they eat the bait they get vaccinated (Brochier et al., 1995). Similarly enteric bacteria such as *Salmonella* can also be used as vectors for carrying a variety of vaccine antigens. Although the safety issues prior to release of these vectors into the environment are very thoroughly investigated there is always a concern, possibly unwarranted, about the potential consequences of release of recombinant agents into the environment. To overcome this concern, interest is developing in oral delivery systems for nonliving antigens. This is possible by encapsulating the antigen in novel polymeric microparticles. These particles can be produced from a range of polymers, which protect the antigen from degradation in the gastrointestinal or respiratory tract and target these vaccines to the tonsilar crypt cells or Peyer's patches of the gastrointestinal or respiratory tract, respectively.

For microparticles to be effective, they must be engulfed by the M cells of the Peyer's patches. Significant work has focused on investigating the optimal size of particles for effective uptake. Uptake can also be modulated by changing the charge, hydrophobicity and linking of specific ligands to the particles to target the vaccine to M cells. For example, the size of polylactide-coglycolide (PLG) microparticles has been studied extensively with regards to their uptake and transport to lymph nodes. If one could generalize, which is risky, particles of $<10\mu m$ are taken up easier by the Peyer's patches than particles $>10\mu m$. This is the case with both PLG and alginate particles (Eldridge et al., 1993; Kim et al., 2002). The current interest in producing nanoparticles for various drug-delivery systems is also being exploited for vaccine delivery and has shown that smaller particles are better than large particles for uptake. Thus, it is clear that size is very important for the efficient uptake of microparticles carrying vaccine antigens. Furthermore, since the mucosa has a negative charge, positive-charged particles are more effective in uptake than negative-charged particles. Another approach for enhancing uptake of microparticles is to link various lectins to the microparticles which then can specifically target M cells (Pappo et al., 1991; Mantis et al., 2002). In addition to being able to directly target microparticles to the oral mucosa, interest is also increasing in being able to use these microparticles to develop single dose vaccines, whereby microparticles of different composition are employed with some particles releasing antigen immediately upon uptake and others releasing antigen at 1–2 weeks after uptake (Cleland et al., 1998; Tamber et al., 2005). This approach is especially attractive since it would provide a priming of the immune system upon vaccination and a booster later on. This has recently been demonstrated with an HIV vaccine in baboons (Cleland et al., 1998). To further enhance the efficacy of microparticles as delivery vehicles it is possible to incorporate adjuvants with the antigen in the microparticles and not only deliver the antigen but also provide the immune system with a boost. Incorporation of saponin-derived adjuvants (QS-21) or CpG ODNs into microparticles have been shown to enhance immune responses to the antigens incorporated in the microparticles (Hunter et al., 2001; Diwan et al., 2002). Although PLG microparticles have been the most extensively studied oral delivery microparticle systems, numerous other microparticles have also been investigated. These include polycapralactone (PEC), starch-based microparticles, alginate microparticles, and most recently polyphosphazenes (Payne et al., 1995; Benoit et al., 1999; McNeal et al., 1999; Wikingsson and Sjoholm, 2002; Rydell and Sjoholm, 2004). The reason for investigating other formulations is that the generation of PLG microparticles generally requires organic solvents and upon degradation in the intracellular environment generates an acidic environment. In some cases this is very detrimental to preservation of different epitopes. This is possibly the reason that incorporation of *Brucella ovis* in PEC microparticles was much more effective in protecting animals from challenge with live *B. ovis* versus encapsulation of the same antigen in PLG which did not confer protection (Murillo et al., 2001). It is also interesting to note that the quality of the immune response induced by the PEC microparticles was associated with a Th1-like response, whereas those induced by PLG encapsulated antigen was associated with a Th2-like immune response.

Significant studies have also been conducted with alginate microparticles in various animal species including cattle, rabbits, and chickens (Bowersock et al., 1999). In each of these cases, oral delivery of the vaccines was enhanced when the antigens were encapsulated in the alginate particles and further enhancement was detectable if the microparticle also contained adjuvants. Due to the mild nature of producing alginate microparticles, it is also possible to encapsulate live microorganisms, thereby protecting them as they transit the stomach to the gastrointestinal tract. This was clearly demonstrated by the ability to encapsulate bovine adenovirus in alginate microspheres and develop immune responses following oral administration of the virus (Mittal et al., 2000).

A most recent class of synthetic biodegradable polymers for vaccine delivery as well as acting as an adjuvant in itself is the polyphosphazenes. Numerous reports have documented that polyphosphazenes perform better than conventional adjuvants when incorporated with vaccines (Andrianov et al., 1998; Payne and Andrianov, 1998). Furthermore, it is possible to formulate polyphosphazenes into microparticles with no organic solvents. Thus, the procedure is extremely mild and one captures the benefit of microencapsulation as well as adjuvanticity all in one type of particle. Although these polyphosphazene microparticles have proven to be very effective in preliminary studies in mice, they require further investigation for application to livestock and companion animals. Although we have identified the areas of current interest and research in microencapsulation and delivery of vaccines in microparticles, it must be emphasized that there are very few studies in large animals demonstrating the application of microparticle formulated antigens for oral immunization. Clearly, much more work is required in large animals but the knowledge that is gained regarding the needs to stimulate the GALT, the size of microparticles, and charges to enhance immune responses combined with preliminary studies in a number of veterinary species demonstrating their potential utility is encouraging and needs to continue to be pursued.

Due to the rather poor immunogenicity of most inactivated and subunit vaccines, they almost always require combination of the vaccine antigen with adjuvants to be effective. The current adjuvants can be classified into: (1) mineral salts, (2) synthetic adjuvants, (3) oil emulsions, (4) plant products, (5) lipid-based vehicles, (6) bacterial products, and, finally (7) molecular adjuvants or cytokines or any combination of the above (Cox and Coulter, 1997). In humans, only aluminum salts or MF-59 are licensed, whereas in veterinary medicine many different adjuvants are in current use. Unfortunately, the exact mechanism by which most adjuvants work is not fully understood and different adjuvants probably have different mechanisms of action. These range from creating a depot of the antigen to attract antigen-presenting cells to the site of vaccination to creating a cytokine environment at the vaccination site to help expand the immune reactive cells. Regardless of the mechanism of action, most licensed adjuvants do not possess all the required features of an ideal adjuvant, therefore, the quest continues for better adjuvants, which not only induce a balanced immune response, but also are safe and do not induce any injection site reactions.

Since many of the adjuvants attract cells of the immune system to the site of injection, one requires a fine balance between attracting the appropriate cells required for induction of immunity, but not cause overstimulation resulting in unwarranted injection site reactions. Indeed, it has been estimated that in cattle, these injection site reactions cost approximately $8/injection site as a result of the need for trimming (van Donkersgoed et al., 1999). It is for this reason that most injections are now given subcutaneously in the neck where meat cuts are of less quality. In the case of cats, aluminum adjuvants have been implicated in fibrosarcomas (Burton and Mason, 1997) and, therefore, the trend is to move to nonadjuvanted vaccines for cats. The major problem with most adjuvants is they generally induce a strong Th2-like immune response, but very little Th1-like immune responses. Because of these concerns, the quest continues for novel adjuvants that do not cause any adverse reactions, but still induce the desired response.

Currently, the most often used adjuvants include aluminum salts, which adhere antigen to their surface and act as a depot for the antigen which is then presented to the antigen-presenting cells infiltrating the injection site. Since the antigen needs to be absorbed to the aluminum salts, it is not surprising that all antigens do not absorb efficiently and require precise pH's for their absorption (Claesson et al., 1988). Another disadvantage of aluminum salts is that they generally induce only Th2-like responses with little cell-mediated immune responses.

Oil emulsions are another very commonly used family of adjuvants in veterinary vaccines. These oil emulsions vary from mineral to plant oils and can be oil in water or water in oil emulsions. The major problem with mineral oils is that they are nonmetabolizable, leading to residues and injection site reactions. To overcome these problems, metabolizable oils are replacing mineral oils. Many of these metabolizable oils are plant-based. In addition to using plant-derived oils, other plant components with immunomodulatory properties are also used, either in combination with oils or alone (Kensil et al., 1991; Wu et al., 1992). Possibly the best known plant-derived adjuvant is saponin (Quil A) or various fractions of Quil A, best known as QS21. One of the mechanisms of action of QS21 is its ability to stimulate cytokines, which help create the environment required to drive the immune response. To further enhance the activity of QS21, it is possible to combine them with other lipid components to form ISCOMs or cage-like structures containing the antigen of interest (Cox et al., 1998; Morein et al., 2004). Thus, ISCOMs act both as delivery vehicles and enhance immunity to the antigen incorporated into the ISCOMs. Currently, most ISCOM-based vaccines incorporate viral glycoproteins into the ISCOM. These vaccines can be delivered either parenterally or mucosally, and can induce both antibody and cellular responses (Takahashi et al., 1990; Morein et al., 2004). One of the disadvantages of ISCOMs is the poor loading capacity and the difficulty of incorporating nonmembranous antigens into ISCOMs (Reid, 1992). Similar to ISCOMs, liposomes have been used experimentally with excellent results. In addition to use with subunit vaccines, lipid-based delivery systems are also efficient in delivery of DNA-based vaccines (Zelphati and Szoka, 1996).

Early vaccines containing whole killed bacteria demonstrated that these vaccines induced significant reactions, which led to the suggestion that many of these components could be good adjuvants. Indeed, this was proven to be a correct assumption. Many of these bacterial products, whether they are LPS or various toxins (cholera toxin, *Escherichia coli* LT, etc.) or bacterial DNA generally induce cell signaling, primarily through Toll-like receptors (TLRs), and production of cytokines or chemokines. It is because of this activity, excessive stimulation can lead to adverse reactions. To overcome these excessive reactions, less toxic analogs of each of these bacterial components are being developed (Ulrich and Myers, 1995).

One of the least reactogenic bacterial components currently showing significant promise as an adjuvant is CpG (Krieg et al., 1999). These are specific synthetic oligonucleotide sequences that signal through TLR-9 receptors. Significant work in the last decade has identified specific CpG motifs that can stimulate both Th1 and Th2-like immune responses (Weeratna et al., 2000). Furthermore, they are active both with recombinant and conventional vaccines (Ioannou et al., 2002). In addition, they are active following in ovo delivery (Gomis et al., 2004).

Since many of the above adjuvants act through their ability to induce cytokines at the injection site, and since cytokines are critical for induction of immunity, numerous investigators have attempted to incorporate cytokines into vaccine formulation with limited success (Heath, 1995). The greatest concern with cytokines as adjuvants is the difficulty of maintaining the current concentration of the cytokine that will drive the immune response without overstimulation with toxic consequences. A recent report of using killed pseudomonas bacteria containing interferon gamma as an adjuvant demonstrated excellent adjuvanticity with no adverse effects (Gaertner et al., 2005). Thus, there is a high probability of eventually using such molecules as adjuvants.

ENHANCING FOOD SAFETY WITH IMPROVED VACCINES

The use of vaccines in animal health is not restricted to the protection of morbidity and mortality of the hosts themselves, but they are regularly employed as key elements in public health programs. Among the first widely and successfully applied applications of vaccines in the animal health were for rabies, a principle zoonotic disease for a variety of domestic and wildlife hosts (see Chapter 33). Classical inactivated rabies vaccines have been used to control this devastating disease in domestic animals and wildlife targets have been addressed with attenuated live approaches and most recently and successfully with a recombinant vectored vaccine. Of perhaps even more profound potential impact on public health is in the control of H5N1 avian influenza, where the use of conventional and recombinant vaccines have been demonstrated to significantly decrease the spread and virus within infected flocks, thereby reducing the spread of the virus vertically into the human population. Other zoonotic diseases for which vaccines have real and potential impact on incidence in the human population are brucellosis in ruminants, Nipah virus and Norwalk Agent in swine.

Focus will be on those diseases for which the exposure to the zoonotic is typically through the food chain. These vaccines carry their own special issues in that the agent may not always be a principal disease in the host, rather a commensal, or part of the host natural flora. *Campylobacter* in poultry and *E. coli* O157 are typical examples of these, with the disease agent residing as part of the host normal flora and only becomes of concern when humans are exposed to contaminated food.

Safety of the water and food supply has clearly led to greatest improvements in public safety in the last century and a half. Statistics from the British home office indicate that in 1900, 140 per 100,000 in the population of the UK died annually from food-related illnesses. Improvements along the food chain reduced that number to approximately 0.4 per 100,000 by 1980, where it has remained constant since then.

Before moving on to animal diseases and the vaccines used to control them, it must be noted that the very vaccines used in food producing animals must in themselves not negatively affect the quality of the food. In the United States, production animals typically include beef, pork and poultry with minor inputs from the sheep and goat industry. The advent of modern vaccine technologies has provided tremendous benefits to large-scale production capabilities for these species and resulted in a positive economic impact with regard to cost of production for the primary producer and cost of product to the consumer. The use of vaccines in production animal medicine has translated into one aspect of the overall production scenario under the concept of "herd health management." The herd health management concept transcends prevention of disease, animal housing and environment, nutrition, and management issues; all of which have contributed to the low cost, bountiful, high quality food supply that this nation has enjoyed for many decades.

There are issues routinely faced by producers with regard to the application of vaccine technologies to production animals. These include the efficiency of mass vaccination, the impact on food quality due to local reactions at the vaccination site, and the rare but serious issue of broken needles at the site of injection. The application of mass vaccination technologies has been most successful in the commercial poultry industry with products that are typically administered by aerosol exposure or via drinking water systems. These technologies have been more difficult to apply to swine and beef primarily because of the design of those production facilities and the formulations currently required to achieve effective vaccine response.

The issue of injection site reactions is primarily associated with swine and beef production because most vaccines are still administered with a traditional syringe and needle approach. These issues are being addressed by targeting specific "low meat value" sites for injections, particularly the lateral neck region. This preserves the integrity and value of the high quality meat cuts associated with the leg and loin regions. In addition, vaccine formulations have improved with better selection of less reactive adjuvants, lower dose volume and targeting efficacy with subcutaneous injections instead of requiring intramuscular deposition of the antigen.

The issue of broken needles is more difficult to address, and the potential for a break always exists during herd vaccination procedures. Though rare, the discovery of a needle in a meat product typically makes headlines and results in damage to the reputation and marketing potential of the processor. Some export customers, specifically the Japan market, set a zero tolerance for broken needles that result in the use of metal detectors to screen meat shipments. If a positive signal is detected in a single sample at the port of entry, the entire shipment is subject to rejection. This is a major expense to the processor. One relatively new technology that holds promise to address the broken needle issue as well as the injection site meat quality issue is the administration of vaccines via needle-free devices.

Several companies are now producing and distributing needle-free devices for the administration of vaccines to swine and cattle. These devices either utilize existing vaccine formulations or more preferably utilize formulations specifically designed to take advantage of the needle-free route of administration. One advantage exploited by needle-free technology is the ability to target the most efficient population of antigen-presenting cells (plasmacytoid dendritic cells) that are concentrated in the dermis. These cells are typically bypassed with a needle injection. Targeting dendritic cells has the potential to reduce antigen quantity, reduce or eliminate the need for adjuvants, and significantly reduce the volume of injection. All of these factors address both the injection site meat quality and broken needle issues.

Other vaccine technologies that have emerged in recent years include the use of recombinant virus vectors and nucleic acid based vaccines. These technologies offer several potential advantages. They allow for the use of a vaccine that allows differentiation of infection from vaccine antigen exposure. This is critical in disease control situations, particularly when dealing with the introduction of exotic disease agents. In some instances, a recombinant vaccine will achieve a better immunologic response than can be obtained with the native antigen and the use of adjuvants may be reduced or eliminated. Nucleic acid vaccines offer a major advantage with regard to purity of the vaccine, elimination of adjuvants, the ability to overcome interference from maternal antibodies in young animals, and the opportunity to prime the immune system for a subsequent boost with a recombinant or native antigen formulation.

Prevention and control of infectious diseases in food animals will continue to be an issue faced by farmers and producers. Some agents are endemic in the respective species and others are highly contagious requiring fast diagnosis and rapid response to achieve control in the absence of vaccination. The other aspect of food animal vaccination that has come to the forefront in recent years is the potential for exotic agents to enter the United States either through natural or incidental routes or via intentional, terrorist introduction into the domestic herds or flocks.

One example of such a threat is influenza virus. The close proximity of people to swine and poultry in some Southeast Asia countries will continue to provide the opportunity for coinfection of hosts with multiple influenza viruses. The potential for reassortment of the virus genome and emergence of strains highly virulent to livestock and human populations will always exist. Vaccination is a leading defense against such a threat.

Vaccines for Food Borne Disease Protection

Salmonella and *Campylobacter* in Poultry

Among the most successfully and broadly applied commercial vaccines used in animal health are targeted to the poultry industry. An entire range of viral and bacterial vaccines are applied throughout the course of production and have allowed diseases such as Newcastle Virus and Marek's disease, which would normally decimate a flock, to be strictly controlled. These vaccines may be applied in the drinking water in the poultry house, in aerosolized spray cabinets in the hatchery, and recently in mass administration in ovo devices. Poultry is also where we find the most consistently applied programs for food safety vaccines in animal health.

According to the Food-Net; bacterial pathogens with the highest relative incidence for causing bacterial food-borne illness in humans during the period 1996–2003 were campylobacter, salmonella, and shigella (Anonymous, 2005c). Most of the campylobacter and salmonella infections are acquired through the handling and consumption of contaminated poultry products. Although more than 2400 different serotypes of salmonella exist, only *Salmonella enterica* subspecies *enterica* (*S. enteriditis*, *S. oranienburg*, *S. Montevideo*, *S. Newport*, *S. typhimurium*, *S. Anatum*, *S. Derby*, and *S. bredeney*) are responsible for gastrointestinal diseases in humans, and they are grouped as paratyphoid (PT) *Salmonellae*. *S. enteriditis*, *S. typhimurium*, and *S. Montevideo* are the most frequently isolated in poultry.

Usually an acute, self-limited gastrointestinal illness, characterized by diarrhea, fever, and abdominal cramps is most common presentation of campylobacter and salmonella infection in humans. Poultry infected with salmonella present minimal clinical signs and there are no clinical signs attributable to campylobacter in chickens of any ages. Both salmonella and campylobacter colonize chicken intestinal tract and large numbers of bacteria shed through feces.

Intestinal contamination is the main source of contamination in the processing plant. Campylobacter colonization (detectable levels) occurs at approximately 2–3 weeks of age and once established in a flock it spreads rapidly. The horizontal transmission between chickens has been well documented, the recent reports suggest that vertical transmission may play a role in spread of salmonella and campylobacter in poultry flocks as well (Gast, 2003; Shane and Stern, 2003).

Preharvest, harvest, and postharvest measures are used in an effort to reduce salmonella and campylobacter incidence in poultry. Preharvest measures include use of feed additives (antibiotics, prebiotics, probiotics, synbiotics, and short-chain organic acids), competitive exclusion, and vaccination. Several vaccines are available for *S. typhimurium* and *S. enteriditis* whereas there are no vaccines for campylobacter for chickens.

Salmonella infection is known to elicit strong humoral responses in chickens whereas, little is known about cellular immune responses. However, it is widely accepted that cell-mediated immunity plays a more important role than humoral immunity for tissue clearance of virulent strains while mucosal antibody responses and promyeolcytes seem to be important in intestinal clearance (Gast, 2003). Indeed, recent studies indicate oral inoculation of chickens with *Salmonella typhimurium* results in early expression of chemokines in spleen and liver, followed by increased expression of gamma interferon and increased number of CD4 and CD8 cells (Withanage et al., 2005). These responses correlate with Th-1 responses to systemic infections in other species. Responses to ST infection in the gut are not as clear, while in a one study an increase in proinflammatory cytokine expression in another there was no substantial increase in cytokine responses (Gast, 2003; Withanage et al., 2005). Vaccination has been widely used to control salmonella infections in poultry. Both live and inactivated vaccines have been utilized though the former have been shown to induce stronger and long-lasting immune responses (Gast, 2003; van Immerseel et al., 2005). However, neither vaccine type can induce complete protection against colonization of the intestines and therefore control programs remain and integrated approach of vaccination, proper preventative measures in the processing plant, handling in the transportation chain, and finally by the end consumer.

Currently no vaccines are available for the control of campylobacter in chickens though several experimental approaches have been shown to induce humoral immune responses and some degree of protection against challenge (Stern et al., 1990; Cawthraw et al., 1994; Khoury and Meinersmann, 1995; Noor et al., 1995; Widders et al., 1996, 1998; Rice et al., 1997; Shane and Stern, 2003). However, there is no evidence of complete protection against intestinal colonization in these studies (Stern et al., 1990; Cawthraw et al., 1994; Khoury and Meinersmann, 1995; Noor et al., 1995; Widders et al., 1996, 1998; Rice et al., 1997). One important aspect of campylobacter colonization is that campylobacter is not usually detected in chickens less than 2–3 weeks of age under commercial broiler production conditions (Jacobs-Reitsma et al., 1995; Evans and Sayers, 2000). However, experimentally it has been shown that chickens less than 2–3 weeks of age are susceptible to campylobacter infection resulting in colonization of the gut suggesting that there is no genetic factors of age-related resistance to infection (Hald et al., 2001; Shane and Stern, 2003). The observed age-related resistance to colonization by campylobacter has recently been attributed to presence of high levels of maternal antibodies (Sahin et al., 2003) in which the decline of maternal antibodies associated with an increase in colonization rate. In addition, maternal antibodies were shown to opsonize campylobacter in in vitro experiments. However, it is well documented that chickens with maternal antibodies can be infected and colonized with campylobacter if higher doses of challenge inoculums' were used (Noor et al., 1995; Hald et al., 2001).

Immune intervention for nondisease causing gut colonizing bacteria is indeed a challenge. One has to consider accessibility of specific immune components into gut content where the bacteria are located. This immediately excludes activity of immune cells in the content though there may be some activity at submucosal level. Although soluble-specific components such as immunoglobulins can be plausible to access to bacteria, their activity in the absence of other components such as complement and phagocytic cells, may not be sufficient to kill or eliminate the bacteria from the gut. One option is to prevent colonization in the first place using immune intervention, and in order to accomplish this, a better understanding of the factors leading to colonization may provide venues to explore vaccine development.

E. coli in Cattle

E. coli is an ubiquitous pathogen of both humans and animals. It is responsible for a wide range of diseases including enteric, septicemic, and urinary track infections. In recent years a significant increase of bloody diarrhea or hemorrhagic colitis caused by a particular strain of *E. coli* (*E. coli* O157:H7) has been observed in most industrialized countries and especially the United States. It has made headlines as hemolytic uremic syndrome, causing death, often in clusters, after consumption of undercooked ground meat, often referred to as "hamburger disease."

The disease manifests in its most common form as severe bloody diarrhea accompanied by abdominal cramps with little or no fever which may resolve in 5–10 days (Anonymous, 2006). In a complication affecting mostly children and the elderly, 2–7% of infections will go to acute renal failure (hemolytic uremic syndrome) and may lead to death (3–5% overall death rate) or long-term complications.

E. coli O157 has an animal reservoir, the primary host being ruminants and specifically cattle. The great majority of cases can be linked back through the handling or consumption of contaminated meat, plants, vegetables, or water. Direct contamination on the farm or in other settings such as petting zoos is responsible for a significant number of cases. The incidence in the United States is estimated at 73,000 cases per annum, including 61 deaths (Anonymous, 2005b, 2006). The strain is highly pathogenic with as little as 1000 bacteria capable of causing the infection.

This makes infection by *E. coli* O157 one of the most frequently reported and severe food-borne illnesses in developed countries. The human disease, its etiology and management has been reviewed recently (Tarr et al., 2005).

While the genus and species *E. coli* is a very important pathogen of all animal species causing mainly enteric diseases such as neonatal diarrhea caused by enterotoxinogenic *E. coli* (ETEC) in calves and piglets, *E. coli* O157 is not pathogenic for cattle. The bacterium colonizes the intestine of the ruminant very efficiently, causing a nonsymptomatic infection. The bacterium is then shed, often intermittently but in high numbers in the feces. Fecal contamination of the meat in the packing plant, bedding, water, fruits, or vegetables on the farm leads to human exposure. The asymptomatic infection is very widespread in many herds across the country. Within a herd, the individual rate is highly heterogeneous from less than 10% to the majority of animals shedding in the environment. Importantly some particular animals shed at very high levels (Matthews et al., 2006).

The genus and species *E. coli* is a very diverse group that comprises approximately 175 O antigen serotypes (e.g., O157). The O antigen is carried by the polysaccharide side chain of the LPS, a component of the bacterial outer membrane. Significant variation is also noted at the chromosomic level (Anonymous, 2005a). More importantly *E. coli*, for the purpose of this discussion are characterized by their virulence factors. Attachment: *E. coli* O157 is part of a group of strains called enteropathogenic *E. coli* (EPEC). Their attachment to the enterocyte causes a lesion that is characterized as effacing because they destroy the cytoskeleton (attaching/effacing *E. coli*). The molecular mechanism has been well described and necessitates at least four major proteins: Esp A, Esp B, Tir, and Intimin. Intimin (a bacterial outer membrane protein) and its receptor

Tir are the attaching molecules. Over 33 virulence factors have been described in this complex process (Deng et al., 2004). An important finding for the understanding of disease transmission, epidemiology, and vaccine studies was the description of preferential adhesion sites in the rectum of the cattle (Naylor et al., 2003). Attachment is a key component of colonization in cattle and the first step of the disease in humans. Toxins: the toxic manifestation of the disease is caused by the secretion of two toxins, called Shiga toxin 1 and 2. The genes for these toxins are carried by bacteriophages. These toxins are also called verocytotoxins and the *E. coli* that secrete them verotoxigenic *E. coli* (STEC or VTEC). They are implicated in the apparition of the hemolytic uremic syndrome in humans (Karch et al., 2005). The resistance of cattle to the effect of these toxins provides an explanation for the absence of disease manifestation in these animals.

Many approaches have been suggested for the development of vaccines using many of the newer technologies as well as more traditional approaches. The target of the vaccination of cattle is the reduction or elimination of infection and excretion, and not disease prevention (as there are no symptoms). This is a particularly difficult target as it must reach a colonization site on the mucosal surface and eliminate or prevent colonization by bacteria that have developed exquisite solutions to parasitize this ecological niche. The most advanced approach is based on the use of *E. coli* secreted antigens that contain several of the proteins involved in the attachment (e.g., Esps and Tir). This approach uses traditional technology but is built on the very precise understanding of the pathogenic mechanisms. It is administered three times by the subcutaneous route. It has led to significant decreases in shedding levels both in challenge models and field trials (Potter et al., 2004). In a more recent test, however, performed under field conditions, the vaccine failed to induce a reduction in prevalence of infection (van Donkersgoed et al., 2005).

Plant (tobacco cell) produced intimin caused a decreased duration of shedding when the vaccine was administered parenterally followed by a booster by the oral route (Judge et al., 2004). Antibodies against intimin passively administered to piglets provided significant protection against a challenge (Dean-Nystrom et al., 2002). A live modified strain of *E. coli* containing a truncated intimin molecule also provided protection against clinical disease in an A/E *E. coli* model in rabbits (Agin et al., 2005). Finally genetically modified toxoids of the shiga toxins will provide both neutralizing antibodies and protection against a toxin challenge in an homologous fashion leading to the conclusions that Stx1 and Stx2 are not cross protective (Wen et al., 2006).

It should also be noted that several other, nonvaccinal approaches have also been proposed to reduce infection at the farm level including nutrition and probiotics. The principal challenges are twofold: first, increase the intrinsic efficacy of the vaccination protocol through optimized formulation or administration regimes, and second, demonstrate the overall efficiency of on farm interventions strategies including infection in reducing the risk for humans. While this has not yet been demonstrated there are many reasons to believe that reduction at the source would provide significant benefits (Matthews et al., 2006).

Considering the highly sporadic as well as the food-borne nature of the disease there is little interest in developing a vaccine for humans. In humans, especially in the light of the increasing levels of antibiotic resistance observed in these strains, improving the treatment tools available, including immunological tools, will be a priority.

Streptococcus suis

Many streptococcal infections have been described in humans. Infection by *S. suis* is a rare zoonotic disease, usually involving specific risk groups consisting of individuals in close occupational contact with infected swine: farmers, slaughterhouse workers, and hunters (of feral swine). Recently, a rare occurrence of a large cluster of patients with a high level of mortality was observed in China (Huang et al., 2005; Normile, 2005). A small number of patients were exposed to the infectious meat (as opposed to the animals or their carcasses). The investigation did not point to a new variant of the bacteria. Thus, the human disease caused by *S. suis* remains a relatively rare but severe zoonotic disease and is not properly speaking a food-borne infection (as opposed to *E. coli* O157, salmonella or listeria). Clinical manifestations are most frequently meningitis (3/4 cases). Sepsis and other very diverse manifestations, including endocarditis, may be observed. Deafness is a very common long-term complication of this streptococcal meningitis (MacInnes and Desrosiers, 1999). Death can occur but the disease is rarely fatal.

The *S. suis* infection in animals is widespread in pig populations across the world. It most frequently colonizes the upper respiratory tract of the swine without causing disease. It can, however, induce diseases either as a primary pathogen (perhaps because of increased virulence of certain strains) or become pathogenic secondarily to another factor: stress, immunodepression, concurrent infections (e.g., viral infections). The disease manifests in many different forms: death, fever, lameness, septicemia, and endocarditis (Martinez et al., 2002).

Immunocompromised individuals frequently are observed with the cattle genotypes of *Cryptosporidium* (Pieniazek et al., 1999; Morgan et al., 2000). In the United States and Australia, most cases appear to be of human origin: of 14 food and waterborne outbreaks, 10 were of human genotype, 3 of bovine genotype, and 1 with both organisms. However, in Europe the bovine genotype is responsible for more infections than the human genotype (McLauchlin et al., 2000; Alves et al., 2001; Guyot et al., 2001). In the United Kingdom, zoonotic transmission has been considered to be the major route of infections in humans (Hunter et al., 2003).

As the sole mechanism for transmission, oocysts have evolved to be dispersed and survive in harsh environments for long periods of time. The oocysts are unusually resistant to natural stresses and many man-made disinfectants. To control infection in animal populations the current best strategy is to move animals to clean areas. For human populations, disinfection procedures are used.

Because all infections with *Cryptosporidium* are initiated by ingestion or inhalation of the oocysts, measures to prevent or limit the spread of infection must be targeted to eliminate or reduce oocysts in the environment. Despite the fact that more than 200 drugs have been tested for activity against *C. parvum*, in vitro or in vivo, there is still no effective treatment (Rehg, 1993; Fayer and Fetterer, 1995; Woods et al., 1996; Armson et al., 1999). A cattle vaccine against *C. parvum* might reduce the environmental contamination and consequently the frequency of waterborne cryptosporidiosis outbreaks in humans. Since cattle may be exposed to *C. parvum* from their day of birth, before their immune system is mature, an active vaccination is not likely to mediate any significant protection. Thus, vaccination against *C. parvum* is being attempted for the purpose of generating hyperimmune colostrum-containing antibodies that may be effective in passive immunotherapy against cryptosporidiosis in calves. Calves fed hyperimmune colostrums, prepared against oocysts, and challenged had significantly reduced patent periods, oocyst shedding and days of diarrhea compared to calves fed normal colostrums (Fayer and Fetterer, 1995). A number of recombinant *C. parvum* surface or internal antigens have been expressed by DNA-cloning technology. Immune colostrum specific for several recombinant *C. parvum* proteins have demonstrated efficacy in murine and ruminant models against cryptosporidiosis. The most promising candidate proteins for vaccine development are surface (CP15/60, CP15, and P23) and micronemal proteins (TRAP-C1, GP900) of sporozoites. Both CP15/60 and P23 are detected by Western blot analyses of cryptosporidium proteins using fecal antibodies and serum from cryptosporidium-infected humans and animals (Lumb et al., 1988; Mead et al., 1988; Hill et al., 1990; Reperant et al., 1994). CP15/60 is defined by the monoclonal antibody (MAb) 5C3. CP15/60 IgA MAb provided protection against infection when administered orally to suckling mice (Tilley et al., 1991). The comparison of two recombinant forms of CP15/60, one produced in a prokaryotic expression system and the other in a eukaryotic expression system, indicated that the most potent immune response was obtained using the eukaryotic form, possibly due to posttranslational modifications (Sagodira et al., 1999). It has been reported that preparturient cows immunized with DNA encoding CP15/60 in the mammary gland produced sporozoite-specific antibodies in the colostrums (Jenkins et al., 1999). Immune colostrum induced a 50% reduction against *C. parvum* development in mice.

P23 is defined by the MAb C6B6 (Mead et al., 1988). IgG1 (Perryman et al., 1996) and IgA (Enriquez and Riggs, 1998) MAbs reactive with P23 were shown to have significant anticryptosporidial activity in mice. Perryman et al. (1999) reported that late gestation Holstein cows immunized with rC7, an immunodominant epitope of P23, developed specific antibodies. Calves receiving immune colostrum were significantly protected against diarrhea and were shedding significantly fewer oocysts than control calves. Antibodies directed against GP900 were shown to inhibit sporozoite in vitro, suggesting a direct role of the molecule in host–parasite interaction (Barnes et al., 1998). The reported involvement of TRAP-C1 homologs in substrate-dependent locomotion of sporozoite (Sultan et al., 1997), as well as in host cell attachment and invasion (Muller et al., 1993), suggest that TRAP-C1 should be considered as a promising vaccine candidate against cryptosporidiosis.

In summary, use of vaccines as a public health tool in the control of food-borne illnesses is a complex and important field. Reducing bio burdens in food producing species, where the target species normally is carrying, or shedding at an already very low rate, or where the human pathogen is part of the animal's normal host flora, is a significant challenge for current vaccine technologies. In the case of developed market places, the threat of food-borne illness is already at impressively low levels, though public expectations are that they will drop still further. Therefore, it must be understood that for successful control, these programs must be components of fully integrated programs, including husbandry, meat processing, and handling at the retail and customer levels.

OPTIMIZING VACCINATION PROTOCOLS

Over the last decade there have been tremendous advances in our understanding of the immune systems of our major domestic animal species. Nonetheless, we still lag behind the depth of knowledge that exists for the immune systems of mice and men. This has not stopped successful vaccination programs from being developed. Much of vaccinology that is practiced in veterinary medicine, and the animal health industry in general, is empirical. It is driven by the realities that exist in raising production animals or working in veterinary practice, where making a living depends on keeping the animals healthy. It is an industry where vaccines are like insurance policies—protection from events that one hopes never happen, and where tenths of pennies can make or break the economics of a business. When used optimally, vaccines have been shown to prevent disease, reduce the need for pharmaceutical intervention (e.g., reduced antibiotic usage) and improve the health and welfare of animals (Knott et al., 1985; McIlroy et al., 1992).

Then there is the definition of protection. The perception that vaccines provide sterilizing immunity, where the disease agent does not establish an infection, while widely held, is generally unfounded and largely unrealistic. It is debatable if such immunity is even in the animal's best interest as exposure to, and transient infection with, pathogens helps keep the immune response active. Vaccines usually provide protection on one of three levels. First, at the lowest level, protection is simply a reduction of clinical signs associated with the infection. The reduction may be decreased disease severity and/or duration, or simply a delay in the onset of the disease. The infection still becomes established and in most cases the animal will be contagious to other in-contact animals, although there may be a reduction in the level of shedding of the infectious agent. Such immunity is not usually capable of stopping an outbreak from progressing.

Second, vaccines may induce immunity that can prevent clinical signs associated with the infection, where the animal shows no overt signs of being infected. This level of protection may actually be associated with a reduction of shedding of the challenge organism, which can reduce spread during a disease outbreak. Unfortunately, there is no general rule to rely upon in this regard, as reduction in shedding tends to be specific for the strains involved—both of the vaccine and the infection. It must be recognized that vaccine protection that prevents clinical signs of disease being identified can lead to a false assumption that the pathogen is not present in a vaccinated population. In such circumstances, especially where economic pressures exist, there may be a tendency to decrease the use of vaccines with disastrous results.

Third, vaccine protection may prevent the infection from becoming established. While this is what many people think all vaccines do, it is a claim that rarely will be found on product labels. The United States Department of Agriculture (USDA), which regulates vaccines in the US, requires that for this claim to be granted the vaccine must be demonstrated to be able to "prevent all colonization or replication of the challenge organism in vaccinated and challenged animals" and "this must be supported with a very high degree of confidence by convincing data" (DeHaven, 2002).

The USDA recognizes these varying levels of protection in the way they allow label claims. The above three categories are often written on labels as "aids in disease control," "for the prevention of disease," and "for the prevention of infection." The vaccine manufacturer has generated a great deal of data, which has been independently reviewed by a government agency, that become the basis for the label claims. Expecting a vaccine to do something that is unrealistic given the data and a label indication is perhaps the greatest cause of reported "vaccine failures." To paraphrase the old adage, if in doubt always read the label.

While on the topic of the level of protection, it should be noted that vaccines can provide efficacy, or protection, that may come from both direct and indirect effects. The direct effect can be relied upon to provide individual immunity—that is the animal that receives the protective dose of vaccine benefits directly. Additionally there may be indirect protection, or herd immunity, that results from vaccination of sufficient numbers of animals in a given population. Herd immunity protects nonvaccinated animals within a population, whether a flock, herd, or group. It is the result of the reduction of the ability of a disease to transmit through the vaccinated individuals. For such immunity to occur, the vaccine should be capable of reducing the shed of the disease causing agent following infection of a vaccinated animal.

The challenge in developing an optimal vaccination program is in dealing with the great diversity that exists within the animal world, and as such there probably is no single optimal program for all occasions. The multitude of factors that need to be considered in choosing a vaccination program are complex—species/breed differences, and within these the genetic selection that has occurred, some of it for resistance to select diseases; the diverse husbandry practices employed, some dating back centuries, some mandated by governments; and the disparate prevalence of diseases has resulted in the adoption of tailored vaccination programs. Such programs generally are designed to consider the risks (e.g., disease prevalence and exposure likelihood), husbandry (e.g., animal-to-animal contact, nutrition), and environmental factors (e.g., wildlife vectors, weather) and of course the animals (e.g., age, previous disease experience).

The conditions that influence the tailored vaccination program change rapidly over the life of the animal. Indeed the time period over which protection is required varies greatly. Chickens, unless they are breeding animals, normally go to market by 6 weeks of age, pigs become pork when just 6 months old, and cattle and sheep frequently become someone's dinner by 18 months of age. Even our beloved pets rarely live beyond their teen years. As we touch upon some of the more unique challenges that exist in optimizing vaccination protocols and practical veterinary vaccinology, the following pages are divided up by life-stage.

Prenatal Vaccination

Strengths

The protection that is provided by the mother to the unborn animal is critical in ensuring the delivery of healthy offspring. In livestock, producing new animals is what generates the income for the farm and as a result the use of vaccines during pregnancy has been widely adopted to enhance this protection. In utero, the fetus is protected from the environment and its pathogens. This protection is provided both physically and immunologically by the dam. The physical part is easy to see—fetal membranes and fluids shield the fetus behind a muscular uterine wall inside a multilayered abdomen. The immunological protection likewise is on several levels—innate and acquired immunity of both the dam and fetus—but unlike physical protection, immunological protection may continue beyond the in utero period into early life.

The immunity of the mother may prevent infections from reaching the fetus or may enable the dam to provide immune cells or antibodies to the fetus. Vaccination of the mother is commonly practiced to accomplish both. In species where litters are common, for example, pigs, vaccination of the dam is an effective way to protect the entire litter with a single vaccine dose. Even in single offspring dams, for example, cattle, vaccination is recognized as providing added insurance to getting a healthy newborn. It can also be an effective way for neonatal animals to gain early immunity against the plethora of infectious challenges they face in the first weeks of life. There is evidence that active vaccination of the fetus may occur with some antigens crossing the placenta and stimulating fetal immunity.

Sows, with a gestation period of 113 days, are often vaccinated near the time of breeding and again 2 weeks prior to farrowing. Such vaccinations have proven effective in increasing litter size and growth performance (Descamps et al., 1990). The vaccination in early pregnancy is aimed at protecting the litter by stimulating active immunity in the sow against common pathogens that cause abortion, metritis, or stillbirths, for example, *Leptospira* spp., parvovirus, *Erysipelothrix rhusiopathiae*. These vaccines, typically given by injection, induce an active immune response in the sow, preventing the viremia or bacteremia from occurring and thus protecting the unborn piglets. Vaccinations in late pregnancy are aimed at boosting immunity to maximize the level of passive immunity (transplacental and colostral antibodies) that a dam passes on to her offspring. Typically these vaccinations are targeted at neonatal diseases causing enteric (e.g., rotavirus, *E. coli*) or respiratory problems (e.g., *Pasteurella multocida, Bordetella bronchiseptica*). Piglets born of vaccinated sows generally have performed significantly better than those of nonvaccinated sows when evaluated on a variety of parameters such as growth and average daily weight gain (Riising et al., 2002).

Similar vaccination protocols are used in dairy cattle. As with the sows, vaccination during pregnancy prevents abortion and helps ensure the birth of a healthy calf. In turn, the successful birth of a calf also ensures another lactation cycle, which after all is what a dairy is all about.

The late pregnancy vaccinations are expected to generate high levels of circulating antibodies that may be passed onto the newborn animal via colostrum. Colostrum is the first milk that is produced by mammals for their babies. Typically this is produced for 2–4 days and its consumption by the newborn is seen as very important for a variety of reasons. Research has shown that maternal antibodies may be concentrated in the colostrum to levels higher than that found in the dam's circulation (Pfeffer et al., 2005). Through the use of dam vaccination, the specificity and level of these antibodies can be manipulated to tailor protection that is most appropriate for the geographical region or environment where the birth occurs. Maternal antibodies can be directly absorbed by the newborn and enter into circulation providing passive immunity against otherwise fatal infections. There is now good evidence that cells and cytokines of the dam's immune system also pass via colostrum to the newborn. These components of colostrum significantly aid the development of the neonatal immune

TABLE 16.1 In utero transplacental transfer of antibodies

Species	Type of placentation	Number of layers	Ig via placenta	Ig via colostrum
Pig, horse, donkey	Epitheliochorial	6	0	+++
Ruminants	Syndesmochorial	5	0	+++
Dog, cat	Endotheliochorial	4	+	+++
Primates	Hemochorial	3	++	+
Rodents	Hemendothelial	1	+++	+

Source: Tizard (2004).

system (Muneta et al., 2005; Reber et al., 2005). The corollary of this is that colostrum deprived animals typically have numerous health problems and frequent infections.

Some maternal antibody crosses the placenta to protect the unborn animal. The level of this transfer depends on the nature of the placenta. Typically of our domestic species, dogs and cats provide the greatest in utero transplacental transfer of antibodies, while pigs, horses, and cattle provide the greatest colostral antibody transfer (Tizard, 2004) (Table 16.1). Even poultry get the benefits of prenatal vaccination. In ovo vaccination is now common. Vaccines of various types are injected directly into the egg at between 9 and 14 days of age, depending on the organism being used as a vaccine. The whole process is automated enabling many thousands of eggs to be vaccinated in a day. The end result is that at hatching the young birds are protected from select pathogens. Research has shown that in ovo vaccination may be more advantageous than posthatch vaccination (Rautenschlein and Haase, 2005).

Challenges

Of course, there are downsides to prenatal vaccination. Not all vaccines are safe to be given to pregnant animals. The administration of live or adjuvanted vaccines during pregnancy carries the risk of inducing a transient febrile reaction and subsequent abortions or stillbirths that may follow a fever. Typically, live bovine herpesvirus vaccines in naïve cows have been associated with abortions due to fetal infection with the vaccine strain and similarly live porcine reproductive and respiratory virus vaccine has been found to reduce litter size and cause abortions in sows. The choice of which vaccines to use in pregnant animals should, as always, come after reading the label and understanding the risks involved.

The timing of administration can play a large part in influencing the safety and the effectiveness of the vaccine. Vaccination soon after breeding has been shown to reduce the conception rate. This is probably a direct result of the transient febrile reaction decreasing implantation of the embryo. Vaccination in late pregnancy, close to parturition, tends to be less effective. Around the time of parturition, there is a well-described phenomenon of reduced immune system activity (Houdijk et al., 2000). Vaccination may not deliver the expected efficacy if given under these suboptimal conditions.

Even when vaccines work well, inducing strong immune responses in the dam, there may be a downside. Passive antibodies, whether transplacental or colostral, can inhibit active immunization of the newborn. In some cases this blocking effect can last months. Unfortunately, the level of passive immunity that blocks vaccine strains may not be sufficient to protect against virulent pathogens. This "gap" between lack of protection and earliest ability to vaccinate has been well described for canine parvovirus (McGavin, 1985). Vaccination of the dam does not eliminate the problem and can actually shift the problem to a time period when the risk of exposure to disease is greater, for example, young animals taken to market.

Opportunities

As Albert Einstein said "In the middle of difficulty lies opportunity." So it is with prenatal vaccination where every challenge becomes an opportunity and vaccine producers are continually seeking safer, more effective vaccines.

The diversity of approaches to improved safety is enormous but ever increasing regulatory hurdles, rather than true safety issues, make more and more vaccines carry the warning "Not for Use in Pregnant Animals." The true benefits of prenatal vaccination are to enable consistent active immunity to develop in the unborn animal. Many neonatal diseases would disappear, if active immunity was present at birth. The resulting increase in health benefits would be significant. Such vaccines are not readily available at present. In ovo vaccination of chickens being perhaps the exception, although even there not all pathogens can be covered by safe and effective in ovo vaccines.

Likewise improved efficacy would be highly desirable to help overcome the periparturient relaxation of immunity that occurs in dams. Such increased efficacy would result in greater consistency of transplacental and colostral immunity transfer. The development of products that could enhance the transfer of passive immunity to the neonates would be greatly welcomed.

Neonatal and Juvenile Vaccination

Strengths

After leaving the protection of the uterus, the neonate begins dealing with a plethora of new antigens, both environmental and infectious. Unfortunately, most of our domestic species are born with relatively immature immune systems inclined toward a Th2-mediated response (Morein et al., 2002). It is during these early weeks of life after birth that newborn animals are the most vulnerable and this is where vaccines can be most beneficially used. The husbandry practices that we use with many of our animals species often entails mixing of animals from a variety of sources, for example, the combining of 3 week old pigs from perhaps hundreds of litters when they move to the nursery, or mixing of hundreds of puppies at 6 weeks of age when they are transported to the pet shop. The combining of very young animals from diverse disease backgrounds when their immune systems are not yet fully developed and not yet primed by vaccination is a huge challenge for maintaining good health. The need for vaccination before, during, and after such commingling of animals is essential to reduce the suffering that disease outbreaks can cause.

No longer are we trying to vaccinate litters, or trying to vaccinate through the dam. The animal is now an individual, albeit perhaps still part of a larger group. It is possible to directly administer vaccines to the animal in accordance with an assessment of its individual needs. As most animal breeders expect to make money from selling the offspring, each individual animal has a monetary value. This enables a fairly objective view to be taken of risk benefit analysis—the benefits arising from the use of vaccines versus the cost of disease. Most animal breeders have tailored vaccination programs to the needs of their animals, with multiple injections of different vaccines being common. Typically neonatal animals receive multiple vaccinations covering the major diseases that their owners expect they may become exposed to during their lives. A kitten could receive eight different vaccines on two or three separate occasions, although many are given in combination—feline herpesvirus, feline calicivirus, feline panleukopenia, feline leukemia virus, rabies virus, and Chlamydia are common; with feline immunodeficiency virus, *B. bronchiseptica* and *Toxoplasma gondii* also available for cats. Puppy vaccinations often involve combinations containing protection against 11 different disease agents, for example, canine parvovirus, distemper, canine adenovirus types 1 and 2, canine parainfluenza, rabies virus, four different *Leptospira* species and *B. bronchiseptica*. These vaccines are usually given 2 or 3 times at intervals of 3–4 weeks starting at 8 weeks of life.

One consequence of now having to vaccinate the individual animals is that in large numbers the task can prove to be daunting. The labor and logistics of vaccinating 20,000 chickens or 10,000 piglets are quite involved. Fortunately, vaccines are available for mass administrations that are optimal for dealing with large flocks or herds. Several options exist with aerosol spray and drinking water being the most common. The vaccines typically are live agents that are diluted in water containing a stabilizer to prolong the in-use viability of the agent. Aerosol vaccination can be tailored to the particular vaccine by adjusting the droplet size that in turn alters the location within the respiratory tract where the droplet eventually lodges (Gough and Allan, 1973; Gomez and Correa, 1978). While originally developed for poultry vaccination, aerosol vaccination has been used in swine (Nielsen et al., 1990) and cattle (Kita et al., 1982) with the same factors as for poultry needing to be considered—droplet size, type of diluent, environmental conditions (e.g., wind and temperature) and strain, and dose of the vaccine organism.

Drinking water, as a vehicle to mass vaccinate animals, while pioneered for poultry, has been used in other species. The addition of the vaccine to regular tap water can be an effective way to vaccinate provided the chlorine is neutralized by the addition of 0.25% skim milk (Kim and Spradbrow, 1978). Like aerosol vaccination the number of doses required for adequate protection will be contingent on the type of agent in the vaccine and the age of the animal.

Challenges

Vaccination of very young animals poses numerous challenges but none as difficult as that of overcoming the interference of maternal immunity. The maternally derived antibodies that most young are born with, or acquire through colostrum consumption, can be protective but as the level wanes there is a period where it is no longer protective. What proves to be worse for the health of the young animal is that these residual antibodies may block effective immunization.

The traditional solution is to administer multiple doses around the period where the maternally derived protection is waning. As the timing cannot be accurately known even this does not guarantee that there is not a susceptible period. As numerous papers testify to, during the canine parvovirus epidemic of the 1980s, many puppies failed to be protected despite receiving vaccinations at weekly intervals (McGavin, 1985). Similar situations exist in other domesticated species, with the result that many doses of vaccine given to young animals are ineffective, not because of a problem with the vaccine, but with the timing of the delivery.

The use of injections to administer vaccines has proven a very effective delivery method but this too has caused problems. Transmission of disease through the use of contaminated needles or vaccine vials has been documented (Witter and Fadly, 2001; Niskanen and Lindberg, 2003). While good hygiene should prevent such problems, the field conditions under which animal vaccinations frequently occur are often not conducive to maintaining sanitary conditions. The presence of blood, mud, feces, and flies is not unexpected. In meat-producing animals the use of needles to deliver products can lead to another problem—broken needles in meat products. The inability to find a broken needle quite often results in the condemnation of the carcass to keep it out of the food chain. While this is a problem with any product, including medicines, delivered by injection, vaccines represent the most common reason for sticking a needle into a meat-producing animal.

The relative immaturity of the newborn's immune system is another challenge that must be factored into vaccination of the neonatal or juvenile animal. The bias for a Th2 response in the neonate (Morein et al., 2002) is not always the preferred type of immune response for protection against disease. The use of adjuvants that drive Th1 immune responses may be required to confirm effective protection. It has been demonstrated that neonatal calves may be able to develop a cell-mediated response comparable to adults, although antibody responses are markedly lower than seen in adult heifers (Nonnecke et al., 2005). This is in normal, healthy, young animals but there are a variety of situations where the young immune system is even less ready to respond to vaccination.

In colostrum deprived animals the maturation of the immune system may be delayed because of a lack of exposure to specific components in this first milk. Vaccination of colostrum-deprived animals should be undertaken with care to recognize that their responsiveness may be compromised compared to that of animals receiving colostrum. Likewise, stress has been associated with poor immune responsiveness and there is little doubt that the newborn goes through many stresses in the first few weeks of life. There, of course, are the routine environmental stressors such as temperature, air quality, etc., but often more critical for animals is the dietary change including weaning at an early age that is routine in most of our domestic species. Pigs, for example, may be weaned at 2–3 weeks of age, and it has been shown that the composition of their starter diets, in particular the levels of protein, lysine, methionine, and threonine, may be critical in the development of their immune system (YongQing et al., 2001; Nonnecke et al., 2005). The simple event of handling the young animals to enable vaccination has been shown to be stressful and potentially reduces the response to the vaccine. Even a car trip to the veterinarian's office may have the same stress-related effect on the immune system. Unfortunately, when simply looking at a young animal it is not possible to determine its responsiveness to vaccination. It is nearly always a guess, which often results in the administration of multiple doses of vaccine to cover the poor responders.

Opportunities

The largest opportunity to improve vaccination of neonatal and juvenile animals is undoubtedly for vaccines that would consistently work in the face of maternal immunity. The health and financial benefits would be immense if we could remove the need to revaccinate young animals because active immunity was blocked by residual but waning passive immunity. The convenience of knowing that you could take your pup or kitten to the veterinarian once to have them immunized, or the need to only vaccinate a herd of 10,000 piglets once, are immense. Likewise sustained release, or long-acting vaccines could facilitate less handling of the animals and hence less stress. Such long-lasting vaccines would increase the likelihood that the vaccine antigens are present when the animal is able to respond to them.

The ability to test for the responsiveness of an animal to a vaccine before actually administering the vaccine could have huge practical benefits for animal owners, especially livestock producers who have large numbers of animals to vaccinate. Currently laboratory testing for antibody levels is possible, but the cost and time make this relatively impractical. It has proven cheaper to vaccinate and hope for a response than to bleed and test. The ability to determine, perhaps via a simple animal-side test, whether the animal has responded to vaccination would be beneficial.

Considerable progress has been made in avoiding needle delivery of vaccines, and not just through alternative routes such as aerosol and drinking water. In the last 10 years, the development of needle-less injector has been an active area of research. These devices use a variety of approaches but most commonly involve high pressures to force the liquid vaccine directly through the skin without a needle. While most of these injectors are effective to some extent, the challenges of dealing with superficial skin contamination and the problem posed by the presence of a dense hair coat to obtain good skin contact for the injection, have hampered their adoption. It is likely that reliable and highly effective needle-less injection devices will be available within the next few years.

Adult Vaccination

Strengths

Many of the issues that surround vaccination of neonatal and juvenile animals tend to resolve with age. Adulthood generally brings a mature immune system including fully functioning innate immunity. The time of this varies by species but, in general, by 1 year of age most or our domestic species are nearing breeding age and can be considered young adults. As stated earlier, chickens and pigs may well have gone to slaughter before 1 year of age but breeding animals of these species are kept beyond 1 year of age. The value of an adult, having been reared and cared for, perhaps for years, means that the economics of vaccination is generally less of a concern. Multiple vaccinations are often given, considering specific disease risks and environmental factors, as part of a tailored vaccination program.

Adult animals, with their mature immune systems, absence of maternally derived immunity, and adjustment to their environmental stressors, are normally the best candidates for vaccination. Even in these near ideal candidates for vaccination, the number of doses that are required for efficacy will depend on the nature of the vaccine. Vaccines fall broadly into one of two categories—live or killed.

Live vaccines typically replicate in the host and hence deliver a large antigenic mass to the immune system resulting in rapid and strong immunity. Live vaccines have several advantages. Formulated without specific immune system stimulants, they generally cause less injection site reactions. Even when not being injected they have the advantage that most retain the infectivity of the wild-type pathogen enabling effective delivery by routes such as aerosol or orally. One potential negative is that, because they replicate, there is the risk that in an immunocompromised individual they may induce some signs of the disease they were meant to prevent.

Killed vaccines might be better termed nonreplicating to cover vaccines that never were alive, for example, subunit-containing vaccines. Most vaccines in this category are made from the culture of live organisms that are inactivated, or killed, using one of a variety of chemicals, with formaldehyde the commonest. The inactivated organisms on there own will often only stimulate a weak or transient immune response. While there are numerous causes of this phenomenon, the common explanations are that inactivation alters the conformation and hence presentation of some antigen epitopes and the inactivated antigens do not signal quite the same danger to the animal's immune system. The "danger signal" model of immunity moves away from the self versus nonself based view and considers the idea that the immune system is more concerned about organisms that damage the animal rather than just being foreign (Gallucci and Matzinger, 2001). To overcome this shortcoming, the addition of an adjuvant is classically done to stimulate a greater immune response, or to drive a particular immune response, for example, Th2 versus Th1. Killed vaccines are safe, in that, without live agents in the final product, they cannot be the source of an infection. However, the presence of the adjuvant often results in greater injection site reactions than is seen with live vaccines. Injection site reactions can range from merely discomforting to quite painful, but in livestock they often end up as imperfections in the meat resulting in trim losses on the slaughter line. The route of vaccination is often dictated by the nature of the injection site reaction. In beef cattle most vaccines are given subcutaneously to avoid reactions damaging the meat. In show animals, for example, horses, intramuscular injections are preferred to avoid any blemishes that subcutaneous injections may cause.

The typical adult will fall into one of two categories with respect to vaccination—naïve or preexposed. The true naïve adult will not have a problem with maternally derived antibodies blocking effective vaccination and normally responds well to vaccinations. In fact the responsiveness to vaccination is much more predictable (strength of protection engendered) and reliable (percentage of animals responding) in these naïve adult animals. This is critical as the adult animal tends to be more adventurous, often traveling and having greater contact with other animals. This is particularly true of the companion animals' species where dogs, cats, and horses may go to shows, boarding facilities, and sporting events. At these venues their opportunity to interact with animals from a diverse range of geographies, and consequently disease backgrounds, is enormous. Even as our livestock species grow older they will have greater exposure risks—stock movement is common and introduced animals whether for breeding or to add production capacity frequently come from markets where mixing of animals is commonplace and disease transmission is prevalent. The onset and duration of immunity that occurs after vaccination of naïve animals will generally be representative of a primary immune response. Such responses tend to be slower in reaching protective levels and the memory may be shorter following a single vaccine dose.

The preexposed adult may have already experienced the disease or have a prior vaccination history. Often these two events will be indistinguishable without an accurate history of the animal. Fortunately when it comes to vaccination it does not matter, both recovery from infection and effective previous vaccination will have engendered an immune memory provided the exposure was not too long ago. The duration of immunity that follows exposure of the immune system to a disease agent, even a vaccine strain, depends on the translation of at least part of the initial response into a memory response. The presence of specific memory T cells enables an amnestic response regardless of the source of the agent inducing the original response. This phenomenon is leveraged in adult vaccination programs to decrease the frequency of revaccination. Traditionally, dog and cats were revaccinated on an annual basis throughout their lifetimes, although it was anticipated that routine vaccines were likely to provide far greater than 1 year of protection. Recent work has proven that under field conditions there is evidence of immunity that can persist for 48 months in some animals (Mouzin et al., 2004a, 2004b). Evidence in horses and cattle would indicate that the duration of immunity following vaccination may be only as long as 1 year but varies with the disease agent (Hannant et al., 1988; Peters et al., 2004). The vaccine manufacturer normally provides guidance as to the duration of immunity that may be anticipated following successful vaccination of animals. Depending on the individual circumstances the duration of immunity may be longer or shorter than the label indicates, especially in environments where exposure to the disease agent occurs. This results in natural boosting of immunity and essentially lifelong immunity.

Challenges

Like their younger counterparts, there are a variety of problems that may be associated with vaccination, for example, injection site reactions and mild transient signs of disease, which are a concern to owners. There are some issues however that tend to be seen more commonly in adults. The reason may be easily understood by remembering that the longer an animal lives the more vaccination events it is likely to have lived through. While this is good with respect to developing immunity, there may be downsides.

Transient lameness may occur following vaccination when the dose is administered into a leg or neck muscle and may be more noticeable in adults simply because of the greater weight of the animals compared to their neonatal or juvenile counterparts. Prior exposure to vaccines may induce a hypersensitivity to components, such as bovine serum albumin, used in manufacturing the product with the result that larger injection site reactions may occur. Other reactions such as arthropathy, vasculitis, neurological dysfunction, and thrombocytopenia have been associated with repeated vaccine exposure (Schattner, 2005). Fibrosarcomas in cats have been linked to vaccine reactions and usually have fatal outcomes (Hershey et al., 2000). Nonetheless the incidence of postvaccinal reactions is relatively rare. In one study, the reaction rate in over 1.2 million dogs receiving 3.4 million doses of vaccine was found to be less than 1 in 250 dogs (Moore et al., 2005).

The use of vaccines during periods of lactation has been found occasionally to cause transient decreases in milk production (Scott et al., 2001). The milk drop appears to be associated with a febrile response that in particular follows the administration of live viral vaccines. In cattle, the milk production loss typically is small but readily noticeable by diligent dairy staff.

Adult animals are often subject to stressful conditions that younger animals are spared. Whether it is the stress of lactation, or the stress of performance in racing or pulling wagons, the result is the same—potential down-regulation of the immune system and reduced responsiveness to vaccination.

Opportunities

The move in companion animals toward longer duration of immunity is a trend that is likely to expand into vaccines for livestock. The labor costs of handling mature livestock are considerable and yet the cost of a disease outbreak can financially devastate a producer. Several promising areas of vaccine research raise the likelihood of longer duration of immunity for all animal species.

Vaccine researchers are tapping into the trends in human health to improve the effectiveness of animal vaccines. The use of DNA vaccines with specific cytotoxic lymphocyte antigens and CpG motifs has been found to induce more potent memory responses (Kennedy et al., 2005). Another promising area of research involves TLRs. These receptors are a family of transmembrane proteins that can interact with a variety of antigens from microbial organisms. Once activated by the interaction, the TLRs can induce an intracellular cascade resulting in select gene expression associated with the immune system. The discovery of TLR agonists or ligands may result in new classes of adjuvants that are more specific and yet safer.

The incorporation in vaccines of interleukins, or genes coding for them, has been a popular area of veterinary research. The intent is that coadministration of antigens with cytokines will result in enhanced immunity. Such products, should they come to market, are likely to also benefit neonatal animals as cytokines may help the maturation of the immune system.

In summary, there is no magic solution to optimizing vaccination programs for animals. A solid understanding of the animal's innate and environmental risk factors as well as the variables such as stress will enable the development of tailored vaccination schedules that best meets the needs of the animal. While no vaccine will be 100% effective under all circumstances, careful consideration of what and when to use, can certainly enhance the chances of successfully protecting animals. Finally, if all else fails, read the label and follow the manufacturer's recommendations. They are based on extensive research and are provided to help optimize the effectiveness of vaccines.

STRATEGIZING THE USE OF ANIMAL VACCINES FOR PROTECTING THE PUBLIC

Health, Modeling, and National Economies

The direct and indirect impacts of infectious diseases, particularly highly contagious diseases, on human and animal health may become rapidly massive and erode a national economy within a few hours or days. These effects may be exponential in nature if the disease is zoonotic and trade sensitive. For example, the recent international crisis of outbreaks of high pathogenicity of the Avian influenza H5N1 (Schudel and Lombard, 2006) reduced trade (Hall, 2004) and negatively impacted the economies of several nations across the globe from Southeast Asia to Western Europe (McQuiston et al., 2005; Vallat and Mallet, 2006) in addition to human morbidity and mortality and serious social disruption (Gust et al., 2001). Other diseases and especially trans-boundary diseases including Nipah virus, West Nile virus, severe acute respiratory syndrome (SARS) (von Overbeck, 2003), classical swine fever, porcine reproductive and respiratory syndrome (PRRS) (Garner et al., 2001), rinderpest, peste des petits ruminants (PPR), Rift Valley fever (RVF), brucellosis, and bovine tuberculosis have been identified as having profound effects on international trade and national economies if introduced into countries free of these diseases (Domenech et al., 2006). Modeling analyses of the immunization against tuberculosis, HIV, Chlamydia trachomatis, and malaria with effective vaccines predicts significant health and economic benefits (Bishai et al., 2001; Bishai and Mercer, 2001; Frick et al., 2004; Maire et al., 2006).

The classic example of a highly contagious animal disease severely impacting international trade, causing social disruption, and having profound repercussions on national and global economies is foot and mouth disease (FMD) (McCauley et al., 1977; Krystynak and Charlebois, 1987; Perry et al., 1999; Paarlberg et al., 2002; Randolph et al., 2002; Belton, 2004; Vallat and Mallet, 2006), which unfortunately occurred in the United Kingdom in 2002 (Samuel and Knowles, 2001; Thompson et al., 2002). The losses attributed to FMD to agriculture and the food chain in the United Kingdom amounted to about £3.1 billion. FMD is an ongoing crisis in several countries and continuously poses a serious threat to FMD-free countries. Based upon the significant risks of an accidental or intentional introduction of zoonotic and nonzoonotic diseases, newer strategies are being developed as countermeasures.

Use of Models for Preemptive Vaccination and Identification of Emerging Disease Vaccine Targets

In response to the growing global crises associated with known, emerging, and unknown infectious diseases of man and animals, more robust disease models and surveillance systems are now being developed to assess the risks and identify potential targets for vaccine research and development; however, more comprehensive interoperable datasets, middleware, and improved micro- and macrolevel models will be required to make the system more efficient and effective (Gust et al., 2001). Several models indicate that vaccination strategies for FMD and respiratory pathogens are justified (Bates et al., 2003a, 2003b, 2003c; Pourbohloul et al., 2005). Even with more sophisticated FMD predictive modeling methods (Perez et al., 2005, 2006; Branscum et al., 2007; Shiilegdamba et al., 2008), the accompanying implementable governmental policies will need to be in place to take tactical and strategic advantage of newer FMD surveillance and vaccine technologies (Marshall and Roger, 2004; Kobayashi et al., 2007a, 2007b), or the control and eradication programs may be prolonged (Whiting, 2003; Perez et al., 2004). Additionally, for purposes of biopreparedness, predicting which vaccines and what quantity of vaccine should be stockpiled becomes a daunting task (Rossides, 2002); however, modeling approaches have been developed to facilitate the decision-making process given the uncertainties and complexities of naturally occurring much less intentional infectious disease outbreaks (Medema et al., 2004).

Reduction and Control of Zoonoses through Strategic Application of Animal Vaccines

In reviewing data from several sources, Lutticken et al. (2007) found that with the increasing demand for food, the enlarging scale of world food production, the augmented transportation of animals and food products, and simultaneous contact of animals with the environment, several microorganisms have established themselves in farmed animals, which although may be relatively harmless to animals are pathogenic to man. These investigators propose options for reducing the risk of transferring zoonotic agents from animals (particularly farm animals) to man by specifically applying veterinary vaccines against viral and bacterial diseases (Lutticken et al., 2007). Other investigations documented that avian influenza, West Nile virus, *Bartonella henselae*, rabies and anthrax vaccines which protect animals reduce or prevent transmission of zoonoses to humans (Capua, 2007; Lutticken et al., 2007; Marano et al., 2007; Miguens, 2007). Vaccination of vector/reservoir species, when efficacious vaccines are available, offers will offer significant advantages to combating zoonotic human disease (Zinsstag and Weiss, 2001; Marano et al., 2007). In summary, when appropriate biopreparedness, management strategies and contingency plans of the future (Westergaard, 2008) are linked with (1) protective DIVA vaccines (Clavijo et al., 2004; Selke et al., 2007) against zoonoses, (2) effective predictive modeling (Branscum et al., 2007), and (3) deployable implementation policies (Kobayashi et al., 2007b), control and prevention of serious zoonotic diseases of man and animals will become more achievable at local, state, and national levels.

References

Agin, T.S., Zhu, C., Johnson, L.A., Thate, T.E., Yang, Z. and Boedeker, E.C. Protection against hemorrhagic colitis in an animal model by oral immunization with isogeneic rabbit enteropathogenic *Escherichia coli* attenuated by truncating intimin. Infect. Immun. 2005; 73(10):6608–6619.

Alves, M., Matos, O. and Antunes, F. Multilocus PCR-RFLP analysis of *Cryptosporidium* isolates from HIV-infected patients from Portugal. Ann. Trop. Med. Parasitol. 2001; 95(6):627–632.

Andrianov, A.K., Chen, J. and Payne, L.G. Preparation of hydrogel microspheres by coacervation of aqueous polyphosphazene solutions. Biomaterials 1998; 19(1–3):109–115.

Anonymous Compendium of measures to prevent disease associated with animals in public settings. MMWR 2005b:1277–1280.

Anonymous. Preliminary FoodNet data on the incidence of infection with pathogens transmitted commonly through food. CDC, 2005c; 54:352–356.

Anonymous. Division of Bacterial and Mycotic Diseases. CDC. http://www.cdc.gov/ncidod/dbmd/diseaseinfo/escherichiacoli_g.htm. 2006.

Armson, A., Meloni, B.P., Reynoldson, J.A. and Thompson, R.C. Assessment of drugs against *Cryptosporidium parvum* using a simple in vitro screening method. FEMS Microbiol. Lett. 1999; 178(2):227–233.

Babiuk, L.A., Babiuk, S.L. and Baca-Estrada, M.E. Novel vaccine strategies. Adv. Virus Res. 2002; 58:29–80.

Babiuk, L.A., Babiuk, S.L., Loehr, B.I. and van Drunnen Littel-van den, H. Nucleic acid vaccines: research tool or commercial reality. Vet. Immunol. Immunopathol. 2000; 76(1–2):1–23.

Barnes, D.A., Bonnin, A., Huang, J.X., Gousset, L., Wu, J., Gut, J., Doyle, P., Dubremetz, J.F., Ward, H. and Petersen, C. A novel multi-domain mucin-like glycoprotein of *Cryptosporidium parvum* mediates invasion. Mol. Biochem. Parasitol. 1998; 96(1–2):93–110.

Bates, T.W., Carpenter, T.E. and Thurmond, M.C. Benefit-cost analysis of vaccination and preemptive slaughter as a means of eradicating foot-and-mouth disease. Am. J. Vet. Res. 2003a; 64(7):805–812.

Bates, T.W., Thurmond, M.C. and Carpenter, T.E. Description of an epidemic simulation model for use in evaluating strategies to control an outbreak of foot-and-mouth disease. Am. J. Vet. Res. 2003b; 64(2):195–204.

Bates, T.W., Thurmond, M.C. and Carpenter, T.E. Results of epidemic simulation modeling to evaluate strategies to control an outbreak of foot-and-mouth disease. Am. J. Vet. Res. 2003c; 64(2):205–210.

Belton, D.J. The macro-economic impact of a foot-and-mouth disease incursion in New Zealand. Dev. Biol. (Basel) 2004; 119:457–461.

Benoit, M.A., Baras, B. and Gillard, J. Preparation and characterization of protein-loaded poly(epsilon-caprolactone) microparticles for oral vaccine delivery. Int. J. Pharm. 1999; 184(1):73–84.

Berthelot-Herault, F., Gottschalk, M., Morvan, H. and Kobisch, M. Dilemma of virulence of *Streptococcus suis*: Canadian isolate 89-1591 characterized as a virulent strain using a standardized experimental model in pigs. Can. J. Vet. Res. 2005; 69(3):236–240.

Bishai, D., Lin, M.K. and Kiyonga, C.W. Modeling the economic benefits of an AIDS vaccine. Vaccine 2001; 20(3–4):526–531.

Bishai, D.M. and Mercer, D. Modeling the economic benefits of better TB vaccines. Int. J. Tuberc. Lung. Dis. 2001; 5(11):984–993.

Bowersock, T.L., HogenEsch, H., Suckow, M., Guimond, P., Martin, S., Borie, D., Torregrosa, S., Park, H. and Park, K. Oral vaccination of animals with antigens encapsulated in alginate microspheres. Vaccine 1999; 17(13–14):1804–1811.

Branscum, A.J., Perez, A.M., Johnson, W.O. and Thurmond, M.C. Bayesian spatiotemporal analysis of foot-and-mouth disease data from the Republic of Turkey. Epidemiol. Infect. 2008; 136(6):833–842.

Brochier, B., Costy, F. and Pastoret, P.P. Elimination of fox rabies from Belgium using a recombinant vaccinia-rabies vaccine: an update. Vet. Microbiol. 1995; 46(1–3):269–279.

Brochier, B., Kieny, M.P., Costy, F., Coppens, P., Bauduin, B., Lecocq, J.P., Languet, B., Chappuis, G., Desmettre, P., Afiademanyo, K. et al. Large-scale eradication of rabies using recombinant vaccinia-rabies vaccine. Nature 1991; 354(6354): 520–522.

Brown, F. Review of accidents caused by incomplete inactivation of viruses. Dev. Biol. Stand. 1993; 81:103–107.

Burton, G. and Mason, K.V. Do postvaccinal sarcomas occur in Australian cats? Aust. Vet. J. 1997; 75(2):102–106.

Busque, P., Higgins, R., Caya, F. and Quessy, S. Immunization of pigs against *Streptococcus suis* serotype 2 infection using a live avirulent strain. Can. J. Vet. Res. 1997; 61(4):275–279.

Capua, I. Vaccination for notifiable avian influenza in poultry. Rev. Sci. Tech. 2007; 26(1):217–227.

Cawthraw, S., Ayling, R., Nuijten, P., Wassenaar, T. and Newell, D.G. Isotype, specificity, and kinetics of systemic and mucosal antibodies to *Campylobacter jejuni* antigens, including flagellin, during experimental oral infections of chickens. Avian Dis. 1994; 38(2):341–349.

Claesson, B.A., Trollfors, B., Lagergard, T., Taranger, J., Bryla, D., Otterman, G., Cramton, T., Yang, Y., Reimer, C.B., Robbins, J.B. et al. Clinical and immunologic responses to the capsular polysaccharide of *Haemophilus influenzae* type b alone or conjugated to tetanus toxoid in 18- to 23-month-old children. J. Pediatr. 1988; 112(5):695–702.

Clavijo, A., Wright, P. and Kitching, P. Developments in diagnostic techniques for differentiating infection from vaccination in foot-and-mouth disease. Vet. J. 2004; 167(1):9–22.

Cleland, J.L., Lim, A., Daugherty, A., Barron, L., Desjardin, N., Duenas, E.T., Eastman, D.J., Vennari, J.C., Wrin, T., Berman, P., Murthy, K.K. and Powell, M.F. Development of a single-shot subunit vaccine for HIV-1. 5. Programmable in vivo autoboost and long lasting neutralizing response. J. Pharm. Sci. 1998; 87(12):1489–1495.

Cox, J.C. and Coulter, A.R. Adjuvants: a classification and review of their modes of action. Vaccine 1997; 15(3):248–256.

Cox, J.C., Sjolander, A. and Barr, I.G. ISCOMs and other saponin based adjuvants. Adv. Drug Deliv. Rev. 1998; 32(3):247–271.

de Greeff, A., Buys, H., Verhaar, R., Dijkstra, J., van Alphen, L. and Smith, H.E. Contribution of fibronectin-binding protein to pathogenesis of *Streptococcus suis* serotype 2. Infect. Immun. 2002; 70(3):1319–1325.

Dean-Nystrom, E.A., Gansheroff, L.J., Mills, M., Moon, H.W. and O'Brien, A.D. Vaccination of pregnant dams with intimin(O157) protects suckling piglets from *Escherichia coli* O157:H7 infection. Infect. Immun. 2002; 70(5):2414–2418.

DeHaven, R. Veterinary Services Memorandum No. 800.202—General Licensing Considerations: Efficacy Studies (U.S.D.o. Agriculture, Ed.), 2002.

Deneer, H.G. and Potter, A.A. Effect of iron restriction on the outer membrane proteins of *Actinobacillus* (*Haemophilus*) *pleuropneumoniae*. Infect. Immun. 1989a; 57(3):798–804.

Deneer, H.G. and Potter, A.A. Iron-repressible outer-membrane proteins of *Pasteurella haemolytica*. J. Gen. Microbiol. 1989b; 135 (Pt 2):435–443.

Deng, W., Puente, J.L., Gruenheid, S., Li, Y., Vallance, B.A., Vazquez, A., Barba, J., Ibarra, J.A., O'Donnell, P., Metalnikov, P., Ashman, K., Lee, S., Goode, D., Pawson, T. and Finlay, B.B. Dissecting virulence: systematic and functional analyses of a pathogenicity island. Proc. Natl. Acad. Sci. U.S.A. 2004; 101(10): 3597–3602.

Descamps, J., Charlier, P., Jolie, R. and Roose, P.D. International Pig Veterinary Society, 11th Congress, Lausanne, Switzerland. 1990.

Diwan, M., Tafaghodi, M. and Samuel, J. Enhancement of immune responses by co-delivery of a CpG oligodeoxynucleotide and tetanus toxoid in biodegradable nanospheres. J. Control. Release 2002; 85(1–3):247–262.

Domenech, J., Lubroth, J., Eddi, C., Martin, V. and Roger, F. Regional and international approaches on prevention and control of animal transboundary and emerging diseases. Ann. N.Y. Acad. Sci. 2006; 1081:90–107.

Donnenberg, M.S. and Kaper, J.B. Construction of an eae deletion mutant of enteropathogenic *Escherichia coli* by using a positive-selection suicide vector. Infect. Immun. 1991; 59(12): 4310–4317.

Eldridge, J.H., Staas, J.K., Chen, D., Marx, P.A., Tice, T.R. and Gilley, R.M. New advances in vaccine delivery systems. Semin. Hematol. 1993; 30(Suppl. 4):16–24, discussion 25.

Enriquez, F.J. and Riggs, M.W. Role of immunoglobulin A monoclonal antibodies against P23 in controlling murine *Cryptosporidium parvum* infection. Infect. Immun. 1998; 66(9): 4469–4473.

Evans, S.J. and Sayers, A.R. A longitudinal study of campylobacter infection of broiler flocks in Great Britain. Prev. Vet. Med. 2000; 46(3):209–223.

Fayer, R. and Fetterer, R. Activity of benzimidazoles against cryptosporidiosis in neonatal BALB/c mice. J. Parasitol. 1995; 81(5):794–795.

Ferguson, M., Wood, D.J. and Minor, P.D. Antigenic structure of poliovirus in inactivated vaccines. J. Gen. Virol. 1993; 74(Pt 4): 685–690.

Frick, K.D., Colchero, M.A. and Dean, D. Modeling the economic net benefit of a potential vaccination program against ocular infection with *Chlamydia trachomatis*. Vaccine 2004; 22(5–6):689–696.

Gaertner, F.H., Babiuk, L.A., Van Moorlehem, E.A., Beskorwayne, T.K., Lee, S.L., Shutter, R.W., Armstrong, J.M. and Griebel, P.J. Amended recombinant cells (ARCs): an economical and surprisingly effective production and delivery vehicle for recombinant bovine IFN-gamma. J. Control. Release 2005; 107(2):189–202.

Gallucci, S. and Matzinger, P. Danger signals: SOS to the immune system. Curr. Opin. Immunol. 2001; 13(1):114–119.

Garner, M.G., Whan, I.F., Gard, G.P. and Phillips, D. The expected economic impact of selected exotic diseases on the pig industry of Australia. Rev. Sci. Tech. 2001; 20(3):671–685.

Garrity, G.M. Bergey's Manual of Systematic Bacteriology. In: (Boone, D.R. and Castenholz, R.W., Eds.), 2nd, New York: Springer, 2005.

Gast, R.K. Paratyphoid infections. In: Diseases of Poultry (Y.M. Saif, H.J. Barnes, J.R. Glisson, A.M. Fadly, L.R. McDougald and D.E. Swayne, Eds.), pp. 583–613. Ames. IA: Iowa State Press, 2003.

Gerdts, V., Mutwiri, G.K., Tikoo, S.K. and Babiuk, L.A. Mucosal delivery of vaccines in domestic animals. Vet. Res. 2006; 37(3):487–510.

Giavedoni, L., Ahmad, S., Jones, L. and Yilma, T. Expression of gamma interferon by simian immunodeficiency virus increases attenuation and reduces postchallenge virus load in vaccinated rhesus macaques. J. Virol. 1997; 71(2):866–872.

Gomez, M. and Correa, J.M. Haemagglutination-inhibiting antibodies and resistance to challenge in fowls following aerosol vaccination against Newcastle disease. Arch. Med. Vet. 1978; 10(2):128–135.

Gomis, S., Babiuk, L., Allan, B., Willson, P., Waters, E., Ambrose, N., Hecker, R. and Potter, A. Protection of neonatal chicks against a lethal challenge of *Escherichia coli* using DNA containing cytosine-phosphodiester-guanine motifs. Avian Dis. 2004; 48(4):813–822.

Gough, R.E. and Allan, W.H. Aerosol vaccination against Newcastle disease: the influence of vaccine diluent. Vet. Rec. 1973; 93(17): 458–461.

Greensfelder, L. Infectious diseases. Polio outbreak raises questions about vaccine. Science 2000; 290(5498):1867–1869.

Gruening, P., Fulde, M., Valentin-Weigand, P. and Goethe, R. Structure, regulation, and putative function of the arginine deiminase system of *Streptococcus suis*. J. Bacteriol. 2006; 188(2):361–369.

Gust, I.D., Hampson, A.W. and Lavanchy, D. Planning for the next pandemic of influenza. Rev. Med. Virol. 2001; 11(1):59–70.

Guyot, K., Follet-Dumoulin, A., Lelievre, E., Sarfati, C., Rabodonirina, M., Nevez, G., Cailliez, J.C., Camus, D. and Dei-Cas, E. Molecular characterization of *Cryptosporidium* isolates obtained from humans in France. J. Clin. Microbiol. 2001; 39(10):3472–3480.

Haesebrouck, F., Pasmans, F., Chiers, K., Maes, D., Ducatelle, R. and Decostere, A. Efficacy of vaccines against bacterial diseases in swine: what can we expect? Vet. Microbiol. 2004; 100(3–4): 255–268.

Hald, B., Knudsen, K., Lind, P. and Madsen, M. Study of the infectivity of saline-stored *Campylobacter jejuni* for day-old chicks. Appl. Environ. Microbiol. 2001; 67(5):2388–2392.

Hall, C. Impact of avian influenza on U.S. poultry trade relations-2002: H5 or H7 low pathogenic avian influenza. Ann. N.Y. Acad. Sci. 2004; 1026:47–53.

Hanlon, L., Argyle, D., Bain, D., Nicolson, L., Dunham, S., Golder, M.C., McDonald, M., McGillivray, C., Jarrett, O., Neil, J.C. and Onions, D.E. Feline leukemia virus DNA vaccine efficacy is enhanced by coadministration with interleukin-12 (IL-12) and IL-18 expression vectors. J. Virol. 2001; 75(18):8424–8433.

Hannant, D., Mumford, J.A. and Jessett, D.M. Duration of circulating antibody and immunity following infection with equine influenza virus. Vet. Rec. 1988; 122(6):125–128.

Heath, A.W. Cytokines as immunological adjuvants. Pharm. Biotechnol. 1995; 6:645–658.

Hershey, A.E., Sorenmo, K.U., Hendrick, M.J., Shofer, F.S. and Vail, D.M. Prognosis for presumed feline vaccine-associated sarcoma after excision: 61 cases (1986–1996). J. Am. Vet. Med. Assoc. 2000; 216(1):58–61.

Hill, B.D., Blewett, D.A., Dawson, A.M. and Wright, S. Analysis of the kinetics, isotype and specificity of serum and coproantibody in lambs infected with *Cryptosporidium parvum*. Res. Vet. Sci. 1990; 48(1):76–81.

Houdijk, J.G.M., Kyriazakis, I., Jackson, F., Huntley, J.F. and Coop, R.L. Can an increased intake of metabolizable protein affect the periparturient relaxation in immunity against *Teladorsagia circumcincta* in sheep? Vet. Parasitol. 2000; 91(1–2):43–62.

Huang, Y.T., Teng, L.J., Ho, S.W. and Hsueh, P.R. *Streptococcus suis* infection. J. Microbiol. Immunol. Infect. 2005; 38(5):306–313.

Hunter, P.R., Chalmers, R.M., Syed, Q., Hughes, L.S., Woodhouse, S. and Swift, L. Foot and mouth disease and cryptosporidiosis: possible interaction between two emerging infectious diseases. Emerg. Infect. Dis. 2003; 9(1):109–112.

Hunter, S.K., Andracki, M.E. and Krieg, A.M. Biodegradable microspheres containing group B Streptococcus vaccine: immune response in mice. Am. J. Obstet. Gynecol. 2001; 185(5): 1174–1179.

Ioannou, X.P., Griebel, P., Hecker, R., Babiuk, L.A. and van Drunen Littel-van den Hurk, S. The immunogenicity and protective efficacy of bovine herpesvirus 1 glycoprotein D plus Emulsigen are increased by formulation with CpG oligodeoxynucleotides. J. Virol. 2002; 76(18):9002–9010.

Jacobs-Reitsma, W.F., van de Giessen, A.W., Bolder, N.M. and Mulder, R.W. Epidemiology of *Campylobacter* spp. at two Dutch broiler farms. Epidemiol. Infect. 1995; 114(3):413–421.

Jenkins, M.C., O'Brien, C., Trout, J., Guidry, A. and Fayer, R. Hyperimmune bovine colostrum specific for recombinant *Cryptosporidium parvum* antigen confers partial protection against cryptosporidiosis in immunosuppressed adult mice. Vaccine 1999; 17(19):2453–2460.

Jones, S.M., Feldmann, H., Stroher, U., Geisbert, J.B., Fernando, L., Grolla, A., Klenk, H.D., Sullivan, N.J., Volchkov, V.E., Fritz, E.A., Daddario, K.M., Hensley, L.E., Jahrling, P.B. and Geisbert, T.W. Live attenuated recombinant vaccine protects nonhuman primates against Ebola and Marburg viruses. Nat. Med. 2005; 11(7): 786–790.

Judge, N.A., Mason, H.S. and O'Brien, A.D. Plant cell-based intimin vaccine given orally to mice primed with intimin reduces time of *Escherichia coli* O157:H7 shedding in feces. Infect. Immun. 2004; 72(1):168–175.

Karch, H., Tarr, P.I. and Bielaszewska, M. Enterohaemorrhagic *Escherichia coli* in human medicine. Int. J. Med. Microbiol. 2005; 295(6–7):405–418.

Kennedy, N.J., Spithill, T.W., Tennent, J., Wood, P.R. and Piedrafita, D. DNA vaccines in sheep: CTLA-4 mediated targeting and CpG motifs enhance immunogenicity in a DNA prime/protein boost strategy. Vaccine 2006; 24(7):970–979.

Kensil, C.R., Patel, U., Lennick, M. and Marciani, D. Separation and characterization of saponins with adjuvant activity from *Quillaja saponaria* Molina cortex. J. Immunol. 1991; 146(2): 431–437.

Khoury, C.A. and Meinersmann, R.J. A genetic hybrid of the *Campylobacter jejuni* flaA gene with LT-B of *Escherichia coli* and assessment of the efficacy of the hybrid protein as an oral chicken vaccine. Avian Dis. 1995; 39(4):812–820.

Kim, B., Bowersock, T., Griebel, P., Kidane, A., Babiuk, L.A., Sanchez, M., Attah-Poku, S., Kaushik, R.S. and Mutwiri, G. K. Mucosal immune responses following oral immunization with rotavirus antigens encapsulated in alginate microspheres. J. Control. Release 2002; 85(1–3):191–202.

Kim, S.J. and Spradbrow, P.B. Administration of a vaccine prepared from the Australian V4 strain of Newcastle disease virus by aerosol and drinking water. Aust. Vet. J. 1978; 54(10):486–489.

Kita, J., Oyrzanowska, J., Prandota, J. and Banbura, M. XIIth World Congress on Diseases of Cattle, Utrecht, Netherlands. 1982.

Knott, G.K.L., Erwin, B.G. and Classick, L.G. Benefits of a clostridial vaccination program in feedlot cattle. Vet. Med. 1985; 80(6): 95–97.

Kobayashi, M., Carpenter, T.E., Dickey, B.F. and Howitt, R.E. A dynamic, optimal disease control model for foot-and-mouth disease: I. Model description. Prev. Vet. Med. 2007a; 79(2–4):257–273.

Kobayashi, M., Carpenter, T.E., Dickey, B.F. and Howitt, R.E. A dynamic, optimal disease control model for foot-and-mouth-disease: II. Model results and policy implications. Prev. Vet. Med. 2007b; 79(2–4):274–286.

Kowalski, J., Gilbert, S.A., van Drunen-Littel-van den Hurk, S., van den Hurk, J., Babiuk, L.A. and Zamb, T.J. Heat-shock promoter-driven synthesis of secreted bovine herpesvirus glycoproteins in transfected cells. Vaccine 1993; 11(11): 1100–1107.

Krieg, A.M., Yi, A.K. and Hartmann, G. Mechanisms and therapeutic applications of immune stimulatory cpG DNA. Pharmacol. Ther. 1999; 84(2):113–120.

Krystynak, R.H. and Charlebois, P.A. The potential economic impact of an outbreak of foot-and-mouth disease in Canada. Can. Vet. J. 1987; 28(8):523–527.

Llull, D., Lopez, R. and Garcia, E. Genetic bases and medical relevance of capsular polysaccharide biosynthesis in pathogenic streptococci. Curr. Mol. Med. 2001; 1(4):475–491.

Lumb, R., Lanser, J.A. and O'Donoghue, P.J. Electrophoretic and immunoblot analysis of *Cryptosporidium* oocysts. Immunol. Cell Biol. 1988; 66(Pt 5–6):369–376.

Lutticken, D., Segers, R.P. and Visser, N. Veterinary vaccines for public health and prevention of viral and bacterial zoonotic diseases. Rev. Sci. Tech. 2007; 26(1):165–177.

MacInnes, J.I. and Desrosiers, R. Agents of the "suis-ide diseases" of swine: *Actinobacillus suis*, *Haemophilus parasuis*, and *Streptococcus suis*. Can. J. Vet. Res. 1999; 63(2):83–89.

Mac Kenzie, W.R., Hoxie, N.J., Proctor, M.E., Gradus, M.S., Blair, K.A., Peterson, D.E., Kazmierczak, J.J., Addiss, D.G., Fox, K.R., Rose, J.B. et al. A massive outbreak in Milwaukee of cryptosporidium infection transmitted through the public water supply. N. Engl. J. Med. 1994; 331(3):161–167.

Maire, N., Aponte, J.J., Ross, A., Thompson, R., Alonso, P., Utzinger, J., Tanner, M. and Smith, T. Modeling a field trial of the RTS,S/AS02A malaria vaccine. Am. J. Trop. Med. Hyg. 2006; 75(2 Suppl.): 104–110.

Mantis, N.J., Cheung, M.C., Chintalacharuvu, K.R., Rey, J., Corthesy, B. and Neutra, M.R. Selective adherence of IgA to murine Peyer's patch M cells: evidence for a novel IgA receptor. J. Immunol. 2002; 169(4):1844–1851.

Marano, N., Rupprecht, C. and Regnery, R. Vaccines for emerging infections. Rev. Sci. Tech. 2007; 26(1):203–215.

Marshall, M. and Roger, P. Policy and science of FMD control: the stakeholders' contribution to decision making. A call for integrated animal disease management. J. Clin. Lab. Immunol. 2004; 53:27–38.

Martinez, G., Harel, J., Lacouture, S. and Gottschalk, M. Genetic diversity of *Streptococcus suis* serotypes 2 and 1/2 isolates recovered from carrier pigs in closed herds. Can. J. Vet. Res. 2002; 66(4):240–248.

Matthews, L., Low, J.C., Gally, D.L., Pearce, M.C., Mellor, D.J., Heesterbeek, J.A., Chase-Topping, M., Naylor, S.W., Shaw, D.J., Reid, S.W., Gunn, G.J. and Woolhouse, M.E. Heterogeneous shedding of *Escherichia coli* O157 in cattle and its implications for control. Proc. Natl. Acad. Sci. U.S.A. 2006; 103(3):547–552.

McCauley, E.H., Aulaqi, N.A., Sundquist, W.B., New, J.C., Jr. and Miller, W.M. A study of the potential economic impact of foot and mouth disease in the United States. Proc. Annu. Meet. U.S. Anim. Health Assoc. 1977; 81:284–296.

McGavin, D. Passive immunity and canine parvovirus. Websters Vet. Digest 1985; 19(1):2–3.

McIlroy, S.G., Goodall, E.A., Bruce, D.W., McCracken, R.M. and McNulty, M.S. The cost benefit of vaccinating broiler flocks against subclinical infectious bursal disease. Avian Pathol. 1992; 21(1):65–76.

McLauchlin, J., Amar, C., Pedraza-Diaz, S. and Nichols, G.L. Molecular epidemiological analysis of *Cryptosporidium* spp. in the United Kingdom: results of genotyping *Cryptosporidium* spp. in 1,705 fecal samples from humans and 105 fecal samples from livestock animals. J. Clin. Microbiol. 2000; 38(11):3984–3990.

McNeal, M.M., Rae, M.N. and Ward, R.L. Effects of different adjuvants on rotavirus antibody responses and protection in mice following intramuscular immunization with inactivated rotavirus. Vaccine 1999; 17(11–12):1573–1580.

McQuiston, J.H., Garber, L.P., Porter-Spalding, B.A., Hahn, J.W., Pierson, F.W., Wainwright, S.H., Senne, D.A., Brignole, T.J., Akey, B.L. and Holt, T.J. Evaluation of risk factors for the spread of low pathogenicity H7N2 avian influenza virus among commercial poultry farms. J. Am. Vet. Med. Assoc. 2005; 226(5):767–772.

Mead, J.R., Arrowood, M.J. and Sterling, C.R. Antigens of *Cryptosporidium* sporozoites recognized by immune sera of infected animals and humans. J. Parasitol. 1988; 74(1):135–143.

Medema, J.K., Zoellner, Y.F., Ryan, J. and Palache, A.M. Modeling pandemic preparedness scenarios: health economic implications of enhanced pandemic vaccine supply. Virus Res. 2004; 103(1–2):9–15.

Miguens, L. The opinion of the production sector on the role of vaccines in the control and eradication of livestock diseases in Argentina. Rev. Sci. Tech. 2007; 26(2):479–483, 485–488.

Mittal, S.K., Aggarwal, N., Sailaja, G., van Olphen, A., HogenEsch, H., North, A., Hays, J. and Moffatt, S. Immunization with DNA, adenovirus or both in biodegradable alginate microspheres: effect of route of inoculation on immune response. Vaccine 2000; 19(2–3):253–263.

Moore, G.E., Guptill, L.F., Ward, M.P., Glickman, N.W., Faunt, K.K., Lewis, H.B. and Glickman, L.T. Adverse events diagnosed within three days of vaccine administration in dogs. J. Am. Vet. Med. Assoc. 2005; 227(7):1102–1108.

Morein, B., Abusugra, I. and Blomqvist, G. Immunity in neonates. Vet. Immunol. Immunopathol. 2002; 87(3–4):207–213.

Morein, B., Hu, K.F. and Abusugra, I. Current status and potential application of ISCOMs in veterinary medicine. Adv. Drug Deliv. Rev. 2004; 56(10):1367–1382.

Morgan, U., Weber, R., Xiao, L., Sulaiman, I., Thompson, R.C., Ndiritu, W., Lal, A., Moore, A. and Deplazes, P. Molecular characterization of *Cryptosporidium* isolates obtained from human immunodeficiency virus-infected individuals living in Switzerland, Kenya, and the United States. J. Clin. Microbiol. 2000; 38(3):1180–1183.

Morgan-Ryan, U.M., Fall, A., Ward, L.A., Hijjawi, N., Sulaiman, I., Fayer, R., Thompson, R.C., Olson, M., Lal, A. and Xiao, L. *Cryptosporidium hominis* n. sp. (Apicomplexa: Cryptosporidiidae) from *Homo sapiens*. J. Eukaryot. Microbiol. 2002; 49(6): 433–440.

Mouzin, D.E., Lorenzen, M.J., Haworth, J.D. and King, V.L. Duration of serologic response to five viral antigens in dogs. J. Am. Vet. Med. Assoc. 2004a; 224(1):55–60.

Mouzin, D.E., Lorenzen, M.J., Haworth, J.D. and King, V.L. Duration of serologic response to three viral antigens in cats. J. Am. Vet. Med. Assoc. 2004b; 224(1):61–66.

Muller, H.M., Reckmann, I., Hollingdale, M.R., Bujard, H., Robson, K.J. and Crisanti, A. Thrombospondin related anonymous protein (TRAP) of *Plasmodium falciparum* binds specifically to sulfated glycoconjugates and to HepG2 hepatoma cells suggesting a role for this molecule in sporozoite invasion of hepatocytes. EMBO J. 1993; 12(7):2881–2889.

Muneta, Y., Yoshihara, K., Minagawa, Y., Nagata, R., Yasuyuki, M., Yamaguchi, T. and Takehara, K. Bovine IL-18 ELISA: detection of IL-18 in sera of pregnant cow and newborn calf, and in colostrum. J. Immunoassay Immunochem. 2005; 26(3):203–213.

Murillo, M., Grillo, M.J., Rene, J., Marin, C.M., Barberan, M., Goni, M.M., Blasco, J.M., Irache, J.M. and Gamazo, C. A *Brucella ovis* antigenic complex bearing poly-epsilon-caprolactone microparticles confer protection against experimental brucellosis in mice. Vaccine 2001; 19(30):4099–4106.

Naylor, S.W., Low, J.C., Besser, T.E., Mahajan, A., Gunn, G.J., Pearce, M.C., McKendrick, I.J., Smith, D.G. and Gally, D.L. Lymphoid follicle-dense mucosa at the terminal rectum is the principal site of colonization of enterohemorrhagic *Escherichia coli* O157:H7 in the bovine host. Infect. Immun. 2003; 71(3):1505–1512.

Nielsen, R., Loftager, M. and Eriksen, L. International Pig Veterinary Society, 11th Congress, Lausanne, Switzerland. 1990.

Niskanen, R. and Lindberg, A. Transmission of bovine viral diarrhea virus by unhygienic vaccination procedures, ambient air, and from contaminated pens. Vet. J. 2003; 165(2):125–130.

Nonnecke, B.J., Waters, W.R., Foote, M.R., Palmer, M.V., Miller, B.L., Johnson, T.E., Perry, H.B. and Fowler, M.A. Development of an adult-like cell-mediated immune response in calves after early vaccination with *Mycobacterium bovis* bacillus Calmette-Guerin. J. Dairy Sci. 2005; 88(1):195–210.

The disease occurs primarily in swine after they enter the fattening barn (beyond 12 weeks of age). The disease has a very significant impact on the economic performances of the herd. The agent *S. suis* is a gram-positive cocci and part of the broad genus *Streptococcus*. A very important feature of the organism is its significant diversity. At the serotype level, 35 different capsular types have been described. Serotype 2 is the most common isolate but its frequency varies significantly from country to country and between North America and Europe. Serotype 1/2 is the next most frequent isolate (Martinez et al., 2002). At the genetic level, significant diversity is observed between herds and even within herds. Interestingly, the pathogenic strains emerging within a herd seem to separate more distinctly from the isolates obtained in healthy swine (Berthelot-Herault et al., 2005). At the molecular level, differences are also observed in expression of putative virulence factors (Smith et al., 1996).

Despite best efforts to understand, the exact mechanisms of pathogenicity of the virulent strains as well as the ability to predict the virulence of isolate strains based on molecular models remain elusive. Various proteins, such as muramidase-released protein (MRP), extracellular protein factor (EPF), and a hemolysin (suilysine) have been suggested (Martinez et al., 2002). Mutants lacking these genes did not demonstrate attenuation (de Greeff et al., 2002). It has been suggested that the simultaneous presence of the three factors could make strains more virulent. More recently additional proteins have been proposed as virulence factors: fibronectin-binding protein (Gruening et al., 2006), and arginine deiminase system (ADS) (Okwumabua and Chinnapapakkagari, 2005). A 38 kDa protein of unknown function was found reactive with serum from infected animals and demonstrated protection (Llull et al., 2001). The capsular polysaccharides of *S. suis* as with other streptococci may also play an important role (Busque et al., 1997).

It follows that the molecular basis of virulence remains unknown and that antigenic variation exists within the putative virulence molecules identified that vaccines for this disease have not been able to use modern technologies, such as recombinant expression, which require a defined antigen. Essentially two types of vaccines have been made available, live attenuated strains (Wisselink et al., 2002; Haesebrouck et al., 2004) or inactivated whole cell bacterins. Neither has met with significant success. The majority of the market remains occupied by autogenous bacterins. Autogenous bacterins could provide a means to avoid the significant antigenic variation of the strain by providing a vaccine derived from the particular strain infecting the herd. Considering the variability of strains within herds, one should expect to change strains regularly to maintain efficacy. *S. suis* induced disease remains a very significant and costly disease for the swine industry. In addition, it is responsible for an infrequent but severe and potentially deadly zoonosis in humans. The lack of a detailed understanding of the virulence has precluded the development of efficacious modern vaccines.

Cryptosporidiosis in Cattle

Cryptosporidiosis is a disease of humans, livestock, companion animals, wildlife, birds, reptiles, and fish infected with one or more species of the protozoan parasite *Cryptosporidium*. The clinical outcome of the disease varies greatly from species to species and is also dependent on virulence factors on the parasite and the host immune response to them. In immunocompetent humans, the infection can be asymptomatic even though the parasite replicates and is excreted. When an illness develops the outcome ranges from mild enteric symptoms without diarrhea to a profuse watery diarrhea. Additional symptoms can include abdominal cramps, vomiting, low-grade fever, fatigue, anorexia (Ungar, 1990; Mac Kenzie et al., 1994). The disease is self limiting and typically resolves within 2 weeks. Organisms usually infect the lower small intestine, but in immunosuppressed patients, it can extend from the esophagus, throughout the intestinal tract to the rectum. In some cases the appendix, gall bladder, bile ducts, pancreas, and occasionally even the respiratory tract have been infected (Ungar, 1990). In humans, the most frequently found organism, designated *Cryptosporidium parvum* genotype 1 (or human genotype), appears host-specific, infecting, and spreading only among humans. This organism is now recognized as a separate species, *Cryptosporidium hominis* (Morgan-Ryan et al., 2002). Another organism, designated *C. parvum* genotype 2 (or cattle genotype), is zoonotic, capable of infecting and spreading among many mammalian hosts. The main reservoir of *C. parvum* genotype 2 is cattle. Virtually all cattle are exposed to the parasite during the first few months of life. Calves will shed large number of oocysts (up to 10^9/g of feces) for approximately 10 days. An active immune response will develop and control the infection. However, evidence suggests that most infected animals will carry *C. parvum* for life, shedding few oocysts intermittently. Collectively, data suggest that vast numbers of oocysts are passed into the environment and that cattle contribute significantly to these large numbers. Zoonotic transmission may be important in cases of direct exposure to cattle feces (farms, undeveloped countries) or in waterborne outbreaks (developed countries).

Noor, S.M., Husband, A.J. and Widders, P.R. In ovo oral vaccination with *Campylobacter jejuni* establishes early development of intestinal immunity in chickens. Br. Poult. Sci. 1995; 36(4): 563–573.

Normile, D. Infectious diseases. WHO probes deadliness of China's pig-borne disease. Science 2005; 309(5739):1308–1309.

Okwumabua, O. and Chinnapapakkagari, S. Identification of the gene encoding a 38-kilodalton immunogenic and protective antigen of *Streptococcus suis*. Clin. Diagn. Lab. Immunol. 2005; 12(4):484–490.

Paarlberg, P.L., Lee, J.G. and Seitzinger, A.H. Potential revenue impact of an outbreak of foot-and-mouth disease in the United States. J. Am. Vet. Med. Assoc. 2002; 220(7):988–992.

Pappo, J., Ermak, T.H. and Steger, H.J. Monoclonal antibody-directed targeting of fluorescent polystyrene microspheres to Peyer's patch M cells. Immunology 1991; 73(3):277–280.

Payne, L.G. and Andrianov, A.K. Protein release from polyphosphazene matrices. Adv. Drug Deliv. Rev. 1998; 31(3):185–196.

Payne, L.G., Jenkins, S.A., Andrianov, A. and Roberts, B.E. Water-soluble phosphazene polymers for parenteral and mucosal vaccine delivery. Pharm. Biotechnol. 1995; 6:473–493.

Perez, A.M., Thurmond, M.C. and Carpenter, T.E. Spatial distribution of foot-and-mouth disease in Pakistan estimated using imperfect data. Prev. Vet. Med. 2006; 76(3–4):280–289.

Perez, A.M., Thurmond, M.C., Grant, P.W. and Carpenter, T.E. Use of the scan statistic on disaggregated province-based data: foot-and-mouth disease in Iran. Prev. Vet. Med. 2005; 71(3–4):197–207.

Perez, A.M., Ward, M.P. and Carpenter, T.E. Control of a foot-and-mouth disease epidemic in Argentina. Prev. Vet. Med. 2004; 65(3–4):217–226.

Perry, B.D., Kalpravidh, W., Coleman, P.G., Horst, H.S., McDermott, J.J., Randolph, T.F. and Gleeson, L.J. The economic impact of foot and mouth disease and its control in South-East Asia: a preliminary assessment with special reference to Thailand. Rev. Sci. Tech. 1999; 18(2):478–497.

Perryman, L.E., Jasmer, D.P., Riggs, M.W., Bohnet, S.G., McGuire, T.C. and Arrowood, M.J. A cloned gene of *Cryptosporidium parvum* encodes neutralization-sensitive epitopes. Mol. Biochem. Parasitol. 1996; 80(2):137–147.

Perryman, L.E., Kapil, S.J., Jones, M.L. and Hunt, E.L. Protection of calves against cryptosporidiosis with immune bovine colostrum induced by a *Cryptosporidium parvum* recombinant protein. Vaccine 1999; 17(17):2142–2149.

Peters, A.R., Thevasagayam, S.J., Wiseman, A. and Salt, J.S. Duration of immunity of a quadrivalent vaccine against respiratory diseases caused by BHV-1, PI3V, BVDV, and BRSV in experimentally infected calves. Prev. Vet. Med. 2004; 66(1–4):63–77.

Pfeffer, A., Shaw, R.J., Green, R.S. and Phegan, M.D. The transfer of maternal IgE and other immunoglobulins specific for *Trichostrongylus colubriformis* larval excretory/secretory product to the neonatal lamb. Vet. Immunol. Immunopathol. 2005; 108(3–4): 315–323.

Pieniazek, N.J., Bornay-Llinares, F.J., Slemenda, S.B., da Silva, A.J., Moura, I.N., Arrowood, M.J., Ditrich, O. and Addiss, D.G. New cryptosporidium genotypes in HIV-infected persons. Emerg. Infect. Dis. 1999; 5(3):444–449.

Potter, A.A., Klashinsky, S., Li, Y., Frey, E., Townsend, H., Rogan, D., Erickson, G., Hinkley, S., Klopfenstein, T., Moxley, R.A., Smith, D.R. and Finlay, B.B. Decreased shedding of *Escherichia coli* O157:H7 by cattle following vaccination with type III secreted proteins. Vaccine 2004; 22(3–4):362–369.

Pourbohloul, B., Meyers, L.A., Skowronski, D.M., Krajden, M., Patrick, D.M. and Brunham, R.C. Modeling control strategies of respiratory pathogens. Emerg. Infect. Dis. 2005; 11(8): 1249–1256.

Raggo, C., Fitzpatrick, D.R., Babiuk, L.A. and Liang, X. Expression of bovine interleukin-1 beta in a bovine herpesvirus-1 vector: in vitro analysis. Virology 1996; 221(1):78–86.

Randolph, T.F., Perry, B.D., Benigno, C.C., Santos, I.J., Agbayani, A.L., Coleman, P., Webb, R. and Gleeson, L.J. The economic impact of foot and mouth disease control and eradication in the Philippines. Rev. Sci. Tech. 2002; 21(3):645–661.

Rautenschlein, S. and Haase, C. Differences in the immunopathogenesis of infectious bursal disease virus (IBDV) following in ovo and post-hatch vaccination of chickens. Vet. Immunol. Immunopathol. 2005; 106(1–2):139–150.

Rautenschlein, S., Sharma, J.M., Winslow, B.J., McMillen, J., Junker, D. and Cochran, M. Embryo vaccination of turkeys against Newcastle disease infection with recombinant fowlpox virus constructs containing interferons as adjuvants. Vaccine 1999; 18(5–6):426–433.

Reber, A.J., Hippen, A.R. and Hurley, D.J. Effects of the ingestion of whole colostrum or cell-free colostrum on the capacity of leukocytes in newborn calves to stimulate or respond in one-way mixed leukocyte cultures. Am. J. Vet. Res. 2005; 66(11): 1854–1860.

Reddy, P.S., Idamakanti, N., Chen, Y., Whale, T., Babiuk, L.A., Mehtali, M. and Tikoo, S.K. Replication-defective bovine adenovirus type 3 as an expression vector. J. Virol. 1999a; 73(11): 9137–9144.

Reddy, P.S., Idamakanti, N., Hyun, B.H., Tikoo, S.K. and Babiuk, L.A. Development of porcine adenovirus-3 as an expression vector. J. Gen. Virol. 1999b; 80(Pt 3):563–570.

Rehg, J.E. Anticryptosporidial activity of lasalocid and other ionophorous antibiotics in immunosuppressed rats. J. Infect. Dis. 1993; 168(6):1566–1569.

Reid, G. Soluble proteins incorporate into ISCOMs after covalent attachment of fatty acid. Vaccine 1992; 10(9):597–602.

Reperant, J.M., Naciri, M., Iochmann, S., Tilley, M. and Bout, D.T. Major antigens of *Cryptosporidium parvum* recognised by serum antibodies from different infected animal species and man. Vet. Parasitol. 1994; 55(1–2):1–13.

Rice, B.E., Rollins, D.M., Mallinson, E.T., Carr, L. and Joseph, S.W. *Campylobacter jejuni* in broiler chickens: colonization and humoral immunity following oral vaccination and experimental infection. Vaccine 1997; 15(17–18):1922–1932.

Riising, H.J., van Empel, P. and Witvliet, M. Protection of piglets against atrophic rhinitis by vaccinating the sow with a vaccine against *Pasteurella multocida* and *Bordetella bronchiseptica*. Vet. Rec. 2002; 150(18):569–571.

Rossides, S.C. A farming perspective on the 2001 foot and mouth disease epidemic in the United Kingdom. Rev. Sci. Tech. 2002; 21(3):831–838.

Rydell, N. and Sjoholm, I. Oral vaccination against diphtheria using polyacryl starch microparticles as adjuvant. Vaccine 2004; 22(9–10):1265–1274.

Sagodira, S., Iochmann, S., Mevelec, M.N., Dimier-Poisson, I. and Bout, D. Nasal immunization of mice with *Cryptosporidium parvum* DNA induces systemic and intestinal immune responses. Parasite Immunol. 1999; 21(10):507–516.

Sahin, O., Luo, N., Huang, S. and Zhang, Q. Effect of Campylobacter-specific maternal antibodies on *Campylobacter jejuni* colonization in young chickens. Appl. Environ. Microbiol. 2003; 69(9):5372–5379.

Samuel, A.R. and Knowles, N.J. Foot-and-mouth disease virus: cause of the recent crisis for the UK livestock industry. Trends Genet. 2001; 17(8):421–424.

Schattner, A. Consequence or coincidence? The occurrence, pathogenesis and significance of autoimmune manifestations after viral vaccines. Vaccines 2005; 23(30):3876–3886.

Schudel, A. and Lombard, M. OIE/FAO International Scientific Conference on Avian Influenza. Emerg. Inf. Dis. 2006; 12(11):1.

Schultz-Cherry, S., Dybing, J.K., Davis, N.L., Williamson, C., Suarez, D.L., Johnston, R. and Perdue, M.L. Influenza virus (A/HK/156/97) hemagglutinin expressed by an alphavirus replicon system protects chickens against lethal infection with Hong Kong-origin H5N1 viruses. Virology 2000; 278(1):55–59.

Scott, H.M., Atkins, G., Willows, B. and McGregor, R. Effects of 2 commercially-available 9-way killed vaccines on milk production and rectal temperature in Holstein-Friesian dairy cows. Can. Vet. J. 2001; 42(10):793–798.

Selke, M., Meens, J., Springer, S., Frank, R. and Gerlach, G.F. Immunization of pigs to prevent disease in humans: construction and protective efficacy of a *Salmonella enterica* serovar *Typhimurium* live negative-marker vaccine. Infect. Immun. 2007; 75(5):2476–2483.

Shane, S.M. and Stern, N.J. Campylobacter infection. In: Diseases of Poultry (Y.M. Saif, H.J. Barnes, J.R. Glisson, A.M. Fadly, L.R. McDougald and D.E. Swayne, Eds.), pp. 615–630. Ames, IA: Iowa State Press, 2003.

Shewen, P.E. and Wilkie, B.N. Cytotoxin of *Pasteurella haemolytica* acting on bovine leukocytes. Infect. Immun. 1982; 35(1):91–94.

Shiilegdamba, E., Carpenter, T.E., Perez, A.M. and Thurmond, M. C. Temporal-spatial epidemiology of foot-and-mouth disease outbreaks in Mongolia, 2000–2002. Vet. Res. Commun. 2008; 32(3):201–207.

Smith, H.E., Vecht, U., Wisselink, H.J., Stockhofe-Zurwieden, N., Biermann, Y. and Smits, M.A. Mutants of *Streptococcus suis* types 1 and 2 impaired in expression of muramidase-released protein and extracellular protein induce disease in newborn germfree pigs. Infect. Immun. 1996; 64(10):4409–4412.

Stern, N.J., Meinersmann, R.J. and Dickerson, H.W. Influence of antibody treatment of *Campylobacter jejuni* on the dose required to colonize chicks. Avian Dis. 1990; 34(3):595–601.

Sultan, A.A., Thathy, V., Frevert, U., Robson, K.J., Crisanti, A., Nussenzweig, V., Nussenzweig, R.S. and Menard, R. TRAP is necessary for gliding motility and infectivity of plasmodium sporozoites. Cell 1997; 90(3):511–522.

Takahashi, H., Takeshita, T., Morein, B., Putney, S., Germain, R.N. and Berzofsky, J.A. Induction of CD8+ cytotoxic T cells by immunization with purified HIV-1 envelope protein in ISCOMs. Nature 1990; 344(6269):873–875.

Tamber, H., Johansen, P., Merkle, H.P. and Gander, B. Formulation aspects of biodegradable polymeric microspheres for antigen delivery. Adv. Drug Deliv. Rev. 2005; 57(3):357–376.

Tarr, P.I., Gordon, C.A. and Chandler, W.L. Shiga-toxin-producing *Escherichia coli* and haemolytic uraemic syndrome. Lancet 2005; 365(9464):1073–1086.

Thompson, D., Muriel, P., Russell, D., Osborne, P., Bromley, A., Rowland, M., Creigh-Tyte, S. and Brown, C. Economic costs of the foot and mouth disease outbreak in the United Kingdom in 2001. Rev. Sci. Tech. 2002; 21(3):675–687.

Tilley, M., Upton, S.J., Fayer, R., Barta, J.R., Chrisp, C.E., Freed, P.S., Blagburn, B.L., Anderson, B.C. and Barnard, S.M. Identification of a 15-kilodalton surface glycoprotein on sporozoites of *Cryptosporidium parvum*. Infect. Immun. 1991; 59(3):1002–1007.

Tizard, I. Veterinary Immunology: An Introduction, 7th ed. New York: W. B. Saunders, 2004.

Townsend, H.G., Penner, S.J., Watts, T.C., Cook, A., Bogdan, J., Haines, D.M., Griffin, S., Chambers, T., Holland, R.E., Whitaker-Dowling, P., Youngner, J.S. and Sebring, R.W. Efficacy of a cold-adapted, intranasal, equine influenza vaccine: challenge trials. Equine Vet. J. 2001; 33(7):637–643.

Ulrich, J.T. and Myers, K.R. Monophosphoryl lipid A as an adjuvant. Past experiences and new directions. Pharm. Biotechnol. 1995; 6:495–524.

Ungar, B.L.P. Cryptosporidiosis in humans (*Homo sapiens*). In: Cryptosporidiosis of Man and Animals (J.P. Dubey, C.A. Speer and R. Fayer, Eds.), pp. 60–82. Boca Raton, FL: CRC Press, 1990.

Vallat, B. and Mallet, E. Ensuring good governance to address emerging and re-emerging animal disease threats: supporting the veterinary services of developing countries to meet OIE international standards on quality. Rev. Sci. Tech. 2006; 25(1):389–401.

van Donkersgoed, J., Dubeski, P.L., Aalhus, J.L., VanderKop, M., Dixon, S. and Starr, W.N. The effect of vaccines and antimicrobials on the formation of injection site lesions in subprimals of experimentally injected beef calves. Can. Vet. J. 1999; 40(4):245–251.

van Donkersgoed, J., Hancock, D., Rogan, D. and Potter, A.A. *Escherichia coli* O157:H7 vaccine field trial in 9 feedlots in Alberta and Saskatchewan. Can. Vet. J. 2005; 46(8):724–728.

van Drunen Littel-van den Hurk, S., Babiuk, S.L. and Babiuk, L.A. Strategies for improved formulation and delivery of DNA vaccines to veterinary target species. Immunol. Rev. 2004; 199:113–125.

van Immerseel, F., Methner, U., Rychlik, I., Nagy, B., Velge, P., Martin, G., Foster, N., Ducatelle, R. and Barrow, P.A. Vaccination and early protection against non-host-specific Salmonella serotypes in poultry: exploitation of innate immunity and microbial activity. Epidemiol. Infect. 2005; 133(6):959–978.

van Oirschot, J.T. Classical attenuated vaccines. In: Veterinary Vaccinology (J.B.P.P. Pastoret, P. Vannier and C. Verschueren, Eds.), pp. 260–262. New York: Elsevier, 1997.

van Oirschot, J.T., Rziha, H.J., Moonen, P.J., Pol, J.M. and van Zaane, D. Differentiation of serum antibodies from pigs vaccinated or infected with Aujeszky's disease virus by a competitive enzyme immunoassay. J. Gen. Virol. 1986; 67(Pt 6):1179–1182.

van Rooijen, G.J. and Moloney, M.M. Plant seed oil-bodies as carriers for foreign proteins. Biotechnology (N.Y.) 1995; 13(1):72–77.

von Overbeck, J. Insurance and epidemics: SARS, West Nile virus and Nipah virus. J. Insur. Med. 2003; 35(3–4):165–173.

Weeratna, R.D., McCluskie, M.J., Xu, Y. and Davis, H.L. CpG DNA induces stronger immune responses with less toxicity than other adjuvants. Vaccine 2000; 18(17):1755–1762.

Wen, S.X., Teel, L.D., Judge, N.A. and O'Brien, A.D. Genetic toxoids of Shiga toxin types 1 and 2 protect mice against homologous but not heterologous toxin challenge. Vaccine 2006; 24(8): 1142–1148.

Westergaard, J.M. Preparedness for major OIE-listed epidemics: implementation of simulation exercises. Zoonoses Public Health 2008; 55(1):37–41.

Whiting, T.L. Foreign animal disease outbreaks, the animal welfare implications for Canada: risks apparent from international experience. Can. Vet. J. 2003; 44(10):805–815.

Widders, P.R., Perry, R., Muir, W.I., Husband, A.J. and Long, K.A. Immunisation of chickens to reduce intestinal colonisation with *Campylobacter jejuni*. Br. Poult. Sci. 1996; 37(4):765–778.

Widders, P.R., Thomas, L.M., Long, K.A., Tokhi, M.A., Panaccio, M. and Apos, E. The specificity of antibody in chickens immunised to reduce intestinal colonisation with *Campylobacter jejuni*. Vet. Microbiol. 1998; 64(1):39–50.

Wikingsson, L. and Sjoholm, I. Polyacryl starch microparticles as adjuvant in oral immunisation, inducing mucosal and systemic immune responses in mice. Vaccine 2002; 20(27–28):3355–3363.

Wisselink, H.J., Stockhofe-Zurwieden, N., Hilgers, L.A. and Smith, H.E. Assessment of protective efficacy of live and killed vaccines based on a non-encapsulated mutant of *Streptococcus suis* serotype 2. Vet. Microbiol. 2002; 84(1–2):155–168.

Withanage, G.S., Wigley, P., Kaiser, P., Mastroeni, P., Brooks, H., Powers, C., Beal, R., Barrow, P., Maskell, D. and McConnell, I. Cytokine and chemokine responses associated with clearance of a primary *Salmonella enterica* serovar *Typhimurium* infection in the chicken and in protective immunity to rechallenge. Infect. Immun. 2005; 73(8):5173–5182.

Witter, R.L. and Fadly, A.M. Reduction of horizontal transmission of avian leukosis virus subgroup J in broiler breeder chickens hatched and reared in small groups. Avian Pathol. 2001; 30(6):641–654.

Woods, K.M., Nesterenko, M.V. and Upton, S.J. Efficacy of 101 antimicrobials and other agents on the development of *Cryptosporidium parvum* in vitro. Ann. Trop. Med. Parasitol. 1996; 90(6):603–615.

Wu, J.Y., Gardner, B.H., Murphy, C.I., Seals, J.R., Kensil, C.R., Recchia, J., Beltz, G.A., Newman, G.W. and Newman, M.J. Saponin adjuvant enhancement of antigen-specific immune responses to an experimental HIV-1 vaccine. J. Immunol. 1992; 148(5):1519–1525.

YongQing, H., YuMing, G., YuPing, Z., Cheng, J. and JingYang, Z. Effect of different dietary levels of protein, lysine, methionine and threonine on immune function in early weaned piglets. Chin. J. Anim. Sci. 2001; 37(4):18–20.

Zelphati, O. and Szoka, F.C., Jr. Intracellular distribution and mechanism of delivery of oligonucleotides mediated by cationic lipids. Pharm. Res. 1996; 13(9):1367–1372.

Zinsstag, J. and Weiss, M.G. Livestock diseases and human health. Science 2001; 294(5542):477.

CHAPTER

17

Misinformation about Vaccines

Martin G. Myers[1,2] *and Diego Pineda*[2]

[1]Sealy Center for Vaccine Development, University of Texas Medical Branch, Galveston, TX, USA

[2]National Network for Immunization Information, Galveston, TX, USA

OUTLINE

ABSTRACT

Erroneous information about vaccine safety (misinformation) has existed since the dawn of vaccines and its dissemination is permitted by the freedom to express an opinion—no matter how incorrect.

Because of the increasing number and timing of vaccines, they are easy scapegoats for adverse events which occur coincident with vaccine administration, particularly those outcomes that are incompletely understood. In addition, vaccine-preventable disease activity has been greatly reduced by immunizations, causing parents to fear what they perceive might be complications of vaccines more than the actual diseases that the vaccines prevent.

It may require years to collect the data necessary to meet the rigorous standards that permit scientists to emphatically reject a vaccine safety concern. During the time needed to collect the evidence, however, those claims can evolve into misinformation.

Misinformers may be unintentional (they are uninformed) or intentional, in which case many have profit incentives. The most effective misinformers, however, are parents who truly believe that their child has been injured by vaccines.

Misinformation about vaccines can—and is—aggressively disseminated. Presented as fact by prominent individuals, it is often featured in publications, expounded on by the news media, discussed on popular TV and radio talk shows, and made the subject of TV dramas and docudramas. Because these compelling stories often feature prominent people, it is little wonder that other parents become confused. Unfortunately, there are barriers in responding to misinformation including innumeracy on the part of the public and poor communication skills on the part of vaccine spokespersons.

Confused parents may delay or refuse immunizations for their child. Many parents have many misperceptions about disease risks and vaccine safety, both those parents who permit their child to be immunized and those who do not. Health professionals can also become confused and some have helped perpetuate misinformation about vaccine safety.

The consequences of misinformation about vaccine safety can be declines in immunization acceptance and coverage followed by disease outbreaks with global consequences.

Countering misinformation about vaccines has become an urgent priority to assure the continued success of immunization programs. Fortunately, parents continue to seek information as well as the means to validate what they find. They also are seeking to identify trusted health providers to give them more information and guidance.

INTRODUCTION

In the 20th century, infectious diseases in the United States and other developed countries began to be better controlled as a consequence of clean water, pest control, and especially because of vaccines (Table 17.1). National immunization programs were successful because of effective childhood vaccines and a broad societal consensus that rare serious side effects—such as vaccine-associated poliomyelitis—were tragic but were tolerable for the benefit of the general public health. Of course, serious vaccine misadventures during the past century led to the vaccine industry being one of the most closely regulated of all US industries (Parkman, 2002).

While the immunization programs have been remarkably successful, they have had to deal with many challenges some of which are consequences of misinformation about vaccines and vaccine safety (Gust et al., 2004, 2005). For example, there have been recent measles outbreaks in the UK (Jansen et al., 2003; Ashmore et al., 2007), the Netherlands (Lernout et al., 2007), Germany (van Treeck, 2006; Bernard et al., 2007; Steart-Freedman and Kovalsky, 2007), and Switzerland (Richard and Masserey Spicker, 2007), attributable to inadequate rates of immunization coverage, at least in part related to the impact of misinformation about the safety of measles-containing vaccines. Similarly, recent outbreaks of mumps in the US seem likely to have been imported from the epidemic in the UK (Centers for Disease Control and Prevention, 2006a), also related to decreased vaccine coverage due to the measles misinformation.

Seventy seven percent of US children 19–35 months of age are fully immunized with all the universally recommended vaccines (Centers for Disease Control and Prevention, 2007)—of course, that also means that 33% of children are under-immunized. In 2001, it was estimated that 3 children per 1000 had never received any vaccines (Smith et al., 2004); thus many others must be incompletely immunized. Under-vaccinated children are consequences of limited health-care access as well as financial and other barriers to immunization, although some are also a consequence of confusion about the importance of vaccines and misperceptions about vaccine safety (Gust et al., 2004). In contrast to under-immunized children, unvaccinated children's families tend to oppose immunization, cluster in communities, and their parents tend not to consider the advice of physicians or other health professionals (Smith et al., 2004; Salmon et al., 1999). Many parents are uninformed about the risks of vaccine-preventable diseases as well as misinformed about the safety of vaccines (Gust et al., 2004).

People opposed to immunization have been present since immunizations began. Cotton Mather's home was bombed (fortunately the bomb did not detonate) for preaching support for variolation, having convinced Zabediel Boylston, a physician, to variolate people in Boston in 1721 (Best et al., 2004). Admittedly, variolation was a practice with severe side effects, including 2% of patients dying, but the risk was less

TABLE 17.1 Vaccines and vaccine-preventable diseases—a remarkable outcome

Diseases	Number of cases before vaccine was widely used[1]	Year vaccine recommended for routine childhood vaccination	Number of cases	
			2005	2006
Smallpox	48,164	Early 1900s	0	0
Diphtheria	175,885	Mid-1940s	0	0
Pertussis	147,271	Mid-1940s	25,616[a]	15,632[a]
Tetanus	1314	Mid-1940s	27	41
Paralytic poliomyelitis	16,316	1955	0[c]	0
Measles	503,282	1963	66	58
Mumps	152,209	1967	314	6584
Rubella	47,745	1969	11	11
Congenital rubella	823		1	1
Invasive *H. influenzae*, type b[b]	20,000	1985	9	29
Invasive *S. pneumoniae*[b]	17,240	2000	1495	1861
Hepatitis A (acute)	26,796	1995	4488	3579
Hepatitis B (acute)	26,107	1986	5119	4713
Varicella	About 4,000,000	1995	32,242	48,445
Deaths	105		3	ND[d]

Source: Adapted from Myers and Pineda (2008) with permission of the publisher.
[a] Numbers of cases of pertussis were at a historic low of 1010 in 1976.
[b] Children less than 6 years of age.
[c] In 2005, there was one case of imported, vaccine-associated paralytic poliomyelitis.
[d] ND, no data are available for 2006 yet.

than the mortality of smallpox in the unimmunized that developed smallpox (14.9%). Initial concerns about the safety of variolation led Benjamin Franklin to decide to not have his 4-year-old son, Franky, immunized by variolation, with tragic consequences:

> In 1736 I lost one of my sons, a fine boy of 4 years old, by the Small Pox... I long regretted bitterly and I still regret that I had not given it to him by inoculation; This I mention for the sake of parents, who omit that operation on the supposition that they should never forgive themselves if a child died under it; my example showing that the regret may be the same either way, and that therefore the safer should be chosen (Hastings Weld, 1859).

MISSING INFORMATION

Missing Community Fear

How can communities knowingly put their children at risk from vaccine-preventable diseases? Of course, communities do not realize that their children may be at risk. In the absence of the vaccine-preventable diseases, parents do not recognize that their communities remain at risk of these infections and that these diseases can be introduced and quickly spread within communities. This loss of societal fear represents one of the extraordinary ironies of the success of immunization programs. Except for smallpox, the vaccine-preventable diseases are still present. Tetanus spores persist in soil and, with global air travel, diseases like measles, mumps, rubella, diphtheria, and poliomyelitis remain just a plane ride away.

Decades ago, when thousands of children (and adults) in the United States contracted these diseases, parents feared these diseases. Successful immunization programs, however, have remarkably lowered disease rates for many of these illnesses. Presently, most parents (and many physicians) have never seen a child with paralytic poliomyelitis, asphyxiated by a diphtheric membrane, or brain damaged by measles, congenital rubella, or *Haemophilus influenzae*, type b (Hib). Some parents even believe that some of the vaccine preventable diseases are "not so bad" (Benin et al., 2006).

Although antivaccine opinions have been around for a long time, the majority of parents have maintained a collective consensus—largely driven by fear of the vaccine-preventable diseases—that vaccines and vaccine mandates are essential to the public health and the safety of children in a given community. In the absence of disease visibility the collective consensus has begun to erode.

The challenge for vaccine policy makers is to communicate that the risks of disease introductions are real—even if they are largely unpredictable.

Missing Data or "The Data are Insufficient to Accept or Reject the Hypothesis"

The timing of immunizations—beginning shortly after birth and then repetitively during the first 2 years of life with continuing immunizations throughout childhood—occurs at the ages when developmental and other problems are being recognized for the first time. Thus, the timing of vaccination—plus their widespread use—make vaccines the perfect "scapegoats" to be blamed for causing diseases of unknown or poorly understood causes.

When a vaccine concern is first suggested—particularly when the association is to a disease about which there is little understanding about etiology—there are often little or no data available to permit an honest scientist to state categorically that "vaccines do not cause that particular disease." And then, as data accumulate, the lack of a demonstrated association does not assure coincidence, requiring multiple studies before scientific consensus is reached that the data "favor rejection of the hypothesized association."

Compounding the issue of the time it takes to collect the needed data is the fact that many well-educated and intelligent people are innumerate—that is, they are unable to process information about risk assessments and measurements (Paulos, 1988). In addition, many suffer from availability bias (Poland and Jacobson, 2001). That is, they make intuitive judgments using readily available information. They base their estimates of how likely an event is based upon how easily they can imagine an example as well as its emotional impact.

Finally, when trying to communicate the complexities of demonstrating coincidence to an innumerate public, vaccinologists and public health officials often "speak in tongues"— using technical jargon as well as using expressions that have very different technical meanings than when they are used in everyday conversational English (Table 17.2). For example, when discussing vaccine safety, the vaccinologist uses the term "adverse event" to describe something that occurred temporally related to vaccine administration whereas many misconstrue that term to mean a "vaccine side effect." For instance, fever is a common side effect of many vaccines, but febrile adverse events after vaccine receipt are not always caused by the vaccine. Similarly, when some serious condition is recognized at about the time vaccines have been administered, the temporal association is difficult for parents to grasp as likely having been coincidental; particularly when public health officials only reassure with expressions such as "the data favor rejection of the theory."

The Uniformed as Misinformer

The uninformed or incompletely informed person may unintentionally disseminate misinformation. Parents rely on many sources for health information (Gellin et al., 2000; Paulussen et al., 2006) including family and friends but, unfortunately, many of these sources may also be uninformed misinformers.

MISSING INFORMATION CAN EVOLVE INTO MISINFORMATION

Once evidence favors rejection of a vaccine safety hypotheses—such as has occurred with thimerosal- or measles-containing vaccines and autism (Institute of Medicine, 2004)—assertions about that vaccine safety concern are misinformation. The evolution of a vaccine safety concern into misinformation is complex but seems to follow a pattern which includes both unintentional and intentional misinformers (Table 17.3).

Consistently, there is a period of scientific uncertainty; a period of missing information where the scientist has difficulty being emphatic that the vaccine and the adverse event are unrelated, even if (s)he believes that the association is coincidental. People respond differently during these periods. Some parents do not immunize their children, perhaps believing that public health and health provider information was "one-sided" or "poor" (Smith et al., 2004); other parents express distrust of the medical community and perhaps a conspiracy by those advocating vaccines (Mills et al., 2005); while the majority, fortunately, trust that their physician or nurse will know what to do (Gellin et al., 2000).

A person's perception of risk is based upon their experiences and knowledge. Someone who thinks that their child—or someone else's child with whom they can empathize—had an adverse outcome because of a vaccine would likely think that vaccines are riskier than a person who has not. Conversely, a parent whose child has had a vaccine-preventable disease—or a physician who has treated that disease—will likely advocate for vaccines.

People respond better to some types of perceived risk than others (Reynolds, 2002). For example, natural

TABLE 17.2 Technical and conversational expressions that may confuse the public

Expression/word	Technical meaning	Conversational meaning
Vaccine adverse event	Something that occurred at about the same time as vaccination, which may or may not be caused by the vaccine	Something caused by the vaccine (vaccine reaction or side effect)
Bias	Systematic error that could lead to the wrong conclusion	Not having an open mind
Controversy	There are different but plausible interpretations of the same data within the scientific community	There is a difference in opinion
The patient " denies xyz"	The patient says she does not have them	The patient reacts defensively to an accusation
Epidemiology	The study of how disease is distributed in a population and of the factors that influence that distribution	Number crunching
Favors rejection of the hypothesis	The data suggest that the hypothesis should be rejected (but you can not prove a negative)	They still do not know
The finding would not go away	We could not find an alternative explanation	They are fudging the data
Inadequate to accept or reject the hypothesis	The data do not allow a definitive statement	They do not know
Naive	The person or animal has not previously been exposed to a particular infection, drug, or vaccine	Unsophisticated, lacking experience, or training
Paralysis	Loss of the ability to move a body part usually as a result of nerve damage	Inability to act, helpless inactivity
Plausible	Theoretically possible	Appearing worthy of belief, factual
Positive	The results of the study concur with the hypothesis; having a value greater than zero	Good
Power	The likelihood of a study finding an effect, if there was one; or, the number of times a number is multiplied by itself	Energy, strength, control
Probably a small risk	The association is likely real but very infrequent; or, uncertainty	Likely that it is a risk
Relative risk	The ratio of two rates of risks, often used to compare risks	The risks are related
Safe	Remote or insignificant risk	No risk or zero risk
Significant	This may not be a chance difference	Important
Not significant	Likely due to chance	Not important
Uncertainty	When the available scientific information is not sufficient to prove a relationship or is not sufficient to favor rejection of a relationship	They do not know

Source: Adapted from Myers and Pineda (2008) with permission of the publisher.

risks (such as infectious diseases) are better tolerated than man made risks (such as vaccine side effects) and risks that affect adults are better tolerated than risks affecting children. Risks that are perceived to have unclear benefits are less tolerated than risks where the benefits are clearly understood. For example, some parents believe that the risk of contracting measles could be lower than the risk of their child possibly experiencing a serious side effect being suggested by other parents who are convinced that their child was harmed following immunization. If they think that there is little benefit from immunizing their child, they may conclude that there is no reason to take the risk of a possible adverse event, even if public health officials try to reassure them. If they are confused as to the risks, they may fail to immunize their child.

Stories about bad things happening to children after a vaccination circulate widely on the internet, are discussed on radio and TV, and described in magazines and newspapers. Despite the fact that serious vaccine safety risks are rare, perceived health risks are the center of attraction to the media, make effective sound bites, and may be egregiously inaccurate (Pribble et al., 2006).

TABLE 17.3 The evolution of vaccine safety concerns into misinformation

A vaccine safety concern is suggested—usually by case reports of possible temporal associations of adverse events with vaccine administration or by an increase of reports of possible adverse events following vaccine administration.
- In the absence of any data making an association, and despite reassurances from the public health authorities, the media may declare that there "may be a problem."[a]
- Some parents become frightened.

Vaccinologists respond that there are insufficient data to be able to state that the vaccine caused the adverse event or that it was just coincidental in time.
- This creates a sense of uncertainty which in turn causes some parents to become confused.

Public health officials compare the possible risk from the adverse event with the known risk of the disease, based on whatever available data there are.
- They make a recommendation.
- Some parents perceive a "cover-up."

The media reports the scientific "uncertainty."
- Parents who are convinced their child was harmed by vaccines want to warn other parents.
- The media often describe a controversy (difference of opinion) between public health officials and parents. True scientific controversy is rarely reported.
- Pseudoscience is promulgated; it is discounted by scientists but is often widely reported by the media.
- Intentional misinformers seek publicity.

Many parents become confused about vaccine benefits and risks.
- Parents see no disease, and reason "why take a risk?"

As data are collected, scientists reach a consensus that the data "favor rejection of the hypothesis."
- Misinformers deny the data.
- Misinformers attempt to discredit the scientists.
- Those who are convinced, make up ad hoc hypotheses.

If community immunity has declined, outbreaks of disease may occur.

Ultimately—perhaps decades later—a cause for the coincidental disease is described.

[a] For example a BBC news headline reported (BBC News Online, 1998, http://news.bbc.co.uk/2/low/uk_news/60510.stm) "child vaccine linked to autism" following the initial case reports by Wakefield et al. (1998) about measles-containing vaccine and some cases of autism.

Media reports about the vaccines-autism theories, for example, continue to be published frequently. Although scientists say "the data suggest no association" (or, that "the data favor rejection of the theory"), parents want to hear a trusted source say "that vaccine does not cause that." In contrast, some who believe that their child (or one that they know) was harmed by vaccine(s) insist that the theory has been established. Unfortunately, media reports often cite both opinions with equal credibility and emphasis; generating uncertainty among parents, affecting their perceptions of risk, and their decision to vaccinate their child (Dannetun et al., 2005).

IMMUNIZATION MANDATES

States require vaccines because they have a responsibility to protect both individuals and the entire population of their state. All states have immunization laws. However, other countries that also have very high levels of immunization coverage—such as Canada and Mexico—do not (WHO, 2007a).

School immunization laws were first established to control outbreaks of smallpox and in recent decades have been widely used to increase vaccine coverage and reduce outbreaks of vaccine-preventable diseases. Immunization requirements are set by states and currently all states have school immunization laws—although there are differences in what may be required in different states (see www.immunizationinfo.org for individual states' requirements and exemptions).

Measles vaccine was licensed in the US in 1963; there was a rapid decline in the incidence of measles from about 438,000 cases/year (1960–1964) to about 42,700/year (1967–1971) in the United States (Centers for Disease Control and Prevention, 1995) (Fig. 17.1A). Both national and local epidemics continued to occur, however (Fig. 17.1A, B).

The constitutional basis for "a community to protect itself against an epidemic of disease" by immunization was established in 1903 (Parmet et al., 2005). This provided a means for states to expand immunization coverage. In the late 1960s and the 1970s immunization laws for school entry were enacted in many states largely on the demands of parents, because of continuing measles epidemics. For example, in Iowa, the creation of immunization requirements for school entry in 1977 (Iowa Administrative Code, 1977) was largely a response of parents to the recurring measles epidemics with the attendant mortality, cases of encephalitis, and overall morbidity (Myers, 1977, unpublished observations); the impact on measles activity was rapid (Fig. 17.1B) with cases dropping from 4333 in 1977 to none in 1981 through 1985 (Iowa Department of Health, 2007).

Parents who oppose vaccines for their children actively seek means to avoid immunizations, including home schooling and lobbying legislators for easier exemptions from immunization laws. Some parents oppose immunization mandates as a violation of their civil liberties. Both groups of parents try to recruit other parents to their "cause."

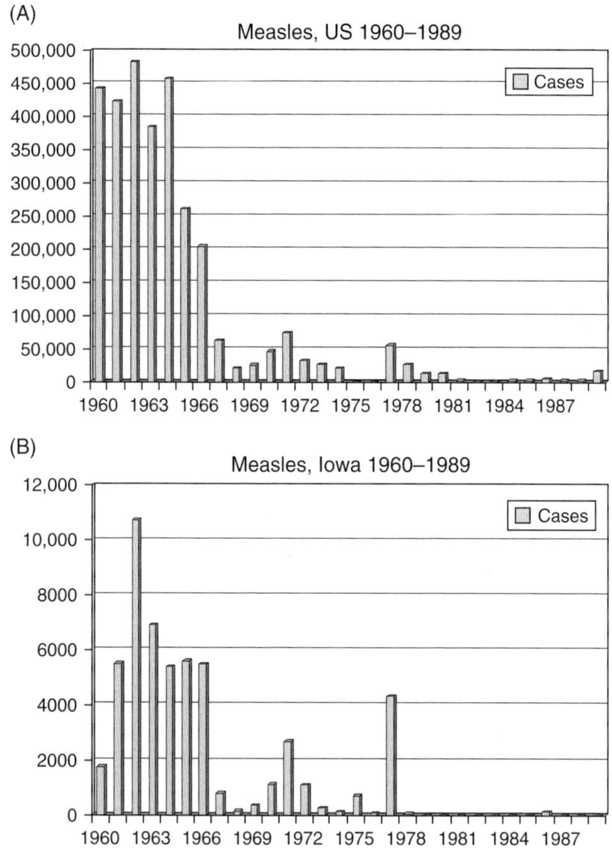

FIGURE 17.1 Measles in (A) the United States (Centers for Disease Control and Prevention, 1995) and (B) Iowa (Iowa Department of Health, 2007, www.idph.state.ia.us/adper/pdf/cade/decades/pdf) 1960–1989. Measles vaccine was licensed in 1963. Mandatory immunization laws were enacted widely by states in the late 1960s and 1970s. Iowa enacted its immunization law in 1977 (Iowa Administrative Code, 1977).

All US states permit exemptions for children who have medical contraindications to immunization. However, most states also permit religious or philosophical exemptions. States permitting easy exemptions have many more unvaccinated children than do states that do not permit philosophical exemptions and unvaccinated children tend to cluster in communities (Smith et al., 2004; Salmon et al., 1999). Children who received exemptions were found to be 35 times more likely to contract measles and were more likely to spread measles to others (Salmon et al., 1999; Parker et al., 2006).

Washington State had school entry immunization exemption rates exceeding 5% in 2005–2006, six counties exceeding 10%, a substantial increase over 1999 (Bardi, 2007). Texas reported a school district with 19.05% conscientious exemptions for the school year 2003–2004, the first year after enactment of the exemption (Immunization Branch, Texas Department of State Health Services, 2003–2004, http://www.dshs.state.tx.us/immunize/coverage/conscientious.shtm).

MISINFORMERS

Misinformation comes from many sources. Some who are uniformed or misinformed unwittingly misinform. Others are parents who honestly believe that their child was harmed by vaccines and they want to warn other parents. Still others are antivaccine activists many of whom may gain by disseminating and popularizing misinformation. For example, profits may come from contributions to organizations, book sales, movie rights, political support, tort precedents, court testimonies, speech making, inches of news space, or minutes of news time.

Conflicts of Interest

Many organizations, government agencies, and universities require their employees (and their immediate families) to divulge potential or perceived financial and other conflicts of interest. Because of the nature of what vaccine researchers do, it is not surprising that some hold patents for their intellectual property, that they may be asked to consult for vaccine companies, or that a portion of their salary may come from grants or contracts with vaccine companies. While it is hard to imagine how an individual could otherwise become a vaccine expert, these types of relationships when active do in fact represent perceived—and at times real—conflicts of interest.

Misinformers make much ado about vaccinologists' real and perceived conflicts of interest. But they are aware of these conflicts of interest because the vaccinologists divulge these associations.

Misinformers may also have conflicts of interest, some obvious others less so. They have no obligation to divulge their own or their families' conflicts of interest. Book authors do not need to divulge their book sales income; investigative reporters (or their spouses) do not need to expose their retirement portfolios; pseudoscientists do not need to divulge their funding sources nor the income they receive from court testimony; and politicians only divulge conflicts of interest that are required by law.

False Experts or "Experts" who Lack Expertise

No credible expert vaccinologist would claim to be expert in *all* the fields that relate to vaccine safety. For example, the IOM Committee on Immunization Safety included experts in pediatrics, neurology, immunology, internal medicine, infectious diseases, genetics, epidemiology, biostatistics, risk perception and communication, decision analysis, public health, nursing, and ethics (Institute of Medicine, 2004).

Some who claim to be experts make claims outside their field of expertise. Such is the case of Dr. Mark Geier:

> Dr. Geier's expertise, training, and experience is [sic] in genetics and obstetrics. He is however a professional witness in areas for which he has no training, expertise, and experience (U.S. Court of Federal Claims, 2003).

Another example would be "...Dr. Eric Ryndland who is considered one of the nations foremost experts on autism..." according to the reporter (Gillen, 2007). He[1] has no publications listed on the National Library of Medicine's Pub Med. It is unclear what, therefore, the reporter utilized to form her opinion of him as an "expert" in her article "To vaccinate or not to vaccinate."

Journalists who have Written Articles or Books about Vaccine Safety

Although many journalists do a credible job writing about scientific and medical issues in ways that makes this information understandable and informative to the general public, they are usually specialized medical or science reporters at large news organizations. Good science journalists let the experts speak through their articles instead of purporting themselves to be experts.

Often times, however, stories—particularly published by smaller news organization and local TV stations—have not been checked for the reliability of the facts and present "controversies" that pit parents or others with an opinion against scientific experts, as if opinions and facts had similar value.

Others simply misinform. Media exposure and "controversies," of course, sell books and attract readers. For example, the journalist David Kirby—with apparently no scientific background or expertise in any field relating to vaccine safety or developmental disorders—wrote a book, *Evidence of Harm*, about thimerosal and autism which misinterprets scientific evidence and quotes out of context, attempting to imply a cover-up by respected scientists. Nevertheless, this book received an award for investigative reporting (Investigative Reporters and Editors, 2006, http://www.ire.org/history/pr/2005ireawards.html). Kirby is a frequent speaker at events for antivaccine and antithimerosal activists.

[1] Neither this spelling, nor Rindland, nor Rynland.

Dan Olmsted, the author of *The Age of Autism*, once a regular columnist for United Press International (UPI), reported on his "research" that autism is less frequent among unvaccinated Amish children in Pennsylvania in contrast to other children. For a time, his articles attracted media attention despite the lack of any scientific content. His research methods included discussing his theories with a man who sells water purifiers in Amish communities (Olmstead, 2005).

Politicians who Claim to be Experts

Many politicians bring together groups with differing interests and expertise—such as parents with opinions and scientists with expertise—in order to collect information and to understand the issues. However, there is little in a politician's training—including those with advanced education including a medical degree—that would qualify them as experts. Of course, politicians often seek support by taking on the causes of different interest groups.

Between 1999 and 2004 the Committee on Government Reform of the US House of Representatives held more than 10 hearings about vaccines, thimerosal, and autism. Rep. Dan Burton chaired this Committee from 1997 to 2002; his claim to expertise being that he is the grandfather of a child whom he believes has autism as a consequence of vaccines. When the IOM Vaccine Safety Committee said in 2004 that the body of scientific evidence favored rejection of a causal relationship between thimerosal-containing vaccines and autism (Institute of Medicine, 2004), Rep. Burton said in a press release:

> Unfortunately, I believe the findings announced in the May 18th IOM report are heavily biased, and unrepresentative of all the available scientific and medical research (Burton, 2004).

Politicians, lawyers, journalists, and others such as parents with an opinion contribute in a very positive way to the public discussion of immunization policy and vaccine safety but they are not authoritative scientific experts. Nonetheless, they are often quoted as if they were.

DISSEMINATION OF MISINFORMATION

Unfortunately, misinformers can disseminate misinformation easily, especially employing the internet and other media.

More than two-thirds of US adults use the internet (Fox, 2005) and 80% of them use the Internet to

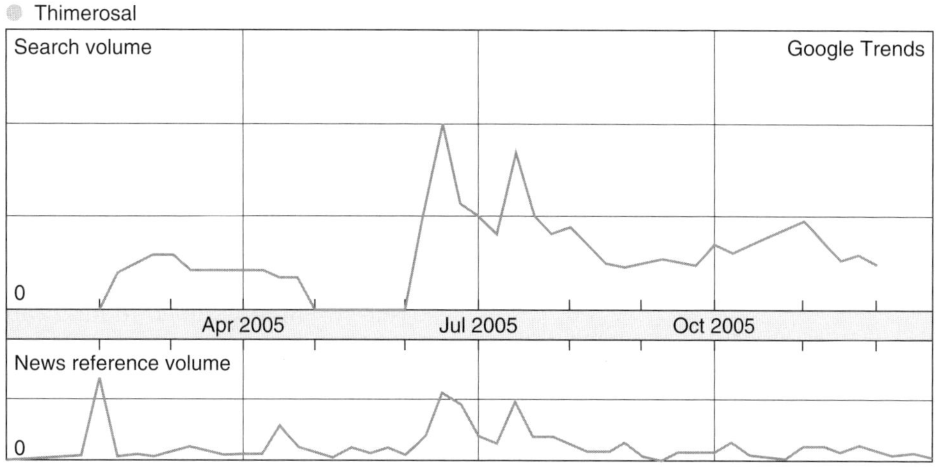

FIGURE 17.2 Thimerosal misinformation events in March and June, 2005 were associated with increased news and search volume for thimerosal. "News reference volume" relates to news stories containing the search word that were accessed on Google News©. "Search volume" describes the volume of internet searches made using that search term.

find health information, including information about vaccines and immunization (Fox, 2006). There are disparities of access to the internet as well as the quality of internet access but those who access the internet (Fox, 2005) appear to have similar demographics to those who are antivaccine (Smith et al., 2004). Those who seek information online usually start at a search engine and do not often check on the source of the information (Fox, 2006). Internet search engines do not distinguish information from misinformation and most people are unable to distinguish reliable from unreliable information. In addition, one-third or more of streaming video on www.YouTube.com were critical of immunizations, with those videos higher rated than video clips favorable to immunizations (Keelan et al., 2007).

An example of how people use the internet in response to misinformation was shown by surges in search engine queries that included the word "thimerosal" following extensive publicity surrounding two misinformation "events": the initial release of the book "Evidence of Harm" by David Kirby (St. Martin's Press) in March, and the publication of the article "Deadly Immunity" by Robert Kennedy Jr. in Rolling Stone and Salon.com in mid-June, 2005. Google Trends© demonstrated both increased news volume and search engine searches for thimerosal in March (book release) and mid-June (article published) (Fig. 17.2) (Myers and Pineda, 2007).

The mainstream media reaches and influences all segments of society. For example, health stories must attract viewers because these account for more than 10% of air time in nightly TV news broadcasts. Unfortunately, these stories are usually unrelated to public health priorities, they may be inaccurate, and provide misinformers with exposure. These reports may contain erroneous information (Pribble et al., 2006).

Unfortunately journalists may feel compelled to present a balance between opinions, reporting "controversies" (or "debates") between the scientific community and misinformers, giving misinformers extensive media exposure and possibly credibility in the minds of the public. For example, on reporting the findings from a large comprehensive study that did *not* suggest any causal association between increasing exposure to mercury in thimerosal—neither prenatal or early in life exposure—and subsequent neuropsychological outcomes at 7–10 years of age (Thompson et al., 2007), Time Magazine in October, 2007 (Time Magazine, 2007), headlined an article "the vaccine debate goes on" and opened the article "Latest Findings. Thimerosal, a mercury-based vaccine preservative, has long been associated with neurological disabilities like autism." The article continued by outlining Centers for Disease Control and Prevention and their "critics" positions.

MISINFORMATION CONTENT

Misinformation content often consists of many of the following:

Pseudoscience

Pseudoscience ("false knowledge") is often presented to the media as "science" by misinformers. It is usually ambiguous, employs hard to comprehend methods, may not be quantifiable, and the results can usually not be duplicated. In many cases, these reports are not subjected to peer review. Often, data may be represented to show one outcome when another is the case. Other times the methods that are used may be likely to give a predetermined outcome. Only data purporting to support the author's claims are presented while conflicting data are ignored or dismissed.

The peer review mechanism employed by most respected journals is intended to ensure that only carefully conducted science is published; but sometimes poorly run studies slip by the most careful editor. Because of the vetting process, however, most pseudoscientific claims about vaccine safety are not published in respected journals but are often published in obscure journals or in the "alternative medicine" press.

A number of studies implying a link between vaccines and neurodevelopmental disorders have been published by Mark Geier and his son David Geier. They often utilize data from the Vaccine Adverse Event Reporting System (VAERS) to try to establish causality (Geier and Geier, 2004). VAERS data cannot be used to establish causality. The purpose of VAERS is to tabulate possible adverse events in order to look for "signals" that should be explored systematically (Varricchio et al., 2004). In addition, the Geiers' reports also often failed to describe methods, important statistical figures were not defined accurately, and data sets overlapped (American Academy of Pediatrics, 2003, http://www.aap.org/profed/thimaut-may03.htm).

In addition to pseudoscience, misinformers often misquote legitimate research claiming that studies support their views, when in fact the original data do not. For example, the November 14, 2005 issue of the New York Times included a full-page advertisement by the group Generation Rescue thanking the researchers who did "groundbreaking research on the connection between mercury and autism" (The New York Times, 2005, http://www.generationrescue.org/images/051114.gif). The ad listed 19 citations of articles—including one by the Geiers. None of the 19 articles shows that mercury causes autism. Although some of them did look for a possible link between mercury and autism, others were about autism in general with no mention of mercury. One of the papers cited was by Andrew Wakefield, about MMR (which does not contain thimerosal) and autism.

Invalid Assumptions

Some invalid assumptions often underlying misinformation claims are that illness in a vaccinated person proves that the vaccine does not work, temporal associations establish causation, and epidemiologic studies are not real science.

Logical Fallacies

Misinformers often use logical fallacies in their arguments. For example, the post hoc argument is common; that is, the temporal association of an adverse event is assumed to infer causality. It is especially seen online, where people report that a vaccine harmed a child because certain symptoms appeared a few days or weeks after vaccination. Also, VAERS reports are purported to be descriptions of vaccine side effects. Often they also make an *argument from ignorance*—claiming that a statement is true only because it has not been proven false, or that it is false only because it has not been proven true.

The ad hominem fallacy—attacking the person—takes many forms in discussions of vaccine safety, alleging cover-ups and conspiracy theories by which misinformers try to discredit those who do not agree with them. Indeed ad hominem arguments are recognized to be a frequent attribute of antivaccination Web sites (Davies et al., 2002) as well as media coverage about antivaccination topics (Leask and Chapman, 1998). Attacks of this type on members of the Institute of Medicine Vaccine Safety Committee and other reputable scientists were so serious that Senator Enzi (2007) filed an investigative report with the US Senate Committee of Health, Education, Labor and Pensions in 2007 refuting the charges. Similarly, the *guilt by association* fallacy claims that a theory or an argument must be false simply because of who else supports the argument.

Misinformers often use the ad misericordiam fallacy to frighten other parents, by featuring anecdotes about children that they claim have been killed or maimed by vaccines (Davies et al., 2002; Wolfe et al., 2002).

Ad hoc Hypotheses

When misinformers find themselves in the position of believing something that becomes unsupportable based on the data, they will often resort to making an ad hoc hypothesis. Attorneys litigating vaccine injury claims do this not infrequently. When multiple studies found no causal association between measles- or thimerosal-containing vaccines with the subsequent development of autism, a new hypothesis was generated that together measles-containing vaccine and the preservative thimerosal in other vaccines cause autism (Office of Special Masters, 2007).

THE CONSEQUENCES OF MISINFORMATION

Outbreaks of vaccine-preventable diseases often begin among the unimmunized and underimmunized and then spread to the fully immunized in the community (Salmon et al., 1999; Parker et al., 2006)—a portion of whom remain susceptible because no vaccine is 100% effective. Sustained transmission of infections within communities occurs when a sufficiently large proportion of the population is susceptible, the proportion differing among the different vaccine preventable communicable diseases (Fine, 2004). Immunization coverage is often utilized as a surrogate marker for the proportion who is immune.

Following periods of intense misinformation, decreases in immunization coverage may occur and disease outbreaks may occur. For example, because of the concerns about whole-cell pertussis vaccine possibly being associated with severe neurologic developmental problems, immunization rates in Great Britain in 1978 had fallen from 80 to 30%. Epidemic pertussis soon followed: between 1977 and 1979 there were 102,500 cases of whooping cough with 36 deaths (Cherry et al., 1988).

In Japan, because of concerns about the whole-cell pertussis vaccine's safety and claims that it was no longer necessary to immunize because pertussis was not present in the community, the age for immunization was changed and immunization coverage for infants fell from about 85% in 1974 to 13.6% in 1976 (Cherry et al., 1988). In 1979, Japan reported 13,105 cases of pertussis with 41 deaths. In the early 1980s Japan re-introduced acellular pertussis-containing vaccines with a reduction in the number of cases (Gangarosa et al., 1998). Similarly, after discontinuing pertussis vaccine, rates of whooping cough returned to the levels seen in the prevaccine era in Sweden. Of 2282 who were hospitalized for whooping cough in 1981–1983, 4% had brain injury from the illness (Cherry et al., 1988).

The specific contribution that misinformation played on these pertussis outbreaks is hard to define because it is not clear when missing information about whole-cell pertussis vaccine safety evolved into misinformation. However, countries whose immunization programs were disrupted by whole-cell pertussis antivaccine movements (Sweden, Japan, the UK, Ireland, Italy, Australia, the former West Germany, and the Russian Federation) experienced ten to 100 times higher pertussis incidence than did countries in which the antivaccine movement had a limited impact on pertussis vaccine coverage (Hungary, the former East Germany, Poland, and the US) (Gangarosa et al., 1998).

Two decades later, another vaccine scare began in the UK. In 1998, Wakefield and others published a series of case reports suggesting that some children who had received measles-containing vaccine suffered bowel injury, permitting absorption of substances that caused brain injury and thus autism (Wakefield et al., 1998). Although Wakefield's report did not provide any evidence of a link between measles vaccine and autism, media coverage was extensive, giving his hypothesis credibility in the eyes of many in the public; the headline in London's *Daily Telegraph* read, "Vaccination may trigger disease linked to autism" (Daily Telegraph, 1998, http://www.telegraph.co.uk/htmlContent.jhtml?html=/archive/1998/02/27/nmmr27.html). Many national and international news outlets carried the same story. Both the initial report and subsequent claims of evidence for measles virus in children with autism have now been dismissed (D'Souza et al., 2006; Doja and Roberts, 2006; Murch et al., 2004; Horton, 2004) but many frightened parents, not surprisingly, feared the vaccine more than the diseases. In the following years MMR vaccination rates fell from 93 to 79% by 2003, with measles (Gust et al., 2005) and mumps (Centers for Disease Control and Prevention, 2006b; WHO, 2007b) (Fig. 17.3) outbreaks as a consequence. The outbreak of mumps in 11 states in 2006 and the epidemic in the UK in 2005 were both caused by mumps virus, genotype G (Centers for Disease Control and Prevention, 2006a), suggesting they may have been epidemiologically related.

Misinformation confuses many. For example, in an unpublished random digit dial telephone survey, almost half the parents with children less than 6 years of age reported that they believed vaccines can cause

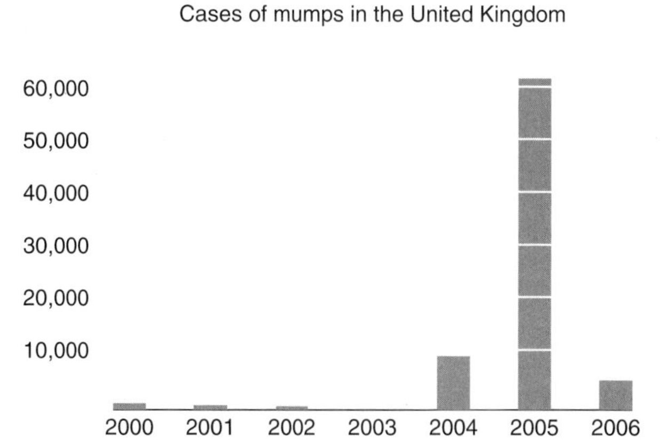

FIGURE 17.3 In 2005, there were more than 63,500 cases of Mumps in the United Kingdom (WHO, 2007b) (Adapted from Myers and Pineda (2008) with permission of the publisher.)

autism (APCO insight, 2006). Misinformation can also confuse health providers, including school personnel overseeing immunization programs (Salmon et al., 2004). In 2006, a 23-month-old asthmatic child with influenza A pneumonia was admitted to a hospital in Texas requiring ventilator support. Two weeks previously she had been denied influenza immunization because the provider believed the child should only receive influenza vaccine without thimerosal preservative, which the practice did not have in stock (Patel, 2006, personal communications).

Misinformation impacts parental decision-making. For example, 13 of 33 mothers interviewed postpartum were vaccine hesitant or outright rejected vaccine for their newborns (Benin et al., 2006). A couple in Tennessee, confused about vaccine safety because of what a friend had told them, decided to delay their daughter's vaccinations. The girl developed Hib meningitis (Snyder, 2000).

Misinformation can have serious consequences for individual families as well as for communities.

VACCINE MISINFORMATION IS A GLOBAL PROBLEM

Misinformation about vaccine safety is a problem that affects developing as well as developed countries. False allegations in Nigeria that oral polio vaccine could cause sterility—thought to be intended to limit the Muslim population size—or that it could cause AIDS, resulted in a resurgence of wild-type polio viruses transmission, cases of polio, and the spread of wild-type virus into 19 previously polio-free countries (Centers for Disease Control and Prevention, 2005–2006). Similar rumors have plagued the introduction of other vaccine programs in other countries. For example, the introduction of a maternal and infant tetanus immunization program necessitated Egypt to train community guides to dispel these types of rumors—and one immunizer had to immunize herself to prove the vaccine safe (U.S. Fund for UNICEF, 2004, www.who.int/entity/immunization_monitoring/diseases/06_fall_2004.pdf).

In contrast to developed countries, antivaccine activities in developing countries focus on national immunization days, religious and political arguments against vaccines, and concerns about "western plots," especially that vaccines could cause involuntary contraception or sterilization (Eastern and Southern Africa Regional Office, United Nations Childrens Fund, 2001).

Part of the difficulty of dealing with misinformation about vaccines in developing countries is resistance to change (Regional Office for South Asia, United Nations Childrens Fund, 2005, http://www.unicef.org/rosa/Immunisation_report_17May_05(final_editing_text).pdf) as well as the recognition that some vaccines and immunization practices have in fact in the past been unsafe (Simonsen et al., 1999). Because the foundations of an effective immunization program are safe and effective vaccines which are delivered in a safe manner, the WHO has put in place systems to try to ensure the safety of the vaccines used in all national programs (Duclso, 2004; Clements et al., 2004).

To measure the impact of a vaccine and to be able to evaluate adverse events following immunization require surveillance of vaccine coverage and disease activity as well as the monitoring of adverse events. Many developing countries have not integrated these components into their national immunization programs (Duclso, 2004). Countries that do have these tools in place, such as the US, still have difficulty dispelling concerns about vaccine-associated anecdotes; in the absence of such an infrastructure, it is little wonder that health officials have difficulty reassuring the public about coincidental associations.

The historical impact of antivaccination rumors on the introduction of immunization programs have led to the development of strategic communication plans for national immunization programs that include the assumption that misinformation and negative reactions will occur as part of the introduction of new vaccines (Regional Office for South Asia, United Nations Childrens Fund, 2005, http://www.unicef.org/rosa/Immunisation_report_17May_05(final_editing_text).pdf).

FIGHTING WITH THE BOOGEYMAN: COUNTERING MISINFORMATION ABOUT VACCINE SAFETY

The vast majority of parents would not opt out of any vaccine and they also understand that school immunization laws protect their children (Gellin et al., 2000). However, there are also many parents who have great misperceptions about the risk and severity of vaccine preventable diseases as well as about the safety of vaccines (Mills et al., 2005; Dannetun et al., 2005; APCO insight, 2006; Salmon et al., 2004; Freed et al., 2004; Fredrickson et al., 2004). Misperceptions about disease risks and vaccine safety are similar among parents who refuse vaccines for their child and among those parents who permit their child to be immunized, although the proportions of parents with misperceptions are greater among those who withhold some or all vaccines from their children (Gust et al., 2004, 2005).

Parental refusal (either refusal or delay) of vaccine for their child is not uncommon (Benin et al., 2006; Fredrickson et al., 2004; Diekma and Committee on Bioethics, 2005; Flanagan-Klygis et al., 2005), most physicians who care for children have to deal with this problem at least once a year—the rate of refusal is estimated to be 7.2/1000 children less than 18 years of age (Fredrickson et al., 2004). Fear of pain, fear of serious side effects, plus the belief that the diseases are not harmful are common reasons for vaccine refusal. These are also common concerns among parents who do immunize their children.

Misinformation about vaccine safety comes in many forms and from many sources. Fortunately, most parents still obtain information about immunizations from trusted health professionals (Gellin et al., 2000). Unfortunately, some health professionals may be uninformed, some themselves may be confused by misinformation, and some (particularly alternative medicine providers) may be "anti-vaccine."

Parents are actively seeking more information (Gust et al., 2004). The most important intervention for countering misinformation may be that a trusted provider addresses the parents' lack of knowledge about the diseases as well as their concerns about vaccine safety. But this approach is limited by the time available to busy clinicians: discussions about vaccines comprising about 15 s during the average visit (Davis et al., 2004).

We have observed increasing use of the NNii Website www.immunizationinfo.org over time—from 17,677 visitors per month in October 2005, to 32,553 in October, 2007 (Myers and Pineda, 2007). The entrance pages to the Web site are usually the Homepage or, if accessed by search engine, often directly to specific vaccines/diseases featured in the section "Vaccines and the Diseases they Prevent." The next most popular section is Immunization Issues, consisting of essays about topical issues on vaccine safety (many suggested by viewers). We also have observed that during times of increased media activity surrounding misinformation events (Fig. 17.2), there is a corresponding surge in readership of archived essays on related topics (see Fig. 17.4A on our Web site. At about the same time, many visitors follow links to information about assessing the reliability of information found on the internet (Fig. 17.4B) (Myers and Pineda, 2007).

A large proportion of parents—whether vaccine hesitant or not—have been confused by misinformation publicized about vaccine safety. Although some parents have made up their mind to not immunize their children—and disregard data that do not support their views (Meszaros et al., 1996)—the majority of parents desire more information. They have many information resources available; unfortunately, they may frequently locate misinformation. However, in addition to seeking information, they also seem to want to confirm the reliability of what they found.

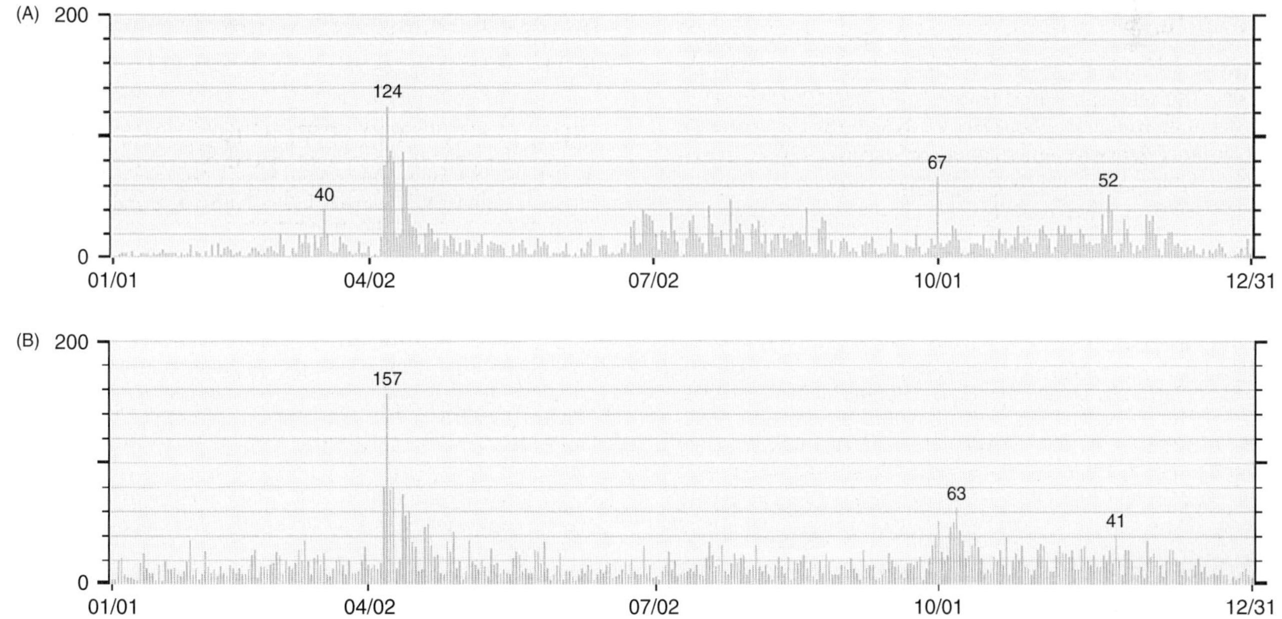

FIGURE 17.4 Search activity at www.immunizationinfo.org coincident with increased media activity surrounding thimerosal-containing vaccines. (A) Daily pageviews of archived essays about thimerosal-containing vaccines. (B) Daily pageviews of archived essays about misinformation about vaccine safety, viewed largely by hyper links (Myers and Pineda, 2008). A pageview is a request for a page, enumerated by www.urchin.org.

Parents need education about the risks and severities of the vaccine-preventable diseases; most do not recognize that they are uninformed.

There seem to be a number of favorable factors to counter misinformation: Most parents want to establish trusting relationships with health professionals; they are seeking more information; they are willing to utilize guidance; and many seek to validate information they locate. These openings would seem to provide the most effective means to sustain and expand confidence in immunization programs in the face of burgeoning misinformation about vaccines and vaccine safety.

In considering the introduction of new vaccines—including vaccines for biothreats and emerging infectious diseases—as well as vaccines developed employing new technologies and new delivery strategies, it will be important to employ communication plans for all the target populations, analogous to those being utilized for the introduction of new vaccines in developing countries (Regional Office for South Asia, United Nations Childrens Fund, 2005, http://www.unicef.org/rosa/Immunisation_report_17May_05(final_editing_text).pdf).

References

American Academy of Pediatrics. Study fails to show a connection between thimerosal and autism. May 16, 2003. http://www.aap.org/profed/thimaut-may03.htm Accessed November 27, 2007.

APCO insight. Understanding the opinion environment on vaccines and thimerosal: a survey among parents of children 6 and under. April, 2006. Unpublished.

Ashmore, J., Addiman, S., Cordery, R., et al. Measles in North East Central London, England: a situation report. Euro. Surveill. 2007; 12(9). www.eurosurveillance.org/ew/2007/070920.asp#2.

Bardi, J. Immunization Program Data, Washington State Department of Health. Personal communication, November 26, 2007.

BBC News Online: UK. Child vaccine linked to autism. Friday, February 27, 1998. http://news.bbc.co.uk/2/low/uk_news/60510.stm.

Benin, A.L., Wisler, D.J., Colson, E. et al. Qualitative analysis of mothers' decision-making about vaccines for infants: the importance of trust. Pediatrics 2006; 117(5):1532–1541.

Bernard, H., Santibanez, S., Siedler, A., et al. An outbreak of measles in lower Bavaria, Germany, January–June 2007. Euro. Surveill. 2007; 12(10). www.eurosurveillance.org/ew/2007/071004.asp#1.

Best, M., Neuhauser, D. and Slavin, L. "Cotton Mather, you dog, dam you! I'll inoculate you with this; with a pox to you": smallpox inoculation, Boston, 1721. (Heroes and martyrs of quality and safety). Qual. Saf. Health Care 2004; 13(1):82–83.

Burton, D. Refuting the IOM's latest vaccines and autism report. Press release, June 2, 2004. http://www.house.gov/burton/pr/2004/pr060204.htm.

Centers for Disease Control and Prevention. Summary of notifiable diseases—United States, 2005. 2006; 55(32):883–894.

Centers for Disease Control and Prevention. Summary of notifiable diseases—United States, 2006. 2007; 56 (33):853–863.

Centers for Disease Control and Prevention. Summary of notifiable diseases, United States, 1994. MMWR 1995; 43(53):1–74.

Centers for Disease Control and Prevention. Progress toward poliomyelitis eradication—Nigeria. MMWR 2005–2006; 56(12):278–281.

Centers for Disease Control and Prevention. Mumps epidemic—Iowa, 2006. MMWR 2006a; 55(13):366–368.

Centers for Disease Control and Prevention. Mumps epidemic—United Kingdom, 2004–2005. MMWR 2006b; 55(07):173–175.

Centers for Disease Control and Prevention. National, State and local area vaccination coverage among children aged 19–35 months—United States, 2006. MMWR 2007; 56(34):880–885.

Cherry, J.D., Brunell, P.A., Golden, G.S. et al. Report of the task force on pertussis and pertussis immunization—1988. Pediatrics 1988; 81:933–984.

Clements, J., Larsen, G. and Jodar, L. Technologies that make administration of vaccines safer. Vaccine 2004; 22:2054–2058.

Dannetun, E., Tegnell, A., Hermansson, G. et al. Parents' reported reasons for avoiding MMR vaccination. Scand. J. Primary Health Care 2005; 23:149–153.

Davies, P., Chapman, S. and Leask, J. Antivaccination activists on the world wide web. Arch. Dis. Child. 2002; 87:22–25.

Davis, T.C., Fredrickson, D.D., Kennen, E.M. et al. Childhood vaccine risk/benefit communication among public health clinics: a time-motion study. Public Health Nursing 2004; 21(3):228–236.

Diekma, D.S. Committee on Bioethics Responding to parental refusals of immunization of children. Pediatrics 2005; 115(5): 1428–1431.

Doja, A. and Roberts, W. Immunizations and autism: a review of the literature. Can. J. Neurol. Sci. 2006; 33:341–346.

D'Souza, Y.D., Fombonne, E. and Ward, B.J. No evidence of persisting measles virus in peripheral blood mononuclear cells from children with autism spectrum disorder. Pediatrics 2006; 118: 1664–1675.

Duclso, P. A global perspective on vaccine safety. Vaccine 2004; 22:2059–2063.

Eastern and Southern Africa Regional Office, United Nations Childrens Fund. Combating antivaccination rumours: lessons learned from case studies in East Africa. 2001. UNICEF, Nairobi, Kenya. http://www.path.org/vaccineresources/files/Combatting_Atnivac_Rumors_UNICEF.pdf. Accessed December 4, 2007.

Enzi, M.B. Thimerosal and autism spectrum disorders: alleged misconduct by government agencies and private entities. Report of the Ranking Member, United States Senate Committee on Health, Education, Labor and Pensions, September 28, 2007.

Fine, P.E. Community immunity [Chapter 56]. In: Vaccines (S.A. Plotkin and W.A. Orenstein, Eds.), 4th ed. Philadelphia, PA: W.B. Saunders Company, 2004.

Flanagan-Klygis, E.A., Sharp, L. and Frader, J. Dismissing the family who refuses vaccines. A study of pediatrician attitudes. Arch. Pediatr. Adolesc. Med. 2005; 159:929–934.

Fox, S. Digital divisions: there are clear differences among those with broadband connections, dial-up connections, and no connections at all to the internet. Pew Internet & American Life Project, October 5, 2005. Washington, DC. http://www.pewinternet.org/report_display.asp?r=165. Accessed November 26, 2007.

Fox, S. Online Health Search 2006: Most internet users start at a search engine when looking for health information online. Very few check the source and date of the information they find. Pew Internet & American Life Project, October 29, 2006. Washington, DC. http://www.pewinternet.org/PPF/r/190/report_display.asp. Accessed November 26, 2007.

Fredrickson, D.D., Davis, T.C., Arnold, C.L. et al. Childhood immunization refusal: provider and parent perceptions. Fam. Med. 2004; 36(6):431–439.

Freed, G.L., Clark, S.J., Hibbs, B.F. et al. Parental vaccine safety concerns. The experiences of pediatricians and family physicians. Am. J. Prev. Med. 2004; 26(1):11–14.

Gangarosa, E.J., Galazka, A.M., Wolfe, C.R., Phillips, L.M., Gangarosa, R.E., Miller, E. and Chen, R.T. Impact of anti-vaccine movements on pertussis control: the untold story. Lancet 1998; 351(9099):356–361.

Geier, D.A. and Geier, M.R. Neurodevelopmental disorders following thimerosal-containing childhood immunizations: a follow-up analysis. Int. J. Toxicol. 2004; 23:369–376.

Geier, M.R. and Geier, D.A. Thimerosal in childhood vaccines, neurodevelopment disorders and heart disease in the United States. J. Am. Physicians Surg. 2003; 8:6–11.

Gellin, B.G., Maibach, E.W. and Marcuse, E.K. Do parents understand immunizations? A national telephone survey. Pediatrics 2000; 106:1097–1102.

Gillen, M. Autism: to vaccinate or not to vaccinate. November 29, 2007. http://cbs4.com/iteam/vaccine.autism.kids.2.598341.html. Accessed November 29, 2007.

Gust, D.A., Kennedy, A., Shui, I. et al. Parent attitudes toward immunizations and healthcare providers. The role of information. Am. J. Prev. Med. 2005; 29(2):105–112.

Gust, D.A., Strine, T.W., Maurice, E. et al. Underimmunization among children: effects of vaccine safety concerns on immunization status. Pediatrics 2004; 114(1):e16–e22.

Hastings Weld, H. Benjamin Franklin: His Autobiography; With A Narrative of His Public Life and Services. Derby & Jackson, 1859.

Horton, R. A statement by the editors of The Lancet. Lancet 2004; 363(9411):820–821.

Immunization Branch, Texas Department of State Health Services. Conscientious exemption: results of interim survey to determine number of conscientious exemption affidavits submitted to schools for school-year 2003–2004. http://www.dshs.state.tx.us/immunize/coverage/conscientious.shtm. Accessed December 6, 2007.

Institute of Medicine Immunization Safety Review: Vaccines and Autism. Washington, DC: National Academies Press, 2004.

Investigative Reporters and Editors, Inc. Toxic dumping, public corruption investigations among winners of 2005 IRE Awards. March 28, 2006. http://www.ire.org/history/pr/2005ireawards.html.

Iowa Administrative Code. Immunization and immunization education: persons attending elementary or secondary schools, licensed child care centers or institutions of higher education. 641-7.10 (139), 1977, as amended.

Iowa Department of Health. Reportable diseases. www.idph.state.ia.us/adper/pdf/cade/decades/pdf. Accessed November 15, 2007.

Jansen, V.A.A., Stollenwerk, N., Jensen, H.J. et al. Measles outbreaks in a population with declining vaccine uptake. Science 2003; 301:804.

Keelan, J., Pavri-Garcia, V., Tomlinson, G. et al. YouTube as a source of information on immunization: a content analysis. JAMA 2007; 298(21):2482–2484.

Leask, J.-A. and Chapman, S. An attempt to swindle nature: press anti-immunisation reportage 1993–1997. Aust. N.Z. J. Pub. Health 1998; 22(1):17–26.

Lernout, T., Kissling, E., Hutse, V., et al. Clusters of measles in Jewish orthodox communities in Antwerp, epidemiologically linked to the United Kingdom. A preliminary report. Euro. Surveill. 2007; 12(11). www.eurosurveillance.org/ew/2007/071115.asp#0.

Meszaros, J.R., Asch, D.A., Baron, J. et al. Cognitive processes and the decisions of some parents to forego pertussis vaccination for their children. J. Clin. Epidemiol. 1996; 49(6):697–703.

Mills, E., Jadad, A.R., Ross, C. et al. Systematic review of qualitative studies exploring parental beliefs and attitudes toward childhood vaccination identifies common barriers to vaccination. J. Clin. Epidemiol. 2005; 58:1081–1088.

Murch, S.H., Anthony, A., Cassen, D.H. et al. Retraction of an interpretation. Lancet 2004; 363:750.

Myers, M.G. and Pineda, D.I. Widely publicized misinformation about vaccine safety causes searches for information. Pediatr. Acad. Soc. Abstract 6293.25, 2007.

Myers, M.G. and Pineda, D.I. Do Vaccines Cause That? A Guide for Evaluating Vaccine Safety Concerns. Galveston, TX: I4PH Press, 2008.

Office of Special Masters. Autism update—September 28, 2007. US Court of Federal Claims, Washington, DC, 2007.

Olmstead, D. The age of autism: a glimpse of the Amish UPI: June 2, 2005. http://www.upi.com/NewsTrack/Science/2005/06/02/the_age_of_autism_a_glimpse_of_the_amish/5801/. Accessed November 26, 2007.

Parker, A.A., Staggs, W., Dayan, G.H. et al. Implications of a 2005 measles outbreak in Indiana for sustained elimination of measles in the United States. N. Engl. J. Med. 2006; 355(5): 447–455.

Parkman, P. The Harry Meyer Jr. MD Lecture. CBER Centennial. 2002. http://www.fda.gov/cber/summaries/cent092302pp.htm. Accessed November 21, 2007.

Parmet, W.E., Goodman, R.A. and Farber, A. Individual rights versus the public's health—100 years after Jacoson v. Masachusetts. N. Engl. J. Med. 2005; 352(7):652–654.

Paulos, J.A. Innumeracy: Mathematical Illiteracy and its Consequences. New York, NY: Hill and Wang, 1988.

Paulussen, T.G.W., Hoekstra, F., Lanting, C.I. et al. Determinants of Dutch parents' decisions to vaccinate their child. Vaccine 2006; 24:644–651.

Poland, G.A. and Jacobson, R.M. Understanding those who do not understand: a brief review of the anti-vaccine movement. Vaccine 2001; 19:2440–2445.

Pribble, J.M., Goldstein, K.M., Fowler, E.F., Greenberg, M.J., Noel, S.K. and Howell, J.D. Medical news for the public to use? What's on local TV news. Am. J. Manag. Care 2006; 12:170–176.

Regional Office for South Asia, United Nations Childrens Fund Building Trust and Responding to Adverse Events Following Immunization in South Asia: Using Strategic Communication. Kathmandu, Nepal: UNICEF, 2005.

Reynolds, B. Crisis and Emergency Risk Communication. Atlanta, GA: Centers for Disease Control and Prevention, 2002.

Richard, J.L. and Masserey Spicker, V. Ongoing measles outbreak in Switzerland: results from November 2006 to July 2007. Euro. Surveill. 2007; 12(7). www.eurosurveillance.org/ew/2007/070726.asp#1.

Salmon, D.A., Haber, M., Gangarosa, E.J. et al. Health consequences of religious and philosophical exemptions from immunization laws: individual and societal risk of measles. JAMA 1999; 282:47–53.

Salmon, D.A., Moulton, L.H., Omer, S.B. et al. Knowledge, attitudes and beliefs of school nurses and personnel and associations with nonmedical immunization exemptions. Pediatrics 2004; 113(6):e552–e559.

Simonsen, L., Kane, A., Lloyd, J. et al. Unsafe injections in the developing world and transmission of bloodborne pathogens: a review. Bull. WHO 1999; 77:789–800.

Smith, P.J., Chu, S.Y. and Barker, L.E. Children who have received no vaccines: who are they and where do they live? Pediatrics 2004; 114(1):187–195.

Snyder B. Vaccine safety morality hit home for girl's parents. The Tennesean, September 17, 2000.

Steart-Freedman, B. and Kovalsky, N. An ongoing outbreak of measles linked to the United Kingdom in an ultra-orthodox Jewish community in Israel. Euro. Surveill. 2007; 12(9). www.eurosurveillance.org/ew/2007/070920.asp#1.

The New York Times. November 14, 2005. http://www.generationrescue.org/images/051114.gif.

Thompson, W.W., Price, C., Goodson, B. et al. Early thimerosal exposure and neuropsychological outcomes at 7 to 10 years. N. Engl. J. Med. 2007; 357(13):1281–1292.

Time Magazine. October 15, 2007.

U.S. Court of Federal Claims. Office of Special Masters. October 9, 2003.

U.S. Fund for UNICEF. The campaign to save mothers and babies from tetanus. MNT Monitor. Fall, 2004. www.who.int/entity/immunization_monitoring/diseases/06_fall_2004.pdf.

Vaccination may trigger disease linked to autism. Celia Hall, Medical Editor. Daily Telegraph, Friday, February 27, 1998, Issue 1008. http://www.telegraph.co.uk/htmlContent.jhtml?html=/archive/1998/02/27/nmmr27.html.

van Treeck, U. Measles outbreak in Germany: over 1000 cases now reported in Nordhein Westfalen. Euro. Surveill. 2006; 11(5). www.eurosurveillance.org/ew/2006/060511.asp#1.

Varricchio, F., Iskander, J., Destefano, F. et al. Understanding vaccine safety information from the vaccine adverse event reporting system. Pediatr. Infect. Dis. J. 2004; 23(4):287–294.

Wakefield, A.J., Murch, S.H., Anthony, A. et al. Ileal-lymphoid-nodular hyperplasia, non-specific colitis, and pervasive developmental disorder in children. Lancet 1998; 351:637–641.

Wolfe, R.M., Sharp, L.K. and Lipsky, M.S. Content and design attributes of antivaccination web sites. JAMA 2002; 287: 3245–3248.

WHO. Country reported immunization coverage. 2007a. http://www.who.int/vaccines/globalsummary/immunization/timeseries/tscoveragebycountry.cfm? Accessed November 21, 2007.

WHO. United Kingdom reported cases. 2007b. http://www.who.int/vaccines/globalsummary/immunization/timeseries/tsincidencebycountry.cfm?country=United%20Kingdom. Accessed November 21, 2007.

CHAPTER

18

Vaccines for Military Application

Andrew W. Artenstein[1,2]

[1]Department of Medicine and Center for Biodefense and Emerging Pathogens, Memorial Hospital of Rhode Island, Pawtucket, RI, USA

[2]The Warren Alpert Medical School of Brown University, Providence, RI, USA

OUTLINE

ABSTRACT

Vaccines have long been used by military forces in order to prevent communicable diseases and thereby fulfill their primary mission: to preserve the fighting force. A tradition that began with the mass vaccination of the Continental Army against smallpox during the War of the American Revolution in the late 18th century continues today with routine and deployment-based vaccination of military forces against potential pathogens of nature and biological weapon threats. As the role of militaries has expanded in recent years to include humanitarian and peacekeeping missions, military applications of vaccines against infectious diseases have concomitantly broadened to include civilian populations worldwide. In many cases involving infectious diseases of military significance, the military has either directly, through the work of force researchers, or indirectly, through investments and resource provision, promoted the development of vaccines. This pattern is expected to persist as new threats are recognized and military forces accept additional challenges.

INTRODUCTION

Military imperatives necessitate force readiness; disease prevention is a critical component of this doctrine. While the ratio of disease-associated deaths to battlefield-associated deaths in U.S. military deployments has progressively decreased over the past century, disease still accounts for the majority of morbidity in military venues (Murray and Horvath, 2007). It is widely recognized that for many diseases of potential military significance, prevention is best accomplished through the deployment of vaccines. Vaccines, considered by many to top the historical list of the ten greatest public health achievements of the 20th century (Centers for Disease Control and Prevention, 1999), have therefore been a cornerstone of military preventive medicine since before the first systematic study of their utility by Edward Jenner in the late 18th century.

Throughout history, the military has been a major force in both the development and the use of vaccines against infectious diseases (Artenstein et al., 2005). The U.S. military, for example, has contributed directly and significantly to numerous vaccines throughout modern history (Table 18.1). In some instances military exigencies have stimulated vaccine research on infectious diseases of mission-specific importance; in others, the military has functioned as a high-volume consumer of routine immunizations.

Aside from their traditional involvement in combat operations, modern military forces have additional roles in global geopolitics including duties as peacekeepers, police, humanitarian relief workers, and disaster assistance personnel (Ryals and Baker, 1996; Baker and Ryals, 1999). These diverse roles beyond their traditional combat duties potentially expose military personnel to a wide array of environments over a broad range of geographic settings throughout the world. Such duties may alter the primary mission from force preservation to the protection and treatment of civilians, notably women and children (Sharp et al., 2001). This evolving role of global militaries since the latter part of the 20th century potentially predisposes forces to an expanded array of pathogens. Vaccines in these settings are used to prevent routine childhood diseases among troops and civilians and for health maintenance of civilians at risk. Military personnel are also at high risk to encounter such infectious diseases during recruit training, a setting in which numerous young adults reside in confined living conditions for extended periods of time and are subjected to intense physical, emotional, and immunologic stressors, increasing their risk of communicable diseases, especially those transmitted by respiratory routes (Benenson, 1984).

Some of the pathogens that threaten military forces cause common infectious diseases that are prevalent in developing nations lacking widespread routine childhood vaccination programs; others are endemic to only select regions of the world and tend to cause sporadic infections or outbreaks in deployed military personnel; while others may cause outbreaks of common diseases when occurring in enclosed populations. To guard against the plethora of infectious diseases that may assail a military in any of its many expanded modern functions, and to fulfill their supplementary roles as humanitarian relief workers in civilian emergencies, a variety of vaccines is widely deployed. Table 18.2 provides a general listing of the current vaccines, not differentiated based on individual service branches, used by the U.S. military under various conditions. As extensive coverage of the individual pathogens and their associated vaccines is provided in other chapters of this text, the present chapter summarizes vaccines as applied to a variety of transmissible diseases of military importance.

HISTORY REDUX: SMALLPOX

For historical reasons smallpox is an appropriate departure point for a discussion of vaccine applications in the military as this disease led to the precedent-setting mass deployment of immunization in troops more than 20 years before the term "vaccination" was first proposed. Smallpox has the dubious distinction of being the greatest historical disease scourge of mankind. Although it was commonly observed from ancient times through the 18th century that survivors of smallpox were protected against further episodes of the disease (Moloo and Artenstein, in press), the history of vaccination from a scientific standpoint is traditionally dated from the 1798 publication of Edward Jenner's landmark experiments with cowpox in which he inoculated a neighbor's boy with purulent material from a milkmaid's hand lesion in Berkeley, England, thus protecting the child against a smallpox challenge (Parish, 1965). However, long before Jenner's human experiments other forms of inoculation were used throughout the world to protect against smallpox.

Variolation, derived from the Latin word "variola" meaning "mark on the skin" (Parish, 1965), describes the technique in which material from the pustules of smallpox-afflicted persons is inoculated into the arm of a healthy person through multiple scratches or

TABLE 18.1 U.S. military contributions to vaccines against infectious diseases (as defined in text)

Disease/agent	Dates	Military relevance	Military contribution	Key personnel
Smallpox	1777	Highly contagious scourge of troops	First large scale inoculation of an army	Gen. George Washington, Dr. Benjamin Rush
Yellow fever	1900	Epidemic disease associated with high mortality in American-occupied areas	Demonstrated that the etiologic agent was a filterable virus transmitted by *A. aegypti* mosquitoes, leading to disease control through vector eradication and eventually, through vaccination	Maj. Walter Reed, Maj. James Carroll, Aristide Agramonte, Jesse Lazaer, Col. William Gorgas
Typhoid	1909	Communicable cause of outbreaks among deployed troops	Development of killed typhoid vaccine that became mandatory for all Army and Navy personnel in 1911 and greatly reduced the morbidity and mortality of typhoid among military personnel	Maj. Frederick F. Russell
Pneumococcus	1945	Prominent respiratory pathogen among troops	Tested first multivalent polysaccharide vaccine at the Army Air Base, Sioux Falls, S.D. (under auspices of the Armed Forces Epidemiological Board), reduced incidence of pneumonia and the pneumococcal carrier state	Dr. Colin MacLeod, Dr. Michael Heidelberger, Lt. Richard Hodges
Hepatitis A	1945; 1985–1990	Cause of major outbreaks among troops in setting of military conflict	1945, demonstrated that passive immunization with pooled normal human immunoglobulin could prevent or attenuate disease; 1980s–1990, investigated safety and immunogenicity of inactivated vaccines and directed a pivotal efficacy study of inactivated HAV vaccine in ~40,000 Thai children	Col. Charles Hoke, Lt. Col. Bruce Innis, Dr. Len Binn, Dr. Stanley Lemon
Adenovirus	1952–1969	Major cause of acute respiratory disease in recruits, disruptions in training, economic losses	Isolated causative agent, later named adenovirus, at Ft. Leonard Wood; described epidemiology and clinical spectrum of adenovirus infections; developed killed bivalent vaccine; recognition of SV40 contamination; developed oral, attenuated, multivalent vaccine	Dr. Maurice Hilleman, Col. Edward Buescher, Maj. Franklin Top, Col. Phillip Russell
Influenza	1957	Epidemic and pandemic disease impede military readiness and fighting force	Described antigenic drift and shift; developed surveillance system for epidemic disease	Dr. Maurice Hilleman
Rubella	1961	Consequences of disease in pregnancy affects military families	Isolated the causative virus after noting interference with enteroviral growth in African green monkey kidney cell cultures; led to development of a safe and effective vaccine using this virologic technique	Cpt. Paul Parkman, Cpt. Malcolm Artenstein, Lt. Col. Edward Buescher
Meningococcal disease	1966–1972	Epidemic meningitis in basic training settings causes significant morbidity and mortality among troops	Described immunologic responses to the bacteria and identified protective responses; developed first polysaccharide immunogen and proved its efficacy in clinical trials	Dr. Malcolm Artenstein, Dr. Irving Goldschneider, Dr. Emil Gotschlich, Maj. Ronald Gold
Japanese encephalitis	1950s; 1980s	Epidemic neurologic disease in Asia; outbreaks in American forces in Korea; potential for impacting deployed troops throughout Asia	Early attempts at vaccination (WWII); significant contributions to the epidemiology and ecology of the virus in the 1950s–1960s; pivotal field trial of inactivated vaccine in Thailand	Dr. Joseph Smadel, Dr. Albert Sabin, Cpt. Edward Buescher, Cpt. William Scherer, Col. Charles Hoke
Hepatitis B	1970s–early 1980s	Blood borne and sexually transmitted disease	Demonstrated protective effect of antibodies	Dr. Saul Krugman, Col. William Bancroft, Dr. Maurice Hilleman

Source: Artenstein et al. (2005).

punctures. The method was introduced into early 18th century Europe via travelers returning from Istanbul and brought to England in 1721 by Lady Mary Wortley Montague, the wife of the British ambassador to the Ottoman court (Barquet and Domingo, 1997). After her daughter was inoculated under the public scrutiny of physicians of the royal court, variolation caught the interest of the royal family, thus gaining

TABLE 18.2 Vaccines for military application under various circumstances

Routine use in recruits	
Tetanus	
Diphtheria	
Pertussis	
Measles	
Mumps	
Rubella	
Influenza	
Meningococcus	
Polio	
Varicella	
Hepatitis A	
Hepatitis B	
Adenovirus	
Deployment to high-risk areas; alert forces	
Anthrax	
Smallpox	
Typhoid	
Yellow fever	
Japanese encephalitis	
Rabies	
Tick-borne encephalitis	
Deployment-based vaccines in development	
Malaria	
Dengue	
Viral encephalitides	
HIV-1	
Plague	
Tularemia	
Viral hemorrhagic fevers	

widespread acceptance in England; the practice ultimately disseminated throughout Europe by the 1740s (Barquet and Domingo, 1997).

Despite an observed mortality rate of 2–3% related to variolation, the procedure still offered better odds than the 15–30% mortality from naturally acquired smallpox. In 1721, Reverend Cotton Mather successfully advocated for its use to abort a smallpox epidemic in Boston (Moloo and Artenstein, in press). Variolation subsequently gained the support of American thought leaders such as Benjamin Franklin and became widespread in the colonies. In 1777, with his army decimated by smallpox and inoculated British troops holding a significant tactical advantage, General George Washington, in consultation with one of colonial America's preeminent physicians and statesman, Dr. Benjamin Rush of Philadelphia, adopted the bold and unprecedented plan to inoculate all susceptible members of the Continental Army, a strategy that is believed to have tipped the balance in favor of the Americans (Artenstein et al., 2005) and to have represented the first large-scale inoculation of a military force.

Jenner published *An Inquiry into the Causes and Effects of the Variolae Vaccinae, a disease discovered in some of the western counties of England, particularly Gloucestershire and Known by the Name of Cow Pox* at his own expense in 1798. In it he coined the term "vaccination" (from the Latin word vacca, for cow) to describe the procedure and distinguish it from variolation (Barquet and Domingo, 1997). The work, though initially met with considerable skepticism, was subsequently confirmed by others and led to the dissemination of vaccination against smallpox in the Western world by the early part of the 19th century (Moloo and Artenstein, in press). The U.S. War Department ordered mandatory vaccination with Jenner's cowpox product, in lieu of variolation, beginning in 1812 (Grabenstein et al., 2006). Smallpox vaccination with vaccinia products distantly related to Jenner's continued to be routinely administered to new recruits until early 1990, nearly 20 years after the cessation of routine civilian use of the vaccine, when it had become clear that the risk outweighed any potential benefits (Redfield et al., 1987).

Smallpox was eradicated as a cause of natural human disease after an intensive global campaign by the World Health Organization and sponsoring countries; the last naturally acquired case occurred in 1977. However, in December 2002 after more than a 12-year hiatus, the U.S. government reinitiated large-scale military vaccination using live vaccinia in response to the perceived threat of bioterrorism involving smallpox. The military program vaccinated over 730,000 individuals as of early 2005; their data reveal that most adverse events occurred at rates below historically anticipated levels (Poland et al., 2005). The relative dearth of "expected" vaccine complications in the recent U.S. vaccination programs was perhaps due to a combination of factors including more rigorous screening for vaccine contraindications than during the previous era of routine vaccine use.

Despite the low incidence of "expected" serious toxicities associated with smallpox vaccination in the military program, a number of unexpected, serious cardiac complications of the use of live vaccinia were noted. Cardiac complications of smallpox vaccines had been reported, albeit infrequently, during the previous era of routine use. Through the 1960s and 1970s, data from Finland and Australia demonstrated rates of myopericarditis as high as 1 per 10,000 vaccinees (Karjalainen et al., 1983); up to 3% of Swedish recruits in the 1960s were found to have nonspecific, asymptomatic T-wave changes on electrocardiogram following smallpox vaccination (Ahlborg et al., 1969).

The major cardiac complications of smallpox vaccination observed in the recent U.S. programs were of two varieties: cardiac ischemic events or myopericarditis (Poland et al., 2005). While a causal relationship of ischemic cardiac events to traditional vaccinia virus vaccination has not been clearly established, their temporal association resulted in a series of government recommendations regarding cardiac screening, new vaccine contraindications, and surveillance for cardiac events that may impact future clinical studies involving newer smallpox vaccine approaches. Additionally, the military program identified 86 cases of myopericarditis during its first two years (Poland et al., 2005), a rate of approximately 1.2 per 10,000, similar to rates observed in Finnish conscripts described 24 years ago (Karjalainen et al., 1983). These rates were clearly higher than age-matched, nonvaccinated individuals, and since cases presented in close temporal association to vaccination, the appropriate conclusion was that smallpox vaccination using live vaccinia virus is associated with an increased risk of myopericarditis.

Currently, smallpox vaccine is given to select U.S. military units that will potentially deploy to areas considered to be at "high risk" due to the threat of biological weapons exposures (Grabenstein et al., 2006). Other developed nations, such as Canada and Great Britain, have not adopted policies of vaccinating large groups of military personnel against smallpox.

VACCINES USED ROUTINELY AND IN RECRUIT TRAINING SETTINGS

Routine childhood immunizations, including those against tetanus, diphtheria, pertussis, measles, mumps, rubella, varicella, and polio are consistently applied to service members early in recruit training to ensure complete penetrance of immune protection against these preventable pathogens. As most young adults have received basic immunizing doses in childhood, these vaccines are generally given either as booster doses to sustain immunity or are given based on the results of serologic screening; subsequent doses are given at intervals as recommended by the Advisory Committee on Immunization Practices (ACIP) (Grabenstein et al., 2006).

Tetanus

Wound-associated tetanus, an enormous disease burden and major cause of battlefield-related mortality throughout history, was effectively controlled by passive immunization of the wounded during World War I and virtually eliminated as a significant risk with the advent of active immunization using formalin-inactivated toxin, tetanus toxoid, in the 1930s (Benenson, 1984). Only 16 cases of tetanus were reported among all U.S. forces serving in World War II following the first large-scale deployment of tetanus toxoid, and most of these occurred in incompletely immunized individuals (Benenson, 1984). Recently, the tetanus, diphtheria, and acellular pertussis combined vaccine product has been widely adopted by the military for use in primary immunization series or for booster doses for service members in recruit settings (Grabenstein et al., 2006). The military deployment of tetanus, diphtheria, and pertussis vaccines may be most beneficial in humanitarian operations in developing countries where routine vaccination is either unavailable or inconsistently applied.

Measles, Mumps, Rubella, Varicella

Classic viral respiratory diseases of childhood, measles, mumps, rubella, and varicella, represented major causes of nonbattlefield morbidity throughout military history (Ottolini and Burnett, 2005). During World War II alone, these infectious diseases accounted for hundreds of thousands of hospital admissions and training days lost to illness among U.S. armed forces (Grabenstein et al., 2006). The practical and economic impact of these viral respiratory diseases in military settings was noted as recently as the 1970s (Meiklejohn, 1983).

Live-virus vaccines against measles (Katz et al., 1960), mumps (Hilleman et al., 1968), and rubella (Meyer and Parkman, 1971) were developed in the 1960s and licensed by the end of the decade. Rubella vaccine, developed shortly after the isolation of the causative agent by military researchers (Fig. 18.1) (Parkman et al., 1962), and measles vaccine were relatively rapidly adopted for routine use in military trainees; mumps vaccine was not uniformly used until 1991, based on the infrequent occurrence of outbreaks of this disease in recruit settings (Grabenstein et al., 2006). Because recent serosurveys have documented that more than 85% of basic trainees are immune to these viruses by virtue of childhood immunization, serologic testing and vaccination of only the nonimmune has become standard (Clardy, 1993; Smoak et al., 1994; Grabenstein et al., 2006). These products, as with other routine childhood vaccines, may be most useful in civilian populations during military humanitarian duties. Varicella vaccine, developed in the 1970s by Takahashi but not licensed in the U.S. until 1995 (Moloo and Artenstein, in press), is also currently used as indicated by the results of individual serologic screening (Burnham et al., 1998).

FIGURE 18.1 Edward Buescher, Paul Parkman, Malcolm Artenstein. (*Source*: Courtesy of Andrew W. Artenstein, MD.) Used with permission.

Hepatitis A

Epidemic jaundice, now known to be due to the hepatitis A virus, has long accompanied armies at war, from Napoleon's campaigns in Russia and North Africa to the U.S. Civil War to World War II and the Korean conflict (Dooley, 2005). Hepatitis A outbreaks among combat forces in North Africa and Italy during World War II accounted for more days of training and force strength lost than any other cause of disease and may have delayed the D-day landing by one month (Dooley, 2005). While investigations clearly showed the infection to be directly related to poor sanitation and hygiene, risk was inherent to military operations because of the unsanitary nature of combat theaters and associated exposures.

Initial forays into the prevention of hepatitis A involved passive immunization with pooled gamma globulin (Dooley, 2005). The short-term efficacy of such an approach was shown by Stokes, Gellis and others in a series of clinical trials, some of which were performed in military populations (Gellis et al., 1945; Stokes and Neefe, 1945). Passive immunization remained the main mode of protection against hepatitis A until viral isolation and propagation in nonhuman primate cell lines in the mid-1980s led to a prototype, formalin-inactivated vaccine (Dooley, 2005). Innis et al. (1994) performed a pivotal clinical trial of hepatitis A vaccine in the early 1990s involving more than 40,000 school children in rural Thailand and demonstrated vaccine safety with efficacy of 94%. Inactivated vaccines were licensed in 1995 and were immediately adopted as a component of the vaccine regimen for recruits and other military personnel (Grabenstein et al., 2006).

Hepatitis B

Hepatitis B, "serum" hepatitis, gained initial notoriety in a military setting as the etiology of a 1942 epidemic among U.S. soldiers who received contaminated lots of yellow fever vaccine manufactured with tainted human albumin (Dooley, 2005). More than 300,000 individuals were likely infected; 50,000 developed jaundice (Norman et al., 1993). First-generation, plasma-derived, hepatitis B subunit vaccines were licensed in 1981. The recombinant subunit hepatitis B vaccine supplanted the plasma-derived product in 1986 and has been applied to military cohorts based upon occupational (e.g., medical personnel) or geographic (e.g., Korean assignment) risk. As hepatitis B vaccination has become routine in childhood, penetrance of immunity is becoming widespread among service members.

Influenza

Influenza, a highly transmissible acute respiratory virus infection, tends to amplify in certain settings, especially crowded, enclosed environments.

In military training camps and aboard naval troop vessels (Earhart et al., 2001), environments in which young adults of varying susceptibility backgrounds and exposure histories are intermixing under cramped conditions, the opportunities for epidemic spread are pronounced. The largest influenza pandemic in recorded history, that of 1918, is believed to have ignited in U.S. training camps and to have disseminated via movement of global military forces in the context of World War I (Taubenberger and Morens, 2006). During the 1918 pandemic, it is estimated that between 50 and 100 million people perished worldwide (Taubenberger and Morens, 2006), including 1.5% of the entire American military force (Grabenstein et al., 2006).

Excessive morbidity and mortality related to seasonal epidemics of influenza and the threat of recurrent pandemic disease have significant impact on military forces and their state of readiness. The Armed Forces Epidemiological Board (AFEB), an advisory group of civilian physicians and scientists, was established in 1941, initially as a response to the influenza threat; the AFEB charter was subsequently expanded to respond to a variety of infectious disease threats (Woodward, 1990). Influenza vaccination among select U.S. military forces began during World War II and became widespread and routinely used by the early 1950s (Grabenstein et al., 2006). Annual immunization is currently deployed to protect against seasonal influenza outbreaks in military settings.

FIGURE 18.2 Maurice Hilleman. (*Source*: Courtesy of the Hilleman family. Used with permission.)

Adenovirus

Acute respiratory disease (ARD) caused by a variety of pathogens has consistently been a leading cause of morbidity among military recruits probably as a result of the crowded living conditions and physical stressors inherent to training camps (Gray et al., 1999). In 1954, Hilleman (Fig. 18.2) and others isolated the virus responsible for the majority of ARD and primary atypical pneumonia in noncombat military settings (Hilleman and Werner, 1954); the agent was later named "adenovirus" based upon the isolation of biologically related organisms from the tonsils and adenoids of children (Enders et al., 1956).

Adenoviruses were noted to have an enormous impact on troop health in recruit environments: attack rates as high as 80% existed in some camps with up to 20% requiring hospitalization; these numbers translated into hundreds of weekly admissions at some of the larger recruiting centers in the U.S. in the mid-20th century (Dudding et al., 1973). Because of the impact on troops and their training, medical resources, and overall economic costs to the military, vaccine development was a high priority. Once it became clear that two serotypes of adenovirus, types 4 and 7, accounted for the majority of ARD in recruits, a formalin-inactivated bivalent vaccine for adenovirus was developed within three years of the initial description of the infectious agent (Hilleman, 1999). Pivotal, large, clinical efficacy trials were performed at multiple large recruit centers; the vaccine proved to be greater than 90% effective and was licensed in 1958 (Hilleman, 1999; Artenstein et al., 2005). Manufacturing problems, however, were identified early and included partial loss of antigenicity in large-scale production and seed lot contamination by Simian virus 40, an oncogenic virus that also contaminated oral and inactivated poliovirus vaccines during the same time period; the license for the parenteral product was revoked in 1963 (Gaydos and Gaydos, 2004).

Second generation, live adenovirus vaccines targeting types 4 and 7 virus, given as enteric-coated capsules, were developed in the mid-1960s under the leadership of Colonel Edward Buescher (Fig. 18.1) and extensively studied at military facilities (Pierce et al., 1968). Through induction of an asymptomatic gastrointestinal infection that is unassociated with the transmission of vaccine virus to close contacts, the vaccines stimulate serum neutralizing antibody in more than 95% of recipients (Top et al., 1971a). The products, either given singly or in combination, resulted in a 50% decrease in hospitalization rates due to ARD in recruits and a greater than 95% reduction in illness due to type 7 adenovirus (Top et al., 1971b). Beginning in 1971 both vaccines were routinely administered to military recruits within hours of their arrival at basic training. A cost-benefit analysis of the U.S. military's adenovirus vaccination program demonstrated a favorable ratio, even accounting for the one-time costs related to vaccine development (Collis et al., 1973).

As a result of adenovirus vaccine use during the period from 1971 to 1996, there were essentially no outbreaks of ARD reported in U.S. military units related to either of the adenovirus serotypes used in the vaccines. However, the sole manufacturer of the vaccines ceased production in 1996; thus a program of reduced vaccine use was instituted prior to the depletion of stocks in 1999 (Gaydos and Gaydos, 2004). The unavailability of adenovirus vaccines has resulted in a recrudescence of ARD due to these agents. Recent surveillance data in sentinel military populations reveal that 55% of throat cultures obtained from recruits with ARD from 1996 to 1998 yielded adenoviruses, primarily types 4 and 7 (Gray et al., 1999). Unvaccinated personnel with ARD were significantly more likely to be infected with adenovirus than their vaccinated counterparts. Outbreaks of adenovirus infection causing ARD have been reported from a variety of military centers since the cessation of routine vaccination and have generally resulted from a large pool of susceptible hosts; nearly 90% of new recruits were seronegative to either type 4 or 7 adenovirus in one study (Gray et al., 1999). Recognizing the benefit of adenovirus vaccine, the U.S. government has taken steps to reinstitute the program by contracting with industry to develop and produce the product for the future.

Meningococcal Disease

Epidemic meningococcal meningitis has been a well-described complication of troop mobilizations in training camps and combat theaters since the latter part of the 19th century. The U.S. forces experienced outbreaks during World War I, prompting the formation of the Armed Forces Epidemiological Board Commission on Meningococcal Meningitis in 1941 to accelerate scientific study and control of the disease (Woodward, 1990). From 1941 to 1943, the Commission documented over 5,000 domestic cases among U.S. Army personnel (Woodward, 1990). The epidemiology in this population clearly showed that the incidence of meningococcal disease was inversely associated with duration of military service; the greatest risk for recruits was during their first three months of military service (Artenstein et al., 2005).

Although sulfa drugs, discovered in 1937, were effective as therapy and prophylaxis for meningococcal disease, reports of sulfa-resistant organisms began to emerge by 1945. By the early 1960s, the majority of clinical isolates of meningococci was resistant to sulfa agents (Artenstein and Gold, 1970), findings confirmed by epidemics of meningococcal disease at multiple military facilities during that time frame (Alexander et al., 1968). The rapid emergence of sulfa-resistance and later rifampin-resistance led the field to focus efforts on vaccine development in the belief that a meningococcal vaccine represented the best hope for long-term control of the problem in military populations.

In 1969, Dr. Malcolm Artenstein (Fig. 18.1), the Chief of the Department of Bacteriology at the Walter Reed Army Institute of Research (WRAIR) and his associates, Emil Gotschlich and Irving Goldschneider, published a series of five classic papers that described in detail the human immunologic response to *Neisseria meningitidis* (Goldschneider et al., 1969a, 1969b; Gotschlich et al., 1969a, 1969b, 1969c). In this work they demonstrated the protective role of serum bactericidal antibody against systemic disease (Goldschneider et al., 1969a); identified the group-specific polysaccharide as the key antigenic determinant (Goldschneider et al., 1969b); and described novel methods for the preparation of high molecular weight group A and C polysaccharides as immunogens (Gotschlich et al., 1969a). In preclinical studies they proved the safety of such preparations in animals and proceeded to show immunogenicity in humans in a phase I clinical trial, with high concentrations of group-specific bactericidal antibody induced after a single, intradermal injection and significant reduction of the nasopharyngeal carriage rates of group C meningococcus (Gotschlich et al., 1969b, 1969c).

Armed with their basic science and preclinical findings, the WRAIR investigators initiated a large clinical trial of a group C meningococcal polysaccharide vaccine. Because there was no method for predicting the location of epidemics, the study involved multiple recruit training centers; and, because of the known rapid acquisition of the carrier state among newly arrived recruits, informed consent and vaccination had to be accomplished within days of arrival (Artenstein et al., 1970). In the pivotal efficacy study 13,763 men at 5 different military posts were vaccinated with a group C polysaccharide, and more than 53,000 served as unimmunized controls. The vaccine proved safe, reduced the acquisition rate of group C meningococcus (although not that of other meningococcal groups), and significantly reduced the rate of systemic meningococcal disease caused by group C organisms during an eight-week period (Artenstein et al., 1970). The results of a second, large field trial, with sample sizes similar to the first, confirmed the protective effect (Gold and Artenstein, 1971).

Based on the aforementioned data, the military began administering meningococcal C polysaccharide vaccine to all recruits in late 1970, resulting in the virtual elimination of meningococcal C disease as a major military health problem (Granoff et al., 2004). The current, tetravalent (groups A, C, Y, and W135) vaccine, licensed in 1981, is indicated for military recruits, asplenics, individuals with terminal complement deficiencies, travelers to areas where meningococcal disease is hyperendemic or epidemic, and subgroups of college students (Granoff et al., 2004). Military personnel deployed to meningococcal-endemic areas such as sub-Saharan Africa receive the vaccine. In the United Kingdom and several other countries, group C meningococcal polysaccharide conjugated to carrier proteins such as tetanus toxoid or diphtheria toxin derivatives has been used with remarkable success in reducing the burden of disease in children, a group in whom polysaccharides are traditionally poorly immunogenic (Granoff et al., 2004). The U.S. military is currently transitioning to the routine use of meningococcal conjugate vaccines for its forces (Grabenstein et al., 2006). The development of a vaccine targeting serogroup B meningococci, the cause of a significant minority of epidemic disease worldwide, remains elusive (Granoff et al., 2004).

VACCINES USEFUL FOR SPECIFIC DEPLOYMENTS AND ALERT FORCES

Selected vaccines have military applications based on deployments or potential deployments of troops to areas with extant vaccine-preventable endemic risks or in members of specialized military forces that are involved in alert settings, such as rapid response or special operations personnel. Similarly, the perceived and ubiquitous threat of biological weapons in military settings extends the repertoire of vaccine-preventable diseases that must be considered in military planning.

Yellow Fever

Yellow fever, currently endemic to sub-Saharan Africa and spotty areas in South America, caused epidemic disease along the eastern seaboard of the U.S. and Central America until the early part of the 20th century. "Yellow Jack" halted Napoleon's armies in New Orleans in 1801 and led to the French abandonment of the Panama Canal Zone in 1889 (Artenstein, et al., 2005). The disease was rampant in Cuba, the island acquired by the U.S. in the Spanish-American War, generating U.S. government–sponsored research into the epidemiology, microbiology, and prevention of yellow fever led by Major Walter Reed. Reed and his group of young army preventive medicine officers elucidated the epidemiology and transmission characteristics of the pathogen; their work directly led to the successful vector control efforts by Major William Gorgas both in Cuba and the Canal Zone, eventually eliminating yellow fever from the latter and allowing the completion of the Panama Canal.

The seminal investigations of Reed and colleagues led to the identification of the viral etiology of yellow fever and, in the 1930s, to the successful attenuation of the virus by serial passage in chick embryos by Max Theiler and colleagues at the Rockefeller Institute (Theiler and Smith, 1937). The resultant live, attenuated strain, 17D, was shown to induce long-term immunity in Rhesus macaques and human volunteers and continues to be used more than 60 years later to protect military forces deploying to yellow fever–endemic regions (Artenstein, et al., 2005; Grabenstein et al., 2006).

Japanese Encephalitis

Japanese encephalitis virus (JE), a mosquito-borne flavivirus that is endemic throughout the continent of Asia, represents a significant risk for deployed troops who are bitten by infected insects as an incidental branch of the viral life cycle (Hoke, 2005). More than 300 cases, 6% of them fatal, were identified in an outbreak among U.S. military personnel assigned to the Korean peninsula in 1950 (Lincoln and Siverson, 1952). Military research in the 1960s confirmed persistent risk in areas of active or planned operations, thus laying the theoretical and epidemiological foundations for vaccine development.

Although vaccine development by the U.S. military and Japanese public health authorities was undertaken to satisfy military exigencies, the ultimate product benefited mostly Asian children who were at the greatest risk from JE (Hoke, 2005). The pivotal vaccine study in the late 1980s enrolled approximately 66,000 subjects in rural Thailand and demonstrated 91% efficacy (Hoke et al., 1988), resulting in licensure by the U.S. Food and Drug Administration in 1992. The JE vaccine, now in transition to a cell culture–based production method, is given to military forces either deploying or on alert to regions with endemic disease.

Rabies

Military forces deployed worldwide are potentially at risk for bites from rabies-infected animals, and preventive approaches to manage the risks are required (Hoke, 2005). Preexposure immunization of forces may be useful in troops assigned to remote or forward areas with ongoing risks of animal exposures (Grabenstein et al., 2006).

Typhoid

Typhoid was a well-recognized scourge of military forces throughout modern history; U.S. troops training for entry into the Spanish-American War experienced multiple large outbreaks of disease as a result of unsanitary conditions at the camps (Grabenstein et al., 2006). Early generations of killed typhoid vaccines, developed by Sir Almroth Wright, were used with equivocal success to protect British troops deployed in the Boer War in 1899 (Benenson, 1984). Major Frederick Russell of the U.S. Army Medical School adapted the British vaccine for subcutaneous use and orchestrated field trials of the new product; mandatory vaccination of American military forces was instituted in 1911 and continued through the Vietnam era, and in concert with improvements in sanitation, significantly limited the impact of typhoid on deployed forces (Grabenstein et al., 2006). Late in the 20th century, whole-cell vaccines were supplanted by subunit and live attenuated typhoid vaccines, and these products are currently applied to military forces in high-risk deployment settings.

Tick-Borne Encephalitis

Virologically similar to JE, tick-borne encephalitis (TBE) is endemic to most areas of Europe and throughout much of Asia; most reported cases derive from Russia (Barrett, et al., 2004). TBE has the potential to be a significant disease threat in military deployments to endemic regions. During the U.S. military involvement in Bosnia-Herzegovinia in the mid-1990s it was estimated that as many as 1% of U.S. forces operating in the region may be at risk for infection, prompting the use of TBE vaccine in 4000 soldiers under an investigational new drug (IND) status (Craig et al., 1999).

Anthrax

Anthrax continues to occupy a position atop the category A agent bioterrorism threat list because of its potential for large-scale morbidity and mortality when used as an aerosolized weapon (Artenstein, 2007). "Proof of principle" of this agent's significance as both a military and civilian threat was amply demonstrated in two events: the unintentional release of weaponized spores in Sverdlovsk in 1979 and the intentional anthrax attacks in the U.S. in 2001 that killed 5, sickened 17 others, and terrorized millions of people around the world (Inglesby et al., 2002).

Due to its obvious military significance, anthrax has historically been mission-relevant for military medical research, resulting in many important scientific contributions deriving from this quarter (Christopher et al., 2005). The current anthrax vaccine, licensed in 1970, is a subunit product that employs protective antigen (PA), one of three protein exotoxins secreted by germinating bacteria in vivo, as the immunogen; vaccine-induced anti-PA antibodies are known to be highly protective in animal models of inhalational disease and human trials involving largely cutaneous anthrax (Inglesby et al., 2002). Approximately 150,000 American military forces were vaccinated during the Persian Gulf War in anticipation of the use of biological weapons by the Iraqi regime (Berezuk and McCarty, 1992; Grabenstein et al., 2006). Beginning in the late 1990s and accelerating since the events of 9/11, the U.S. military embarked on an expanded program of anthrax vaccination in certain deployable forces; to date over 1.5 million individuals felt to be at potential risk because of assignment location have received in excess of 5 million doses of anthrax vaccine in this program (Grabenstein et al., 2006). Numerous cohort studies and data reviews in the aftermath of the expanded use of anthrax vaccine by the military have failed to reveal a causal relationship between anthrax vaccine and chronic comorbid medical conditions (Sever et al., 2004; Grabenstein et al., 2006).

Alphaviruses

In addition to JE and yellow fever, a number of other vector-borne viral pathogens pose risk to military forces either as environmental threats related to specific geographic deployments or as potential agents of biological warfare. The alphaviruses, while antigenically and molecularly distinct from one another, share a mosquito-borne transmission pattern and cause similar clinical syndromes: viral encephalitis, seen with Venezuelan equine encephalitis (VEE), Eastern equine encephalitis (EEE), and Western equine encephalitis (WEE); or febrile rash with serositis in Chikungunya, Ross River virus, and O'nyong nyong (Pittman and Plotkin, 2004). The U.S. military has developed IND vaccine products targeting the viral encephalitides and Chikungunya. These vaccines have been limited in their scope of use largely to at-risk laboratory personnel and have shown promise in reducing laboratory-acquired infections with these agents (Pittman and Plotkin, 2004; Hoke, 2005). Because the use of these vaccines has only been implemented on a small scale, safety has not been assessed on a broad scale. Immunologic interference related to the co-administration of these vaccines has also been noted (Pittman and Plotkin, 2004).

Viral Hemorrhagic Fevers

These agents, deriving from multiple families of RNA viruses, are generally arthropod- or rodent-borne infections that result in clinical syndromes involving systemic symptoms, disseminated hemorrhagic manifestations, and vascular collapse (Thomas et al., 2005). Hemorrhagic fever viruses are considered to be high-priority threat agents for bioterrorism (Borio et al., 2002) and are threats of nature based upon incursions into their geographic and ecologic niches by military forces (Thomas et al., 2005; Grabenstein et al., 2006). The U.S. military has developed safe and immunogenic vaccines against Rift Valley fever and Argentine hemorrhagic fever (AHF) caused by the Junin arenavirus (Pittman and Plotkin, 2004). The latter live product has been found to be 95% effective in preventing disease in endemic settings in formal phase II testing in humans (Maiztegui et al., 1998); it is currently used to protect civilians in Argentina.

Viral hemorrhagic fevers of significant military importance but for which vaccines remain largely elusive and in active development include dengue, hantaviruses, and disease caused by the filoviruses Marburg and Ebola. Novel strategies, based on advances in molecular biology and genetics, are being applied to these pathogens; much of the research is being driven by or involves collaboration with military scientists, a testament to the relevance of these pathogens to the military's mission (Thomas et al., 2005).

CONCLUSIONS

The development of safe and effective vaccines targeting diseases related to other biological weapons, such as plague and tularemia in addition to improved anthrax and smallpox vaccines, is a high priority and of obvious military application. Similarly, based upon the sheer global burden of disease, malaria continues to pose a significant threat to military forces deployed in many parts of the world. While a vaccine solution to this enormous public health problem remains elusive, substantial resources are engaged to achieve this goal. Military medical research has been a leader in the development of malaria vaccines to date and will likely be a major consumer of future effective products for both combat and humanitarian operations (Ockenhouse et al., 2005).

HIV-1 also represents a global threat to militaries. As with other sexually transmitted and blood-borne pathogens, HIV-1 may represent a direct threat to military forces in sporadic fashion (Artenstein et al., 1995; Brodine et al., 1995). However, the most significant threat posed by the worldwide HIV epidemic to deployed military forces is its deleterious impact on global security and the stability of governments, especially those in developing nations. The destabilizing influence of the HIV-1 epidemic serves to create or exacerbate humanitarian crises that subsequently involve military forces in expanded roles. Thus, an effective vaccine against HIV-1 may have important military application.

A variety of infectious threats continually face military forces as occupational hazards related either to the epidemiology of deployments, wartime vulnerabilities, or based upon the epidemiology of troop encampments and training environments. While many such threats have been controlled or prevented with vaccines, others remain. Vigorous research continues toward the goal of developing safe and effective products against these incompletely addressed pathogens.

References

Ahlborg, B., Linroth, K. and Nrodgren, B. ECG-changes without subjective symptoms after smallpox vaccination of military personnel. Acta Med. Scand. 1969; 464(Suppl.):113–126.

Alexander, C.E., Sanborn, W.R., Cherriere, G., Crocker, W.H., Jr., Ewald, P.E. et al. Sulfadiazine-resistant group A Neisseria meningitides. Science 1968; 161:1019.

Artenstein, A.W. Anthrax: from antiquity to answers. J. Infect. Dis. 2007; 195:471–473.

Artenstein, A.W., Coppola, J., Brown, A.E., Carr, J.K., Sanders-Buell, E. et al. Multiple introductions of HIV-1 subtype E into the Western Hemisphere. Lancet 1995; 346:1197–1198.

Artenstein, A.W., Opal, J.M., Opal, S.M., Tramont, E.C., Peter, G. et al. History of U.S. military contributions to the study of vaccines against infectious diseases. Mil. Med. 2005; 170:3–11.

Artenstein, M.S. and Gold, R. Current status of prophylaxis of meningococcal disease. Mil. Med. 1970; 135:735–739.

Artenstein, M.S., Gold, R., Zimmerly, J.G., Wyle, F.A., Schneider, H. et al. Prevention of meningococcal disease by group C polysaccharide vaccine. N. Engl. J. Med. 1970; 282:417–420.

Baker, M.S. and Ryals, P.A. The medical department in military operations other than war. Part I. Planning for deployment. Mil. Med. 1999; 164:572–579.

Barquet, N. and Domingo, P. Smallpox: the triumph over the most terrible of ministers of death. Ann. Intern. Med. 1997; 127:635–642.

Barrett, P.N., Dorner, F., Ehrlich, H. and Plotkin, S.A. Tick-borne encephalitis virus vaccine. In: Vaccines (S.A. Plotkin and W.A. Orenstein, Eds.), 4th ed., pp. 1039–1055. Philadelphia, PA: Elsevier, Inc., 2004.

Benenson, A.S. Immunization and military medicine. Rev. Infect. Dis. 1984; 6:1–12.

Berezuk, G.P. and McCarty, G.E. Investigational drugs and vaccines fielded in support of Operation Desert Storm. Mil. Med. 1992; 157:404–406.

Borio, L., Inglesby, T., Peters, C.J., Schmaljohn, A.L., Hughes, J.M. et al. Hemorrhagic fever viruses as biological weapons: medical and public health management. J. Am. Med. Assoc. 2002; 287:2391–2405.

Brodine, S.K., Mascola, J.R., Weiss, P.J., Ito, S., Porter, K.R. et al. Detection of diverse HIV-1 subtypes in the United States. Lancet 1995; 346:1198–1199.

Burnham, B.R., Wells, T.S. and Riddle, J.R. A cost-benefit analysis of a routine varicella vaccination program for United States Air Force Academy cadets. Mil. Med. 1998; 163:631–634.

Centers for Disease Control and Prevention Achievements in public health, 1900–1999 impact of vaccines universally recommended for children—United States, 1990–1998. Morb. Mortal. Wkly. Rep. 1999; 48:243–248.

Christopher, G.W., Agan, B.K., Cieslak, T.J. and Olson, P.E. History of U.S. military contributions to the study of bacterial zoonoses. Mil. Med. 2005; 170:39–48.

Clardy, W.F. Susceptibility in USAF recruits to vaccine-preventable diseases. Vaccine 1993; 11:573–575.

Collis, P.B., Dudding, B.A., Winter, P.E., Russell, P.K. and Buescher, E.L. Adenovirus vaccines in military recruit populations: a cost-benefit analysis. J. Infect. Dis. 1973; 128:745–752.

Craig, S.C., Pittman, P.R., Lewis, T.E., Rossi, C.A., Henchal, E.A. et al. An accelerated schedule for tick-borne encephalitis vaccine: the American military experience in Bosnia. Am. J. Trop. Med. Hyg. 1999; 61:874–878.

Dooley, D.P. History of U.S. military contributions to the study of viral hepatitis. Mil. Med. 2005; 170:71–76.

Dudding, B.A., Top, F.H. and Winter, P. Acute respiratory disease in military trainees: the adenovirus surveillance program, 1966–1971. Am. J. Epidemiol. 1973; 97:187–198.

Earhart, K.C., Beadle, C., Miller, L.K., Pruss, M.W., Gray, G.C. et al. Outbreak of influenza in highly vaccinated crew of U.S. Navy ship. Emerg. Infect. Dis. 2001; 7:463–465.

Enders, J.F., Bell, J.A., Dingle, J.H., Francis, T., Hilleman, M.R. et al. "Adenoviruses": group name proposed for new respiratory tract viruses. Science 1956; 124:119–120.

Gaydos, C.A. and Gaydos, J.C. Adenovirus vaccine. In: Vaccines (S.A. Plotkin and W.A. Orenstein, Eds.), 4th ed., pp. 863–885. Philadelphia, PA: Elsevier, Inc., 2004.

Gellis, S.S., Stokes, J., Jr., Brother, G.M., Hall, W.M., Gilmore, H.R. et al. The use of human immune serum globulin (gamma-globulin) in infectious (epidemic) hepatitis in the Mediterranean theater of operations. I. Studies on prophylaxis in two epidemics of infectious hepatitis. J. Am. Med. Assoc. 1945; 128:1062–1063.

Gold, R. and Artenstein, M.S. Meningococcal infections 2. Field trial of group C meningococcal polysaccharide vaccine in 1969–1970. Bull. World Health Organ. 1971; 45:279–282.

Goldschneider, I., Gotschlich, E.C. and Artenstein, M.S. Human immunity to the meningococcus I. The role of humoral antibodies. J. Exp. Med. 1969a; 129:1307–1326.

Goldschneider, I., Gotschlich, E.C. and Artenstein, M.S. Human immunity to the meningococcus II. The development of natural immunity. J. Exp. Med. 1969b; 129:1327–1348.

Gotschlich, E.C., Liu, T.Y. and Artenstein, M.S. Human immunity to the meningococcus III. Preparation and immunochemical properties of the group A, group B, and group C meningococcal polysaccharides. J. Exp. Med. 1969a; 129:1349–1365.

Gotschlich, E.C., Liu, T.Y. and Artenstein, M.S. Human immunity to the meningococcus IV. Immunogenicity of group A and group C meningococcal polysaccharides in human volunteers. J. Exp. Med. 1969b; 129:1367–1384.

Gotschlich, E.C., Liu, T.Y. and Artenstein, M.S. Human immunity to the meningococcus V. The effect of immunization with meningococcal group C polysaccharide on the carrier state. J. Exp. Med. 1969c; 129:1385–1395.

Grabenstein, J.D., Pittman, P.R., Greenwood, J.T. and Engler, R.J.M. Immunization to protect the U.S. armed forces: heritage, current practice, and prospects. Epidemiol. Rev. 2006; 28:3–26.

Granoff, D.M., Feavers, I.M. and Borrow, R. Meningococcal vaccines. In: Vaccines (S.A. Plotkin and W.A. Orenstein, Eds.), 4th ed., pp. 959–987. Philadelphia, PA: Elsevier, Inc., 2004.

Gray, G.C., Callahan, J.D., Hawksworth, A.W., Fisher, C.A. and Gaydos, J.C. Respiratory diseases among U.S. military personnel: countering emerging threats. Emerg. Infect. Dis. 1999; 5:379–387.

Hilleman, M.R. Personal historical chronicle of six decades of basic and applied research in virology, immunology, and vaccinology. Immunol. Rev. 1999; 170:7–27.

Hilleman, M.R., Bunyak, E.B., Weibel, R.E. and Stokes, J., Jr. Live attenuated mumps-virus vaccine. N. Engl. J. Med. 1968; 278:227–232.

Hilleman, M.R. and Werner, J.H. Recovery of new agent from patients with acute respiratory illness. Proc. Soc. Exp. Biol. Med. 1954; 85:183–188.

Hoke, C.H. History of U.S. military contributions to the study of viral encephalitis. Mil. Med. 2005; 170:92–105.

Hoke, C.H., Nisalak, A., Sangawhipa, N., Jatanasen, S., Laorakopongse, T. et al. Protection against Japanese encephalitis by inactivated vaccines. N. Engl. J. Med. 1988; 19:608–614.

Inglesby, T.V., O'Toole, T., Henderson, D.A., Bartlett, J.G., Ascher, M.S. et al. Anthrax as a biological weapon, 2002: updated recommendations for management. J. Am. Med. Assoc. 2002; 287:2236–2252.

Innis, B.L., Snitbhan, R., Kunasol, P., Laorakpongse, T., Poopatanakool, W. et al. Protection against hepatitis A by an inactivated vaccine. J. Am. Med. Assoc. 1994; 271:1328–1334.

Karjalainen, J., Heikkila, J., Nieminen, M.D., Jalanko, H., Kleemola, M. et al. Etiology of mild acute infectious myocarditis. Relation to clinical features. Acta Med. Scand. 1983; 213:65–73.

Katz, S.L., Kempe, C.H., Black, F.L., Lepow, M.L., Krugman, S. et al. Studies on an attenuated measles virus vaccine VIII. General summary and evaluation of results of vaccine. N. Engl. J. Med. 1960; 263:180–184.

Lincoln, A.F. and Siverson, S.E. Acute phase of Japanese B encephalitis: two hundred and one cases in American soldiers, Korea, 1950. J. Am. Med. Assoc. 1952; 150:268–273.

Maiztegui, J.I., McKee, K.T., Jr., Barrera Oro, J.G., Harrison, L.H., Gibbs, P.H. et al. Protective efficacy of a live attenuated vaccine against Argentine hemorrhagic fever. AHF Study Group. J. Infect. Dis. 1998; 177:277–283.

Meiklejohn, G. Viral respiratory disease at Lowry Air Force Base in Denver, 1952–1982. J. Infect. Dis. 1983; 148:775–784.

Meyer, H.M. and Parkman, P.D. Rubella vaccination: a review of practical experience. J. Am. Med. Assoc. 1971; 215:613–619.

Moloo, H. and Artenstein, A.W. History of vaccines. In: Encyclopedia of Public Health (C.J. Murray, Ed.), Elsevier, Oxford, UK, 2008; in press.

Murray, C.K. and Horvath, L.L. An approach to prevention of infectious diseases during military deployments. Clin. Infect. Dis. 2007; 44:424–430.

Norman, J.E., Beebe, G.W., Hoofnagle, J.H. and Seeff, L.B. Mortality follow-up of the 1942 epidemic of hepatitis B in the U.S. Army. Hepatology 1993; 18:790.

Ockenhouse, C.F., Magill, A., Smith, D. and Milhous, W. History of U.S. military contributions to the study of malaria. Mil. Med. 2005; 170:12–16.

Ottolini, M.G. and Burnett, M.W. History of U.S. military contributions to the study of respiratory infections. Mil. Med. 2005; 170:66–70.

Parish, H.J. A history of immunization. London: E & S Livingston Ltd., 1965.

Parkman, P.D., Buescher, E.L. and Artenstein, M.S. Recovery of rubella virus from Army recruits. Proc. Soc. Exp. Biol. Med. 1962; 111:225–230.

Pierce, W.E., Rosenbaum, M.J. and Edwards, E.A. Live and inactivated adenovirus vaccines for the prevention of acute respiratory illness in naval recruits. Am. J. Epidemiol. 1968; 87:237–246.

Pittman, P. and Plotkin, S.A. Miscellaneous limited-use vaccines. In: Vaccines (S.A. Plotkin and W.A. Orenstein, Eds.), 4th ed., pp. 989–998. Philadelphia, PA: Elsevier, Inc., 2004.

Poland, G.A., Grabenstein, J.D. and Neff, J.M. The US smallpox vaccination program: a review of a large modern era smallpox vaccination implementation program. Vaccine 2005; 23:2078–2081.

Redfield, R.R., Wright, D.C., James, W.D., Jones, T.S., Brown, C. et al. Disseminated vaccinia in a military recruit with human immunodeficiency virus (HIV) disease. N. Engl. J. Med. 1987; 316:673–676.

Ryals, P.A. and Baker, M.S. Military medicine in operations other than war part II: humanitarian relief missions for naval reserve fleet hospitals. Mil. Med. 1996; 161:502–504.

Sever, J.L., Brenner, A.I., Gale, A.D., Lyle, J.M., Ward, B.J. et al. Safety of anthrax vaccine: an expanded review and evaluation of adverse events reported to the Vaccine Adverse Event Reporting System (VAERS). Pharmacoepidemiol. Drug Saf. 2004; 13:825–840.

Sharp, T.W., Wightman, J.M., Davis, M.J., Sherman, S.S. and Burkle, F.M. Military assistance in complex emergencies: what have we learned since the Kurdish relief effort? Prehosp. Disaster Med. 2001; 16:197–208.

Smoak, B.L., Novakoski, W.L., Mason, C.J. and Erickson, R.L. Evidence for a recent decrease in measles susceptibility among young American adults. J. Infect. Dis. 1994; 170:216–219.

Stokes, J., Jr. and Neefe, J.R. The prevention and attenuation of infectious hepatitis by gamma-globulin: preliminary note. J. Am. Med. Assoc. 1945; 127:144–145.

Taubenberger, J.K. and Morens, D.M. 1918 Influenza: the mother of all pandemics. Emerg. Infect. Dis. 2006; 12:15–22.

Theiler, M. and Smith, H.H. The use of yellow fever virus modified by in vitro cultivation for human immunization. J. Exp. Med. 1937; 65:787–800.

Thomas, S.J., Lawler, J.V. and Endy, T.P. History of U.S. military contributions to the study of viral hemorrhagic fevers. Mil. Med. 2005; 170:77–91.

Top, F.H., Buescher, E.L., Bancroft, W.H. and Russell, P.K. Immunization with live types 7 and 4 adenovirus vaccines. II. Antibody response and protective effect against acute respiratory disease due to adenovirus type 7. J. Infect. Dis. 1971b; 124:155–160.

Top, F.H., Grossman, R.A., Bartelloni, P.J., Segal, H.E., Dudding, B.A. et al. Immunization with live types 7 and 4 adenovirus vaccines. I. Safety, infectivity, antigenicity, and potency of adenovirus type 7 vaccine in humans. J. Infect. Dis. 1971a; 124:148–154.

Woodward, T.E. The Armed Forces Epidemiological Board: its first fifty years 1940–1990. In: Textbooks of Military Medicine (R. Zajtchuk and D.P. Jenkins, Eds.), pp. 1–150. Washington, DC: Department of the Army, Office of the Surgeon General and Borden Institute, 1990, Section 1.

SECTION III

VIRAL VACCINES

CHAPTER

19

Dengue

David W. Vaughn[1], *Stephen S. Whitehead*[2] *and Anna P. Durbin*[3]

[1]Military Infectious Diseases Research Program, US Army Medical Research and Development Command, Fort Detrick, MD, USA

[2]Laboratory of Infectious Diseases, National Institute of Allergy and Infectious Diseases, Bethesda, MD, USA

[3]Centre for Immunization Research, Department of International Health, Johns Hopkins Bloomberg School of Public Health, Baltimore, MD, USA

OUTLINE

ABSTRACT

Molecular evolution studies suggest that dengue virus (DENV) evolved 1000 years ago and entered a sustained human–mosquito cycle between 125 and 320 years ago. While it is unlikely that DENV would be used as a biothreat agent, DENV has emerged since World War II as the most important mosquito-borne *viral* pathogen infecting an estimated 100 million persons each year. Infection with any of the four DENV serotypes (DENV-1, 2, 3, and 4) can be inapparent, result in classic dengue fever with high fever, headache, eye pain and muscle ache, or progress at the time of defervescence to dengue hemorrhagic fever (DHF) characterized by hemorrhagic manifestations and plasma leakage that can lead to shock and death. The immunopathological mechanisms by which DENV causes the clinical features of DHF are intricate and include aberrant humoral and cellular immune responses. Previous DENV infections may predispose to more severe disease by the induction of enhancing antibody and cross-reactive T cells. Treatment is supportive relying upon careful fluid management which can be lifesaving. Prevention currently depends on vector control which has been largely unsuccessful. Several dengue vaccine candidates have advanced to clinical trials to include classically and molecularly attenuated live virus vaccines, chimeric vaccines using dengue and yellow fever virus backbones, and DNA vaccines. Whole virus inactivated and recombinant subunit vaccine candidates should soon enter into clinical testing. Live attenuated DENV vaccines offer the most promise in terms of broad, long-lasting protection, and although they are economical to produce, they may carry the risks of enhanced reactogenicity in recipients with preexisting flavivirus antibodies, adverse events among the immunocompromised, vaccine virus transmission by vector mosquitoes, and the risk of adventitious agents. Chimeric vaccine approaches use the same gene sequence for the nonstructural proteins for each of the four DENV serotypes avoiding the need for attenuating mutations in the structural genes and potentially reducing interference in the replication of the four virus vaccine components within the multivalent vaccine recipient. DNA vaccines, in combination with other approaches, may increase the complexity and effectiveness of the immune response. Whole inactivated virus approaches have been used successfully for other viral diseases, reduce interference issues in multivalent vaccines, and have protected nonhuman primates from viremia following challenge with wild-type virus. Recombinant subunit vaccines have also protected nonhuman primates from viremia and offer a more focused approach in an attempt to tailor antibody and cell-mediated immure responses. While pathogenesis studies seek to dissect immune responses in an attempt to avoid vaccine-related disease enhancement, these risks, which apply to all vaccine approaches, may need to be evaluated empirically.

INTRODUCTION

Despite the fact that approximately 2.5 billion people live in regions where dengue virus (DENV) transmission by mosquito is possible, and despite the resurgence of dengue disease over the last 25 years, dengue remains classified as a neglected tropical disease by the World Health Organization. This neglect may be compounded by the low likelihood that DENV would be used as an agent of bioterrorism. Nevertheless, the disease burden created by DENV is significant, especially among children living in endemic regions of Southeast Asia, the Indian subcontinent, and Central and South America. Because the vast majority of DENV infections are asymptomatic, many would consider these infections unimportant. However, previous infection with DENV or the presence of maternally derived DENV antibodies is the most recognized risk factor for the progression to serious dengue disease. Therefore, all DENV infections are significant, whether they are inapparent or result in dengue fever (DF) or the more severe dengue hemorrhagic fever (DHF). Although vector control is currently the most effective measure to reduce the impact of DENV, its effectiveness is dependent on the intensity and sustainability of such programs, which are tenuous at best. Recent advances in our understanding of dengue pathogenesis and in vaccine development show promise to deliver effective and long-lasting dengue control.

HISTORY OF DENGUE DISEASE

Epidemics of DF-like illness have been identified in the medical literature as far back as 992 CE or 1014 years ago. A Chinese "encyclopedia of disease symptoms and remedies" described an illness called "water poison" associated with flying insects and water that included rash, fever, eye pain, arthralgias, myalgias, and hemorrhagic manifestations (Nobuchi, 1979; Gubler, 1997). If this was truly a description of a dengue epidemic then recent molecular evolution studies of DENV suggest that this was one of the first. According to Twiddy et al. (2003), DENV may have evolved approximately 1000 years ago, transferred from solely a monkey–mosquito sylvatic cycle to a sustained human–mosquito cycle between 125 and

320 years ago, and diversified into the four DENV serotypes only in the last 100 years. While the name "dengue" probably comes from Africa (see below), the virus itself most likely originated in Asia as all four DENV serotypes have been identified in sylvatic cycles in Asia while only DENV serotype 2 (DENV-2) has been identified in forest cycles in Africa (Gubler, 1997).

More commonly cited as the first clinical description of dengue was that provided by Dr. Benjamin Rush, a signer of the United States Declaration of Independence and Surgeon in Charge of Hospitals during the American Revolutionary War under General George Washington (Rush, 1789). Following the epidemic of fever in Philadelphia in the summer of 1780, Dr. Rush provided the following description:

> This fever generally came on with rigor, but seldom with a regular chilly fit.... The pains which accompanied this fever were exquisitely severe in the head, back, and limbs. The pains in the head were sometimes in the back parts of it, and at other times they occupied only the eyeballs.... A few complained of their flesh being sore to the touch, in every part of the body. From these circumstances, the disease was sometimes believed to be a rheumatism. But its more general name among all classes of people was, the *Break-bone fever*.

Dr. Rush did not use the term "dengue" but rather "bilious remitting fever" and "break-bone fever" and was probably describing febrile illnesses with multiple etiologies. Donald Carey agrees that Rush was, for the most part, describing what is currently referred to as DF (illness associated with DENV infection). However, the disease referred to as dengue throughout most of the 19th century more closely resembled the clinical syndrome now associated with chikungunya virus infection (Carey, 1971) (see Chapter 32 for Chikungunya). The term dengue was first used in Cuba in 1928 meaning "affected." The Spanish term may have been based on the Swahili term "dinga" or "dyenga" meaning "a sudden cramp-like seizure" used during chikungunya-like epidemics on the East Coast of Africa in 1823 (Rigau-Perez, 1998). The term dengue and the virus (chikungunya) may have traveled to the Caribbean as a part of the slave trade. With infrequent epidemics the terms "break-bone fever" and "dengue" were confused. Toward the end of the 19th century, the term dengue was increasingly applied to Rush's break-bone fever including outbreaks in Australia reported in 1898 (Hare, 1898). The disease previously known as dengue became chikungunya meaning "that which bends up" following outbreaks in East Africa in 1952 with the isolation of that alphavirus (Robinson, 1955).

In 1906, US Army physicians Ashburn and Craig, working in the Philippines, demonstrated that dengue was caused by an "ultramicroscopic nonfilterable agent" or virus by injecting healthy volunteers with filtered serum from febrile dengue patients using syringe and mosquitoes (Ashburn and Craig, 1907). This dispelled earlier contentions that dengue was caused by a bacteria or parasite. This was only the second human viral pathogen described following the yellow fever virus (YFV) described by Reed et al. (1901). Their work also confirmed the work of Graham (1903) at the Syrian Protestant College in Beruit (now American University) that DENV was transmitted by mosquito vectors. With controlled experimental infections and careful clinical descriptions the clinical label of dengue became fixed.

The Japanese and United States accelerated dengue research during World War II leading to DENV isolation by intracranial injection of mice with patient serum. Dr. Hotta and Dr. Kumura isolated DENV-1 from specimens collected between 1942 and 1945 (Hotta, 1952) and Dr. Sabin and Dr. Schlesinger isolated DENV-1 and DENV-2 from specimens collected in Hawaii/India and New Guinea, respectively, in 1944 (Sabin and Schlesinger, 1945; Sabin, 1952). In 1956, a DHF epidemic occurred in Manila resulting in the identification and naming of DENV-3 and DENV-4 viruses by Dr. Hammon (Hammon et al., 1960).

ETIOLOGIC AGENT

DENV is a member of the *Flavivirus* genus of the virus family Flaviviridae. The virus is a spherical particle containing a single-strand, positive-sense RNA genome about 11 kb in length. The genome of the virus contains a single open reading frame which is expressed as a large polyprotein subsequently processed by virus-encoded and host cell proteases into three structural and at least seven nonstructural (NS) proteins. The termini of the genome contain untranslated regions (UTRs) that are important for viral replication. The 5′ UTR is relatively short (approximately 100 nucleotides) and has a type I cap structure (m^7GpppA), while the 3′ UTR (approximately 450 nucleotides) contains a number of conserved RNA structures and lacks a terminal poly(A) tract. The overall genome organization is 5′UTR-C-prM-E-NS1-NS2A-NS2B-NS3-NS4A-NS4B-NS5-3′UTR (C, capsid; prM, membrane precursor; and E, envelope). The infectious virion is approximately 50 nm in diameter and has a relatively smooth surface consisting of a compact arrangement of 180 copies of the E glycoprotein on the virus lipid envelope (Fig. 19.1). Immunity to the virus is mediated primarily by neutralizing

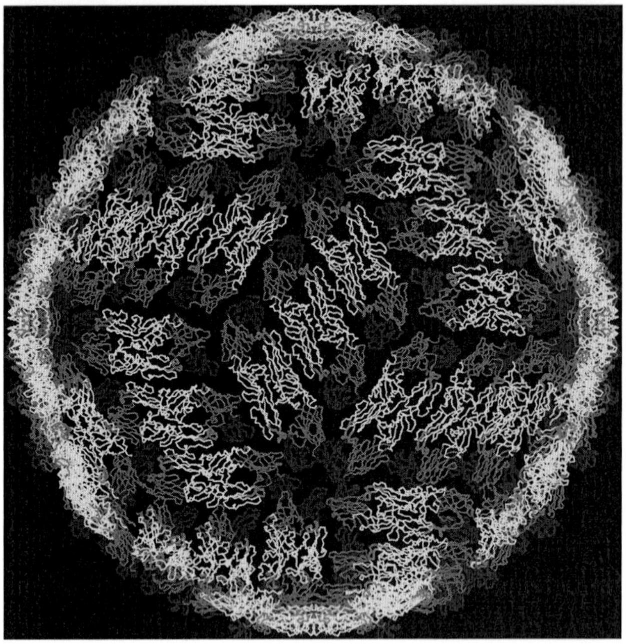

FIGURE 19.1 Arrangement of the E proteins on the surface of the mature virus (Kuhn et al., 2002). The virus surface is covered with 30 protein "rafts," each containing three anti-parallel dimers of E protein. Each monomer is colored as follows: red, domain I (central region); yellow, domain II (dimerization and fusion region); blue, domain III; green (receptor-binding region) (see color plate section).

antibodies to the E glycoprotein which also defines each of the four DENV serotypes (DENV-1, DENV-2, DENV-3, and DENV-4).

Recent crystallographic and cryo-electron microscopic techniques have greatly advanced the understanding of the E glycoprotein structure and its organization on the surface of the virus (Modis et al., 2005; Mukhopadhyay et al., 2005). The molecular conformation of E protein is highly dependent on the pH of the environment: at relatively alkaline pH, the E protein molecules are arranged in a dimeric conformation and lay flat along the virus surface in a herringbone pattern, giving the virus its characteristically smooth appearance; at the relatively acidic pH of the endosomal compartment, the E protein molecules shift to a trimeric conformation and protrude from the virus surface during the process of viral and endosomal membrane fusion which facilitates release of the viral genome into the cell cytoplasm. It is the inherent flexibility of the E protein structure which allows for these considerable conformational changes required during virus maturation and infection. Recent studies have demonstrated that glycosylation sites on neighboring E proteins are bound by a single carbohydrate recognition site of DC-SIGN, a proposed receptor of DENV (Pokidysheva et al., 2006). As the sole protein exposed on the virus surface, the E glycoprotein must function to bind cellular receptors, interact with host cell membranes, direct viral assembly and budding, as well as display antigenic determinants that elicit immune responses.

The primary antigen of DENV is the E glycoprotein, although antibody responses directed toward prM and NS1 have been identified. The E glycoprotein monomer can be divided into three structural/functional domains, I, II, and III (Fig. 19.1). Recent crystallographic studies have suggested that epitopes located among each of the three domains may impart different functionality to the elicited immune response (Modis et al., 2005). Serotype-specific, neutralizing antibodies are generally elicited by epitopes on the surface of domain III, which has also been implicated in cell–surface receptor binding. Although antibodies against epitopes in domain I are generally nonneutralizing, antibodies that recognize epitopes in domain II, which contains the fusion peptide region, have been shown to neutralize virus, probably by inhibition of membrane fusion. Therefore, it is possible that neutralization of DENV can occur not only at the level of virus attachment, but also at the level of membrane fusion and viral entry, as first demonstrated for West Nile virus (Gollins and Porterfield, 1986).

PROTECTIVE IMMUNE RESPONSE

Antibody-Mediated Protection

Humoral immunity is thought to play both a protective and a permissive role in DENV infection. Early studies in mice found that antibody, but not cell-mediated immunity, was critical to the recovery from acute DENV infection (Chaturvedi et al., 1977, 1978). In addition, epidemiologic data suggests that maternal antibody protects infants from DENV infection for the first few months of life (Halstead et al., 2002). Solid protection against a second infection with a homotypic DENV has been demonstrated for as long as 18 months after primary infection but is believed to be much longer, perhaps even lifelong (Sabin, 1952). The evidence for long-lasting homotypic immunity is derived primarily from in vitro studies that detected neutralizing antibody in the sera of persons infected with DENV more than four decades earlier (Halstead, 1974; Papaevangelou and Halstead, 1977; Okuno et al., 1983). Sabin (1952) suggested that short-lived heterotypic immunity may exist resulting in milder illness up to 9 months following primary infection. Animal studies in nonhuman primates have also found little evidence of long-lived cross protection between heterologous DENV (Halstead et al., 1973a; Whitehead et al., 1970; Scherer et al., 1972).

The kinetics of antibody production following primary and secondary DENV infections are quite different (Russell et al., 1967, 1969; Innis et al., 1989). In primary DENV infection, IgM antibodies appear first, generally between the 4th and the 7th day of illness, peak within 2 weeks, and then decay to undetectable levels within 6 months after infection (Innis et al., 1989; Vaughn et al., 1997). In contrast to the high levels of IgM antibody that are present early after primary infection, only low levels of IgG antibodies against DENV are detected during the acute phase of illness. IgM antibody levels greatly exceed IgG levels for 2–4 weeks following primary infection. Both IgG and IgM antibodies have the ability to neutralize virus and to inhibit the hemagglutination (HI) of gander red blood cells but only IgG antibody has complement fixation (CF) activity. Secondary DENV infection produces an anamnestic response as the result of common antigens shared by DENV. Characteristic of an anamnestic response, IgG is the primary antibody produced; IgM antibody is markedly reduced. High levels of IgG antibody are detected early and rise dramatically in contrast to IgM antibody levels, which are elevated only modestly. Secondary infection is associated with high neutralizing, HI, and CF antibody titers that have broad cross-reactivity and are comprised almost entirely of IgG. For this reason, the hemagglutination-inhibition titer has been used to classifying DENV infections as primary or secondary, and a patient with an HI titer of ≥1:2560 in an acute specimen is defined as having a secondary flavivirus infection (Anonymous, 1997).

Different methods of detecting dengue antibody vary in their specificity and their sensitivity over time. HI antibodies are usually cross-reactive, even in primary DENV infections, for months to years (Whitehead et al., 1970; Scherer et al., 1978; Scott et al., 1976; Vaughn et al., 1997). The HI serotype specificity response is variable over years; becoming monotypic in some patients, remaining heterotypic in some patients, and disappearing altogether in others (Hammon et al., 1966; Scott et al., 1976; Papaevangelou and Halstead, 1977). For this reason, the absence of HI antibody is an unreliable criterion for defining a dengue-naive individual. Of the methods used to measure dengue antibody, the neutralizing antibody test exhibits the greatest dengue serotype specificity (Scherer et al., 1978, Russell et al., 1967). Neutralizing antibody titers to homologous virus following primary infection are usually below 1:1000 and cross-reactions tend to be low titered. In addition, neutralizing antibody titer to homologous virus can increase progressively over a period of at least 6 months. Neutralizing antibody produced following secondary infection is broadly cross-reactive and can remain so for years (Russell et al., 1967; Papaevangelou and Halstead, 1977). The specificity may narrow over time but this is variable among individuals (Innis, 1997). Despite the induction of heterologous antibody with secondary infection, DENV have been shown to demonstrate the phenomenon of original antigenic sin (Halstead et al., 1983). During secondary DENV infection, Halstead demonstrated that neutralizing antibody responses were detected against both homotypic and heterotypic viruses but the highest titers were detected to the original infecting DENV serotype. In general, the presence of neutralizing antibody to more than one DENV serotype in late convalescent serum indicates a history of multiple prior DENV infections, or at least prior flavivirus infection, and only the absence of neutralizing antibody to all four DENV serotypes is evidence that no previous DENV infection has occurred.

Inhibition of virus attachment and virus-mediated membrane fusion are thought to be the mechanisms of DENV neutralization (Gollins and Porterfield, 1986; Roehrig et al., 1998). The DENV E protein is considered to be the major antigen participating in antibody-mediated neutralization of virus. In primary DENV infection in humans, the E protein elicits the first immune responses and the ones with the longest duration (Churdboonchart et al., 1991). Monoclonal antibody (MAb) studies have corroborated the E protein as a target for neutralizing antibodies (He et al., 1995; Kaufman et al., 1987, 1989). MAb against DENV-2 E and prM proteins protected mice against challenge with both heterologous and homologous DENV. Neutralization activity correlated well with protective ability for E monoclonal antibodies. Although prM monoclonal antibodies were also protective, their protective ability did not correlate as well with neutralizing activity. Antibodies to the NS proteins can also be detected following primary and secondary DENV infection. NS1, NS3, and NS5 are detected following primary infection and antibodies to C, prM, NS1, NS3, NS4A, and NS5 can be detected in the convalescent sera of patients with secondary DENV infection (Kuno et al., 1990; Churdboonchart et al., 1991; Huang et al., 1999). Although NS1 antibodies do not have neutralizing activity, NS1 MAbs were able to partially protect passively immunized mice from intracerebral challenge with DENV-2 virus (Henchal et al., 1988). In these experiments solid protection equivalent to that conferred by polyclonal neutralizing antibody was achieved with one pair of DENV-2 NS1 MAbs given as a combination. NS1 is a soluble, complement-fixing antigen that is expressed on the surface of

DENV-infected cells. Studies have suggested that NS1 antibody may be protective by promoting complement-mediated lysis of infected cells (Schlesinger et al., 1986, 1987; Falgout et al., 1990). What role, if any, antibody to the other NS proteins plays in the immune response to DENV has not been clarified.

Cell-Mediated Immunity

Although virus-specific cytotoxic T lymphocytes (CTL) have been reported to be important in the recovery from certain viral infections such as influenza (Wells et al., 1981; Yap et al., 1978), the role of CTL in recovery from DENV infection is less defined and has not been studied to the same extent as the antibody response. Both $CD4^+$ $CD8^-$ and $CD4^-$ $CD8^+$ DENV-reactive memory T cells are generated after primary DENV infection. In a study of volunteers who developed DF in response to infection with a candidate live dengue vaccine, activation of $CD4^+$ T cells occurred during the presence of viremia and activation of $CD8^+$ T cells followed $CD4^+$ T cell activation (Kurane et al., 1995). Activation of T cells preceded the disappearance of viremia and T cell activation gradually subsided once viremia ended, suggesting that T cell activation contributed to the control of DENV infection. The presence of DENV-specific $CD4^+$ $CD8^-$ T lymphocytes has been detected in peripheral blood mononuclear cells (PBMC) collected from persons naturally infected with DENV or immunized with live attenuated candidate dengue vaccines (Kurane et al., 1989a, 1989b; Dharakul et al., 1994). The proliferative responses of $CD4^+$ $CD8^-$ T lymphocytes in bulk culture after primary infection are both serotype-specific and serotype-cross-reactive memory responses, however when examined at the clonal level, the responses are mainly dengue serotype-cross-reactive (Kurane et al., 1991a). DENV-specific $CD4^+$ T lymphocyte clones were established from a volunteer who had been immunized with yellow fever vaccine 2 years prior to being infected with DENV-3.

Although $CD4^+$ $CD8^-$ T lymphocytes from this volunteer were primarily serotype-specific in bulk culture, at the clonal level there were serotype-cross-reactive and flavivirus-cross-reactive responses. DENV-specific $CD4^+$ T cell clones established from other DENV recipients also demonstrated a heterogenous response (Kurane and Ennis, 1994b), indicating that both serotype-specific and serotype-cross-reactive $CD4^+$ memory CTL are present in most individuals after primary DENV infection. Activated $CD4^+$ $CD8^-$ T lymphocytes produce a predominantly Th-1 response consisting of high titers of IFN-γ and IL-2, and are effective at lysing DENV-infected cells expressing the MHC class II phenotype (Kurane et al., 1990). $CD4^+$ T cells can efficiently lyse both target cells expressing DENV antigen as well as uninfected bystander cells, perhaps contributing to the hepatocyte damage observed during acute dengue illness (Gagnon et al., 1999). In addition, the expression of IFN-γ by activated $CD4^+$ $CD8^-$ T lymphocytes may also contribute to the pathogenesis of DHF/DSS (see below).

Although cytolytic responses to DENV have been detected in $CD4^+$ T cell populations, $CD8^+$ cytotoxic T cells are thought to contribute the majority of the cytolytic activity (Mathew et al., 1996). Following primary DENV infection, $CD8^+$ T lymphocytes exhibit significant serotype-specific proliferation with a variable level of serotype-cross-reactivity (Gwinn et al., 2003; Mathew et al., 1996; Dharakul et al., 1994; Bukowski et al., 1989). Although proliferative responses are generated to one or more heterologous DENV, the most significant proliferative and cytolytic responses are nearly always directed toward the homologous serotype. The cross-reactivity among DENV memory T cells is contributed to by the relatively high degree of sequence homology between the viruses. Studies of T cell clones from DENV-infected individuals have identified clones that not only recognize DENV, but also recognize other flaviviruses such as YFV, West Nile virus, Kunjin virus, and Japanese encephalitis virus (JEV) (Spaulding et al., 1999; Mathew et al., 1998; Kurane et al., 1991a). The NS3 protein is highly conserved across the flaviviruses and NS3 has been identified as an important target of $CD4^+$ and $CD8^+$ T cells (Zivna et al., 2002; Spaulding et al., 1999; Mathew et al., 1998). NS3 was found to be the most frequently recognized viral antigen by the T cells of Vietnamese adults experiencing a secondary DENV infection (Simmons et al., 2005). These studies have demonstrated that most DENV NS3-specific T cells are serotype-cross-reactive. The E protein, capsid protein, and NS proteins other than NS3 have also been identified as targets of $CD4^+$ and $CD8^+$ T cells from DENV-infected individuals (Bukowski et al., 1989; Simmons et al., 2005; Mathew et al., 1996; Green et al., 1997; Livingston et al., 1994; Gagnon et al., 1996).

T cell activation appears to contribute to the resolution of DENV infection and IFN-γ receptor-dependent immune responses have been demonstrated to mediate early DENV clearance in a mouse model, although IFNα/β receptor pathways were more crucial in limiting initial DENV replication (Shresta et al., 2004b). Although activation of T cells may contribute to the control of DENV infection it may also contribute to the immunopathogenesis of DHF/DSS. Markers of

T cell activation such as sIL-2R, sCD4, sCD8, IL-2, CD69, and IFN-γ have been measured during acute DENV infection with wild-type or vaccine candidate viruses (Kurane et al., 1991b, 1995; Green et al., 1999a). A significant increase in the percentage of $CD8^+$ and NK cells expressing CD69 during the febrile period of illness was found in children with DHF compared to those with DF or other nondengue viral illnesses. Evidence of marked T cell activation, including higher levels of sIL2-R, sCD4, IL-2, and IFN-γ, were found in the sera of Thai children with DF or DHF compared with sera from healthy Thai children. In addition, higher levels of sCD8 were found in the sera of DHF patients, but not patients with DF, suggesting greater T cell activation in DHF. Hospitalized children suffering from acute secondary DENV infection had higher circulating numbers of NS3-specific $CD8^+$ T cells than did normal DENV-immune individuals. During acute infection, these $CD8^+$ T cells exhibited a huge proliferative response balanced by massive apoptosis (Mongkolsapaya et al., 2003). The clones that proliferated demonstrated a relatively lower affinity for the DENV causing the secondary infection and a higher affinity for serotypes that were presumed to be from a prior DENV infection. These $CD8^+$ T cells may be less effective at clearing DENV encountered in secondary infection. In addition, patients infected with DENV-1 or DENV-4 mounted IFN-γ responses to DENV-2 peptides that were equivalent in strength to those of patients infected with DENV-2. This suggests that, in contrast to primary infection, serotype-cross-reactive rather than serotype-specific T cells dominate the acute response during secondary infection. The role of T cell activation in producing the clinical signs and symptoms of DHF/DSS is discussed below and schematized in Fig. 19.2.

EPIDEMIOLOGY

Significance as Public Health Problem

More than 2.5 billion persons in 100 countries are at risk for dengue by virtue of living in areas with the

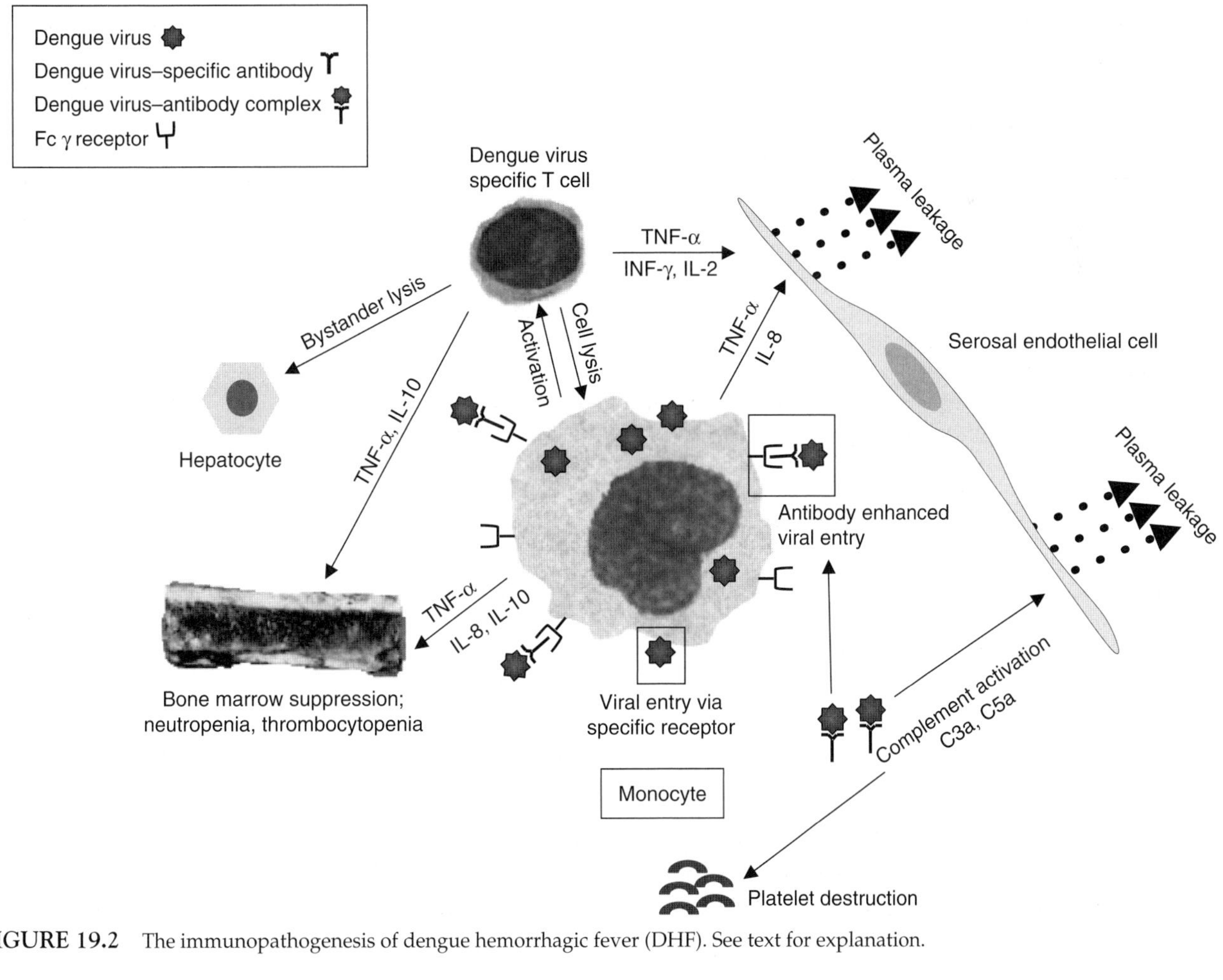

FIGURE 19.2 The immunopathogenesis of dengue hemorrhagic fever (DHF). See text for explanation.

necessary vector mosquitoes (primarily *Aedes agypti*, *A. albopictus*, and *A. polynesiensis*) and a tropical or subtropical climate (Fig. 19.3). It has been estimated that DENV infects 100 million people each year. Most infections are subclinical or result in self-limited febrile illness though up to 500,000 cases of DHF occur each year and more than 60,000 deaths due to dengue have been reported to date (Thomas et al., 2003; Gubler, 1998). DHF, a severe form of DENV infection, has been widely recognized since DENV infection was associated with Philippine hemorrhagic fever and Thai hemorrhagic fever in the 1950s (Hammon et al., 1960; Halstead et al., 1963). However, the literature documents dengue outbreaks with hemorrhagic manifestations including epistaxis, hematemesis, and gingival bleeding occurring as early as the late 1890s in North Queensland (Hare, 1898). DHF was reported sporadically from the Americas (Honduras, Jamaica, Puerto Rico, Curacao) in the late 1960s and 1970s (Anonymous, 1990). The Cuban DHF outbreak in 1981 marked the start of severe dengue disease in the Americas (Kouri et al., 1989).

Dengue epidemics can be intense both in terms of clinical attack rates when large numbers of nonimmune individuals are exposed (e.g., 20,000 cases among US forces on Saipan during World War II over a 3-month period) and morbidity and mortality (e.g., 344,203 reported cases, 10,312 severe cases, and 158 deaths in the Cuba outbreak of 1981 (Pinheiro and Corber, 1997) or 711,919 cases of DF and 2229 cases of DHF in Brazil in 2002). While dramatic outbreaks have occurred in Central and South America, India, Southeast Asia, and the Pacific Islands, there have been few reports of dengue for Africa (Thomas et al., 2003). In fact, no dengue cases were reported to the WHO between 1998 and 2001 (Anonymous, 2001) though there are rare reports of clinical disease (Durand et al., 2000). This may be due to limited circulation of DENV, under reporting (limited surveillance and laboratory capabilities) or reduced virus virulence in genetically African populations (Halstead et al., 2001). Many of the reported dengue outbreaks in Africa are among foreigners (Thomas et al., 2003).

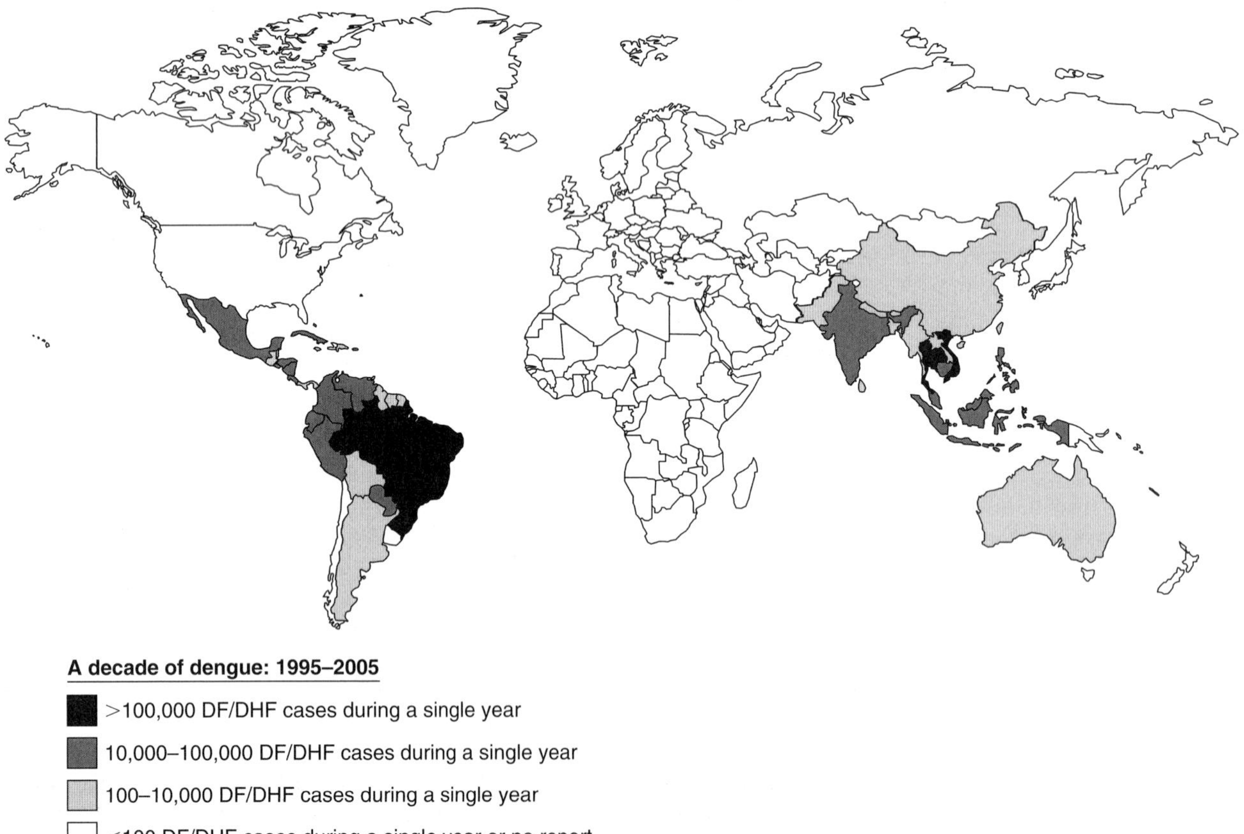

FIGURE 19.3 Countries reporting dengue fever (DF) and dengue hemorrhagic fever (DHF) from 1995 through 2005. Source of information: WHO DengueNet (www.who.int/csr/disease/dengue/denguenet) and PAHO (www.paho.org/english/ad/dpc/cd/dengue.htm).

When dengue first enters a community it affects all age groups. In hyperendemic areas where all four DENV serotypes circulate, adults become immune and dengue becomes a disease of children. In the past, dengue was considered to be an urban disease where large numbers of susceptibles could sustain transmission. However, with modern intensity and rapidity of transportation, incidence rates in more rural areas equal those in urban centers (Cummings et al., 2004). A person infected with DENV can travel to any population center worldwide with dengue vectors during the asymptomatic incubation period.

Potential as Biothreat Agent

Viruses causing viral hemorrhagic fevers (VHFs) have been classified by both the CDC and NIAID as Category A priority pathogens. Although DENV can cause hemorrhagic fever, it has relatively low potential for use as a biothreat agent due to its very low incidence of person-to-person spread and inability to be transmitted by the aerosol route. Initially the NIAID listed DENV as a Category A pathogen, but recommended in its 2003 Research Agenda for Category B and C Priority Pathogens that DENV be transferred from Category A to Category C (http://biodefense.niaid.nih.gov). However, according to the 2006 Progress Report of the NIAID Biodefense Research Agenda for Category A Agents, DENV is still included as a Category A pathogen. Currently, the CDC does not include DENV on the Bioterrorism Agents/Diseases list (http://www.bt.cdc.gov/agent/agentlist-category.asp). In its 2002 Consensus Statement, the Working Group on Civilian Biodefense excluded DENV from the list of hemorrhagic fever viruses that pose serious risk as biological weapons because it is not transmissible by small-particle aerosol, and primary DENV infection rarely causes hemorrhagic fever (Borio et al., 2002).

CLINICAL DISEASE

Infection with DENV may result in a spectrum of outcomes ranging from clinically inapparent infection to life-threatening hemorrhagic fever accompanied by vascular compromise and shock (Guzman et al., 2000; Anonymous, 1997; Burke et al., 1988; Endy et al., 2002). The WHO has provided case definitions for the major clinical syndromes caused by DENV: classic DF, DHF, and dengue shock syndrome (DSS) (Anonymous, 1997). All four serotypes of DENV can cause the full clinical spectrum of illness; it is primarily the degree of vascular leak that differentiate these syndromes.

Classic Dengue Fever

Most symptomatic cases of dengue present as classic DF, the clinic course and presentation of which has been well described in both experimental and natural infection studies (Simmons, 1931; Eram et al., 1979; Siler et al., 1926; Sabin, 1952; Liu et al., 1989; Halstead et al., 1969). The incubation period of dengue can range from 3 to 14 days but is generally 4–7 days. Dengue presents with a sudden onset of fever accompanied by headache; pain in the eyes, muscles, back, and joints; flushing of the face; anorexia; abdominal pain; nausea; occasionally vomiting; and rash. The epithet *break-bone fever* was given to dengue due to the back, limb, and joint pains that can be so severe as to be incapacitating. The fever that accompanies dengue generally does not exceed 40°C and typically has a duration 3–5 days, but may last a week or longer. Clinicians have described a two-peaked fever pattern in some dengue patients: the saddle-backed curve (Simmons, 1931; Siler et al., 1926). In these instances, fever begins with the onset of symptoms and may subside on the 3rd or 4th day only to return 1–3 days later. The saddle-backed fever pattern was noted in 40–85% of patients infected experimentally with a DENV. Rash is a common physical finding in dengue with a reported incidence in some studies of greater than 80%, however it is often evanescent and polymorphic in its appearance (Siler et al., 1926; Sabin, 1952; Halstead et al., 1969, 1970; Liu et al., 1989; Lai et al., 2004). The rash is most commonly seen on the trunk, inside of the arms and thighs, and the plantar and palmar surfaces. It may present as macular, maculopapular, morbilliform, scarlatiniform, or petechial in character. Intense pruritus and desquamation of the rash have been reported. DENV infection tends to present as a milder illness in younger children compared with older children and adults who more typically manifest the symptoms of classic DF described above (Halstead et al., 1969, 1970; Innis, 1995). In young children, DENV infection is often subclinical or a mild undifferentiated febrile illness. Respiratory symptoms, including cough; pharyngeal injection; sore throat; and rhinitis are frequently present in children with primary or secondary DENV infection (Halstead et al., 1969). Hepatomegaly may be present in children, especially in those with the more severe DHF or DSS (Eram et al., 1979; Nimmannitya, 1987; Nimmannitya et al., 1969; Kalayanarooj et al., 1997; Wichmann et al., 2004). Common laboratory abnormalities reported with DENV infection include leukopenia, thrombocytopenia, and elevation in serum transaminases (Siler et al., 1926; Sabin, 1952; Kuo et al., 1992; Kalayanarooj et al., 1997; Simmons, 1931; Souza et al., 2004). These laboratory abnormalities are transient. Leukopenia is most marked during the febrile period and is thought to be due to bone marrow suppression by the virus (La Russa and Innis, 1995). Full recovery from DF can be expected, however it may be prolonged with many patients complaining of weakness and depression for many weeks following recovery from the acute phase illness.

Dengue Hemorrhagic Fever/Shock Syndrome

DHF is a more severe form of DENV infection that is characterized by fever, thrombocytopenia, hemorrhagic manifestations, and evidence of altered vascular permeability with leakage of intravascular fluid into interstitial spaces (vascular leak syndrome) (Carlos et al., 2005; Nimmannitya, 1987; Kalayanarooj et al., 1997; Gubler, 1998; Hayes et al., 1988). DHF is primarily a disease of children less than 15 years of age in hyperendemic areas (Anonymous, 1997; Eram et al., 1979). The presence of plasma leakage and thrombocytopenia is the critical feature that distinguishes DHF from DF. The hemorrhagic manifestations of DHF include capillary fragility; petechiae, ecchymoses, or purpura; bleeding from the mucosa, gastrointestinal tract, or other site; and hematemesis or melena. Capillary fragility can be documented by a positive tourniquet test. This test is performed by inflating the blood pressure cuff midway between the diastolic and systolic blood pressure for 5 min and then counting the number of petechiae in a defined area. Twenty or more petechiae in a 2.5 cm^2 area is considered a positive test. The WHO has classified DHF into four severity grades (Table 19.1) where Grades III and IV are designated DSS (Anonymous, 1997). The clinical course of DHF resembles classical DF in its initial presentation of abrupt onset of high fever and other nonspecific constitutional signs of a few days duration. Shortly before, during, or after the time of defervescence, the patient's condition suddenly deteriorates with the advent of hemorrhagic manifestations and symptoms of hypovolemia due to plasma leakage. Hepatomegaly may become evident in a high percentage of children, particularly those in Thailand where up to 98% of children diagnosed with DHF were reported to have liver enlargement (Nimmannitya, 1987; Eram et al., 1979; Wichmann et al., 2004). In the study by Nimmannitya, liver enlargement was not necessarily accompanied by elevated transaminases as 74% of pediatric dengue cases in that study had normal ALT levels despite clinical hepatomegaly. However, mild-to-moderate elevations in both AST and ALT have been commonly reported in dengue patients, with more severe enzyme elevations observed in DHF and DSS cases (Nguyen et al., 1997; Souza et al., 2004). Increased vascular permeability allows the loss of plasma into the interstitial spaces. Pleural effusion, especially right-sided, is nearly always present in DHF patients (Nimmannitya, 1987; Pramulijo and Harun, 1991; Venkata Sai et al., 2005; Kalayanarooj et al., 1997). Ascites is also common, with a reported incidence of more than 50% in DHF patients examined by ultrasound (Venkata Sai et al., 2005; Pramulijo and Harun, 1991). Hemoconcentration occurs as a result of plasma leakage and an increase in the hematocrit of $\geq$20% over the course of illness or recovery is a diagnostic criterion for DHF. Hypovolemic shock ensues with sufficient leakage of plasma into the interstitial spaces. In mild cases of DHF (Grades I and II), the degree of plasma leakage is such that there is little effect on the blood pressure. In Grades III and IV, the shock is profound; the patient is often restless and irritable with cool, clammy extremities and a rapid, weak pulse. The course of shock is short but life-threatening; patients usually succumb or recover within 24 h of the onset of shock. The tourniquet test may be negative in patients who present in frank shock but will generally turn positive once the patient is hemodynamically stable (Innis, 1995).

Fulminant liver failure and neurological manifestations have been described in patients with dengue (Subramanian et al., 2005; Lawn et al., 2003; Nimmannitya et al., 1987; Patey et al., 1993; Solomon et al., 2000). There have been case reports of children and adults with confirmed DENV infection whose clinical presentation included fever, jaundice, altered sensorium, and liver transaminase levels in the thousands. Histological examination of the liver has shown marked steatosis, hepatitis with necrosis, and Councilman bodies in some of these cases. Reye's syndrome associated with dengue has also been reported (Kho et al., 1981; Alvarez and Ramirez Ronda, 1985). Encephalopathy, convulsions, meningismus, coma, and transverse myelitis have been reported neurological complications of DENV infection, although they are infrequent in occurrence and difficult to distinguish from co-morbidities that include DENV infection (Solomon et al., 2000).

Early diagnosis of dengue can be difficult as the symptom complex produced by DENV infection is similar to that of many other infectious diseases. In addition, the diagnosis of dengue can be definitively proven only by recovery of the virus or viral antigen

TABLE 19.1 Dengue hemorrhagic fever case definition adapted from WHO (1)

	Platelet count	Hemorrhagic manifestations	Plasma leakage[1]	Circulatory compromise
Dengue fever	Variable	Variable, includes petechiae, epistaxis	Absent	No
Dengue hemorrhagic fever				
Grade 1	< 100,000/mm^3	Positive tourniquet test and/or easy bruising	Present	No
Grade II	< 100,000/mm^3	Spontaneous bleeding in addition to manifestations of grade I	Present	No
Dengue shock syndrome				
Grade III	< 100,000/mm^3	May be hemorrhage	Present	Rapid, weak pulse and narrow pulse pressure or hypotension, with the presence of cold, clammy skin and restlessness
Grade IV	< 100,000/mm^3	May be hemorrhage	Present	Profound shock with undetectable blood pressure

[1]Objective evidence of plasma leakage includes fluctuation in the hematocrit ≥ 20% during the course of illness or clinical signs such as pleural effusion, ascites, or hypoproteinemia.

from blood or tissue of a patient, or by serological confirmation. The combination of a positive tourniquet test, depressed total white cell count, depressed absolute neutrophil count, and elevated AST level were found to be predictive in discerning DF from other febrile illnesses in children (Kalayanarooj et al., 1997). Even when dengue is suspected, it is difficult to predict which patients will deteriorate to DHF/DSS. Frequency of abdominal pain, restlessness, epistaxis, thrombocytopenia, and increased hematocrit have been found to be higher in those children hospitalized with DHF compared with those hospitalized with DF (Carlos et al., 2005; Eram et al., 1979; Hayes et al., 1988; Cohen and Halstead, 1966). Individual risk factors that may predispose to severe disease include underlying chronic disease, age, and race (Bravo et al., 1987; Guzman et al., 2002).

TREATMENT

Typical DF is a self-limited illness that can be managed at home. However, DF is often managed in hospital as prior to defervescence, DF and DHF are indistinguishable and the onset of shock with DHF can be rapid. If managed at home, patients or their parents should be instructed on the early signs and symptoms of plasma leakage, and have a means of quick transport to the hospital. Antipyretics may be useful to reduce the risk of febrile convulsions in young children. Acetaminophen, dosed according to age, is preferred over salicylates which may increase the risk of bleeding complications (Anonymous, 1997).

The successful management of DHF relies upon a high index of suspicion to promptly recognize the onset of plasma leakage followed by frequent assessment of intravascular volume status with monitoring for hemorrhagic complications. A rise in hematocrit with concomitant drop in platelet count may precede the onset of clinically apparent shock. A treatment algorithm developed at the Bangkok Children's Hospital (now the Queen Sirikit Institute of Child Health) outlines the importance of and approach to careful fluid management (Vaughn and Green, 2000). Careful and rapid replacement of plasma losses can reverse or prevent shock and prevent complications due to shock and electrolyte imbalances such as disseminated intravascular coagulation. Over hydration must be avoided. Colloid solutions (e.g., Dextran 70) appear to restore cardiac index and blood pressure and normalize hematocrit more rapidly than crystalloids (e.g., Ringer's lactate) (Dung et al., 1999; Soni et al., 2001; Khongphatthanayothin et al., 2003). Drug interventions have yet to be shown to improve outcomes over careful fluid management (Tassniyom et al., 1993; Leyssen et al., 2000; Atrasheuskaya et al., 2003).

A lack of improvement may indicate inadequate fluid resuscitation, internal bleeding, or iatrogenic fluid overload. Fresh whole-blood transfusions may be required or fluid restriction and treatment with furosemide. Fresh frozen plasma or platelets may also be considered in cases where disseminated intravascular coagulation (DIC) or other coagulaopathies complicate the clinical course (Anonymous, 1997; Mairuhu et al., 2003). The Queen Sirikit Institute of Child Health has seen a steady decline in case fatality rates, from 10% in 1970 to 2% in 1984 to 0.2% in 1990 (Nimmannitya, 1997). Mortality is usually linked to delayed provision of supportive treatment and/or premorbid chronic illness (Anonymous, 1997).

PATHOGENESIS

DHF is defined by symptoms of DF accompanied by vascular leakage and evidence of capillary fragility and bleeding. Fluid extravasation secondary to vascular leak may become so severe as to result in hemodynamic compromise and shock (DSS). Plasma leakage into the serous spaces is evident in most cases of DHF. As described previously, the pleural spaces are those most commonly involved, but the pericardial and peritoneal effusions are not uncommon. In addition, serous membranes of involved spaces are notably edematous. Microscopic studies of autopsy specimens show pericapillary edema in soft tissues and swelling of capillary endothelial cells (EC). In general, the tissue damage observed is not very severe compared to the severity of illness (Bhamarapravati et al., 1967).

The immunopathologic mechanisms driving these clinical manifestations are complex and not entirely understood. Components of both the humoral and cellular immune systems play a role in precipitating events that result in severe disease (Green and Rothman, 2006). Antibody-dependent enhancement (ADE) and cross-reactive T cell responses are thought to be important immunologic mediators of DHF. Secondary DENV infection with a serotype different from that causing primary infection has been identified as a major epidemiologic risk factor for DHF (Sangkawibha et al., 1984; Burke et al., 1988; Halstead and Yamarat, 1965). ADE of infection is produced upon secondary DENV infection when heterologous antibody present in the host binds to the infecting DENV but is not able to neutralize the virus. These DENV–antibody complexes then attach to the Fcγ receptors on circulating monocytes aiding entry of the DENV into this target cell. Halstead demonstrated that peripheral blood leukocytes (PBL) from both nonimmune rhesus monkeys and humans were nonpermissive for DENV infection, in contrast to PBLs from immune rhesus monkeys and humans which supported the replication of DENV to high titer (Halstead et al., 1976; Marchette et al., 1976). He later demonstrated nonpermissive cultured PBLs could become permissive to DENV infection when nonneutralizing antibody was added to the culture medium (Halstead and O'Rourke, 1977a, 1977b). The ability of passively transferred DENV antibody to enhance DENV infection has been demonstrated in vivo in rhesus macaques, substantiating the theory of ADE. Titers of DENV were significantly higher in nonimmune rhesus monkeys given dengue-immune human cord blood serum than in those monkeys that received normal cord blood serum (Halstead, 1979). In addition, a role for ADE in the development of severe dengue disease in endemic areas is suggested by the timing of DHF in young infants 6–12 months of age when maternally derived DENV antibody titers decline below protective levels (Kliks et al., 1988; Pengsaa et al., 2006). It is during this critical time period that infants are at increased risk for severe disease following DENV infection despite the fact that they have never been previously infected with DENV and lack DENV-specific cellular immunity. Eventually with the complete degradation of maternal antibody, infants emerge from this period of enhanced susceptibility to DHF.

Investigators have found a correlation between the level of viremia and the severity of dengue disease, as higher levels of viremia were found in DHF/DSS patients than in those with DF (Libraty et al., 2002; Murgue et al., 2000; Vaughn et al., 2000). ADE may provide a mechanism for increased viral entry into monocytes leading to higher levels of viral replication in these patients. As discussed previously, during secondary DENV infection activated T cells produce IFN-γ. It has been reported that IFN-γ increases the number of Fcγ receptors on human monocytes (Perussia et al., 1983; Guyre et al., 1983). DENV infection of U937 cells, a human monocytic cell line, can be augmented by IFN-γ in the presence of anti-DENV antibody (Kontny et al., 1988). In addition, the percentage of DENV-infected cells appears to correlate with the number of Fcγ receptors on the U937 cells. Elevated levels of IFN-γ have been documented during the acute phase of DHF/DSS and in one study, peak IFN-γ levels occurred within 2 days of peak viremia (Nguyen et al., 2004; Libraty et al., 2002). Induction of Fcγ receptors on monocytes could potentially lead to increased viral entry and replication within monocytes, leading to higher levels of viremia and potentially, more severe disease.

Increased vascular permeability with resultant plasma leakage into the serous cavities is the clinical hallmark of DHF/DSS. The lack of structural damage to EC demonstrated in autopsy studies and the short-lived nature of the vascular leak syndrome implicate serum inflammatory mediators as inducers of altered vascular permeability in serous tissues (Bhamarapravati, 1961; Bhamarapravati et al., 1967). A number of different cytokines as well as anaphylatoxins have been identified as putative effectors of the vascular leak syndrome. Complement activation has also been implicated in the pathogenesis of DHF/DSS (Russell et al., 1969; Churdboonchart et al., 1983; Suvatte et al., 1973). Depressed levels of complement; particularly C3, C4, and C5, have been documented in patients with DHF/DSS. The nadir of C3 and C4 coincided with, or occurred within 24h of, the onset of shock. The nonstructural protein NS1 has been identified as a possible precipitator of complement activation in patients with DHF (Avirutnan et al., 2006). The authors demonstrated that NS1 purified from DEN-infected cells activated complement and that significantly higher levels of NS1 were found in patients with DHF compared with DF. In addition, serum and pleural fluid levels of the terminal complement complex SC5b-9 were highly correlated with disease severity in this study. The role of complement activation in the vascular leak syndrome of dengue is further evidenced by the finding of elevated levels of C3a and C5a (anaphylatoxins) in DSS patients, which occurred at or around the time of defervescence (Bokisch et al., 1973; Malasit, 1987). Elevated plasma levels of C3a correlated with severity of disease in these patients. Both C3a and C5a cause the release of histamine, increase vascular permeability, and produce vasodilation providing a mechanism for the extravasation of plasma into the serous cavities.

It is becoming evident that an excessive cellular immune response to DENV infection, driven by activated, cross-reactive memory T lymphocytes, is critical to the immunopathology associated with DHF/DSS. Markers of T cell activation such as CD69, IFN-γ, interleukin-2 (IL-2), soluble interleukin -2 receptor (sIL-2R), soluble CD4 (sCD4), and soluble CD8 (sCD8) have been detected in the blood of patients with DHF/DSS (Kurane et al., 1991b; Green et al., 1999a), suggesting that such activation may play a role in the pathogenesis of severe dengue disease. In addition, elevated levels of tumor necrosis factor alpha (TNF-α), soluble TNF-α (sTNF-α), IL-2, interleukin 6 (IL-6), interleukin-8 (IL-8), and interleukin-10 (IL-10) measured in the blood of patients with DHF/DSS have correlated with severity of disease (Kurane and Ennis, 1994a; Hober et al., 1993, 1996; Bethell et al., 1998; Green et al., 1999b, 1999c; Raghupathy et al., 1998; Juffrie et al., 2000; Kuno and Bailey, 1994; Nguyen et al., 2004). Significantly elevated levels of TNF-α and/or sTNFR have been detected in the blood of patients DHF/DSS compared to both DF patients and normal controls (Hober et al., 1993, 1996, 1998; Gagnon et al., 2002; Nguyen et al., 2004; Iyngkaran et al., 1995). TNF-α is known to activate cultured EC and increase vascular permeability (Pober and Cotran, 1990). In in vitro studies, TNF-α has been demonstrated to increase the permeability of DENV-infected human EC, perhaps by inducing a morphological change in the cells (Dewi et al., 2004). Additionally, serum collected from four dengue patients during the acute phase of their illness induced the activation of human EC, with sera taken on the day of defervescence having the strongest activating effect (Cardier et al., 2005). TNF-α levels were markedly elevated in these samples and pretreatment of the dengue sera with anti-TNF-α antibody inhibited the EC activating effect, suggesting that TNF-α was critical for activation. Additionally, TNF-α may play a role in the hemostatic defects associated with DHF/DSS. Elevated TNF-α concentrations were significantly more common in the plasma of DF patients presenting with hemorrhagic manifestations than in those patients without hemorrhagic manifestations (Azeredo et al., 2001). Individuals with a TNF-α polymorphism that is associated with higher levels of TNF-α were 25 times more likely to develop DHF than those individuals without the polymorphism (Fernandez-Mestre et al., 2004).

IL-8 has also been implicated as a mediator of increased vascular permeability in DHF/DSS patients. It is secreted by a variety of cells including macrophages, monocytes, EC, and neutrophils. IL-8 is known to promote neutrophil adherence to vascular endothelium as well as to cause morphological changes to the EC that directly affects the permeability of the EC (Talavera et al., 2004). DENV infection of EC and monocytes in vitro induces the production of IL-8 and significantly higher levels of IL-8 have been found in DHF/DSS patients (Raghupathy et al., 1998; Juffrie et al., 2000; Bosch et al., 2002; Huang et al., 2000c). In addition, elevated levels of IL-8 and RANTES were detected in the pleural fluid of DHF/DSS patients (Avirutnan et al., 1998).

The underlying mechanisms responsible for the bleeding manifestations in DHF are not well understood. Postmortem studies of DHF/DSS cases have demonstrated the presence of hemorrhage and fibrin thrombi in the various tissues, suggesting the occurrence of DIC, at least in the most severe cases (Fresh et al., 1969; Srichaikul et al., 1975). Disturbances of coagulation and fibrinolysis have been demonstrated in all grades of DHF, with some correlation between the magnitude of derangement and the severity of disease (Srichaikul et al., 1977; Bhamarapravati, 1989; Isarangkura et al., 1987; Huang et al., 2001; Krishnamurti et al., 2001; Carlos et al., 2005). The degree of DIC observed in most cases of DHF is not severe and is not thought to be the major cause of bleeding and shock in these patients (Srichaikul et al., 1977; Mitrakul et al., 1973). The most frequent laboratory findings of coagulopathy reported in DHF are markers of coagulation activation and include an elevated partial thromboplastin time (PTT), elevated thrombin time (TT), decreased fibrinogen level, and decreased platelet count. When compared to DIC induced by bacterial sepsis, fibrinogen is only moderately depleted and fibrin degradation products are not as elevated in DHF (Suvatte et al., 1973). Activation of the coagulation system occurs during the acute febrile phase of illness and is generally more pronounced than activation of the fibrinolytic system (Van Gorp et al., 2002; Srichaikul et al., 1977; Avila-Aguero et al., 2004). It is thought to occur primarily by the stimulation of the intrinsic pathway as the PTT is abnormal in nearly all cases of coagulopathy and the prothrombin time (PT) is often normal. Tissue factor, a critical procoagulant protein responsible for initiating the coagulation cascade, is synthesized predominantly by monocytes and EC. Levels of tissue factor and plasminogen activator inhibitor type 1 (PAI-1), a major inhibitor of fibrinolysis, were shown to be elevated in children admitted to hospital with DSS (Wills et al., 2002). Elevated levels of these procoagulant factors in the blood of DSS patients suggests that factors produced by activated monocytes, EC, or platelets may be responsible for initiation of the coagulation system. A variety of cytokines, including IL-6, have been described as activators of the coagulation system (van der Poll et al., 2001; Huang et al., 2003b) and may contribute to the coagulopathy observed in DHF. Elevated levels of IL-6 have been documented in the blood of patients with DHF/DSS (Kuno and Bailey, 1994; Nguyen et al., 2004; Juffrie et al., 2001; Avila-Aguero et al., 2004; Huang et al., 2000c, 2003b) and may be associated with the hemostatic defects observed in severe dengue. DENV-infected EC produce IL-6 and tPA, a mediator of fibrinolysis and increased levels of IL-6 were shown to correlate with elevated levels of tissue plasminogen activator (tPA) in the blood of DHF/DSS patients (Huang et al., 2003b). In addition, anti-IL-6-antibody inhibited the production of tPA by cultured EC suggesting that IL-6 can regulate DENV-induced tPA production by EC. TNF-α may indirectly influence coagulation activation by its effect on IL-6 and may play a direct role in fibrinolysis. Suharti et al. (2002) reported that TNF-α levels were significantly associated with d-dimer formation, an activation marker of fibrinolysis, in patients with DSS .

Thrombocytopenia is a hallmark of DHF and is present, to some degree, in all patients with DHF. The etiology of thrombocytopenia in dengue is twofold: increased destruction and decreased production of platelets. Platelet kinetic studies done in DHF patients have demonstrated hepatic sequestration and increased destruction of platelets, possible due to antigen–antibody immune complex formation on the platelet surface (Boonpucknavig et al., 1979; Mitrakul et al., 1977). Aspirates of bone marrow taken from DENV-infected patients early in infection documented a hypocellular marrow with reduced megakaryocytopoiesis (Nelson et al., 1964; Putintseva et al., 1986). Reduced granulocytopoiesis and erythrocytopoiesis in the bone marrow were also evident early in infection, indicating that the cause of the hypocellularity was at the level of a pluripotent hematopoietic stem cell or that DENV was infecting all cells lines. Studies using long-term marrow culture (LTMC) as a surrogate for bone marrow established that DENV-induced suppression of hematopoiesis occurred and was most consistent with altered cytokine production by stromal cells (La Russa and Innis, 1995; Murgue et al., 1997). Bone marrow stromal cells are very permissive for DENV infection and have been demonstrated to produce cytokines including IL-8, macrophage inflammatory protein-1α, and TNF-α, which can induce myelosuppression. Numerous studies have correlated elevated IL-10 levels with the severity of thrombocytopenia in DHF patients (Perez et al., 2004; Libraty et al., 2002; Azeredo et al., 2001). Although the role of IL-10 in causing thrombocytopenia in DHF is not yet known, IL-10 has been demonstrated to reduce the platelet count in normal healthy volunteers, primarily by reducing platelet production (Sosman et al., 2000). In DHF patients, IL-10 levels peaked at or near the time of defervescence and also correlated with the degree of plasma leakage. Because IL-10 is a known inhibitor of proinflammatory cytokines, it is thought that IL-10 may increase in response to elevated levels of INF-γ and TNF-α as a negative feedback mechanism (Green et al., 1999c).

Hepatic enlargement and elevation of liver transaminases are common clinical findings in children with DHF. Histopathological evaluation of liver biopsies has demonstrated a variety of abnormal findings including foci of midzonal hepatocellular necrosis, microvesicular steatosis, and the presence of Councilman bodies (Bhamarapravati et al., 1967; Huerre et al., 2001). Although the type of cell infected by DENV is controversial, DENV antigen has been found in the hepatocytes and Kupffer cells (Jessie et al., 2004; Huerre et al., 2001; Hall et al., 1991; Lucia and Kangwanpong, 1994). Despite the presence of antigen in hepatic cells, it remains unclear whether or not viral replication occurs in the liver and whether or not DENV is directly hepatotoxic. Although DENV readily infects primary Kupffer cells in culture, viral replication was aborted and the infected Kupffer cells underwent rapid cell death by apoptosis (Marianneau et al., 1999). Libraty et al. (2002) reported that hepatic transaminase elevation was associated with sIL-2R, a marker of immune activation, and not with virus titer. Immune activation may be the primary mechanism of hepatocellular necrosis as $CD4^+$ T cell clones from a DENV-4-infected donor induced the lysis of cognate cells as well as bystander HepG2 cells (Gagnon et al., 1999). It is possible that hepatocyte damage is incurred when DENV-specific CTL are activated by DENV-infected Kupffer cells and then lyse hepatocytes via a bystander effect.

In summary, the immunopathological mechanisms by which DENV causes the clinical features of DHF are intricate and include aberrant humoral and cellular immune responses. Previous DENV infections may predispose to more severe disease by the induction of enhancing antibody and cross-reactive T cells. Upon secondary infection, DENV utilizes enhancing antibodies to gain entry to its primary target cell, the monocyte, leading to higher viral titers in the host. In addition, $CD4^+$ and $CD8^+$ T cells recognizing cross-reactive DENV epitopes become activated and induce the production of proinflammatory cytokines initiating the cytokine cascade. Dysregulation of these cytokines culminates in the clinical features of DHF/DSS; plasma leakage, hemoconcentration, and hemorrhagic manifestations.

VACCINES

History

Blanc and Caminopetros (1929) were the first to publish an approach to dengue vaccine development. They attempted to produce an attenuated living virus that would give rise to low-grade "symptomless" dengue with subsequent immunity. It was thought that DENV in blood could be attenuated by mixing it for 5 min with one-twentieth of its volume of ox bile, while the virus was killed by similar exposure to one-fifteenth of its volume of bile. Twenty-two persons were doubly inoculated, first with the 1:15 bile virus and later with the 1:20 bile virus; and when tested for immunity by challenge with wild-type virus 3 weeks later only two of the individuals developed dengue. It is not known which DENV serotype or serotypes were being used in this work.

Simmons et al. (1931) developed the US military's first dengue vaccine in 1929. These researchers conducted investigations in which 100% of 35 individuals experimentally infected with dengue were later found to be resistant to a second attack 400 days later. This suggested the possibility of durable immunity and a prophylactic vaccine. They benefited from the work of Siler et al. (1926) who confirmed that DENV was transmitted by *Aedes aegypti*, determined the period of infectivity of dengue patients, and determined the extrinsic incubation time for virus replication in *A. aegypti*. Simmons and colleagues prepared a "mosquito" vaccine by grinding 2010 DENV-infected *A. aegypti* mosquitoes in a sterile porcelain mortar with a salt solution and chemically pure phenol and formalin. After the suspension had stood for 48 h, it was transferred to a sterile bottle with the addition of 10 cc of sterile physiologic salt solution to further dilute the preservatives. It stood another 8 days to destroy bacteria and 20 cc of salt solution was added. This solution was slowly centrifuged to eliminate remaining insoluble matter. Before use, the vaccine was examined for bacteria (10 day aerobic and anaerobic cultures), bacterial toxins, and safety tested in rabbits, guinea pigs, and white mice, all of which remained normal. One milliliter doses were given to each of the three authors and then two volunteers. While these injections failed to prevent subsequent infection and disease, some of the volunteers developed only mild cases of dengue. In retrospect, the weakness of their approach was to give two doses of vaccine just 4 days apart and to challenge just 1 week after the second dose. Based on the work of Blanc and Caminopetros, the authors started to pursue a live attenuated vaccine derived by serial passage from mosquito to mosquito.

Both the Japanese and Americans made significant contributions to dengue vaccinology during WWII (Kimura and Hotta, 1944; Sabin, 1952). Sabin (1952) used a dermal neutralization test to show that there were two different DENV serotypes, the Hawaii and New Guinea strains B, C, and D later designated as DENV-1 and DENV-2 respectively. The test was performed by mixing and then incubating convalescent serum with acute serum containing known virus. The mixture was then injected intracutaneously into a human volunteer. The virus was considered neutralized if the subject did not develop local or systemic signs of infection. He confirmed that serotype-specific antibody was protective and observed that heterologous cross protection lasted for at least 2 months. He adapted the Hawaii strain virus by mouse brain passage for use as a vaccine in man. Tests conducted in human subjects showed that the virus became attenuated in humans after the 7th mouse brain passage (Sabin, 1952). Schlesinger et al. (1956) similarly adapted the New Guinea strain in mouse brain and tested it as a monovalent vaccine or in combination with the attenuated Hawaiian strain and yellow fever 17D. The Sabin DENV-1 mouse brain adapted virus underwent a further seven passages in suckling mouse brain. Fourteen human subjects received this strain, designated MD-1, and were immunized without disease (Wisseman et al., 1963). In 1963 Wisseman and colleagues (Bellanti et al., 1966) conducted a field trial of a mouse brain adapted DENV-1 virus among 1113 adolescent males in Puerto Rico during a DENV-3 epidemic. Among the 561 vaccine recipients, the attack rate of DF was 39% of the rate observed for the placebo control group. Further work in mouse brain derived live virus vaccines were curtailed due to concerns with host cell contaminants and adventitious agents (Wisseman, 1966).

Animal Models

An immense hurdle to understanding the role of the immune system in the control of DENV infection, as well as its contribution to the immunopathology of DHF/DSS, is the lack of an animal model that reliably reproduces human disease. Dengue viruses have only three known natural hosts: humans, various nonhuman primates, and mosquitoes. In rural areas with a low prevalence of human DENV infection, sylvatic transmission of DENV has been documented (Rico-Hesse, 1990; de Silva et al., 1999; Fagbami et al., 1977; Wolfe et al., 2001) and nonhuman primates are thought to be the primary mammalian hosts (Rudnick, 1966). For these reasons, several species of nonhuman primates have been evaluated as animal models for both primary and secondary DENV infection (Halstead et al., 1973b, 1973c; Whitehead et al., 1970; Scherer et al., 1972, 1978; Schiavetta et al., 2003). Rhesus macaques have been the most extensively used nonhuman primate model for dengue. In studies done in the early 1970s, macaques were infected with each of the four DENV serotypes at doses ranging from as low as 8–50 pfu to as high as 50,000 pfu (Halstead et al., 1973b). As in human infection, monocytes were determined to be the cells in the peripheral blood of monkeys that were permissive for DENV infection (Halstead et al., 1977). These studies determined that the onset and duration of viremia in macaques after primary DENV infection was similar to that seen in humans, however peak virus titers were generally lower. In addition, the serological response of macaques was very similar to that described in humans. Despite the confirmation of viremia in the monkeys, they developed few clinical signs of dengue, even after secondary DENV infection. Modest axillary and inguinal lymphadenopathy was reported 4–10 days after infection and some animals developed leukopenia. In studies of heterologous DENV infection in more than 100 macaques, no animal developed a classic DHF/DSS syndrome (Halstead et al., 1973c). However, there was a significant increase in peak viremia titer in secondary DENV-2 infection compared to primary infection. One animal that received DENV-2 challenge 3 months after a primary DENV-4 infection developed some laboratory abnormalities associated with DHF including thrombocytopenia, elevated PT, decrease in complement level, and an early unsustained elevation in hematocrit. Additionally, the tissue distribution of DENV was studied after primary and heterologous secondary infection in these macaques (Marchette et al., 1973). Prior to the period of viremia, virus could be recovered only from the skin at the inoculation site and regional lymph nodes. During the period of viremia, virus was recovered primarily from the skin at the inoculation site and less frequently, from skin at remote sites and axillary and inguinal lymph nodes. Virus was only occasionally recovered from other sites such as the thymus, spleen, lung, and gastrointestinal tract. Interestingly, there was suggestive evidence of a greater amount and wider distribution of virus in the tissues after secondary infection.

Other nonhuman primates that have been demonstrated to support DENV infection include chimpanzees, gibbons, and owl monkeys. Chimpanzees experimentally inoculated with low-passaged clinical isolates of DENV strains 1–4 had the onset of viremia around day 2 postinfection and a duration ranging from 5 to 9 days (Scherer et al., 1978). The mean peak titers in the animals were lower than that generally seen in human infection and none of the animals developed clinical or laboratory evidence of dengue. Secondary challenge with a heterotypic virus was done in a subset of animals, none of which developed any clinical illness. Gibbons infected sequentially with heterologous DENV all became viremic after primary infection (Whitehead et al., 1970). Gibbons were challenged at 6 and 12 weeks after primary infection and low-level viremia was detected in some of the monkeys, primarily in those challenged at 12 weeks. No animal developed illness that was dengue-related after primary or secondary infection. Owl monkeys infected with 20,000 pfu of DENV became viremic but the susceptibility to the different DENV serotypes was inconsistent and the duration of viremia quite short (Schiavetta et al., 2003). Six of the 20 infected owl monkeys developed neutropenia. Overt illness, described as extreme lethargy and markedly decreased appetite, was reported in two of the infected owl monkeys, however one of these monkeys was never viremic.

The prohibitive cost and restricted availability of nonhuman primates has prompted investigators to evaluate other less expensive, more available species as potential animal models for dengue. A number of different inbred, gene knockout, and immunodeficient strains of mice have been used in the evaluation of potential dengue vaccines as well as to study the immune response to wild-type DENV infection. Suckling mice can be infected intracerebrally (i.c.) with nonadapted DENV and although the animals develop fatal neurological symptoms, they do not become viremic and do not develop clinical or laboratory evidence of DHF/DSS (Raut et al., 1996). Passaging of DENV in suckling mouse brain was an early method of developing candidate dengue vaccines, as some viruses would become attenuated with multiple passages (Schlesinger et al., 1956; Hotta, 1969; Sabin, 1952). In addition, the attenuation phenotype could be assessed in this model by the amount of neurological disease that developed after i.c. inoculation. Adult mice, in contrast, do not develop any symptoms of illness when infected peripherally with nonadapted DENV. To produce symptoms of encephalitis in adult mice requires challenge with highly mouse-adapted strains of DENV. DENV-2 antigen could be recovered from young adult mice inoculated intraperitoneally (i.p.) with a nonadapted DENV-2 virus, however no replicating virus was recovered from the sera or organs of infected animals and none of the animals became symptomatic (Boonpucknavig et al., 1981).

Because nonmouse adapted DENV do not replicate well, if at all, in older mice, more sophisticated mouse models have subsequently been developed. Numerous models utilizing severe combined immunodeficient (SCID) mice have been developed to improve DENV replication in the mouse. SCID mice reconstituted with human peripheral blood lymphocytes (PBL) were inoculated with DENV-1 to determine if these mice would support DENV replication (Wu et al., 1995). The infection rate of the hu-PBL-SCID mice was quite low, possibly due to a small number of human target cells in the reconstituted mice, and none of the mice produced antibody. In another experiment, investigators engrafted SCID mice with various hemopoietic cell lines to determine if mice could be infected peripherally with DENV (Lin et al., 1998). Only those SCID mice engrafted with K562 cells, an erythroleukemia cell line, were susceptible to DENV infection. Mice did not become infected with DENV unless the virus was injected directly into the K562 tumor mass. Infected mice exhibited limb weakness and paralysis and died approximately 4 weeks postinfection. High titers of virus were recovered from the brain and peripheral blood of infected mice and the titer of virus appeared to correlate with the extent of neurological symptoms. To improve the peripheral infectivity of DENV in a SCID mouse model, SCID mice were transplanted with a human hepatocarcinoma cell line, HepG2, which is known to support DENV replication (An et al., 1999). Although the HepG2 cells were injected into the spleen of the SCID mice, the majority of cells translocated to the liver soon after injection. At approximately 2 months after transplantation of the HepG2 cells, mice were infected i.p. with DENV-2. Approximately 80% of the infected mice became viremic by day 8 postinfection and developed titers ranging from 3.2 $\log_{10}$ to 6.8 $\log_{10}$ pfu/ml. Virus was detected in the livers of all infected animals with the HepG2 cells appearing to be the major target of infection, as the majority of these cells were destroyed during infection. The infected mice developed anorexia, weight loss, and eventually hindleg paralysis. Significant laboratory abnormalities that developed in infected mice included reduced platelet counts, elevated PTT, and an elevated serum TNF-α that developed with the onset of paralysis. Mice also developed an elevated HCT but this could be attributed to dehydration caused by the paralysis. The HepG2-engrafted SCID mouse model has been used to evaluate the role of DENV-specific $CD8^+$ cells in secondary DENV infection (An et al., 2004). The DENV-2 specific T cell clone 2D42 was isolated from infected BALB/c mice. 2D42 is a $CD8^+$ T cell clone that recognizes epitopes on the NS3 protein of DENV-2. Two days after HepG2-SCID mice were inoculated with 2D42 cells, they were infected with DENV-2 virus. The HepG2-SCID mice inoculated with 2D42 cells prior to infection developed dramatically lower platelet counts and died sooner than HepG2-SCID that did not receive the $CD8^+$ T cells. HepG2-SCID mice that received normal thymocytes prior to DENV-2 infection died earlier than did the mice that received the 2D42 cells indicating that DENV-specific $CD8^+$ may provide both a protective and pathogenic role in secondary DENV infection. Whether this model can be expanded to other DENV and other T cell clones has yet to be determined.

The nonobese diabetic (NOD)/SCID mouse has recently been described as a model for DF (Bente et al., 2005). Mice transplanted with human cord blood hematopoietic progenitor ($CD34^+$) cells developed a full range of dendritic cells, defined as $CD11^+$ or $CD123^+$ lineage negative, HLA-DR^+ cells. Eight or more weeks posttransplant the mice were infected subcutaneously with DENV-2 strain K0049, a strain selected for its ability to replicate well in human dendritic cells in vitro. The humanized NOD/SCID mice developed a significant decrease in platelets on day 8 postinfection and a marked erythema of the skin in the majority of mice on days 2–4 postinfection that continued through day 14 in some mice. Viremia was detected by RT-PCR and peaked on day 2 and lasted until day 14 postinfection. Nonhumanized NOD/SCID control mice demonstrated lower viremias and did not develop erythema or decrease in platelet count. Virus could be detected in the spleen, liver, and skin of some of the hu-NOD/SCID mice but not in any of the control mice.

Because SCID mice do not have an intact immune system, other mouse models are being developed to help identify which components of the immune system are critical for the control of DENV infection, as well as those that may contribute to the immunopathology of DHF/DSS. Knockout mice deficient in both the IFN-α/β and IFN-γ receptor genes (AG129) were infected i.p. with both mouse-adapted and non-adapted strains of DENV-2 (Johnson and Roehrig, 1999). Only AG129 mice infected with the mouse-adapted strain became symptomatic indicating that both IFN-α/β and IFN-γ systems are important for the control of DENV. These mice developed neurological symptoms by day 7 that progressed to death by day 21 postinfection. Virus was recovered from the serum of AG129 mice, peaking at day 3 and persisting at low levels until day 9. Virus was also recovered from the spleens and brain tissue of the infected AG129 mice. In contrast, none of the wild-type 129 (WT129) mice developed any symptoms and no virus was recovered from tissues or serum. Additional studies with the IFN-α/β receptor deficient (A129), IFN-γ receptor deficient (G129) and AG129 mice determined that the activities of both IFN-α/β and IFN-γ are vital for protection against primary DENV infection (Shresta et al., 2004b). IFN-α/β may be important in the control of early DENV dissemination to CNS as well as other tissues. AG129 mice were more susceptible to both mouse-adapted DENV-1 and DENV-2 viruses, indicating that IFN-γ action may be more critical for protection against DENV than IFN-α/β. The DENV strain PL046 was further adapted to AG129 mice by alternating passages in AG129 mice and C6/36 cells (Shresta et al., 2006). Although the adapted virus, designated D2S10 did not cause significantly higher viremia titers in AG129 mice than did PL046, it was more virulent. Mice infected with D2S10 developed hunched posture, ruffled fur, and lethargy by day 3 or 4 postinfection. Interestingly, there was evidence of increased vascular permeability of the large and small intestine in these mice, although no evidence of increased vascular permeability of the pleural, pericardial, or peritoneal membranes was detected. TNF-α was implicated in these studies as a mediator of disease.

Unlike the IFN receptor knockout mice described above, the A/J mouse strain is an inbred immunocompetent strain that can be infected with high titers of a nonmouse adapted DENV-2 strain (Huang et al., 2000b). DENV-2 virus causes a nonfatal paraplegia in approximately 70% of infected A/J mice 2–3 weeks after infection. In addition, 10–14 days after infection, mice developed thrombocytopenia thought to be the result of anti-platelet antibody. Indeed, anti-platelet antibody was detected in the serum of infected A/J mice as early as 4 days postinfection and antibody levels were maintained for 2–3 weeks. Based on these results, other investigators have characterized the primary immune response to DENV-2 in this model (Shresta et al., 2004a). They confirmed that A/J mice can be infected peripherally with a nonmouse adapted strain of DENV-2 and that a proportion of mice develop paralysis. DENV-2 virus was recovered from CNS tissue of infected mice but not from the serum or any other organs. DENV-infected mice were found to have an increased number of NK cells expressing CD69, an early activation marker, at 3 days postinfection. In addition, there were a significant number of activated B cells detected in the spleens of DENV-infected mice compared with controls. Although an increase in activated T cells was also noted, it was not as significant. The only other DENV strain that was tested in this model was DENV-2 16881 and no mice developed paralysis after infection with the 16881 strain. This suggests this model may be restricted in its ability to evaluate other DENV.

As stated above, the time course of DENV replication and antibody response in nonhuman primates is comparable to what occurs in humans. For this reason, the attenuation phenotype and immunogenicity of experimental dengue vaccines can be evaluated in monkeys to identify the best candidate to be taken into human clinical trials. In addition, refinement of this model may prove useful for the study of the immunopathology of DHF/DSS. Mouse models have become helpful in dissecting out important components of the immune response to dengue and have provided some clues to which factors may be involved in DHF/DSS. In addition, mouse models that support virus replication of different nonmouse adapted DENV play an important role in the early evaluation of potential vaccine candidate viruses.

Vaccines in Development

While several candidate dengue vaccines have entered clinical trials in recent years, the development of safe and effective dengue vaccines faces many challenges. Dengue vaccine development proceeds

TABLE 19.2 A partial list of dengue vaccines under active development including the general approach, developers, number of dengue virus (DENV) genes in the vaccine for each serotype, and the clinical phase of development

Vaccine approach	Developer(s)	DENV genes	Status
Live-attenuated (traditional)	Sanofi-pasteur/Mahidol University	10	Tetravalent Phase 2
Live-attenuated (traditional)	GlaxoSmithKline Biologicals/Walter Reed Army Institute of Research (WRAIR)	10	Tetravalent Phase 2b
Molecularly attenuated	US National Institutes of Health	10	DENV-1 and 4 Phase 1
Molecularly attenuated	US Food and Drug Administration/WRAIR	10	Preclinical
Live chimeric (YFV)	Acambis/sanofi-pasteur	2	Tetravalent Phase 2
Live chimeric (DENV-4)	US National Institutes of Health	2 + 8	DENV-2 and 3 Phase 1
Live chimeric (DEN-2)	InViragen/CDC	2 + 8	Preclinical
DNA	Naval Medical Research Center/WRAIR	2	Preclinical
Purified inactivated virus	WRAIR	3	Preclinical
Recombinant subunit	Hawaii Biotech, Inc.	2	Preclinical

without the benefit of a full understanding of the pathogenesis of severe dengue disease or an adequate animal disease model. And unlike other multivalent vaccines where vaccines can be developed and fielded as they are developed, four dengue vaccines must be developed and combined as a single vaccine to avoid the theoretical risk of antibody-dependent immune enhancement of disease severity with subsequent natural exposure if a broadly protective immune response is not elicited with vaccination. Table 19.2 outlines nine leading approaches to vaccine development. The order of vaccine candidates reflects the bias that the presentation of more genes offers a greater opportunity for adequate immune responses across a genetically diverse human population and that live approaches are superior to inactive approaches.

Live Attenuated Virus Vaccines (Traditional Approach)

Most licensed human vaccines in the United States are empirically derived attenuated live viral vaccines [measles, mumps, rubella, adenovirus 4, adenovirus 7, varicella, rotavirus, smallpox, yellow fever, and influenza]. Note that one of these vaccines protects against another flavivirus (YFV) and that there is a live attenuated Japanese encephalitis vaccine available in China and elsewhere raising expectations for the development of a successful dengue vaccine. In 1971, a Dengue Task Force was formed by the Virus Commission, United States Armed Forces Epidemiology Board to foster cooperative efforts to develop dengue vaccines using traditional methods (Halstead, 1978). Subsequently, research teams from Mahidol University (Bangkok, Thailand) and the Walter Reed Army Institute of Research (WRAIR, Silver Spring, MD) independently developed monovalent, bivalent, trivalent, and tetravalent dengue vaccine candidates.

The Mahidol vaccine virus candidates, DENV-1 16007 PDK-13, DENV-2 16681 PDK-53, DENV-3 16562 PGMK 30/FRhL3, and DENV-4 1036 PDK-48, were clinical isolates grown in either primary dog kidney (PDK) cells (DENV-1, 2, and 4) or primary African green monkey kidney (PGMK) cells and then in fetal rhesus lung (FRhL) cells (DENV-3). Similarly the WRAIR candidates, DENV-1 45AZ5 PDK-27/FRhL3, DENV-2 16803 PDK-50/FRhL3, DENV-3 CH53489 PDK-20/FRhL3, DENV-4 341750 PDK-6/FRhL3 were all grown in PDK cells with terminal passages in FRhL cells (Bhamarapravati and Yoksan, 2000).

The Mahidol University monovalent vaccines were well-tolerated in adult flavivirus-naive volunteers in Thailand (Bhamarapravati et al., 1987; Bhamarapravati and Yoksan, 1997, 2000) and the United States (Vaughn et al., 1996; Kanesa-thasan et al., 2001b). Phase 1 clinical testing identified passage levels with minimal reactogenicity while inducing 100% seroconversion for all four monovalent components (some with a booster dose) (Bhamarapravati and Yoksan, 1997). Bivalent and trivalent formulations using DENV-1, 2, and 4 vaccine candidates elicited uniform seroconversions in Thai subjects (Bhamarapravati and Yoksan, 1989). However, when all four serotypes were combined into a tetravalent vaccine and given in a single dose, only DENV-3 viremia and neutralizing antibody were elicited (Kanesa-thasan et al., 2001b). Studies evaluating formulations using lower doses of DENV-3 with one or two boosters were subsequently pursued to overcome this apparent viral interference phenomenon. New formulations were evaluated in adults (Sabchareon et al., 2002) and children (Sabchareon et al., 2004). While immunogenicity remained high following multiple dosing (up to 100% seroconversion following three doses of tetravalent vaccine over 12 months), moderate reactogenicity, which included one child who had a week long dengue-like illness and nearly universal systemic reactogenicity among Australian adult volunteers, prompted development of a replacement DENV-3 component (Kitchener et al., 2006). However, a new biologically cloned DENV-3 component was also highly reactogenic (Sanchez et al., 2006) and this vaccine development approach is not currently active.

While the WRAIR vaccine demonstrated protective efficacy as a tetravalent formulation in rhesus macaques (Sun et al., 2006), like the Mahidol vaccine, it has primarily been developed through a series of Phase 1 evaluations of monovalent and tetravalent vaccine candidates (Sun et al., 2003; Edelman et al., 2003; Kanesa-thasan et al., 2002). Among the initially selected passage levels, the seroconversion rates were 100, 92, 46, and 58% for single dose of DENV-1-4, respectively. The WRAIR DENV-2, 3, and 4 vaccine viruses were well-tolerated by volunteers. However, the DENV-1 PDK-20/FRhL3 monovalent candidate was associated with increased reactogenicity with 40% developing fever and generalized rash (Edelman et al., 1994; Mackowiak et al., 1994). In an effort to reduce reactogenicity, the DENV-1 component was replaced with a further passaged virus (PDK-27 rather than PDK-20) and to improve immunogenicity, the DENV-4 component was replaced with a lower passaged virus (PDK-6 rather than PDK-20). Current tetravalent formulations are in Phase 2 testing in North America and Asia.

There are important safety issues for live dengue vaccines. Foremost, is the theoretical risk of enhanced disease due to natural DENV exposure in endemic regions following DENV vaccination. Less than tetravalent dengue vaccines have not been fielded to date due to the generally accepted need for a vaccine to induce primary-type immune responses to all four DENV serotypes simultaneously. The simultaneous production of neutralizing antibodies to all four serotypes of DENV is predicted to minimize the risk of disease enhancement following natural infection. ADE appears to occur with neutralizing antibodies at subneutralizing concentrations, so a vaccine that induces protection for a period of time may later increase the risk for enhanced disease. This is particularly a concern for vaccines that induce low levels of neutralizing antibodies (Burton et al., 2000), but could be a problem for any vaccine given enough time. The lessons learned with inactivated respiratory syncyntial virus (RSV) and measles vaccines suggest that low-level neutralizing antibody responses, lack of induction of neutralizing antibody to important epitopes (e.g., G protein in RSV), high $CD4^+$ proliferation, and absent $CD8^+$ memory responses are associated with exaggerated disease with subsequent natural infection (Murphy et al., 1986). Skewing toward a Th-2 type response by vaccination may also play a role in disease enhancement (Griffin et al., 1994; Graham et al., 1993). It is anticipated that tetravalent seroconversion with adequate $CD8^+$ memory responses will preclude enhanced disease even many years following vaccination. Broad Th-1 responses seem most likely with live virus approaches to vaccine development.

A second concern is that enhanced vaccine reactogenicity is possible when a live tetravalent DENV vaccine is administered to persons with preexisting anti-flavivirus antibody. Specifically, preexisting monovalent anti-DENV antibody could neutralize one component of a live tetravalent vaccine and enhance the replication of heterologous viral components to cause DF or DHF. To date, enhanced disease has not been observed in yellow fever immune vaccine recipients (Bancroft et al., 1984), volunteers receiving closely spaced sequential DENV of different serotypes, or in volunteers screened for flavivirus antibody yet experiencing anamnestic antibody response patterns following vaccination (Kanesa-Thasan et al., 2003). In this last group, volunteers were screened using the plaque reduction neutralization assay for DENV. Preexisting flavivirus antibody was probably nondengue and perhaps represents a lesser risk. Empiric administration of tetravalent vaccines to volunteers with documented preexisting partial DENV immunity has yet to demonstrate an increased risk though the numbers of volunteers tested is small to date (Kanesa-thasan et al., 2002). There have been recent suggestions that wild-type DENV-4 and DENV-2 viruses may be naturally attenuated in many flavivirus-naive hosts (Vaughn, 2000). Safety data for DENV-4 and DENV-2 component vaccines administered to flavivirus-naive volunteers must be considered in this context. Given the observation that secondary DENV infections are also largely subclinical (Burke et al., 1988; Endy et al., 2002), large numbers of partial dengue-immune volunteers will need to receive candidate vaccines before this risk can be adequately assessed. However, since live attenuated vaccine viruses replicate to very low levels in vaccinees, it is unlikely that ADE would be capable of increasing titers to levels associated with serious disease (Vaughn et al., 2000).

Other safety concerns with live attenuated vaccines include cell culture-derived adventitious agents, community spread of vaccine virus by local vector mosquitoes, vaccine virus neurovirulence, and the effects of vaccine administration to immunocompromised hosts.

Live Virus Vaccines Attenuated by Engineered Mutations

Identification of mutations capable of appropriately attenuating DENV for vaccine use has been a fundamental goal of dengue vaccine development. As described above, passage of DENV in PDK cells has lead to the accumulation of mutations associated with an attenuation phenotype, however, it was not until 2000 that specific mutations derived by this empirical approach were identified as contributing to the attenuation of the DENV-2 PDK-53 vaccine candidate (Butrapet et al., 2000). In a separate vaccine strategy developed at the Laboratory of Infectious Diseases (NIAID, NIH, Bethesda, MD), the DENV-4 full-length cDNA clone was used to engineer deletion mutations in the 5′ UTR for structure/function analysis of this region (Cahour et al., 1995). Although several mutant constructs yielded virus, the resulting viruses were not always genetically stable and were not further developed as vaccine candidates. However, deletion mutations introduced into the 3′ UTR of DENV-4 were more stable and conferred varying levels of attenuation in rhesus monkeys compared to the wild-type parent virus (Men et al., 1996). Among the deletions created, which ranged from 30 to 262 nucleotides, the 3′ 172-143 deletion mutation, later referred to as Δ30, showed a desirable balance between level of attenuation and immunogenicity in monkeys. The DENV-4 virus containing the Δ30 mutation was subsequently evaluated in adult human volunteers and was shown to be safe, asymptomatic, and immunogenic at all doses administered (10^1–10^5 pfu) (Durbin et al., 2001, 2005).

The favorable evaluation of the DEN4Δ30 vaccine in humans supported a unique strategy to create vaccine candidates for the other three DENV serotypes. Because the structure of the DENV 3′ UTR is well conserved among all four serotypes, it was reasoned that deletion of nucleotides analogous to the Δ30 mutation in each serotype would likely result in attenuation. Introduction of the Δ30 mutation into DENV-1 resulted in a vaccine candidate attenuated to levels similar to that observed in monkeys for DEN4Δ30 (Whitehead et al., 2003a) and well tolerated and immunogenic in humans (Durbin et al., 2006a). Although introduction of the Δ30 mutation into DENV-2 conferred only a modest level of attenuation (Blaney et al., 2004b), introduction of the Δ30 mutation into DENV-3 failed to attenuate the resulting virus (Blaney et al., 2004a). Nevertheless, the Δ30 mutation has been shown to contribute to the attenuation observed for chimeric viruses DEN2/4Δ30 and DEN3/4Δ30 (Blaney et al., 2004b; Whitehead et al., 2003b) and tetravalent formulations containing these viruses along with DEN1Δ30 and DEN4Δ30 have been tested in monkeys and shown to be attenuated and elicit balanced antibody responses (Blaney et al., 2005).

Other strategies to identify attenuating mutations for DENV have included the generation of point mutations throughout the virus genome. Chemical mutagenesis has been successfully used to identify point mutations exhibiting a range of useful phenotypes, including temperature sensitivity, small plaque size, enhanced replication in Vero cells, reduced replication in mouse brain, reduced replication in SCID mice transplanted with human liver cells, and reduced infectivity for mosquitoes (Blaney et al., 2002, 2003; Hanley et al., 2003). Paired charge-to-alanine mutagenesis of the DENV-4 NS5 coding region has yielded a menu of attenuating mutations (Hanley et al., 2002), and viruses bearing several of these mutations have been successfully tested in monkeys (Hanley et al., 2004) and are currently being evaluated in humans. Substitution or deletion of noncontiguous nucleotides in the 3′-terminal stem-loop structure of DENV genome has yielded viruses DEN2mutF, with reduced replication in tissue culture mosquito cells (Zeng et al., 1998), and DEN1mutF, with reduced replication in mosquito cells and attenuated replication in monkeys (Markoff et al., 2002). It is possible that the mutF set of mutations could be introduced into the remaining DENV serotypes to develop a tetravalent vaccine in a manner analogous to that proposed for the Δ30 mutation.

Live Attenuated Virus Chimeric Virus Vaccines

Recombinant DNA technology has made it possible to expedite the development of several unique attenuated DENV vaccine candidates. For more than a decade it has been possible to produce chimeric DENV in which the structural protein coding region of

TABLE 19.3 Intertypic and interspecific chimeric viruses expressing dengue structural genes

Structural genes	Genetic background	Tested in monkeys	Tested in humans	References	GenBank accession no.
DENV-1 PR/94	DENV-4Δ30	X		Blaney et al. (2007)	
DENV-2 NGC proto	DENV-4Δ30	X	X	Whitehead et al. (2003b)	AY243467
DENV-3 Sleman/78	DENV-4Δ30	X	X	Blaney et al. (2004a)	AY656168
DENV-1 WP	DENV-4 814669	X		Bray and Lai (1991)	
DENV-2 NGC proto	DENV-4 814669	X		Bray et al. (1996)	
DENV-2 NGC mouse	DENV-4 814669			Bray and Lai (1991)	
DENV-3 CH53489	DENV-4 814669			Chen et al. (1995)	
DENV-1 16007	DENV-2 PDK-53	X		Huang et al. (2000a)	AF180817
DENV-3 16562	DENV-2 PDK-53	X		Huang et al. (2003a)	
DENV-4 1036	DENV-2 PDK-53	X		Huang et al. (2003a)	
DENV-1 PUO359	YF-17D	X	X	Guirakhoo et al. (2001)	
DENV-2 PUO218	YF-17D	X	X	Guirakhoo et al. (2000)	AF038402
DENV-3 PaH881/88	YF-17D	X	X	Guirakhoo et al. (2001)	AF349753
DENV-4 1228	YF-17D	X	X	Guirakhoo et al. (2001)	
DENV-2 PR159	YF-17D			van Der Most et al. (2000)	M19197
DENV-2 NGC	YF-17DD			Caufour et al. (2001)	M29095
DENV-2 NGC/40247	YF-17DD			Caufour et al. (2001)	
DENV-2 PUO218	YF-17D			Chambers et al. (2003)	AF038402
DENV-4 VR-1257	YF-17D			Chambers et al. (2003)	
DENV-2 PUO218	JEV Nakayama			Chambers et al. (2006)	
DENV-4 S-14	JEV JaOArS982			Mathenge et al. (2004)	AY559316

a flavivirus is replaced by that from a specific DENV serotype. The generation of chimeric DENV was first reported by scientists at the NIH in 1991 (Bray and Lai, 1991) shortly after their recovery of DENV-4 from a full-length cDNA clone (Lai et al., 1991). Since that beginning, numerous chimeric viruses expressing DENV structural proteins have been created, as listed in Table 19.3. Intertypic DENV representing all four serotypes have been developed using various DENV-2 and DENV-4 viruses as the genetic background. In addition, interspecific chimeric DENV have been generated using flavivirus species such as YFV and JEV as the genetic background. Although the genetic materials for each chimeric virus approach have differed, the goal has remained the same: use chimerization to bring together the immunogenic structural genes of DENV and the attenuated NS genes of a related flavivirus to create safe vaccine candidates for each of the four DENV serotypes.

For the generation of chimeric viruses, it is generally accepted that the genetic backbone should be attenuated. However, experience has shown that chimerization itself can lead to attenuation. Intertypic chimeric viruses created with a wt DENV-4 background and wt DENV-2 structural genes were shown to be significantly attenuated compared to either wt parent virus in mice, mosquitoes, or rhesus monkeys (Whitehead et al., 2003b). Nevertheless, the use of an attenuated background further augmented the level of attenuation of the DENV-4/DENV-2 chimeric viruses (Whitehead et al., 2003b). Several attenuated vaccine viruses or vaccine candidate viruses have been used as the genetic background for creating chimeric dengue vaccine candidates. These include the use of YFV vaccine strain 17D (van Der Most et al., 2000; Guirakhoo et al., 2001; Caufour et al., 2001; Chambers et al., 2003), PDK passaged DENV-2 vaccine candidate PDK-53 (Huang et al., 2003a), and genetically modified vaccine candidate DEN4Δ30 (Whitehead et al., 2003b; Blaney et al., 2004b). For the most part, wild-type, unaltered dengue structural genes have been used with these attenuated backgrounds, which should ensure that infectivity and immunospecificity is preserved.

For chimeric viruses expressing DENV structural genes, it is clear that the level of attenuation observed for the background virus is maintained in the chimeric virus without the need for attenuating mutations in the structural genes. However, this has not been the case for certain other chimeric flaviviruses, such as chimeric YFV/JEV in which it was necessary to use structural genes from an attenuated strain of JEV in order to create an attenuated chimeric virus, even though the genetic background was YF vaccine strain 17D (Chambers et al., 1999). It is unclear why chimerization itself leads to attenuation of the resulting virus. It is reasonable to assume that tropism, attachment, and infectivity of the dengue chimeric viruses is not altered compared to the parent virus from which the structural gene is derived since both viruses are coated on their surface with the same authentic E glycoprotein, an attractive feature of the chimeric virus vaccine approach. The basis of the observed attenuation is likely the result of decreased virus replication due to reductions in gene expression, genome replication, or packaging efficiency. The suggestion has been made that attenuation may reflect the relative incompatibility of the structural and NS proteins such as proteases (Bray and Lai, 1991). Chimeric DENV resulting from the substitution of the capsid C protein along with the prM and E proteins have been more attenuated than substitution of only the prM and E regions. In fact, substitution of the C, prM, and E region of DENV-2 into DEN4Δ30 resulted in a chimeric virus that replicated in tissue culture, yet demonstrated minimal replication in mice and undetectable replication in monkeys, possibly the result of an inefficient, host-mediated interaction between the DENV-2 C protein and the DEN4Δ30 genome during virus assembly (Whitehead et al., 2003b).

Several promising chimeric dengue vaccine candidates have been successfully evaluated in preclinical studies and in human studies. In general, the chimeric vaccine candidates that have been evaluated in nonhuman primates have demonstrated decreased levels of viremia and a reduced number of viremic days compared to parental viruses from which the structural genes were derived (Whitehead et al., 2003b; Bray et al., 1996; Guirakhoo et al., 2002, 2004; Blaney et al., 2004a). Based on the success of these nonhuman primate studies, selected vaccine candidates have been evaluated in human clinical trials. The ChimeriVax platform, developed by Acambis, Inc. (Cambridge, MA), is based on YFV 17D, and has been used to create chimeric vaccine candidates for each of the dengue serotypes. The monovalent ChimeriVax-DEN2 vaccine candidate was successfully evaluated in humans and shown to be safe and immunogenic (Guirakhoo et al., 2006). Early reports of the Phase 1 testing of the ChimeriVax tetravalent vaccine indicate that it is safe without any serious adverse side effects. Phase 1 testing of the NIH chimeric vaccine candidate DEN2/4Δ30 has shown it to be safe and immunogenic at a dose of 10^3 pfu (Durbin et al., 2006b), and clinical evaluation of DEN3/4Δ30 is currently underway.

Live Vectored Vaccines

Although numerous recombinant vector systems have been used to express DENV antigens, the resulting vaccine candidates have achieved very limited success. Early vaccine candidates including recombinant vaccinia viruses (VV) expressing C-prM-E-NS1-NS2A (Zhao et al., 1987), C-prM-E (Deubel et al., 1988), or E (Hahn et al., 1990) failed to induce significant DENV-specific antibodies in rodents or monkeys, although 3-week-old offspring born to immunized mice were resistant to dengue encephalitis following intracerebral challenge with wt DENV (Bray et al., 1989). Since mice were also protected against encephalitis following immunization with VV expressing nonstructural protein NS1 (Falgout et al., 1990), it would appear that cell-mediated immunity in the absence of a neutralizing antibody response to E protein plays a significant role in protection afforded by VV vectors. Recombinant VV vectors expressing full-length E protein have been shown to express only intracellular E protein, therefore, VV expressing C-terminal truncated E proteins lacking the hydrophobic membrane anchor region and subsequently secreted from the cell have been shown to be more immunogenic (Men et al., 1991). By analogy to other flavivirus expression systems, it has been suggested that expression of DENV prM and E leads to the formation of empty virus-like particles, and that co-expression of the C protein may destabilize or prevent secretion of these particles, thereby decreasing overall immunogenicity (Fonseca et al., 1994). In an updated approach, the replication-deficient modified vaccinia Ankara (MVA) was used as a vector to express the C-terminal truncated E protein of DENV-2 (Men et al., 2000). For studies in monkeys, two doses of this vector conferred relatively low levels of neutralizing antibody and only partial protection upon virus challenge, while three doses resulted in higher levels of neutralizing antibody and complete protection following challenge. Construction of MVA vectors expressing E protein of the other three dengue serotypes has not been reported.

Because of their safety and low pathogenicity in humans, replication-defective adenoviruses vectors have been used to express the E protein of DENV-2 (Jaiswal et al., 2003). Three doses of this vector were required to achieve a significant antibody response in mice. Because of the ability of the adenovirus genome to accommodate lengthy insertions, it is envisioned that a single recombinant adenovirus vector can be constructed to express the E proteins from all four dengue serotypes, and such work is reportedly underway (Jaiswal et al., 2003). Prime-boost strategies using adenovirus expressing domain III of the DENV-2 envelope protein or DNA vectors expressing the same domain have been shown to be immunogenic in mice (Khanam et al., 2006).

The rationale behind the live vectored vaccine approach is the proven safety of the vectors used to generate the various vaccine candidates. In addition, poxviruses and adenoviruses have been shown to be potent inducers of both humoral and cellular immunity. However, there is widespread immunity in the human population to both of these vectors. Recent evaluation in humans of a vaccinia-JEV vaccine candidate demonstrated that while vaccinia-naive volunteers mounted a neutralizing antibody response, vaccinia-immune volunteers failed to develop neutralizing antibodies (Kanesa-thasan et al., 2001a), indicating that preexisting antivector immunity may have rendered the recombinant vaccine ineffective. Nevertheless, several solutions have been proposed to overcome the observed immune inactivation. Priming with a DNA expression vector prior to boosting with the virus vector has been shown to circumvent antivector immunity (Yang et al., 2003). In addition, adenoviruses and poxviruses from other animal species have been suggested for use as vectors.

Nucleic Acid Vaccines

DNA vaccines consist of DNA constructs that are introduced into cells and subsequently translated into immunogenic proteins. The DNA construct is a plasmid that contains a eukaryotic promoter and termination sequences to drive transcription of DENV target genes in the vaccine recipient (Whalen, 1996). After the DNA vaccine enters cells, transcribed RNA is translated to produce proteins to be processed and presented to the immune system in the context of MHC molecules and generate humoral and cell-mediated immune responses. A variety of adjuvants have been studied in conjunction with the DNA vaccines to enhance immune responses. In addition, various routes of administration and methods of delivery have been investigated.

DNA vaccines afford advantages over conventional vaccines including ease of production, stability, and transport at room temperature, decreased likelihood of replication interference as can been seen with live attenuated approaches, and it is possible to vaccinate against multiple pathogens in a single vaccination (Gurunathan et al., 2000).

Preclinical evaluation of the first dengue DNA vaccine candidate expressing the PrM and E genes was published in 1997 (Kochel et al., 1997) following the early successes in animals with DNA vaccines targeting HIV (Wang et al., 1993), influenza (Ulmer et al., 1993), and malaria (Sedegah et al., 1994). Work by Kochel et al. at the Naval Medical Research Center (NMRC, Bethesda, MD) evaluated two eukaryotic plasmid expression vector systems expressing the PrM protein and 92% of the E protein for DENV-2 virus (New Guinea C strain). Both constructs induced neutralizing antibody in all mice (Kochel et al., 1997), with a subsequent improvement seen with the addition of immunostimulatory CpG motifs (Porter et al., 1998). In subsequent experiments using genes from the West Pacific 74 strain of DENV-1 virus and the pVR1012 plasmid, it was determined that the full-length E gene with PrM served as a better immunogen (Raviprakash et al., 2000) and was shown to reduced the frequency and duration of viremia in rhesus macaques following challenge with wild-type virus (Raviprakash et al., 2006). Konishi et al. (2000) successfully immunized mice with a similar DENV-2 vaccine construct using C-PrM-E genes, Purdy and Chang (2005) have expressed virus-like particles expressing PrM and E proteins in Chinese hamster ovary cell culture, and Costa et al. (2006) have shown protection in mice using a DENV-2 NS-1 construct. The NMRC has also pursued "DNA shuffling" and screening with some success in mice (Apt et al., 2006). At present, the DNA approach has produced modest neutralizing antibody levels in nonhuman primates with only a portion of animals fully protected from viremia (Putnak et al., 2003; Raviprakash et al., 2001, 2003). With the recognition that skin dendritic cells are high permissive for DENV replication (Wu et al., 2000), ongoing efforts are focused on vaccine delivery systems to target these cells.

DNA vaccines used in combination with other approaches, such as inactivated virus vaccines discussed below may increase the complexity and effectiveness of the immune response (Eo et al., 2001; Doolan and Hoffman, 2001; Tellier et al., 1998). Preliminary studies in mice and monkeys at the NMRC suggest that this approach may be effective for dengue (Simmons et al., 2001b, 2006). In the mice, DNA vaccination alone elicited primarily an IgG2a antibody response while the recombinant protein used elicited an exclusive IgG1 antibody response. Vaccination with different products in sequence or simultaneously elicited a more balanced distribution of both antibody subclasses. However, for a vaccine targeted for 1 billion children in the tropics, an 8-component vaccine will present obstacles of cost and complexity. The NMRC DNA vaccine is currently undergoing Phase 1 clinical testing to be followed by the assessment of prime-boost strategies.

The DNA approach also carries its own risks (Klinman et al., 1997). These include the theoretical risk of nucleic acid integration into the host's chromosomal

DNA to potentially inactivate tumor suppressor genes or activate oncogenes. This risk appears to be well below the spontaneous mutation frequency for mammalian cells (Nichols et al., 1995; Martin et al., 1999). However, if a mutation due to DNA integration is a part of a multiple hit phenomenon leading to carcinogenesis, it could take many years before this problem became evident. Another concern is that foreign DNA might induce anti-DNA antibodies leading to autoimmune diseases such as systemic lupus erythematosis. However, studies in lupus-prone mice, normal mice, rabbits, and people to date have not validated this concern (Parker et al., 1999; Mor et al., 1997) and, in fact, DNA vaccines are being proposed as an approach to the management of autoimmune diseases (Prud'homme et al., 2001; von Herrath and Whitton, 2000; Karin, 2000).

Infectious virus RNA derived directly from cDNA has been used to successfully immunize mice with nanogram amounts of tick-borne encephalitis virus (Mandl et al., 1998; Dubensky et al., 1998). This approach mimics natural infection with advantages of increased vaccine purity and stability over traditional cell culture-based approaches. It also eliminates the risk of DNA integration and anti-DNA autoimmune disease.

Whole Virus Inactivated Vaccines

Inactivated whole virus vaccines are the second most common type of licensed vaccine in the United States after live attenuated viral vaccines. Inactivated vaccines have at least two advantages over live attenuated vaccines; inactivated vaccines cannot revert to a more pathogenic phenotype and they are unlikely to interfere with each other in combination. Moreover, induction of cell-mediated and humoral immune responses has been demonstrated with inactivated flavivirus vaccines (Aihara et al., 2000). On the other hand, inactivated or killed vaccines express only the part of the viral genome that encodes structural proteins resulting in a less broad immune response. Other potential disadvantages of these vaccines are the difficulty in manufacturing preparations of sufficient titer, their increased cost per dose, and current requirements for multiple immunizations. Even so, two successful inactivated flavivirus vaccines are safe, effective, and licensed for use to prevent Japanese encephalitis (Hoke et al., 1988) and tick-borne encephalitis (Craig et al., 1999). Inactivated dengue vaccine may be useful as a traveler's vaccine or as a part of a prime-boost strategy with live or replicating vaccines.

A purified, inactivated DENV-2 vaccine compliant with good manufacturing practices has been manufactured by the Department of Biologics Research (WRAIR, Silver Spring, MD) and a DENV-1 equivalent will soon enter clinical trials (Putnak et al., 1996a, 1996b). For preparation of DENV-2-PIV, DENV-2 strain S16803, Vero cell passage three, was propagated in certified Vero cells. Virus from the culture supernatant fluid was concentrated by ultrafiltration and purified on sucrose gradients. The high-titer purified virus (approximately 9 $\log_{10}$pfu/ml) was inactivated with formalin at 22°C for 10 days. The DENV-2 vaccine with alum and other adjuvants induced high neutralizing antibody levels and protection against viremia in a primate model (Putnak et al., 1996a, 2005; Simmons et al., 2006).

Recombinant Subunit Vaccines

Dengue antigens, primarily prM and E proteins, can be expressed in bacteria, yeast, mosquito cells, or mammalian cells, and then purified to prepare nonreplicating, subunit vaccines. In general, a truncated E protein is expressed that lacks its carboxy-terminal hydrophobic membrane anchor region in an effort to maximize the secretion and solubility of the protein. These modified E proteins were shown in 1991 to be immunogenic and protective in mice following inoculation with recombinant vaccinia virus expressing the truncated E protein (Men et al., 1991) or with baculovirus-expressed truncated E protein (Putnak et al., 1991). Since that time, a number of expression systems have been used to generate E proteins: *E. coli* expressing domain III of the E protein of DENV-2 as a fusion protein with the maltose-binding protein (Simmons et al., 2001a), E-NS1 protein of DENV-2 as a fusion protein with staphylococcal protein A (Srivastava et al., 1995); *Pichia pastoris* expressing E protein of DENV-4 (Guzman et al., 2004), DENV-2/hepatitis B surface antigen hybrid (Bisht et al., 2002), or subviral particles of DENV-1 (Sugrue et al., 1997); baculovirus expression in *Spodoptera frugiperda* cells of full-length E protein aggregates of DENV-2 (Kelly et al., 2000), His-tagged E protein of DENV-2 (Staropoli et al., 1997), hybrid E protein of DENV-2/DENV-3 (Bielefeldt-Ohmann et al., 1997), or E proteins of DENV-2 and DENV-3 (Delenda et al., 1994); *Drosophila melanogaster* cells (Schneider-2) stably expressing E protein of DENV-2 (Putnak et al., 2005); and CHO-K1 cells stably expressing subviral particles of DENV-2 (Konishi and Fujii, 2002). Not surprisingly, all the subunit preparations listed above elicited moderate to high levels of antibody following immunization of mice. Although such vaccines are anticipated to achieve a high level of safety, it is generally accepted that multiple doses will be required and that adjuvants will be needed to increase immunogenicity.

To date, a DENV subunit vaccine has not been tested in humans. However, two studies in macaques have recently been completed using monovalent DENV-2 or DENV-4 truncated E proteins. Guzman et al. (2003) immunized macaques with four doses of 100 μg of E protein (DENV-4), using alum as an adjuvant. DENV antibody titers were barely detectable by ELISA and hemagglutination-inhibition following all four doses and at the time of virus challenge, neutralizing antibody titers were only 1:30. Although partial protection against wt DENV-2 challenge viremia was demonstrated, the researchers considered the results insufficient to recommend the subunit preparation as a vaccine candidate and suggested that factors including conformational differences between recombinant and native E protein, antigen presentation, and immunization schedule could have influenced the lack of full protection.

In collaboration with Hawaii Biotech, Inc. (Alea, Hawaii), Putnak et al. (2005) immunized macaques with two doses of DENV-2 truncated E protein produced in drosophila cells and formulated with each of five different adjuvants containing alum salts with or without the addition of 3D-MPL (Corixa, Inc.) or QS-21 (Antigenics, Inc.) adjuvants. Additional macaques were immunized with experimental preparations of live attenuated vaccine virus or formalin inactivated vaccine virus. DENV neutralizing antibody titers prior to challenge varied widely (1:60–1:5300) among the subunit vaccinated monkeys and several exhibited significant challenge virus replication even in the presence of relatively high levels of virus neutralizing antibody. Nevertheless, one group of monkeys receiving the highest dose of antigen along with both 3D-MPL and QS-21 were completely protected, an improvement over previous results with baculovirus-expressed structural proteins of DENV-4 (Eckels et al., 1994). In all of the reported macaque studies, there was no evidence of enhanced challenge virus replication, even in the presence of minimally protective antibody levels. It is likely that additional vaccination regimens including higher antigen doses or more frequent boosting will be required to maintain antibody titers at protective levels.

PROSPECTS FOR THE FUTURE

In recent years there has been heightened awareness of the global morbidity and mortality related to dengue. As a result, there has been a resurgence in dengue research, particularly in the area of vaccine development. Vaccination is thought to have the greatest potential to reduce the incidence of dengue in endemic areas. As discussed in previous sections, the prospect of an effective vaccine for the prevention of dengue is closer to becoming a reality. The number of candidate dengue vaccines has expanded to include live attenuated dengue vaccines, both recombinant and biologically derived; subunit protein vaccines; DNA vaccines; vectored vaccines; and whole virus inactivated vaccines. The pace of vaccine development has increased but the challenges to the implementation of a successful dengue vaccine program are still significant. Prior to the initiation of large-scale vaccine trials in endemic regions, the infrastructure must be in place to determine not only the efficacy of the vaccine but to also evaluate the impact of vaccination on disease severity in dengue-exposed as well as dengue-naive individuals. It will be essential that accurate case definitions for DF/DHF/DSS be consistently applied during these trials. The WHO classification system of dengue disease, although useful for decades, may need to be refined to more effectively identify and triage which dengue patients will go on to develop severe disease (Deen et al., 2006). Efforts to this end are ongoing and it is expected that the WHO will provide guidelines for the design and conduct of dengue vaccine trials in endemic areas.

An expanding area of dengue research is that of therapeutics. As discussed above, the treatment for dengue is limited to supportive measures only. Currently there are no licensed antiviral or other therapeutic agents available for the treatment acute dengue illness. Several possible targets for antiviral therapy have been identified including viral attachment, the viral polymerase, and other components of viral replication (Takhampunya et al., 2006; Vasudevan et al., 2001; Altmeyer, 2004). In addition, further understanding of the aberrant immune response may provide opportunities for the use of immune modulating drugs to prevent or treat disease progression in dengue.

A third area of intensive research in dengue is the development of an animal model that reproducibly mimics the clinical picture seen in humans. Such a model would be extremely useful in determining the components of the immune response that contribute to the hemorrhagic and vascular leak syndromes so characteristic of severe dengue. Novel diagnostic and therapeutic agents could then be evaluated and optimized in preclinical testing, improving the ability to identify the most useful agents for clinical practice. Vaccine development would also benefit from an animal model which reproducibly mimics human disease; to not only identify attenuation characteristics, but to also assess the risk of DHF/DSS conferred by vaccination to populations of dengue-endemic regions. Progress continues to be made toward the development of a useful animal model and, with the ability to genetically manipulate specific components of the immune response in mice expanding, such a prospect may not be far off.

KEY ISSUES

- DENV is the most important mosquito-borne viral pathogen infecting more than 100 million persons annually.
- There are four serotypes of DENV, each capable of causing the full spectrum of clinical disease.
- Epidemiological studies have demonstrated the greatest risk factor for severe dengue disease is secondary infection with a DENV serotype different from that which caused primary infection.
- The immunopathologic mechanisms of severe or enhanced dengue are complex and involve elements of both the humoral immune system, via ADE, and the cellular immune system.
- Neutralizing antibody directed against the structural protein E is thought to be the most important mediator of protection from DENV.
- Long-lasting, perhaps lifelong, homotypic immunity is generated after primary DENV infection.
- Only short-term heterotypic immunity is generated after primary infection.
- There are currently no antiviral treatments for DENV; treatment is supportive only.
- There are currently no licensed vaccines for the prevention of dengue disease, however numerous candidate vaccines are in development.
- A viable vaccine for the prevention of dengue will have to protect against all four serotypes of DENV.
- Future areas of DENV research include continued vaccine development and implementation, development of novel diagnostic and therapeutic agents, and further characterization of aberrant immune response seen in severe dengue.

ACKNOWLEDGMENT

The authors wish to thank John Bisbing for his management of the reference database used in the writing of this chapter.

References

Vaccines. Philadelphia: W.B. Saunders Company, 1999.

Aihara, H., Takasaki, T., Toyosaki-Maeda, T., Suzuki, R., Okuno, Y. and Kurane, I. T-cell activation and induction of antibodies and memory T cells by immunization with inactivated Japanese encephalitis vaccine. Viral. Immunol. 2000; 13:179–186.

Altmeyer, R. Virus attachment and entry offer numerous targets for antiviral therapy. Curr. Pharm. Des. 2004; 10:3701–3712.

Alvarez, M.E. and Ramirez Ronda, C.H. Dengue and hepatic failure. Am. J. Med. 1985; 79:670–674.

An, J., Kimura-Kuroda, J., Hirabayashi, Y. and Yasui, K. Development of a novel mouse model for dengue virus infection. Virology 1999; 263:70–77.

An, J., Zhou, D.S., Zhang, J.L., Morida, H., Wang, J.L. and Yasui, K. Dengue-specific CD8+ T cells have both protective and pathogenic roles in dengue virus infection. Immunol. Lett. 2004; 95:167–174.

Anonymous Dengue haemorrhagic fever (DHF): increase in number of cases in the Americas 1980–1987. Wkly. Epidemiol. Rec. 1990; 65:13–15.

Anonymous Dengue Haemorrhagic Fever: Diagnosis, Treatment, Prevention and Control, 2nd ed. Geneva: World Health Organization, 1997.

Anonymous. Dengue update. TDR News, 2001.

Apt, D., Raviprakash, K., Brinkman, A., Semyonov, A., Yang, S., Skinner, C., Diehl, L., Lyons, R., Porter, K. and Punnonen, J. Tetravalent neutralizing antibody response against four dengue serotypes by a single chimeric dengue envelope antigen. Vaccine 2006; 24:335–344.

Ashburn, P.M. and Craig, C.F. Experimental investigations regarding the etiology of dengue fever. J. Infect. Dis. 1907; 4:440–475.

Atrasheuskaya, A., Petzelbauer, P., Fredeking, T.M. and Ignatyev, G. Anti-TNF antibody treatment reduces mortality in experimental dengue virus infection. FEMS Immunol. Med. Microbiol. 2003; 35:33–42.

Avila-Aguero, M.L., Avila-Aguero, C.R., Um, S.L., Soriano-Fallas, A., Canas-Coto, A. and Yan, S.B. Systemic host inflammatory and coagulation response in the dengue virus primo-infection. Cytokine 2004; 27:173–179.

Avirutnan, P., Malasit, P., Seliger, B., Bhakdi, S. and Husmann, M. Dengue virus infection of human endothelial cells leads to chemokine production, complement activation, and apoptosis. J. Immunol. 1998; 161:6338–6346.

Avirutnan, P., Punyadee, N., Noisakran, S., Komoltri, C., Thiemmeca, S., Auethavornanan, K., Jairungsri, A., Kanlaya, R., Tangthawornchaikul, N., Puttikhunt, C., Pattanakitsakul, S.N., Yenchitsomanus, P.T., Mongkolsapaya, J., Kasinrerk, W., Sittisombut, N., Husmann, M., Blettner, M., Vasanawathana, S., Bhakdi, S. and Malasit, P. Vascular leakage in severe dengue virus infections: a potential role for the nonstructural viral protein NS1 and complement. J. Infect. Dis. 2006; 193:1078–1088.

Azeredo, E.L., Zagne, S.M., Santiago, M.A., Gouvea, A.S., Santana, A.A., Neves-Souza, P.C., Nogueira, R.M., Miagostovich, M.P. and Kubelka, C.F. Characterisation of lymphocyte response and cytokine patterns in patients with dengue fever. Immunobiology 2001; 204:494–507.

Bancroft, W.H., Scott, R.M., Eckels, K.H., Hoke, C.H., Jr., Simms, T.E., Jesrani, K.D., Summers, P.L., Dubois, D.R., Tsoulos, D. and Russell, P.K. Dengue virus type 2 vaccine: reactogenicity and immunogenicity in soldiers. J. Infect. Dis. 1984; 149:1005–1010.

Bellanti, J.A., Bourke, R.T.A., Rosenzweig, E.C. and Eylar, O.R. Report of dengue vaccine trial in the Caribbean, 1963: a collaborative study. Bull. World Health Organ. 1966; 35:93.

Bente, D.A., Melkus, M.W., Garcia, J.V. and Rico-Hesse, R. Dengue fever in humanized NOD/SCID mice. J. Virol. 2005; 79:13797–13799.

Bethell, D.B., Flobbe, K., Cao, X.T., Day, N.P., Pham, T.P., Buurman, W.A., Cardosa, M.J., White, N.J. and Kwiatkowski, D. Pathophysiologic and prognostic role of cytokines in dengue hemorrhagic fever. J. Infect. Dis. 1998; 177:778–782.

Bhamarapravati, N. The spectrum of pathological changes in Thai haemorrhagic fever. SEATO Med. Res. Monogr. Symp. Haemorrhagic Fever 1961; 2:76–84.

Bhamarapravati, N. Hemostatic defects in dengue hemorrhagic fever. Rev. Infect. Dis. 1989; 11:S826–S829.

Bhamarapravati, N., Tuchinda, P. and Boonyapaknavik, V. Pathology of Thailand haemorrhagic fever: a study of 100 autopsy cases. Ann. Trop. Med. Parasitol. 1967; 61:500–510.

Bhamarapravati, N. and Yoksan, S. Study of bivalent dengue vaccine in volunteers. Lancet 1989; 1:1077.

Bhamarapravati, N. and Yoksan, S. Live attenuated tetravalent dengue vaccine. In: Dengue and Dengue Hemorrhagic Fever (D.J. Gubler and G. Kuno, Eds.). New York: CAB International, 1997.

Bhamarapravati, N. and Yoksan, S. Live attenuated tetravalent dengue vaccine. Vaccine 2000; 18(Suppl. 2):44–47.

Bhamarapravati, N., Yoksan, S., Chayaniyayothin, T., Angsubphakorn, S. and Bunyaratvej, A. Immunization with a live attenuated dengue-2-virus candidate vaccine (16681-PDK 53): clinical, immunological and biological responses in adult volunteers. Bull. World Health Organ. 1987; 65:189–195.

Bielefeldt-Ohmann, H., Beasley, D.W., Fitzpatrick, D.R. and Aaskov, J.G. Analysis of a recombinant dengue-2 virus-dengue-3 virus hybrid envelope protein expressed in a secretory baculovirus system. J. Gen. Virol. 1997; 78:2723–2733.

Bisht, H., Chugh, D.A., Raje, M., Swaminathan, S.S. and Khanna, N. Recombinant dengue virus type 2 envelope/hepatitis B surface antigen hybrid protein expressed in *Pichia pastoris* can function as a bivalent immunogen. J. Biotechnol. 2002; 99:97–110.

Blanc, G. and Caminopetros, J. Contributions to the study of vaccination against dengue. Bull. Acad. Med. 1929; 102:40–47.

Blaney, J., Johnson, D., Manipon, G., Firestone, C., Hanson, C., Murphy, B. and Whitehead, S. Genetic basis of attenuation of dengue virus type 4 small plaque mutants with restricted replication in suckling mice and in SCID mice transplanted with human liver cells. Virology 2002; 300:125.

Blaney, J.E., Jr., Hanson, C.T., Firestone, C.Y., Hanley, K.A., Murphy, B.R. and Whitehead, S.S. Genetically modified, live attenuated dengue virus type 3 vaccine candidates. Am. J. Trop. Med. Hyg. 2004a; 71:811–821.

Blaney, J.E., Jr., Hanson, C.T., Hanley, K.A., Murphy, B.R. and Whitehead, S.S. Vaccine candidates derived from a novel infectious cDNA clone of an American genotype dengue virus type 2. BMC Infect. Dis. 2004b; 4:39.

Blaney, J.E., Jr., Manipon, G.G., Firestone, C.Y., Johnson, D.H., Hanson, C.T., Murphy, B.R. and Whitehead, S.S. Mutations which enhance the replication of dengue virus type 4 and an antigenic chimeric dengue virus type 2/4 vaccine candidate in Vero cells. Vaccine 2003; 21:4317–4327.

Blaney, J.E., Jr., Matro, J.M., Murphy, B.R. and Whitehead, S.S. Recombinant, live-attenuated tetravalent dengue virus vaccine formulations induce a balanced, broad, and protective neutralizing antibody response against each of the four serotypes in rhesus monkeys. J. Virol. 2005; 79:5516–5528.

Blaney, J.E., Jr., Sathe, N.S., Hanson, C.T., Firestone, C.Y., Murphy, B.R. and Whitehead, S.S. Vaccine candidates for dengue virus type 1 (DEN1) generated by replacement of the structural genes of rDEN4 and rDEN4Delta30 with those of DEN1. Virol. J. 2007; 4:23.

Bokisch, V.A., Muller-Eberhard, H.J. and Dixon, F.J. The role of complement in hemorrhagic shock syndrome (dengue). Trans. Assoc. Am. Physicians 1973; 86:102–110.

Boonpucknavig, S., Vuttiviroj, O. and Boonpucknavig, V. Infection of young adult mice with dengue virus type 2. Trans. R. Soc. Trop. Med. Hyg. 1981; 75:647–653.

Boonpucknavig, S., Vuttivirojana, O., Bunnag, C., Bhamarapravati, N. and Nimmannitya, S. Demonstration of dengue antibody complexes on the surface of platelets from patients with dengue hemorrhagic fever. Am. J. Trop. Med. Hyg. 1979; 28:881–884.

Borio, L., Inglesby, T., Peters, C.J., Schmaljohn, A.L., Hughes, J.M., Jahrling, P.B., Ksiazek, T., Johnson, K.M., Meyerhoff, A., O'Toole, T., Ascher, M.S., Bartlett, J., Breman, J.G., Eitzen, E.M., Jr., Hamburg, M., Hauer, J., Henderson, D.A., Johnson, R.T., Kwik, G., Layton, M., Lillibridge, S., Nabel, G.J., Osterholm, M.T., Perl, T.M., Russell, P. and Tonat, K. Hemorrhagic fever viruses as biological weapons: medical and public health management. JAMA 2002; 287:2391–2405.

Bosch, I., Xhaja, K., Estevez, L., Raines, G., Melichar, H., Warke, R.V., Fournier, M.V., Ennis, F.A. and Rothman, A.L. Increased production of interleukin-8 in primary human monocytes and in human epithelial and endothelial cell lines after dengue virus challenge. J. Virol. 2002; 76:5588–5597.

Bravo, J.R., Guzman, M.G. and Kouri, G.P. Why dengue haemorrhagic fever in Cuba? 1. Individual risk factors for dengue haemorrhagic fever/dengue shock syndrome (DHF/DSS). Trans. R. Soc. Trop. Med. Hyg. 1987; 81:816–820.

Bray, M. and Lai, C.J. Construction of intertypic chimeric dengue viruses by substitution of structural protein genes. Proc. Natl. Acad. Sci. U.S.A. 1991; 88:10342–10346.

Bray, M., Men, R. and Lai, C.J. Monkeys immunized with intertypic chimeric dengue viruses are protected against wild-type virus challenge. J. Virol. 1996; 70:4162–4166.

Bray, M., Zhao, B.T., Markoff, L., Eckels, K.H., Chanock, R.M. and Lai, C.J. Mice immunized with recombinant vaccinia virus expressing dengue 4 virus structural proteins with or without nonstructural protein NS1 are protected against fatal dengue virus encephalitis. J. Virol. 1989; 63:2853–2856.

Bukowski, J.F., Kurane, I., Lai, C.J., Bray, M., Falgout, B. and Ennis, F.A. Dengue virus-specific cross-reactive CD8+ human cytotoxic T lymphocytes. J. Virol. 1989; 63:5086–5091.

Burke, D.S., Nisalak, A., Johnson, D.E. and Scott, R.M. A prospective study of dengue infections in Bangkok. Am. J. Trop. Med. Hyg. 1988; 38:172–180.

Burton, D.R., Williamson, R.A. and Parren, P.W. Antibody and virus: binding and neutralization. Virology 2000; 270:1–3.

Butrapet, S., Huang, C.Y., Pierro, D.J., Bhamarapravati, N., Gubler, D.J. and Kinney, R.M. Attenuation markers of a candidate dengue type 2 vaccine virus, strain 16681 (PDK-53), are defined by mutations in the 5′ noncoding region and nonstructural proteins 1 and 3. J. Virol. 2000; 74:3011–3019.

Cahour, A., Pletnev, A., Vazeillefalcoz, M., Rosen, L. and Lai, C.J. Growth-restricted dengue virus mutants containing deletions in the 5′ noncoding region of the RNA genome. Virology 1995; 207:68–76.

Cardier, J.E., Marino, E., Romano, E., Taylor, P., Liprandi, F., Bosch, N. and Rothman, A.L. Proinflammatory factors present in sera from patients with acute dengue infection induce activation and apoptosis of human microvascular endothelial cells: possible role of TNF-alpha in endothelial cell damage in dengue. Cytokine 2005; 30:359–365.

Carey, D.E. Chikungunya and dengue: a case of mistaken identity? J. Hist. Med. Allied Sci. 1971; 26:243–262.

Carlos, C.C., Oishi, K., Cinco, M.T., Mapua, C.A., Inoue, S., Cruz, D.J., Pancho, M.A., Tanig, C.Z., Matias, R.R., Morita, K., Natividad, F.F.,

Igarashi, A. and Nagatake, T. Comparison of clinical features and hematologic abnormalities between dengue fever and dengue hemorrhagic fever among children in the Philippines. Am. J. Trop. Med. Hyg. 2005; 73:435–440.

Caufour, P.S., Motta, M.C., Yamamura, A.M., Vazquez, S., Ferreira, I.I., Jabor, A.V., Bonaldo, M.C., Freire, M.S. and Galler, R. Construction, characterization and immunogenicity of recombinant yellow fever 17D-dengue type 2 viruses. Virus Res. 2001; 79:1–14.

Chambers, T.J., Jiang, X., Droll, D.A., Liang, Y., Wold, W.S. and Nickells, J. Chimeric Japanese encephalitis virus/dengue 2 virus infectious clone: biological properties, immunogenicity and protection against dengue encephalitis in mice. J. Gen. Virol. 2006; 87:3131–3140.

Chambers, T.J., Liang, Y., Droll, D.A., Schlesinger, J.J., Davidson, A.D., Wright, P.J. and Jiang, X. Yellow fever virus/dengue-2 virus and yellow fever virus/dengue-4 virus chimeras: biological characterization, immunogenicity, and protection against dengue encephalitis in the mouse model. J. Virol. 2003; 77:3655–3668.

Chambers, T.J., Nestorowicz, A., Mason, P.W. and Rice, C.M. Yellow fever/Japanese encephalitis chimeric viruses: construction and biological properties. J. Virol. 1999; 73:3095–3101.

Chaturvedi, U.C., Tandon, P. and Mathur, A. Effect of immunosuppression on dengue virus infection in mice. J. Gen. Virol. 1977; 36:449–458.

Chaturvedi, U.C., Tandon, P., Mathur, A. and Kumar, A. Host defence mechanisms against dengue virus infection of mice. J. Gen. Virol. 1978; 39:293–302.

Chen, W., Kawano, H., Men, R., Clark, D. and Lai, C.J. Construction of intertypic chimeric dengue viruses exhibiting type 3 antigenicity and neurovirulence for mice. J. Virol. 1995; 69:5186–5190.

Churdboonchart, V., Bhamarapravati, N. and Futrakul, P. Crossed immunoelectrophoresis for the detection of spit products of the third complement in dengue hemorrhagic observations in patients' plasma. Am. J. Trop. Med. Hyg. 1983; 32:569–576.

Churdboonchart, V., Bhamarapravati, N., Peampramprecha, S. and Sirinavin, S. Antibodies against dengue viral proteins in primary and secondary dengue hemorrhagic fever. Am. J. Trop. Med. Hyg. 1991; 44:481–493.

Cohen, S.N. and Halstead, S.B. Shock associated with dengue infection. I. Clinical and physiologic manifestations of dengue hemorrhagic fever in Thailand, 1964. J. Pediatrics 1966; 68:448–456.

Costa, S.M., Paes, M.V., Barreto, D.F., Pinhao, A.T., Barth, O.M., Queiroz, J.L., Armoa, G.R., Freire, M.S. and Alves, A.M. Protection against dengue type 2 virus induced in mice immunized with a DNA plasmid encoding the non-structural 1 (NS1) gene fused to the tissue plasminogen activator signal sequence. Vaccine 2006; 24:195–205.

Craig, S.C., Pittman, P.R., Lewis, T.E., Rossi, C.A., Henchal, E.A., Kuschner, R.A., Martinez, C., Kohlhase, K.F., Cuthie, J.C., Welch, G.E. and Sanchez, J.L. An accelerated schedule for tick-borne encephalitis vaccine: the American military experience in Bosnia. Am. J. Trop. Med. Hyg. 1999; 61:874–878.

Cummings, D.A., Irizarry, R.A., Huang, N.E., Endy, T.P., Nisalak, A., Ungchusak, K. and Burke, D.S. Travelling waves in the occurrence of dengue haemorrhagic fever in Thailand. Nature 2004; 427:344–347.

Deen, J.L., Harris, E., Wills, B., Balmaseda, A., Hammond, S.N., Rocha, C., Dung, N.M., Hung, N.T., Hien, T.T. and Farrar, J.J. The WHO dengue classification and case definitions: time for a reassessment. Lancet 2006; 368:170–173.

Delenda, C., Staropoli, I., Frenkiel, M.P., Cabanie, L. and Deubel, V. Analysis of C-terminally truncated dengue 2 and dengue 3 virus envelope glycoproteins: processing in insect cells and immunogenic properties in mice. J. Gen. Virol. 1994; 75:1569–1578.

de Silva, A.M., Dittus, W.P., Amerasinghe, P.H. and Amerasinghe, F.P. Serologic evidence for an epizootic dengue virus infecting toque macaques (*Macaca sinica*) at Polonnaruwa, Sri Lanka. Am. J. Trop. Med. Hyg. 1999; 60:300–306.

Deubel, V., Kinney, R.M., Esposito, J.J., Cropp, C.B., Vorndam, A.V., Monath, T.P. and Trent, D.W. Dengue 2 virus envelope protein expressed by a recombinant vaccinia virus fails to protect monkeys against dengue. J. Gen. Virol. 1988; 69:1921–1929.

Dewi, B.E., Takasaki, T. and Kurane, I. In vitro assessment of human endothelial cell permeability: effects of inflammatory cytokines and dengue virus infection. J. Virol. Methods 2004; 121:171–180.

Dharakul, T., Kurane, I., Bhamarapravati, N., Yoksan, S., Vaughn, D.W., Hoke, C.H., Jr. and Ennis, F.A. Dengue virus-specific memory T cell responses in human volunteers receiving a live attenuated dengue virus type 2 candidate vaccine. J. Infect Dis. 1994; 170:27–33.

Doolan, D.L. and Hoffman, S.L. DNA-based vaccines against malaria: status and promise of the Multi-Stage Malaria DNA Vaccine Operation. Int. J. Parasitol. 2001; 31:753–762.

Dubensky, T.W., Jr., Polo, J.M. and Liu, M.A. Live virus vaccines: something old, something new, something borrowed... [news; comment]. Nat. Med. 1998; 4:1357–1358.

Dung, N.M., Day, N.P., Tam, D.T., Loan, H.T., Chau, H.T., Minh, L.N., Diet, T.V., Bethell, D.B., Kneen, R., Hien, T.T., White, N.J. and Farrar, J.J. Fluid replacement in dengue shock syndrome: a randomized, double-blind comparison of four intravenous-fluid regimens. Clin. Infect Dis. 1999; 29:787–794.

Durand, J.P., Vallee, L., de Pina, J.J. and Tolou, H. Isolation of a dengue type 1 virus from a soldier in West Africa (Cote d'Ivoire). Emerg. Infect. Dis. 2000; 6:83–84.

Durbin, A.P., Karron, R.A., Sun, W., Vaughn, D.W., Reynolds, M.J., Perreault, J.R., Thumar, B., Men, R., Lai, C.J., Elkins, W.R., Chanock, R.M., Murphy, B.R. and Whitehead, S.S. Attenuation and immunogenicity in humans of a live dengue virus type-4 vaccine candidate with a 30 nucleotide deletion in its 3′-untranslated region. Am. J. Trop. Med. Hyg. 2001; 65:405–413.

Durbin, A.P., McArthur, J., Marron, J.A., Blaney, J.E., Jr., Thumar, B., Wanionek, K., Murphy, B.R. and Whitehead, S.S. The live attenuated dengue serotype 1 vaccine rDEN1Delta30 is safe and highly immunogenic in healthy adult volunteers. Hum. Vaccines 2006a; 2:167–173.

Durbin, A.P., Mcarthur, J.H., Marron, J.A., Blaney, J.E., Thumar, B., Wanionek, K., Murphy, B.R. and Whitehead, S.S. rDEN2/4Delta30(ME), a live attenuated chimeric dengue serotype 2 vaccine is safe and highly immunogenic in healthy dengue-naive adults. Hum.Vaccines 2006b; 2:255–260.

Durbin, A.P., Whitehead, S.S., McArthur, J., Perreault, J.R., Blaney, J.E., Jr., Thumar, B., Murphy, B.R. and Karron, R.A. rDEN4 Delta 30, a live attenuated dengue virus type 4 vaccine candidate, is safe, immunogenic, and highly infectious in healthy adult volunteers. J. Infect. Dis. 2005; 191:710–718.

Eckels, K.H., Dubois, D.R., Summers, P.L., Schlesinger, J.J., Shelly, M., Cohen, S., Zhang, Y.M., Lai, C.J., Kurane, I., Rothman, A., Hasty, S. and Howard, B. Immunization of monkeys with baculovirus-dengue type-4 recombinants containing envelope and nonstructural proteins: evidence of priming and partial protection. Am. J. Trop. Med. Hyg. 1994; 50:472–478.

Edelman, R., Tacket, C.O., Wasserman, S.S., Vaughn, D.W., Eckels, K.H., Dubois, D.R., Summers, P.L. and Hoke, C.H., Jr. A live attenuated dengue-1 vaccine candidate (45AZ5) passaged in primary dog kidney cell culture is attenuated and immunogenic for humans. J. Infect. Dis. 1994; 170:1448–1455.

Edelman, R., Wasserman, S.S., Bodison, S.A., Putnak, R.J., Eckels, K.H., Tang, D., Kanesa-Thasan, N., Vaughn, D.W., Innis, B.L. and Sun, W. Phase I trial of 16 formulations of a tetravalent live-attenuated dengue vaccine. Am. J. Trop. Med. Hyg. 2003; 69:48–60.

Endy, T.P., Chunsuttiwat, S., Nisalak, A., Libraty, D.H., Green, S., Rothman, A.L., Vaughn, D.W. and Ennis, F.A. Epidemiology of inapparent and symptomatic acute dengue virus infection: a prospective study of primary school children in Kamphaeng Phet, Thailand. Am. J. Epidemiol. 2002; 156:40–51.

Eo, S.K., Gierynska, M., Kamar, A.A. and Rouse, B.T. Prime-boost immunization with DNA vaccine: mucosal route of administration changes the rules. J. Immunol. 2001; 166:5473–5479.

Eram, S., Setyabudi, Y., Sadono, I., Sutrisno, D.S., Gubler, D.J. and Saroso, J.S. Epidemic dengue hemorrhagic fever in rural Indonesia: II. Clinical studies. Am. J. Trop. Med. Hyg. 1979; 28:711–716.

Fagbami, A.H., Monath, T.P. and Fabiyi, A. Dengue virus infections in Nigeria: a survey for antibodies in monkeys and humans. Trans. R. Soc. Trop. Med. Hyg. 1977; 71:60–65.

Falgout, B., Bray, M., Schlesinger, J.J. and Lai, C.J. Immunization of mice with recombinant vaccinia virus expressing authentic dengue virus nonstructural protein NS1 protects against lethal dengue virus encephalitis. J. Virol. 1990; 64:4356–4363.

Fernandez-Mestre, M.T., Gendzekhadze, K., Rivas-Vetencourt, P. and Layrisse, Z. TNF-alpha-308A allele, a possible severity risk factor of hemorrhagic manifestation in dengue fever patients. Tissue Antigens 2004; 64:469–472.

Fonseca, B.A.L., Pincus, S., Shope, R.E., Paoletti, E. and Mason, P.W. Recombinant vaccinia viruses co-expressing dengue-1 glycoproteins prM and E induce neutralizing antibodies in mice. Vaccine 1994; 12:279–285.

Fresh, J.W., Reyes, V., Clarke, E.J. and Uylangco, C.V. Philippine hemorrhagic fever: a clinical, laboratory, and necropsy study. J. Lab. Clin. Med. 1969; 73(3):451–458.

Gagnon, S.J., Ennis, F.A. and Rothman, A.L. Bystander target cell lysis and cytokine production by dengue virus-specific human CD4(+) cytotoxic T-lymphocyte clones. J. Virol. 1999; 73:3623–3629.

Gagnon, S.J., Mori, M., Kurane, I., Green, S., Vaughn, D.W., Kalayanarooj, S., Suntayakorn, S., Ennis, F.A. and Rothman, A.L. Cytokine gene expression and protein production in peripheral blood mononuclear cells of children with acute dengue virus infections. J. Med. Virol. 2002; 67:41–46.

Gagnon, S.J., Zeng, W., Kurane, I. and Ennis, F.A. Identification of two epitopes on the dengue 4 virus capsid protein recognized by a serotype-specific and a panel of serotype- cross-reactive human CD4+ cytotoxic T-lymphocyte clones. J. Virol. 1996; 70:141–147.

Gollins, S.W. and Porterfield, J.S. A new mechanism for the neutralization of enveloped viruses by antiviral antibody. Nature 1986; 321:244–246.

Graham, B.S., Henderson, G.S., Tang, Y.W., Lu, X., Neuzil, K.M. and Colley, D.G. Priming immunization determines T helper cytokine mRNA expression patterns in lungs of mice challenged with respiratory syncytial virus. J. Immunol. 1993; 151:2032–2040.

Graham, H. The dengue: a study of its pathology and mode of propagation. J. Trop. Med. 1903; 6:209–214.

Green, S., Kurane, I., Pincus, S., Paoletti, E. and Ennis, F.A. Recognition of dengue virus NS1-NS2a proteins by human CD4+ cytotoxic T lymphocyte clones. Virology 1997; 234:383–386.

Green, S., Pichyangkul, S., Vaughn, D.W., Kalayanarooj, S., Nimmannitya, S., Nisalak, A., Kurane, I., Rothman, A.L. and Ennis, F.A. Early CD69 expression on peripheral blood lymphocytes from children with dengue hemorrhagic fever. J. Infect. Dis. 1999a; 180:1429–1435.

Green, S., Vaughn, D.W., Kalayanarooj, S., Nimmannitya, S., Suntayakorn, S., Nisalak, A., Lew, R., Innis, B.L., Kurane, I., Rothman, A.L. and Ennis, F.A. Early immune activation in acute dengue illness is related to development of plasma leakage and disease severity. J. Infect. Dis. 1999b; 179:755–762.

Green, S., Vaughn, D.W., Kalayanarooj, S., Nimmannitya, S., Suntayakorn, S., Nisalak, A., Rothman, A.L. and Ennis, F.A. Elevated plasma interleukin-10 levels in acute dengue correlate with disease severity. J. Med. Virol. 1999c; 59:329–334.

Green, S. and Rothman, A. Immunopathological mechanisms in dengue and dengue hemorrhagic fever. Curr. Opin. Infect. Dis. 2006; 19:429–436.

Griffin, D.E., Ward, B.J. and Esolen, L.M. Pathogenesis of measles virus infection: an hypothesis for altered immune responses. J. Infect. Dis. 1994; 170(Suppl. 1):S24–S31.

Gubler, D.J. Dengue and dengue hemorrhagic fever: its history and resurgence as a global public health problem. In: Dengue and Dengue Hemorrhagic Fever (D.J. Gubler and G. Kuno, Eds.). New York: CAB International, 1997.

Gubler, D.J. Dengue and dengue hemorrhagic fever. Clin. Microbiol. Rev. 1998; 11:480–496.

Guirakhoo, F., Arroyo, J., Pugachev, K.V., Miller, C., Zhang, Z.X., Weltzin, R., Georgakopoulos, K., Catalan, J., Ocran, S., Soike, K., Ratterree, M. and Monath, T.P. Construction, safety, and immunogenicity in nonhuman primates of a chimeric yellow fever-dengue virus tetravalent vaccine. J. Virol. 2001; 75:7290–7304.

Guirakhoo, F., Kitchner, S., Morrison, D., Forrat, R., Mccarthy, K., Nichols, R., Yoksan, S., Duan, X., Ermak, T., Kanesa-Thasan, N., Bedford, P., Lang, J., Quentin-Millet, M. and Monath, T. Hum. Vaccines 2006; 2:60–67.

Guirakhoo, F., Pugachev, K., Arroyo, J., Miller, C., Zhang, Z.X., Weltzin, R., Georgakopoulos, K., Catalan, J., Ocran, S., Draper, K. and Monath, T.P. Viremia and immunogenicity in nonhuman primates of a tetravalent yellow fever-dengue chimeric vaccine: genetic reconstructions, dose adjustment, and antibody responses against wild-type dengue virus isolates. Virology 2002; 298:146–159.

Guirakhoo, F., Pugachev, K., Zhang, Z., Myers, G., Levenbook, I., Draper, K., Lang, J., Ocran, S., Mitchell, F., Parsons, M., Brown, N., Brandler, S., Fournier, C., Barrere, B., Rizvi, F., Travassos, A., Nichols, R., Trent, D. and Monath, T. Safety and efficacy of chimeric yellow fever-dengue virus tetravalent vaccine formulations in nonhuman primates. J. Virol. 2004; 78:4761–4775.

Guirakhoo, F., Weltzin, R., Chambers, T.J., Zhang, Z., Soike, K., Ratterree, M., Arroyo, J., Georgakopoulos, K., Catalan, J. and Monath, T.P. Recombinant chimeric yellow fever-dengue type 2 virus is immunogenic and protective in nonhuman primates. J. Virol. 2000; 74:5477–5485.

Gurunathan, S., Klinman, D.M. and Seder, R.A. DNA vaccines: immunology, application, and optimization*. Annu. Rev. Immunol. 2000; 18:927–974.

Guyre, P.M., Morganelli, P.M. and Miller, R. Recombinant immune interferon increases immunoglobulin G Fc receptors on cultured human mononuclear phagocytes. J. Clin. Invest. 1983; 72:393–397.

Guzman, M.G., Kouri, G., Bravo, J., Valdes, L., Vazquez, S. and Halstead, S.B. Effect of age on outcome of secondary dengue 2 infections. Int. J. Infect. Dis. 2002; 6:118–124.

Guzman, M.G., Kouri, G., Valdes, L., Bravo, J., Alvarez, M., Vazquez, S., Delgado, I. and Halstead, S.B. Epidemiologic studies on dengue in Santiago de Cuba, 1997. Am. J. Epidemiol. 2000; 152:793–799.

Guzman, M.G., Mune, M. and Kouri, G. Dengue vaccine: priorities and progress. Expert Rev. Anti Infect. Ther. 2004; 2:895–911.

Guzman, M.G., Rodriguez, R., Hermida, L., Alvarez, M., Lazo, L., Mune, M., Rosario, D., Valdes, K., Vazquez, S., Martinez, R., Serrano, T., Paez, J., Espinosa, R., Pumariega, T. and Guillen, G. Induction of neutralizing antibodies and partial protection from viral challenge in *Macaca fascicularis* immunized with recombinant dengue 4 virus envelope glycoprotein expressed in *Pichia pastoris*. Am. J. Trop. Med. Hyg. 2003; 69:129–134.

Gwinn, W., Sun, W., Innis, B.L., Caudill, J. and King, A.D. Serotype-specific T(H)1 responses in recipients of two doses of candidate live-attenuated dengue virus vaccines. Am. J. Trop. Med. Hyg. 2003; 69:39–47.

Hahn, Y.S., Lenches, E.M., Galler, R., Rice, C.M., Dalrymple, J. and Strauss, J.H. Expression of the structural proteins of dengue 2 virus and yellow fever virus by recombinant vaccinia viruses. Arch. Virol. 1990; 115:251–265.

Hall, W.C., Crowell, T.P., Watts, D.M., Barros, V.L., Kruger, H., Pinheiro, F. and Peters, C.J. Demonstration of yellow fever and dengue antigens in formalin-fixed paraffin-embedded human liver by immunohistochemical analysis. Am. J. Trop. Med. Hyg. 1991; 45:408–417.

Halstead, S.B. Etiologies of the experimental dengues of Siler and Simmons. Am. J. Trop. Med. Hyg. 1974; 23:974–982.

Halstead, S.B. Studies on the attenuation of dengue 4. Asian J. Infect. Dis. 1978; 2:112–117.

Halstead, S.B. In vivo enhancement of dengue virus infection in Rhesus monkeys by passively transferred antibody. J. Infect. Dis. 1979; 140:527–533.

Halstead, S.B., Casals, J., Shotwell, H. and Palumbo, N. Studies on the immunization of monkeys against Protection derived from single and sequential virus infections. Am. J. Trop. Med. Hyg. 1973a; 22:365–374.

Halstead, S.B., Shotwell, H. and Casals, J. Studies on the pathogenesis of dengue infection in monkeys. I. Clinical laboratory responses to primary infection. J. Infect. Dis. 1973b; 128:7–14.

Halstead, S.B., Shotwell, H. and Casals, J. Studies on the pathogenesis of dengue infection in monkeys. II. Clinical laboratory responses to heterologous infection. J. Infect. Dis. 1973c; 128:15–22.

Halstead, S.B., Lan, N.T., Myint, T.T., Shwe, T.N., Nisalak, A., Kalyanarooj, S., Nimmannitya, S., Soegijanto, S., Vaughn, D.W. and Endy, T.P. Dengue hemorrhagic fever in infants: research opportunities ignored. Emerg. Infect. Dis. 2002; 8:1474–1479.

Halstead, S.B., Marchette, N.J., Chow, J.S. and Lolekha, S. Dengue virus replication enhancement in peripheral blood leukocyte from immune human beings (39160). Soc. Exp. Biol. Med. 1976; 151:136–139.

Halstead, S.B., Nimmannitya, S. and Margiotta, M.R. Dengue and Chikungunya virus infection in man in Thailand, 1962–1964: II. Observations on disease in outpatients. Am. J. Trop. Med. Hyg. 1969; 18:972–983.

Halstead, S.B. and O'Rourke, E.J. Antibody-enhanced dengue virus infection in primate leukocytes. Nature 1977a; 265:739–741.

Halstead, S.B. and O'Rourke, E.J. Dengue viruses and mononuclear phagocytes. I. Infection enhancement by non-neutralizing antibody. J. Exp. Med. 1977b; 146:201–217.

Halstead, S.B., O'Rourke, E.J. and Allison, A.C. Dengue viruses and mononuclear phagocytes: II. Identity of blood and tissue leukocytes supporting in vitro infection. J. Exp. Med. 1977; 146:218–228.

Halstead, S.B., Rojanasuphot, S. and Sangkawibha, N. Original antigenic sin in dengue. Am. J. Trop. Med. Hyg. 1983; 32:154–156.

Halstead, S.B., Streit, T.G., Lafontant, J.G., Putvatana, R., Russell, K., Sun, W., Kanesa-Thasan, N., Hayes, C.G. and Watts, D.M. Haiti: absence of dengue hemorrhagic fever despite hyperendemic dengue virus transmission. Am. J. Trop. Med. Hyg. 2001; 65:180–183.

Halstead, S.B., Udomsakdi, S., Simasathien, P., Singharaj, P., Sukhavachana, P. and Nisalak, A. Observations related to pathogenesis of dengue hemorrhagic fever. I. Experience with classification of dengue viruses. Yale J. Biol. Med. 1970; 42:261–275.

Halstead, S.B. and Yamarat, C. Recent epidemics of hemorrhagic fever in Thailand. Observations related to pathogenesis of a "new" dengue disease. J. Am. Pub. Health Assoc. 1965; 55:1386–1395.

Halstead, S.B., Yamarat, C. and Scanlon, J.E. The Thai hemorrhagic fever epidemic of 1962 (a preliminary report). J. Med. Assoc. Thai. 1963; 46:449–462.

Hammon, W.M., Rudnick, A. and Sather, G.E. Viruses associated with epidemic hemorrhagic fevers of the Philippines and Thailand. Science 1960; 131:1102–1103.

Hammon, W.M., Sather, G.E., Bond, J.O. and Lewis, F.Y. Effect of previous dengue infection and yellow fever vaccination on St. Louis encephalitis virus serological surveys in Tampa Bay area of Florida. Am. J. Epidemiol. 1966; 83:571–585.

Hanley, K.A., Lee, J.J., Blaney, J.E., Jr., Murphy, B.R. and Whitehead, S.S. Paired charge-to-alanine mutagenesis of dengue virus type 4 NS5 generates mutants with temperature-sensitive, host range, and mouse attenuation phenotypes. J. Virol. 2002; 76:525–531.

Hanley, K.A., Manlucu, L.R., Gilmore, L.E., Blaney, J.E., Jr., Hanson, C.T., Murphy, B.R. and Whitehead, S.S. A trade-off in replication in mosquito versus mammalian systems conferred by a point mutation in the NS4B protein of dengue virus type 4. Virology 2003; 312:222–232.

Hanley, K.A., Manlucu, L.R., Manipon, G.G., Hanson, C.T., Whitehead, S.S., Murphy, B.R. and Blaney, J.E., Jr. Introduction of mutations into the non-structural genes or 3′ untranslated region of an attenuated dengue virus type 4 vaccine candidate further decreases replication in rhesus monkeys while retaining protective immunity. Vaccine 2004; 22:3440–3448.

Hare, F.E. The 1897 epidemic of dengue in North Queensland. Aust. Med. Gaz. 1898; 21 March: 17:98–107.

Hayes, C.G., Manaloto, C.R., Gonzalez, A. and Ranoa, C.P. Dengue infections in the Philippines: clinical and virological findings in 517 hospitalized patients. Am. J. Trop. Med. Hyg. 1988; 39:110–116.

He, R.T., Innis, B.L., Nisalak, A., Usawattanakul, W., Wang, S.L., Kalayanarooj, S. and Anderson, R. Antibodies that block virus attachment to vero cells are a major component of the human neutralizing antibody response against dengue virus type 2. J. Med. Virol. 1995; 45:451–461.

Henchal, E.A., Henchal, L.S. and Schlesinger, J.J. Synergistic interactions of anti-NS1 monoclonal antibodies protect passively immunized mice from lethal challenge with dengue 2 virus. J. Gen. Virol. 1988; 69:2101–2107.

Hober, D., Delannoy, A.S., Benyoucef, S., De Groote, D. and Wattre, P. High levels of sTNFR p75 and TNFα in dengue-infected patients. Microbiol. Immunol. 1996; 40:569–573.

Hober, D., Nguyen, T.L., Shen, L., Ha, D.Q., Huong, V.T., Benyoucef, S., Nguyen, T.H., Bui, T.M., Loan, H.K., Le, B.L., Bouzidi, A., De Groote, D., Drouet, M.T., Deubel, V. and Wattre, P. Tumor necrosis factor alpha levels in plasma and whole-blood culture in dengue-infected patients: relationship between virus detection and pre-existing specific antibodies. J. Med. Virol. 1998; 54:210–218.

Hober, D., Poli, L., Roblin, B., Gestas, P., Chungue, E., Granic, G., Imbert, P., Pecarere, J.L., Vergezpascal, R., Wattre, P. and Maniez Montreuil, M. Serum levels of tumor necrosis factor-alpha (TNF-alpha), interleukin-6 (IL-6), and interleukin-1 beta (IL-1 beta) in dengue-infected patients. Am. J. Trop. Med. Hyg. 1993; 48:324–331.

Hoke, C.H., Jr., Nisalak, A., Sangawhipa, N., Jatanasen, S., Laorakpongse, T., Innis, B.L., Kotchasenee, S., Gingrich, J.B.,

Latendresse, J., Fukai, K. and Al, E. Protection against Japanese encephalitis by inactivated vaccines. N. Engl. J. Med. 1988; 319:608–614.

Hotta, S. Experimental studies on dengue. I. Isolation, identification and modification of the virus. J. Infect. Dis. 1952; 90:1–9.

Hotta, S. Dengue and Related Hemorrhagic Diseases. St. Louis: Warren H. Green, Inc., 1969.

Huang, C.Y., Butrapet, S., Pierro, D.J., Chang, G.J., Hunt, A.R., Bhamarapravati, N., Gubler, D.J. and Kinney, R.M. Chimeric dengue type 2 (vaccine strain PDK-53)/dengue type 1 virus as a potential candidate dengue type 1 virus vaccine. J. Virol. 2000a; 74:3020–3028.

Huang, C.Y., Butrapet, S., Tsuchiya, K.R., Bhamarapravati, N., Gubler, D.J. and Kinney, R.M. Dengue 2 PDK-53 virus as a chimeric carrier for tetravalent dengue vaccine development. J. Virol. 2003a; 77:11436–11447.

Huang, J.H., Wey, J.J., Sun, Y.C., Chin, C., Chien, L.J. and Wu, Y.C. Antibody responses to an immunodominant nonstructural 1 synthetic peptide in patients with dengue fever and dengue hemorrhagic fever. J. Med. Virol. 1999; 57:1–8.

Huang, K.J., Li, S.Y., Chen, S.C., Liu, H.S., Lin, Y.S., Yeh, T.M., Liu, C.C. and Lei, H.Y. Manifestation of thrombocytopenia in dengue-2-virus-infected mice. J. Gen. Virol. 2000b; 81:2177–2182.

Huang, Y.H., Lei, H.Y., Liu, H.S., Lin, Y.S., Chen, S.H., Liu, C.C. and Yeh, T.M. Tissue plasminogen activator induced by dengue virus infection of human endothelial cells. J. Med. Virol. 2003b; 70:610–616.

Huang, Y.H., Lei, H.Y., Liu, H.S., Lin, Y.S., Liu, C.C. and Yeh, T.M. Dengue virus infects human endothelial cells and induces IL-6 and IL-8 production. Am. J. Trop. Med. Hyg. 2000c; 63:71–75.

Huang, Y.H., Liu, C.C., Wang, S.T., Lei, H.Y., Liu, H.L., Lin, Y.S., Wu, H.L. and Yeh, T.M. Activation of coagulation and fibrinolysis during dengue virus infection. J. Med. Virol. 2001; 63:247–251.

Huerre, M.R., Lan, N.T., Marianneau, P., Hue, N.B., Khun, H., Hung, N.T., Khen, N.T., Drouet, M.T., Huong, V.T., Ha, D.Q., Buisson, Y. and Deubel, V. Liver histopathology and biological correlates in five cases of fatal dengue fever in Vietnamese children. Virchows Arch. 2001; 438:107–115.

Innis, B.L. Dengue and dengue hemorrhagic fever. In: Kass Handbook of Infectious Diseases: Exotic Virus Infections (J.S. Porterfield, Ed.), 1st ed. London: Chapman & Hall Medical, 1995.

Innis, B.L. Antibody responses to dengue virus infection. In: Dengue and Dengue Hemorrhagic Fever (D.J. Gubler and G. Kuno, Eds.). Cambridge: CAB International, 1997.

Innis, B.L., Nisalak, A., Nimmannitya, S., Kusalerdchariya, S., Chongswasdi, V., Suntayakorn, S., Puttisri, P. and Hoke, C.H., Jr. An enzyme-linked immunosorbent assay to characterize dengue infections where dengue and Japanese encephalitis co-circulate. Am. J. Trop. Med. Hyg. 1989; 40:418–427.

Isarangkura, P.B., Pongpanich, B., Pintadit, P., Phanichyakarn, P. and Valyasevi, A. Hemostatic derangement in dengue haemorrhagic fever. Southeast Asian J. Trop. Med. Public Health 1987; 18:331–339.

Iyngkaran, N., Yadav, M. and Sinniah, M. Augmented inflammatory cytokines in primary dengue infection progressing to shock. Singapore Med. J. 1995; 36:218–221.

Jaiswal, S., Khanna, N. and Swaminathan, S. Replication-defective adenoviral vaccine vector for the induction of immune responses to dengue virus type 2. J. Virol. 2003; 77:12907–12913.

Jessie, K., Fong, M.Y., Devi, S., Lam, S.K. and Wong, K.T. Localization of dengue virus in naturally infected human tissues, by immunohistochemistry and in situ hybridization. J. Infect. Dis. 2004; 189:1411–1418.

Johnson, A.J. and Roehrig, J.T. New mouse model for dengue virus vaccine testing. J. Virol. 1999; 73:783–786.

Juffrie, M., Meer, G.M., Hack, C.E., Haasnoot, K., Sutaryo, Veerman, A.J. and Thijs, L.G. Inflammatory mediators in dengue virus infection in children: interleukin-6 and its relation to C-reactive protein and secretory phospholipase A2. Am. J. Trop. Med. Hyg. 2001; 65:70–75.

Juffrie, M., Van der Meer, G.M., Hack, C.E., Haasnoot, K., Sutaryo, Veerman, A.J. and Thijs, L.G. Inflammatory mediators in dengue virus infection in children: interleukin-8 and its relationship to neutrophil degranulation. Infect. Immun. 2000; 68:702–707.

Kalayanarooj, S., Vaughn, D.W., Nimmannitya, S., Green, S., Suntayakorn, S., Kunentrasai, N., Viramitrachai, W., Ratanachu-Eke, S., Kiatpolpoj, S., Innis, B.L., Rothman, A.L., Nisalak, A. and Ennis, F.A. Early clinical and laboratory indicators of acute dengue illness. J. Infect. Dis. 1997; 176:313–321.

Kanesa-Thasan, N., Hernandez , L., Lyons, A., Putnak, R., Sun, W., Mckinney, D. and Vaughn, D.W. A phase I study of the WRAIR tetravalent live attenuated dengue vaccine in flavivirus-immune adult volunteers. 5th Annual Conference on Vaccine Research. Baltimore, MD, 2002.

Kanesa-Thasan, N., Smucny, J.J., Hoke, C.H., Marks, D.H., Konishi, E., Kurane, I., Tang, D.B., Vaughn, D.W., Mason, P.W. and Shope, R.E. Safety and immunogenicity of NYVAC-JEV and ALVAC-JEV attenuated recombinant Japanese encephalitis virus-poxvirus vaccines in vaccinia-nonimmune and vaccinia-immune humans. Vaccine 2001a; 19:483–491.

Kanesa-Thasan, N., Sun, W., Kim-Ahn, G., van Albert, S., Putnak, J.R., King, A., Raengsakulsrach, B., Christ-Schmidt, H., Gilson, K., Zahradnik, J.M., Vaughn, D.W., Innis, B.L., Saluzzo, J.F. and Hoke, C.H., Jr. Safety and immunogenicity of attenuated dengue virus vaccines (Aventis Pasteur) in human volunteers. Vaccine 2001b; 19:3179–3188.

Kanesa-Thasan, N., Sun, W., Ludwig, G.V., Rossi, C., Putnak, J.R., Mangiafico, J.A., Innis, B.L. and Edelman, R. Atypical antibody responses in dengue vaccine recipients. Am. J. Trop. Med. Hyg. 2003; 69:32–38.

Karin, N. Gene therapy for T cell-mediated autoimmunity: teaching the immune system how to restrain its own harmful activities by targeted DNA vaccines. Isr. Med. Assoc. J. 2000; 2(Suppl.):63–68.

Kaufman, B.M., Summers, P.L., Dubois, D.R., Cohen, W.H., Gentry, M.K., Timchak, R.L., Burke, D.S. and Eckels, K.H. Monoclonal antibodies for dengue virus prM glycoprotein protect mice against lethal dengue infection. Am. J. Trop. Med. Hyg. 1989; 41:576–580.

Kaufman, B.M., Summers, P.L., Dubois, D.R. and Eckels, K.H. Monoclonal antibodies against dengue 2 virus E-glycoprotein protect mice against lethal dengue infection. Am. J. Trop. Med. Hyg. 1987; 36:427–434.

Kelly, E.P., Greene, J.J., King, A.D. and Innis, B.L. Purified dengue 2 virus envelope glycoprotein aggregates produced by baculovirus are immunogenic in mice. Vaccine 2000; 18:2549–2559.

Khanam, S., Khanna, N. and Swaminathan, S. Induction of neutralizing antibodies and T cell responses by dengue virus type 2 envelope domain III encoded by plasmid and adenoviral vectors.Vaccine 2006; 24(42–43): 6513–6525.

Kho, L.K., Sumarmo, W.H., Jahja, E. and Gubler, D.J. Dengue hemorrhagic fever accompanied by encephalopathy in Jakarta. Southeast Asian J. Trop. Med. Public Health 1981; 12:83–86.

Khongphatthanayothin, A., Suesaowalak, M., Muangmingsook, S., Bhattarakosol, P. and Pancharoen, C. Hemodynamic profiles of patients with dengue hemorrhagic fever during toxic stage: an echocardiographic study. Intensive Care Med. 2003; 29:570–574.

Kimura, R. and Hotta, S. Studies on dengue: anti-dengue active immunization experiments in mice. Jpn. J. Bacteriol. 1944; 1:96–99.

Kitchener, S., Nissen, M., Nasveld, P., Forrat, R., Yoksan, S., Lang, J. and Saluzzo, J.F. Immunogenicity and safety of two live-attenuated tetravalent dengue vaccine formulations in healthy Australian adults. Vaccine 2006; 24:1238–1241.

Kliks, S.C., Nimmannitya, S., Nisalak, A. and Burke, D.S. Evidence that maternal dengue antibodies are important in the development of dengue hemorrhagic fever in infants. Am. J. Trop. Med. Hyg. 1988; 38:411–419.

Klinman, D.M., Takeno, M., Ichino, M., Gu, M., Yamshchikov, G., Mor, G. and Conover, J. DNA vaccines: safety and efficacy issues. Springer Semin. Immunopathol. 1997; 19:245–256.

Kochel, T., Wu, S.J., Raviprakash, K., Hobart, P., Hoffman, S., Porter, K. and Hayes, C. Inoculation of plasmids expressing the dengue-2 envelope gene elicit neutralizing antibodies in mice. Vaccine 1997; 15:547–552.

Konishi, E. and Fujii, A. Dengue type 2 virus subviral extracellular particles produced by a stably transfected mammalian cell line and their evaluation for a subunit vaccine. Vaccine 2002; 20:1058–1067.

Konishi, E., Yamaoka, M., Kurane, I. and Mason, P.W. A DNA vaccine expressing dengue type 2 virus premembrane and envelope genes induces neutralizing antibody and memory B cells in mice. Vaccine 2000; 18:1133–1139.

Kontny, U., Kurane, I. and Ennis, F.A. Gamma interferon augments Fc gamma receptor-mediated dengue virus infection of human monocytic cells. J. Virol. 1988; 62:3928–3933.

Kouri, G.P., Guzman, M.G., Bravo, J.R. and Triana, C. Dengue haemorrhagic fever/dengue shock syndrome: lessons from the Cuban epidemic, 1981. Bull. World Health Organ. 1989; 67:375–380.

Krishnamurti, C., Kalayanarooj, S., Cutting, M.A., Peat, R.A., Rothwell, S.W., Reid, T.J., Green, S., Nisalak, A., Endy, T.P., Vaughn, D.W., Nimmannitya, S. and Innis, B.L. Mechanisms of hemmorrhage in dengue without circulatory collapse. Am. J. Trop. Med. Hyg. 2001; 65:840–847.

Kuhn, R.J., Zhang, W., Rossmann, M.G., Pletnev, S.V., Corver, J., Lenches, E., Jones, C.T., Mukhopadhyay, S., Chipman, P.R., Strauss, E.G., Baker, T.S. and Strauss, J.H. Structure of dengue virus. Implications for flavivirus organization, maturation, and fusion. Cell 2002; 108:717–725.

Kuno, G. and Bailey, R.E. Cytokine responses to dengue infection among Puerto Rican patients. Mem. Inst. Oswaldo Cruz. (Rio de Janeiro) 1994; 89:179–182.

Kuno, G., Vorndam, A.V., Gubler, D.J. and Gomez, I. Study of anti-dengue NS1 antibody by western blot. J. Med. Virol. 1990; 32:102–108.

Kuo, C.H., Tai, D.I., Chang Chien, C.S., Lan, C.K., Chiou, S.S. and Liaw, Y.F. Liver biochemical tests and dengue fever. Am. J. Trop. Med. Hyg. 1992; 47:265–270.

Kurane, I., Brinton, M.A., Samson, A.L. and Ennis, F.A. Dengue virus-specific, human CD4+ CD8-cytotoxic T-cell clones: multiple patterns of virus cross-reactivity recognized by NS3-specific T-cell clones. J. Virol. 1991a; 65:1823–1828.

Kurane, I., Innis, B.L., Nimmannitya, S., Nisalak, A., Meager, A., Janus, J. and Ennis, F.A. Activation of T lymphocytes in dengue virus infections. High levels of soluble interleukin 2 receptor, soluble CD4, soluble CD8, interleukin 2, and interferon-gamma in sera of children with dengue. J. Clin. Invest. 1991b; 88:1473–1480.

Kurane, I. and Ennis, F.A. Cytokines in dengue virus infections: role of cytokines in the pathogenesis of dengue hemorrhagic fever. Semin. Virol. 1994a; 5:443–448.

Kurane, I. and Ennis, F.A. Cytotoxic T lymphocytes in dengue virus infection. Curr. Top. Microbiol. Immunol. 1994b; 189:93–108.

Kurane, I., Innis, B.L., Hoke, C.H., Jr., Eckels, K.H., Meager, A., Janus, J. and Ennis, F.A. T cell activation in vivo by dengue virus infection. J. Clin. Lab. Immunol. 1995; 46:35–40.

Kurane, I., Innis, B.L., Nisalak, A., Hoke, C.H., Jr., Nimmannitya, S., Meager, A. and Ennis, F.A. Human T cell responses to dengue virus antigens. Proliferative responses and interferon gamma production. J. Clin. Invest. 1989a; 83:506–513.

Kurane, I., Meager, A. and Ennis, F.A. Dengue virus-specific human T cell clones. Serotype crossreactive proliferation, interferon gamma production, and cytotoxic activity. J. Exp. Med. 1989b; 170:763–775.

Kurane, I., Rothman, A.L., Bukowski, J.F., Kontny, U., Janus, J., Innis, B.L., Nisalak, A., Nimmannitya, S., Meager, A. and Ennis, F.A. T-lymphocyte responses to dengue viruses. In: New Aspects of Positive-Strand RNA viruses (M.A. Brinton and F.X. Heinz, Eds.). Washington, DC: American Society for Microbiology, 1990.

La Russa, V.F. and Innis, B.L. Mechanisms of dengue virus-induced bone marrow suppression. Baillieres Clin. Haematol. 1995; 8:249–270.

Lai, C.J., Zhao, B.T., Hori, H. and Bray, M. Infectious RNA transcribed from stably cloned full-length cDNA of dengue type 4 virus. Proc. Natl. Acad. Sci. U.S.A. 1991; 88:5139–5143.

Lai, P.C., Lee, S.S., Kao, C.H., Chen, Y.S., Huang, C.K., Lin, W.R., Wann, S.R., Lin, H.H., Yen, M.Y. and Liu, Y.C. Characteristics of a dengue hemorrhagic fever outbreak in 2001 in Kaohsiung. J. Microbiol. Immunol. Infect. 2004; 37:266–270.

Lawn, S.D., Tilley, R., Lloyd, G., Finlayson, C., Tolley, H., Newman, P., Rice, P. and Harrison, T.S. Dengue hemorrhagic fever with fulminant hepatic failure in an immigrant returning to Bangladesh. Clin. Infect. Dis. 2003; 37:e1–e4.

Leyssen, P., De Clercq, E. and Neyts, J. Perspectives for the treatment of infections with Flaviviridae. Clin. Microbiol. Rev. 2000; 13:67–82, table of contents.

Libraty, D.H., Endy, T.P., Houng, H.S., Green, S., Kalayanarooj, S., Suntayakorn, S., Chansiriwongs, W., Vaughn, D.W., Nisalak, A., Ennis, F.A. and Rothman, A.L. Differing influences of virus burden and immune activation on disease severity in secondary dengue-3 virus infections. J. Infect. Dis. 2002; 185:1213–1221.

Lin, Y.L., Liao, C.L., Chen, L.K., Yeh, C.T., Liu, C.I., Ma, S.H., Huang, Y.Y., Huang, Y.L., Kao, C.L. and King, C.C. Study of Dengue virus infection in SCID mice engrafted with human K562 cells. J. Virol. 1998; 72:9729–9737.

Liu, H.W., Ho, T.L., Hwang, C.S. and Liao, Y.H. Clinical observations of virologically confirmed dengue fever in the 1987 outbreak in southern Taiwan. Kao-Hsiung I Hsueh Ko Hsueh Tsa Chih (Kaohsiung J. Med. Sci.) 1989; 5:42–49.

Livingston, P.G., Kurane, I., Lai, C.J., Bray, M. and Ennis, F.A. Recognition of envelope protein by dengue virus serotype-specific human CD4+ CD8-cytotoxic T-cell clones. J. Virol. 1994; 68:3283–3288.

Lucia, H.L. and Kangwanpong, D. Identification of dengue virus-infected cells in paraffin-embedded tissue using in situ polymerase chain reaction and DNA hybridization. J. Virol. Methods 1994; 48:1–8.

Mackowiak, P.A., Wasserman, S.S., Tacket, C.O., Vaughn, D.W., Eckels, K.H., Dubois, D.R., Hoke, C.H., Jr. and Edelman, R. Quantitative relationship between oral temperature and severity of illness following inoculation with candidate attenuated dengue virus vaccines. Clin. Infect. Dis. 1994; 19:948–950.

Mairuhu, A.T., Mac Gillavry, M.R., Setiati, T.E., Soemantri, A., Ten Cate, H., Brandjes, D.P. and Van Gorp, E.C. Is clinical outcome of dengue-virus infections influenced by coagulation and fibrinolysis? A critical review of the evidence. Lancet Infect. Dis. 2003; 3:33–41.

Malasit, P. Complement and dengue haemorrhagic fever/shock syndrome. Southeast Asian J. Trop. Med. Public Health 1987; 18:316–320.

Mandl, C.W., Aberle, J.H., Aberle, S.W., Holzmann, H., Allison, S.L. and Heinz, F.X. In vitro-synthesized infectious RNA as an attenuated live vaccine in a flavivirus model [see comments]. Nat. Med. 1998; 4:1438–1440.

Marchette, N.J., Halstead, S.B. and Chow, J.S. Replication of dengue viruses in cultures of peripheral blood leukocytes from dengue-immune Rhesus monkeys. J. Infect. Dis. 1976; 133:274–282.

Marchette, N.J., Halstead, S.B., Falkler, W.A., Jr., Stenhouse, A. and Nash, D. Studies on the pathogenesis of dengue infection in monkeys. III. Sequential distribution of virus in primary and heterologous infections. J. Infect. Dis. 1973; 128:23–30.

Marianneau, P., Steffan, A.M., Royer, C., Drouet, M.T., Jaeck, D., Kirn, A. and Deubel, V. Infection of primary cultures of human Kupffer cells by Dengue virus: no viral progeny synthesis, but cytokine production is evident. J. Virol. 1999; 73:5201–5206.

Markoff, L., Pang, X., Houng, H.S., Falgout, B., Olsen, R., Jones, E. and Polo, S. Derivation and characterization of a dengue type 1 host range-restricted mutant virus that is attenuated and highly immunogenic in monkeys. J. Virol. 2002; 76:3318–3328.

Martin, T., Parker, S.E., Hedstrom, R., Le, T., Hoffman, S.L., Norman, J., Hobart, P. and Lew, D. Plasmid DNA malaria vaccine: the potential for genomic integration after intramuscular injection. Hum. Gene Ther. 1999; 10:759–768.

Mathenge, E.G., Parquet Mdel, C., Funakoshi, Y., Houhara, S., Wong, P.F., Ichinose, A., Hasebe, F., Inoue, S. and Morita, K. Fusion PCR generated Japanese encephalitis virus/dengue 4 virus chimera exhibits lack of neuroinvasiveness, attenuated neurovirulence, and a dual-flavi immune response in mice. J. Gen. Virol. 2004; 85:2503–2513.

Mathew, A., Kurane, I., Green, S., Stephens, H.A., Vaughn, D.W., Kalayanarooj, S., Suntayakorn, S., Chandanayingyong, D., Ennis, F.A. and Rothman, A.L. Predominance of HLA-restricted cytotoxic T-lymphocyte responses to serotype-cross-reactive epitopes on nonstructural proteins following natural secondary dengue virus infection. J. Virol. 1998; 72:3999–4004.

Mathew, A., Kurane, I., Rothman, A.L., Zeng, L.L., Brinton, M.A. and Ennis, F.A. Dominant recognition by human CD8+ cytotoxic T lymphocytes of dengue virus nonstructural proteins NS3 and NS1.2a. J. Clin. Invest. 1996; 98:1684–1691.

Men, R., Bray, M., Clark, D., Chanock, R.M. and Lai, C.J. Dengue type 4 virus mutants containing deletions in the 3′ noncoding region of the RNA genome: analysis of growth restriction in cell culture and altered viremia pattern and immunogenicity in rhesus monkeys. J. Virol. 1996; 70:3930–3937.

Men, R., Wyatt, L., Tokimatsu, I., Arakaki, S., Shameem, G., Elkins, R., Chanock, R., Moss, B. and Lai, C.J. Immunization of rhesus monkeys with a recombinant of modified vaccinia virus Ankara expressing a truncated envelope glycoprotein of dengue type 2 virus induced resistance to dengue type 2 virus challenge. Vaccine 2000; 18:3113–3122.

Men, R.H., Bray, M. and Lai, C.J. Carboxy-terminally truncated dengue virus envelope glycoproteins expressed on the cell surface and secreted extracellularly exhibit increased immunogenicity in mice. J. Virol. 1991; 65:1400–1407.

Mitrakul, C., Futrakul, P., Vulayasevee, A., Pongpanich, B., Bhamarapravati, N., Bhanchet, P., Phanichyakarn, P., Pariyanonda, A., Udomsakdi, S., Sirisinha, S., Calcott, M.A., Mcconahey, P.J., Dixon, F.J., Muller-Eberhard, H.J., Tuchinda, P., Bukkavesa, S., Suwatte, V., Top, F.H., Jr., Russell, P.K., Mansuwan, P. and Nimmanitya, S. Pathogenetic mechanisms in dengue haemorrhagic fever: report of an international collaborative study. Bull. World Health Organ. 1973; 48:117–133.

Mitrakul, C., Poshyachinda, M., Futrakul, P., Sangkawibha, N. and Ahandrik, S. Hemostatic and platelet kinetic studies in dengue hemorrhagic fever. Am. J. Trop. Med. Hyg. 1977; 26:975–984.

Modis, Y., Ogata, S., Clements, D. and Harrison, S.C. Variable surface epitopes in the crystal structure of dengue virus type 3 envelope glycoprotein. J. Virol. 2005; 79:1223–1231.

Mongkolsapaya, J., Dejnirattisai, W., Xu, X.N., Vasanawathana, S., Tangthawornchaikul, N., Chairunsri, A., Sawasdivorn, S., Duangchinda, T., Dong, T., Rowland-Jones, S., Yenchitsomanus, P.T., Mcmichael, A., Malasit, P. and Screaton, G. Original antigenic sin and apoptosis in the pathogenesis of dengue hemorrhagic fever. Nat. Med. 2003; 9:921–927.

Mor, G., Singla, M., Steinberg, A.D., Hoffman, S.L., Okuda, K. and Klinman, D.M. Do DNA vaccines induce autoimmune disease? Hum. Gene Ther. 1997; 8:293–300.

Mukhopadhyay, S., Kuhn, R.J. and Rossmann, M.G. A structural perspective of the flavivirus life cycle. Nat. Rev. Microbiol. 2005; 3:13–22.

Murgue, B., Cassar, O., Guigon, M. and Chungue, E. Dengue virus inhibits human hematopoietic progenitor growth in vitro. J. Infect. Dis. 1997; 175:1497–1501.

Murgue, B., Roche, C., Chungue, E. and Deparis, X. Prospective study of the duration and magnitude of viraemia in children hospitalised during the 1996–1997 dengue-2 outbreak in French Polynesia. J. Med. Virol. 2000; 60:432–438.

Murphy, B.R., Prince, G.A., Walsh, E.E., Kim, H.W., Parrott, R.H., Hemming, V.G., Rodriguez, W.J. and Chanock, R.M. Dissociation between serum neutralizing and glycoprotein antibody responses of infants and children who received inactivated respiratory syncytial virus vaccine. J. Clin. Microbiol. 1986; 24:197–202.

Nelson, E.R., Bierman, H.R. and Chulajata, R. Hematologic findings in the 1960 hemorrhagic fever epidemic (dengue) in Thailand. Am. J. Trop. Med. Hyg. 1964; 13:642–649.

Nguyen, T.H., Lei, H.Y., Nguyen, T.L., Lin, Y.S., Huang, K.J., Le, B.L., Lin, C.F., Yeh, T.M., Do, Q.H., Vu, T.Q., Chen, L.C., Huang, J.H., Lam, T.M., Liu, C.C. and Halstead, S.B. Dengue hemorrhagic fever in infants: a study of clinical and cytokine profiles. J. Infect. Dis. 2004; 189:221–232.

Nguyen, T.L., Nguyen, T.H. and Tieu, N.T. The impact of dengue haemorrhagic fever on liver function. Res. Virol. 1997; 148:273–277.

Nichols, W.W., Ledwith, B.J., Manam, S.V. and Troilo, P.J. Potential DNA vaccine integration into host cell genome. Ann. N.Y Acad. Sci. 1995; 772:30–39.

Nimmannitya, S. Clinical spectrum and management of dengue haemorrhagic fever. Southeast Asian J. Trop. Med. Public Health 1987; 18:392–397.

Nimmannitya, S. Dengue hemorrhagic fever: diagnosis and management. In: Dengue and Dengue Hemorrhagic Fever (D.J. Gubler and G. Kuno, Eds.). Cambridge: CAB International, 1997.

Nimmannitya, S., Halstead, S.B., Cohen, S. and Margiotta, M.R. Dengue and chikungunya virus infection in man in Thailand, 1962–1964. I. Observations on hospitalized patients with hemorrhagic fever. Am. J. Trop. Med. Hyg. 1969; 18:954–971.

Nimmannitya, S., Thisyakorn, U. and Hemsrichart, V. Dengue haemorrhagic fever with unusual manifestations. Southeast Asian J. Trop. Med. Public Health 1987; 18:398–406.

Nobuchi, H. The symptoms of a dengue-like illness recorded in a Chinese medical encyclopedia. Kanpo no Rinsho 1979; 26:422–425.

Okuno, Y., Fukunaga, T., Tadano, M., K., F., Ikeda, T., Sekii, K. and Ariyoshi, H. Serological studies on volunteers inoculated experimentally with a dengue virus strain in 1943. Biken J. 1983; 26:161–163.

Papaevangelou, G. and Halstead, S.B. Infections with two dengue viruses in Greece in the 20th century. Did dengue hemorrhagic fever occur in the 1928 epidemic? Am. J. Trop. Med. Hyg. 1977; 80:46–51.

Parker, S.E., Borellini, F., Wenk, M.L., Hobart, P., Hoffman, S.L., Hedstrom, R., Le, T. and Norman, J.A. Plasmid DNA malaria vaccine: tissue distribution and safety studies in mice and rabbits. Hum. Gene Ther. 1999; 10:741–758.

Patey, O., Ollivaud, L., Breuil, J. and Lafaix, C. Unusual neurologic manifestations occurring during dengue fever infection. Am. J. Trop. Med. Hyg. 1993; 48:793–802.

Pengsaa, K., Luxemburger, C., Sabchareon, A., Limkittikul, K., Yoksan, S., Chambonneau, L., Chaovarind, U., Sirivichayakul, C., Lapphra, K., Chanthavanich, P. and Lang, J. Dengue virus infections in the first 2 years of life and the kinetics of transplacentally transferred dengue neutralizing antibodies in Thai children. J. Infect. Dis. 2006; 194:1570–1576.

Perez, A.B., Garcia, G., Sierra, B., Alvarez, M., Vazquez, S., Cabrera, M.V., Rodriguez, R., Rosario, D., Martinez, E., Denny, T. and Guzman, M.G. IL-10 levels in dengue patients: some findings from the exceptional epidemiological conditions in Cuba. J. Med. Virol. 2004; 73:230–234.

Perussia, B., Dayton, E.T., Lazarus, R., Fanning, V. and Trinchieri, G. Immune interferon induces the receptor for monomeric IgG1 on human monocytic and myeloid cells. J. Exp. Med. 1983; 158:1092–1113.

Pinheiro, F.P. and Corber, S.J. Global situation of dengue and dengue haemorrhagic fever, and its emergence in the Americas. World Health Stat. Q 1997; 50:161–169.

Pober, J.S. and Cotran, R.S. Cytokines and endothelial cell biology. Physiol. Rev. 1990; 70:427–451.

Pokidysheva, E., Zhang, Y., Battisti, A.J., Bator-Kelly, C.M., Chipman, P.R., Xiao, C., Gregorio, G.G., Hendrickson, W.A., Kuhn, R.J. and Rossmann, M.G. Cryo-EM reconstruction of dengue virus in complex with the carbohydrate recognition domain of DC-SIGN. Cell 2006; 124:485–493.

Porter, K.R., Kochel, T.J., Wu, S.J., Raviprakash, K., Phillips, I. and Hayes, C.G. Protective efficacy of a dengue 2 DNA vaccine in mice and the effect of CpG immuno-stimulatory motifs on antibody responses. Arch. Virol. 1998; 143:997–1003.

Pramulijo, H.S. and Harun, S.R. Ultrasound findings in dengue haemorrhagic fever. Pediatr. Radiol. 1991; 21:100–102.

Prud'homme, G.J., Lawson, B.R., Chang, Y. and Theofilopoulos, A.N. Immunotherapeutic gene transfer into muscle. Trends Immunol. 2001; 22:149–155.

Purdy, D.E. and Chang, G.J. Secretion of noninfectious dengue virus-like particles and identification of amino acids in the stem region involved in intracellular retention of envelope protein. Virology 2005; 333:239–250.

Putintseva, E., Vega, G. and Fernandez, L. Alterations in thrombopoiesis in patients with thrombocytopenia produced by dengue hemorrhagic fever. Nouv. Rev. Fr. Hematol. 1986; 28:269–273.

Putnak, R., Barvir, D.A., Burrous, J.M., Dubois, D.R., D'Andrea, V.M., Hoke, C.H., Jr., Sadoff, J.C. and Eckels, K.H. Development of a purified, inactivated, dengue-2 virus vaccine prototype in Vero cells: immunogenicity and protection in mice and rhesus monkeys. J. Infect. Dis. 1996a; 174:1176–1184.

Putnak, R., Cassidy, K., Conforti, N., Lee, R., Sollazzo, D., Truong, T., Ing, E., Dubois, D., Sparkuhl, J., Gastle, W. and Hoke, C.H., Jr. Immunogenic and protective response in mice immunized with a purified, inactivated, dengue-2 virus vaccine prototype made in fetal rhesus lung cells. Am. J. Trop. Med. Hyg. 1996b; 55:504–510.

Putnak, R., Coller, B.A., Voss, G., Vaughn, D.W., Clements, D., Peters, I., Bignami, G., Houng, H.S., Chen, R.C., Barvir, D.A., Seriwatana, J., Cayphas, S., Garcon, N., Gheysen, D., Kanesa-Thasan, N., Mcdonell, M., Humphreys, T., Eckels, K.H., Prieels, J.P. and Innis, B.L. An evaluation of dengue type-2 inactivated, recombinant subunit, and live-attenuated vaccine candidates in the rhesus macaque model. Vaccine 2005; 23:4442–4452.

Putnak, R., Feighny, R., Burrous, J., Cochran, M., Hackett, C., Smith, G. and Hoke, C.H., Jr. Dengue-1 virus envelope glycoprotein gene expressed in recombinant baculovirus elicits virus-neutralizing antibody in mice and protects them from virus challenge. Am. J. Trop. Med. Hyg. 1991; 45:159–167.

Putnak, R., Fuller, A., Vanderzanden, L., Innis, B.L. and Vaughn, D.W. Vaccination of rhesus macaques against dengue-2 virus with a plasmid DNA vaccine encoding the viral prM and E genes. Am. J. Trop. Med. Hyg. 2003; 68:469–476.

Raghupathy, R., Chaturvedi, U.C., Al-Sayer, H., Elbishbishi, E.A., Agarwal, R., Nagar, R., Kapoor, S., Misra, A., Mathur, A., Nusrat, H., Azizieh, F., Khan, M.A. and Mustafa, A.S. Elevated levels of IL-8 in dengue hemorrhagic fever. J. Med. Virol. 1998; 56:280–285.

Raut, C.G., Deolankar, R.P., Kolhapure, R.M. and Goverdhan, M.K. Susceptibility of laboratory-bred rodents to the experimental infection with dengue virus type 2. Acta. Virol. 1996; 40:143–146.

Raviprakash, K., Apt, D., Brinkman, A., Skinner, C., Yang, S., Dawes, G., Ewing, D., Wu, S.J., Bass, S., Punnonen, J. and Porter, K. A chimeric tetravalent dengue DNA vaccine elicits neutralizing antibody to all four virus serotypes in rhesus macaques. Virology 2006; 353:166–173.

Raviprakash, K., Ewing, D., Simmons, M., Porter, K.R., Jones, T.R., Hayes, C.G., Stout, R. and Murphy, G.S. Needle-free biojector injection of a dengue virus type 1 DNA vaccine with human immunostimulatory sequences and the GM-CSF gene increases immunogenicity and protection from virus challenge in Aotus monkeys. Virology 2003; 315:345–352.

Raviprakash, K., Kochel, T.J., Ewing, D., Simmons, M., Phillips, I., Hayes, C.G. and Porter, K.R. Immunogenicity of dengue virus type 1 DNA vaccines expressing truncated and full length envelope protein. Vaccine 2000; 18:2426–2434.

Raviprakash, K., Marques, E., Ewing, D., Lu, Y., Phillips, I., Porter, K.R., Kochel, T.J., August, T.J., Hayes, C.G. and Murphy, G.S. Synergistic neutralizing antibody response to a dengue virus type 2 DNA vaccine by incorporation of lysosome-associated membrane protein sequences and use of plasmid expressing GM-CSF. Virology 2001; 290:74–82.

Reed, W., Carroll, J. and Agramonte, A. The etiology of yellow fever: an additional note. JAMA 1901; 36:431–440.

Rico-Hesse, R. Molecular evolution and distribution of dengue viruses type 1 and 2 in nature. Virology 1990; 174:479–493.

Rigau-Perez, J.G. The early use of break-bone fever (Quebranta huesos, 1771) and dengue (1801) in Spanish. Am. J. Trop. Med. Hyg. 1998; 59:272–274.

Robinson, M.C. An epidemic of virus disease in Southern Province, Tanganyika Territory, in 1952–53 I. Clinical features. Trans. R. Soc. Trop. Med. Hyg. 1955; 49:28–32.

Roehrig, J.T., Bolin, R.A. and Kelly, R.G. Monoclonal antibody mapping of the envelope glycoprotein of the dengue 2 virus, Jamaica. Virology 1998; 246:317–328.

Rudnick, A. Studies of the ecology of dengue in Malaysia. Bull. World Health Organ. 1966; 35:78–79.

Rush, B. An Account of the Bilious Remitting Fever, as it Appeared in Philadelphia, in the Summer and Autumn of the Year 1780. Medical Inquires and Observations, 1st ed. Philadelphia: Prichard and Hall, 1789.

Russell, P.K., Intavivat, A. and Kanchanapilant, S. Anti-dengue immunoglobulins and serum beta 1 c/a globulin levels in dengue shock syndrome. J. Immunol. 1969; 102:412–420.

Russell, P.K., Udomsakdi, S. and Halstead, S.B. Antibody response in dengue and dengue hemorrhagic fever. Jpn. J. Med. Sci. Biol. 1967; 20:103–108.

Sabchareon, A., Lang, J., Chanthavanich, P., Yoksan, S., Forrat, R., Attanath, P., Sirivichayakul, C., Pengsaa, K., Pojjaroen-Anant, C., Chambonneau, L., Saluzzo, J.F. and Bhamarapravati, N. Safety and immunogenicity of a three dose regimen of two tetravalent live-attenuated dengue vaccines in five- to twelve-year-old Thai children. Pediatr. Infect. Dis. J. 2004; 23:99–109.

Sabchareon, A., Lang, J., Chanthavanich, P., Yoksan, S., Forrat, R., Attanath, P., Sirivichayakul, C., Pengsaa, K., Pojjaroen-Anant, C., Chokejindachai, W., Jagsudee, A., Saluzzo, J.F. and Bhamarapravati, N. Safety and immunogenicity of tetravalent live-attenuated dengue vaccines in Thai adult volunteers: role of serotype concentration, ratio, and multiple doses. Am. J. Trop. Med. Hyg. 2002; 66:264–272.

Sabin, A.B. Research on dengue during World War II. Am. J. Trop. Med. Hyg. 1952; 1:30–50.

Sabin, A.B. and Schlesinger, R.W. Production of immunity to dengue with virus modified by propagation in mice. Science 1945; 101:640–642.

Sanchez, V., Gimenez, S., Tomlinson, B., Chan, P.K., Thomas, G.N., Forrat, R., Chambonneau, L., Deauvieau, F., Lang, J. and Guy, B. Innate and adaptive cellular immunity in flavivirus-naive human recipients of a live-attenuated dengue serotype 3 vaccine produced in Vero cells (VDV3). Vaccine 2006; 24:4914–4926.

Sangkawibha, N., Rojanasuphot, S., Ahandrik, S., Viriyapongse, S., Jatanasen, S., Salitul, V., Phanthumachinda, B. and Halstead, S.B. Risk factors in dengue shock syndrome: a prospective epidemiologic study in Rayong, Thailand. I. The 1980 outbreak. Am. J. Epidemiol. 1984; 120:653–669.

Scherer, W.F., Breakenridge, F.A. and Dickerman, R.W. Cross-protection studies and search for subclinical disease in new world monkeys infected sequentially with different immunologic types of dengue viruses. Am. J. Epidemiol. 1972; 95:67–79.

Scherer, W.F., Russell, P.K., Rosen, L., Casals, J. and Dickerman, R.W. Experimental infection of chimpanzees with dengue viruses. Am. J. Trop. Med. Hyg. 1978; 27:590–599.

Schiavetta, A.M., Harre, J.G., Wagner, E., Simmons, M. and Raviprakash, K. Variable susceptibility of the owl monkey (*Aotus nancymae*) to four serotypes of dengue virus. Contemp. Top. Lab. Anim. Sci. 2003; 42:12–20.

Schlesinger, J.J., Brandriss, M.W., Cropp, C.B. and Monath, T.P. Protection against yellow fever in monkeys by immunization with yellow fever virus nonstructural protein NS1. J. Virol. 1986; 60:1153–1155.

Schlesinger, J.J., Brandriss, M.W. and Walsh, E.E. Protection of mice against dengue 2 virus encephalitis by immunization with the dengue 2 virus non-structural glycoprotein NS1. J. Gen. Virol. 1987; 68:853–857.

Schlesinger, R.W., Gordon, I., Frankel, J.W., Winter, J.W., Patterson, P.R. and Dorrance, W.R. Clinical and serologic response of man to immunization with attnuated dengue and yellow fever viruses. J. Immunol. 1956; 77:352–364.

Scott, R.M., Nimmannitya, S., Bancroft, W.H. and Mansuwan, P. Shock syndrome in primary dengue infections. Am. J. Trop. Med. Hyg. 1976; 25:866–874.

Sedegah, M., Hedstrom, R., Hobart, P. and Hoffman, S.L. Protection against malaria by immunization with plasmid DNA encoding circumsporozoite protein. Proc. Natl. Acad. Sci. U.S.A. 1994; 91:9866–9870.

Shresta, S., Kyle, J.L., Robert Beatty, P. and Harris, E. Early activation of natural killer and B cells in response to primary dengue virus infection in A/J mice. Virology 2004a; 319:262–273.

Shresta, S., Kyle, J.L., Snider, H.M., Basavapatna, M., Beatty, P.R. and Harris, E. Interferon-dependent immunity is essential for resistance to primary dengue virus infection in mice, whereas T- and B-cell-dependent immunity is less critical. J. Virol. 2004b; 78:2701–2710.

Shresta, S., Sharar, K.L., Prigozhin, D.M., Beatty, P.R. and Harris, E. Murine model for dengue virus-induced lethal disease with increased vascular permeability. J. Virol. 2006; 80:10208–10217.

Siler, J.F., Hall, M.W. and Hitchens, A.P. Dengue: its history, epidemilogy, mechanism of transmission, etiology, clinical manifestations, immunity, and prevention. Philippine J. Sci. 1926; 29:1–304.

Simmons, C.P., Dong, T., Chau, N.V., Dung, N.T., Chau, T.N., Thao Le, T.T., Dung, N.T., Hien, T.T., Rowland-Jones, S. and Farrar, J. Early T-cell responses to dengue virus epitopes in Vietnamese adults with secondary dengue virus infections. J. Virol. 2005; 79:5665–5675.

Simmons, J.S. Dengue fever. Am. J. Trop. Med. Hyg. 1931; 11:77–102.

Simmons, J.S., St John, J.H. and Reynolds, F.H.K. Experimental studies of dengue. Philippine J. Sci. 1931; 44:1–252.

Simmons, M., Murphy, G.S. and Hayes, C.G. Short report: antibody responses of mice immunized with a tetravalent dengue recombinant protein subunit vaccine. Am. J. Trop. Med. Hyg. 2001a; 65:159–161.

Simmons, M., Murphy, G.S., Kochel, T., Raviprakash, K. and Hayes, C.G. Characterization of antibody responses to combinations of a dengue-2 DNA and dengue-2 recombinant subunit vaccine. Am. J. Trop. Med. Hyg. 2001b; 65:420–426.

Simmons, M., Porter, K.R., Hayes, C.G., Vaughn, D.W. and Putnak, R. Characterization of antibody responses to combinations of a dengue virus type 2 DNA vaccine and two dengue virus type 2 protein vaccines in rhesus macaques. J. Virol. 2006; 80:9577–9585.

Solomon, T., Dung, N.M., Vaughn, D.W., Kneen, R., Thao, L.T., Raengsakulrach, B., Loan, H.T., Day, N.P., Farrar, J., Myint, K.S., Warrell, M.J., James, W.S., Nisalak, A. and White, N.J. Neurological manifestations of dengue infection. Lancet 2000; 355:1053–1059.

Soni, A., Chugh, K., Sachdev, A. and Gupta, D. Management of dengue fever in ICU. Indian J. Pediatr. 2001; 68:1051–1055.

Sosman, J.A., Verma, A., Moss, S., Sorokin, P., Blend, M., Bradlow, B., Chachlani, N., Cutler, D., Sabo, R., Nelson, M., Bruno, E., Gustin, D., Viana, M. and Hoffman, R. Interleukin 10-induced thrombocytopenia in normal healthy adult volunteers: evidence for decreased platelet production. Br. J. Haematol. 2000; 111:104–111.

Souza, L.J., Alves, J.G., Nogueira, R.M., Gicovate Neto, C., Bastos, D.A., Siqueira, E.W., Souto Filho, J.T., Cezario Tde, A., Soares, C.E. and Carneiro Rda, C. Aminotransferase changes and acute hepatitis in patients with dengue fever: analysis of 1,585 cases. Braz. J. Infect. Dis. 2004; 8:156–163.

Spaulding, A.C., Kurane, I., Ennis, F.A. and Rothman, A.L. Analysis of murine CD8(+) T-cell clones specific for the dengue virus NS3 protein: flavivirus cross-reactivity and influence of infecting serotype. J. Virol. 1999; 73:398–403.

Srichaikul, T., Nimmannitya, S., Artchararit, N., Siriasawakul, T. and Sungpeuk, P. Fibrinogen metabolism and disseminated intravascular coagulation in dengue hemorrhagic fever. Am. J. Trop. Med. Hyg. 1977; 26:525–532.

Srichaikul, T., Punyagupta, S., Nitiyanant, P. and Alkarawong, K. Disseminated intravascular coagulation in adult dengue haemorrhagic fever: report of three cases. Southeast Asian J. Trop. Med. Public Health 1975; 6:106–114.

Srivastava, A.K., Putnak, J.R., Warren, R.L. and Hoke, C.H., Jr. Mice immunized with a dengue type 2 virus E and NS1 fusion protein made in *Escherichia coli* are protected against lethal dengue virus infection. Vaccine 1995; 13:1251–1258.

Staropoli, I., Frenkiel, M.P., Megret, F. and Deubel, V. Affinity-purified dengue-2 virus envelope glycoprotein induces neutralizing

antibodies and protective immunity in mice. Vaccine 1997; 15:1946–1954.
Subramanian, V., Shenoy, S. and Joseph, A.J. Dengue hemorrhagic fever and fulminant hepatic failure. Dig. Dis. Sci. 2005; 50:1146–1147.
Sugrue, R.J., Fu, J., Howe, J. and Chan, Y.C. Expression of the dengue virus structural proteins in *Pichia pastoris* leads to the generation of virus-like particles. J. Gen. Virol. 1997; 78:1861–1866.
Suharti, C., Van Gorp, E.C., Setiati, T.E., Dolmans, W.M., Djokomoeljanto, R.J., Hack, C.E., Ten, C.H. and Van der Meer, J.W. The role of cytokines in activation of coagulation and fibrinolysis in dengue shock syndrome. Thromb. Haemost. 2002; 87:42–46.
Sun, W., Edelman, R., Kanesa-Thasan, N., Eckels, K.H., Putnak, J.R., King, A.D., Houng, H.S., Tang, D., Scherer, J.M., Hoke, C.H., Jr. and Innis, B.L. Vaccination of human volunteers with monovalent and tetravalent live-attenuated dengue vaccine candidates. Am. J. Trop. Med. Hyg. 2003; 69:24–31.
Sun, W., Nisalak, A., Gettayacamin, M., Eckels, K.H., Putnak, J.R., Vaughn, D.W., Innis, B.L., Thomas, S.J. and Endy, T.P. Protection of rhesus monkeys against dengue virus challenge after tetravalent live attenuated dengue virus vaccination. J. Infect. Dis. 2006; 193:1658–1665.
Suvatte, V., Pongpipat, D., Tuchinda, S., Ratanawongs, A., Tuchinda, P. and Bukkavesa, S. Studies on serum complement C3 and fibrin degradation products in Thai hemorrhagic fever. J. Med. Assoc. Thai. 1973; 56:24–32.
Takhampunya, R., Padmanabhan, R. and Ubol, S. Antiviral action of nitric oxide on dengue virus type 2 replication. J. Gen. Virol. 2006; 87:3003–3011.
Talavera, D., Castillo, A.M., Dominguez, M.C., Gutierrez, A.E. and Meza, I. IL8 release, tight junction and cytoskeleton dynamic reorganization conducive to permeability increase are induced by dengue virus infection of microvascular endothelial monolayers. J. Gen. Virol. 2004; 85:1801–1813.
Tassniyom, S., Vasanawathana, S., Chirawatkul, A. and Rojanasuphot, S. Failure of high-dose methylprednisolone in established dengue shock syndrome: a placebo-controlled, double-blind study. Pediatrics 1993; 92:111–115.
Tellier, M.C., Pu, R., Pollock, D., Vitsky, A., Tartaglia, J., Paoletti, E. and Yamamoto, J.K. Efficacy evaluation of prime-boost protocol: canarypoxvirus-based feline immunodeficiency virus (FIV) vaccine and inactivated FIV-infected cell vaccine against heterologous FIV challenge in cats. AIDS 1998; 12:11–18.
Thomas, S.J. ,Strickman, D. and Vaughn, D.W. Dengue epidemiology: virus epidemiology, ecology, and emergence. Adv. Virus Res. 2003; 61:235–289.
Twiddy, S.S., Holmes, E.C. and Rambaut, A. Inferring the rate and time-scale of dengue virus evolution. Mol. Biol. Evol. 2003; 20:122–129.
Ulmer, J.B., Donnelly, J.J., Parker, S.E., Rhodes, G.H., Felgner, P.L., Dwarki, V.J., Gromkowski, S.H., Deck, R.R., Dewitt, C.M. and Friedman, A.etal. Heterologous protection against influenza by injection of DNA encoding a viral protein. Science 1993; 259:1745–1749.
van Der Most, R.G., Murali-Krishna, K., Ahmed, R. and Strauss, J.H. Chimeric yellow fever/dengue virus as a candidate dengue vaccine: quantitation of the dengue virus-specific CD8 T-cell response. J. Virol. 2000; 74:8094–8101.
van der Poll, T., De Jonge, E. and Levi, M. Regulatory role of cytokines in disseminated intravascular coagulation. Semin. Thromb. Hemost. 2001; 27:639–651.
Van Gorp, E.C., Setiati, T.E., Mairuhu, A.T., Suharti, C., Cate HT, H., Dolmans, W.M., Van der Meer, J.W., Hack, C.E. and Brandjes, D.P. Impaired fibrinolysis in the pathogenesis of dengue hemorrhagic fever. J. Med. Virol. 2002; 67:549–554.
Vasudevan, S.G., Johansson, M., Brooks, A.J., Llewellyn, L.E. and Jans, D.A. Characterisation of inter- and intra-molecular interactions of the dengue virus RNA dependent RNA polymerase as potential drug targets. Farmaco 2001; 56:33–36.
Vaughn, D.W. Invited commentary: dengue lessons from Cuba. Am. J. Epidemiol. 2000; 152:800–803.
Vaughn, D.W. and Green, S. Dengue and dengue hemorrhagic fever. In: Hunter's Textbook of Tropical Medicine and Emerging Infectious Diseases (G.T. Strickland, Ed.), 8th ed. Philadelphia: W.B. Saunders Company, 2000.
Vaughn, D.W., Green, S., Kalayanarooj, S., Innis, B.L., Nimmannitya, S., Suntayakorn, S., Endy, T.P., Raengsakulrach, B., Rothman, A.L., Ennis, F.A. and Nisalak, A. Dengue viremia titer, antibody response pattern, and virus serotype correlate with disease severity. J. Infect. Dis. 2000; 181:2–9.
Vaughn, D.W., Green, S., Kalayanarooj, S., Innis, B.L., Nimmannitya, S., Suntayakorn, S., Rothman, A.L., Ennis, F.A. and Nisalak, A. Dengue in the early febrile phase: viremia and antibody responses. J. Infect. Dis. 1997; 176:322–330.
Vaughn, D.W., Hoke, C.H., Jr., Yoksan, S., Lachance, R., Innis, B.L., Rice, R.M. and Bhamarapravati, N. Testing of a dengue 2 live-attenuated vaccine (strain 16681 PDK 53) in ten American volunteers. Vaccine 1996; 14:329–336.
Venkata Sai, P.M., Dev, B. and Krishnan, R. Role of ultrasound in dengue fever. Br. J. Radiol. 2005; 78:416–418.
Von Herrath, M.G. and Whitton, J.L. DNA vaccination to treat autoimmune diabetes. Ann. Med. 2000; 32:285–292.
Wang, B., Boyer, J., Srikantan, V., Coney, L., Carrano, R., Phan, C., Merva, M., Dang, K., Agadjanan, M., Gilbert, L. et al. DNA inoculation induces neutralizing immune responses against human immunodeficiency virus type 1 in mice and nonhuman primates. DNA Cell Biol. 1993; 12:799–805.
Wells, M.A., Ennis, F.A. and Albrecht, P. Recovery from a viral respiratory infection II. Passive transfer of immune spleen cells to mice with influenza pneumonia. J. Immunol. 1981; 126:1042–1046.
Whalen, R.G. DNA vaccines for emerging infectious diseases: what if? Emerg. Infect. Dis. 1996; 2:168–175.
Whitehead, R.H., Chaicumpa, V., Olson, L.C. and Russell, P.K. Sequential dengue virus infections in the white-handed gibbon (*Hylobates lar*). Am. J. Trop. Med. Hyg. 1970; 19:94–102.
Whitehead, S.S., Falgout, B., Hanley, K.A., Blaney, J.E., Jr., Markoff, L. and Murphy, B.R. A live, attenuated dengue virus type 1 vaccine candidate with a 30-nucleotide deletion in the 3′ untranslated region is highly attenuated and immunogenic in monkeys. J. Virol. 2003a; 77:1653–1657.
Whitehead, S.S., Hanley, K.A., Blaney, J.E., Jr., Gilmore, L.E., Elkins, W.R. and Murphy, B.R. Substitution of the structural genes of dengue virus type 4 with those of type 2 results in chimeric vaccine candidates which are attenuated for mosquitoes, mice, and rhesus monkeys. Vaccine 2003b; 21:4307–4316.
Wichmann, O., Hongsiriwon, S., Bowonwatanuwong, C., Chotivanich, K., Sukthana, Y. and Pukrittayakamee, S. Risk factors and clinical features associated with severe dengue infection in adults and children during the 2001 epidemic in Chonburi, Thailand. Trop. Med. Int. Health 2004; 9:1022–1029.
Wills, B.A., Oragui, E.E., Stephens, A.C., Daramola, O.A., Dung, N.M., Loan, H.T., Chau, N.V., Chambers, M., Stepniewska, K., Farrar, J.J. and Levin, M. Coagulation abnormalities in dengue hemorrhagic fever: serial investigations in 167 Vietnamese children with dengue shock syndrome. Clin. Infect. Dis. 2002; 35:277–285.
Wisseman, C.L., Jr. Prophylaxis of dengue, with special reference to live virus vaccine. Proc. Japan Soc. Trop. Med. 1966; 7:51–56.
Wisseman, C.L., Jr., Sweet, B.H., Rosenzweig, E.C. and Eylar, O.R. Attenuated living type 1 dengue vaccines. Am. J. Trop. Med. Hyg. 1963; 12:620–623.

Wolfe, N.D., Kilbourn, A.M., Karesh, W.B., Rahman, H.A., Bosi, E.J., Cropp, B.C., Andau, M., Spielman, A. and Gubler, D.J. Sylvatic transmission of arboviruses among Bornean orangutans. Am. J. Trop. Med. Hyg. 2001; 64:310–316.

Wu, S.J., Grouard-Vogel, G., Sun, W., Mascola, J.R., Brachtel, E., Putvatana, R., Louder, M.K., Filgueira, L., Marovich, M.A., Wong, H.K., Blauvelt, A., Murphy, G.S., Robb, M.L., Innis, B.L., Birx, D.L., Hayes, C.G. and Frankel, S.S. Human skin Langerhans cells are targets of dengue virus infection. Nat. Med. 2000; 6:816–820.

Wu, S.J.L., Hayes, C.G., Dubois, D.R., Windheuser, M.G., Kang, Y.H., Watts, D.M. and Sieckmann, D.G. Evaluation of the severe combined immunodeficient (SCID) mouse as an animal model for dengue viral infection. Am. J. Trop. Med. Hyg. 1995; 52:468–476.

Yang, Z.Y., Wyatt, L.S., Kong, W.P., Moodie, Z., Moss, B. and Nabel, G.J. Overcoming immunity to a viral vaccine by DNA priming before vector boosting. J. Virol. 2003; 77:799–803.

Yap, K.L., Ada, G.L. and Mckenzie, I.F. Transfer of specific cytotoxic T lymphocytes protects mice inoculated with influenza virus. Nature 1978; 273:238–239.

Zeng, L., Falgout, B. and Markoff, L. Identification of specific nucleotide sequences within the conserved 3′-SL in the dengue type 2 virus genome required for replication. J. Virol. 1998; 72:7510–7522.

Zhao, B.T., Prince, G., Horswood, R., Eckels, K., Summers, P., Chanock, R. and Lai, C.J. Expression of dengue virus structural proteins and nonstructural protein NS1 by a recombinant vaccinia virus. J. Virol. 1987; 61:4019–4022.

Zivna, I., Green, S., Vaughn, D.W., Kalayanarooj, S., Stephens, H.A., Chandanayingyong, D., Nisalak, A., Ennis, F.A. and Rothman, A.L. T cell responses to an HLA-B*07-restricted epitope on the dengue NS3 protein correlate with disease severity. J. Immunol. 2002; 168:5959–5965.

CHAPTER

20

Ebola and Marburg

Gary P. Kobinger[1,2], *Maria Croyle*[3] *and Heinz Feldmann*[1,2]
[1]Special Pathogens Program, National Microbiology Laboratory, Public Health Agency of Canada, Winnipeg, MB, Canada
[2]Department of Medical Microbiology, University of Manitoba, Winnipeg, MB, Canada
[3]Division of Pharmaceutics, College of Pharmacy, The University of Texas at Austin, Austin, TX, USA

OUTLINE

ABSTRACT

Marburg virus (MARV) and Ebola virus (EBOV), family Filoviridae, were discovered in 1967 and 1976, respectively, but remained largely unknown until 1995 when EBOV reemerged in Kikwit, Democratic Republic of the Congo (DRC). Since then Angola, DRC, Gabon, Republic of Congo, Sudan, and Uganda have experienced outbreaks of EBOV and MARV hemorrhagic fever, which varied from a few cases to several hundreds with case fatality rates ranging from 50–90%. In addition, EBOV has emerged as a devastating pathogen for the great apes in some of these countries leading to a dramatic reduction of these endangered species. More recently, different fruit and insectivore bats have been implicated as potential reservoirs for MARV and EBOV. Being highly pathogenic for human and nonhuman primates, and the subject of former weapon programs, makes filoviruses one of the most feared pathogens worldwide today. The lack of pre- and post-exposure intervention makes the development of rapid diagnostics, new antivirals, and protective vaccines a priority for the preparedness of many nations. Further insight into the immunology and pathogenesis of filoviruses will help to establish the urgently needed immune protective markers to evaluate experimental vaccine candidates. This chapter reviews and evaluates the current status of filovirus vaccine research, a field that has made tremendous progress in the past decade leading to several promising vaccine platforms of which one is currently in phase I clinical trials. All these platforms show excellent pre-exposure protective efficacies in macaques, the gold standard animal model, but in particular one candidate vaccine has proven promising in post-exposure treatment. It is expected that within the next decade a vaccine for MARV and EBOV will be available.

INTRODUCTION

Marburg virus (MARV) was identified as the causative agent of several human cases of hemorrhagic fever that occurred in Marburg, Germany, in 1967 (Siegert et al., 1967). The first recorded emergence of Ebola virus (EBOV) was in Central Africa, in southern Sudan and northern Zaire, now the Democratic Republic of the Congo, in the summer of 1976. Both viruses were classified in the order of Mononegavirales, as they are using a non-segmented, negative-strand RNA genome as a genetic template for replication. The family of Filoviridae was proposed in 1982 to segregate EBOV and MARV on the basis of newly characterized morphological, physiochemical, and biological features (Feldmann et al., 2003; Kiley et al., 1982). The family of Filoviridae is divided into the Marburgvirus and Ebolavirus genera which are further subdivided into distinct species such as *Lake Victoria marburgvirus* (LVMARV) and *Zaire*, *Reston*, *Sudan* or *Cote d'Ivoire ebolaviruses* (ZEBOV, REBOV, SUBOV, or CIEBOV, respectively) (Feldmann et al., 2004).

ANTIGENS ENCODED BY AGENT

Filoviruses are long, cylindrical, filamentous, enveloped particles that can be straight, curved, or coiled and of variable lengths although with a consistent diameter of approximately 80 nm (Fig. 20.1). Their negative-strand RNA genome is non-segmented and measures approximately 19 kb in length (Feldmann et al., 2003; Sanchez et al., 2007). The genomes of filoviruses encode for seven genes, although EBOV has the distinctive feature of expressing an additional protein through RNA editing (Sanchez et al., 1996; Volchkov et al., 1995). MARV and EBOV express a nucleoprotein (NP), a transmembrane glycoprotein (GP), an RNA-dependent RNA polymerase (L), and four additional virion structural proteins: VP24, VP30, VP35, and VP40 (Feldmann and Kiley, 1999; Sanchez et al., 2007) (Fig. 20.1). For EBOV, RNA editing occurs in the GP gene and results in expression of a truncated version of the glycoprotein, designated soluble glycoprotein (sGP), which is secreted into the extracellular environment of infected cells. NP, VP30, VP35, and the L protein are associated with the virus genomic RNA to form the ribonucleoprotein complex (RNP). NP, VP35, and L are essential and sufficient for the transcription and replication of MARV, whereas EBOV also require the presence of VP30. The inside of the viral membrane is coated with the matrix protein VP40 and a second membrane-associated protein VP24. The external surface of the virus membrane is spiked with homotrimers of $GP_{1,2}$ which are processed through the endosomal pathway including folding, disulfide bridge formation, glycozylation, and cleavage of the precursor GP into GP_1 and GP_2 by furin or a furin-like endoprotease. Trimers of $GP_{1,2}$ spikes mediate receptor binding and fusion and are thus the major target for neutralizing host immune responses (Feldmann et al., 2001).

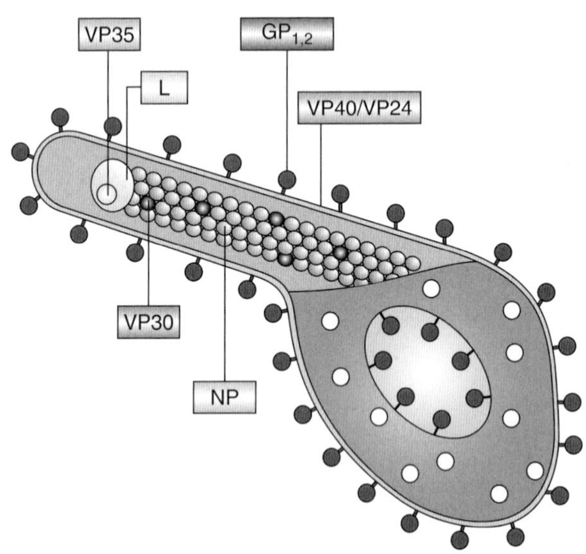

FIGURE 20.1 Schematic illustration of a filovirus particle. Four proteins form the RNP: polymerase or large (L) protein, NP, virion structural protein 30 (VP30), and virion structural protein 35 (VP35). The glycoprotein (GP) is anchored with the carboxy-terminal part in the virion membrane. Homotrimers of GP form the spikes on the surface of the virion. VP40 is the matrix protein and VP24 another membrane-associated protein. Adapted from Feldmann et al. (2003) (see color plate section).

EPIDEMIOLOGY

Outbreaks of filovirus infections have primarily been localized to Central Africa (Fig. 20.2) and despite their dramatic consequences on affected communities they are of relatively low impact on global human health (Sanchez et al., 1995). However, filovirus infections are feared worldwide as imported cases or the result of an intentional release (Borio et al., 2002). With mortality rates ranging from 23–90% and documented human-to-human transmission, the introduction or intentional release of adequately formulated filoviruses in a densely populated area, such as a major city, could be devastating. No clinical modalities are currently available to the general public. However, several strategies, including vaccines and antiviral drugs, are being evaluated and a few promising candidates have successfully passed advanced preclinical testing.

CLINICAL DISEASE, PATHOGENESIS, AND TREATMENT

Filoviruses cause hemorrhagic fever with a survival rate that can be as low as 10% among infected individuals, listing them among the most lethal viruses known to man (Wilson et al., 2001a, 2001b). The mode of primary transmission to humans remains evasive. Fruit bats are suspected to be a natural reservoir of EBOV and MARV, and could also act as a primary source of infection to the great apes, which in turn have transmitted infections to humans in the past (Leroy et al., 2005; Towner et al., 2007). Filoviruses are extremely infectious in experimental animal models, although natural spread among apes and humans is less efficient and normally requires direct contact. However, there is experimental and epidemiologic evidence indicating that filoviruses can also be transmitted through mucosal exposure (Jaax et al., 1996; Johnson et al., 1995). The disease course of a filovirus infection in human or nonhuman primates is dramatic, leading to a fulminant syndrome characterized by systemic inflammation, multi-organ failure, and coagulopathy with hemorrhages (Borio et al., 2002; Feldmann et al., 2003; Sanchez et al., 2007). The incubation period may vary from 6 to 21 days. Initial symptoms are nonspecific, such as fever, headache, nausea, diarrhea, and abdominal pain. Severe clinical cases rapidly progress over a 7–14-day period to the fully blown hemorrhagic fever syndrome and death (Baskerville et al., 1978; Bowen et al., 1978; Fisher-Hoch, 1983; Sanchez et al., 2007). The virus appears to initially infect monocytes, macrophages, and dendritic cells leading to dysregulation of the innate immunity and a systemic inflammatory response syndrome. The end result is massive destruction of critical organs (e.g., liver), vascular damage, and hemorrhages (Feldmann et al., 2003). While its clinical course is well described, the specific mechanisms underlying the pathogenicity of filovirus infections have not been defined due to difficulties in obtaining clinical specimens, studying the disease in remote areas where outbreaks occur, and the level of

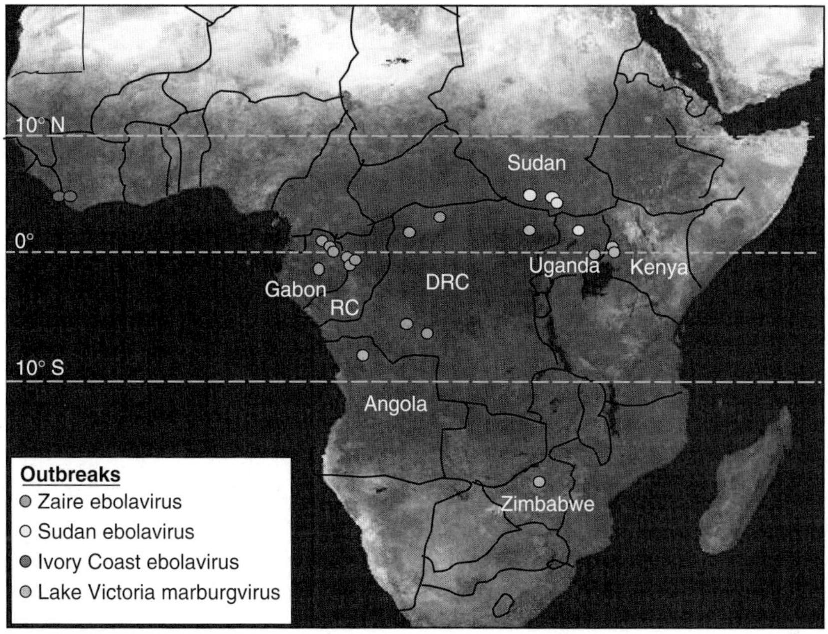

FIGURE 20.2 Summary of location, timeline, and types of filovirus outbreaks in Africa. RC, Republic of Congo; DRC, Democratic Republic of Congo. Adapted from Groseth et al. (2007).

biohazard containment required for clinical analysis. In addition, the rapid progression of the disease has afforded little opportunity to develop acquired immunity and has hindered identification of the components of the immune system that may protect against filovirus infections.

VACCINES

Attempts to develop vaccines for filoviruses began shortly after MARV and EBOV were discovered. Immunization regimens of formalin-fixed or heat-inactivated virion preparations were first evaluated in guinea pigs and nonhuman primates. Unfortunately, the level of protection achieved was inconsistent among different studies. However, all studies demonstrated partial protection of vaccinated guinea pigs or nonhuman primates suggesting that complete protection could be generated with a more efficient immunization strategy (Bausch and Geisbert, 2007; Chupurnov et al., 1995; Geisbert et al., 2002, 2003; Ignatyev et al., 1994, 1996; Lupton et al., 1980; Mikhailov et al., 1994).

More recently, the focus of filovirus vaccine development has been more on EBOV (Table 20.1). Protocols have mainly concentrated on the use of subunit vaccines based on a single or a combination of viral structural proteins to induce protective immunity against EBOV. The GP, NP, and the viral structural proteins VP24, VP30, VP35, and VP40 were evaluated as potential antigenic determinants of protective immune responses in different immunization strategies (Table 20.1). Initially, vaccine carriers of viral antigens included naked DNA, liposomes, vaccinia, alphavirus replicons, and virus-like particles (VLPs) (Geisbert et al., 2002; Pushko et al., 2000, 2001; Vanderzanden et al., 1998; Warfield et al., 2003; Wilson et al., 2001a; Xu et al., 1998). DNA immunization of mice with both EBOV GP and NP constructs stimulated detectable cytotoxic T cell responses to both antigens and some degree of protection (Vanderzanden et al., 1998; Xu et al., 1998). However, the optimal protocol of DNA-based vaccine with GP and NP in guinea pigs resulted only in partial protection against EBOV (Chupurnov et al., 1995; Vanderzanden et al., 1998; Xu et al., 1998). Immune sera from vaccinated guinea pigs did not inhibit EBOV replication in vitro and passive transfer in naïve animals failed to protect them against EBOV. Recombinant vaccinia viruses expressing EBOV NP, VP35, VP40, or VP24 failed to protect guinea pigs from a lethal challenge with EBOV. Vaccinia-encoding GP was the only construct capable of inducing some levels of protection in guinea pigs but failed to protect *Cynomolgus macaques* (Geisbert et al., 2002). Serum antibody titers by ELISA and endpoint titers in the 80% plaque reduction neutralization test ($PRNT_{80}$), which detect neutralizing antibody (NAB) remained low following immunization with vaccinia expressing EBOV GP. Immunization with VEEV replicons expressing GP resulted in low but detectable endpoint titers of NAB in $PRNT_{80}$. Specific, non-neutralizing antibodies were also detected by ELISA in response to immunization with EBOV VP24, VP30, VP35, VP40, and NP but it is not clear what role these antibodies may play in protection against the lethal challenge. The best immunization strategy with VEEV-based vaccine resulted in mixed protection in mice and guinea pigs and failed to protect nonhuman primates (Pushko et al., 2000).

Complete protection of nonhuman primates against EBOV was first reported in 2000 with the use of naked DNA in combination with replication incompetent adenoviruses (Ads) based on serotype 5 (Adhu5) administered in a prime-boost regimen (Sullivan et al., 2000) (Table 20.1). This approach was later improved by a single immunization with recombinant Adhu5 expressing GP and NP (Sullivan et al., 2003). A vaccine based on vesicular stomatitis virus (VSV) was also shown to protect nonhuman primates against a lethal challenge with EBOV (Jones et al., 2005). Recently, a recombinant human parainfluenza virus type 3 (HPIV3) vector expressing ZEBOV GP also showed full protection of guinea pigs and nonhuman primates against a homologous ZEBOV challenge (Bukreyev et al., 2006, 2007) (Table 20.1).

However, the first success in fully protecting nonhuman primates against a lethal challenge of filovirus was demonstrated against MARV following immunization with a VEEV-based vaccine expressing GP and NP (Hevey et al., 1998). Several immunization strategies were also evaluated in parallel against MARV including DNA, Adhu5, alphavirus replicons, VSV, VLPs, vaccinia, and bacculovirus vaccine platforms (Table 20.2) (Becker et al., 1996; Daddario-DiCaprio et al., 2006a; Hevey et al., 1997, 1998, 2001; Jones et al., 2005; Riemenschneider et al., 2003; Swenson et al., 2005; Warfield et al., 2004). Similarly to EBOV vaccinated animals, MARV infection resulted in variable protection of immunized animals depending on the vaccine strategy and animal model selected. Remarkably, the first demonstration of post-exposure filovirus vaccine efficacy in nonhuman primates was demonstrated against MARV with 100% survival of VSV-MARV vaccinated animals (Daddario-DiCaprio et al., 2006b). This was also done more recently for EBOV with 50% survival among VSV-EBOV vaccinated animals (Feldmann et al., 2007).

TABLE 20.1 Vaccine strategies against Ebola virus

Vector	Antigen	Animal model	Protection	References
DNA	GP	Mouse	Partial	Xu et al. (1998)
		Guinea pig	Partial	Vanderzanden et al. (1998)
	NP	Mouse	Partial	Chupurnov et al. (1995)
		Guinea pig	Partial	Xu et al. (1998), Vanderzanden et al. (1998), Chupurnov et al. (1995)
	GP + NP	Guinea pig	Full	Sullivan et al. (2000)
	sGP	Guinea pig	Partial	Xu et al. (1998)
Vaccinia	VP24	Guinea pig	No	Geisbert et al. (2002)
	VP35	Guinea pig	No	Geisbert et al. (2002)
	VP40	Guinea pig	No	Geisbert et al. (2002)
	GP	Guinea pig	Partial	Geisbert et al. (2002)
		Macaque	No	Geisbert et al. (2002)
VEEV replicon	GP	Mouse	Full	Pushko et al. (2000)
		Guinea pig	Full	Pushko et al. (2000)
		Macaque	No	Geisbert et al. (2002)
	NP	Mouse	Full	Geisbert et al. (2002)
		Guinea pig	Partial	Pushko et al. (2001)
		Macaque	No	Wilson and Hart (2001)
	GP + NP	Mouse	Full	Pushko et al. (2001)
		Guinea pig	Full	Pushko et al. (2001)
		Macaque	No	Geisbert et al. (2002)
	VP24	Mouse	Partial	Wilson et al. (2001b)
	VP30	Mouse	Partial	Wilson et al. (2001b)
	VP35	Mouse	Partial	Wilson et al. (2001b)
	VP40	Mouse	Partial	Wilson et al. (2001b)
DNA + Adhu5	GP + NP	Macaque	Full	Sullivan et al. (2000, 2003)
VSV	ΔGP	Macaque	Full	Jones et al. (2005)
HPIV3	GP	Guinea pig	Full	Bukreyev et al. (2006, 2007)
		Macaque	Full	Bukreyev et al. (2006, 2007)
Baculovirus	GP	Guinea pig	No	Riemenschneider et al. (2003)
DNA + Baculovirus	ΔGP	Guinea pig	Partial	Riemenschneider et al. (2003)

Note: GP: glycoprotein; ΔGP: recombinant GP protein; sGP: secreted GP; NP: nucleoprotein; AdHu5: recombinant adenovirus serotype 5; VSV: recombinant vesicular stomatitis virus; VEEV: Venezuelan equine encephalitis virus; HPIV3: human parainfluenza virus type 3; VP: viral protein.

VACCINES IN DEVELOPMENT

Adenovirus as Vaccine Vectors

Vectors based on recombinant adenoviruses (Ads) were initially mostly considered in the setting of gene therapy. A number of different serotypes of Ads have been isolated from humans causing a spectrum of illnesses ranging from upper respiratory tract infection to diarrheal diseases (Horwitz, 2001). These viruses are non-enveloped with linear double-stranded genomes of approximately 36 kb in length (Shenk and Williams, 1984).

The initial applications of Ads as vectors for gene therapy were based on recombinants in which the immediate early gene E1 is deleted (Rosenfeld et al., 1991; Stratford-Perricaudet et al., 1990). This substantially attenuated the virus, diminishing its toxicity and allowing for expression of heterologous genes. Virtually all experiments performed to date with adenoviral vectors utilized a construct based on Adhu5 for adenovirus serotype 5 (Adhu5). Ad vectors were shown to elicit vibrant B and T cell responses to both the Ad capsid proteins as well as the transgene products (Yang et al., 1994a, 1994b). Consequently, re-administration of the vector is ineffective because of neutralizing antibodies fiber capsid protein. In addition, target cells expressing the transgene product are quickly eliminated as a result of rapid re-activation of cytotoxic T lymphocytes (CTL) to both the transgene product as well as residually expressed viral genes. The intensity of the immune response is promoted mainly through the transduction of antigen-presenting cells in vivo (Jooss et al., 1998).

The strength and amplitude of the immune response following in vivo administration of Ads compelled several scientists to consider them as vaccine

TABLE 20.2 Vaccine strategies against Marburg virus

Vector	Antigen	Animal model	Protection	References
DNA	GP (Musoke)	Guinea pig	Partial	Hevey et al. (1998)
		Guinea pig	Partial	Riemenschneider et al. (2003)
		Macaque	Partial	Riemenschneider et al. (2003)
	GP (Ravn)	Guinea pig	Full	Riemenschneider et al. (2003)
DNA and bacculovirus	GP (Musoke) + GPΔTM	Guinea pig	Full	Hevey et al. (2001)
Vaccinia	GP (Musoke)	Guinea pig	No	Becker et al. (1996)
VLP	GP + VP40 (Musoke)	Guinea pig	Full	Warfield et al. (2004); Swenson et al. (2005)
VLP	GP (Musoke) + VP40 (Ebola)	Guinea pig	Full	Swenson et al. (2005)
VLP	GP (Ebola) + VP40 (Musoke)	Guinea pig	Partial	Swenson et al. (2005)
VEEV replicon	GP (Musoke)	Guinea pig	Partial	Hevey et al. (1998)
	GP (Musoke)	Guinea pig	Full	Hevey et al. (2001)
	NP (Musoke)	Guinea pig	Full	Hevey et al. (1998)
	VP40 (Musoke)	Guinea pig	Partial	Hevey et al. (1998)
	GPΔTM (Musoke)	Guinea pig	Full	Hevey et al. (1998)
	VP24 (Musoke)	Guinea pig	Partial	Hevey et al. (1998)
	VP30 (Musoke)	Guinea pig	No	Hevey et al. (1998)
	VP35 (Musoke)	Guinea pig	Partial	Hevey et al. (1998)
	GP (Musoke)	Macaque	Full	Hevey et al. (1998)
	GP + NP (Musoke)	Macaque	Full	Hevey et al. (1998)
	NP (Musoke)	Macaque	Partial	Hevey et al. (2001)
VSV	GP	Macaque	Full	Jones et al. (2005)

Note: GP: glycoprotein; GPΔTM: recombinant GP protein; NP: nucleoprotein; VSV: recombinant vesicular stomatitis virus; VEEV: Venezuelan equine encephalitis virus; VLP: virus-like particle; VP: viral protein.

carriers. In recent years adenoviral vectors have been shown to elicit important immune responses to various antigens including bacterial, viral, and even self-antigens in the context of cancer vaccines. Antigens derived from pathogens such as anthrax, plague, hepatitis C, rabies, and SARS were successful at stimulating strong immune responses often correlating with protection of susceptible animals against challenge with the infectious agent (Gao et al., 2003; Makimura et al., 1996; Tan et al., 2003; Vos et al., 2001).

In conjunction with naked DNA, Ad vector was the first vaccine capable of protecting nonhuman primates against ZEBOV challenge (Sullivan et al., 2000). In that study, cynomologus monkeys were immunized three times with a mixture of plasmid DNAs encoding the NP of ZEBOV and GPs of different EBOVs (ZEBOV, SEBOV, CIEBOV) followed by a booster of Ad expressing ZEBOV GP five months after priming with DNA. The monkeys were challenged with a low dose of ZEBOV (6 plaque-forming units (pfu) per animal) and only vaccinated animals survived the challenge. Antibody responses, T-cell proliferation, and CTL responses indicated that antibody and T-helper CD4+ cells were important for protection and that cell-mediated immunity although possibly important might not have been an absolute requirement. This would be in agreement with observations that passive transfer of EBOV immune serum can be protective in different animal models normally susceptible to infection. This approach was further improved to obtain complete protection of cynomolgus macaques to ZEBOV only 28 days post-vaccination with a single dose of Adhu5 expressing GP mixed with Ad expressing NP (Sullivan et al., 2003). This work led to a Phase I clinical trial by the U.S. National Institutes of Health (NIH) with early results demonstrating that the DNA component is well tolerated and capable of eliciting an immune response against EBOV GP in normal adults (Martin et al., 2006). More recently, the NIH initiated another Phase I clinical trial in normal adults immunized with the first-generation Adhu5 expressing the EBOV GP with the objective of evaluating safety and immune responses to the vaccine (http://clinicaltrials.gov/show/NCT00374309).

An important problem in the use of currently available adenoviral vector for human vaccines is natural pre-exposure to the virus, which leads to a long-lasting immunity capable of compromising vaccine efficacy. Infection with natural Ads is frequent during the first quarter of life resulting in high levels of serum NAB to Ad serotype. Approximately 30–50% of the human population in North America and over

90% in developing countries has levels of NAB capable of compromising Ad vaccine efficacy against EBOV in different animal models (Kobinger et al., 2006). Several studies have shown that these levels of NABs can also neutralize relatively high doses of different vaccine vectors based on Adhu5 (Chirmule et al., 1999; Fitzgerald et al., 2003; Xiang et al., 2002). Experimental studies demonstrated that several animal models including mice, guinea pigs, and nonhuman primates could no longer generate a protective immune response in the presence of circulating NAB against the Ad particles (Kobinger et al., 2006; Nanda et al., 2005). Results from the STEP HIV clinical trial reported an increased number of HIV seroconversionin Ad-vaccinated individuals with prior immunity to adenovirus (Moorel et al., 2008). Consequently, alternate strategies such as the evaluation of different vaccine vectors based on distinct serotypes of human or simian Ads that are not commonly circulating in the human population as well as different routes of immunization are currently being investigated (Jones et al., 2007; Kobinger et al., 2006; Nanda et al., 2005; Patel et al., 2007; Roy et al., 2004). Human Ad serotype 6, against which the prevalence of NAB in humans is less than 5%, has shown comparable immunogenicity to Adhu5 in nonhuman primates (Capone et al., 2006). A vaccine regimen based on chimpanzee Ads were also shown to be protective against ZEBOV and SARS viruses in different animal models pretreated with human NABs capable of inhibiting the efficacy of Ad serotype 5-based vaccine (Kobinger et al., 2006).

Vesicular Stomatitis Virus as Vaccine Vectors

VSV is a non-segmented negative-stranded RNA virus and belongs to the family Rhabdoviridae, genus *Vesiculovirus* (Lyles and Rupprecht, 2007). Its simple structure and rapid high-titered growth in mammalian and many other cell types have made it a preferential tool for molecular and cell biologists in the past 30 years (Lyles and Rupprecht, 2007). Naturally occurring human infections with VSV are rare. However, several cases of VSV infection have been reported in individuals directly exposed to infected livestock and researchers directly exposed within laboratory environments (Lyles and Rupprecht, 2007). Most VSV infections are thought to be asymptomatic in humans; however, febrile illness, including chills, myalgia, and nausea, has been reported in infected individuals (Roberts et al., 1999).

The VSV genome is simple and well characterized at the molecular level, which makes the manipulation and production of VSV vaccine vector relatively easy. Replication of the virus occurs within the cytoplasm of infected cells and is not known to undergo genetic recombination or integration into the cellular genome. Similar to Ad-based vector, recombinant VSV is able to accommodate relatively large inserts and multiple genes in its genome. This ability to incorporate multiple inserts of significant sizes also offers advantages over other RNA virus vectors, such as those based on alphaviruses and poliovirus (Davis et al., 1996; Liljestrom and Garoff, 1991; Morrow and Rixon, 1994; Xiong et al., 1989).

VSV can stimulate robust immune responses against self as well as foreign viral antigens making this virus a potent vaccine carrier. Several candidate vaccines based on VSV were shown to elicit immune protection correlating with strong humoral and cellular immune responses in different animal models (Dietzchold and Rupprecht, 1996; Zinkernagel et al., 1978). The VSV vaccine platform was the first to prevent AIDS in nonhuman primates infected with SHIV, a hybrid human and simian immunodeficiency virus that is lethal to macaques (Li et al., 1992). VSV-based vaccine has since then demonstrated efficacy against a number of pathogens, such as herpes simplex 2, tuberculosis, *Listeria monocytogenes*, SARS, Lassa virus, EBOV, and MARV viruses, in different animal models including mouse, guinea pigs, and nonhuman primates (Geisbert et al., 2005; Jones et al., 2005; Kapadia et al., 2005; Klas et al., 2006; Natuk et al., 2006; Xing and Lichty, 2006). Neutralization of VSV-based vaccines is unlikely to be problematic in humans since there is a low rate of preexisting immunity to VSV as demonstrated by the low seroprevalence of VSV antibodies in the general population (Roberts et al., 1999). Several of these VSV vaccine vectors remain replication-competent, which probably explains, at least in part, the substantial immunogenicity inherent to this vaccine carrier. A unique and remarkable ability of VSV-based vaccine is the rapidity by which immunization induces protection. Nonhuman primates can survive a lethal challenge when vaccinated 30 minutes after exposure to EBOV (Feldmann et al., 2007) and MARV (Daddario-DiCaprio et al., 2006a, 2006b). Cross-protection was also demonstrated with the VSV-MARV vaccine in nonhuman primates against a closely related strain of MARV. However, cross-protection against EBOV was not achieved following immunization with the VSV-MARV vaccine (Daddario-DiCaprio et al., 2006a).

Routes for Vaccine Delivery

The route of vaccine administration can have a significant effect on the strength of the immune response against a given pathogen. In general, parenterally administered vaccines largely stimulate systemic responses, whereas vaccines administered by a mucosal route may lead to efficient mucosal and systemic responses through different mechanisms (Lajeunesse et al., 2004; Boyaka et al., 2003). Mucosal vaccination can also overcome barriers of parenteral immunization caused by preexisting systemic immunity to the vaccine carrier from previous exposure (e.g., natural infection, vaccination) (Basset et al., 2003; Zhou and Ertl, 2006). Rapid progression of filovirus infections and the level of biohazard containment required for analysis of clinical samples have significantly hindered identification of the components of the immune system necessary for effective protection against infection. A prominent feature of fatal EBOV infection is the relative lack of a virus-specific immune response (Baize et al., 1999, 2002; Heffernan et al., 2005). In contrast, survivors have exhibited strong IgM antibody responses, clearance of viral antigen, and sustained T cell responses (Ksiazek et al., 1999; Leroy et al., 2001; Sanchez et al., 2004). This strongly suggests that both cellular and humoral immune responses are necessary for protection against EBOV infection. Administration of filovirus vaccines currently in development by various routes may provide some insight to the types of immune responses necessary for protection against infection.

To date, the primary method of administration of filovirus vaccine candidates is by intramuscular injection (Bramwell et al., 2005; Jones et al., 2005; Sullivan et al., 2000), the same route by which many newly approved vaccines are being administered (Arvin and Greenberg, 2006). Although this route of delivery ensures that the entire vaccinating dose and formulation enters the body intact, it poses significant safety risks for healthcare providers, patients, and the community. This is especially important in developing countries where not only is the demand for these vaccines high, but also the rate of occupational needle-stick injuries (Pruss-Ustun et al., 2005). As a result, significant training programs must be in place prior to mass vaccination campaigns, which may be necessary in the event of a sudden outbreak of infection. In addition, actions necessary to prevent disease transmission significantly reduce the number of people that can be vaccinated in a reasonable amount of time, and limit the utility of this delivery method when national vaccination campaigns are required.

Needle-free vaccine delivery is an attractive vaccination strategy for many reasons. Vaccines administered in this manner will offer improved safety with respect to disease transmission and reduce the pain associated with vaccination. In many cases, self-vaccination may be possible. This approach also increases the potential for the vaccine to be formulated and produced in a manner that may reduce or abolish the necessity for maintaining the "cold chain" required for proper storage of vaccines to remote areas, which has been estimated to burden worldwide programs (Das, 2004). For these reasons, this method of delivery is supported by many public health organizations, including the World Health Organization, the Global Alliance for Vaccines and Immunization and the Centers for Disease Control and Prevention (Giudice and Campbell, 2006). Currently, needle-free vaccine delivery approaches focus on administration under the skin (the subcutaneous route), the external surface of the skin (the transdermal route) and mucosal surfaces, specifically oral and nasal immunization. To date, mucosal immunization is the only method of needle-free vaccine delivery that has been commonly and successfully used in vaccination programs and which may be useful in the design of vaccination strategies against filoviruses (Bramwell et al., 2005; Levine and Sztein, 2004; Neutra and Kozlowski, 2006).

Several Ad or VSV-based vaccine candidates designed to afford protection against infection with *Mycobacterium tuberculosis*, influenza, papilloma virus, rabies, measles, respiratory syncytial virus, and HIV have also been administered by the intranasal route (Jiang et al., 2006; Kahn et al., 2001; Publicover et al., 2005; Reuter et al., 2002; Roberts et al., 1998; Schlereth et al., 2003; Xing and Lichty, 2006; Zhou et al., 2006). A recombinant HPIV3 vector expressing the EBOV GP could fully protect guinea pigs following intranasal vaccination (Bukreyev et al., 2006). Two doses of this vector with granulocyte-macrophage colony-stimulating factor could also afford complete protection to nonhuman primates vaccinated through the respiratory route (Bukreyev et al., 2007). Results from many studies strongly suggest that both Ad and VSV vectors are capable of crossing the respiratory tract and inducing strong cellular and humoral immune responses after a single immunization. Recently, reports indicated that nasal administration of Ad or VSV-based Ebola vaccine induces full protection of mice against a lethal challenge with ZEBOV (Jones et al., 2007; Patel et al., 2007). In the future, intranasal vaccination strategies against filovirus infection may prove a useful alternative to intramuscular immunization. This may become extremely relevant in the context of sudden outbreaks and malicious intentional release.

PROSPECTS: PHARMACOLOGY AND TOXICOLOGY OF VACCINES AND VACCINE VECTORS

In general, the primary component of a vaccine is either the pathogen of interest in an inactivated or an attenuated form, a purified protein subunit of the pathogen, or a plasmid or other vector (virus or bacterial) capable of eliciting endogenous production of antigenic components of the target pathogen (Barouch, 2006; Clark and Cassidy-Hanley, 2005; Dudek and Knipe, 2006; Mayr et al., 2005). Other components of vaccine products, which cannot be discounted during preclinical development, are adjuvants to promote and strengthen the immune response, excipients for stabilization of antigens and vectors, and the delivery device/container closure system. The pharmacology and toxicology of filovirus vaccines is not well studied and relevant studies are currently in progress.

PROSPECTS FOR THE FUTURE

Despite several promising results, vaccine development is still in the experimental stages. The process of moving an experimental vaccine candidate into clinical trials is a time-consuming and expensive process. It has become clear that the rodent models (mouse and guinea pig), although important for the development of antivirals, therapeutics, and vaccines, are not necessarily predictive for the efficacy of the same drugs in nonhuman primates, which are the gold standard for predicting efficacy in humans. A lack in proper biocontainment, in particular animal space, has been a major drawback in the process of vaccine development against filoviruses and related high-containment pathogens. This lack will most likely be properly addressed in the future with the current construction of several facilities in the United States and worldwide. Vaccine development also needs industrial partners, which are largely missing today. Programs need to be established that will make it interesting for companies to contribute to the development of approved vaccines for human use. Nevertheless, it is expected that a first-generation MARV and EBOV vaccine will be available over the next decade at least for medical, personal, first responder, and other high-risk professional groups.

KEY ISSUES

- Ebola and Marburg hemorrhagic fevers are considered highly communicable diseases.
- Bats might serve as reservoirs for EBOV and MARV.
- Currently there is no licensed treatment for or vaccine against EBOV and MARV.
- Progress has been made to understand the pathogenesis of EBOV and MARV in experimentally infected animals, and massive virus replication, immune suppression, vascular leakage, and organ destruction have been identified as key determinants.
- Several vaccine candidates have shown promising results in the protection against EBOV and MARV challenges in different animal models. The most promising candidates are based on replication-deficient Ad vectors, replication-competent vesicular stomatitis and HPIV3 vectors, and VLP preparations.
- There is little understanding about protective parameters in the host and correlates of protective immunity have not been well defined.
- To have utility in outbreak and biodefense settings as well as for laboratory exposures, the ideal preventive vaccines against EBOV and MARV need to confer protection in a single-dose regime and/or work in post-exposure prophylaxis.
- It is expected that first-generation vaccines will be available against EBOV and MARV over the next decade at least for high-risk exposure groups and post-exposure treatment.

References

Arvin, A.M. and Greenberg, H.B. New viral vaccines. Vaccine 2006; 344(1):240–249.

Baize, S., Leroy, E.M., Georges, A.J., Georges-Courbot, M.C., Capron, M., Bedjabaga, I., Lansoud-Soukate, J. and Mavoungou, E. Inflammatory responses in Ebola virus-infected patients. Clin. Exp. Immunol. 2002; 128(1):163–168.

Baize, S., Leroy, E.M., Georges-Courbot, M.C., Capron, M., Lansoud-Soukate, J., Debre, P., Fisher-Hoch, S.P., McCormick, J.B. and Georges, A.J. Defective humoral responses and extensive intravascular apoptosis are associated with fatal outcome in Ebola virus-infected patients. Nat. Med. 1999; 5(4):423–426.

Barouch, D.H. Rational design of gene-based vaccines. J. Pathol. 2006; 208(2):283–289.

Baskerville, A., Bowen, E.T., Platt, G.S., McArdell, L.B. and Simpson, D.I. The pathology of experimental Ebola virus infection in monkeys. J. Pathol. 1978; 125(3):131–138.

Basset, C., Holton, J., O'Mahony, R. and Roitt, I. Innate immunity and pathogen-host interaction. Vaccine 2003; 21(Suppl. 2):S12–S23.

Bausch, D.G. and Geisbert, T.W. Development of vaccines for Marburg hemorrhagic fever. Expert Rev. Vaccines 2007; 6(1):57–74.

Becker, S., Volchkov, V.E., Muhleberger, E. and Al, E. A recombinant vaccinia virus expressing the surface glycoprotein of Marburg virus does not protect guinea pigs against Marburg virus-infection. Xth International Congress of Virology. Jerusalem, Israel, 1996.

Borio, L., Inglesby, T., Peters, C.J., Schmaljohn, A.L., Hughes, J.M., Jahrling, P.B., Ksiazek, T., Johnson, K.M., Meyerhoff, A., O'Toole, T., Ascher, M.S., Bartlett, J., Breman, J.G., Eitzen, E.M., Jr., Hamburg, M., Hauer, J., Henderson, D.A., Johnson, R.T., Kwik, G., Layton, M., Lillibridge, S., Nabel, G.J., Osterholm, M.T., Perl, T.M., Russell, P. and Tonat, K. Hemorrhagic fever viruses as biological weapons: medical and public health management. JAMA 2002; 287(18):2391–2405.

Bowen, E.T., Platt, G.S., Simpson, D.I., McArdell, L.B. and Raymond, R.T. Ebola haemorrhagic fever: experimental infection of monkeys. Trans. R. Soc. Trop. Med. Hyg. 1978; 72(2):188–191.

Boyaka, P.N., Tafaro, A., Fischer, R., Leppla, S.H., Fujihashi, K. and McGhee, J.R. Effective mucosal immunity to anthrax: neutralizing antibodies and Th cell responses following nasal immunization with protective antigen. J. Immunol. 2003; 170(11):5636–5643.

Bramwell, V.W., Eyles, J.E. and Oya Alpar, H. Particulate delivery systems for biodefense subunit vaccines. Adv. Drug Deliv. Rev. 2005; 57(9):1247–1265.

Bukreyev, A., Rollin, P.E., Tate, M.K., Yang, L., Zaki, S.R., Shieh, W.J., Murphy, B.R., Collins, P.L. and Sanchez, A. Successful topical respiratory tract immunization of primates against Ebola virus. J. Virol. 2007; 81(12):6379–6388.

Bukreyev, A., Yang, L., Zaki, S.R., Shieh, W.J., Rollin, P.E., Murphy, B.R., Collins, P.L. and Sanchez, A. A single intranasal inoculation with a paramyxovirus-vectored vaccine protects guinea pigs against a lethal-dose Ebola virus challenge. J. Virol. 2006; 80(5):2267–2279.

Capone, S., Meola, A., Ercole, B.B., Vitelli, A., Pezzanera, M., Ruggeri, L., Davies, M.E., Tafi, R., Santini, C., Luzzago, A., Fu, T.M., Bett, A., Colloca, S., Cortese, R., Nicosia, A. and Folgori, A. A novel adenovirus type 6 (Ad6)-based hepatitis C virus vector that overcomes preexisting anti-ad5 immunity and induces potent and broad cellular immune responses in rhesus macaques. J. Virol. 2006; 80(4):1688–1699.

Chirmule, N., Propert, K., Magosin, S., Qian, Y., Qian, R. and Wilson, J. Immune responses to adenovirus and adeno-associated virus in humans. Gene Ther. 1999; 6(9):1574–1583.

Chupurnov, A.A., Chernukhin, I.V., Ternovoi, V.A., Kudoiarova, N.M., Makhova, N.M., Azaev, M. and Smolina, M.P. Attempts to develop a vaccine against Ebola fever. Vopr. Virusol. 1995; 40(6):257–260.

Clark, T.G. and Cassidy-Hanley, D. Recombinant subunit vaccines: potentials and constraints. Dev. Biol. (Basel) 2005; 121:153–163.

Daddario-DiCaprio, K.M., Geisbert, T.W., Geisbert, J.B., Stroher, U., Hensley, L.E., Grolla, A., Fritz, E.A., Feldmann, F., Feldmann, H. and Jones, S.M. Cross-protection against Marburg virus strains by using a live, attenuated recombinant vaccine. J. Virol. 2006a; 80(19):9659–9666.

Daddario-DiCaprio, K.M., Geisbert, T.W., Stroher, U., Geisbert, J.B., Grolla, A., Fritz, E.A., Fernando, L., Kagan, E., Jahrling, P.B., Hensley, L.E., Jones, S.M. and Feldmann, H. Postexposure protection against Marburg haemorrhagic fever with recombinant vesicular stomatitis virus vectors in non-human primates: an efficacy assessment. Lancet 2006b; 367(9520):1399–1404.

Das, P. Revolutionary vaccine technology breaks the cold chain. Lancet Infect. Dis. 2004; 4(12):719.

Davis, N.L., Brown, K.W. and Johnston, R.E. A viral vaccine vector that expresses foreign genes in lymph nodes and protects against mucosal challenge. J. Virol. 1996; 70(6):3781–3787.

Dietzchold, B. and Rupprecht, C.E. Rhabdoviruses. 3rd ed. In: Fields Virology, Vol. 1, pp. 1137–1159. Philadelphia: Lippincott-Raven Publishers, 1996.

Dudek, T. and Knipe, D.M. Replication-defective viruses as vaccines and vaccine vectors. Virology 2006; 344(1):230–239.

Feldmann, H., Geisberg, T.W., Jarling, P.B., Klenk, H.D., Netesov, S.V., Peters, C.J., Sanchez, A., Swanepoel, R. and Volchkov, V.E. Virus Taxonomy, Eighth Report of the International Committee on Taxonomy of Viruses. In: Fauquet, C.M., Mayo, C.M., Maniloff, J., Desselberger, U. and Ball, L.A. (eds.), Elsevier Academic Press, San Diego, 2004, pp. 645.

Feldmann, H., Jones, S.M., Daddario-DiCaprio, K.M., Geisbert, J.B., Stroher, U., Grolla, A., Bray, M., Fritz, E.A., Fernando, L., Feldmann, F., Hensley, L.E. and Geisbert, T.W. Effective post-exposure treatment of Ebola infection. PLoS Pathog 2007; 3(1):e3.

Feldmann, H., Jones, S., Klenk, H.D. and Schnittler, H.J. Ebola virus: from discovery to vaccine. Nat. Rev. Immunol. 2003; 3(8):677–685.

Feldmann, H. and Kiley, M.P. Classification, structure, and replication of filoviruses. Curr. Top. Microbiol. Immunol. 1999; 235:1–21.

Feldmann, H., Volchkov, V.E., Volchkova, V.A., Stroher, U. and Klenk, H.D. Biosynthesis and role of filoviral glycoproteins. J. Gen. Virol. 2001; 82(Pt 12):2839–2848.

Fisher-Hoch, S.P. The haemostatic defect in viral haemorrhagic fevers. Br. J. Haematol. 1983; 55(4):565–571.

Fitzgerald, J.C., Gao, G.P., Reyes-Sandoval, A., Pavlakis, G.N., Xiang, Z.Q., Wlazlo, A.P., Giles-Davis, W., Wilson, J.M. and Ertl, H.C. A simian replication-defective adenoviral recombinant vaccine to HIV-1 gag. J. Immunol. 2003; 170(3):1416–1422.

Gao, W., Tamin, A., Soloff, A., D'Aiuto, L., Nwanegbo, E., Robbins, P.D., Bellini, W.J., Barratt-Boyes, S. and Gambotto, A. Effects of a SARS-associated coronavirus vaccine in monkeys. Lancet 2003; 362(9399):1895–1896.

Geisbert, T.W., Hensley, L.E., Larsen, T., Young, H.A., Reed, D.S., Geisbert, J.B., Scott, D.P., Kagan, E., Jahrling, P.B. and Davis, K.J. Pathogenesis of Ebola hemorrhagic fever in cynomolgus macaques: evidence that dendritic cells are early and sustained targets of infection. Am. J. Pathol. 2003; 163(6):2347–2370.

Geisbert, T.W., Jones, S., Fritz, E.A., Shurtleff, A.C., Geisbert, J.B., Liebscher, R., Grolla, A., Stroher, U., Fernando, L., Daddario, K.M., Guttieri, M.C., Mothe, B.R., Larsen, T., Hensley, L.E., Jahrling, P.B. and Feldmann, H. Development of a new vaccine for the prevention of Lassa fever. PLoS Med. 2005; 2(6):e183.

Geisbert, T.W., Pushko, P., Anderson, K., Smith, J., Davis, K.J. and Jahrling, P.B. Evaluation in nonhuman primates of vaccines against Ebola virus. Emerg. Infect. Dis. 2002; 8(5):503–507.

Giudice, E.L. and Campbell, J.D. Needle-free vaccine delivery. Adv. Drug Deliv. Rev. 2006; 58(1):68–89.

Groseth, A., Feldmann, H. and Strong, J.E. The ecology of Ebola virus. Trends Microbiol. 2007; 15(9):408–416.

Heffernan, R.T., Pambo, B., Hatchett, R.J., Leman, P.A., Swanepoel, R. and Ryder, R.W. Low seroprevalence of IgG antibodies to Ebola virus in an epidemic zone: Ogooue-Ivindo region, Northeastern Gabon, 1997. J. Infect. Dis. 2005; 191(6):964–968.

Hevey, M., Negley, D., Geisbert, J., Jahrling, P. and Schmaljohn, A. Antigenicity and vaccine potential of Marburg virus glycoprotein expressed by baculovirus recombinants. Virology 1997; 239(1):206–216.

Hevey, M., Negley, D., Pushko, P., Smith, J. and Schmaljohn, A. Marburg virus vaccines based upon alphavirus replicons protect guinea pigs and nonhuman primates. Virology 1998; 251(1):28–37.

Hevey, M., Negley, D., VanderZanden, L., Tammariello, R.F., Geisbert, J., Schmaljohn, C., Smith, J.F., Jahrling, P.B. and Schmaljohn, A.L. Marburg virus vaccines: comparing classical and new approaches. Vaccine 2001; 20(3–4):586–593.

Horwitz, M.S. Adenovirus immunoregulatory genes and their cellular targets. Virology 2001; 279(1):1–8.

Ignatyev, G.M., Agafonov, A.P., Streltsova, M.A. and Kashentseva, E.A. Inactivated Marburg virus elicits a nonprotective immune response in Rhesus monkeys. J. Biotechnol. 1996; 44(1–3):111–118.

Ignatyev, G.M., Strel'tsova, M.A., Agafionov, A.P., Zhukova, N.A., Kashentseva, E.A. and Vorobeva, M.S. The immunity indices of animals immunized with the inactivated Marburg virus after infection with homologous virus. Vopr. Virusol. 1994; 39(1):13–17.

Jaax, N.K., Davis, K.J., Geisbert, T.J., Vogel, P., Jaax, G.P., Topper, M. and Jahrling, P.B. Lethal experimental infection of rhesus monkeys with Ebola-Zaire (Mayinga) virus by the oral and conjunctival route of exposure. Arch. Pathol. Lab. Med. 1996; 120(2):140–155.

Jiang, P., Liu, Y., Yin, X., Yuan, F., Nie, Y., Luo, M., Aihua, Z., Liyin, D., Ding, M. and Deng, H. Elicitation of neutralizing antibodies by intranasal administration of recombinant vesicular stomatitis virus expressing human immunodeficiency virus type 1 gp120. Biochem. Biophys. Res. Commun. 2006; 339(2):526–532.

Johnson, E., Jaax, N., White, J. and Jahrling, P. Lethal experimental infections of rhesus monkeys by aerosolized Ebola virus. Int. J. Exp. Pathol. 1995; 76(4):227–236.

Jones, S.M., Feldmann, H., Stroher, U., Geisbert, J.B., Fernando, L., Grolla, A., Klenk, H.D., Sullivan, N.J., Volchkov, V.E., Fritz, E.A., Daddario, K.M., Hensley, L.E., Jahrling, P.B. and Geisbert, T.W. Live attenuated recombinant vaccine protects nonhuman primates against Ebola and Marburg viruses. Nat. Med. 2005; 11(7): 786–790.

Jones, S.M., Stroher, U., Fernando, L., Qiu, X., Alimonti, J., Melito, P., Bray, M., Klenk, H.-D. and Feldmann, H. Assessment of a vesicular stomatitis virus-based vaccine using the mouse model of Ebola virus hemorrhagic fever. J. Infect. Dis. 2007, Nov 15; 196 Suppl 2:S404–12.

Jooss, K., Yang, Y., Fisher, K.J. and Wilson, J.M. Transduction of dendritic cells by DNA viral vectors directs the immune response to transgene products in muscle fibers. J. Virol. 1998; 72(5):4212–4223.

Kahn, J.S., Roberts, A., Weibel, C., Buonocore, L. and Rose, J.K. Replication-competent or attenuated, nonpropagating vesicular stomatitis viruses expressing respiratory syncytial virus (RSV) antigens protect mice against RSV challenge. J. Virol. 2001; 75(22):11079–11087.

Kapadia, S.U., Rose, J.K., Lamirande, E., Vogel, L., Subbarao, K. and Roberts, A. Long-term protection from SARS coronavirus infection conferred by a single immunization with an attenuated VSV-based vaccine. Virology 2005; 340(2):174–182.

Kiley, M.P., Bowen, E.T., Eddy, G.A., Isaacson, M., Johnson, K.M., McCormick, J.B., Murphy, F.A., Pattyn, S.R., Peters, D., Prozesky, O.W., Regnery, R.L., Simpson, D.I., Slenczka, W., Sureau, P., van der Groen, G., Webb, P.A. and Wulff, H. Filoviridae: a taxonomic home for Marburg and Ebola viruses? Intervirology 1982; 18(1–2):24–32.

Klas, S.D., Lavine, C.L., Whitt, M.A. and Miller, M.A. IL-12-assisted immunization against Listeria monocytogenes using replication-restricted VSV-based vectors. Vaccine 2006; 24(9):1451–1461.

Kobinger, G.P., Feldmann, H., Zhi, Y., Schumer, G., Gao, G., Feldmann, F., Jones, S. and Wilson, J.M. Chimpanzee adenovirus vaccine protects against Zaire Ebola virus. Virology 2006; 346(2):394–401.

Ksiazek, T.G., Rollin, P.E., Williams, A.J., Bressler, D.S., Martin, M.L., Swanepoel, R., Burt, F.J., Leman, P.A., Khan, A.S., Rowe, A.K., Mukunu, R., Sanchez, A. and Peters, C.J. Clinical virology of Ebola hemorrhagic fever (EHF): virus, virus antigen, and IgG and IgM antibody findings among EHF patients in Kikwit, Democratic Republic of the Congo, 1995. J. Infect. Dis. 1999; 179(Suppl. 1):S177–S187.

Lajeunesse, M., Zhang, Q. and Finn, A. Mucosal immunity to infections and its importance in future vaccinology. Adv. Exp. Med. Biol. 2004; 549:13–22.

Leroy, E.M., Baize, S., Debre, P., Lansoud-Soukate, J. and Mavoungou, E. Early immune responses accompanying human asymptomatic Ebola infections. Clin. Exp. Immunol. 2001; 124(3):453–460.

Leroy, E.M., Kumulungui, B., Pourrut, X., Rouquet, P., Hassanin, A., Yaba, P., Delicat, A., Paweska, J.T., Gonzalez, J.P. and Swanepoel, R. Fruit bats as reservoirs of Ebola virus. Nature 2005; 438(7068):575–576.

Levine, M.M. and Sztein, M.B. Vaccine development strategies for improving immunization: the role of modern immunology. Nat. Immunol. 2004; 5(5):460–464.

Li, J., Lord, C.I., Haseltine, W., Letvin, N.L. and Sodroski, J. Infection of cynomolgus monkeys with a chimeric HIV-1/SIVmac virus that expresses the HIV-1 envelope glycoproteins. J. Acquir. Immune Defic. Syndr. 1992; 5(7):639–646.

Liljestrom, P. and Garoff, H. A new generation of animal cell expression vectors based on the Semliki Forest virus replicon. Biotechnology (N.Y.) 1991; 9(12):1356–1361.

Lupton, H.W., Lambert, R.D., Bumgardner, D.L., Moe, J.B. and Eddy, G.A. Inactivated vaccine for Ebola virus efficacious in guinea pig model. Lancet 1980; 2(8207):1294–1295.

Lyles, D.S. and Rupprecht, C.E. Rhabdoviridae. In: Fields Virology, Vol. 1, pp. 1363–1408. 2 vols, 2007.

Makimura, M., Miyake, S., Akino, N., Takamori, K., Matsuura, Y., Miyamura, T. and Saito, I. Induction of antibodies against structural proteins of hepatitis C virus in mice using recombinant adenovirus. Vaccine 1996; 14(1):28–36.

Martin, J.E., Sullivan, N.J., Enama, M.E., Gordon, I.J., Roederer, M., Koup, R.A., Bailer, R.T., Chakrabarti, B.K., Bailey, M.A., Gomez, P.L., Andrews, C.A., Moodie, Z., Gu, L., Stein, J.A., Nabel, G.J. and Graham, B.S. A DNA vaccine for Ebola virus is safe and immunogenic in a phase I clinical trial. Clin. Vaccine Immunol. 2006; 13(11):1267–1277.

Mayr, U.B., Walcher, P., Azimpour, C., Riedmann, E., Haller, C. and Lubitz, W. Bacterial ghosts as antigen delivery vehicles. Adv. Drug Deliv. Rev. 2005; 57(9):1381–1391.

Mikhailov, V.V., Borisevich, I.V., Chernikova, N.K., Potryvaeva, N.V. and Krasnianskii, V.P. The evaluation in hamadryas baboons of the possibility for the specific prevention of Ebola fever. Vopr. Virusol. 1994; 39(2):82–84.

Morrow, J.A. and Rixon, F.J. Analysis of sequences important for herpes simplex virus type 1 latency-associated transcript promoter activity during lytic infection of tissue culture cells. J. Gen. Virol. 1994; 75(Pt 2):309–316.

Nanda, A., Lynch, D.M., Goudsmit, J., Lemckert, A.A., Ewald, B.A., Sumida, S.M., Truitt, D.M., Abbink, P., Kishko, M.G., Gorgone, D.A., Lifton, M.A., Shen, L., Carville, A., Mansfield, K.G., Havenga, M.J. and Barouch, D.H. Immunogenicity of recombinant fiber-chimeric adenovirus serotype 35 vector-based vaccines in mice and rhesus monkeys. J. Virol. 2005; 79(22): 14161–14168.

Natuk, R.J., Cooper, D., Guo, M., Calderon, P., Wright, K.J., Nasar, F., Witko, S., Pawlyk, D., Lee, M., DeStefano, J., Tummolo, D., Abramovitz, A.S., Gangolli, S., Kalyan, N., Clarke, D.K., Hendry, R.M., Eldridge, J.H., Udem, S.A. and Kowalski, J. Recombinant vesicular stomatitis virus vectors expressing herpes simplex virus type 2 gD elicit robust CD4+ Th1 immune responses and are protective in mouse and guinea pig models of vaginal challenge. J. Virol. 2006; 80(9):4447–4457.

Neutra, M.R. and Kozlowski, P.A. Mucosal vaccines: the promise and the challenge. Nat. Rev. Immunol. 2006; 6(2):148–158.

Patel, A., Zhang, Y., Croyle, M., Tran, K., Gray, M., Strong, J., Feldmann, H., Wilson, J.M. and Kobinger, G.P. Mucosal delivery of adenovirus-based vaccine protects against Ebola virus infection in mice. J. Infect. Dis. 2007,Nov 15; 196 Suppl 2:S413–20.

Pruss-Ustun, A., Rapiti, E. and Hutin, Y. Estimation of the global burden of disease attributable to contaminated sharps injuries among health-care workers. Am. J. Ind. Med. 2005; 48(6): 482–490.

Publicover, J., Ramsburg, E. and Rose, J.K. A single-cycle vaccine vector based on vesicular stomatitis virus can induce immune responses comparable to those generated by a replication-competent vector. J. Virol. 2005; 79(21):13231–13238.

Pushko, P., Bray, M., Ludwig, G.V., Parker, M., Schmaljohn, A., Sanchez, A., Jahrling, P.B. and Smith, J.F. Recombinant RNA replicons derived from attenuated Venezuelan equine encephalitis virus protect guinea pigs and mice from Ebola hemorrhagic fever virus. Vaccine 2000; 19(1):142–153.

Pushko, P., Geisbert, J., Parker, M., Jahrling, P. and Smith, J. Individual and bivalent vaccines based on alphavirus replicons protect guinea pigs against infection with Lassa and Ebola viruses. J. Virol. 2001; 75(23):11677–11685.

Reuter, J.D., Vivas-Gonzalez, B.E., Gomez, D., Wilson, J.H., Brandsma, J.L., Greenstone, H.L., Rose, J.K. and Roberts, A. Intranasal vaccination with a recombinant vesicular stomatitis virus expressing cottontail rabbit papillomavirus L1 protein provides complete protection against papillomavirus-induced disease. J. Virol. 2002; 76(17):8900–8909.

Riemenschneider, J., Garrison, A., Geisbert, J., Jahrling, P., Hevey, M., Negley, D., Schmaljohn, A., Lee, J., Hart, M.K., Vanderzanden, L., Custer, D., Bray, M., Ruff, A., Ivins, B., Bassett, A., Rossi, C. and Schmaljohn, C. Comparison of individual and combination DNA vaccines for *B. anthracis*, Ebola virus, Marburg virus and Venezuelan equine encephalitis virus. Vaccine 2003; 21(25–26):4071–4080.

Roberts, A., Buonocore, L., Price, R., Forman, J. and Rose, J.K. Attenuated vesicular stomatitis viruses as vaccine vectors. J. Virol. 1999; 73(5):3723–3732.

Roberts, A., Kretzschmar, E., Perkins, A.S., Forman, J., Price, R., Buonocore, L., Kawaoka, Y. and Rose, J.K. Vaccination with a recombinant vesicular stomatitis virus expressing an influenza virus hemagglutinin provides complete protection from influenza virus challenge. J. Virol. 1998; 72(6):4704–4711.

Rosenfeld, M.A., Siegfried, W., Yoshimura, K., Yoneyama, K., Fukayama, M., Stier, L.E., Paakko, P.K., Gilardi, P., Stratford-Perricaudet, L.D., Perricaudet, M. et al. Adenovirus-mediated transfer of a recombinant alpha 1-antitrypsin gene to the lung epithelium in vivo. Science 1991; 252(5004):431–434.

Roy, S., Gao, G., Clawson, D.S., Vandenberghe, L.H., Farina, S.F. and Wilson, J.M. Complete nucleotide sequences and genome organization of four chimpanzee adenoviruses. Virology 2004; 324(2):361–372.

Sanchez, A., Geisbert, J. B. and Feldmann, H. Filoviridae: Marburg and Ebola Viruses. 5th ed, Vol. 1, pp. 1409–1448. 2 vols. Philadelphia: Wolters Kluwer, 2007.

Sanchez, A., Ksiazek, T.G., Rollin, P.E., Peters, C.J., Nichol, S.T., Khan, A.S. and Mahy, B.W. Reemergence of Ebola virus in Africa. Emerg. Infect. Dis. 1995; 1(3):96–97.

Sanchez, A., Lukwiya, M., Bausch, D., Mahanty, S., Sanchez, A.J., Wagoner, K.D. and Rollin, P.E. Analysis of human peripheral blood samples from fatal and nonfatal cases of Ebola (Sudan) hemorrhagic fever: cellular responses, virus load, and nitric oxide levels. J. Virol. 2004; 78(19):10370–10377.

Sanchez, A., Trappier, S.G., Mahy, B.W., Peters, C.J. and Nichol, S.T. The virion glycoproteins of Ebola viruses are encoded in two reading frames and are expressed through transcriptional editing. Proc. Natl. Acad. Sci. U.S.A. 1996; 93(8):3602–3607.

Schlereth, B., Buonocore, L., Tietz, A., Meulen Vt, V., Rose, J.K. and Niewiesk, S. Successful mucosal immunization of cotton rats in the presence of measles virus-specific antibodies depends on degree of attenuation of vaccine vector and virus dose. J. Gen. Virol. 2003; 84(Pt 8):2145–2151.

Shenk, T. and Williams, J. Genetic analysis of adenoviruses. Curr. Top. Microbiol. Immunol. 1984; 111:1–39.

Siegert, R., Shu, H.L., Slenczka, W., Peters, D. and Muller, G. On the etiology of an unknown human infection originating from monkeys. Dtsch. Med. Wochenschr. 1967; 92(51):2341–2343.

Stratford-Perricaudet, L.D., Levrero, M., Chasse, J.F., Perricaudet, M. and Briand, P. Evaluation of the transfer and expression in mice of an enzyme-encoding gene using a human adenovirus vector. Hum. Gene. Ther. 1990; 1(3):241–256.

Sullivan, N.J., Geisbert, T.W., Geisbert, J.B., Xu, L., Yang, Z.Y., Roederer, M., Koup, R.A., Jahrling, P.B. and Nabel, G.J. Accelerated vaccination for Ebola virus haemorrhagic fever in non-human primates. Nature 2003; 424(6949):681–684.

Sullivan, N.J., Sanchez, A., Rollin, P.E., Yang, Z.Y. and Nabel, G.J. Development of a preventive vaccine for Ebola virus infection in primates. Nature 2000; 408(6812):605–609.

Swenson, D.L., Warfield, K.L., Negley, D.L., Schmaljohn, A., Aman, M.J. and Bavari, S. Virus-like particles exhibit potential as a pan-filovirus vaccine for both Ebola and Marburg viral infections. Vaccine 2005; 23(23):3033–3042.

Tan, Y., Hackett, N.R., Boyer, J.L. and Crystal, R.G. Protective immunity evoked against anthrax lethal toxin after a single intramuscular administration of an adenovirus-based vaccine encoding humanized protective antigen. Hum. Gene Ther. 2003; 14(17):1673–1682.

Towner, J.S., Pourrut, X., Albarino, C.G., Nkogue, C.N., Bird, B.H., Grard, G., Ksiazek, T.G., Gonzalez, J.P., Nichol, S.T. and Leroy, E.M. Marburg virus infection detected in a common African bat. PLoS ONE 2007; 2(1):e764.

Vanderzanden, L., Bray, M., Fuller, D., Roberts, T., Custer, D., Spik, K., Jahrling, P., Huggins, J., Schmaljohn, A. and Schmaljohn, C. DNA vaccines expressing either the GP or NP genes of Ebola virus protect mice from lethal challenge. Virology 1998; 246(1):134–144.

Volchkov, V.E., Becker, S., Volchkova, V.A., Ternovoj, V.A., Kotov, A.N., Netesov, S.V. and Klenk, H.D. GP mRNA of Ebola virus is edited by the Ebola virus polymerase and by T7 and vaccinia virus polymerases. Virology 1995; 214(2):421–430.

Vos, A., Neubert, A., Pommerening, E., Muller, T., Dohner, L., Neubert, L. and Hughes, K. Immunogenicity of an E1-deleted recombinant human adenovirus against rabies by different routes of administration. J. Gen. Virol. 2001; 82(Pt 9):2191–2197.

Warfield, K.L., Bosio, C.M., Welcher, B.C., Deal, E.M., Mohamadzadeh, M., Schmaljohn, A., Aman, M.J. and Bavari, S. Ebola virus-like particles protect from lethal Ebola virus infection. Proc. Natl. Acad. Sci. U.S.A. 2003; 100(26):15889–15894.

Warfield, K.L., Swenson, D.L., Negley, D.L., Schmaljohn, A.L., Aman, M.J. and Bavari, S. Marburg virus-like particles protect guinea pigs from lethal Marburg virus infection. Vaccine 2004; 22(25–26):3495–3502.

Wilson, J.A., Bosio, C.M. and Hart, M.K. Ebola virus: the search for vaccines and treatments. Cell Mol. Life Sci. 2001a; 58(12–13):1826–1841.

Wilson, J.A., Bray, M., Bakken, R. and Hart, M.K. Vaccine potential of Ebola virus VP24, VP30, VP35, and VP40 proteins. Virology 2001b; 286(2):384–390.

Wilson, J.A. and Hart, M.K. Protection from Ebola virus mediated by cytotoxic T lymphocytes specific for the viral nucleoprotein. J. Virol. 2001; 75(6):2660–2664.

Xiang, Z., Gao, G., Reyes-Sandoval, A., Cohen, C.J., Li, Y., Bergelson, J.M., Wilson, J.M. and Ertl, H.C. Novel, chimpanzee serotype 68-based adenoviral vaccine carrier for induction of antibodies to a transgene product. J. Virol. 2002; 76(6):2667–2675.

Xing, Z. and Lichty, B.D. Use of recombinant virus-vectored tuberculosis vaccines for respiratory mucosal immunization. Tuberculosis (Edinb.) 2006; 86(3–4):211–217.

Xiong, C., Levis, R., Shen, P., Schlesinger, S., Rice, C.M. and Huang, H.V. Sindbis virus: an efficient, broad host range vector for gene expression in animal cells. Science 1989; 243(4895):1188–1191.

Xu, L., Sanchez, A., Yang, Z., Zaki, S.R., Nabel, E.G., Nichol, S.T. and Nabel, G.J. Immunization for Ebola virus infection. Nat. Med. 1998; 4(1):37–42.

Yang, Y., Nunes, F.A., Berencsi, K., Furth, E.E., Gonczol, E. and Wilson, J.M. Cellular immunity to viral antigens limits E1-deleted adenoviruses for gene therapy. Proc. Natl. Acad. Sci. U.S.A. 1994a; 91(10):4407–4411.

Yang, Y., Nunes, F.A., Berencsi, K., Gonczol, E., Engelhardt, J.F. and Wilson, J.M. Inactivation of E2a in recombinant adenoviruses improves the prospect for gene therapy in cystic fibrosis. Nat. Genet. 1994b; 7(3):362–369.

Zhou, D., Cun, A., Li, Y., Xiang, Z. and Ertl, H.C. A chimpanzee-origin adenovirus vector expressing the rabies virus glycoprotein as an oral vaccine against inhalation infection with rabies virus. Mol. Ther. 2006; 14(5):662–672.

Zhou, D. and Ertl, H.C. Therapeutic potential of adenovirus as a vaccine vector for chronic virus infections. Expert Opin. Biol. Ther. 2006; 6(1):63–72.

Zinkernagel, R.M., Adler, B. and Holland, J.J. Cell-mediated immunity to vesicular stomatitis virus infections in mice. Exp. Cell. Biol. 1978; 46(1–2):53–70.

CHAPTER

21

Alphaviral Encephalitides

Scott C. Weaver and Slobodan Paessler

Department of Pathology, Centers for Tropical Diseases, and Biodefense and Emerging Infectious Diseases, and Sealy Center for Vaccine Development, University of Texas Medical Branch, Galveston, TX, USA

OUTLINE

ABSTRACT

Alphaviruses that typically cause encephalitis include eastern (EEEV), Venezuelan (VEEV), and western equine encephalitis viruses (WEEV). All are important emerging viral pathogens of humans, equids, and other domesticated animals. Although these viruses do not exhibit the high human fatality rates characteristic of some more virulent biothreat viruses, all are highly infectious via the aerosol route and highly incapacitating, making them biothreat agents. Despite vaccine development efforts dating back many decades, licensed human vaccines are not available. Investigational new drug vaccines against VEEV and EEEV are offered to at-risk laboratory workers through the U.S. Army Special Immunizations program. However, these vaccines suffer from reactogenicity and poor rates of seroconversion (VEEV) and poor immunogenicity requiring multiple doses and frequent boosters (EEEV). Veterinary vaccines are widely available in America but most suffer from poor and short-lived immunogenicity, and the risk of residual, virulent virus in inactivated preparations. Recent advances in understanding of molecular aspects of alphavirus pathogenesis and genetics promise improvements in live-attenuated vaccines against all three encephalitis alphaviruses, as well as safer virus strains for inactivated vaccine preparations. Virus-like particles that express either alphavirus or other antigens, based on alphavirus replicon genomes, are also being widely developed as vaccine platforms.

INTRODUCTION

Alphaviral encephalitides comprise three groups of viruses in the family *Togaviridae*, genus *Alphavirus*: the eastern (EEEV), western (WEEV), and Venezuelan equine encephalitis viruses (VEEV) (Weaver et al., 2005). Like most alphaviruses, they are all mosquito-borne, relying on biological (requiring replication in the mosquito vector), and horizontal transmission in nature (Griffin, 2007). All are also zoonotic viruses that rely on birds or small mammals as reservoir hosts within enzootic foci of transmission. These viruses regularly cause encephalitis in humans and equids in many regions of the tropical, subtropical, and temperate Americas (Tsai et al., 2002). Although other alphaviruses, including Sindbis and Semliki Forest viruses, cause encephalitis in laboratory mice, and have been used as models of human encephalitis, only EEEV, VEEV, and WEEV cause encephalitis during natural transmission. All three were isolated during the early part of the 20th century and are relatively well understood from an epidemiologic and ecologic standpoint. However, little progress has been made in preventing or controlling human disease. Effective veterinary vaccines have contributed to reducing disease in equids and other domestic animals, but no licensed human vaccines have been produced. Furthermore, no effective antivirals have been identified and protection of the human population remains focused on vector control and limiting exposure to mosquito bites (Tsai et al., 2002).

SHORT HISTORY OF THE DISEASES

Alphaviral encephalitides were first recognized in the 19th century in the northeastern United States, where outbreaks of equine encephalitis consistent with EEE were noted (Hanson, 1957). Equine outbreaks consistent with VEE were first recognized in northern South America during the 1920s (Walton and Grayson, 1988), and WEE outbreaks were first recognized in California in 1930 (Reisen and Monath, 1988). Virus isolations were made shortly after the recognition of a likely viral etiology for these epizootics; EEEV was first isolated from horses in Argentina in 1930 (not identified until later) and in the eastern United States in 1933 (Ten Broeck and Merrill, 1933); VEEV was first isolated from a horse in Venezuela in 1938 (Kubes and Rios, 1939; Beck and Wyckoff, 1938), and WEEV was isolated from a horse in California in 1930 (Meyer et al., 1931). Human isolates of all three viruses followed soon thereafter, although VEEV was not recognized as an important human pathogen until the 1960s (Weaver et al., 2004). These three different alphaviruses were quickly recognized as distinct using in vivo neutralization tests, and later characterized in greater detail using complement fixation, hemagglutination inhibition, plaque reduction neutralization, and other serological tests.

ETIOLOGIC AGENTS

Classification

The alphaviruses were initially categorized as group A arboviruses based on antigenic relationships and virion morphology (Calisher et al., 1980). Although originally grouped with the flaviviruses as members of the family *Togaviridae*, genetic studies later revealed fundamental differences in genome organization, and other genetic and replication characteristics. Eventually, the alphaviruses became one of two genera in the family *Togaviridae*, the other being Rubella virus, the sole member of the genus *Rubivirus* (Weaver et al., 2005). Currently, the genus *Alphavirus* includes 29 species; several include antigenically or genetically based subtypes, which can have fundamental differences in epidemiological and virulence properties. All alphaviruses are arthropod vector borne except for

Salmon pancreatic disease and Southern elephant seal viruses, which have not been demonstrated to have vectors.

Antigens Encoded by Agent

The alphavirus genome encodes seven major proteins; four of these are called nonstructural because they are not incorporated into virus particles (Fig. 21.1). The nonstructural proteins are translated from genomic, positive strand RNA in the cytoplasm and participate in various ways in genome replication and modifications, and in their own nonstructural polyprotein processing. Differential temporal cleavage patterns of the nonstructural polyprotein regulate the synthesis of plus and minus strand RNA during replication. The structural proteins are encoded by a separate open reading frame that is transcribed as a subgenomic message and synthesized in excess of the genomic RNA. The structural polyprotein is translated in the cytoplasm in the following order; the capsid, followed by the E2 and E1 envelope glycoprotein. The capsid possesses a serine protease activity, cleaves itself from the nascent polyprotein, and remains in the cytoplasm. The remainder is processed through the secretory apparatus with cellular proteases responsible for further cleavages. Both E2 and E1 are glycosylated, and become embedded into the plasma membrane as heterodimers. Virions mature when 240 copies of the capsid protein combine in the cytoplasm with 1 molecule of genomic RNA to form a nucleocapsid; the nucleocapsid then recognizes the cytoplasmic tail of the E2 protein to initiate budding, which results in envelopment of the virion by the plasma membrane and 240 copies of the embedded E2/E1 dimer. The final arrangement includes 80 trimers of the E2/E1 dimers, which form spikes on the virion surface (Fig. 21.1). The E2 proteins protrude to form the outer part of these spikes, while the E1 protein lies primarily underneath, parallel with the envelope (Pletnev et al., 2001). Of the structural proteins, only the E1 structure has been solved to atomic resolution, and is homologous to the flavivirus E protein (Lescar et al., 2001).

Preparation of alphavirus antigens for serological tests can rely on cell cultures or animal tissues (Beaty et al., 1989). In either case, antigens extracted are composed primarily of the structural proteins, which are produced in much greater quantities than the nonstructural proteins in infected cells. Antigens produced from purified virus include only the structural proteins. Hemagglutination relies primarily on the E1 protein, which contains a highly conserved fusion motif thought to mediate endosome membrane fusion at low pH for entry into the cytoplasm following receptor-mediated endocytosis. However, E2 also has some involvement. This E1 fusion motif is exposed upon a reduction to acidic pH (Gibbons et al., 2004). Most neutralization epitopes, determined by the binding sites of monoclonal antibodies that neutralize infectivity, are found on the E2 protein, with a few mapped to E1. The E2 protein is also thought to be the most important for interactions with cellular receptors, which are poorly understood for alphaviruses. The E1 protein contains more epitopes conserved among alphaviruses than E2, and this probably explains the greater cross-reactivity seen among alphaviruses in E1-dependent hemagglutination inhibition tests than in neutralization. Specific E2 and/or E1 epitopes have been mapped and characterized in some detail for VEEV, WEEV, and EEEV (Roehrig and Bolin, 1997; Roehrig, 1993; Pereboev et al., 1996; Roehrig et al., 1990).

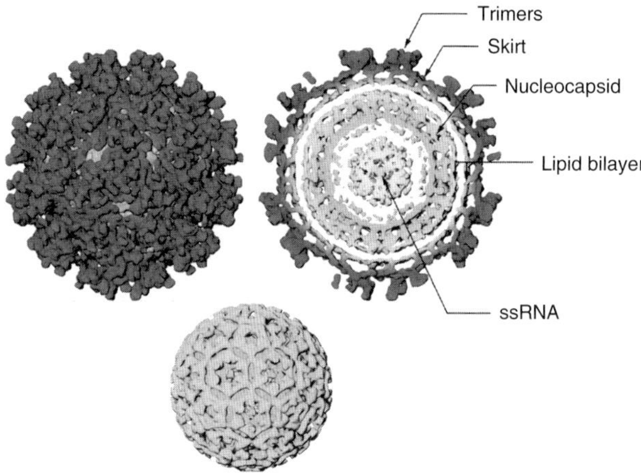

FIGURE 21.1 Cryo-electron microscopic reconstruction of the VEE virion at 14Å resolution. Upper left, external view of the virion showing the envelope glycoproteins on the surface; upper right, cross section of the virion showing the RNA genome at the center surrounded by 240 copies of the capsid protein, the bilipid envelope derived from the host cell plasma membrane, the E1 envelope glycoprotein lying parallel to the envelope, and the E2 envelope glycoprotein forming spikes on the surface; lower, nucleocapsid showing the capsid proteins in T = 4 icosahedral symmetry. Courtesy of Angel Paredes and Wah Chiu, Baylor College of Medicine, and Stanley Watowich and Scott Weaver, University of Texas Medical Branch (see color plate section).

PROTECTIVE IMMUNE RESPONSE

Antibody

Alphavirus infection induces a typical humoral immune response, with IgM appearing within one week of infection and persisting for a few months. The IgG response usually appears after 1–2 weeks and persists for years. A variety of evidence indicates that titers of serum-neutralizing antibodies are strongly predictive of protection against alphaviral disease (Griffin, 2007; Tsai et al., 2002). The end of viremia often coincides with the first appearance of IgM, and passive transfer studies have demonstrated that antibodies can protect against VEE and WEE, including both neutralizing and non-neutralizing, and those directed against E2 and E1 (Hunt and Roehrig, 1985; Roehrig, 1993). Protection from VEEV infection is apparently mediated primarily by antibody. However, the correlation between protection and neutralizing serum antibody titers is inconsistent when infection is aerosol-mediated (Elvin et al., 2002); induction of local respiratory tract immunity, including specific IgG and IgA, potentially correlates with protection against aerosol exposure.

Cell-Mediated Immunity

Alphaviruses induce cell-mediated immunity including CD4 and CD8 T lymphocytes, and macrophages. The role of cell-mediated immunity in recovery and protection from disease is not well established (Griffin, 2007). In fact, some studies have failed to identify any cell-mediated immunity after VEEV (Jones et al., 2003) and Sindbis infection of mice (Rowell and Griffin, 2002). However, cell-mediated immunity in combination with innate and humoral immunity may be required to prevent the development of encephalitis and/or for clearance from the brain. Mice with selective immune deficiencies can be protected by vaccination against VEEV, with the α/β T cell subset playing a more important role than γ/δ T cells (Paessler et al., 2007). Sindbis virus RNA has been shown to persist in the brains of infected mice, and can be cleared noncytolytically by a combination of antibodies directed at the E2 protein and interferon gamma (Griffin et al., 1997; Binder and Griffin, 2001). Attenuated variants of VEEV replicate persistently in the brains of some immunodeficient animals after challenge with wild-type virus (Paessler et al., 2007).

EPIDEMIOLOGY

All three alphaviral encephalitides are zoonotic, mosquito-borne viruses that are maintained in enzootic cycles by small mammalian or avian reservoir hosts. VEEV occurs in forest or swamp habitats ranging from North to South America (Fig. 21.2). VEEV emerges to cause equine epizootics and epidemics when enzootic variants, generally in subtype ID that occur in northern South America, mutate to become equine virulent and viremia competent (Weaver et al., 2004; Greene et al., 2005b; Anishchenko et al., 2006). Equine amplification in agricultural settings results in widespread spillover to humans who reside near horses, donkeys, and mules. In contrast, EEEV and WEEV do not rely on amplification in secondary hosts, but spill over directly from enzootic, avian cycles into humans and domestic animals that reside near these habitats. In the case of EEEV, enzootic circulation in North America occurs in hardwood swamp habitats inhabited by the enzootic vector, *Culiseta melanura*, and possibly other *Culex* spp. vectors (Cupp et al., 2004; Weaver, 2001). In Central and South America, forest habitats are the sites of enzootic foci and *Culex* (*Melanoconion*) spp. are the most common vectors. In Latin America, EEEV is associated with little human disease, probably due to inherently reduced virulence compared to North American strains (Aguilar et al., 2005, 2007a). WEEV occurs in avian cycles associated with agroecosystems in western North America, South America, and probably Central America. In North America, *Cx. tarsalis* is the principal vector, while other mosquitoes probably transmit in Latin America (Reisen, 2001; Reisen and Monath, 1988).

Significance as Public Health Problem

Of the three alphaviral encephalitides, VEEV is by far the most important pathogen both from a human and equine standpoint. Epidemics and equine epizootics have typically involved up to hundreds of thousands of cases in both hosts, with high mortality rates in equids and high morbidity in humans. Attack rates in both hosts often exceed 50% in affected areas, and most human infections are apparent (Bowen and Calisher, 1976). Mortality in horses, usually due to overt encephalitis, is typically 30–80%. Humans usually develop a severe, highly debilitating febrile illness, and some patients, especially children, develop neurological disease with convulsions, disorientation, and drowsiness, occasionally culminating in coma and death. Although overall mortality rarely exceeds 1% in humans, rates of neurological disease and permanent sequelae are usually 5–15%. Epidemiologic studies have also identified teratogenic effects of human VEEV infection with many premature abortions and stillborn births during outbreaks (Rivas et al., 1997; Briceno Rossi, 1967), and nonhuman primate studies have confirmed these findings (London et al., 1977).

Historically, EEEV and WEEV have caused massive equine epizootics involving thousands to tens of thousands of cases in North and South America.

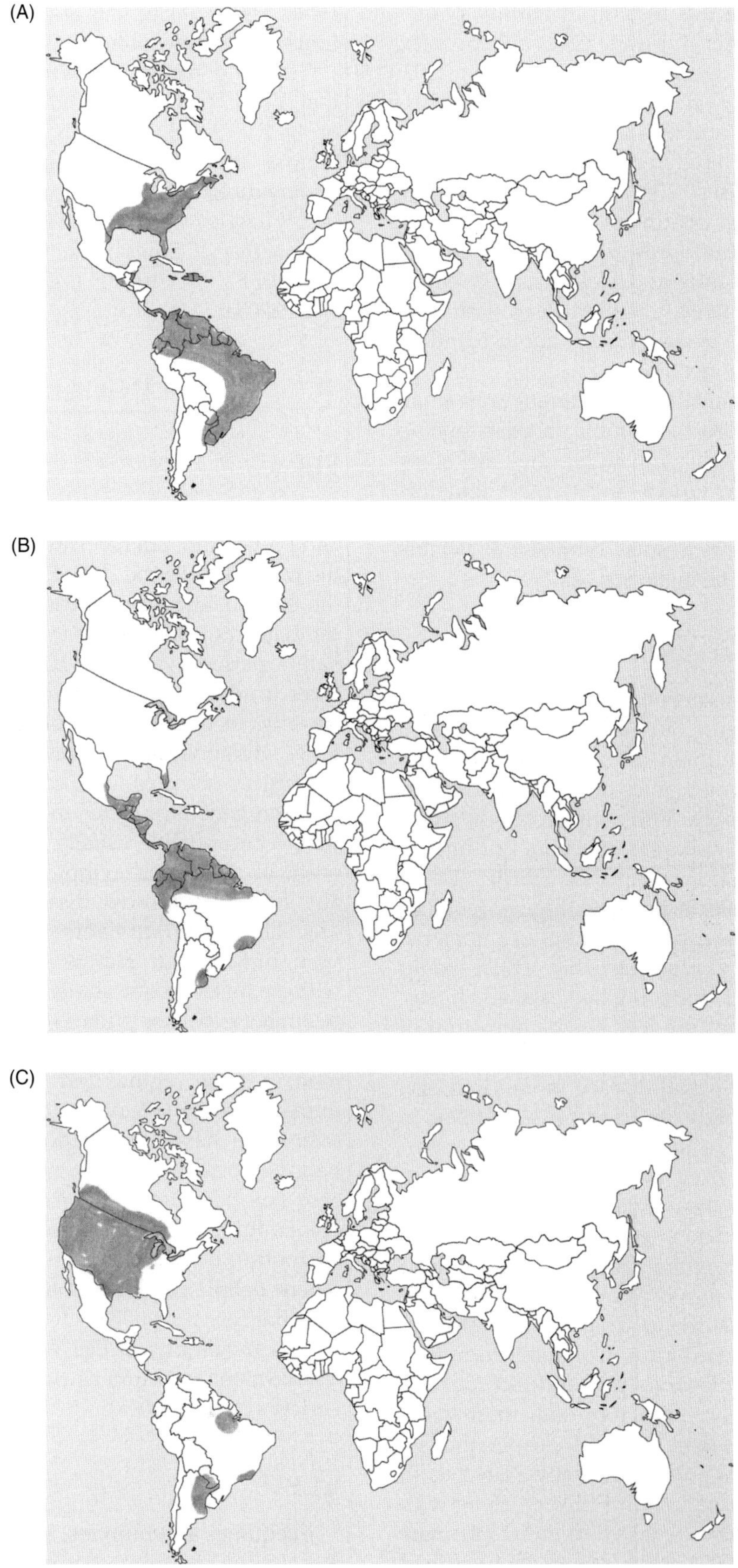

FIGURE 21.2 Map showing the distribution of the encephalitic alphaviruses: (a) VEEV, (b) EEEV, and (c) WEEV.

In North America, EEEV is also an important pathogen of emus, pheasants, turkeys, pigs, sheep, deer, and a variety of bird species (Beckwith et al., 2002; Gottdenker et al., 2003; Tate et al., 2005; Weaver, 2001; Bauer et al., 2005; Day and Stark, 1996). In recent decades, WEE epizootics have declined dramatically for unknown reasons (Reisen, 2001; Reisen and Monath, 1988), while EEEV has continued to cause sporadic epizootics involving hundreds of horses, primarily in states along the Atlantic and Gulf coasts (Weaver, 2001; Morris, 1988). However, human cases with both EEEV and WEEV have averaged only about 5 and 15 annually, respectively, since 1964.

In summary, EEEV and WEEV generate only small numbers of human cases and there is probably no potential for a commercial human vaccine market for these biothreats. VEEV causes considerable endemic and epidemic disease, but principally in developing countries of Latin America, so a commercially viable vaccine would need to be inexpensive.

Potential as Biothreat Agents

All three alphavirus encephalitides are considered potential biothreat agents (Sidwell and Smee, 2003). All can cause severe human disease, culminating in fatal encephalitis. However, rates of attack and apparent infection vary widely. VEEV causes high attack rates and high rates of apparent infection, but human cases are rarely fatal. The nonspecific febrile disease that precedes encephalitis is, however, highly debilitating as well as immunosuppressive. These characteristics of typical human VEE suggest that VEEV would be highly effective at immobilizing armed forces. Although mortality rates are low, the combination of extremely efficient aerosol infectivity and moderate (5–15%) rates of neurological disease in children also suggests that VEEV might be used for terrorism. In contrast to VEEV, EEEV causes high rates of inapparent human infections during natural epidemics; during a 1959 epidemic in New Jersey, the apparent: ratio was estimated at 1:23 (Goldfield et al., 1968). However, also in contrast to VEEV, apparent EEE cases usually lead to neurological disease that is often fatal. Survivors of EEEV infection often suffer permanent neurological sequelae, making it a feared virus in endemic locations. Although laboratory infections probably attributable to aerosol exposure have been reported, these cases are far fewer in number than for VEEV despite the fact that EEEV is isolated regularly in many state health laboratories in North America. This history, along with higher aerosol EEEV doses required to initiate primate disease (Reed et al., 2007) compared to VEEV (Reed et al., 2004, 2005b), suggests that EEEV is less infectious via the aerosol route.

The long history of laboratory infections suggests that VEEV is more infectious for humans than EEEV or WEEV (SALS, 1980; U.S. Department of Health and Human Services, 1999). Experimental aerosol infections of cynomolgous macaques corroborate these data; the VEEV ID_{50} is 7×10^4 PFU (Reed et al., 2004), while EEEV is more than 10 times higher (10^6 PFU) (Reed et al., 2007), as is WEEV (2×10^6 PFU, LD is $>7.3 \log_{10}$ PFU) (Reed et al., 2005a).

CLINICAL DISEASE

EEEV

According to the U.S. Centers for Disease Control and Prevention, 220 confirmed cases of EEE occurred in the U.S. from 1964 to 2004. EEEV is the most virulent of the encephalitic alphaviruses, with a case-fatality rate estimated at >33%. After a very short (4–10 days) incubation period symptoms begin with sudden onset of fever, headache, and general muscle pains of increasing severity. In human cases of encephalitis, fever, headache, vomiting, respiratory symptoms, leucocytosis, hematuria, seizures, and coma may occur. Surviving children generally suffer more severe sequelae. Death due to encephalitis usually occurs 2–10 days after the onset of signs and symptoms, and survivors generally suffer progressive, disabling mental and physical sequelae. These survivors usually live a normal life span, but without gainful employment, and a patient with sequelae costs an average of US$3 million in treatment expenses (Villari et al., 1995).

In Latin America, equine EEE outbreaks are common in some regions but there is little evidence of human disease (Weaver, 2001). EEEV-specific human antibodies have been reported in many regions of Latin America. However, no evidence of human disease has been found in Argentina despite intensive epidemiological studies following extensive equine epizootics, or in an enzootic focus near Iquitos, Peru where people are undoubtedly exposed to the virus (Aguilar et al., 2007a). It is likely that the South American variety of EEEV is less infectious and/or less virulent for humans than is the North American variety (Weaver, 2001).

VEEV

In equines and humans, VEEV causes a spectrum of disease ranging from unapparent to acute encephalitis in all age groups, with no sex bias observed. However, children are more likely to develop fatal encephalitis and suffer permanent neurological sequelae than adults. VEEV also infects the fetus in pregnant women and causes birth defects as well as spontaneous abortions and stillborn births. In humans, the incubation period is usually 2–5 days (Johnson and Martin, 1974). Most infections are apparent with high attack rates during epidemics. Symptoms appear abruptly, including malaise, fever, chills, and severe retroorbital or occipital headache. Myalgia typically centers in the thighs and lumbar region of the back. Signs usually include leukopenia, tachycardia, and fever, frequently accompanied by nausea, vomiting, and diarrhea. Signs and symptoms of central nervous system (CNS) disorders occur less frequently and typically include convulsions, somnolence, confusion, and photophobia. Acute disease usually subsides 4–6 days after onset, followed by asthenia that can last for several weeks. The illness is sometimes biphasic with recurrence 4–8 days after onset. A small proportion of human cases proceed to stupor and coma, sometimes followed by death (generally less than 1% of cases).

WEEV

Following an incubation period of about 2–10 days, WEEV causes a broad spectrum of human disease ranging from unapparent to fatal encephalitis. Illness typically begins with the sudden onset of headache, followed by dizziness, chills, fever, myalgias, and malaise (Tsai et al., 2002). Prodromal symptoms intensify over several days, often with an acute exacerbation of dizziness and vomiting, increasing somnolence or confusion. Neck stiffness is present in about 50% of cases, and neurologic signs usually include weakness and generalized tremulousness, especially of the hands, lips, and tongue. Stupor or coma develop in <10% of cases. General improvement begins several days after the defervescence of fever, typically within a week to 10 days after the onset. In infants, the initial presentation and clinical progression is more rapid, evolving from a nonspecific illness of fever, irritability, and diarrhea to convulsions and coma. Overall, mortality is 4% and is highest among the elderly, although serious sequelae are more common in recovered infants and children, including extensive brain damage resulting in mental retardation, developmental delays, convulsions, spasticity, and extrapyramidal movement disorders. Perinatal illness following late trimester or postpartum maternal infection have also been reported.

TREATMENT

Treatment of VEE, EEE, and WEE is limited to supportive therapies to maintain fluid balance and prevent intracranial pressure. Little can be done to prevent neurological disease and death in infected patients (Tsai et al., 2002; Weaver, 2001). Intracranial pressure, which contributes to morbidity and mortality, can be treated, as well as convulsions. Airway protection is needed in patients with loss of consciousness, along with hyperventilation accompanied by anesthesia and sedation. Cerebral blood flow can be maintained via $PaCO_2$ regulation. Brain swelling can be minimized by regulating serum sodium and osmolarity.

Because EEEV, VEEV, and WEEV are maintained in zoonotic mosquito-borne transmission cycles often found in swamps and forests, little can be done in an ecologically acceptable manner to control virus circulation (Weaver, 2001). Control usually relies on early detection of enzootic amplification in birds or rodents, followed by mosquito control to reduce vector populations. Sentinel birds used for detection of seroconversion, as well as adult female mosquito collections for virus isolation, can detect high levels of EEEV and WEEV circulation. Equine disease usually precedes human VEE disease. Personal protection against mosquito bites is usually the most effective means of VEE, EEE, and WEE prevention, especially for individuals who reside, work, or recreate near equids (VEE) and enzootic habitats (EEE, WEE). Equine vaccination is effective in preventing VEE epizootics and epidemics, but must be maintained continuously, especially in northern South America where the progenitors of major outbreaks occur (Weaver et al., 2004).

Although Ribavirin has been shown to have some inhibitory effect on alphaviruses in vitro, no human trials have been described. Because neurological disease follows a prodromal illness for at least several days to about a week, human cases are often diagnosed when acquired immunity is already detected and virus is cleared from the periphery. At this time, it is unlikely that antiviral treatment alone can control CNS disease. Stimulators of innate immunity, such as poly-IC, or interferons administered directly are known to prevent alphaviral disease in animals when administered before or soon after infection (Worthington et al., 1973; Worthington and Baron, 1971; Finter, 1966; Bradish and Titmuss, 1981; Aguilar et al., 2005). However, early treatment is generally impossible because the alphavirus incubation period is usually 24 h or longer. Based on previous studies

(Coppenhaver et al., 1995), combination treatments of poly-IC and alphavirus antibodies (Phillpotts et al., 2002) may protect against disease even if administered after invasion of the CNS, and should be explored further.

PATHOGENESIS

Description of the Disease Processes

EEEV

Pathogenesis of EEEV infections leading to invasion of the human CNS is poorly understood. Clinical studies using magnetic resonance imaging and computed tomography show changes in the basal ganglia and thalami suggesting brain edema, ischemia, and hypoperfusion in the early stage of disease (Deresiewicz et al., 1997). Gross pathological investigations of fatal human cases report brain edema with necrosis, facial or generalized edema, vascular congestion and hemorrhage in the brain and visceral organs (Farber et al., 1940; Femster, 1938, 1957; Getting, 1941; Azimi and Grossman, 1997; Jordan et al., 1965; Bastian et al., 1975). Loss of neurons, perivascular cuffing, microglial proliferation, vasculitis, and focal inflammatory infiltration are the most typical findings of human encephalitis. Vascular lesions are also common with thrombi and extravasation of red cells, and necrosis and demyelinization also occur (Rozdilsky et al., 1968). In the brain, antigen is localized to the perikaryon and neuronal dendrites, with little detected in glial cells (Garen et al., 1999). Cell death by apoptosis is conspicuous, primarily in glial and inflammatory cells. Neuronal death is associated with cytoplasmic swelling or eosinophilia and nuclear pyknosis. The degree of inflammation is not correlated with the presence of antigen, especially in cerebral cortex and spinal cord, where infected cells already have presumably been cleared before death of the patient. EEEV infection is also accompanied by visceral and pulmonary congestion. Apoptosis occurs in glial and inflammatory cells, but is less commonly associated with neuronal cell death (Garen et al., 1999). Similar lesions occur in primates after intracerebral inoculation (Nathanson et al., 1969).

Mice (Liu et al., 1970), hamsters (Paessler et al., 2004), guinea pigs (Walder et al., 1980), and rhesus (Nathanson et al., 1969) and Cynomolgus macaques (Reed et al., 2007) have been used for experimental EEE pathogenesis studies. Murine models for EEEV (Liu et al., 1970) reproduce the lymphoid involvement and cerebral pathology. Most strains of EEEV produce high rates of fatal encephalitis in peripherally infected mice (Schoepp et al., 2002; Liu et al., 1970); hamsters more faithfully reproduce the vasculitis associated with microhemorrhages in the brain that dominates the pathological picture in fatal human EEE (Paessler et al., 2004). In mice, initial replication is detected in fibroblasts, osteoblasts, and skeletal muscle myocytes near the site of subcutaneous infection (Vogel et al., 2005). Virus is first detected in the brain one day after infection, with rapid interneuronal spread leading to direct neuronal death by day 4. Invasion of the CNS by EEEV probably occurs by a vascular route, rather than via peripheral nerves or the olfactory bulb like VEEV. In hamsters, neuronal cell death is detectable only in late stages of disease after EEEV replicates in a variety of visceral organs, produces viremia, and penetrates the brain (Paessler et al., 2004). In general, the virus is capable of establishing productive infection in all parts of the mouse or hamster brain. Experimental studies have suggested early infection of periventricular and perivascular neuronal cells in the basal ganglia and hippocampus (Paessler et al., 2004). In contrast to VEEV, EEEV appears to rapidly infect the brain of animals via the blood, and the first antigen-positive neuronal cells are located in the basal ganglia and brain stem in the hamster model (Paessler et al., 2004). Certain findings in animal models, such as the hamster, are similar to early clinical manifestations of EEEV infections in humans. The inflammatory response can be very prominent in the brain if animals survive at least five days and is usually produced by macrophages, lymphocytes, and neutrophils. Some of the animal models also display the peripheral pathological changes described in fatal human EEE cases, including congestion and numerous microhemorrhages in the liver, spleen, and lung (Fig. 21.3).

VEEV

The infection of humans mostly produces a "flu-like" disease, and neurological disease (VEE) may be characterized by disorientation, ataxia, mental depression, and convulsions, which can be detected in up to 14% of infected individuals, especially children (Johnson and Martin, 1974). Overall mortality rates during epidemics rarely exceed 1%. Sequelae following VEE-related neurological disease in humans and rats is common and described elsewhere (Leon, 1975; Garcia-Tamayo et al., 1979). The typical complication of VEEV infections that is reported is a long-term immunosuppression in patients upon clinical recovery. The predominant pathological findings in fatal human VEE cases reveal: (a) in the CNS: edema, congestion,

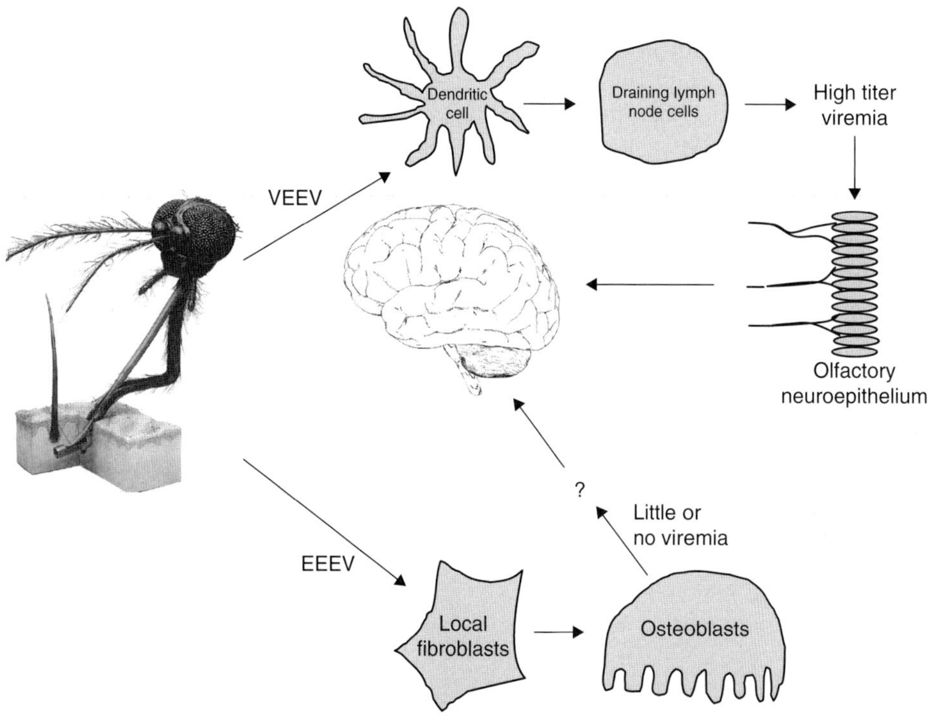

FIGURE 21.3 Illustration of VEEV and EEEV pathogenesis based on findings in the mouse models. WEEV pathogenesis is less well understood, but probably more closely resembles that of EEEV because there is little evidence of extensive lymphoid involvement. Most studies of peripheral infection have used needle inoculation of animals, but mosquitoes deposit alphavirus-infectious saliva both intra- and extravascularly in the dermis (Smith et al., 2006; Turell and Spielman, 1992), so natural, mosquito-borne pathogenesis could differ.

hemorrhages, vasculitis, meningitis, and encephalitis; (b) in the lungs: interstitial pneumonia, alveolar hemorrhage, congestion, and edema; (c) in lymphoid tissue: follicular necrosis and lymphocyte depletion; and (d) in the liver: diffuse hepatocellular degeneration (de la Monte et al., 1985; Ehrenkranz and Ventura, 1974; Johnson et al., 1968). Laboratory mice, hamsters, and guinea pigs are well developed as animal models for VEE, with a murine model for VEEV-induced encephalitis and lymphotropism being established (Davis et al., 1994, 1995; Kundin, 1966; Jackson et al., 1991). Previous studies have demonstrated that the murine model is characterized by biphasic disease, which starts with the productive infection of lymphoid tissue and ends in the destruction of the CNS by viral replication and a "toxic" neuroinflammatory response, which is uniformly lethal (Gorelkin, 1973; Grieder et al., 1995, 1997; Schoneboom et al., 1999, 2000; Garcia-Tamayo et al., 1979; Ryzhikov et al., 1991, 1995). By the time encephalitis has developed in an infected mouse, the infectious virus is usually absent from the peripheral organs and blood. However, virus replicates to high titers in the brain and mice die 5–7 days after infection due to fatal encephalitis (Anishchenko et al., 2004; Paessler et al., 2003). The route of VEEV invasion of the CNS has been studied extensively, and appears to rely on entry into the brain via the olfactory nerve and possibly the dental nerves (Ryzhikov et al., 1995; Charles et al., 1995; MacDonald and Johnston, 2000; Aronson et al., 2000). All VEEV strains including enzootic and epizootic produce similar disease in a variety of mouse strains, both inbred and outbred (Wang et al., 1999).

WEEV

Like the other human alphaviral encephalitides, WEE pathogenesis is poorly understood. In postmortem examinations the brain appears normal or exhibits a moderate degree of vascular congestion (Tsai et al., 2002). Perivascular infiltrates of lymphocytes, plasma cells, and neutrophils are characteristic, along with vascular necrosis in some cases. Lesions are distributed mainly in the subcortical white matter, the internal capsule, thalami, basal ganglia, substantia nigra, dentate nucleus, and molecular layer of the cerebellum and gray matter of the brainstem and spinal cord. Neurons are found in various stages of degeneration. Widely scattered focal areas of demyelination are seen.

Although pathogenesis studies of WEEV in small animal models are quite limited, the mouse has received most of the focus. With many WEEV strains, only partial mortality occurs following peripheral infection; higher levels of mortality accompany intracranial and intranasal infection (Liu et al., 1970; Julander et al., 2007). Bianchi et al. (1993) studied three epizootic (associated with equine epizootics) WEEV strains (McMillan, CBA87, and CBACIV 180) and an enzootic strain (AG80-646). Neurovirulence and neuroinvasiveness for adult mice correlated with viremia and virus replication in the brain. Epizootic viruses were neurovirulent and neuroinvasive, whereas the enzootic strain was attenuated. Intranasal infectivity of eight WEEV strains revealed that all were 100% lethal for adult BALB/c mice, with some variation in the time to death (Nagata et al., 2006). Based on the time of death the high-virulence phenotype (rapid death) included strains California, Fleming, and McMillan, and the low-virulence group included CBA87, Mn548, B11, Mn520, and 71V-1658.

Histopathological findings of WEEV infection in baby mice include lesions in muscle, cartilage, and bone marrow (acute necrotizing degeneration), while infection in three-week-old mice causes diffuse meningoencephalitis and lesions in the heart, lungs, kidney, liver, and brown fat (Aguilar, 1970; Monath et al., 1978). In addition to mice, guinea pigs, rabbits, and newborn chickens have also been evaluated as models for WEE along with hamsters. Guinea pigs suffer fatal disease from epizootic strains but survive enzootic AG80-646 strain infection. However, rabbits showed the inverse pattern of disease, with the enzootic strain lethal (Bianchi et al., 1997). Viremia level correlated with the virulence in all of these models. In hamsters, (Julander et al., 2007) infected intraperitoneally, disease progression is rapid, with viremia and virus replication in the brain, liver, and kidney 2–4 days after inoculation. Foci of virus replication occur in neurons of the cerebral cortex and midbrain. In cynomolgous macaques, WEEV produces partial fatality when administered in high doses via the aerosol route (Reed et al., 2005a). Three to four days of fever occur four to six days after infection. Clinical signs of encephalitis begin coincident with fever onset, and progress rapidly, accompanied by elevated white cell counts (predominantly neutrophils and monocytes). Immunohistochemical evidence of WEEV antigen is found in the brain and CNS, along with a marked lymphocytic infiltrate (Reed et al., 2005a).

Immune Response to Infection

IFN-α and -β are important in protecting against alphaviral disease (Grieder and Vogel, 1999; Spotts et al., 1998). Previous studies showed a correlation between alphavirus virulence and resistance to IFN-α and -β. Some studies with VEEV suggested that the IFN-α and -β resistance or sensitivity phenotype correlated with epizootic potential and equine virulence (Spotts et al., 1998; Jahrling et al., 1976), although others observed little or no difference (Anishchenko, 2004). In contrast, IFN-γ is believed to have a lesser role in protecting against alphavirus infection.

The appearance of neutralizing IgM usually coincides with the clearance of peripheral infection, but replication in the CNS continues. Noncytolytic clearance of SINV from the murine CNS involves immune-mediated, noncytolytic clearance from neurons, an activity that requires both CD4+ and CD8+ T cells, and that T cells mainly require IFN-γ for clearing virus from certain neuron populations (Binder and Griffin, 2001). However, RNA can persist after infectious virus is no longer detected. The early lymphoid phase of VEEV infection ends with the appearance of IgM about 3–4 days postinfection, and the CNS pathology in a VEEV-infected mouse has a significant immunologic component in addition to direct killing of neurons via apoptosis (Charles et al., 2001). Although neutralizing antibodies are believed to be the principal determinants of immune protection, recent studies of vaccinated mice with immune deficiencies indicate that antibodies are not absolutely required for protection, and that α/β T cells can also protect against fatal disease (Paessler et al., 2006, 2007). IFN-γ receptor KO mice did not show any major difference in viremia or mortality compared to wild-types, supporting the conclusions of previous studies that showed no difference in disease outcome in IFN-γ-receptor deficient mice infected with SFV (Muller et al., 1994).

Detailed studies of the immune response to VEEV strain TC-83 demonstrate immune responses in draining lymph nodes 24h following murine vaccination, with increased levels of activated B cells and both CD4(+) and CD8(+) T cells (Bennett et al., 2000). Activated monocyte/macrophages and natural killer cells are also seen between 6h and 7 days postvaccination. VEEV-specific IgG appears in the serum on day 5 postvaccination, with titers rising on days 7, 14, and 21. Isotypes of IgG measured on days 7 and 21 are predominantly of the IgG2a subclass, indicating a Th1-mediated response. Cytokine mRNAs detected in the serum include the Th1 cytokine IFN-γ and the inflammatory cytokine TNF-α, whereas the Th2 cytokine IL4 is not detected (Bennett et al., 2000).

In studies with Sindbis virus as a model of alphaviral encephalitis, cellular genes such as *bax*, *bak*, and *bcl-2*, which regulate the apoptotic pathway, can modulate the outcome of infection (Jan et al., 2000). During Semliki Forest virus–induced murine encephalitis, infiltrating leukocytes and neural precursor cells undergo apoptosis, while productively infected neurons undergo necrosis (Sammin et al., 1999). Immature mouse neurons infected with Sindbis or Semliki Forest viruses die of caspase-dependent apoptosis, while mature neurons survive by producing factors inhibiting viral-induced apoptosis. Apoptosis is induced at the time of alphavirus fusion with the cell membrane, and virus replication is not required (Jan and Griffin, 1999). Virus fusion apparently initiates the apoptotic cascade by inducing sphingomyelin degradation and ceramide release. In mature mice, antibody clears virus from the infected cells, however, viral RNA persists in infected neurons. The interruption of continued antibody production leads to viral reactivation. In contrast, T cells are not protective, although CD8(+) cells may act indirectly, possibly via cytokines, to assist in the clearance of viral RNA from neurons.

VACCINES

History

VEEV Vaccines

Vaccines to protect equines and humans against VEEV were first produced soon after the virus was isolated in 1938 (Beck and Wyckoff, 1938; Kubes and Rios, 1939). These formalin-inactivated preparations were initially made from mouse brain and other animal tissues following infection with virulent, unmodified subtype IAB strains (Randall et al., 1949). Genetic studies showing an extremely high degree of sequence conservation among IAB strains isolated from 1938 to 1973 (Kinney et al., 1992; Weaver et al., 1999), as well as the isolation of live virus from at least one human vaccine preparation (Sutton and Brooke, 1954). The lack of IAB outbreaks since 1973, coincident with the replacement of inactivated VEEV vaccines by the live-attenuated TC-83 strain, supports the vaccine origin hypothesis for all IAB outbreaks during the 1960s and 1970s.

The TC-83 attenuated VEEV vaccine strain was produced in 1961 by 83 serial passages of the subtype IAB stain Trinidad donkey in guinea pig heart cells (Berge et al., 1961), and soon thereafter was administered to humans at risk for laboratory infection (McKinney et al., 1963). Reverse genetic studies indicate that two mutations are the principal determinants of attenuation: genome nucleotide position 3 in the 5′ untranslated region, and amino acid position 12 in the E2 envelope glycoprotein (Kinney et al., 1993). TC-83 is generally administered in live form prepared from chicken embryo cells (Cole et al., 1976) as U.S. Food and Drug Administration (FDA) Investigational New Drug No. 142 (Pittman et al., 1996). A formalin-inactivated version called C84 has also been used in humans since the 1970s as FDA IND No. 914 (Pittman et al., 1996; Edelman et al., 1979). Evaluation of 821 vaccinated laboratory workers, including some boosted with C-84, showed 82% seroconversion (80% plaque reduction neutralization titers ≥1:20). Side effects occurred in 23% of vaccinees but no long-term sequelae were noted (Pittman et al., 1996).

In the case of contraindications or lack of seroconversion to TC-83, a formalin-inactivated version, C-84 is used (Cole et al., 1974). Unlike TC-83, this vaccine produces only occasional, mild, local, and systemic reactions; it augments preexisting VEEV antibody titers in seropositive recipients of TC-83, and induces high titers of cross-reactive antibodies in nonimmune subjects after one primary and two booster vaccinations, with circulating antibody persisting for at least 14 months (Edelman et al., 1979). However, some rodent studies indicate that TC-83 protects mice better against aerosol challenge (Jahrling and Stephenson, 1984). Other studies using cynomolgous macaques indicate that TC-83 (single dose) and C-84 (3 doses) elicit similar neutralizing antibody responses and protect against peripheral infection, but neither protects all animals against aerosol exposure (Pratt et al., 1998). Neither pre-challenge antibody titers nor lymphoproliferative responses predicted well the outcome of challenge. Microspheres composed of poly (DL-lactide-co-glycolide) augmenting IgG antibody levels and neutralization titers to the C84 inactivated VEEV vaccine and afford enhanced protection against VEEV challenge (Greenway et al., 1995).

The TC-83 vaccine induces VEEV-specific IgM and IgG in the respiratory tracts of immunized mice, and passive intranasal immunization with IgG protects against aerosol challenge (Phillpotts, 1999). The mucosal lymphotropism of attenuated VEEV strains probably explains their efficacy against aerosol challenge of mice (Charles et al., 1997), although a role for cell-mediated immunity in protection has also been suggested (Elvin et al., 2002). Some studies have shown that TC-83 induces limited cross-immunity against heterologous VEEV subtypes (Fillis and Calisher, 1979); others showed breakthrough infections and disease only after aerosol challenge. The risk of breakthroughs appears to increase with time after vaccination, and is more likely with epizootic, homologous serogroup IA/B viruses than with enzootic, heterologous strains (Phillpotts and Wright, 1999).

Another issue with live alphavirus vaccines is immunologic interference with a second vaccination.

Persons previously vaccinated with VEEV strain TC-83 exhibit poor neutralizing antibody responses to a live-attenuated chikungunya virus (CHIKV) vaccine, and vice versa. ELISA antibody responses demonstrate cross-reactive IgG to VEEV after primary CHIKV immunization, followed by an anamnestic response upon subsequent VEEV vaccination (McClain et al., 1998).

EEEV and WEEV Vaccines

Although live-attenuated WEEV (Hughes and Johnson, 1967; Binn et al., 1966) and EEEV (Brown and Officer, 1975; Brown et al., 1975) vaccine strains were tested in the 1970s, only multivalent, formalin-inactivated vaccines against VEEV, EEEV, and WEEV are currently administered to horses. In equids, these vaccines protect against challenge after two initial doses (Barber et al., 1978) but require frequent, life-long boosters (Waldridge et al., 2003). The EEEV vaccine is also immunogenic and efficacious in some but not all avian species (Tengelsen et al., 2001; Snoeyenbos et al., 1978; Olsen et al., 1997; Clark et al., 1987). Both immunologic cross-reactivity among alphaviruses and extensive strain variation are potential issues with alphavirus vaccines. For example, the presence of preexisting EEEV or WEEV antibodies appears to interfere with TC-83 immunogenicity in horses (Vanderwagen et al., 1975; Ferguson et al., 1978). Due to antigenic differences among EEEV strains, the inactivated vaccine prepared from a North American strain elicits high-neutralizing titers against most or all North American isolates, but not against South American strains (Strizki and Repik, 1995). These data suggest that the current equine vaccines produced in the U.S. may not be efficacious in Latin America. The risks of using wild-type, virulent alphavirus strains in inactivated vaccine preparations continue despite indications that this virus was responsible for initiation of VEE epizootics from the 1940s to 1970s (Weaver et al., 1999). The occurrence of EEE in a California horse shortly after vaccination, also suggesting residual live virus, also underscores the risks associated with these vaccines (Franklin et al., 2002).

Veterinary Vaccines

The TC-83 vaccine strain was initially used in live form during the 1969–1971 epizootic that swept through much of Central America, Mexico, and into Texas. Extensive field testing in horses showed live TC-83 vaccine to be safe and efficacious (Walton et al., 1972) with long-lasting immunity (Walton and Johnson, 1972). Viremia titers were generally below the predicted infection threshold of epizootic vectors (Baker et al., 1978), vaccine virus was isolated from mosquitoes in Louisiana (Pedersen et al., 1972), raising environmental concerns. In Texas, TC-83 proved minimally reactive with no documented abortions or deaths attributable to vaccination, and most antibody titers were maintained for at least 11 months. Vaccine virus was isolated 1–14 days after vaccination and antibody appeared about a week after vaccination. The lack of VEEV isolations from horses with antibodies against EEEV and WEEV suggested some degree of cross-protection (Calisher and Maness, 1975), as did some serologic (Ferguson et al., 1977) and animal model studies (Cole and McKinney, 1971).

Duration of Immunity

Human and murine immunity generated by TC-83 against homologous subtype IAB strains lasts at least 7–9 years, while cross-reactive immunity against other subtypes is not as long lived (Burke et al., 1977; Fillis and Calisher, 1979). Most human TC-83 vaccinees retained neutralization titers for at least 6 years, and many TC-83 responders who were also boosted with C-84 remained seropositive for over 10 years. However, many TC-83 nonresponders who were boosted with C-84 failed to retain titers ≥1:20 for more than 1–2 years (Pittman et al., 1996).

Contraindications of Vaccination, Including Special Risk Groups

Pregnant women, persons infected with human immunodeficiency virus (HIV), or persons in any way immunosuppressed or immunocompromised are excluded from vaccination under the U.S. Army's Special Immunization Program. A known allergy to egg products also grounds for exclusion. Pancreatic involvement in hamster infections with VEEV (Gorelkin and Jahrling, 1974) is the rationale for exclusion of diabetic persons from vaccination with strain TC-83.

Adverse Events

Adverse reactions to the TC-83 live VEEV vaccine strain have included a mild, febrile illness often including malaise, headache, fever, chills, myalgia, sore throat, and occasionally, nausea, arthralgia, anorexia, vomiting, diarrhea, erythema, and rash. Adverse events to the inactivated VEEV and EEEV vaccines include typical local and systemic allergic

TABLE 21.1 New alphavirus vaccines under development

Agent	Vaccine type	Attenuation or immunogenicity strategy	References	Stage of development
VEEV	Live virus	Ablation of E3-E2 cleavage site, compensatory mutation in E1 protein gene	Davis et al. (1995), Pratt et al. (2003), Reed et al. (2005b)	Phase I human studies completed
		SINV/VEEV nonstructural/structural protein gene chimeras	Paessler et al. (2003, 2006)	Preclinical
EEEV		WEEV/EEEV nonstructural/structural protein gene chimeras	Schoepp et al. (2002)	Preclinical
		SINV/EEEV nonstructural/structural protein gene chimeras	Wang et al. (2007)	Preclinical
Various	Replicon particle	Defective genome expressing some critical antigenic proteins, packaged by envelope glycoproteins supplied in trans	Frolov et al. (1997), Pushko et al. (2000)	Preclinical (for expression of alphavirus antigens)

reactions. Long-term sequelae following alphavirus vaccination have not been reported. A case of hydrops fetalis and fetal demise resulting from TC-83 vaccination of a pregnant woman indicated that this vaccine should not be used during pregnancy (Casamassima et al., 1987).

Current Licensed Vaccines

Although EEEV, VEEV, and WEEV vaccines have been administered to persons at risk by the U.S. Army under IND protocols since the 1960s, there are no alphavirus vaccines licensed for human use. Human vaccines against alphaviruses have never been licensed.

VACCINES IN DEVELOPMENT

Past alphavirus vaccine development has focused on the use of live, attenuated viruses or inactivated, whole virus. Although some recent development strategies have utilized DNA that encodes alphavirus structural proteins (Riemenschneider et al., 2003) most recent efforts have focused on improved strategies for alphavirus attenuation and the use of replicon particles.

Need for Improved Alphavirus Vaccines

As with other live RNA virus vaccines including polio where attenuation was derived via cell culture adaptation, resulting in a small number of attenuating point mutations, the TC-83 VEEV strain is probably unstably attenuated due to selection in humans or domestic animals for reversions. This inherent problem, probably reflected in the high rates of reactogenicity among human vaccinees (Pittman et al., 1996), was the impetus for the development of new live vaccine strains using genetic engineering strategies for improved stability of attenuation. Currently, two basic attenuation strategies are under development, along with the use of replicon particles that are limited to a single cycle of replication in vaccinees (Table 21.1).

Discovery/Basic Science

Antigen Discovery

Little effort has been reported on attempts to discover and exploit new alphavirus antigens for vaccines.

Preclinical Development

Adjuvant Development

The IND alphavirus vaccines described above are generally used with only Alum adjuvant. However, microspheres composed of poly (DL-lactide-co-glycolide) enhance IgG and neutralizing antibody responses as well as mucosal immunity to the VEEV C-84 vaccine (Greenway et al., 1995, 1998) and thus show promise for further development. Recent studies indicate that the alphavirus replicon particles have intrinsic systemic and mucosal adjuvant activity due to their RNA replication activity (Thompson et al., 2006).

Delivery Technology

Delivery of alphavirus structural proteins as DNA vaccines has shown promise in mice (Riemenschneider et al., 2003), but results with primates or large animals have not been reported.

Development of Animal Models

As described above, mice, hamsters, and guinea pigs have been used most commonly as animal models for human and equine EEE, VEE, and WEE. Recently, Reed and colleagues have conducted detailed pathogenesis studies using cynomolgous macaques exposed to alphavirus aerosols. These primates develop fever 4–6 days after WEEV exposure, with a duration of 3–4 days (Reed et al., 2005a). Clinical signs of encephalitis begin after the cessation of fever, with elevated white blood cell counts dominated by neutrophils and monocytes. In animals that die, immunohistochemical evidence of viral antigen is found in the brain and CNS with marked lymphocytic infiltrates. Experimental aerosol infections of cynomolgous macaques with VEEV subtypes IE or VEE complex subtype IIIA Mucambo virus develop fever, viremia, and lymphopenia in a dose-dependent manner (Reed et al., 2004). Most develop clinical signs of encephalitis, including loss of balance and hypothermia. Following aerosol exposure to EEEV, cynomolgus macaques develop encephalitis that corresponds with the onset of fever and tachycardia (Reed et al., 2007). Viremia is transient or undetectable and the onset of illness was dose dependent. A prominent leukocytosis occurs, with neutrophils predominating.

New Alphavirus Vaccine Platforms

Although a bacterially expressed E1 envelope glycoprotein from WEEV was shown to be immunogenic in mice, it exhibited poor protection (Das et al., 2004). The entire structural protein region (C-E3-E2-6K-E1), or the complete glycoprotein region (E3-E2-6K-E1) of VEEV expressed by baculoviruses are highly immunogenic in mice and protected these animals against challenge, while individual glycoprotein regions (E3-E2 and 6K-E1) are incompletely processed but nevertheless provide complete protection (Hodgson et al., 1999). Vaccinia-based VEEV vaccines have also been produced and shown to be immunogenic and protective in the mouse model (Kinney et al., 1988a, 1988b; Bennett et al., 1998). Various vaccinia constructs including all or only some of the structural proteins protect mice well against peripheral challenge but provide only partial protection against aerosol exposure (Phillpotts et al., 2000). The inability of these vaccines to elicit a sufficient IgA response may be responsible for the lack of efficacy against airborne challenge.

Recombinant, defective type 5 adenoviruses expressing the E3-E2-6K structural genes of VEEV protect against homologous challenge virus but less so against enzootic VEE complex strains (Phillpotts et al., 2005). A DNA vaccine against VEEV has also been shown to be efficacious in mice (Riemenschneider et al., 2003).

Purified envelope glycoproteins extracted from VEEV, EEEV, and WEEV are also immunogenic and protective (Pedersen, 1976). Novel gene-shuffling techniques are also under development to produce multivalent alphavirus vaccines (Locher et al., 2005). DNA vaccines against WEEV has shown partial efficacy in mice (Nagata et al., 2005).

New Alphavirus Vaccine Concept Development

Two main discoveries that led to the novel alphavirus attenuation strategies are outlined in Table 21.1. First, the discovery that WEEV is the descendent of a recombinant alphavirus (Hahn et al., 1988; Weaver et al., 1997) led to the interest in generating by recombinant DNA technology artificial recombinants to study interactions among alphavirus genes (Kuhn et al., 1996). These studies determined that chimeric alphaviruses involving two different species are highly attenuated, although the mechanism of attenuation is not known. The lack of attenuation in chimeras derived from different strains of the same alphavirus, for example, different subtypes of VEEV (Greene et al., 2005a; Powers et al., 2000), suggests that incompatibilities between more distantly related sequences or proteins results in suboptimal interactions and replication efficiency, although this is manifested more in vivo than in vitro. This attenuation has been exploited to develop several chimeric vaccine candidates against VEEV and EEEV (Table 21.1). The second basic discovery that led to a new alphavirus attenuation strategy was the identification of VEEV mutants that result in a lack of cleavage between the E3 and E2 proteins during secretory processing of the PE2 polyprotein (Davis et al., 1995). RNA transcripts derived from cDNA clones containing deletions in this cleavage site are nonviable but give rise to a small proportion of viable revertants with second-site suppressor mutations and incorporate unprocessed PE2 into virions. The combination of the lethal PE2 cleavage mutation and a suppressor mutation at E1 position 253 is highly attenuated and immunogenic (Pratt et al., 2003; Reed et al., 2005b; Hart et al., 2000).

Another basic alphavirology development that has translated into new vaccine strategies is the replicon particle. Because only the nonstructural proteins and cis-acting RNA sequences are required for alphavirus genome replication, the structural protein genes can be replaced by foreign antigens that are expressed at high levels in infected cells (Frolov et al., 1997; Liljestrom and Garoff, 1991). However, for these replicon genomes to enter the cytoplasm for replication, they must be either introduced by chemical or electromagnetic transfection methods, or packaged into virus-like particles by capsid and envelope proteins provided in trans from a second genetic construct. Often, two separate, transcribed "helper" plasmids encode the structural packaging proteins to avoid the generation of complete viral genomes via recombination (Pushko et al., 1997). Another approach is to develop packaging cell lines that constitutively express the structural proteins (Polo et al., 1999). Because the replicon genomes are defective and cannot produce structural proteins, they express the foreign genes during only one round of replication and cannot spread to other cells, an important safety consideration. The alphavirus replicon systems have been used primarily to express immunogens from other viruses or proteins (Zhou et al., 1994; Davis et al., 2002), but can also be used to immunize against homologous or heterologous alphaviruses without the safety concerns of a live replicating virus.

Another recent discovery that will probably lead to new, live alphavirus vaccine development approaches is the molecular basis of the shutoff of vertebrate host cell gene expression that occurs during alphavirus infections which undoubtedly contributes to virulence. Unlike the Old World Sindbis and Semliki Forest viruses that regulate this shutoff via their nsP2 (Frolova et al., 2002; Garmashova et al., 2006), EEEV (Aguilar et al., 2007b), and VEEV (Garmashova et al., 2007) use the capsid protein for this purpose. The mechanism of WEEV host cell gene expression shutoff is unknown, but the derivation of its capsid gene from an EEEV ancestor suggests it also uses the capsid protein. Inactivation of this capsid function could therefore be exploited for alphavirus attenuation.

Attenuated VEEV Strain V3526

Because of the human reactogenicity of the TC-83 vaccine and other concerns from animal studies such as glucose intolerance and reduced insulin release after vaccination (Rayfield et al., 1976), next generation, live-attenuated VEEV vaccine strains have been developed using infectious cDNA clones derived from the Trinidad donkey strain. Attenuation was achieved by inserting either attenuating mutations or a PE2 cleavage-signal mutation combined with an E1 gene resuscitating mutation. The latter strain, called V3526, is immunogenic for mice and nonhuman primates (Pratt et al., 2003). V3526 induces systemic and mucosal protection in mice more efficiently than TC-83 (Hart et al., 2000) and is stably attenuated after cell culture and mouse passages, remaining less neurovirulent in mice than TC-83 (Ludwig et al., 2001). V3526 also appears to have a lower risk for environmental transmission and distribution (Turell et al., 1999; Rao et al., 2004). Although this vaccine strain elicits neutralizing antibodies in macaques only against homologous subtype IAB VEEV strains, it protects against both IAB and IE strains in aerosol challenges (Reed et al., 2005b).

Chimeric alphaviruses have recently been developed as vaccine candidates. A chimera derived from a cDNA clone including the 5′ 2/3 of the WEEV genome, encoding the nonstructural proteins, and the 3′ 1/3 of the EEEV genome, encoding structural proteins, is highly attenuated in mice (Schoepp et al., 2002). This chimera protects against EEEV but not WEEV challenge. A chimeric clone encoding the nonstructural proteins of Sindbis virus and the structural proteins of the VEEV TC-83 strain is also highly attenuated and protects against wild-type VEEV challenge (Paessler et al., 2003).

Clinical Trials

Phase I

Clinical Phase I trials of the VEEV TC-83 and C-84 vaccines are described above (Pittman et al., 1996). Results of Phase I trials of the U.S. Army's inactivated EEEV and WEEV vaccines have not been published. A Phase I trial of the V3526 attenuated VEEV vaccine was completed in 2006, but the results have not yet been reported. No Phase II or III studies have been undertaken for any alphavirus vaccines.

POSTEXPOSURE IMMUNOPROPHYLAXIS

No studies of alphavirus immunoprophylaxis have been reported.

PROSPECTS FOR THE FUTURE

Several new strategies for generating robustly and stably attenuated alphaviruses have been described recently, and offer promise for the generation of improved, live virus vaccines. However, the stability of the attenuation phenotypes will require close scrutiny due to the potentially fatal consequences of reversion to wild-type virulence. Additional studies of the determinants of attenuation will be required to predict phenotype stability, as well as experiments to test stability empirically. These new attenuation strategies should also lead to improvements in the efficiency and safety of inactivated vaccine production for both human and veterinary use, which currently rely on the propagation of wild-type, virulent, select agent strains in some cases. Replicon particles, which combine most of the safety of inactivated vaccines with improved immunogenicity, also offer promise not only for alphaviral disease, but for other antigenic targets using alphavirus vectors. Further effort is needed to better define the critical determinants of both humoral and cellular immunity, so that other new strategies such as virus-like particles or DNA-based immunization can be exploited for alphaviruses.

KEY ISSUES

- No human alphavirus vaccine suitable for licensure is currently in clinical development.
- An ideal alphavirus vaccine for biodefense purposes would require only one dose, generate long-lasting immunity, be relatively inexpensive to produce, and be stable for long-term storage.
- An ideal human vaccine for protection against endemic and epidemic VEE in Latin America would have the same properties, but cost would be an overriding issue.
- Because EEEV and WEEV typically cause only small numbers of human cases in the U.S., there is probably not a sufficient market for commercial vaccine development.
- A safe and effective equine vaccine is critical for preventing epidemics of VEE, which rely on equine amplification for transmission to large numbers of people.
- The current equine vaccine against VEEV relies on only two point attenuating mutations, suggesting the possibility that reversion could generate a VEE epidemic involving large numbers of human cases.
- Further identification and characterization of critical alphavirus antigens is needed to exploit newer vaccine platform technologies for biodefense.

References

Aguilar, M.J. Pathological changes in brain and other target organs of infant and weanling mice after infection with non-neuroadapted western equine encephalitis virus. Infect. Immun. 1970; 2:533–542.

Aguilar, P.V., Paessler, S., Carrara, A.S., Baron, S., Poast, J., Wang, E., Moncayo, A.C., Anishchenko, M., Watts, D., Tesh, R.B. and Weaver, S.C. Variation in interferon sensitivity and induction among strains of eastern equine encephalitis virus. J. Virol. 2005; 79(17): 11300–11310.

Aguilar, P.V., Robich, R.M., Turell, M.J., O'Guinn, M.L., Klein, T.A., Huaman, A., Guevara, C., Rios, Z., Tesh, R.B., Watts, D.M., Olson, J. and Weaver, S.C. Endemic eastern equine encephalitis in the Amazon region of Peru. Am. J. Trop. Med. Hyg. 2007a; 76:293–298.

Aguilar, P.V., Weaver, S.C. and Basler, C.F. Capsid protein of eastern equine encephalitis virus inhibits host cell gene expression. J. Virol. 2007b; 81:3866–3876.

Anishchenko, M., Bowen, R.A., Paessler, S., Austgen, L., Greene, I.P. and Weaver, S.C. Venezuelan encephalitis emergence mediated by a phylogenetically predicted viral mutation. Proc. Natl. Acad. Sci. U.S.A. 2006; 103:4994–4999.

Anishchenko, M., Paessler, S., Greene, I., Aguilar, P.V., Carrara, A.S. and Weaver, S.C. Generation and characterization of closely related epizootic and enzootic infectious cDNA clones for studying interferon sensitivity and emergence mechanisms of Venezuelan equine encephalitis. J. Virol. 2004; 78:1–8.

Aronson, J.F., Grieder, F.B., Davis, N.L., Charles, P.C., Knott, T., Brown, K. and Johnston, R.E. A single-site mutant and revertants arising in vivo define early steps in the pathogenesis of Venezuelan equine encephalitis virus. Virology 2000; 270:111–123.

Azimi, P. and Grossman, M. Encephalitis in boy from South Carolina. Pediatr. Infect. Dis. J. 1997; 16:824–825.

Baker, E.F., Jr., Sasso, D.R., Maness, K., Prichard, W.D. and Parker, R.L. Venezuelan equine encephalomyelitis vaccine (strain TC-83): a field study. Am. J. Vet. Res. 1978; 39:1627–1631.

Barber, T.L., Walton, T.E. and Lewis, K.J. Efficacy of trivalent inactivated encephalomyelitis virus vaccine in horses. Am. J. Vet. Res. 1978; 39:621–625.

Bastian, F.O., Wende, R.D., Singer, D.B. and Zeller, R.S. Eastern equine encephalomyelitis. Histopathologic and ultrastructural changes with isolation of the virus in a human case. Am. J. Clin. Pathol. 1975; 64:10–13.

Bauer, R.W., Gill, M.S., Poston, R.P. and Kim, D.Y. Naturally occurring eastern equine encephalitis in a Hampshire weather. J. Vet. Diagn. Invest. 2005; 17:281–285.

Beaty, B.J., Calisher, C.H. and Shope, R.E. Arboviruses. In: Diagnostic procedures for viral, rickettsial and chlamydial infections (N.J. Schmidt and R.W. Emmons, Eds.), 6th ed. Washington, DC: American Public Health Association, 1989.

Beck, C.E. and Wyckoff, R.W.G. Venezuelan equine encephalomyelitis. Science 1938; 88:530.

Beckwith, W.H., Sirpenski, S., French, R.A., Nelson, R. and Mayo, D. Isolation of eastern equine encephalitis virus and West Nile virus from crows during increased arbovirus surveillance in Connecticut, 2000. Am. J. Trop. Med. Hyg. 2002; 66:422–426.

Bennett, A.M., Elvin, S.J., Wright, A.J., Jones, S.M. and Phillpotts, R.J. An immunological profile of Balb/c mice protected from airborne challenge following vaccination with a live attenuated Venezuelan equine encephalitis virus vaccine. Vaccine 2000; 19:337–347.

Bennett, A.M., Lescott, T. and Phillpotts, R.J. Improved protection against Venezuelan equine encephalitis by genetic engineering of a recombinant vaccinia virus. Viral. Immunol. 1998; 11:109–117.

Berge, T.O., Banks, I.S. and Tigertt, W.D. Attenuation of Venezuelan equine encephalomyelitis virus by in vitro cultivation in guinea pig heart cells. Am. J. Hyg. 1961; 73:209–218.

Bianchi, T.I., Aviles, G., Monath, T.P. and Sabattini, M.S. Western equine encephalomyelitis: virulence markers and their epidemiologic significance. Am. J. Trop. Med. Hyg. 1993; 49:322–328.

Bianchi, T.I., Aviles, G. and Sabattini, M.S. Biological characteristics of an enzootic subtype of western equine encephalomyelitis virus from Argentina. Acta Virol. 1997; 41:13–20.

Binder, G.K. and Griffin, D.E. Interferon-gamma-mediated site-specific clearance of alphavirus from CNS neurons. Science 2001; 293:303–306.

Binn, L.N., Sponseller, M.L., Wooding, W.L., Mcconnell, S.J., Spertzel, R.O. and Yager, E.H. Efficacy of an attenuated western encephalitis vaccine in equine animals. Am. J. Vet. Res. 1966; 27:1599–1604.

Bowen, G.S. and Calisher, C.H. Virological and serological studies of Venezuelan equine encephalomyelitis in humans. J. Clin. Microbiol. 1976; 4:22–27.

Bradish, C.J. and Titmuss, D. The effects of interferon and double-stranded RNA upon the virus-host interaction: studies with Togavirus strains in mice. J. Gen. Virol. 1981; 53:21–30.

Briceno Rossi, A.L. Rural epidemic encephalitis in Venezuela caused by a group A arbovirus (VEE). Progr. Med. Virol. 1967; 9:176–203.

Brown, A. and Officer, J.E. An attenuated variant of Eastern encephalitis virus: biological properties and protection induced in mice. Arch. Virol. 1975; 47:123–138.

Brown, A., Vosdingh, R. and Zebovitz, E. Attenuation and immunogenicity of ts mutants of Eastern encephalitis virus for mice. J. Gen. Virol. 1975; 27:111–116.

Burke, D.S., Ramsburg, H.H. and Edelman, R. Persistence in humans of antibody to subtypes of Venezuelan equine encephalomyelitis (VEE) virus after immunization with attenuated (TC-83) VEE virus vaccine. J. Infect. Dis. 1977; 136:354–359.

Calisher, C.H. and Maness, K.S. Laboratory studies of Venezuelan equine encephalitis virus in equines, Texas, 1971. J. Clin. Microbiol. 1975; 2:198–205.

Calisher, C.H., Shope, R.E., Brandt, W., Casals, J., Karabatsos, N., Murphy, F.A., Tesh, R.B. and Wiebe, M.E. Proposed antigenic classification of registered arboviruses I. Togaviridae, *Alphavirus*. Intervirology 1980; 14:229–232.

Casamassima, A.C., Hess, L.W. and Marty, A. TC-83 Venezuelan equine encephalitis vaccine exposure during pregnancy. Teratology 1987; 36:287–289.

Charles, P.C., Brown, K.W., Davis, N.L., Hart, M.K. and Johnston, R.E. Mucosal immunity induced by parenteral immunization with a live attenuated Venezuelan equine encephalitis virus vaccine candidate. Virology 1997; 228:153–160.

Charles, P.C., Trgovcich, J., Davis, N.L. and Johnston, R.E. Immunopathogenesis and immune modulation of Venezuelan equine encephalitis virus-induced disease in the mouse. Virology 2001; 284:190–202.

Charles, P.C., Walters, E., Margolis, F. and Johnston, R.E. Mechanism of neuroinvasion of Venezuelan equine encephalitis virus in the mouse. Virology 1995; 208:662–671.

Clark, G.G., Dein, F.J., Crabbs, C.L., Carpenter, J.W. and Watts, D.M. Antibody response of sandhill and whooping cranes to an eastern equine encephalitis virus vaccine. J. Wildl. Dis. 1987; 23:539–544.

Cole, F.E., Jr., May, S.W. and Eddy, G.A. Inactivated Venezuelan equine encephalomyelitis vaccine prepared from attenuated (TC-83 strain) virus. Appl. Microbiol. 1974; 27:150–153.

Cole, F.E., Jr. and McKinney, R.W. Cross-protection in hamsters immunized with group A arbovirus vaccines. Infect. Immun. 1971; 4:37–43.

Cole, F.E., Jr., Pedersen, C.E., Jr., Robinson, D.M. and Eddy, G.A. Improved method for production of attenuated Venezuelan equine encephalomyelitis (TC-83 strain) vaccine. J. Clin. Microbiol. 1976; 3:460–462.

Coppenhaver, D.H., Singh, I.P., Sarzotti, M., Levy, H.B. and Baron, S. Treatment of intracranial alphavirus infections in mice by a combination of specific antibodies and an interferon inducer. Am. J. Trop. Med. Hyg. 1995; 52:34–40.

Cupp, E.W., Tennessen, K.J., Oldland, W.K., Hassan, H.K., Hill, G.E., Katholi, C.R. and Unnasch, T.R. Mosquito and arbovirus activity during 1997–2002 in a wetland in northeastern Mississippi. J. Med. Entomol. 2004; 41:495–501.

Das, D., Gares, S.L., Nagata, L.P. and Suresh, M.R. Evaluation of a Western Equine Encephalitis recombinant E1 protein for protective immunity and diagnostics. Antiviral. Res. 2004; 64:85–92.

Davis, N.L., Brown, K.W., Greenwald, G.F., Zajac, A.J., Zacny, V.L., Smith, J.F. and Johnston, R.E. Attenuated mutants of Venezuelan equine encephalitis virus containing lethal mutations in the PE2 cleavage signal combined with a second-site suppressor mutation in E1. Virology 1995; 212:102–110.

Davis, N.L., Grieder, F.B., Smith, J.F., Greenwald, G.F., Valenski, M.L., Sellon, D.C., Charles, P.C. and Johnston, R.E. A molecular genetic approach to the study of Venezuelan equine encephalitis virus pathogenesis. Arch. Virol. Suppl. 1994; 9:99–109.

Davis, N.L., West, A., Reap, E., Macdonald, G., Collier, M., Dryga, S., Maughan, M., Connell, M., Walker, C., Mcgrath, K., Cecil, C., Ping, L.H., Frelinger, J., Olmsted, R., Keith, P., Swanstrom, R., Williamson, C., Johnson, P., Montefiori, D. and Johnston, R.E. Alphavirus replicon particles as candidate HIV vaccines. IUBMB Life 2002; 53:209–211.

Day, J.F. and Stark, L.M. Eastern equine encephalitis transmission to emus (*Dromaius novaehollandiae*) in Volusia County, Florida: 1992 through 1994. J. Am. Mosq. Control. Assoc. 1996; 12: 429–436.

De la Monte, S., Castro, F., Bonilla, N.J., Gaskin de Urdaneta, A. and Hutchins, G.M. The systemic pathology of Venezuelan equine encephalitis virus infection in humans. Am. J. Trop. Med. Hyg. 1985; 34:194–202.

Deresiewicz, R.L., Thaler, S.J., Hsu, L. and Zamani, A.A. Clinical and neuroradiographic manifestations of eastern equine encephalitis. N. Engl. J. Med. 1997; 336:1867–1874.

Edelman, R., Ascher, M.S., Oster, C.N., Ramsburg, H.H., Cole, F.E. and Eddy, G.A. Evaluation in humans of a new, inactivated vaccine for Venezuelan equine encephalitis virus (C-84). J. Infect. Dis. 1979; 140:708–715.

Ehrenkranz, N.J. and Ventura, A.K. Venezuelan equine encephalitis virus infection in man. Annu. Rev. Med. 1974; 25:9–14.

Elvin, S.J., Bennett, A.M. and Phillpotts, R.J. Role for mucosal immune responses and cell-mediated immune functions in protection from airborne challenge with Venezuelan equine encephalitis virus. J. Med. Virol. 2002; 67:384–393.

Farber, S., Hill, A., Connerly, M.L. and Dingle, J.H. Encephalitis in infants and children caused by a virus of the eastern variety of equine encephalitis. JAMA 1940; 114:1725–1731.

Femster, R. Outbreak of encephalitis in man due to the eastern virus of equine encephalomyelitis. N. Engl. J. Med. 1938; 28: 1403–1410.

Femster, R. Equine encephalitis in Massachusetts. N. Engl. J. Med. 1957; 257:701–704.

Ferguson, J.A., Reeves, W.C. and Hardy, J.L. Antibody studies in ponies vaccinated with Venezuelan equine encephalomyelitis (strain TC-83) and other alphavirus vaccines. Am. J. Vet. Res. 1977; 38:425–430.

Ferguson, J.A., Reeves, W.C., Milby, M.M. and Hardy, J.L. Study of homologous and heterologous antibody response in California horses vaccinated with attenuated Venezuelan equine encephalomyelitis vaccine (strain TC-83). Am. J. Vet. Res. 1978; 39:371–376.

Fillis, C.A. and Calisher, C.H. Neutralizing antibody responses of humans and mice to vaccination with Venezuelan encephalitis (TC-83) virus. J. Clin. Microbiol. 1979; 10:544–549.

Finter, N. Interferon as an antiviral agent in vivo: quantitative and temporal aspects of the production of mice against Semliki Forest virus. Br. J. Exp. Pathol. 1966; 47:361–371.

Franklin, R.P., Kinde, H., Jay, M.T., Kramer, L.D., Green, E.G., Chiles, R.E., Ostlund, E., Husted, S., Smith, J. and Parker, M.D. Eastern equine encephalomyelitis virus infection in a horse from California. Emerg. Infect. Dis. 2002; 8:283–288.

Frolov, I., Frolova, E. and Schlesinger, S. Sindbis virus replicons and Sindbis virus: assembly of chimeras and of particles deficient in virus RNA. J. Virol. 1997; 71:2819–2829.

Frolova, E.I., Fayzulin, R.Z., Cook, S.H., Griffin, D.E., Rice, C.M. and Frolov, I. Roles of nonstructural protein nsP2 and alpha/beta interferons in determining the outcome of Sindbis virus infection. J. Virol. 2002; 76:11254–11264.

Garcia-Tamayo, J., Carreno, G. and Esparza, J. Central nervous system alterations as sequelae of Venezuelan equine encephalitis virus infection in the rat. J. Pathol. 1979; 128:87–91.

Garen, P.D., Tsai, T.F. and Powers, J.M. Human eastern equine encephalitis: immunohistochemistry and ultrastructure. Mod. Pathol. 1999; 12:646–652.

Garmashova, N., Gorchakov, R., Frolova, E. and Frolov, I. Sindbis virus nonstructural protein nsP2 is cytotoxic and inhibits cellular transcription. J. Virol. 2006; 80:5686–5696.

Garmashova, N., Gorchakov, R., Volkova, E., Paessler, S., Frolova, E. and Frolov, I. The Old World and New World alphaviruses use different virus-specific proteins for induction of transcriptional shutoff. J. Virol. 2007; 81:2472–2484.

Getting, V. Equine encephalomyelitis in Massachusetts. N. Engl. J. Med. 1941; 224:999–1006.

Gibbons, D.L., Vaney, M.C., Roussel, A., Vigouroux, A., Reilly, B., Lepault, J., Kielian, M. and Rey, F.A. Conformational change and protein-protein interactions of the fusion protein of Semliki Forest virus. Nature 2004; 427:320–325.

Goldfield, M., Welsh, J.N. and Taylor, B.F. The 1959 outbreak of eastern encephalitis in New Jersey. 5. The inapparent infection: disease ratio. Am. J. Epidemiol. 1968; 87:32–33.

Gorelkin, L. Venezuelan equine encephalomyelitis in an adult animal host. An electron microscopic study. Am. J. Pathol. 1973; 73:425–442.

Gorelkin, L. and Jahrling, P.B. Pancreatic involvement by Venezuelan equine encephalomyelitis virus in the hamster. Am. J. Pathol. 1974; 75:349–362.

Gottdenker, N.L., Howerth, E.W. and Mead, D.G. Natural infection of a great egret (*Casmerodius albus*) with eastern equine encephalitis virus. J. Wildl. Dis. 2003; 39:702–706.

Greene, I.P., Paessler, S., Anishchenko, M., Smith, D.R., Brault, A.C., Frolov, I. and Weaver, S.C. Venezuelan equine encephalitis virus in the guinea pig model: evidence for epizootic virulence determinants outside the e2 envelope glycoprotein gene. Am. J. Trop. Med. Hyg. 2005a; 72:330–338.

Greene, I.P., Paessler, S., Austgen, L., Anishchenko, M., Brault, A.C., Bowen, R.A. and Weaver, S.C. Envelope glycoprotein mutations mediate equine amplification and virulence of epizootic Venezuelan equine encephalitis virus. J. Virol. 2005b; 79:9128–9133.

Greenway, T.E., Eldridge, J.H., Ludwig, G., Staas, J.K., Smith, J.F., Gilley, R.M. and Michalek, S.M. Enhancement of protective immune responses to Venezuelan equine encephalitis (VEE) virus with microencapsulated vaccine. Vaccine 1995; 13:1411–1420.

Greenway, T.E., Eldridge, J.H., Ludwig, G., Staas, J.K., Smith, J.F., Gilley, R.M. and Michalek, S.M. Induction of protective immune responses against Venezuelan equine encephalitis (VEE) virus aerosol challenge with microencapsulated VEE virus vaccine. Vaccine 1998; 16:1314–1323.

Grieder, F.B., Davis, B.K., Zhou, X.D., Chen, S.J., Finkelman, F.D. and Gause, W.C. Kinetics of cytokine expression and regulation of host protection following infection with molecularly cloned Venezuelan equine encephalitis virus. Virology 1997; 233:302–312.

Grieder, F.B., Davis, N.L., Aronson, J.F., Charles, P.C., Sellon, D.C., Suzuki, K. and Johnston, R.E. Specific restrictions in the progression of Venezuelan equine encephalitis virus-induced disease resulting from single amino acid changes in the glycoproteins. Virology 1995; 206:994–1006.

Grieder, F.B. and Vogel, S.N. Role of interferon and interferon regulatory factors in early protection against Venezuelan equine encephalitis virus infection. Virology 1999; 257:106–118.

Griffin, D., Levine, B., Tyor, W., Ubol, S. and Despres, P. The role of antibody in recovery from alphavirus encephalitis. Immunol. Rev. 1997; 159:155–161.

Griffin, D.E. Alphaviruses. In: Fields' Virology (D.M. Knipe and P.M. Howley, Eds.), 5th ed. New York: Lippincott, Williams and Wilkins, 2007.

Hahn, C.S., Lustig, S., Strauss, E.G. and Strauss, J.H. Western equine encephalitis virus is a recombinant virus. Proc. Natl. Acad. Sci. U.S.A. 1988; 85:5997–6001.

Hanson, R.P. An epizootic of equine encephalomyelitis that occurred in Massachusetts in 1831. Am. J. Trop. Med. Hyg. 1957; 6:858–862.

Hart, M.K., Caswell-Stephan, K., Bakken, R., Tammariello, R., Pratt, W., Davis, N., Johnston, R.E., Smith, J. and Steele, K. Improved mucosal protection against Venezuelan equine encephalitis virus is induced by the molecularly defined, live-attenuated V3526 vaccine candidate. Vaccine 2000; 18:3067–3075.

Hodgson, L.A., Ludwig, G.V. and Smith, J.F. Expression, processing, and immunogenicity of the structural proteins of Venezuelan equine encephalitis virus from recombinant baculovirus vectors. Vaccine 1999; 17:1151–1160.

Hughes, J.P. and Johnson, H.N. A field trial of a live-virus western encephalitis vaccine. J. Am. Vet. Med. Assoc. 1967; 150:167–171.

Hunt, A.R. and Roehrig, J.T. Biochemical and biological characteristics of epitopes on the E1 glycoprotein of western equine encephalitis virus. Virology 1985; 142:334–346.

Jackson, A.C., Sengupta, S.K. and Smith, J.F. Pathogenesis of Venezuelan equine encephalitis virus infection in mice and hamsters. Vet. Pathol. 1991; 28:410–418.

Jahrling, P.B., Navarro, E. and Scherer, W.F. Interferon induction and sensitivity as correlates to virulence of Venezuelan encephalitis viruses for hamsters. Arch. Virol. 1976; 51:23–35.

Jahrling, P.B. and Stephenson, E.H. Protective efficacies of live attenuated and formaldehyde-inactivated Venezuelan equine encephalitis virus vaccines against aerosol challenge in hamsters. J. Clin. Microbiol. 1984; 19:429–431.

Jan, J.T., Chatterjee, S. and Griffin, D.E. Sindbis virus entry into cells triggers apoptosis by activating sphingomyelinase, leading to the release of ceramide. J. Virol. 2000; 74:6425–6432.

Jan, J.T. and Griffin, D.E. Induction of apoptosis by Sindbis virus occurs at cell entry and does not require virus replication. J. Virol. 1999; 73:10296–10302.

Johnson, K.M. and Martin, D.H. Venezuelan equine encephalitis. Adv. Vet. Sci. Comp. Med. 1974; 18:79–116.

Johnson, K.M., Shelokov, A., Peralta, P.H., Dammin, G.J. and Young, N.A. Recovery of Venezuelan equine encephalomyelitis virus in Panama. A fatal case in man. Am. J. Trop. Med. Hyg. 1968; 17:432–440.

Jones, L.D., Bennett, A.M., Moss, S.R., Gould, E.A. and Phillpotts, R.J. Cytotoxic T-cell activity is not detectable in Venezuelan equine encephalitis virus-infected mice. Virus Res. 2003; 91:255–259.

Jordan, R.A., Wagner, J.A. and Mccrumb, F.R. Eastern equine encephalitis: report of a case with autopsy. Am. J. Trop. Med. Hyg. 1965; 14:470–474.

Julander, J.G., Siddharthan, V., Blatt, L.M., Schafer, K., Sidwell, R.W. and Morrey, J.D. Effect of exogenous interferon and an interferon inducer on western equine encephalitis virus disease in a hamster model. Virology 2007; 360:454–460.

Kinney, R.M., Chang, G.J., Tsuchiya, K.R., Sneider, J.M., Roehrig, J.T., Woodward, T.M. and Trent, D.W. Attenuation of Venezuelan equine encephalitis virus strain TC-83 is encoded by the 5′-noncoding region and the E2 envelope glycoprotein. J. Virol. 1993; 67:1269–1277.

Kinney, R.M., Esposito, J.J., Johnson, B.J., Roehrig, J.T., Mathews, J.H., Barrett, A.D. and Trent, D.W. Recombinant vaccinia/Venezuelan equine encephalitis (VEE) virus expresses VEE structural proteins. J. Gen. Virol. 1988a; 69:3005–3013.

Kinney, R.M., Esposito, J.J., Mathews, J.H., Johnson, B.J., Roehrig, J.T., Barrett, A.D. and Trent, D.W. Recombinant vaccinia virus/Venezuelan equine encephalitis (VEE) virus protects mice from peripheral VEE virus challenge. J. Virol. 1988b; 62:4697–4702.

Kinney, R.M., Tsuchiya, K.R., Sneider, J.M. and Trent, D.W. Molecular evidence for the origin of the widespread Venezuelan equine encephalitis epizootic of 1969 to 1972. J. Gen. Virol. 1992; 73:3301–3305.

Kubes, V. and Rios, F.A. The causitive agent of infectious equine encephalomyelitis in Venezuela. Science 1939; 90:20–21.

Kuhn, R.J., Griffin, D.E., Owen, K.E., Niesters, H.G. and Strauss, J.H. Chimeric Sindbis-Ross River viruses to study interactions between alphavirus nonstructural and structural regions. J. Virol. 1996; 70:7900–7909.

Kundin, W.D. Pathogenesis of Venezuelan equine encephalomyelitis virus. II. Infection in young adult mice. J. Immunol. 1966; 96:49–58.

Leon, C.A. Sequelae of Venezuelan equine encephalitis in humans: a four year follow-up. Int. J. Epidemiol. 1975; 4:131–140.

Lescar, J., Roussel, A., Wien, M.W., Navaza, J., Fuller, S.D., Wengler, G. and Rey, F.A. The Fusion glycoprotein shell of Semliki Forest virus: an icosahedral assembly primed for fusogenic activation at endosomal pH. Cell 2001; 105:137–148.

Liljestrom, P. and Garoff, H. A new generation of animal cell expression vectors based on the Semliki Forest virus replicon. Biotechnology (N.Y.) 1991; 9:1356–1361.

Liu, C., Voth, D.W., Rodina, P., Shauf, L.R. and Gonzalez, G. A comparative study of the pathogenesis of western equine and eastern equine encephalomyelitis viral infections in mice by intracerebral and subcutaneous inoculations. J. Infect. Dis. 1970; 122:53–63.

Locher, C.P., Paidhungat, M., Whalen, R.G. and Punnonen, J. DNA shuffling and screening strategies for improving vaccine efficacy. DNA Cell Biol. 2005; 24:256–263.

London, W.T., Levitt, N.H., Kent, S.G., Wong, V.G. and Sever, J.L. Congenital cerebral and ocular malformations induced in rhesus monkeys by Venezuelan equine encephalitis virus. Teratology 1977; 16:285–295.

Ludwig, G.V., Turell, M.J., Vogel, P., Kondig, J.P., Kell, W.K., Smith, J.F. and Pratt, W.D. Comparative neurovirulence of attenuated and non-attenuated strains of Venezuelan equine encephalitis virus in mice. Am. J. Trop. Med. Hyg. 2001; 64:49–55.

MacDonald, G.H. and Johnston, R.E. Role of dendritic cell targeting in Venezuelan equine encephalitis virus pathogenesis. J. Virol. 2000; 74:914–922.

McClain, D.J., Pittman, P.R., Ramsburg, H.H., Nelson, G.O., Rossi, C.A., Mangiafico, J.A., Schmaljohn, A.L. and Malinoski, F.J. Immunologic interference from sequential administration of live attenuated alphavirus vaccines. J. Infect. Dis. 1998; 177:634–641.

McKinney, R.W., Berge, T.O., Sawyer, W.D., Tigertt, W.D. and Crozier, D. Use of an attenuated strain of Venezuelan equine encephalomyelitis virus for immunization in man. Am. J. Trop. Med. Hyg. 1963; 12:597–603.

Meyer, K.F., Haring, C.M. and Howitt, B. The etiology of epizootic encephalomyelitis of horses in the San Joachin Valley, 1930. Science 1931; 74:227–228.

Monath, T.P., Kemp, G.E., Cropp, C.B. and Chandler, F.W. Necrotizing myocarditis in mice infected with Western equine encephalitis virus: clinical, electrocardiographic, and histopathologic correlations. J. Infect. Dis. 1978; 138:59–66.

Morris, C.D. Eastern equine encephalomyelitis. In: The Arboviruses: Epidemiology and Ecology (T.P. Monath, Ed.), Vol. III. Boca Raton, Florida: CRC Press, 1988.

Muller, U., Steinhoff, U., Reis, L.F., Hemmi, S., Pavlovic, J., Zinkernagel, R.M. and Aguet, M. Functional role of type I and type II interferons in antiviral defense. Science 1994; 264:1918–1921.

Nagata, L.P., Hu, W.G., Masri, S.A., Rayner, G.A., Schmaltz, F.L., Das, D., Wu, J., Long, M.C., Chan, C., Proll, D., Jager, S., Jebailey, L., Suresh, M.R. and Wong, J.P. Efficacy of DNA vaccination against western equine encephalitis virus infection. Vaccine 2005; 23:2280–2283.

Nagata, L.P., Hu, W.G., Parker, M., Chau, D., Rayner, G.A., Schmaltz, F.L. and Wong, J.P. Infectivity variation and genetic diversity among strains of Western equine encephalitis virus. J. Gen. Virol. 2006; 87:2353–2361.

Nathanson, N., Stolley, P.D. and Boolukos, P.J. Eastern equine encephalitis. Distribution of central nervous system lesions in man and Rhesus monkey. J. Comp. Pathol. 1969; 79:109–115.

Olsen, G.H., Turell, M.J. and Pagac, B.B. Efficacy of eastern equine encephalitis immunization in whooping cranes. J. Wildl. Dis. 1997; 33:312–315.

Paessler, S., Aguilar, P., Anishchenko, M., Wang, H.Q., Aronson, J., Campbell, G., Cararra, A.S. and Weaver, S.C. The hamster as an animal model for eastern equine encephalitis—and its use in studies of virus entrance into the brain. J. Infect. Dis. 2004; 189:2072–2076.

Paessler, S., Fayzulin, R.Z., Anishchenko, M., Greene, I.P., Weaver, S.C. and Frolov, I. Recombinant Sindbis/Venezuelan equine encephalitis virus is highly attenuated and immunogenic. J. Virol. 2003; 77:9278–9286.

Paessler, S., Ni, H., Petrakova, O., Fayzulin, R.Z., Yun, N., Anishchenko, M., Weaver, S.C. and Frolov, I. Replication and clearance of Venezuelan equine encephalitis virus from the brains of animals vaccinated with chimeric SIN/VEE viruses. J. Virol. 2006; 80:2784–2796.

Paessler, S., Yun, N.E., Judy, B.M., Dziuba, N., Zacks, M.A., Grund, A.H., Frolov, I., Campbell, G.A., Weaver, S.C. and Estes, D.M. Alpha-beta T cells provide protection against lethal encephalitis in the murine model of VEEV infection. Virology 2007; 367(2):307–323.

Pedersen, C.E., Jr. Preparation and testing of vaccines prepared from the envelopes of Venezuelan, eastern, and western equine encephalomyelitis viruses. J. Clin. Microbiol. 1976; 3:113–118.

Pedersen, C.E., Jr., Robinson, D.M. and Cole, F.E., Jr. Isolation of the vaccine strain of Venezuelan equine encephalomyelitis virus from mosquitoes in Louisiana. Am. J. Epidemiol. 1972; 95:490–496.

Pereboev, A.V., Razumov, I.A., Svyatchenko, V.A. and Loktev, V.B. Glycoproteins E2 of the Venezuelan and eastern equine encephalomyelitis viruses contain multiple cross-reactive epitopes. Arch. Virol. 1996; 141:2191–2205.

Phillpotts, R.J. Immunity to airborne challenge with Venezuelan equine encephalitis virus develops rapidly after immunization with the attenuated vaccine strain TC-83. Vaccine 1999; 17:2429–2435.

Phillpotts, R.J., Jones, L.D. and Howard, S.C. Monoclonal antibody protects mice against infection and disease when given either before or up to 24h after airborne challenge with virulent Venezuelan equine encephalitis virus. Vaccine 2002; 20:1497–1504.

Phillpotts, R.J., Lescott, T.L. and Jacobs, S.C. Vaccinia virus recombinants encoding the truncated structural gene region of Venezuelan equine encephalitis virus (VEEV) give solid protection against peripheral challenge but only partial protection against airborne challenge with virulent VEEV. Acta Virol. 2000; 44:233–239.

Phillpotts, R.J., O'Brien, L., Appleton, R.E., Carr, S. and Bennett, A. Intranasal immunisation with defective adenovirus serotype 5 expressing the Venezuelan equine encephalitis virus E2 glycoprotein protects against airborne challenge with virulent virus. Vaccine 2005; 23:1615–1623.

Phillpotts, R.J. and Wright, A.J. TC-83 vaccine protects against airborne or subcutaneous challenge with heterologous mouse-virulent strains of Venezuelan equine encephalitis virus. Vaccine 1999; 17:982–988.

Pittman, P.R., Makuch, R.S., Mangiafico, J.A., Cannon, T.L., Gibbs, P.H. and Peters, C.J. Long-term duration of detectable neutralizing antibodies after administration of live-attenuated VEE vaccine and following booster vaccination with inactivated VEE vaccine. Vaccine 1996; 14:337–343.

Pletnev, S.V., Zhang, W., Mukhopadhyay, S., Fisher, B.R., Hernandez, R., Brown, D.T., Baker, T.S., Rossmann, M.G. and Kuhn, R.J. Locations of carbohydrate sites on alphavirus glycoproteins show that E1 forms an icosahedral scaffold. Cell 2001; 105:127–136.

Polo, J.M., Belli, B.A., Driver, D.A., Frolov, I., Sherrill, S., Hariharan, M.J., Townsend, K., Perri, S., Mento, S.J., Jolly, D.J., Chang, S.M., Schlesinger, S. and Dubensky, T.W., Jr. Stable alphavirus packaging cell lines for Sindbis virus and Semliki Forest virus-derived vectors. Proc. Natl. Acad. Sci. U.S.A. 1999; 96:4598–4603.

Powers, A.M., Brault, A.C., Kinney, R.M. and Weaver, S.C. The use of chimeric Venezuelan equine encephalitis viruses as an approach for the molecular identification of natural virulence determinants. J. Virol. 2000; 74:4258–4263.

Pratt, W.D., Davis, N.L., Johnston, R.E. and Smith, J.F. Genetically engineered, live attenuated vaccines for Venezuelan equine encephalitis: testing in animal models. Vaccine 2003; 21:3854–3862.

Pratt, W.D., Gibbs, P., Pitt, M.L. and Schmaljohn, A.L. Use of telemetry to assess vaccine-induced protection against parenteral and aerosol infections of Venezuelan equine encephalitis virus in non-human primates. Vaccine 1998; 16:1056–1064.

Pushko, P., Bray, M., Ludwig, G.V., Parker, M., Schmaljohn, A., Sanchez, A., Jahrling, P.B. and Smith, J.F. Recombinant RNA replicons derived from attenuated Venezuelan equine encephalitis virus protect guinea pigs and mice from Ebola hemorrhagic fever virus. Vaccine 2000; 19:142–153.

Pushko, P., Parker, M., Ludwig, G.V., Davis, N.L., Johnston, R.E. and Smith, J.F. Replicon-helper systems from attenuated Venezuelan equine encephalitis virus: expression of heterologous genes in vitro and immunization against heterologous pathogens in vivo. Virology 1997; 239:389–401.

Randall, R., Maurer, F.D. and Smadel, J.E. Immunization of laboratory workers with purified Venezuelan equine encephalomyelitis vaccine. J. Immunol. 1949; 63:313–318.

Rao, V., Hinz, M.E., Roberts, B.A. and Fine, D. Environmental hazard assessment of Venezuelan equine encephalitis virus vaccine candidate strain V3526. Vaccine 2004; 22:2667–2673.

Rayfield, E.J., Gorelkin, L., Curnow, R.T. and Jahrling, P.B. Virus-induced pancreatic disease by Venezuelan encephalitis virus. Alterations in glucose tolerance and insulin release. Diabetes 1976; 25:623–631.

Reed, D.S., Lackemeyer, M.G., Garza, N.L., Norris, S., Gamble, S., Sullivan, L.J., Lind, C.M. and Raymond, J.L. Severe encephalitis in cynomolgus macaques exposed to aerosolized eastern equine encephalitis virus. J. Infect. Dis. 2007; 196:441–450.

Reed, D.S., Larsen, T., Sullivan, L.J., Lind, C.M., Lackemeyer, M.G., Pratt, W.D. and Parker, M.D. Aerosol exposure to western equine encephalitis virus causes fever and encephalitis in cynomolgus macaques. J. Infect. Dis. 2005a; 192:1173–1182.

Reed, D.S., Lind, C.M., Lackemeyer, M.G., Sullivan, L.J., Pratt, W.D. and Parker, M.D. Genetically engineered, live, attenuated vaccines protect nonhuman primates against aerosol challenge with a virulent IE strain of Venezuelan equine encephalitis virus. Vaccine 2005b; 23:3139–3147.

Reed, D.S., Lind, C.M., Sullivan, L.J., Pratt, W.D. and Parker, M.D. Aerosol infection of cynomolgus macaques with enzootic strains of venezuelan equine encephalitis viruses. J. Infect. Dis. 2004; 189:1013–1017.

Reisen, W.K. Western equine encephalitis. In: The Encyclopedia of Arthropod-transmitted Infections (M.W. Service, Ed.). Wallingford, U.K.: CAB International, 2001.

Reisen, W.K. and Monath, T.P. Western equine encephalomyelitis. In: The Arboviruses: Epidemiology and Ecology (T.P. Monath, Ed.), Vol. V. Boca Raton, Florida: CRC Press, 1988.

Riemenschneider, J., Garrison, A., Geisbert, J., Jahrling, P., Hevey, M., Negley, D., Schmaljohn, A., Lee, J., Hart, M.K., Vanderzanden, L., Custer, D., Bray, M., Ruff, A., Ivins, B., Bassett, A., Rossi, C. and Schmaljohn, C. Comparison of individual and combination DNA vaccines for B. anthracis, Ebola virus, Marburg virus and Venezuelan equine encephalitis virus. Vaccine 2003; 21:4071–4080.

Rivas, F., Diaz, L.A., Cardenas, V.M., Daza, E., Bruzon, L., Alcala, A., De la Hoz, O., Caceres, F.M., Aristizabal, G., Martinez, J.W., Revelo, D., De la Hoz, F., Boshell, J., Camacho, T., Calderon, L., Olano, V.A., Villarreal, L.I., Roselli, D., Alvarez, G., Ludwig, G. and Tsai, T. Epidemic Venezuelan equine encephalitis in La Guajira, Colombia, 1995. J. Infect. Dis. 1997; 175:828–832.

Roehrig, J.T. Immunogens of encephalitis viruses. Vet. Microbiol. 1993; 37:273–284.

Roehrig, J.T. and Bolin, R.A. Monoclonal antibodies capable of distinguishing epizootic from enzootic varieties of Subtype I Venezuelan equine encephalitis viruses in a rapid indirect immunofluorescence assay. J. Clin. Microbiol. 1997; 35:1887–1890.

Roehrig, J.T., Hunt, A.R., Chang, G.J., Sheik, B., Bolin, R.A., Tsai, T.F. and Trent, D.W. Identification of monoclonal antibodies capable of differentiating antigenic varieties of eastern equine encephalitis viruses. Am. J. Trop. Med. Hyg. 1990; 42:394–398.

Rowell, J.F. and Griffin, D.E. Contribution of T cells to mortality in neurovirulent Sindbis virus encephalomyelitis. J. Neuroimmunol. 2002; 127:106–114.

Rozdilsky, B., Robertson, H.E. and Chorney, J. Western encephalitis: report of eight fatal cases. Saskatchewan epidemic, 1965. Can. Med. Assoc. J. 1968; 98:79–86.

Ryzhikov, A.B., Ryabchikova, E.I., Sergeev, A.N. and Tkacheva, N.V. Spread of Venezuelan equine encephalitis virus in mice olfactory tract. Arch. Virol. 1995; 140:2243–2254.

Ryzhikov, A.B., Tkacheva, N.V., Sergeev, A.N. and Ryabchikova, E.I. Venezuelan equine encephalitis virus propagation in the olfactory tract of normal and immunized mice. Biomed. Sci. 1991; 2:607–614.

Sammin, D.J., Butler, D., Atkins, G.J. and Sheahan, B.J. Cell death mechanisms in the olfactory bulb of rats infected intranasally with Semliki forest virus. Neuropathol. Appl. Neurobiol. 1999; 25:236–243.

Schoepp, R.J., Smith, J.F. and Parker, M.D. Recombinant chimeric western and eastern equine encephalitis viruses as potential vaccine candidates. Virology 2002; 302:299–309.

Schoneboom, B.A., Fultz, M.J., Miller, T.H., Mckinney, L.C. and Grieder, F.B. Astrocytes as targets for Venezuelan equine encephalitis virus infection. J. Neurovirol. 1999; 5:342–354.

Schoneboom, B.A., Lee, J.S. and Grieder, F.B. Early expression of IFN-alpha/beta and iNOS in the brains of Venezuelan equine encephalitis virus-infected mice. J. Interferon Cytokine Res. 2000; 20:205–215.

Sidwell, R.W. and Smee, D.F. Viruses of the Bunya- and Togaviridae families: potential as bioterrorism agents and means of control. Antiviral. Res. 2003; 57:101–111.

Smith, D.R., Aguilar, P.V., Coffey, L.L., Gromowski, G.D., Wang, E. and Weaver, S.C. Venezuelan equine encephalitis virus transmission and effect on pathogenesis. Emerg. Infect. Dis. 2006; 12:1190–1196.

Snoeyenbos, G.H., Weinack, O.M. and Rosenau, B.J. Immunization of pheasants for eastern encephalitis. Avian Dis. 1978; 22:386–390.

Spotts, D.R., Reich, R.M., Kalkhan, M.A., Kinney, R.M. and Roehrig, J.T. Resistance to alpha/beta interferons correlates with the epizootic and virulence potential of Venezuelan equine encephalitis viruses and is determined by the 5′ noncoding region and glycoproteins. J. Virol. 1998; 72:10286–10291.

Strizki, J.M. and Repik, P.M. Differential reactivity of immune sera from human vaccinees with field strains of eastern equine encephalitis virus. Am. J. Trop. Med. Hyg. 1995; 53:564–570.

Sutton, L.S. and Brooke, C.C. Venezuelan equine encephalomyelitis due to vaccination in man. J. Am. Med. Assoc. 1954; 155:1473–1476.

Tate, C.M., Howerth, E.W., Stallknecht, D.E., Allison, A.B., Fischer, J.R. and Mead, D.G. Eastern equine encephalitis in a free-ranging white-tailed deer (*Odocoileus virginianus*). J. Wildl. Dis. 2005; 41:241–245.

Ten Broeck, C. and Merrill, M. A serological difference between eastern and western equine encephalomyelitis virus. Proc. Soc. Exp. Biol. Med. 1933; 31:217–220.

Tengelsen, L.A., Bowen, R.A., Royals, M.A., Campbell, G.L., Komar, N. and Craven, R.B. Response to and efficacy of vaccination against eastern equine encephalomyelitis virus in emus. J. Am. Vet. Med. Assoc. 2001; 218:1469–1473.

The Subcommittee on Arbovirus Laboratory Safety of the American Committee on Arthropod-Borne Viruses Laboratory safety for arboviruses and certain other viruses of vertebrates. Am. J. Trop. Med. Hyg. 1980; 29:1359–1381.

Thompson, J.M., Whitmore, A.C., Konopka, J.L., Collier, M.L., Richmond, E.M., Davis, N.L., Staats, H.F. and Johnston, R.E. Mucosal and systemic adjuvant activity of alphavirus replicon particles. Proc. Natl. Acad. Sci. U.S.A. 2006; 103:3722–3727.

Tsai, T.F., Weaver, S.C. and Monath, T.P. Alphaviruses. In: Clinical Virology (D.D. Richman, R.J. Whitley and F.G. Hayden, Eds.). Washington, DC: ASM Press, 2002.

Turell, M.J., Ludwig, G.V., Kondig, J. and Smith, J.F. Limited potential for mosquito transmission of genetically engineered, live-attenuated Venezuelan equine encephalitis virus vaccine candidates. Am. J. Trop. Med. Hyg. 1999; 60:1041–1044.

Turell, M.J. and Spielman, A. Nonvascular delivery of Rift Valley fever virus by infected mosquitoes. Am. J. Trop. Med. Hyg. 1992; 47:190–194.

U.S. Department of Health and Human Services Biosafety in Microbiological and Biomedical Laboratories, 4th ed. Washington, DC: U.S. Government Printing Office, 1999.

Vanderwagen, L.C., Pearson, J.L., Franti, C.E., Tamm, E.L., Riemann, H.P. and Behymer, D.E. A field study of persistence of antibodies in California horses vaccinated against western, eastern, and Venezuelan equine encephalomyelitis. Am. J. Vet. Res. 1975; 36:1567–1571.

Villari, P., Spielman, A., Komar, N., Mcdowell, M. and Timperi, R.J. The economic burden imposed by a residual case of eastern encephalitis. Am. J. Trop. Med. Hyg. 1995; 52:8–13.

Vogel, P., Kell, W.M., Fritz, D.L., Parker, M.D. and Schoepp, R.J. Early events in the pathogenesis of eastern equine encephalitis virus in mice. Am. J. Pathol. 2005; 166:159–171.

Walder, R., Jahrling, P.B. and Eddy, G.A. Differentiation markers of eastern equine encephalitis (EEE) viruses and virulence. In: Arboviruses in the Mediterranean Countries (J. Vesenjak-Hirjan, Ed.). Stuttgart: Gustav Fischer Verlag, 1980.

Waldridge, B.M., Wenzel, J.G., Ellis, A.C., Rowe-Morton, S.E., Bridges, E.R., D'Andrea, G. and Wint, R. Serologic responses to eastern and western equine encephalomyelitis vaccination in previously vaccinated horses. Vet. Ther. 2003; 4:242–248.

Walton, T.E., Brautigam, F.E., Ferrer, J.A. and Johnson, K.M. Epizootic Venezuelan equine encephalomyelitis in Central America. Disease pattern and vaccine evaluation in Nicaragua, 1969–1970. Am. J. Epidemiol. 1972; 95:247–254.

Walton, T.E. and Grayson, M.A. Venezuelan equine encephalomyelitis. In: The Arboviruses: Epidemiology and Ecology (T.P. Monath, Ed.), Vol. IV. Boca Raton, FL: CRC Press, 1988.

Walton, T.E. and Johnson, K.M. Persistence of neutralizing antibody in Equidae vaccinated with Venezuelan equine encephalomyelitis vaccine strain TC-83. J. Am. Vet. Med. Assoc. 1972; 161:916–918.

Wang, E., Barrera, R., Boshell, J., Ferro, C., Freier, J.E., Navarro, J.C., Salas, R., Vasquez, C. and Weaver, S.C. Genetic and phenotypic changes accompanying the emergence of epizootic subtype IC Venezuelan equine encephalitis viruses from an enzootic subtype ID progenitor. J. Virol. 1999; 73:4266–4271.

Wang, E., Petrakova, O., Adams, A.P., Aguilar, P.V., Kang, W., Paessler, S., Volk, S.M., Frolov, I. and Weaver, S.C. Chimeric Sindbis/eastern equine encephalitis vaccine candidates are highly attenuated and immunogenic in mice. Vaccine 2007; 25(43):7573–7581.

Weaver, S.C. Eastern equine encephalitis. In: The Encyclopedia of Arthropod-transmitted Infections (M.W. Service, Ed.). Wallingford, U.K.: CAB International, 2001.

Weaver, S.C., Ferro, C., Barrera, R., Boshell, J. and Navarro, J.C. Venezuelan equine encephalitis. Annu. Rev. Entomol. 2004; 49:141–174.

Weaver, S.C., Frey, T.K., Huang, H.V., Kinney, R.M., Rice, C.M., Roehrig, J.T., Shope, R.E. and Strauss, E.G. Togaviridae. In: Virus Taxonomy, VIIIth Report of the ICTV (C.M. Fauquet, M.A. Mayo, J. Maniloff, U. Desselberger and L.A. Ball, Eds.). London: Elsevier/Academic Press, 2005.

Weaver, S.C., Kang, W., Shirako, Y., Rumenapf, T., Strauss, E.G. and Strauss, J.H. Recombinational history and molecular evolution of western equine encephalomyelitis complex alphaviruses. J. Virol. 1997; 71:613–623.

Weaver, S.C., Pfeffer, M., Marriott, K., Kang, W. and Kinney, R.M. Genetic evidence for the origins of Venezuelan equine encephalitis virus subtype IAB outbreaks. Am. J. Trop. Med. Hyg. 1999; 60:441–448.

Worthington, M. and Baron, S. Late therapy with an interferon stimulator in an arbovirus encephalitis in mice. Proc. Soc. Exp. Biol. Med. 1971; 136:323–327.

Worthington, M., Levy, H. and Rice, J. Late therapy of an arbovirus encephalitis in mice with interferon and interferon stimulators. Proc. Soc. Exp. Biol. Med. 1973; 143:638–643.

Zhou, X., Berglund, P., Rhodes, G., Parker, S.E., Jondal, M. and Liljestrom, P. Self-replicating Semliki Forest virus RNA as recombinant vaccine. Vaccine 1994; 12:1510–1514.

CHAPTER

22

Foot-and-Mouth Disease

Peter W. Mason[1,2] *and Marvin J. Grubman*[3]

[1]Department of Pathology, University of Texas Medical Branch, Galveston, TX, USA

[2]Sealy Center for Vaccine Development, University of Texas Medical Branch, Galveston, TX, USA

[3]U.S. Department of Agriculture, Agricultural Research Service, Plum Island Animal Disease Center, Greenport, NY, USA

OUTLINE

ABSTRACT

Foot-and-mouth disease (FMD) is a devastating disease of livestock that has had a significant impact on world economies and public health. Its importance to mankind is confirmed by the fact that FMD virus (FMDV) was the first animal virus discovered, and that FMD was among the first animal diseases for which vaccines were developed.

Due to its rapid spread and the high cost of eradication campaigns, countries that are free of the disease are susceptible to huge economic losses if the disease were to be introduced either accidentally or deliberately in an act of agroterrorism. Vaccines developed in the 1950s and 1960s have helped to control this scourge in many nations, and these products have been further developed and refined over the last 30 years. Although these vaccines are very useful as part of eradication campaigns in countries where FMDV is enzootic, it is not feasible to use these products in prophylactic vaccination of entire regions. In addition, these vaccines are not ideally suited to control outbreaks in disease-free countries. Thus, for the portions of the world that are currently FMD free, there is a need for new vaccines that can be applied in the areas surrounding an outbreak to rapidly dampen the spread of infection. Although the market for this type of vaccine is small, there are several groups trying to satisfy this important need. This chapter will detail the history of FMD vaccine development emphasizing recent advances and prospects for future development.

INTRODUCTION

The first reports of foot-and-mouth disease (FMD) date from 1514, when a livestock disease compatible with the signs of FMD was described. These signs include the formation of fluid-filled vesicles on the mouth, feet, nares, and teats of the affected animals, producing severe discomfort but rarely death. Although this gruesome disease is limited to cloven-hoofed animals, it has severely affected mankind through its impact on animal agriculture. In areas of the world where food and draft animals are essential for subsistence agriculture, FMD can affect nutrition. In countries that have highly developed animal industry and free trade, outbreaks cause economic devastation. This latter aspect of the disease has raised concern that the etiologic agent of the disease, FMD virus (FMDV), could be used as a bioterrorist weapon.

Due to its economic importance, high virulence, and utility as a model for studying virus evolution and speciation, FMD has been extensively examined and is the subject of many reviews (Brown, 2003; Doel, 2003; Domingo et al., 2003; Grubman and Baxt, 2004; Jackson et al., 2003; Knowles and Samuel, 2003; Mason et al., 2003; Sutmoller et al., 2003; Thomson et al., 2003). The purpose of this chapter is to briefly recount the development of vaccines that can be used to effectively reduce the impact of FMD in enzootic countries and to describe the development of second-generation vaccines with enhanced safety and/or efficacy that may be particularly well suited to use in the face of an outbreak as "emergency vaccines."

Brief History of FMD

FMDV was responsible for thousands of outbreaks of the disease in Europe at the end of the 1800s, prompting the German government to provide funding to Frederick Loeffler to identify the etiologic agent of the disease and develop methods to control it. Loeffler applied state-of-the-art ceramic filter technology that had been used to separate disease-causing bacteria from their toxins to demonstrate that the infectious agent or "instigator" in the lymph from FMD vesicles was smaller than the smallest known bacteria. Since he reasoned that the disease-causing agent could have been a toxin, he undertook serial animal-to-animal passages with diluted filtered preparations to demonstrate that the "instigator" was infectious. The results of these studies, which were published in 1898 (Loeffler, 1898; Loeffler and Frosch, 1898), heralded the start of animal virology. A summary of Loeffler's achievements with FMD was published several years later (Loeffler, 1909). This document recounted his studies demonstrating that FMD induces a protective immunity to reinfection, and that convalescent serum could be used as a "highly beneficial prophylactic." In the same work, Loeffler reported how convalescent serum could be successfully administered in the field and indicated that administration of serum followed by the "instigator" present in the lymph of diseased animals could induce active immunity to FMD. In this report, Loeffler emphasized the value of developing both serum and active immunity to control FMD; however, his death a few years later prevented his pursuit of more effective products to control FMD.

During the first few decades of the 1900s, studies on FMD helped to further characterize the nature of the agent and the immune response it elicited, demonstrating among other things that there were multiple serotypes of the virus, providing a preview for one of the largest obstacles to controlling the disease, namely, the need for multivalent vaccines.

In the first third of the 20th century, the United States experienced multiple outbreaks of FMD. Most of these outbreaks appeared to result from direct importation of virus-contaminated products. In the case of the 1908 outbreak, retrospective analyses ascribed the mode of introduction as an FMDV-contaminated seed stock of vaccinia virus that was imported from Japan for production of the human smallpox vaccine (Mohler and Rosenau, 1909). The outbreak in 1914, the most severely experienced in U.S. history, affected 22 states and the District of Columbia. Control of this outbreak, which lasted for 20 months, included the destruction of over 170,000 animals (Mohler, 1924). The last outbreak in the United States, which occurred in 1929, was ascribed to importation of infected garbage offloaded from ships and fed to pigs, in clear violation of the U.S. Department of Agriculture (USDA) regulations (Mohler, 1929). Ironically, this mechanism of transmission is still repeating itself 75 years later, as evidenced by a similar episode that introduced the Pan-Asian strain of FMDV into the port of Durban, South Africa, in 2000.

The United States has remained FMD free since 1929. This situation can be attributed to a number of factors, including the extensive efforts by the United States, from 1946 to 1954, in assisting Mexico to eliminate FMD by a combination of slaughter and ring vaccination and the absence of land routes connecting the United States to FMD-affected areas (the United States helps to maintain an FMD vaccination-free "sentinel" zone along the northern border of Colombia, permitting rapid detection of FMD that could threaten disease-free Central America). The USDA also maintains strict import restrictions on biological products (including foods, medicines, and animal-derived research products) and live animals, thus helping to reduce the likelihood of accidental introduction of FMD. Recently, the Department of Homeland Security has initiated programs to help prevent deliberate introduction of FMD into the United States.

Economic Impact and International Agencies that Control FMD

In the early 20th century, the devastating nature of outbreaks of animal diseases imported into Europe resulted in the formation of an international agency, the Office International des Épizooties (OIE). Although founded originally to control rinderpest, which had been introduced into Europe by trade with Africa, FMD soon became one of the most important diseases that the OIE tracked and controlled. Today, the OIE receives voluntary reports of the activity of FMD and several other animal diseases and has the ability to declare countries "free" of FMD, following a rigorous investigation of a country's surveillance data and infrastructure. The FMD "disease-free" status, which is used by trading partners to control importation of live animals, can add considerable value to a country's animal products. As a result, countries that cannot prove they are free of FMD have a severe economic disadvantage, and when outbreaks occur in countries that wish to quickly regain their disease-free status, national agricultural authorities may implement draconian slaughter policies to eliminate the disease. In this context, the fact that FMD vaccination can actually mask the circulation of live FMDV can provide an economic disincentive for vaccination (see below). Figure 22.1, which shows the

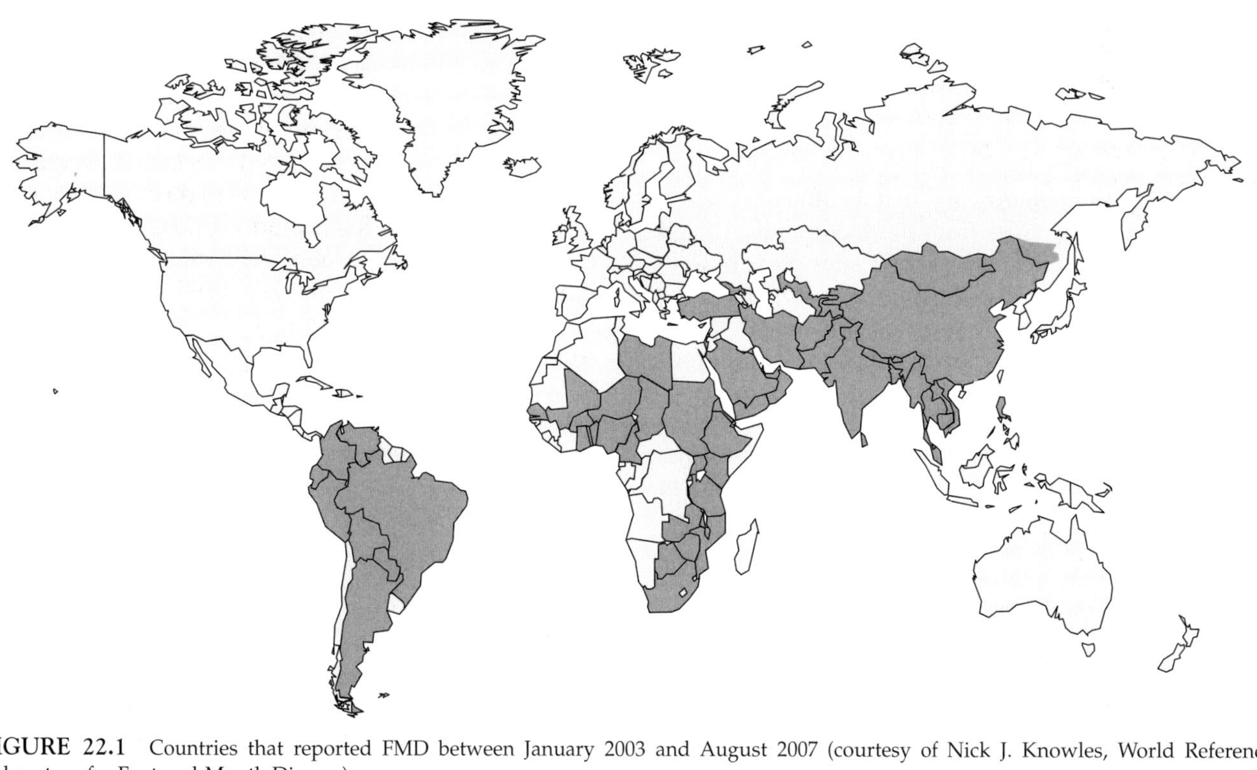

FIGURE 22.1 Countries that reported FMD between January 2003 and August 2007 (courtesy of Nick J. Knowles, World Reference Laboratory for Foot-and-Mouth Disease).

current distribution of FMD in the world, demonstrates the presence of the disease in many developing countries and absence from large industrial countries that pay the highest prices for animal products.

Due to these economic factors, as well as its potential to directly affect animal productivity (and hence human nutrition), the Food and Agriculture Organization (FAO) of the United Nations (UN), the World Health Organization (WHO), and the Pan American Health Organization (PAHO) all maintain programs for tracking and controlling FMD.

DESCRIPTION OF THE DISEASE

Disease Presentation and Pathogenesis

The signs of FMD are variable, depending on both host and virus strains. Susceptible hosts include essentially all cloven-hoofed animals, but the majority of research efforts have been focused on economically important livestock species: cattle, pigs, and sheep. Among these three species, disease signs may vary substantially, and altered pathogenic properties have been associated with individual isolates. In susceptible hosts infected with virulent isolates, FMD is characterized by fever, vesicular lesions, and erosions of epithelium of the mouth, tongue, nares, muzzle, feet, and teats of affected animals (Grubman and Baxt, 2004). Signs of disease can appear within 24 h of contact with the agent, but the time of onset of the disease and the severity of infection depend on the infectious dose, the individual isolate, the host species, and the individual animal's age and health status. Recovery of large amounts of virus from vesicular lesions indicates that FMDV replicates to high titers in oral and pedal epithelial tissues, but recovery of the virus from the blood and exhaled breath indicates replication in other sites contribute to pathogenesis and spread.

FMDV can spread by many different methods; direct inoculation by multiple routes can reproduce the disease, and contact transmission among animals occurs quite readily. The virus can also persist on surfaces in animal containment facilities and can be transferred by workers or equipment. These methods of spread are of major importance in outbreak propagation, but airborne spread is perhaps the most dreaded mechanism since it is the least likely to be controlled by sanitary measures, even those that involve destruction of affected herds. Once FMDV enters a susceptible animal, it rapidly reproduces, permitting shedding to surrounding animals by highly efficient contact transmission. Early studies on the development of FMD lesions in a guinea pig model for FMD demonstrated that following establishment of foci of infected cells in the *stratum spinosum* layer of the epithelium, lesions progressed, filling with vesicular fluid and ballooning cells that are a hallmark of the infection (Galloway and Nicolau, 1928). Twenty years later, an "explant" tissue culture system was used to demonstrate a similar progression of infection in fragments of bovine lingual epithelium (Frenkel, 1949).

Carrier State

Almost a century ago, Loeffler predicted that animals that had recovered from FMD could become chronic carriers of the agent of the disease, and that the existence of such carriers could explain some outbreaks of FMD (Loeffler, 1909). However, it was not until the late 1950s that a chronic infection defined as the "carrier state" was definitively demonstrated (van Bekkum et al., 1959). In this work, van Bekkum and coworkers demonstrated that infectious FMDV could be recovered from the oral/pharyngeal tract of convalescent animals, using an apparatus (a cup probang) designed for the recovery of tuberculosis bacilli. Subsequent studies on the carrier state demonstrated that up to 50% of convalescent bovines become carriers (as defined by recovery of live virus 28 days or later postinfection), and that these animals can carry the virus for many years. It has also been demonstrated that vaccinated animals exposed to live virus can become carriers (Sutmoller et al., 2003). In addition, Cape buffalo (an important wildlife reservoir of FMDV in sub-Saharan Africa) appear to become carriers for life, and sheep but not pigs also become carriers (Sutmoller et al., 2003).

To date, there has been no documented experimental transmission of FMDV from carrier cattle to naive livestock animals in the laboratory; however, laboratory transmission has been observed from Cape buffalo to domestic cattle (Thomson et al., 2003). Further, molecular epidemiological analyses of an outbreak in northern Zimbabwe in 1991 strongly suggest a role for carrier buffalo in disease transmission to livestock (Thomson et al., 2003). Despite the paucity of data on a role for carriers in sustaining or generating outbreaks, the fact that carriers can be readily generated among apparently healthy animals that have been vaccinated and then deliberately infected (Sutmoller et al., 2003) has led to important rules for certification of "FMD disease-free status" by the OIE. Specifically, the possibility that vaccinated animals could become subclinically infected and subsequently become carriers has contributed to the development of OIE guidelines that permit a country to be declared free of FMD only after a year without evidence of infection if animals vaccinated to control the outbreak are not destroyed. On the other hand, OIE guidelines for disease-free status require only 3 months without evidence of infection if vaccination is *not* employed. These guidelines were responsible, in part, for the fact that all vaccinated animals were destroyed in the aftermath of the 2001 outbreaks in the Netherlands (Bouma et al., 2003). After the 2001 FMD outbreak in Europe, the 1-year period required in countries that have vaccinated was reduced to 6 months to help reduce pressure on countries to implement widespread slaughter of apparently healthy vaccinated animals.

MOLECULAR BIOLOGY OF FMDV

Genome Structure and Polyprotein Synthesis

FMDV is the type member of the genus *Aphthovirus* of the family Picornaviridae and is an antigenically variable virus consisting of seven serotypes: A, O, C, South African Territories (SATs) 1–3, and Asia-1. Although the FMDV genome shares many features with other picornavirus genomes, it contains several features that are either unique to FMDV, its genus, or in other cases only present in a subset of picornavirus genomes. Like all picornaviruses, FMDV has a positive-sense RNA genome that is covalently linked via a 5′-terminal uridine to a small protein, 3B (also known as VPg). The genome encodes a single long polyprotein that is co- and posttranslationally processed into a large number of mature products by viral-encoded proteinases. Figure 22.2 shows a schematic diagram of the genome with the "mature" protein products identified; however, it is likely that several partially cleaved precursor products serve essential functions in viral replication (Mason et al., 2003).

The FMDV genome begins with a long 5′-untranslated region (5′UTR) that has a highly conserved secondary structure containing several sub-features. At the 5′-end of the genomic RNA is a 370-base domain originally identified by enzymatic digestion with nucleases that cleaved this short portion of the genome from the downstream poly(C) tract (Mason et al., 2003; Fig. 22.2). This highly structured RNA, known as the S-fragment, can be folded into a long stem-loop structure (Fig. 22.2). The poly(C) tract is followed by a series of pseudoknots of unknown function. Naturally occurring strains can have three or four pseudoknots, but viruses with as few as two pseudoknots have also been identified (Mason et al., 2003). The pseudoknots are followed by a short stem loop, the *cre* (*cis*-acting replicative element), that is an essential RNA replication element (Mason et al., 2003). The *cre* is followed by a complex RNA structure that was defined almost two

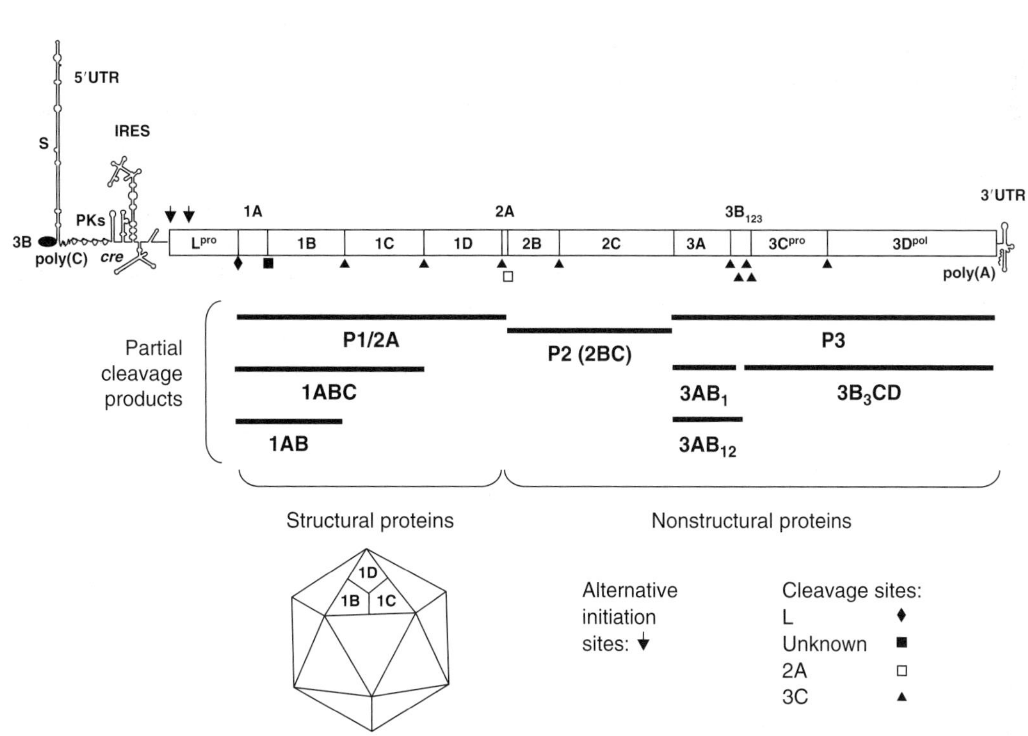

FIGURE 22.2 Schematic diagram of the FMDV genome showing the position of the genetic elements. Open boxes indicate protein-encoding regions; lines indicate RNA structures; UTR = untranslated region; S-fragment = short fragment of the genome; poly(C) = poly cytosine tract; PK = pseudoknot; *cre* = *cis*-acting replicative element; IRES = internal ribosome entry site. Reprinted with permission from Mason et al. (2003).

decades ago as the internal ribosome entry site (IRES) (Mason et al., 2003). The IRES directs host cell ribosomes to the initiation codons at the start of the viral polyprotein-coding region, which in the case of FMDV are found over 1200 bases from the 5′-end of the viral genome. Translation initiation from an IRES, differs from translation initiation of the majority of cellular mRNAs, which initiate protein synthesis following ribosome binding to the 7-methyl-guanosine (MeG) cap found at the 5′-end of the mRNAs. IRES function is critical to the pathobiology of FMDV and other picornaviruses since the utilization of an alternative translation initiation system permits the virus to shut off translation of critical host cell factors, allowing the virus to outrun its host (Mason et al., 2003).

Unlike most picornaviruses, the aphthovirus polyprotein does not initiate with its structural proteins; rather, it initiates with the leader proteinase (L^{pro}). L^{pro} cleaves the polyprotein between its C-terminus and the N-terminus of the capsid precursor protein P1-2A. This cleavage releases the growing polypeptide chain with a free N-terminal glycine on protein 1A, which is subsequently myristylated by cellular enzymes. In addition to its role in maturation of viral proteins, L^{pro} cleaves cellular proteins—including eIF-4G, an essential protein in recruitment of RNAs to the ribosome—and has been identified as an important virulence factor (Mason et al., 2003).

The P1-2A capsid precursor is cleaved at its C-terminus by 2A, a short peptide that autocatalytically removes itself from the P2 polyprotein and remains associated with P1 until subsequent processing (Mason et al., 2003). The P2 and P3 precursor polypeptides code for a number of viral nonstructural (NS) proteins, including the 3C proteinase, $3C^{pro}$, and the RNA-dependent RNA polymerase, $3D^{pol}$. $3C^{pro}$ is responsible for most of the cleavages of the viral polyprotein, including the processing of P1-2A into the capsid proteins 1AB, 1C, and 1D (Mason et al., 2003). The processing of capsid protein 1AB occurs by an unknown mechanism, is autocatalytic, and is required for generation of infectious virus (Mason et al., 2003).

Virion Structure

The FMDV virion is composed of 60 copies of each of four different proteins encapsidating the 3B-linked genome. Of these four proteins, three—1B, 1C, and 1D—are externally exposed while the fourth, 1A, is entirely interior in the virion. X-ray crystallographic analyses have revealed that like other picornaviruses, the FMDV surface proteins have a conserved eight-stranded β-barrel structure (Acharya et al., 1989). Further, two of these proteins, 1C and 1D, contain internal N-terminal extensions that appear to be missing on 1B. However, protein 1A appears to serve as the N-terminal extension of 1B during folding, consistent with the fact that 1AB is present at the early stages of virion morphogenesis (Jackson et al., 2003). Although much of the impetus for determining the structure of the FMDV capsid came from the pursuit of better vaccines, the first crystallography studies on a type O virus (Acharya et al., 1989) failed to reveal an ordered structure for the loop between the G and H β-barrels of protein 1D, which had been shown by a variety of methods to contain strongly neutralizing epitopes (Brown, 1992). Interestingly, this loop is also disordered in the structures for type C (Lea et al., 1994) and type A (Fry et al., 2005) viruses but was present on forms of a type O virus crystallized under reducing conditions (Logan et al., 1993). These data are consistent with a large number of epidemiology-, peptide-, and epitope-mapping studies indicating that this portion of the capsid has genetic and structural flexibility that appears to contribute to the ability of the virus to rapidly undergo antigenic variation (Domingo et al., 2003), complicating vaccine-based control of FMD (see Antigenic Variation).

Antigenic Variation

Early work on FMDV discovered the existence of multiple serotypes by demonstration that convalescent animals could not be reinfected by the same isolate but were susceptible to diseases caused by isolates from a different location (Vallée and Carré, 1922). These studies were repeated in a second laboratory (Waldmann and Trautwein, 1926) that used similar techniques to show that there were three different serotypes of virus circulating in Europe in the early 1900s (originally designated A, B, and C, but later renamed A, O, and C to be consistent with the A and O strains identified by Vallée and Carré). The ramifications of these finding are very significant. Despite the ability of animals to be resistant to reinfection with the same serotype, they can be infected with a different serotype, so naive animals can suffer multiple episodes of FMD. Thus, countries with multiple serotypes of FMDV need multivalent vaccines, and countries that want vaccines to protect their animals from all strains of FMDV need to vaccinate with a heptavalent vaccine.

An additional problem with vaccination coverage is that FMDV can evolve rapidly in the field, and as a result, vaccines for individual serotypes, particularly type A, can rapidly become out-of-date, as field strains evolve (Doel, 2003). A portion of the reason for this rapid evolution appears to be the flexibility of the G–H loop of the capsid structure (see Virion Structure), which has also appeared to limit the utility of peptide-based vaccines (see Protein and Synthetic Peptide Vaccines) targeted to this region of the virion surface (Domingo et al., 2003).

The ability of FMDV to undergo extensive genetic variation in cell culture as well as in the field has made it a model for the study of viral quasispecies and evolution (Domingo et al., 2003). In addition, this flexibility, most easily recognized by sequencing the 1D portion of the genome, has revolutionized the study of FMD outbreaks, allowing readily comparable sequence data generated from multiple laboratories to be utilized in place of serology to identify serotypes and subtypes, and even track viral changes during individual outbreaks (Knowles and Samuel, 2003).

DEVELOPMENT OF TRADITIONAL VACCINES FOR FMD

Animal-Derived Vaccines

In the 1930s, as technology progressed in the development of vaccines for both animal and human diseases, the methods utilized to formalin-inactivate toxins were applied to material derived from FMDV-infected animals. This procedure involved the growth of FMDV in the tongues of live animals, followed by slaughter 2 days later, and the collection of the diseased epithelium. The epithelial material containing large quantities of viral antigen was then inactivated with formaldehyde-containing solutions and then formulated into a vaccine using an aluminum hydroxide adjuvant. Using this method, only 200 doses of vaccine could be made from each animal (Sutmoller et al., 2003), thus requiring a large number of animals for commercial vaccine production. To obtain this large number of animals, these vaccines were produced in animals housed near slaughterhouses, and for economic reasons, the meat from the infected animals was then sold in the market. Given our knowledge of how easily FMD outbreaks can be propagated, these combination vaccine plant and slaughterhouse facilities were likely to have generated their own vaccine business since they could have been a source for outbreaks in the surrounding community. Despite the somewhat barbaric process of intentionally causing disease in food animals, the vaccines were efficacious. Nevertheless, the need to infect live animals, the large volume of material needed, and the presence of a significant amount of bovine antigens in the preparation left considerable room for further development.

Cell Culture-Derived Inactivated Vaccines

The development of Frenkel's tongue epithelium method to cultivate FMDV produced a method of vaccine generation that was safer and more economically effective than vaccines derived from infected live animals. The basis of the vaccine was an "organ culture" produced by sterile removal of the epithelium of the tongue from freshly slaughtered cows, followed by mincing and suspension in a nutrient broth in large fermenters (Frenkel, 1951). Virus harvests from these fermenters were formaldehyde-inactivated using methods similar to those applied to the homogenates of diseased epithelium prepared directly from infected cattle (see Animal-Derived Vaccines). These vaccines were among the first to be used extensively in Europe, and their application in the Netherlands coincided with a dramatic drop in the number of FMD outbreaks (Brown, 1992).

The next generation of vaccines for FMDV followed the rapid acceleration of eukaryotic cell cultivation technology developed for basic research and human vaccine production in the 1950s and early 1960s. Work at the USDA's Plum Island laboratory demonstrated that primary cultures of embryonic bovine kidney cells were an effective substrate for virus growth and vaccine preparation (Bachrach et al., 1955). Later work with the baby hamster kidney (BHK) cell line demonstrated that this cell line could be used to grow virus (Mowat and Chapman, 1962), and BHK cells have become the "workhorse" for FMDV antigen production system being used at large manufacturing plants throughout the world to produce essentially all of the antigens now used to prepare FMD vaccines. In some cases, vaccines are prepared using suspended cultures of BHK cells, but there is clear evidence that growth of cells in suspension can result in selection of viruses with altered antigenic properties (Bolwell et al., 1989). More recent studies indicate that tissue culture propagation can select viruses with altered receptor utilization, antigenic structure, and virulence (Sa-Carvalho et al., 1997), probably because selection of viruses in cell culture to permit utilization of different receptors can result in changes in the virion structure that affect both tropism in vivo and antigenic structure. Antigenic variants selected in vitro can be as important to FMD control as the evolution of variants in the field since efforts to produce vaccine strains with improved growth properties in cell culture can select variants with altered cell-binding properties and antigenic structures yielding vaccine antigens with inferior immunogenicity.

A major advance in the production of safe and effective FMD vaccines came from the realization that formaldehyde-inactivated virus harvests could contain residual infectivity, and thus FMD vaccines could, in some cases, produce outbreaks. Fears that FMD vaccines themselves could cause outbreaks have contributed to the "curious" resistance to FMD vaccination that is perceived by many practitioners of human health as an inability of the animal health community to understand basic vaccination strategies.

The problem with formaldehyde inactivation is not limited to FMD vaccines. In the case of poliovirus, formalin inactivation was essential to the production of the first inactivated polio vaccine (IPV). However, soon after IPV went into widespread use, the investigations of the "Cutter incident" demonstrated that certain lots of formaldehyde-inactivated IPV made by Cutter Laboratories were linked to cases of poliomyelitis, and that live virus could be recovered from these lots of IPV (Nathanson and Langmuir, 1963). These findings attributed the infection of vaccinees to the inability of formaldehyde to penetrate aggregates of poliovirus in the viral harvests used as the source of antigen for IPV. This problem in vaccine manufacture was eventually remedied by introduction of a filtration step, whereas in the case of FMD vaccines, an alternative inactivant replaced formaldehyde (see below). In the case of polio, the relatively low infectivity of the virus (less than 1% of those infected were expected to develop symptoms of the disease) made the incidence of disease due to "incomplete" inactivation rare. In the case of FMDV, which is much more virulent, the incompletely inactivated virus in vaccine preparations is much more likely to produce disease (and outbreaks). However, during early stages of FMD vaccination campaigns, when outbreaks caused by "naturally" circulating strains were quite high, the relatively small numbers of outbreaks that could have been caused by incompletely inactivated vaccines would have been difficult to detect. However, the application of molecular epidemiology to outbreaks in the late stages of the FMD eradication campaign in Europe revealed that many outbreaks could be linked to formaldehyde-inactivated vaccines prepared from subtypes of FMDV that were not currently circulating (Beck and Strohmaier, 1987). This seminal work helped to prompt the switch to a different class of chemical inactivants, the imines. Imines were introduced as inactivants in the late 1950s in experimental products (Brown and Crick, 1959), and unlike aldehyde-based inactivants, imines display linear inactivation kinetics, allowing for greater certainty in obtaining a completely inactivated product (Bahnemann, 1990). Today, all FMD vaccines produced in European and South American manufacturing facilities utilize imine inactivation and, in most cases, employ partial purification of the viral antigen to remove BHK cell antigens. The resulting viral antigens are then formulated into vaccines by the addition of aluminum hydroxide/saponin or a mineral oil emulsion (Doel, 2003).

Live-Attenuated Vaccines (LAVs)

Following adaptation of FMD to grow in suckling mice (Skinner, 1951), it was soon appreciated that mouse-passaged viruses were attenuated in some animal species. This prompted efforts to develop these adapted viruses as more reliable (and potentially less expensive) vaccines to replace the animal-tissue-derived inactivated products available at the time (see Cell Culture-Derived Inactivated Vaccines). Thus, live-attenuated products were developed in the United Kingdom, but large-scale testing in South Africa revealed their incomplete attenuation, resulting in disease in the vaccinated population, and only provided protection for about 70% of the vaccinated animals (Martin and Edwards, 1965). At the same time, mouse adaptation was pursued in the Soviet Union as a method to produce LAVs for FMD, and the technology was transferred to the People's Republic of China, which utilized mouse-adapted viruses that were amplified in rabbits (to produce larger volumes of inocula) until quite recently (J. Callis and Q. Zhao, personal communication). Simultaneously with these animal-based attenuation and propagation methods, researchers in South America pursued adaptation of FMDV to chicken embryos to attenuate the virus. These vaccines also suffered from incomplete attenuation, and as vaccine plants in the Europe and South America developed better methods to grow a viral antigen for inactivated vaccines (notably the adoption of the BHK cell culture system) and reliably inactivated it, live-attenuated products were largely dropped.

Emergency Vaccines

Although FMD outbreaks in disease-free nations are devastating, continuous vaccination with the existing products is not cost effective. This is due to multiple factors, including the high cost of vaccination for a rapidly turned over livestock population, and the need to vaccinate against all circulating strains of virus. Furthermore, if the intent is to protect against deliberate introduction of FMD, the breadth of vaccine coverage would need to be extended to historical strains that are currently not in circulation. There is also a concern that vaccination might facilitate undetected spread of infection since animals vaccinated with existing vaccines can become infected (Orsel et al., 2007). These and other factors, many of which are based on economic concerns that vaccination confounds ready detection of a disease, preventing rapid return of countries suffering outbreaks to trade in animal products, have led many countries, including the United Kingdom, to never implement vaccination in response to an outbreak. In the case of the United States, the last FMD outbreak occurred before the availability of useful vaccines, so the precise criteria that will be applied in the face of an outbreak remains unclear. Nevertheless, North America, Europe, and Japan maintain vaccine banks that could be utilized to slow outbreaks. Due to the above-mentioned concerns that vaccines might produce a population of animals that could become asymptomatic carriers or shedders, the major utility of these banked vaccines might be to slow spread of the infection, and all vaccinated animals might be destroyed as part of outbreak control. This strategy was utilized in the Netherlands in 2001, where the "vaccinate and destroy" policy was implemented, in part, to allow for more orderly destruction of all animals in the zones surrounding the confirmed outbreaks. Interestingly, there were infections detected on some farms shortly after vaccines were applied, suggesting that vaccines that induced immunity more quickly could have utility in outbreak situations.

These latter findings reinforce the need for "outbreak" vaccines that provide a rapid immunity, a criterion that is less important in a countrywide eradication campaign or prophylactic vaccination of a border region with a country that has enzootic FMD. Nevertheless, the horrific outbreak in the United Kingdom in 2001 has caused several researchers to focus on the rapid acquisition of immunity for a small, but potentially profitable, vaccine bank market where high-priced vaccines might be justified to protect critical genetic stock or rapidly induce immunity in the ring around an outbreak. Studies to explore the ability of traditional vaccines to provide the short-term immunity needed for ring vaccination have noted that swine (which are potent excreters of virus in outbreaks) cannot mount a protective immune response within the first week of vaccination (Doel et al., 1994). However, more recent studies, prompted by the catastrophic depopulation of animals in the United Kingdom in 2001, have demonstrated in cattle that vaccination can reduce shedding and disease within 4 days of administration (Cox et al., 2005; Golde et al., 2005).

SECOND-GENERATION VACCINES FOR FMD

Although the traditional cell culture-derived inactivated vaccine is effective, there are a number of concerns with its use, particularly during outbreaks in disease-free countries. These include difficulty in differentiating vaccinated animals from the infected ones, the need to induce rapid protection in the event of an outbreak in a disease-free country, the inability of inoculation with this vaccine to block the development of the carrier state, the need for expensive high-containment facilities for vaccine production, and so forth. Despite the excellent safety record of these high-containment facilities, there was an outbreak of FMD in the United Kingdom in August 2007 on farms near the Pirbright high-containment laboratory caused by a strain of virus present in both an FMD vaccine-manufacturing facility and the Institute of Animal Health located at this site. This occurrence may necessitate further scrutiny of a manufacturing process that requires the growth of thousands of liters of infectious FMDV that contains millions of cattle-infectious units per milliliter (Doel, 2003). Moreover, in terms of manufacture of FMD vaccines in the United States, federal law only allows work with infectious FMDV at the Plum Island laboratory. Therefore, the United States is completely dependent on foreign manufacturers for vaccines, a situation that has received additional attention following the September 11 terrorist attacks. Over the past 25–30 years, investigators have attempted to address these concerns by using new technologies to develop alternative vaccines.

Genetically Engineered Live-Attenuated or Inactivated Vaccines

As described above, in the 1950s and the 1960s, attempts were made to develop live-attenuated FMD vaccines by classical procedures. Although this approach was successfully used by Sabin to develop an oral poliovirus vaccine, it was not successful for FMDV (see Live-Attenuated Vaccines). With the advent of recombinant DNA techniques and the construction of full-length infectious FMDV cDNA clones (Rieder et al., 1993; Zibert et al., 1990), efforts were made to design genetically engineered infectious clones that either had specific mutations in selected regions of the genome (McKenna et al., 1995) or deletion of an entire protein-coding region (Piccone et al., 1995).

Mason and coworkers engineered an infectious clone in which a portion of the G–H loop of 1D required for virus binding to susceptible cells was deleted and virus particles were produced that were unable to bind to cells (McKenna et al., 1995). This virus did not cause disease in either suckling mice or swine and cattle inoculated with deleted virus in an oil emulsion developed an FMDV-specific neutralizing antibody response and were protected from clinical disease when challenged with virulent homologous virus (McKenna et al., 1995). These workers also engineered an infectious clone that contained a serotype

A virus in which the G–H loop of 1D was substituted with the homologous region from serotype O or C (Rieder et al., 1994). Chemically inactivated vaccines prepared from the chimeric viruses induced antibodies in guinea pigs that neutralized both type A and either type O or type C viruses. Swine inoculated with the inactivated A/C chimera were protected from challenge with type A virus and partially protected against challenge with type C.

In an alternative approach, Piccone et al. (1995) constructed a full-length clone from a laboratory strain of type A12 virus that lacked the complete coding region of an NS protein, L^{pro}. The advantage of this approach as compared to the classical method of attenuation is that the risk of reversion to virulence is significantly reduced. Virus derived from this clone, leaderless virus, was infectious in BHK-21 cells demonstrating that L^{pro} was not essential for virus growth in this cell culture system. Subsequent studies demonstrated that leaderless virus was attenuated in both swine and cattle (Chinsangaram et al., 1998b; Mason et al., 1997). Based on the above observations, leaderless virus was examined as an LAV candidate (Chinsangaram et al., 1998b; Mason et al., 1997). Both swine and cattle inoculated with this virus developed a significant neutralizing antibody response but did not develop clinical disease. Challenge of vaccinated cattle, by both direct inoculation in the tongue and by contact with a naive infected animal in the same room, and challenge of vaccinated swine, by exposure to an infected animal in the same room, resulted in less severe and delayed clinical disease as compared to naive animals, but the vaccinated animals were not completely protected. These results indicate the potential for the rational design of LAVs by deleting a particular coding region but demonstrate the difficulty in designing these viruses, so that they do not cause clinical disease and yet can replicate sufficiently to induce a protective immune response.

Subsequently, a leaderless virus was constructed with a capsid-coding region from a field strain of serotype O in the genetic background of strain A12 (Almeida et al., 1998). This virus was avirulent in cattle but caused a mild disease in swine and was transmitted to a naive animal in the same room. Thus, although L^{pro} is a major virulence factor, it is clear that there are multiple regions of the FMDV genome involved in pathogenesis, and a more complete understanding of these factors is necessary in order to develop a useful LAV.

Nevertheless, attenuated FMDVs including leaderless virus and genetically engineered viruses lacking or with altered cell-binding sites could be used as a source of antigen for traditional inactivated vaccines (Chinsangaram et al., 1998b; McKenna et al., 1995). The use of these attenuated strains of viruses rather than highly virulent field strains as a source of antigen for vaccine production could reduce the risks associated with virus escape or incomplete virus inactivation. Moreover, chimeric technology in which capsid-coding regions of difficult-to-cultivate viruses (such as some SAT types) are inserted into other backbones may provide alternative methods for the production of seed viruses for traditional inactivated vaccine manufacture (van Rensburg et al., 2004).

Protein and Synthetic Peptide Vaccines

Early analyses of the FMDV capsid proteins by enzymatic treatment identified a protein on the virus surface, 1D, which was sensitive to trypsin and absorbed neutralizing antibody [see Mason et al. (2003) for a review]. Inoculation of this protein into animals induced both a neutralizing antibody response and protected swine when subsequently challenged with virus [see Grubman and Baxt (2004) for a review]. Soon after these discoveries, Bachrach and coworkers used recombinant DNA techniques to create an *Escherichia coli*-expressed 1D vaccine candidate, and demonstrated that swine and cattle given two inoculations of this purified recombinant protein developed high levels of neutralizing antibody and protection against challenge (Kleid et al., 1981). Subsequent work demonstrated that fragments of 1D, particularly the C-terminal half of the molecule, could elicit a neutralizing antibody response (Strohmaier et al., 1982). Based on the sequence of the 1D protein from a number of FMDV serotypes, Brown and collaborators identified a region of the protein that was highly variable (amino acids 141–160) and, using chemically synthesized peptides, demonstrated that this region could induce neutralizing antibodies and protect guinea pigs against virus challenge (Bittle et al., 1982). Subsequent studies by a number of laboratories have used chemically synthesized 1D peptides in combination with peptides containing T cell epitopes (to enhance the humoral immune response) to produce high-titer antibodies to these 1D peptides (DiMarchi et al., 1986; Francis et al., 1991; Nargi et al., 1999; Pfaff et al., 1982). Other groups have used viral vectors, naked DNA plasmids, transgenic plants, or plants infected with recombinant viruses to express 1D-coding regions (particularly the 141–160 region) to produce vaccine candidates with the ability to generate strong anti-FMDV titers (Kit et al., 1991; Kitson et al., 1991; Wigdorovitz et al., 1999a, 1999b; Wong et al., 2000, 2002). While in some cases these immunogens induced a significant neutralizing antibody response, they either did not induce protection in livestock species or required multiple inoculations to induce protection (Grubman and Baxt, 2004). All of these approaches present a limited number of viral epitopes that represent only continuous regions on the viral capsid to the inoculated animal. However, some of the antigenic sites on the FMDV capsid are composed of discontinuous epitopes corresponding to regions from more than one structural protein or from different loops on the same protein (Mason et al., 2003). Thus, with the known variability and quasispecies nature of FMDV (Domingo et al., 2003), vaccination of animals with only a subset of epitopes representing a small number of the continuous epitopes present on the virus could result in the selection of antigenic variants that could cause outbreaks if these animals were exposed to an infectious virus. This supposition was supported by the results of a large-scale synthetic peptide vaccination study that revealed limited protection against challenge, with disease in the unprotected cattle arising from antigenic variants with alterations in the regions corresponding to the immunizing peptide (Taboga et al., 1997). Thus, it appears that development of an efficacious FMD peptide vaccine will require: (1) new technology to construct peptides that present discontinuous epitopes and (2) inclusion of peptides that represent multiple epitopes of the virus.

Empty Capsid Vaccines

An alternative approach to the use of immunogens containing only a limited subset of viral capsid epitopes is the construction of an immunogen that contains the entire repertoire of immunogenic sites present on intact FMDV. Empty viral capsids, virus particles lacking nucleic acid, are naturally produced in FMDV-infected cell cultures, are antigenically similar to infectious virus, and are as immunogenic as virions in animals (Rowlands, 1974; Rweyemamu et al., 1979). Utilizing recombinant DNA technology, FMDV constructs that contain the portions of the viral genome required for empty capsid synthesis, processing, and assembly have been produced (Belsham et al., 1991; Lewis et al., 1991; Roosien et al., 1990). These constructs include the coding region for the viral structural protein precursor (P1-2A) and $3C^{pro}$ (the NS protein required for capsid precursor processing and assembly) but lack the coding regions for most of the other NS proteins (see Fig. 22.2). Furthermore, this immunogen lacks FMDV nucleic acid and cannot synthesize infectious viral nucleic acid when inoculated into animals. Thus, this type of vaccine is not infectious, does not require high-containment facilities for production, and can be produced in the United States. In addition, vaccines derived from such products could be easily used as "marker" vaccines since companion diagnostic assays could be designed to detect antibodies against FMDV NS proteins not present in the vaccine; thus vaccinated animals can be distinguished from infected animals.

A number of approaches to deliver this type of empty capsid immunogen to animals have been examined; these include inoculation of proteins expressed in *E. coli* (Grubman et al., 1993; Lewis et al., 1991) or recombinant baculovirus-infected cells (Grubman et al., 1993; Lewis et al., 1991; Saiz et al., 1994), direct inoculation of DNA encoding the immunogen (Beard et al., 1999; Benvenisti et al., 2001; Cedillo-Barron et al., 2001; Chinsangaram et al., 1998a), or direct inoculation of recombinant viral vectors encoding the immunogen (Abrams et al., 1995; Berinstein et al., 2000; Mayr et al., 1999, 2001; Moraes et al., 2002; Sanz-Parra et al., 1999a, 1999b). Thus far, the most effective delivery system is a recombinant, replication-defective human adenovirus type 5 (Ad5) vector (Mayr et al., 1999, 2001; Moraes et al., 2002). This vector is derived from the replication-competent form of Ad5 by deletion of a portion of the genome; the resulting replication-defective virus has the capacity to incorporate 5–8 kbp of foreign DNA and is only able to grow in cell cultures that provide the missing Ad5 proteins (Graham et al., 1977). This vector has a number of positive features including infectivity in multiple animal species, including swine and bovine (ensuring uptake and expression of foreign genes in these species) (Prevec et al., 1989), a lack of pathogenicity in animals, and a predicted high level of safety in humans (since replication-competent viruses were used extensively to immunize U.S. military recruits) (Couch et al., 1963; Top et al., 1971). In addition, since the empty capsids are produced in the inoculated animal, it can, like FMDV infection, induce both humoral and cell-mediated immune responses.

Despite their inability to productively replicate in animals, immunization of a number of animal species with replication-defective Ad5 vectors containing transgenes from other pathogens has resulted in induction of an immune response against the transgene and protection from challenge (Moraes et al., 2002; Pacheco et al., 2005; Russell, 2000). Furthermore, in the case of FMDV, the faithful replication of Ad5 and transgene DNA sidesteps the problem of selection of antigenic variants during cell passage of FMDV for traditional vaccine production (see Antigenic Variation and Cell Culture-Derived Inactivated Vaccines).

An Ad5-FMDV vector encoding the capsid and 3C^{pro} regions of a laboratory strain of FMDV, type A12, induced an FMDV-specific neutralizing antibody response in swine, and after two inoculations, five of six animals challenged by contact were protected from clinical disease (Mayr et al., 1999, 2001). The animal that was not protected had significantly less severe disease than the animals in the control group. It has also been shown that processing of the P1-2A capsid precursor protein was essential for induction of a neutralizing antibody response and protection since Ad5 vectors either containing an inactive 3C^{pro} (Mayr et al., 2001) or lacking the 3C^{pro}-coding region (Sanz-Parra et al., 1999b) did not protect animals from challenge. In subsequent studies, an Ad5 vector was constructed that contained the P1-2A-coding region from an FMDV field strain, A24 Cruzeiro, from South America (Moraes et al., 2002). In these studies, it was demonstrated that a single inoculation of a 10-fold higher dose than in the earlier study induced a neutralizing antibody response. The titers of antibodies obtained were low but detectable at 1 week postvaccination and increased by the second week. When these swine were challenged by direct inoculation at 7, 14, or 42 days postvaccination, all vaccinated animals were protected from clinical disease, and no virus was detected in samples of blood or nasal swabs. Furthermore, there was no evidence of challenge virus replication in the vaccinated animals since none of these animals developed antibodies against the viral NS proteins as assayed by radioimmunoprecipitation.

The Ad5-A24 vaccine was also tested in cattle since these are the most economically important animals susceptible to FMDV (Pacheco et al., 2005). Cattle were administered a single high dose of vaccine and co-housed with animals given the same dose of a control Ad5 vector. Seven days later, all animals were challenged by intradermal inoculation on the tongue. The control animals developed fever, viremia for 3 days, and lesions on all four feet. None of the Ad5-A24-vaccinated animals developed fever or viremia, and only one animal had a single lesion on the dental pad. However, this latter animal showed no evidence of systemic spread of the virus. Sera from these animals were tested post-challenge, and all five animals showed serological evidence of limited challenge virus replication.

In a preliminary study in cattle, animals were inoculated with two high doses of Ad5-A24 9 weeks apart and developed FMDV-specific neutralizing antibody titers of greater than 1000, but no antibodies against viral NS proteins were detected (Grubman, 2005). These results clearly demonstrate that animals inoculated with two doses of this subunit vaccine can be differentiated from infected animals indicating the utility of this immunogen as a "marker" vaccine.

Cytokine-Enhanced Vaccines

To aid in the control of outbreaks of FMD in disease-free regions, it is essential to have vaccines that rapidly protect animals from infection, thus limiting virus spread (see Emergency Vaccines). In these regions, both traditional vaccines and genetically derived vaccine candidates, such at the Ad5-vectored empty capsids, suffer from their inability to induce protective immunity in less than 7 days. The economically devastating FMD outbreaks in Taiwan in 1997, and in the United Kingdom and the Netherlands in 2001, stimulated interest in developing protective measures to rapidly block or limit virus replication during outbreaks in disease-free countries and thereby contain disease spread.

The use of cytokines to modulate responses against immunization with candidate FMD vaccines based on DNA and recombinant virus vectors, as well as traditional vaccines, has been investigated by several groups. Interleukin 1 (IL-1) and IL-2 proteins have been shown to enhance the humoral immune response to inactivated FMD vaccines in a mouse model (McCullough et al., 1992). Cedillo-Barron et al. (2001) demonstrated that the inclusion of the gene for porcine granulocyte-macrophage colony-stimulating factor (pGM-CSF) with a DNA empty capsid vaccine resulted in a significant increase in antibody levels against FMDV and some improvement in protection from FMD challenge. However, in these studies, three DNA inoculations were given prior to challenge. Interestingly, Caron et al. (2005) found that co-delivery of an Ad5 vector encoding pGM-CSF and an Ad5-FMDV empty capsid was less effective than an empty capsid construct alone. However, studies with other immunogens indicate that the adjuvant effect of GM-CSF is most pronounced when this cytokine is co-expressed in the same vector (Wang et al., 2002).

Type I interferon (IFNα/β) is a cytokine that provides one of the first lines of host defense against virus infection (Biron and Sen, 2001) and can rapidly protect cell culture against infection by all FMDV serotypes (Chinsangaram et al., 2003; Grubman and Baxt, 2004). IFN has been used clinically to reduce or block virus replication, most notably in the treatment of patients with hepatitis C or B (Grubman, 2003). However, the half-life of IFN in the blood is approximately 5h, and thus clinical use requires multiple, high-dose inoculations that can have adverse systemic effects (Grubman,

2003). Thus, Chinsangaram et al. (2003) designed experiments to deliver the IFN gene directly to animals by using a replication-defective Ad5 vector designed to produce IFN endogenously, with the hope that this expression system could deliver the compound in a manner that would overcome rapid clearance. In these experiments, swine were inoculated with a single dose of Ad5-porcine IFNα (pIFNα) (or a control Ad5 vector) and challenged a day later by direct inoculation with virulent FMDV A24. The control Ad5-inoculated animals developed typical clinical signs of disease as well as viremia and antibody response to viral NS proteins. All animals given a low dose of Ad5-pIFNα developed clinical disease, but the disease was delayed and less severe as compared to the control group. All animals given a high dose of Ad5-pIFNα were completely protected from clinical disease, had no viremia, and did not develop antibodies to FMDV NS proteins (Chinsangaram et al., 2003). There was a direct correlation between protection and the level of pIFNα in the blood of these animals. These studies also showed that Ad5-pIFNα produced detectable IFN within 4 h of administration, and subsequent studies have shown that the protection observed at 24 h postinoculation lasted for 3–5 days (Moraes et al., 2003). Furthermore, administration of Ad5-pIFNα a day after challenge was shown to be capable of reducing viremia, virus shedding, and disease severity (Moraes et al., 2003). The combination of Ad5-pIFNα and the Ad5-A24 vaccine appears to afford swine exposed to FMDV shortly after treatment both rapid protection and a more robust FMDV-specific neutralizing antibody response than either approach alone (Moraes et al., 2003). More recent experiments have shown that, in addition to inducing a rapid antiviral response, IFNα can function as an adjuvant and enhance the long-term protective response of the Ad5-A24 vaccine in pigs (de Avila Botton et al., 2006).

In studies in cattle, administration of Ad5-pIFNα did not completely protect animals from FMDV challenge, although the disease was delayed and less severe as compared to control animals (Wu et al., 2003). The incomplete protection in cattle correlated with the reduced levels of IFN detected in their blood and was very similar to the results in swine given the low-dose Ad5-pIFNα that was not protective (see above) (Chinsangaram et al., 2003). Thus, it appears that successful application of this approach to cattle will require an enhancement of the level of IFN activity by: (1) using promoters that produce higher levels of IFN, (2) altering the capsid of human Ad5 so that it can more efficiently infect bovine cells, and/or (3) targeting the Ad5 vector to cells or tissues that would increase the effectiveness of the treatment. Other approaches to enhance early innate or adaptive immune responses to FMD vaccines include the addition of multiple cytokines that can act synergistically or combining vaccine antigens with other immune stimulating products such as CpG or dsRNA mimetics to enhance protection. In support of this latter approach, Moraes et al. (2007) have recently demonstrated that the combination of IFNα and type II IFN (IFNγ) can synergistically inhibit FMDV replication in cell culture and sterilely protect swine from virus challenge.

PROSPECTS FOR THE FUTURE

The natural or deliberate release of FMDV in many areas of the world would be devastating. In developed countries with large naive populations of livestock, an outbreak would have immediate adverse economic consequences, and the application of the current control strategy of large-scale slaughter of infected and in-contact susceptible animals would be met with massive public resistance. Thus, despite the availability of vaccines that can be successfully used to control FMD in long-term eradication campaigns, developed countries have begun to invest in alternative control measures that are better suited to emergency vaccination and vaccination-to-live policies. These latter concerns have refocused FMD vaccine development on the production of immunogens that lack one or more of the virus-encoded antigens that can be use as part of "marker" vaccines, which allow for unequivocal distinction between infected and vaccinated animals during post-outbreak serological surveys required to demonstrate freedom from disease. Vaccines with a higher antigen load or augmented with adjuvants or cytokines that stimulate a rapid immune response are also being investigated. Finally, the demonstration that delivery of IFNα can induce protection more rapidly than a vaccine has produced a push for development of either antivirals that do not have vaccine antigens included in them (and hence eliminate the need for differentiating serological tests post-outbreak) or vaccines that are augmented by cytokines designed to activate the innate immune system while vaccine-induced immunity is being established. The development of this latter technology into a useable combination vaccine product will require a concerted collaborative effort by governments of developed countries and veterinary vaccine manufacturers. Recently, the U.S. Department of Homeland Security and the USDA entered into a collaborative agreement with GenVec, Inc., a U.S. biotechnology company, to produce Ad5-vectored FMD vaccines for the USDA-APHIS National Veterinary Stockpile.

While the focus on new technologies may produce products that can limit the spread of FMD outbreaks in disease-free countries, control of FMD in countries where the disease is enzootic will continue to rely on existing vaccines and the diligence of local veterinary authorities in rapidly identifying outbreaks, ensuring compliance with mandatory vaccination policies, and controlling trade and animal movement. These latter activities are likely to become more difficult, as globalization and economic development increase the number of animals being produced and the speed at which animals and their products reach marketplaces in distant lands. However, increased transparency in trade and sharing of resources could help to control this devastating disease worldwide, improving the economic situation in all countries.

KEY ISSUES

- FMD is the most highly regulated agricultural disease in the world and its presence in countries with active animal-product export economies can be devastating.
- Infections with the etiologic agent of the disease, FMDV, are not usually fatal and can cause difficult-to-detect diseases in certain species, complicating outbreak control.
- FMDV is one of the most rapidly mutating viruses known and exists in the form of multiple subtypes and serotypes.
- FMDV vaccines are currently being manufactured using chemically inactivated cell culture-derived virus. These vaccines prevent disease, but not infection, and have demonstrated utility in campaigns to eradicate disease where the circulating serotypes are known. However, application in an outbreak of unknown origin is more difficult due to the need to match the vaccine to the serotype(s) causing the outbreak.
- Vaccination campaigns in the face of spreading outbreaks are complicated by the fact that the disease spreads so rapidly from host to host that the outbreak can arrive to locations before vaccination has produced immunity sufficient to prevent viral shedding and/or disease.
- Existence of an FMDV carrier state in a portion of vaccinated animals that have been infected with the virus and the ability of vaccinated animals to shed the virus if they become infected have been used as arguments against vaccination, especially in the early stages of outbreaks when it appears that slaughter and quarantine alone can control spread of the disease.
- New vaccines as well as antivirals/cytokines that provide immunity more rapidly and/or can be produced without the need for production of the tremendous amounts of infectious material required for manufacture of the current vaccine are being developed and may provide additional tools to help control FMD.

ACKNOWLEDGMENTS

We thank Nick J. Knowles from the Institute of Animal Health's Pirbright laboratory, Surrey, U.K., for supplying Fig. 22.1.

References

Abrams, C.C., King, A.M. and Belsham, G.J. Assembly of foot-and-mouth disease virus empty capsids synthesized by a vaccinia virus expression system. J. Gen. Virol. 1995; 76(Pt 12): 3089–3098.

Acharya, R., Fry, E., Stuart, D., Fox, G., Rowlands, D. and Brown, F. The three-dimensional structure of foot-and-mouth disease virus at 2.9 Å resolution. Nature 1989; 337(6209):709–716.

Almeida, M.R., Rieder, E., Chinsangaram, J., Ward, G., Beard, C., Grubman, M.J. and Mason, P.W. Construction and evaluation of an attenuated vaccine for foot-and-mouth disease: difficulty adapting the leader proteinase-deleted strategy to the serotype O1 virus. Virus Res. 1998; 55(1):49–60.

Bachrach, H.L., Hess, W.R. and Callis, J. Foot-and-mouth disease virus: its growth and cytopathogenicity in tissue culture. Science 1955; 122:1269–1270.

Bahnemann, H.G. Inactivation of viral antigens for vaccine preparation with particular reference to the application of binary ethylenimine. Vaccine 1990; 8(4):299–303.

Beard, C., Ward, G., Rieder, E., Chinsangaram, J., Grubman, M.J. and Mason, P.W. Development of DNA vaccines for foot-and-mouth disease, evaluation of vaccines encoding replicating and non-replicating nucleic acids in swine. J. Biotechnol. 1999; 73(2–3):243–249.

Beck, E. and Strohmaier, K. Subtyping of European foot-and-mouth disease virus strains by nucleotide sequence determination. J. Virol. 1987; 61(5):1621–1629.

Belsham, G.J., Abrams, C.C., King, A.M., Roosien, J. and Vlak, J.M. Myristoylation of foot-and-mouth disease virus capsid protein precursors is independent of other viral proteins and occurs in both mammalian and insect cells. J. Gen. Virol. 1991; 72 (Pt 3):747–751.

Benvenisti, L., Rogel, A., Kuznetzova, L., Bujanover, S., Becker, Y. and Stram, Y. Gene gun-mediate DNA vaccination against foot-and-mouth disease virus. Vaccine 2001; 19(28–29):3885–3895.

Berinstein, A., Tami, C., Taboga, O., Smitsaart, E. and Carrillo, E. Protective immunity against foot-and-mouth disease virus induced by a recombinant vaccinia virus. Vaccine 2000; 18(21):2231–2238.

Biron, C. and Sen, G.C. Interferons and other cytokines. In: Field's Virology (P.M. Howley and D.M. Knipe, Eds.), pp. 321–351. Philadelphia: Lippincott Williams & Wilkins, 2001.

Bittle, J.L., Houghten, R.A., Alexander, H., Shinnick, T.M., Sutcliffe, J.G., Lerner, R.A., Rowlands, D.J. and Brown, F.

Protection against foot-and-mouth disease by immunization with a chemically synthesized peptide predicted from the viral nucleotide sequence. Nature 1982; 298(5869):30–33.

Bolwell, C., Brown, A.L., Barnett, P.V., Campbell, R.O., Clarke, B.E., Parry, N.R., Ouldridge, E.J., Brown, F. and Rowlands, D.J. Host cell selection of antigenic variants of foot-and-mouth disease virus. J. Gen. Virol. 1989; 70(Pt 1):45–57.

Bouma, A., Elbers, A.R., Dekker, A., de Koeijer, A., Bartels, C., Vellema, P., van der Wal, P., van Rooij, E.M., Pluimers, F.H. and de Jong, M.C. The foot-and-mouth disease epidemic in the Netherlands in 2001. Prev. Vet. Med. 2003; 57(3):155–166.

Brown, F. New approaches to vaccination against foot-and-mouth disease. Vaccine 1992; 10(14):1022–1026.

Brown, F. The history of research in foot-and-mouth disease. Virus Res. 2003; 91(1):3–7.

Brown, F. and Crick, J. Application of gel diffusion analysis to a study of the antigenic structure of inactivated vaccines prepared from the virus of foot-and-mouth disease. J. Immunol. 1959; 82:444–447.

Caron, L., Brum, M.C.S., Moraes, M.P., Golde, W.T., Arns, C.W. and Grubman, M.J. The effect of granulocyte-macrophage colony-stimulating factor on the potency and efficacy of a foot-and-mouth disease virus subunit vaccine. Pesqui. Vet. Bras. 2005; 25:150–158.

Cedillo-Barron, L., Foster-Cuevas, M., Belsham, G.J., Lefevre, F. and Parkhouse, R.M. Induction of a protective response in swine vaccinated with DNA encoding foot-and-mouth disease virus empty capsid proteins and the 3D RNA polymerase. J. Gen. Virol. 2001; 82(Pt 7):1713–1724.

Chinsangaram, J., Beard, C., Mason, P.W., Zellner, M.K., Ward, G. and Grubman, M.J. Antibody response in mice inoculated with DNA expressing foot-and-mouth disease virus capsid proteins. J. Virol. 1998a; 72(5):4454–4457.

Chinsangaram, J., Mason, P.W. and Grubman, M.J. Protection of swine by live and inactivated vaccines prepared from a leader proteinase-deficient serotype A12 foot-and-mouth disease virus. Vaccine 1998b; 16(16):1516–1522.

Chinsangaram, J., Moraes, M.P., Koster, M. and Grubman, M.J. Novel viral disease control strategy: adenovirus expressing alpha interferon rapidly protects swine from foot-and-mouth disease. J. Virol. 2003; 77:1621–1625.

Couch, R.B., Chanock, R.M., Cate, T.R., Lang, D.J., Knight, V. and Heubner, R.J. Immunization with types 4 and 7 adenovirus by selective infection of the intestinal tract. Am. Rev. Respir. Dis. 1963; 88:394–403.

Cox, S.J., Voyce, C., Parida, S., Reid, S.M., Hamblin, P.A., Paton, D.J. and Barnett, P.V. Protection against direct-contact challenge following emergency FMD vaccination of cattle and the effect on virus excretion from the oropharynx. Vaccine 2005; 23(9):1106–1113.

de Avila Botton, S., Brum, M.C., Bautista, E., Koster, M., Weiblen, R., Golde, W.T. and Grubman, M.J. Immunopotentiation of a foot-and-mouth disease virus subunit vaccine by interferon alpha. Vaccine 2006; 24(17):3446–3456.

DiMarchi, R., Brooke, G., Gale, C., Cracknell, V., Doel, T. and Mowat, N. Protection of cattle against foot-and-mouth disease by a synthetic peptide. Science 1986; 232(4750):639–641.

Doel, T.R. FMD vaccines. Virus Res. 2003; 91(1):81–99.

Doel, T.R., Williams, L. and Barnett, P.V. Emergency vaccination against foot-and-mouth disease: rate of development of immunity and its implications for the carrier state. Vaccine 1994; 12(7):592–600.

Domingo, E., Escarmis, C., Baranowski, E., Ruiz-Jarabo, C.M., Carrillo, E., Nunez, J.I. and Sobrino, F. Evolution of foot-and-mouth disease virus. Virus Res. 2003; 91(1):47–63.

Francis, M.J., Hastings, G.Z., Brown, F., McDermed, J., Lu, Y.A. and Tam, J.P. Immunological evaluation of the multiple antigen peptide (MAP) system using the major immunogenic site of foot-and-mouth disease virus. Immunology 1991; 73(3):249–254.

Frenkel, H.S. Histological changes in explanted bovine epithelial tongue tissue infected with the virus of foot-and-mouth disease. Am. J. Vet. Res. 1949; 10:142–145.

Frenkel, H.S. Research on foot-and-mouth disease. II. The cultivation of the virus in explanations of tongue epithelium of bovine animals. Am. J. Vet. Res. 1950; 11:371–373.

Fry, E.E., Newman, J.W., Curry, S., Najjam, S., Jackson, T., Blakemore, W., Lea, S.M., Miller, L., Burman, A., King, A.M. and Stuart, D.I. Structure of foot-and-mouth disease virus serotype $A10_{61}$ alone and complexed with oligosaccharide receptor: receptor conservation in the face of antigenic variation. J. Gen. Virol. 2005; 86(Pt 7):1909–1920.

Galloway, I.A. and Nicolau, S. Histological study of the development of lesions of foot-and-mouth disease in the tongue of guinea-pig, rabbit, and ferret. In: Third Progress Report of the British Foot-and-Mouth Disease Research Committee. Great Britain Ministry of Agriculture and Fisheries, Vol. Appendix III, 1928:104–115.

Golde, W.T., Pacheco, J.M., Duque, H., Doel, T., Penfold, B., Ferman, G.S., Gregg, D.R. and Rodriguez, L.L. Vaccination against foot-and-mouth disease virus confers complete clinical protection in 7 days and partial protection in 4 days: use in emergency outbreak response. Vaccine 2005; 23(50):5775–5782.

Graham, F.L., Smiley, J., Russell, W.C. and Nairn, R. Characteristics of a human cell line transformed by DNA from human adenovirus 5. J. Gen. Virol. 1977; 36:59–74.

Grubman, M.J. New approaches to rapidly control foot-and-mouth disease outbreaks. Expert Rev. Anti Infect. Ther. 2003; 1(4):579–586.

Grubman, M.J. Development of novel strategies to control foot-and-mouth disease: marker vaccines and antivirals. Biologicals 2005; 33:227–234.

Grubman, M.J. and Baxt, B. Foot-and-mouth disease. Clin. Microbiol. Rev. 2004; 17(2):465–493.

Grubman, M.J., Lewis, S.A. and Morgan, D.O. Protection of swine against foot-and-mouth disease with viral capsid proteins expressed in heterologous systems. Vaccine 1993; 11(8):825–829.

Jackson, T., King, A.M., Stuart, D.I. and Fry, E. Structure and receptor binding. Virus Res. 2003; 91(1):33–46.

Kit, M., Kit, S., Little, S.P., Di Marchi, R.D. and Gale, C. Bovine herpesvirus-1 (infectious bovine rhinotracheitis virus)-based viral vector which expresses foot-and-mouth disease epitopes. Vaccine 1991; 9(8):564–572.

Kitson, J.D., Burke, K.L., Pullen, L.A., Belsham, G.J. and Almond, J.W. Chimeric polioviruses that include sequences derived from two independent antigenic sites of foot-and-mouth disease virus (FMDV) induce neutralizing antibodies against FMDV in guinea pigs. J. Virol. 1991; 65(6):3068–3075.

Kleid, D.G., Yansura, D., Small, B., Dowbenko, D., Moore, D.M., Grubman, M.J., McKercher, P.D., Morgan, D.O., Robertson, B.H. and Bachrach, H.L. Cloned viral protein vaccine for foot-and-mouth disease: responses in cattle and swine. Science 1981; 214(4525):1125–1129.

Knowles, N.J. and Samuel, A.R. Molecular epidemiology of foot-and-mouth disease virus. Virus Res. 2003; 91(1):65–80.

Lea, S., Hernandez, J., Blakemore, W., Brocchi, E., Curry, S., Domingo, E., Fry, E., Abu-Ghazaleh, R., King, A., Newman, J. et al. The structure and antigenicity of a type C foot-and-mouth disease virus. Structure 1994; 2(2):123–139.

Lewis, S.A., Morgan, D.O. and Grubman, M.J. Expression, processing, and assembly of foot-and-mouth disease virus capsid structures in heterologous systems: induction of a neutralizing antibody response in guinea pigs. J. Virol. 1991; 65(12):6572–6580.

Loeffler, F. IV. Bericht der kommission zur erforschung der maul- und klauenseuche bei dem institut für infektionskrankheiten in Berlin. Centralblatt Bakteriologie Parasitenkunde Infektionskrankheiten 1898; 24:569–574.

Loeffler, F. Die serotherapie, die seroprophylaxe und die impfung bei maul- und klauenseuche und deren wert für die veterinärpolizei. Dtsch. Med. Wochenschr. 1909; 48:2097–2101.

Loeffler, F. and Frosch, P. Berichte der kommission zur erforschung der maul- und klauenseuche bei dem institut für infektionskrankheiten in Berlin. Centralblatt Bakteriologie Parasitenkunde Infektionskrankheiten 1898; 23:371–391.

Logan, D., Abu-Ghazaleh, R., Blakemore, W., Curry, S., Jackson, T., King, A., Lea, S., Lewis, R., Newman, J., Parry, N. et al. Structure of a major immunogenic site on foot-and-mouth disease virus. Nature 1993; 362(6420):566–568.

Martin, W.B. and Edwards, L.T. A field trial in South Africa of an attenuated vaccine against foot-and-mouth disease. Res. Vet. Sci. 1965; 36:196–201.

Mason, P.W., Grubman, M.J. and Baxt, B. Molecular basis of pathogenesis of FMDV. Virus Res. 2003; 91(1):9–32.

Mason, P.W., Piccone, M.E., McKenna, T.S., Chinsangaram, J. and Grubman, M.J. Evaluation of a live-attenuated foot-and-mouth disease virus as a vaccine candidate. Virology 1997; 227(1):96–102.

Mayr, G.A., Chinsangaram, J. and Grubman, M.J. Development of replication-defective adenovirus serotype 5 containing the capsid and 3C protease coding regions of foot-and-mouth disease virus as a vaccine candidate. Virology 1999; 263(2):496–506.

Mayr, G.A., O'Donnell, V., Chinsangaram, J., Mason, P.W. and Grubman, M.J. Immune responses and protection against foot-and-mouth disease virus (FMDV) challenge in swine vaccinated with adenovirus-FMDV constructs. Vaccine 2001; 19:2152–2162.

McCullough, K.C., Pullen, L. and Parkinson, D. The immune response against foot-and-mouth disease virus: influence of the T lymphocyte growth factors IL-1 and IL-2 on the murine humoral response in vivo. Immunol. Lett. 1992; 31(1):41–46.

McKenna, T.S., Lubroth, J., Rieder, E., Baxt, B. and Mason, P.W. Receptor binding site-deleted foot-and-mouth disease (FMD) virus protects cattle from FMD. J. Virol. 1995; 69(9):5787–5790.

Mohler, J. The 1929 outbreak of foot-and-mouth disease in California. J. Am. Vet. Med. Assoc. 1929; 75:309–318.

Mohler, J.R. Foot-and-mouth disease, with special reference to the outbreak of 1914. United States Department of Agriculture, Department Circular, 1924:325.

Mohler, J.R. and Rosenau, M.J. The origin of the recent outbreak of foot-and-mouth disease in the United States. United States Department of Agriculture, Department Circular, 1909:147.

Moraes, M.P., Chinsangaram, J., Brum, M.C.S. and Grubman, M.J. Immediate protection of swine from foot-and-mouth disease: a combination of adenoviruses expressing interferon alpha and a foot-and-mouth disease virus subunit vaccine. Vaccine 2003; 22:268–279.

Moraes, M.P., de los Santos, T., Koster, M., Turecek, T., Wang, H., Andreyev, V.G. and Grubman, M.J. Enhanced antiviral activity against foot-and-mouth disease virus by a combination of type I and II porcine interferons. J. Virol. 2007; 81(13):7124–7135.

Moraes, M.P., Mayr, G.A., Mason, P.W. and Grubman, M.J. Early protection against homologous challenge after a single dose of replication-defective human adenovirus type 5 expressing capsid proteins of foot-and-mouth disease virus (FMDV) strain A24. Vaccine 2002; 20(11–12):1631–1639.

Mowat, G.N. and Chapman, W.G. Growth of foot-and-mouth disease virus in a fibroblastic cell line derived from hamster kidneys. Nature 1962; 194:253–255.

Nargi, F., Kramer, E., Mezencio, J., Zamparo, J., Whetstone, C., Van Regenmortel, M.H., Briand, J.P., Muller, S. and Brown, F. Protection of swine from foot-and-mouth disease with one dose of an all-D retro peptide. Vaccine 1999; 17(22):2888–2893.

Nathanson, N. and Langmuir, A.D. The Cutter incident: poliomyelitis following formaldehyde-inactivated poliovirus vaccination in the United States during the spring of 1955. II Relationship of poliomyelitis to Cutter vaccine. Am. J. Hyg. 1963; 78:29–60.

Orsel, K., de Jong, M.C., Bouma, A., Stegeman, J.A. and Dekker, A. Foot and mouth disease virus transmission among vaccinated pigs after exposure to virus shedding pigs. Vaccine 2007; 25(34):6381–6391.

Pacheco, J.M., Brum, M.C., Moraes, M.P., Golde, W.T. and Grubman, M.J. Rapid protection of cattle from direct challenge with foot-and-mouth disease virus (FMDV) by a single inoculation with an adenovirus-vectored FMDV subunit vaccine. Virology 2005; 337(2):205–209.

Pfaff, E., Mussgay, M., Bohm, H.O., Schulz, G.E. and Schaller, H. Antibodies against a preselected peptide recognize and neutralize foot and mouth disease virus. EMBO J. 1982; 1(7):869–874.

Piccone, M.E., Rieder, E., Mason, P.W. and Grubman, M.J. The foot-and-mouth disease virus leader proteinase gene is not required for viral replication. J. Virol. 1995; 69(9):5376–5382.

Prevec, L., Schneider, M., Rosenthal, K.L., Belbeck, L.W., Derbyshire, J.B. and Graham, F.L. Use of human adenovirus-based vectors for antigen expression in animals. J. Gen. Virol. 1989; 70:429–434.

Rieder, E., Baxt, B., Lubroth, J. and Mason, P.W. Vaccines prepared from chimeras of foot-and-mouth disease virus (FMDV) induce neutralizing antibodies and protective immunity to multiple serotypes of FMDV. J. Virol. 1994; 68(11):7092–7098.

Rieder, E., Bunch, T., Brown, F. and Mason, P.W. Genetically engineered foot-and-mouth disease viruses with poly(C) tracts of two nucleotides are virulent in mice. J. Virol. 1993; 67(9):5139–5145.

Roosien, J., Belsham, G.J., Ryan, M.D., King, A.M. and Vlak, J.M. Synthesis of foot-and-mouth disease virus capsid proteins in insect cells using baculovirus expression vectors. J. Gen. Virol. 1990; 71(Pt 8):1703–1711.

Rowlands, D.J., Sangar, D.V. and Brown, F. A comparative chemical and serological study of the full and empty particles of foot-and-mouth disease virus. J. Gen. Virol. 1974; 26:227–238.

Russell, W.C. Update on adenovirus and its vectors. J. Gen. Virol. 2000; 81:2573–2604.

Rweyemamu, M.M., Terry, G. and Pay, T.W. Stability and immunogenicity of empty particles of foot-and-mouth disease virus. Arch. Virol. 1979; 59(1–2):69–79.

Sa-Carvalho, D., Rieder, E., Baxt, B., Rodarte, R., Tanuri, A. and Mason, P.W. Tissue culture adaptation of foot-and-mouth disease virus selects viruses that bind to heparin and are attenuated in cattle. J. Virol. 1997; 71(7):5115–5123.

Saiz, J.C., Cairo, J., Medina, M., Zuidema, D., Abrams, C., Belsham, G.J., Domingo, E. and Vlak, J.M. Unprocessed foot-and-mouth disease virus capsid precursor displays discontinuous epitopes involved in viral neutralization. J. Virol. 1994; 68(7):4557–4564.

Sanz-Parra, A., Jimenez-Clavero, M.A., Garcia-Briones, M.M., Blanco, E., Sobrino, F. and Ley, V. Recombinant viruses expressing the foot-and-mouth disease virus capsid precursor polypeptide (P1) induce cellular but not humoral antiviral immunity and partial protection in pigs. Virology 1999a; 259(1):129–134.

Sanz-Parra, A., Vazquez, B., Sobrino, F., Cox, S.J., Ley, V. and Salt, J.S. Evidence of partial protection against foot-and-mouth disease in cattle immunized with a recombinant adenovirus vector expressing the precursor polypeptide (P1) of foot-and-mouth disease virus capsid proteins. J. Gen. Virol. 1999b; 80:671–679.

Skinner, H.H. Propagation of strains of foot-and-mouth disease virus in unweaned white mice. Proc. R. Soc. Med. 1951; 44:1041–1044.

Strohmaier, K., Franze, R. and Adam, K.H. Location and characterization of the antigenic portion of the FMDV immunizing protein. J. Gen. Virol. 1982; 59:295–306.

Sutmoller, P., Barteling, S.S., Olascoaga, R.C. and Sumption, K.J. Control and eradication of foot-and-mouth disease. Virus Res. 2003; 91(1):101–144.

Taboga, O., Tami, C., Carrillo, E., Nunez, J.I., Rodriguez, A., Saiz, J.C., Blanco, E., Valero, M.L., Roig, X., Camarero, J.A., Andreu, D., Mateu, M.G., Giralt, E., Domingo, E., Sobrino, F. and Palma, E.L. A large-scale evaluation of peptide vaccines against foot-and-mouth disease: lack of solid protection in cattle and isolation of escape mutants. J. Virol. 1997; 71(4):2606–2614.

Thomson, G.R., Vosloo, W. and Bastos, A.D. Foot and mouth disease in wildlife. Virus Res. 2003; 91(1):145–161.

Top, F.H., Buescher, E.L., Bancroft, W.H. and Russell, P.K. Immunization with live types 7 and 4 adenovirus vaccines. II. Antibody response and protective effect against acute respiratory disease due to adenovirus type 7. J. Infect. Dis. 1971; 124:155–160.

Vallée, H. and Carré, H. Sur la pluralite des virus aphteux. C. R. Acad. Sci. 1922; 174:1498.

van Bekkum, J.G., Frenkel, H.S., Frederiks, H.H.J. and Frenkel, S. Observations on the carrier state of cattle exposed to foot-and-mouth disease virus. Tijdschr. Diergeneeskd. 1959; 84:1159–1164.

van Rensburg, H.G., Henry, T.M. and Mason, P.W. Studies of genetically defined chimeras of a European type A virus and a South African Territories type 2 virus reveal growth determinants for foot-and-mouth disease virus. J. Gen. Virol. 2004; 85(Pt 1):61–68.

Waldmann, O. and Trautwein, K. Experimentelle Untersuchungen, über die Pluralität des Maul- und Klauenseuchevirus. Berl. Tierarztl. Wochenschr. 1926; 42:569–571.

Wang, J., Zganiacz, A. and Xing, Z. Enhanced immunogenicity of BCG vaccine by using a viral-based GM-CSF transgene adjuvant formulation. Vaccine 2002; 20(23–24):2887–2898.

Wigdorovitz, A., Carrillo, C., Dus Santos, M.J., Trono, K., Peralta, A., Gomez, M.C., Rios, R.D., Franzone, P.M., Sadir, A.M., Escribano, J.M. and Borca, M.V. Induction of a protective antibody response to foot and mouth disease virus in mice following oral or parenteral immunization with alfalfa transgenic plants expressing the viral structural protein VP1. Virology 1999a; 255(2):347–353.

Wigdorovitz, A., Perez Filgueira, D.M., Robertson, N., Carrillo, C., Sadir, A.M., Morris, T.J. and Borca, M.V. Protection of mice against challenge with foot and mouth disease virus (FMDV) by immunization with foliar extracts from plants infected with recombinant tobacco mosaic virus expressing the FMDV structural protein VP1. Virology 1999b; 264(1):85–91.

Wong, H.T., Cheng, S.C., Chan, E.W., Sheng, Z.T., Yan, W.Y., Zheng, Z.X. and Xie, Y. Plasmids encoding foot-and-mouth disease virus VP1 epitopes elicited immune responses in mice and swine and protected swine against viral infection. Virology 2000; 278(1):27–35.

Wong, H.T., Cheng, S.C., Sin, F.W., Chan, E.W., Sheng, Z.T. and Xie, Y. A DNA vaccine against foot-and-mouth disease elicits an immune response in swine which is enhanced by co-administration with interleukin-2. Vaccine 2002; 20(21–22):2641–2647.

Wu, Q., Brum, M.C.S., Caron, L., Koster, M. and Grubman, M.J. Adenovirus-mediated type I interferon expression delays and reduces disease signs in cattle challenged with foot-and-mouth disease virus. J. Int. Cyt. Res. 2003; 23:359–368.

Zibert, A., Maass, G., Strebel, K., Falk, M.M. and Beck, E. Infectious foot-and-mouth disease virus derived from a cloned full-length cDNA. J Virol 1990; 64(6):2467–2473.

CHAPTER

23

Hantavirus

Christopher D. Hammerbeck, Victoria Wahl-Jensen and Jay W. Hooper

Virology Division, U.S. Army Medical Research Institute of Infectious Diseases, Fort Detrick, MD, USA

OUTLINE

ABSTRACT

Hantaviruses are known to cause diseases of military and public health concern. The association of hantaviruses and war can be attributed to the fact that these viruses are carried by rodents and are transmitted to humans via rodent excreta. During sustained ground combat operations, human–rodent contact inevitably increases due to an increased number of personnel working and sleeping in the rodents' habitat, environmental disruption, and a breakdown in hygiene. During the Korean War, casualties caused by hantaviruses impacted military operations and spurred research aimed at identifying the etiological agent responsible for what was then known as Korean hemorrhagic fever, and developing countermeasures to prevent and treat the disease. This research resulted in the identification of multiple hantaviruses responsible for hemorrhagic fever with renal syndrome (HFRS), including Hantaan virus, Seoul virus, Dobrava virus, and Puumala virus. An ongoing effort to develop vaccines against these viruses has resulted in inactivated-virus vaccines that have been licensed in Asia. In addition, efforts to develop safe, effective molecular subunit vaccines against these viruses have shown promise. The only drug to show any clinical efficacy against HFRS is ribavirin used soon after the first symptoms of the disease. Ribavirin has not been licensed by the Food and Drug Administration (FDA) for use as a treatment for HFRS. The viruses associated with HFRS pose a relatively low threat as biological weapons because the viruses are difficult to grow to high titer, HFRS is not contagious, and the case-fatality rates range between 0.1% and 15%. However, starting in 1993, several previously unknown, highly pathogenic hantaviruses were discovered in the Americas. At least one of these viruses poses a substantial bioterrorism threat. Andes virus grows to levels ten times higher than the HFRS-associated hantaviruses, can spread from person to person under close-contact conditions, and is highly pathogenic (30–50% cases are fatal). Andes virus and several other newly emerging hantaviruses in the Americas are associated with rapid onset of pulmonary edema followed by cardiogenic shock. There are no vaccines or effective drugs to prevent or treat hantavirus pulmonary syndrome (HPS). Here, we review the viruses associated with HFRS and HPS with an emphasis on ongoing efforts to develop vaccines against these dangerous pathogens.

INTRODUCTION

Hantaviruses are rodent-borne viruses that cause hemorrhagic fever in humans. These viruses are associated with diverse disease syndromes with varying degrees of severity. Diseases caused by hantaviruses are generally manifested as either hemorrhagic fever with renal syndrome (HFRS) or hantavirus pulmonary syndrome (HPS). The most prevalent and lethal HFRS-associated hantavirus is Hantaan virus (HTNV) (20,000–50,000 cases per year, mostly in Asia) with a case-fatality rate of 5–15% (Fang et al., 2006). The most prevalent hantavirus in Central Europe, northern Europe/Scandinavia, and western Russia is Puumala virus (PUUV); whereas in southern Europe and the Balkans, both PUUV and Dobrava virus (DOBV) cause disease. The most prevalent and lethal HPS-associated hantaviruses are Andes virus (ANDV) in South America and Sin Nombre virus (SNV) in North America. There have been close to 2000 cases of HPS between 1993 and 2007 with an overall mortality rate of approximately 40%, despite state-of-the-art treatment in modern intensive care units. The mechanisms underlying the pathogenesis of the vascular leak syndromes associated with HFRS and HPS remain unknown. There are no vaccines or specific antiviral drugs licensed by the U.S. Food and Drug Administration to treat or prevent HFRS or HPS.

SHORT HISTORY OF THE DISEASE

The Western world became aware of hantaviruses for the first time during an outbreak of HFRS in 1951 during the Korean War. This disease affected nearly 3000 United Nations' troops by 1954 and exhibited a mortality rate of 7% (Johnson, 2001). Despite extensive investigations and reports on this outbreak, the etiologic agent was not identified at that time. Indeed, the mystery agent for what was then called Korean hemorrhagic fever remained elusive until HTNV, named for the Hantaan river that ran near the border between North and South Korea where human cases were endemic, was isolated in 1978 by Lee et al. (1978a, 1978b). Shortly after the discovery of HTNV, other hantaviruses were identified in Eurasia (Brummer-Korvenkontio et al., 1982, 1980; Gresikova et al., 1984). It is widely believed that other wartime epidemics of nephritis and hemorrhagic fever were caused by hantaviruses. For example, 10,000 German troops developed epidemic nephritis in Finland during World War II (Stuhlfault, 1943). More recently, epidemics of HFRS occurred in military units engaged in the conflicts associated with the breakup of the former Yugoslavia (Markotic et al., 2002; Hukic et al., 1996).

Hantaviruses did not burst onto the American scene until an outbreak of fulminate pulmonary illness occurred in the southwest of the United States in 1993 (Nichol et al., 1993). Unlike HFRS, this syndrome largely affected the lungs and was aptly termed HPS. Evidence that cardiac dysfunction plays an important role in the disease caused by the newly discovered hantaviruses led some groups to call the disease hantavirus cardiopulmonary syndrome (HCPS). It is interesting that while it took decades to identify the causative agent of HFRS, the etiology of HPS was determined in approximately 30 days (Johnson, 2001). The agent responsible for this initial outbreak, as well as the majority of disease in North America since then, is SNV. Retrospective studies have shown that HPS was present but unrecognized in North America before the 1993 outbreak in the Four Corners region (Frampton et al., 1995; Schwarcz et al., 1996; Zaki et al., 1996).

After the discovery of pathogenic hantaviruses in North America, numerous hantaviruses that cause HPS were identified in Central and South America. As was the case in North America, HPS was not recognized in Central or South America until outbreaks of fatal disease occurred. The reader is referred to a review of recent advances in research on hantaviruses of Latin America (Pini, 2004). Unlike HFRS and even HPS in North America, HPS in Chile and Argentina has occurred in clusters, and there is evidence that the causative agent, ANDV, can spread from one person to the other (Lazaro et al., 2007; Toro et al., 1998; Vial et al., 2006; Wells et al., 1997). This development is a serious concern, especially as there have been instances where persons have contracted the disease in South America and then traveled internationally before developing symptoms, and fatal disease (Reynolds et al., 2007). Although person-to-person transmission and the export of cases are relatively rare occurrences, they are important events in the history of hantavirus disease because they demonstrate that the disease is capable of spreading beyond the initial human–rodent interactions.

ETIOLOGIC AGENTS

Hantaviruses represent a diverse group of viruses within a separate genus of the family Bunyaviridae, each carried by a specific rodent host. The viruses that have been associated with human disease are listed in Table 23.1. Viruses not yet associated with disease are listed in Table 23.2. Members of this genus are pleomorphic (70–210 nm), generally spherical particles with a distinctive grid-like appearance (Goldsmith et al., 1995; McCormick et al., 1982; White et al., 1982). Virions possess a host-derived lipid envelope that is studded with oligomers of the surface glycoproteins G1 and G2, now known as G_n and G_c, respectively (Fig. 23.1). The genome of hantaviruses is tripartite, negative-sense, single-stranded RNA that encodes three structural proteins (Schmaljohn and Hooper, 2001). The small RNA segment (S) encodes the nucleocapsid (N) protein, the medium segment (M) encodes the glycoprotein precursor (GPC) that is processed to G_n and G_c, and the large segment (L) encodes the viral RNA-dependent RNA polymerase (RdRp). Each RNA segment is complexed with multiple copies of N protein to form ribonucleocapsids. It is likely that most virions contain single copies of the S, M, and L ribonucleocapsids; however, reassortant studies report that it is possible for virions to contain multiple copies of one or more genome segments (Rodriguez et al., 1998). Hantavirus virions are relatively stable in the environment. For example, PUUV has been shown to remain infectious up to 11 days at room temperature (Kallio et al., 2006).

PROTECTIVE IMMUNE RESPONSES

Humoral Immune Responses

During hantavirus infection, B cells upregulate production of all immunoglobin (Ig) isotypes including IgA, IgE, IgG, and IgM (Alexeyev et al., 1994; Bostik et al., 2000; Groen et al., 1994; Lundkvist et al., 1993). Increases in IgA and IgE antibody titers are typically seen before and during the acute phase of infection, while production of IgM and IgG usually occur during and after the acute phase with IgG titers continuing to increase early into the convalescent phase. While IgM antibodies specific for hantaviral N, G_n, and G_c proteins can be found during infection with both HPS and HFRS viruses, anti-N-protein-specific IgM predominates. Similarly, N-protein is the most common viral antigen recognized by IgG antibodies, which are also the most abundant Ig produced during hantavirus infection. Neutralizing antibodies target the hantavirus surface glycoproteins G_n and G_c, but there is no convincing evidence that antibodies to the N-protein or RdRp neutralize virus. The ability of N-protein-specific Ig to neutralize virus cannot be ruled out entirely, however, as passive transfer of N-protein-specific antibodies delayed time-to-death after experimental infection of suckling mice (Yoshimatsu et al., 1993). High levels of neutralizing antibodies upon hospital admission have been correlated with a milder course of SNV-associated HPS, suggesting that the production of high titers of neutralizing antibodies can mitigate and possibly protect against disease (Bharadwaj et al., 2000). In these cases, mild disease was associated with neutralizing antibody titers less than 1:800 while antibody titers during severe disease were greater than 1:800. The conclusion that neutralizing antibodies are sufficient to protect against hantavirus is supported by numerous passive transfer experiments involving either monoclonal antibodies, convalescent sera, or sera from vaccinated animals (see "Postexposure Immunoprophylaxis"). Still, while vaccine studies in animals indicate that neutralizing antibodies are sufficient to protect, they are not necessary for protection. For example, hamsters

TABLE 23.1 Hantaviruses known to disease in humans

Disease	Virus species	Virus strain	Abbreviation	Rodent reservoir
A. Old World				
HFRS[a]	*Hantaan virus*	Amur virus	AMRV	*Apodemus peninsulae*
		Hantaan virus-CUMB-B11 virus[b]	HTNV	Unknown (*human isolate*)
		Hantaan virus-76-118 virus		*Apodemus agrarius*
		Hantaan virus-HoJo virus[b]		Unknown (*human isolate*)
		Hantaan virus-HV114 virus		*Apodemus agrarius*
		Hantaan virus-Lee virus[b]		Unknown (*human isolate*)
	Seoul virus	Seoul virus-HR80-39	SEOV	*Rattus norvegicus; Rattus rattus*
		Seoul virus-Biken-1 virus[b]		
		Seoul virus-HB55 virus[b]		
		Seoul virus-KI-83-262 virus[b]		
		Seoul virus-KI-85-1 virus[b]		
		Seoul virus-KI-88-15 virus[b]		
		Seoul virus-L99 virus		*Rattus losea*
		Seoul virus-R22 virus[b]		*Rattus norvegicus*
		Seoul virus-SR-11 virus		
	Dobrava-Belgrade virus	Dobrava-Aa virus	DOBV	*Apodemus agrarius*
		Dobrava-Af virus[b]		*Apodemus flavicollis*
		Saaremaa virus	SAAV	*Apodemus agrarius*
	Puumala virus	1324Cg/79 virus[b]	PUUV	*Clethrionomys glareolus*
		Bashkiria Cg18-20 virus[b]		
		K27 virus[b]		Unknown (*human isolate*) suggested *Clethrionomys sp.*
		Kazan virus[b]		*Clethrionomys rufocanus*
		p360 virus[b]		Unknown (*human isolate*) suggested *Clethrionomys sp.*
		Sotkamo virus		*Clethrionomys glareolus; Clethrionomys rufocanus*
		Tobetsu-60Cr-93 virus[b]		*Clethrionomys rufocanus*
		Urmurtia/338Cg/92 virus		*Clethrionomys glareolus*
		Vindeln/L20Cg/83 virus[b]		
		Vranica virus[b]		*Clethrionomys rufocanus*
B. New World				
HPSc	*Andes virus*	Andes virus	ANDV	*Oligoryzomys longicauddatus*
		Bermejo virus	BMJV	*Oligoryzomys chacoensis*
		Lechiquanas virus	LECV	*Oligoryzomys flavescens*
		Maciel virus	MCLV	*Necromys benefactus*
		Orán virus	ORNV	*Oligoryzomys longicaudatus*
	Sin Nombre virus	Monongahela virus	MGLV	*Peromyscus maniculatus nubiterrae*
		Sin Nombre virus-NMH10	SNV	*Peromyscus maniculatus*
		Sin Nombre-Convict Creek 107		*Peromycus maniculatus*
	Laguna Negra virus	Laguna Negra virus	LNV	*Calomys laucha*
	New York virus	New York virus-RI-1 virus	NYV	*Peromyscus leucopus*
	Black Creek Canal virus	Black Creek Canal virus	BCCV	*Sigmodon hispidus*
	Bayou virus	Bayou virus	BAYV	*Oryzomys palustris*
	Unknown	Araraquara virus[b]	ARAV	*Bolomys lasiurus*
	Unknown	Castelo dos Sonhos virus[b]	CASV	Unknown
	Unknown	Choclo virus[b]	CHOV	*Oligoryzomys fulvescens*
	Unknown	Human 39694 virus[b]	Hu39694	Unknown
	Unknown	Juquitiba virus[b]	JUQV	*Oligoryzomys nigripes*

Source: Derived from Fauquet et al., 2005; Kariwa et al., 2007; Klein and Calisher, 2007; Muranyi et al., 2005.

[a] Hemorrhagic fever with renal syndrome.

[b] Virus not listed in Fauquet et al., 2005.

[c] Hantavirus pulmonary syndrome.

TABLE 23.2 Other hantaviruses not yet associated with human disease

Disease	Virus species	Virus strain	Abbreviation	Rodent reservoir
Unknown	*Andes virus*	Pergamino virus	PRGV	*Akadon azarae*
	Cano Delgadito virus	Cano Delgadito virus[b]	CADV	*Sigomdon alstoni*
	El Moro Canyon virus	El Moro Canyon virus-RM-97 virus	ELMCV	*Reithrodontomys megalotis*
	Hantaan virus	Da Bie Shan	DBSV	*Niviventer confucianus*
	Isla Vista virus	Isla Vista virus	ISLAV	*Microtus californicus*
	Khabarovsk virus	Khabarovsk virus	KHAV	*Microtus fortis*
	Muleshoe virus	Muleshoe virus	MULV	*Sigmodon hispidus*
	Prospect Hill virus	Bloodland Lake virus	BLLV	*Microtus ochrogaster*
		Prospect Hill virus	PHV	*Microtus pennsylvanicus*
	Puumala virus	Hokkaido virus-Kamiiso-8Cr-95	HOKV	*Clethrionomys rufocanus*
		Muju virus	MUJV	*Eothenomys regulus*
	Rio Mamoré virus	Rio Mamoré virus[c]	RIOMV	*Oligoryzomys microtis*
	Rio Segundo virus	Rio Segundo virus	RIOS	*Reithrodontomys mexicanus*
	Sin Nombre virus	Blue River virus	BRV	*Peromycus leucopus*
	Thailand virus	Thailand virus[d]	THAIV	*Bandicota indica; Bandicota savilei*
	Thottapolayam virus	Thottapalayam virus[d]	TPMV	*Suncus murinus*
	Topografov virus	Topografov virus	TOPV	*Lemmus sibiricus*
	Tula virus[d,e]	Tula virus-Moravia/Ma5302V	TULV	*Microtus arvalis; Microtus rossiaemeridionalis*
		Tula virus-Tula/Ma76/87		*Microtus rossiaemeridionalis*
		Tula virus-Malacky/Ma32/94[a]		*Microtus arvalis*
	Unknown	Calabazo virus[a,b]	na	*Zygodontomys brevicauda*
	Unknown	Limestone Canyon virus[a]	LSCV	*Peromyscus boylii*
	Unknown	Maporal virus[a]	MAPV	*Oligoryzomys fulvescens*
	Unknown	Sangassou virus[a]	na	*Hylomyscus simus*
	Unknown	Soochong virus[a]	SOOV	*Apodemus peninsulae*
	Unknown	Alto Paraguay[a]	ALP	*Holochilus chacarius*
	Unknown	Ape Aime-Itapúa virus[a]	AAI	*Akodon montensis*
	Unknown	Itapúa37 virus[a]	IP37	*Oligoryzomys nigripes*
	Unknown	Itapúa38 virus[a]	IP38	*Oligoryzomys nigripes*

Source: Derived from Fauquet et al., 2005; Kariwa et al., 2007; Klein and Calisher, 2007; Muranyi et al., 2005.
[a] Virus not listed in Fauquet et al. (2005).
[b] Positive seroconversion without confirmed clinical disease.
[c] Emerging evidence that virus may cause symptoms consistent with HPS.
[d] Emerging evidence that virus may cause symptoms consistent with HFRS.
[e] HFRS disease included pulmonary involvement.

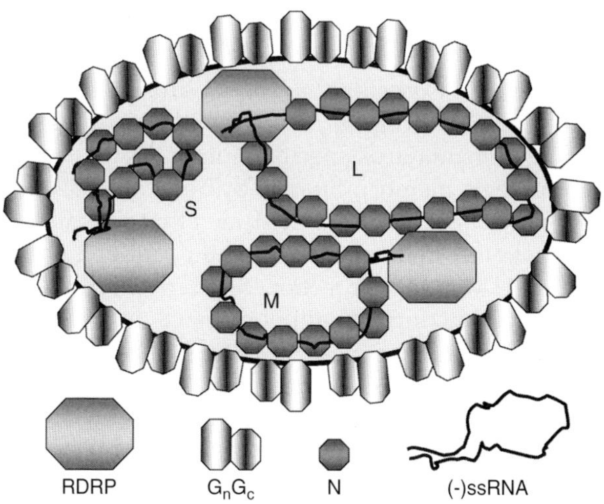

FIGURE 23.1 Hantaviruses are pleomorphic, generally spherical viruses with a diameter of 80–120 nm. They contain a host-derived lipid membrane that contains oligomers of the G_n/G_c glycoproteins. The three single-strand, negative-sense RNA segments are associated with nucleocapsid (N) protein and the RNA-dependent RNA polymerase (RdRp).

previously infected with HTNV, Seoul virus (SEOV), DOBV, PUUV, or SNV were protected against lethal HPS caused by ANDV despite the absence of detectable ANDV cross-neutralizing antibodies (Hooper et al., 2001b). It is likely that cellular responses to the N-protein and/or glycoproteins contribute to protection. In this respect, vaccination with purified N-protein can effectively stimulate T-cell responses and has been shown to protect in several animal models in the absence of neutralizing antibodies (Bharadwaj et al., 2002; de Carvalho Nicacio et al., 2002; Woo et al., 2005). Immune complexes can be found in the sera and on platelets of HFRS patients (Kanerva et al., 1998a) but the role that complement and Fc receptor-bearing cells play in antibody-mediated protection in vivo remains unknown. In vitro experiments have demonstrated that adding complement can enhance the neutralizing activity in sera several fold as measured by plaque reduction neutralization (Takenaka et al., 1985); however, activating complement by the classical pathway has also been correlated with a more severe course of nephropathia epidemica (NE) after PUUV infection (Paakkala et al., 2000).

Cellular Immune Responses

The role of the human cellular immune response to hantavirus infection is not well defined. During infection in laboratory mice, $CD4^+$ and $CD8^+$ T cells have been shown to protect against HTNV infection in immunocompromised animals (Asada et al., 1987), and in other models, a strong correlation was observed between the number of HTNV-specific, $IFN\gamma^+$ $CD8^+$ T cells and the amount of viral antigen such that mice with large numbers of $IFN\gamma^+$ $CD8^+$ T cells had little to no detectable viral N-protein (Araki et al., 2003). Unlike human infection, however, hantavirus infection of *Mus* species does not cause notable disease (Araki et al., 2004). Hantavirus infection of humans results in the induction of strong immune responses. Elevated numbers of circulating T cells are found during HFRS (Huang et al., 1994), and biopsies of acute PUUV-infected patients reveal high numbers of activated and, predominantly $CD8^+$, T cells (Mustonen et al., 1994; Temonen et al., 1996). Similarly, increased numbers of cytokine-producing cells are seen in the lungs of patients with HPS (Mori et al., 1999). Both cytotoxic $CD8^+$ T cells and helper $CD4^+$ T cells (T_H1 and T_H2) appear important for controlling hantavirus infection in humans. $CD8^+$ T cell epitopes can be found in G_n, G_c, and N-proteins, but like the distribution of serum Ig, N-protein-specific $CD8^+$ T cells appear to be the most abundant (Van Epps et al., 1999; de Carvalho Nicacio et al., 2001; Van Epps et al., 2002). Patients diagnosed with PUUV and HTNV infection have elevated T_H1 (TNFα, IFNγ, and IgG3) and T_H2 (IL-6, IL-10, and IgG1) responses and elevated levels of proinflammatory cytokines such as interferon (IFN)α and interleukin (IL)-1 (Groen et al., 1994; Krakauer et al., 1994, 1995; Linderholm et al., 1996; Lundkvist et al., 1993; Temonen et al., 1996). The same is true during human infection with HPS-causing hantaviruses, where elevated levels of proinflammatory cytokines (IL-1α, IL-1β, and IFNα) and T_H1 (IL-2, IFNγ, and tumor necrosis factor (TNF)) and T_H2 cytokines (IL-4, IL-6, and TNFβ) are seen during fatal HPS cases (Mori et al., 1999). The correlation between the intensity of the immune response and the severity of clinical disease has further led to the suggestion that hantavirus disease pathogenesis is mediated by the immune system (see "Pathogenesis"). A more complete synopsis of all the immunological changes known to occur during hantavirus infection was recently reviewed by Klein and Calisher (2007).

Both T_H1 and T_H2 responses are characteristic of the immune responses elicited during the acute phase of infection of rodent reservoirs such as deer mice (SNV) (Schountz et al., 2007) and Norway rats (SEOV) (Klein et al., 2001; Tanishita et al., 1986). Yet as viral persistence is established, T cells from persistently infected deer mice predominantly expressed TGF-β, and most expressed FoxP3, consistent with inducible and T_H3 regulatory T cells (Schountz et al., 2007), and increased percentages of $CD4^+CD25^+FoxP3^+$ regulatory T cells and TGFβ mRNA expression are found during persistent SEOV infection of Norway rats (Easterbrook et al., 2007), supporting the hypothesis that suppressing the rodent host immune response is necessary to achieve long-term persistence (Meyer and Schmaljohn, 2000). Little is known about the regulatory T-cell response during human hantavirus infection, though. The presence of IL-10, IL-4, and TGFβ during the course of hantavirus infection could indicate a potential role for subsets of regulatory cells (Mills, 2004) in controlling the immune response to and immunopathology of human hantavirus infection. An analysis of acute dengue virus infection revealed increased numbers of $CD4^+CD25^{high}FoxP3^+$ T regulatory cells that are capable of suppressing vasoactive cytokine production, but due to relatively low frequencies, are thought to be insufficient to control immunopathology (Luhn et al., 2007). The extent of regulatory T-cell responses during hantavirus infection is not known, and what role these cells may play in limiting or contributing to hantavirus disease pathology will need to be explored.

Immunologic Memory

Secondary exposures to hantaviruses have not been reported, possibly owing to the formation of long-lived hantavirus-specific humoral and cellular immunologic memory. Studies by Ye et al. (2004) and Valdivieso et al. (2006) demonstrated that convalescent HPS patients maintain neutralizing antibody titers through 11 years postinfection. Similarly, high titers of PUUV neutralizing IgG antibody have been detected in the serum of convalescent patients up through 20 years postinfection (Horling et al., 1992; Lundkvist et al., 1993). Moreover, HTNV- and PUUV-specific memory $CD4^+$ and $CD8^+$ T cells, with frequencies as high as 100–300 per 10^6 for a single epitope, can persist for decades after infection (Van Epps et al., 1999, 2002). Furthermore, given the high degree of cross-reactivity of some T-cell N-protein epitopes (Van Epps et al., 1999) and the suggestion that cross-neutralizing antibody responses can be generated during infection and vaccination (de Carvalho Nicacio et al., 2002;

Hooper et al., 2001a, 2006), it is possible that exposure to one species of hantavirus could afford protection from a subsequent exposure to multiple species of hantaviruses.

EPIDEMIOLOGY

As stated above, hantaviruses that cause disease in humans are carried by rodents. To date, hantaviruses carried by insectivores, such as shrews and moles, have not been associated with human disease, although there is anecdotal evidence that these animals may also harbor pathogenic hantaviruses (Okumura et al., 2007). In general, infection of the natural rodent host results in a chronic carrier state without pronounced pathology or signs of disease. However, in-depth histological studies identified lesions within the lungs of deer mice infected with SNV and white-footed mice infected with NY-1 virus (Netski et al., 1999; Lyubsky et al., 1996). There is also a report that PUUV-infected animals are less likely to survive the winter, suggesting that infection has a negative effect on host fitness (Kallio et al., 2007). Disease in humans occurs when persons are exposed to contaminated rodent feces, urine, or saliva. The most common mode of transmission is thought to be the inhalation of aerosolized rodent droppings; however, contact with open wounds, rodent bites, and ingestion of contaminated material are also possible modes of transmission. Ingestion of virus as a mode of infection is not well documented; however, laboratory hamsters can be readily infected through the gut (gavage needle) with ANDV, supporting the possibility that hantaviruses could be transmitted by the ingestion of contaminated food (Hooper et al., 2008).

In humans, hantaviruses cause disease in the young and old, male and female. In most studies, HFRS and HPS occurred predominantly in working-age males. The preponderance of disease in this population is likely related to occupational exposure. Epidemiologic studies of HFRS and HPS report increased incidence of hantavirus disease in persons working or sleeping in environments inhabited by rodents, which include agricultural workers, forest workers, and soldiers (Mulic and Ropac, 2002; Sinclair et al., 2007; Vapalahti et al., 2003; Abu Sin et al., 2007). In many regions, hantavirus disease has a seasonal peak. For example, most of the cases in the 2005 outbreaks in Europe occurred in June and July (Heyman et al., 2007). The geographic range of hantavirus disease is shown in Fig. 23.2. The geographic range of the rodent hosts of pathogenic hantavirus are shown in Fig. 23.3. Although disease has not been detected in all of these regions, the presence of the rodent reservoir indicates that the potential for hantavirus disease exists. Endemic disease occurs in China, the Korean peninsula, Russia, Europe, and the Americas. A scan of the literature indicates that approximately 40 countries have reported hantavirus disease, the presence of virus, or serological evidence of infection (Figure 23.2 and Table 23.3), while several other countries have reported rare and sporadic HFRS in port cities that can probably be attributed to SEOV infections spread by rats carried port-to-port on ships.

China, Korea, and Far Eastern Russia

HFRS is a major health problem in China. It is estimated that from 1950 to 1997, there were 1.25 million cases of HFRS with 44,300 deaths (Yan et al., 2007; Fang et al., 2006; Kariwa et al., 2007). Liaoning province is one area with a particularly high incidence of HFRS, 141.7 cases/100,000 (Lin et al., 2007). South Korea reported 3039 cases of HFRS between 1997 and 2006 (DisWeb, 2003). During that time, there has been a trend of increased frequencies of cases with more than 400 cases per year for the last 3 years. Based on its geographic location, it is very likely that North Korea has a significant number of HFRS cases; however, as is the case for many countries, the number of recent HFRS cases was not readily available. One review from 1996 reported 316 HFRS cases in North Korea from 1961 to 1997 [sic] (Lee, 1996). In Far Eastern Russia, there were 4442 cases of HFRS between 1978 and 1997 (Tkachenko et al., 1998). Several other countries in Asia, including Australia, Fiji, Hong Kong, India, Indonesia, Japan, Malaysia, Mongolia, Myanmar, Singapore, Sri Lanka, Taiwan, Thailand, and Vietnam, have reported rare or sporadic cases of HFRS, or seroepidemiological evidence that hantaviruses exist and can cause infections (reviewed in Kariwa et al., 2007). The major viruses that cause HFRS in the Far East are HTNV, carried by *Apodemus agrarius* and SEOV carried by *Ratttus norvegicus*. Inactivated-virus vaccines against HFRS are licensed for use in China and South Korea (see "Vaccines").

Western Russia and Eastern Europe

HFRS has been a reportable disease in Russia since 1978. In a review of HFRS in Russia, (Tkachenko et al. (1998) reports that between 1978 and 1997, there were 109,082 cases in western Russia. Specific regions of Russia have reported relatively large outbreaks. For example, the Bashkiria region has consistently

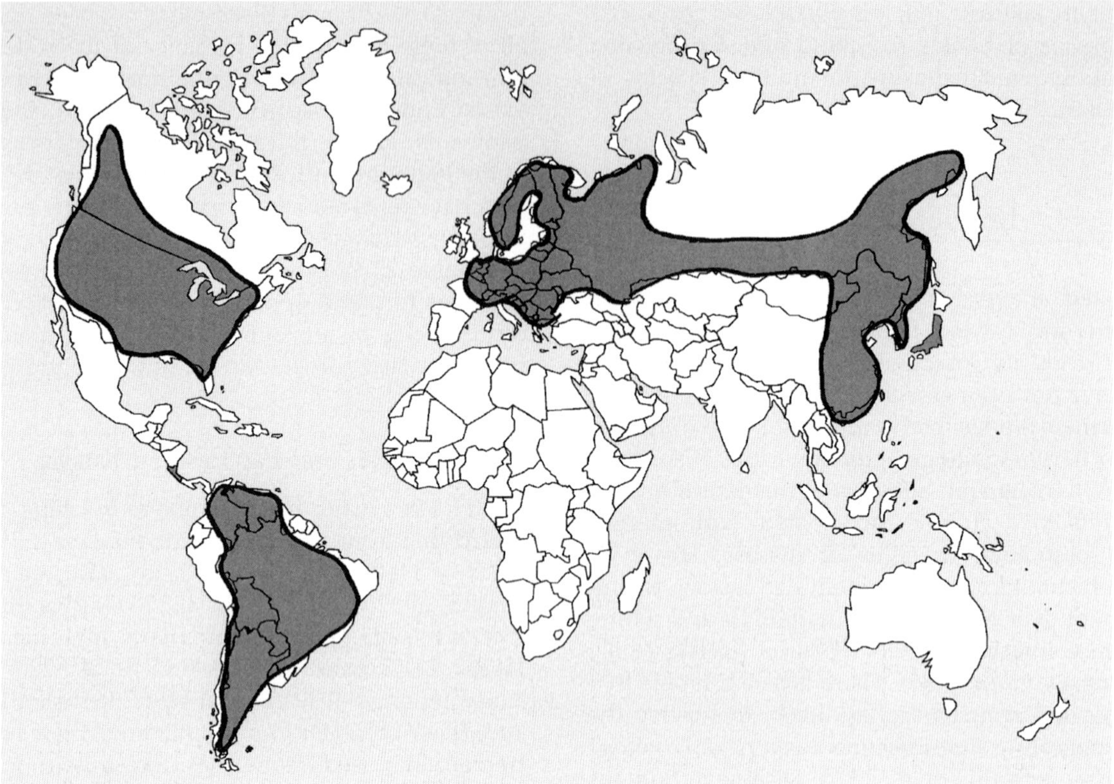

FIGURE 23.2 Distribution of hantavirus disease worldwide. Hantaviruses associated with HPS are found exclusively in North, Central, and South America. HFRS-associated viruses are widespread throughout Europe, the Balkans, Western Russia, China, and the Korean peninsula, Asia.

reported high number of cases, including epidemics in 1993 and 1997, where the number of cases approached 150 and 287 cases/100,000 persons, respectively (Niklasson et al., 1993; Tkachenko et al., 1998). More recently (in 2007), there were outbreaks (more than 3000 cases) in the vicinity of the towns of Voronezh and Lipetsk (Dybas, 2007). As in China and Korea, most of the HFRS cases in Russia occur in rural areas; however, there can be epidemics in urban areas, as was the situation in the 1997 Bashkiria outbreak (Tkachenko et al., 1998). In western Russia, most of the HFRS is caused by PUUV, carried by bank voles, *Clethrionomys glareolus*. Case-fatality rates indicate that strains of PUUV in Russia, such as strain K27 isolated from a fatal human case, cause a more severe form of disease (0.4% case fatality) than the strains of PUUV found in Finland and Sweden (0.1% case fatality) (Tkachenko et al., 1998). Several other countries in Eastern Europe, including Byelorus, Estonia, Georgia, Latvia, Lithuania, Poland, Romania, and Ukraine, have reported rare or sporadic cases of HFRS or serologic evidence that pathogenic hantaviruses are endemic (reviewed in Tkachenko et al., 1999; Vapalahti et al., 2003; Avsic-Zupanc, 1998). There is no HFRS vaccine licensed for use in Russia or Eastern Europe.

Northern and Central Europe/Scandinavia

A review published in 2003 indicated that 12 countries in Europe reported five or more cases of HFRS per year (Vapalahti et al., 2003). Other countries in Europe with rare or sporadic cases of HFRS or seroepidemiologic evidence of infection include Austria, Czech Republic, Great Britain, Portugal, and Switzerland (Lee, 1996; Vapalahti et al., 2003). It is likely that HFRS in Europe is underdiagnosed because the disease is relatively mild as compared to HFRS in Asia or HPS in the Americas. A study in Belgium reported that the seroprevalence was at 3.8%, well above the number of diagnosed cases (van der Groen et al., 1983). Sweden and Finland have the most cases (300 and 1000 per year, respectively) followed by

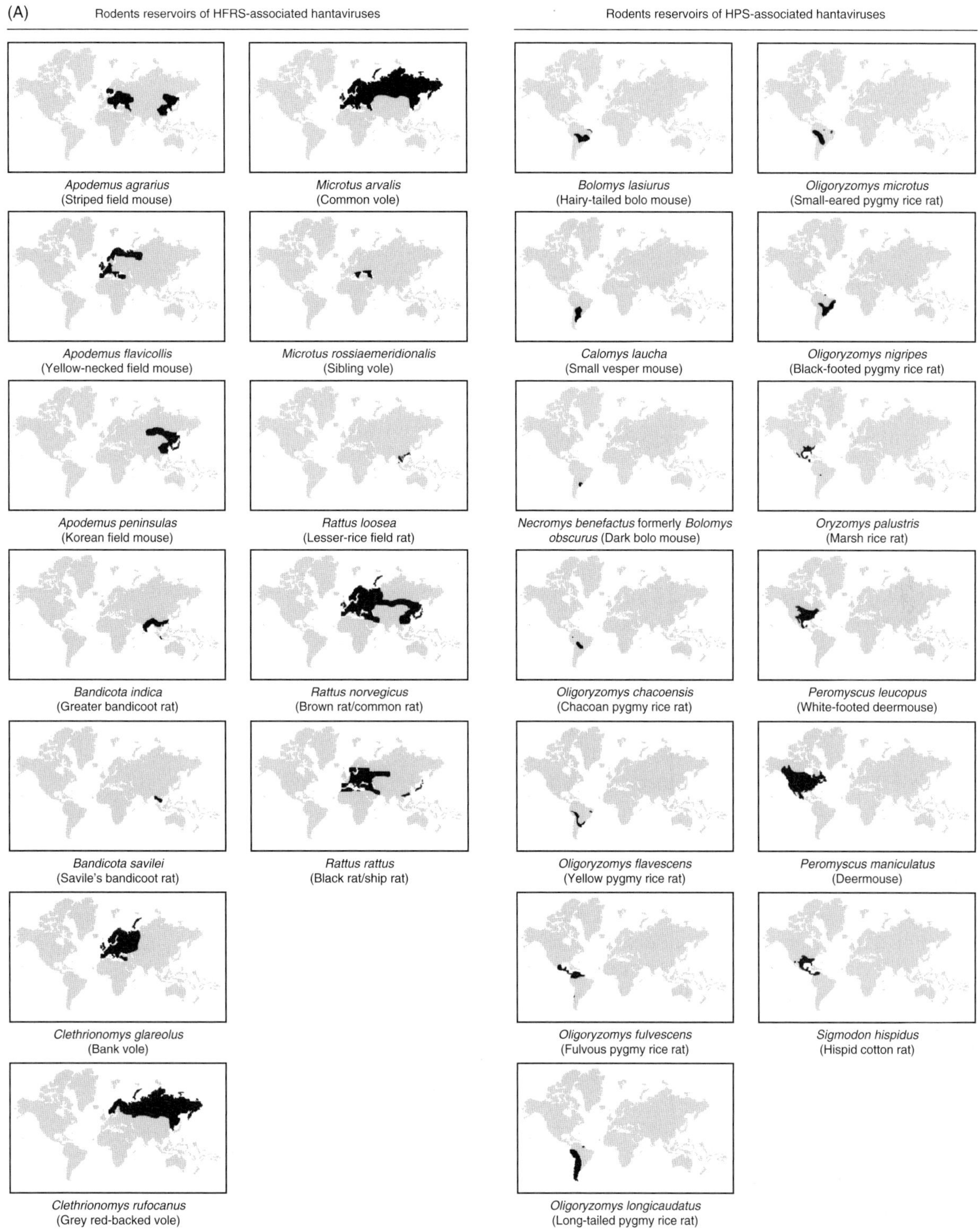

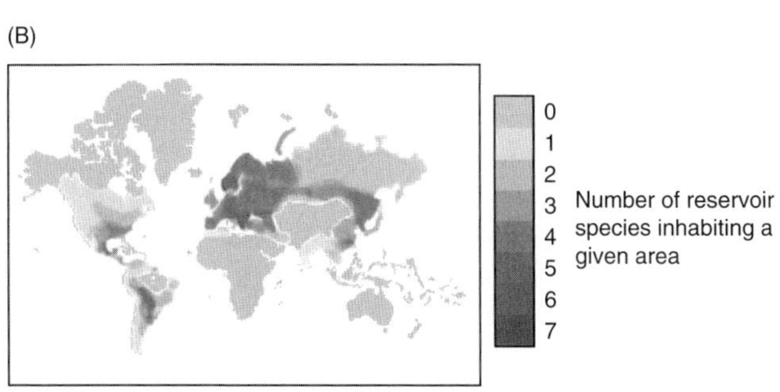

FIGURE 23.3 Distribution of hantavirus rodent reservoirs. (A) Schematic representation of the distribution of individual rodent species that are reservoirs of known pathogenic hantaviruses. (B) Estimate of the number of different rodent species in a given area that are reservoirs for hantaviruses known to cause human disease (see color plate section).

Germany, France, and Belgium, each with ~100 cases per year. These national numbers mask the health problem caused by HFRS in specific geographic regions. For example, the number of cases in northern Sweden is much higher than in the south. Most of the cases are in Vasterbotten County where the incidence is normally ~23.5 cases/100,000 (Settergren et al., 1988). In Finland, the number of cases in the province of Mikkeli was 70 cases/100,000, mostly in farmers (Vapalahti et al., 1999). PUUV is carried by the red bank vole, *C. glareolus*, and DOBV is carried by the yellow-necked mouse, *A. flavicollis*. These viruses are believed to cause most of the cases of HFRS in Northern Europe and Scandinavia.

Recently, HFRS has become more of a health concern in Europe. In 2005, there was a relatively large series of outbreaks in Belgium, France, the Netherlands, Luxembourg, and Germany (1114 cases) caused by PUUV (Heyman et al., 2007). In Germany, many of the cases occurred in relatively large cities, including Osnabuck and Cologne where the annual incidence was 8.5 and 4.2 cases/100,000, respectively (Abu Sin et al., 2007). In 2007, Sweden experienced an almost 10-fold increase in the annual number of HFRS (nephropathica epidemica) cases (Pettersson et al., 2008). Vasterbotten County accounted for 800 of the 2200 cases, with two fatalities. In Europe and elsewhere, there is a suspected link between climate change, food-source production, reservoir population, and incidence of human disease. Climatic conditions that result in increased production of food source (e.g., mast years resulting in high yields of oak and beech nuts) can, in turn, result in increase in reservoir population and incidence of human disease (Heyman et al., 2007; Vapalahti et al., 2003). Although HFRS in Europe has not resulted in high numbers of fatalities, this disease does have a substantial impact on the healthcare system because many patients are hospitalized and some require hemodialysis. There is no HFRS vaccine licensed for use anywhere in Northern and Central Europe or Scandinavia.

Southern Europe/Balkans

A review in 1999 indicated that there were approximately 100 HFRS cases per year in the Balkans (Avsic-Zupanc, 1998). Essentially, all of the countries in the Balkans have reported cases of HFRS (Table 23.3). There is also evidence of HFRS in Albania (Antoniadis et al., 1996). The military conflicts in the former Yugoslavia resulted in increased numbers of HFRS cases, as would be expected based on the increased numbers of personnel working and sleeping outdoors (Avsic-Zupanc, 1998). Severe cases of HFRS in the Balkans are usually caused by DOBV carried by the yellow-necked field mouse, *A. flavicollis* (Avsic-Zupanc et al., 1992). PUUV is also circulating in this region and has been associated with disease (Lundkvist et al., 1997). There is no HFRS vaccine licensed for use in Southern Europe or the Balkans.

North America

HPS in the Americas is relatively rare, but the disease is among the most pathogenic of any acute viral disease. Like HFRS, HPS afflicts mostly working-age males in rural areas (CDC, 2007). Since the first outbreak of HPS in the "Four Corners region" of the southwestern United States, there have been approximately 465 cases of HPS with 35% case fatality. In Canada, there have been ~100 cases. The hantavirus responsible for most of the HPS cases in the United States is SNV carried by deer mice, *Peromyscus maniculatus*. Other hantavirues that have caused disease in the United States include Black Creek Canal virus (BCCV), Bayou virus (BAYV), and New York virus (NYV). There is no HPS vaccine licensed for use in North America.

Central America

HPS is not widespread in Central America. The only country that has reported significant outbreaks is Panama (Bayard et al., 2004). Specific regions, such as the Azuero peninsula, have seroprevalance rates as high as 30% (Armien et al., 2004). A recent serial serosurvey revealed a very high seroprevalence against hantaviruses in certain communities where changes in agribusiness may have brought hantavirus-carrying rodents into more frequent contact with humans. The hantavirus associated with human disease in Panama is Choclo virus, carried by *Oligoryomys fulvescens* (Vincent et al., 2000). There is no HPS vaccine licensed for use in Central America.

South America

Several countries in South America have reported hantavirus disease including Argentina, Bolivia, Brazil, Chile, Columbia, Paraguay, Uruguay, and Venezuela (PAHO, 2004). A list of the viruses that cause HPS in South America is shown in Table 23.1, and a map of the rodent geographic range is shown in Fig. 23.3. ANDV, carried by the long-tailed pygmy rice rat, *O. longicauddatus*, and related viruses cause

TABLE 23.3 Worldwide distribution of HPS and HFRS cases

Country[a]	Regions of interest	Number of cases	Years	Reference
East Asia				
China		20,000–50,000	per year	Fang et al. (2006)
	Liaoning province	141.7/100,000	per year	Lin et al. (2007)
North Korea		316	1961–1996	Lee (1996)
South Korea		100–400	per year	DisWeb (2006)
Russia and Eastern Europe				
Russia		3000	per year	Vapalahti et al. (2003)
	Bashkiria region	44.6/100,000	per year	Tkachenko et al. (1999)
Hungary		136	1952–1980	Lee (1996)
Northern and Central Europe/Scandinavia				
Belgium		372	2005[b]	Heyman et al. (2007)
	Luxenburg rovince	33.8/100,000	2005[b]	Heyman et al. (2007)
Denmark		10	per year	Vapalahti et al. (2003)
Finland		1000	per year	Vapalahti et al. (2003)
	Mikkeli province	70/100,000	per year	Vapalahti et al. (2003)
France		253	2005[b]	Heyman et al. (2007)
	Ardennes district	32.7/100,000	2005[b]	Heyman et al. (2007)
Germany		448	2005[b]	Heyman et al. (2007)
	Osnabuck city	8.5/100,000	2005[b]	Heyman et al. (2007)
Italy		14	1984–1987	Lee (1996)
Luxembourg		14	2005[b]	Heyman et al. (2007)
Netherlands		27	2005[b]	Heyman et al. (2007)
Norway		50	per year	Vapalahti et al. (2003)
Sweden		300	per year	Vapalahti et al. (2003)
	Vasterbotten county	23.5/100,000	per year	Settergren et al. (1988)
Southern Europe/Balkans				
Bosnia-Herzegovina		487	1952–1995	Avsic-Zupanc (1998)
Bulgaria		399	1952–1995	Avsic-Zupanc (1998)
Croatia		134	1952–1995	Avsic-Zupanc (1998)
Greece		21	1952–1995	Avsic-Zupanc (1998)
Macedonia		10	1952–1995	Avsic-Zupanc (1998)
Montenegro		129	1952–1995	Avsic-Zupanc (1998)
Serbia		228	1952–1995	Avsic-Zupanc (1998)
Slovakia		10	per year	Vapalahti et al. (2003)
Slovenia		106	1952–1995	Avsic-Zupanc (1998)
North America				
Canada		88	1993–2004	PAHO (2004)
United States		465	1993–2007	http://www.cdc.gov
Central America				
Panama		35	1993–2004	PAHO (2004)
South America				
Argentina		592	1993–2004	PAHO (2004)
Bolivia		36	1993–2004	PAHO (2004)
Brazil		423	1993–2004	PAHO (2004)
Chile		331	1993–2004	PAHO (2004)
Columbia		8	2006	Xinhua news service 14 March 2006
Paraguay		99	1993–2004	PAHO (2004)
Uruguay		48	1993–2004	PAHO (2004)
Venezuela		2	1993–2004	PAHO (2004)

[a] Additional countries reporting hantavirus disease, or seroepidemiological evidence of pathogenic hantaviruses, are listed in the text.
[b] Epidemic year.

most of the cases in Chile, Argentina. Juquitiba virus, carried by the black-footed pygmy rice rat, *O. nigripes*, is associated with most of the cases in Brazil. Interestingly, ANDV is the only hantavirus associated with person-to-person transmission, and many of the cases in Chile and Argentina have occurred in clusters (Lazaro et al., 2007; Toro et al., 1998; Vial et al., 2006; Wells et al., 1997). There is no HPS vaccine licensed for use in South America.

CLINICAL DISEASE

HPS, HFRS, and NE are three clinical diseases caused by hantaviruses. The diseases are similar in many respects, and share a number of clinical manifestations which have been the focus of several recent reviews (Jonsson et al., 2007; Mertz et al., 2006; Muranyi et al., 2005). Clinical disease, which is often divided into four to five distinct phases, occurs after an incubation period varying from about 7 to 36 days (Muranyi et al., 2005; Vial et al., 2006; Young et al., 2000) and results from the infection of endothelial cells and subsequent increase in vascular permeability.

Hantavirus Pulmonary Syndrome

After the virus incubation period, the onset of HPS begins with a prodromal, febrile period that typically lasts 3–5 days but can last as long as 10 days (Hjelle and Glass, 2000; Peters and Khan, 2002). Symptoms during the onset of HPS are nonspecific and flu-like and include fever, myalgia, malaise, and may also consist of headache, dizziness, anorexia, abdominal pain, nausea, vomiting, and diarrhea (Enria et al., 2001). Other reported early symptoms included a transient skin rash and conjunctival suffusion. Hantavirus-specific IgG and IgM antibodies can also be detected during this phase. At the end of this phase, however, more serious symptoms may present, such as the abrupt onset of a nonproductive cough, and shortness of breath and tachypnea that mark the beginning of the cardiopulmonary phase (Hallin et al., 1996; Levy and Simpson, 1994). It is at this point that most patients seek medical attention and, in severe cases, may rapidly develop pulmonary edema and respiratory failure, usually within 4–24h. Whereas chest radiographs of patients may be normal during the prodromal febrile phase, interstitial edema and the presence of bilateral pulmonary infiltrates usually develop within the first few hours of the cardiopulmonary phase (Ketai et al., 1994). A presumptive diagnosis may be made during this phase based on the presence of pulmonary edema and an abnormal peripheral blood smear, including thrombocytopenia, myelocytosis, a lack of significant toxic granulation in neutrophils, hemoconcentration, and >10% immunoblastic lymphocytes (Mertz et al., 2006). Furthermore, the cardiogenic shock syndrome leads to arrhythmias, lactic acidosis, hypotension, and oliguria, while pulmonary edema contributes to tachypnea, exertional dyspnea, and nonproductive cough with hypoxemia as a common feature. Hypovolemia occurs as a result of high-protein-fluid leakage from the blood into the lung interstitium and alveoli, and this further aggravates shock symptoms. The cardiopulmonary phase usually lasts 2–4 days, but in severe cases, approximately half of the patients succumb to hypoxia and circulatory compromise within 24–48h of developing pulmonary edema. Fatality rates for the most common HPS hantaviruses, SNV and ANDV, range from 30% to 50%, respectively, while deaths from other HPS strains occur in approximately 15% of the cases (Doyle et al., 1998). Patients that survive the cardiopulmonary phase enter the third or diuretic phase, resulting in the rapid clearance of pulmonary edema and spontanteous diuresis followed by cessation of fever and shock. The final convalescent phase may last up to 2 months and may be marked by weakness, fatigue, and abnormal pulmonary function including abnormal diffusion capacity, but usually results in full recovery.

While New World hantaviruses are largely associated with pulmonary dysfunction, cases of renal involvement have been associated with BCCV and BYV, and renal failure was a common feature of Oran-virus-associated HPS infection (Levis et al., 1998; Nichol, 2001). Other clinical symptoms may include conjunctival injection, facial flushing, pharyngeal congestion, petechiae, abnormal urine analysis, and/or frank hemorrhages (Lazaro et al., 2000).

Hemorrhagic Fever with Renal Symptoms

The early clinical manifestations of HFRS are similar to HPS with an initial febrile phase that manifest with acute flu-like symptoms such as high fever, backache, abdominal pain, chills, myalgia, malaise, and bradycardia. The febrile phase, which can last 3–5 days, is also characterized by photophobia, pharynx enathema, diffuse reddening of the face, development of petechia on the palate, hematuria, atypical gross proteinuria, and can include conjunctival hemorrhage and transient visual impairment. After the onset of fever, patients enter a 3–6-day hypotensive phase with symptoms of varying degrees including leukocytosis, thrombocytopenia, elevated serum creatine, tubulinterstitial nephritis, necrotizing glomerulonephritis, IgA nephropathy, petechiae, and nose bleeds to more severe complications that can include intracranial hemorrhages (Linderholm and Elgh, 2001). Nausea and vomiting with severe clinical shock and hypotension, likely due to massive capillary leakage, have also been reported during the hypotensive phase (Warner, 1996). Renal complications of HFRS, acute renal failure caused by interstitial hemorrhage and interstitial infiltrates (Beers and Berkow, 2005; Sirotin and Keiser, 2001), will typically manifest in the oliguric phase stage of illness. During this phase that begins approximately 8 days after the onset of symptoms, hemorrhagic manifestations become prominent, patients have reported abdominal and/or back pain, and have reduced urinary output. After this, the polyuric or diuretic phase, beginning approximately on day 11, is characterized by a dramatic increase in urinary output, as much as 3–6 L/day, and considerable fluid and electrolyte shifts are observed. The subsequent convalescent phase can last from weeks to several months. Sequelae such as chronic renal failure and hypertension (Makela et al., 2000) can occur during this phase but are rare. Approximately 10% to 15% of HFRS cases are clinically severe with 6–15% of cases becoming fatal. Note that for all phases of HFRS, the overall number of clinical phases and severity may differ considerably on a patient-to-patient basis. Other clinical manifestations that may be involved during the course of HFRS include visual impairment; acute myopia; CNS involvement, resulting in seizures, myocarditis, gastrointestinal hemorrhages; and thyroid, liver, pancreas, and lung involvement.

Nephropathia Epidemica

NE is a less severe form of disease similar to HFRS that is typically caused by PUUV and occurs most commonly in Scandinavia and Central Europe. The majority of NE cases are asymptomatic (~90%), and the overall case-fatality rate of severe NE with acute renal failure is far lower than HFRS (approximately 0.1% to 1% for NE versus 5% for HRFS) (Warner, 1996; Linderholm and Elgh, 2001). Clinically, NE presents similarly to HFRS, while severe neurological complications such as paralysis of the bladder or seizures are reported in approximately 1% of NE patients. After a 3–4-day febrile phase and a 3-day oliguric phase, patients develop polyuria. The convalescent phase may extend over several weeks, and sequelae are rare.

TREATMENT

Treatment of patients with HPS or HFRS remains largely supportive. With respect to HFRS, fluid restriction during the hypotensive and oliguric phases reduced mortality with KHF (Steer, 1955). Nonsteroid anti-inflammatory drugs and any drug that is toxic to the kidneys should be avoided, and close management of electrolyte balance is critical (Linderholm and Elgh, 2001). Dialysis may be appropriate in severe cases of HFRS but is rarely needed in cases of NE.

Intensive care for HPS patients includes careful fluid management, monitoring of cardiac output, and ventilatory support (Castillo et al., 2001; Hallin et al., 1996; Levy and Simpson, 1994). Fluid balance must be vigilantly managed, but not overloaded, due to complication with pulmonary edema (Warner, 1996). Cardiac output and function should be monitored closely and maintained. Supplemental oxygen may be sufficient to treat patients with mild HPS, but in severe cases, disease progression and respiratory failure will occur rapidly and may require venoarterial extracorporeal membrane oxygenation (ECMO) therapy to provide cardiopulmonary support. The efficacy of ECMO therapy during early HPS cases is debatable (Hallin et al., 1996; Crowley et al., 1998; Enria et al., 2001), but more recent studies involving ECMO treatment in the United States (Mertz et al., 2006) and Chile (Tomicic et al., 2005) have been more promising, with a survival rate of approximately 85%. Much of this recent success is attributable to advances in diagnostic tools (e.g., ELISA, PCR, Western blot, strip immunoblot assay) that provide a more rapid diagnosis resulting in more rapid and aggressive treatment.

The use of steroid therapy has also been reported to have some effect in influencing the positive outcome of hantavirus diseases. In two separate cases, patients suffering from PUUV infection improved after extended administration of corticosteroids (Dunst et al., 1998; Seitsonen et al., 2006). The use of steroids to treat hantavirus disease is not without drawbacks, though, as immunosuppression has resulted in the occurrence of secondary infections (Siqueira et al., 2007). Given the apparent relationship between the immune response and pathogenesis of hantavirus disease, the use of methylprednisolone in the treatment of HPS is being investigated in a controlled study in Chile (Mertz et al., 2006).

There is currently no Food and Drug Administration (FDA) approved antiviral drug or immunotherapeutic agent approved for the treatment of hantaviral diseases. The antiviral drug ribavirin has shown success in reducing viral titers, increasing survival rates, and reducing the severity of HFRS in patients in China (Huggins et al., 1991) and in HTNV-infected suckling mice (Huggins et al., 1986). The use of ribavirin as a treatment for HPS has not proved successful (Chapman et al., 1999; Mertz et al., 2004); however, the number of HPS patients to test the efficacy of ribavirin has been inadequate. In these studies, HPS patients treated with ribavirin did not begin until after the onset of the cardiopulmonary phase of disease. In contrast, the HFRS patients treated with ribavirin began treatment before the onset of renal complications, suggesting that the efficacy of ribavirin treatment may depend largely on the phase of infection and the severity of disease when ribavirin is administered. While ribavirin may not be effective in treating patients after the onset of clinical symptoms, it may be possible to use ribavirin as a prophylactic treatment for people at high risk of contracting disease such as recent household contacts of ANDV index cases where there is the threat of person-to-person transmission (Martinez et al., 2005; Padula et al., 1998). Additionally, evidence that patients with lower titers of neutralizing antibodies often develop a more severe disease (Bharadwaj et al., 2000) had led to the speculation that administration of high-dose neutralizing antibodies as a postexposure immunoprophylatic may reduce disease severity and improve survival (see "Postexposure Immunoprophylaxis").

PATHOGENESIS

In rodents, hantavirus infection results in a largely nonpathogenic, persistent infection (Meyer and Schmaljohn, 2000). In humans, however, clinically productive hantavirus infection results in one of the two diseases, HPS or HFRS, that are both characterized by changes in endothelial cell permeability and vascular edema (Cosgriff, 1991; Lee, 1991; Nolte et al., 1995; Zaki et al., 1995). While the mechanisms underlying the pathogenesis of HPS and HFRS are not completely understood, the direct effects of viral infection of endothelial cells and the immune response to hantavirus infection have been proposed as playing important roles.

Possible Direct Role of Viral Infection of Endothelial Cells on Pathogenesis

Analysis of hantavirus-infected patients (Huang et al., 1994; Zaki et al., 1995; Nolte et al., 1995) and experimentally infected hamsters (Hooper et al., 2001b; Milazzo et al., 2002; Wahl-Jensen et al., 2007) suggests that endothelial cells are the primary targets of hantaviruses; an observation validated by multiple in vitro studies (Gavrilovskaya et al., 1998, 1999; Geimonen et al., 2002; Yanagihara and Silverman, 1990). Infection of endothelial cells appears to be mediated by the expression of β1 and β3 integrins on the surface of endothelial cells as forced expression of these integrins renders cells permissive to hantavirus infection while antibody blockade of these integrins prevents infection (Gavrilovskaya et al., 1998, 1999). Hemorrhagic diseases, in general, are caused by the dysregulation of permeability of the vascular endothelium. Integrins expressed by endothelial cells, such as αvβ3, play key roles in maintaining vascular integrity and regulating permeability (Hynes et al., 1999; Kevil et al., 1998; Leavesley et al., 1993; Sugimori et al., 1997; Tsukada et al., 1995). In this respect, several human hemorrhagic diseases, including Glanzmann's disease and Goodpasture's syndrome, and at least three autoimmune hemorrhagic diseases result from the dysregulation of β3-integrin function (Borza et al., 2003; Gunwar et al., 1991; Hodivala-Dilke et al., 1999; Joutsi-Korhonen et al., 2004; Kalluri et al., 1994; Watkins et al., 2002). Moreover, β3-integrin-deficient mice have vascular permeability defects and mucocutaneous hemorrhage similar to that seen during HFRS (Hodivala-Dilke et al., 1999). Accordingly, infection of endothelial cells by nonpathogenic hantaviruses such as Prospect Hill (PH) virus involves the β1 integrin, whereas infection by pathogenic hantaviruses is associated with β3 integrin (Gavrilovskaya et al., 1998, 1999). Pathogenic hantaviruses bind inactive αvβ3 integrin structures (Raymond et al., 2005), suggesting that infection may interfere with the formation of active αvβ3 conformations which subsequently would inhibit normal β3-integrin function, as seen by Gavrilovskaya et al. (2002). This has lead to the suggestion that virus binding to αvβ3 integrin may disrupt the integrity of endothelial cell junctional complexes leading to increased permeability. Endothelial cells are held together by three major types of junctions, i.e., tight junctions, gap junctions, and adherens junctions, along with numerous other adhesion molecules including PECAM that can interact with αvβ3 integrin (reviewed by Bazzoni and Dejana, 2004; Wallez and Huber, 2007). Inhibition of αvβ3-integrin signaling has been shown to induce the formation of incomplete adherens junctions (Ozaki et al., 2007); however, there is little direct visual evidence that hantavirus infection in vivo results in gap formation between endothelial cells (Wahl-Jensen et al., 2007; Zaki et al., 1995). Nevertheless, in vivo disruption of adherens junctions by targeting VE-cadheren can result in increased permeability without the formation of visible gaps (reviewed by Bazzoni and Dejana, 2004). Still, other in vitro experiments have demonstrated that infection in and of itself does not affect the integrity of adherens junctions (Sundstrom et al., 2001) or increase permeability (Niikura et al., 2004; Sundstrom et al., 2001). Alternatively, hantavirus may influence permeability by altering the function of endothelial cell cation channels by promoting the influx of Ca^{2+} through αvβ3-integrin signaling (Bhattacharya et al., 2000; Kawasaki et al., 2004). TNFα has been shown to amplify Ca^{2+} influx and endothelial cell permeability by increasing the expression of the TRPC1 Ca^{2+} channel (Paria et al., 2004). Moreover, Ca^{2+} signaling can increase endothelial cell permeability by disassembling VE-cadherin junctions (Sandoval et al., 2001). Intriguingly, TNFα increased the permeability of hantavirus-infected cells and further prolonged permeability even after TNFα was removed (Niikura et al., 2004). Whether this is the result of increased Ca^{2+} signaling is unknown, but it may reflect the importance of endothelial cell and/or adaptive antiviral responses to elicit full hantavirus pathogenesis.

Possible Role of Innate Antiviral Responses on Pathogenesis

Nonpathogenic hantaviruses are not associated with human disease even though, like pathogenic species, they can infect human endothelial cells (Yanagihara and Silverman, 1990). Early experiments demonstrated that HTNV infection of endothelial cells can be blocked by adding exogenous type I IFN (Pensiero et al., 1992), and Geimonen et al. (2002) demonstrated that there is a distinct temporal difference in the induction of the interferon response between pathogenic and nonpathogenic hantavirus strains. Within 24 h of infection, nonpathogenic hantaviruses upregulated numerous interferon-inducible genes (ISG), whereas pathogenic strains largely failed to induce ISG expression. Both pathogenic and nonpathogenic viruses induced IFNβ late after infection (4 days), yet both can inhibit signaling from exogenous type I interferon via STAT-1/2 (Spiropoulou et al., 2007). How pathogenic hantaviruses regulate endothelial cell interferon responses is unknown, but it may relate to portions of the G_n cytoplasmic tail (Alff et al., 2006) and the presence of immuno-tyrosine activation motifs (ITAMs) (Geimonen et al., 2003), as has been demonstrated for other viruses (Beaufils et al., 1993; Dehghani et al., 2002; Lee et al., 1998a; Miller et al., 1995; Willems et al., 1995; Xu et al., 1999).

Aside from forming a barrier within the vasculature and regulating vascular permeability, endothelial cells also secrete cytokines and chemokines, and upregulate adhesion molecules and costimulatory molecules that can prevent infection and can recruit and stimulate cells of the immune system (Choi et al., 2004). Infection of endothelial cells does not appear to directly affect the expression of many cytokines, adhesion and costimulatory molecules (Sundstrom et al., 2001). Rather, expression of these molecules appears to be regulated by inflammatory cytokines such as IFNγ and TNFα. Both pathogenic and nonpathogenic hantaviruses have been shown to induce the translocation of NF-κB and interferon regulatory factors to the nucleus and the subsequent upregulation of leukocyte chemokines such as RANTES, IL-8, and IP-10 (Sundstrom et al., 2001), albeit somewhat species-specifically. At later times after infection, endothelial cells infected by some pathogenic hantaviruses have also been shown to upregulate cytokines such as IL-6, which, along with IL-1 from infected or activated macrophages, may cause increases in endothelial permeability (Kerkar et al., 2006; Maruo et al., 1992; Nooteboom et al., 2006) and have been implicated in the pathogenesis of HFRS (Linderholm et al., 1996). If cytokines mediate vascular permeability during hantavirus disease in vivo, then histological evidence from human cases or experimentally infected hamsters suggests that it does not necessarily involve the restructuring of endothelial gap junctions (Wahl-Jensen et al., 2007; Zaki et al., 1995). Alternatively, IL-1 and IL-6 may synergistically induce vascular endothelial growth factor (VEGF) expression by endothelial cells (Loeffler et al., 2005; Sironi et al., 1989) and the VEGF induced under those conditions could lead to the formation of vesiculo-vacuolar organelles (VVOs) within endothelial cells (reviewed in Dvorak and Feng, 2001; Feng et al., 1999). These cytoplasmic, grape-like clusters provide a direct link between the luminal and ablumenal sides of endothelial cells and have been shown to provide the major route of macromolecular extravasation at sites of increased vascular permeability in veins associated with experimental tumors and in animal models of allergic inflammatory eye disease. Further studies are needed to determine if the plasma leakage in HPS involves a paracellular or transcellular mechanism. Inducible nitric oxide synthase (iNOS), which can be produced by endothelial cells, neutrophils, and macrophages and is known to alter alveolar epithelial and endothelial cell permeability, is upregulated during HPS (Davis et al., 2002). There is further evidence that nitric oxide can upregulate αVβ3-integrin expression on endothelial cells, in vitro (Lee et al., 2000), suggesting a mechanism by which the immune response to hantavirus may propagate hantavirus infection. Additionally, several cytokines produced by lung endothelial cells, epithelial cells, and macrophages are known to influence the deposition of fibrin that is characteristic of HPS pathology in both hamsters (Wahl-Jensen et al., 2007) and humans (Zaki et al., 1995). Permeability of the vascular endothelium is accompanied by an influx of plasma proteins including fibrinogen, which is converted to fibrin at sites of inflammation. TNFα, possibly from macrophages or T cells, can induce the expression of TGFβ from lung fibroblasts (Sullivan et al., 2005) and epithelial cells (Warshamana et al., 2001), which can subsequently act in concert with TNFα to enhance the formation of fibrin by inhibiting mechanisms of fibrinolysis (Idell et al., 1992). The presence of fibrin can further enhance vascular permeability, possibly by directly inducing the secretion of vasoactive IL-1 from mononuclear cells such as macrophages (Perez and Roman, 1995). Whether regulatory cytokines such as IL-10, which can inhibit the synthesis of inflammatory cytokines and chemokines and decrease fibrin deposition by increasing fibrinolysis (Okada et al., 2000), can improve the course of HPS has not been investigated.

These data suggest that the ability of pathogenic hantaviruses to regulate aspects of the endothelial cell antiviral response may contribute to hantavirus disease pathogenesis. The rapid induction of type I IFN and leukocyte chemokine responses in response to nonpathogenic hantaviruses may contribute to decreased viral replication and dissemination, allowing for more rapid viral clearance which may ultimately lead to a lack of disease. In contrast, the delayed type I IFN responses stimulated by pathogenic hantaviruses may lead to increased viral replication and dissemination, resulting in large numbers of infected endothelial cells secreting vasoactive cytokines and attracting activated leukocytes.

Possible Role of Adaptive Immune Response in Pathogenesis

The noticeable lack of cytopathic effect after hantavirus infection of endothelial cells and the presence of activated T cells during hantavirus disease pathogenesis have led to the speculation that hantavirus disease pathology is mediated, at least in part, by hantavirus-specific effector T cells. Consistent with this idea, increased numbers of activated $CD8^+$ T cells are observed during the acute phase of HFRS (Huang et al., 1994) and acute PUUV infections (Mustonen et al., 1994; Temonen et al., 1996) with disproportionate CD4 to CD8 T-cell ratios (Chen and Yang, 1990). Similarly, large numbers of $CD4^+$ and $CD8^+$ T cells and large numbers of cytokine-producing cells are present in the lungs, spleens, and hearts of HPS patients (Mori et al., 1999; Nolte et al., 1995; Zaki et al., 1995), and the numbers of T cells present in the blood of acute HPS patients can correlate with the clinical severity of the disease (Kilpatrick et al., 2004). In fatal HPS cases, mononuclear cell infiltrates in these tissues were found to include a mixed population of macrophages and T cells which consisted of a disproportionately large number of activated $CD8^+$ T cells. Further studies suggest a genetic linkage between the severity of disease, and MHC haplotype has been observed in patients with NE and HPS, further linking $CD8^+$ T cell responses with disease. Makela et al. (2002) and Mustonen et al. (1996) showed that the HLA-B8-DR3 haplotype was associated with a more severe NE disease outcome) and that the HLA-B27 haplotype was associated with milder disease (Mustonen et al., 1998). Other data suggest a link between the HLA-B35 haplotype and a more severe form of HPS caused by SNV (Kilpatrick et al., 2004).

The extent to which T cells may mediate the pathogenesis of hantavirus infection may also be influenced by infected endothelial cells themselves (reviewed in Choi et al., 2004). Human endothelial cells constitutively express both MHC I and MHC II complexes, allowing them to directly present viral peptides to both activated $CD4^+$ and $CD8^+$ T cells. Moreover, endothelial cells also express a variety of early and late costimulatory molecules and secrete cytokines that can augment or prolong T-cell responses, including IL-6, which enhances T cell–endothelial cell interactions (Watson et al., 1996). $CD8^+$ T cells typically eliminate virus-infected cells using either a perforin/granzyme- or Fas/FasL-mediated death pathway. However, the lungs of HPS patients or infected hamsters reveal little obvious damage to endothelial cells (Hooper et al., 2001b; Nolte et al., 1995; Wahl-Jensen et al., 2007; Zaki et al., 1995), suggesting that if T cells contribute to endothelial cell permeability, then vascular leakage is caused by the release of vasoactive cytokines, rather that endothelial cell apoptosis. Why infected endothelial cells are not subject to elimination by $CD8^+$ T cells and natural killer (NK) cells is still not clear, but elevated levels of perforin, granzyme B, and the epithelial cell apoptosis marker caspase-cleaved cytokeratin-18 in the serum of patients infected with PUUV argues against a global inhibition of perforin and granzyme synthesis. This suggests that infected epithelial cells do not benefit from the same type of protection as endothelial cells (Klingstrom et al., 2006). Increased levels of TNFα and transforming growth factor beta (TGFβ) have been found in kidney biopsies from acute NE patients (Temonen et al., 1996). Moreover, genetic predisposition to produce high amounts of TNF has been associated with severe NE (Kanerva et al., 1998b). Systemic levels of inflammatory cytokines have also been reported in plasma of HFRS patients and include TNFα, soluble TNFα receptor, IFNα, IFNγ, IL-6, and IL-10 (Linderholm et al., 1996). Similarly, Mori et al. (1999) described an increased prevalence of cytokines such as IL-1α, IL-1β, IL-2, Il-4, IL-6, TNFα, TNFβ, and IFNγ in the lungs and spleens of fatal HPS cases. Many of these cytokines are produced by activated T cells, and all of them are implicated in increased vascular permeability (Beynon et al., 1993; Kerkar et al., 2006; Kotowicz et al., 2004; Maruo et al., 1992; Nooteboom et al., 2006). Consistent with a role for TNFα in immunopathology, mouse models of influenza virus (Xu et al., 2004) and respiratory syncytial virus infection (Rutigliano and Graham, 2004) have shown that the lung injury caused during infection is primarily due to CD8$^+$ T-cell-derived TNFα.

Hantaviral antigens have been detected in the macrophages, dendritic cells, and, to a lesser extent, lymphocytes (Huang et al., 1994; Zaki et al., 1995; Nolte et al., 1995). Whether interactions between T cells and hantavirus-infected, antigen-presenting cells or direct infection of T cells can lead to a dysregulated T-cell responses and overproduction of vasoactive cytokines or can cause the T-cell apoptosis and lymphopenia as was observed shortly after infection in both humans and experimentally infected hamsters (Wahl-Jensen et al., 2007; Zaki et al., 1995) is not known.

Taken together, these data suggest multiple ways that the innate and adaptive immune responses may contribute to hantavirus disease pathology. How these mechanisms contribute to disease will need to be explored further in an animal model that mimics hantavirus pathology in humans (e.g., Syrian hamsters). Ultimately, many of these mechanisms may work synergistically to promote disease, as shown in Fig. 23.4. Endothelial cells infected by pathogenic strains of hantaviruses, likely are unable to upregulate early type I IFN responses, allowing for efficient viral replication and dissemination. As T cells become activated, they migrate to the site of infection in response to endothelial-cell-derived chemokines, and encounter infected endothelial cells expressing viral antigens. The continued interactions between activated T cells and endothelial cells and vasoactive cytokines secreted by both may then act synergistically to induce changes in vascular permeability and edema consistently seen in hantavirus disease pathology.

There is a large body of circumstantial evidence implicating vasoactive cytokines in the pathogenesis of hantaviral diseases. Whether T cells are the primary source of these cytokines has yet to be determined. Still, it is likely that disease pathogenesis is a product of prolonged exposure to these cytokines and not merely the presence of these cytokines. Vaccines are likely to play an important role in limiting the production of these cytokines in the event of subsequent exposure to hantaviruses by either preventing infection altogether by neutralizing antibodies or by stimulating a rapid and robust memory T-cell response that will eliminate virus quickly before prolonged exposure to cytokines can cause disease.

VACCINES

The identification and isolation of the etiologic agent of HFRS by Lee et al. (1978a) provided a target for the development of vaccines to prevent HFRS. In general, two approaches have been used to develop candidate hantavirus vaccines: inactivated-virus vaccines and subunit molecular vaccines. To our knowledge, an attenuated live-virus vaccine approach has not been attempted. Inactivated hantavirus vaccines involve the replication of virus in rodent brains or cell culture. The virus is then chemically inactivated (e.g., formalin or β-propiolactone), combined with adjuvant (e.g., aluminum hydroxide), and delivered intramuscularly. In contrast, the subunit molecular vaccine approach avoids a requirement to replicate, purify, and inactivate highly pathogenic hantaviruses. Rather, modern molecular techniques are used to produce target immunogens delivered as proteins or as genes. Both protein-based and gene-based candidate hantavirus vaccines have been prepared and tested in animals. One gene-based molecular vaccine has been tested in phase 1 and 2 clinical trials, and several candidate gene-based vaccines have been tested in nonhuman primates. A summary of the published candidate inactivated-virus vaccines and subunit molecular vaccines is provided below.

Inactivated-Virus Vaccines

Efforts to produce inactivated hantavirus vaccines to protect against HPS have not been reported. Inactivated-virus HFRS vaccines based on HTNV, SEOV, and PUUV have been produced and tested in humans (reviewed in Hooper and Li, 2001; Cho et al., 2002). Versions of these vaccines are licensed for use in the Republic of Korea and China. The Korean vaccine, marketed as Hantavax®, is produced using formalin-inactivated HTNV grown in mouse brains, precipitated, purified, combined with aluminum hydroxide as an adjuvant, and administered intramuscularly twice at a 1-month interval. Hantavax® and a bivalent vaccine (HTNV and PUUV produced in hamster brains, inactivated with formalin, combined with aluminum hydroxide, and administered subcutaneously three times at 1-month intervals), were able to elicit antibodies in humans, as measured by immunofluorescence and, to a lesser extent, by plaque reduction neutralization test (Cho and Howard, 1999; Cho et al., 2002; Lee et al., 1998b). A vaccine similar to Hantavax®, but produced in Vero cells, was produced and tested in mice (Choi et al., 2003). That vaccine was more immunogenic than Hantavax®, eliciting higher neutralizing antibody responses, and inducing a

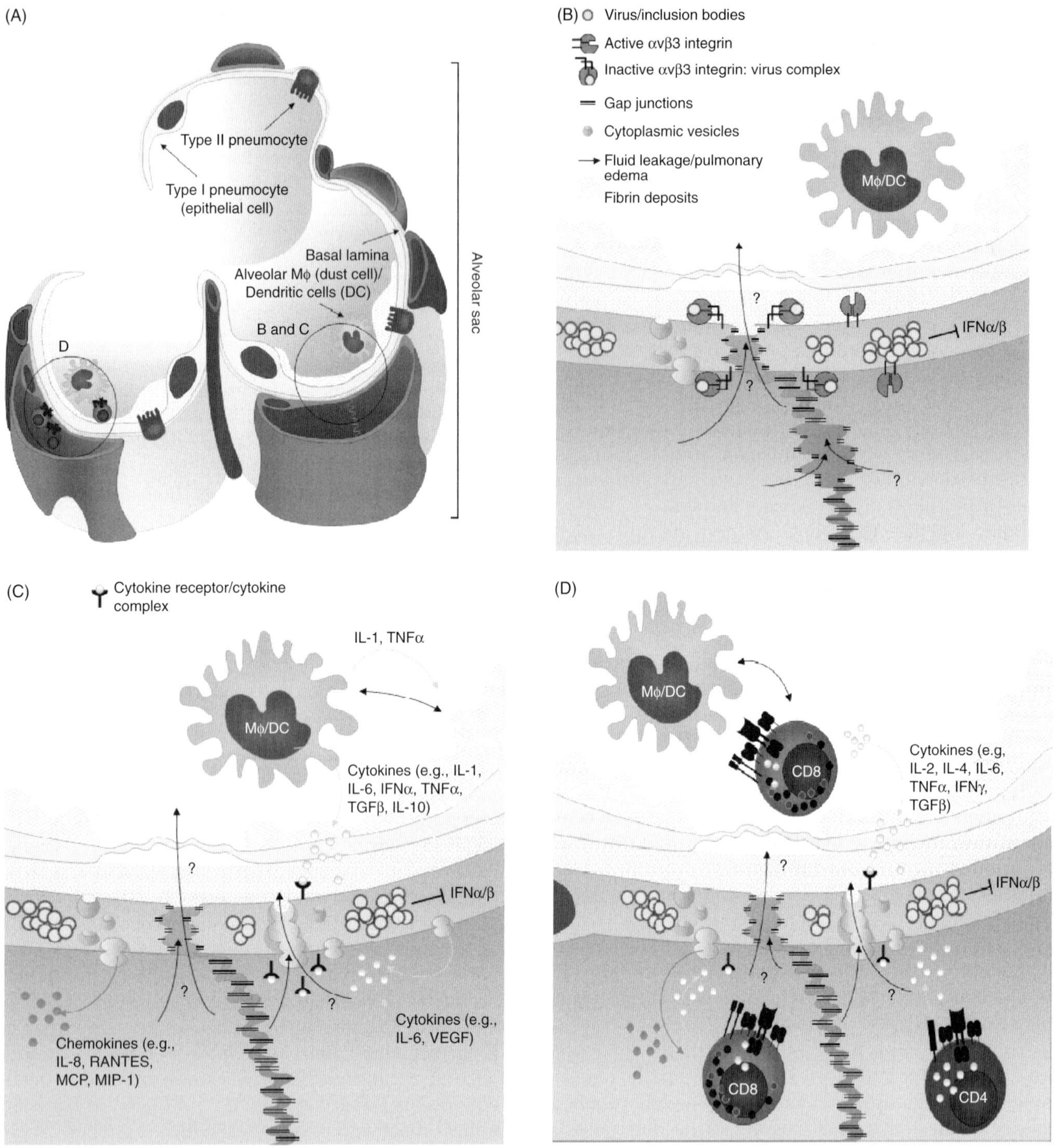

FIGURE 23.4 Possible mechanisms of hantvirus pulmonary syndrome disease pathogensis. (A) Gross schematic diagram of lung alveoli, associated cell types, and vasculature. (B) Vascular permeability as a direct result of virus binding. During infection of endothelial cells, hantaviruses may bind to inactive (bent) αvβ3 integrin, inhibiting normal integrin function, which may cause endothelial cell junctions to open, allowing fluid to pass between endothelial cells and into the alveolar sac. Viral inhibition of type I IFN responses for efficient viral replication and dissemination. (C) Vascular permeability as a result of endothelial cell responses to infection. Infection of endothelial cells may induce the upregulation of cytokines and chemokines involved in promoting permeability via endothelial cell junctions and/or transcellular pathways, initiating fibrin deposition and attracting leukocytes,. Alternatively, infection may directly interfere with normal endothelial cell processes leading to abnormal regulation of vascular integrity. (D) Vascular permeability as a result of T-cell activation. Chemokines secreted by infected endothelial cells may attract activated $CD4^+$ and $CD8^+$ T cells. As T cells interact with viral antigens presented by infected endothelial cells and macrophages and dendritic cells, they secrete large amounts of cytokines that promote permeability via transcellular pathways or by interfering with the normal function of endothelial cell junctional proteins, contribute to fibrin deposition, and further enhance the activation state of endothelial cells and antigen-presenting cells.

greater level of protection in a mouse infection model when lower doses were administered.

In China, the inactivated-virus vaccines have been produced in cell culture, not in rodent brains. There are at least three vaccines licensed for use in China that are described in the literature. One vaccine involved the growth of Seoul-like virus (strain L99) in golden hamster kidney cells (GHKC) and inactivated using 0.025% formalin (Song et al., 1992). Another vaccine involved the growth of HTNV (strain Z10) in Mongolian gerbil kidney cells (MGKC) and inactivation by 0.025% β-propiolactone. A third bivalent vaccine manufactured by Tianyuan Corporation involved the replication of HTNV (strain Z10) and SEOV (strain Z5) in MGKC and inactivation by β-propiolactone. All of these vaccines included 0.5 mg/ml of aluminum hydroxide as adjuvant (Yongxin et al., 1998). In general, the vaccination schedules were an initial series of three vaccinations at 4- and 2-week intervals followed by a 1-year booster vaccination. All of these vaccines were similar to Hantavax® in that they elicited good antibodies as measured by IFAT and ELISA. Neutralizing antibody responses were detected with lower frequencies than IFAT or ELISA; however, after three vaccinations the neutralizing antibody titers could be detected in up to 100% of the vaccine recipients (Song et al., 1992; Zhu et al., 1994; Yongxin et al., 1998). The inactivated vaccines appear to be safe with minimal adverse effects, and there has been no evidence of antibody-dependent enhancement of disease.

The protective efficacy of these inactivated-virus vaccines remains ambiguous. For example, there are reports that the Chinese MGKC and GHKC vaccines were 94–98% protective in large-scale (tens of thousands) trials (Yongxin et al., 1998). In contrast, there are more recent reports published by Chinese researchers indicating that there is still no effective prophylactic vaccine directed at HFRS (Zhang et al., 2007). Epidemiologic data suggest that there are fewer cases of HFRS in areas where the vaccine has been used; however, there are still 20,000–50,000 cases of HFRS in China per year (Fang et al., 2006). It remains unclear as to how much of a role change in hantavirus-related education and/or changes in living conditions have affected the rates of HFRS in Asia. One epidemiologic study on the efficacy of Hantavax® reports that the apparent protective effects in Korea could be due to chance (Park et al., 2004).

Subunit Molecular Vaccines

Regardless of their efficacy, inactivated hantavirus vaccines have many drawbacks, including difficulty in replicating the virus to high titer, danger of working with high concentrations of a biosafety level 3 or 4 agent, difficulties ensuring complete inactivation of the virus, and the loss of protective epitopes by the inactivation process. To avoid these caveats, researchers around the world have embarked on efforts to identify protective immunogens and use those subunits of the virus as candidate vaccines. A summary of some of the efforts aimed at ultimately developing subunit molecular hantavirus vaccines are provided in Tables 23.4 and 23.5.

Protein-Based Vaccines

Several groups have investigated the immunogenicity and protective efficacy of recombinant hantavirus proteins. Technologies to produce the proteins include recombinant baculovirus, *E. coli*, yeast, and most recently, transgenic plants. In some studies, the N-protein elicited immune responses in vaccinated mice, hamsters, and bank voles (Dargeviciute et al., 2002; de Carvalho Nicacio et al., 2002; Geldmacher et al., 2004a, 2004b; Klingstrom et al., 2004; Lundkvist et al., 1996; Schmaljohn et al., 1990). In most cases, anti-N-protein antibodies were detected by ELISA, while in some studies, the activation of the cellular immune response was detected (de Carvalho Nicacio et al., 2002; Klingstrom et al., 2004). In all of these studies, vaccination with the purified N-protein by the subcutaneous or intramuscular route elicited partial or complete protection against infection. An experiment testing the oral delivery of PUUV N-proteins produced in transgenic plants concluded that N-protein was not immunogenic by that route because it was degraded by proteases in the digestive tract (Khattak et al., 2004). Neutralizing antibodies are not directed at the N-protein; thus, it is likely that cell-mediated immunity plays a critical role in protection afforded by vaccination with purified N-protein.

There is only one published study examining the immunogenicity and protective efficacy of purified recombinant hantavirus glycoproteins as subunit vaccines. In that study, HTNV G_n or G_c or G_n/G_c expressed in insect cells infected with recombinant baculovirus was used to vaccinate hamsters (Schmaljohn et al., 1990). The highest level of neutralizing antibodies, and protection against infection, was elicited when the vaccine consisted of both G_n and G_c. In another report, SNV G_n or G_c peptides plus adjuvant were used as vaccine immunogen in a deer mouse infection model (Bharadwaj et al., 2002). The highest level of protection was elicited by G_n peptides. To date, there are no published reports of protein-based subunit hantavirus vaccines being tested in lethal disease models, and no studies have been performed in nonhuman primates.

Vaccines with Virus-Like Particles

It is often possible to generate excellent immunity against viruses when the vaccine consists of particles containing protective immunogens in the form of virus-like particles (VLP). One of the most exciting recent examples of a successful VLP vaccine is Merk's Gardasil®, a VLP vaccine against human papilloma viruses (Stanley, 2007; Villa et al., 2006). VLP approaches aimed at hantaviruses have been investigated. Ulrich et al. (1998, 1999) have described animal studies involving chimeric hepatitis B VLP containing N-protein peptides. More recently, a chimeric-polyoma VLP approach was used to elicit anti-N-protein antibodies in mice (Gedvilaite et al., 2004). In another approach, Lee et al. (2006) demonstrated that vesicular stomatitis virus (VSV) pseudotypes with Hantaan G_n/G_c on their surface elicited neutralizing antibodies, and protective immunity, in mice. Hantavirus VLP have not yet been tested in nonhuman primates or in animal disease models.

Live-Virus-Vectored Vaccines

There has been only one candidate subunit molecular vaccine tested in humans. The vaccine is a recombinant vaccinia virus expressing the HTNV S- and M-gene products (i.e., nucleocapsid protein, G_n and G_c). The HTNV G_c/G_n virus-based vaccines, and a similar SEOV-based version, protected against infection in a hamster or gerbil infection model, respectively (Chu et al., 1995; Schmaljohn et al., 1990; Xu et al., 1992). This live-virus vaccine was administered either subcutaneously or intramuscularly in phase 1and 2 clinical trials (McClain et al., 2000). HTNV neutralizing antibodies were elicited in vaccinia-naïve individuals; however, the level of neutralizing antibodies elicited in persons who had previously been vaccinated with vaccinia virus (i.e., the smallpox vaccine) was unacceptably low. Poor immunogenicity in persons previously vaccinated with the smallpox vaccine, coupled with the dangers associated with live vaccinia virus-based vaccines (e.g., progressive vaccinia, eczema vaccinatum) (see Chapter 37), discouraged advanced development of the vaccinia-vectored HFRS vaccine.

A far more experimental live-virus approach involving recombinant cytomegalovirus expressing SNV G_n was tested in a deer mouse model and was shown to elicit a detectable antibody response against G_n (Rizvanov et al., 2006, 2003). To our knowledge, other live-virus-vectored molecular vaccines, including adenovirus-vectored or live-VSV-vectored vaccines, have not been reported.

Packaged Replicons

Vaccine platforms involving viruses that infect but do not replicate have also been used to investigate as putative approaches to develop hantavirus vaccines. Sindbis virus replicons expressing the SEOV N-protein or G_n/G_c were tested in hamsters (Kamrud et al., 1999). These alphavirus replicons elicited antibody responses, and in the case of G_n/G_c, neutralizing antibodies were produced. Hamsters vaccinated with G_n/G_c replicons were partially protected against SEOV infection. Replicons based on another alphavirus, Semliki Forest virus, expressing the PUUV N, G_n, G_c, or G_n/G_c expressed protein in vitro; however, vaccination studies were not reported (Billecocq et al., 2003; Kallio-Kokko et al., 2001).

DNA Vaccines

A great deal of work involving the delivery of plasmid DNA as a vaccine has been performed in recent years. A summary of many of the studies is provided in Table 23.4. In general, DNA vaccines have been delivered as plasmids either by intramuscular injection or by particle-mediated epidermal delivery (gene gun). In addition, there is one report of an ANDV M-gene-based DNA vaccine delivered by electroporation (Hooper et al., 2008).

Vaccines involving the delivery of the S gene as naked DNA or as DNA-launched replicons have elicited antibody responses against SEOV, PUUV, and SNV (Bharadwaj et al., 2002; Bucht et al., 2001; Hooper et al., 1999; Kamrud et al., 1999; Lindkvist et al., 2007). In those experiments, immune responses were detected (e.g., anti-N-protein antibodies by ELISA) and, in some instances, partial protection was observed in rodent infection models (Bharadwaj et al., 2002; Bucht et al., 2001; Kamrud et al., 1999).

Vaccines involving the delivery of the fragments (i.e., peptides) of the M gene have shown some efficacy (Bharadwaj et al., 2002); however, much greater levels of neutralizing antibodies and protection were observed when the vaccine involved delivery of the full-length M gene (Custer et al., 2003; Hooper et al., 1999, 2001a, 2006, 2008). The first DNA vaccine to elicit high-titer neutralizing antibodies against a hantavirus was a gene-gun-delivered plasmid containing the SEOV M gene (Hooper et al., 1999). This was followed by similar vaccines containing the HTNV M gene, ANDV M gene, and a plasmid containing

TABLE 23.4 Subunit molecular vaccines against hantaviruses tested in small animal models

Vaccine	Source of immunogen	Route of administration; no. of vaccinations; interval	Measure of immune response	Animal model	Protection[a]	Reference
Protein						
Baculovirus-derived N_P	HTNV	i.m.; 2; 4-wk	N_P-ELISA	Hamster	Yes	Schmaljohn et al. (1990)
Baculovirus-derived N_P + adjuvant	PUUV	s.c.; 3; 3-wk	N_P-ELISA	Bank vole	Yes	Lundkvist et al. (1996)
Baculovirus-derived G_n	HTNV	i.m.; 2; 4-wk	(−)PRNT	Hamster	Most	Schmaljohn et al. (1990)
Baculovirus –derived G_c	HTNV	i.m.; 2; 4-wk	(−)PRNT	Hamster	Some	Schmaljohn et al. (1990)
Baculovirus-derived G_n/G_c	HTNV	i.m.; 2; 4-wk	+PRNT	Hamster	Yes	Schmaljohn et al. (1990)
E. coli-derived N_P + adjuvant	PUUV	s.c.; 3; 3-wk	N_P-ELISA	Bank vole	Yes	Lundkvist et al. (1996)
	PUUV	s.c.; 3; 3-wk	N_P-ELISA	Bank vole	Yes (PUUV)	de Carvalho Nicacio et al. (2002)
	TOPV	s.c.; 3; 3-wk	N_P-ELISA	Bank vole	Yes (PUUV)	de Carvalho Nicacio et al. (2002)
	ANDV	s.c.; 3; 3-wk	N_P-ELISA	Bank vole	Some (PUUV)	de Carvalho Nicacio et al. (2002)
	DOBV	s.c.; 3; 3-wk	N_P-ELISA	Bank vole	Some (PUUV)	de Carvalho Nicacio et al. (2002)
	DOBV	s.c.; 3; 3-10wk	N_P-ELISA	Mice	Some	Klingstrom et al. (2004)
Yeast-derived N_P + adjuvant	PUUV	i.m.; 3; 1-wk	N_P-ELISA	Bank vole	Yes	Dargeviciute et al. (2002)
	DOBV	s.c.; 3; 1-wk	N_P-ELISA	Mice	nd	Geldmacher et al. (2004a)
Plant-derived N_P	PUUV	p.o.; 4; 1, 15, 14 days	(−) N_P-ELISA	Mice	nd	Khattak et al. (2004)
Virus-like particle						
Chimeric HBV VLP (N_P peptides) + adjuvant	PUUV	s.c.; 3; 3-wk	N_P-ELISA	Bank vole	Some	Ulrich et al. (1998, 1999)
	PUUV	i.p. + s.c.; 2; 10-days	N_P-ELISA	Mice	nd	Geldmacher et al. (2004b)
	DOBV	i.p. + s.c.; 2; 10-days	N_P-ELISA	Mice	nd	Geldmacher et al. (2004b)
	HTNV	i.p. + s.c.; 2; 10-days	N_P-ELISA	Mice	nd	Geldmacher et al. (2004b)
Chimeric polyoma VLP (N_P peptides) + adjuvant	PUUV	s.c.; 3; 4-wk	N_P-ELISA	Mice	nd	Gedvilaite et al. (2004b)
VSV pseudotype (G_n/G_c) + adjuvant	HTNV	s.c.; 3; 3-2 wk	+PRNT	Mice	yes	Lee et al. (2006)
Live-virus-vectored						
Recombinant-vaccinia N_P	HTNV	1-Scarified; 1-i.p; 4-wk	N_P-ELISA	Hamster	No	Schmaljohn et al. (1990)
	SEOV	s.c.; 2; 4-wk	N_P-ELISA	Gerbil	Most	Xu et al. (1992)
Recombinant-vaccinia G_n	HTNV	1-Scarified; 1-i.p; 4-wk	(−)PRNT	Hamster	No	Schmaljohn et al. (1990)
Recombinant-vaccinia G_c	HTNV	1-Scarified; 1-i.p; 4-wk	(−)PRNT	Hamster	Most	Schmaljohn et al. (1990)
Recombinant-vaccinia G_n/G_c	HTNV	1-Scarified; 1-i.p; 4-wk	+PRNT	Hamster	Yes	Schmaljohn et al. (1990)
	SEOV	s.c.; 2; 4-wk	+PRNT	Gerbil	Yes	Xu et al. (1992)
Recombinant-vaccina $N_P/G_n/G_c$	HTNV	i.m.; 2; 30-day	+PRNT	Hamster	Yes	Chu et al. (1995)
Recombinant-cytomegalovirus (G_n)	SNV	i.p; 2; 1 year	G_n ELISA	Deer mice	nd	Rizvanov et al. (2003)

(Continued)

TABLE 23.4 (Continued)

Vaccine	Source of immunogen	Route of administration; no. of vaccinations; interval	Measure of immune response	Animal model	Protection[a]	Reference
Packaged Replicons						
Packaged Sindbis replicon N_P	SEOV	s.c.; 3; 4-wk	N_P-ELISA	Hamster	No	Kamrud et al. (1999)
Packaged Sindbis replicon G_n/G_c	SEOV	s.c.; 3; 4-wk	+PRNT	Hamster	Some	Kamrud et al. (1999)
PCR-generated linear fragments						
Linear DNA modified N_P peptide	PUUV	Gene gun; 4; 2-wk	N_P-ELISA	Mice	nd	Johansson et al. (2002)
Plasmid DNA						
DNA-launched Sindbis replicon N_P	SEOV	Gene gun; 3; 3-wk	N_P-ELISA	Hamster	Some	Kamrud et al. (1999)
DNA-launched Sindbis replicon G_n/G_c	SEOV	Gene gun; 3; 3-wk	+PRNT	Hamster	Yes	Kamrud et al. (1999)
DNA vaccine N_P (truncated, secreted)	PUUV	i.m.; 4; 3-wk	N_P-ELISA	Mice	Some	Bucht et al. (2001)
DNA vaccine N_P	SEOV	Gene gun; 3; 4-wk	N_P-ELISA	Hamster	No	Hooper et al. (1999)
	SNV	i.m.; 3; 3-wk	Proliferation	Deer mice	Some	Bharadwaj et al. (2002)
	PUUV	i.m.; 4; 3-wk	N_P-ELISA	Mice	Some	Bucht et al. (2001)
	SEOV	Gene gun; 5; 2-wk	N_P-ELISA	Mice	Nd	Lindkvist et al. (2007)
	PUUV	Gene gun; 5; 2-wk	N_P-ELISA	Mice	nd	Lindkvist et al. (2007)
	SNV	Gene gun; 5; 2-wk	N_P-ELISA	Mice	nd	Lindkvist et al. (2007)
DNA vaccine G_n or G_c peptides	SNV	i.m.; 3; 3-wk	Proliferation	Deer mice	No/yes	Bharadwaj et al. (2002)
DNA vaccine G_n/G_c	SEOV	Gene gun; 3; 4-wk	+PRNT	Hamster	Yes	Hooper et al. (1999)
	HTNV	Gene gun; 3; 3-wk	+PRNT	Hamster	Yes	Hooper et al. (2001)
	ANDV	Gene gun; 3; 3-wk	None	Hamster	No[b]	Custer et al. (2003)
	ANDV	Electroporation; 2; 4-wk	+PRNT	Rabbit	nd	Hooper et al. (2008)
	HTNV/ ANDV	Gene gun; 4; 2-1-5-wk	None	Hamster	No[b]	Hooper et al. (2006)
	PUUV	Gene gun; 3; 3-wk	+PRNT	Hamster	Yes	Hooper et al. (unpublished)

Notes: N_P, N-protein; G_n, G_n glycoprotein; G_c, G_c glycoprotein; Proliferation, splenocyte proliferation; s.c., subcutaneous; i.m., intramuscular; i.p., intraperitoneal; p.o., per oral; wk, week; nd, not determined; PRNT, plaque reduction neutralization test; HBV, hepatitis B virus; VLP, virus-like particle; VSV, vesicular stomatitis virus.

[a]Protection against *infection* after homologous virus challenge, unless otherwise indicated (virus); for ANDV, protection against *lethal disease*.

[b]Same vaccine was immunogenic in nonhuman primates (see Table 23.5).

both the HTNV and ANDV M gene (Custer et al., 2003; Hooper et al., 2001a, 2006). The SEOV and HTNV M-gene-based DNA vaccines elicited immune responses in hamsters that cross-protected against SEOV, HTNV, and DOBV, but not PUUV (Hooper et al., 2001a).

Several of the full-length M-gene-based DNA vaccines have been tested in nonhuman primates (Table 23.5). Rhesus macaques vaccinated with the SEOV or HTNV DNA vaccines by a gene gun produced high-titer neutralizing antibodies that exhibited

TABLE 23.5 Subunit molecular vaccines against hantaviruses tested in nonhuman primates

Vaccine	Route of administration; no. of vaccinations; interval	Vaccine	Neutralizing antibody: First detected	Neutralizing antibody: Maximum $PRNT_{50}$ titer reported	Reference
Recombinant vaccina $N/G_n/G_c$ (HTNV)	i.m.; 2; 6-wk	Rhesus	After 2 vacc.	640	Hooper et al. (2001a)
SEOV M-gene-based DNA vaccine	Gene gun; 3; 3-wk	Rhesus	After 2 vacc.	1280	Hooper et al. (2001a)
HTNV M-gene-based DNA vaccine	Gene gun; 3; 3-wk	Rhesus	After 1 vacc.	10,240	Hooper et al. (2001a)
	Gene gun; 4; 3–6-wk	Rhesus	After 2 vacc.	20,480	Custer et al. (2003)
ANDV M-gene-based DNA vaccine	Gene gun; 4; 3–6-wk	Rhesus	After 2 vacc.	20,480	Custer et al. (2003)
	Gene gun; 4; 3-wk	Rhesus	After 2 vacc.	10,240	Custer et al. (2003)
	Gene gun; 4	Cynomolgus	nd	1280	Hooper et al. (2006)
HTNV/ANDV M-gene-based DNA vaccine	Gene gun; 4; 3-wk	Rhesus	After 2, 4 vacc.	640	Hooper et al. (2006)
PUUV M-gene-based DNA vaccine	Gene gun; 3; 3-wk	Rhesus	After 2 vacc.	1016	Hooper et al. (unpublished)

Notes: i.m., intramuscular; $PRNT_{50\ titer}$, reciprocal of the highest serum dilution reducing plaque number by 50% as measured by plaque reduction neutralization test; nd, not determined.

cross-reactivity with SEOV, HTNV, and DOBV, but less cross-reactivity against PUUV (Hooper et al., 2001a). Recently, a PUUV M-gene-based DNA vaccine was constructed and tested in nonhuman primates. This vaccine delivered by gene gun elicited neutralizing antibodies against PUUV in nonhuman primates (Hooper et al., unpublished observations). This same vaccine elicited PUUV neutralizing antibodies in hamsters and protected against infection with PUUV. A combined vaccine consisting of the HTNV and PUUV M-gene-based DNA vaccines, theoretically, should cross-neutralize against the full spectrum of major HFRS-associated hantaviruses.

To develop an HPS vaccine, the full-length ANDV M gene was cloned into a DNA vaccine plasmid and tested in animal models (Custer et al., 2003). Interestingly, this vaccine was not immunogenic in hamsters, but was highly immunogenic in nonhuman primates (i.e., rhesus and cynomolgus macaques) and rabbits (Custer et al., 2003; Hooper et al., 2006, 2008). The neutralizing antibody levels produced in the macaques were very high (Table 23.5). In one study, $PRNT_{50}$ titers as high as 1:20,480 were produced (Custer et al., 2003). Sera from the rhesus macaques vaccinated with the ANDV DNA vaccine using the gene gun cross-neutralized SNV and BCCV (Custer et al., 2003); however, sera from the rabbits vaccinated by electroporation failed to cross-neutralize (Hooper et al., 2008). Thus, it might be necessary to construct additional DNA vaccine plasmids (e.g., SNV M-gene-based DNA vaccine) to produce a vaccine that cross-protects against the many different hantaviruses that cause HPS.

A DNA vaccine containing both the HTNV and ANDV full-length M genes, pWRG/HA-M, was not immunogenic in hamsters but was immunongenic in nonhuman primates (Hooper et al., 2006). The reason why DNA vaccines containing the ANDV M gene are poorly immunogenic in hamsters, but are immunogenic in nonhuman primates and rabbits, remains unknown. It is tempting to speculate that this phenomenon could be related to the high virulence of ANDV in hamsters. A more in-depth discussion of the ANDV/hamster lethal HPS model is provided in the "Prospects for the Future" section below.

POSTEXPOSURE IMMUNOPROPHYLAXIS

There is an unquestioned need for the development of hantavirus vaccines for use in regions of the world where there are high numbers of HFRS or HPS cases. Similarly, vaccines are needed for persons in high-risk groups, including soldiers deployed to endemic regions. However, HPS and HFRS outbreaks can occur suddenly in areas previously void of disease where pretreatment vaccines are unavailable or impractical. In such scenarios (including biological weapons attack), the emergency use of postexposure immunoprophylactic products could be used to prevent disease in potentially exposed individuals, mitigate panic, and contain the spread of disease. Postexposure immunoprophylactic products might also be used in combination with emergency vaccination (e.g., as in the case of rabies), or as part of a therapeutic treatment. Passive antibody has been used to prevent and treat other viral diseases including Argentine hemorrhagic fever (Enria et al., 1984) and has been proposed as a means of defending against biological weapons threats (Casadevall, 2002).

There is reason to believe that immunotherapeutic products could be effective in preventing and treating hantavirus disease. Numerous animal studies have demonstrated that monoclonal or polyclonal antibodies protected rodents from infection with hantaviruses (Arikawa et al., 1992; Medina et al., 2007; Schmaljohn et al., 1990; Xu et al., 2002; Yoshimatsu et al., 1993; Zhang et al., 1989). One study also found passive transfer of neutralizing monoclonal antibodies before challenge could delay the effects of PUUV on the blood chemistry of one of the two infected nonhuman primates (Klingstrom et al., 2005).

We used the ANDV/hamster lethal HPS model (see "Prospects for the Future") to test whether antibodies could protect before and/or after exposure to a lethal dose of ANDV (Custer et al., 2003). Serum from nonhuman primates vaccinated with the ANDV M-gene-based DNA vaccine containing high levels of neutralizing antibodies protected hamsters against an intramuscular challenge with ANDV when given 1 day before challenge. More importantly, passive transfer of the sera up to 5 days after challenge protected the hamsters. This was an exciting finding because it demonstrated that a postexposure prophylactic approach could be a viable way to prevent HPS in persons exposed or potentially exposed to the virus. Recently, we produced ANDV-neutralizing antibodies in rabbits using the ANDV M DNA vaccine and confirmed that pre- or postchallenge passive transfer of the serum could confer protection against a lethal mucosal challenge with ANDV (Hooper et al., 2008).

The aforementioned animal experiments suggest that postexposure use of antibodies could be a way to prevent disease in exposed individuals. There are clinical data supporting this concept. For example, high levels of neutralizing antibodies in HPS patients at hospital admission correlated with higher survival rates (Bharadwaj et al., 2000). Likewise, low levels of viremia at hospital admission are associated with improved clinical outcome (Xiao et al., 2006). Clinical studies testing the capacity of human convalescent serum, or laboratory-derived neutralizing antibodies, to effectively prevent and treat hantavirus diseases are needed. Examples of possible test sites include Chile and Argentina, where HPS can occur in clusters (Lazaro et al., 2007; Toro et al., 1998; Vial et al., 2006; Wells et al., 1997).

PROSPECTS FOR THE FUTURE

As climatic changes affect the environment, we can expect to see changes in rodent populations and distributions. It is likely that these changes will alter the ways in which humans come in contact with hantaviruses, and these changes could result in changes in the rates and distribution of HPS cases in the Americas and HFRS in Eurasia. Moreover, changes in interactions between different species of rodents could, theoretically, increase the possibility that different hantaviruses might coinfect the rodent host and produce reassortant progeny viruses. There is evidence that hantaviruses have reassorted in nature (Henderson et al., 1995; Li et al., 1995). Reassortant viruses could exhibit altered biological properties including virulence, as is the case for influenza virus (genetic shift). Thus, the threat posed by naturally occurring hantavirus disease is dynamic and must be carefully monitored.

The threat posed by unnaturally occurring hantavirus disease caused by their intentional use as biological weapons should not be ignored. ANDV is of particular concern for several reasons including its availability in nature, high lethality, rapid disease onset and progression, transmission properties (e.g., person-to-person transmission), and replication characteristics (e.g., ANDV replicates to titers as high as 10^{7-8} in cell culture and hamster organs).

The threat posed by hantaviruses is of particular concern when one considers the paucity of effective medical countermeasures to prevent and treat HPS and HFRS. There is no HPS vaccine available anywhere in the world. There are no HFRS vaccines, except the inactivated-virus vaccines in China and South Korea. There are no antiviral drugs licensed to specifically treat hantavirus disease. Finally, medical treatments that due exist such as ECMO for HPS and hemodialysis for HFRS are extraordinarily expensive and require a medical infrastructure not available in many parts of the world.

Providing education on steps that can be taken to avoid exposure to hantaviruses (e.g., avoid inhalation or ingestion of rodent excreta) can greatly reduce the incidence of HPS and HFRS; however, medical countermeasures to prevent and treat these diseases remain desperately needed. The development of animal models that accurately mimic hantavirus disease in humans should accelerate the development of vaccines, treatments, and an understanding of hantavirus disease pathogenesis. Such a model was discovered in 2001. When injected into Syrian hamsters, ANDV causes a lethal vascular-leak disease that closely resembled HPS in humans (Hooper et al., 2001b). Similarities include (1) the incubation time, (2) rapid disease onset after the first symptoms, (3) infection of endothelial cells and the absence of necrosis, (4) severe pulmonary edema and pleural effusion, (5) mild thrombocytopenia, (6) neutrophilia, and (7) cardiogenic impairment and hypotension (Campen et al., 2006; Hooper et al., 2001b; Wahl-Jensen et al., 2007). A second hantavirus, Maporal virus, was reported to cause similar disease manifestations in Syrian hamsters (Milazzo et al., 2002). ANDV is highly pathogenic in adult Syrian hamsters when administered by the intramuscular ($LD_{50} = 8$ pfu, mean-time-death = 13 days), subcutaneous ($LD_{100} \leq 2000$, mean-time-death = 16 days), intranasal ($LD_{50} = 95$ pfu, mean-time-to-death = 17.5 days), or intragastric routes ($LD_{50} = 225$ pfu, mean-time-death = 19 days) (Hooper et al., 2001b, 2008). The sex of the animal does not alter the disease course (Hooper et al., 2008). This lethal disease model of HPS is currently being used to develop candidate hantavirus vaccines, immunoprophylactics, and therapeutics. In addition, experiments using this model are underway to test hypotheses aimed at understanding the pathogenesis of hantavirus vascular-leak syndromes. One limitation on the ANDV/hamster model work is that it requires the highest level of containment (biosafety level 4). This constraint will be alleviated as more high containment laboratories become operational around the world. Another constraint facing the development of medical countermeasures to HPS and HFRS is limited funding. International collaboration and coordination, including cost-sharing, will be a key factor in the future development of vaccines and cures for diseases caused by hantaviruses.

KEY ISSUES

- Pathogenic hantaviruses pose a danger to persons working in environments populated by rodents that carry the virus (i.e., Eurasia and the Americas).
- Certain hantaviruses (i.e., ANDV) pose an increased danger because they are highly pathogenic and can spread from one person to the other (close contacts).
- There is a small animal model for HPS, but there is still no small animal model for HFRS.
- The pathogenesis of HPS and HFRS is not understood; and it remains unclear as to what role the immune response plays in the disease.
- There are no FDA-licensed vaccines or drugs to prevent or cure HPS or HFRS.

ACKNOWLEDGMENTS

The authors of this review were supported by the Military Infectious Disease Research Program (Area T), U.S. Army Medical Research and Material Command. Opinions, interpretations, conclusions, and recommendations are those of the authors and are not necessarily endorsed by the U.S. Army or the Department of Defense.

References

Abu Sin, M., Stark, K., Van Treeck, U., Dieckmann, H., Uphoff, H., Hautmann, W., Bornhofen, B., Jensen, E., Pfaff, G. and Koch, J. Risk factors for hantavirus infection in Germany, 2005. Emerg. Infect. Dis. 2007; 13:1364–1366.

Alexeyev, O.A., Ahlm, C., Billheden, J., Settergren, B., Wadell, G. and Juto, P. Elevated levels of total and Puumala virus-specific immunoglobulin E in the Scandinavian type of hemorrhagic fever with renal syndrome. Clin. Diagn. Lab. Immunol. 1994; 1:269–272.

Alff, P.J., Gavrilovskaya, I.N., Gorbunova, E., Endriss, K., Chong, Y., Geimonen, E., Sen, N., Reich, N.C. and Mackow, E.R. The pathogenic NY-1 hantavirus G1 cytoplasmic tail inhibits RIG-I- and TBK-1-directed interferon responses. J. Virol. 2006; 80:9676–9686.

Antoniadis, A., Stylianakis, A., Papa, A., Alexiou-Daniel, S., Lampropoulos, A., Nichol, S.T., Peters, C.J. and Spiropoulou, C.F. Direct genetic detection of Dobrava virus in Greek and Albanian patients with hemorrhagic fever with renal syndrome. J. Infect. Dis. 1996; 174:407–410.

Araki, K., Yoshimatsu, K., Lee, B.H., Kariwa, H., Takashima, I. and Arikawa, J. Hantavirus-specific CD8(+)-T-cell responses in newborn mice persistently infected with Hantaan virus. J. Virol. 2003; 77:8408–8417.

Araki, K., Yoshimatsu, K., Lee, B.H., Okumura, M., Kariwa, H., Takashima, I. and Arikawa, J. Age-dependent hantavirus-specific CD8(+) T-cell responses in mice infected with Hantaan virus. Arch. Virol. 2004; 149:1373–1382.

Arikawa, J., Yao, J.S., Yoshimatsu, K., Takashima, I. and Hashimoto, N. Protective role of antigenic sites on the envelope protein of Hantaan virus defined by monoclonal antibodies. Arch. Virol. 1992; 126:271–281.

Armien, B., Pascale, J.M., Bayard, V., Munoz, C., Mosca, I., Guerrero, G., Armien, A., Quiroz, E., Castillo, Z., Zaldivar, Y., Gracia, F., Hjelle, B. and Koster, F. High seroprevalence of hantavirus infection on the Azuero peninsula of Panama. Am. J. Trop. Med. Hyg. 2004; 70:682–687.

Asada, H., Tamura, M., Kondo, K., Okuno, Y., Takahashi, Y., Dohi, Y., Nagai, T., Kurata, T. and Yamanishi, K. Role of T lymphocyte subsets in protection and recovery from Hantaan virus infection in mice. J. Gen. Virol. 1987; 68(Pt. 7):1961–1969.

Avsic-Zupanc, T. Hantaviruses and hemorrhagic fever with renal syndrome in the Balkans. In: Emergence and Control of Rodent-borne Viral Diseases (Hantaviral and Arenal Diseases) (J.F. Saluzzo and B. Dodet, Eds.). Veyrier-du-lac, Annecy, France: Elsevier, 1998.

Avsic-Zupanc, T., Xiao, S.Y., Stojanovic, R., Gligic, A., Van Der Groen, G. and Leduc, J.W. Characterization of Dobrava virus: a hantavirus from Slovenia, Yugoslavia. J. Med. Virol. 1992; 38:132–137.

Bayard, V., Kitsutani, P.T., Barria, E.O., Ruedas, L.A., Tinnin, D.S., Munoz, C., De Mosca, I.B., Guerrero, G., Kant, R., Garcia, A., Caceres, L., Gracio, F.G., Quiroz, E., DE Castillo, Z., Armien, B., Libel, M., Mills, J.N., Khan, A.S., Nichol, S.T., Rollin, P.E., Ksiazek, T.G. and Peters, C.J. Outbreak of hantavirus pulmonary syndrome, Los Santos, Panama, 1999–2000. Emerg. Infect. Dis. 2004; 10:1635–1642.

Bazzoni, G. and Dejana, E. Endothelial cell-to-cell junctions: molecular organization and role in vascular homeostasis. Physiol. Rev. 2004; 84:869–901.

Beaufils, P., Choquet, D., Mamoun, R.Z. and Malissen, B. The (YXXL/I)2 signalling motif found in the cytoplasmic segments of the bovine leukaemia virus envelope protein and Epstein-Barr virus latent membrane protein 2A can elicit early and late lymphocyte activation events. Embo J. 1993; 12:5105–5112.

Beers, M.H. and Berkow, R. Infectious diseases; Viral Diseases. In: *The Merck Manual of Diagnosis and Therapy*, 17th Ed., Indianapolis Wiley Publishers, 2005.

Beynon, H.L., Haskard, D.O., Davies, K.A., Haroutunian, R. and Walport, M.J. Combinations of low concentrations of cytokines and acute agonists synergize in increasing the permeability of endothelial monolayers. Clin. Exp. Immunol. 1993; 91: 314–319.

Bharadwaj, M., Mirowsky, K., Ye, C., Botten, J., Masten, B., Yee, J., Lyons, C.R. and Hjelle, B. Genetic vaccines protect against Sin Nombre hantavirus challenge in the deer mouse (Peromyscus maniculatus). J. Gen. Virol. 2002; 83:1745–1751.

Bharadwaj, M., Nofchissey, R., Goade, D., Koster, F. and Hjelle, B. Humoral immune responses in the hantavirus cardiopulmonary syndrome. J. Infect. Dis. 2000; 182:43–48.

Bhattacharya, S., Ying, X., Fu, C., Patel, R., Kuebler, W., Greenberg, S. and Bhattacharya, J. Alpha(v)beta(3) integrin induces tyrosine phosphorylation-dependent Ca(2+) influx in pulmonary endothelial cells. Circ. Res. 2000; 86:456–462.

Billecocq, A., Coudrier, D., Boue, F., Combes, B., Zeller, H., Artois, M. and Bouloy, M. Expression of the nucleoprotein of the Puumala virus from the recombinant Semliki Forest virus replicon: characterization and use as a potential diagnostic tool. Clin. Diagn. Lab. Immunol. 2003; 10:658–663.

Borza, D.B., Neilson, E.G. and Hudson, B.G. Pathogenesis of Goodpasture syndrome: a molecular perspective. Semin. Nephrol. 2003; 23:522–531.

Bostik, P., Winter, J., Ksiazek, T.G., Rollin, P.E., Villinger, F., Zaki, S.R., Peters, C.J. and Ansari, A.A. Sin nombre virus (SNV) Ig isotype antibody response during acute and convalescent phases of hantavirus pulmonary syndrome. Emerg. Infect. Dis. 2000; 6:184–187.

Brummer-Korvenkontio, M., Henttonen, H. and Vaheri, A. Hemorrhagic fever with renal syndrome in Finland: ecology and virology of nephropathia epidemica. Scand. J. Infect. Dis. Suppl. 1982; 36:88–91.

Brummer-Korvenkontio, M., Vaheri, A., Hovi, T., Von Bonsdorff, C.H., Vuorimies, J., Manni, T., Penttinen, K., Oker-Blom, N. and Lahdevirta, J. Nephropathia epidemica: detection of antigen in bank voles and serologic diagnosis of human infection. J. Infect. Dis. 1980; 141:131–134.

Bucht, G., Sjolander, K.B., Eriksson, S., Lindgren, L., Lundkvist, A. and Elgh, F. Modifying the cellular transport of DNA-based vaccines alters the immune response to hantavirus nucleocapsid protein. Vaccine 2001; 19:3820–3829.

Campen, M.J., Milazzo, M.L., Fulhorst, C.F., Obot Akata, C.J. and Koster, F. Characterization of shock in a hamster model of hantavirus infection. Virology 2006; 356:45–49.

Casadevall, A. Antibodies for defense against biological attack. Nat. Biotechnol. 2002; 20:114.

Castillo, C., Naranjo, J., Sepulveda, A., Ossa, G. and Levy, H. Hantavirus pulmonary syndrome due to Andes virus in Temuco, Chile: clinical experience with 16 adults. Chest 2001; 120:548–554.

CDC, S.P.B. All about Hantaviruses. Centers for Disease Control and Prevention, 2007. http://www.cdc.gov/ncidod/diseases/hanta/hps/. Accessed in 2007.

Chapman, L.E., Mertz, G.J., Peters, C.J., Jolson, H.M., Khan, A.S., Ksiazek, T.G., Koster, F.T., Baum, K.F., Rollin, P.E., Pavia, A.T., Holman, R.C., Christenson, J.C., Rubin, P.J., Behrman, R.E., Bell, L.J., Simpson, G.L. and Sadek, R.F. Intravenous ribavirin for hantavirus pulmonary syndrome: safety and tolerance during 1 year of open-label experience. Ribavirin Study Group. Antivir. Ther. 1999; 4:211–219.

Chen, L.B. and Yang, W.S. Abnormalities of T cell immunoregulation in hemorrhagic fever with renal syndrome. J. Infect. Dis. 1990; 161:1016–1019.

Cho, H.W. and Howard, C.R. Antibody responses in humans to an inactivated hantavirus vaccine (Hantavax). Vaccine 1999; 17:2569–2575.

Cho, H.W., Howard, C.R. and Lee, H.W. Review of an inactivated vaccine against hantaviruses. Intervirology 2002; 45:328–333.

Choi, J., Enis, D.R., Koh, K.P., Shiao, S.L. and Pober, J.S. T lymphocyte-endothelial cell interactions. Annu. Rev. Immunol. 2004; 22:683–709.

Choi, Y., Ahn, C.J., Seong, K.M., Jung, M.Y. and Ahn, B.Y. Inactivated Hantaan virus vaccine derived from suspension culture of Vero cells. Vaccine 2003; 21:1867–1873.

Chu, Y.K., Jennings, G.B. and Schmaljohn, C.S. A vaccinia virus-vectored Hantaan virus vaccine protects hamsters from challenge with Hantaan and Seoul viruses but not Puumala virus. J. Virol. 1995; 69:6417–6423.

Cosgriff, T.M. Mechanisms of disease in hantavirus infection: pathophysiology of hemorrhagic fever with renal syndrome. Rev. Infect. Dis. 1991; 13:97–107.

Crowley, M.R., Katz, R.W., Kessler, R., Simpson, S.Q., Levy, H., Hallin, G.W., Cappon, J., Krahling, J.B. and Wernly, J. Successful treatment of adults with severe hantavirus pulmonary syndrome with extracorporeal membrane oxygenation. Crit. Care. Med. 1998; 26:409–414.

Custer, D.M., Thompson, E., Schmaljohn, C.S., Ksiazek, T.G. and Hooper, J.W. Active and passive vaccination against hantavirus pulmonary syndrome with Andes virus M genome segment-based DNA vaccine. J. Virol. 2003; 77:9894–9905.

Dargeviciute, A., Brus Sjolander, K., Sasnauskas, K., Kruger, D.H., Meisel, H., Ulrich, R. and Lundkvist, A. Yeast-expressed Puumala hantavirus nucleocapsid protein induces protection in a bank vole model. Vaccine 2002; 20:3523–3531.

Davis, I.C., Zajac, A.J., Nolte, K.B., Botten, J., Hjelle, B. and Matalon, S. Elevated generation of reactive oxygen/nitrogen species in hantavirus cardiopulmonary syndrome. J. Virol. 2002; 76:8347–8359.

De Carvalho Nicacio, C., Gonzalez Della Valle, M., Padula, P., Bjorling, E., Plyusnin, A. and Lundkvist, A. Cross-protection against challenge with Puumala virus after immunization with nucleocapsid proteins from different hantaviruses. J. Virol. 2002; 76:6669–6677.

De Carvalho Nicacio, C., Sallberg, M., Hultgren, C. and Lundkvist, A. T-helper and humoral responses to Puumala hantavirus nucleocapsid protein: identification of T-helper epitopes in a mouse model. J. Gen. Virol. 2001; 82:129–138.

Dehghani, H., Brown, C.R., Plishka, R., Buckler-White, A. and Hirsch, V.M. The ITAM in Nef influences acute pathogenesis of AIDS-inducing simian immunodeficiency viruses SIVsm and SIVagm without altering kinetics or extent of viremia. J. Virol. 2002; 76:4379–4389.

Disweb, C.D.I.S. Communicable diseases statistics, Major statistics, Reported cases by year. National Institutes of Health, 2003.

Disweb, C.D.I.S. Republic of Korea: Korea Center for Disease Control and Prevention. http://dis.cdc.go.kr/eng_statistics/statistics.asp, National Institute of Health, 2003, accessed in 2007.

Doyle, T.J., Bryan, R.T. and Peters, C.J. Viral hemorrhagic fevers and hantavirus infections in the Americas. Infect. Dis. Clin. North Am. 1998; 12:95–110.

Dunst, R., Mettang, T. and Kuhlmann, U. Severe thrombocytopenia and response to corticosteroids in a case of nephropathia epidemica. Am. J. Kidney Dis. 1998; 31:116–120.

Dvorak, A.M. and Feng, D. The vesiculo-vacuolar organelle (VVO). A new endothelial cell permeability organelle. J. Histochem. Cytochem. 2001; 49:419–432.

Dybas, C. L. Russia Sees Ill Effects of 'General Winter's' Retreat. Washington Post, 2007.

Easterbrook, J.D., Zink, M.C. and Klein, S.L. Regulatory T cells enhance persistence of the zoonotic pathogen Seoul virus in its reservoir host. Proc. Natl. Acad. Sci. U.S.A. 2007; 104:15502–15507.

Enria, D.A., Briggiler, A.M., Fernandez, N.J., Levis, S.C. and Maiztegui, J.I. Importance of dose of neutralising antibodies in treatment of Argentine haemorrhagic fever with immune plasma. Lancet 1984; 2:255–256.

Enria, D.A., Briggiler, A.M., Pini, N. and Levis, S. Clinical manifestations of New World hantaviruses. Curr. Top. Microbiol. Immunol. 2001; 256:117–134.

Fang, L., Yan, L., Liang, S., De Vlas, S.J., Feng, D., Han, X., Zhao, W., Xu, B., Bian, L., Yang, H., Gong, P., Richardus, J.H. and Cao, W. Spatial analysis of hemorrhagic fever with renal syndrome in China. BMC Infect. Dis. 2006; 6:77.

Fauquet, C.M., Mayo, M.A., Maniloff, J., Desselberger, U. and Ball, L.A. (Eds.) Virus Taxonomy. Eighth Report of the International Committee on Taxonomy of Viruses. San Diego, CA: Elsevier Academic Press, 2005.

Feng, D., Nagy, J.A., Pyne, K., Hammel, I., Dvorak, H.F. and Dvorak, A.M. Pathways of macromolecular extravasation across microvascular endothelium in response to VPF/VEGF and other vasoactive mediators. Microcirculation 1999; 6:23–44.

Frampton, J.W., Lanser, S. and Nichols, C.R. Sin Nombre virus infection in 1959. Lancet 1995; 346:781–782.

Gavrilovskaya, I.N., Brown, E.J., Ginsberg, M.H. and Mackow, E.R. Cellular entry of hantaviruses which cause hemorrhagic fever with renal syndrome is mediated by beta3 integrins. J. Virol. 1999; 73:3951–3959.

Gavrilovskaya, I.N., Peresleni, T., Geimonen, E. and Mackow, E.R. Pathogenic hantaviruses selectively inhibit beta3 integrin directed endothelial cell migration. Arch. Virol. 2002; 147:1913–1931.

Gavrilovskaya, I.N., Shepley, M., Shaw, R., Ginsberg, M.H. and Mackow, E.R. Beta3 integrins mediate the cellular entry of hantaviruses that cause respiratory failure. Proc. Natl. Acad. Sci. U.S.A. 1998; 95:7074–7079.

Gedvilaite, A., Zvirbliene, A., Staniulis, J., Sasnauskas, K., Kruger, D.H. and Ulrich, R. Segments of puumala hantavirus nucleocapsid protein inserted into chimeric polyomavirus-derived virus-like particles induce a strong immune response in mice. Viral Immunol. 2004; 17:51–68.

Geimonen, E., Lamonica, R., Springer, K., Farooqui, Y., Gavrilovskaya, I.N. and Mackow, E.R. Hantavirus pulmonary syndrome-associated hantaviruses contain conserved and functional ITAM signaling elements. J. Virol. 2003; 77:1638–1643.

Geimonen, E., Neff, S., Raymond, T., Kocer, S.S., Gavrilovskaya, I.N. and Mackow, E.R. Pathogenic and nonpathogenic hantaviruses differentially regulate endothelial cell responses. Proc. Natl. Acad. Sci. U.S.A. 2002; 99:13837–13842.

Geldmacher, A., Schmaler, M., Kruger, D.H. and Ulrich, R. Yeast-expressed hantavirus Dobrava nucleocapsid protein induces a strong, long-lasting, and highly cross-reactive immune response in mice. Viral Immunol. 2004a; 17:115–122.

Geldmacher, A., Skrastina, D., Petrovskis, I., Borisova, G., Berriman, J.A., Roseman, A.M., Crowther, R.A., Fischer, J., Musema, S., Gelderblom, H.R., Lundkvist, A., Renhofa, R., Ose, V., Kruger, D.H., Pumpens, P. and Ulrich, R. An amino-terminal segment of hantavirus nucleocapsid protein presented on hepatitis B virus core particles induces a strong and highly cross-reactive antibody response in mice. Virology 2004b; 323:108–119.

Goldsmith, C.S., Elliott, L.H., Peters, C.J. and Zaki, S.R. Ultrastructural characteristics of Sin Nombre virus, causative agent of hantavirus pulmonary syndrome. Arch. Virol. 1995; 140:2107–2122.

Gresikova, M., Rajcani, J., Sekeyova, M., Brummer-Korvenkontio, M., Kozuch, O., Labuda, M., Turek, R., Weismann, P., Nosek, J. and Lysy, J. Haemorrhagic fever virus with renal syndrome in small rodents in Czechoslovakia. Acta Virol. 1984; 28:416–421.

Groen, J., Gerding, M., Jordans, J.G., Clement, J.P. and Osterhaus, A.D. Class and subclass distribution of hantavirus-specific serum antibodies at different times after the onset of nephropathia epidemica. J. Med. Virol. 1994; 43:39–43.

Gunwar, S., Bejarano, P.A., Kalluri, R., Langeveld, J.P., Wisdom, B.J., Jr., Noelken, M.E. and Hudson, B.G. Alveolar basement membrane: molecular properties of the noncollagenous domain (hexamer) of collagen IV and its reactivity with Goodpasture autoantibodies. Am. J. Respir. Cell Mol. Biol. 1991; 5:107–112.

Hallin, G.W., Simpson, S.Q., Crowell, R.E., James, D.S., Koster, F. T., Mertz, G.J. and Levy, H. Cardiopulmonary manifestations of hantavirus pulmonary syndrome. Crit. Care. Med. 1996; 24: 252–258.

Henderson, W.W., Monroe, M.C., ST Jeor, S.C., Thayer, W.P., Rowe, J.E., Peters, C.J. and Nichol, S.T. Naturally occurring Sin Nombre virus genetic reassortants. Virology 1995; 214:602–610.

Heyman, P., Cochez, C., Ducoffre, G., Mailles, A., Zeller, H., ABU Sin, M., Koch, J., Van Doornum, G., Koopmans, M., Mossong, J. and Schneider, F. Haemorrhagic fever with renal syndrome: an analysis of the outbreaks in Belgium, France, Germany, the Netherlands and Luxembourg in 2005. Euro Surveill. 2007; 12:E15–E16.

Hjelle, B. and Glass, G.E. Outbreak of hantavirus infection in the Four Corners region of the United States in the wake of the 1997–1998 El Nino-southern oscillation. J. Infect. Dis. 2000; 181:1569–1573.

Hodivala-Dilke, K.M., Mchugh, K.P., Tsakiris, D.A., Rayburn, H., Crowley, D., Ullman-Cullere, M., Ross, F.P., Coller, B.S., Teitelbaum, S. and Hynes, R.O. Beta3-integrin-deficient mice are a model for Glanzmann thrombasthenia showing placental defects and reduced survival. J. Clin. Invest. 1999; 103:229–238.

Hooper, J.W., Custer, D.M., Smith, J. and Wahl-Jensen, V. Hantaan/Andes virus DNA vaccine elicits a broadly cross-reactive neutralizing antibody response in nonhuman primates. Virology 2006; 347:208–216.

Hooper, J.W., Custer, D.M., Thompson, E. and Schmaljohn, C.S. DNA vaccination with the Hantaan virus M gene protects hamsters against three of four HFRS hantaviruses and elicits a high-titer neutralizing antibody response in Rhesus monkeys. J. Virol. 2001a; 75:8469–8477.

Hooper, J.W., Ferro, A.M. and Wahl-Jensen, V. Immune serum produced by DNA vaccination protects hamsters against lethal respiratory challenge with Andes virus. J. Virol. 2008; 82:1332–1338.

Hooper, J.W., Kamrud, K.I., Elgh, F., Custer, D. and Schmaljohn, C.S. DNA vaccination with hantavirus M segment elicits neutralizing antibodies and protects against seoul virus infection. Virology 1999; 255:269–278.

Hooper, J.W., Larsen, T., Custer, D.M. and Schmaljohn, C.S. A lethal disease model for hantavirus pulmonary syndrome. Virology 2001b; 289:6–14.

Hooper, J.W. and Li, D. Vaccines against hantaviruses. Curr. Top. Microbiol. Immunol. 2001; 256:171–191.

Horling, J., Lundkvist, A., Huggins, J.W. and Niklasson, B. Antibodies to Puumala virus in humans determined by neutralization test. J. Virol. Methods 1992; 39:139–147.

Huang, C., Jin, B., Wang, M., Li, E. and Sun, C. Hemorrhagic fever with renal syndrome: relationship between pathogenesis and cellular immunity. J. Infect. Dis. 1994; 169:868–870.

Huggins, J.W., Hsiang, C.M., Cosgriff, T.M., Guang, M.Y., Smith, J.I., Wu, Z.O., Leduc, J.W., Zheng, Z.M., Meegan, J.M., Wang, Q.N. et al. Prospective, double-blind, concurrent, placebo-controlled clinical trial of intravenous ribavirin therapy of hemorrhagic fever with renal syndrome. J. Infect. Dis. 1991; 164:1119–1127.

Huggins, J.W., Kim, G.R., Brand, O.M. and Mckee, K.T., Jr. Ribavirin therapy for Hantaan virus infection in suckling mice. J. Infect. Dis. 1986; 153:489–497.

Hukic, M., Kurt, A., Torstensson, S., Lundkvist, A., Wiger, D. and Niklasson, B. Haemorrhagic fever with renal syndrome in north-east Bosnia. Lancet 1996; 347:56–57.

Hynes, R.O., Bader, B.L. and Hodivala-Dilke, K. Integrins in vascular development. Braz. J. Med. Biol. Res. 1999; 32:501–510.

Idell, S., Zwieb, C., Boggaram, J., Holiday, D., Johnson, A.R. and Raghu, G. Mechanisms of fibrin formation and lysis by human lung fibroblasts: influence of TGF-beta and TNF-alpha. Am. J. Physiol. 1992; 263:L487–L494.

Johansson, P., Lindgren, T., Lunström, M., Holmström, A., Elgh, F. and Bucht, G. PCR-generated linear DNA fragments utilized as a hantavirus DNA vaccine. Vaccine. 2002; Sep 10; 20(27–28):3379–3388.

Johnson, K.M. Hantaviruses: history and overview. Curr. Top. Microbiol. Immunol. 2001; 256:1–14.

Jonsson, C.B., Hooper, J. and Mertz, G. Treatment of hantavirus pulmonary syndrome. Antiviral Res. 2008; Apr; 78(1):162–169.

Joutsi-Korhonen, L., Preston, S., Smethurst, P.A., Ijsseldijk, M., Schaffner-Reckinger, E., Armour, K.L., Watkins, N.A., Clark, M.R., De Groot, P.G., Farndale, R.W., Ouwehand, W.H. and Williamson, L.M. The effect of recombinant IgG antibodies against the leucine-33 form of the platelet beta3 integrin (HPA-1a) on platelet function. Thromb. Haemost. 2004; 91:743–754.

Kallio, E.R., Klingstrom, J., Gustafsson, E., Manni, T., Vaheri, A., Henttonen, H., Vapalahti, O. and Lundkvist, A. Prolonged survival of Puumala hantavirus outside the host: evidence for indirect transmission via the environment. J. Gen. Virol. 2006; 87:2127–2134.

Kallio, E.R., Voutilainen, L., Vapalahti, O., Vaheri, A., Henttonen, H., Koskela, E. and Mappes, T. Endemic hantavirus infection impairs the winter survival of its rodent host. Ecology 2007; 88:1911–1916.

Kallio-Kokko, H., Leveelahti, R., Brummer-Korvenkontio, M., Lundkvist, A., Vaheri, A. and Vapalahti, O. Human immune response to Puumala virus glycoproteins and nucleocapsid protein expressed in mammalian cells. J. Med. Virol. 2001; 65:605–613.

Kalluri, R., Gattone, V.H., 2nd., Noelken, M.E. and Hudson, B.G. The alpha 3 chain of type IV collagen induces autoimmune Goodpasture syndrome. Proc. Natl. Acad. Sci. U.S.A. 1994; 91:6201–6205.

Kamrud, K.I., Hooper, J.W., Elgh, F. and Schmaljohn, C.S. Comparison of the protective efficacy of naked DNA, DNA-based Sindbis replicon, and packaged Sindbis replicon vectors expressing hantavirus structural genes in hamsters. Virology 1999; 263:209–219.

Kanerva, M., Mustonen, J. and Vaheri, A. Pathogenesis of puumala and other hantavirus infections. Rev. Med. Virol. 1998a; 8:67–86.

Kanerva, M., Vaheri, A., Mustonen, J. and Partanen, J. High-producer allele of tumour necrosis factor-alpha is part of the susceptibility MHC haplotype in severe puumala virus-induced nephropathia epidemica. Scand. J. Infect. Dis. 1998b; 30:532–534.

Kariwa, H., Yoshimatsu, K. and Arikawa, J. Hantavirus infection in East Asia. Comp. Immunol. Microbiol. Infect. Dis. 2007; 30:341–356.

Kawasaki, J., Davis, G.E. and Davis, M.J. Regulation of Ca^{2+}-dependent K^+ current by alphavbeta3 integrin engagement in vascular endothelium. J. Biol. Chem. 2004; 279:12959–12966.

Kerkar, S., Williams, M., Blocksom, J.M., Wilson, R.F., Tyburski, J.G. and Steffes, C.P. TNF-alpha and IL-1beta increase pericyte/endothelial cell co-culture permeability. J. Surg. Res. 2006; 132:40–45.

Ketai, L.H., Williamson, M.R., Telepak, R.J., Levy, H., Koster, F.T., Nolte, K.B. and Allen, S.E. Hantavirus pulmonary syndrome: radiographic findings in 16 patients. Radiology 1994; 191: 665–668.

Kevil, C.G., Payne, D.K., Mire, E. and Alexander, J.S. Vascular permeability factor/vascular endothelial cell growth factor-mediated permeability occurs through disorganization of endothelial junctional proteins. J. Biol. Chem. 1998; 273:15099–15103.

Khattak, S., Darai, G. and Rosen-Wolff, A. Puumala virus nucleocapsid protein expressed in transgenic plants is not immunogenic after oral administration. Virus Genes 2004; 29:109–116.

Kilpatrick, E.D., Terajima, M., Koster, F.T., Catalina, M.D., Cruz, J. and Ennis, F.A. Role of specific CD8+ T cells in the severity of a fulminant zoonotic viral hemorrhagic fever, hantavirus pulmonary syndrome. J. Immunol. 2004; 172:3297–3304.

Klein, S.L., Bird, B.H. and Glass, G.E. Sex differences in immune responses and viral shedding following Seoul virus infection in Norway rats. Am. J. Trop. Med. Hyg. 2001; 65:57–63.

Klein, S.L. and Calisher, C.H. Emergence and persistence of hantaviruses. Curr. Top. Microbiol. Immunol. 2007; 315:217–252.

Klingstrom, J., Falk, K.I. and Lundkvist, A. Delayed viremia and antibody responses in Puumala hantavirus challenged passively immunized cynomolgus macaques. Arch. Virol. 2005; 150:79–92.

Klingstrom, J., Hardestam, J., Stoltz, M., Zuber, B., Lundkvist, A., Linder, S. and Ahlm, C. Loss of cell membrane integrity in puumala hantavirus-infected patients correlates with levels of epithelial cell apoptosis and perforin. J. Virol. 2006; 80:8279–8282.

Klingstrom, J., Maljkovic, I., Zuber, B., Rollman, E., Kjerrstrom, A. and Lundkvist, A. Vaccination of C57/BL6 mice with Dobrava hantavirus nucleocapsid protein in Freund's adjuvant induced partial protection against challenge. Vaccine 2004; 22:4029–4034.

Kotowicz, K., Callard, R.E., Klein, N.J. and Jacobs, M.G. Interleukin-4 increases the permeability of human endothelial cells in culture. Clin. Exp. Allergy 2004; 34:445–449.

Krakauer, T., Leduc, J.W. and Krakauer, H. Serum levels of tumor necrosis factor-alpha, interleukin-1, and interleukin-6 in hemorrhagic fever with renal syndrome. Viral Immunol. 1995; 8:75–79.

Krakauer, T., Leduc, J.W., Morrill, J.C., Anderson, A.O. and Krakauer, H. Serum levels of alpha and gamma interferons in hemorrhagic fever with renal syndrome. Viral Immunol. 1994; 7:97–101.

Lazaro, M.E., Cantoni, G.E., Calanni, L.M., Resa, A.J., Herrero, E.R., Iacono, M.A., Enria, D.A. and Gonzalez Cappa, S.M. Clusters of hantavirus infection, southern Argentina. Emerg. Infect. Dis. 2007; 13:104–110.

Lazaro, M.E., Resa, A.J., Barclay, C.M., Calanni, L., Samengo, L., Martinez, L., Padula, P.J., Pini, N., Lasala, M.B., Elsner, B. and Enria, D.A. Hantavirus pulmonary syndrome in southern Argentina. Medicina (B. Aires) 2000; 60:289–301.

Leavesley, D.I., Schwartz, M.A., Rosenfeld, M. and Cheresh, D.A. Integrin beta 1- and beta 3-mediated endothelial cell migration is triggered through distinct signaling mechanisms. J. Cell Biol. 1993; 121:163–170.

Lee, B.H., Yoshimatsu, K., Araki, K., Okumura, M., Nakamura, I. and Arikawa, J. A pseudotype vesicular stomatitis virus containing Hantaan virus envelope glycoproteins G1 and G2 as an alternative to hantavirus vaccine in mice. Vaccine 2006; 24:2928–2934.

Lee, H., Guo, J., Li, M., Choi, J.K., Demaria, M., Rosenzweig, M. and Jung, J.U. Identification of an immunoreceptor tyrosine-based activation motif of K1 transforming protein of Kaposi's sarcoma-associated herpes virus. Mol. Cell. Biol. 1998a; 18:5219–5228.

Lee, H.W. Hantaviruses: an emerging disease. Phil. J. Microbiol. Infect. Dis. 1996; 25:S19–S24.

Lee, H.W., Chu, Y.K., Woo, Y.D., An, C.N., Kim, H., Tkachenko, E. and Gligic, A. Vaccines against hemorrhagic fever with renal syndrome. In: Emergence and Control of Rodent-borne Viral Diseases (Hantaviral and Arenal Diseases) (J.F. Saluzzo and B. Dodet, Eds.). Veyrier-du-lac, Annecy, France: Elsevier, 1998b.

Lee, H.W., Lee, P.W. and Johnson, K.M. Isolation of the etiologic agent of Korean hemorrhagic fever. J. Infect. Dis. 1978a; 137:298–308.

Lee, H.W., Lee, P.W. and Johnson, K.M. Isolation of the etiologic agent of Korean hemorrhagic fever. J. Infect. Dis. 1978b; 137:298–308.

Lee, J.S. Clinical features of hemorrhagic fever with renal syndrome in Korea. Kidney Int. Suppl. 1991; 35:S88–S93.

Lee, P.C., Kibbe, M.R., Schuchert, M.J., Stolz, D.B., Watkins, S.C., Griffith, B.P., Billiar, T.R. and Shears, L.L., 2nd. Nitric oxide induces angiogenesis and upregulates alpha(v)beta(3) integrin expression on endothelial cells. Microvasc. Res. 2000; 60:269–280.

Levis, S., Morzunov, S.P., Rowe, J.E., Enria, D., Pini, N., Calderon, G., Sabattini, M. and ST Jeor, S.C. Genetic diversity and epidemiology of hantaviruses in Argentina. J. Infect. Dis. 1998; 177:529–538.

Levy, H. and Simpson, S.Q. Hantavirus pulmonary syndrome. Am. J. Respir. Crit. Care Med. 1994; 149:1710–1713.

Li, D., Schmaljohn, A.L., Anderson, K. and Schmaljohn, C.S. Complete nucleotide sequences of the M and S segments of two hantavirus isolates from California: evidence for reassortment in nature among viruses related to hantavirus pulmonary syndrome. Virology 1995; 206:973–983.

Lin, H., Liu, Q., Guo, J., Zhang, J., Wang, J. and Chen, H. Analysis of the geographic distribution of HFRS in Liaoning province between 2000 and 2005. BMC Public Health 2007; 7:207.

Linderholm, M., Ahlm, C., Settergren, B., Waage, A. and Tarnvik, A. Elevated plasma levels of tumor necrosis factor (TNF)-alpha, soluble TNF receptors, interleukin (IL)-6, and IL-10 in patients with hemorrhagic fever with renal syndrome. J. Infect. Dis. 1996; 173:38–43.

Linderholm, M. and Elgh, F. Clinical characteristics of hantavirus infections on the Eurasian continent. Curr. Top. Microbiol. Immunol. 2001; 256:135–151.

Lindkvist, M., Lahti, K., Lilliehook, B., Holmstrom, A., Ahlm, C. and Bucht, G. Cross-reactive immune responses in mice after genetic vaccination with cDNA encoding hantavirus nucleocapsid proteins. Vaccine 2007; 25:1690–1699.

Loeffler, S., Fayard, B., Weis, J. and Weissenberger, J. Interleukin-6 induces transcriptional activation of vascular endothelial growth factor (VEGF) in astrocytes in vivo and regulates VEGF promoter activity in glioblastoma cells via direct interaction between STAT3 and Sp1. Int. J. Cancer 2005; 115:202–213.

Luhn, K., Simmons, C.P., Moran, E., Dung, N.T., Chau, T.N., Quyen, N.T., Thao Le, T.T., Van Ngoc, T., Dung, N.M., Wills, B., Farrar, J., Mcmichael, A.J., Dong, T. and Rowland-Jones, S. Increased frequencies of CD4+ CD25(high) regulatory T cells in acute dengue infection. J. Exp. Med. 2007; 204:979–985.

Lundkvist, A., Bjorsten, S. and Niklasson, B. Immunoglobulin G subclass responses against the structural components of Puumala virus. J. Clin. Microbiol. 1993; 31:368–372.

Lundkvist, A., Hukic, M., Horling, J., Gilljam, M., Nichol, S. and Niklasson, B. Puumala and Dobrava viruses cause hemorrhagic fever with renal syndrome in Bosnia-Herzegovina: evidence of highly cross-neutralizing antibody responses in early patient sera. J. Med. Virol. 1997; 53:51–59.

Lundkvist, A., Vapalahti, O., Plyusnin, A., Sjolander, K.B., Niklasson, B. and Vaheri, A. Characterization of Tula virus antigenic determinants defined by monoclonal antibodies raised against baculovirus-expressed nucleocapsid protein. Virus Res. 1996; 45:29–44.

Lyubsky, S., Gavrilovskaya, I., Luft, B. and Mackow, E. Histopathology of *Peromyscus leucopus* naturally infected with pathogenic NY-1 hantaviruses: pathologic markers of HPS viral infection in mice. Lab. Invest. 1996; 74:627–633.

Makela, S., Ala-Houhala, I., Mustonen, J., Koivisto, A.M., Kouri, T., Turjanmaa, V., Vapalahti, O., Vaheri, A. and Pasternack, A. Renal function and blood pressure five years after puumala virus-induced nephropathy. Kidney Int. 2000; 58:1711–1718.

Makela, S., Mustonen, J., Ala-Houhala, I., Hurme, M., Partanen, J., Vapalahti, O., Vaheri, A. and Pasternack, A. Human leukocyte antigen-B8-DR3 is a more important risk factor for severe Puumala hantavirus infection than the tumor necrosis factor-alpha(-308) G/A polymorphism. J. Infect. Dis. 2002; 186:843–846.

Markotic, A., Nichol, S.T., Kuzman, I., Sanchez, A.J., Ksiazek, T.G., Gagro, A., Rabatic, S., Zgorelec, R., Avsic-Zupanc, T., Beus, I. and Dekaris, D. Characteristics of Puumala and Dobrava infections in Croatia. J. Med. Virol. 2002; 66:542–551.

Martinez, V.P., Bellomo, C., San Juan, J., Pinna, D., Forlenza, R., Elder, M. and Padula, P.J. Person-to-person transmission of Andes virus. Emerg. Infect. Dis. 2005; 11:1848–1853.

Maruo, N., Morita, I., Shirao, M. and Murota, S. IL-6 increases endothelial permeability in vitro. Endocrinology 1992; 131:710–714.

Mcclain, D.J., Summers, P.L., Harrison, S.A., Schmaljohn, A.L. and Schmaljohn, C.S. Clinical evaluation of a vaccinia-vectored Hantaan virus vaccine. J. Med. Virol. 2000; 60:77–85.

Mccormick, J.B., Sasso, D.R., Palmer, E.L. and Kiley, M.P. Morphological identification of the agent of Korean haemorrhagic fever (Hantaan virus) as a member of the Bunyaviridae. Lancet 1982; 1:765–768.

Medina, R.A., Mirowsky-Garcia, K., Hutt, J. and Hjelle, B. Ribavirin, human convalescent plasma and anti-beta3 integrin antibody inhibit infection by Sin Nombre virus in the deer mouse model. J. Gen. Virol. 2007; 88:493–505.

Mertz, G.J., Hjelle, B., Crowley, M., Iwamoto, G., Tomicic, V. and Vial, P.A. Diagnosis and treatment of new world hantavirus infections. Curr. Opin. Infect. Dis. 2006; 19:437–442.

Mertz, G.J., Miedzinski, L., Goade, D., Pavia, A.T., Hjelle, B., Hansbarger, C.O., Levy, H., Koster, F.T., Baum, K., Lindemulder, A.,

Wang, W., Riser, L., Fernandez, H. and Whitley, R.J. Placebo-controlled, double-blind trial of intravenous ribavirin for the treatment of hantavirus cardiopulmonary syndrome in North America. Clin. Infect. Dis. 2004; 39:1307–1313.

Meyer, B.J. and Schmaljohn, C.S. Persistent hantavirus infections: characteristics and mechanisms. Trends Microbiol. 2000; 8:61–67.

Milazzo, M.L., Eyzaguirre, E.J., Molina, C.P. and Fulhorst, C.F. Maporal viral infection in the Syrian golden hamster: a model of hantavirus pulmonary syndrome. J. Infect. Dis. 2002; 186: 1390–1395.

Miller, C.L., Burkhardt, A.L., Lee, J.H., Stealey, B., Longnecker, R., Bolen, J.B. and Kieff, E. Integral membrane protein 2 of Epstein-Barr virus regulates reactivation from latency through dominant negative effects on protein--tyrosine kinases. Immunity 1995; 2:155–166.

Mills, K.H. Regulatory T cells: friend or foe in immunity to infection? Nat. Rev. Immunol. 2004; 4:841–855.

Mori, M., Rothman, A.L., Kurane, I., Montoya, J.M., Nolte, K.B., Norman, J.E., Waite, D.C., Koster, F.T. and Ennis, F.A. High levels of cytokine-producing cells in the lung tissues of patients with fatal hantavirus pulmonary syndrome. J. Infect. Dis. 1999; 179:295–302.

Mulic, R. and Ropac, D. Epidemiologic characteristics and military implications of hemorrhagic fever with renal syndrome in Croatia. Croat. Med. J. 2002; 43:581–586.

Muranyi, W., Bahr, U., Zeier, M. and Van Der Woude, F.J. Hantavirus infection. J. Am. Soc. Nephrol. 2005; 16:3669–3679.

Mustonen, J., Helin, H., Pietila, K., Brummer-Korvenkontio, M., Hedman, K., Vaheri, A. and Pasternack, A. Renal biopsy findings and clinicopathologic correlations in nephropathia epidemica. Clin. Nephrol. 1994; 41:121–126.

Mustonen, J., Partanen, J., Kanerva, M., Pietila, K., Vapalahti, O., Pasternack, A. and Vaheri, A. Genetic susceptibility to severe course of nephropathia epidemica caused by Puumala hantavirus. Kidney Int. 1996; 49:217–221.

Mustonen, J., Partanen, J., Kanerva, M., Pietila, K., Vapalahti, O., Pasternack, A. and Vaheri, A. Association of HLA B27 with benign clinical course of nephropathia epidemica caused by Puumala hantavirus. Scand. J. Immunol. 1998; 47:277–279.

Netski, D., Thran, B.H. and St Jeor, S.C. Sin Nombre virus pathogenesis in *Peromyscus maniculatus*. J. Virol. 1999; 73:585–591.

Nichol, S.T. Chapter 49: Bunyaviruses, In: Fields Virology, 4th ed. (D.M. Knipe, P.M. Howley., D.E. Griffin, R.A. Lamb, M.A. Martin, B. Roizman, S.E. Straus Ed.), Lippincott Williams and Wilkins, Philadelphia, 2001.

Nichol, S.T., Spiropoulou, C.F., Morzunov, S., Rollin, P.E., Ksiazek, T.G., Feldmann, H., Sanchez, A., Childs, J., Zaki, S. and Peters, C.J. Genetic identification of a hantavirus associated with an outbreak of acute respiratory illness. Science 1993; 262:914–917.

Niikura, M., Maeda, A., Ikegami, T., Saijo, M., Kurane, I. and Morikawa, S. Modification of endothelial cell functions by Hantaan virus infection: prolonged hyper-permeability induced by TNF-alpha of hantaan virus-infected endothelial cell monolayers. Arch. Virol. 2004; 149:1279–1292.

Niklasson, B., Hornfeldt, B., Mullaart, M., Settergren, B., Tkachenko, E., Myasnikov Yu, A., Ryltceva, E.V., Leschinskaya, E., Malkin, A. and Dzagurova, T. An epidemiologic study of hemorrhagic fever with renal syndrome in Bashkirtostan (Russia) and Sweden. Am. J. Trop. Med. Hyg. 1993; 48:670–675.

Nolte, K.B., Feddersen, R.M., Foucar, K., Zaki, S.R., Koster, F.T., Madar, D., Merlin, T.L., Mcfeeley, P.J., Umland, E.T. and Zumwalt, R.E. Hantavirus pulmonary syndrome in the United States: a pathological description of a disease caused by a new agent. Hum. Pathol. 1995; 26:110–120.

Nooteboom, A., Bleichrodt, R.P. and Hendriks, T. Modulation of endothelial monolayer permeability induced by plasma obtained from lipopolysaccharide-stimulated whole blood. Clin. Exp. Immunol. 2006; 144:362–369.

Okada, K., Fujita, T., Minamoto, K., Liao, H., Naka, Y. and Pinsky, D.J. Potentiation of endogenous fibrinolysis and rescue from lung ischemia/reperfusion injury in interleukin (IL)-10-reconstituted IL-10 null mice. J. Biol. Chem. 2000; 275:21468–21476.

Okumura, M., Yoshimatsu, K., Kumperasart, S., Nakamura, I., Ogino, M., Taruishi, M., Sungdee, A., Pattamadilok, S., Ibrahim, I.N., Erlina, S., Agui, T., Yanagihara, R. and Arikawa, J. Development of serological assays for Thottapalayam virus, an insectivore-borne hantavirus. Clin. Vaccine Immunol. 2007; 14:173–181.

Ozaki, M., Ogita, H. and Takai, Y. Involvement of integrin-induced activation of protein kinase C in the formation of adherens junctions. Genes Cells 2007; 12:651–662.

Paakkala, A., Mustonen, J., Viander, M., Huhtala, H. and Pasternack, A. Complement activation in nephropathia epidemica caused by Puumala hantavirus. Clin. Nephrol. 2000; 53:424–431.

Padula, P.J., Edelstein, A., Miguel, S.D., Lopez, N.M., Rossi, C.M. and Rabinovich, R.D. Hantavirus pulmonary syndrome outbreak in Argentina: molecular evidence for person-to-person transmission of Andes virus. Virology 1998; 241:323–330.

PAHO. Number of Cases and Deaths from Hantavirus Pulmonary Syndrome (HPS), 2004.

Paria, B.C., Vogel, S.M., Ahmmed, G.U., Alamgir, S., Shroff, J., Malik, A.B. and Tiruppathi, C. Tumor necrosis factor-alpha-induced TRPC1 expression amplifies store-operated Ca2+ influx and endothelial permeability. Am. J. Physiol. Lung Cell. Mol. Physiol. 2004; 287:L1303–L1313.

Park, K., Kim, C.S. and Moon, K.T. Protective effectiveness of hantavirus vaccine. Emerg. Infect. Dis. 2004; 10:2218–2220.

Pensiero, M.N., Sharefkin, J.B., Dieffenbach, C.W. and Hay, J. Hantaan virus infection of human endothelial cells. J. Virol. 1992; 66:5929–5936.

Perez, L.E., Chandrasekar, B., Saldarriaga, O.A., Zhao, W., Arteaga, L.T., Travi, B.L. and Melby, P.C. Reduced nitric oxide synthase 2 (NOS2) promoter activity in the Syrian hamster renders the animal functionally deficient in NOS2 activity and unable to control an intracellular pathogen. J. Immunol. 2006; 176:5519–5528.

Perez, R.L. and Roman, J. Fibrin enhances the expression of IL-1 beta by human peripheral blood mononuclear cells. Implications in pulmonary inflammation. J. Immunol. 1995; 154:1879–1887.

Peters, C.J. and Khan, A.S. Hantavirus pulmonary syndrome: the new American hemorrhagic fever. Clin. Infect. Dis. 2002; 34:1224–1231.

Pettersson, L.K.J., Hardestam, J., Lundkvist, Å., Ahlm, C. and Evander M. Hantavirus RNA in saliva from patients with hemorrhagic fever with renal syndrome. Emerg. Infect. Dis. [Epub ahead of print] 2008.

Pini, N. Hantavirus pulmonary syndrome in Latin America. Curr. Opin. Infect. Dis. 2004; 17:427–431.

Raymond, T., Gorbunova, E., Gavrilovskaya, I.N. and Mackow, E.R. Pathogenic hantaviruses bind plexin-semaphorin-integrin domains present at the apex of inactive, bent alphavbeta3 integrin conformers. Proc. Natl. Acad. Sci. U.S.A. 2005; 102:1163–1168.

Reynolds, S., Galanis, E., Krajden, M., Morshed, M., Bowering, D., Abelson, W. and Kollmann, T.R. Imported fatal hantavirus pulmonary syndrome. Emerg. Infect. Dis. 2007; 13:1424–1425.

Rizvanov, A.A., Khaiboullina, S.F., Van Geelen, A.G. and St Jeor, S.C. Replication and immunoactivity of the recombinant *Peromyscus maniculatus* cytomegalovirus expressing hantavirus G1 glycoprotein in vivo and in vitro. Vaccine 2006; 24:327–334.

Rizvanov, A.A., Van Geelen, A.G., Morzunov, S., Otteson, E.W., Bohlman, C., Pari, G.S. and St Jeor, S.C. Generation of a recombinant cytomegalovirus for expression of a hantavirus glycoprotein. J. Virol. 2003; 77:12203–12210.

Rodriguez, L.L., Owens, J.H., Peters, C.J. and Nichol, S.T. Genetic reassortment among viruses causing hantavirus pulmonary syndrome. Virology 1998; 242:99–106.

Rutigliano, J.A. and Graham, B.S. Prolonged production of TNF-alpha exacerbates illness during respiratory syncytial virus infection. J. Immunol. 2004; 173:3408–3417.

Sandoval, R., Malik, A.B., Minshall, R.D., Kouklis, P., Ellis, C.A. and Tiruppathi, C. Ca(2+) signalling and PKCalpha activate increased endothelial permeability by disassembly of VE-cadherin junctions. J. Physiol. 2001; 533:433–445.

Schmaljohn, C.S., Chu, Y.K., Schmaljohn, A.L. and Dalrymple, J.M. Antigenic subunits of Hantaan virus expressed by baculovirus and vaccinia virus recombinants. J. Virol. 1990; 64:3162–3170.

Schmaljohn, C.S. and Hooper, J.W. *Bunyaviridae*: the viruses and their replication. In: Fields Virology, 4th ed. (D.M. Knipe, P.M. Howley., D.E. Griffin, R.A. Lamb, M.A. Martin, B. Roizman, and S.E. Straus Ed.), Lippincott Williams & Wilkins, 2001.

Schountz, T., Prescott, J., Cogswell, A.C., Oko, L., Mirowsky-Garcia, K., Galvez, A.P. and Hjelle, B. Regulatory T cell-like responses in deer mice persistently infected with Sin Nombre virus. Proc. Natl. Acad. Sci. U.S.A. 2007; 104:15496–15501.

Schwarcz, S.K., Shefer, A.M. and Zaki, S.R. Retrospective diagnosis of a fatal case of the hantavirus pulmonary syndrome, 1980. West J. Med. 1996; 164:348–350.

Seitsonen, E., Hynninen, M., Kolho, E., Kallio-Kokko, H. and Pettila, V. Corticosteroids combined with continuous venovenous hemodiafiltration for treatment of hantavirus pulmonary syndrome caused by Puumala virus infection. Eur. J. Clin. Microbiol. Infect. Dis. 2006; 25:261–266.

Settergren, B., Juto, P., Wadell, G., Trollfors, B. and Norrby, S.R. Incidence and geographic distribution of serologically verified cases of nephropathia epidemica in Sweden. Am. J. Epidemiol. 1988; 127:801–807.

Sinclair, J.R., Carroll, D.S., Montgomery, J.M., Pavlin, B., Mccombs, K., Mills, J.N., Comer, J.A., Ksiazek, T.G., Rollin, P.E., Nichol, S.T., Sanchez, A.J., Hutson, C.L., Bell, M. and Rooney, J.A. Two cases of hantavirus pulmonary syndrome in Randolph County, West Virginia: a coincidence of time and place? Am. J. Trop. Med. Hyg. 2007; 76:438–442.

Siqueira, R.C., Jorge, R. and Figueiredo, L.T. Bilateral acute acquired toxoplasmic retinochoroiditis after steroid therapy for hantavirus pulmonary syndrome: case report. Arq. Bras. Oftalmol. 2007; 70:513–516.

Sironi, M., Breviario, F., Proserpio, P., Biondi, A., Vecchi, A., Van Damme, J., Dejana, E. and Mantovani, A. IL-1 stimulates IL-6 production in endothelial cells. J. Immunol. 1989; 142:549–553.

Sirotin, B.Z. and Keiser, N.P. On the history of the study of haemorrhagic fever with renal syndrome in eastern Russia. Nephrol. Dial. Transplant. 2001; 16:1288–1290.

Song, G., Huang, Y.C., Hang, C.S., Hao, F.Y., Li, D.X., Zheng, X.L., Liu, W.M., Li, S.L., Huo, Z.W., Huei, L.J. et al. Preliminary human trial of inactivated golden hamster kidney cell (GHKC) vaccine against haemorrhagic fever with renal syndrome (HFRS). Vaccine 1992; 10:214–216.

Spiropoulou, C.F., Albarino, C.G., Ksiazek, T.G. and Rollin, P.E. Andes and Prospect Hill hantaviruses differ in early induction of interferon although both can downregulate interferon signaling. J. Virol. 2007; 81:2769–2776.

Stanley, M. Prevention strategies against the human papillomavirus: the effectiveness of vaccination. Gynecol. Oncol. 2007; 107:S19–S23.

Steer, A. Pathology of hemorrhagic fever: a comparison of the findings; 1951 and 1952. Am. J. Pathol. 1955; 31:201–221.

Stuhlfault, K. Bericht über ein neues schlammfieberähnliches Krankenheitsbild bei Deutschen Truppen in Lappland. Dtsch. Med. Wochenschr. 1943; 69:439–443, 474–477.

Sugimori, T., Griffith, D.L. and Arnaout, M.A. Emerging paradigms of integrin ligand binding and activation. Kidney Int. 1997; 51:1454–1462.

Sullivan, D.E., Ferris, M., Pociask, D. and Brody, A.R. Tumor necrosis factor-alpha induces transforming growth factor-beta1 expression in lung fibroblasts through the extracellular signal-regulated kinase pathway. Am. J. Respir. Cell. Mol. Biol. 2005; 32:342–349.

Sundstrom, J.B., Mcmullan, L.K., Spiropoulou, C.F., Hooper, W.C., Ansari, A.A., Peters, C.J. and Rollin, P.E. Hantavirus infection induces the expression of RANTES and IP-10 without causing increased permeability in human lung microvascular endothelial cells. J. Virol. 2001; 75:6070–6085.

Takenaka, A., Gibbs, C.J., Jr. and Gajdusek, D.C. Antiviral neutralizing antibody to Hantaan virus as determined by plaque reduction technique. Arch. Virol. 1985; 84:197–206.

Tanishita, O., Takahashi, Y., Okuno, Y., Tamura, M., Asada, H., Dantas, J.R., Jr., Yamanouchi, T., Domae, K., Kurata, T. and Yamanishi, K. Persistent infection of rats with haemorrhagic fever with renal syndrome virus and their antibody responses. J. Gen. Virol. 1986; 67(Pt. 12):2819–2824.

Temonen, M., Mustonen, J., Helin, H., Pasternack, A., Vaheri, A. and Holthofer, H. Cytokines, adhesion molecules, and cellular infiltration in nephropathia epidemica kidneys: an immunohistochemical study. Clin. Immunol. Immunopathol. 1996; 78:47–55.

Tkachenko, E., Dekonenko, A., Ivanov, A., Dzaguova, T., Ivanov, L., Slonova, R., Nurgaleeva, R., Stepanenko, A., Ivanidze, E. and Zagidullin, I. Hemorrhagic fever with renal syndrome and hantaviruses in Russia. In: Emergence and Control of Rodent-Borne Viral Diseases (Hantaviral and Arenal Diseases) (J.F. Saluzzo and B. Dodet, Eds.). Veyrier-du-lac, Annecy, France: Elsevier, 1998.

Tkachenko, E., Ivanidze, E., Zagidullin, I. and Dekonenko, A. HFRS in Eurasia: Russia and the Republics of the former USSR. In: Manual of Hemorrhagic Fever with Renal Syndrome and Hantavirus Pulmonary Syndrome (H.W. Lee, C. C. Calisher and C. Schmaljohn Eds.). Seoul, WHO Collaborating Center for Virus Reference and Research (Hantaviruses), Asian Instititue for Life Sciences, 1999.

Tomicic, V., Espinoza, M., Torres, J., Abarca, J., Montes, J.M., Luppi, M., Concha, A., Reynolds, E., Laporte, A., Canals, C. and Vial, P. Addition of an arterio-venous shunt during veno-arterial extracorporeal life support in a patient with hantavirus pulmonary syndrome. Rev. Med. Chil. 2005; 133:817–822.

Toro, J., Vega, J.D., Khan, A.S., Mills, J.N., Padula, P., Terry, W., Yadon, Z., Valderrama, R., Ellis, B.A., Pavletic, C., Cerda, R., Zaki, S., Shieh, W.J., Meyer, R., Tapia, M., Mansilla, C., Baro, M., Vergara, J.A., Concha, M., Calderon, G., Enria, D., Peters, C.J. and Ksiazek, T.G. An outbreak of hantavirus pulmonary syndrome, Chile, 1997. Emerg. Infect. Dis. 1998; 4:687–694.

Tsukada, H., Ying, X., Fu, C., Ishikawa, S., Mckeown-Longo, P., Albelda, S., Bhattacharya, S., Bray, B.A. and Bhattacharya, J. Ligation of endothelial alpha v beta 3 integrin increases capillary hydraulic conductivity of rat lung. Circ. Res. 1995; 77:651–659.

Ulrich, R., Koletzki, D., Lachmann, S., Lundkvist, A., Zankl, A., Kazaks, A., Kurth, A., Gelderblom, H.R., Borisova, G., Meisel, H. and Kruger, D.H. New chimaeric hepatitis B virus core particles carrying hantavirus (serotype Puumala) epitopes: immunogenicity and protection against virus challenge. J. Biotechnol. 1999; 73:141–153.

Ulrich, R., Lundkvist, A., Meisel, H., Koletzki, D., Sjolander, K.B., Gelderblom, H.R., Borisova, G., Schnitzler, P., Darai, G.

and Kruger, D.H. Chimaeric HBV core particles carrying a defined segment of Puumala hantavirus nucleocapsid protein evoke protective immunity in an animal model. Vaccine 1998; 16:272–280.

Valdivieso, F., Vial, P., Ferres, M., Ye, C., Goade, D., Cuiza, A. and Hjelle, B. Neutralizing antibodies in survivors of Sin Nombre and Andes hantavirus infection. Emerg. Infect. Dis. 2006; 12:166–168.

van der Groen, G., Piot, P., Desmyter, J., Colaert, J., Muylle, L., Tkachenko, E.A., Ivanov, A.P., Verhagen, R. and van Ypersele de Strihou, C. Seroepidemiology of Hantaan-related virus infections in Belgian populations. Lancet 1983; 2:1493–1494.

Van Epps, H.L., Schmaljohn, C.S. and Ennis, F.A. Human memory cytotoxic T-lymphocyte (CTL) responses to Hantaan virus infection: identification of virus-specific and cross-reactive CD8(+) CTL epitopes on nucleocapsid protein. J. Virol. 1999; 73:5301–5308.

Van Epps, H.L., Terajima, M., Mustonen, J., Arstila, T.P., Corey, E.A., Vaheri, A. and Ennis, F.A. Long-lived memory T lymphocyte responses after hantavirus infection. J. Exp. Med. 2002; 196:579–588.

Vapalahti, K., Paunio, M., Brummer-Korvenkontio, M., Vaheri, A. and Vapalahti, O. Puumala virus infections in Finland: increased occupational risk for farmers. Am. J. Epidemiol. 1999; 149:1142–1151.

Vapalahti, O., Mustonen, J., Lundkvist, A., Henttonen, H., Plyusnin, A. and Vaheri, A. Hantavirus infections in Europe. Lancet Infect. Dis. 2003; 3:653–661.

Vial, P.A., Valdivieso, F., Mertz, G., Castillo, C., Belmar, E., Delgado, I., Tapia, M. and Ferres, M. Incubation period of hantavirus cardiopulmonary syndrome. Emerg. Infect. Dis. 2006; 12:1271–1273.

Villa, L.L., Costa, R.L., Petta, C.A., Andrade, R.P., Paavonen, J., Iversen, O.E., Olsson, S.E., Hoye, J., Steinwall, M., RIIS-Johannessen, G., Andersson-Ellstrom, A., Elfgren, K., Krogh, G., Lehtinen, M., Malm, C., Tamms, G.M., Giacoletti, K., Lupinacci, L., Railkar, R., Taddeo, F.J., Bryan, J., Esser, M.T., Sings, H.L., Saah, A.J. and Barr, E. High sustained efficacy of a prophylactic quadrivalent human papillomavirus types 6/11/16/18 L1 virus-like particle vaccine through 5 years of follow-up. Br. J. Cancer 2006; 95:1459–1466.

Vincent, M.J., Quiroz, E., Gracia, F., Sanchez, A.J., Ksiazek, T.G., Kitsutani, P.T., Ruedas, L.A., Tinnin, D.S., Caceres, L., Garcia, A., Rollin, P.E., Mills, J.N., Peters, C.J. and Nichol, S.T. Hantavirus pulmonary syndrome in Panama: identification of novel hantaviruses and their likely reservoirs. Virology 2000; 277:14–19.

Wahl-Jensen, V., Chapman, J., Asher, L., Fisher, R., Zimmerman, M., Larsen, T. and Hooper, J.W. Temporal analysis of Andes virus and Sin Nombre virus infections of Syrian hamsters. J. Virol. 2007; 81:7449–7462.

Wallez, Y. and Huber, P. Endothelial adherens and tight junctions in vascular homeostasis, inflammation and angiogenesis. *Biochim. Biophys. Acta.* 2008; Mar; 1778(3):794–809.

Warner, G.S. Hantavirus illness in humans: review and update. South Med. J. 1996; 89:264–271.

Warshamana, G.S., Corti, M. and Brody, A.R. TNF-alpha, PDGF, and TGF-beta(1) expression by primary mouse bronchiolar-alveolar epithelial and mesenchymal cells: tnf-alpha induces TGF-beta(1). Exp. Mol. Pathol. 2001; 71:13–33.

Watkins, N.A., Smethurst, P.A., Allen, D., Smith, G.A. and Ouwehand, W.H. Platelet alphaIIbbeta3 recombinant autoantibodies from the B-cell repertoire of a post-transfusion purpura patient. Br. J. Haematol. 2002; 116:677–685.

Watson, C., Whittaker, S., Smith, N., Vora, A.J., Dumonde, D.C. and Brown, K.A. IL-6 acts on endothelial cells to preferentially increase their adherence for lymphocytes. Clin. Exp. Immunol. 1996; 105:112–119.

Wells, R.M., Sosa Estani, S., Yadon, Z.E., Enria, D., Padula, P., Pini, N., Mills, J.N., Peters, C.J. and Segura, E.L. An unusual hantavirus outbreak in southern Argentina: person-to-person transmission? Hantavirus Pulmonary Syndrome Study Group for Patagonia. Emerg. Infect. Dis. 1997; 3:171–174.

White, J.D., Shirey, F.G., French, G.R., Huggins, J.W., Brand, O.M. and Lee, H.W. Hantaan virus, aetiological agent of Korean haemorrhagic fever, has Bunyaviridae-like morphology. Lancet 1982; 1:768–771.

Willems, L., Gatot, J.S., Mammerickx, M., Portetelle, D., Burny, A., Kerkhofs, P. and Kettmann, R. The YXXL signalling motifs of the bovine leukemia virus transmembrane protein are required for in vivo infection and maintenance of high viral loads. J. Virol. 1995; 69:4137–4141.

Woo, G.J., Chun, E.Y., Kim, K.H. and Kim, W. Analysis of immune responses against nucleocapsid protein of the Hantaan virus elicited by virus infection or DNA vaccination. J. Microbiol. 2005; 43:537–545.

Xiao, R., Yang, S., Koster, F., Ye, C., Stidley, C. and Hjelle, B. Sin Nombre viral RNA load in patients with hantavirus cardiopulmonary syndrome. J. Infect. Dis. 2006; 194:1403–1409.

Xinhua News Service. Columbia reports new outbreak of hanta virus. Poeple's Daily Online. March 14, 2006. http://english.people.com.cn/200603/14/eng20060314_250492.html

Xu, L., Yoon, H., Zhao, M.Q., Liu, J., Ramana, C.V. and Enelow, R.I. Cutting edge: pulmonary immunopathology mediated by antigen-specific expression of TNF-alpha by antiviral CD8+ T cells. J. Immunol. 2004; 173:721–725.

Xu, X., Ruo, S.L., Mccormick, J.B. and Fisher-Hoch, S.P. Immunity to hantavirus challenge in Meriones unguiculatus induced by vaccinia-vectored viral proteins. Am. J. Trop. Med. Hyg. 1992; 47:397–404.

Xu, X.N., Laffert, B., Screaton, G.R., Kraft, M., Wolf, D., Kolanus, W., Mongkolsapay, J., Mcmichael, A.J. and Baur, A.S. Induction of Fas ligand expression by HIV involves the interaction of Nef with the T cell receptor zeta chain. J. Exp. Med. 1999; 189:1489–1496.

Xu, Z., Wei, L., Wang, L., Wang, H. and Jiang, S. The in vitro and in vivo protective activity of monoclonal antibodies directed against Hantaan virus: potential application for immunotherapy and passive immunization. Biochem. Biophys. Res. Commun. 2002; 298:552–558.

Yan, L., Fang, L.Q., Huang, H.G., Zhang, L.Q., Feng, D., Zhao, W.J., Zhang, W.Y., Li, X.W. and Cao, W.C. Landscape elements and Hantaan virus-related hemorrhagic fever with renal syndrome, People's Republic of China. Emerg. Infect. Dis. 2007; 13:1301–1306.

Yanagihara, R. and Silverman, D.J. Experimental infection of human vascular endothelial cells by pathogenic and nonpathogenic hantaviruses. Arch. Virol. 1990; 111:281–286.

Ye, C., Prescott, J., Nofchissey, R., Goade, D. and Hjelle, B. Neutralizing antibodies and Sin Nombre virus RNA after recovery from hantavirus cardiopulmonary syndrome. Emerg. Infect. Dis. 2004; 10:478–482.

Yongxin, Y., Zhiyong, A., Zhihui, Y. and Guanmu, D. Inactivated cell-culture hantavirus vaccine developed in China. In: Emergence and Control of Rodent-borne Viral Diseases (Hantaviral and Arenal Diseases) (J.F. Saluzzo and B. Dodet, Eds.). Veyrier-du-lac, Annecy, France: Elsevier, 1998.

Yoshimatsu, K., Yoo, Y.C., Yoshida, R., Ishihara, C., Azuma, I. and Arikawa, J. Protective immunity of Hantaan virus nucleocapsid and envelope protein studied using baculovirus-expressed proteins. Arch. Virol. 1993; 130:365–376.

Young, J.C., Hansen, G.R., Graves, T.K., Deasy, M.P., Humphreys, J.G., Fritz, C.L., Gorham, K.L., Khan, A.S., Ksiazek, T.G., Metzger, K.B. and Peters, C.J. The incubation period of hantavirus pulmonary syndrome. Am. J. Trop. Med. Hyg. 2000; 62:714–717.

Zaki, S.R., Greer, P.W., Coffield, L.M., Goldsmith, C.S., Nolte, K.B., Foucar, K., Feddersen, R.M., Zumwalt, R.E., Miller, G.L., Khan,

A.S. et al. Hantavirus pulmonary syndrome. Pathogenesis of an emerging infectious disease. Am. J. Pathol. 1995; 146:552–579.

Zaki, S.R., Khan, A.S., Goodman, R.A., Armstrong, L.R., Greer, P. W., Coffield, L.M., Ksiazek, T.G., Rollin, P.E., Peters, C.J. and Khabbaz, R.F. Retrospective diagnosis of hantavirus pulmonary syndrome, 1978–1993: implications for emerging infectious diseases. Arch. Pathol. Lab. Med. 1996; 120:134–139.

Zhang, F.L., Wu, X.A., Luo, W., Bai, W.T., Liu, Y., Yan, Y., Wang, H.T. and Xu, Z.K. The expression and genetic immunization of chimeric fragment of Hantaan virus M and S segments. Biochem. Biophys. Res. Commun. 2007; 354:858–863.

Zhang, X.K., Takashima, I. and Hashimoto, N. Characteristics of passive immunity against hantavirus infection in rats. Arch. Virol. 1989; 105:235–246.

Zhu, Z.Y., Tang, H.Y., Li, Y.J., Weng, J.Q., Yu, Y.X. and Zeng, R.F. Investigation on inactivated epidemic hemorrhagic fever tissue culture vaccine in humans. Chin. Med. J. (Engl.) 1994; 107:167–170.

CHAPTER

24

Hepatitis C

Meital Gal-Tanamy[1], *Christopher Walker*[2], *Steven Foung*[3] *and Stanley M. Lemon*[1]

[1]Center for Hepatitis Research, Institute for Human Infections and Immunity, and the Department of Microbiology and Immunology, University of Texas Medical Branch, Galveston, TX, USA

[2]The Research Institute at Nationwide Children's Hospital, Columbus, OH, USA

[3]Department of Pathology, Stanford University School of Medicine, Palo Alto, CA, USA

OUTLINE

ABSTRACT

Hepatitis C virus (HCV) is a major public health problem, with over 180 million persons infected worldwide and at risk for developing cirrhosis or liver cancer. Current interferon-based treatment options are expensive, often associated with considerable toxicity, and moreover are of limited efficacy. Thus, vaccines capable of either preventing new infections or exerting a therapeutic effect on established infections would be of value. Both types of vaccines remain elusive goals, however, in part because immunity against the virus is very poorly understood. An impressive degree of genetic (and likely antigenic) variation among HCV strains is another barrier to vaccine development. In addition, there are no readily available animal models of hepatitis C, which greatly complicates vaccine development. Protective immunity has been correlated mainly with virus-specific CD4+ and CD8+ T-cell responses, while the role played by antibodies to the virus is less clear but still potentially important. Immunization with recombinant viral envelope proteins induces primarily antibody and CD4+ Th-cell responses. While not preventing acute infection, such a vaccine may have prevented persistent infection in chimpanzees challenged with a virus closely related to the vaccine strain. Other vaccine strategies that have been explored include the use of synthetic peptides, genetic immunization, recombinant vectors, virus-like particles, or combinations of these. While many of these approaches have induced CD8+ T-cell responses, as well as CD4+ Th and B-cell responses, few have shown clear evidence of providing protective immunity against HCV challenge. Other efforts have focused on ex vivo stimulation of dendritic cells, and passive immunization with polyclonal and monoclonal antibodies. This chapter summarizes what is known of the immune responses that control HCV infection and different strategies that have been pursued for HCV vaccine development, and outlines the current status of vaccine candidates in pre-clinical and clinical development.

INTRODUCTION

Hepatitis C virus (HCV) was identified and shown to be the cause of almost all cases of post-transfusion "non-A, non-B" viral hepatitis in the late 1980s (Choo et al., 1989; Kuo et al., 1989). The virus has a unique capacity to establish long-term, persistent infections in the majority of acutely infected persons, placing such individuals at significant risk for progressive hepatic fibrosis, cirrhosis, and liver cancer. This occurs despite the activation of virus-specific CD4+ and CD8+ T cells and the presence of virus-neutralizing antibodies in infected persons. HCV has the capacity to disrupt early events in the innate immune response to infection, but the specific mechanisms underlying viral persistence and the failure of virus-specific B- and T-cell responses to eliminate the infection remain incompletely defined. HCV infection currently accounts for less than 15% of acute viral hepatitis cases within the United States. However, it is by far the leading cause of chronic viral hepatitis and is present in over 40% of persons with chronic liver disease. Efforts at vaccine development, yet to succeed despite almost two decades of work, are thus aimed at both prevention of infection (prophylactic vaccines) and treatment of established infections (therapeutic vaccines).

HEPATITIS C VIRUS

Genome Organization and Virus Structure

HCV is classified within the genus *Hepacivirus* of the family *Flaviviridae*, and it shares a distant phylogenetic relationship with yellow fever virus and other classical flaviviruses. However, HCV has many features that distinguish it from other members of this virus family, including the organization of its polyprotein that contains two envelope glycoproteins, E1 and E2 (Fig. 24.1). Much has been learned about the molecular virology of HCV and its interactions with host cells since the discovery of the virus. A complete description of the agent is well beyond the scope of this chapter, and for this the reader is referred to detailed reviews elsewhere (Lemon et al., 2006; Lindenbach and Rice, 2005). Only a brief summary is provided here.

Similar to other members of the *Flaviviridae*, the HCV genome is comprised of a single RNA molecule that is single-stranded and messenger sense (Choo et al., 1989). The genomic RNA has a relatively lengthy 5′ nontranslated region of approximately 342 bases, containing an internal ribosome entry site (IRES) that directs the 5′ cap-independent initiation of viral translation (Lemon and Honda, 1997) (Fig. 24.1). This is followed by a long open reading frame encoding a single polyprotein that undergoes co-translational processing events directed by both host cell and virus-encoded proteases resulting in the production of ten mature proteins (Fig. 24.1). Proteolytic cleavages directed by host cell signal peptidases produce a series of structural proteins which include the core, and E1 and E2 envelope proteins, as well as a small membrane-associated viroporin, p7. Virus-encoded proteases are responsible for additional processing events that generate the nonstructural proteins of HCV.

There is no high-resolution model of the structure of the envelope proteins, and only a rudimentary

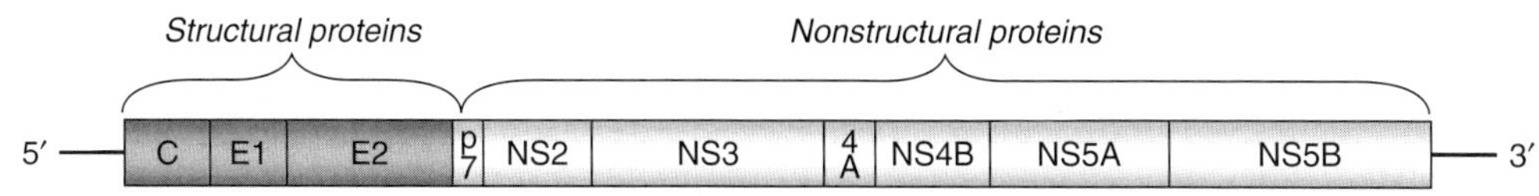

FIGURE 24.1 Organization of the positive-sense RNA genome of HCV. The noncoding RNA sequences are shown as solid lines, while the polyprotein coding region is shown as a box. The 5′ nontranslated segment of the genome contains an IRES that controls translation of the polyprotein. The segment of the polyprotein that contains structural proteins (core, E1, and E2) known to be present within the virus particle is heavily shaded, while nonstructural proteins are lightly shaded; p7 and NS2 are presumed to be not present within the virus particle, but are also not required for replication of virus RNA.

understanding of their membrane topology. Signal sequences at the carboxy termini of the core and E1 coding regions direct the secretion of E1 and E2 into the lumen of the ER; both remain anchored to the membrane by hydrophobic C-terminal sequences as type 1 transmembrane proteins. The core protein is released from the membrane by an intramembranous cleavage carried out by signal peptide peptidase (McLauchlan et al., 2002), and remains within the cytoplasm where it associates with lipid droplets and mitochondria. E1 and E2 are massively glycosylated within the ER and the Golgi, with about 50% of their mass made up of sugars in their mature form. This is reminiscent of the "glycan shield" presented by human immunodeficiency virus (HIV), and may contribute to the relatively poor effectiveness of virus neutralization by antibodies in vivo. It also poses challenges for vaccine development. E1 and E2 proteins have strong ER retention signals, and details of the secretion process remain obscure (Penin et al., 2004; Dubuisson and Rice, 1996). However, the recent advent of cell culture-infectious virus systems is providing much new information about viral assembly, which appears to be dependent upon the association of the core with lipid droplets and involves multiple nonstructural proteins (Miyanari et al., 2007; Gastaminza et al., 2008).

At least six nonstructural proteins are derived from the remainder of the polyprotein, each of which likely has multiple functions, many only partly understood. These include NS2, which contains in its C-terminal half a unique cysteine protease activity that spans the NS2/NS3 junction and cleaves in *cis* at this site. Crystallographic studies suggest that NS2 functions as a dimer with composite active sites (Lorenz et al., 2006). The NS3 protein contains in its amino terminus residues that function as part of the *cis*-active NS2/NS3 protease and, after cleavage from NS2 and the intercalation of NS4A, form the mature NS3/4A serine protease that is responsible for all of the other downstream cleavage events within the polyprotein (Kim et al., 1996). The carboxy two-thirds of the NS3 protein folds into a separate functional domain-containing NTPase and RNA helicase activities that are essential for virus replication (Yao et al., 1997). In addition to its role as an NS3 protease accessory factor, NS4A contains a membrane anchor that fixes the NS3/4A complex to cellular membranes (Lin et al., 1995). NS4B is a hydrophobic protein; it appears to be responsible for reorganizing intracellular membranes into a "membranous web" that appears to be the site of replication of viral RNA (Moradpour et al., 2003). NS5A is a multi-functional protein that contributes to interferon (IFN)-resistance and also plays an essential role in replication of viral RNA as well as assembly of the virus particle (Enomoto et al., 1996; Miyanari et al., 2007; Tellinghuisen et al., 2005); Finally, NS5B, which is derived from the carboxy terminal portion of the polyprotein, is an RNA-dependent RNA polymerase that forms the catalytic core of the replicase (Bressanelli et al., 1999).

Many, if not all, of these proteins have important interactions with cellular proteins that contribute directly or indirectly to viral RNA replication (Lemon et al., 2006). Prominent among these interactions is the ability of the mature NS3/4A protease to proteolytically target key cellular signaling molecules required for the induction of IFN-mediated antiviral cellular defenses through both the Toll-like receptor 3 (TLR3) and retinoic acid-inducible gene I (RIG-I) pathways (Li et al., 2005a; Meylan et al., 2005).

Virus Structure

Little is known about the structure of the virus particle, and only a few reliable images of HCV particles obtained from infected humans or chimpanzees exist (Kaito et al., 1994). These show a rather ill-defined amorphous membrane enveloping what appears to be a more ordered nucleocapsid structure. Evidence suggests that the E1 and E2 glycoproteins form a noncovalent heterodimer that is presented on the surface of the virion. E2 contains a highly variable domain near its amino terminus (HVR-1 domain), which is likely to form an immunogenic loop on the surface of the virion and which may interact with neutralizing antibodies (Kato et al., 1993; Bartosch et al., 2003a).

As such, it may be analogous to the V3 loop of HIV-1. It may also play a role in determining the interaction of the E2 protein with SR-B1, a putative co-receptor important for viral entry into cells (Bartosch et al., 2003c; Roccasecca et al., 2003). However, this loop can be deleted without complete loss of infectivity (Forns et al., 2000b). E2 is considered to be the major target of neutralizing antibodies. Further details of the interactions of antibodies with the virus are described below.

Growing evidence supports the involvement of apolipoproteins (ApoB and ApoE) produced within the hepatocyte as an integral component of the viral structure. Virus found in the serum of infected chimpanzees or humans generally demonstrates a significantly lesser buoyant density than infectious virus produced in cell culture (see below) (Lindenbach et al., 2006). Moreover, serum-derived virus can be precipitated efficiently with antibodies to ApoB and ApoE, and progressive detergent disruption of these particles has shown that ApoB remains tightly associated with the virus even after treatment with 0.5% deoxycholic acid or 0.18% NP-40 (Nielsen et al., 2006; Diaz et al., 2006). In addition, the production of virus particles by infected Huh7 cells is blocked by siRNA knockdown of ApoB as well as inhibition of microsomal transport protein, which participates in the secretion of triglyceride-rich very-low-density lipoproteins from liver cells (VLDL) (Huang et al., 2007; Gastaminza et al., 2008). Thus, HCV particles have been described as "lipoviroparticles" based on a high content of host lipoproteins and the possible involvement of the apolipoprotein secretory apparatus in their assembly and release. The implications of this feature of the virus structure for the success of vaccines remain uncertain. However, the immunoprecipitation of virus from an infected patient was notably less efficient with antibodies to E2 (~25% precipitation) than antibodies to ApoB or ApoE (>95%) (Nielsen et al., 2006). Further studies are needed to understand the significance of this finding, but it is possible that the ability of neutralizing antibodies to access E2 might be limited by the association of infectious virus particles with host lipoproteins.

Cell Culture Replication of Virus

Wild-type HCV replicates inconsistently in cultures of primary chimpanzee or human hepatocytes; it has also been shown to be capable of very low level, persistent replication in some B- and T-cell derived lymphoid cell lines (Shimizu et al., 1992; Shimizu and Yoshikura, 1994; Shi and Lai, 2001; Lanford et al., 1994). More permissive in vitro cell culture systems supporting the replication of HCV would be helpful for gaining a better understanding of the virus, as well for vaccine development. However, such systems remain generally elusive.

Infectious cDNA clones of the genome were constructed following identification of HCV, and the infectivity of RNA transcribed from several of these was established by direct intrahepatic injection of synthetic RNA in chimpanzees (Kolykhalov et al., 1997; Yanagi et al., 1997). These early clones benefited many aspects of HCV research, including the development of subgenomic HCV RNA replicons in many laboratories. These RNAs encode selectable markers and a minimal polyprotein segment extending from NS3 to NS5B. They contain authentic 5′ and 3′ termini, and undergo efficient autonomous amplification following transfection into permissive cell cultures, typically human hepatoma (Huh7) cells (Lohmann et al., 1999). Replicons propelled the field forward and led to many studies that have usefully characterized various aspects of viral RNA replication. In general, efficient replication of these replicon RNAs is associated with the acquisition of specific cell culture-adaptive mutations, often involving the NS5A protein. Importantly, these mutations may be highly attenuating when introduced back into the wild-type virus (Bukh et al., 2002).

In recent years, a limited number of HCV strains that are capable of reproducing the entire replication cycle in cultured cells have become available (Lindenbach et al., 2005; Yi et al., 2006; Wakita et al., 2005; Zhong et al., 2005). Most notable among these viruses is the JFH-1 strain, a remarkable genotype 2a virus that for reasons still unknown is capable of moderately efficient replication in Huh7 cells (Wakita et al., 2005). These cell culture-produced viruses are referred to as "HCVcc." Important from the perspective of vaccine development, the JFH-1 strain has been used to construct chimeric viruses with envelope proteins from different virus genotypes, opening up new approaches to assessing the presence of serum neutralizing antibodies (Yi et al., 2007; Pietschmann et al., 2006).

Sequence Diversity among Different HCV Strains

Different strains of HCV have been classified into six distinct major "genotypes" based on nucleotide sequence divergence (Simmonds, 2004; Simmonds et al., 2005). The genetic distance between some genotypes is large enough to suggest that there are likely

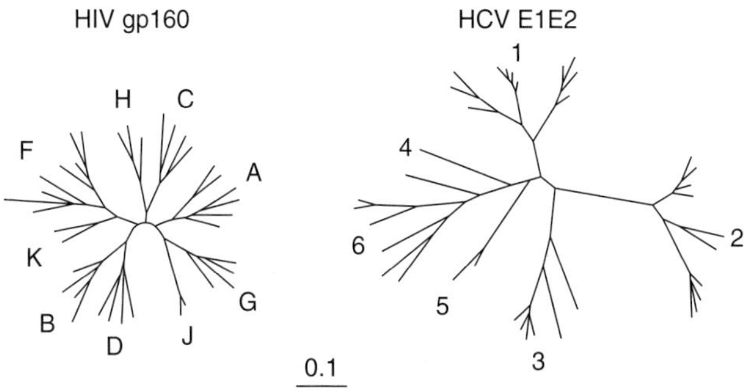

FIGURE 24.2 Comparative phylogenetic analyses of sequence diversity in the envelope protein sequences of HCV and HIV. Trees, displayed here on the same scale, were generated separately for both viruses using a maximum-likelihood search method and amino acid sequences translated from aligned nucleotide sequences available in the Los Alamos databases. HIV subtypes and HCV genotypes are labeled. The figure is provided courtesy of Dr. Stuart Ray, Johns Hopkins University.

to be significant serotypic differences. This is reflected in substantial differences in the amino acid sequences of the E1 and E2 glycoproteins of different genotypes, which are considered to be the major, if not exclusive, target for neutralizing antibodies. The extent of diversity among these envelope protein sequences exceeds that existing among the gp160 proteins of different subtypes of the HIV (Fig. 24.2). Nonetheless, although this presents a very large hurdle to be overcome for the development of a successful vaccine, monoclonal antibodies have been isolated which are capable of broad recognition of viruses from different genotypes (Owsianka et al., 2005).

Genotype 1 HCV is considerably more refractory to IFN therapy than nongenotype 1 strains, with sustained response rates of only about 50% with the best-available therapy, a combination of pegylated IFN and ribavirin. In addition, spontaneous resolution of acute infection appears to occur with greater frequency in genotype 3 infections than in infections with other genotypes (Lehmann et al., 2004). Otherwise, there is little evidence for differences in the pathogenicity of various genotypes.

CLINICAL DISEASE AND PATHOGENESIS

Most acute HCV infections are associated with minimal symptoms and only rarely with identifiable jaundice (Feinstone et al., 1975). Thus, patients presenting to physicians with acute hepatitis C are much more likely to be icteric but represent only a small fraction of all acute infections (Alter et al., 1992). The incubation period extending from exposure to onset of acute liver disease, when it occurs, is of the order of 8 weeks. Virus replication is maximal shortly after infection. The level of viremia declines subsequently with the appearance of antibody to viral proteins and T-cell mediated immunity to HCV, and in most persons remains constant at a relatively fixed level for many years thereafter (Alter et al., 1992; Thimme et al., 2002). However, in as many as 20–40% of infected persons, these immune responses clear the infection and viral persistence is not established.

Approximately two-thirds of persons who progress to persistent infection go on to develop persistent or intermittent elevations in serum alanine aminotransferase (ALT) levels, a marker of hepatic inflammation (Alter et al., 1992). At liver biopsy, these individuals may have fibrosis, the extent of which does not correlate with the magnitude of serum enzyme elevations. As many as a third of these patients will ultimately develop cirrhosis (Kage et al., 1997; Alter et al., 1992; Seeff et al., 1992). Cirrhosis may develop within as little as 5 years of the initial infection, but typically is found in persons who have been infected for several decades. Factors associated with disease progression include an older age at infection, regular alcohol consumption, coinfection with HIV or hepatitis B virus, and obesity and insulin resistance (Strader et al., 2004; Consensus Development Panel, 2002). Once cirrhosis is established, there is a substantial risk (perhaps as high as 3–4% per year) of developing primary hepatocellular carcinoma (Saito et al., 1991). In one US series, 42% of patients with hepatocellular carcinoma had detectable HCV genomic RNA in liver tissue (Abe et al., 1998). Liver injury most likely results from the direct action of virus-specific cytotoxic T-cells, as well as from secondary mediators of inflammation, such as IFN-γ, and other cytokines and chemokines, that are released as a result of a virus-specific T-cell response (Rehermann and Nascimbeni, 2005; Thimme et al., 2002). CD8+ T-cell mediated responses may suppress virus replication and lower the magnitude of viremia, as described below. However, in the absence of eliminating the virus altogether, these T-cell responses may be destructive and result in significant inflammation within the liver.

Extrahepatic manifestations of HCV infection are relatively common, and include glomerulonephritis and type II mixed cryoglobulinemia associated with vasculitis (Johnson et al., 1993; Palekar and Harrison, 2005; Agnello et al., 1992; Marcellin et al., 1993). Such individuals have circulating immune complexes containing viral RNA. Other potentially associated clinical conditions include porphyria cutanea tarda, sicca syndrome, and a variety of autoimmune diseases as well as B-cell lymphoma. The pathogenesis of these conditions is poorly understood (Lemon et al., 2006).

ANIMAL MODELS OF HEPATITIS C

In addition to the limited availability of fully permissive cell culture systems that are capable of supporting the entire HCV life-cycle, the lack of a small animal model has bedeviled the field since its inception. Chimpanzees remain the only well-validated animal model for HCV infections in humans (Lanford et al., 2001). Chimpanzees recapitulate many features of hepatitis C in humans (Fig. 24.3), but there are significant differences in the course of infection in chimpanzees and humans. Persistent infections develop in only 30–50% of experimentally challenged chimpanzees, less than the 60–80% persistence rate in humans, and moreover are not usually associated with significant liver disease (Bassett et al., 1998). Chimpanzees also demonstrate a very restricted antibody response to the envelope proteins

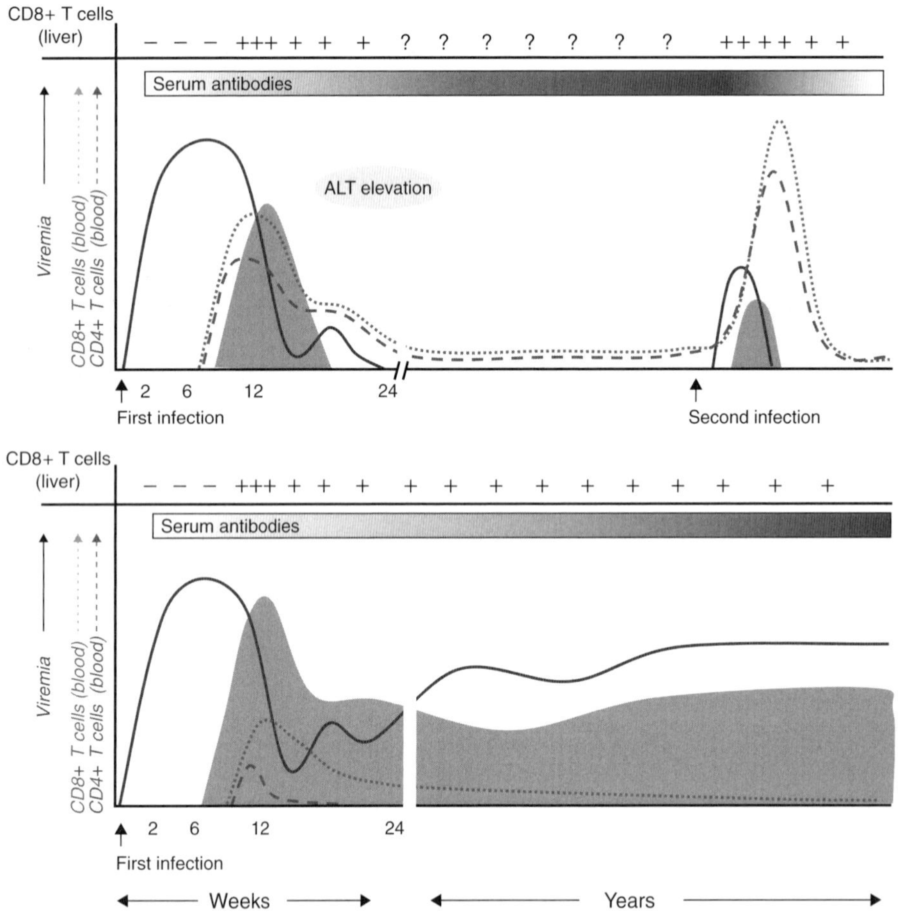

FIGURE 24.3 Typical course of acute resolving and persisting HCV infections in humans and chimpanzees. Top panel: Viremia usually peaks at 8–12 weeks after infection in acute resolving hepatitis C. Expansion of virus-specific CD4+ and CD8+ T cells in blood and their infiltration into liver is temporally associated with a decline in viremia and rise in serum ALT. Successful T-cell responses fully contract only after apparent resolution of infection. Whether memory T cells reside in the liver is unknown (indicated by "?"). Serum antibody responses to structural and nonstructural proteins develop within the first month of infection and titers decline in the years following virus clearance; the kinetics of neutralizing antibody are not known. Spontaneous resolution of acute hepatitis C generally confer only partial protection against re-infection. Second infections are more likely to resolve and are of shorter duration, at least in part due to rapid expansion of memory CD4+ and CD8+ T cells. Bottom panel: HCV infections that persist are characterized by variable T-cell responses that often target fewer epitopes and reach lower peak frequencies. HCV-specific CD4+ T-cell responses either never develop or fail before virus is cleared. CD8+ T cells are sometimes detected in blood for weeks to months following infection but can persist in the chronically infected liver for decades. Durable serum antibody responses to viral proteins develop early in infection. Neutralizing antibodies also appear but at least initially have a limited capacity for recognition of rapidly emerging HCV variants.

compared to humans (Bassett et al., 1999), which should raise concerns regarding their use for evaluation of vaccines. Moreover, the availability of this endangered animal species is limited, and its use in vaccine studies extraordinarily expensive. As an alternative, mice with chimeric human livers have been created by transplanting human hepatocytes into SCID mice bearing a lethal plasminogen activator transgene (SCID-Alb-uPa) (Mercer et al., 2001). While difficult to produce, such animals are susceptible to infection with HCV. However, their underlying immune deficiency makes them unsuitable for most studies of vaccine immunogenicity or protective immunity.

TREATMENT

Current treatment options are limited to recombinant IFN-α-based therapies, which are suboptimal. The best-available therapies, combinations of pegylated IFN-α and ribavirin, result in sustained viral clearance in only ~50% in those infected with genotype 1 viruses, which are most prevalent (Davis et al., 1998; McHutchison and Fried, 2003). Cures are possible, and sustained clearance of the virus is associated with improved clinical outcomes. However, a recent multi-center, NIH-sponsored study (HALT-C) indicated that attempts at long-term suppression of infection with IFN are not beneficial in those who fail to clear the virus. A number of candidate antiviral compounds with specific activity against HCV are currently under development. Many of these target the NS3/4A protease or NS5B polymerase enzymatic activities that are required for viral replication. Early emergence of resistance has been documented in phase I/II clinical trials of such compounds, however, and current concepts call for IFN-α to remain a major component of therapy in the foreseeable future (Chung et al., 2008).

EPIDEMIOLOGY

Worldwide, over 170 million people are estimated to be infected with HCV, reflecting the unique capacity of this virus to establish long standing, persistent infections. The overall prevalence of active infection is approximately 2.2% (Fig. 24.4) (Global Burden of Disease for Hepatitis C, 2004). Within the United States, HCV infection is the leading cause of chronic hepatitis and cirrhosis, and an increasingly important factor in the etiology of hepatocellular carcinoma (Alter et al., 1999; El Serag, 2004; Vong and Bell, 2004). The age-adjusted risk for liver cancer increased by 226% between 1993 and 1999 among Americans over 65 years of age, most likely reflecting prior increases in prevalence of HCV infection (Davila et al., 2004). Liver cancer is increasingly found in younger individuals, with African-American and Hispanic males at particularly high risk. This disparity closely mirrors racial and gender differences in HCV infection prevalence (Alter et al., 1999). Nonetheless, the incidence of new HCV infections is relatively low. Only 671 confirmed cases of acute hepatitis C were reported to the US Centers for Disease Control in 2005, leading (after corrections for under-reporting) to estimates of about 20,000 new infections nationally. This low estimated

FIGURE 24.4 Global prevalence of HCV infection. Prevalence is based on serological reports available to the World Health Organization (Global Burden of Disease for Hepatitis C, 2004).

incidence rate has led some to question the need for preventive vaccines.

Transmission occurs primarily by percutaneous exposure. While relatively common previously, transmission by blood and plasma-derived products has been virtually eliminated by donor testing and chemical inactivation procedures in well-developed countries. However, patient-to-patient transmission of HCV can result from unsafe therapeutic injections and other health-care procedures, and contributes substantially to the disease burden in the developing world (Hauri et al., 2004; Williams et al., 2004). In contrast, the overwhelming risk factor in industrialized nations is illicit injection drug use (Shepard et al., 2005). Active HCV infections are highly prevalent among injection drug users, and transmission is facilitated by sharing paraphernalia for the preparation of drugs as well as syringes and needles. Percutaneous exposures are a minor source of infection among health-care workers in industrialized nations, and do not result in an increased prevalence of infection in this group above the general population. Sexual activity appears to be an inefficient means for transmission of this virus. However, some studies have shown correlations between sexual practices or numbers of sex partners and the risk of infection with HCV. Recent outbreaks of hepatitis C have been identified among homosexual males who were co-infected with HIV (Gotz et al., 2005; Rauch et al., 2005). Maternal–infant transmission of HCV also occurs at very low rates, but is increased when the mother is co-infected with HIV (Mast et al., 2005; Roberts and Yeung, 2002).

Importantly, HCV can remain viable on surfaces under ambient environmental conditions for at least 16 h (Kamili et al., 2006). For a variety of reasons, however, HCV poses little risk as an agent of bioterrorism.

ANTIGENS EXPRESSED BY HCV

Antibodies directed against many of the mature HCV proteins develop during the course of infection with the virus, including most notably NS3, core, and E2. Current diagnostic assays use recombinant antigens derived from the core and nonstructural (NS3, NS4, and NS5) regions of the HCV genome (Centers for Disease Control and Prevention, 2003). These tests do not measure immunity to the virus, but are rather indicators of previous infection with HCV. Antibodies to E2, the larger of the two envelope glycoproteins, may possess substantial neutralizing activity (see below), while there is no evidence that antibodies to core or to any of the nonstructural proteins are capable of providing protection.

A number of epitopes recognized by virus-specific T cells have been identified. These map in a distributed fashion across the polyprotein in a class I histocompatibility antigen-restricted manner. A compendium of these epitopes is maintained on the web by the Immune Epitope Database and Analysis Resource (http://www.immuneepitope.org/). No single HCV protein has been implicated as dominant in stimulating either CD4+ or CD8+ T-cell subsets (Ward et al., 2002).

HOST IMMUNE RESPONSES

While early reports suggested the absence of protection against rechallenge in chimpanzees that had been experimentally infected with HCV (Farci et al., 1992), the existence of protective immunity following the resolution of HCV infection is now well established in the chimpanzee model (Lanford et al., 2004; Bassett et al., 2001; Weiner et al., 2001). Such immunity is typically not sterilizing in nature, but can broadly extend across genotypes, last for many years, and lead to prevention of long-term viral persistence if not short-term, acute re-infection. Protection from persistent infection is associated with a vigorous T-cell response, and has been shown to be dependent upon both memory CD4+ T cell and CD8+ T-cell subsets (Bassett et al., 2001; Grakoui et al., 2003; Shoukry et al., 2003). As indicated below, the role of antibody in such protection is less certain. Rechallenge of previously infected chimpanzees leads to marked increases in serum antibodies to the envelope proteins, but these responses generally do not correlate well with the outcome of the infection (Bassett et al., 2001). The existence of this nonsterilizing protective immunity, and in particular its ability to protect against persistent infection, provides perhaps the best evidence that it might be possible to develop broadly protective vaccines.

Innate Immune Responses to Infection

The disruption of innate immune signaling pathways that regulate induction of IFN, cytokine, and chemokine responses could be an important factor underlying the failure of virus-specific T cells to control the infection. The NS3/4A protease disrupts signal transduction pathways that normally lead to the synthesis of type 1 IFN α/β and hundreds of IFN-inducible genes (ISGs) in response to virus infection. TLR3 is expressed on the surface of cells, or within intracellular vesicles, activates IFN regulatory factor 3 (IRF-3), and induces IFN-β synthesis when bound by extracellular dsRNA. Signaling through this pathway is abrogated by NS3/4A-mediated cleavage of the TLR3 adaptor protein, TRIF (Toll-IL1 receptor domain-containing adaptor inducing IFN-β, otherwise known as TICAM-1), which links TLR3 to a TRAF3 complex containing noncanonical and canonical kinases of the IKK complex that are responsible for activation of IRF-3 and NF-κB (Li et al., 2005a; Ferreon et al., 2005). In contrast to TLR3, RIG-I is expressed within the cytoplasm. Upon binding its viral RNA ligand, RIG-I is recruited to mitochondria, where it interacts with MAVS (mitochondrial antiviral signaling protein, also known as IPS-1, VISA, or Cardiff), and initiates formation of a signalosome complex leading to activation of IRF-3 and NF-κB (Seth et al., 2005; Kawai et al., 2005; Meylan et al., 2005). The RIG-I pathway is also disrupted by NS3/4A, which directs the specific cleavage of MAVS (Loo et al., 2006; Meylan et al., 2005; Li et al., 2005b). The disruption of these signaling pathways may serve to delay or prevent the synthesis of IFNs and other cytokines in response to HCV infection. Hypothetically, this could both impair T-cell activation as well as reduce the recruitment of virus-specific CD8+ cells to the liver (Lang et al., 2006; Kabelitz and Medzhitov, 2007; Schulz et al., 2005). However, despite these findings, microarray studies of liver tissues from infected chimpanzee and humans show abundant expression of a large number of ISGs (Bigger et al., 2001, 2004; Su et al., 2002; Lau et al., 2008). The cell type(s) expressing these ISGs remain undefined, and many questions remain concerning the disruption of these innate signaling pathways and how this might contribute to the lack of host control of HCV infection.

Adaptive T-Cell Responses

Studies in chimpanzees and humans suggest that a sustained T-cell response targeting multiple HCV epitopes is usually associated with a successful outcome of infection. The importance of the T-cell response is also supported by a temporal relationship in the kinetics of virus-specific T-cell expansion and reductions in the abundance of circulating virus during acute infection (Cooper et al., 1999; Shoukry et al., 2003; Lechner et al., 2000; Thimme et al., 2001). In addition, antibody-mediated depletion of memory CD4+ helper or CD8+ cytotoxic subsets prior to rechallenge with HCV leads to prolonged viremia in chimpanzees (Shoukry et al., 2003; Grakoui et al., 2003). Other studies have documented substantial reductions in the magnitude of the primary viremia in chimpanzees immunized with nonstructural HCV proteins that are not thought to be targets of neutralizing antibodies (Folgori et al., 2006). Interestingly, the T-cell response to virus challenge is typically delayed in acute infections in chimpanzees. Even when the infection is ultimately cleared, expansion of virus-specific T-cell subsets and the intrahepatic expression of IFN-γ does not occur until 6–8 weeks after acute infection (Thimme et al., 2002; Woollard et al., 2003). Similar delays in acute phase adaptive T-cell immunity have also been described in humans (Shoukry et al., 2004).

Despite a great deal of research over the past 20 years, no clear cut explanation has yet emerged for the failure of host T-cell responses to clear HCV infections. It is likely that the high frequency of persistent virus infections results from a combination of several distinct mechanisms affecting CD4+ helper and CD8+ cytotoxic T-cell subsets. Failure to sustain a CD4+ T-cell response past the point of apparent resolution of infection is perhaps the most reliable predictor of HCV persistence (Shoukry et al., 2004). An inability to generate a durable CD4+ or CD8+ T-cell response may be caused at least in part by disruption of innate signaling pathways, as described above, and consequently impaired cytokine responses to infection. The liver is a classic, immunoprivileged organ, and studies in animal models show that even well-primed liver-specific cytotoxic T-lymphocyte responses may result in little if any liver damage in the absence of TLR3 signaling and appropriate cytokine and chemokine expression in the liver (Lang et al., 2006).

Beyond the possibility of weak antigen priming, reasons for impaired CD4+ T-cell immunity are still poorly understood, in part because these cells are difficult to visualize or detect using functional assays in chronic hepatitis C. Defects underlying failed CD8+ T-cell responses are better defined. HCV can persist despite acute phase CD8+ T-cell responses that are sometimes robust (Lauer et al., 2005). The abundance of circulating IFN-γ secreting CD8+ T cells recognizing HCV-specific peptides can peak as late as 6–12 months after acute infection, and is eventually followed by the loss of some or all epitope reactivities from blood (Cox et al., 2005a). Nonetheless, HCV-specific CD8+ T-cell populations can survive for decades in the persistently infected liver. Failure to recognize and eliminate HCV-infected hepatocytes is probably caused in part by the accumulation of amino acid substitutions within MHC class I-restricted HCV epitopes (Bowen and Walker, 2005; Cox et al., 2005b). These mutations may lead to viral escape from cytotoxic T-lymphocyte responses, thereby compounding potentially inadequate T-cell priming or memory responses. The capacity of the virus to escape from these controlling T-cell responses represents a significant challenge for vaccine development. However, it is important to emphasize that escape mutations do not emerge within all MHC class I epitopes even when cognate CD8+ T cells are visualized within the liver with MHC class I tetramers.

HCV-specific T cells are frequently defective in proliferative responses, as well as IFN-γ production and cytotoxic effector functions (Wedemeyer et al., 2002). Markers associated with T-cell memory can be reduced or absent, including the interleukin-7 receptor alpha (CD127) receptor required for IL-7 driven homeostatic proliferation. Impaired function may be associated with expression of molecules like programmed death-1 (PD-1) that can transmit inhibitory signals to antigen-specific CD8+ T cells upon engagement of PD-1 ligand(s) (Radziewicz et al., 2007). Strategies to interrupt PD-1 signaling, perhaps in combination with an HCV vaccine, represent promising new approaches for restoration of T-cell function and therapy of chronic hepatitis C (Grakoui et al., 2006).

Humoral Immune Responses

The role of B-cell immunity in resolution of acute infection and prevention of re-infection also remains incompletely defined. Neutralizing antibodies are typically delayed in appearance in acute HCV infection, generally do not confer sterile immunity, and are usually present in chronically infected persons (Farci et al., 1992; Netski et al., 2005; Helle et al., 2007; Lai et al., 1994; Prince, 1994). Nonetheless, substantial data suggest that virus-specific antibodies exert a level of control over HCV infection. This has been demonstrated in chimpanzee challenge studies in which antibodies isolated from a patient with chronic hepatitis C were able to neutralize the infectivity of an inoculum collected from the same subject 2 years previously (Farci et al., 1994). Replication of HCV has also been blocked in chimpanzees by prior in vitro neutralization of an inoculum with antibodies directed against a hypervariable region located near the amino terminus of the E2 protein (HVR-1, see Fig. 24.5, Farci et al., 1996), or by incubation with pooled immunoglobulins prepared from infected individuals (Yu et al., 2004). In other studies, infection was delayed, but not prevented, by passive transfer of the immunoglobulin preparation to animals prior to challenge with infectious virus (Krawczynski et al., 1996). Human monoclonal antibodies (HMAbs) that are capable of broadly neutralizing genetically diverse HCV strains, and protecting against infection with heterologous HCV in transgenic mice possessing chimeric human livers have been also generated recently (Law et al., 2008).

The functional activities of antibodies have been assessed in several different in vitro, cell-based assays. Antibodies to E2 can block the binding of soluble recombinant E2 glycoproteins to CD81, which is widely accepted to be a co-receptor for the virus (Allander et al., 2000; Ishii et al., 1998; Bartosch et al., 2003c; Pileri et al., 1998). More recently, retrovirus particles have been pseudotyped with the E1 and E2 proteins (Bartosch et al., 2003b; Hsu et al., 2003) to produce HCV pseudoparticles ("HCVpp"), which have been used as surrogates for assessing the functional activities of antibodies. Antibodies that limit the ability of HCVpp to attach to and enter cells are present in very low abundance and are very strain-specific early in the course of infection, when antibodies first appear (Lavillette et al., 2005; Meunier et al., 2005; Logvinoff et al., 2004; Netski et al., 2005). However, their titer increases and functional activity broadens after several months of infection so as to include the ability to neutralize heterologous HCV genotypes (Lavillette et al., 2005; Meunier et al., 2005; Logvinoff et al., 2004; Netski et al., 2005). An inverse correlation has been demonstrated between the magnitude of viremia and serum HCVpp neutralizing antibody titers in

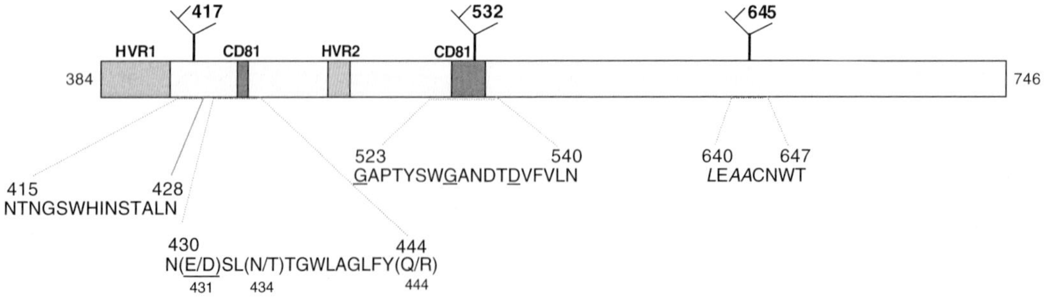

FIGURE 24.5 Mapping of neutralization epitopes on the HCV E2 protein sequence. Regions that are associated with CD81 binding hypervariable regions, and glycosylation sites that affect neutralization by domain B-reactive monoclonal human antibodies are labeled. The underlined residues are in the epitope recognized by CBH-2, a domain B-reactive human monoclonal antibody. Leu640, Ala642, and Ala643 (shown in italics) are necessary for overall E2 structure. Substitution of Ala431 with Glu results in escape from CBH-2 neutralization (Keck et al., submitted).

one study of chronically infected subjects (Lavillette et al., 2005).

Infectious virus produced in cell culture (HCVcc) has also been used to document the neutralizing activity of antibodies (Lindenbach et al., 2005; Yi et al., 2006). Although such data remain limited, they are largely consistent with earlier studies using pseudotyped retrovirus particles, including the limited genotype specificity of early antibody. In one study, sera collected from two of three patients about 1 year after acute infection with genotype 1a virus neutralized only genotype 1a virus, and not genotype 2a virus, while the third patient had a much greater titer of neutralizing antibody against the genotype 1a virus (Yi et al., 2007).

All in vitro neutralization studies should be viewed with some caution, however, given present uncertainties concerning the structure of HCV described above, and significant differences in the buoyant densities of HCVcc and HCV found in serum or plasma (Lindenbach et al., 2006).

VIRAL NEUTRALIZATION EPITOPES

As indicated above, there are no detailed crystallographic or cryoelectron microscopic reconstructions of the HCV envelope, and thus current understanding of structure–function relationships within the envelope are rudimentary and based largely on mutational analyses. Moreover, the envelope glycoproteins display some of the highest levels of genetic diversity among the HCV proteins, with E2 displaying greater variability than E1. The hypervariable HVR1 region found at the amino terminus of E2 (Fig. 24.5) is highly immunogenic, and is a major determinant for isolate-specific neutralizing antibody responses (Farci et al., 1996; Shimizu et al., 1994). However, such antibody provides little protection against the virus, as the HVR1 sequence continuously evolves during the course of infection, apparently in response to selective forces exerted by HVR1-specific neutralizing antibodies on the viral quasispecies population. This was clearly demonstrated in a study of sequential HCV sequences isolated from one patient over a 26-year period, in which serum antibodies failed to neutralize the concurrent dominant E1E2 species present in the blood (von Hahn et al., 2007). Escape was associated with mutations within the HVR1 that led to decreased binding and neutralization by MAbs that were produced to the HVR1 sequence present in the earliest isolate obtained from the patient.

More broadly neutralizing antibodies are usually directed against conformational epitopes within E2 (Allander et al., 2000; Hadlock et al., 2000; Bugli et al., 2001; Habersetzer et al., 1998; Ishii et al., 1998). Cross-competition analyses with HMAbs recognizing conformational epitopes on the HCV E2 molecules have delineated at least three immunogenic clusters of overlapping epitopes with distinct functions and properties (Keck et al., 2004, 2005). Non-neutralizing antibodies recognize one cluster, which has been designated as domain A, while neutralizing HMAbs can be segregated into two distinct groups recognizing two different clusters of epitopes, designated domains B and C respectively (Keck et al., 2007).

HMAbs recognizing E2 domain B epitopes demonstrate varying abilities to neutralize HCVpp containing glycoproteins derived from the six different HCV genotypes (Owsianka et al., 2008). One such antibody, CBH-5, reduced the infectivity of HCVpp derived from all genotypes by more than 60%. Importantly, CBH-5 and other domain B antibodies block the binding of E2 to CD81. Alanine scanning mutagenesis has shown that four highly conserved E2 residues (Gly523, Pro525, Gly530, and Asp535) are required for CBH-5 binding (Owsianka et al., 2008). Two of these residues (Gly530 and Asp535) have been shown to be essential for the interaction of E2 with CD81. Another residue (Gly523) also appears to have involved in the interaction with CD81, since an alanine substitution at this position lessens CD81 binding by 80% and HCVpp infectivity by 90% (Owsianka et al., 2006). The data thus suggest that CBH-5 exerts its potent and broad neutralizing effect on HCV by competing with CD81 for binding to conserved residues on E2 that are important for viral entry. Consistent with this, broadly neutralizing human antibodies derived from combinatorial libraries isolated from individuals with chronic hepatitis C also recognize epitopes containing Gly523, Trp529, Gly530, and Asp535 (Johansson et al., 2007; Law et al., 2008; Perotti et al., 2008). In addition, HCVpp bearing an alanine substitution at Asn532 demonstrate enhanced neutralization by sera from individuals infected with genotype 1a, 1b, 2b, 3, 4, and 5 viruses, suggesting that all of these sera contain antibodies directed at this region (Falkowska et al., 2007). Taken together, these data suggest that domain B represents a conserved, immunodominant site recognized by broadly reactive, neutralizing antibodies.

Of particular concern from a vaccine perspective, single amino acid substitutions within domain B antibodies can lead to virus escape from some domain B-reactive antibodies, thus recapitulating the escape observed with antibodies directed against the HVR1 domain. CBH-2 is a HMAb that recognizes a broadly conserved epitope in domain B that contains residues Gly523, Gly530, and Asp535 (similar to CBH-5).

These are also contact residues for E2 binding to CD81, as described above. However, a single amino acid substitution at residue 431, which is located at a considerable distance within the linear sequence of E2 from other domain B residues (Fig. 24.5), results in complete escape from CBH-2-mediated neutralization in a genotype 1a virus (Keck et al., 2008, submitted). These recent results highlight the substantial challenges inherent in developing HCV vaccines, and show that an effective vaccine will need to induce antibodies to both conserved as well as invariant epitopes to lessen the probabilities of virus escape.

APPROACHES TO VACCINE DEVELOPMENT

Sterilizing immunity is not essential to the success of a hepatitis C vaccine. The public health consequences of HCV arise from its ability to cause liver disease in the context of persistent infection, and thus the ability to prevent persistent infection should be the goal in vaccine development. As described in greater detail below, the pre-clinical development of a recombinant envelope protein vaccine was pursued for more than a decade by Chiron prior to its acquisition by Novartis, reaching phase I/II safety and immunogenicity trials in humans (Houghton and Abrignani, 2005). Vaccines comprised on recombinant envelope proteins appear generally safe and capable of eliciting both cellular and humoral immune responses that may be at least partially protective (Leroux-Roels et al., 2004; Houghton and Abrignani, 2005; Stamataki et al., 2007). Immunization with both recombinant envelope glycoproteins may prevent or delay the onset of viremia following infectious virus challenge (Puig et al., 2004; Choo et al., 1994; Houghton and Abrignani, 2005). However, it is far from certain that recombinant envelope protein vaccines, which are designed primarily to elicit neutralizing antibodies and CD4+ T cells, will prove to be sufficiently immunogenic for prevention of persistent human infections.

As indicated above, there are no well defined in vitro correlates of immunity, either for protection against acute infection or for prevention of progression of acute infections to long-term persistence of the virus. Moreover, vaccine developers are severely handicapped in not having adequate tools available to study the potential protection afforded by immunization. A number of pre-clinical studies have been carried out in chimpanzees, but the availability of this animal model is severely limited and the numbers of animals included in individual studies typically very small. In most cases, these studies have involved challenge with relatively low-titer infectious inocula that were closely related to the virus from which the vaccine antigens were taken. The extreme genetic heterogeneity of the envelope glycoproteins of HCV, described above, is a major challenge to be overcome. Moreover, chimpanzees appear to have a remarkably high rate of spontaneous resolution of HCV infection (Bassett et al., 1998), making determination of the effects of a vaccine on virus persistence very difficult.

However, as noted above, chimpanzees and possibly also humans who have successfully resolved previous infections with HCV may have a level of protective immunity that, while not capable of preventing acute re-infection, may prevent subsequent progression to persistent infection when rechallenged with the virus. Substantial data suggest that such immunity is mediated by CD4+ and CD8+ T cells (Bassett et al., 2001; Grakoui et al., 2003; Major et al., 2002; Shoukry et al., 2003). Thus, the most successful vaccine strategies may be those designed to stimulate the T-cell responses elicited during naturally resolving HCV infections. A number of approaches have been developed for priming class I and II restricted T cells against HCV (Houghton and Abrignani, 2005; Inchauspe and Feinstone, 2003). Examples include the use of virus-like particles (VLP) comprised of the HCV nucleocapsid and envelope proteins (Lechmann et al., 2001), recombinant viral vectors (Wuest et al., 2004; Vidalin et al., 2000; Abraham et al., 2004), and bacterial vectors (Simon et al., 2003; Wedemeyer et al., 2001; Abraham et al., 2004), as well as genetic immunization with plasmid sequences encoding segments of both structural and nonstructural proteins of the virus (Satoi et al., 2001; Pancholi et al., 2003; Encke et al., 1998). Some studies have also combined two of these modalities, such as priming with DNA immunization followed by boosting with a recombinant poxvirus vector (Pancholi et al., 2000), or priming with recombinant adenovirus, then boosting with DNA (Folgori et al., 2006). Although relatively few of these candidate approaches to immunization have been tested in the chimpanzee model, many have proven capable of priming specific CD4+ and CD8+ T-cell responses against HCV in mice. Several of these vaccine candidates have also been shown to be capable of stimulating HCV-specific T cells in nonhuman primates such as baboons (Jeong et al., 2004) and rhesus macaques (Rollier et al., 2005; Polakos et al., 2001; Capone et al., 2006), which may be more reliable models of the immunogenic response in humans.

In general, vaccines that are capable of priming virus-specific T-cell responses have been found to delay or diminish the magnitude of viremia in immunized chimpanzees following virus challenge, compared to unimmunized control animals (Folgori et al., 2006; Youn et al., 2005; Rollier et al., 2004; Forns et al., 2000a). Suppression of viremia was particularly profound with one vaccine that expressed only nonstructural proteins of the virus and thus was incapable of inducing virus-neutralizing activities (Folgori et al., 2006). Protection in this case was almost certainly mediated exclusively by T cells. However, despite evidence of exerting some control over the acute infection, in many studies there has been no evidence of a reduction in rates of progression to long-term, persistent infection (Forns et al., 2000a; Rollier et al., 2004). All of these studies have been limited by the small numbers of animals available, but this is nonetheless a critical distinction in that the significance of HCV as a human pathogen relates to its ability to establish persistent infection and cause chronic liver disease. There is no good explanation for the failure of such T-cell vaccines to protect against persistent infection, but it may reflect deficiencies in the frequency and perhaps breadth of the memory T-cell populations they elicit.

In summary, while there are no clearly established in vitro correlates of protective immunity against infection with HCV, strong indications from multiple sources suggest that CD4+ and CD8+ T-cell responses, while critical for control of acute infection, are insufficient for prevention of long-term persistent infection, at least in chimpanzees. At the same time, some evidence exists to support the importance of neutralizing antibody to the virus, and the ability of B-cell responses to modify the course of infection. Thus, an ideal vaccine should elicit both robust T-cell as well as B-cell responses. However, despite intensive efforts on the part of many, and despite continued progress in the field, no candidate vaccine has yet been shown to produce the level and/or type of immunity required for prevention of HCV-associated liver disease (Table 24.1).

RECOMBINANT PROTEIN VACCINES

Chiron Recombinant gpE1/gpE2 Vaccine

Early work at Chiron, the company that first identified the virus responsible for hepatitis C, focused on the development of recombinant HCV envelope glycoproteins as vaccine candidates (Choo et al., 1994). The E1 and E2 glycoproteins were expressed in mammalian cells with a recombinant vaccinia vector and found to associate with each other as a noncovalent heterodimer that bound antibody present in many serum samples from patients with chronic hepatitis C (Ralston et al., 1993). Immunization with this "gpE1/gpE2" preparation induced virus-specific antibodies and Th-cell responses in naïve chimpanzees (Houghton and Abrignani, 2005; Choo et al., 1994). In a landmark series of experiments, chimpanzees were immunized over a period of 6 months with three doses of the gpE1/gpE2 vaccine (generally 30–80 μg each) in an oil/water adjuvant, then challenged intravenously with 10–100 chimpanzee-infectious doses (CID_{50}) of virus 2 months later. Remarkably, 5 out of 12 immunized animals were completely protected against homologous challenge with the strain of virus from which the vaccine antigen was derived (HCV-1), and only two of the infected animals progressed to chronic infection (Houghton and Abrignani, 2005). In contrast, ten of ten unimmunized control animals (including nonconcurrently studied historic controls) became infected, seven of them persistently (Houghton and Abrignani, 2005). This difference in the proportion of animals sustaining a chronic infection was statistically significant ($p < 0.03$), but then the rate of persistent infections in the unimmunized animals was higher than that reported in other studies (Bassett et al., 1998). Similar protection against persistent infection was observed following challenge of immunized chimpanzees with a different strain of genotype 1a virus (HCV-H): nine of nine immunized animals became infected, but only one of the nine animals progressed to chronicity. In contrast, 8 of 14 nonimmunized animals developed chronic infection (all 14 were infected). It is not possible to predict how protective this genotype 1a vaccine might be against genotype 1b or genotype 2 or 3 infections.

A significant limitation to the chimpanzee studies of the gpE1/gpE2 vaccine is the lack of understanding of the significance of protection against an intravenous challenge with 10–100 CID_{50} of virus. Although this has become a standard in the field, it is not clear how it relates to the real-world situation. While a contaminated blood transfusion would provide a challenge many orders of magnitude greater than this, such an event would be very unlikely today because of screening procedures in place to protect the blood supply. The typical inoculum size in the setting of community-acquired infections is not known and can only be guessed at. Nonetheless, the data obtained in these chimpanzee challenge experiments suggest that the gpE1/gpE2 vaccine may be capable of protecting against persistent infection with at least genotype 1a viruses (Houghton and Abrignani, 2005). This is likely to be accomplished this via the induction of neutralizing antibodies, since the existence of sterilizing immunity in some animals correlated at least roughly with the magnitude of the antibody response

TABLE 24.1 HCV vaccine candidates that have entered clinical trials

Type	Primary clinical goal	Immunogen	Developer	Significant findings	Stage
Recombinant protein	Prevention	gpE1/gpE2 in oil/water adjuvant	Chiron/Novartis	Induces humoral and cellular immune responses and may protect against persistent infection with closely related virus in chimpanzees	Phase 1b
	Therapy	gpE1 in alum adjuvant (INO-101)	Innogenetics	Primes humoral and cellular immune responses in humans, but did not reduce liver injury in patients with chronic hepatitis C	Discontinued after Phase 2
	Therapy	Core protein in ISCOMATRIX® adjuvant	Chiron/Novartis and CSL	Primes T-cell responses in macaques	? Phase 1
	Therapy	NS3–NS4–NS5–Core polyprotein in ISCOMATRIX® adjuvant	Chiron/Novartis and CSL	Primes T-cell responses in chimpanzees	? Phase 1
Synthetic peptide	Therapy	Core, NS3, and NS4 peptides with poly-arginine adjuvant (IC41)	Intercell	Induced CD4+ and CD8+ T-cell responses in HLA-A2+ HCV-infected persons	Phase 2
	Therapy	HCV peptides formulated with influenza virosomes	Pevion Biotech	Induces virus-specific CTL and Th response in mice	Phase 1
	Therapy	"Personalized" peptide cocktail derived from E1, E2, NS3, and NS5A	Kurume University	Induced humoral and cellular immune responses in HCV-infected patients	Phase 1
DNA	Therapy	ChronVac-C®	Tripep and Inovio	Not available	Phase 1
Recombinant vectors	Prevention	MVA expressing NS3, NS4, and NS5B (TG 4040)	Transgene	Induces T-cell response in mice	Phase 1
	Therapy	Heat-killed yeast expressing Core and NS3 (GI-5005 Tarmogen)	GlobeImmune	Primes T-cell response in mice	Phase 1/2
Antibody	Therapy	HCIG human polyclonal antibody (Civacir®)	NABI/Biotest	Delayed infection in chimpanzees, but no impact on HCV viremia in liver transplant recipients	Phase 2
	Therapy	Ab68 + Ab65 monoclonal antibodies (XTL-6865)	XTL Biopharmaceuticals	No statistically significant changes in HCV viremia in patients with chronic hepatitis C	Discontinued after Phase 1

to E2 as assessed by inhibition of E2 binding to CD81 and neutralization of HCVpp entry (Houghton and Abrignani, 2005).

Encouraged by these results, a randomized phase I clinical study has been carried out to evaluate the safety, tolerability, and immunogenicity of the Chiron gpE1/gpE2 vaccine given at doses of 4, 20, or 100 μg with MF59 adjuvant versus a placebo in a total of 60 HCV-naïve subjects. The study was completed in August, 2005. Results have not been published as of February 2008, but accounts indicate that all 45 vaccine recipients developed neutralizing antibodies. Potentially important to the future of the gpE1/gpE2 vaccine, Chiron was acquired by Novartis in 2006, leading to significant changes in the management of the development program.

Innogenetics Recombinant E1 Vaccine

Recombinant E1 (in the absence of E2) has also been evaluated as an immunogen for therapeutic vaccination. The vaccine, developed by Innogenetics, consisted of a 135 amino acid polypeptide representing carboxy-terminally truncated E1 from a genotype 1b virus, adjuvanted with alum (Nevens et al., 2003). The underlying rationale behind the selection of E1 as a vaccine immunogen appears to have been based on correlations observed between high anti-E1 antibody levels and response to IFN therapy, as well as the absence of high titer anti-E1 antibodies in many persons with chronic HCV infection (Leroux-Roels et al., 2004; Nevens et al., 2003). Preliminary studies in chronically infected chimpanzees immunized with E1 protein reportedly resulted in improvements in disease (Maertens et al., 2000). In a phase I immunogenicity and tolerability study involving 20 healthy, HCV-negative volunteers, a series of three doses of this vaccine, 20 μg each, given at 3-week intervals, was well tolerated, and resulted in strong Th1 and anti-E1 antibody responses that were increased following a fourth dose given at 26 weeks (Leroux-Roels et al., 2004). In a subsequent pilot study, 35 patients with chronic hepatitis C were randomized to receive either five doses of this vaccine (at weeks 0, 4, 8, 12, and 24, $n = 26$) or placebo ($n = 9$) (Nevens et al., 2003). By week 48–50, serum ALT levels had declined significantly in the immunized group, but they were not different statistically at this point in time from the ALT values observed in the smaller number of control patients. Beyond 50 weeks, all of the patients enrolled in the study were offered an additional series of five doses of open-label vaccine. Follow-up liver biopsies suggested improvements in histopathology in 9 of 24 patients, but it is difficult to interpret these data in the absence of a bone fide control group. There were no significant changes noted in the level of viremia (Nevens et al., 2003). The only conclusion to be drawn from this study was that further analysis of the vaccine might be warranted.

Results of a follow-up phase IIb trial of the E1 therapeutic vaccine candidate were reported by Innogenetics in June, 2005, and demonstrated no differences in disease progression in groups of patients with chronic hepatitis C who were treated with vaccine versus placebo. In September of 2007, Innogenetics announced the results of a longer, 3 year, prospective, multicenter, randomized, double-blind, placebo-controlled phase IIb trial involving 122 patients. Those in the vaccine group had received four courses of vaccine consisting of recombinant E1 produced in yeast at doses of 50 μg each. Liver biopsies taken at the beginning and end of the study were examined for the primary clinical end-point which was a reduction in liver tissue damage. Patients treated with the vaccine were reported to show strong immune responses against the E1 protein, but failed to demonstrate a significant reduction in liver tissue damage assessed by histopathologic examination. Following this study, the E1 therapeutic vaccine program was discontinued.

Other Recombinant Protein Vaccine Candidates

In studies distinct from the Chiron program, E1E2, expressed from a recombinant Sindbis virus, has also been evaluated for its ability to elicit protective immunity in a small number of chimpanzees. Two animals were immunized: one who had recovered from a prior infection with HCV, and another who received priming doses of a DNA plasmid and a peptide immunogen prior to the recombinant E1E2 (Puig et al., 2004). The very small number of animals studied, compounded by differences in the pre-immunization history of the animals as well as the adjuvants and challenge inocula used, preclude any specific conclusions about the efficacy of the immunization. However, viremia appeared to have been delayed by several weeks in the two immunized animals following challenge with 100 CID_{50} of homologous virus, when compared with an unimmunized control animal who was challenged with only 3.5 CID_{50} (Puig et al., 2004). The naïve immunized animal that received the priming dose of the peptide immunogen progressed to persistent infection, while the other two animals resolved the infection. Importantly, both immunized animals became infected despite having developed robust antibody responses (including to the HVR1 domain of E2) and proliferative T-cell responses.

Other HCV proteins and several different adjuvants have also been evaluated as potential vaccine immunogens. Investigators at Chiron and CSL collaborated on evaluation of a prototype core protein-ISCOMATRIX vaccine that elicited high titers of antibody to core and primed strong CD4+ and CD8+ T-cell responses in rhesus macaques (Polakos et al., 2001). As many as 0.30–0.71 and 0.32–2.21% of the circulating CD8+ and CD4+ T cells, respectively, recognized core peptides by intracellular cytokine staining. An ISCOMATRIX-adjuvanted NS3-4-5–Core polyprotein is also reported to have elicited broad Th1-like CD4+ and CD8+ T-cell responses in chimpanzees; it may have ameliorated disease but did not prevent persistent infection following heterologous virus challenge (Houghton and Abrignani, 2005).

SYNTHETIC PEPTIDE VACCINES

Several different candidate vaccines under development incorporate synthetic peptides that represent conserved epitopes that have been shown to be recognized by HCV-specific T cells. These vaccines are for the most part being developed for therapeutic rather than preventive purposes, and are either cocktails of multiple peptides, or synthetic multi-epitope polypeptides. Several have entered clinical trials. All are potentially limited by the HLA-restricted nature of epitopes recognized by CD8+ T cells, as well as the sequence diversity existing among different HCV strains.

Intercell IC41 Peptide Vaccine

This therapeutic HCV vaccine candidate contains peptides representing three Th epitopes and five conserved HLA-A2 restricted cytotoxic T-cell epitopes derived from the core, NS3, and NS4 sequences, complexed with a proprietary poly-arginine adjuvant (Schlaphoff et al., 2007; Firbas et al., 2006). Clinical trials have been confined to HLA-A2+ individuals, who comprise about 40–50% of all Caucasians. In a phase I clinical trial involving ascending doses of the vaccine, 50% and 40% of healthy HCV-naïve adults who were immunized with 5.0 μg doses of the peptide vaccine developed positive IFN-γ Eli-spot tests for HCV-reactive CD4+ and CD8+ T cells, respectively (Firbas et al., 2006). Similar responses were not seen in placebo recipients. Comparable results were obtained in tetramer and proliferation assays, supporting the immunogenicity of the vaccine. In a second clinical study, HLA-A2 patients with chronic hepatitis C received a total of six doses of the vaccine at 4-week intervals. Modest increases in the frequency of HCV tetramer-positive CD8+ cells were observed in about one quarter of these patients, but there were no changes noted in circulating HCV RNA levels (Schlaphoff et al., 2007). This vaccine remains in development. In February, 2008, Intercell AG released the results of a study involving 50 patients with chronic genotype 1 infection. Study subjects who received eight intradermal injections of the IC41 vaccine at bi-weekly intervals were said to have demonstrated a minimal (40%) but statistically significant reduction in the abundance of circulating virus.

Pevion Biotech Virosome-Formulated Peptide Vaccine

Pevion is currently enrolling patients in a phase I clinical trial of virosome-formulated HCV peptides as a candidate therapeutic immunogen. The concept underlying this vaccine is that reconstituted influenza virosomes (IRIVs), composed of phospholipids and influenza surface glycoproteins, are capable of efficiently delivering peptide epitopes to the MHC class I antigen presentation pathway due to their fusogenic activity. Virosomes containing a 132-amino acid long peptide fragment of the HCV core protein were capable of inducing virus-specific cytotoxic and IFN-γ-producing T cells in HLA-A2.1 transgenic mice (Amacker et al., 2005).

Other Peptide-Based Vaccine Candidates

An academic group has reported the development of humoral and cellular immune responses to HCV peptides that were used to immunize HCV-infected HLA-A24+ patients who had failed IFN therapy (Yutani et al., 2007). Specific peptide sequences were selected to immunize each study subject based on pre-immunization stimulation of CD8+ cytotoxic T cells, making this a "personalized" vaccine. Reductions in viral load and improvements in serum ALT levels were noted in some of these vaccine recipients in this uncontrolled clinical study.

Several other studies have evaluated different strategies for presenting peptide epitopes to the immune system in mouse models. A library containing HVR1 peptide variants and a multi-epitope polypeptide immunogen containing HVR1 sequences both induced strong and specific antibody responses (Roccasecca et al., 2001; Torresi et al., 2007). A conceptually different approach, involving a synthetic multi-epitope polypeptide containing triple tandem repeats of five conserved epitopes from HCV was capable of inducing antibody as well as cytotoxic lymphocytes in BALB/c mice (Chen et al., 2007). The antibody elicited by this peptide immunogen was reported to have antiviral activity in infected transgenic mice harboring chimeric human livers. In a separate study, multi-epitope peptides containing sequences from NS3, NS4, and NS5B were capable of eliciting both primary and memory T-cell responses (Fournillier et al., 2006). Whether such immunogens will eventually make it to clinical trials remains to be seen.

GENETIC IMMUNIZATION

A number of studies have evaluated different approaches for DNA immunization against HCV in mice, including the use of DNA encoding a multi-genotype E1 library (Encke et al., 2007), envelope proteins (Nakano et al., 1997; Fournillier et al., 1998; Tedeschi et al., 1997), a combination of E1 and core epitopes (Shi et al., 2006), and NS3 (Gao et al., 2006). In general, the immunization of mice with DNA results in robust induction of mostly Th1- and CD8+ T-cell responses. Efforts to enhance DNA delivery have involved use of a gene gun (Nakano et al., 1997; Fournillier et al., 1998; Tedeschi et al., 1997), as well as in vivo electroporation (Zucchelli et al., 2000; Ahlen et al., 2007).

In general, naked DNA is less immunogenic in larger animals than in mice. However, in a small trial involving two chimpanzees, immunization with DNA encoding a truncated E2 protein resulted in an anti-E2 antibody response (including to the HVR1 domain). However, there was little evidence of protective immunity as both chimpanzees became infected following challenge with 100 CID_{50} of homologous virus (Forns et al., 2000a). The immunized chimpanzees resolved the infection, while a control animal challenged with 3 CID_{50} of virus did not. Unfortunately, it is not possible to draw any conclusions from this study concerning the potential efficacy of DNA immunization given the small number of animals, as well as differences in the magnitude of the inocula used to challenge the control versus the immunized animals. In another study, DNA encoding the E1E2 envelope proteins was adsorbed onto cationic poly (lactide co-glycolide) (PLG) microparticles (O'Hagan et al., 2004). Administered as a series of 3–4 doses, this DNA vaccine generated antibody responses comparable to MF59-adjuvanted gpE1/gpE2 protein in rhesus macaques, and also was capable of priming for a response to a protein booster (O'Hagan et al., 2004).

Replication-defective adenoviruses are very effective for delivery of genetic vaccines (Fattori et al., 2006), and strong immune responses have been observed following delivery of DNA to chimpanzees by this method (Folgori et al., 2006). Recombinant replication-deficient adenovirus vectors derived from two different adenovirus subtypes were used to deliver the first three genetic immunizations against genotype 1b NS3–NS5B proteins, at 0, 4, and 24 weeks, and were then followed by three doses of DNA given between weeks 35 and 39. The animals were challenged at week 49 with 100 CID_{50} of a heterologous, genotype 1a inoculum (Folgori et al., 2006). Since there are no neutralization determinants within the NS3–NS5B segment of the polyprotein, this vaccination approach was designed to elicit T-cell immunity. Proliferative CD4+ and CD8+ cell responses were observed in five of five immunized chimpanzees, while potent cytotoxic activity was induced in four animals. Importantly, cross-reactive T cells recognizing heterologous genotype 1a epitopes were detected in four of the five immunized animals, while intrahepatic HCV-specific CD8+ T cells were detected in all (Folgori et al., 2006). On challenge with a heterologous genotype 1a virus, viremia was delayed by at least a week and reduced 100-fold in magnitude, compared to five control, nonimmunized chimpanzees. The infection resolved spontaneously in four of the five immunized chimpanzees, but this also occurred in three of the five control animals. Strikingly, none of the immunized animals demonstrated biochemical evidence of acute hepatitis in association with HCV infection, while ALT elevations were noted in all of the controls (Folgori et al., 2006).

Although the immunization regimen used in this study is not one that would be amenable to incorporation into clinical practice, this study is important because it demonstrated that T-cell immunity, elicited by genetic immunization, can provide protection against acute liver disease and also limit virus replication following challenge with a heterologous strain of HCV (Folgori et al., 2006). This latter finding is particularly worthy of emphasis, because the genotype 1b and 1a viruses used as the basis for immunization and challenge in this study, respectively, differed from each other close to the limits of genetic diversity observed among genotype 1 viruses of both subtypes (Fig. 24.2) (Folgori et al., 2006). Although the study involved a relatively large number of animals, including five nonimmunized animals studied as contemporary controls, no conclusions can be drawn from it regarding the ability of such immunization to prevent viral persistence.

Similar results were obtained in another study, in which uninfected chimpanzees received three priming doses of DNA encoding genotype 1b core, E1E2, and NS3-5 proteins, followed by a recombinant adenovirus booster expressing the same proteins. This immunization regimen induced HCV-specific antibody and T-cell proliferation and IFN-γ Eli-spot responses, and resulted in a 100-fold reduction in the peak viremia following challenge with a different genotype 1b virus (Youn et al., 2005). A prime-boost strategy was also tested in a study in which two chimpanzees were primed with plasmid DNA encoding four HCV proteins (core, NS3, and truncated forms of E1 and E2), followed by a protein boost with the same recombinant proteins. Both of the immunized animals were infected on heterologous virus challenge, although the magnitude of viremia was reduced in comparison to an unimmunized control animal (Rollier et al., 2004). One of the immunized animals rapidly resolved the infection. Interestingly, immunization in this animal resulted in a bias towards Th1-type cytokine responses for E1 and NS3. In yet another study, immunization of naïve chimpanzees with DNA encoding core–E1E2 and NS3 followed by a poxvirus booster immunization using the modified vaccinia Ankara (MVA) vector resulted in robust HCV-specific immune responses as well as marked reduction in the magnitude of acute viremia on challenge with a heterologous virus (Rollier et al., 2007). However, despite good evidence for partial control of the acute infection, three of the four immunized animals in this study progressed to persistent infection.

Taken together, these studies show that genetic immunization, used either for priming or boosting immune responses to HCV, can result in robust virus-specific CD4+ and CD8+ T-cell responses that are capable of limiting the extent of virus replication upon challenge with even distantly related heterologous viruses of the same genotype. However, the bulk of the evidence would suggest that such immunization approaches do not provide a level or type of immunity that is sufficient for protection against persistent infection. This is in contrast with other evidence summarized above that suggests that immunization with recombinant gpE1/gpE2, while also failing to provide sterilizing immunity, might provide some protection against the development of subsequent persistent infection with closely related virus. Unfortunately, however, spontaneous resolution of acute HCV infections is a frequent occurrence in nonimmunized, HCV-naïve chimpanzees (Bassett et al., 1998). It is likely that compelling evidence for vaccine-induced protection against persistent infections will come only from clinical trials carried out in human subjects.

Genetic immunization is also being considered as an approach to therapy for persons with established, chronic HCV infection. A candidate therapeutic DNA vaccine, ChronVac-C (TriPep), is currently being evaluated in a nonrandomized, open-label phase I study of safety and immunogenicity. Based on pre-clinical results obtained in mice (Ahlen et al., 2007), this study will evaluate the use of electroporation to enhance delivery of the ChronVac-C vaccine in treatment-naïve patients with low virus loads.

RECOMBINANT VECTORED VACCINES

A variety of strategies for expression of HCV proteins by viral vectors have been explored extensively in mice. Viral vectors that have been utilized for the expression of both structural and nonstructural proteins of HCV include adenovirus (Arribillaga et al., 2002; Frelin et al., 2004), poxviruses, including canarypox virus and the MVA strain of vaccinia (Pancholi et al., 2003; Abraham et al., 2004), vesicular stomatitis virus (Majid et al., 2006), alphaviruses (Perri et al., 2003; Vidalin et al., 2000; Brinster et al., 2002), and ovine atadenovirus (Wuest et al., 2004). While of interest, however, such studies are generally not predictive of the immune responses to such immunization in larger animals, including humans. Nonetheless, a candidate virus-vectored vaccine, TG4040, is being developed by Transgene for use in humans using the MVA strain of vaccinia virus as an expression platform. MVA is highly attenuated, and not capable of replication in mammalian cells. The TG4040 vaccine is designed to express the nonstructural NS3, NS4, and NS5B proteins, and is intended to stimulate a therapeutic T-cell response. In transgenic HLA class I mice, an accelerated schedule of immunization with the vaccine was capable of inducing virus-specific cytolytic CD8+ T cells as well as specific CD4+ T responses (Fournillier et al., 2007). The TG4040 vaccine advanced to early phase I clinical trials to assess its safety and immunogenicity in 2007.

Bacterial and yeast-derived vectors have also been evaluated for expression of HCV proteins. The use of transformed attenuated *Salmonella typhimurium* $AroA^-$ strains expressing core or NS3 has been shown to efficiently prime cytotoxic T cells in mice, and to provide the ability to immunize mice orally (Liao et al., 2007; Wedemeyer et al., 2001). Yet another approach is the GlobeImmune GI-5005 Tarmogen, which is composed of whole, heat-inactivated recombinant *Saccharomyces cerevisiae* expressing a NS3–core fusion protein. In mice, GI-5005 induced antigen-specific proliferative and cytotoxic T-cell responses associated with Th1-type cytokine secretion, and led to eradication of tumor cells that expressed the NS3 protein (Haller et al., 2007). GI-5005 is being developed as a potential therapeutic vaccine. In November, 2007, GlobeImmune announced plans to advance GI-2005 to a phase II clinical trial after reporting results of a phase Ib trial in which six patients were said to have achieved reductions in circulating viral load of 0.75–1.25 $\log_{10}$.

INACTIVATED HCV AND VIRUS-LIKE PARTICLE VACCINES

The development of a conventional cell culture-propagated, whole virus vaccine became feasible with discovery of the genotype 2a JFH-1 virus that is capable of replication in cultured Huh7 human hepatoma cells (Lindenbach et al., 2006; Wakita et al., 2005). Inter-genotypic virus chimeras, constructed in the JFH-1 background but in which the structural proteins are derived from more prevalent genotype 1 viruses, are also replication competent and produce as many if not more virus particles (Pietschmann et al., 2006; Yi et al., 2007). Based on reported proprietary enhancements in the growth of virus and the ability of inactivated cell culture-produced virus to elicit HCVcc-neutralizing antibodies in mice, Toray Industries, Inc. announced plans in August, 2007, to develop an inactivated whole virus hepatitis C vaccine in collaboration with the National Institute for Infectious Diseases in Tokyo. So far, this work has not been published in the scientific literature, and numerous regulatory obstacles will need to be overcome to produce an acceptable product. Whether inactivated, cell culture-produced virus would have advantages over the use of recombinant proteins as immunogens is unknown, as is the ability of inactivated virus to induce protective immunity in animals.

HCV-like particles (HCV-LPs), composed of HCV structural proteins expressed in insect cells and assembled into particulate structures (Baumert et al., 1998), represent an interesting alternative to HCV vaccine development. These particles have a virus-like structure, and so could express conformational determinants that exist naturally but are not present in soluble recombinant proteins. HCV-LPs have been produced in insect cells with a recombinant baculovirus expressing cDNA encoding the structural proteins of HCV (core, E1, and E2, with or without p7) (Baumert et al., 1998). Morphologic and biophysical studies suggest that such particles may resemble virions present in infected humans (Baumert et al., 1998). In mice, HCV-LPs induced both humoral and cytotoxic and Th-cell responses directed against the core, E1, and E2 proteins (Baumert et al., 1999; Lechmann et al., 2001; Murata et al., 2003). Furthermore, immunization with HCV-LPs was able to protect mice against challenge with a recombinant vaccinia virus expressing the structural proteins of HCV (Murata et al., 2003). In another series of studies, HCV-LPs were combined with a novel adjuvant (AS01B, monophosphoryl lipid A [MPL], and QS21) to induce increased titers of antibodies directed against E1 and E2 (Qiao et al., 2003). However, HCV-specific humoral and cellular immune responses induced by HCV-LPs in baboons were only marginally enhanced by the use of adjuvants (Jeong et al., 2004). In a recent study, chimpanzees immunized with HCV-LPs or HCV-LPs adjuvanted with AS01B (two animals each) developed IFN-γ^+, IL-2+, CD4+, and CD8+ T-cell proliferative responses specific for core, E1, and E2. Surprisingly, however, there was little if any antibody response detected. The immunized chimpanzees were challenged with 100 CID_{50} of homologous infectious virus. Each of the immunized animals became infected, but the magnitude and duration of viremia was less than in historic control animals that had been similarly challenged in the past (Elmowalid et al., 2007). These early pre-clinical results are promising with regard to the cellular and humoral immune responses induced by HCV-LPs. Chimeric hepatitis B particles have also been studied as a potential expression platform for HCV antigens (Vietheer et al., 2007).

DENDRITIC CELL IMMUNIZATION

Ex vivo stimulation of dendritic cells (DC) has been considered based on the ability of DCs to potently present antigen and stimulate T-cell responses, and a large but conflicting body of evidence suggesting that DC function may be impaired in patients with chronic HCV infection (Decalf et al., 2007; Kanto and Hayashi, 2007; Kanto et al., 2004; Auffermann-Gretzinger et al., 2001). To date, this strategy has been studied only in vitro and in mouse models. In one study, considerably stronger cytotoxic T-lymphocyte responses were primed by injecting mice with DCs transduced with a recombinant adenovirus expressing the core protein, resulting in partial protection against challenge with a recombinant core protein-expressing vaccinia virus (Matsui et al., 2002). A second research group has also demonstrated the induction of humoral and cellular immune responses, and similar partial protection in a mouse challenge model, after adoptive immunization with DCs pulsed with the core protein in vitro (Encke et al., 2005). Immunization with DCs transduced to express nonstructural proteins has also been shown to generate strong virus-specific cytotoxic CD8+ and CD4+ Th1 responses (Racanelli et al., 2004; Zabaleta et al., 2008). Moreover, in recently published work, adoptive immunization with DCs transduced with adenoviral vectors encoding NS3 protein resulted in protection against the growth of NS3-expressing tumor cells, and also down-regulated intrahepatic expression of HCV RNA following transfection by hydrodynamic injection (Zabaleta et al., 2008). In other studies, DCs from healthy, HCV-naive human subjects which were transduced to express nonstructural proteins with recombinant adenovirus were able to prime for antigen-specific autologous T-cell responses in vitro (Li et al., 2008). Taken together, these studies show that adoptive immunization with DCs transduced ex vivo to express HCV proteins has considerable potential to stimulate virus-specific immune responses. These studies need to be expanded to include the chimpanzee model in order to determine whether DC immunization has significant advantages over other immunization strategies for control of hepatitis C. In the long run, if it proves to be of practical value, DC immunization is likely be more useful for therapy than for prevention.

PASSIVE IMMUNIZATION

Early support for the potential efficacy of passive immunization against HCV came from a clinical trial, carried out prior to the discovery of HCV, in which the administration of pooled human γ-globulin protected against the development of chronic hepatitis following acute, transfusion-acquired non-A, non-B hepatitis acquired by blood transfusion (Knodell et al., 1976). However, since this study was carried out, screening for HCV has been introduced into transfusion medicine, and all donor units that test positive for HCV (and might contain significant levels of neutralizing antibodies to HCV) are discarded and no longer included in the manufacture of pooled human immune globulin (IG). Thus, an HCV-specific IG, anti-HCV immunoglobulin (HCIG), was developed that was enriched in antibodies to the virus (Piazza et al., 1999). The administration of HCIG to chimpanzees 1 h after challenge with infectious virus was found to extend the incubation period leading to acute hepatocellular disease, but did not delay or otherwise modify HCV infection (Krawczynski et al., 1996). The greatest clinical need for such a product would be in prevention of graft infection following liver transplantation in patients with chronic hepatitis C. However, in a randomized, open-label study involving such patients, the administration of 17 doses of HCIG (Civacir, Nabi Biopharmaceuticals) over a period of 14 weeks failed to suppress HCV infection or reduce viral RNA levels in transplant recipients (Davis et al., 2005). An expanded phase II study of Civacir is currently in progress.

HMAbs developed by XTL Biopharmaceuticals (Ab-XTL68 and Ab-XTL65, formulated as XTL-6865) recognize different conformational epitopes on the E2 glycoprotein. These antibodies were capable of inhibiting HCV infection and reducing viral load in infected mice with chimeric human livers (Eren et al., 2006). The administration of high doses of Ab-XTL68 also resulted in transient reductions in HCV RNA in a randomized, double-blind, placebo-controlled trial involving chronically infected patients undergoing liver transplantation (Schiano et al., 2006). However, in March, 2007, XTL announced that no statistically significant changes were observed in viral RNA levels in a phase I, ascending dose study of XTL-6865, which contains both antibodies. The current development status of XTL-6865 is uncertain. Genmab is currently pursuing the pre-clinical development of another human antibody with potential neutralizing activity against HCV (Hu-Max-HepC).

KEY ISSUES

- Immunity to HCV is very poorly understood. Acute infections progress to chronicity despite the induction of virus-specific CD4+ and CD8+ T cells and virus-neutralizing antibodies. However, prior infection protects against long-term persistence on rechallenge of chimpanzees, and this is dependent on CD4+ and CD8+ memory subsets. Although the role of antibodies in such protection is not well defined, a successful vaccine will likely need to induce both robust T- and B-cell responses.
- The capacity of HCV to persist for years in many infected persons may stem in part from its ability to mutate and thereby escape established CD8+ T cell responses and antibody-mediated neutralization.
- The six major genotypes of HCV have diverse genetic sequences. Evidence that immunity of any type can confer broad protection against infection with different genotypes is limited.
- Vaccine development is hindered by the absence of a small animal model of hepatitis C, and by the high costs and lack of availability of the chimpanzee model.
- Significant HCV-related disease (hepatic fibrosis, cirrhosis, and liver cancer) is typically associated with chronic infection. Acute infections are generally not severe. Thus sterilizing immunity is not required, and a vaccine that is capable of preventing persistent infection would be of substantial value.
- As current IFN-based treatment options for hepatitis C are inadequate, a therapeutic vaccine would be of value. Some candidates have entered clinical trials, but none has yet achieved the desired outcome of controlling or modifying the course of HCV infection.

ACKNOWLEDGMENTS

The work was supported in parts by grants from the National Institutes of Health, U19-A140035 (Stanley M. Lemon), R37-AI47367, and U19-AI48231 (Christopher Walker), and R01-HL079381 (Steven Foung). Meital Gal-Tanamy is a McLaughlin Postdoctoral Research Fellow.

References

Abe, K., Edamoto, Y., Park, Y.N., Nomura, A.M., Taltavull, T.C., Tani, M. and Thung, S.N. In situ detection of hepatitis B, C, and G virus nucleic acids in human hepatocellular carcinoma tissues from different geographic regions. Hepatology 1998; 28:568–572.

Abraham, J.D., Himoudi, N., Kien, F., Berland, J.L., Codran, A., Bartosch, B., Baumert, T., Paranhos-Baccala, G., Schuster, C., Inchauspe, G. and Kieny, M.P. Comparative immunogenicity analysis of modified vaccinia Ankara vectors expressing native or modified forms of hepatitis C virus E1 and E2 glycoproteins. Vaccine 2004; 22:3917–3928.

Agnello, V., Chung, R.T. and Kaplan, L.M. A role for hepatitis C virus infection in type II cryoglobulinemia. N. Engl. J. Med. 1992; 327:1490–1495.

Ahlen, G., Soderholm, J., Tjelle, T., Kjeken, R., Frelin, L., Hoglund, U., Blomberg, P., Fons, M., Mathiesen, I. and Sallberg, M. In vivo electroporation enhances the immunogenicity of hepatitis C virus nonstructural 3/4A DNA by increased local DNA uptake, protein expression, inflammation, and infiltration of CD3+ T cells. J. Immunol. 2007; 179:4741–4753.

Allander, T., Drakenberg, K., Beyene, A., Rosa, D., Abrignani, S., Houghton, M., Widell, A., Grillner, L. and Persson, M.A. Recombinant human monoclonal antibodies against different conformational epitopes of the E2 envelope glycoprotein of hepatitis C virus that inhibit its interaction with CD81. J. Gen. Virol. 2000; 81:2451–2459.

Alter, M.J., Kruszon-Moran, D., Nainan, O.V., McQuillan, G.M., Gao, F., Moyer, L.A., Kaslow, R.A. and Margolis, H.S. The prevalence of hepatitis C virus infection in the United States, 1988 through 1994. N. Engl. J. Med. 1999; 341:556–562.

Alter, M.J., Margolis, H.S., Krawczynski, K., Judson, F.N., Mares, A., Alexander, W.J., Hu, P.Y., Miller, J.K., Gerber, M.A., Sampliner, R.E., Meeks, E.L. and Beach, M.J. The natural history of community-acquired hepatitis C in the United States. N. Engl. J. Med. 1992; 327:1899–1905.

Amacker, M., Engler, O., Kammer, A.R., Vadrucci, S., Oberholzer, D., Cerny, A. and Zurbriggen, R. Peptide-loaded chimeric influenza virosomes for efficient in vivo induction of cytotoxic T cells. Int. Immunol. 2005; 17:695–704.

Arribillaga, L., de Cerio, A.L., Sarobe, P., Casares, N., Gorraiz, M., Vales, A., Bruna-Romero, O., Borras-Cuesta, F., Paranhos-Baccala, G., Prieto, J., Ruiz, J. and Lasarte, J.J. Vaccination with an adenoviral vector encoding hepatitis C virus (HCV) NS3 protein protects against infection with HCV-recombinant vaccinia virus. Vaccine 2002; 21:202–210.

Auffermann-Gretzinger, S., Keeffe, E.B. and Levy, S. Impaired dendritic cell maturation in patients with chronic, but not resolved, hepatitis C virus infection. Blood 2001; 97:3171–3176.

Bartosch, B., Bukh, J., Meunier, J.C., Granier, C., Engle, R.E., Blackwelder, W.C., Emerson, S.U., Cosset, F.L. and Purcell, R.H. In vitro assay for neutralizing antibody to hepatitis C virus: evidence for broadly conserved neutralization epitopes. Proc. Natl. Acad. Sci. USA 2003a; 100:14199–14204.

Bartosch, B., Dubuisson, J. and Cosset, F.L. Infectious hepatitis C virus pseudo-particles containing functional E1–E2 envelope protein complexes. J. Exp. Med. 2003b; 197:633–642.

Bartosch, B., Vitelli, A., Granier, C., Goujon, C., Dubuisson, J., Pascale, S., Scarselli, E., Cortese, R., Nicosia, A. and Cosset, F.L. Cell entry of hepatitis C virus requires a set of co-receptors that include the CD81 tetraspanin and the SR-B1 scavenger receptor. J. Biol. Chem. 2003c; 278:41624–41630.

Bassett, S.E., Brasky, K.M. and Lanford, R.E. Analysis of hepatitis C virus-inoculated chimpanzees reveals unexpected clinical profiles. J. Virol. 1998; 72:2589–2599.

Bassett, S.E., Guerra, B., Brasky, K., Miskovsky, E., Houghton, M., Klimpel, G.R. and Lanford, R.E. Protective immune response to hepatitis C virus in chimpanzees rechallenged following clearance of primary infection. Hepatology 2001; 33:1479–1487.

Bassett, S.E., Thomas, D.L., Brasky, K.M. and Lanford, R.E. Viral persistence, antibody to E1 and E2, and hypervariable region 1 sequence stability in hepatitis C virus-inoculated chimpanzees. J. Virol. 1999; 73:1118–1126.

Baumert, T.F., Ito, S., Wong, D.T. and Liang, T.J. Hepatitis C virus structural proteins assemble into virus like particles in insect cells. J. Virol. 1998; 72:3827–3836.

Baumert, T.F., Vergalla, J., Satoi, J., Thomson, M., Lechmann, M., Herion, D., Greenberg, H.B., Ito, S. and Liang, T.J. Hepatitis C virus-like particles synthesized in insect cells as a potential vaccine candidate. Gastroenterology 1999; 117:1397–1407.

Bigger, C.B., Brasky, K.M. and Lanford, R.E. DNA microarray analysis of chimpanzee liver during acute resolving hepatitis C virus infection. J. Virol. 2001; 75:7059–7066.

Bigger, C.B., Guerra, K.M., Brasky, K.M., Hubbard, G., Beard, M.R., Luxon, B., Lemon, S.M. and Lanford, R.E. Intrahepatic gene expression during chronic hepatitis C virus infection in chimpanzees. J. Virol. 2004; 78:13779–13792.

Bowen, D.G. and Walker, C.M. Mutational escape from CD8+ T cell immunity: HCV evolution, from chimpanzees to man. J. Exp. Med. 2005; 201:1709–1714.

Bressanelli, S., Tomei, L., Roussel, A., Incitti, I., Vitale, R.L., Mathieu, M., De Francesco, R. and Rey, F.A. Crystal structure of the RNA-dependent RNA polymerase of hepatitis C virus. Proc. Natl. Acad. Sci. USA 1999; 96:13034–13039.

Brinster, C., Chen, M., Boucreux, D., Paranhos-Baccala, G., Liljestrom, P., Lemmonier, F. and Inchauspe, G. Hepatitis C virus non-structural protein 3-specific cellular immune responses following single or combined immunization with DNA or recombinant Semliki Forest virus particles. J. Gen. Virol. 2002; 83:369–381.

Bugli, F., Mancini, N., Kang, C.Y., Di, C.C., Grieco, A., Manzin, A., Gabrielli, A., Gasbarrini, A., Fadda, G., Varaldo, P.E., Clementi, M. and Burioni, R. Mapping B-cell epitopes of hepatitis C virus E2 glycoprotein using human monoclonal antibodies from phage display libraries. J. Virol. 2001; 75:9986–9990.

Bukh, J., Pietschmann, T., Lohmann, V., Krieger, N., Faulk, K., Engle, R.E., Govindarajan, S., Shapiro, M., St Claire, M. and Bartenschlager, R. Mutations that permit efficient replication of hepatitis C virus RNA in Huh-7 cells prevent productive replication in chimpanzees. Proc. Natl. Acad. Sci. USA 2002; 99:14416–14421.

Capone, S., Meola, A., Ercole, B.B., Vitelli, A., Pezzanera, M., Ruggeri, L., Davies, M.E., Tafi, R., Santini, C., Luzzago, A., Fu, T.M., Bett, A., Colloca, S., Cortese, R., Nicosia, A. and Folgori, A. A novel adenovirus type 6 (Ad6)-based hepatitis C virus vector that overcomes preexisting anti-ad5 immunity and induces potent and broad cellular immune responses in rhesus macaques. J. Virol. 2006; 80:1688–1699.

Centers for Disease Control and Prevention. Guidelines for laboratory testing and result reporting of antibody to hepatitis C virus. Centers for Disease Control and Prevention. Morbidity and Mortality Weekly Report 52, 2003; 1–13, 15.

Chen, L., Gao, T., Yang, N., Huang, J., Chen, Y., Gao, T., Li, Q. and Ren, D. Immunization with a synthetic multiepitope antigen induces humoral and cellular immune responses to hepatitis C virus in mice. Viral Immunol. 2007; 20:170–179.

Choo, Q.-L., Kuo, G., Ralston, R., Weiner, A., Chien, D., Van Nest, G., Han, J., Berger, K., Thudium, K., Kuo, C., Kansopon, J., McFarland, J., Tabrizi, A., Ching, K., Moss, B., Cummins, L.B., Houghton, M. and Muchmore, E. Vaccination of chimpanzees against infection by the hepatitis C virus. Proc. Natl. Acad. Sci. USA 1994; 91:1294–1298.

Choo, Q.-L., Kuo, G., Weiner, A.J., Overby, L.R., Bradley, D.W. and Houghton, M. Isolation of a cDNA clone derived from a blood-borne non-A, non-B viral hepatitis genome. Science 1989; 244:359–362.

Chung, R.T., Gale, M., Jr., Polyak, S.J., Lemon, S.M., Liang, T.J. and Hoofnagle, J.H. Mechanisms of action of interferon and ribavirin in chronic hepatitis C: summary of a workshop. Hepatology 2008; 47:306–320.

Consensus Development Panel NIH Consensus Statement on Management of Hepatitis C: 2002. NIH Consens. State Sci. Statements. 2002; 19:1–46.

Cooper, S., Erickson, A.L., Adams, E.J., Kansopon, J., Weiner, A.J., Chien, D.Y., Houghton, M., Parham, P. and Walker, C.M. Analysis of a successful immune response against hepatitis C virus. Immunity 1999; 10:439–449.

Cox, A.L., Mosbruger, T., Lauer, G.M., Pardoll, D., Thomas, D.L. and Ray, S.C. Comprehensive analyses of CD8+ T cell responses during longitudinal study of acute human hepatitis C. Hepatology 2005a; 42:104–112.

Cox, A.L., Mosbruger, T., Mao, Q., Liu, Z., Wang, X.H., Yang, H.C., Sidney, J., Sette, A., Pardoll, D., Thomas, D.L. and Ray, S.C. Cellular immune selection with hepatitis C virus persistence in humans. J. Exp. Med. 2005b; 201:1741–1752.

Davila, J.A., Morgan, R.O., Shaib, Y., McGlynn, K.A. and El Serag, H.B. Hepatitis C infection and the increasing incidence of hepatocellular carcinoma: a population-based study. Gastroenterology 2004; 127:1372–1380.

Davis, G.L., Esteban-Mur, R., Rustgi, V., Hoefs, J., Gordon, S.C., Trepo, C., Shiffman, M.L., Zeuzem, S., Craxi, A., Ling, M.H. and Albrecht, J. Interferon alfa-2b alone or in combination with ribavirin for the treatment of relapse of chronic hepatitis C. International Hepatitis Interventional Therapy Group [see comments]. N. Engl. J. Med. 1998; 339:1493–1499.

Davis, G.L., Nelson, D.R., Terrault, N., Pruett, T.L., Schiano, T.D., Fletcher, C.V., Sapan, C.V., Riser, L.N., Li, Y., Whitley, R.J. and Gnann, J.W., Jr. A randomized, open-label study to evaluate the safety and pharmacokinetics of human hepatitis C immune globulin (Civacir) in liver transplant recipients. Liver Transpl. 2005; 11:941–949.

Decalf, J., Fernandes, S., Longman, R., Ahloulay, M., Audat, F., Lefrerre, F., Rice, C.M., Pol, S. and Albert, M.L. Plasmacytoid dendritic cells initiate a complex chemokine and cytokine network and are a viable drug target in chronic HCV patients. J. Exp. Med. 2007; 204:2423–2437.

Diaz, O., Delers, F., Maynard, M., Demignot, S., Zoulim, F., Chambaz, J., Trepo, C., Lotteau, V. and Andre, P. Preferential association of Hepatitis C virus with apolipoprotein B48-containing lipoproteins. J. Gen. Virol. 2006; 87:2983–2991.

Dubuisson, J. and Rice, C.M. Hepatitis C virus glycoprotein folding: disulfide bond formation and association with calnexin. J. Virol. 1996; 70:778–786.

El Serag, H.B. Hepatocellular carcinoma: recent trends in the United States. Gastroenterology 2004; 127:S27–S34.

Elmowalid, G.A., Qiao, M., Jeong, S.H., Borg, B.B., Baumert, T.F., Sapp, R.K., Hu, Z., Murthy, K. and Liang, T.J. Immunization with hepatitis C virus-like particles results in control of hepatitis C virus infection in chimpanzees. Proc. Natl. Acad. Sci. USA 2007; 104:8427–8432.

Encke, J., Findeklee, J., Geib, J., Pfaff, E. and Stremmel, W. Prophylactic and therapeutic vaccination with dendritic cells against hepatitis C virus infection. Clin. Exp. Immunol. 2005; 142:362–369.

Encke, J., Radunz, W., Eisenbach, C., Geib, J., Gehrke, S., Pfaff, E. and Stremmel, W. Development of a heterologous, multigenotype vaccine against hepatitis C virus infection. Eur. J. Clin. Invest. 2007; 37:396–406.

Encke, J., zu Putlitz, J., Geissler, M. and Wands, J.R. Genetic immunization generates cellular and humoral immune responses against the nonstructural proteins of the hepatitis C virus in a murine model. J. Immunol. 1998; 161:4917–4923.

Enomoto, N., Sakuma, I., Asahina, Y., Kurosaki, M., Murakami, T., Yamamoto, C., Ogura, Y., Izumi, N., Marumo, F. and Sato, C. Mutations in the nonstructural protein 5A gene and response to interferon in patients with chronic hepatitis C virus 1b infection. N. Engl. J. Med. 1996; 334:77–81.

Eren, R., Landstein, D., Terkieltaub, D., Nussbaum, O., Zauberman, A., Ben-Porath, J., Gopher, J., Buchnick, R., Kovjazin, R., Rosenthal-Galili, Z., Aviel, S., Ilan, E., Shoshany, Y., Neville, L., Waisman, T., Ben-Moshe, O., Kischitsky, A., Foung, S.K., Keck, Z.Y., Pappo, O., Eid, A., Jurim, O., Zamir, G., Galun, E. and Dagan, S. Preclinical evaluation of two neutralizing human monoclonal antibodies against hepatitis C virus (HCV): a potential treatment to prevent HCV reinfection in liver transplant patients. J. Virol. 2006; 80:2654–2664.

Falkowska, E., Kajumo, F., Garcia, E., Reinus, J. and Dragic, T. Hepatitis C virus envelope glycoprotein E2 glycans modulate entry, CD81 binding, and neutralization. J. Virol. 2007; 81:8072–8079.

Farci, P., Alter, H.J., Govindarajan, S., Wong, D.C., Engle, R., Lesniewski, R.R., Mushahwar, I.K., Desai, S.M., Miller, R.H., Ogata, N. and Purcell, R.H. Lack of protective immunity against reinfection with hepatitis C virus. Science 1992; 258:135–140.

Farci, P., Alter, H.J., Wong, D.C., Miller, R.H., Govindarajan, S., Engle, R., Shapiro, M. and Purcell, R.H. Prevention of hepatitis C virus infection in chimpanzees after antibody-mediated in vitro neutralization. Proc. Natl. Acad. Sci. USA 1994; 91:7792–7796.

Farci, P., Shimoda, A., Wong, D., Cabezon, T., De Gioannis, D., Strazzera, A., Shimizu, Y., Shapiro, M., Alter, H.J. and Purcell, R.H. Prevention of hepatitis C virus infection in chimpanzees by hyperimmune serum against the hypervariable region 1 of the envelope 2 protein. Proc. Natl. Acad. Sci. USA 1996; 93:15394–15399.

Fattori, E., Zampaglione, I., Arcuri, M., Meola, A., Ercole, B.B., Cirillo, A., Folgori, A., Bett, A., Cappelletti, M., Sporeno, E., Cortese, R., Nicosia, A. and Colloca, S. Efficient immunization of rhesus macaques with an HCV candidate vaccine by heterologous priming-boosting with novel adenoviral vectors based on different serotypes. Gene Ther. 2006; 13:1088–1096.

Feinstone, S.M., Kapikian, A.Z., Purcell, R.H., Alter, H.J. and Holland, P.V. Transfusion-associated hepatitis not due to viral hepatitis type A or B. N. Engl. J. Med. 1975; 292:767–770.

Ferreon, J.C., Ferreon, A.C.M., Li, K. and Lemon, S.M. Molecular determinants involved in TRIF proteolysis by the hepatitis C virus NS3/4A protease. J. Biol. Chem. 2005; 280:20483–20492.

Firbas, C., Jilma, B., Tauber, E., Buerger, V., Jelovcan, S., Lingnau, K., Buschle, M., Frisch, J. and Klade, C.S. Immunogenicity and safety of a novel therapeutic hepatitis C virus (HCV) peptide vaccine: a randomized, placebo controlled trial for dose optimization in 128 healthy subjects. Vaccine 2006; 24:4343–4353.

Folgori, A., Capone, S., Ruggeri, L., Meola, A., Sporeno, E., Ercole, B.B., Pezzanera, M., Tafi, R., Arcuri, M., Fattori, E., Lahm, A., Luzzago, A., Vitelli, A., Colloca, S., Cortese, R. and Nicosia, A. A T-cell HCV vaccine eliciting effective immunity against heterologous virus challenge in chimpanzees. Nat. Med. 2006; 12:190–197.

Forns, X., Payette, P.J., Ma, X., Satterfield, W., Eder, G., Mushahwar, I.K., Govindarajan, S., Davis, H.L., Emerson, S.U., Purcell, R.H. and Bukh, J. Vaccination of chimpanzees with plasmid DNA encoding the hepatitis C virus (HCV) envelope E2 protein modified the infection after challenge with homologous monoclonal HCV. Hepatology 2000a; 32:618–625.

Forns, X., Thimme, R., Govindarajan, S., Emerson, S.U., Purcell, R.H., Chisari, F.V. and Bukh, J. Hepatitis C virus lacking the hypervariable region 1 of the second envelope protein is infectious and causes acute resolving or persistent infection in chimpanzees. Proc. Natl. Acad. Sci. USA 2000b; 97:13318–13323.

Fournillier, A., Dupeyrot, P., Martin, P., Parroche, P., Pajot, A., Chatel, L., Fatmi, A., Gerossier, E., Bain, C., Lone, Y.C., Trepo, C. and Inchauspe, G. Primary and memory T cell responses induced by hepatitis C virus multiepitope long peptides. Vaccine 2006; 24:3153–3164.

Fournillier, A., Gerossier, E., Evlashev, A., Schmitt, D., Simon, B., Chatel, L., Martin, P., Silvestre, N., Balloul, J.M., Barry, R. and Inchauspe, G. An accelerated vaccine schedule with a poly-antigenic hepatitis C virus MVA-based candidate vaccine induces potent, long lasting and in vivo cross-reactive T cell responses. Vaccine 2007; 25:7339–7353.

Fournillier, A., Nakano, I., Vitvitski, L., Depla, E., Vidalin, O., Maertens, G., Trepo, C. and Inchauspe, G. Modulation of immune responses to hepatitis C virus envelope E2 protein following injection of plasmid DNA using single or combined delivery routes. Hepatology 1998; 28:237–244.

Frelin, L., Ahlen, G., Alheim, M., Weiland, O., Barnfield, C., Liljestrom, P. and Sallberg, M. Codon optimization and mRNA amplification effectively enhances the immunogenicity of the hepatitis C virus nonstructural 3/4A gene. Gene Ther. 2004; 11:522–533.

Gao, M., Wang, H.P., Wang, Y.N., Zhou, Y. and Wang, Q.L. HCV–NS3 Th1 minigene vaccine based on invariant chain CLIP genetic substitution enhances CD4(+) Th1 cell responses in vivo. Vaccine 2006; 24:5491–5497.

Gastaminza, P., Cheng, G., Wieland, S., Zhong, J., Liao, W. and Chisari, F.V. Cellular determinants of hepatitis C virus assembly, maturation, degradation and secretion. J. Virol. 2008; 82:2241–2249.

Global Burden of Disease (GBD) for Hepatitis C J. Clin. Pharmacol. 2004; 44:20–29.

Gotz, H.M., van, D.G., Niesters, H.G., den Hollander, J.G., Thio, H.B. and de, Z.O. A cluster of acute hepatitis C virus infection among men who have sex with men—results from contact tracing and public health implications. AIDS 2005; 19:969–974.

Grakoui, A., John, W.E., Hanson, H.L., Walker, C. and Ahmed, R. Turning on the off switch: regulation of anti-viral T cell responses in the liver by the PD-1/PD-L1 pathway. J. Hepatol. 2006; 45:468–472.

Grakoui, A., Shoukry, N.H., Woollard, D.J., Han, J.H., Hanson, H.L., Ghrayeb, J., Murthy, K.K., Rice, C.M. and Walker, C.M. HCV persistence and immune evasion in the absence of memory T cell help. Science 2003; 302:659–662.

Habersetzer, F., Fournillier, A., Dubuisson, J., Rosa, D., Abrignani, S., Wychowski, C., Nakano, I., Trepo, C., Desgranges, C. and Inchauspe, G. Characterization of human monoclonal antibodies specific to the hepatitis C virus glycoprotein E2 with in vitro binding neutralization properties. Virology 1998; 249:32–41.

Hadlock, K.G., Lanford, R.E., Perkins, S., Rowe, J., Yang, Q., Levy, S., Pileri, P., Abrignani, S. and Foung, S.K. Human monoclonal antibodies that inhibit binding of hepatitis C virus E2 protein to CD81 and recognize conserved conformational epitopes. J. Virol. 2000; 74:10407–10416.

Haller, A.A., Lauer, G.M., King, T.H., Kemmler, C., Fiolkoski, V., Lu, Y., Bellgrau, D., Rodell, T.C., Apelian, D., Franzusoff, A. and Duke, R.C. Whole recombinant yeast-based immunotherapy induces potent T cell responses targeting HCV NS3 and Core proteins. Vaccine 2007; 25:1452–1463.

Hauri, A.M., Armstrong, G.L. and Hutin, Y.J. The global burden of disease attributable to contaminated injections given in health care settings. Int. J. STD AIDS 2004; 15:7–16.

Helle, F., Goffard, A., Morel, V., Duverlie, G., McKeating, J., Keck, Z.Y., Foung, S., Penin, F., Dubuisson, J. and Voisset, C. The neutralizing activity of anti-hepatitis C virus antibodies is modulated by specific glycans on the E2 envelope protein. J. Virol. 2007; 81:8101–8111.

Houghton, M. and Abrignani, S. Prospects for a vaccine against the hepatitis C virus. Nature 2005; 436:961–966.

Hsu, M., Zhang, J., Flint, M., Logvinoff, C., Cheng-Mayer, C., Rice, C.M. and McKeating, J.A. Hepatitis C virus glycoproteins mediate pH-dependent cell entry of pseudotyped retroviral particles. Proc. Natl. Acad. Sci. USA 2003; 100:7271–7276.

Huang, H., Sun, F., Owen, D.M., Li, W., Chen, Y., Gale, M., Jr. and Ye, J. Hepatitis C virus production by human hepatocytes dependent on assembly and secretion of very low-density lipoproteins. Proc. Natl. Acad. Sci. USA 2007; 104:5848–5853.

Inchauspe, G. and Feinstone, S. Development of a hepatitis C virus vaccine. Clin. Liver Dis. 2003; 7:243–259.

Ishii, K., Rosa, D., Watanabe, Y., Katayama, T., Harada, H., Wyatt, C., Kiyosawa, K., Aizaki, H., Matsuura, Y., Houghton, M., Abrignani, S. and Miyamura, T. High titers of antibodies inhibiting the binding of envelope to human cells correlate with natural resolution of chronic hepatitis C. Hepatology 1998; 28:1117–1120.

Jeong, S.H., Qiao, M., Nascimbeni, M., Hu, Z., Rehermann, B., Murthy, K. and Liang, T.J. Immunization with hepatitis C virus-like particles induces humoral and cellular immune responses in nonhuman primates. J. Virol. 2004; 78:6995–7003.

Johansson, D.X., Voisset, C., Tarr, A.W., Aung, M., Ball, J.K., Dubuisson, J. and Persson, M.A. Human combinatorial libraries yield rare antibodies that broadly neutralize hepatitis C virus. Proc. Natl. Acad. Sci. USA 2007; 104:16269–16274.

Johnson, R.J., Gretch, D.R., Yamabe, H., Hart, J., Bacchi, C.E., Hartwell, P., Couser, W.G., Corey, L., Wener, M.H., Alpers, C.A. and Willson, R. Membranoproliferative glomerulonephritis associated with hepatitis C virus infection. N. Engl. J. Med. 1993; 328:465–470.

Kabelitz, D. and Medzhitov, R. Innate immunity—cross-talk with adaptive immunity through pattern recognition receptors and cytokines. Curr. Opin. Immunol. 2007; 19:1–3.

Kage, M., Shimamatu, K., Nakashima, E., Kojiro, M., Inoue, O. and Yano, M. Long-term evolution of fibrosis from chronic hepatitis to cirrhosis in patients with hepatitis C: morphometric analysis of repeated biopsies. Hepatology 1997; 25:1028–1031.

Kaito, M., Watanabe, S., Tsukiyama-Kohara, K., Yamaguchi, K., Kobayashi, Y., Konishi, M., Yokoi, M., Ishida, S., Suzuki, S. and Kohara, M. Hepatitis C virus particle detected by immunoelectron microscopic study. J. Gen. Virol. 1994; 75:1755–1760.

Kamili, S., Krawczynski, K., McCaustland, K.A., Li, X. and Alter, M.J. The infectivity of hepatitis C virus after drying and storing at room temperature. Infect. Control Hospital Epidemiol. 2007; 28:519–524.

Kanto, T. and Hayashi, N. Innate immunity in hepatitis C virus infection: interplay among dendritic cells, natural killer cells and natural killer T cells. Hepatol. Res. 2007; 37(Suppl 3):S319–S326.

Kanto, T., Inoue, M., Miyatake, H., Sato, A., Sakakibara, M., Yakushijin, T., Oki, C., Itose, I., Hiramatsu, N., Takehara, T., Kasahara, A. and

Hayashi, N. Reduced numbers and impaired ability of myeloid and plasmacytoid dendritic cells to polarize T helper cells in chronic hepatitis C virus infection. J. Infect. Dis. 2004; 190:1919–1926.

Kato, N., Sekiya, H., Ootsuyama, Y., Nakazawa, T., Hijikata, M., Ohkoshi, S. and Shimotohno, K. Humoral immune response to hypervariable region 1 of the putative envelope glycoprotein (gp70) of hepatitis C virus. J. Virol. 1993; 67:3923–3930.

Kawai, T., Takahashi, K., Sato, S., Coban, C., Kumar, H., Kato, H., Ishii, K.J., Takeuchi, O. and Akira, S. IPS-1, an adaptor triggering RIG-I- and Mda5-mediated type I interferon induction. Nat. Immunol. 2005; 6:981–988.

Keck, Z.Y., Li, T.K., Xia, J., Bartosch, B., Cosset, F.L., Dubuisson, J. and Foung, S.K. Analysis of a highly flexible conformational immunogenic domain A in hepatitis C virus E2. J. Virol. 2005; 79:13199–13208.

Keck, Z.Y., Op De, B.A., Hadlock, K.G., Xia, J., Li, T.K., Dubuisson, J. and Foung, S.K. Hepatitis C virus E2 has three immunogenic domains containing conformational epitopes with distinct properties and biological functions. J. Virol. 2004; 78:9224–9232.

Keck, Z.Y., Xia, J., Cai, Z., Li, T.K., Owsianka, A.M., Patel, A.H., Luo, G. and Foung, S.K. Immunogenic and functional organization of hepatitis C virus (HCV) glycoprotein E2 on infectious HCV virions. J. Virol. 2007; 81:1043–1047.

Kim, J.L., Morgenstern, K.A., Lin, C., Fox, T., Dwyer, M.D., Landro, J.A., Chambers, S.P., Markland, W., Lepre, C.A., O'Malley, E.T., Harbeson, S.L., Rice, C.M., Murcko, M.A., Caron, P.R. and Thomson, J.A. Crystal structure of the hepatitis C virus NS3 protease domain complexed with a synthetic NS4A cofactor peptide. Cell 1996; 87:343–355.

Knodell, R.G., Conrad, M.E., Ginsberg, A.L. and Bell, C.J. Efficacy of prophylactic gamma-globulin in preventing non-A, non-B post-transfusion hepatitis. Lancet 1976; 1:557–561.

Kolykhalov, A.A., Agapov, A.A., Blight, K.J., Mihalik, K., Feinstone, S.M. and Rice, C.M. Transmission of hepatitis C by intrahepatic inoculation with transcribed RNA. Science 1997; 277:570–574.

Krawczynski, K., Alter, M.J., Tankersley, D.L., Beach, M., Robertson, B.H., Lambert, S., Kuo, G., Spelbring, J.E., Meeks, E., Sinha, S. and Carson, D.A. Effect of immune globulin on the prevention of experimental hepatitis C virus infection. J. Infect. Dis. 1996; 173:822–828.

Kuo, G., Choo, Q.-L., Alter, H.J., Gitnick, G.L., Redeker, A.G., Purcell, R.H., Miyamura, T., Dienstag, J.L., Alter, M.J., Stevens, C.E., Tegtmeier, G.E., Bonino, F., Colombo, M., Lee, W.-S., Kuo, C., Berger, K., Shuster, J.R., Overby, L.R., Bradley, D.W. and Houghton, M. An assay for circulating antibodies to a major etiologic virus of human non-A, non-B hepatitis. Science 1989; 244:362–364.

Lai, M.E., Mazzoleni, A.P., Argiolu, F., De, V.S., Balestrieri, A., Purcell, R.H., Cao, A. and Farci, P. Hepatitis C virus in multiple episodes of acute hepatitis in polytransfused thalassaemic children. Lancet 1994; 343:388–390.

Lanford, R.E., Bigger, C., Bassett, S. and Klimpel, G. The chimpanzee model of hepatitis C virus infections. ILAR J. 2001; 42:117–126.

Lanford, R.E., Guerra, B., Chavez, D., Bigger, C., Brasky, K.M., Wang, X.H., Ray, S.C. and Thomas, D.L. Cross-genotype immunity to hepatitis C virus. J. Virol. 2004; 78:1575–1581.

Lanford, R.E., Sureau, C., Jacob, J.R., White, R. and Fuerst, T.R. Demonstration of in vitro infection of chimpanzee hepatocytes with hepatitis C virus using strand-specific RT/PCR. Virology 1994; 202:606–614.

Lang, K.S., Georgiev, P., Recher, M., Navarini, A.A., Bergthaler, A., Heikenwalder, M., Harris, N.L., Junt, T., Odermatt, B., Clavien, P.A., Pircher, H., Akira, S., Hengartner, H. and Zinkernagel, R.M. Immunoprivileged status of the liver is controlled by Toll-like receptor 3 signaling. J. Clin. Invest. 2006; 116:2456–2463.

Lau, D.T., Fish, P.M., Sinha, M., Owen, D.M., Lemon, S.M. and Gale, M. Jr. Patterns of IRF-3 activation, hepatic interferon-stimulated gene expression and immune cell infiltration in patients with chronic hepatitis C. Hepatology, 2008; 47:799–809.

Lauer, G.M., Lucas, M., Timm, J., Ouchi, K., Kim, A.Y., Day, C.L., Schulze Zur, W.J., Paranhos-Baccala, G., Sheridan, I., Casson, D.R., Reiser, M., Gandhi, R.T., Li, B., Allen, T.M., Chung, R.T., Klenerman, P. and Walker, B.D. Full-breadth analysis of CD8+ T-cell responses in acute hepatitis C virus infection and early therapy. J. Virol. 2005; 79:12979–12988.

Lavillette, D., Morice, Y., Germanidis, G., Donot, P., Soulier, A., Pagkalos, E., Sakellariou, G., Intrator, L., Bartosch, B., Pawlotsky, J.M. and Cosset, F.L. Human serum facilitates hepatitis C virus infection, and neutralizing responses inversely correlate with viral replication kinetics at the acute phase of hepatitis C virus infection. J. Virol. 2005; 79:6023–6034.

Law, M., Maruyama, T., Lewis, J., Giang, E., Tarr, A.W., Stamataki, Z., Gastaminza, P., Chisari, F.V., Jones, I.M., Fox, R.I., Ball, J.K., McKeating, J.A., Kneteman, N.M. and Burton, D.R. Broadly neutralizing antibodies protect against hepatitis C virus quasispecies challenge. Nat. Med. 2008; 14:25–27.

Lechmann, M., Murata, K., Satoi, J., Vergalla, J., Baumert, T.F. and Liang, T.J. Hepatitis C virus-like particles induce virus-specific humoral and cellular immune responses in mice. Hepatology 2001; 34:417–423.

Lechner, F., Wong, D.K., Dunbar, P.R., Chapman, R., Chung, R.T., Dohrenwend, P., Robbins, G., Phillips, R., Klenerman, P. and Walker, B.D. Analysis of successful immune responses in persons infected with hepatitis C virus. J. Exp. Med. 2000; 191:1499–1512.

Lehmann, M., Meyer, M.F., Monazahian, M., Tillmann, H.L., Manns, M.P. and Wedemeyer, H. High rate of spontaneous clearance of acute hepatitis C virus genotype 3 infection. J. Med. Virol. 2004; 73:387–391.

Lemon, S.M. and Honda, M. Internal ribosome entry sites within the RNA genomes of hepatitis C virus and other flaviviruses. Semin. Virol. 1997; 8:274–288.

Lemon, S.M., Walker, C., Alter, M.J. and Yi, M.K. Hepatitis C virus. In: Field's Virology (D.M. Knipe, P.M. Howley, D.E. Griffin, R.A. Lamb, M.A. Malcolm, B. Roizman and S.E. Straus, Eds.). Lippincott Williams & Wilkins, 2007; pp.1253–1304.

Leroux-Roels, G., Depla, E., Hulstaert, F., Tobback, L., Dincq, S., Desmet, J., Desombere, I. and Maertens, G. A candidate vaccine based on the hepatitis C E1 protein: tolerability and immunogenicity in healthy volunteers. Vaccine 2004; 22:3080–3086.

Li, K., Foy, E., Ferreon, J.C., Nakamura, M., Ferreon, A.C.M., Ikeda, M., Ray, S.C., Gale, M., Jr. and Lemon, S.M. Immune evasion by hepatitis C virus NS3/4A protease-mediated cleavage of the TLR3 adaptor protein TRIF. Proc. Natl. Acad. Sci. USA 2005a; 102:2992–2997.

Li, W., Krishnadas, D.K., Kumar, R., Tyrrell, D.L. and Agrawal, B. Priming and stimulation of hepatitis C virus-specific CD4+ and CD8+ T cells against HCV antigens NS4, NS5a or NS5b from HCV-naive individuals: implications for prophylactic vaccine. Int. Immunol. 2008; 20:89–104.

Li, X.D., Sun, L., Seth, R.B., Pineda, G. and Chen, Z.J. Hepatitis C virus protease NS3/4A cleaves mitochondrial antiviral signaling protein off the mitochondria to evade innate immunity. Proc. Natl. Acad. Sci. USA 2005b; 102:17717–17722.

Liao, X.L., Ren, H., Zhao, P., Zhu, S.Y., Cao, J., Chen, Z.H., Zhao, L.J., Pan, W., Feitelson, M.A. and Qui, Z.T. Efficient induction of mouse immune responses to hepatitis C virus by viral core protein-carrying attenuated *Salmonella typhimurium*. Viral Immunol. 2007; 20:469–478.

Lin, C., Thomson, J.A. and Rice, C.M. A central region in the hepatitis C virus NS4A protein allows formation of an active

NS3–NS4A serine proteinase complex in vivo and in vitro. J. Virol. 1995; 69:4373–4380.
Lindenbach, B.D., Evans, M.J., Syder, A.J., Wolk, B., Tellinghuisen, T.L., Liu, C.C., Maruyama, T., Hynes, R.O., Burton, D.R., McKeating, J.A. and Rice, C.M. Complete replication of hepatitis C virus in cell culture. Science 2005; 309:623–626.
Lindenbach, B.D., Meuleman, P., Ploss, A., Vanwolleghem, T., Syder, A.J., McKeating, J.A., Lanford, R.E., Feinstone, S.M., Major, M.E., Leroux-Roels, G. and Rice, C.M. Cell culture-grown hepatitis C virus is infectious in vivo and can be recultured in vitro. Proc. Natl. Acad. Sci. USA 2006; 103:3805–3809.
Lindenbach, B.D. and Rice, C.M. Unravelling hepatitis C virus replication from genome to function. Nature 2005; 436:933–938.
Logvinoff, C., Major, M.E., Oldach, D., Heyward, S., Talal, A., Balfe, P., Feinstone, S.M., Alter, H., Rice, C.M. and McKeating, J.A. Neutralizing antibody response during acute and chronic hepatitis C virus infection. Proc. Natl. Acad. Sci. USA 2004; 101:10149–10154.
Lohmann, V., Korner, F., Koch, J., Herian, U., Theilmann, L. and Bartenschlager, R. Replication of subgenomic hepatitis C virus RNAs in a hepatoma cell line. Science 1999; 285:110–113.
Loo, Y.M., Owen, D.M., Li, K., Erickson, A.L., Johnson, C.L., Fish, P., Carney, D.S., Wang, T., Ishida, H., Yoneyama, M., Fujita, T., Saito, T., Lee, W.M., Hagedorn, C.H., Lau, D.T., Weinman, S.A., Lemon, S.M. and Gale, M., Jr. Viral and therpeutic control of interferon beta promoter stimulator 1 during hepatitis C virus infection. Proc. Natl. Acad. Sci. USA 2006; 103:6001–6006.
Lorenz, I.C., Marcotrigiano, J., Dentzer, T.G. and Rice, C.M. Structure of the catalytic domain of the hepatitis C virus NS2-3 protease. Nature 2006; 442:831–835.
Maertens, G., Priem, S., Ducatteeuw, A., Verschoorl, E., Verstrepen, B., Roskams, T., Desmet, V., Fuller, S., Van, H.K., Vandeponseele, P., Bosman, F., Buyse, M.A., Van Doorn, L.J., Heeney, J., Kos, A. and Depla, E. Improvement of chronic active hepatitis C in chronically infected chimpanzees after therapeutic vaccination with the HCV E1 protein. Acta Gastroenterol. Belg. 2000; 63:203.
Majid, A.M., Ezelle, H., Shah, S. and Barber, G.N. Evaluating replication-defective vesicular stomatitis virus as a vaccine vehicle. J. Virol. 2006; 80:6993–7008.
Major, M.E., Mihalik, K., Puig, M., Rehermann, B., Nascimbeni, M., Rice, C.M. and Feinstone, S.M. Previously infected and recovered chimpanzees exhibit rapid responses that control hepatitis C virus replication upon rechallenge. J. Virol. 2002; 76:6586–6595.
Marcellin, P., Descamps, V., Martinot-Peignoux, M., Larzul, D., Xu, L., Boyer, N., Pham, B.-N., Crickx, B., Guillevin, L., Belaich, S., Erlinger, S. and Benhamou, J.-P. Cryoglobulinemia with vasculitis associated with hepatitis C virus infection. Gastroenterology 1993; 104:272–277.
Mast, E.E., Hwang, L.Y., Seto, D.S., Nolte, F.S., Nainan, O.V., Wurtzel, H. and Alter, M.J. Risk factors for perinatal transmission of hepatitis C virus (HCV) and the natural history of HCV infection acquired in infancy. J. Infect. Dis. 2005; 192:1880–1889.
Matsui, M., Moriya, O., bdel-Aziz, N., Matsuura, Y., Miyamura, T. and Akatsuka, T. Induction of hepatitis C virus-specific cytotoxic T lymphocytes in mice by immunization with dendritic cells transduced with replication-defective recombinant adenovirus. Vaccine 2002; 21:211–220.
McHutchison, J.G. and Fried, M.W. Current therapy for hepatitis C: pegylated interferon and ribavirin. Clin. Liver Dis. 2003; 7:149–161.
McLauchlan, J., Lemberg, M.K., Hope, G. and Martoglio, B. Intramembrane proteolysis promotes trafficking of hepatitis C virus core protein to lipid droplets. EMBO J. 2002; 21:3980–3988.
Mercer, D.F., Schiller, D.E., Elliott, J.F., Douglas, D.N., Hao, C., Rinfret, A., Addison, W.R., Fischer, K.P., Churchill, T.A., Lakey, J.R., Tyrrell, D.L. and Kneteman, N.M. Hepatitis C virus replication in mice with chimeric human livers. Nat. Med. 2001; 7:927–933.
Meunier, J.C., Engle, R.E., Faulk, K., Zhao, M., Bartosch, B., Alter, H., Emerson, S.U., Cosset, F.L., Purcell, R.H. and Bukh, J. Evidence for cross-genotype neutralization of hepatitis C virus pseudo-particles and enhancement of infectivity by apolipoprotein C1. Proc. Natl. Acad. Sci. USA 2005; 102:4560–4565.
Meylan, E., Curran, J., Hofmann, K., Moradpour, D., Binder, M., Bartenschlager, R. and Tschopp, J. Cardif is an adaptor protein in the RIG-I antiviral pathway and is targeted by hepatitis C virus. Nature 2005; 437:1167–1172.
Miyanari, Y., Atsuzawa, K., Usuda, N., Watashi, K., Hishiki, T., Zayas, M., Bartenschlager, R., Wakita, T., Hijikata, M. and Shimotohno, K. The lipid droplet is an important organelle for hepatitis C virus production. Nat. Cell Biol. 2007; 9:1089–1097.
Moradpour, D., Gosert, R., Egger, D., Penin, F., Blum, H.E. and Bienz, K. Membrane association of hepatitis C virus nonstructural proteins and identification of the membrane alteration that harbors the viral replication complex. Antiviral Res. 2003; 60:103–109.
Murata, K., Lechmann, M., Qiao, M., Gunji, T., Alter, H.J. and Liang, T.J. Immunization with hepatitis C virus-like particles protects mice from recombinant hepatitis C virus-vaccinia infection. Proc. Natl. Acad. Sci. USA 2003; 100:6753–6758.
Nakano, I., Maertens, G., Major, M.E., Vitvitski, L., Dubuisson, J., Fournillier, A., De, M.G., Trepo, C. and Inchauspe, G. Immunization with plasmid DNA encoding hepatitis C virus envelope E2 antigenic domains induces antibodies whose immune reactivity is linked to the injection mode. J. Virol. 1997; 71:7101–7109.
Netski, D.M., Mosbruger, T., Depla, E., Maertens, G., Ray, S.C., Hamilton, R.G., Roundtree, S., Thomas, D.L., McKeating, J. and Cox, A. Humoral immune response in acute hepatitis C virus infection. Clin. Infect. Dis. 2005; 41:667–675.
Nevens, F., Roskams, T., Van, V.H., Horsmans, Y., Sprengers, D., Elewaut, A., Desmet, V., Leroux-Roels, G., Quinaux, E., Depla, E., Dincq, S., Vander, S.C., Maertens, G. and Hulstaert, F. A pilot study of therapeutic vaccination with envelope protein E1 in 35 patients with chronic hepatitis C. Hepatology 2003; 38:1289–1296.
Nielsen, S.U., Bassendine, M.F., Burt, A.D., Martin, C., Pumeechockchai, W. and Toms, G.L. Association between hepatitis C virus and very-low-density lipoprotein (VLDL)/LDL analyzed in iodixanol density gradients. J. Virol. 2006; 80:2418–2428.
O'Hagan, D.T., Singh, M., Dong, C., Ugozzoli, M., Berger, K., Glazer, E., Selby, M., Wininger, M., Ng, P., Crawford, K., Paliard, X., Coates, S. and Houghton, M. Cationic microparticles are a potent delivery system for a HCV DNA vaccine. Vaccine 2004; 23:672–680.
Owsianka, A., Tarr, A.W., Juttla, V.S., Lavillette, D., Bartosch, B., Cosset, F.L., Ball, J.K. and Patel, A.H. Monoclonal antibody AP33 defines a broadly neutralizing epitope on the hepatitis C virus E2 envelope glycoprotein. J. Virol. 2005; 79:11095–11104.
Owsianka, A.M., Tarr, A.W., Keck, Z.Y., Li, T.K., Witteveldt, J., Adair, R., Foung, S.K., Ball, J.K. and Patel, A.H. Broadly neutralizing human monoclonal antibodies to the hepatitis C virus E2 glycoprotein. J. Gen. Virol. 2008; 89:653–659.
Owsianka, A.M., Timms, J.M., Tarr, A.W., Brown, R.J., Hickling, T.P., Szwejk, A., Bienkowska-Szewczyk, K., Thomson, B.J., Patel, A.H. and Ball, J.K. Identification of conserved residues in the E2 envelope glycoprotein of the hepatitis C virus that are critical for CD81 binding. J. Virol. 2006; 80:8695–8704.
Palekar, N.A. and Harrison, S.A. Extrahepatic manifestations of hepatitis C. South. Med. J. 2005; 98:1019–1023.

Pancholi, P., Liu, Q., Tricoche, N., Zhang, P., Perkus, M.E. and Prince, A.M. DNA prime-canarypox boost with polycistronic hepatitis C virus (HCV) genes generates potent immune responses to HCV structural and nonstructural proteins. J. Infect. Dis. 2000; 182:18–27.

Pancholi, P., Perkus, M., Tricoche, N., Liu, Q. and Prince, A.M. DNA immunization with hepatitis C virus (HCV) polycistronic genes or immunization by HCV DNA priming-recombinant canarypox virus boosting induces immune responses and protection from recombinant HCV-vaccinia virus infection in HLA-A2.1-transgenic mice. J. Virol. 2003; 77:382–390.

Penin, F., Dubuisson, J., Rey, F.A., Moradpour, D. and Pawlotsky, J.M. Structural biology of hepatitis C virus. Hepatology 2004; 39:5–19.

Perotti, M., Mancini, N., Diotti, R.A., Tarr, A.W., Ball, J.K., Owsianka, A., Adair, R., Patel, A.H., Clementi, M. and Burioni, R. Identification of a broadly cross-reacting and neutralizing human monoclonal antibody directed against the hepatitis C virus E2 protein. J. Virol. 2008; 82:1047–1052.

Perri, S., Greer, C.E., Thudium, K., Doe, B., Legg, H., Liu, H., Romero, R.E., Tang, Z., Bin, Q., Dubensky, T.W., Jr., Vajdy, M., Otten, G.R. and Polo, J.M. An alphavirus replicon particle chimera derived from Venezuelan equine encephalitis and sindbis viruses is a potent gene-based vaccine delivery vector. J. Virol. 2003; 77:10394–10403.

Piazza, M., Sagliocca, L., Tosone, G., Guadagnino, V., Stazi, M.A., Orlando, R., Borgia, G., Rosa, D., Abrignani, S., Palumbo, F., Manzin, A. and Clementi, M. Prophylaxis of hepatitis C with intramuscular immunoglobulin: clinical and economic appraisal. BioDrugs 1999; 12:291–300.

Pietschmann, T., Kaul, A., Koutsoudakis, G., Shavinskaya, A., Kallis, S., Steinmann, E., Abid, K., Negro, F., Dreux, M., Cosset, F.L. and Bartenschlager, R. Construction and characterization of infectious intragenotypic and intergenotypic hepatitis C virus chimeras. Proc. Natl. Acad. Sci. USA 2006; 103:7408–7413.

Pileri, P., Uematsu, Y., Campagnoli, S., Galli, G., Falugi, F., Petracca, R., Weiner, A.J., Houghton, M., Rosa, D., Grandi, G. and Abrignani, S. Binding of hepatitis C virus to CD81. Science 1998; 282:938–941.

Polakos, N.K., Drane, D., Cox, J., Ng, P., Selby, M.J., Chien, D., O'Hagan, D.T., Houghton, M. and Paliard, X. Characterization of hepatitis C virus core-specific immune responses primed in rhesus macaques by a nonclassical ISCOM vaccine. J. Immunol. 2001; 166:3589–3598.

Prince, A.M. Immunity in hepatitis C virus infection. Vox Sang. 1994; 67(Suppl 3):227–228.

Puig, M., Major, M.E., Mihalik, K. and Feinstone, S.M. Immunization of chimpanzees with an envelope protein-based vaccine enhances specific humoral and cellular immune responses that delay hepatitis C virus infection. Vaccine 2004; 22:991–1000.

Qiao, M., Murata, K., Davis, A.R., Jeong, S.H. and Liang, T.J. Hepatitis C virus-like particles combined with novel adjuvant systems enhance virus-specific immune responses. Hepatology 2003; 37:52–59.

Racanelli, V., Behrens, S.E., Aliberti, J. and Rehermann, B. Dendritic cells transfected with cytopathic self-replicating RNA induce crosspriming of CD8+ T cells and antiviral immunity. Immunity 2004; 20:47–58.

Radziewicz, H., Ibegbu, C.C., Fernandez, M.L., Workowski, K.A., Obideen, K., Wehbi, M., Hanson, H.L., Steinberg, J.P., Masopust, D., Wherry, E.J., Altman, J.D., Rouse, B.T., Freeman, G.J., Ahmed, R. and Grakoui, A. Liver-infiltrating lymphocytes in chronic human hepatitis C virus infection display an exhausted phenotype with high levels of PD-1 and low levels of CD127 expression. J. Virol. 2007; 81:2545–2553.

Ralston, R., Thudium, K., Berger, K., Kuo, C., Gervase, B., Hall, J., Selby, M., Kuo, G., Houghton, M. and Choo, Q.L. Characterization of hepatitis C virus envelope glycoprotein complexes expressed by recombinant vaccinia viruses. J. Virol. 1993; 67:6753–6761.

Rauch, A., Rickenbach, M., Weber, R., Hirschel, B., Tarr, P.E., Bucher, H.C., Vernazza, P., Bernasconi, E., Zinkernagel, A.S., Evison, J. and Furrer, H. Unsafe sex and increased incidence of hepatitis C virus infection among HIV-infected men who have sex with men: the Swiss HIV Cohort Study. Clin. Infect. Dis. 2005; 41:395–402.

Rehermann, B. and Nascimbeni, M. Immunology of hepatitis B virus and hepatitis C virus infection. Nat. Rev. Immunol. 2005; 5:215–229.

Roberts, E.A. and Yeung, L. Maternal–infant transmission of hepatitis C virus infection. Hepatology 2002; 36:S106–S113.

Roccasecca, R., Ansuini, H., Vitelli, A., Meola, A., Scarselli, E., Acali, S., Pezzanera, M., Ercole, B.B., McKeating, J., Yagnik, A., Lahm, A., Tramontano, A., Cortese, R. and Nicosia, A. Binding of the hepatitis C virus E2 glycoprotein to CD81 is strain specific and is modulated by a complex interplay between hypervariable regions 1 and 2. J. Virol. 2003; 77:1856–1867.

Roccasecca, R., Folgori, A., Ercole, B.B., Puntoriero, G., Lahm, A., Zucchelli, S., Tafi, R., Pezzanera, M., Galfre, G., Tramontano, A., Mondelli, M.U., Pessi, A., Nicosia, A., Cortese, R. and Meola, A. Mimotopes of the hyper variable region 1 of the hepatitis C virus induce cross-reactive antibodies directed against discontinuous epitopes. Mol. Immunol. 2001; 38:485–492.

Rollier, C., Depla, E., Drexhage, J.A., Verschoor, E.J., Verstrepen, B.E., Fatmi, A., Brinster, C., Fournillier, A., Whelan, J.A., Whelan, M., Jacobs, D., Maertens, G., Inchauspe, G. and Heeney, J.L. Control of heterologous hepatitis C virus infection in chimpanzees is associated with the quality of vaccine-induced peripheral T-helper immune response. J. Virol. 2004; 78:187–196.

Rollier, C., Verschoor, E.J., Paranhos-Baccala, G., Drexhage, J.A., Verstrepen, B.E., Berland, J.L., Himoudi, N., Barnfield, C., Liljestrom, P., Lasarte, J.J., Ruiz, J., Inchauspe, G. and Heeney, J.L. Modulation of vaccine-induced immune responses to hepatitis C virus in rhesus macaques by altering priming before adenovirus boosting. J. Infect. Dis. 2005; 192:920–929.

Rollier, C.S., Paranhos-Baccala, G., Verschoor, E.J., Verstrepen, B.E., Drexhage, J.A., Fagrouch, Z., Berland, J.L., Komurian-Pradel, F., Duverger, B., Himoudi, N., Staib, C., Meyr, M., Whelan, M., Whelan, J.A., Adams, V.C., Larrea, E., Riezu, J.I., Lasarte, J.J., Bartosch, B., Cosset, F.L., Spaan, W.J., Diepolder, H.M., Pape, G.R., Sutter, G., Inchauspe, G. and Heeney, J.L. Vaccine-induced early control of hepatitis C virus infection in chimpanzees fails to impact on hepatic PD-1 and chronicity. Hepatology 2007; 45:602–613.

Saito, I., Miyamura, T., Ohbayashi, A., Harada, H., Katayama, T., Kikuchi, S., Watanabe, Y., Koi, S., Onji, M., Ohta, Y., Choo, Q.-L., Houghton, M. and Kuo, G. Hepatitis C virus infection is associated with the development of hepatocellular carcinoma. Proc. Natl. Acad. Sci. USA 1991; 87:6547–6549.

Satoi, J., Murata, K., Lechmann, M., Manickan, E., Zhang, Z., Wedemeyer, H., Rehermann, B. and Liang, T.J. Genetic immunization of wild-type and hepatitis C virus transgenic mice reveals a hierarchy of cellular immune response and tolerance induction against hepatitis C virus structural proteins. J. Virol. 2001; 75:12121–12127.

Schiano, T.D., Charlton, M., Younossi, Z., Galun, E., Pruett, T., Tur-Kaspa, R., Eren, R., Dagan, S., Graham, N., Williams, P. V. and Andrews, J. Monoclonal antibody HCV-AbXTL68 in patients undergoing liver transplantation for HCV: results of a phase 2 randomized study. Liver Transpl. 2006; 12:1381–1389.

Schlaphoff, V., Klade, C.S., Jilma, B., Jelovcan, S.B., Cornberg, M., Tauber, E., Manns, M.P. and Wedemeyer, H. Functional and phenotypic characterization of peptide-vaccine-induced HCV-specific CD8+ T cells in healthy individuals and chronic hepatitis C patients. Vaccine 2007; 25:6793–6806.

Schulz, O., Diebold, S.S., Chen, M., Naslund, T.I., Nolte, M.A., Alexopoulou, L., Azuma, Y.T., Flavell, R.A. and Liljestrom, P.Reis e Sousa Toll-like receptor 3 promotes cross-priming to virus-infected cells. Nature 2005; 433:887–892.

Seeff, L.B., Buskell-Bales, Z., Wright, E.C., Durako, S.J., Alter, H.J., Iber, F.L., Hollinger, F.B., Gitnick, G., Knodell, R.G., Perrillo, R.P., Stevens, C.E. and Hollingsworth, C.G.NHLBI Study Group Long-term mortality after transfusion-associated non-A, non-B hepatitis. N. Engl. J. Med. 1992; 327:1906–1911.

Seth, R.B., Sun, L., Ea, C.K. and Chen, Z.J. Identification and characterization of MAVS, a mitochondrial antiviral signaling protein that activates NF-kappaB and IRF 3. Cell 2005; 122:669–682.

Shepard, C.W., Finelli, L. and Alter, M.J. Global epidemiology of hepatitis C virus infection. Lancet Infect. Dis. 2005; 5:558–567.

Shi, L., Liu, S., Fan, G.X., Sheng, L., Ren, H.X. and Yuan, Y.K. Effective induction of type 1 cytotoxic T cell responses in mice with DNA vaccine encoding two hepatitis C virus cytotoxic T lymphocyte epitopes. Viral Immunol. 2006; 19:702–711.

Shi, S.T. and Lai, M.M. Hepatitis C viral RNA: challenges and promises. Cell Mol. Life Sci. 2001; 58:1276–1295.

Shimizu, Y.K., Hijikata, M., Iwamoto, A., Alter, H.J., Purcell, R.H. and Yoshikura, H. Neutralizing antibodies against hepatitis C virus and the emergence of neutralization escape mutant viruses. J. Virol. 1994; 68:1494–1500.

Shimizu, Y.K., Iwamoto, A., Hijikata, M., Purcell, R.H. and Yoshikura, H. Evidence for in vitro replication of hepatitis C virus genome in a human T-cell line. Proc. Natl. Acad. Sci. USA 1992; 89:5477–5481.

Shimizu, Y.K. and Yoshikura, H. Multicycle infection of hepatitis C virus in cell culture and inhibition by alpha and beta interferons. J. Virol. 1994; 68:8406–8408.

Shoukry, N.H., Cawthon, A.G. and Walker, C.M. Cell-mediated immunity and the outcome of hepatitis C virus infection. Annu. Rev. Microbiol. 2004; 58:391–424.

Shoukry, N.H., Grakoui, A., Houghton, M., Chien, D.Y., Ghrayeb, J., Reimann, K.A. and Walker, C.M. Memory CD8+ T cells are required for protection from persistent hepatitis C virus infection. J. Exp. Med. 2003; 197:1645–1655.

Simmonds, P. Genetic diversity and evolution of hepatitis C virus—15 years on. J. Gen. Virol. 2004; 85:3173–3188.

Simmonds, P., Bukh, J., Combet, C., Deleage, G., Enomoto, N., Feinstone, S., Halfon, P., Inchauspe, G., Kuiken, C., Maertens, G., Mizokami, M., Murphy, D.G., Okamoto, H., Pawlotsky, J.M., Penin, F., Sablon, E., Shin, I., Stuyver, L.J., Thiel, H.J., Viazov, S., Weiner, A.J. and Widell, A. Consensus proposals for a unified system of nomenclature of hepatitis C virus genotypes. Hepatology 2005; 42:962–973.

Simon, B.E., Cornell, K.A., Clark, T.R., Chou, S., Rosen, H.R. and Barry, R.A. DNA vaccination protects mice against challenge with *Listeria monocytogenes* expressing the hepatitis C virus NS3 protein. Infect. Immun. 2003; 71:6372–6380.

Stamataki, Z., Coates, S., Evans, M.J., Wininger, M., Crawford, K., Dong, C., Fong, Y.L., Chien, D., Abrignani, S., Balfe, P., Rice, C.M., McKeating, J.A. and Houghton, M. Hepatitis C virus envelope glycoprotein immunization of rodents elicits cross-reactive neutralizing antibodies. Vaccine 2007; 25:7773–7784.

Strader, D.B., Wright, T., Thomas, D.L. and Seeff, L.B. Diagnosis, management, and treatment of hepatitis C. Hepatology 2004; 39:1147–1171.

Su, A.I., Pezacki, J.P., Wodicka, L., Brideau, A.D., Supekova, L., Thimme, R., Wieland, S., Bukh, J., Purcell, R.H., Schultz, P.G. and Chisari, F.V. Genomic analysis of the host response to hepatitis C virus infection. Proc. Natl. Acad. Sci. USA 2002; 99:15669–15674.

Tedeschi, V., Akatsuka, T., Shih, J.W., Battegay, M. and Feinstone, S.M. A specific antibody response to HCV E2 elicited in mice by intramuscular inoculation of plasmid DNA containing coding sequences for E2. Hepatology 1997; 25:459–462.

Tellinghuisen, T.L., Marcotrigiano, J. and Rice, C.M. Structure of the zinc-binding domain of an essential component of the hepatitis C virus replicase. Nature 2005; 435:374–379.

Thimme, R., Bukh, J., Spangenberg, H.C., Wieland, S., Pemberton, J., Steiger, C., Govindarajan, S., Purcell, R.H. and Chisari, F.V. Viral and immunological determinants of hepatitis C virus clearance, persistence, and disease. Proc. Natl. Acad. Sci. USA 2002; 99:15661–15668.

Thimme, R., Oldach, D., Chang, K.M., Steiger, C., Ray, S.C. and Chisari, F.V. Determinants of viral clearance and persistence during acute hepatitis C virus infection. J. Exp. Med. 2001; 194:1395–1406.

Torresi, J., Stock, O.M., Fischer, A.E., Grollo, L., Drummer, H., Boo, I., Zeng, W., Earnest-Silveira, L. and Jackson, D.C. A self-adjuvanting multiepitope immunogen that induces a broadly cross-reactive antibody to hepatitis C virus. Hepatology 2007; 45:911–920.

Vidalin, O., Fournillier, A., Renard, N., Chen, M., Depla, E., Boucreux, D., Brinster, C., Baumert, T., Nakano, I., Fukuda, Y., Liljestrom, P., Trepo, C. and Inchauspe, G. Use of conventional or replicating nucleic acid-based vaccines and recombinant Semliki forest virus-derived particles for the induction of immune responses against hepatitis C virus core and E2 antigens. Virology 2000; 276:259–270.

Vietheer, P.T., Boo, I., Drummer, H.E. and Netter, H.J. Immunizations with chimeric hepatitis B virus-like particles to induce potential anti-hepatitis C virus neutralizing antibodies. Antivir. Ther. 2007; 12:477–487.

von Hahn, T., Yoon, J.C., Alter, H., Rice, C.M., Rehermann, B., Balfe, P. and McKeating, J.A. Hepatitis C virus continuously escapes from neutralizing antibody and T-cell responses during chronic infection in vivo. Gastroenterology 2007; 132:667–678.

Vong, S. and Bell, B.P. Chronic liver disease mortality in the United States, 1990–1998. Hepatology 2004; 39:476–483.

Wakita, T., Pietschmann, T., Kato, T., Date, T., Miyamoto, M., Zhao, Z., Murthy, K., Habermann, A., Krauslich, H.-G., Mizokami, M., Bartenschlager, R. and Liang, T.J. Production of infectious hepatitis C virus in tissue culture from a cloned viral genome. Nat. Med. 2005; 11:791–796.

Ward, S., Lauer, G., Isba, R., Walker, B. and Klenerman, P. Cellular immune responses against hepatitis C virus: the evidence base 2002. Clin. Exp. Immunol. 2002; 128:195–203.

Wedemeyer, H., Gagneten, S., Davis, A., Bartenschlager, R., Feinstone, S. and Rehermann, B. Oral immunization with HCV-NS3-transformed *Salmonella*: induction of HCV-specific CTL in a transgenic mouse model. Gastroenterology 2001; 121:1158–1166.

Wedemeyer, H., He, X.S., Nascimbeni, M., Davis, A.R., Greenberg, H.B., Hoofnagle, J.H., Liang, T.J., Alter, H. and Rehermann, B. Impaired effector function of hepatitis C virus-specific CD8+ T cells in chronic hepatitis C virus infection. J. Immunol. 2002; 169:3447–3458.

Weiner, A.J., Paliard, X., Selby, M.J., Medina-Selby, A., Coit, D., Nguyen, S., Kansopon, J., Arian, C.L., Ng, P., Tucker, J., Lee, C.T., Polakos, N.K., Han, J., Wong, S., Lu, H.H., Rosenberg, S., Brasky, K.M., Chien, D., Kuo, G. and Houghton, M. Intrahepatic genetic inoculation of hepatitis C virus RNA confers cross-protective immunity. J. Virol. 2001; 75:7142–7148.

Williams, I.T., Perz, J.F. and Bell, B.P. Viral hepatitis transmission in ambulatory health care settings. Clin. Infect. Dis. 2004; 38:1592–1598.

Woollard, D.J., Grakoui, A., Shoukry, N.H., Murthy, K.K., Campbell, K.J. and Walker, C.M. Characterization of HCV-specific Patr class II restricted CD4+ T cell responses in an acutely infected chimpanzee. Hepatology 2003; 38:1297–1306.

Wuest, T., Both, G.W., Prince, A.M., Hofmann, C. and Loser, P. Recombinant ovine atadenovirus induces a strong and sustained T cell response against the hepatitis C virus NS3 antigen in mice. Vaccine 2004; 22:2717–2721.

Yanagi, M., Purcell, R.H., Emerson, S.U. and Bukh, J. Transcripts from a single full-length cDNA clone of hepatitis C virus are infectious when directly transfected into liver of a chimpanzee. Proc. Natl. Acad. Sci. USA 1997; 97:8738–8743.

Yao, N.H., Hesson, T., Cable, M., Hong, Z., Kwong, A.D., Le, H.V. and Weber, P.C. Structure of the hepatitis C virus RNA helicase domain. Nat. Struct. Biol. 1997; 4:463–467.

Yi, M., Ma, Y., Yates, J. and Lemon, S.M. Compensatory mutations in E1, p7, NS2 and NS3 enhance yields of cell culture-infectious inter-genotypic chimeric hepatitis C virus. J. Virol. 2007; 81:629–638.

Yi, M., Villanueva, R.A., Thomas, D.L., Wakita, T. and Lemon, S.M. Production of infectious genotype 1a hepatitis C virus (Hutchinson strain) in cultured human hepatoma cells. Proc. Natl. Acad. Sci. USA 2006; 103:2310–2315.

Youn, J.W., Park, S.H., Lavillette, D., Cosset, F.L., Yang, S.H., Lee, C.G., Jin, H.T., Kim, C.M., Shata, M.T., Lee, D.H., Pfahler, W., Prince, A.M. and Sung, Y.C. Sustained E2 antibody response correlates with reduced peak viremia after hepatitis C virus infection in the chimpanzee. Hepatology 2005; 42:1429–1436.

Yu, M.Y., Bartosch, B., Zhang, P., Guo, Z.P., Renzi, P.M., Shen, L.M., Granier, C., Feinstone, S.M., Cosset, F.L. and Purcell, R.H. Neutralizing antibodies to hepatitis C virus (HCV) in immune globulins derived from anti-HCV-positive plasma. Proc. Natl. Acad. Sci. USA 2004; 101:7705–7710.

Yutani, S., Yamada, A., Yoshida, K., Takao, Y., Tamura, M., Komatsu, N., Ide, T., Tanaka, M., Sata, M. and Itoh, K. Phase I clinical study of a personalized peptide vaccination for patients infected with hepatitis C virus (HCV) 1b who failed to respond to interferon-based therapy. Vaccine 2007; 25:7429–7435.

Zabaleta, A., Llopiz, D., Arribillaga, L., Silva, L., Riezu-Boj, J.I., Lasarte, J.J., Borras-Cuesta, F., Prieto, J. and Sarobe, P. Vaccination against hepatitis C virus with dendritic cells transduced with an adenovirus encoding NS3 protein. Mol. Ther. 2008; 16:210–217.

Zhong, J., Gastaminza, P., Cheng, G., Kapadia, S., Kato, T., Burton, D.R., Wieland, S.F., Uprichard, S.L., Wakita, T. and Chisari, F.V. Robust hepatitis C virus infection in vitro. Proc. Natl. Acad. Sci. USA 2005; 102:9294–9299.

Zucchelli, S., Capone, S., Fattori, E., Folgori, A., Di, M.A., Casimiro, D., Simon, A.J., Laufer, R., La, M.N., Cortese, R. and Nicosia, A. Enhancing B- and T-cell immune response to a hepatitis C virus E2 DNA vaccine by intramuscular electrical gene transfer. J. Virol. 2000; 74:11598–11607.

CHAPTER

25

Human Immunodeficiency Virus

Farah Cassis-Ghavami, Marcel Curlin, Robert Geise and Ann Duerr
HIV Vaccine Trials Network, Seattle, WA, USA

OUTLINE

ABSTRACT

Human immunodeficiency virus (HIV) is one of the greatest worldwide public health challenges of the last century. Since being identified over 20 years ago, HIV has claimed an estimated 25 million lives. Currently, an estimated 33 million individuals are living with HIV/AIDS. Although it causes infections worldwide, this virus has especially targeted areas of the developing world, with prevalence rates nearing 50% among women of child-bearing age in some areas of sub-Saharan Africa.

Primary infection may be characterized by an acute viral syndrome or may be entirely asymptomatic, and individuals are often unaware of their infection. Symptomatic illness usually occurs several years after infection, and is manifested by significant-to-severe immune suppression. Although antiretroviral therapy (ART) is generally effective at suppressing viral replication, treatment is not universally available and is often associated with serious side effects. Also, due to the high rate of mutation during viral replication, ART may become ineffective in noncompliant individuals.

The structure, genetics, and replication characteristics of HIV make it a challenging pathogen. HIV is a remarkably diverse virus, with two major types, and multiple subtypes and recombinant forms circulating worldwide. The viral envelope varies considerably from isolate to isolate, and has few conserved regions that can be effectively targeted by host antibody responses. Glycosylation of protein structures on the envelope coating hinder access by neutralizing antibodies, and widespread mutational change within the genome permits escape from cellular immune mechanisms. HIV preferentially infects activated host immune cells, which are diverted from their normal cellular biosynthetic pathways to produce virus particles, and undergo premature apoptosis. However, infected CD4+ T cells may also remain transcriptionally silent, leaving the incorporated proviral HIV genome dormant for many years. This results in a reservoir of infected cells that persists despite apparently effective therapy.

The development of an HIV vaccine that is protective and easily and economically deliverable is a daunting endeavor for scientists, public health officials, and government agencies. The field of HIV vaccine development has met with a number of recent disappointments. Both the VAXGEN antibody-based vaccine and the Merck adenovirus T-cell-stimulating vaccine showed no efficacy in protecting from infection or reducing viral loads. In fact, the Merck product, tested in the Americas and South Africa, may have led to an increased susceptibility to HIV infection in individuals with evidence of preexisting serological immunity to the adenovirus vector.

A new paradigm of HIV vaccine effectiveness may need to be considered. This paradigm includes vaccines that may: (1) prevent infection; (2) allow infection that is subsequently cleared without clinical disease; (3) delay clinical progression in the vaccinated individual; or (4) minimally impact disease in the infected individual, but reduce infection of others. Several new approaches are actively being tested in HIV vaccine development. DNA and peptide-based vaccines, heterologous prime-boost regimens, and alternative viral vector are under consideration and development. Scientists continue to use many different methodologies to optimize immunogenic HIV insert sequences in order to overcome the tremendous variability presented by potential infecting viruses. Other approaches seek to increase the recognition of viral antigens through the use of adjuvants and optimized modes of immunogen delivery. The next decade will provide opportunities for these hurdles to be overcome, and will likely see the emergence of new challenges as second- and third-generation vaccines are developed. Multidisciplinary approaches to vaccination may ultimately lead to complete control of this pandemic.

INTRODUCTION

The ongoing worldwide epidemic caused by the human immunodeficiency virus type 1 (HIV-1) remains a major global health challenge. HIV-1 causes the acquired immunodeficiency syndrome (AIDS), which is responsible for tremendous human suffering and economic loss throughout the world. According to The Joint United Nations Program on HIV/AIDS (UNAIDS), in 2007 the AIDS epidemic claimed more than 2 million lives, and an estimated 2.5 million people acquired HIV (UNAIDS, 2007). The number of people living with the virus globally is approximately 33 million. Without wider access to antiretroviral therapy (ART), it is likely that nearly all of these people will die of AIDS in the next two decades. While recent efforts are moving forward to provide treatment and HIV-related care to those most in need in southern Africa, the pandemic continues to spread, encompassing such areas as India, China, and the former Soviet Union. Current projections suggest that an additional 45 million people will become infected with HIV between 2004 and 2010 unless the world succeeds in mounting a drastically expanded global prevention effort (UNAIDS, 2004). AIDS has become the leading infectious disease killer, and the fourth leading cause of death overall. In severely affected countries, life expectancy has fallen by more than 10 years (Dorling et al., 2006).

Since 1996, potent new ARTs, including combination regimens with protease inhibitors, have shown that HIV-1 infection can become a chronic, manageable disease among individuals with access to these medications. However, an effective vaccine remains our best chance of ultimately curtailing this epidemic.

HISTORY OF HIV

Discovery of HIV

AIDS was first recognized in the United States in 1981 following a sudden outbreak of opportunistic infections, including Pneumocystis carinii pneumonia and Kaposi's sarcoma (KS) among homosexual men (Durack, 1981; Gottlieb et al., 1981; Masur et al., 1981). Subsequent epidemiologic studies implicated an infectious agent that was transmitted during sexual intercourse, through intravenous drug abuse, by therapies utilizing blood and blood products, and vertically from mother to child (Broder et al., 1994).

In 1983, Barre-Sinoussi, Chermann, and Montagier at the Pasteur Institute in Paris isolated a retrovirus from lymph node cells of a patient with lymphadenopathy; accordingly, this virus was designated lymphadenopathy-associated virus (LAV) (Wain-Hobson et al., 1991; Barre-Sinoussi et al., 1983). Several other laboratories were also searching for the agent responsible for AIDS. In 1984, Gallo and associates reported the characterization of another human retrovirus distinct from human T-lymphotropic virus (HTLV) that they called HTLV-III (Gallo et al., 1984; Popovic et al., 1984; Sarngadharan et al., 1984; Schupbach et al., 1984). Levy et al. (1984) simultaneously reported the identification of retroviruses that they named the AIDS-associated retroviruses (ARVs). Within a short time, the three prototype viruses (LAV, HTLV-III, and ARV) were recognized as members of the same group of retroviruses, and their properties identified them as lentiviruses. Their proteins were all distinct from those of HTLV (Rabson and Martin, 1985). The AIDS viruses had many properties distinguishing them from HTLV. For all these reasons, in 1986, the International Committee on Taxonomy of Viruses recommended giving the AIDS virus a separate name, human immunodeficiency virus (Coffin et al., 1986).

HIV isolates were subsequently recovered from the blood of many patients with AIDS, AIDS-related complex, and neurologic syndromes, as well as from the peripheral blood mononuclear cells (PBMCs) of several clinically healthy individuals (Salahuddin et al., 1985; Levy and Shimabukuro, 1985; Levy et al., 1985). The widespread transmission of this agent was appreciated and its close association with AIDS and related illnesses strongly supported its role in these diseases. Soon after the discovery of HIV-1, a separate epidemic caused by a related human retrovirus, HIV-2, was identified in West Africa (Clavel et al., 1987). Both HIV subtypes can lead to AIDS, although disease progression due to HIV-2 infection appears to be slower (Reeves and Doms, 2002; Whittle et al., 1994; Kanki et al., 1992).

HIV CLASSIFICATION AND STRUCTURE

HIV Structure

HIV is a member of the genus *Lentivirus* in the Retroviridae family. Retroviruses are so called

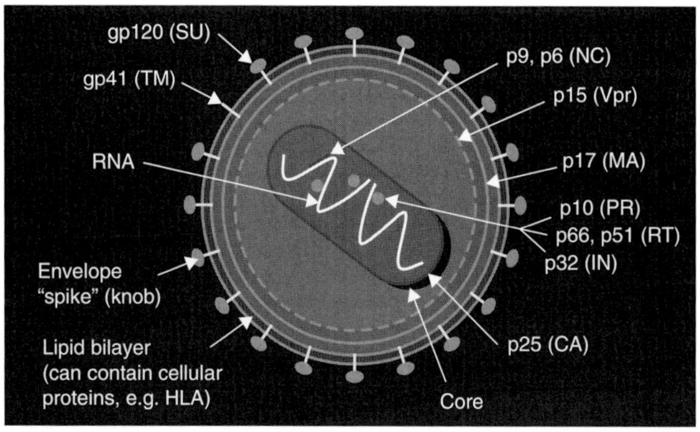

FIGURE 25.1 HIV virus structure. Reproduced with permission from Levy (2007).

because their RNA genome is transcribed into DNA within the cell using the viral enzyme reverse transcriptase (RT). This DNA then enters the nucleus and integrates into the cellular chromosome. A mature virion has a spherical shape and consists of an outer lipid bilayer membrane or envelope that surrounds the cone-shaped nucleocapsid (Fig. 25.1). The core appears to be attached to the viral outer envelope at its narrow end (Hoglund et al., 1992). The outer membrane contains approximately 10–20 spiked knobs (Zhu et al., 2003), which are assembled as trimers of the outer envelope protein gp120 bound to the transmembrane portion, gp41. The viral membrane is cholesterol-rich (Liao et al., 2001) and includes cellular proteins (Hoglund et al., 1992).

Each mature virion is composed of two molecules of single-stranded RNA surrounded by three *gag* gene cleavage products: the p17 outer matrix protein; the p24 major capsid protein, which forms the capsid shell; and the p7 nucleoprotein, which binds tightly to the viral RNA. The nucleocapsid is composed of a p17 myristoylated outer-matrix protein sandwiched between the membrane-bound envelope and the p24 core capsid protein. The p7 protein binds the two positive-strand copies of complete viral RNA attached at the packaging site, and it also binds to p24. A number of other viral proteins required for the early phases of infection are incorporated with the virion. These include protease, which is essential for viral assembly, and reverse transcriptase and integrase, which are needed after entry for viral DNA synthesis and integration (Mandell et al., 2005).

HIV Genome

The HIV genome is approximately 9.8 kb in length (Muesing et al., 1985). Both ends of the provirus are

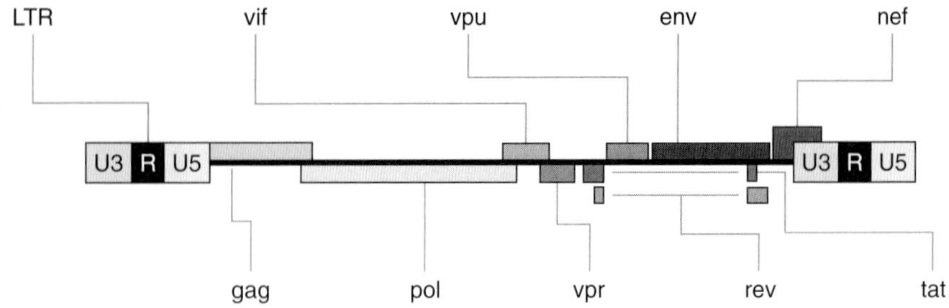

FIGURE 25.2 HIV genome. An overview of the organization of the approximately 9-kb genome of the HIV provirus. Reprinted with permission from Greene and Matija (2003).

flanked by a repeated sequence known as the long terminal repeat (LTR). The HIV proviral genome encodes at least nine proteins (Fig. 25.2) (Gallo et al., 1988). Table 25.1 lists the viral proteins and their functions.

HIV Phylogeny

Currently, based on full-length viral genome sequencing, there are three major HIV-1 groups called M (main), O (outlier), and N (non M or O) (Fig. 25.3). Eight HIV-2 groups have been identified (Peeters et al., 2003). HIV-1 group M has been divided into nine clades (or subtypes), designated A to D, F to H, J, and K (Peeters et al., 2003; Robertson et al., 2000). Some subtypes have been distinguished for groups O, but only a few isolates of group N have been identified. No subtypes of the HIV-2 groups have yet been identified (Robertson et al., 2000). Each HIV-1 M clade differs from the others in amino acid composition by at least 20% in the envelope region and 15% in the Gag region (Robertson et al., 2000).

In the HIV-1 M group, subtype A is found primarily in Central Africa, clade B in North America and Europe, clade C in South Africa and India, and clade D in Central Africa. Subtype F includes a few isolates from Brazil (Potts et al., 1993) and all of the viruses thus far characterized from children in Romania (Dumitrescu et al., 1994). Other group M subtypes include viruses from Russia (G) (Bobkov et al., 1994), Africa and Taiwan (H) (Los Alamos National Laboratory, 2001), Zaire (clade J) (Laukkanen et al., 1999), and Cameroon (clade K) (Roques et al., 1999). The M group also includes many circulating recombinant forms (CRF) (Korber et al., 2001; Thomson et al., 2002), which encode genetic structures from two or more subtypes (e.g., A/E = CRF01AE; A/G = CRF02AG). Presently there are 16 recognized CRFs derived from HIV-1 group M isolates (Peeters et al., 2003; Robertson et al., 2000).

PATHOGENESIS

Overview

HIV transmission to an uninfected host occurs across mucosal surfaces during sexual contact, parturition, and early infancy (Ruprecht et al., 1999), and by traumatic introduction of virus particles across the intact dermis during percutaneous exposure. Infectious virus particles attach to a restricted set of host target cells through specific receptor/ligand mediated interactions between viral and cell membrane surfaces. Following attachment, fusion of the viral lipid membrane and the host cell membrane results in the release of viral nucleocapsid contents into the cytoplasm. The host RNA genome undergoes reverse transcription during transit to the cell nucleus, and subsequently integration into the host genome. Transcriptionally active cells produce new virus particles by nonlytic mechanisms to complete the life cycle. Viral genes may also remain transcriptionally silent within long-lived lymphocytes cells, resulting in the presence of a latent reservoir. Host humoral and cellular responses mediate the lysis of infected cells and contribute to the rapid clearance of virus particles from blood, but generally do not result in suppression of viremia or elimination of infected cells. Infected CD4+ T lymphocytes undergo cell death by direct and indirect mechanisms, resulting in depletion of CD4 T-lymphocytes and profound immune dysregulation over a period ranging from years to decades. Host demise is ultimately precipitated by onset of opportunistic infections.

Transmission across the Epithelium

Horizontal transmission of HIV to an uninfected host generally occurs through either percutaneous exposure, or across a mucosal surface. During percutaneous infection, a source of replication-competent virus particles is introduced into the systemic circulation,

TABLE 25.1 HIV proteins and their functions

Proteins[a]	Size (KD[a])	Function
Gag	p24	Capsid (CA), structural protein
	p17	Matrix (MA) protein, myristoylated
	p7	Nucleocapsid (NC) protein; helps in reverse transcription
	p6	Role in budding (L domain)
Polymerase (pol)	p66, p51	Reverse transcriptase (RT); RNAse H-inside core
Protease (PR)	p10	Post-translation processing of viral proteins
Integrase (IN)	p32	Viral cDNA integration
Envelope	gp120	Envelope surface (SU) protein
	gp41	Envelope transmembrane (TM) protein
Tat[b]	p14	Transactivation
Rev[b]	p19	Regulation of viral mRNA expression
Nef	p27	Pleiotropic, can increase or decrease virus replication
Vif	p23	Increases virus infectivity and cell:cell transmission; helps in proviral DNA synthesis and/or in virion assembly
Vpr	p15	Helps in virus replication; transactivation
Vpu[b,c]	p16	Helps in virus release; disrupts gp160:CD4 complexes
Vpx[d]	p15	Helps in entry and infectivity
Tev[b]	p26	Tat/Rev activities

Source: Reprinted with permission from Levy (2007).

[a] See Fig. 25.2 for location of the viral genes on the HIV genome.

[b] Not found associated within the virion.

[c] Only present with HIV-1.

[d] Only encoded by HIV-2; may be a duplication of Vpr.

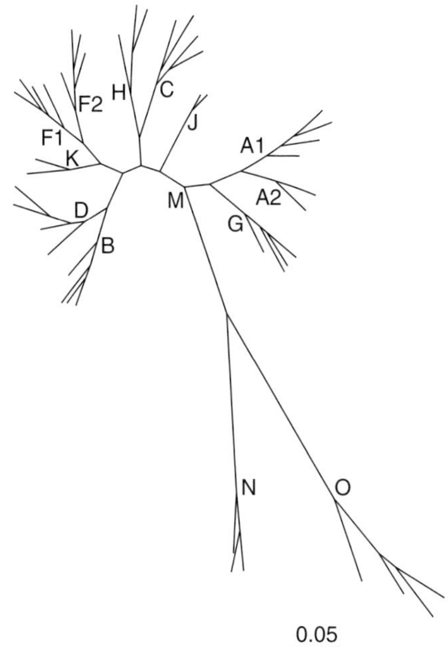

FIGURE 25.3 HIV-1 phylogenetic tree. HIV gene sequence phylogeny showing HIIV-1 major groups M, N, and O, and group M subtypes. Reprinted with permission from Levy (2007).

where direct contact between host CD4+ T lymphocytes and infectious virions presumably results in productive infection. Infection across mucosal surfaces of the genitourinary and gastrointestinal (GI) tracts generally takes place through sexual contact, during parturition, and after exposure through breastfeeding. Viral translocation across mucosal surfaces may occur through several mechanisms. Viruses are taken up and transported across the epithelium by intraepithelial dendritic cells (DC), which may either be directly infected, or trap intact virions on the cell surface in a form suitable for antigen presentation (Geijtenbeek et al., 2000; Kwon et al., 2002). HIV has also been shown to cross the epithelium through mechanical breaches caused by trauma or infection, allowing direct contact with and infection of local CD4+ T-cells and tissue macrophages. Infected DC (Kwon et al., 2002), T-cells, and macrophages present at the site of transmission travel via afferent lymphatic channels to regional lymph nodes (Pope et al., 1995), where surface-expressed and newly synthesized virus particles establish new rounds of infection. In this way, HIV subverts normal antigen-processing pathways to access suitable targets for viral replication within the host.

The Replication Cycle

Host Cell Binding

HIV replication cycle occurs through a series of partially defined steps from viral entry to the budding of new particles from the infected cell surface. Infectious virus particles attach to the host target cell through the interaction between a trimeric complex of three surface gp120 molecules (Earl et al., 1990) and three transmembrane gp41 molecules within the viral envelope, and CD4 expressed on the host plasma membrane. Binding of viral gp120 with CD4 triggers a series of conformational changes which expose binding sites to an additional host cell coreceptor that is required for entry (Sullivan et al., 1998). Although 12 distinct chemokine receptor molecules have been shown to serve as chemokine coreceptors for HIV entry in vitro, only two appear to be important for viral pathogenesis in humans (Berson et al., 1996; Deng et al., 1996; Feng et al., 1996). Primary infection and chronic disease are generally characterized by viral populations tropic for CCR5, while viruses tropic for CXCR4 are observed in late infection in some individuals (Scarlatti et al., 1997), and may herald profound immune dysfunction.

Membrane Fusion

Transmembrane gp41 is present as a six helix bundle in a noncovalent association with gp120 in the viral envelope (Chan et al., 1997). Full binding of gp120 to CD4 and either CCR5 or CXCR4 induces major conformational changes within g41 (Chan and Kim, 1998), which dissociates from gp120, and inserts a hydrophobic fusion peptide "harpoon" into the host cell membrane. Gp41 subsequently folds back upon itself into a hairpin-like structure, drawing the viral and cell membranes into close proximity, immediately prior to fusion (Chan and Kim, 1998). Fusion of the viral and cell membranes is followed by release of the contents of the nucleocapsid into the cytoplasm. Following entry, both strands of the viral RNA genome are associated with reverse transcriptase, integrase, viral protein R (Vpr), lysine transfer RNA (tRNA^{LYS}), host cell actin filaments, and other host and viral proteins (Karageorgos et al., 1993; Bukrinskaya et al., 1998). This aggregation constitutes the reverse transcriptase complex which effectuates reverse transcription of the viral genomic RNA into DNA during transit to the cell nucleus (Karageorgos et al., 1993).

Entry and Reverse Transcription

Reverse transcription of the diploid single-stranded (+) RNA genome into double-stranded DNA is accomplished by the retroviral enzyme reverse transcriptase. Reverse transcription of the RNA template occurs through a series of well-defined molecular steps, beginning with the priming of an RNA strand by tRNA^{LYS}, proceeding through two-strand-transfer jumps in which a nascent DNA segment jumps to a complementary region of the template strand, and ultimately resulting in the polymerization of two complementary proviral DNA strands. During this process, the RNA template undergoes degradation catalyzed by the RNAse H portion of reverse transcriptase. Because RNAse does not have a proofreading function, reverse transcription is extremely error-prone, resulting in the frequent introduction of point mutations during each generation. In addition, strand-transfer errors may result in insertions, deletions, and recombination in the newly transcribed viral genome (reviewed in Gotte et al., 1999). These changes underlie the enormous genetic variability characteristic of HIV.

Nuclear Entry and Integration

After reverse transcription, the HIV preintegration complex (PIC), composed of double-stranded proviral DNA, integrase, and several additional proteins (Miller et al., 1997), enters the host cell nucleus. In contrast with other retroviruses, the HIV PIC may enter the nucleus of both dividing and nondividing cells (Weinberg et al., 1991). The size of the PIC is considerably larger than the 25 nm diameter of nuclear pores (Pemberton et al., 1998), and the precise mechanisms by which entry takes place are unclear (Dvorin et al., 2002). This process may involve binding of a "DNA flap" domain to host nuclear targeting proteins (Zennou et al., 2000), or transient herniations of the nuclear membrane with intravasation of cytoplasmic contents into the nucleus. Once in the nucleus, the viral integrase gene and poorly characterized cellular cofactors catalyze proviral DNA integration into the host chromosome, preferentially targeting active host gene regions. Proviral DNA may also undergo autointegration, resulting in episomic, replication-incompetent LTR circles (Li et al., 2001).

Transcription

The integrated HIV genome may reside in a dormant state within resting cells (Adams et al., 1994), or become transcriptionally active. The former circumstance accounts for the persistence of HIV within latent reservoirs despite therapy. Transcriptional activation of the integrated proviral genome appears to be dependent on several host cellular factors that are upregulated following cell activation, and is greatly enhanced by the presence of the viral tat protein (Kao et al., 1987). Tat acts in part to recruit transcriptional factors to the HIV LTR region, a process that is inefficient in the absence of tat (Wei et al., 1998). The HIV genome contains numerous donor (5′) and acceptor (3′) splice sites, allowing at least 30 alternative mRNA molecules to be made from a single HIV proviral gene. Transcription products include the full-length unspliced primary transcript (~9 kb) and a variety of partially and fully spliced mRNAs encoding the viral *env* protein and the six accessory gene products of the viral genome (Saltarelli et al., 1996). Viral replication depends on a balanced production of unspliced and spliced transcriptional products.

Viral Assembly and Budding

Viral mRNA translation products (gag, gag-pol, vpr, and vif gene products) and two full-length viral genomic RNA molecules are assembled at the plasma membrane for packaging into a new virus particle (Fig. 25.4) (Wilk et al., 2001). Assembly and budding appear to require the action of both viral proteins and host cytosolic proteins (Zimmerman et al., 2002), which are diverted from their normal cytosolic functions. Host proteins involved in viral assembly and budding include tumor suppressor gene 101, which normally functions to direct products to incorporation within multivesicular bodies for degradation (Martin-Serrano et al., 2003; Bieniasz, 2006). Budding of virus particles takes place preferentially near cholesterol-rich rafts in the plasma membrane (Liao et al., 2001; Ono and Freed, 2001), resulting in enrichment of cholesterol within the viral membrane, a factor that may be important for subsequent fusion events during infection of a new host cell (Liao et al., 2001).

Viral Latency

Most T cells that become infected during successive rounds of replication are activated CD4+ T lymphocytes which go on to become productively infected cells producing progeny viruses, and ultimately die after several days due to viral cytopathic effects or host immune mechanisms. However, in some cases, T cells destined to become long-lived, quiescent memory T lymphocytes express sufficient CCR5 on the cell surface to become infected. These cells progress to a state of transcriptional inactivity, but contain one or more HIV proviral genomes, resulting in the maintenance of a long-lived latent viral reservoir (Finzi et al., 1997; Chun et al., 1997). Latently infected cells may undergo reactivation at a later time by exposure to antigen, resulting in upregulation of transcriptional activity and virus production. The size of the latent HIV reservoir in humans is unclear, but has been estimated to be approximately 10^6 cells, and may have a half-life in excess of 43 months (Finzi et al., 1997; reviewed in Pierson et al., 2000). The presence of the latent reservoir is responsible for persistent infection despite long-term effective therapy, and ensures that decades of continuous and complete inhibition of viral replication would be required to cure HIV through ART alone. "Flushing out" latently infected cellular reservoirs of HIV continues to be investigated as an approach to eradicating HIV from the host (Kulkosky et al., 2002; Stellbrink et al., 2002; Fraser et al., 2000).

Death of Infected Target Cells

A wide range of host cells are susceptible to HIV infection. However, the cell populations most directly involved in the process of transmission and virus production during all stages of illness are CD4-bearing cells of the host immune system, including DC,

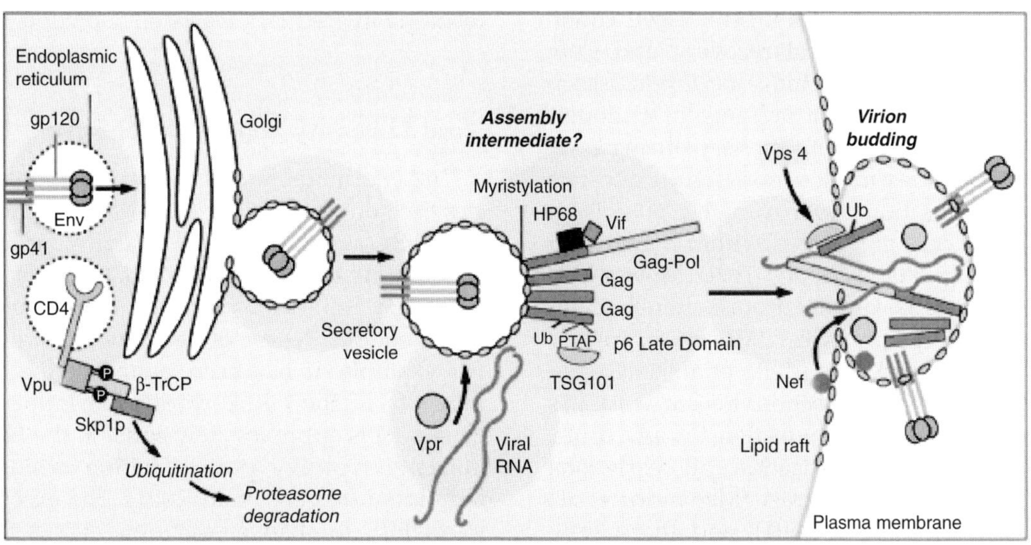

FIGURE 25.4 Viral assembly and budding. The figure depicts assembly of new virus particles, and budding from a cholesterol-rich raft in the plasma membrane. Adapted with permission from HIV InSite (see color plate section).

activated and resting memory CD4+ T lymphocytes, and macrophages. Of these, the activated memory CD4+ T cell, and in particular, the HIV-specific memory T cell (Douek et al., 2002), is preferentially infected in vivo. Other host cells may also support infection, including CD8+ T cells (Roederer et al., 1995) astrocytes and other cells of the central nervous system (Rothenaigner et al., 2007), capillary endothelial cells, columnar cells of the GI tract, pulmonary fibroblasts, renal tubular cells, and several other cell types. The full pathophysiologic significance of these other infected cell populations is unclear. Productively infected cells typically have a shortened life span. For example, the activated CD4+ T cell typically dies within several days of infection (Patki et al., 2000; Bolton et al., 2002; Lenardo et al., 2002). Infected tissue macrophages and DC may persist for weeks to months (Smith et al., 2001), respectively. Numerous mechanisms have been advanced to account for the profound loss of CD4+ T cells that occurs during primary HIV infection, including direct cell killing of host cells (Mattapallil et al., 2005; Li et al., 2005), cytopathic effects of HIV transcriptional products (Bolton and Lenardo, 2007; Sakai et al., 2006; Jones et al., 2007; Weinhold et al., 1989), host cytolytic activity, and bystander effects causing apoptosis of uninfected neighboring cells (Holm and Gabuzda, 2005). The relative contribution of direct and indirect mechanisms to CD4+ cell loss has been a matter of significant debate (Anderson et al., 1998; Grossman et al., 2002; Veazey and Lackner, 2005).

CD4 T-Lymphocyte Homeostasis

Primary HIV infection is typically associated with an abrupt fall in the peripheral CD4+ T-cell count, which is followed by a partial recovery and slow decline over years before reaching the levels associated with AIDS. However, the presence of a modest decline in CD4+ T cells in the periphery during early and chronic disease may mask a more significant state of immune dysregulation. Approximately 60–65% of all body CD4+ T cells are located within the GI tract (Cerf-Bensussan and Guy-Grand, 1991; Mowat and Viney, 1997), either as lamina propria lymphocytes or intraepithelial lymphocytes. Moreover, nearly all of these are activated cells (Zeitz et al., 1988) and thus ideal targets for HIV infection. Recent work has demonstrated that following infection by any route, there is early massive viral replication and nearly total CD4+ cell loss within the GI tract (Mehandru et al., 2004, 2006; Brenchley et al., 2004), and that CD4+ cell recovery in this site during chronic infection is very poor despite ART (Mehandru et al., 2006). The profound immunologic disruption at this site may underlie the immunologic changes leading to failure of the immune system during disease progression.

The gradual failure of CD4 T-cell homeostasis that occurs during chronic disease progression has been ascribed to a state of heightened immune activation (Giorgi and Detels, 1989; Bouscarat et al., 1996; Lederman et al., 2000; Hazenberg et al., 2000; Sousa et al., 2002), increased T-cell production and T-cell loss (De Boer et al., 2003). In the uninfected host, naïve CD4+ thymocytes may occasionally become activated in response to antigens, and then may enter the resting memory CD4+ pool, but more frequently undergo apoptosis (Patki et al., 2000). CD4+ T-cell loss is balanced by new thymic production of naïve CD4+ cells. In the otherwise healthy adult, abrupt depletion of CD4 cells is followed by gradual recovery that is limited by thymic output, and is attenuated in comparison with CD8 recovery (Atkinson et al., 1982; Watanabe et al., 1997; Mackall et al., 1997). During chronic HIV infection, after the profound CD4+ T-cell loss of primary infection, recovery of CD4+ T cells is further hampered by abnormally upregulated activation and apoptotic death of CD4+ cells, histological and functional disruption of lymphatic tissue, and impaired output from the marrow (Louache et al., 1992; Marandin et al., 1996) and thymus (Kalayjian et al., 2003). Thus, CD4 T-cell homeostasis during chronic infection is characterized by abnormally high cellular turnover (De Boer et al., 2003). The precise dynamic mechanisms leading to T-cell depletion over time remain a matter of some conjecture. According to one widely held view, failure is likened to a "tap and drain" model (Amadori et al., 1996), in which CD4 T-cell loss slightly exceeds production capacity, resulting in gradual CD4 T-cell depletion (Fig. 25.5).

Viral Diversity and Pathogenesis

The enormous rate of virus production (10^{10} particles each day) (Ho et al., 1995; Ramratnam et al., 1999) and high rate of error (10^{-5} changes per base pair per generation) (Mansky and Temin, 1995) characteristic of HIV replication permit the virus to accumulate a tremendous degree of genomic variation over time. For example, it has been estimated that the genetic diversity of the entire worldwide influenza A virus gene pool is approximately equal to the HIV-1 genetic diversity typically present within a single infected individual (Korber et al., 2001). The exceptional ability of HIV to change over time allows the infecting viral population to effectively adapt to environmental selective pressures. Through antigenic variation,

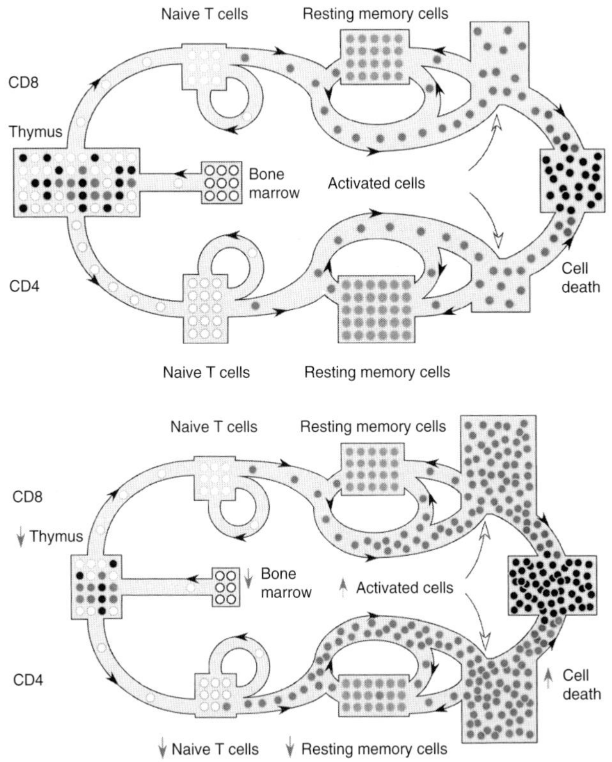

FIGURE 25.5 Disruption of normal T-cell homeostasis due to HIV infection. (A) Under normal conditions, losses from the resting memory CD4+ and CD8+ T lymphocyte pools through activation, and apoptosis are balanced by the input of new naïve T cells from the marrow and thymus. (B) During chronic HIV infection, increased activation of both CD4+ and CD8+ T cells results in increased cell turnover. CD4+ T-cell homeostasis is relatively impaired because of reduced output from marrow and thymus, resulting in a smaller pool of naïve T cells and resting memory CD4+ T cells. Naïve cells, clear center; resting memory cells, light gray center; activated cells, hatched center; dying cells, black center; HIV-infected cells, speckled center. Processes/cell populations affected by HIV infection are highlighted by upward and downward arrows Adapted with permission from Douek (2003) (see color plate section).

HIV is able to evade host humoral and cell-mediated immunity. Similarly, HIV readily evolves resistance mutations to nearly all known antiretroviral agents, leading to therapeutic failure during clinical management (Fournier et al., 2005; Haddad et al., 2000; Orkin et al., 2005; Raguin et al., 2004; Saag, 2004; Sabin et al., 2005). Finally, HIV is able to exploit a wide range of cellular niches. For example, by expanding its coreceptor tropism during disease progression to allow CXCR4 usage, HIV may infect a wider range of cells including naïve T cells, thereby further exacerbating host CD4 depletion (Scarlatti et al., 1997; Connor et al., 1997; de Roda Husman et al., 1997; Berkowitz et al., 1998). Thus, viral evolution drives many of the processes resulting in disease progression and poor clinical outcome in the human host.

PROTECTIVE IMMUNE RESPONSE

Overview

Both humoral and cell-mediated adaptive immune responses are elicited in response to HIV infection. Because cytotoxic T lymphocytes (CTL) can only recognize infected cells, cytolytic mechanisms cannot prevent infection, but are thought to be the primary immunologic mechanism in response to a variety of viral infections in mammals. Cytotoxic CD8+ T-cell responses appear to be the earliest meaningful responses following HIV infection, and are credited with reducing plasma viral load to steady-state levels from the peak levels seen during primary infection. Humoral immunity arises within weeks of primary HIV infection, and the detection of host antibodies (Ab) to HIV antigens forms the basis of the HIV enzyme linked immunosorbent assay (ELISA), the serological test that is most widely used to screen for HIV infection. Host Ab are capable of rendering virus particles inactive, and theoretically could provide sterilizing immunity from infection. However, neutralizing antibodies (nAb) are not believed to have a major role in controlling viral replication during chronic HIV infection. Less information is available on the role of innate immune mechanisms in controlling HIV.

Antibody Response to HIV

Humoral immunity constitutes one of the two major limbs of the adaptive immune system in mammals. Ab are elicited in response to most infections in humans, and are capable of providing protection from challenge by a wide range of bacterial and viral pathogens after vaccination or natural infection (Robbins et al., 1995). HIV-specific Ab are primarily directed against epitopes within the viral gp120 and gp41 proteins that are expressed on the surface of the viral plasma membrane (Hansen et al., 1990; Scanlan et al., 2002; Ugolini et al., 1997; Valenzuela et al., 1997; Sattentau and Moore, 1995). The antibody response to HIV is therefore confined to targets within a relatively narrow portion of the viral genome. HIV-specific Ab have been shown to interfere with HIV replication through several mechanisms. nAb, by definition, abrogate the ability of the virus to infect new host cells. Anti-HIV nAb generally function by binding to conserved epitopes on gp120 and gp41 necessary for engagement with CD4 (Ugolini et al., 1997; Zhou et al., 2007) or the host cell chemokine coreceptor (Decker et al., 2005; Trkola et al., 1996), or to domains necessary for fusion (Fig. 25.6) (Golding et al., 2002;

Gorny and Zolla-Pazner, 2000). Non-nAb may also impact viral replication, through antibody-dependent cellular cytotoxicity (ADCC) (Tyler et al., 1990; Koup et al., 1991; Jewett and Bonavida, 1990). Ab binding to viral peptides expressed on the surface of the infected cell stimulate engagement of cytotoxic cells bearing the Fc receptor, including NK cells, macrophages, and other cell types, resulting in cell-mediated lysis.

HIV has evolved several static and dynamic strategies to evade host humoral response. In its functional form, gp120 is present on the viral surface in a noncovalently bound trimeric configuration (Earl et al., 1990), in which sensitive epitopes are shielded from exposure to Ab. Binding to CD4 induces conformational changes which temporarily expose chemokine coreceptor binding sites (Sullivan et al., 1998; Sattentau and Moore, 1995) to allow completion of attachment and fusion. Because several important nAb epitopes are located within these binding regions, important targets for the nAb response are only transiently exposed to neutralization (Labrijn et al., 2003). In addition, monomeric gp120 and other fragmentary degradation products are copiously produced during the process of viral replication (Moore et al., 2006; Wyatt and Sodroski, 1998). While monomeric gp120 is highly immunogenic, Ab generated in response to monomeric gp120 are typically nonneutralizing (Sattentau and Moore, 1995). Immune responses are thereby diverted to decoy targets and are rendered ineffective in clearing circulating viruses. HIV has also been shown to evolve under selective pressure imposed by humoral immunity. The HIV *env* genes contain five hypervariable loops, V1–V5, which are the most highly variable regions of the genome. Extensive length polymorphism, particularly within V1 and V2, and point mutations throughout *env* permit viral escape from circulating Ab (Moog et al., 1997; Deeks et al., 2006; Albert et al., 1990; Geffin et al., 2003). Studies have also demonstrated that bulky glycan sidechains inserted at N-linked glycosylation sites (NLGS) may protect important epitopes within gp120, and that virus escape from nAb in vivo and in vitro can be directly attributed to the evolution of new NLGS within env under antibody-mediated selective pressure (Wei et al., 2003). Thus, the virus particle may be protected from nAb by an "evolving glycan shield."

Following natural infection in humans, anti-HIV Ab arise within weeks and continue to persist and evolve throughout chronic infection (Wei et al., 2003), before waning in intensity in some individuals with the onset of AIDS. Most Ab produced by the host are ineffective at neutralizing circulating virus particles. Host plasma has been shown in some laboratory studies to neutralize viruses that had been circulating at earlier times, but most studies suggest that the host-specific nAb

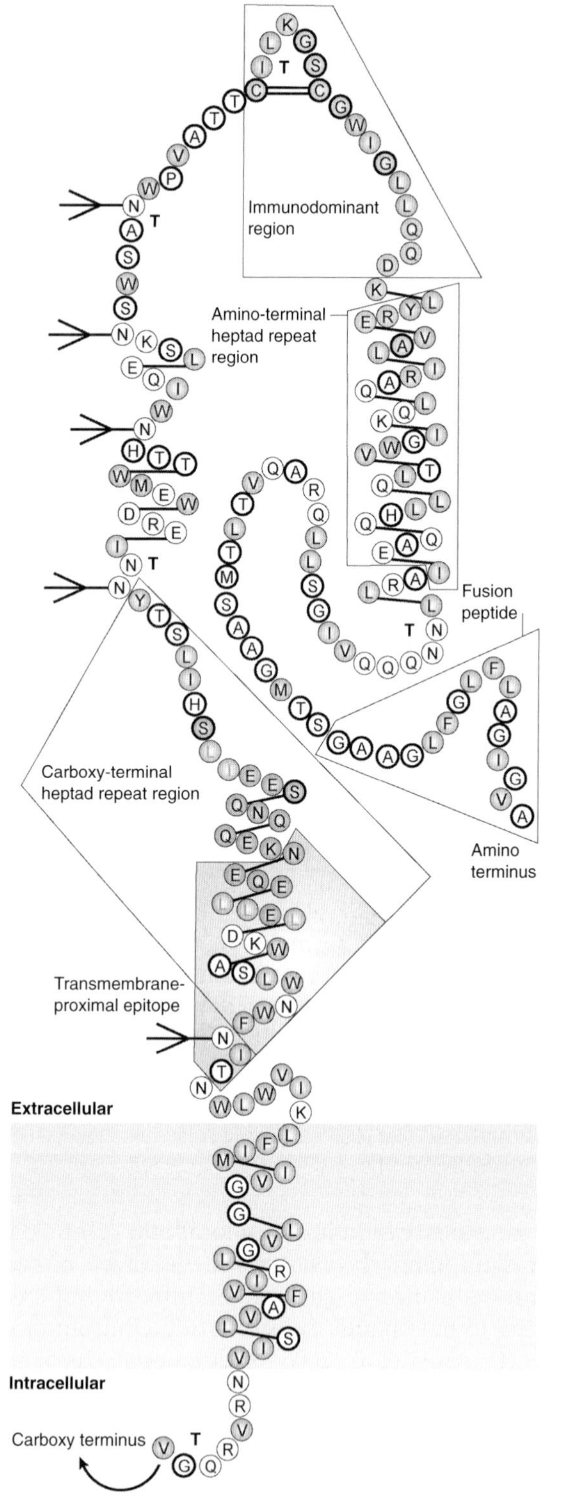

FIGURE 25.6 HIV gp41. Schematic diagram of HIV transmembrane protein gp41, depicting functional regions important for virus entry into the host cell, and sites important for immune responses. Hydrophobic amino acids shown in gray, charged amino acids shown as unfilled circles, neutral amino acids outlined in bold, potential N-linked glycosylation sites shown as branched stick figures. Important antibody epitopes are presented in turquoise, pink, and yellow. Adapted from Zolla-Pazner (2004) (see color plate section).

response lags significantly behind viral evolutionary changes (Moog et al., 1997; Albert et al., 1990; Wei et al., 2003), and is thus poorly able to neutralize contemporaneous virus populations. Passive immunity studies have shown that antibody responses can block infection in macaques and chimpanzees (Emini et al., 1992; Gauduin et al., 1995; Shibata et al., 1999; Mascola et al., 2000). However, there is relatively little evidence for a relationship between nAb responses and clinical outcome during natural infection. Some, but not all, studies in long-term nonprogressors have suggested an association between the nAb response and slowed disease progression (Harrer et al., 1996; Montefiori et al., 1996; Binley et al., 1998). Two large phase III efficacy trials based on eliciting Ab to *env* epitopes failed to demonstrate protection from infection. The impact of humoral immunity on infection and disease progression and the importance of Ab in providing protection from HIV challenge after vaccination continue to be a matter of debate.

Cell-Mediated Immunity to HIV

Cytotoxic CD8+ T cells recognize infected host cells through a highly specific interaction with viral proteins that are coexpressed on the surface of infected cells in association with major histocompatibility (MHC) class I molecules. These complexes arise when viral proteins produced during HIV replication are cleaved into short peptides of 8–11 amino acids by cell peptidases, and redirected to the endoplasmic reticulum for binding to MHC molecules and eventual transportation to the plasma membrane. The CD8+ T cell engages infected cells through binding of the T-cell receptor (TCR) with viral antigen–MHC complexes. TCR binding is epitope-specific, and CD8+ T cells are restricted to a single epitope based on the configuration of the TCR. After binding, several mechanisms mediate killing of the infected target cell (Kagi et al., 1994). Following the release of lysis granules, perforin may insert into the target cell plasma membrane, creating ion channels which cause osmotic dysregulation of and lysis of the target cell. Binding of Fas ligand (Fas-L) with Fas expressed on the target cell surface causes apoptotic cell death through a series of programmed cytoplasmic and nuclear events (Nagata and Golstein, 1995). Activated CD8+ cells also release interferon gamma (INF-γ), as well as the beta chemokines MIP-1α, MIP-1β, and RANTES, the natural ligands of CCR5 (Cocchi et al., 1995). Release of these soluble products cause neighboring cells to become intrinsically more refractory to viral infections, or less susceptible to HIV by reducing the surface expression and/or binding availability of CCR5 (Cocchi et al., 1995).

A number of factors act to limit the effectiveness of the CD8+ cytotoxic response in humans. CD8+ T cells from HIV-1 infected individuals have shown numerous defects, including dysregulated cytokine expression (Tanaka et al., 1999; Garba et al., 2002; Barker et al., 1995), abnormal skewing towards preterminally differentiated maturation states (Champagne et al., 2001), upregulation of immunomodulatory surface receptors (Trautmann et al., 2006), impairment of granule-dependent lysis (perforin and granzyme A) (Yang et al., 2005; Andersson et al., 2002; Chiang et al., 1996; Trabattoni et al., 2004), anergy in the presence of HIV antigens (Lichterfeld et al., 2004), and increased susceptibility to apoptotic signals (Petrovas et al., 2004). Finally, HIV-1 may interfere with CD4 helper function by causing depleting CD4 cells (Douek et al., 2002; Brenchley et al., 2006), or by inducing an inappropriate skewing of the CD4 helper function toward a Th2 rather than a Th1 pattern of cytokine response, favoring antibody over cellular immune mechanisms. Because virus-specific CD4+ T-helper function appears to be required to maintain maximally effective CD8 responses (Matloubian et al., 1994; Battegay et al., 1994; Hasenkrug et al., 1998; Kalams et al., 1999), dysregulation of CD4 T-cell function is likely to further weaken cytotoxic immunity during disease progression.

Vigorous CTL responses to HIV infection are typical (Betts et al., 2001; Migueles and Connors, 2001; Gea-Banacloche et al., 2000; Pantaleo et al., 1994), and several lines of evidence indirectly suggest that cytolytic mechanisms are of primary importance in response to HIV infection. In animal models, depletion of CD8+ T cells abrogates the usual dip in viral load from peak viremia following primary infection (Matano et al., 1998) and interferes with host ability to maintain steady-state viral loads (Jin et al., 1999; Schmitz et al., 1999), suggesting that control is mediated in part by cytolytic mechanisms. In the SIV model of HIV infection, vaccination studies have shown that robust CTL responses may result in some degree of control of infection (Amara et al., 2001; Shiver et al., 2002; Barouch et al., 2000). In humans, a dip in peak viremia is also typically observed at some time following primary infection, and this has been correlated with the appearance of HIV-specific CTL (Borrow et al., 1994; Koup et al., 1994). Furthermore, disease progression has been shown to be profoundly influenced by host human leukocyte antigen (HLA) class 1 genotype (Carrington and O'Brien, 2003; Kaslow et al., 1996; Carrington et al., 1999; Gao et al., 2001). Because of the prominent role of HLA in antigen presentation and CTL lysis, this observation implies that CTL mechanisms play a significant role in the host response to infection. Several studies have shown relatively strong CTL responses in long-term nonprogressors, suggesting that CTL mechanisms may be a significant determinant of disease progression (Harrer et al., 1996; Rinaldo et al., 1995; Betts et al., 1999). Moreover, finding a direct link between the targeting of particular CTL epitopes and disease progression has been elusive.

EPIDEMIOLOGY

Since the first cases of AIDS were reported in 1981, HIV infection has grown to pandemic proportions, resulting thus far in an estimated 65 million infections and 25 million deaths (CDC, 1981; UNAIDS, 2006). An estimated 33.2 (30.6–36.1) million people worldwide live with HIV-1 (UNAIDS, 2007). In 2007, there were 2.5 million new HIV-1 infections and 2.1 million AIDS deaths (UNAIDS, 2007). Today, this pandemic has left no region of the world untouched (Inciardi and Williams, 2005) (Fig. 25.7). Currently, more than 6800 persons become infected with HIV and more than 5700 persons die from AIDS each day, mostly because of inadequate access to HIV prevention and treatment services. The HIV pandemic remains the most serious infectious disease challenging public health (UNAIDS, 2007).

Worldwide, heterosexual contact is the most common mode of transmission, accounting for about 85% of all HIV-1 infections (Simon et al., 2006). In the United States, sexual contact between men remains the most common means of sexual transmission, accounting for approximately 40% of the new cases (UNAIDS, 2006). Intravenous drug use is an increasing mode of HIV-1 transmission, particularly in Eastern Europe and Southeast Asia (UNAIDS, 2006). The rapid spread of infection of HIV-1 in these regions through injecting drug use is important, since it provides a bridge for rapid establishment of more generalized epidemics (Simon et al., 2006).

Sub-Saharan Africa remains the most seriously affected world region, with AIDS established as the leading cause of death there (UNAIDS, 2007). With only about 10% of the world's population, this region contains more than two-thirds (68%) of all people who are HIV positive, and more than three-quarters (76%) of all AIDS deaths in 2007 occurred here (UNAIDS, 2007). Transmission in this region is primarily through heterosexual contact. Southern Africa is the epicenter of the AIDS epidemic. This subregion accounts for 35% of all people living with HIV and almost one-third (32%) of all new HIV infections and AIDS deaths globally (UNAIDS, 2007). All countries in the region except Angola have an estimated adult (i.e., aged 15–49 years) HIV prevalence exceeding 10% (UNAIDS, 2006). Although sub-Saharan Africa continues to bear a disproportionate burden of HIV-1 infections, an increasing number of countries now report stabilization or decline in prevalence (e.g., Zambia, Tanzania, Kenya, Ghana, Rwanda, Burkina Faso, and Zimbabwe) (UNAIDS, 2006). There is some evidence to attribute these reductions to changes in sexual behavior, such as postponement of sexual debut, reduction in casual relationships, and more consistent condom use in casual relationships (Halperin and Epstein, 2004; Cates, 2005).

Unlike other regions, the majority of people living with HIV in sub-Saharan Africa (61%) are

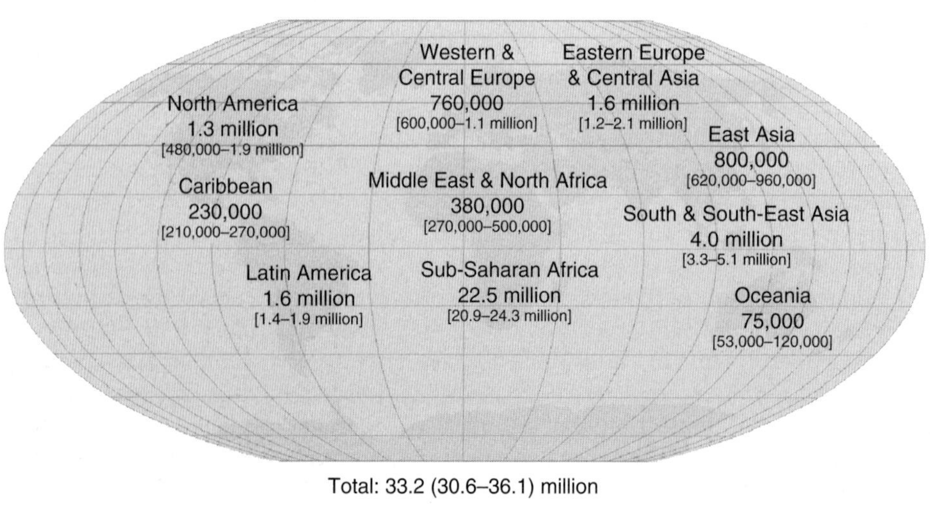

FIGURE 25.7 Global HIV epidemiology. Adults and children estimated to be living with HIV, 2007. Reproduced with permission from UNAIDS (http://www.unaids.org).

women (UNAIDS, 2007). Overall, a quarter of all new HIV-1 infections occur in persons younger than 25 years (UNAIDS, 2006). HIV-1 infection rates are three to six times higher in female adolescents than in their male counterparts (UNAIDS, 2006 Abdool-Karim and Abdool-Karim, 2002; Shisana and Davids, 2004; Pettifor et al., 2005); this difference is attributed to widespread sexual coupling of young women with older men. The increasing burden of HIV infections in women in this region has important implications for mother-to-child transmission as well as the large number of orphaned children. An estimated 11.4 million children in this region have been orphaned due to HIV and AIDS (Marston et al., 2005).

CLINICAL DISEASE

Clinical signs and symptoms of HIV may not become apparent until several years after infection. People who are HIV infected through sexual exposure can be viremic in 4–11 days (Kahn and Walker, 1998). During this early time period, a clinical "acute-retroviral" syndrome is reported in 40–80% of the infected individuals. Symptoms include fever, rash, lymphadenoapthy, headache, pharyngitis, aseptic menegitis, arthralgias, and myalgias. Symptoms are often mistakenly attributed to another viral illness, and usually resolve within 14 days. In one study, 94% of individuals sought medical attention during this period, but only one in four received the diagnosis of acute HIV infection (Schacker et al., 1996). During primary infection, individuals typically have very high plasma virus levels. In fact, one study showed that individuals with acute infection were the source for half of all new infections in a large urban setting (Brenner et al., 2007).

After this initial viremic period, an infected individual that goes unrecognized and/or untreated may be without symptoms from their infection. The median time from infection to progression to AIDS is 8–10 years, with a general decline of CD4 counts of 50–75 cells/year and increasing plasma viral load in untreated individuals (Lyles et al., 2000). However, irreversible immunologic changes appear to occur soon after infection, with severe depletion of CD4+ lymphocytes particularly in lymphoid tissue of the GI tract (Brenchley et al., 2004). The pace of disease progression is generally influenced by an individual's viral load, general health, and other concomitant infections and illnesses. The risk of developing an AIDS-defining illness in a 3-year period is correlated with plasma viral load and peripheral CD4+ T-cell count (Mellors et al., 1997).

Individuals with CD4 counts greater than 500 cells/μl are generally asymptomatic, but may experience lymphadenopathy, a variety of dermatologic conditions, and appear to be at increased risk of acquiring hematologic malignancies such as non-Hodgkins lymphoma. Individuals with a CD4 count between 200 and 500 may continue to be asymptomatic or may present with mild disease. Recurrent mild viral disease (herpes simplex, zoster infections, oral hairy leukoplakia), fungal infections (vaginal or oral candidiasis), and bacterial infections (sinusitis, recurrent pneumonias) may manifest at this point. Complaints of fatigue, weight loss, and various myalgias and arthralgias are common. As CD4 counts decline to between 50 and 200, untreated individuals often begin to experience more significant manifestations of immune compromise. These may include pneumocystis pneumonia (which remains the most common opportunistic infection in AIDS patients), toxplasmosis encephalitis, parasitic diarrheal illnesses, and oncologic conditions such as invasive carcinoma of the cervix, rectum, or anus.

Severely immunocompromised untreated HIV-infected individuals (CD4 counts less than 50) are at risk for additional opportunistic infections such as disseminated Mycobacterium avium complex disease (manifested as fever, weight loss, bone marrow involvement, as well as a number of less common organ involvements), cryptococcal meningitis, a number of diarrheal illnesses (bacterial, viral, and parasitic), disseminated endemic mycoses, and HIV wasting syndrome and encephalopathies.

MANAGEMENT AND TREATMENT OF HIV

Initial management of a newly diagnosed HIV patient includes taking a detailed medical history and performing a thorough physical examination. Key elements in evaluating an HIV-infected individual include information regarding social and sexual history to identify the risk factors associated not only with HIV acquisition but also HIV transmission. This history should also include information on patient sexual practices including type and number of sexual partners and use of barrier methods of protection, history of sexually transmitted infections (STIs), drug and alcohol use, the exchange of sex for money or drugs, and intravenous or other drug use. This history should be an opportunity to counsel patients on ways to prevent the acquisition of concurrent STIs and decrease the risk of HIV transmission. It is also a time to address barriers to successful management and treatment of HIV infection, and to identify current medical issues that may limit use of antiretroviral drugs. Certain laboratory, radiologic, and clinical studies will also assist with management (Table 25.2). Genotypic viral resistance assays are currently part of the recommendation for all HIV-infected individuals as they enter care, regardless of how long they have been infected, to assess for transmission of resistant viral isolates (Hammer, 2005). These assays are not universally available. Infected individuals should be screened annually for tuberculosis, STIs, and age-appropriate health screenings as recommended by a number of professional societies including the American Cancer Society, the Infectious Disease Society of America, the Centers for Disease Control, and the American College of Physicians.

Prior to the availability of effective therapy in the mid-1990s, HIV infection was predictably associated with gradual host decline leading to death many years after infection in most individuals. This pattern is still widespread in many developing countries with high prevalence of HIV infection and limited access to HIV therapy. The advent of highly active ART in certain areas has a marked impact on survival from HIV disease. In eight large US cities, mortality declined from 29.4 per 100 person years in 1995 to 8.8 per 100 person years in the second quarter of 1997 (Palella et al., 1998). Over the past decade, ART has continued to improve with less toxic medications and better dosing schedules. Also, the number of drugs and number of classes of available agents dramatically increased. There are presently five classes of FDA-approved medications encompassing 23 different individual drugs (Table 25.3). The five classes (entry inhibitors, nucleoside/nucleotide reverse transcriptase inhibitors, nonnucleoside reverse transcriptase inhibitors, protease inhibitors, and integrase inhibitors) interrupt different stages of the HIV life cycle, both at extracellular and intracellular stages in viral production.

Presently, standard therapy includes using at least three different drugs from two classes of medications. Current recommendations from a number of international bodies recommend starting therapy prior to significant immune compromise (Hammer et al., 2006). This approach should be balanced against the long-term side effects from medications, which may include metabolic abnormalities, increased cardiovascular events, dysmorphic body changes, and significant GI intolerance. Also, in developing countries, there are significant limitations to access that impact the choice and timing of ARTs. In general, individuals with symptomatic HIV disease or CD4 counts less than 200 should be initiated for appropriate ART. Individuals with CD4 counts greater than 350 do not require immediate initiation of therapy, although those with very high viral loads (greater than 100,000) should consider therapy soon, as such a high viral load predicts a more rapid decline in CD4 count and faster progression to AIDS. For those with CD4 counts between 200 and 350, the decision to commence therapy should be made on an individual basis. Issues to consider in starting therapy include concomitant medical issues and medications, social situations that might impact compliance, and drug or alcohol issues that may create a barrier to successful therapy.

TABLE 25.2 Baseline recommended testing in newly diagnosed HIV patient

General evaluations	HIV-specific testing
Complete blood count	CD 4 cell counts
Electrolytes, glucose, blood urea nitrogen, creatinine	Plasma HIV-1 RNA
Bilirubin, alkaline phosphatase, transaminases (AST/ALT)	Genotypic test for antiretroviral resistance
Creatine kinase	
Amylase, lipase	
Fasting lipid profile	
Serology for syphilis	
Sexually transmitted infection screen	
Serologies for Hepatitis A, B and C	
Toxoplasmosis serologies	
CMV serologies	
Cervical papanicolaou smear	
Anal screening for HPV and atypia	
Tuberculin skin test	
EKG	
Chest X-ray	

Note: Additional screening tests may be added depending on local guidelines and potential patient exposures.
Source: Adapted from Hammer (2005).

Another aspect of HIV infection management is monitoring for and treating opportunistic infections and giving appropriate prophylaxis when patients are immune compromised. Pathogens for which primary prophylaxis is recommended include: *Pneumocystis jiroveci* (CD4 counts less than 200—prophylaxis with trimethoprim/sulfamethoxazole [TMP/SMX]), *Toxoplasma gondii* (CD4 counts less than 100—prophylaxis with TMP/SMX), *Mycobacterium avium* complex (CD4 counts less than 50—prophylaxis with azithromycin), and *Mycobacterium tuberculosis* (any individual with a positive PPD without evidence of disease should receive 9 months of isoniazid (INH) therapy, as well as proper vaccine management against influenza, pneumococcus, and hepatitis A and B). In some cases, prophylaxis can

TABLE 25.3 FDA-approved antiretrovirals by class (as of January 2008)

Nucleoside/nucleotide reverse transcriptase inhibitor	Emtricitabine Lamivudine Zidovudine Tenofovir disoproxil fumarate Didanosine Stavudine Abacavir
Nonnucleoside reverse transcriptase inhibitor	Etravirine Delavirdine Efavirenz Nevirapine
Protease inhibitor	Amprenavir Tipranavir Indinavir Saquinavir Lopinavir Ritonavir Fosamprenavir Darunavir Atazanavir Nelfinavir
Fusion inhibitor/entry inhibitor	Enfuvirtide Maraviroc
Integrase inhibitors	Raltegravir
Combination drugs: trade name (constituent medications)	Kaletra (lopinavir/ritonavir) Atripla (tenofovir/emtracitibine/efavirenz) Combivir (zidovudine/lamivudine) Epzicom (abacavir/lamivudine) Trizivir (zidovudine/lamivudine/abacavir) Truvada (tenofovir/emtricitabine)

be discontinued when the immune system recovers after ART.

VACCINES

HIV vaccine trial research received impetus from the US government as early as 1984. At that time, Margaret Heckler, Secretary of Health and Human Services, predicted that an HIV vaccine would be in human trials within 2 years. Perhaps the greatest political push came from President Bill Clinton, who in 1997 challenged researchers to find an effective vaccine within 10 years, evoking the imagery of John F. Kennedy and his challenge to the NASA science program to put a man on the moon. A wide variety of HIV vaccine design approaches have been pursued, and over 50 candidates have been tested in clinical trials. Of these candidates, only few have advanced to phase 2B or 3 clinical efficacy testing (Table 25.4).

The first HIV vaccine efficacy trials, initiated in the 1990s, focused on the attempt to elicit neutralizing antibody responses using recombinant HIV envelop glycoprotein (rgp120) antigens produced from cell-line adapted HIV strains. In experiments in animal models and in the early-phase clinical trials that preceded efficacy studies, the glycoprotein candidate immunogens (rgp120 and rgp160) were safe and immunogenic (Keefer et al., 1997; Migasena et al., 2000; Wright et al., 1999; McElrath et al., 2000; Dolin et al., 1991). Experiments in nonhuman primates indicated that this approach elicited nAb, although the antibody response was narrow in most cases. In experimental systems, the elicited response was capable of neutralizing only HIV strains that closely resembled the vaccine antigen, and did not protect against challenge with primary or heterologous challenge virus (Arthur et al., 1989; Berman et al., 1988). Despite these results, two large efficacy trials were launched by the VaxGen Company, fueled in part by uncertainty about whether animal studies would accurately predict efficacy in humans. The trials were huge undertakings and proved that HIV efficacy trials can be successfully conducted at multiple international sites with good community support. The VAX004 study tested a bivalent, subtype B rgp120 vaccine (AIDSVAX B/B) among 5403 volunteers in the US, Canada, and Europe. The vaccine did not protect against infection (vaccine efficacy = 6%, 95% CI = [−17%, 24%]) and did not affect HIV disease course (HIV viral load or rate of ART) in participants who became infected (Flynn et al., 2005). The second trial tested AIDSVAX B/E, which included both subtype B and subtype E rgp120 antigens, among 2546 injecting drug users in Thailand. Vaccine efficacy was estimated at 0.1% (95% CI, −30.8% to 23.8%; $P = .99$, log-rank test) and no significant effects on disease course were seen (Pitisuttithum et al., 2006).

The failure of the VaxGen trials underscored the difficulty of eliciting broadly nAb that could block infection with HIV. Most subsequent trials of HIV vaccine candidates have involved products that aim to induce HIV-specific cytotoxic T-cell responses. These cells recognize HIV epitopes displayed on cell surfaces in conjunction with HLA and act against infected cells but not free virus. They limit the spread of infection by destroying infected cells via apoptosis, or by secreting chemokines and cytokines that interfere with subsequent rounds of infection. It is anticipated, based on studies in nonhuman primates, that this type of

TABLE 25.4 Ongoing phase 2 and phase 3 HIV preventive vaccine trials (as of November 1, 2007)

Trial name and description	Organizer/manufacturer	Site locations	Description of vaccine
Phase 3			
RV 144: A trial of live recombinant AVLAC –HIV priming with VaxGen gp120 B/E boost—enrollment complete and in follow-up	Walter Reed Army Institute of Research (WRAIR), Armed Forces Research Inst., Thai MoH, Thai AIDS Vaccine Evaluation Group. Manufacturers: Aventis Pasteur and VaxGen	Thailand	Live canary pox with a clade B env and gag-pol fusion boosted with a subunit clade B/E gp120
Phase 2 and 2b (test of concept)			
STEP and Phambili (HVTN 502 and 503): Multicenter test of concept trial using 3-dose regimen. Enrolment and further vaccination suspended September 2007	Merck, US Division of AIDS (DAIDS), HIV Vaccine Trials Network (HVTN), South African AIDS Vaccine Initiative (SAAVI)	North and South America, Australia, and South Africa	Replicative incompetent Ad5 with clade B gag, pol, and nef inserts
Phase 2			
HVTN 204: Clinical trial evaluating the safety and immunogenicity of DNA plasmid vaccine followed by adenovirus boost. Enrolment completed and in followup	DAIDS, HVTN. Manufacturer: VRC	North and South America, South Africa	VRC 6 plasmid DNA of ENV A, B and C plus clade B gag, pol, and nef. Boosted with Ad5 vector with a ENV A, B, and C and a gag-pol fusion insert
IAVI A002: Placebo-control double-blind trial evaluating the safety and immunogenicity of AAV vaccine	International AIDS Vaccine Initiative (IAVI). Manufacturer: Targeted Genetics	South Africa, Uganda, and Zambia	Adeno-associated virus 2 with a gag, pol, and ΔRT
ANRS VAC 18: Randomized double-blind trial evaluating the safety and immunogenicity of LIPO-5 versus placebo	Agence Nationale de Recherches sur le SIDA (ANRS). Manufacturer: Aventis Pasteur	France	5 lipopeptides containing CTL epitopes from gag, pol, nef

Source: IAVI Database (2007); AIDS Vaccine Clearing House (2007); Duerr et al. (2006).

response will not prevent infection but will ameliorate disease course by limiting viral replication.

In the second wave of HIV vaccine trials, a variety of "T cell" vaccine candidates were tested; the first major candidates employed pox-virus vectors. The prototype for this approach is vaccinia, the poxvirus that was used as a vector in the global smallpox campaign. Vaccinia itself was abandoned as an HIV vaccine vector due to safety concerns raised by a case of fatal disseminated vaccinia in an immune-compromised, HIV-infected individual (Redfield et al., 1987). A number of attenuated pox-virus vectors have been tested. Of these, constructs based on canarypox were studied most extensively in humans, including efficacy testing. These viral vectors are host-range-restricted and undergo abortive replication in human cells. The HIV insert sequences are expressed and presented without replication of the vector itself. In clinical trials involving over 400 HIV-infected participants and more than 1500 uninfected participants (Franchini et al., 2004), canarypox (ALVAC) vaccine candidates were safe and immunogenic (Gupta et al., 2002; de Bruyn et al., 2004). While nAb responses were limited, initial trials found T-cell responses in 30–40% of trial participants when multiple time points were tested (Duerr et al., 2008). In a pivotal trial, the second-generation ALVAC prime (vCP1452) and recombinant gp120 (AIDSVAX™ B/B subunit) boost regimen failed to meet the preset immunogenicity criteria and did not move forward to efficacy testing (Russell et al., 2007). A similar regimen (the canarypox vector vCP 1521 containing the HIV-1 clade B *env*, *gag*, and protease genes, in combination with AIDSVAX clade B and E rgp120) is being tested in a 16,000-person efficacy trial in Thailand. The trial's primary objective is to evaluate the regimen's ability to prevent HIV infection; results are expected in 2009. The rationale for this trial has been questioned (Burton et al., 2004) due to poor performance of the components (or closely related ones) in earlier trials. The sponsors of the trial defended the decision to advance the regimen to efficacy testing based on responses to the combined regimen seen in nonhuman primate studies and earlier clinical trials that met predefined milestones, as well as the extensive review process prior to initiation of the efficacy trial (McNeil et al., 2004). Additional pox-virus-vectored HIV vaccines are also being moved forward in human trials. A modified vaccinia ankara product (MVA) is scheduled to enter phase 2 studies late in 2008.

The most recent wave of HIV vaccine testing has been dominated by replication-defective adenovirus serotype 5 (Ad5) vectors, used alone or in combination with DNA. One such product, produced by Merck, uses an admixture of three adenoviruses containing codon-optimized subtype B *gag*, *pol*, or *nef* genes, respectively. This product was tested as a three-dose regimen. Early clinical testing indicated excellent safety and immunogenicity (the vaccine elicits a robust T-cell response, which is somewhat diminished among individuals with preexisting immunity to Ad5), and experiments in nonhuman primates showed protection against SHIV, but not SIV, challenge (Casimiro et al., 2005). This vaccine was evaluated in two large efficacy trials. The first, called the STEP Study, was conducted in North and South America, the Caribbean, and Australia among 3000 at-risk volunteers, half of whom had evidence of prior exposure to Ad5. The second trial, which began 2 years after the first, was scheduled to enroll 3000 at-risk heterosexual men and women in South Africa. These companion studies sought to evaluate vaccine efficacy among men who have sex with men and heterosexuals, and in individuals living in countries with HIV subtypes B and C epidemics. A joint endpoint of efficacy for prevention of HIV acquisition and efficacy for amelioration of disease course (measured as reduction of set-point viral load) was evaluated in both trials. Vaccinations were discontinued in both trials in September 2007, when an interim analysis of data from the STEP study showed no efficacy for either outcome. In the study population examined in this interim analysis (participants with baseline Ad5 < 200), there were 24 infections among 741 vaccinees and 21 infections among 762 placebo recipients; the geometric mean plasma vRNA level was similar in infected vaccine vs. placebo recipients (~40,000 vs. ~26,000 copies/ml, respectively). Among all men who participated in this trial (women were excluded from further analysis as only one infection occurred among a woman), 49 HIV infections occurred in 914 vaccine recipients and 33 in 922 placebo recipients (Merck Research and Development News, 2007). Initial multivariate analyses suggested that any increased risk associated with vaccination was limited to men with evidence of prior exposure to Ad5, and uncircumcised men. The treatment hazard ratio (vaccine vs. placebo) was 0.8 (0.4, 1.6) in circumcised men with baseline Ad5 titers ≤ 18, 1.4 (0.6, 3.2) in circumcised men with baseline Ad5 titers > 18, 2.5 (0.8, 8.0) in uncircumcised men with baseline Ad5 titers ≤ 18 and 4.3 (1.7, 11.0) in uncircumcised men with baseline Ad5 titers > 18 (Merck Research and Development News, 2007). The reasons for this apparent increase in susceptibility to HIV associated with vaccination among men with prior Ad immunity and uncircumcised men are presently unclear, but continue to be investigated.

The second Ad5 candidate, developed by the US National Institutes of Health's (NIH) Vaccine Research Center (VRC), is designed to provide multiclade protection. The vaccine regimen consists of a six-plasmid DNA vaccine expressing subtype B *gag*, *pol*, *nef*, and subtype A, B, C envelop genes, respectively, given as a prime at 0, 1, and 3 months. This priming regimen is followed by a boost at 6 months consisting of a mixture of four adenoviruses containing inserts comprised of a subtype B *gag-pol* gene fusion product, and envelope genes from HIV-1 subtypes A, B, and C. In animal models, this vaccine regimen was immunogenic (Shiver et al., 2002); it did not prevent infection but was associated with decreased viral load, less frequent progression to AIDS, and preservation of central memory cells (Letvin et al., 2006; Mascola et al., 2005; Santra et al., 2005). The regimen was found to be well tolerated and immunogenic in phase 1 and 2 trials and is scheduled for efficacy testing in early 2009.

The extensive previous experimental use of adenovirus vectors in gene therapy and vaccinology, and the established safety and immunogenicity properties of these vectors continue to make adenoviruses attractive for use in vaccine development. However, in light of the STEP study, alternatives to adenovirus 5 vector constructs are being investigated, using serotypes with lower incidence of natural infection. Less frequently occurring adenovirus subtypes, such as adenovirus serotype 26 and adenovirus serotype 35, as well as chimeric products of several adenoviruses, can be used as vectors to elicit an immune response and are entering clinical trials in humans (Abbink et al., 2007; Lemckert et al., 2005). Other potential vectors include adeno-associated virus (AAV) and vesicular stomatitis virus (VSV). AAV requires a helper virus for replication, and is not known to cause human disease. Studies using an SIV insert in macaques that were later challenged with SIV showed that the vaccinated animals had better virologic suppression and control than placebo recipients (Johnson et al., 2005). AAV has been employed as a gene vector for the treatment of other conditions including cystic fibrosis, hemophilia, and rheumatologic disorders. However, the use of this vector suffered a setback when a participant receiving AAV-based therapy for inflammatory arthritis died from disseminated histoplasmosis. Although later found not likely to be associated with the AAV product, this event has slowed the development of AAV as a vaccine vector. A recombinant rabies-VSV product with an SIV insert has also been used in macaques that were later challenged with a pathogenic SHIV, and those that were vaccinated also had attenuated disease progression (McKenna et al., 2007). These potential vectors are in late-phase preclinical or early clinical testing and will be used alone or in prime-boost regimens.

In the field of HIV vaccine research, a number of alternative approaches have been studied to increase immunogenicity and delivery of antigen. One of these methods, a heterologous prime-boost vaccine strategy is being used in about half of the presently ongoing trials. This technique uses a "prime" vaccine that initially stimulates the immune system generating memory T and B cells. Later, a boost is given with a different vector, a different insert, or sometimes both. Conceptually, this approach may work by circumventing immunity that might have developed to the initial vector. However, there is limited data to show that this approach elicits better immune responses.

Additionally, novel delivery methods have mechanically improved vaccine immunogenicity and antigen delivery. The Biojector® is one such device. It uses a single, needleless, syringe-based device that injects a vaccine product under high pressure by compressed carbon dioxide cartridges. This device is used primarily to inject DNA vaccine product into the tissues, theoretically allowing for increased cellular uptake. However, immunogenicity results obtained with this technique have been inconsistent (Brave et al., 2007; Meseda et al., 2006; Rao et al., 2006). Electroporation is a second method that is being developed to help improve the immunogenicity of DNA vaccines. Already used experimentally in gene therapy and cancer vaccines, electroporation increases the permeability of cell membranes and theoretically enhances the uptake of the immunogen into cells. In nonhuman primates (Luckay et al., 2007), electroporation has increased immune response 10- to 40-fold at one-fifth the dose of immunogen compared to conventional needle-and-syringe delivery.

Adding adjuvants to HIV vaccine candidates is another area of ongoing research. Many of the current vaccine strategies do not produce immune responses that are robust, long-lived, and appropriately focused on production of nAb or cytotoxic responses. Adjuvants could overcome this problem by targeting the antigen to antigen-presenting cells, or increasing immune response by stimulating production of cytokines and costimulatory molecules, or both. Adjuvants such as polymeric microspheres (e.g., polylactide-coglycolide or PLG) have been tested with HIV vaccine candidates to increase immune response through facilitating interactions with antigen-presenting cells (Singh and Srivastava, 2003). Other adjuvants being tested in conjunction with HIV vaccine candidates attempt to increase induction of relevant cytokines and upregulate costimulatory molecules. CpG, unmethylated cytosine-guanine dinucleotides, which acts as a ligand for toll-like receptor 9 (TLR9), is one example (Kojima et al., 2002). Stimulation of TLRs, in turn, enhances and directs the immune response. A related approach is the administration of vaccine candidates with costimulatory molecules such as interleukin-12 (IL-12), IL-15, or granulocyte-macrophage stimulating factor (GM-CSF) in an attempt to manipulate the immune response and increase cell-mediated immunity to the coadministered HIV vaccine antigen (Calarota and Weiner, 2004).

When potential genetic inserts are considered for vectors, the biological diversity of HIV becomes a significant issue. One the major challenges of the HIV epidemic is the significant viral diversity of the target pathogen. The rate of mutation of HIV is so rapid that the diversity of virus circulating within an infected individual's blood may be broader than the genetic diversity of influenza circulating worldwide (Fig. 25.8) (Korber et al., 2001). It is also unclear that inserts based on a single clade or several limited clades will provide protection against genetically diverse isolates. Although the responses generated by natural infection have been shown in vitro to have cross-clade reactivity, this protection has not been demonstrated in humans (Coplan et al., 2005; Cao et al., 1997).

Several approaches have been used to select genetic inserts for vaccines. Gene segments from one or more "representative" naturally-occurring viral sequences may be selected. Alternatively, artificial insert sequences may be designed by a variety of methods, to maximize the chances of inducing broadly reactive immune responses of protective value following heterologous challenge. Consensus sequences are constructed by simply using the most common amino acid at each position in the genome among naturally occurring strains (Gaschen et al., 2002). Ancestral sequences are developed using methods based on phylogenetic analysis of existing sequences: maximum likelihood analysis of circulating sequences is used to infer the most likely genetic makeup of an "ancestor" for currently circulating viruses (Gaschen et al., 2002). Center-of-tree design approaches seek to find a phylogenetically determined ancestral sequence which minimizes the genetic distance to current isolates (Gaschen et al., 2002; Mullins et al., 2004). Finally, epitope-based methods include a series of nucleic acid sequences encoding immunogenic epitopes of high frequency within the circulating HIV genome.

CONCLUSIONS

HIV is one of the greatest public health challenges faced worldwide over the last century. Despite advances in therapy that have made HIV a manageable chronic illness in areas where ART is available, the best approach to stemming the epidemic globally will be to develop a vaccine that is capable of reducing the incidence of transmission and/or attenuate disease progression in infected individuals. Unique aspects of the epidemiology and pathophysiology of HIV infection have made the development of an effective and economically feasible vaccine a challenging endeavor for scientists, public health officials, governments, and populations at risk. In the face of these difficulties, the research community may be required to consider vaccines that meet only some of the properties of an ideal vaccine. As we move forward, several questions will need to be addressed, including: (1) What would be the impact of a vaccine that mitigates disease but does not directly prevent infection in vaccinees? What data would be necessary to support the development and distribution of such a vaccine? (2) What would be the effect of such a vaccine on mucosal viral shedding and the subsequent risk of HIV transmission from the vaccinated individual? (3) What role would such a vaccine have in multifaceted prevention programs? and (4) Could slower disease progression due to administration of a "therapeutic" vaccine aid the implementation of treatment programs in resource-limited settings, by making the number of individuals needing therapy more manageable? The next decade will provide the opportunities for these and other questions to be answered, and will likely see the proposal of a number of new questions as second- and third-generation vaccines are developed. Multidisciplinary approaches with vaccination as one element of a combined strategy may ultimately lead to complete control of this pandemic.

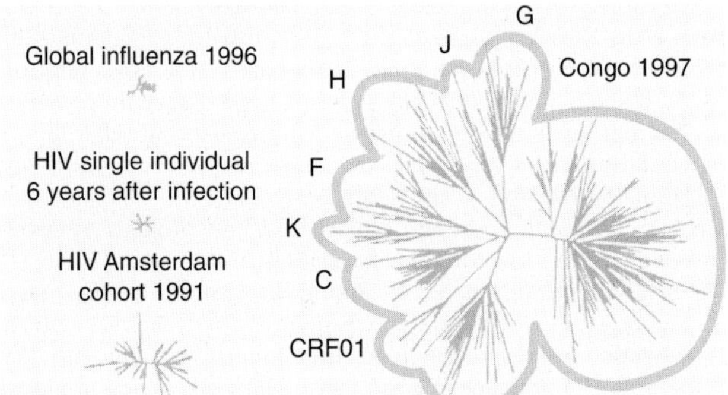

FIGURE 25.8 HIV genetic diversity. Comparison of genetic diversity of global influenza in a calendar year, HIV in an individual, a single city cohort in 1991, and a single sub-Saharan country in 1997. Reproduced with permission from Duerr et al. (2008) and Garber et al. (2004).

KEY ISSUES

- HIV has claimed an estimated 25 million lives, and there are currently 33 million infected persons.
- The HIV epidemic has had a significant socioeconomic impact in virtually all developing and developed nations, but disproportionately afflicts populations at its epicenter in sub-Saharan Africa.
- The HIV epidemic continues to expand into large population centers in Asia and other regions around the world.
- In most untreated individuals, HIV infection eventually leads to collapse of the immune system and death due to opportunistic infections.
- Rare individuals may remain entirely asymptomatic, but there have been no documented cases of spontaneous cure from infection.
- While ART may arrest disease progression in some, treatment is costly and unavailable to most infected individuals.
- There is presently no cure for HIV, nor a preventative vaccine.
- An effective vaccine will be critical to controlling the epidemic.
- Current vaccine efforts seek to find a vaccine that can either reduce the likelihood of infection, or attenuate disease progression in infected individuals.
- Vaccine design is impeded by the tremendous antigenic variation characteristic of HIV.

- Early vaccine development efforts sought to elicit nAb responses; recent vaccine efforts have focused on eliciting a protective T-cell response.

ACKNOWLEDGMENT

The authors thank Katie Skibinski for editorial assistance in the preparation of this chapter.

References

Abbink, P., Lemckert, A.A., Ewald, B.A., Lynch, D.M., Denholtz, M., Smits, S., Holterman, L., Damen, I., Vogels, R., Thorner, A.R., O'Brien, K.L., Carville, A., Mansfield, K.G., Goudsmit, J., Havenga, M.J. and Barouch, D.H. Comparative seroprevalence and immunogenicity of six rare serotype recombinant adenovirus vaccine vectors from subgroups B and D. J. Virol. 2007; 81(9):4654–4663.

Abdool-Karim, Q. and Abdool-Karim, S.S. The evolving HIV epidemic in South Africa. Int. J. Epidemiol. 2002; 31(1):37–40.

Adams, M., Sharmeen, L., Kimpton, J., Romeo, J.M., Garcia, J. V., Peterlin, B.M., Groudine, M. and Emerman, M. Cellular latency in human immunodeficiency virus-infected individuals with high CD4 levels can be detected by the presence of promoter-proximal transcripts. Proc. Natl. Acad. Sci. U.S.A. 1994; 91(9):3862–3866.

AIDS Vaccine Clearing House. Accessed Nov. 1, 2007. http://aidsvaccineclearinghouse.org/.

Albert, J., Abrahamsson, B., Nagy, K., Aurelius, E., Gaines, H., Nystrom, G. and Fenyo, E.M. Rapid development of isolate-specific neutralizing antibodies after primary HIV-1 infection and consequent emergence of virus variants which resist neutralization by autologous sera. Aids 1990; 4(2):107–112.

Amadori, A., Zamarchi, R. and Chieco-Bianchi, L. CD4: CD8 ratio and HIV infection: the 'tap-and-drain' hypothesis. Immunol. Today 1996; 17(9):414–417.

Amara, R.R., Villinger, F., Altman, J.D., Lydy, S.L., O'Neil, S.P., Staprans, S.I., Montefiori, D.C., Xu, Y., Herndon, J.G., Wyatt, L. S., Candido, M.A., Kozyr, N.L., Earl, P.L., Smith, J.M., Ma, H.L., Grimm, B.D., Hulsey, M.L., Miller, J., McClure, H.M., McNicholl, J.M., Moss, B. and Robinson, H.L. Control of a mucosal challenge and prevention of AIDS by a multiprotein DNA/MVA vaccine. Science 2001; 292(5514):69–74.

Anderson, R.W., Ascher, M.S. and Sheppard, H.W. Direct HIV cytopathicity cannot account for CD4 decline in AIDS in the presence of homeostasis: a worst-case dynamic analysis. J. Acquir. Immune Defic. Syndr. Hum. Retrovirol. 1998; 17(3):245–252.

Andersson, J., Kinloch, S., Sonnerborg, A., Nilsson, J., Fehniger, T.E., Spetz, A.L., Behbahani, H., Goh, L.E., McDade, H., Gazzard, B., Stellbrink, H., Cooper, D. and Perrin, L. Low levels of perforin expression in CD8+ T lymphocyte granules in lymphoid tissue during acute human immunodeficiency virus type 1 infection. J. Infect. Dis. 2002; 185(9):1355–1358.

Arthur, L.O., Bess, J.W., Jr., Waters, D.J., Pyle, S.W., Kelliher, J.C., Nara, P.L., Krohn, K., Robey, W.G., Langlois, A.J., Gallo, R.C. et al. Challenge of chimpanzees (Pan troglodytes) immunized with human immunodeficiency virus envelope glycoprotein gp120. J. Virol. 1989; 63(12):5046–5053.

Atkinson, K., Hansen, J.A., Storb, R., Goehle, S., Goldstein, G. and Thomas, E.D. T-cell subpopulations identified by monoclonal antibodies after human marrow transplantation. I. Helper-inducer and cytotoxic-suppressor subsets. Blood 1982; 59(6):1292–1298.

Barker, E., Mackewicz, C.E. and Levy, J.A. Effects of TH1 and TH2 cytokines on CD8+ cell response against human immunodeficiency virus: implications for long-term survival. Proc. Natl. Acad. Sci. U.S.A. 1995; 92(24):11135–11139.

Barouch, D.H., Santra, S., Schmitz, J.E., Kuroda, M.J., Fu, T.M., Wagner, W., Bilska, M., Craiu, A., Zheng, X.X., Krivulka, G. R., Beaudry, K., Lifton, M.A., Nickerson, C.E., Trigona, W. L., Punt, K., Freed, D.C., Guan, L., Dubey, S., Casimiro, D., Simon, A., Davies, M.E., Chastain, M., Strom, T.B., Gelman, R. S., Montefiori, D.C., Lewis, M.G., Emini, E.A., Shiver, J.W. and Letvin, N.L. Control of viremia and prevention of clinical AIDS in rhesus monkeys by cytokine-augmented DNA vaccination. Science 2000; 290(5491):486–492.

Barre-Sinoussi, F., Chermann, J.C., Rey, F., Nugeyre, M.T., Chamaret, S., Gruest, J., Dauguet, C., Axler-Blin, C., Vezinet-Brun, F., Rouzioux, C., Rozenbaum, W. and Montagnier, L. Isolation of a T-lymphotropic retrovirus from a patient at risk for acquired immune deficiency syndrome (AIDS). Science 1983; 220(4599):868–871.

Battegay, M., Moskophidis, D., Rahemtulla, A., Hengartner, H., Mak, T.W. and Zinkernagel, R.M. Enhanced establishment of a virus carrier state in adult CD4+ T-cell-deficient mice. J. Virol. 1994; 68(7):4700–4704.

Berkowitz, R.D., Alexander, S., Bare, C., Linquist-Stepps, V., Bogan, M., Moreno, M.E., Gibson, L., Wieder, E.D., Kosek, J., Stoddart, C.A. and McCune, J.M. CCR5- and CXCR4-utilizing strains of human immunodeficiency virus type 1 exhibit differential tropism and pathogenesis in vivo. J. Virol. 1998; 72(12):10108–10117.

Berman, P.W., Groopman, J.E., Gregory, T., Clapham, P.R., Weiss, R. A., Ferriani, R., Riddle, L., Shimasaki, C., Lucas, C., Lasky, L.A. et al. Human immunodeficiency virus type 1 challenge of chimpanzees immunized with recombinant envelope glycoprotein gp120. Proc. Natl. Acad. Sci. U.S.A. 1988; 85(14):5200–5204.

Berson, J.F., Long, D., Doranz, B.J., Rucker, J., Jirik, F.R. and Doms, R.W. A seven-transmembrane domain receptor involved in fusion and entry of T-cell-tropic human immunodeficiency virus type 1 strains. J. Virol. 1996; 70(9):6288–6295.

Betts, M.R., Ambrozak, D.R., Douek, D.C., Bonhoeffer, S., Brenchley, J.M., Casazza, J.P., Koup, R.A. and Picker, L.J. Analysis of total human immunodeficiency virus (HIV)-specific CD4(+) and CD8(+) T-cell responses: relationship to viral load in untreated HIV infection. J. Virol. 2001; 75(24):11983–11991.

Betts, M.R., Krowka, J.F., Kepler, T.B., Davidian, M., Christopherson, C., Kwok, S., Louie, L., Eron, J., Sheppard, H. and Frelinger, J. A. Human immunodeficiency virus type 1-specific cytotoxic T lymphocyte activity is inversely correlated with HIV type 1 viral load in HIV type 1-infected long-term survivors. AIDS Res. Hum. Retroviruses 1999; 15(13):1219–1228.

Bieniasz, P.D. Late budding domains and host proteins in enveloped virus release. Virology 2006; 344(1):55–63.

Binley, J.M., Jin, X., Huang, Y., Zhang, L., Cao, Y., Ho, D.D. and Moore, J.P. Persistent antibody responses but declining cytotoxic T-lymphocyte responses to multiple human immunodeficiency virus type 1 antigens in a long-term nonprogressing individual with a defective p17 proviral sequence and no detectable viral RNA expression. J. Virol. 1998; 72(4):3472–3474.

Bobkov, A., Cheingsong-Popov, R., Garaev, M., Rzhaninova, A., Kaleebu, P., Beddows, S., Bachmann, M.H., Mullins, J. I., Louwagie, J., Janssens, W. et al. Identification of an env G subtype and heterogeneity of HIV-1 strains in the Russian Federation and Belarus. Aids 1994; 8(12):1649–1655.

Bolton, D.L., Hahn, B.I., Park, E.A., Lehnhoff, L.L., Hornung, F. and Lenardo, M.J. Death of CD4(+) T-cell lines caused by human immunodeficiency virus type 1 does not depend on caspases or apoptosis. J. Virol. 2002; 76(10):5094–5107.

Bolton, D.L. and Lenardo, M.J. Vpr cytopathicity independent of G2/M cell cycle arrest in human immunodeficiency virus type 1-infected CD4+ T cells. J. Virol. 2007; 81(17):8878–8890.

Borrow, P., Lewicki, H., Hahn, B.H., Shaw, G.M. and Oldstone, M.B. Virus-specific CD8+ cytotoxic T-lymphocyte activity associated with control of viremia in primary human immunodeficiency virus type 1 infection. J. Virol. 1994; 68(9):6103–6110.

Bouscarat, F., Levacher-Clergeot, M., Dazza, M.C., Strauss, K. W., Girard, P.M., Ruggeri, C. and Sinet, M. Correlation of CD8 lymphocyte activation with cellular viremia and plasma HIV RNA levels in asymptomatic patients infected by human immunodeficiency virus type 1. AIDS Res. Hum. Retroviruses 1996; 12(1):17–24.

Brave, A., Boberg, A., Gudmundsdotter, L., Rollman, E., Hallermalm, K., Ljungberg, K., Blomberg, P., Stout, R., Paulie, S., Sandstrom, E., Biberfeld, G., Earl, P., Moss, B., Cox, J.H. and Wahren, B. A new multi-clade DNA prime/recombinant MVA boost vaccine induces broad and high levels of HIV-1-specific CD8(+) T-cell and humoral responses in mice. Mol. Ther. 2007; 15(9):1724–1733.

Brenchley, J.M., Ruff, L.E., Casazza, J.P., Koup, R.A., Price, D.A. and Douek, D.C. Preferential infection shortens the life span of human immunodeficiency virus-specific CD4+ T cells in vivo. J. Virol. 2006; 80(14):6801–6809.

Brenchley, J.M., Schacker, T.W., Ruff, L.E., Price, D.A., Taylor, J.H., Beilman, G.J., Nguyen, P.L., Khoruts, A., Larson, M., Haase, A.T. and Douek, D.C. CD4+ T cell depletion during all stages of HIV disease occurs predominantly in the gastrointestinal tract. J. Exp. Med. 2004; 200(6):749–759.

Brenner, B.G., Roger, M., Routy, J.P., Moisi, D., Ntemgwa, M., Matte, C., Baril, J.G., Thomas, R., Rouleau, D., Bruneau, J., Leblanc, R., Legault, M., Tremblay, C., Charest, H. and Wainberg, M.A. High rates of forward transmission events after acute/early HIV-1 infection. J. Infect. Dis. 2007; 195(7):951–959.

Broder, S., Merigan, T.C. and Bolognesi, D. Textbook of AIDS Medicine. Baltimore: Williams Wilkins, 1994, xix, 951 p., [14] p. of plates.

Bukrinskaya, A., Brichacek, B., Mann, A. and Stevenson, M. Establishment of a functional human immunodeficiency virus type 1 (HIV-1) reverse transcription complex involves the cytoskeleton. J. Exp. Med. 1998; 188(11):2113–2125.

Burton, D.R., Desrosiers, R.C., Doms, R.W., Feinberg, M.B., Gallo, R.C., Hahn, B., Hoxie, J.A., Hunter, E., Korber, B., Landay, A., Lederman, M.M., Lieberman, J., McCune, J.M., Moore, J.P., Nathanson, N., Picker, L., Richman, D., Rinaldo, C., Stevenson, M., Watkins, D.I., Wolinsky, S.M. and Zack, J.A. Public health. A sound rationale needed for phase III HIV-1 vaccine trials. Science 2004; 303(5656):316.

Calarota, S.A. and Weiner, D.B. Enhancement of human immunodeficiency virus type 1-DNA vaccine potency through incorporation of T-helper 1 molecular adjuvants. Immunol. Rev. 2004; 199:84–99.

Cao, H., Kanki, P., Sankale, J.L., Dieng-Sarr, A., Mazzara, G.P., Kalams, S.A., Korber, B., Mboup, S. and Walker, B.D. Cytotoxic T-lymphocyte cross-reactivity among different human immunodeficiency virus type 1 clades: implications for vaccine development. J. Virol. 1997; 71(11):8615–8623.

Carrington, M., Nelson, G.W., Martin, M.P., Kissner, T., Vlahov, D., Goedert, J.J., Kaslow, R., Buchbinder, S., Hoots, K. and O'Brien, S.J. HLA and HIV-1: heterozygote advantage and B*35-Cw*04 disadvantage. Science 1999; 283(5408):1748–1752.

Carrington, M. and O'Brien, S.J. The influence of HLA genotype on AIDS. Annu. Rev. Med. 2003; 54:535–551.

Casimiro, D.R., Wang, F., Schleif, W.A., Liang, X., Zhang, Z.Q., Tobery, T.W., Davies, M.E., McDermott, A.B., O'Connor, D. H., Fridman, A., Bagchi, A., Tussey, L.G., Bett, A.J., Finnefrock, A.C., Fu, T.M., Tang, A., Wilson, K.A., Chen, M., Perry, H.C., Heidecker, G.J., Freed, D.C., Carella, A., Punt, K.S., Sykes, K. J., Huang, L., Ausensi, V.I., Bachinsky, M., Sadasivan-Nair, U., Watkins, D.I., Emini, E.A. and Shiver, J.W. Attenuation of simian immunodeficiency virus SIVmac239 infection by prophylactic immunization with dna and recombinant adenoviral vaccine vectors expressing Gag. J. Virol. 2005; 79(24):15547–15555.

Cates, W., Jr. Review of non-hormonal contraception (condoms, intrauterine devices, nonoxynol-9 and combos) on HIV acquisition. J. Acquir. Immune Defic. Syndr. 2005; 38(Suppl. 1):S8–S10.

Cerf-Bensussan, N. and Guy-Grand, D. Intestinal intraepithelial lymphocytes. Gastroenterol. Clin. North Am. 1991; 20(3):549–576.

Champagne, P., Ogg, G.S., King, A.S., Knabenhans, C., Ellefsen, K., Nobile, M., Appay, V., Rizzardi, G.P., Fleury, S., Lipp, M., Forster, R., Rowland-Jones, S., Sekaly, R.P., McMichael, A.J. and Pantaleo, G. Skewed maturation of memory HIV-specific CD8 T lymphocytes. Nature 2001; 410(6824):106–111.

Chan, D.C. and Kim, P.S. HIV entry and its inhibition. Cell 1998; 93(5):681–684.

Chan, D.C., Fass, D., Berger, J.M. and Kim, P.S. Core structure of gp41 from the HIV envelope glycoprotein. Cell 1997; 89(2):263–273.

Chiang, C.F., Tenenbaum, S.A., Verret, C.R., Leissinger, C.A. and Garry, R.F. Activity of granzyme A, a serine protease in the killing granules of cytotoxic T lymphocytes, is reduced in cells from HIV-infected hemophiliacs. AIDS Res. Hum. Retroviruses 1996; 12(3):235–239.

Chun, T.W., Carruth, L., Finzi, D., Shen, X., DiGiuseppe, J.A., Taylor, H., Hermankova, M., Chadwick, K., Margolick, J., Quinn, T. C., Kuo, Y.H., Brookmeyer, R., Zeiger, M.A., Barditch-Crovo, P. and Siliciano, R.F. Quantification of latent tissue reservoirs and total body viral load in HIV-1 infection. Nature 1997; 387(6629):183–188.

Clavel, F., Mansinho, K., Chamaret, S., Guetard, D., Favier, V., Nina, J., Santos-Ferreira, M.O., Champalimaud, J.L. and Montagnier, L. Human immunodeficiency virus type 2 infection associated with AIDS in West Africa. N. Engl. J. Med. 1987; 316(19):1180–1185.

Cocchi, F., DeVico, A.L., Garzino-Demo, A., Arya, S.K., Gallo, R.C. and Lusso, P. Identification of RANTES, MIP-1 alpha, and MIP-1 beta as the major HIV-suppressive factors produced by CD8+ T cells. Science 1995; 270(5243):1811–1815.

Coffin, J., Haase, A., Levy, J.A., Montagnier, L., Oroszlan, S., Teich, N., Temin, H., Toyoshima, K., Varmus, H., Vogt, P. et al. Human immunodeficiency viruses. Science 1986; 232(4751):697.

Connor, R.I., Sheridan, K.E., Ceradini, D., Choe, S. and Landau, N. R. Change in coreceptor use coreceptor use correlates with disease progression in HIV-1--infected individuals. J. Exp. Med. 1997; 185(4):621–628.

Coplan, P.M., Gupta, S.B., Dubey, S.A., Pitisuttithum, P., Nikas, A., Mbewe, B., Vardas, E., Schechter, M., Kallas, E.G., Freed, D.C., Fu, T.M., Mast, C.T., Puthavathana, P., Kublin, J., Brown Collins, K., Chisi, J., Pendame, R., Thaler, S.J., Gray, G., McIntyre, J., Straus, W.L., Condra, J.H., Mehrotra, D.V., Guess, H.A., Emini, E. A. and Shiver, J.W. Cross-reactivity of anti-HIV-1 T cell immune responses among the major HIV-1 clades in HIV-1-positive individuals from 4 continents. J. Infect. Dis. 2005; 191(9):1427–1434.

De Boer, R.J., Mohri, H., Ho, D.D. and Perelson, A.S. Turnover rates of B cells, T cells, and NK cells in simian immunodeficiency virus-infected and uninfected rhesus macaques. J. Immunol. 2003; 170(5):2479–2487.

de Bruyn, G., Rossini, A.J., Chiu, Y.L., Holman, D., Elizaga, M. L., Frey, S.E., Burke, D., Evans, T.G., Corey, L. and Keefer, M.C. Safety profile of recombinant canarypox HIV vaccines. Vaccine 2004; 22(5-6):704–713.

de Roda Husman, A.M., Koot, M., Cornelissen, M., Keet, I.P., Brouwer, M., Broersen, S.M., Bakker, M., Roos, M.T., Prins, M., de Wolf, F., Coutinho, R.A., Miedema, F., Goudsmit, J. and Schuitemaker, H. Association between CCR5 genotype and the clinical course of HIV-1 infection. Ann. Intern. Med. 1997; 127(10):882–890.

Decker, J.M., Bibollet-Ruche, F., Wei, X., Wang, S., Levy, D.N., Wang, W., Delaporte, E., Peeters, M., Derdeyn, C.A., Allen, S., Hunter, E., Saag, M.S., Hoxie, J.A., Hahn, B.H., Kwong, P.D., Robinson, J.E. and Shaw, G.M. Antigenic conservation and immunogenicity of the HIV coreceptor binding site. J. Exp. Med. 2005; 201(9):1407–1419.

Deeks, S.G., Schweighardt, B., Wrin, T., Galovich, J., Hoh, R., Sinclair, E., Hunt, P., McCune, J.M., Martin, J.N., Petropoulos, C.J. and Hecht, F.M. Neutralizing antibody responses against autologous and heterologous viruses in acute versus chronic human immunodeficiency virus (HIV) infection: evidence for a constraint on the ability of HIV to completely evade neutralizing antibody responses. J. Virol. 2006; 80(12):6155–6164.

Deng, H., Liu, R., Ellmeier, W., Choe, S., Unutmaz, D., Burkhart, M., Di Marzio, P., Marmon, S., Sutton, R.E., Hill, C.M., Davis, C.B., Peiper, S.C., Schall, T.J., Littman, D.R. and Landau, N.R. Identification of a major co-receptor for primary isolates of HIV-1. Nature 1996; 381(6584):661–666.

Dolin, R., Graham, B.S., Greenberg, S.B., Tacket, C.O., Belshe, R.B., Midthun, K., Clements, M.L., Gorse, G.J., Horgan, B.W., Atmar, R.L. et al. The safety and immunogenicity of a human immunodeficiency virus type 1 (HIV-1) recombinant gp160 candidate vaccine in humans. NIAID AIDS Vaccine Clinical Trials Network. Ann. Intern. Med. 1991; 114(2):119–127.

Dorling, D., Shaw, M. and Davey Smith, G. Global inequality of life expectancy due to AIDS. BMJ 2006; 332(7542):662–664.

Douek, D.C. Disrupting T-cell homeostasis: how HIV-1 infection causes disease. AIDS Rev. 2003; 5(3):172–177.

Douek, D.C., Brenchley, J.M., Betts, M.R., Ambrozak, D.R., Hill, B.J., Okamoto, Y., Casazza, J.P., Kuruppu, J., Kunstman, K., Wolinsky, S., Grossman, Z., Dybul, M., Oxenius, A., Price, D.A., Connors, M. and Koup, R.A. HIV preferentially infects HIV-specific CD4+ T cells. Nature 2002; 417(6884):95–98.

Duerr, A., Corey, L., and Wasserheit, J.N. HIV vacccines. In King K. Holmes, Frederick P. Sparling, Walter E. Stamm, Peter Poit, Judith N. Wasserheit, Lawrence Corey, Myron S. cohen, and D. Heather Watts (Eds.), *Sexually Transmitted Diseases* (4th edition) New York: McGraw-Hill, Health Professions Division, 2008; 1937–1954.

Duerr, A., Wasserheit, J. and Corey, L. HIV vaccines: new frontiers in vaccine development. Clin. Infect. Dis. 2006; 43:500–511.

Dumitrescu, O., Kalish, M.L., Kliks, S.C., Bandea, C.I. and Levy, J.A. Characterization of human immunodeficiency virus type 1 isolates from children in Romania: identification of a new envelope subtype. J. Infect. Dis. 1994; 169(2):281–288.

Durack, D.T. Opportunistic infections and Kaposi's sarcoma in homosexual men. N. Engl. J. Med. 1981; 305(24):1465–1467.

Dvorin, J.D., Bell, P., Maul, G.G., Yamashita, M., Emerman, M. and Malim, M.H. Reassessment of the roles of integrase and the central DNA flap in human immunodeficiency virus type 1 nuclear import. J. Virol. 2002; 76(23):12087–12096.

Earl, P.L., Doms, R.W. and Moss, B. Oligomeric structure of the human immunodeficiency virus type 1 envelope glycoprotein. Proc. Natl. Acad. Sci. U.S.A. 1990; 87(2):648–652.

Emini, E.A., Schleif, W.A., Nunberg, J.H., Conley, A.J., Eda, Y., Tokiyoshi, S., Putney, S.D., Matsushita, S., Cobb, K.E., Jett, C. M. et al. Prevention of HIV-1 infection in chimpanzees by gp120 V3 domain-specific monoclonal antibody. Nature 1992; 355(6362):728–730.

Centers for Disease Control (CDC) Epidemiology of HIV/AIDS – United States, 1981–2005. MMWR Morb. Mortal. Wkly. Rep. 2006; 55(21):589–592.

Feng, Y., Broder, C.C., Kennedy, P.E. and Berger, E.A. HIV-1 entry cofactor: functional cDNA cloning of a seven-transmembrane, G protein-coupled receptor. Science 1996; 272(5263):872–877.

Finzi, D., Hermankova, M., Pierson, T., Carruth, L.M., Buck, C., Chaisson, R.E., Quinn, T.C., Chadwick, K., Margolick, J., Brookmeyer, R., Gallant, J., Markowitz, M., Ho, D.D., Richman, D.D. and Siliciano, R.F. Identification of a reservoir for HIV-1 in patients on highly active antiretroviral therapy. Science 1997; 278(5341):1295–1300.

Flynn, N.M., Forthal, D.N., Harro, C.D., Judson, F.N., Mayer, K.H. and Para, M.F. Placebo-controlled phase 3 trial of a recombinant glycoprotein 120 vaccine to prevent HIV-1 infection. J. Infect. Dis. 2005; 191(5):654–665.

Fournier, S., Chaffaut, C., Maillard, A., Loze, B., Lascoux, C., Gerard, L., Timsit, J., David, F., Bergmann, J.F., Oksenhendler, E., Sereni, D., Chevret, S. and Molina, J.M. Factors associated with virological response in HIV-infected patients failing antiretroviral therapy: a prospective cohort study. HIV. Med. 2005; 6(2):129–134.

Franchini, G., Gurunathan, S., Baglyos, L., Plotkin, S. and Tartaglia, J. Poxvirus-based vaccine candidates for HIV: two decades of experience with special emphasis on canarypox vectors. Expert. Rev. Vaccines 2004; 3(4 Suppl):S75–sS88.

Fraser, C., Ferguson, N.M., Ghani, A.C., Prins, J.M., Lange, J.M., Goudsmit, J., Anderson, R.M. and de Wolf, F. Reduction of the HIV-1-infected T-cell reservoir by immune activation treatment is dose-dependent and restricted by the potency of antiretroviral drugs. Aids 2000; 14(6):659–669.

Gallo, R., Wong-Staal, F., Montagnier, L., Haseltine, W.A. and Yoshida, M. HIV/HTLV gene nomenclature. Nature 1988; 333(6173):504.

Gallo, R.C., Salahuddin, S.Z., Popovic, M., Shearer, G.M., Kaplan, M., Haynes, B.F., Palker, T.J., Redfield, R., Oleske, J., Safai, B. et al. Frequent detection and isolation of cytopathic retroviruses (HTLV-III) from patients with AIDS and at risk for AIDS. Science 1984; 224(4648):500–503.

Gao, X., Nelson, G.W., Karacki, P., Martin, M.P., Phair, J., Kaslow, R., Goedert, J.J., Buchbinder, S., Hoots, K., Vlahov, D., O'Brien, S.J. and Carrington, M. Effect of a single amino acid change in MHC class I molecules on the rate of progression to AIDS. N. Engl. J. Med. 2001; 344(22):1668–1675.

Garber, D.A., Silverstri, G. and Feinberg, M.B. Prospects for an AIDS vaccine: Three big questions, no easy answers. Lancet Infect Dis. 2004; 4:397–413.

Garba, M.L., Pilcher, C.D., Bingham, A.L., Eron, J. and Frelinger, J.A. HIV antigens can induce TGF-beta(1)-producing immunoregulatory CD8+ T cells. J. Immunol. 2002; 168(5):2247–2254.

Gaschen, B., Taylor, J., Yusim, K., Foley, B., Gao, F., Lang, D., Novitsky, V., Haynes, B., Hahn, B.H., Bhattacharya, T. and Korber, B. Diversity considerations in HIV-1 vaccine selection. Science 2002; 296(5577):2354–2360.

Gauduin, M.C., Safrit, J.T., Weir, R., Fung, M.S. and Koup, R.A. Pre- and postexposure protection against human immunodeficiency virus type 1 infection mediated by a monoclonal antibody. J. Infect. Dis. 1995; 171(5):1203–1209.

Gea-Banacloche, J.C., Migueles, S.A., Martino, L., Shupert, W.L., McNeil, A.C., Sabbaghian, M.S., Ehler, L., Prussin, C., Stevens, R., Lambert, L., Altman, J., Hallahan, C.W., de Quiros, J.C. and

Connors, M. Maintenance of large numbers of virus-specific CD8+ T cells in HIV-infected progressors and long-term non-progressors. J. Immunol. 2000; 165(2):1082–1092.

Geffin, R., Hutto, C., Andrew, C. and Scott, G.B. A longitudinal assessment of autologous neutralizing antibodies in children perinatally infected with human immunodeficiency virus type 1. Virology 2003; 310(2):207–215.

Geijtenbeek, T.B., Kwon, D.S., Torensma, R., van Vliet, S.J., van Duijnhoven, G.C., Middel, J., Cornelissen, I.L., Nottet, H.S., KewalRamani, V.N., Littman, D.R., Figdor, C.G. and van Kooyk, Y. DC-SIGN, a dendritic cell-specific HIV-1-binding protein that enhances trans-infection of T cells. Cell 2000; 100(5):587–597.

Giorgi, J.V. and Detels, R. T-cell subset alterations in HIV-infected homosexual men: NIAID multicenter AIDS cohort study. Clin. Immunol. Immunopathol. 1989; 52(1):10–18.

Golding, H., Zaitseva, M., de Rosny, E., King, L.R., Manischewitz, J., Sidorov, I., Gorny, M.K., Zolla-Pazner, S., Dimitrov, D.S. and Weiss, C.D. Dissection of human immunodeficiency virus type 1 entry with neutralizing antibodies to gp41 fusion intermediates. J. Virol. 2002; 76(13):6780–6790.

Gorny, M.K. and Zolla-Pazner, S. Recognition by human monoclonal antibodies of free and complexed peptides representing the prefusogenic and fusogenic forms of human immunodeficiency virus type 1 gp41. J. Virol. 2000; 74(13):6186–6192.

Gotte, M., Li, X. and Wainberg, M.A. HIV-1 reverse transcription: a brief overview focused on structure–function relationships among molecules involved in initiation of the reaction. Arch. Biochem. Biophys. 1999; 365(2):199–210.

Gottlieb, M.S., Schroff, R., Schanker, H.M., Weisman, J.D., Fan, P.T., Wolf, R.A. and Saxon, A. Pneumocystis carinii pneumonia and mucosal candidiasis in previously healthy homosexual men: evidence of a new acquired cellular immunodeficiency. N. Engl. J. Med. 1981; 305(24):1425–1431.

Grossman, Z., Meier-Schellersheim, M., Sousa, A.E., Victorino, R. M. and Paul, W.E. CD4+ T-cell depletion in HIV infection: are we closer to understanding the cause? Nat. Med. 2002; 8(4):319–323.

Greene, Warner, C. and Peterlinn, B. Matija. Molecualr insights into HIV Biology. HIV Insite Knowledge Base Chapter February 2006. http://hivinsite.ucsf.edu/Insite?page=kb-02-01-01.

Gupta, K., Hudgens, M., Corey, L., McElrath, M.J., Weinhold, K., Montefiori, D.C., Gorse, G.J., Frey, S.E., Keefer, M.C., Evans, T.G., Dolin, R., Schwartz, D.H., Harro, C., Graham, B., Spearman, P. W., Mulligan, M. and Goepfert, P. Safety and immunogenicity of a high-titered canarypox vaccine in combination with rgp120 in a diverse population of HIV-1-uninfected adults: AIDS Vaccine Evaluation Group Protocol 022A. J. Acquir. Immune. Defic. Syndr. 2002; 29(3):254–261.

Haddad, D.N., Birch, C., Middleton, T., Dwyer, D.E., Cunningham, A.L. and Saksena, N.K. Evidence for late stage compartmentalization of HIV-1 resistance mutations between lymph node and peripheral blood mononuclear cells. Aids 2000; 14(15):2273–2281.

Halperin, D.T. and Epstein, H. Concurrent sexual partnerships help to explain Africa's high HIV prevalence: implications for prevention. Lancet 2004; 364(9428):4–6.

Hammer, S.M., Saag, M.S., Schechter, M., Montaner, J.S., Schooley, R.T., Jacobsen, D.M., Thompson, M.A., Carpenter, C.C., Fischl, M.A., Gazzard, B.G., Gatell, J.M., Hirsch, M.S., Katzenstein, D.A., Richman, D.D., Vella, S., Yeni, P.G. and Volberding, P. A. Treatment for adult HIV infection: 2006 recommendations of the International AIDS Society-USA panel. JAMA 2006; 296(7):827–843.

Hammer, S.M. Clinical practice. Management of newly diagnosed HIV infection. N. Engl. J. Med. 2005; 353(16):1702–1710.

Hansen, J.E., Clausen, H., Nielsen, C., Teglbjaerg, L.S., Hansen, L. L., Nielsen, C.M., Dabelsteen, E., Mathiesen, L., Hakomori, S.I. and Nielsen, J.O. Inhibition of human immunodeficiency virus (HIV) infection in vitro by anticarbohydrate monoclonal antibodies: peripheral glycosylation of HIV envelope glycoprotein gp120 may be a target for virus neutralization. J. Virol. 1990; 64(6):2833–2840.

Harrer, T., Harrer, E., Kalams, S.A., Elbeik, T., Staprans, S.I., Feinberg, M.B., Cao, Y., Ho, D.D., Yilma, T., Caliendo, A.M., Johnson, R.P., Buchbinder, S.P. and Walker, B.D. Strong cytotoxic T cell and weak neutralizing antibody responses in a subset of persons with stable nonprogressing HIV type 1 infection. AIDS Res. Hum. Retroviruses 1996; 12(7):585–592.

Hasenkrug, K.J., Brooks, D.M. and Dittmer, U. Critical role for CD4(+) T cells in controlling retrovirus replication and spread in persistently infected mice. J. Virol. 1998; 72(8):6559–6564.

Hazenberg, M.D., Stuart, J.W., Otto, S.A., Borleffs, J.C., Boucher, C. A., de Boer, R.J., Miedema, F. and Hamann, D. T-cell division in human immunodeficiency virus (HIV)-1 infection is mainly due to immune activation: a longitudinal analysis in patients before and during highly active antiretroviral therapy (HAART). Blood 2000; 95(1):249–255.

Ho, D.D., Neumann, A.U., Perelson, A.S., Chen, W., Leonard, J.M. and Markowitz, M. Rapid turnover of plasma virions and CD4 lymphocytes in HIV-1 infection. Nature 1995; 373(6510):123–126.

Hoglund, S., Ofverstedt, L.G., Nilsson, A., Lundquist, P., Gelderblom, H., Ozel, M. and Skoglund, U. Spatial visualization of the maturing HIV-1 core and its linkage to the envelope. AIDS Res. Hum. Retroviruses 1992; 8(1):1–7.

Holm, G.H. and Gabuzda, D. Distinct mechanisms of CD4+ and CD8+ T-cell activation and bystander apoptosis induced by human immunodeficiency virus type 1 virions. J. Virol. 2005; 79(10):6299–6311.

IAVI Database. Accessed Nov. 1, 2007http://www.iavireport.org/trialsdb/.

Inciardi, J.A. and Williams, M.L. Editor's introduction: the global epidemiology of HIV and AIDS. AIDS Care 2005; 17(Suppl. 1):S1–S8.

Jewett, A. and Bonavida, B. Peripheral blood monocytes derived from HIV+ individuals mediate antibody-dependent cellular cytotoxicity (ADCC). Clin. Immunol. Immunopathol. 1990; 54(2):192–199.

Jin, X., Bauer, D.E., Tuttleton, S.E., Lewin, S., Gettie, A., Blanchard, J., Irwin, C.E., Safrit, J.T., Mittler, J., Weinberger, L., Kostrikis, L.G., Zhang, L., Perelson, A.S. and Ho, D.D. Dramatic rise in plasma viremia after CD8(+) T cell depletion in simian immunodeficiency virus-infected macaques. J. Exp. Med. 1999; 189(6):991–998.

Johnson, P.R., Schnepp, B.C., Connell, M.J., Rohne, D., Robinson, S., Krivulka, G.R., Lord, C.I., Zinn, R., Montefiori, D.C., Letvin, N.L. and Clark, K.R. Novel adeno-associated virus vector vaccine restricts replication of simian immunodeficiency virus in macaques. J. Virol. 2005; 79(2):955–965.

Jones, G.J., Barsby, N.L., Cohen, E.A., Holden, J., Harris, K., Dickie, P., Jhamandas, J. and Power, C. HIV-1 Vpr causes neuronal apoptosis and in vivo neurodegeneration. J. Neurosci. 2007; 27(14):3703–3711.

Kagi, D., Vignaux, F., Ledermann, B., Burki, K., Depraetere, V., Nagata, S., Hengartner, H. and Golstein, P. Fas and perforin pathways as major mechanisms of T cell-mediated cytotoxicity. Science 1994; 265(5171):528–530.

Kahn, J.O. and Walker, B.D. Acute human immunodeficiency virus type 1 infection. N. Engl. J. Med. 1998; 339(1):33–39.

Kalams, S.A., Buchbinder, S.P., Rosenberg, E.S., Billingsley, J.M., Colbert, D.S., Jones, N.G., Shea, A.K., Trocha, A.K. and Walker,

B.D. Association between virus-specific cytotoxic T-lymphocyte and helper responses in human immunodeficiency virus type 1 infection. J. Virol. 1999; 73(8):6715–6720.

Kalayjian, R.C., Landay, A., Pollard, R.B., Taub, D.D., Gross, B. H., Francis, I.R., Sevin, A., Pu, M., Spritzler, J., Chernoff, M., Namkung, A., Fox, L., Martinez, A., Waterman, K., Fiscus, S.A., Sha, B., Johnson, D., Slater, S., Rousseau, F. and Lederman, M.M. Age-related immune dysfunction in health and in human immunodeficiency virus (HIV) disease: association of age and HIV infection with naive CD8+ cell depletion, reduced expression of CD28 on CD8+ cells, and reduced thymic volumes. J. Infect. Dis. 2003; 187(12):1924–1933.

Kanki, P., M'Boup, S., Marlink, R., Travers, K., Hsieh, C.C., Gueye, A., Boye, C., Sankale, J.L., Donnelly, C., Leisenring, W. et al. Prevalence and risk determinants of human immunodeficiency virus type 2 (HIV-2) and human immunodeficiency virus type 1 (HIV-1) in west African female prostitutes. Am. J. Epidemiol. 1992; 136(7):895–907.

Kao, S.Y., Calman, A.F., Luciw, P.A. and Peterlin, B.M. Anti-termination of transcription within the long terminal repeat of HIV-1 by tat gene product. Nature 1987; 330(6147):489–493.

Karageorgos, L., Li, P. and Burrell, C. Characterization of HIV replication complexes early after cell-to-cell infection. AIDS Res. Hum. Retroviruses 1993; 9(9):817–823.

Kaslow, R.A., Carrington, M., Apple, R., Park, L., Munoz, A., Saah, A.J., Goedert, J.J., Winkler, C., O'Brien, S.J., Rinaldo, C., Detels, R., Blattner, W., Phair, J., Erlich, H. and Mann, D.L. Influence of combinations of human major histocompatibility complex genes on the course of HIV-1 infection. Nat. Med. 1996; 2(4):405–411.

Keefer, M.C., Wolff, M., Gorse, G.J., Graham, B.S., Corey, L., Clements-Mann, M.L., Verani-Ketter, N., Erb, S., Smith, C. M., Belshe, R.B., Wagner, L.J., McElrath, M.J., Schwartz, D. H. and Fast, P. Safety profile of phase I and II preventive HIV type 1 envelope vaccination: experience of the NIAID AIDS Vaccine Evaluation Group. AIDS Res. Hum. Retroviruses 1997; 13(14):1163–1177.

Kojima, Y., Xin, K.Q., Ooki, T., Hamajima, K., Oikawa, T., Shinoda, K., Ozaki, T., Hoshino, Y., Jounai, N., Nakazawa, M., Klinman, D. and Okuda, K. Adjuvant effect of multi-CpG motifs on an HIV-1 DNA vaccine. Vaccine 2002; 20(23–24):2857–2865.

Korber, B., Gaschen, B., Yusim, K., Thakallapally, R., Kesmir, C. and Detours, V. Evolutionary and immunological implications of contemporary HIV-1 variation. Br. Med. Bull. 2001; 58:19–42.

Koup, R.A., Robinson, J.E., Nguyen, Q.V., Pikora, C.A., Blais, B., Roskey, A., Panicali, D. and Sullivan, J.L. Antibody-dependent cell-mediated cytotoxicity directed by a human monoclonal antibody reactive with gp120 of HIV-1. Aids 1991; 5(11):1309–1314.

Koup, R.A., Safrit, J.T., Cao, Y., Andrews, C.A., McLeod, G., Borkowsky, W., Farthing, C. and Ho, D.D. Temporal association of cellular immune responses with the initial control of viremia in primary human immunodeficiency virus type 1 syndrome. J. Virol. 1994; 68(7):4650–4655.

Kulkosky, J., Nunnari, G., Otero, M., Calarota, S., Dornadula, G., Zhang, H., Malin, A., Sullivan, J., Xu, Y., DeSimone, J., Babinchak, T., Stern, J., Cavert, W., Haase, A. and Pomerantz, R. J. Intensification and stimulation therapy for human immunodeficiency virus type 1 reservoirs in infected persons receiving virally suppressive highly active antiretroviral therapy. J. Infect. Dis. 2002; 186(10):1403–1411.

Kwon, D.S., Gregorio, G., Bitton, N., Hendrickson, W.A. and Littman, D.R. DC-SIGN-mediated internalization of HIV is required for trans-enhancement of T cell infection. Immunity 2002; 16(1):135–144.

Labrijn, A.F., Poignard, P., Raja, A., Zwick, M.B., Delgado, K., Franti, M., Binley, J., Vivona, V., Grundner, C., Huang, C.C., Venturi, M., Petropoulos, C.J., Wrin, T., Dimitrov, D.S., Robinson, J., Kwong, P.D., Wyatt, R.T., Sodroski, J. and Burton, D.R. Access of antibody molecules to the conserved coreceptor binding site on glycoprotein gp120 is sterically restricted on primary human immunodeficiency virus type 1. J. Virol. 2003; 77(19):10557–10565.

Laukkanen, T., Albert, J., Liitsola, K., Green, S.D., Carr, J.K., Leitner, T., McCutchan, F.E. and Salminen, M.O. Virtually full-length sequences of HIV type 1 subtype J reference strains. AIDS Res. Hum. Retroviruses 1999; 15(3):293–297.

Lederman, M.M., Kalish, L.A., Asmuth, D., Fiebig, E., Mileno, M. and Busch, M.P. 'Modeling' relationships among HIV-1 replication, immune activation and CD4+ T-cell losses using adjusted correlative analyses. Aids 2000; 14(8):951–958.

Lemckert, A.A., Sumida, S.M., Holterman, L., Vogels, R., Truitt, D.M., Lynch, D.M., Nanda, A., Ewald, B.A., Gorgone, D.A., Lifton, M.A., Goudsmit, J., Havenga, M.J. and Barouch, D.H. Immunogenicity of heterologous prime-boost regimens involving recombinant adenovirus serotype 11 (Ad11) and Ad35 vaccine vectors in the presence of anti-ad5 immunity. J. Virol. 2005; 79(15):9694–9701.

Lenardo, M.J., Angleman, S.B., Bounkeua, V., Dimas, J., Duvall, M. G., Graubard, M.B., Hornung, F., Selkirk, M.C., Speirs, C.K., Trageser, C., Orenstein, J.O. and Bolton, D.L. Cytopathic killing of peripheral blood CD4(+) T lymphocytes by human immunodeficiency virus type 1 appears necrotic rather than apoptotic and does not require env. J. Virol. 2002; 76(10):5082–5093.

Letvin, N.L., Mascola, J.R., Sun, Y., Gorgone, D.A., Buzby, A.P., Xu, L., Yang, Z.Y., Chakrabarti, B., Rao, S.S., Schmitz, J.E., Montefiori, D.C., Barker, B.R., Bookstein, F.L. and Nabel, G.J. Preserved CD4+ central memory T cells and survival in vaccinated SIV-challenged monkeys. Science 2006; 312(5779):1530–1533.

Levy, J.A. HIV and the Pathogenesis of AIDS, 3rd ed. Washington, DC: ASM Press, 2007, xvii, 644 p., [16] p. of plates.

Levy, J.A., Hoffman, A.D., Kramer, S.M., Landis, J.A., Shimabukuro, J.M. and Oshiro, L.S. Isolation of lymphocytopathic retroviruses from San Francisco patients with AIDS. Science 1984; 225(4664):840–842.

Levy, J.A., Shimabukuro, J., McHugh, T., Casavant, C., Stites, D. and Oshiro, L. AIDS-associated retroviruses (ARV) can productively infect other cells besides human T helper cells. Virology 1985; 147(2):441–448.

Levy, J.A. and Shimabukuro, J. Recovery of AIDS-associated retroviruses from patients with AIDS or AIDS-related conditions and from clinically healthy individuals. J. Infect. Dis. 1985; 152(4):734–738.

Li, L., Olvera, J.M., Yoder, K.E., Mitchell, R.S., Butler, S.L., Lieber, M., Martin, S.L. and Bushman, F.D. Role of the non-homologous DNA end joining pathway in the early steps of retroviral infection. Embo. J. 2001; 20(12):3272–3281.

Li, Q., Duan, L., Estes, J.D., Ma, Z.M., Rourke, T., Wang, Y., Reilly, C., Carlis, J., Miller, C.J. and Haase, A.T. Peak SIV replication in resting memory CD4+ T cells depletes gut lamina propria CD4+ T cells. Nature 2005; 434(7037):1148–1152.

Liao, Z., Cimakasky, L.M., Hampton, R., Nguyen, D.H. and Hildreth, J.E. Lipid rafts and HIV pathogenesis: host membrane cholesterol is required for infection by HIV type 1. AIDS Res. Hum. Retroviruses 2001; 17(11):1009–1019.

Lichterfeld, M., Kaufmann, D.E., Yu, X.G., Mui, S.K., Addo, M.M., Johnston, M.N., Cohen, D., Robbins, G.K., Pae, E., Alter, G., Wurcel, A., Stone, D., Rosenberg, E.S., Walker, B.D. and Altfeld, M. Loss of HIV-1-specific CD8+ T cell proliferation after acute HIV-1 infection and restoration by vaccine-induced HIV-1-specific CD4+ T cells. J. Exp. Med. 2004; 200(6):701–712.

Los Alamos National Laboratory HIV Sequence Compendium. Los Alamos, NM: Theoretical Biology and Biophysics Group T-10, Los Alamos National Laboratory, 2001, p.v.

Louache, F., Henri, A., Bettaieb, A., Oksenhendler, E., Raguin, G., Tulliez, M. and Vainchenker, W. Role of human immunodeficiency virus replication in defective in vitro growth of hematopoietic progenitors. Blood 1992; 80(12):2991–2999.

Luckay, A., Sidhu, M.K., Kjeken, R., Megati, S., Chong, S.Y., Roopchand, V., Garcia-Hand, D., Abdullah, R., Braun, R., Montefiori, D.C., Rosati, M., Felber, B.K., Pavlakis, G.N., Mathiesen, I., Israel, Z.R., Eldridge, J.H. and Egan, M.A. Effect of plasmid DNA vaccine design and in vivo electroporation on the resulting vaccine-specific immune responses in rhesus macaques. J. Virol. 2007; 81(10):5257–5269.

Lyles, R.H., Munoz, A., Yamashita, T.E., Bazmi, H., Detels, R., Rinaldo, C.R., Margolick, J.B., Phair, J.P. and Mellors, J.W. Natural history of human immunodeficiency virus type 1 viremia after seroconversion and proximal to AIDS in a large cohort of homosexual men. Multicenter AIDS Cohort Study. J. Infect. Dis. 2000; 181(3):872–880.

Mackall, C.L., Fleisher, T.A., Brown, M.R., Andrich, M.P., Chen, C. C., Feuerstein, I.M., Magrath, I.T., Wexler, L.H., Dimitrov, D. S. and Gress, R.E. Distinctions between CD8+ and CD4+ T-cell regenerative pathways result in prolonged T-cell subset imbalance after intensive chemotherapy. Blood 1997; 89(10):3700–3707.

Mandell, G.L., Douglas, R.G., Bennett, J.E. and Dolin, R. Mandell, Douglas, and Bennett's principles and practice of infectious diseases. Elsevier/Churchill Livingstone [Book review (E-STREAMS)]. Available at: http://www.e-streams.com/es0806_7/es0867_4196.html, 2005.

Mansky, L.M. and Temin, H.M. Lower in vivo mutation rate of human immunodeficiency virus type 1 than that predicted from the fidelity of purified reverse transcriptase. J. Virol. 1995; 69(8):5087–5094.

Marandin, A., Katz, A., Oksenhendler, E., Tulliez, M., Picard, F., Vainchenker, W. and Louache, F. Loss of primitive hematopoietic progenitors in patients with human immunodeficiency virus infection. Blood 1996; 88(12):4568–4578.

Marston, M., Zaba, B., Salomon, J.A., Brahmbhatt, H. and Bagenda, D. Estimating the net effect of HIV on child mortality in African populations affected by generalized HIV epidemics. J. Acquir. Immune. Defic. Syndr. 2005; 38(2):219–227.

Martin-Serrano, J., Zang, T. and Bieniasz, P.D. Role of ESCRT-I in retroviral budding. J. Virol. 2003; 77(8):4794–4804.

Mascola, J.R., Sambor, A., Beaudry, K., Santra, S., Welcher, B., Louder, M.K., Vancott, T.C., Huang, Y., Chakrabarti, B.K., Kong, W.P., Yang, Z.Y., Xu, L., Montefiori, D.C., Nabel, G.J. and Letvin, N.L. Neutralizing antibodies elicited by immunization of monkeys with DNA plasmids and recombinant adenoviral vectors expressing human immunodeficiency virus type 1 proteins. J. Virol. 2005; 79(2):771–779.

Mascola, J.R., Stiegler, G., VanCott, T.C., Katinger, H., Carpenter, C.B., Hanson, C.E., Beary, H., Hayes, D., Frankel, S.S., Birx, D. L. and Lewis, M.G. Protection of macaques against vaginal transmission of a pathogenic HIV-1/SIV chimeric virus by passive infusion of neutralizing antibodies. Nat. Med. 2000; 6(2):207–210.

Masur, H., Michelis, M.A., Greene, J.B., Onorato, I., Stouwe, R.A., Holzman, R.S., Wormser, G., Brettman, L., Lange, M., Murray, H.W. and Cunningham-Rundles, S. An outbreak of community-acquired Pneumocystis carinii pneumonia: initial manifestation of cellular immune dysfunction. N. Engl. J. Med. 1981; 305(24):1431–1438.

Matano, T., Shibata, R., Siemon, C., Connors, M., Lane, H.C. and Martin, M.A. Administration of an anti-CD8 monoclonal antibody interferes with the clearance of chimeric simian/human immunodeficiency virus during primary infections of rhesus macaques. J. Virol. 1998; 72(1):164–169.

Matloubian, M., Concepcion, R.J. and Ahmed, R. CD4+ T cells are required to sustain CD8+ cytotoxic T-cell responses during chronic viral infection. J. Virol. 1994; 68(12):8056–8063.

Mattapallil, J.J., Douek, D.C., Hill, B., Nishimura, Y., Martin, M. and Roederer, M. Massive infection and loss of memory CD4+ T cells in multiple tissues during acute SIV infection. Nature. 2005; 434(7037):1093–1097.

McElrath, M.J., Corey, L., Montefiori, D., Wolff, M., Schwartz, D., Keefer, M., Belshe, R., Graham, B.S., Matthews, T., Wright, P., Gorse, G., Dolin, R., Berman, P., Francis, D., Duliege, A.M., Bolognesi, D., Stablein, D., Ketter, N. and Fast, P. A phase II study of two HIV type 1 envelope vaccines, comparing their immunogenicity in populations at risk for acquiring HIV type 1 infection. AIDS Vaccine Evaluation Group. AIDS Res. Hum. Retroviruses. 2000; 16(9):907–919.

McKenna, P.M., Koser, M.L., Carlson, K.R., Montefiori, D.C., Letvin, N.L., Papaneri, A.B., Pomerantz, R.J., Dietzschold, B., Silvera, P., McGettigan, J.P. and Schnell, M.J. Highly attenuated rabies virus-based vaccine vectors expressing simian-human immunodeficiency virus89.6P Env and simian immunodeficiency virusmac239 Gag are safe in rhesus macaques and protect from an AIDS-like disease. J. Infect. Dis. 2007; 195(7):980–988.

McNeil, J.G., Johnston, M.I., Birx, D.L. and Tramont, E.C. Policy rebuttal. HIV vaccine trial justified. Science 2004; 303(5660):961.

Mehandru, S., Poles, M.A., Tenner-Racz, K., Jean-Pierre, P., Manuelli, V., Lopez, P., Shet, A., Low, A., Mohri, H., Boden, D., Racz, P. and Markowitz, M. Lack of mucosal immune reconstitution during prolonged treatment of acute and early HIV-1 infection. PLoS Med. 2006; 3(12):e484.

Mehandru, S., Poles, M.A., Tenner-Racz, K., Horowitz, A., Hurley, A., Hogan, C., Boden, D., Racz, P. and Markowitz, M. Primary HIV-1 infection is associated with preferential depletion of CD4+ T lymphocytes from effector sites in the gastrointestinal tract. J. Exp. Med. 2004; 200(6):761–770.

Mellors, J.W., Munoz, A., Giorgi, J.V., Margolick, J.B., Tassoni, C.J., Gupta, P., Kingsley, L.A., Todd, J.A., Saah, A.J., Detels, R., Phair, J.P. and Rinaldo, C.R., Jr. Plasma viral load and CD4+ lymphocytes as prognostic markers of HIV-1 infection. Ann. Intern. Med. 1997; 126(12):946–954.

Meseda, C.A., Stout, R.R. and Weir, J.P. Evaluation of a needle-free delivery platform for prime-boost immunization with DNA and modified vaccinia virus ankara vectors expressing herpes simplex virus 2 glycoprotein D. Viral. Immunol. 2006; 19(2):250–259.

Merck Research and Development News. Data from STEP Study Presented at Open Scientific Session Confirm Merck's Investigational HIV Vaccine was Not Effective. Nov. 7, 2007. [cited Feb. 6, 2008] http://www.merck.com/newsroom/press_releases/research_and_development/2007_1107.html.

Migasena, S., Suntharasamai, P., Pitisuttithum, P., Kitayaporn, D., Wasi, C., Huang, W., Vanichseni, S., Koompong, C., Kaewkungwal, J., Raktham, S., Ippolito, T., Hanson, C., Gregory, T., Heyward, W.L., Berman, P. and Francis, D. AIDSVAX (MN) in Bangkok injecting drug users: a report on safety and immunogenicity, including macrophage-tropic virus neutralization. AIDS. Res. Hum. Retroviruses. 2000; 16(7):655–663.

Migueles, S.A. and Connors, M. Frequency and function of HIV-specific CD8(+) T cells. Immunol. Lett. 2001; 79(1–2):141–150.

Miller, M.D., Farnet, C.M. and Bushman, F.D. Human immunodeficiency virus type 1 preintegration complexes: studies of organization and composition. J. Virol. 1997; 71(7):5382–5390.

Montefiori, D.C., Baba, T.W., Li, A., Bilska, M. and Ruprecht, R.M. Neutralizing and infection-enhancing antibody responses do not correlate with the differential pathogenicity of SIVmac239delta3 in adult and infant rhesus monkeys. J. Immunol. 1996; 157(12):5528–5535.

Moog, C., Fleury, H.J., Pellegrin, I., Kirn, A. and Aubertin, A.M. Autologous and heterologous neutralizing antibody responses following initial seroconversion in human immunodeficiency virus type 1-infected individuals. J. Virol. 1997; 71(5):3734–3741.

Moore, P.L., Crooks, E.T., Porter, L., Zhu, P., Cayanan, C.S., Grise, H., Corcoran, P., Zwick, M.B., Franti, M., Morris, L., Roux, K.H., Burton, D.R. and Binley, J.M. Nature of nonfunctional envelope proteins on the surface of human immunodeficiency virus type 1. J. Virol. 2006; 80(5):2515–2528.

Mowat, A.M. and Viney, J.L. The anatomical basis of intestinal immunity. Immunol. Rev. 1997; 156:145–166.

Muesing, M.A., Smith, D.H., Cabradilla, C.D., Benton, C.V., Lasky, L.A. and Capon, D.J. Nucleic acid structure and expression of the human AIDS/lymphadenopathy retrovirus. Nature. 1985; 313(6002):450–458.

Mullins, J.I., Nickle, D.C., Heath, L., Rodrigo, A.G. and Learn, G.H. Immunogen sequence: the fourth tier of AIDS vaccine design. Expert. Rev. Vaccines. 2004; 3(4 Suppl):S151–s159.

Nagata, S. and Golstein, P. The Fas death factor. Science 1995; 267(5203):1449–1456.

Ono, A. and Freed, E.O. Plasma membrane rafts play a critical role in HIV-1 assembly and release. Proc. Natl. Acad. Sci. U.S.A. 2001; 98(24):13925–13930.

Orkin, C., Stebbing, J., Nelson, M., Bower, M., Johnson, M., Mandalia, S., Jones, R., Moyle, G., Fisher, M. and Gazzard, B. A randomized study comparing a three- and four-drug HAART regimen in first-line therapy (QUAD study). J. Antimicrob. Chemother. 2005; 55(2):246–251.

Palella, F.J., Jr., Delaney, K.M., Moorman, A.C., Loveless, M.O., Fuhrer, J., Satten, G.A., Aschman, D.J. and Holmberg, S.D. Declining morbidity and mortality among patients with advanced human immunodeficiency virus infection. HIV Outpatient Study Investigators. N. Engl. J. Med. 1998; 338(13):853–860.

Pantaleo, G., Demarest, J.F., Soudeyns, H., Graziosi, C., Denis, F., Adelsberger, J.W., Borrow, P., Saag, M.S., Shaw, G.M., Sekaly, R. P. et al. Major expansion of CD8+ T cells with a predominant V beta usage during the primary immune response to HIV. Nature. 1994; 370(6489):463–467.

Patki, A.H., Zielske, S.P., Sieg, S.F. and Lederman, M.M. Preferential S phase entry and apoptosis of CD4(+) T lymphocytes of HIV-1-infected patients after in vitro cultivation. Clin. Immunol. 2000; 97(3):241–247.

Peeters, M., Toure-Kane, C. and Nkengasong, J.N. Genetic diversity of HIV in Africa: impact on diagnosis, treatment, vaccine development and trials. Aids 2003; 17(18):2547–2560.

Pemberton, L.F., Blobel, G. and Rosenblum, J.S. Transport routes through the nuclear pore complex. Curr. Opin. Cell Biol. 1998; 10(3):392–399.

Petrovas, C., Mueller, Y.M. and Katsikis, P.D. HIV-specific CD8+ T cells: serial killers condemned to die? Curr. HIV Res. 2004; 2(2):153–162.

Pettifor, A.E., Rees, H.V., Kleinschmidt, I., Steffenson, A.E., MacPhail, C., Hlongwa-Madikizela, L., Vermaak, K. and Padian, N.S. Young people's sexual health in South Africa: HIV prevalence and sexual behaviors from a nationally representative household survey. Aids. 2005; 19(14):1525–1534.

Pierson, T., McArthur, J. and Siliciano, R.F. Reservoirs for HIV-1: mechanisms for viral persistence in the presence of antiviral immune responses and antiretroviral therapy. Annu. Rev. Immunol. 2000; 18:665–708.

Pitisuttithum, P., Gilbert, P., Gurwith, M., Heyward, W., Martin, M., van Griensven, F., Hu, D., Tappero, J.W. and Choopanya, K. Randomized, double-blind, placebo-controlled efficacy trial of a bivalent recombinant glycoprotein 120 HIV-1 vaccine among injection drug users in Bangkok, Thailand. J. Infect. Dis. 2006; 194(12):1661–1671.

Centers for Disease Control (CDC) Pneumocystis pneumonia–Los Angeles. MMWR Morb. Mortal. Wkly Rep. 1981; 30(21): 250–252.

Pope, M., Gezelter, S., Gallo, N., Hoffman, L. and Steinman, R.M. Low levels of HIV-1 infection in cutaneous dendritic cells promote extensive viral replication upon binding to memory CD4+ T cells. J. Exp. Med. 1995; 182(6):2045–2056.

Popovic, M., Sarngadharan, M.G., Read, E. and Gallo, R.C. Detection, isolation, and continuous production of cytopathic retroviruses (HTLV-III) from patients with AIDS and pre-AIDS. Science. 1984; 224(4648):497–500.

Potts, K.E., Kalish, M.L., Lott, T., Orloff, G., Luo, C.C., Bernard, M. A., Alves, C.B., Badaro, R., Suleiman, J., Ferreira, O. et al. Genetic heterogeneity of the V3 region of the HIV-1 envelope glycoprotein in Brazil. Brazilian Collaborative AIDS Research Group. Aids 1993; 7(9):1191–1197.

Rabson, A.B. and Martin, M.A. Molecular organization of the AIDS retrovirus. Cell 1985; 40(3):477–480.

Raguin, G., Chene, G., Morand-Joubert, L., Taburet, A.M., Droz, C., Le Tiec, C., Clavel, F. and Girard, P.M. Salvage therapy with amprenavir, lopinavir and ritonavir 200 mg/d or 400 mg/d in HIV-infected patients in virological failure. Antivir. Ther. 2004; 9(4):615–625.

Ramratnam, B., Bonhoeffer, S., Binley, J., Hurley, A., Zhang, L., Mittler, J.E., Markowitz, M., Moore, J.P., Perelson, A.S. and Ho, D.D. Rapid production and clearance of HIV-1 and hepatitis C virus assessed by large volume plasma apheresis. Lancet 1999; 354(9192):1782–1785.

Rao, S.S., Gomez, P., Mascola, J.R., Dang, V., Krivulka, G.R., Yu, F., Lord, C.I., Shen, L., Bailer, R., Nabel, G.J. and Letvin, N.L. Comparative evaluation of three different intramuscular delivery methods for DNA immunization in a nonhuman primate animal model. Vaccine 2006; 24(3):367–373.

Redfield, R.R., Wright, D.C., James, W.D., Jones, T.S., Brown, C. and Burke, D.S. Disseminated vaccinia in a military recruit with human immunodeficiency virus (HIV) disease. N. Engl. J. Med. 1987; 316(11):673–676.

Reeves, J.D. and Doms, R.W. Human immunodeficiency virus type 2. J. Gen. Virol. 2002; 83(Pt. 6):1253–1265.

Rinaldo, C., Huang, X.L., Fan, Z.F., Ding, M., Beltz, L., Logar, A., Panicali, D., Mazzara, G., Liebmann, J., Cottrill, M. et al. High levels of anti-human immunodeficiency virus type 1 (HIV-1) memory cytotoxic T-lymphocyte activity and low viral load are associated with lack of disease in HIV-1-infected long-term non-progressors. J. Virol. 1995; 69(9):5838–5842.

Robbins, J.B., Schneerson, R. and Szu, S.C. Perspective: hypothesis: serum IgG antibody is sufficient to confer protection against infectious diseases by inactivating the inoculum. J. Infect. Dis. 1995; 171(6):1387–1398.

Robertson, D.L., Anderson, J.P., Bradac, J.A., Carr, J.K., Foley, B., Funkhouser, R.K., Gao, F., Hahn, B.H., Kalish, M.L., Kuiken, C., Learn, G.H., Leitner, T., McCutchan, F., Osmanov, S., Peeters, M., Pieniazek, D., Salminen, M., Sharp, P.M., Wolinsky, S. and Korber, B. HIV-1 nomenclature proposal. Science 2000; 288(5463):55–56.

Roederer, M., Dubs, J.G., Anderson, M.T., Raju, P.A., Herzenberg, L.A. and Herzenberg, L.A. CD8 naive T cell counts decrease progressively in HIV-infected adults. J. Clin. Invest. 1995; 95(5):2061–2066.

Roques, P., Menu, E., Narwa, R., Scarlatti, G., Tresoldi, E., Damond, F., Mauclere, P., Dormont, D., Chaouat, G., Simon, F. and

Barre-Sinoussi, F. An unusual HIV type 1 env sequence embedded in a mosaic virus from Cameroon: identification of a new env clade. European Network on the study of in utero transmission of HIV-1. AIDS Res. Hum. Retroviruses. 1999; 15(17):1585–1589.

Rothenaigner, I., Kramer, S., Ziegler, M., Wolff, H., Kleinschmidt, A. and Brack-Werner, R. Long-term HIV-1 infection of neural progenitor populations. Aids 2007; 21(17):2271–2281.

Ruprecht, R.M., Baba, T.W., Liska, V., Ray, N.B., Martin, L.N., Murphey-Corb, M., Rizvi, T.A., Bernacky, B.J., Keeling, M.E., McClure, H.M. and Andersen, J. Oral transmission of primate lentiviruses. J. Infect. Dis. 1999; 179(Suppl 3):S408–s412.

Russell, N.D., Graham, B.S., Keefer, M.C., McElrath, M.J., Self, S. G., Weinhold, K.J., Montefiori, D.C., Ferrari, G., Horton, H., Tomaras, G.D., Gurunathan, S., Baglyos, L., Frey, S.E., Mulligan, M.J., Harro, C.D., Buchbinder, S.P., Baden, L.R., Blattner, W.A., Koblin, B.A. and Corey, L. Phase 2 study of an HIV-1 canarypox vaccine (vCP1452) alone and in combination with rgp120: negative results fail to trigger a phase 3 correlates trial. J. Acquir. Immune. Defic. Syndr. 2007; 44(2):203–212.

Saag, M.S. Initiation of antiretroviral therapy: implications of recent findings. Top. HIV Med. 2004; 12(3):83–88.

Sabin, C.A., Hill, T., Lampe, F., Matthias, R., Bhagani, S., Gilson, R., Youle, M.S., Johnson, M.A., Fisher, M., Scullard, G., Easterbrook, P., Gazzard, B. and Phillips, A.N. Treatment exhaustion of highly active antiretroviral therapy (HAART) among individuals infected with HIV in the United Kingdom: multicentre cohort study. BMJ. 2005; 330(7493):695.

Sakai, K., Dimas, J. and Lenardo, M.J. The Vif and Vpr accessory proteins independently cause HIV-1-induced T cell cytopathicity and cell cycle arrest. Proc. Natl. Acad. Sci. U.S.A. 2006; 103(9):3369–3374.

Salahuddin, S.Z., Markham, P.D., Popovic, M., Sarngadharan, M. G., Orndorff, S., Fladagar, A., Patel, A., Gold, J. and Gallo, R. C. Isolation of infectious human T-cell leukemia/lymphotropic virus type III (HTLV-III) from patients with acquired immunodeficiency syndrome (AIDS) or AIDS-related complex (ARC) and from healthy carriers: a study of risk groups and tissue sources. Proc. Natl. Acad. Sci. U.S.A. 1985; 82(16):5530–5534.

Saltarelli, M.J., Hadziyannis, E., Hart, C.E., Harrison, J.V., Felber, B.K., Spira, T.J. and Pavlakis, G.N. Analysis of human immunodeficiency virus type 1 mRNA splicing patterns during disease progression in peripheral blood mononuclear cells from infected individuals. AIDS Res. Hum. Retroviruses. 1996; 12(15):1443–1456.

Santra, S., Seaman, M.S., Xu, L., Barouch, D.H., Lord, C.I., Lifton, M.A., Gorgone, D.A., Beaudry, K.R., Svehla, K., Welcher, B., Chakrabarti, B.K., Huang, Y., Yang, Z.Y., Mascola, J.R., Nabel, G. J. and Letvin, N.L. Replication-defective adenovirus serotype 5 vectors elicit durable cellular and humoral immune responses in nonhuman primates. J. Virol. 2005; 79(10):6516–6522.

Sarngadharan, M.G., Popovic, M., Bruch, L., Schupbach, J. and Gallo, R.C. Antibodies reactive with human T-lymphotropic retroviruses (HTLV-III) in the serum of patients with AIDS. Science 1984; 224(4648):506–508.

Sattentau, Q.J. and Moore, J.P. Human immunodeficiency virus type 1 neutralization is determined by epitope exposure on the gp120 oligomer. J. Exp. Med. 1995; 182(1):185–196.

Scanlan, C.N., Pantophlet, R., Wormald, M.R., Ollmann Saphire, E., Stanfield, R., Wilson, I.A., Katinger, H., Dwek, R.A., Rudd, P.M. and Burton, D.R. The broadly neutralizing anti-human immunodeficiency virus type 1 antibody 2G12 recognizes a cluster of alpha1-- > 2 mannose residues on the outer face of gp120. J. Virol. 2002; 76(14):7306–7321.

Scarlatti, G., Tresoldi, E., Bjorndal, A., Fredriksson, R., Colognesi, C., Deng, H.K., Malnati, M.S., Plebani, A., Siccardi, A.G., Littman, D.R., Fenyo, E.M. and Lusso, P. In vivo evolution of HIV-1 coreceptor usage and sensitivity to chemokine-mediated suppression. Nat. Med. 1997; 3(11):1259–1265.

Schacker, T., Collier, A.C., Hughes, J., Shea, T. and Corey, L. Clinical and epidemiologic features of primary HIV infection. Ann. Intern. Med. 1996; 125(4):257–264.

Schmitz, J.E., Kuroda, M.J., Santra, S., Sasseville, V.G., Simon, M.A., Lifton, M.A., Racz, P., Tenner-Racz, K., Dalesandro, M., Scallon, B.J., Ghrayeb, J., Forman, M.A., Montefiori, D.C., Rieber, E. P., Letvin, N.L. and Reimann, K.A. Control of viremia in simian immunodeficiency virus infection by CD8+ lymphocytes. Science 1999; 283(5403):857–860.

Schupbach, J., Popovic, M., Gilden, R.V., Gonda, M.A., Sarngadharan, M.G. and Gallo, R.C. Serological analysis of a subgroup of human T-lymphotropic retroviruses (HTLV-III) associated with AIDS. Science 1984; 224(4648):503–505.

Shibata, R., Igarashi, T., Haigwood, N., Buckler-White, A., Ogert, R., Ross, W., Willey, R., Cho, M.W. and Martin, M.A. Neutralizing antibody directed against the HIV-1 envelope glycoprotein can completely block HIV-1/SIV chimeric virus infections of macaque monkeys. Nat. Med. 1999; 5(2):204–210.

Shisana, O. and Davids, A. Correcting gender inequalities is central to controlling HIV/AIDS. Bull. World Health Organ. 2004; 82(11):812.

Shiver, J.W., Fu, T.M., Chen, L., Casimiro, D.R., Davies, M.E., Evans, R.K., Zhang, Z.Q., Simon, A.J., Trigona, W.L., Dubey, S.A., Huang, L., Harris, V.A., Long, R.S., Liang, X., Handt, L., Schleif, W.A., Zhu, L., Freed, D.C., Persaud, N.V., Guan, L., Punt, K. S., Tang, A., Chen, M., Wilson, K.A., Collins, K.B., Heidecker, G.J., Fernandez, V.R., Perry, H.C., Joyce, J.G., Grimm, K.M., Cook, J.C., Keller, P.M., Kresock, D.S., Mach, H., Troutman, R. D., Isopi, L.A., Williams, D.M., Xu, Z., Bohannon, K.E., Volkin, D.B., Montefiori, D.C., Miura, A., Krivulka, G.R., Lifton, M.A., Kuroda, M.J., Schmitz, J.E., Letvin, N.L., Caulfield, M.J., Bett, A. J., Youil, R., Kaslow, D.C. and Emini, E.A. Replication-incompetent adenoviral vaccine vector elicits effective anti-immunodeficiency-virus immunity. Nature 2002; 415(6869):331–335.

Simon, V., Ho, D.D. and Abdool Karim, Q. HIV/AIDS epidemiology, pathogenesis, prevention, and treatment. Lancet 2006; 368(9534):489–504.

Singh, M. and Srivastava, I. Advances in vaccine adjuvants for infectious diseases. Curr. HIV Res. 2003; 1(3):309–320.

Smith, B.A., Gartner, S., Liu, Y., Perelson, A.S., Stilianakis, N.I., Keele, B.F., Kerkering, T.M., Ferreira-Gonzalez, A., Szakal, A.K., Tew, J.G. and Burton, G.F. Persistence of infectious HIV on follicular dendritic cells. J. Immunol. 2001; 166(1):690–696.

Sousa, A.E., Carneiro, J., Meier-Schellersheim, M., Grossman, Z. and Victorino, R.M. CD4 T cell depletion is linked directly to immune activation in the pathogenesis of HIV-1 and HIV-2 but only indirectly to the viral load. J. Immunol. 2002; 169(6):3400–3406.

Stellbrink, H.J., van Lunzen, J., Westby, M., O'Sullivan, E., Schneider, C., Adam, A., Weitner, L., Kuhlmann, B., Hoffmann, C., Fenske, S., Aries, P.S., Degen, O., Eggers, C., Petersen, H., Haag, F., Horst, H.A., Dalhoff, K., Mocklinghoff, C., Cammack, N., Tenner-Racz, K. and Racz, P. Effects of interleukin-2 plus highly active antiretroviral therapy on HIV-1 replication and proviral DNA (COSMIC trial). Aids 2002; 16(11):1479–1487.

Sullivan, N., Sun, Y., Sattentau, Q., Thali, M., Wu, D., Denisova, G., Gershoni, J., Robinson, J., Moore, J. and Sodroski, J. CD4-Induced conformational changes in the human immunodeficiency virus type 1 gp120 glycoprotein: consequences for virus entry and neutralization. J. Virol. 1998; 72(6):4694–4703.

Tanaka, M., Hirabayashi, Y., Gatanaga, H., Aizawa, S., Hachiya, A., Takahashi, Y., Tashiro, E., Kohsaka, T., Oyamada, M., Ida, S. and Oka, S. Reduction in interleukin-2-producing cells but not Th1 to Th2 shift in moderate and advanced stages of human immunodeficiency virus type-1-infection: direct analysis of intracellular cytokine concentrations in CD4+ CD8- T cells. Scand. J. Immunol. 1999; 50(5):550–554.

Thomson, M.M., Perez-Alvarez, L. and Najera, R. Molecular epidemiology of HIV-1 genetic forms and its significance for vaccine development and therapy. Lancet Infect. Dis. 2002; 2(8):461–471.

Trabattoni, D., Piconi, S., Biasin, M., Rizzardini, G., Migliorino, M., Seminari, E., Boasso, A., Piacentini, L., Villa, M.L., Maserati, R. and Clerici, M. Granule-dependent mechanisms of lysis are defective in CD8 T cells of HIV-infected, antiretroviral therapy-treated individuals. Aids 2004; 18(6):859–869.

Trautmann, L., Janbazian, L., Chomont, N., Said, E.A., Gimmig, S., Bessette, B., Boulassel, M.R., Delwart, E., Sepulveda, H., Balderas, R.S., Routy, J.P., Haddad, E.K. and Sekaly, R. P. Upregulation of PD-1 expression on HIV-specific CD8+ T cells leads to reversible immune dysfunction. Nat. Med. 2006; 12(10):1198–1202.

Trkola, A., Dragic, T., Arthos, J., Binley, J.M., Olson, W.C., Allaway, G.P., Cheng-Mayer, C., Robinson, J., Maddon, P.J. and Moore, J.P. CD4-dependent, antibody-sensitive interactions between HIV-1 and its co-receptor CCR-5. Nature 1996; 384(6605):184–187.

Tyler, D.S., Stanley, S.D., Zolla-Pazner, S., Gorny, M.K., Shadduck, P.P., Langlois, A.J., Matthews, T.J., Bolognesi, D.P., Palker, T. J. and Weinhold, K.J. Identification of sites within gp41 that serve as targets for antibody-dependent cellular cytotoxicity by using human monoclonal antibodies. J. Immunol. 1990; 145(10):3276–3282.

Ugolini, S., Mondor, I., Parren, P.W., Burton, D.R., Tilley, S.A., Klasse, P.J. and Sattentau, Q.J. Inhibition of virus attachment to CD4+ target cells is a major mechanism of T cell line-adapted HIV-1 neutralization. J. Exp. Med. 1997; 186(8):1287–1298.

UNAIDS. AIDS Epidemic Update: December 2006. http://data.unaids.org/pub/EpiReport/2006/2006_EpiUpdate_en.pdf.

UNAIDS, WHO. AIDS Epidemic Update: December 2004. http://whqlibdoc.who.int/unaids/2004/9291733903.pdf.

UNAIDS, WHO. AIDS Epidemic Update: December 2007. http://data.unaids.org/pub/EPISlides/2007/2007_epiupdate_en.pdf.

Valenzuela, A., Blanco, J., Krust, B., Franco, R. and Hovanessian, A.G. Neutralizing antibodies against the V3 loop of human immunodeficiency virus type 1 gp120 block the CD4-dependent and -independent binding of virus to cells. J. Virol. 1997; 71(11):8289–8298.

Veazey, R.S. and Lackner, A.A. HIV swiftly guts the immune system. Nat. Med. 2005; 11(5):469–470.

Wain-Hobson, S., Vartanian, J.P., Henry, M., Chenciner, N., Cheynier, R., Delassus, S., Martins, L.P., Sala, M., Nugeyre, M.T., Guetard, D. et al. LAV revisited: origins of the early HIV-1 isolates from Institut Pasteur. Science 1991; 252(5008):961–965.

Watanabe, N., De Rosa, S.C., Cmelak, A., Hoppe, R., Herzenberg, L.A. and Roederer, M. Long-term depletion of naive T cells in patients treated for Hodgkin's disease. Blood 1997; 90(9):3662–3672.

Wei, P., Garber, M.E., Fang, S.M., Fischer, W.H. and Jones, K.A. A novel CDK9-associated C-type cyclin interacts directly with HIV-1 Tat and mediates its high-affinity, loop-specific binding to TAR RNA. Cell 1998; 92(4):451–462.

Wei, X., Decker, J.M., Wang, S., Hui, H., Kappes, J.C., Wu, X., Salazar-Gonzalez, J.F., Salazar, M.G., Kilby, J.M., Saag, M.S., Komarova, N.L., Nowak, M.A., Hahn, B.H., Kwong, P.D. and Shaw, G.M. Antibody neutralization and escape by HIV-1. Nature 2003; 422(6929):307–312.

Weinberg, J.B., Matthews, T.J., Cullen, B.R. and Malim, M.H. Productive human immunodeficiency virus type 1 (HIV-1) infection of nonproliferating human monocytes. J. Exp. Med. 1991; 174(6):1477–1482.

Weinhold, K.J., Lyerly, H.K., Stanley, S.D., Austin, A.A., Matthews, T.J. and Bolognesi, D.P. HIV-1 GP120-mediated immune suppression and lymphocyte destruction in the absence of viral infection. J. Immunol. 1989; 142(9):3091–3097.

Whittle, H., Morris, J., Todd, J., Corrah, T., Sabally, S., Bangali, J., Ngom, P.T., Rolfe, M. and Wilkins, A. HIV-2-infected patients survive longer than HIV-1-infected patients. Aids 1994; 8(11):1617–1620.

Wilk, T., Gross, I., Gowen, B.E., Rutten, T., de Haas, F., Welker, R., Krausslich, H.G., Boulanger, P. and Fuller, S.D. Organization of immature human immunodeficiency virus type 1. J. Virol. 2001; 75(2):759–771.

Wright, P.F., Lambert, J.S., Gorse, G.J., Hsieh, R.H., McElrath, M. J., Weinhold, K., Wara, D.W., Anderson, E.L., Keefer, M.C., Jackson, S., Wagner, L.J., Francis, D.P., Fast, P.E. and McNamara, J. Immunization with envelope MN rgp120 vaccine in human immunodeficiency virus-infected pregnant women. J. Infect. Dis. 1999; 180(4):1080–1088.

Wyatt, R. and Sodroski, J. The HIV-1 envelope glycoproteins: fusogens, antigens, and immunogens. Science 1998; 280(5371):1884–1888.

Yang, O.O., Lin, H., Dagarag, M., Ng, H.L., Effros, R.B. and Uittenbogaart, C.H. Decreased perforin and granzyme B expression in senescent HIV-1-specific cytotoxic T lymphocytes. Virology 2005; 332(1):16–19.

Zeitz, M., Greene, W.C., Peffer, N.J. and James, S.P. Lymphocytes isolated from the intestinal lamina propria of normal nonhuman primates have increased expression of genes associated with T-cell activation. Gastroenterology 1988; 94(3):647–655.

Zennou, V., Petit, C., Guetard, D., Nerhbass, U., Montagnier, L. and Charneau, P. HIV-1 genome nuclear import is mediated by a central DNA flap. Cell 2000; 101(2):173–185.

Zhou, T., Xu, L., Dey, B., Hessell, A.J., Van Ryk, D., Xiang, S.H., Yang, X., Zhang, M.Y., Zwick, M.B., Arthos, J., Burton, D.R., Dimitrov, D.S., Sodroski, J., Wyatt, R., Nabel, G.J. and Kwong, P. D. Structural definition of a conserved neutralization epitope on HIV-1 gp120. Nature 2007; 445(7129):732–737.

Zhu, P., Chertova, E., Bess, J., Lifson, J.D., Arthur, L.O., Liu, J., Jr., Taylor, K.A. and Roux, K.H. Electron tomography analysis of envelope glycoprotein trimers on HIV and simian immunodeficiency virus virions. Proc. Natl. Acad. Sci. U.S.A. 2003; 100(26):15812–15817.

Zimmerman, C., Klein, K.C., Kiser, P.K., Singh, A.R., Firestein, B. L., Riba, S.C. and Lingappa, J.R. Identification of a host protein essential for assembly of immature HIV-1 capsids. Nature 2002; 415(6867):88–92.

Zolla-Pazner, S. Identifying epitopes of HIV-1 that induce protective antibodies. Nat. Rev. Immunol. 2004; 4(3):199–210.

CHAPTER

26

Human Papillomavirus

William Bonnez
School of Medicine and Dentistry, University of Rochester, Rochester, NY, USA

OUTLINE

ABSTRACT

Human papillomaviruses (HPVs) include over 100 different genotypes that infect the stratified squamous epithelia and may cause proliferative lesions. Some of the HPVs are responsible for cutaneous warts, while one subset of HPVs is found in the skin lesions of epidermodysplasia verruciformis, a condition that leads to the development of cutaneous squamous cell carcinomas. A third subset of HPVs is associated with infections that can cause benign, premalignant, and malignant lesions not only of the genital tract, the anus, but also of the oropharynx. These lesions include external genital warts, intraepithelial neoplasias of the penis (PIN), anus (AIN), vulva (VIN), vagina (VAIN), and cervix (CIN), as well as cancer of the cervix. Cervical cancer is the second most commonly acquired cancer in women worldwide. This significant disease burden explains the great importance of genital (or mucosal) HPVs. The availability since 2006 of highly effective vaccines to prevent a large fraction of these morbidities has been a major advance whose promises are growing. The two vaccines available, Gardasil and Cervarix, are based on the expression of the major viral capsid protein in yeast or insect cells, and the spontaneous assembly of this protein into a noninfectious capsid called a virus-like particle (VLP), because it retains the appearance, size, and immunologic properties of the native, infectious viral particle.

Gardasil is made of the VLPs of HPV types 6 and 11 that account for 90 of external genital warts, and of HPV types 16 and 18 that account for over 70% of cervical cancers worldwide. Cervarix is a bivalent vaccine that includes only HPV VLP of types 16 and 18. Trials have shown that these vaccines protect virtually 100% of the subjects against the immediate precursor to cancer, high-grade CIN, caused by the vaccine HPV types the subjects are naive to. Gardasil has received indications in the United States not only for the prevention in 9–26-year-old women of cervical cancer and CIN of all grades, but also for the prevention of VIN and VAIN grades 2 and 3, as well as of external genital warts. The most effective use of the vaccine is when administered before sexual debut, at a time the subject is naive for the infections caused by the HPV types included in the vaccine. The efficacy of the HPV vaccine has now been demonstrated up to the age of 45 years in women. Trials in men for the prevention of external genital warts are in progress, and could serve as a basis for the immunization of both males and females. The HPV vaccine, especially when it covers a broad range of viral types, holds the promise of not only affecting cervical cancer and its screening, but also of reducing the burden of the other cancers attributable to HPV, such as those of the vulva, vagina, penis, and anus, as well as a majority of oropharyngeal cancers. To attain these goals, maintenance of safety, favorable cost–benefit ratios, and broad delivery and accessibility of the vaccine will need to be achieved.

INTRODUCTION

Papillomaviruses are small (55nm in diameter), non-enveloped, icosahedral viruses, containing a double-stranded, circular DNA genome about 8000 base pairs long. They are found throughout higher vertebrates, mostly mammals and birds, causing cutaneous and mucosal tumors. The over-100 distinct human papillomavirus (HPV) genotypes are restricted to man and the squamous stratified epithelia. The tumors they cause range from benign (warts, papillomas, condylomas) to malignant (mostly squamous cell carcinomas). A subset of about 40 of these viruses have assumed a great medical and public health importance for the infections and diseases they cause in the anogenital tract. Cervical cancer is the best-established example of the causal link between HPV and cancer. The availability since 2006 of effective vaccines to prevent this disease and others has been a major public health advance whose scope of application is likely to grow.

SHORT HISTORY OF THE DISEASES

Cutaneous and anogenital warts, as well as anogenital cancers, have been recognized since antiquity (Routh et al., 1997). The transmissible nature of warts was established by Jadassohn in 1895, the first of several investigators to perform experimental inoculations in man of infectious materials derived from warts (Rowson and Mahy, 1967). Ciuffo reported in 1907 that the agent was filterable, and the viral nature was further confirmed in 1949 with the demonstration by electron microscopy of viral particles in warts (Ciuffo, 1907; Strauss et al., 1949). The DNA of the first HPV—type 1—was cloned in 1975 from a plantar wart, and the existence of more than one type of HPVs was established in 1977 (Favre et al., 1975; Orth et al., 1977; Gissmann et al., 1977). HPV6 and 11, the main agents of genital warts, were identified in 1980 and 1982, respectively (Gissmann and zur Hausen, 1980; Gissmann et al., 1982). HPV-11 was first isolated from a laryngeal papilloma, a condition that gained recognition in 1871, and for which a papillomavirus etiology was first suspected in 1973 (Mackenzie, 1871; Boyle et al., 1973).

In 1935, Rous and Beard had shown that benign papillomas induced by the cottontail rabbit papillomavirus (CRPV) (Shope rabbit papillomavirus) could progress to invasive carcinomas (Rous and Beard, 1935). This observation, as well as the description in 1922 by Lewandowsky and Lutz of epidermodysplasia verruciformis, followed by the identification in 1966 of papillomavirus particles in the disease's lesions, which later in life may convert to a squamous cell carcinoma, provided the original insight into the potential oncogenicity of HPV (Lewandowsky and Lutz, 1922; Ruiter and van Mullem, 1966). A relationship between HPV and cancer of the cervix was first clearly proposed in 1975 (zur Hausen et al., 1975; Munoz and De The, 1975). The isolation in cervical carcinomas of HPV-16 (Durst et al., 1983) in 1983 and of HPV-18 (Boshart et al., 1984) a year later marked the beginning of the experimental and epidemiologic quest that eventually established the causal link between a subset of HPVs and various malignancies of the anogenital tract, as well as of the respiratory and upper digestive tracts. An important landmark in that respect was the demonstration that 99.7% of cervical carcinomas worldwide contain HPV DNA (Walboomers et al., 1999).

ETIOLOGIC AGENTS

Classification

HPVs belong to the genuses *Alpha, Beta, Gamma, Mu,* or *Nupapillomavirus* of the family Papovaviridae (see http://www.ncbi.nlm.nih.gov/ICTVdb/Ictv/index.htm). Some of the genuses are further divided into species, each represented by a species type. For example, HPV type 16, an Alphapapillomavirus, is the representative type of species 9, which also contains types 31, 33, 35, 52, 58, and 67 (de Villiers et al., 2004). In contrast to many viruses, the taxonomy of HPVs is based on genotyping, not on serotyping. A new type is one that differs from others by more than 10% in the nucleotide sequence of the L1 open reading frame (ORF), which encodes for the major capsid protein. Subtypes differ from one another by 5–10%, while variants do by more than 2%, but less than 5% (de Villiers et al., 2004). As of mid-2007, 107 types have been assigned, but this is a classification in progress with some type assignments that have been changed (e.g., types 46, 55, and 64).

From a clinical standpoint, it is practical to identify three groups of diseases that approximate the delineation of different HPV genuses. The first group is made mostly of cutaneous warts (most notably hand, plantar, flat warts), corresponding essentially to the Gamma, Mu, and Nupapillomaviruses. The second is the rare genodermatose epidermodysplasia verruciformis that is associated with the Betapapillomaviruses. The third is the large group of genital and/or mucosal HPVs made of the *Alphapapillomaviruses*.

Antigens Encoded by Agent

The antigenicity of HPV is a complex and unsettled field, particularly as it relates to the immune response during natural infections (Egelkrout and Galloway, 2007). This section is limited to the salient findings.

The papillomavirus major capsid protein when heat-denatured under reducing conditions harbors an antigen shared by all papillomaviruses tested (Jenson et al., 1980). This has been exploited for the development of a cross-specific antibody (sometimes referred as the "DAKO" antibody) that is useful, although not very sensitive, for the histologic diagnosis of papillomavirus infection by immunocytochemistry (Jenson et al., 1985). Although results have been discordant, it does not appear that this antigen is recognized during a natural infection (Dillner et al., 1990; Kirnbauer et al., 1994).

The study of the HPV capsid antigens was long hampered by the difficulty of obtaining HPV virions by either culture or lesion extraction. This obstacle was lifted in the 1990s by the possibility of creating virus-like particles (VLP) by expressing the major papillomavirus capsid protein in a suitable, usually eukaryotic, expression system—the same technology that was the basis for the development of the current HPV vaccines. The investigation of the antigenic properties of VLPs from different HPV genotypes demonstrated a generally good agreement between serotype and genotype (Rose et al., 1994a, 1994b; Giroglou et al., 2001). This was not entirely surprising since both entities are directly or indirectly dependent on the L1 nucleotide sequence. For the same reason, at least partial antigenic cross-specificity was noted for closely related genotypes, such as types 6 and 11 (Bonnez et al., 1991; Touze et al., 1998; Wang et al., 2003a). Several hypervariable regions are present on L1. Some of these regions, which form exposed loops, bear some of the conformational type-specific antigenic determinants of the capsid. Although the number and locations of these determinants are not completely defined, it appears that for HPV-6 the main type-specific determinant is borne by loops EF and BC, by loops FG and HI for HPV-16, and by loop EF for HPV-31 (Christensen et al., 2001; McClements et al., 2001; Carter et al., 2003, 2006; Wang et al., 2003b). Most of these conformational epitopes are neutralizing. For example, for HPV-16 such epitopes are found in loops DE, FG, and HI (Carter et al., 2003; White et al., 1999; Varsani et al., 2006; Culp et al., 2007; Ryding et al., 2007; Day et al., 2007). Although neutralizing linear epitopes have been described, they are probably less important than the conformational ones (Christensen et al., 1996a, 1996b; Combita et al., 2002). HPV L1 VLPs are currently the best antigens for diagnostic serology (Egelkrout and Galloway, 2007; Dillner, 1999). However, the sensitivity of adequately specific immunoassays based on these antigens is approximately 50–80%, an unsatisfactory standard that has limited the use of these tests to seroepidemiology.

The L2 minor capsid protein contains subdominant, broadly cross-neutralizing epitopes (Roden et al., 2000; Slupetzky et al., 2007; Kondo et al., 2007). These epitopes are linear, and for HPV-16, are located in the amino-terminus portion of the polypeptide (Gambhira et al., 2007).

Full or partial polypeptides encoded by the non-structural ORFs E2, E4, E6, and E7 have been used for serological testing. Immunoassays based on these antigens are insensitive, particularly if one tries to preserve specificity, and often lead to variable results from study to study (Egelkrout and Galloway, 2007; Galloway, 1992). Nevertheless, E6 and E7 polypeptides seem to generate a consistent antibody response in about half of patients with cervical cancer (Müller et al., 1992; Meschede et al., 1998).

PROTECTIVE IMMUNE RESPONSE

Antibody

The humoral response to the viral capsid through the generation of neutralizing antibodies appears to be the sole source of protective immunity against HPV infection. However, as detailed later, this protective immune response requires immunization with HPV L1 VLPs to fully develop. During a natural infection, as shown, for instance, with HPV-6, -11, and -16, there is a seroconversion to HPV L1 VLP in about half of the patients in whom HPV DNA is detected, but this antibody response is insufficient to provide protection from either a future infection or disease (Kirnbauer et al., 1994; Heim et al., 1995; Eisemann et al., 1996; Wikstrom et al., 1997; Studentsov et al., 2003; Viscidi et al., 2004; Tachezy et al., 2006). The reason is likely related to the weak neutralizing antibody titers generated by a natural infection.

Cross-specific neutralizing antibodies to the L2 polypeptides can be generated in animal models of papillomavirus infection (Roden et al., 2000; Slupetzky et al., 2007; Kondo et al., 2007; Gambhira et al., 2007). Whether the magnitude of this antibody response would be sufficient to prevent human HPV infection or disease has not yet been established.

CMI

Although cell-mediated immunity is important in the control of established HPV-related tumors and the resolution of HPV infections, it appears to play no role in their prevention (Eiben Lyons et al., 2007; Kanodia et al., 2007; Sheu et al., 2007).

EPIDEMIOLOGY

HPV infections are extremely ubiquitous, more so than was ever appreciated. Prevalence studies of cutaneous HPV infections have shown rates of 80% in immunocompetent hosts, and approaching 95% in immunosuppressed patients (Antonsson et al., 2000). These infections also tend to persist, as half of infected individuals will retain the same infection when sampled 6 years later (Hazard et al., 2006).

Genital or mucosal HPV infections are less prevalent than cutaneous HPV infections. It has been estimated that worldwide 10.4% of women with normal cervical cytology have a genital HPV infection (Burchell et al., 2006). Nevertheless, prevalence varies considerably according to age. No more than 1% of newborn babies have oral or genital HPV infections (Smith et al., 2004). However, by 20 years of age, at the peak-prevalence, about 23% of women worldwide who have normal cervical cytology have a cervical infection (Burchell et al., 2006). This reflects the sexual mode of transmission of genital HPV. The prevalence thereafter steadily decreases with age, but there are clearly geographic variations in the age–prevalence profiles, whereby the curve may stay relatively flat, or decrease, but present a second smaller peak in older age (Burchell et al., 2006; Smith et al., 2006). Genital HPV infections are rapidly acquired. For example, in a cohort of university women who had a negative HPV DNA cervicovaginal lavage at baseline, 2 years later the cumulative incidence of infection was 32%. In a study of 18–20-year-old male students, the cumulative incidence of any HPV infection of the genitalia after 2 years was 62% (Partridge et al., 2007).

These HPV infections can lead to clinical manifestations. Cutaneous warts, in at least one of their three main manifestations—common (hand and feet), plantar, and flat warts—eventually affect a majority of the population with prevalence rates reported to be as high as 20% (Williams et al., 1993; Larsson and Lidén, 1980). However, they peak during childhood and adolescence, and thereafter eventually disappear in virtually all subjects. One exception to this pattern is the hand warts observed in butchers, meat packers, and fish handlers—lesions that are strongly associated to HPV-7 (Bonnez, 2002a).

The peak age for genital warts is 20–29 years. In the United States, in a large population survey of 18–59-year-olds, 5.7% of subjects (7.4% women and 3.9% men) reported having had genital warts (Dunne et al., 2005). In a study of about 70,000 women aged 18–45 years from Finland, Norway, Sweden, and Denmark, the equivalent figure was 10.6% (Kjaer et al., 2007). Although data are limited, genital wart incidence appears to be increasing (Centers for Disease Control and Surveillance, 2003; Koshiol et al., 2004). A related condition, recurrent respiratory papillomatosis, is rare, with a prevalence of 11 cases per 100,000 for the more common juvenile-onset variant (Derkay, 2001).

HPV infection is a necessary, if not sufficient, condition for the development of cervical cancer (Schiffman et al., 2007). In 2002, with 493,343 new cases and 273,505 deaths, cancer of the uterine cervix ranked second in incidence after breast, and third in mortality after breast and colon, in women worldwide (Parkin et al., 2005). As shown in Fig. 26.1, the geographic distribution of this cancer is unequal, with the developing countries most severely affected (Parkin et al., 2005). In the United States, 2007 estimates place cervical cancer 14th in incidence (11,150 new cases)

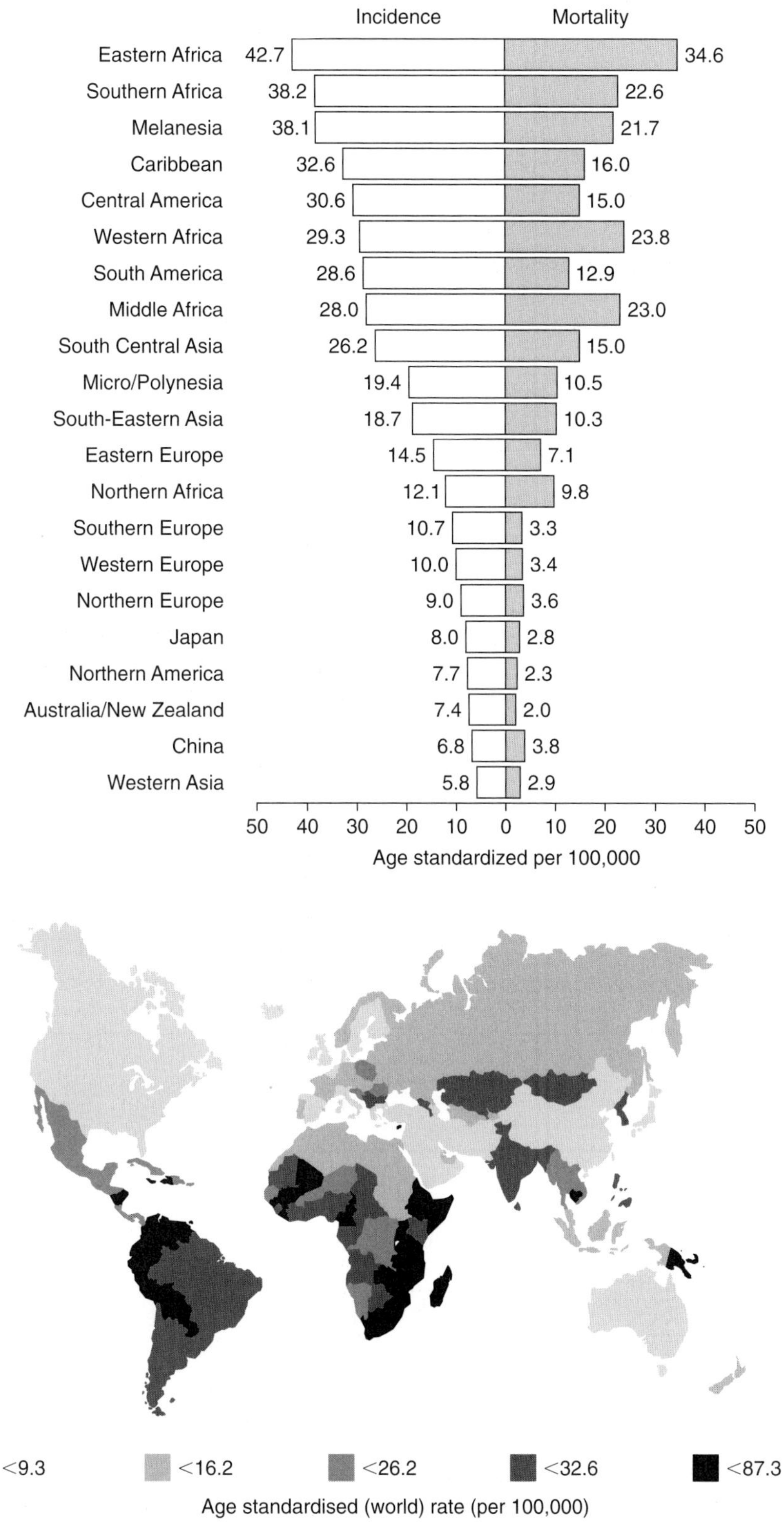

FIGURE 26.1 Age-standardized incidence and mortality rates for cervical cancer (per 100,000) (reproduced from Parkin et al. (2005). Data shown per 100,000.

and 16th in mortality (3670 deaths) among all female cancers (Jemal et al., 2007). Given that cervical cancer only represents approximately 1% of the HPV-related precursor lesions detected on cervical cytology, it is obvious that the public health impact of HPV is much larger that the one limited to cervical cancer (Stoler, 2000). In the United States, the economic burden of genital HPV infection has been estimated to be $3.9 billion in 2000, with cervical cancer representing only 5% of these costs (Chesson et al., 2004). The causal link between some HPVs and cancer is not limited to the cervix, for which the evidence is the strongest, but also extends to some histologic variants of cancers of the vulva, vagina, penis, and anus. HPV accounts for 40–90% of the cancers in these locations (Parkin and Bray, 2006). Altogether, these other cancers represent in the United States about one-third of the HPV attributable cancer burden (Jemal et al., 2007; Parkin and Bray, 2006). It has also recently become clear that HPV plays a dominant role in a majority of oropharyngeal cancers distinct from those related to alcohol and tobacco use (D'Souza et al., 2007). Altogether, the incidence of these non-cervical HPV-associated cancers is as high in women as that of cervical cancer (Jemal et al., 2007) (Fig. 26.2).

Although HPVs are pathogenic, and for some types even oncogenic, they are ubiquitous and not potential biothreat agents. The reasons are that their transmission requires direct contact with the susceptible anatomic areas of the target host, their incubation is long, and only a small percentage of the infections ultimately lead to cancers, which take years to develop.

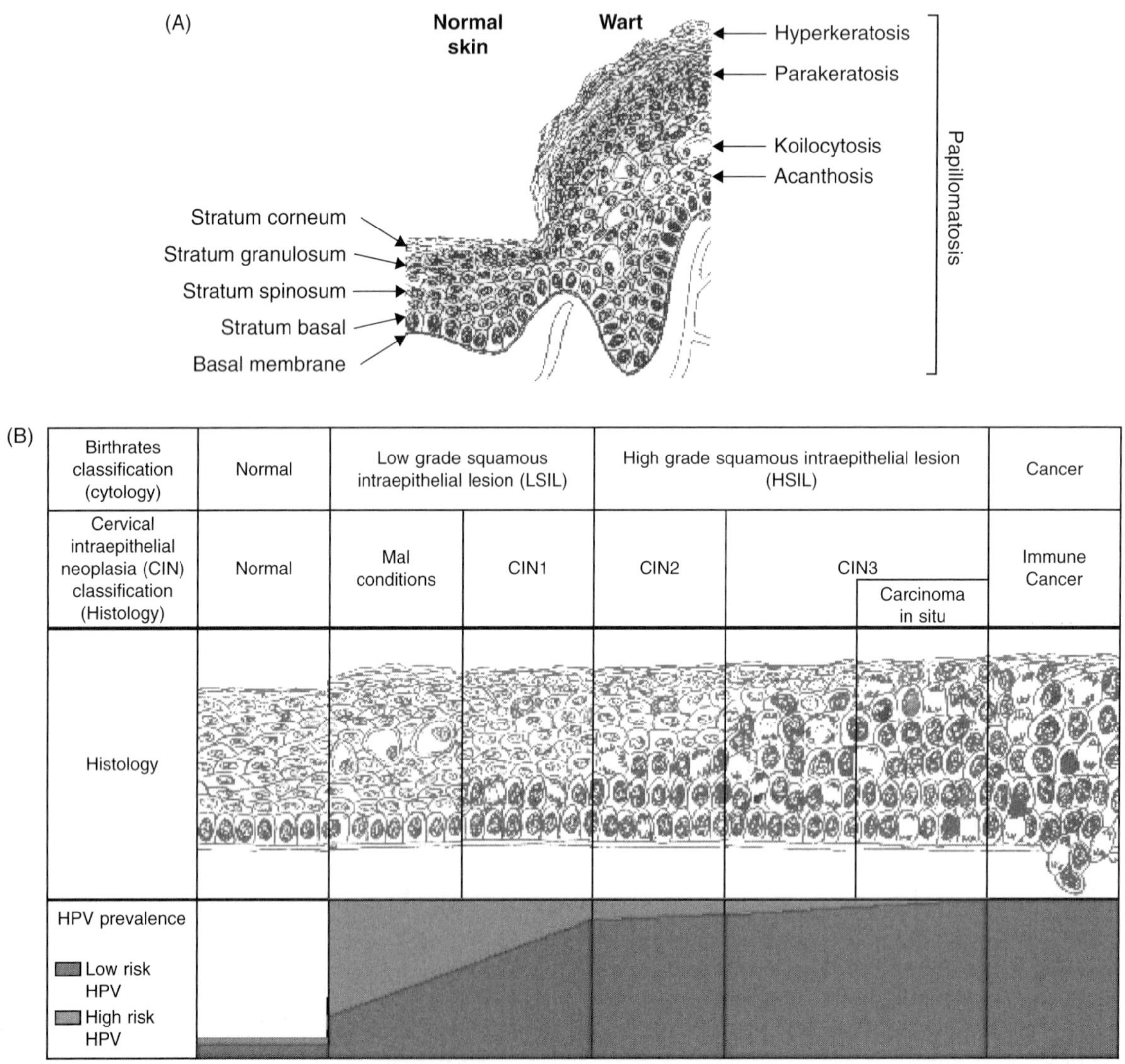

FIGURE 26.2 (A) Histologic features of a genital/cutaneous wart. (B) Histopathogenesis of cervical HPV-associated lesions (see text for details).

CLINICAL DISEASE

Most HPV infections are asymptomatic, latent, or subclinical, and involve the squamous, stratified epithelia of the body as well as the glandular epithelium of the uterine cervix. Consequently, the clinical manifestations arise in those tissues and range from benign proliferative lesions, warts, papillomas, and condylomas, to malignant tumors, namely squamous cell carcinomas and cervical adenocarcinoma.

Cutaneous warts are typically distributed over the acral extremities and the face. They are mostly caused by HPV types 1, 2, and 4. They include common warts (verrucae vulgaris), the most usual form, which are found on the dorsum of the hands, between the fingers, and around the nails. They can also be found on the foot soles where they tend to aggregate and form flat, mosaic warts. Distinct in appearance is the deep plantar wart (verruca plantaris) that is usually solitary, raised, with a ragged surface and a deep base. Plane warts (verrucae planae, juvenile warts) are typically found on the face of children, but can also be found on the extremities.

Epidermodysplasia verruciformis is a rare genodermatosis involving the whole body skin, which is remarkable not only for the HPV types that appear predominantly associated with it such as types 5, 8, 9, 12, 14, 15, and 17, but also for the high propensity for the resulting lesions, which range from plane warts to brown-to-red papules, to progress, by the fifth decade, into squamous cell carcinomas in the sun-exposed areas. This disease has been the earliest model of an association between HPV and cancer (Orth, 2006).

Anogenital warts [condyloma(-ta) acuminatum (-ta), venereal warts] develop on the external genitalia and perineal area. Over 95% of them are caused by either HPV-6 or -11 (Brown et al., 1999). Commonly, a patient will present with 5–6 lesions, each rarely exceeding 1 cm in diameter. In the uncircumcised male, the preputial cavity is the site of predilection, in contrast to the circumcised male, in whom the penile shaft is most often involved. The pubic area, the groin, and the urethral meatus can be involved, but the scrotum much less commonly. In females, lesions are typically present on the posterior introitus extending toward the perineum, but the labia majora and minora are also often involved. Some lesions can be hard to discern either because of their small area or for being very flat. The aid of a magnifying instrument, such as a colposcope, with the prior application of 3–5% acetic acid (white vinegar), facilitates the identification of these lesions. There is evidence that penile flat papules contribute to disease transmission (Bleeker et al., 2002, 2003). In contrast, genital warts can form very large, giant condylomas that tend to have a local invasive behavior, thus forming verrucous carcinomas. *Buschke-Löwenstein* tumors are related, but distinct tumors.

In both males and females, warts can be observed around the anus, sometimes extending into the anal canal, up to the pectinate line. Anal sexual play, including anal intercourse, contributes to this occurrence. Different HPV-related lesions can arise on the vaginal wall in the form of either flat or spiked condylomas. Condylomas can also be observed on the cervix. Papillomas may also grow in the larynx, on the vocal cords causing airway obstruction. This unusual condition, called recurrent respiratory papillomatosis, is caused mostly by the same HPVs causing genital warts, types 6 and 11 (Derkay, 2001; Derkay and Darrow, 2006). Its age distribution is bi-modal, with a more common juvenile-onset and an adult-onset form.

Different HPV-related lesions can be found in the oral cavity (Bonnez, 2002b). Oral squamous cell papillomas (also called squamous papillomas) are the most common, and with oral condylomata acuminata—a different histologic entity—they are largely caused by the genital HPV types 6, 11, and 16. Oral verruca vulgaris are histologically distinct and caused by cutaneous HPV, such as types 2, 4, 57. Focal epithelial hyperplasia of the oral cavity (Heck's disease or Archard's disease) is a rather unique, diffuse, and usually transient oral disease caused predominantly by HPV-3 and -13 (Syrjanen, 2003).

TREATMENT

The benign and self-limited nature of most HPV lesions, combined with the limitations of the current treatment options, makes observation without treatment an acceptable approach in many instances of HPV disease. Nevertheless, physical or cosmetic discomfort, psychological burden, and reduction of the transmission risk are treatment considerations. The treatment options rely on the physical or chemical destruction of the lesions, as well as on immunomodulation. Their use rests on evidence of widely varied quality.

Cutaneous warts are treated mostly with the topical application of preparations containing salicylic acid, a keratolytic agent (Micali et al., 2004; Gibbs and Harvey, 2006). These preparations can be found over the counter, often in combination with lactic acid and collodion. Glutaraldehyde, silver nitrate, and formaldehyde are some of many other chemicals that have been used. A major adjunctive measure is the removal prior to treatment of any superficial cornified layers by soaking the area and using a pumice stone, sandpaper, or an emery board. Cryotherapy is an alternate option that can be applied with over-the-counter cryogenics, although better freezing is obtained in the medical office with liquid nitrogen delivered by spray, probes, or cotton applicators. The therapeutic application on the wart of an occlusive dressing with duct tape has received a lot of attention, especially after the results of a randomized clinical trial (Focht et al., 2002). However, two subsequent trials failed to replicate these results (Wenner et al., 2007; de Haen et al., 2006).

Imiquimod, an imidazoquinoline amine compound that stimulates the local production of interferon and other cytokines, has been a useful addition to the treatment of genital warts, as it allows self-treatment (Garland, 2003; Skinner, 2003; Chang et al., 2005). It is available as a 5% cream (Aldara) that is applied three times a week on alternate days for 8 weeks, and up to 16 weeks. Complete response rates are about 60%, better for females than for males (Garland et al., 2001). Another self-administered treatment is podofilox, as a 5% solution or gel (Condylox) that is applied three consecutive days per week, for up to four cycles (Longstaff and von Krogh, 2001; Lacey et al., 2003). The recurrence rate is high, resulting in a net complete response rate of 30–50%. Among the practitioner's applied treatments, podophyllotoxin, a mitotic spindle inhibitor, has fallen out of favor because of its poor efficacy–toxicity profile (von Krogh and Longstaff, 2001). Podofilox, one of the active components of podophyllotoxin, has replaced it. Bi- or trichloracetic acid, cryotherapy, laser surgery, and electrosurgery are among the treatment methods for genital warts most commonly employed by health practitioners (Bonnez, 2002c; Lacey, 2005; Workowski and Berman, 2007). There is no firm basis to compare their respective efficacy, but they vary substantially in cost, and their indications do not completely overlap (Alam and Stiller, 2001). The selection is usually guided by the technique's availability to the practitioner.

The treatment of intraepithelial neoplasias outside the cervix is not well codified, and is generally based on the use of observation, imiquimod, cryotherapy, electrosurgery, laser therapy, or local blade excision (Lacey, 2005). Elaborate consensus guidelines have been developed for the management and treatment of the different stages of cervical intraepithelial neoplasia (http://www.asccp.org/) (Wright et al., 2007). They incorporate observation, ablation (laser surgery, cryotherapy), and excisional (electrosurgery, cold-blade surgery) treatment. The treatment of recurrent respiratory papillomatosis is mostly by surgical debridement, with the use of a microdebrider device that is preferred to laser surgery or cold steel mircrosurgery (Derkay and Darrow, 2006). Adjuvant therapy includes systemic interferon and intralesional cidofovir injections. A therapeutic vaccine, HspE7, based on the fusion protein of the heat-shock protein 65 of *Mycobacterium bovis* BCG and E7 of HPV-16, has shown encouraging results in an open-label trial (Derkay et al., 2005).

PATHOGENESIS

Description of Disease Process

HPV diseases are the consequence of the proliferation of the epidermal layers. In a benign lesion, the strata spinosum, granulosum, and corneum are thicker than in the normal epithelium, resulting in acanthosis, parakeratosis, and hyperkeratosis, respectively. In addition, within the stratum spinosum the presence of koilocytes is quasi-pathognonomic of HPV. Koilocytes are large keratinocytes with a perinuclear halo. By histology, the nuclei may appear shriveled, but on cytology they are more easily revealed to be binucleated, and typically larger and with a more effaced chromatin than normal nuclei. The relative magnitude of the different histologic features, and the position and abundance of the koilocytosis are features that are related to the HPV type causing the infection and the nature of the lesion, for instance a common wart as opposed to a genital wart. The infection also leads to a change of the cytoarchitecture within the lesion, with a deepening of the rete ridges in the basement membranes and the protrusion of the lesion above the normal epithelium, a phenomenon called papillomatosis. In an inverted papilloma, such as encountered in the nasal passages, this proliferative process extends toward the dermis rather than above the surface.

In malignant lesions, some of the above features may be retained in the milder histologic grades, but the main change is the proliferation of the basal layer with cells that acquire malignant characteristic with a higher nuclear to cytoplasmic ratio and frequent mitoses, some abnormal. According to the fraction of the epithelium this dyskaryotic proliferation occupies, three grades are recognized in what is called an intraepithelial neoplasia (also improperly called dysplasia). In grade 1, the mildest, the lower third of the epithelium is involved; in grade 2, more than one-third, but less than two-thirds is involved; and in grade 3 more than two-thirds is involved. If the whole epithelium is involved, as in the ultimate grade 3, this is carcinoma in situ, the breach of the basement membrane signaling invasive carcinoma. Cervical intraepithelial neoplasia is shortened as CIN, while the vulvar, vaginal, anal, and penile forms go by the acronyms of VIN, VAIN, AIN, and PIN, respectively. The cytologic diagnosis of these lesions in the cervix has been based on the Bethesda classification (Solomon et al., 2002). Low-grade squamous intraepithelial lesions (LSIL) regroup cervical condyloma and CIN1, while high-grade squamous intraepithelial lesions (HSIL) correspond to CIN2 and 3. The same dichotomic division between low- and high-grade intraepithelial neoplasias is recommended by some for the purpose of histologic classification.

These pathologic changes are the result of viral transcription (Hebner and Laimins, 2006; Snijders et al., 2006). This interaction starts in the dividing basal keratinocytes where, under the control of an early promoter located in the upstream regulatory region of the viral genome, transcripts that are produced encode the ORFs of E6, E7, E1, and E2. E1 and E2 proteins are involved in viral DNA replication, which has to harness the replicative enzymes of the host cell. E6 and E7 proteins are involved with oncogenicity for the high-risk HPVs. As one of the daughter cells ascends and differentiates through the epithelium, viral transcripts are then made from a late promoter located in the E7 ORF. This leads to the expression of the E1, E2, E4, E5, as well as L1 and L2 ORFs. The role of the E4 and E5 proteins is poorly understood, but they associate with the cytokeratin network and the Golgi apparatus, respectively. L1 and L2 proteins are the major and minor capsid proteins, respectively, and are thus necessary for the production of viral particles. The viral capsid is composed of 72 capsomeres, each being made of five L1 and one L2 proteins.

Among the best-understood aspects of HPV molecular pathogenesis are the functions of the E6 and E7 oncoproteins. They interfere with two important cellular tumor-suppressor proteins, p53 and the retinoblastoma (Rb) proteins, respectively. Both of these cellular proteins control the G1/S transition of the cell cycle. E6 also interacts with the function of the PSD-95/discs large/ZO-1 (PDZ) proteins that are important in cell signaling and cell-to-cell adhesion. It also activates the telomerase, thus increasing cell longevity, and in concert with E7 immortalizes keratinocytes. E7 itself can derepress the class I histone deacetylases (HDACs) that are important in blocking the transition of the cell cycle into the S phase. It can also promote chromosomal instability. Methylation of selected sites in the upstream regulatory region of the viral genome and chromosomal alterations (polyploidy, aneuploidy, deletions) are some of the signaling events that mark the transition from benign to malignant (Garnett and Duerksen-Hughes, 2006). Integration of the viral genome into the host genome is another such event. Even if its true significance is in question, integration causes the derepression of E6 and E7. E5, E6, and E7 individually and together, contribute to the inhibition of apoptosis, a process that would normally eliminate abnormal cells (Garnett and Duerksen-Hughes, 2006).

Immune Response to Infection

HPV infection itself can blunt the immune response (Eiben Lyons et al., 2007; Kanodia et al., 2007; Sheu et al., 2007). HPV-16 E5 down-regulates the expression of HLA-A and HLA-B, the major histocompatibility class I proteins responsible for the presentation of viral peptides to cytotoxic T lymphocytes (Ashrafi et al., 2006). Nevertheless, the host is capable of containing, even eliminating HPV infection, by opposing these mechanisms of immune evasion. Cytokines such as interleukin 2 (IL-2) and interferon gamma (type II interferon), as well as interferons alpha and beta (type I interferons) and tumor necrosis factor (TNF) alpha, are important defense mechanisms that are mediated by natural killer cells. In addition to this innate immune response, an adaptative cytotoxic immune response driven by a CD4T cell response can be noted, and is mostly directed at E6 and E7. This response is not HPV type specific, and its effectiveness is unclear, but it appears important enough to account for the greater HPV disease burden encountered in immunodeficient or immunosuppressed subjects (Scott et al., 2001; Nicol et al., 2005). A humoral immune response can probably develop to most of the viral HPV proteins, but it is usually in a minority of subjects. The most important response is to L1. As seen with HPV-6, -11, and -16, only half of the patients with a new HPV infection will develop antibodies to HPV L1 VLP, but the magnitude of this antibody response is insufficient to provide protection from a future HPV infection or disease (Kirnbauer et al., 1994; Heim et al., 1995; Eisemann et al., 1996; Wikstrom et al., 1997; Studentsov et al., 2003; Viscidi et al., 2004; Tachezy et al., 2006).

VACCINES

History

It is in man that the first vaccines against HPV disease were developed (Biberstein, 1944). These vaccines were aimed to be therapeutic and were autogenous—the vaccine was made from the patient's own genital warts to eradicate the remaining lesions (Biberstein, 1944; Abcarian and Sharon, 1982). The results of several case series were very good, but they failed to be confirmed in a double-blind, cross-over, placebo-controlled, randomized study (Malison et al., 1982). The trend favored the vaccine. Unfortunately, the study was underpowered to show a difference. The prophylactic efficacy of such vaccines, or vaccines prepared directly from a lesion, was never studied in man, but some protection was observed in the cow with bovine papillomavirus (BPV) and in the domestic rabbit with the cottontail rabbit papillomavirus (Jarrett et al., 1990; Evans et al., 1962a, 1962b). Nevertheless, as these animal models are sufficiently different from HPV disease in man caution is always required in extrapolating the results. For example, BPV types 1 and 2 infect the epidermis and dermis (Jarrett, 1985), BPV-4 causes disease in the alimentary tract (Jarrett et al., 1978), and CRPV causes a different pathology in its natural host than in the domestic rabbit (Kreider and Bartlett, 1985). It is likely that these limitations are an important reason why none of the vaccine concepts tried in these models made it into the clinic until the advent of the VLP-based HPV vaccine. An example of these inconsequential leads were the positive results obtained with BPV-4 and CRPV full L1 fusion proteins (Pilacinski et al., 1986; Jin et al., 1990; Lin et al., 1992, 1993). Conflicting results were obtained using a full L2 fusion protein, depending on the type of BPV used, or when using CRPV (Lin et al., 1992; Jarrett et al., 1991; Campo et al., 1993). Similarly, immunization with a fusion protein containing the C-terminus of L2 or the full E2 gave different results with BPV-4 and CRPV (Campo et al., 1993; Christensen et al., 1991; Chandrachud et al., 1995; Selvakumar et al., 1995). Favorable results were seen with a full unfused L1 protein (Pilacinski et al., 1986), a combination of L1 C and N-terminus fusion proteins (Jarrett et al., 1991), a L2 N-terminus fusion protein (Chandrachud et al., 1995), a full E1 fusion protein (Selvakumar et al., 1995), a combination of full E1 and E2 fusion proteins (Selvakumar et al., 1995), and E7 and combined E7 and L2 fusion proteins (Campo et al., 1993). No protection was conferred by immunization with fusion proteins made from CRPV L1 C-terminus, middle segment, N-terminus, or a combination of the C-terminus and middle segment (Lin et al., 1993; Christensen et al., 1991).

Several groups of investigators independently pursued the idea of an HPV vaccine with the conviction that it would be based on the expression of the major capsid protein in its native conformation. Although it had been shown that HPV-1 virions purified from plantar warts were the best antigens to detect a specific serologic response in patients with cutaneous warts, it then was not possible to either grow or purify genital HPV virions (Kienzler et al., 1983). With the description of the human xenograft nude mouse model, it became possible to grow HPV-11 virions (Kreider et al., 1985). Using them as antigen, one could detect with great specificity antibodies in genital or recurrent respiratory patients (Bonnez et al., 1991, 1992a, 1992b). The nude mouse model also allowed to demonstrate that the HPV-11 virions could induce strong neutralizing antibodies in the rabbit (Christensen and Kreider, 1990; Bonnez et al., 1990, 1992). This gave the impetus to find a noninfectious substitute as a vaccine. Intracellular, native L1 protein appeared to be made when HPV-1 L1 was expressed in simian cells, using an SV-40 vector (Ghim et al., 1992). In 1991, the synthesis and purification of HPV-16 VLPs with a vaccinia vector was reported (Zhou et al., 1991). This was promoted as a potential vaccine, but other than the fact that no immunologic evidence was given to this effect, the authors asserted that VLPs could be made only with L1 + L2 alone. This was amply proven wrong. Moreover, when other investigators tried to replicate that work with what should have been the same HPV-16 strain, VLPs failed to materialize (Kirnbauer et al., 1993). This was shown to be due to a single point mutation in that particular strain. This explained why these original 'VLPs' did not have the size and appearance of the papillomavirus VLPs made by other investigators, then and since (Kirnbauer et al., 1992; Rose et al., 1993a). It was then shown that HPV VLPs could be made with L1 alone, and that they had the same immunologic properties as the intact virions, in particular that immunized rabbits could generate neutralizing antibodies against an HPV-11 skin infection (Rose et al., 1993a, 1993b, 1994). These experiments done entirely with human viruses and tissues offered the most coherent basis for the current vaccine. Subsequent work revealed that a human HPV vaccine would have to be multivalent, as different HPV genotypes also corresponded to different serotypes (Rose et al., 1994a).

Papillomavirus VLPs were then tested in different animals, and excellent protection was seen against challenge with the homologous virus in a canine oral papillomavirus (COPV) model (Suzich et al., 1995), as well as in CRPV and BPV-4 models (Christensen et al., 1996b; Kirnbauer et al., 1996). Passive immunization was also protective in the COPV and CRPV models, indicating the key role of neutralizing antibodies in the vaccine efficacy (Suzich et al., 1995; Breitburd et al., 1995).

The first clinical trials began in 1997 with the successful demonstration of the initial safety and immunogenicity of monovalent HPV L1 vaccines against HPV-11 and HPV-16 (Evans et al., 2001; Harro et al., 2001). Production of very high neutralizing antibody titers to the immunizing VLPs was demonstrated. It should be noted that the cellular immune response induced by HPV VLPs, as measured by lymphocyte transformation assays, was not type specific (Evans et al., 2001; Pinto et al., 2006). This work opened the clinical development of the current vaccines.

Current Licensed Vaccines

There are currently two vaccines, both containing HPV L1 VLPs that have been licensed for administration in at least some countries: Gardasil and Cervarix. The approval for Cervarix is very recent, and does not concern the United States yet. The major significant differences at this point between these two vaccines are that Gardasil is quadrivalent, directed at HPV types 6, 11, 16, and 18, and contains an alum adjuvant whereas Cervarix is bivalent, including only HPV-16 and -18 and contains alum and monophosphyl lipid A adjuvant system. This difference means that only Gardasil is approved for the prevention of external genital warts. Otherwise, for HPV-16 and -18 cervical infections and diseases, the two vaccines appear at this point to have very similar immunogenicity, safety, and efficacy profiles. Less published information is available for Cervarix than for Gardasil, which will be covered first.

Gardasil

Clinical Trials

Phase I The development of Gardasil was preceded by dose-escalation studies that evaluated the immunogenicity and tolerance of monovalent HPV11 and HPV-16 L1 VLP vaccines (Fife et al., 2004; Brown et al., 2004; Poland et al., 2005). An initial study enrolled HPV DNA negative and HPV seronegative subjects for the type tested and found good neutralization titers with intramuscular doses equal to or higher than 40 μg for HPV-16 and 50 μg for HPV-11 VLP (Fife et al., 2004). Neutralization antibodies in this study and all subsequent studies were measured by competition for VLP binding against a labeled mouse monoclonal with known specificity for the neutralizing epitopes of the HPV type tested (Opalka et al., 2003; Dias et al., 2005). The retention of the antibody response was high after 3 years. A post-hoc assessment of the vaccine efficacy for the prevention of persistent HPV-16 infection was conducted in one of the population originating from this trial (Brown et al., 2004). There were 5 cases per 100 person-years at risk in the control group compared to none in the vaccine group. A dose ranging study (0, 10, 20, 40, and 80 μg) of HPV-16 L1 VLP was conducted in women regardless of their baseline cervical HPV DNA and HPV serostatus (Poland et al., 2005). It showed no difference in local or systemic reactions among treatment groups. The titers after immunization in the subjects who were seronegative at baseline were 36–78-fold higher than those in the subjects who had mounted a serologic response to HPV-16 after a natural infection. Moreover, post-immunization titers were 1.1–2.4-fold higher in the subjects who were seropositive at baseline, compared to those who were seronegative.

Gardasil, the quadrivalent vaccine preparation including HPV-6, -11, -16, and -18 VLPs, was evaluated in a dose-escalating study, followed by a dose-ranging study in 16–23-year-old women (Villa et al., 2006a). This study led to the selection of the current vaccine dosing, and confirmed the high immunogenicity for each of the HPV types of the vaccine preparation, with serologic responses being sustained for up to 3 years.

Phase II It was the study by Koutsky et al. (2002) that firmly established the clinical value of the HPV VLP vaccine. The investigators enrolled 2392 women aged 16–26 years who had had no more than five lifetime sexual partners to be randomized to receive either HPV-16 L1 VLP (40 μg) formulated with amorphous aluminum hydroxyphosphate sulfate adjuvant or adjuvant alone given intramuscularly at day 0, month 2, and month 6. Genital samples to test for the presence of HPV DNA were obtained at enrollment, one month after the last immunization, and every six months thereafter. The patients were monitored with cytology, and colposcopy if indicated by protocol. The endpoint of the study was the presence, starting one month after the third immunization, of HPV-16 infection on two or more consecutive visits (persistent infection). A predetermined number of cases of persistent infection was reached after a median follow-up of 17.4 months. Excluding from the analysis those women who had been HPV-16 DNA positive or HPV-16 seropositive at baseline, the analysis showed that none of the 768 vaccinees, as opposed to 41 of the 765 placebo recipients, had developed a persistent infection. In the secondary analysis, 10 of the cases of persistent HPV-16 infection had developed related CIN, thus pointing to a disease prevention effect as well. Although pain at the injection site was the most common side effect, there was no difference in the rates of adverse reactions between the two study groups.

The analysis after 48 months of follow-up of the study populations showed 7 cases of HPV-16 persistent infections in the 755 vaccine recipients, but 111 cases in the placebo group, yielding an efficacy of 94% (95% CI, 88–98%) (Mao et al., 2006). Three cases each of CIN2 and CIN3 were noted in the placebo recipients, but none in the vaccine group. As in the phase I studies, the HPV-16 VLP serum titers peaked after the third immunization, then declined, but by month 18 reached a plateau still many folds higher than the levels attained after a natural infection.

An extension of the dose-ranging study that led to the dosing selection for Gardasil offered an opportunity to look up to 5 years post-enrollment at the combined incidence of HPV-6, -11, -16, and -18 infection and disease (CIN and genital warts) in placebo (adjuvant) and vaccine (the formulation that was later approved as Gardasil) recipients who were seronegative and HPV DNA negative for the vaccine types analyzed (Villa et al., 2006b). This endpoint was reached in 2 of 235cases in the vaccine group and in 46 of 233 in the placebo groups—a vaccine efficacy of 96%. The efficacy was 100% when looking at disease only, with no cases of CIN or genital warts in the vaccine recipients, but 6 in the placebo recipients. In order to study the immune memory response, those subjects whose study had been extended beyond 3 years received a fourth injection at month 60 (Olsson et al., 2007). This fourth immunization with the vaccine induced a strong anamnestic response. The antibody titers reached stable levels higher than one month after the third immunization.

The clinical development of the HPV vaccine presented two particular problems. It was anticipated that in order to achieve maximal efficacy, the vaccine would have to be administered before the acquisition of HPV infection. Therefore, HPV persistent infection or disease could not be used as an endpoint of vaccine efficacy in an essentially sexually naive population. A protective neutralizing titer after immunization was regarded as an adequate surrogate to bridge the results in 16–26-year-olds to 9–15-year-olds. The second difficulty was the complete or near-complete efficacy of the vaccine in 16–26-year-olds. This meant that it was impossible to determine a neutralizing antibody level below which either persistent HPV infection or, more importantly, disease would start breaking through. The vaccine had to be shown to induce at least equivalent antibody titers in pre-adolescents as in adolescent or adult women. This "bridging" efficacy was established in a randomized, double-blind trial of Gardasil immunization in 506 girls (10–15-year-old), 510 boys (10–15-year-old)—boys and girls being sexually naive—and 513 women (16–23-year-old) (Block et al., 2006). The resulting antibody titers to HPV-6, -11, -16, or -18 VLPs in boys and girls were non-inferior to those in women. They were in fact 1.7–2.7-fold higher, with a trend showing higher titers in boys. These results were repeated in a study done in boys and girls 9–15 years old that compared the vaccine against a non-aluminum adjuvant (Reisinger et al., 2007). In the first study, boys and girls were more likely to have fevers after vaccination, 13.8 and 12.8%, respectively, than women, 7.3%. In the second study, vaccine recipients reported one or more injection site–related adverse events more often than the placebo recipients, 75.3% vs. 50.0%.

Phase III The primary endpoint of the phase III trials was the demonstration that the vaccine not just prevented infection, but that it also reduced disease incidence. In contrast to cervical HPV infection, CIN2/3 was regarded as the proper surrogate endpoint for cervical cancer because of the following factors: (a) it was the immediate precursor for cervical cancer; (b) it would be amenable to secondary prevention measures when found; (c) its detection and removal had been shown to prevent cancer; and (d) it was fully treatable (Kang and Lagakos, 2004). The trial results are summarized in Table 26.1 (Garland et al., 2007; Koutsky, 2007; Ault, 2007; Joura et al., 2007; Luna et al., 2007). They clearly show that the greatest benefit of the vaccine was in subjects that were seronegative and HPV DNA negative for the vaccine HPV types, indicating that maximum benefit of vaccination occurs before sexual debut (Hildesheim and Herrero, 2007). Nevertheless, even in women with virological (serology or PCR) evidence of preexisting vaccine-type HPV infection, the vaccine was 100% effective at preventing incident CIN2/3 or cervical adenocarcinoma in situ (AIS) by vaccine HPV types for which the women were negative at enrollment (FUTURE II Study Group, 2007). Similarly, vaccination was 94% effective in preventing vaginal or vulvar HPV-associated lesions, including external genital warts. In this combined study of 15–26-year-old women, 26.8% of the enrollees had current or past evidence of at least one vaccine-type HPV infection.

HPV-16 and -18 account for only 70% of cervical cancers and a smaller fraction of CIN, especially low grade. The vaccine has little impact on the infections and diseases caused by the other HPV types. The "intention-to-treat population—any lesion regardless of type" analysis (Table 26.1), which also includes women not naive for the vaccine types, provides a look at the overall effectiveness rather than just efficacy of the vaccine. The results are substantially worse, and it has been pointed out that the vaccine efficacy of 17% noted in that analysis of protocol 017 implies that "129 women would need to be vaccinated in order to prevent one case of CIN2/3 or AIS" (Koutsky, 2007; Sawaya and Smith-McCune, 2007).

TABLE 26.1 Summary of the efficacy results of Gardasil against HPV-related lesions

Lesion prevented					Caused by HPV						
Condylomas (external and vaginal)	VIN/VAIN 1	VIN/VAIN 2/3	CIN 1	CIN 2/3 & AIS	6/11	16/18	Others	Number of subjects[a]	Vaccine efficacy (%) (95% confidence interval)	Merck protocol	References
Per-protocol susceptible population[b]											
•					•	•		4540	100 (92–100)	013	Garland et al. (2007)
	•				•	•		4540	100 (49–100)	013	Garland et al. (2007)
		•				•		15,596	100 (72–100)	007, 013, 015	Joura et al. (2007)
			•	•	•	•		4499	100 (94–100)	013	Garland et al. (2007)
				•		•		17,129	99 (93–100)	005, 007, 013, 015	Ault (2007)
			•	•	•	•		3819	100 (61–100)	019[c]	Luna et al. (2007)
Unrestricted susceptible population[d]											
•					•	•		5351	96 (86–99)	013	Garland et al. (2007)
	•				•	•		5351	82 (16–98)	013	Garland et al. (2007)
		•				•		17,531	97 (79–100)	007, 013, 015	Joura et al. (2007)
			•	•	•	•		4951	98 (92–100)	013	Garland et al. (2007)
				•		•		19,466	98 (93–100)	005, 007, 013, 015	Ault (2007)
Intention-to-treat population[e]											
•					•	•		5455	76 (61–86)	013	Garland et al. (2007)
	•				•	•		5455	63 (<0-88)	013	Garland et al. (2007)
		•				•		18,174	71 (37–88)	007, 013, 015	Joura et al. (2007)
			•	•	•	•		5455	55 (40–66)	013	Garland et al. (2007)
				•		•		20,583	44 (31–55)	005, 007, 013, 015	Ault (2007)

(Continued)

TABLE 26.1 (Continued)

Lesion prevented: Condylomas (external and vaginal)	VIN/VAIN 1	VIN/VAIN 2/3	CIN 1	CIN 2/3 & AIS	Caused by HPV 6/11	Caused by HPV 16/18	Caused by HPV Others	Number of subjects[a]	Vaccine efficacy (%) (95% confidence interval)	Merck protocol	References
Intention-to-treat population - Any lesion regardless of type											
•					•	•	•	5455	51 (32–65)	013	Garland et al. (2007)
	•				•	•	•	5455	18 (<0–46)	013	Garland et al. (2007)
		•				•	•	18,174	49 (18–69)	007, 013, 015	Joura et al. (2007)
			•	•	•	•	•	5455	20 (8–31)	013	Garland et al. (2007)
				•		•	•	20,583	18 (7–29)	005, 007, 013, 015	Ault (2007)

[a] The subjects were randomized 1:1 between the vaccine and placebo groups, and the two groups were tightly balanced. Except where noted (Protocol 019), the subjects were 16–26-year-old women, and results are based on 3 years of follow-up.
[b] This was the population used for the primary endpoint analysis. Subjects were seronegative and HPV DNA negative at day 1 for the HPV types included in the vaccine. They had a normal Pap smear, received all 3 immunizations, and cases were counted starting 1 month after the third immunization.
[c] Protocol 019 enrolled 24–45-year-old women, and the disease endpoint was assessed by cytology at a mean 1.65 years of follow-up. In that study, the vaccine was 91% effective preventing persistent infection, external genital warts, or CIN caused by HPV-6, -11, -16, or -18.
[d] This population included subjects who were seronegative and HPV DNA negative at day 1 for the HPV types included in the vaccine, but it included protocol violators, individuals who had received less than 3 doses, who had developed HPV DNA positivity during the vaccination phase, and who had an abnormal Pap smear at day 1. Case counting began 1 month after the first immunization. About a quarter of the subjects were positive by HPV DNA or serology at day 1 for at least one of the vaccine types.
[e] This population was identical to the Unrestricted Susceptible Population, except that it included subjects who were HPV seropositive and/or HPV DNA positive for the vaccine types at day 1.

This is a pessimistic and restrictive interpretation for several reasons. These low-efficacy figures are those observed at 3 years of follow-up. As long as the vaccine is effective, they should improve with further follow-up, because the placebo recipients continue to acquire lesions at a higher rate than the vaccinees. This increase in efficacy in the general population has been verified with Cervarix with 5-year follow-up (Harper et al., 2006). The figures also vary according to the geographic area, possibly reflecting different HPV type distributions. The vaccine efficacy for CIN2/3 and AIS in North America in the "intention-to-treat" population was double that of Asia (Ault, 2007). HPV-16 and -18 are more oncogenic than the other HPVs, and preventing one case of CIN2/3 caused by one of these two types should have more impact on cervical cancer than preventing one case of CIN2/3 caused by one of the non-vaccine types (Schiffman et al., 2005). Finally, vaccination cannot be dismissed when it is the only preventive measure available for HPV-associated cancers other than those of the cervix. Even with cancer of the cervix, half of the affected women are those who have never been screened, and another 10% has not been screened in the previous 5 years ((NIH) NIoH, 1996; Leyden et al., 2005). Given that the socioeconomic barriers to cervical cancer screening are different from those applying to vaccination, vaccination might be the only preventative option for some women (Hewitt et al., 2004).

A further analysis of the phase II/III trials examined the impact of baseline characteristics on the immunogenicity of the vaccine (Giuliano et al., 2007). As already observed, age at vaccination was inversely proportional to the antibody titers induced by immunization. Preexisting HPV antibodies resulted in a higher serologic response than in the absence of baseline antibodies. Otherwise, there was no detectable effect of race/ethnicity, region of residence, lactation status, hormonal contraceptive usage, smoking status, Pap smear diagnosis, immunosuppressant, or anti-inflammatory drug use, and number of sex partners. The absence of effect of anti-inflammatory drug use on vaccine immunogenicity is preliminary, but reassuring in light of animal and human in vitro experiments that had shown an inhibitory effect of cyclooxygenase-2 inhibitors (Ryan et al., 2006).

Agent/Strain Used in Vaccine, How Produced, Number of Doses Administered to Give Protective Immunity Gardasil (also sold in some countries as Silgard) is made by Merck and is distributed in European countries by Sanofi Pasteur, a joint venture of Merck and Sanofi-Aventis. It was first marketed in June 2006. This quadrivalent vaccine is made in a *Saccharomyces cerevisiae* yeast expression system, and contains L1 VLP of HPV types 6, 11, 16, and 18 at concentrations of 20, 40, 40, and 20 μg, respectively. A 0.5 ml dose also includes approximately 225 μg of an amorphous aluminum hydroxyphosphate sulfate proprietary adjuvant. The preparation is sterile and does not contain a preservative or antibiotics. The vaccine is administered as three doses at months 0, 2, and 6. It is given intramuscularly, preferably in the deltoid area of the upper arm or in the higher anterolateral area of the thigh.

Protective immunity is believed to be exclusively provided by the neutralizing antibodies induced by immunization. A protective antibody threshold has not yet been established given the very high efficacy of the vaccine. It is therefore unclear after how many doses immunity is established. It was however noted that the vaccine retained a very high efficacy, even when the analysis included patients who had not received all three injections (see Table 26.1). Clinical studies are in progress to examine the effectiveness of abbreviated immunization schedules (Advisory Committee on Immunization Practices, 2007, www.cdc.gov/vaccines/recs/acip/downloads/acip_min_feb07.pdf). Gardasil package insert (https://www.merckvaccines.com/gardasilProductPage_frmst.html) indicates that neutralizing antibody titers measured one month after the third dose are the highest for all the vaccine HPV types if the second dose is administered on time or early (36–50 days after dose 1) and/or if the third dose is administered on time or late (138–160 days after dose 2).

Clinical and serological endpoints have been followed up to 5 years at this point , and there is still no evidence that a booster immunization is or will be necessary (Villa et al., 2006b; Olsson et al., 2007). Nevertheless, a booster immunization was evaluated 5 years after the first immunization in a subgroup of 104 16–23-year-old female vaccine recipients (Olsson et al., 2007). For each of the HPV types, the neutralization titers were superior to titers observed one month after the third immunization, demonstrating a very strong memory response. In that study, subjects who had received the placebo for their first three immunizations were also boosted, and among those who had been seropositive but PCR HPV DNA negative at enrollment, the titers after this single vaccine administration were comparable to those measured after three initial immunizations in the seronegative subjects.

Vaccine Recommendations, Including Potential to Administer with Other Vaccines The Advisory Committee on Immunization Practices (ACIP), which is part of Centers for Disease Control and Prevention and a leading group on immunization recommendations in the United States, recommends the vaccine to 11–12-year-old girls, with a catch-up immunization for 13–26-year-old girls (Markowitz et al., 2007) (Table 26.2). The Vaccine for Children Fund for free vaccination, a federally mandated program administered by each state and targeted to the underprivileged population, is applicable to the 9–18-year-old female population.

In the 11–12-year-old group for which the vaccine is indicated, two other vaccines are recommended, the tetanus and diphtheria toxoids and acellular pertussis vaccine (Tdap) and the meningococcal vaccine (MCV4) (FUTURE II Study Group, 2007). At this date the ACIP and the American Academy of Pediatrics (AAP) recommend administering these single-dose vaccines with one of the HPV vaccine doses. It is recommended to use a different site for the injection of each vaccine. Although there are no theoretical reasons for concern, there are no data currently available evaluating the safety and efficacy of this practice. This should change by the end of 2008, as a study is currently underway (ClinicalTrials.gov Identifier: NCT00325130). A catch-up immunization with the hepatitis B vaccine may also be necessary in the same age group, and a study has shown no effect on the safety and immunogenicity of the HPV vaccine (Markowitz et al., 2007; Gardasil Package Insert, 2007, www.fda.gov/cber/label/hpvmer040307LB.pdf).

Several American professional organizations have endorsed the use of the vaccine with recommendations usually similar to those of the ACIP. They include the AAP (American Academy of Pediatrics Committee on Infectious Diseases, 2007, http://www.cispimmunize.org/ill/pdf/HPVprovisional.pdf), the American College of Obstetricians and Gynecologists (ACOG) (ACOG Committee, 2006), the American Association of Family Practitioners (AAFP) (Armstrong, 2006), the Society for Adolescent Medicine (SAM) (Friedman et al., 2006), the American College Health Association (ACHA), (ACHA guidelines, 2006) and the American Cancer Society (ACS) (Saslow et al., 2007). The ACS guidelines do not recommend the vaccine for the 19–26-year-old female population.

It is important to note that HPV immunization does not change in any way as the current guidelines for cervical cancer screening (Pap smear) (Saslow et al., 2002; ACOG Practice Bulletin, 2003; Screening for cervical cancer, 2003, http://www.ahrq.gov/clinic/3rduspstf/cervcan/cervcanrr.pdf).

TABLE 26.2 HPV quadrivalent vaccine, Gardasil, recommendations (United States)

Indications

- 11–12-year-old girls
- Catch-up immunization for 13–26-year-old girls
- Can be administered to girls as young as 9 years
- Most effective if administered before sexual debut
- Given as 3 intramuscular doses, at months 0, 2, and 6
- May be given at the same time as the hepatitis B (Hep B), the tetanus and diphtheria (Td), the tetanus, diphtheria, acellular pertussis (Tdap), or the meningococcal (MCV4) vaccines
- Can be given to patients with genital warts, an equivocal or abnormal Pap smear, or the presence of cervical high-risk HPV DNA (Hybrid Capture II assay), but the subject should be advised that the vaccine is not therapeutic and does not protect against diseases by already established HPV types
- May be given to lactating women
- May be given to immunosuppressed or immunodeficient patients, but efficacy may be reduced
- Cervical cancer screening (Pap smear) recommendations are unchanged for vaccine recipients and should be reminded to the subject
- Protective sexual behaviors (including abstinence, monogamy, limiting the number of sexual partners, using condoms) should be continued since the vaccine does not protect against all HPV infections and does not affect other sexually transmitted infections
- An interrupted immunization series should be resumed (not restarted at dose 1) as soon as it becomes possible

Contraindications

- Immediate hypersensitivity to yeast or any of the vaccine components
- Pregnancy (if vaccinated during pregnancy report the case to the vaccine pregnancy registry, tel. 800-986-8999)
- Moderate or severe acute illness
- The vaccine does not contain thimerosal or mercury

Countries Where Vaccine is Licensed, Including Manufacturers As of the end of October 2007, Gardasil has been approved in 87 different markets, including the 3 North American countries (Canada, United States, and Mexico), 15 Caribbean and Central American countries, 6 South American countries, 38 European countries, 15 Middle Eastern and African countries, and 11 Asian or Pacific countries, and more applications are pending.

Indications for Vaccination/Target Populations (Who is Vaccinated?) In the United States, Gardasil has been approved for administration to females ages 9 through 26 years for the prevention of cancer of the cervix; cervical AIS; CIN grades 1, 2, and 3; VAIN grades 2 and 3; VIN grades 2 and 3; and genital warts.

The vaccine is recommended irrespective of cervical HPV DNA or Pap smear status at the time of immunization. It should be noted that some countries, such as Australia, have allowed for Gardasil to be administered to males, and in other countries the recommended age of immunization differs. Most states in the United States do not plan to make the vaccine mandatory for school attendance, a very sensitive and controversial issue to date.

Duration of Immunity The efficacy of Gardasil has been established up to 3 years in over 10,000 subjects (see Table 26.1). More limited efficacy data on 241 subjects are available at 5 years demonstrating complete protection against cervical persistent infection and disease (Villa et al., 2006b). Sustained neutralizing titers for the different HPV types included in the vaccine have also been documented for at least 5 years (Olsson et al., 2007). The absence of events indicating vaccine failure has precluded the establishment of a neutralizing antibody titer threshold below which protection starts waning. A mass-screening program with central registries in place in the Nordic countries (Denmark, Sweden, Norway, and Finland) should allow to determine not only the duration of vaccine effectiveness, but also the impact on cervical cancer screening, and long-term safety (Lehtinen et al., 2006; Barnaba et al., 2006). Modeling of the decay of the antibody response would indicate neutralizing titers lasting for at least 12 years above those following a natural infection (Fraser et al., 2007).

Contraindications of Vaccination, Including Special Risk Groups The vaccine is contraindicated in subjects with a known history of allergy to yeast, or to any of the active substances or excipients included in the vaccine. Individuals who develop symptoms indicative of allergic reaction should not receive further doses.

The vaccine is also contraindicated during pregnancy and is classified as a pregnancy category B. During the clinical trials 1115 of 10,418 vaccine recipients and 1151 of 9120 placebo recipients reported at least one pregnancy. The outcome of these pregnancies was known for 996 in the vaccine group and for 1018 in the placebo group, and the rates of fetal loss were comparable, 38 and 40%, respectively (Miller, 2006). There were 15 cases of congenital anomalies, 5 of them occurring within 30 days of immunization in the Gardasil recipients and 16 in the placebo recipients (Miller, 2006). The anomalies were those expected to occur in women aged 16–26 years. The manufacturer maintains a pregnancy registry.

Although breast-feeding is not a contraindication to the vaccine, and 995 nursing mothers received the vaccine during the clinical trials, caution is advised as it is not known whether the vaccine antigens or the antibodies induced by immunization are excreted in human milk (Gardasil Package Insert, 2007, www.fda.gov/cber/label/hpvmer040307LB.pdf).

The vaccine is not intended to be used for the treatment of existing genital warts, cervical cancer, CIN, VAIN, or VIN. No therapeutic efficacy has been demonstrated on the cervical HPV infections already present at onset of vaccination (Koutsky, 2007). It will be important to examine whether HPV seropositive women who have cleared their vaccine-HPV infection are less likely to acquire another vaccine type-related disease (Advisory Committee on Immunization Practices, 2007, www.cdc.gov/vaccines/recs/acip/downloads/acip_min_feb07.pdf).

In general, the vaccine does not protect against diseases due to non-vaccine types, but a recent analysis demonstrates a 45% efficacy against CIN2/3 and AIS caused by either HPV-31 or -45, virus types related to HPV-16 and -18, respectively (Brown and The FUTURE Study Group, 2007). These two types account for 2.9 and 6.7%, respectively, of all cervical cancers (Munoz et al., 2004).

The vaccine does not contain any HPV DNA and is noninfectious. Therefore, it is not contraindicated in immunosuppressed or immunodeficient individuals. However, its efficacy in these populations has not been established.

Adverse Events Local adverse reactions were noted in about 85% of the Gardasil recipients, and this was a rate higher only by 6.5–9.4% than seen in the placebo recipients (Garland et al., 2007; Koutsky, 2007). They include pain at the injection site (85%) and swelling (26%) both of which are seen 10% more commonly than with the placebo. Fever in the vaccine and placebo recipients was seen in 13.5 and 10.2% of the subjects, respectively (Garland et al., 2007). Girls and boys younger than 16 years were 7–16% less prone than older females to experience pain at the injection site. They were, however, more likely to have fever after immunization (Block et al., 2006). In a combined population of over 17,000 subject events, serious adverse reactions were seen only once in each vaccine and placebo equal-sized groups. Similarly, there were no differences in the rates of other adverse effects.

As of June 30, 2007, more than 7 million doses of Gardasil had been distributed. The Vaccine Adverse Event Reporting System (VAERS) had received a total of 2531 reports (Anon, 2007, http://www.cdc.gov/vaccines/vpd-vac/hpv/downloads/hpv-gardasil-gbs.pdf). Less than 6% of the reports were classified as serious events defined according to the Code of Federal Regulations. This represents less than half the rate of serious adverse reactions for any vaccines reported to VAERS. Thirteen cases of Guillain-Barré syndrome (GBS) had been reported by that date. Two met the case definition of GBS, occurring within 6 weeks after immunization to Gardasil alone, but both subjects had also received the meningococcal vaccine (Menectra). The number of observed GBS is consistent with the number expected by chance alone.

Thromboembolic events were rare in the prelicensure studies. One case was reported in both vaccine and placebo groups, and after licensure two additional cases were reported to VAERS. None of them, however, could be attributed to the vaccine (Iskander, 2007).

As of May 8, 2007, VAERS had registered 11 reports of syncopal episodes (Iskander, 2007). In the studies that led to licensure, 4.0% of 5088 vaccine recipients experienced dizziness as opposed to 3.7% of 3790 of placebo recipients (Advisory Committee on Immunization Practices, 2007, www.cdc.gov/vaccines/recs/acip/downloads/acip_min_feb07.pdf). This led the ACIP to recommend a period of 15 min of observation after vaccination (Advisory Committee on Immunization Practices, 2007, www.cdc.gov/vaccines/recs/acip/downloads/acip_min_feb07.pdf).

Cervarix

Clinical Trials

Phase I Cervarix is a bivalent vaccine, containing 20 μg each of HPV-16 and -18 L1 VLP formulated with an AS04 adjuvant made of 500 μg aluminum hydroxide and 50 μg 3-*O*-deacyl-4′- monophosphoryl lipid A (MPL). An initial study in human, mice, and monkey subjects demonstrated the ability for the vaccine given intramuscularly at 0, 1, and 6 months to induce higher binding and neutralizing antibody titers (1.6–8.5-fold) when formulated with AS04 than with aluminum hydroxide alone (Giannini et al., 2006). This antibody response persisted for at least 3.5 years.

Phase II The immunogenicity, tolerance, and efficacy of Cervarix was demonstrated in a phase II study enrolling 15–25-year-old women with no more than 6 lifetime sexual partners who were randomized 1:1 to receive either the vaccine or the adjuvant alone by intramuscular injection at months 0, 1, and 6 (Harper et al., 2004). The prevention of transient or persistent incident cervical infection with the vaccine HPV types in women that were HPV-16 and -18 DNA negative and seronegative at entry was the primary endpoint. A secondary endpoint was the prevention of persistent HPV-16 and -18 infection, or of cytologic or histologic atypical cell of unknown significance (ASCUS) or CIN. In the cohort analyzed according to protocol, at up to 18 months of follow-up, the vaccine efficacy in the 366 vaccinees and 355 placebo recipients was 91.6% (95% CI, 64.5–98.0%) for incident cervical HPV16 or 18 infections, 100% (95% CI, 47–100%) for persistent infections, and 92.9% (95% CI, 70–98.3%) for cytologic ASCUS or higher. When analyzed at up to 4.5 years, these results were still excellent, with similar vaccine efficacy figures of 96.9% (95% CI, 81.3–99.9%), 94.3% (95% CI, 63.2–99.9%), and 95.7% (95% CI, 83.5–99.5%), respectively (Harper et al., 2006). When CIN2/3 by cytology or histology was the endpoint, the efficacy was 100% (95% CI, −07.7 to 100%). In that study, Cervarix was shown to induce cross-protection against cervical infections caused by some non-vaccine types, with an efficacy of 54.5% against HPV-31, and 94.2% against HPV-45 (Harper et al., 2006).

The immunogenicity of Cervarix was demonstrated in 10–25-year-old females, with 100% seroconversion for binding antibodies to HPV-16 and -18 VLPs (Pedersen et al., 2007). As with Gardasil, the group of 10–14-year-old females had titers higher by about twofold than the titers in the 15–25-year-old females.

Phase III At present, only interim results of an ongoing phase III study of Cervarix (protocol HPV-008) have been presented (Dubin, 2007). This study has enrolled 18,644 women aged 15–25 years, unscreened for either cervical HPV DNA or cytology, to be randomized 1:1 to receive intramuscularly at least one of three vaccine doses (at months 0, 1, and 6) of either Cervarix or GSK hepatitis A virus (HAV) vaccine (control) that contains the same adjuvant as Cervarix. Cases are counted after the first vaccine dose. The endpoint is histologically proven CIN2 or higher (CIN2+) associated with HPV-16 or -18. The interim analysis was triggered when 23 cases of HPV-16/18 CIN2+ accumulated in the total vaccinated cohort. The per-protocol interim analysis carried out in women who were HPV-16 or -18 negative at baseline showed a vaccine efficacy of 99.3% (95% CI, 53.4–99.3%) at 15-month follow-up. When adjusted to include only those cases in which the HPV infection was present in preceding visits, the efficacy increased to 100% (95%, 74.2–100%). In this study, statistically significant cross-protective vaccine efficacy was also observed against 6-month persistent infections caused by HPV types 45 (vaccine efficacy of 59.9%), 31 (36.1%), and 52 (31.6%), a type related to HPV-16. The vaccine was well tolerated with serious adverse events noted in three patients in the Cervarix and five in the HAV group. General and local symptoms were slightly more common with Cervarix, 85.0 and 91.2%, respectively, than with HAV, 81.6 and 79.9%.

The vaccine reactogenicity profile was similar in all groups. There were also no differences between the two vaccine groups in the rates of abnormal or premature infants from the pregnancies that occurred during the study. The rates of unsolicited adverse events, including chronic and autoimmune diseases were comparable in both groups.

Agent/Strain Used in Vaccine, How Produced, Number of Doses Administered to Give Protective Immunity Cervarix is manufactured by GlaxoSmithKline Biologicals, and was first approved in Australia in May 2007. It is a bivalent vaccine made in insect cells (*Spodoptera frugiperda* Sf-9 and *Trichoplusia ni* Hi-5 cell lines), using a baculovirus expression system. It contains in a 0.5 ml dose, 20 μg each of HPV-16 and HPV-18 L1 VLP, mixed with an AS04 adjuvant (500 μg aluminum hydroxide and 50 μg MPL). The vaccine is given in three doses injected intramuscularly at months 0, 1, and 6.

Vaccine Recommendations, Including Potential to Administer with Other Vaccines It is unknown at present what recommendations will be issued by ACIP for Cervarix use in the United States, but they are likely to be very similar to those of Gardasil for cervical HPV diseases given the similitudes in content and results of the two vaccines. However, the absence of HPV-6 and -11 VLPs in Cervarix is likely to affect specific recommendations regarding the prevention against genital warts. Furthermore, no data have yet been presented on the impact of Cervarix on vulvar and vaginal intraepithelial neoplasias caused by HPV-16 and -18. Price is likely to be a consideration in the recommendations offered by third-payer agencies and insurance companies.

Countries Where Vaccine is Licensed Cervarix has been approved in Australia, Europe, the United Arab Emirates, Kenya, the Philippines, and Japan. Other approvals are expected, including in the United States.

Indications for Vaccination/Target Populations (Who is Vaccinated?) The data supporting the regulatory authorizations that led to Cervarix approval in different countries are not available at present. However, efficacy and safety data from the ongoing phase III study (HPV-008) done on a 15–25-year-old population of women unscreened for current or prior high-risk HPV infections and for cervical cytology have been presented to ACIP, and are a basis of approval for this population (Dubin, 2007). It should be noted that Australia, for example, has approved the vaccine in 10–45-year-old females, as well as in males. As with Gardasil, Pap smear or cervical HPV DNA status are not contraindications to vaccination with Cervarix.

Duration of Immunity It is unknown to what extent higher post-immunization titers predict the durability of protection. Higher titers of binding and neutralizing antibodies to HPV-16 and -18 were seen after immunization with AS04 than with aluminum hydroxide adjuvants (Giannini et al., 2006). The adjuvants in Gardasil and Cervarix differ in formulation, dose, and incorporation process into the vaccine. GSK is currently conducting a double-blind, comparative trial of the two vaccines to evaluate their respective immunogenicity (ClinicalTrials.gov Identifier NCT00423046).

Contraindications of Vaccination, Including Special Risk Groups No data are available on Cervarix contraindications, but it is anticipated that they will resemble those listed for Gardasil. In that respect, Cervarix is not intended for therapeutic use. A large-scale study has clearly demonstrated the absence of effect on existing cervical infection (Hildesheim et al., 2007). We do not yet know the impact of vaccination with Cervarix on the natural history of preexisting disease.

Adverse Events Cervarix causes the same type of adverse reactions as Gardasil. To the extent comparisons are legitimate, local pain and redness might be slightly more common, and redness was as common as swelling (Harper et al., 2004). The rates of systemic reactions did not differ between the vaccine and placebo groups (Harper et al., 2004, 2006). In the interim analysis of the HPV-008 study, general and local symptoms were 3.4 and 1.3 percentage points higher, respectively, with Cervarix than with the HAV vaccine control (Dubin, 2007). There were otherwise no differences between the two groups in the rates of unsolicited and serious adverse events. Likewise, no differences were observed in the rates of abnormal or premature births that occurred in the approximately 670 pregnancies recorded in each group.

Vaccines in Development

The yeast and insect cell expression systems are expensive biotechnologies, and this has stimulated the search for many alternative solutions to reduce the production costs. Two types of vaccines seem to be close to entering the clinical phase of development.

The first one is based on the production of HPV VLP and capsomeres in *Escherichia coli* (Li et al., 1997; Chen et al., 2001; Garcea and Gissmann, 2004; Schaedlich et al., 2007). A different vaccine approach relies on the use of HPV L2 polypeptides, particularly those derived from the N-terminus, which offer broad cross-neutralization among HPV types and other animal papillomaviruses (Roden et al., 2000; Slupetzky et al., 2007; Kondo et al., 2007). Solutions are being developed to augment the weaker immunogenicity of these types of vaccines compared to HPV VLPs (Karanam et al., 2007).

Rationale of Second/Third/New-Generation Vaccines

Other than the important issue of vaccine cost, an area of improvement for future prophylactic vaccine is the broadening of protection against HPV types non-included in the current vaccines. Such augmented vaccines could give not only added protection against cervical cancer, but also have a much more important impact on any intraepithelial neoplasias than the present vaccines. However, expanded vaccine will be more costly to manufacture.

PROSPECTS FOR THE FUTURE

Several ongoing studies should lead to an extension of the indications for the HPV vaccine. Gardasil has already demonstrated its immunogenicity, efficacy, and safety in women older than 26 and up to 45 years old (Luna et al., 2007). Cervarix is also being evaluated in the same age population. A randomized, double-blind, placebo-controlled trial (ClinicalTrials.gov Identifier NCT00423046) is evaluating its immunogenicity in 18–45-year-old women, while another study is focused on the 18–35-year-old female population (ClinicalTrials.gov NCT00306241). A third trial is examining the safety, immunogenicity, and efficacy of the vaccine in women older than 26 years (ClinicalTrials.gov Identifier NCT00456807).

The use of the vaccine in special populations is also being evaluated. The U.S. National Institute of Allergy and Infectious Diseases is sponsoring a study of the safety and immunogenicity of Gardasil in HIV-infected male and female 7–12-year-old children (ClinicalTrials.gov NCT00339040). Men who are infected or are the recipients of a solid organ or bone marrow transplant are at high risk for HPV infections and diseases. They are being considered for clinical trials of Gardasil (Advisory Committee on Immunization Practices, 2007, www.cdc.gov/vaccines/recs/acip/downloads/acip_min_feb07.pdf).

One issue of great importance is the place of male HPV immunization in the prevention of cervical cancer. Several studies have examined favorably the epidemiologic consequences and even the cost-effectiveness of this strategy (Hughes et al., 2002; Elbasha et al., 2007; Dasbach et al., 2006). An indication for male immunization with Gardasil will hinge on the results due in 2008 of a randomized, double-blind, placebo-controlled study in males for the prevention of genital warts (ClinicalTrials.gov Identifier NCT00090285).

It is essential to eventually demonstrate that HPV vaccination actually reduces the rates of cervical cancer. Both vaccine manufacturers have put in place in the Nordic countries studies that are either time extensions of current studies, or an examination of cancer registries with data matched to centralized immunization registries. A corollary question is to evaluate the impact vaccination might have on screening (Kulasingam and Myers, 2003). There are large costs to be saved if the vaccine, particularly one that includes more types than the current ones, has a large impact not only on cancer, but also on CIN of any stage. This could eventually lead to not only the recommendation of the vaccine in young males, but also to a more vigorous pursuit of immunization in general. If cervical HPV infection and disease prevalence decline, the screening approach will have to be redesigned to address the resulting increase in false positive results.

The current focus on the cervix, vagina, and vulva should not hide the potential importance HPV immunization may have in curbing diseases such as anal, oropharyngeal, even some penile cancers, as well as recurrent respiratory papillomatosis (Freed and Derkay, 2006). These possibilities are likely to fuel a great deal of interest in these vaccines in the foreseeable future.

KEY ISSUES

- Vaccination impact on cervical cancer—it remains to be demonstrated that vaccination reduces the rate of cervical cancer. The evidence for the efficacy of the HPV vaccine rests at the moment on the reduction of the rate of CIN2/3 and AIS caused by the HPV types included in the vaccine. However, an overwhelming body of concordant biologic and epidemiologic evidence indicates that an HPV infection by a high-risk type is the necessary, if not sufficient, condition for the development of CIN2/3 and then cervical cancer. Prevention of cervical cancer by the vaccine is the last proof needed to irrefutably establish the causal link between HPV and cervical cancer. Failure of the vaccine to prevent cervical cancer would invalidate this paradigm. The realistic challenge, which will probably take more than a decade to be met, is to be able, with the means already in place, to demonstrate the beneficial effect of the vaccine on cancer. The same questions of course apply to vulvar and vaginal cancers, but they will be more difficult to answer because of the lower incidence of these tumors.

- Vaccination effectiveness—the vaccine efficacy is now well demonstrated in women, but vaccine effectiveness needs to be better defined, i.e., what is the impact of the vaccine in the general population on the HPV-associated diseases, regardless of the cervical HPV DNA status, serologic status, or the prevalent and incident HPV type exposures of the vaccinees. A high effectiveness of the vaccine could have a major impact on cervical cancer screening protocols and procedures. A related issue is the evaluation of HPV vaccination on herd immunity, as this might affect the efforts at disseminating the vaccination.
- Breadth of HPV coverage—vaccine effectiveness is related to the HPV types that are covered by HPV vaccination. In addition to the types included in the current vaccines, protection extends to some of the phylogenetically related HPV types. Because this protection is only partial, one may have to rely on the addition of other types to the vaccine to significantly increase its effectiveness.
- Duration of efficacy—the duration of efficacy of the current three-dose immunization is unknown. Although a booster immunization at 5 years has already been shown to induce a strong anamnestic antibody response, and could be used whenever protection wanes with time, this represents a significant cost that one hopes to avoid. Fortunately, current modeling of the antibody kinetics appears to suggest a long durability.
- Efficacy in men—the efficacy of the vaccine in men will not be known for Gardasil until the second part of 2008 in all likelihood. The endpoint of the current studies is the prevention of genital warts. The vaccine was already shown to be effective at preventing external genital warts in women. Given that vulvar warts occur on the mucosal surfaces as well as on the skin, there is strong reason to expect that the vaccine will also be effective in preventing warts on the male genitalia skin.
- Efficacy and effectiveness in special populations—now that early results indicate that Gardasil is effective in preventing CIN in women aged 26–45 years, vaccine effectiveness in that age group will have to be ascertained before broad guidelines could be developed and endorsed by insurance companies for the use of the vaccine in that age group. Other groups in whom HPV vaccination needs to be evaluated include immunodeficient (e.g., HIV) and immunosuppressed (e.g., transplant recipients) hosts, men who have sex with men, and the sexual partners of subjects with active HPV disease.
- Impact of the HPV vaccine disease natural history and HPV transmission—HPV vaccination has clearly no impact on established cervical HPV infection, but it remains to be seen whether it alters the natural history of HPV disease by, for instance, reducing the development of new warts, or decreasing the relapse rate of treated disease. Does vaccination reduce the risk of HPV transmission from a diseased subject to his sexual partner or from a pregnant mother with genital warts to her offspring at risk of developing recurrent respiratory papillomatosis?
- HPV type replacement—one concern is that the eradication of HPV types included in the vaccine would lead to the accelerated replacement by other oncogenic HPV types. Some found possible evidence of this in a study of Gardasil, but a subgroup randomization imbalance was a more likely explanation (Sawaya and Smith-McCune, 2007; FDA, 2006, www.fda.gov/cber/sba/hpvmer060806s.pdf). No such evidence of HPV replacement was found when looked for in a study of Cervarix (Harper et al., 2006). This question will undoubtedly remain, but several observations make this scenario unlikely. First, HPV-16 and -18 are the most oncogenic of all HPV types; hence, given the slow evolutionary clock of papillomaviruses, it is unlikely that the types not included in the vaccine would quickly acquire an increased oncogenic potential (Schiffman et al., 2005; Bernard, 2005). Second, multiple HPV infections show that infections with different types are not independent, but rather synergistic events. Consequently, the eradication of HPV-16 and -18, should lead to a decrease of the infections by the other HPV types that tend to be associated with types 16 and 18 (Elbasha and Galvani, 2005). A surge in other HPV infections could be artifactual. The eradication of the vaccine types might allow the PCR to pick up weak HPV DNA viral copy numbers heretofore undetected.
- Impact of vaccination on other HPV-associated diseases—the same HPV types responsible for

cervical cancer and included in the HPV vaccines, are even more common in anal cancers and variants of penile cancers. They are also responsible for up to 70% of oropharyngeal cancers. It should become a priority to try to establish at least for the anus and the oropharynx if HPV vaccination is beneficial. For these cancers, as well as for those of the vulva and vagina, in the absence of screening vaccination is the only potential preventative measure. Although recurrent respiratory papillomatosis is a relatively rare condition, it causes significant morbidity. Gardasil administered to the mother might prevent the development of laryngeal papillomas in the child.

- Vaccine safety—the HPV vaccines appear to be remarkably safe, and no area of genuine concern has emerged so far. Nevertheless, continued safety is essential for the sustained acceptability of the vaccine, and vigilance is necessary as with any novel biological product.
- Vaccine accessibility—the current HPV vaccines are among the most expensive ones. For example, each dose of Gardasil is charged $120 by the manufacturer in the United States. Prices of Gardasil or Cervarix in other countries are at this time generally similar or higher. This is clearly a major obstacle at the moment for the dissemination of HPV vaccination where it is most needed, especially the developing world. However, past experience with other vaccines is encouraging, and prices are likely to come down.

References

Abcarian, H. and Sharon, N. Long-term effectiveness of the immunotherapy of anal condyloma acuminatum. Dis. Colon Rectum 1982; 25:648–651.

ACHA guidelines Recommendations for institutional prematriculation immunizations. J. Am. Coll. Health Assoc. 2006; 55: 126–132.

ACOG Committee Opinion No. 344: human papillomavirus vaccination. Obstet. Gynecol. 2006; 108:699–705.

ACOG Practice Bulletin Cervical cytology screening. Number 45, August 2003. Int. J. Gynaecol. Obstet. 2003; 83:237–247.

Advisory Committee on Immunization Practices. February 21–22, 2007, Atlanta, Georgia—Record of the Proceedings. Centers for Disease Control and Prevention. 2007, www.cdc.gov/vaccines/recs/acip/downloads/acip_min_feb07.pdf (accessed June 15, 2007).

Alam, M. and Stiller, M. Direct medical costs for surgical and medical treatment of condylomata acuminata. Arch. Dermatol. 2001; 137:337–341.

American Academy of Pediatrics Committee on Infectious Diseases. Recommended immunization schedules for children and adolescents—United States, 2007. Pediatrics. 2007; 119:207–208, 3 p following 208.

American Academy of Pediatrics Committee on Infectious Diseases. Prevention of human papillomavirus infection: provisional recommendations for immunization of females with quadrivalent human papillomavirus vaccine. American Academy of Pediatrics. 2007, http://www.cispimmunize.org/ill/pdf/HPVprovisional.pdf. (accessed October 31, 2007).

Anon. HPV: Gardasil and GBS. CDC/Office of the Chief Science Officer. 2007, http://www.cdc.gov/vaccines/vpd-vac/hpv/downloads/hpv-gardasil-gbs.pdf (accessed November 11, 2007).

Antonsson, A., Forslund, O., Ekberg, H., Sterner, G. and Göran Hansson, B. The ubiquity and impressive genomic diversity of human skin papillomaviruses suggest a commensalic nature of these viruses. J. Virol. 2000; 74:11636–11641.

Armstrong, C. Recommended Adult Immunization Schedule, United States, 2007. Am. Fam. Physician 2006; 74:2115.

Ashrafi, G.H., Brown, D.R., Fife, K.H. and Campo, M.S. Down-regulation of MHC class I is a property common to papillomavirus E5 proteins. Virus Res. 2006; 120:208–211.

Ault, K.A. Effect of prophylactic human papillomavirus L1 virus-like-particle vaccine on risk of cervical intraepithelial neoplasia grade 2, grade 3, and adenocarcinoma in situ: a combined analysis of four randomised clinical trials. Lancet 2007; 369:1861–1868.

Barnabas, R.V., Laukkanen, P., Koskela, P., Kontula, O., Lehtinen, M. and Garnett, G.P. Epidemiology of HPV 16 and cervical cancer in Finland and the potential impact of vaccination: mathematical modelling analyses. PLoS Med. 2006; 3:e138.

Bernard, H.U. The clinical importance of the nomenclature, evolution and taxonomy of human papillomaviruses. J. Clin. Virol. 2005; 32(Suppl. 1):S1–S6.

Biberstein, H. Immunization therapy of warts. Arch. Dermatol. Syphil. 1944; 50:12–22.

Bleeker, M.C., Hogewoning, C.J., Van Den Brule, A.J., Voorhorst, F.J., Van Andel, R.E., Risse, E.K. et al. Penile lesions and human papillomavirus in male sexual partners of women with cervical intraepithelial neoplasia. J. Am. Acad. Dermatol. 2002; 47:351–357.

Bleeker, M.C., Hogewoning, C.J., Voorhorst, F.J., van den Brule, A.J., Snijders, P.J., Starink, T.M. et al. Condom use promotes regression of human papillomavirus-associated penile lesions in male sexual partners of women with cervical intraepithelial neoplasia. Int. J. Cancer 2003; 107:804–810.

Block, S.L., Nolan, T., Sattler, C., Barr, E., Giacoletti, K.E., Marchant, C.D. et al. Comparison of the immunogenicity and reactogenicity of a prophylactic quadrivalent human papillomavirus (types 6, 11, 16, and 18) L1 virus-like particle vaccine in male and female adolescents and young adult women. Pediatrics 2006; 118:2135–2145.

Bonnez, W. A comment on "Butcher's warts: dermatological heritage or testable misinformation?". Arch. Dermatol. 2002a; 138:411.

Bonnez, W. Issues with HIV and oral human papillomavirus infections. AIDS Reader 2002b; 12:174–176.

Bonnez, W. Papillomaviruses. In: Antimicrobial Therapy and Vaccines (V.L. Yu, R. Weber and D. Raoult, Eds.), 2nd ed. New York, NY: Apple Tree Productions, LLC, 2002c.

Bonnez, W., DaRin, C., Rose, R.C. and Reichman, R.C. Use of human papillomavirus type 11 virions in an ELISA to detect specific antibodies in humans with condylomata acuminata. J. Gen. Virol. 1991; 72:1343–1347.

Bonnez, W., Kashima, H.K., Leventhal, B., Mounts, P., Rose, R.C., Reichman, R.C. et al. Antibody response to human papillomavirus (HPV) type 11 in children with juvenile-onset recurrent respiratory papillomatosis (RRP). Virology 1992a; 188:384–387.

Bonnez, W., Rose, R.C. and Reichman, R.C. Neutralization of HPV-11 infection by specific polyclonal antiserum directed against HPV-11 viral particles. Ninth International Papillomavirus Workshop. Heidelberg, Germany 1990:354.

Bonnez, W., Rose, R.C. and Reichman, R.C. Antibody-mediated neutralization of human papillomavirus type 11 (HPV-11) infection in the nude mouse. Detection of HPV-11 mRNAs. J. Infect. Dis. 1992b; 165:376–380.

Boshart, M., Gissmann, L., Ikenberg, H., Kleinheinz, A., Scheurlen, W. and zur Hausen, H. A new type of papillomavirus DNA, its presence in genital cancer biopsies and in cell lines derived from cervical cancer. EMBO J. 1984; 3:1151–1157.

Boyle, W.F., Riggs, J.L., Oshiro, L.S. and Lennette, E.H. Electron microscopic identification of papova virus in laryngeal papilloma. Laryngoscope 1973; 83:1102–1108.

Breitburd, F., Kirnbauer, R., Hubbert, N.L., Nonnenmacher, B., Trin-Dinh-Desmarquet, C., Schiller, J.T. et al. Immunization with viruslike particles from cottontail rabbit papillomavirus (CRPV) can protect against experimental CRPV infection. J. Virol. 1995; 69:3959–3963.

Brown, D. and The FUTURE Study Group. HPV type 6/11/16/18 vaccine: first analysis of cross-protection against persistent infection, cervical intraepithelial neoplasia (CIN), and adenocarcinoma in situ (AIS) caused by oncogenic HPV types in addition to 16/18. In: 47th Interscience Conference on Antimicrobial Agents and Chemotherapy, September 17–20, 2007; Chicago, IL; p. G-1720b.

Brown, D.R., Fife, K.H., Wheeler, C.M., Koutsky, L.A., Lupinacci, L.M., Railkar, R. et al. Early assessment of the efficacy of a human papillomavirus type 16 L1 virus-like particle vaccine. Vaccine 2004; 22:2936–2942.

Brown, D.R., Schroeder, J.M., Bryan, J.T., Stoler, M.H. and Fife, K.H. Detection of multiple human papillomavirus types in condylomata acuminata lesions from otherwise healthy and immunosuppressed patients. J. Clin. Microbiol. 1999; 37:3316–3322.

Burchell, A.N., Winer, R.L., de Sanjose, S. and Franco, E.L. Chapter 6: epidemiology and transmission dynamics of genital HPV infection. Vaccine 2006; 24(Suppl. 3):S52–S61.

Campo, M.S., Grindlay, G.J., O'Neil, B.W., Chandrachud, L.M., McGarvie, G.M. and Jarrett, W.F.H. Prophylactic and therapeutic vaccination against a mucosal papillomavirus. J. Gen. Virol. 1993; 74:945–953.

Carter, J.J., Wipf, G.C., Benki, S.F., Christensen, N.D. and Galloway, D.A. Identification of a human papillomavirus type 16-specific epitope on the C-terminal arm of the major capsid protein L1. J. Virol. 2003; 77:11625–11632.

Carter, J.J., Wipf, G.C., Madeleine, M.M., Schwartz, S.M., Koutsky, L.A. and Galloway, D.A. Identification of human papillomavirus type 16 L1 surface loops required for neutralization by human sera. J. Virol. 2006; 80:4664–4672.

Centers for Disease Control and Surveillance. Sexually Transmitted Disease Surveillance, 2002. Atlanta, GA: U.S. Department of Health and Human Services, Centers for Disease Control and Prevention, September 2003.

Chandrachud, L.M., Grindlay, G.J., McGarvie, G.M., O'Neil, B.W., Wagner, E.R., Jarrett, W.F.H. et al. Vaccination of cattle with the N-terminus of L2 is necessary and sufficient for preventing infection by bovine papillomavirus-4. Virology 1995; 211:204–208.

Chang, Y.C., Madkan, V., Cook-Norris, R., Sra, K. and Tyring, S. Current and potential uses of imiquimod. South Med. J. 2005; 98:914–920.

Chen, X.S., Casini, G., Harrison, S.C. and Garcea, R.L. Papillomavirus capsid protein expression in *Escherichia coli*: purification and assembly of HPV11 and HPV16 L1. J. Mol. Biol. 2001; 307:173–182.

Chesson, H.W., Blandford, J.M., Gift, T.L., Tao, G. and Irwin, K.L. The estimated direct medical cost of sexually transmitted diseases among American youth, 2000. Perspect. Sex. Reprod. Health 2004; 36:11–19.

Christensen, N.D., Cladel, N.M., Reed, C.A., Budgeon, L.R., Embers, M.E., Skulsky, D.M. et al. Hybrid papillomavirus L1 molecules assemble into virus-like particles that reconstitute conformational epitopes and induce neutralizing antibodies to distinct HPV types. Virology 2001; 291:324–334.

Christensen, N.D., Dillner, J., Eklund, C., Carter, J.J., Wipf, G.C., Reed, C.A. et al. Surface conformational and linear epitopes on HPV-16 and HPV-18 L1 virus-like particles as defined by monoclonal antibodies. Virology 1996a; 223:174–184.

Christensen, N.D., Reed, C.A., Cladel, N.M., Han, R. and Kreider, J.W. Immunization with viruslike particles induces long-term protection of rabbits against challenge with cottontail rabbit papillomavirus. J. Virol. 1996b; 70:960–965.

Christensen, N.D. and Kreider, J.W. Antibody-mediated neutralization in vivo of infectious papillomavirus. J. Virol. 1990; 64:3151–3156.

Christensen, N.D., Kreider, J.W., Kan, N.C. and DiAngelo, S.L. The open reading frame L2 of cottontail rabbit papillomavirus contains antibody-inducing neutralizing epitopes. Virology 1991; 181:572–579.

Ciuffo, G. Imnesto positivo con filtrato di verruca volgare. Giorn. Ital. Mal. Venere Pelle 1907; 48:12–17.

Combita, A.L., Touze, A., Bousarghin, L., Christensen, N.D. and Coursaget, P. Identification of two cross-neutralizing linear epitopes within the L1 major capsid protein of human papillomaviruses. J. Virol. 2002; 76:6480–6486.

Culp, T.D., Spatz, C.M., Reed, C.A. and Christensen, N.D. Binding and neutralization efficiencies of monoclonal antibodies, Fab fragments, and scFv specific for L1 epitopes on the capsid of infectious HPV particles. Virology 2007; 361:435–446.

Dasbach, E.J., Elbasha, E.H. and Insinga, R.P. Mathematical models for predicting the epidemiologic and economic impact of vaccination against human papillomavirus infection and disease. Epidemiol. Rev. 2006; 28:88–100.

Day, P.M., Thompson, C.D., Buck, C.B., Pang, Y.Y., Lowy, D.R. and Schiller, J.T. Neutralization of human papillomavirus with monoclonal antibodies reveals different mechanisms of inhibition. J. Virol. 2007; 81:8784–8792.

de Haen, M., Spigt, M.G., van Uden, C.J., van Neer, P., Feron, F.J. and Knottnerus, A. Efficacy of duct tape vs placebo in the treatment of verruca vulgaris (warts) in primary school children. Arch. Pediatr. Adolesc. Med. 2006; 160:1121–1125.

Derkay, C.S. Recurrent respiratory papillomatosis. Laryngoscope 2001; 111:57–69.

Derkay, C.S. and Darrow, D.H. Recurrent respiratory papillomatosis. Ann. Otol. Rhinol. Laryngol. 2006; 115:1–11.

Derkay, C.S., Smith, R.J., McClay, J., van Burik, J.A., Wiatrak, B.J., Arnold, J. et al. HspE7 treatment of pediatric recurrent respiratory papillomatosis: final results of an open-label trial. Ann. Otol. Rhinol. Laryngol. 2005; 114:730–737.

de Villiers, E.M., Fauquet, C., Broker, T.R., Bernard, H.U. and zur Hausen, H. Classification of papillomaviruses. Virology 2004; 324:17–27.

Dias, D., Van Doren, J., Schlottmann, S., Kelly, S., Puchalski, D., Ruiz, W. et al. Optimization and validation of a multiplexed luminex assay to quantify antibodies to neutralizing epitopes on human papillomaviruses 6, 11, 16, and 18. Clin. Diagn. Lab. Immunol. 2005; 12:959–969.

Dillner, J. The serological response to papillomaviruses. Semin. Cancer Biol. 1999; 9:423–430.

Dillner, L., Moreno-Lopez, J. and Dillner, J. Serological responses to papillomaviruse group-specific antigens in women with neoplasia of the cervix uteri. J. Clin. Microbiol. 1990; 28:624–627.

D'Souza, G., Kreimer, A.R., Viscidi, R., Pawlita, M., Fakhry, C., Koch, W.M. et al. Case-control study of human papillomavirus and oropharyngeal cancer. N. Engl. J. Med. 2007; 356:1944–1956.

Dubin, G. HPV-008: Phase II study of the efficacy of GSK's cervical cancer candidate vaccine in 15 to 25 year old women—planned interim analysis results. Advisory Committee on Immunization Practices, Atlanta, Georgia. 2007, http://www.cdc.gov/vaccines/

recs/acip/downloads/mtg-slides-jun07/34-hpv2-dubin.pdf (accessed November 11, 2007).
Dunne, E.F., Thep-Amnuay, S, Whitehead, S.J. and Markowitz, L.E. Genital warts among 18–59 year olds in the U.S. National Health and Nutrition Examination Survey (NHANES), 1999–2002. In: 16th Meeting of the International Society for Sexually Transmitted Diseases Research; 2005 July 10–13; Amsterdam, The Netherlands; 2005. p. MP-123.
Durst, M., Gissmann, L., Ikenberg, H. and zur Hausen, H. A papillomavirus DNA from a cervical carcinoma and its prevalence in cancer biopsy samples from different geographic regions. Proc. Natl. Acad. Sci. U.S.A. 1983; 80:3812–3815.
Egelkrout, E.M. and Galloway, D.A. The humoral immune response to human papillomavirus. In: The Papillomaviruses (R.L. Garcea and D. DiMaio, Eds.), pp. 277–312. New York, NY: Springer, 2007.
Eiben Lyons, G., Nishimura, M.I. and Kast, W.M. Cell-mediated immune responses to human papillomavirus. In: The Papillomaviruses (R.L. Garcea and D. DiMaio, Eds.), pp. 313–335. New York, NY: Springer, 2007.
Eisemann, C., Fisher, S.G., Gross, G., Muller, M. and Gissmann, L. Antibodies to human papillomavirus type 11 virus-like particles in sera of patients with genital warts and in control groups. J. Gen. Virol. 1996; 77:1799–1803.
Elbasha, E.H., Dasbach, E.J. and Insinga, R.P. Model for assessing human papillomavirus vaccination strategies. Emerg. Infect. Dis. 2007; 13:28–41.
Elbasha, E.H. and Galvani, A.P. Vaccination against multiple HPV types. Math. Biosci. 2005; 197:88–117.
Evans, T.G., Bonnez, W., Rose, R.C., Koenig, S., Demeter, L., Suzich, J.A. et al. A phase 1 study of a recombinant viruslike particle vaccine against human papillomavirus type 11 in healthy adult volunteers. J. Infect. Dis. 2001; 183:1485–1493.
Evans, C.A., Gorman, L.R., Ito, Y. and Weiser, R.S. Antitumor immunity in the Shope papilloma-carcinoma complex of rabbits. I. Papilloma regression induced by homologous and autologous tissue vaccines. J. Natl. Cancer Inst. 1962a; 29:277–285.
Evans, C.A., Gorman, L.R., Ito, Y. and Weiser, R.S. A vaccination procedure which increases the frequency of regression of Shope papillomas of rabbits. Nature 1962b; 193:289–290.
Favre, M., Orth, G., Croissant, O. and Yaniv, M. Human papillomavirus DNA: physical map. Proc. Natl. Acad. Sci. U.S.A. 1975; 72:4810–4814.
FDA Executive Summary—GARDASIL Licensing Action. 2006, www.fda.gov/cber/sba/hpvmer060806s.pdf (accessed June 15, 2007).
Fife, K.H., Wheeler, C.M., Koutsky, L.A., Barr, E., Brown, D.R., Schiff, M.A. et al. Dose-ranging studies of the safety and immunogenicity of human papillomavirus Type 11 and Type 16 virus-like particle candidate vaccines in young healthy women. Vaccine 2004; 22:2943–2952.
Focht 3rd., D.R., Spicer, C. and Fairchok, M.P. The efficacy of duct tape vs cryotherapy in the treatment of verruca vulgaris (the common wart). Arch. Pediatr. Adolesc. Med. 2002; 156:971–974.
Fraser, C., Tomassini, J.E., Xi, L., Golm, G., Watson, M., Giuliano, A.R. et al. Modeling the long-term antibody response of a human papillomavirus (HPV) virus-like particle (VLP) type 16 prophylactic vaccine. Vaccine 2007; 25:4324–4333.
Freed, G.L. and Derkay, C.S. Prevention of recurrent respiratory papillomatosis: role of HPV vaccination. Int. J. Pediatr. Otorhinolaryngol. 2006; 70:1799–1803.
Friedman, L.S., Kahn, J., Middleman, A.B., Rosenthal, S.L. and Zimet, G.D. Human papillomavirus (HPV) vaccine: a position statement of the Society for Adolescent Medicine. J. Adolesc. Health 2006; 39:620.
FUTURE II Study Group. Prophylactic efficacy of a quadrivalent human papillomavirus (HPV) vaccine in women with virological evidence of HPV infection. J. Infect. Dis. 2007; 196:1438–1446.
Galloway, D.A. Serological assays for the detection of HPV antibodies. In: The Epidemiology of Cervical Cancer and Human Papillomavirus (N. Mu¤oz, F.X. Bosch, K.V. Shah and A. Meheus, Eds.), pp. 147–161. Lyon, France: International Agency for Research on Cancer, 1992.
Gambhira, R., Jagu, S., Karanam, B., Gravitt, P.E., Culp, T.D., Christensen, N.D., et al. Protection of rabbits against challenge with rabbit papillomaviruses by immunization with the N-terminus of HPV16 minor capsid antigen L2. J. Virol. 2007; 81(21):11585–11592.
Garcea, R.L. and Gissmann, L. Virus-like particles as vaccines and vessels for the delivery of small molecules. Curr. Opin. Biotechnol. 2004; 15:513–517.
Gardasil Package Insert. FDA. 2007, www.fda.gov/cber/label/hpvmer040307LB.pdf (accessed Jun 15, 2007).
Garland, S.M. Imiquimod. Curr. Opin. Infect. Dis. 2003; 16:85–89.
Garland, S.M., Hernandez-Avila, M., Wheeler, C.M., Perez, G., Harper, D.M., Leodolter, S. et al. Quadrivalent vaccine against human papillomavirus to prevent anogenital diseases. N. Engl. J. Med. 2007; 356:1928–1943.
Garland, S.M., Sellors, J.W., Wikstrom, A., Petersen, C.S., Aranda, C., Aractingi, S. et al. Imiquimod 5% cream is a safe and effective self-applied treatment for anogenital warts—results of an open-label, multicentre Phase IIIB trial. Int. J. STD AIDS 2001; 12:722–729.
Garnett, T.O. and Duerksen-Hughes, P.J. Modulation of apoptosis by human papillomavirus (HPV) oncoproteins. Arch. Virol. 2006; 151:2321–2335.
Ghim, S.J., Jenson, A.B. and Schlegel, R. HPV-1 L1 protein expressed in cos cells displays conformational epitopes found on intact virions. Virology 1992; 190:548–552.
Giannini, S.L., Hanon, E., Moris, P., Van Mechelen, M., Morel, S., Dessy, F. et al. Enhanced humoral and memory B cellular immunity using HPV16/18 L1 VLP vaccine formulated with the MPL/aluminium salt combination (AS04) compared to aluminium salt only. Vaccine 2006; 24:5937–5949.
Gibbs, S. and Harvey, I. Topical treatments for cutaneous warts. [update of Cochrane Database Syst Rev. 2003; (3):CD001781; PMID: 12917913]. Cochrane Database Syst. Rev. 2006.
Giroglou, T., Sapp, M., Lane, C., Fligge, C., Christensen, N.D., Streeck, R.E. et al. Immunological analyses of human papillomavirus capsids. Vaccine 2001; 19:1783–1793.
Gissmann, L., Diehl, V., Schultz-Coulon, H.J. and zur Hausen, H. Molecular cloning and characterization of human papillomavirus DNA derived from a laryngeal papilloma. J. Virol. 1982; 44:393–400.
Gissmann, L., Pfister, H. and zur Hausen, H. Human papillomaviruses: characterization of 4 different isolates. Virology 1977; 76:569–580.
Gissmann, L. and zur Hausen, H. Partial characterization of viral DNA from human genital warts (condylomata acuminata). Int. J. Cancer 1980; 25:605–609.
Giuliano, A.R., Lazcano-Ponce, E., Villa, L., Nolan, T., Marchant, C., Radley, D. et al. Impact of baseline covariates on the immunogenicity of a quadrivalent (types 6, 11, 16, and 18) human papillomavirus virus-like-particle vaccine. J. Infect. Dis. 2007; 196:1153–1162.
Harper, D.M., Franco, E.L., Wheeler, C., Ferris, D.G., Jenkins, D., Schuind, A. et al. Efficacy of a bivalent L1 virus-like particle vaccine in prevention of infection with human papillomavirus types 16 and 18 in young women: a randomised controlled trial. Lancet 2004; 364:1757–1765.

Harper, D.M., Franco, E.L., Wheeler, C.M., Moscicki, A.B., Romanowski, B., Roteli-Martins, C.M. et al. Sustained efficacy up to 4.5 years of a bivalent L1 virus-like particle vaccine against human papillomavirus types 16 and 18: follow-up from a randomised control trial. Lancet 2006; 367:1247–1255.

Harro, C.D., Pang, Y.-YS., Roden, R.B.S., Hildesheim, A., Wang, Z., Reynolds, M.J. et al. Safety and immunogenicity trial in adult volunteers of a human papillomavirus 16 L1 virus-like particle vaccine. J. Natl. Cancer Inst. 2001; 93:284–292.

Hazard, K., Karlsson, A., Andersson, K., Ekberg, H., Dillner, J. and Forslund, O. Cutaneous human papillomaviruses persist on healthy skin. J. Invest. Dermatol. 2007; 127(1):116–119.

Hebner, C.M. and Laimins, L.A. Human papillomaviruses: basic mechanisms of pathogenesis and oncogenicity. Rev. Med. Virol. 2006; 16:83–97.

Heim, K., Christensen, N.D., Hoepfl, R., Wartusch, B., Pinzger, G., Zeimet, A. et al. Serum IgG, IgM, and IgA reactivity to human papillomavirus types 11 and 6 virus-like particles in different gynecologic patient groups. J. Infect. Dis. 1995; 172:395–402.

Hewitt, M., Devesa, S.S. and Breen, N. Cervical cancer screening among U.S. women: analyses of the 2000 National Health Interview Survey. Prev. Med. 2004; 39:270–278.

Hildesheim, A. and Herrero, R. Human papillomavirus vaccine should be given before sexual debut for maximum benefit. J. Infect. Dis. 2007; 196:1431–1432.

Hildesheim, A., Herrero, R., Wacholder, S., Rodriguez, A.C., Solomon, D., Bratti, M.C. et al. Effect of human papillomavirus 16/18 L1 viruslike particle vaccine among young women with preexisting infection: a randomized trial. JAMA 2007; 298:743–753.

Hughes, J.P., Garnett, G.P. and Koutsky, L. The theoretical population-level impact of a prophylactic human papilloma virus vaccine. Epidemiology 2002; 13:631–639.

Iskander, J. Quadrivalent human papillomavirus vaccine (HPV4): United States post-licensure safety update. Advisory Committee on Immunization Practices. Atlanta, Georgia. 2007, http://www.cdc.gov/vaccines/recs/acip/downloads/mtg-slides-jun07/35-hpv3-iskander.pdf (accessed November 11, 2007).

Jarrett, W.F.H. Bovine papillomaviruses. Clin. Dermatol. 1985; 3:8–19.

Jarrett, W.F.H., Murphy, J., O'Neil, B.W. and Laird, H.M. Virus-induced papillomas of the alimentary tract of cattle. Int. J. Cancer 1978; 22:323–328.

Jarrett, W.F., O'Neil, B.W., Gaukroger, J.M., Laird, H.M., Smith, K.T. and Campo, M.S. Studies on vaccination against papillomaviruses: a comparison of purified virus, tumor extract, and transformed cells in prophylactic vaccination. Vet. Record 1990; 126:449–452.

Jarrett, W.F.H., Smith, K.T., O'Neil, B.W., Gaukroger, J.M., Chandrachud, L.M., Grindlay, G.J. et al. Studies on vaccination against papillomaviruses: prophylactic and therapeutic vaccination with recombinant structural proteins. Virology 1991; 184:33–42.

Jemal, A., Siegel, R., Ward, E., Murray, T., Xu, J. and Thun, M.J. Cancer statistics, 2007. CA Cancer J. Clin. 2007; 57:43–66.

Jenson, A.B., Kurman, R.J. and Lancaster, W.D. Detection of papillomavirus common antigens in lesions of skin and mucosa. Clin. Dermatol. 1985; 3:56–63.

Jenson, A.B., Rosenthal, J.D., Olson, C., Pass, F., Lancaster, W.D. and Shah, K. Immunologic relatedness of papillomaviruses from different species. J. Natl. Cancer Inst. 1980; 64:495–500.

Jin, X.W., Cowsert, L., Marshall, D., Reed, D., Pilacinski, W., Lim, L.Y. et al. Bovine serological response to a recombinant BPV-1 major capsid protein vaccine. Intervirology 1990; 31:345–354.

Joura, E.A., Leodolter, S., Hernandez-Avila, M., Wheeler, C.M., Perez, G., Koutsky, L.A. et al. Efficacy of a quadrivalent prophylactic human papillomavirus (types 6, 11, 16, and 18) L1 virus-like-particle vaccine against high-grade vulval and vaginal lesions: a combined analysis of three randomised clinical trials. Lancet 2007; 369:1693–1702.

Kang, M. and Lagakos, S.W. Evaluating the role of human papillomavirus vaccine in cervical cancer prevention. Stat. Methods Med. Res. 2004; 13:139–155.

Kanodia, S., Fahey, L.M. and Kast, W.M. Mechanisms used by human papillomaviruses to escape the host immune response. Curr. Cancer Drug Targets 2007; 7:79–89.

Karanam, B., Gambhira, R., Peng, S., Jagu, S., Kim, D.-J., Ketner, G., et al. Vaccination with TA-CIN (L2E6E7) and a semi-synthetic quillaja saponin analog GPI-0100 elicits protective humoral and tumor immunity. In: 24th International Papillomavirus Conference and Clinical Workshop, November 3–9, 2007; Beijing, China; 2007. p. 6A-04.

Kienzler, J.L., Lemoine, M.T., Orth, G., Jibard, N., Blanc, D., Laurent, R. et al. Humoral and cell-mediated immunity to human papillomavirus type 1 (HPV-1) in human warts. Br. J. Dermatol. 1983; 108:665–672.

Kirnbauer, R., Booy, F., Cheng, N., Lowy, D.R. and Schiller, J.T. Papillomavirus L1 major capsid protein self-assembles into virus-like particles that are highly immunogenic. Proc. Natl. Acad. Sci. U.S.A. 1992; 89:12180–12184.

Kirnbauer, R., Chandrachud, L.M., O'Neil, B.W., Wagner, E.R., Griondlay, G.J., Armstrong, A. et al. Virus-like particles of bovine papillomavirus type 4 in prophylactic and therapeutic immunization. Virology 1996; 219:37–44.

Kirnbauer, R., Hubbert, N.L., Wheeler, C.M., Becker, T.M., Lowy, D.R. and Schiller, J.T. A virus-like particle enzyme-linked immunosorbent assay detects serum antibodies in a majority of women infected with human papillomavirus type 16. J. Natl. Cancer Inst. 1994; 86:494–499.

Kirnbauer, R., Taub, J., Greenstone, H., Roden, R., Dürst, M., Gissmann, L. et al. Efficient self-assembly of human papillomavirus type 16 L1 and L1-L2 into virus-like particles. J. Virol. 1993; 67:6929–6936.

Kjaer, S.K., Tran, T.N., Sparen, P., Tryggvadottir, L., Munk, C., Dasbach, E. et al. The burden of genital warts: a study of nearly 70,000 women from the general female population in the 4 Nordic countries. J. Infect. Dis. 2007; 196:1447–1454.

Kondo, K., Ishii, Y., Ochi, H., Matsumoto, T., Yoshikawa, H. and Kanda, T. Neutralization of HPV16, 18, 31, and 58 pseudovirions with antisera induced by immunizing rabbits with synthetic peptides representing segments of the HPV16 minor capsid protein L2 surface region. Virology 2007; 358:266–272.

Koshiol, J.E., Laurent, S.A. and Pimenta, J.M. Rate and predictors of new genital warts claims and genital warts-related healthcare utilization among privately insured patients in the United States. Sex. Transm. Dis. 2004; 31:748–752.

Koutsky, L.A. Quadrivalent vaccine against human papillomavirus to prevent high-grade cervical lesions. N. Engl. J. Med. 2007; 356:1915–1927.

Koutsky, L.A., Ault, K.A., Wheeler, C.M., Brown, D.R., Barr, E., Alvarez, F.B. et al. A controlled trial of a human papillomavirus type 16 vaccine. N. Engl. J. Med. 2002; 347:1645–1651.

Kreider, J.W. and Bartlett, G.L. Shope rabbit papilloma-carcinoma complex. a model system of HPV infections. Clin. Dermatol. 1985; 3:20–26.

Kreider, J.W., Howett, M.K., Wolfe, S.A., Bartlett, G.L., Zaino, R.J., Sedlacek, T.V. et al. Morphological transformation *in vivo* of human uterine cervix with papillomavirus from condylomata acuminata. Nature 1985; 317:639–641.

Kulasingam, S.L. and Myers, E.R. Potential health and economic impact of adding a human papillomavirus vaccine to screening programs. JAMA 2003; 290:781–789.

Lacey, C.J. Therapy for genital human papillomavirus-related disease. J. Clin. Virol. 2005; 32(Suppl. 1):S82–S90.

Lacey, C.J., Goodall, R.L., Tennvall, G.R., Maw, R., Kinghorn, G.R., Fisk, P.G. et al. Randomised controlled trial and economic evaluation of podophyllotoxin solution, podophyllotoxin cream, and podophyllin in the treatment of genital warts. Sex. Transm. Infect. 2003; 79:270–275.

Larsson, P.A. and Lidén, S. Prevalence of skin diseases among adolescents 12–16 years of age. Acta Derm. Venereol. (Stockh.) 1980; 60:415–423.

Lehtinen, M., Herrero, R., Mayaud, P., Barnabas, R., Dillner, J., Paavonen, J. et al. Chapter 28: studies to assess the long-term efficacy and effectiveness of HPV vaccination in developed and developing countries. Vaccine 2006; 24(Suppl. 3):S233–S241.

Lewandowsky, F. and Lutz, W. Ein Fall einer bisher nicht beschriebenen Hauterkrankung (Epidermodysplasia verruciformis). Arch. Dermatol. Syph. (Berl.) 1922; 141:193–203.

Leyden, W.A., Manos, M.M., Geiger, A.M., Weinmann, S., Mouchawar, J., Bischoff, K. et al. Cervical cancer in women with comprehensive health care access: attributable factors in the screening process. J. Natl. Cancer Inst. 2005; 97:675–683.

Li, M., Cripe, T.P., Estes, P.A., Lyon, M.K., Rose, R.C. and Garcea, R.L. Expression of the human papillomavirus type 11 L1 capsid protein in Escherichia coli: characterization of protein domains involved in DNA binding and capsid assembly. J. Virol. 1997; 71:2988–2995.

Lin, Y.-L., Borenstein, L.A., Ahmed, R. and Wettstein, F.O. Cottontail rabbit papillomavirus L1 protein-based vaccines: protection is achieved only with a full-length, nondenatured product. J. Virol. 1993; 67:4154–4162.

Lin, Y.-L., Borenstein, L.A., Selvakumar, R., Ahmed, R. and Wettstein, F.O. Effective vaccination against papilloma development by immunization with L1 or L2 structural protein of cottontail rabbit papillomavirus. Virology 1992; 187:612–619.

Longstaff, E. and von Krogh, G. Condyloma eradication: self-therapy with 0.15–0.5% podophyllotoxin versus 20–25% podophyllin preparations—an integrated safety assessment. Regul. Toxicol. Pharmacol. 2001; 33:117–137.

Luna, J., Saah, A., Hood, S., Bautista, O.M. and Barr, E. Investigators, F.I., Safety, efficacy, and immunogenicity of quadrivalent HPV vaccine (Gardasil) in women aged 24–45. In: 24th International Papillomavirus Conference and Clinical Workshop, November 3–9, 2007; Beijing, China; 2007. p. PA1-04.

Mackenzie, M. Essay on Growths in the Larynx: With Reports, and an Analysis of One Hundred Consecutive Cases Treated by the Author. Philadelphia: Lindsay & Blakiston, 1871.

Malison, M.D., Morris, R. and Jones, L.W. Autogenous vaccine therapy for condyloma acuminatum. A double-blind controlled study. Br. J. Vener. Dis. 1982; 58:62–65.

Mao, C., Koutsky, L.A., Ault, K.A., Wheeler, C.M., Brown, D.R., Wiley, D.J. et al. Efficacy of human papillomavirus-16 vaccine to prevent cervical intraepithelial neoplasia: a randomized controlled trial. Obstet. Gynecol. 2006; 107:18–27.

Markowitz, L.E., Dunne, E.F., Saraiya, M., Lawson, H.W., Chesson, H. and Unger, E.R. Quadrivalent human papillomavirus vaccine: recommendations of the Advisory Committee on Immunization Practices (ACIP). MMWR. Morb. Mortal. Wkly. Rep. 2007; 56:1–24.

McClements, W.L., Wang, X.M., Ling, J.C., Skulsky, D.M., Christensen, N.D., Jansen, K.U. et al. A novel human papillomavirus type 6 neutralizing domain comprising two discrete regions of the major capsid protein L1. Virology 2001; 289:262–268.

Meschede, W., Zumbach, K., Braspenning, J., Scheffner, M., Benitez-Bribiesca, L., Luande, J. et al. Antibodies against early proteins of human papillomaviruses as diagnostic markers for invasive cervical cancer. J. Clin. Microbiol. 1998; 36:475–480.

Micali, G., Dall'Oglio, F., Nasca, M.R. and Tedeschi, A. Management of cutaneous warts: an evidence-based approach. Am. J. Clin. Dermatol. 2004; 5:311–317.

Miller, N.B. Gardasil™ : quadrivalent human papillomavirus 6, 11, 16, 18 L1 VLP vaccine. CBER, FDA. 2006, www.fda.gov/ohrms/dockets/AC/06/slides/2006-4222S-2.ppt (accessed October 31, 2007).

Müller, M., Viscidi, R.P., Sun, Y., Guerrero, E., Hill, P.M., Shah, F. et al. Antibodies to HPV-16 E6 and E7 proteins as markers for HPV-16-associated invasive cervical cancer. Virology 1992; 187:508–514.

Munoz, N., Bosch, F.X., Castellsague, X., Diaz, M., de Sanjose, S., Hammouda, D. et al. Against which human papillomavirus types shall we vaccinate and screen? The international perspective. Int. J. Cancer 2004; 111:278–285.

Munoz, N. and De The, G. Virus papillomateux et cancer du col chez la femme. In: Biomedical Aspects of Human Wart Virus Infection; 1975 December 15–16 (M. Prunieras, Ed.), p. 101. Lyon, France: Fondation Mérieux, 1975.

Nicol, A.F., Fernandes, A.T. and Bonecini-Almeida Mda, G. Immune response in cervical dysplasia induced by human papillomavirus: the influence of human immunodeficiency virus-1 co-infection—review. Mem. Inst. Oswaldo Cruz 2005; 100:1–12.

(NIH) NIoH. NIH Consensus Statement: Cervical Cancer. 1996. pp. 1–38.

Olsson, S.E., Villa, L.L., Costa, R.L., Petta, C.A., Andrade, R.P., Malm, C. et al. Induction of immune memory following administration of a prophylactic quadrivalent human papillomavirus (HPV) types 6/11/16/18 L1 virus-like particle (VLP) vaccine. Vaccine 2007; 25:4931–4939.

Opalka, D., Lachman, C.E., MacMullen, S.A., Jansen, K.U., Smith, J.F., Chirmule, N. et al. Simultaneous quantitation of antibodies to neutralizing epitopes on virus-like particles for human papillomavirus types 6, 11, 16, and 18 by a multiplexed luminex assay. Clin. Diagn, Lab. Immunol. 2003; 10:108–115.

Orth, G. Genetics of epidermodysplasia verruciformis: insights into host defense against papillomaviruses. Semin. Immunol. 2006; 18:362–374.

Orth, G., Favre, M. and Croissant, O. Characterization of a new type of human papillomavirus that causes skin warts. J. Virol. 1977; 24:108–120.

Parkin, D.M. and Bray, F. Chapter 2: the burden of HPV-related cancers. Vaccine 2006; 24(Suppl. 3):S11–S25.

Parkin, D.M., Bray, F., Ferlay, J. and Pisani, P. Global cancer statistics, 2002. CA Cancer J. Clin. 2005; 55:74–108.

Partridge, J.M., Hughes, J.P., Feng, Q., Winer, R.L., Weaver, B.A., Xi, L.-F. et al. Genital human papillomavirus infection in men: incidence and risk factors in a cohort of university students. J. Infect. Dis. 2007; 196:1128–1136.

Pedersen, C., Petaja, T., Strauss, G., Rumke, H.C., Poder, A., Richardus, J.H. et al. Immunization of early adolescent females with human papillomavirus type 16 and 18 L1 virus-like particle vaccine containing AS04 adjuvant. J. Adolesc. Health 2007; 40:564–571.

Pilacinski, W.P., Glassman, D.L., Glassman, K.F., Reed, D.E., Lum, M.A., Marshall, R.F. et al. Immunization against bovine papillomavirus infection. In: Papillomaviruses (D. Evered and S. Clark, Eds.), pp. 136–156. Chichester, UK: Wiley, 1986.

Pinto, L.A., Viscidi, R., Harro, C.D., Kemp, T.J., Garcia-Pineres, A.J., Trivett, M. et al. Cellular immune responses to HPV-18, -31, and -53 in healthy volunteers immunized with recombinant HPV-16 L1 virus-like particles. Virology 2006; 353:451–462.

Poland, G.A., Jacobson, R.M., Koutsky, L.A., Tamms, G.M., Railkar, R., Smith, J.F. et al. Immunogenicity and reactogenicity of a novel vaccine for human papillomavirus 16: a 2-year randomized controlled clinical trial. Mayo Clin. Proc. 2005; 80:601–610.

Reisinger, K.S., Block, S.L., Lazcano-Ponce, E., Samakoses, R., Esser, M.T., Erick, J. et al. Safety and persistent immunogenicity of a quadrivalent human papillomavirus types 6, 11, 16, 18 L1 virus-like particle vaccine in preadolescents and adolescents: a randomized controlled trial. Pediatr. Infect. Dis. J. 2007; 26:201–209.

Roden, R.B., Yutzy 4th., W.H., Fallon, R., Inglis, S., Lowy, D.R. and Schiller, J.T. Minor capsid protein of human genital papillomaviruses contains subdominant, cross-neutralizing epitopes. Virology 2000; 270:254–257.

Rose, R.C., Bonnez, W., Da Rin, C., McCance, D.J. and Reichman, R.C. Serological differentiation of human papillomavirus types 11, 16 and 18 using recombinant virus-like particles. J. Gen. Virol. 1994a; 75:2445–2449.

Rose, R.C., Bonnez, W., Reichman, R.C. and Garcea, R.L. Expression of human papillomavirus type 11 L1 protein in insect cells: in vivo and in vitro assembly of viruslike particles. J. Virol. 1993a; 67:1936–1944.

Rose, R.C., Bonnez, W. and Reichman, R.C. Human papillomavirus type 11 (HPV-11) recombinant virus-like particles (VLPs) induce the formation of neutralizing antibodies. In: 12th International Papillomavirus Conference, September 26–October 1, 1993b; Baltimore, Maryland. p. 64.

Rose, R.C., Reichman, R.C. and Bonnez, W. Human papillomavirus type 11 (HPV-11) recombinant virus-like particles (VLPs) induce the formation of neutralizing antibodies and detect HPV-specific antibodies in human sera. J. Gen. Virol. 1994b; 75:2075–2079.

Rous, P. and Beard, J.W. The progression to carcinoma of virus-induced rabbit papillomas (Shope). J. Exp. Med. 1935; 62:523–548.

Routh, H.B., Bhowmik, K.R. and Parish, L.C. Myths, fables and even truths about warts and human papillomavirus. Clin. Dermatol. 1997; 15:305–307.

Rowson, K.E.K. and Mahy, B.W.J. Human papova (wart) virus. Bacteriol. Rev. 1967; 31:110–131.

Ruiter, M. and van Mullem, P.J. Demonstration by electron microscopy of an intranuclear virus in epidermodysplasia verruciformis. J. Invest. Dermatol. 1966; 47:247–252.

Ryan, E.P., Malboeuf, C.M., Bernard, M., Rose, R.C. and Phipps, R.P. Cyclooxygenase-2 inhibition attenuates antibody responses against human papillomavirus-like particles. J. Immunol. 2006; 177:7811–7819.

Ryding, J., Dahlberg, L., Wallen-Ohman, M. and Dillner, J. Deletion of a major neutralizing epitope of human papillomavirus type 16 virus-like particles. J. Gen. Virol. 2007; 88:792–802.

Saslow, D., Castle, P.E., Cox, J.T., Davey, D.D., Einstein, M.H., Ferris, D.G. et al. American Cancer Society guideline for human papillomavirus (HPV) vaccine use to prevent cervical cancer and its precursors. CA Cancer J. Clin. 2007; 57:7–28.

Saslow, D., Runowicz, C.D., Solomon, D., Moscicki, A.B., Smith, R.A., Eyre, H.J. et al. American Cancer Society guideline for the early detection of cervical neoplasia and cancer. CA Cancer J. Clin. 2002; 52:342–362.

Sawaya, G.F. and Smith-McCune, K. HPV vaccination—more answers, more questions. N. Engl. J. Med. 2007; 356:1991–1993.

Schaedlich, L., Garcea, R.L. and Gissmann L. HPV 16 chimeric capsomeres: candidate vaccines against persistent HPV infection. In: 24th International Papillomavirus Conference and Clinical Workshop, November 3–9, 2007; Beijing, China; 2007. p. 6A-05.

Schiffman, M., Castle, P.E., Jeronimo, J., Rodriguez, A.C. and Wacholder, S. Human papillomavirus and cervical cancer. Lancet 2007; 370:890–907.

Schiffman, M., Herrero, R., Desalle, R., Hildesheim, A., Wacholder, S., Rodriguez, A.C. et al. The carcinogenicity of human papillomavirus types reflects viral evolution. Virology 2005; 337:76–84.

Scott, M., Nakagawa, M. and Moscicki, A.B. Cell-mediated immune response to human papillomavirus infection. Clin. Diagn. Lab. Immunol. 2001; 8:209–220.

Screening for cervical cancer—recommendations and rationale. U.S. Preventive Services Task Force. AHRQ. 2003, http://www.ahrq.gov/clinic/3rduspstf/cervcan/cervcanrr.pdf (accessed October 31, 2007).

Selvakumar, R., Borenstein, L.A., Lin, Y.L., Ahmed, R. and Wettstein, F.O. Immunization with nonstructural proteins E1 and E2 of cottontail rabbit papillomavirus stimulates regression of virus-induced papillomas. J. Virol. 1995; 69:602–605.

Sheu, B.C., Chang, W.C., Lin, H.H., Chow, S.N. and Huang, S.C. Immune concept of human papillomaviruses and related antigens in local cancer milieu of human cervical neoplasia. J. Obstet. Gynaecol. Res. 2007; 33:103–113.

Skinner, R.B., Jr. Imiquimod. Dermatol. Clin. 2003; 21:291–300.

Slupetzky, K., Gambhira, R., Culp, T.D., Shafti-Keramat, S., Schellenbacher, C., Christensen, N.D. et al. A papillomavirus-like particle (VLP) vaccine displaying HPV16 L2 epitopes induces cross-neutralizing antibodies to HPV11. Vaccine 2007; 25:2001–2010.

Smith, J.S., Rana, R.K. and Pimenta, J.M. Global review of age-specific prevalence of human papillomavirus (types 16/18/overall). In: 23rd International Papillomavirus Conference and Workshop, September 1–7, 2006; Prague, Czech Republic; 2006. p. PS15-6.

Smith, E.M., Ritchie, J.M., Yankowitz, J., Wang, D., Turek, L.P. and Haugen, T.H. HPV prevalence and concordance in the cervix and oral cavity of pregnant women. Infect. Dis. Obstet. Gynecol. 2004; 12:45–56.

Snijders, P.J., Steenbergen, R.D., Heideman, D.A. and Meijer, C.J. HPV-mediated cervical carcinogenesis: concepts and clinical implications. J. Pathol. 2006; 208:152–164.

Solomon, D., Davey, D., Kurman, R., Moriarty, A., O'Connor, D., Prey, M. et al. The 2001 Bethesda System: terminology for reporting results of cervical cytology. JAMA 2002; 287:2114–2119.

Stoler, M.H. Advances in cervical screening technology. Mod. Pathol. 2000; 13:275–284.

Strauss, M.J., Shaw, E.W., Bunting, H. and Melnick, J.L. "Crystalline" virus-like particles from skin papillomas characterized by intranuclear inclusion bodies. Proc. Soc. Exp. Biol. Med. 1949; 72:46–50.

Studentsov, Y.Y., Ho, G.Y., Marks, M.A., Bierman, R. and Burk, R.D. Polymer-based enzyme-linked immunosorbent assay using human papillomavirus type 16 (HPV16) virus-like particles detects HPV16 clade-specific serologic responses. J. Clin. Microbiol. 2003; 41:2827–2834.

Suzich, J.A., Ghim, S.-J., Palmer-Hill, F.J., White, W.I., Tamura, J.K., Bell, J.A. et al. Systemic immunization with papillomavirus L1 protein completely prevents the development of viral mucosal papillomas. Proc. Natl. Acad. Sci. U.S.A. 1995; 92:11553–11557.

Syrjanen, S. Human papillomavirus infections and oral tumors. Med. Microbiol. Immunol. (Berl.) 2003; 192:123–128.

Tachezy, R., Mikyskova, I., Ludvikova, V., Rob, L., Kucera, T., Slavik, V. et al. Longitudinal study of patients after surgical treatment for cervical lesions: detection of HPV DNA and prevalence of HPV-specific antibodies. Eur. J. Clin. Microbiol. Infect. Dis. 2006; 25:492–500.

Touze, A., Dupuy, C., Mahe, D., Sizaret, P.Y. and Coursaget, P. Production of recombinant virus-like particles from human papillomavirus types 6 and 11, and study of serological reactivities between HPV 6, 11, 16 and 45 by ELISA: implications for papillomavirus prevention and detection. FEMS Microbiol. Lett. 1998; 160:111–118.

Varsani, A., Williamson, A.L., Jaffer, M.A. and Rybicki, E.P. A deletion and point mutation study of the human papillomavirus type 16 major capsid gene. Virus Res. 2006; 122:154–163.

Villa, L.L., Ault, K.A., Giuliano, A.R., Costa, R.L., Petta, C.A., Andrade, R.P. et al. Immunologic responses following administration of a vaccine targeting human papillomavirus Types 6, 11, 16, and 18. Vaccine 2006a; 24:5571–5583.

Villa, L.L., Costa, R.L., Petta, C.A., Andrade, R.P., Paavonen, J., Iversen, O.E. et al. High sustained efficacy of a prophylactic quadrivalent human papillomavirus types 6/11/16/18 L1 virus-like particle vaccine through 5 years of follow-up. Br. J. Cancer 2006b; 95:1459–1466.

Viscidi, R.P., Schiffman, M., Hildesheim, A., Herrero, R., Castle, P.E., Bratti, M.C. et al. Seroreactivity to human papillomavirus (HPV) types 16, 18, or 31 and risk of subsequent HPV infection: results from a population-based study in Costa Rica. Cancer Epidemiol. Biomarkers Prev. 2004; 13:324–327.

von Krogh, G. and Longstaff, E. Podophyllin office therapy against condyloma should be abandoned. Sex. Transm. Infect. 2001; 77:409–412.

Walboomers, J.M., Jacobs, M.V., Manos, M.M., Bosch, F.X., Kummer, J.A., Shah, K.V. et al. Human papillomavirus is a necessary cause of invasive cervical cancer worldwide. J. Pathol. 1999; 189:12–19.

Wang, X.M., Cook, J.C., Lee, J.C., Jansen, K.U., Christensen, N.D., Ludmerer, S.W. et al. Human papillomavirus type 6 virus-like particles present overlapping yet distinct conformational epitopes. J. Gen. Virol. 2003a; 84:1493–1497.

Wang, X., Wang, Z., Christensen, N.D. and Dillner, J. Mapping of human serum-reactive epitopes in virus-like particles of human papillomavirus types 16 and 11. Virology 2003b; 311:213–221.

Wenner, R., Askari, S.K., Cham, P.M., Kedrowski, D.A., Liu, A. and Warshaw, E.M. Duct tape for the treatment of common warts in adults: a double-blind randomized controlled trial. Arch. Dermatol. 2007; 143:309–313.

White, W.I., Wilson, S.D., Palmer-Hill, F.J., Woods, R.M., Ghim, S.-J., Hewitt, L.A. et al. Characterization of a major neutralizing epitope on human papillomavirus type 16 L1. J. Virol. 1999; 73:4882–4889.

Wikstrom, A., Eklund, C., von Krogh, G., Lidbrink, P. and Dillner, J. Antibodies against human papillomavirus type 6 capsids are elevated in men with previous condylomas. APMIS 1997; 105:884–888.

Williams, H.C., Pottier, A. and Strachan, D. The descriptive epidemiology of warts in British schoolchildren. Br. J. Dermatol. 1993; 128:504–511.

Workowski, K.A. and Berman, S.M. Centers for Disease Control and Prevention sexually transmitted diseases treatment guidelines. Clin. Infect. Dis. 2007; 44(Suppl. 3):S73–S76.

Wright, T.C., Jr., Massad, L.S., Dunton, C.J., Spitzer, M., Wilkinson, E.J. and Solomon, D. 2006 consensus guidelines for the management of women with cervical intraepithelial neoplasia or adenocarcinoma in situ. Am. J. Obstet. Gynecol. 2007; 197:340–345.

Zhou, J., Sun, X.Y., Stenzel, D.J. and Frazer, I.H. Expression of vaccinia recombinant HPV-16 L1 and L2 ORF proteins in epithelial cells is sufficient for assembly of HPV virion-like particles. Virology 1991; 185:251–257.

zur Hausen, H., Gissmann, L., Steiner, W., Dippold, W. and Dreger, I. Human papilloma viruses and cancer. Bibl. Haematol. 1975; 43:569–571.

CHAPTER

27

Influenza

Joan E. Nichols[1] *and James W. LeDuc*[2]

[1]Department of Internal Medicine, Division of Infectious Diseases, Department of Microbiology and Immunology, and Sealy Center for Vaccine Development, University of Texas Medical Branch, Galveston, TX, USA

[2]Institute for Human Infections and Immunity, Department of Microbiology and Immunology, University of Texas Medical Branch, Galveston, TX, USA

OUTLINE

ABSTRACT

Influenza virus is a globally important respiratory pathogen which is associated with a high degree of morbidity and mortality annually. The rapid evolution of influenza A and B viruses contributes to the annual seasonal epidemics (localized outbreaks) in humans as well as occasional pandemic (worldwide) outbreaks. Despite improvements in development of antiviral therapies during the last decade, vaccination remains the most effective method of prophylaxis. For those at risk of developing complications from influenza infection, annual vaccination is recommended as it induces a good degree of protection and is generally well tolerated by the recipient. Currently there are two types of influenza vaccines in use, the live-attenuated vaccine (LAV) given intranasally/orally, and the inactivated vaccine (IV) delivered subcutaneously or intramuscularly. The available trivalent IV (TIV) elicits good serum antibody responses but induces poor mucosal IgA antibody and cell-mediated immunity. In comparison, the LAV elicits a more complete (humoral and cellular) immune response, similar to the type of immune response produced in the development of natural immunity to a specific strain. IV and LAV vaccines for seasonal influenza are effective in preventing and curbing the spread of influenza-related disease, but new technologies such as reverse genetics, production of recombinant proteins and use of cell culture to manufacture vaccine stocks could be used to improve the production of seasonal vaccines and to develop pandemic vaccines in the future. These new methods of virus production should help to develop vaccines that would be safe, cross-protective against variant influenza strains, require less virus per dose than current vaccines, and take less time to prepare the large amounts of virus necessary for vaccine production. Furthermore, pandemic vaccines against highly virulent strains, such as the H5N1 virus, can be generated much faster using reverse genetics approaches. Other technological breakthroughs such as production of vaccines that promote the induction of innate immune responses by signaling through Toll-like receptors or the use of adjuvants that boost the immunogenicity of current vaccine formulations would decrease the amount of antigen required to promote a response in recipients. Finally, development of "universal influenza virus vaccines," which target conserved components of the virus that are not subject to change, would allow us to produce both seasonal and pandemic vaccines that would be protective against more strains than the strains circulating in any given year.

INTRODUCTION

There are three types of influenza virus termed A, B, and C. Influenza types A or B viruses cause epidemics which are localized outbreaks of disease every year. Type C influenza virus infection causes either a very mild respiratory illness or no symptoms at all and does not cause epidemics. It is estimated that influenza is responsible for an average of 51,000 deaths annually in the United States, with a range from about 7000 to about 72,000 deaths from localized epidemic outbreaks in any given year (Kilbourne, 2006). Three pandemics (worldwide outbreaks) have occurred in the previous century: "Spanish influenza" in 1918, "Asian influenza" in 1957, and "Hong Kong influenza" in 1968 (Simonsen et al., 1998a; Kilbourne, 2006). The 1918 pandemic killed an estimated 40–50 million people worldwide (Simonsen et al., 1998a; Bridges et al., 2002). It is because of this the 1918 influenza pandemic is considered one of the deadliest disease events in human history. Subsequent pandemics have been less severe causing an estimated 2 million deaths in 1957 and 1 million deaths in 1968 (Simonsen et al., 1998a; Bridges et al., 2002). Anyone can become infected with the influenza virus, and serious problems related to influenza can happen at any age. The most vulnerable population includes people aged 65 years and older, people of any age with chronic medical conditions, and very young children (Simonsen et al., 1998a; Bridges et al., 2002). Complications include pneumonia, bronchitis, and sinus and ear infections. Use of influenza vaccine has been shown to reduce influenza-related deaths and hospitalizations in all age groups. Because of public health concerns vaccination efforts have focused on influenza types A and B. The best way to limit effects of influenza in the general population is through annual vaccination. Both inactivated (i.e., killed virus) influenza vaccines (IV) and live-attenuated vaccines (LAV) are currently licensed and available for use in the United States (Nicas et al., 2005).

Influenza A virus is highly infectious, with high transmission rate, and respiratory secretions in people infected with influenza A virus have a large quantity of virus, which has the potential to become aerosolized when the patient sneezes or coughs (Nicas et al., 2005). Virus titers measured in nasopharyngeal washes peak approximately 2–3 days after infection and can reach very high levels [up to 10^7 50% tissue culture infective dose ($TCID_{50}$)/mL] (Knight, 1973; Nicas et al., 2005). Influenza viruses can be transmitted through aerosols, large droplets, or direct contact with secretions (or fomites) and these three modes are not mutually exclusive (Knight, 1973; Murphy et al., 1973;

Douglas, 1975; Nicas et al., 2005). The virus has a short-incubation period (1–2 days) with a 4–6-h replication cycle followed by release of virus for hours before cell death resulting in infection of adjacent cells by newly synthesized virus. The decay of virus infectivity increases rapidly at relative humidity >40%.[8] Increased survival of influenza virus in aerosols at low relative humidity has been suggested as a factor that accounts for the seasonality of influenza (Loosli et al., 1943; Hemmes et al., 1960, 1962; Alford et al., 1966; Lowen et al., 2007). Infection of animal models with influenza virus in controlled heat and humidity environments show that cold temperatures and low relative humidity are favorable to the spread of influenza virus (Hemmes et al., 1960; Alford et al., 1966; Knight, 1973).

SHORT HISTORY OF THE DISEASE

Of all the pandemics caused by the influenza viruses in the 20th century, the 1918 pandemic, or "Spanish influenza," was by far the most devastating. It was associated with an exceptionally high mortality rate, killing an estimated 40–50 million people worldwide (Alford et al., 1966; Kilbourne, 1977; Crosby, 1989; Glezen, 1996; Reid et al., 2001). The unusually high mortality among the 15- to 34-year-old population significantly lowered the average life expectancy in the United States (Kilbourne, 1977; Crosby, 1989; Glezen, 1996; Reid et al., 2001). This degree of morbidity and mortality was not seen in the later influenza pandemics of 1957 and 1968.

Influenza was first noticed as a respiratory disease in swine. In 1918, J.S. Koen, a veterinarian, observed an outbreak of a respiratory disease in pigs at a local state fair that was similar symptomatically to the disease seen in humans during the "Spanish" influenza pandemic of 1918 (Koen, 1919). Attempts to isolate the cause of the infection in pigs by Peter Olitsky and Frederick Gates by filtration of respiratory isolates was unsuccessful due to evaluation of the filtered material alone as a possible cause of the respiratory disease with no examination of material that passed through the filters (Hill, 1942). Eventually Richard Shope, working at the Rockefeller Institute for Comparative Pathology in Princeton, New Jersey, repeated the filtration experiments inoculating either the filtered material or the material that passed through the filter into pigs (Shope and Lewis, 1931). Shope demonstrated that pigs became ill only when the respiratory filtrate was used for inoculation and the disease could be transferred repeatedly between pigs (Shope and Lewis, 1931). Using techniques similar to Shope's, Wilson Smith, Christopher Andrewes, and Patrick Laidlaw at the National Institute for Medical Research in the United Kingdom soon isolated the virus from humans (Laidlaw, 1935). Both groups later demonstrated that sera from people infected with the 1918 virus could neutralize the effects of the pig virus suggesting that the swine virus was related to the 1918 human pandemic virus (Shope, 1936).

ETIOLOGIC AGENTS OF HUMAN INFLUENZA

Seasonal influenza or "common influenza" is caused by influenza A and B, two types of influenza viruses that cause epidemic human disease (Wright and Webster, 2001). Influenza type C viruses have been shown to cause mild illness in humans and do not cause epidemics or pandemics (Francis et al., 1950; Minuse et al., 1954; Wright and Webster, 2001; Coiras et al., 2003). Subtypes of influenza A cause disease in humans and are classified by the types of surface hemagglutinin (HA) or neuraminidase (NA) glycoprotein present. For influenza A there are 16 known HA subtypes and 9 known NA subtypes and many different combinations of HA and NA proteins are possible. There are three subtypes of influenza A virus that have been known to be transmitted from person-to-person and to circulate in humans, H1N1, H2N2, and H3N2 (Wright and Webster, 2001). Influenza B viruses are usually found only in humans. They can cause morbidity and mortality among humans, but in general, are associated with less severe epidemics than influenza A viruses and although influenza type B viruses can cause human epidemics, they have not caused pandemics (Baine et al., 1980; Wright et al., 1980; Wright and Webster, 2001). There have been reports of lower respiratory infections in children suggesting that influenza C virus infection in infants can be associated with high fever and respiratory symptoms severe enough to require hospitalization (Moriuchi et al., 1991; Calvo et al., 2006).

Avian influenza ("bird flu") is caused by influenza viruses that occur naturally among wild birds. There is no preexisting human immunity to these viruses and due to this lack of immunity avian viruses have the potential to become pandemic influenza viruses. Subtypes of influenza which have been documented to cause disease in humans include H5, H7, and H9 viruses (Katz et al., 1999; Alexander and Brown, 2000; Wright and Webster, 2001; Peiris et al., 2004; ; Kobasa and Kawaoka, 2005; Chen et al., 2006; Wong and Yuen, 2006; WHO, 2005, 2007).

The H5N1 virus is deadly to domestic fowl and can be transmitted from birds to humans. H5 infections, such as high pathogenicity avian influenza (HPAI) H5N1 viruses currently circulating in Asia and Europe, have been documented among humans and sometimes cause severe illness or death (Alexander and Brown, 2000; Wright and Webster, 2001; Peiris et al., 2004; Kobasa and Kawaoka, 2005; Chen et al., 2006; WHO, 2005, 2007). H7 infection in humans is rare but can occur among persons who have direct contact with infected birds (Wright et al., 1980; Alexander and Brown, 2000; Kobasa and Kawaoka, 2005; The World Health Organization, 2005; Wong and Yuen, 2006; WHO, 2007). Symptoms have included conjunctivitis and/or upper respiratory symptoms. H7 viruses have been associated with both low-pathogenicity avian influenza (LPAI) (e.g., H7N2, H7N7) and HPAI (e.g., H7N3, H7N7), and these viruses have caused mild-to-severe and fatal illness in humans (Alexander and Brown, 2000; Wright and Webster, 2001; Peiris et al., 2004; Kobasa and Kawaoka, 2005; The World Health Organization, 2005; Chen et al., 2006; Wong and Yuen, 2006; WHO, 2007). Influenza A H9 has rarely been reported to infect humans. However, this subtype has been documented in some cases in a low pathogenic form (Alexander and Brown, 2000; Wright and Webster, 2001; Kobasa and Kawaoka, 2005; WHO, 2005, 2007).

CLASSIFICATION

Influenza virus, a member of the Orthomyxoviridae family, consists of a segmented, negative-stranded RNA genome. Influenza virus types A and B have eight RNA segments. Subtype C has seven RNA segments (Lamb, 1983; Lamb and Krug, 2001; Tyler and Nathanson, 2001). For influenza A the multipartite genome is encapsidated, with each segment in a separate nucleocapsid [i.e., a ribonuclear protein complex (RNP)] (Fig. 27.1). Eight different segments of negative-sense single-stranded RNA are present; this allows for genetic reassortment in single cells infected with more than one virus and may result in multiple strains that are different from the parental viruses that infected the cell. The genome of influenza A consists of 10 genes encoding for different proteins. These include the following: three transcriptases (PB2, PB1, and PA) which form the viral polymerase, two surface glycoproteins HA and NA, two matrix proteins (M1 and M2), one nucleocapsid protein (NP), and two nonstructural proteins (NS1 and NS2) (Fig. 27.1). The virus envelope glycoproteins (HA and NA) are distributed evenly over the virion surface, forming characteristic

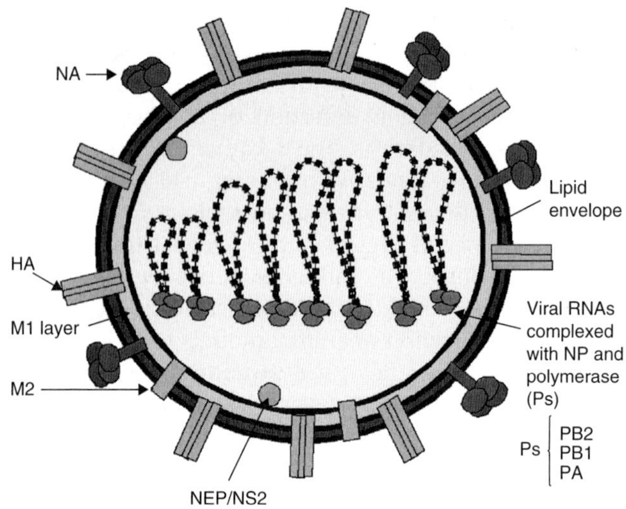

FIGURE 27.1 Schematic diagram of an influenza A particle. HA, hemagglutinin; NA, neuraminidase; M2, ion channel; NS2/NEP, nonstructural protein 2. Influenza virus RNA replication takes place in the cell nucleus and NS2/NEP is involved in the export of RNPs out of the nucleus. M, matrix protein; NP, nucleoprotein; Ps, the viral polymerase which is formed of three components PB1, PB2, and PA. RNP, the ribonucleoprotein complex containing the viral RNA; NP, nucleoproteins and the viral polymerase.

spike-shaped structures. Antigenic variation in these proteins is used as part of the influenza A virus subtype definition (but not used for influenza B or C viruses). HA is found in the virion as a trimer and NA as well as M2 as tetramers. The matrix protein (M) underlies the lipid envelope (Lamb, 1983; Lamb and Krug, 2001; Tyler and Nathanson, 2001). For influenza A virus the HA, NA, and M2 proteins are inserted into the host cell-derived lipid envelope.

Influenza viruses are named according to (a) type of virus (A, B, or C); (b) where the virus was isolated; (c) the specimen number; and (d) the year the virus was isolated. For example, A/PR/8/34 indicates that the virus was type A, isolated from specimen number 8 obtained from a patient in Puerto Rico in 1934. This is a specific strain of influenza virus. An "H7N2 virus" designates influenza A subtype that has an HA 7 protein and an NA 2 protein. Similarly an "H5N1" virus has an HA 5 protein and an NA 1 protein.

ANTIGENS ENCODED BY THE AGENT

The structure of influenza A and B viruses are similar but influenza C is different. The HA of influenza A virus is responsible for the attachment of virus to the sialic acid-containing receptors on the host cell surface and also for fusion between the virus and endosome membranes resulting in release of viral RNPs into the cytoplasm (Lamb, 1983; Lamb and Krug, 2001; Tyler and Nathanson, 2001). HA is a homotrimer of approximately 220kDa and is a type 1 membrane protein that is anchored in the viral membrane by its C-terminus. The head of HA contains the receptor-binding site at the tip of each monomer and the five antigenic sites that bind neutralizing antibodies (Lamb, 1983; Gerhard, 2001; Lamb and Krug, 2001; Tyler and Nathanson, 2001). HA is the main determinant of virulence of the virus and plays an important role in the viral life cycle. As the major antigenic determinant, it is the key ingredient of influenza vaccines for both influenza A and B viruses.

The influenza A NA is a tetrameric glycoprotein of approximately 240kDa and is a type 2 membrane protein that is anchored in the viral membrane by its N-terminus. The NA consists of a hydrophobic stalk and a globular head that contains the enzymatic and antigenic sites (Lamb, 1983; Lamb and Krug, 2001; Tyler and Nathanson, 2001). NA promotes release of virus progeny from infected cells, prevents viral aggregation and aids movement of virus through the respiratory tract. Neuraminidase sialidase activity seems to be conserved across virus subtypes and could potentially be an important target for antiviral therapy. Another influenza A membrane protein, the M2 ion channel, is a tetrameric membrane channel important for release of viral RNAs from the virion, is conserved across virus types, and also has the potential to be a target for development of antiviral therapy and vaccine development (Lamb, 1983; Lamb and Krug, 2001; Tyler and Nathanson, 2001).

PROTECTIVE IMMUNE RESPONSE

Immunity to primary infection with influenza is mediated by antibodies produced by the host to the major viral surface antigens, HA and NA (Gerhard, 2001; Lamb and Krug, 2001; Tyler and Nathanson, 2001). Antibody against one influenza virus type or subtype confers limited or no protection against another type or subtype of influenza. Furthermore, antibody to one antigenic variant of influenza virus often does not protect against a new antigenic variant of the same type or subtype (Lamb, 1983; Sugaya et al., 1994; Gerhard, 2001; Lamb and Krug, 2001; Tyler and Nathanson, 2001). Frequent development of antigenic variants through "antigenic drift" is the virologic basis for seasonal epidemics or localized influenza A outbreaks and the reason for the usual incorporation of one or more strains in each year's influenza vaccine. More dramatic antigenic changes, or "shifts," occur less frequently and can result in the emergence of a novel influenza virus with the potential to cause a pandemic (Lamb and Krug, 2001; Tyler and Nathanson, 2001).

Protection from primary infection with influenza virus is mediated by antibodies, cytotoxic T lymphocytes (CTL) and T-helper CD4+ cells, but protection from reinfection is largely mediated by antibodies directed against the surface glycoproteins, principally HA but NA also (Biddison et al., 1981; McMichael et al., 1983; Sugaya et al., 1994; Jameson et al., 1998, 1999). In humans little epidemiological evidence has been shown for the involvement of T cell-mediated immunity in protection from reinfection with influenza virus although some partial protection may occur (Biddison et al., 1981; McMichael et al., 1983; Sugaya et al., 1994; Jameson et al., 1998, 1999).

Most vaccinated children and adults develop post-vaccination hemagglutination inhibition (HAI) antibody titers (Nichol et al., 1995; Lamb and Krug, 2001; Tyler and Nathanson, 2001; Jefferson et al., 2005; Ashkenazi et al., 2006; Lewis et al., 2007; Belshe et al., 2007; Manzoli et al., 2007). These antibody titers are protective against the illness caused by strains similar to those included in the vaccine. The effectiveness of influenza vaccine depends primarily on the age and immunocompetence of the vaccine recipient and the degree of similarity between the viruses in the vaccine and those in circulation (Bell et al., 1978; Couch, 1999; de Bruijn et al., 1999; Jefferson et al., 2005; Yamanaka et al., 2005; Ashkenazi et al., 2006). Although response to vaccine may produce lower anti-HA antibody than healthy individuals some protection is provided (Bell et al., 1978; Couch, 1999; de Bruijn et al., 1999; Jefferson et al., 2005; Yamanaka et al., 2005; Ashkenazi et al., 2006). Influenza vaccination is highly effective in HIV-1-infected persons and does not seem to be associated with substantial changes in viral load or CD4 cell count (Couch, 1999; Yamanaka et al., 2005).

EPIDEMIOLOGY

Influenza is a globally important respiratory pathogen that causes nearly annual epidemics and occasional pandemics. Influenza epidemics generally occur in the winter months in the Northern hemisphere and in May–September in the Southern hemisphere (Lamb and Krug, 2001; Tyler and Nathanson, 2001; Wright and Webster, 2001). Antigenic variants of influenza A viruses emerge every year due to changes in the virus itself and this requires that updated vaccine formulations be prepared annually based on ongoing international surveillance of circulating strains of influenza virus by the World Health Organization (Wright and Webster, 2001).

Influenza A viruses constantly evolve by two mechanisms (a) antigenic drift, which involves point mutations in the antigenic sites of the two major virion surface proteins HA and NA and (b) antigenic shift, the sudden replacement of HA or NA of one subtype by another subtype (Lamb and Krug, 2001; Tyler and Nathanson, 2001; Wright and Webster, 2001; Ferguson et al., 2003; Webby et al., 2007). Gradual changes that occur through mutations in the two genes that code for the main surface proteins, HA and NA are referred to as antigenic drift. These point mutations occur unpredictably and result in minor changes to these surface proteins. Antigenic drift produces new virus strains that may not be recognized by antibodies to earlier influenza strains. Influenza B viruses undergo antigenic drift less rapidly than influenza A viruses. Antigenic shift refers to the occurrence of an abrupt, major change in HA and/or NA to produce a novel influenza A virus subtype in humans that were not currently circulating among people. Antigenic shift can occur either through direct animal (poultry)-to-human transmission or through mixing of human influenza A and animal influenza A virus genes to create a new human influenza A subtype virus through a process called genetic reassortment (Lamb and Krug, 2001; Tyler and Nathanson, 2001; Wright and Webster, 2001). Since 1977, H1N1 and H3N2 viruses, and influenza B viruses have circulated globally in the human population. In 2001, influenza A (H1N2) viruses emerged after what was probably a genetic reassortment between human A (H1N1) and A (H3N2) viruses (Lamb and Krug, 2001; Tyler and Nathanson, 2001; Wright and Webster, 2001). The current hypothesis is that new influenza viruses emerge by genetic reassortment (shift), or exchange of genetic material between two influenza viruses. For example, a large number of influenza subtypes are found commonly in animal species and all subtypes have been isolated from birds (Webster et al., 2006; Webby et al., 2007). Aquatic birds, particularly wild ducks, are primary reservoirs of influenza viruses (Sugaya et al., 1994; Lamb and Krug, 2001; Wright and Webster, 2001; Webster et al., 2006; Webby et al., 2007). The virus multiplies in the gastrointestinal (GI) tract and usually causes no illness. It is excreted in the feces, and the droppings contaminate the water. Thus, aquatic birds and domestic fowl become infected in nature by coming into contact with contaminated water.

Human viruses preferentially bind to sialic acid (SA) linked to galactose by α-2,6 linkages that are the main type found on the epithelial cells of the human respiratory tract, while avian viruses tend to bind to α-2,3 linkages that are found on duck intestinal epithelium (Webster et al., 2006). The specificity for different receptors has long been one of the explanations for the species barrier between avian and human influenza viruses (Couceiro et al., 1993; Ito et al., 1998; Wright and Webster, 2001; Gambaryan et al., 2003; Matrosovich et al., 2004). The presence of both SA α-2,3 and α-2,6 linkages in the pig tracheal epithelium is the reason why pigs may serve as the "mixing vessel" for the genesis of new viral types through coinfection (Ito et al., 1998). Chickens may have a similar role, in that their lung and intestinal epithelia contain both types of linkages (Gambaryan et al., 2003; Matrosovich et al., 2004). In the human respiratory epithelium, it has been shown that α-2,3 and α-2,6 linkages are found on ciliated and nonciliated cells, respectively, thereby allowing human infection by avian influenza viruses (Matrosovich et al., 2004). Pigs are considered intermediate hosts for influenza viruses and they possess receptors for human and avian influenza viruses. They can be infected by coming into contact with infected birds or water contaminated by them. Avian viruses may also spread to other mammals such as whales, seals, and horses and occasionally may cause disease outbreaks (Lamb and Krug, 2001; Tyler and Nathanson, 2001; Wright and Webster, 2001). Pigs, however, may become infected with human influenza viruses if they come into contact with infected individuals. If an avian H5N1 virus infects a domestic pig, and concurrently the pig acquires H3N2 infection from a farm worker there is potential that a new virus subtype, H5N2, may emerge as a result of the genetic reassortment between the two viruses. The new H5N2 variant then could spread to humans. Since the human population does not possess antibodies to this new virus, it is possible that widespread epidemics may occur.

To become a pandemic virus capable of causing a global influenza outbreak with world wide spread three conditions must be met: (1) a new subtype of influenza A virus is introduced into the human population produced through genetic reassortment resulting in an antigenic shift, (2) the virus causes serious illness in humans, and (3) the virus can spread easily from person-to-person in a sustained manner. Influenza type A viruses can infect people, birds, pigs, horses, and other animals, but wild birds are considered the natural hosts for these viruses. There are substantial genetic differences between the influenza A subtypes that typically infect birds and those that infect both people and birds.

Avian influenza A virus strains are further classified as low pathogenic (LPAI) or highly pathogenic (HPAI) on the basis of specific molecular genetic and pathogenesis criteria that require specific testing. Most avian influenza A viruses are LPAI viruses that are generally associated with mild disease symptoms in poultry. In contrast, HPAI viruses can cause severe illness that has a high mortality in poultry. Avian influenza A viruses of the subtypes H5 and H7, including H5N1, H7N7, and H7N3 viruses, have been associated with HPAI, and human infections with these viruses have ranged from mild (H7N3, H7N7) to severe and fatal disease (H7N7, H5N1) (Webster et al., 1981; Kurtz et al., 1996; Zhou et al., 1996; Yuen et al., 1998; Peiris et al., 1999, 2004; Tweed et al., 2004; Butt et al., 2005; Puzelli et al., 2005; The World Health Organization, 2005; Chen et al., 2006; Wong and Yuen, 2006; WHO, 2007). Human illness due to infection with LPAI viruses vary from very mild symptoms (e.g., conjunctivitis) to influenza-like illness (Webster et al., 1981; Kurtz et al., 1996; Zhou et al., 1996; Yuen et al., 1998; Peiris et al., 1999; Taubenberger et al., 2001; Tweed et al., 2004; Puzelli et al., 2005; The World Health Organization, 2005; Wong and Yuen, 2006). Examples of LPAI viruses that have infected humans include H7N7, H9N2, and H7N2. In general, direct human infection with avian influenza viruses occurs very infrequently, and has been associated with direct contact (e.g., touching) with infected sick or dead infected birds (domestic fowl) (Webster et al., 1981; The World Health Organization, 2005; WHO, 2007).

SIGNIFICANCE AS A PUBLIC HEALTH PROBLEM

Between 1889 and 1977 six pandemics of influenza have been recognized. Based on studies using sera obtained from persons who lived through the 1918–1919 pandemic period, the pandemic of 1918–1919 was caused by an H1N1 virus (WHO, 2007). Recently, the 1918 virus was further identified as H1N1 by complete sequencing of HA and NA from formalin-fixed, paraffin-embedded, lung tissue samples from patients who died of influenza during the 1918 pandemic, as well as from a frozen sample obtained by in situ biopsy of the lung of a victim buried in permafrost in Alaska since 1918 (Sims et al., 2003). Phylogenetic analysis suggested that the H1N1 virus had circulated between mammals and birds, and might have become adapted in humans before the 1918 pandemic (WHO, 2007). H1N1 was replaced by H2N2; the new subtype emerged to cause the Asian flu pandemic of 1957 (Simonsen et al., 1998a; Bridges et al., 2002; Nicas et al., 2005; Kilbourne, 2006). Eleven years later, another new influenza virus, H3N2 emerged to cause the Hong Kong flu pandemic of 1968 (Simonsen et al., 1998a; Bridges et al., 2002; Nicas et al., 2005; Kilbourne, 2006). The H3N2 subtype has persisted up to the present time. Currently both H3N2 and H1N1 viruses circulate together in the human population. Human H5N1 infections have been confirmed by WHO in Thailand, Vietnam, Indonesia, Cambodia, China, Turkey, and Iraq. A total of 385 cases with 243 deaths have been reported to the WHO as of June 19, 2008 with a mortality rate of greater than 50% (Sims et al., 2003; Peiris et al., 2004; The World Health Organization, 2005; Chen et al., 2006). Additional suspected human cases are under investigation in countries affected by agricultural outbreaks. Human H5N1 cases have been associated with direct exposure to infected poultry, and there is little evidence of human-to-human transmission (Sims et al., 2003; Peiris et al., 2004; The World Health Organization, 2005; Chen et al., 2006). The acquisition of characteristics enabling and sustaining transmission between people of H5N1 virus has been of great concern due to the potential to produce a pandemic with devastating global consequences.

Global Surveillance for Influenza

Since 1952 the WHO has coordinated global surveillance for influenza viruses in circulation worldwide. The objective of the surveillance system is to isolate and characterize currently circulating influenza strains in order to provide optimum antigenic and genetic matches for influenza vaccine production. Surveillance is accomplished through a network of 118 National Influenza Centers in 89 countries, coupled with 4 WHO Collaborating Centers located in the United Kingdom, Japan, Australia, and the United States. Recent concerns regarding the threat of pandemic influenza is facilitating the establishment of new National Influenza Centers in additional countries. Activities of the National Influenza Centers are guided by Terms of Reference as established by the WHO and agreed to by national Ministries of Health. Centers collect appropriate clinical specimens from patients throughout the year, but especially during influenza season. They report outbreaks to WHO and network partners, and they isolate and provide preliminary characterization of viruses in circulation. A representative sample of isolates made throughout the year, as well as unusual isolates and isolates obtained late in the transmission season, are forwarded to the four Collaborating Centers for more in depth characterization and sequencing, if deemed appropriate. The National Influenza Centers also notify WHO immediately of any unusual outbreaks that might suggest the emergence of a novel influenza virus with pandemic potential. Weekly descriptive reports are submitted to WHO electronically throughout the influenza season, as well as in annual reports, and this data is posted on FLUNET, an electronic platform that is part of the WHO Communicable Diseases Global Atlas (http://gamapserver.WHO.INT/GlobalAtlas). The Centers provide training and provide technical advice within their countries and, if appropriate, regionally. WHO provides reagents annually to all Centers to facilitate standardized identification of influenza viruses, and reports results of antigenic and genetic characterization of viruses in circulation, and publishes reports on global influenza activity.

Representatives of the four WHO Influenza Collaborating Centers and key national laboratories meet at WHO Headquarters in Geneva twice a year, in February for the Northern hemisphere and in September for the Southern hemisphere, to recommend vaccine composition for the coming influenza season. Information regarding the antigenic and genetic characterization of viruses collected by the network of National Influenza Centers is shared along with epidemiological summaries of transmission patterns seen. Through these meetings an attempt is made to predict the characteristics of influenza viruses likely to be in circulation during the coming season. At the conclusion of these deliberations, the information is provide to global influenza vaccine manufacturers and published in WHO's Weekly Epidemiological Record and other publications. Specific strains of H1N1, H3N2, and influenza B are suggested for inclusion in the vaccine (Table 27.1). WHO also collaborates with national vaccine licensing agencies to provide viruses for vaccine production and vaccine potency testing.

Increasing global concern brought about by H5N1 avian influenza and other outbreaks in which avian influenza viruses have resulted in human infections has led to the creation by WHO of a working group on influenza research at the human–animal interface. This initiative provides a linkage between WHO and the two principle international agencies addressing animal health, the World Organization for Animal Health (OIE) and the Food and Agriculture Organization of the United Nations (FAO). In 2005 the FAO and OIE announced the creation of a new joint scientific network to support veterinary services in the control of avian influenza under the acronym OFFLU. The objectives of the new network are to collaborate with the WHO human surveillance networks on issues that affect both humans and animals, to include early definition of candidate vaccines for possible human use and to coordinate research activities on influenza viruses. Results of these efforts and other focused surveillance specific to H5N1 avian influenza human infections are periodically posted on the WHO website.

POTENTIAL AS A BIOTHREAT AGENT

Influenza virus is considered an NIAID category C priority biothreat agent. Agents in this category are emerging, readily available, and easily disseminated. In addition, these organisms are capable of causing high morbidity and mortality rates. Infection from many biothreat agents (e.g., *Bacillus anthracis* that causes anthrax, Lassa virus that causes Lassa fever) also produce flu-like symptoms. Accordingly, in the event of a bioterrorist attack, one of the problems facing first responders and health-care workers is the ability to distinguish bioagent infection from natural influenza infection. Without the quick and specific identification of influenza infection, a patient would likely be sent home only to die days later when infection with a biothreat agent moved beyond the treatable stage.

CLINICAL DISEASE

Influenza illness is characterized by the abrupt onset of respiratory signs and symptoms (fever, myalgia, headache, malaise, nonproductive cough, sore throat, and rhinitis) (Glezen and Couch, 1997; Wright and Webster, 2001). In children, otitis media, nausea, and vomiting have also been reported with influenza illness. Uncomplicated influenza illness typically resolves after 3–7 days, although cough and malaise can persist for more than 2 weeks (Monto and Kioumehr, 1975; Monto et al., 1985; Glezen and Couch, 1997). In people with underlying medical conditions such as pulmonary or cardiac disease, influenza infection can lead to the development of secondary bacterial pneumonia or primary influenza viral pneumonia, and can occur as a coinfection with other respiratory pathogens. Influenza-related deaths from pneumonia are usually caused by exacerbations of cardiopulmonary conditions or other chronic diseases. Infants and young children have the highest influenza infection and hospitalization rates in pediatrics (Piedra and Glezen, 1991; Belay et al., 1999; Heikkinen et al., 1999; Neuzil et al., 2000, 2002; Munoz et al., 2002; Wang et al., 2003). The immaturity of the infant's immune system and the absence of prior immunity and exposure to the virus are potential contributors. Although most children that suffer from influenza infection are otherwise healthy, underlying

TABLE 27.1 Influenza vaccine manufacturers (from WHO website, October 2007—www.who.int)

Country	Producer	Web site
Australia	CSL	http://www.csl.com.au
Austria, Czech Republic	Baxter-Immuno AG	http://www.baxtervaccines.com
Belgium	GSK	http://www.gsk.com
Brazil	Butantan	http://www.butantan.gov.br
Canada	GSK-ID Biomedical	http://www.gsk.com
China	Zhejiang Tianyuan-BioPharmaceutical Co. Ltd.	http://www.ty-pharm.com
China	Sinovac Kexing Biological Product Co. Ltd.	http://www.sinovac.com
China	Beijing Institute of Biological Products (BIBP)	http://www.ccibp.com/
China	Shanghai Institute of Biological Products (SIBP)	http://www.siobp.com
China	Changchun Institute of Biological Products (CCIBP)	http://www.ccibp.com/
China	Changchun Changsheng Life Sciences Ltd.	cssy@public.cc.j1.cn
China	Lanzhou Institute of Biological Products (LIBP)	http://www.vacmic.com
China	Shenzhen-Sanofi Pasteur (Filler)	http://sanofipasteur.com
China	Changzhou Yanshen Biotechnology Co. Ltd.	Not available
China	Liaoning Tiancheng Bio-pharmacy Institute Co. Ltd.	Not available
France	Sanofi Pasteur	http://www.sanofipasteur.com
Germany	Novartis	http://www.Novartis-behring.de
Hungary	Omninvest Ltd.	http://www.omninvest.hu
Italy	Novartis	http://www.novartis-behring.de
Japan	Denka Seiken Co. Ltd.	http://www.denka-seiken.co.jp
Japan	Chemo-Sero-Therapeutic Research Institute	http://www.kaketsuken.or.jp
Japan	Kitasato Institute	http://www.kitasato.or.jp
Japan	Biken	http://www.biken.osaka-u.ac.jp
Korea	Dong Shin Pharmaceuticals	http://www.dong-shin.com
Korea	Korea Vaccine Co. Ltd.	http://www.kovax-v.co.kr
Korea	Korea Green Cross	http://www.kuraish.com, http://www.rheinbiotech.com, http://www.crucell-asia.com
Netherlands	Solvay Healthcare	http://www.solvaypharmaceuticals.nl
Romania	Cantacuzino Institute	http://www.cantacuzino.ro
Russia	Immunopreparat Research Productive Association, Ufa, Products Immunologicals and Drugs. Irkutsk, RIVS, Saint Petersburg	http://www.microgen.ru
Serbia	Torlak Institute	http://www.torlakinstitut.com
Switzerland	Crucell	http://www.crucell.com
UK	Novartis	http://www.Novartis-behring.de
USA	Medimmune-Avirion	http://www.medimmune.com
USA	Sanofi Pasteur	http://www.sanofipasteur.com
USA	Wyeth Lederle	http://www.wyeth.com/

chronic medical conditions further increase the risk for complications (Piedra and Glezen, 1991; Belay et al., 1999; Heikkinen et al., 1999; Neuzil et al., 2000, 2002; Munoz et al., 2002; Wang et al., 2003). Epidemics of influenza A are also accompanied by mortality, mostly among those 65 years and older. In the history of influenza pandemics, the only exception was the 1918 pandemic, which caused high mortality among young adults as well as the elderly.

Influenza caused by H5N1 shares some features with the virus responsible for the Spanish influenza pandemic of 1918 (Webster et al., 1981; Wright and Webster,

2001; Tyler and Nathanson, 2001; The World Health Organization, 2005; Chen et al., 2006; WHO, 2007). H5N1 or avian influenza has some of the standard seasonal influenza symptoms such as high fever, headache, and muscle aches, but people quickly become seriously ill and pneumonia, difficulty breathing, and multiorgan failure have been common among people who have been stricken with H5N1 influenza (Peiris et al., 2004; The World Health Organization, 2005; Chen et al., 2006; Wong and Yuen, 2006; WHO, 2007). Morbidity and mortality in those people with H5N1 are severe in previously healthy, young, and middle-aged persons. Diarrhea occurs more frequently than with influenza caused by the seasonal human-adapted subtypes that commonly circulate. The frequency of pneumonia and diarrhea distinguish avian from seasonal influenza. Over 50% of reported H5N1 influenza cases have been fatal. Death is primarily due to respiratory or multiorgan failure (Peiris et al., 2004; The World Health Organization, 2005; Chen et al., 2006; WHO, 2007). Unlike human-adapted subtypes, H5N1 is found in high titers in lower respiratory tract specimens, throat swabs and stool (Peiris et al., 2004; Puzelli et al., 2005; The World Health Organization, 2005; Chen et al., 2006; Wong and Yuen, 2006; WHO, 2007). The current H5N1 virus has also been isolated from serum (Puzelli et al., 2005). Although influenza caused by H5N1 is generally known for its aggressive disease course and associated high mortality rates, there is evidence that mild disease and asymptomatic infections occur.

TREATMENT

Vaccination is the best way to limit and control influenza infection and yearly vaccination is recommended (Fiore et al., 2007). In 1976, the US Food and Drug Administration (FDA) approved amantadine to both treat and prevent influenza type A in adults and children 1-year-old or older (Hayden and Pavia, 2006). FDA-approved rimantadine—a derivative of amantadine—in 1993 to treat and prevent influenza infection in adults and prevent influenza in children. These two drugs act against influenza A viruses but not against influenza B viruses. In 1999, FDA-approved two additional drugs to fight the flu: Relenza (zanamivir) and Tamiflu (oseltamivir), the first of a new class of antiviral drugs called NA inhibitors (Hayden and Pavia, 2006; Antiviral Agents for Seasonal Influenza: Information for Health Professionals). Zanamivir and oseltamivir inhibit both influenza A and B viruses. Studies have shown that all four drugs can reduce the duration of flu symptoms by 1 day if taken within 2 days of the onset of the illness. Amantadine is approved for treating and preventing uncomplicated influenza A virus infection in adults and children who are 1 year of age or older (Hayden and Pavia, 2006; Antiviral Agents for Seasonal Influenza: Information for Health Professionals). Rimantadine is approved for treating and preventing uncomplicated influenza virus A infection in adults and for preventing, but not treating, such infections in children. Zanamivir is approved for preventing influenza A and B virus infections in people 5 years of age and older, and for treating uncomplicated influenza A and B virus infections in people 7 years of age and older who have not had symptoms for more than 2 days. Oseltamivir is approved for treating uncomplicated influenza virus infection in people 1 year of age or older who have not had symptoms for more than 2 days (Hayden and Pavia, 2006; Antiviral Agents for Seasonal Influenza: Information for Health Professionals). Amantadine and rimantadine are taken orally in pill form. Zanamivir, available as a dry powder, can be inhaled using a device known as a "Diskhaler." The recommended dosage is two inhalations twice a day, morning and night, for 5 days. Oseltamivir is available as a pill and for adults is taken twice daily for 5 days. For children, the dose of oseltamivir depends on the child's weight. A liquid suspension of oseltamivir can be taken by children or adults who cannot swallow a capsule. Both amantadine and rimantadine have been shown to cause CNS and GI side effects. Amantadine has been associated with more frequent CNS adverse reactions than rimantadine, especially in the elderly (Couch, 2000; Hayden and Pavia, 2006; de Jong et al., 2005; Antiviral Agents for Seasonal Influenza: Information for Health Professionals). Since zanamivir is taken by inhalation, it should not be given to persons with underlying airway disease. The primary side effect of oseltamivir is nausea and vomiting; thus it should be taken with food.

Viral resistance to adamantanes has the ability to emerge rapidly during treatment because a single point mutation at amino acid positions 26, 27, 30, 31, or 34 of the M2 protein can confer cross-resistance to both amantadine and rimantadine (Couch, 2000; de Jong et al., 2005). Due to the high proportion of circulating influenza viruses in the US in recent years that have been resistant to the adamantanes, the CDC currently recommends that neither amantadine nor rimantadine be used for the treatment or chemoprophylaxis of influenza in the United States. Early therapy with NA inhibitors, either oseltamivir or zanamivir, reduces the duration of symptoms, the duration of disability, and the risk of lower respiratory tract complications (Couch, 2000; Hayden and Pavia, 2006).

Current WHO treatment guidelines for influenza A/H5N1 infection, the strain of concern in current pandemic planning, recommend NA inhibitors for treatment and prophylaxis. The guidelines also include a soft recommendation for the use of M2 inhibitors (amantadine and rimantadine) for strains of the virus still susceptible to these drugs. Recent isolates of influenza A H5N1 show varying resistance to the adamantanes (amantadine and rimantadine): the NA inhibitors, oseltamivir and zanamivir, are active against influenza A H5N1. However, the emergence of high-level resistance to oseltamivir during oseltamivir treatment has been demonstrated in some patients with influenza A H5N1 infections.

PATHOGENESIS

Influenza A is highly contagious causing infection of the upper and lower respiratory tract. Primary viral pneumonia is the most severe complication observed during influenza infection and is associated with high mortality (Bender and Small, 1992). Evaluation of the pathogenesis of influenza is influenced by both the physical and immunological status of the host as well as the virulence of the infecting strain of virus. A great deal of information regarding influenza pathogenesis has been learned from the study of animal models or from evaluation of highly pathogenic avian influenza viruses. However, the precise details of the pathogenesis in humans remains largely unknown at this time.

Respiratory epithelial cells are the primary targets for influenza virus infection and once the virus enters the respiratory tract where virus replicates in bronchial and alveolar epithelial cells (Matrosovich et al., 2004) infection can spread to adjacent resident alveolar macrophages (AM). Although the disease is often associated with systemic symptoms the majority of virus replication in uncomplicated influenza infection is thought to occur in the respiratory tract (Bender and Small, 1992).

Following infection, AM die predominantly by apoptosis, whereas epithelial cells have been shown to undergo necrotic death (Fesq et al., 1994). Cell infection as well as necrosis/apoptosis of infected cells triggers immune responses with production of cytokines such as tumor necrosis factor (TNF-α), interleukin (IL)-1, and chemokines such as monocyte chemoattractant protein (MCP)-1, RANTES, MIP-1α and β, interferon-induced protein (IP)-10 and IL-8 (Virelizier et al., 1979; Fujisawa et al., 1987; Bender and Small, 1992; Meerschaert and Furie, 1995; Braciak et al., 1996; Doherty, 1996; Friedland, 1996; Lehmann et al., 1996; Sprenger et al., 1996; Adachi et al., 1997; Hofmann et al., 1997; Kuziel et al., 1997; Bussfeld et al., 1998; Matsukura et al., 1998; Cox and Subbarao, 1999; Dawson et al., 2000; Julkunen et al., 2000; Van Reeth, 2000; Julkunen et al., 2001; Maus et al., 2002; Chan et al., 2005). Initial release of inflammatory mediators by infected lung epithelium, resident tissue macrophages, and AMs results in the initial extravasation of macrophages, and neutrophils and then later NK cells and T cells, from the peripheral blood into infected lung tissue within days after influenza A virus infection (Virelizier et al., 1979; Fujisawa et al., 1987; Bender and Small, 1992; Meerschaert and Furie, 1995; Braciak et al., 1996; Doherty, 1996; Lehmann et al., 1996; Sprenger et al., 1996; Adachi et al., 1997; Hofmann et al., 1997; Kuziel et al., 1997; Cox and Subbarao, 1999; Dawson et al., 2000; Julkunen et al., 2000, 2001; Maus et al., 2002; Chan et al., 2005). Although polymorphonuclear leukocytes and alveolar macrophages provide some early protection, resident tissue macrophages can become infected and induced to produce high levels of inflammatory cytokines (Fujisawa et al., 1987; Bender and Small, 1992; Doherty, 1996; Lehmann et al., 1996; Hofmann et al., 1997; Dawson et al., 2000; Julkunen et al., 2000). There is evidence that this early inflammatory pathology can contribute to severe disease, especially with the highly lethal H5N1 viruses (Chan et al., 2005).

Studies of isolates of avian influenza A (H5N1) from patients revealed that viral virulence factors include the presence of a highly cleavable HA that can be activated by multiple cellular proteases, a specific substitution in the polymerase basic protein 2 (Glu627Lys) which can enhance replication (Shinya et al., 2004; Guan et al., 2002; Perkins and Swayne, 2002; Zitzow et al., 2002; Li et al., 2004; Avian Influenza A, 2004; Horimoto et al., 2004; Sturm-Ramirez et al., 2004; Govorkova et al., 2004), and a substitution in nonstructural protein 1 (Asp92Glu) that confers increased resistance to inhibition by interferons and TNF-α in vitro and prolonged replication in swine, (Seo et al., 2002; Cheung et al., 2002) as well as greater production of cytokines, particularly TNF-α, in human macrophages exposed to the virus (Cheung et al., 2002).

DESCRIPTION OF DISEASE PROCESS

Cytokines and chemokines have long been thought to play a role in the extreme pathology associated with highly pathogenic influenza viruses (Cretescu et al., 1978; Akashi et al., 1994; Imashuku et al., 1996;

Taubenberger et al., 1997; Hayden et al., 1998; Fisman, 2000; Kaiser et al., 2001; Seo et al., 2002; Cheung et al., 2002; Perkins and Swayne, 2002; Zitzow et al., 2002; Govorkova et al., 2004; Kobasa et al., 2004; Sturm-Ramirez et al., 2004; Beigel et al., 2005; Tumpey et al., 2005; Buchy et al., 2007). While there is little direct evidence from the human pathology caused by the 1918 pandemic virus, it is thought that both the 1918 virus and avian H5N1 viruses cause a dysregulated and exaggerated production of inflammatory cytokines or "cytokine storm" that, in turn, results in pulmonary edema, primary and secondary pneumonias with acute bronchopneumonia. Recently, recombinant viruses generated by reverse genetics expressing the HA and NA of the 1918 influenza virus were shown to induce significantly higher levels of IFN-δ, TNF-α, MIP-2, and MIP-1α and MCP-1, MIP-1β, MIP-2, MIP-3α, IL-1, IL-6, IL-12, IL-18, and G-CSF in the lungs of infected mice (Akashi et al., 1994; Imashuku et al., 1996; Taubenberger et al., 1997; Hayden et al., 1998; Fisman, 2000; Kaiser et al., 2001; Cheung et al., 2002; Kobasa et al., 2004; Beigel et al., 2005; Tumpey et al., 2005; de Jong et al., 2006), providing some support that elevated chemokine and cytokine levels were a hallmark of the severe human disease caused by the 1918 virus (Kobasa et al., 2004; Beigel et al., 2005; Tumpey et al., 2005).

High virus levels were found in those infected with H5N1 and correlated with fatal outcomes (Peiris et al., 2004). High serum levels of IL-6, TNF-α, IFN-γ, and sIL-2R were found in patients during outbreaks in Hong Kong in 1997 of H5N1 (Akashi et al., 1994; Imashuku et al., 1996; Taubenberger et al., 1997 Fisman, 2000; Mou et al., 2006), while those infected in 2003 and 2004–2005 were shown to have had elevated serum concentrations of the chemokines IP-10, MCP-1, and monokine induced by IFN-γ (MIG), compared to patients with uncomplicated seasonal influenza infections. Studies have also demonstrated higher levels of induction of proinflammatory cytokines such as TNF-α and IFN-β, by H5N1 viruses compared to human-adapted H3N2 and H1N1 viruses (Kaiser et al., 2001). Generally, disease severity has correlated strongly with cytokine levels (Cretescu et al., 1978; Kaiser et al., 2001; Buchy et al., 2007) and highest levels of chemokines, and the neutrophil chemoattractant IL-8, were detected in individuals who died (Cretescu et al., 1978; Kaiser et al., 2001; Buchy et al., 2007). Abnormally high levels of IP-10 and MIG were also observed in other fatal cases of H5N1 influenza infection (Cretescu et al., 1978; Buchy et al., 2007) and other studies have highlighted the strong correlation between IL-6, IFN-α, and TNF-α levels and the severity of disease symptoms (Cretescu et al., 1978; Hayden et al., 1998; Kaiser et al., 2001; Buchy et al., 2007).

IMMUNE RESPONSE TO INFECTION

The humoral immune system, including both the mucosal and systemic responses, plays a major role in the development of immunity to influenza. During infection of humans, antibodies are produced to all the major viral proteins (Murphy et al., 1973; Potter and Oxford, 1979; Askonas et al., 1982) Antibodies to the surface glycoproteins, HA and NA, are associated with resistance to infection, whereas antibodies to the conserved internal antigens, M and NP, have not been shown to be protective (Gerhard, 2001). The CTL response is mainly directed against the matrix and nucleocapsid proteins (Biddison et al., 1981; McMichael et al., 1983; Jameson et al., 1998). Although the CTL response does not confer protection to infection, it is important for the clearance of virus and recovery from illness.

ANTIBODY

The mucosal tissues of the upper respiratory tract are the main portal of entry for influenza virus, and the mucosal immune system provides the first line of defense against infection. Both mucosal and systemic antibody production contributes to resistance to influenza infection and disease (Murphy et al., 1973; Askonas et al., 1982; Clements et al., 1983). Antibodies secreted locally in the upper respiratory tract are a major factor in resistance to natural infection (Artenstein et al., 1964; Murphy et al., 1973; Clements et al., 1983, 1986; Mazanec et al., 1995) and after natural infection secretory immunoglobulin A (SIgA) and serum IgG are involved in protection of the respiratory tract against reinfection with the same virus or an antigenically similar strain of virus. Nasal secretions have been shown to contain neutralizing antibodies to influenza HA and NA, which are primarily of the IgA isotype (Rossen et al., 1970; Jurgensen et al., 1973; Murphy and Clements, 1989; Yewdell and Hackett, 1989; Mazanec et al., 1995). During primary infection, all three major Ig classes (IgG, IgA, and IgM) specific to HA can be detected by enzyme-linked immunosorbent assay in nasal washings, although IgA and IgM are more frequently detected than IgG (Rossen et al., 1970; Jurgensen et al., 1973; Murphy and Clements, 1989; Yewdell and Hackett, 1989; Mazanec et al., 1995).

The humoral immune system produces antibodies against different influenza antigens, of which the HA-specific antibody is the most important for neutralization of the virus and thus prevention of illness. The NA-specific antibodies are less effective in preventing infection, but they lessen the release of virus from infected cells.

Cell-Mediated Immunity

Influenza-specific cellular lymphocytes have been detected in the blood and the lower respiratory tract secretions of infected subjects. Cytolysis of influenza-infected cells is mediated by CTLs in concert with influenza-specific antibodies and complement (Quinnan et al.,1980; Ruben et al., 1980; Biddison et al., 1981; McMichael et al., 1983, 1986; Fleischer et al., 1985; Gerhard and Mozdanowska, 2001). The primary cytotoxic response is detectable in blood after 6–14 days (Ruben et al., 1980; McMichael et al., 1983, 1986). CTLs that recognize both surface HA, NA, and M2 protein as well as the internal nonglycosylated proteins of influenza M, NP, and PB2 proteins, have been isolated (Biddison et al., 1981; Fleischer et al., 1985). Cell-mediated immunity plays a role in recovery from influenza infection and may also prevent influenza-associated complications, but it does not seem to contribute significantly in preventing infection.

The level of memory CTLs to influenza does not correlate with susceptibility to infection or illness following experimental influenza infection, but it does correlate with an increased rate of clearance of the virus from the respiratory tract. There is a limited amount of data on human CTL responses to influenza available in the literature, but information from influenza infections using the murine model shows that a very strong T-helper response, is induced (McMichael et al., 1986). The CTL response has been shown to be cross-reactive between influenza A strains (Fleischer et al., 1985; Gerhard and Mozdanowska, 2001; Nguyen et al., 2001) and is very important in minimizing the spread of virus in the respiratory system (Fleischer et al., 1985; McMichael et al., 1986; Gerhard and Mozdanowska, 2001).

VACCINES

Vaccination induces a good degree of protection (60–90% efficacy) and is well tolerated by the recipient (Parkman et al., 1977; Couch and Kasel, 1983; La Montagne et al., 1983; Couch et al., 1984, 1986, 1996; Maassab and DeBorde, 1985; Kaiser et al., 1999; Simonsen et al., 1998b; Couch, 2000; Bridges et al., 2000; Halperin et al., 2002). For those at risk of complications from influenza, annual vaccination is recommended due to the antigenic changes in circulating strains. The current trivalent IV (TIV) and LAV are both safe and efficacious for healthy adults, elderly, and children (Barker and Mullooly, 1980; Couch and Kasel, 1983; Couch et al., 1984, 1986; Maassab and DeBorde, 1985; Belshe et al., 1998; Kaiser et al., 1999; Simonsen et al., 1998b; Couch, 2000; CDC, 2000; Bridges et al., 2000; Halperin et al., 2002). Vaccine is available for all persons, including school age children, who run the risk of becoming ill with influenza or of transmitting influenza to others and all children aged 6 months–4 years (Couch and Kasel, 1983; Couch et al., 1984; Kaiser et al., 1999; Simonsen et al., 1998b; Couch, 2000; Bridges et al., 2000). Vaccination is also recommended for adults aged 50 years or older; individuals aged 6 months–18 years at increased risk for Reye's syndrome because of long-term treatment with aspirin; women who will be pregnant during the influenza season; persons with chronic pulmonary (including asthma), cardiovascular (excluding hypertension), renal, hepatic, hematological, or metabolic disorders (including diabetes); immunosuppressed persons; those with any condition that can compromise respiratory function (Couch and Kasel, 1983; Couch et al., 1984; Kaiser et al., 1999; Simonsen et al., 1998b; Couch, 2000; Bridges et al., 2000). Vaccine is recommended for residents of nursing homes and other chronic care facilities; health care personnel; healthy household contacts and caregivers of children younger than 5 years; and adults aged 50 years or older (especially contacts of infants younger than 6 months); and persons with medical conditions increasing their risk for severe complications from influenza (Couch and Kasel, 1983; Couch et al., 1984; Kaiser et al., 1999; Simonsen et al., 1998b; Couch, 2000; Bridges et al., 2000).

Currently available IV as well as LAV are prepared from virus grown in embryonated hen's eggs (Palese, 2006). Most influenza virus vaccines used in the United States and Europe consist of embryonated egg-grown and formaldehyde-inactivated preparations, which, after purification, are chemically disrupted with a non-ionic detergent (e.g., Triton X-100). Split virus preparations show lower pyrogenicity than whole virus vaccines (Couch et al., 1996; Keitel et al., 1997; Palese, 2006). In general, single dose for adults contains the equivalent of 45 μg HA (15 μg HA for each of the three antigenic components). This dose is approximately the amount of purified virus obtained from the allantoic fluid of one infected embryonated egg.

Live-attenuated influenza virus vaccines for strains of type A and type B influenza were developed by serial passage in chick-embryo cells at 25°C until they became cold-adapted (Murphy et al., 1973; Palese, 2006). Vaccine strains are prepared by the reassortment of gene segments from a master strain and a putative epidemic strain, followed by selection for a strain with genes for HA and NA from the putative epidemic strain and the remaining genes from the cold-adapted parent. The vaccine is given intranasally by spray and induces mild upper respiratory tract symptoms in about 10–15% of persons (Maassab and DeBorde, 1985; Couch et al., 1986; Belshe et al., 1998).

Table 27.2 shows a list of the influenza strains used in vaccines for the flu seasons from 1989–2008. Currently all vaccine stocks are produced in eggs, but not all new antigenic variants replicate well in this system as was seen in development of vaccine to protect against the A/Fujian virus in 2000. The US FDA rejected the use of the most appropriate H3N2 strain, A/Fujian/411/2002, and instead used the same strain as in the 2002 formulation (Palese, 2006). This decision was made primarily because the A/Fujian/411/2002 strain had first been isolated in Madin–Darby canine kidney (MDCK) cells rather than in embryonated eggs. Use of MDCK cells for virus isolation is not allowed by FDA's rules, which do not yet encompass advanced technologies or scientifically sound purification procedures based on limiting dilutions or cloning with DNA. Because no egg-culture suitable A/Fujian-like virus was available, the vaccine was made with the A/Panama/2007/99 virus as the H3 component (Palese, 2006). It is important to realize that even though this strain was not an optimal match for the A/Fujian epidemic virus, it provided some protection, particularly in adults.

The 2007–2008 influenza vaccine formulation contains an A/Solomon Islands/3/2006 (H1N1)-like virus; an A/Wisconsin/67/2005 (H3N2)-like virus; and a B/Malaysia/2506/2004-like virus. The three strains for the new influenza vaccine formulation were confirmed by the US FDAs Vaccines and Related Biological Products Advisory Committee in February 2007 and correspond with recommendations made by the WHO in February. The A/Solomon Islands/3/2006 virus replaced the previous A/New Caledonia/99 virus which had circulated between 2000 and 2007.

HISTORY

The first influenza vaccine developed was a monovalent preparation which means that the vaccine was formulated from a single strain of influenza virus. In the 1960s and 1970s the influenza vaccines were bivalent or composed of two strains. Since 1978 influenza vaccines have been trivalent with the inclusion of a strain of influenza B and two strains of

TABLE 27.2 Strains used in vaccines for the flu seasons shown—1989–2007

Season	H1N1	H3N2	Type B
89–90	A/Taiwan/86	A/Shanghai/87	B/Yamagata/88
90–91	A/Taiwan/86	A/Shanghai/89	B/Yamagata/88
91–92	A/Taiwan/86	A/Beijing/89	B/Panama/90
92–93	A/Texas/91	A/Beijing/89	B/Panama/90
93–94	Unchanged	Unchanged	Unchanged
94–95	A/Texas/91	A/Shandong/93	B/Panama/90
95–96	A/Texas/91	A/Johannesburg/94	B/Harbin/94
96–97	A/Texas/91	A/Nanchang/95	B/Harbin/94
97–98	A/Johannesburg/96	A/Nanchang/95	B/Harbin/94
98–99	A/Beijing/95	A/Sydney/97	B/Beijing/93
99–00	A/Beijing/95	A/Sydney/97	B/Yamanashi/98
00–01	A/New Caledonia/99	A/Panama/99	B/Yamanashi/98
01–02	A/New Caledonia/99	A/Panama/99	B/Victoria/00 or similar
02–03	A/New Caledonia/99	A/Moscow/99	B/Hong Kong/2001
03–04	A/New Caledonia/99	A/Moscow/99	B/Hong Kong/2001
04–05	A/New Caledonia/99	A/Fujian/2002	B/Shanghai/2002
05–06	A/New Caledonia/99	A/California/2004	B/Shanghai/2002
06–07	A/New Caledonia/1999	A/Wisconsin/2005	B/Malaysia/2004
07–08	A/Solomon Islands/2006	A/Wisconsin/2005	B/Malaysia/2004
08–09	A/Brisbane/59/2007	A/Brisbane/10/2007	B/Florida/4/2006

influenza A. From the 1970s on the vaccine has been changed to parallel the changes in currently circulating strains of virus with occasional changes in one or all of the strains included in the vaccine preparation. Surveillance by the WHO influences the strain selection included in each year's vaccine preparation and production of vaccine occurs year round.

Influenza virus vaccines first developed in the 1940s consisted of partially purified preparations of influenza viruses grown in embryonated hen's eggs. Because of substantial contamination by egg-derived components, these killed (formaldehyde-treated) vaccines were highly pyrogenic and lacking in efficacy. A major breakthrough in vaccine manufacture was the development of the zonal ultracentrifuge in the 1960s (Gerin and Anderson, 1969). This technology, which originated from uses for military purposes, revolutionized the purification process and industrial production of many viruses for vaccines including influenza virus. It remains the basis for the manufacturing process of current influenza virus vaccines. Most prototype seed strains are provided to the manufacturers by government agencies, which create high-yielding strains through classical reassortment with a high-yielding laboratory strain, A/PR/8/34 (Palese, 2006).

Whole inactivated virus vaccines were the first influenza vaccines to be produced. The currently circulating strain of influenza was inoculated into embryonated eggs, harvested 2–3 days later and inactivated. Because of the high incidence of reactions seen in individuals given whole, inactivated virus vaccine, attempts were made to produce a vaccine which was less reactogenic but conserved immunogenicity. Split vaccines can be prepared from inactivated particles disrupted with detergents. These vaccines have been shown to induce fewer side effects and were found to be as immunogenic as whole virus vaccine. Subunit vaccines which contain only the HA and NA antigen can also be produced. Subunit vaccines are used in aqueous suspension or may be absorbed to carriers. Volunteers given subunit vaccines experienced fewer reactions than those given whole virus vaccines (La Montagne et al., 1983; Couch et al., 1996).

Recently licensed LAVs induce immunity that is similar to that provided from natural infection with induction of both humoral and cellular immunity. Normal methods for attenuation include use of repeated passages and temperature adaptation (Palese, 2006). Attenuated strains of virus can also be mixed with wild-type virus to produce reassortants that contain the RNA fragments which code for wild-type HA and NA, and all the other genetic material form the attenuated strain (Kilbourne, 1969). When given intranasally, attenuated vaccines produce few side effects (Couch et al., 1986; Margolis et al., 1990; Nichol et al., 1996).

CURRENT LICENSED VACCINES

Influenza vaccination is the primary method for preventing seasonal influenza and its complications. Vaccines need to be evaluated and, if necessary, updated to provide coverage for the circulating strains of influenza every year and, consequently, the vaccine must also be administered every year. Table 27.3 lists the vaccines currently licensed for use in the US Both IV and LAV formulations of the vaccine are available. Both LAV and IV contain strains of influenza viruses that are antigenically equivalent to the annually recommended strains: one influenza A (H3N2) virus, one A (H1N1) virus, and one B virus. Each year, one or more virus strains can potentially be changed on the basis of global surveillance for influenza viruses and the emergence and spread of new strains. Both vaccines are administered annually to provide optimal protection against influenza virus infection. Although both types of vaccines are effective, the vaccines differ in several aspects. Inactivated influenza vaccine contains killed viruses, and thus cannot produce signs or symptoms of influenza virus infection. In contrast, LAV contains live, attenuated viruses and, therefore, has a potential to produce mild signs or symptoms related to influenza virus infection. LAV is administered intranasally by sprayer, whereas inactivated influenza vaccine is administered intramuscularly by injection.

The US FDA announced the licensure of the first vaccine in the United States for humans against the avian influenza H5N1 virus in 2007. This inactivated influenza virus vaccine is indicated for immunization of persons 18 through 64 years of age who are at increased risk of exposure to the H5N1 influenza virus subtype contained in the vaccine. This vaccine was derived from the A/Vietnam/1203/2004 influenza virus (Table 27.3).

Thimerosal is a very effective preservative that has been used since the 1930s to prevent contamination in some multidose vials of vaccines (preservatives are not required for vaccines in single dose vials). Thimerosal contains approximately 49% ethyl mercury. There is no convincing evidence of harm caused by the low doses of thimerosal in vaccines, except for minor reactions like redness and swelling at the injection site. However, in July 1999 the Public Health Service (PHS) agencies, the American Academy of Pediatrics (AAP), and vaccine manufacturers agreed that thimerosal should be reduced or eliminated in vaccines as a precautionary measure.

TABLE 27.3 Currently licensed vaccines*

Vaccine	Manufacturer	Form of vaccine	Dosage	Thimerisol preservative	Age indication for administration
Fluzone® seasonal flu vaccine	Sanofi Pasteur Inc.	Inactivated: One dose required 15 μg each of HA of two influenza A & one influenza B	Multidose vial	Present 0.01% or (25 μg mercury/dose)	6 months and older
			Single dose: 0.25 ml syringe	None	6–35 months
			Single dose: 0.5 ml syringe or vial	None	36 months and older
Fluvirin® seasonal flu vaccine	Novartis Vaccine	Inactivated: One dose required 15 μg each of HA of two influenza A & one influenza B	Multidose vial	Present 0.01% or (25 μg mercury/dose)	4 years and older
			Single dose: Prefilled 0.5 ml syringe	Preservative free: (1 mcg or less mercury/0.5 ml dose)	4 years and older
Fluvarix® seasonal flu vaccine	Glaxo Smith Kline Biologicals	Inactivated: One dose required 15 μg each of HA of two influenza A & one influenza B	Single dose: Prefilled 0.5 ml syringe	Preservative free (1 mcg or less mercury/0.5 ml dose)	18 years and older
FluMist® seasonal flu vaccine	MedImmune Vaccines, Inc.	Live Attenuated: One dose required 15 μg each of HA of two influenza A & one influenza B	Single dose sprayer	None	Healthy immunocompetent persons 2–49 years
AFluria® seasonal flu vaccine	CSL Biotherapies	Inactivated: One dose required 15 μg each of HA of two influenza A & one influenza B	Single dose	None	18 years and older
Flulaval® seasonal flu vaccine	ID Biomedical Corporation	Inactivated: One dose required 15 μg each of HA of two influenza A & one influenza B	Multidose vial	Yes 0.01% or (25 μg mercury/dose)	18 years and older
Influenza virus vaccine, H5N1 pandemic flu vaccine	Sanofi Pasteur Inc.	Inactivated: 90 μg/hemagglutinin (HA) of the influenza virus strain A/Vietnam/1203/2004 (H5N1, clade 1) two doses required			18–64 years of age

*as of July 2007.

DURATION OF IMMUNITY

A critical aspect of influenza vaccine development is the demonstration that immunization is capable of inducing a protective immune response. In individuals who have been immunologically primed by exposure to related viruses by infection or by immunization, a single dose of 15 μg HA per strain, is considered to give high levels of protective immunity in younger adults and to prevent severe consequences of infection in the elderly. Although vaccines are usually prepared from whole virus, split products or from purified subunits and occasionally from whole virions, the immune responses to immunization in primed populations are considered equivalent for each type of vaccine. Secretory antibody induced by inactivated viral vaccines, is thought to persist for about a year, after having reached a peak level at about 4–6 weeks following immunization or infection (Kilbourne, 1969). Inactivated virus vaccine induced greater serum IgA, IgG, and IgM HA antibody responses than did live virus vaccine (Artenstein et al., 1964; Kilbourne, 1969; Rossen et al., 1970; Couch et al., 1984; Clements et al., 1986; Murphy and Clements, 1989; Kroon et al., 1994; Mazanec et al., 1995). Although vaccination with both types resulted in significantly increased levels of nasal-wash HA antibodies, there was a striking difference in the predominant isotype of local antibody induced by each type of vaccine (Waldman and Ganguly, 1975). Recipients of live virus vaccine developed IgA antibody more frequently whereas recipients of inactivated influenza vaccine developed IgG antibody more frequently (Artenstein et al., 1964; Rossen et al., 1970; Clements and Murphy, 1986; Clements et al., 1986; Murphy and Clements, 1989; Mazanec et al., 1995).

CONTRAINDICATION OF VACCINATION, INCLUDING SPECIAL RISK GROUPS

Current vaccine components are grown in embryonated chicken eggs and are rendered noninfectious. The vaccine is safe. Soreness at the site of injection is the most common complaint. Fever, malaise, and myalgia are infrequent (Margolis et al., 1990; ACIP, 1999). The vaccine is recommended to be given each year prior to the influenza season. Immediate—presumably allergic—reactions (e.g., hives, angioedema, allergic asthma, and systemic anaphylaxis) occur rarely after influenza vaccination (Couch et al., 1986; Margolis et al., 1990; Nichol et al., 1996). These reactions probably result from hypersensitivity to certain vaccine components; the majority of reactions probably are caused by residual egg protein. Although current influenza vaccines contain only a limited quantity of egg protein, this protein can induce immediate hypersensitivity reactions among persons who have severe egg allergy (Wright et al., 1977; Couch et al., 1986; Margolis et al., 1990; Nichol et al., 1996; James et al., 1998). Influenza vaccine should, therefore, not be administered to anyone with a history of hypersensitivity to any vaccine component, including eggs, egg products, or thimerosal.

The vaccine should not be administered to individuals who have a prior history of Guillain–Barré syndrome (Schonberger et al., 1979; Hurwitz et al., 1981; Kaplan et al., 1982). People who have severely weakened immune systems or people who are in contact with others with severely weakened immune systems should not be given live vaccine formulations. The estimated risk of becoming infected with the current vaccine virus formulation after close contact with a person vaccinated with the nasal-spray LAV is low (0.6–2.4%) (Maassab and DeBorde, 1985; Couch et al., 1986; Belshe et al., 1998). Because the viruses are weakened, infection is unlikely to result in influenza illness symptoms. Immune compromised individuals should not be given the AV since it contains live although attenuated virus.

ADVERSE EVENTS

There are risks associated with all vaccines including current influenza vaccine formulations. The most common local and systemic adverse reactions to vaccine include soreness at the vaccination site that can last up to 2 days, pain, and swelling, fever, malaise, and myalgia (Margolis et al., 1990; ACIP, 1999). Other adverse reactions may occur. The most frequent side effects of the killed vaccine after IM injection are pain, redness, and swelling at the injection site lasting 1–2 days, and systemic side effects such as headache, fever, malaise, and myalgia in about 5% of people receiving vaccines (Margolis et al., 1990; Nichol et al., 1996). These side effects are largely due to a local immune response, with interferon production leading to systemic effects. Local side effects are more common with whole virus vaccines than subunit or split vaccines, and also more common with intradermal vaccination than intramuscular (IM) vaccination.

The inactivated vaccine does not contain live virus, and cannot cause influenza infection. Live-attenuated virus vaccines do contain live virus; however, side effects are rare, with a runny nose, congestion, sore throat, and headache being the most commonly reported symptoms, with occasional abdominal pain, vomiting, and myalgia (Couch et al., 1986). They are not recommended for children less than 2 years.

Inactivated influenza virus vaccine was associated with the occurrence of the Guillain–Barré syndrome during the program of immunization against swine influenza in 1976 (Schonberger et al., 1979; Hurwitz et al., 1981; Kaplan et al., 1982). However, studies in subsequent years did not confirm such a relationship. A similar relation was reported again recently, but the calculated risk was only one additional case of Guillain–Barré syndrome per million people vaccinated, a risk lower than that of severe influenza in all age groups and particularly among people at high risk for complications (Schonberger et al., 1979; Hurwitz et al., 1981; Kaplan et al., 1982). Concern about the Guillain–Barré syndrome should not deter people from receiving an inactivated influenza virus vaccine.

PRECLINICAL, PHASE I, PHASE II, AND PHASE III TRIALS

The current egg-based technology for producing influenza vaccine was created in the 1950s. This current process of production of reassortant virus stocks followed by repeated passaging of the strain in embryonated hen's eggs to allow for egg adaptation and growth enhancement is time-consuming. A number of novel methods of production of vaccine virus are currently in process and include cell-culture based culture or production of recombinant viral proteins. Table 27.4 lists a number of vaccine formulations that are currently in preclinical or phase I–III trials in the US. Many of the candidate vaccines listed in Table 27.4 were formulated using methods that do not rely on embryonated hen's egg culture and hold great promise for the development of future influenza vaccines.

RATIONALE OF SECOND/THIRD/NEW GENERATION VACCINES

Use of Cell Lines to Produce Vaccine

Cell culture-derived influenza vaccines are currently in development as an option to egg-based production, which uses embryonated hen's egg as the standard method for production (Patriarca, 2007). There are a number of problems associated with this process, which include a dependency on the timely availability of embryonated hen's eggs as well as the selection of specific changes in the viral proteins due to egg culture (Robertson et al., 1985; Katz et al., 1987; Harmon et al., 1988; Katz and Webster, 1989; Robertson, 1993; Hardy et al., 1995; Ito et al., 1997; Patriarca, 2007; Reisinger et al., 2007) The H5 avian influenza virus strains responsible for recent epizoonotic outbreaks in Asia are lethal to chicken embryos (Palese, 2006; Patriarca, 2007). Also, due to the high pathogenicity of avian influenza virus strains, the conventional production of avian influenza virus vaccines would require the use of biosafety level 3 containment facilities for virus production. Vaccine development and production using the standard egg-based method often takes several months following the identification of new potential strains using current egg-based culture methods and typically require reassortment with a high-yield strain. Therefore, a culture method that can rapidly produce new influenza virus for vaccine production is needed as a priority for seasonal and especially for pandemic influenza preparedness.

The tissue culture process is easily expanded for large scale production of virus using large bioreactor systems to produce human influenza antigens, avian influenza antigens, or other animal influenza antigens as necessary (Patriarca, 2007). Several tissue culture cell types are being used to produce virus stocks for vaccine trials. These include MDCK (Ito et al., 1997; Patriarca, 2007), Vero cells, PER.C6 (patented human cell line) and an avian embryonic stem cell-derived cell line, EBx (a cell line is derived from chicken embryo cells, treated with a chemical mutagen to render the cell line immortal) (Patriarca, 2007). For production of cell-culture derived vaccines, live influenza virus is used to infect cells in culture. Once the viral infection has propagated through the cells, the live virus is harvested and inactivated in much the same way as in conventional egg-based influenza vaccines.

Several important regulatory issues will need to be addressed before a switch to cell culture-based production systems can occur. These include proof that the vaccine product is free of cell-derived contaminants as well as testing for any other contaminating viruses. Tissue culture-based influenza A vaccines have currently been approved for use throughout the European Union. Large scale safety and immunogenicity studies have been conducted in Europe, and trials of safety and immunogenicity are currently taking place in the United States as well. These studies show that the cell-culture based vaccine is of comparable safety and immunogenicity as compared to egg culture-based vaccines (Harmon et al., 1988).

Recombinant DNA-Based Approaches

The influenza A virus (Fig. 27.1) has ten gene products: HA and NA on the surface; the M2 ion channel and a variety of intraviral proteins; the M1 matrix proteins; and various RNA packaging proteins that make up the virion. HA and NA, which are on the surface of the virus, are available for attack by the immune system, but they are subject to immune pressures and antigenic drift (mutations) in their genetic material over time. HA protein has been synthesized by a variety of recombinant DNA (rDNA)-based approaches for vaccine development (Subbarao and Katz, 2004). HA or other influenza proteins can be made alone, can

TABLE 27.4 Preclinical, phase I, II, and III trials*

Virus growth conditions and formulation of vaccine	Indevelopment seasonal	Indevelopment pandemic
Egg culture inactivated	Improved products are produced periodically by most major companies	Inactivated whole virion: • CSL Biotherapies, H5N1 (phase I) • Solvay Pharmaceuticals, H5N1 (phase I) • Baxter H5N1 (phase I & II) • Berna Biotech, H9N2 (phase II) • Glaxo Smith Kline Biologicals, H9N2, H5N1 (phase I) • Biken, H5N1 with alum (phase I & II) Inactivated split virus: • Sanofi Pasteur, H5N1, H7N1 (phase I) • Glaxo Smith Kline Biologicals (GSK), H5N1 (phase I) • CSL Limited, H5N1 (phase II) • Biken , H5N1 (phase I and I & II) Inactivated subunit: • Novartis Vaccines and Diagnostics, H9N2, H5N1, H5N3 (phase I, II, & III) • Solvay Pharmaceuticals, H5N1 (phase I) Berna Biotech, H9N2, Virosome-based (phase II).
Cell culture inactivated	Nobilon International (phase I, grown in MDCK) Solvay Pharmaceuticals (phase I & II, grown in PER.C_6 line) Sanofi Pasteur (phase I, grown in MDCK) Glaxo Smith Kline (in I & II, grown in MDCK) Novartis (phase III, grown in MDCK)	Baxter, H5N1 (start phase III wild-type in Vero) Nobilon International, H5N1 (phase I, grown in MDCK) Nobilon/Vivalis H5N1, (preclinical in cell, preclinical in embryonic chicken embryo fibroblast cells, EBx® cell line) Sanofi Pasteur H5N1, H7N7, H7N1 (phase I or I & II) Solvay Pharmaceuticals, H5N1 (phase I, grown in human Per.C_6® line) GSK H5N1 (preclinical, grown in MDCK)
Recombinant or DNA vaccine	Merck & Co., Inc., M2 protein Universal Vaccine (pretrial) Protein Sciences (phase III in 2007) Novavax VLP (preclinical) Acambis (phase I), recombinant Hep-B core fusion with M2 and grown in *E. coli* VaxInnate Inc. (phase I), flagellin fused with M2 and grown in *E. coli.* Protein Sciences (phase I), FluBIØk® recombinant HA, NA and M proteins grown in caterpillar cells with baculovirus to produce the recombinant protein. *E. coli* (phase I & II)	Merck & Co., Inc., M2 protein universal (pretrial) Acambis ACAM + Flu A® possible pandemic vaccine (phase I) M2e protein universal Protein Sciences, H5N1 (phase III in 2007) Novavax H5N1, VLP (phase I) Powermed, H5N1, DNA vaccine (phase I)
Live attenuated egg or cell culture	Medimmune MDCK (cell-based) (phase I)	Medimmune, H5N1, H9N2 (currently in phase I for both egg- and cell-based)

*as of Octorber 2007.

be co-expressed with other proteins to boost immunogenicity (Prel et al., 2007; Treanor et al., 2007), or HA and NA can be expressed together with the matrix protein that underlies the membrane to make a virus-like particle (VLP) (Galarza et al., 2005a, 2005b; Ionescu et al., 2006; Pushko et al., 2007; Bright et al., 2007; Quan et al., 2007). Recombinant HA produced in a baculovirus system (insect tissue culture expression of HA followed by purification of the HA) has been successfully formulated into trivalent vaccine, and this trivalent vaccine has demonstrated safety and efficacy in clinical trials in the United States (Treanor et al., 2007).

VLPs mimic the three-dimensional structure of a virus but do not contain genetic material, so they cannot replicate or cause infection. As VLPs maintain functional properties of both influenza surface proteins (HA and NA), they have been shown to activate multiple arms of the immune system to generate a broadly protective immune response (Galarza et al., 2005a, 2005b; Ionescu et al., 2006; Pushko et al., 2007; Bright et al., 2007; Quan et al., 2007). VLPs apparently have a high propensity for being taken up by the immune system; an effective dose of a VLP vaccine in a mouse model is about one-tenth the dose necessary for HA itself (Pushko et al., 2007; Bright et al., 2007; Quan et al., 2007). The potential advantage here is the reduction in the amount of vaccine necessary to elicit an effective immune response.

Antigens other than HA are also good candidates for the production of influenza vaccines. The NA protein, although one possible candidate, is susceptible to antigenic drift and immune pressure, and because of this does not offer an advantage over HA as a surface target. NA is also not a strong neutralizing antigen and blocking the enzymatic activity with antibody may confer some benefit. Conserved influenza antigens are the NS1 protein (a nonstructural nuclear protein), M1 (the matrix protein), NP, and M2 (which is the ion channel). In the context of the influenza virus, the M2 ion channel protein is particularly important at this time for the production of a "universal vaccine." The ion channel is important in virus replication, and during the uncoating of the virus a pH change must take place before the viral genome is released for the purpose of replication (Lamb et al., 1985; Neirynck et al., 1999). The M2e peptide appears to be relatively conserved and is a relatively stable target that in theory should not change year after year (Holsinger and Lamb, 1991; Pinto et al., 1992; Black et al., 1993; Neirynck et al., 1999). Different methods of making the influenza M2e based vaccine are in development and these include: (1) hooking four copies of the 24 amino acid M2 protein to flagellin produced in *Escherichia coli*; and (2) hooking a copy of M2e to Pam3Cys (e.g., tripalmitoyl cysteine, a synthetic adjuvant linked to M2e, the antigenic protein) to make Pam3Cys.M2e (Lamb et al., 1985). The flagellin fusion is thought to be a better method, since the M2e 24 amino acid peptide alone does not illicit an immune response.

Another option being developed is to use M2e fused at the N-terminus of the hepatitis B core protein (HBc) subunit or inserted in the immunodominant loop as a vaccine (Neirynck et al., 1999). Evidence indicates that these types of "universal" influenza A vaccines are promising and a number of candidate vaccines based on this methodology are currently being tested in human clinical trials.

DNA Vaccines

DNA vaccines are expressed in situ after DNA vaccination and these vaccines induce immune responses in the form of both humoral and CMI responses, including CTL (Fynan et al., 1993a; Deck et al., 1997; Kodihalli et al., 1999, 2000; Drape et al., 2006). DNA vaccines are comprised simply of *E. coli*-derived plasmid DNA, are not infectious, do not replicate, and encode only the proteins of interest without production of bacterial proteins. The functional components of the plasmid include a strong promoter system (such as the immediate early promoter of cytomegalovirus), a convenient cloning site for insertion of a gene of interest, a polyadenylation termination sequence, a prokaryotic origin of replication for production in *E. coli*, and a selectable marker (e.g., ampicillin resistance gene) to facilitate selection of bacterial cells containing the plasmid. Some of the earliest work on DNA vaccines was conducted using influenza as a model system. These studies at least in animals including nonhuman primates, demonstrated production of hemagglutination inhibiting (HI) antibodies and CTL. The potency of DNA vaccines encoding influenza virus antigens in many of these animal studies was demonstrated by the low amounts of DNA required to induce immunity, using either IM injection of the vaccine or by use of the gene gun (Fynan et al., 1993b). Combined immunization with DNA encoding HA (which can generate neutralizing antibodies) and DNA encoding NP and M1 (which can induce broad T-cell responses) may provide a greater breadth of protection than can be obtained with conventional inactivated influenza vaccines and may be useful in humans against influenza (Fynan et al., 1993a; Kodihalli et al., 1999). The first human trial of a DNA vaccine designed to prevent H5N1 avian influenza infection began on December 21, 2006. The vaccine was delivered with a particle-mediated epidermal delivery (PMED) needle-free injection system, which delivers DNA to the epidermal layer of the skin where it enters the cells of the immune network, creating immunity and facilitating both treatment and prevention of disease (Fynan et al., 1993b).

Adjuvants

Adjuvants are agents or drugs that have few or no antigenic effects or properties but may have the potential to increase the efficacy or potency of vaccines or other drugs when given at the same time (see Chapter 9). Some of the most common adjuvants used in vaccine preparations are aluminum salts and monophosphoryl lipid (MPL)/QS21 cocktails. The response to potential pandemic flu vaccines has been shown to be much improved by the addition of MF59 (Nicholson et al., 2001; Stephenson et al., 2003; Atmar et al., 2006; Bresson et al., 2006; Baldo et al., 2007). MF59 also allows significant dose sparing, which is essential to increase the pandemic vaccine production capacity. Unfortunately, traditional adjuvants, such as Alum, do not appear to be potent for pandemic influenza vaccines. Although poorly understood scientifically, the successful development of adjuvants is important because they can greatly increase the efficacy of existing vaccine supplies. Most of the current IV and LAV virus vaccines do not contain an adjuvant. To stretch the available supply, antigen-sparing adjuvant approaches should be considered and many new approaches are under development.

Alternative Delivery of Antigens

Vaccines can be delivered in ways other than by direct IM injection. Any method can be used to introduce the vaccine directly to dendritic cells (the cells that carry antigen to the lymph nodes) located in the intradermal compartment of the skin, whence they will be effectively taken into the immune system. Thus, vaccines can be delivered through the skin using patches, microabrasion, or microtines coated with antigen. They can also be delivered through the intranasal system; Flu-Mist® administers a LAV by spray into the intranasal compartment to create local mucosal immunity.

PRECLINICAL DEVELOPMENT INCLUDING RELEVANT ANIMAL MODELS

There are a number of animal models presently available for the study of the pathogenesis of influenza. Influenza A virus will infect a number of Old World and New World primates (Kalter et al., 1969; London et al., 1972; Berendt and Hall, 1977). The gibbon and baboon develop clinical illness with nasal application of the virus, and the squirrel, cynomolgus and rhesus monkeys develop illness when the virus is inoculated intratracheally (Berendt, 1974; Berendt and Hall, 1977). Primate models have a number of disadvantages, including a lack of commercial availability of the reagents necessary to characterize the host response in detail. Mammals such as horses and pigs that are natural hosts for influenza have also been used experimentally. However, their large size and the limited availability of reagents available also preclude their use as good models in the laboratory.

Small-animal models that have been used to study influenza virus pathogenesis include the ferret, in which human influenza virus was originally isolated (Marois et al., 1971; Smith and Sweet, 1988; Herlocher et al., 2001, 2004; Yen et al., 2005; Lipatov et al., 2006; Govorkova et al., 2007; Belser et al., 2007). Adult ferrets become ill after infection with influenza A viruses, exhibiting fever, lethargy, and weight loss (Marois et al., 1971; Smith and Sweet, 1988; Herlocher et al., 2001, 2004; Yen et al., 2005; Lipatov et al., 2006; Govorkova et al., 2007; Belser et al., 2007). The ferret model has been used in recent studies of H5N1 viruses (Lipatov et al., 2006; Govorkova et al., 2007), the transmission of influenza and the development of resistance to antiviral therapy (Herlocher et al., 2001; Yen et al., 2005). Unfortunately, as is seen for other models, ferrets are outbred and again reagents are not readily available for evaluating the correlates of protective immunity. In contrast, laboratory strains of mice are inbred and there is an abundance of reagents to characterize both innate and adaptive immune responses. However, mice are not naturally infected with influenza viruses, and therefore most studies are performed using strains that have been adapted by extensive serial passage to replicate efficiently in this mammal (Hernandez et al., 2003; Evseenko et al., 2007; Joseph et al., 2007).

POSTEXPOSURE IMMUNE IMMUNOPROPHYLAXIS

Little information related to immunoprophylaxis for influenza infection is currently available. Limited trials of human monoclonal or polyclonal antibodies in animals have yielded promising results but large scale trials of passive antibody use for influenza have not yet been done (Krieg, 1999; Hanson et al., 2006). The ability of some nucleic acid-based drugs to elicit broad-spectrum antiviral immunity is of particular interest in the development of products that have the potential for prevention and treatment of influenza infection. Immunomodulating products that enhance innate and/or cellular immunity and provide some level of antiviral resistance may offer the potential of protection against a number of seasonal and avian strains of influenza viruses, regardless of genetic mutations, reassortments, recombinations, zoonotic origin, or development of drug-resistance. Nucleic acid-based drugs currently in development that can stimulate the host's immune responses against viral infections include CpG containing oligonucleotides and dsRNA, such as poly ICLC (Levy et al., 1975; Wong et al., 1995). These drugs are Toll-like receptor (TLR) agonists, and are in various stages of clinical development. Liposomal poly ICLC provided protection against both seasonal and avian influenza viruses in experimental studies in animals (Wong et al., 1999). Mounting evidence in animal studies suggest synthetic dsRNA, such as poly ICLC can elicit effective and protective antiviral immune responses to influenza infection. Antiviral immune responses induced by these drugs have been shown to confer complete protection in mice against multiple challenge doses of influenza A viruses, including influenza A/PR/8/34 (H1N1) and A/Aichi (H3N2) strains (Wong et al., 1999). The observation that liposomal poly ICLC has been shown to be effective against various subtypes of influenza A viruses (H1N1, H3N2) may well suggest it is particularly well suited to deal with constantly evolving viruses, such as the influenza A viruses (Wong et al., 1999, 2005).

PROSPECTS FOR THE FUTURE

Influenza remains a serious disease despite the availability of good antivirals as well as IV and LAV vaccines, which are effective for most recipients. Current vaccines are safe and efficacious but there are problems with the current methodologies for vaccine production which can delay manufacture of sufficient vaccine stock viruses in pandemic situations. The human devastation caused by past influenza pandemics as well as recent concerns with avian influenza are a reminder that the threat of a future pandemic is real and that new approaches for vaccine production are necessary to meet the needs of the future.

KEY ISSUES

- Current vaccines are produced by growing virus in chicken egg–embryo culture. Problems associated with adequacy of egg supply, biological safety, killing of chick embryos by virulent avian strains and appropriateness of virus stocks used for inoculation make egg–embryo culture problematic and suggest that the development of nonegg-based culture systems should be a priority for future of influenza vaccine development.
- Development of better methods of selection and production of viral antigens using reverse genetics for both seasonal and pandemic strains of virus are needed if we are to produce adequate virus for the development of vaccines which reduce morbidity and mortality and improve influenza virus protection.
- Advances in vaccine research which show great promise for the future include evaluation of pure subunit vaccines, development of new adjuvants, development of DNA vaccines, enhancing methods for mucosal immunization, and the use of immunomodulators for enhancing immune responses.
- Development of better methods of selection and production of viral antigens for both seasonal and pandemic strains of virus are needed if we are to produce adequate virus to reduce morbidity and mortality and improve influenza virus protection in the future.

References

ACIP: Advisory Committee on Immunization Practices. Prevention and control of influenza: recommendations of the (ACIP). MMWR Recomm. Rep. 1999; 48:1–28.

Adachi, M., Matsukura, S., Tokunaga, H. and Kokubu, F. Expression of cytokines on human bronchial epithelial cells induced by influenza virus A. Int. Arch. Allergy Immunol. 1997; 113:307–311.

Akashi, K., Hayashi, S., Gondo, H., Mizuno, S., Harada, M., Tamura, K. et al. Involvement of interferon-gamma and macrophage colony-stimulating factor in pathogenesis of haemophagocytic lymphohistiocytosis in adults. Br. J. Haematol. 1994; 87:243–250.

Alexander, D.J. and Brown, I.H. Recent zoonoses caused by influenza A viruses. Revue Sci. Tech. 2000; 19(1):197–225.

Alford, R.H., Kasel, J.A., Gerone, P.J. and Knight, V. Human influenza resulting from aerosol inhalation. Proc. Soc. Exp. Biol. Med. 1966; 122:800–804.

Antiviral Agents for Seasonal Influenza: Information for Health Professionals. Available at http://www.cdc.gov/flu/professionals/antivirals/index.htm.

Artenstein, M.S., Bellanti, J.A. and Buescher, E.L. Identification of the antiviral substances in nasal secretions. Proc. Soc. Exp. Biol. Med. 1964; 117:558–564.

Ashkenazi, S., Vertruyen, A., Arístegui, J., Esposito, S., McKeith, D.D., Klemola, T., Biolek, J., Kühr, J., Bujnowski, T., Desgrandchamps, D., Cheng, S.M., Skinner, J., Gruber, W.C. and Forrest, B.D.CAIV-T Study Group Superior relative efficacy of live attenuated influenza vaccine compared with inactivated influenza vaccine in young children with recurrent respiratory tract infections. Pediatr. Infect. Dis. J. 2006; 25(10):870–879.

Askonas, B.A., McMichael, A.J. and Webster, R.G. The immune response to influenza viruses and the problem of protection against infection. In: Basic and Applied Influenza Research (A.S. Beare, Ed.), pp. 159–188. Boca Raton, FL: CRC Press, 1982.

Atmar, R.L., Keitel, W.A., Patel, S.M., Katz, J.M., She, D., El Sahly, H., Pompey, J., Cate, T.R. and Couch, R.B. Safety and immunogenicity of nonadjuvanted and MF59-adjuvanted influenza A/H9N2 vaccine preparations. Clin. Infect. Dis. 2006; 43(9):1135–1142.

Avian Influenza A (H5N1). Weekly Epidemiological Record. 2004; 79(7):65–70. (Also available at Sims, L.D., Ellis, T.M., Liu, K.K., et al. Avian influenza in Hong Kong 1997–2002. Avian Dis. 2003; 47(Suppl.):832–838.

Baine, W.B., Luby, J.P. and Martin, S.M. Severe illness with influenza B. Am. J. Med. 1980; 68(2):181–189.

Baldo, V., Baldovin, T., Floreani, A., Carraro, A.M. and Trivello, R. MF59-adjuvanted influenza vaccine confers superior immunogenicity in adult subjects (18–60 years of age) with chronic diseases who are at risk of post-influenza complications. Vaccine 2007; 25(20):3955–3961.

Barker, W.H. and Mullooly, J.P. Impact of epidemic type A influenza in a defined adult population. Am. J. Epidemiol. 1980; 112:798–811.

Beigel, J.H., Farrar, J., Han, A.M., Hayden, F.G., Hyer, R., de Jong, M.D. et al. Avian influenza A (H5N1) infection in humans. N. Engl. J. Med. 2005; 353:1374–1385.

Belay, E.D., Bresee, J.S., Holman, R.C., Khan, A.S., Shahriari, A. and Schonberger, L.B. Reye's syndrome in the United States from 1981 through 1997. N. Engl. J. Med. 1999; 340:1377–1382.

Bell, T.D., Chai, H., Berlow, B. and Daniels, G. Immunization with killed influenza virus in children with chronic asthma. Chest 1978; 73:140–145.

Belser, J.A., Lu, X., Maines, T.R., Smith, C., Li, Y., Donis, R.O., Katz, J.M. and Tumpey, T.M. Pathogenesis of avian influenza (H7) virus infection in mice and ferrets: enhanced virulence of Eurasian H7N7 viruses isolated from humans. J. Virol. 2007; 81(20):11139–11147.

Belshe, R.B., Edwards, K.M., Vesikari, T., Black, S.V., Walker, R.E., Hultquist, M., Kemble, G. and Connor, E.M. CAIV-T Comparative Efficacy Study Group Live attenuated versus inactivated influenza vaccine in infants and young children. N. Engl. J. Med. 2007; 356(7):685–696, Erratum in: 356(12), p. 1283.

Belshe, R.B., Mendelman, P.M., Treanor, J. et al. The efficacy of live attenuated, cold-adapted, trivalent, intranasal influenza virus vaccine in children. N. Engl. J. Med. 1998; 338:1405–1412.

Bender, B.S. and Small, P.A., Jr. Influenza: pathogenesis and host defense. Sem. Respir. Infect. 1992; 7:38–45.

Berendt, R. Simian model for the evaluation of immunity to influenza. Infect. Immun. 1974; 9:101–105.

Berendt, R. and Hall, W. Reaction of squirrel monkeys to intratracheal inoculation with influenza/A/New Jersey/76 (Swine) virus. Infect. Immun. 1977; 16:476–479.

Biddison, W.E., Sharrow, S.O. and Shearer, G.M. T cell subpopulations required for the human cytotoxic T lymphocyte response to influenza virus: evidence for T cell help. J. Immunol. 1981; 127(2):487–491.

Black, R.A., Rota, P.A., Gorodkova, N., Klenk, H.D. and Kendal, A.P. Antibody response to the M2 protein of influenza A virus expressed in insect cells. J. Gen. Virol. 1993; 74:143–146.

Braciak, T.A., Bacon, D., Xing, Z., Torry, D.J., Graham, F.L., Schall, T.J., Richards, C.D., Croitoru, K. and Gauldie, J. Overexpression of RANTES using a recombinant adenovirus vector induces the tissue-directed recruitment of monocytes to the lung. J. Immunol. 1996; 157:5076–5084.

Bresson, J.L., Perronne, C., Launay, O., Gerdil, C. and Saville, M.W. Safety and immunogenicity of an inactivated split-virion influenza A/Vietnam/1194/2004 (H5N1) vaccine: phase I randomised trial. Lancet 2006; 367:1657–1664.

Bridges, C.B., Fukuda, K., Uyeki, T.M., Cox, N.J. and Singleton, J.A. Centers for Disease Control and Prevention, Advisory Committee on Immunization Practices. Prevention and control of influenza. Recommendations of the Advisory Committee on Immunization Practices (ACIP). MMWR Recomm. Rep. 2002; 51(RR-3):1–31.

Bridges, C.B., Winquist, A.G., Fukuda, K., Cox, N.J., Singleton, J.A. and Strikas, R.A. Advisory Committee on Immunization Practices (ACIP), Centers for Disease Control and Prevention. Prevention and control of influenza. Recommendations of the Advisory Committee on Immunization Practices (ACIP). MMWR Recomm. Rep. 2000; 49:1–38.

Bright, R.A., Carter, D.M., Daniluk, S., Toapanta, F.R., Ahmad, A., Gavrilov, V., Massare, M., Pushko, P., Mytle, N., Rowe, T., Smith, G. and Ross, T.M. Influenza virus-like particles elicit broader immune responses than whole virion inactivated influenza virus or recombinant hemagglutinin. Vaccine 2007; 25(19):3871–3878.

Buchy, P., Mardy, S., Vong, S., Toyoda, T., Aubin, J.T., Miller, M., Touch, S., Sovann, L., Dufourcq, J.B., Richner, B., Tu, P.V., Tien, N.T., Lim, W., Peiris, J.S. and Van der Werf, S. Influenza A/H5N1 virus infection in humans in Cambodia. J. Clin. Virol. 2007; 39(3):164–168, Epub 2007 May 29.

Bussfeld, D., Kaufmann, A., Meyer, R.G., Gemsa, D. and Sprenger, H. Differential mononuclear leukocyte attracting chemokine production after stimulation with active and inactivated influenza A virus. Cell. Immunol. 1998; 186:1–7.

Butt, K.M., Smith, G.J., Chen, H., Zhang, L.J., Leung, Y.H., Xu, K.M., Lim, W., Webster, R.G., Yuen, K.Y., Peiris, J.S. and Guan, Y. Human infection with an avian H9N2 influenza A virus in Hong Kong in 2003. J. Clin. Microbiol. 2005; 43(11):5760–5767.

Calvo, C., García-García, M.L., Centeno, M., Pérez-Breña, P. and Casas, I. Influenza C virus infection in children, Spain. Emerg. Infect. Dis. 2006; 12(10):1621–1622.

CDC: Centers for Disease Control and Prevention. Delayed supply of influenza vaccine and adjunct ACIP influenza vaccine recommendations for the 2000–01 influenza season. MMWR Morb. Mortal. Wkly. Rep. 2000; 49:619–622.

Chan, M.C., Cheung, C.Y., Chui, W.H., Tsao, S.W., Nicholls, J.M., Chan, Y.O. et al. Proinflammatory cytokine responses induced by influenza A (H5N1) viruses in primary human alveolar and bronchial epithelial cells. Respir. Res. 2005; 6:135.

Chen, H., Smith, G.J., Li, K.S. et al. Establishment of multiple sublineages of H5N1 influenza virus in Asia: Implications for pandemic control. Proc. Natl. Acad. Sci. U.S.A. 2006; 103:2845–2850.

Cheung, C.Y., Poon, L.L., Lau, A.S., Luk, W., Lau, Y.L., Shortridge, K.F. et al. Induction of proinflammatory cytokines in human macrophages by influenza A (H5N1) viruses: a mechanism for the unusual severity of human disease? Lancet 2002; 360:1831–1837.

Clements, M.L., Betts, R.F., Tierney, E.L. and Murphy, B.R. Serum and nasal wash antibodies associated with resistance to experimental challenge with influenza A wild-type virus. J. Clin. Microbiol. 1986; 24:157–160.

Clements, M.L. and Murphy, B.R. Development and persistence of local and systemic antibody responses in adults given live attenuated or inactivated influenza A virus vaccine. J. Clin. Microbiol. 1986; 23(1):66–72.

Clements, M.L., O'Donnell, S., Levine, M.M., Chanock, R.M. and Murphy, B.R. Dose response of A/Alaska/6/77 (H3N2) cold-adapted reassortant vaccine virus in adult volunteers: role of local antibody in resistance to infection with vaccine virus. Infect. Immun. 1983; 40:1044–1051.

Coiras, M.T., Perez-Brena, P., Garcia, M.L. and Casas, I. Simultaneous detection of influenza A, B, and C viruses, respiratory syncytial virus, and adenoviruses in clinical samples by multiplex reverse transcription nested-PCR assay. J. Med. Virol. 2003; 69:132–144.

Couceiro, J.N., Paulson, J.C. and Baum, L.G. Influenza virus strains selectively recognize sialyloligosaccharides on human respiratory epithelium: the role of the host cell in selection of hemagglutinin receptor specificity. Virus Res. 1993; 29:155–165.

Couch, R.B. Influenza, influenza virus vaccine, and human immunodeficiency virus infection. Clin. Infect. Dis. 1999; 28:548–551.

Couch, R.B. Prevention and treatment of influenza. N. Engl. J. Med. 2000; 343(24):1778–1787.

Couch, R.B. and Kasel, J.A. Immunity to influenza in man. Ann. Rev. Microbiol. 1983; 37:529–549.

Couch, R.B., Kasel, J.A., Six, H.R., Cate, T.R. and Zahradnik, J.M. Immunological reactions and resistance to infection with influenza virus. In: Molecular Virology and Epidemiology of Influenza, Beecham Symposium, London, 1983 (C. Stuart-Harris and C. Potter, Eds.), pp. 119–153. London: Academic Press, 1984.

Couch, R.B., Keitel, W.A., Cate, T.R., Quarles, J.A., Taber, L.A. and Glezen, W.P. Prevention of influenza virus infections by current inactivated influenza virus vaccines. In: Options for the Control of Influenza III (L.E. Brown, A.W. Hampson and R.B. Webster, Eds.), pp. 97–106. Amsterdam: Elsevier Science B.V., 1996.

Couch, R.B., Quarles, J.M., Cate, T.R. and Zahradnik, J.M. Clinical trials with live cold-reassortant influenza virus vaccines. In: Options for the Control of Influenza (A.P. Kendal and P.A. Patriarca, Eds.), Vol. 36 of UCLA symposia on molecular and cellular biology, pp. 223–241. New York: Alan R. Liss, 1986.

Cox, N.J. and Subbarao, K. Influenza. Lancet 1999; 354:1277–1282.

Cretescu, L., Beare, A.S. and Schild, G.C. Formation of antibody to matrix protein in experimental human influenza A virus infections. Infect. Immun. 1978; 22:322–327.

Crosby, A. America's Forgotten Pandemic. Cambridge, UK: Cambridge University Press, 1989.

Dawson, T.C., Beck, M.A., Kuziel, W.A., Henderson, F. and Maeda, F. Contrasting effects of CCR5 and CCR2 deficiency in the pulmonary inflammatory response to influenza A virus. Am. J. Pathol. 2000; 156:1951–1959.

de Bruijn, I.A., Remarque, E.J., Jol-van der Zijde, C.M., van Tol, M.J., Westendorp, R.G. and Knook, D.L. Quality and quantity of the humoral immune response in healthy elderly and young subjects after annually repeated influenza vaccination. J. Infect. Dis. 1999; 179(1):31–36.

Deck, R.R., DeWitt, C.M., Donnelly, J.J., Liu, M.A. and Ulmer, J.B. Characterization of humoral immune responses induced by an influenza hemagglutinin DNA vaccine. Vaccine 1997; 15(1):71–78.

de Jong, M.D., Simmons, C.P., Thanh, T.T., Hien, V.M., Smith, G.J., Chau, T.N. et al. Fatal outcome of human influenza A (H5N1) is associated with high viral load and hypercytokinemia. Nat. Med. 2006; 12:1203–1207.

de Jong, M.D., Tran, T.T., Truong, H.K., Vo, M.H., Smith, G.J., Nguyen, V.C., Bach, V.C., Phan, T.Q., Do, Q.H., Guan, Y., Peiris, J.S., Tran, T.H. and Farrar, J. Oseltamivir. Resistance during treatment of influenza A (H5N1) infection. N. Engl. J. Med. 2005; 353(25):2667–2672.

Doherty, P.C. Immune responses to viruses. In: Clinical Immunology, Principles and Practice (R.R. Rich, Jr., T.A. Fleisher, Jr., B.D. Schwarz, Jr., W.T. Shearer, Jr. and W. Strober, Jr., Eds.), pp. 535–549. St. Louis, MO: Mosby, 1996.

Douglas, R.G. Influenza in man. In: The Influenza Viruses and Influenza (E.D. Kilbourne, Ed.), pp. 375–447. New York: Academic Press, 1975.

Drape, R.J., Macklin, M.D., Barr, L.J. et al. Epidermal DNA vaccine for influenza is immunogenic in humans. Vaccine 2006; 24(21):4475–4481.

Evseenko, V.A., Bukin, E.K., Zaykovskaya, A.V., Sharshov, K.A., Ternovoi, V.A., Ignatyev, G.M. and Shestopalov, A.M. Experimental infection of H5N1 HPAI in BALB/c mice. Virol. J. 2007; 4:77.

Ferguson, N.M., Galvani, A.P. and Bush, R.M. Ecological and immunological determinants of influenza evolution. Nature 2003; 422(6930):428–433.

Fesq, H., Bacher, M., Nain, M. and Gemsa, D. Programmed cell death (apoptosis) in human monocytes infected by influenza A virus. Immunobiology 1994; 190:175–182.

Fiore, A.E., Shay, D.K., Haber, P., Iskander, J.K., Uyeki, T.M., Mootrey, G., Bresee, J.S. and Cox, N.J. Advisory Committee on Immunization Practices (ACIP), Centers for Disease Control and Prevention. Prevention and control of influenza. Recommendations of the Advisory Committee on Immunization Practices (ACIP). MMWR Recomm. Rep. 2007; 56(RR-6):1–54.

Fisman, D.N. Hemophagocytic syndromes and infection. Emerg. Infect. Dis. 2000; 6:601–608.

Fleischer, B., Becht, H. and Rott, R. Recognition of viral antigens by human influenza A virus-specific T lymphocyte clones. J. Immunol. 1985; 135:2800–2804.

Francis, T., Jr., Quilligan, J.J., Jr. and Minuse, E. Identification of another epidemic respiratory disease. Science 1950; 112(2913):495–497.

Friedland, J.S. Chemokines in viral disease. Res. Virol. 1996; 147:131–138.

Fujisawa, H., Tsuru, S., Taniguchi, M., Zinnaka, Y. and Nomoto, K. Protective mechanisms against pulmonary infection with influenza virus: I. Relative contribution of polymorphonuclear leukocytes and of alveolar macrophages to protection during the early phase of intranasal infection. J. Gen. Virol. 1987; 68:425–432.

Fynan, E.F., Robinson, H.L. and Webster, R.G. Use of DNA encoding influenza hemagglutinin as an avian influenza vaccine. DNA Cell Biol. 1993a; 12(9):785–789.

Fynan, E.F., Webster, R.G., Fuller, D.H., Haynes, R.J., Santoro, J.C. and Robinson, H.L. DNA vaccines: protective immunizations by parenteral, mucosal, and gene-gun inoculations. Proc. Natl. Acad. Sci. U.S.A. 1993b; 90(24):11478–11482.

Galarza, J.M., Latham, T. and Cupo, A. Virus-like particle (VLP) vaccine conferred complete protection against a lethal influenza virus challenge. Viral Immunol. 2005a; 18(1):244–251.

Galarza, J.M., Latham, T. and Cupo, A. Virus-like particle vaccine conferred complete protection against a lethal influenza virus challenge. Viral Immunol. 2005b; 18(2):365–372.

Gambaryan, A.S., Tuzikov, A.B., Bovin, N.V. et al. Differences between influenza virus receptors on target cells of duck and chicken and receptor specificity of the 1997 H5N1 chicken and human influenza viruses from Hong Kong. Avian Dis. 2003; 47 (3 Suppl.):1154–1160.

Gerhard, W. The role of the antibody response in influenza virus infection. Curr. Top. Microbiol. Immunol. 2001; 260:171.

Gerhard, W. and Mozdanowska, K. Roles of CD4+ T and B cells in influenza virus infection. In: Options for the Control of Influenza IV (A. Osterhaus, N.J. Cox, and A.W. Hampson, Eds.), pp. 311–318. New York, NY: Elsevier Science, 2001.

Gerin, J.L. and Anderson, N.G. Purification of influenza virus in the K-II zonal centrifuge. Nature 1969; 221:1255–1256.

Glezen, W.P. Emerging infections: pandemic influenza. Epidemiol. Rev. 1996; 18:64–76.

Glezen, W.P. and Couch, R.B. Influenza viruses. In: Viral Infections of Humans: Epidemiology and Control (A.S. Evans and R.A. Kaslow, Eds.), 4th ed., pp. 473–505. New York: Plenum Medical Book, 1997.

Govorkova, E.A., Ilyushina, N.A., Boltz, D.A., Douglas, A., Yilmaz, N. and Webster, R.G. Efficacy of oseltamivir therapy in ferrets inoculated with different clades of H5N1 influenza virus. Antimicrob. Agents Chemother. 2007; 51(4):1414–1424.

Govorkova, E.A., Rehg, J.E., Krauss, S. et al. Lethality to ferrets of H5N1 influenza viruses isolated from humans and poultry in 2004. J. Virol. 2004; 79:2191–2198.

Guan, Y., Peiris, J.S.M., Lipatov, A.S. et al. Emergence of multiple genotypes of H5N1 avian influenza viruses in Hong Kong SAR. Proc. Natl. Acad. Sci. U.S.A. 2002; 99:8950–8955.

Halperin, S.A., Smith, B., Mabrouk, T., Germain, M., Trepanier, P., Hassell, T. et al. Safety and immunogenicity of trivalent, inactivated, mammalian cell culture-derived influenza vaccine in healthy adults, seniors, and children. Vaccine 2002; 20:1240–1247.

Hanson, B.J., Boon, A.C., Lim, A.P., Webb, A., Ooi, E.E. and Webby, R.J. Passive immunoprophylaxis and therapy with humanized monoclonal antibody specific for influenza A H5 hemagglutinin in mice. Respir. Res. 2006; 7:126.

Hardy, C.T., Young, S.A., Webster, R.G., Naeve, C.W. and Owens, R.J. Egg fluids and cells of the chorioallantoic membrane of embryonated chicken eggs can select different variants of influenza A (H3N2) viruses. Virology 1995; 211:302–306.

Harmon, M.W., Rota, P.A., Walls, H.H. and Kendal, A.P. Antibody response in humans to influenza virus type B host-cell-derived variants after vaccination with standard (egg-derived) vaccine or natural infection. J. Clin. Microbiol. 1988; 26:333–337.

Hayden, F.G., Fritz, R., Lobo, M.C., Alvord, W., Strober, W. and Straus, S.E. Local and systemic cytokine responses during experimental human influenza A virus infection. Relation to symptom formation and host defense. J. Clin. Invest. 1998; 101:643–649.

Hayden, F.G. and Pavia, A.T. Antiviral management of seasonal and pandemic influenza. J. Infect. Dis. 2006; 194(Suppl. 2):S119–S126.

Heikkinen, T., Thint, M. and Chonmaitree, T. Prevalence of various respiratory viruses in the middle ear during acute otitis media. N. Engl. J. Med. 1999; 340:260–264.

Hemmes, J.H., Winkler, K.C. and Kool, S.M. Virus survival as a seasonal factor in influenza and poliomyelitis. Nature 1960; 188:430–431.

Hemmes, J.H., Winkler, K. and Kool, S.M. Virus survival as a seasonal factor in influenza and poliomyelitis. Antonie Van Leeuwenhoek 1962; 28:221–233.

Herlocher, M.L., Elias, S., Truscon, R., Harrison, S., Mindell, D., Simon, C. and Monto, A.S. Ferrets as a transmission model for influenza: sequence changes in HA1 of type A (H3N2) virus. J. Infect. Dis. 2001; 184(5):542–546.

Herlocher, M.L., Truscon, R., Elias, S., Yen, H.L., Roberts, N.A., Ohmit, S.E. and Monto, A.S. Influenza viruses resistant to the antiviral drug oseltamivir: transmission studies in ferrets. J. Infect. Dis. 2004; 190(9):1627–1630.

Hernandez, E., Ramisse, F., Lhonneux, A., Noury, J., Bazin, H. and Cavallo, J.D. Compared protective effect of nasal immunoprophylaxis using a new human monoclonal IgM antibody, human polyclonal antibodies, F(ab')2, amantadine, and zanamivir for prophylaxis of influenza A virus pneumonia in mice. Mil. Med. 2003; 168(3):246–251.

Hill, J. Influenza, or ferrets to the front. In: Silent Enemies: The Story of Diseases of War and their Control. G.P. Putnam and Sons, 1942, Books for Libraries Press, Freeport, NY. Reprinted 1970. pp. 127–136.

Hofmann, P., Sprenger, H., Kaufmann, A., Bender, A., Hasse, C., Nain, M. and Gemsa, D. Susceptibility of mononuclear phagocytes to influenza A virus infection and possible role in the antiviral response. J. Leukoc. Biol. 1997; 61:408–414.

Holsinger, L.J. and Lamb, R.A. Influenza virus M2 integral membrane protein is a homotetramer stabilized by formation of disulfide bonds. Virology 1991; 183:32–43.

Horimoto, T., Fukuda, N., Iwatsuki-Horimoto, K. et al. Antigenic differences between H5N1 human influenza viruses isolated in 1997 and 2003. J. Vet. Med. Sci. 2004; 66:303–305.

Hurwitz, E.S., Schonberger, L.B., Nelson, D.B. and Holman, R.C. Guillain–Barré syndrome and the 1978–1979 influenza vaccine. N. Engl. J. Med. 1981; 304:1557–1561.

Imashuku, S., Hibi, S., Fujiwara, F. and Todo, S. Hyper-interleukin (IL)-6-naemia in haemophagocytic lymphohistiocytosis. Br. J. Haematol. 1996; 93:803–807.

Ionescu, R.M., Przysiecki, C.T., Liang, X., Garsky, V.M., Fan, J., Wang, B., Troutman, R., Rippeon, Y., Flanagan, E., Shiver, J. and Shi, L. Pharmaceutical and immunological evaluation of human papillomavirus virus like particle as an antigen carrier. J. Pharm. Sci. 2006; 95(1):70–79.

Ito, T., Suzuki, Y., Takada, A., Kawamoto, A., Otsuki, K., Masuda, H., Yamada, M., Suzuki, T., Kida, H. and Kawaoka, Y. Differences in sialic acid–galactose linkages in the chicken egg amnion and allantois influence human influenza virus receptor specificity and variant selection. J. Virol. 1997; 71:3357–3362.

Ito, T., Couceiro, J.N., Kelm, S. et al. Molecular basis for the generation in pigs of influenza A viruses with pandemic potential. J. Virol. 1998; 72:7367–7373.

James, J.M., Zeiger, R.S., Lester, M.R. et al. Safe administration of influenza vaccine to patients with egg allergy. J. Pediatr. 1998; 133:624–628.

Jameson, J., Cruz, J. and Ennis, F.A. Human cytotoxic T-lymphocyte repertoire to influenza A viruses. J. Virol. 1998; 72:8682.

Jameson, J., Cruz, J., Terajima, M. and Ennis, F.A. Human CD8+ and CD4+ T lymphocyte memory to influenza A viruses of swine and avian species. J. Immunol. 1999; 162(12):7578–7583.

Jefferson, T., Rivetti, D., Rivetti, A., Rudin, M., Di Pietrantonj, C. and Demicheli, V. Efficacy and effectiveness of influenza vaccines in elderly people: a systematic review. Lancet 2005; 366(9492):1165–1174, Epub 2005 Sept. 22. Review. Erratum in: Lancet 2006 Mar. 25; 367(9515):986.

Joseph, T., McAuliffe, J., Lu, B., Jin, H., Kemble, G. and Subbarao, K. Evaluation of replication and pathogenicity of avian influenza a H7 subtype viruses in a mouse model. J. Virol. 2007; 81(19):10558–10566.

Julkunen, I., Melen, K., Nyqvist, M., Pirhonen, J., Sareneva, T. and Matikainen, S. Inflammatory responses in influenza A virus infection. Vaccine 2000; 19(Suppl. 1):S32–S37.

Julkunen, I., Sareneva, T., Pirhonen, J., Ronni, T., Melen, K. and Matikainen, S. Molecular pathogenesis of influenza A virus infection and virus-induced regulation of cytokine gene expression. Cytokine Growth Factor Rev. 2001; 12:171–180.

Jurgensen, P.F., Olsen, G.N., Johnson, J.E., Jr. et al. Immune response of the human respiratory tract. II. Cell-mediated immunity in the lower respiratory tract to tuberculin and mumps and influenza viruses. J. Infect. Dis. 1973; 128:730–735.

Kaiser, L., Couch, R.B., Galasso, G.J. et al. First International Symposium on Influenza and Other Respiratory Viruses: summary and overview: Kapalua, Maui, Hawaii, December 4–6, 1998. Antiviral Res. 1999; 42:149–175.

Kaiser, L., Fritz, R.S., Straus, S.E., Gubareva, L. and Hayden, F.G. Symptom pathogenesis during acute influenza: interleukin-6 and other cytokine responses. J. Med. Virol. 2001; 64:262–268.

Kalter, S., Heberling, R., Vice, T., Lief, F. and Rodriguez, A. Influenza (A2/Hong Kong/68) in the baboon (*Papio* sp.). Proc. Soc. Exp. Biol. Med. 1969; 132:357–361.

Kaplan, J.E., Katona, P., Hurwitz, E.S. and Schonberger, L.B. Guillain–Barré syndrome in the United States 1979–1980 and 1980–1981: lack of an association with influenza vaccination. J. Am. Med. Assoc. 1982; 248:698–700.

Katz, J.M., Lim, W., Buxton-Bridges, C. et al. Antibody responses in individuals infected with avian influenza A (H5N1) viruses and detection of anti-H5 antibody among household and social contacts. J. Infect. Dis. 1999; 180(6):1763–1770.

Katz, J.M., Naeve, C.W. and Webster, R.G. Host cell-mediated variation in H3N2 influenza viruses. Virology 1987; 156:386–395.

Katz, J.M. and Webster, R.G. Efficacy of inactivated influenza A virus (H3N2) vaccines grown in mammalian cells or embryonated eggs. J. Infect. Dis. 1989; 160:191–198.

Keitel, W.A., Cate, T.R., Couch, R.B., Huggins, L.L. and Hess, K.R. Efficacy of repeated annual immunization with inactivated influenza virus vaccines over a five year period. Vaccine 1997; 15:1114–1122.

Kilbourne, E.D. Future influenza vaccines and the use of genetic recombinants. Bull. World Health Organ. 1969; 41:643–645.

Kilbourne, E.D. Influenza pandemics in perspective. J. Am. Med. Assoc. 1977; 237:1225–1228.

Kilbourne, E.D. Influenza pandemics of the 20th century. Emerg. Infect. Dis. 2006; 12(1):9–14, www.cdc.gov/eid.

Knight, V. Airborne transmission and pulmonary deposition of respiratory viruses. In: Airborne Transmission and Airborne Infections (J.F. Hers and K.C. Winkler, Eds.), VIth International Symposium on Aerobiology, pp. 175–182. New York: Wiley, 1973.

Kobasa, D. and Kawaoka, Y. Emerging influenza viruses: past and present. Curr. Mol. Med. 2005; 5(8):791–803.

Kobasa, D., Takada, A., Shinya, K., Hatta, M., Halfmann, P., Theriault, S. et al. Enhanced virulence of influenza A viruses with the haemagglutinin of the 1918 pandemic virus. Nature 2004; 431:703–707.

Kodihalli, S., Goto, H., Kobasa, D.L., Krauss, S., Kawaoka, Y. and Webster, R.G. DNA vaccine encoding hemagglutinin provides protective immunity against H5N1 influenza virus infection in mice. J. Virol. 1999; 73(3):2094–2098.

Kodihalli, S., Kobasa, D.L. and Webster, R.G. Strategies for inducing protection against avian influenza A virus subtypes with DNA vaccines. Vaccine 2000; 18(23):2592–2599.

Koen, J.S. A practical method for field diagnosis of swine diseases. Am. J. Vet. Med. 1919; 14:468–470.

Krieg, M. Mechanisms and applications of immune stimulatory CpG oligodeoxynucleotides. Biochim. Biophys. Acta 1999; 1489:107–116.

Kroon, F.P., van Dissel, J.T., de Jong, J.C. and van Furth, R. Antibody response to influenza, tetanus and pneumococcal vaccines in HIV-seropositive individuals in relation to the number of $CD4^+$ lymphocytes. AIDS 1994; 8:469–476.

Kurtz, J., Manvell, R.J. and Banks, J. Avian influenza virus isolated from a woman with conjunctivitis. Lancet 1996; 348:901–902.

Kuziel, W.A., Morgan, S.J., Dawson, T.C., Griffin, S., Smithies, O., Ley, K. and Maeda, N. Severe reduction in leukocyte adhesion and monocyte extravasation in mice deficient in CC chemokine receptor 2. Proc. Natl. Acad. Sci. U.S.A. 1997; 94:12053–12058.

La Montagne, J.R., Noble, G.R., Quinnan, G.V. et al. Summary of clinical trials of inactivated influenza vaccine—1978. Rev. Infect. Dis. 1983; 5:723–736.

Laidlaw, P.P. Epidemic influenza: a virus disease. Lancet 1935:118–123.

Lamb, R.A. The influenza virus RNA segments and their encoded proteins. In: Genetics of Influenza Viruses (P. Palese and D.W. Kingsbury, Eds.), pp. 26–69. Vienna: Springer, 1983.

Lamb, R.A. and Krug, R.M. Orthomyxoviridae: the viruses and their replication. In: Fields Virology (D.M. Knipe and P.M. Howley, Eds.), 4th ed., pp. 1487–1531. Philadelphia, PA: Lippincott, Williams and Wilkins, 2001.

Lamb, R.A., Zebedee, S.L. and Richardson, C.D. Influenza virus M2 protein is an integral membrane protein expressed on the infected-cell surface. Cell 1985; 40:627–633.

Lehmann, C., Sprenger, H., Nain, M., Bacher, M. and Gemsa, D. Infection of macrophages by influenza A virus: characteristics of tumour necrosis factor-α (TNF-α) gene expression. Res. Virol. 1996; 147:123–130.

Levy, H.B., Baer, G., Baron, S., Buckler, C.E., Gibbs, C.J., Iadorola, M.J. et al. A modified polyriboinosinic-polyribocytidylic acid complex that induces interferon in monkeys. J. Infect. Dis. 1975; 132:434–439.

Lewis, E.N., Griffin, M.R., Szilagyi, P.G., Zhu, Y., Edwards, K.M. and Poehling, K.A. Childhood influenza: number needed to vaccinate to prevent 1 hospitalization or outpatient visit. Pediatrics 2007; 120(3):467–472.

Li, K.S., Guan, Y., Wang, J. et al. Genesis of a highly pathogenic and potentially pandemic H5N1 influenza virus in eastern Asia. Nature 2004; 430:209–213.

Lipatov, A.S., Hoffmann, E., Salomon, R., Yen, H.L. and Webster, R.G. Cross-protectiveness and immunogenicity of influenza A/Duck/Singapore/3/97(H5) vaccines against infection with A/Vietnam/1203/04(H5N1) virus in ferrets. J. Infect. Dis. 2006; 194(8):1040–1043.

London, W.T., Alter, H.J., Lander, J. and Purcell, R.H. Serial transmission in rhesus monkeys of an agent related to hepatitis-associated antigen. J. Infect. Dis. 1972; 125:382–389.

Loosli, C., Lemon, H., Robertson, O. and Appel, E. Experimental airborne influenza infection. 1. Influence of humidity on survival of virus in air. Proc. Soc. Exp. Biol. Med. 1943; 53:205–206.

Lowen, A.C., Mubareka, S., Steel, J., Palese, P. and Plo, S. Influenza virus transmission is dependent on relative humidity and temperature. Pathogens 2007; 3(10):1470–1476, e151, doi: 10.1371/journal.ppat.0030151.

Maassab, H.F. and DeBorde, D.C. Development and characterization of cold-adapted viruses for use as live virus vaccines. Vaccine 1985; 3:355–369.

Manzoli, L., Schioppa, F., Boccia, A. and Villari, P. The efficacy of influenza vaccine for healthy children: a meta-analysis evaluating potential sources of variation in efficacy estimates including study quality. Pediatr. Infect. Dis. J. 2007; 26(2):97–106.

Margolis, K.L., Nichol, K.L., Poland, G.A. and Pluhar, R.E. Frequency of adverse reactions to influenza vaccine in the elderly: a randomized, placebo-controlled trial. J. Am. Med. Assoc. 1990; 264:1139–1141.

Marois, P., Boudreault, A., DiFranco, E. and Pavilanis, V. Response of ferrets and monkeys to intranasal infection with human, equine and avian influenza viruses. Can. J. Comp. Med. 1971; 35:71–76.

Matrosovich, M.N., Matrosovich, T.Y., Gray, T. et al. Human and avian influenza (AI) viruses target different cell types in cultures of human airway epithelium. Proc. Natl. Acad. Sci. U.S.A. 2004; 101:4620–4624.

Matsukura, S., Kokubu, F., Kubo, H., Tomita, T., Tokunaga, H., Kadokura, M., Yamamoto, T., Kuroiwa, Y., Ohno, T., Suzaki, H. and Adachi, M. Expression of RANTES by normal airway epithelial cells after influenza virus A infection. Am. J. Respir. Cell Mol. Biol. 1998; 18:255–264.

Maus, U., von Grote, K., Kuziel, W.A., Mack, M., Miller, E.J., Cihak, J., Stangassinger, M., Maus, R., Schlondorff, D., Seeger, W. and Lohmeyer, J. The role of CC chemokine receptor 2 in alveolar monocyte and neutrophil immigration in intact mice. Am. J. Respir. Crit. Care Med. 2002; 166:268–273.

Mazanec, M.B., Coudret, C.L. and Fletcher, D.R. Intracellular neutralization of influenza virus by immunoglobulin A anti-hemagglutinin monoclonal antibodies. J. Virol. 1995; 69:1339–1344.

McMichael, A.J., Gotch, F.M., Noble, G.R. and Beare, P.A. Cytotoxic T-cell immunity to influenza. N. Engl. J. Med. 1983; 309:13–17.

McMichael, A.J., Michie, C.A., Gotch, F.M., Smith, G.L. and Moss, B. Recognition of influenza A virus nucleoprotein by human cytotoxic T lymphocytes. J. Gen. Virol. 1986; 67(Pt 4):719–726.

Meerschaert, J. and Furie, M.B. The adhesion molecules used by monocytes for migration across endothelium include CD11a/CD18: CD11b/CD18, and VLA-4 on monocytes and ICAM-1, VCAM-1, and other ligands on endothelium. J. Immunol. 1995; 154:4099–4112.

Minuse, E., Quilligan, J.J., Jr. and Fransis, T., Jr. Type C influenza virus. I. Studies of the virus and its distribution. J. Lab. Clin. Med. 1954; 43(1):31–42.

Monto, A.S. and Kioumehr, F. The Tecumseh Study of respiratory illness. IX. Occurrence of influenza in the community, 1966–1971. Am. J. Epidemiol. 1975; 102:553–563.

Monto, A.S., Koopman, J.S. and Longini, I.M., Jr. Tecumseh Study of illness. XII. Influenza infection and disease, 1976–1981. Am. J. Epidemiol. 1985; 121:811–822.

Moriuchi, H., Katsushima, N., Nishimura, H., Nakamura, K. and Numazaki, Y. Community-acquired influenza C virus infection in children. J. Pediatr. 1991; 118:235–238.

Mou, S.S., Nakagawa, T.A., Riemer, E.C., McLean, T.W., Hines, M.H. and Shetty, A.K. Hemophagocytic lymphohistiocytosis complicating influenza A infection. Pediatrics 2006; 118:e216–e219.

Munoz, F.M., Demmler, G.J. and Glezen, W.P. Influenza in children in Houston, Texas. Pediatr. Res. 2002; 47:272A.

Murphy, B.R. and Clements, M.L. The systemic and mucosal immune response of humans to influenza A virus. In: Current Topics in Microbiology and Immunology (J. Mestecky and J.R. McGhee, Eds.), pp. 107–116. New York, NY: Springer Verlag, 1989.

Murphy, B.R., Chalhub, E.G., Nusinoff, S.R., Kasel, J. and Chanock, R.M. Temperature-sensitive mutants of influenza virus. 3. Further characterization of the ts-1(E) influenza a recombinant (H3N2) virus in man. J. Infect. Dis. 1973; 128:479–487.

Neirynck, S., Deroo, T., Saelens, X., Vanlandschoot, P., Min Jou, W. and Fiers, W. A universal influenza A vaccine based on the extracellular domain of the M2 protein. Nat. Med. 1999; 5:1119–1120.

Neuzil, K.M., Mellen, B.G., Wright, P.F., Mitchel, E.F., Jr. and Griffin, M.R. The effect of influenza on hospitalisations, outpatient visits, and courses of antibiotics in children. N. Engl. J. Med. 2000; 342:225–231.

Neuzil, K.M., Reed, G.W., Mitchel, E.F., Simonsen, L. and Griffin, M.R. Impact of influenza on acute cardiopulmonary hospitalizations in pregnant women. Am. J. Epidemiol. 1998; 148:1094–1102.

Neuzil, K.M., Zhu, Y., Griffin, M.R. et al. Burden of interpandemic influenza in children younger than 5 years: a 25-year prospective study. J. Infect. Dis. 2002; 185(2):147–152.

Nguyen, H.H., van Ginkel, F.W., Vu, H.L., McGhee, J.R. and Mestecky, J. Mechanism of heterosubtypic immunity to influenza A virus infection. In: Options for the Control of Influenza IV (A. Osterhaus, N.J. Cox and A.W. Hampson, Eds.), pp. 333–340. New York, NY: Elsevier Science, 2001.

Nicas, M., Nazaroff, W.W. and Hubbard, A. Toward understanding the risk of secondary airborne infection: emission of respirable pathogens. J. Occup. Environ. Hyg. 2005; 2:143–154.

Nichol, K.L., Lind, A., Margolis, K.L., Murdoch, M., McFadden, R., Hauge, M. et al. The effectiveness of vaccination against influenza in healthy, working adults. N. Engl. J. Med. 1995; 333:889–893.

Nichol, K.L., Margolis, K.L., Lind, A. et al. Side effects associated with influenza vaccination in healthy working adults: a randomized, placebo-controlled trial. Arch. Intern. Med. 1996; 156:1546–1550.

Nicholson, K.G., Stephenson, I., Glück, R., Mischler, R., Newman, R.M., Palache, A.M., Verlander, N.Q., Warburton, F., Wood, J.M. and Zambon, M.C. Safety and antigenicity of non-adjuvanted and MF59-adjuvanted influenza A/Duck/Singapore/97 (H5N3) vaccine: a randomised trial of two potential vaccines against H5N1 influenza. Lancet 2001; 357:1937–1943.

Palese, P. Making better influenza virus vaccines? Emerg. Infect. Dis. 2006; 12(1):61–65.

Parkman, P.D., Hopps, H.E., Rastogi, S.C. and Meyer, H.M., Jr. Summary of clinical trials of influenza virus vaccines in adults. J. Infect. Dis. 1977; 136(Suppl.):S722–S730.

Patriarca, P.A. Use of cell lines for the production of influenza virus vaccines: an appraisal of technical, manufacturing, and regulatory considerations. World Health Organization, 2007 (http://www.who.int/vacine_research/diseases/influenza/WHO_FluCell_Substrate_Version8.pdg). Updated 10 April 2007, accessed 12 Oct. 2007.

Peiris, J.S., Yu, W.C., Leung, C.W., Cheung, C.Y., Ng, W.F., Nicholls, J.M. et al. Re-emergence of fatal human influenza A subtype H5N1 disease. Lancet 2004; 363:617–619.

Peiris, M., Yuen, K.Y., Leung, C.W. et al. Human infection with influenza H9N2. Lancet 1999; 354:916–917.

Perkins, L.E. and Swayne, D.E. Pathogenicity of a Hong Kong-origin H5N1 highly pathogenic avian influenza virus for emus, geese, ducks, and pigeons. Avian Dis. 2002; 46:53–63.

Piedra, P.A. and Glezen, W.P. Influenza in children: epidemiology, immunity and vaccines. Semin. Pediatr. Infect. Dis. 1991; 2:140–146.

Pinto, L.H., Holsinger, L.J. and Lamb, R.A. Influenza virus M2 protein has ion channel activity. Cell 1992; 69:517–528.

Potter, C.W. and Oxford, J.S. Determinants of immunity to influenza infection in man. Br. Med. Bull. 1979; 35:69–75.

Prel, A., Le Gall-Reculé, G., Cherbonnel, M., Grasland, B., Amelot, M. and Jestin, V. Assessment of the protection afforded by triple baculovirus recombinant coexpressing H5, N3, M1 proteins against a homologous H5N3 low-pathogenicity avian influenza virus challenge in Muscovy ducks. Avian Dis. 2007; 51 (1 Suppl.):484–489.

Pushko, P., Tumpey, T.M., Van Hoeven, N., Belser, J.A., Robinson, R., Nathan, M., Smith, G., Wright, D.C. and Bright, R.A. Evaluation of influenza virus-like particles and Novasome adjuvant as candidate vaccine for avian influenza. Vaccine 2007; 25(21):4283–4290, Epub 2007 Mar. 9.

Puzelli, S., di Trani, L., Fabiani, C. et al. Serological analysis of serum samples from humans exposed to avian H7 influenza viruses in Italy between 1999 and 2003. J. Infect. Dis. 2005; 192:1318–1322.

Quan, F.S., Huang, C., Compans, R.W. and Kang, S.M. Virus-like particle vaccine induces protective immunity against homologous and heterologous strains of influenza virus. J. Virol. 2007; 81(7):3514–3524, Epub 2007 Jan. 24.

Quinnan, G.V., Ennis, F.A., Tuazon, C.U. et al. Cytotoxic lymphocytes and antibody-dependent complement-mediated cytotoxicity induced by administration of influenza vaccine. Infect. Immun. 1980; 30:362–369.

Reid, A.H., Taubenberger, J.K. and Fanning, T.G. The 1918 Spanish influenza: integrating history and biology. Microbes Infect. 2001; 3:81–87.

Reisinger, K.S., Block, S.L., Bugarini, R. and Holmes, S. Safety, tolerability and immunogenicity of a Madin–Darby canine kidney cell-based influenza vaccine compared to Fluvirin®: a phase II study. Abstract presented at Options for the Control of Influenza VI, Toronto, ON, Canada, 2007; no. 2007; P717, 17–23.

Robertson, J.S. Clinical influenza virus and the embryonated hen's egg. Rev. Med. Virol. 1993; 3:97–106.

Robertson, J.S., Naeve, C.W., Webster, R.G., Bootman, J.S., Newman, R. and Schild, G.C. Alterations in the hemagglutinin associated with adaptation of influenza B virus to growth in eggs. Virology 1985; 143:166–174.

Rossen, R.D., Butler, W.T., Waldman, R.H. et al. The proteins in nasal secretion. II. A longitudinal study of IgA and neutralizing antibody levels in nasal washings from men infected with influenza virus. J. Am. Med. Assoc. 1970; 211:1157–1161.

Ruben, F.L., Bachmayer, H. and Fireman, P. Parameters of humoral and cellular immunity to influenza A/USSR/90/77(H1N1): virus in various age groups. J. Infect. Dis. 1980; 142:699–703.

Saravolac, E.G., Sabuda, D., Crist, C., Blasetti, K., Schnell, G., Yang, H., Kende, M., Levy, H.B. and Wong, J.P. Immunoprophylactic strategies against respiratory influenza virus infection. Vaccine 2001; 19(17–19):2227–2232.

Schonberger, L.B., Bregman, D.J., Sullivan-Bolyai, J.Z. et al. Guillain–Barré syndrome following vaccination in the National Influenza Immunization Program, United States, 1976–1977. Am. J. Epidemiol. 1979; 110:105–123.

Seo, S.H., Hoffman, E. and Webster, R.G. Lethal H5N1 influenza viruses escape host anti-viral cytokine responses. Nat. Med. 2002; 8:950–954.

Shinya, K., Hamm, S., Hatta, M., Ito, H., Ito, T. and Kawaoka, Y. PB2 amino acid at position 627 affects replicative efficiency, but not cell tropism, of Hong Kong H5N1 influenza A viruses in mice. Virology 2004; 320:258–266.

Shope, R.E. The incidence of neutralizing antibodies for swine influenza in the sera of human beings of different ages. J. Exp. Med. 1936; 63:669–684.

Shope, R.E. and Lewis, P.A. Swine influenza. J. Exp. Med. 1931; 54:349–359.

Simonsen, L., Clarke, M.J., Schonberger, L.B. et al. Pandemic versus epidemic influenza mortality: a pattern of changing age distribution. J. Infect. Dis. 1998a; 178:53–60.

Simonsen, L., Conn, L.A., Pinner, R.W. and Teutsch, S. Trends in infectious disease hospitalizations in the United States, 1980–1994. Arch. Intern. Med. 1998b; 158:1923–1928.

Sims, L.D., Ellis, T.M., Liu, K.K. et al. Avian influenza in Hong Kong 1997–2002. Avian Dis. 2003; 47:832–838.

Smith, H. and Sweet, C. Lessons for human influenza from pathogenicity studies with ferrets. Rev. Infect. Dis. 1988; 10(1):56–75.

Sprenger, H., Meyer, R.G., Kaufmann, A., Bussfeld, D., Rischkowsky, E. and Gemsa, D. Selective induction of monocyte and not neutrophil-attracting chemokines after influenza A virus infection. J. Exp. Med. 1996; 184:1191–1196.

Stephenson, I., Nicholson, K.G., Colegate, A., Podda, A., Wood, J., Ypma, E. and Zambon, M. Boosting immunity to influenza H5N1 with MF59-adjuvanted H5N3 A/Duck/Singapore/97 vaccine in a primed human population. Vaccine 2003; 21:1687–1693.

Sturm-Ramirez, K.M., Ellis, T., Bousfield, B. et al. Reemerging H5N1 influenza viruses in Hong Kong in 2002 are highly pathogenic to ducks. J. Virol. 2004; 78:4892–4901.

Subbarao, K. and Katz, J.M. Influenza vaccines generated by reverse genetics. Curr. Top. Microbiol. Immunol. 2004; 283:313–342.

Sugaya, K., Nerome, M., Ishida, M., Matsumoto, K., Mitamura, M. and Nirasawa, M. Efficacy of inactivated influenza vaccine in preventing antigenically drifted influenza type A and well-matched type B. J. Am. Med. Assoc. 1994; 272:1122–1126.

Taubenberger, J.K., Reid, A.H., Fanning, T.G. and Janczewski, T.A. Integrating historical, clinical and molecular genetic data in order to explain the origin and virulence of the 1918 Spanish influenza virus. Philos. Trans. R. Soc. Lond. B Biol. Sci. 2001; 356:1829–1839.

Taubenberger, J.K., Reid, A.H., Krafft, A.E., Bijwaard, K.E. and Fanning, T.G. Initial genetic characterization of the 1918 "Spanish" influenza virus. Science 1997; 275:1793–1796.

The World Health Organization Global Influenza Program Surveillance Network. Evolution of H5N1 avian influenza viruses in Asia. Emerg. Infect. Dis. 2005; 11:1515–1522.

To, K.F., Chan, P.K., Chan, K.F., Lee, W.K., Lam, W.Y., Wong, K.F. et al. Pathology of fatal human infection associated with avian influenza A H5N1 virus. J. Med. Virol. 2001; 63:242–246.

Treanor, J.J., Schiff, G.M., Hayden, F.G. et al. Safety and immunogenicity of a baculovirus-expressed hemagglutinin influenza vaccine. J. Am. Med. Assoc. 2007; 297(14):1577–1582.

Tumpey, T.M., Garcia-Sastre, A., Taubenberger, J.K., Palese, P., Swayne, D.E., Pantin-Jackwood, M.J. et al. Pathogenicity of influenza viruses with genes from the 1918 pandemic virus: functional roles of alveolar macrophages and neutrophils in limiting virus replication and mortality in mice. J. Virol. 2005; 79:14933–14944.

Tweed, S.A., Skowronski, D.M., David, S.T. et al. Human illness from avian influenza H7N3, British Columbia. Emerg. Infect. Dis. 2004; 10:2196–2199.

Tyler, K.L. and Nathanson, N. Pathogenesis of viral infections. In: Fields Virology (D.M. Knipe and P.M. Howley, Eds.), 4th ed., pp. 199–2431. Philadelphia, PA: Lippincott, Williams and Wilkins, 2001.

Van Reeth, K. Cytokines in the pathogenesis of influenza. Vet. Microbiol. 2000; 74:104–109.

Virelizier, J.L., Allison, A.C. and Schild, G.C. Immune responses to influenza virus in the mouse, and their role in control of the infection. Br. Med. Bull. 1979; 35:65–68.

Waldman, R.H. and Ganguly, R. Duration of circulating and secretory antibody and cell-mediated immunity following immunization. Dev. Biol. Stand. 1975; 28:273–284.

Wang, Y.H., Huang, Y.C., Chang, L.Y., Kao, H.T., Lin, P.Y., Huang, C.G. and Lin, T.Y. Clinical characteristics of children with influenza A virus infection requiring hospitalization. J. Microbiol. Immunol. Infect. 2003; 36(2):111–116.

Webby, R.J., Webster, R.G. and Richt, J.A. Influenza viruses in animal wildlife populations. Curr. Top. Microbiol. Immunol. 2007; 315:67–83.

Webster, R.G., Geraci, J., Petursson, G. et al. Conjunctivitis in human beings caused by influenza A virus of seals. N. Engl. J. Med. 1981; 304:911.

Webster, R.G., Peiris, M., Chen, H. and Guan, Y. H5N1 outbreaks and enzootic influenza. Emerg. Infect. Dis. 2006; 12(1):3–8.

WHO. Epidemic and Pandemic Alert and Response (EPR): Situation Updates—Avian Influenza, 2008. http://www.who.int/csr/disease/avian_influenza/updates/en/index.html, accessed on 13 Aug, 2008.

Wong, J.P., Nagata, L.P., Christopher, M.E., Salazar, A.M. and Dale, R.M. Prophylaxis of acute respiratory virus infection using nucleic acid-based drugs. Vaccine 2005; 23:2266–2268.

Wong, J.P., Saravolac, E.G., Sabuda, D., Levy, H.B. and Kende, M. Prophylactic and herapeutic efficacies of poly(IC.LC) against respiratory influenza A virus infection in mice. Antimicrob. Agents Chemother. 1995; 39:2574–2576.

Wong, J.P., Yang, H., Nagata, L., Kende, M., Levy, H., Schnell, G. et al. Liposome-mediated immunotherapy against respiratory influenza virus infection using double stranded RNA poly ICLC. Vaccine 1999; 17:1788–1795.

Wong, S.S. and Yuen, K.Y. Avian influenza virus infections in humans. Chest 2006; 129(1):156–168.

Wright, P.F., Bryant, J.D. and Karzon, D.T. Comparison of influenza B/Hong Kong virus infections among infants, children, and young adults. J. Infect. Dis. 1980; 141(4):430–435.

Wright, P.F., Thompson, J., Vaughn, W.K., Folland, D.S., Sell, S.H. and Karzon, D.T. Trials of influenza A/New Jersey/76 virus vaccine in normal children: an overview of age-related antigenicity and reactogenicity. J. Infect. Dis. 1977; 136(Suppl.):S731–S741.

Wright, P.F. and Webster, R. Orthomyxoviruses. In: Fields Virology (D.M. Knipe and P.M. Howley, Eds.), 4th ed., pp. 1533–1568. Philadelphia, PA: Lippincott, Williams and Wilkins, 2001.

Yamanaka, H., Teruya, K., Tanaka, M., Kikuchi, Y., Takahashi, T., Kimura, S. and Oka, S. HIV/Influenza Vaccine Study Team. Efficacy and immunologic responses to influenza vaccine in HIV-1-infected patients. J. Acquir. Immune Defic. Syndr. 2005; 39(2):167–173.

Yen, H.L., Herlocher, L.M., Hoffmann, E., Matrosovich, M.N., Monto, A.S., Webster, R.G. and Govorkova, E.A. Neuraminidase inhibitor-resistant influenza viruses may differ substantially in fitness and transmissibility. Antimicrob. Agents Chemother. 2005; 49(10):4075–4084.

Yewdell, J.W. and Hackett, C.J. The specificity and function of T lymphocytes induced by influenza A viruses. In: The Influenza Viruses (R. Krug, Ed.). New York, NY: Plenum Press, 1989.

Yuen, K.Y., Chan, P.K.S., Peiris, M. et al. Clinical features and rapid viral diagnosis of human disease associated with avian influenza A H5N1 virus. Lancet 1998; 351:467–471.

Zhou, N., He, S., Zhang, T., Zou, W., Shu, L., Sharp, G.B. and Webster, R.G. Influenza infection in humans and pigs in southeastern China. Arch. Virol. 1996; 141(3–4):649–661.

Zitzow, L.A., Rowe, T., Morken, T., Shieh, W.J., Zaki, S. and Katz, J.M. Pathogenesis of avian influenza A (H5N1) viruses in ferrets. J. Virol. 2002; 76:4420–4429.

CHAPTER

28

Japanese Encephalitis

Ichiro Kurane

Department of Virology, National Institute of Infectious Diseases, Tokyo, Japan

OUTLINE

ABSTRACT

Japanese encephalitis (JE) is the most important viral encephalitis in Asia. JE occurs in annual epidemics or endemically in many Asian countries. Nearly 3 billion people, approximately 60% of the world's population, are at risk from JE virus infection. JE virus is maintained in nature between vector mosquitoes and vertebrate animals, especially pigs. Pigs also play a role as an amplifier for JE virus. Humans are infected by the bite of infected mosquitoes. Neutralizing antibody provides the best evidence that protective immunity has been established, and the biological assay of neutralization shows correlation with protection. Vaccination is the most effective measure of prevention of JE in humans. Vaccine coverage must be maintained indefinitely in all individuals who are potentially exposed to the virus. However, the vaccines are not available to all the populations at risk in the world.

Four types of JE vaccines are currently used in the world: three types of inactivated vaccines and one type of live-attenuated vaccine. Mouse brain-derived inactivated JE vaccine is produced from infected mouse brains by inactivation and extensive purification of the virus. This type of vaccine is produced with Nakayama or Beijing-1 strain, and that produced with Nakayama strain has been used as the international JE vaccine. Primary hamster kidney (PHK) cell-derived inactivated JE vaccine and Vero cell-derived inactivated JE vaccine have been used in China. Live-attenuated JE vaccine derived from PHK cells also have been used in China and some other countries. Furthermore, additional Vero cell-derived inactivated vaccines and chimeric vaccines are under development, and phase II and III clinical trials are being undertaken in humans. Multiple types of JE vaccines will be available in the near future. The governments of the countries where JE is endemic or epidemic will need to determine which types of vaccine will be used in respective countries.

INTRODUCTION

Japanese encephalitis (JE) virus is the most important cause of viral encephalitis in Asia (Monath and Tsai, 2002). Nearly 3 billion people, approximately 60% of the world's population, are at risk from JE virus infection. In the last 20 years, JE virus transmission has intensified in certain countries. Further, the disease has extended its geographical range to previously unaffected areas of Asia and northern Australia. JE is a vaccine-preventable disease. All the populations at risk of JE virus infection in JE-endemic or -epidemic countries should be immunized with JE vaccine.

JE VIRUS

Genome and Structure

JE virus belongs to the family Flaviviridae (genus *Flavivirus*). The genome of JE virus is a single and plus-stranded RNA, nearly 11 kb in length (Sumiyoshi et al., 1987). The JE virus genome encodes an open reading frame flanked by about 100- and 600-nucleotideuntranslated regions at the 5′- and 3′-ends, respectively. The uninterrupted open reading frame codes for three structural proteins—the capsid (C), preM [which is the precursor to the membrane (M) protein], and the envelope (E) protein—and seven nonstructural proteins: NS1, NS2A, NS2B, NS3, NS4A, NS4B, and NS5 (Kimura-Kuroda et al., 1990). JE virion has a spherical shape about 40–50 nm in diameter. The nucleocapsid is about 30 nm in diameter and covered by the envelope. The envelope contains the E and M proteins. The E protein is glycosylated and responsible for attachment to still unidentified cellular receptors and fusion with cell membrane. The E protein contains the main epitopes recognized by neutralizing antibodies. It is believed that the E protein plays the most important role in induction of protective immunity (Kimura-Kuroda et al., 1990; Konishi et al., 1999). JE virus is grouped into the JE-serocomplex group in flaviviruses. JE virus demonstrates a high level of cross-reactivity with West Nile virus, St. Louis encephalitis virus, and Murray Valley encephalitis virus.

Ecology and Transmission

JE virus is maintained in nature between vector mosquitoes and vertebrate animals, especially pigs (Monath and Tsai, 2002). The principal vector of JE virus is culex mosquitoes, mainly *Culex tritaeniorhynchus* in Asia (Rosen, 1986). Wild birds probably play a role in the maintenance of JE virus in nature and in carrying the virus to new regions.

Humans are infected by the bite of infected mosquitoes. Pigs play a role as an amplifier for JE virus. Infectivity titers of JE virus reach high levels in pigs. Many mosquitoes are infected with the virus by biting infected pigs. The infected mosquitoes then transmit JE virus to humans. However, humans do not transmit virus to biting mosquitoes because of low titers of viremia; thus, humans are considered to be dead-end hosts.

JAPANESE ENCEPHALITIS

Clinical Symptoms

JE virus infection can be asymptomatic or symptomatic, demonstrating clinical illness of encephalitis, meningitis, or febrile illness in humans. One to three in 1000 infected humans develop clinical illness. JE is an inflammatory disease of the central nervous system, which consists of cerebrum, cerebellum, and spinal cord.

The incubation period of JE is 5–15 days. The illness starts with fever and headache. Other early symptoms include vomiting, diarrhea, and muscle aches. Meningeal irritation becomes apparent on the second day, and then cerebral symptoms, including altered consciousness ranging from apathy to coma, rapidly develop. The case fatality rate is about 20–30%. Approximately 30–50% of the surviving patients have neuropsychiatric sequelae. Complete recovery occurs in only one-third of the patients.

Pathogenesis

The pathogenesis of JE is not completely understood. JE virus is transmitted to humans by mosquito bite. It first multiplies in the local lymph nodes or other organs and then enters blood vessels, causing viremia. The virus reaches neuronal cells in the central nervous system by crossing the blood-brain barrier. It is generally believed that in most JE virus infections, the virus does not enter the central nervous system, which is protected by the blood-brain barrier, but results in inapparent infection.

Laboratory Tests

Clinical diagnosis is a prerequisite for early and proper treatment. Laboratory tests are essential for confirmatory diagnosis of JE. Virus isolation from blood or cerebrospinal fluid of patients can be successful at a very early stage of the illness but is usually very difficult. Recently, the IgM-capture ELISA has been widely used for detection of specific IgM (Martin et al., 2000). Hemagglutination inhibition (HI) test is still used for serodiagnosis in many laboratories.

EPIDEMIOLOGY

Global Distribution of JE

Approximately 20,000 JE cases with 6000 deaths are reported annually in the world. The actual number, however, is estimated to be a few-fold greater. A surveillance standard has been published, but implementation of a surveillance system specifically for JE is incomplete in many Asian countries, as encephalitis is not differentiated by etiology.

JE occurs in annual epidemics or endemically in many Asian countries, including Japan, Korea, Taiwan, China, Vietnam, Thailand, Malaysia, Myanmar, India, Nepal, and Sri Lanka (Vaughn and Hoke, 1992; Basnyat et al., 2001). The endemic regions are surrounded by Japan, China, and Korea in the north, Indonesia and Australia in the south, India and Pakistan in the west, and the Philippines and Guam in the east (Fig. 28.1). Nearly 3 billion people, approximately 60% of the world's population, are believed to be at risk of JE virus infection. The first JE patient in Papua New Guinea was reported in 1995 and then again in 1997 (Mackenzie, 1999; Hanson et al., 2004; Rubin and Baird, 2004). In Australia, three JE patients were first reported in Badu Island in the Torres Strait

FIGURE 28.1 The areas where Japanese encephalitis is naturally found.

in 1995 (Hanna et al., 1996). In 1998, patients were again reported in Badu Island and, for the first time, in Cape York Peninsula in the Australian Subcontinent (Hanna et al., 1999). Thus, JE is no more a disease restricted to Asia only but is also found in Oceania.

Seasonal Prevalence

JE virus transmission season differs among endemic regions. It is mainly due to numbers and activities of vector mosquitoes. The climate, especially temperature and rainfall, affects the number and activity of vector mosquitoes. The changes in agricultural procedures, including rice cultivation, also affect the mosquito density and eventually the size and frequency of the transmission cycle of JE virus. Levels of immunity among pig population also influence the transmission cycle.

There is a year-round transmission of JE virus in tropical countries (Monath and Tsai, 2002). In the Southeast Asian countries, JE virus transmission mainly occurs between May and December and throughout the year in limited sites. However, the transmission pattern in the temperate zone (e.g., Japan, Korea, and northern China) is rather well defined (Monath and Tsai, 2002). The epidemic season in these areas usually starts in May or June and ends in September or October.

CONTROL OF JE

Protective Immunity against JE Virus

Protective immunity against JE virus has been long studied using mouse models. It is accepted that neutralizing antibodies play a key role in the protection. Passive transfer of neutralizing antibodies before virus inoculation protected mice from lethal challenge (Oya, 1988; Kimura-Kuroda and Yasui, 1988). Thus, preexisting neutralizing antibodies are important for protection of the infection. There was a linear relationship between the levels of neutralizing antibodies and the titers of challenge viruses that were resisted by mice. It was also demonstrated that when antibody was passively transferred after infection, partial protection was observed. Although complement-mediated cell lysis and antibody-dependent cell-mediated cytotoxicity were attributed to the protective activity, the mechanism of protection by anti-NS1 antibodies is not completely understood. The role of cytotoxic T cells (CTLs) in the recovery from JE virus infection has not been well established. However, it has been reported that passive transfer of virus-specific CTLs also protects mice from lethal challenge by JE virus (Murali-Krishna et al., 1998).

Control of JE in Nature

Considering the transmission cycle of JE virus to humans, there are several ways that are theoretically effective for preventing JE virus infection of humans. The control of mosquitoes and the vaccination of pigs are effective in certain circumstances; however, these measures are not practical as a means of preventing human illness. At this point, vaccination of humans is the most effective measure of prevention of JE in humans. This has been demonstrated epidemiologically in Japan, Korean and Taiwan in East Asia. Humans are incidental hosts; thus, coverage must be maintained indefinitely in all the people who are potentially exposed to the virus.

JE VACCINES

Effect of JE Vaccine on JE Epidemics

The total number of reported JE cases has been recently decreased, according to a report of the World Health Organization (WHO). China accounted for the largest number of JE cases in the world. In the last few decades, however, China achieved a significant reduction in the number of JE cases, mainly because of a nationwide implementation of JE vaccination (Yu, 1995). In Japan, the annual JE case number is less than 10 since the 1990s. Similar trends are observed in Korea and Taiwan.

The age distribution of JE patients varies in countries. In countries where JE virus is hyperendemic, half of the patients are children under 4 years of age, and most of the patients are children under 10 years of age (Gajanana et al., 1995). However, in Japan, Korea, and Taiwan, where mass JE vaccination has been extensively implemented, JE occurs among adult populations rather than children, the majority of the patients being at over 50 years of age (Matsunaga et al., 1999; Petersen and Marfin, 2005). As stated above, most of the JE patients are children, and the children at risk should receive safe and efficacious JE vaccine to acquire protective immunity. However, this goal has not been achieved in most of the countries in JE-endemic and -epidemic regions of the world, as JE vaccines are not distributed to all at risk. Insufficient doses of vaccines are produced, and some of the currently available vaccines are still expensive for use in developing countries.

Currently Produced JE Vaccines (Table 28.1)

Four types of JE vaccines are currently used in the world: three types of inactivated vaccines and one

TABLE 28.1 Currently produced JE vaccines

Type of vaccine	Organs or cells	Virus strains	Countries
Inactivated	Mouse brain	Beijing-1	Japan, Thailand
	Mouse brain	Nakayama	Korea, Taiwan, Vietnam, India, Japan (for export only)
	PHK cells	P3	China
	Vero cells	P3	China
Live attenuated	PHK cells	SA 14-14-2	China

Note: PHK: primary hamster kidney

type of live-attenuated vaccine (Halstead and Tsai, 2004). These are mouse brain-derived inactivated JE vaccine, primary hamster kidney (PHK) cell-derived inactivated JE vaccine, Vero cell-derived inactivated JE vaccine, and live-attenuated JE vaccine derived from PHK cells.

Inactivated JE Vaccine

Mouse Brain-Derived Inactivated JE Vaccine Nakayama and Beijing-1 strains are used as production strains. The vaccine is produced from infected mouse brains by inactivation and extensive purification of the virus. Mouse brain-derived inactivated JE vaccine produced with Nakayama strain was first licensed in Japan in 1954 and has been used as the international JE vaccine. This type of vaccine is currently produced in Japan, Korea, Taiwan, Thailand, India, and Vietnam. The strain for vaccine production was changed to Beijing-1 strain in Japan in 1989. Beijing-1 strain is used as the virus seed in Japan and Thailand, and Nakayama strain in Korea, Vietnam, and India. The efficacy of mouse brain-derived inactivated JE vaccine was evaluated in two field trials. In a study in Taiwan in 1966, the efficacy of the Nakayama vaccine was 80% after two immunizations (Hsu et al., 1971) while other studies in Thailand demonstrated that the efficacies of monovalent Nakayama vaccine and Nakayama and Beijing-1 bivalent vaccines were both 91% (Hoke et al., 1988).

PHK Cell-Derived Inactivated JE Vaccine The vaccine is produced from JE virus propagated in PHK by inactivation and purification (Halstead and Tsai, 2004). PHK cell-derived inactivated JE vaccine has been produced exclusively in China using P3 strain since 1968. Approximately 70 million doses are distributed annually in China, and this vaccine had been the principal JE vaccine in China until the live-attenuated vaccine was produced. Regional trials in China showed protection against JE with efficacies of 85–87%.

Vero Cell-Derived Inactivated JE Vaccine The vaccine is produced from JE virus propagated in Vero cells by inactivation and purification. Vero cell-derived inactivated JE vaccine using P3 strain has been licensed in China since 1998. Approximately 10 million doses have been distributed in China up to 2006. Regional trials in China showed protection against JE with efficacies of over 90%.

Live-Attenuated JE Vaccine

Live-Attenuated JE Vaccine Derived from PHK Cells A PHK cell-derived, live-attenuated JE vaccine was developed using the SA 14-14-2 strain and licensed in 1988 in China (Tsai, 2000). The efficacy of the vaccine has been reported more than 99% (Bista et al., 2001). It has been reported that more than 100 million children received the SA 14-14-2 vaccine with no serious adverse events. Two-dose immunization 1 week apart is recommended as a routine immunization, although single-dose immunization showed a comparable efficacy. The vaccine has also been licensed in Korea (Sohn et al., 1999), Nepal, and India.

Adverse Events

Adverse events due to the mouse brain-derived inactivated JE vaccine have been studied. Local reactions at the injection sites and fevers occur in approximately 10% of vaccinated children after vaccination with mouse brain-derived inactivated JE vaccine in Japan. In addition, severe allergic symptoms have been reported recently (Plesner et al., 2000; Marfin et al., 2005). The incidence of systemic reactions characterized by generalized urticaria, respiratory symptoms, and cardiovascular symptoms have been noted. Some showed serum IgE against gelatin included as a stabilizer; however, many have remained obscure in etiology (Sakaguchi and Inouye, 1998).

Since JE vaccine used infected mouse brain as a source, concerns existed regarding the possible incidence of severe adverse reaction in nervous systems of vaccinees. Severe neurological disorders, including acute disseminated encephalomyelitis (ADEM), were reported in relation to JE vaccination (Fukuda et al., 1997). The incidence rate is about one–two in 1,000,000 doses in Japan, but the etiological relationship has not been confirmed in most of these cases. Adverse events

TABLE 28.2 JE vaccines under development

Type of vaccine	Organs or cells	Virus strains	Countries
Inactivated	Vero cells	Beijing-1	Japan
	Vero cells	SA 14-14-2	Austria
Live (chimeric vaccine)	Vero cells	SA 14-14-2	U.S.A.

of other types of JE vaccines have not been extensively analyzed or reported.

JE VACCINES UNDER DEVELOPMENT

These are summarized in Table 28.2.

Inactivated Vaccines

Vero Cell-Derived Inactivated JE Vaccine

Mouse brain-derived inactivated JE vaccine has been considered as the internationally accepted vaccine for a long period of time. The contribution of this vaccine to the prevention of JE in the past is well appreciated, but the fact that the vaccine is produced from infected mouse brain raises some concerns. It has been expected that the production of inactivated JE vaccines will eventually be shifted from infected animal brains to cell culture.

New Vero cell-derived inactivated JE vaccines have been developed using Beijing-1 and SA 14-14-2 strains as seed viruses in Japan and Europe, respectively (Sugawara et al., 2002). These candidate vaccines demonstrated high levels of immunogenicity and safety in phase I and II studies, and phase III clinical trials are being performed. On the other hand, the changes of the source of JE virus from mouse brain to cell culture will require new quality control measures to be implemented.

Live Vaccine

Chimeric JE Vaccine

Chambers et al. (1997) developed a chimera virus by replacing preM and E genes of yellow fever virus 17D strain with those of JE virus SA 14-14-2 strain. This chimera virus induced JE virus-specific antibodies in the vaccinees. The chimera vaccine proved efficacious in yellow fever-immune individuals. The vaccinees showed viremia for 4–7 days after inoculation, and some developed fever and malaise as those who had received yellow fever 17D vaccination. The yellow fever–JE chimeric vaccine is expected to be as potent as yellow fever vaccine in the duration of specific antibody for years, even after a single injection (Monath et al., 2002). Currently, the phase II studies have been completed for the development program of the yellow fever–JE chimeric vaccine.

EVALUATION OF JE VACCINE

Vaccine Potency

Neutralizing antibody provides the best evidence that protective immunity has been established, and the biological assay of neutralization shows correlation with protection (Hombach et al., 2005). A linear titer-protection relationship exists, and data from efficacy trials corroborate the role of neutralizing antibody in protection. Moreover, neutralizing antibody-mediated protection has been demonstrated for homotypic and heterotypic JE strains. It is accepted that a reasonable threshold antibody level for protection is 1:10 dilution of serum in a 50% plaque reduction neutralization test (PRNT). Roles of immunological memory and cellular immunity in the protection have also been considered, but details are not known.

Assessment of Protective Potency of New Vaccine

The potency of the vaccine should be determined by titration of neutralizing antibody induced in immunized mice by the PRNT. Mice are usually immunized twice via the intraperitoneal route. Neutralization antibody titers should be calculated as a 50% PRNT, which represents the neutralizing potency most reliably. The candidate vaccine needs to be tested in parallel with a reference vaccine (standard). The challenge JE virus strain should be homologous to the vaccine strain.

Evaluation in Clinical Studies

The new vaccine should assure the protection against JE infection. As stated above, a neutralizing antibody titer of not less than 1:10 dilution is considered to be the level that assures protective immunity. In clinical studies, two endpoints are considered (Hombach et al., 2005). The first endpoint is the seroconversion rate using a 1:10 dilution of serum. The second endpoint is geometric mean titers (GMT) of induced neutralizing antibodies. The comparison of protective efficacy between the tested and control vaccines in the field study is theoretically possible; however, such a trial will require an enormous number of volunteers to obtain statistically significant data and is extremely difficult.

Further, in the current situation where JE vaccines are actually used, it is unethical to perform double-blind phase III studies including a placebo group. Thus, head-to-head comparison in neutralizing antibody responses between a new candidate vaccine and the currently licensed mouse brain-derived vaccines is the test protocol to be considered. When a new candidate vaccine is produced with a strain different from that used for currently licensed vaccine(s), selection of the challenge virus strains in neutralization assays will be a key issue. It has been reported that neutralizing antibody titers are usually higher against homologous strains than against heterologous strains. In particular, the titers are lower against those belonging to other virus genotypes.

Thus, the question will be: which strains of JE virus will be used as a challenge virus in neutralization assay? The possible strains are: (i) the new vaccine-producing strain(s), (ii) the current vaccine strain, and (iii) a third strain that is antigenically (and genetically) between the new vaccine strain and the current vaccine strain. There is no agreement yet on which approach will be followed. The selection of the third neutral strain is scientifically quite difficult. Therefore, the use of the strains homologous to the respective vaccine-producing strains is likely to be most reasonable.

FURTHER STUDIES

International References

JE vaccine reference is important to assure the potency of the vaccines, that is, levels of induced neutralizing antibodies to JE virus. However, there is no international vaccine reference at this point, although the reference vaccines have been prepared with Nakayama and Beijing-1 strains for domestic use in Japan. Currently, the countries that produce JE vaccines use Japanese vaccine references or prepare their own references based on the Japanese reference. At present, there is a trial of preparing a regional JE reference among Southeast Asian countries. Thus, designation of international reference will be the first step for the harmonization of the quality control of JE vaccine in the world.

It is expected that the vaccine reference will possess a potency similar to that used in field phase III trials and demonstrated sufficient potencies. Further, the vaccine reference is expected to be prepared for each respective strain of JE vaccine. This suggests that development of new reference vaccine(s) will be required because vaccine development using new strains of JE virus is being pursued.

Effect of Other Flaviviruses

It is known that there is an immunological cross-reaction among flaviviruses. The regions where JE virus is endemic are overlapped with areas where other flaviviruses are endemic; especially dengue viruses. If vaccinees are immune to other flaviviruses, the induction of neutralizing antibodies to JE virus will be modulated. Thus, the vaccinees' immune status to JE virus should be carefully screened in clinical trials. In studies to evaluate JE vaccine potency for the induction of JE virus neutralizing antibodies, the immune status of the candidates to other flaviviruses as well as JE virus must be carefully screened.

Cross-Reactive Protection against JE Virus in Heterologous Genotypes

JE virus strains are grouped into five genotypes, based on the nucleotide sequence of the E gene (Solomon et al., 2003). However, JE virus strains comprise only one serotype (Tsarev et al., 2000). Current JE vaccines are produced using Beijing-1 or Nakayama strains (mouse brain-derived inactivated JE vaccine), Beijing P3 (PHK cell-derived inactivated JE and Vero cell-derived inactivated JE vaccines in China), and SA14-14-2 (live attenuated vaccine in China). New Vero cell-derived inactivated JE vaccines under development use Beijing-1 and SA 14-14-2 strains. All these strains belong to genotype 3. In neutralizing antibody titrations of the sera obtained from JE vaccine-immunized mice, titers against homologous virus strains are generally higher than those against heterologous JE strains within the same genotype. The neutralizing antibody titers are further lower against the strains belonging to other genotypes. This raises a question whether the level of protective immunity against heterologous strains may not be as high as against homologous strains. The protective efficacy of 91% determined from clinical trials in Thailand may have been higher if JE virus strains very closely related to the prevalent strains in the region had been used as a production strain. Thus, selection of strains for vaccine production is another issue to be carefully addressed in the future.

In terms of using neutralizing antibody levels as a correlate of protection, it should be further confirmed whether JE vaccine produced using genotype 3 JE virus strains induce similar levels of protective immunity against all the genotypes of JE virus in humans. It should be, however, noted that it is generally accepted that JE vaccines produced with genotype 3 strains induce protective immunity to JE virus in other genotypes as well as those in the same genotype.

KEY ISSUES

- Four types of JE vaccines are currently used in the world. Furthermore, additional Vero cell-derived inactivated vaccine and a new chimeric vaccine are under development.
- Multiple types of JE vaccines will be available in the near future.
- The governments of the countries where JE is endemic or epidemic will need to determine which types of vaccine to be used.
- There are still many questions to be addressed on JE vaccines, and further studies are still needed to develop third-generation JE vaccines.
- Although JE is a vaccine-preventable disease, availability of the vaccine is a practical issue.
- The vaccines should be distributed and available to all the populations at risk.

References

Basnyat, B., Zimmerman, M.D., Shrestha, Y. et al. Persistent Japanese encephalitis in Kathmandu: the need for immunization. J. Travel Med. 2001; 8:270–271.

Bista, M.B., Banerjee, M.K., Shin, S.H. et al. Efficacy of single-dose SA 14-14-2 vaccine against Japanese encephalitis: a case control study. Lancet 2001; 358:791–795.

Chambers, T.J., Nestorowicz, A., Mason, P.W. et al. Yellow fever/Japanese encephalitis chimaeric viruses: construction and biological properties. J. Virol. 1997; 73:3095–3101.

Fukuda, H., Umehara, F., Kawahigashi, N. et al. Acute disseminated myelitis after Japanese B encephalitis vaccination. J. Neurol. Sci. 1997; 148:113–115.

Gajanana, A., Thenmozi, V., Samuel, P. et al. A community-based study of subclinical flavivirus infections in children in an area of Tamil Nadu, India, where Japanese encephalitis is endemic. Bull. World Health Organ. 1995; 73:237–244.

Halstead, S.B. and Tsai, T.F. Japanese encephalitis vaccine. In: Vaccine (S. Plotin and W.A. Orenstein, Eds.), 5th ed., pp. 919–957. Philadelphia: The W.B. Saunders Co., 2004.

Hanna, J.N., Ritchie, S.A., Phillips, D.A. et al. An outbreak of Japanese encephalitis in the Torres Strait, Australia, 1995. Med. J. Aust. 1996; 165:256–260.

Hanna, J.N., Ritchie, S.A., Phillips, D.A. et al. Japanese encephalitis in north Queensland, Australia, 1998. Med. J. Aust. 1999; 170(11):533–536.

Hanson, J.P., Taylor, C.T., Richads, A.R. et al. Japanese encephalitis acquired near Port Moresby: implications for residents and travellers to Papua New Guinea. Med. J. Aust. 2004; 181:282–283.

Hoke, C.H., Nisalak, A., Sangawhipa, N. et al. Protection against Japanese encephalitis by inactivated vaccines. N. Engl. J. Med. 1988; 319:608–614.

Hombach, J., Solomon, T., Kurane, I., Jacobson, J. and Wood, D. Report on a WHO consultation on immunological endpoints for evaluation of new Japanese encephalitis vaccines, WHO, Geneva, 2–3 September, 2004. Vaccine 2005; 23(45): 5205–5211.

Hsu, T.C., Chow, L.P., Wei, H.Y., Chen, C.L., Hsu, S.T., Huang, C.T., Kitaoka, M. and Sunaga, H. A completed field trial for an evaluation of the effectiveness of mouse-brain Japanese encephalitis vaccine. In: Immunization for Japanese encephalitis (W.M. Hammon, M. Kitaoka and W.G. Downs, Eds.), pp. 258–265. Tokyo, Japan: Jgaku Shoin Ltd., 1971.

Kimura-Kuroda, J., Ogimoto, M. et al. Induction of protective immunity in animals vaccinated with recombinant vaccinia viruses that express PreM and E glycoproteins of Japanese encephalitis virus. J. Virol. 1990; 64:2788–2795.

Kimura-Kuroda, J. and Yasui, K. Protection of mice against Japanese encephalitis virus by passive administration of monoclonal antibodies. J. Immunol. 1988; 141:3606–3610.

Konishi, E., Yamaoka, M., Khin, S.W. et al. The anamnestic neutralizing antibody response is critical for protection of mice from challenge following vaccination with a plasmid encoding the Japanese encephalitis virus premembrane and envelope genes. J. Virol. 1999; 73:5527–5534.

Mackenzie, J.S. The ecology of Japanese encephalitis virus in the Australian region. Clin. Virol. 1999; 27:1–17.

Marfin, A.A., Eidex, R.S., Kozarsky, P.E. et al. Yellow fever and Japanese encephalitis vaccines: indications and complications. Infect. Dis. Clin. North Am. 2005; 19:151–168.

Martin, D.A., Muth, D.A., Brown, T. et al. Standardization of immunoglobulin M capture enzyme-linked immunosorbent assays for routine diagnosis of arboviral infections. J. Clin. Microbiol. 2000; 38:1823–1826.

Matsunaga, Y., Yabe, S., Taniguchi, K. et al. Current status of Japanese encephalitis in Japan. J. Jpn. Assoc. Infect. Dis. 1999; 73:97–103, In Japanese.

Monath, T.P., McCarthy, K., Bedford, P. et al. Clinical proof of principle for ChimeriVax™: recombinant live, attenuated vaccines against flavivirus infections. Vaccine 2002; 20:1004–1018.

Monath, T.P. and Tsai, T.F. Flaviviruses. In: Clinical Virology (D.D. Richman, R.J. Whitley and F.G. Hayden, Eds.), pp. 1097–1151. Washington, D.C: ASM Press, 2002.

Murali-Krishna, K., Ravi, V. and Manjunath, R. Protection of adult but not newborn mice against lethal intracerebral challenge with Japanese encephalitis virus by adaptively transferred virus-specific cytotoxic T lymphocytes: requirement for L3T4 cells. J. Gen. Virol. 1998; 77:705–714.

Oya, A. Japanese encephalitis vaccine. Acta Paediatr. Jpn. 1988; 30:175–184.

Petersen, L.R. and Marfin, A.A. Shifting epidemiology of Flaviviridae. J. Travel Med. 2005(Suppl. 1):S3–11.

Plesner, A., Ronne, T. and Wachmann, H. Case-control study of allergic reactions to Japanese encephalitis vaccine. Vaccine 2000; 18:1830–1836.

Rosen, L. The natural history of Japanese encephalitis virus. Annu. Rev. Microbiol. 1986; 40:76–79.

Rubin, G. and Baird, J. New recommendations on Japanese encephalitis vaccination for travelers to Papua New Guinea. Med. J. Aust. 2004; 181:283.

Sakaguchi, M. and Inouye, S. Two patterns of systemic immediate-type reactions to Japanese encephalitis vaccines. Vaccine 1998; 16:68–69.

Sohn, Y.M., Park, M.S., Rho, H.O. et al. Primary and booster immune responses to SA14-14-2 Japanese encephalitis vaccine in Korean infants. Vaccine 1999; 17:2259–2264.

Solomon, T., Ni, H., Beasley, D.W.C. et al. Origin and evolution of Japanese encephalitis virus in Southeast Asia. J. Virol. 2003; 77:3091–3098.

Sugawara, K., Nishiyama, K., Ishikawa, Y. et al. Development of Vero cell-derived inactivated Japanese encephalitis vaccine. Biologicals 2002; 30:303–314.

Sumiyoshi, H., Mori, C., Fuke, I. et al. Complete nucleotide sequence of the Japanese encephalitis virus genome RNA. Virology 1987; 161:497–510.

Tsai, T.F. New initiatives for the control of Japanese encephalitis by vaccination. Mitutes of a WHO/CVI meeting in Bangkok, Thailand, 13–15 October 1998. Vaccine 2000; 18:1–25.

Tsarev, S.A., Sanders, M.L., Vaughn, D.W. et al. Phylogenetic analysis suggests only one serotype of Japanese encephalitis virus. Vaccine 2000; 18(Suppl. 2):36–43.

Vaughn, D.W. and Hoke, C.H., Jr. The epidemiology of Japanese encephalitis: prospects for prevention. Epidemiol. Rev. 1992; 14:197–221.

Yu, Y.X. Japanese encephalitis in China. Southeast Asian J. Trop. Med. Public Health 1995; 26(Suppl. 3):17–21.

CHAPTER

29

Junin (Argentine Hemorrhagic Fever)

Kelly T. McKee, Jr.[1], *Delia A. Enria*[2] *and Julio G. Barrera Oro*[3]

[1]Quintiles Public Health and Government Services, Quintiles Transnational Corporation, Durham, NC, USA

[2]Instituto Nacional de Enfermedades Virales Humanas (INEVH) "Dr. Julio I. Maiztegui", Pergamino, Buenos Aires Province, Argentina

[3]GoldbeltRaven LLC/USAMRIID, Fort Detrick, MD, USA

OUTLINE

ABSTRACT

Argentine hemorrhagic fever (AHF) is a rodent-borne viral hemorrhagic fever endemic to north-central Argentina. The etiologic agent, Junin virus, has caused significant morbidity and mortality since its recognition approximately 50 years ago. Junin is a Clade B arenavirus, and possesses many of the biological and physicochemical characteristics for potential use as a biological weapon. Consequently, the virus is regarded as a major (CDC Category A) biothreat agent.

Numerous attempts were made during the 1950s, 1960s, and 1970s by Argentine investigators to develop inactivated and live vaccines; for a variety of reasons, all ultimately proved unsuitable. In 1981, a collaborative venture involving the Argentine government, the U.S. Department of Defense, the United Nations Development Program, and the Pan American Health Organization was initiated to develop an AHF vaccine. This program resulted in the creation of Candid #1, a live-attenuated Junin vaccine initially produced at the Salk Institute in Swiftwater, PA. Candid #1 was proven safe and effective in a series of clinical trials conducted in both the United States and Argentina. In the wake of technology transfer, Candid #1 was manufactured and licensed for use in Argentina. The vaccine has had substantial impact in reducing the incidence of AHF, and is currently incorporated into the Argentine National Vaccine Program. Candid #1 remains the only vaccine proven effective against a disease caused by an arenavirus.

INTRODUCTION AND HISTORY

Argentine hemorrhagic fever (AHF) is a severe, rodent-borne zoonosis endemic to the fertile grasslands, or "humid pampas," of north-central Argentina (Fig. 29.1). AHF was first recognized as a clinical entity during the 1950s, following reports of a hemorrhagic fever syndrome among farmers and agricultural workers (Arribalzaga, 1955; Alvarez Ambrosetti et al., 1959; Rugiero et al., 1959b). Many epidemiologic features of AHF (e.g., its seasonality, demographics, and patterns of spread) were noted in the initial report of the disease (Arribalzaga, 1955; Sabattini and Maiztegui, 1970). Virus recovered in 1958 from patients with various clinical manifestations (Parodi et al., 1958; Pirosky et al., 1959a) was confirmed as the cause of AHF via volunteer human inoculation (Pirosky et al., 1959a, 1959b; Rugiero et al., 1959a). Although this agent, subsequently called Junin virus, was recovered consistently from the organs and body fluids of several rodent species, the drylands vesper mouse, *Calomys musculinus*, was ultimately established as its principal natural reservoir (Mills et al., 1991; Sabattini et al., 1967, 1977; Sabattini and Gonzales, 1967).

FIGURE 29.1 Geographic distribution of Junin virus in nature.

ETIOLOGIC AGENT

Junin virus is a member of the Tacaribe complex of New World arenaviruses (Mettler et al., 1963). Phylogenetic reconstruction of New World arenaviruses based on N gene sequences has allowed grouping into three lineages (Clades A, B, and C). Junin shares membership in Clade B with three other agents causing hemorrhagic fever in the Americas: Machupo, Guanarito, and Sabia viruses (causative agents of Bolivian hemorrhagic fever, Venezuelan hemorrhagic fever, and hemorrhagic fever in Brazil, respectively), suggesting radiation of this highly pathogenic phenotype from a common ancestor (Bowen et al., 1996).

Junin is an enveloped virus containing two single-stranded RNA molecules. Virions are pleomorphic; particles varying in diameter from about 40 to 200nm, with evenly spaced glycoprotein spikes protruding 7–10nm from the surface (Buchmeier et al., 2006). Electron-dense structures representing host-cell ribosomes are frequently seen within the viral envelope. The genome consists of two RNA segments, termed L (7.2kb) and S (3.4kb), that serve as ambisense templates for mRNA synthesis (Meyer et al., 2002). The L RNA segment encodes the viral RNA-dependent RNA polymerase (L protein), and a small zinc-binding (Z) matrix protein that plays a regulatory role in virus replication. The S RNA segment contains genes for the nucleocapsid protein (NP) and the viral glycoprotein precursor (GP-C). The NP is the principal structural protein of the viral nucleocapsid, and complexes with genomic RNA. Translation of GP-C sequences yields the GP-C precursor protein and a small signal peptide (SP) important in targeting the nascent polypeptide to the rough endoplasmic reticulum for glycosylation. The GP-C is processed by a cellular enzyme (subtilase SKI-1/S1P) to produce the two mature surface glycoproteins GP-1 (40–46kDa) and GP-2 (35kDa) (Buchmeier et al., 2006; York et al., 2004). The GP-1 and GP-2 associate noncovalently with myristoylated SP to form the mature glycoprotein complex, visible as the spikes on the virion surface (York et al., 2004). GP-1 is situated at the top of the spike, and functions as the virion attachment protein mediating binding with cell surface receptors. GP-2 forms the stalk of the spike; it promotes the acid-dependent fusion of viral and cellular membranes within internalized endosomes that is essential to viral entry (Buchmeier et al., 2006; Kunz et al., 2002). The role of the small residue SP associated with the complex is unclear, but it appears to be required for efficient pH-dependent membrane fusion (York et al., 2004). Serum neutralizing antibodies are directed primarily against the viral glycoprotein (Sanchez et al., 1989).

Host-cell receptors are important to virus pathogenicity and tropism. It has been shown that α-dystroglycan serves as the major cellular receptor for Old World and Clade C New World arenaviruses, but not Clade A and B viruses (Spiropoulou et al., 2002). Recently, transferin receptor-1 (TfR1) was identified as a cellular binding site for Clade B arenaviruses (including Junin). The behavior and expression of this receptor on macrophages, activated lymphocytes, and endothelial cells are important to viral replication and disease pathogenesis (Radoshitzky et al., 2007).

PROTECTIVE IMMUNE RESPONSE

It is likely that both cellular and humoral immunity are involved in protection from AHF. The use of immune plasma therapy and the relationship between therapeutic efficacy and neutralizing antibody content have established the importance of neutralizing antibody in recovery from disease (Enria et al., 1984, 1986; Maiztegui et al., 1979). That neutralizing antibody is similarly critical to protection is strongly inferred. In a study demonstrating efficacy of the Candid #1 live-attenuated Junin virus vaccine (see below), the vast majority (91.1%) demonstrated seroconversion (Maiztegui et al., 1998). Junin virus-specific IgG subclass patterns were similar between patients convalescing from AHF and subjects receiving Candid #1 vaccine, suggesting an important role for IgG_1 in recovery and protection (del Carmen Saavedra et al., 2003). Patterns of neutralizing antibodies observed among nonhuman primates receiving the Candid #1 vaccine and subsequently surviving challenge with virulent wild-type Junin virus indicate that humoral immunity plays a significant role in protection (McKee et al., 1992). While human AHF immune plasma protected guinea pigs from experimental Junin infection, the $F(ab')_2$ fraction alone did not (Kenyon et al., 1990). Although evidence strongly supports a pivotal role for elimination of virus by neutralizing antibody, observations in model systems argue for an important role of antibody in elimination of virus-infected cells as well. In particular, resistance to Junin infection develops in conjunction with appearance of virus-specific antibody-dependent cellular cytotoxicity (ADCC), and anti-Junin virus antibodies can sensitize cells to complement-mediated lysis, (Kenyon and Peters, 1989; Peters et al., 1987).

A role for cellular immunity in protection has also been inferred, but the evidence is less clear. In some Candid #1-vaccinated individuals who do not develop measurable antibody, Junin virus-specific lymphoproliferative responses have been observed

(Maiztegui et al., 1998). Rarely, experimental animals may be protected from virulent Junin virus challenge in the absence of detectable neutralizing antibody (McKee et al., 1992). Vaccination of guinea pigs with recombinant vaccinia virus expressing Junin glycoproteins protected against disease, but not necessarily infection, in the face of low or undetectable levels of neutralizing antibody (Lopez et al., 2000). Therefore, it is likely that a balanced immune response, involving humoral, cellular, and perhaps other, as yet undefined, factors, is involved in protection from serious illness and death from AHF.

EPIDEMIOLOGY

When AHF was initially recognized, the endemic area comprised about 16,000 km^2 within rural portions of Buenos Aires province, where the population at risk was estimated to be 270,000 persons (Maiztegui, 1975). Over the ensuing years, however, a progressive expansion of the disease to the north, east, and west has been documented (Maiztegui et al., 1986; Maiztegui and Sabattini, 1977). By end of the 20th century, the endemic zone occupied portions of four provinces (Buenos Aires, Cordoba, Santa Fe, and La Pampa), encompassing an area more than 9 times that originally described, and placing approximately 5,000,000 persons at risk (Garcia et al., 2000) (Fig. 29.2).

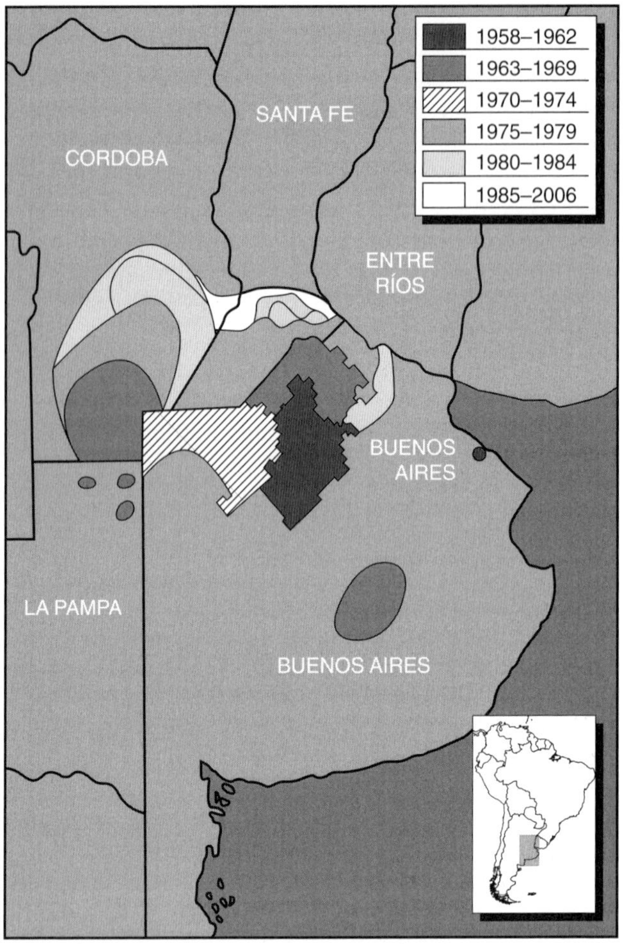

FIGURE 29.2 Progressive expansion of AHF region of endemnicity.

As the AHF-endemic area has expanded, temporal and geographic patterns have emerged. Throughout the endemic area, Junin virus infection in both rodents and humans tends to be focal in distribution. The incidence of human disease (AHF) has been observed to be highest during the first 5–10 years following it recognition in newly affected areas, with subsequent decline with the passage of time. However, pockets of disease have reemerged in some of these historically active areas. The basis for these spatial and temporal patterns is undefined, but subtle changes in reservoir rodent genetics and susceptibility to infection, the dynamics of virus reintroduction, and other factors are likely involved.

In recent years, the geographic expansion of the endemic area appears to have slowed (Enria and Barrera Oro, 2002). Whether this signals an approach to the ultimate extent of spread of this disease is unknown; the natural distribution of *C. musculinus* remains larger than the disease-endemic area, and, when sought, virus can be recovered from rodents outside the disease-endemic zone (Mills et al., 1991; Mills and Childs, 1998).

Prior to the availability of a preventive vaccine, between 100 and 800 cases of AHF were diagnosed each year in the AHF endo-epidemic area (Arribalzaga, 1955; Maiztegui, 1975). Cases occurred year-round in all age groups and both genders, but most occurred in 15–60-year-old male agricultural workers during a March–June epidemic season coinciding with the annual grain harvests and higher rodent population densities. Beginning in 1992, in conjunction with selective vaccination of high-risk adults using the Candid #1 live-attenuated Junin virus vaccine, a dramatic decrease in AHF incidence along with a shift in case demographics toward women, children under 15 years of age, and individuals without rural contact has been observed (Enria and Barrera Oro, 2002; Enria et al., 1999).

An association between AHF and indigenous rodents was made shortly after the disease was recognized (Pirosky et al., 1959a). Subsequently, the epidemiology of AHF and distribution of Junin infections among humans were found to be linked closely to the ecology and patterns of infection in these rodents (Mills and Childs, 1998).

C. musculinus has been established as the principal reservoir for Junin virus (Mills et al., 1991; Sabattini et al., 1977; Carballal et al., 1986). This species has comprised the vast majority of virus and/or antigen-positive rodent captures in the endemic area, and persistent viremia as well as salivary shedding have been documented in naturally and laboratory-infected animals. However, the virus has also been recovered from organs and body fluids of other rodent species inhabiting the central Argentine pampas, including *Calomys laucha* and *Akodon azarae*, and, less frequently *Mus musculus*, *Bolomys obscurus*, and *Oligoryzomys flavescens* (Mills et al., 1991; Sabattini et al., 1967, 1977; Sabattini and Gonzales, 1967).

The dynamics of Junin infection in rodent reservoir species is complex. The virus traditionally has been thought to establish asymptomatic, persistent infections, and to be maintained in nature by a balance between horizontal and vertical transmission (Vitullo and Merani, 1990). Carrier males may infect virus-negative females, producing vertically infected offspring of both genders (Buchmeier et al., 2006). However, chronically infected females infected as newborns are generally sterile as adults, presumably limiting their role in producing chronically viremic offspring (Vitullo and Merani, 1988), and Junin infections acquired vertically have been observed to result in growth retardation, decreased survivorship, and impaired reproductive ability (Vitullo et al., 1987). Field observations implicate horizontal transmission as the predominant mode of Junin virus spread among *C. musculinus* (Mills et al., 1992). Under experimental conditions, about half of *C. musculinus* infected as adults become chronic carriers with persistent viremia, while the remainder develop antibody and clear the infection (Vitullo and Merani, 1990).

Shedding of Junin in secreta and excreta by chronically infected *C. musculinus* underlies acquisition of the virus by humans. It is thought that infection occurs incidentally, through exposure to aerosols and/or fomites during the course of routine occupational (e.g., farming) or other activities that intrude on the natural habitat of infected rodents (Maiztegui, 1975). *C. musculinus* is an opportunistic species, widely distributed across the pampas. Although numerically prevalent in farm fields, particularly during the harvest season, this rodent appears to have a preference for linear habitats such as roadsides and fencerows (Mills et al., 1992). Thus, although transmission of Junin virus to humans may occur to laborers or others working in fields, contact along roadsides before or after entering the fields (e.g., as the individual walks or drives into roadside vegetation; or rests, eats, or engages in other activities along fences or roadways) may represent a greater risk of exposure to the virus (Mills et al., 1992). This feature of habitat preference could also explain disease acquisition among nontraditional demographic groups.

BIOTHREAT AGENT POTENTIAL

There is no direct evidence that Junin virus has ever been weaponized or used intentionally to infect unsuspecting persons. It was extensively studied, however, by scientists working in the well-documented offensive biological weapons programs of the former Soviet Union (Alibek, 1999). Junin possesses multiple epidemiological, clinical, and physiochemical characteristics that make it highly suitable for use in this capacity. The virus is stable and highly infectious by aerosol (inhalation of infectious aerosols and/or fomites being principle mode of acquisition of AHF in nature), is relatively readily accessible, causes significant morbidity and mortality among those infected, appears on the basis of animal studies to be infectious at low doses, and, despite having limited potential for casual person-to-person spread, is capable of instilling fear and panic in at-risk populations. In conjunction with other pathogenic arenaviruses (e.g., Lassa, Machupo) it is grouped with other viral hemorrhagic fever agents by the U.S. Centers for Disease Control and Prevention as a Category A (highest threat level) pathogen (Centers for Disease Control and Prevention, 2000); as such, it is a focus of priority resource allocation for countermeasure development.

CLINICAL DISEASE

Details of the evolution of the AHF clinical syndrome derive from field observations as well as two meticulously chronicled human volunteer inoculations (Pirosky et al., 1959a; Rugiero et al., 1959a). Human infection with Junin virus usually results in symptomatic illness, although subclinical infections have been reported (Maiztegui, 1975). The onset of disease is insidious. Following an incubation period of 1–2 weeks after natural exposure (precisely 39.25h in one volunteer receiving a total of 3.47×10^6 suckling mouse IC LD_{50} by sequential intramuscular and subcutaneous injections), malaise, myalgia, anorexia, and fever gradually develop. Within 3–4 days, severe prostration accompanied by back pain, headache, dizziness, and gastrointestinal disturbances appear. On physical examination, facial flushing and edema, conjunctival injection, and petechiae on the palate and axillary surfaces are observed. The gums frequently appear congested, and bleeding is commonly seen, particularly among those with poor dentition. Hemorrhage along the gingival margins is a characteristic finding. Cardiovascular effects include electrocardiographic abnormalities, moderate bradycardia, and orthostatic hypotension. Neurological changes are commonly seen, the most frequent of which are tremors of the tongue and hands, diminished deep tendon reflexes, lethargy, and cutaneous hyperaesthesia; meningeal signs are generally not evident (Maiztegui, 1975; Mills and McKee, 2000).

The acute febrile period generally lasts for 8–10 days, and most patients then go on to improve and recover. Convalescence is typically prolonged, but sequelae are uncommon (Maiztegui, 1975; Mills and McKee, 2000). In up to a third of cases, however, progression to a more serious form of disease occurs. In these patients, evolution along one of three lines is seen: a predominantly hemorrhagic syndrome with diffuse ecchymoses, bleeding from venipuncture sites, and oozing from mucous membranes; a progressive neurologic deterioration characterized by delirium, coma, and convulsions; or a mixed hemorrhagic-neurologic syndrome with features common to each, and accompanied by shock (Maiztegui, 1975; Mills and McKee, 2000). Case-fatality rates may approach 30% among untreated patients with severe disease, with death generally occurring during the second week of illness (Enria and Barrera Oro, 2002). Pregnant women, particularly in the third trimester, are at increased risk for death from AHF, and fetal mortality is high (Briggiler et al., 1990). Among survivors of severe AHF, convalescence may be protracted, 1–3 months or more, with weight loss, fatigue, autonomic instability, and hair loss commonly seen.

During the acute phase of AHF, leucopenia (total white blood cell count $<4000/mm^3$) and thrombocytopenia (platelet count $<100{,}000/mm^3$) are almost always evident. In a retrospective case-control study, these two parameters in combination had a sensitivity of 100% and specificity of 71% for the disease (Harrison et al., 1999). Proteinuria and oliguria are frequently observed, although anuria (except in terminal cases) is rare (Maiztegui, 1975). Hyaline casts, inclusion-containing cells, and red blood cells may be seen in urinary sediment. Abnormalities in routine clotting studies may be seen in severely affected cases. Serum aspartate aminotransferase and alanine aminotransferase values are generally normal or only slightly elevated; hyperglycemia and glycosuria have been occasionally seen (Maiztegui, 1975; Mills and McKee, 2000).

TREATMENT

Supportive care, with attention to hydration, osmotic shifts, and electrolyte alterations, is the cornerstone of management for viral hemorrhagic fevers, including AHF (Buchmeier et al., 2006; Centers for Disease Control and Prevention, 1995). Measures to minimize microvascular trauma (e.g., avoidance of travel, gentle sedation, minimizing use of intramuscular injections and drugs that affect clotting [such as aspirin]) should be implemented. Interventions to counter specific pathologic conditions (e.g., bleeding diathesis) should be applied according to clinical judgment. Management of shock can be challenging, due to the increased vascular permeability seen with the disease. Person-to-person spread of Junin virus is uncommon, but has been documented in sexual partners (Briggiler et al., 1987).

Administration of immune plasma obtained from convalescent AHF patients has significantly improved outcome in patients with this disease. When administered within 8 days of onset of illness, mortality was reduced from 20–30% to 1–2% (Maiztegui et al., 1979). Therapeutic efficacy correlates with the content of neutralizing antibody contained in plasma; at least 3000 therapeutic units (Σ[neutralizing antibody titer $\times$ plasma volume]/body weight [kg]), generally equivalent to 2–3 convalescent plasma units, is the optimal dosage (Enria et al., 1984, 1986). About 10% of treated survivors will develop a curious late neurologic syndrome characterized by fever, headache, tremors, and ataxia (Maiztegui et al., 1979). The cause of this condition that develops 4–6 weeks after receipt of plasma therapy is unknown, but it is typically benign and self-limited.

PATHOGENESIS

In many respects, AHF is a classical viral hemorrhagic fever, presenting nonspecifically as a febrile illness, manifesting clinical signs and symptoms of vascular injury and dysregulation, and with relatively unimpressive gross pathology. Much of our understanding of AHF pathogenesis has been informed by studies of Junin and related arenaviruses in experimental animal systems, particularly guinea pigs (Guerrero et al., 1977; Oubina et al., 1984; Weissenbacher et al., 1975b) and nonhuman primates (Gonzales et al., 1983; Green et al., 1987; McKee et al., 1985, 1987; Weissenbacher et al., 1979). Studies of blood and tissues samples from humans with the disease have also yielded valuable insights into the mechanisms underlying the physiologic and pathologic changes observed.

Junin viral replication is thought to occur at the initial site of infection, generally the lungs (Buchmeier et al., 2006). Although pneumonic infiltrates are not evident early, it is felt that local pulmonary infection and spread to regional (usually hilar) lymphatics occurs, with subsequent dissemination to other parenchymal tissues. As infection progresses, a wide variety of other organs may become affected, notably vascular endothelium, myocardium, kidneys, and the central nervous system. The gross pathological changes to these tissues tend to be nonspecific and often relatively unimpressive (Maiztegui, 1975). Destruction of splenic white pulp and the cortical areas of lymph nodes may be prominent, but frank necrosis is otherwise infrequent (Buchmeier et al., 2006; Gonzalez et al., 1980).

The genesis of bleeding abnormalities common to patients with AHF is beginning to be elucidated. Although the cause of quantitative alterations observed in platelet and white blood cell counts of patients with AHF is poorly understood, bone marrow hypoplasia and decreased erythropoietin levels are likely related (Marta et al., 2000). There is evidence for low-level but persistent coagulation and fibrinolytic system activation (Heller et al., 1995). In conjunction with thrombocytopenia, platelet function appears to be compromised; a circulating inhibitor of platelet aggregation, and platelet activation inhibitors (nitric oxide [NO] and prostacyclin [PGI_2]) have been described (Gomez et al., 2003, Cummins et al., 1990). Although total serum complement activity, as well as levels of some complement proteins (specifically, C2, C3, and C5) may be decreased, immune complex formation seems not to be important to disease pathogenesis (de Bracco et al., 1978).

Tropism of Junin virus for macrophages and endothelium, and subsequent release of immune mediators from these cells are believed to be central to AHF pathogenesis. Macrophages and activated lymphocytes, both of which express the recently characterized New World hemorrhagic fever arenavirus receptor, TfR1, to high levels, appear to be the principal targets of the virus (Radoshitzky et al., 2007). This receptor is also highly expressed on endothelial cells. Patients with AHF demonstrate high circulating levels of interferon-α and tumor necrosis factor (TNF)-α correlating with disease severity (Heller et al., 1992; Levis et al., 1984, 1985). Elevated levels of interleukins (IL) 6, 8, and 10 are also seen, and correlations have been observed between levels of TNF-α and levels of both IL-8 and IL-10, and between IL-8 and elastase α1 antitrypsin complex (E-α1AT) (a marker of neutrophil lysosomal degranulation) (Marta et al., 1999). Therefore, in addition to direct effects of viral infection on endothelium and immune cells, mediators released from, and/or acting on, these cells are likely stimuli for endothelial cell dysfunction, coagulation defects, and neutrophil activation.

A possible role for virus-specific factors in the varying clinical syndromes observed in patients with severe AHF is unresolved. Patterns of infection seen in humans (hemorrhagic, neurologic, or mixed) and, in some cases, virulence, were replicated when patient isolates were inoculated experimentally into rhesus macaques (McKee et al., 1985, 1987; Peters et al., 1987). However, fidelity of disease patterns with clinical isolates was not seen to the same degree in guinea pigs (Kenyon et al., 1986). Moreover, sequence analysis of GP-1 and GP-2 genes from seven Junin virus isolates recovered from patients with varying presentations failed to yield a genetic marker corresponding to clinical syndrome (Garcia et al., 2000).

VACCINES

History

Shortly after the isolation of Junin virus in 1958, Argentine scientists began the quest to develop a vaccine against AHF. Efforts were made to develop both inactivated and live-attenuated vaccines, and involved both homologous (Junin) and heterologous (Tacaribe) viruses (Barrera Oro and McKee, 1991).

Initial attempts to produce a killed virus vaccine focused on formalin-inactivation of Junin grown in mouse brain. These products were reported to induce neutralizing antibodies in mice and guinea pigs, and to protect against challenge in these animals with virulent Junin virus strains (Barrera Oro et al., 1967; Pirosky et al., 1959a). One formalin-inactivated mouse brain Junin vaccine was assessed for efficacy in humans (Agnese, 1999), then administered to more than 15,000 inhabitants of the AHF-endemic area between 1959 and 1962; however, no formal evaluation of immunogenicity or efficacy was undertaken (Mettler, 1969). Later, a formalin-inactivated Junin vaccine was produced in cell culture. While this candidate immunogen-induced neutralizing antibodies in laboratory animals, it was unsuccessful in protecting against challenge with virulent strains of Junin virus (Videla et al., 1989).

Alternative inactivation strategies included photoinactivation of Junin virus following exposure to neutral red dye, ultraviolet radiation, heating, and acetone (de Martinez Segovia et al., 1980; Parodi et al., 1965; Carballal et al., 1985; D'Aiutolo et al., 1979). Guinea pigs immunized with mouse brain pools of two different Junin virus strains inactivated using neutral red and ultraviolet light were challenged with a virulent Junin virus strain; 50% of the animals were reported to survive the challenge. This vaccine was subsequently assessed in 11 human volunteers, but was successful in eliciting complement fixing antibodies in only 2 previously immune individuals (de Guerrero et al., 1977).

Attempts to produce a viral subunit vaccine were made as well. One such candidate, a purified G38 capsular glycoprotein emulsified in Freund's complete adjuvant, induced specific neutralizing antibodies in rabbits, and protected guinea pigs from challenge with virulent Junin virus strains (Cresta et al., 1980). However, further evaluation of this product was not undertaken.

In the wake of a fortuitous observation that the prototype XJ strain of Junin virus lost virulence for guinea pigs after approximately 40 serial passages in newborn mouse brain (Casals, 1965; Mettler et al., 1963), enthusiasm for developing a live-attenuated vaccine ensued. A virus pool was created at the U.S. National Institutes of Health by passing a plaque-purified isolate from mouse brain passage 14 of XJ strain virus into newborn hamster brain. This strain, designated XJ Clone 3, was further passed in continuous cell culture and newborn mouse brain by Argentine scientists for use as a vaccine (Barrera Oro and McKee, 1991; Rugiero et al., 1981). The XJ Clone 3 vaccine proved attenuated for, and immunogenic in, guinea pigs (De Guerrero et al., 1969), and, beginning in 1968, human clinical trials with this product were initiated (Rugiero et al., 1969; Weissenbacher et al., 1969). By 1970, a total of 636 persons of both sexes were vaccinated. Although clinical reactions were recorded in about 75% of intensively followed individuals, most signs and symptoms were of relatively mild-to-moderate severity, and immunogenicity, measured by neutralizing antibodies, was reported to be 96% (Rugiero et al., 1974). The persistence of the immune response was established in 90% of a sample of vaccinees studied between 7 and 9 years after inoculation (Rugiero et al., 1981). In 1971, however, an expert panel from the National Academy of Medicine of Argentina, recommended that vaccination with XJ Clone 3 be suspended due to the absence of sound documentation of viral passage history, and disclosure of passages in heteroploid cell lines and mouse-brain substrate (Weissenbacher et al., 1987). Further characterization of this product led to the demonstration of long-term viral persistence, neurotropism, and neurovirulence in guinea pigs and nonhuman primates. Similar problems plagued additional attempts to develop attenuated viruses from prototype XJ strain Junin virus (Barrera Oro and McKee, 1991).

The realization that laboratory animals infected with Tacaribe, a New World arenavirus related serologically and chemically to Junin, developed neutralizing antibodies cross-reactive against Junin virus (Parodi and Coto, 1964; Tauraso and Shelokov, 1965; Weissenbacher et al., 1975a) prompted its examination as a potential heterologous vaccine. While immunization of guinea pigs and nonhuman primates with Tacaribe virus evoked persistent antibodies against Junin, and cross protection against Junin virus challenge could be demonstrated (Coto et al., 1980; Damonte et al., 1981; Samoilovich et al., 1988, 1984; Weissenbacher et al., 1982, 1987; Weissenbacher and Damonte, 1983), further development has not been pursued.

Current Licensed Vaccines

Discussions held in 1976 during the First International Seminar on Arenaviral Hemorrhagic Fevers (Eddy and Barrera Oro, 1977) prompted formation of an international collaboration to develop a safe and effective vaccine against AHF. This project, which was initiated in 1979, brought together experts in vaccine development representing the governments of Argentina and the United States under the auspices of the United Nations Development Program and the Pan American Health Organization (PAHO). Utilizing high-containment laboratories at the United States Army Medical Research Institute of Infectious Diseases (USAMRIID) at Fort Detrick, Maryland, a program was established under the direction of Dr. Julio G. Barrera Oro from Argentina's Malbran Institute to produce an experimental Junin vaccine to be tested in human volunteers under the supervision of U.S., Argentine, and PAHO regulatory agencies. The initiative culminated in the production of Candid #1, a live-attenuated Junin vaccine derived from a well-documented descendent of prototype XJ strain Junin virus that was passaged and "cloned" in certified diploid mammalian cells (Barrera Oro and McKee, 1991).

Candid #1 ($XJGP_2$ MB_{44} $FRhL_{19}$), was derived from the 44th mouse brain passage (XJ#44) of XJ strain Junin virus. It was determined to be at least as attenuated as the previous Argentine live vaccine candidate (XJ Clone 3) in guinea pigs, and was assay-negative for murine viruses. Later studies of nucleotide and amino acid sequences of L and Z genes from Candid #1 and its progenitors demonstrated 12-point mutations in the L protein unique to the vaccine strain, suggesting a molecular basis for the attenuation phenotype (Goñi et al., 2006). Preclinical studies at USAMRIID demonstrated the safety, immunogenicity, and protective efficacy of Candid #1 in guinea pigs and rhesus macaques; the median protective dose of the vaccine was less than 34 plaque-forming units (pfu) for guinea pigs, and less than 16 pfu for rhesus macaques (Barrera Oro and Eddy, 1982; Barrera Oro et al., 1985; Kenyon et al., 1989; McKee et al., 1992).

Additionally, the vaccine protected guinea pigs and rhesus macaques against Machupo virus, the etiologic agent of Bolivian Hemorrhagic Fever; a dose of 3 pfu was sufficient to protect rhesus from challenge with a virulent strain of Machupo (Barrera Oro and Eddy, 1982; Barrera Oro et al., 1988; Jahrling et al., 1988; Lupton et al., 1988).

Using lots of vaccine produced in the United States at the Salk Institute in Swiftwater, PA, clinical studies with Candid #1 were initiated in 1985. A series of phase I and II clinical trials performed in both the U.S. and Argentina demonstrated the safety and immunogenicity of the vaccine in more than 200 human volunteers: none of the vaccinated subjects developed significant clinical, hematological, biochemical, or urinary abnormalities, and >91% of vaccinees developed specific neutralizing antibodies (Barrera Oro et al., 1987; MacDonald et al., 1988, 1987; Maiztegui et al., 1988; Maiztegui and McKee, 1989). Between 1988 and 1990, a prospective, double-blind, placebo-controlled phase III efficacy trial was performed in 6500 human volunteers living and/or working in a high-risk region of Santa Fe Province, Argentina. No significant adverse events attributable to the vaccine were recorded, and 91.1% of vaccinees developed Junin-specific neutralizing antibodies. Vaccine efficacy during the 2–3 year follow-up proved excellent; 95% for protection against AHF (95% CI: 82–99%), and 84% for protection against infection with any clinical symptoms (95% CI: 60–94%) (Maiztegui et al., 1998).

Between 1994 and 2003, the remaining limited quantities of vaccine produced at the Salk Institute were used to vaccinate (under protocols approved by Argentine regulatory authorities) more than 240,000 highly at-risk individuals 15 years of age and older in the AHF-endemic zone. No significant adverse events attributable to Candid #1 were reported, and Junin virus-specific neutralizing antibodies were detected in 88.5–95.7% of recipients studied. Moreover, neutralizing antibodies at titers ≥1:10 were shown to persist for more than 10 years in >90% of vaccinees. With the gradual introduction of Candid #1, the annual incidence of AHF has fallen dramatically (Fig. 29.3). The estimated effectiveness of Candid #1 vaccine for protection against AHF for the period 1992–1999 was 98.1% (95% CI: 96.6–99%) (Enria and Barrera Oro, 2002, Enria et al., 1999).

Following completion of phase III efficacy trials, efforts to transfer the technology for producing Candid #1 from the U.S. to Argentina were undertaken. The Ministry of Health of Argentina determined that vaccine production could best be accomplished at a nascent facility colocated with the Instituto Nacional de Enfermedades Virales Humans (INEVH) in Pergamino.

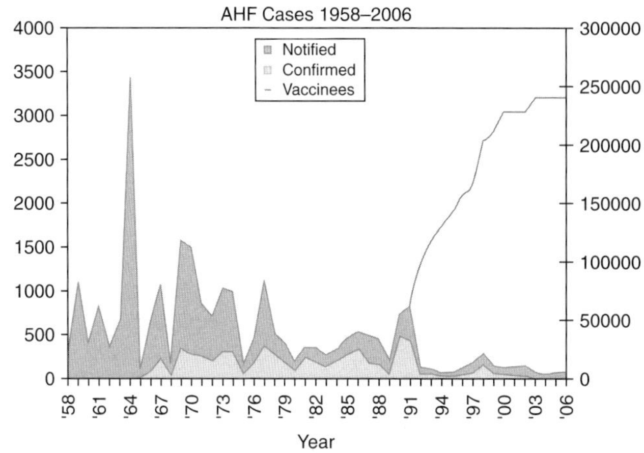

FIGURE 29.3 Incidence of AHF in relation to introduction of Candid #1 Junin vaccine in Argentina. "Notified" indicates cases of human disease clinically compatible with AHF, and reported as suspected AHF, to authorities; "Confirmed" indicates AHF cases confirmed virologically and/or serologically. Source: Argentine National Program for AHF Control, Pergamino, Argentina (Enria, personal communication).

A program for training INEVH personnel in good manufacturing practices (GMP) and quality control practices at the Salk Institute in Pennsylvania (the location where vaccine for use in all clinical trials had been manufactured) was established, and efforts commenced to manufacture Candid #1 in Argentina (Ambrosio et al., 2006). The INEVH facility was certified in 2001 by the Argentine regulatory authority (ANMAT) for production of attenuated viral vaccines.

During 2005–2006, a bridging study to establish the equivalence of Candid #1 vaccine produced in the U.S. with that produced in Argentina was conducted in 946 human subjects. During a 3-month period of active follow-up, no severe adverse events linked to the vaccination were recorded. Vaccine immunogenicity, measured by titration of neutralizing antibodies against Junin virus, was more than 95% (Enria et al., in press). In 2006, Candid #1 vaccine was licensed by the regulatory agency of Argentina (ANMAT, Disposicion No. 4812, August 29, 2006).

Vaccine Administration, Contraindications, and Precautions

Candid #1 is administered as a single intramuscular injection. Requirements for booster immunization have not been determined.

Contraindications for receipt of Candid #1 are similar to those for receipt of live-attenuated vaccines in general, and include the existence of malignancy, chronic, or uncontrolled illnesses that could affect the immune response, as well as the concurrent use of immunosuppressive therapies. Deferral of vaccination should be considered for those with acute febrile illnesses.

Specific contraindications include:

- Congenital or acquired immune deficit.
- Pregnancy: the effects that this vaccine could produce during pregnancy are unknown, and for this reason unpredictable.
- Lactation.
- Allergies to vaccine components.

Candid #1 should not be administered concurrently with other vaccines, as there is no data on possible interactions. Candid #1 has not been evaluated in children less than 15 years of age.

Manufacturing and Licensure

Candid #1 is manufactured exclusively at the Instituto Nacional de Enfermedades Virales Humanas "Dr. J.I. Maiztegui," ANLIS, Ministerio de Salud de la Nación. Monteagudo 2510 (2700) Pergamino, Pcia. de Buenos Aires, Argentina. The vaccine is licensed for use only in Argentina.

Vaccine Indications

Candid #1 is indicated for the prevention of AHF. In Argentina, Candid #1 is included in the National Immunization Agenda (Program). The vaccine is approved for use in adults and children 15 years of age and older that are living and/or working in the endemic area for AHF; in general, recipients derive from the provinces of Santa Fe, Córdoba, Buenos Aires, and La Pampa. The total population at risk for AHF in the Argentine endemic area is estimated currently to be approximately 5,000,000 persons.

Candid #1 is an orphan vaccine (Lang and Wood, 1999). It is destined to be used in a restricted area of Argentina, where AHF is endemic. It may also be used by investigators around the world working with Junin virus. Preclinical data suggest that Candid #1 might be useful for the protection against Bolivian hemorrhagic fever (Barrera Oro et al., 1988; Jahrling et al., 1988; Lupton et al., 1988); however, no clinical trails to address this indication have been performed.

Duration of Immunity

Persistence of neutralizing antibodies against Junin virus have been demonstrated for more than 10 years following receipt of a single dose of Candid #1 vaccine (Enria and Barrera Oro, 2002). Neutralizing antibodies directed against Junin virus are considered a surrogate marker for efficacy (Enria et al., in press).

Adverse Events Associated with Vaccination

Local reactions following receipt of Candid #1 vaccine include mild pain, discomfort, itching, erythema, and induration at the inoculation site. Systemic adverse events that have been reported in temporal association (within 3 weeks) with vaccination include headache, lassitude, myalgias, fever, nausea and/or vomiting, retroocular pain, dizziness, low-back pain, and exanthema. Mild leucopenia (<4000 white blood cells/mm^3) and thrombocytopenia (<150,000 platelets/mm^3) have also been reported. All events reported have been mild and have disappeared either spontaneously or with symptomatic treatment (Enria et al., in press).

As Candid #1 is a live-attenuated Junin virus vaccine, the possibility exists that a recipient could develop signs and symptoms of AHF. If a diagnosis of AHF is made, administration of AHF immune plasma should be considered within the first week from onset of symptoms.

Vaccines in Development

There are currently no efforts underway to develop a next-generation vaccine for AHF. However, technologies and molecular engineering methodologies applied to developing and/or producing vaccines for other pathogenic arenaviruses could prove useful in future efforts to control or eliminate AHF (Auperin et al., 1988; Clegg and Lloyd, 1987; Morrison et al., 1989; Fisher-Hoch et al., 1989; Lopez et al., 2000; Rico-Hesse, 1999; Bredenbeek et al., 2006; Giacalone et al., 2007; Pushko et al., 2001).

Postexposure Immunoprophylaxis

There is no recognized role for Candid #1 vaccine in prophylaxis after exposure to Junin virus. Plasma obtained from AHF survivors that contains anti-Junin neutralizing antibody is effective in treating individuals with AHF if administered within 8 days of disease onset (Enria et al., 1984, 1986a; Maiztegui et al., 1979). Therefore, it is reasonable to consider immunoprophylaxis following high-risk exposure to Junin virus (e.g., needlestick) with immune plasma (Peters and Shelokov, 1990).

PROSPECTS FOR THE FUTURE

AHF is an endemic disease for which effective control can be anticipated. Therapy to reduce the case-fatality rate is proven and available, and Candid #1 vaccine has been demonstrated to reduce the incidence of the illness. However, as in other zoonotic diseases with a nonhuman reservoir, viral eradication is not likely. For this reason, even with good vaccine coverage, isolated cases and small outbreaks of AHF can be expected to continue.

KEY ISSUES

- AHF a rodent-borne zoonosis endemic to north-central Argentina, is a major public health problem.
- The etiologic agent of AHF, Junin virus, is a CDC Category A biothreat agent.
- Candid #1, a safe and effective live-attenuated Junin virus vaccine, is licensed in Argentina for prevention of AHF.
- Introduction of Candid #1 into Argentina's National Immunization Program has resulted in a significant reduction in AHF incidence.
- Candid #1 is currently the only vaccine proven effective against an arenavirus.

References

Agnese, G. Historia de una enfermedad, imaginario y espacio rural: La fiebre hemorrágica argentina—1943–1963. Tesis para optar al grado de Licenciatura en Historia. 5.2. Experiencias en el Hospital Neuropsiquiátrico. Pontificia Universidad Católica Argentina. Facultad de derecho y Ciencias Sociales. Argentina: Instituto de Historia. Rosario, 1999.

Alibek, K. Biohazard. New York, NY: Random House, 1999.

Alvarez Ambrosetti, E., Cintora, F.A., Locisero, R., Magnoni, C., Milani, H. and Vaccarezza, R., Fiebre hemorragica epidemica; observaciones clinicas. Dia Medico 1959; 31:232–235.

Ambrosio, A.M., del Carmen Saavedra, M., Riera, L.M. and Fassio, R.M. La producción nacional de vacuna a virus Junin vivo atenuado (Candid d1) anti-fiebre hemorrágica argentina. Acta Bioquím. Clín. Latinoam. 2006; 40:5–17.

Arribalzaga, R.A. Una nueva enfermedad epidémica a germen desconocido: Hipertérmia, nefrotóxica, leucopénica y enantemática. Día Méd. 1955; 27:1204–1210.

Auperin, D.D., Esposito, J.J., Lange, J.V., Bauer, S.P., Knight, J., Sasso, D.R. and Mccormick, J.B. Construction of a recombinant vaccinia virus expressing the Lassa virus glycoprotein gene and protection of guinea pigs from a lethal Lassa virus infection. Virus Res. 1988; 9:233–248.

Barrera Oro, J.G. and Eddy, G.A. Characteristics of candidate live attenuated Junin virus vaccine. IV International Conference on Comparative Virology. Banff, Alberta, Canada. Abstract S4–S10, 1982.

Barrera Oro, J.G., Girola, R.A. and Gutman Frugone, L. Immunological studies with the Junin virus. II. Acquired immunity in guinea pigs with a formol-inactivated virus. Medicina (B. Aires) 1967; 27:279–282.

Barrera Oro, J.G., Lupton, H.W., Jahrling, P.B., Meegan, J., Kenyon, R.H. and Peters, C.J. Cross protection against Machupo virus with Candid d1 live-attenuated Junin virus vaccine. I. The postvaccination prechallenge immune response. Second International Conference on the Impact of Viral Diseases on the Development of Latin American Countries and the Caribbean Region. Mar del Plata, Argentina. Abstract P.E.1, 1988.

Barrera Oro, J., Macdonald, C., Kenyon, R.H., Meegan, J.M., Cole, F. Jr., Lupton, H.W., and Peters, C.J. Virus isolation and immune response in humans inoculated with a live-attenuated Junin virus (JV) vaccine. VII International Congress of Virology. Edmonton, Alberta, Canada. Abstract R3.45, 1987.

Barrera Oro, J.G. and McKee, K.T., Jr. Toward a vaccine against Argentine hemorrhagic fever. Bull. Pan Am. Health Organ. 1991; 25:118–126.

Barrera Oro, J.G., McKee Jr., K.T., Kuehne, A.I., Moe, J.B., Green, D.E., Cole, F.E. and Lupton, H.W. Preclinical trials of a live, attenuated Junin virus vaccine in rhesus macaques. Institut Pasteur Symposium on Vaccines and Vaccinations. Paris, France. Abstract P28, 1985.

Bowen, M.D., Peters, C.J. and Nichol, S.T. The phylogeny of New World (Tacaribe complex) arenaviruses. Virology 1996; 219:285–290.

Bredenbeek, P.J., Molenkamp, R., Spaan, W.J., Deubel, V., Marianneau, P., Salvato, M.S., Moshkoff, D., Zapata, J., Tikhonov, I., Patterson, J., Carrion, R., Ticer, A., Brasky, K. and Lukashevich, I.S. A recombinant Yellow Fever 17D vaccine expressing Lassa virus glycoproteins. Virology 2006; 345:299–304.

Briggiler, A.M., Enria, D.A., Feuillade, M.R. and Maiztegui, J.I. Contagio interhumano e infección inaparente por virus Junin en matrimonios del área endémica de fiebre hemorrágica Argentina. Medicina (B. Aires) 1987; 47:565.

Briggiler, A.M., Levis, S., Enria, D.A. and Maiztegui, J.I. Argentine hemorrhagic fever in pregnant women. Medicina (B. Aires) 1990; 50:443.

Buchmeier, M.J., De La Torre, J.-C. and Peters, C.J. Arenaviridae: the viruses and their replication. In: Field's Virology (D.M. Knipe, P.M. Howley, D.E. Griffin, R.A. Lamb, M.A. Martin, B. Roizman and S.E. Straus, Eds.), 5th ed. Philadelphia, PA: Lippincott, Williams, and Wilkins, 2006.

Carballal, G., Videla, C., Dulout, F., Cossio, P.M., Acuna, A.M. and Bianchi, N.O. Experimental infection of *Akodon molinae* (Rodentia, Cricetidae) with Junin virus. J. Med. Virol. 1986; 19:47–54.

Carballal, G., Videla, C., Oubina, J.R. and Frigerio, M.J. Inactivated Junin virus antigens. Medicina (B. Aires) 1985; 45: 153–158.

Casals, J. Serological studies on Junin and Tacaribe viruses. Am. J. Trop. Med. Hyg. 1965; 14:794–796.

Centers for Disease Control and Prevention. Notice to readers update: management of patients with suspected viral hemorrhagic fever—United States. MMWR Morb. Mortal. Wkly. Rep. 1995; 44:475–479.

Centers for Disease Control and Prevention. Biological and chemical terrorism: strategic plan for preparedness and response. MMWR Morb. Mortal. Wkly. Rep. 2000; 49:1–14.

Clegg, J.C. and Lloyd, G. Vaccinia recombinant expressing Lassa-virus internal nucleocapsid protein protects guineapigs against Lassa fever. Lancet 1987; 2:186–188.

Coto, C.E., Damonte, E.B., Catello, M.A. and Weissenbacher, M.C. Protection of guinea pigs inoculated with Tacaribe virus against lethal doses of Junin virus. J. Infect. Dis. 1980; 141:389–393.

Cresta, B., Padula, P. and de Martinez Segovia, M. Biological properties of Junin virus proteins. I. Identification of the immunogenic glycoprotein. Intervirology 1980; 13:284–288.

Cummins, D., Molinas, F.C., Lerer, G., Maiztegui, J.I., Faint, R. and Machin, S.J. A plasma inhibitor of platelet aggregation in patients with Argentine hemorrhagic fever. Am. J. Trop. Med. Hyg. 1990; 42:470–475.

D'aiutolo, A.C., Lampuri, J.S. and Coto, C.E. Reactivacion del virus Junin inactivado por la luz ultravioleta. Medicina (B. Aires) 1979; 39:801.

Damonte, E.B., Calello, M.A., Coto, C.E. and Weissenbacher, M.C. Immunization of guinea pigs against Argentinian hemorrhagic fever with Tacaribe virus replicated in human diploid cells. Medicina (B. Aires) 1981; 41:467–470.

de Bracco, M.M., Rimoldi, M.T., Cossio, P.M., Rabinovich, A., Maiztegui, J.I., Carballal, G. and Arana, R.M. Argentine hemorrhagic fever. Alterations of the complement system and anti-Junin-virus humoral response. N. Engl. J. Med. 1978; 299:216–221.

de Guerrero, L.B., Boxaca, M., Weissenbacher, M. and Frigerio, M.J. Experimental infection of the guinea pig with Junin virus. Clinical picture, dissemination, and elimination of the virus. Medicina (B. Aires) 1977; 37:271–278.

De Guerrero, L.B., Weissenbacher, M.C. and Parodi, A.S. Immunization against Argentinian hemorrhagic fever using an attenuated strain of Junin virus. I. Modified strain of Junin virus. Immunization of guinea pigs. Medicina (B. Aires) 1969; 29:1–5.

de Martinez Segovia, Z.M., Arguelles, G. and Tokman, A. Antigenic capacity of Junin virus inactivated by photodynamic oxidation. Medicina (B. Aires) 1980; 40:156–160.

del Carmen Saavedra, M., Sottosanti, J.M., Riera, L. and Ambrosio, A.M. IgG subclasses in human immune response to wild and attenuated (vaccine) Junin virus infection. J. Med. Virol. 2003; 69:447–450.

Eddy, G.A. and Barrera Oro, J.G. Discussion general. Medicina (B. Aires) 1977; 37(Suppl. 3):257–259.

Enria, D., Franco, S.G., Ambrosio, A., Vallejos, K., Levis, S. and Maiztegui, J.I. Current status of the treatment of Argentine hemorrhagic fever. Med. Microbiol. Immunol. 1986; 175:173–176.

Enria, D.A. and Barrera Oro, J.G. Junin virus vaccines. Curr. Top. Microbiol. Immunol. 2002; 263:239–261.

Enria, D.A., Briggiler, A.M., Fernandez, N.J., Levis, S.C. and Maiztegui, J.I. Importance of dose of neutralizing antibodies in treatment of Argentine hemorrhagic fever with immune plasma. Lancet 1984; 2:255–256.

Enria, D.A., Briggiler, A.M., Feuillade, M.R., Ambrosio, A.M., Crivelli, E. y el grupo de estudio de la vacuna contra la Fiebre Hemorragica Argentina. Estudio clínico puente, en fase III, en voluntarios humanos sanos, de entre 15 y 65 años de edad, de ambos géneros, a riesgo de adquirir Fiebre Hemorrágica Argentina, aleatorio, a doble ciego, para evaluar la comparabilidad entre la vacuna Candid d1 producida en la Argentina y la producida en los Estados Unidos. Medicina (B. Aires) (in press).

Enria, D.A., Feuillade, M.R., Levis, S.C., Briggiler, A.M., Ambrosio, A.M., Saavedra, M.C., Becker, J.L., Riera, L., Calderon, G., Pini, N., Sottosanti, J., Aviles, G., Grarcia, J. and Sabattini, M. Impact of vaccination of a high risk population for Argentine hemorrhagic fever with a live attenuated Junin virus vaccine. In: Factors in the Emergence and Control of Rodent-Borne Viral Diseases (J.F. Saluzzo and B. Dodet, Eds.). Paris, France: Elsevier SAS, 1999.

Fisher-Hoch, S.P., Mccormick, J.B., Auperin, D., Brown, B.G., Castor, M., Perez, G., Ruo, S., Conaty, A., Brammer, L. and Bauer, S. Protection of rhesus monkeys from fatal Lassa fever by vaccination with a recombinant vaccinia virus containing the Lassa virus glycoprotein gene. Proc. Natl. Acad. Sci. U.S.A. 1989; 86:317–321.

Garcia, J.B., Morzunov, S.P., Levis, S., Rowe, J., Calderon, G., Enria, D., Sabattini, M., Buchmeier, M.J., Bowen, M.D. and St Jeor, S.C. Genetic diversity of the Junin virus in Argentina: geographic and temporal patterns. Virology 2000; 272:127–136.

Giacalone, M.J., Zapata, J.C., Berkley, N.L., Sabbadini, R.A., Chu, Y.L., Salvato, M.S. and Mcguire, K.L. Immunization with non-replicating *E. coli* minicells delivering both protein antigen and DNA protects mice from lethal challenge with lymphocytic choriomeningitis virus. Vaccine 2007; 25:2279–2287.

Gomez, R.M., Pozner, R.G., Lazzari, M.A., D'atri, L.P., Negrotto, S., Chudzinski-Tavassi, A.M., Berria, M.I. and Schattner, M. Endothelial cell function alteration after Junin virus infection. Thromb. Haemost. 2003; 90:326–333.

Goñi, S.E., Iserte, J.A., Ambrosio, A.M., Romanowski, V., Ghiringhelli, P.D. and Lozano, M.E. Genomic features of attenuated Junin virus vaccine strain candidate. Virus Genes 2006; 32:37–41.

Gonzales, P.H., Laguens, R.P., Frigerio, M.J., Calello, M.A. and Weissenbacher, M.C. Junin virus infection of *Callithrix jacchus*: pathologic features. Am. J. Trop. Med. Hyg. 1983; 32:417–423.

Gonzalez, P.H., Cossio, P.M., Arana, R., Maiztegui, J.I. and Laguens, R.P. Lymphatic tissue in Argentine hemorrhagic fever. Pathologic features. Arch. Pathol. Lab. Med. 1980; 104:250–254.

Green, D.E., Mahlandt, B.G. and McKee, K.T.J. Experimental Argentine hemorrhagic fever in rhesus macaques: virus-specific variations in pathology. J. Med. Virol. 1987; 22:113–133.

Guerrero, L.B., Boxaca, M.C., Weissenbacher, M.C. and Frigerio, M.J. Infeccion experimental del cobayo con virus Junin. Cuadro clínico, diseminación y eliminación del virus. Medicina (B. Aires) 1977; 37:271–278.

Harrison, L.H., Halsey, N.A., McKee, K.T.J., Peters, C.J., Barrera Oro, J.G., Briggiler, A.M., Feuillade, M.R. and Maiztegui, J.I. Clinical case definitions for Argentine hemorrhagic fever. Clin. Infect. Dis. 1999; 28:1091–1094.

Heller, M.V., Marta, R.F., Sturk, A., Maiztegui, J.I., Hack, C.E., Cate, J.W. and Molinas, F.C. Early markers of blood coagulation and fibrinolysis activation in Argentine hemorrhagic fever. Thromb. Haemost. 1995; 73:368–373.

Heller, M.V., Saavedra, M.C., Falcoff, R., Maiztegui, J.I. and Molinas, F.C. Increased levels of tumor necrosis factor alpha in Argentine hemorrhagic fever. J. Infect. Dis. 1992; 166:1203–1204.

Jahrling, P.B., Trotter, R.W., Barrera Oro, J.G., Lupton, H.W., Cosgriff, T.M., Lewis, R.W., Parrish, D.B., Smith, S.B. and Peters, C.J. Cross-protection against Machupo virus with Candid d1 Junin virus vaccine: III. Post-challenge clinical findings. Second International Conference on the Impact of Viral Diseases on the Development of Latin American Countries and the Caribbean Region. Mar del Plata, Argentina. Abstract P.E.3, 1988.

Kenyon, R.H., Condie, R.M., Jahrling, P.B. and Peters, C.J. Protection of guinea pigs against experimental Argentine hemorrhagic fever by purified human IgG: importance of elimination of infected cells. Microb. Pathog. 1990; 9:219–226.

Kenyon, R.H., McKee, K., Barrera Oro, J., Ragland, D., Crabbs, C. and Higgins, Y. Protective efficacies in rhesus monkeys of Candid 1 strain Junin virus (JV) and Tacaribe virus against aerosol challenge with virulent JV. 38th Annual Meeting of the American Society of Tropical Medicine and Hygiene. Honolulu, Hawaii. Abstract 136, 1989.

Kenyon, R.H., McKee, K.T.J., Maiztegui, J.I., Green, D.E. and Peters, C.J. Heterogeneity of Junin virus strains. Med. Microbiol. Immunol. 1986; 175:169–172.

Kenyon, R.H. and Peters, C.J. Actions of complement on Junin virus. Rev. Infect. Dis. 1989; 11(Suppl. 4):S771–S776.

Kunz, S., Borrow, P. and Oldstone, M.B. Receptor structure, binding, and cell entry of arenaviruses. Curr. Top. Microbiol. Immunol. 2002; 262:111–137.

Lang, J. and Wood, S.C. Development of orphan vaccines: an industry perspective. Emerg. Infect. Dis. 1999; 5:749–756.

Levis, S.C., Saavedra, M.C., Ceccoli, C., Falcoff, E., Feuillade, M.R., Enria, D.A., Maiztegui, J.I. and Falcoff, R. Endogenous interferon in Argentine hemorrhagic fever. J. Infect. Dis. 1984; 149:428–433.

Levis, S.C., Saavedra, M.C., Ceccoli, C., Feuillade, M.R., Enria, D.A., Maiztegui, J.I. and Falcoff, R. Correlation between endogenous interferon and the clinical evolution of patients with Argentine hemorrhagic fever. J. Interferon Res. 1985; 5:383–389.

Lopez, N., Scolaro, L., Rossi, C., Jacamo, R., Candurra, N., Pujol, C., Damonte, E.B. and Franze-Fernandez, M.T. Homologous and heterologous glycoproteins induce protection against Junin virus challenge in guinea pigs. J. Gen. Virol. 2000; 81:1273–1281.

Lupton, H.W., Jahrling, P.B., Barrera Oro, J.G. and Peters, C.J. Cross-protection against Machupo virus with Candid d1 live-attenuated Junin virus vaccine: II. Post-challenge virological and immunological findings. Second International Conference on the Impact of Viral Diseases on the Development of Latin American Countries and the Caribbean Region. Mar del Plata, Argentina. Abstract P.E.2, 1988.

MacDonald, C., McKee, K., Meegan, J., Kenyon, R., Parrish, D., Maiztegui, J., Lupton, H., Peters, C. and Barrera Oro, J. Initial evaluation in humans of a live-attenuated vaccine against Argentine hemorrhagic fever. XVI Army Science Conference. Hampton, VA, 1988.

MacDonald, C., McKee, K., Peters, C., Feinsod, F., Cosgriff, T. and Barrera Oro, J. Initial clinical assessment of humans inoculated with a live-attenuated Junin virus (JV) vaccine. VII International Congress of Virology. Edmonton, Alberta, Canada. Abstract R3.27, 1987.

Maiztegui, J.I. Clinical and epidemiological patterns of Argentine haemorrhagic fever. Bull. World Health Organ. 1975; 52:567–575.

Maiztegui, J., Levis, S., Enria, D., Feuillade, M.R., Cavanagh, P., Briggiler, A., Conti, O., Vallejos, D., Tiano, E., Jaitovich, I., Gamboa, G., Becker, J.L., Saavedra, C., Ambrosio, A., McKee, K. T. and Barrera Oro, J. Inocuidad e inmunogenicidad en seres humanos de la cepa Candid 1 de virus Junin. Medicina (B. Aires) 1988; 6:660.

Maiztegui, J.I., Fernandez, N.J. and De Damilano, A.J. Efficacy of immune plasma treatment of Argentine hemorrhagic fever and association between treatment and a late neurological syndrome. Lancet 1979; 2:1216–1217.

Maiztegui, J.I., Feuillade, M.R. and Briggiler, A.M. Progressive extension of the endemic area and changing incidence of AHF. Med. Microbiol. Immunol. 1986; 175:149–152.

Maiztegui, J.I. and McKee, K.T.J. Inoculation of human volunteers with a vaccine against Argentine hemorrhagic fever. VI International Conference on Comparative and Applied Virology. Banff, Alberta, Canada. Abstract S4, 1989.

Maiztegui, J.I., McKee, K.T.J., Barrera Oro, J.G., Harrison, L. H., Gibbs, P.H., Feuillade, M.R., Enria, D.A., Briggiler, A.M., Levis, S.C., Ambrosio, A.M., Halsey, N.A., Peters, C.J. and Group, A.T.A.S. Protective efficacy of a live-attenuated vaccine against Argentine hemorrhagic fever. J. Infect. Dis. 1998; 177:277–283.

Maiztegui, J.I. and Sabattini, M.S. Extensión progresiva del área endémica de fiebre hemorrágica argentina. Medicina (B. Aires) 1977; 37(Suppl. 3):162–166.

Marta, R.F., Enria, D. and Molinas, F.C. Relationship between hematopoietic growth factors levels and hematological parameters in Argentine hemorrhagic fever. Am. J. Hematol. 2000; 64:1–6.

Marta, R.F., Montero, V.S., Hack, C.E., Sturk, A., Maiztegui, J.I. and Molinas, F.C. Proinflammatory cytokines and elastase-alpha-1-antitrypsin in Argentine hemorrhagic fever. Am. J. Trop. Med. Hyg. 1999; 60:85–89.

McKee, K.T.J., Barrera Oro, J.G., Kuehne, A.I., Spisso, J.A. and Mahlandt, B.G. Candid No. 1 Argentine hemorrhagic fever vaccine protects against lethal Junin virus challenge in rhesus macaques. Intervirology 1992; 34:154–163.

McKee, K.T.J., Mahlandt, B.G., Maiztegui, J.I., Eddy, G.A. and Peters, C.J. Experimental Argentine hemorrhagic fever in rhesus macaques: virus strain-dependent clinical response. J. Infect. Dis. 1985; 45:218–221.

McKee, K.T.J., Mahlandt, B.G., Maiztegui, J.I., Green, D.E. and Peters, C.J. Virus-specific factors in experimental Argentine hemorrhagic fever in rhesus macaques. J. Med. Virol. 1987; 22:99–111.

Mettler, N.E. Argentine hemorrhagic fever: current knowledge. PAHO Scientific Publication 183. Washington DC: Pan American Health Organization, 1969.

Mettler, N.E., Casals, J. and Shope, R.E. Study of the antigenic relationships between Junin virus, the etiologic agent of Argentinian hemorrhagic fever, and other arthropod-borne viruses. Am. J. Trop. Med. Hyg. 1963; 12:647–652.

Meyer, B.J., De La Torre, J.C. and Southern, P.J. Arenaviruses: genomic RNAs, transcription, and replication. Curr. Top. Microbiol. Immunol. 2002; 262:139–157.

Mills, J.N. and Childs, J.E. Ecologic studies of rodent reservoirs: their relevance for human health. Emerg. Infect. Dis. 1998; 4:529–537.

Mills, J.N., Ellis, B.A., McKee, K.T.J., Calderon, G.E., Maiztegui, J.I., Nelson, G.O., Ksiazek, T.G., Peters, C.J. and Childs, J.E. A longitudinal study of Junin virus activity in the rodent reservoir of Argentine hemorrhagic fever. Am. J. Trop. Med. Hyg. 1992; 47:749–763.

Mills, J.N., Ellis, B.A., McKee, K.T.J., Ksiazek, T.G., Barrera Oro, J.G., Maiztegui, J.I., Calderon, G.E., Peters, C.J. and Childs, J.E. Junin virus activity in rodents from endemic and nonendemic loci in central Argentina. Am. J. Trop. Med. Hyg. 1991; 44:589–597.

Mills, J.N. and McKee, K.T.J. South American hemorrhagic fevers. In: Hunter's Tropical Medicine and Emerging Infectious Diseases (G.T. Strickland, Ed.), 8th ed. Philadelphia, PA: W.B. Saunders, Co., 2000.

Morrison, H.G., Bauer, S.P., Lange, J.V., Esposito, J.J., Mccormick, J.B. and Auperin, D.D. Protection of guinea pigs from Lassa fever by vaccinia virus recombinants expressing the nucleoprotein or the envelope glycoproteins of Lassa virus. Virology 1989; 171:179–188.

Oubina, J.R., Carballal, G., Videla, C.M. and Cossio, P.M. The guinea pig model for Argentine hemorrhagic fever. Am. J. Trop. Med. Hyg. 1984; 33:1251–1257.

Parodi, A.S. and Coto, C.E. Immunization of guinea pigs against the Junin virus by inoculation with Tacaribe virus. Medicina (B. Aires) 1964; 24:151–153.

Parodi, A.S., De Guerrero, L.B. and Weissenbacher, M. Fiebre hemorrágica argentina: vacunación con virus Junin inactivado. Cienc. Investigación 1965; 21:132–133.

Parodi, A.S., Greenway, D., Rugiero, H.R., Rivero, E., Frigerio, M.J., de la Barrera, J.M., Mettler, N., Garzon, F., Boxaca, M., Guerrero, L., and Nota, R. Sobre la etiología del brote epidémico en Junín. Día Méd. 1958; 30:2300–2302.

Peters, C.J., Jahrling, P.B., Liu, C.T., Kenyon, R.H., McKee, K.T.J. and Barrera Oro, J.G. Experimental studies of arenaviral hemorrhagic fevers. Curr. Top. Microbiol. Immunol. 1987; 134:5–68.

Peters, C.J. and Shelokov, A. Viral hemorrhagic fever. Curr. Ther. Infect. Dis. 1990; 3:355–360.

Pirosky, I., Zuccarini, J., Molinelli, E.A., Di Pietro, A., Barrera Oro, J.G., Martini, P. and Copello, A.R. Virosis hemorrágica del noroeste bonaerense (endemoepidémica, febril, enantémática y leucopénica). Buenos Aires: Ministerio de Asistencia Social y Salud Pública, Instituto Nacional de Microbiología, 1959a.

Pirosky, I., Zuccarini, J., Montinelli, E.A., Di Pietro, A., Martini, P., Ferreyra, B., Gutman Frugone, L.F. and Vazquez, T. Virosis hemorrágica del noroeste bonaerense. Endemoepidémica, febril, enantemática y leucopénica. I. La primera inoculación experimental al hombre. Orientación Méd. 1959b; 8:144–148.

Pushko, P., Geisbert, J., Parker, M., Jahrling, P. and Smith, J. Individual and bivalent vaccines based on alphavirus replicons protect guinea pigs against infection with Lassa and Ebola viruses. J. Virol. 2001; 75:11677–11685.

Radoshitzky, S.R., Abraham, J., Spiropoulou, C.F., Kuhn, J.H., Nguyen, D., Li, W., Nagel, J., Schmidt, P.J., Nunberg, J.H., Andrews, N.C., Farzan, M. and Choe, H. Transferrin receptor 1 is a cellular receptor for New World haemorrhagic fever arenaviruses. Nature 2007; 446:92–96.

Rico-Hesse, R. Vaccines for emergent American arenaviruses. In: Factors in the emergence and control of rodent-borne viral diseases (J.B. Saluzzo and B. Dodet, Eds.), Paris, France: Elsevier SAS, 1999.

Rugiero, H.A., Magnoni, C. and Cintora, F.A. Immunización contra la fiebre hemorrágica argentina con una cepa atenuada de virus Junín: análisis de 636 vacunados. Pren. Méd. Argent. 1974; 61:231–240.

Rugiero, H.A., Magnoni, C., De Guerrero, L.B., Milani, H.A., Izquierdo, F.P., Milani, H.L. and Weber, E.L. Immunogenicity of A/USSR (H1N1) subunit vaccine in unprimed young adults (Persistence of antibodies and clinical evaluation in volunteers 7 to 9 years following the vaccination against Argentine hemorrhagic fever). J. Med. Virol. 1981; 7:227–232.

Rugiero, H.R., Astarloa, L., Gonzalez Cambaceres, C., Maglio, F. and Squassi, G. Immunization against Argentina hemorrhagic fever using a strain attenuated Junin virus. II. Immunization of volunteers, clinical and laboratory analysis. Medicina (B. Aires) 1969; 29:81–87.

Rugiero, H.R., Greenway, D.J., Parodi, A.S., Lomban, F.R., Frigerio, M.J., Precerutti, A.C. and Boxaca, M. Inoculación voluntaria del virus de la fiebre hemorrágica epidémica. Día Méd. 1959a; 31:218–224.

Rugiero, H.R., Greenway, D.J., Parodi, A.S., Rivero, E., Frigerio, M., Mettler, N.E., Boxaca, M., Nota, N., de Guerrero, L. and Garzon, F. Fiebre hemorragica epidemica; sintesis clinico-evolutiva. Dia Med. 1959b; 31:236–243.

Sabattini, M.S. and Gonzales, L.E. Identificación directa de virus Junin en roedores silvestres infectados en la naturaleza. Rev. Soc. Argent. Biol. 1967; 43:252–260.

Sabattini, M.S., Gonzales, L.E. and Crespo, J.A. Aislamiento de virus Junin en roedores silvestres de la Provincia de Cordoba. Seg. Jorn. Entomoepidemiol. Arg. 1967; 3:365–373.

Sabattini, M.S., Gonzales De Rios, L.E., Diaz, G. and Vega, V.R. Infección natural y experimental de roedores con virus Junin. Medicina (B. Aires) 1977; 37(Suppl. 3):149–161.

Sabattini, M.S. and Maiztegui, J.I. Fiebre hemorrágica Argentina. Medicina (B. Aires) 1970; 30(Suppl. 1):111–128.

Samoilovich, S.R., Calello, M.A., Laguens, R.P. and Weissenbacher, M.C. Long-term protection against Argentine hemorrhagic fever in Tacaribe virus infected marmosets: virologic and histopathologic findings. J. Med. Virol. 1988; 24:229–236.

Samoilovich, S.R., Pecci Saavedra, J., Frigerio, M.J. and Weissenbacher, M.C. Nasal and intrathalamic inoculations of primates with Tacaribe virus: protection against Argentine hemorrhagic fever and absence of neurovirulence. Acta Virol. 1984; 28:277–281.

Sanchez, A., Pifat, D.Y., Kenyon, R.H., Peters, C.J., Mccormick, J.B. and Kiley, M.P. Junin virus monoclonal antibodies: characterization and cross-reactivity with other arenaviruses. J. Gen. Virol. 1989; 70:1125–1132.

Spiropoulou, C.F., Kunz, S., Rollin, P.E., Cambell, K.P. and Oldstone, M.B.A. New World arenavirus Clade C, but not Clade A and B viruses, utilizes α-dystroglycan as its major receptor. J. Virol. 2002; 76:5140–5146.

Tauraso, N. and Shelokov, A. Protection against Junin virus by immunization with live Tacaribe virus. Proc. Soc. Exp. Biol. Med. 1965; 119:608–611.

Videla, C., Carballal, G., Remorini, P. and La Torre, J. Formalin inactivated Junin virus: immunogenicity and protection assays. J. Med. Virol. 1989; 29:215–220.

Vitullo, A.D., Hodara, V.L. and Merani, M.S. Effect of persistent infection with Junin virus on growth and reproduction of its natural reservoir, *Calomys musculinus*. Am. J. Trop. Med. Hyg. 1987; 37:663–669.

Vitullo, A.D. and Merani, M.S. Is vertical transmission sufficient to maintain Junin virus in nature? J. Gen. Virol. 1988; 69:1437–1440.

Vitullo, A.D. and Merani, M.S. Vertical transmission of Junin virus in experimentally infected adult *Calomys musculinus*. Intervirology 1990; 31:339–344.

Weissenbacher, M., De Guerrero, L.B., Help, G. and Parodi, A.S. Immunization against Argentina hemorrhagic fever using a strain of attenuated Junin virus. 3. Serological reactions in volunteers. Medicina (B. Aires) 1969; 29:88–92.

Weissenbacher, M.C., Calello, M.A., Colillas, O.J., Rondinone, S.N. and Frigerio, M.M. Argentine hemorrhagic fever: a primate model. Intervirology 1979; 11:363–365.

Weissenbacher, M.C., Coto, C.E. and Calello, M.A. Cross-protection between Tacaribe complex viruses. Presence of neutralizing antibodies against Junin virus (Argentine hemorrhagic fever) in guinea pigs infected with Tacaribe virus. Intervirology 1975a; 6:42–49.

Weissenbacher, M.C., Coto, C.E., Calello, M.A., Rondinone, S.N., Damonte, E.B. and Frigerio, M.J. Cross-protection in nonhuman primates against Argentine hemorrhagic fever. Infect. Immun. 1982; 35:425–430.

Weissenbacher, M.C. and Damonte, E.B. Fiebre hemorrágica argentina. Adel. Microbiol. Enferm. Infecc. 1983; 2:119–171.

Weissenbacher, M.C., Guerrero, L.B. and Boxaca, M.C. Experimental biology and pathogenesis of Junin virus infection in animals and man. Bull. World Health Organ. 1975b; 52:507–515.

Weissenbacher, M.C., Laguens, R.P. and Coto, C.E. Argentine hemorrhagic fever. Curr. Top. Microbiol. Immunol. 1987; 134:79–116.

York, J., Romanowski, V., Lu, M. and Nunberg, J.H. The signal peptide of the Junin arenavirus envelope glycoprotein is myristoylated and forms an essential subunit of the mature G1-G2 complex. J. Virol. 2004; 78:10783–10792.

CHAPTER

30

Measles

William J. Moss[1], *Diane E. Griffin*[2] *and W. Harry Feinstone*[2]

[1]Department of Epidemiology, International Health and Molecular Microbiology and Immunology, Johns Hopkins Bloomberg School of Public Health, Baltimore, MD, USA

[2]Department of Molecular Microbiology and Immunology, John Hopkins Bloomberg School of Public Health, Baltimore, MD, USA

OUTLINE

ABSTRACT

Measles is a highly contagious disease characterized by a prodromal illness of fever, cough, coryza, and conjunctivitis followed by the appearance of a generalized maculopapular rash. Measles virus is a nonsegmented, single-stranded, negative-sense RNA virus and a member of the *Morbillivirus* genus in the family of Paramyxoviridae. Although RNA viruses have high mutation rates, measles virus is an antigenically monotypic virus and the surface proteins responsible for inducing protective immunity have retained their antigenic structure. The public health significance is that measles vaccines developed decades ago from a single measles virus strain remain protective worldwide. Prior to the development and widespread use of measles vaccine, 30 million cases of measles were estimated to occur each year, resulting in more than 1 million deaths. Several live, attenuated measles vaccines are available, either as single-antigen vaccines or in combination with rubella and mumps vaccines (MR and MMR vaccines). Most of the currently used measles vaccines were derived from the Edmonston strain of measles virus that was isolated by Enders and Peebles in 1954.

Measles vaccines are recommended for all susceptible children and adults for whom the vaccine is not contraindicated. Despite progress in reducing measles mortality, measles remains a major cause of vaccine-preventable death and an important cause of morbidity and mortality in children, particularly sub-Saharan Africa and in Asia. The ideal measles vaccine would be inexpensive, safe, heat-stable, immunogenic in neonates or very young infants, and administered as a single dose without needle or syringe. A number of vaccine candidates with some of these characteristics are undergoing preclinical studies, including DNA vaccines and various viral and bacterial vectored vaccines. The high infectivity of measles virus is a characteristic suitable to a biothreat agent. However, increasingly high levels of measles vaccination coverage throughout the world as part of accelerated measles control efforts would protect many from the deliberate use of measles virus as a biothreat agent. Genetic engineering of a measles virus strain that was not neutralized by antibodies induced by the current attenuated measles vaccines would likely have reduced infectivity, as suggested by the fact that wild-type measles viruses have not mutated to alter their neutralizing epitopes. Measles virus meets many of the biological criteria for disease eradication. Measles virus has no nonhuman reservoir, can be accurately diagnosed, and measles vaccination is a highly effective intervention. Where measles virus differs from smallpox and polio viruses is that it is more highly infectious, necessitating higher levels of population immunity to interrupt transmission. It remains unclear whether the threat from bioterrorism precludes stopping measles vaccination after eradication, but provision of a second opportunity for measles vaccination likely could be stopped following eradication.

INTRODUCTION

Measles is a highly contagious disease characterized by a prodromal illness of fever, cough, coryza, and conjunctivitis followed by the appearance of a generalized maculopapular rash. Deaths from measles are due largely to an increased susceptibility to secondary bacterial and viral infections, attributed to a prolonged state of immune suppression. Despite the development of an effective attenuated vaccine, measles remains a leading vaccine-preventable cause of childhood mortality worldwide, particularly sub-Saharan Africa and in Asia, and continues to cause outbreaks in communities with low vaccination coverage rates in industrialized nations.

HISTORY OF MEASLES

Measles is one of the most important infectious diseases of humans and has caused millions of deaths

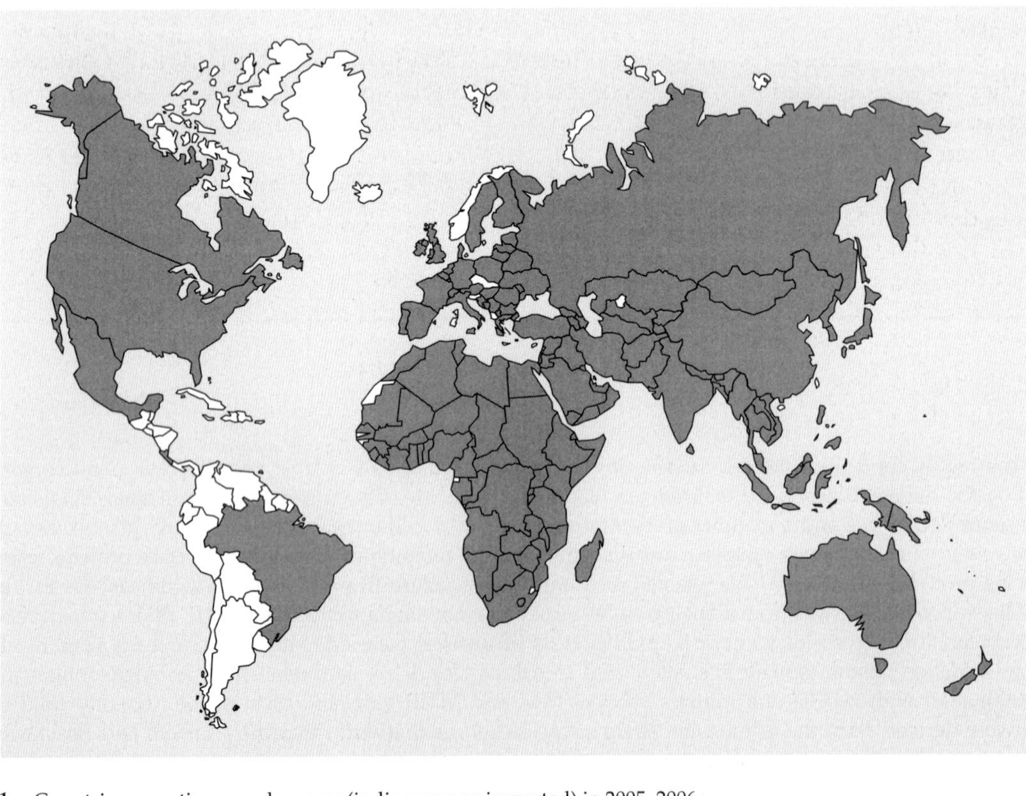

FIGURE 30.1 Countries reporting measles cases (indigenous or imported) in 2005–2006

since its emergence thousands of years ago. Measles virus most closely resembles rinderpest virus, a pathogen of cattle, and likely evolved as a zoonotic infection in communities where cattle and humans lived in close proximity. Measles virus is believed to have become established in human populations about 5000–10,000 years ago when human populations achieved sufficient size in Middle Eastern river valley civilizations to maintain virus transmission.

Abu Becr, an Arab physician also known as Rhazes, is generally credited with distinguishing smallpox from measles in the 9th century. He dated the first description of measles to the 6th century. However, epidemics identified as measles were not recorded until the 11th and 12th centuries, and measles was first mentioned as a childhood disease in 1224. The name "morbilli" was derived from the Italian meaning "little diseases" to distinguish it from plague, "il morbo." Sanvages in 1763 defined morbilli as measles, but called it rubeola, leading to confusion with rubella.

Introduction of measles into previously unexposed populations has been associated with high mortality. One quarter of the population on the Fiji Islands died after the introduction of measles virus in 1875. Millions died as a result of European exploration of the New World, largely due to the introduction of diseases such as smallpox and measles into native Amerindian populations. The high mortality from these diseases facilitated European conquest of the Americas (McNeill, 1976).

Many of the basic principles of measles epidemiology and infection were elucidated by Peter Panum, a Danish physician who was sent to the Faroe Islands in 1846 during a large measles epidemic (Panum, 1938). Panum deduced the highly contagious nature of the disease, the 14-day incubation period, the lifelong immunity following infection, and postulated a respiratory route of transmission. Measles virus first was isolated from the blood of David Edmonston and others by Enders and Peebles (1954). The development of vaccines against measles soon followed.

MEASLES VIRUS

Measles virus is a spherical, nonsegmented, single-stranded, negative-sense RNA virus and a member of the *Morbillivirus* genus in the family of Paramyxoviridae. Other members of the *Morbillivirus* genus are rinderpest virus and canine distemper virus. Although RNA viruses have high mutation rates, measles virus is an antigenically monotypic virus, meaning that the surface proteins responsible for inducing protective immunity have retained their antigenic structure. The public health significance is that measles vaccines developed decades ago from a single measles virus strain remain protective worldwide. Measles virus is killed by ultraviolet light and heat. Attenuated measles vaccine viruses retain these characteristics, necessitating a cold chain for transportation and storage.

The measles virus RNA genome consists of approximately 16,000 nucleotides and is enclosed in a lipid-containing envelope derived from the host cell. The genome encodes eight proteins, two of which (V and C) are nonstructural proteins and are transcribed from the phosphoprotein (P) gene. Of the six structural proteins, P, large protein (L), and nucleoprotein (N) form the nucleocapsid housing the viral RNA. The hemagglutinin protein (H), fusion protein (F), and matrix protein (M), together with lipids from the host cell membrane, form the viral envelope.

The H protein interacts with F to mediate fusion of the viral envelope with the host cell membrane (Malvoisin and Wild, 1993). The primary function of the H protein is to bind to the host cellular receptors for measles virus. The two identified receptors are CD46 and CD150 (SLAM). CD46 is a complement regulatory molecule expressed on all nucleated cells in humans. SLAM, an acronym for signaling lymphocyte activation molecule, is expressed on activated T and B lymphocytes and antigen-presenting cells. The binding sites on H for these receptors overlap and strains of measles virus differ in the efficiency with which each is used. Wild-type measles virus binds to cells primarily through the cellular receptor SLAM whereas most vaccine strains bind to CD46; however, most measles virus strains can use both CD46 and SLAM as receptors during acute infection (Schneider et al., 2002). Additional, as yet unidentified receptors for measles virus exist on human endothelial and epithelial cells (Andres et al., 2003).

Other measles virus proteins are involved in viral replication. The P protein regulates transcription, replication, and the efficiency with which the N assembles into nucleocapsids (Spehner et al., 1997). The M protein links ribonucleoproteins with envelope proteins during virion assembly. The functions of V and C proteins have not been clearly defined, but both appear to contribute to the virulence of measles virus by regulating transcription and sensitivity to the antiviral effects of interferon (IFN) α/β (Valsamakis et al., 1998; Patterson et al., 2000).

Variability within the genome is sufficient to allow for molecular epidemiologic investigation. Genetic characterization of wild-type measles viruses is based on sequence analysis of the genes encoding the

N and H proteins. One of the most variable regions of the measles virus genome is the 450-nucleotide sequence at the carboxy-terminal of the N protein, with up to 12% variability between wild-type viruses. The World Health Organization (WHO) recognizes 8 clades of measles virus (designated A through H) and 23 genotypes (World Health Organization, 2006). New genotypes likely will be identified with enhanced surveillance and molecular characterization. As measles control efforts intensify, molecular surveillance of circulating measles virus strains can be used to document interruption of measles virus transmission and to identify the source and transmission pathways of measles virus outbreaks (Rota and Bellini, 2003). Molecular epidemiologic tools also would be important in documenting deliberate bioterrorist introductions of wild-type or genetically modified measles virus strains.

IMMUNE RESPONSES TO MEASLES VIRUS

Host immune responses to measles virus are essential for viral clearance, clinical recovery, and the establishment of long-term immunity. The early nonspecific (innate) immune responses that occur during the prodromal phase of the illness include activation of natural killer (NK) cells and production of IFN-α and β. These innate immune responses contribute to the control of measles virus replication before the onset of more specific adaptive immune responses. The protective efficacy of antibodies to measles virus is illustrated by the immunity conferred to infants from passively acquired maternal antibodies and the protection of exposed, susceptible individuals following administration of antimeasles virus immune globulin (Black and Yannet, 1960). The first measles virus-specific antibodies produced after infection are of the IgM subtype, followed by a switch to predominantly IgG3 and then to IgG1 and IgG4 isotypes (Isa et al., 2002). IgA antibodies to measles virus are found in mucosal secretions. The most abundant and rapidly produced antibodies are against N, and the absence of antibodies to N is the best indicator of seronegativity to measles virus. Antibodies to H and F proteins contribute to virus neutralization and are sufficient to provide protection against measles virus infection.

Evidence for the importance of cellular immunity to measles virus is demonstrated by the ability of children with agammaglobulinemia to fully recover from measles, whereas children with severe defects in T-lymphocyte function often develop severe or fatal disease (Good and Zak, 1956). Monkeys depleted of $CD8^+$ T-lymphocytes and challenged with wild-type measles virus had a more extensive rash, higher measles virus loads, and longer duration of viremia than control animals, further confirming the importance of cellular immunity in measles virus clearance (Permar et al., 2003). $CD4^+$ T-lymphocytes also are activated in response to measles virus infection and secrete cytokines capable of modulating the humoral and cellular immune responses. Plasma cytokine profiles show increased levels of IFN-γ during the acute phase, followed by a shift to high levels of interleukin (IL)-4 and IL-10 during convalescence (Moss et al., 2002). The initial predominant Th1 response (characterized by IFN-γ) is presumed to be essential for viral clearance, while the later Th2 response (characterized by IL-4) promotes the development of measles virus-specific antibodies.

The duration of protective immunity following wild-type measles virus infection is generally thought to be lifelong. The immunologic mechanisms involved in sustaining high levels of neutralizing antibody to measles virus are not completely understood, although general principles of immunologic memory probably govern this process. Immunologic memory to measles virus includes both continued production of measles virus-specific IgG antibodies and the circulation of measles virus-specific $CD4^+$ and $CD8^+$ T-lymphocytes (Ovsyannikova et al., 2003). Although immune protection is assessed by measurement of antimeasles virus antibodies, long lasting cellular immune responses almost certainly play an important role in protection from infection and disease.

Young infants in the first months of life are protected against measles by maternally acquired IgG antibodies. An active transport mechanism in the placenta is responsible for the transfer of IgG antibodies from the maternal circulation to the fetus starting at about 28 weeks gestation and continuing until birth (Crowe, 2001). Three factors determine the degree and duration of protection in the newborn: (1) the level of maternal antimeasles antibodies; (2) the efficiency of placental transfer; and (3) the rate of catabolism in the child. Although providing passive immunity to young infants, maternally acquired antibodies can interfere with the immune responses to the attenuated measles vaccine by inhibiting replication of vaccine virus. In general, maternally acquired antibodies are no longer present in the majority of children by 9 months of age, the time of routine measles vaccination in many countries. Women with vaccine-induced immunity tend to have lower antimeasles virus antibody titers than women with naturally acquired immunity, and their children may be susceptible to measles at an earlier age.

EPIDEMIOLOGY

Prior to the development and widespread use of measles vaccine, 30 million cases of measles were estimated to occur each year, resulting in more than 1 million deaths. Despite progress in reducing measles mortality, measles remains the most frequent cause of vaccine-preventable death and an important cause of morbidity and mortality in children, particularly in sub-Saharan Africa and in Asia (Henao-Restrepo et al., 2003).

The disease burden due to measles decreased over the past several decades due to a number of factors. Measles mortality declined in developed countries in association with economic development, improved nutritional status and supportive care, particularly antibiotic therapy for secondary bacterial pneumonia. Remarkable progress in reducing measles incidence and mortality has been, and continues to be, made in resource-poor countries as a consequence of increasing measles vaccine coverage, provision of a second opportunity for measles vaccination through supplementary immunization activities, and efforts by the WHO, the United Nations Children's Fund (UNICEF) and their partners to target 45 countries for accelerated and sustained measles mortality reduction. Specifically, this targeted strategy aims to achieve >90% measles vaccination coverage in every district of the 45 countries and to ensure that all children receive a second opportunity for measles immunization (World Health Organization, 2001). Provision of vitamin A through polio and measles vaccination campaigns has contributed further to the reduction in measles mortality (World Health Organization, 2005).

In 2003, the World Health Assembly endorsed a resolution urging member countries to reduce the number of deaths attributed to measles by 50% by the end of 2005 compared with 1999 estimates. Overall global measles mortality in 2005 was estimated to be 345,000 deaths (uncertainty bounds 247,000 and 458,000 deaths) (Wolfson et al., 2007). This estimate represents a 60% decrease from 1999, when the global number of measles deaths was estimated to be 873,000 (uncertainty bounds 634,000 and 1,140,000 deaths). The largest decrease in measles mortality was in Africa, contributing 72% of the global reduction in measles mortality.

Measles incidence has a typical temporal pattern characterized by annual, seasonal epidemics superimposed upon longer epidemic cycles of 2–5 years or more. In temperate climates, annual measles outbreaks typically occur in the late winter and early spring. These annual outbreaks are likely the result of social networks facilitating transmission (e.g., congregation of children at school) and environmental factors favoring the viability and transmission of measles virus (Fine and Clarkson, 1982). Measles cases continue to occur during the interepidemic period in densely populated communities but at low incidence. The longer cycles occurring every several years result from the accumulation of susceptible persons over successive birth cohorts and the subsequent decline in the number of susceptibles following an outbreak. In the absence of a vaccination program these longer epidemic cycles tend to occur every 2–4 years. Measles vaccination programs that achieve coverage rates in excess of 80% extend the interepidemic period to 4–8 years by reducing the number of susceptible individuals.

Humans are the only reservoir for measles virus, a characteristic important for the potential eradication of measles. Nonhuman primates may be infected with measles virus and develop an illness similar to measles in humans, with rash, coryza, and conjunctivitis. However, populations of wild monkeys are not of sufficient size to maintain measles virus transmission.

Measles virus is transmitted primarily by respiratory droplets small enough to traverse several feet but too large to remain suspended in the air for long periods of time. The symptoms induced during the prodrome, particularly sneezing and coughing, enhance transmission. Measles virus also may be transmitted by the airborne route, suspended on small particles for a prolonged time. Direct contact with infected secretions can transmit measles virus, but the virus does not survive long on fomites as it is quickly killed by heat and ultraviolet radiation. The average incubation period for measles, the time from infection to clinical disease, is approximately 10 days to the onset of fever and 14 days to the onset of rash (range 7–18 days). The incubation period may be shorter in infants and following a large inoculum of virus, and longer in adults. During this seemingly quiescent period, the virus is rapidly replicating and infecting target tissues.

Measles virus is one of the most highly contagious infectious agents and outbreaks can occur in populations in which less than 10% of persons are susceptible. Chains of transmission commonly occur among household contacts, school children, and health care workers. Generally, persons with measles are infectious for several days before and after the onset of rash, when titers of measles virus in the blood and body fluids are highest. As with many other acute viral infections (SARS-coronavirus being an exception), the fact that measles virus is contagious prior to the onset of recognizable disease hinders the effectiveness of quarantine measures. Measles virus can be isolated in tissue culture from the urine as late as 1 week after rash onset. Detection of measles virus in body fluids by a variety of means, including identification of multinucleated giant cells in nasal secretions or the use of reverse transcriptase-polymerase chain reaction (RT-PCR), suggests the potential for a prolonged infectious period in persons immunocompromised by severe malnutrition or human immunodeficiency virus type 1 (HIV-1) infection (Dossetor et al., 1977; Permar et al., 2001; Riddell et al., 2007). However, whether detection of measles virus by these methods indicates prolonged contagiousness is unclear.

In densely populated urban settings with low vaccination coverage rates, measles is a disease of young children. The cumulative distribution can reach 50% by 1 year of age, with a significant proportion of children acquiring measles virus infection before 9 months, the age of routine vaccination. As measles vaccine coverage increases, or population density decreases, the age distribution shifts toward older children. In such situations, measles cases predominate in school-age children. Infants and younger children, although susceptible if not protected by immunization, are not exposed to measles virus at a rate sufficient to cause a large disease burden in this age group. As vaccination coverage increases further, the age distribution of cases may be shifted into adolescence and young adulthood, as seen in measles outbreaks in the United States, Brazil, and Australia (Hutchins et al., 1996; de Quadros et al., 1998; Lambert et al., 2000), necessitating targeted measles vaccination programs for these older age groups.

The high infectivity of measles virus is a characteristic suitable to a biothreat agent. However, increasingly high levels of measles vaccination coverage throughout the world as part of accelerated measles control efforts would protect many from the deliberate use of measles virus as a biothreat agent. Genetic engineering of a measles virus strain that was not neutralized by antibodies induced by the current attenuated measles vaccines would likely have reduced infectivity, as suggested by the fact that wild-type measles viruses have not mutated to alter their neutralizing epitopes.

CLINICAL DISEASE

Clinically apparent measles begins with a prodrome characterized by fever, cough, coryza, and conjunctivitis. Koplik's spots, small white lesions on the buccal mucosa inside the mouth, may be visible during the prodrome and allow the astute clinician to diagnose measles prior to the onset of rash. The prodromal symptoms intensify several days before the onset of rash. The characteristic erythematous and maculopapular rash appears first on the face and behind the ears, and then spreads in a centrifugal fashion to the trunk and extremities. The rash lasts for 3–4 days and fades in the same manner as it appeared.

In uncomplicated measles, clinical recovery begins soon after appearance of the rash. Complications occur in up to 40% of measles cases, and the risk of complication is increased by extremes of age, malnutrition, and vitamin A deficiency (Morley, 1969). Complications of measles have been described in almost every organ system. The respiratory tract is a frequent site of complication, with pneumonia accounting for most measles-associated deaths (Duke and Mgone, 2003). Pneumonia is caused by secondary viral or bacterial infections, or by measles virus itself. Pathologically, measles virus infection of the lung is characterized by multinucleated giant cells, formed when measles virus proteins on the cell surface allow cells to fuse. Other respiratory complications include laryngotracheobronchitis and otitis media. Mouth ulcers, or stomatitis, may hinder children from eating or drinking. Many children with measles develop diarrhea, which further contributes to malnutrition. Keratoconjunctivitis is common after measles, particularly in children with vitamin A deficiency, and was a frequent cause of blindness.

Because the rash of measles is a consequence of the cellular immune response, persons with impaired cellular immunity, such as those with the acquired immunodeficiency syndrome (AIDS), may not develop the characteristic measles rash. These persons have a high case fatality and may develop a giant cell pneumonitis due to measles virus (Moss et al., 1999). T-lymphocyte defects due to causes other than HIV-1 infection, such as cancer chemotherapy, are also associated with increased severity of measles.

Rare but serious complications of measles involve the central nervous system. Postmeasles encephalomyelitis complicates approximately 1 in 1000 cases, mainly older children and adults. Encephalomyelitis occurs within 2 weeks of the onset of rash and is characterized by fever, seizures, and a variety of neurological abnormalities. The finding of periventricular demyelination, the induction of immune responses to myelin basic protein, and the absence of measles virus in the brain suggest that postmeasles encephalomyelitis is an autoimmune disorder triggered by measles virus infection. Other central nervous system complications that occur months to years after acute infection are measles inclusion body encephalitis (MIBE) and subacute sclerosing panencephalitis (SSPE). In contrast to postmeasles encephalomyelitis, MIBE and SSPE are caused by persistent measles virus infection. MIBE is a rare but fatal complication that affects individuals with defective cellular immunity and typically occurs months after infection. SSPE is a slowly progressive disease characterized by seizures, progressive deterioration of cognitive and motor functions followed by death that occurs 5–15 years after measles virus infection, most often in persons infected with measles virus before 2 years of age.

There are conflicting and inconclusive data suggesting that measles virus infection causes or contributes to the development of chronic diseases, including multiple sclerosis, Paget's disease, inflammatory bowel disease, and otosclerosis (Perry and Halsey, 2004). However, no causal association has been established between measles and these conditions.

The characteristic clinical features of measles are of sufficient sensitivity and specificity to have high predictive value in regions where measles is endemic. However, laboratory diagnosis is necessary where measles virus transmission rates are low or in immunocompromised persons who may not have the characteristic clinical manifestations. Infection with rubella, parvovirus B19, human herpes virus 6, and dengue viruses may mimic measles. Detection of IgM antibodies to measles virus by enzyme immunoassay (EIA) is the standard method of diagnosing acute measles (Bellini and Helfand, 2003). Alternatively, seroconversion using IgG-specific EIA, hemagglutinin inhibition, complement fixation, or virus neutralization assays can be used to diagnose acute measles based on testing serum or plasma obtained during the acute and convalescent phases.

Measles virus can be isolated in tissue culture from white blood cells, respiratory tract secretions, and urine, although the ability to isolate measles virus diminishes quickly after rash onset. Amplification and detection of measles virus RNA by RT-PCR from blood, urine, and nasal discharge is highly sensitive in detecting measles virus RNA and allows sequencing of the measles virus genome for molecular epidemiologic studies.

TREATMENT

No specific antiviral drug is used routinely to treat measles virus infection, although the broad antiviral agent ribavirin has been used to treat immunocompromised persons or persons with SSPE, alone or in combination with IFN-α or intravenous immunoglobulin (Forni et al., 1994; Solomon et al., 2002). The major components of case management include provision of vitamin A, prompt treatment of secondary bacterial infections, and nutritional support. Several placebo-controlled trials have demonstrated marked reductions in morbidity and mortality in hospitalized children with measles treated with vitamin A. The WHO recommends administration of two daily doses of 200,000 IU of vitamin A to all children with measles 12 months of age or older. Lower doses (100,000 IU) are recommended for children less than 12 months of age. Overall, this regimen resulted in a 64% reduction in the risk of mortality (RR = 0.36, 95% CI 0.14–0.82) (D'Souza and D'Souza, 2002). Pneumonia-specific mortality is reduced, and the impact is greatest in children less than 2 years of age (D'Souza and D'Souza, 2002).

Secondary bacterial infections are a major cause of morbidity and mortality following measles (Duke and Mgone, 2003), and effective case management involves prompt treatment with antibiotics. Various strategies have been used to guide antibiotic therapy in children with measles. Antibiotics are indicated for children with measles who have clinical evidence of bacterial infection, including pneumonia, otitis media, skin infection, eye infection, and severe mouth ulcers. *Streptococcus pneumoniae* and *Haemophilus influenza* type b are the most common causes of bacterial pneumonia following measles, and antibiotic therapy should be directed against these pathogens. Whether all children with measles, or all hospitalized children with measles, should be given prophylactic antibiotics remains controversial. Limited evidence suggests that antibiotics administered as prophylaxis to all children presenting with measles may reduce the incidence of pneumonia but not mortality (Duke and Mgone, 2003). The potential benefits of antibiotic prophylaxis need to be weighed against the risks of accelerating antibiotic resistance.

Vitamin A has been widely distributed through polio and measles supplemental immunization activities as well as through routine child health services. Pooled analysis of community-based studies of vitamin A supplementation of apparently healthy children resulted in a 39% reduction in measles-associated mortality (Villamor and Fawzi, 2000). Thus, vitamin A is not only effective in reducing mortality when used to treat hospitalized children with measles but community-based supplementation programs also can result in measles mortality reduction.

PATHOGENESIS

Respiratory droplets from infected persons serve as vehicles of transmission by carrying infectious virus to epithelial cells of the respiratory tract of susceptible hosts. During the 10–14 day incubation period between infection and the onset of clinical signs and symptoms, measles virus replicates and spreads within the infected host. Initial viral replication occurs in epithelial cells at the portal of entry in the upper respiratory tract and the virus then spreads to local lymphatic tissue. Replication in local lymph nodes is followed by viremia (the presence of virus in the blood) and the dissemination of measles virus to many organs, including lymph nodes, skin, kidney, gastrointestinal tract and liver, where the virus replicates in epithelial and endothelial cells as well as monocytes and macrophages.

Although measles virus infection is clinically inapparent during the incubation period, the virus is actively replicating and the host immune responses are developing. Evidence of these processes can be detected. During the incubation period, the number of circulating lymphocytes is reduced (lymphopenia) (Auwaerter et al., 1999). Measles virus can be isolated from the nasopharynx and blood during the later part of the incubation period and during the several days prior to the onset of rash when levels of viremia are highest. The prodrome ends with the appearance of the measles rash. The rash results from measles virus-specific cellular immune responses and marks the beginning of viral clearance from blood and tissue. Histologic examination of the rash reveals infected capillary endothelial cells and a mononuclear cell infiltrate (Kimura et al., 1975). Clearance of infectious virus from the blood and other tissues occurs within the first week after the appearance of the rash, although measles virus RNA can be detected in body fluids of some children for at least several months using a RT-PCR-based assay (Permar et al., 2001; Riddell et al., 2007).

The intense immune responses induced by measles virus infection are paradoxically associated with depressed responses to unrelated (nonmeasles virus) antigens, lasting for several weeks to months beyond resolution of the acute illness. This state of immune suppression enhances susceptibility to secondary bacterial and viral infections causing pneumonia and diarrhea, and is responsible for much measles-related morbidity and mortality (Beckford et al., 1985; Greenberg et al., 1991). Delayed-type hypersensitivity (DTH) responses to recall antigens, such as tuberculin, are suppressed (Tamashiro et al., 1987) and cellular and humoral responses to new antigens are impaired following measles virus infection (Coovadia et al., 1978). Reactivation of tuberculosis and remission of autoimmune diseases have been described after measles and are attributed to this state of immune suppression.

Abnormalities of both the innate and adaptive immune responses have been described following measles virus infection. Transient lymphopenia with a reduction in $CD4^+$ and $CD8^+$ T-lymphocytes occurs in children following measles virus infection. Functional abnormalities of immune cells have also been detected, including decreased lymphocyte proliferative responses (Hirsch et al., 1984). Dendritic cells, major antigen-presenting cells, mature poorly, lose the ability to stimulate proliferative responses in lymphocytes, and undergo cell death when infected with measles virus in vitro (Servet-Delprat et al., 2000). The dominant Th2 response in children recovering from measles can inhibit Th1 responses and increase susceptibility to intracellular pathogens (Griffin et al., 1985; Griffin and Ward, 1993). The production of IL-12, important for the generation of Th1-type immune response, decreases following binding of the CD46 receptor (Karp et al., 1996) and is low for several weeks in children with measles (Atabani et al., 2001). This diminished ability to produce IL-12 could further result in a limited Th1 immune response to other pathogens. Furthermore, engagement of CD46 and CD3 on monocytes induces production of high levels of IL-10 and transforming growth factor (TGF)-β, an immunomodulatory and immunosuppressive cytokine profile characteristic of regulatory T cells (Kemper et al., 2003). The role of these cytokines in the immune suppression following measles is supported by in vivo evidence of elevated levels of IL-10 in the plasma of children after measles virus infection (Moss et al., 2002).

MEASLES VACCINES

History of Measles Vaccines

Attenuation of measles virus was achieved primarily by serial passage in chick embryo cells. The first attenuated measles vaccine licensed in the United States was Edmonston B. This vaccine was immunogenic and widely used between 1963 and 1975, but was frequently associated with fever and rash. The Schwarz and Moraten (more attenuated) strains were derived from the original Edmonston strain but further attenuated through additional passage in chick embryo fibroblasts. Despite differences in their passage history, these vaccine strains have identical genomic sequences (Parks et al., 2001). The Moraten vaccine (Merck) is the only measles vaccine used in the United States, whereas the Schwarz vaccine is used in many countries throughout the world. Other attenuated measles vaccines have been produced from locally derived wild-type strains, particularly in Russia, China, and Japan. One vaccine strain, the Edmonston-Zagreb vaccine, was also passaged in human diploid cells after attenuation in chick embryo fibroblasts, which may account for its increased immunogenicity and reactogenicity. Measles vaccines are relatively heat stable in the lyophilized (dry) form, but rapidly lose potency when exposed to heat after reconstitution.

In the 1960s, a formalin-inactivated, alum-precipitated measles vaccine (FIMV) was licensed and administered to children in the United States. Three doses of inactivated vaccine elicited a protective antibody response but antibody titers waned within months (Carter et al., 1962). Up to 60% of immunized children exposed to measles developed atypical measles, characterized by high fever, pneumonitis, and a petechial rash on the extremities (Fulginiti et al., 1967; Nader et al., 1968), leading to withdrawal of the FIMV in 1967. In a rhesus macaque model, atypical measles was shown to be associated with immune complex deposition in affected tissues and a systemic and pulmonary eosinophilia (Polack et al., 1999). The antibody response consisted of high levels of complement fixing antibodies with low avidity for measles virus, characteristics that may have promoted exaggerated immune complex formation and disease.

Seroconversion rates with attenuated measles vaccines in young infants are low because of immunologic immaturity and the interference of transplacentally acquired maternal antibodies with replication of vaccine virus (Gans et al., 1998). To protect young infants against measles, high-titer preparations containing 10–100 times the standard dose of vaccine virus were evaluated in several countries. Seroconversion rates in 4–6 month old infants immunized with high-titer measles vaccine were comparable to those of 9–15 month old children vaccinated with standard-titer measles vaccine, and the protective antibody response persisted for over 2 years. However, high-titer measles vaccine resulted in a poorly understood increase in mortality in immunized girls 1–2 years after vaccination compared to girls immunized with standard-titer vaccine at 9 months of age (Holt et al., 1993; Aaby et al., 1996). The increased mortality was attributable to infections commonly associated with measles, such as diarrhea and pneumonia. Although these studies were carried out in countries with different levels of socioeconomic development, excess mortality was observed only in countries with poor socioeconomic conditions and frequent malnutrition (Senegal, Haiti, and Guinea Bissau). The basis for the increased mortality in girls is not understood.

Current Licensed Measles Vaccines

Several live, attenuated measles vaccines are available worldwide, either as single-antigen vaccines or in combination with rubella and mumps vaccines (MR and MMR vaccines). Recently, a combined measles-mumps-rubella-varicella vaccine was licensed by the United States Food and Drug Administration (ProQuad, Merck and Co., Inc.). Licensed combination vaccines do not reduce the immunogenicity of the measles component. Most of the currently used measles vaccines were derived from the Edmonston strain of measles virus that was isolated by Enders and Peebles in 1954. These vaccines have undergone different passage histories in cell culture, but nucleotide sequence analysis shows less than 0.6% genetic differences between the vaccine strains. Vaccines in widespread use that were derived from the Edmonston measles virus stain include the Schwarz, Edmonston-Zagreb, and Moraten strains. Vaccines derived from other measles virus strains include CAM-70, Leningrad-16, and Shanghai 191.

The live, attenuated measles vaccines are typically cultured in primary chick embryo or human diploid (e.g., Edmonston-Zagreb) cells for several days. The supernatant fluid is harvested and frozen. Vaccine stocks are quality and safety tested before they are lyophilized. Measles vaccines may contain sorbitol or gelatin as stabilizers and the antibiotic neomycin, but do not contain thimerosal. Traces of the reverse transcriptase of an avian retrovirus can be found in vaccines cultured in chick embryo fibroblasts but there is no evidence that this is harmful to vaccine recipients. Prior to use, the vaccine must be reconstituted in sterile diluent. Measles vaccines lose about half of their potency after reconstitution if stored at 20°C for 1h, and lose almost all potency if stored at 37°C for 1h.

Measles vaccines are recommended for all susceptible children and adults for whom the vaccine is not contraindicated (Table 30.1). Adolescents and young adults may constitute a susceptible population,

TABLE 30.1 Recommendations for measles vaccination

Recommendations of the Advisory Committee on Immunization Practices for the United States

- A two-dose schedule with MMR vaccine is recommended.
- The first dose is recommended at 12–15 months of age.
- The second dose is recommended at 4–6 years of age.
- Adults at increased risk should receive special consideration for measles vaccination, including international travelers, students attending colleges and other post-high school educational institutions, and persons who work at health care facilities.
- Measles vaccine is not indicated for persons with severe hypersensitivity (anaphylaxis) to neomycin or gelatin, or who are severely immunocompromised.

Recommendations of the World Health Organization

- Measles vaccine should be administered at 9 months of age.
- A second opportunity for measles vaccination should be provided through mass measles vaccination campaigns.
- HIV-infected children should receive measles vaccine at 6 and 9 months of age, unless severely immunocompromised, because of their increased risk of severe measles.

including university students, health care workers, and military recruits. A single dose of measles vaccine provides lifelong immunity in the majority of vaccine recipients. However, a second opportunity for measles vaccination provides protection to those with primary vaccine failure and a chance to reach unvaccinated children. The second opportunity may be delivered through routine health services at school entry or through supplementary immunization activities such as national immunization days.

Measles vaccines are usually injected subcutaneously but can be administered intramuscularly. Each 0.5 mL dose of vaccine contains at least 1000 infective units of measles vaccine virus. The optimal age of measles vaccination is determined by consideration of the age-dependent increase in seroconversion rates following measles vaccination and the average age of infection. In regions of intense measles virus transmission, the average age of infection is low and the optimal strategy is to vaccinate against measles as young as possible. However, both maternally acquired antibodies and immunologic immaturity reduce the protective efficacy of measles vaccination in early infancy (Gans et al., 1998). In many parts of the world, 9 months is considered the optimal age of measles vaccination (Halsey, 1983), and is the age recommended by the Expanded Program on Immunization (EPI). Most countries following the EPI schedule administer measles vaccine alone, although more countries are introducing combined measles and rubella vaccines as rubella control programs expand. Under some circumstances, provision of an early dose of measles vaccine at 6 months of age (e.g., in outbreaks or to HIV-infected children) is appropriate. In regions that have achieved measles control or elimination, where the risk of measles in infants is low, the age of measles vaccination is increased to ensure that a higher proportion of children develop protective immunity. For example, in the United States the first dose of measles vaccine is administered at 12–15 months of age, as the combined MMR vaccine.

Measles vaccine induces both humoral and cellular immune responses. Antibodies first appear between 12 and 15 days after vaccination and peak at 21–28 days. IgM antibodies appear transiently in blood, IgA antibodies are predominant in mucosal secretions, and IgG antibodies persist in blood for years. Vaccination also induces measles virus-specific T-lymphocytes (Ovsyannikova et al., 2003; Wong-Chew et al., 2004). Although both humoral and cellular responses can be induced by measles vaccine, they are of lower magnitude and shorter duration compared to those following wild-type measles virus infection (Ward et al., 1995).

The proportion of children who develop protective antibody titers following measles vaccination depends on the presence of inhibitory maternal antibodies and the immunologic maturity of the vaccine recipient, as well as the dose and strain of vaccine virus. Frequently cited figures are that approximately 85% of children develop protective antibody titers when given measles vaccine at 9 months of age and 90–95% respond when vaccinated at 12 months of age (Cutts et al., 1995). Concurrent acute infections may interfere with the immune response to measles vaccine, although this is probably uncommon (Scott et al., 1999). Polymorphisms in human immune response genes also influence immune responses to measles vaccine (Ovsyannikova et al., 2004).

The duration of immunity following measles vaccination is more variable and shorter than following wild-type measles virus infection, with an estimated 5% of children developing secondary vaccine failure at 10–15 years after vaccination (Anders et al., 1996). Immunologic boosting from repeated exposure to measles virus may play a role in maintaining protective antibody levels in communities with measles virus transmission (Whittle et al., 1999).

The WHO encourages measles vaccination of all susceptible children and adults. Because measles can be more severe in HIV-infected persons and the risk of serious adverse events appears to be small, the WHO recommends that asymptomatic HIV-infected persons, and even those who are symptomatic but not severely immunocompromised, receive measles vaccine. The American Academy of Pediatrics recommends that severely immunocompromised children and adults, defined by low $CD4^+$ T-lymphocyte cell counts or percentage, should not receive measles vaccine because of the potential risk of vaccine-related pneumonitis. Measles vaccine is contraindicated in persons with a history of anaphylactic reaction to neomycin, gelatin, or other vaccine components.

Fever occurs in approximately 5–15% of recipients 6–12 days following measles vaccination, and a rash occurs in approximately 5% of recipients. These signs and symptoms are a consequence of the host immune response to replicating measles vaccine virus, but do not result in serious morbidity or mortality. Rarely, thrombocytopenia may occur.

Although assumed to be rare, the risk of disease caused by attenuated measles vaccine virus in HIV-infected persons is unknown. The only documented case of disease induced by vaccine virus in an HIV-infected person was in a 20-year-old man who died 15 months after receiving his second dose of measles vaccine (Angel et al., 1998). He had a very low $CD4^+$ T-lymphocyte cell count but no HIV-related symptoms at the time of vaccination. Ten months later he developed giant cell pneumonitis and measles vaccine virus was identified in his lung. Fatal, disseminated infection with measles vaccine virus has been reported rarely in persons with other impairments of immune function (Monafo et al., 1994), and MIBE due to vaccine virus was reported in a child with an uncharacterized immune deficiency (Bitnun et al., 1999).

Much public attention has focused on a purported association between MMR vaccine and autism following publication of a report in 1998 hypothesizing that MMR vaccine may cause a syndrome of autism and intestinal inflammation (Wakefield et al., 1998). The events that followed, and the public concern over the safety of MMR vaccine, led to diminished vaccine coverage in the United Kingdom and provide important lessons in the misinterpretation of epidemiologic evidence and the communication of scientific results to the public (Offit and Coffin, 2003). The publication that incited the concern was a case series describing 12 children with a regressive developmental disorder and chronic enterocolitis. Nine of the children had autism. Onset of the developmental delay was associated by the parents with MMR vaccination in eight children. This simple temporal association was misinterpreted and misrepresented as a possible causal relationship, first by the lead author of the study and then by the media and public. Subsequently, several comprehensive reviews and additional epidemiological studies rejected evidence of a causal relationship between MMR vaccination and autism (DeStefano and Thompson, 2004). One of the most conclusive studies was a large retrospective cohort study of over half a million Danish children that found the relative risk of MMR vaccine for autistic disorder to be 0.92 (95% confidence interval, 0.68–1.24) (Madsen et al., 2002).

Measles Vaccines in Development

Aerosol administration of measles vaccine was first evaluated in the early 1960s in several countries, including the former Soviet Union and the United States. More recent studies in South Africa (Dilraj et al., 2000) and Mexico (Bennett et al., 2002) have shown that aerosol administration of measles vaccine is highly effective in boosting antibody titers, although the primary immune response to aerosol measles vaccine is lower than following subcutaneous administration (Wong-Chew et al., 2004). Administration of measles vaccine by aerosol has the potential to greatly facilitate measles vaccination during mass campaigns, and the WHO plans to test and bring to licensure an aerosol measles vaccine by 2009.

The ideal measles vaccine would be inexpensive, safe, heat-stable, immunogenic in neonates or very young infants, and administered as a single dose without needle or syringe. The age at vaccination should coincide with the EPI schedule to maximize compliance and share resources. Finally, a new vaccine should not elicit atypical measles upon exposure of immunized individuals to wild-type measles virus and should not be associated with prolonged immunosuppression, adversely affecting immune responses to subsequent infections.

A number of vaccine candidates with some of these characteristics are undergoing preclinical studies (Table 30.2). Naked cDNA vaccines are thermostable, inexpensive, and could theoretically elicit antibody responses in the presence of passively acquired maternal antibody. DNA vaccines encoding either or both the measles H and F proteins are safe, immunogenic, and protective against measles challenge in naïve, juvenile rhesus macaques (Polack et al., 2000). A different

TABLE 30.2 Measles vaccine candidates

Vaccine type	Viral proteins	Model	Delivery	References
DNA	H, F, H/F	Macaques	Intradermal gene-gun	Polack et al. (2000)
DNA	H/F/N	Macaques	Intradermal	Premenko-Lanier et al. (2003, 2004)
Sindbis virus replicon particles	H, F, H/F	Macaques	Intradermal	Pan et al. (2005)
Sindbis virus replicon DNA	H, F, H/F	Mice	Intramuscular	Song et al. (2005)
Chimeric human-bovine parainfluenza virus type 3	H	Macaques	Intranasal	Skiadopoulos et al. (2001)
Salmonella enterica serovar Typhi CVD 908-*htrA*	H	Cotton rat	Intranasal	Pasetti et al. (2003)
Shigella flexneri 2a CVD 1208	H	Cotton rat	Intranasal	Pasetti et al. (2003)
Transgenic plant protein	H	Mice	Intraperitoneal, oral	Huang et al., (2001); Webster et al. (2002)

construct, containing H, F, and N genes and an IL-2 molecular adjuvant, provided protection to infant macaques in the presence of neutralizing antibody (Premenko-Lanier et al., 2003, 2004). Alternative techniques for administering measles DNA, such as alphavirus (Pan et al., 2005), parainfluenza virus (Skiadopoulos et al., 2001), or enteric bacterial (Pasetti et al., 2003) vectors, also are under investigation. Immune responses to intranasal administration of measles virus vaccines is enhanced by the use of adjuvants (Chabot et al., 2005). Novel oral immunization strategies have been developed using plant-based expression of the measles virus H protein in tobacco (Webster et al., 2005). These vaccines induced humoral immune responses following both primary immunization and as a booster dose after a DNA prime. In addition, attenuated recombinant measles virus vaccines have been used as vectors for the delivery of genes from other pathogens (Tangy and Naim, 2005).

POSTEXPOSURE IMMUNOPROPHYLAXIS

Immune globulin can be given intramuscularly to susceptible persons within 6 days of exposure to prevent or lessen the severity of disease. Postexposure immunoprophylaxis is indicated for those at high risk of severe disease, including susceptible household contacts between 6 months and 1 year of age, pregnant women, and immunocompromised persons. Where resources are not limited, HIV-infected children, or those exposed to HIV but whose infection status is unknown, should receive postexposure immunoprophylaxis. Persons who are not immunocompromised and who have received at least one dose of measles vaccine at 12 months of age or older should not receive postexposure immunoprophylaxis. If given with 72h of exposure, measles vaccine may provide protection against disease.

PROSPECTS FOR THE FUTURE

The possibility of measles eradication has been discussed for almost 40 years (Sencer et al., 1967). Serious consideration of measles eradication began in the late 1960s as smallpox eradication was nearing completion and the effective, long-term immunity induced by measles vaccine became apparent. Measles virus meets many of the biologic criteria for disease eradication (Moss and Griffin, 2006). Measles virus has no nonhuman reservoir; infection can be accurately diagnosed, and measles vaccination is a highly effective intervention. Although measles virus displays sufficient genetic variability to conduct molecular epidemiologic analyses, the antigenic epitopes against which protective antibodies develop have remained stable. Measles virus differs from smallpox and polio viruses, however, in that it is more highly infectious, necessitating much higher levels of population immunity to interrupt transmission.

Potential barriers to measles eradication include: (1) lack of political will; (2) difficulties of measles control in densely populated urban environments; (3) the HIV epidemic; (4) waning immunity and the potential transmission from subclinical cases; (5) transmission among susceptible adults; (6) the risk of unsafe injections; and (7) unfounded fears of disease due to measles vaccine (Orenstein et al., 2000). Whether the threat from bioterrorism precludes stopping measles vaccination after eradication is a topic of debate, but, at the least, a single-dose rather than a two-dose measles vaccination strategy could be adopted (Meissner et al., 2004). The elimination of endemic measles virus transmission in large geographic areas, such as the Americas, suggests that global eradication is feasible with current vaccination strategies (de Quadros, 2004). Many believe this to be a realistic and morally imperative goal, but, as polio eradication efforts have shown, the endgame may be full of challenges.

KEY ISSUES

- Measles is highly contagious viral infection.
- Great progress has been made in measles control and elimination through accelerated efforts, including increased routine vaccination coverage rates and supplementary immunization activities.
- Despite the reduction in measles cases and deaths, measles remains a leading cause of vaccine-preventable death worldwide.
- The high infectivity of measles virus is a characteristic suitable to a biothreat agent.
- Increasingly high levels of measles vaccination coverage throughout the world would protect many from the deliberate use of measles virus as a biothreat agent.
- Measles vaccine virus can be used as a vector to deliver genes derived from other pathogens.
- Genetic engineering of a measles virus strain that was not neutralized by antibodies induced by the current attenuated measles vaccines would likely have reduced infectivity, although this is not certain.

References

Aaby, P., Samb, B., Simondon, F., Knudsen, K., Seck, A.M., Bennett, J. et al. A comparison of vaccine efficacy and mortality during routine use of high-titre Edmonston-Zagreb and Schwarz standard measles vaccines in rural Senegal. Trans. R. Soc. Trop. Med. Hyg. 1996; 90:326–330.

Anders, J.F., Jacobson, R.M., Poland, G.A., Jacobsen, S.J. and Wollan, P.C. Secondary failure rates of measles vaccines: a metaanalysis of published studies. Pediatr. Infect. Dis. J. 1996; 15:62–66.

Andres, O., Obojes, K., Kim, K.S., ter Meulen, V. and Schneider-Schaulies, J. CD46- and CD150-independent endothelial cell infection with wild-type measles viruses. J. Gen. Virol. 2003; 84:1189–1197.

Angel, J.B., Walpita, P., Lerch, R.A., Sidhu, M.S., Masuredar, M., DeLellis, R.A. et al. Vaccine-associated measles pneumonitis in an adult with AIDS. Ann. Intern. Med. 1998; 129:104–106.

Atabani, S.F., Byrnes, A.A., Jaye, A., Kidd, I.M., Magnusen, A.F., Whittle, H. et al. Natural measles causes prolonged suppression of interleukin-12 production. J. Infect. Dis. 2001; 184:1–9.

Auwaerter, P.G., Rota, P.A., Elkins, W.R., Adams, R.J., DeLozier, T., Shi, Y. et al. Measles virus infection in rhesus macaques: altered immune responses and comparison of the virulence of six different virus strains. J. Infect. Dis. 1999; 180:950–958.

Beckford, A.P., Kaschula, R.O. and Stephen, C. Factors associated with fatal cases of measles. A retrospective autopsy study. S. Afr. Med. J. 1985; 68:858–863.

Bennett, J.V., de Castro, F., Valdespino-Gomez, J.L., Garcia-Garcia, M.L., Islas-Romero, R., Echaniz-Aviles, G. et al. Aerosolized measles and measles-rubella vaccines induce better measles antibody booster responses than injected vaccines: randomized trials in Mexican schoolchildren. Bull. World Health Organ. 2002; 80:806–812.

Bellini, W.J. and Helfand, R.F. The challenges and strategies for laboratory diagnosis of measles in an international setting. J. Infect. Dis. 2003; 187(Suppl. 1):S283–S290.

Bitnun, A., Shannon, P., Durward, A., Rota, P.A., Bellini, W.J., Graham, C. et al. Measles inclusion-body encephalitis caused by the vaccine strain of measles virus. Clin. Infect. Dis. 1999; 29:855–861.

Black, F.L. and Yannet, H. Inapparent measles after gamma globulin administration. JAMA 1960; 173:1183–1188.

Carter, C.H., Conway, T.J., Cornfeld, D., Iezzoni, D.G., Kempe, C.H., Moscovici, C. et al. Serologic response of children to inactivated measles vaccine. JAMA 1962; 179:848–853.

Chabot, S., Brewer, A., Lowell, G., Plante, M., Cyr, S., Burt, D.S. et al. A novel intranasal Protollin-based measles vaccine induces mucosal and systemic neutralizing antibody responses and cell-mediated immunity in mice. Vaccine 2005; 23:1374–1383.

Coovadia, H.M., Wesley, A., Henderson, L.G., Brain, P., Vos, G.H. and Hallett, A.F. Alterations in immune responsiveness in acute measles and chronic post-measles chest disease. Int. Arch. Allergy Appl. Immunol. 1978; 56:14–23.

Crowe, J.E.J. Influence of maternal antibodies on neonatal immunization against respiratory viruses. Clin. Infect. Dis. 2001; 33(10):1720–1727.

Cutts, F.T., Grabowsky, M. and Markowitz, L.E. The effect of dose and strain of live attenuated measles vaccines on serological responses in young infants. Biologicals 1995; 23:95–106.

de Quadros, C.A. Can measles be eradicated globally? Bull. World Health Organ. 2004; 82:134–138.

de Quadros, C.A., Hersh, B.S., Nogueira, A.C., Carrasco, P.A. and da Silveira, C.M. Measles eradication: experience in the Americas. Bull. World Health Organ 1998; 76(Suppl. 2):47–52.

DeStefano, F. and Thompson, W.W. MMR vaccine and autism: an update of the scientific evidence. Expert. Rev. Vaccines 2004; 3:19–22.

Dilraj, A., Cutts, F.T., de Castro, J.F., Wheeler, J.G., Brown, D., Roth, C. et al. Response to different measles vaccine strains given by aerosol and subcutaneous routes to schoolchildren: a randomised trial. Lancet 2000; 355:798–803.

Dossetor, J., Whittle, H.C. and Greenwood, B.M. Persistent measles infection in malnourished children. BMJ 1977; 1:1633–1635.

D'Souza, R.M. and D'Souza, R. Vitamin A for the treatment of children with measles: a systematic review. J. Trop. Pediatr. 2002; 48:323–327.

Duke, T. and Mgone, C.S. Measles: not just another viral exanthem. Lancet 2003; 361:763–773.

Enders, J.F. and Peebles, T.C. Propagation in tissue cultures of cytopathic agents from patients with measles. Proc. Soc. Exp. Biol. Med. 1954; 86:277–286.

Fine, P.E. and Clarkson, J.A. Measles in England and Wales—I: an analysis of factors underlying seasonal patterns. Int. J. Epidemiol. 1982; 11:5–14.

Forni, A.L., Schluger, N.W. and Roberts, R.B. Severe measles pneumonitis in adults: evaluation of clinical characteristics and therapy with intravenous ribavirin. Clin. Infect. Dis. 1994; 19:454–462.

Fulginiti, V.A., Eller, J.J., Downie, A.W. and Kempe, C.H. Altered reactivity to measles virus: atypical measles in children previously immunized with inactivated measles virus vaccines. JAMA 1967; 202:1075.

Gans, H.A., Arvin, A.M., Galinus, J., Logan, L., DeHovitz, R. and Maldonado, Y. Deficiency of the humoral immune response to measles vaccine in infants immunized at age 6 months. JAMA 1998; 280:527–532.

Good, R.A. and Zak, S.J. Disturbances in gamma globulin synthesis as "experiments of nature". Pediatrics 1956:109–149.

Greenberg, B.L., Sack, R.B., Salazar-Lindo, E., Budge, E., Gutierrez, M., Campos, M. et al. Measles-associated diarrhea in hospitalized children in Lima, Peru: pathogenic agents and impact on growth. J. Infect. Dis. 1991; 163:495–502.

Griffin, D.E., Cooper, S.J., Hirsch, R.L., Johnson, R.T., Soriano, I.L., Roedenbeck, S. et al. Changes in plasma IgE levels during complicated and uncomplicated measles virus infections. J. Allergy Clin. Immunol. 1985; 76:206–213.

Griffin, D.E. and Ward, B.J. Differential CD4T cell activation in measles. J. Infect. Dis. 1993; 168:275–281.

Halsey, N.A. The optimal age for administering measles vaccine in developing countries. In: Recent Advances in Immunization: A Bibliographic Review (N.A. Halsey and C.A. de Quadros, Eds.), pp. 4–17. Washington, DC: Pan American Health Organization, 1983.

Henao-Restrepo, A.M., Strebel, P., Hoekstra, E.J., Birmingham, M. and Bilous, J. Experience in global measles control, 1990–2001. J. Infect. Dis. 2003; 187(Suppl. 1):S15–S21.

Hirsch, R.L., Griffin, D.E., Johnson, R.T., Cooper, S.J., Lindo de Soriano, I., Roedenbeck, S. et al. Cellular immune responses during complicated and uncomplicated measles virus infections of man. Clin. Immunol. Immunopathol. 1984; 31:1–12.

Holt, E.A., Moulton, L.H., Siberry, G.K. and Halsey, N.A. Differential mortality by measles vaccine titer and sex. J. Infect. Dis. 1993; 168:1087–1096.

Huang, Z., Dry, I., Webster, D., Strugnell, R. and Wesselingh, S. Plant-derived measles virus hemagglutinin protein induces neutralizing antibodies in mice. Vaccine 2001; 19:2163–2171.

Hutchins, S., Markowitz, L., Atkinson, W., Swint, E. and Hadler, S. Measles outbreaks in the United States, 1987 through 1990. Pediatr. Infect. Dis. J. 1996; 15(1):31–38.

Isa, M.B., Martinez, L., Giordano, M., Passeggi, C., de Wolff, M.C. and Nates, S. Comparison of immunoglobulin G subclass profiles induced by measles virus in vaccinated and naturally infected individuals. Clin. Diagn. Lab. Immunol. 2002; 9:693–697.

Karp, C.L., Wysocka, M., Wahl, L.M., Ahearn, J.M., Cuomo, P.J., Sherry, B. et al. Mechanism of suppression of cell-mediated immunity by measles virus. Science 1996; 273:228–231.

Kemper, C., Chan, A.C., Green, J.M., Brett, K.A., Murphy, K.M. and Atkinson, J.P. Activation of human $CD4^+$ cells with CD3 and CD46 induces a T-regulatory cell 1 phenotype. Nature 2003; 421:388–392.

Kimura, A., Tosaka, K. and Nakao, T. Measles rash. I. Light and electron microscopic study of skin eruptions. Arch. Virol. 1975; 47:295–307.

Lambert, S.B., Morgan, M.L., Riddell, M.A., Andrews, R.M., Kelly, H.A., Leydon, J.A. et al. Measles outbreak in young adults in Victoria, 1999. Med. J. Aust. 2000; 173:467–471.

Madsen, K.M., Hviid, A., Vestergaard, M., Schendel, D., Wohlfahrt, J., Thorsen, P. et al. A population-based study of measles, mumps, and rubella vaccination and autism. N. Engl. J. Med. 2002; 347:1477–1482.

Malvoisin, E. and Wild, T.F. Measles virus glycoproteins: studies on the structure and interaction of the haemagglutinin and fusion proteins. J. Gen. Virol. 1993; 74(Pt. 11):2365–2372.

McNeill, W.H. Plagues and Peoples. London: Penguin, 1976.

Meissner, H.C., Strebel, P.M. and Orenstein, W.A. Measles vaccines and the potential for worldwide eradication of measles. Pediatrics 2004; 114:1065–1069.

Monafo, W.J., Haslam, D.B., Roberts, R.L., Zaki, S.R., Bellini, W.J. and Coffin, C.M. Disseminated measles infection after vaccination in a child with a congenital immunodeficiency. J. Pediatr. 1994; 124:273–276.

Morley, D. Severe measles in the tropics. Br. Med. J. 1969; 1:297–300.

Moss, W.J., Cutts, F. and Griffin, D.E. Implications of the human immunodeficiency virus epidemic for control and eradication of measles. Clin. Infect. Dis. 1999; 29:106–112.

Moss, W.J. and Griffin, D.E. Global measles elimination. Nat. Rev. Microbiol. 2006; 4:900–908.

Moss, W.J., Ryon, J.J., Monze, M. and Griffin, D.E. Differential regulation of interleukin (IL)-4, IL-5, and IL-10 during measles in Zambian children. J. Infect. Dis. 2002; 186:879–887.

Nader, R.R., Horwitz, M.S. and Rousseau, J. Atypical exanthem following exposure to natural measles: eleven cases in children previously inoculated with killed vaccine. J. Pediatr. 1968; 72:22–28.

Offit, P.A. and Coffin, S.E. Communicating science to the public: MMR vaccine and autism. Vaccine 2003; 22:1–6.

Orenstein, W.A., Strebel, P.M., Papania, M., Sutter, R.W., Bellini, W.J. and Cochi, S.L. Measles eradication: is it in our future? Am. J. Public Health 2000; 90:1521–1525.

Ovsyannikova, I.G., Dhiman, N., Jacobson, R.M., Vierkant, R.A. and Poland, G.A. Frequency of measles virus-specific CD4+ and CD8+ T cells in subjects seronegative or highly seropositive for measles vaccine. Clin. Diagn. Lab. Immunol. 2003; 10:411–416.

Ovsyannikova, I.G., Jacobson, R.M. and Poland, G.A. Variation in vaccine response in normal populations. Pharmacogenomics 2004; 5:417–427.

Pan, C.H., Valsamakis, A., Colella, T., Nair, N., Adams, R.J., Polack, F.P. et al. Inaugural article: modulation of disease, T cell responses, and measles virus clearance in monkeys vaccinated with H-encoding alphavirus replicon particles. Proc. Natl. Acad. Sci. U.S.A. 2005; 102:11581–11588.

Panum, P. Observations made during the epidemic of measles on the Faroe Islands in the year 1846. Med. Classics 1938; 3:829–886.

Parks, C.L., Lerch, R.A., Walpita, P., Wang, H.P., Sidhu, M.S. and Udem, S.A. Comparison of predicted amino acid sequences of measles virus strains in the Edmonston vaccine lineage. J. Virol. 2001; 75:910–920.

Pasetti, M.F., Barry, E.M., Losonsky, G., Singh, M., Medina-Moreno, S.M., Polo, J.M. et al. Attenuated *Salmonella enterica* serovar Typhi and *Shigella flexneri* 2a strains mucosally deliver DNA vaccines encoding measles virus hemagglutinin, inducing specific immune responses and protection in cotton rats. J. Virol. 2003; 77:5209–5217.

Patterson, J.B., Thomas, D., Lewicki, H., Billeter, M.A. and Oldstone, M.B. V and C proteins of measles virus function as virulence factors in vivo. Virology 2000; 267:80–89.

Permar, S.R., Klumpp, S.A., Mansfield, K.G., Kim, W.K., Gorgone, D.A., Lifton, M.A. et al. Role of CD8(+) lymphocytes in control and clearance of measles virus infection of rhesus monkeys. J. Virol. 2003; 77:4396–4400.

Permar, S.R., Moss, W.J., Ryon, J.J., Monze, M., Cutts, F., Quinn, T.C. et al. Prolonged measles virus shedding in human immunodeficiency virus-infected children, detected by reverse transcriptase-polymerase chain reaction. J. Infect. Dis. 2001; 183:532–538.

Perry, R.T. and Halsey, N.A. The clinical significance of measles: a review. J. Infect. Dis. 2004; 189(Suppl. 1):S4–S16.

Polack, F.P., Auwaerter, P.G., Lee, S.H., Nousari, H.C., Valsamakis, A., Leiferman, K.M. et al. Production of atypical measles in rhesus macaques: evidence for disease mediated by immune complex formation and eosinophils in the presence of fusion-inhibiting antibody. Nat. Med. 1999; 5:629–634.

Polack, F.P., Lee, S.H., Permar, S., Manyara, E., Nousari, H.G., Jeng, Y. et al. Successful DNA immunization against measles: neutralizing antibody against either the hemagglutinin or fusion glycoprotein protects rhesus macaques without evidence of atypical measles. Nat. Med. 2000; 6:776–781.

Premenko-Lanier, M., Rota, P.A., Rhodes, G., Verhoeven, D., Barouch, D.H., Lerche, N.W. et al. DNA vaccination of infants in the presence of maternal antibody: a measles model in the primate. Virology 2003; 307:67–75.

Premenko-Lanier, M., Rota, P.A., Rhodes, G.H., Bellini, W.J. and McChesney, M.B. Protection against challenge with measles virus (MV) in infant macaques by an MV DNA vaccine administered in the presence of neutralizing antibody. J. Infect. Dis. 2004; 189:2064–2071.

Riddell, M.A., Moss, W.J., Hauer, D., Monze, M. and Griffin, D.E. Slow clearance of measles virus RNA after acute infection. J. Clin. Virol. 2007; 39:312–317.

Rota, P.A. and Bellini, W.J. Update on the global distribution of genotypes of wild type measles viruses. J. Infect. Dis. 2003; 187(Suppl. 1):S270–S276.

Schneider, U., von Messling, V., Devaux, P. and Cattaneo, R. Efficiency of measles virus entry and dissemination through different receptors. J. Virol. 2002; 76:7460–7467.

Scott, S., Cutts, F.T. and Nyandu, B. Mild illness at or after measles vaccination does not reduce seroresponse in young children. Vaccine 1999; 17:837–843.

Sencer, D.J., Dull, H.B. and Langmuir, A.D. Epidemiologic basis for eradication of measles in 1967. Public Health Rep. 1967; 82:253–256.

Servet-Delprat, C., Vidalain, P.O., Azocar, O., Le Deist, F., Fischer, A. and Rabourdin-Combe, C. Consequences of Fas-mediated human dendritic cell apoptosis induced by measles virus. J. Virol. 2000; 74:4387–4393.

Skiadopoulos, M.H., Surman, S.R., Riggs, J.M., Collins, P.L. and Murphy, B.R. A chimeric human-bovine parainfluenza virus type 3 expressing measles virus hemagglutinin is attenuated for replication but is still immunogenic in rhesus monkeys. J. Virol. 2001; 75:10498–10504.

Solomon, T., Hart, C.A., Vinjamuri, S., Beeching, N.J., Malucci, C. and Humphrey, P. Treatment of subacute sclerosing panencephalitis with interferon-alpha, ribavirin, and inosiplex. J. Child. Neurol. 2002; 17:703–705.

Song, M.K., Vindurampulle, C.J., Capozzo, A.V., Ulmer, J., Polo, J.M., Pasetti, M.F. et al. Characterization of immune responses induced by intramuscular vaccination with DNA vaccines encoding measles virus hemagglutinin and/or fusion proteins. J. Virol. 2005; 79:9854–9861.

Spehner, D., Drillien, R. and Howley, P.M. The assembly of the measles virus nucleoprotein into nucleocapsid-like particles is modulated by the phosphoprotein. Virology 1997; 232:260–268.

Tamashiro, V.G., Perez, H.H. and Griffin, D.E. Prospective study of the magnitude and duration of changes in tuberculin reactivity during uncomplicated and complicated measles. Pediatr. Infect. Dis. J. 1987; 6:451–454.

Tangy, F. and Naim, H.Y. Live attenuated measles vaccine as a potential multivalent pediatric vaccination vector. Viral Immunol. 2005; 18:317–326.

Valsamakis, A., Schneider, H., Auwaerter, P.G., Kaneshima, H., Billeter, M.A. and Griffin, D.E. Recombinant measles viruses with mutations in the C, V, or F gene have altered growth phenotypes in vivo. J. Virol. 1998; 72:7754–7761.

Villamor, E. and Fawzi, W.W. Vitamin A supplementation: implications for morbidity and mortality in children. J. Infect. Dis. 2000; 182(Suppl. 1):S122–S133.

Wakefield, A.J., Murch, S.H., Anthony, A., Linnell, J., Casson, D.M., Malik, M. et al. Ileal-lymphoid-nodular hyperplasia, non-specific colitis, and pervasive developmental disorder in children. Lancet 1998; 351:637–641.

Ward, B., Boulianne, N., Ratnam, S., Guiot, M.C., Couilard, M. and De Serres, G. Cellular immunity in measles vaccine failure: demonstration of measles antigen-specific lymphoproliferative responses despite limited serum antibody production after revaccination. J. Infect. Dis. 1995; 172:1591–1595.

Webster, D.E., Cooney, M.L., Huang, Z., Drew, D.R., Ramshaw, I.A., Dry, I.B. et al. Successful boosting of a DNA measles immunization with an oral plant-derived measles virus vaccine. J. Virol. 2002; 76:7910–7912.

Webster, D.E., Thomas, M.C., Huang, Z. and Wesselingh, S.L. The development of a plant-based vaccine for measles. Vaccine 2005; 23:1859–1865.

Whittle, H.C., Aaby, P., Samb, B., Jensen, H., Bennett, J. and Simondon, F. Effect of subclinical infection on maintaining immunity against measles in vaccinated children in West Africa. Lancet 1999; 353:98–101.

Wolfson, L.J., Strebel, P.M., Gacic-Dobo, M., Hoekstra, E.J., McFarland, J.W. and Hersh, B.S. Has the 2005 measles mortality reduction goal been achieved? A natural history modelling study. Lancet 2007; 369:191–200.

Wong-Chew, R.M., Islas-Romero, R., Garcia-Garcia, M.L., Beeler, J.A., Audet, S., Santos-Preciado, J.I. et al. Induction of cellular and humoral immunity after aerosol or subcutaneous administration of Edmonston-Zagreb measles vaccine as a primary dose to 12-month-old children. J. Infect. Dis. 2004; 189:254–257.

World Health Organization Progress in reducing measles mortality—worldwide 1999–2003. Wkly. Epidemiol. Rec. 2005: 78–81.

World Health Organization Global distribution of measles and rubella genotypes—update. Wkly. Epidemiol. Rec. 2006; 81:474–479.

World Health Organization United Nations Children's Fund. Measles mortality reduction and regional elimination strategic plan 2001–2005. Geneva, Switzerland: World Health Organization, 2001.

CHAPTER

31

Hendra and Nipah

Vincent Deubel[1] *and Fabian T. Wild*[2]

[1]Institut Pasteur of Shanghai, Chinese Academy of Sciences, Shanghai, China

[2]Inserm, U758, IFR128 BioSciences Lyon-Gerland, Lyon, France

OUTLINE

ABSTRACT

Among viruses that can be called "emerging viruses," Hendra and Nipah viruses, together with SARS coronavirus and avian influenza viruses, are on the top of those listed for the last 15 years. These two viruses were new genetically close-related viruses, clustering in a new genus *Henipavirus*, in the family *Paramyxoviridae*. They emerged in different animals and crossed the species barrier to infect humans, in Australia in 1994 and in Malaysia in 1998, causing different clinical outcomes with Nipah outbreaks reported later on in India and Bangladesh.

Then, both spread geographically showing pathological and epidemiological features similar to other endemic zoonotic viruses, each outbreak challenging public health services to control the disease. Very rapidly, scientists identified the reservoir hosts of the viruses by serology and by isolating viruses in *Pteropus* fruit bats. Due to the high mortality rate caused by henipaviruses in domestic animals and in humans, and their possible transmission by aerosol, they were immediately categorized to be handled in Biosafety Level-4 containment. Thus, only a few teams have been able to undertake research on the viruses, and many aspects of the virus life cycle, host responses to infection, pathogenesis, and prophylaxis still need to be solved. From the review presented in this chapter, it is obvious that these viruses represent potential agents for bioterrorism against which protection should be rapidly implemented.

INTRODUCTION

Based on unique genetic characteristics, two newly identified highly pathogenic agents, Hendra virus (HeV) and Nipah virus (NiV) have been classified in the genus *Henipavirus* in the family *Paramyxoviridae* (Wang et al., 2000; Mayo, 2002). These zoonotic viruses have caused fatal diseases in animals, HeV in horses, and NiV in pigs, and have crossed the species barrier to infect humans with mortality rates up to 75% (Eaton et al., 2007). They emerged from *Pteropus* fruit bats. They must be handled in Biosafety Level 4 containment. Due to the high pathogenicity of the viruses and cultural reasons, only few necropsy samples are available for pathogenesis studies. However, several animal models have been examined for further studies on vaccine and therapy development.

HISTORY OF THE DISEASE

HeV virus first occurred in 1994 in Brisbane, Australia, as the causative agent of a highly fatal disease of race horses. During the outbreak, 14 of 21 infected horses died of an acute respiratory disease, and one of two infected humans, the trainer, also died (Murray et al., 1995; Selvey et al., 1995). A second incident in which two horses and a farmer died came to light a year later, although infection actually occurred prior to the Brisbane outbreak. The farmer developed a mild meningitis illness after he participated to the necropsies of the two horses, but then recovered. Thirteen months later, he became ill and died from a severe encephalitis, which was found to be caused by HeV (O'Sullivan et al., 1997). In total, only four human cases of HeV infection have been recorded and two of them died (Murray et al., 1995; O'Sullivan et al., 1997). Retrospective analysis of equine specimens confirmed that they too had been infected with HeV indicating that the horses were the probable source of infection. Thus, the virus reactivated after a period of latency. Two more horses died of HeV infection in 1999 and 2004.

NiV, closely related to HeV, emerged in 1998 in Peninsular Malaysia, as a highly pathogenic infection of pigs and humans (Chua et al., 1999) (Fig. 31.1). It resulted in 265 human cases of whom 105 died, and a further 11 cases and 1 death among abattoir workers in Singapore (Paton et al., 1999) (Table 31.1). About 90% of human cases were among pig farmers or others associated with the transport or culling of pigs (Goh et al., 2000; Chua et al., 1999). Only 8% of patients reported no contact with pigs suggesting other sources of infection, such as infected dogs and cats (Parashar et al., 2000).

The disease in humans was predominantly encephalitic (Lee et al., 1999) whereas that in pigs was an acute respiratory disease with or without neurological signs (Mohd Nor et al., 2000). No human-to-human transmission was observed, but the disease in pigs was highly contagious. The majority of patients who survived acute NiV infection made a full recovery, but about 20% had residual neurological sequelae. The outbreak was controlled by the culling of over 1.1 million pigs and by implementing strict quarantine measures.

More recently, outbreaks of NiV infection have been reported from Bangladesh and India (Fig. 31.1). Five outbreaks have occurred in Bangladesh between 2001 and 2005 with a case fatality rate over 75% (Hsu et al., 2001; Eaton et al, 2007; Field et al., 2007; Harcourt et al., 2005; WHO, 2004, ICDDR, 2004b) (Table 31.1). The epidemics in Faridpur district showed specific features not observed during previous NiV epidemics with patients who developed acute respiratory distress syndrome. Human-to-human transmission was also reported during this epidemic (WHO, 2004). The incubation period of contaminated people through close contact with sick patients was from 6 to 11 days and 6 members of one family and community died after direct contact or subsequent contact with one index case (ICDDR, 2004b). Epidemiological investigations of the epidemic in Goalanda failed to establish any link with pigs, but rather suggested that contamination could have occurred by eating fruits spoiled by urine or saliva of fruit bats during the night or by drinking palm juice (Luby et al., 2006). One outbreak has been reported in Siliguri from West Bengal in

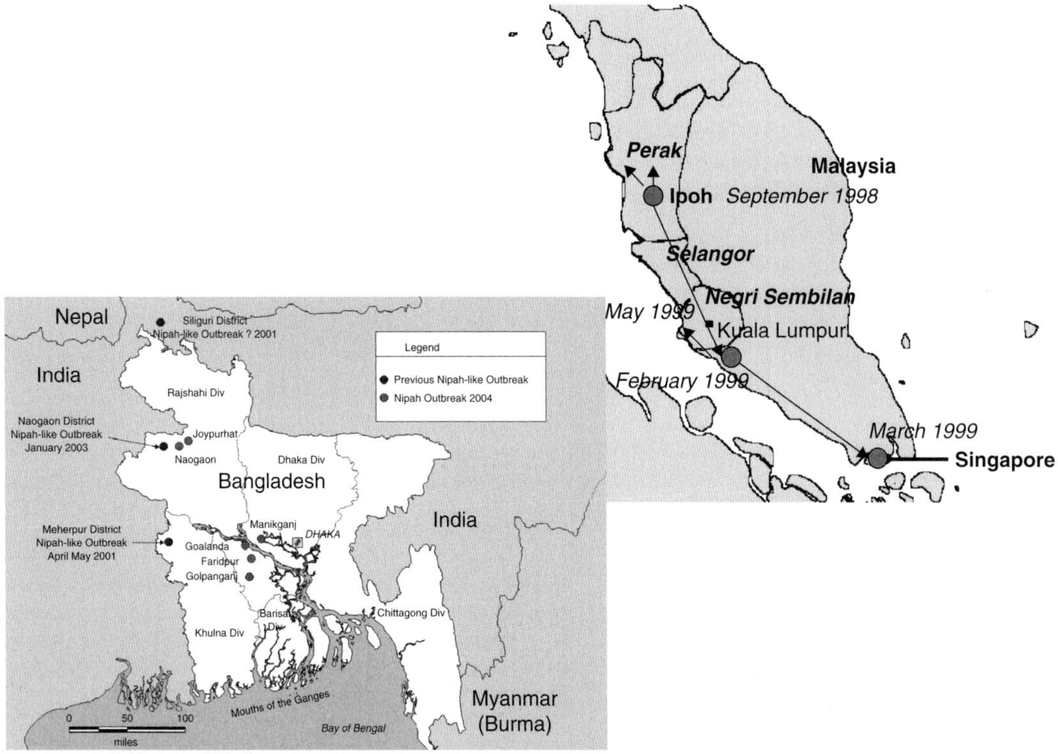

FIGURE 31.1 Geographic distribution of Nipah virus outbreaks, 1998–2005 (see Table 31.1).

TABLE 31.1 Nipah virus outbreaks, 1998–2005

Dates	Location	Number of cases	Number of deaths	CFR (%)
September–May 1998	Malaysia	265	105	40
March 1999	Singapore	11	1	9
February 2001	Siliguri, India	66	45	68
April–May 2001	Meherpur, Bangladesh	13	9	69
January 2003	Naogaon, Bangladesh	12	8	67
January 2004	Goalando, Bangladesh	29	22	76
April 2004	Faridpur, Bangladesh	36	27	75
January–March 2005	Tangail, Bangladesh	12	11	92

India (Harit et al., 2006). A new genetic variant was identified as responsible for the epidemics (Harcourt et al., 2005). These outbreaks have been characterized by very high fatality rates. In Tangail District, Bangladesh, from December 15, 2004, through January 31, 2005, 12 persons were infected with NiV and developed fever with new onset of seizures or altered mental status. Eleven (92%) died. The strain was genetically related to NiV but showed overall 91.2% nucleotide homology and >92% amino acid homology with the NiV strain isolated in Malaysia (Harcourt et al., 2005). An outbreak of febrile illness associated with central nervous system (CNS) defects was notified in the two first months of 2001 in Siliguri, West Bengal, India (Chadha et al., 2006). Retrospective serological analysis detected NiV-specific IgM and IgG antibodies in 9 of 18 patients and RT-PCR detected RNA from NiV in urine samples from 5 of them.

ETIOLOGIC AGENT

Virus Morphology

Henipaviruses are pleiomorphic with spherical or filamentous structures ranging from 40 to 2000 nm in size (Fig. 31.2A). Nucleocapsids are visible in electron microscopy with a diameter of about 18 nm (Hyatt

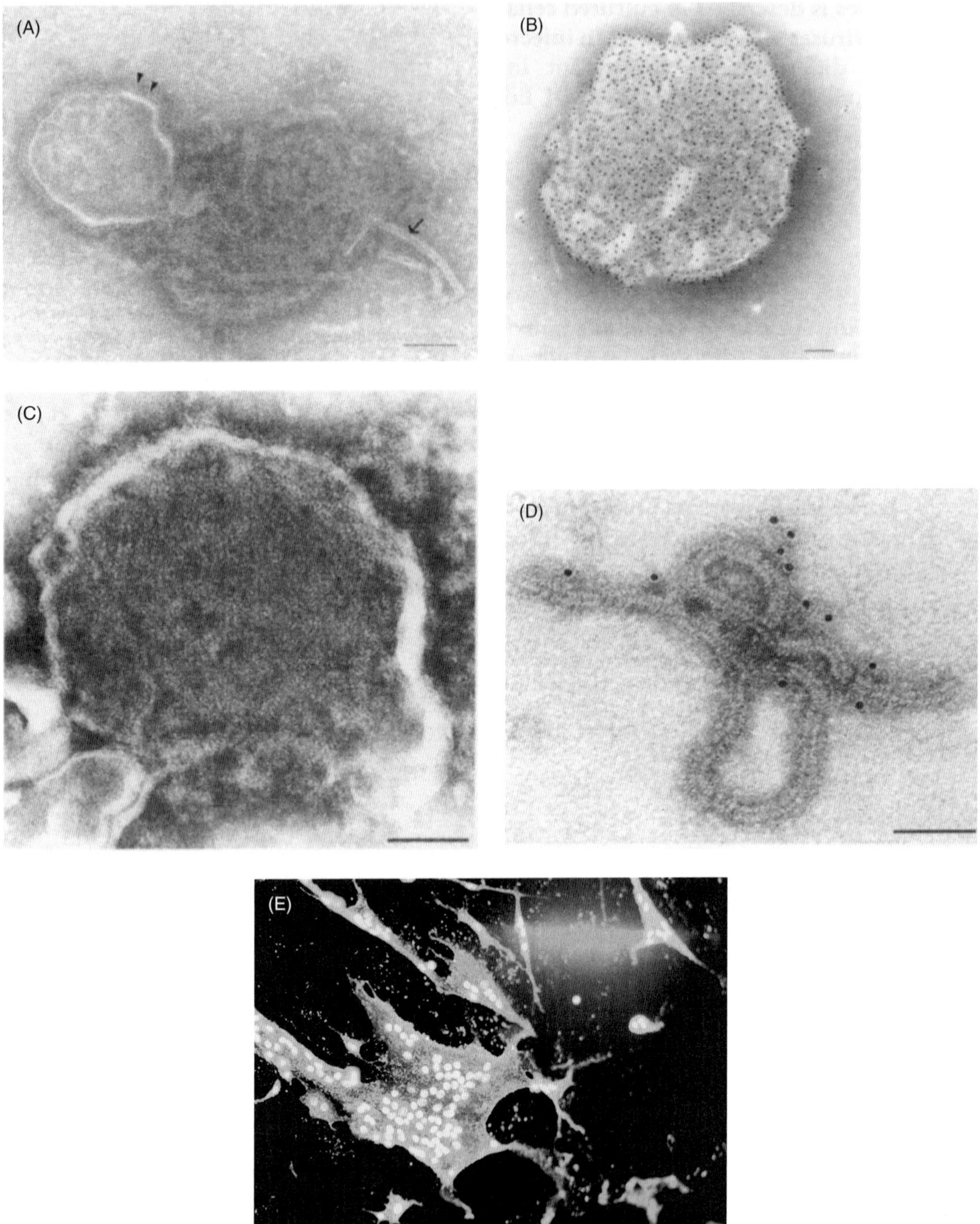

FIGURE 31.2 Microscopy of Nipah virus. (A–D) Electron microscopy showing enveloped virus (A–C) and nucleocapsids (D) (from Hyatt et al., 2001); (B) Syncytia of BHK-21 cells infected with Nipah virus (m.o.i. 0.025, 24 h postinfection) and immunostained with anti-Nipah antibodies and counterstained with Hoechst (original magnification: 20×) (see color plate section).

et al., 2001). Viruses form characteristic aggregation of nucleocapsids in the cytoplasm of infected cells (Goldsmith et al., 2003) (Fig. 31.2B). Infected cells form large syncitia (Fig. 31.2E) with nucleus disposed at the periphery (NiV) or more concentrated inside the cell (HeV). NiV form larger syncitia than HeV or Tioman virus, another paramyxovirus isolated from bats. Virus replication leads to virus release in the cell supernatant within 5–6 h with titers higher than 10^8 PFU/ml. Vero cells rapidly fuse, within 1–3 days, with

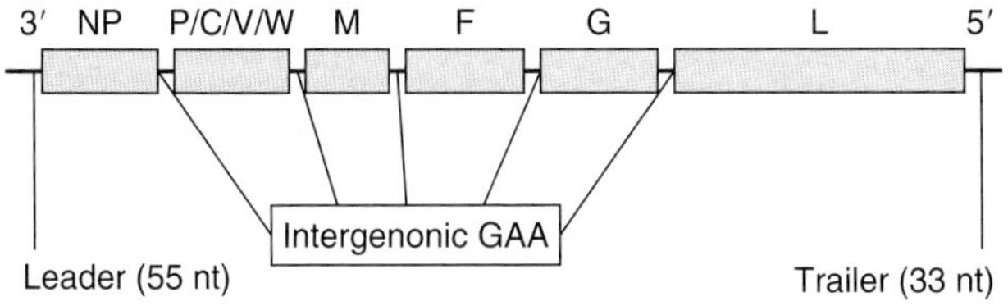

FIGURE 31.3 Structure of henipavirus genome.

a multiplicity of infection ranging from 1 to 0.001, respectively.

Virus Classification

The *Paramyxoviridae* family is subdivided based on the nature of the encoded genes and their organization within the genome. NiV and HeV are closely related to the morbilliviruses and respiroviruses coding for six proteins with similar functions. However, they differ from the respiroviruses in that they lack a functional neuraminidase. They can be further differentiated by the size of their genome, 18 kb, compared with 15.1–15.9 kb for other paramyxoviruses. They have now been classified in a separate genus, *Henipavirus*.

The genome of HeV is 18,234 nucleotides, while that of NiV isolated in the Malaysian epidemic in 1998 is 18,246 nucleotides (Harcourt et al., 2000, 2001; Wang et al., 2001; Chan et al., 2001), whereas that isolated in Bangladesh in 2004 contains an extra six nucleotides (Harcourt et al., 2005). Thus, the genome size follows the rule of six as was proposed from in vitro studies (Halpin et al., 2004). The 3′ leader (55 nucleotides) and the 5′ trailer (33 nucleotides) sequences function as promoters for the positive and negative strands, respectively. The 12-terminal 3′ and 5′ nucleotides are highly conserved and complementary (Wang et al., 2001). The organization of the different genes within the genome is shown in Fig. 31.3. The intergenic sequence GAA is the same as in morbilliviruses and respiroviruses.

Molecular Epidemiology

Several strains of HeV were isolated from horses, humans, and bats, in different epidemic locations in Australia (1994–1996) and NiV was isolated from pigs, humans, and bats in Malaysia, India, Bangladesh, and Cambodia (1999–2004). Full-length sequences of HeV and NiV genomes share up to 75% nucleotide sequence similarity but only 40–45% with other viruses in *Paramyxoviridae* family (Wang et al., 2000).

Partial sequence analysis of HeV showed high similarity between horse, human, and bat isolates over a 2-year period. However, full-length or partial sequencing of NiV has shown significant divergence, in particular between Malaysian/Cambodian and India/Bangladesh clusters which differed by about 5% in nucleotide sequence. Clusters shared more than 98% nucleotide sequence similarity (Reynes et al., 2005; Chua et al., 2002a; Harcourt et al., 2005; Chadha et al., 2006; Luby et al., 2006; Abubakar et al., 2004). Higher divergence was noticed in Bangladesh/India outbreaks, suggesting different introductions of the virus by different colonies of Pteroptera. No correlation between amino acid change and epidemic or virulence features has been identified up to now.

Antigens Encoded by Agent

The *nucleoprotein gene* codes for a protein of 532 amino acids having 92.1% homology between the two viruses. In keeping with other Paramyxoviruses, the variable region is at the –COOH terminus where 22 of the 42 amino acid differences between the N proteins of NiV and HeV are located in a 28-amino-acid portion.

The *P gene* is transcribed into an mRNA (2704 nucleotides) coding for a P protein of 707 (HeV) or 709 (NiV) amino acids. This protein is approximately 100 amino acids larger than the P protein of other paramyxoviruses (Lamb and Kolakofsky, 2001). The overall homology between NiV and HeV is 67.6%, but the amino- (137 amino acids) and carboxy- (321 amino acids) terminals contain approximately 80% identity. Like other paramyxoviruses, a second protein, C is translated from a different open reading frame (+23 or +26 nucleotides) to give a protein of 166 amino acids, which has 83.2% identity for NiV and HeV. Interestingly, HeV, but not NiV, has the potential to code for a second protein in the same frame as C, but this has so far not been identified in virus-infected cells. Similar to other paramyxoviruses, henipaviruses have found a mechanism of increasing their genetic potential. During transcription of the *P gene*, one or two G residues may be inserted at a specific editing site. This leads to the synthesis of a fusion-protein in which the N-terminal P protein contains 406 amino acids fused to (+1G) 55 (HeV) or 52 (NiV) amino acid C-terminal giving a V protein or (+2Gs) leading to a 47 amino acid C-terminal W protein (Fig. 31.4).

The *matrix (M) gene* codes for a protein of 352 amino acids with a homology of 89% between NiV and HeV, although 12 of the changes occur within the first 13 amino acids. In paramyxoviruses, the M protein

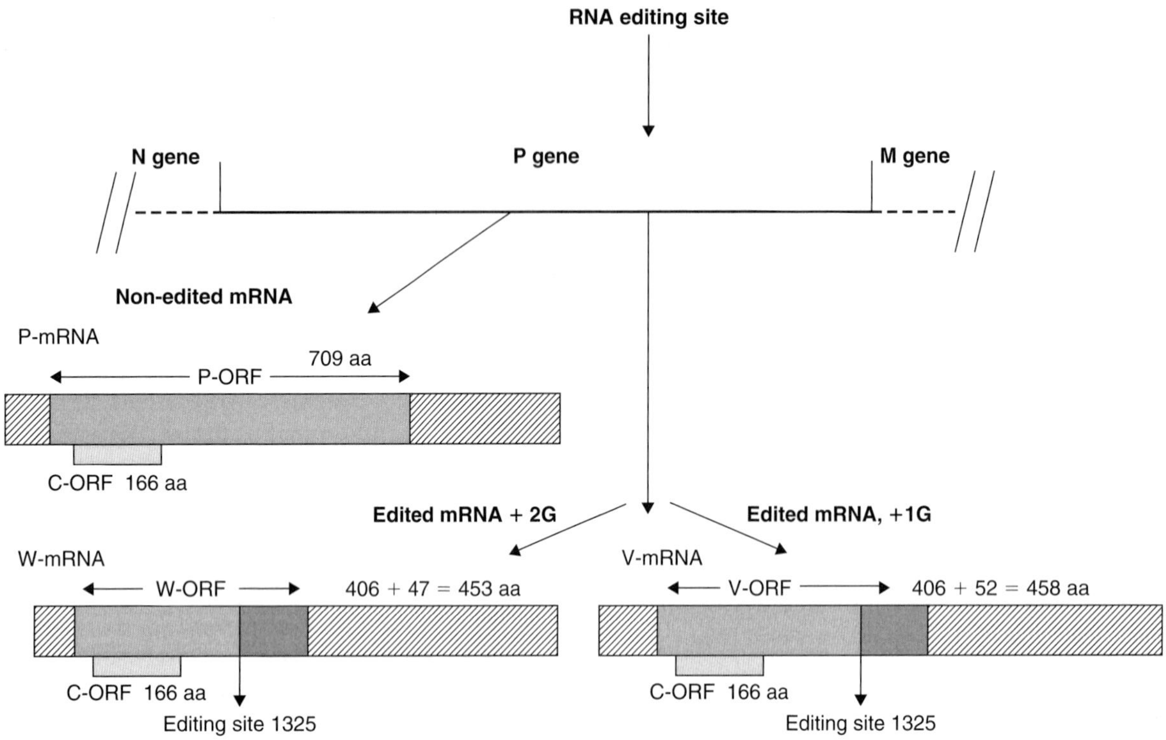

FIGURE 31.4 Translation strategy of the *P gene* of henipaviruses.

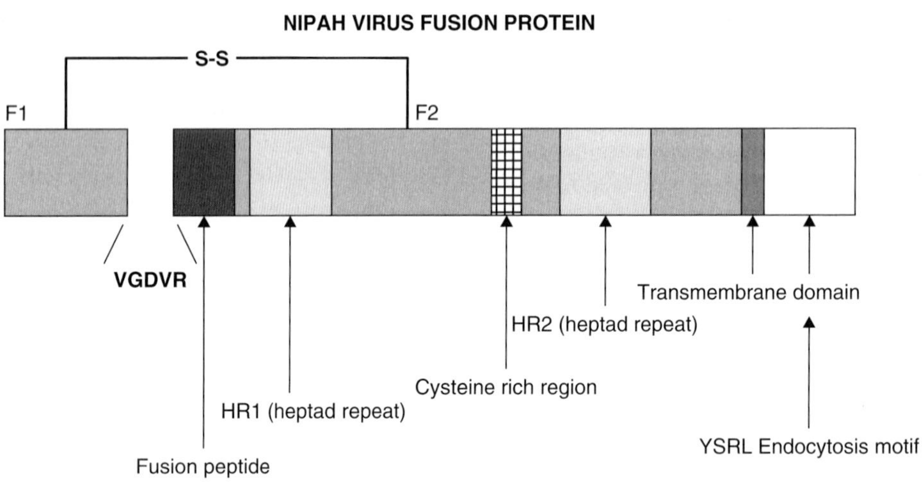

FIGURE 31.5 Henipavirus fusion F protein (see color plate section).

participates in the budding process. Expression of the NiV M protein is sufficient to produce budding virus-like particles that are physically and morphologically similar to NiV. A "late domain-like" sequence YMYL within the NiV M protein plays a role in this function. Deletion of this sequence causes the M to relocate to the nucleus (Ciancanelli and Basler, 2006).

The *fusion (F) gene* codes for a protein for both viruses which is 546 amino acids containing 88.1% homology (Ni/HeV). Similar to other paramyxoviruses, it is a type 1 transmembrane protein with a membrane-spanning region close to the carboxy-terminal, with two α-helical regions (HR1, HR2) involved in the fusion process and a cysteine-rich zone implicated in the interaction with the virus attachment protein (Fig. 31.5). The F protein is synthesized as an inactive precursor, F0, that is subsequently cleaved into two subunits F1 + F2 liberating at the amino terminus of F1 a hydrophobic peptide, which can interact with the host cell membrane. The cleavage site has a unique single basic residue (lysine in HeV and arginine in NiV) and is not cleaved by the protease furin, as in other paramyxoviruses (Moll et al., 2004a; Craft and Dutch, 2005). It has been proposed that the inactive uncleaved form is transported to the surface and subsequently internalized within the endosomal compartment, where it is cleaved and recycled to the cell surface as a fusion-active protein (Pager et al., 2004; Diederich et al., 2005; Meulendyke et al. 2005). Endocytosis is normally mediated by tyrosine dileucine–dependent signals in the cytoplasmic tails of transmembrane proteins. Mutation of these motifs (YXXF) in the NiV F glycoprotein inhibited endocytosis and fusion (Diederich et al., 2005).

HeV mutated in the motif at position 525 (Y525A) reached the plasma membrane rapidly, but was subsequently endocytosed much slower compared to the wild-type version (Meulendyke et al., 2005). After endocytosis both wild-type and mutated versions were cleaved. Pager and Dutch (2005) have shown that the endosomal/lysosomal protease cathepsin L is involved in the proteolytic maturation of HeV F. Cathepsin L–specific inhibitors reduced cleavage and the purified enzyme could cleave F0 into the F1 + F2 subunits. The cleavage of HeV F was initially assumed to be after a monobasic residue in the GDVK/R sequence. However, mutation of the basic residue or the upstream amino acids failed to eliminate cleavage (Craft and Dutch, 2005; Moll et al., 2004a). The apparent lack of specificity for cleavage is in keeping with the cleavage properties of cathepsin L.

These studies suggest that the henipavirus F protein is transported to the plasma membrane in an uncleaved form (F0), where it is endocytosed and subsequently cleaved by cathepsin L before returning to the membrane. HeV virions were found to contain both uncleaved and cleaved forms of F (Michalski et al., 2000). This may imply that during infection the viruses may use either membrane fusion or endocytosis during virus entry.

There are six potential N-linked glycosylation sites (three in F1 and three in F2), but only two sites in the F2 subunit (N67 and N99) and two in F1 (N414 and N464) have been shown to be glycosylated. Mutation of the sites in the F2 moiety does not affect cleavage, transport, or fusion activity. In contrast, mutation of the F1 sites leads to a greater than 80% reduction of expression of F at the cell surface. N414 is particularly critical for the folding and transport of the protein (Moll et al., 2004a, 2004b; Carter et al., 2005).

The *G gene* codes for the host-cell attachment protein, which is a class 2, type transmembrane protein of 602 amino acids HeV or 600, NiV (83.3% homology). This membrane protein spans the membrane leading to a membrane-proximal stalk domain with a large C-terminal globular head. This latter domain contains the receptor-binding region and by analogy with the structures determined for other paramyxoviruses is organized into a six-blade β-sheet propeller structure. The henipavirus G protein shares structural features characteristic of viruses in the genus *Respirovirus*. The locations of all seven proposed disulfide bonds are conserved. The overall canonical folding pattern of six parallel sheets, which is conserved in viral, bacterial, and eukaryotic neuraminidases, is also retained in henipavirus G. Among the seven active site residues known to be important for neuraminidase activity (Langedijk et al., 1997) only one is present. Structural models of the G protein suggest that the sites of interaction on G with the cell receptor are on the globular head. Mapping studies using escape mutants of either NiV or HeV propose that there is a major discontinuous epitope, which commences near the shallow depression in the center of the top surface of the globular head and extends to the rim of the barrel-like structure on the top loops of β-sheet 5 (White et al., 2005; Guillaume et al., 2006a).

The G glycoprotein is expressed at the plasma membrane, but unlike the F glycoprotein there are no motifs in the cytoplasmic tail, which would aid endocytosis. However, it appears to be constitutively internalized with the bulk flow during membrane turnover (Vogt et al., 2005). In several Paramyxoviruses, the two glycoproteins have been shown to be noncovalently linked (Malvoisin and Wild, 1993). The NiV glycoproteins can be co-immunoprecipitated and mutational studies have established that the avidity of the interactions is inversely related to their fusogenic activity (Aguilar et al., 2006, 2007). The interaction of the two viral glycoproteins would be a means by which G could be co-endocytosed even though it does not contain the appropriate motif.

The *L gene* in both viruses codes for a protein of 2244 amino acids with 86.8% homology at the protein level. In the region predicted to contain the catalytic site, both viruses contain the sequence GDNE that varies from other sequenced nonsegmented negative strand viruses (GDNQ). Mutation of E→Q in NiV did not affect its efficiency (Magoffin et al., 2007).

VIRUS–CELL INTERACTIONS

The Host Cell Receptor

Studies on tissues from humans and infected animals have shown that henipaviruses infected a number of tissues including lungs, endothelial cells, and neurons (Hooper et al., 2001; Wong et al., 2002). In vitro, the viruses infect a wide range of cells of different species. The lack of an active neuraminidase site makes it unlikely that sialic acid can act as a cellular receptor, especially as neuraminidase treatment of cells failed to effect virus infection (Eaton et al., 2004).

Two groups have identified a cellular protein, ephrin-B2, that permits the attachment and penetration of henipaviruses into host cells (Negrete et al., 2005; Bonaparte et al., 2005). Ephrin-B2 is a critical gene involved in embryonic development, and has established roles in vasculogenesis and axonal guidance. Ephrin genes are highly conserved and have been found in all animal species examined. Both ephrin-B2 and its cognate receptor EphB4 have tyrosine signaling and PDZ binding motifs in their cytoplasmic domains (Kullander and Klein, 2002). Forward signaling mediated by EphB4 facilitates anti-adhesive and repulsive behavior on contact with ephrin-B2 expressing cells, while ephrin-B2 "reverse" signaling facilitates adhesion upon contact with EphB4-expressing cells (Sakamoto et al., 2004). Bennett et al. (1995) identified ephrin-B2 mRNA in most human tissues examined with the exception of blood leukocytes and the pancreas. Cerretti et al. (1995) confirmed its expression in human adult lung and kidney. Ephrin-B2 persists in adult arteries where it extends into some of the smallest diameter microvessels. It is expressed in vascular smooth muscle of arteries, but not veins (Shin et al., 2001). The ephrin-B2 is expressed in neurons and is thought to act postsynaptically by modulating glutamate receptors (Grunwald et al., 2001; Henderson et al., 2001; Contractor et al., 2002) and thereby modulates Ca^{2+} influx into the postsynaptic cells. Thus, the expression of ephrin-B2 in different tissues is similar to the tropism observed for henipaviruses.

The crystal structure of the extracellular domain of mouse ephrin-B2 has been determined (Toth et al., 2001). It has a β-barrel globular domain composed of eight strands arranged in two sheets around a hydrophobic core. In the crystal structure, the ephrin-B2 ectodomains are associated as extensively interacting dimers, which is likely a functional form in vivo. There are two surface regions on the proposed structure, I—a small concave pocket of hydrophobic nature surrounded by charged amino acids, represents a site critical for Eph receptor binding, whereas a second area (II) is implicated in dimerization.

The interactive regions between ephrin-B2 and the henipavirus attachment protein have yet to be mapped. However, soluble NiV G protein with deletions in the globular head domains (372–393, 437–464, or 559–587) fails to interact with ephrin-B2 (Negrete et al., 2005). This would suggest that the conformational folding of the viral glycoprotein is important in its interaction with the receptor. Using soluble forms of NiV or HeV and ephrin-B2, it was shown that there was a high-affinity interaction, which could be specifically blocked with monoclonal antibodies, which neutralized both NiV and HeV.

In a search for alternative receptors for NiV, it was shown that soluble NiV-G could also bind to ephrin-B3, but not to any of the other ephrins. The binding affinity to cell surface B3 was lower than ephrin-B2, but was sufficient to permit viral entry of virus (Negrete et al., 2006). Competition studies with soluble ephrins B2 and B3 suggest that they interact with an overlapping site. Mutational analysis showed that the Leu-Trp residues in the solvent exposed G-H loop of ephrins B2 and B3 were critical determinants of NiV binding and entry. Replacement of the Tyr-Met residues in the homologous positions in ephrin-B1 with Leu-Trp conferred NiV receptor activity to ephrin-B1.

In contrast to ephrin-B2, ephrin-B3 does not appear to be expressed in the endothelium, but is found in the CNS in overlapping and distinct patterns with ephrin-B2 (Flenniken et al., 1996; Bergemann et al., 1998). Presumably in cells where both receptors are expressed, the virus would use ephrin-B2 due to its greater affinity. However, in regions such as the corpus callosum (Liebl et al., 2003) and the spinal cord (Yokoyama et al. 2001), ephrin-B3 is distinctly expressed and could be the major receptor.

The Fusion Process

In paramyxovirus infections, the virus and cell membranes undergo fusion during the initiation of virus entry into the cell and then during in late infection when the infected cell membranes interact with their neighboring uninfected cells. To initiate this phenomenon, the virus G attachment protein interacts with the cell receptor. The F protein, which is noncovalently linked to the G protein, dissociates and the hydrophobic region at the N-terminal of F1 is introduced into the cell membrane. The two α-helical regions (HR1, HR2) associate in the form of a six helical bundle (Xu et al., 2004a, 2004b), which in turn brings the viral and cell membranes together to fuse. The whole process can be analyzed by expressing the two virus glycoproteins (G and F) in cells that possess the appropriate host cell receptor. In the majority of the paramyxovirus systems, it has been shown that the two glycoproteins must be homologous, i.e., from the same virus. In the case of henipaviruses, heterologous pairs of F and G proteins from NiV and HeV can induce fusion (Bossart et al., 2001, 2002; Tamin et al., 2002). The proportion of the two viral glycoproteins at the cell membrane can influence fusion. Optimum fusion was observed when the G and F were expressed at the membrane in equal amounts. Further, the HeV G protein levels are modulated by both a more rapid protein turnover and a slower protein trafficking than that for HeV F (Whitman and Dutch, 2007).

Mutational studies on the cytoplasmic tail region of NiV F showed that they could affect the avidities of the NiV–F and –G interactions. Mutations decreasing these interactions increased the fusion activity (Aguilar et al., 2007). Similar observations have been previously shown with measles virus glycoproteins (Plemper et al., 2002).

In the second stage of fusion, the two α-helical regions of the F glycoprotein associate to bring the viral and cell membranes together. It has been shown that for a number of the paramyxoviruses synthetic polypeptides corresponding to the HR2 region inhibit the association of the two α-helical regions and thus block fusion (Lambert et al., 1996; Wild and Buckland, 1997). Similar studies with henipaviruses have shown that peptides corresponding to the HR2 region also block fusion (Bossart et al., 2002). The peptides, which efficiently block fusion, are relatively long (42 mer) and show hydrophobic properties, which would limit their in vivo use. More recent studies have developed a second generation of peptides which are shorter (32 mer) and include a capped peptide via amidation and acetylation and two poly (ethylene glycol)-linked (PEGylated) peptides, one PEG moiety at the C-terminus and the other at the N-terminus. These peptides are highly soluble and inhibited fusion and virus infection at similar concentrations to the nonmodified peptides (Bossart et al., 2005b). From studies with other peptides it can be assumed that such modifications will increase their half-life in vivo.

The amino acid sequences in this region of the F protein are somewhat conserved among the paramyxoviruses, fusion inhibition by the HR2 peptides has generally been found to be virus specific. Peptides specific for NiV or HeV F proteins equally inhibited the other virus, but not morbilliviruses (Bossart et al., 2001, 2002). An exception to this rule appears to be human Parainfluenza virus 3. Peptides to its HR2 region block HeV-induced fusion more efficiently than the homologous virus (Porotto et al., 2006). Further studies to elucidate the mechanism involved would help in the design of more efficient inhibitors.

EPIDEMIOLOGY

Significance as Public Health and Agricultural Problems

The first epidemic episode of NiV emerged in a piggery causing primarily respiratory disease in 1998 through 1999 and then progressed to pig farmers causing lethal encephalitis. In 1998, in Perak State located North from Kuala Lumpur, the causative agent was initially thought to be Japanese encephalitis virus but high mortality in humans and acute respiratory disease in pigs and death did not meet the case definition of a mosquito-borne flavivirus (Chua et al., 1999; Mohd Nor et al., 2000). Indeed, vaccination of the population in the epidemic area against Japanese encephalitis virus and mosquito control did not stop the spread of the disease. From February to April 1999, the virus was probably imported by infected pigs to Negeri Sembilan where intensive pig farms were affected. More than one million pigs were culled to stop the epidemics. The virus isolated from the cerebrospinal fluid (CSF) of a patient was identified as closely related to HeV and was called Nipah, the name of a river in the village from where the patient originated (Chua et al., 2000a). Eleven cases of encephalitis and one death were then identified in pig abattoir workers in Singapore (Paton et al., 1999). Since then no more cases have been reported in this region.

Megacheroptera Fruit Bats as the Natural Host of Henipaviruses and Source of Infection in Animals and Humans

HeV and NiV are the only zoonotic paramyxoviruses known to be highly pathogenic for humans. Serological investigations rapidly showed that fruit bats (flying foxes) of genus *Pteropus* in the suborder *Megachiroptera* could be involved in virus transmission of HeV to horses (Young et al., 1996). HeV was then isolated from two species of *Pteropus* (Halpin et al., 2000). Transmission from bats to horses may have occurred through ingestion of pastures contaminated with infected urine, placenta or aborted fetal material, or from partially eaten fruit or fruit pulp spat out by bats. Transmission between horses appeared to be a rare event, and there was no evidence of direct transmission from bats to humans. One study collecting samples from nasal and throat swabs and urine showed NiV in a significant amount of patients with an acute disease, but no seroconversion was observed in health workers (Chua et al., 2001).

Close phylogenetic and serological relationships with HeV facilitated the rapid identification of Pteropid bat species as the reservoir hosts of NiV (Mohd Yob et al., 2001) and isolation from their urine in Tioman Island (Chua et al., 2002a). It is believed that transmission from bats to pigs was through fruit spats, placental or aborted fetal material, or other contaminated material falling into pig pens. The virus was further isolated from urine of *Pteropus Lylei* in Cambodia (Reynes et al.,

2005). Antibody prevalence was observed in dogs of farmers where the NiV epidemics had emerged (Mohd Yob et al., 2001), but the role of dogs in the virus life cycle has not been identified.

In Bangladesh, most people may have been infected by drinking raw date palm sap contaminated by fruit bats *(Pteropus giganteus)* (Luby et al., 2006). These outbreaks have been characterized by very high fatality rates, by sometimes lack of contact with either bats or a domestic spillover host, and by the probability of human-to-human transmission (ICDDR,B, 2004a, 2004b). This latter feature is of very significant global concern; if correct, future outbreaks may spread to other areas where fruit bats are not part of the normal fauna. It may also have a significant impact in the biothreat potential.

The geographic range of Pteropid fruit bats extends as overlapping populations from Pacific Islands through southern and southeastern Asia, and Australasia to the Middle East. Additional species occur in the southern Indian Ocean, including Pembe Island and Madagascar. There are epidemiological and serological evidence that fruit bats are the major reservoir for HeV and that NiV is widespread in *Pteropus* in Southeast Asia (Mohd Yob et al., 2001; Olson et al., 2002; Wacharapluesadee et al., 2005; Lehle et al, 2007). However, it cannot be excluded that neutralizing antibodies found in several species of fruit bats in Cambodia, Thailand, Malaysia, Madagascar, and Vietnam (Huong VTQ et al., unpublished results) may be cross-neutralizing antibodies to another related but to date not identified virus in bats. However, NiV RNA was found in bat saliva and urine of bats in Thailand, suggesting the persistence of similar NiV infection in these animals (Wacharapluesadee et al., 2005).

Pteropus poliocephalus (gray headed fruit bats) experimentally infected with NiV showed subclinical manifestation, transient systemic virus, and episodic low-level virus excretion in urine in the presence of neutralizing antibodies (Middleton et al., 2007). Low and intermittent excretion of NiV in the urine of infected bats may be sufficient to maintain virus survival among bats. This strain of bats was also infected with HeV and showed subclinical symptoms and virus transfer to the fetus (Williamson et al., 2000). This suggests that henipaviruses may be maintained in nature through transplacental transmission from the mother to the fetus.

Cats, horses, or pigs are natural hosts of HeV and NiV. Recovery of viruses from urine as well as from the nasal discharge or saliva of these viruses may favor viral shedding and possibly field transmission of the viruses between domestic animals and between those animals and humans.

These results confirm that bats carrying henipavirus represent a public health hazard. However, how the virus can be transferred to intermediate host or directly to human remains puzzling.

Since bats are natural hosts of the viruses, epidemiological investigations have observed dramatic changes in the habitats of these animals to explain the installation of bat colonies in new environments planted with fruit trees and occupied by humans and pig farms. In 1997/1998, deforestation due to fires in Borneo resulted in the migration of fruit bats over several regions of Southeast Asia in the months directly preceding the NiV disease outbreak. This was exacerbated by a unusual drought event the same years (Chua et al., 2002b). In Malaysia, NiV found new ideal ecological situation to be transmitted from its reservoir host to pigs and ultimately to the human population (Chua et al., 2002b). In China and Southeast Asia, countless numbers of villages where fruit trees are spoiled by fruit bats, pigs roam freely, and only poor sanitary conditions are available, provide ideal environments for natural transmission of NiV from infected bats.

Potential as Biothreat Agent

NiV and HeV are highly virulent pathogens and are on priority lists of potential bioterrorism agents because of their capacity to rapidly spread in pig farms and stud farms, incurring high transmission and lethality rates in humans, and potentially causing medical and economical disasters (Lam, 2003). Whether preferential neurotropism, or lower virus doses in urine of patients, or other factors may account for low human-to-human transmission remains unknown. However, a strain of NiV has recently been identified in Bangladesh with ability to cause acute respiratory distress syndrome and to be transmitted from person-to-person (ICDDR,B, 2004b).

CLINICAL DISEASE

Clinical Symptoms

There have been only four human cases besides horses infected with HeV, limiting information on clinical and pathophysiological features of the disease. Pathological manifestation of HeV infection in horses and humans is that of flu-like syndrome with severe pneumonitis and respiratory failure, and renal failure and arterial thrombosis (Selvey et al., 1995). Lungs show lesions of congestion, hemorrhages, and edema

(Hooper et al., 2001). Chest radiography may show mild interstitial shadowing. In NiV infection, after an average of 10 days of incubation, patients presented with nonspecific systemic symptoms of fever, desorientation/confusion, guidiness, headache, myalgia, and sore throat. The more severe cases showed drowsiness and altered level of consciousness that required ventilation (Chong et al., 2002). In another study, 40% of them had cough and respiratory syndromes, but only 28% had convulsions (Wong et al., 2002). Duration of illness was quite variable and lasted from less than 5 days to more than 35 days. Most common clinical signs were reduced consciousness, hyporeflexia or areflexia, segmental myoclonus, and seizure. These symptoms reflected brain stem dysfunction (Chua et al., 1999; Goh et al., 2000; Wong et al., 2002). The general status of patients showing hypotension and tachycardia deteriorated rapidly to death. High viral replication in the CNS may be an important factor for high mortality (Chua et al., 2000b). Magnetic resonance imaging (MRI) shows focal and scattered areas of increased signal intensity with a cortical distribution within the white matter in T2-weighted images. Transient punctate cortical hyperintensities were seen on T1-weighted images (Paton et al., 1999; Lim et al., 2002).

Some patients showed late onset of the disease, up to 4 months after the presumed infection episode, and in more than 10% of patients who showed mild encephalitis a few days after infection, relapsed few months to 4 years with severe meningo-encephalitis causing death in up to 18% of the cases. Symptoms were similar compared to acute manifestation, with increased manifestations of seizure and focal cortical signs. The overall mortality was as high as 40%. On the other hand, up to 50% recovered fully. As for the others, the residual neurological signs were mostly mild but some patients retained permanent sequelae (Ng et al., 2004; Sejvar et al., 2007).

Atypical Symptoms

A study of a 24 month follow up after the NiV outbreak in Malaysia observed 12 survivors (7.5%) of acute encephalitis who had relapsed encephalitis and 10 patients (3.4%) had late-onset encephalitis (Tan et al., 2002). The mean interval between the first neurological episode and the time of initial infection was 8.4 months. Patients showed common clinical features observed in primary of NiV infection. Four of the 22 relapsed and late-onset encephalitis patients (18%) died. Cortical lesion were observed by MRI with high intensities of signals (Lim et al., 2002). Focal encephalitis with NiV antigens in neurons was observed in brains of deceased patients.

In Bangladesh, similar symptoms were observed, but respiratory distress and cough were symptoms characteristic of this disease. Chest radiographs showed bilateral infiltrates consistent with acute respiratory distress syndrome. One of the two fatal cases of HeV infection in Australia also showed respiratory infection and lesions in the lung (Selvey et al., 1995).

TREATMENT

An open-label trial of ribavirin was carried out during the NiV epidemics in Malaysia (Chong et al., 2001). The study suggested that ribavirin was able to reduce the mortality of acute NiV encephalitis by 36%. However, reduction of NiV replication by Ribavirin was observed in Vero cells infected with the virus but not in vivo in hamster or guinea pig animal models (Georges-Courbot et al., 2006). The difference of protective efficacy of Ribavirin in humans and in hamsters may have several explanations: (1) the study in humans was not randomized and performed at posteriori, and grouping treated and nontreated patients may have introduced some bias; (2) the dose and mode of infection of NiV used for the challenge or the virus replication in hamsters may be higher and different than in humans, and may have not been blocked by the drug in vivo; (3) Ribavirin may be differently metabolized in humans than in hamsters. Several other drugs known to be active against paramyxoviruses like inhibitors of either S-adenosylhomocysteine hydrolase or of IMP deshydrogenase, or of OMP decarboxylase, or of CTP synthetase, were tested in vitro. A few of them were active in vitro but inactive against NiV infection when tested in the hamster model (Georges-Courbot et al., 2006).

Many viruses are susceptible to interferon (IFN). Viral-derived double-stranded RNA is one of the activating factors of the IFN cascade. One treatment strategy is to inject in infected individuals external IFN to cure or at least to maintain low doses of viruses. Another alternative is to inject an inducer of IFN response. Poly(I)-poly(C12U) (Ampligen), a strong IFN-α and IFN-β inducer, showed anti-HIV activity in clinical trials and in several animal models of positive- and negative-strand RNA viruses. This compound was tested in the hamster model of NiV infection and substantial protection of treated animals was observed if injected for 10 consecutive days after infection (Georges-Courbot et al., 2006). Although no infectious virus was detected, RT-PCR positive viral sequences were recovered in organs of hamsters surviving from the challenge. These results deserve further studies, in particular to investigate if any virus relapse would occur in animals treated with Ampligen and protected against a lethal challenge.

Galectin-1 is a lectin produced by several cell types which modulates cytokine synthesis and increases IL-6 production by dendritic cells. It was also shown to inhibit cell fusion provoked by NiV infection (Levroney et al., 2005). The combined activities of this endogenous lectin may provide a potent drug candidate to treat patients against henipavirus infection.

PATHOGENESIS

Description of Disease Process

HeV and NiV showed similar pathogenesis and vascular endothelium tropism as pathognomonic of the disease, with resulting syndrome of interstitial pneumonia, meningitis, and atrophic glomeruli (Selvey et al., 1995; O'Sullivan et al., 1997; Wong et al., 2002). In both viral diseases, syncytia in microvascular endothelium were observed. In one case of HeV infection, syncytia developed particularly in the blood vessels of the lung (Murray et al., 1995; Selvey et al., 1995) and in the other case, they were more prominent in brain microvascular (O'Sullivan et al., 1997) (Fig. 31.6). Systemic vaculitis, thrombosis, ischemia, and parenchymal necrosis were the main histopathological hallmarks (Fig. 31.7). Pathological lesions with plaques of necrosis were visible in the white and grey matter in CNS and in spinal cord. Inflammatory cells in perivascular cuffing consisted of macrophages, lymphocytes, neutrophils, and microglia cells (Fig. 31.7). Neuronophagia was active. Multiple eosinophilic viral inclusions were visible in the cytoplasm of few neurons, but only in 63% of the cases (Wong et al., 2002). These inclusions characteristics of paramyxoviruses also identified by electron microscopy (Hyatt et al., 2001; Wong et al., 2002) correspond to inclusion bodies of nucleocapsids. Vasculitis, endothelial cell damage, and thrombosis lead to microinfarctus and microvessel rupture which appear to be the mechanism for the virus to cross the blood–brain barrier and invade the neuronal parenchyma.

Histology of necropsies of one case of HeV and one case NiV relapse were performed (O'Sullivan et al., 1997; Wong et al., 2002). In HeV patient, discrete necrosis was observed in the cortex, brainstem, and cerebellum. Multinucleated cells and viral inclusions were observed in the brain, and other organs. Inflammatory cells were scattered in the brain parenchyma. In one

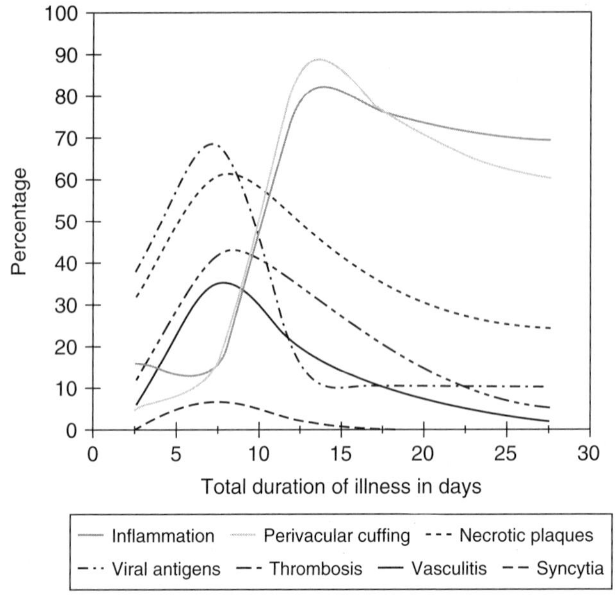

FIGURE 31.6 Temporal distribution of microscopic findings and viral antigen in the CNS in human fatal Nipah virus infection (from Wong et al., 2002).

NiV patient, parenchymal lesions were larger with numerous macrophage infiltrates. Intracellular inclusion in neurons were more prominent. No vasculitis or necrotic plaques were observed. HeV genome was amplified from the serum and CSF of the patient and the sequence of 500 nucleotides in the matrix gene was identical to that of the virus that circulated 13 months before. No NiV nucleotide sequence was available from the NiV case. Neither HeV nor NiV was isolated from the brain of either patient.

Animal Models

As in the human disease, NiV and HeV cause systemic infection in several animals. Guinea pigs infected with high doses of HeV develop encephalitis (Williamson et al., 2001), whereas NiV receiving 10^7pfu only showed transient fever and weight loss (Wong et al., 2003). In contrast, both NiV and HeV induce an acute infection in golden hamsters that appears to reproduce the pathology and pathogenesis of the acute human NiV infection, with primarily vascular infection, perivacular cuffing and thrombosis, and neurotropism with numerous lesions in the brain (Wong et al., 2003) (Fig. 31.8). This model would be useful to study pathological processes of each virus, and in particular the causes and features of delayed onset and relapse characterized in humans for both viruses. In a feline model, infection of cats

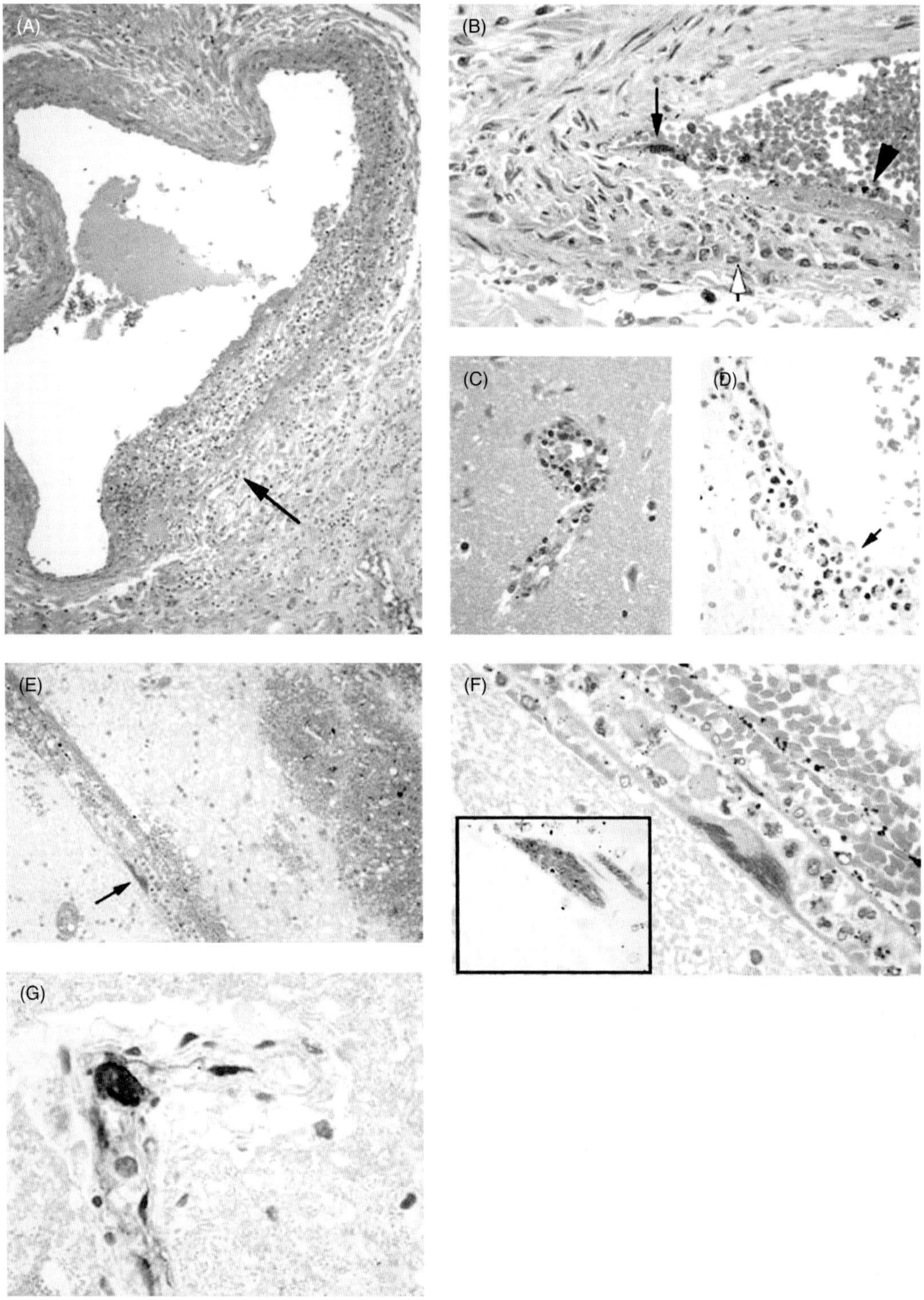

FIGURE 31.7 Vascular pathology and viral immunolocalization in Nipah virus infection in human. (A) Vasculitis in a lung artery. Focal and transmural mixed inflammatory infiltrate is visible (arrow). (B) High-power magnification of another pulmonary vessel showing endothelial syncytium (arrow) and ulceration (arrowhead). Note focal transmural inflammation (open arrow). (C) Vasculitis in a small cerebral vessel. (D) A cerebral venule showing endothelial ulceration (arrow) associated with inflammatory cellular debris. (E) Cerebral hemorrhage adjacent to a vessel with a multinucleated giant cell (arrow). (F) Higher-power magnification showing endothelial origin of syncytium seen in (E). (Inset) Intranuclear immunostaining of viral antigens in same cell. (G) Positive immunostaining is also seen in the cytoplasm of an endothelial syncytial cell protruding in lumen of a cerebral vessel. H&E stain: (A–F and inset in F) immunoalkaline phosphatase with napthol fast red substrate and hematoxylin counterstain (G). Original magnifications: 25× (A); 100× (B–D); 50× (E); 250× (F, inset); 158× (F, G) (from Wong et al., 2002) (see color plate section).

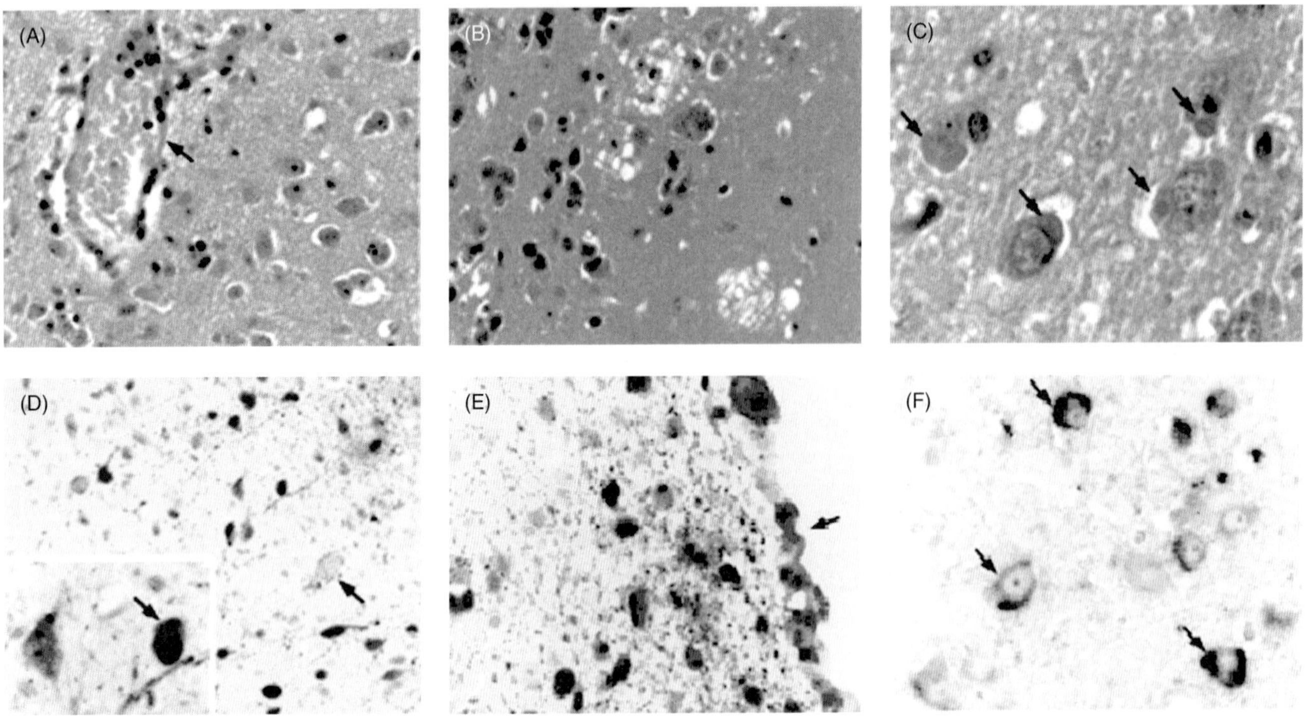

FIGURE 31.8 Cerebral pathology in acute Nipah infection in a hamster model. (A) Small vessel vasculitis (arrow) characterized by mild inflammation in the vicinity of infected neurons. (B) Focal areas of parenchymal ischemia, infarction, and edema. (C) Neurons with eosinophilic viral inclusions (arrows). (D) Immunolocalization of viral antigens to neurons in the nucleus (inset, arrow), cytoplasm, and processes near a vasculitic vessel (arrow). (E) Viral antigens localized to ependymal lining (arrow) and neurons. (F) Neurons demonstrating viral RNA in the cytoplasm (arrows). (A–C) Hemalin-phloxine-safranin stain; (D, E) IHC, hematoxylin counterstain; (F) in situ hybridization, hematoxylin counterstain. Original magnifications: 40× (A, B, E, D (inset); 20× (D); 100× (C, F) (from Wong et al., 2003) (see color plate section).

with NiV induces an acute febrile reaction followed by severe respiratory disease. There was an apparent lack of neurological signs, despite the demonstration of NiV genome in brain tissue (Mungall et al., 2006). Detection of NiV in placenta and fetal tissues from pregnant cats infected with NiV suggests that both vertical and horizontal transmission of NiV could play a role in spillover events (Mungall et al., 2007). NiV infection of pigs leads to a 10–15% mortality rate with few animals showing neurological signs (Mohd Nor et al., 2000). Experimental infection of pigs showed that following the initial replication in the upper respiratory tract; the virus invades the CNS directly via the cranial nerves and by crossing the blood–brain barrier. NiV also replicated in the respiratory system, including the lung, concurrently with the spread through the nervous and lymphoreticular systems. The respiratory tract is one of the main targets early in the infection (Weingartl et al., 2005).

Saimiri monkeys were highly susceptible to NiV and showed clinical symptoms of CNS infection and histopathological lesions similar to those described in humans or animal models (Contamin et al., to be published).

Inoculation of fertile eggs via yolk sac with NiV led to embryo's death with severe pathogenesis in CNS and tropism for vascular endothelial cells and neurons (Tanimura et al., 2006). This model has been proposed for virus titration and for study of the vascular and neuronal tropisms of NiV.

IMMUNE RESPONSE TO INFECTION

Innate Immunity

The IFN response is one of the host's primary defense mechanisms against viral infection. IFN-α/β is synthesized and secreted by cells in direct response to viral infection. Viruses have devised various strategies to circumvent this IFN-induced antiviral state through the expression of viral proteins that inhibit specific components of the IFN system. The IFN antagonist functions of the *Paramyxovirinae* are all coded by the *P gene* and in most cases are carried out by one of the accessory proteins: C, V, or W. These proteins all target the IFN signaling, or JAK/STAT pathway, but do so through distinct mechanisms. The Henipavirus V

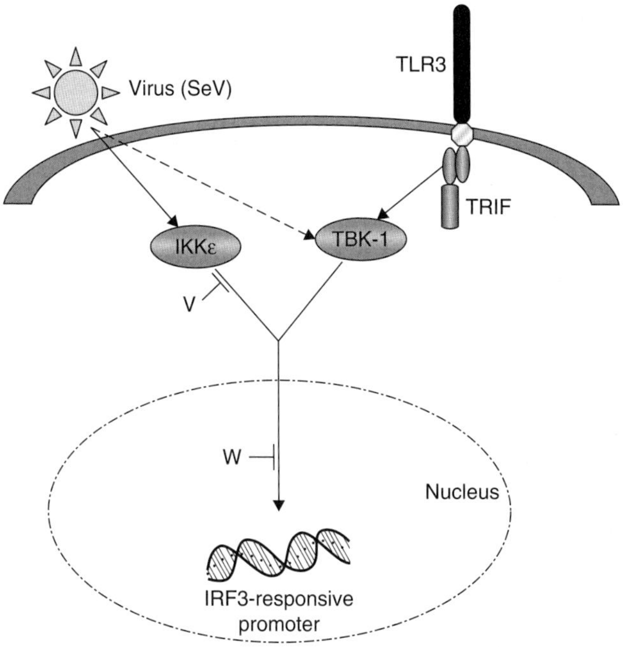

FIGURE 31.9 Illustration of the virus and TLR-3 activated signaling pathways and the points where NiV V and W proteins are predicted to exert their inhibitory activity (from Shaw et al., 2005).

and W proteins localize to the cytoplasm and nucleus, respectively, and have both been shown to function as inhibitors of JAK/STAT signaling (Rodriguez et al., 2002, 2003; Shaw et al., 2004, 2005). The V and W proteins also block viral activation of the IFN-β promoter and the IFN regulatory factor 3-response IFN-stimulated gene 54 promoter. Only the W protein shows strong inhibition of promoter activation in response to stimulation of Toll-like receptor (TLR) 3. The nuclear localization of the W protein enables it to target both virus and TLR pathways, whereas the cytoplasmic V protein is restricted to inhibiting the virus pathway (Fig. 31.9). The V protein of NiV has been shown to antagonize the IFN response in human cells via the sequestration of STAT1 and STAT2. This activity can be inactivated by the mutation of a single amino acid (Hagmaier et al., 2006).

The henipavirus C protein by analogy with other paramyxoviruses has a dual role. It inhibits viral RNA transcription and replication, but also acts by blocking IFN action. Park et al. (2003) have shown that NiV C expression prevents the establishment of an IFN-induced antiviral state.

Protective Immune Responses

Protective immunity to paramyxoviruses has been attributed to both cellular and humoral responses. Children suffering from hypogammaglobulinemia recover normally from measles infections, whereas those suffering from defects in cellular immune responses have complications leading to encephalitis (Good and Zak, 1956). On the other hand, antimeasles hyperimmune serum protects children from infection (Gallagher, 1935). Thus, cellular responses are important in protection, but once in place, the antibodies provide a first line of defense. Studies with a number of paramyxoviruses have shown that both of the viral glycoproteins (the attachment and fusion proteins) can induce antibodies that neutralize the virus in vitro (Giraudon and Wild, 1985) and protect animals from a challenge after passive administration (Malvoisin and Wild, 1990).

Little is known about humoral response against Henipaviruses and nothing about the cellular response. Virus can persist in humans hidden in some cells in the brain or other unknown sites, even in the presence of neutralizing antibodies. The time course of viremia and immune responses have been described by Wong et al. (2002) and are described in Fig. 31.6. Envelope glycoproteins G and F of Henipavirus are highly antigenic and induce antibodies that are able to neutralize virus infection in vitro and in vivo (see below). Zhu et al. (2006) developed human monoclonal antibodies to HeV. They panned a naïve human antibody library with soluble HeV-G. The Fab fragments were selected and showed that one had a high HeV neutralization and also had a cross reactivity with NiV. The conversion of this Fab fragment into an IgG1 immunoglobulin significantly increased its anti-fusion and neutralization activity.

VACCINATION STRATEGIES

The development of inactivated vaccines against paramyxoviruses in the 1960s showed that they induced virus-neutralizing antibodies, which were protective. The decline in these antibodies levels with time and subsequent exposure to virus led to a disease, which was more severe than if the children had not been vaccinated (Nader et al., 1968). Much later studies, in monkeys showed that vaccination with the inactivated measles vaccine resulted in a priming for a nonprotective type 2 CD4 response, in contrast to the responses induced by the wild-type or attenuated viruses (Polack et al., 1999). In both monkeys and humans, measles infections initially induce a Th1 type response, which converges into a Th-2 response with the induction of antiviral antibodies. These observations led to the conclusion that new vaccines against paramyxoviruses should induce similar immune responses to that in natural infection.

Few studies have been conducted to develop anti-henipavirus vaccines. To define the antigens inducing neutralizing antibodies, the two NiV glycoproteins have been expressed as vaccinia recombinants and used to immunize mice or guinea pigs. Both G and F glycoproteins induced neutralizing antibodies, although G was more efficient. (Tamin et al., 2002; Guillaume et al., 2004a). Immunization of rabbits with a soluble form of HeV G (lacking the cytoplasmic tail) induced neutralizing antibodies to both HeV and NiV (Bossart et al., 2005a). In contrast, murine monoclonal antibodies against either HeV or NiV G did not cross neutralize although NiV anti-F monoclonal antibodies do (White et al., 2005; Guillaume et al., 2006b). Thus, potential vaccines should contain at least one of the viral glycoproteins. In a hamster model, immunization with either of the NiV glycoproteins (G or F) using a vaccinia vector, protected the animals against a lethal challenge (Guillaume et al., 2004a). The protective mechanisms involved are in part humoral. Hamsters given either monospecific polyclonal or monoclonal anti-G or F antibodies are protected from a subsequent challenge with a lethal dose of NiV (Guillaume et al., 2004a, 2006b). Further, enhanced forms of the disease were not observed when lower levels of antibody were used.

In the pig model, animals immunized with the recombinant canarypox virus (ALVAC) expressing either of the glycoproteins were protected against a challenge. Virus could not be detected after challenge and no virus was shed in the infected animals. These vaccines stimulated both type 1 and type 2 cytokine responses (Weingartl et al., 2006). In the cat model, animals that were vaccinated with an adjuvented preparation of either soluble NiV-G or HeV-G were protected against a subsequent virus challenge (Mungall et al., 2006). The different studies show that protection by vaccination is possible using different preparations of the two viral glycoproteins. However, there is no information on long-term immunity provided by such vaccines.

LABORATORY DIAGNOSIS OF HENIPAVIRUS

Virus identification is a key element in the control of the disease. Henipaviruses grow in a number of human and animal cell lines to high titers, inducing cell syncitia. Due to the lack of BSL-4 facilities in endemic regions for NiV, cell cultures could be handled in BSL-3, but then transferred to BSL-4 containment if any further work has to be carried out. Virus identification is performed either by indirect immunofluorescence of virus-infected cells using group- and type-specific monoclonal antibodies, or RT-PCR on extracts of cell supernatants using specific primers or probes. Several RT-PCR and real-time PCR protocols have been developed for both HeV and NiV (Smith et al., 2001; Guillaume et al., 2004b; Chang et al., 2006; Wacharapluesadee and Hemachudha, 2007). TaqMan assay is usually used for virus genome detection in serum, CSF or urine of patients and animals. Several genes have been targeted depending of the specificity required. The relatively conserved matrix protein M has been used for HeV and NiV gene amplification. But comparative nucleotide sequences available for most of henipaviruses would be valuable to synthesize a set of primers with broad specificity to detect all known viruses in the Henipavirus genus. However, the specificity of each current technique should be validated with NiV variants from Bangladesh, since the probe used in Taqman technique may by too discriminative.

Histology and immunohistology of clinical samples have great value when polyclonal and monoclonal antibodies are available. In particular, antibodies against nucleoprotein of henipaviruses can identify intracellular inclusions in fixed tissues (Wong et al., 2002; Tanimura et al., 2004).

Electron microscopy is one element essential to identify new emerging viruses and should be performed on supernatant and virus-infected cells (Murray et al., 1995; Chua et al., 2000a).

IgM and IgG titration of henipaviruses can be performed by ELISA and neutralization tests. Clinical samples are serum or CSF. Usually, antigens used for ELISA tests are crude extracts from virus-infected Vero cells (Daniels et al., 2001) or recombinant proteins expressed in *E. coli* (Yu et al., 2006; Chen et al., 2006; Kashiwazaki et al., 2004). Reactions should be validated using recombinant F, G, and M proteins prepared in baculovirus-infected insect cells (Esheragi et al., 2005). However, these in-house tests are limited and no commercial kit has been developed up to now. The reference technique remains the seroneutralization test with microplaques, which provides an accurate and specific way to calculate antibody dilution in serum or CSF that neutralize 50% of cell toxicity caused by NiV or HeV ($TCID_{50}$). This test is restricted to BSL-4 laboratories in which live virus can be manipulated. However, production of virus NiV pseudotyped particles was recently made from VSV-ΔG-Luc or VSV-ΔG-GFP viruses, recombinant VSV derived from a full-length complementary DNA clone of VSV in which the G-protein envelope has been replaced with Luciferase or EGFP reporter genes

(Negrete et al., 2006; Patch et al. 2007; Porotto et al., 2007). NiV or HeV envelope proteins F and G were provided in trans, and the VSV-ΔG-NiV and VSV-ΔG-NiV pseudotyped viruses collected from transfected cells were used to transduce new cell's cell lines. Positive neutralization of the pseudotypes by anti-NiV or anti-HeV can then be followed by inhibition of transduction recorded by absence of luciferase or fluorescence (GFP) activity in cells 1 day posttransduction. This very sensitive and rapid test can be used for large series of seroneutralization and epidemiological investigations in BSL-2 containment. Another recent technique does not use live virus but rather the binding inhibition of protein G to ephrin-2 receptor fixed on beads as a new test of anti-NiV neutralizing antibodies (Bossart et al., 2007).

PROSPECTS FOR THE FUTURE

Henipavirus circulation in Southeast Asia should be monitored on a large scale in bats by ELISA and seroneutralization using recombinant technologies. These techniques should be included in any serodiagnosis of encephalitis in all potential endemic countries for NiV where fatal cases are recorded. Epidemiological investigations should address the cause of NiV emergence and the mechanisms of virus transmission from the reservoir to intermediary host and finally to humans (Epstein et al., 2006). The possible role of vectors, such as ticks that feed on bats and occasionally on other animals should be clarified. Countermeasures to prevent virus transmission to livestock and humans should be implemented and information to targeted populations should be diffused. Epidemiologists and health workers should be prepared to identify any unusual clinical manifestation where a bat could be involved. Better prophylaxis and treatments available at low cost to hamper any spread of virus in an epidemic area should be developed in research laboratories and tested in preclinical and clinical trials.

KEY ISSUES

- Endemicity of NiV in Southeast Asia and recurrent outbreaks in Bangladesh with possible human-to-human transmission (Gurley et al., 2007) highlight the importance and urgency of establishing a strong surveillance system supported by a network of laboratories equipped to handle and diagnose highly pathogenic viruses and including patient intensive care units, use of appropriate personal protective equipment, barrier nursing, and safe disposal of potentially infected material in the prevention and control measures for NiV/HeV virus infection.
- Although the major impact of Henipaviruses is human illness, the release of those zoonotic agents would have consequences for both humans and animals because infected and non-infected animals might need to be sacrificed to control the spread of infection. Continued research is necessary to develop effective strategies to limit the impact of these biological threats. Although it is possible to protect animals by vaccination and also by passive administration of antibodies, the main question is if a vaccine is developed should it be for humans or aimed to create a barrier in the domestic animal population such as pigs? The latter policy would be relevant in Malaysia were the pig was an intermediate and amplifying host, but in countries such as Bangladesh, man was infected possibly after drinking contaminated date palm sap (Luby et al., 2006). Before developing a vaccine for humans, it is necessary that the number of cases due to NiV infection be evaluated. It is obvious that until now due to a lack of viral reagents, the number of encephalitis cases due to NiV has been largely underestimated.
- It is highly probable that flying foxes harboring viruses like henipaviruses will continue to be displaced from their native habitat by natural or human forces and will come in close contact with human population or spillover hosts. Accidental or deliberate spray of NiV in pig farms may have a huge economical impact but could be limited if the virus is rapidly identified. Its transmission by aerosol, either by contaminated spillover host or through human-to-human transmission, could be rapidly blocked. But if it is like avian flu or SARS, one henipa-like virus becomes highly transmissible to humans by infecting cells of the upper respiratory tract, it would potentially cause a worldwide threat. Thus, henipaviruses should be considered as potential agents of bioterrorism (Lam, 2003) and handled in strict biosafety condition and security containments.

References

AbuBakar, S., Chang, L.Y., Ali, A.R., Sharifah, S.H. and Yusoff Zamrod, Z. Isolation and molecular identification of Nipah virus from pigs. Emerg. Infect. Dis. 2004; 10:2228–2230.

Aguilar, H.C., Matreyek, K.A., Choi, D.Y., Filone, C.M., Hashimi, S.T., Levroney Negrete, O.A., Bertolotti-Ciarlet, A., Choi, D.Y.,

McHardy, I., Fulcher, J.A., Su, S.V., Wolf, M.C., Kohatsu, L., Baum, L.G. and Lee, B. N-glycans on Nipah virus fusion protein protect against neutralization but reduce membrane fusion and viral entry. J. Virol. 2006; 80:4878–4889.

Aguilar, H.C., Matreyek, K.A., Choi, D.Y., Filone, C.M., Young, S. and Lee, B. Polybasic KKR motifs in the cytoplasmic tail of Nipah virus fusion protein modulates membrane fusion by inside-out signalling. J. Virol. 2007; 81:4520–4532.

Bennett, B.D., Zeigler, F.C., Gu, Q., Fendly, B., Goddard, A.D., Gillett, N. and Matthews, W. Molecular cloning of a ligand for the EPH-related receptor protein-tyrosine kinase Htk. Proc. Natl. Acad. Sci. U. S. A. 1995; 92:1866–1870.

Bergemann, A.D., Zhang, L., Chiang, M.K., Brambilla, R., Klein, R. and Flanagan, J.G. Ephrin-B3, a ligand for the receptor EphB3, expressed at the midline of the developing neural tube. Oncogene 1998; 16:471–480.

Bonaparte, M.I., Dimitrov, A.S., Bossart, K.N., Crameri, G., Mungall, B.A., Bishop, K.A., Choudhry, V., Dimitrov, D.S., Wang, L.F., Eaton, B.T. and Broder, C.C. Ephrin-B2 ligand is a functional receptor for Hendra virus and Nipah virus. Proc. Natl. Acad. Sci. U. S. A. 2005; 102:10652–10657.

Bossart, K.N., Crameri, G., Dimitrov, A.S., Mungall, B.A., Feng, Y.R., Patch, J.R., Choudhary, A., Wang, L.F., Eaton, B.T. and Broder, C.C. Receptor binding, fusion inhibition, and induction of cross-reactive neutralizing antibodies by a soluble G glycoprotein of Hendra virus. J. Virol. 2005a; 79:6690–6702.

Bossart, K.N., McEachern, J.A., Hickey, A.C., Choudhry, V., Dimitrov, D.S., Eaton, B.T. and Wang, L.F. Neutralization assays for differential henipavirus serology using Bio-Plex protein array systems. J. Virol. Methods 2007; 142:29–40.

Bossart, K.N., Mungall, B.A., Crameri, G., Wang, L.F., Eaton, B.T. and Broder, C.C. Inhibition of Henipavirus fusion and infection by heptad-derived peptides of the Nipah virus fusion glycoprotein. Virol. J. 2005b; 18:2:57.

Bossart, K.N., Wang, L.F., Eaton, B.T. and Broder, C.C. Functional expression and membrane fusion tropism of the envelope glycoproteins of Hendra virus. Virology 2001; 290:121–135.

Bossart, K.N., Wang, L.F., Flora, M.N., Chua, K.B., Lam, S.K., Eaton, B.T. and Broder, C.C. Membrane fusion tropism and heterotypic functional activities of the Nipah virus and Hendra virus envelope glycoproteins. J. Virol. 2002; 76:11186–11198.

Carter, J.R., Pager, C.T., Fowler, S.D. and Dutch, R.E. Role of N-linked glycosylation of the Hendra virus fusion protein. J. Virol. 2005; 79:7922–7925.

Cerretti, D.P., Vanden Bos, T., Nelson, N., Kozlosky, C.J., Reddy, P., Maraskovsky, E., Park, L.S., Lyman, S.D., Copeland, N.G., Gilbert, D.J., Jenkins, N.A. and Fletcher, F.A. Isolation of LERK-5: a ligand of the eph-related receptor tyrosine kinases. Mol. Immunol. 1995; 32:1197–1205.

Chadha, .M.S., Comer, J.A., Lowe, L., Rotra, P.A., Rollin, P.E., Bellini, W.J., Ksiazek, T.G. and Mishra, A. Nipah virus-associated encephalitis outbreak, Siliguri, India. Emerg. Infect. Dis. 2006; 12:235–239.

Chan, Y.P., Chua, K.B., Koh, C.L., Lim, M.E. and Lam, S.K. Complete nucleotide sequences of Nipah virus isolates from Malaysia. J. Gen. Virol. 2001; 82:2151–2155.

Chang, L.Y., Ali, A.R., Hassan, S.S. and AbuBakar, S. Quantitativ estimation of Nipah virus replication kinetics in vitro. Virol. J. 2006; 3:47.

Chen, J.M., Yu, M., Morrissy, C., Zhao, Y.G., Meehan, G., Sun, Y.X., Wang, Q.H., Zhang, W., Wang, L.F. and Wang, Z.L. A comparative indirect ELISA for the detection of henipavirus antibodies based on a recombinant nucleocapsid protein expressed in *Escherichia coli*. J. Virol. Methods 2006; 136:273–276.

Chong, H.T., Kamarulzaman, A., Tan, C.T., Goh, K.J., Thayaparan, T., Kunjapan, S.R., Chew, N.K., Chua, K.B. and Lam, S.K. Treatment of acute Nipah encephalitis with ribavirin. Ann. Neurol. 2001; 49:810–813.

Chong, H.T., Kunjapan, S.R., Thayaparan, T., Tong, J., Petharunam, V., Jusoh, M.R. and Tan, C.T. Nipah encephalitis outbreak in Malaysia, clinical features in patients from Seremban. Can. J. Neurol. Sci. 2002; 29:83–87.

Chua, K.B., Bellini, W.J., Rota, P.A., Harcourt, B.H., Tamin, A., Lam, S.K., Ksiazek, T.G., Rollin, P.E., Zaki, S.R., Shieh, W., Goldsmith, C.S., Gubler, D.J., Roehrig, J.T., Eaton, B., Gould, A.R., Olson, J., Field, H., Daniels, P., Ling, A.E., Peters, C.J., Anderson, L.J. and Mahy, B.W. Nipah virus: a recently emergent deadly paramyxovirus. Science 2000a; 288:1432–1435.

Chua, K.B., Chua, B.H. and Wang, C.W. Anthropogenic deforestation, El Nino and the emergence of Nipah virus in Malaysia. Malaysian J. Pathol. 2002c; 24:15–21.

Chua, K.B., Goh, K.J., Wong, K.T., Kamarulzaman, A., Tan, P.S., Ksiazek, T.G., Zaki, S.R., Paul, G., Lam, S.K. and Tan, C.T. Fatal encephalitis due to Nipah virus among pig-farmers in Malaysia. Lancet 1999; 354:1257–1259.

Chua, K.B., Koh, C.L., Hooi, P.S., Wee, K.F., Khong, J.H., Chua, B.H., Chan, Y.P., Lim, M.E. and Lam, S.K. Isolation of Nipah virus from Malaysian Island flying-foxes. Microbes Infect. 2002b; 4:145–151.

Chua, K.B., Lam, S.K., Goh, K.J., Hooi, P.S., Ksiazek, T.G., Kamarulzaman, A., Olson, J. and Tan, C.T. The presence of Nipah virus in respiratory secretions and urine of patients during an outbreak of Nipah virus encephalitis in Malaysia. J. Infect. Dis. 2001; 42:40–43.

Chua, K.B., Lam, S.K., Tan, C.T., Hooi, P.S., Goh, K.J., Chew, N.K., Tan, K.S., Kamarulzaman, A. and Wong, K.T. High mortality in Nipah encephalitis is associated with presence of virus in cerebrospinal fluid. Ann. Neurol. 2000b; 48:802–805.

Chua, K.B., Wang, L.F., Lam, S.K. and Eaton, B.T. Full length genome sequence of Tioman virus, a novel paramyxovirus in the genus Rubulavirus isolated from fruit bats in Malaysia. Arch. Virol. 2002a; 147:1323–1348.

Ciancanelli, M.J. and Basler, C.F. Mutation of YMYL in the Nipah virus matrix protein abrogates budding and alters subcellular localization. J. Virol. 2006; 80:12070–12078.

Contractor, A., Rogers, C., Maron, C., Henkemeyer, M., Swanson, G.T. and Heinemann, S.F. Trans-synaptic Eph receptor-ephrin signalling in hippocampal mossy fiber. Science 2002; 296:1864–1869.

Craft, W.W., Jr. and Dutch, R.E. Sequence motif upstream of the Hendra virus fusion protein cleavage site is not sufficient to promote efficient proteolytic processing. Virology 2005; 341:130–140.

Daniels, P., Ksiazek, T. and Eaton, B.T. Laboratory diagnosis of Nipah and Hendra virus infections. Microbes Infect. 2001; 3:289–295.

Diederich, S., Moll, M., Klenk, H.D. and Maisner, A. The Nipah virus fusion protein is cleaved within the endosomal compartment. J. Biol. Chem. 2005; 280:29899–29903.

Eaton, B.T., MacKenzie, J.S. and Wang, L.F. Henipavirus. In: Fields Virology (D.M. Knipe and P.M. Howley, Eds.), Vol. 2, pp. 1587–1600. Philadelphia: Lippincott, Williams & Williams, 2007.

Eaton, B.T., Wright, P.J., Wang, L.F., Sergeyev, O., Michalski, W.P., Bossart, K.N. and Broder, C.C. Henipaviruses: recent observations on regulation of transcription and the nature of the cell receptor. Arch. Virol. Suppl. 2004; 18:122–131.

Epstein, J.H., Field, H.E., Luby, S., Pulliam, J.R. and Daszak, P. Nipah virus: impact, origins, and causes of emergence. Curr. Infect. Dis. Rep. 2006; 8:59–65.

Eshaghi, M., Tan, W.S., Chin, W.K. and Yusoff, K. Purification of the extra-cellular domain of Nipah virus glycoprotein produced in *Escherichia coli* and possible application in diagnosis. J. Biotechnol. 2005; 116:221–226.

Field, H.E., Mackenzie, J.S. and Daszak, P. Henipaviruses: emerging paramyxoviruses associated with fruit bats. Curr. Top. Microbiol. Immunol. 2007; 315:133–159.

Flenniken, A.M., Gale, N.W., Yancopoulos, G.D. and Wilkinson, D.G. Distinct and overlapping expression patterns of ligands for Eph-related receptor tyrosine kinases during mouse embryogenesis. Dev. Biol. 1996; 179:382–401.

Gallagher, J.R. Use of convalescent measles serum to control measles in a preparatory school. Am. J. Public Health 1935; 25:595–598.

Georges-Courbot, M.C., Contamin, H., Faure, C., Loth, P., Baize, S., Leyssen, P., Neyts, J. and Deubel, V. Poly(I)-poly(C12U) but not ribavirin prevents death in a hamster model of Nipah infection. Antimicrob. Agents Chemother. 2006; 50:1768–1772.

Giraudon, P. and Wild, T.F. Correlation between epitopes on hemagglutinin of measles virus and biological activities: passive protection by monoclonal antibodies is related to their hemagglutination inhibiting activity. Virology 1985; 144:46–58.

Goh, K.J., Tan, C.T., Chew, N.K., Tan, P.S., Kamarulzaman, A., Sarji, S.A., Wong, K.T., Abdullah, B.J., Chua, K.B. and Lam, S.K. Clinical features of Nipah virus encephalitis among pig farmers in Malaysia. N. Engl. J. Med. 2000; 342:1229–1235.

Goldsmith, C.S., Whistler, T., Rollin, P.E., Ksiazek, T.G., Rota, P.A., Bellini, W.J., Daszak, P., Wong, K.T., Shieh, W.J. and Zaki, S.R. Elucidation of Nipah virus morphogenesis and replication using ultrastructural and molecular approaches. Virus Res. 2003; 92:89–98.

Good, R.A. and Zak, S.J. Disturbances in gamma globulin synthesis as "experiments of nature". Pediatrics 1956; 28:21–28.

Grunwald, I.C., Korte, M., Wolfer, D., Wilkinson, G.A., Unsicker, K., Lipp, H.P., Bonhoeffer, T. and Klein, R. Kinase-independent requirement of EphB2 receptors in hippocampal synaptic plasticity. Neuron 2001; 32:1027–1040.

Guillaume, V., Aslan, H., Ainouze, M., Guerbois, M., Wild, T.F., Buckland, R. and Langedijk, J.P.M. Evidence of a potential receptor-binding site on the Nipah virus G protein (NiV-G): identification of globular head residues with a role in fusion promotion and their localisation on an NiV-G structural model. J. Virol. 2006a; 80:7546–7554.

Guillaume, V., Contamin, H., Loth, P., Georges-Courbot, M.C., Lefeuvre, A., Marianneau, P., Chua, K.B., Lam, S.K., Buckland, R., Deubel, V. and Wild, T.F. Nipah virus: vaccination and passive protection studies in a hamster model. J. Virol. 2004a; 78:834–840.

Guillaume, V., Contamin, H., Loth, P., Grosjean, I., Georges-Courbot, M.C., Deubel, V., Buckland, R. and Wild, T.F. Antibody prophylaxis and therapy against Nipah virus infection in hamsters. J. Virol. 2006b; 80:1972–1978.

Guillaume, V., Lefeuvre, A., Faure, C., Marianneau, P., Buckland, R., Lam, S.K., Wild, T.F. and Deubel, V. Specific detection of Nipah Virus using real-time RT-PCR (TaqMan). J. Virol. Methods 2004b; 120:229–237.

Gurley, E.S., Montgomery, J.M., Woodward, J.L. and Breiman, R.F. Person-to-person transmission of Nipah virus in a Bangladeshi community. Emerg. Infect. Dis. 2007; 13:1031–1037.

Hagmaier, K., Stock, N., Goodbourn, S., Wang, L.F. and Randall, R. A single amino acid substitution in the V protein of Nipah virus alters its ability to block interferon signalling in cells from different species. J. Gen. Virol. 2006; 87:3649–3653.

Halpin, K., Bankamp, B., Harcourt, B.H., Bellini, W.J. and Rota, P.A. Nipah virus conforms to the rule of six in a minigenome replication assay. J. Gen. Virol. 2004; 85:701–707.

Halpin, K., Young, P.L., Field, H.E. and Mackenzie, J.S. Isolation of Hendra virus from pteropid bats: a natural reservoir of Hendra virus. J. Gen. Virol. 2000; 81:1927–1932.

Harcourt, B.H., Lowe, L., Tamin, A., Liu, X., Bankamp, B., Bowden, N., Rollin, P.E., Comer, J.A., Ksiazek, T.G., Hossain, M.J., Gurley, E.S., Breiman, R.F., Bellini, W.J. and Rota, P.A. Genetic characterization of Nipah virus, Bangladesh, 2004. Emerg. Infect. Dis. 2005; 11:1594–1597.

Harcourt, B.H., Tamin, A., Halpin, K., Ksiazek, T.G., Rollin, P.E., Bellini, W.J. and Rota, P.A. Molecular characterization of the polymerase gene and genomic termini of Nipah virus. Virology 2001; 287:192–201.

Harcourt, B.H., Tamin, A., Ksiazek, T.G., Rollin, P.E., Anderson, L. J., Bellini, W.J. and Rota, P.A. Molecular characterization of Nipah virus, a newly emergent paramyxovirus. Virology 2000; 271:334–349.

Harit, A.K., Ichhpujani, R.L., Gupta, S., Gill, K.S., Lal, S., Ganguly, N.K. and Agarwal, S.P. Nipah/Hendra virus outbreak in Siliguri, West Bengal, India. Indian J. Med. Res. 2006; 123:553–560.

Henderson, J.T., Georgiou, J., Jia, Z., Robertson, J., Elowe, S., Roder, J.C. and Pawson, T. The receptor tyrosine kinase EphB2 regulates NMDA-dependent synaptic function. Neuron 2001; 32:1041–1056.

Hooper, P., Zaki, S., Daniels, P. and Middleton, D. Comparative pathology of the diseases caused by Hendra and Nipah viruses. Microbes Infect. 2001; 3:315–322.

Hsu, V.P., Hossain, M.J., Parashar, U.D., Ali, M.M., Ksiazek, T.G., Kuzmin, I., Niezgoda, M., Rupprecht, C., Bresee, J. and Breiman, R.F. Nipah virus encephalitis reemergence, Bangladesh in, 2001. Emerg. Infect. Dis. 2001; 12:2082–2087.

Hyatt, A.D., Zaki, S.R., Goldsmith, C.S., Wise, T.G. and Hengstberger, S.G. Ultrastructure of Hendra virus and Nipah virus within cultured cells and host animals. Microbes Infect. 2001; 3:297–306.

ICDDR,B Nipah virus encephalitis outbreak over wide area of western Bangladesh, 2004. Health Sci. Bull. 2004a; 2(1):7–11.

ICDDR,B Person-to-person transmission of Nipah virus during outbreak in Faridpur District, 2004. Health Sci. Bull. 2004b; 2(2):5–9.

Kashiwazaki, Y., Na, Y.N., Tanimura, N. and Imada, T. A solid phase blocking ELISA for detection of antibodies to Nipah virus. J. Virol. Method 2004; 121:259–261.

Kullander, K. and Klein, R. Mechanisms and functions of Eph and ephrin signalling. Nat. Rev. Mol. Cell Biol. 2002; 3:475–486.

Lam, S.K. Nipah virus—a potential agent of bioterrorism? Antiviral Res. 2003; 57:113–119.

Lamb, R.A. and Kolakofsky, D. Paramyxoviridae: the viruses and their replication. In: Fields Virology (D.M. Knipe, P.M. Howley, D.E. Griffin et al., Eds.), 4th ed., pp. 1305–1340. Philadelphia: Lippincott, Williams & Williams, 2001.

Lambert, D.M., Barney, S., Lambert, A.L., Guthrie, K., Medinas, R., Davis, D.E., Bucy, T., Erickson, J., Merutka, G. and Petteway, S.R., Jr. Peptides from conserved regions of paramyxovirus fusion (F) proteins are potent inhibitors of viral fusion. Proc. Natl. Acad. Sci. U. S. A. 1996; 93:2186–2191.

Langedijk, J.P., Daus, F.J. and van Oirschot, J.T. Sequence and structure alignment of Paramyxoviridae attachment proteins and discovery of enzymatic activity for a morbillivirus hemagglutinin. J. Virol. 1997; 71:6155–6167.

Lee, K.E., Umapathi, T., Tan, C.B., Tjia, H.T., Chua, T.S., Oh, H.M., Fock, K.M., Kurup, A., Das, A., Tan, A.K. and Lee, W.L. The neurological manifestations of Nipah virus encephalitis, a novel paramyxovirus. Ann. Neurol. 1999; 46:428–432.

Lehle, C., Razafitrimo, G., Razainirina, J., Andriaholinirina, N., Goodman, S.M., Faure, C., Georges-Courbot, M.C., Rousset, D.

and Reynes, J.M. Henipavirus and Tioman virus antibodies in pteropodid bats, Madagascar. Emerg. Infect. Dis. 2007; 13:159–161.

Levroney, E.L., Aguilar, H.C., Fulcher, J.A., Kohatsu, L., Pace, K.E., Pang, M., Gurney, K.B., Baum, L.G. and Lee, B. Novel innate immune functions for galectin-1: galectin-1 inhibits cell fusion by Nipah virus envelope glycoproteins and augments dendritic cell secretion of proinflammatory cytokines. J. Immunol. 2005; 175:413–420.

Liebl, D.J., Morris, C.J., Henkemeyer, M. and Parada, L.F. mRNA expression of ephrins and eph receptor tyrosine kinases in the neonatal and adult mouse central nervous system. J. Neurosci. Res. 2003; 71:7–22.

Lim, C.C., Lee, K.E., Lee, W.L., Tambyah, P.A., Lee, C.C., Sitoh, Y.Y., Auchus, A.P., Lin, B.K. and Hui, F. Nipah virus encephalitis: serial MR study of an emerging disease. Radiology 2002; 222:219–226.

Luby, S.P., Rahman, M., Hossain, M.J., Blum, L.S., Husain, M.M., Gurley, E., Khan, R., Ahmed, B.N., Rahman, S., Nahar, N., Kenah, E., Comer, J.A. and Ksiazek, T.G. Foodborne transmission of Nipah virus, Bangladesh. Emerg. Infect. Dis. 2006; 12:1888–1894.

Magoffin, D.E., Halpin, K., Rota, P.A. and Wang, L.F. Effects of single amino acid substitutions at the E residue in the conserved GDNE motif of the Nipah virus polymerase (L) protein. Arch. Virol. 2007; 152:827–832.

Malvoisin, E. and Wild, T.F. Contribution of measles virus fusion protein in protective immunity: anti-F monoclonal antibodies neutralize virus infectivity and protect mice against challenge. J. J. Virol. 1990; 64:5160–5162.

Malvoisin, E. and Wild, T.F. Measles virus glycoproteins: studies on the structure and interaction of the haemagglutinin and fusion proteins. J. Gen. Virol. 1993; 74:2365–2372.

Mayo, M.A. Virus taxonomy—Houston, 2002. Arch. Virol. 2002; 147:1071–1076.

Meulendyke, K.A., Wurth, M.A., McCann, R.O. and Dutch, R.E. Endocytosis plays a critical role in proteolytic processing of the Hendra virus fusion protein. J. Virol. 2005; 79:12643–12649.

Michalski, W.P., Crameri, G., Wang, L.F., Shiell, B.J. and Eaton, B. The cleavage activation and sites of glycosylation in the fusion protein of Hendra virus. Virus Res. 2000; 69:83–93.

Middleton, D.J., Morrissy, C.J., van der Heide, B.M., Russell, G.M., Braun, M.A., Westbury, H.A., Halpin, K. and Daniels, P.W. Experimental Nipah virus infection in pteropid bats (*Pteropus poliocephalus*). J. Comp. Pathol. 2007; 136:266–272.

Mohd Nor, M.N., Gan, C.H. and Ong, B.L. Nipah virus infection of pigs in peninsular Malaysia. Rev. - Off. Int. Epizoot. 2000; 19:160–165.

Mohd Yob, J., Field, H., Rashdi, A.M., Morrissy, C., van der Heide, B., Rota, P., bin Adzhar, A., White, J., Daniels, P., Jamaluddin, A. and Ksiazek, T. Nipah virus infection in bats (order Chiroptera) in peninsular Malaysia. Emerg. Infect. Dis. 2001; 7:439–441.

Moll, M., Diederich, S., Klenk, H.D., Czub, M. and Maisner, A. Ubiquitous activation of the Nipah virus fusion protein does not require a basic amino acid at the cleavage site. J. Virol. 2004a; 78:9705–9712.

Moll, M., Kaufmann, A. and Maisner, A. Influence of N-glycans on processing and biological activity of the Nipah virus fusion protein. J. Virol. 2004b; 78:7274–7278.

Mungall, B.A., Middleton, D., Crameri, G., Bingham, J., Halpin, K., Russel, G., Green, D., McEachern, J., Pritchard, L.I., Eaton, B.T., Wang, L.F., Bossart, K.N. and Broder, C.C. Feline model of acute Nipah virus infection and protection with a soluble glycoprotein-based subunit vaccine. J. Virol. 2006; 80:12293–12302.

Mungall, B.A., Middleton, D., Crameri, G., Halpin, K., Bingham, J., Eaton, B.T. and Broder, C.C. Vertical transmission and fetal replication of Nipah virus in an experimentally infected cat. J. Infect. Dis. 2007; 196:812–816.

Murray, K., Selleck., P., Hooper, P., Hyatt, A., Gould, A., Gleeson, L., Westbury, H., Hiley, L., Selvey, L. and Rodwell, B. A morbillivirus that caused fatal disease in horses and humans. Science 1995; 268:94–97.

Nader, P.R., Horwitz, M.S. and Rousseau, J. Atypical exanthem following exposure to natural measles: eleven cases in children previously inoculated with killed vaccine. J. Pediatr. 1968; 72:22–28.

Negrete, O.A., Levroney, E.L., Aguilar, H.C., Bertolotti-Ciarlet, A., Nazarian, R., Tajyar, S. and Lee, B. EphrinB2 is the entry receptor for Nipah virus, an emergent deadly paramyxovirus. Nature 2005; 436:401–405.

Negrete, O.A., Wolf, M.C., Aguilar, H.C., Enterlein, S., Wang, W., Mühlberger, E., Su, S.V., Bertolotti-Ciarlet, A., Flick, R. and Lee, B. Two key residues in Ephrin B3 are critical for its use as an alternative receptor for Nipah virus. PLOS Pathog. 2006; 2:1–9.

Ng, B.Y., Lim, C.C., Yeoh, A. and Lee, W.L. Neuropsychiatric sequelae of Nipah virus encephalitis. J. Neuropsychiatry Clin. Neurosci. 2004; 16:500–504.

Olson, J.G., Rupprecht, C., Rollin, P.E., An, U.S., Niezgoda, M., Clemins, T., Walston, J. and Ksiazek, T.G. Antibodies to Nipah-like virus in bats (*Pteropus lylei*), Cambodia. Emerg. Infect. Dis. 2002; 8:987–988.

O'Sullivan, J.D., Allworth, A.M., Paterson, D.L., Snow, T.M., Boots, R., Gleeson, L.J., Gould, A.R., Hyatt, A.D. and Bradfield, J. Fatal encephalitis due to novel paramyxovirus transmitted from horses. Lancet 1997; 349:93–95.

Pager, C.T. and Dutch, R.E. Cathepsin L is involved in proteolytic processing of the Hendra virus fusion protein. J. Virol. 2005; 79:12714–12720.

Pager, C.T., Wurth, M.A. and Dutch, R.E. Subcellular localization and calcium and pH requirements for proteolytic processing of the Hendra virus fusion protein. J. Virol. 2004; 78:9154–9163.

Parashar, U.D., Sunn, L.M., Ong, F., Mounts, A.W., Arif, M.T., Ksiazek, T.G., Kamaluddin, M.A., Mustafa, A.N., Kaur, H., Ding, L.M., Othman, G., Radzi, H.M., Kitsutani, P.T., Stockton, P.C., Arokiasamy, J., Gary, H.E., Jr. and Anderson, L.J. Case-control study of risk factors for human infection with a new zoonotic paramyxovirus, Nipah virus, during a 1998–1999 outbreak of severe encephalitis in Malaysia. J. Infect. Dis. 2000; 181:1755–1759.

Park, M.S., Shaw, M.L., Munoz-Jordan, J., Cros, J.F., Nakaya, T., Bouvier, N., Palese, P., Garcia-Sastre, A. and Basler, C.F. Newcastle disease virus (NDV)-based assay demonstrates interferon-antagonist activity for the NDV V protein and the Nipah virus V, W, and C proteins. J. Virol. 2003; 77:1501–1511.

Patch, J.R., Crameri, G., Wang, L.F., Eaton, B.T. and Broder, C.C. Quantitative analysis of Nipah virus proteins released as virus-like particles reveals central role for the matrix protein. Virol. J. 2007; 4:1–14.

Paton, N.I., Leo, Y.S., Zaki, S.R., Auchus, A.P., Lee, K.E., Ling, A.E., Chew, S.K., Ang, B., Rollin, P.E., Umapathi, T., Sng, I., Lee, C.C., Lim, E. and Ksiazek, T.G. Outbreak of Nipah virus infection among abattoir workers in Singapore. Lancet 1999; 354:1253–1256.

Plemper, R.K., Hammond, A.L., Gerlier, D., Fielding, A.K. and Cattaneo, R. Strength of envelope protein interaction modulates cytopathicity of measles virus. J. Virol. 2002; 76:5051–5061.

Polack, F.P., Auwaerter, P.G., Lee, S.H., Nousari, H.C., Valsamakis, A., Leiferman, K.M., Diwan, A., Adams, R.J. and Griffin, D.E. Production of atypical measles in rhesus macaques: evidence for disease mediated by immune complex formation and eosinophils in the presence of fusion-inhibiting antibody. Nat. Med. 1999; 15:629–634.

Porotto, M., Carta, P., Deng, Y., Kellogg, G.E., Whitt, M., Lu, M., Mungall, B.A. and Moscona, A. Molecular determinants of antiviral potency of paramyxovirus entry inhibitors. J. Virol. 2007; 81:10567–10574.

Porotto, M., Doctor, L., Carta, P., Fornabaio, M., Greengard, O., Kellog, G.E. and Moscona, A. Inhibition of Hendra virus fusion. J. Virol. 2006; 80:9837–9849.

Reynes, JM., Counor, D., Ong, S., Faure, C., Seng, V., Molia, S., Walston, J., Georges-Courbot, M.C., Deubel, V. and Sarthou, J.L. Nipah virus in Lyle's flying foxes, Cambodia. Emerg. Infect. Dis. 2005; 11:1042–1047.

Rodriguez, J.J., Cruz, C.D. and Horvath, C.M. Identification of the nuclear export signal and STAT-binding domains of the Nipah virus V protein reveals mechanisms underlying interferon evasion. J. Virol. 2004; 78:5358–5367.

Rodriguez, J.J. and Horvath, C.M. Host evasion by emerging paramyxoviruses: Hendra virus and Nipah virus v proteins inhibit interferon signaling. Viral Immunol. 2004; 17:210–219.

Rodriguez, J.J., Parisien, J.P. and Horvath, C.M. Nipah virus V protein evades alpha and gamma interferons by preventing STAT1 and STAT2 activation and nuclear accumulation. J. Virol. 2002; 76:11476–11483.

Rodriguez, J.J., Wang, L.F. and Horvath, C.M. Hendra virus V protein inhibits interferon signaling by preventing STAT1 and STAT2 nuclear accumulation. J. Virol. 2003; 77:11842–11845.

Sakamoto, H., Zhang, X.Q., Suenobu, S., Ohbo, K., Ogawa, M. and Suda, T. Cell adhesion to ephrinb2 is induced by EphB4 independently of its kinase activity. Biochem. Biophys. Res. Commun. 2004; 321:681–687.

Sejvar, J.J., Hossain, J., Saha, S.K., Gurley, E.S., Banu, S., Hamadani, J.D., Faiz, M.A., Siddiqui, F.M., Mohammad, Q.D., Mollah, A.H., Uddin, R., Alam, R., Rahman, R., Tan, C.T., Bellini, W., Rota, P., Breiman, R.F. and Luby, S.P. Long-term neurological and functional outcome in Nipah virus infection. Ann. Neurol. 2007; 62:235–242.

Selvey, L.A., Wells, R.M., McCormack, J.G., Ansford, A.J., Murray, K., Rogers, R.J., Lavercombe, P.S., Selleck, P. and Sherindan, J.W. Infection of humans and horses by a newly described morbillivirus. Med. J. Aust. 1995; 162:642–645.

Shaw, M.L., Cardenas, W.B., Zamarin, D., Palese, P. and Basler, C.F. Nuclear localization of the Nipah virus W protein allows for inhibition of both virus- and toll-like receptor 3-triggered signaling pathways. J. Virol. 2005; 79:6078–6088.

Shaw, M.L., Garcia-Sastre, A., Palese, P. and Basler, C.F. Nipah virus V and W proteins have a common STAT1-binding domain yet inhibit STAT1 activation from the cytoplasmic and nuclear compartments, respectively. J. Virol. 2004; 78:5633–5641.

Shin, D., Garcia-Cardena, G., Hayashi, S., Gerety, S., Asahara, T., Stavrakis, G., Isner, J., Folkman, J., Gimbrone, M.A., Jr. and Anderson, D.J. Expression of ephrinB2 identifies a stable genetic difference between arterial and venous vascular smooth muscle as well as endothelial cells, and marks subsets of microvessels at sites of adult neovascularization. Dev. Biol. 2001; 230:139–150.

Smith, I.L., Halpin, K., Warrilow, D. and Smith, G.A. Development of a fluorogenic RT-PCR assay (TaqMan) for the detection of Hendra virus. J. Virol. Methods 2001; 98:33–40.

Tamin, A., Harcourt, B.H., Ksiazek, T.G., Rollin, P.E., Bellini, W.J. and Rota, P.A. Functional properties of the fusion and attachment glycoproteins of Nipah virus. Virology 2002; 296:190–200.

Tan, C.T., Goh, K.J., Wong, K.T., Sarji, S.A., Chua, K.B., Chew, N.K., Murugasu, P., Loh, Y.L., Chong, H.T., Tan, K.S., Thayaparan, T., Kumar, S. and Jusoh, M.R. Relapsed and late-onset Nipah encephalitis. Ann. Neurol. 2002; 51:703–708.

Tanimura, N., Imada, T., Kashiwazaki, Y., Shamusudin, S., Syed Hassan, S., Jamaluddin, A., Russell, G. and White, J. Reactivity of anti-Nipah virus monoclonal antibodies to formalin-fixed, paraffin-embedded lung tissues from experimental Nipah and Hendra virus infections. J. Veterinarian Med. Sci. 2004; 66:1263–1266.

Tanimura., N., Imada, T., Kashiwazaki, Y. and Sharifah, S.H. Distribution of viral antigens and development of lesions in chicken embryos inoculated with Nipah virus. J. Comp. Pathol. 2006; 135:74–82.

Toth, J., Cutforth, T., Gelinas, A.D., Bethoney, K.A., Bard, J. and Harrison, C.J. Crystal structure of an ephrin ectodermain. Dev. Cell 2001; 1:83–92.

Vogt, C., Eickmann, M., Diederich, S., Moll, M. and Maisner, A. Endocytosis of the Nipah virus glycoproteins. J. Virol. 2005; 79:3865–3872.

Wacharapluesadee, S. and Hemachudha, T. Duplex nested RT-PCR for detection of Nipah virus RNA from urine specimens of bats. J. Virol. Methods 2007; 141:97–101.

Wacharapluesadee, S., Lumlertdacha, B., Boongird, K., Wanghongsa, S., Chanhome, L., Rollin, P., Stockton, P., Rupprecht, C.E., Ksiazek, T.G. and Hemachudha, T. Bat Nipah virus, Thailand. Emerg. Infect. Dis. 2005; 11:1949–1951.

Wang, L., Harcourt, B.H., Yu, M., Tamin, A., Rota, P.A., Bellini, W.J. and Eaton, B.T. Molecular biology of Hendra and Nipah viruses. Microbes Infect. 2001; 3:279–287.

Wang, L.F., Yu, M., Hansson, E., Pritchard, L.I., Shiell, B., Michalski, W.P. and Eaton, B.T. The exceptionally large genome of Hendra virus: support for creation of a new genus within the family Paramyxoviridae. J. Virol. 2000; 74:9972–9979.

Weingartl, H., Czub, S., Copps, J., Berhane, Y., Middleton, D., Marszal, P., Gren, J., Smith, G., Ganske, S., Manning, L. and Czub, M. Invasion of the central nervous system in a porcine host by Nipah virus. J. Virol. 2005; 79:7528–7534.

Weingartl, H.M., Berhane, Y., Caswell, J.L., Loosmore, S., Audonnet, J-C., Roth, J.A. and Czub, M. Recombinant Nipah virus vaccines protect pigs against challenge. J. Virol. 2006; 80:7929–7938.

White, J.R., Boyd, V., Crameri, G.S., Duch, C.J., van Laar, R.K., Wang, L.F. and Eaton, B.T. Location of, immunogenicity of and relationships between neutralization epitopes on the attachment protein (G) of Hendra virus. J. Gen. Virol. 2005; 86:2839–2848.

Whitman, S.D. and Dutch, R.E. Surface density of the Hendra G protein modulates Hendra F protein-promoted membrane fusion : role for Hendra G protein trafficking and degradation. Virology 2007; 363:419–429.

Wild, T.F. and Buckland, R. Inhibition of measles virus infection and fusion with peptides corresponding to the leucine zipper region of the fusion protein. J. Gen. Virol. 1997; 78:107–111.

Williamson, M.M., Hooper, P.T., Selleck, P.W., Westbury, H.A. and Slocombe, R.F. Experimental Hendra virus infection in pregnant guinea-pigs and fruit bats (*Pteropus poliocephalus*). J. Comp. Pathol. 2000; 122:201–207.

Williamson, M.M., Hooper, P.T., Selleck, P.W., Westbury, H.A. and Slocombe, R.F.S. A guinea-pig model of Hendra virus encephalitis. J. Comp. Pathol. 2001; 124:273–279.

Wong, K.T., Grosjean, I., Brisson, C., Blanquier, B., Fevre-Montange, M., Bernard, A., Loth, P., Georges-Courbot, M.C., Chevallier, M., Akaoka, H., Marianneau, P., Lam, S.K., Wild, T.F. and Deubel, V. A golden hamster model for human acute Nipah virus infection. Am. J. Pathol. 2003; 163:2127–2137.

Wong, K.T., Shieh, W.J., Kumar, S., Norain, K., Abdullah, W., Guarner, J., Goldsmith, C.S., Chua, K.B., Lam, S.K., Tan, C.T., Goh, K.J., Chong, H.T., Jusoh, R., Rollin, P.E., Ksiazek, T.G. and Zaki, S.R.Nipah Virus Pathology Working Group Nipah virus infection: pathology and pathogenesis of an emerging paramyxoviral zoonosis. Am. J. Pathol. 2002; 161:2153–2167.

World Health Organization Nipah virus outbreak(s) in Bangladesh, January–April 2004. Wkly. Epidemiol. Rec. 2004; 79:168–171.

Xu, Y., Gao, S., Cole, D.K., Zhu, J., Su, N., Wang, H., Gao, G.F. and Rao, Z. Basis for fusion inhibition by peptides: analysis of the heptad repeat regions of the fusion proteins from Nipah and Hendra viruses, newly emergent zoonotic paramyxoviruses. Biochem. Biophys. Res. Commun. 2004a; 315:664–670.

Xu, Y., Lou, Z., Liu, Y., Cole, D.K., Su, N., Qin, L., Li, X., Bai, Z., Rao, Z. and Gao, G.F. Crystallization and preliminary crystallographic analysis of the fusion core from two new zoonotic paramyxoviruses, Nipah virus and Hendra virus. Acta Crystallogr. Section D, Biol. Crystallogr. 2004b; 60:1161–1164.

Yokoyama, N., Romero, M.I., Cowan, C.A., Galvan, P., Helmbacher, F., Charnay, P., Parada, L.F. and Henkemeyer, M. Forward signalling mediated by ephrinB3 prevents contraleral cortiico-spinal axons from recrossing the spinal cord midline. Neuron 2001; 29:85–97.

Young, P.L., Halpin, K., Selleck, P.W., Field, H., Gravel, J.L., Kelly, M.A. and Mackenzie, J.S. Serologic evidence for the presence in Pteropus bats of a Paramyxovirus related to equine morbillivirus. Emerg. Infect. Dis. 1996; 2:239–240.

Yu, F., Khairullah, N.S., Inoue, S., Balasubramaniam, V., Berendam, S.J., The, L.K., Ibrahim, N.S., Abdul Rahman, S., Hassan, S.S., Hasebe, F., Sinniah, M. and Morita, K. Serodiagnosis using recombinant Nipah virus nucleocapsid protein expressed in *Escherichia coli*. J. Clin. Microbiol. 2006; 44:3134–3138.

Zhu, Z., Dimitrov, A.S., Bossart, K.N., Crameri, G., Bishop, K.A., Choudhry, V., Mungall, B.A., Feng, Y-R., Choudry, A., Zhang, M-Y., Feng, Y., Wang, L-F., Xiao, X., Eaton, B.T., Broder, C.C. and Dimitrov, D.S. Potent neutralization of Hendra and Nipah viruses by human monoclonal antibodies. J. Virol. 2006; 80:891–899.

CHAPTER

32

O'nyong-nyong and Chikungunya

Ann M. Powers[1] *and Aaron C. Brault*[2]

[1]Division of Vector-Borne Infectious Diseases, Centers for Disease Control and Prevention, Rampart Road, Fort Collins, CO, USA

[2]Departments of Pathology and Microbiology and Immunology, Center for Vector-Borne Diseases, School of Veterinary Medicine, University of California, Davis, CA, USA

OUTLINE

ABSTRACT

At the individual level, vaccines have the potential to modulate the severity of illness as well as eliminate or reduce the likelihood of transmission of the pathogen to others making vaccination one of the most effective disease intervention tools available. Several vaccines against arthropod-borne viruses (arboviruses), including yellow fever virus and Japanese encephalitis virus, have been developed and are in widespread use. As there are approximately 600 known arboviruses, many of which cause severe human disease, additional vaccine development is warranted.

Among the arboviruses pathogenic to humans are chikungunya (CHIKV) and o'nyong-nyong (ONNV) viruses. These viruses cause explosive outbreaks and induce a prolonged, debilitating arthralgic illness that can have a dramatic impact on the affected communities. "Despite these facts, there are no therapeutics or vaccines commercially available for either virus". Preliminary vaccine development and characterization has been initiated for CHIKV leading to Phase 2 human clinical trials. Results from these studies have been promising to date. Unfortunately, the marketability of a vaccine against this virus is uncertain thus limiting the rapid development of a commercial product. As epidemics of illness caused by both CHIKV and ONNV continue to occur with greater intensity and frequency, the possibility of producing a vaccine for widespread distribution may also be more likely. This chapter describes the biology, ecology, and epidemiology of these viruses providing a framework from which to propose and support alphavirus vaccine development.

INTRODUCTION

Arthropod-borne viruses (arboviruses) are the causative agents of some of the most important emerging infectious diseases that are significant, global, public health problems (Gubler, 2001). One group of these viruses transmitted by invertebrate vectors belongs to the family *Togaviridae* and cause diseases ranging from mild febrile illness, polyarthritis, or encephalitis. Two of these viruses, chikungunya virus (CHIKV) and o'nyong-nyong virus (ONNV) are geographically widespread, are associated with severe morbidity, and can cause explosive outbreaks. The lack of specific therapeutics, prophylaxis, or antiviral treatments against these agents clearly highlights the need for an effective vaccine as a first-line defense that can quickly generate strong immunity. However, to accomplish this objective, for CHIKV and ONNV, there is first a need to develop a deeper understanding of the basic pathogenesis and biology of these viruses. A more complete study of the ecology, epidemiology, host immunology, and virus–vector interactions, as well as development of appropriate animal models for disease studies all require additional research for the design and implementation of an appropriate and efficacious vaccine program. Significant progress toward identifying a highly protective vaccine against CHIKV has already been well initiated as described below and further knowledge will continue to improve the likelihood of success.

HISTORY OF THE DISEASE

Chikungunya

CHIKV was first isolated from the serum of a febrile human in Tanganyika (Tanzania) in 1953 during an epidemic of dengue-like illness (Robinson, 1955; Ross, 1956). The highest disease incidence occurred on the Makonde plateau, an area devoid of woodland habitat, and was documented as a disease completely new to the area (Lumsden, 1955; Casals and Whitman, 1957). While the initial assessment was that the outbreak was due to a dengue virus (a flavivirus), serological and antigenic characterization of the isolates from the outbreak indicated that it was an alphavirus closely related to Mayaro and Semliki Forest viruses (Casals and Whitman, 1957; Spence and Thomas, 1959). Since that original epidemic and identification of the etiological agent, a significant period of CHIKV activity occurred between the 1960s and 1980s when the virus was isolated repeatedly from numerous countries in central and southern Africa as well as in Senegal and Nigeria in western Africa. Some of the outbreaks that occurred in Africa during this period included those in Senegal in 1962, Uganda's Entebbe area in 1968 (McCrae et al., 1971), and South Africa from 1975 to 1977 (Jupp and McIntosh, 1988). The virus also spread from these African origins to the Asian continent where frequent outbreaks have been reported covering Malaysia, Indonesia, India, Cambodia, and Thailand from 1960s to 2003 (Chastel, 1963; Jadhav et al., 1965; Myers et al., 1965; Rao et al., 1965; Munasinghe et al., 1966; Rao, 1966; Halstead et al., 1969a, 1969b, 1969c, 1969d; Nimmannitya et al., 1969; Thaung et al., 1975; Marchette et al., 1978; Thein et al., 1992; Thaikruea et al., 1997; Lam et al., 2001; Mourya et al., 2001; Kit, 2002; Laras et al., 2005). Numerous large cities in Southeast Asia including Calcutta and Bangkok have been identified as particularly active sites of transmission and disease (Pavri, 1964; Sarkar et al., 1965b; Halstead et al., 1969c; Burke et al., 1985). Recently, there has been a resurgence of CHIKV outbreaks with major disease clusters documented in Democratic Republic of the Congo (Nur et al., 1999; Pastorino et al., 2004), Ivory Coast in 1996/1997 (Thonnon et al., 1999), Senegal in 1986 and 1996/1997 (Diallo et al., 1999), Indonesia in 2003 (Porter et al., 2004; Laras et al., 2005), Comoros in 2005 (Sergon et al., 2007), Reunion in 2006 (Schuffenecker et al., 2006), and India in 2006/2007 (Arankalle et al., 2007). Retrospective reviews have also suggested that CHIKV epidemics occurred as early as 1779 but were frequently documented inaccurately as dengue outbreaks (Carey, 1971). Overall, CHIKV has caused numerous well-documented outbreaks in both Africa and Southeast Asia, involving hundreds of

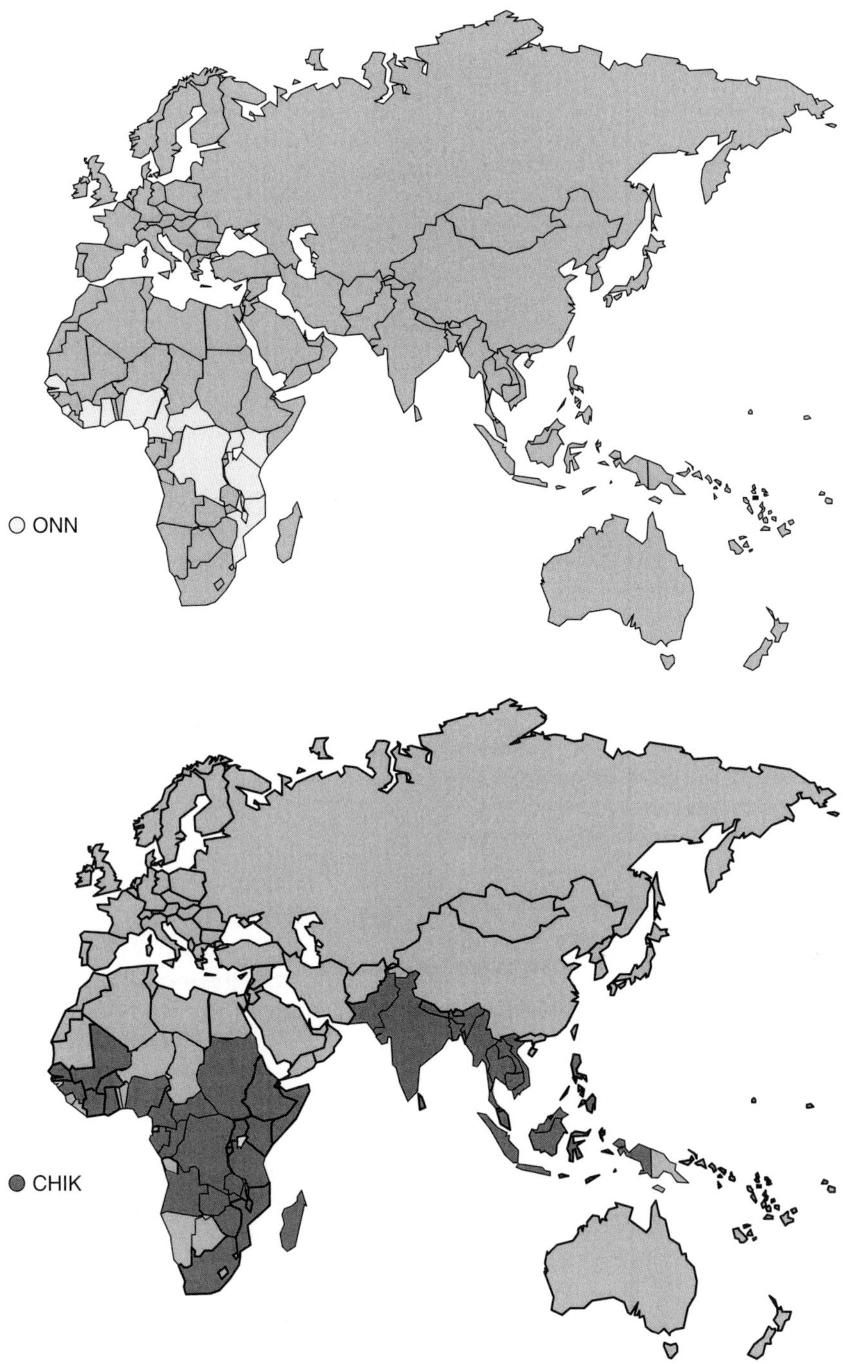

FIGURE 32.1 Geographic distribution of (a) ONNV and (b) CHIKV. Viral activity has been documented by viral isolation or serological detection of virus-specific antibodies.

thousands of people (Fig. 32.1). Outbreaks continue to occur with increased frequency and new geographic areas are likely to be involved (Jupp and Kemp, 1996).

O'nyong-nyong

In 1959, one of the largest arboviral epidemics ever documented began in the Acholi region of Uganda and subsequently swept across East Africa, involving over 2 million reported cases and lasting over 3 years (Haddow et al., 1960; Johnson, 1988). The clinical features of the disease resembled those of both chikungunya and dengue with primary symptoms including fever, severe joint and back pain, and rash. During the course of the outbreak, affected regions were estimated to have over 50% of the population affected (Williams et al., 1965b). A viral isolate was obtained from the serum of a febrile patient and was determined to be closely related to but distinct from CHIKV. The virus was given the name o'nyong-nyong virus (ONNV) after the Acholi word for "very painful and weak" that had come to be associated with the disease (Williams and Woodall, 1961). The virus effectively disappeared until 1996 when the second documented ONNV epidemic occurred in southern Uganda (Rwaguma et al., 1997; Lanciotti et al., 1998; Kiwanuka et al., 1999; Sanders et al., 1999). This outbreak began in the rural south-central regions of Uganda and, over a period of several months, extended into several adjacent areas focusing primarily near lakes and swamps. The outbreak was so extensive that the infection rate in affected areas was estimated to be nearly 70% (Sanders et al., 1999). Between 1962 and 1996, only sporadic evidence of ONNV transmission was documented with some of these infections determined to be caused by Igbo Ora virus, a serological subtype of ONNV (Moore et al., 1975; Woodruff et al., 1978; Marshall et al., 1982; Lhuillier et al., 1988; Olaleye et al., 1988, 1990). During 2003, another outbreak of febrile illness found to be caused by ONNV was reported among Liberian refugees in western Cote d'Ivoire (Posey et al., 2005). While ONNV is extremely closely related to CHIKV, the outbreak patterns and distribution of the two viruses are quite distinct with ONNV causing only periodic massive epidemics in contrast to the more frequent CHIKV outbreaks.

ETIOLOGIC AGENT(S): CLASSIFICATION AND ANTIGENS ENCODED BY AGENT

Both CHIKV and ONNV are members of the *Alphavirus* genus in the family *Togaviridae*. The only other genus within the *Togaviridae* is the *Rubivirus* genus whose sole species is Rubella virus (Weaver et al., 2005). The *Alphavirus* genus consists of 29 distinct species that either cause encephalitis, febrile illness with arthralgia, or are not known to cause disease in humans. The members are primarily vector-borne with nearly all utilizing mosquitoes as their invertebrate vectors.

Like all alphaviruses, CHIKV and ONNV have a genome consisting of a linear, positive sense, single-stranded RNA molecule of approximately 11.8 kb. The genomes of CHIKV and ONNV have both been sequenced, Khan et al. (2002) and Strauss et al. (1988), respectively and are 11,826 and 11,835 nucleotides in length, respectively. The nonstructural proteins, required for viral replication, are encoded in the 5′ two thirds of the genome, while the structural genes are collinear with the 3′one-third (Fig. 32.2). The structural proteins are produced by translation of an mRNA that is generated from an internal, subgenomic promoter immediately downstream of the nonstructural open reading frame. The 5′ end of the genome has a 7-methylguanosine cap while the 3′ end is polyadenylated. CHIKV and ONNV have 3′ noncoding regions of 503 and 425 nucleotides, respectively; the repeat sequence elements present in this region of the two virus species are distinct and generate different predicted secondary structures (Levinson et al., 1990; Pfeffer et al., 1998; Khan et al., 2002). Additionally, an internal polyadenylation site [I-poly (A)] of variable length has been identified in the 3′ noncoding region of one CHIKV genome (Khan et al., 2002) but no comparable I-poly (A) has been found in ONNV or other CHIKV strains.

There are four nonstructural genes, designated nsP1-4. All of these proteins are necessary for viral replication and each has specific functions in this process. The nsP1 protein is required for initiation of synthesis of minus-strand RNA (Sawicki et al., 1981; Wang et al., 1991; Sawicki and Sawicki, 1998) and has an additional function as a methyltransferase to cap the genomic and subgenomic RNAs during transcription (Durbin and Stollar, 1985; Mi and Stollar, 1991; Strauss and Strauss, 1994). The nsP2 protein has RNA helicase activity in its N-terminus (Hodgman, 1988; Gorbalenya et al., 1990) while the C-terminal domain functions as a proteinase for the alphavirus nonstructural polyprotein (Ding and Schlesinger, 1989; Hahn et al., 1989; Hardy and Strauss, 1989). There are three sites in the nonstructural polyprotein that are cleaved by both nsP2 and polyproteins containing nsP2, but the cleavage-site preferences of the different polyproteins vary. Differential cleavage results in alternative processing pathways for the nonstructural polyprotein at different times during the infections process (de Groot et al., 1990). For example, early during infection, P123 and P1234 (made by read-through translation of a stop codon present between

FIGURE 32.2 Arrangement of CHIKV and ONNV genes and genetic elements.

the nsP3 and nsP4 genes) are produced by translation of the genomic mRNA. As the infection proceeds, P123 cleaves between nsP1 and nsP2 to produce nsP1 and P23. The P23 polyprotein is autoproteolytic; it cleaves itself to generate the individual nsP2 and nsP3 proteins. It is likely that the nsP2 proteinase activity has a regulatory function by modulating the rate of viral replication. The functions of the nsP3 gene are not fully understood. This protein consists of two domains, a widely conserved N-terminal domain and a hypervariable carboxy terminus (Strauss et al., 1988; LaStarza et al., 1994b). The C-terminal region has been shown to tolerate numerous mutations, including large deletions, and still produce viable viral particles in vertebrate cells (Davis et al., 1989; Li et al., 1990). Interestingly, some nsP3 mutations in one alphavirus, Sindbis, while doing little to impair replication in mammalian cells, significantly reduced the replicative ability of the virus in mosquito cells (Lastarza et al., 1994a) suggesting that this gene may aid in defining cell type and invertebrate vector specificity. The final nonstructural protein, nsP4, is presumably the viral RNA-dependent RNA polymerase due to the characteristic GDD motif (Kamer and Argos, 1984).

The structural gene products are generated by translation of a subgenomic message to produce a polyprotein that is processed to produce a capsid protein, the two major envelope surface glycoproteins (E1 and E2) as well as two small peptides, E3 and 6K (Simizu et al., 1984; Strauss and Strauss, 1994). The highly conserved capsid protein associates with the nascent positive sense viral RNA to generate nucleocapsids. E1 and E2 are posttranscriptionally modified in the endoplasmic reticulum and Golgi apparatus before being transported to the plasma membrane (Schlesinger and Schlesinger, 1986) where they maintain a close association with each other forming a dimeric spike structure (Anthony and Brown, 1991; Paredes et al., 1993). As virion formation proceeds, the cytoplasmic nucleocapsids are trafficked to the cell membrane where they bind to the surface glycoproteins before budding from the cell (Garoff and Simons, 1974; Ekstrom et al., 1994; Garoff et al., 1998). The E2 protein, in particular, has been found to be an important determinant of antigenicity and cell receptor binding in both the vertebrate host and the insect vector for other alphaviruses. In mice, a single mutation in the Venezuelan equine encephalitis virus (VEEV) E2 glycoprotein delayed replication of the mutant by almost 2 days and significantly reduced the pathogenicity of the virus (Davis et al., 1994), while in the insect, one VEEV strain, containing a single mutation in the E2, was restricted in its ability to disseminate from the midgut following oral infection (Woodward et al., 1991). Additionally, the E2 gene has been implicated in increased vector infection of epizootic mosquito vectors (Brault et al., 2002, 2004).

PROTECTIVE IMMUNE RESPONSE

Both CHIKV and ONNV are highly immunogenic and generate strong antibody responses that are hallmarks of the infection. Antibodies are generated quite early in infection as documented by serological assays using either serum or cerebral spinal fluid (CSF). Serum samples from CHIKV infections collected within 5 days of symptom onset are typically free of neutralizing (N), complement-fixing (CF), and hemagglutination inhibiting (HI) antibodies. However, samples collected as early as 2 weeks postinfection already contain substantial levels of both N and HI antibodies (Carey et al., 1969; Nimmannitya et al., 1969). Additional information on N antibodies against CHIK and ONN viruses has been obtained from serosurveys following outbreaks and during interepidemic periods (Salim and Porterfield, 1973; Rodhain et al., 1989; Adesina and Odelola, 1991). For example, after the 1996–1997 ONNV outbreak in Uganda, 45% of the people surveyed in affected areas had ONNV-specific antibodies (Sanders et al., 1999). An interepidemic serosurvey, conducted in the early 1970s in Sudan, demonstrated 13% and 23% of those tested had neutralizing antibodies against CHIKV and ONNV, respectively (Salim and Porterfield, 1973). High-titered CHIKV neutralizing antibody levels have also been detected in laboratory workers infected by occupational exposure (Tomori et al., 1981).

While both viruses generate high titered N antibodies, these antibodies are not equally cross-neutralizing even though the viruses are extremely closely related. This is and will be an important consideration for vaccine development against the species pair. Single dose (infection immune) antisera reveal a one-way difference in neutralization of homologous versus heterologous viruses. Early studies indicated that CHIKV immune serum could neutralize both CHIKV and ONNV while ONNV antisera neutralized only the homologous virus (Porterfield, 1961; Karabatsos, 1975; Chanas et al., 1976). One study designed to further characterize this one-way relationship compared small and large plaque variants of CHIKV with one strain of ONNV (Chanas et al., 1979). This study revealed that antiserum against ONNV did not neutralize CHIKV to detectable levels but that nonneutralizing antibodies did, in fact, bind to the CHIK virions. At the time, it was suggested that the arrangement of the antigens on the CHIKV virions was sufficiently different to keep heterologous antibodies from effectively neutralizing the virus. Further studies utilized panels of monoclonal antibodies (mabs) to elucidate the interactions between CHIKV, ONNV, and antibodies against the two viruses. In this study, the authors again found that the majority of CHIKV-specific mabs would react with ONNV but as few of 12% of the mabs generated against ONNV were reactive against certain strains of CHIKV. The results of this study suggested that ONNV had certain antigenic epitopes that were unique to the envelope proteins of this virus and thus, would not bind the closely related CHIKV (Blackburn et al., 1995). The changes in the ONNV glycoprotein epitopes that are reflected by the inability of ONNV mabs to bind CHIKV may correlate with the distinctive ability of ONNV to infect anopheline mosquitoes while the other alphaviruses lack this capacity. Development of vaccines against CHIKV and ONNV would need to consider these epitope differences if broad cross-protection were to be obtained.

The role of cell-mediated immunity has not been extensively studied for either CHIKV or ONNV infections. Due to the lack of mortality associated with these viruses and the frequent misdiagnosis of dengue for CHIKV or ONNV infections, these alphaviruses have not been considered high priority agents until recently. However, another alphavirus in the Semliki Forest antigenic complex, Ross River virus (RRV), was shown to induce the proliferation of T-lymphocytes in infected mouse spleen cells (Aaskov et al., 1983). Curiously, little virus-specific cytotoxic activity was detected in these cells. Studies with other alphaviruses (Semliki Forest and Sindbis) demonstrated that cytotoxic T-lymphocyte (CTL) activity correlated with clearance from peripheral blood (Peck et al., 1979; Blackman and Morris, 1984); however, this CTL activity was not believed to protect against acute alphavirus viremia (Kagi and Hengartner, 1996). Recent work describing the characterization of an alphavirus CD8 CTL epitope that localizes to the capsid protein found that the epitope was conserved across many of the Semliki Forest complex viruses including CHIKV and correlated with CTL cross-reactivity from mice that had been infected with these viruses (Linn et al., 1998). Further examination of the role of CMI in CHIKV and ONNV infections is certainly warranted.

EPIDEMIOLOGY

For two viruses that are so closely related, both serologically and genetically, CHIKV and ONNV have curiously distinct transmission cycles and epidemiology, a condition that could impact any broad control or vaccine strategies for these agents. Geographically, ONNV is fairly restricted with its range limited to select portions of Africa (Fig. 32.1); the virus has been isolated in samples collected from Uganda, Kenya, Tanzania, Malawi, Mozambique, Cameroon, and the Central African Republic (Woodall et al., 1963; Johnson, 1988). The antigenic subtype Igbo Ora has additionally been identified in the Central African Republic and Nigeria, extending the geographic range of ONNV to western Africa (Moore et al., 1974, 1975; Olaleye et al., 1989). In contrast, CHIKV is prevalent throughout Africa as well as Southeast Asia and parts of Oceania. The virus causes not only large outbreaks throughout this distribution but appears to be enzootic across tropical regions of Africa. Molecular epidemiological studies have suggested that CHIKV originated in Africa and subsequently moved to Asia and the Indian Ocean areas (Powers et al., 2000) and that lineages are continually circulating even in the absence of identified vertebrate hosts (Yadav et al., 2003b).

The major distinction between the two viruses is that they are transmitted not only by different species of mosquitoes but by vectors that are in different families. In fact, ONNV is distinct among alphaviruses in that it is the only species to be principally vectored by anopheline mosquitoes; all the other alphaviruses utilize culicine species in their transmission cycles. The majority of mosquitoes collected in both the 1959–1962 and 1996–1997 outbreaks were found to be primarily endophilic and the epidemic transmission of ONNV is most certainly between man and infected *Anopheles gambiae* and *Anopheles funestus* mosquitoes (Corbet et al., 1961; Williams et al., 1965a; Johnson, 1988; Lutwama et al., 1999). During both outbreaks, *An. funestus* was the most abundant mosquito collected and isolates of ONNV were obtained during each epidemic from this species. An isolate was also obtained from *An. funestus* during the interepidemic period in 1978 suggesting that this mosquito may also be involved in the enzootic maintenance cycle (Johnson et al., 1981). *An. gambiae* was incriminated as a vector during the first outbreak but there were too few of this species collected during the second outbreak to confirm its role as an epidemic vector. Additionally, an isolate of ONNV was recovered from a pool of *Mansonia uniformis* during the 1996–1997 outbreak and previous literature has suggested that *Mansonia* species mosquitoes may be ONNV vectors (Haddow et al., 1960; Lutwama et al., 1999). The 1997 isolate was the first ever recorded from a *Mansonia* species so the epidemic and maintenance potential of mosquitoes in this genus remains to be determined.

In sharp contrast to ONNV, CHIKV in the West and Central Africa range appears to be maintained in a sylvatic cycle involving wild nonhuman primates and forest-dwelling *Aedes* spp. mosquitoes. The virus has been isolated from sylvatic mosquito species in several countries including Senegal, Cote d'Ivoire, Central African Republic, and South Africa. The mosquito species involved vary geographically and with ecological conditions; however, the major species involved in sylvatic cycles are *Aedes furcifer*, *Aedes taylori*, *Aedes luteocephalus*, *Aedes africanus*, and *Aedes neoafricanus* (McIntosh et al., 1977; Jupp and McIntosh, 1988; Diallo et al., 1999). Isolates of CHIKV have been obtained from all these species but the principal vectors during epidemics in these geographic regions appear to be members of the *Ae. furcifer-taylori* group (Jupp and McIntosh, 1990). In these primarily rural regions the outbreaks have tended to be of smaller scale and appear to be heavily dependent upon the sylvatic mosquito densities that increase with periods of heavy rainfall (Lumsden, 1955).

While forest-dwelling mosquito species are primary vectors in West and Central Africa, the urban mosquito *Aedes aegypti* has been found to be the most significant vector in Asia with virtually all Asian mosquito isolates coming from this species. These are urban and peridomestic, anthropophilic mosquitoes that maintain close associations with humans and thus, are likely to be responsible for the much larger outbreaks documented there. Urban or large outbreaks in East Africa are also likely associated with the presence of *Ae. aegypti* mosquitoes. Numerous laboratory studies have examined distinct populations of *Ae. aegypti* to understand the variable susceptibilities of this species in the transmission of CHIKV (Soekiman, 1987; Banerjee et al., 1988; Mourya et al., 1994). Other common peridomestic species, including *Aedes albopictus*, *Aedes vittatus*, and *Anopheles stephensi*, that have been found in abundance in CHIKV endemic areas have also been assessed for their vectorial capacity (Tesh et al., 1976; Soekiman et al., 1986a, 1986b; Mourya, 1987; Mourya and Banerjee, 1987; Turell et al., 1992, Yadav et al., 2003a). All three of these species were found to be competent vectors in the laboratory; their role as primary vectors in urban outbreaks of CHIKV disease should be considered.

Numerous field and laboratory studies have been conducted on the mosquito vectors involved in the transmission of CHIKV and ONNV. Unfortunately, there is far less information on the vertebrate hosts involved in viral maintenance. Most of the speculation regarding potential vertebrate reservoirs has been derived from serosurveys and laboratory animal studies that have demonstrated the presence of antibodies against these viruses (McIntosh, 1961; Paul and Singh, 1968; Bedekar and Pavri, 1969; Johnson et al., 1977; Marshall et al., 1982). Both humans and wild nonhuman primates throughout the humid forests and semiarid savannas of Africa have been found to have significant antibody levels against CHIKV. It is postulated that the appearance of CHIKV epidemics follows a 3–4 year cyclical pattern that coincides with the repopulation of susceptible, nonimmune, wild primates (Jupp and McIntosh, 1988). To date, a vertebrate reservoir or sylvan transmission cycle for CHIKV has not been identified outside Africa. Similarly, a vertebrate host for ONNV has not been determined although antibody has been detected in cattle, goats, lizards, and several species of rodents (Johnson et al., 1977).

The two major ONNV epidemics were both extensively characterized epidemiologically and revealed that the virus moved extremely rapidly through the affected areas and infected a significant proportion of the population (Lumsden, 1955; Ross, 1956; Williams et al., 1965b; Sanders et al., 1999). In 1959–1961, over 2 million people were infected during the 2-year outbreak. Similarly, during the 1996 outbreak infection rates were estimated to be as high as 68% with attack rates over 40% (Sanders et al., 1999). CHIKV outbreaks have varied dramatically in magnitude depending upon whether the outbreak was localized to urban or rural settings. However, in both settings, the attack rate has been found to be significant and comparable; morbidity rates between 30% and 70% are typical with postoutbreak seroprevalence rates commonly ranging from 50% to 80% (Halstead et al., 1969c; Padbidri and Gnaneswar, 1979; Jupp and McIntosh, 1988). The seropositivity rate has been found to be slightly lower in rural outbreaks suggesting that the vectors transmitting the virus in rural ecological settings have insufficient human contact or are less efficient at transmission to human hosts (McCrae et al., 1971). Rates of immunity are also lower in young children as is the percentage of people over 60 years of age demonstrating clinical illness during outbreaks. This indicates that the virus produces long-lived antibody that is presumably protective for life (Kokernot et al., 1965a, 1965b; Reuben, 1967). Occupational exposure differences to mosquitoes are likely the cause of any higher incidence rates that may be associated with males compared with females.

Although mortality is not a major concern of either ONNV or CHIKV infection, these viruses are still significant public health problems. The acute phase of the illness causes fever and joint pain that is so incapacitating that inactivity for several weeks is not uncommon. With the accompanying high attack rate, this can lead to crippling absenteeism rates as well as a significant drain on local health care systems. For example, in the initial ONNV outbreak, up to 10% of the local workforce was ill at any given time with 25% being incapacitated for 5 days or more (Williams et al., 1965b).

POTENTIAL BIOTHREAT AGENT

Neither CHIKV nor ONNV has been classified as potential biothreat agents as the encephalitic alphaviruses VEEV, eastern equine encephalitis virus (EEEV), and western equine encephalitis virus (WEEV) have been. However, CHIKV has been documented to cause laboratory infections that could be associated with aerosol route of transmission so the possibility of using this virus as an aerosolized biothreat agent cannot be completely excluded.

CLINICAL DISEASE

A classic triad of signs for CHIKV and ONNV infection includes fever, arthralgia, and rash that may or may not be accompanied by other indicators of disease (Deller and Russell, 1967; Tesh, 1982; McGill, 1995; Adebajo, 1996). The overall clinical illness is virtually identical between these two alphaviruses but there are occasional distinctions that may be suggestive markers for differential diagnosis in regions where the viruses overlap (Fig. 32.1 for geographic distributions).

CHIK virus infection produces an illness in humans that typically begins with a sudden onset of fever reaching as high as 104°F (Deller and Russell, 1967) that may last as long as 10 days. The fever almost always precedes the rash and joint pain and has occasionally been reported as biphasic with the recurrence noted on the 4th or 5th day of illness (Robinson, 1955; Jadhav et al., 1965). Additionally, there have been reported cases of febrile convulsions in young children (Moore et al., 1974). Fever is also documented with ONNV infection; however, it usually ranges from 99°F to 101°F and lasts for only 2–3 days. No biphasic fevers have been reported for ONNV.

The rash associated with infection is maculopapular and erythematous in character, is visible starting 2–5 days postinfection, and may last up to 10 days. The nonpruritic rash is distributed primarily on the face, limbs, and trunk of the body. There has also been infrequent documentation of hemorrhagic manifestations including hematemesis and melena due to CHIKV infection in Southeast Asia (Sarkar et al., 1965a) where hemorrhagic disease is typically attributed to infection with dengue viruses.

Perhaps the most significant manifestation of the illness associated with these viruses is the severe joint pain that occurs with virtually every clinical case (Deller and Russell, 1968; Fourie and Morrison, 1979). Even the names given to the virus indicate the degree of discomfort caused: "chikungunya" is Makonde meaning "that which bends up" and "o'nyong-nyong" is an Acholi word for a similar sentiment. The arthralgia is most commonly symmetrical and peripheral being noted in the ankles, toes, fingers, elbows, wrists, and knees. The joints exhibit extreme tenderness and swelling with the patients frequently reporting incapacitating pain that lasts for weeks or months. Paresthesia of the skin over the affected joints is common with CHIKV infection suggesting neurological involvement but no evidence of neurologic sequelae have been observed. Most infections completely resolve at that time but there have been documented cases of CHIKV-induced arthralgia persisting for several years (Brighton et al., 1983; McGill, 1995) with up to 12% of patients with CHIKV disease developing chronic joint problems (Calisher, 1999). In a rare instance, the rheumatic manifestations resulted in joint destruction before resolution after numerous years (Brighton and Simson, 1984).

Other symptoms that have been reported with CHIKV or ONNV infection include headache, retro-orbital pain, photophobia, lumbar back pain, chills, weakness, malaise, nausea, vomiting, and myalgia (Shore, 1961; Brighton, 1981; Johnson, 1988; McGill, 1995; Calisher, 1999). Some combination of these symptoms is generally reported by all patients but their presence and frequency are variable. Additionally, signs including conjunctivitis, pharyngitis, and lymphadenopathy may be observed. In particular, enlargement and tenderness of the cervical lymph node is frequently quite severe and a distinctive characteristic of ONNV infection. Other nodes including the postoccipital, axillary, and inguinal may also be enlarged.

Occasionally, unusual clinical presentation due to CHIKV infection has been observed. For example, one study revealed evidence of myocarditis after acute febrile illness and serologic evidence of CHIKV infection (Obeyesekere and Hermon, 1972, 1973). Presentation included sinus tachycardia, ventricular ectopics, cardiomegaly, and abnormal electrocardiograms. Signs of congestive heart failure were documented several months after initial illness. Cases of mortality, neurological disease, and intrauterine transmission were also reported in the 2006 Reunion outbreak (Schuffenecker et al., 2006; Powers and Logue, 2007). While clinical cases of this nature are not typical, they serve as a reminder that the viruses may have diverse and perhaps evolving patterns of virulence.

Laboratory parameters for CHIKV or ONNV infections can be quite variable and frequently do not aid in diagnosis. Results include a normal hematocrit (varying from 40% to 50%), thrombocytopenia or alternatively, no obvious platelet deficiencies, normal urinanalysis, and low to normal WBC counts of 4000–7000 although leukocytosis has been documented (Shore, 1961; Jadhav et al., 1965; Deller and Russell, 1968; Tesh, 1982). The erythrocyte sedimentation rate may be slightly elevated which is suggestive of rheumatoid arthritis and may confound diagnosis of isolated CHIKV infections. Overall, standard laboratory testing is of little value and diagnosis is dependent upon isolation of virus or confirmation of virus-specific antibodies using acute and convalescent sera. Diagnosis is facilitated during large epidemics while small rural outbreaks or individual cases are often never diagnosed. Unfortunately, the symptoms of both CHIKV and ONNV infection are quite similar to those of many other pathogens affecting people in the endemic areas. One particular difficulty in identifying the widespread nature of CHIKV infection is its overlapping distribution with dengue viruses. Because the clinical symptoms of CHIKV infection often mimic those of dengue fever and because CHIKV circulates in regions where dengue virus is endemic, it has been postulated that many cases of dengue virus infection are misdiagnosed and that the incidence of CHIKV infection is much higher than previously reported (Myers and Carey, 1967; Carey, 1971).

TREATMENT

Because both CHIKV and ONNV induce nonfatal, self-limiting diseases, treatment is entirely supportive for the symptoms. With the primary signs including fever and joint pain with swelling, analgesics, antipyretics, and anti-inflammatory agents are the most appropriate treatment. Commonly, these agents would include aspirin but due to the reports of possible hemorrhagic manifestations, aspirin should be avoided in favor of other options (Tesh, 1982) including acetaminophen, ibuprofen, steroid therapy, and nonsteroidal agents such as indomethacin. However, as some of these therapies can have serious side effects, the need for prolonged use should be a consideration in treatment regimens. For chronic cases of arthritis due to CHIKV, chloroquine phosphate was shown in one study to provide relief to patients that had limited response to nonsteroidal anti-inflammatory drugs (Brighton, 1984). Additional actions including bed rest and fluids are recommended.

The lack of specific treatment for CHIKV and ONNV infection has prompted laboratory studies to identify antiviral agents effective against these viruses. Compounds including ribavirin, sulfated polysaccharides (iota carrageenan, fucoidan, and dextran sulfate), 6-azauridine, glycyrrhizin, and interferon-α have been evaluated for their ability to inhibit replication of CHIKV in cell culture. With the exception of the polysaccharides, all were found to have both potent and selective antiviral activity (Briolant et al., 2004). Additionally, these four compounds have demonstrated efficacy in the treatment of other diseases (Deneau and Farber, 1975; Andrei and De Clercq, 1993, Grieder and Vogel, 1999; van Rossum et al., 1999; Ryman et al., 2000) warranting additional testing for their use against these arthralgic alphaviruses. In particular, multidrug therapies may prove useful as each treatment has distinct pharmacological actions that may produce synergistic effects as well as reduce the likelihood of resistance.

PATHOGENESIS

While numerous field and laboratory studies have investigated serological outcomes of CHIKV or ONNV infection, virtually no effort has been placed on direct examination of the mechanism(s) of pathogenesis due to infection with these viruses. However, several recent papers elucidating the molecular and cellular aspects of RRV disease development may also be applicable to CHIKV and ONNV infections. RRV is another alphavirus in the Semliki Forest antigenic complex and is closely related to both CHIKV and ONNV genetically. Additionally, RRV causes epidemic polyarthritis generating a pattern of illness including rheumatic symptoms, rash, and fever (Harley et al., 2001) similar to that seen with CHIKV and ONNV. Also, like CHIKV disease, the symptoms may persist for several months with prolonged painful arthralgia suggesting the development of disease may result from similar pathways.

RRV patients with chronic joint pain were found to have viral antigen in the synovial fluid and this virus attracted inflammatory infiltrates to the affected joints (Rulli et al., 2005). CHIKV patients with prolonged joint pain demonstrated high titers of specific antibodies suggesting persistent antigen presence, but no virus could be cultured from synovial fluid in a small cohort of subjects (Brighton et al., 1983). In addition, RRV has been found to persistently infect macrophages and evidence suggests that the ability of the virus to be maintained in these cells for extensive periods contributes to their pathogenesis (Way et al., 2002). Macrophages were found to be directly responsible for muscle tissue destruction as treatment of animals with macrophage toxic compounds prior to RRV infection completely prevented clinical diseases and clearance of macrophages from infected tissue was associated with a decrease in symptoms (Lidbury et al., 2000). Because macrophages were found to be infected by and responsible for RRV disease, it was postulated that chemokines may also be involved in pathogenesis (Lidbury et al., 2000; Mahalingam et al., 2003) and some evidence has been accumulated supporting this idea (Mateo et al., 2000). The proposed model is that RRV-infected macrophages secrete chemokines that are attractive for additional macrophage. The newly immigrated macrophages are then susceptible to further infection thus propagating the cycle. Additionally, RRV can bind specific antibodies resulting in complexes that can bind and infect monocytes and macrophage via their cellular Fc receptors (Rulli et al., 2005). This antibody-dependent enhancement (ADE) pathway could have implications for vaccination potential including suppression of the host cell antiviral responses (Lidbury and Mahalingam, 2000; Suhrbier and La Linn, 2003). Due to the similarities in clinical symptomology between RRV and CHIKV, these mechanisms need to be closely examined in the case of CHIKV infections.

The T-cell response may also be a factor in CHIKV/ONNV pathogenesis as it is with RRV. CD8+ T cells are the major immunological cell type associated with epidemic polyarthritis patients that quickly recover from the febrile rash in contrast to the CD4+ cells that predominate in the synovial fluid of patients with chronic disease (Fraser and Becker, 1984). When productively and persistently infected *in vitro* macrophages are treated with CD8+ T cells generated by vaccination of mice with RRV capsid protein, the infection is completely cleared (Linn et al., 1998). This suggests that a defective cell-mediated immune response may also play a role in chronic disease and viral persistence (Fraser, 1986). Again, these studies need to be extended to the other alphavirus arthritides to determine if this is a broad-spectrum occurrence.

VACCINES: *HISTORY*

Due to the significant infection and attack rates during outbreaks, the extensive geographic distribution of these viruses, and the severe morbidity associated with clinical disease due to CHIKV or ONNV infection, a vaccine is highly desirable for control of disease caused by these agents. In addition, there have been numerous documented cases of laboratory acquired infections of CHIKV (U.S. Department of Health and Human Services: Biosafety in Microbiological and Biomedical Laboratories) indicating that a vaccine would also be beneficial to personnel with occupational risk. However, while there has been extensive work in vaccinology for several other alphaviruses (Rayner et al., 2002; Nalca et al., 2003; Johnston and Davis, 2004), the history of vaccine development for ONNV and CHIKV is short. This is particularly true for ONNV where no vaccine development has ever been documented. For CHIKV, scientists at the Walter Reed Army Institute of Research, The Salk Institute, and the U.S. Army Medical Research Institute of Infectious Diseases have been the primary investigators to undertake vaccine development and characterization. Some additional work focusing on potential CHIKV vaccines has been performed at the Japanese National Institute of Health (Kitaoka, 1967) and the Kobe University School of Medicine (Nakao and Hotta, 1973). However, none of these efforts have yet resulted in a licensed vaccine. Some human clinical trials have been performed with promising results but, to date, all CHIKV vaccine candidates must be considered "vaccines in development."

One of the first published reports of a possible vaccine against CHIKV was the description of a formalin-killed, CHIKV-infected, suckling mouse brain (SMB) preparation. This project began as an attempt by researchers in Japan to protect children against Thai hemorrhagic fever that was found to be caused by dengue and chikungunya viruses (Halstead and Buescher, 1961; Halstead et al., 1963). The SMB preparation of CHIKV vaccine was derived from 2- to 5-day-old mice that had been infected with the prototype strain of virus. The material was formalin inactivated and assessed for the ability to induce N and HI antibodies in immunized mice. At 15 days postinoculation, mice were found to have both HI and protective N antibodies suggesting that a CHIKV vaccine could be protective (Kitaoka, 1967).

A second early examination of a potential CHIKV vaccine was the use of an African strain of virus, one that had previously been through 177 SMB passages, and harvested from baby hamster kidney (BHK) cell culture. The recovered virus was either ultraviolet (UV) or formalin inactivated to determine if inactivation method would impact protective efficacy. The purified preparations were injected subcutaneously (s.c.) into monkeys using a 3-dose schedule at days 0, 14, and 63. Both induced N antibodies with the UV-inactivated material proving to be the better approach (Nakao and Hotta, 1973). For both this study and the SMB vaccine screened above, no additional information was published as follow up to this initial developmental work.

VACCINES IN DEVELOPMENT

The most extensive work performed in the development of a human CHIKV vaccine was initiated by investigators at Walter Reed Army Institute of Research.

Their first efforts involved comparative testing of vaccines prepared in both avian and mammalian cells. This early work utilized CHIKV strain 168 that was obtained from the 1952–1953 outbreak in Tanganyika (Mason and Haddow, 1957). The virus was used to infect chick embryo (CE), SMB, and African green-monkey kidney cell (GMKC) cultures to ascertain the most appropriate production system (Harrison et al., 1967). Virus harvested from each of these preparations was formalin inactivated using the standard techniques that had been developed for a Rift Valley fever vaccine (Randall et al., 1962). Potency testing for the products consisted of two intraperitoneal (i.p.) inoculations of the vaccine into adult mice at 0 and 7 days. Challenge of the vaccinated animals occurred 14 days later by the intracerebral (i.c.) route. This first round of potency tests suggested that the CE preparation would be unsuitable for further development, as serological responses and protective capacity in the mice were poor. A separate study comparing CE and GMKC products indicated that CE cultures were quite efficacious generating 4–5 times the level of antibodies developed in response to the GMKC vaccine and providing more protection against homologous challenge virus (White et al., 1972). The distinction between these two disparate results was the infectivity titer of the virus prior to formalin inactivation; to generate suitable CE vaccine, highly concentrated CE suspension cultures as described for EEE vaccine production were required (White et al., 1971). While the SMB product did induce antibodies and was found to be protective, the possibility of allergic reactions in humans limited enthusiasm for further development of this material. The GMKC vaccine was also found to induce high levels of antibodies and was protective against i.c. challenge with homologous virus; therefore, these cells were used for further characterization of their vaccine potential. Further tests included cross-strain challenge in both mice and monkeys, viremia profile determinations, and antibody development assessment. The immunogenic potential of the GMKC vaccine was demonstrated in all these tests with no detectable viremia and good protection in monkeys after challenge with four strains of CHIKV (Harrison et al., 1967) and indicated that further development of this vaccine was warranted.

The next step in development of the GMKC CHIKV vaccine examined alternate methods of inactivation of the virus. The first products were prepared by the long used approach of formalin inactivation. Because CHIKV virions are lipid-containing particles, the methodology being tested for experimental influenza and measles vaccines (Davenport et al., 1964; Norrby et al., 1966) of extracting the virus with surfactant (Tween 80) and ether was investigated (Eckels et al., 1970). The approach had been shown previously to produce an immunogenic vaccine that lacked any infectivity and that same phenomenon was observed with GMKCs infected with CHIKV. Additionally, this vaccine stimulated the development of CF, HI, and N antibodies as well as protecting mice against challenge with homologous virus. In fact, the hemagglutination (HA) activity of the tween-ether vaccine was maintained or even slightly enhanced while the formalin treatment destroyed the HA activity. A lack of knowledge as to the character of the immunizing antigen of the CHIKV particles precluded an explanation for this phenomenon. No further development of CHIKV vaccine using this method of preparation has been published.

The continuation of work by these investigators focused on safety and immunogenicity by looking for immune response, adverse events, and possible side effects of the formalin-inactivated product from GMKC culture. In this phase, the strain of virus used was changed from the highly passaged strain originating from the first outbreak in Africa to an isolate obtained directly from the serum of an infected patient in Thailand. The strain was designated 15561 and was used for all further development by these investigators. Extensive actions to provide purity assurances were undertaken to confirm the elimination of hepatitis agents and simian and murine adventitious agents from the starting material and all culture materials used. A vaccine lot was produced, formalin-inactivated, lyophilized, and used in safety and sterility tests. Potency tests were conducted in mice and were followed up with a human volunteer study consisting of 16 subjects. Subjects were given two doses of vaccine s.c. 28 days apart and examined for local and systemic reactions during the course of the study. No adverse events were noted in any of the volunteers and most developed N antibody by day 14 with all subjects demonstrating significant levels by day 42. CF and HI antibody levels remained low throughout the study in all subjects (Harrison et al., 1971). However, the initial results demonstrated both safety and immunogenicity and were promising enough to continue with further serologic studies examining long-term persistence of antibody. This antibody persistence study was undertaken using 12 laboratory or office workers (all employed in the building housing the research facilities) who were immunized with 4 distinct alphavirus vaccines (DeMeio et al., 1979). Subjects received the CHIKV vaccine after all other alphavirus vaccines; this occurred at 7 years 9 months after immunization with the VEE live-attenuated vaccine, 3 years and 1 month post-EEE vaccination, and 11 months after the WEE vaccine. Unfortunately, the CHIKV vaccine failed to elicit any significant N antibody levels in any of the participants. The lack of efficiency was attributed to the low potency of the CHIKV vaccine lot used as the ED_{50} value of 0.33 found in mice may not have been expected to generate significant levels of antibodies. Curiously, this vaccine did induce a significant rise in titer against EEEV in one subject and two other subjects developed CHIKV N antibody in response to immunization with the VEEV vaccine.

The disappointing immunogenicity demonstrated by the formalin-inactivated CHIKV vaccine in the long-term antibody persistence study led to the development of a second generation CHIKV vaccine. The 15561 strain from human serum was used as seed material to generate vaccine CHIK 181/clone 25 resulting from a series of 18 plaque to plaque passages in MRC-5 cells of the starting virus which had been passaged 11 times in GMKC culture (Levitt et al., 1986). The CHIK 181/clone 25 strain exhibited small plaques, was temperature sensitive (ts), had decreased virulence for suckling mice, and showed a reduced level of viremia in monkeys. All of these characteristics have been associated with attenuation in other host systems (Eckels et al., 1980; Halstead et al., 1984) suggesting that 181/25 would have reduced virulence as well. The protective efficacy of this vaccine was evaluated in both mouse and monkey model systems. Mice were inoculated with a single 0.5 mL dose, i.p. and challenged i.c. with a virulent, heterologous strain of virus (168) 2 weeks following immunization. Monkeys received an intramuscular (i.m.) dose of pilot lot of vaccine 37 days prior to i.m. challenge with 5.0 $\log_{10}$ plaque-forming units (pfu) of nonattenuated parental virus (15561). The CD-1 mice were protected completely when immunized with 4.5–6.5 $\log_{10}$pfu of vaccine and 50% of the mice survived when vaccinated with only 2.5 $\log_{10}$pfu. In the monkey study, all the animals developed neutralizing antibody and were completely aviremic upon challenge indicting strong protective immunity. Additionally, the 181/clone 25 seed generated either no viremia or significantly reduced levels with a marked delay in onset compared with the parental virus in monkeys. Presumably, this reduced virulence in monkeys would correlate with virulence patterns in humans.

Because this was a live virus, additional safety testing included examining the potential of the vaccine to be transmitted by mosquitoes in the case a viremic vaccinee was bit by a mosquito (Turell and Malinoski, 1992). This phenomenon has been reported for the live attenuated VEEV vaccine (Pedersen et al., 1972). Of concern was the possibility that the vaccine characteristics could be ablated during replication of the virus in the mosquito and that the virus would revert to virulence before being transmitted to another vertebrate host. The vaccine was examined in both *Ae. aegypti* and *Ae. albopictus*, the two mosquitoes known to be competent urban vectors of the virus (Tesh et al., 1976; Banerjee et al., 1988; Turell et al., 1992). While the vaccine strain was able to replicate in both species and even be transmitted by the mosquitoes that received the virus via intrathoracic inoculation, there was no evidence of reversion to virulence. Based upon the low viremias the vaccine strain generated in test subjects, it was postulated that it would be unlikely for either mosquito vector to become infected from feeding upon a vaccinated human (Turell and Malinoski, 1992). The promising results generated by these studies led to the submission of an investigational new drug (IND) application for the CHIK 181/clone 25 vaccine strain to provide a route for the initiation of human clinical trials.

CLINICAL TRIALS

Phase 1 clinical trials consisted of both alphavirus naïve individuals (38) as well as 36 additional subjects that had previously been immunized with a live attenuated VEEV vaccine (McClain et al., 1998). Vaccines were delivered s.c. at a dose of 4.4 $\log_{10}$pfu/0.5 mL. Mild symptoms including headache, fever, myalgia, and injection site tenderness were reported with equal frequency for both the placebo and CHIKV vaccine cohorts in the naïve subject study. Approximately one-third of these subjects had transient but detectable levels of viremia. However, the levels were too low to be detected directly from the serum but were only documented after cell culture amplification. All CHIKV vaccine subjects seroconverted as noted by the appearance of IgM and IgG antibody detected in ELISA as well as N antibody with a peak titer occurring at 28 days. Of those with preexisting alphavirus vaccine history, no local or systemic symptoms were reported after immunization. Antibody responses were poor in this group with only 36% of the subjects generating detectable levels of CHIKV-specific antibodies. However, there was some cross-reactivity as subjects vaccinated against CHIKV generated antibodies that were detected in a VEEV IgG ELISA with an anamnestic response upon subsequent immunization with VEEV vaccine. This information suggested that the live attenuated CHIKV vaccine was both safe and efficacious but that sequential immunization with live-attenuated alphavirus vaccines could hinder the development of protective immune responses to subsequent alphavirus vaccines or infections.

The positive results derived from the Phase 1 clinical trial were sufficient to progress to a Phase 2 randomized, double-blind, placebo-controlled CHIKV vaccine trial to further assess safety and immunogenicity. A total of 73 subjects was enrolled in this study with

59 receiving a single s.c. inoculation of ~10^5 pfu of TSI-GSD-218 vaccine (CHIK 181/clone 25 strain) in 0.5mL (Edelman et al., 2000). After clinical and serological evaluation of both vaccine and placebo groups, the incidence of local or systemic reactions was found to be comparable in both groups; this indicated safety of the vaccine. The only differences noted were the descriptions of minor joint pain by five subjects in the vaccine group with all symptoms lasting less than 24h. Serologically 98% of those receiving the CHIKV vaccine developed N antibodies that peaked in titer by day 28 postimmunization and over 85% were still seropositive at 1 year. This vaccine was found to be highly immunogenic, well-tolerated, and safe. It appears to be a very promising alphavirus vaccine for those at risk and additional long-term studies of duration of immunity are certainly warranted.

POSTEXPOSURE IMMUNOPROPHYLAXIS

No studies have examined any possibilities for postexposure immunoprophylaxis to either CHIKV or ONNV. Potential approaches may include passive transfer of antibodies from sera of previously infected patients or produced from specific donors, monoclonal antibody therapy, or immediate postexposure vaccination. Because these viruses have a short incubation period and can produce illness in just a few days, some of these options are not particularly attractive. Early identification of outbreaks would be an essential component of disease control to insure that people in affected areas but not yet exposed or infected could receive any vaccine that should become available.

PROSPECTS FOR THE FUTURE

While the preliminary basic science studies and clinical human trials provide promise for a CHIKV vaccine that would be extremely beneficial from a public health perspective, the lack of mortality and marketability are certainly factors that may limit rapid development of any licensed vaccine. Additionally, continued advancement in the understanding of the basic biology of the viruses themselves and their ability to induce immunity in the vertebrate hosts is necessary to aid in the design of the most appropriate vaccine strategy. Ultimately, given the areas where the viruses are found, any vaccine distribution must be used in concert with a comprehensive mosquito control program that would employ a diverse group of measures to be implemented for both short-term (during outbreak situations) and long-term (sustained) control of vector mosquitoes. A successful program will be a partnership among the local population, government and health ministry officials, the scientific community, and the members of the communication network providing accurate and timely information. Examples of short-term vector control measures include covering of open water breeding sites or applying larvicidal products if coverage is not feasible, elimination of artificial breeding sites such as discarded waste or used tires, and the use of adulticide products for immediate control of adult mosquito populations early in an outbreak. However, these temporary and short-term vector control measures must be used in conjunction with long-term solutions including monitoring of levels of vector mosquitoes, developing a system of communication between disease surveillance activities and vector control activities to insure that control measures are implemented rapidly when clusters of vector-borne disease are identified, and education of the public on measures to maintain a mosquito-free environment at their homes and places of work. Hopefully, these immediately available options for mosquito control can be implemented while the scientific, political, and social barriers to CHIKV and ONNV vaccine development and distribution can be resolved. Given the intraglobal atmosphere currently present and the potential for any disease to suddenly appear in unexpected areas such as West Nile virus has demonstrated, vaccination strategies for CHIKV and ONNV should be seriously considered.

KEY ISSUES

- CHIKV and ONNV cause significant morbidity, are widely distributed, cause explosive outbreaks, are responsible for numerous laboratory acquired infections, and can easily be imported into naïve regions due to travel from endemic areas. These traits clearly demonstrate the need for a highly immunogenic vaccine.
- Distinctly different ecological and epidemiological patterns exist for these closely related viruses potentially impacting the utility of a single vaccination strategy.
- No licensed vaccines are currently available. Various formulations of potential CHIKV vaccines have been examined and subjected to human clinical trials. No ONNV vaccine development has ever been published.
- There is significant cross-reactivity among alphaviruses and the antibodies developed against

them. This has led to concerns including the possibility of ADE of infection as well as impaired development of neutralizing serologic responses and protective immunity. Vaccine strategies such as the use of live, attenuated preparations must consider these possibilities.

References

Aaskov, J.G., Dalglish, D.A., Davies, C.E., Tucker, M. and Donaldson, M.D. Cell-mediated immune response to Ross River virus in mice: evidence for a defective effector cell response. Aust. J. Exp. Biol. Med. Sci. 1983; 61(Pt 5):529–540.

Adebajo, A.O. Rheumatic manifestations of tropical diseases. [Review] [30 refs]. Curr. Opin. Rheumatol. 1996; 8(1):85–89.

Adesina, O.A. and Odelola, H.A. Ecological distribution of Chikungunya haemagglutination inhibition antibodies in human and domestic animals in Nigeria. Trop. Geogr. Med. 1991; 43(3):271–275.

Andrei, G. and De Clercq, E. Molecular approaches for the treatment of hemorrhagic fever virus infections. Antiviral Res. 1993; 22(1):45–75.

Anthony, R.P. and Brown, D.T. Protein–protein interactions in an alphavirus membrane. J. Virol. 1991; 65(3):1187–1194.

Arankalle, V.A., Shrivastava, S., Cherian, S., Gunjikar, R.S., Walimbe, A.M., Jadhav, S.M., Sudeep, A.B. and Mishra, A.C. Genetic divergence of Chikungunya viruses in India (1963–2006) with special reference to the 2005–2006 explosive epidemic. J. Gen. Virol. 2007; 88(7):1967–1976.

Banerjee, K., Mourja, D.T. and Malunjkar, A.S. Susceptibility & transmissibility of different geographical strains of *Aedes aegypti* mosquitoes to Chikungunya virus. Indian J. Med. Res. 1988; 87:134–138.

Bedekar, S.D. and Pavri, K.M. Studies with Chikungunya virus. II. Serological survey of humans and animals in India. Indian J. Med. Res. 1969; 57(7):1193–1197.

Blackburn, N.K., Besselaar, T.G. and Gibson, G. Antigenic relationship between chikungunya virus strains and o'nyong nyong virus using monoclonal antibodies. Res. Virol. 1995; 146(1):69–73.

Blackman, M.J. and Morris, A.G. Gamma interferon production and cytotoxicity of spleen cells from mice infected with Semliki Forest virus. J. Gen. Virol. 1984; 65(Pt 5):955–961.

Brault, A.C., Powers, A.M., Ortiz, D., Estrada-Franco, J.G., Navarro-Lopez, R. and Weaver, S.C. Venezuelan equine encephalitis emergence: enhanced vector infection from a single amino acid substitution in the envelope glycoprotein. Proc. Natl. Acad. Sci. U.S.A. 2004; 101(31):11344–11349.

Brault, A.C., Powers, A.M. and Weaver, S.C. Vector infection determinants of Venezuelan equine encephalitis virus reside within the E2 envelope glycoprotein. J. Virol. 2002; 76(12):6387–6392.

Brighton, S.W. Chikungunya virus infections. S. Afr. Med. J. 1981; 59(16):552.

Brighton, S.W. Chloroquine phosphate treatment of chronic Chikungunya arthritis. An open pilot study. S. Afr. Med. J. 1984; 66(6):217–218.

Brighton, S.W., Prozesky, O.W. and de la Harpe, A.L. Chikungunya virus infection. A retrospective study of 107 cases. S. Afr. Med. J. 1983; 63(9):313–315.

Brighton, S.W. and Simson, I.W. A destructive arthropathy following Chikungunya virus arthritis: a possible association. Clin. Rheumatol. 1984; 3(2):253–258.

Briolant, S., Garin, D., Scaramozzino, N., Jouan, A. and Crance, J.M. In vitro inhibition of Chikungunya and Semliki Forest viruses replication by antiviral compounds: synergistic effect of interferon-alpha and ribavirin combination. Antiviral Res. 2004; 61(2):111–117.

Burke, D.S., Nisalak, A. and Nimmannitya, S. Disappearance of Chikungunya virus from Bangkok. Trans. R. Soc. Trop. Med. Hyg. 1985; 79(3):419–420.

Calisher, C.H. Chikungunya, O'nyong nyong and Mayaro viruses (Togaviridae). In: Encyclopedia of Virology (A. Granoff and R.G. Webster, Eds.), Vol. 3, 2nd ed. pp. 236–241. London: Elsevier Science & Technology Books, 1999.

Carey, D.E. Chikungunya and dengue: a case of mistaken identity? J. Hist. Med. Allied Sci. 1971; 26(3):243–262.

Carey, D.E., Myers, R.M., DeRanitz, C.M., Jadhav, M. and Reuben, R. The 1964 chikungunya epidemic at Vellore, South India, including observations on concurrent dengue. Trans. R. Soc. Trop. Med. Hyg. 1969; 63(4):434–445.

Casals, J. and Whitman, L. Mayaro virus: a new human disease agent. I. Relationship to other arbor viruses. Am. J. Trop. Med. Hyg. 1957; 6(6):1004–1011.

Chanas, A.C., Hubalek, Z., Johnson, B.K. and Simpson, D.I. A comparative study of o'nyong nyong virus with chikungunya virus and plaque variants. Arch. Virol. 1979; 59(3):231–238.

Chanas, A.C., Johnson, B.K. and Simpson, D.I. Antigenic relationships of alphaviruses by a simple micro-culture cross-neutralization method. J. Gen. Virol. 1976; 32(2):295–300.

Chastel, C. Human infections in Cambodia by the Chikungunya virus or an apparently closely related agent. I. Clinical aspects. Isolations and identification of the viruses. Serology. Bull. Soc. Pathol. Exot. Filiales 1963; 56:892–915.

Corbet, P.S., Williams, M.C. and Gillett, J.D. O'nyong-nyong fever: an epidemic virus disease in East Africa. IV. Vector studies at epidemic sites. Trans. R. Soc. Trop. Med. Hyg. 1961; 55:463–480.

Davenport, F.M., Hennessy, A.V., Brandon, F.M., Webster, R.G., Barrett, C.D., Jr. and Lease, G.O. Comparisons of serologic and febrile responses in humans to vaccination with influenza A viruses or their hemagglutinins. J. Lab. Clin. Med. 1964; 63:5–13.

Davis, N.L., Grieder, F.B., Smith, J.F., Greenwald, G.F., Valenski, M.L., Sellon, D.C., Charles, P.C. and Johnston, R.E. A molecular genetic approach to the study of Venezuelan equine encephalitis virus pathogenesis. Arch. Virol. 1994; 9:99–109.

Davis, N.L., Willis, L.V., Smith, J.F. and Johnston, R.E. In vitro synthesis of infectious Venezuelan equine encephalitis virus RNA from a cDNA clone: analysis of a viable deletion mutant. Virology 1989; 171(1):189–204.

de Groot, R.J., Hardy, W.R., Shirako, Y. and Strauss, J.H. Cleavage-site preferences of Sindbis virus polyproteins containing the non-structural proteinase. Evidence for temporal regulation of polyprotein processing in vivo. EMBO J. 1990; 9(8):2631–2638.

Deller, J.J., Jr. and Russell, P.K. An analysis of fevers of unknown origin in American soldiers in Vietnam. Ann. Intern. Med. 1967; 66(6):1129–1143.

Deller, J.J., Jr. and Russell, P.K. Chikungunya disease. Am. J. Trop. Med. Hyg. 1968; 17(1):107–111.

DeMeio, J.L., DeSanctis, A.N. and Thomas, W.J. Persistence in humans of antibody after immunization with four alphavirus vaccines. Asian J. Infect. Dis. 1979; 3(3):119–124.

Deneau, D.G. and Farber, E.M. The treatment of psoriasis with azaribine. Dermatologica 1975; 151(3):158–163.

Diallo, M., Thonnon, J., Traore-Lamizana, M. and Fontenille, D. Vectors of chikungunya virus in Senegal: current data and transmission cycles. Am. J. Trop. Med. Hyg. 1999; 60(2):281–286.

Ding, M.X. and Schlesinger, M.J. Evidence that Sindbis virus NSP2 is an autoprotease which processes the virus nonstructural polyprotein. Virology 1989; 171(1):280–284.

Durbin, R.K. and Stollar, V. Sindbis virus mutants able to replicate in methionine-deprived *Aedes albopictus* cells. Virology 1985; 144(2):529–533.

Eckels, K.H., Harrison, V.R. and Hetrick, F.M. Chikungunya virus vaccine prepared by Tween-ether extraction. Appl. Microbiol. 1970; 19(2):321–325.

Eckels, K.H., Harrison, V.R., Summers, P.L. and Russell, P.K. Dengue-2 vaccine: preparation from a small-plaque virus clone. Infect. Immun. 1980; 27(1):175–180.

Edelman, R., Tacket, C.O., Wasserman, S.S., Bodison, S.A., Perry, J.G. and Mangiafico, J.A. Phase II safety and immunogenicity study of live chikungunya virus vaccine TSI-GSD-218. Am. J. Trop. Med. Hyg. 2000; 62(6):681–685.

Ekstrom, M., Liljestrom, P. and Garoff, H. Membrane protein lateral interactions control Semliki Forest virus budding. EMBO J. 1994; 13(5):1058–1064.

Fourie, E.D. and Morrison, J.G. Rheumatoid arthritic syndrome after chikungunya fever. S. Afr. Med. J. 1979; 56(4):130–132.

Fraser, J.R. Epidemic polyarthritis and Ross River virus disease. Clin. Rheum. Dis. 1986; 12(2):369–388.

Fraser, J.R. and Becker, G.J. Mononuclear cell types in chronic synovial effusions of Ross River virus disease. Aust. N.Z. J. Med. 1984; 14(4):505–506.

Garoff, H., Hewson, R. and Opstelten, D.J.E. Virus maturation by budding. Microbiol. Mol. Biol. Rev. (Washington, DC) 1998; 62(4):1171–1190.

Garoff, H. and Simons, K. Location of the spike glycoproteins in the Semliki Forest virus membrane. Proc. Natl. Acad. Sci. U.S.A. 1974; 71(10):3988–3992.

Gorbalenya, A.E., Koonin, E.V. and Wolf, Y.I. A new superfamily of putative NTP-binding domains encoded by genomes of small DNA and RNA viruses. FEBS Lett. 1990; 262(1):145–148.

Grieder, F.B. and Vogel, S.N. Role of interferon and interferon regulatory factors in early protection against Venezuelan equine encephalitis virus infection. Virology 1999; 257(1):106–118.

Gubler, D.J. Human arbovirus infections worldwide. Ann. N.Y. Acad. Sci. 2001; 951:13–24.

Haddow, A.J., Davies, C.W. and Walker, A.J. O'nyong-nyong fever: an epidemic virus disease in east Africa I. Introduction. Trans. R. Soc. Trop. Med. Hyg. 1960; 54:517.

Hahn, Y.S., Strauss, E.G. and Strauss, J.H. Mapping of RNA-temperature-sensitive mutants of Sindbis virus: assignment of complementation groups A, B, and G to nonstructural proteins. J. Virol. 1989; 63(7):3142–3150.

Halstead, S.B. and Buescher, E.L. Hemorrhagic disease in rodents infected with virus associated with Thai hemorrhagic fever. Science 1961; 134:475–476.

Halstead, S.B., Eckels, K.H., Putvatana, R., Larsen, L.K. and Marchette, N.J. Selection of attenuated dengue 4 viruses by serial passage in primary kidney cells. IV. Characterization of a vaccine candidate in fetal rhesus lung cells. Am. J. Trop. Med. Hyg. 1984; 33(4):679–683.

Halstead, S.B., Nimmannitya, S. and Margiotta, M.R. Dengue and chikungunya virus infection in man in Thailand, 1962–1964. II. Observations on disease in outpatients. Am. J. Trop. Med. Hyg. 1969a; 18(6):972–983.

Halstead, S.B., Scanlon, J.E., Umpaivit, P. and Udomsakdi, S. Dengue and chikungunya virus infection in man in Thailand, 1962–1964. IV. Epidemiologic studies in the Bangkok metropolitan area. Am. J. Trop. Med. Hyg. 1969b; 18(6):997–1021.

Halstead, S.B., Udomsakdi, S., Scanlon, J.E. and Rohitayodhin, S. Dengue and chikungunya virus infection in man in Thailand, 1962–1964. V. Epidemiologic observations outside Bangkok. Am. J. Trop. Med. Hyg. 1969c; 18(6):1022–1033.

Halstead, S.B., Udomsakdi, S., Singharaj, P. and Nisalak, A. Dengue chikungunya virus infection in man in Thailand, 1962–1964. 3. Clinical, epidemiologic, and virologic observations on disease in non-indigenous white persons. Am. J. Trop. Med. Hyg. 1969d; 18(6):984–996.

Halstead, S.B., Yammerat, C. and Scanlon, J.E. The Thai hemorrhagic fever epidemic of 1962, a preliminary report. J. Med. Assoc. Thailand 1963; 46:449–462.

Hardy, W.R. and Strauss, J.H. Processing the nonstructural proteins of Sindbis virus: Nonstructural proteinase is in the C-terminal half of nsP2 and functions both in cis and in trans. J. Virol. 1989; 63:4653–4664.

Harley, D., Sleigh, A. and Ritchie, S. Ross River virus transmission, infection, and disease: a cross-disciplinary review. Clin. Microbiol. Rev. 2001; 14(4):909–932.

Harrison, V.R., Binn, L.N. and Randall, R. Comparative immunogenicities of chikungunya vaccines prepared in avian and mammalian tissues. Am. J. Trop. Med. Hyg. 1967; 16(6):786–791.

Harrison, V.R., Eckels, K.H., Bartelloni, P.J. and Hampton, C. Production and evaluation of a formalin-killed Chikungunya vaccine. J. Immunol. 1971; 107(3):643–647.

Hodgman, T.C. A new superfamily of replicative proteins [letter] [published erratum appears in Nature 1988;333(6173):578]. Nature 1988; 333(6168):22–23.

Jadhav, M., Namboodripad, M., Carman, R.H., Carey, D.E. and Myers, R.M. Chikungunya disease in infants and children in Vellore: a report of clinical and haematological features of virologically proved cases. Indian J. Med. Res. 1965; 53(8):764–776.

Johnson, B.K. O'nyong-nyong virus disease. In: The Arboviruses: Epidemiology and Ecology (T.P. Monath, Ed.), Vol. III, pp. 217–223. Boca Raton, FL: CRC Press, 1988.

Johnson, B.K., Chanas, A.C., Shockley, P., Squires, E.J., Gardner, P., Wallace, C., Simpson, D.I., Bowen, E.T., Platt, G.S., Way, H., Parsons, J. and Grainger, W.E. Arbovirus isolations from, and serological studies on, wild and domestic vertebrates from Kano Plain, Kenya. Trans. R. Soc. Trop. Med. Hyg. 1977; 71(6):512–517.

Johnson, B.K., Gichogo, A., Gitau, G., Patel, N., Ademba, G., Kirui, R., Highton, R.B. and Smith, D.H. Recovery of o'nyong-nyong virus from *Anopheles funestus* in Western Kenya. Trans. R. Soc. Trop. Med. Hyg. 1981; 75(2):239–241.

Johnston, R.E. and Davis, N.L. Future vaccines against emerging encephalitides. Arch. Virol. Suppl. 2004; (18):207–220.

Jupp, P.G. and Kemp, A. What is the potential for future outbreaks of chikungunya, dengue and yellow fever in southern Africa? [editorial]. S. Afr. Med. J. 1996; 86(1):35–37.

Jupp, P.G. and McIntosh, B.M. Chikungunya virus disease. In: The Arbovirus: Epidemiology and Ecology (T.P. Monath, Ed.), Vol. II, pp. 137–157. Boca Raton, FL: CRC Press, 1988.

Jupp, P.G. and McIntosh, B.M. *Aedes furcifer* and other mosquitoes as vectors of chikungunya virus at Mica, northeastern Transvaal, South Africa. J. Am. Mosquito Control Assoc. 1990; 6(3): 415–420.

Kagi, D. and Hengartner, H. Different roles for cytotoxic T cells in the control of infections with cytopathic versus noncytopathic viruses. Curr. Opin. Immunol. 1996; 8(4):472–477.

Kamer, G. and Argos, P. Primary structural comparison of RNA-dependent polymerases from plant, animal and bacterial viruses. Nucleic Acids Res. 1984; 12(18):7269–7282.

Karabatsos, N. Antigenic relationships of group A arboviruses by plaque reduction neutralization testing. Am. J. Trop. Med. Hyg. 1975; 24(3):527–532.

Khan, A.H., Morita, K., Parquet Md Mdel, C., Hasebe, F., Mathenge, E.G. and Igarashi, A. Complete nucleotide sequence of chikungunya virus and evidence for an internal polyadenylation site. J. Gen. Virol. 2002; 83(Pt 12):3075–3084.

Kit, L.S. Emerging and re-emerging diseases in Malaysia. Asia Pac. J. Public Health 2002; 14(1):6–8.

Kitaoka, M. Japanese encephalitis vaccine including a preliminary report on dengue fever and Chikungunya vaccines. Jpn. J. Med. Sci. Biol. 1967; 20:41–56.

Kiwanuka, N., Sanders, E.J., Rwaguma, E.B., Kawamata, J., Ssengooba, F.P., Najjemba, R., Were, W.A., Lamunu, M., Bagambisa, G., Burkot, T.R., Dunster, L., Lutwama, J.J., Martin, D.A., Cropp, C.B., Karabatsos, N., Lanciotti, R.S., Tsai, T.F. and Campbell, G.L. O'nyong-nyong fever in south-central Uganda, 1996–1997: clinical features and validation of a clinical case definition for surveillance purposes. Clin. Infect. Dis. 1999; 29(5):1243–1250.

Kokernot, R.H., Casaca, V.M., Weinbren, M.P. and McIntosh, B.M. Survey for antibodies against arthropod-borne viruses in the sera of indigenous residents of Angola. Trans. R. Soc. Trop. Med. Hyg. 1965a; 59(5):563–570.

Kokernot, R.H., Szlamp, E.L., Levitt, J. and McIntosh, B.M. Survey for antibodies against arthropod-borne viruses in the sera of indigenous residents of the Caprivi Strip and Bechuanaland Protectorate. Trans. R. Soc. Trop. Med. Hyg. 1965b; 59(5):553–562.

Lam, S.K., Chua, K.B., Hooi, P.S., Rahimah, M.A., Kumari, S., Tharmaratnam, M., Chuah, S.K., Smith, D.W. and Sampson, I.A. Chikungunya infection: an emerging disease in Malaysia. Southeast Asian J. Trop. Med. Public Health 2001; 32(3):447–451.

Lanciotti, R.S., Ludwig, M.L., Rwaguma, E.B., Lutwama, J.J., Kram, T.M., Karabatsos, N., Cropp, B.C. and Miller, B.R. Emergence of epidemic o'nyong-nyong fever in Uganda after a 35-year absence: genetic characterization of the virus. Virology 1998; 252(1):258–268.

Laras, K., Sukri, N.C., Larasati, R.P., Bangs, M.J., Kosim, R., Djauzi Wandra, T., Master, J., Kosasih, H., Hartati, S., Beckett, C., Sedyaningsih, E.R., Beecham, H.J., III and Corwin, A.L. Tracking the re-emergence of epidemic chikungunya virus in Indonesia. Trans. R. Soc. Trop. Med. Hyg. 2005; 99(2):128–141.

Lastarza, M.W., Grakoui, A. and Rice, C.M. Deletion and duplication mutations in the C-terminal nonconserved region of Sindbis virus nsP3: effects on phosphorylation and on virus replication in vertebrate and invertebrate cells. Virology 1994a; 202(1):224–232.

LaStarza, M.W., Lemm, J.A. and Rice, C.M. Genetic analysis of the nsP3 region of Sindbis virus: evidence for roles in minus-strand and subgenomic RNA synthesis. J. Virol. 1994b; 68(9):5781–5791.

Levinson, R.S., Strauss, J.H. and Strauss, E.G. Complete sequence of the genomic RNA of O'nyong-nyong virus and its use in the construction of alphavirus phylogenetic trees. Virology 1990; 175(1):110–123.

Levitt, N.H., Ramsburg, H.H., Hasty, S.E., Repik, P.M., Cole, F.E., Jr. and Lupton, H.W. Development of an attenuated strain of chikungunya virus for use in vaccine production. Vaccine 1986; 4(3):157–162.

Lhuillier, M., Cunin, P., Mazzariol, M.J., Monteny, N., Cordellier, R. and Bouchite, B. A rural epidemic of Igbo Ora virus (with interhuman transmission) in the Ivory Coast 1984–1985. Bull. Soc. Pathol. Exot. Filiales 1988; 81(3):386–395.

Li, G.P., La Starza, M.W., Hardy, W.R., Strauss, J.H. and Rice, C.M. Phosphorylation of Sindbis virus nsP3 in vivo and in vitro. Virology 1990; 179(1):416–427.

Lidbury, B.A. and Mahalingam, S. Specific ablation of antiviral gene expression in macrophages by antibody-dependent enhancement of Ross River virus infection. J. Virol. 2000; 74(18):8376–8381.

Lidbury, B.A., Simeonovic, C., Maxwell, G.E., Marshall, I.D. and Hapel, A.J. Macrophage-induced muscle pathology results in morbidity and mortality for Ross River virus-infected mice. J. Infect. Dis. 2000; 181(1):27–34.

Linn, M.L., Mateo, L., Gardner, J. and Suhrbier, A. Alphavirus-specific cytotoxic T lymphocytes recognize a cross-reactive epitope from the capsid protein and can eliminate virus from persistently infected macrophages. J. Virol. 1998; 72(6): 5146–5153.

Lumsden, W.H.R. An epidemic of virus disease in Southern Province, Tanganyika Territory, in 1952–53. II. General description and epidemiology. Trans. R. Soc. Trop. Med. Hyg. 1955; 49:33–57.

Lutwama, J.J., Kayondo, J., Savage, H.M., Burkot, T.R. and Miller, B.R. Epidemic O'Nyong-Nyong fever in southcentral Uganda, 1996–1997: entomologic studies in Bbaale village, Rakai District. Am. J. Trop. Med. Hyg. 1999; 61(1):158–162.

Mahalingam, S., Friedland, J.S., Heise, M.T., Rulli, N.E., Meanger, J. and Lidbury, B.A. Chemokines and viruses: friends or foes? Trends Microbiol. 2003; 11(8):383–391.

Marchette, N.J., Rudnick, A., Garcia, R. and MacVean, D.W. Alphaviruses in Peninsular Malaysia: I. Virus isolations and animal serology. Southeast Asian J. Trop. Med. Public Health 1978; 9(3):317–329.

Marshall, T.F., Keenlyside, R.A., Johnson, B.K., Chanas, A.C. and Smith, D.H. The epidemiology of O'nyong-nyong in the Kano Plain, Kenya. Ann. Trop. Med. Parasitol. 1982; 76(2):153–158.

Mason, P.J. and Haddow, A.J. An epidemic of virus disease in Southern Province, Tanganyika Territory, in 1952–53; an additional note on Chikungunya virus isolations and serum antibodies. Trans. R. Soc. Trop. Med. Hyg. 1957; 51(3):238–240.

Mateo, L., La Linn, M., McColl, S.R., Cross, S., Gardner, J. and Suhrbier, A. An arthrogenic alphavirus induces monocyte chemoattractant protein-1 and interleukin-8. Intervirology 2000; 43(1):55–60.

McClain, D.J., Pittman, P.R., Ramsburg, H.H., Nelson, G.O., Rossi, C.A., Mangiafico, J.A., Schmaljohn, A.L. and Malinoski, F.J. Immunologic interference from sequential administration of live attenuated alphavirus vaccines. J. Infect. Dis. 1998; 177(3):634–641.

McCrae, A.W., Henderson, B.E., Kirya, B.G. and Sempala, S.D. Chikungunya virus in the Entebbe area of Uganda: isolations and epidemiology. Trans. R. Soc. Trop. Med. Hyg. 1971; 65(2):152–168.

McGill, P.E. Viral infections: alpha-viral arthropathy. [Review] [31 refs]. Baillieres Clin. Rheumatol. 1995; 9(1):145–150.

McIntosh, B.M. Susceptibility of some African wild rodents to infection with various arthropod-borne viruses. Transfusion (Paris) 1961; 55:63–68.

McIntosh, B.M., Jupp, P.G. and Dos Santos, I. Rural epidemic of chikungunya in South Africa with involvement of *Aedes* (Diceromyia) *furcifer* (Edwards) and baboons. S. Afr. J. Sci. 1977; 73:267–269.

Mi, S. and Stollar, V. Expression of Sindbis virus nsP1 and methyltransferase activity in *Escherichia coli*. Virology 1991; 184(1):423–427.

Moore, D.L., Causey, O.R., Carey, D.E., Reddy, S., Cooke, A.R., Akinkugbe, F.M., David-West, T.S. and Kemp, G.E. Arthropod-borne viral infections of man in Nigeria, 1964–1970. Ann. Trop. Med. Parasitol. 1975; 69(1):49–64.

Moore, D.L., Reddy, S., Akinkugbe, F.M., Lee, V.H., David-West, T.S., Causey, O.R. and Carey, D.E. An epidemic of chikungunya fever at Ibadan, Nigeria, 1969. Ann. Trop. Med. Parasitol. 1974; 68(1):59–68.

Mourya, D.T. Absence of transovarial transmission of Chikungunya virus in *Aedes aegypti* & *Ae. albopictus* mosquitoes. Indian J. Med. Res. 1987; 85:593–595.

Mourya, D.T. and Banerjee, K. Experimental transmission of chikungunya virus by *Aedes vittatus* mosquitoes. Indian J. Med. Res. 1987; 86:269–271.

Mourya, D.T., Gokhale, M.D., Malunjkar, A.S., Bhat, H.R. and Banerjee, K. Inheritance of oral susceptibility of *Aedes aegypti* to chikungunya virus. Am. J. Trop. Med. Hyg. 1994; 51(3):295–300.

Mourya, D.T., Thakare, J.R., Gokhale, M.D., Powers, A.M., Hundekar, S.L., Jayakumar, P.C., Bondre, V.P., Shouche, Y.S. and Padbidri, V.S. Isolation of chikungunya virus from *Aedes aegypti* mosquitoes collected in the town of Yawat, Pune District, Maharashtra State, India. Acta Virol. 2001; 45(5–6):305–309.

Munasinghe, D.R., Amarasekera, P.J. and Fernando, C.F. An epidemic of dengue-like fever in Ceylon (chikungunya—a clinical and haematological study. Ceylon Med. J. 1966; 11(4):129–142.

Myers, R.M. and Carey, D.E. Concurrent isolation from patient of two arboviruses, chikungunya and dengue type 2. Science 1967; 157(794):1307–1308.

Myers, R.M., Carey, D.E., Reuben, R., Jesudass, E.S., De Ranitz, C. and Jadhav, M. The 1964 epidemic of dengue-like fever in South India: isolation of chikungunya virus from human sera and from mosquitoes. Indian J. Med. Res. 1965; 53(8):694–701.

Nakao, E. and Hotta, S. Immunogenicity of purified, inactivated chikungunya virus in monkeys. Bull. World Health Organ. 1973; 48(5):559–562.

Nalca, A., Fellows, P.F. and Whitehouse, C.A. Vaccines and animal models for arboviral encephalitides. Antiviral Res. 2003; 60(3):153–174.

Nimmannitya, S., Halstead, S.B., Cohen, S.N. and Margiotta, M.R. Dengue and chikungunya virus infection in man in Thailand, 1962–1964. I. Observations on hospitalized patients with hemorrhagic fever. Am. J. Trop. Med. Hyg. 1969; 18(6):954–971.

Norrby, E., Lagercrantz, R. and Gard, S. Measles vaccination. V. The booster effect of purified hemagglutinin in children previously immunized with this product or formalin-killed vaccine. Acta Paediatr. Scand. 1966; 55(1):73–78.

Nur, Y.A., Groen, J., Heuvelmans, H., Tuynman, W., Copra, C. and Osterhaus, A.D. An outbreak of West Nile fever among migrants in Kisangani, Democratic Republic of Congo. Am. J. Trop. Med. Hyg. 1999; 61(6):885–888.

Obeyesekere, I. and Hermon, Y. Myocarditis and cardiomyopathy after arbovirus infections (dengue and chikungunya fever). Br. Heart J. 1972; 34(8):821–827.

Obeyesekere, I. and Hermon, Y. Arbovirus heart disease: myocarditis and cardiomyopathy following dengue and chikungunya fever—a follow-up study. Am. Heart J. 1973; 85(2):186–194.

Olaleye, O.D., Oladosu, L.A., Omilabu, S.A., Baba, S.S. and Fagbami, A.H. Complement fixing antibodies against arboviruses in horses at Lagos, Nigeria. Revue d Elevage et de Medecine Veterinaire des Pays Tropicaux 1989; 42(3):321–325.

Olaleye, O.D., Omilabu, S.A. and Baba, S.S. Growth of Igbo-Ora virus in some tissue cultures. Acta Virol. 1990; 34(4):367–371.

Olaleye, O.D., Omilabu, S.A. and Fagbami, A.H. Igbo-Ora virus (an alphavirus isolated in Nigeria): a serological survey for haemagglutination inhibiting antibody in humans and domestic animals. Trans. R. Soc. Trop. Med. Hyg. 1988; 82(6):905–906.

Padbidri, V.S. and Gnaneswar, T.T. Epidemiological investigations of chikungunya epidemic at Barsi, Maharashtra state, India. J. Hyg. Epidemiol. Microbiol. Immunol. 1979; 23(4):445–451.

Paredes, A.M., Brown, D.T., Rothnagel, R., Chiu, W., Schoepp, R.J., Johnston, R.E. and Prasad, B.V. Three-dimensional structure of a membrane-containing virus. Proc. Natl. Acad. Sci. U.S.A. 1993; 90(19):9095–9099.

Pastorino, B., Muyembe-Tamfum, J.J., Bessaud, M., Tock, F., Tolou, H., Durand, J.P. and Peyrefitte, C.N. Epidemic resurgence of Chikungunya virus in democratic Republic of the Congo: identification of a new central African strain. J. Med. Virol. 2004; 74(2):277–282.

Paul, S.D. and Singh, K.R. Experimental infection of *Macaca radiata* with Chikungunya virus and transmission of virus by mosquitoes. Indian J. Med. Res. 1968; 56(6):802–811.

Pavri, K.M. Presence ff Chikungunya antibodies in human sera collected from Calcutta and Jamshedpur before 1963. Indian J. Med. Res. 1964; 52:698–702.

Peck, R., Brown, A. and Wust, C.J. In vitro heterologous cytotoxicity by T effector cells from mice immunized with Sindbis virus. J. Immunol. 1979; 123(4):1763–1766.

Pedersen, C.E., Jr., Robinson, D.M. and Cole, F.E., Jr. Isolation of the vaccine strain of Venezuelan equine encephalomyelitis virus from mosquitoes in Louisiana. Am. J. Epidemiol. 1972; 95(5):490–496.

Pfeffer, M., Kinney, R.M. and Kaaden, O.R. The alphavirus 3′-nontranslated region: size heterogeneity and arrangement of repeated sequence elements. Virology 1998; 240(1):100–108.

Porter, K.R., Tan, R., Istary, Y., Suharyono, W., Sutaryo Widjaja, S., Ma'Roef, C., Listiyaningsih, E., Kosasih, H., Hueston, L., McArdle, J. and Juffrie, M. A serological study of Chikungunya virus transmission in Yogyakarta, Indonesia: evidence for the first outbreak since 1982. Southeast Asian J. Trop. Med. Public Health 2004; 35(2):408–415.

Porterfield, J.S. Cross-neutralization studies with group A arthropod-borne viruses. Bull. World Health Organ. 1961; 24:735.

Posey, D.L., O'Rourke, T., Roehrig, J.T., Lanciotti, R.S., Weinberg, M. and Maloney, S. O'Nyong-nyong fever in West Africa. Am. J. Trop. Med. Hyg. 2005; 73(1):32.

Powers, A.M., Brault, A.C., Tesh, R.B. and Weaver, S.C. Re-emergence of chikungunya and o'nyong-nyong viruses: evidence for distinct geographical lineages and distant evolutionary relationships. J. Gen. Virol. 2000; 81(Pt 2):471–479.

Powers, A.M. and Logue, C.H. Changing patterns of chikungunya virus: re-emergence of a zoonotic arbovirus. J. Gen. Virol. 2007; 88(9):2363–2377.

Randall, R., Gibbs, C.J., Jr., Aulisio, C.G., Binn, L.N. and Harrison, V.R. The development of a formalin-killed Rift Valley fever virus vaccine for use in man. J. Immunol. 1962; 89:660–671.

Rao, T.R. Recent epidemics caused by chikungunya virus in India, 1963–1965. Sci. Cult. 1966; 32:215.

Rao, T.R., Carey, D.E. and Pavri, K.M. Preliminary isolation and identification of chikungunya virus from cases of dengue-like illness in Madras city. Indian J. Med. Res. 1965; 53(8):689–693.

Rayner, J.O., Dryga, S.A. and Kamrud, K.I. Alphavirus vectors and vaccination. Rev. Med. Virol. 2002; 12(5):279–296.

Reuben, R. Some entomological and epidemiological observations on the 1964 outbreak of Chikungunya fever in South India. Indian J. Med. Res. 1967; 55(1):1–12.

Robinson, M.C. An epidemic of virus disease in Southern Province, Tanganyika Territory, in 1952–53. Trans. R. Soc. Trop. Med. Hyg. 1955; 49:28.

Rodhain, F., Gonzalez, J.P., Mercier, E., Helynck, B., Larouze, B. and Hannoun, C. Arbovirus infections and viral haemorrhagic fevers in Uganda: a serological survey in Karamoja district, 1984. Trans. R. Soc. Trop. Med. Hyg. 1989; 83(6):851–854.

Ross, R.W. The Newala epidemic. III. The virus: isolation, pathogenic properties and relationship to the epidemic. J. Hyg. 1956; 54:177–191.

Rulli, N.E., Suhrbier, A., Hueston, L., Heise, M.T., Tupanceska, D., Zaid, A., Wilmes, A., Gilmore, K., Lidbury, B.A. and Mahalingam, S. Ross River virus: molecular and cellular aspects of disease pathogenesis. Pharmacol. Ther. 2005; 107(3):329–342.

Rwaguma, E.B., Lutwama, J.J., Sempala, S.D., Kiwanuka, N., Kamugisha, J., Okware, S., Bagambisa, G., Lanciotti, R., Roehrig, J.T. and Gubler, D.J. Emergence of epidemic O'nyong-nyong fever in southwestern Uganda, after an absence of 35 years. Emerg. Infect. Dis. 1997; 3(1):77.

Ryman, K.D., Klimstra, W.B., Nguyen, K.B., Biron, C.A. and Johnston, R.E. Alpha/beta interferon protects adult mice from fatal Sindbis virus infection and is an important determinant of cell and tissue tropism. J. Virol. 2000; 74(7):3366–3378.

Salim, A.R. and Porterfield, J.S. A serological survey on arbovirus antibodies in the Sudan. Trans. R. Soc. Trop. Med. Hyg. 1973; 67(2):206–210.

Sanders, E.J., Rwaguma, E.B., Kawamata, J., Kiwanuka, N., Lutwama, J.J., Ssengooba, F.P., Lamunu, M., Najjemba, R., Were, W.A., Bagambisa, G. and Campbell, G.L. O'nyong-nyong fever in south-central Uganda, 1996–1997: description of the epidemic and results of a household-based seroprevalence survey. J. Infect. Dis. 1999; 180(5):1436–1443.

Sarkar, J.K., Chatterjee, S.N., Chakravarti, S.K. and Mitra, A.C. Chikungunya virus infection with haemorrhagic manifestations. Indian J. Med. Res. 1965a; 53(10):921–925.

Sarkar, J.K., Chatterjee, S.N., Chakravarty, S.K. and Mitra, A.C. The causative agent of Calcutta haemorrhagic fever: chikungunya or dengue. Bull. Calcutta School Trop. Med. 1965b; 13(2):53–54.

Sawicki, D.L. and Sawicki, S.G. Role of the nonstructural polyproteins in alphavirus RNA synthesis. Adv. Exp. Med. Biol. 1998; 440:187–198.

Sawicki, S.G., Sawicki, D.L., Kaariainen, L. and Keranen, S. A Sindbis virus mutant temperature-sensitive in the regulation of minus-strand RNA synthesis. Virology 1981; 115(1):161–172.

Schlesinger, M. and Schlesinger, S. Formation and assembly of alphavirus glycoproteins. In: The Togaviridae and Flaviviridae (S. Schelsinger and M.J. Schlesinger, Eds.), pp. 121–148. New York: Plenum Publishing Corp., 1986.

Schuffenecker, I., Iteman, I., Michault, A., Murri, S., Frangeul, L., Vaney, M.C., Lavenir, R., Pardigon, N., Reynes, J.M., Pettinelli, F., Biscornet, L., Diancourt, L., Michel, S., Duquerroy, S., Guigon, G., Frenkiel, M.P., Bréhin, A.C., Cubito, N., Desprès, P., Kunst, F., Rey, F.A., Zeller, H. and Brisse, S. Genome microevolution of chikungunya viruses causing the Indian Ocean outbreak. PLoS Med. 2006; 3(7):1058–1070.

Sergon, K., Yahaya, A.A., Brown, J., Bedja, S.A., Mlindasse, M., Agata, N., Allaranger, Y., Ball, M.D., Powers, A.M., Ofula, V., Onyango, C., Konongoi, L.S., Sang, R., Njenga, M.K. and Breiman, R.F. Seroprevalence of Chikungunya virus infection on Grande Comore Island, Union of the Comoros, 2005. Am. J. Trop. Med. Hyg. 2007; 76(6):1189–1193.

Shore, H. O'nyong-nyong fever: an epidemic virus disease in East Africa. III. Some clinical and epidemiological observations in the Northern Province of Uganda. Trans. R. Soc. Trop. Med. Hyg. 1961; 55:361.

Simizu, B., Yamamoto, K., Hashimoto, K. and Ogata, T. Structural proteins of Chikungunya virus. J. Virol. 1984; 51(1):254–258.

Soekiman, S. A study on susceptibility of Indonesia colonies of *Aedes aegypti* and *Aedes albopictus* mosquitoes to experimental infection with dengue type 3 and chikungunya viruses. Kobe J. Med. Sci. 1987; 33(1):19–34.

Soekiman, S., Konishi, E.J. and Matsumura, T. Susceptibility of Indonesia colonies of *Aedes aegypti* and *Aedes albopictus* mosquitoes to experimental infection with chikungunya virus. Kobe J. Med. Sci. 1986a; 32(4):127–132.

Soekiman, S., Matsumura, T. and Yamanishi, H. Multiplication of chikungunya virus in salivary glands of *Aedes albopictus* (Oahu strain) mosquitoes: an electron microscopic study. Jpn. J. Med. Sci. Biol. 1986b; 39(5–6):207–211.

Spence, L.P. and Thomas, L. Application of haemagglutination and complement fixation techniques to the identification and serological classification of arthropod-borne viruses; studies on Chikungunya and Makonde viruses. Trans. R. Soc. Trop. Med. Hyg. 1959; 53(3):248–255.

Strauss, E.G., Levinson, R., Rice, C.M., Dalrymple, J. and Strauss, J.H. Nonstructural proteins nsP3 and nsP4 of Ross River and O'Nyong-nyong viruses: sequence and comparison with those of other alphaviruses. Virology 1988; 164(1):265–274.

Strauss, J.H. and Strauss, E.G. The alphaviruses: gene expression, replication, and evolution [published erratum appears in Microbiol. Rev. 1994;58(4):806]. [Review] [629 refs]. Microbiol. Rev. 1994; 58(3):491–562.

Suhrbier, A. and La Linn, M. Suppression of antiviral responses by antibody-dependent enhancement of macrophage infection. Trends Immunol. 2003; 24(4):165–168.

Tesh, R.B. Arthritides caused by mosquito-borne viruses. Annu. Rev. Med. 1982; 33:31–40.

Tesh, R.B., Gubler, D.J. and Rosen, L. Variation among geographic strains of *Aedes albopictus* in susceptibility to infection with chikungunya virus. Am. J. Trop. Med. Hyg. 1976; 25(2):326–335.

Thaikruea, L., Charearnsook, O., Reanphumkarnkit, S., Dissomboon, P., Phonjan, R., Ratchbud, S., Kounsang, Y. and Buranapiyawong, D. Chikungunya in Thailand: a re-emerging disease? Southeast Asian J. Trop. Med. Public Health 1997; 28(2):359–364.

Thaung, U., Ming, C.K., Swe, T. and Thein, S. Epidemiological features of dengue and chikungunya infections in Burma. Southeast Asian J. Trop. Med. Public Health 1975; 6(2):276–283.

Thein, S., La Linn, M., Aaskov, J., Aung, M.M., Aye, M., Zaw, A. and Myint, A. Development of a simple indirect enzyme-linked immunosorbent assay for the detection of immunoglobulin M antibody in serum from patients following an outbreak of chikungunya virus infection in Yangon, Myanmar. Trans. R. Soc. Trop. Med. Hyg. 1992; 86(4):438–442.

Thonnon, J., Spiegel, A., Diallo, M., Diallo, A. and Fontenille, D. Chikungunya virus outbreak in Senegal in 1996 and 1997. Bull. Soc. Pathol. Exot. 1999; 92(2):79–82.

Tomori, O., Monath, T.P., O'Connor, E.H., Lee, V.H. and Cropp, C.B. Arbovirus infections among laboratory personnel in Ibadan, Nigeria. Am. J. Trop. Med. Hyg. 1981; 30(4):855–861.

Turell, M.J., Beaman, J.R. and Tammariello, R.F. Susceptibility of selected strains of *Aedes aegypti* and *Aedes albopictus* (Diptera: Culicidae) to chikungunya virus. J. Med. Entomol. 1992; 29(1):49–53.

Turell, M.J. and Malinoski, F.J. Limited potential for mosquito transmission of a live, attenuated chikungunya virus vaccine. Am. J. Trop. Med. Hyg. 1992; 47(1):98–103.

van Rossum, T.G., Vulto, A.G., Hop, W.C., Brouwer, J.T., Niesters, H.G. and Schalm, S.W. Intravenous glycyrrhizin for the treatment of chronic hepatitis C: a double-blind, randomized, placebo-controlled phase I/II trial. J. Gastroenterol. Hepatol. 1999; 14(11):1093–1099.

Wang, Y.F., Sawicki, S.G. and Sawicki, D.L. Sindbis virus nsP1 functions in negative-strand RNA synthesis. J. Virol. 1991; 65(2):985–988.

Way, S.J., Lidbury, B.A. and Banyer, J.L. Persistent Ross River virus infection of murine macrophages: an in vitro model for the study of viral relapse and immune modulation during long-term infection. Virology 2002; 301(2):281–292.

Weaver, S.C., Frey, T.K., Huang, H.V., Kinney, R.M., Rice, C.M., Roehrig, J.T., Shope, R.E. and Strauss, E.G. Togaviridae. In: Virus Taxonomy: Eighth Report of the International Committee on Taxonomy of Viruses (C.M. Fauquet, M.A. Mayo, J. Maniloff, U. Desselberger and L.A. Ball, Eds.), pp. 999–1008. Amsterdam: Elsevier Academic Press, 2005.

White, A., Berman, S. and Lowenthal, J.P. Inactivated eastern equine encephalomyelitis vaccines prepared in monolayer and concentrated suspension chick embryo cultures. Appl. Microbiol. 1971; 22(5):909–913.

White, A., Berman, S. and Lowenthal, J.P. Comparative immunogenicities of Chikungunya vaccines propagated in monkey kidney monolayers and chick embryo suspension cultures. Appl. Microbiol. 1972; 23(5):951–952.

Williams, M.C. and Woodall, J.P. O'nyong-nyong fever: an epidemic virus disease in East Africa. II. Isolation and some properties of the virus. Trans. R. Soc. Trop. Med. Hyg. 1961; 55:135–141.

Williams, M.C., Woodall, J.P., Corbet, P.S. and Gillett, J.D. O'nyong-nyong fever: an epidemic virus disease in East Africa. VIII. Virus isolations from *Anopheles* mosquitoes. Trans. R. Soc. Trop. Med. Hyg. 1965a; 59:300–306.

Williams, M.C., Woodall, J.P. and Gillett, J.D. O'nyong-Nyong fever: an epidemic virus disease in East Africa. VII. Virus isolations from man and serological studies up to July 1961. Trans. R. Soc. Trop. Med. Hyg. 1965b; 59:186–197.

Woodall, J.P., Williams, M.C. and Lule, M. O'nyong-nyong. East Afr. Virus Res. Inst. Rep. 1963; 13:31.

Woodruff, A.W., Bowen, E.T. and Platt, G.S. Viral infections in travellers from tropical Africa. Br. Med. J. 1978; 1(6118):956–958.

Woodward, T.M., Miller, B.R., Beaty, B.J., Trent, D.W. and Roehrig, J.T. A single amino acid change in the E2 glycoprotein of Venezuelan equine encephalitis virus affects replication and dissemination in *Aedes aegypti* mosquitoes. J. Gen. Virol. 1991; 72(10):2431–2435.

Yadav, P., Gokhale, M.D., Barde, P.V., Singh, D.K., Mishra, A.C. and Mourya, D.T. Experimental transmission of Chikungunya virus by *Anopheles stephensi* mosquitoes. Acta Virol. 2003a; 47(1):45–47.

Yadav, P., Shouche, Y.S., Munot, H.P., Mishra, A.C. and Mourya, D.T. Genotyping of Chikungunya virus isolates from India during 1963–2000 by reverse transcription-polymerase chain reaction. Acta Virol. 2003b; 47(2):125–127.

CHAPTER

33

Rabies

Anthony R. Fooks[1], *Nicholas Johnson*[1] *and Charles E. Rupprecht*[2]
[1]Rabies and Wildlife Zoonoses Group, Veterinary Laboratories Agency (Weybridge), Addlestone, Surrey, UK
[2]Centers for Disease Control and Prevention, Atlanta, GA, USA

OUTLINE

ABSTRACT

Rabies is unique among the viral diseases of man in that it kills virtually every individual it infects. The illness associated with the disease, in previous times referred to as hydrophobia, is particularly unpleasant for the victim but also for health care workers and relatives who have to witness it. Global estimates of death from rabies suggest that one person dies from the disease every 10 min, and more than 300 others are exposed. This claim is supported by a study of Malawian children with viral encephalitis, where 3 (11.5%) of 26 children originally clinically diagnosed with cerebral malaria were later laboratory-confirmed as having rabies suggesting that human disease is underreported in some rabies-endemic countries. The disease is distributed worldwide and is endemic in many countries causing an estimated 50,000–70,000 human deaths each year, although the true burden of the disease is unknown due to underreporting and poor surveillance systems in many areas of the world (Meltzer and Rupprecht, 1998; Fooks, 2005). Following exposure, the virus invades the peripheral nervous system before entering the CNS leading to malfunction, organ failure, and ultimately death. Two clinical manifestations, "furious" and "dumb" (paralytic) rabies are observed following infection although the mechanisms of both syndromes are still not well understood. Rabies virus belongs to the *Lyssavirus* genus, which contains seven recognized genotypes. The genus can be divided into two broad groups or phylogroups. Genotype 1 ("classical" rabies virus), 4, 5, 6, and 7 belong to phylogroup I, while genotypes 2 and 3 have been placed in phylogroup II on the basis of glycoprotein sequence, pathogenicity, and immunogenicity. Additionally, there are four newly described lyssavirus species: Aravan, Khujand, Irkut, and West Caucasian Bat Virus that have been recently isolated from bats in Eurasia. All lyssaviruses are capable of causing clinical "rabies" in mammals.

Throughout the world, greater than 10 million people, many of whom are unvaccinated, have to endure extreme anxiety of "suspect" rabies following exposure from a rabid animal. Although preexposure vaccination and postexposure prophylaxis (PEP) for rabies is highly effective, ignorance about the disease, especially in developing countries, impedes timely intervention. In rare circumstances, vaccine failures have been reported. Highly effective vaccines have been developed for both animal and human use that induce long-term immunity against virus challenge; however the rabies virus vaccines were derived from isolates obtained between 50 and 100 years ago. Such strains are unlikely to take into account the genetic variation of the virus in the field accompanying the spatiotemporal fluctuation of reservoir species. Current evidence suggests that human vaccines, based on inactivated rabies virus, are not fully protective against all of the lyssavirus genotypes.

The ease of obtaining live virus from an animal reservoir is straightforward although propagating the virus requires specific skills and specialist laboratory facilities. Rabies virus is poorly transmitted by the airborne route, but does present a substantial hazard to personnel exposed to aerosols of rabies virus. The virus is unlikely to be an effective bioweapon if the aim of any terrorist act is to achieve rapid dissemination of the agent with maximum human exposure, largely due to the unpredictability of transmission and long incubation periods. However, the ability of the disease to provoke fear makes the virus a biothreat agent that could result in considerable public concern if deliberately reintroduced into a rabies-free region, i.e., Western Europe. In addition, governments confronted with a deliberate reintroduction would be obliged to instigate costly elimination strategies, protection of domestic livestock, and public information programs suggesting that the virus could be used as a means of "agroterroism." As a result the virus is classified as a "category C emerging infectious disease threat."

Future candidate vaccines need to be designed with broad-spectrum activity against all lyssaviruses, and ease of delivery using either oral or parenteral routes focusing on the creation of herd immunity among species. Furthermore, they must elicit long-term immunity, inducing both adaptive and innate immunity in a single-dose and produced at a cost that will allow widespread use throughout the world but especially in developing countries where rabies is still endemic. No antiviral agents are effective against the rabies virus and there is an urgent need for biologicals that inhibit the virus and provide relief from this devastating disease.

INTRODUCTION

Rabies is an ancient, if not one of the oldest reported diseases of man and has been reported in the writings of most civilizations for more than three millennia (Blaisdell, 1994; Neville, 2004; Swabe, 2004). The Greek word for rabies is "lyssa," a word of unknown origin although the earliest documented use of the word *lyssa* is from Homer's Iliad where it refers to a "raging dog" (Neville, 2004), and has been adopted as the genus name (*Lyssavirus*) for the group of viruses to which rabies belongs. This observation suggests that rabies did exist in Homer's time (8th–9th century BC). It is likely however, that rabies existed a long time before Homer first used the word *lyssa* to describe the disease. In the 3rd century BC, Andreas of Carystos wrote a treatise on rabies making him the first author of a script devoted entirely to rabies (Neville, 2004). Some of the first reports of animal rabies may date to the *Laws of Eshunna* in Mesopotamia, in the 23rd century BC, where the Babylonian word *Segû* meant both *to be mad* and *to become rabid*, and where the word *kaduh-hu*, which meant *to have the mouth open*, was synonymous with canine rabies. In the 17th century, the ancient word *rabiem* ("the madness") was used to describe the disease. Canine rabies was also reported in the 6th century BC, in the *Avesta* (Persia), in the 4th century BC by Aristotle (Greece), in the 1st century BC in the *Susrutasamhita* (India), in the 1st century AD by Rhazes and Avicenna (Arabia), in the 4th century in the *Talmud* and in Egyptian texts, then by Roman authors (Dioscorides, Galen), by the Byzantine

Hippiastrians (5th century), by Arab authors (10th century) (Neville, 2004; Swabe, 2004). In the 4th century BC, the absence of any mention of rabies by the "Father of Medicine," Hippocrates, led historians to deduce that rabies had not reached Europe by the 5th century BC. However, Hippocrates did refer to "frenetics" who drink little, are disturbed by every sound and afflicted with trembling suggesting that rabies was indeed present.

Lacking scientific knowledge of the disease, the proposed methods at the time for preventing rabies were largely improbable without any real prospect for success. In the 1st century BC, the considered cause and prescribed treatment that was used until the 17th century for rabies was defined. Grattius Faliscus wrote a poem describing a "little grub" supposed *small worm* (*lyssa*, in Greek), that was located under the tongue of rabid dogs. While the dogs were still young, the part of the tongue thought to be responsible for the malady was removed in an attempt to prevent the dog contracting the disease. It was during the reign of Tiberius (14–37 AD) that detailed instructions from an encyclopedia compiled by Aulus Cornelius Celsus for the treatment of bites from rabid dogs was reported (Neville, 2004). Celsus concluded that the bites from rabid animals contained a poisonous substance which, unless treated, would prove fatal. The "poison" that Celsus referred to was from the Latin word for "virus." Throughout the period, many diagnostic and control methods were also proposed. Two influential medical authorities were Rufus and Soranus. Rufus believed that rabies was sexually transmitted and Soranus reported a baby afflicted with rabies as being terrified of its mother's breast. A treatise, probably written by Soranus, described the contagious nature of rabies, not just from a bite, but also from the breath or claws of a rabid dog. The accurate recording of detailed clinical symptoms and the logic behind observing and recording the results from different treatments for rabies was undertaken by Galen in the 2nd century AD. He reported that clinical signs from a rabid dog included having saliva flowing from its mouth, irrationally running and then stopping, and biting with furious rage (Neville, 2004). Some of the earliest reported descriptions of the clinical signs of rabid dogs and subsequent treatments that were based on scientific evidence came from Aretaeus (2nd century AD), Philumenis (2nd century AD), Felix (2nd century AD), Aphrodisiensis (3rd century AD), Oribasius (4th century AD), Pelagonius (4th century AD), Vegetius (4th century AD), Aetius of Amida (5th/6th century AD), and Paul of Aegina (7th century AD).

In 1576, George Tuberville reported that the time of an animals' breeding influenced the animals' susceptibility to madness. A decade later, in 1587, Conrad Gesner described several variations of "madness" followed by an overview of the characteristics, habits, and diseases of numerous animal species by Edward Topsell in 1607. The latter identified rabies as being the most important disease of dogs. In 1613, Thomas Spackman, an English physician, devoted a whole book to the description of rabies. In the 17th century, religious popularity endorsed the faith in miracles, not least of all from St. Hubert, the patron saint of hunters, who by the custom of visiting his shrine could cure the afflicted from rabies (Swabe, 2004).

The "furious" stage of rabies is characterized by radical changes in the patient's behavior; changes that would be of grave concern to all those in attendance. In addition, the "madness" that embodies the final stages of the disease has a deep social and metaphysical significance, especially in ancient cultures (Swabe, 2004). The fear of a rabid person is therefore deep-rooted in history, hence the universal dread of the disease. Following infection, the human patient is undoubtedly reduced to a wild, uncontrolled animalistic state. A rabid person breaches the gap between humans and other animals thus opening the realms of reality to fantasy, myth, and folklore. The result to onlookers was to promote their prejudices and inspire their imaginations, sense of danger, and disgust. The domesticated dog was also reduced to an uncivilized and highly dangerous state, shocking to humans who were unaccustomed to witnessing savage canine behavior.

Popular fantasies about vampirism and werewolves were also deeply entrenched beliefs in medieval folklore and have accounted for the demonism of the disease that reduces perfectly healthy individuals to an "inhuman state" (Gomez-Alonso, 1998). It became apparent that the earliest medical practitioners were prepared to attempt any possible outlandish practice for rabies whether the treatment used was based on accurate scientific and medical knowledge or not. Many writers of the time had recognized the "hydrophobia" in patients and also the changes in behavior. In addition, the link between a "poisonous substance" in the saliva of dogs and in human disease had been reported in many ancient texts. The practice of "worming" young puppies has been described previously. Celsus in the 1st century AD stressed the need to attend to the wound afflicted by the bite of a rabid animal as soon as possible, suggesting cauterization to draw out the poison. An alternative approach that was practiced in ancient times was to immerse someone in water, usually a well. The patient was left in the well until they had drunk sufficient water to overcome the hydrophobia. More bizarre and tenuous treatments ranged from the healing power from a piece of hyena skin wrapped in cloth (Scribonius Largus) as a supposed prophylactic against rabies to curative substances including one remedy that was derived from the skull of a hanged man to treat patients afflicted from the bite of a mad dog (Antaeus). Numerous antidotes for rabies were "herbal remedies," often adopted as a "pick-and-mix" approach from previous medical writers. These included the root of the dog rose, a bitch's menstrual fluid, the ash of sea-crabs, garlic, hellebore, castoreum, rose oil, honey, laurel berries with latter supposed improvements including chamomile, sorrel, saffron, and nut kernel. Although many of these remedies were reputed to be efficacious including the mixture of river-crab ash and gentian, which was reported to never fail, yet in fact, not one remedy was ever proven to be effective (Neville, 2004). If any animal survived any of the practices/remedies that were on offer, it was certain that the animal was not suffering from rabies (Swabe, 2004).

The 19th century saw the application of scientific method in attempting to understand the nature of rabies and effective preventative measures. In 1804, the German scientist, Zinke, demonstrated the infectious nature of saliva from rabid dogs. Galtier succeeded in protecting sheep and goats by intravenous inoculation with virulent saliva, with a series of experiments between 1879 and 1881. However, no effective prophylaxis in animals was available before Pasteur's discovery in 1885. It should be recalled that the original goal of Pasteur's research was to develop a vaccine for use in dogs before exposure and that his decision to use this vaccine in man, after exposure, was only taken under pressure from his peers. Pasteur was, moreover, working on *virulent matter* (nerve tissues) and had not isolated the pathogen. It was not until the advent of the concept of "filterable agents," that in 1903, Remlinger isolated and identified the etiology of rabies as a virus. Some of the first vaccines used extensively in exposed animals consisted of Hogyes' method, using six injections of fixed virus diluted in saline. Thereafter in the early 20th century, many domestic animal vaccines contained the attenuated virus (Flury), or were viruses partially inactivated by phenol, as described initially by Fermi, and later by Semple. Effective adjuvanted vaccines containing inactivated viruses were not available largely until the latter half of the 20th century.

ETIOLOGIC AGENT(S)

Classification

Rabies virus is single-stranded, negative-sense, nonsegmented RNA virus within the *Lyssavirus* genus, which along with three other recognized animal virus genera form the family Rhabdoviridae within the order Mononegavirales. Four families comprise the Mononegavirales including: Filoviridae (e.g., Ebola virus), Paramyxoviridae (e.g., measles virus), Bornaviridae (Borna disease virus), and Rhabdoviridae.

The *Lyssavirus* genus is subdivided into four serotypes and seven major species, or genotypes, all adapted to replication in the mammalian CNS (Bourhy et al., 1993; Badrane et al., 2001). The archetype virus of the genus is classical rabies virus (RABV) (Fig. 33.1). Prior to the wider availability of genetic analysis, strains of RABV were identified by serotype analysis using hyperimmune sera and later by panels of antinucleocapsid monoclonal antibodies. These panels were able to differentiate lyssaviruses into four distinct serotypes. The first serotype (serotype 1) contained classical rabies virus and Australian bat lyssavirus (ABLV); serotype 2 included Lagos bat virus (LBV); serotype 3

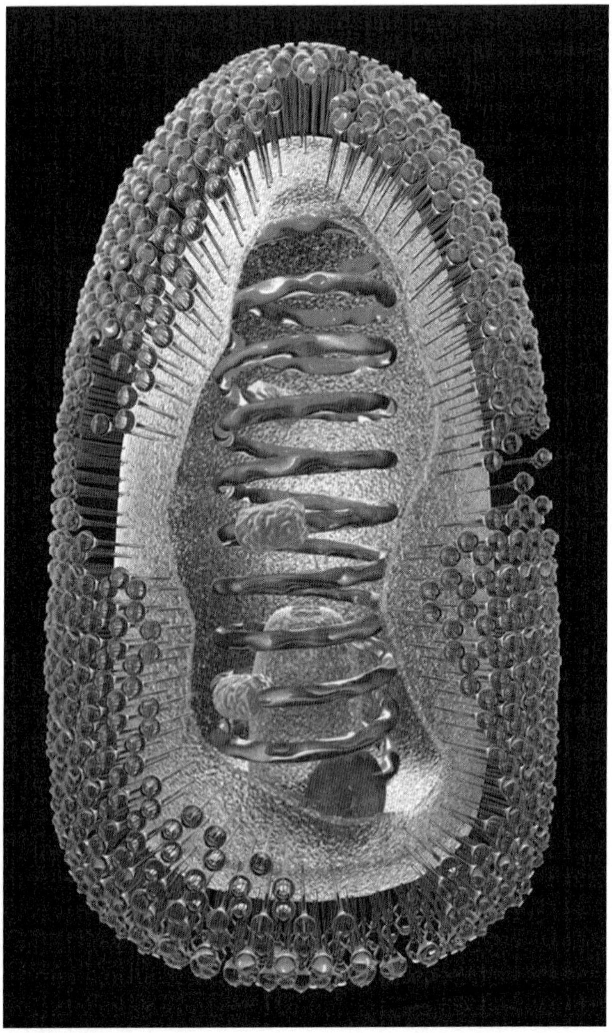

FIGURE 33.1 Schematic diagram of rabies virus (see color plate section).

included Mokola virus (MOKV); and serotype 4 consisting of Duvenhage virus (DUVV), European Bat Lyssavirus type-1 (EBLV-1) from Denmark and European Bat Lyssavirus type-2 (EBLV-2) from Finland.

With the development of molecular genetic analysis the *Lyssavirus* genus was differentiated into seven genetically divergent lineages. These include, RABV (genotype 1), LBV (genotype 2), MOKV (genotype 3), DUVV (genotype 4), EBLV type-1 (genotype 5), EBLV type-2 (genotype 6), and ABLV (genotype 7). With one exception (MOKV), all genotypes have been isolated from bats (Shope et al., 1970; Kuzmin et al., 2005). Four additional rabies-related viruses have recently been isolated from bats in Eurasia (Botvinkin et al., 2003; Fooks, 2004): Aravan virus, Khujand virus, Irkut virus, and West Caucasian Bat Virus (WCBV). These have been proposed as new members of the *Lyssavirus* genus. Aravan virus was isolated from the brain of a lesser mouse-eared bat (*Myotis blythi*) in Southern Kyrgyzstan in 1991 (Arai et al., 2003). A decade later, Khujand virus was isolated from the brain of a whiskered bat (*Myotis mystacinus*) in Northern Tajikistan (Kuzmin et al., 2003). In 2002, Irkut virus was isolated from a greater tube-nosed bat (*Murina leucogaster*) in Eastern Siberia. Later that year, WCBV was isolated from a bent-winged bat (*Miniopterus schreibersi*) in the Caucasus Mountains, approximately 100 km southeast of the town of Krasnodar and 150 km from the Eastern Turkish border. Subsequent studies have segregated the various genotypes into larger groups or phylogroups based on a number of genotypic and phenotypic characteristics (Badrane et al., 2001). Genotypes 1, 4, 5, 6, and 7, Aravan virus, Khujand virus, and Irkut virus comprise phylogroup I, while phylogroup II contains the two divergent African viruses, genotypes 2 and 3 and the most divergent, WCBV (which may be considered eventually in its own phylogroup).

Within genotypes, and especially for classical rabies, variants or biotypes are associated with different hosts (dog, fox, bat) and represent coevolution of particular lineages of virus with a specific reservoir.

EPIDEMIOLOGY

Rabies is a fatal zoonotic disease of mammals and is endemic on all continents with the exception of Antarctica and several islands (Hawaii, New Zealand) and, at least in regards to rabies in carnivores, an increasing number of European countries (Fig. 33.2). However, maintaining a rabies-free status

FIGURE 33.2 Global distribution of rabies virus variants.

incurs considerable costs and there is a continual risk of reimportation. Although targeted vaccination programs have been successful in eliminating the disease from carnivores over large areas, the virus still causes 30,000–50,000 human deaths per year. Fortunately the virus is limited in the range of species that can serve as suitable hosts. Those taxa that support cycles of infection are termed reservoir species and are limited to members of the Carnivora and Chiroptera orders. Factors such as the capacity to inflict a skin puncturing bite, high mobility and density and the degree of within-species interactions appear to favor certain species in their ability to transmit virus from one member to the next. All other species, including man, are considered spillover hosts in which the virus infects individuals, but is rarely transmitted further. Principal among the reservoir hosts is the domestic dog (*Canis familiaris*). Its proximity to humans as a companion animal, make it a key link in the transmission of rabies to man. Where data are recorded, the main source of human infection are bites from rabid dogs. Cats are often implicated in the transmission of rabies but due to their solitary nature act as a spillover host. A range of wildlife species also harbor the virus including the fox, raccoon, skunk, mongoose, and many species of bats (Bourhy et al., 1992; McColl et al., 2000; Rupprecht et al., 2002), and all are capable of transmitting rabies virus to humans.

Until recently, epidemiological studies were limited to recording the prevalence of cases and associating rabies with particular hosts. The ability to distinguish individual rabies viruses from the same host species has greatly increased our understanding of the interaction of viruses with particular reservoir hosts and has enabled retrospective investigations in human and domestic animal cases based on the particular rabies virus variant identified. This capacity to analyze the rabies virus variant has been dependent on developments in the field of molecular biology and as a result of the large variation observed within the viral genome. The gene structure of RABV, and the related lyssaviruses, is highly conserved but within these genes there appears to be variation in the length of coding and noncoding sequences and enormous sequence variation that often translates to amino acid variation in the expressed proteins. This has enabled researchers to measure viral variation and link this to geographical distribution, host reservoir, and transmission by spillover hosts. Originally, virus typing using panels of monoclonal antibodies was used to discriminate antigenic variants between other lyssaviruses and RABV (Smith, 1989). This approach has largely been superceded by genome sequence analysis. Many regions of the genome have been used to measure virus variation and as yet there appears to be no consensus as to which is the most appropriate sequence to use. Where comparative studies have been attempted, similar relationships between virus variants have been obtained (Bourhy et al., 1999; Johnson et al., 2002a). However, to date, surveys of GenBank indicate that most information is available for the viral nucleoprotein gene. Early studies attempted to compare rabies viruses from throughout the world (Smith et al., 1992; Kissi et al. 1995). A large number of later studies have focused on more meaningful panels of viruses based on rabies-endemic developing countries. For example, a molecular epidemiological study of rabies in Turkey demonstrates a clear distinction between rabies in the west of the country and viruses in the east (Fig. 33.3). Similar work has been conducted in Latin America, including Colombia and Mexico (Pãez et al., 2005; Velasco-Villa

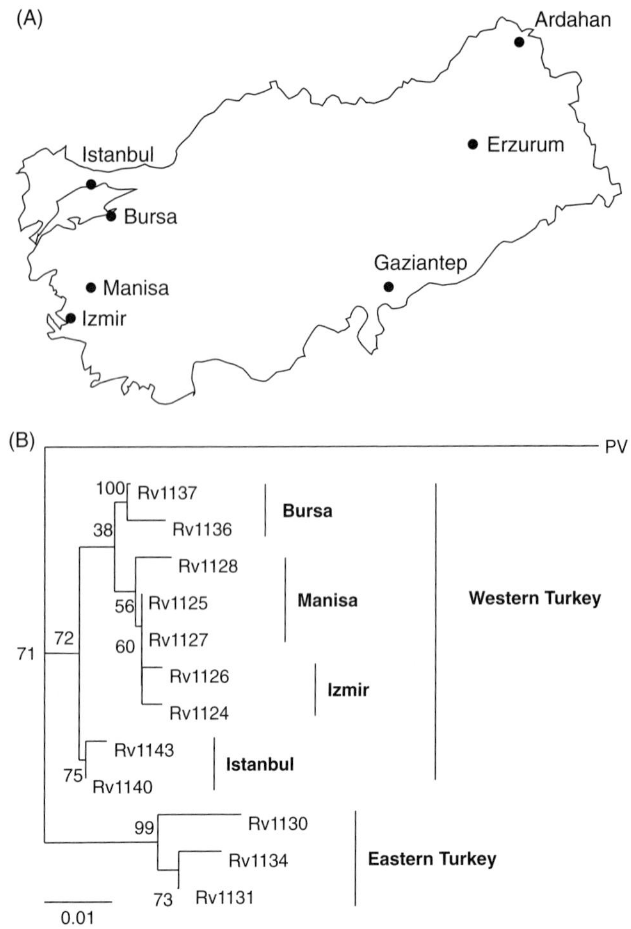

FIGURE 33.3 (A) A map of Turkey showing the principal cities that have reported rabies in recent years. (B) A phylogenetic tree demonstrating the division of rabies viruses in Turkey between those found in the western cities including the capital, Istanbul, and those found in eastern cities.

et al., 2005). Such studies could form the basis of rabies elimination programs by allowing investigators to distinguish between endemic variants and reintroductions. A further application has been the ability to confirm the source of human rabies cases where little information is available from the case history (Smith et al., 2003; Willoughby et al., 2005; Johnson et al., 2002b, Fooks et al., 2003).

SPECIES AFFECTED

In practice, it is possible to infect experimentally any mammalian species with rabies virus, although the degree of resistance of each species varies considerably according to the isolate and dose of rabies virus inoculated, the age of the animal, and the route of inoculation. This variation in resistance (species barrier) could explain why natural infection is unevenly distributed in different species, the majority of which are simply victims and not vectors of rabies virus (Rupprecht et al., 2002). Birds are also susceptible, but appear more naturally resistant to rabies (Schneider and Burstler, 1967). The range of species which is susceptible to rabies virus is large and differs according to continent.

Circumstantial evidence suggests that lyssaviruses may have evolved from the African continent and since became established outside of Africa as a result of zoogeographic spread (Rupprecht et al., 1991). Urban (dog) rabies predominates in Africa. In sub-Saharan Africa, the domestic dog is the principal reservoir whereas in the countries of southern Africa, a rabies cycle circulates in specific hosts, which is independent of the canine cycle. Different biotypes are maintained in the side-striped jackal (*Canis adustus*), the yellow mongoose (*Cynictis penicillata*) (Nel et al., 2005b), and the bat-eared fox (*Otocyon megalotis*) (Sabeta et al., 2007). Infections into occasional hosts such as kudu antelopes in Namibia as a result of spillover from the jackal population have been reported (Mansfield et al., 2006).

The domestic dog is also the principal reservoir in the Indian sub-continent (including India, Pakistan, and Bangladesh), where human fatalities from rabies are the highest in the world.

In contrast to most of the globe, no wild carnivore vectors/reservoirs of rabies have been identified definitively in southeast Asia, the dog being the only species responsible for enzootic rabies in this region. Insectivorous and frugivorous bats are also hosts of lyssavirus variants throughout Asia and have independent cycles of infection (Tacal and Boado, 1966; Pal et al., 1980; Arguin et al., 2002; Lumlertdacha et al., 2005; Tang et al., 2005). Although Australia is free of terrestrial rabies, several species of bats (frugivorous, insectivorous) are also responsible for perpetuating rabies (Gould et al., 1998; Samaratunga et al., 1998; McColl et al., 2000).

In the United States, in the early part of the 20th century, there were scores of cases of human rabies each year but this has decreased to between one and three cases in the 1980s. In 2004, rabies was responsible for eight cases, including four people who died following receipt of infected organs and tissues from an Arkansas donor bitten by a bat (Srinivasan et al., 2005). In addition, a Wisconsin teenager contracted rabies from a bat bite and survived rabies (Willoughby et al., 2005). Widespread vaccination of companion animals post World War II has been largely responsible for the decrease in human cases. Vaccination campaigns were implemented in the 1940s and greatly reduced canine variants of rabies virus by the 1960s. In 2005, wild animals accounted for the majority of all reported cases (greater than 92%), whereas domestic species accounted for 8%. Raccoons are the most commonly reported wildlife species (40%) followed by skunks and bats (23 and 22%, respectively) (Blanton et al., 2006).

In the Americas, all cases of bat rabies are caused by a variant of genotype 1. For example, the most common species found in the US include: the big brown bat (*Eptesicus fuscus*); the Brazilian (Mexican) free-tailed bat (*Tadarida brasiliensis mexicana*); the little brown bat (*Myotis lucifugus*); the hoary bat (*Lasiurus cinereus*); and the western pipistrelle bat (*Pipistrellus hesperus*). The variants of rabies virus that are predominantly reported to cause cases in humans are associated with the silver-haired (*Lasionycteris noctivagans*) and the eastern pipistrelle (*Pipistrellus subflavus*) bats. Bats account for an increasingly larger proportion of human rabies incidents in the US (Messenger et al., 2002). As the majority of cases of indigenous human rabies in the US since the 1950s was caused by insectivorous-bat variants of rabies virus, it led to conclusions that the silver-haired bat variant demonstrated a higher neuroinvasiveness and subsequent virulence in man. Although the route of entry into nerves in the epithelium leading to invasion of the CNS has never been established, the silver-haired bat variant does replicate more efficiently and at lower temperatures than dog variants of rabies virus (Dietzschold et al., 2000).

In contrast, the dog is also the main vector/reservoir in many countries of Central and South America, and a totally independent rabies cycle exists in hematophagous bats (vampire bats), chiefly the species *Desmodus rotundus* (Lord et al., 1975). This species is still responsible for the infection of thousands of domestic livestock and many cases in humans. Insectivorous bats also act as a reservoir for rabies but are rarely the cause of spillover infections.

Since 1989, the increased use of oral rabies vaccines (ORV) has been instrumental in successfully eliminating sylvatic rabies from large areas within Europe, Canada, and the US (Slate et al., 2005; MacInnes et al., 2001). For example, from 1990 onwards, we have witnessed the elimination of rabies from terrestrial mammals (principally the red fox) in many Western European countries: the Netherlands (1991), Switzerland (1999), France (2000), Belgium and Luxembourg (2001), and the Czech Republic (2004) resulting in these countries being declared "rabies-free." The last recorded case of terrestrial rabies in Germany occurred in February 2006 close to the city of Mainz (Thomas Muller, personal communication). Oral rabies vaccination field trials were first reported in Switzerland in 1978 using a live-attenuated rabies virus strain (Street Alabama Dufferin, SAD). The use of a genetically modified vaccine (vaccinia recombinant expressing the rabies virus glycoprotein, VRG) has also been widely used in recent years. While the threat in Europe from rabies in foxes has diminished in western Europe as reported by Bourhy et al. (2005), dog rabies in some eastern European countries, fox rabies in central and eastern Europe, and raccoon dog rabies in northeastern Europe continue to pose a risk. Similar ORV efforts focusing upon raccoons, foxes, and coyotes are operative throughout North America (Slate et al., 2005).

Among bats, rabies virus perpetuates itself in a cycle that is usually independent of rabies in terrestrial mammals, even though the latter, especially man, may be the victim of bites by rabid bats, from time to time. However, on rare occasions in the US, "spillover" cases of bat variants to terrestrial mammals have resulted in infection, and may occasionally lead to local enzootic establishment (Daoust et al., 1996; Leslie et al., 2006).

Animal rabies is distributed virtually worldwide. However, some areas (particularly island countries, such as the British Isles, Hawaii, Australia, New Zealand, Japan, Taiwan) have succeeded in protecting themselves against further incursion of the disease in terrestrial carnivores. Other countries have progressively eliminated rabies by disease prevention measures or prophylaxis: this is the case in Western Europe, and in some countries in the Southern Cone of Latin America. However, maintaining a rabies-free status incurs considerable costs and there is a continual risk of reimportation as was the case in July 2004, when a 4-month-old female dog was introduced to France via Spain and caused possible exposure to several dogs, cats, and human beings (Servas et al., 2005). Moreover, given their ability to fly, rabies in bats creates special circumstances in considering the indigenous global occurrence of disease (Constantine, 2003).

PATHOGENESIS

Rabies is transmitted through the saliva of rabid animals (Fig. 33.4A). Bites are the most common means of transmission. Occasionally less traumatic transfer of saliva such as the licking of a wound or broken skin can lead to deposition of virus. Deeper bites that lead to exposure of muscle tissue appear to be the most effective at transmitting the virus. As a result, the World Health Organization (2005) categorizes wounds for humans into three groups of increasing severity from category 1, which involves licking of intact skin and requires no action, to category 2, licking of broken skin, and category 3 single or multiple transdermal bites or scratches that warrant postexposure prophylaxis (PEP). The virus appears to persist at the bite site that leads to the long incubation times associated with rabies. Several unusual case records document that years elapse between a bite and the development of symptoms, although in humans the time between exposure and disease is usually measured in weeks to months. What the virus is doing, if anything is unclear, mainly because it is a very complex scenario to model with any accuracy. One theory suggests that some pathogenic strains are more adapted to replication at lower temperatures that may enhance persistence in the dermal layers (Dietzschold et al., 2000). Alternatively the virus persists in a dormant state or replicates at low levels in muscle or other tissues. However, it is clearly capable of evading the immune system for a considerable length of time. At some point, possibly a random event, RABV encounters a nerve cell, either directly or at nerve–muscle junctions where it enters the neuron through interaction of the viral glycoprotein with one of a number of protein receptors (Fig. 33.4B). At least three have been identified to date, the acetylcholine receptor (Lentz et al., 1982; Lewis et al., 2000; Gastka et al., 1996) shared with muscle cells, the neuronal cell adhesion molecule (Thoulouze et al., 1998) and the p75 neutrophin receptor (p75NTR) (Tuffereau et al., 1998; Langevin et al., 2002). The relative importance of each of these receptors is unclear but could be interpreted as a sign that the virus glycoprotein has adapted to bind a range of receptors to maximize the chance of entering a nerve cell. Once within the peripheral nervous system, virus migrates

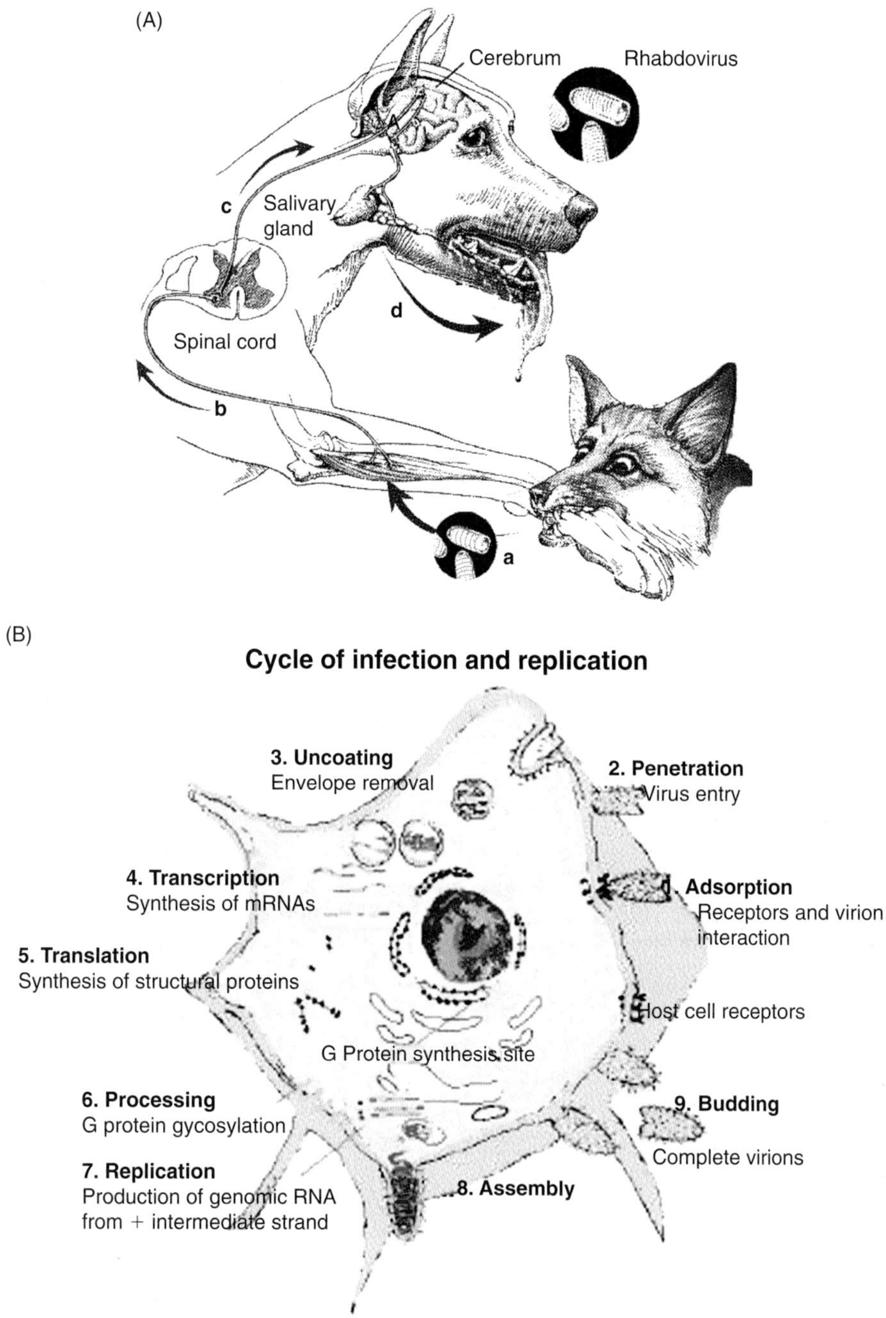

FIGURE 33.4 The infectious path of rabies virus. (A) From animal bite to CNS. Adapted from Infectious Diseases of the Dog and Cat, C. Greene, 3rd ed, Saunders. (B) Intraneuronal: from reception to assembly. Developed by L. Orciari, CDC. (see color plate section).

along peripheral nerves by retrograde movement toward the CNS at approximately 50–100 mm per day. Virus replication is detectable within nerve cell bodies (Tsiang, 1993). In humans, this often leads to the first overt symptoms of disease with paresthesia, a tingling sensation at the site of the bite. Once within the CNS the virus appears to replicate rapidly with associated developments of disease symptoms. The host response to infection varies. The immune-privileged status of the brain may limit the ability of the adaptive immune system to react to infection. More recent data suggest that alternative innate-mechanisms of immunity within the CNS do respond to infection, such as the generation of interferon, although this does not appear to effectively limit the infection and may lead to further damage to the brain (Stewart and Sulkin, 1966; Turner, 1972; Marcovistz et al., 1984, 1994; Prosniak et al., 2001; Saha and Rangarajan,

2003; McKimmie et al., 2005; Johnson et al., 2006a). Subsequently, the virus disseminates centrifugally out from the brain to a wide range of organs. Critically for the virus, one of the major portals of exit is the salivary glands through which the virus is shed, leading to further cycles of infection. Although not proven, it is possible that the virus targets areas of the brain that modulate behavior thus making the victim more likely to bite further individuals (Fig. 33.4A).

QUASISPECIES

The concept of quasispecies has been difficult to confirm in virus samples. One difficulty in molecular epidemiology and phylogeny is that the methods used select a single (presumably dominant) isolate from a virus strain, therefore population variants would be excluded. In addition, nearly all virus studies are selective in that they examine a subset of viruses that lead to disease fatalities. Currently no field isolate, particularly from wild animals, which are clinically silent and fail to kill the host have been detected. If they exist they have been overlooked in these analyses.

Benmansour et al. (1992) have suggested that quasispecies exist in RABV using the glycoprotein sequence. A high level of intrinsic heterogeneity has been directly observed during the adaptation of a human RABV isolate to growth in cell culture (Benmansour et al., 1992). Using G gene sequence, a street RABV isolate from a dog was compared with that of CVS displaying a 10% divergence in overall amino acid composition. Only 6% divergence was noted in the ectodomain, suggesting that structural constraints are exerted in this portion of the glycoprotein. A human RABV isolate cultured from saliva had only five amino acids difference (1%) from a canine isolate, an indication of their close relatedness. These differences could have originated during transmission from dog to dog, dog to human, or during isolation on cell culture; nonetheless they do demonstrate genetic evolution of street RABV. This evolution was further evidenced by the selection of cell-adapted virus variants, which displayed new amino acid substitutions in the glycoprotein, one of which was associated with neurovirulence in mice (Arg333) (Benmansour et al., 1992). This intrinsic heterogeneity was further demonstrated by sequencing molecular clones of the G gene, which revealed that only one-third of the viral genomes present in the brain of a rabid animal (dog) had the consensus sequence. Two-thirds of the clones analyzed displayed from one to three amino acid substitutions. These heterogeneous populations provide the virus with the capacity to adapt quickly and easily to new environmental conditions.

The occurrence of RABV quasispecies has also been shown to have implications in disease pathogenesis. Two virus variants of the ("fixed") CVS-24 stain have been described that differ genotypically and phenotypically. One is dominant in neuronal cells, the other in nonneuronal cells, both in vitro and in vivo (murine model) (Morimoto et al., 1998). The 10 amino acid differences in the G protein between these two variants have been shown to produce differences in pathogenesis. Both are virulent in neonatal mice but only the neurotropic variant retained the ability to infect adult mice. This suggests that certain aspects of host development, such as nervous and immune system maturation may be important in the virus variant selection process (Morimoto et al., 1998). Other factors that may be involved in generating sequence heterogeneity in RABV, include duration of infection, route of transmission, virus load, virus and host protein cooperation (Kissi et al., 1999).

Kissi et al. (1999) examined the potential for variation in a wild RABV isolate from a fox and examined mutation in several genes following passage in mice, dogs, cats, and cell culture. Again substitutions in the G protein were the most common. However, passage (2–3) in animals other than mice was not sufficient to produce a new dominant variant from the quasispecies population. It has been suggested that two mechanisms of genomic evolution of the RABV quasispecies exist during adaptation to environmental change. Firstly, a limited accumulation of mutations with no replacement of the dominant sequence (conservative changes) and secondly, a less frequent but rapid selective enrichment of favored variants where advantageous mutation has occurred (Kissi et al., 1999).

In the field, genetic variability has been especially evident at wave fronts of disease spread, an observation that reflects and supports the quasispecies concept, i.e., the hypothesis that virus populations evolve most rapidly in a new environment, but that after a few "passages" the most fit variants are stabilized rapidly (Domingo et al., 1985; Holland et al., 1991). Despite the obvious potential for random mutation, overall high levels of genetic conservation are found in wild RABV. This suggests that substantial selective pressures are operative in vivo, during productive infections in various "host-adaptive landscapes." RABV genomes appear to become trapped at local fitness optima, adaptations to new hosts, or the adoption of other transmission strategies may be difficult due to structural and functional constraints on the virus. Several simultaneous coadaptive changes may be required to affect host species transfer (Wandeler et al., 1994) as appears to have occurred with the wolf to dog and fox to raccoon dog.

POTENTIAL AS BIOTHREAT AGENT

Lyssaviruses are potential biothreat agents because confirmation of rabies in a population would spread "terror"; a principal goal of any terrorist. One reason for rabies being used as a weapon is for "notoriety" and "fear," largely related to the public familiarity with the disease. In addition, media attention associated with the presence of the disease in a population would promote "event" and "purpose." The viruses are capable of causing disease in all mammals including man resulting in widespread disruption, especially to the agricultural industry and related economies such as tourism or disruption in food supply. In addition, the disease has the capacity to confer a severe burden to human health. Although lyssaviruses are naturally fragile, are easily inactivated, and do not persist in the environment (Rupprecht et al., 2002), this group of viruses is capable of genetic change and new and emerging strains are being identified. Currently used human vaccines are not fully protective against all of the members of this genus. Limitations of biologicals could exacerbate a scenario with even conventional rabies viruses, as even a single rabid animal can result in hundreds of human exposures (Rotz et al., 1998). Lyssaviruses are primarily spread by the bite of an infected animal. However, recent data support the hypothesis that under optimum environmental conditions, virus can be spread by the airborne route (Johnson et al., 2006b). Rabies has a worldwide distribution ensuring the relative ease of obtaining a source of virus via an infected animal. The most important, and probably intangible, facet of rabies is the "fear" the disease provokes. Although the disease is usually vaccine preventable and there is a need for an appropriate transmission vector to cause infection in man, any perceived threat in the form of rabies used as a bioweapon is realistic. Deliberate release of virus in the form of a reservoir host in a previously rabies-free area could result in economic losses due to livestock death and subsequent costs of elimination. It could also invoke "fear" in a human population without requiring large-scale human fatalities. For these reasons, rabies could be perceived as a both an economic and psychological bioweapon. To date, there are no substantial documentations that lyssaviruses have been weaponized. Nevertheless, with adequate technology, concentrations in cell culture can be obtained that exceed 1×10^8 $TCID_{50}$ without a loss in virulence.

CLINICAL DISEASE

The first characteristic signs of rabies usually appear after an incubation period of 15–30 days. This period is only an approximate indication, corresponding to an exposure of average intensity in a species of average receptivity to a given virus. The incubation period can be much longer and may extend from weeks to months and even years (Rupprecht et al., 2002). The clinical signs of rabies may vary according to the species; "furious" (encephalitic) or "paralytic (dumb)" forms of the disease being observed in almost all species affected (Warrell and Warrell, 2004). The majority of patients (70%) demonstrate rabies in the "furious" form. In humans, the onset of clinical rabies is characterized by nonspecific prodromal features consisting of headache, fever, anorexia, nausea, vomiting, and diarrhea. This may be associated with symptoms such as anxiety, aggression, insomnia, hallucinations, and nightmares. Clinical symptoms progress rapidly especially as the virus spreads through the central nervous system (CNS) and the neurological impairment increases. Symptoms often fluctuate with paroxysms of hyperventilation and dysphasia, usually provoked by sight or sound of water, which are punctuated by periods of relative lucidity. The patient may also complain of intolerance to tactile, auditory, and visual stimuli. Paraesthesiae or pain at the bite site is observed in about 45% of cases, which may be followed by progressive pain or numbness in the limbs. Usually within 4–7 days, from the onset of symptoms, pain is reported in the neck region, which increases in intensity before becoming a permanent headache.

Hyperexcitability and hydrophobia are characteristic of the furious (encephalitic) form of rabies. These patients usually die within 7 days (average 5 days) after the clinical onset (Hemachuda and Phuapradit, 1997). In patients suffering from "paralytic" rabies, the average length of survival is 13 days. The initial clinical features of "furious" rabies are fever and hyperactivity. The latter is in response to internal stimuli such as thirst and nervousness, and to external stimuli including light and noise. The typical "furious" clinical symptoms then include: aerophobia, hydrophobia, fluctuating consciousness, spasms, and autonomic dysfunctions (hypersalivation and piloerection).

The initial clinical features of "paralytic" rabies are similar to those of "furious" rabies albeit later and milder. Inspiratory spasms occur in all patients with "paralytic" rabies during the preterminal phase. The bitten limb will usually present with weakness, which eventually spreads to all of the limbs, pharyngeal, facial, and respiratory muscles. Other symptoms include hypersalivation and altered perception. In many cases, a fear of death is continually expressed leading to confusion, urinary incontinence, and atrial fibrillation followed by respiratory spasms, breathing difficulties, convulsions, profuse salivation, and ultimately cardiorespiratory arrest (Hemachuda and Phuapradit, 1997; Solomon et al., 2005). In contrast, an ascending paralysis, usually from the bitten limb, is characteristic of the paralytic or "dumb" form of the disease. The paralysis then spreads rapidly and symmetrically with a loss of sphincter control and paralysis of the muscles of deglutination and respiration. The latter occurs as a terminal event.

MORBIDITY AND MORTALITY

Rabies is almost always fatal, once symptoms appear (Hemachuda and Phuapradit, 1997; Rupprecht et al., 2002; Warrell and Warrell, 2004). It has been reported that five vaccinated patients with "atypical" rabies have survived as none of these patients showed the "usual" clinical symptoms of rabies (Hemachuda, 1997). In 2005, a single case of survival from rabies in a teenager from the US following rabies virus transmission from a bat bite was reported (Willoughby et al., 2005). In this case, the patient showed "typical" clinical symptoms of rabies and although she had not received pre or postexposure vaccination against rabies, her treatment consisted of an innovative and largely unproven regimen of drug-induced coma plus a combination of antiviral therapy. Although premature in understanding, the success in treating this patient, and the history of prior survivors, does suggest that the clinical course of rabies can be altered in a small number of cases. Attempts at duplicating the original achievement of the "Milwaukee protocol" from different laboratories in patients suffering rabies have been unsuccessful, leading to the supposition that additional fundamental studies and development of relevant animal models are still required before this protocol is universally accepted (Hemachudha et al., 2006; Jackson et al., 2007; Willoughby 2007). Therefore, until the treatment used is substantiated, once clinical symptoms develop rabies remains a fatal and not a curable disease.

DIAGNOSTIC METHODS

A diagnosis of rabies is usually made on clinical grounds in the first instance. In areas free of rabies, a history of travel to a rabies-endemic country and/or evidence of a bite wound support a rabies diagnosis. A preliminary diagnosis should be verified through the use of specific laboratory tests. Rabies is diagnosed in many laboratories following positive histological examination of brain tissue by the direct fluorescent antibody test (FAT), which employs the detection of the nucleocapsid protein (Goldwasser and Kissling, 1958; Dean et al., 1996). This is a rapid, sensitive, and specific test (2–3h) with a high specificity. Histopathological preparation on formalin fixed paraffin-embedded brain preparations, stained with Hemotoxylin and Eosin (H&E), can be performed as well, but take several hours longer than the FAT. Aggregates of viral protein ("Negri bodies") often accumulate within certain neurons (particularly in the brain stem, cerebellum, and hippocampus) and can be observed under the light microscope, however false-positive results have been reported due to nonspecific inclusion bodies. Although the older method of examination for Negri bodies may be rapid, it has a low specificity and sensitivity compared to other methods (Nietfeld et al., 1989). Virus isolation can be achieved by the mouse inoculation test (MIT). This is performed by preparing homogenized suspension of suspect tissue (normally brain tissue) that is administered to mice by intracranial injection (Koprowski, 1996). The mice are monitored up to 28 days postinfection. The rabies tissue culture infection test (RTCIT) is comparable to the MIT and the results are obtained within 2–4 days (Rudd and Trimarchi, 1987; Crick and King, 1988). Laboratories with access to highly specific monoclonal antibodies have the opportunity to employ immunohistochemical techniques in detecting antigen in fixed tissues. Rapid immunohistochemistry protocols are under development for enhanced field surveillance (Lembo et al., 2007). Molecular techniques including the reverse transcriptase-polymerase chain reaction (RT-PCR) can be used to identify the presence of viral RNA in a range of samples (Sacramento et al., 1991), including saliva, thus providing the ability to attempt antemortem diagnosis in humans (Johnson et al., 2002b; Smith et al., 2003). An advantage of molecular diagnostic techniques, especially RT-PCR, is the ability to amplify template in preparation for sequencing. This may be required if more than one lyssavirus genotype is present in the area when the infection occurred. In general, the highly variable regions of the genome are the most suitable for differentiating closely related isolates. In contrast, the conserved regions are suitable for typing distantly related lyssaviruses. Given the cost, time, and opportunity for both false positives from laboratory contamination, as well as false negatives from primer mismatches, these variables, among others, usually limit the diagnostic utility of PCR to national research laboratories.

TREATMENT

Although there is no proven treatment that exists for rabies once clinical signs have appeared in an animal, postexposure prophylaxis is highly effective in man before clinical symptoms are observed and is also likely to be effective in animals. For this reason the actions taken following a bite from a potentially rabid animal are critically important to ensure prevention of disease. In the event of clinical suspicion, it is advisable to isolate the suspect animal, and to euthanize it (Rupprecht and Gibbons, 2004). Previous studies have demonstrated that thorough cleansing of the wound or wounds, with copious amounts of soap and water can increase survival by 50% (Kaplan et al., 1962). Subsequently, medical advice should be sought and PEP initiated as soon as possible. This consists of administering several injections of an anti-rabies vaccine according to a protocol defined by the medical authorities, principally the WHO. Vaccine is supplemented by the administration of rabies immune globulins (RIGs) infiltrated around the wounds. In some exceptional circumstances, recovery from clinical rabies has been documented both in humans and animals (Hattwick et al., 1972; Porras et al., 1976; Willoughby et al., 2005) but as yet the reason for these recoveries has not been elucidated.

PHYSICAL RESISTANCE OF THE VIRUS

Lyssaviruses are regarded as relatively fragile and are susceptible to a range of physical agents: virions are rapidly destroyed by heat, and heating to 50°C for 15 min completely inactivates the virus. Lyssaviruses are also inactivated by light and ultraviolet rays, and by different chemical agents including lipid solvents (ether, chloroform), quaternary ammoniums, bleach, and soapy solutions (practical application to the washing of bite wounds and their disinfection) (Kaplan, 1996). Phenol, formaldehyde, β-propiolactone, acetylethyleneimine also destroy the virus (practical application for inactivation of the virus during vaccine manufacture). In addition, lyssaviruses are preserved by 50% glycerine (application to the preservation of samples for dispatch to the laboratory in the event of lengthy transport) and quite resistant to putrefaction under certain circumstances, and virus has been identified in the carcass of a rabid animal that had been buried for 9 days (Lewis and Thacker, 1974). Rabies virus has been shown to survive from naturally infected foxes, in the form of a suspension of salivary gland tissue in physiological saline containing 10% calf serum. At temperatures of 37 and 20°C there was a rapid decline in the amount of virus over the first 24 h with complete inactivation after 96 h. The effect of temperature was marked, with virus surviving 144 h at 5°C when exposed on a flat surface but being inactivated within 42 h at 20°C (Matouch et al., 1987). Solar radiation increased the rate of virus destruction; inactivation was complete after one and a half hours when exposed to intense sunshine at 30°C, but inactivation took 20 h at the same temperature in the absence of sunshine.

IMMUNOLOGY

Antigens Encoded by Agent

The *Lyssavirus* genome is composed of a single strand of negative-sense, nonsegmented RNA (around 12 kb) encoding for five separate proteins and comprising a vestigial intergenic region (psi) between the G and L cistrons. The genes are ordered from 3′ to 5′ end of the genome as follows: nucleoprotein (N), phosphoprotein (P, NS, or M1), matrix protein (M or M2), glycoprotein (G), and RNA-dependent RNA polymerase (L) (Fig. 33.5). Structurally, the virion is composed of a nucleocapsid incorporating the N, P, and L proteins in tight association with the genomic RNA. The M protein forms a capsid structure which is in turn surrounded by a host-derived envelope into which the G protein is inserted. As a result the G protein is the only surface exposed viral protein. The G- and N-gene products are the major stimuli involved in eliciting an immune response in vaccinated animals (Wunner et al., 1988; Rupprecht et al., 1991) and therefore have been the focus of much research.

The G-genes of fixed strains of RABV and that of wild-type strains are drastically different (Lafon, 2005) and as such suggest that the glycoprotein is responsible for degrees of virulence and plays a fundamental role in both pathogenicity and neuroinvasiveness in different hosts (Faber et al., 2007). Of critical importance to neurovirulence is the presence of an arginine at position 333 of the glycoprotein. A change of residue at this position severely attenuates the virus variant through a loss of neuroinvasiveness (Dietzschold et al., 1983). Differences in pathogenicity between lyssavirus strains might also be dependent on the length of the genomic region between the L-gene start methionine and the transcription termination and polyadenylation (TTP) motif, which might affect transcriptional activity and RNA stability. All available lyssavirus sequences to date, including the EBLVs, use the second TTP site with the exception being Pasteur Virus (PV) and PV-derived isolates (Marston et al., 2007).

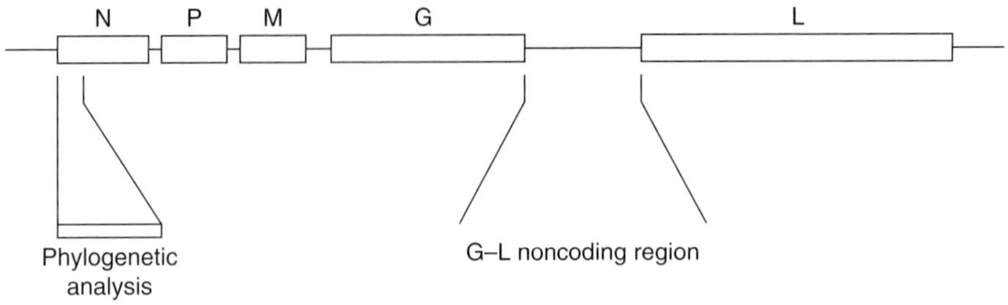

FIGURE 33.5 A schematic representation of the rabies virus genome showing the 5 coding regions for the nucleoprotein (N), phosphoprotein (P), matrix protein (M), glycoprotein (G), and the RNA polymerase (L). Also shown is the 400 base pair region of the nucleoprotein used for phylogenetic analysis of panels of rabies viruses (see Fig. 33.3) and the 500 bp G–L intergenic region with no known coding function.

PROTECTIVE IMMUNE RESPONSES

The principal property of RABV is its ability to infect neurons and gain access to the CNS, an immune-privileged site. If the virus can gain access to the CNS before an immune response is generated, the fate of the host is largely sealed. The principal correlate of protection against rabies is a strong virus neutralizing antibody response (Hemachuda, 1994). This response is used as a convenient marker to demonstrate the utility of vaccination in animals (Fooks et al., 2002). Antibodies directed to the viral glycoprotein are believed to provide the most effective immunological means as this is the only surface exposed antigen (Celis et al., 1986) possibly by the inhibition of cell-to-cell spread of infectious virus particles (Hooper et al., 1998). However, the glycoprotein of different RABV strains have different immunogenic properties, possibly related to the ability of certain strains to induce apoptosis (Morimoto et al., 1999). It is hypothesized that fixed, laboratory adapted strains of RABV overproduce glycoprotein and trigger the host cell to apoptose. This in turn makes the virus more immunogenic and reduces virulence. By contrast, the street or wild-type strains of RABV do not overproduce glycoprotein and maintain virulence for the host without apoptosis (Theerasurakarn and Ubol, 1998; Bahloul and Lafon, 2003).

The three-dimensional structure of the glycoprotein has not been fully elucidated, and as yet no crystal structure for this protein has been resolved. However, it is postulated to form trimeric structures within the virion envelope, which may have frustrated efforts to create recombinant vaccines against rabies. A number of epitopes have been identified using monoclonal antibody mapping. The principal epitopes are conformational (Seif et al., 1985; Prehaud et al., 1988) although many linear, nonconformational epitopes have been demonstrated (Dietzschold et al., 1990: Luo et al., 1997: Ni et al., 1995: Raux et al., 1995: Mansfield et al., 2004). However, it is unclear whether naturally infected or vaccinated individuals produce antibodies against these sites. Studies of RABV sequences encoding for the glycoprotein have not demonstrated great variation at suspected epitopes or for any selective pressure at particular glycoprotein sites driven either by the need to evade the host immune system or to adapt to new hosts (Holmes et al., 2002).

RABV-specific cell-mediated responses have been demonstrated (Celis et al., 1986; Cho et al., 1987) and may play a role in protection but this may be limited to assisting in the development of the antibody response. Gene knockout experiments in mice suggest that the loss of the cell-mediated arm of the immune system lowers but does not preclude resistance to rabies, whereas loss of antibody production invariably leads to fatal infection (Hooper, 2005).

Although the contribution of cell-mediated and innate immune responses to protection against rabies virus replication has not been fully defined, it is considered that they are not as important as the antibody response. As a result only a small number of studies have sought to identify T cell epitopes on particular virus-encoded proteins. Such epitopes have been mapped to the phosphoprotein (Celis et al., 1990), nucleoprotein (Ertl et al., 1991), and glycoprotein (Desmezieres et al., 1999). The entry of primed cytotoxic T cells into the CNS would lead to the destruction of virally infected cells, which in turn could contribute to the manifestations of disease (Sugamata et al., 1992). They may also play a role in the expression of proinflammatory cytokines, especially type 1 interferons. The expression of interferon-γ, TNF-α, IL-1α, and TLR3 are markedly increased following rabies infection and correlate with disease symptoms (Marquette et al., 1996; McKimmie et al., 2005). However, it is unclear if T cell infiltration is a common feature of late stage rabies virus infection (Wang et al., 2005). Indeed, RABV-specific cytotoxic T cells, in association with the production of inflammatory cytokines such as tumor necrosis factor-α, may be detrimental to the host and induce paralytic disease. Human cases of infection do not always show evidence of an antibody response and although it is often used diagnostically, serological tests for rabies are often inconclusive (Crepin et al., 1998). In contrast, vaccination is highly effective either given before exposure or prophylactically.

VACCINES

The development of vaccination against rabies began early in the 19th century with the pioneering work of Louis Pasteur. Without the benefit of our current understanding of the pathogen or the immune system, he developed a vaccine that prevented the development of disease following exposure. The first vaccines used the desiccated spinal cord of rabbits infected with rabies and required a long course of injections (which Pasteur as a scientist, not a clinician, was unable to administer himself). Although nerve tissue-derived vaccines are still in use today, WHO (2005) does not recommend their use due to side effects induced by components of the vaccine preparation. A 20th century milestone has been the development of cell-culture derived vaccines. These vaccines, such as human-diploid cell vaccine (HDCV: Imovax®, Sanofi), are considerably safer but are more expensive to produce. However, they are now being superceded by the purified chick embryo cell vaccine (PCECV; Rabipur®, RabAvert, Chiron Behring Vaccines, USA), and Vero cell vaccines (PVRV; Verorab, Sanofi). Such vaccines can be used for both pre and postexposure prophylaxis. Preexposure vaccination is recommended for those at high risk, including research workers, veterinarians, animal handlers, and travelers to rabies-endemic areas. The vaccination schedule consists of three intramuscular injections at day 0, 7 and 21 or 28. Postexposure vaccination is more complex and its use may depend on the level of exposure. A number of regimens have been developed (Dreesen, 1997) that also include the use of RIG. The RIG is derived from equine or human sources, although the cost of its production has cast doubt on its future availability (WHO, 2005). There are little data on the persistence of protection although one study demonstrated that 14 of 22 vaccine recipients possessed neutralizing antibody levels above 0.5 IU/ml five years after immunization (Rodrigues et al., 1987). By current WHO recommendations, those that are at continuous risk of exposure to the virus are recommended to monitor serum antibody levels at 6 monthly intervals and to obtain a booster vaccination if the antibody level drops below 0.5 IU/ml.

Parenteral vaccination of animals against rabies is undertaken by the subcutaneous or intramuscular route at doses recommended by the vaccine manufacturer. The vaccines most frequently used in tropical countries until recently were modified live-virus vaccines obtained by passage in laboratory animals or in cell culture (ERA/SAD strains or derivatives) and in chicken embryos (Flury strain, high or low egg passages: HEP or LEP). These vaccines had the advantage of being easy to produce, were inexpensive, and provided protection for 2–3 years. They had the disadvantage, however, of being heat-sensitive and dangerous for some species (e.g., the LEP vaccine in the cat). These live-virus vaccines have therefore gradually been replaced by inactivated virus vaccines produced in cell culture, which had the advantage of lacking the potential threat of vaccine-induced rabies. The length of immunity conferred can also equal or exceed that offered by live-virus vaccines, especially when an adjuvant is added. Such vaccination should be given to companion animals in rabies-endemic areas and is a requirement for many animals being transferred from a rabies-endemic area to a rabies-free country (Fooks et al., 2002).

Japan is believed to be the first country to use canine vaccination on a large scale to control rabies. A Semple-type vaccine of rabbit brain origin was developed in the early 1920s, and its success aroused interest in the US, where research on methods to assess the potency of this type of vaccine, and on establishing the duration of immunity in vaccinated dogs, was undertaken.

A further development in the vaccination of animals has been the development of live-attenuated rabies vaccines that can be administered orally (Baer et al., 1971). This approach was pioneered in the field in Switzerland (Wandeler et al., 1988) and has been used extensively in Europe and North America. The three vaccines used most frequently are based on the ERA/SAD strain, the SAG2 strain, and recombinant vaccinia-rabies glycoprotein vaccine (VRG). These vaccines are usually offered in liquid form, 1–2 ml of which are placed in a vaccine container, such as a plastic capsule, which is encased in a meatmeal or fishmeal-based bait. Often the content and mode of presentation of the bait needs to be refined to target particular species and provide sufficient uptake to ensure a high level of vaccine coverage. These baits can either be distributed in the wild (manually or by aerial means) in the case of wild carnivores or stray dogs, or distributed individually by hand outs directly in the case of domestic dogs. In view of the potential danger that these vaccines present to nontarget species (notably man), who could ingest them, the distribution of oral vaccines should be administered by authorized persons and must be undertaken within a regulated national vaccination program. This approach has led to the virtual elimination of rabies from the fox population in much of Europe and has contributed to the control of rabies in a range of wildlife species in North America.

Modern, cell culture-based rabies vaccines with their origin from genotype 1/serotype 1 strains, are composed of, and elicit protective immunity against, strains within phylogroup I. It is unclear, however, that vaccines based on genotype 1/serotype 1 strains will protect against infection with viruses from the other phylogroups. Studies to assess the ability of rabies vaccines to protect against the other lyssaviruses continue as results vary widely depending on the vaccine virus strain, challenge method, and host species. There are three main vaccine strains: PV, Pitman-Moore (PM), or Flury low egg passage (LEP). Hanlon and coworkers have shown that RABV-based vaccines provide little to no crossprotection against the recently identified lyssavirus, West Caucasian Bat virus. The inability of conventional preexposure immunization against rabies and postexposure prophylaxis against these newly emerging lyssaviruses has serious public health implications (Hanlon et al., 2005). As conventional prophylaxis does not exist for these new viruses, there is an urgent need for the development of a broad-spectrum vaccine and an appropriate immune globulin with cross-neutralizing ability against all of the viruses within the *Lyssavirus* genus.

Future DNA vaccines or recombinant viral vectors expressing immune-dominant regions may be able to confer protective immunity against all of the lyssaviruses. These prototype vaccines should be simple to produce, cost-effective, safe from genetic reassortment, and able to elicit a balanced Th1/Th2 immune response. In this respect, both the G- and N-proteins have been expressed in prokaryotic and eukaryotic expression systems and used as vaccine-delivery systems for both parenteral and oral application (Rupprecht et al., 1986; Prehaud et al., 1990; Yarosh et al., 1996). Despite numerous research studies, DNA vaccination against rabies has not proved sufficiently efficacious to warrant phase I clinical trials (Ertl, 2003). Edible plants expressing rabies virus proteins have also been considered for the large-scale production of antigens. Plant-derived proteins could provide an inexpensive delivery vehicle for subunit vaccines in developing countries to vaccinate both animals and man in mass vaccination campaigns (Yusibov et al., 1997; Ma et al., 2005). Transgenic tomatoes have been engineered to express the RABV glycoprotein, freeze-dried, and fed to mice resulting in the generation of G-specific antibodies (McGarvey et al., 1995). However, major issues about antigen concentration and delivery remain.

Although initially complex, the ability to produce recombinant negative-strand viruses is now very common. This has included the reverse genetics of rabies virus (Schnell et al., 1994) and has lead to a revolution in our understanding of the relationship between particular viral components and disease pathogenesis (Faber et al., 2004). This technology now offers a great many opportunities to design and develop a range of live-attenuated vaccine candidates tailored to the particular virus variant being targeted. For example, in the United States this could find many applications where current oral delivery vaccines have not proven effective at reducing rabies among a number of wildlife vectors (Dietzschold and Schnell, 2002). Indeed, the level of sophistication achieved by this approach has enabled rationally attenuated rabies viruses to be redesigned for use as a vaccine against other pathogens and also incorporating a beneficial cytokine gene (McGettigan et al., 2006).

PUBLIC HEALTH SIGNIFICANCE

WHO estimates that there are in excess of 50,000 rabies deaths in humans each year. The highest number of human rabies cases is reported in Africa and Asia, particularly the Indian sub-continent. These figures however, are considered to be conservative estimates due to underreporting of the disease in developing countries. This is largely a result of poor health care infrastructures and also because of insufficient control measures confronting significant reservoirs present in domestic animals and wildlife species. For these reasons, it is unlikely that rabies will be truly eradicated globally, especially considering the role of bats. Control efforts are focused on virus elimination in animals that might have frequent contact with man, particularly companion animals.

The Pan American Health Organization demonstrated in the late 1990s that human rabies could be effectively reduced through a program of parenteral vaccination of domestic animals in conjunction with rabies education campaigns, especially in rural rabies-endemic regions of developing countries of the Americas. In this way, the transmission cycle from animals to humans was disrupted, acutely limiting virus spread and reducing the public health risk. However, this approach does not yet address elimination or reduction of rabies in wildlife reservoirs. This provides a compelling argument for an understanding of the epidemiological cycle of rabies in a specific region so that appropriate control policies can be used to target particular animal reservoirs. Furthermore, bat rabies has never been successfully eliminated, and it is not clear what strategies would be successful, so true eradication of the disease is not possible in many parts of the world. Population controls used to reduce transmission of RABV from vampire bats in Latin America using nonselective destructive methods are considered to be obsolete and offer only a short-term respite from the problem. In many cases, the problem can be exacerbated where habitat is destroyed or disrupted. Bats incubating the disease can travel up to 100 km and transmit rabies to new colonies resulting in further dissemination of the virus in geographical locations that may have been previously rabies-free.

Apart from infection by bites, exceptional cases of infection by aerosols have been reported (in caves inhabited by large colonies of infected bats) or by the ingestion of infected animal carcasses. However, these natural modes of infection are not considered to be of critical public health importance in epidemiological terms at the present time (Constantine, 1962).

PROSPECTS FOR THE FUTURE

With the advent and widespread use of molecular biology techniques, it is now possible to construct lyssaviruses from cloned cDNA (Schnell et al., 1994). This pivotal report has led to an explosion in "reverse genetics" allowing genome rearrangements and the selection and isolation of "rescued" viruses that do not occur naturally. Possible drawbacks in these technologies is their use in developing viruses with altered phenotypes that demonstrate a broader cell-tropism with enhanced targeting moieties. A genetically engineered strain of rabies virus could potentially be generated with a high neuroinvasiveness, the potential ability to change tropisms, to be airborne, and to evade immunity from current rabies vaccines. A more simplistic and less technologically challenging scenario would be a strain of virus that targeted feral cats in urban areas. This would also cause widespread disruption and risks to public health. Lastly, opportunistic scenarios and "timing" of an event could be utilized for terrorist acts. In 2004, a shortage in the supply of Imovax® rabies vaccine (Sanofi Pastuer) in Canada led a second vaccine, RabAvert® (Chiron), to be licensed for use in Canada in a relatively short time. In the US and Europe, manufacturing problems in the shortage of Imovax® rabies vaccine led to concerns over available vaccine for PEP. The global shortage of immune globulins for human use as part of a PEP regimen has led to numerous approaches, some coordinated by WHO, to provide an alternative product for use throughout the world. Additionally, the need for national strategic stockpiles of these orphan biologicals has received recent attention.

It is necessary for governments and international organizations to combat any terrorist use of biological agents, such as rabies, by implementing a global, integrated surveillance program that has an "early detection" signal with open communication channels for the free-flow of data and information between cooperating countries, before such events are documented.

KEY ISSUES

- Lyssaviruses are zoonotic.
- Lyssaviruses are present in all continents of the world.
- Lyssaviruses must be notifiable in all countries.
- Lyssaviruses should be considered as potential biothreat agents.
- Transmission of a lyssavirus from one host to another is principally by the dissemination of virus-laden saliva from a biting incident.
- Lyssaviruses are readily inactivated in the environment and are poorly transmitted by the airborne route.
- Detection of the virus in a previously "rabies-free region" would result in widespread disorder, especially if used for agroterrorism.
- Misconceptions concerning rabies still avail. It is largely the "notoriety" and "fear" of the disease, especially in countries that are free from terrestrial disease in animals, in which the disease would cause the most disruption.
- Efficacious animal (live-attenuated and recombinant viral-vectored) vaccines are available for both parenteral and oral use.
- The large-scale control of rabies in wildlife has been successfully demonstrated with the use of oral vaccines.
- Fully efficacious and safe (inactivated) cell-culture produced human vaccines are available.
- There is no reliable treatment for rabies once clinical symptoms appear.
- Postexposure prophylaxis against rabies is only effective in humans before the onset of clinical symptoms.
- Any PEP regimen must include both a course of vaccination and the administration of immune globulin to be fully effective.

- New vaccine candidates and alternative immune globulins are under development in order to neutralize and cross-protect against new and emerging lyssavirus strains.

As part of the emergency preparedness to deal rapidly with any deliberate release of rabies virus or a related lyssavirus, an incident report plan (IRP) has been prepared. The IRP will be coordinated from national veterinary/medical authorities in alliance with international organizations, especially the WHO and OIE and used to quickly identify an infected area and to galvanize a rapid response team comprised of designated experts involved with emergency planning in consultation with rabies experts representing the World Health Organization Communicable Disease Surveillance and Response Collaborating Centre network and the OIE rabies reference laboratory network. Additional support from international specialists in veterinary public health, epidemiologists, clinicians, and representatives from other government departments and nongovernmental organizations will be provided in such an emergency.

ACKNOWLEDGMENT

Anthony R. Fooks and Nicholas Johnson are supported by the UK Department for Environment, Food and Rural Affairs (Defra) [Grant SEV3500].

References

Arai, Y.T., Kuzmin, I.V., Kameoka, Y. and Botvinkin, A.D. New Lyssavirus genotype from the lesser mouse-eared bat (*Myotis blythi*), Kyrghyzstan. Emerg. Infect. Dis. 2003; 9(3):1–9.

Arguin, P.M., Murray-Lillibridge, K., Miranda, M.E., Smith, J.S., Calaor, A.B. and Rupprecht, C.E. Serologic evidence of Lyssavirus infections among bats, the Philippines. Emerg. Infect. Dis. 2002; 8:258–262.

Badrane, H., Bahloul, C., Perrin, P. and Tordo, N. Evidence of two Lyssavirus phylogroups with distinct pathogenicity and immunogenicity. J. Virol. 2001; 75:3268–3276.

Baer, G.M., Abelseth, M.K. and Debbie, J.G. Oral vaccination of foxes against rabies. Am. J. Epidemiol. 1971; 93:487–490.

Bahloul, L. and Lafon, M. Apoptosis and rabies virus neuroinvasion. Biochimie 2003; 85:777–788.

Benmansour, A., Brahimi, M., Tuffereau, C., Coulon, P., Lafay, F. and Flamand, A. Rapid sequence evolution of street rabies glycoprotein is related to the highly heterogeneous nature of the viral population. Virology 1992; 187:33–45.

Blaisdell, J.D. The deadly bite of ancient animals: written evidence for rabies, or the lack thereof, in the ancient Egyptian and Mesopotamian texts. Vet. Hist. 1994; 8:22–28.

Blanton, J.D., Krebs, J.W., Hanlon, C.A. and Rupprecht, C.E. Rabies surveillance in the United States during 2005. J. Am. Vet. Med. Assoc. 2006; 229:1897–1911.

Botvinkin, A.D., Poleschuk, E.M., Kuzmin, I.V., Borisova, T.I., Gazaryan, S.V., Yager, P. and Rupprecht, C.E. Novel lyssaviruses isolated from bats in Russia. Emerg. Infect. Dis. 2003; 9:1623–1625.

Bourhy, H., Dacheux, L., Strady, C. and Mailles, A. Rabies in Europe in 2005. Eurosurveillance 2005; 10(11):213–216.

Bourhy, H., Kissi, B., Audry, L., Smreczak, M., Sadkowska-Todys, M., Kulonen, K., Tordo, N., Zmudzinski, J.F. and Holmes, E.C. Ecology and evolution of rabies virus in Europe. J. Gen. Virol. 1999; 80:2545–2557.

Bourhy, H., Kissi, B., Lafon, M., Sacramento, D. and Tordo, N. Antigenic and molecular characterization of bat rabies virus in Europe. J. Clin. Microbiol. 1992; 30:2419–2426.

Bourhy, H., Kissi, B. and Tordo, N. Molecular diversity of the Lyssavirus genus. Virology 1993; 194:70–81.

Brookes, S.M., Parsons, G., Johnson, N., McElhinney, L.M. and Fooks, A.R. Rabies human diploid cell vaccine elicits cross-neutralising and cross-protecting immune responses against European and Australian bat lyssaviruses. Vaccine 2005; 23(32):4101–4109.

Celis, E., Larson, J., Otvos, L., Jr. and Wunner, W.H. Identification of a rabies virus T cell epitope on the basis of its similarity with a hepatitis B surface antigen peptide presented to T cells by the same MHC molecule (HLA-DPw4). J. Immunol. 1990; 145:305–310.

Celis, E., Miller, R.W., Wiktor, T.J., Dietzshold, B. and Koprowski, H. Isolation and characterization of human T cell lines and clones reactive to rabies virus: antigen-specificity and production of interferon-gamma. J. Immunol. 1986; 136:692–697.

Cho, S.N., Nahara, S., Mifune, K. and Kawai, A. Murine T cell clones directed to rabies virus: isolation and some of their properties. J. Gen. Virol. 1987; 68:1115–1123.

Constantine, D.G. Rabies transmission by the non-bite route. Public Health Rep. 1962; 77:287.

Constantine, D.G. Geographic translocation of bats: known and potential problems. Emerg. Infect. Dis. 2003; 9:17–21.

Constantine, D.G., Emmons, R.W. and Woodie, J.D. Rabies virus in nasal mucosa of naturally infected bats. Science 1972; 175:1255–1256.

Constantine, D.G., Tierkel, E.S., Kleckner, M.D. and Hawkins, D.M. Rabies in New Mexico cavern bats. Public Health Rep. 1968; 83:303–316.

Crepin, P., Audry, L., Rotivel, Y., Gacoin, A., Caroff, C. and Bourhy, H. Intravitam diagnosis of human rabies by PCR using saliva and cerebrospinal fluid. J. Clin. Microbiol. 1998; 36:1117–1121.

Crick, J. and King, A. Culture of rabies virus in vitro. In: Rabies (J.B. Campbell and K.M. Charlton, Eds.), pp. 47–66. Boston, MA: Kluwer Academic, 1988.

Daoust, P.Y., Wandeler, A.I. and Casey, G.A. Cluster of rabies cases of probable bat origin among red foxes in Prince Edward Island, Canada. J. Wildl. Dis. 1996; 32(2):403–406.

Dean, D.J., Abelseth, M.K. and Atanasiu, P. The fluorescent antibody test. In: Laboratory Techniques in Rabies (F.-X. Meslin, M.M. Kaplan and H. Kowprowski, Eds.), 4th ed., pp. 88–93. Geneva: World Health Organization, 1996.

Desmezieres, E., Jacob, Y., Saron, M.F., Delpeyroux, F., Tordo, N. and Perrin, P. Lyssavirus glycoproteins expressing immunologically potent foreign B cell and cytotoxic T lymphocyte epitopes as prototypes for multivalent vaccines. J. Gen. Virol. 1999; 80:2343–2351.

Dietzschold, B., Gore, M., Marchadier, D., Niu, H.S., Bunschoten, H.M., Otvos, L., Wunner, W.H., Ertl, H.C.J., Osterhaus, A.D.M.E. and Koprowski, H. Structural and immunological characterization of a linear virus-neutralizing epitope of the rabies virus glycoprotein and its possible use in a synthetic vaccine. J. Virol. 1990; 64:3804–3809.

Dietzschold, B., Morimoto, K., Hooper, D.C., Smith, J.S., Rupprecht, C.E. and Koprowski, H. Genotypic and phenotypic diversity of rabies virus variants involved in human rabies: implications for postexposure prophylaxis. J. Hum. Virol. 2000; 3:50–57.

Dietzschold, B. and Schnell, M.J. New approaches to the development of live attenuated rabies vaccines. Hybrid. Hybridomics 2002; 21:129–134.

Dietzschold, B., Wunner, W.H., Wiktor, T.J., Lopes, D., Lafon, M., Smith, C.L. and Koprowski, H. Characterization of an antigenic determinant of the glycoprotein that correlates with pathogenicity of rabies virus. Proc. Natl. Acad. Sci. U.S.A. 1983; 80:70–74.

Domingo, E., Martinez-Salas, E., Sobrino, F., de la Torre, J.C., Portela, A., Ortin, J., Lopez-Galindez, C., Perez-Brena, P., Villanueva, N., Najera, R. et al. The quasispecies (extremely heterologous) nature of viral RNA genome populations: biological relevance—a review. Gene 1985; 49:1–8.

Dreesen, D.W. A global review of rabies vaccines for human use. Vaccine 1997; 15(Suppl.):S2–S6.

Ertl, H.C. Rabies DNA vaccines for protection and therapeutic treatment. Expert Opin. Biol. Ther. 2003; 3:639–644.

Ertl, H.C., Dietzschold, B. and Otvos, L., Jr. T helper cell epitope of rabies virus nucleoprotein defined by tri- and tetrapeptides. Eur. J. Immunol. 1991; 21:1–10.

Faber, M., Faber, M.L., Li, J., Preuss, M.A., Schnell, M.J. and Dietzschold, B. Dominance of a nonpathogenic glycoprotein gene over a pathogenic glycoprotein gene in rabies virus. J. Virol. 2007; 81(13):7041–7047.

Faber, M., Pulmanausahakul, R., Nagao, K., Prosniak, M., Rice, A.B., Koprowski, H., Schnell, M.J. and Dietzschold, B. Identification of viral genomic elements responsible for rabies virus neuroinvasiveness. Proc. Natl. Acad. Sci. U.S.A. 2004; 101:16328–16332.

Fooks, A.R. The challenge of emerging *lyssaviruses*. Expert Rev. Vaccines 2004; 3(4):89–92.

Fooks, A.R. Rabies in Europe—editorial. Eurosurveillance 2005; 10(11):1–4.

Fooks, A.R., McElhinney, L.M., Brookes, S.B., Johnson, N., Keene, V., Parsons, G. and Soldan, A. Rabies antibody testing and the UK Pet Travel Scheme. Vet. Rec. 2002; 150:428–430.

Fooks, A.R., McElhinney, L.M., Pounder, D., Finnegan, C., Mansfield, K., Brookes, S.M., Parsons, G., White, K., McIntyre, P.G. and Nathwani, D. Case report: isolation of a European Bat Lyssavirus type-2a from a fatal case of rabies encephalitis. J. Med. Virol. 2003; 71:281–289.

Gastka, M., Horvath, J. and Lentz, T.L. Rabies virus binding to the nicotinic acetylcholine receptor alpha subunit demonstrated by virus overlay protein binding assay. J. Gen. Virol. 1996; 77:2437–2440.

Goldwasser, R.A. and Kissling, R.E. Fluorescent antibody staining of street and fixed rabies virus antigens. Proc. Soc. Exp. Biol. Med. 1958; 98:219–223.

Gomez-Alonso, J. Rabies: a possible explanation for the vampire legend. Neurology 1998; 51:856–859.

Gould, A.R., Hyatt, A.D., Lunt, R., Kattenbelt, J.A., Hengstberger, S. and Blacksell, S.D. Characterisation of a novel lyssavirus isolated from Pteropid bats in Australia. Virus Res. 1998; 54(2):165–187.

Hanlon, C.A., Kuzmin, I.V., Blanton, J.D., Weldon, W.C., Manangan, J.S. and Rupprecht, C.E. Efficacy of rabies biologics against new lyssaviruses from Eurasia. Virus Res. 2005; 111:44–54.

Hattwick, M.A., Weis, T.T., Strechshulte, C.J., Baer, G.M. and Gregg, M.B. Recovery from rabies: a case report. Ann. Intern. Med. 1972; 76:931–942.

Hemachuda, T. Human rabies: clinical aspects, pathogenesis and potential therapy. In: Lyssaviruses (C.E. Rupprecht, B. Dietzschold and H. Kowprowski, Eds.), pp. 121–143. Berlin: Springer-Verlag, 1994.

Hemachuda, T. Rabies. In: Central Nervous System Infectious Diseases and Therapy (K.L. Roos, Ed.), pp. 573–600. New York: Marcel Dekker Inc., 1997.

Hemachudha, T., Mitrabhakdi, E., Wilde, H., Vejabhuti, A., Siripataravanit, S. and Kingnate, D. Additional reports of failure to responds to treatment after rabies exposure in Thailand. Clin. Infect. Dis. 1999; 28:143–144.

Hemachuda, T. and Phuapradit, P. Rabies. Curr. Opin. Neurol. 1997; 10:260–267.

Hemachudha, T., Sunsaneewitayakul, B., Desudchit, T., Suankratay, C., Sittipunt, C., Wacharapluesadee, S., Khawplod, P., Wilde, H. and Jackson, A.C. Failure of therapeutic coma and ketamine for therapy of human rabies. J. Neurovirol. 2006; 12(5):407–409.

Holland, J.J., de la Torre, J.C., Clarke, D.K. and Duarte, E. Quantitation of relative fitness and great adaptability of clonal populations of RNA viruses. J. Virol. 1991; 65:2960–2967.

Holmes, E.C., Woelk, C.H., Kassis, R. and Bourhy, H. Genetic constraints and the adaptive evolution of rabies virus in nature. Virology 2002; 292:247–257.

Hooper, D.C. The role of immune responses in the pathogenesis of rabies. J. Neurovirol. 2005; 11:88–92.

Hooper, D.C., Morimoto, K., Bette, M., Weihe, E., Koprowski, H. and Dietzschold, B. Collaboration of antibody and inflammation in clearance of rabies virus from the central nervous system. J. Virol. 1998; 72:3711–3719.

Jackson, A.C., Scott, C.A., Owen, J., Weli, S.C. and Rossiter, J.P. Therapy with minocycline aggravates experimental rabies in mice. J. Virol. 2007; 81(12):6248–6253.

Johnson, N., McElhinney, L.M., Smith, J., Lowings, P. and Fooks, A.R. Phylogenetic comparison of the genus *Lyssavirus* using distal coding sequences of the glycoprotein and nucleoprotein genes. Arch. Virol. 2002a; 147:2111–2123.

Johnson, N., McKimmie, C.S., Mansfield, K.L., Wakeley, P.R., Brookes, S.M., Fazakerley, J.K. and Fooks, A.R. Lyssavirus infection activates interferon gene expression in the brain. J. Gen. Virol. 2006a; 87:2663–2667.

Johnson, N., Phillpotts, R. and Fooks, A.R. Airborne transmission of Lyssaviruses. J. Med. Microbiol. 2006b; 55(6):785–790.

Johnson, N., Wade Lipscombe, D., Stott, R., Gopal Rao, G., Mansfield, K., Smith, J., McElhinney, L.M. and Fooks, A.R. Investigation of a human case of rabies in the United Kingdom. J. Clin. Virol. 2002b; 25:351–356.

Kaplan, M.M. Safety precautions in handling rabies virus. In: Laboratory Techniques in Rabies (M.M. Kaplan and H. Kowprowski, Eds.), 3rd ed., pp. 3–8. Geneva: World Health Organization, 1996.

Kaplan, M.M., Cohen, D., Koprowski, H., Dean, D. and Ferrigan, L. Studies on the local treatment of wounds for the prevention of rabies. Bull. World Health Organ. 1962; 26:765–775.

Kissi, B., Badrane, H., Audry, L., Lavenu, A., Tordo, N., Brahimi, M. and Bourhy, H. Dynamics of rabies virus quasispecies during serial passages in heterologous hosts. J. Gen. Virol. 1999; 80:2041–2050.

Kissi, B., Tordo, N. and Bourhy, H. Genetic polymorphism in the rabies virus nucleoprotein. Virology 1995; 209:526–537.

Koprowski, H. The mouse inoculation test. In: Laboratory Techniques in Rabies (F.X. Meslin, M. Kaplan and H. Koprowski, Eds.), 4th ed., pp. 80–87. Geneva: World Health Organisation, 1996.

Kuzmin, I.V., Hughes, G.J., Botvinkin, A.D., Orciari, L.A. and Rupprecht, C.E. Phylogenetic relationships of Irkut and West Caucasian bat viruses within the Lyssavirus genus and suggested quantitative criteria based on the N gene sequence for lyssavirus genotype definition. Virus Res. 2005; 111:28–43.

Kuzmin, I.V., Orciari, L.A., Arai, Y.T., Smith, J.S., Hanlon, C.A., Kameoka, Y. and Rupprecht, C.E. Bat lyssaviruses (Aravan and Khujand) from Central Asia: phylogenetic relationships according to N, P and G gene sequences. Virus Res. 2003; 97:65–79.

Lafon, M. Bat rabies—the Achilles heel of a viral killer? Lancet 2005; 366:876–877.

Langevin, C., Jaaro, H., Bressanelli, S., Fainzilber, M. and Tuffereau, C. Rabies virus glycoprotein (RVG) is a trimeric ligand for the N-terminal cysteine-rich domain of the mammalian p75 neurotrophin receptor. J. Biol. Chem. 2002; 277:37655–37662.

Larghi, O.P. and Diaz, A.M. Cross protection of mice against different rabies virus isolates. Zentralbl. Bakteriol. Mikrobiol. Hyg. A 1985; 259(2):268–274.

Lembo, T., Haydon, D.T., Velasco-Villa, A., Rupprecht, C.E., Packer, C., Brandão, P.E., Kuzmin, I.E., Fooks, A.R., Barrat, J. and Cleaveland, S. Molecular epidemiology identifies only a single rabies virus variant circulating in complex carnivore communities of the Serengeti. Proc. Biol. Sci. 2007; 274(1622):2123–2130.

Lentz, T.L., Burrage, T.G., Smith, A.L., Crisk, J. and Tignor, G.H. Is the acetylcholine receptor a rabies virus receptor? Science 1982; 215:182–184.

Leslie, M.J., Messenger, S., Rohde, R.E., Smith, J., Cheshier, R., Hanlon, C. and Rupprecht, C.E. Bat-associated rabies virus in Skunks. Emerg. Infect. Dis. 2006; 12(8):1274–1277.

Lewis, P., Fu, Y. and Lentz, T.L. Rabies virus entry at the neuromuscular junction in nerve-muscle cocultures. Muscle Nerve 2000; 23:720–730.

Lewis, V.J. and Thacker, W.L. Limitations of deteriorated tissue for rabies diagnosis. Health Lab. Sci. 1974; 11:8–12.

Lord, R.D., Fuenzalida, E., Delpietro, H., Larghi, O.P., de Díaz, A.M. and Lãzaro, L. Observations on the epizootiology of vampire bat rabies. Bull. Pan Am. Health Organ. 1975; 9(3):189–195.

Lumlertdacha, B., Boongird, K., Wanghongsa, S., Wacharapluesadee, S., Chanhome, L., Khawplod, P., Hemachudha, T., Kuzmin, I. and Rupprecht, C.E. Survey for bat lyssaviruses, Thailand. Emerg. Infect. Dis. 2005; 11:232–236.

Luo, T.R., Minamoto, N., Ito, H., Goto, H., Hiraga, S., Ito, N., Sugiyama, M. and Kinjo, T. A virus neutralizing eptiope on the glycoprotein of rabies virus that contains Trp251 is a linear epitope. Virus Res. 1997; 51:35–41.

Ma, J.K., Chikwamba, R., Sparrow, P., Fischer, R., Mahoney, R. and Twyman, R.M. Plant-derived pharmaceuticals—the road forward. Trends Plant Sci. 2005; 10:580–585.

MacInnes, C.D., Smith, S.M., Tinline, R.R., Ayers, N.R., Bachmann, P., Ball., D.G., Calder, L.A., Crosgrey, S.J., Fielding, C., Hauschilt, P., Honig, J.M., Johnston, D.H., Lawson, K.F., Nunan, C.P., Pedde, M.A., Pond, B., Stewart, R.B. and Voigt, D.R. Elimination of rabies from red foxes in eastern Ontario. J. Wildl. Dis. 2001; 37: 119–132.

Mallawa, M., Fooks, A.R., Banda, D., Chikungwa, P., Mankhambo, L., Molyneux, E., Molyneux, M.E. and Solomon, T. Rabies encephalitis in a malaria-endemic area of Malawi, Africa. Emerg. Infect. Dis. 2007; 13(1):136–139.

Mansfield, K., McElhinney, L.M., Hübschle, O., Mettler, F., Sabeta, C., Nel, L.H. and Fooks, A.R. A molecular epidemiological study of rabies epizootics in Kudu (*Tragelaphus strepsiceros*) in Namibia. BMC Vet. Res. 2006; 2:2.

Mansfield, K.L., Johnson, N. and Fooks, A.R. Identification of a conserved linear epitope at the N terminus of the rabies glycoprotein. J. Gen. Virol. 2004; 85:3279–3283.

Marcovistz, R., Hovanessian, A.G. and Tsiang, H. Distribution of rabies virus, interferon and interferon-mediated enzymes in the brains of virus-infected rats. J. Gen. Virol. 1984; 65:995–997.

Marcovistz, R., Leal, E.C., Matos, D.C. and Tsiang, H. Interferon production and immune response induction in apathogenic rabies virus-infected mice. Acta Virol. 1994; 38:193–197.

Marquette, C., Ceccaldi, P.-E., Ban, E., Weber, P., Tsiang, H. and Haour, F. Alteration of interleukin-1 alpha production and interleukin-1 alpha binding sites in mouse brain during rabies infection. Arch. Virol. 1996; 141:573–585.

Marston, D., McElhiney, L.M., Johnson, N., Müller, T., Conzelmann, K.K., Tordo, N. and Fooks, A.R. Comparative analysis of the full genome sequence for European Bat Lyssavirus type-1 and type-2 with other lyssaviruses and evidence for a conserved transcription termination and polyadenylation motif in the G-L 3′-nontranslated region. J. Gen. Virol. 2007; 88:1302–1314.

Matouch, O., Jaros, J. and Pohl, P. Survival of rabies virus under external conditions. Vet. Med. 1987; 32(LX):669–674.

McColl, K.A., Tordo, N. and Aguilar Setien, A.A. Bat lyssavirus infections. Rev. Sci. Tech. 2000; 19:177–196.

McGarvey, P.G., Hammond, J., Dienelt, M.M., Hooper, D.C., Michaels, F.H., Fu, Z.F., Dietzschold, B. and Koprowski, H. Expression of the rabies virus glycoprotein in transgenic tomatoes. Bio/Technology 1995; 13:1484–1487.

McGettigan, J.P., Koser, M.L., McKenna, P.M., Smith, M.E., Marvin, J.M., Eisenlohr, L.C., Dietzschold, B. and Schnell, M.J. Enhanced humoral HIV-1-specifc immune responses generated from recombinant rhabdoviral-based vaccine vectors co-expressing HIV-1 proteins and IL-2. Virology 2006; 344:363–377.

McKimmie, C.S., Johnson, N., Brown, A.R., Fooks, A.R. and Fazakerley, J.K. Viruses selectively upregulate Toll-like receptors in the central nervous system. Biochem. Biophys. Res. Commun. 2005; 336(3):925–933.

Meltzer, M.I. and Rupprecht, C.E. A review of the economics of the prevention and control of rabies: part I, global impact and rabies in humans. Pharmacoeconomics 1998; 14:365–383.

Meslin, F.X. and Stohr, K. Prospects for immunisation against rabies in developing countries. In: Rabies Control in Asia (B. Dodet and F.-X. Meslin, Eds.), pp. 15–18. Paris: Elsevier, 1997.

Messenger, S.L., Smith, J.S. and Rupprecht, C.E. Emerging epidemiology of bat-associated cryptic cases of rabies in humans in the United States. Clin. Infect. Dis. 2002; 35:738–747.

Morimoto, K., Hooper, D.C., Carbaugh, H., Fu, Z., Koprowski, H. and Dietzschold, B. Rabies virus quasispecies: implications for pathogenesis. Proc. Natl. Acad. Sci. U.S.A. 1998; 95: 3152–3156.

Morimoto, K., Hooper, D.C., Spitsin, S., Koprowski, H. and Dietzschold, B. Pathogenicity of different rabies virus variants inversely correlates with apoptosis and rabies virus glycoprotein expression in infected primary neuron cultures. J. Virol. 1999; 73:510–518.

Nel, L.H. Vaccines for lyssaviruses other than rabies. Expert Rev. Vaccines 2005a; 4:533–540.

Nel, L.H., Sabeta, C.T., von Teichman, B., Jaftha, J.B., Rupprecht, C.E. and Bingham, J. Mongoose rabies in southern Africa: re-evaluation based on molecular epidemiology. Virus Res. 2005b; 109:165–173.

Neville, J. Chapter 1—rabies in the ancient world. In: Historical Perspective of Rabies in Europe and the Mediterranean Basin (A.A. King, A.R. Fooks, M. Aubert and A.I. Wanderler, Eds.), pp. 1–12. OIE Publications, Paris, France, 2004.

Ni, Y., Tominga, Y., Honda, Y., Morimoto, K., Sakamoto, S. and Kawai, A. Mapping and characterization of a sequential epitope on the rabies virus glycoprotein which is recognized by a neutralizing monoclonal antibody Rg719. Microbiol. Immunol. 1995; 39:693–702.

Nietfeld, J.C., Rakich, P.M., Tyler, D.E. and Bauer, R.W. Rabies-like inclusions in dogs. J. Vet. Diag. Invest. 1989; 1:333–338.

Pãez, A., Saad, C., Nũnez, C. and Bóshell, J. Molecular epidemiology of rabies in northern Colombia 1994–2003. Evidence for human and fox rabies associated with dogs. Epidemiol. Infect. 2005; 133:529–536.

Pal, S.R., Arora, B., Chhuttani, P.N., Broor, S., Choudhury, S., Joshi, R.M. and Ray, S.D. Rabies virus infection of a flying fox bat, *Pteropus policephalus* in Chandigarh, Northern India. Trop. Geogr. Med. 1980; 32:265–267.

Porras, C., Barboza, J.J., Fuenzalida, E., Adaros, H.L., Oviedo, A.M. and Furst, J. Recovery from rabies in man. Ann. Intern. Med. 1976; 8:44–48.

Prehaud, C., Coulon, P., Lafay, F., Thiers, C. and Flamand, A. Antigenic site II of rabies virus glycoprotein: structure and role in viral virulence. J. Virol. 1988; 62:1–7.

Prehaud, C., Harris, R.D., Fulop, V., Koh, C.L., Wong, J., Flamand, A. and Bishop, D.H. Expression, characterization, and purification of a phosphorylated rabies nucleoprotein synthesized in insect cells by baculovirus vectors. Virology 1990; 178:486–497.

Prosniak, M., Hooper, D.C., Dietzschold, B. and Koprowski, H. Effect of rabies virus infection on gene expression in mouse brain. Proc. Natl. Acad. Sci. U.S.A. 2001; 98:2758–2763.

Raux, H., Coulon, P., Lafay, F. and Flamand, A. Monoclonal antibodies which recognize the acidic configuration of the rabies virus glycoprotein at the surface of the virion can be neutralizing. Virology 1995; 210:400–408.

Rodrigues, F.M., Mandke, V.B., Roumiantzeff, M., Rao, C.V., Mehta, J.M., Pavri, K.M. and Poonwalla, C. Persistence of rabies antibody 5 years after pre-exposure prophylaxis with human diploid cell antirabies vaccine and antibody response to a single booster dose. Epidemiol. Infect. 1987; 99:91–95.

Rotz, L.D., Hensley, J.A., Rupprecht, C.E. and Childs, J.E. Large-scale human exposures to rabid or presumed rabid animals in the United States: 22 cases (1990–1996). J. Am. Vet. Med. Assoc. 1998; 212:1198–1200.

Rudd, R.J. and Trimarchi, C.V. Comparison of sensitivity of BHK-21 and murine neuroblastoma cells in the isolation of a street strain rabies virus. J. Clin. Microbiol. 1987; 25:1456–1458.

Rupprecht, C.E., Dietzschold, B., Wunner, W.H. and Koprowski, H. Antigenic relationships of lyssaviruses. In: The Natural History of Rabies (G.M. Baer, Ed.), 2nd ed., pp. 69–100. Boca Raton, FL: CRC, 1991.

Rupprecht, C.E. and Gibbons, R.V. Clinical practice. Prophylaxis against rabies. N. Engl. J. Med. 2004; 351(25):2626–2635.

Rupprecht, C.E., Hanlon, C.A. and Hemachudha, T. Rabies re-examined. Lancet Infect. Dis. 2002; 2:327–343.

Rupprecht, C.E., Wiktor, T.J., Johnston, D.H., Hamir, A.N., Dietzschold, B., Wunner, W.H., Glickman, L.T. and Koprowski, H. Oral immunization and protection of raccoons (*Procyon lotor*) with a vaccinia-rabies glycoprotein recombinant virus vaccine. Proc. Natl. Acad. Sci. U.S.A. 1986; 8:7947–7950.

Sabeta, C., Mansfield, K., McElhinney, L.M., Fooks., A.R. and Nel, L.H. Molecular epidemiology of rabies in bat-eared foxes (*Otocyon megalotis*) in South Africa. Virus Res. 2007; 129:1–10.

Sacramento, D., Bourhy, H. and Tordo, N. PCR technique as an alternative method for diagnosis and molecular epidemiology of rabies virus. Mol. Cell. Probes 1991; 5:229–240.

Saha, A. and Rangarajan, P. Common host genes are activated in mouse brain by Japanese encephalitis and rabies viruses. J. Gen. Virol. 2003; 84:1729–1735.

Samaratunga, H., Searle, J.W. and Hudson, N. Non-rabies Lyssavirus human encephalitis from fruit bats: Australian bat lyssavirus infection (pteropid Lyssavirus) infection. Neoropathol. Appl. Neurobiol. 1998; 24:331–335.

Schneider, L. and Burstler, H. Studies on the pathogenesis of rabies in fowls after intracerebral infection. Zentralbl. Veterinarmed B 1967; 14:598–624.

Schnell, M.J., Mebatsion, T. and Conzelmann, K.K. Infectious rabies viruses from cloned cDNA. EMBO J. 1994; 13:4195–4203.

Seif, I., Coulon, P., Rollin, P.E. and Flamand, A. Rabies virulence: effects on pathogenicity and sequence characterization of rabies virus mutations affecting antigenic site III of the glycoprotein. J. Virol. 1985; 53:926–934.

Selimov, M.A., Mikhailovskii, E.M. and Gribencha, S.V. Epidemiology of rabies in the USSR between 1964 and 1971. Simpozium Po Beshenstvu (Symp. Rabies) 1972:79–82.

Servas, V., Mailles, A., Neau, D., Castor, C., Manetti, A., Fouquet, E., Ragnaud, J.-M., Bourhy, H., Paty, M., Melik, N., Astoul, J., Cliquet, F., Moiton, M., Francois, C., Coustillas, M., Minet, J. C., Parriaud, P., Capek, I. and Filleul, L. An imported case of canine rabies in Aquitaine: investigation and management of the contacts at risk, August 2004–March 2005. Eurosurveillance 2005; 10(11):222–225.

Shill, M., Baynes, R.D. and Miller, S.D. Fatal encephalitis despite appropriate post-exposure prophylaxis. A case report. N. Engl. J. Med. 1987; 316:1257–1258.

Shope, R.E., Murphy, F.A., Harrison, A.K., Causey, O.R., Kemp, G.E., Simpson, D.I.H. and Moore, D.L. Two African viruses serologically and morphologically related to rabies virus. J. Virol. 1970; 6:690–692.

Slate, D., Rupprecht, C.E., Rooney, J.A., Donovan, D., Lein, D.H. and Chipman, R.B. Status of oral rabies vaccination in wild carnivores in the United States. Virus Res. 2005; 111:68–76.

Smith, J.S. Rabies virus epitopic variation: use in ecologic studies. Adv. Virus Res. 1989; 36:215–307.

Smith, J., McElhinney, L.M., Parsons, G., Brink, N., Doherty, T., Agranoff, D., Miranda, M.E. and Fooks, A.R. Case report: rapid antemortem diagnosis of a human case of rabies imported into the UK from the Philippines. J. Med. Virol. 2003; 69:150–155.

Smith, J.S., Orciari, L.A., Yager, P.A., Seidel, H.D. and Warner, C.K. Epidemiologic and historical relationships among 87 rabies virus isolates determined by limited sequence analysis. J. Infect. Dis. 1992; 166:296–307.

Solomon, T., Marston, D., MacPherson, M., Felton, T., Shaw, S., McElhinney, L.M., Das, K., Mansfield, K., Wainwright, J., Kwong, G.N.M. and Fooks, A.R. Paralytic rabies following a 2 week holiday in India. BMJ 2005; 331:469–470.

Srinivasan, A., Burton, E.C., Kuehnert, A.J., Ruprecht, C., Sutker, W.L., Ksiazek, T.G., Paddock, C.D., Guarner, J., Shieh, W.J., Goldsmith, C., Hanlon Anlon, C.A., Zoretic, J., Fischbach, B., Niezgoda, M., El-Feky, W.H., Orciari, L., Sanchez, E.Q., Likos, A., Klintmalm, G.B., Cardo, D., Leduc, J., Chamberland, M.E., Jernigan, D.B. and Zaki, S.R. Rabies in Transplant Recipients Investigation Team Transmission of rabies virus from an organ donor to four transplant recipients. N. Engl. J. Med. 2005; 352:1103–1111.

Stewart, W.E., II and Sulkin, S.E. Interferon production in hamsters experimentally infected with rabies virus. Proc. Soc. Exp. Biol. Med. 1966; 123:650–654.

Sugamata, M., Miyazawa, M., Mori, S., Spangrude, G.J., Ewalt, L.C. and Lodmell, D.L. Paralysis of street rabies virus-infected mice is dependent on T-lymphocytes. J. Virol. 1992; 66:1252–1260.

Swabe, J. Chapter 22—folklore, perceptions, science and rabies. In: Historical Perspective of Rabies in Europe and the Mediterranean Basin (A.A. King, A.R. Fooks, M. Aubert and A.I. Wanderler, Eds.), pp. 311–322. OIE Publications, Paris, France, 2004.

Tacal, J.V., Jr. and Boado, P.A., Jr. Examination for negri bodies of the brains of a species of Philippine bat (*Scotophilus temminckii* Horsfield). Acta Med. Philipp. 1966; 3:103–106.

Tang, X., Luo, M., Zhang, S., Fooks, A.R., Hu, R. and Tu, C. Pivotal role of dogs in rabies transmission, China. Emerg. Infect. Dis. 2005; 11(12):1970–1972.

Theerasurakarn, S. and Ubol, S. Apoptosis induction in brain during the fixed strain of rabies virus infection correlates with onset and severity of illness. J. Neurovirol. 1998; 4:407–414.

Thoulouze, M.I., Lafage, M., Schachner, M., Hartmann, U., Cremer, H. and Lafon, M. The neural cell adhesion molecule is a receptor for rabies virus. J. Virol. 1998; 72:7181–7190.

Tsiang, H. Pathophysiology of rabies virus infection of the nervous system. Adv. Virus Res. 1993; 42:375–412.

Tuffereau, C., Benejean, J., Blondel, D., Kieffer, B. and Flamand, A. Low-affinity nerve-growth factor receptor (p75NTR) can serve as a receptor for rabies virus. EMBO J. 1998; 17:7250–7259.

Turner, G.S. Rabies vaccines and interferon. J. Hyg. 1972; 70:445–453.

Velasco-Villa, A., Orciari, L.A., Souza, V., JuÃrez-Islas, V., Gomez-Sierra, M., Castillo, A., Flisser, A. and Rupprecht, C.E. Molecular epizootiology of rabies associated with terrestrial carnivores in Mexico. Virus Res. 2005; 111:13–27.

Wandeler, A.I., Capt, S., Kappeler, A. and Hauser, R. Oral immunization of wildlife against rabies: concept and first field experiments. Rev. Infect. Dis. 1988; 10(Suppl.):S649–S653.

Wandeler, A.I., Nadin-Davis, S.A., Tinline, R.R. and Rupprecht, C.E. Rabies epidemiology: some ecological and evolutionary perspectives. Curr. Top. Microbiol. Immunol. 1994; 187:297–324.

Wang, Z.W., Sarmento, L., Wang, Y., Li, X.-Q., Dhingra, V., Tseggai, T., Jiang, B. and Fu., Z.F. Attenuated rabies virus activates, while pathogenic rabies virus evades, the host innate immune responses in the central nervous system. J. Virol. 2005; 79:12554–12565.

Warrell, M.J. and Warrell, D.A. Rabies and other lyssavirus diseases. Lancet 2004; 363:959–969.

Wilde, H., Sirikawin, S., Sabcharoen, A., Kingnate, D., Tantawichien, T., Harischandra, P.A., Chaiyabutr, N., de Silva, D.G., Liyanage, J.B. and Sitprija, V. Failure of postexposure treatment of rabies in children. Clin. Infect. Dis. 1996; 22:228–232.

Willoughby, R.E., Jr. A cure for a rabies? Sci. Am. 2007; 296:88–95.

Willoughby, R.E., Jr., Tieves, K.S., Hoffman, G.M., Ghanayem, N.S., Amlie-Lefond, C.M., Schwabe, M.J., Chusid, M.J. and Rupprecht, C.E. Survival after treatment of rabies with induction of coma. N. Engl. J. Med. 2005; 352:2508–2514.

Winkler, W.G., Baker, E.F. and Hopkins, C.C. An outbreak of non-bite transmitted rabies in a laboratory animal colony. Am. J. Epidemiol. 1972; 95:267–277.

World Health Organization. WHO expert consultation on rabies. WHO technical Report Series 931. 1st report, Geneva, Switzerland, 2005.

Wunner, W.H., Larson, J.K., Dietzschold, B. and Smith, C.L. The molecular biology of rabies viruses. Rev. Infect. Dis. 1988; 10(Suppl. 4):S771–S781.

Yarosh, O.K., Wanderler, A.I., Graham, F.L., Campbell, J.B. and Prevec, L. Human adenovirus type 5 vectors expressing rabies glycoprotein. Vaccine 1996; 14:1257–1264.

Yusibov, V., Modelska, A., Steplewski, K., Agadjanyan, M., Weiner, D., Hooper, D.C. and Koprowski, H. Antigens produced in plants by infection with chimeric plant viruses immunize against rabies virus and HIV-1. Proc. Natl. Acad. Sci. U.S.A. 1997; 94:5784–5788.

CHAPTER

34

Ross River Virus: Epidemic Polyarthritis

John Aaskov

Institute of Health and Biomedical Innovation; World Health Organisation, Collaborating Centre for Arbovirus Reference and Research, Queensland University of Technology, Brisbane, Australia

OUTLINE

ABSTRACT

Ross River virus is a mosquito-borne Alphavirus which is endemic in Australia and Papua New Guinea and which causes approximately 5000 cases of a disease known as epidemic polyarthritis each year. Horses are the only hosts, other than humans, known to develop clinical signs or symptoms following a natural infection. The dramatic increase in the number of cases reported between 1980 and 1990, and thereafter, reflects the development and commercialization of ELISA-based tests to diagnose this disease and the requirement for it to be notified to health authorities. Patients develop arthritis in the small joints of the hands and feet and also may experience fever, rash, and a range of other nonspecific signs and symptoms. Disability lasts from a few weeks to a year or more. The disease has been estimated to cost Australia more than US$10 million each year in direct medical costs. The epidemic potential of the virus was demonstrated in 1979–1980 when it spread to Fiji, New Caledonia, Samoa, Tonga, and the Cook Islands causing thousands of clinical infections. Interestingly, this outbreak by-passed New Zealand. After an interval of 25 years, RRV transmission was detected again in Fiji in 2004. Isolated cases of epidemic polyarthritis also are diagnosed among visitors returning from Australia to Europe and to areas of the southern United States where there are mosquitoes which are able to transmit the virus. Vector control is an expensive and inefficient mechanism for controlling this disease because of the wide range of mosquito and vertebrate hosts for RRV and the wide range of different habitats employed by these hosts. Natural infection in humans appears to provide lifelong immunity against a second clinical infection so this virus appears to be an ideal target for a vaccine. Experimental animals immunized with RRV inactivated with binary ethyleneimine or with formalin developed no adverse signs or symptoms and did not develop a viremia when challenged intravenously with large doses of virus. The titers of neutralizing antibody elicited by these killed vaccines in experimental animals were similar to those in humans following a natural infection.

INTRODUCTION

Ross River virus (RRV) causes approximately 5000 cases of epidemic polyarthritis in Australia and an unquantitated number in Papua New Guinea each year. While the emergence of epidemic polyarthritis in Australia is probably a reflection of the emergence of a capacity by laboratories in this country to diagnose RRV infection in a simple and reliable manner, the epidemic of RRV infection in the Pacific in 1979–1980 was a clear demonstration of the epidemic potential of this virus in nonimmune human populations. It remains to be seen whether the apparent absence of RRV from most areas of the Pacific involved in the 1979–1980 outbreak has allowed a large enough population of nonimmune hosts to develop for another outbreak to occur in this region. There also may be value in monitoring the return of epidemic polyarthritis patients to areas of the United States where the local mosquitoes are known to be competent to transmit the virus. RRV infection is difficult to prevent through vector control programs alone because of the large number of potential hosts and mosquito vectors. However, natural infection with this virus appears to elicit lifelong immunity against clinical disease.

HISTORY OF THE DISEASE

Shope and Anderson (1960) recognized that the outbreaks of polyarthritis that had been observed throughout eastern Australia since 1928 (Nimmo, 1928; Halliday and Horan, 1943; Harris, 1944; Sibree, 1944; Dowling, 1946; Goswell, 1946; Short, 1949; Anderson and French, 1957; Fuller and Warner, 1957; Wilson, 1957) resembled that in Tanganyika (Tanzania) in 1952 (Robinson, 1955) from which one of Casals' (Casals, 1957) Group A arboviruses (Chikungunya) had been isolated (see Chapter 32). Using a panel of Group A arboviruses (Semliki Forest, Mayaro, Chikungunya, Bebaru, and Getah) to test acute and convalescent sera from patients who developed rash and polyarthritis during an outbreak of "epidemic polyarthritis" in Mildura and surrounding towns in the Murray Valley (Fig. 34.1) in 1956, Shope and Anderson (1960) were able to show diagnostic rises in hemagglutination inhibiting (HI) antibody titers to Semliki Forest virus (three patients) and Bebaru virus (two additional patients). A diagnostic rise in neutralizing antibody titers against Bebaru virus also was detected in four of these five patients.

Anderson et al. (1961) then detected antibody that neutralized Bebaru virus in sera from humans from Queensland, Victoria, the Northern Territory, and Papua New Guinea and in horses, pigs, and cattle from throughout Queensland. Stanley and Choo (1964) also detected antibody that reacted with Bebaru virus in human sera from central and Western Australia and in horses from south-west Western Australia. Together, these studies suggested that epidemic polyarthritis was caused by a Group A arbovirus (Alphavirus) and that infection with this (these) agents was widespread in Australia.

Isolation of RRV by Doherty et al. (1963) enabled serological tests to be used to confirm that it, rather than Bebaru virus, was the causative agent of epidemic polyarthritis (Doherty et al., 1963, 1964).

FIGURE 34.1 Locations in Australia mentioned in the text.

Subsequent serological studies have found the subclinical infection rates with RRV in humans to vary from approximately 0.5% per annum in southern Australia (Echuca-Melbourne Collaborative Group, 1976; Fraser et al., 1986) to 1.5% per annum, and greater, in northern Australia (Doherty et al., 1966; Aaskov et al., 1981e).

ETIOLOGICAL AGENT

Classification

Family: Togaviridae. Genus: *Alphavirus* (Fig. 34.2).

Doherty et al. (1963) used suckling mice to isolate the prototype strain of RRV (T48) from a pool of 88 female *Aedes vigilax* mosquitoes which had been collected in a mangrove habitat near Ross River in Townsville in northeastern Australia between January 27 and February 6, 1959.

Although Doherty et al. (1972) recovered RRV from a child with headache, fever, and chills in 1971, using suckling mice, the first isolate of RRV from an epidemic polyarthritis patient was made, again using suckling mice, by Mataika and Miles during the outbreak of RRV infection in Fiji in 1979 (Aaskov et al., 1981c). Subsequently, Tesh et al. (1981) recovered a single isolate of RRV from epidemic polyarthritis patients in Samoa and Rosen et al. (1981) recovered 49 isolates of RRV from acute phase sera from epidemic polyarthritis patients in the Cook Islands and demonstrated viremia in these patients for up to a week following onset of symptoms. Despite more than two decades of attempts to isolate RRV from epidemic polyarthritis patients in Australia using a variety of avian and mammalian systems, the first isolates were not made until 1983—using C6-36 mosquito cells (Aaskov et al., 1985). Subsequent experience in Australia suggested that virus could be recovered from approximately 10% of acute phase seronegative serum samples from epidemic polyarthritis patients using cultures of C6-36 cells. This rate is lower than that reported by Rosen et al. (1981) in the Cook Islands but few Australian patients present as early after onset of symptoms as did those in the Cook Islands.

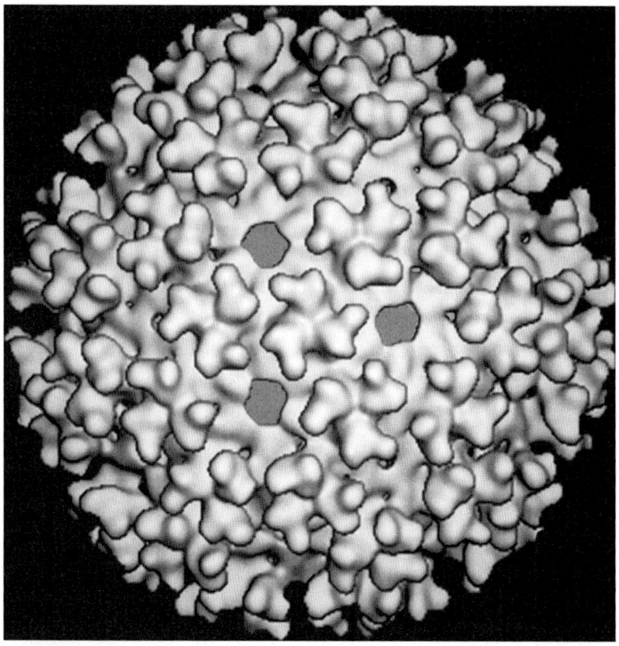

FIGURE 34.2 Cryo-EM reconstruction of RRV showing the clusters of trimeric E1-E2 heterodimers over the surface of the virion. The dark areas in the centre of the virion are regions in which the lipid envelope is exposed.

Shope and Anderson (1960) provided the first serological evidence that the causative agent of epidemic polyarthritis was a "Group A" arbovirus and the subsequent serological studies with RRV (Doherty et al., 1963, 1964) confirmed its classification in this family. However, it was not until the studies by Martin et al. (1979) and Richardson et al. (1980) and the limited nucleotide sequencing by Dalgarno et al. (1983) that there was biochemical evidence to support the subsequent classification of RRV as an *Alphavirus* in the family Togaviridae.

The genomes of the prototype strain of RRV (T48) and that of a mouse avirulent strain (NB5092) have been sequenced (Faragher et al., 1988).

By analogy with other alphaviruses, the genome of RRV (Faragher et al., 1988) is believed to code for four nonstructural proteins (nsP1-4) and five structural proteins (capsid [C], envelope protein 3 [E3], envelope protein 2 [E2], 6K protein, and envelope protein 1 [E1]) although only C, E2, and E1 can be detected in mature virions. RRV falls into the subgroup of Alphaviruses with an OPAL stop codon (UGA) at the 3′ end of the nsP3 protein. However, recent studies with O'nyong-nyong (Lanciotti et al., 1998) and RRVes (Aaskov, unpublished) suggest that Alphavirus populations may be polymorphic at this site having genomes with the OPAL codon and genomes with a codon for Arginine (CGA).

Cryo-electron microscopic reconstructions of the RRV virion (Cheng et al., 1995; Smith et al., 1995; Mukhopadhyay et al., 2002; Zhang et al., 2005) (Fig. 34.2) revealed a genome encased in a nucleocapsid composed of 240 copies of the capsid protein in a $T = 4$ icosahedral lattice and surrounded by an host cell-derived lipid bilayer approximately 40Å thick. Anchored in the lipid bilayer were 80 spikes each composed of three E1-E2 envelope protein heterodimers.

There are no reports of the E3 protein being detected in mature RRV virions. Amino acid residues around E218 of the E2 protein may play a role in attachment to host cells (Kerr et al., 1993; Heil et al., 2001) and the region ^{80}VYTGVUPFMWGGAYCFCD^{97} in the E1 glycoprotein of RRV shows extensive homology with the same region of other alphaviruses and with other viral proteins which are involved in fusion of virions with host cell membranes (Strauss and Strauss, 1994) and may play a similar role in the infection of cells by RRV.

Antigens Encoded by the Agent

While all the structural and nonstructural proteins of RRV are likely to be immunogenic, evidence of humans producing antibodies, or cell-mediated responses, against most of them is lacking.

Sera from epidemic polyarthritis patients and from experimental animals infected with RRV react with the E1 and E2 structural proteins (Kistner et al., 2007). IgG antibody in serum from epidemic polyarthritis patients reacts with both reduced and nonreduced RRV E1 and E2 proteins in western blots, albeit more strongly with the nonreduced proteins. IgM antibody from epidemic polyarthritis patients reacts with nonreduced E1 and E2 proteins in western blots, but rarely, if at all with reduced proteins. The reaction of human anti-RRV IgM with the E2 protein invariably is stronger than the reaction with E1 (unpublished observations). Sera from experimentally infected animals react with the C protein as well as E1 and E2 (Kistner et al., 2007) while that from patients has not been observed to react with C (unpublished observations). There have been no systematic studies of the reaction of sera from patients with nonstructural proteins.

As with other alphaviruses, many of the determinants involved in antibody-mediated neutralization, cell tropism, and mouse virulence have been identified in the E2 glycoprotein of RRV (Kerr et al., 1992, 1993; Meek et al., 1989; Vrati et al., 1986, 1988, 1996). Three, and possibly four, epitopes recognized by monoclonal antibodies which neutralized infection of RRV in vitro have been identified in the E2 glycoprotein of RRV. These involve amino acids at positions E4 (Kerr et al., 1993), E216, E232-234 and E246-251 (Vrati et al., 1988). However, polyclonal sera raised against peptides corresponding to the three putative epitopes between amino acids 216 and 251 of the RRV E2 protein failed to neutralize virus in vitro (Kerr et al., 1992). One possible explanation for these data was that the epitopes recognized by the neutralizing monoclonal antibodies in the RRV E2 protein were conformational or discontinuous and that the polyclonal sera raised against the linear peptides bound to virions with insufficient affinity to neutralize infection.

It is not known whether sera from humans recognize the same epitopes in the E2 protein as these neutralizing monoclonal antibodies but, in an epitope blocking ELISA, Oliveira et al. (2006) found sera from epidemic polyarthritis patients, which inhibits the binding of an anti-E2 mouse monoclonal antibody to Ross River virions. No serological epitopes have been identified in the E1 glycoprotein despite the extensive serological responses to this protein by both humans and experimentally infected animals.

Class switching of anti-RRV antibodies from IgM to IgG occurs readily in both humans and experimental animals indicating the virus is eliciting an effective T_H (CD4) response. Furthermore, Fraser and Becker (1984) observed predominantly CD4 lymphocytes in the synovial effusion of epidemic polyarthritis patients, although the role or function of these cells could not be identified. In contrast, these authors (Fraser et al., 1983) reported predominantly CD8 lymphocytes in the exanthem associated with many RRV infections. Again, neither the function nor the epitopes recognized by these cells could be determined.

An H-2K^d restricted T-cell epitope (QYSGGRFTI) has been identified in the capsid protein of RRV (Linn et al., 1998). Murine CD8 T-cells recognizing this epitope were able to destroy RAW264.7 cells (a murine macrophage cell line) infected with RRV in vitro but had no effect on viremia in mice.

PROTECTIVE IMMUNE RESPONSE

Epidemic polyarthritis patients do not develop a second clinical infection and in a prospective serological study (Aaskov et al., 1981e) a fourfold rise in preexisting anti-RRV antibody titers was observed in several participants suggesting they had experienced a subclinical anamnestic infection. While these observations suggest a natural infection elicits a lifelong immunity, they do not identify which arm(s) of the immune response are responsible for this protection. Also lacking are any data for the outcome of RRV infections in humans who are immunosuppressed or who have defined immunodeficiencies.

Early laboratory studies (Doherty et al., 1963) in which mixtures of RRV and antibody were injected into newborn mice demonstrated that antibody alone could prevent infection in vivo—at least in experimental animals. There have been no definitive studies which demonstrate that T cells (CD4 or CD8) alone can protect against an infection with RRV.

FIGURE 34.3 Geographic distribution of confirmed transmission of RRV.

EPIDEMIOLOGY

Although the rate of infection of humans with RRV at any given locality appears to be independent of age (Doherty et al., 1966; Echuca-Melbourne Collaborative Group, 1976), clinical disease in children is rare (Aaskov et al., 1981c) and when it does occur, arthritis is mild or nonexistent (Doherty et al., 1972; unpublished observations) (Fig. 34.3). The incidence of disease in those over 40 years of age decreases, presumably due to immunity elicited by previous subclinical and clinical infections. There have been no confirmed reports of an epidemic polyarthritis patient developing a second clinical infection and there is no other epidemiological evidence that the antibody-dependent enhancement of monocyte/macrophage cell lines by RRV observed in vitro (Linn et al., 1996; Lidbury and Mahalingam, 2000) plays any role the epidemiology of RRV infection.

While there was extensive serological evidence of human infection with RRV prior to 1979 (Doherty, 1972), disease appeared to occur as outbreaks in geographically discrete regions. In areas where epidemic polyarthritis was endemic, almost all cases were reported following the summer rains (December–April)—new cases in the July–October interval were rare (Doherty et al., 1971). Subsequent expansion of RRV diagnostic facilities and epidemic polyarthritis becoming a notifiable disease in 1990 showed this to be an incomplete picture of human disease (Fig. 34.4).

The detection of anti-RRV antibodies in a wide range of native and domestic animals raised the possibility of a zoonotic cycle of transmission of RRV with humans as an incidental host. RRV has been recovered from asymptomatic native birds (*Grallina cyanoleuca* and *Microeca fascinans*; Whitehead et al., 1968) and horses (Pascoe et al., 1978), and from horses with musculoskeletal disease (Azoulas et al., 2003). Horses and little corellas (*Cacatua sanguinea*) have been shown to develop viremias sufficient to infect *Culex annulirostris* mosquitoes (Kay et al., 1986). *Aedes aegypti* (Taylor and Marshall, 1975) and *Aedes polynesiensis* (Gubler, 1981) mosquitoes could be infected by feeding on viremic baby laboratory mice, and *Aedes vigilax* mosquitoes

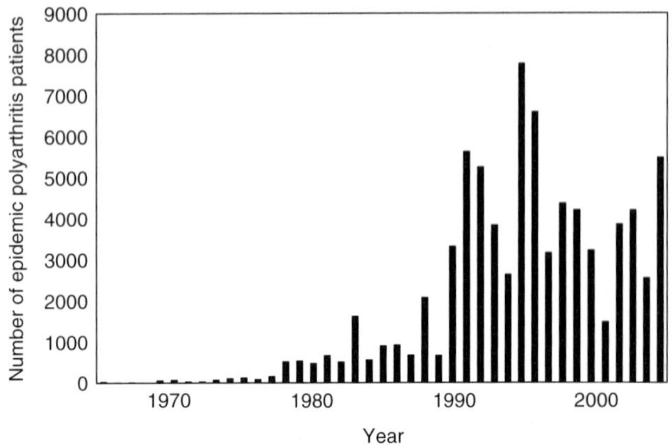

FIGURE 34.4 Number of cases of epidemic polyarthritis reported each year in Australia from 1966 to 2006. Notification became compulsory in 1990.

could be infected by feeding on viremic gray-headed flying foxes (*Pteropus poliocephalus*) (Ryan et al., 1997). A range of other native, domestic, and feral animals have been shown to develop viremia following experimental infection with RRV (Whitehead, 1969; Marshall and Miles, 1984; Kay and Aaskov, 1989) but they have not been assessed for their capacity to infect known vectors of RRV.

Although, RRV has been isolated from more than 30 species of mosquitoes (Russell, 1995, 2002), the competence of most to transmit this virus has not been assessed. The logistics of these studies are magnified by the need to assess the competence of mosquito species collected from different localities to transmit a range of RRV genotypes (Sammels et al., 1995).

Two events have directed attention to the possibility that RRV may be able to maintain itself in human–mosquito–human cycles. The first was the epidemic of RRV infection in the Pacific in 1979–1980 and the second was the realization, with the provision of comprehensive diagnostic RRV serology and national reporting of cases of epidemic polyarthritis that human infection occurred year round and that many patients were urban dwellers (Aaskov et al., 1981e) who lived nowhere near significant numbers of most of the native animals proposed as likely amplifying hosts for RRV.

Although there are anecdotal accounts of the first cases of epidemic polyarthritis appearing in the Murray Valley with the arrival, each year, of sugarcane cutters from north Queensland to pick stone fruit, the first definitive evidence of movement of RRV over significant distances came with the outbreak of RRV infection in the Pacific in 1979–1980. Evidence to support the concept that the virus was introduced into the Pacific with a viremic human came from the subsequent reports of tourists developing epidemic polyarthritis within days of returning to Australia from Fiji (Bennett et al., 1980).

The first cases of epidemic polyarthritis were recognized, in Fiji, in late April 1979 near Nadi International Airport. Cases then began to appear throughout the island of Viti Levu, peaking in late May or early June. Cases subsequently appeared on the second largest island, Vanua Levu, and continued for more than a year with virus being isolated from two patients from Labasa in January 1981 (Marshall and Miles, 1984). In August 1979, epidemic polyarthritis appeared in American Samoa and new cases continued to be recognized until at least January 1980 (Tesh et al., 1981). By November 1979, cases of epidemic polyarthritis were recognized in the Wallis and Futuna Islands (Fauran et al., 1984) and in January 1980 they were being reported from Tonga (Kay, 1981), the Cook Islands (Rosen et al., 1981), and New Caledonia (Fauran et al., 1984). Serological testing of preepidemic sera from Fiji (Aaskov et al., 1981c) suggested RRV transmission may have occurred there in the years preceding the outbreak but was not sustained. Similar testing of sera from Western Samoa suggested that RRV infection also may have occurred there prior to 1980 although no cases of epidemic polyarthritis were recognized (Marshall and Miles, 1984). However, in the absence of travel histories for the individuals involved, it is possible they were infected with RRV during a visit to Australia.

Taken together, these data suggested that RRV employs multiple host–vector cycles, each adapted to a particular ecological niche. The observation by Hawkes et al. (1985) that the incidence of epidemic polyarthritis could vary almost 10-fold between two very similar communities (237 per 10^5 residents in Forbes and 27 per 10^5 residents in Parkes), only 30 km apart, during the same outbreak, in the same geophysical region of New South Wales, suggested that these niches may be quite small. It seems possible that multiple niches could exist even within a single large metropolitan center. It also is likely that these niches are dynamic environments which constantly change in response to human (development, habitation, vector control programs) or natural (tides, temperature, rain, wind) intervention.

Significance as a Public Health Problem

Approximately 5000 cases of epidemic polyarthritis are reported in Australia each year (Fig. 34.4) and the direct medical costs of the disease have been estimated to be US$ 10 million annually (Aaskov et al., 1998). Given the scores of possible vertebrate hosts and the scores of potential mosquito vectors, vector control programs can hope only to reduce the number of cases of epidemic polyarthritis in urban areas. Efforts to motivate individuals to take adequate personal protective measures have fallen, largely, on deaf ears.

The cost of vector control programs designed to limit this disease have not been quantitated and it is probably unreasonable to do so because a significant motivation for mosquito control programs in Australia is nuisance reduction rather than disease prevention. The social costs of epidemic polyarthritis have not been quantitated.

No estimate has been made of the cost of the epidemic of RRV infection in 1979–1980, which resulted in tens of thousands of cases of clinical disease across the Pacific (Aaskov et al., 1981c; Rosen et al., 1981; Tesh et al., 1981), or of the disease burden in Papua New Guinea (Scrimgeour et al., 1987). Each year small numbers of tourists from Europe and the US also develop epidemic polyarthritis after visiting Australia as do foreign military personnel who participate in exercises in northern Australia.

Potential as a Biothreat

While RRV would not appear to have potential as a conventional biological weapon, its remarkable capacity to cause epidemics, as demonstrated during the outbreak in the Pacific in 1979–1980 (Aaskov et al., 1981c; Rosen et al., 1981; Tesh et al., 1981; Kay, 1981), marks this virus as a biothreat. The recent reports of cases of epidemic polyarthritis originating in Fiji (Klapsing et al., 2005; and unpublished) point to the possibility of a repeat of the earlier epidemic in the Pacific. The ability of rapid air travel to put viremic humans into the southern states of the United States where mosquitoes have been shown to be competent to transmit this virus (Mitchell et al., 1987) suggest RRV also poses a threat to that region.

CLINICAL DISEASE

The term "epidemic polyarthritis" was coined by Dowling (1946) and now is used to describe the disease caused by RRV in humans. The recent use by some of the term "epidemic polyarthritis" to describe infections caused by other mosquito-borne viruses is without foundation and causes confusion.

The interval from infection with RRV to onset of symptoms has been estimated to be from 3 to 21 days (Anderson and French, 1957; Fraser and Cunningham, 1980; Rosen et al., 1981). Disease occurs most commonly in adults (Mudge and Aaskov, 1983; Hawkes et al., 1985; Curran et al., 1997) with symptoms being milder and of shorter duration in children. There is a decreasing incidence of disease with age, presumably due to many long-term residents being immune as a result of previous clinical or subclinical infections (Aaskov and Doherty, 1994). The ratio of subclinical to clinical RRV infections varies from approximately 20:1 in areas like Queensland where infection is endemic to 4:1 (in Fiji; Aaskov et al., 1981c) or even lower (1:2 in Griffith, New South Wales; Hawkes et al., 1985) during outbreaks of disease. Those with the HLA DR7 haplotype are 3.3 times more likely to develop clinical disease following an infection with RRV than those with other haplotypes, while HLA DR3 appeared to provide significant protection against clinical disease (Fraser et al., 1980).

Epidemic polyarthritis is characterized by arthritis, particularly in the knees and the small joints of the hands and feet and may be accompanied by a maculopapular rash and or fever/chills. Patients also may develop a range of other nonspecific signs and symptoms (e.g., paresthesia, lethargy, depression, headache, nausea; Aaskov et al., 1981c; Mudge and Aaskov, 1983; Hawkes et al., 1985; Fraser, 1986; Condon and Rouse, 1995; Seldon and Cameron, 1996; Harley et al., 2001; Mylonas et al., 2002).

The clinical presentation of what is believed to have been epidemic polyarthritis appears to have changed with both time and location (Keary et al., 1997). While to Nimmo (1928), Halliday and Horan (1943), Goswell (1946), and Shope and Anderson (1960) arthritis was the most noticeable feature of this disease, but to Dowling (1946), Weber (1946), Wilson (1957), and Fuller and Warner (1957) it was the rash. In a survey by Mudge and Aaskov (1983), rash was reported from 78% of epidemic polyarthritis patients from South Australia but from only 40% of patients from the remainder of the country. Even with careful observation, rash was detected in only 20% of Fijian patients (Aaskov et al., 1981c). Seglenieks and Moore (1974) reported that in the same area of South Australia rash was the most prominent feature of the 1956 outbreak of RRV infection while in 1971 it was arthritis. It is unclear whether the pathological processes leading to the rash are the same as those that lead to arthritis (see below). There is no evidence that different strains of RRV cause different symptoms, or symptoms of varying severity, in humans. Another feature of epidemic polyarthritis that appears to be changing is the duration of symptoms and, perhaps related to this, more frequent reports of lethargy and depression. It is difficult to quantitate these changes because of the differing methodology used to obtain the data but they might be influenced by diagnosis and reporting of more patients with milder disease now that laboratory confirmation is more readily available. While two early studies (Sibree, 1944; Goswell, 1946) reported hospitalization of significant numbers of patients for from 1 to 4 weeks, there were few reports of long-term sequelae. In contrast, hospitalization of epidemic polyarthritis patients is now uncommon but long-term sequelae are being reported more often. More recent studies suggest epidemic polyarthritis patients are absent from work for an average of 2 days (Mylonas et al., 2002) and unable to perform normal duties for, on average, 2–6 weeks (Hawkes et al., 1985; Condon and Rouse, 1995; Seldon and Cameron, 1996; Westley-Wise et al., 1996). Symptoms usually resolve over 3–6 months (Aaskov et al., 1981b, 1987; Harley et al., 2002; Mylonas et al., 2002) but some patients may have symptoms lasting a year or more (Fraser, 1986).

There is a body of evidence suggesting that strains of RRV T48 and NB5092 vary in their tropism for cardiac and neurological tissue (Mims et al., 1973; Murphy et al., 1973) but the issue of the tissue tropism of virus populations that have not been passaged extensively in mice has not been explored, and not in human tissues.

There also are isolated reports of involvement of the renal and central nervous systems in RRV disease. Davies et al. (1982) postulated on the basis of retrospective serology that RRV might cause segmental necrotizing glomerulonephritis but it was not until two later studies (Fraser et al., 1988b; Anstey et al., 1991) that a clearer association between RRV infection and glomerulonephritis and hematuria was made. Changes observed in the renal biopsy made by Fraser et al. (1988b) were compatible with a serological or immune-complex mediated pathogenesis. However, no viral antigen was detected in this tissue.

RRV is highly neurotropic in young mice (Doherty et al., 1963; Mims et al., 1973; Murphy et al., 1973) and possibly in horses (Gard et al., 1977), but evidence for infection of the human central nervous system has been less convincing. Perhaps the first such report was that of the death of a soldier of encephalitis in Katherine (Northern Territory) in 1942 or 1943, several months after developing what was believed to be epidemic polyarthritis (Short, 1984). Other tenuous examples include two from Queensland (Aaskov, 1979), one from the Northern Territory (Lucas and Qiao, 1999), one from Papua New Guinea (Scrimgeour, 1999), and one of a focal encephalitis in a 58-year-old male German tourist on return from a visit to Queensland. This patient had developed a febrile illness and prolonged arthritis while in Australia where a RRV infection was confirmed by serology. Eight weeks after onset of symptoms, he presented to a hospital in Germany complaining of weakness, disturbed temperature sensation, and paresthesia in his left forearm. Cerebrospinal fluid was normal as were routine blood tests and blood chemistry. A cerebal magnetic resonance tomography with gadolinium IV showed three high-density subcortical plaques. It was felt that one of these, in the right hemisphere, could have explained the transient neurological symptoms (Aaskov, unpublished data). Perhaps the strongest evidence of meningitis following a RRV infection was provided by Penna and Irving (1993). A patient with symptoms (fever, arthralgia) and serology (HI titer 80; anti-RRV IgM positive) compatible with a diagnosis of epidemic polyarthritis presented with the additional symptoms of headache and neck stiffness during the 1993 outbreak of RRV infection in South Australia. Virus isolation could not be attempted but cerebrospinal fluid contained 100 leukocytes/μl consisting of 80% lymphocytes and 20% neutrophils. No fungal or bacterial pathogens could be cultured and no serological evidence of infection with other pathogens was obtained.

Just as RRV can be extremely neurotropic in young mice but rarely causes central nervous system symptoms in humans, it readily crosses the placenta of pregnant mice (Aaskov et al., 1981a; Milner and Marshall, 1984) but rarely does so in humans. Following the outbreak of RRV infection in Fiji in 1979, cord blood from approximately 5% (11/198) of children born to mothers who were 11–19 weeks pregnant when infected, contained anti-RRV IgM antibody suggesting in utero exposure to virus or viral antigen (Aaskov et al., 1981d). These children, with anti-RRV IgM in their cord blood, appeared normal at birth and following a physical examination several years later. In a smaller retrospective study of children born to women in the Cook Islands who were infected with RRV during the first 21–22 weeks of pregnancy (39 children), Aleck et al. (1983) found no evidence of in utero infection nor of increased rates of abnormalities in children born to infected mothers. However, the sample size (39 subjects) in the study by Aleck et al. (1983) was too small to detect an in utero infection, if the incidence of transplacental transmission of virus was similar to that reported by Aaskov et al. (1981d) in Fiji.

TREATMENT

The results of the only formal trial of treatment for epidemic polyarthritis suggested recovery was faster in patients taking corticosteroids (Mylonas et al., 2004). However, most clinicians employ symptomatic treatment with nonsteroidal anti-inflammatory medication (Fraser, 1986). A number of patients with prolonged, severe, arthritis have gained relief when taking prednisolone.

PATHOGENESIS

Titers of anti-RRV antibodies and of autoantibodies do not correlate with the presence or duration of arthritic symptoms of epidemic polyarthritis and complement activation has not been detected in the joints of patients (Clarris et al., 1975; Aaskov et al., 1981b, 1981c; Fraser et al., 1981, 1987, 1988a). Together with the predominantly mononuclear nature of the inflammatory response in the arthritic joints of these patients (see below), these observations point to a cell-mediated rather than an antibody-mediated pathogenesis for this disease.

As outlined above, the feature common to all epidemic polyarthritis patients is arthritis in the small joints, particularly the hands and feet. Biopsies of joint synovium from a small number of patients revealed mild hyperplasia, mononuclear cell infiltration, and some fibrin deposition (Hazelton et al., 1985; Soden et al., 2000). The cell infiltrate consisted of approximately equal numbers of monocytes/macrophages and lymphocytes. Both B and T lymphocytes were detected with approximately 60% of the T lymphocytes being $CD4^+$. The cell exudate in the joints of epidemic polyarthritis patients consisted predominantly of mononuclear leukocytes with reports of the relative proportions of macrophages and lymphocytes in the exudate varying extensively (Halliday and Horan, 1943; Clarris et al., 1975; Fraser et al., 1981; Fraser and Becker, 1984; Hazelton et al., 1985). The lymphocyte populations in the exudate consisted of B lymphocytes, natural killer (NK) cells and $CD4^+$ and $CD8^+$ T lymphocytes (Fraser and Becker, 1984; Hazelton et al., 1985).

T lymphocytes from the peripheral blood of epidemic polyarthritis patients proliferated in a virus-specific manner when cultured with purified RRV in vitro (Aaskov et al., 1981b). However, the interval over which lymphocytes recovered from patients would proliferate when reexposed to virus in vitro did not correlate with the duration of symptoms. A better, but inverse, correlation was observed between peripheral blood NK cell activity and symptoms (Aaskov et al., 1987). NK cell activity was depressed soon after onset of symptoms and remained depressed for as long as arthritis persisted, months after virus could no longer be detected in joints or in the peripheral blood. Significant depression of NK cell activity also occurred when patients had severe relapses of arthritis. In the study reported by Hazelton et al. (1985), functional NK cells were recovered from the knee of an epidemic polyarthritis patient and subsequently Aaskov et al. (1987) were able to demonstrate killing of uninfected autologous synovial cells by NK cells.

Although there are no reports of recovery of live virus from the joints of epidemic polyarthritis patients, Fraser et al. (1981) were able to detect viral antigen in mononuclear leukocytes in the exudate from the joints (knee or ankle) of five patients for up to 7 days after onset of symptoms. This is the same interval over which it has been possible to isolate RRV from the peripheral blood of patients (Rosen et al., 1981; Aaskov et al., 1985). Soden et al. (2000) were able to detect viral RNA in inflamed synovial tissue from 2 of 12 patients for up to 5 weeks after onset of symptoms but could not recover live virus from the joints of any of these patients.

Journeaux et al. (1987) were able to establish prolonged (up to 5 weeks), but not persistent, infections with RRV in human synovial cells in vitro by lowering the temperature of the cultures from 37°C to 35°C—the approximate temperature of small peripheral joints. This interval correlates reasonably well with the duration of intense arthritis experienced by many patients early after onset of symptoms (Fraser, 1986a; Seldon and Cameron, 1996; Westley-Wise et al., 1996). At the end of this interval, only fibroblast-like (type B) synovial cells remained in the virus-infected cultures. Superficially, these data could be taken as suggesting that the macrophage-like (type A) synovial cells supported virus replication and that when these poorly replicating cells had been destroyed, the infection terminated. However, we, and others (Cunningham and Fraser, 1985) have shown human fibroblast-like synovial cells to be able to be infected with RRV in vitro so it is difficult to understand how they survived exposure to RRV in this system. The levels of extracellular RRV produced in vitro by the primary synovial cell cultures (Journeaux et al., 1987) was very low as evidenced by its detection with the highly sensitive C6-36 mosquito cell line but not with Vero cells. The earlier failure of Cunningham and Fraser (1985) to detect prolonged infection of human synovial cells with RRV in vitro may have been due to their use of relatively insensitive virus detection assays.

Linn et al. (1996) and Way et al. (2002) have been able to establish persistent RRV infections in murine macrophage cell lines—but not in human macrophage/monocyte cell lines or in human tissues. Way et al. (2002) were able to detect viral RNA in their cells when levels of extracellular virus were undetectable by conventional cell culture techniques. Furthermore, they were able to manipulate culture conditions to induce these cells to again produce infectious extracellular virus—mirroring the relapses reported by some epidemic polyarthritis patients.

Although not all epidemic polyarthritis patients develop a rash, many of the histological changes associated with the rash mirror those seen in joints. A perivascular infiltrate of mononuclear leukocytes (monocytes/macrophages and CD8$^+$ T-cells) was observed in the dermis adjacent to the rash. RRV antigen was detected in this region as well as in basal epidermal and eccrine duct cells. No immunoglobulin or complement (Clq, C3) deposition was detected (Fraser et al., 1983).

While these data point to a possible role for T lymphocytes in the pathogenesis of epidemic polyarthritis, only one T-lymphocyte clone (CD4$^+$) was recovered after more than 2 years work by a group with more than a decade of experience producing T-cell clones from patients with other viral infections (Suhrbier, unpublished observations).

There is an extensive literature describing the pathogenesis of RRV infection in mice beginning with that of Mims et al. (1973) (reviewed by Suhrbier and Linn 2004 and Rulli et al., 2005). This points to a central role for monocytes/macrophages and cytokines/chemokines in the pathogenesis of epidemic polyarthritis. However, many of these observations remain to be confirmed in humans or in human tissue (Mateo et al., 1999).

VACCINES

Epidemic polyarthritis is an ideal target for a vaccine. There are no confirmed reports of an epidemic polyarthritis patient ever having a second clinical infection (i.e., immunity appears to be lifelong) and the limited genetic analyses which have been performed suggest there is very little genetic variation between isolates of RRV (Faragher et al., 1985; Sammels et al., 1995).

In the absence of any markers of attenuation of virulence of RRV for humans, a decision was made to attempt to develop a killed vaccine. Irradiation with ultraviolet light had been used previously to inactivate RRV (Aaskov and Tucker, 1979) while retaining critical epitopes recognized by IgG and IgM antibodies in sera from epidemic polyarthritis patients (Oseni et al., 1983). However, at that time, there was no equipment or facilities available that would have enabled this procedure to be undertaken on an industrial scale. The inactivation of viruses by formalin or beta-propiolactone has nonlinear kinetics which makes determination of the time required for inactivation of all virions difficult (Harrison et al., 1971; Cole et al., 1974; Barrett et al., 1984; Klockmann et al., 1989; Barteling and Vreeswijk, 1991). Furthermore, both these agents affect the antigenic structure of proteins. Formalin cross-links adjacent proteins and beta-propiolactone alters the chemical structure of amino acids such as methionine, cysteine, histidine, praline, and lysine (Taubman and Atassi, 1968; Boutwell et al., 1969). Inactivation of RRV with beta-propiolactone resulted in a significant reduction of binding of anti-RRV IgM antibody to virions in indirect ELISAs but had less effect on the attachment of IgG (unpublished observations).

RRV recovered from a child with epidemic polyarthritis and who had no evidence of other infectious or chronic diseases (Yu and Aaskov, 1994) was inactivated with binary ethylenimine (BEI), purified by centrifugation on a sucrose gradient, and evaluated as a vaccine. This treatment had no detectable effect on the binding to the BEI-inactivated virions of human polyclonal anti-RRV IgM or IgG. It has been our experience that recognition of RRV virions by IgM antibodies is extremely sensitive to any change in the conformation of the E1 and E2 glycoproteins. Furthermore, RNA extracted from the BEI-inactivated virus was at least one thousand times less infectious than RNA extracted from live virus, i.e., 1.5 μg RNA from the vaccine caused neither morbidity nor mortality when injected intracerebally into newborn mice (Aaskov et al., 1997).

Neither vaccine nor vaccine with Alhydrogel adjuvant caused any adverse reactions when administered intramuscularly to mice. Biopsies of the sites of injection revealed a strong inflammatory response but no significant tissue damage (Yu and Aaskov, 1994).

Mice immunized with one dose of vaccine elicited lower total antibody titers (indirect ELISA) than mice immunized with vaccine plus Alhydrogel adjuvant, but the levels of anti-RRV neutralizing antibodies in mice immunized with vaccine alone were higher than in mice given vaccine plus adjuvant. Furthermore, the levels of protection against in vivo challenge with live virus were higher in mice given one dose of vaccine alone than in those given vaccine plus adjuvant. Mice given two doses of 0.2–30 μg vaccine or two doses of 0.2–30 μg vaccine with adjuvant produced no detectable viremia when challenged intravenously with 10^5 TCID of RRV 75 days after the second immunization (Yu and Aaskov, 1994).

Because of an observed association between HLA DR7 and clinical disease following human infection with RRV and a claimed increased prevalence of clinical RRV infections in females, the responses of male and female, inbred and outbred, mice were evaluated. Levels of neutralizing antibodies were similar in all groups 7 weeks after two injections with 2 μg vaccine. While sera from these animals neutralized a panel of contemporary strains of RRV to varying degrees ($\log_{10}$ neutralization indices from 2.0 to >3.5), these titers are similar to those of former epidemic polyarthritis patients who are immune to a second clinical infection (see below).

Despite the promise of a BEI-inactivated RRV vaccine in animal experiments, BEI has not been employed for the production of a human vaccine and so there are no protocols to satisfy regulatory authorities of its safety or at least convince them that the BEI could be completely removed or converted to a harmless metabolite. Prior to committing to these processes, studies were conducted to determine whether or not standard formalin-inactivation protocols could be employed to produce a safe immunogenic RRV vaccine (Kistner et al., 2007).

The strain of RRV used to prepare the BEI-inactivated vaccine was grown serum-free in vaccine-certified Vero cells in commercial-scale fermenters, inactivated with formalin and any contaminating Vero cell DNA destroyed with Pronase. The virus was filtered through 1.2 and 0.2μm filters throughout the formalin treatment process and finally purified by centrifugation on a sucrose gradient. Examination of this material by electron microscopy revealed few examples of clumps of two or more virions. SDS-PAGE of the formalin-inactivated virus preparation revealed capsid (C) and envelope proteins (E1, E2) of anticipated sizes as well as several bands of protein of larger molecular weight which were recognized by polyclonal anti-RRV antisera. These were presumed to be oligomers of E1 and E2 cross-linked by formalin. Human anti-RRV IgM antibodies reacted as well with the E1 and E2 proteins from formalin-inactivated virus as they did with these proteins from native virus suggesting that the prolonged treatment with formalin had not damaged immunodominant epitopes. In contrast to the response to BEI-inactivated vaccine, formalin-inactivated RRV elicited both lower levels of antibody and lower levels of neutralization in vitro and in vivo than vaccine with $Al(OH)_3$ adjuvant.

Two doses of 0.625μg formalin-inactivated virus with adjuvant elicited titers of neutralizing antibodies ($\log_{10}$ 2.7) similar to those in humans who were immune following a natural infection with RRV ($\log_{10}$ 1.1–3.5, median 2.4).

Two different inactivation protocols have produced killed RRV vaccines that, at clinically relevant doses, elicited strong neutralizing responses in experimental animals with no apparent adverse effects, even after challenge with live virus. The virus grows to high titer in serum-free cultures in industrial scale fermenters and can be purified using conventional industrial protocols. A clinical trial of a formalin-inactivated, U.V.-irradiated, RRV vaccine is in progress. It is likely to be a cost-effective alternative to the burden of 5000 cases of epidemic polyarthritis in Australia each year and it also could have a place as a "tourist vaccine" and for military personnel coming to Australia for joint exercises. It also would be a very useful tool in the event of the introduction of RRV into the southern United States or if there is another outbreak of RRV infection in the Pacific.

KEY ISSUES

- Ross River virus causes approximately 5000 cases of epidemic polyarthritis in Australia each year and caused a major outbreak in Pacific countries in 1979–1980. Patients commonly are symptomatic for up to 9 months and the health costs associated with this disease are estimated at US$10 million each year.
- Because so many different species of mosquitoes are potential vectors of Ross River virus and the virus has so many vertebrate hosts, mosquito control programs are not cost effective and will not be able to prevent this disease.
- The virus grows to high titer in cell culture and retains immunogenicity when inactivated by a number of methods. This makes Ross River virus an ideal candidate for an inactivated vaccine.

References

Aaskov, J.G. RRV disease. In: Arbovirus Research in Australia. Proceedings of the Second Symposium (T.D. St. George and E.L. French, Eds.), Vol. 2, pp. 164–167. Brisbane: CSIRO, 1979.

Aaskov, J.G., Chen, J.Y., Hanh, T.H. and Dennington, P.M. Surveillance for RRV infection using blood donors. Am. J. Trop. Med. Hyg. 1998; 58:726–730.

Aaskov, J.G., Dalglish, D.A., Harper, J.J., Douglas, J.F., Donaldson, M.D. and Hertzog, P.J. Natural killer cells in viral arthritis. Clin. Exp. Immunol. 1987; 68:23–32.

Aaskov, J.G., Davies, C.E.A., Tucker, M. and Dalglish, D.A. Effect on mice of infection during pregnancy with three Australian arboviruses. Am. J. Trop. Med. Hyg. 1981a; 30:198–203.

Aaskov, J.G. and Doherty, R.L. Arboviral zoonoses of Australasia. In: Handbook of Zoonoses (W. Beran, Ed.), 2nd ed., pp. 301–317. Boca Raton, FL: CRC Press, 1994.

Aaskov, J.G., Fraser, J.R.E. and Dalglish, D.A. Specific and non-specific immunological changes in epidemic polyarthritis patients. Aust. J. Exp. Biol. Med. Sci. 1981b; 59:599–608.

Aaskov, J.G., Mataika, J.U., Lawrence, G.W., Rabukawaqa, V., Tucker, M.M., Miles, J.A.R. and Dalglish, D.A. An epidemic of RRV infection in Fiji, 1979. Am. J. Trop. Med. Hyg. 1981c; 30:1053–1059.

Aaskov, J.G., Nair, K., Lawrence, G.W., Dalglish, D.A. and Tucker, M. Evidence for transplacental transmission of RRV in humans. Med. J. Aust. 1981d; 2:20–21.

Aaskov, J.G., Ross, P., Davies, C.E.A., Innis, M.D., Guard, R.W., Stallman, N.D. and Tucker, M. Epidemic polyarthritis in north-eastern Australia, 1978–1979. Med. J. Aust. 1981e; 2:17–19.

Aaskov, J.G., Ross, P.V., Harper, J.J. and Donaldson, M.D. Isolation of RRV from epidemic polyarthritis patients in Australia. Aust. J. Exp. Biol. Med. Sci. 1985; 63:587–597.

Aaskov, J. and Tucker, M. Ultraviolet inactivation of RRV for use in haemagglutination inhibition studies. Aust. J. Med. Technol. 1979; 10:109–111.

Aaskov, J.G., Williams, L. and Yu, S. A candidate RRV vaccine: pre-clinical evaluation. Vaccine 1997; 15:1396–1404.

Aleck, K.A., Rosen, L., Pettitt, D.J., Boveington, C. and Bennett, P.H. Absence of intrauterine infection following RRV infection during pregnancy. Am. J. Trop. Med. Hyg. 1983; 32:618–620.

Anderson, S.G., Doherty, R.L. and Carley, J.G. Epidemic polyarthritis: antibody to a group A arthropod borne virus in Australia and the island of New Guinea. Med. J. Aust. 1961; 1:273–276.

Anderson, S.G. and French, E.L. An epidemic exanthem associated with polyarthritis in the Murray Valley, 1956. Med. J. Aust. 1957; 2:113–117.

Anstey, N., Currie, B. and Tai, K.S. RRV disease presenting with hematuria. Southeast Asian J. Trop. Med. Public Health 1991; 22:281–283.

Azoulas, J.K., Wishart, E., Bibby, S. and Ainsworth, C. Isolation of RRV from mosquitoes and from horses with signs of musculoskeletal disease. Aust. Vet. J. 2003; 81:344–347.

Barrett, A.D.T., Hunt, N. and Dimmock, N.J. A rapid method for the inactivation of virus infectivity prior to assay for interferons. J. Virol. Methods 1984; 8:349–351.

Barteling, S.J. and Vreeswijk, J. Developments in foot and mouth vaccines. Vaccine 1991; 9:75–88.

Bennett, N.M., Cunningham, A.L., Fraser, J.R.E. and Speed, B.R. Epidemic polyarthritis acquired in Fiji. Med. J. Aust. 1980; 1:316–317.

Boutwell, R.K., Colburn, N.H. and Muckerman, C.C. In vivo reactions of beta propiolactone. Ann. N.Y. Acad. Sci. 1969; 163:751–764.

Casals, J. The arthropod borne group of animal viruses. Trans. N.Y. Acad. Sci. 1957; 19:219–235.

Cheng, R.H., Kuhn, R.J., Olson, N.H., Rossman, M.G., Choi, H.K., Smith, T.J. and Baker, T.S. Nucleocapsid and glycoprotein organisation in an enveloped virus. Cell 1995; 80:621–630.

Clarris, B.J., Doherty, R.L., Fraser, J.R.E., French, E.L. and Muirden, K.D. Epidemic polyarthritis: a cytological, virological and immunological study. Aust. N.Z. J. Med. 1975; 5:450–457.

Cole, F.E., May, S.W. and Eddy, G.A. Inactivated Venezeulan equine encephalomyelilitis vaccine prepared from attenuated (TC-83 strain) virus. Appl. Microbiol. 1974; 27:150–153.

Condon, R.J. and Rouse, I.L. Acute symptoms and sequelae or RRV infection in South-Western Australia: a follow up study. Clin. Diag. Virol. 1995; 3:273–284.

Cunningham, A.L. and Fraser, J.R.E. RRV infection of human synovial cells in vitro. Aust. J. Exp. Biol. Med. Sci. 1985; 63:197–204.

Curran, M., Harvey, B., Crerar, S., Oliver, G., D'Souza, R., Myint, H., Rann, C. and Andrews, R. Australia's notifiable diseases status, 1996. Annual Report of the National Notifiable Diseases Surveillance System. Commun. Dis. Intell. 1997; 21:281–307.

Dalgarno, L., Rice, C.M. and Strauss, J.H. RRV 26s RNA: complete nucleotide sequence and deduced sequence of the encoded structural proteins. Virology 1983; 129:170–187.

Davies, D.J., Moran, J.E., Niall, J.F. and Ryan, G.B. Segmental necrotising glomerulonephritis with anti-neutrophil antibody: possible arbovirus aetiology. Br. Med. J. 1982; 285:606.

Doherty, R.L. Arboviruses of Australia. Aust. Vet. J. 1972; 48:172–180.

Doherty, R.L., Barrett, W.J., Gorman, B.M. and Whitehead, R.H. Epidemic polyarthritis in eastern Australia. Med. J. Aust. 1971; 1:5–8.

Doherty, R.L., Carley, J.G. and Best, J.C. Isolation of RRV from man. Med. J. Aust. 1972; 1:1083–1084.

Doherty, R.L., Gorman, B.M., Whitehead, R.H. and Carley, J.G. Studies of epidemic polyarthritis: the significance of three group A arboviruses isolated from mosquitoes in Queensland. Australas. Ann. Med. 1964; 13:322–327.

Doherty, R.L., Gorman, B.M., Whitehead, R.H. and Carley, J.G. Studies of arthropod borne virus infections in Queensland. V. Survey of antibodies to group A arboviruses in man and other animals. Aust. J. Exp. Biol. Med. Sci. 1966; 44:365–378.

Doherty, R.L., Whitehead, R.H., Gorman, B.M. and O'Gower, A.K. The isolation of a third Group A arbovirus in Australia with preliminary observations on its relationship to epidemic polyarthritis. Aust. J. Sci. 1963; 26:183–184.

Dowling, P.G. Epidemic polyarthritis. Med. J. Aust. 1946; 1:245–246.

Echuca-Melbourne Collaborative Group Arbovirus infection in a Murray Valley community. 1. Prevalence of antibodies, December, 1974. Med. J. Aust. 1976; 1:257–259.

Faragher, S.G., Marshall, I.D. and Dalgarno, L. RRV genetic variants in Australia and the Pacific Islands. Aust. J. Exp. Biol. Med. Sci. 1985; 63:473–488.

Faragher, S.G., Meek, A.D.J., Rice, C.M. and Dalgarno, L. Genome sequences of a mouse avirulent and a mouse virulent strain of RRV. Virology 1988; 163:509–526.

Fauran, P., Donaldson, M., Harper, J., Oseni, R.A. and Aaskov, J.G. Characterisation of RRVes isolated from patients with polyarthritis in New Caledonia and Wallis and Futuna Islands. Am. J. Trop. Med. Hyg. 1984; 33:1228–1231.

Fraser, J.R.E. Epidemic polyarthritis and RRV disease. Clin. Rheum. Dis. 1986; 12:369–388.

Fraser, J.R.E. and Becker, G.J. Mononuclear cell types in chronic synovial effusions of RRV disease. Aust. N.Z. J. Med. 1984; 14:505–506.

Fraser, J.R.E., Christie, D.G., Gust, I.D., White, J., Leach, R., Macaulay, E.D., Ahern, A.P., Alexander, J., Jones, J.M., Lung, D.Y.L., Moore, F.M., Moysey, C.D., Phillips, P.J. and Pryor, I. Arbovirus infection in a Murray Valley community. Aust. N.Z. J. Med. 1986; 16:52–57.

Fraser, J.R.E. and Cunningham, A.L. Incubation time of epidemic polyarthritis. Med. J. Aust. 1980; 1:550–551.

Fraser, J.R.E., Cunningham, A.L., Clarris, BJ., Aaskov, J.G. and Leach, R. Cytology of synovial effusions in epidemic polyarthritis. Aust. N.Z. J. Med. 1981; 11:168–173.

Fraser, J.R.E., Cunningham, A.L., Mathews, J.D. and Riglar, A. Immune complexes and RRV disease (epidemic polyarthritis). Rheumatol. Int. 1988a; 8:113–117.

Fraser, J.R.E., Cunningham, A.L., Muller, H.K., Sinclair, R.A. and Standish, H.G. Glomerulonephritis in the acute phase of RRV disease (epidemic polyarthritis). Clin. Nephrol. 1988b; 29:149–152.

Fraser, J.R.E., Ratnamohan, V.M., Dowling, J.P.G., Becker, C.J. and Varigos, C.A. The exanthem of RRV infection: histology, location of virus antigen and nature of inflammatory infiltrate. J. Clin. Pathol. 1983; 36:1256–1263.

Fraser, J.R.E., Rowley, M.J. and Tait, B. Collagen antibodies in RRV disease (epidemic polyarthritis). Rheumatol. Int. 1987; 7:267–269.

Fraser, J.R.E., Tait, B., Aaskov, J.G. and Cunningham, A.L. Possible genetic determinants in epidemic polyarthritis caused by RRV infection. Aust. N.Z. J. Med. 1980; 10:597–603.

Fuller, C.O. and Warner, P. Some epidemiological and laboratory observations on an epidemic rash and polyarthritis occurring in the upper Murray region of south Australia. Med. J. Aust. 1957; 2:117–120.

Gard, G.P., Marshall, I.D., Walker, K.H., Ackland, H.M. and De Sarem, W.G. Association of Australian arboviruses with nervous disease in horses. Aust. Vet. J. 1977; 53:61–66.

Goswell, G. A further report on an epidemic of acute polyarthritis. Med. J. Aust. 1946; 2:861–863.

Gubler, D.J. Transmission of RRV by *Aedes polynesiensis* and *Aedes aegypti*. Am. J. Trop. Med. Hyg. 1981; 30:1303–1306.

Halliday, J.H. and Horan, J.P. An epidemic of polyarthritis in the Northern Territory. Med. J. Aust. 1943; 2:293–295.

Harley, D., Sleigh, A. and Ritchie, S. RRV transmission, infection and disease: a cross-disciplinary review. Clin. Microbiol. Rev. 2001; 14:909–932.

Harris, L. Polyarthritis in the Northern Territory. Med. J. Aust. 1944; 1:546.
Harrison, V.R., Eckels, K.H., Bartelloni, P.J. and Hampton, C. Production and evaluation of a formalin killed chikungunya vaccine. J. Immunol. 1971; 107:643–647.
Harley, D., Sleigh, A. and Ritchie, S. Ross River virus transmission, infection, and disease: a disciplinary review. Clin. Microbiol. Rev. 2001; 14:909–932.
Hawkes, R.A., Boughton, C.R., Nairn, H.M. and Stallman, N.D. A major outbreak of epidemic polyarthritis in New South Wales during the summer of 1983/1984. Med. J. Aust. 1985; 143:330–333.
Hazelton, R.A., Hughes, C. and Aaskov, J.G. The inflammatory response in the synovium of a patient with Ross River arbovirus infection. Aust. N.Z. J. Med. 1985; 15:336–339.
Heil, M.L., Albee, A., Strauss, J.H. and Kuhn, R.J. An amino acid substitution in the coding region of the E2 glycoprotein adapts RRV to utilise heparin sulphate as an attachment moiety. J. Virol. 2001; 75:6303–6309.
Journeaux, S.F., Brown, W.G. and Aaskov, J.G. Prolonged infection of human synovial cells with RRV. J. Gen. Virol. 1987; 68:3165–3169.
Kay, B.H. Assignment Report, World Health Organisation, Regional Office for the South Pacific, Manila, 1981.
Kay, B.H. and Aaskov, J.G. RRV disease (epidemic polyarthritis). In: The Arboviruses: Epidemiology and Ecology (T.P. Monath, Ed.), Vol. IV, pp. 93–112. Boca Raton, FL: CRC Press, 1989.
Kay, B.H., Hall, R.A., Fanning, I.D., Mottram, P., Young, P.L. and Pollitt, C.C. Experimental infection of vertebrates with Murray Valley encephalitis and RRVes. In: Arbovirus Research in Australia: Proceedings, Fourth Symposium (T.D. St George, B.H. Kay and J. Blok, Eds.), pp. 71–75. Brisbane: CSIRO, 1986.
Keary, P.J., Soden, M.A.M. and Mudge, P.R. The changing epidemiology of RRV disease in South Australia. Med. J. Aust. 1997; 166:333.
Kerr, P.J., Fitzgerald, S., Tregear, G.W., Dalgarno, L. and Weir, R.C. Characterisation of a major neutralisation domain of RRV using anti-viral and anti-peptide antibodies. Virology 1992; 187:338–342.
Kerr, P.J., Weir, R.C. and Dalgarno, L. RRV variants selected during passage in chick embryo fibroblasts: serological, genetic and biological changes. Virology 1993; 193:446–449.
Kistner, O., Barrett, N., Bruhmann, A., Reiter, M., Mundt, W., Savidis-Dacho, H., Schrober-Bendixen, S., Dorner, F. and Aaskov, J.G. The pre-clinical testing of a formaldehyde inactivated RRV vaccine designed for use in humans. Vaccine 2007; 25:4845–4852.
Klapsing, P., MacLean, J.D., Glaze, S., McClean, K.L., Drebot, M.A., Lanciotti, R.S. and Campbell, G.J. RRV disease re-emergence, Fiji, 2003–2004. Emerg. Infect. Dis. 2005; 11:613–615.
Klockmann, U., Bock, H.L., Franke, V., Hein, B., Reiner, G. and Hilfenhaus, J. Preclinical investigations of the safety, immunogenicity and efficacy of a purified inactivated tick-borne encephalitis vaccine. J. Biol. Standards 1989; 17:331–342.
Lanciotti, R.S., Ludwig, M.L., Rwaguma, E.B., Lutwama, J.J., Kram, T.M., Karabatsos, N., Cropp, B.C. and Miller, B.R. Emergence of epidemic O'nyong-nyong fever in Uganda after a 35 year absence: genetic characterisation of the virus. Virology 1998; 252:258–268.
Lidbury, B.A. and Mahalingam, S. Specific ablation of anti-viral gene expression in macrophages by antibody-dependent enhancement of RRV infection. J. Virol. 2000; 74:8376–8381.
Linn, M.L., Aaskov, J.G. and Suhrbier, A. Antibody dependent enhancement and persistence in macrophages of an arbovirus associated with arthritis. J. Gen. Virol. 1996; 77:407–411.
Linn, M.L., Mateo, L., Gardiner, J. and Suhrbier, A. Alphavirus-specific cytotoxic T lymphocytes recognise a cross-reactive epitope from the capsid protein and can eliminate virus from persistently infected macrophages. J. Virol. 1998; 72:5146–5153.
Lucas, R.E. and Qiao, M. A case of encephalitis in central Australia due to RRV? Aust. N.Z. J. Med. 1999; 29:268–270.
Marshall, I.D. and Miles, J.A.R. RRV and epidemic polyarthritis. Curr. Top. Vector Res. 1984; 2:31–56.
Martin, J.H.B., Weir, R.C. and Dalgarno, L. Replication of standard and defective RRV in BHK cells: patterns of viral RNA and polypeptide synthesis. Arch. Virol. 1979; 61:87–103.
Mateo, L., Linn, M.L., McColl, S.R., Cross, S., Gardiner, J. and Suhrbier, A. An arthrogenic alphavirus induces monocyte chemoattractant protein-1 and interlukin-8. Intervirology 1999; 43:55–60.
Meek, A.D., Faragher, S.G., Weir, R.C. and Dalgarno, L. Genetic and phenotypic studies on RRV variants of enhanced virulence selected during mouse passage. Virology 1989; 172:399–407.
Milner, A.R. and Marshall, I.D. Pathogenesis of in utero infections with abortogenic and non abortogenic alphaviruses in mice. J. Virol. 1984; 50:66–72.
Mims, C.A., Murphy, F.A., Taylor, W.P. and Marshall, I.D. Pathogenesis of RRV infection in mice. I. Ependymal infection, cortical thinning, and hydrocephalus. J. Infect. Dis. 1973; 127:121–128.
Mitchell, C.J., Miller, B.R. and Gubler, D.J. Vector competence of *Aedes albopictus* from Houston, Texas, for dengue serotypes 1 to 4, yellow fever and RRVes. J. Am. Mosquito Control Assoc. 1987; 3:460–465.
Mudge, P.R. and Aaskov, J.G. Epidemic polyarthritis in Australia, 1980–1981. Med. J. Aust. 1983; 2:269–273.
Mukhopadhyay, S., Chipman, P.R., Hong, E.M., Kuhn, R.J. and Rossmans, M.G. In vitro assembled alphavirus core like particles maintain a structure similar to that of nucleocapsid cores in mature viruses. J. Virol. 2002; 76:11128–11132.
Murphy, F.A., Taylor, W.P., Mims, C.A. and Marshall, I.D. Pathogenesis of RRV infection in mice. II. Muscle and brown fat lesions. J. Infect. Dis. 1973; 127:129–138.
Mylonas, A.D., Brown, A.M., Carthew, T.L., McGrath, B., Purdie, D.M., Pandeya, N., Vecchio, P.C., Collins, L.G., Gardiner, I.D., de Looze, F.J., Reymond, E.J. and Suhrbier, A. Natural history of RRV-induced epidemic polyarthritis. Med. J. Aust. 2002; 177:356–360.
Mylonas, A.D., Harley, D., Purdie, D.M., Pandeya, N., Vecchio, P.C., Farmer, J.F. and Suhrbier, A. Corticosteroid therapy in an alphavirus arthritis. J. Clin. Rheumatol. 2004; 10:326–330.
Nimmo, J.R. An unusual epidemic. Med. J. Aust. 1928; 1:549–550.
Oliveira, N.M.M., Broom, A.K., Mackenzie, J.S., Smith, D.W., Lindsay, M.D.A., Kay, B.H. and Hall, R.A. Epitope blocking enzyme linked immunosorbent assay for detection of antibodies to RRV in vertebrate sera. Clin. Vaccine Immunol. 2006; 13:814–817.
Oseni, R.A., Donaldson, M.D., Dalglish, D.A. and Aaskov, J.G. Detection by ELISA of IgM antibodies to RRV in serum from patients with suspected epidemic polyarthritis. Bull. World Health Organ. 1983; 61:703–708.
Pascoe, R.R.R., St George, T.D. and Cybinski, D.H. The isolation of RRV from a horse. Aust. Vet. J. 1978; 54:600.
Penna, J.E. and Irving, L.G. Evidence for meningitis in RRV infection. Med. J. Aust. 1993; 159:492–493.
Richardson, M.A., Boulton, R.W., Raghow, R.S. and Dalgarno, L. Polypeptide synthesis in Alphavirus-infected *Aedes albopictus* cells during the establishment of persistent infection. Arch. Virol. 1980; 63:263–274.
Robinson, M.C. An epidemic of virus disease in Southern Province, Tanganyika Territory, in 1952–53. I. Clinical features. Trans. R. Soc. Trop. Med. Hyg. 1955; 49:28–32.
Rosen, L., Gubler, D.J. and Bennett, P.H. Epidemic polyarthritis (Ross River) virus infection in the Cook Islands. Am. J. Trop. Med. Hyg. 1981; 30:1294–1302.
Rulli, N.E., Suhrbier, A., Hueston, L., Heise, M.T., Tupanceska, D., Zaid, A., Wilmes, A., Gilmore, K., Lidbury, B.A. and

Mahalingam, S. RRV: molecular and cellular aspects of disease pathogenesis. Pharmacol. Ther. 2005; 107:324–329.

Russell, R.C. Arboviruses and their vectors in Australia: an update on the ecology and epidemiology of some mosquito-borne arboviruses. Rev. Med. Vet. Entomol. 1995; 83:141–158.

Russell, R.C. RRV: ecology and distribution. Annu. Rev. Entomol. 2002; 47:1–31.

Ryan, P.A., Martin, L., Mackenzie, J.S. and Kay, B.H. Investigation of grey-headed flying foxes (*Pteropus poliocephalus*) (Megachiroptera: Pteropodidae) and mosquitoes in the ecology of RRV in Australia. Am. J. Trop. Med. Hyg. 1997; 57:476–482.

Sammels, L.M., Coelen, R.J., Lindsay, M.D. and Mackenzie, J.S. Geographic distribution and evolution of RRV in Australia and the Pacific Islands. Virology 1995; 212:20–29.

Scrimgeour, E.M. Suspected RRV encephalitis in Papua New Guinea. Aust. N.Z. J. Med. 1999; 29:559.

Scrimgeour, E.M., Aaskov, J.G. and Matz, L.R. RRV in Papua New Guinea. Trans. R. Soc. Trop. Med. Hyg. 1987; 81:833–834.

Seglenieks, Z. and Moore, B.W. Epidemic polyarthritis in South Australia: report of an outbreak in 1971. Med. J. Aust. 1974; 2:552–556.

Seldon, S.M. and Cameron, A.S. Changing epidemiology of RRV disease in South Australia. Med. J. Aust. 1996; 165:313–317.

Shope, R.E. and Anderson, S.G. The virus aetiology of epidemic exanthem and polyarthritis. Med. J. Aust. 1960; 1:156–158.

Short, B. Polyarthritis of unknown origin. Med. J. Aust. 1949; 2:762.

Short, B. Polyarthritis of unknown origin. Med. J. Aust. 1984; 1:112–119.

Sibree, E.W. Acute polyarthritis in Queensland. Med. J. Aust. 1944; 2:565–567.

Smith, T.J., Cheng, R.H., Olsen, N.H., Peterson, P., Chase, E., Kuhn, R.J. and Baker, T.S. Putative receptor binding sites on Alphaviruses as visualised by cryoelectron microscopy. Proc. Natl. Acad. Sci. U.S.A. 1995; 92:10648–10652.

Soden, M., Vasudevan, H., Roberst, B., Coelen, R., Hamlin, G., Vasudevan, S. and La Brooy, J. Detection of viral ribonucleic acid and histologic analysis of inflamed synovium in RRV infection. Arthritis Rheum. 2000; 43:365–369.

Stanley, N.F. and Choo, S.B. Studies of arboviruses in Western Australia. Serological epidemiology. Bull. World Health Organ. 1964; 30:221–226.

Strauss, J.H. and Strauss, E.G. The alphaviruses: gene expression, replication and evolution. Microbiol. Rev. 1994; 58:491–562.

Suhrbier, A. and Linn, M.L. Clinical and pathological aspects of arthritis due to RRV and other Alphaviruses. Curr. Opin. Rheumatol. 2004; 16:374–379.

Taubman, M.A. and Atassi, M.Z. Reaction of beta-propiolactone with amino acids and its specificity for methionine. Biochem. J. 1968; 106:829–834.

Taylor, W.P. and Marshall, ID. Adaptation studies with RRV: retention of field level virulence. 1. Gen. Virol. 1975; 28:73–83.

Tesh, R.E., McLean, R.G., Shroyer, D.A., Calisher, C.H. and Rosen, L. RRV (Togaviridae: Alphavirus) infection (epidemic polyarthritis) in American Samoa. Trans. R. Soc. Trop. Med. Hyg. 1981; 75:426–431.

Vrati, S., Faragher, S.G., Weir, R.C. and Dalgarno, L. RRV mutants with a deletion in the E2 gene: properties of the virion, virus specific macromolecule synthesis and attenuation of virulence for mice. Virology 1986; 151:222–232.

Vrati, S.C., Fernon, C.A., Dalgarno, L. and Weir, R.C. Location of a major antigenic site involved in RRV neutralisation. Virology 1988; 162:346–353.

Vrati, S., Kerr, P.J., Weir, R.C. and Dalgarno, L. Entry kinetics and mouse virulence of RRV mutants altered in neutralisation epitopes. J. Virol. 1996; 70:1745–1750.

Way, S.J.R., Lidbury, B.A. and Banyer, J.L. Persistent RRV infection of murine macrophages: an in vitro model for the study of viral relapse and immune modulation during long term infection. Virology 2002; 301:281–292.

Weber, F.C., Oppel, T. and Raymond, R.W. A mild exanthematous disease seen in the Schouten Islands. Amer. J. Trop. Med. 1946; 26:489–495.

Westley-Wise, V.J., Beard, J.R., Sladden, T.J., Dunn, T.M. and Simpson, J. RRV infection on the North Coast of New South Wales. Aust. N.Z. J. Public Health 1996; 20:87–92.

Whitehead, R.H. Experimental infection of vertebrates with Ross River and Sindbis viruses, two group A arboviruses isolated in Australia. Aust. J. Exp. Biol. Med. Sci. 1969; 47:11–15.

Whitehead, R.H., Doherty, R.L., Domrow, R., Standfast, H.A. and Wetters, E.J. Studies of the epidemiology of arthropod-borne virus infections at Mitchell River Mission, Cape York Peninsula, North Queensland. Trans. R. Soc. Trop. Med. Hyg. 1968; 62:439–445.

Wilson, J.G. The Murray Valley rash. Med. J. Aust. 1957; 2:120–122.

Yu, S. and Aaskov, J.G. Development of a candidate vaccine against RRV infection. Vaccine 1994; 12:1118–1124.

Zhang, W., Heil, M., Kuhn, R.J. and Baker, T.S. Heparin binding sites on RRV revealed by electron cryo-microscopy. Virology 2005; 332:511–518.

CHAPTER

35

Rotavirus

Monica Malone McNeal and David I. Bernstein
Division of Infectious Diseases, Cincinnati Children's Hospital Medical Center, University of Cincinnati, Cincinnati, OH, USA

OUTLINE

ABSTRACT

The most common cause of severe diarrhea in children under 5 years of age is rotavirus. Rotavirus infections are responsible for over 600,000 deaths worldwide, representing 5% of all childhood deaths. Ninety percent of these deaths occur in developing countries. In addition, it is estimated that societal costs, including doctor's visits and hospitalizations, are approximately 1 billion dollars per year. Since the 1990s, The World Health Organization (WHO) has given priority to finding a safe and effective rotavirus vaccine. This quest has been going on for over 30 years. Natural infection with rotavirus has been shown to provide protection from subsequent severe disease, although the exact mechanism of protection is unclear. Both live and non live vaccines have been developed but only live oral rotavirus vaccines have progressed to human trials. One vaccine, Rotashield, was licensed in the United States in 1998 but was later withdrawn due to safety concerns because of its association with intussusception, a serious condition that causes bowel obstruction. After several years, and the completion of the largest vaccine trials ever conducted for a vaccine, two new live attenuated vaccines became available. Rotarix®, an attenuated human strain became available in 2004 in Mexico and now is approved in over 100 countries. In 2006, another vaccine, RotaTeq™, a live five component bovine human reassortant was licensed in the United States. In 2006, the WHO issued a recommendation that the two vaccines could be introduced in regions where they had been evaluated in clinical trials. The licensure of these two vaccines should dramatically decrease the burden of disease in countries where it is approved but the true impact of the vaccines will not be realized until it is available in the countries with the highest disease burden.

INTRODUCTION

Rotaviruses are recognized as the single most important cause of severe infantile gastroenteritis worldwide and have been estimated to be responsible for >600,000 deaths annually (Parashar et al., 2003, 2006). Transmission of rotaviruses occurs primarily by the fecal–oral route, providing a highly efficient mechanism for universal exposure that circumvents differences in national cultural practices and public health standards (Kapikian et al., 2001). Nearly all children less than 5 years of age have experienced at least one rotavirus infection (Glass et al., 2006; Rodriguez et al., 1987; Kapikian, 1993).

The most common symptoms associated with rotavirus disease are diarrhea and vomiting accompanied by fever (Kapikian, 1993; Staat et al., 2002). Rotavirus illness can be mild and of short duration or produce severe dehydration leading to hospitalization and mortality if not treated. Severe disease occurs primarily in young children, most commonly between 6 and 24 months of age and treatment of rotavirus illness is limited to supportive measures such as oral or intravenous rehydration (Kapikian, 1993; Staat et al., 2002; Rodriguez et al., 1977). The only specific method to combat rotavirus disease is the development of a vaccine. Fortunately, between 2004 and 2006, two new rotavirus vaccine candidates have become available and their licensure is proceeding in countries around the world.

The reduced severity of illness, typically associated with sequential rotavirus infections, provide the basis for the development of rotavirus vaccines (Velazquez et al., 1996). Thus, a realistic goal for a rotavirus vaccine would be to protect against severe disease and hospitalization. The purpose of this review is to present a summary of the development of rotavirus vaccines including early vaccine candidates, the current licensed vaccines, and other vaccine candidates under development. In addition, the epidemiology, pathogenesis, molecular virology, and immunology of rotavirus will be briefly discussed.

DISEASE BURDEN

In the United States, nearly every child is infected with rotavirus by the age of 5 and rotavirus illness results in 20–70 deaths annually (Fischer et al., 2007; Glass et al., 1996). Infection with rotavirus often results in more severe illness than other pathogens and therefore rotavirus accounts for a higher percentage of gastroenteritis episodes requiring medical intervention (Clark and McKendrick, 2004; Wilhelmi et al., 2003). During peak rotavirus seasons, 70% of all gastrointestinal hospitalizations may be due to rotavirus-associated gastroenteritis (Parashar et al., 1998). It is estimated that 1 child in 7 will require a doctor's visit at a clinic or emergency room and 1 in 67 to 1 in 87 will be hospitalized because of a rotavirus illness (Malek et al., 2006; Fischer et al., 2007). Based on these estimates, rotaviruses are responsible for 5–10% of all gastroenteritis episodes in children under 5 years of age leading to over 50,000 hospitalizations. Depending on how hospital surveillance studies were conducted, these estimates may be lower than the actual numbers of rotavirus-related hospitalizations (Hsu et al., 2005). Similar estimates of the disease burden due to rotavirus were recently reported in a large study conducted in seven European countries (Van Damme et al., 2007a; Giaquinto et al., 2007a), demonstrating similar patterns of rotavirus disease in developed countries.

Globally, rotavirus disease has a much more dramatic and significant impact on infant health. Although rates of rotavirus illness among children are similar throughout the world, the resulting mortality differs substantially. It is estimated that rotavirus illness is responsible for over 600,000 deaths annually, representing 5% of all deaths in children younger than 5 years of age worldwide (Parashar et al., 2003, 2006). More than 90% of these deaths occur in Africa and Asia (Fig. 35.1). Greater than 100,000 deaths occur in India and sub-Saharan Africa and 35,000 occur in China (Parashar et al., 2003; Glass et al., 2005). Therefore, safe and effective rotavirus vaccines are urgently needed, especially in countries

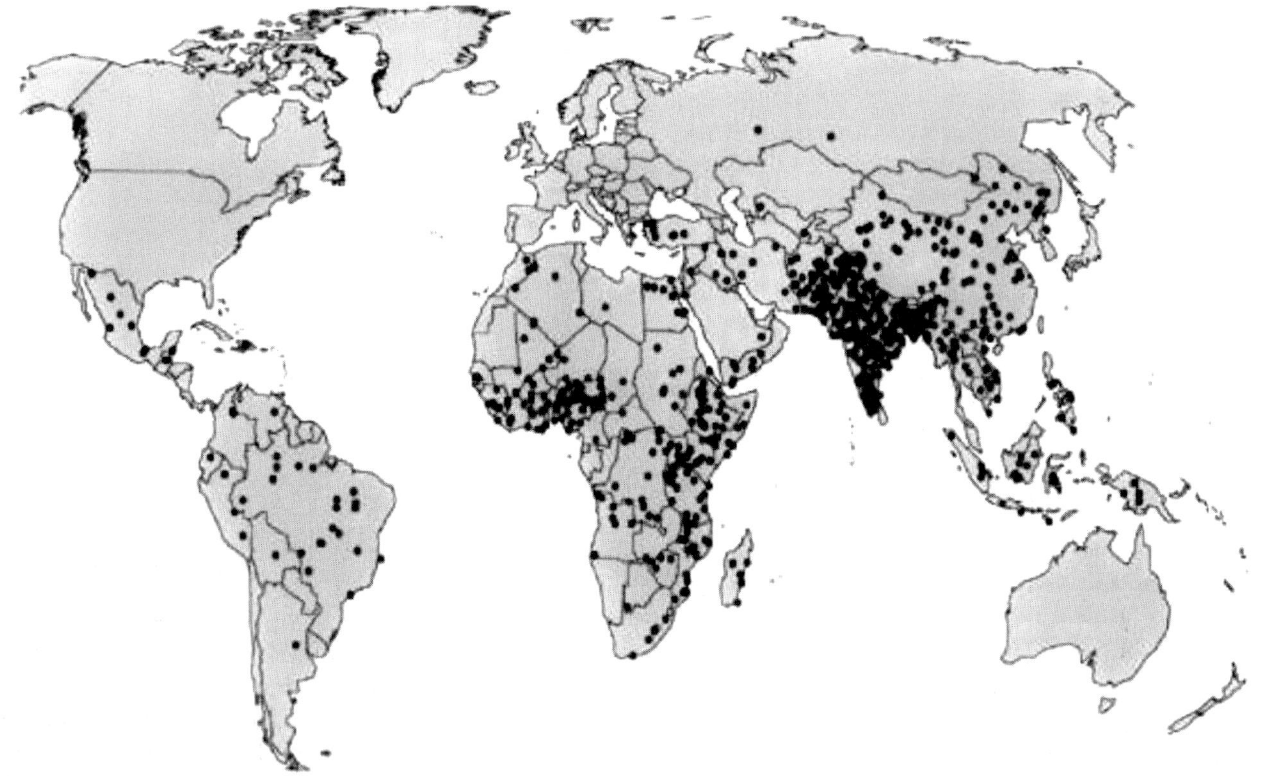

FIGURE 35.1 Distribution of deaths due to rotavirus. Each dot represents 1000 deaths. (Image courtesy of Umesh Parashar, Center for Disease Control and Prevention, Atlanta, GA.)

with the highest rotavirus mortality and the most limited resources available to treat diarrheal illnesses (Girard et al., 2006).

HISTORY OF ROTAVIRUS

The first rotaviruses to be described, based on pathology and epidemiology, were murine strains which were classified under the general description as the agents responsible for "epizootic diarrhea of infant mice," i.e., EDIM (Cheever and Mueller, 1947; Pappenheimer and Enders, 1947). Murine rotaviruses were also among the first to be visualized by electron microscopy (Adams and Kraft, 1963), along with viruses obtained from the rectal swabs of monkeys that contained viruses with comparable morphologic features (Malherbe and Harwin, 1963). These agents were described as 70-nm particles that had a wheel-like appearance and were later designated "rota" viruses from the Latin word for wheel (Flewett et al., 1974; Wyatt et al., 1978). Mebus et al. (1969) demonstrated the presence of these particles in stools of calves with diarrhea, thus associating these viruses with a diarrheal disease.

The correlation between these viruses and severe diarrhea in young children was reported first in 1973 by Bishop and colleagues, who used electron microscopy to examine biopsy specimens of duodenal mucosa from children with acute gastroenteritis (Bishop et al., 1973, 1974). Within a short time other investigators confirmed the association between the presence of rotavirus in feces and acute gastroenteritis.

In addition to their distinctive morphologic features, human rotaviruses along with their animal counterparts, share a group antigen (Kapikian et al., 1976; Woode et al., 1976) and are classified as members of the *Rotavirus* genus within the Reoviridae family (Matthews, 1979). In 1980, particles that were morphologically indistinguishable from established rotavirus strains but lacked the common group antigen were discovered in pigs (Bridger, 1980; Saif et al., 1980). This finding led to the identification of rotaviruses belonging to six additional groups (B–G) based on common group antigens, with the original rotavirus strains classified as group A (Saif and Jiang, 1994). Only groups A–C have been associated with human diseases, and most known cases of rotavirus gastroenteritis are caused by group A strains. Although, there have been large outbreaks reported in China and Japan associated with

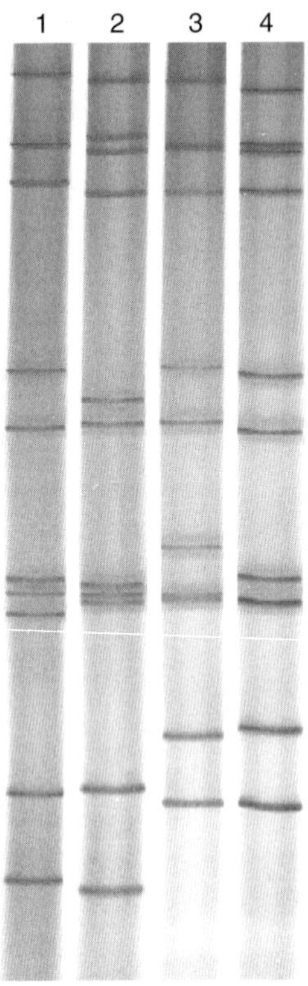

FIGURE 35.2 Polyacrylamide gel electrophoretic patterns of genomic RNAs obtained from group A human rotaviruses and visualized by silver staining. The patterns demonstrate the characteristic four size classes of RNA separated into groups of four, two, three, and two segments each. Human rotavirus strains included (from left to right) lane 1, Wa; lane 2, ST-3; lane 3, DS-1; lane 4, natural isolate containing a short pattern similar to DS-1.

nongroup A strains, particularly among adults (Hung, 1988; Kuzuya et al., 1998; Matsumoto et al., 1989), group A rotaviruses are the strains to which vaccine development has been directed.

PROPERTIES OF THE VIRUS

Rotavirus is a double-stranded RNA virus. The genome segments of rotavirus can be extracted from viral particles and separated by polyacrylamide gel electrophoresis into 11 bands visualized by ethidium bromide or silver staining as shown in Fig. 35.2. Each rotavirus strain has a characteristic RNA profile or electropherotype, a property that has been used extensively in epidemiologic studies of these viruses. The characteristic RNA electrophoretic pattern of group A rotaviruses consists of four size classes containing segments 1–4, 5 and 6, 7–9, and 10 and 11 (Kapikian et al., 2001). A short pattern is seen when the segment 11 runs slower than segment 10. RNA segments of strains belonging to less well characterized rotavirus groups (i.e., groups B–G) also can be separated into four size classes, but the distribution of segments within these classes differs from group to group (Saif, 1990).

The genome of rotavirus encodes the six structural proteins—VP1–VP4, VP6, and VP7—and six nonstructural proteins designated NSP1–NSP6 (Estes and Kapikian, 2007; Estes, 2001). Each segment encodes one known rotavirus protein except segment 11 which encodes both NSP5 and NSP6 using alternative open reading frames (Mattion et al., 1991). The genome segments range in size from approximately 660 to 3300 base pairs, and their encoded proteins, whose known functions are briefly described in Table 35.1, have molecular weights of approximately 12,000–125,000 (Estes and Kapikian, 2007).

Computer-generated images of rotavirus particles obtained by cryoelectron microscopy show it is ca. 100nm in diameter and is composed of three concentric protein layers depicted in Fig. 35.3 (Pesavento et al., 2006; Prasad and Chiu, 1994; Shaw et al., 1993). The outer layer contains 780 molecules of the VP7 glycoprotein and 60 dimers or trimers of the VP4 protein (Dormitzer et al., 2004) which forms spike-like projections that extend through and 11–12nm beyond the VP7 layer (Pesavento et al., 2006; Prasad and Chiu, 1994). The VP4 protein is anchored to the intermediate layer of the particle composed of 780 molecules of the VP6 protein. The innermost layer contains 120 molecules of the VP2 protein that interact with 12 molecules each of the viral transcriptase (VP1) and the guanylyltransferase (VP3) along with the 11 segments of the double-stranded RNA genome (Valenzuela et al., 1991; Estes, 2001; Liu et al., 1992; Estes and Kapikian, 2007).

Replication

Rotaviruses are activated by cleavage of the outer capsid VP4 protein by trypsin-like proteases into proteins termed VP5 and VP8 which remain associated with the virus (Lopez et al., 1985; Arias et al., 1996; Zarate et al., 2000; Estes et al., 1981). After attachment to receptors on the cytoplasmic membrane, through association with the VP8 protein, the activated virion either passes directly through this membrane or is taken into the cytoplasm within a vesicle (Hewish et al., 2000; Graham et al., 2005, 2006; Nava et al., 2004; Blanchard et al., 2007; Lopez and Arias, 2006). During membrane penetration or soon thereafter, the outer capsid proteins are removed (Cuadras et al., 1997). The RNA-dependent RNA polymerase (i.e., the VP1 transcriptase) associated with the inner shell is stimulated to synthesize the 11 viral mRNAs that are capped by VP3 (Patton, 1994; Patton and Spencer, 2000; Patton et al., 2006). The mRNAs are extruded from the virus cores through channels

TABLE 35.1 Rotavirus gene segments and properties of encoded proteins

RNA segment	No. of base pairs	Encoded protein	Molecular weight of protein (H 10^{-4})	Properties of proteins
1	3300	VP1	12.5	Inner core protein, RNA-dependent RNA polymerase, binds to ssRNA, complexes with VP3
2	2700	VP2	9.4	Inner core protein, RNA binding, needed for VP1 replicase activity
3	2600	VP3	8.8	Inner core protein, guanylyltransferase, methyltransferase, complexes with VP1
4	2360	VP4	8.7	Outer capsid protein, hemagglutinin, receptor binding, virulence, neutralization protein, fusogenic protein
5	1600	NSP1	5.9	Nonstructural protein, RNA binding, IRF regulatory protein
6	1360	VP6	4.5	Intermediate capsid protein, group and subgroup antigen, required for transcription
7	1100	NSP3	3.5	Nonstructural protein, mRNA binding, inhibits host translation
8	1060	NSP2	3.7	Nonstructural protein, RNA and NSP5 binding, virosome formation
9	1060	VP7	3.7	Outer capsid glycoprotein, neutralization protein
10	750	NSP4	2.0	Nonstructural glycoprotein, transmembrane protein, enterotoxin
11	660	NSP5	2.2	Nonstructural protein, phosphorylated, NSP2 binding, VP2 and NSP6 interaction
		NSP6	1.2	Nonstructural protein, NSP5 binding

in the VP2 and VP6 protein layers at the 12 vertices of the viral particles, and subsequently translated into viral proteins (Prasad et al., 1996; Lawton et al., 1997). Once viral proteins accumulate within the cytoplasm, large inclusions or viroplasms are formed in which the assembly of virion precursors is initiated (Patton et al., 2003, 2006; Fabbretti et al., 1999). Particle assembly may be initiated by the formation of complexes within the viroplasm that contain plus-strand RNAs from the 11 genome segments along with VP1 and VP3 and RNA-binding nonstructural proteins NSP2, NSP5, and NSP6 (Patton et al., 2003; Taraporewala et al., 2006; Taraporewala and Patton, 2004).

Although the mechanism is not completely defined, the virus assembles and packages one of each of the plus-strand RNAs within individual precursor viral complexes. These complexes eventually lose their nonstructural proteins, evolve into double-layered viral particles with the sequential addition of VP2 and VP6, and convert their single-stranded RNAs into double-stranded genome segments (Silvestri et al., 2004; Patton et al., 2004; Tortorici et al., 2006).

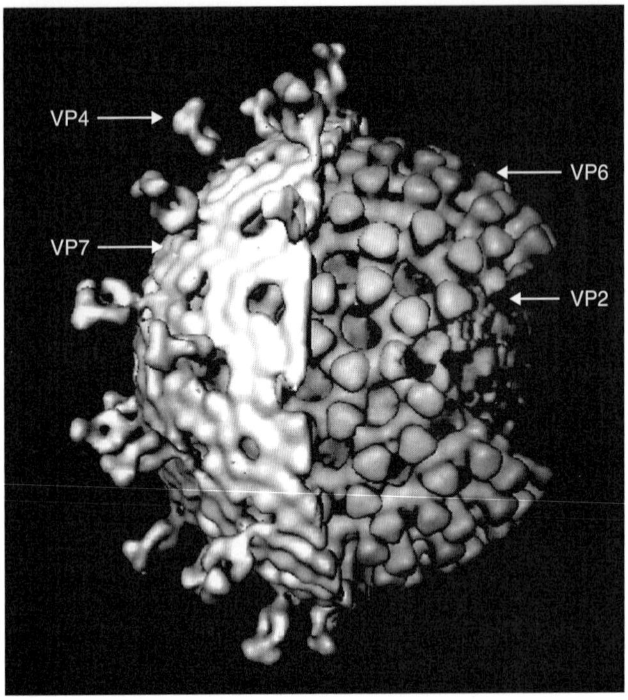

FIGURE 35.3 Image obtained by cryoelectron microscopy. (Image courtesy of B.V.V. Prasad, Baylor College of Medicine, Houston, TX.)

The double-layered particles then bud into the rough endoplasmic reticulum (ER) after their association with the NSP4 transmembrane glycoprotein. VP4 is either added prior to entry into the ER or sometime thereafter. The other rotavirus glycoprotein, VP7, which becomes sequestered within the rough ER, is added to complete the formation of mature viral particles (Estes, 2001; Patton et al., 2007). These mature viruses accumulate within the lumen of the rough ER until cell lysis occurs (Estes and Kapikian, 2007).

Serotypes

The two outer capsid proteins, VP7 and VP4, contain neutralization epitopes (Greenberg et al., 1983; Hoshino et al., 1984). Early studies using sera from hyperimmunized animals in cross-neutralization assays described a number of different serotypes found to infect humans and animals based on the VP7 protein (Hoshino et al., 1984; Wyatt et al., 1983). Viruses were given a G type referring to the glycosylated structure of this protein. Researchers later found that most of the neutralizing antibody induced when an animal was hyperimmunized, by injecting live preparations, was directed against the VP7, but that in some cases, i.e., after oral infection, VP4 could be the dominant neutralization protein (Clark et al., 1990; Perez-Schael et al., 1990; Ward et al., 1988, 1993). A dual serotyping scheme was therefore developed.

VP7 serotypes are determined by cross-neutralization and by using panels of monoclonal antibodies (Ward et al., 1991; Coulson et al., 1987; Taniguchi et al., 1987), VP4 serotype determination however, is more difficult (Mackow et al., 1988; Padilla-Noriega et al., 1992; Taniguchi et al., 1988). Two numeric systems were established to classify the VP4 protein, or P type referring to the protease sensitivity of this protein. The P serotype is based on neutralization assays using antisera against recombinant expressed VP4 proteins or viruses with specific VP4 genes (Gorziglia et al., 1990; Snodgrass et al., 1992). The second classification is based on comparative nucleic hybridization and sequence analyses or genotypes (Gentsch et al., 1992; Gouvea et al., 1994). The P serotype is indicated by an open number and the genotype is noted with a bracketed number. For example, the most common P type worldwide belongs to serotype P1A and genotype 8 and is therefore designated P1A[8]. Most recently, rotaviruses are often only classified by genotyping for the P protein. To date, 15 G types and 25 P types have been identified on the basis of sequence analyses from animal and human isolates (Estes and Cohen, 1989; Gentsch et al., 1996; Liprandi et al., 2003; McNeal et al., 2005; Rahman et al., 2005). However, as discussed below only a few G and P type combinations account for the vast majority of human infections.

Rotaviruses, similar to other viruses with segmented genomes, have the ability to form reassortants. During replication, if a coinfection has occurred, gene segments from each parent virus can be incorporated into new progeny viruses. The gene segments from different viruses can segregate independently and the new viruses produced can contain genes from either parent. Rotaviruses have an extremely wide host range, but natural cross-species infections, particularly between animals and humans, may be rare. Nevertheless, numerous human isolates appear to be animal strains or animal–human rotavirus reassortants as determined by sequence analyses (Cunliffe et al., 2000; Das et al., 1993a, 1993b, Matthijnssens et al., 2006). This property of reassortment has contributed to the diversity of G and P types found throughout the world. In addition, as will be discussed, a number of reassortant viruses have been used as vaccine candidates.

GLOBAL DISTRIBUTION OF HUMAN ROTAVIRUS SEROTYPES

Early studies using cross-neutralization assays and monoclonal antibodies to determine serotypes found that four main G types (G1, G2, G3, and G4) accounted for 90% of the stains isolated from humans. However, it was common in these early studies to have a high percentage of strains that were not typed by these methods. With the development of surveillance studies using reverse-transcription polymerase chain reaction (RT-PCR) genotyping and automated nucleotide sequencing of VP7, for G type, and VP4, for P type, more strains have been identified and genotyped. Over 42 different P–G combinations of strains infecting humans have been described, containing 10 different G types and 11 different P types (Gentsch et al., 2005). The most common G types still remain G1 through G4 but now also include G5, G8, G9, G10, and G12 (especially G9) in various regions around the world. The most common types for the P protein are P[8] and P[4] in addition to the more recently described P[6].

The global distribution of G and P serotypes can vary greatly by location and time, or, as was found in some studies, vary little over sequential seasons (Ward et al., 1991; Pipittajan et al., 1991). A review by Santos and Hoshino (2005) has summarized rotavirus serotypes from 124 studies performed between 1989 and 2004 and included data on 45,571 isolates collected from 52 countries on 5 continents. The four common G types, G1, G2, G3, and G4, with either P[8] or P[4], represented over 88% of the strains identified worldwide during this period. Serotype G9, containing either P[8] or P[6], was found to be an emerging strain in a number of different locations. Overall, the most common strains responsible for rotavirus illness have been found to be G1P[8] strains (Gentsch et al., 1996; Santos and Hoshino, 2005). In North America, Europe, and Australia the G1P[8] strain was responsible for over 70% of rotavirus infections. In South America and Asia the incidence was only 30% while in Africa it was only 23%. Africa has a G8 frequency as high as that of G3 or G4, and P[6] was found in a third of all isolates. In addition, in Africa, 27% of the strains identified had unusual combinations, such as G8 with P[4] or P[6]. Viruses containing G5 have been shown to be increasing in South America (Santos and Hoshino, 2005).

In another study by Gentsch et al. (2005), data on studies conducted in 35 countries, involving over 20,000 isolates were analyzed. The four common strains, G1P[8], G2P[4], G3P[8], and G4P[8] again represented 72% of all strains with G9 containing either P[6] or P[8] identified in over 2% of the isolates. Although P[6] was first identified in combination with G9 strains, recent studies have found that the G9P[8] strain may be more predominant (Iturriza-Gomara et al., 2000; Kirkwood et al., 2003, 2006a; Montenegro et al., 2007). It was also found that G types including G5, G6, G8, and G10 in various combinations with P[4], P[6], P[8], P[9], and P[14] were found in approximately 1.2% of the total strains identified. Latin America, Africa, and Asia also had a higher incidence of P[6] stains than Europe or North America and developing countries were found to have a higher incidence of unusual strains (Santos and Hoshino, 2005; Gentsch et al., 2005). In European countries, G1, G2, G3, G4, and G9 were the predominant G serotypes found, although the incidence varied by country (Van Damme et al., 2007b). Similar findings were reported from Central and South-Eastern Europe (Tcheremenskaia et al., 2007). A review of studies from China showed that G1 was the overall predominate strain but that G3 viruses were becoming more common in recent studies (Orenstein et al., 2007; Wang et al., 2007). G12 viruses, either with P[6] or P[8], appear to be the new emerging strains in India (Ramani et al., 2007) and Australia (Kirkwood et al., 2006b). The diversity of rotavirus serotypes, especially in developing countries where mortality due to rotavirus is greatest, has had an effect on vaccine development, although the extent of cross protection remains unclear as discussed below.

EPIDEMIOLOGY OF ROTAVIRUS

Severe human rotavirus disease occurs most commonly between 6 and 24 months of age (Kapikian and Chanock, 1990; Rodriguez et al., 1987; Ward et al., 1991), but the age may be lower in less developed countries (Haffejee and Moosa, 1990; Raul Velazquez et al., 1993). Neonatal rotavirus infections, which are often asymptomatic, are common and appear to be endemic in some newborn nurseries (Bishop et al., 1983; Perez-Schael et al., 1984). The asymptomatic nature of neonatal rotavirus infections may be due, at least partially, to protection from transplacental antibody that can persist for the first months of life (Bernstein et al., 1990). The onset of rotavirus disease in infants has been reported to coincide with the decline of maternal IgG antibody (Zheng et al., 1989).

Older children and adults, including elderly patients, are susceptible to reinfections with rotavirus, which can cause mild but rarely severe disease (Nakajima et al., 2001; Griffin et al., 2002; Marshall et al., 2003; Iijima et al., 2006). The reduced severity of rotavirus disease in older children and adults is due primarily to the immunity induced by previous rotavirus infections.

As with other respiratory and enteric viruses, distinct seasonality is associated with rotavirus disease (Estes and Kapikian, 2007; Cook et al., 1990; Haffejee, 1995; Kapikian et al., 1974; Turcios et al., 2006). This seasonality is particularly evident in temperate climates, where rotaviruses are responsible for the large increase in hospitalizations and deaths due to diarrheal diseases found during the winter season (Parashar et al., 1998). The seasonality of rotavirus disease is less apparent in tropical climates but disease is still more prevalent in the drier, cooler months (Haffejee, 1995). The cause for the seasonality of rotavirus disease remains unknown.

PATHOGENESIS

The incubation period for rotavirus is approximately 1–3 days (Kapikian et al., 1983). In children with symptoms, the onset is often abrupt, with fever and vomiting followed by explosive, watery diarrhea. Vomiting may precede the diarrhea in approximately half the cases (Haffejee, 1991). Fever occurs commonly during rotavirus illness, with estimates of between 45 and 84% of patients (Rodriguez et al., 1977; Uhnoo et al., 1986). The disease is usually self-limited, lasting 4–8 days, although the duration of symptoms ranged between 2 and 22 days in a Guatemalan study (Wyatt et al., 1979). In a study conducted in the United States, 63% of hospitalized children with rotavirus infection had fever, vomiting, and diarrhea at presentation, but children also presented with combinations of only one or two of these symptoms (Staat et al., 2002).

After fecal–oral transmission of rotavirus, infection is initiated in the upper intestine and typically leads to a series of histologic and physiologic changes. Studies in calves revealed that rotavirus infection caused the villus epithelium to change from columnar to cuboidal, which resulted in a shortening and stunting of the villi. The cells at the villus tips became denuded, while in the underlying lamina propria, the numbers of reticulum-like cells increased and mononuclear cell infiltration was observed. The infection started at the proximal end of the small intestine and advanced distally (Mebus et al., 1974; Mebus, 1976). From the few studies examining the pathologic changes in the intestines of humans, the resulted changes appeared to be similar to those found in animals (Holmes et al., 1975; Suzuki and Konno, 1975).

Although rotaviruses cause severe diarrhea in numerous species, including humans, the mechanisms responsible for the illness have not been completely determined and may be due to multiple factors. From various studies in animals, malabsorption of carbohydrates, sodium and calcium fluxes, retarded differentiation of uninfected enterocytes, or crypt cell secretions have been found to be involved in causing diarrhea in specific animal studies (Davidson et al., 1977; Graham et al., 1984; Collins et al., 1988). Diarrhea also has been induced in infant mice and rats by intraperitoneal inoculation with the rotavirus NSP4 protein as well as with a 22 amino acid peptide derived from this protein (Ball et al., 1996; Morris et al., 1999). Investigators observed that this protein and its peptide increased the levels of intracellular calcium by activating a calcium-dependent signal transduction pathway that mobilizes transport of this ion from the ER (Tian et al., 1994, 1995). Further reports suggest that NSP4 possesses membrane destabilization activity that may result from increased intracellular calcium concentrations, resulting in cytoskeleton disorganization and cell death (Tian et al., 1996; Browne et al., 2000; Brunet et al., 2000a, 2000b). Thus, binding NSP4 to intestinal epithelium after its release from infected cells may contribute to altered ion transport and diarrhea. Another factor that may have a role in rotavirus-induced diarrhea is the enteric nervous system which lies under the villus epithelium. It has been reported that rotavirus infection can activate this system in mice and drugs that block nerve activity attenuate rotavirus-induced fluid secretion in vitro and attenuate diarrhea in vivo (Lundgren et al., 2000). Whether this is a major mechanism of diarrhea occurring after rotavirus infection in humans remains to be determined.

The tissue tropism of rotavirus infection in humans was thought initially to be restricted to the villi of the small intestine. Because no consistent evidence of extraintestinal replication of rotavirus had been shown, the general assumption was that rotavirus pathology is strictly intestinal. However, instances of nongastrointestinal rotavirus-associated disease, including the association with abnormal liver function, as well as respiratory and nervous system involvement were reported (Lewis et al., 1979; Matsuno et al., 1983; Hongou et al., 1998; Nigro, 1991; Honeyman et al., 2000; Hattori et al., 1992). This spread was explained when it was reported that both antigenemia (presence of rotavirus protein in the blood) and viremia (presence of live rotavirus in the blood) were found during rotavirus infection in several strains of animals and in humans (Blutt et al., 2003). Since that time, this observation has been confirmed by a number of studies in humans (Chiappini et al., 2005, 2006; Fischer et al., 2005; Blutt et al., 2006, 2007; Ray et al., 2006).

IMMUNITY

After over 30 years of research involving studies of natural rotavirus infection, vaccine studies, and animal studies, the mechanisms of immunity to rotavirus infection remain unclear. Humoral antibody responses include an early IgM response followed by the production of rotavirus IgG and IgA (Riepenhoff-Talty et al., 1981; Grimwood et al., 1988). Infection also induces local, intestinal antibodies that are predominantly IgA (Bernstein et al., 1989; Coulson et al., 1990; Matson et al., 1993; Brown et al., 2000). T cell responses have been difficult to measure in humans, but several rotavirus proteins have been identified that contain T cell epitopes and have been used to measure T cell responses after rotavirus infection (Offit et al., 1993; Jaimes et al., 2002; Franco et al., 1994).

As stated before, virus neutralization epitopes are found on VP4 and VP7 proteins located on the surface of the virion, and neutralizing antibody to the infecting virus can be detected after infection or vaccination. However, VP6 appears to be the most immunogenic protein and antibody made against VP6 is not neutralizing. One of the most controversial questions concerning the development of immunity to rotavirus infection is whether or not neutralizing or serotype-specific antibody is necessary for protection. As will be discussed, this has had a dramatic effect on vaccine development.

Studies dealing with natural rotavirus infections were used initially to understand rotavirus immunity and the role of serotype specific antibodies. Many investigators reported that natural rotavirus infections produce incomplete protection, but many studies have shown that previous infections protect against severe disease associated with reinfection (O'Ryan et al., 1990; Bernstein et al., 1991; Clemens et al., 1992, Ward and Bernstein, 1994; Velazquez et al., 1996; Franco et al., 2006). In a large study conducted in Mexico, protection from both rotavirus reinfection and rotavirus diarrhea increased with each new infection (Velazquez et al., 1996). However, sequential infections even with the same serotype have been reported. In an early study reporting rotavirus disease with reinfection by the same serotype, the investigators noted that protection of young children in a Japanese orphanage lasted 6 months then declined after 1 year. This study noted a close correlation between titers of serotype-specific antibody and protection (Chiba et al., 1986). This study is often cited to support the idea that neutralizing antibody is necessary for protection. While it is generally accepted that if present at a sufficient titer, neutralizing antibody will protect, it is still unclear whether only neutralizing antibody is necessary for protection.

Although, reinfections with rotavirus can be common, other studies have shown that protection lasts at least 1 year. Neonates infected within the first 2 weeks of life were protected against severe disease but not against reinfection in one study (Bishop et al., 1983). In another study, infants who developed a symptomatic or an asymptomatic rotavirus infection during the first year of the study were protected against contracting a subsequent rotavirus illness or even an asymptomatic reinfection during the following year (Bernstein et al., 1991). Similarly, in another study, a natural rotavirus infection in the first year was found to be 93% protective against a symptomatic reinfection in the second year. This protection occurred even though the G1 strains that circulated during the first year were responsible for only 66% of rotavirus disease in the second year (Ward and Bernstein, 1994).

Correlations of serotype-specific neutralizing antibody and protection have not been supported in other larger studies. In the largest study, conducted in Bangladesh during a 2-year period when the four major G serotypes circulated, the titers of both preexisting homologous and heterologous neutralizing antibody were significantly lower in patients with acute rotavirus disease than in matched control subjects. Further analysis, however, could not find a correlation with serotype-specific neutralizing antibody, and protection seemed to correlate better with the magnitude of the response rather than with specific neutralizing responses (Ward et al., 1992). Differences in these studies evaluating the role of serotype specific antibody and the protection provided by natural infection may be due to the variation in circulating strains, the inoculum size, or the duration of protection.

Due to the difficulties associated with human studies, animal models using mice, rabbits, and gnotobiotic pigs have been developed (Ward et al., 1990; Conner et al., 1988; Saif et al., 1996). But again results from these studies have not completely identified the mechanisms of protection against rotavirus illness or infection possibly due to the use of different species. Questions remain regarding the relevance of these animal models to human rotavirus disease. Early studies in mice showed that while CD8 T cells are involved in resolution of an infection, antibody is necessary for protection against reinfection (McNeal et al., 1995; Franco and Greenberg, 1995; Franco et al., 1997). In a recent study, mice given an oral immunization of a fully heterotypic rotavirus were almost completely protected from shedding when later challenged, and this protection was dependent on the ability of rotavirus-specific IgA to be transported through intestinal epithelial cells or be present at the intestinal mucosa (VanCott et al., 2006). If these studies mirror what is needed in humans, then local antibody may be the effector of protection and non-neutralzing antibody that is not serotype-specific can also be involved in protection. Thus, serum levels of rotavirus-specific IgA may be a good correlate of protection, because it appears to parallel antibody levels in the gut (Brown et al., 2000; Jiang et al., 2002). Our understanding of protective immune responses will improve with further animal evaluations and should be aided by studies of the two new vaccines. The absence of a reliable immunologic marker of protection, however, continues to render vaccine trials more difficult.

ROTAVIRUS VACCINES

Public health institutions around the world have given the development of an effective rotavirus vaccine a high priority (Parashar and Glass, 2006). Numerous cost-effectiveness analyses in many countries have evaluated the need and benefit of a rotavirus vaccine. Although studies in the United States have shown mixed results, rotavirus vaccination is considered a cost-effective intervention (Widdowson et al., 2007) and a recent study has indicated pediatricians in this country are willing to implement the recently licensed vaccine (Kempe et al., 2007). Studies in other countries, especially countries that have high mortality due to rotavirus disease, clearly have shown that a rotavirus immunization program would be cost-effective from the perspective of society and the health care system (Glass et al., 2005; Podewils et al., 2005; Giaquinto et al., 2007b; Lorgelly et al., 2007). Because natural rotavirus infections can induce excellent protection, at least against severe rotavirus disease, vaccine efforts have been directed mostly at the development of live attenuated rotavirus vaccines given orally.

Early efforts concentrated on the use of animal rotavirus strains, labeled the Jennerian approach because it relies on the natural attenuation of animal viruses in humans for safety and largely heterotypic immune responses for protection. The initial efforts with animal rotavirus vaccines yielded inconsistent results. Therefore, in an attempt to make the vaccines more closely related to human strains, human rotavirus genes coding for the proteins that induce neutralizing antibody, VP4 and VP7, were introduced into these animal strains by creating reassortant viruses as described earlier. This approach is often labeled the modified Jennerian approach. Another approach has been to use human rotaviruses, either isolated from asymptomatic infections in human neonates or isolated from a child with rotavirus illness and attenuated by cell culture adaptation and passage.

Animal Strains

The first vaccine trial was conducted 10 years after the identification of rotavirus as an agent of severe diarrhea. This vaccine candidate was based on an animal strain, a bovine rotavirus termed RIT4237 that was derived from the Lincoln isolate, NCDV, and was a G6P[1] virus (Hoshino and Kapikian, 1994). The initial studies of the RIT4237 vaccine produced variable results. The vaccine was safe and effective in Finland, providing a protective efficacy of approximately 50% against all disease and greater than 80% protection against severe disease. This trial, thus supported the hypothesis that heterotypic protection was possible since this virus shared neither G nor P serotypes with the predominant human types. Antibody responses after vaccination showed a seroconversion rate of 53% as measured by serum IgG (IgA antibody was not determined in these early studies) but protection against severe diarrhea was detected even in vaccinees who did not seroconvert, suggesting that IgG was not a good measurement of vaccine take (Vesikari et al., 1983, 1984, 1985). However, later studies in developing countries were disappointing, showing little or no efficacy (DeMol et al., 1986; Hanlon et al., 1987; Lanata et al., 1989). Similarly, the vaccine failed to provide protection in a study performed on a Navajo reservation in Arizona (Santosham et al., 1991) and ultimately development of this vaccine was discontinued.

Another bovine rotavirus based vaccine, WC3, also appeared to be safe but replicated poorly in humans with between 5 and 17% of vaccinees shedding detectable vaccine virus (Clark et al., 1988). Immunized infants developed a neutralizing antibody response to WC3 in addition to rotavirus IgA and IgG antibody responses (Bernstein et al., 1990). The initial studies conducted in Philadelphia using one dose of the vaccine appeared promising, showing a protection rate of 76% against rotavirus diarrhea and 100% protection against more severe cases during a season that had predominantly G1 strains (Clark et al., 1986). Unfortunately, later trials in Cincinnati (Bernstein et al., 1990) and in less developed countries (Georges-Courbot et al., 1991) showed no significant protection. As will be discussed, WC3 was later used to develop reassortant vaccines and is indeed the backbone of the RotaTeq™ vaccine now licensed in the USA (Merck Research Company).

The next major vaccine candidate was a simian rotavirus isolated from the diarrheal stool of a rhesus monkey and termed MMU18006 or more commonly known as RRV. This vaccine was a G3P[3] strain and thus shared the G type with a strain that is common in humans. This strain replicated better in humans than the bovine strains, perhaps because of the sharing of the G3 serotype. RRV vaccination was associated with mild side effects, including low-grade fever and mild diarrhea, especially when it was given to older children who had lost maternal antibodies (Vesikari et al., 1986; Perez-Schael et al., 1987; Losonsky et al.,

1986). RRV was found to induce serum rotavirus IgG and IgA, intestinal rotavirus IgA, and RRV neutralizing antibody responses, but no neutralizing antibody against non-G3 human rotaviruses could be detected (Losonsky et al., 1986). Protection with this vaccine was inconsistent, ranging from greater than 50%, even in developing countries, to moderate (20–50%) to nonexistent (Midthun and Kapikian, 1996). Some evidence suggested that protection was more effective in G3 than G1 outbreaks but the evidence was not conclusive (Flores et al., 1990; Conner et al., 1994). This virus was later used to make the reassortants containing human rotavirus VP7 genes that became Rotashield™ vaccine (Wyeth).

One additional animal strain that has been developed into a vaccine is a G10P[12] strain isolated from a lamb, designated LLR. In 2000 and 2001, a vaccine based on this strain and produced by the Lanzhou Institute of Biological Products was introduced in China (Ling-Qiao, 2001). The vaccine was not tested in a controlled trial against a placebo group, therefore the efficacy is not known.

A summary of the various vaccines based on animal strains is presented in Table 35.2.

Reassortant Vaccines

Because of the belief that homotypic immunity, or the development of serotype-specific antibody responses, might increase the protection seen with rotavirus vaccines and because of a lack of consistent protection after vaccination with heterotypic animal strains, the next vaccines developed were reassortant vaccines (Table 35.3). These vaccines contained the genes encoding VP7 or the VP7 and VP4 proteins of human rotavirus strains, with the remainder of the genes from an animal strain. As discussed earlier, these proteins were chosen because they induce neutralizing antibody. One goal of this strategy was to create a multivalent vaccine containing viruses with human rotavirus genes representing the main human serotypes, initially G1–4.

Rotashield™

The simian strain, RRV, was used as the background to make three different reassortant viruses containing the VP7 gene from human strains D (G1 serotype), DS-1 (G2 serotype), and ST-3 (G4 serotype) with the

TABLE 35.2 Animal rotavirus strains used as vaccine candidates

Candidate	Origin	G and P type	Comments
RIT4237	Bovine	G6P[1]	In small trials, safe but variable results; no efficacy in developing countries; discontinued development
WC3	Bovine	G6P[5]	Safe but variable results; virus used to make reassortants containing human genes
MU18006 (RRV)	Simian	G3P[3]	Some incidence of fever in vaccinees; inconsistent protection; virus used to make reassortants containing human genes
LLR	Lamb	G10P[12]	Used in China, efficacy not known

TABLE 35.3 Animal human reassortant rotavirus strains used as vaccine candidates

Candidate	Virus background	Human gene substitutions	Comments
RRV-TV (Rotashield)	RRV G3P[3]	G1, G2, G4	Some fever in vaccinees; 49–64% protective against rotavirus; 80% protective against severe disease; licensed in United States in 1998; withdrawn due to association with intussusception
RotaTeq™	WC3 G6P[5]	G1, G2, G3, G4, P[8]	Safe, no association with intussusception after trials involving over 60,000 infants; highly protective, especially against severe disease; three doses given to infants starting at 6–12 weeks of age; licensed in the United States in 2006
UK reassortants	UK G6P[5]	G1, G2, G3, G4	In small studies, safe and protective; under development in several countries; may add more reassortants, possibly for G9 and G8

G3 serotype represented by RRV itself. The reassortant viruses were initially tested individually in various trials for safety, immunogenicity, reactogenicity, and efficacy. Each RRV reassortant had a safety profile comparable to RRV, with low-grade fevers occurring in about 20% of the vaccinees. Protection rates varied with the different monovalent viruses depending on the study, but generally heterotypic and homotypic protection were measured (Midthun and Kapikian, 1996). Therefore, all four strains were incorporated into a tetravalent vaccine, RRV-TV which later was licensed as Rotashield™.

Vaccination with RRV-TV produced a rotavirus IgG and IgA serum antibody response, but neutralizing antibodies were produced predominantly to RRV rather than to the human serotypes (Bernstein et al., 1995; Rennels et al., 1996). This finding appears to be consistent with data suggesting that VP4 rather than VP7 is more immunogenic after live oral infection (Ward et al., 1988, 1993). Extensive evaluations of the RRV-TV were completed before licensure. In two large trials conducted at centers across the United States, the RRV-TV vaccine was found to be safe, with the subjects developing a slight increase in temperature after the first dose. The efficacy of the vaccine against any rotavirus disease for the first year was 49–64%, whereas during a 2-year follow-up, it was 57%. Protection against severe disease was 80% (Bernstein et al., 1995; Rennels et al., 1996). Vaccination also significantly decreased the number of medical office visits for rotavirus gastroenteritis or dehydration. In a postlicensure vaccine effectiveness study, it was determined that three doses of the vaccine were 100% effective in preventing hospitalization due to rotavirus (Staat et al., 2006). Vaccination appeared to provide protection against both G1 and G3 serotypes (Rennels et al., 1996; Bernstein, 2000). Similar results were reported from Finnish studies except that fever occurred somewhat more commonly and efficacy was somewhat enhanced (Joensuu et al., 1997). Protection similar to that seen in the US studies was also provided by the vaccine when it was evaluated in a less developed country, Venezuela (Perez-Schael et al., 1997).

As a result of these studies, RRV-TV or Rotashield™, was licensed in the United States in 1998 and recommended for general use in all infants (American Academy of Pediatrics, 1998). However, less than 1 year after licensing, 15 cases of intussusception that occurred shortly after vaccination were reported to the Vaccine Adverse Events Reporting System (VAERS). Intussusception is a form of intestinal blockage caused when a segment of the bowel prolapses into a more distal segment of the intestine (Centers for Disease Control, 1999a). Further analysis by the Advisory Committee on Immunization Practices led to the conclusion that an association existed between receipt of the vaccine and development of intussusception. Their recommendation for the use of Rotashield™ was withdrawn and the vaccine was withdrawn by the manufacturer (Centers for Disease Control, 1999b).

Much debate and controversy occurred after the use of Rotashield™ was halted. Initial publications describing the association reported an increased risk of the development of intussusception from 3 to 14 days after receipt of the first dose of Rotashield™ (adjusted odds ratio 21.7) and a smaller risk after administration of the second dose of vaccine (Murphy et al., 2001). The initial estimate was that one case of intussusception attributable to vaccination with RRV would occur for every 4670–9474 infants vaccinated. Therefore, the decision was made to discontinue the use of Rotashield™ (Murphy et al., 2001, 2002, 2003). In more recent analysis, the consensus was that the risk was approximately 1 in 10,000 (Peter and Myers, 2002). Further, when the data were further analyzed, it appeared that the risk of intussusception was age related and increased substantially in children greater than 90 days of age, leading some researchers to suggest that vaccination with any live rotavirus vaccine should be restricted to younger children (Simonsen et al., 2005; Rothman et al., 2006). The Food and Drug Administration (FDA) also recommended that future trials using live oral rotavirus vaccines should be conducted with more than 60,000 children to ensure that the risk of intussusception from a new vaccine would be less than was attributed to Rotashield™ (Glass et al., 2004).

The pathogenic mechanism underlying the association of the Rotasheild™ vaccine and intussusception has not been defined (Lynch et al., 2006). Several studies have investigated the possible infectious etiology of intussusception, but these studies have had a small sample size and several different pathogens, including adenoviruses, have been identified (Bhisitkul et al., 1992; Hsu et al., 1998). It does not appear that wild-type rotavirus causes intussusception as there is a lack of seasonal variation for intussusception that would be expected to occur if rotavirus was a major cause of intussusception (Rennels et al., 1998; Parashar et al., 2000; Chang et al., 2002).

RotaTeq™

After the bovine WC3 vaccine, discussed previously, was shown to be safe but not consistently effective, a monovalent vaccine containing the VP7 gene of a human G1 rotavirus (WI79-9) and the remainder of the genes from WC3, was developed. This vaccine was reported to be effective especially against more severe disease during a predominantly serotype G1 outbreak in Rochester (Clark et al., 1990). Next, because of the idea that serotype-specific protection was necessary, a WC3-based reassortant quadrivalent vaccine containing three viruses with gene substitutions of the VP7 gene with human G1, G2, or G3, and one virus containing a VP4 gene substitution with a human P[8] on a WC3 background was evaluated. In studies conducted at multiple centers in the United States, it was shown to be safe and effective against all cases of rotavirus gastroenteritis (75%) and especially against severe cases (100%) (Clark et al., 2004). Finally by adding a reassortant virus containing a human G4 VP7 to the above four viruses, a pentavalent vaccine, RotaTeq™, was developed by Merck Research Laboratories (Clark et al., 2006).

Because of the association of Rotashield™ with intussusception, the large pivotal trial of RotaTeq™ included over 70,000 infants from 11 countries. The vaccine was shown to be safe and, unlike Rotashield™, it did not induce fever (Vesikari et al., 2006d). Most importantly there was no association with intussusception. Intussusception occurred in 12 vaccine recipients and 15 placebo recipients within 1 year after the first dose of vaccine including six vaccine recipients and five placebo recipients within 42 days after any dose (relative risk, 1.6; 95% confidence interval, 0.4–6.4). The vaccine was also highly effective, reducing all G1–G4 rotavirus gastroenteritis by 74.0%, severe gastroenteritis by 98.0%, and hospitalizations and emergency room visits by 94.5% (Vesikari et al., 2006d). On February 3, 2006 RotaTeq™ was licensed by the US FDA for use in infants. The American Academy of Pediatrics recommended the vaccine which is administered in a three-dose schedule between 6 and 32 weeks of age, with a first dose administered between 6 and 12 weeks of age and all three doses administered by 32 weeks of age (American Academy of Pediatrics, C.O.I.D., 2007a, 2007b). In the United States, postmarketing safety of RotaTeq™ is being monitored by the CDC and the FDA. As of February, 2007 there was no suggestion of an association of RotaTeq™ vaccination with intussusception and CDC reaffirmed its recommendation for routine vaccination of infants with RotaTeq™ (Centers for Disease Control, 2007).

Bovine UK Reassortant Vaccine

The bovine UK strain, which is also a G6P[5] similar to WC3, has also been used to develop a reassortant tetravalent vaccine. The bovine UK reassortant vaccine was developed in the laboratory of Albert Kapikian of the National Institutes of Health in Bethesda, MD where RRV-TV was also developed. The tetravalent UK reassortant vaccine contains single VP7 gene substitutions from G1, G2, G3, or G4 human rotaviruses on a 10-gene UK background. Initially, monoreassortants were developed and evaluated. They were found to be safe and immunogenic in infants under 6 months of age and also did not induce fever (Clements-Mann et al., 1999). The tetravalent UK vaccine was then evaluated for safety and immunogenicity in the USA and later for safety/immunogenicity/efficacy in Finland. Although the studies were small, protection was significant (60% against any rotavirus disease and 90% against severe rotavirus disease; $P < 0.02$) (Vesikari et al., 2006c). In order to facilitate the production and commercialization of this vaccine, the NIH Office of Technology Transfer has granted licenses to at least eight groups throughout the world, seven of which are in developing nations, including vaccine manufacturers in Brazil, China, and India. There has been additional work done to develop UK-based reassortants containing additional VP7 genes for some of the emerging viruses such as G8 or G9 strains, thereby, proposing a hexavalent reassortant vaccine (Kapikian et al., 2005).

Human Strains

The use of human strains for vaccine candidates is based on the data that show that natural infection can provide protection against subsequent infection or disease. The human strains considered as vaccine candidates are either naturally attenuated, as thought to be the case with neonatal strains, or attenuated by culture adaptation and multiple passage. The candidates discussed below and listed in Table 35.4 are each composed of one strain and therefore, rely on homotypic and heterotypic mechanisms of protection.

Neonatal Strains

The first natural history study showed that asymptomatically infected neonates had a subsequent reduction in the frequency and severity of rotavirus diarrhea, thereby, generating interest in the use of neonatal strains as vaccine candidates (Bishop et al., 1983) Therefore, one of the first human strains to be tested as a rotavirus vaccine was a neonatal strain isolated from an asymptomatic newborn infant in Venezula. The strain, M37 is a G1P[6]. Several small studies with this candidate showed that the vaccine was safe but only moderately immunogenic, inducing neutralizing antibody responses to the

TABLE 35.4 Human strains developed as vaccine candidates

Candidate	Origin	G and P type	Comments
M37	Neonatal	G1P[6]	Isolated in Venezuela; safe but not very immunogenic; no protection in small trials; discontinued
RV3	Neonatal	G3P[6]	Isolated in Australia; safe, and small trials showed limited protection at low dose; plans for Phase II and III trials at higher doses
116E	Neonatal	G9P[11]	Isolated in India; natural bovine human reassortant; small trials have shown immune responses; development continues in India
I321	Neonatal	G10P[11]	Isolated in India; natural bovine human reassortant; small trials have shown very poor immune responses; development discontinued
Rotarix®	Child with gastroenteritis	G1P[8]	Isolated in Cincinnati, OH; safe, no association with intussusception after trials involving over 63,000 infants; highly protective, especially against severe disease; given in two doses; licensed in Mexico in 2004 and in over 90 countries worldwide to date

M37 strain and not to other G1 strains such as Wa (Midthun et al., 1991). In an efficacy study of 203 Finnish infants, there was no protection against predominantly G1 strains and this vaccine candidate was not pursued (Vesikari et al., 1991).

RV3, another vaccine candidate, is a G3P[6] human rotavirus that was isolated in an obstetric nursery in Melbourne, Australia, where it caused endemic, asymptomatic infections in newborn infants in the 1970s. As stated above, neonates infected with this virus were 100% protected against severe rotavirus disease, caused primarily by heterotypic G2P[4] strains, for their first 3 years of life (Bishop et al., 1983). Once RV3 was developed into a vaccine candidate, Phase I and early Phase II trials were conducted. Although an immune response to the vaccine (6.5×10^5 ffu) was detected in only 46% of subjects, protection in the responders against primarily serotype G1 circulating rotaviruses was 54% ($P = 0.08$) (Barnes et al., 1997, 2002). Future evaluations will utilize a higher dose of vaccine to improve the immune response and protection.

Two other, more recent vaccine candidates, were derived from two neonatal strains isolated from asymptomatically infected neonates in India in the mid-1980s, one from New Delhi and the other from the Bangalore region of India. Both of these strains are natural reassortants between bovine and human strains. The strain from New Delhi, 116E, a G9P[11] strain, contains a VP4 gene segment from a bovine rotavirus and the other 10 genes are from a human strain (Das et al., 1993a). The other strain I321, is a G10P[11] strain and contains nine genes of bovine origin and only gene segments for two nonstructural proteins from a human strain (Dunn et al., 1993). Early studies suggested that neonates infected with viruses similar to these two strains were protected against subsequent rotavirus illness (Bhan et al., 1993; Aijaz et al., 1996), but a recent report by another group suggests otherwise for G10P[11] or I321-like viruses (Banerjee et al., 2007). Both I321 and 116E vaccine candidates were tested in a small, placebo-controlled trial, that showed only the 116E strain elicited significant immune responses (Bhandari et al., 2006). Therefore, only the 116E strain is being pursued as a vaccine candidate in India under a consortium with partners from the United States including the Children's Vaccine Program at PATH (Dennehy, 2007).

Rotarix®

The second marketed vaccine, Rotarix, was first licensed in Mexico in 2005 and has since been approved in over 100 countries worldwide including developed, less developed, and developing nations (Bernstein, 2006). This vaccine is based on the attenuated human strain, 89-12. This strain was obtained from an infant with rotavirus gastroenteritis in Cincinnati, OH and is a G1P[8] strain, representing the most common strain worldwide (Santos and Hoshino, 2005). The isolate was attenuated by multiple passages in tissue culture and then developed into a vaccine. Early studies with 89-12 showed that it was safe but a low-grade fever did develop in 19% of recipients. Results of a multicenter efficacy trial showed that two doses of this vaccine provided 89% protection against any rotavirus disease and 100% protection from severe disease (Bernstein et al., 1999).

The 89-12 strain was further purified by limiting dilution and passage in tissue culture by GlaxoSmithKline. The final product was called RIX4414 and is now marketed as Rotarix (GlaxoSmithKline), a two dose oral vaccine. Initial safety testing in Finland showed that the vaccine was safe and more attenuated than the parent vaccine strain as it did not produce the fevers that had been detected with the parent 89-12 (Vesikari et al., 2004a). Subsequent studies in Singapore, Finland, Latin America, and the US confirmed that it was safe, not associated with fever and still remained highly immunogenic (Phua et al., 2005; Salinas et al., 2005). Other studies showed the vaccine did not interfere with other concomitantly administered childhood vaccines (Dennehy et al., 2005; Ruiz-Palacios et al., 2007). In the initial efficacy trial of RIX4414 conducted in Finland over two rotavirus seasons, the vaccine was 73% protective against all rotavirus gastroenteritis and 90% protective against severe illness even though a relatively low dose was used (Vesikari et al., 2004b).

Similar to RotaTeq™, and because of the concerns about intussusception, the pivotal vaccine trial with RIX4414 also involved large numbers. Over 63,000 infants from several countries in Latin America and Finland were enrolled in a safety trial. The RIX4414 vaccine ($10^{6.5}$ median cell culture infective dose) was given in a two dose schedule at approximately 2 and 4 months of age (compared to the three dose schedule of RotaTeq™). This large study showed the vaccine was safe, did not induce fever and again, most importantly, was not associated with intussusception. During the 31-day period after each dose, six vaccine recipients and seven placebo recipients had confirmed intussusception with no evidence of clustering (difference in risk, −0.32 per 10,000 infants, 95% confidence interval, −2.91 to 2.18; $P = 0.78$). During the entire study, there were 25 cases of intussusception, 16 in the placebo group and 9 in the vaccine group (Ruiz-Palacios et al., 2006).

Efficacy data from a subset of 20,000 infants from this trial showed 85% protection against severe rotavirus diarrhea and hospitalization. Protection against more severe gastroenteritis was 100%. It was also demonstrated that protection was high (86%) not only against severe rotavirus diarrhea caused by G1P[8] strains but also against G3P[8], G4P[8] and G9P[8] strains which all shared the VP4 P[8] genotype. Efficacy against G2P[4] strains (strains that are not matched for either VP4 or VP7) was 41%. However, in a meta-analysis efficacy was calculated to be 67–71%, indicating that the vaccine will provide substantial protection even against strains that do not have either G1 or P[8] specificity (Ruiz-Palacios et al., 2006).

In a more recent trial, conducted in six European countries, in a population more similar to that used in trials for RotaTeq™ over 4000 infants were evaluated. Protection in this study was 87% against any rotavirus gastroenteritis, 96% against severe disease and 100% against hospitalization due to rotavirus. In this study, efficacy against G3, G4, and G9 strains was similar to that against G1 strains and was over 95%. Efficacy against the unrelated G2 strains was 75%, suggesting that heterotypic or non-neutralizing antibody responses are also involved in protection (Vesikari et al., 2006a, 2006b). Of importance, protection against hospitalization for gastroenteritis of any etiology was 75%, demonstrating the importance of rotavirus as a pathogen of severe gastroenteritis in these countries.

One of the best correlations between rotavirus antibody induced by vaccination and protection was shown in studies using Rotarix. If a vaccinee did not develop a measurable serum rotavirus IgA response after two doses of Rotarix, they were significantly more likely to develop rotavirus disease when compared to subjects that did develop an IgA response (De Vos et al., 2004). However, protection rates against severe rotavirus illness were higher than the rate of vaccinees who had a measurable IgA response after Rotarix immunization (Salinas et al., 2005).

Similar to RotaTeq™, postlicensure safety of Rotarix is being monitored in the countries where use has begun. Additional trials are ongoing in developing countries in Asia and Africa to determine safety and efficacy in these areas where rotavirus disease burden is greatest. In addition, an application for licensure in the United States was accepted for review by the FDA in August, 2007.

NONLIVING ROTAVIRUS VACCINE CANDIDATES

In addition to live oral rotavirus vaccines, a number of nonliving vaccine candidates have been developed and evaluated in animal models. Nonliving vaccine candidates were developed in attempt to formulate a vaccine that would be more effective than live oral vaccines, since even natural infection does not always result in 100% protection from reinfection. In addition, these candidates were often studied in an attempt to better define mechanisms of protection, including determinations regarding the protective efficacy of various viral proteins. Finally after the association of Rotashield with intussusception, possible safety concerns with live virus vaccines added to the possible need for nonliving vaccine candidates. The nonliving candidates studied include DNA vaccines (Chen et al., 1997; Choi et al., 2007), inactivated purified triple and double-layered virus particles (McNeal et al., 1998, 1999), recombinant virus-like particles (VLPs) containing VP2 and VP6, with or without VP4 and VP7 (O'Neal et al., 1998), and recombinant expressed VP6 protein (Choi et al., 2002a). To date none of these candidates have been tested in humans.

Using animal models, each of the above candidates have shown significant levels of protection, especially in the adult mouse model which uses reduction of viral shedding as its measure of protection. Of note, VLPs did not induce protection from illness when used in the gnotobiotic pig model, currently the only illness animal model (Yuan et al., 2000). Except for DNA-based vaccines, the other candidates often require the use of an effective adjuvant, usually recombinant formulations of bacterial toxins, to induce protection (Choi et al., 2002b; Bertolotti-Ciarlet et al., 2003). However, the use of this type of adjuvant in humans greatly increases the safety concerns associated with these vaccine approaches. If the present licensed live oral vaccines, RotaTeq™ and Rotarix, or other live oral rotavirus vaccines prove to be ineffective or have safety issues, the present nonliving vaccines may be further developed.

CONCLUSIONS

Worldwide, rotavirus infections are responsible for over 600,000 deaths, representing 5% of all childhood deaths in children less than 5 years of age. A safe and effective rotavirus vaccine is the best hope of combating this disease. After the largest vaccine trials in history, two live oral vaccines, RotaTeq™ and Rotarix, have been licensed or will soon be licensed in many countries around the world. Both vaccines were shown to be safe and effective, decreasing severe disease and hospitalization due to rotavirus by 85–100%. In addition, neither vaccine has been associated with intussusception in either the large trials or postmarketing surveillance. However, neither vaccine has been evaluated, to date, in the poorest countries where it is most needed. Assuming the vaccines are also effective in these countries, the next hurdle will be to make the vaccines available to these populations. The GAVI Alliance, formerly known as the Global Alliance for Vaccines and Immunization, has committed to helping with the financial cost of the vaccine in countries which qualify for support. In addition, in February, 2007, Rotarix was granted WHO prequalification status, an important first step in getting a rotavirus vaccine to poor countries.

KEY ISSUES

- Rotavirus is the most common cause of gastroenteritis in young children in all parts of the world. It is responsible for over 600,000 deaths around the world, and over 50,000 hospitalizations in the US.
- Rotavirus is a segmented double-stranded RNA virus. Serotyping is based on the two outer capsid proteins that induce neutralizing antibodies VP7 (G type) and VP4 (P type).
- The mechanism of protection against rotavirus is not completely clear but neutralizing antibodies play a role. The role of non-neutralizing antibodies and T cells are less clear but are probably important.
- Rotavirus vaccine development has largely concentrated on the use of live attenuated vaccines administered by the oral route in order to generate local, intestinal immunity.
- The initial approach of using animal rotavirus strains that were naturally attenuated for humans was replaced by the use of reassorted strains, i.e., replacement of animal rotavirus genes with human genes (most commonly encoding VP7 or VP7 and VP4 proteins) or using an attenuated human strain.
- The first licensed rotavirus vaccine Rotashield™ used a rhesus monkey rotavirus strain as the backbone for a quadrivalent vaccine. This vaccine was associated with intussusception and was quickly withdrawn from the market.
- There are now two rotavirus vaccines available in the US, EU, Latin America as well as many other countries. Rotateq™ is a pentavalent vaccine based on reassortants of a bovine rotavirus while Rotarix™ is an attenuated human strain.
- Both vaccine have undergone rigorous testing in large human trials of over 60,000 infants and appear safe with neither being associated with intussusception.
- Both vaccines also appear to protect infants from the most common human serotypes.

- The remaining challenge for both vaccines is to show they can impact disease in the poorest countries of the world where mortality is the highest.

References

Adams, W.R. and Kraft, L.M. Epizootic diarrhea of infant mice: identification of the etiologic agent. Science 1963; 141:359–360.

Aijaz, S., Gowda, K., Jagannath, H.V., Reddy, R.R., Maiya, P.P., Ward, R.L., Greenberg, H.B., Raju, M., Babu, A. and Rao, C.D. Epidemiology of symptomatic human rotaviruses in Bangalore and Mysore, India, from 1988 to 1994 as determined by electropherotype, subgroup and serotype analysis. Arch. Virol. 1996; 141:715–726.

American Academy of Pediatrics, C.O.I.D. Prevention of rotavirus disease: guidelines for use of rotavirus vaccine. Pediatrics 1998; 1–2:1483–1491.

American Academy of Pediatrics, C.O.I.D. Recommended immunization schedules for children and adolescents—United States, 2007. Pediatrics 2007a; 119:207–208, 3 p following 208.

American Academy of Pediatrics, C.O.I.D. Prevention of rotavirus disease: guidelines for use of rotavirus vaccine. Pediatrics 2007b; 119:171–182.

Arias, C.F., Romero, P., Alvarez, V. and Lopez, S. Trypsin activation pathway of rotavirus infectivity. J. Virol. 1996; 70:5832–5839.

Ball, J.M., Tian, P., Zeng, C.Q., Morris, A.P. and Estes, M.K. Age-dependent diarrhea induced by a rotaviral nonstructural glycoprotein. Science 1996; 272:101–104.

Banerjee, I., Gladstone, B.P., Le Fevre, A.M., Ramani, S., Iturriza-Gomara, M., Gray, J.J., Brown, D.W., Estes, M.K., Muliyil, J.P., Jaffar, S. and Kang, G. Neonatal infection with G10P[11] rotavirus did not confer protection against subsequent rotavirus infection in a community cohort in Vellore, South India. J. Infect. Dis. 2007; 195:625–632.

Barnes, G.L., Lund, J.S., Adams, L., Mora, A., Mitchell, S.V., Caples, A. and Bishop, R.F. Phase 1 trial of a candidate rotavirus vaccine (RV3) derived from a human neonate. J. Paediatr. Child. Health 1997; 33:300–304.

Barnes, G.L., Lund, J.S., Mitchell, S.V., De Bruyn, L., Piggford, L., Smith, A.L., Furmedge, J., Masendycz, P.J., Bugg, H.C., Bogdanovic-Sakran, N., Carlin, J.B. and Bishop, R.F. Early phase II trial of human rotavirus vaccine candidate RV3. Vaccine 2002; 20:2950–2956.

Bernstein, D.I. Rotavirus vaccine, current status and future prospects. BioDrugs 2000; 14:275–311.

Bernstein, D.I. Live attenuated human rotavirus vaccine, Rotarix. Semin. Pediatr. Infect. Dis. 2006; 17:188–194.

Bernstein, D.I., Glass, R.I., Rodgers, G., Davidson, B.L. and Sack, D.A. Evaluation of rhesus rotavirus monovalent and tetravalent reassortant vaccines in US children. US Rotavirus Vaccine Efficacy Group. JAMA 1995; 273:1191–1196.

Bernstein, D.I., Kacica, M.A., McNeal, M.M., Schiff, G.M. and Ward, R.L. Local and systemic antibody response to rotavirus WC3 vaccine in adult volunteers. Antiviral Res. 1989; 12:293–300.

Bernstein, D.I., Sack, D.A., Rothstein, E., Reisinger, K., Smith, V.E., O'Sullivan, D., Spriggs, D.R. and Ward, R.L. Efficacy of live, attenuated, human rotavirus vaccine 89-12 in infants: a randomised placebo-controlled trial. Lancet 1999; 354:287–290.

Bernstein, D.I., Sander, D.S., Smith, V.E., Schiff, G.M. and Ward, R.L. Protection from rotavirus reinfection: 2-year prospective study. J. Infect. Dis. 1991; 164:277–283.

Bernstein, D.I., Smith, V.E., Sander, D.S., Pax, K.A., Schiff, G.M. and Ward, R.L. Evaluation of WC3 rotavirus vaccine and correlates of protection in healthy infants. J. Infect. Dis. 1990; 162: 1055–1062.

Bertolotti-Ciarlet, A., Ciarlet, M., Crawford, S.E., Conner, M.E. and Estes, M.K. Immunogenicity and protective efficacy of rotavirus 2/6-virus-like particles produced by a dual baculovirus expression vector and administered intramuscularly, intranasally, or orally to mice. Vaccine 2003; 21:3885–3900.

Bhan, M.K., Lew, J.F., Sazawal, S., Das, B.K., Gentsch, J.R. and Glass, R.I. Protection conferred by neonatal rotavirus infection against subsequent rotavirus diarrhea. J. Infect. Dis. 1993; 168:282–287.

Bhandari, N., Sharma, P., Glass, R.I., Ray, P., Greenberg, H., Taneja, S., Saksena, M., Rao, C.D., Gentsch, J.R., Parashar, U., Maldonado, Y., Ward, R.L. and Bhan, M.K. Safety and immunogenicity of two live attenuated human rotavirus vaccine candidates, 116E and I321, in infants: results of a randomised controlled trial. Vaccine 2006; 24:5817–5823.

Bhisitkul, D.M., Todd, K.M. and Listernick, R. Adenovirus infection and childhood intussusception. Am. J. Dis. Child. 1992; 146:1331–1333.

Bishop, R.F., Barnes, G.L., Cipriani, E. and Lund, J.S. Clinical immunity after neonatal rotavirus infection. A prospective longitudinal study in young children. N. Engl. J. Med. 1983; 309:72–76.

Bishop, R.F., Davidson, G.P., Holmes, I.H. and Ruck, B.J. Virus particles in epithelial cells of duodenal mucosa from children with acute non-bacterial gastroenteritis. Lancet 1973; 2:1281–1283.

Bishop, R.F., Davidson, G.P., Holmes, I.H. and Ruck, B.J. Detection of a new virus by electron microscopy of faecal extracts from children with acute gastroenteritis. Lancet 1974; 1:149–151.

Blanchard, H., Yu, X., Coulson, B.S. and Von Itzstein, M. Insight into host cell carbohydrate-recognition by human and porcine rotavirus from crystal structures of the virion spike associated carbohydrate-binding domain (VP8*). J. Mol. Biol. 2007; 367:1215–1226.

Blutt, S.E., Fenaux, M., Warfield, K.L., Greenberg, H.B. and Conner, M.E. Active viremia in rotavirus-infected mice. J. Virol. 2006; 80: 6702–6705.

Blutt, S.E., Kirkwood, C.D., Parreno, V., Warfield, K.L., Ciarlet, M., Estes, M.K., Bok, K., Bishop, R.F. and Conner, M.E. Rotavirus antigenaemia and viraemia: a common event? Lancet 2003; 362:1445–1449.

Blutt, S.E., Matson, D.O., Crawford, S.E., Staat, M.A., Azimi, P., Bennett, B.L., Piedra, P.A. and Conner, M.E. Rotavirus antigenemia in children is associated with viremia. PLoS Med. 2007; 4:e121.

Bridger, J.C. Detection by electron microscopy of caliciviruses, astroviruses and rotavirus-like particles in the faeces of piglets with diarrhoea. Vet. Rec. 1980; 107:532–533.

Brown, K.A., Kriss, J.A., Moser, C.A., Wenner, W.J. and Offit, P.A. Circulating rotavirus-specific antibody-secreting cells (ASCs) predict the presence of rotavirus-specific ASCs in the human small intestinal lamina propria. J. Infect. Dis. 2000; 182: 1039–1043.

Browne, E.P., Bellamy, A.R. and Taylor, J.A. Membrane-destabilizing activity of rotavirus NSP4 is mediated by a membrane-proximal amphipathic domain. J. Gen. Virol. 2000; 81:1955–1959.

Brunet, J.P., Cotte-Laffitte, J., Linxe, C., Quero, A.M., Geniteau-Legendre, M. and Servin, A. Rotavirus infection induces an increase in intracellular calcium concentration in human intestinal epithelial cells: role in microvillar actin alteration. J. Virol. 2000a; 74:2323–2332.

Brunet, J.P., Jourdan, N., Cotte-Laffitte, J., Linxe, C., Geniteau-Legendre, M., Servin, A. and Quero, A.M. Rotavirus infection induces cytoskeleton disorganization in human intestinal epithelial cells: implication of an increase in intracellular calcium concentration. J. Virol. 2000b; 74:10801–10806.

Centers for Disease Control Intussusception among recipients of rotavirus vaccine—United States, 1998–1999. MMWR 1999a; 48:577–581.

Centers for Disease Control Withdrawal of rotavirus vaccine recommendation. JAMA 1999b; 282:2113–2114.

Centers for Disease Control Postmarketing monitoring of intussusception after RotaTeq vaccination—United States, February 1, 2006–February 15, 2007. MMWR 2007; 56:218–222.

Chang, E.J., Zangwill, K.M., Lee, H. and Ward, J.I. Lack of association between rotavirus infection and intussusception: implications for use of attenuated rotavirus vaccines. Pediatr. Infect. Dis. J. 2002; 21:97–102.

Cheever, F.S. and Mueller, J.H. Epidemic diarrheal disease of suckling mice. I. Manifestations, epidemiology, and attempts to transmit the disease. J. Exp. Med. 1947; 85:405–416.

Chen, S.C., Fynan, E.F., Robinson, H.L., Lu, S., Greenberg, H.B., Santoro, J.C. and Herrmann, J.E. Protective immunity induced by rotavirus DNA vaccines. Vaccine 1997; 15:899–902.

Chiappini, E., Azzari, C., Moriondo, M., Galli, L. and De Martino, M. Viraemia is a common finding in immunocompetent children with rotavirus infection. J. Med. Virol. 2005; 76:265–267.

Chiappini, E., Galli, L. and De Martino, M. Viremia and clinical manifestations in children with rotavirus infection. J. Infect. Dis. 2006; 193:1333.

Chiba, S., Yokoyama, T., Nakata, S., Morita, Y., Urasawa, T., Taniguchi, K., Urasawa, S. and Nakao, T. Protective effect of naturally acquired homotypic and heterotypic rotavirus antibodies. Lancet 1986; 2:417–421.

Choi, A.H., McNeal, M.M., Basu, M., Flint, J.A., Stone, S.C., Clements, J.D., Bean, J.A., Poe, S.A., Vancott, J.L. and Ward, R.L. Intranasal or oral immunization of inbred and outbred mice with murine or human rotavirus VP6 proteins protects against viral shedding after challenge with murine rotaviruses. Vaccine 2002a; 20:3310–3321.

Choi, A.H., McNeal, M.M., Flint, J.A., Basu, M., Lycke, N.Y., Clements, J.D., Bean, J.A., Davis, H.L., Mccluskie, M.J., Vancott, J.L. and Ward, R.L. The level of protection against rotavirus shedding in mice following immunization with a chimeric VP6 protein is dependent on the route and the coadministered adjuvant. Vaccine 2002b; 20:1733–1740.

Choi, A.H., Smiley, K., Basu, M., McNeal, M.M., Shao, M., Bean, J.A., Clements, J.D., Stout, R.R. and Ward, R.L. Protection of mice against rotavirus challenge following intradermal DNA immunization by Biojector needle-free injection. Vaccine 2007; 25: 3215–3218.

Clark, B. and McKendrick, M. A review of viral gastroenteritis. Curr. Opin. Infect. Dis. 2004; 17:461–469.

Clark, H.F., Bernstein, D.I., Dennehy, P.H., Offit, P., Pichichero, M., Treanor, J., Ward, R.L., Krah, D.L., Shaw, A., Dallas, M.J., Laura, D., Eiden, J.J., Ivanoff, N., Kaplan, K.M. and Heaton, P. Safety, efficacy, and immunogenicity of a live, quadrivalent human-bovine reassortant rotavirus vaccine in healthy infants. J. Pediatr. 2004; 144:184–190.

Clark, H.F., Borian, F.E., Bell, L.M., Modesto, K., Gouvea, V. and Plotkin, S.A. Protective effect of WC3 vaccine against rotavirus diarrhea in infants during a predominantly serotype 1 rotavirus season. J. Infect. Dis. 1988; 158:570–587.

Clark, H.F., Borian, F.E., Modesto, K. and Plotkin, S.A. Serotype 1 reassortant of bovine rotavirus WC3, strain WI79-9, induces a polytypic antibody response in infants. Vaccine 1990; 8:327–332.

Clark, H.F., Furukawa, T., Bell, L.M., Offit, P.A., Perrella, P.A. and Plotkin, S.A. Immune response of infants and children to low-passage bovine rotavirus (strain WC3). Am. J. Dis. Child. 1986; 140:350–356.

Clark, H.F., Offit, P.A., Plotkin, S.A. and Heaton, P.M. The new pentavalent rotavirus vaccine composed of bovine (strain WC3)-human rotavirus reassortants. Pediatr. Infect. Dis. J. 2006; 25:577–583.

Clemens, J.D., Ward, R.L., Rao, M.R., Sack, D.A., Knowlton, D.R., Van Loon, F.P., Huda, S., McNeal, M., Ahmed, F. and Schiff, G. Seroepidemiologic evaluation of antibodies to rotavirus as correlates of the risk of clinically significant rotavirus diarrhea in rural Bangladesh. J. Infect. Dis. 1992; 165:161–165.

Clements-Mann, M.L., Makhene, M.K., Mrukowicz, J., Wright, P.F., Hoshino, Y., Midthun, K., Sperber, E., Karron, R. and Kapikian, A.Z. Safety and immunogenicity of live attenuated human-bovine (UK) reassortant rotavirus vaccines with VP7-specificity for serotypes 1, 2, 3 or 4 in adults, children and infants. Vaccine 1999; 17:2715–2725.

Collins, J., Starkey, W.G., Wallis, T.S., Clarke, G.J., Worton, K.J., Spencer, A.J., Haddon, S.J., Osborne, M.P., Candy, D.C. and Stephen, J. Intestinal enzyme profiles in normal and rotavirus-infected mice. J. Pediatr. Gastroenterol. Nutr. 1988; 7:264–272.

Conner, M.E., Estes, M.K. and Graham, D.Y. Rabbit model of rotavirus infection. J. Virol. 1988; 62:1625–1633.

Conner, M.E., Matson, D.O. and Estes, M.K. Rotavirus vaccines and vaccination potential. Curr. Top. Microbiol. Immunol. 1994; 185:285–337.

Cook, S.M., Glass, R.I., Lebaron, C.W. and Ho, M.S. Global seasonality of rotavirus infections. Bull. World Health Organ. 1990; 68:171–177.

Coulson, B.S., Grimwood, K., Masendycz, P.J., Lund, J.S., Mermelstein, N., Bishop, R.F. and Barnes, G.L. Comparison of rotavirus immunoglobulin A coproconversion with other indices of rotavirus infection in a longitudinal study in childhood. J. Clin. Microbiol. 1990; 28:1367–1374.

Coulson, B.S., Unicomb, L.E., Pitson, G.A. and Bishop, R.F. Simple and specific enzyme immunoassay using monoclonal antibodies for serotyping human rotaviruses. J. Clin. Microbiol. 1987; 25:509–515.

Cuadras, M.A., Arias, C.F. and Lopez, S. Rotaviruses induce an early membrane permeabilization of MA104 cells and do not require a low intracellular Ca^{2+} concentration to initiate their replication cycle. J. Virol. 1997; 71:9065–9074.

Cunliffe, N.A., Gentsch, J.R., Kirkwood, C.D., Gondwe, J.S., Dove, W., Nakagomi, O., Nakagomi, T., Hoshino, Y., Bresee, J.S., Glass, R.I., Molyneux, M.E. and Hart, C.A. Molecular and serologic characterization of novel serotype G8 human rotavirus strains detected in Blantyre, Malawi. Virology 2000; 274:309–320.

Das, B.K., Gentsch, J.R., Hoshino, Y., Ishida, S., Nakagomi, O., Bhan, M.K., Kumar, R. and Glass, R.I. Characterization of the G serotype and genogroup of New Delhi newborn rotavirus strain 116E. Virology 1993a; 197:99–107.

Das, M., Dunn, S.J., Woode, G.N., Greenberg, H.B. and Rao, C.D. Both surface proteins (VP4 and VP7) of an asymptomatic neonatal rotavirus strain (I321) have high levels of sequence identity with the homologous proteins of a serotype 10 bovine rotavirus. Virology 1993b; 194:374–379.

Davidson, G.P., Gall, D.G., Petric, M., Butler, D.G. and Hamilton, J.R. Human rotavirus enteritis induced in conventional piglets. Intestinal structure and transport. J. Clin. Invest. 1977; 60: 1402–1409.

Demol, P., Zissis, G., Tubzler, J.P. et al. Failure of live attenuated oral rotavirus vaccine. Lancet 1986; 2:108.

Dennehy, P.H. Rotavirus vaccines: an update. Vaccine 2007; 25:3137–3141.

Dennehy, P.H., Brady, R.C., Halperin, S.A., Ward, R.L., Alvey, J.C., Fischer, F.H., Jr., Innis, B.L., Rathfon, H., Schuind, A. and De Vos, B.

Comparative evaluation of safety and immunogenicity of two dosages of an oral live attenuated human rotavirus vaccine. Pediatr. Infect. Dis. J. 2005; 24:481–488.

De Vos, B., Vesikari, T., Linhares, A.C., Salinas, B., Perez-Schael, I., Ruiz-Palacios, G.M., Guerrero Mde, L., Phua, K.B., Delem, A. and Hardt, K. A rotavirus vaccine for prophylaxis of infants against rotavirus gastroenteritis. Pediatr. Infect. Dis. J. 2004; 23:S179–S182.

Dormitzer, P.R., Nason, E.B., Prasad, B.V. and Harrison, S.C. Structural rearrangements in the membrane penetration protein of a non-enveloped virus. Nature 2004; 430:1053–1058.

Dunn, S.J., Greenberg, H.B., Ward, R.L., Nakagomi, O., Burns, J.W., Vo, P.T., Pax, K.A., Das, M., Gowda, K. and Rao, C.D. Serotypic and genotypic characterization of human serotype 10 rotaviruses from asymptomatic neonates. J. Clin. Microbiol. 1993; 31:165–169.

Estes, M.K. Rotaviruses and their replication. In: Fields Virology (D. Knipe, P. Howley, D. Griffin, R. Lamb, M. Martin, B. Roizman and S. Strauss, Eds.), 4th ed. Philadelphia: Lippincott, Williams and Wilkins, 2001.

Estes, M.K. and Cohen, J. Rotavirus gene structure and function. Microbiol. Rev. 1989; 53:410–449.

Estes, M.K., Graham, D.Y. and Mason, B.B. Proteolytic enhancement of rotavirus infectivity: molecular mechanisms. J. Virol. 1981; 39:879–888.

Estes, M.K. and Kapikian, A.Z. Rotaviruses. In: Fields Virology (D. Knipe and P. Howley, Eds.), 5th ed. Philadelphia: Lippincott Williams & Wilkins, 2007.

Fabbretti, E., Afrikanova, I., Vascotto, F. and Burrone, O.R. Two non-structural rotavirus proteins, NSP2 and NSP5, form viroplasm-like structures in vivo. J. Gen. Virol. 1999; 80(Pt. 2):333–339.

Fischer, T.K., Ashley, D., Kerin, T., Reynolds-Hedmann, E., Gentsch, J., Widdowson, M.A., Westerman, L., Puhr, N., Turcios, R.M. and Glass, R.I. Rotavirus antigenemia in patients with acute gastroenteritis. J. Infect. Dis. 2005; 192:913–919.

Fischer, T.K., Viboud, C., Parashar, U., Malek, M., Steiner, C., Glass, R. and Simonsen, L. Hospitalizations and deaths from diarrhea and rotavirus among children <5 years of age in the United States, 1993–2003. J. Infect. Dis. 2007; 195:1117–1125.

Flewett, T.H., Bryden, A.S., Davies, H., Woode, G.N., Bridger, J.C. and Derrick, J.M. Relation between viruses from acute gastroenteritis of children and newborn calves. Lancet 1974; 2:61–63.

Flores, J., Perez-Schael, I., Blanco, M., White, L., Garcia, D., Vilar, M., Cunto, W., Gonzalez, R., Urbina, C., Boher, J. et al. Comparison of reactogenicity and antigenicity of M37 rotavirus vaccine and rhesus-rotavirus-based quadrivalent vaccine. Lancet 1990; 336:330–334.

Franco, M.A., Angel, J. and Greenberg, H.B. Immunity and correlates of protection for rotavirus vaccines. Vaccine 2006; 24:2718–2731.

Franco, M.A. and Greenberg, H.B. Role of B cells and cytotoxic T lymphocytes in clearance of and immunity to rotavirus infection in mice. J. Virol. 1995; 69:7800–7806.

Franco, M.A., Lefevre, P., Willems, P., Tosser, G., Lintermanns, P. and Cohen, J. Identification of cytotoxic T cell epitopes on the VP3 and VP6 rotavirus proteins. J. Gen. Virol. 1994; 75(Pt 3): 589–596.

Franco, M.A., Tin, C. and Greenberg, H.B. CD8+ T cells can mediate almost complete short-term and partial long-term immunity to rotavirus in mice. J. Virol. 1997; 71:4165–4170.

Gentsch, J.R., Glass, R.I., Woods, P., Gouvea, V., Gorziglia, M., Flores, J., Das, B.K. and Bhan, M.K. Identification of group A rotavirus gene 4 types by polymerase chain reaction. J. Clin. Microbiol. 1992; 30:1365–1373.

Gentsch, J.R., Laird, A.R., Bielfelt, B., Griffin, D.D., Banyai, K., Ramachandran, M., Jain, V., Cunliffe, N.A., Nakagomi, O., Kirkwood, C.D., Fischer, T.K., Parashar, U.D., Bresee, J.S., Jiang, B. and Glass, R.I. Serotype diversity and reassortment between human and animal rotavirus strains: implications for rotavirus vaccine programs. J. Infect. Dis. 2005; 192(Suppl. 1):S146–S159.

Gentsch, J.R., Woods, P.A., Ramachandran, M., Das, B.K., Leite, J.P., Alfieri, A., Kumar, R., Bhan, M.K. and Glass, R.I. Review of G and P typing results from a global collection of rotavirus strains: implications for vaccine development. J. Infect. Dis. 1996; 174(Suppl. 1):S30–S36.

Georges-Courbot, M.C., Monges, J., Siopathis, M.R., Roungou, J.B., Gresenguet, G., Bellec, L., Bouquety, J.C., Lanckriet, C., Cadoz, M., Hessel, L. et al. Evaluation of the efficacy of a low-passage bovine rotavirus (strain WC3) vaccine in children in Central Africa. Res. Virol. 1991; 142:405–411.

Giaquinto, C., Van Damme, P., Huet, F., Gothefors, L., Maxwell, M., Todd, P. and Da Dalt, L. Clinical consequences of rotavirus acute gastroenteritis in Europe, 2004–2005: the REVEAL study. J. Infect. Dis. 2007a; 195(Suppl. 1):S26–S35.

Giaquinto, C., Van Damme, P., Huet, F., Gothefors, L. and Van der Wielen, M. Costs of community-acquired pediatric rotavirus gastroenteritis in 7 European countries: the REVEAL Study. J. Infect. Dis. 2007b; 195(Suppl. 1):S36–S44.

Girard, M.P., Steele, D., Chaignat, C.L. and Kieny, M.P. A review of vaccine research and development: human enteric infections. Vaccine 2006; 24:2732–2750.

Glass, R.I., Bresee, J.S., Parashar, U.D., Jiang, B. and Gentsch, J. The future of rotavirus vaccines: a major setback leads to new opportunities. Lancet 2004; 363:1547–1550.

Glass, R.I., Bresee, J.S., Turcios, R., Fischer, T.K., Parashar, U.D. and Steele, A.D. Rotavirus vaccines: targeting the developing world. J. Infect. Dis. 2005; 192(Suppl. 1):S160–S166.

Glass, R.I., Kilgore, P.E., Holman, R.C., Jin, S., Smith, J.C., Woods, P.A., Clarke, M.J., Ho, M.S. and Gentsch, J.R. The epidemiology of rotavirus diarrhea in the United States: surveillance and estimates of disease burden. J. Infect. Dis. 1996; 174 (Suppl. 1):S5–S11.

Glass, R.I., Parashar, U.D., Bresee, J.S., Turcios, R., Fischer, T.K., Widdowson, M.A., Jiang, B. and Gentsch, J.R. Rotavirus vaccines: current prospects and future challenges. Lancet 2006; 368:323–332.

Gorziglia, M., Larralde, G., Kapikian, A.Z. and Chanock, R.M. Antigenic relationships among human rotaviruses as determined by outer capsid protein VP4. Proc. Natl. Acad. Sci. U.S.A. 1990; 87:7155–7159.

Gouvea, V., Santos, N. and Timenetsky Mdo, C. VP4 typing of bovine and porcine group A rotaviruses by PCR. J. Clin. Microbiol. 1994; 32:1333–1337.

Graham, D.Y., Sackman, J.W. and Estes, M.K. Pathogenesis of rotavirus-induced diarrhea. Preliminary studies in miniature swine piglet. Dig. Dis. Sci. 1984; 29:1028–1035.

Graham, K.L., Fleming, F.E., Halasz, P., Hewish, M.J., Nagesha, H.S., Holmes, I.H., Takada, Y. and Coulson, B.S. Rotaviruses interact with alpha4beta7 and alpha4beta1 integrins by binding the same integrin domains as natural ligands. J. Gen. Virol. 2005; 86:3397–3408.

Graham, K.L., Takada, Y. and Coulson, B.S. Rotavirus spike protein VP5* binds alpha2beta1 integrin on the cell surface and competes with virus for cell binding and infectivity. J. Gen. Virol. 2006; 87:1275–1283.

Greenberg, H.B., Valdesuso, J., Van Wyke, K., Midthun, K., Walsh, M., Mcauliffe, V., Wyatt, R.G., Kalica, A.R., Flores, J. and Hoshino, Y. Production and preliminary characterization of monoclonal antibodies directed at two surface proteins of rhesus rotavirus. J. Virol. 1983; 47:267–275.

Griffin, D.D., Fletcher, M., Levy, M.E., Ching-Lee, M., Nogami, R., Edwards, L., Peters, H., Montague, L., Gentsch, J.R. and Glass, R.I.

Outbreaks of adult gastroenteritis traced to a single genotype of rotavirus. J. Infect. Dis. 2002; 185:1502–1505.

Grimwood, K., Lund, J.C., Coulson, B.S., Hudson, I.L., Bishop, R.F. and Barnes, G.L. Comparison of serum and mucosal antibody responses following severe acute rotavirus gastroenteritis in young children. J. Clin. Microbiol. 1988; 26:732–738.

Haffejee, I.E. The pathophysiology, clinical features and management of rotavirus diarrhoea. Q. J. Med. 1991; 79:289–299.

Haffejee, I.E. The epidemiology of rotavirus infections: a global perspective. J. Pediatr. Gastroenterol. Nutr. 1995; 20:275–286.

Haffejee, I.E. and Moosa, A. Rotavirus studies in Indian (Asian) South African infants with acute gastro-enteritis: II. Clinical aspects and outcome. Ann. Trop. Paediatr. 1990; 10:245–254.

Hanlon, P., Hanlon, L., Marsh, V., Byass, P., Shenton, F., Hassan-King, M., Jobe, O., Sillah, H., Hayes, R., M'Boge, B.H. et al. Trial of an attenuated bovine rotavirus vaccine (RIT 4237) in Gambian infants. Lancet 1987; 1:1342–1345.

Hattori, H., Torii, S., Nagafuji, H., Tabata, Y. and Hata, A. Benign acute myositis associated with rotavirus gastroenteritis. J. Pediatr. 1992; 121:748–749.

Hewish, M.J., Takada, Y. and Coulson, B.S. Integrins alpha2beta1 and alpha4beta1 can mediate SA11 rotavirus attachment and entry into cells. J. Virol. 2000; 74:228–236.

Holmes, I.H., Ruck, B.J., Bishop, R.F. and Al, E. Infantile enteritis viruses: morphogenesis and morphology. J. Virol. 1975; 16:937–943.

Honeyman, M.C., Coulson, B.S., Stone, N.L., Gellert, S.A., Goldwater, P.N., Steele, C.E., Couper, J.J., Tait, B.D., Colman, P.G. and Harrison, L.C. Association between rotavirus infection and pancreatic islet autoimmunity in children at risk of developing type 1 diabetes. Diabetes 2000; 49:1319–1324.

Hongou, K., Konishi, T., Yagi, S., Araki, K. and Miyawaki, T. Rotavirus encephalitis mimicking afebrile benign convulsions in infants. Pediatr. Neurol. 1998; 18:354–357.

Hoshino, Y. and Kapikian, A.Z. Rotavirus antigens. Curr. Top. Microbiol. Immunol. 1994; 185:179–227.

Hoshino, Y., Wyatt, R.G., Greenberg, H.B., Flores, J. and Kapikian, A.Z. Serotypic similarity and diversity of rotaviruses of mammalian and avian origin as studied by plaque-reduction neutralization. J. Infect. Dis. 1984; 149:694–702.

Hsu, H.Y., Kao, C.L., Huang, L.M., Ni, Y.H., Lai, H.S., Lin, F.Y. and Chang, M.H. Viral etiology of intussusception in Taiwanese childhood. Pediatr. Infect. Dis. J. 1998; 17:893–898.

Hsu, V.P., Staat, M.A., Roberts, N., Thieman, C., Bernstein, D.I., Bresee, J., Glass, R.I. and Parashar, U.D. Use of active surveillance to validate international classification of diseases code estimates of rotavirus hospitalizations in children. Pediatrics 2005; 115:78–82.

Hung, T. Rotavirus and adult diarrhea. Adv. Virus Res. 1988; 35.

Iijima, Y., Iwamoto, T., Nukuzuma, S., Ohishi, H., Hayashi, K. and Kobayashi, N. An outbreak of rotavirus infection among adults in an institution for rehabilitation: long-term residence in a closed community as a risk factor for rotavirus illness. Scand. J. Infect. Dis. 2006; 38:490–496.

Iturriza-Gomara, M., Cubitt, D., Steele, D., Green, J., Brown, D., Kang, G., Desselberger, U. and Gray, J. Characterisation of rotavirus G9 strains isolated in the UK between 1995 and 1998. J. Med. Virol. 2000; 61:510–517.

Jaimes, M.C., Rojas, O.L., Gonzalez, A.M., Cajiao, I., Charpilienne, A., Pothier, P., Kohli, E., Greenberg, H.B., Franco, M.A. and Angel, J. Frequencies of virus-specific CD4(+) and CD8(+) T lymphocytes secreting gamma interferon after acute natural rotavirus infection in children and adults. J. Virol. 2002; 76: 4741–4749.

Jiang, B., Gentsch, J.R. and Glass, R.I. The role of serum antibodies in the protection against rotavirus disease: an overview. Clin. Infect. Dis. 2002; 34:1351–1361.

Joensuu, J., Koskenniemi, E., Pang, X.L. and Vesikari, T. Randomised placebo-controlled trial of rhesus-human reassortant rotavirus vaccine for prevention of severe rotavirus gastroenteritis. Lancet 1997; 350:1205–1209.

Kapikian, A.Z. Viral gastroenteritis. JAMA 1993; 269:627–630.

Kapikian, A.Z. and Chanock, R.M. Rotaviruses. In: Virology (B.N. Fields, D.M. Knipe and R.M. Chanock, Eds.). New York: Raven Press, 1990.

Kapikian, A.Z., Cline, W.L., Kim, H.W., Kalica, A.R., Wyatt, R.G., Vankirk, D.H., Chanock, R.M., James, H.D., Jr. and Vaughn, A.L. Antigenic relationships among five reovirus-like (RVL) agents by complement fixation (CF) and development of new substitute CF antigens for the human RVL agent of infantile gastroenteritis. Proc. Soc. Exp. Biol. Med. 1976; 152:535–539.

Kapikian, A.Z., Hoshino, Y. and Chanock, R.M. Rotaviruses. In: Fields Virology (D.M. Knipe, P. Howley and D. Griffin, Eds.), 2nd ed. Philadelphia: Lippincott, Williams and Wilkins, 2001.

Kapikian, A.Z., Kim, H.W., Wyatt, R.G., Rodriguez, W.J., Ross, S., Cline, W.L., Parrott, R.H. and Chanock, R.M. Reoviruslike agent in stools: association with infantile diarrhea and development of serologic tests. Science 1974; 185:1049–1053.

Kapikian, A.Z., Simonsen, L., Vesikari, T., Hoshino, Y., Morens, D.M., Chanock, R.M., La Montagne, J.R. and Murphy, B.R. A hexavalent human rotavirus-bovine rotavirus (UK) reassortant vaccine designed for use in developing countries and delivered in a schedule with the potential to eliminate the risk of intussusception. J. Infect. Dis. 2005; 192(Suppl. 1):S22–S29.

Kapikian, A.Z., Wyatt, R.G., Levine, M.M., Yolken, R.H., Vankirk, D.H., Dolin, R., Greenberg, H.B. and Chanock, R.M. Oral administration of human rotavirus to volunteers: induction of illness and correlates of resistance. J. Infect. Dis. 1983; 147:95–106.

Kempe, A., Daley, M.F., Parashar, U.D., Crane, L.A., Beaty, B.L., Stokley, S., Barrow, J., Babbel, C., Dickinson, L.M., Widdowson, M.A., Alexander, J.P. and Berman, S. Will pediatricians adopt the new rotavirus vaccine? Pediatrics 2007; 119:1–10.

Kirkwood, C., Bogdanovic-Sakran, N., Palombo, E., Masendycz, P., Bugg, H., Barnes, G. and Bishop, R. Genetic and antigenic characterization of rotavirus serotype G9 strains isolated in Australia between 1997 and 2001. J. Clin. Microbiol. 2003; 41:3649–3654.

Kirkwood, C.D., Bogdanovic-Sakran, N., Cannan, D., Bishop, R.F. and Barnes, G.L. National Rotavirus Surveillance Program annual report, 2004–05. Commun. Dis. Intell. 2006a; 30:133–136.

Kirkwood, C.D., Cannan, D., Bogdanovic-Sakran, N., Bishop, R.F. and Barnes, G.L. National Rotavirus Surveillance Program annual report, 2005–06. Commun. Dis. Intell. 2006b; 30:434–438.

Kuzuya, M., Fujii, R., Hamano, M., Yamada, M., Shinozaki, K., Sasagawa, A., Hasegawa, S., Kawamoto, H., Matsumoto, K., Kawamoto, A., Itagaki, A., Funatsumaru, S. and Urasawa, S. Survey of human group C rotaviruses in Japan during the winter of 1992 to 1993. J. Clin. Microbiol. 1998; 36:6–10.

Lanata, C.F., Black, R.E., Del Aguila, R., Gil, A., Verastegui, H., Gerna, G., Flores, J., Kapikian, A.Z. and Andre, F.E. Protection of Peruvian children against rotavirus diarrhea of specific serotypes by one, two, or three doses of the RIT 4237 attenuated bovine rotavirus vaccine. J. Infect. Dis. 1989; 159:452–459.

Lawton, J.A., Estes, M.K. and Prasad, B.V. Three-dimensional visualization of mRNA release from actively transcribing rotavirus particles. Nat. Struct. Biol. 1997; 4:118–121.

Lewis, H.M., Parry, J.V., Davies, H.A., Parry, R.P., Mott, A., Dourmashkin, R.R., Sanderson, P.J., Tyrrell, D.A. and Valman, H.B.

A year's experience of the rotavirus syndrome and its association with respiratory illness. Arch. Dis. Child. 1979; 54:339–346.

Ling-Qiao, Z. A rotavirus vaccine licensed in China. Health News 2001; 31:1.

Liprandi, F., Gerder, M., Bastidas, Z., Lopez, J.A., Pujol, F.H., Ludert, J.E., Joelsson, D.B. and Ciarlet, M. A novel type of VP4 carried by a porcine rotavirus strain. Virology 2003; 315:373–380.

Liu, M., Mattion, N.M. and Estes, M.K. Rotavirus VP3 expressed in insect cells possesses guanylyltransferase activity. Virology 1992; 188:77–84.

Lopez, S. and Arias, C.F. Early steps in rotavirus cell entry. Curr. Top. Microbiol. Immunol. 2006; 309:39–66.

Lopez, S., Arias, C.F., Bell, J.R., Strauss, J.H. and Espejo, R.T. Primary structure of the cleavage site associated with trypsin enhancement of rotavirus SA11 infectivity. Virology 1985; 144:11–19.

Lorgelly, P.K., Joshi, D., Iturriza Gomara, M., Gray, J. and Mugford, M. Exploring the cost effectiveness of an immunization programme for rotavirus gastroenteritis in the United Kingdom. Epidemiol. Infect. 2007:1–12.

Losonsky, G.A., Rennels, M.B., Kapikian, A.Z., Midthun, K., Ferra, P.J., Fortier, D.N., Hoffman, K.M., Baig, A. and Levine, M.M. Safety, infectivity, transmissibility and immunogenicity of rhesus rotavirus vaccine (MMU 18006) in infants. Pediatr. Infect. Dis. 1986; 5:25–29.

Lundgren, O., Peregrin, A.T., Persson, K., Kordasti, S., Uhnoo, I. and Svensson, L. Role of the enteric nervous system in the fluid and electrolyte secretion of rotavirus diarrhea. Science 2000; 287:491–495.

Lynch, M., Shieh, W.J., Bresee, J.S., Tatti, K.M., Gentsch, J.R., Jones, T., Jiang, B., Hummelman, E., Zimmerman, C.M., Zaki, S.R. and Glass, R.I. Intussusception after administration of the rhesus tetravalent rotavirus vaccine (Rotashield): the search for a pathogenic mechanism. Pediatrics 2006; 117:e827–e832.

Mackow, E.R., Shaw, R.D., Matsui, S.M., Vo, P.T., Dang, M.N. and Greenberg, H.B. The rhesus rotavirus gene encoding protein VP3: location of amino acids involved in homologous and heterologous rotavirus neutralization and identification of a putative fusion region. Proc. Natl. Acad. Sci. U.S.A. 1988; 85:645–649.

Malek, M.A., Curns, A.T., Holman, R.C., Fischer, T.K., Bresee, J.S., Glass, R.I., Steiner, C.A. and Parashar, U.D. Diarrhea- and rotavirus-associated hospitalizations among children less than 5 years of age: United States, 1997 and 2000. Pediatrics 2006; 117:1887–1892.

Malherbe, H. and Harwin, R. The cytopathic effects of vervet monkey viruses. S. Afr. Med. J. 1963; 37:407–411.

Marshall, J., Botes, J., Gorrie, G., Boardman, C., Gregory, J., Griffith, J., Hogg, G., Dimitriadis, A., Catton, M. and Bishop, R. Rotavirus detection and characterisation in outbreaks of gastroenteritis in aged-care facilities. J. Clin. Virol. 2003; 28:331–340.

Matson, D.O., O'Ryan, M.L., Herrera, I., Pickering, L.K. and Estes, M.K. Fecal antibody responses to symptomatic and asymptomatic rotavirus infections. J. Infect. Dis. 1993; 167:577–583.

Matsumoto, K., Hatano, M., Kobayashi, K., Hasegawa, A., Yamazaki, S., Nakata, S., Chiba, S. and Kimura, Y. An outbreak of gastroenteritis associated with acute rotaviral infection in schoolchildren. J. Infect. Dis. 1989; 160:611–615.

Matsuno, S., Utagawa, E. and Sugiura, A. Association of rotavirus infection with Kawasaki syndrome. J. Infect. Dis. 1983; 148:177.

Matthews, R.E. The classification and nomenclature of viruses. Summary of results of meetings of the International Committee on Taxonomy of Viruses in The Hague, September 1978. Intervirology 1979; 11:133–135.

Matthijnssens, J., Rahman, M., Martella, V., Xuelei, Y., De Vos, S., De Leener, K., Ciarlet, M., Buonavoglia, C. and Van Ranst, M. Full genomic analysis of human rotavirus strain B4106 and lapine rotavirus strain 30/96 provides evidence for interspecies transmission. J. Virol. 2006; 80:3801–3810.

Mattion, N.M., Mitchell, D.B., Both, G.W. and Estes, M.K. Expression of rotavirus proteins encoded by alternative open reading frames of genome segment 11. Virology 1991; 181:295–304.

McNeal, M.M., Barone, K.S., Rae, M.N. and Ward, R.L. Effector functions of antibody and CD8+ cells in resolution of rotavirus infection and protection against reinfection in mice. Virology 1995; 214:387–397.

McNeal, M.M., Rae, M.N., Bean, J.A. and Ward, R.L. Antibody-dependent and -independent protection following intranasal immunization of mice with rotavirus particles. J. Virol. 1999; 73:7565–7573.

McNeal, M.M., Rae, M.N., Conner, M.E. and Ward, R.L. Stimulation of local immunity and protection in mice by intramuscular immunization with triple- or double-layered rotavirus particles and QS-21. Virology 1998; 243:158–166.

McNeal, M.M., Sestak, K., Choi, A.H., Basu, M., Cole, M.J., Aye, P.P., Bohm, R.P. and Ward, R.L. Development of a rotavirus-shedding model in rhesus macaques, using a homologous wild-type rotavirus of a new P genotype. J. Virol. 2005; 79:944–954.

Mebus, C.A. Reovirus-like calf enteritis. Am. J. Dig. Dis. 1976; 21:592–598.

Mebus, C.A., Stair, E.L., Underdahl, N.R. and Al, E. Pathology of neonatal calf diarrhea induced by a reo-like virus. Vet. Pathol. 1974; 8:490–505.

Mebus, C.A., Underdahl, N.R., Rhodes, M.B. and Al, E. Calf diarrhea (scours): reproduced with a virus from a field outbreak. Univ. Nebraska Res. Bull. 1969; 233:1–16.

Midthun, K., Halsey, N.A., Jett-Goheen, M., Clements, M.L., Steinhoff, M., King, J.C., Karron, R., Wilson, M., Burns, B., Perkis, V. et al. Safety and immunogenicity of human rotavirus vaccine strain M37 in adults, children, and infants. J. Infect. Dis. 1991; 164:792–796.

Midthun, K. and Kapikian, A.Z. Rotavirus vaccines: an overview. Clin. Microbiol. Rev. 1996; 9:423–434.

Montenegro, F.M., Correia, J.B., Rodrigues Falbo, A., Dove, W., Nakagomi, T., Nakagomi, O., Cuevas, L.E., Cunliffe, N.A. and Hart, C.A. Anticipating rotavirus vaccines in Brazil: detection and molecular characterization of emerging rotavirus serotypes G8 and G9 among children with diarrhoea in Recife, Brazil. J. Med. Virol. 2007; 79:335–340.

Morris, A.P., Scott, J.K., Ball, J.M., Zeng, C.Q., O'Neal, W.K. and Estes, M.K. NSP4 elicits age-dependent diarrhea and Ca(2+)mediated I(−) influx into intestinal crypts of CF mice. Am. J. Physiol. 1999; 277:G431–G444.

Murphy, B.R., Morens, D.M., Simonsen, L., Chanock, R.M., La Montagne, J.R. and Kapikian, A.Z. Reappraisal of the association of intussusception with the licensed live rotavirus vaccine challenges initial conclusions. J. Infect. Dis. 2003; 187:1301–1308.

Murphy, T.V., Gargiullo, P.M., Massoudi, M.S., Nelson, D.B., Jumaan, A.O., Okoro, C.A., Zanardi, L.R., Setia, S., Fair, E., Lebaron, C.W., Wharton, M. and Livengood, J.R. Intussusception among infants given an oral rotavirus vaccine. N. Engl. J. Med. 2001; 344:564–572.

Murphy, T.V., Gargiullo, P.M. and Wharton, M. More on rotavirus vaccination and intussusception. N. Engl. J. Med. 2002; 346: 211–212.

Nakajima, H., Nakagomi, T., Kamisawa, T., Sakaki, N., Muramoto, K., Mikami, T., Nara, H. and Nakagomi, O. Winter seasonality and rotavirus diarrhoea in adults. Lancet 2001; 357:1950.

Nava, P., Lopez, S., Arias, C.F., Islas, S. and Gonzalez-Mariscal, L. The rotavirus surface protein VP8 modulates the gate and fence function of tight junctions in epithelial cells. J. Cell. Sci. 2004; 117:5509–5519.

Nigro, G. Pancreatitis with hypoglycemia-associated convulsions following rotavirus gastroenteritis. J. Pediatr. Gastroenterol. Nutr. 1991; 12:280–282.

Offit, P.A., Hoffenberg, E.J., Santos, N. and Gouvea, V. Rotavirus-specific humoral and cellular immune response after primary, symptomatic infection. J. Infect. Dis. 1993; 167:1436–1440.

O'Neal, C.M., Clements, J.D., Estes, M.K. and Conner, M.E. Rotavirus 2/6 virus like particles administered intranasally with cholera toxin, *Escherichia coli* heat-labile toxin (LT), and LT-R192G induce protection from rotavirus challenge. J. Virol. 1998; 72:3390–3393.

Orenstein, E.W., Fang, Z.Y., Xu, J., Liu, C., Shen, K., Qian, Y., Jiang, B., Kilgore, P.E. and Glass, R.I. The epidemiology and burden of rotavirus in China: a review of the literature from 1983 to 2005. Vaccine 2007; 25:406–413.

O'Ryan, M.L., Matson, D.O., Estes, M.K., Bartlett, A.V. and Pickering, L.K. Molecular epidemiology of rotavirus in children attending day care centers in Houston. J. Infect. Dis. 1990; 162:810–816.

Padilla-Noriega, L., Fiore, L., Rennels, M.B., Losonsky, G.A., Mackow, E.R. and Greenberg, H.B. Humoral immune responses to VP4 and its cleavage products VP5* and VP8* in infants vaccinated with rhesus rotavirus. J. Clin. Microbiol. 1992; 30:1392–1397.

Pappenheimer, A.W. and Enders, J.F. An epidemic diarrheal disease of suckling mice. II. Inclusions in the intestional epithelial cells. J. Exp. Med. 1947; 85:417–422.

Parashar, U.D., Gibson, C.J., Bresse, J.S. and Glass, R.I. Rotavirus and severe childhood diarrhea. Emerg. Infect. Dis. 2006; 12:304–306.

Parashar, U.D. and Glass, R.I. Public health. Progress toward rotavirus vaccines. Science 2006; 312:851–852.

Parashar, U.D., Holman, R.C., Clarke, M.J., Bresee, J.S. and Glass, R.I. Hospitalizations associated with rotavirus diarrhea in the United States, 1993 through 1995: surveillance based on the new ICD-9-CM rotavirus-specific diagnostic code. J. Infect. Dis. 1998; 177:13–17.

Parashar, U.D., Holman, R.C., Cummings, K.C., Staggs, N.W., Curns, A.T., Zimmerman, C.M., Kaufman, S.F., Lewis, J.E., Vugia, D.J., Powell, K.E. and Glass, R.I. Trends in intussusception-associated hospitalizations and deaths among US infants. Pediatrics 2000; 106:1413–1421.

Parashar, U.D., Hummelman, E.G., Bresee, J.S., Miller, M.A. and Glass, R.I. Global illness and deaths caused by rotavirus disease in children. Emerg. Infect. Dis. 2003; 9:565–572.

Patton, J., Kearney, K. and Taraporewala, Z.F. Rotavirus genome replication: role of the RNA-binding protein. In: Viral Gastroenteritis (U. Desselberger and J. Gray, Eds.), Vol. 9, pp. 165–183. The Netherlands: Elsevier Science, 2003.

Patton, J.T. Rotavirus replication. Curr. Top. Microbiol. Immunol. 1994; 185:107–127.

Patton, J.T., Silvestri, L.S., Tortorici, M.A., Vasquez-del Carpio, R. and Taraporewala, Z.F. Rotavirus genome replication and morphogenesis: role of the viroplasm. Curr. Top. Microbiol. Immunol. 2006; 309:169–187.

Patton, J.T. and Spencer, E. Genome replication and packaging of segmented double-stranded RNA viruses. Virology 2000; 277:217–225.

Patton, J.T., Vasquez-del Carpio, R. and Spencer, E. Replication and transcription of the rotavirus genome. Curr. Pharm. Des. 2004; 10:3769–3777.

Patton, J.T., Vasquez-del Carpio, R., Tortorici, M.A. and Taraporewala, Z.F. Coupling of rotavirus genome replication and capsid assembly. Adv. Virus Res. 2007; 69:167–201.

Perez-Schael, I., Blanco, M., Vilar, M., Garcia, D., White, L., Gonzalez, R., Kapikian, A.Z. and Flores, J. Clinical studies of a quadrivalent rotavirus vaccine in Venezuelan infants. J. Clin. Microbiol. 1990; 28:553–558.

Perez-Schael, I., Daoud, G., White, L., Urbina, G., Daoud, N., Perez, M. and Flores, J. Rotavirus shedding by newborn children. J. Med. Virol. 1984; 14:127–136.

Perez-Schael, I., Gonzalez, M., Daoud, N., Perez, M., Soto, I., Garcia, D., Daoud, G., Kapikian, A.Z. and Flores, J. Reactogenicity and antigenicity of the rhesus rotavirus vaccine in Venezuelan children. J. Infect. Dis. 1987; 155:334–338.

Perez-Schael, I., Guntinas, M.J., Perez, M., Pagone, V., Rojas, A.M., Gonzalez, R., Cunto, W., Hoshino, Y. and Kapikian, A.Z. Efficacy of the rhesus rotavirus-based quadrivalent vaccine in infants and young children in Venezuela. N. Engl. J. Med. 1997; 337:1181–1187.

Pesavento, J.B., Crawford, S.E., Estes, M.K. and Prasad, B.V. Rotavirus proteins: structure and assembly. Curr. Top. Microbiol. Immunol. 2006; 309:189–219.

Peter, G. and Myers, M.G. Intussusception, rotavirus, and oral vaccines: summary of a workshop. Pediatrics 2002; 110:e67.

Phua, K.B., Quak, S.H., Lee, B.W., Emmanuel, S.C., Goh, P., Han, H.H., De Vos, B. and Bock, H.L. Evaluation of RIX4414, a live, attenuated rotavirus vaccine, in a randomized, double-blind, placebo-controlled phase 2 trial involving 2464 Singaporean infants. J. Infect. Dis. 2005; 192(Suppl. 1):S6–S16.

Pipittajan, P., Kasempimolporn, S., Ikegami, N., Akatani, K., Wasi, C. and Sinarachatanant, P. Molecular epidemiology of rotaviruses associated with pediatric diarrhea in Bangkok, Thailand. J. Clin. Microbiol. 1991; 29:617–624.

Podewils, L.J., Antil, L., Hummelman, E., Bresee, J., Parashar, U.D. and Rheingans, R. Projected cost-effectiveness of rotavirus vaccination for children in Asia. J. Infect. Dis. 2005; 192 (Suppl. 1): S133–S145.

Prasad, B.V. and Chiu, W. Structure of rotavirus. Curr. Top. Microbiol. Immunol. 1994; 185:9–29.

Prasad, B.V., Rothnagel, R., Zeng, C.Q., Jakana, J., Lawton, J.A., Chiu, W. and Estes, M.K. Visualization of ordered genomic RNA and localization of transcriptional complexes in rotavirus. Nature 1996; 382:471–473.

Rahman, M., Matthijnssens, J., Nahar, S., Podder, G., Sack, D.A., Azim, T. and Van Ranst, M. Characterization of a novel P[25],G11 human group a rotavirus. J. Clin. Microbiol. 2005; 43:3208–3212.

Ramani, S., Banerjee, I., Gladstone, B.P., Sarkar, R., Selvapandian, D., Le Fevre, A.M., Jaffar, S., Iturriza-Gomara, M., Gray, J.J., Estes, M.K., Brown, D.W. and Kang, G. Geographic information systems and genotyping in identification of rotavirus G12 infections in residents of an urban slum with subsequent detection in hospitalized children: emergence of G12 genotype in South India. J. Clin. Microbiol. 2007; 45:432–437.

Raul Velazquez, F., Calva, J.J., Lourdes Guerrero, M., Mass, D., Glass, R.I., Pickering, L.K. and Ruiz-Palacios, G.M. Cohort study of rotavirus serotype patterns in symptomatic and asymptomatic infections in Mexican children. Pediatr. Infect. Dis. J. 1993; 12:54–61.

Ray, P., Fenaux, M., Sharma, S., Malik, J., Subodh, S., Bhatnagar, S., Greenberg, H., Glass, R.I., Gentsch, J. and Bhan, M.K. Quantitative evaluation of rotaviral antigenemia in children with acute rotaviral diarrhea. J. Infect. Dis. 2006; 194:588–593.

Rennels, M.B., Glass, R.I., Dennehy, P.H., Bernstein, D.I., Pichichero, M.E., Zito, E.T., Mack, M.E., Davidson, B.L. and Kapikian, A.Z. Safety and efficacy of high-dose rhesus-human reassortant rotavirus vaccines: report of the National Multicenter Trial. United States Rotavirus Vaccine Efficacy Group. Pediatrics 1996; 97:7–13.

Rennels, M.B., Parashar, U.D., Holman, R.C., Le, C.T., Chang, H.G. and Glass, R.I. Lack of an apparent association between intussusception and wild or vaccine rotavirus infection. Pediatr. Infect. Dis. J. 1998; 17:924–925.

Riepenhoff-Talty, M., Bogger-Goren, S., Li, P., Carmody, P.J., Barrett, H.J. and Ogra, P.L. Development of serum and intestinal antibody response to rotavirus after naturally acquired rotavirus infection in man. J. Med. Virol. 1981; 8:215–222.

Rodriguez, W.J., Kim, H.W., Arrobio, J.O., Brandt, C.D., Chanock, R.M., Kapikian, A.Z., Wyatt, R.G. and Parrott, R.H. Clinical features of acute gastroenteritis associated with human reovirus-like agent in infants and young children. J. Pediatr. 1977; 91:188–193.

Rodriguez, W.J., Kim, H.W., Brandt, C.D., Schwartz, R.H., Gardner, M.K., Jeffries, B., Parrott, R.H., Kaslow, R.A., Smith, J.I. and Kapikian, A.Z. Longitudinal study of rotavirus infection and gastroenteritis in families served by a pediatric medical practice: clinical and epidemiologic observations. Pediatr. Infect. Dis. J. 1987; 6:170–176.

Rothman, K.J., Young-Xu, Y. and Arellano, F. Age dependence of the relation between reassortant rotavirus vaccine (RotaShield) and intussusception. J. Infect. Dis. 2006; 193:898, author reply 898-9.

Ruiz-Palacios, G.M., Guerrero, M.L., Bautista-Marquez, A., Ortega-Gallegos, H., Tuz-Dzib, F., Reyes-Gonzalez, L., Rosales-Pedraza, G., Martinez-Lopez, J., Castanon-Acosta, E., Cervantes, Y., Costa-Clemens, S. and Devos, B. Dose response and efficacy of a live, attenuated human rotavirus vaccine in Mexican infants. Pediatrics 2007; 120:e253–e261.

Ruiz-Palacios, G.M., Perez-Schael, I., Velazquez, F.R., Abate, H., Breuer, T., Clemens, S.C., Cheuvart, B., Espinoza, F., Gillard, P., Innis, B.L., Cervantes, Y., Linhares, A.C., Lopez, P., Macias-Parra, M., Ortega-Barria, E., Richardson, V., Rivera-Medina, D.M., Rivera, L., Salinas, B., Pavia-Ruz, N., Salmeron, J., Ruttimann, R., Tinoco, J.C., Rubio, P., Nunez, E., Guerrero, M.L., Yarzabal, J.P., Damaso, S., Tornieporth, N., Saez-Llorens, X., Vergara, R.F., Vesikari, T., Bouckenooghe, A., Clemens, R., De Vos, B. and O'Ryan, M. Safety and efficacy of an attenuated vaccine against severe rotavirus gastroenteritis. N. Engl. J. Med. 2006; 354:11–22.

Saif, L.J. Nongroup A rotaviruses. In: Viral Diarrheas of Man and Animals (L.J. Saif and K.W. Theil, Eds.). Boca Raton: CRC Press, 1990.

Saif, L.J., Bohl, E.H., Theil, K.W., Cross, R.F. and House, J.A. Rotavirus-like, calicivirus-like, and 23-nm virus-like particles associated with diarrhea in young pigs. J. Clin. Microbiol. 1980; 12:105–111.

Saif, L.J. and Jiang, B. Nongroup A rotaviruses of humans and animals. Curr. Top. Microbiol. Immunol. 1994; 185:339–371.

Saif, L.J., Ward, L.A., Yuan, L., Rosen, B.I. and To, T.L. The gnotobiotic piglet as a model for studies of disease pathogenesis and immunity to human rotaviruses. Arch. Virol. Suppl. 1996; 12:153–161.

Salinas, B., Perez Schael, I., Linhares, A.C., Ruiz Palacios, G.M., Guerrero, M.L., Yarzabal, J.P., Cervantes, Y., Costa Clemens, S., Damaso, S., Hardt, K. and De Vos, B. Evaluation of safety, immunogenicity and efficacy of an attenuated rotavirus vaccine, RIX4414: a randomized, placebo-controlled trial in Latin American infants. Pediatr. Infect. Dis. J. 2005; 24:807–816.

Santos, N. and Hoshino, Y. Global distribution of rotavirus serotypes/genotypes and its implication for the development and implementation of an effective rotavirus vaccine. Rev. Med. Virol. 2005; 15:29–56.

Santosham, M., Letson, G.W., Wolff, M., Reid, R., Gahagan, S., Adams, R., Callahan, C., Sack, R.B. and Kapikian, A.Z. A field study of the safety and efficacy of two candidate rotavirus vaccines in a Native American population. J. Infect. Dis. 1991; 163:483–487.

Shaw, A.L., Rothnagel, R., Chen, D., Ramig, R.F., Chiu, W. and Prasad, B.V. Three-dimensional visualization of the rotavirus hemagglutinin structure. Cell 1993; 74:693–701.

Silvestri, L.S., Taraporewala, Z.F. and Patton, J.T. Rotavirus replication: plus-sense templates for double-stranded RNA synthesis are made in viroplasms. J. Virol. 2004; 78:7763–7774.

Simonsen, L., Viboud, C., Elixhauser, A., Taylor, R.J. and Kapikian, A.Z. More on RotaShield and intussusception: the role of age at the time of vaccination. J. Infect. Dis. 2005; 192 (Suppl. 1):S36–S43.

Snodgrass, D.R., Hoshino, Y., Fitzgerald, T.A., Smith, M., Browning, G.F. and Gorziglia, M. Identification of four VP4 serological types (P serotypes) of bovine rotavirus using viral reassortants. J. Gen. Virol. 1992; 73(Pt 9):2319–2325.

Staat, M.A., Azimi, P.H., Berke, T., Roberts, N., Bernstein, D.I., Ward, R.L., Pickering, L.K. and Matson, D.O. Clinical presentations of rotavirus infection among hospitalized children. Pediatr. Infect. Dis. J. 2002; 21:221–227.

Staat, M.A., Cortese, M.M., Bresee, J.S., Begue, R.E., Vitek, C., Rhodes, P., Zhang, R., Gentsch, J., Roberts, N.E., Jaeger, J.L., Ward, R., Bernstein, D.I. and Dennehy, P.H. Rhesus rotavirus vaccine effectiveness and factors associated with receipt of vaccine. Pediatr. Infect. Dis. J. 2006; 25:1013–1018.

Suzuki, H. and Konno, T. Reovirus-like particles in jejunal mucosa of a Japanese infant with acute infectious non-bacterial gastroenteritis. Tohoku J. Exp. Med. 1975; 115:199–221.

Taniguchi, K., Maloy, W.L., Nishikawa, K., Green, K.Y., Hoshino, Y., Urasawa, S., Kapikian, A.Z., Chanock, R.M. and Gorziglia, M. Identification of cross-reactive and serotype 2-specific neutralization epitopes on VP3 of human rotavirus. J. Virol. 1988; 62:2421–2426.

Taniguchi, K., Urasawa, T., Morita, Y., Greenberg, H.B. and Urasawa, S. Direct serotyping of human rotavirus in stools by an enzyme-linked immunosorbent assay using serotype 1-, 2-, 3-, and 4-specific monoclonal antibodies to VP7. J. Infect. Dis. 1987; 155:1159–1166.

Taraporewala, Z.F., Jiang, X., Vasquez-del Carpio, R., Jayaram, H., Prasad, B.V. and Patton, J.T. Structure-function analysis of rotavirus NSP2 octamer by using a novel complementation system. J. Virol. 2006; 80:7984–7994.

Taraporewala, Z.F. and Patton, J.T. Nonstructural proteins involved in genome packaging and replication of rotaviruses and other members of the Reoviridae. Virus Res. 2004; 101:57–66.

Tcheremenskaia, O., Marucci, G., De Petris, S., Ruggeri, F.M., Dovecar, D., Sternak, S.L., Matyasova, I., Dhimolea, M.K., Mladenova, Z. and Fiore, L. Molecular epidemiology of rotavirus in Central and South-Eastern Europe. J. Clin. Microbiol. 2007; 45: 2197–2204.

Tian, P., Ball, J.M., Zeng, C.Q. and Estes, M.K. The rotavirus nonstructural glycoprotein NSP4 possesses membrane destabilization activity. J. Virol. 1996; 70:6973–6981.

Tian, P., Estes, M.K., Hu, Y., Ball, J.M., Zeng, C.Q. and Schilling, W.P. The rotavirus nonstructural glycoprotein NSP4 mobilizes Ca^{2+} from the endoplasmic reticulum. J. Virol. 1995; 69: 5763–5772.

Tian, P., Hu, Y., Schilling, W.P., Lindsay, D.A., Eiden, J. and Estes, M.K. The nonstructural glycoprotein of rotavirus affects intracellular calcium levels. J. Virol. 1994; 68:251–257.

Tortorici, M.A., Shapiro, B.A. and Patton, J.T. A base-specific recognition signal in the 5′ consensus sequence of rotavirus plus-strand RNAs promotes replication of the double-stranded RNA genome segments. RNA 2006; 12:133–146.

Turcios, R.M., Curns, A.T., Holman, R.C., Pandya-Smith, I., Lamonte, A., Bresee, J.S. and Glass, R.I. Temporal and geographic trends of rotavirus activity in the United States, 1997–2004. Pediatr. Infect. Dis. J. 2006; 25:451–454.

Uhnoo, I., Olding-Stenkvist, E. and Kreuger, A. Clinical features of acute gastroenteritis associated with rotavirus, enteric adenoviruses, and bacteria. Arch. Dis. Child. 1986; 61:732–738.

Valenzuela, S., Pizarro, J., Sandino, A.M., Vasquez, M., Fernandez, J., Hernandez, O., Patton, J. and Spencer, E. Photoaffinity labeling of rotavirus VP1 with 8-azido-ATP: identification of the viral RNA polymerase. J. Virol. 1991; 65:3964–3967.

Vancott, J.L., Prada, A.E., McNeal, M.M., Stone, S.C., Basu, M., Huffer, B., Jr., Smiley, K.L., Shao, M., Bean, J.A., Clements, J.D., Choi, A.H. and Ward, R.L. Mice develop effective but delayed protective immune responses when immunized as neonates either intranasally with nonliving VP6/LT(R192G) or orally with live rhesus rotavirus vaccine candidates. J. Virol. 2006; 80:4949–4961.

Van Damme, P., Giaquinto, C., Huet, F., Gothefors, L., Maxwell, M. and Van der Wielen, M. Multicenter prospective study of the burden of rotavirus acute gastroenteritis in Europe, 2004–2005: the REVEAL study. J. Infect. Dis. 2007a; 195 (Suppl. 1):S4–S16.

Van Damme, P., Giaquinto, C., Maxwell, M., Todd, P. and Van der Wielen, M. Distribution of rotavirus genotypes in Europe, 2004–2005: the REVEAL Study. J. Infect. Dis. 2007b; 195 (Suppl. 1): S17–S25.

Velazquez, F.R., Matson, D.O., Calva, J.J., Guerrero, L., Morrow, A.L., Carter-Campbell, S., Glass, R.I., Estes, M.K., Pickering, L.K. and Ruiz-Palacios, G.M. Rotavirus infections in infants as protection against subsequent infections. N. Engl. J. Med. 1996; 335:1022–1028.

Vesikari, T., Isolauri, E., Delem, A., D'Hondt, E., Andre, F.E., Beards, G.M. and Flewett, T.H. Clinical efficacy of the RIT 4237 live attenuated bovine rotavirus vaccine in infants vaccinated before a rotavirus epidemic. J. Pediatr. 1985; 107:189–194.

Vesikari, T., Isolauri, E., Delem, A., D'Hondt, E., Andre, F.E. and Zissis, G. Immunogenicity and safety of live oral attenuated bovine rotavirus vaccine strain RIT 4237 in adults and young children. Lancet 1983; 2:807–811.

Vesikari, T., Isolauri, E., D'Hondt, E., Delem, A., Andre, F.E. and Zissis, G. Protection of infants against rotavirus diarrhoea by RIT 4237 attenuated bovine rotavirus strain vaccine. Lancet 1984; 1:977–981.

Vesikari, T., Kapikian, A.Z., Delem, A. and Zissis, G. A comparative trial of rhesus monkey (RRV-1) and bovine (RIT 4237) oral rotavirus vaccines in young children. J. Infect. Dis. 1986; 153:832–839.

Vesikari, T., Karvonen, A., Korhonen, T., Espo, M., Lebacq, E., Forster, J., Zepp, F., Delem, A. and De Vos, B. Safety and immunogenicity of RIX4414 live attenuated human rotavirus vaccine in adults, toddlers and previously uninfected infants. Vaccine 2004a; 22:2836–2842.

Vesikari, T., Karvonen, A., Korhonen, T., Espo, M., Lebacq, E., Forster, J., Zepp, F., Delem, A. and de Vos, B. Human rotavirus vaccine Rotarix^{TM} (RIX4414) is highly efficacious in Europe. 24th Annual Meeting of the European Society for Pediatric Infectious Diseases, Basel, Switzerland, 2006a.

Vesikari, T., Karvonen, A., Puustinen, L., Szakal, E.D., Zeng, S.Q., Delem, A. and De Vos, B. A short report on highlights of worldwide development of RIX4414: an European experience. Vaccine 2006b; 24:3779.

Vesikari, T., Karvonen, A., Puustinen, L., Zeng, S.Q., Szakal, E.D., Delem, A. and De Vos, B. Efficacy of RIX4414 live attenuated human rotavirus vaccine in Finnish infants. Pediatr. Infect. Dis. J. 2004b; 23:937–943.

Vesikari, T., Karvonen, A.V., Majuri, J., Zeng, S.Q., Pang, X.L., Kohberger, R., Forrest, B.D., Hoshino, Y., Chanock, R.M. and Kapikian, A.Z. Safety, efficacy, and immunogenicity of 2 doses of bovine-human (UK) and rhesus-rhesus-human rotavirus reassortant tetravalent vaccines in Finnish children. J. Infect. Dis. 2006c; 194:370–376.

Vesikari, T., Matson, D.O., Dennehy, P., Van Damme, P., Santosham, M., Rodriguez, Z., Dallas, M.J., Heyse, J.F., Goveia, M.G., Black, S.B., Shinefield, H.R., Christie, C.D., Ylitalo, S., Itzler, R.F., Coia, M.L., Onorato, M.T., Adeyi, B.A., Marshall, G.S., Gothefors, L., Campens, D., Karvonen, A., Watt, J.P., O'Brien, K.L., Dinubile, M.J., Clark, H.F., Boslego, J.W., Offit, P.A. and Heaton, P.M. Safety and efficacy of a pentavalent human-bovine (WC3) reassortant rotavirus vaccine. N. Engl. J. Med. 2006d; 354:23–33.

Vesikari, T., Ruuska, T., Koivu, H.P., Green, K.Y., Flores, J. and Kapikian, A.Z. Evaluation of the M37 human rotavirus vaccine in 2- to 6-month-old infants. Pediatr. Infect. Dis. J. 1991; 10:912–917.

Wang, Y.H., Kobayashi, N., Zhou, D.J., Yang, Z.Q., Zhou, X., Peng, J.S., Zhu, Z.R., Zhao, D.F., Liu, M.Q. and Gong, J. Molecular epidemiologic analysis of group A rotaviruses in adults and children with diarrhea in Wuhan city, China, 2000–2006. Arch. Virol. 2007; 152:669–685.

Ward, R.L. and Bernstein, D.I. Protection against rotavirus disease after natural rotavirus infection. US Rotavirus Vaccine Efficacy Group. J. Infect. Dis. 1994; 169:900–904.

Ward, R.L., Clemens, J.D., Knowlton, D.R., Rao, M.R., Van Loon, F.P., Huda, N., Ahmed, F., Schiff, G.M. and Sack, D.A. Evidence that protection against rotavirus diarrhea after natural infection is not dependent on serotype-specific neutralizing antibody. J. Infect. Dis. 1992; 166:1251–1257.

Ward, R.L., Clemens, J.D., Sack, D.A., Knowlton, D.R., McNeal, M.M., Huda, N., Ahmed, F., Rao, M. and Schiff, G.M. Culture adaptation and characterization of group A rotaviruses causing diarrheal illnesses in Bangladesh from 1985 to 1986. J. Clin. Microbiol. 1991; 29:1915–1923.

Ward, R.L., Knowlton, D.R., Schiff, G.M., Hoshino, Y. and Greenberg, H.B. Relative concentrations of serum neutralizing antibody to VP3 and VP7 proteins in adults infected with a human rotavirus. J. Virol. 1988; 62:1543–1549.

Ward, R.L., McNeal, M.M., Sander, D.S., Greenberg, H.B. and Bernstein, D.I. Immunodominance of the VP4 neutralization protein of rotavirus in protective natural infections of young children. J. Virol. 1993; 67:464–468.

Ward, R.L., McNeal, M.M. and Sheridan, J.F. Development of an adult mouse model for studies on protection against rotavirus. J. Virol. 1990; 64:5070–5075.

Widdowson, M.A., Meltzer, M.I., Zhang, X., Bresee, J.S., Parashar, U.D. and Glass, R.I. Cost-effectiveness and potential impact of rotavirus vaccination in the United States. Pediatrics 2007; 119:684–697.

Wilhelmi, I., Roman, E. and Sanchez-Fauquier, A. Viruses causing gastroenteritis. Clin. Microbiol. Infect. 2003; 9:247–262.

Woode, G.N., Bridger, J.C., Jones, J.M., Flewett, T.H., Davies, H.A., Davis, H.A. and White, G.B. Morphological and antigenic relationships between viruses (rotaviruses) from acute gastroenteritis of children, calves, piglets, mice, and foals. Infect. Immun. 1976; 14:804–810.

Wyatt, R.G., James, H.D., Jr., Pittman, A.L., Hoshino, Y., Greenberg, H.B., Kalica, A.R., Flores, J. and Kapikian, A.Z. Direct isolation in cell culture of human rotaviruses and their characterization into four serotypes. J. Clin. Microbiol. 1983; 18:310–317.

Wyatt, R.G., Kalica, A.R., Mebus, C.A. and Al, E. Reovirus-like agents (rotaviruses) associated with diarrheal illness in animals and man. In: Perspectives in Virology (M. Pollard, Ed.). New York: Raven Press, 1978.

Wyatt, R.G., Yolken, R.H., Urrutia, J.J., Mata, L., Greenberg, H.B., Chanock, R.M. and Kapikian, A.Z. Diarrhea associated with rotavirus in rural Guatemala: a longitudinal study of 24 infants and young children. Am. J. Trop. Med. Hyg. 1979; 28:325–328.

Yuan, L., Geyer, A., Hodgins, D.C., Fan, Z., Qian, Y., Chang, K.O., Crawford, S.E., Parreno, V., Ward, L.A., Estes, M.K., Conner, M.E. and Saif, L.J. Intranasal administration of 2/6-rotavirus-like particles with mutant *Escherichia coli* heat-labile toxin (LT-R192G) induces antibody-secreting cell responses but not protective immunity in gnotobiotic pigs. J. Virol. 2000; 74:8843–8853.

Zarate, S., Espinosa, R., Romero, P., Mendez, E., Arias, C.F. and Lopez, S. The VP5 domain of VP4 can mediate attachment of rotaviruses to cells. J. Virol. 2000; 74:593–599.

Zheng, B.J., Lo, S.K., Tam, J.S., Lo, M., Yeung, C.Y. and Ng, M.H. Prospective study of community-acquired rotavirus infection. J. Clin. Microbiol. 1989; 27:2083–2090.

CHAPTER

36

SARS

Bart L. Haagmans and Albert D.M.E. Osterhaus

Department of Virology, Erasmus MC, Rotterdam, The Netherlands

OUTLINE

ABSTRACT

Five years after the first severe acute respiratory syndrome (SARS) outbreak, several candidate SARS-coronavirus (CoV) vaccines are at various stages of preclinical and clinical development. Based on the observation that SARS-CoV infection is efficiently controlled upon passive transfer of antibodies directed against the spike (S) protein of SARS-CoV, vaccines containing the S protein have been formulated. Animals immunized with inactivated whole virus vaccines or live-recombinant vaccines expressing the SARS-CoV S protein (e.g., using rabies virus, vesicular stomatitis virus, bovine parainfluenza virus type 3, adenovirus, or attenuated vaccinia virus MVA as a vector), as well as mice immunized with DNA vaccines expressing the S protein gene all developed neutralizing antibodies to SARS-CoV and were protected against SARS-CoV challenge.

Although much effort has been focused on developing a SARS vaccine, the commercial viability of such a vaccine for SARS-CoV will ultimately depend on whether the virus re-emerges in the near future. This vaccine should induce highly cross-reactive neutralizing antibodies to protect against newly emerging viruses related to SARS-CoV and protect both the gastrointestinal and respiratory tract in the absence of significant side effects. Given the fact that in the previous outbreak mainly the elderly succumbed to the infection, special attention should be given to vaccines that are able to efficiently protect aged individuals.

INTRODUCTION

Severe acute respiratory syndrome coronavirus (SARS-CoV) first emerged in the human population in November 2002. Phylogenetic analysis of SARS-CoV isolates from animals indicated that this virus most probably originated from bats, was transmitted first to palm civets and subsequently to humans at the wet markets in southern China. Subsequent outbreaks occurred early 2003 in Hong Kong, Hanoi, Toronto, and Singapore, and could be directly traced back to one index patient who acquired the infection in Guangdong and traveled to Hong Kong. A worldwide epidemic was halted through the efforts of the World Health Organization, which responded rapidly to this threat by issuing a global alert, rigorous local containment efforts, warning against unnecessary travel to affected areas, and by creating a network of international experts to combat this virus. In the end only 8096 people became ill, and 774 people died in this first SARS epidemic. Because SARS-CoV could re-emerge and cause another epidemic at any time, development of effective vaccines remains of vital importance.

ETIOLOGIC AGENT

Although several infectious agents, including chlamydia, influenza A subtype H5N1, and human metapneumovirus, were considered as a possible cause of SARS, three groups independently reported the isolation of a previously unrecognized CoV from clinical specimens of SARS patients (Peiris et al., 2003; Rota et al., 2003; Drosten et al., 2003).

Through electron microscopy, serology, and reverse-transcription PCR with consensus- and random-primers, and subsequent sequencing of the replicase gene, its identity could be revealed and consistently demonstrated in clinical specimens from patients with the disease but not in healthy controls. To conclusively establish a causal role for this CoV, cynomolgous macaques were inoculated with a SARS-CoV isolate. Because the disease in macaques caused by SARS-CoV infection was pathologically similar to that seen in human patients with SARS, and since the virus was successfully re-isolated from the nasal swabs and lung lesions of these animals, and a specific antibody response to the virus was shown in the infected animals, SARS-CoV proved to be the causative agent of this infectious disease (Fouchier et al., 2003; Kuiken et al., 2003).

CLASSIFICATION AND ANTIGENS ENCODED

SARS-CoV is a single-stranded, positive-sense RNA virus, phylogentically related to coronaviruses from group 2 despite the fact that it does not encode a hemagglutinin-esterase protein (Snijder et al., 2003). The genome is packaged together with the nucleocapsid protein, at least five membrane proteins (M, E, 3a, 7a, and 7b) and the spike (S) protein (Fig. 36.1). The S1 region within the S protein, and more specifically a 193-amino acid fragment of the S protein (corresponding to residues 318–510), has been identified as the region that interacts with the cell receptor, angiotensin-converting enzyme 2 (Li et al., 2005a). The majority of neutralizing antibodies are directed against this region of the S protein. Antibodies raised against the N-terminal region of 3a protein or the M

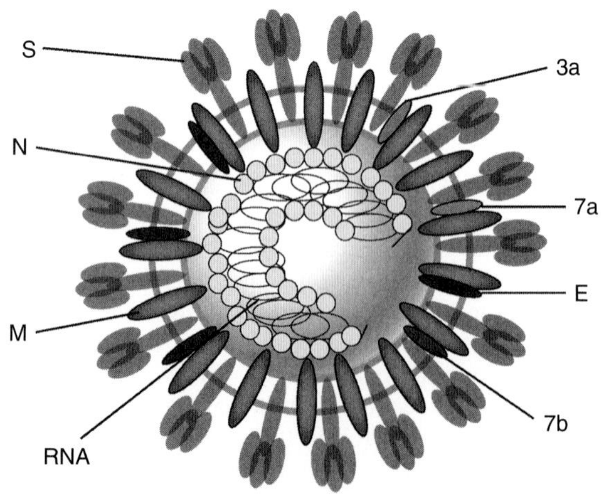

FIGURE 36.1 Schematic diagram of the SARS-CoV particle. S, spike protein; M, membrane protein, E, envelope protein; N, nucleocapsid protein; 3a, 7a, and 7b; structural proteins of SARS-CoV.

protein also inhibit SARS-CoV replication in vitro but their relevance in protection remains unclear. The genome also encodes two large poly-proteins with diverse enzymatic activities needed for efficient replication and several accessory proteins with unknown function (3b, 6, 8a, 8b, and 9b).

EPIDEMIOLOGY

At the end of 2004, 30 countries reported a total of 8096 probable cases of SARS (Fig. 36.2). Pathogenic SARS-CoVs do not circulate in the human population at the moment, but their re-emergence from animal reservoirs may likely occur in the future. Because many of the early SARS patients in Guandong had epidemiological links to the live-animal market trade, different animal species were tested for the presence of SARS-like viruses. Soon after the outbreak, a SARS-like coronavirus, which had more than 99% homology with human SARS-CoV, was detected by RT-PCR in the nasal and fecal swabs of palm civets (*Paguma larvata*) and a raccoon dog (*Nyctereutes procyonoides*) (Guan et al., 2003). More recent studies indicate that bats may potentially act as natural reservoirs for SARS-like CoVs (Li et al., 2005b; Lau et al., 2005). However, sequence comparison of the S protein genes from bat SARS-like CoV and palm civet SARS-like CoV revealed only 64% genetic homology. Subsequent studies by Tang et al. (2006) have demonstrated that approximately 6% of bats sampled in China were positive for CoVs. Interestingly, these CoVs are genetically diverse and many bat CoVs clustered with existing group 1 viruses, while others formed a separate lineage that included only viruses from bats (putative group 5). Other SARS-CoV like viruses clustered in a putative group 4 consisting of two subgroups, one of bat CoVs and another of SARS-CoVs from humans and other mammalian hosts. However, from these studies the direct progenitor of the SARS-CoV isolated from palm civets has not been identified. Major genetic variations in the S protein gene of these viruses from civet cats, seemed essential for the transition from animal-to-human transmission

FIGURE 36.2 Reported suspected SARS cases from November 1, 2002 to July 31, 2003 (data from the World Health Organization, http://www.who.int/csr/sars/country/table2004_04_21/en/index.html). Countries with >5 suspected cases are indicated.

to human-to-human transmission, which eventually caused the SARS outbreak of 2002–2003.

SIGNIFICANCE AS PUBLIC HEALTH PROBLEM AND POTENTIAL AS BIOTHREAT AGENT

There is at present no evidence for the virus persisting in the human population. Possible options for the re-emergence of SARS include the escape of the virus from laboratories, which has already occurred on three occasions. The re-emergence of the virus from its animal reservoir remains possible, given that the virus is detectable in the feces and respiratory secretions of some animals. Indeed, SARS-CoV re-emerged in four patients in Guangdong in December 2003, although these SARS-like CoVs caused milder clinical disease (Liang et al., 2004). The US National Institute of Allergy and Infectious Diseases Biodefense Network classified SARS-CoV as a category C priority pathogen pointing out that SARS-CoV could be a potential biothreat agent.

CLINICAL DISEASE

The clinical symptoms of SARS-CoV infection are those of lower respiratory tract disease and include fever, malaise, peripheral T cell lymphocytopenia, decreased platelet counts, prolonged coagulation profiles, and mildly elevated serum hepatic enzymes (Peiris et al., 2004; Li et al., 2004). Chest radiography reveals infiltrates with subpleural consolidation or "ground glass" changes compatible with viral pneumonitis. Around 20–30% of individuals with SARS require management in intensive care units and the overall case:fatality rate reached approximately 10%.

Although the main clinical symptoms are those of severe respiratory illness, SARS-CoV actually also causes a gastrointestinal and urinary tract infection; SARS-CoV can be detected in the feces and urine of patients, and electron microscopic studies of biopsies of the upper and lower intestinal mucosae of patients with SARS confirmed the presence of the virus in these tissues (Peiris et al., 2004). Fecal transmission proved to be important in at least one major community outbreak in Hong Kong (Amoy Gardens), in which over 300 patients were infected within a few days.

Three features of SARS may be relevant for intervention strategies. First, progressive age dependence in mortality and disease severity is observed in SARS patients. In fact, none of the SARS-CoV-infected children aged below 12 years in Hong Kong required intensive care or mechanical ventilation (Ng et al., 2004). This is not totally explained by comorbid factors but similar age dependence in mortality is seen in patients with other (nonviral) causes of acute respiratory stress syndrome (Rubenfeld et al., 2005). Second, virus transmission is low in the first days of illness and peaks around day 10 after disease onset (Chu et al., 2004). Finally, several studies revealed that high viral load in the nasopharyngeal aspirate was found to be an independent predictor of mortality (Hung et al., 2004; Chu et al., 2004). Therefore, vaccine strategies aimed at reducing the viral load may suffice to provide clinical benefit.

TREATMENT

The first efforts to treat SARS patients were mainly based on the use of ribavirin and corticosteroids. Ribavirin, which targets IMP dehydrogenase, has been known a long time as a broad-spectrum antiviral agent. However, current data do not support the use of ribavirin for SARS treatment; in vitro studies did not show significant antiviral activity (Cinatl et al., 2003) and ribavirin enhanced the infectivity of SARS-CoV in mice (Barnard et al., 2006). On the other hand, a protective effect of interferon (IFN)-α has been observed in a preliminary study during the SARS outbreak (Loutfy et al., 2003). These results are in concordance with several studies that noted antiviral activity in vitro (Cinatl et al., 2003; Hensley et al., 2004) and animal studies showing that pegylated IFN-α effectively reduced SARS-CoV replication and excretion, viral antigen expression by type 1 pneumocytes and the pulmonary damage in cynomolgous macaques that were infected experimentally with SARS-CoV (Haagmans et al., 2004). However, despite an extensive literature reporting on SARS treatments, it is not possible to determine whether treatments benefited patients during the SARS outbreak. Because of variation in treatment regimens—particularly the wide range in doses, duration of therapy, and route of administration of ribavirin and corticosteroids, no clear conclusion can be drawn regarding the efficacy of the drugs tested (Stockman et al., 2006). In the event of a future outbreak of SARS-CoV or another novel agent, attempts should be made to develop treatment protocols, organize randomized trials and to collect and contribute information for a standardized minimum dataset that could facilitate analysis of treatment outcomes among different settings (Stockman et al., 2006).

PATHOGENESIS

The major sources of transmission in humans are droplets that deposit on the respiratory epithelium. Unlike the situation in several other respiratory viral infections, viral load of SARS-CoV in the upper respiratory tract peaked around day 10 after disease onset (Peiris et al., 2004). Therefore, virus transmission may be less efficient in the first days of illness, a finding supported by epidemiological observations. Real-time PCR assays detect SARS-CoV during the first week in specimens of the lower respiratory tract (e.g., bronchoalveolar lavage, sputum, endotracheal aspirates), nasopharyngeal aspirate, throat swabs, and/or serum (Chan et al., 2004), whereas fecal samples may show very high viral loads toward the end of the first week and second week of illness. In typical cases, which were largely confined to adult and elderly individuals, SARS presented with acute respiratory distress syndrome, characterized by the presence of diffuse alveolar damage and multiorgan dysfunction upon autopsy (Nicholls et al., 2003). The pathological changes in lung alveoli most likely follow a common pathway characterized by an acute phase of protein-rich alveolar fluid influx into the alveolar lumina as a consequence of the injury to the alveolar wall. Subsequently type-2 pneumocyte hyperplasia takes place to replace the loss of infected type-1 pneumocytes and to cover the denuded epithelial basement membrane, resulting in restoration of the normal alveolar architecture. Severe alveolar injury may lead to fibrosis with loss of alveolar function in more protracted cases (Fig. 36.3).

INNATE IMMUNE RESPONSE TO INFECTION

It has been hypothesized that the pathological changes observed in the lungs are initiated by a disproportional innate immune response, illustrated by elevated levels of inflammatory cytokines and chemokines, such as CXCL10 (IP-10), CCL2 (MCP-1), IL-6, IL-8, IL-12, IL-1β, and IFN-γ (Huang et al., 2005; Wong et al., 2004). These in vivo data have been confirmed in vitro, demonstrating that SARS-CoV infection induces a range of cytokines and chemokines in diverse cell types (Cheung et al., 2005; Law et al., 2005). Although in vitro studies argued that production of type I IFNs is inhibited or delayed by SARS-CoV, in SARS-CoV-infected macaques, and also early during SARS in humans, type I IFNs can be readily demonstrated (De Lang et al., 2007). Because prophylactic treatment of macaques with pegylated IFN-α reduces SARS-CoV replication in the lungs, regulation of the production of IFNs may be important in controlling SARS (Haagmans et al., 2004).

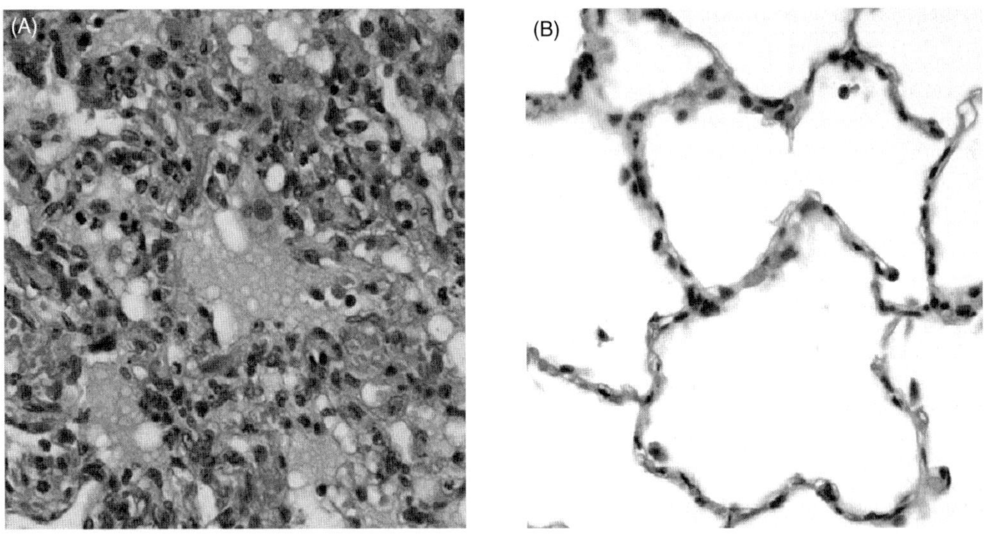

FIGURE 36.3 Acute lung injury after experimental infection of nonhuman primates with SARS-CoV. Shown are the histological changes in the lungs after SARS-CoV infection (A) as compared to the normal lung tissue (B). Viral infection of bronchiolar epithelial cells, type 1 and type 2 pneumocytes results in the influx of macrophages and neutrophils (A). The subsequent loss of type 1 cells may cause flooding of the alveolus by edema fluid (A) and in later stages fibrosis of the alveolar wall (deposition of collagen and influx of fibroblasts).

HUMORAL IMMUNE RESPONSE

Seroconversion usually occurs in weeks 2 or 3 of illness and virus-neutralizing antibodies can be detected in convalescent human serum. In patients who had recovered from SARS-neutralizing antibody titers peaked at month 4 but were undetectable in 16% of patients at month 36 (Cao et al., 2007). Neutralizing antibodies may be directed against different regions of the S protein (S1 and S2) as monoclonal antibodies against these epitopes exert potent neutralization of SARS-CoV in vitro (Sui et al., 2004; Lip et al., 2006). Conversely, peptides which are located in these regions were able to induce neutralizing antibodies (Bisht et al., 2005; Keng et al., 2005; Zhang et al., 2004).

Although experiments in diverse animal models have revealed that relatively low levels of neutralizing antibodies exert potent protection against lower respiratory tract infection, the neutralizing antibody titer necessary to achieve protection in humans exposed to SARS-CoV is not known. A concern in case of re-emergence of SARS is the possible absence of cross protection against these viruses. However, recent studies by He et al. (2006) have shown that the some neutralizing epitopes of SARS-CoV have been maintained during cross-species transmission, suggesting that receptor binding domain-based vaccines may induce broad protection against both human and animal SARS-CoV variants.

CELLULAR IMMUNE RESPONSE

In most SARS autopsies, extensive necrosis of the spleen and atrophy of the white pulp with severe lymphocyte depletion have been observed. On the other hand, rapid phase in peripheral lymphocyte recovery usually coincided with improved clinical conditions of SARS patients (Peiris et al., 2004). Long-lived memory T cell responses against SARS-CoV nucleocapsid and S protein have been demonstrated in recovered SARS patients, although their relevance in antiviral protection is not well understood (Yang et al., 2006; Peng et al., 2006; Li et al., 2006). However, despite potent immune responses and clinical recovery, peripheral lymphocyte counts in the recovered patients were not restored to normal levels (Li et al., 2006).

Interestingly, mice that lack NK-T cells, or NK cells, or T and B cells all cleared the virus by day 9 after infection (Glass et al., 2004). These data argue that cell-mediated immune responses are not essential to control virus clearance.

VACCINES

Based on the observation that SARS-CoV infection is efficiently controlled upon passive transfer of antibodies directed against the S protein of SARS-CoV, a range of vaccines containing the S protein/gene has been developed. Animals immunized with inactivated whole virus vaccines or live-recombinant vaccines expressing the SARS-CoV S protein (e.g., using rabies virus, vesicular stomatitis virus, bovine parainfluenza virus type 3, adenovirus, or attenuated vaccinia virus MVA as a vector), as well as mice immunized with DNA vaccines expressing the S protein gene all developed neutralizing antibodies to SARS-CoV and were protected against SARS-CoV challenge. Table 36.1 displays an overview of vaccines that have been tested for efficacy in animal models.

Inactivated Whole Virus and Subunit Vaccines

Inactivated SARS vaccines have been reported to elicit high titers of S protein-specific neutralizing antibodies. Few studies, however, have addressed whether inactivated whole SARS-CoV virions confer protection from virus challenge. Mice that were immunized twice with a candidate SARS-CoV vaccine, produced through a two-step inactivation procedure involving sequential formaldehyde and U.V. inactivation developed high-antibody titers against the SARS-CoV S protein and high levels of neutralizing antibodies (Spruth et al., 2006) (see Chapter 11). Moreover, the vaccine conferred protective immunity as demonstrated by prevention of SARS-CoV replication in the respiratory tract of mice after intranasal challenge with SARS-CoV. Protection of mice was correlated to the antibody titer against the SARS-CoV S protein and neutralizing antibody titer. Similar results have been obtained using a beta-propiolactone inactivated SARS-CoV vaccine in mice (Stadler et al., 2005). In addition, two Chinese groups have demonstrated protective efficacy of inactivated SARS vaccines in rhesus monkeys (Qin et al., 2006; Zhou et al., 2005). A soluble recombinant polypeptide containing the N-terminal segment of the S glycoprotein may suffice to induce neutralizing antibodies and protective immunity in mice (Bisht et al., 2005). In addition, a trimeric recombinant S protein was able to elicit an efficacious protective immune response in hamsters (Kam et al., 2007).

One of the most promising vaccine candidates is based on the combination of recombinant S protein with the Protollin adjuvant. In both young and aged mice, an intranasal Protollin-formulated S protein

TABLE 36.1 SARS-CoV vaccines[a]

Author	Vaccine[b]	Animal species	Immunogenicity[c]	Protection
Spruth et al. (2006)	Inactivated whole virus	Mice and macaques	Neutralizing Abs	Yes
Stadler et al. (2005)	Inactivated whole virus	Mice	Neutralizing Abs	Yes
Qin et al. (2006)	Inactivated whole virus	Macaques	Neutralizing Abs	Yes
Zhou et al. (2005)	Inactivated whole virus	Macaques	Neutralizing Abs	Yes
Bisht et al. (2005)	Subunit	Mice	Neutralizing Abs	Yes
Kam et al. (2007)	Subunit	Mice	Neutralizing Abs	Yes
Hu et al. (2007)	Subunit	Mice	Neutralizing Abs/ lung IgA	Yes
Yang et al. (2004)	Plasmid DNA vector	Mice	Neutralizing Abs	Yes
See et al. (2006)	Adenovirus vector	Mice	Neutralizing Abs	Yes
Bisht et al. (2004)	MVA vector	Mice	Neutralizing Abs	Yes
Chen et al. (2005)	MVA vector	Macaques	Neutralizing Abs	Yes
Weingartl et al. (2004)	MVA vector	Ferrets	No neutralizing Abs	No
Buchholz et al. (2004)	Parainfluenza virus vector	Hamsters	Neutralizing Abs	Yes
Bukreyev et al. (2004)	Parainfluenza virus vector	Macaques	Neutralizing Abs	Yes
Kapadia et al. (2005)	VSV vector	Mice	Neutralizing Abs	Yes
Vogel et al. (2007)	VSV vector	Aged mice	Neutralizing Abs	Yes
DiNapoli et al. (2007)	NDV vector	African green monkeys	Neutralizing Abs	Yes
Deming et al. (2006)	VEEV vector	Aged mice	Neutralizing Abs	Yes
		Aged mice and heterologous challenge	No neutralizing Abs	No

[a] Only those SARS-CoV vaccines containing the spike protein/gene and tested for protection against a SARS-CoV challenge are listed.
[b] MVA, modified vaccinia virus Ankara; VSV, vesicular stomatitis virus; NDV, Newcastle disease virus; VEEV, Venezuelan equine encephalitis virus.
[c] Presence of neutralizing antibodies at time of challenge.

vaccine elicited high levels of antigen-specific IgG in serum and significant levels of antigen-specific lung IgA (Hu et al., 2007). In contrast, mice immunized intramuscularly with Alum absorbed S protein did not develop detectable IgA responses. Following virus challenge of the aged mice, no virus was detected in the lungs of mice vaccinated intranasally, whereas intramuscularly immunized mice did not show significant control of virus replication compared to controls (Hu et al., 2007).

Vectored Vaccines

A DNA vaccine encoding the S glycoprotein of the SARS-CoV induces T cell, neutralizing antibody responses, and protective immunity in a mouse model (Yang et al., 2004). These authors also demonstrated that antibody responses in mice vaccinated with an expression vector encoding a form of S protein that includes its transmembrane domain elicited neutralizing antibodies. Viral replication was reduced by more than six orders of magnitude in the lungs of mice vaccinated with these S protein plasmid DNA expression vectors, and protection was mediated by a humoral, but not a T-cell-dependent, immune mechanism. Subsequent studies using a prime-boost combination of DNA and whole killed SARS-CoV vaccines elicited higher antibody responses than DNA or whole killed virus vaccines alone (Kong et al., 2005). Apart from this study, several other groups have analyzed the immunogenicity of SARS DNA vaccines but none of these challenged the vaccinated animals with SARS-CoV.

Adenovirus-vector based vaccination strategies against SARS-CoV were employed early on after the SARS outbreak to demonstrate that vaccinated rhesus macaques developed virus-neutralizing antibody responses against fragment S1 of the S protein and T cell responses against the nucleocapsid (Gao et al., 2003). More recently, See et al. (2006) demonstrated that vaccination of C57B/L6 mice with adenovirus type 5-expressing S and nucleocapsid administered intranasally, but not intramuscularly, significantly limited SARS-CoV replication in the lungs.

The highly attenuated modified vaccinia virus Ankara (MVA) has been used to express the S glycoprotein of SARS-CoV in vaccination experiments using mouse, ferret, and rhesus monkey models (Bisht et al., 2004; Chen et al., 2005; Weingartl et al., 2004). Intranasal and intramuscular administration of MVA encoding the SARS-CoV S protein led to the induction of a humoral immune response in BALB/c mice, as well as reduced viral titers in the respiratory tract (Bisht et al., 2005).

Recombinant bovine-human parainfluenza virus type 3 vector (BHPIV3) is being developed as a live attenuated, intranasal pediatric vaccine against human parainfluenza virus type 3. Immunization of African green monkeys with a single dose of BHPIV3 expressing SARS-CoV S protein administered via the respiratory tract induced the production of SARS-CoV neutralizing antibodies (Bukreyev et al., 2004). A recombinant BHPIV3 expressing SARS-CoV structural protein (S, M, and N) individually or in combination has been evaluated for immunogenicity and protective efficacy in hamsters (Buchholz et al., 2004). In the absence of S protein, expression of M, N, or E did not induce a detectable serum SARS-CoV-neutralizing antibody response and no protection against SARS-CoV challenge in the respiratory tract, whereas the vectors expressing the S protein induced neutralizing antibody responses and protection.

Recombinant rabies virus expressing the S protein of SARS-CoV induced a neutralizing antibody response in mice (Faber et al., 2005). Similarly, an attenuated vesicular stomatitis virus vector that encodes the SARS-CoV S protein may be used to induce neutralizing antibody responses (Kapadia et al., 2005). Mice vaccinated with recombinant vesicular stomatitis virus expressing S protein developed SARS-CoV-neutralizing antibody and were able to control a challenge with SARS-CoV performed at either 1 month or 4 months after a single vaccination. In addition, by passive antibody transfer experiments these authors demonstrated that the antibody response induced by the vaccine was sufficient to control SARS-CoV infection.

The efficacy of these vectors was further demonstrated in studies using aged mice. In aged mice, vaccinated with recombinant vesicular stomatitis virus expressing S protein, antibody titers induced were sufficient to protect them against subsequent challenge with SARS-CoV (Vogel et al., 2007).

African green monkeys immunized via the respiratory tract with two doses of a recombinant Newcastle disease virus encoding the S protein developed a relatively high titer of SARS-CoV neutralizing antibodies and upon challenge demonstrated a 1000-fold reduction in pulmonary SARS-CoV titer compared with control animals (DiNapoli et al., 2007).

Finally, Venezuelan equine encephalitis virus based vaccines have been tested extensively in young and aged mice. Most importantly, different recombinant SARS-CoV bearing epidemic and zoonotic S protein variants were used to challenge the vaccinated mice. Venezuelan equine encephalitis virus replicon particles expressing the 2003 epidemic Urbani SARS-CoV strain S glycoprotein but not particles containing the nucleocapsid protein from the same strain provided complete short- and long-term protection against homologous strain challenge in young and senescent mice (Deming et al., 2006). Although the S protein encoding vaccine provided complete short-term protection against heterologous (strain GD03) challenge in young mice, only limited protection was seen in vaccinated senescent animals. Interestingly, nucleocapsid-encoding vaccines not only failed to protect from homologous or heterologous challenge but also resulted in enhanced immunopathology with eosinophilic infiltrates within the lungs of SARS-CoV-challenged mice (Deming et al., 2006).

BASIC SCIENCE AND RATIONALE OF NEW GENERATION VACCINES

A new generation of vaccines may be obtained from manipulating the full-length infectious cDNA clone of SARS-CoV. One approach would be to delete the ORFs 3a, 3b, 6, 7a, 7b, 8a, 8b, or 9b similar to other coronavirus mutants generated previously; some mouse hepatitis or feline infectious peritonitis deletion viruses replicate to the same extent as wild-type viruses in vitro but are severely attenuated in vivo making them potential vaccine candidates. However, SARS-CoV deletion mutants lacking ORFs 3a, 3b, 6, 7a, or 7b, grew similar to that of the parental wild-type virus in the mouse model (Yount et al., 2005). On the other hand, a recombinant SARS-CoV that lacks the E gene was attenuated both in vitro and in vivo (DeDiego et al., 2007). Viable recombinant virus with the E gene deleted was recovered in Vero cells with a titer around 10^6 pfu/ml but titers in the respiratory tract of hamsters were 100–1000-fold reduced compared to wild-type SARS-CoV replication, suggesting that this mutant is attenuated. Multiplication of these viruses in packaging cell lines would provide the missing protein in trans and would make a promising SARS-CoV vaccine candidate that has the E gene deleted.

Live attenuated virus vaccines may revert to wild-type and recombine with other circulating human or zoonotic coronaviruses. In order to prevent this, it has been proposed to delete an essential gene, located in a position distant from gene E, and the relocation of the deleted gene to the position previously occupied by gene E (Enjuanes et al., 2008). A potential recombination leading to the rescue of gene E would lead to the loss of the essential gene. Alternatively, the transcriptional regulatory sequences (TRS) of a vaccine virus could be genetically manipulated to a sequence incompatible with the TRS of any known circulating coronavirus as described by Yount et al. (2006). This virus could be further modified by building attenuating mutations on the genetic backbone of the recombination resistant TRS rewired virus either for use as a safe high titer seed stock for making killed vaccines or as a live virus vaccine. One such attenuating mutation could be targeted to the nonstructural protein 1. Recombinant Mouse Hepatitis Virus (MHV) encoding a deletion in the nsp1-coding sequence grew normally in tissue culture but was severely attenuated in vivo (Züst et al., 2007). Low doses of nsp1 mutant MHV elicited potent cytotoxic T cell responses and protected mice against homologous and heterologous virus challenge. This attenuation strategy provides a new paradigm for the development of highly efficient coronavirus vaccines.

PRECLINICAL DEVELOPMENT, INCLUDING RELEVANT ANIMAL MODELS

Although several types of vectored vaccines have been developed, several companies favored the classical approach using inactivated whole virus to develop a vaccine to be used for preclinical testing (see Chapter 11). Methods in place for the production of available vaccines could be easily used using well-established technologies. In the preclinical development stage, it is preferable that different animal species are used to evaluate the safety and efficacy of candidate vaccines.

Overall, a wide range of animal species, including rodents (mice and hamsters), carnivores (ferrets and cats), and nonhuman primates (cynomolgus and rhesus macaques, common marmosets, and African green monkeys) can be experimentally infected with SARS-CoV (Subbarao et al., 2004; Roberts et al., 2005b; Martina et al., 2003; Kuiken et al., 2003; McAuliffe et al., 2004; Haagmans and Osterhaus, 2006). Most species show no clinical signs of disease, although the virus replicates efficiently in respiratory tissues. Aged mice and ferrets on the other hand, show signs of clinical disease, albeit in the absence of the typical lung lesions seen in humans with SARS (Roberts et al., 2005a). In contrast, inoculation of SARS-CoV in the respiratory tract of cynomolgus macaques causes infection of bronchial epithelial cells and type-1 pneumocytes 1–4 days postinfection, followed by extensive type-2 pneumocyte hyperplasia in the lungs at 4–6 days postinfection (Kuiken et al., 2003; Haagmans and Osterhaus, 2006). The lesions, consisting of multiple foci of acute diffuse alveolar damage and characterized by flooding of alveoli with protein-rich edema fluid mixed with variable numbers of neutrophils, are quite similar to those observed in humans in the acute stages of SARS.

Remarkably, vaccine candidates tested in ferrets showed reduced efficacy. Vaccination with MVA encoding the S protein induced only moderate antibody responses and consequently did not protect against intranasal SARS-CoV infection and even resulted in an inflammatory response in the livers of the vaccinated ferrets (Weingartl et al., 2004). Whether these aberrant responses resulted from immunopathological mechanisms, like antibody-dependent enhancement of infection, or represented recall responses to viral antigen in the liver is not clear at the moment but deserves further investigation. In addition, limited protection from SARS-CoV challenge was observed in ferrets vaccinated with inactivated whole virus (Darnell et al., 2007). Two out of four ferrets showed little or no neutralizing antibody even after the second immunization and none of the vaccinated ferrets were able to reduce virus excretion at day 2 after challenge but subsequently cleared the virus more rapidly compared to control animals. Adenovirus-based vaccines tested in ferrets seemed more powerful as they protected the lower respiratory tract efficiently but had less effect on virus excretion in the upper respiratory tract (Kobinger et al., 2007).

CLINICAL TRIALS

Five years after the first SARS outbreak a range of candidate vaccines have been developed. Early in 2006 some companies in China and the US initiated phase 1 trials. The first clinical trial has been initiated by a Chinese company, Sinovac Biotech of Beijing in collaboration with the Chinese academy of Medical Sciences using an inactivated whole virus vaccine. To evaluate the safety and immunogenicity of this vaccine, 36 subjects received two doses of vaccine or placebo control. On day 42, all individuals showed seroconversion and peak titers of neutralizing antibodies were reached 2 weeks after the second vaccination followed by a significant decline 4 weeks later (Lin et al., 2007). Several other candidate SARS vaccines are at various stages of preclinical and clinical development.

POSTEXPOSURE IMMUNOPROPHYLAXIS

In SARS patients who recover, high levels of neutralizing antibody responses are observed, suggesting that antibody responses play a role in determining the ultimate disease outcome of SARS-CoV-infected patients (Zhang et al., 2006). Although attempts have been made to test the efficacy of serum preparations from seroconvalescent SARS patients in the acute phase of SARS, no conclusive evidence has been obtained regarding their efficacy. In mice, on the other hand, SARS-CoV infection is efficiently controlled upon passive transfer of convalescent immunoglobulines (Subbarao et al., 2004). The concept that antibodies protect against SARS has been further explored through the generation of human monoclonal antibodies against SARS-CoV. Prophylactic administration of a human monoclonal antibody reduced replication of SARS-CoV in the lungs of infected ferrets by 1000-fold, completely prevented the development of SARS-CoV-induced macroscopic lung pathology, and abolished shedding of virus in pharyngeal secretions (Ter Meulen et al., 2004). In subsequent studies, several other monoclonal antibodies were evaluated for their efficacy in mouse and hamster models (Sui et al., 2005; Traggiai et al., 2004).

PROSPECTS FOR THE FUTURE

The importance of assessing immunogenicity of candidate SARS-CoV vaccines using virus neutralization assays is well acknowledged, but the variety of these tests in use is a significant problem since there is at this time no consensus on the most sensitive, specific, and reproducible assay system. To compare data from each of the candidate vaccines requires international standardization of the immunological assays and the availability of an antibody standard used for the evaluation of these vaccines. To test cross reactivity of antibodies generated by vaccination, murine leukemia virus was used to generate infectious particles containing different S proteins (Giroglou et al., 2004).

Enhanced disease and mortality have been observed in kittens immunized against or infected with a type I coronavirus, feline infectious peritonitis virus (FIPV), when subsequently exposed to FIPV infection (Weiss and Scott, 1981). Macrophages are able to take up feline coronavirus-antibody complexes more efficiently causing the virus to replicate to higher titers. Interestingly, one study also demonstrated that antibodies against human SARS-CoV isolates enhance entry of pseudo-typed viruses expressing the civet cat SARS-like CoV S protein into cells, but not replication (Yang et al., 2005). To date, there is no evidence for enhanced replication following SARS-CoV challenge in previously immunized animals.

One other problem which may arise after vaccination with whole inactivated virus when absorbed with certain adjuvants such as alum, could relate to the induction of skewed Th2 recall responses similar to what has been observed in children vaccinated with inactivated respiratory syncytial and measles virus vaccines.

Although much effort has been focused on developing a SARS vaccine, the commercial viability of developing a vaccine for SARS-CoV will ultimately depend on whether the virus re-emerges in the near future. It is questionable whether possible future outbreaks will cause major outbreaks but vaccines, antivirals, or passive immunization would be relevant in the context of protecting high-risk individuals such as laboratory and health-care workers.

KEY ISSUES

- Five years after the first SARS outbreak, several candidate SARS-CoV vaccines are at various stages of preclinical and clinical development.
- Based on the observation that SARS-CoV infection is efficiently controlled upon passive transfer of antibodies directed against the S protein of SARS-CoV, vaccines encoding this protein have been developed.
- Animals immunized with diverse vaccines containing the S protein or S protein gene developed SARS-CoV-neutralizing antibodies and were protected against SARS-CoV challenge.
- Future challenges for the development of SARS-CoV vaccines are linked to the potential re-emergence of this virus. Vaccines able to induce highly cross-reactive antibodies which efficiently protect the gastrointestinal and respiratory tract are needed. Given the fact that in the previous outbreak mainly elderly humans succumbed to the infection, special attention should be given to protect specifically these individuals.

References

Barnard, D.L., Day, C.W., Bailey, K., et al. Enhancement of the infectivity of SARS-CoV in BALB/c mice by IMP dehydrogenase inhibitors, including ribavirin. Antiviral Res. Mar 24; Epub, 2006.

Bisht, H., Roberts, A., Vogel, L. et al. Severe acute respiratory syndrome coronavirus spike protein expressed by attenuated vaccinia virus protectively immunizes mice. Proc. Natl. Acad. Sci. U.S.A. 2004; 101(17):6641–6646.

Bisht, H., Roberts, A., Vogel, L. et al. Neutralizing antibody and protective immunity to SARS coronavirus infection of mice induced by a soluble recombinant polypeptide containing an N-terminal segment of the spike glycoprotein. Virology 2005; 334(2):160–165.

Buchholz, U.J., Bukreyev, A., Yang, L. et al. Contributions of the structural proteins of severe acute respiratory syndrome coronavirus to protective immunity. Proc. Natl. Acad. Sci. U.S.A. 2004; 101(26):9804–9809.

Bukreyev, A., Lamirande, E.W., Buchholz, U.J. et al. Mucosal immunisation of African green monkeys (*Cercopithecus aethiops*) with an attenuated parainfluenza virus expressing the SARS coronavirus spike protein for the prevention of SARS. Lancet 2004; 363(9427):2122–2127.

Cao, W.C., Liu, W., Zhang, P.H. et al. Disappearance of antibodies to SARS-associated coronavirus after recovery. N. Engl. J. Med. 2007; 357(11):1162–1163.

Chan, K.H., Poon, L.L., Cheng, V.C. et al. Detection of SARS coronavirus in patients with suspected SARS. Emerg. Infect. Dis. 2004; 10:294–299.

Che, X.Y., Hao, W., Wang, Y. et al. Nucleocapsid protein as early diagnostic marker for SARS. Emerg. Infect. Dis. 2004; 10(11):1947–1949.

Chen, Z., Zhang, L., Qin, C. et al. Recombinant modified vaccinia virus Ankara expressing the spike glycoprotein of severe acute respiratory syndrome coronavirus induces protective neutralizing antibodies primarily targeting the receptor binding region. J. Virol. 2005; 79(5):2678–2688.

Cheung, C.Y., Poon, L.L., Ng, I.H. et al. Cytokine responses in severe acute respiratory syndrome coronavirus-infected macrophages in vitro: possible relevance to pathogenesis. J. Virol. 2005; 79:7819–7826.

Chu, C.M., Poon, L.L., Cheng, V.C. et al. Initial viral load and the outcomes of SARS. CMAJ 2004; 171(11):1349–1352.

Cinatl, J., Morgenstern, B., Bauer, G. et al. Treatment of SARS with human interferons. Lancet 2003; 362:293–294.

Darnell, M.E., Plant, E.P., Watanabe, H., Byrum, R., St. Claire, M., Ward, J.M. and Taylor, D.R. Severe acute respiratory syndrome coronavirus infection in vaccinated ferrets. J. Infect. Dis. 2007; 196(9):1329–1338, Epub 2007 September 27.

DeDiego, M.L., Alvarez, E., Almazán, F. et al. A severe acute respiratory syndrome coronavirus that lacks the E gene is attenuated in vitro and in vivo. J. Virol. 2007; 81(4):1701–1713, Epub 2006 November 15.

De Lang., A., Baas, T., Teal, T., Leijten, L.M., Rain, B., Osterhaus, A.D., Haagmans, B.L. and Katze, M.G. Functional genomics highlights differential induction of antiviral pathways in the lungs of SARS-CoV-infected macaques. PLos Pathog 2007; Aug 10; 3(8): e112.

Deming, D., Sheahan, T., Heise, M. et al. Vaccine efficacy in senescent mice challenged with recombinant SARS-CoV bearing epidemic and zoonotic spike variants. PLoS Med. 2006; 3(12):e525, Erratum in: PLoS Med. 2007 February; 4(2):e80.

DiNapoli, J.M., Kotelkin, A., Yang, L. et al. Newcastle disease virus, a host range-restricted virus, as a vaccine vector for intranasal immunization against emerging pathogens. Proc. Natl. Acad. Sci. U.S.A. 2007; 104(23):9788–9793, Epub 2007 May 29.

Drosten, C., Gunther, S., Preiser, W. et al. Identification of a novel coronavirus in patients with severe acute respiratory syndrome. N. Engl. J. Med. 2003; 348(20):1967–1976.

Enjuanes, L., Dediego, M.L., Alvarez, E., et al. Vaccines to prevent severe acute respiratory syndrome coronavirus-induced disease. Virus Res. 2008; 133(1):45–62. [Epub 2007, April 9].

Faber, M., Lamirande, E.W., Roberts, A. et al. A single immunization with a rhabdovirus-based vector expressing severe acute respiratory syndrome coronavirus (SARS-CoV) S protein results in the production of high levels of SARS-CoV-neutralizing antibodies. J. Gen. Virol. 2005; 86:1435–1440.

Fouchier, R.A., Kuiken, T., Schutten, M. et al. Aetiology: Koch's postulates fulfilled for SARS virus. Nature 2003; 423(6937):240.

Gao, W., Tamin, A., Soloff, A. et al. Effects of a SARS-associated coronavirus vaccine in monkeys. Lancet 2003; 362(9399):1895–1896.

Giroglou, T., Cinatl, J., Jr., Rabenau, H. et al. Retroviral vectors pseudotyped with severe acute respiratory syndrome coronavirus S protein. J. Virol. 2004; 78(17):9007–9015.

Glass, W.G., Subbarao, K., Murphy, B. et al. Mechanisms of host defense following severe acute respiratory syndrome-coronavirus (SARS-CoV) pulmonary infection of mice. J. Immunol. 2004; 173(6):4030–4039.

Guan, Y., Zheng, B.J., He, Y.Q. et al. Isolation and characterization of viruses related to the SARS coronavirus from animals in southern China. Science 2003; 302(5643):276–278.

Haagmans, B.L., Kuiken, T., Martina, B.E. et al. Pegylated interferon-alpha protects type 1 pneumocytes against SARS coronavirus infection in macaques. Nat. Med. 2004; 10:290–293, Epub 2004 February 22.

Haagmans, B.L. and Osterhaus, A.D. Nonhuman primate models for SARS. PLoS Med. 2006; 3(5):e194, Epub 2006 April 18.

He, Y., Li, J., Li, W. et al. Cross-neutralization of human and palm civet severe acute respiratory syndrome coronaviruses by antibodies targeting the receptor-binding domain of spike protein. J. Immunol. 2006; 176(10):6085–6092.

Hensley, L.E., Fritz, L.E., Jahrling, P.B. et al. Interferon-beta 1a and SARS coronavirus replication. Emerg. Infect. Dis. 2004; 10:317–319.

Hu, M.C., Jones, T., Kenney, R.T. et al. Intranasal Protollin-formulated recombinant SARS S-protein elicits respiratory and serum neutralizing antibodies and protection in mice. Vaccine 2007; 25(34):6334–6340, Epub 2007 June 29.

Huang, K.J., Su, I.J., Theron, M. et al. An interferon gamma-related cytokine storm in SARS patients. J. Med. Virol. 2005; 75:185–194.

Hung, I.F., Cheng, V.C., Wu, A.K. et al. Viral loads in clinical specimens and SARS manifestations. Emerg. Infect. Dis. 2004; 9:1550–1557.

Kam, Y.W., Kien, F., Roberts, A. et al. Antibodies against trimeric S glycoprotein protect hamsters against SARS-CoV challenge despite their capacity to mediate FcgammaRII-dependent entry into B cells in vitro. Vaccine 2007; 25(4):729–740, Epub 2006 August 22.

Kapadia, S.U., Rose, J.K., Lamirande, E. et al. Long-term protection from SARS coronavirus infection conferred by a single immunization with an attenuated VSV-based vaccine. Virology 2005; 340(2):174–182.

Keng, C.T., Zhang, A. and Shen, S. Amino acids 1055 to 1192 in the S2 region of severe acute respiratory syndrome coronavirus S protein induce neutralizing antibodies: implications for the development of vaccines and antiviral agents. J. Virol. 2005; 79(6):3289–3296.

Kobinger, G.P., Figueredo, J.M., Rowe, T. et al. Adenovirus-based vaccine prevents pneumonia in ferrets challenged with the SARS coronavirus and stimulates robust immune responses in macaques. Vaccine 2007; 25(28):5220–5231, Epub 2007 May 7.

Kong, W.P., Xu, L., Stadler, K. et al. Modulation of the immune response to the severe acute respiratory syndrome spike glycoprotein by gene-based and inactivated virus immunization. J. Virol. 2005; 79(22):13915–13923.

Kuiken, T., Fouchier, R.A., Schutten, M. et al. Newly discovered coronavirus as the primary cause of severe acute respiratory syndrome. Lancet 2003; 362(9380):263–270.

Lau, S.K., Woo, P.C., Li, K.S. et al. Severe acute respiratory syndrome coronavirus-like virus in Chinese horseshoe bats. Proc. Nat .Acad. Sci. U.S.A. 2005; 102(39):14040–14045.

Law, H.K., Cheung, C.Y., Ng, H.Y. et al. Chemokine up-regulation in SARS-coronavirus-infected, monocyte-derived human dendritic cells. Blood 2005; 106:2366–2374.

Li, F., Li, W., Farzan, M. et al. Structure of SARS coronavirus spike receptor-binding domain complexed with receptor. Science 2005a; 309(5742):1864–1868.

Li, T., Qiu, Z., Zhang, L. et al. Significant changes of peripheral T lymphocyte subsets in patients with severe acute respiratory syndrome. J. Infect. Dis. 2004; 189(4):648–651.

Li, T., Xie, J., He, Y. et al. Long-term persistence of robust antibody and cytotoxic T cell responses in recovered patients infected with SARS coronavirus. PLoS ONE 2006; 1:e24.

Li, W., Shi, Z., Yu, M. et al. Bats are natural reservoirs of SARS-like coronaviruses. Science 2005b; 310(5748):676–679.

Liang, G., Chen, Q., Xu, J. et al. Laboratory diagnosis of four recent sporadic cases of community-acquired SARS, Guangdong province, China. Emerg. Infect. Dis. 2004; 10:1774–1781.

Lin, J.T., Zhang, J.S., Su, N. et al. Safety and immunogenicity from a phase I trial of inactivated severe acute respiratory syndrome coronavirus vaccine. Antivir. Ther. 2007; 12(7):1107–1113.

Lip, K.M., Shen, S. and Yang, X. Monoclonal antibodies targeting the HR2 domain and the region immediately upstream of the HR2 of the S protein neutralize in vitro infection of severe acute respiratory syndrome coronavirus. J. Virol. 2006; 80(2):941–950.

Liu, L., Fang, Q., Deng, F. et al. Natural mutations in the receptor binding domain of spike glycoprotein determine the reactivity of cross-neutralization between palm civet coronavirus and severe acute respiratory syndrome coronavirus. J. Virol. 2007; 81(9):4694–4700, Epub 2007 February 21.

Loutfy, M.R., Blatt, L.M., Siminovitch, K.A. et al. Interferon alfacon-1 plus corticosteroids in severe acute respiratory syndrome: a preliminary study. JAMA 2003; 290(24):3222–3228.

Martina, B.E., Haagmans, B.L., Kuiken, T. et al. Virology: SARS virus infection of cats and ferrets. Nature 2003; 425(6961):915.

McAuliffe, J., Vogel, L., Roberts, A. et al. Replication of SARS coronavirus administered into the respiratory tract of African Green, rhesus and cynomolgus monkeys. Virology 2004; 330(1):8–15.

Ng, P.C., Leung, C.W., Chiu, W.K. et al. SARS in newborns and chil dren. Biol. Neonate 2004; 85(4):293–298.

Nicholls, J.M., Poon, L.L., Lee, K.C. et al. Lung pathology of fatal severe acute respiratory syndrome. Lancet 2003; 361(9371):1773–1778.

Peiris, J.S., Guan, Y. and Yuen, K.Y. Severe acute respiratory syndrome. Nat. Med. 2004; 10(12 Suppl):S88–S97.

Peiris, J.S., Lai, S.T., Poon, L.L. et al. Coronavirus as a possible cause of severe acute respiratory syndrome. Lancet 2003; 361(9366):1319–1325.

Peng, H., Yang, L.T., Wang, L.Y. et al. Long-lived memory T lymphocyte responses against SARS coronavirus nucleocapsid protein in SARS-recovered patients. Virology 2006; 351(2):466–475, Epub 2006 May 11, 2006.

Qin, E., Shi, H., Tang, L. et al. Immunogenicity and protective efficacy in monkeys of purified inactivated Vero-cell SARS vaccine. Vaccine 2006; 24(7):1028–1034, Epub 2005 September 12, 2006.

Roberts, A., Paddock, C., Vogel, L. et al. Aged BALB/c mice as a model for increased severity of severe acute respiratory syndrome in elderly humans. J. Virol. 2005a; 79(9):5833–5838.

Roberts, A., Vogel, L., Guarner, J. et al. Severe acute respiratory syndrome coronavirus infection of golden Syrian hamsters. J. Virol. 2005b; 79(1):503–511.

Rota, P.A., Oberste, M.S., Monroe, S.S. et al. Characterization of a novel coronavirus associated with severe acute respiratory syndrome. Science 2003; 300(5624):1394–1399.

Rubenfeld, G.D., Caldwell, E., Peabody, E. et al. Incidence and outcomes of acute lung injury. N. Engl. J. Med. 2005; 353:1685–1693.

See, R.H., Zakhartchouk, A.N., Petric, M. et al. Comparative evaluation of two severe acute respiratory syndrome (SARS) vaccine candidates in mice challenged with SARS coronavirus. J. Gen. Virol. 2006; 87(Pt 3):641–650.

Snijder, E.J., Bredenbeek, P.J., Dobbe, J.C. et al. Unique and conserved features of genome and proteome of SARS-coronavirus, an early split-off from the coronavirus group 2 lineage. J. Mol. Biol. 2003; 331(5):991–1004.

Spruth, M., Kistner, O., Savidis-Dacho, H. et al. A double-inactivated whole virus candidate SARS coronavirus vaccine stimulates neutralising and protective antibody responses. Vaccine 2006; 24(5):652–661, Epub 2005 August 26.

Stadler, K., Roberts, A., Becker, S. et al. SARS vaccine protective in mice. Emerg. Infect. Dis. 2005; 11(8):1312–1314.

Stockman, L.J., Bellamy, R. and Garner, P. SARS: systematic review of treatment effects. PLoS Med. 2006; 3(9):e343.

Subbarao, K., McAuliffe, J., Vogel, L. et al. Prior infection and passive transfer of neutralizing antibody prevent replication of severe acute respiratory syndrome coronavirus in the respiratory tract of mice. J. Virol. 2004; 78(7):3572–3577.

Sui, J., Li, W. and Murakami, A. Potent neutralization of severe acute respiratory syndrome (SARS) coronavirus by a human mAb to S1 protein that blocks receptor association. Proc. Natl. Acad. Sci. U.S.A. 2004; 101:2536–2541.

Sui, J., Li, W., Roberts, A. et al. Evaluation of human monoclonal antibody 80R for immunoprophylaxis of severe acute respiratory syndrome by an animal study, epitope mapping, and analysis of spike variants. J. Virol. 2005; 79(10):5900–5906.

Tang, X.C., Zhang, J.X., Zhang, S.Y. et al. Prevelance and genetic diversity of coronaviruses in bats from China. J. Virol. 2006; 80:7481–7490.

Ter Meulen, J., Bakker, A.B., van den Brink, E.N. et al. Human monoclonal antibody as prophylaxis for SARS coronavirus infection in ferrets. Lancet 2004; 363(9427):2139–2141.

Traggiai, E., Becker, S., Subbarao, K. et al. An efficient method to make human monoclonal antibodies from memory B cells: potent neutralization of SARS coronavirus. Nat. Med. 2004; 10(8):871–875.

Vogel, L.N., Roberts, A., Paddock, C.D. et al. Utility of the aged BALB/c mouse model to demonstrate prevention and control strategies for severe acute respiratory syndrome coronavirus (SARS-CoV). Vaccine 2007; 25(12):2173–2179, Epub 2006 December 11.

Weingartl, H., Czub, M., Czub, S. et al. Immunization with modified vaccinia virus Ankara-based recombinant vaccine against severe acute respiratory syndrome is associated with enhanced hepatitis in ferrets. J. Virol. 2004; 78(22):12672–12676.

Weiss, R.C. and Scott, F.W. Antibody-mediated enhancement of disease in feline infectious peritonitis: comparisons with dengue hemorrhagic fever. Comp. Immunol. Microbiol. Infect. Dis. 1981; 4(2):175–189.

Wong, C.K., Lam, C.W., Wu, A.K. et al. Plasma inflammatory cytokines and chemokines in severe acute respiratory syndrome. Clin. Exp. Immunol. 2004; 136:95–103.

Yang, L.T., Peng, H., Zhu, Z.L. et al. Long-lived effector/central memory T-cell responses to severe acute respiratory syndrome coronavirus (SARS-CoV) S antigen in recovered SARS patients. Clin. Immunol. 2006; 120(2):171–178, Epub 2006 June 16.

Yang, Z.Y., Kong, W.P., Huang, Y. et al. A DNA vaccine induces SARS coronavirus neutralization and protective immunity in mice. Nature 2004; 428(6982):561–564.

Yang, Z.Y., Werner, H.C., Kong, W.P. et al. Evasion of antibody neutralization in emerging severe acute respiratory syndrome coronaviruses. Proc. Natl. Acad. Sci. U.S.A. 2005; 102:797–801.

Yount, B., Roberts, R.S., Lindesmith, L. and Baric, R.S. Rewiring the severe acute respiratory syndrome coronavirus (SARS-CoV) transcription circuit: engineering a recombination-resistant genome. Proc. Natl. Acad. Sci. U.S.A. 2006; 103(33):12546–12551, Epub 2006 August 4.

Yount, B., Roberts, R.S., Sims, A.C. et al. Severe acute respiratory syndrome coronavirus group-specific open reading frames encode nonessential functions for replication in cell cultures and mice. J. Virol. 2005; 79(23):14909–14922.

Zhang, H., Wang, G., Li, J. et al. Identification of an antigenic determinant on the S2 domain of the severe acute respiratory syndrome coronavirus spike glycoprotein capable of inducing neutralizing antibodies. J. Virol. 2004; 78(13):6938–6945.

Zhang, L., Zhang, F., Yu, W. et al. Antibody responses against SARS coronavirus are correlated with disease outcome of infected individuals. J. Med. Virol. 2006; 78:1–8.

Zhou, J., Wang, W., Zhong, Q. et al. Immunogenicity, safety, and protective efficacy of an inactivated SARS-associated coronavirus vaccine in rhesus monkeys. Vaccine 2005; 23(24):3202–3209.

Zou, Z., Yang, Y., Chen, J. et al. Prognostic factors for severe acute respiratory syndrome: a clinical analysis of 165 cases. Clin. Infect. Dis. 2004; 38(4):483–489, Epub 2004 January 28.

Züst, R., Cervantes-Barragán, L., Kuri, T. et al. Coronavirus nonstructural protein 1 is a major pathogenicity factor: implications for the rational design of coronavirus vaccines. PLoS Pathog. 2007; 3(8):e109.

CHAPTER

37

Smallpox

Rick Kennedy[1] *and Gregory A. Poland*[1,2]

[1]Mayo Vaccine Research Group, Mayo Clinic College of Medicine, Rochester, MN, USA

[2]Program in Translational Immunovirology and Biodefense, Mayo Clinic College of Medicine, Rochester, MN, USA

OUTLINE

ABSTRACT

Throughout history mankind has been ravaged by smallpox, a devastating disease which touched every corner of the earth and was capable of destroying entire civilizations. Through repeated epidemics and pandemics smallpox altered the course of history and is thought to have killed more people than any other infectious disease. Although attempts to control and mitigate the disease had been practiced for thousands of years, it still took nearly two centuries after Jenner's initial experiments with cowpox and vaccination before smallpox was brought under control. The eventual eradication of smallpox in 1980 is undoubtedly one of mankind's greatest medical achievements. It is a sad commentary on human nature that, only a few decades after this dreadful scourge was eradicated, smallpox is once again a subject of international concern, due to its potential as a biological weapon. Smallpox has been used as a biological weapon in the past and history has clearly and repeatedly shown that whether intentional or not, outbreaks of this disease in a susceptible population are devastating. Given the morbidity and mortality of this disease it is not surprising that the early part of the 21st century has seen millions of dollars being spent on preparedness, prevention, and therapeutic measures to combat the smallpox threat.

INTRODUCTION

Smallpox, a deadly disease with telltale clinical symptoms and signs, has plagued mankind for millennia. Before its eradication it was endemic (with the exception of Australia) throughout the world and claimed hundreds of millions of lives (Table 37.1). In the last century alone, it is estimated to have killed 500 million people (Fenner, 1988). It has wiped out entire civilizations and altered the course of empires and yet its origin remains obscure (Dixon, 1962; Hopkins, 1983; Hays, 2005).

Smallpox has been eradicated world-wide and variola-virus is currently found only in secure repositories in the US and Russia.

This chapter deals with not only the history of the disease and the organism that causes it, but will also focus on the vaccines and other strategies used to control the disease. This is especially important in the context of being the only infectious disease which has been successfully eradicated, and yet remains capable of inordinate morbidity, mortality, and societal disruption. This chapter will also discuss current issues as well as future prospects concerning smallpox and related poxviruses.

TABLE 37.1 Smallpox epidemics throughout history

Year	Location	Details
BC		
1000–2000	Egypt	Smallpox signs found in several Egyptian mummies
1350	Egypt	Epidemic among the Hittites thought to be smallpox
568	Mecca	Ethiopian army suffer outbreak during the "Elephant War"
430	Athens	Plague of Athens likely caused by smallpox
395	Sicily	Plague among Carthaginian army thought to be smallpox
165	Rome	Antonine Plague thought to be caused by smallpox
AD		
710–731	Europe	Islamic armies bring smallpox to Spain, Portugal, France
735	Japan	Epidemic in Nara kills thousands
1241	Iceland	First epidemic in Iceland
1368	Siam	Smallpox outbreak kills the king of Burma
1507	Caribbean	Epidemic wipes out entire tribes on Hispaniola
1519	Mexico	Cortez brings smallpox to Aztecs, death toll ~3.5 million
1524	Peru	Smallpox follows Pizarro to the Incan empire
1617–1619	Massachusetts	Severe outbreak eliminates many of the indigenous tribes
1623	Russia	First recorded occurrences of smallpox in Russia
1636–1697	Massachusetts	Boston suffers major epidemics about once every 10 years
1702	Canada	A severe outbreak in Quebec kills 1/4 of the inhabitants
1713	South Africa	Epidemic in Cape Town started by contaminated clothing
1763	Ottawa	Deliberate infection of Indians with smallpox by the British
1776	Quebec	An epidemic wipes out a portion of the Continental Army
1789	Australia	Severe outbreak among aborigines
1795	Germany	Epidemic in Berlin caused by variolation
1804–1810	Russia	Severe epidemics kill ~1 million Russians
1824	Europe	European smallpox pandemic
1868–1884	India	Five major outbreaks in India kill over 2.5 million people
1870	Europe	Pandemic during Franco-Prussian war kills over 500,000
1896/1904	US/Africa	Variola minor occurs in Florida and South Africa

EARLY HISTORY

The causative agent of the disease, variola major (Latin for mottled (*varius*) or pimple (*varus*)), is likely to have evolved by adaptation to humans from a virus which infected animals who lived in close proximity to the human populations. It appears to have developed in concert with civilizations reaching sufficient density to support a disease which only infected humans and required face-to-face contact for reliable transmission. Well-settled and thriving agricultural societies provided the close proximity and numbers of susceptible victims necessary for smallpox to establish itself. It is thought that smallpox originated between 5000 and 10,000 BC in the Nile river valley and Fertile Crescent areas of the Near East (Fenner, 1988). Perhaps the best early evidence of smallpox comes from ancient mummies, such as that of Rhamses V (died 1157 BC) shown in Fig. 37.1. These mummies bear scars characteristic of smallpox (Fenner, 1988; Dixon, 1962). From this base, the disease spread to Asia, Europe, and northern Africa. The earliest definitive mention of smallpox in India is by the physician Vagbhata in the 7th century AD and in China there are unmistakable descriptions of smallpox by the physician Ko Hung in 340 AD (Dixon, 1962). Europe likely had intermittent importations of smallpox throughout the early part of the first millennium AD until its population reached sufficient size to support endemic disease. Smallpox was periodically reintroduced to Europe by Muslim Armies in the 6th and 7th centuries and by knights returning from the Crusades of the 11th and 12th centuries (Dixon, 1962).

In Africa, smallpox spread across the Mediterranean coast with the expansion of Islam in the 8th century. Exploration and trade brought the disease to the coastal towns and then into the interior of Africa where it remained endemic until the 1970s.

The Western hemisphere remained free of smallpox until the European explorations and conquests in the 15th–18th centuries (Patterson and Runge, 2002; Eyler, 2003). In 1519 Cortez's army brought the disease into central Mexico where it killed millions of Aztec Indians. Less than a decade later, smallpox infected the Incan empire 2000 miles to the south, killing hundreds of thousands of individuals. Early settlers

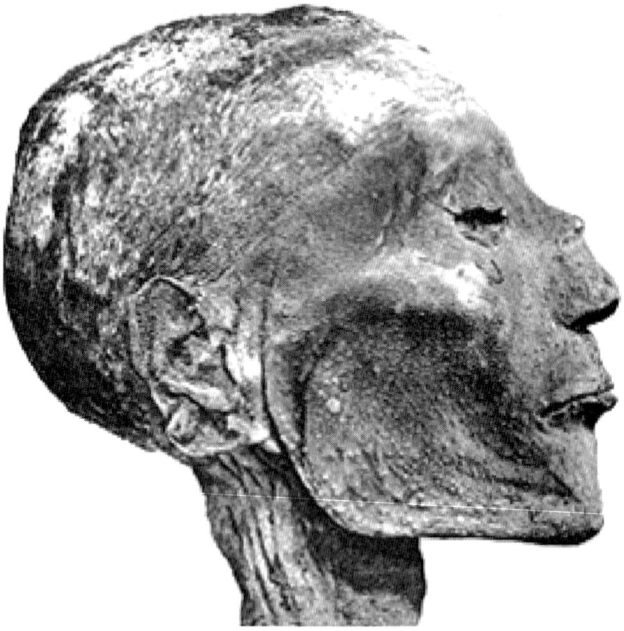

FIGURE 37.1 The mummified head of Rhamses V with typical scarring from variola infection.

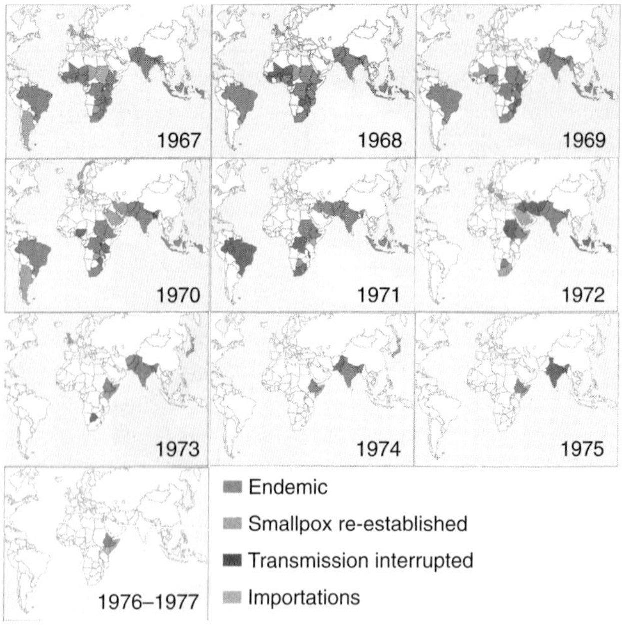

FIGURE 37.2 Worldwide smallpox prevalence during the Intensified Eradication Effort (1967–1977). Reproduced with permission from Fenner et al. (1980).

on the eastern coast of North America brought smallpox into contact with the local Indian tribes with the same tragic consequences.

In the 17th and 18th centuries, smallpox was endemic throughout the world with periodic deadly epidemics. Smallpox is estimated to have killed 400,000 people per year and to have caused more than one-third of all blindness in Europe during this period (Fenner, 1988). Smallpox never became established in Australia but it did cause major epidemics with massive numbers of casualties, mainly among the aborigines.

In the 1890s a milder version of smallpox was identified in South Africa and at about the same time in the southeastern USA. This disease, called variola minor, amaas, or alastrim, was considerably less severe than variola major, with a mortality rate of ~1% (Fenner, 1988; Shchelkunov et al., 2000; Bray and Buller, 2004). By the 1930s alastrim had replaced variola major in the US, Canada, the United Kingdom, and parts of Central America (Bray and Buller, 2004) while variola major continued to circulate in Asia and Africa.

ERADICATION

In spite of having achieved local control, smallpox remained a significant public health issue worldwide, even in countries without endemic cases. The World Health Organization (WHO) officially adopted the smallpox eradication program in 1959, but little progress was made until the Intensified Smallpox Eradication Program began in 1967. At this time smallpox was endemic in 31 countries in South America, Africa, India, and Asia. The eradication program consisted of a twofold strategy (Behbehani, 1980; Fenner, 1988). (1) Massive vaccination campaigns in areas with endemic smallpox and in the neighboring countries were quickly implemented. It was hoped that by achieving 80% vaccine coverage, sufficient herd immunity would be obtained to disrupt transmission and halt the spread of the disease. (2) Sophisticated disease surveillance systems were developed to allow early detection and containment of outbreaks. Once an outbreak was detected, all known or possible contacts with infected victims were vaccinated (Eichner, 2003).

Figure 37.2 tracks the worldwide reduction in endemic smallpox areas from 1967 to 1977. Within 3 years South America and most of Africa had been cleared of the disease. By 1975 clearance of smallpox from Asia occurred as well as the end of variola major infections. Two years later, on October 26, 1977, the last natural case of smallpox occurred in Somalia. The victim was Ali Maow Maalin a hospital worker who, ironically, was afraid to receive the vaccine.

On May 8, 1980 the 33rd World Health Assembly declared that smallpox had been eradicated from the world. During the eradication effort the WHO had spent approximately $112 million and enlisted hundreds of workers, while the international community at large had donated vast amounts of vaccine, supplies, cash, and volunteer work. It is estimated that eradication of smallpox directly saved over $1 billion in global health care costs annually (Behbehani, 1983). While it is hard to put a price on a single case of a disease, a Pakistani traveler introduced smallpox to Great Britain in 1961. The resulting epidemic involved 67 smallpox cases, the vaccination of 5.5 million people, and cost $3.6 million (Behbehani, 1983).

During the Intensified Eradication Program, it was also decided that all samples of the virus would be destroyed or sent to one of the two approved repositories, the Centers for Disease Control and Prevention in Atlanta, GA, and the Research Institute of Viral Preparations in Moscow (virus specimens were subsequently moved to the Russian State Research Center of Virology and Biotechnology in Koltsovo) (Alibek, 2004). A fatal laboratory-acquired case of smallpox occurred in 1978 in Birmingham, England and no doubt spurred many of the remaining smallpox laboratories to destroy or transfer their samples to the two official repositories.

For the next 30 years smallpox was no longer a public health problem and the only attention paid to the disease were the efforts to comply with the WHO recommendations regarding vaccine stockpiles and the limited research on variola conducted at the two reference laboratories. This research initially focused on gene sequencing and later moved into developing animal models and diagnostic and detection assays (Massung et al., 1992, 1994; Antoine et al., 1998; Shchelkunov et al., 2000, 2001; Esposito et al., 2006; Li et al., 2006, 2007). Most basic research into variola virus and smallpox disease stopped, although work with other poxviruses, particularly vaccinia virus continued. In fact, shortly after the eradication, vaccinia began to show great promise as a vector for other vaccines.

There remains considerable debate over the fate of the last remaining variola specimens (Roizman et al., 1994; Henderson and Fenner, 2001; Breman et al., 2003) which are slated to be destroyed in 2011 (WHO, 2006). Arguments both for and against the destruction of the variola stocks center on whether or not continued research and testing is necessary in order to prepare for possible future biological attacks or outbreaks of emerging poxviruses.

POXVIRUS BIOLOGY

The etiologic agents responsible for smallpox are variola major and variola minor. Both are members of the Poxviridae family which are large, complex, DNA viruses that replicate in the cytoplasm of the host cell (Fields et al., 1996). Two subfamilies exist: Entomopoxvirinae, which infect insects and Chordopoxvirinae, which infect vertebrate animals (Table 37.2).

TABLE 37.2 Poxvirus families and representative species

Genus	Example	Reservoir	Zoonotic hosts
Avipoxvirus	Fowlpox virus	Birds, arthropods	
Capripoxvirus	Goatpox	Ungulates, arthropods	
	Lumpy skin disease virus	Ungulates, arthropods	
Leporipoxvirus	Myxoma	Rabbits, squirrels	
	Shope fibroma virus	Rabbits, squirrels	
Molluscipoxvirus	Molluscum contagiosum	Humans	
Orthopoxvirus	Buffalopox virus	Unknown	Buffaloes, humans
	Camelpox virus	Camels	Camels, humans
	Cowpox virus	Rodents	Wide host range
	Ectromelia virus	Rodents	
	Monkeypox virus	Rodents, squirrels	Monkeys, humans
	Uasin Gishu disease virus	Unknown	Horses
	Vaccinia virus	Unknown	Wide host range
	Variola virus	Humans	None
Parapoxvirus	Bovine papular stomatitis virus	Cattle	Humans
	Orf	Ungulates	Humans, cats
Suipoxvirus	Swinepox virus	Swine	
Tanapoxvirus	Yaba monkey tumor virus	Monkeys	Humans
Yatapoxvirus	Tanapox virus	Rodents, insects	Humans, monkeys

As shown in Fig. 37.3, Poxviruses have linear, double-stranded, DNA genomes 130–300 kb pairs in size (Fields et al., 1996). At each end of the genome is a hairpin loop and inverted terminal repeats which covalently join the two strands into one continuous chain (Gubser et al., 2004). Orthopoxvirus genomes contain 200–250 open reading frames (ORF), with most of the essential conserved genes located in the central region of the genome (Wittek, 1982; McCraith et al., 2000; Upton et al., 2003; Randall et al., 2004). An excellent

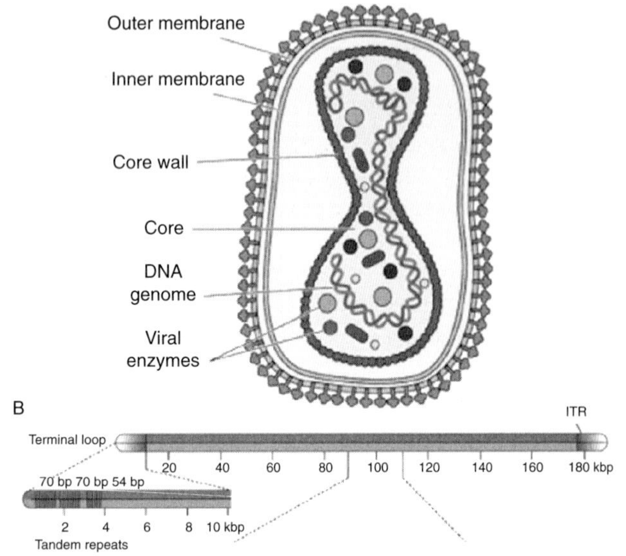

FIGURE 37.3 Diagram of (A) poxvirus virion and (B) genome structure. Reproduced with permission from Harrison et al. (2004).

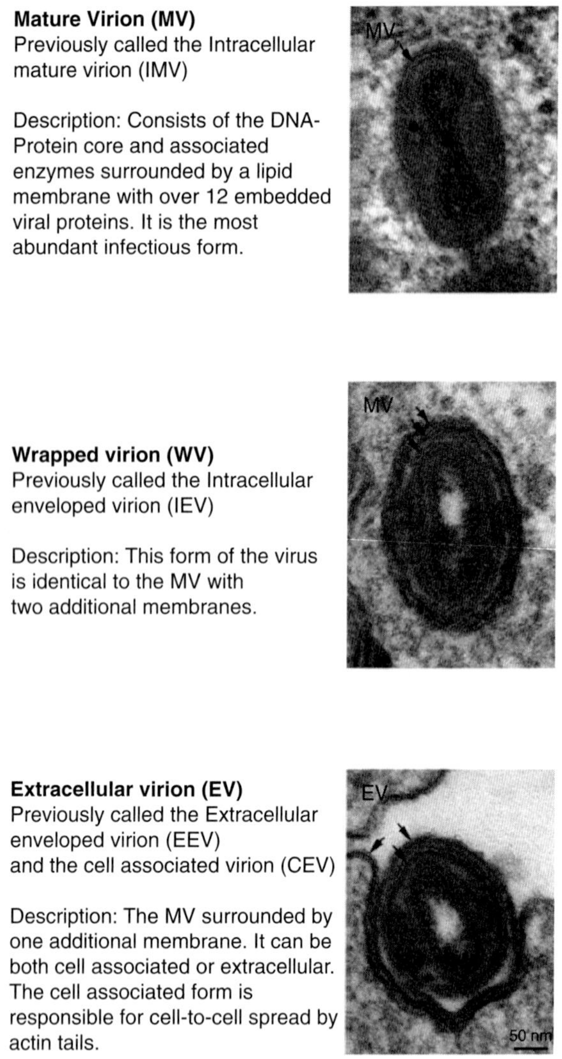

FIGURE 37.4 Infectious forms of poxviruses. Used with permission from Moss (2006).

online database with regularly updated genomic and proteomic information can be found at http://www.poxvirus.org.

Figure 37.3 also shows the brick shaped, or rounded, rectangular virions about 350 × 270 nm. The outer lipid membrane is studded with protein tubule structures as well as cellular antigens (Krauss et al., 2002). Enclosed within this membrane is a dumbbell shaped nucleoprotein core and two lateral bodies. This virion, consisting of the core, lateral bodies, and membrane is fully infectious. Many strains of poxviruses can exist in other forms containing additional lipid bilayers (Fig. 37.4) which likely play distinct roles in viral spreading within the host (Buller and Palumbo, 1991; Smith et al., 2002; McFadden, 2005b; Moss, 2006).

The viral life cycle is shown in Fig. 37.5 and can be divided into several discrete stages. The timeframe for one complete viral life cycle (Entry →Assembly→ Release of progeny virus) depends heavily on the viral strain and the type of host cell infected. For vaccinia virus, maximal yields are obtained 12–24 h after infection, although progeny virus can be seen in just 4–6 h postinfection (Buller and Palumbo, 1991).

Like most viruses poxviruses are able to quickly shutdown and/or control critical cellular processes such as DNA translation and transcription, protein synthesis, cell cycle, apoptosis, intracellular transport, and cytoskeletal machinery. Poxviruses also possess a large number of genes aimed at subverting a wide variety of key host defense mechanisms (Palumbo et al., 1994; Alcami et al., 1999, 2000; Bowie et al., 2000; Moss and Shisler, 2001; Seet et al., 2001; Shisler and Moss, 2001; Fruh et al., 2002; Katze et al., 2002; Symons et al., 2002; Dunlop et al., 2003; Reading et al., 2003; Seet et al., 2003a, 2003b; Li et al., 2004a, 2005; McFadden, 2004; Haga and Bowie, 2005; Stack et al., 2005; Alejo et al., 2006; Taylor and Barry, 2006; Webb et al., 2006; Humrich et al., 2007; Nuara et al., 2007).

Both vaccinia and variola viruses are stable for several days at room temperature and for months at 4°C, however they are inactivated relatively quickly at temperatures above 50°C (Harper, 1961). Viral suspensions in most liquids are more labile than dried virus such as scabs or freeze-dried vaccine (Cardone, 1953, 1956; Ferrier et al., 2004), in fact variola major has been recovered from scabs kept in a laboratory cupboard for over 13 years. Both variola and vaccinia are sensitive to ultraviolet light, and most common cleaning and disinfecting agents such as alcohols, bleach, ordinary

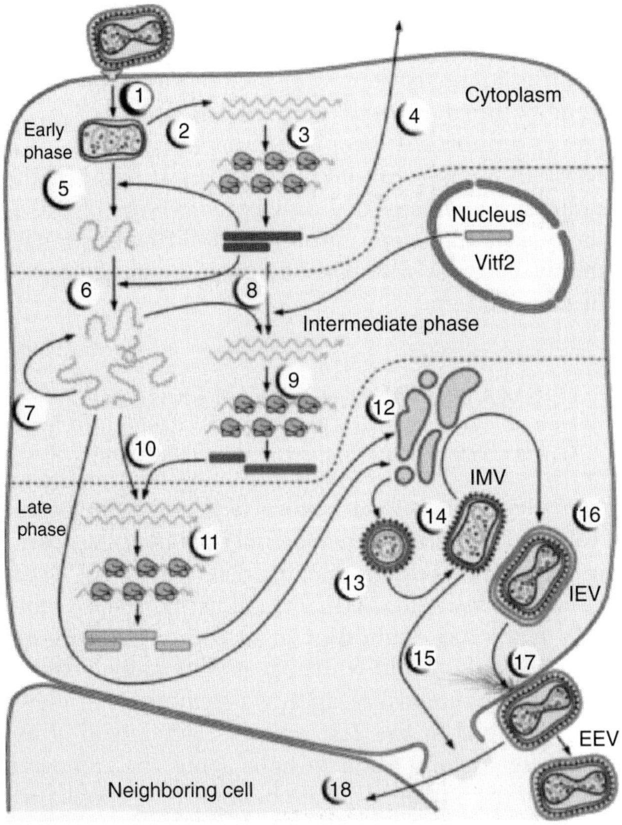

FIGURE 37.5 Poxvirus life cycle. Description of steps. (1) Entry. (2) Early mRNA synthesis. (3) Early gene transcription. (4) Secretion of modulatory proteins. (5) Release of genome. (6) Replication of viral DNA. (7) Additional viral genomes made. (8) Intermediate mRNA synthesis. (9) Intermediate gene transcription. (10) Late mRNA synthesis. (11) Late gene transcription. (12) Early virion assembly. (13) Formation of immature virion. (14) Brick shaped IMV. (15) Cell lysis releases IMV. (16) Formation of IEV. (17) Transport to cell surface (CEV). (18) Direct transfer to neighbor cell or release as EEV. Used with permission from Flint et al. (2003).

detergents, and quaternary ammonium compounds (Tanabe and Hotta, 1976; Ferrier et al., 2004).

THE IMMUNE RESPONSE TO POXVIRUSES

Smallpox was eradicated well before the advent of modern immunologic techniques for assaying cell-mediated immunity; consequently much of what we know about the immune response to variola major is confined to humoral immunity (Fenner, 1988). In ordinary cases of smallpox antibody titers rose 6–10 days after illness (approximately 3 weeks after infection) and persisted for several years following infection. The magnitude and speed with which the humoral immune response developed correlated with both disease severity and survival. Infection left survivors with lifelong immunity. Prominent physicians in the 19th and 20th centuries estimated that only 1 in 1000 subjects would suffer a second attack, and field reports from the Intensified Eradication Effort revealed that rarely would one see a case of variola major in a pockmarked person (Dixon, 1962; Fenner, 1988).

Immune responses to vaccination show a slightly different picture. Vaccination produces robust antibody responses far more quickly than infection (~10 days after immunization) and is likely the reason that early postexposure vaccination reduced or even eliminated the clinical symptoms of smallpox. Historically, protection from disease was thought to wane 3–10 years after vaccination (Behbehani, 1983) and the CDC recommends revaccination every 10 years for those working with vaccinia and every 3 years for those working with pathogenic Orthopoxviruses such as monkeypox.

More recent work shows clear and distinct roles for both humoral and cellular immune responses as well as innate effector mechanisms such as $\gamma\delta$ T cells, NK cells, complement, and defensins (Haga and Bowie, 2005; Stack et al., 2005). Like antibody levels both CD4 and CD8T cell responses also rise dramatically in the few weeks following vaccination. CD8 responses steadily decline with a half-life of 8–15 years, while antibodies and CD4 responses persist for decades (Hammarlund et al., 2003; Kan et al., 2007; Sivapalasingam et al., 2007) but it is not known how much protection against disease this residual immunity may provide (Nishiura and Eichner, 2006).

Subjects with T or B cell deficiencies are at increased risk of serious adverse events following immunization with vaccinia virus, and may have had higher rates of severe smallpox disease (Fulginiti, 2003; Fulginiti et al., 2003b; Wollenberg and Engler, 2004). In some animal studies, the adoptive transfer of immune serum or virus-specific T cells confers full protection against subsequent vaccinia challenge indicating that either arm of the immune response is sufficient (Ramirez et al., 2002; Snyder et al., 2004; Chaudhri et al., 2006), whereas other studies show that while humoral responses provide protection against disease, $CD8^+$ T cell responses prolong survival but do not offer protection against death (Edghill-Smith et al., 2005; Fang and Sigal, 2005). The emerging picture is that during primary infection cellular immune responses function early to limit viral spreading while antibody responses function at later time points to clear the pathogen and provide protection from subsequent reinfection (Panchanathan et al., 2006).

Following the events of September 11 and renewed interest in smallpox, a number of researchers focused on the identification of immunogenic proteins, peptides, and epitopes from vaccinia. The first CD8+ T cell epitopes were described in 2003, and a few years later, vaccinia-specific MHC I epitopes numbered in the hundreds (Drexler et al., 2003; Terajima et al., 2003, 2006; Snyder et al., 2004; Dong and Denny, 2006; Moutaftsi et al., 2006; Terajima and Ennis, 2006; Kennedy and Poland, 2007). MHC II restricted epitopes are also beginning to be identified (Tang et al., 2006; Jing et al., 2007; Mitra-Kaushik et al., 2007; Moutaftsi et al., 2007). The identification of numerous B cell epitopes has also been accomplished, and unlike T cells, B cells preferentially target structural and membrane proteins (Bell et al., 2004; Davies et al., 2005a, 2005b; Jones-Trower et al., 2005).

EPIDEMIOLOGY

Smallpox is a human disease for which there are no known animal or insect vectors (Fenner, 1988). In spite of the fact that viral particles are relatively resistant to environmental conditions, transmission almost always occurs via close physical contact with an infected subject who has the characteristic rash or with infected bedding or clothing (Breman and Henderson, 2002). Although rare, airborne transmission did occur under certain conditions (Meiklejohn et al., 1961; Downie et al., 1965; Wehrle et al., 1970). Outbreaks spread slowly and were easy to trace given that: (1) transmission required prolonged, close physical contact, (2) the incubation period was long, and (3) patients were infectious only when bedridden with a highly visible rash. Infection conferred lifelong immunity to the survivors, and there was no evidence of latent infection, recurrences, or a carrier state. Subclinical infections occurred but were rare and confined to vaccinated individuals. These subjects rarely shed virus and had little impact on disease transmission (Fenner, 1988).

Historical accounts indicate that there were two separate epidemiological patterns to the disease. When smallpox was introduced into a naive population it spread rapidly, infecting nearly every susceptible individual. These outbreaks affected all age groups and had very high mortality rates. The introduction of the disease to the Native Americans and Australian aborigines are the best documented examples of smallpox's destructive power, however there is no reason to believe that it was any less deadly among the naive populations in China, India, or Europe hundreds or thousands of years earlier.

In geographical areas with endemic smallpox the disease continually circulated through the community with seasonal outbreaks mostly in winter and spring when conditions (i.e., temperature and humidity) allowed the virus to persist longer in the environment. These larger outbreaks would occur when sufficient numbers of uninfected people accumulated in a town or city to allow a full-fledged epidemic. Most of the adults in the affected region were survivors of previous attacks and were immune to the disease, consequently disease outbreaks in these areas primarily affected children.

SMALLPOX AS A BIOLOGICAL WEAPON

Smallpox has a long history of use as a biological weapon, with most of the documented incidents coming from the Western Hemisphere (Hopkins, 1983; Bhalla and Warheit, 2004; Frischknecht, 2003) (Table 37.3). After the eradication it was thought that the likelihood of smallpox being intentionally reintroduced as a biological weapon was remote. That viewpoint changed in the 21st century (Henderson et al., 1999) after details came to light about the extensive nature of the Soviet and Iraqi biological weapons programs (Alibek, 2004; Roffey et al., 2002). The anthrax attacks in the US in 2001, as well as the large number of conventional terrorist attacks shortly thereafter

TABLE 37.3 Historical uses of smallpox as a biological weapon

Date	Description
1500s	Spanish conquistadors give Native Americans contaminated clothing. Resulting outbreaks contribute to the downfall of the Aztec and Inca empires.
1760s	French and Indian war—The British army gives Indians tribes loyal to the French contaminated blankets. Outbreaks estimated to have ~50% mortality.
1776	Revolutionary war—British send infected civilians out of Boston in hope of infecting continental Army. Washington institutes mandatory vaccination of soldiers.
1800s	Blankets from smallpox victims deliberately given to Native Americans causing widespread epidemics.
1860s	Civil War—Former Kentucky Governor sells contaminated clothing to Union troops.
1980s–?	The Soviet Union develops a bioweapons program which includes large quantities of smallpox. Iraqi biological weapons program attempts to develop camelpox as a biological weapon. International concern that other countries are also involved in biological weapons development and testing.

were grim reminders that individuals and groups were willing to use such weapons. The CDC created a categorized list of potential biological agents with variola listed as one of the most likely pathogens to be used as a biological weapon (Henderson et al., 1999; Bhalla and Warheit, 2004). Smallpox has several key features which make it an excellent choice for a biological weapon (Fig. 37.6) (Tegnell et al., 2002; Blendon et al., 2003; Mahy, 2003; Pennington, 2003; Whitley, 2003; Cleri et al., 2006).

- Deadly, disfiguring illness
- Extremely low infectious dose
- High mortality rate
- Contagious pathogen
- Long incubation time allows for the perpetrator to escape
- Would cause widespread panic and disruption
- Modern populations largely susceptible
- No effective treatment
- Insufficient vaccine to deal with large epidemics
- Odorless and tasteless, making it difficult to detect

FIGURE 37.6 Characteristics of smallpox which make it an ideal biological weapon.

CLINICAL DISEASE

The clinical course of the disease can be divided into several discrete phases (Fenner, 1988; Breman and Henderson, 2002) shown in Fig. 37.7. The average incubation period is 12–14 days (range: 7–17 days) during which time the subject is asymptomatic and noninfectious. This is followed by a prodromal phase accompanied by the abrupt onset of severe headache, backache, high fever (>40°C), prostration, and sometimes vomiting. Symptoms may last for 2–4 days. The rash begins with an enanthema involving the mouth and oropharynx followed a day later by small reddish macules which develop into papules over a period of 1–2 days. One to two days later the papules become vesicles 2–5 mm in size. The lesions initially occur on the face and extremities and gradually cover the entire body. Round, hard, fluid-filled pustules soon develop and over the next 5–8 days begin to crust over. All of the lesions typically progress with the same kinetics.

FIGURE 37.7 Smallpox disease progression and characteristics. Used with permission from Dr. David Heymann, World Health Organization and Breman and Henderson (2002).

Recovery occurs as the scabs fall off, leaving 65–80% of the victims with extensive scarring typically concentrated on the face (Fenner, 1988). Blindness from viral keratitis or secondary infection occurs in 1% of cases (Behbehani, 1983). Encephalitis is rare but can occur (1% of cases) (Dixon, 1962). Rarely, disease occurring in children can lead to arthritis due to infection and disruption of the growing bone and joint tissues (Fenner, 1988). Variola minor cases resemble mild cases of ordinary type smallpox while disease caused by variola major can be classified into five main types based on clinical severity (Table 37.4).

TABLE 37.4 Classification of smallpox

Designation	Symptoms	Mortality
Ordinary	Fever, backache, malaise, headache, delirium	~30%
	The extent of the rash can vary	
	Discrete pocks	<10%
	Semiconfluent	25–50%
	Confluent	50–75%
Less severe (far more common in vaccinated individuals)		
Modified type	Rash has discrete, smaller lesions which rapidly resolve	Rare
Variola sine eruptione	Asymptomatic or brief fever, influenza like symptoms; patients may or may not present with rash	0%
More severe (account for ~10% of all cases)		
Flat type	Lesions are flat, evolve slowly, and eventually coalesce; rash may be discrete or confluent	80% immune 99% naive
Hemorrhagic	Prolonged, severe prodrome; dusky rash, bleeding from mucosal membranes	97%
Variola minor		
Alastrim or amaas	Similar to ordinary type variola major but far more mild symptoms	1–2%

Adapted from: Fenner (1988), Breman and Henderson (2002), and Cleri et al. (2006).

The two milder forms are called modified type and variola sine eruptione and are more commonly seen in vaccinated individuals as they have preexisting immunity which lessens the severity of disease symptoms. Ten percent of the cases are much more severe: flat type or hemorrhagic smallpox. Death rates are extremely high for both of these classes of disease. In addition, victims of hemorrhagic smallpox act as "super-spreaders" of disease by virtue of the higher viral titers, prolonged prodrome, and profuse bleeding. It was noted early on that the extent and distribution of the lesions correlated with mortality. Patients with fewer discrete lesions were more likely to recover, while victims with dense or confluent lesions typically had a poor prognosis (Fenner, 1988; Breman and Henderson, 2002).

TREATMENT

There remains no effective treatment for smallpox. In the past, patients were typically isolated and then provided with supportive care such as intravenous (IV) fluids, antibiotics to reduce secondary infections, and medications for pain relief. If an outbreak were to occur today our options would be essentially unchanged.

Data from numerous outbreaks during the 20th century indicate that postexposure immunization can reduce the severity of symptoms. Immunization within 3–4 days of smallpox infection is thought to provide nearly complete protection from disease, while immunization later than this may provide partial protection and reduce the disease severity (Breman and Henderson, 2002; Massoudi et al., 2003; Mortimer, 2003a; Nishiura and Eichner, 2007). This is likely due to the long incubation period of variola virus which allows time for the crossprotective antivaccinia responses to develop. Consistent with this data, the CDC recommends vaccination within 3 days of exposure.

Historically, numerous drugs were tried and although initial reports of thiosemicarbazones and arabinoside derivatives showed therapeutic effects, further studies could not confirm these results and showed high levels of toxicity (Neyts and De Clercq, 2003; Sliva and Schnierle, 2007). Cidofovir, a nucleoside analog DNA polymerase inhibitor has shown antiviral activity in vitro and in animal models with several poxviruses (Smee et al., 2001b; De Clercq, 2002; De Clercq and Neyts, 2004; Neyts et al., 2004; Quenelle et al., 2004a; Knorr et al., 2006). Although it may be used to treat smallpox or vaccine complications under an investigational new drug application its utility is limited by its severe renal toxicity

(Bray and Roy, 2004). Several groups have focused on modifications of cidofovir which enhance its oral availability, activity against poxviruses, and reduced nephrotoxicity with promising results (Smee et al., 2001a; Bray et al., 2000; Buller et al., 2004; Painter and Hostetler, 2004; Quenelle et al., 2004b). Other drugs showing promising activity against poxviruses include: ribivirin, trifluridine, a variety of tetrapyrrole compounds (Chen-Collins et al., 2003), and small molecule inhibitors of essential viral enzymes (Byrd et al., 2004; McFadden, 2005a; Reeves et al., 2005; De Clercq, 2001) with limited homology to human enzymes (Smee and Sidwell, 2003; Yang et al., 2005; Sliva and Schnierle, 2007).

Vaccinia immune globulin (VIG) is concentrated gamma-globulin from pooled plasma of recently vaccinated individuals. It was initially developed in the 1950s to combat vaccine-related complications (Fenner, 1988; Bray, 2004). There is extensive literature attesting to its efficacy in treating disease and vaccine complications and while there are no controlled therapeutic trials, the United States and other countries have once again begun making and stockpiling VIG (Goldsmith et al., 2004; Hopkins and Lane, 2004; Hopkins et al., 2004).

Yet another area of exploration is the use of immunotherapeutics to combat smallpox infections (Atrasheuskaya et al., 2004; Liu et al., 2004; Law et al., 2005). There are a few reports indicating that combined vaccination and VIG administration is more effective than postexposure vaccination alone. One avenue of current research is to develop humanized monoclonal antibodies with similar or enhanced efficacy compared to VIG (Casadevall, 2002; Chen et al., 2006). Along a similar vein, one can envision the use of agents which block viral virulence factors, or molecules such as TNFα antagonists to reduce immunopathology, or even the use of adjuvants to redirect and shape a developing immune response to increase protective immunity while decreasing inflammation and its associated side effects (Ahmed et al., 2005; Staib et al., 2005).

In summary, there is currently no treatment for smallpox infection other than immediate postexposure vaccination and perhaps VIG. This picture is likely to change in the future, especially given the research impetus and the results in this field so far.

PATHOGENESIS

Much of our understanding of smallpox pathogenesis comes from studies using related poxviruses such as mousepox, rabbitpox, and monkeypox (Fenner et al., 1957; Marshall and Fenner, 1960, 2000; Wallace and Buller, 1985; Esteban and Buller, 2005; Levine et al., 2007). The virus enters the respiratory tract and rapidly enters the local lymph nodes. Within a few days, a brief, low-level viremia deposits the virus into the liver and spleen, where it grows to extremely high titers over the next 4–14 days (Breman and Henderson, 2002). A secondary viremia marks the beginning of the prodromal phase when large quantities of the virus enter the bloodstream and infects the mucous membranes of the mouth and pharynx followed by infection of the dermal layer of the skin. The characteristic lesions form on the mouth and tongue 24h before the rash appears on the skin. All of the lesions contain large quantities of virus (Fenner, 1988). The virus is also present in the urine, blood, and other body fluids as well as internal organs such as the liver, spleen, lymph nodes, kidneys, and bone marrow (Sarkar et al., 1973a, 1973b). Virus is present in respiratory secretions and can be found on clothing, linen, and bedding. Viral shedding is greatest during the first 10 days of the rash however patients remain infectious until all scabs and crusts are shed (Mitra et al., 1974; Sarkar et al., 1974; Breman and Henderson, 2002).

The immune response develops within a week of the first symptoms of disease and consists of both humoral and cellular responses. Survival depends on the ability of both arms of the immune system to curb viral replication and eliminate virally infected cells. These immune responses are a two-edged sword as deaths after infection were frequently due to immunopathology.

In addition to variola virus there are a number of other poxviruses which can cause human disease. These are shown in Table 37.5. Of particular concern for public health officials is monkeypox, which has been associated with sporadic outbreaks in Africa (Ligon, 2004; Levine et al., 2007), and a 2003 outbreak in the Midwestern United States (Kile et al., 2005) related to the unregulated importation of exotic pets. There are also recent reports from South America describing outbreaks of novel poxviruses in cattle with documented transmission to humans (Damaso et al., 2000; Schatzmayr et al., 2000; da Fonseca et al., 2002; de Souza Trindade et al., 2003; Nagasse-Sugahara et al., 2004) and an outbreak of buffalopox in India (Singh et al., 2006). While usually less severe than smallpox, these other poxviruses can spread from human-to-human and are considered important emerging zoonotic pathogens (Di Giulio and Eckburg, 2004; Jezek et al., 1986; Fine et al., 1988; Frey and Belshe, 2004; Fleischauer et al., 2005; Learned

TABLE 37.5 Poxviruses that infect humans and cause disease

Genus and species (disease)	Primary reservoir	Geographic region	Mode of transmission	Protection provided by vaccinia vaccination
Orthopoxvirus				
Cowpox	Rodents	Europe, Africa, central and northern Africa	Direct contact	Yes
Monkeypox	Rodents	Central and West Africa	Direct contact, respiratory droplets	Yes
Vaccinia	Unknown[a]		Direct contact	
Variola (smallpox)b	Humans	US, Russia	Direct contact, respiratory droplets	Yes
Yatapoxvirus				
Tanapox	Nonhuman primates	Kenya, Zaire	Direct contact	No
Yabapox	Nonhuman primates	Central Africa	Direct contact	No
Parapoxvirus				
Pseudocowpox (milker's nodules and paravaccinia)	Ungulates	Worldwide	Direct contact	No
Bovine papular stomatitis	Ungulates	US, Canada, Africa, Australia, New Zealand, Great Britain, Europe	Direct contact	No
Orf	Ungulates	North America, Europe, New Zealand	Direct contact	No
Sealpox	Seals	North Sea, Pacific Ocean, Atlantic	Direct contact	No
Molluscipoxvirus				
Molluscipox (molluscum contagiosum)	Humans	Worldwide	Direct contact	No

[a] Genetic analysis of vaccinia reveals it to be different from cowpox; its origin is unknown, but it may be an extinct horsepox virus.
[b] Smallpox was declared eradicated in 1980.

et al., 2005). Of further concern is camelpox, the closest genetic relative to variola virus (Afonso et al., 2002b), which was a component of the Iraqi biological weapons program (Zilinskas, 1997).

HISTORY OF THE SMALLPOX VACCINE

Variolation

Well before the development of microbiology and the germ theory the idea that physical separation from the ill would protect the healthy was independently developed and utilized by most civilizations (Dixon, 1962; Fenner, 1988). Separate huts or even villages were designated for those harboring infection and ships entering port cities were frequently quarantined.

China, India, and several Middle Eastern countries had developed a procedure to protect themselves from the effects of smallpox (Dixon, 1962; Behbehani, 1983; Fenner, 1988). In Asia, people who had undergone less severe disease provided infected scab material which was ground to a fine powder and inhaled through the nose by healthy individuals. This procedure was termed insufflation and resulted in a severe, generalized rash which resembled a mild case of smallpox. In Persia, a modification of this procedure was used whereby pus from a smallpox lesion was scraped into the skin of a healthy individual, who soon developed a severe local lesion frequently accompanied by satellite pustules. The cutaneous method of variolation was thought to be somewhat less severe than the inhalation method. Both practices were later called "variolation" to distinguish them from "vaccination" with cowpox virus and were characterized by a much less severe form of smallpox with mortality rates of about 1–2% in contrast to smallpox's 30% death rate. An unfortunate side effect of variolation was that the subjects were contagious, and could and did cause further outbreaks of smallpox. Variolation was practiced in Asia and the Middle East and was slowly brought to Europe by travelers to the Middle East. The main advocate of the practice was Lady Mary Wortley Montagu, who was introduced to the technique in Turkey (Dinc and Ulman, 2007) and helped to establish the practice of variolation in England (Behbehani, 1983).

Vaccination

Contrary to popular belief, the idea that cowpox (an orthopox virus related to variola) infection brought with it immunity to smallpox did not originate with Edward Jenner. Numerous reports describing this phenomenon have been found dating back to the 1760s, over 30 years before Jenner published his work. While practicing variolation in his country, Jenner noticed that he was unable to variolate many of his patients who had previously contracted cowpox. The potential to develop a safe and effective protective measure against smallpox held great appeal for him and he began investigating this phenomenon by inoculating his 10-month-old son and two neighborhood children with cowpox taken from the hands of his son's nurse. When he tried to variolate them some time later, both his son and the nurse were unable to be infected with smallpox. Over the next 10 years he carefully performed numerous additional experiments. The most famous of his subjects was James Phipps, who, at the age of 8, was inoculated with cowpox taken from an infected nursemaid, Sarah Nelmes on May 14, 1796. Despite vocal opposition, the success of his "inoculation" was clear and within 10 years of its inception, large-scale vaccination was being practiced throughout the world. As vaccination was safer, less painful, and did not spread smallpox many European nations began to outlaw variolation and impose compulsory vaccination. It soon became apparent that inoculation with cowpox did not provide lifelong protection against smallpox and regular revaccination was instituted. Widespread vaccination began to control the number, frequency, and mortality of smallpox epidemics throughout the 19th century and had even eliminated smallpox in many of the more developed countries by the 1940s and 1950s.

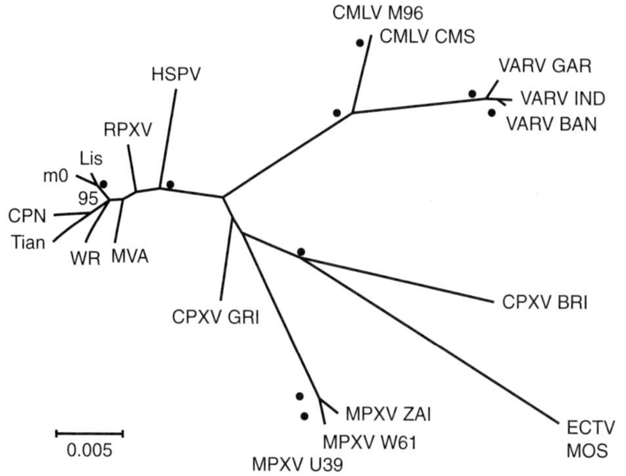

FIGURE 37.8 Sequence diversity in poxviruses. Conserved OPV central genomic nucleotide sequences were aligned using DIALIGN, and gapped regions were realigned with CLUSTAL W and trimmed with Gblocks. The scale indicates estimated distance. For additional details on the generation of the tree see the reference. Reproduced with permission from Tulman et al. (2007).

SMALLPOX VACCINE

In the early years of vaccination, a wide variety of poxviruses were utilized including: cowpox, sheeppox, goatpox, horsepox, swinepox, and buffalopox. Over time physicians began to use vaccinia virus for immunization against smallpox. The source of vaccinia virus is, to this day, unknown. While various explanations for its origins have been postulated, the most common are: (1) a derivative smallpox virus, altered by years of passage in laboratories and domesticated animals, (2) a derivative cowpox virus, similarly altered by laboratory passage, and (3) a remnant strain of now extinct horsepox virus. As shown in Fig. 37.8 numerous poxviruses have been sequenced (Yau and Rouhandeh, 1973; Bhat et al., 1991; Afonso et al., 2000, 2002a, 2002b, 2005, 2006; Tulman et al., 2001, 2004) and genetic analysis shows that although closely related to rabbitpox, horsepox, and cowpox, vaccinia is a distinct viral species (Shchelkunov et al., 2000; Esposito et al., 2006; Tulman et al., 2006). Regardless of its origin, vaccinia virus proved to be a relatively safe and effective vaccine, and was the basis for all of the smallpox vaccines in widespread use during the eradication efforts.

Throughout the 1800s several key improvements were made in vaccine production and usage. Italian physicians began using calves as a source of the vaccine in 1805 and commenced large-scale vaccine production in calves in 1840 (Dixon, 1962). Another key development was the use of glycerol as a stabilizer and phenol as a bactericidal agent which occurred during the same time period (Fenner, 1988). Numerous different virus strains and species were used, grown using varied methodologies, had different titers, take rates, side effects, and efficacy (Table 37.6).

The advent of the Intensified Eradication Program brought with it standardized production methods, reporting of vaccine take rates, and quality control measures including: seed lots for virus production, maximal acceptable bacterial counts, minimum acceptable viral titers, and defined heat stability (Fenner, 1988). The vast majority of the vaccine strains were produced using large healthy animals (mainly calves but also sheep, and water buffaloes) as *vaccinifers*. The belly or the flanks of these animals were shaved, disinfected, and rinsed with sterile water.

TABLE 37.6 Vaccinia strains used in smallpox vaccines

Strain	Locations
EM-63	USSR
New York City Board of Health	North and South America
Lister	UK, Africa, North and South America, Asia, Europe, India
Temple of Heaven	China
BIEM	USSR
Bern	Germany, Yugoslavia, Austria
B-15	USSR
Patwadangar	India
Copenhagen	Denmark
Paris	France, Brazil, Syria, Turkey
Budapest	Hungary
Darien	Japan
Ikeda	Japan
Gam	USSR
MRIVP	USSR
Per	USSR
Tashkent	USSR
Tom	USSR
Williamsport	USA
LMC	UK
Bordeaux	Africa, Portugal
Massachusetts 999	Argentina
Bohemia	Czechoslovakia
Hamburg	Germany
Aosta	Italy
Sweden	Sweden
Finland	Finland
Vienna	Bulgaria
Spain	Spain
Tom	USSR

Two perpendicular series of parallel scratches 1mm apart were made on the prepared area. High titer seed virus was rubbed into the scarified skin and incubated for several days. The vaccine *pulp*, a mixture of epidermal cells, leukocytes, plasma, hair, bacteria, and the virus, was then scraped from the animal's skin and further processed. The vaccine lymph was prepared by grinding and homogenizing the pulp and adding glycerol (50% final concentration) and phenol (0.4% final concentration). The resulting liquid vaccine required constant refrigeration and, despite being satisfactory for countries with cold chain transportation, it was not acceptable in tropical and developing countries. Freeze-dried vaccines were more stable and could survive higher temperatures than the liquid counterparts and from the 1950s onward freeze-dried vaccines were used almost exclusively.

Vaccination techniques varied widely, but by the 1950s all of the methods had the same basic principle: to introduce the virus into the epidermis. Vaccination sites included the external side of the upper arm over the deltoid muscle, the flexor surface of the forearm, a small portion of the back, and various locations on the legs. A number of different instruments were used to introduce the virus including: needles, lancets, small knives to scratch the virus into the skin, the rotary lancet to abrade the epidermis, jet injectors which used compressed air to force the virus into the subcutaneous tissue, and the bifurcated needle, developed by Benjamin Rubin at Wyeth Laboratories in 1965. The rotary lancet was frequently associated with more severe local reactions and was phased out for safer methods. The jet injectors were quite effective but were rapidly replaced by the cheap and simple-to-use bifurcated needle (Fenner, 1988).

One major advance was the replacement of higher pathogenicity strains with safer ones which caused fewer and less severe side effects (Kretzschmar et al., 2006). By the end of the eradication effort the two most utilized commonly strains were the New York City Board of Health (NYCBOH) and Lister strains (Fenner, 1988).

In order to further reduce vaccine complications a number of laboratories developed attenuated vaccinia strains. These included CVI-78 and CVII strains, which were less reactogenic than the NYCBOH strain (Kempe et al., 1968), the modified vaccinia Ankara (MVA) strain developed through 572 serial passages in chick embryo fibroblasts (Mayr et al., 1978; Blanchard et al., 1998), and LC16m8, a Japanese strain attenuated through passage in rabbit kidney cells at 30°C (Sugimoto and Yamanouchi, 1994; Kenner et al., 2006). MVA and LC16m8 were used on tens of thousands of subjects with fewer adverse reactions than the nonattenuated strains, however as these attenuated strains were produced at the very end of the eradication effort efficacy data is lacking (Mayr, 2003; Slifka, 2005). Modern studies of these attenuated strains have refocused primarily on MVA, LC16m8, and NYVAC (McCurdy et al., 2004; Tartaglia et al., 1992). They have been safely tested in immunocompromised animal models (Earl et al., 2004; Wyatt et al., 2004), but unfortunately appear to be less immunogenic than their nonattenuated cousins and it is not clear whether or not they would provide protection against smallpox (Coulibaly et al., 2005; Puissant and Combadiere, 2006; Staib et al., 2006; Parrino et al., 2007).

In the 1960s many researchers tried to develop inactivated smallpox vaccine in an effort to reduce complication rates. A variety of inactivation methods were tested including: heat killing, formaldehyde, UV or gamma irradiation, photodynamic inactivation, and others. For reasons that are still not understood, none of these vaccines produced satisfactory results (Fenner, 1988). One possibility is that they consist almost entirely of nonenveloped viral particles and hence may not have primed for immune recognition and clearance of the enveloped virions, a necessary requirement for protective immunity.

Currently, the only licensed vaccine is Dryvax, made by Wyeth Laboratories. The existing available doses were manufactured in the 1970s and early 1980s and stored frozen at −20°C. The smallpox vaccine is the most reactogenic of all FDA approved vaccines and has a large number of contraindications (Table 37.7), limiting its usefulness in the general population (Lane and Fulginiti, 2003). These include anyone with a compromised immune system (HIV, cancer, radiation therapy, organ transplant recipients), with skin disease such as eczema or atopic dermatitis, with a variety of heart conditions, pregnancy, and others. Many of these conditions are so widespread that it is estimated that as much as 50% of the US population today could not receive the vaccine (Kemper et al., 2002). This situation adds further urgency to the need to develop safer vaccine candidates.

The current vaccination technique is essentially unchanged from what was done during the global smallpox eradication effort (Wyeth, 2002). The typical site for vaccination is the deltoid area of the upper arm. The use of alcohol to prep the area is contraindicated as this can inactivate the vaccine virus. The multiuse vaccine vial is reconstituted with water and a sterile bifurcated needle is dipped into the vaccine. The needle will hold ~2.5 μl of vaccine containing 250,000 pock forming units of virus. The needle is then used to make several insertions in a small area of the arm (5 mm in diameter). Percutaneous immunization has been shown to promote more robust immune responses than other vaccination routes (Fenner, 1988; Fulginiti et al., 2003a). For primary vaccination three insertions are made, and if no blood is visible another three insertions are made. For revaccination 15 insertions are made. The needle is then discarded into an appropriately labeled biohazard container. Excess vaccine is then absorbed with sterile gauze and the vaccine site is covered to prevent spread of the virus. The bandage should be changed every 1–3 days and the site should be kept dry. Care should be taken to wash hands after direct contact with the site or potentially contaminated objects, and to wash clothing, towels, and bedding separately with detergent and/or bleach and to dispose of the scab properly once it falls off.

TABLE 37.7 Contraindications for smallpox vaccination

Cancer
Organ/bone marrow transplant
Steroid therapy
Disease affecting immune system (HIV, lymphoma, leukemia)
Primary immune deficiency disorder
Lupus or other severe autoimmune disease
Darier's disease
Active skin disease/condition (burns, impetigo, psoriasis, acne, poison ivy, chickenpox, herpes, severe rashes)
Current or past atopic dermatitis (eczema)
Current history of heart problems (congestive heart failure, heart attack, cardiomyopathy, stroke, chest pain)
Three or more of the following • High blood pressure • High cholesterol • Diabetes or high blood sugar • Current smoker • First degree relative with heart condition before the age of 50 Pregnancy or breastfeeding
Allergies to polymyxin B, streptomycin, chlortetracycline, neomycin, or latex

After vaccination a pock will form at the site after 3–4 days (Fig. 37.9). Over the first week, this vesicle typically grows to 5–15 mm in diameter, fills with pus, and begins to drain. After 2–3 weeks the blister will dry up and a scab will form. The scab will fall off a short time later leaving a small scar. For individuals receiving a repeat vaccination the vesicle will be smaller and resolve more quickly. Individuals receiving the vaccine should be in good general health and not have any symptoms of illness when they receive the vaccine.

Like all immunizations, the smallpox vaccine can induce a number of side effects and adverse events. These are listed in Table 37.8 and can range from common annoyances to life-threatening conditions (Lane et al., 1969, 1970b, 1970a; Goldstein et al., 1975; Fenner, 1988; Fulginiti et al., 2003b; Lane and Goldstein, 2003; Casey et al., 2005; Kretzschmar et al., 2006). Most individuals receiving the vaccine can expect to experience one or more of the following reactions: fever, headache, muscle and joint aches, chills, nausea, swelling, soreness and itching at the vaccine site, and regional lymphadenopathy. Satellite lesions may form near the vaccine site. These are typically mild and last for several days. Studies of the large-scale vaccination programs started after September 11 showed a lower number of most adverse events than expected based on historical data (Casey et al., 2005). However, multiple cases of myo- and pericarditis, inflammation of the heart tissue or the pericardium respectively,

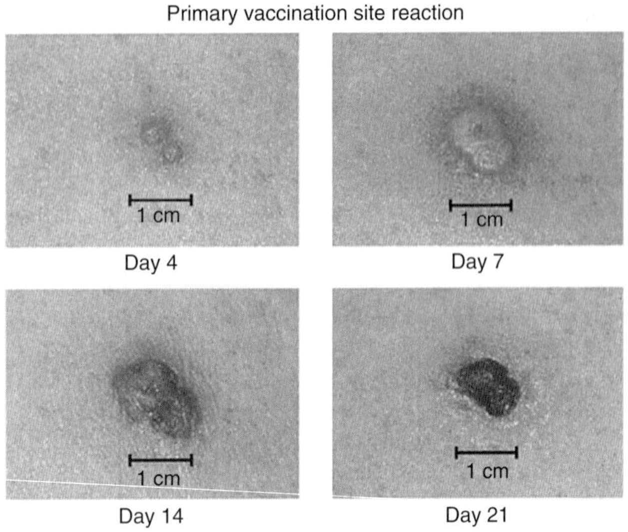

FIGURE 37.9 Characteristic smallpox vaccine "Take." Courtesy of the CDC. http://www.bt.cdc.gov/agent/smallpox/vaccination/index.asp.

TABLE 37.8 Frequencies of adverse events following smallpox vaccination (numbers represent events per million vaccinations)

	1963 National Survey[a]	1968 National Survey[a]	2003 Vaccination Programs[b]
Postvaccinial encephalitis	1.9	2.9	2.6
Progressive vaccinia	1.1	0.9	0
Eczema vaccinatum	8.7	10.4	0
Generalized vaccinia	20.8	23.4	59.7
Accidental inoculation	13.6	25.4	221.8
Myopericarditis	ND[c]	ND[c]	133.6
Death	1.1	1.3	3[d]

[a] Data from 1963 and 1968 are taken from nationwide surveys conducted by the CDC (Neff et al., 1967; Lane et al., 1969). The first number represents the incidence rate among primary vaccinees.
[b] Data from reports on the two US Smallpox Vaccination Programs that began in 2002 among civilian (n = 40,422) and military (n = 730,580) personnel (Grabenstein and Winkenwerder, 2003; Casey et al., 2005; Poland et al., 2005).
[c] ND, no data.
[d] The three deaths listed, occurred among the civilian personnel and were a result of myocardial infarctions. It is not known if they are directly related to the vaccine. No deaths have been reported among the 730,000+ military personnel receiving the vaccine.

were identified at a much higher rate than historically recorded (Halsell et al., 2003; Poland and Neff, 2003; Poland et al., 2005). Although a causal link between smallpox vaccination and myopericarditis has not been conclusively proven it remains a concern. Although the majority of these subjects recovered promptly it remains to be seen whether or not they suffer any long-term sequelae and consequently the CDC included a variety of heart conditions on the list of contraindications (Arness et al., 2004; Cassimatis et al., 2004; Gami et al., 2005).

The most common side effects of vaccination are: (1) inadvertent autoinoculation which occurs when vaccinia virus from the vaccination site is transferred to other sites on the body or to other individuals (Sepkowitz, 2003) and (2) erythematous or urticarial skin rashes, collectively termed generalized vaccinia. These rashes are fairly frequent and typically resolve spontaneously requiring no special care (Fulginiti et al., 2003b; Wollenberg and Engler, 2004). Ocular vaccinia can result in keratitis and the scarring may result in impaired vision or blindness. Topical antiviral agents can be used to treat the symptoms and in some cases VIG may be recommended. Bacterial superinfection of the vaccination site can also occur, typically by staphylococci or β-hemolytic streptococci (Fulginiti et al., 2003b). These types of infections are readily treated by appropriate antibiotic therapy and their incidence is greatly reduced by proper care of the vaccination site (Vellozzi et al., 2004).

Other more serious complications occur in patients with some degree of immunologic deficit and include: eczema vaccinatum, progressive vaccinia (vaccinia necrosum), and postvaccinal encephalitis (Fenner, 1988; Fulginiti et al., 2003b; Rock et al., 2004; Vellozzi et al., 2005).

Eczema vaccinatum can be either a localized or systemic spread of the infection. The rash progression mimics that of the normal vaccination site, however, the lesions are typically confluent and cover the areas of the skin with current or past eczema. If left untreated, patients can develop septic shock and die, therefore prompt treatment with VIG is recommended as soon as possible after diagnosis. Given the predisposition of this complication for eczematous patients it is important to refrain from vaccinating these individuals unless absolutely necessary (Engler et al., 2002).

Progressive vaccinia (also called vaccinia necrosum), is a severe illness with local progressive necrosis at the vaccination site. Inflammatory responses are absent and the lesion fails to resolve. This is soon followed by secondary lesions elsewhere on the body. Nearly all cases involved individuals with defects in cell-mediated immunity (O'Connell et al., 1964; Redfield et al., 1987). Lymphocyte transfer through blood transfusion allows for viral clearance but also results in lethal graft-versus-host disease. Before the use of VIG and antiviral agents the disease was uniformly fatal and even the current treatment regimen (prompt hospitalization and supportive care, massive doses of VIG, and/or cidofovir) is not effective unless the T cell defect is minor or reversible.

Postvaccinial encephalitis is another rare complication with high mortality and morbidity rates. Historically the death rate was 15–25% and 1/3 of the survivors were left with permanent neurological damage. Hospitalization and supportive care are required but there is no specific therapy and VIG is ineffective and not recommended.

While rare, smallpox vaccine can cause fetal complications, hence the contraindication for pregnant women (Hassett, 2003). There have been approximately 50 reported cases of fetal vaccinia, which results in stillbirth or death of the infant shortly after delivery (Fulginiti et al., 2003b).

Historically, for every million primary vaccine recipients several hundred people would experience complications severe enough to require hospitalization and one to two people would die, typically the result of either postvaccinial encephalitis or progressive vaccinia. After the eradication of smallpox, even this small risk was deemed unacceptable and routine vaccination stopped. The complication rate in individuals receiving revaccination is far lower (about 1/10) for all adverse events.

SMALLPOX VACCINES IN DEVELOPMENT

Two obstacles hinder the testing of new smallpox vaccines. The first is that there are very few historical studies that quantitate protective immunity and there are no defined immune correlates of protection. The second point is that testing the clinical efficacy of new vaccines is not possible in the absence of endemic smallpox. Consequently, next generation vaccines are being compared side-by-side with the already licensed vaccines in order to demonstrate equivalent immunogenicity in humans and protective efficacy in animal models.

One of the first priorities has been to reexamine the licensed vaccines using modern immunological techniques and involves several large-scale vaccination programs (Stienlauf et al., 1999; Grabenstein and Winkenwerder, 2003; Poland and Neff, 2003; Combadiere et al., 2004; Kennedy et al., 2004; Orr et al., 2004; Poland et al., 2005; Amanna et al., 2006; Treanor et al., 2006; Simpson et al., 2007). Another priority was to determine if the existing Dryvax vaccine could be diluted without a loss of immunogenicity. The ability to dilute a vaccine with proven efficacy would substantially increase the utility and number of persons who could be protected given the limited vaccine stock available. These studies found that "take rates" and immune markers (antibody levels and cellular immune responses) did not differ between groups receiving undiluted, a 1:5 dilution or a 1:10 dilution of Dryvax (Frey et al., 2002; Talbot et al., 2004; Couch et al., 2007), and in fact, reducing the vaccine dose also reduced vaccine-associated side effects (Couch et al., 2007). Studies have also been conducted with the Aventis Pastuer and Lister-based vaccines with similar results (Kim et al., 2005; Hsieh et al., 2006).

New smallpox vaccines are also being developed and tested. In September 2000, Acambis Inc. received a contract to create a stockpile of 40 million doses of a new vaccine. The next year they received a second contract to create an additional 155 million doses. Acambis created two tissue culture-based vaccines using the NYCBOH strain of vaccinia ACAM1000 and ACAM2000, grown in MRC-5 or Vero cells respectively (Weltzin et al., 2003; Monath et al., 2004; Artenstein et al., 2005). Aventis Pastuer produced a vaccine similar to Dryvax which was approved for use until 1997. This vaccine is undergoing additional testing as an Investigational New Drug (IND) (Rock et al., 2006). Additionally, NIH is supporting research into MVA-based vaccines including ACAM3000 (Acambis) and IMVAMUNE-BN (Bavarian Nordic) (Vollmar et al., 2006). These vaccines are currently undergoing clinical trials and are destined to join the US strategic national stockpile which in 2004 contained over 400 million doses of smallpox vaccines. Other groups are developing vaccines based on other attenuated strains such as the Japanese LC16m8 strain of vaccinia or on the NYVAC (Kidokoro et al., 2005; Empig et al., 2006; Najera et al., 2006). An updated list of smallpox vaccine clinical trials can be found at http://www.clinicaltrials.gov.

A number of groups are working on developing DNA, protein, and peptide-based subunit vaccines for smallpox (Otero et al., 2005; Thornburg et al., 2007). Several promising candidates have demonstrated protective efficacy in multiple animal models (Hooper et al., 2000, 2003, 2004, 2007; Fogg et al., 2004, 2007; Meseda et al., 2005; Heraud et al., 2006; Sakhatskyy et al., 2006; Xiao et al., 2007). Another subunit vaccine produced in plants protects mice from lethal poxvirus infection and may provide a stable, easily administered vaccine (Golovkin et al., 2007). Theoretically these vaccines will be safer to administer as they would contain no live virus. While such vaccines work well in animal models they do not induce robust immune responses in humans. This lack of immunogenicity is currently the focus of intense research and in the future we may see these types of vaccines not only for smallpox but also other pathogens as well (Bonilla-Guerrero and Poland, 2003; Poland, 2005).

Mice, rabbits, and nonhuman primates are typically used as animal models for studying poxvirus pathogenesis and immunity (Turner, 1967; Wallace and Buller, 1985; Williamson et al., 1990; Martinez et al., 2000; Smith and Kotwal, 2002; Esteban and Buller, 2005). Just as important as the choice of animal host is the selection of the virus strain to be used in the studies. As mentioned earlier, studies using an animal and its natural pathogen (i.e., mice and mousepox, or nonhuman primates, and monkeypox) provide different results from studies using vaccinia virus. It should be noted that these are, at best, close approximations of the human disease. For example, researchers have been able to mimic certain symptoms (characteristic rash, high mortality) of the human disease by the IV inoculation of extremely high titers (10^9 pfu) of variola (Rubins et al., 2004) into nonhuman primates which are typically resistant to variola infection.

As mentioned before, vaccinia virus is highly amenable to the insertion of exogenous DNA sequences and has shown promise as a vector for the development of vaccines for other pathogens including HIV, HBV, rabies, RSV, malaria, TB, leishmania, and numerous cancer antigens as well (Xing et al., 2005; Lindsey et al., 2006; Bejon et al., 2006; Perez-Jimenez et al., 2006; Puissant and Combadiere, 2006; Weyer et al., 2007; Wang et al., 2007; Hutchings et al., 2005).

PROSPECTS FOR THE FUTURE

In the aftermath of September 11 the United States took stock of preparedness in the event of an intentional release of smallpox (or any other biological agent). A smallpox scenarios is of concern because (1) there was no effective treatment for the disease, (2) the US population under 35 years of age was not immunized and those over 35 had not received a vaccine for decades and were likely to have waning immunity, (3) the United States had a stockpile of only 7–15 million doses of vaccine and the WHO only had 500,000 doses, (4) VIG was an unproven product and was in short supply, (5) vaccine production facilities were nonexistent and the production methods would not meet FDA requirements, (6) there was only one antiviral drug approved for human use that might be effective against smallpox—Cidofovir, (7) most physicians had never seen a case of smallpox, making early detection unlikely, and (8) public health infrastructure was inadequate and unprepared for the mass casualty situations likely to occur with a biological attack. The United States allocated hundreds of millions of dollars into biodefense spending in an effort to address these issues. In 2004, President Bush authorized Project Bioshield which appropriated \$5.2 billion over 10 years to address these issues (Schuler, 2004, 2005; Lam et al., 2006; Franco and Deitch, 2007). Several years later in 2007, considerable progress had been made (LeDuc and Jahrling, 2001; Breman et al., 2003; Janoff and Lynfield, 2003; Mahalingam et al., 2004; Arita, 2005; Bossi et al., 2006; Ruef, 2006). The strategic national stockpile now holds over 400 million doses of live smallpox vaccine, with additional doses of attenuated strains currently being tested and produced (Rosenthal et al., 2001; Wiser et al., 2007). An important question now before us is what type of vaccination policy, if any, to promote (Bicknell and James, 2003; Mortimer, 2003b). Universally safe subunit vaccines are under development. Numerous promising drug candidates have been discovered with activity against poxviruses (Baker et al., 2003; Bray and Roy, 2004). A wide range of new methods for detection, identification, and analysis of poxviruses and immune responses to them have been developed (Espy et al., 2002; Earl et al., 2003; Leparc-Goffart et al., 2003; Newman et al., 2003; Cosma et al., 2004; Li et al., 2004b; Rotz and Hughes, 2004; Leparc-Goffart et al., 2005). First responders and public health infrastructure at the state and local levels has been bolstered in order to cope with mass casualty events (Meltzer et al., 2001; Hull et al., 2003; Kaplan et al., 2003). Emergency management plans have been formulated and tested using advanced computer algorithms to predict both smallpox transmission and preventive measures (Gani and Leach, 2001; O'Toole et al., 2002; Eichner, 2003; Eichner and Dietz, 2003; Grais et al., 2003; Kiang and Krathwohl, 2003). Details regarding the US emergency response plans and vaccination policy in the event of smallpox release can be found at the CDC website (http://www.bt.cdc.gov/agent/smallpox/index.asp).

This remarkable progress has come with a large price-tag; nevertheless we are far better prepared for an outbreak of smallpox now than we were a few years ago. We have a greater understanding of viral biology, infection and immunology, useful knowledge for the creation of vaccines not just for smallpox but also for SARS, avian influenza or any of the other emerging viral infectious diseases and threats we may encounter. We have developed antiviral drugs with activity against poxviruses as well as other virus families. We have sequenced poxvirus genomes and have begun to harness viral proteins as therapeutic agents for other

Key elements to eradication effort
No animal reservoir
Long incubation period
Transmission required close contact
Patients only infectious during rash
Readily recognizable rash
Severe disease with high mortality and morbidity
Infectious persons bedridden
Cheap, stable and effective vaccine
International cooperation and political will

FIGURE 37.10 Key elements of the successful smallpox eradication effort.

diseases and conditions (De Clercq, 2001). We are closer to developing peptide and protein-based vaccines with far better safety profiles than live viral vaccines. There has been research into immunologic adjuvants with the potential to be used in numerous vaccines. Public health infrastructure is now better able to cope with not only biological attacks, but also natural disasters such as hurricanes, earthquakes, tsunamis, and natural outbreaks. Continuing progress in all of these areas will be necessary in order for us to be prepared for biological threats of either man-made or natural origin.

CONCLUSIONS

The eradication of smallpox is one of the greatest triumphs of modern medicine and illustrates the full potential an effective vaccination program can realize. The eradication effort cost hundreds of millions of dollars, widespread international cooperation, countless hours of effort by thousands of health care workers worldwide, and required the wide availability of an effective vaccine (Fig. 37.10). While large, these costs are dwarfed by the savings in human life, not to mention the elimination of health care, quarantine, surveillance, and prevention measure costs associated with smallpox. Several other diseases are now being considered for eradication including measles and polio. The eradication of these diseases will no doubt require efforts of a similar magnitude as those needed to eliminate smallpox yet the end results will be equally impressive.

KEY ISSUES

- Smallpox has probably killed more people than any other disease in history.
- Edward Jenner proved that cowpox can protect against smallpox infection.
- The use of vaccinia virus as a smallpox vaccine led to the eradication of smallpox in 1980.
- Smallpox has been used as a biological weapon for centuries.
- The Soviet Union had a massive bioweapons program beginning in the 1980s.
- The threat of smallpox being used as a biological weapon today is small but would have a devastating impact on society.
- The currently licensed vaccine can cause severe complications.
- New vaccines are needed but cannot be tested for efficacy against smallpox.
- There is still no effective treatment for smallpox.

References

Afonso, C.L., Delhon, G., Tulman, E.R., Lu, Z., Zsak, A., Becerra, V.M., Zsak, L., Kutish, G.F. and Rock, D.L. Genome of deerpox virus. J. Virol. 2005; 79:966–977.

Afonso, C.L., Tulman, E.R., Delhon, G., Lu, Z., Viljoen, G.J., Wallace, D.B., Kutish, G.F. and Rock, D.L. Genome of crocodilepox virus. J. Virol. 2006; 80:4978–4991.

Afonso, C.L., Tulman, E.R., Lu, Z., Zsak, L., Kutish, G.F. and Rock, D.L. The genome of fowlpox virus. J. Virol. 2000; 74:3815–3831.

Afonso, C.L., Tulman, E.R., Lu, Z., Zsak, L., Osorio, F.A., Balinsky, C., Kutish, G.F. and Rock, D.L. The genome of swinepox virus. J. Virol. 2002a; 76:783–790.

Afonso, C.L., Tulman, E.R., Lu, Z., Zsak, L., Sandybaev, N.T., Kerembekova, U.Z., Zaitsev, V.L., Kutish, G.F. and Rock, D.L. The genome of camelpox virus. Virology 2002b; 295:1–9.

Ahmed, C.M., Burkhart, M.A., Subramaniam, P.S., Mujtaba, M.G. and Johnson, H.M. Peptide mimetics of gamma interferon possess antiviral properties against vaccinia virus and other viruses in the presence of poxvirus B8R protein. J. Virol. 2005; 79:5632–5639.

Alcami, A., Khanna, A., Paul, N.L. and Smith, G.L. Vaccinia virus strains Lister, USSR and Evans express soluble and cell-surface tumour necrosis factor receptors. J. Gen. Virol. 1999; 80(Pt 4):949–959.

Alcami, A., Symons, J.A. and Smith, G.L. The vaccinia virus soluble alpha/beta interferon (IFN) receptor binds to the cell surface and protects cells from the antiviral effects of IFN. J. Virol. 2000; 74:11230–11239.

Alejo, A., Ruiz-Arguello, M.B., Ho, Y., Smith, V.P., Saraiva, M. and Alcami, A. A chemokine-binding domain in the tumor necrosis factor receptor from variola (smallpox) virus. Proc. Natl. Acad. Sci. U. S. A. 2006; 103:5995–6000.

Alibek, K. Smallpox: a disease and a weapon. Int. J. Infect. Dis. 2004; 8(Suppl 2):S3–S8.

Amanna, I.J., Slifka, M.K. and Crotty, S. Immunity and immunological memory following smallpox vaccination. Immunol. Rev. 2006; 211:320–337.

Antoine, G., Scheiflinger, F., Dorner, F. and Falkner, F.G. The complete genomic sequence of the modified vaccinia Ankara strain: comparison with other orthopoxviruses. Virology 1998; 244:365–396.

Arita, I. Smallpox vaccine and its stockpile in 2005. Lancet Infect. Dis. 2005; 5:647–652.

Arness, M.K., Eckart, R.E., Love, S.S., Atwood, J.E., Wells, T.S., Engler, R.J., Collins, L.C., Ludwig, S.L., Riddle, J.R., Grabenstein, J.D. and Tornberg, D.N. Myopericarditis following smallpox vaccination. Am. J. Epidemiol. 2004; 160:642–651.

Artenstein, A.W., Johnson, C., Marbury, T.C., Morrison, D., Blum, P.S., Kemp, T., Nichols, R., Balser, J.P., Currie, M. and Monath, T.P. A novel, cell culture-derived smallpox vaccine in vaccinia-naive adults. Vaccine 2005; 23:3301–3309.

Atrasheuskaya, A.V., Bukin, E.K., Fredeking, T.M. and Ignatyev, G.M. Protective effect of exogenous recombinant mouse interferon-gamma and tumour necrosis factor-alpha on ectromelia virus infection in susceptible BALB/c mice. Clin. Exp. Immunol. 2004; 136:207–214.

Baker, R.O., Bray, M. and Huggins, J.W. Potential antiviral therapeutics for smallpox, monkeypox and other orthopoxvirus infections. Antiviral Res. 2003; 57:13–23.

Behbehani, A.M. The smallpox story: man finally defeats an old adversary. J. Kans. Med. Soc. 1980; 81:447–456, 474.

Behbehani, A.M. The smallpox story: life and death of an old disease. Microbiol. Rev. 1983; 47:455–509.

Bejon, P., Peshu, N., Gilbert, S.C., Lowe, B.S., Molyneux, C.S., Forsdyke, J., Lang, T., Hill, A.V. and Marsh, K. Safety profile of the viral vectors of attenuated fowlpox strain FP9 and modified vaccinia virus Ankara recombinant for either of 2 preerythrocytic malaria antigens, ME-TRAP or the circumsporozoite protein, in children and adults in Kenya. Clin. Infect. Dis. 2006; 42:1102–1110.

Bell, E., Shamim, M., Whitbeck, J.C., Sfyroera, G., Lambris, J.D. and Isaacs, S.N. Antibodies against the extracellular enveloped virus B5R protein are mainly responsible for the EEV neutralizing capacity of vaccinia immune globulin. Virology 2004; 325:425–431.

Bhalla, D.K. and Warheit, D.B. Biological agents with potential for misuse: a historical perspective and defensive measures. Toxicol. Appl. Pharmacol. 2004; 199:71–84.

Bhat, P.P., Mishra, B.P. and Bhat, P.N. Identification and size variation of terminal fragments of sheep pox virus genome. Indian J. Exp. Biol. 1991; 29:434–436.

Bicknell, W. and James, K. The new cell culture smallpox vaccine should be offered to the general population. Rev. Med. Virol. 2003; 13:5–15.

Blanchard, T.J., Alcami, A., Andrea, P. and Smith, G.L. Modified vaccinia virus Ankara undergoes limited replication in human cells and lacks several immunomodulatory proteins: implications for use as a human vaccine. J. Gen. Virol. 1998; 79(Pt 5):1159–1167.

Blendon, R.J., DesRoches, C.M., Benson, J.M., Herrmann, M.J., Taylor-Clark, K. and Weldon, K.J. The public and the smallpox threat. N. Engl. J. Med. 2003; 348:426–432.

Bonilla-Guerrero, R. and Poland, G.A. Smallpox vaccines: current and future. J. Lab. Clin. Med. 2003; 142:252–257.

Bossi, P., Garin, D., Guihot, A., Gay, F., Crance, J.M., Debord, T., Autran, B. and Bricaire, F. Bioterrorism: management of major biological agents. Cell. Mol. Life Sci. 2006; 63:2196–2212.

Bowie, A., Kiss-Toth, E., Symons, J.A., Smith, G.L., Dower, S.K. and O'Neill, L.A. A46R and A52R from vaccinia virus are antagonists of host IL-1 and toll-like receptor signaling. Proc. Natl. Acad. Sci. U. S. A. 2000; 97:10162–10167.

Bray, M. Henry Kempe and the birth of vaccinia immune globulin. Clin. Infect. Dis. 2004; 39:767–769.

Bray, M. and Buller, M. Looking back at smallpox. Clin. Infect. Dis. 2004; 38:882–889.

Bray, M., Martinez, M., Smee, D.F., Kefauver, D., Thompson, E. and Huggins, J.W. Cidofovir protects mice against lethal aerosol or intranasal cowpox virus challenge. J. Infect. Dis. 2000; 181:10–19.

Bray, M. and Roy, C.J. Antiviral prophylaxis of smallpox. J. Antimicrob. Chemother. 2004; 54:1–5.

Breman, J.G., Arita, I. and Fenner, F. Preventing the return of smallpox. N. Engl. J. Med. 2003; 348:463–466.

Breman, J.G. and Henderson, D.A. Diagnosis and management of smallpox. N. Engl. J. Med. 2002; 346:1300–1308.

Buller, R.M., Owens, G., Schriewer, J., Melman, L., Beadle, J.R. and Hostetler, K.Y. Efficacy of oral active ether lipid analogs of cidofovir in a lethal mousepox model. Virology 2004; 318:474–481.

Buller, R.M. and Palumbo, G.J. Poxvirus pathogenesis. Microbiol. Rev. 1991; 55:80–122.

Byrd, C.M., Bolken, T.C., Mjalli, A.M., Arimilli, M.N., Andrews, R.C., Rothlein, R., Andrea, T., Rao, M., Owens, K.L. and Hruby, D.E. New class of orthopoxvirus antiviral drugs that block viral maturation. J. Virol. 2004; 78:12147–12156.

Cardone, E. The stability of the viral potency of smallpox vaccine when prepared in calf globulin, dehydrated, and stored without refrigeration. J. Lab. Clin. Med. 1953; 42:693–699.

Cardone, E. The stability of the viral potency of smallpox vaccine when prepared in the globulin and other protein fractions of calf plasma, dehydrated, and stored at tropical temperatures. J. Lab. Clin. Med. 1956; 48:117–123.

Casadevall, A. Passive antibody administration (immediate immunity) as a specific defense against biological weapons. Emerg. Infect. Dis. 2002; 8:833–841.

Casey, C.G., Iskander, J.K., Roper, M.H., Mast, E.E., Wen, X.J., Torok, T.J., Chapman, L.E., Swerdlow, D.L., Morgan, J., Heffelfinger, J.D., Vitek, C., Reef, S.E., Hasbrouck, L.M., Damon, I., Neff, L., Vellozzi, C., McCauley, M., Strikas, R.A. and Mootrey, G. Adverse events associated with smallpox vaccination in the United States, January-October 2003. Jama 2005; 294:2734–2743.

Cassimatis, D.C., Atwood, J.E., Engler, R.M., Linz, P.E., Grabenstein, J.D. and Vernalis, M.N. Smallpox vaccination and myopericarditis: a clinical review. J. Am. Coll. Cardiol. 2004; 43:1503–1510.

Chaudhri, G., Panchanathan, V., Bluethmann, H. and Karupiah, G. Obligatory Requirement for Antibody in Recovery from a Primary Poxvirus Infection. J. Virol. 2006; 80:6339–6344.

Chen, Z., Earl, P., Americo, J., Damon, I., Smith, S.K., Zhou, Y.H., Yu, F., Sebrell, A., Emerson, S., Cohen, G., Eisenberg, R.J., Svitel, J., Schuck, P., Satterfield, W., Moss, B. and Purcell, R. Chimpanzee/human mAbs to vaccinia virus B5 protein neutralize vaccinia and smallpox viruses and protect mice against vaccinia virus. Proc. Natl. Acad. Sci. U. S. A. 2006; 103:1882–1887.

Chen-Collins, A.R., Dixon, D.W., Vzorov, A.N., Marzilli, L.G. and Compans, R.W. Prevention of poxvirus infection by tetrapyrroles. BMC Infect. Dis. 2003; 3:9.

Cleri, D.J., Porwancher, R.B., Ricketti, A.J., Ramos-Bonner, L.S. and Vernaleo, J.R. Smallpox as a bioterrorist weapon: myth or menace? Infect. Dis. Clin. North. Am. 2006; 20:329–357, ix.

Combadiere, B., Boissonnas, A., Carcelain, G., Lefranc, E., Samri, A., Bricaire, F., Debre, P. and Autran, B. Distinct time effects of vaccination on long-term proliferative and IFN-gamma-producing T cell memory to smallpox in humans. J. Exp. Med. 2004; 199:1585–1593.

Cosma, A., Buhler, S., Nagaraj, R., Staib, C., Hammarin, A.L., Wahren, B., Goebel, F.D., Erfle, V. and Sutter, G. Neutralization assay using a modified vaccinia virus Ankara vector expressing the green fluorescent protein is a high-throughput method to monitor the humoral immune response against vaccinia virus. Clin. Diagn. Lab. Immunol. 2004; 11:406–410.

Couch, R.B., Winokur, P., Edwards, K.M., Black, S., Atmar, R. L., Stapleton, J.T., Kissner, J.M., Shinefield, H., Denny, T.N., Bybel, M.J., Newman, F.K. and Yan, L. Reducing the dose of smallpox vaccine reduces vaccine-associated morbidity without reducing vaccination success rates or immune responses. J. Infect. Dis. 2007; 195:826–832.

Coulibaly, S., Bruhl, P., Mayrhofer, J., Schmid, K., Gerencer, M. and Falkner, F.G. The nonreplicating smallpox candidate vaccines defective vaccinia Lister (dVV-L) and modified vaccinia

Ankara (MVA) elicit robust long-term protection. Virology 2005; 341:91–101.

da Fonseca, F.G., Trindade, G.S., Silva, R.L., Bonjardim, C.A., Ferreira, P.C. and Kroon, E.G. Characterization of a vaccinia-like virus isolated in a Brazilian forest. J. Gen. Virol. 2002; 83:223–228.

Damaso, C.R., Esposito, J.J., Condit, R.C. and Moussatche, N. An emergent poxvirus from humans and cattle in Rio de Janeiro State: Cantagalo virus may derive from Brazilian smallpox vaccine. Virology 2000; 277:439–449.

Davies, D.H., Liang, X., Hernandez, J.E., Randall, A., Hirst, S., Mu, Y., Romero, K.M., Nguyen, T.T., Kalantari-Dehaghi, M., Crotty, S., Baldi, P., Villarreal, L.P. and Felgner, P.L. Profiling the humoral immune response to infection by using proteome microarrays: high-throughput vaccine and diagnostic antigen discovery. Proc. Natl. Acad. Sci. U. S. A. 2005a; 102:547–552.

Davies, D.H., McCausland, M.M., Valdez, C., Huynh, D., Hernandez, J.E., Mu, Y., Hirst, S., Villarreal, L., Felgner, P.L. and Crotty, S. Vaccinia virus H3L envelope protein is a major target of neutralizing antibodies in humans and elicits protection against lethal challenge in mice. J. Virol. 2005b; 79:11724–11733.

De Clercq, E. Vaccinia virus inhibitors as a paradigm for the chemotherapy of poxvirus infections. Clin. Microbiol. Rev. 2001; 14:382–397.

De Clercq, E. Cidofovir in the treatment of poxvirus infections. Antiviral Res. 2002; 55:1–13.

De Clercq, E. and Neyts, J. Therapeutic potential of nucleoside/nucleotide analogues against poxvirus infections. Rev. Med. Virol. 2004; 14:289–300.

de Souza Trindade, G., da Fonseca, F.G., Marques, J.T., Nogueira, M.L., Mendes, L.C., Borges, A.S., Peiro, J.R., Pituco, E.M., Bonjardim, C.A., Ferreira, P.C. and Kroon, E.G. Aracatuba virus: a vaccinialike virus associated with infection in humans and cattle. Emerg. Infect. Dis. 2003; 9:155–160.

Di Giulio, D.B. and Eckburg, P.B. Human monkeypox: an emerging zoonosis. Lancet Infect. Dis. 2004; 4:15–25.

Dinc, G. and Ulman, Y.I. The introduction of variolation 'A La Turca' to the West by Lady Mary Montagu and Turkey's contribution to this. Vaccine 2007; 25:4261–4265.

Dixon, C.W. Smallpox. London: J. & A. Churchill, 1962.

Dong, Y. and Denny, T.N. HLA-A2-Restricted human CD8+ cytotoxic T lymphocyte responses to a novel epitope in vaccinia virus that is conserved among orthopox viruses. J. Infect. Dis. 2006; 194:168–175.

Downie, A.W., Meiklejohn, M., St Vincent, L., Rao, A.R., Sundara Babu, B.V. and Kempe, C.H. The recovery of smallpox virus from patients and their environment in a smallpox hospital. Bull. World Health Organ. 1965; 33:615–622.

Drexler, I., Staib, C., Kastenmuller, W., Stevanovic, S., Schmidt, B., Lemonnier, F.A., Rammensee, H.G., Busch, D.H., Bernhard, H., Erfle, V. and Sutter, G. Identification of vaccinia virus epitope-specific HLA-A*0201-restricted T cells and comparative analysis of smallpox vaccines. Proc. Natl. Acad. Sci. U. S. A. 2003; 100:217–222.

Dunlop, L.R., Oehlberg, K.A., Reid, J.J., Avci, D. and Rosengard, A.M. Variola virus immune evasion proteins. Microbes Infect. 2003; 5:1049–1056.

Earl, P.L., Americo, J.L. and Moss, B. Development and use of a vaccinia virus neutralization assay based on flow cytometric detection of green fluorescent protein. J. Virol. 2003; 77:10684–10688.

Earl, P.L., Americo, J.L., Wyatt, L.S., Eller, L.A., Whitbeck, J.C., Cohen, G.H., Eisenberg, R.J., Hartmann, C.J., Jackson, D.L., Kulesh, D.A., Martinez, M.J., Miller, D.M., Mucker, E.M., Shamblin, J.D., Zwiers, S.H., Huggins, J.W., Jahrling, P.B. and Moss, B. Immunogenicity of a highly attenuated MVA smallpox vaccine and protection against monkeypox. Nature 2004; 428:182–185.

Edghill-Smith, Y., Bray, M., Whitehouse, C.A., Miller, D., Mucker, E., Manischewitz, J., King, L.R., Robert-Guroff, M., Hryniewicz, A., Venzon, D., Meseda, C., Weir, J., Nalca, A., Livingston, V., Wells, J., Lewis, M.G., Huggins, J., Zwiers, S.H., Golding, H. and Franchini, G. Smallpox vaccine does not protect macaques with AIDS from a lethal monkeypox virus challenge. J. Infect. Dis. 2005; 191:372–381.

Eichner, M. Case isolation and contact tracing can prevent the spread of smallpox. Am. J. Epidemiol. 2003; 158:118–128.

Eichner, M. and Dietz, K. Transmission potential of smallpox: estimates based on detailed data from an outbreak. Am. J. Epidemiol. 2003; 158:110–117.

Empig, C., Kenner, J.R., Perret-Gentil, M., Youree, B.E., Bell, E., Chen, A., Gurwith, M., Higgins, K., Lock, M., Rice, A.D., Schriewer, J., Sinangil, F., White, E., Buller, R.M., Dermody, T.S., Isaacs, S.N. and Moyer, R.W. Highly attenuated smallpox vaccine protects rabbits and mice against pathogenic orthopoxvirus challenge. Vaccine 2006; 24:3686–3694.

Engler, R.J., Kenner, J. and Leung, D.Y. Smallpox vaccination: Risk considerations for patients with atopic dermatitis. J. Allergy Clin. Immunol. 2002; 110:357–365.

Esposito, J.J., Sammons, S.A., Frace, A.M., Osborne, J.D., Olsen-Rasmussen, M., Zhang, M., Govil, D., Damon, I.K., Kline, R., Laker, M., Li, Y., Smith, G.L., Meyer, H., Leduc, J.W. and Wohlhueter, R.M. Genome sequence diversity and clues to the evolution of variola (smallpox) virus. Science 2006; 313:807–812.

Espy, M.J., Cockerill, I.F., Meyer, R.F., Bowen, M.D., Poland, G.A., Hadfield, T.L. and Smith, T.F. Detection of smallpox virus DNA by LightCycler PCR. J. Clin. Microbiol. 2002; 40:1985–1988.

Esteban, D.J. and Buller, R.M. Ectromelia virus: the causative agent of mousepox. J. Gen. Virol. 2005; 86:2645–2659.

Eyler, J.M. Smallpox in history: the birth, death, and impact of a dread disease. J. Lab. Clin. Med. 2003; 142:216–220.

Fang, M. and Sigal, L.J. Antibodies and CD8+ T Cells are complementary and essential for natural resistance to a highly lethal cytopathic virus. J. Immunol. 2005; 175:6829–6836.

Fenner, F. Smallpox and its eradication. Geneva: World Health Organization, 1988.

Fenner, F. Adventures with poxviruses of vertebrates. FEMS Microbiol. Rev. 2000; 24:123–133.

Fenner, F., Poole, W.E., Marshall, I.D. and Dyce, A.L. Studies in the epidemiology of infectious myxomatosis of rabbits. VI. The experimental introduction of the European strain of myxoma virus into Australian wild rabbit populations. J. Hyg. (Lond.) 1957; 55:192–206.

Ferrier, A., Garin, D. and Crance, J.M. Rapid inactivation of vaccinia virus in suspension and dried on surfaces. J. Hosp. Infect. 2004; 57:73–79.

Fields, B.N., Knipe, D.M. and Howley, P.M. Fields virology. Philadelphia: Lippincott-Raven, 1996.

Fine, P.E., Jezek, Z., Grab, B. and Dixon, H. The transmission potential of monkeypox virus in human populations. Int. J. Epidemiol. 1988; 17:643–650.

Fleischauer, A.T., Kile, J.C., Davidson, M., Fischer, M., Karem, K.L., Teclaw, R., Messersmith, H., Pontones, P., Beard, B.A., Braden, Z. H., Cono, J., Sejvar, J.J., Khan, A.S., Damon, I. and Kuehnert, M.J. Evaluation of human-to-human transmission of monkeypox from infected patients to health care workers. Clin. Infect. Dis. 2005; 40:689–694.

Flint, S.J. Principles of virology: molecular biology, pathogenesis, and control of animal viruses. Washington, DC.: ASM Press, 2004.

Fogg, C., Lustig, S., Whitbeck, J.C., Eisenberg, R.J., Cohen, G.H. and Moss, B. Protective immunity to vaccinia virus induced by vaccination with multiple recombinant outer membrane proteins of intracellular and extracellular virions. J. Virol. 2004; 78:10230–10237.

Fogg, C.N., Americo, J.L., Lustig, S., Huggins, J.W., Smith, S.K., Damon, I., Resch, W., Earl, P.L., Klinman, D.M. and Moss, B. Adjuvant-enhanced antibody responses to recombinant proteins correlates with protection of mice and monkeys to orthopoxvirus challenges. Vaccine 2007; 25:2787–2799.

Franco, C. and Deitch, S. Billions for biodefense: federal agency biodefense funding, FY2007-FY2008. Biosecur. Bioterror. 2007; 5:117–133.

Frey, S.E. and Belshe, R.B. Poxvirus zoonoses – putting pocks into context. N. Engl. J. Med. 2004; 350:324–327.

Frey, S.E., Couch, R.B., Tacket, C.O., Treanor, J.J., Wolff, M., Newman, F.K., Atmar, R.L., Edelman, R., Nolan, C.M. and Belshe, R.B. Clinical responses to undiluted and diluted smallpox vaccine. N. Engl. J. Med. 2002; 346:1265–1274.

Frischknecht, F. The history of biological warfare. Human experimentation, modern nightmares and lone madmen in the twentieth century. EMBO Rep. 2003; 4(Spec No):S47–S52.

Fruh, K., Bartee, E., Gouveia, K. and Mansouri, M. Immune evasion by a novel family of viral PHD/LAP-finger proteins of gamma-2 herpesviruses and poxviruses. Virus Res. 2002; 88:55–69.

Fulginiti, V.A. Risks of smallpox vaccination. Jama 2003; 290:1452, author reply 1452.

Fulginiti, V.A., Papier, A., Lane, J.M., Neff, J.M. and Henderson, D.A. Smallpox vaccination: a review, part I. Background, vaccination technique, normal vaccination and revaccination, and expected normal reactions. Clin. Infect. Dis. 2003a; 37:241–250.

Fulginiti, V.A., Papier, A., Lane, J.M., Neff, J.M. and Henderson, D.A. Smallpox vaccination: a review, part II. Adverse events. Clin. Infect. Dis. 2003b; 37:251–271.

Gami, A.S., Murphy, J. and Cooper, L.T., Jr. Cautious interpretation of data regarding myopericarditis associated with smallpox vaccination. J. Am. Coll. Cardiol. 2005; 45:160, author reply 160-1.

Gani, R. and Leach, S. Transmission potential of smallpox in contemporary populations. Nature 2001; 414:748–751.

Goldsmith, J.C., Eller, N., Mikolajczyk, M., Manischewitz, J., Golding, H. and Scott, D.E. Intravenous immunoglobulin products contain neutralizing antibodies to vaccinia. Vox. Sang. 2004; 86:125–129.

Goldstein, J.A., Neff, J.M., Lane, J.M. and Koplan, J.P. Smallpox vaccination reactions, prophylaxis, and therapy of complications. Pediatrics 1975; 55:342–347.

Golovkin, M., Spitsin, S., Andrianov, V., Smirnov, Y., Xiao, Y., Pogrebnyak, N., Markley, K., Brodzik, R., Gleba, Y., Isaacs, S.N. and Koprowski, H. Smallpox subunit vaccine produced in Planta confers protection in mice. Proc. Natl. Acad. Sci. U. S. A. 2007; 104:6864–6869.

Grabenstein, J.D. and Winkenwerder, W., Jr. US military smallpox vaccination program experience. Jama 2003; 289:3278–3282.

Grais, R.F., Ellis, J.H. and Glass, G.E. Forecasting the geographical spread of smallpox cases by air travel. Epidemiol. Infect. 2003; 131:849–857.

Gubser, C., Hue, S., Kellam, P. and Smith, G.L. Poxvirus genomes: a phylogenetic analysis. J. Gen. Virol. 2004; 85:105–117.

Haga, I.R. and Bowie, A.G. Evasion of innate immunity by vaccinia virus. Parasitology 2005; 130(Suppl):S11–S25.

Halsell, J.S., Riddle, J.R., Atwood, J.E., Gardner, P., Shope, R., Poland, G.A., Gray, G.C., Ostroff, S., Eckart, R.E., Hospenthal, D.R., Gibson, R.L., Grabenstein, J.D., Arness, M.K. and Tornberg, D.N. Myopericarditis following smallpox vaccination among vaccinia-naive US military personnel. Jama 2003; 289:3283–3289.

Hammarlund, E., Lewis, M.W., Hansen, S.G., Strelow, L.I., Nelson, J.A., Sexton, G.J., Hanifin, J.M. and Slifka, M.K. Duration of antiviral immunity after smallpox vaccination. Nat. Med. 2003; 9:1131–1137.

Harper, G.J. Airborne micro-organisms: survival tests with four viruses. J. Hyg. (Lond.) 1961; 59:479–486.

Harrison, S.C., Alberts, B., Ehrenfeld, E., Enquist, L., Fineberg, H., McKnight, S.L., Moss, B., O'Donnell, M., Ploegh, H., Schmid, S.L., Walter, K.P. and Theriot, J. Discovery of antivirals against smallpox. Proc Natl. Acad. Sci. U. S. A. 2004; 101:11178–11192.

Hassett, D.E. Smallpox infections during pregnancy, lessons on pathogenesis from nonpregnant animal models of infection. J. Reprod. Immunol. 2003; 60:13–24.

Hays, J.N. Epidemics and pandemics : their impacts on human history. Santa Barbara, Calif.: ABC-CLIO, 2005.

Henderson, D. and Fenner, F. Recent Events and Observations Pertaining to Smallpox Virus Destruction in 2002. Clinical Infectious Diseases 2001; 33:1057–1059.

Henderson, D.A., Inglesby, T.V., Bartlett, J.G., Ascher, M.S., Eitzen, E., Jahrling, P.B., Hauer, J., Layton, M., McDade, J., Osterholm, M.T., O'Toole, T., Parker, G., Perl, T., Russell, P.K. and Tonat, K. Smallpox as a biological weapon: medical and public health management. Working Group on Civilian Biodefense. Jama 1999; 281:2127–2137.

Heraud, J.M., Edghill-Smith, Y., Ayala, V., Kalisz, I., Parrino, J., Kalyanaraman, V.S., Manischewitz, J., King, L.R., Hryniewicz, A., Trindade, C.J., Hassett, M., Tsai, W.P., Venzon, D., Nalca, A., Vaccari, M., Silvera, P., Bray, M., Graham, B.S., Golding, H., Hooper, J.W. and Franchini, G. Subunit recombinant vaccine protects against monkeypox. J. Immunol. 2006; 177:2552–2564.

Hooper, J.W., Custer, D.M., Schmaljohn, C.S. and Schmaljohn, A.L. DNA vaccination with vaccinia virus L1R and A33R genes protects mice against a lethal poxvirus challenge. Virology 2000; 266:329–339.

Hooper, J.W., Custer, D.M. and Thompson, E. Four-gene-combination DNA vaccine protects mice against a lethal vaccinia virus challenge and elicits appropriate antibody responses in nonhuman primates. Virology 2003; 306:181–195.

Hooper, J.W., Golden, J.W., Ferro, A.M. and King, A.D. Smallpox DNA vaccine delivered by novel skin electroporation device protects mice against intranasal poxvirus challenge. Vaccine 2007; 25:1814–1823.

Hooper, J.W., Thompson, E., Wilhelmsen, C., Zimmerman, M., Ichou, M.A., Steffen, S.E., Schmaljohn, C.S., Schmaljohn, A.L. and Jahrling, P.B. Smallpox DNA vaccine protects nonhuman primates against lethal monkeypox. J. Virol. 2004; 78:4433–4443.

Hopkins, D.R. Princes and peasants: smallpox in history. Chicago: University of Chicago Press, 1983.

Hopkins, R.J., Kramer, W.G., Blackwelder, W.C., Ashtekar, M., Hague, L., Winker-La Roche, S.D., Berezuk, G., Smith, D. and Leese, P.T. Safety and pharmacokinetic evaluation of intravenous vaccinia immune globulin in healthy volunteers. Clin. Infect. Dis. 2004; 39:759–766.

Hopkins, R.J. and Lane, J.M. Clinical efficacy of intramuscular vaccinia immune globulin: a literature review. Clin. Infect. Dis. 2004; 39:819–826.

Hsieh, S.M., Chen, S.Y., Sheu, G.C., Hung, M.N., Chou, W. H., Chang, S.C. and Hsu, K.H. Clinical and immunological responses to undiluted and diluted smallpox vaccine with vaccinia virus of Lister strain. Vaccine 2006; 24:510–515.

Hull, H.F., Danila, R. and Ehresmann, K. Smallpox and bioterrorism: public-health responses. J. Lab. Clin. Med. 2003; 142:221–228.

Humrich, J.Y., Thumann, P., Greiner, S., Humrich, J.H., Averbeck, M., Schwank, C., Kampgen, E., Schuler, G. and Jenne, L. Vaccinia virus impairs directional migration and chemokine receptor switch of human dendritic cells. Eur. J. Immunol. 2007; 37:954–965.

Hutchings, C.L., Gilbert, S.C., Hill, A.V. and Moore, A.C. Novel protein and poxvirus-based vaccine combinations for simultaneous induction of humoral and cell-mediated immunity. J. Immunol. 2005; 175:599–606.

Janoff, E.N. and Lynfield, R. Smallpox: remembrance of things past, or the coming plague? J. Lab. Clin. Med. 2003; 142:211–215.

Jezek, Z., Arita, I., Mutombo, M., Dunn, C., Nakano, J.H. and Szczeniowski, M. Four generations of probable person-to-person transmission of human monkeypox. Am. J. Epidemiol. 1986; 123:1004–1012.

Jing, L., Chong, T.M., Byrd, B., McClurkan, C.L., Huang, J., Story, B. T., Dunkley, K.M., Aldaz-Carroll, L., Eisenberg, R.J., Cohen, G.H., Kwok, W.W., Sette, A. and Koelle, D.M. Dominance and diversity in the primary human CD4 T cell response to replication-competent vaccinia virus. J. Immunol. 2007; 178: 6374–6386.

Jones-Trower, A., Garcia, A., Meseda, C.A., He, Y., Weiss, C., Kumar, A., Weir, J.P. and Merchlinsky, M. Identification and preliminary characterization of vaccinia virus (Dryvax) antigens recognized by vaccinia immune globulin. Virology 2005; 343:128–140.

Kan, V.L., Manischewitz, J., King, L.R. and Golding, H. Durable neutralizing antibodies after remote smallpox vaccination among adults with and without HIV infection. Aids 2007; 21:521–524.

Kaplan, E.H., Craft, D.L. and Wein, L.M. Analyzing bioterror response logistics: the case of smallpox. Math. Biosci. 2003; 185:33–72.

Katze, M.G., He, Y. and Gale, M., Jr. Viruses and interferon: a fight for supremacy. Nat. Rev. Immunol. 2002; 2:675–687.

Kempe, C.H., Fulginiti, V., Minamitani, M. and Shinefield, H. Smallpox vaccination of eczema patients with a strain of attenuated live vaccinia (CVI-78). Pediatrics 1968; 42:980–985.

Kemper, A.R., Davis, M.M. and Freed, G.L. Expected adverse events in a mass smallpox vaccination campaign. Eff. Clin. Pract. 2002; 5:84–90.

Kennedy, J.S., Frey, S.E., Yan, L., Rothman, A.L., Cruz, J., Newman, F.K., Orphin, L., Belshe, R.B. and Ennis, F.A. Induction of human T cell-mediated immune responses after primary and secondary smallpox vaccination. J. Infect. Dis. 2004; 190:1286–1294.

Kennedy, R. and Poland, G.A. T-Cell epitope discovery for variola and vaccinia viruses. Rev. Med. Virol. 2007; 17:93–113.

Kenner, J., Cameron, F., Empig, C., Jobes, D.V. and Gurwith, M. LC16m8: an attenuated smallpox vaccine. Vaccine 2006; 24:7009–7022.

Kiang, K.M. and Krathwohl, M.D. Rates and risks of transmission of smallpox and mechanisms of prevention. J. Lab. Clin. Med. 2003; 142:229–238.

Kidokoro, M., Tashiro, M. and Shida, H. Genetically stable and fully effective smallpox vaccine strain constructed from highly attenuated vaccinia LC16m8. Proc. Natl. Acad. Sci. U. S. A. 2005; 102:4152–4157.

Kile, J.C., Fleischauer, A.T., Beard, B., Kuehnert, M.J., Kanwal, R.S., Pontones, P., Messersmith, H.J., Teclaw, R., Karem, K.L., Braden, Z.H., Damon, I., Khan, A.S. and Fischer, M. Transmission of monkeypox among persons exposed to infected prairie dogs in Indiana in 2003. Arch. Pediatr. Adolesc. Med. 2005; 159:1022–1025.

Kim, S.H., Yeo, S.G., Jang, H.C., Park, W.B., Lee, C.S., Lee, K.D., Kim, H.B., Kim, N.J., Kim, Y.T., Jee, Y., Cho, H., Oh, M.D. and Choe, K.W. Clinical responses to smallpox vaccine in vaccinia-naive and previously vaccinated populations: undiluted and diluted Lancy-Vaxina vaccine in a single-blind, randomized, prospective trial. J. Infect. Dis. 2005; 192:1066–1070.

Knorr, C.W., Allen, S.D., Torres, A.R. and Smee, D.F. Effects of cidofovir treatment on cytokine induction in murine models of cowpox and vaccinia virus infection. Antiviral Res. 2006; 72:125–133.

Krauss, O., Hollinshead, R., Hollinshead, M. and Smith, G.L. An investigation of incorporation of cellular antigens into vaccinia virus particles. J. Gen. Virol. 2002; 83:2347–2359.

Kretzschmar, M., Wallinga, J., Teunis, P., Xing, S. and Mikolajczyk, R. Frequency of adverse events after vaccination with different vaccinia strains. PLoS Med. 2006; 3:e272.

Lam, C., Franco, C. and Schuler, A. Billions for biodefense: Federal Agency biodefense funding, FY2006-FY2007. Biosecur. Bioterror. 2006; 4:113–127.

Lane, J.M. and Fulginiti, V.A. Transmission of vaccinia virus and rationale for measures for prevention. Clin. Infect. Dis. 2003; 37:281–284.

Lane, J.M. and Goldstein, J. Adverse events occurring after smallpox vaccination. Semin. Pediatr. Infect. Dis. 2003; 14:189–195.

Lane, J.M., Ruben, F.L., Abrutyn, E. and Millar, J.D. Deaths attributable to smallpox vaccination, 1959 to 1966, and 1968. Jama 1970a; 212:441–444.

Lane, J.M., Ruben, F.L., Neff, J.M. and Millar, J.D. Complications of smallpox vaccination, 1968. N. Engl. J. Med. 1969; 281:1201–1208.

Lane, J.M., Ruben, F.L., Neff, J.M. and Millar, J.D. Complications of smallpox vaccination, 1968: results of ten statewide surveys. J. Infect. Dis. 1970b; 122:303–309.

Law, M., Putz, M.M. and Smith, G.L. An investigation of the therapeutic value of vaccinia-immune IgG in a mouse pneumonia model. J. Gen. Virol. 2005; 86:991–1000.

Learned, L.A., Reynolds, M.G., Wassa, D.W., Li, Y., Olson, V.A., Karem, K., Stempora, L.L., Braden, Z.H., Kline, R., Likos, A., Libama, F., Moudzeo, H., Bolanda, J.D., Tarangonia, P., Boumandoki, P., Formenty, P., Harvey, J.M. and Damon, I.K. Extended interhuman transmission of monkeypox in a hospital community in the Republic of the Congo, 2003. Am. J. Trop. Med. Hyg. 2005; 73:428–434.

LeDuc, J.W. and Jahrling, P.B. Strengthening national preparedness for smallpox: an update. Emerg. Infect. Dis. 2001; 7:155–157.

Leparc-Goffart, I., Poirier, B., El Zaouk, A., Tissier, M.H. and Fuchs, F. New generation of cell culture assay for smallpox vaccine potency. J. Clin. Microbiol. 2003; 41:3687–3689.

Leparc-Goffart, I., Poirier, B., Garin, D., Tissier, M.H., Fuchs, F. and Crance, J.M. Standardization of a neutralizing anti-vaccinia antibodies titration method: an essential step for titration of vaccinia immunoglobulins and smallpox vaccines evaluation. J. Clin. Virol. 2005; 32:47–52.

Levine, R.S., Peterson, A.T., Yorita, K.L., Carroll, D., Damon, I.K. and Reynolds, M.G. Ecological niche and geographic distribution of human monkeypox in Africa. PLoS ONE 2007; 2:e176.

Li, G., Chen, N., Feng, Z., Buller, R.M., Osborne, J., Harms, T., Damon, I., Upton, C. and Esteban, D.J. Genomic sequence and analysis of a vaccinia virus isolate from a patient with a smallpox vaccine-related complication. Virol. J. 2006; 3:88.

Li, P., Wang, N., Zhou, D., Yee, C.S.K., Chang, C.-H., Brutkiewicz, R.R. and Blum, J.S. Disruption of MHC Class II-Restricted Antigen Presentation by Vaccinia Virus. J. Immunol. 2005; 175:6481–6488.

Li, W.X., Li, H., Lu, R., Li, F., Dus, M., Atkinson, P., Brydon, E.W., Johnson, K.L., Garcia-Sastre, A., Ball, L.A., Palese, P. and Ding, S.W. Interferon antagonist proteins of influenza and vaccinia viruses are suppressors of RNA silencing. Proc. Natl. Acad. Sci. U. S. A. 2004a; 101:1350–1355.

Li, Y., Ropp, S.L., Zhao, H., Damon, I.K. and Esposito, J.J. Orthopoxvirus pan-genomic DNA assay. J. Virol. Methods 2007; 141:154–165.

Li, Z., Rubin, S.A., Taffs, R.E., Merchlinsky, M., Ye, Z. and Carbone, K.M. Mouse neurotoxicity test for vaccinia-based smallpox vaccines. Vaccine 2004b; 22:1486–1493.

Ligon, B.L. Monkeypox: a review of the history and emergence in the Western hemisphere. Semin. Pediatr. Infect. Dis. 2004; 15:280–287.

Lindsey, K.R., Gritz, L., Sherry, R., Abati, A., Fetsch, P.A., Goldfeder, L.C., Gonzales, M.I., Zinnack, K.A., Rogers-Freezer, L., Haworth, L., Mavroukakis, S.A., White, D.E., Steinberg, S.M., Restifo, N.P., Panicali, D.L., Rosenberg, S.A. and Topalian, S.L. Evaluation of prime/boost regimens using recombinant poxvirus/tyrosinase vaccines for the treatment of patients with metastatic melanoma. Clin. Cancer Res. 2006; 12:2526–2537.

Liu, G., Zhai, Q., Schaffner, D.J., Wu, A., Yohannes, A., Robinson, T.M., Maland, M., Wells, J., Voss, T.G., Bailey, C. and Alibek, K. Prevention of lethal respiratory vaccinia infections in mice with interferon-alpha and interferon-gamma. FEMS Immunol. Med. Microbiol. 2004; 40:201–206.

Mahalingam, S., Damon, I.K. and Lidbury, B.A. 25 years since the eradication of smallpox: why poxvirus research is still relevant. Trends Immunol. 2004; 25:636–639.

Mahy, B.W. An overview on the use of a viral pathogen as a bioterrorism agent: why smallpox? Antiviral Res. 2003; 57:1–5.

Marshall, I.D. and Fenner, F. Studies in the epidemiology of infectious myxomatosis of rabbits. VII. The virulence of strains of myxoma virus recovered from Australian wild rabbits between 1951 and 1959. J. Hyg. (Lond.) 1960; 58:485–488.

Martinez, M.J., Bray, M.P. and Huggins, J.W. A mouse model of aerosol-transmitted orthopoxviral disease: morphology of experimental aerosol-transmitted orthopoxviral disease in a cowpox virus-BALB/c mouse system. Arch. Pathol. Lab. Med. 2000; 124:362–377.

Massoudi, M.S., Barker, L. and Schwartz, B. Effectiveness of postexposure vaccination for the prevention of smallpox: results of a delphi analysis. J. Infect. Dis. 2003; 188:973–976.

Massung, R.F., Liu, L.I., Qi, J., Knight, J.C., Yuran, T.E., Kerlavage, A.R., Parsons, J.M., Venter, J.C. and Esposito, J.J. Analysis of the complete genome of smallpox variola major virus strain Bangladesh-1975. Virology 1994; 201:215–240.

Massung, R.F., McFadden, G. and Moyer, R.W. Nucleotide sequence analysis of a unique near-terminal region of the tumorigenic poxvirus, Shope fibroma virus. J. Gen. Virol. 1992; 73(Pt 11):2903–2911.

Mayr, A. Smallpox vaccination and bioterrorism with pox viruses. Comp. Immunol. Microbiol. Infect. Dis. 2003; 26:423–430.

Mayr, A., Stickl, H., Muller, H.K., Danner, K. and Singer, H. [The smallpox vaccination strain MVA: marker, genetic structure, experience gained with the parenteral vaccination and behavior in organisms with a debilitated defence mechanism (author's transl)]. Zentralbl. Bakteriol. [B] 1978; 167:375–390.

McCraith, S., Holtzman, T., Moss, B. and Fields, S. Genome-wide analysis of vaccinia virus protein-protein interactions. Proc. Natl. Acad. Sci. U. S. A. 2000; 97:4879–4884.

McCurdy, L.H., Larkin, B.D., Martin, J.E. and Graham, B.S. Modified vaccinia Ankara: potential as an alternative smallpox vaccine. Clin. Infect. Dis. 2004; 38:1749–1753.

McFadden, G. Smallpox: an ancient disease enters the modern era of virogenomics. Proc. Natl. Acad. Sci. U. S. A. 2004; 101:14994–14995.

McFadden, G. Gleevec casts a pox on poxviruses. Nat. Med. 2005a; 11:711–712.

McFadden, G. Poxvirus tropism. Nat. Rev. Microbiol. 2005b; 3:201–213.

Meiklejohn, G., Kempe, C.H., Downie, A.W., Berge, T.O., St Vincent, L. and Rao, A.R. Air sampling to recover variola virus in the environment of a smallpox hospital. Bull. World Health Organ. 1961; 25:63–67.

Meltzer, M.I., Damon, I., LeDuc, J.W. and Millar, J.D. Modeling potential responses to smallpox as a bioterrorist weapon. Emerg. Infect. Dis. 2001; 7:959–969.

Meseda, C.A., Garcia, A.D., Kumar, A., Mayer, A.E., Manischewitz, J., King, L.R., Golding, H., Merchlinsky, M. and Weir, J.P. Enhanced immunogenicity and protective effect conferred by vaccination with combinations of modified vaccinia virus Ankara and licensed smallpox vaccine Dryvax in a mouse model. Virology 2005; 339:164–175.

Mitra, A.C., Sarkar, J.K. and Mukherjee, M.K. Virus content of smallpox scabs. Bull. World Health Organ. 1974; 51:106–107.

Mitra-Kaushik, S., Cruz, J., Stern, L.J., Ennis, F.A. and Terajima, M. Human Cytotoxic CD4+ T Cells Recognize HLA-DR1-Restricted Epitopes on Vaccinia Virus Proteins A24R and D1R Conserved among Poxviruses. J. Immunol. 2007; 179:1303–1312.

Monath, T.P., Caldwell, J.R., Mundt, W., Fusco, J., Johnson, C.S., Buller, M., Liu, J., Gardner, B., Downing, G., Blum, P.S., Kemp, T., Nichols, R. and Weltzin, R. ACAM2000 clonal Vero cell culture vaccinia virus (New York City Board of Health strain) – a second-generation smallpox vaccine for biological defense. Int. J. Infect. Dis. 2004; 8(Suppl 2):S31–s44.

Mortimer, P.P. Can postexposure vaccination against smallpox succeed? Clin. Infect. Dis. 2003a; 36:622–629.

Mortimer, P.P. The new cell culture smallpox vaccine should not be offered to the general population. Rev. Med. Virol. 2003b; 13:17–20.

Moss, B. Poxvirus entry and membrane fusion. Virology 2006; 344:48–54.

Moss, B. and Shisler, J.L. Immunology 101 at poxvirus U: immune evasion genes. Semin. Immunol. 2001; 13:59–66.

Moutaftsi, M., Bui, H.H., Peters, B., Sidney, J., Salek-Ardakani, S., Oseroff, C., Pasquetto, V., Crotty, S., Croft, M., Lefkowitz, E.J., Grey, H. and Sette, A. Vaccinia virus-specific CD4+ T cell responses target a set of antigens largely distinct from those targeted by CD8+ T cell responses. J. Immunol. 2007; 178:6814–6820.

Moutaftsi, M., Peters, B., Pasquetto, V., Tscharke, D.C., Sidney, J., Bui, H.H., Grey, H. and Sette, A. A consensus epitope prediction approach identifies the breadth of murine T(CD8+)-cell responses to vaccinia virus. Nat. Biotechnol. 2006; 24:817–819.

Nagasse-Sugahara, T.K., Kisielius, J.J., Ueda-Ito, M., Curti, S.P., Figueiredo, C.A., Cruz, A.S., Silva, M.M., Ramos, C.H., Silva, M.C., Sakurai, T. and Salles-Gomes, L.F. Human vaccinia-like virus outbreaks in Sao Paulo and Goias States, Brazil: virus detection, isolation and identification. Rev. Inst. Med. Trop. Sao Paulo 2004; 46:315–322.

Najera, J.L., Gomez, C.E., Domingo-Gil, E., Gherardi, M.M. and Esteban, M. Cellular and biochemical differences between two attenuated poxvirus vaccine candidates (MVA and NYVAC) and role of the C7L gene. J. Virol. 2006; 80:6033–6047.

Newman, F.K., Frey, S.E., Blevins, T.P., Mandava, M., Bonifacio, A., Jr., Yan, L. and Belshe, R.B. Improved assay to detect neutralizing antibody following vaccination with diluted or undiluted vaccinia (Dryvax) vaccine. J. Clin. Microbiol. 2003; 41:3154–3157.

Neyts, J. and De Clercq, E. Therapy and short-term prophylaxis of poxvirus infections: historical background and perspectives. Antiviral Res. 2003; 57:25–33.

Neyts, J., Leyssen, P., Verbeken, E. and De Clercq, E. Efficacy of cidofovir in a murine model of disseminated progressive vaccinia. Antimicrob. Agents Chemother. 2004; 48:2267–2273.

Nishiura, H. and Eichner, M. Estimation of the duration of vaccine-induced residual protection against severe and fatal smallpox based on secondary vaccination failure. Infection 2006; 34:241–246.

Nishiura, H. and Eichner, M. Interpreting the epidemiology of postexposure vaccination against smallpox. Int. J. Hyg. Environ. Health 2007.

Nishiura, H. and Eichner, M. Interpreting the epidemiology of postexposure vaccination against smallpox. Int. J. Hyg. Environ. Health 2008; 211:219–226.

Nuara, A.A., Buller, R.M. and Bai, H. Identification of residues in the ectromelia virus gamma interferon-binding protein involved in expanded species specificity. J. Gen. Virol. 2007; 88:51–60.

O'Connell, C.J., Karzon, D.T., Barron, A.L., Plaut, M.E. and Ali, V.M. Progressive Vaccinia with Normal Antibodies. A Case Possibly Due to Deficient Cellular Immunity. Ann. Intern. Med. 1964; 60:282–289.

Orr, N., Forman, M., Marcus, H., Lustig, S., Paran, N., Grotto, I., Klement, E., Yehezkelli, Y., Robin, G., Reuveny, S., Shafferman, A. and Cohen, D. Clinical and immune responses after revaccination of israeli adults with the Lister strain of vaccinia virus. J. Infect. Dis. 2004; 190:1295–1302.

Otero, M., Calarota, S.A., Dai, A., De Groot, A.S., Boyer, J.D. and Weiner, D.B. Efficacy of novel plasmid DNA encoding vaccinia antigens in improving current smallpox vaccination strategy. Vaccine 2005.

Otero, M., Calarota, S.A., Dai, A., De Groot, A.S., Boyer, J.D. and Weiner, D.B. Efficacy of novel plasmid DNA encoding vaccinia antigens in improving current smallpox vaccination strategy. Vaccine 2006; 24:4461–4470.

O'Toole, T., Mair, M. and Inglesby, T.V. Shining light on "Dark Winter". Clin. Infect. Dis. 2002; 34:972–983.

Painter, G.R. and Hostetler, K.Y. Design and development of oral drugs for the prophylaxis and treatment of smallpox infection. Trends Biotechnol. 2004; 22:423–427.

Palumbo, G.J., Buller, R.M. and Glasgow, W.C. Multigenic evasion of inflammation by poxviruses. J. Virol. 1994; 68:1737–1749.

Panchanathan, V., Chaudhri, G. and Karupiah, G. Protective Immunity against Secondary Poxvirus Infection Is Dependent on Antibody but Not on CD4 or CD8 T-Cell Function. J. Virol. 2006; 80:6333–6338.

Parrino, J., McCurdy, L.H., Larkin, B.D., Gordon, I.J., Rucker, S.E., Enama, M.E., Koup, R.A., Roederer, M., Bailer, R.T., Moodie, Z., Gu, L., Yan, L. and Graham, B.S. Safety, immunogenicity and efficacy of modified vaccinia Ankara (MVA) against Dryvax challenge in vaccinia-naive and vaccinia-immune individuals. Vaccine 2007; 25:1513–1525.

Patterson, K.B. and Runge, T. Smallpox and the Native American. Am. J. Med. Sci. 2002; 323:216–222.

Pennington, H. Smallpox and bioterrorism. Bull. World Health Organ. 2003; 81:762–767.

Perez-Jimenez, E., Kochan, G., Gherardi, M.M. and Esteban, M. MVA-LACK as a safe and efficient vector for vaccination against leishmaniasis. Microbes Infect. 2006; 8:810–822.

Poland, G.A. Smallpox vaccines: from first to second to third generation. Lancet 2005; 365:362–363.

Poland, G.A., Grabenstein, J.D. and Neff, J.M. The US smallpox vaccination program: a review of a large modern era smallpox vaccination implementation program. Vaccine 2005; 23:2078–2081.

Poland, G.A. and Neff, J.M. Smallpox vaccine: problems and prospects. Immunol. Allergy Clin. North Am. 2003; 23:731–743.

Puissant, B. and Combadiere, B. Keeping the memory of smallpox virus. Cell. Mol. Life Sci. 2006; 63:2249–2259.

Quenelle, D.C., Collins, D.J. and Kern, E.R. Cutaneous infections of mice with vaccinia or cowpox viruses and efficacy of cidofovir. Antiviral Res. 2004a; 63:33–40.

Quenelle, D.C., Collins, D.J., Wan, W.B., Beadle, J.R., Hostetler, K.Y. and Kern, E.R. Oral treatment of cowpox and vaccinia virus infections in mice with ether lipid esters of cidofovir. Antimicrob. Agents Chemother. 2004b; 48:404–412.

Ramirez, J.C., Tapia, E. and Esteban, M. Administration to mice of a monoclonal antibody that neutralizes the intracellular mature virus form of vaccinia virus limits virus replication efficiently under prophylactic and therapeutic conditions. J. Gen. Virol. 2002; 83:1059–1067.

Randall, A.Z., Baldi, P. and Villarreal, L.P. Structural proteomics of the poxvirus family. Artif. Intell. Med. 2004; 31:105–115.

Reading, P.C., Moore, J.B. and Smith, G.L. Steroid hormone synthesis by vaccinia virus suppresses the inflammatory response to infection. J. Exp. Med. 2003; 197:1269–1278.

Redfield, R.R., Wright, D.C., James, W.D., Jones, T.S., Brown, C. and Burke, D.S. Disseminated vaccinia in a military recruit with human immunodeficiency virus (HIV) disease. N. Engl. J. Med. 1987; 316:673–676.

Reeves, P.M., Bommarius, B., Lebeis, S., McNulty, S., Christensen, J., Swimm, A., Chahroudi, A., Chavan, R., Feinberg, M.B., Veach, D., Bornmann, W., Sherman, M. and Kalman, D. Disabling poxvirus pathogenesis by inhibition of Abl-family tyrosine kinases. Nat. Med. 2005; 11:731–739.

Rock, M.T., Yoder, S.M., Talbot, T.R., Edwards, K.M. and Crowe, J.E., Jr. Adverse events after smallpox immunizations are associated with alterations in systemic cytokine levels. J. Infect. Dis. 2004; 189:1401–1410.

Rock, M.T., Yoder, S.M., Talbot, T.R., Edwards, K.M. and Crowe, J.E., Jr. Cellular immune responses to diluted and undiluted aventis pasteur smallpox vaccine. J. Infect. Dis. 2006; 194:435–443.

Roffey, R., Tegnell, A. and Elgh, F. Biological warfare in a historical perspective. Clin. Microbiol. Infect. 2002; 8:450–454.

Roizman, B., Joklik, W., Fields, B. and Moss, B. The destruction of smallpox virus stocks in national repositories: a grave mistake and a bad precedent. Infect. Agents Dis. 1994; 3:215–217.

Rosenthal, S.R., Merchlinsky, M., Kleppinger, C. and Goldenthal, K.L. Developing new smallpox vaccines. Emerg. Infect. Dis. 2001; 7:920–926.

Rotz, L.D. and Hughes, J.M. Advances in detecting and responding to threats from bioterrorism and emerging infectious disease. Nat. Med. 2004; 10:S130–S136.

Rubins, K.H., Hensley, L.E., Jahrling, P.B., Whitney, A.R., Geisbert, T.W., Huggins, J.W., Owen, A., Leduc, J.W., Brown, P.O. and Relman, D.A. The host response to smallpox: analysis of the gene expression program in peripheral blood cells in a nonhuman primate model. Proc. Natl. Acad. Sci. U. S. A. 2004; 101:15190–15195.

Ruef, C. Five years after 9/11-fearing smallpox, ... and the vaccine against it. Infection 2006; 34:239–240.

Sakhatskyy, P., Wang, S., Chou, T.H. and Lu, S. Immunogenicity and protection efficacy of monovalent and polyvalent poxvirus vaccines that include the D8 antigen. Virology 2006; 355:164–174.

Sarkar, J.K., Mitra, A.C. and Mukherjee, M.K. Duration of virus excretion in the throat of asymptomatic household contacts of smallpox patients. Indian J. Med. Res. 1974; 62:1800–1803.

Sarkar, J.K., Mitra, A.C., Mukherjee, M.K. and De, S.K. Virus excretion in smallpox. 2. Excretion in the throats of household contacts. Bull. World Health Organ. 1973a; 48:523–527.

Sarkar, J.K., Mitra, A.C., Mukherjee, M.K., De, S.K. and Mazumdar, D.G. Virus excretion in smallpox. 1. Excretion in the throat, urine, and conjunctiva of patients. Bull. World Health Organ. 1973b; 48:517–522.

Schatzmayr, H.G., Lemos, E.R., Mazur, C., Schubach, A., Majerowicz, S., Rozental, T., Schubach, T.M., Bustamante, M.C. and Barth, O.M. Detection of poxvirus in cattle associated with human cases in the State of Rio de Janeiro: preliminary report. Mem. Inst. Oswaldo Cruz 2000; 95:625–627.

Schuler, A. Billions for biodefense: federal agency biodefense funding, FY2001-FY2005. Biosecur. Bioterror. 2004; 2:86–96.

Schuler, A. Billions for biodefense: federal agency biodefense budgeting, FY2005-FY2006. Biosecur. Bioterror. 2005; 3:94–101.

Seet, B.T., Johnston, J.B., Brunetti, C.R., Barrett, J.W., Everett, H., Cameron, C., Sypula, J., Nazarian, S.H., Lucas, A. and McFadden, G. Poxviruses and immune evasion. Annu. Rev. Immunol. 2003a; 21:377–423.

Seet, B.T., McCaughan, C.A., Handel, T.M., Mercer, A., Brunetti, C., McFadden, G. and Fleming, S.B. Analysis of an orf virus chemokine-binding protein: Shifting ligand specificities among a family of poxvirus viroceptors. Proc. Natl. Acad. Sci. U. S. A. 2003b; 100:15137–15142.

Seet, B.T., Singh, R., Paavola, C., Lau, E.K., Handel, T.M. and McFadden, G. Molecular determinants for CC-chemokine recognition by a poxvirus CC-chemokine inhibitor. Proc. Natl. Acad. Sci. U. S. A. 2001; 98:9008–9013.

Sepkowitz, K.A. How contagious is vaccinia? N. Engl. J. Med. 2003; 348:439–446.

Shchelkunov, S.N., Totmenin, A.V., Babkin, I.V., Safronov, P.F., Ryazankina, O.I., Petrov, N.A., Gutorov, V.V., Uvarova, E.A., Mikheev, M.V., Sisler, J.R., Esposito, J.J., Jahrling, P.B., Moss, B. and Sandakhchiev, L.S. Human monkeypox and smallpox viruses: genomic comparison. FEBS Lett. 2001; 509:66–70.

Shchelkunov, S.N., Totmenin, A.V., Loparev, V.N., Safronov, P.F., Gutorov, V.V., Chizhikov, V.E., Knight, J.C., Parsons, J.M., Massung, R.F. and Esposito, J.J. Alastrim smallpox variola minor virus genome DNA sequences. Virology 2000; 266:361–386.

Shisler, J.L. and Moss, B. Immunology 102 at poxvirus U: avoiding apoptosis. Semin. Immunol. 2001; 13:67–72.

Simpson, E.L., Hercher, M., Hammarlund, E.K., Lewis, M.W., Slifka, M.K. and Hanifin, J.M. Cutaneous responses to vaccinia in individuals with previous smallpox vaccination. J. Am. Acad. Dermatol. 2007; 57:442–444.

Singh, R.K., Hosamani, M., Balamurugan, V., Satheesh, C.C., Shingal, K.R., Tatwarti, S.B., Bambal, R.G., Ramteke, V. and Yadav, M.P. An outbreak of buffalopox in buffalo (Bubalus bubalis) dairy herds in Aurangabad, India. Rev. Sci. Tech. 2006; 25:981–987.

Sivapalasingam, S., Kennedy, J.S., Borkowsky, W., Valentine, F., Zhan, M.X., Pazoles, P., Paolino, A., Ennis, F.A. and Steigbigel, N.H. Immunological memory after exposure to variola virus, monkeypox virus, and vaccinia virus. J. Infect. Dis. 2007; 195:1151–1159.

Slifka, M.K. The Future of Smallpox Vaccination: is MVA the key? Med. Immunol. 2005; 4:2.

Sliva, K. and Schnierle, B. From actually toxic to highly specific – novel drugs against poxviruses. Virol. J. 2007; 4:8.

Smee, D.F., Bailey, K.W. and Sidwell, R.W. Treatment of lethal vaccinia virus respiratory infections in mice with cidofovir. Antivir Chem. Chemother. 2001a; 12:71–76.

Smee, D.F., Bailey, K.W., Wong, M.H. and Sidwell, R.W. Effects of cidofovir on the pathogenesis of a lethal vaccinia virus respiratory infection in mice. Antiviral Res. 2001b; 52:55–62.

Smee, D.F. and Sidwell, R.W. A review of compounds exhibiting anti-orthopoxvirus activity in animal models. Antiviral Res. 2003; 57:41–52.

Smith, G.L., Vanderplasschen, A. and Law, M. The formation and function of extracellular enveloped vaccinia virus. J. Gen. Virol. 2002; 83:2915–2931.

Smith, S.A. and Kotwal, G.J. Immune response to poxvirus infections in various animals. Crit. Rev. Microbiol. 2002; 28:149–185.

Snyder, J.T., Belyakov, I.M., Dzutsev, A., Lemonnier, F. and Berzofsky, J.A. Protection against lethal vaccinia virus challenge in HLA-A2 transgenic mice by immunization with a single CD8+ T-cell peptide epitope of vaccinia and variola viruses. J. Virol. 2004; 78:7052–7060.

Stack, J., Haga, I.R., Schroder, M., Bartlett, N.W., Maloney, G., Reading, P.C., Fitzgerald, K.A., Smith, G.L. and Bowie, A.G. Vaccinia virus protein A46R targets multiple Toll-like-interleukin-1 receptor adaptors and contributes to virulence. J. Exp. Med. 2005; 201:1007–1018.

Staib, C., Kisling, S., Erfle, V. and Sutter, G. Inactivation of the viral interleukin 1beta receptor improves CD8+ T-cell memory responses elicited upon immunization with modified vaccinia virus Ankara. J. Gen. Virol. 2005; 86:1997–2006.

Staib, C., Suezer, Y., Kisling, S., Kalinke, U. and Sutter, G. Short-term, but not post-exposure, protection against lethal orthopoxvirus challenge after immunization with modified vaccinia virus Ankara. J. Gen. Virol. 2006; 87:2917–2921.

Stienlauf, S., Shoresh, M., Solomon, A., Lublin-Tennenbaum, T., Atsmon, Y., Meirovich, Y. and Katz, E. Kinetics of formation of neutralizing antibodies against vaccinia virus following re-vaccination. Vaccine 1999; 17:201–204.

Sugimoto, M. and Yamanouchi, K. Characteristics of an attenuated vaccinia virus strain, LC16m0, and its recombinant virus vaccines. Vaccine 1994; 12:675–681.

Symons, J.A., Adams, E., Tscharke, D.C., Reading, P.C., Waldmann, H. and Smith, G.L. The vaccinia virus C12L protein inhibits mouse IL-18 and promotes virus virulence in the murine intranasal model. J. Gen. Virol. 2002; 83:2833–2844.

Talbot, T.R., Stapleton, J.T., Brady, R.C., Winokur, P.L., Bernstein, D.I., Germanson, T., Yoder, S.M., Rock, M.T., Crowe, J.E., Jr. and Edwards, K.M. Vaccination success rate and reaction profile with diluted and undiluted smallpox vaccine: a randomized controlled trial. Jama 2004; 292:1205–1212.

Tanabe, I. and Hotta, S. Effect of disinfectants on variola virus in cell culture. Appl. Environ. Microbiol. 1976; 32:209–212.

Tang, J., Murtadha, M., Schnell, M., Eisenlohr, L.C., Hooper, J. and Flomenberg, P. Human T-cell responses to vaccinia virus envelope proteins. J. Virol. 2006; 80:10010–10020.

Tartaglia, J., Perkus, M.E., Taylor, J., Norton, E.K., Audonnet, J.C., Cox, W.I., Davis, S.W., van der Hoeven, J., Meignier, B., Riviere, M. et al. NYVAC: a highly attenuated strain of vaccinia virus. Virology 1992; 188:217–232.

Taylor, J.M. and Barry, M. Near death experiences: poxvirus regulation of apoptotic death. Virology 2006; 344:139–150.

Tegnell, A., Wahren, B. and Elgh, F. Smallpox--eradicated, but a growing terror threat. Clin. Microbiol. Infect. 2002; 8:504–509.

Terajima, M., Cruz, J., Leporati, A., Demkowicz, W.E., Jr., Kennedy, J.S. and Ennis, F.A. Identification of Vaccinia CD8+ T-Cell Epitopes Conserved among Vaccinia and Variola Viruses Restricted by Common MHC Class I Molecules, HLA-A2 or HLA-B7. Human Immunology 2006a; In Press.

Terajima, M., Cruz, J., Leporati, A.M., Demkowicz, W.E., Jr., Kennedy, J.S. and Ennis, F.A. Identification of vaccinia CD8+ T-cell epitopes conserved among vaccinia and variola viruses restricted by common MHC class I molecules, HLA-A2 or HLA-B7. Hum. Immunol. 2006b; 67:512–520.

Terajima, M., Cruz, J., Raines, G., Kilpatrick, E.D., Kennedy, J.S., Rothman, A.L. and Ennis, F.A. Quantitation of CD8+ T cell responses to newly identified HLA-A*0201-restricted T cell epitopes conserved among vaccinia and variola (smallpox) viruses. J. Exp. Med. 2003; 197:927–932.

Terajima, M. and Ennis, F.A. Using HLA-transgenic mice to identify immunodominant human CD8(+) T cell epitopes-Does (genome) size matter? Immunol. Lett. 2006; 105:97–98.

Thornburg, N.J., Ray, C.A., Collier, M.L., Liao, H.X., Pickup, D.J. and Johnston, R.E. Vaccination with Venezuelan equine encephalitis replicons encoding cowpox virus structural proteins protects mice from intranasal cowpox virus challenge. Virology 2007; 362:441–452.

Treanor, J., Wu, H., Liang, H. and Topham, D.J. Immune responses to vaccinia and influenza elicited during primary versus recent or distant secondary smallpox vaccination of adults. Vaccine 2006; 24:6913–6923.

Tulman, E.R., Afonso, C.L., Lu, Z., Zsak, L., Kutish, G.F. and Rock, D.L. Genome of lumpy skin disease virus. J. Virol. 2001; 75:7122–7130.

Tulman, E.R., Afonso, C.L., Lu, Z., Zsak, L., Kutish, G.F. and Rock, D.L. The genome of canarypox virus. J. Virol. 2004; 78:353–366.

Tulman, E.R., Delhon, G., Afonso, C.L., Lu, Z., Zsak, L., Sandybaev, N.T., Kerembekova, U.Z., Zaitsev, V.L., Kutish, G.F. and Rock, D.L. Genome of horsepox virus. J. Virol. 2006; 80:9244–9258.

Tulman, E.R., Delhon, G., Afonso, C.L., Lu, Z., Zsak, L., Sandybaev, N.T., Kerembekova, U.Z., Zaitsev, V.L., Kutish, G.F. and Rock, D.L. Genome of Horsepox Virus. J. Virol. 2007; 80(18):9244–9258.

Turner, G.S. Respiratory infection of mice with vaccinia virus. J. Gen. Virol. 1967; 1:399–402.

Upton, C., Slack, S., Hunter, A.L., Ehlers, A. and Roper, R.L. Poxvirus orthologous clusters: toward defining the minimum essential poxvirus genome. J. Virol. 2003; 77:7590–7600.

Vellozzi, C., Averhoff, F., Lane, J.M., Wen, X., Moore, A.C., Santibanez, S., Kroger, A., Hasbrouck, L.M., Kennedy, A. and Casey, C.G. Superinfection following smallpox vaccination (Vaccinia), United States: surveillance January 2003 through January 2004. Clin. Infect. Dis. 2004; 39:1660–1666.

Vellozzi, C., Lane, J.M., Averhoff, F., Maurer, T., Norton, S., Damon, I. and Casey, C. Generalized vaccinia, progressive vaccinia, and eczema vaccinatum are rare following smallpox (vaccinia) vaccination: United States surveillance, 2003. Clin. Infect. Dis. 2005; 41:689–697.

Vollmar, J., Arndtz, N., Eckl, K.M., Thomsen, T., Petzold, B., Mateo, L., Schlereth, B., Handley, A., King, L., Hulsemann, V., Tzatzaris, M., Merkl, K., Wulff, N. and Chaplin, P. Safety and immunogenicity of IMVAMUNE, a promising candidate as a third generation smallpox vaccine. Vaccine 2006; 24:2065–2070.

Wallace, G.D. and Buller, R.M. Kinetics of ectromelia virus (mousepox) transmission and clinical response in C57BL/6j, BALB/cByj and AKR/J inbred mice. Lab. Anim. Sci. 1985; 35:41–46.

Wang, Z., Rosa, C.L., Li, Z., Ly, H., Krishnan, A., Martinez, J., Britt, W.J. and Diamond, D.J. Vaccine properties of a novel marker gene-free recombinant modified vaccinia Ankara expressing immunodominant CMV antigens pp65 and IE1. Vaccine 2007; 25:1132–1141.

Webb, T.J., Litavecz, R.A., Khan, M.A., Du, W., Gervay-Hague, J., Renukaradhya, G.J. and Brutkiewicz, R.R. Inhibition of CD1d1-mediated antigen presentation by the vaccinia virus B1R and H5R molecules. Eur. J. Immunol. 2006; 36:2595–2600.

Wehrle, P.F., Posch, J., Richter, K.H. and Henderson, D.A. An airborne outbreak of smallpox in a German hospital and its significance with respect to other recent outbreaks in Europe. Bull. World Health Organ. 1970; 43:669–679.

Weltzin, R., Liu, J., Pugachev, K.V., Myers, G.A., Coughlin, B., Blum, P.S., Nichols, R., Johnson, C., Cruz, J., Kennedy, J.S., Ennis, F.A. and Monath, T.P. Clonal vaccinia virus grown in cell culture as a new smallpox vaccine. Nat. Med. 2003; 9:1125–1130.

Weyer, J., Rupprecht, C.E., Mans, J., Viljoen, G.J. and Nel, L.H. Generation and evaluation of a recombinant modified vaccinia virus Ankara vaccine for rabies. Vaccine 2007; 25:4213–4222.

Whitley, R.J. Smallpox: a potential agent of bioterrorism. Antiviral Res. 2003; 57:7–12.

WHO (2006), Vol. A59/10.

Williamson, J.D., Reith, R.W., Jeffrey, L.J., Arrand, J.R. and Mackett, M. Biological characterization of recombinant vaccinia viruses in mice infected by the respiratory route. J. Gen. Virol. 1990; 71(Pt 11):2761–2767.

Wiser, I., Balicer, R.D. and Cohen, D. An update on smallpox vaccine candidates and their role in bioterrorism related vaccination strategies. Vaccine 2007; 25:976–984.

Wittek, R. Organization and expression of the poxvirus genome. Experientia 1982; 38:285–297.

Wollenberg, A. and Engler, R. Smallpox, vaccination and adverse reactions to smallpox vaccine. Curr. Opin. Allergy Clin. Immunol. 2004; 4:271–275.

Wyatt, L.S., Earl, P.L., Eller, L.A. and Moss, B. Highly attenuated smallpox vaccine protects mice with and without immune deficiencies against pathogenic vaccinia virus challenge. Proc. Natl. Acad. Sci. U. S. A. 2004; 101:4590–4595.

Wyeth Smallpox vaccine Dried, Calf Lymph Type. Dryvax insert, 2002.

Xiao, Y., Aldaz-Carroll, L., Ortiz, A.M., Whitbeck, J.C., Alexander, E., Lou, H., Davis, H.L., Braciale, T.J., Eisenberg, R.J., Cohen, G.H. and Isaacs, S.N. A protein-based smallpox vaccine protects mice from vaccinia and ectromelia virus challenges when given as a prime and single boost. Vaccine 2007; 25:1214–1224.

Xing, Z., Santosuosso, M., McCormick, S., Yang, T.C., Millar, J., Hitt, M., Wan, Y., Bramson, J. and Vordermeier, H.M. Recent advances in the development of adenovirus- and poxvirus-vectored tuberculosis vaccines. Curr. Gene Ther. 2005; 5:485–492.

Yang, H., Kim, S.K., Kim, M., Reche, P.A., Morehead, T.J., Damon, I.K., Welsh, R.M. and Reinherz, E.L. Antiviral chemotherapy facilitates control of poxvirus infections through inhibition of cellular signal transduction. J. Clin. Invest. 2005; 115:379–387.

Yau, T. and Rouhandeh, H. Characterization of monkey pox virus DNA genome. Biochim. Biophys. Acta 1973; 299:210–217.

Zilinskas, R.A. Iraq's biological weapons. The past as future? Jama 1997; 278:418–424.

CHAPTER

38

Tick-Borne Encephalitis

Axel T. Lehrer[1] and Michael R. Holbrook[2]

[1]PanThera Biopharma, Aiea, HI, USA

[2]Department of Pathology, Center for Biodefense and Emerging Diseases, Sealy Center for Vaccine Development, and Institute for Human Infections and Immunity, University of Texas Medical Branch, Galveston, TX, USA

OUTLINE

ABSTRACT

Tick-borne encephalitis (TBE) is a disease caused by a group of closely related tick-borne flaviviruses with an endemic range across Asia from Japan into Western Europe. TBE can range from a subclinical or mild disease to a severe disease with a high incidence of encephalitis and long-term sequelae. Related viruses have been found in North America, India, Saudi Arabia, and Malaysia, and several are associated with viral hemorrhagic fever rather than encephalitis.

There are currently four inactivated virus vaccines available for the prevention of TBE—two produced in Russia and two licensed for use in Europe. Although effective, there are some limitations with the use of inactivated virus, primarily the need for multiple doses for development of a protective immune response. There are also limitations on the availability of the vaccines and questions about the efficacy of these vaccines in protecting against tick-borne flaviviruses that cause hemorrhagic fever. Although the current vaccines have made a significant impact in parts of Western Europe, the overall incidence of disease is still increasing in many countries. A number of modern approaches are currently being examined for development of a new generation of TBE vaccines. These include the use of live-attenuated viruses, chimeric viruses, DNA vaccines, and protein subunit vaccines. Each technology has shown some promise in animal models, but clinical trials have not been reported for any of the novel vaccine candidates. With current licensure requirements, the active use of any new vaccine is still several years away; however, with disease incidence increasing, the need for improved vaccines will continue.

INTRODUCTION

Tick-borne encephalitis (TBE) is a disease that is endemic from Western Europe, across Asia, and into parts of Japan. TBE accounts for 7000–10,000 cases of illness annually with nearly 4000 cases reported in Europe alone in 2006 [Tick-Borne Encephalitis International Scientific Working Group (TBE-ISW) http://www.tbe-info.com]. The causative agent for TBE is TBE virus (TBEV), a member of the TBE serocomplex of flaviviruses that are related serologically and genetically. TBEV causes two clinically distinct diseases, depending on the subtype of virus. The disease found in Europe is typically biphasic with a relatively low mortality rate and few sequelae in survivors. The disease in eastern Asia is much more severe with a higher mortality rate, significant sequelae in survivors, and the potential to develop chronic or persistent disease. To date there are no effective therapeutics for the treatment of TBE.

There are currently four known vaccines available for prevention of TBEV and one that is in use in India for Kyasanur Forest disease virus (KFDV). Two of the TBEV vaccines are licensed for use in Europe while the remaining two are manufactured for use in Russia. All four TBEV vaccines are inactivated virus vaccines that are manufactured following similar processes. The principal variation between the four vaccines is the choice of virus strain used. Two of the manufacturers use a Western subtype TBEV as the basis for their vaccine while both of the Russian vaccines have been developed using a Far Eastern subtype virus.

There is currently a significant effort to develop TBEV vaccines that are not based on the use of an inactivated virus. The concerns about inactivated virus vaccines center around the potential for incomplete inactivation, the need for multiple vaccinations and boosters, and relatively poor immunogenicity in some people. Modern vaccine development focuses on the use of subunit vaccines, live-attenuated vaccines, or DNA-based vaccines. The benefit of subunit vaccines is that there is no virus present in the vaccine; hence, the likelihood of virus reactivation or reversion is nil. The limitations of subunit vaccines include the need for multiple vaccinations and boosters, and their efficacy may be determined by the quality of the adjuvant used in the vaccine formulation. Live-attenuated vaccines include naturally attenuated viruses, artificially attenuated viruses, and recombinant virus vaccines. There have been a wide variety of candidates developed for vaccine trials and all have the advantage of requiring a single vaccination with periodic boosters. The limitations of these vaccines are the significant concern about vaccine reversion and the use of genetically modified viruses in humans. DNA-based vaccines have shown some promise as most are a combination of live-attenuated and subunit vaccines that are capable of replicating, but only contain a portion of the TBEV genome and cannot generate infectious TBEV. Most DNA vaccines use alternative virus vectors, such as adenoviruses, to express TBEV proteins and induce a protective immune response. However, the use of DNA vaccines has met with some resistance, as there is concern about the production of replication-competent highly pathogenic viruses, an unlikely event given the construction of the vaccines.

As with vaccine development for many viruses, the development of vaccines for TBEV has entered the modern age where the demands for safety and efficacy sometimes collide.

HISTORY OF TBE

A neurologic disease with high mortality was first recognized around 1932 in Far Eastern regions of the former Union of Soviet and Socialist Republics (USSR) (Silber and Soloviev, 1946). Initially, this disease was thought to be Japanese summer encephalitis. The virus was thought to be transmitted through interpersonal contact and movement of the disease between endemic regions was believed to be due to movement of forest workers (Silber and Soloviev, 1946). The first published description of TBE was in 1936 by Silber et al. and described the disease in far-eastern Russia (then the USSR) (Silber and Soloviev, 1946). Laborers in forested regions frequently became ill from a severe neurologic disease that often resulted in death. In 1937, L.A. Silber led a special expedition with a mandate to study TBE in the eastern USSR. Silber et al. determined that ticks were the primary vector for the contagion and that *Ixodes persulcatus* ticks were the most likely vector (Silber and Soloviev, 1946). Subsequently, Chumakov et al. demonstrated experimentally that *I. persulcatus* could transmit the disease (Chumakov and Gladkikh, 1939). In 1937, two groups were able to isolate the virus and clearly demonstrate that the isolated virus was responsible for the disease (Silber and Soloviev, 1946). Subsequently, the virus was named Far Eastern encephalitis virus (Silber and Soloviev, 1946). The isolation of the virus also led to the first effort to generate a vaccine, as Smorodintsev developed a formalin-inactivated vaccine that was used in forest workers in endemic regions (Smorodintsev et al., 1941). A number of subsequent studies characterized the clinical aspects of Far Eastern encephalitis including the first descriptions of chronic and progressive forms of the disease (Silber and Soloviev, 1946). In the early 1940s Far Eastern encephalitis was described in Siberia and the European part of the USSR (Chumakov et al., 1944). It was noted at that time that the virus in Siberia and European USSR caused a different clinical disease than Far Eastern encephalitis, and it was proposed by Chumakov et al. (1944) that the diseases be termed Far Eastern and Western encephalitis to reflect their clinical differences.

Outside of the USSR, TBE was first recognized in epidemics in Czechoslovakia in 1948 and 1949 and the virus was isolated in 1949 from both humans and ticks (Gallia et al., 1949; Hloucal, 1960). This virus was found to be serologically related to the Far Eastern and Western encephalitis viruses, and the milder clinical disease in the Czechoslovakian epidemics suggested that these viruses were of the Western subtype. In 1951–1952, a large epidemic of Western encephalitis was reported in the Roznava district of Czechoslovakia. In this outbreak, it was determined that transmission of the virus in goat milk played a major role in the epidemic in which more than 600 people were affected (Blaskovic, 1958). Smorodintsev had previously shown that Far Eastern encephalitis virus could be transmitted by drinking raw goat's milk (Silber and Soloviev, 1946; Smorodintsev, 1958).

Since its discovery, TBE has now been recognized in many countries in Europe, including most of Eastern Europe and parts of Western Europe (Fig. 38.1). The endemic range for Western subtype TBE reaches east to the Ural mountains while the endemic range for the Far Eastern subtype ranges from Eastern Europe to far-eastern Russia and Hokkaido island in Japan with some reports of TBEV in far-eastern China. The endemic areas of the viruses are essentially coincident with the endemic range of their primary tick vectors. *Ixodes ricinus*, the principle vector for Western subtype TBEV, is found primarily in Europe, parts of Asia, and North Africa. *I. persulcatus*, the vector for Far Eastern subtype TBEV, is found across northern Asia (Brown et al., 2005).

Since the discovery and identification of TBEV, a number of related viruses have been identified (Table 38.1). All of these viruses are transmitted by ticks and there is a certain degree of serological cross-reactivity among the viruses. These viruses include Omsk hemorrhagic fever virus (OHFV) that was first isolated in 1947 in Siberia from human blood (Chumakov, 1948); Langat virus (LGTV) isolated in Malaysia in 1956 from tick pools (Gordon Smith, 1956); Kyasanur Forest disease virus (KFDV) that was isolated in India in 1957 from humans, monkeys, and ticks (Work, 1958); Powassan virus (POWV) isolated in Canada in 1958 from human brain (McLean and Donohue, 1958); and Alkhurma virus (ALKV) isolated in Saudi Arabia in 1996 from human blood (Zaki, 1997). OHFV, KFDV, and ALKV are all associated with viral hemorrhagic fever, and KFDV and ALKV have also been shown to produce encephalitis in humans (Zaki, 1997). POWV appears to cause primarily subclinical disease but can result in extremely severe and lethal encephalitis. LGTV does not typically cause disease in humans and in the 1970s, the E5 strain of LGTV was developed and tested as a potential live-attenuated vaccine for TBEV (Mayer, 1975; Mayer et al., 1976; Price and Thind, 1973; Price et al., 1970).

Since the discovery of these viruses, the tick-borne flaviviruses in the TBE serocomplex have been shown to be closely related serologically. Genetic analysis of these viruses has subsequently confirmed serological data. The TBEV have since been divided into three subtypes (Far Eastern, Siberian, and Western) and genetic variation between the viruses correlates well with differences in clinical presentation described when Far Eastern and Western encephalitis were first recognized.

ETIOLOGIC AGENTS

Classification

The genus *Flavivirus* in the family *Flaviviridae* consists of approximately 70 related viruses that are typically transmitted by mosquitoes or ticks, yet there are several flaviviruses that are classified as "no known

FIGURE 38.1 Worldwide distribution of tick-borne flaviviruses with significant human health issue.

vector" as there does not appear to be an arthropod host associated with virus transmission (Calisher and Gould, 2003). The mosquito-borne flaviviruses consist of viruses such as dengue virus (DENV), yellow fever virus (YFV), and viruses in the Japanese encephalitis virus serocomplex including Japanese encephalitis (JEV) and West Nile viruses (WNV). The mosquito-borne viruses can also be delineated by their arthropod vectors, as DENV and YFV are typically transmitted by *Aedes spp.* mosquitoes while JEV serocomplex viruses are transmitted by *Culex spp.* mosquitoes. There is also a significant amount of serological and genetic diversity among the mosquito-borne flaviviruses (Gould et al., 2001, 2004; Zanotto et al., 1995).

There are two subsets of tick-borne flaviviruses: those that are associated with seabirds and those with mammals (Gould et al., 2001, 2004). Mammal-associated tick-borne flaviviruses subset contains the only viruses known to cause disease in humans and is commonly referred to as the TBE serocomplex of viruses. The TBE serocomplex is genetically divided into four major clades representing Far Eastern subtype TBEV, Siberian subtype TBEV, Western subtype TBEV, and Louping Ill virus (LIV) (Gould et al., 2001, 2004; Zanotto et al., 1995). Other viruses in the serocomplex have very few species and form their own clades.

The progenitor virus for the TBE serocomplex is thought to be POWV (Gould et al., 2001, 2004; Zanotto et al., 1995). The evaluation of evolutionary trends of the TBE serocomplex is that the virus migrated from eastern Asia across Asia and into Europe with Western subtype TBE and louping ill viruses (LIV) representing the most recently evolved virus subtypes (Gould et al., 2001, 2004; Zanotto et al., 1995) (Fig. 38.2).

The level of serologic and genetic diversity among the tick-borne flaviviruses is much less than among the mosquito-borne viruses. There is approximately 30% nucleotide variation between POWV and LIV, and Western subtype TBEV, the most distant relatives. There is 16% nucleotide variation between Far Eastern subtype and Western subtype TBEV. The diversity at the amino acid level is about 23% between POWV and LIV, the least related viruses (Lin et al., 2003). There is less than 4% amino acid diversity between Far Eastern and Western subtypes of TBEV (Lin et al., 2003). This high level of similarity at the nucleotide and amino acid

TABLE 38.1 Members of the mammalian tick-borne encephalitis group, including major stains

Virus	Subtype	Strain	Site of isolation	Source	Year of isolation	References
TBEV	Far Eastern	Sofjin[a]	Far-eastern USSR	Human brain	1937	Silber and Soloviev (1946)
	Siberian	Vasilchenko	Novosibirsk			
	Western	Byelorussia strain 256	Western USSR	*Ixodes ricinus*	1940	
		Stillerova	Czechoslovakia	Human brain	1950	
		Absettarov	USSR	Human blood	1951	
		Kumlinge A52	Finland	*I. ricinus*	1951	Oker-Blom et al. (1962)
		Graz	Austria	Human CNS	1953	
		Hypr	Czechoslovakia	Human blood	1953	
		Neudörfl[a]	Austria	Ticks	~1971	Kunz (2003)
Powassan		Byers[a]	Canada	Human brain	1958	McLean and Donohue (1958)
Omsk hemorrhagic fever		Kubrin[a]	USSR	Human blood	1947	Chumakov (1948)
		Guriev	USSR	Human blood	?	
Kyasanur Forest Disease		P9605[a]	India	Human blood	1957	Work (1958)
		W377	India	Monkey brain	1957	
Alkhurma			Saudi Arabia	Human	1996	Zaki (1997)
Langat		TP21[a]	Malaysia	*Ixodes granulatus*	1956	Gordon Smith (1956)

Source: Adapted from Theiler and Downs, The arthropod-borne viruses of vertebrates, 1973.
[a] Prototype strain.

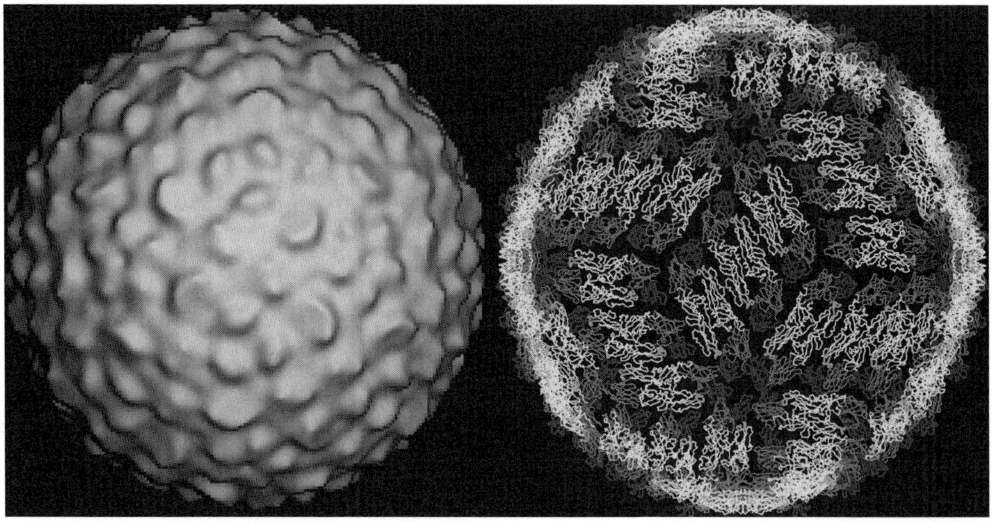

FIGURE 38.2 A cladogram showing the evolutionary cline of TBE serocomplex viruses across Asia. FETBE: Far Eastern TBEV; TSE: Turkish sheep encephalitis virus; GGE: Greek goat encephalitis virus; WTBE: Western subtype TBEV; SSE: Spanish sheep encephalitis virus; LI: Louping Ill virus. Reprinted from Zanotto et al. (1995), with permission from Elsevier (see color plate section).

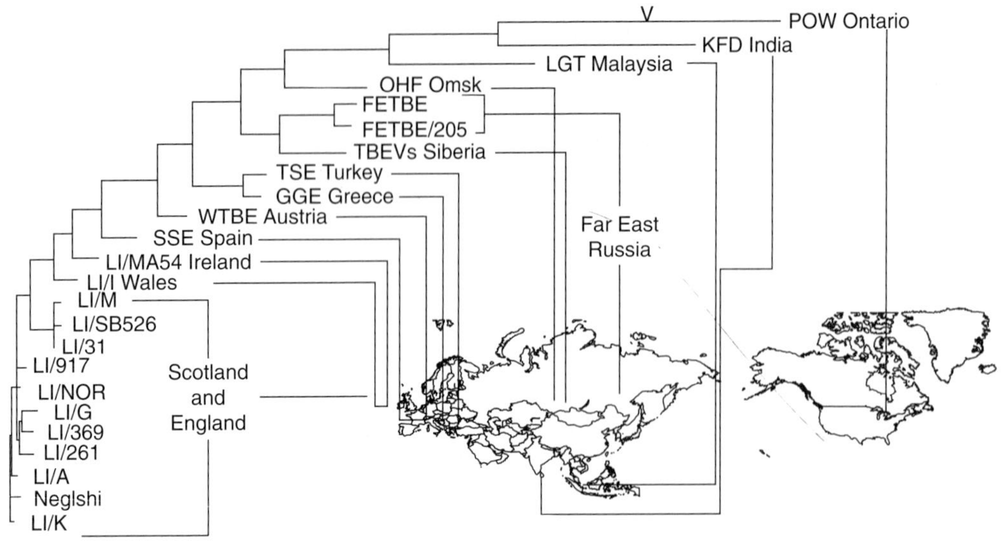

FIGURE 38.3 Cryo-electron microscopy surface representation of dengue-2 virus, a mosquito-borne flavivirus (left). The incorporation of the viral envelope protein into electron densities is observed by cryo-electron microscopy. The arrangement of E protein dimers are shown along with putative carbohydrate moieties. Adapted from Kuhn and Strauss (2003), with permission from Elsevier.

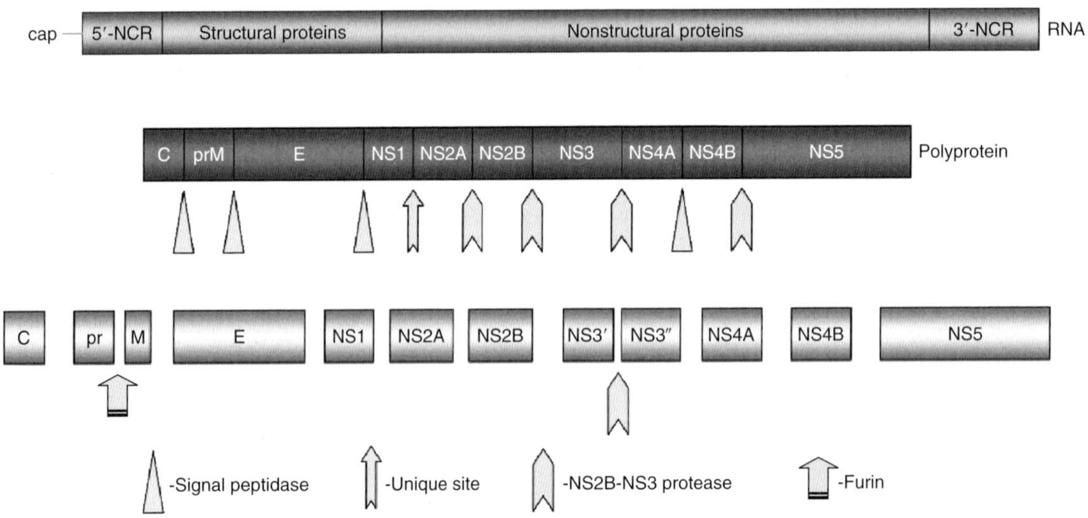

FIGURE 38.4 A diagrammatic representation of the flavivirus genome and polyprotein. Protease cleavage sites are also indicated.

level is further borne out in structural similarities of the viral envelope (E) protein, the primary immunogen on the surface of the virion, as these viruses are difficult to distinguish serologically. The high level of similarity further suggests that vaccines raised against one member of the TBEV serocomplex will show a high level of cross-reactivity against other members of the complex.

Flavivirus Genome Organization

The flaviviruses are small enveloped viruses with a diameter of approximately 50 nm. The virus particle is composed of 90 envelope (E) protein dimers arranged in a herringbone type array (Kuhn et al., 2002) (Fig. 38.3). Interspersed among the E protein lattice is the viral membrane (M) protein, though the precise location of the M protein is not clear. The viral particle contains the single-stranded positive-sense RNA viral genome that is approximately 11 kb in size. The viral genome is translated as a single polyprotein and is co- and post-translationally cleaved to generate 10 individual proteins, three structural and seven nonstructural proteins (Fig. 38.4). The structural proteins consist of the capsid (C), pre-membrane (prM), and E proteins while the nonstructural proteins include NS1, NS2A, NS2B, NS3, NS4A, NS4B, and NS5. NS2B and NS3 associate to produce a viral protease and NS5 functions as a helicase and the viral RNA polymerase. The flavivirus nonstructural proteins serve a number of functions, but are primarily associated with viral replication (e.g., NS1, NS3, and NS5) and polyprotein processing (e.g., NS2B and NS3), although the function of some of the nonstructural proteins is poorly understood. Several nonstructural proteins have been associated with stimulating or moderating the host response to viral infection by interferon induction or inhibition (e.g., NS5) and induction of apoptosis (e.g., M protein).

The structural proteins make up the total protein component of the flavivirus virions as none of the nonstructural proteins are included in the virion. The C protein is associated with the viral RNA although the role of the C protein in RNA organization in the virion is still unclear. The prM protein functions as a chaperone and facilitates folding of the E protein. This is an important role that cannot be overlooked, as a number of vaccine development strategies have included DNA vectors that encode the prM and E proteins to generate subviral particles. The lack of the prM protein results in poor protein expression and poor immunogenicity. During virion organization the prM protein is assembled along with the E protein, but the protein is cleaved during virus maturation by a furin-like protein leaving only the M protein in the mature virion with the E protein. The flavivirus E protein is the major virus protein on the surface of the virion. The E protein is the primary immunogen for the flaviviruses and serves as a receptor-binding protein and fusion protein (Chen et al., 1997; Chu et al., 2005; Hung et al., 2004; Lee and Lobigs, 2000; Mandl et al., 2000). The structures of the E protein from several flaviviruses have been solved and it has been shown to have three distinct domains—I, II, and III (Modis et al., 2003; Rey et al., 1995). Domain I is the central domain and serves as a hinge between domains II and III. Domain II is the dimerization domain and also contains the viral fusion peptide. Domain III is the receptor-binding domain (Chu et al., 2005; Hung et al., 2004; Lee et al., 2006) and is also a principal target for neutralizing antibodies (Beasley and Barrett, 2002; Crill and Roehrig, 2001; Lin and Wu, 2003; Sanchez et al., 2005; Thullier et al., 2001). During viral attachment and entry, the virus binds to an as-yet-unidentified cell surface receptor, is internalized in clathrin-coated pits, and undergoes fusion with endosomal membranes during a low pH-induced conformational rearrangement (Bressanelli et al., 2004; Heinz and Allison, 2000, 2001; Kuhn et al., 2002; Modis et al., 2004; Stiasny et al., 2002, 2005; Zhang et al., 2004). Following fusion, the viral RNA is released into the cell cytoplasm where replication occurs.

The flavivirus entry provides potential intervention points for neutralizing antibodies in a protective response to viral infection. The first is at the point of viral attachment. A neutralizing antibody could bind to the E protein domain III and prevent the virus from binding to cell surface receptors. A number of neutralizing monoclonal antibodies are known to interact with domain III to prevent infection by different flaviviruses (Beasley and Barrett, 2002; Crill and Roehrig, 2001; Kaufmann et al., 2006; Lin and Wu, 2003; Sanchez et al., 2005; Volk et al., 2004; Wu et al., 2003; Wu and Lin, 2001). A second point of inhibition is prevention of membrane fusion. Antibodies can bind to the viral E protein and inhibit the pH-induced rearrangement of the E protein that occurs in cellular endosomes to allow fusion of the viral envelope protein with the endosomal membrane. Several monoclonal antibodies are thought to neutralize flavivirus infection in this manner. Epitopes for these antibodies have been mapped to several regions of the viral E protein including domains I and II (Beasley and Aaskov, 2001; Oliphant et al., 2006).

EPIDEMIOLOGY

Significance as Public Health Problem

TBE is a widespread arboviral encephalitis that is of significant human heath importance. As discussed earlier, TBEV was first isolated in far-eastern Russia and has since been found across Asia with related viruses found throughout Europe, the British Isles, and also in North America (Fig. 38.1). TBE is a significant human health issue in Europe where 2000–4000 cases are reported annually with the total number of cases increasing (TBE-ISW http://www.tbe-info.com) (Table 38.2). The current case numbers are reflective of a significant vaccination effort in some countries in Europe, particularly Austria, where total case numbers have been reduced by 90% of their pre-vaccine totals. Prior to the development and effective use of TBE vaccines there were more than 600 cases annually in Austria. Although TBE is a notifiable disease in many countries, most countries in Europe were not reporting consistently prior to the mid-1970s.

However, despite a reduction in the number of annual TBE cases due to improved vaccination coverage and changes in personal behavior in some countries, the range of TBE endemicity appears to be increasing and therefore the total case number is also on the rise. Recent reports of virus isolation or amplification of TBE RNA from tick pools has provided evidence of TBEV circulation in locations where this virus has not been previously found (Haglund, 2002; Skarpaas et al., 2004). The increase in the endemic range of TBE is thought to be associated with the increased range of *I. ricinus*, the principal vector for TBEV in Europe and *I. persulcatus*, the vector for Far Eastern subtype TBEV

TABLE 38.2 Incidence of TBEV in Europe and Russia (1990–2006)

	Year																
Country	**1990**	**1991**	**1992**	**1993**	**1994**	**1995**	**1996**	**1997**	**1998**	**1999**	**2000**	**2001**	**2002**	**2003**	**2004**	**2005**	**2006**
Austria	89	128	84	102	178	109	128	99	62	41	60	54	60	82	54	100	84
Belarus			2	20	50	66	97	67	78	26	23	61	18	25			
Croatia	23	60	27	76	87	59	57	25	24	26	18	27	30				20
Czech Republic	193	356	338	629	613	744	571	415	422	490	719	411	647	606	500	642	1113
Denmark									1	4	3	1	1	4	8	4	
Estonia	37	68	163	166	177	175	177	404	387	185	272	215	90	237	182	164	171
Finland	9		14	25	16	23	10	19	17	12	41	33	38	16	31	17	18
France	2	1	2	5	4	6	1	1	2	5	0	0	2	6	7	0	
Germany		44	142	118	306	226	114	211	148	115	133	253	226	278	274	431	547
Hungary	222	288	206	329	258	234	224	99	84	51	45	76	80	114	59	90	115
Italy			2	2	8	6	8	8	11	5	15	19	6	14	23	22	14
Latvia	122	227	287	791	1366	1341	716	874	1029	350	544	303	153	365	251	142	171
Lithuania	9	14	17	198	284	426	309	645	548	171	419	298	168	763	425	242	462
Norway									1	1	2	1	2	1	3	0	5
Poland	8	4	8	249	181	267	257	201	209	101	170	205	126	339	262	174	316
Slovak Republic	14	24	16	51	60	89	101	76	54	57	92	76	62	74	70	28	
Slovenia	235	245	210	194	492	260	406	274	136	150	190	260	262	275	204	297	372
Sweden	54	75	83	51	116	68	44	76	64	53	133	128	105	105	160	130	163
Switzerland	26	37	66	44	97	60	62	123	68	112	91	107	53	116	138	206	259
Total-Europe	1043	1571	1667	3050	4293	4159	3282	3617	3345	1955	2970	2528	2129	3420	2651	2689	3810
Russia	6529	6796	7968	10,943	9889	10,141	12,830	10,156	10,332	11,910	8901	8867	7279	8190	6886	7140	7320

Source: Adapted from: http://www.tbe-info.com/upload/medialibrary/Reported_Cases.pdf.

(Haglund, 2002; Johan et al., 2006; Lindgren and Gustafson, 2001; Skarpaas et al., 2004, 2006; Skarphedinsson et al., 2005). However, studies of climate change and its correlation with incidence of TBE suggest that climate change specifically may not be associated with the northern migration of TBE (Randolph, 2004; Rogers and Randolph, 2006). However, the increasing range of TBEV is important as new endemic foci have naïve populations and do not yet have extensive vaccination programs in place.

TBEV is transmitted to humans via the bite of infected ticks. The tick life cycle consists of three distinct phases: larvae, nymph, and adult. Ticks in all three phases have been shown to have the capacity to transmit TBEV although only the larvae and nymphs are likely to play a significant role in the TBEV life cycle (Nuttall and Labuda, 2005). TBEV has also been shown to be transmitted transstadially (Nuttall and Labuda, 2005) and transovarially (Rehacek, 1962). TBEV transmission cycles between infected ticks and small mammals or birds, which are the principal hosts for immature ticks. Adult ticks typically feed on goats, deer, or other ungulates that are not typically susceptible to TBEV and, subsequently, do not play a significant role in virus transmission (Nuttall and Labuda, 2005).

Potential as Biothreat Agent

TBEV is considered a potential biothreat agent largely due to its pathogenicity, ability to be transmitted via aerosol, lack of therapeutic treatment, and limited vaccine coverage. While these parameters make TBEV a highly dangerous virus, the inability of TBEV to be transmitted by human-to-human contact, the instability of the virus in UV light, and the lack of a suitable arthropod vector in many parts of the world would make extensive outbreaks of this virus difficult to engineer. TBEV is relatively stable when kept in cool conditions and it appears to also remain active in liquids for extended periods of time (Kharitonova and Leonov, 1985). There has also been evidence, based on accidental laboratory exposures, of aerosol transmission of TBEV (Scherer et al., 1980); however, conclusive studies have not been published. TBEV can also clearly survive the stomach and gut since the ingestion of contaminated milk has long been recognized as a means of transmission (Gritsun et al., 2003b; Moritsch and Kovac, 1960; Pogodina, 1958, 1960).

CLINICAL DISEASE

As discussed earlier, TBEV infection can result in diseases with varying severity depending on the subtype of TBEV involved. Western subtype TBEV generally causes a milder biphasic disease while Far Eastern subtype typically causes a much more severe disease with paresis and neurological sequelae more common and a higher mortality rate. The typical course of TBE can last up to 8 weeks from the time of infection with more severe cases lasting considerably longer. There has also been evidence of latent or persistent infections due to Far Eastern or Siberian subtype TBEV in both humans and nonhuman primates (Frolova and Pogodina, 1984; Gritsun et al., 2003a, 2003b; Pogodina et al., 1981a).

Western subtype TBEV and Far Eastern subtype TBEV cause clinically distinct diseases. Western subtype TBEV is typically characterized as a biphasic disease. However, in approximately 65% of persons infected with Western subtype TBEV, the infection remains inactive or ends after the first phase of disease. Of the remaining 35% of patients, the majority have a disease that follows the typical biphasic disease profile and progresses to the much more severe second stage characterized by neurological symptoms. In some cases, the first phase of disease is very mild, or asymptomatic, with the onset of clinical illness corresponding to the second, severe, phase of disease (Gritsun et al., 2003b; Reisner, 1981). Following infection, Western subtype TBEV typically has an incubation period that lasts from 2 to 18 days (Gritsun et al., 2003b). There is no indication of disease during the incubation period. Following the incubation period, disease onset can be rapid with nonspecific systemic signs of disease predominating. These signs include headache, fatigue, myalgia, nausea, general malaise, and fever with body temperatures rising to more than 38°C. This first stage of disease generally lasts for 1–8 days and corresponds to the viremic phase. The first phase of disease is followed by an afebrile period that can last from 1 to 20 days. Symptoms of disease generally subside during this period. The second phase of disease begins with a rapid onset of symptoms following the afebrile phase. Manifestations are generally similar to the first phase of disease, but are more severe with extremely high fevers and significant evidence of neurological involvement. Patients typically develop meningitis, meningoencephalitis, meningoencephalomyelitis, or meningoencephaloradiculitis (Gritsun et al., 2003b). The meningitis usually resolves and the case fatality rate is 1–2%. Some patients who have more severe disease can develop long-term sequelae; however, long-term health effects of the infection are generally minor. There is also evidence of age-dependence in disease severity whereby children have less severe disease and sequelae than adults, especially the elderly (Krausler, 1981; Kunz, 1974; Logar et al., 2000; Messner, 1981).

In persons infected with a Far Eastern strain of TBEV, the incubation period is similar and onset is sudden. Symptoms can include headache, high fever, vomiting, myalgia, photophobia, and sensorial and visual changes. Prodromal symptoms are rare. Patients also develop signs of neurological involvement such as paresis, paralysis, seizures, and coma over the course of disease (Burke and Monath, 2001; Clarke and Casals, 1965). Paralysis normally begins in the upper extremities including the shoulders and neck and is accompanied by hyperesthesia. Disease can be complicated by flaccid lower motor neuron paralysis where evidence of ascending paralysis or hemiparesis is a development that results in a poor prognosis (Clarke and Casals, 1965). A poliomyelitis-like illness has also been described (Silber and Soloviev, 1946). The case fatality rate is estimated at 20–30%. Unlike infection with Western subtype TBEV, neurological sequelae are much more common in individuals infected with the Far Eastern subtype virus. These include atrophy and paresis of the brachial plexus and neck muscles, paresis of the lower extremities, and other sequelae resembling poliomyelitis. As mentioned above, TBEV has also been shown to cause persistent or chronic disease, although only the intermediately pathogenic Siberian subtype has clearly been associated with this phenomenon (Gritsun et al., 2003a, 2003b; Pogany et al., 2003; Pogodina et al., 1981a, 1981b, 1981c).

TREATMENT

There are currently no approved therapeutic interventions for TBE. The onset of encephalitis or meningitis requires supportive care in an intensive care ward where treatment protocols are general and depend on the severity of disease. Corticosteroids may be used in certain circumstances to reduce inflammation while specific treatments to reduce or prevent seizures may also be provided if symptoms arise. General pain medications and sedatives may be used. Intubation and ventilation support may also be required.

A few experimental therapies for TBE have been attempted in animal model systems. These include the use of phosprenyl, an immunomodulatory compound (Danilov et al., 1996), the interferon inducers Poly I: C and tilorone hydrochloride (Hofmann and Kunz, 1972) and amiphenosine (Libikova, 1968). None of these therapies have been included in clinical trials.

PATHOGENESIS

Description of Disease Process

The exact mechanism of the disease process during TBEV infection has not yet been determined, but general hypotheses can be drawn based on clinical data, and data from other neurotropic flaviviruses and arboviruses.

Virus is introduced via tick bite and thought to replicate locally where it may either infect or be phagocytosed by resident macrophages or Langerhans cells. These cells are then believed to migrate to draining lymph nodes where further replication occurs. Replication in the lymph nodes leads to release of the virus into the lymphatic system and subsequent release into venous blood flow where the virus is distributed systemically. It is also possible that virus could penetrate endothelial cells lining lymph vessels to allow infection of surrounding tissues.

TBEV clearly targets neural tissues, but it is not clear how the virus penetrates the blood-brain barrier (BBB). Several hypotheses have been proposed for penetration of the BBB by neurotropic flaviviruses, but none have been conclusively proven. These include migration across or through endothelial cells (Dropulic and Masters, 1990; Liou and Hsu, 1998), the use of toll-like receptors (Wang et al., 2004), and migration across the olfactory epithelium (McMinn et al., 1996; Monath et al., 1983). There have also been suggestions for the point of viral penetration of the BBB through damaged endothelial cells (Lustig et al., 1992). Given the presence of encephalitis or meningitis in afflicted individuals, it seems reasonable to assume that damage to vascular endothelial cells and migration of inflammatory cells may at least contribute to viral introduction into the brain; however, it is not clear whether inflammation is a response to virus that is already in the CNS.

As discussed above, Far Eastern subtype TBEV frequently causes neurological damage to peripheral neurons, particularly in the brachial plexus. These data suggest the possibility of peripheral penetration of the CNS and subsequent migration of virus to the brain in cerebral spinal fluid. The use of magnetic resonance imaging (MRI) has recently proven useful in the detection of damage to the spinal cord and brain during some acute TBEV infections. MRI has identified damage to a range of tissues including anterior horns of cervical and thoracic vertebrae, lumbar nerve roots, basal ganglia, and thalamus in more severe cases (Alkadhi and Kollias, 2000; Bender et al., 2005; Pfefferkorn et al., 2007; Stich et al., 2007), but has also shown that the majority of afflicted persons have normal MRI results (Alkadhi and Kollias, 2000).

Studies examining the localization of Western subtype TBEV viral antigen in historical postmortem samples have identified a fairly specific distribution of viral antigen in the brain. TBEV appears to target large neurons of the anterior pons, medulla oblongata, dentate nucleus, Purkinje cells, and striatum (Gelpi et al., 2005). This study also found that infiltration of cytotoxic T cells was a common result of infection. These data and other evidence suggested that the CD8 positive T cells may contribute to the loss of nerve cells (Gelpi et al., 2005).

PROTECTIVE IMMUNE RESPONSE

The immune response to TBEV infection includes all aspects of host immunity from a strong innate inflammatory response to activation of cell-mediated immunity (CMI) and humoral immunity (HI). Evidence suggests that a cell-mediated response is required for effective clearance of virus, but CMI is also thought to be at least partially responsible for neural damage during severe cases of disease (Gelpi et al., 2005, 2006). HI also plays a protective role during TBEV infection as anti-TBEV IgM and IgG levels in both serum and the CNS are elevated during infection (Gunther et al., 1997; Kaiser and Holzmann, 2000). The vast majority of studies examining the immune response to TBEV have been performed in vaccinated individuals. In these studies, both CMI and HI play a significant role during vaccination. The majority of TBEV vaccines tested to date are non-replicating vaccines consisting of inactivated virus, subviral particles, or viral subunits. In all cases, it is evident that the stimulation of neutralizing antibodies is correlated with protective immunity. As with other virus diseases, the stimulation of CMI is more closely associated with viral replication; hence, a significant cell-mediated response is not usually seen following vaccination with inactivated virus.

Humoral Immunity

The role of humoral immunity in the protection against infection by flaviviruses, including TBEV, is well established. All currently licensed vaccines have been evaluated for their efficacy by determination of the neutralizing antibody titer induced by the vaccine. The development of neutralizing antibodies is the clinical correlate of protection of a successful vaccine. What is less well understood is the role of humoral immunity in the protection of naïve individuals during a natural infection with TBEV. Very few studies have examined this aspect of disease carefully in humans with studies on convalescent patients making up the majority of these studies. It is clear that the production of virus-specific IgM and IgG occurs during manifest infections and that both are found in serum and the central nervous system (CNS). IgM levels are presumably high during the prodromal stage of disease (8–14 days following infection). IgM levels wane over the course of disease, but are still detectable in some patients at 4–10 months post-infection (Hofmann et al., 1983). It is not clear why some patients would maintain elevated IgM levels.

A majority of patients with TBE have been shown to have elevated IgM and IgG levels in their CSF at presentation with patients demonstrating intrathecal production of IgM and IgG at 15 days post-presentation (Kaiser and Holzmann, 2000). However, it has been shown that the presence of IgM or IgG in the CSF is not a prognostic indicator for the outcome of disease (Gunther et al., 1997; Kaiser and Holzmann, 2000). Patients with low levels of serum IgG (Kaiser and Holzmann, 2000) or CSF IgM (Gunther et al., 1997) early in disease may develop a more severe illness due to an inability to rapidly clear the virus and allow an increased viremia. Individuals with manifest TBE develop high neutralizing titers of IgG with a peak toward the end of the disease phase. IgG titers wane somewhat, but remain elevated (Klockmann et al., 1991). As there are no documented cases of secondary infections with TBEV it is assumed that protective antibody titers remain elevated for life.

The demonstration of virus-specific IgM or IgG is the principal diagnostic tool used for demonstrating a positive TBEV infection. ELISAs represent the primary assays used for the determination of TBEV infection while hemagglutination inhibition assays have also been used as a diagnostic tool. However, neutralization assays are considered the best method for demonstrating the induction of a protective antibody response against TBEV as previous exposure to flaviviruses (e.g., yellow fever vaccine) can confound results from ELISA and hemagglutination inhibition assays (Holzmann et al., 1996). The limitation of neutralization assays is the necessity of working with live virus. TBEV is considered a BSL-3 or BSL-4 virus in most parts of the world; hence, it is not readily available in clinical diagnostic laboratories.

The development of protective antibodies is the basis for successful vaccine development (see below for more extensive discussion). Studies in humans have measured IgM and IgG levels in vaccinated mice and found that neutralizing antibodies are produced following the initial vaccination, but that the titer is significantly increased following secondary and tertiary vaccination. The levels of neutralizing antibodies in humans remain high for several years post-vaccination with early studies indicating that positive hemagglutination titers fell below 70% at 3–4 years after the final boost (Kunz et al., 1991). A more recent study found that some vaccinated individuals were seropositive more than 8 years following their final vaccination, suggesting that the 3-year requirement for boosters could be extended, at least in healthy adults under 50 years old (Hainz et al., 2005; Rendi-Wagner et al., 2004b). However, in elderly persons serum antibody titers wane more quickly, so that shorter booster intervals are indicated in order to maintain protection (Hainz et al., 2005).

Cell-Mediated Immunity

The role of CMI in TBE is unclear at the present time. There have been very few studies that have examined CMI in TBE patients and most point to the CMI as being more detrimental to the host, via CD8-mediated killing of neurons, than beneficial. One study examined the proliferative ability of T-cells isolated from infected individuals and found that a depression of the proliferative response of T-cells led to more severe disease (Shilov and Ryzhaenkov, 2000). Studies examining the CSF of TBE patients have found that an increased cell counts in the CSF are common and that these cells are predominantly lymphocytes (Logar et al., 2000; Ziebarth-Schroth, 1972). There is also evidence that increased CNS cell counts are correlated with more severe disease (Kaiser and Holzmann, 2000). Histopathology studies have also suggested that CD8+ cells may be associated with damage to neurons (Gelpi et al., 2005). Indirect examination of the role of CMI in the CNS suggests that a significant T cell response is likely, as the chemokines CXCL10 and CXCL11 were both elevated in infected individuals (Lepej et al., 2007). CXCL10 and CXCL11 are IFN-γ responsive chemokines that are associated with T-cell recruitment.

As discussed above, the production of a strong neutralizing antibody response to vaccination is the hallmark of successful vaccination. The cell-mediated response is typically less robust as viral replication appears to be required for a strong T-cell response and current vaccines are inactivated non-replicating viruses. In order to stimulate a more profound T-cell response, a number of replication competent vaccines have been developed. Examination of the cellular response to replication-competent vaccines found induction of Th1 type (CD8+)-mediated immunity (Aberle et al., 2005) that may be associated with viral nonstructural proteins (Gomez et al., 2003) while viral C and E proteins are associated with induction of a CD4+ response (Gomez et al., 2003). There remains a number of unknowns regarding the induction of a strong cell-mediated response to vaccination, including the development of a memory response and the potential induction of a detrimental inflammatory response.

While replication-competent vaccines appear to be the best approach for generating a more complete and long-lasting immune response to vaccination, there remains a significant concern about the potential production of infectious virus from these systems.

HISTORY OF TBE VACCINES

Tissue culture–derived TBE vaccines were first developed in 1960 in Russia (Levkovich et al., 1960), and field trials have thereafter demonstrated the efficacy (Chumakov et al., 1963). In 1971, a collaborative effort between the Institute of Virology at the University of Vienna, Austria, and the Microbiological Research Establishment at Porton Down, United Kingdom, started with the aim of developing a vaccine to protect against TBE. An Austrian isolate (strain Neudörfl) of the Western TBEV subtype, subcloned in specific pathogen-free (SPF) chicken embryo cells was used to produce the vaccine antigen. The initial vaccine contained virus grown in suspensions of primary SPF chicken embryo cells, clarified by centrifugation and purified by hydroxylapatite chromatography after formalin inactivation (Kunz, 2003). The vaccine was stabilized with human albumin and further contained aluminum hydroxide as adjuvant. Between 1973 and 1975, this candidate vaccine was administered to about 30,000 people. The Austrian company Immuno took over the vaccine from that point and started commercial marketing of the product in 1976 under the name FSME-IMMUN®. More than 400,000 people received the first generation of this vaccine in Austria and high seroconversion rates (greater than 90% by hemagglutination inhibition test) were observed after two doses (Kunz et al., 1976). Due to a relatively quick decline in antibodies after the second dose, a third immunization was indicated after 9–12 months (Barrett et al., 2003). This vaccine showed a concerning rate of local and systemic side effects, such as headache, malaise, and pyrexia. Since contaminating proteins were assumed to be their cause, continuous-flow zonal centrifugation of the inactivated virus was introduced in 1979 to purify the antigen (Heinz et al., 1980). The new product had a significantly improved purity level, showed good immunogenicity and a reduced level of side effects (Kunz et al., 1980). In the late 1990s the manufacturing process was further modified (Barrett et al., 2003) by changing the seed virus used to inoculate the primary chicken embryo cells from mouse brain–derived material to seed stocks passed twice in chicken embryo cell culture. This passage history eliminated the potential risk of carrying contaminating mouse brain protein into the final product. Further modifications at that point were the elimination of thiomersal and human serum albumin (used as stabilizer) from the formulation. This new formulation was named TicoVac®. Unexpectedly, the elimination of stabilizer led to an increase in adverse events, mostly high fever in infants and young children (Marth and Kleinhappl, 2001). Subsequently, FSME-IMMUN® containing human serum albumin as stabilizer was reintroduced in 2001 in dosages for adult and pediatric applications.

A second European product (Encepur) was developed by Behring-Werke, Marburg (Germany), in the late 1980s using the K-23 strain, a Western subtype isolate from Southern Germany, and reached the market in 1992 (Girgsdies and Rosenkranz, 1996). A pediatric formulation was developed and released in 1995 due to the high incidence of adverse reactions observed in children, especially after the first dose. The Encepur vaccines are produced using essentially the same process used for production of FSME-IMMUN®, the only differences are that virus seed of K-23 was always produced in chicken embryo cells and different stabilizers were applied.

The first vaccine tested in the early 1960s to control KFD in the area of Shimoga district, Mysore state of India, was a 5–10% suspension of formalin-inactivated RSSEV (Far Eastern subtype) (mouse brain preparation) produced by Walter Reed Army Institute of Research laboratory, Washington, D.C., United States. It was injected subcutaneously, two doses a week apart followed by a third dose five weeks after the second. About 4000 vaccinees displayed no unwanted reaction such as allergic and febrile reactions (Aniker et al., 1962). The vaccine induced weak hemagglutination inhibition antibody response but stimulated no complement fixation antibody response. The vaccine also failed to evoke boost responses in many individuals with previous KFDV infection. The RSSEV-based vaccine was therefore found to be ineffective in reducing the attack rate of KFD or in modifying the disease course (Pavri et al., 1962; Shah et al., 1962). The first effort to produce a KFDV-based vaccine was made by growing KFDV in brains of infant mice and subsequent inactivation by formalin. The vaccine retained potency up to six months if it was stored in a refrigerator and induced neutralizing antibodies in mice (Mansharamani et al., 1965). Another experimental KFDV vaccine was also produced by growing the virus in chick embryo cells, but the product was poorly immunogenic and failed to evoke neutralizing antibody responses in mice (Dandawate et al., 1965a, 1965b). In 1966, another formalized experimental vaccine was successfully prepared by growing KFDV in chick embryo fibroblast cell cultures. The vaccine was found to be immunogenic, potent, stable, and safe (Mansharamani and Dandawate, 1967; Mansharamani et al., 1967). Field trials with this tissue culture vaccine yielded acceptable results, as they induced neutralizing antibodies in 50% of the vaccinees and stimulated protective responses in ~23% of vaccinees (Banerjee et al., 1969). Efforts were also made to develop a live-attenuated KFDV vaccine by attenuating the strain P9605 through serial tissue culture passages. Langurs (*Presbytis entellus*) vaccinated with this attenuated KFDV generated neutralizing antibodies and survived challenge (Bhatt and Anderson, 1971). In another experiment, two dosages of formalin-inactivated KFDV administered to langurs by subcutaneous route induced neutralizing antibodies. The antibodies, though transient, could be detected up to 15 months after administration of the first dose of vaccine. Although the vaccine did not completely protect the langurs from challenge, it prevented death (Bhatt and Dandawate, 1974). A surveillance study indicated that the formalized KFDV vaccine has some, but not absolute, protective effect on vaccinees in Sagar-Sorab taluks of Shimoga district (Upadhyaya et al., 1979).

LGTV is a member of the TBE serocomplex isolated in Malaysia, which is a BSL-2 virus. An attenuated LGTV vaccine based on the E5 strain was also tested and showed protection against KFDV (Thind, 1981). Kayser et al. (1985) studied the human antibody response to immunization with 17D YFV and inactivated TBEV vaccine. They reported that vaccines producing hemagglutination inhibition titers >20 against TBEV showed cross-reaction with KFD. The formalin-inactivated KFDV vaccine produced in chick embryo fibroblasts has been licensed and is currently in use in the endemic areas in Karnataka state of India. The locations for vaccination were selected on the basis of prevalence of KFDV activity in the previous years, including the villages from which mortality in monkeys was reported and those adjacent to the KFDV-affected areas. The efficacy of the vaccine was satisfactory, exerting a highly significant protective effect (Dandawate, 1994). In outbreak situations, almost all individuals, including children, are being routinely vaccinated by the local government authorities. However, the occurrence of KFD cases, despite vaccination, suggests some changes in virus antigenic determinants have occurred. The KFDV strain currently used for vaccine preparation was isolated late in the 1950s. Thereafter, KFDV strains have not been well characterized at the molecular level; hence, possible antigenic drift and diversity since its first introduction in India during the late 1950s remain poorly understood. The trend of increasing KFD cases in Karnataka state warrants development of a new vaccine preparation that includes currently circulating KFDV. Improper storage of vaccine and lack of maintenance of cold chain resulting in inactivation of the vaccine could be another reason for the vaccine failures.

CURRENT LICENSED VACCINES

At least four formalin-inactivated TBE vaccines are currently available, derived from virus grown in chick embryo fibroblasts, inactivated by formaldehyde and purified by either chromatography or continuous flow centrifugation (Table 38.3). These vaccines are administered along a two-dose schedule followed by a booster immunization at one year and recommended booster injections every 3–5 years. They are produced in Austria (Baxter Vaccine; previously Immuno AG), Germany (Novartis, previously Beringwerke), and the Russian Federation (Institute of Poliomyelitis and Viral Encephalitides and Virion Company). Active surveillance in Austria has demonstrated a dramatic decline in the incidence of TBE in vaccinated groups, with an estimated vaccine efficacy of at least 96% (Kunz, 2003). The main reported side effect with currently available vaccines is post-vaccination fever and allergic reactions in children. A series of improvements to the available vaccines were introduced to reduce their reactogenicity. In Germany, the vaccine is widely used to immunize children in high-risk areas. The Russian vaccines also induce high seroconversion rates and are believed to be highly effective, albeit safety data are not readily available (Table 38.3).

An inactivated virus vaccine for KFD was developed many years ago for use in India (see above)

TABLE 38.3 Current Licensed TBE vaccines

	FSME-IMMUN®	Encepur ™	Encevir	TBE vaccine
Manufacturer	Baxter Vaccines, Vienna, Austria	Novartis Vaccines and Diagnostics, Germany	Virion Corporation, Tomsk, Russia	Chumakov Institute for Poliomyelitis and Viral Encephalitides, Moscow, Russia
Virus strain (subtype)	Neudörfl (Western subtype)	K-23 (Western subtype)	205 (Far Eastern subtype)	Sofjin (Far Eastern subtype)
Production method	Cultured in primary chicken embryo cells, purified after formaldehyde inactivation by continuous-flow zonal centrifugation	Cultured in primary chicken embryo cells, purified after formaldehyde inactivation by continuous-flow zonal centrifugation	Cultured in primary chicken embryo cells, purified and concentrated after formaldehyde inactivation	Cultured in primary chicken embryo cells, purified and concentrated after formaldehyde inactivation
Excipients	Aluminum hydroxide, human serum albumin	Aluminum hydroxide, sucrose	?	?
Countries licensed	Austria, Germany, Switzerland, Hungary, Czech Republic, Baltic States, UK, Canada	Germany, Austria, Switzerland, Czech Republic, Baltic States, Russia	Russia	Russia
Pediatric vaccine	1–<16 years old	1–<12 years old	Same as adults	Same as adults
Vaccination schedule (conventional)	0, 1–3 months, 6–15 months Booster doses for adults as per 2005 Austrian Immunization Plan <60 years: first booster after 3 years, subsequently 5-year intervals ≥60 years: 3-year-interval booster doses for children per manufacturer: 3-year intervals	0, 1–3 months, 9–12 months Booster doses for adults as per 2005 Austrian Immunization Plan <60 years: first booster after 3 years, subsequently 5-year intervals ≥60 years: 3-year-interval booster doses for children per manufacturer: 3-year intervals	3 doses	3 doses
Vaccination schedule (accelerated)	Day 0, day 14, 6–15 months Booster doses: as above	Day 0, 7, 21 First booster at 12–18 months, subsequent booster doses as above	NA	NA
Immunogenicity (seroconversion) (conventional schedule)	Adults: 92.9–97% after second dose, 100% after third dose Children: 98.5–100% after second dose, 100% after third dose	Adults: 100% after second dose Children: clinical trials not available	NA	NA
Immunogenicity (seroconversion) (accelerated schedule)	Adults: clinical trials not available Children: 95% after second dose	Adults: 100% after primary series (3 doses) Children: 100% after primary series (3 doses)	NA	NA
Safety	Mild–moderate systemic and local reactions common Fever in very young children common Fever in older children occasional Fever in adults infrequent Severe neurologic reactions rare		NA	NA

(Dandawate et al., 1965a, 1965b, 1980; Mansharamani and Dandawate, 1967; Mansharamani et al., 1965, 1967; Upadhyaya et al., 1979). The vaccine called anti-KFD is produced by the Virus Diagnostic Laboratory (VDL) of the health department of the State government of Karnataka (Shimoga, India). This vaccine is not particularly efficacious and boosters are required annually. Annual epidemics still occur in the endemic area with several hundred cases reported every year (Pattnaik, 2006).

VACCINES IN DEVELOPMENT

Rationale for Next-Generation Vaccines

Issues with Current Vaccines

Safety Since the development of the first TBE vaccines in Europe in the early 1970s (Barrett et al., 2003), a number of safety issues have been observed and new formulations introduced aimed at addressing these issues. As stated above, in 2000, a new, albumin-free formulation of TBE vaccine (Ticovac) was released by Baxter. This formulation elicited severe vaccination side effects in the form of high fever in more than 1% of vaccinated children. As a consequence, in Germany the vaccine use was restricted to individuals 12 years and older. Recently, a lower dosage (50% of the adult dose) with human albumin as a stabilizer was approved in children under the age of 12 after studies showed that this formulation does not induce fever (Marth and Kleinhappl, 2001). A study of the other commercial TBE vaccine Encepur®, established that doses of antigen administered to children should be at least twofold lower than those administered to adults to reduce the risk of fever above 38°C (Girgsdies and Rosenkranz, 1996). Most of the side effects seem to be attributable to the ability of inactivated TBEV particles in combination with alum to stimulate a strong cytokine response (TNF-α and IL-1β) (Marth and Kleinhappl, 2001). Polygeline was used as a proteineous stabilizer in former vaccines causing a relatively high rate of allergic reactions, especially in children. A new formulation has recently proven to show a lower incidence of side effects (Zent and Hennig, 2004). In a study using current (adult) formulations, local reactions were observed in 33–45, systemic reactions in 13–38, and fever in 0.2–9% of the individuals in study groups after receiving their first dose of vaccine (Loew-Baselli et al., 2006). Thus, while the new formulations appear to be safer than their previous counterparts, safety remains a concern. Due to the risks associated with TBE vaccinations [between 2001 and 2005, approximately 1% of vaccinees experienced serious adverse reactions (source: Baxter Product monograph FSME-IMMUN)], vaccination and re-vaccination rates in some endemic areas are lower than one would expect based on virus prevalence, disease severity, and incidence rates, especially in children.

In addition to clinical safety concerns, there are also issues related to vaccine production. Large-scale growth of pathogenic viruses requiring BSL-3 containment adds significantly to the challenges and cost of manufacturing the existing inactivated vaccines.

Multiple Doses All current vaccines require at least three doses to achieve sufficient protective efficacy, followed by another booster after 1–3 years (Beran, 2005; Schondorf et al., 2007). Even when the recommended vaccination schedule is followed, not all subjects develop protective titers of neutralizing antibodies (Ehrlich et al., 2003). Cases of vaccine breakthrough are well documented in the literature, showing the importance of sufficient neutralizing antibody titers (Bender et al., 2004; Kleiter et al., 2007). Shortening of the time intervals between initial vaccinations (accelerated schedule) compromises the magnitude of maximal antibody levels and the duration of protective immunity (Zent and Broker, 2005). Thus, induction of protective efficacy with the currently available products requires several doses and an extended period of time—an abbreviated dosing schedule would therefore represent a significant improvement.

Frequent Boosters The currently available vaccines normally require boosters every 3 years. In addition, clinical data suggest that severe disease outcome in TBE patients can be correlated with lower neutralizing titers (Kaiser and Holzmann, 2000) showing the importance of maintaining sufficient neutralizing antibody titers against TBE viruses. Passive protection by serum transfer was previously used as a treatment of TBE infections in humans, but is not available anymore since there were indications that this treatment can lead to severe disease outcome (Waldvogel et al., 1996). This might be linked to the fact that success with serum transfer is only successful if it occurs before the virus enters the CNS (Nalca et al., 2003). This suggests that in order to properly protect individuals over the long term with the vaccines currently available, regular boosting is required to maintain appropriate serum titers of neutralizing antibodies. Increased durability of immune response and protective efficacy would represent a distinct improvement to the existing vaccines.

Lower Efficacy in the Elderly Population Recent data indicate that it may be acceptable for individuals up to 49 years old to extend the booster interval with the currently available vaccines to 5 years (Beran, 2005). For older persons boosters are still required every 3 years (Rendi-Wagner et al., 2004a). This is due to lower virus neutralizing titers observed in elderly persons, especially after only completing the initial vaccination course (3 doses) (Rendi-Wagner et al., 2004a). Hainz et al. (2005) described a study of vaccine efficacy, which showed that 1–2 years after the last vaccination, 16% in a group of persons over 60 years old had TBE-specific ELISA titers below the recommended levels (<100VIE units). Among individuals of the same age group who were last vaccinated 2–3 years previously, 27% tested negative. Waning antibody titers potentially compromising efficacy, seem to be particularly problematic in the immunosenescent population.

Protection Range The commercially available vaccines are only indicated for protection against a single TBE subtype (CEEV or RSSEV). While cross-neutralizing activity against Far Eastern strains has been demonstrated with current vaccines, the cross-reactive virus-neutralizing titers are lower (Hayasaka et al., 2001; Leonova et al., 2007). No current product claims protective efficacy against the even more distantly related viruses of the TBE complex such as POWV, OHFV, and KFDV. A vaccine which could protect against the spectrum of TBE complex viruses would provide a significant advantage over existing products—especially since no products of satisfactory quality are available for protection against KFDV and OHFV.

Other Approaches to Vaccine Development

Recent experimental approaches to produce novel TBEV vaccines include DNA vaccines (Schmaljohn et al., 1997) and recombinant vaccinia viruses (Dmitriev et al., 1996) (Table 38.4). Vaccinia strains expressing the segment of proteins from E to NS1 provided the best level of protection. Novel vaccines could also be based on attenuated strains of TBE viruses. Studies have shown that deletions in the 3′ noncoding region can lead to highly attenuated mutants in the mouse model (Mandl et al., 1998). Large deletions in the capsid protein seem to lead to attenuation by disturbing virus assembly (Kofler et al., 2002). An approach using chimeric viruses based on swapping the prM and E structural protein genes into a dengue-4 virus backbone with a 30 nucleotide deletion in the 3′ noncoding region (TBEV/DEN4Δ30 and LGT/DEN4) has shown good safety in monkey trials (Rumyantsev et al., 2006); vaccine efficacy was demonstrated against LGTV challenge, but suggests that further improvement might be necessary. Candidate vaccines based on purified, insect cell–expressed recombinant subunits have demonstrated broad protective potential, which included protection against OHFV in the mouse model.

CLINICAL TRIALS

At this time no clinical trials for novel vaccines are being reported. Post-marketing surveillance continues on the European formulations.

POST-EXPOSURE IMMUNOPROPHYLAXIS

In Russia, prevention of TBE with the use of specific immunoglobulins is still available (via a product from the Chumakov Institute of Poliomyelitis and Viral Encephalitides, Moscow). This treatment is effective when administered to tick-bitten persons within 3 days post-bite. This therapy is reported to be effective in 98% of cases in a curative context and 100% in a prophylactic context (i.e., administered to individuals before visiting natural foci of TBEV) (Charrel et al., 2004). In Europe, the two products based on TBE immunoglobulin [Encegam® (Behring) and FSME-bulin (Baxter)] were produced. However, they initially lost their license for administration to children aged up to 14 years because of a suspected association between post-exposure application of immunoglobulin and severe forms of the disease (Arras et al., 1996; Kluger et al., 1995; Waldvogel et al., 1996). Subsequently, both products were withdrawn completely from the market and are no longer available, even though the suspected antibody-dependent enhancement (demonstrated in vitro) could not be shown in vivo in the mouse model (Kreil and Eibl, 1997). It is likely that the remaining Russian product will have little future because of safety regulations regarding the use of human-derived products for treating humans.

Theoretically, active TBE vaccination could be considered as an alternative to post-exposure prophylaxis—but since increased reactogenicity has been associated with higher antigen doses, particularly at the beginning of the primary immunization course, it is unlikely that currently vaccines will be recommended for this use.

TABLE 38.4 TBE vaccines in development

Candidate	Description	Stage of development	Developer
DNA vaccine	DNA vectors expressing CEEV E or prM+E region of virus polyprotein	Some of the candidates were successful in achieving protective efficacy in mice. Gene gun application was more successful than injection	University of Vienna, Austria
DNA vaccine	DNA vectors expressing either CEEV or RSSEV prM+E region of virus polyprotein	Proof-of-concept in mouse and nonhuman primates successful with vaccine administered using particle bombardment technique	USAMRIID, Frederick, USA
DNA vaccine	DNA vectors expressing CEEV NS1 protein	DNA vaccine alone protected a significant portion of vaccinated mice from live virus challenge (Timofeev, 2004). A prime-boost approach applying a recombinant vaccinia vector expressing NS1 protein also showed a high level of protection against RSSEV (Aleshin, 2005)	Chumakov Institute of Poliomyelitis and Viral Encephalitides, Moscow, Russia
Recombinant TBE/vaccinia viruses	Recombinant vaccinia viruses were created incorporating various polyprotein segments from the capsid protein C to NS3	Candidates were successful in protecting mice from live virus challenge	VECTOR, Novosibirsk, Russia
Recombinant TBE/vaccinia viruses	Recombinant vaccinia vector was created expressing the prM+E section of CEEV	Good immunogenicity and efficacy was seen with sufficient doses in the mouse model (Holzer, 1999)	Baxter, Orth, Austria
Recombinant TBE/adenovirus	Adenovirus (Ad51) expressing NS1 protein	Vaccine candidate protected mice against challenge and potentiated immune responses observed if administered together with killed virus vaccine	Chumakov Institute of Poliomyelitis and Viral Encephalitides, Moscow, Russia
Recombinant TBE/dengue 4 viruses	Attenuated dengue-4 vaccine strain expressing Langat or RSSEV prM+E	Recombinant viruses are safe in immunodeficient mice and induce neutralizing antibody titers in mice and nonhuman primates. Protective efficacy against Langat virus was demonstrated in nonhuman primates	NIH/NIAID, Bethesda, USA
Attenuated Langat virus	Attenuated version selected after serial passages of LGT E5 in embryonated chicken eggs—strain LGT E5	A previous candidate (LGT TP21) caused neurological disease in 1/18,570 vaccinees in a Russian clinical trial (Smorodincev, 1986). The new candidate vaccine strain shows markedly reduced neurovirulence in SCID mice (Rumyantsev et al., 2006)	NIH/NIAID, Bethesda, USA
Attenuated TBE virus	Attenuation is achieved by intentional amino acid deletions in the capsid protein region	Attenuated virus has shown safety and immunogenicity in mice	University of Vienna, Austria
Peptides from TBE NS1 protein	Various peptides identified in the NS1 protein of RSSEV representing linear epitopes were tested	One peptide (aa 37–55) provided partial protection in the mouse model of TBE virus infection	Chumakov Institute of Poliomyelitis and Viral Encephalitides and Institute of Bioorganic chemistry (both Moscow, Russia)
Recombinant E subunits	Envelope subunit proteins of RSSEV and CEEV expressed by insect cells	Candidates have shown efficacy against RSSEV, CEEV, and OHFV in the mouse model	Hawaii Biotech, Aiea, USA
Recombinant subviral particles and other recombinant proteins	Various forms of recombinantly expressed or virus-derived subunits from CEEV	Recombinant subviral particles and rosettes showed good immunogenicity and efficacy in the mouse model, while soluble forms were much less immunogenic (Heinz et al., 1995)	University of Vienna, Austria

PROSPECTS FOR THE FUTURE

All current novel TBE vaccine approaches are still of mainly academic interest and have not reached the industrial development stage. It is unlikely that a recombinant subunit or live-attenuated TBEV vaccine will be licensed within the next 5 years.

Post-exposure treatment using TBE Ig is no longer recommended in almost all European countries and it is expected that during the next 5 years the recommendation to use Igs for post-exposure treatment will be revoked for the remaining European Union (EU) member states. Recommendations for post-exposure prophylaxis on the basis of active TBE immunization are not expected in the near future.

KEY ISSUES

- The endemic range of TBE viruses is increasing in multiple regions across the world.
- The increase in TBE case numbers is believed to be largely associated with an increase in outdoor activity and increased contact with the tick vectors.
- Although mortality rates associated with TBEV infections are not very high, the most serious effect is the high prevalence of long-term sequelae caused especially by the Far Eastern strains.
- Due to considerable stability, members of the TBE complex are considered to be serious candidates to be used as biowarfare agents that could be disseminated via various routes of exposure.
- The quality and quantity of available vaccine in many Eastern European countries is poor. Vaccine coverage in those endemic areas is well below the desirable range.
- No commercial vaccines are available to protect against "exotic" members of the TBE complex (Kyasanur Forest Disease and Alkhurma viruses).
- The cost of well-established and effective "Western" vaccines is high and as recently as in the 2007 season there has been a shortage in available doses.
- Vaccines require routine boosts to maintain efficacy, especially in the elderly population.

References

Aberle, J.H., Aberle, S.W., Kofler, R.M. and Mandl, C.W. Humoral and cellular immune response to RNA immunization with flavivirus replicons derived from tick-borne encephalitis virus. J. Virol. 2005; 79:15107–15113.

Aleshin, S.E., Timofeev, A.V., Khoretonenko, M.V., Zakharova, L.G., Pashvykina, G.V., Stephenson, J.R., Shneider, A.M., Altstein, A.D. Combined prime-boost vaccination against tick-borne encephalitis (TBE) using a recombinant vaccinia virus and a bacterial plasmid both expressing TBE virus non-structural NS1 protein. BMC Microbiol. 2005; 5:45.

Alkadhi, H. and Kollias, S.S. MRI in tick-borne encephalitis. Neuroradiology 2000; 42:753–755.

Aniker, S.P., Work, T.H., Chandrasekharaiya, T., Murthy, D.P., Rodrigues, F.M., Ahmed, R., Kulkarni, K.G., Rahman, S.H., Mansharamani, H. and Prasanna, H.A. The administration of formalin-inactivated RSSE virus vaccine in the Kyasanur Forest disease area of Shimoga District, Mysore State. Indian J. Med. Res. 1962; 50:147–152.

Arras, C., Fescharek, R. and Gregersen, J.P. Do specific hyperimmunoglobulins aggravate clinical course of tick-borne encephalitis? Lancet 1996; 347:1331.

Banerjee, K., Dandawate, C.N., Bhatt, P.N. and Rao, T.R. Serological response in humans to a formolized Kyasanur Forest disease vaccine. Indian J. Med. Res. 1969; 57:969–974.

Barrett, P.N., Schober-Bendixen, S. and Ehrlich, H.J. History of TBE vaccines. Vaccine 2003; 21(Suppl. 1):S41–S49.

Beasley, D.W. and Aaskov, J.G. Epitopes on the dengue 1 virus envelope protein recognized by neutralizing IgM monoclonal antibodies. Virology 2001; 279:447–458.

Beasley, D.W. and Barrett, A.D. Identification of neutralizing epitopes within structural domain III of the West Nile virus envelope protein. J. Virol. 2002; 76:13097–13100.

Bender, A., Jager, G., Scheuerer, W., Feddersen, B., Kaiser, R. and Pfister, H.W. Two severe cases of tick-borne encephalitis despite complete active vaccination—the significance of neutralizing antibodies. J. Neurol. 2004; 251:353–354.

Bender, A., Schulte-Altedorneburg, G., Walther, E.U. and Pfister, H.W. Severe tick borne encephalitis with simultaneous brain stem, bithalamic, and spinal cord involvement documented by MRI. J. Neurol. Neurosurg. Psychiatry 2005; 76:135–137.

Beran, J. Immunisation against tick-borne encephalitis by widely used vaccines: short-term history and current recommendations. Clin. Microbiol. Infect. 2005; 11:424–426.

Bhatt, P.N. and Anderson, C.R. Attenuation of a strain of Kyasanur Forest disease virus for mice. Indian J. Med. Res. 1971; 59:199–205.

Bhatt, P.N. and Dandawate, C.N. Studies on the antibody response of a formalin inactivated Kyasanur Forest Disease Virus vaccine in langurs "Presbytis entellus". Indian J. Med. Res. 1974; 62:820–826.

Blaskovic, D. Tick-borne encephalitis in Europe. Some aspects of the epidemiology and control of the disease. Ann. Soc. Belge Med. Trop. 1958; 38:867–883.

Bressanelli, S., Stiasny, K., Allison, S.L., Stura, E.A., Duquerroy, S., Lescar, J., Heinz, F.X. and Rey, F.A. Structure of a flavivirus envelope glycoprotein in its low-pH-induced membrane fusion conformation. EMBO J. 2004; 23:728–738.

Brown, R.N., Lane, R.S. and Dennis, D.T. Geographic distribution of tick-borne diseases and their vectors. In: Tick-Borne Diseases of Humans (J.L. Goodman, D.T. Dennis and D.E. Sonenshine, Eds.), pp. 363–392. Washington, D.C.: ASM Press, 2005.

Burke, D.S. and Monath, T.P. Flaviviruses. In: Field's Virology (D.M. Knipe and P.M. Howley, Eds.), pp. 1043–1125. Philadelphia: Lippincott Williams & Wilkins, 2001.

Calisher, C.H. and Gould, E.A. Taxonomy of the virus family Flaviviridae. Adv. Virus Res. 2003; 59:1–19.

Charrel, R.N., Attoui, H., Butenko, A.M., Clegg, J.C., Deubel, V., Frolova, T.V., Gould, E.A., Gritsun, T.S., Heinz, F.X., Labuda, M., Lashkevich, V.A., Loktev, V., Lundkvist, A., Lvov, D.V., Mandl, C.W., Niedrig, M., Papa, A., Petrov, V.S., Plyusnin, A., Randolph, S., Suss, J., Zlobin, V.I. and de Lamballerie, X. Tick-borne virus diseases of human interest in Europe. Clin. Microbiol. Infect. 2004; 10:1040–1055.

Chen, Y., Maguire, T., Hileman, R.E., Fromm, J.R., Esko, J.D., Linhardt, R.J. and Marks, R.M. Dengue virus infectivity depends on envelope protein binding to target cell heparan sulfate. Nat. Med. 1997; 3:866–871.

Chu, J.J., Rajamanonmani, R., Li, J., Bhuvanakantham, R., Lescar, J. and Ng, M.L. Inhibition of West Nile virus entry by using a recombinant domain III from the envelope glycoprotein. J. Gen. Virol. 2005; 86:405–412.

Chumakov, M.P. Results of the study made on Omsk hemorrhagic fever (OH) by an expedition of the Institute of Neurology. Vestn Akad Med Navk SSR 1948; 2:19–26.

Chumakov, M.P., Gagarina, A.V., Vilner, L.M., Khanina, M.K., Rodin, I.M., Vasenovich, M.I., Lakina, V.I. and Finogenova, E.V. Experience in the experimental production and control of tissue culture vaccine against tick encephalitis. Vopr. Virusol. 1963; 29:415–420.

Chumakov, M.P. and Gladkikh, S. On the role of Ixodidae in communicating spring-summer encephalitis. Bull. de Biol. et de Med. Exper. de l'URSS 1939; 7:229–238.

Chumakov, M.P., Zeitlenok, N.A., Varobieva, N.N., Shifrin, I.A., Orlova, N.N. and Beliaeva, A.P. Study of virus encephalitides. Geographical distribution and certain characteristics of the epidemiology of tick-borne encephalitis in man in Europe and part of the USSR, Siberia, Kazakhstan. Neuropathol. Psychiatry 1944; 13:20–23.

Clarke, D.H. and Casals, J. Arboviruses: Group B. In: Viral and Rickettsial Infections of Man (F.L. Horsfall, Jr. and J.B. I. Tamm, Eds.). Philadelphia: Lippincott, 1965.

Crill, W.D. and Roehrig, J.T. Monoclonal antibodies that bind to domain III of the dengue virus E glycoprotein are the most efficient blockers of virus adsorption to vero cells. J. Virol. 2001; 75:7769–7773.

Dandawate, C.N., Desai, G.B., Achar, T.R. and Banerjee, K. Field evaluation of formalin inactivated Kyasanur forest disease virus tissue culture vaccine in three districts of Karnataka state. Indian J. Med. Res. 1994; 99:152–158.

Dandawate, C.N., Mansharamani, H.J. and Jhala, H.I. Experimental vaccine against Kyasanur Forest Disease (KFD) virus from embryonated eggs. I. Adaptation of the virus to developing chick embryo and preparation of formolised vaccines. Indian J. Pathol. Bacteriol. 1965a; 8:241–260.

Dandawate, C.N., Mansharamani, H.J. and Jhala, H.I. Experimental vaccine against Kyasanur Forest Disease (KFD) virus from embryonated eggs. II. Comparative immunogenicity of vaccines inactivated with formalin and beta-propio-lactone (BPL). Indian J. Pathol. Bacteriol. 1965b; 8:261–270.

Dandawate, C.N., Upadhyaya, S. and Banerjee, K. Serological response to formolized Kyasanur forest disease virus vaccine in humans at Sagar and Sorab Talukas of Shimoga district. J. Biol. Stand. 1980; 8:1–6.

Danilov, L.L., Maltsev, S.D., Deyeva, A.V., Narovlyansky, A.N., Sanin, A.V., Ozherelkov, S.V. and Pronin, A.V. Phosprenyl: a novel drug with antiviral and immunomodulatory activity. Arch. Immunol. Ther. Exp. (Warsz) 1996; 44:395–400.

Dmitriev, I.P., Khromykh, A.A., Ignatyev, G.M., Gainullina, M.N., Ageenko, V.A., Dryga, S.A., Vorobyeva, M.S. and Sandakhchiev, L.S. Immunization with recombinant vaccinia viruses expressing structural and part of the nonstructural region of tick-borne encephalitis virus cDNA protect mice against lethal encephalitis. J. Biotechnol. 1996; 44:97–103.

Dropulic, B. and Masters, C.L. Entry of neurotropic arboviruses into the central nervous system: an in vitro study using mouse brain endothelium. J. Infect. Dis. 1990; 161:685–691.

Ehrlich, H.J., Pavlova, B.G., Fritsch, S., Poellabauer, E.M., Loew-Baselli, A., Obermann-Slupetzky, O., Maritsch, F., Cil, I., Dorner, F. and Barrett, P.N. Randomized, phase II dose-finding studies of a modified tick-borne encephalitis vaccine: evaluation of safety and immunogenicity. Vaccine 2003; 22:217–223.

Frolova, M.P. and Pogodina, V.V. Persistence of tick-borne encephalitis virus in monkeys. VI. Pathomorphology of chronic infection in central nervous system. Acta Virol. 1984; 28:232–239.

Gallia, F., Rampas, J. and Hollender, L. Laboratory infection with encephalitis virus. Cas. Lek. Ces. 1949; 88:224.

Gelpi, E., Preusser, M., Garzuly, F., Holzmann, H., Heinz, F.X. and Budka, H. Visualization of Central European tick-borne encephalitis infection in fatal human cases. J. Neuropathol. Exp. Neurol. 2005; 64:506–512.

Gelpi, E., Preusser, M., Laggner, U., Garzuly, F., Holzmann, H., Heinz, F.X. and Budka, H. Inflammatory response in human tick-borne encephalitis: analysis of postmortem brain tissue. J. Neurovirol. 2006; 12:322–327.

Girgsdies, O.E. and Rosenkranz, G. Tick-borne encephalitis: development of a paediatric vaccine. A controlled, randomized, double-blind and multicentre study. Vaccine 1996; 14:1421–1428.

Gomez, I., Marx, F., Saurwein-Teissl, M., Gould, E.A. and Grubeck-Loebenstein, B. Characterization of tick-borne encephalitis virus-specific human T lymphocyte responses by stimulation with structural TBEV proteins expressed in a recombinant baculovirus. Viral Immunol. 2003; 16:407–414.

Gordon Smith, C.E. A virus resembling Russian spring-summer encephalitis virus from an *Ixodid* in Malaya. Nature 1956; 178:581–582.

Gould, E.A., de Lamballerie, X., Zanotto, P.M. and Holmes, E.C. Evolution, epidemiology, and dispersal of flaviviruses revealed by molecular phylogenies. Adv. Virus Res. 2001; 57:71–103.

Gould, E.A., Moss, S.R. and Turner, S.L. Evolution and dispersal of encephalitic flaviviruses. Arch. Virol. Suppl. 2004; 18:65–84.

Gritsun, T.S., Frolova, T.V., Zhankov, A.I., Armesto, M., Turner, S.L., Frolova, M.P., Pogodina, V.V., Lashkevich, V.A. and Gould, E.A. Characterization of a Siberian virus isolated from a patient with progressive chronic tick-borne encephalitis. J. Virol. 2003a; 77:25–36.

Gritsun, T.S., Lashkevich, V.A. and Gould, E.A. Tick-borne encephalitis. Antiviral Res. 2003b; 57:129–146.

Gunther, G., Haglund, M., Lindquist, L., Skoldenberg, B. and Forsgren, M. Intrathecal IgM, IgA and IgG antibody response in tick-borne encephalitis. Long-term follow-up related to clinical course and outcome. Clin. Diagn. Virol. 1997; 8:17–29.

Haglund, M. Occurrence of TBE in areas previously considered being non-endemic: Scandinavian data generate an international study by the International Scientific Working Group for TBE (ISW-TBE). Int. J. Med. Microbiol. 2002; 291(Suppl. 33):50–54.

Hainz, U., Jenewein, B., Asch, E., Pfeiffer, K.P., Berger, P. and Grubeck-Loebenstein, B. Insufficient protection for healthy elderly adults by tetanus and TBE vaccines. Vaccine 2005; 23:3232–3235.

Hayasaka, D., Goto, A., Yoshii, K., Mizutani, T., Kariwa, H. and Takashima, I. Evaluation of European tick-borne encephalitis virus vaccine against recent Siberian and far-eastern subtype strains. Vaccine 2001; 19:4774–4779.

Heinz, F.X. and Allison, S.L. Structures and mechanisms in flavivirus fusion. Adv. Virus Res. 2000; 55:231–269.

Heinz, F.X. and Allison, S.L. The machinery for flavivirus fusion with host cell membranes. Curr. Opin. Microbiol. 2001; 4:450–455.

Heinz, F.X., Allison, S.L., Schalich, J., Holzmann, H., Mandl, C.W. and Kunz, C. Recombinant and virion-derived soluble and particulate immunogens for vaccination against tick-borne encephalitis. Vaccine. 1995; 13(17):1636–1642.

Heinz, F.X., Kunz, C. and Fauma, H. Preparation of a highly purified vaccine against tick-borne encephalitis by continuous flow zonal ultracentrifugation. J. Med. Virol. 1980; 6:213–221.

Hloucal, L. Tick-borne encephalitis as observed in Czechoslovakia. J. Trop. Med. 1960; 63:293–296.

Hofmann, H. and Kunz, C. The protective effect of the interferon inducers tilorone hydrochloride and Poly I:C on experimental tick-borne encephalitis in mice. Arch. Gesamte Virusforsch 1972; 37:262–266.

Hofmann, H., Kunz, C., Heinz, F.X. and Dippe, H. Detectability of IgM antibodies against TBE virus after natural infection and after vaccination. Infection 1983; 11:164–166.

Holzer, G.W., Remp, G., Pfleiderer, M., Enzersberger, O.M., Emsenhuber, W., Hämmerle, T., Gruber, F., Urban, C., Falkner, F. G. and Dorner, F. Highly efficient induction of protective immunity by a vaccinia virus vector defective in late gene expression. J. Virol. 1999; 73(6):4536–4542.

Holzmann, H., Kundi, M., Stiasny, K., Clement, J., McKenna, P., Kunz, C. and Heinz, F.X. Correlation between ELISA, hemagglutination inhibition, and neutralization tests after vaccination against tick-borne encephalitis. J. Med. Virol. 1996; 48:102–107.

Hung, J.J., Hsieh, M.T., Young, M.J., Kao, C.L., King, C.C. and Chang, W. An external loop region of domain III of dengue virus type 2 envelope protein is involved in serotype-specific binding to mosquito but not mammalian cells. J. Virol. 2004; 78:378–388.

Johan, F., Asa, L., Rolf, A., Barbro, C., Ingvar, E., Mats, H., Ake, L., Sirkka, V. and Bo, S. Tick-borne encephalitis (TBE) in Skane, southern Sweden: a new TBE endemic region? Scand. J. Infect. Dis. 2006; 38:800–804.

Kaiser, R. and Holzmann, H. Laboratory findings in tick-borne encephalitis—correlation with clinical outcome. Infection 2000; 28:78–84.

Kaufmann, B., Nybakken, G.E., Chipman, P.R., Zhang, W., Diamond, M.S., Fremont, D.H., Kuhn, R.J. and Rossmann, M.G. West Nile virus in complex with the Fab fragment of a neutralizing monoclonal antibody. Proc. Natl. Acad. Sci. U.S.A. 2006; 103:12400–12404.

Kayser, M., Klein, H., Paasch, I., Pilaski, J., Blenk, H. and Heeg, K. Human antibody response to immunization with 17D yellow fever and inactivated TBE vaccine. J. Med. Virol. 1985; 17:35–45.

Kharitonova, N.N. and Leonov, Y.A. Omsk Hemorrhagic Fever. New Delhi: Amerind Publishing Co., 1985.

Kleiter, I., Jilg, W., Bogdahn, U. and Steinbrecher, A. Delayed humoral immunity in a patient with severe tick-borne encephalitis after complete active vaccination. Infection 2007; 35:26–29.

Klockmann, U., Bock, H.L., Kwasny, H., Praus, M., Cihlova, V., Tomkova, E. and Krivanec, K. Humoral immunity against tick-borne encephalitis virus following manifest disease and active immunization. Vaccine 1991; 9:42–46.

Kluger, G., Schottler, A., Waldvogel, K., Nadal, D., Hinrichs, W., Wundisch, G.F. and Laub, M.C. Tickborne encephalitis despite specific immunoglobulin prophylaxis. Lancet 1995; 346:1502.

Kofler, R.M., Heinz, F.X. and Mandl, C.W. Capsid protein C of the tick-borne encephalitis virus tolerates large internal deletions and is a favorable target for attenuation of virulence. J. Virol. 2002; 76:3534–3543.

Krausler, J. 23 years of TBE in the district of Neunkirchen (Austria). In: Tick-Borne Encephalitis (C. Kunz, Ed.), pp. 6–12. Wein: Facultas Verlag, 1981.

Kreil, T.R. and Eibl, M.M. Pre- and postexposure protection by passive immunoglobulin but no enhancement of infection with a flavivirus in a mouse model. J. Virol. 1997; 71:2921–2927.

Kuhn, R.J. and Strauss, J.H. Enveloped viruses. Adv. Protein Chem. 2003; 64:363–377.

Kuhn, R.J., Zhang, W., Rossman, M.G., Pletnev, S.V., Corver, J., Lenches, E., Jones, C.T., Mukhopadhyay, S., Chipman, P.R., Strauss, E.G., Baker, T.S. and Strauss, J.H. Structure of dengue virus: implications for flavivirus organization, maturation and fusion. Cell 2002; 108:717–725.

Kunz, C. Fruhsommer-Meningoenzehalitis. Pediatr. Prax. 1974; 14:189–192.

Kunz, C. TBE vaccination and the Austrian experience. Vaccine 2003; 21(Suppl. 1):S50–S55.

Kunz, C., Heinz, F.X. and Hofmann, H. Immunogenicity and reactogenicity of a highly purified vaccine against tick-borne encephalitis. J. Med. Virol. 1980; 6:103–109.

Kunz, C., Hofmann, H. and Dippe, H. Early summer meningoencephalitis vaccination, a preventive medicine measure with high acceptance in Austria. Wien Med Wochenschr 1991; 141:273–276.

Kunz, C., Hofmann, H. and Stary, A. Field studies with a new tick-borne encephalitis (TBE) vaccine. Zentralbl Bakteriol [Orig A] 1976; 234:141–144.

Lee, E. and Lobigs, M. Substitutions at the putative receptor binding site of an encephalitic flavivirus alter virulence and host cell tropism and reveal a role for glycosaminoglycans in entry. J. Virol. 2000; 74:8867–8875.

Lee, J.W., Chu, J.J. and Ng, M.L. Quantifying the specific binding between West Nile virus envelope domain III protein and the cellular receptor alphaVbeta3 integrin. J. Biol. Chem. 2006; 281:1352–1360.

Leonova, G.N., Ternovoi, V.A., Pavlenko, E.V., Maistrovskaya, O.S., Protopopova, E.V. and Loktev, V.B. Evaluation of vaccine Encepur Adult for induction of human neutralizing antibodies against recent Far Eastern subtype strains of tick-borne encephalitis virus. Vaccine 2007; 25:895–901.

Lepej, S.Z., Misic-Majerus, L., Jeren, T., Rode, O.D., Remenar, A., Sporec, V. and Vince, A. Chemokines CXCL10 and CXCL11 in the cerebrospinal fluid of patients with tick-borne encephalitis. Acta Neurol. Scand. 2007; 115:109–114.

Levkovich, E.N., Zasukhina, G.D., Chumakov, M.P., Lashkevich, V.A. and Gagarina, A.V. Tissue culture vaccine against tick-borne encephalitis. Vopr. Virusol. 1960; 5:233–236.

Libikova, H. Treatment of experimental tick-borne encephalitis in mice with amiphenosine (5-(p-aminophenyl) cytosine). Acta Virol. 1968; 12:384.

Lin, C.W. and Wu, S.C. A functional epitope determinant on domain III of the Japanese encephalitis virus envelope protein interacted with neutralizing-antibody combining sites. J. Virol. 2003; 77:2600–2606.

Lin, D., Li, L., Dick, D., Shope, R.E., Feldmann, H., Barrett, A.D.T. and Holbrook, M.R. Analysis of the complete genome of the tick-borne flavivirus Omsk hemorrhagic fever virus. Virology 2003; 313:81–90.

Lindgren, E. and Gustafson, R. Tick-borne encephalitis in Sweden and climate change. Lancet 2001; 358:16–18.

Liou, M.-L. and Hsu, C.-Y. Japanese encephalitis virus is transported across the cerebral blood vessels by endocytosis in mouse brain. Cell Tissue Res. 1998; 293:389–394.

Loew-Baselli, A., Konior, R., Pavlova, B.G., Fritsch, S., Poellabauer, E., Maritsch, F., Harmacek, P., Krammer, M., Barrett, P.N. and Ehrlich, H.J. Safety and immunogenicity of the modified adult tick-borne encephalitis vaccine FSME-IMMUN: results of two large phase 3 clinical studies. Vaccine 2006; 24:5256–5263.

Logar, M., Arnez, M., Kolbl, J., Avsic-Zupanc, T. and Strle, F. Comparison of the epidemiological and clinical features of tick-borne encephalitis in children and adults. Infection 2000; 28:74–77.

Lustig, S., Danenberg, H.D., Kafri, Y., Kobiler, D. and Ben-Nathan, D. Viral neuroinvasion and encephalitis induced by lipopolysaccharide and its mediators. J. Exp. Med. 1992; 176:707–712.

Mandl, C.W., Allison, S.L., Holzmann, H., Meixner, T. and Heinz, F.X. Attenuation of tick-borne encephalitis virus by structure-based site-specific mutagenesis of a putative flavivirus receptor binding site. J. Virol. 2000; 74:9601–9609.

Mandl, C.W., Holzmann, H., Meixner, T., Rauscher, S., Stadler, P.F., Allison, S.L. and Heinz, F.X. Spontaneous and engineered deletions in the 3′ noncoding region of tick-borne encephalitis virus: construction of highly attenuated mutants of a flavivirus. J. Virol. 1998; 72:2132–2140.

Mansharamani, H.J. and Dandawate, C.N. Experimental vaccine against Kyasanur Forest disease (KFD) virus from tissue culture source. II. Safety testing of the vaccine in cortisone sensitized Swiss albino mice. Indian J. Pathol. Bacteriol. 1967; 10:25–32.

Mansharamani, H.J., Dandawate, C.N. and Krishnamurthy, B.G. Experimental vaccine against Kyasanur Forest disease (KFD) virus from tissue culture source. I. Some data on the preparation and antigenicity tests of vaccines. Indian J. Pathol. Bacteriol. 1967; 10:9–24.

Mansharamani, H.J., Dandawate, C.N., Krishnamurthy, B.G., Nanavati, A.N. and Jhala, H.I. Experimental vaccine against Kyasanur Forest Disease (KFD) virus from mouse brain source. Indian J. Pathol. Bacteriol. 1965; 12:159–177.

Marth, E. and Kleinhappl, B. Albumin is a necessary stabilizer of TBE-vaccine to avoid fever in children after vaccination. Vaccine 2001; 20:532–537.

Mayer, V. A live vaccine against tick-borne encephalitis: integrated studies. I. Basic properties and behavior of the E5 "14" clone (Langat virus). Acta Virol. 1975; 19:209–218.

Mayer, V., Orolin, D., Pogady, J., Starek, M., Kubistova, K., Gado-Sova, E. and Buran, I. Experimental live tick-borne encephalitis vaccine (Langat E5 "14" virus clone): volunteers 1 and 2 years after single-dose immunization. Acta Virol. 1976; 20:215–225.

McLean, D.M. and Donohue, W.L. Powassan virus: Isolation of virus from a fatal case of encephalitis. Can. Med. Assoc. J. 1958; 80:708–711.

McMinn, P.C., Dalgarno, L. and Weir, R.C. A comparison of the spread of Murray Valley encephalitis viruses of high or low neuroinvasiveness in the tissues of Swiss mice after peripheral inoculation. Virology 1996; 220:414–423.

Messner, H. Pediatric problems of TBE. In: Tick-Borne Encephalitis (C. Kunz, Ed.), pp. 25–27. Wein: Facultas Verlag, 1981.

Modis, Y., Ogata, S., Clements, D. and Harrison, S.C. A ligand-binding pocket in the dengue virus envelope glycoprotein. Proc. Natl. Acad. Sci. U.S.A. 2003; 100:6986–6991.

Modis, Y., Ogata, S., Clements, D. and Harrison, S.C. Structure of the dengue virus envelope protein after membrane fusion. Nature 2004; 427:313–319.

Monath, T.P., Cropp, C.B. and Harrison, A.K. Mode of entry of a neurotropic arbovirus into the central nervous system. Reinvestigation of an old controversy. Lab. Invest. 1983; 48:399–410.

Moritsch, H. and Kovac, W. Investigations on pathogenesis of alimentary infection with tick-borne encephalitis virus in mice. In: Biology of Viruses of the Tick-Borne Encephalitis Complex (H. Libikova, Ed.), pp. 283–285. Smolenice: Academic Press, 1960.

Nalca, A., Fellows, P.F. and Whitehouse, C.A. Vaccines and animal models for arboviral encephalitides. Antiviral Res. 2003; 60:153–174.

Nuttall, P.A. and Labuda, M. Tick-borne encephalitis. In: Tick-Borne Diseases of Humans (J.L. Goodman, D.T. Dennis and D.E. Sonenshine, Eds.), pp. 150–163. Washington, D.C.: ASM Press, 2005.

Oker-Blom, N., Kaariainen, L., Brummer-Korvenkontio, M. and Weckstrom, P. Isolation and occurrence of the viruses of the tick-borne encephalitis complex in Finland. In: Biology of Viruses of the Tick-Borne Encephalitis Complex (H. Libikova, Ed.), pp. 423–429. New York: Academic Press, 1962.

Oliphant, T., Nybakken, G.E., Engle, M., Xu, Q., Nelson, C.A., Sukupolvi-Petty, S., Marri, A., Lachmi, B.E., Olshevsky, U., Fremont, D.H., Pierson, T.C. and Diamond, M.S. Antibody recognition and neutralization determinants on domains I and II of West Nile Virus envelope protein. J. Virol. 2006; 80:12149–12159.

Pattnaik, P. Kyasanur forest disease: an epidemiological view in India. Rev. Med. Virol. 2006; 16:151–165.

Pavri, K.M., Gokhale, T. and Shah, K.V. Serological response to Russian spring-summer encephalitis virus vaccine as measured with Kyasanur Forest disease virus. Indian J. Med. Res. 1962; 50:153–161.

Pfefferkorn, T., Feddersen, B., Schulte-Altedorneburg, G., Linn, J. and Pfister, H.W. Tick-borne encephalitis with polyradiculitis documented by MRI. Neurology 2007; 68:1232–1233.

Pogany, J., Fabian, M.R., White, K.A. and Nagy, P.D. A replication silencer element in a plus-strand RNA virus. EMBO J. 2003; 22:5602–5611.

Pogodina, V.V. Resistance of tick-borne encephalitis virus to gastric juice. Vopr. Virusol. 1958; 3:271–275.

Pogodina, V.V. The course of alimentary infection and development of immunity in tick-borne encephalitis. In: Biology of Viruses of the Tick-Borne Encephalitis Complex (H. Libikova, Ed.), pp. 275–281. Smolenice: Academic Press, 1960.

Pogodina, V.V., Frolova, M.P., Malenko, G.V., Fokina, G.I., Levina, L.S., Mamonenko, L.L., Koreshkova, G.V. and Ralf, N.M. Persistence of tick-borne encephalitis virus in monkeys. I. Features of experimental infection. Acta Virol. 1981a; 25:337–343.

Pogodina, V.V., Levina, L.S., Fokina, G.I., Koreshkova, G.V., Malenko, G.V., Bochkova, N.G. and Rzhakhova, O.E. Persistence of tic-borne encephalitis virus in monkeys. III. Phenotypes of the persisting virus. Acta Virol. 1981b; 25:352–360.

Pogodina, V.V., Malenko, G.V., Fokina, G.I., Levina, L.S., Koreshkova, G.V., Rzhakhova, O.E., Bochkova, N.G. and Mamonenko, L.L. Persistence of tick-borne encephalitis virus in monkeys. II. Effectiveness of methods used for virus detection. Acta Virol. 1981c; 25:344–351.

Price, W.H. and Thind, I.S. Immunization of mice against Russian spring-summer virus complex and monkeys against Powassan virus with attenuated Langat E5 virus. Am. J. Trop. Med. Hyg. 1973; 22:100–108.

Price, W.H., Thind, I.S., Teasdale, R.D. and O'Leary, W. Vaccination of human volunteers against Russian spring-summer (RSS) virus complex with attenuated Langat E5 virus. Bull. World Health Organ. 1970; 42:89–94.

Randolph, S.E. Evidence that climate change has caused 'emergence' of tick-borne diseases in Europe? Int. J. Med. Microbiol. 2004; 293(Suppl. 37):5–15.

Rehacek, J. Transovarial transmission of tick-borne encephalitis virus by ticks. Acta Virol. 1962; 6:220–226.

Reisner, H. Clinic and treatment of tick-borne encephalitis (TBE): introduction. In: Tick-Borne Encephalitis (C. Kunz, Ed.), pp. 1–5. Wein: Facultas Verlag, 1981.

Rendi-Wagner, P., Kundi, M., Zent, O., Banzhoff, A., Jaehnig, P., Stemberger, R., Dvorak, G., Grumbeck, E., Laaber, B. and Kollaritsch, H. Immunogenicity and safety of a booster vaccination against tick-borne encephalitis more than 3 years following the last immunisation. Vaccine 2004a; 23:427–434.

Rendi-Wagner, P., Kundi, M., Zent, O., Dvorak, G., Jaehnig, P., Holzmann, H., Mikolasek, A. and Kollaritsch, H. Persistence of protective immunity following vaccination against tick-borne encephalitis—longer than expected? Vaccine 2004b; 22:2743–2749.

Rey, F.A., Heinz, F.X., Mandl, C.W., Kunz, C. and Harrison, S.C. The envelope glycoprotein from tick-borne encephalitis at 2Å resolution. Nature 1995; 375:291–298.

Rogers, D.J. and Randolph, S.E. Climate change and vector-borne diseases. Adv. Parasitol. 2006; 62:345–381.

Rumyantsev, A.A., Chanock, R.M., Murphy, B.R. and Pletnev, A.G. Comparison of live and inactivated tick-borne encephalitis virus vaccines for safety, immunogenicity and efficacy in rhesus monkeys. Vaccine 2006; 24:133–143.

Sanchez, M.D., Pierson, T.C., McAllister, D., Hanna, S.L., Puffer, B.A., Valentine, L.E., Murtadha, M.M., Hoxie, J.A. and Doms, R.W. Characterization of neutralizing antibodies to West Nile virus. Virology 2005; 336:70–82.

Scherer, W.F., Eddy, G.A., Monath, T.P. and Walton, T.E. Laboratory safety for arboviruses and certain other viruses of vertebrates. Am. J. Trop. Med. Hyg. 1980; 29:1359–1381.

Schmaljohn, C., Vanderzanden, L., Bray, M., Custer, D., Meyer, B., Li, D., Rossi, C., Fuller, D., Fuller, J., Haynes, J. and Huggins, J. Naked DNA vaccines expressing the prM and E genes of Russian spring summer encephalitis virus and Central European encephalitis virus protect mice from homologous and heterologous challenge. J. Virol. 1997; 71:9563–9569.

Schondorf, I., Beran, J., Cizkova, D., Lesna, V., Banzhoff, A. and Zent, O. Tick-borne encephalitis (TBE) vaccination: applying the most suitable vaccination schedule. Vaccine 2007; 25:1470–1475.

Shah, K.V., Aniker, S.P., Murthy, D.P., Rodrigues, F.M., Jayadeviah, M.S. and Prasanna, H.A. Evaluation of the field experience with formalin-inactivated mouse brain vaccine of Russian spring-summer encephalitis virus against Kyasanur Forest disease. Indian J. Med. Res. 1962; 50:162–174.

Shilov, Y.I. and Ryzhaenkov, V.G. Alterations in the proliferative response of peripheral lymphocytes in tick-borne encephalitis. Dokl. Biol. Sci. 2000; 375:553–555.

Silber, L.A. and Soloviev, V.D. Far eastern tick-borne spring-summer (spring) encephalitis. Am. Rev. Sov. Med. 1946; 3(Suppl.):1–75.

Skarpaas, T., Golovljova, I., Vene, S., Ljostad, U., Sjursen, H., Plyusnin, A. and Lundkvist, A. Tickborne encephalitis virus, Norway and Denmark. Emerg. Infect. Dis. 2006; 12:1136–1138.

Skarpaas, T., Ljostad, U. and Sundoy, A. First human cases of tick-borne encephalitis, Norway. Emerg. Infect. Dis. 2004; 10:2241–2243.

Skarphedinsson, S., Jensen, P.M. and Kristiansen, K. Survey of tickborne infections in Denmark. Emerg. Infect. Dis. 2005; 11:1055–1061.

Smorodintsev, A.A. Tick-borne spring-summer encephalitis. Prog. Med. Virol. 1958; 1:210–248.

Smorodintsev, A.A. and Dubov, A.V. Tick-borne encephalitis and its vaccine prophylaxis, Meditsina, Leningrad, Russia. 1986; p. 172–211.

Smorodintsev, A.A., Kagan, N.W., Levkovitsch, E.N. and Dankovskij, N.L. Experimenteller und epidemiologischer Beitrag zur activen Immunisierung gegen die Fruhling-Sommer-zecken-encephalitis. Arch. Ges. Virusforsch. Bd. 1941; 2:1–25.

Stiasny, K., Allison, S.L., Schalich, J. and Heinz, F.X. Membrane interactions of the tick-borne encephalitis virus fusion protein E at low pH. J. Virol. 2002; 76:3784–3790.

Stiasny, K., Kossl, C. and Heinz, F.X. Differences in the postfusion conformations of full-length and truncated class II fusion protein E of tick-borne encephalitis virus. J. Virol. 2005; 79:6511–6515.

Stich, O., Reinhard, M. and Rauer, S. MRI scans of cervical cord provide evidence of anterior horn lesion in a patient with tick-borne encephalomyeloradiculitis. Eur. J. Neurol. 2007; 14:e5–e6.

Thind, I.S. Attenuated Langat E5 virus as a live virus vaccine against Kyasanur forest disease virus. Indian J. Med. Res. 1981; 73:141–149.

Thullier, P., Demangel, C., Bedouelle, H., Megret, F., Jouan, A., Deubel, V., Mazie, J. and Lafaye, P. Mapping of a dengue virus neutralizing epitope critical for the infectivity of all serotypes: insight into the neutralization mechanism. J. Gen. Virol. 2001; 82:1885–1892.

Timofeev, A.V., Butenko, V.M. and Stephenson, J.R. Genetic vaccination of mice with plasmids encoding the NS1 non-structural protein from tick-borne encephalitis virus and dengue 2 virus. Virus Genes. 2004; 28(1):85–97.

Upadhyaya, S., Dandawate, C.N. and Banerjee, K. Surveillance of formolized KFD virus vaccine administration in Sagar-Sorab talukas of Shimoga district. Indian J. Med. Res. 1979; 69:714–719.

Volk, D.E., Beasley, D.W., Kallick, D.A., Holbrook, M.R., Barrett, A.D. and Gorenstein, D.G. Solution structure and antibody binding studies of the envelope protein domain III from the New York strain of West Nile virus. J. Biol. Chem. 2004; 279(37): 38755–38761.

Waldvogel, K., Bossart, W., Huisman, T., Boltshauser, E. and Nadal, D. Severe tick-borne encephalitis following passive immunization. Eur. J. Pediatr. 1996; 155:775–779.

Wang, T., Town, T., Alexopoulou, L., Anderson, J.F., Fikrig, E. and Flavell, R.A. Toll-like receptor 3 mediates West Nile virus entry into the brain causing lethal encephalitis. Nat. Med. 2004; 10:1366–1373.

Work, T. Russian spring-summer encephalitis virus in India: Kyasanur Forest disease. Prog. Med. Virol. 1958; 1:248–279.

Wu, K.P., Wu, C.W., Tsao, Y.P., Kuo, T.W., Lou, Y.C., Lin, C.W., Wu, S.C. and Cheng, J.W. Structural basis of a flavivirus recognized by its neutralizing antibody: solution structure of the domain iii of the Japanese encephalitis virus envelope protein. J. Biol. Chem. 2003; 278:46007–46013.

Wu, S.C. and Lin, C.W. Neutralizing peptide ligands selected from phage-displayed libraries mimic the conformational epitope on domain III of the Japanese encephalitis virus envelope protein. Virus Res. 2001; 76:59–69.

Zaki, A.M. Isolation of a flavivirus related to the tick-borne encephalitis complex from human cases in Saudi Arabia. Trans. R. Soc. Trop. Med. Hyg. 1997; 91:179–181.

Zanotto, P.M., Gao, G.F., Gritsun, T., Marin, M.S., Jiang, W.R., Venugopal, K., Reid, H.W. and Gould, E.A. An arbovirus cline across the northern hemisphere. Virology 1995; 210:152–159.

Zent, O. and Broker, M. Tick-borne encephalitis vaccines: past and present. Expert Rev. Vaccines 2005; 4:747–755.

Zent, O. and Hennig, R. Post-marketing surveillance of immediate allergic reactions: polygeline-based versus polygeline-free pediatric TBE vaccine. Vaccine 2004; 23:579–584.

Zhang, Y., Zhang, W., Ogata, S., Clements, D., Strauss, J.H., Baker, T.S., Kuhn, R.J. and Rossmann, M.G. Conformational changes of the flavivirus e glycoprotein. Structure (Camb) 2004; 12:1607–1618.

Ziebarth-Schroth, A. Fruhsommermeningoenzephalitis (FSME). Klinik und besondere Verlaufsformen. Wien Klin Wochenschr 1972; 84:778–781.

CHAPTER

39

West Nile

John T. Roehrig

Arboviral Diseases Branch, Division of Vector-Borne Infectious Diseases, Centers for Diseases Control and Prevention, U.S. Public Health Service, Department of Health and Human Services, Fort Collins, CO, USA

OUTLINE

ABSTRACT

West Nile virus (WNV) is an emerging flavivirus throughout the western hemisphere. Originally isolated in Africa in 1937, WNV jumped continents, first to Europe, and later to North America in 1999. Since its original discovery in the New York City area in 1999, WNV has been on an inexorable march west and south through North, Central, and South America, demonstrating a 20% human morbidity rate and a 30% equine case-fatality rate. By the end of 2007, there have been over 27,000 human cases of WNV with 1086 human case fatalities in the United States alone. The unexpected introduction of WNV into the United States continues to serve as an excellent example of the effects an emerging infectious disease can have on the public health infrastructure of a country, and the subsequent steps taken to track, prevent, and control such events.

INTRODUCTION

Although West Nile virus (WNV) can be considered an "old" virus historically, it is currently an "emerging" viral agent and serves as a model of a bioterror threat. In 1999 WNV was first discovered in the New York City area. How and when WNV made the jump from the Middle East to North America has been and probably will never be adequately explained (Lanciotti et al., 1999). It has been assumed that the virus was introduced in an infected mosquito or a legally or illegally imported bird. In the fall of that year, the U.S. Centers for Disease Control and Prevention's Division of Quarantine undertook a study to identify commerce flow between the two regions, but were never able to determine a likely source for the virus. Regardless of how WNV jumped continents, WNV has had a significant economic and public health impact on the United States. As the virus moved westward with a 20% human morbidity rate and a 30% equine case-fatality rate, considerable resources were invested in surveillance, diagnostics, prevention, and control of this disease. National guidelines for prevention and control of WNV in the United States were drafted in 1999, modified twice, and are still used today (http://www.cdc.gov/ncidod/dvbid/westnile/resources/wnv-guidelines-aug-2003.pdf).

There were a number of lessons learned or relearned from this experience. The state and local public health infrastructure needed to respond to a major vector-borne disease outbreak was in shambles from years of neglect. The last major vector-borne disease outbreak in the United States occurred in the late 1970s and was caused by St. Louis encephalitis (SLE) virus. Even though enzootic/endemic transmission of viruses like LaCrosse and eastern equine encephalitis viruses occur on a yearly basis, public health dollars were spent on vector-borne disease programs in only a handful of jurisdictions. The national collective public health expertise was maintained only at the CDC through their longitudinal programs. Because of that expertise, a national response was mounted rapidly, and programs were initiated and funded to reestablish the infrastructure and knowledge base necessary for vector-borne disease prevention and control programs. These programs are essential since it is apparent that once a mosquito-borne agent takes hold in a new region, it is virtually impossible to eradicate, and that even with relatively low human case-fatality rates, the social, human behavioral, and economic load can be significant. As always, the trick is to develop a creative vision to guide retention of the newly developed expertise as vector-borne diseases again fade from the short collective memories of the public and politicians.

SHORT HISTORY OF THE DISEASE

WNV was first isolated from the serum of a febrile woman in the West Nile district of Uganda in 1937. Since that time, WNV has circulated in endemic and occasionally epidemic transmission cycles throughout Europe, Western Asia, Africa, the Middle East, Australia (as Kunjin virus), and North and Central America (Fig. 39.1). Major outbreaks of WNV have been documented throughout the world (Table 39.1). In 1999 a WNV outbreak was recognized in the United States for the first time. This initial human and animal outbreak was identified in the New York City area. Genetic studies determined that this virus introduction likely occurred from the Middle East, most likely from Israel (Lanciotti et al., 1999). Since that time, WNV has spread westward through the entire continental U.S. and into Canada, Mexico, Central America, and some Caribbean islands (Table 39.2 and Fig. 39.2).

ETIOLOGIC AGENT

Classification

WNV is an arthropod-borne virus (arbovirus) and as such is transmitted to humans and other animals through the bite of a virus-infected mosquito, usually of the genus *Culex*. WNV is a member of the family *Flaviviridae*, genus *Flavivirus*. The WNV genome is a positive-sense, single-stranded RNA of about 11,000 nucleotides (Fig. 39.3). The RNA genome has a single open-reading frame that encodes a polycistronic precursor protein (Rice et al., 1985). Three structural proteins: capsid, (C), premembrane (prM), and envelope (E), and seven nonstructural proteins: (NS1, NS2A, NS2B, NS3, NS4A, NS4B, and NS5) are encoded by the genome (Chambers and Rice, 1987; Rice et al., 1985). The 5′-end of the genome is noncoding, capped, and not polyadenylated. The 3′-end of the genome is noncoding and assumes a stabilized, folded structure that has been well-studied. Serological classification places WNV within the Japanese encephalitis virus (JEV) complex along with other medically important flaviviruses such as SLE, Murray Valley encephalitis (MVE), and Kunjin (KUN) viruses.

Antigens Encoded by the Agent

The WN virion contains a nucleocapsid formed by C proteins that protects the viral genomic RNA. The virion is enveloped. The envelope is derived from host

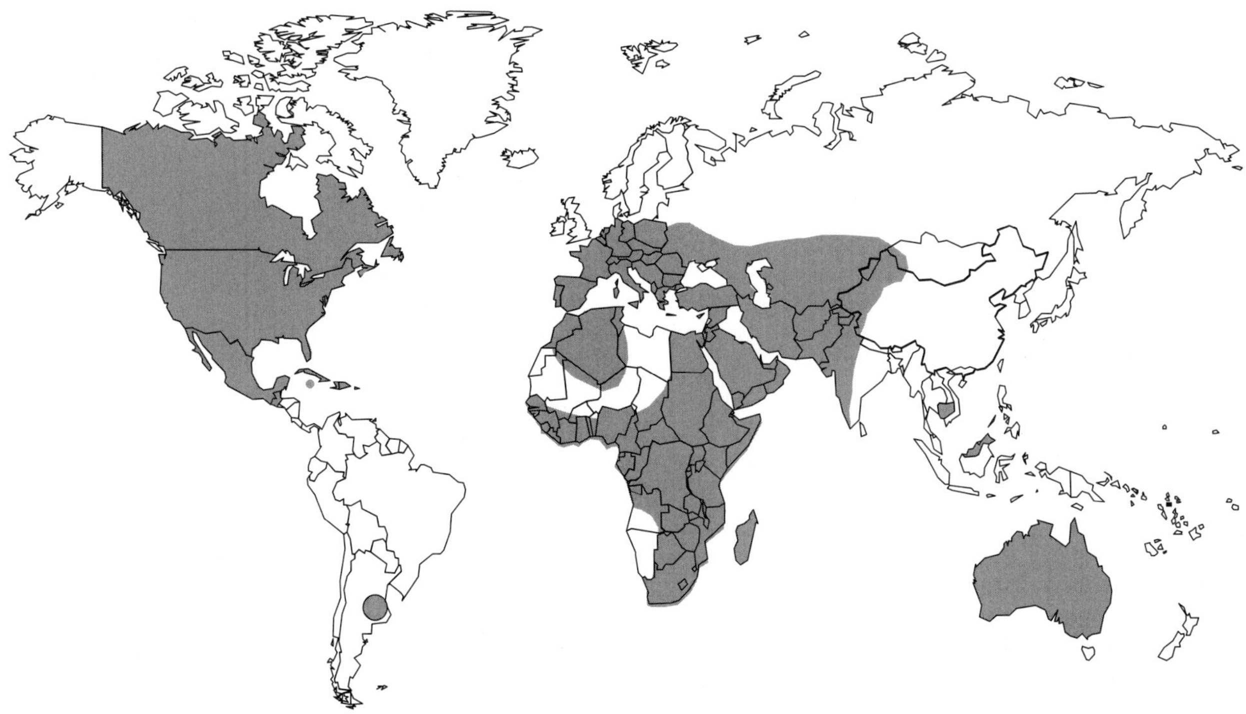

FIGURE 39.1 Geographic distribution of WNV.

TABLE 39.1 Outbreaks of WNV worldwide

Location	Date(s)
Israel	1951–1954, 1957, 1999–2003
South Africa	1974
Algeria	1994
Morocco	1996
Tunisia	1997, 2003
Czech Republic	1997
Italy	1998
France	2003
Romania	1996–2000
Russia	1999–2001
North America and Central America	1999–2006

cell membrane that has been modified by the insertion of the prM and E-proteins. The prM protein is cleaved in a host cell-dependent manner to pr and M protein during late viral maturation. The M protein stays incorporated in the virion envelope while the pr is released into the extracellular fluid. The virion structure has been solved by cryoelectron microscopy (Fig. 39.4). The envelope is highly ordered and is composed of 90 closely packed homodimers of the E-protein (Kuhn et al., 2002; Mukhopadhyay et al., 2003). The primary antigens for all flaviviruses are the E, C, prM, M, NS1, NS3, and NS5 proteins. The immunogenicity of the NS2A, NS2B, NS4A, and NS4B has not been well-studied.

PROTECTIVE IMMUNE RESPONSE

Antibody

The adaptive, protective immune response to WNV infection is driven by the presence or absence of antibodies that neutralize virus infectivity. Virus-neutralizing antibodies are largely elicited by the E-protein (for review, see Roehrig, 2003). There is nothing remarkable about the human antibody response to WNV infection. Upon infection WNV elicits IgM, IgG, and IgA responses. The IgM response begins early, frequently before onset of symptoms, and can persist in serum or cerebrospinal fluid (CSF) in cases of WN neuroinvasive disease (WNND) (Kapoor et al., 2004; Roehrig et al., 2003). IgM is usually always detectable in serological assays by 7–8 days post-onset of symptoms. IgA antibodies are also detectable early in infection and have half-lives similar to IgM (Prince et al., 2005b).

TABLE 39.2 WNV activity in the United States, 1999–2006

Year	States[a]	Humans/fatalities	Birds	Mosquito pools	Veterinary
1999	4	62/7	?	16	25
2000	11 + DC	21/2	4305	515	63
2001	27 + DC	66/9	7338	919	731
2002	44 + DC	4156/284	16,739	6604	14,571
2003	46 + DC	9862/264	12,066	8384	5251
2004	47 + DC + PR	2539/100	7396	8759	1544
2005	48 + DC	3000/119	5393	11,816	1253
2006	48 + DC	4269/177	4106	11,898	1121
2007	47 + DC	3630/124	2182	8215	507

[a]Abbreviations: DC = District of Columbia; PR = Puerto Rico.

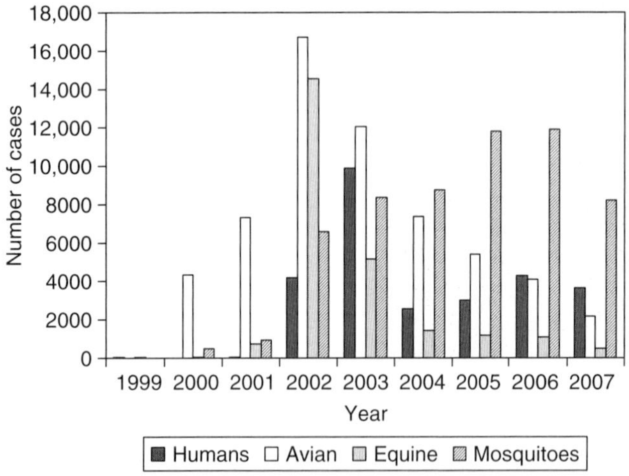

FIGURE 39.2 WNV case distribution by animal type from 1999 to 2005 in the United States.

IgG antibodies are detectable soon after infection and are maintained for years.

Structural Proteins

The molecular structure of the flaviviral E-protein has been solved by x-ray crystallography for WN (Nybakken et al., 2006; Kanai et al., 2006), dengue (DEN) 2, DEN 3, and tick-borne encephalitis (TBE) viruses (Modis et al., 2003, 2005; Rey, 2003; Rey et al., 1995). While the E-protein amino acid sequences vary, the overall E-protein structure appears to be conserved among all flaviviruses studied thus far. The E-protein is the most important viral antigen for eliciting a protective immune response and is a Class II viral fusion protein that undergoes an irreversible oligomeric reorganization at acidic pH (Modis et al., 2004; Rey, 2003; Zhang et al., 2004). Antibodies to the E-protein limit or neutralize viral infectivity by blocking attachment of viruses to cells and blocking virus-mediated cell-membrane fusion, both of which are critical steps in virus infection (Butrapet et al., 1998; Crill and Roehrig, 2001; Gollins and Porterfield, 1986; Roehrig, 2003).

The E-proteins of TBE and DEN viruses have been studied most completely, and flaviviral structure–function relationships that are generally applicable to all flaviviruses have been developed and are reviewed in detail elsewhere (Allison et al., 1999; Guirakhoo et al., 1989; Heinz et al., 1982, 1983, 1984, 1991; Mandl et al., 1989; Rey et al., 1995; Roehrig et al., 1998). Briefly, three structural domains, I, II, and III, have been identified in the E-protein monomer (Fig. 39.5). These structural domains appear to be identical to the functional and antigenic domains C, A, and B previously identified for TBE and DEN virus E-proteins (Rey et al., 1995). Domain II (A-domain) is the dimerization region and contains the fusion peptide. This domain contains flavivirus cross-reactive epitopes. Domain III is an immunoglobulin-like structure that is likely involved in cell attachment and can be rescued from proteolytically cleaved E-protein as a stable 9–10 kDa peptide (approximately amino acids 300–400). Domain III can also be expressed in a variety of expression systems. Domain III elicits strongly neutralizing and virus-specific antibodies. Domain I contains the molecular "hinge" about which the low pH catalyzed oligomeric reorganization occurs (Bressanelli et al., 2004; Modis et al., 2004; Stiasny et al., 2004). It is not surprising that important epitopes in Domains I and II are highly conformational. It is important to note that only DEN viruses are glycosylated at N67, and WNV does not have an N-linked glycosylation motif at this location. In addition, utilization of the N153 N-linked motif by WNV is variable. It has been shown that usage of the N153 glycosylation motif helps define viral neuroinvasion (Beasley et al., 2004, 2005).

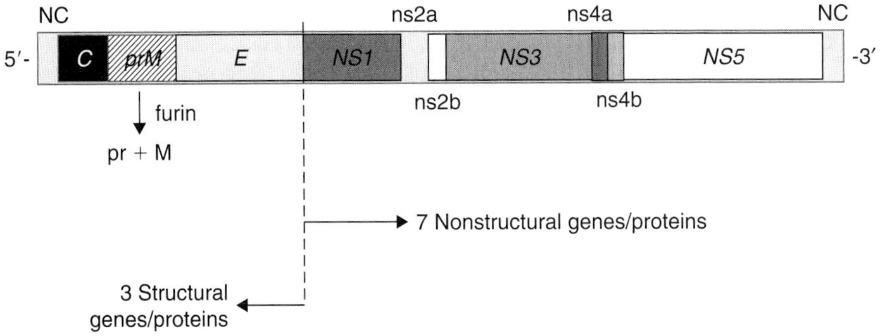

FIGURE 39.3 Flaviviral genome. Abbreviations: NC, noncoding; ns, nonstructural. The serine protease furin cleaves the prM into pr and M protein.

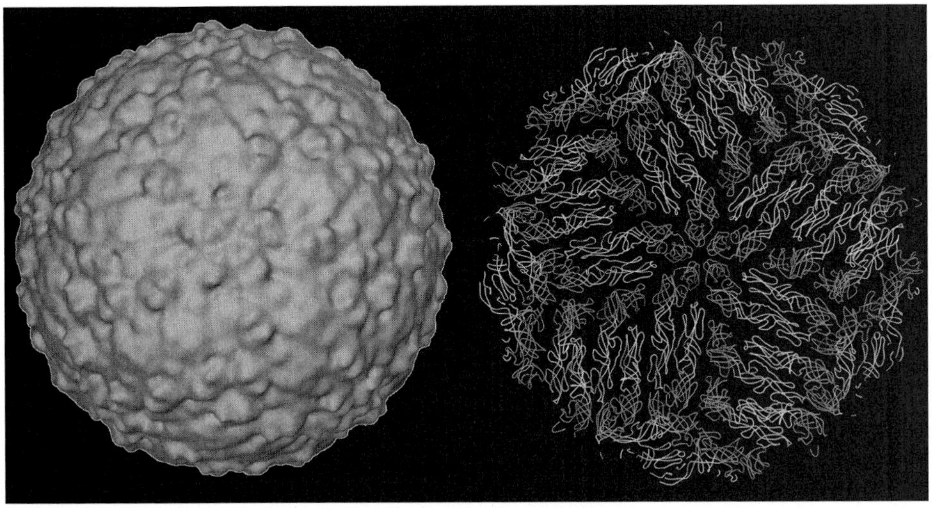

FIGURE 39.4 Cryoelectron microscopic structure of WNV. Panel A, view of the WN virion looking down the fivefold axis. Panel B, C-alpha trace showing the fitted E-protein heterodimers looking down the fivefold axis. Courtesy of Baerbel Kaufmann and Richard Kuhn, Purdue University (see color plate section).

The C and prM proteins can elicit antiviral antibody. It has been suggested that anti-prM and anti-E-protein antibodies can neutralize virus infectivity by interfering with viral attachment to dendritic cell-specific ICAM-3 grabbing nonintegrin (DC-SIGN) bearing dendritic cells (Barba-Spaeth et al., 2005; Kwan et al., 2005; Lozach et al., 2005; Marovich et al., 2001; Modis et al., 2005; Navarro-Sanchez et al., 2003; Palmer et al., 2005; Sakuntabhai et al., 2005; Tassaneetrithep et al., 2003; Wu et al., 2000). The significance of this observation in vivo has yet to be thoroughly explored; however, evidence with both WNV and DEN virus suggest that virus–DC-SIGN interaction may play a pivotal role in capturing virus through mannose residues in their N-linked glycosylation sites onto susceptible cells (Pokidysheva et al., 2006).

Nonstructural Proteins

The most important antibody response to flaviviral nonstructural proteins is elicited by the NS1 protein (Cane and Gould, 1988; Costa et al., 2006; Henchal et al., 1988; Konishi et al., 1991; Macdonald et al., 2005; Putnak and Schlesinger, 1990; Schlesinger et al., 1986, 1987, 1993; Zhang et al., 1988). The anti-NS1 response has been intensely studied in DEN and yellow fever (YF) virus-infected mice. Passive antibody protection studies using murine anti-NS1 monoclonal antibodies demonstrated some protective effects against DEN or YF virus infection. Since the NS1 protein is not expressed in the infectious virion, but rather it is expressed on virus-infected cell surfaces, it is believed that anti-NS1 antibodies mediate protection at the level of the virus-infected cell. Similar studies with viruses causing encephalitis—like WNV—have not been performed. Antibody responses to other NS proteins of WNV have not been well-studied; however, the antibody response to the WNV NS5 protein has served as the target to diagnose WNV infections in humans using a microsphere immunoassay.

T-Cell Epitopes

As with all viruses, both CD4 and CD8 T-cells are required to drive the immune response to flaviviruses. WNV is no exception. WNV T-cell epitopes have not been well-defined; however, results from other

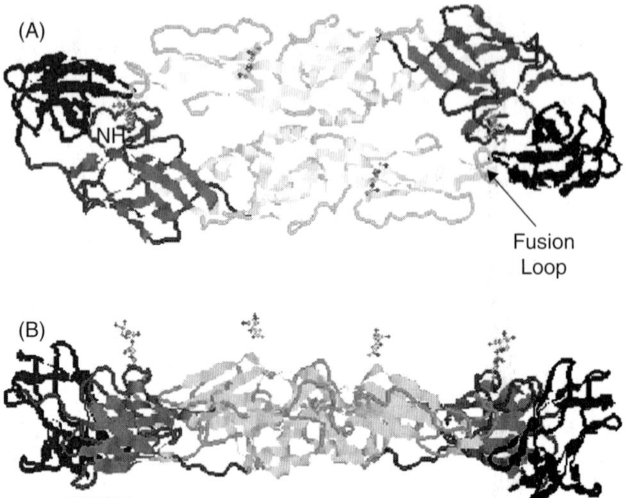

FIGURE 39.5 Crystallographic molecular structure of the DEN2 E-glycoprotein homodimer. Panel A, homodimer top view (aa 1–394). Panel B, homodimer lateral view. Structural domains of monomers identified by color: gray, Domain I; white, Domain II; black, domain III. N-linked glycosylation shown in ball-and-stick notation. Fusion loop (aa 98–110)) marked in DII. Structural coordinates from Modis et al. (2003) (see color plate section).

closely related flaviviruses (e.g., JE and MVE viruses) can likely be extrapolated to WNV. Recent studies in animal models have reaffirmed the requirement for a CD8 T-cell response for viral clearance and recovery (Klein et al., 2005; Shrestha and Diamond, 2004; Shrestha et al., 2006; Wang et al., 2003b). The NS3 protein seems to contain the most important flaviviral cytotoxic T-cell epitopes (Douglas et al., 1994; Hughes, 2001; Kesson et al., 2002; Kurane et al., 1991, 1995, 1998; Lobigs et al., 1994, 2003; Mathew et al., 1996; Okamoto et al., 1998; Rothman et al., 1996; Zivny et al., 1995). Helper T-cell epitopes have been identified on the E-protein as well (Mathews et al., 1991, 1992; Roehrig et al., 1992).

EPIDEMIOLOGY

Significance as Public Health Problem

WNV now accounts for the largest number of cases of viral encephalitis in the United States. Worldwide WNV is an arbovirus, primarily transmitted by the mosquitoes of the genus *Culex* (e.g., *Culex tarsalis*, *Culex pipiens pipiens*, *Culex pipiens quinquefasciatus*, *Culex salinarius*, and *Culex nigirpalpus* in the United States). Mosquito-borne transmission to humans in temperate climates usually peaks in the late summer and early fall (Fig. 39.6). Mosquito-borne transmission to humans in milder or more tropical climates can occur throughout the year, whenever mosquitoes are active.

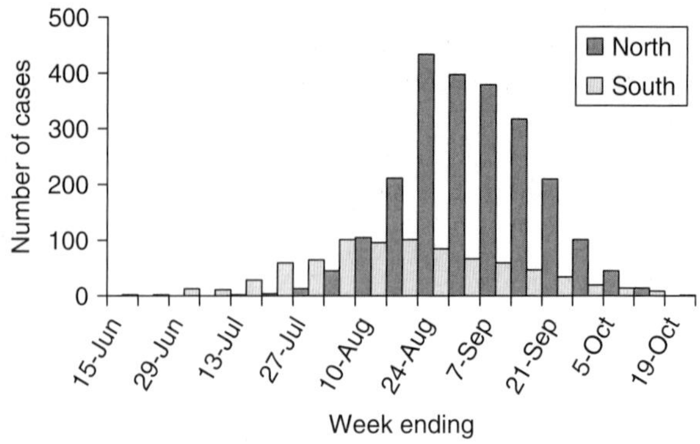

FIGURE 39.6 Distribution of human WNND in temperate (north) and more tropical (south) regions of the United States.

In the United States alone, however, 62 species of mosquitoes have been shown to be infected with WNV. The actual vector-status of many of these mosquito species remains to be determined. WNV is a zoonotic disease with birds being the primary natural reservoir. Over 300 species of birds have been shown to be WNV-infected in the western hemisphere. Humans are primarily infected through the bite of a WNV-infected mosquito. Recently, other modes of WNV transmission have been identified such as blood transfusion, tissue transplantation, percutaneous occupational exposure, breast feeding, and intrauterine transfer (Hayes and O'Leary, 2004). The last two modes of transmission have been documented, but are very rare (O'Leary et al., 2006; Paisley et al., 2006).

Even though WNV has now been in the United States since 1999, molecular epidemiological analysis of current and past strains of the U.S. WNV has demonstrated low-level genetic drift, with remarkable overall phenotypic stability (Davis et al., 2005; Lanciotti et al., 1999, 2002). WNV can be divided into two genetic lineages (1 and 2), with lineage 1 WNVs primarily responsible for major human outbreaks (Fig. 39.7). Lineage 1 strains have been divided into four clades: A, B, Indian, and KUN (Beasley et al., 2004).

Potential as Biothreat Agent

The potential of using WNV as an agent to inflict mass human casualties via aerosol application is small. The experience of the United States following the introduction of WNV has demonstrated, however, that WNV

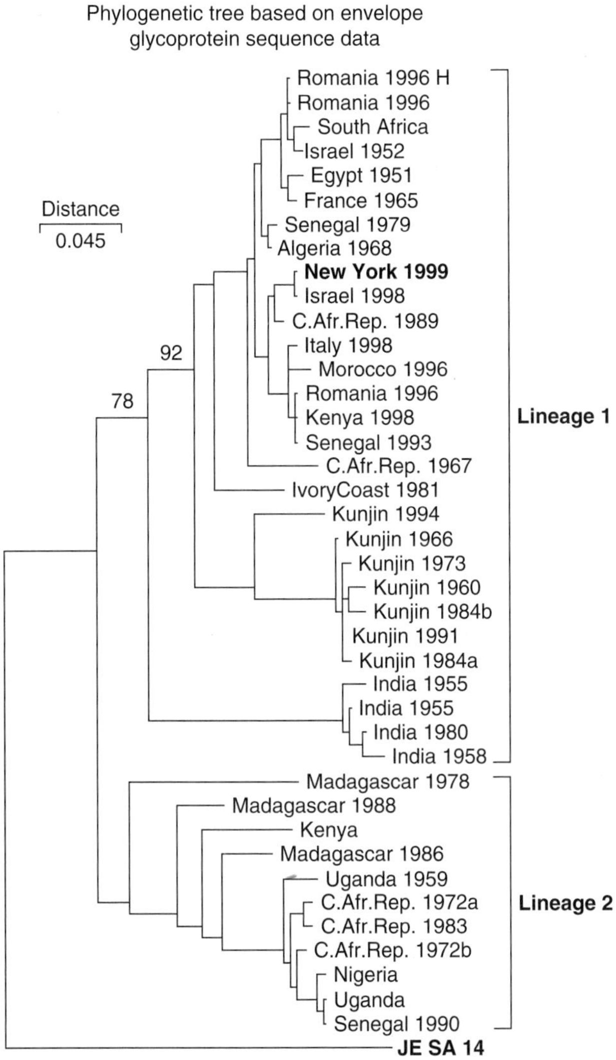

FIGURE 39.7 Phylogenetic tree of WNV based on E-protein gene sequence data. Two major lineages (1 and 2) are shown. Courtesy of R. Lanciotti, CDC. MEGA, distance tree, Kimura 2-parameter, neighbor-joining.

transmission can cause significant human, avian, and equine outbreaks, and can be quite financially costly. The determination that WNV can be transmitted from human-to-human through transfer of contaminated blood products has resulted in the screening of the entire U.S. blood supply for WNV on a continuing basis (Anon 2002; Anon 2003a; Anon 2003b; Busch et al., 2005a; Cameron et al., 2006; Custer et al., 2006; Epstein, 2005; Gallian et al., 2005; Hardinger et al., 2003; Harrington et al., 2003; Hiatt et al., 2003; Humar, 2004; Iwamoto et al., 2003; Kiberd and Forward, 2004; Kleinman et al., 2005; Lee et al., 2005; Macedo de Oliveira et al., 2004; Petersen and Epstein, 2005; Prince et al., 2005a; Rados, 2003; Stramer et al., 2005). This testing is quite costly, but has been shown to have likely averted over 4000 blood-borne WNV infections in the United States since it was initiated. A similar survey for WNV-contaminated blood in Mexico, identified only one WNV positive specimen (Sanchez-Guerrero et al., 2006).

CLINICAL DISEASE

Approximately 20% of human infections with WNV result in clinical disease. Less than 1% of human infections with WNV result in WNND (Mostashari et al., 2001). Incidence of WNND and death increase with age; however, persons of all ages appear equally susceptible to infection. West Nile fever (WNF) is characterized by an acute onset of flu-like symptoms including fever, headache, malaise, and myalgia, and frequently a transient rash. WNND may result in meningitis, encephalitis, and poliomyelitis-like flaccid paralysis resulting from infection of the anterior horn cells of the spinal motor neurons (Hayes et al., 2005; Jeha et al., 2003; Li et al., 2003; Sejvar et al., 2003). Individuals with WNND may suffer from relatively long-term sequellae that may persist for years (Klee et al., 2004). The overall case-fatality rate for WNND is approximately 9–10% (O'Leary et al., 2004).

DIAGNOSIS

The diagnosis of WNV infection in humans is primarily based upon the detection of WNV-specific IgM in serum or CSF specimens from individuals with clinically compatible symptoms of WNF or WNND (Malan et al., 2004; Martin et al., 2002, 2004). WNV-specific IgM is usually detectable by 8 days post-onset of clinical symptoms and is measured in an IgM antibody-capture enzyme-linked immunosorbent assay (MAC-ELISA). There are currently three U.S. Food and Drug Administration approved WNV-reactive MAC-ELISAs commercially available. Individuals with WNV-specific IgM in their CSF are considered confirmed WNV infections. One caveat of IgM testing is the possible long-term persistence of WNV IgM in serum (greater than 12 months with some patients) (Roehrig et al., 2003). This observation could confuse interpretation of the actual time of infection.

Newer platforms for detection of anti-WNV antibody have been developed. These microsphere immunoassays detect IgM or IgG antibody specific for either the E or NS5 proteins (Johnson et al., 2005, 2007; Wong et al., 2003, 2004). The advantages of these tests are the requirement for smaller specimen volumes, less time to perform testing, and the ability to multiplex testing using a variety of arboviral antigens with a single specimen.

Because of the possibility of flaviviral cross-reactivity of serum IgM with other flaviviruses, a second-tier and more virus-specific assay (plaque-reduction neutralization test, PRNT) is frequently performed. The PRNT is more virus-specific that ELISAs, but is usually only performed in state public health laboratories or national reference laboratories. Because of the aforementioned serological cross-reactivities of flaviviral immune sera, it is also important to acquire a complete travel and vaccination history from a suspected WNV patient.

For infected humans, nucleic acid amplification tests (NAATs) such as reverse-transcriptase polymerase chain reaction (RT-PCR) or real-time PCR (e.g., TaqMan) and nucleic acid sequence-based assay (NASBA) are useful in only about 50% of cases maximally (Lanciotti, 2003; Lanciotti and Kerst, 2001; Lanciotti et al., 2000). This low level of sensitivity results from a low WNV viremia at the time of hospital presentation of an ill individual. The NAATs are more useful in screening blood donations for WNV contamination (Macedo de Oliveira et al., 2004; Pealer et al., 2003). WNV NAAT-positive blood donations are usually acquired prior to the onset of disease in the blood donor when viremias are higher. Currently every U.S. blood donation is tested for WNV contamination using FDA approved NAATs (Busch et al., 2005a, 2005b; Custer et al., 2006, 2004; Kleinman et al., 2005; Prince et al., 2005a; Tobler et al., 2005).

With appropriate fresh or formalin-fixed tissue specimens (e.g., brain tissue), WNV infection and can be unambiguously identified by immunohistochemistry or in situ PCR, and occasionally virus isolation (Agamanolis et al., 2003; Iwamoto et al., 2003; Shieh et al., 2000). Unfortunately these specimens are only available upon autopsy of a fatal WNV infection.

TREATMENT

Current treatment for WNF and encephalitis is purely supportive, and the degree of treatment depends on the severity of symptoms. Mild WNV infections may require only medication for pain and liquids for rehydration; however, severe WNND may require ventilator support.

Antivirals

There are no licensed antiviral agents designed for or proven to be effective against WNV infection. Two antivirals, ribavirin and interferon, have been used on a compassionate basis, with very limited success. Ribavirin can inhibit WNV multiplication in vitro in human neural cell culture at high doses (Crance et al., 2003; Day et al., 2005; Jackson, 2004; Jordan et al., 2000; Leyssen et al., 2005). There have been no controlled subhuman primate studies to determine the ability of ribavirin to limit WNV infection.

More is known about the antiviral effects of alpha-interferon; however, no controlled experimental studies of the therapeutic value of interferon during WNV infection of subhuman primates or humans have been reported (Anderson and Rahal, 2002; Guo et al., 2005; Hrnicek and Mailliard, 2004; Kalil et al., 2005; Liu et al., 2005; Morrey et al., 2004; Samuel and Diamond, 2005; Scholle and Mason, 2005; Wang et al., 2003a). In the mouse model of infection, WNV resistance has been linked to the presence or absence of functional 2′,5′-oligoadenylate synthetase and RNase L gene complexes that participate in the interferon response (Kajaste-Rudnitski et al., 2005; Lucas et al., 2003; Mashimo et al., 2002; Perelygin et al., 2002; Silvia et al., 2004; Urosevic, 2003; Yakub et al., 2005). The requirement for interferon in an innate murine response to WNV infection was confirmed by observing enhanced virulence of WNV in interferon-deficient mice. While human trials are underway, to date only anecdotal observations of the antiviral efficacy of alpha-interferon in humans have been reported.

Preexposure Antibody Prophylaxis and Postexposure Antibody Therapy

Because protection from flaviviral infection can be mediated by virus-neutralizing antibody, it seems plausible that preexposure antibody prophylaxis would be a reasonable approach to disease prevention. Studies in mice using polyclonal and monoclonal antibodies have shown that virus-neutralizing antibodies can prevent challenge with WN, SLE, JE, TBE, or MVE viruses (Engle and Diamond, 2003; Gould et al., 2005; Hsieh et al., 1963; Kreil, 2004; Kreil et al., 1997, 1998a, 1998b; Kreil and Eibl, 1997; Mathews and Roehrig, 1984; Roehrig et al., 2001; Roodyn, 1974; Simkova, 1959). The timing of the antibody application is critical for animal survival. Results from most previous investigations suggest that once the virus has entered the brain, any therapeutic effect of antibody is lost. A recent study with WNV in an inbred mouse and hamster models demonstrated that murine or human MAbs or human single-chain variable region antibody fragments human could cure also WNV-infected animals (Morrey et al., 2006, 2007; Gould et al., 2005; Oliphant et al., 2005). Similar data does not exist for humans.

PATHOGENESIS

After being bitten by a WNV-infected mosquito, the virus replicates in local cell populations. Splenic dendritic cells have been implicated as a site of WNV replication (Kulkarni et al., 1991). Virus then spreads to the draining lymph nodes and enters the blood stream. It has been recently hypothesized that virus entry into the central nervous system (CNS) is via increased permeability of the blood-brain barrier (Wang et al., 2004). Previous studies with SLE virus suggested that this virus enters the CNS directly from the blood through exposed neurons of olfactory system (Monath et al., 1983). Other mechanisms of viral entry into the CNS, such as retrograde virus transport via non-olfactory neurons, has not been ruled out (Hunsperger and Roehrig, 2005). WNV will infect a variety of neurons in the brain, brainstem, and spinal cord (Agamanolis et al., 2003; Ceccaldi et al., 2004; Guarner et al., 2004; Hayes et al., 2005; Kleinschmidt-DeMasters et al., 2004; Shieh et al., 2000). There are no well-characterized genetic factors associated with human risk of acquiring WNND; however, deficiencies in the 2′, 5′-oligoadenylate synthetase pathways and the subsequent interferon response has been associated with murine susceptibility to WNV (Bonnevie-Nielsen et al., 1995; Kajaste-Rudnitski et al., 2005; Lucas et al., 2003; Mashimo et al., 2002; Perelygin et al., 2002; Yakub et al., 2005). A generalized skin rash has been associated with WNF; however, it is not known if the virus replicates in the skin.

Preliminary observations in mice and humans indicate that a genetic predisposition to severe WNV disease may be associated with a nonfunctional CCR5 gene product. The significance of these observations await corroborative studies (Glass et al., 2005, 2006).

VACCINES

No WNV vaccines are approved for human use. There are, however, four vaccines approved by the United States Department of Agriculture (USDA) for use in horses (Ng et al., 2003). The vaccine currently commercially available is a formalin-inactivated cell culture-based vaccine requiring two doses and is produced by Fort Dodge Animal Health. The parental virus for this vaccine is a strain of WNV isolated from a horse 1999 WNV outbreak in the New York City area. It is currently administered as a monovalent vaccine; however, there should be no problem with multivalent formulations including other killed equine encephalitis vaccines. Currently annual booster immunizations are recommended for this vaccine.

The second vaccine, developed in a partnership between Fort Dodge Animal Health and researchers at the U.S. Centers for Disease Control and Prevention is a DNA vaccine that expresses the prM and E-protein of WNV in the form of a virus-like particle (Davis et al., 2001). This vaccine has USDA approval and is currently in production for an eventual commercial release. It has also been used to protect successfully the endangered California condor population in the western United States (G.-J. Chang, personal communication). This is the first DNA vaccine ever approved for commercial applications, and is also in Phase I human clinical trials.

The other vaccines are recombinant vaccines. One vaccine is a canarypox virus-vectored WNV vaccine (Minke et al., 2004), and the other is a WNV/YF chimeric virus (Arroyo et al., 2004).

VACCINES IN DEVELOPMENT

Rationale of Second/Third/New Generation Vaccines

Besides inactivated and DNA vaccines, a variety of other WNV are currently in various stages of development (Table 39.3). Any WNV vaccine will require appropriate expression of the E and prM proteins. Three new vaccines: a WNV/YFV chimera (ChimeriVax, Acambis, Ltd.); a WNV/DEN2V chimera (U.S. CDC); and a poxvirus-vectored vaccine, express the WNV prM and E-proteins in heterologous viral backgrounds. An additional candidate is an attenuated WNV vaccine, and as such elicits an authentic adaptive immune response to all WNV proteins. The benefit of including the NS1 protein as part of a recombinant vaccine is still an open issue. From a public health perspective, the additional protection possibly afforded by anti-NS1 antibody has to be balanced by the inability to differentiate current infections with WNV from individuals vaccinated with prM/E-based WNV vaccines based on the presence or absence of anti-NS1 antibody.

Discovery/Basic Science

Flavivirus vaccine development has had a checkered past. The YF vaccine developed in the early

1900s has been one of the safest and most effective live-attenuated vaccines ever produced. Since that time, a variety of approaches have been used for other medically important flaviviruses such and TBEV and JEV, with reasonable success. The identification of approaches to produce vaccines by recombinant DNA technology has been problematic. There were many attempts to develop E-protein based vaccines in the 1980s, most of which were not successful. Problems were associated with our lack of understanding of the requirements for appropriate expression of the E-protein. It was determined that many E-protein epitopes were discontinuous (nonlinear) and, therefore, conformational in nature. It was also hypothesized (and later proven) that the E-protein is a class II fusion protein that undergoes an oligomeric reorganization resulting in cell-membrane fusion. This oligomeric reorganization results in loss of important E-protein epitopes and implies that that vaccine production protocols will require careful control of pH during production runs.

Studies with TBEV and DENV determined that E-protein synthesis required the presence of prM protein. The prM protein likely functions as a chaperone protein for the E-protein maintaining its conformation during maturation in low pH exocytic vesicles. Late in viral maturation the prM protein is cleaved by cellular serine proteases (e.g., furin) activating the E-protein fusion potential (Elshuber et al., 2003). Failure to cleave the prM protein results in altered expression of E-protein epitopes, an E-protein with reduced fusion capacity, and virus with lower infectivity. Since the prM cleavage enzymes are cell-dependent, the virion prM/E composition and virus infectivity will vary depending on the host cell used for vaccine production. For other viruses encoding Class II fusion protein (e.g., alphaviruses) ablation of the multibasic serine protease cleavage motif in the analogous chaperone proteins has actually been used as an attenuating mutation for vaccine development.

The contribution of the nonstructural proteins to protective humoral and cellular immunity needs further evaluation. For example, even though NS1 has been shown to be of some value eliciting protective antibody, a recent study with DENV confirmed that the E-protein drives the protective antibody response when analyzed using replicating WNV/DENV chimeric viruses differing E and NS proteins (Calvert et al., 2006). It is also unclear whether or not chimeric WNV vaccines that carry heterologous nonstructural proteins will be less efficient at eliciting long-term immunity and effective viral clearance mechanisms when a vaccinee is faced with a wild-type viral challenge.

Preclinical Development, Including Relevant Animal Models

Preclinical evaluation of vaccines for encephalitic flaviviruses has the advantage of the availability of a number of relevant animal models. Initial animal testing for attenuation, neuroinvasion, and neuroinvasiveness is usually performed in suckling, weanling, or adult mice. Recently a C57Bl/6J mouse model has been used for WNV studies (Diamond et al., 2003). The limitation of this mouse strain is the variable infectivity of WNV. A hamster model has also recently been developed for WNV. Similar to the C57Bl/6J mouse model, the hamster model shows variable WNV infectivity (Sbrana et al., 2005; Tesh et al., 2005; Tonry et al., 2005; Xiao et al., 2001). Persistent WNV infection has been identified in these hamsters (Tesh et al., 2005; Tonry et al., 2005). This observation should be valuable in studying persistent WNV disease. Subhuman primates and horses serve as large animal models for WNV vaccine testing, and have been used successfully.

Human Vaccine Trials

There are two vaccines in limited Phase I or Phase II human clinical trials at this time—the ChimeriVax WNV (Monath et al., 2006) and the WNV DNA vaccines (Table 39.3). There are no WNV vaccines in Phase III human clinical trials. Adequate Phase I and II human clinical trials for WNV vaccines should be achievable. Due to the inconsistent and unpredictable nature of WNV activity, Phase III human efficacy studies are likely to be problematic and may require a large number of study participants. Since the highest value target for WNV vaccination will be the elderly due to the increased severity of WNV infection in this age group, it will be important to include older individuals in all stages of vaccine evaluation.

PROSPECTS FOR THE FUTURE

The likelihood that WNV or related flaviviruses will be used as a bioterrorism agent is not high. Flaviviruses in general are not overly stable in the environment, nor is there good evidence that they are easily transmitted via aerosol. As was the case with WNV, the real threat would be establishing a stable natural mosquito–host transmission cycle. This would permit natural spread of disease. Once these natural cycles are established, and depending in the reservoir host, the outbreak could be contained if the host is relatively immobile (e.g., rodents) or be widespread if the

TABLE 39.3 WNV vaccines

Vaccine	Status	Target Animal: Humans	Target Animal: Equines	Target Animal: Other animals	References
TME	Preclinical			×	(Malkinson et al., 2001; Samina et al., 2005)
Inactivated WNV	Approved equine		×		(Ng et al., 2003; Samina et al., 2005)
WNV/Canarypoxvirus	Approved equine		×		(Grosenbaugh et al., 2004; Karaca et al., 2005; Minke et al., 2004; Siger et al., 2004)
WNV DNA	Approved equine Phase I human trial	×	×	×	(Chang et al., 2001; Davis et al., 2001; Monath, 2001; Turell et al., 2003)
WNV/DEN chimera	Preclinical	×	×	×	(Hanley et al., 2005; Huang et al., 2005)
Recombinant attenuated WNV	Preclinical	×			(Lustig et al., 2000; Yamshchikov et al., 2004)
WNV/YF chimera	Approved equine Phase II human trial	×	×		(Arroyo et al., 2001, 2004; Johnson et al., 2003; Langevin et al., 2003; Monath et al., 2001)

host is very mobile (e.g., birds). Because reverse genetics works well with flaviviruses, genetic modifications could be made that would result in phenotypic differences that could enhance virus virulence or transmissibility. Whether or not such manipulations would be worth the effort is beyond the scope of this review.

There are no significant scientific impediments to the development of a safe and effective WNV vaccine for humans. The economics of the need for this vaccine will continue to drive its development. If WNV continues to be a public health problem in the United States—and there is no reason to believe it would not—vaccine development will continue to a successful conclusion.

KEY ISSUES

Key issues surrounding WNV that still need to be addressed:

- Continued development of a WNV vaccine.
- Developing more specific serodiagnostics for WNV and other flaviviruses.
- Understanding better the viral virulence and pathogenesis in the human.
- Developing approaches to better predict WNV activity.
- Maintaining vector-borne viral disease expertise and response infrastructure.

ACKNOWLEDGMENTS

The author would like to acknowledge the members of the Arbovirus Diseases Branch, CDC. Without the tireless efforts of these individuals, the CDC would not have been able to adequately respond to the WNV emergency in 1999 and subsequent years.

References

Anon (2002). From the Centers for Disease Control and Prevention. Investigations of West Nile virus infections in recipients of blood transfusions. JAMA 2002; 288(20):2535–2536.

Anon (2002).Detection of West Nile virus in blood donations—United States, 2003. MMWR Morb. Mortal. Wkly. Rep. 2003a; 52(32):769–772.

Anon (2002).Update: detection of West Nile virus in blood donations—United States, 2003. MMWR Morb. Mortal. Wkly. Rep. 2003b; 52(38):916–919.

Agamanolis, D.P., Leslie, M.J., Caveny, E.A., Guarner, J., Shieh, W.J. and Zaki, S.R. Neuropathological findings in West Nile virus encephalitis: a case report. Ann. Neurol. 2003; 54(4):547–551.

Allison, S.L., Stiasny, K., Stadler, K., Mandl, C.W. and Heinz, F.X. Mapping of functional elements in the stem-anchor region of tick-borne encephalitis virus envelope protein E. J. Virol. 1999; 73(7):5605–5612.

Anderson, J.F. and Rahal, J.J. Efficacy of interferon alpha-2b and ribavirin against West Nile virus in vitro. Emerg. Infect. Dis. 2002; 8(1):107–108.

Arroyo, J., Miller, C.A., Catalan, J. and Monath, T.P. Yellow fever vector live-virus vaccines: West Nile virus vaccine development. Trends Mol. Med. 2001; 7(8):350–354.

Arroyo, J., Miller, C., Catalan, J., Myers, G.A., Ratterree, M.S., Trent, D.W. and Monath, T.P. ChimeriVax-West Nile virus live-attenuated vaccine: preclinical evaluation of safety, immunogenicity, and efficacy. J. Virol. 2004; 78(22):12497–12507.

Barba-Spaeth, G., Longman, R.S., Albert, M.L. and Rice, C.M. Live attenuated yellow fever 17D infects human DCs and allows for presentation of endogenous and recombinant T cell epitopes. J. Exp. Med. 2005; 202(9):1179–1184.

Beasley, D.W., Davis, C.T., Whiteman, M., Granwehr, B., Kinney, R.M. and Barrett, A.D. Molecular determinants of virulence of West Nile virus in North America. Arch. Virol. Suppl. 2004(18):35–41.

Beasley, D.W., Whiteman, M.C., Zhang, S., Huang, C.Y., Schneider, B.S., Smith, D.R., Gromowski, G.D., Higgs, S., Kinney, R.M. and

Barrett, A.D. Envelope protein glycosylation status influences mouse neuroinvasion phenotype of genetic lineage 1 West Nile virus strains. J. Virol. 2005; 79(13):8339–8347.

Bonnevie-Nielsen, V., Heron, I., Monath, T.P. and Calisher, C.H. Lymphocytic 2′,5′-oligoadenylate synthetase activity increases prior to the appearance of neutralizing antibodies and immunoglobulin M and immunoglobulin G antibodies after primary and secondary immunization with yellow fever vaccine. Clin. Diagn. Lab. Immunol. 1995; 2(3):302–306.

Bressanelli, S., Stiasny, K., Allison, S.L., Stura, E.A., Duquerroy, S., Lescar, J., Heinz, F.X. and Rey, F.A. Structure of a flavivirus envelope glycoprotein in its low-pH-induced membrane fusion conformation. EMBO J. 2004; 23(4):728–738.

Busch, M.P., Caglioti, S., Robertson, E.F., McAuley, J.D., Tobler, L.H., Kamel, H., Linnen, J.M., Shyamala, V., Tomasulo, P. and Kleinman, S.H. Screening the blood supply for West Nile virus RNA by nucleic acid amplification testing. N. Engl. J. Med. 2005a; 353(5):460–467.

Busch, M.P., Tobler, L.H., Saldanha, J., Caglioti, S., Shyamala, V., Linnen, J.M., Gallarda, J., Phelps, B., Smith, R.I., Drebot, M. and Kleinman, S.H. Analytical and clinical sensitivity of West Nile virus RNA screening and supplemental assays available in 2003. Transfusion 2005b; 45(4):492–499.

Butrapet, S., Kimura-Kuroda, J., Zhou, D.S. and Yasui, K. Neutralizing mechanism of a monoclonal antibody against Japanese encephalitis virus glycoprotein E. Am. J. Trop. Med. Hyg. 1998; 58(4):389–398.

Calvert, A.E., Huang, C.Y., Kinney, R.M. and Roehrig, J.T. Non-structural proteins of dengue 2 virus offer limited protection to interferon-deficient mice after dengue 2 virus challenge. J. Gen. Virol. 2006; 87(Pt 2):339–346.

Cameron, C., Reeves, J., Antonishyn, N., Tilley, P., Alport, T., Eurich, B., Towns, D., Lane, D. and Saldanha, J. West Nile virus in Canadian blood donors. Transfusion 2005; 45(4):487–491.

Cane, P.A. and Gould, E.A. Reduction of yellow fever virus mouse neurovirulence by immunization with a bacterially synthesized non-structural protein (NS1) fragment. J. Gen. Virol. 1988; 69(Pt 6):1241–1246.

Ceccaldi, P.E., Lucas, M. and Despres, P. New insights on the neuropathology of West Nile virus. FEMS Microbiol. Lett. 2004; 233(1):1–6.

Chambers, T.J. and Rice, C.M. Molecular biology of the flaviviruses. Microbiol. Sci. 1987; 4(7):219–223.

Chang, G.J., Davis, B.S., Hunt, A.R., Holmes, D.A. and Kuno, G. Flavivirus DNA vaccines: current status and potential. Ann. N.Y. Acad. Sci. 2001; 951:272–285.

Costa, S.M., Paes, M.V., Barreto, D.F., Pinhao, A.T., Barth, O.M., Queiroz, J.L., Armoa, G.R., Freire, M.S. and Alves, A.M. Protection against dengue type 2 virus induced in mice immunized with a DNA plasmid encoding the non-structural 1 (NS1) gene fused to the tissue plasminogen activator signal sequence. Vaccine 2006; 24(2):195–205.

Crance, J.M., Scaramozzino, N., Jouan, A. and Garin, D. Interferon, ribavirin, 6-azauridine and glycyrrhizin: antiviral compounds active against pathogenic flaviviruses. Antiviral Res. 2003; 58(1):73–79.

Crill, W.D. and Roehrig, J.T. Monoclonal antibodies that bind to domain III of dengue virus E glycoprotein are the most efficient blockers of virus adsorption to Vero cells. J. Virol. 2001; 75(16):7769–7773.

Custer, B., Busch, M.P., Marfin, A.A. and Petersen, L.R. The cost-effectiveness of screening the U.S. blood supply for West Nile virus. Ann. Intern. Med. 2005; 143(7):486–492.

Custer, B., Tomasulo, P.A., Murphy, E.L., Caglioti, S., Harpool, D., McEvoy, P. and Busch, M.P. Triggers for switching from minipool testing by nucleic acid technology to individual-donation nucleic acid testing for West Nile virus: analysis of 2003 data to inform 2004 decision making. Transfusion 2004; 44(11):1547–1554.

Davis, B.S., Chang, G.J., Cropp, B., Roehrig, J.T., Martin, D.A., Mitchell, C.J., Bowen, R. and Bunning, M.L. West Nile virus recombinant DNA vaccine protects mouse and horse from virus challenge and expresses in vitro a noninfectious recombinant antigen that can be used in enzyme-linked immunosorbent assays. J. Virol. 2001; 75(9):4040–4047.

Davis, C.T., Ebel, G.D., Lanciotti, R.S., Brault, A.C., Guzman, H., Siirin, M., Lambert, A., Parsons, R.E., Beasley, D.W., Novak, R.J., Elizondo-Quiroga, D., Green, E.N., Young, D.S., Stark, L.M., Drebot, M.A., Artsob, H., Tesh, R.B., Kramer, L.D. and Barrett, A.D. Phylogenetic analysis of North American West Nile virus isolates, 2001–2004: evidence for the emergence of a dominant genotype. Virology 2005; 342(2):252–265.

Day, C.W., Smee, D.F., Julander, J.G., Yamshchikov, V.F., Sidwell, R.W. and Morrey, J.D. Error-prone replication of West Nile virus caused by ribavirin. Antiviral Res. 2005; 67(1):38–45.

Diamond, M.S., Shrestha, B., Marri, A., Mahan, D. and Engle, M. B cells and antibody play critical roles in the immediate defense of disseminated infection by West Nile encephalitis virus. J. Virol. 2003; 77(4):2578–2586.

Douglas, M.W., Kesson, A.M. and King, N.J. CTL recognition of west Nile virus-infected fibroblasts is cell cycle dependent and is associated with virus-induced increases in class I MHC antigen expression. Immunology 1994; 82(4):561–570.

Elshuber, S., Allison, S.L., Heinz, F.X. and Mandl, C.W. Cleavage of protein prM is necessary for infection of BHK-21 cells by tick-borne encephalitis virus. J. Gen. Virol. 2003; 84(Pt 1):183–191.

Engle, M.J. and Diamond, M.S. Antibody prophylaxis and therapy against West Nile virus infection in wild-type and immunodeficient mice. J. Virol. 2003; 77(24):12941–12949.

Epstein, J.S. Insights on donor screening for West Nile virus. Transfusion 2005; 45(4):460–462.

Gallian, P., De Micco, P., de Lamballerie, X., Levayer, T., Levacon, F., Guntz, P., Mercier, B., Dupond, I., Cornillot, C. and Andreu, G. Prevention of West Nile virus transmission by blood transfusion: a comparison of nucleic acid test screening assays. Transfusion 2005; 45(9):1540–1541.

Glass, W.G., Lim, J.K., Cholera, R., Pletnev, A.G., Gao, J.L. and Murphy, P.M. Chemokine receptor CCR5 promotes leukocyte trafficking to the brain and survival in West Nile virus infection. J. Exp. Med. 2005; 202(8):1087–1098.

Glass, W.G., McDermott, D.H., Lim, J.K., Lekhong, S., Yu, S.F., Frank, W.A., Pape, J., Cheshier, R.C. and Murphy, P.M. CCR5 deficiency increases risk of symptomatic West Nile virus infection. J. Exp. Med. 2006; 203(1):35–40.

Gollins, S.W. and Porterfield, J.S. A new mechanism for the neutralization of enveloped viruses by antiviral antibody. Nature 1986; 321(6067):244–246.

Gould, L.H., Sui, J., Foellmer, H., Oliphant, T., Wang, T., Ledizet, M., Murakami, A., Noonan, K., Lambeth, C., Kar, K., Anderson, J.F., de Silva, A.M., Diamond, M.S., Koski, R.A., Marasco, W.A. and Fikrig, E. Protective and therapeutic capacity of human single-chain Fv-Fc fusion proteins against West Nile virus. J. Virol. 2005; 79(23):14606–14613.

Grosenbaugh, D.A., Backus, C.S., Karaca, K., Minke, J.M. and Nordgren, R.M. The anamnestic serologic response to vaccination with a canarypox virus-vectored recombinant West Nile virus (WNV) vaccine in horses previously vaccinated with an inactivated WNV vaccine. Vet. Ther. 2004; 5(4):251–257.

Guarner, J., Shieh, W.J., Hunter, S., Paddock, C.D., Morken, T., Campbell, G.L., Marfin, A.A. and Zaki, S.R. Clinicopathologic study and laboratory diagnosis of 23 cases with West Nile virus encephalomyelitis. Hum. Pathol. 2004; 35(8):983–990.

Guirakhoo, F., Heinz, F.X. and Kunz, C. Epitope model of tick-borne encephalitis virus envelope glycoprotein E: analysis of structural properties, role of carbohydrate side chain, and conformational changes occurring at acidic pH. Virology 1989; 169(1):90–99.

Guo, J.T., Hayashi, J. and Seeger, C. West Nile virus inhibits the signal transduction pathway of alpha interferon. J. Virol. 2005; 79(3):1343–1350.

Hanley, K.A., Goddard, L.B., Gilmore, L.E., Scott, T.W., Speicher, J., Murphy, B.R. and Pletnev, A.G. Infectivity of West Nile/dengue chimeric viruses for West Nile and dengue mosquito vectors. Vector Borne Zoonotic Dis. 2005; 5(1):1–10.

Hardinger, K.L., Miller, B., Storch, G.A., Desai, N.M. and Brennan, D.C. West Nile virus-associated meningoencephalitis in two chronically immunosuppressed renal transplant recipients. Am. J. Transplant. 2003; 3(10):1312–1315.

Harrington, T., Kuehnert, M.J., Kamel, H., Lanciotti, R.S., Hand, S., Currier, M., Chamberland, M.E., Petersen, L.R. and Marfin, A.A. West Nile virus infection transmitted by blood transfusion. Transfusion 2003; 43(8):1018–1022.

Hayes, E.B. and O'Leary, D.R. West Nile virus infection: a pediatric perspective. Pediatrics 2004; 113(5):1375–1381.

Hayes, E.B., Sejvar, J.J., Zaki, S.R., Lanciotti, R.S., Bode, A.V. and Campbell, G.L. Virology, pathology, and clinical manifestations of West Nile virus disease. Emerg. Infect. Dis. 2005; 11(8):1174–1179.

Heinz, F.X., Berger, R., Majdic, O., Knapp, W. and Kunz, C. Monoclonal antibodies to the structural glycoprotein of tick-borne encephalitis virus. Infect. Immun. 1982; 37(3):869–874.

Heinz, F.X., Berger, R., Tuma, W. and Kunz, C. A topological and functional model of epitopes on the structural glycoprotein of tick-borne encephalitis virus defined by monoclonal antibodies. Virology 1983; 126(2):525–537.

Heinz, F.X., Mandl, C.W., Holzmann, H., Kunz, C., Harris, B.A., Rey, F. and Harrison, S.C. The flavivirus envelope protein E: isolation of a soluble form from tick-borne encephalitis virus and its crystallization. J. Virol. 1991; 65(10):5579–5583.

Heinz, F.X., Tuma, W., Guirakhoo, F., Berger, R. and Kunz, C. Immunogenicity of tick-borne encephalitis virus glycoprotein fragments: epitope-specific analysis of the antibody response. J. Gen. Virol. 1984; 65(Pt 11):1921–1929.

Henchal, E.A., Henchal, L.S. and Schlesinger, J.J. Synergistic interactions of anti-NS1 monoclonal antibodies protect passively immunized mice from lethal challenge with dengue 2 virus. J. Gen. Virol. 1988; 69(Pt 8):2101–2107.

Hiatt, B., DesJardin, L., Carter, T., Gingrich, R., Thompson, C. and de Magalhaes-Silverman, M. A fatal case of West Nile virus infection in a bone marrow transplant recipient. Clin. Infect. Dis. 2003; 37(9):e129–e131.

Hrnicek, M.J. and Mailliard, M.E. Acute West Nile virus in two patients receiving interferon and ribavirin for chronic hepatitis C. Am. J. Gastroenterol. 2004; 99(5):957.

Hsieh, W.C., Wang, S.P. and Rasmussen, A.F. Prophylactic therapy of Japanese encephalitis passive immunization combined with CO_2 inhalation. Proc. Soc. Exp. Biol. Med. 1963; 112:267–269.

Huang, C.Y., Silengo, S.J., Whiteman, M.C. and Kinney, R.M. Chimeric dengue 2 PDK-53/West Nile NY99 viruses retain the phenotypic attenuation markers of the candidate PDK-53 vaccine virus and protect mice against lethal challenge with West Nile virus. J. Virol. 2005; 79(12):7300–7310.

Hughes, A.L. Evolutionary change of predicted cytotoxic T cell epitopes of dengue virus. Infect. Genet. Evol. 2001; 1(2):123–130.

Humar, A. Screening for West Nile virus: more uncertainty. Am. J. Transplant. 2004; 4(8):1217–1218.

Hunsperger, E. and Roehrig, J.T. Characterization of West Nile viral replication and maturation in peripheral neurons in culture. J. Neurovirol. 2005; 11(1):11–22.

Iwamoto, M., Jernigan, D.B., Guasch, A., Trepka, M.J., Blackmore, C.G., Hellinger, W.C., Pham, S.M., Zaki, S., Lanciotti, R.S., Lance-Parker, S.E., DiazGranados, C.A., Winquist, A.G., Perlino, C.A., Wiersma, S., Hillyer, K.L., Goodman, J.L., Marfin, A.A., Chamberland, M.E. and Petersen, L.R. Transmission of West Nile virus from an organ donor to four transplant recipients. N. Engl. J. Med. 2003; 348(22):2196–2203.

Jackson, A.C. Therapy of West Nile virus infection. Can. J. Neurol. Sci. 2004; 31(2):131–134.

Jeha, L.E., Sila, C.A., Lederman, R.J., Prayson, R.A., Isada, C.M. and Gordon, S.M. West Nile virus infection: a new acute paralytic illness. Neurology 2003; 61(1):55–59.

Johnson, A.J., Cheshier, R.C., Cosentino, G., Masri, H.P., Mock, V., Oesterle, R., Lanciotti, R.S., Martin. D.A., Panella, A.j., Kosoy, O. and Biggerstaff. Validation of a microsphere-based immunoassay for detection of anti-West Nile virus and anti-St. Louis encephalitis virus immunoglobulin m antibodies. Clin. Vaccine Immunol. 2007; 14(9):1084–1093.

Johnson, A.J., Noga, A.J., Kosoy, O., Lanciotti, R.S., Johnson, A.A. and Biggerstaff, B.J. Duplex microsphere-based immunoassay for detection of anti-West Nile virus and anti-St. Louis encephalitis virus immunoglobulin m antibodies. Clin. Diagn. Lab. Immunol. 2005; 12(5):566–574.

Johnson, B.W., Chambers, T.V., Crabtree, M.B., Arroyo, J., Monath, T.P. and Miller, B.R. Growth characteristics of the veterinary vaccine candidate ChimeriVax-West Nile (WN) virus in *Aedes* and *Culex* mosquitoes. Med. Vet. Entomol. 2003; 17(3):235–243.

Jordan, I., Briese, T., Fischer, N., Lau, J.Y. and Lipkin, W.I. Ribavirin inhibits West Nile virus replication and cytopathic effect in neural cells. J. Infect. Dis. 2000; 182(4):1214–1217.

Kajaste-Rudnitski, A., Mashimo, T., Frenkiel, M.P., Guenet, J.L., Lucas, M. and Despres, P. The 2′,5′ oligoadenylate synthetase 1b is a potent inhibitor of West Nile virus replication inside infected cells. J. Biol. Chem. 2005; 281(8):4624–4637.

Kalil, A.C., Devetten, M.P., Singh, S., Lesiak, B., Poage, D.P., Bargenquast, K., Fayad, P. and Freifeld, A.G. Use of interferon-alpha in patients with West Nile encephalitis: report of 2 cases. Clin. Infect. Dis. 2005; 40(5):764–766.

Kanai, R., Kar, K., Anthony, K., Gould, L.H., Ledizet, M., Fikrig, E., Marasco, W.A., Koski, R.A. and Modis, Y. Crystal structure of West Nile virus envelope glycoprotein reveals viral surface epitopes. J. Virol. 2006; 80:11000–11008.

Kapoor, H., Signs, K., Somsel, P., Downes, F.P., Clark, P.A. and Massey, J.P. Persistence of West Nile Virus (WNV) IgM antibodies in cerebrospinal fluid from patients with CNS disease. J. Clin. Virol. 2004; 31(4):289–291.

Karaca, K., Bowen, R., Austgen, L.E., Teehee, M., Siger, L., Grosenbaugh, D., Loosemore, L., Audonnet, J.C., Nordgren, R. and Minke, J.M. Recombinant canarypox vectored West Nile virus (WNV) vaccine protects dogs and cats against a mosquito WNV challenge. Vaccine 2005; 23(29):3808–3813.

Kesson, A.M., Cheng, Y. and King, N.J. Regulation of immune recognition molecules by flavivirus, West Nile. Viral Immunol. 2002; 15(2):273–283.

Kiberd, B.A. and Forward, K. Screening for West Nile virus in organ transplantation: a medical decision analysis. Am. J. Transplant. 2004; 4(8):1296–1301.

Klee, A.L., Maidin, B., Edwin, B., Poshni, I., Mostashari, F., Fine, A., Layton, M. and Nash, D. Long-term prognosis for clinical West Nile virus infection. Emerg. Infect. Dis. 2004; 10(8): 1405–1411.

Klein, R.S., Lin, E., Zhang, B., Luster, A.D., Tollett, J., Samuel, M.A., Engle, M. and Diamond, M.S. Neuronal CXCL10 directs CD8+ T-cell recruitment and control of West Nile virus encephalitis. J. Virol. 2005; 79(17):11457–11466.

Kleinman, S., Glynn, S.A., Busch, M., Todd, D., Powell, L., Pietrelli, L., Nemo, G., Schreiber, G., Bianco, C. and Katz, L. The 2003 West Nile virus United States epidemic: the America's Blood Centers experience. Transfusion 2005; 45(4):469–479.

Kleinschmidt-DeMasters, B.K., Marder, B.A., Levi, M.E., Laird, S.P., McNutt, J.T., Escott, E.J., Everson, G.T. and Tyler, K.L. Naturally acquired West Nile virus encephalomyelitis in transplant recipients: clinical, laboratory, diagnostic, and neuropathological features. Arch. Neurol. 2004; 61(8):1210–1220.

Konishi, E., Pincus, S., Fonseca, B.A., Shope, R.E., Paoletti, E. and Mason, P.W. Comparison of protective immunity elicited by recombinant vaccinia viruses that synthesize E or NS1 of Japanese encephalitis virus. Virology 1991; 185(1):401–410.

Kreil, T.R. West Nile virus: recent experience with the model virus approach. Dev. Biol. (Basel) 2004; 118:101–105.

Kreil, T.R., Burger, I., Attakpah, E., Olas, K. and Eibl, M.M. Passive immunization reduces immunity that results from simultaneous active immunization against tick-borne encephalitis virus in a mouse model. Vaccine 1998a; 16(9–10):955–959.

Kreil, T.R., Burger, I., Bachmann, M., Fraiss, S. and Eibl, M.M. Antibodies protect mice against challenge with tick-borne encephalitis virus (TBEV)-infected macrophages. Clin. Exp. Immunol. 1997; 110(3):358–361.

Kreil, T.R. and Eibl, M.M. Pre- and postexposure protection by passive immunoglobulin but no enhancement of infection with a flavivirus in a mouse model. J. Virol. 1997; 71(4):2921–2927.

Kreil, T.R., Maier, E., Fraiss, S. and Eibl, M.M. Neutralizing antibodies protect against lethal flavivirus challenge but allow for the development of active humoral immunity to a nonstructural virus protein. J. Virol. 1998b; 72(4):3076–3081.

Kuhn, R.J., Zhang, W., Rossmann, M.G., Pletnev, S.V., Corver, J., Lenches, E., Jones, C.T., Mukhopadhyay, S., Chipman, P.R., Strauss, E.G., Baker, T.S. and Strauss, J.H. Structure of dengue virus: implications for flavivirus organization, maturation, and fusion. Cell 2002; 108(5):717–725.

Kulkarni, A.B., Mullbacher, A. and Blanden, R.V. Functional analysis of macrophages, B cells and splenic dendritic cells as antigen-presenting cells in West Nile virus-specific murine T lymphocyte proliferation. Immunol. Cell Biol. 1991; 69(Pt 2):71–80.

Kurane, I., Brinton, M.A., Samson, A.L. and Ennis, F.A. Dengue virus-specific, human CD4+ CD8-cytotoxic T-cell clones: multiple patterns of virus cross-reactivity recognized by NS3-specific T-cell clones. J. Virol. 1991; 65(4):1823–1828.

Kurane, I., Okamoto, Y., Dai, L.C., Zeng, L.L., Brinton, M.A. and Ennis, F.A. Flavivirus-cross-reactive, HLA-DR15-restricted epitope on NS3 recognized by human CD4+ CD8-cytotoxic T lymphocyte clones. J. Gen. Virol. 1995; 76(Pt 9):2243–2249.

Kurane, I., Zeng, L., Brinton, M.A. and Ennis, F.A. Definition of an epitope on NS3 recognized by human CD4+ cytotoxic T lymphocyte clones cross-reactive for dengue virus types 2, 3, and 4. Virology 1998; 240(2):169–174.

Kwan, W.H., Helt, A.M., Maranon, C., Barbaroux, J.B., Hosmalin, A., Harris, E., Fridman, W.H. and Mueller, C.G. Dendritic cell precursors are permissive to dengue virus and human immunodeficiency virus infection. J. Virol. 2005; 79(12):7291–7299.

Lanciotti, R.S. Molecular amplification assays for the detection of flaviviruses. Adv. Virus Res. 2003; 61:67–99.

Lanciotti, R.S., Ebel, G.D., Deubel, V., Kerst, A.J., Murri, S., Meyer, R., Bowen, M., McKinney, N., Morrill, W.E., Crabtree, M.B., Kramer, L.D. and Roehrig, J.T. Complete genome sequences and phylogenetic analysis of West Nile virus strains isolated from the United States, Europe, and the Middle East. Virology 2002; 298(1):96–105.

Lanciotti, R.S. and Kerst, A.J. Nucleic acid sequence-based amplification assays for rapid detection of West Nile and St. Louis encephalitis viruses. J. Clin. Microbiol. 2001; 39(12):4506–4513.

Lanciotti, R.S., Kerst, A.J., Nasci, R.S., Godsey, M.S., Mitchell, C.J., Savage, H.M., Komar, N., Panella, N.A., Allen, B.C., Volpe, K.E., Davis, B.S. and Roehrig, J.T. Rapid detection of West Nile virus from human clinical specimens, field-collected mosquitoes, and avian samples by a TaqMan reverse transcriptase-PCR assay. J. Clin. Microbiol. 2000; 38(11):4066–4071.

Lanciotti, R.S., Roehrig, J.T., Deubel, V., Smith, J., Parker, M., Steele, K., Crise, B., Volpe, K.E., Crabtree, M.B., Scherret, J.H., Hall, R.A., MacKenzie, J.S., Cropp, C.B., Panigrahy, B., Ostlund, E., Schmitt, B., Malkinson, M., Banet, C., Weissman, J., Komar, N., Savage, H.M., Stone, W., McNamara, T. and Gubler, D.J. Origin of the West Nile virus responsible for an outbreak of encephalitis in the northeastern United States. Science 1999; 286(5448):2333–2337.

Langevin, S.A., Arroyo, J., Monath, T.P. and Komar, N. Host-range restriction of chimeric yellow fever-West Nile vaccine in fish crows (*Corvus ossifragus*). Am. J. Trop. Med. Hyg. 2003; 69(1):78–80.

Lee, D.H., Mathew, J., Pfahler, W., Ma, D., Valinsky, J., Prince, A.M. and Andrus, L. Individual donor nucleic acid amplification testing for detection of West Nile virus. J. Clin. Microbiol. 2005; 43(10):5111–5116.

Leyssen, P., Balzarini, J., De Clercq, E. and Neyts, J. The predominant mechanism by which ribavirin exerts its antiviral activity in vitro against flaviviruses and paramyxoviruses is mediated by inhibition of IMP dehydrogenase. J. Virol. 2005; 79(3): 1943–1947.

Li, J., Loeb, J.A., Shy, M.E., Shah, A.K., Tselis, A.C., Kupski, W.J. and Lewis, R.A. Asymmetric flaccid paralysis: a neuromuscular presentation of West Nile virus infection. Ann. Neurol. 2003; 53(6):703–710.

Liu, W.J., Wang, X.J., Mokhonov, V.V., Shi, P.Y., Randall, R. and Khromykh, A.A. Inhibition of interferon signaling by the New York 99 strain and Kunjin subtype of West Nile virus involves blockage of STAT1 and STAT2 activation by nonstructural proteins. J. Virol. 2005; 79(3):1934–1942.

Lobigs, M., Arthur, C.E., Mullbacher, A. and Blanden, R.V. The flavivirus nonstructural protein NS3 is a dominant source of cytotoxic T cell peptide determinants. Virology 1994; 202(1):195–201.

Lobigs, M., Mullbacher, A. and Regner, M. MHC class I upregulation by flaviviruses: Immune interaction with unknown advantage to host or pathogen. Immunol. Cell Biol. 2003; 81(3):217–223.

Lozach, P.Y., Burleigh, L., Staropoli, I., Navarro-Sanchez, E., Harriague, J., Virelizier, J.L., Rey, F.A., Despres, P., Arenzana-Seisdedos, F. and Amara, A. Dendritic cell-specific intercellular adhesion molecule 3-grabbing non-integrin (DC-SIGN)-mediated enhancement of dengue virus infection is independent of DC-SIGN internalization signals. J. Biol. Chem. 2005; 280(25):23698–23708.

Lucas, M., Mashimo, T., Frenkiel, M.P., Simon-Chazottes, D., Montagutelli, X., Ceccaldi, P.E., Guenet, J.L. and Despres, P. Infection of mouse neurones by West Nile virus is modulated by the interferon-inducible 2′-5′ oligoadenylate synthetase 1b protein. Immunol. Cell Biol. 2003; 81(3):230–236.

Lustig, S., Olshevsky, U., Ben-Nathan, D., Lachmi, B.E., Malkinson, M., Kobiler, D. and Halevy, M. A live attenuated West Nile virus strain as a potential veterinary vaccine. Viral Immunol. 2000; 13(4):401–410.

Macdonald, J., Tonry, J., Hall, R.A., Williams, B., Palacios, G., Ashok, M.S., Jabado, O., Clark, D., Tesh, R.B., Briese, T. and Lipkin, W.I. NS1 protein secretion during the acute phase of West Nile virus infection. J. Virol. 2005; 79(22):13924–13933.

Macedo de Oliveira, A., Beecham, B.D., Montgomery, S.P., Lanciotti, R.S., Linnen, J.M., Giachetti, C., Pietrelli, L.A., Stramer, S.L. and Safranek, T.J. West Nile virus blood transfusion-related infection despite nucleic acid testing. Transfusion 2004; 44(12):1695–1699.

Malan, A.K., Martins, T.B., Hill, H.R. and Litwin, C.M. Evaluations of commercial West Nile virus immunoglobulin G (IgG) and IgM enzyme immunoassays show the value of continuous validation. J. Clin. Microbiol. 2004; 42(2):727–733.

Malkinson, M., Banet, C., Khinich, Y., Samina, I., Pokamunski, S. and Weisman, Y. Use of live and inactivated vaccines in the control of West Nile fever in domestic geese. Ann. N.Y. Acad. Sci. 2001; 951:255–261.

Mandl, C.W., Guirakhoo, F., Holzmann, H., Heinz, F.X. and Kunz, C. Antigenic structure of the flavivirus envelope protein E at the molecular level, using tick-borne encephalitis virus as a model. J. Virol. 1989; 63(2):564–571.

Marovich, M., Grouard-Vogel, G., Louder, M., Eller, M., Sun, W., Wu, S.J., Putvatana, R., Murphy, G., Tassaneetrithep, B., Burgess, T., Birx, D., Hayes, C., Schlesinger-Frankel, S. and Mascola, J. Human dendritic cells as targets of dengue virus infection. J. Investig. Dermatol. Symp. Proc. 2001; 6(3):219–224.

Martin, D.A., Biggerstaff, B.J., Allen, B., Johnson, A.J., Lanciotti, R.S. and Roehrig, J.T. Use of immunoglobulin m cross-reactions in differential diagnosis of human flaviviral encephalitis infections in the United States. Clin. Diagn. Lab. Immunol. 2002; 9(3):544–549.

Martin, D.A., Noga, A., Kosoy, O., Johnson, A.J., Petersen, L.R. and Lanciotti, R.S. Evaluation of a diagnostic algorithm using immunoglobulin M enzyme-linked immunosorbent assay to differentiate human West Nile Virus and St. Louis encephalitis virus infections during the 2002 West Nile Virus epidemic in the United States. Clin. Diagn. Lab. Immunol. 2004; 11(6):1130–1133.

Mashimo, T., Lucas, M., Simon-Chazottes, D., Frenkiel, M.P., Montagutelli, X., Ceccaldi, P.E., Deubel, V., Guenet, J.L. and Despres, P. A nonsense mutation in the gene encoding 2'-5'-oligoadenylate synthetase/L1 isoform is associated with West Nile virus susceptibility in laboratory mice. Proc. Natl. Acad. Sci. U.S.A. 2002; 99(17):11311–11316.

Mathew, A., Kurane, I., Rothman, A.L., Zeng, L.L., Brinton, M.A. and Ennis, F.A. Dominant recognition by human CD8+ cytotoxic T lymphocytes of dengue virus nonstructural proteins NS3 and NS1.2a. J. Clin. Invest. 1996; 98(7):1684–1691.

Mathews, J.H., Allan, J.E., Roehrig, J.T., Brubaker, J.R., Uren, M.F. and Hunt, A.R. T-helper cell and associated antibody response to synthetic peptides of the E glycoprotein of Murray Valley encephalitis virus. J. Virol. 1991; 65(10):5141–5148.

Mathews, J.H. and Roehrig, J.T. Elucidation of the topography and determination of the protective epitopes on the E glycoprotein of Saint Louis encephalitis virus by passive transfer with monoclonal antibodies. J. Immunol. 1984; 132(3):1533–1537.

Mathews, J.H., Roehrig, J.T., Brubaker, J.R., Hunt, A.R. and Allan, J.E. A synthetic peptide to the E glycoprotein of Murray Valley encephalitis virus defines multiple virus-reactive T- and B-cell epitopes. J. Virol. 1992; 66(11):6555–6562.

Minke, J.M., Siger, L., Karaca, K., Austgen, L., Gordy, P., Bowen, R., Renshaw, R.W., Loosmore, S., Audonnet, J.C. and Nordgren, B. Recombinant canarypoxvirus vaccine carrying the prM/E genes of West Nile virus protects horses against a West Nile virus-mosquito challenge. Arch. Virol. Suppl. 2004; 18:221–230.

Modis, Y., Ogata, S., Clements, D. and Harrison, S.C. A ligand-binding pocket in the dengue virus envelope glycoprotein. Proc. Natl. Acad. Sci. U.S.A. 2003; 100(12):6986–6991.

Modis, Y., Ogata, S., Clements, D. and Harrison, S.C. Structure of the dengue virus envelope protein after membrane fusion. Nature 2004; 427(6972):313–319.

Modis, Y., Ogata, S., Clements, D. and Harrison, S.C. Variable surface epitopes in the crystal structure of dengue virus type 3 envelope glycoprotein. J. Virol. 2005; 79(2):1223–1231.

Monath, T.P. Prospects for development of a vaccine against the West Nile virus. Ann. N.Y. Acad. Sci. 2001; 951:1–12.

Monath, T.P., Arroyo, J., Miller, C. and Guirakhoo, F. West Nile virus vaccine. Curr. Drug Targets Infect. Disord. 2001; 1(1):37–50.

Monath, T.P., Cropp, C.B. and Harrison, A.K. Mode of entry of a neurotropic arbovirus into the central nervous system. Reinvestigation of an old controversy. Lab. Invest. 1983; 48(4):399–410.

Monath, T.P., Liu, J., Kanesa-Thasan, N., Myers, G.A., Nichols, R., Deary, A., McCarthy, K., Johnson, C., Ermak, T., Shin, S., Arroyo, J., Guirakhoo, F., Kennedy, J.S., Ennis, F.A., Green, S. and Bedford, P. A live, attenuated recombinant West Nile virus vaccine. Proc. Natl. Acad. Sci. U.S.A. 2006; 103:6694–6699.

Morrey, J.D., Day, C.W., Julander, J.G., Blatt, L.M., Smee, D.F. and Sidwell, R.W. Effect of interferon-alpha and interferon-inducers on West Nile virus in mouse and hamster animal models. Antivir. Chem. Chemother. 2004; 15(2):101–109.

Morrey, J.D., Siddharthan, V., Olsen, A.L., Roper, G.Y., Wang, H., Baldwin, T.J., Koenig, S., Nordstrom, J.L. and Diamond, M.S. Humanized monoclonal antibody against West Nile virus envelope protein administered after neuronal infection protects against lethal encephalitis in hamsters. J. Infect. Dis. 2006; 194(9):1300–1308.

Morrey, J.D., Siddharthan, V., Olsen, A.L., Wang, H., Julander, J.G., Hall, J.O., Nordstrom, J.L., Koenig, S., Johnson, S. and Diamond, M.S. Defining limits of treatment with humanized neutralizing monoclonal antibody for West Nile virus neurological infection in a hamster model. Antimicrob. Agents Chemother. 2007; 51(7):2396–2402.

Mostashari, F., Bunning, M.L., Kitsutani, P.T., Singer, D.A., Nash, D., Cooper, M.J., Katz, N., Liljebjelke, K.A., Biggerstaff, B.J., Fine, A.D., Layton, M.C., Mullin, S.M., Johnson, A.J., Martin, D.A., Hayes, E.B. and Campbell, G.L. Epidemic West Nile encephalitis, New York, 1999: results of a household-based seroepidemiological survey. Lancet 2001; 358(9278):261–264.

Mukhopadhyay, S., Kim, B.S., Chipman, P.R., Rossmann, M.G. and Kuhn, R.J. Structure of West Nile virus. Science 2003; 302(5643):248.

Navarro-Sanchez, E., Altmeyer, R., Amara, A., Schwartz, O., Fieschi, F., Virelizier, J.L., Arenzana-Seisdedos, F. and Despres, P. Dendritic-cell-specific ICAM3-grabbing non-integrin is essential for the productive infection of human dendritic cells by mosquito-cell-derived dengue viruses. EMBO Rep. 2003; 4(7):723–728.

Ng, T., Hathaway, D., Jennings, N., Champ, D., Chiang, Y.W. and Chu, H.J. Equine vaccine for West Nile virus. Dev. Biol. (Basel) 2003; 114:221–227.

Nybakken, G.E., Nelson, C.A., Chen, B.R., Diamond, M.S. and Fremont, D.H. Crystal structure of the West Nile virus envelope glycoprotein. J. Virol. 2006; 80:11467–11474.

O'Leary, D.R., Kuhn, S., Kniss, K.L., Hinckley, A.F., Rasmussen, S.A., Pape, W.J., Kightlinger, L.K., Beecham, B.D., Miller, T.K., Neitzel, D.F., Michaels, S.R., Campbell, G.L., Lanciotti, R.S. and Hayes, E.B. Birth outcomes following West Nile Virus infection of pregnant women in the United States: 2003–2004. Pediatrics 2006; 117(3):e537–e545.

O'Leary, D.R., Marfin, A.A., Montgomery, S.P., Kipp, A.M., Lehman, J.A., Biggerstaff, B.J., Elko, V.L., Collins, P.D., Jones, J.E. and Campbell, G.L. The epidemic of West Nile virus in the United States, 2002. Vector Borne Zoonotic Dis. 2004; 4(1):61–70.

Oliphant, T., Engle, M., Nybakken, G.E., Doane, C., Johnson, S., Huang, L., Gorlatov, S., Mehlop, E., Marri, A., Chung, K.M., Ebel, G.D., Kramer, L.D., Fremont, D.H. and Diamond, M.S. Development of a humanized monoclonal antibody with therapeutic potential against West Nile virus. Nature Med. 2005; 11(5):522–530.

Okamoto, Y., Kurane, I., Leporati, A.M. and Ennis, F.A. Definition of the region on NS3 which contains multiple epitopes recognized by dengue virus serotype-cross-reactive and flavivirus-cross-reactive, HLA-DPw2-restricted CD4+ T cell clones. J. Gen. Virol. 1998; 79(Pt 4):697–704.

Paisley, J.E., Hinckley, A.F., O'Leary, D.R., Kramer, W.C., Lanciotti, R.S., Campbell, G.L. and Hayes, E.B. West Nile virus infection among pregnant women in a northern Colorado community, 2003 to 2004. Pediatrics 2006; 117(3):814–820.

Palmer, D.R., Sun, P., Celluzzi, C., Bisbing, J., Pang, S., Sun, W., Marovich, M.A. and Burgess, T. Differential effects of dengue virus on infected and bystander dendritic cells. J. Virol. 2005; 79(4):2432–2439.

Pealer, L.N., Marfin, A.A., Petersen, L.R., Lanciotti, R.S., Page, P.L., Stramer, S.L., Stobierski, M.G., Signs, K., Newman, B., Kapoor, H., Goodman, J.L. and Chamberland, M.E. Transmission of West Nile virus through blood transfusion in the United States in 2002. N. Engl. J. Med. 2003; 349(13):1236–1245.

Perelygin, A.A., Scherbik, S.V., Zhulin, I.B., Stockman, B.M., Li, Y. and Brinton, M.A. Positional cloning of the murine flavivirus resistance gene. Proc. Natl. Acad. Sci. U.S.A. 2002; 99(14):9322–9327.

Petersen, L.R. and Epstein, J.S. Problem solved? West Nile virus and transfusion safety. N. Engl. J. Med. 2005; 353(5):516–517.

Pokidysheva, E., Zhang, Y., Battisti, A.J., Bator-Kelly, C.M., Chipman, P.R., Xiao, C., Gregorio, G.G., Hendrickson, W.A., Kuhn, R.J. and Rossmann, M.G. Cryo-EM reconstruction of dengue virus in complex with the carbohydrate recognition domain of DC-SIGN. Cell 2006; 124(3):485–493.

Prince, H.E., Lape-Nixon, M., Busch, M.P., Tobler, L.H., Foster, G.A. and Stramer, S.L. Utilization of follow-up specimens from viremic blood donors to assess the value of West Nile virus immunoglobulin G avidity as an indicator of recent infection. Clin. Diagn. Lab. Immunol. 2005a; 12(9):1123–1126.

Prince, H.E., Tobler, L.H., Lape-Nixon, M., Foster, G.A., Stramer, S.L. and Busch, M.P. Development and persistence of West Nile virus-specific immunoglobulin M (IgM), IgA, and IgG in viremic blood donors. J. Clin. Microbiol. 2005b; 43(9):4316–4320.

Putnak, J.R. and Schlesinger, J.J. Protection of mice against yellow fever virus encephalitis by immunization with a vaccinia virus recombinant encoding the yellow fever virus non-structural proteins, NS1, NS2a and NS2b. J. Gen. Virol. 1990; 71(Pt 8):1697–1702.

Rados, C. First test approved to help detect West Nile virus. FDA Consum. 2003; 37(5):18–19.

Rey, F.A. Dengue virus envelope glycoprotein structure: new insight into its interactions during viral entry. Proc. Natl. Acad. Sci. U.S.A. 2003; 100(12):6899–6901.

Rey, F.A., Heinz, F.X., Mandl, C., Kunz, C. and Harrison, S.C. The envelope glycoprotein from tick-borne encephalitis virus at 2A resolution. Nature 1995; 375(6529):291–298.

Rice, C.M., Lenches, E.M., Eddy, S.R., Shin, S.J., Sheets, R.L. and Strauss, J.H. Nucleotide sequence of yellow fever virus: implications for flavivirus gene expression and evolution. Science 1985; 229(4715):726–733.

Roehrig, J.T. Antigenic structure of flavivirus proteins. Adv. Virus Res. 2003; 59:141–175.

Roehrig, J.T., Bolin, R.A. and Kelly, R.G. Monoclonal antibody mapping of the envelope glycoprotein of the dengue 2 virus, Jamaica. Virology 1998; 246(2):317–328.

Roehrig, J.T., Johnson, A.J., Hunt, A.R., Beaty, B.J. and Mathews, J.H. Enhancement of the antibody response to flavivirus B-cell epitopes by using homologous or heterologous T-cell epitopes. J. Virol. 1992; 66(6):3385–3390.

Roehrig, J.T., Nash, D., Maldin, B., Labowitz, A., Martin, D.A., Lanciotti, R.S. and Campbell, G.L. Persistence of virus-reactive serum immunoglobulin m antibody in confirmed West Nile virus encephalitis cases. Emerg. Infect. Dis. 2003; 9(3):376–379.

Roehrig, J.T., Staudinger, L.A., Hunt, A.R., Mathews, J.H. and Blair, C.D. Antibody prophylaxis and therapy for flavivirus encephalitis infections. Ann. N.Y. Acad. Sci. 2001; 951:286–297.

Roodyn, L. Immunization and other prophylactic measures. Br. Med. J. 1974; 4(5945):648–650.

Rothman, A.L., Kurane, I. and Ennis, F.A. Multiple specificities in the murine CD4+ and CD8+ T-cell response to dengue virus. J. Virol. 1996; 70(10):6540–6546.

Sakuntabhai, A., Turbpaiboon, C., Casademont, I., Chuansumrit, A., Lowhnoo, T., Kajaste-Rudnitski, A., Kalayanarooj, S.M., Tangnararatchakit, K., Tangthawornchaikul, N., Vasanawathana, S., Chaiyaratana, W., Yenchitsomanus, P.T., Suriyaphol, P., Avirutnan, P., Chokephaibulkit, K., Matsuda, F., Yoksan, S., Jacob, Y., Lathrop, G.M., Malasit, P., Despres, P. and Julier, C. A variant in the CD209 promoter is associated with severity of dengue disease. Nat. Genet. 2005; 37(5):507–513.

Samina, I., Khinich, Y., Simanov, M. and Malkinson, M. An inactivated West Nile virus vaccine for domestic geese-efficacy study and a summary of 4 years of field application. Vaccine 2005; 23(41):4955–4958.

Samuel, M.A. and Diamond, M.S. Alpha/beta interferon protects against lethal West Nile virus infection by restricting cellular tropism and enhancing neuronal survival. J. Virol. 2005; 79(21):13350–13361.

Sanchez-Guerrero, S.A., Romero-Estrella, S., Rodriguez-Ruiz, A., Infante-Ramirez, L., Gomez, A., Villanueva-Vidales, E., Garcia-Torres, M., Dominguez, A.M., Vazquez, J.A., Calderon, E.D., Valiente-Banuet, L., Linnen, J.M., Broulik, A., Harel, W. and Marin, Y.L.R.A. Detection of West Nile virus in the Mexican blood supply. Transfusion 2006; 46(1):111–117.

Sbrana, E., Tonry, J.H., Xiao, S.Y., da Rosa, A.P., Higgs, S. and Tesh, R.B. Oral transmission of West Nile virus in a hamster model. Am. J. Trop. Med. Hyg. 2005; 72(3):325–329.

Schlesinger, J.J., Brandriss, M.W., Cropp, C.B. and Monath, T.P. Protection against yellow fever in monkeys by immunization with yellow fever virus nonstructural protein NS1. J. Virol. 1986; 60(3):1153–1155.

Schlesinger, J.J., Brandriss, M.W. and Walsh, E.E. Protection of mice against dengue 2 virus encephalitis by immunization with the dengue 2 virus non-structural glycoprotein NS1. J. Gen. Virol. 1987; 68(Pt 3):853–857.

Schlesinger, J.J., Foltzer, M. and Chapman, S. The Fc portion of antibody to yellow fever virus NS1 is a determinant of protection against YF encephalitis in mice. Virology 1993; 192(1):132–141.

Scholle, F. and Mason, P.W. West Nile virus replication interferes with both poly(I:C)-induced interferon gene transcription and response to interferon treatment. Virology 2005; 342(1):77–87.

Sejvar, J.J., Haddad, M.B., Tierney, B.C., Campbell, G.L., Marfin, A.A., Van Gerpen, J.A., Fleischauer, A., Leis, A.A., Stokic, D.S. and Petersen, L.R. Neurologic manifestations and outcome of West Nile virus infection. JAMA 2003; 290(4):511–515.

Shieh, W.J., Guarner, J., Layton, M., Fine, A., Miller, J., Nash, D., Campbell, G.L., Roehrig, J.T., Gubler, D.J. and Zaki, S.R. The role of pathology in an investigation of an outbreak of West Nile encephalitis in New York, 1999. Emerg. Infect. Dis. 2000; 6(4):370–372.

Shrestha, B. and Diamond, M.S. Role of CD8+ T cells in control of West Nile virus infection. J. Virol. 2004; 78(15):8312–8321.

Shrestha, B., Samuel, M.A. and Diamond, M.S. CD8+ T cells require perforin to clear West Nile virus from infected neurons. J. Virol. 2006; 80(1):119–129.

Siger, L., Bowen, R.A., Karaca, K., Murray, M.J., Gordy, P.W., Loosmore, S.M., Audonnet, J.C., Nordgren, R.M. and Minke, J.M. Assessment of the efficacy of a single dose of a recombinant vaccine against West Nile virus in response to natural challenge with West Nile virus-infected mosquitoes in horses. Am. J. Vet. Res. 2004; 65(11):1459–1462.

Silvia, O.J., Pantelic, L., Mackenzie, J.S., Shellam, G.R., Papadimitriou, J. and Urosevic, N. Virus spread, tissue inflammation and antiviral response in brains of flavivirus susceptible and resistant mice acutely infected with Murray Valley encephalitis virus. Arch. Virol. 2004; 149(3):447–464.

Simkova, A. A study on the prophylactic application of immune gamma-globulin in tick-borne encephalitis. Acta Virol. 1959; 3:240–249.

Stiasny, K., Bressanelli, S., Lepault, J., Rey, F.A. and Heinz, F.X. Characterization of a membrane-associated trimeric low-pH-induced form of the class II viral fusion protein E from tick-borne encephalitis virus and its crystallization. J. Virol. 2004; 78(6):3178–3183.

Stramer, S.L., Fang, C.T., Foster, G.A., Wagner, A.G., Brodsky, J.P. and Dodd, R.Y. West Nile virus among blood donors in the United States, 2003 and 2004. N. Engl. J. Med. 2005; 353(5):451–459.

Tassaneetrithep, B., Burgess, T.H., Granelli-Piperno, A., Trumpfheller, C., Finke, J., Sun, W., Eller, M.A., Pattanapanyasat, K., Sarasombath, S., Birx, D.L., Steinman, R.M., Schlesinger, S. and Marovich, M.A. DC-SIGN (CD209) mediates dengue virus infection of human dendritic cells. J. Exp. Med. 2003; 197(7):823–829.

Tesh, R.B., Siirin, M., Guzman, H., Travassos da Rosa, A.P., Wu, X., Duan, T., Lei, H., Nunes, M.R. and Xiao, S.Y. Persistent West Nile virus infection in the golden hamster: studies on its mechanism and possible implications for other flavivirus infections. J. Infect. Dis. 2005; 192(2):287–295.

Tobler, L.H., Bianco, C., Glynn, S.A., Schreiber, G.B., Dille, B.J., Prince, H.E., Lanciotti, R.S., Linnen, J.M., Gallarda, J., Shyamala, V., Smith, D., Kleinman, S.H. and Busch, M.P. Detection of West Nile virus RNA and antibody in frozen plasma components from a voluntary market withdrawal during the 2002 peak epidemic. Transfusion 2005; 45(4):480–486.

Tonry, J.H., Xiao, S.Y., Siirin, M., Chen, H., da Rosa, A.P. and Tesh, R.B. Persistent shedding of West Nile virus in urine of experimentally infected hamsters. Am. J. Trop. Med. Hyg. 2005; 72(3):320–324.

Turell, M.J., Bunning, M., Ludwig, G.V., Ortman, B., Chang, J., Speaker, T., Spielman, A., McLean, R., Komar, N., Gates, R., McNamara, T., Creekmore, T., Farley, L. and Mitchell, C.J. DNA vaccine for West Nile virus infection in fish crows (*Corvus ossifragus*). Emerg. Infect. Dis. 2003; 9(9):1077–1081.

Urosevic, N. Is flavivirus resistance interferon type I-independent? Immunol. Cell Biol. 2003; 81(3):224–229.

Wang, T., Scully, E., Yin, Z., Kim, J.H., Wang, S., Yan, J., Mamula, M., Anderson, J.F., Craft, J. and Fikrig, E. IFN-gamma-producing gamma delta T cells help control murine West Nile virus infection. J. Immunol. 2003a; 171(5):2524–2531.

Wang, T., Town, T., Alexopoulou, L., Anderson, J.F., Fikrig, E. and Flavell, R.A. Toll-like receptor 3 mediates West Nile virus entry into the brain causing lethal encephalitis. Nat. Med. 2004; 10(12):1366–1373.

Wang, Y., Lobigs, M., Lee, E. and Mullbacher, A. CD8+ T cells mediate recovery and immunopathology in West Nile virus encephalitis. J. Virol. 2003b; 77(24):13323–13334.

Wong, S.J., Boyle, R.H., Demarest, V.L., Woodmansee, A.N., Kramer, L.D., Li, H., Drebot, M., Koski, R.A., Fikrig, E., Martin, D.A. and Shi, P.Y. Immunoassay targeting nonstructural protein 5 to differentiate West Nile virus infection from dengue and St. Louis encephalitis virus infections and from flavivirus vaccination. J. Clin. Microbiol. 2003; 41(9):4217–4223.

Wong, S.J., Demarest, V.L., Boyle, R.H., Wang, T., Ledizet, M., Kar, K., Kramer, L.D., Fikrig, E. and Koski, R.A. Detection of human anti-flavivirus antibodies with a West Nile virus recombinant antigen microsphere immunoassay. J. Clin. Microbiol. 2004; 42(1):65–72.

Wu, S.J., Grouard-Vogel, G., Sun, W., Mascola, J.R., Brachtel, E., Putvatana, R., Louder, M.K., Filgueira, L., Marovich, M.A., Wong, H.K., Blauvelt, A., Murphy, G.S., Robb, M.L., Innes, B.L., Birx, D.L., Hayes, C.G. and Frankel, S.S. Human skin Langerhans cells are targets of dengue virus infection. Nat. Med. 2000; 6(7):816–820.

Xiao, S.Y., Guzman, H., Zhang, H., Travassos da Rosa, A.P. and Tesh, R.B. West Nile virus infection in the golden hamster (*Mesocricetus auratus*): a model for West Nile encephalitis. Emerg. Infect. Dis. 2001; 7(4):714–721.

Yakub, I., Lillibridge, K.M., Moran, A., Gonzalez, O.Y., Belmont, J., Gibbs, R.A. and Tweardy, D.J. Single nucleotide polymorphisms in genes for 2′-5′-oligoadenylate synthetase and RNase L in patients hospitalized with West Nile virus infection. J. Infect. Dis. 2005; 192(10):1741–1748.

Yamshchikov, G., Borisevich, V., Seregin, A., Chaporgina, E., Mishina, M., Mishin, V., Kwok, C.W. and Yamshchikov, V. An attenuated West Nile prototype virus is highly immunogenic and protects against the deadly NY99 strain: a candidate for live WN vaccine development. Virology 2004; 330(1):304–312.

Zhang, Y., Zhang, W., Ogata, S., Clements, D., Strauss, J.H., Baker, T.S., Kuhn, R.J. and Rossmann, M.G. Conformational changes of the flavivirus E glycoprotein. Structure 2004; 12(9):1607–1618.

Zhang, Y.M., Hayes, E.P., McCarty, T.C., Dubois, D.R., Summers, P.L., Eckels, K.H., Chanock, R.M. and Lai, C.J. Immunization of mice with dengue structural proteins and nonstructural protein NS1 expressed by baculovirus recombinant induces resistance to dengue virus encephalitis. J. Virol. 1988; 62(8):3027–3031.

Zivny, J., Kurane, I., Leporati, A.M., Ibe, M., Takiguchi, M., Zeng, L.L., Brinton, M.A. and Ennis, F.A. A single nine-amino acid peptide induces virus-specific, CD8+ human cytotoxic T lymphocyte clones of heterogeneous serotype specificities. J. Exp. Med. 1995; 182(3):853–863.

CHAPTER

40

Yellow Fever

Sareen E. Galbraith[1,2] *and Alan D.T. Barrett*[2]

[1]Brain Infections Group, Divisions of Neurological Science and Medical Microbiology, University of Liverpool, Liverpool, UK

[2]Department of Pathology, Sealy Center for Vaccine Development, Center for Biodefense and Emerging Infectious Diseases, and Institute for Human Infections and Immunity, University of Texas Medical Branch, Galveston, TX, USA

OUTLINE

ABSTRACT

Yellow fever virus (YFV) is a mosquito-borne flavivirus that is the causative agent of yellow fever (YF), which is a disease that ranges in severity from an influenza-like syndrome to severe liver and renal dysfunction, circulatory shock, and hemorrhage. The disease causes substantial morbidity and mortality; case fatality rates often exceed

20% and, despite the availability of a vaccine, it is still a major public health problem for sub-Saharan Africa and tropical South America, and travelers to endemic areas. The virus is maintained in endemic areas of Africa and South America by enzootic transmission between mosquitoes and monkeys, and the epidemiology of the disease reflects the geographical distribution of the mosquito vectors. There are seven genotypes of this virus with five circulating in different geographic areas of Africa and two in South America. Two live-attenuated YF vaccines were simultaneously developed: the French neurotropic vaccine (FNV), or Dakar vaccine, and the 17D vaccine. In the 1980s the manufacture of FNV was completely discontinued because the World Health Organization (WHO) prohibited vaccination of children under 14 years old due to the high incidence of encephalitis. The development of the highly effective, live-attenuated 17D vaccine in the late 1930s is one of the milestones in vaccine research. The vaccine is safe and highly efficacious with one dose of vaccine giving immunity for at least 10 years, and may be lifelong immunity. Neutralizing antibodies are the correlate of protection. There are three substrains of the YF 17D vaccine in production today (17D-204, 17DD, and 17D-213). The WHO has standardized the seed lot and manufacturing processes, and published guidelines for approval of vaccine production and vaccination certificates. There are a number of contraindications for YF vaccination which include age, thymus disease, pregnancy, breast-feeding, immune suppression, and hypersensitivity to eggs. In recent years, a rare but significant number of severe adverse events have been reported involving vaccine-associated viscerotropic and neurotropic disease. Current vaccine development involves utilizing the 17D vaccine virus backbone, via infectious clone technology, to develop a variety of chimeric virus vaccines, some of which have reached clinical trials for human use.

INTRODUCTION

Yellow fever virus (YFV) is a mosquito-borne flavivirus that is the causative agent of yellow fever (YF), which is a disease that ranges in severity from an influenza-like syndrome to severe liver and renal dysfunction, circulatory shock, and hemorrhage. The disease causes substantial morbidity and mortality; case fatality rates often exceed 20% (Monath, 2004) and, despite the availability of a vaccine, is still a major public health problem for sub-Saharan Africa and tropical South America, and travelers to endemic areas (WHO, 2003). The virus is maintained in endemic areas of Africa and South America by enzootic transmission between mosquitoes and monkeys, and it can be transmitted transovarially from one mosquito to another (Marchous and Simond, 1905; Beaty et al., 1980; Monath 2004; Barrett and Monath, 2003b). There is an urban transmission cycle involving *Aedes aegypti* mosquitoes and humans (WHO, 2003; Barrett and Monath, 2003b). A highly effective, live-attenuated vaccine (strain 17D) has been available since the late 1930s (Strokes et al., 1928).

SHORT HISTORY OF THE DISEASE

Although the first mention of YF is in a Mayan manuscript reporting an epidemic that occurred in Guadeloupe and the Yucatan in 1648 (Carter, 1931; Hobson 1963), Griffin Hughes was the first to use the term "yellow fever" to describe the disease in his book in 1750 (Garrison, 1929). It is thought that the virus originated in Africa and then spread to South America due to the slave trade and trading routes (Lepiniec et al., 1994; Mutebi et al., 2001; Bryant et al., 2007). Subsequently, large epidemics occurred throughout the 18th and 19th centuries in the Caribbean islands, the United States, Africa, Europe, West Indies, and South America (Garrison, 1929; Powell, 1949; Taylor, 1951; Duffy, 1966; Coleman, 1984, 1987; Waddell, 1990–1992; Patterson, 1992). The isolation of wild-type YFV in the late 1920s (Strokes et al., 1928; Mathis et al., 1928) led directly to the development of highly effective live-attenuated vaccines in the 1930s. The vaccines, together with mosquito control measures, made epidemic control possible and resulted in the eradication of the virus from Europe and the United States. Currently, YF is considered to be a reemerging disease in areas of sub-Saharan Africa and tropical South America, where it causes regular epidemics. Thirty-three countries (15°N to 10°S of the equator), with a combined population of 508 million, are at risk in Africa. In the Americas, nine South American countries and several Caribbean islands are at risk, with Bolivia, Brazil, Colombia, Ecuador, and Peru at the greatest risk. Unvaccinated ecotourism travelers to endemic areas are also at risk of contracting YF.

ETIOLOGIC AGENT(S)

There were many debates regarding the agent that caused YF and Carlos Findlay was the first to suggest that mosquitoes transmitted the disease (Findlay, 1881; Chastel, 2003). Reed et al. (1901) confirmed that mosquitoes (*Ae. aegypti*) transmitted the agent and demonstrated that it was a virus because it was capable of passing through bacterial proof filters. Thus, YFV was the first animal virus shown to be transmitted by an arthropod host.

Classification

YFV is the prototype member of the genus *Flavivirus*, which is named after the Latin word for yellow, *Flavus*. The *Flavivirus* genus is one of three genera in the *Flaviviridae* family, the others being *Pestivirus* and *Hepacivirus*. The *Flavivirus* genus contains three major ecological subdivisions: tick-borne viruses, mosquito-borne viruses, and viruses with no known arthropod vector. The tick-borne viruses are subdivided into mammalian and seabird groups. The mosquito-borne viruses are subdivided into Aroa, dengue, Japanese encephalitis, Kokobera, Ntaya, Spondweni, and YF groups. The viruses with no known arthropod vector contain the Modoc and Rio Bravo groups. The YF group contains mosquito-borne Banzi, Bouboui, Edge Hill, Jugra, Potiskum, Saboya, Sepik, Uganda S, Wesselsbron, and YF viruses, plus the non-vector-borne Entebbe bat, Sokoluk and Yokose viruses. YFV has a spherical, enveloped nucleocapsid of 30 nm diameter and the entire virion measures approximately 50 nm in diameter. The envelope (E) protein is a dimer that sits parallel to the surface of the virus in association with the membrane (M) protein (Kuhn et al., 2002).

ANTIGENS ENCODED BY AGENT

The genome of YFV is approximately 11,000 nucleotides in length, depending on the strain, with the 17D-204 vaccine virus genome 10,862 nucleotides in length. The 3411 amino acid polyprotein is flanked by a 5′ noncoding region of 118 nucleotides and a 3′ noncoding region of 511 nucleotides (Rice et al., 1985; Dupuy et al., 1989). The 5′ terminus has a type 1 cap followed by two conserved nucleotides (AG) and the 3′ terminus lacks a poly A tract (Rice et al., 1985). The polyprotein encodes 10 proteins: the structural proteins; capsid (C), membrane (M) and envelope (E) are encoded by the N-terminal one-third of the polyprotein and the nonstructural proteins; NS1, NS2A, NS2B, NS3, NS4A, NS4B and NS5 are encoded by the C-terminal two-thirds of the polyprotein. Five YFV proteins (E, M, NS1, NS2B, and NS3) have been shown to be involved in immune responses in the host and these are illustrated in Fig. 40.1. The major immunogen is the E protein, which encodes epitopes inducing neutralizing antibodies that are primarily responsible for the protective immune response. The M protein has also been shown to contain epitopes inducing neutralizing antibodies but these have weak neutralizing activity (Schlesinger et al., 1983). NS1 is unusual, as it is a nonstructural protein that is glycosylated (N_{130} and N_{208}); it induces antibody-dependent cellular cytotoxicity and fixes complement. Human cytotoxic T cell epitopes are found on the E structural protein

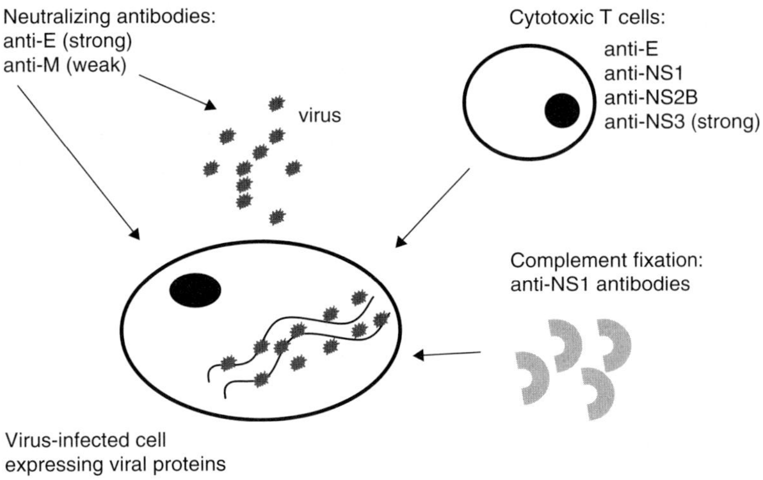

FIGURE 40.1 Yellow fever virus proteins that induce the host immune response (from Monath, 2001). The photomicrograph shows yellow fever virus particles in endoplasmic reticulum (magnification × 280,000). The E glycoprotein, which is inserted in the cell membrane–derived envelope, contains three conformational epitopes that are responsible for virus neutralization. These neutralizing antibody epitopes on the E protein are the principal mediators of protection against reinfection. The E protein and the nonstructural proteins (NS1, NS2B, and NS3) have cytotoxic T cell epitopes which generate cytotoxic T cells that clear virus-infected cells during recovery. Type-specific and cross-reactive complement fixation epitopes are also found on the NS1 protein, which generate production of cytotoxic antibodies that bind complement and help clear virus during the recovery phase.

and the NS1, NS2B, and NS3 nonstructural proteins (Co et al., 2002; Van der Most et al., 2002). Studies with mosquito-borne West Nile virus (WNV) were the first to show that the majority of flavivirus E protein epitopes were conformational (Gollins and Porterfield, 1986). Monoclonal antibodies have identified a number of overlapping epitopes on the E protein (Cammack and Gould, 1986; Schlesinger et al., 1984). Physically these epitopes are either strain-, YF type-, complex-specific or flavivirus genus common, while biologically some of these epitopes are associated with hemagglutination inhibition (HI), which may or may not be associated with neutralization (Schlesinger et al., 1983, 1984; Geske et al., 1983; Buckley and Gould, 1985; Cammack and Gould, 1986; Heinz, 1986; Gould et al., 1985, 1989; Barrett et al., 1989, 1990a, 1990b; Sil et al., 1992; Ledger et al., 1992). Overall, few epitopes are involved in neutralization and very few elicit high titer neutralization (Buckley and Gould, 1985; Cammack and Gould 1986; Ledger et al., 1992). To date, few epitopes have been mapped to specific amino acids on the E protein: two YF type-specific epitopes have been mapped to amino acids 71/72 and 153/155, a wild-type epitope to amino acid 173, and a 17D-204 substrain-specific epitope to amino acids 305 and 325 (Lobigs et al., 1987; Ryman et al., 1997, 1998). Like the E protein, NS1 has been shown to have YF type- and complex-specific, and flavivirus genus common epitopes. Some of these NS1 epitopes are involved in complement fixation but none are involved in neutralization (Putnak and Schlesinger, 1990; Schlesinger et al., 1983, 1986, 1990).

PROTECTIVE IMMUNE RESPONSE

Infection with wild-type YFV results in lifelong immunity and the YF 17D vaccine is also highly effective; immunity occurs within 10 days in 95% of vaccinees and induces protective immunity against all known wild-type strains. YFV induces rapid and specific humoral and cell-mediated immune responses, which will be discussed under the next two subheadings. The humoral response is better characterized than the cell-mediated response, while the innate immune response is an active area of research at the present time.

Antibody Response

IgM antibodies are detected by one week postvaccination, reach a peak by two weeks postvaccination, and then they decline over several months (Lhuillier et al., 1986). The neutralizing and HI antibody responses are rapid and detected by seven days postvaccination (Porterfield, 1954; Theiler and Casals, 1958). The HI antibody peaks 1–2 months post-immunization and then steadily declines, whereas the neutralizing antibody response persists and is the primary mechanism of protection against reexposure. Complement fixation antibodies are detected by two weeks post-wild-type infection and their levels increase rapidly but they are not seen postvaccination. These antibodies bind complement to help clear virus during recovery from infection and their levels drop by 4–12 months post-wild-type infection (Putnak and Schlesinger, 1990; Schlesinger et al., 1986, 1990). All studies to date indicate that neutralizing antibody is the correlate of protection with over 98% of vaccinees being fully protected for at least 10 years and neutralizing antibodies persist for at least 45 years (Sawyer, 1931; Bauer and Hudson, 1930; Groot and Ribeiro 1962; Poland et al., 1981; Lobigs et al., 1987; Brandriss et al., 1990; Niedrig et al., 1999; Monath and Heinz, 1996; Monath et al., 2000).

CMI Response

The17D vaccine strain is a potent inducer of CD4+ and CD8+ cytotoxic T cell responses against the nonstructural proteins NS1, NS2B, NS3, and the E structural protein (Reinhardt et al., 1998; Co et al., 2002; van der Most et al., 2002). A study by Santos et al. (2005) showed increases in CD3+, CD4+, and CD8+ T cells in vaccinees and re-vaccinees at 30 days postvaccination. Studies involving CD4 knockout mice have shown that CD4+ T cells are central to the immune response against YFV through antibody and cytokine production, B cell activation, and posterior plasma cell differentiation (Liu and Chambers 2001). CD8+ T cells also contribute to the protective immune response by mediating viral clearance (Doherty, 1996). The CD8 T cell response to YFV reaches its peak 1–2 weeks postvaccination and is detectable up to 19 months, which is thought to contribute to vaccine efficacy (Reinhardt et al., 1998; Co et al., 2002). There is also a significant increase in tumor necrosis factor (TNF) peaking on days 2 and 7 postvaccination (Hacker et al., 1998), which interferes with virus replication by activating or causing the differentiation of T cells and macrophages (Herbein and O'Brien, 2000). One recent study examining the activation events and modulatory pathways in T cells after vaccination found that CD4+ T cells were activated early (day 7), CD19+ T cells were activated by day 15, and CD8+ T cells were activated late (day 30). This up-regulation of modulatory features on CD4+ and CD8+ cells at day 15 caused a lower level of CD38+ T cells by day 30 postvaccination (Martins et al., 2007).

Innate Immune Response

Recent studies have shown that the innate immune system also plays an important role in determining the strength and quality of adaptive immune responses against viruses such as YF. Toll-like receptors (TLR) are key sensors that recognize double- and single-stranded RNA (Takeda et al., 2003; Beutler, 2004). The 17D vaccine strain replicates minimally (perhaps abortively) in dendritic cells without causing substantial apoptotic cell death (Barba-Spaeth et al., 2005; Palmer et al., 2007) and stimulates TLR 2, 7, 8, and 9 (Querec et al., 2006), which results in the production of type I interferon and other inflammatory cytokines (Wheelock and Sibley, 1965) and dendritic cell activation and maturation. These activated dendritic cells most likely migrate to regional lymph nodes and stimulate both cell-mediated and humoral adaptive immune responses (Iwasaki and Medzhitov, 2004; Barba-Spaeth et al., 2005; Kawai and Akira, 2006; Palmer et al., 2007; Querec and Pulendran, 2007). Viral interaction with alternate TLRs modifies the Th1 and Th2 cytokine balance produced by the activated immune cells, and it is possible that nonviral vaccine components could also influence this balance (Querec et al., 2006; Querec and Pulendran, 2007). The interferon-induced peripheral lymphocyte response from vaccinees may be mediated in part through the action of 2′-5′ oligoadenylate synthetase. Levels of this enzyme are increased in lymphocytes by day 4 postvaccination and reach a maximum at day 7. This enzyme catalyzes the production of 2′-5′ oligoadenylates, which activates latent RNAase, which degrades viral mRNA and impairs cellular protein synthesis (Bonnevie-Nielsen et al., 1989, 1995). Natural killer cells have also been shown to play a role early in virus infection (Fagraeus et al., 1982).

EPIDEMIOLOGY

YF is endemic and intermittently epidemic in tropical South America and sub-Saharan Africa, with an estimated 200,000 cases and 30,000 deaths annually (Figs 40.2A and B; Robertson et al., 1996; Vainio and Cutts, 1998; Monath, 2001; Vasconcelos 2003; Vasconcelos et al., 2004; Tomori, 2004; Gubler 2004; Bryan et al., 2004; WHO, 1998, 2001, 2007a, 2007b, 2007c; WHO/UNICEF, 2005; Roberts, 2007; Barrett and Higgs, 2007). YFV is thought to have originated in Africa and then was subsequently introduced into South America by slave trade travelers (Bryant et al., 2007, Bryant et al., 2004). This is supported by the greater heterogeneity of African YFV genomes compared to those in South America (Lepiniec et al., 1994; Mutebi et al., 2001; Gould et al., 2001; Barrett and Monath, 2003b). Further supporting evidence comes from the phylogenetic studies that show that West African strains are closer to South American strains than the central or East African strains (Lepiniec et al., 1994; Chang et al., 1995; Wang et al., 1996) and there is more genetic diversity in central and West African strains (Barrett and Higgs, 2007). Nucleotide sequencing studies have identified seven genotypes, based on ≥9% nucleotide variation, using the prM and E sequences in a neighbor-joining tree, with 5 circulating in different geographic areas of Africa and two of South America (Lepiniec et al., 1994; Mutebi et al., 2001; Vasconcelos et al., 2004). In general, YFV strains have evolved slowly (Deubel et al., 1986; Lepiniec et al., 1994; Chang et al., 1995). There are two genotypes in West Africa: the West Africa genotype I (occurs in Eastern Ivory Coast, Burkina Faso, and Nigeria to Cameroon) and the West Africa genotype II (occurs in Western Ivory Coast and Mali to Senegal) (Lepiniec et al., 1994; Mutebi et al., 2001). The three other genotypes were identified in Central and Eastern Africa: the East and Central African genotype (occurs in Central African Republic, central Sudan, Ethiopia, and the Democratic Republic of Congo), the East African genotype (occurs in Kenya, Southern Sudan, and Angola), and the Angola genotype (based on a single isolate from Angola) (Mutebi et al., 2001; Onyango et al., 2004b). There are two genotypes in South America: the South American genotype I (occurs in Brazil, Panama, Columbia, Ecuador, Venezuela, and Trinidad) and the South American genotype II (occurs mainly in Peru and Bolivia, with some isolates from Brazil and Trinidad) (Bryant and Barrett, 2003).

The virus has three transmission cycles: the sylvatic (or jungle), the intermediate, and the urban cycle. The transmission cycle involves horizontal and vertical transmission of the virus in the mosquito vectors (Diallo et al., 2000; Vasconcelos et al., 2001b; Mondet et al., 2002). The importance of transovarial transmission (Marchous and Simond, 1905; Beaty et al., 1980) in maintaining the transmission cycle is not known. It is thought that the infected mosquito eggs allow for "over wintering" of the virus because they survive desiccation and hatch the next rainy season (Barrett and Monath, 2003b). In the jungle cycle, which occurs in tropical rainforests, the virus is transmitted only between lower primates by canopy-dwelling wild mosquitoes, usually *Haemagogus* spp. in South

(A)

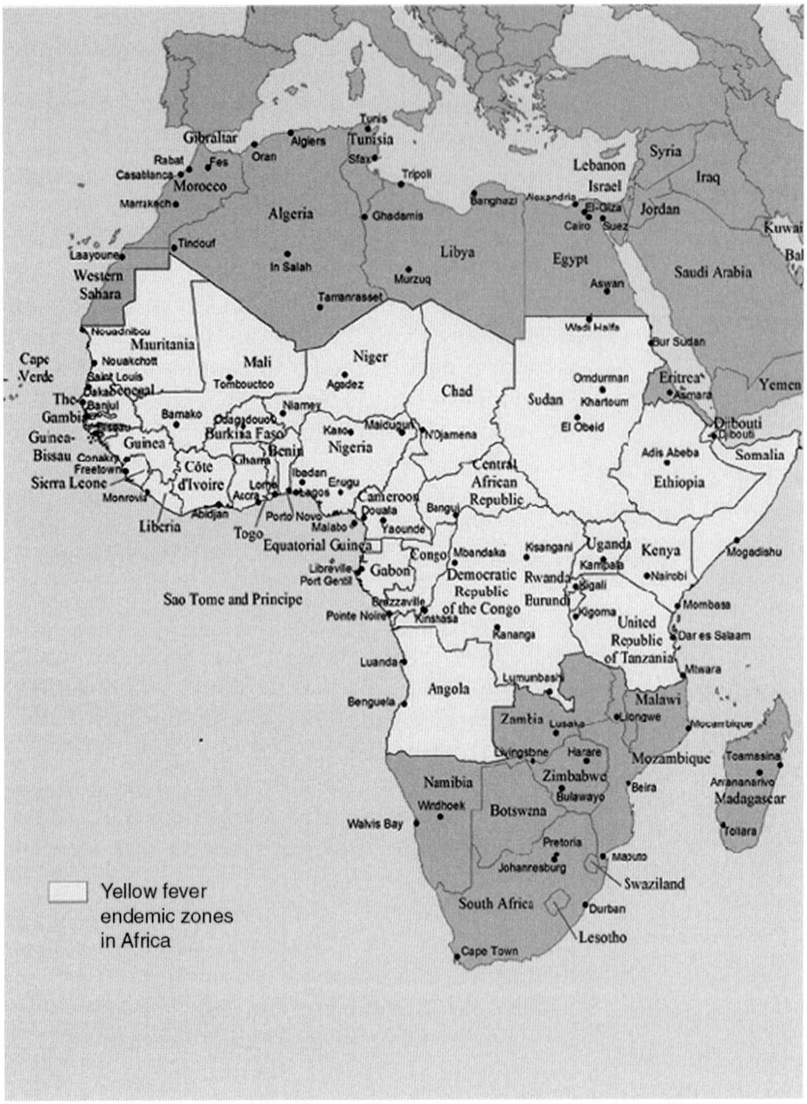

FIGURE 40.2 Yellow Fever endemic zones in Africa (A) and South America (B) as of 2007 (from CDC Web site http://wwwn.cdc.gov/travel/yellowBookCh4-YellowFever.aspx#667). Yellow fever occurs in sub-Saharan Africa and tropical South America, where it is endemic and intermittently epidemic. These endemic areas have evidence of yellow fever transmission to humans and/or its potential, because there are competent mosquito vectors and the virus is present in nonhuman primates.

America (Vasconcelos et al., 2003) and mostly *Aedes africanus* in Africa. The virus is enzootic in the rainforests of Africa and South America. The disease in non-human primates varies with South American primates often succumbing to the disease, whereas African primates often show no signs of infection. Human infections in the jungle cycle are usually occurring in young men who work in the rainforest areas (Barrett and Monath, 2003b). The intermediate, or savannah, cycle only occurs in the jungles of Africa, where there is some human activity on the moist savannah ground, usually farming. A variety of *Aedes* sp. of semidomestic mosquitoes that feed on nonhuman primates are the vectors in this cycle (Germain et al., 1980; Mutebi and Barrett, 2002) and they will often also feed on humans if they come into the savannah areas. It is this cycle in Africa that is regarded as the zone of emergence because this is thought to be where the virus evolved to become an important human pathogen (Barrett and Higgs, 2007). In the urban transmission cycle, the virus is transmitted between humans by domestic *Ae. aegypti* mosquitoes in Africa (Johnson et al., 2002; Barrett and Higgs, 2007). Urban YF has not been seen in South America since 1928 except for small outbreaks in Brazil in 1928, Trinidad in 1954, and Bolivia in 1999 (Monath, 1999; Van der Stuyft et al., 1999; Barrett et al., 2007; WHO, 2006a, 2006b).

Thus, YF epidemiology reflects the geographical distribution of the mosquito vectors: *Haemagogus* spp. in South America and *Aedes* spp. in Africa (Monath, 2004). The vast majority of cases (>80%) are in Africa due to a combination of more people in Africa and more effective vaccination programs in South America.

(B)

FIGURE 40.2 (Continued)

The greatest risk of a large epidemic occurs when infected humans return to urban areas and are fed on by the domestic vector mosquito *Ae. aegypti*, which then transmits the virus to other humans. The world distribution of *Ae. aegypti* extends from 40° North to 40° South latitude (most tropical to subtropical regions) and this distribution is expected to increase with global warming (Fig. 40.3, Gubler, 2002; http://www.searo.who.int/LinkFiles/Dengue_chap-7.pdf).

Significance as Public Health Problem

In the 18th and 19th centuries, YF was a huge public health problem until mosquito control measures and production of an effective vaccine brought the epidemics under control in the 20th century. Yet as we enter the 21st century this virus is once again a significant public health problem (Robertson et al., 1996; WHO, 1998; Barrett and Higgs, 2007) and is classified as a reemerging disease.

Urban YF was eliminated from South America in 1928 (see above), whereas sylvatic YF continues to be transmitted from monkeys to humans via mosquitoes in the rainforest ecosystem. Thus, most South American cases are in residents and those who work in forested or transitional areas of Bolivia, Brazil, Colombia, Ecuador, Venezuela, Guyana, French Guiana, and Peru (http://wwwn.cdc.gov/travel/yellowBookCh4-YellowFever.aspx#667). The disease is only partly controlled due to incomplete vaccination of residents and workers, and provides a potential source of infection to population centers that have been reinfested with *Ae. aegypti*. Sylvatic YF is

(A)

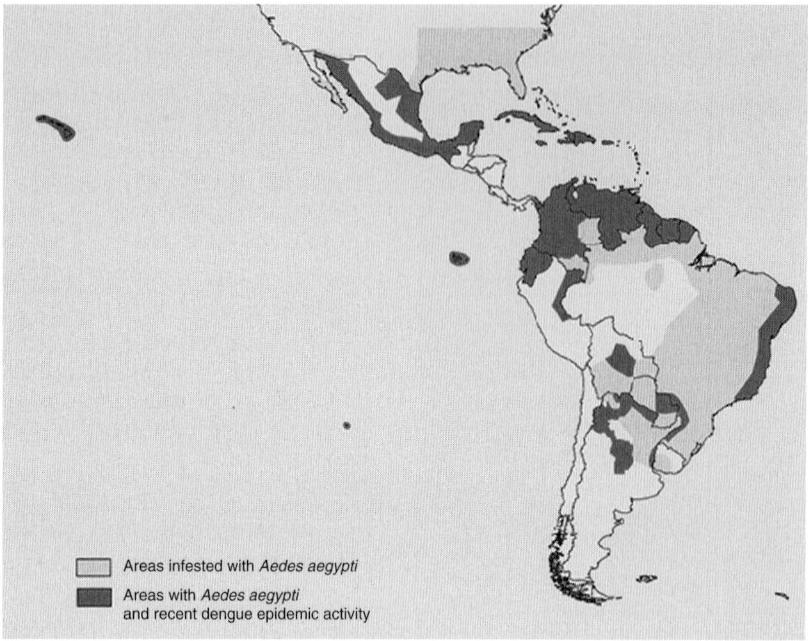

(B)

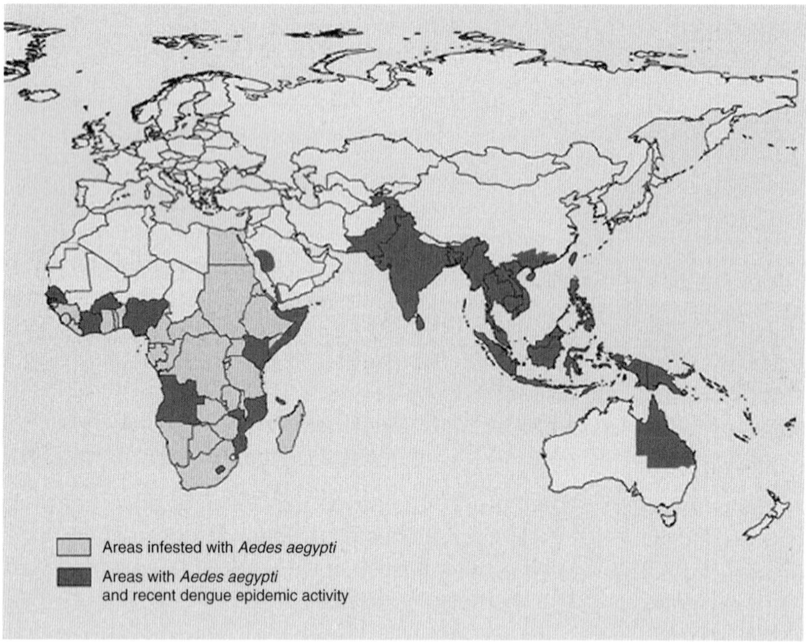

FIGURE 40.3 Distribution of *Aedes aegypti* in the Americas (A) and Eastern Hemisphere (B) (http://www.searo.who.int/LinkFiles/Dengue_chap-7.pdf). *Ae. aegypti* is found in or near human habitations and has two peak periods of biting activity: several hours in the morning after daybreak and several hours before it gets dark. However, this mosquito may feed at any time during the day, especially indoors, in shady areas, or when it is overcast.

endemic in the north and west of Brazil (Mondet et al., 1996; Vasconcelos et al., 1997, 2001c; Codeço et al., 2004) and there is a risk of emergence in the large coastal population centers, such as Sao Paulo and Rio de Janeiro (Filippis et al., 2001). In addition, reinfestation of the country with the *Ae. aegypti* vector in the 1970s greatly increased the risk of YF urbanization in Brazil (Fig. 40.3; Gubler 2002; Lourenco-de-Oliveira et al., 2004). It is hard to quantitate the risk, but it could be as high as 29% during epizootic periods (Codeço et al., 2004; WHO, 2007c). The expanding infestation in South America by *Ae. aegypti* has resulted in large epidemics of dengue, another mosquito-borne flavivirus disease also transmitted by this mosquito, yet there has not been a corresponding increase in YF disease. Many authors believe it is only a matter of time before urban YF returns to South America. Nonetheless, there continues to be periodic epidemics and epizootics of

YF activity in the border area of eastern Colombia and western Venezuela (Rifakis et al., 2006), and the foothills of the eastern Andes mountains.

In Africa, the risk of urban YF epidemics has increased greatly because of the increased risk of YFV transmission due to high mosquito vector densities in areas where there are large nonimmune populations. It is significant that most YF cases are in infants and children, who have the greatest risk of severe disease with case fatality rates as high as 20% (Arguin et al., 2007). The United Nations reported that African urban population growth rates are the most rapid in the world (United Nations, Population Division, 2001), with the proportion of urban to rural populations predicted to reach 63% by the year 2020, including 70 cities that will have more than a million people by 2015 (United Nations, World Urbanization Prospects, the 1999 revision; Meeting the Urban Challenge, Population Reports Series M, number 16 United Nations, World Urbanization prospects, the 1999 revision). Urban cities have large numbers of "shantytowns" with inadequate water supplies and lack of proper public sanitation services; people store drinking water in containers that serve as breeding sites for *Ae. aegypti*. Since 2000, 18 countries in Africa have reported cases of YF (Roberts, 2007). The situation is complicated as high levels of resources, both from national and international support, are needed to effectively control urban outbreaks, which deprive these countries of scarce resources to control other diseases and to improve health services. Today, the World Health Organization (WHO) estimates that 12 countries (Benin, Nigeria, Burkino Faso, Ghana, Liberia, Senegal, Cote d'Ivoire, Cameroon, Gabon, Toga, Guinea, and Sierra Leone) have large nonimmune populations that are at high risk and 100 million people need immunization now (http://www.who.int/csr/don/archive/disease/yellow_fever/en/index.html) and a large catch-up immunization campaign, cosponsored by WHO and Global Alliance for Vaccines and Immunization, is being planned to fill the need. A further problem is that traditionally the countries most severely afflicted by YF have followed a policy of emergency control vaccination, rather than the more effective inclusion of the 17D vaccine in their Expanded Program of Immunization (EPI) (Monath and Nasidi, 1993).There was an epidemic in Cameroon in 1990 (Vicens et al., 1993) and outbreaks in Senegal in 1995 and 1996 (Thonnon et al., 1998), Sudan in 2003 (Onyango et al., 2004a, 2004b), Cote d'Ivoire in 2006 (World Health Organization, 2006b), and Togo in 2007 (World Health Organization, 2007a). For the continent of Africa to effectively bring YF under control it is estimated that it will require 80% vaccination coverage of children under age one, who are at most risk (Tomori, 2002) and that YF should be added to the EPI of all countries where the virus is endemic (Moreau et al., 1999).

In addition to the public health problems for residents of endemic areas in Africa and South America, it is a significant hazard to national and international unvaccinated travelers who visit endemic areas. During 1970–2002, there were 10 cases reported in unimmunized travelers from the United States and Europe with a mortality rate of 89% (Digoutte et al., 1981; World Health Organization, 1986, 1998, 2000; Nolla-Salas and Sadalls-Radresa, 1989; Monath and Cetron, 2002; Barnett, 2007). The number of travelers from the United States to South America and Africa has more than doubled since 1988 (International Trade Administration. U.S. resident travel to Canada, Mexico and overseas countries historical visitation outbound 1988–1999. http://www.tinet.ita.doc.gov/research/reports/basic/national). Currently, 9 million tourists from North America, Europe, and Asia travel each year to countries where YF is endemic (WHO, 1996), and potentially increase the risk of introducing YFV into North and Central America, Caribbean, Middle East, Asia, Australia, and Oceania (Tomori, 2004; Barrett and Higgs, 2007).

POTENTIAL AS BIOTHREAT AGENT

During the 1930s the Japanese experimented with YFV as a biological weapon in China. After World War II, during the 1950s, the American military researched YFV as a biological weapon, including its spread by aerosol. Most countries classify YFV at biosafety level three, or equivalent, due to the availability of a vaccine. YFV was originally on the list of CDC select agents but it has since been removed (http://www.cdc.gov/od/sap/docs/salist.pdf) due to the availability of an effective and safe vaccine. Thus, this virus is not considered a biothreat agent, although there is no effective treatment and probably insufficient worldwide vaccine production (Pugachev et al., 2005; Roberts, 2007) if the virus was used as an agent of bioterrorism. Finally, there is no doubt that YFV would be classified as a biosafety level 4 pathogen if there was no vaccine available.

CLINICAL DISEASE

There is a wide spectrum of the clinical disease from nonspecific abortive illness to a rapid and fatal hemorrhagic fever (Kerr, 1951). The incubation period is 3–6 days and there is an abrupt disease onset with fever, chills, malaise, lower back pain, myalgia, nausea, and dizziness. The patient presents with bradycardia relative to their height of fever, an acute febrile illness, and congestion of face and conjunctivae. On average, the fever lasts 3 days and younger children may show febrile convulsions. During this time the patient is viremic, with infectious virus particles 10^4/ml of blood or greater. Between two and three days after disease onset the "period of infection" begins where the patient shows the first signs of jaundice typical of YF and their serum transaminase levels begin to rise. The "period of infection" is normally followed by a 2–24h "period of remission" where the symptoms disappear and the virus is cleared from the blood by the immune response. At this stage the majority of patients recover without any side effects, although the "convalescent phase" lasts several weeks and is characterized by fatigue and prolonged weakness with abnormal liver function tests that can last more than two months. Approximately 15–25% go on to develop "the period of intoxication" where they develop a more severe illness with vomiting, epigastric pain, renal failure, fever, jaundice, and hemorrhagic diathesis. It is at this stage that the jaundice becomes worse and the serum transaminase levels rise, reflecting the severity of the disease (Oudart and Rey, 1970). Patients exhibit a multifactorial bleeding disorder with hemorrhagic manifestations, such as bleeding from gums and needle puncture sites, epistaxis, ecchymoses, petechiae and, in many cases, there is also melena or metrorrhagia, major bleeding, and coffee grounds hematemesis. Approximately 20–50% of patients with this hepatorenal disease die 7–10 days after onset. These patients become delirious and fall into a stupor followed by coma and death, with symptoms of hypotension, hypothermia, hyperkalaemia, Cheyne-Stokes respirations, hypoglycemia, and metabolic acidosis. At this late disease stage there is also edema and perivascular hemorrhages in the brain (Monath, 2001).

TREATMENT

There is no specific treatment available for patients suffering from wild-type YF, YF vaccine-associated viscerotropic (YEL-AVD), and YF vaccine-associated neurotropic disease (YEL-AND) (CDC, 2008). Ribavirin postexposure treatment showed some benefit when used to treat the hamster model of viscerotrophic disease (Sbrana et al., 2004). A recent study confirmed that treatment with IFN alfacon-1 or ribavirin (starting 2 days postinfection) significantly improved survival of hamsters challenged with YFV. Pre-virus exposure treatment with IFN alfacon-1 also improved YF disease in infected hamsters (Julander et al., 2007b). In another study by the same group, the T-1106 (substituted pyrazine) compound was used to significantly improve survival, again in the hamster model, using two treatment regimes—one starting 4h prior to virus inoculation and continuing twice daily until 7 days post-virus inoculation and the other one beginning 4 days after virus challenge with twice-daily treatment for 8 days (Julander et al., 2007a). Another recent study suggested that corticosteroid therapy could also influence survival of patients suffering from YEL-AVD (Vellozzi et al., 2006), although this needs further investigation to see if it is of proven benefit.

Palliative treatment of disease symptoms has changed little and includes maintenance of patient nutrition; prevention of hypoglycemia; nasogastric suction to prevent gastric distension and aspiration; intravenous cimetidine to prevent gastric bleeding; treatment of hypotension by fluid replacement and vasoactive drugs; administration of oxygen; correction of metabolic acidosis; treatment of bleeding with fresh-frozen plasma; and dialysis if needed due to renal failure (Monath, 1987). Intensive care may improve the outcome for seriously ill patients, although this is not available in poorer developing countries where most cases occur and oral rehydration salts and paracetamol are used to treat the dehydration and fever, and antibiotics are used to treat any superimposed bacterial infection (WHO, 2001).

PATHOGENESIS

Studies of YFV pathogenesis utilize experimental infection of nonhuman primates because this is the only model that shows the viscerotropic disease typical of YF infection in humans and the LD_{50} can be less than 1 plaque-forming unit for some strains in experimentally infected monkeys (Theiler, 1951; Tigertt et al., 1960). European hedgehogs show viscerotropic disease, with lesions very similar to those in nonhuman primates, and neurotropic disease after experimental infection (Findlay and Clarke, 1934a, 1934b) and a hamster model of viscerotropic disease has recently been developed by serial passage of the virus in hamsters (Tesh et al., 2001; Xiao et al., 2001; McArthur et al., 2003, 2005; Sbrana et al., 2006). This hamster model is useful, as it shows many similarities to the disease seen in humans; however, the model requires most wild-type strains to be adapted by passage in hamsters. Wild-type and vaccine viruses cause encephalitis only in various rodent models (Monath, 2001). Animals succumb to infection when virus is inoculated directly into the brain while neurotropic disease is age-related when virus is administered by a peripheral route. For example, 17D vaccines are lethal in mice up to 5 days old, wild-type strains up to approximately 10–12 days old, and the FNV strain up to 3 weeks of age (Fitzgeorge and Bradish, 1980). The intraperitoneal inoculation of 8-day-old mice is often used to study the mouse neuroinvasive phenotype.

Description of Disease Process

Kupffer cells in the liver become infected on day 1 and the virus spreads to the kidney, bone marrow, spleen, and lymph nodes (Theiler, 1951; Tigertt et al., 1960). Fig. 40.4 illustrates the main pathogenic processes of YFV. Spleen germinal centers, tonsils, lymph nodes, and Peyer's patches all show necrosis in both monkeys and human beings, but it is uncertain whether or not it is a direct virus effect (Monath et al., 1981). There are two main areas of disease pathology: renal and hepatic damage. Renal pathology shows eosinophilic degeneration and fatty change of tubular epithelium without inflammation. Renal tubular cells of fatal human cases contain antigen, which suggests direct viral injury plays an important role, although this is not observed in experimentally infected monkeys (Monath, 2001; Theiler, 1951; Tigertt et al., 1960). Histological changes are observed in the basement membrane and cells lining the Bowman's capsule. Antigen is present in glomerulae 2–3 days after infection of monkeys; hence, the marked albuminuria is thought to be due to alteration of glomerular function (Monath et al., 1981). Severe hepatic pathology is a late event characterized by infection and degeneration of hepatocytes in both monkeys and humans (Theiler, 1951; Tigertt et al., 1960; Monath, 2001). Hepatic injury shows a mid-zonal distribution, with sparing of cells around the central vein and portal tracts. Antigen and virus RNA have been demonstrated in these midzonal hepatocytes (Monath et al., 1989; Monath, 2001), which undergo eosinophilic degeneration and have condensed nuclear chromatin that is typical of apoptotic cell death (Marianneau et al., 1998). Hepatic injury

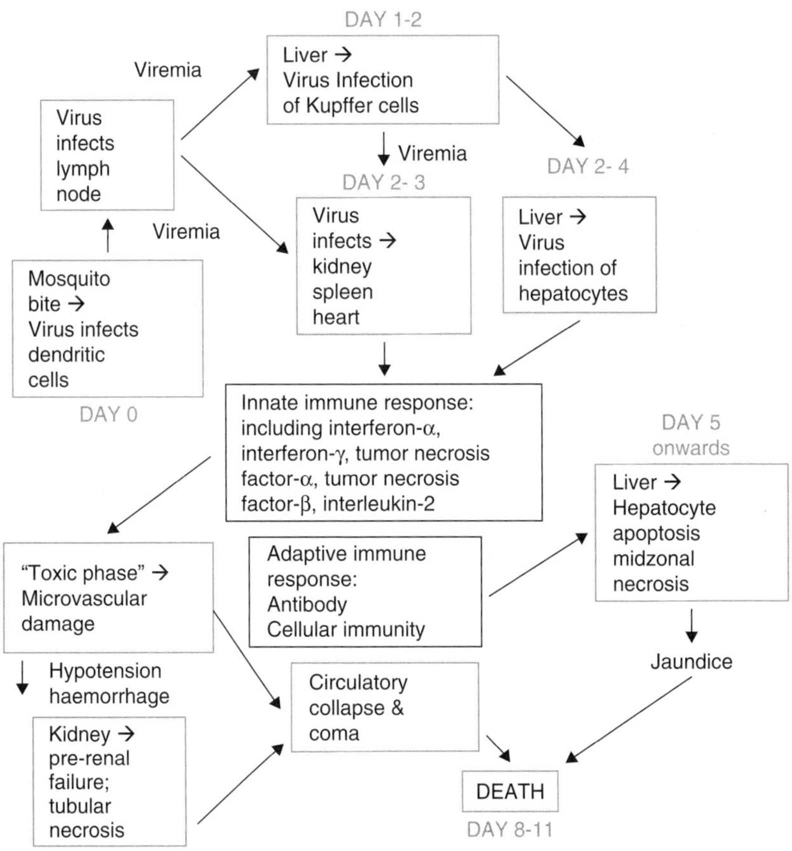

FIGURE 40.4 Pathogenesis of yellow fever virus. This is based on studies in experimentally infected monkeys and human case reports (**bold**). Speculative mechanisms (*italics*) are drawn from in vitro data or reports on other flavivirus infections. CTL: cytotoxic T lymphocyte; DIC: disseminated intravascular coagulation; IL: interleukin (from Monath, 2001).

from apoptosis explains the virtual absence of inflammatory cells, preservation of the reticulin framework, and healing without fibrosis (Monath 2001).

Hypotension and shock, which are thought to be mediated by cytokine dysregulation, followed by death, are the last stages of the disease. Kupffer cells and splenic macrophages that are infected with YFV, or are activated in response to virus injury, produce these cytokines, such as TNFα. Cytotoxic T cells could also be responsible for cell injury, oxygen free radical formation, endothelial damage and microthrombosis, disseminated intravascular coagulation, tissue anoxia, oliguria, and shock. Direct viral injury to the heart may contribute to shock because myocardial fibers have been shown to contain viral antigen and undergo apoptosis (Monath, 2001). Further studies of patients or experimentally infected monkeys are necessary to better understand the disease process, especially the late-stage shock syndrome.

Immune Response to Infection

Infection with YFV results in a rapid and specific immune response against the virus, which is highly effective and induces lifelong protective immunity. The specifics of this response are discussed further in "Protective Immune Response" above. Previous infection with heterologous flaviviruses is also known to affect immune responses to YFV, and consequently disease severity, because prior flavivirus infection can induce a cross-protective immune response. As early as the 1930s, infection with dengue virus (DEN) was thought to partially protect against YF infection and later experiments demonstrated that this was correct (Snijders et al., 1934; Sabin, 1952; Frederiksen, 1955; Ashcroft, 1979; Kurane et al., 1991). To achieve this cross-protective immune response it was necessary to actively immunize monkeys with DEN (Snijders et al., 1934; Theiler and Anderson, 1975). In addition, another study showed that African flaviviruses gave similar partial cross-protection against subsequent challenge with YFV (Henderson et al., 1970), which suggests that cross-protection between flaviviruses (termed group B arboviruses when the studies were undertaken) has an influence on the epidemiology of YF in Africa. Furthermore, epidemiological studies during a YF epidemic in the Gambia showed that humans with prior flavivirus immunity had a lower risk of developing severe YF disease (Monath et al., 1980) and studies using the hamster model of severe YF provide supportive evidence that prior exposure to heterologous flaviviruses reduces subsequent risk of fatal YFV infection (Xiao et al., 2003).

VACCINES

History

There had always been research on developing a YF vaccine but for many years it was not known what caused the disease YF. However, once the virus was isolated by two groups in 1927–1928 (Asibi and French Viscerotropic viruses) (Mathis et al., 1928; Strokes et al., 1928), there were many attempts to develop a vaccine. Several failed attempts were made to produce an inactivated single-dose vaccine by groups in France (Pettit, 1931), Brazil (Aragao, 1929), and the United Kingdom (Hindle, 1928, 1928–1929). Formaldehyde or phenol was used to inactivate virus-infected monkey liver or spleen. Testing of Hindle's vaccine in monkeys gave nonreproducible results (Hindle, 1928–1929; Okell, 1930; Davis 1931), a clinical trial in Brazil gave inconclusive results (Pettit, 1931), and at least one person who received the vaccine became infected with YFV (Burke and Davis, 1930). Subsequent attempts were made to improve the inactivated vaccines with "better" antigens using heat- or ultraviolet light–inactivated tissue cultures or mouse brain cultures. These vaccines were tested in monkeys and the fully inactivated preparations caused no illness, but also gave no protection, whereas vaccines with residual live virus caused illness and stimulated protective immunity (Gordon and Hughes, 1936). The only inactivated vaccine that gave protection against challenge was tested in mice and rabbits, but no further testing was undertaken in primates. This vaccine used multiple doses of phenol-inactivated YFV-infected mouse brain (Sellards and Bennett, 1937).

Once the virus had been isolated, there was research to develop a live-attenuated vaccine by adaptation of virus to nonhuman tissue. Two live-attenuated YF vaccines were simultaneously developed—the French neurotropic vaccine (FNV), or Dakar vaccine, in Senegal, which was attenuated by passaging in mouse brain, and the 17D vaccine, which was attenuated by passaging in mouse tissue and embryonated chicken eggs (Strokes et al., 1928; Mathis et al., 1928). In 1927, the wild-type French viscerotropic virus (FVV), or French virus strain, of YFV was isolated from Francoise Mayali, who was a patient in Dakar, Senegal (Mathis et al., 1928). The work was carried at the Institute Pasteur at Dakar and the virus was attenuated by intracerebral passage in mice. After 128 passages, the FNV was isolated and this virus underwent a further 109 passages to generate the first vaccine at passage 237. This virus was tested in monkeys for virulence and although it showed attenuation of visceral YF when inoculated extraneurally, it also showed increased neurotropic virulence for monkeys and mice (Durieux, 1956a). This led to the first clinical trial in 1931, where a suspension of FNV and a quantity of human immune serum were simultaneously injected into volunteers because the virus was considered too dangerous to administer alone and the immune serum was used to limit the spread of the virus. A year later, FNV was inoculated subcutaneously without the immune serum. The success of this trial led to a compulsory vaccination by scarification order in 1941 for the whole military and civilian population of French West Africa using virus at passage 237–260. A year later, a program of compulsory immunization was initiated for all of colonial French West and Equatorial Africa. The success of this compulsory immunization policy was highlighted by deaths during major YFV epidemics in Nigeria and the Gold Coast, where immunization was not compulsory. Over the course of the next 10 years, the FNV strain was widely used as a vaccine until in 1951–1952, during epidemics in Central America and Nigeria, cases of postvaccine encephalitis came to light. The rate in Nigeria was 3–4/1000 vaccinations and most cases were children (38% case fatality rate). Over 38 million doses of FNV were administered mainly by scarification, sometimes alongside the smallpox vaccine, in the Francophone countries of West Africa from 1939 and 1952 (Durieux, 1956b). A further 10 years passed until in 1961 the FNV was contraindicated for children under 10 years, because of this high incidence of encephalitis in children. In 1966, production of the 17D vaccine was initiated in Dakar, Senegal, but it wasn't until 1980 that the manufacture of the FNV was completely discontinued because the WHO prohibited the vaccination of children under 14 years old (Robertson, 1993).

The 17D vaccine was derived from the Asibi strain, which was isolated from the blood of a patient (Asibi) in Ghana and passaged in monkeys (Strokes et al., 1928). Subsequently, Theiler and Smith (1937a) undertook serial intracerebral passages in mice and, like the French workers, observed that the resulting virus was attenuated in its ability to cause viscerotropic disease but showed increased neurovirulence in monkeys (Sellards, 1931). This Asibi strain of YFV was established in embryonic mouse tissue and passaged 18 times, followed by 58 passages in minced whole chick embryonic tissue. After these 58 passages, the brain and spinal cord were removed from the whole chick embryonic tissue and the virus was passaged a further 100 times and named as 17D at passage 176 (Penna, 1965). Interestingly, Theiler and Smith were working on a number of variants concurrently (17A, 17B, 17C, 17D, 17E, etc.); 17E was partially attenuated and only 17D was fully attenuated. Furthermore, Theiler was unable to repeat the attenuation process to re-derive 17D independently (Theiler and Smith, 1937a). By 1937, the 17D vaccine was being tested in humans, first in YF immune individuals and then in nonimmune individuals (Theiler and Smith, 1937b) and by 1938 initial immunization against YF had started in Brazil. Vaccination of 59,000 people indicated that the vaccine was attenuated in children and women at any stage of pregnancy. Mild reactions occurred in 10–15% of vaccinees and more intense reactions occurred in 1–2% of vaccinees at 5–8 days after vaccination, with 95% of them acquiring immunity as measured by specific antibodies (Penna, 1965). Vaccine batches began to be produced in various laboratories from available subcultures after multiple passaging of the original 17D vaccine. This led to failure of immunization due to over-attenuation of 17D vaccine batches that had been passaged 300–350 times from the initial Asibi virus because they were no longer infectious for humans (Theiler, 1951). In 1942, there were cases of hepatitis in England, Brazil, and the United States due to the practice of adding 10% human serum to the vaccine, which was contaminated with a hepatitis-causing virus, as this was necessary for stability of the vaccine virus (Smith, 1951). In Brazil, there were a few cases of encephalitis in children, depending on which substrain was utilized to vaccinate them; 0.36% of vaccinees showed CNS involvement and one person died (Georges et al., 1985). By 1945 the Standing Technical Committee on Health of the United Nations Relief and Rehabilitation Administration established manufacturing standards to control vaccine production (WHO, 1945; United Nations Relief and Rehabilitation Administration 1946). In addition, the WHO established a seed lot system to resolve these problems and those of over-attenuation or under-attenuation of 17D vaccine by passaging in tissue culture (WHO, 1959). In the early 1950s, 12 million people were successfully vaccinated with 17D in southern Brazil in response to outbreaks of YF (de Sousa Manso, 1956). Following the large epidemic of YF in West Africa in 1965, 1966 saw the decline in use of the FNV and the beginning of the sole use of 17D-derived vaccines, which are still in use today.

Current Licensed Vaccines

There are three substrains of the 17D vaccine that are used in vaccine production today. The 17D-204 vaccine substrain is utilized in all countries, except Brazil, where the 17DD vaccine substrain is used. The 17D-204 vaccine was developed from the original attenuated 17D by continued chick embryo passage (without neuronal tissue) from passage 176 to passage 204. Subsequently, the virus was passaged in embryonated chicken eggs and all currently manufactured vaccines are manufactured at passage levels between 235 and 240. The 17DD vaccine was derived by passage in whole chick embryonic tissue with the neuronal tissue removed from p176-195; however, subsequent passages were undertaken independently in Brazil such that its development differed from 17D-204. This vaccine was passaged in embryonated chicken eggs and all currently manufactured vaccines are at passage levels 286 (Barrett, 1987, 1997). During the 1970s and 1980s it became apparent that some vaccines had been prepared in eggs contaminated with Avian leucosis virus (ALV) and so a number of manufacturers prepared ALV-free seeds of 17D virus in order to remove the endogenous retrovirus. The Robert Koch Institute in Germany, on behalf of WHO, established a new seed lot of 17D-204, termed 213/77 or 17D-213, which was certified free of ALV contamination, and is used at passage 239–240 of 17D-204 virus in embryonated chicken eggs (Post et al., 1992).

Over the years there have been many manufacturers. The 17D-204 substrain has been manufactured in France, Senegal, United States, the Netherlands, United Kingdom, Germany, and India, and 17DD vaccine in Brazil and Colombia. The 17D-213 vaccine has been manufactured in Nigeria and Russia, plus Berna Biotech in Switzerland (Pfister et al., 2006) but never marketed by the latter. At the present time there are only four recognized producers: Sanofi Pasteur in France and United States, Institut Pasteur, Dakar, Senegal, and Bio-Manguinhos, FioCruz, Rio de Janiero, Brazil. All manufacturers are prequalified by the WHO to provide vaccine for use in developing countries. The Chinese also produce a 17D vaccine for internal use, but the substrain is not known.

The genomes of 17DD (Duarte dos Santos et al., 1995; Galler et al., 1998), 17D-204 (Rice et al., 1985, Dupuy et al., 1989), and 17D-213 (Duarte dos Santos et al., 1995; Galler et al., 1998) vaccine viruses and parent wild-type Asibi virus have been sequenced (Hahn et al., 1987). Unfortunately, the original 17D virus is not available. The three substrains differ slightly in sequence, thus justifying their classification as substrains (Galler et al., 1998), but they share 20 amino acid substitutions and four nucleotide changes in the 3′ noncoding region. The capsid gene and 5′ noncoding region of wild-type Asibi and 17D vaccine viruses were identical in sequence (Table 40.1; Monath and Barrett, 2003).

The genomes of FVV and FNV have also been sequenced and found to differ by 35 amino acids and 4 nucleotide changes in the 3′ noncoding region (Table 40.2; Wang et al., 1995). Again the 5′ noncoding regions were identical in sequence. Interestingly, the two attenuation processes (17D substrains and FNV) shared two common amino acid substitutions at M-36 (L → F) and NS4B-95 (I → M; Tables 40.1 and 40.2; Monath and Barrett, 2003; Wang et al., 1995). However, there is no evidence to date that these encode common attenuating mutations. In addition, monoclonal antibodies have been generated against wild-type and vaccine strains. Wild-type (Schlesinger et al., 1983; Sil et al., 1992; Gould et al.,1985, 1989; Cammack and Gould, 1986; Barrett et al., 1990a, 1990b; Sil et al., 1992) and vaccine-specific epitopes (i.e., recognize 17D and FNV viruses only) (Gould et al., 1985, 1989; Schlesinger et al., 1983, 1984; Barrett et al., 1990a, 1990b; Sil et al., 1992), and 17D-204- and 17DD substrain–specific epitopes (Schlesinger et al., 1983; Geske et al., 1983; Gould et al., 1985; Barrett et al., 1989) have all been identified on the E protein.

TABLE 40.1 Comparison of the amino acid sequence of the Asibi virus with a consensus sequence for the 17D substrains (17DD; 17D-204; 17D-213) derived from it

Gene	AA position	AA change (Asibi–17D)
M	36[a]	L–F
E	52	G–R
	170	A–V
	173	T–I
	200	K–T
	299	M–I
	305	S–F
	380	T–R
	407	A–V
NS1	307	I–V
NS2A	61	M–V
	110	T–A
	115	T–A
	126	S–F
NS2B	109	I–L
NS3	485	D–N
NS4A	146	V–A
NS4B	95[a]	I–M
NS5	836	E–K
	900	P–L
3′UTR	–	U–C
	–	U–C
	–	G–A
	–	A–C

Source: Adapted from Monath and Barrett (2003).
[a] These amino acid substitutions are common to all wild-type and vaccine strains, respectively. Reproduced with permission from Barrett, A.D. Advances in Virus Research, Vol. 60. Elsevier, 2003.

TABLE 40.2 Comparison of the nucleotide and amino acid sequences of the French viscerotropic virus (FVV) and the French neurotropic vaccine (FNV) derived from it

NT position	FVV	FNV	Gene	AA position	AA change (FVV–FNV)
357	U	C	C	80	V–A
386	A	G	–	90	S–G
554	G	A	prM	25	V–M
854	C	U	M	35[a]	L–F
1134	C	U	E	54	A–V
1398	A	G	–	142	Q–R
1432	U	A	–	153	N–K
1653	G	A	–	227	G–E
1718	A	G	–	249	N–D
2193	U	C	–	407	V–A
2344	G	A	–	457	M–I
2478	G	C	NS1	9	G–A
3135	C	U	–	228	P–L
3344	A	G	–	298	T–A
3557	A	C	–	369	M–L
3752	A	G	NS2A	25	I–V
3821	A	G	–	48	T–A
4046	C	U	–	123	H–Y
4358	U	G	NS2B	60	S–A
4607	A	G	NS3	13	I–V
4644	C	A	–	25	S–Y
5145	U	A	–	192	M–K
6332	G	A	–	588	A–T
6527	U	C	NS4A	30	F–L
6843	A	U	2K	9	Y–F
6942	U	C	NS4B	19	L–S
7171	A	G	–	95[a]	I–M
7178	G	A	–	98	V–I
7380	C	U	–	165	A–V
7622	G	A	–	246	E–K
7641	G	C	NS5	2	R–T
8409	U	C	–	258	I–T
8640	G	A	–	335	R–K
9615	A	G	–	660	K–R
10,268	A	G	–	878	I–V
10,357	C	U	3′UTR	–	–
10,358	A	G	–	–	–
10,404	G	A	–	–	–
10,798	A	G	–	–	–

Note: The FNV strain used here was obtained from the Institut Pasteur, Paris, and the FVV strain from Senegal, 1927 (a human isolate with the following passage history: monkey p25, smb p1, Vero p1).
Source: Reproduced with permission from Wang, Eryu, et al. Journal of General Virology, 76, part 11, pages 2749–2755, Society for General Microbiology, 1995.
[a] These amino acid substitutions are common to all wild-type and vaccine strains, respectively.

Agent/Strain Used in Vaccine, How Produced, Number of Doses Administered to Give Protective Immunity

In 1957, the WHO published "Requirements for YF Vaccine" which standardized the seed lot and manufacturing processes for the 17D substrain viruses. Subsequently, in 1959, they published guidelines for WHO approval of vaccine production and vaccination certificates, which were updated in 1971, 1976, and 1981 (World Health Organization 1959, 1971, 1976, 1981). Seed viruses are assessed for the potential to be viscerotropic by measuring liver enzyme levels in nonhuman primates. All new primary and secondary seed pools for vaccine production have to be tested for immunogenicity and neurovirulence in groups of 10 or more Cynomologus or Rhesus monkeys (Levenbrook et al., 1987). The criterion is that not more than 10% of monkeys develop severe clinical signs of encephalitis and 90% of monkeys must have viremia that should not exceed 1.65×10^4 pfu/ml (or mouse LD_{50}/ml) and must develop neutralizing antibodies (WHO, 1976). These neutralizing antibodies are measured by either protection of groups of six or more weanling mice against intracerebral mice challenge with the FNV substrain or by plaque reduction assay. In addition, any inflammatory and histological changes associated with pathogenesis are scored in the brains of the monkeys. The WHO also sets guidelines for the manufacture of vaccine, and it should be noted that techniques do vary between the manufacturers; hence, complete standardization is difficult (Barrett, 1987, 1997). The vaccine virus is cultured in embryonated chicken eggs, with the requirement that all embryos must be harvested by 12 days postinoculation (World Heath Organization, 1976), and then the embryoes are minced, aliquoted, and lyophilized using aseptic technique. After the suspension is freeze-dried, it is placed in nitrogen-filled flame-sealed ampoules and stored at 0–5°C temperature. Some producers include stabilizers, such as gelatin, peptone, lactose, sorbitol, histidine, or alanine, with the vaccine to maintain the viability of the vaccine at high temperatures such as that found in tropical countries (De Souza Lopes et al., 1988a; Barmr and Bronnert, 1984; Georges et al., 1985). Once produced, the WHO has designated control tests that must be carried out to determine vaccine potency. This lyophilized vaccine must not lose more than 90% infectivity after incubation at 37°C for 2 weeks and it must be tested for any contaminating agents (WHO, 1945). The virus content of the vaccine must be at least 1000 mouse LD_{50}, or 1000 pfu, and this is quantified by mouse LD_{50} or a cell culture assay (plaque reduction or infectious dose$_{50}$) in porcine or monkey kidney Vero cells (Barrett, 1987, 1997). Since there can be variability in the calculation of virus titers by different manufacturers using one of these two methods (Seagroatt and Magrath, 1983) an international reference standard was established for vaccine potency. Despite this some more recent studies have shown that there are inconsistencies between the amount of virus needed to infect tissue culture cells and to kill a mouse (De Souza Lopes et al., 1988b; Sood et al., 1995). In addition, there is variability in infectivity levels. One study showed that 178 pfu (30 mouse LD_{50}) (Freestone et al., 1977) immunized 85% of volunteers while another showed that 200 pfu (600 mouse LD_{50}) (De Souza Lopes et al., 1988b) immunized all volunteers. In practice, most vaccines are found to contain between 5000 and 200,000 pfu/dose (De Souza Lopes et al., 1987). Control tests have to be performed in three different laboratories in the WHO list of approved laboratories. If the batch passes these tests it is approved by the WHO. The vaccine is reconstituted with 0.5 ml water and one dose injected subcutaneously. Vaccinees receive WHO-approved vaccination certificates for 10 years protective immunity (Barrett, 1987, 1997; CDC, 2008) with booster immunizations given as required.

Vaccine Recommendations, Including Potential to Administer with Other Vaccines

The WHO recommends that this vaccine is not administered to children less than 6 months of age, and 9 months in some countries. In the United States, the CDC recommends that adults and children over 9 months old (Cetron et al., 2002) who are at risk of contracting YFV infection are vaccinated. Tourists traveling to Africa and South America plus adults working there are the main groups receiving vaccination. The recommended interval to the booster dose is 10 years (http://wwwn.cdc.gov/travel/yellowBookCh1-GenRecVaccination.aspx). It should be pointed out that it is likely that a single dose gives at least 20–30 years immunity, maybe lifelong immunity. However, data are only available on long-term immunity for a relatively small group of vaccinees such that a 10-year booster regimen is used to maintain immunity. There are a number of contraindications of vaccination: age, pregnancy, breast-feeding, thymus disease, immune suppression, and hypersensitivity, and these are discussed in the section "Contraindications of Vaccination, Including Special Risk Groups."

In general, commonly administered vaccines can safely and effectively be given on the same day without impairing antibody responses or increasing the rates of adverse reactions. Inactivated vaccines do not interfere with the immune response to other inactivated or live-virus vaccines; hence, they may be administered at the same time, but at separate sites. The immune response to a live-virus vaccine could be impaired if administered within 28 days of another live-virus vaccine. Whenever possible, injected live-virus vaccines should be given at least 28 days apart. If two injected live-virus vaccines are not administered on the same day but less than 28 days apart, the second vaccine should be re-administered at least 4 weeks later (Kroger et al., 2006). Simultaneous administration of all necessary vaccines is encouraged for people who are at the recommended age to receive vaccination and for whom no contraindications exist (http://wwwn.cdc.gov/travel/yellowBookCh1-GenRecVaccination.aspx). Table 40.3 shows vaccines that have been successfully coadministered with the 17D vaccine. Although chloroquine (antimalarial prophylaxis) is known to inhibit the replication of YFV in vitro, it does not seem to adversely affect antibody responses to the 17D vaccine (Tsai et al., 1986). There are no data on coadministration with the inactivated Japanese encephalitis (JE) vaccine, although interference is unlikely because prior infection or immunization with JE does not interfere with YF vaccination (Wisseman et al., 1962; Sweet et al., 1962).

TABLE 40.3 Agents and vaccines that can be coadministered with the 17D vaccine

Agent/vaccine	Target	References
YF immune globulin	U.S. adults	Kaplan et al. (1984)
Chloroquine tablets	U.S. adults	Tsai et al. (1986)
Measles vaccine	Brazilian children Nigerian children	Stefano et al. (1999) Ruben et al. (1973)
Small pox vaccine	Nigerian children U.S. adults	Ruben et al. (1973) Tauraso et al. (1972)
Meningococcal + typhoid vaccine	U.S. adults	Dukes (1996)
Hepatitis A vaccine	French adults Spanish adults	Dumas et al. (1997) Gil et al. (1996)
Hepatitis B vaccine	Senegalese children	Yvonnet et al. (1986)
Typhoid vaccine	French adults	Dumas et al. (1997)
Diptheria-pertussis-tetanus antigen	Nigerian children	Ruben et al. (1973)
Bivalent cholera and typhoid vaccine	Adults	Bovier and Noble (1999)
MMR vaccine	Brazilian children	Collaborative group for Studies with Yellow Fever Vaccine (2007)

Countries Where Vaccine Is Licensed, Including Manufacturers

There are four countries where the vaccine is licensed for manufacture: Brazil (Bio-Manguinhos, Fio-Cruz, Oswaldo Cruz Foundation, Rio de Janeiro); United States (Sanofi Pasteur Inc., Swiftwater, Philadelpia); France (Sanofi Pasteur, Marcy L'Etoile), and Senegal [Institut Pasteur, Dakar (an updated situation from the WHO technical report series no 872, 1988)].

Indications for Vaccination/Target Populations

As early as 1988, the Joint WHO/UNICEF Technical Group on Immunization in Africa recommended incorporation of YF vaccine in routine child immunization programs of countries at risk for YF in Africa and South America. Currently the WHO recommends that the vaccine be administered in countries at risk for outbreaks: 31 African countries, 2 countries in WHO Eastern Mediterranean Region, and 11 South American countries. The following is a summary of the differing vaccination policies of at-risk countries. Angola, Benin, Burkina Faso, Cameroon, Central African Republic, Chad, Congo, Côte d'Ivoire, Democratic Republic of Congo, Gabon, Ghana, Gambia, Guinea Bissau, Liberia, Mali, Niger, Sao Tome & Principe, Senegal, Sierra Leone, Togo, and Kenya (part of country) immunize infants at 9 months old. Namibia vaccinates only regions at risk. Nigeria, Bahamas, and Belize vaccinate infants at 9 months and recommend the vaccine for travelers to endemic countries. Argentina, Panama, Paraguay, and Venezuela immunize infants at 1 year old who reside in tropical areas of the country, at-risk groups, and recommend the vaccine for travelers. Bolivia immunizes all infants at 12–23 months old. Colombia, Equador (part of country), Guyana, Peru, and Suriname (part of the country) immunize infants at 1 year old. Honduras immunizes children at 1 year old and recommends the vaccine for travelers. Brazil immunizes infants at 9 months old, adults every 10 years, and recommends the vaccine for travelers. Jamaica immunizes infants at 1 year old, adults every 10 years, and recommends the vaccine for travelers. Trinidad & Tobago immunize infants at 1 year old and children at 11 years old. The Maldives immunizes people from 18–45 years old in part of the country. France immunizes infants at 6 months old and 11 years old (part of the country) and has mandatory vaccination for French Guyana. In general, other countries outside of sub-Saharan Africa and tropical South America recommend vaccination for travelers and others who are at risk of contracting YF (WHO vaccine-preventable diseases: monitoring system 2006).

Since 2000, the Global Alliance for Vaccines and Immunization (GAVI) has provided support to poorer countries for improvement of these routine immunization policies, to help provide safe injections, and increase YF vaccination coverage. The Yellow Fever Initiative, led by WHO and UNICEF, with the support of GAVI, was also launched to help increase vaccination coverage and reduce the risk of YF outbreaks in Africa (WHO, 2007c).

In May 2007, at the World Health Assembly, the WHO and partners announced the goal is to immunize more than 48 million people in 12 West African countries over the next 5 years as a catch-up campaign to reestablish an immune barrier against YF (Roberts, 2007).

Duration of Immunity

Immunization has long been known to induce a protective antibody response that provides excellent protection against infection with the virus (Sawyer, 1931; Bauer and Hudson, 1930; Groot and Ribeiro, 1962; Rosenzweig et al., 1963; Monath and Heinz, 1996; Monath et al., 2000). Ninety percent of vaccine recipients develop neutralizing antibodies within 10 days after immunization, and 99% within 30 days. Studies have shown that protective immunity lasts as long as 35 years (Poland et al., 1981, Niedrig et al., 1999). Protection may very well be lifelong, as there are no known cases of YF infection in patients who have been vaccinated and had an appropriate initial immune response (Monath et al., 2002a). Although immunity after a single vaccination may be lifelong, international health regulations recommend revaccination at 10-year intervals. The WHO guidelines say a single dose of vaccine provides protection for 10 years and recommend booster vaccinations every 10 years, but this recommendation is not supported by specific clinical evidence (WHO, 2001). Little is known about the effect of one or more booster vaccinations on the duration of immunity. A recent study indicated that revaccination produced more pronounced neutralizing antibody increases (anamnastic response) in people with low antibody titers after the first dose of vaccine (Hepburn et al., 2006). Most vaccinees did not maintain a high neutralizing antibody titer for more than 3–5 years. If a persistent high titer is needed for protection, booster vaccinations with a frequency less than every 10 years would be needed, except in patients who have a high titer already. However, it is unknown whether a high titer is necessary for protection. All available data suggest that the detection of any level of neutralizing antibodies is sufficient for protective immunity. These findings raise new questions as to whether 10-year boosting is the best strategy and demonstrate the need for additional research on optimal dosing schedules (Hepburn et al., 2006).

Contraindications of Vaccination, Including Special Risk Groups

There are a number of contraindications for YF vaccination—the first is age (CDC, 2008). Infants younger than 6 months should not be vaccinated because they are more susceptible to YEL-AND (also known as postvaccinal encephalitis; Cetron et al., 2002). People older than 60 years have an increased risk for adverse events following vaccination (Martin et al., 2001a; Lawrence et al., 2004; Monath et al., 2005; Massad et al., 2005). There is also growing evidence to suggest that YEL-AVD and YEL-AND also appear more frequently in older people (Hayes, 2007; Barwick et al., 2004). Thus, the risk for death due to natural YF infection needs to be balanced against the risk of YEL-AVD and YEL-AND as a result of vaccination in elderly travelers (Khromava et al., 2005).

The second contraindication for YF vaccination is thymus disease. Individuals who have thymus disease or have had a thymectomy should not be vaccinated because this is a risk factor for YEL-AVD (Troillet and Laurencet, 2001; Barwick-Eidex, 2004; Vellozzi et al., 2006; Hayes, 2007; CDC, 2008). Four out of the 40 vaccine recipients with YEL-AVD reported worldwide have had thymus disease. One fatal case in the United States occurred in a 67-year-old woman, who had a thymectomy (Martin et al., 2001b) and a non fatal case occurred in a 70-year-old man who had hyperthyroidism, myasthenia gravis, and had received a thymectomy (Vellozzi et al., 2006). A third nonfatal case in Switzerland occurred in a 50-year-old man (Troillet and Laurencet, 2001) and a fourth fatal case in Colombia occurred in a 44-year-old man, who had both had a thymectomy (Hayes, 2007; Barwick-Eidex, 2004).

The third contraindication for YF vaccination is pregnancy. Currently there are insufficient data to determine whether the vaccine is safe to administer during pregnancy. Therefore, vaccination should only occur if travel to an area with risk of YF is unavoidable. Since pregnancy may affect immunologic function, serologic testing is necessary, and infants born to these women should be monitored for evidence of congenital infection and other possible adverse effects (CDC, 2008). Nonetheless, limited follow-up studies of women and newborn children indicate no abnormal effects suggesting that vaccination of pregnant women may not need to be contraindicated (Nasidi et al., 1993; Tsai et al., 1993; Robert et al., 1999; Suzano et al., 2006; Cavalcanti et al., 2007).

The fourth contraindication for YF vaccination is breast-feeding; vaccination of nursing mothers should be avoided, because of the risk of virus transmission to the baby. It is not known whether YFV is excreted in breast milk, but there have been no reports of adverse events or transmission of the virus to the baby. Nursing mothers should only be vaccinated if travel to an area with risk of YF is unavoidable (CDC, 2008).

The fifth contraindication for YF vaccination is immune suppression; any immunosuppressed person (HIV infection, leukemia, lymphoma, other malignancy, thymus disease, thymectomy, corticosteroid treatment, alkylating drug treatment, antimetabolite treatment, or radiation therapy) who is unable to effectively resist viral infections should not be vaccinated. Low-dose (prednisone or equivalent 20 mg/day), short-term (less than 2 weeks) systemic corticosteroid therapy or intra-articular, bursal, or tendon injections with corticosteroids; and intranasal corticosteroids are not thought to constitute an increased hazard (CDC, 1990, 2008). HIV-infected people who have laboratory verification of T lymphocyte counts and adequate immune system function (e.g., CD4 $> 200/\text{mm}^3$) can be vaccinated and monitored if travel to an area with risk of YF is unavoidable (Tattevin et al., 2004). Limited data indicate that the seroconversion rate is reduced; hence, serologic testing is necessary (Cetron et al., 2002; Sibailly et al., 1997). Family members of an immunosuppressed person, who themselves have no contraindications, should be vaccinated. HIV-infected individuals who have inadequate immune system function should not be vaccinated because fatal YEL-AND occurred in a 53-year-old patient with undiagnosed HIV infection who had a low CD4 count (108 cells/mm^3) (Kengsakul et al., 2002).

The sixth contraindication for YF vaccination is hypersensitivity to eggs; the vaccine is contraindicated because it is produced in chicken embryos. People who are able to eat eggs or egg products may receive the vaccine; however, some egg-sensitive persons are not allergic to cooked eggs. Gelatin, which is used as a stabilizer in some vaccines, might also cause allergic reactions. If vaccination of a person with gelatin hypersensitivity or with a questionable history of egg or chicken hypersensitivity is necessary because travel to an area with risk of YF is unavoidable, an intradermal test dose of vaccine should be administered (Aventis Pasteur, 1996; Arguin et al., 2007).

Adverse Events

The 17D vaccine is considered one of the safest vaccines and normally it is well tolerated with recipients reporting injection site pain, inflammation, mild headaches, myalgia, low-grade fevers, backache, or other minor symptoms that occur 5–11 days after vaccination. In non-placebo-controlled clinical trials, where the events reported could be unrelated to vaccination, the incidence of these minor symptoms varies from 4 to 25% and somewhere between 1 and 7% of vaccinees have to curtail regular activities (Theiler and Smith, 1937b; Smith et al., 1938; Freestone et al., 1977; Moss-Blundell et al., 1981; Digoutte et al., 1995; Lang et al., 1999; Monath et al., 2002a; Dumas et al., 1997; Kelso et al., 1999). First-time vaccinees have mild viremia (not exceeding 10^2 pfu/ml for ≥ 1 day between the third and seventh day after vaccination) which is associated with elevations of IFN-α, TNF-α which probably cause the minor symptoms reported by the vacinees, and markers of T cell activation (Wheelock and Sibley, 1965; Hacker et al., 1998; Reinhardt et al., 1998; Monath, 1999; Monath et al., 2002b). Revaccination after 10 years results in no viremia and an even lower incidence of adverse reactions because the 17D virus is neutralized by preformed antibody (Monath et al., 2002a, 2002b; Reinhardt et al., 1998).

Immediate severe hypersensitivity reactions to vaccination, which are characterized by rash, anaphylaxis, urticaria, or asthma or a combination of these, are rare with an incidence of $<0.8/100{,}000$ vaccinees. The vaccine is contraindicated for persons with a history of oral intolerance to eggs because it is a chicken egg extract. There can also be allergic reactions to other vaccine components in people without egg allergies (Kelso et al., 1999).

Serious adverse reactions to the vaccine are very unusual and there are two types: (1) vaccine-associated neurotropic disease (YEL-AND), encephalitis caused by neuroinvasion of the virus, and (2) vaccine-associated viscerotrophic disease (YEL-AVD), pansystemic infection with hepatitis, which is very similar to wild-type YF infection. The CDC/FDA Working Group on YF vaccines provided diagnostic definitions for YEL-AND and YEL-AVD. To diagnose YEL-AND a patients must show encephalopathy or inflammation on EEG, or CSF, testing and imaging or else they must have clinical disease with symptoms of headache, fever, and focal or general neurological dysfunction. These must occur within one month of vaccination and must be in the absence of other possible etiology. Supportive evidence comes from YFV histopathology and isolation of the vaccine virus from the patient's blood. To diagnose YEL-AVD a patient must have disease symptoms of nausea, vomiting, malaise, fever, diarrhea, myalgia, dyspnea, and they must have evidence of YFV-mediated tissue damage (jaundice, liver dysfunction, renal impairment, tachycardia, rhabdomyolysis, thrombocytopenia, hypotension, myocarditis, disseminated intravascular coagulation, and hemorrhage) in the absence of other etiology within 10 days of vaccination (Cetron et al., 2002).

Yellow Fever Vaccine–Associated Neurotropic Disease (YEL-AND) YEL-AND (previously named postvaccine encephalitis) has been divided into three categories based on recommendations of the Advisory Committee on Immunization Practices: (1) cases with neurotropic involvement, (2) cases with an autoimmune involvement of the central nervous system, and (3) cases with an autoimmune involvement of the peripheral nervous system. All cases have similar profiles, irrespective of the manufacturing source of the vaccine, and they occur in first-time vaccinees with an onset of illness 4–27 days postvaccination and a case fatality rate <5% (Cetron et al., 2002). Historically, YEL-AND was seen as encephalitis in infants, the first cases were in children who received the FNV by scarification in the 1950s during epidemics in Central America and Nigeria only, with the highest incidence in Nigeria (3–4/1000 vaccinations; 38% case fatality rate) (Durieux, 1956b; Stones and Mac Namara, 1955). It has also occasionally been seen in young babies receiving the 17D vaccine (Anonymous, 1966). After vaccine administration was limited to children ≥6 months in 1969 there have been few reports of YEL-AND in the United States (CDC, 1969). Since 1992, six cases of encephalitis among adult vaccinees have been reported to the U.S. Vaccine Adverse Event Reporting System (VAERS) (CDC, 2002; Marfin et al., 2006) and 10 cases of autoimmune neurological disease have been reported (CDC, 2008). A total of 21 cases worldwide had been reported by June 2001 (Monath, 1999b) showing an increase in the rate of YEL-AND, but this could simply be due to the increased surveillance.

It is difficult to estimate an accurate incidence rate for YEL-AND: the earliest ones varied from 50 to 400 cases per 100,000 infants under 6 months (Monath, 2004). During a 1993 Kenyan vaccination campaign, active hospital-based surveillance estimated a rate of 5.8 cases per 100,000 vaccinees (Monath, 2004). By 2005, the reporting rate was estimated to be 0.4 reported cases per 100,000 doses distributed in the United States (Marfin et al., 2005; Khromava et al., 2005) and the rate in the United Kingdom, where another 17D-204-based vaccine was used, was similar (Kitchener, 2004). The reporting rate was highest among older vaccinees, with 1.4 reported cases per 100,000 doses in persons 60 years and older (Khromava et al., 2005). Crude recent estimates of the reported rate in the United States are lower at 0.3–0.5 cases per 100,000 doses of vaccine distributed, although the rate is higher for people over 60 years at 1.8 cases per 100,000 doses (CDC, 2008). In the most recent study, McMahon et al. (2007) give a rate of 0.23 cases per 100,000 distributed vaccine doses in the United States.

Yellow Fever Vaccine–Associated Viscerotropic Disease (YEL-AVD) Yellow fever vaccine–associated viscerotrophic disease (YEL-AVD, previously named febrile multiple organ system failure or postvaccination multiple organ system failure) has been described among recipients of vaccines produced by all 17D-204 and 17DD manufacturers and, like YEL-AND, all cases have occurred in first-time vaccinees (Cetron et al., 2002). The vaccine virus has been cultured from blood, serum, heart, liver, spleen, skin, brain, spinal cord, kidney, lungs, and skeletal muscle. Liver biopsy of a fatal case before death showed minimal periportal inflammation, mild microvesicular fatty change, and focal degeneration of parenchymal tissue without hepatocellular necrosis. YF antigen can be detected in the liver and spleen. Microscopic examination of liver tissue showed mid-zonal necrosis, steatosis, eosinophilic degeneration of hepatocytes, apoptosis with Councilman bodies, microvesicular fatty changes, and minimal inflammation. Myocarditis and tubular necrosis have also been observed and the neutralizing antibody titers against the virus were higher than expected in some patients (Vasconcelos et al., 2001a; Chan et al., 2001; Martin et al., 2001b; Doblas et al., 2006; Hayes, 2007).

YEL-AVD was first reported in the literature in 2001 (Chan et al., 2001; Martin et al., 2001b; Vasconcelos et al., 2001a; Adhiyaman et al., 2001; Troillet and Laurencet, 2001; Werfel and Popp, 2001) and was initially considered to be a recent phenomenon until Engel et al. (2006) reported that a case had occurred in a Brazilian vaccine recipient in 1975 (Vasconcelos et al., 2004). The reports investigated seven cases, and six of them were fatal following administration of three different vaccines, giving a case fatality rate of 85% (Chan et al., 2001; Vasconcelos et al., 2001a; Martin et al., 2001b). Chan et al. (2001) reported the death of a 56-year-old man taking aspirin and atorvastatin who was vaccinated against YF and meningococcus. Vasconcelos et al. (2001a) reported the deaths of a healthy 22-year-old woman and a 5-year-old girl with a history of prematurity and recurrent bouts of bronchitis and diarrhea who was vaccinated against YF, measles, mumps, and rubella. Martin et al. (2001b) reported the deaths of a 79-year-old woman with a history of hypothyroidism, polymyalgia rheumatica, and hypertension; a 67-year-old woman, who had a colectomy, ileostomy, and thymectomy; and a 63-year-old man with history of hypertension who was vaccinated against YF, quadrivalent A/C/Y/W-135 meningococcus, oral polio, and hepatitis A. Martin et al. (2001b) reported the only survivor, a 76-year-old man with a history of Crohns disease, following YF and oral typhoid vaccination. Subsequently, further reports described individuals who survived YEL-AVD, giving a revised case fatality rate of 60% (Adhiyaman et al., 2001; Troillet and Laurencet, 2001; Werfel and Popp, 2001). Adhiyaman et al. (2001) reported a 45-year-old man who was immunized against YF, poliovirus, hepatitis A, and tetanus; Troillet and Laurencet (2001) described a 50-year-old man, who had surgery for thyoma; and Werfel and Popp (2001) reported a 71-year-old man with a history of sigmoid diverticulosis and borreliosis who all survived. The next year, a 25-year-old-man who received YF, influenza, and poliovirus vaccinations and a 70-year-old man who received YF vaccination were identified as having survived YEL-AVD (Levy et al., 2002).

Following these initial 12 reports, an additional 28 cases have been identified worldwide and reported to the CDC as of October 2007, which gives a total of 40 cases thus far (Hayes, 2007). Some of the cases have been detailed in the literature: Doblas et al. (2006) reported the death of a 26-year-old woman who was vaccinated for YF, diphtheria, and tetanus; Zhou et al. (2005) reported a 60-year-old man who had chemotherapy for rectal cancer and was vaccinated against YF and meningococcus, and survived; Gerasimon and Lowry (2005) reported the death of a 22-year-old woman who was vaccinated against YF and influenza; and Kitchener (2004) retrospectively found four cases of YEL-AVD where all of the people had received concurrent tetanus toxoid (either combined or singular formulation) vaccination. The most recent cases are a cluster of four fatal cases in Peru from a total of 42,000 given the same lot of vaccine. At the time of writing this chapter, the cause of this cluster is currently under investigation. Two potential risk factors for the development of YEL-AVD have been identified: advanced age (60 years or older) (Khromava et al., 2005; Lawrence et al., 2004) and a history of thymus disease with a thymectomy (Barwick-Eidex, 2004).

The incidence of YEL-AVD is estimated by dividing the number of reported cases by the number of YF vaccine doses distributed. The estimated frequency of YEL-AVD in the United States is 3–5 cases per 1,000,000 doses distributed, while rates are higher among people 60–69 years old (1.1 cases per 100,000 doses), and higher again among those who are over 70 years (3.2 per 100,000 doses). The estimated incidence in the United Kingdom is similar at 0.25 cases per 100,000 doses (Kitchener et al., 2004). An Australian review of adverse events following YF immunization revealed a similar increasing frequency of serious adverse events with increasing age (Lawrence et al., 2004) and a Brazilian study revealed a similar estimated risk of fatal YEL-AVD ranging from 0.004 to 0.21 fatalities per 100,000 doses (Struchiner et al., 2004).

Vaccines in Development

YF vaccine is used principally in tropical climates and the initial lyophilized vaccines without stabilizers deteriorated rapidly when exposed to temperatures above −20°C. So the development of more heat-stable vaccines became necessary. In 1987, the WHO recommended that each vaccine lot meet the following stability test: maintenance of potency (>1000 mouse intracerebral LD_{50}/human dose) with mean loss of titer ≤10-fold after being held at 37°C for 14 days. At that time only half of the YF vaccines produced worldwide met those stability standards. A number of additives were systematically investigated and the most successful formulation, used by a number of manufacturers, employs sugars, amino acids, and divalent cations [lactose (4%), sorbitol (2%), histidine (0.01 M), alanine (0.01 M), in phosphate buffered saline with Ca^{2+} and Mg^{2+}]. This stabilized vaccine loses only 0.3–0.5 $\log_{10}$ of infectivity after being held at 37°C for 14 days and it is stable for two years at 4–22°C. Despite its stability when freeze-dried, the 17D vaccine has very limited stability after reconstitution and must be discarded after an hour, and as such use of the normal multi-dose vials in tropical climates reduces the vaccine effectiveness. Improvement in vaccine stability after reconstitution is an area for future research (Monath, 1996).

The 17D vaccine is highly effective with 95% of people vaccinated showing immunity after a week and safe because >500 million doses have been administered worldwide with very rare serious side effects. Developments in molecular genetic techniques have utilized what is termed "reverse genetics" where the YFV RNA genome can be recovered from the cDNA infectious clone (Rice et al., 1989). Bray and Lai (1991) used two serotypes of DEN to create the first chimeric flavivirus and show that the structural genes of one serotype could be replaced by that of another. Chambers et al. (1999) created chimeric 17D/JE viruses that were attenuated for mice by swapping the prM and E protein genes from JE virus into the 17D 204 backbone; interestingly swapping the complete structural genes resulted in nonviable virus. This led to the ChimeriVax™ platform developed by Acambis Inc., where the YF 17D-204 vaccine virus backbone is used to create a new generation of chimeric vaccines, against DEN, WNV, and JE by swapping of the prM and E protein genes of the heterologous flaviviruses into the 17D infectious clone backbone (Pugachev et al., 2005). This approach has proved to be very promising with the ChimeriVax-WN in phase I clinical trials (Arroyo et al., 2004; Monath et al., 2006), tetravalent ChimeriVax-DEN (Guirakhoo et al., 2002, 2004; Guirkahoo et al., 2001) in phase 2 clinical trials, and ChimeriVax-JE in phase III clinical trials (Monath et al., 2002b, 2003). Table 40.4 gives a summary of the progress of these vaccines in clinical trials. The sequencing of the 17DD genome (Duarte dos Santos et al., 1995; Galler et al., 1998) allowed the production of an infectious clone from this substrain and chimeric flavivirus vaccines were also constructed by swapping the prM and E protein genes (Bonaldo et al., 2000). A chimeric 17DD/DEN1 vaccine, using the VeMir95 Venezuelan isolate, was not avirulent for mice and was more attenuated for monkeys than the commercial 17DD vaccine (Mateu et al., 2007). A chimeric 17DD/DEN2 vaccine, using the 44-2 wild-type Southeast Asian isolate, was similarly attenuated in monkey tests (Galler et al., 2005). A chimeric 17DD/Den 2 New Guinea C (NGC) vaccine, with the carboxy terminus of an E gene from a Brazilian isolate, was attenuated in mice and gave partial protection against a lethal challenge with DEN2 NGC virus (Caufour et al., 2001). While these 17DD-based vaccines have shown promise in monkey trials, none have progressed to human clinical trials.

TABLE 40.4 17D vaccine candidates that are under development for clinical use

Vaccine (strain)	Company	Clinical trial	References
ARILVAX™ (17D)	Acambis	Phase III	Monath et al. (2002a) Belmusto-Worn et al. (2005)
BERNA-YF	Berna Biotech	Phase III	Pfister et al. (2005)
Chimeri Vax-DEN2 (16681)	Acambis	Phase I	Guirakhoo et al. (2004, 2006)
Tetravalent Chimeri Vax-DEN DEN 1 (PUO359) DEN 2 (16681) DEN 3 (PaH881/88) DEN 4 (1228)	Acambis/ Oravax/Aventis Pasteur	Phase II	Guirkahoo et al. (2001) Guirakhoo et al. (2002) Guirakhoo et al. (2004)
Chimeri Vax-JE (SA14-14-2)	Acambis	Phase II	Monath et al. (2002b, 2003)
Chimeri Vax-WNV (NY99)	Acambis	Phase 1	Arroyo et al. (2004) Monath et al. (2006)

There has been recent interest in using the 17D backbone to develop chimeric YF viruses as a diagnostic tool for flaviviruses that need to be handled at BSL3. For example, Chimerivax-St Louis encephalitis (SLE) virus is being used at BSL2 as a surrogate for SLE virus in neutralization tests (Pugachev et al., 2004).

The 17D backbone has been used to develop a recombinant vaccine against Lassa fever virus; where the 17D 204 infectious clone expressing Lassa fever virus GP1 and GP2 in frame between the E and NS1 genes protected guinea pigs against fatal Lassa fever (Bredenbeek et al., 2006). Both the 17D-204 and the 17DD backbones are being used as prototype malaria vaccines (Tao et al., 2005; Bonaldo et al., 2002, 2005, 2006) and the 17D-204 backbone as a prototype cancer vaccine (McAllister et al., 2000) to express foreign epitopes, rather than complete antigens. The 17D-204 backbone expressing malaria epitopes at the viral serine protease cleavage site between NS2B and NS3 diminished murine liver parasite burden by 70% and provided protection for 4–8 weeks. Boosting with irradiated sporozoites (mimics immune response of vaccine-primed individuals in endemic areas) after vaccination conferred sterile immunity to 90% of mice indicating that vaccination could be sustained and magnified by the bite of infected mosquitoes (Tao et al., 2005). A second chimeric 17DD backbone expressing malaria epitopes, this time in the E protein fg loop, was attenuated for mice and monkey neurovirulence, and mouse tests showed production of antibodies that recognized the parasite protein (Bonaldo et al., 2002, 2005, 2006). To facilitate cancer vaccine studies, ovalbumin epitopes have been inserted, as model systems at different sites in the 17D-204 genome (N terminus of the polypeptide or between proteins C and prM, NS2A and NS2B, NS2B and NS3, NS3 and NS4A, or NS4A and NS4B). Immunization of mice elicited protective immunity against challenge with lethal doses of malignant melanoma cells expressing ovalbumin, and virus immunotherapy induced regression of established solid tumors and pulmonary metastases (McAllister et al., 2000). HIV vaccines are currently being developed using this same strategy (Van Epps, 2005).

Studies are also underway to develop replication-deficient 17D vaccine vectors (replicons) expressing foreign antigens in place of the envelope protein genes, e.g., Hepatitis C virus (Molenkamp et al., 2003). These so-called replicons undergo one round of replication and stimulate strong immune responses without the risk of complications associated with live vaccines, especially in immunosuppressed people (Jones et al., 2005; Mason et al., 2006).

Rationale of Second/Third/New-Generation Vaccines

A recent study evaluated the incremental health and programmatic cost impact of using single-dose lyophilized 17D vaccines as compared to the standard multi-dose vials in Cambodia, Ghana, and Bangladesh. A cost-effectiveness model found that single-dose presentations of the thermostable YF vaccine is a potentially cost-effective way of reducing childhood deaths and disability in low-resource countries in Asia and Africa (Levin et al., 2007).

Clinical Trials

There have been three recent phase III clinical trials involving two new YF vaccines that have been produced by companies who have bought the vaccine production technology rights. The first of these, the ARILVAX™ vaccine (Acambis, United States), was compared with a current commercial vaccine, YF-VAX® (Sanofi Pasteur Inc., USA), in a phase III double blind clinical trial in U.S. adults. There was 99.3% seroconversion of individuals given YF-VAX and 98.6% seroconversion of individuals given ARILVAX. Both vaccines were well tolerated by the recipients and the mean antibody responses elicited were well above minimal levels needed for protection. No serious adverse events were reported, although the YF-VAX recipients had more pain, edema, and inflammation at the site of the injection (Monath et al., 2002a; Barrett, 2004). The first randomized, double-blind, phase III vaccine trial in children (age 9 months–10 years) was carried out in Peru and compared the efficiency of this ARILVAX vaccine with the YF-VAX vaccine. Both vaccines were well tolerated and highly immunogenic, although the seroconversion rates were significantly higher for the ARILVAX vaccine. Any adverse reactions reported were mild and resolved without medical intervention (Belmusto-Worn et al., 2005; Barrett, 2004).

The immunogenicity and safety of another new YF vaccine, the BERNA-YF vaccine (Flavimun®, Berna Biotech Ltd., Switzerland), manufactured after the transfer of the vaccine technology from the Robert Koch Institute in Germany, was compared with the original German vaccine and a current, commercial vaccine, Stamaril® (Sanofi Pasteur, France), in a phase III bridging clinical trial. An effective immune response, with seroprotection in 100% of the individuals in all vaccine groups was elicited by all three vaccines and no adverse events were reported. There were no differences in the immune responses generated between three consecutive production batches of the BERNA-YF vaccine (Pfister et al., 2005).

POSTEXPOSURE IMMUNOPROPHYLAXIS

Early studies showed that passive antibody, the interferon inducer poly(I-poly(C), or interferon treatments are effective only before or within hours of infection, and as such these can be utilized in postexposure prophylaxis after a known exposure event, e.g., a laboratory-acquired infection (Monath,1987).

PROSPECTS FOR THE FUTURE

The new ChimeriVax platform vaccines have reached phase III clinical trials and are manufactured in Vero monkey kidney cell culture, which facilitates rapid vaccine manufacture at short notice to combat outbreaks of disease. One of the great problems with YF vaccine is supply and demand. Large numbers of embryonated chicken eggs are not available at short notice and it takes considerable labor to manufacture the vaccine. This opens up the possibility of changing the manufacture of the current 17D vaccine from embryonated chicken eggs to cell culture (Freire et al., 2005). This approach has not moved forward very far, in part due to the need of obtaining extensive safety and efficacy data on a "new" product compared to the current egg-based vaccine, which has a well-known safety and efficacy profile. Nonetheless, if a chimeric 17D-based vaccine is licensed and used it may provide the impetus to undertake additional research for a 17D Vero cell–derived vaccine. It has also been proposed that a reverse genetics Vero cell vaccine would have the advantage of a vaccine virus of defined sequence and would eliminate the quasi species found in the current vaccine. This, of course, assumes that the heterogeneity in the current vaccine does not contribute to the efficacy of the vaccine.

The reemergence of YFV in North America is a serious threat as evidenced by the West Nile epidemic brought to New York in 1999 (Lanciotti et al., 1999). Unvaccinated tourists who visit YFV endemic areas are at high risk of being infected and air travel would allow these viremic tourists to travel home quickly and be the source of an epidemic. Primary care physicians, in general, are ill prepared to diagnose or treat YF cases due to lack of experience and the *Ae. aegypti* domestic vector is present in large numbers in the Southern states. YF is a zoonosis and therefore vector control is central to the control of the disease. The development of molecular genetic techniques allows for the possibility of control using transgenic mosquitoes, although there are a huge number of ethical issues about the release of transgenic vectors. Mosquitoes have already been experimentally engineered with resistance to YFV (Higgs et al., 1998) and this would allow for competition studies in the field to see if naturally occurring populations could be replaced with genetically engineered resistant populations. This research would open up the whole area of flavivirus control using transgenic resistant mosquito vectors.

KEY ISSUES

- YFV was the first virus demonstrated to be transmitted by mosquitoes.
- Development of live-attenuated YF vaccines is one of the milestones in vaccine research and it is one of the safest and most efficacious vaccines developed to date.
- YF is a reemerging enzootic disease that is still a major public health problem despite the availability of the vaccine.
- More research is necessary to better understand the disease pathology and host immune response.
- In recent years, a rare but significant number of severe adverse events have been reported involving vaccine-associated viscerotropic disease and vaccine-associated neurotropic disease.
- The 17D vaccine virus infectious clone backbone is being successfully used to develop a variety of chimeric virus vaccines, some of which have reached clinical trials for human use.

References

Adhiyaman, V., Oke, A. and Cefai, C. Effects of yellow fever vaccination. Lancet 2001; 358:1907–1908.

Anonymous. Fatal viral encephalitis following 17D yellow fever vaccine inoculation. Report of a case in a 3-year-old child. J. Am. Med. Assoc. 1966; 198:671–672.

Aragao, H.deB. Report upon some researches on yellow fever. Memórias do Instituto Oswaldo Cruz 1929; 2:35–46.

Arguin, P.M., Kozarsky, P.E., Navin, A.W. and Reed, C. CDC Health Information for International Travel 2008: The Yellow Book. Philadelphia: Elsevier Health Sciences Division, 2007.

Arroyo, J., Miller, C., Catalan, J., Myers, G.A., Ratterree, M.S., Trent, D.W. and Monath, T.P. ChimeriVax-West Nile virus live-attenuated vaccine: preclinical evaluation of safety, immunogenicity, and efficacy. J. Virol. 2004; 78:12497–12507.

Ashcroft, M.T. Historical evidence of resistance to yellow fever acquired by residence in India. Trans. R. Soc. Trop. Med. Hyg. 1979; 73:247–248.

Aventis Pasteur YF-VAX Package Insert. Philadelphia: Swiftwater, 1996.

Barba-Spaeth, G., Longman, R.S., Albert, M.L. and Rice, C.M. Live attenuated yellow fever 17D infects human DCs and allows for presentation of endogenous and recombinant T cell epitopes. J. Exp. Med. 2005; 202:1179–1184.

Barmr, M. and Bronnert, C. Thermostabilisation du vaccine antiamaril 17D lyophilize. Essai de substances protectrices. J. Biol. Standardization 1984; 12:435–442.

Barnett, E.D. Yellow fever: epidemiology and prevention. Clin. Infect. Dis. 2007; 44:850–856.

Barrett, A.D.T. Yellow fever vaccines. Bull. Institute Pasteur 1987; 85:103–124.

Barrett, A.D., Mathews, J.H., Miller, B.R., Medlen, A.R., Ledger, T.N. and Roehrig, J.T. Identification of monoclonal antibodies that distinguish between 17D-204 and other strains of yellow fever virus. J. Gen. Virol. 1990a; 71:13–18.

Barrett, A.D. and Monath, T.P. Pathogenesis and pathophysiology of yellow fever. Adv. Virus Res. 2003a; 60:343–395.

Barrett, A.D. and Monath, T.P. Epidemiology and ecology of yellow fever virus. Adv. Virus Res. 2003b; 61:291–315.

Barrett, A.D., Monath, T.P., Barban, V., Niedrig, M. and Teuwen, D.E. 17D yellow fever vaccines: new insights. A report of a workshop held during the World Congress on Medicine and Health in the tropics, Marseille, France, Monday 12 September 2005. Vaccine 2007; 25:2758–2765.

Barrett, A.D., Monath, T.P., Cropp, C.B., Adkins, J.A., Ledger, T.N., Gould, E.A., Schlesinger, J.J., Kinney, R.M. and Trent, D.W. Attenuation of wild-type yellow fever virus by passage in HeLa cells. J. Gen. Virol. 1990b; 71:2301–2306.

Barrett, A.D., Pryde, A., Medlen, A.R., Ledger, T.N., Whitby, J.E., Gibson, C.A., De Silva, M., Groves, D.J., Langley, D.J. and Minor, P.D. Examination of the envelope glycoprotein of yellow fever vaccine viruses with monoclonal antibodies. Vaccine 1989; 7:333–336.

Barrett, A.D.T. Yellow fever vaccines. Biologicals 1997; 25:17–25.

Barrett, A.D.T. Current status of the Arilvax; yellow fever vaccine. Exp. Rev. Vaccines 2004; 3:413–420.

Barrett, A.D.T. and Higgs, S. Yellow fever: a disease that has yet to be conquered. Ann. Rev. Entomol. 2007; 5:209–229.

Barwick, R.S., Marfin, A.A. and Cetron, M.S. Yellow fever vaccine-associated disease. In: Emerging Infections (W.M. Scheld, B.E. Murray and J.M. Hughes, Eds.), 6th ed., pp. 25–34. Washington, D.C.: ASM Press, 2004.

Barwick-Eidex, R. History of thymoma and yellow fever vaccination. Lancet 2004; 364:936.

Bauer, J.H. and Hudson, N.P. Duration of immunity in human yellow fever as shown by protective power of serum. J. Prev. Med. 1930; 4:177.

Beaty, B.J., Tesh, R.B. and Aitken, T.H. Transovarial transmission of yellow fever virus in Stegomyia mosquitoes. Am. J. Trop. Med. Hyg. 1980; 29:125–132.

Belmusto-Worn, V.E., Sanchez, J.L., McCarthy, K., Nichols, R., Bautista, C.T., Magill, A.J., Pastor-Cauna, G., Echevarria, C., Laguna-Torres, V.A., Samame, B.K., Baldeon, M.E., Burans, J.P., Olson, J.G., Bedford, P., Kitchener, S. and Monath, T.P. Randomized, double-blind, phase III, pivotal field trial of the comparative immunogenicity, safety, and tolerability of two yellow fever 17D vaccines (Arilvax and YF-VAX) in healthy infants and children in Peru. Am. J. Trop. Med. Hyg. 2005; 72:189–197.

Beutler, B. Inferences, questions and possibilities in Toll-like receptor signalling. Nature 2004; 430:257–263.

Bonaldo, M.C., Caufour, P.S., Freire, M.S. and Galler, R. The yellow fever 17D vaccine virus as a vector for the expression of foreign proteins: development of new live flavivirus vaccines. Memórias do Instituto Oswaldo Cruz 2000; 95:215–223.

Bonaldo, M.C., Garratt, R.C., Caufour, P.S., Freire, M.S., Rodrigues, M.M., Nussenzweig, R.S. and Galler, R. Surface expression of an immunodominant malaria protein B cell epitope by yellow fever virus. J. Mol. Biol. 2002; 315:873–885.

Bonaldo, M.C., Garratt, R.C., Freire, M.S. and Galler, R. Expression of foreign protein epitopes at the surface of recombinant yellow fever 17D viruses based on three-dimensional modeling of its envelope protein. Cell Biochem. Biophys. 2006; 44:313–324.

Bonaldo, M.C., Garratt, R.C., Marchevsky, R.S., Coutinho, E.S., Jabor, A.V., Almeida, L.F., Yamamura, A.M., Duarte, A.S., Oliveira, P.J., Lizeu, J.O., Camacho, L.A., Freire, M.S. and Galler, R. Attenuation of recombinant yellow fever 17D viruses expressing foreign protein epitopes at the surface. J. Virol. 2005; 79:8602–8613.

Bonnevie-Nielsen, V., Heron, I., Monath, T.P. and Calisher, C.H. Lymphocytic 2′, 5′-oligoadenylate synthetase activity increases prior to the appearance of neutralizing antibodies and immunoglobulin M and immunoglobulin G antibodies after primary and secondary immunization with yellow fever vaccine. Clin. Diagn. Lab. Immunol. 1995; 2:302–306.

Bonnevie-Nielsen, V., Larsen, M.L., Frifelt, J.J., Michelsen, B. and Lernmark, A. Assocation of IDDM and attenuated response of 2′,5′-oligoadenylate synthetase to yellow fever vaccine. Diabetes 1989; 38:1636–1642.

Bosch, I., Herrera, F., Navarro, J.C., Lentino, M., Dupuis, A., Maffei, J., Jones, M., Fernandez, E., Perez, N., Perez-Eman, J., Guimaraes, A.E., Barrera, R., Valero, N., Ruiz, J., Velasquez, G., Martinez, J., Comach, G., Komar, N., Spielman, A. and Kramer, L. West Nile virus, Venezuela. Emerg. Infect. Dis. 2007; 13:651–653.

Bovier, R.H. and Noble, S. Bivalent cholera and typhoid vaccine. Drugs 1999; 58:91–96.

Brandriss, M.W., Schlesinger, J.J. and Walsh, E.E. Immunogenicity of a purified fragment of 17D yellow fever envelope protein. J. Infect. Dis. 1990; 161:1134–1139.

Bray, M. and Lai, C.J. Construction of intertypic chimeric dengue viruses by substitution of structural protein genes. Proc. Natl. Acad. Sci. U.S.A. 1991; 88:10342–10346.

Bredenbeek, P.J., Molenkamp, R., Spaan, W.J., Deubel, V., Marianneau, P., Salvato, M.S., Moshkoff, D., Zapata, J., Tikhonov, I., Patterson, J., Carrion, R., Ticer, A., Brasky, K. and Lukashevich, I.S. A recombinant yellow fever 17D vaccine expressing Lassa virus glycoproteins. Virology 2006; 345:299–304.

Bryan, C.S., Moss, S.W. and Kahn, R.J. Yellow fever in the Americas. Infect. Dis. Clin. North Am. 2004; 18:275–292.

Bryant, J.E. and Barrett, A.D. Comparative phylogenies of yellow fever isolates from Peru and Brazil. FEMS Immunol. Med. Microbiol. 2003; 39:103–118.

Bryant, J.E., Holmes, E.C. and Barrett, A.D. Out of Africa: a molecular perspective on the introduction of yellow fever virus into the Americas. PLoS Pathog. 2007; 3:0668–0673.

Buckley, A. and Gould, E.A. Neutralization of yellow fever virus studied using monoclonal and polyclonal antibodies. J. Gen. Virol. 1985; 66:2523–2531.

Burke, A.W. and Davis, N.C. Notes on laboratory infections with yellow fever. Am. J. Trop. Med. Hyg. 1930(Suppl. 1–10):419–426.

Cammack, N. and Gould, E.A. Topographical analysis of epitope relationships on the envelope glycoprotein of yellow fever 17D vaccine and the wild type Asibi parent virus. Virology 1986; 150:333–341.

Carter, H.R. Yellow Fever: An Epidemiological and Historical Study of Its Place of Origin. Baltimore: The Williams and Wilkins Company, 1931.

Caufour, P.S., Motta, M.C.A., Yamamura, A.M.Y., Vazquez, S., Ferreira, I.I., Jabor, A.V., Bonaldo, M.C., Freire, M.S. and Galler, R. Construction, characterization and immunogenicity of recombinant yellow fever 17D-dengue type 2 viruses. Virus Res. 2001; 79:1–14.

Cavalcanti, D.P., Salomão, M.A., Lopez-Camelo, J. and Pessoto, M.A. Campinas Group of Yellow Fever Immunization during Pregnancy Early exposure to yellow fever vaccine during pregnancy. Trop. Med. Int. Health 2007; 12:833–837.

Centers for Disease Control Recommendations of the Immunization Practices Advisory Committee. Yellow fever vaccine. MMWR Recomm. Rep. 1969; 18:189.

Centers for Disease Control Yellow fever vaccine: recommendations of the Immunization Practices Advisory Committee (ACIP). MMWR Recomm. Rep. 1990; 39:1–6.

Centers for Disease Control Fatal yellow fever in a traveler returning from Venezuela, (1999). MMWR Recomm. Rep. 2000; 49:303–305.

Centers for Disease Control Adverse events associated with 17D-derived yellow fever vaccination—United States, 2001–2002. MMWR Recomm. Rep. 2002; 51:989–993.

Cetron, M.S., Marfin, A.A., Julian, K.G., Gubler, D.J., Sharp, D.J., Barwick, R.S., Weld, L.H., Chen, R., Clover, R.D., Deseda-Tous, J., Marchessault, V., Offit, P.A. and Monath, T.P. Yellow fever vaccine. Recommendations of the Advisory Committee on Immunization Practices (ACIP). MMWR Recomm. Rep. 2002; 51:1–11.

Chambers, T.J., Nestorowicz, A., Mason, P.M. and Rice, C.M. Yellow fever/Japanese encephalitis chimeric viruses: construction and biological properties. J. Virol. 1999; 73:3095–3101.

Chambers, T.J., Tsai, T.F., Pervikov, Y. and Monath, T.P. Vaccine development against dengue and Japanese encephalitis: report of a World Health Organization meeting. Vaccine 1997; 15:1494–1502.

Chan, R.C., Penney, D.J., Little, D., Carter, I.W., Roberts, J.A. and Rawlinson, W.D. Hepatitis and death following vaccination with 17D-204 yellow fever vaccine. Lancet 2001; 358:121–122.

Chang, G.J., Cropp, B.C., Kinney, R.M., Trent, D.W. and Gubler, D.J. Nucleotide sequence variation of the envelope protein gene identifies two distinct genotypes of yellow fever virus. J. Virol. 1995; 69:5773–5780.

Chastel, C. Centenary of the discovery of yellow fever virus and its transmission by a mosquito (Cuba 1900–1901). Bulletin de la Société de Pathologie Exotique (1990) 2003; 96:250–256.

Co, M.D., Terajima, M., Cruz, J., Ennis, F.A. and Rothman, A.L. Human cytotoxic T lymphocyte responses to live attenuated 17D yellow fever vaccine: identification of HLA-B35-restricted CTL epitopes on nonstructural proteins NS1, NS2b, NS3, and the structural protein E. Virology 2002; 293:151–163.

Codeço, C.T., Luz, P.M. and Struchiner, C.J. Risk assessment of yellow fever urbanization in Rio de Janeiro, Brazil. Trans. R. Soc. Trop. Med. Hyg. 2004; 98:702–710.

Coleman, W. Epidemiological method in the 1860s: yellow fever at Saint-Nazaire. Bull. Hist. Med. 1984; 58:145–163.

Coleman, W. Yellow Fever in the North. Madison: University of Wisconsin Press, 1987.

Collaborative Group for Studies with Yellow Fever Vaccine Randomized, double-blind, multicenter study of the immunogenicity and reactogenicity of 17DD and WHO 17D-213/77 yellow fever vaccines in children: implications for the Brazilian National Immunization Program. Vaccine 2007; 25:3118–3123.

Davis, N.C. Uso experimental de uma vaccine cloroformada contra a febre amarela. Brasil-médico 1931; 45:268–273.

De Souza Lopes, O., de Almeida Guimares, S.S.D. and de Carvalho, R. Studies on yellow fever vaccine I. Quality control parameters. J. Biol. Standardization 1987; 15:323–329.

De Souza Lopes, O., de Almeida Guimares, S.S.D. and de Carvalho, R. Studies on yellow fever vaccine II. Stability of the reconstituted product. J. Biol. Standardization 1988a; 16:71–76.

De Souza Lopes, O., de Almeida Guimares, S.S.D. and de Carvalho, R. Studies on yellow fever vaccine III. Dose response in volunteers. J. Biol. Standardization 1988b; 16:77–82.

De Sousa Manso, C. Mass vaccination against yellow fever in Brazil. In: YF Vaccination (K.C. Smithburn, C. Duriex, R. Koerber, H.A. Penna, G.W.A. Dick, G. Courtois, C. de Sousa Manso, G. Stuart and P.H. Bonnel, Eds.), Monograph Series No. 30, pp. 123–139. Geneva: World Health Organization, 1956.

Deubel, V., Digoutte, J.P., Monath, T.P. and Girard, M. Genetic heterogeneity of yellow fever virus strains from Africa and the Americas. J. Gen. Virol. 1986; 67:209–213.

Diallo, M., Thonnon, J. and Fontenille, D. Vertical transmission of the yellow fever virus by *Aedes aegypti* (Diptera, Culicidae): dynamics of infection in F1 adult progeny of orally infected females. Am. J. Trop. Med. Hyg. 2000; 62:151–156.

Digoutte, J.-P., Cornet, M., Deubel, V. and Downs, W.G. Yellow fever. In: Exotic Viral Infections (J.S. Porterfield, Ed.), pp. 67–102. London: Chapman and Hall, 1995.

Digoutte, J.P., Plessart, H., Salaun, J.J., Heme, G., Ferrara, L. and Germain, M. Three cases of yellow fever contracted in Senegal. Bull. WHO 1981; 59:759–766.

Doblas, A., Domingo, C., Bae, H.G., Bohorquez, C.L., De Ory, F., Niedrig, M., Mora, D., Carrasco, F.J. and Tenorio, A. Yellow fever vaccine-associated viscerotropic disease and death in Spain. J. Clin. Virol. 2006; 36:156–158.

Doherty, P.C. Cytotoxic T cell effector and memory function in viral immunity. Curr. Top. Microbiol. Immunol. 1996; 206:1–14.

Duarte dos Santos, C.N., Post, P.R., Carvalho, R., Ferreira, I.I., Rice, C.M. and Galler, R. Complete nucleotide sequence of yellow fever virus vaccine strains 17DD and 17D-213. Virus Res. 1995; 35:35–41.

Duffy, J. Sword of Pestilence. Louisianna: Lousianna State University Press, 1966.

Dukes, C., Froeschle, J., George, J. et al. Safety and immunogenicity of simultaneous administration of Typhim Vi (TV), YF-VAX (YF) and Menomune (MV). 36th ICAAC, 1996, Abstracts.

Dumas, R., Forrat, R., Lang, J., Farinelli, T. and Loutan, L. Safety and immunogenicity of a new inactivated hepatitis A vaccine in concurrent administration with a typhoid fever vaccine or a typhoid fever + yellow fever vaccine. Adv. Ther. 1997; 14:160–167.

Dupuis, A.P., Marra, P.P. and Kramer, L.D. Serologic evidence of West Nile virus transmission, Jamaica, West Indies. Emerg. Infect. Dis. 2003; 9:860–863.

Dupuy, A.P., Despres, P., Cahour, A., Girard, M. and Bouloy, M. Nucleotide sequence comparison of the genome of two 17D-204 yellow fever vaccines. Nucleic Acids Res. 1989; 17:3989–3995.

Durieux, C. Preparation of yellow fever vaccine at Institut Pasteur, Dakar. In: YF Vaccination (K.C. Smithburn, C. Duriex, R. Koerber, H.A. Penna, G.W.A. Dick, G. Courtois, C. de Sousa Manso, G. Stuart and P.H. Bonnel, Eds.), Monograph Series No. 30, pp. 31–32. Geneva: World Health Organization, 1956a.

Durieux, C. Mass yellow fever vaccination in French Africa South of Sahara. In: YF Vaccination (K.C. Smithburn, C. Duriex, R. Koerber, H.A. Penna, G.W.A. Dick, G. Courtois, C. de Sousa Manso, G. Stuart and P.H. Bonnel, Eds.), Monograph Series No. 30, pp. 115–122. Geneva: World Health Organization, 1956b.

Engel, A.R., Vasconcelos, P.F., McArthur, M.A. and Barrett, A.D. Characterization of a viscerotropic yellow fever vaccine variant from a patient in Brazil. Vaccine 2006; 24:2803–2809.

Fagraeus, A., Ehrnst, A., Klein, E., Patarroyo, M. and Goldstein, G. Characterization of blood mononuclear cells reacting with K 562 cells after yellow fever vaccination. Cell. Immunol. 1982; 67:37–48.

Fernandez-Salas, I., Contreras-Cordero, J.F., Blitvich, B.J., Gonzalez-Rojas, J.I., Cavazos-Alvarez, A., Marlenee, N.L., Elizondo-Quiroga, A., Lorono-Pino, M.A., Gubler, D.J., Cropp, B.C., Calisher, C.H. and Beaty, B.J. Serologic evidence of West Nile virus infection in birds, Tamaulipas State, Mexico. Vector Borne Zoonotic Dis. (Larchmont, N.Y.) 2003; 3:209–213.

Filippis, A.M., Schatzmayr, H.G., Nicolai, C., Baran, M., Miagostovich, M.P., Sequeira, P.C. and Nogueira, R.M. Jungle yellow fever, Rio de Janeiro. Emerg. Infect. Dis. 2001; 7:484–485.

Findlay, C. El mosquito hipotéticamente considerado como agente de transmissión de la fiebre amailla. Anuals of Royal Academy. De cien.méd. de la Habana 1881; 18:147–169.

Findlay, G.M. and Clarke, L.P. Susceptibility of hedgehog to yellow fever: viscerotrophic virus. Trans. R. Soc. Trop. Med. Hyg. 1934a; 28:193–200.

Findlay, G.M. and Clarke, L.P. Susceptibility of hedgehog to yellow fever: neurotrophic virus. Trans. R. Soc. Trop. Med. Hyg. 1934b; 28:335–345.

Fitzgeorge, R. and Bradish, C.J. Differentiation of strains of yellow fever virus in gamma-irradiated mice. J. Gen. Virol. 1980; 50:345–356.

Frederiksen, H. Historical evidence for interference between dengue and yellow fever. Am. J. Trop. Med. Hyg. 1955; 4:483–491.

Freestone, D.S., Ferris, R.D., Weinberg, A.L. and Kelly, A. Stabilized 17D strain yellow fever vaccine: dose response studies, clinical reactions and effects on hepatic function. J. Biol. Standardization 1977; 5:181–186.

Freire, M.S., Mann, G.F., Marchevsky, R.S., Yamamura, A.M., Almeida, L.F., Jabor, A.V., Malachias, J.M., Coutinho, E.S. and Galler, R. Production of yellow fever 17DD vaccine virus in primary culture of chicken embryo fibroblasts: yields, thermo and genetic stability, attenuation and immunogenicity. Vaccine 2005; 23:2501–2512.

Galler, R., Marchevsky, R.S., Caride, E., Almeida, L.F., Yamamura, A.M., Jabor, A.V., Motta, M.C., Bonaldo, M.C., Coutinho, E.S. and Freire, M.S. Attenuation and immunogenicity of recombinant yellow fever 17D-dengue type 2 virus for rhesus monkeys. Braz. J. Med. Biol. Res. 2005; 38:1835–1846.

Galler, R., Post, P.R., Santos, C.N. and Ferreira, I.I. Genetic variability among yellow fever virus 17D substrains. Vaccine 1998; 16:1024–1028.

Garrison, F.H. An Introduction to the History of Medicine with Medical Chronology, Suggestions for Study and Bibliographical Data. Philadelphia: W.B. Saunders Company, 1929.

Georges, A.-J., Tible, F., Meunier, D.-M.-Y., Gonzalez, J.-P., Beraud, A.M., Sissoko-Dybdahl, N.-R., Abdul-Wahid, S., Fritzell, B., Girad, M. and Georges-Courbot, M.-C. Thermostability and efficacy in the field of a new stabilised yellow fever virus. Vaccine 1985; 3:313–315.

Gerasimon, G. and Lowry, K. Rare case of fatal yellow fever vaccine-associated viscerotropic disease. South. Med. J. 2005; 98:653–656.

Germain, M., Francy, D.B., Monath, T.P., Ferrara, L., Bryan, J., Salaun, J.J., Heme, G., Renaudet, J., Adam, C. and Digoutte, J.P. Yellow fever in the Gambia, 1978–1979: entomological aspects and epidemiological correlations. Am. J. Trop. Med. Hyg. 1980; 29:929–940.

Geske, T., Nichtila, P., Seethaler, H., Koch, M.A. and L'Agestehr, J. Establishment of hybridomas producing antibodies to viral surface epitopes related to pathogenic properties of yellow fever virus strains. Immunobiology 1983; 165:263.

Gibbons, R.V. and Vaughn, D.W. Dengue: an escalating problem. Br. Med. J. 2002; 324:1563–1566.

Gil, A., González, A., Dal-Ré, R. and Calero, J.R. Interference assessment of yellow fever vaccine with the immune response to a single-dose inactivated hepatitis A vaccine (1440 E.U.): a controlled study in adults. Vaccine 1996; 14:1028–1030.

Gollins, S.W. and Porterfield, J.S. A new mechanism for the neutralization of enveloped viruses by antiviral antibody. Nature 1986; 321(6067):244–246.

Gordon, J.E. and Hughes, T.P. A study of inactivated yellow fever virus as immunizing agent. J. Immunol. 1936; 30:221–234.

Gould, E.A., Buckley, A., Cammack, N., Barrett, A.D., Clegg, J.C., Ishak, R. and Varma, M.G. Examination of the immunological relationships between flaviviruses using yellow fever virus monoclonal antibodies. J. Gen. Virol. 1985; 66:1369–1382.

Gould, E.A., Buckley, A., Cane, P.A., Higgs, S. and Cammack, N. Use of a monoclonal antibody specific for wild-type yellow fever virus to identify wild-type antigenic variants in 17D vaccine pools. J. Gen. Virol. 1989; 70:1889–1894.

Gould, E.A., de Lamballerie, X., Zanotto, P.M. and Holmes, E.C. Evolution, epidemiology, and dispersal of flaviviruses revealed by molecular phylogenies. Adv. Virus Res. 2001; 57:71–103.

Groot, H. and Ribeiro, R.B. Neutralizing and haemagglutination-inhibiting antibodies to YF 17 years after vaccination with 17D vaccine. Bull. WHO 1962; 27:699–707.

Gubler, D.J. Dengue and dengue hemorrhagic fever: its history and resurgence as a global public health problem. In: Dengue and Dengue Hemorrhagic Fever (D.J. Gubler and G. Kuno, Eds.), pp. 1–22. New York: CABI Publishing, 1997.

Gubler, D.J. Epidemic dengue/dengue hemorrhagic fever as a public health, social and economic problem in the 21st century. Trends Microbiol. 2002; 10:100–103.

Gubler, D.J. The changing epidemiology of yellow fever and dengue, 1900 to 2003: full circle? Comp. Immunol. Microbiol. Infect. Dis. 2004; 27:319–330.

Guirkahoo, F., Arroyo, J., Pugachev, K.V., Miller, C., Zhang, Z.-X., Weltzin, R., Georgakopoulos, K., Catalan, J., Ocran, S., Soike, K., Ratterree, M. and Monath, T.P. Construction, safety, and immunogenicity in nonhuman primates of a chimeric yellow fever-dengue virus tetravalent vaccine. J. Virol. 2001; 75:7290–7304.

Guirakhoo, F., Kitchener, S., Morrison, D., Forrat, R., McCarthy, K., Nichols, R., Yoksan, S., Duan, X., Ermak, T.H., Kanesa-Thasan, N., Bedford, P., Lang, J., Quentin-Millet, M.J. and Monath, T.P. Live attenuated chimeric yellow fever dengue type 2 (ChimeriVax-DEN2) vaccine: Phase I clinical trial for safety and immunogenicity: effect of yellow fever pre-immunity in induction of cross neutralizing antibody responses to all 4 dengue serotypes. Hum. Vaccines 2006; 2:60–67.

Guirakhoo, F., Pugachev, K., Zhang, Z., Myers, G., Levenbook, I., Draper, K., Lang, J., Ocran, S., Mitchell, F., Parsons, M., Brown, N., Brandler, S., Fournier, C., Barrere, B., Rizvi, F., Travassos, A., Nichols, R., Trent, D. and Monath, T. Safety and efficacy of chimeric yellow fever-dengue virus tetravalent vaccine formulations in nonhuman primates. J. Virol. 2004; 78:4761–4775.

Guirakhoo, F., Pugachev, K.V., Arroyo, J., Miller, C., Zhang, Z.-X., Weltzin, R., Georgakopoulos, K., Catalan, J., Ocran, S., Draper, K. and Monath, T.P. Viremia and immunogenicity in nonhuman primates of a tetravalent yellow fever-dengue chimeric vaccine: genetic reconstructions, dose adjustment, and antibody responses against wild-type dengue virus isolates. Virology 2002; 298:146–159.

Hacker, U.T., Jelinek, T., Erhardt, S., Eigler, A., Hartmann, G., Nothdurft, H.D. and Endres, S. In vivo synthesis of tumor necrosis factor-alpha in healthy human after yellow fever vaccination. J. Infect. Dis. 1998; 177:774–778.

Hahn, C.S., Dalrymple, J.M., Strauss, J.H. and Rice, C.M. Comparison of the virulent Asibi strain of yellow fever virus with the 17D vaccine strain derived from it. Proc. Natl. Acad. Sci. U.S.A. 1987; 84:2019–2023.

Hayes, E.B. Acute viscerotropic disease following vaccination against yellow fever. Trans. R. Soc. Trop. Med. Hyg. 2007; 101:967–971.

Heinz, F.X. Epitope mapping of flavivirus glycoproteins. Adv. Virus Res. 1986; 31:103–168.

Henderson, B.E., Cheshire, P.P., Kirya, G.B. and Lule, M. Immunologic studies with yellow fever and selected African group B arboviruses in rhesus and vervet monkeys. Am. J. Trop. Med. Hyg. 1970; 19:110–118.

Hepburn, M.J., Kortepeter, M.G., Pittman, P.R., Boudreau, E.F., Mangiafico, J.A., Buck, P.A., Norris, S.L. and Anderson, E.L. Neutralizing antibody response to booster vaccination with the 17D yellow fever vaccine. Vaccine 2006; 24:2843–2849.

Herbein, G. and O'Brien, W.A. Tumor necrosis factor (TNF)-alpha and TNF receptors in viral pathogenesis. Proc. Soc. Exp. Biol. Med. 2000; 223:241–257.

Higgs, S., Rayner, J.O., Olson, K.E., Davis, B.S., Beaty, B.J. and Blair, C.D. Engineered resistance in *Aedes aegypti* to a West African and a South American strain of yellow fever virus. Am. J. Trop. Med. Hyg. 1998; 58:663–670.

Hindle, E. A yellow fever vaccine. Br. Med. J. 1928; 1:976.

Hindle, E. An experimental study of yellow fever. Trans. R. Soc. Trop. Med. Hyg. 1928–1929; 22:405.

Hobson, W. World Health and History. Bristol: Wright, 1963.

Iwasaki, A. and Medzhitov, R. Toll-like receptor control of the adaptive immune responses. Nat. Immunol. 2004; 5:987–995.

Jennings, A.D., Gibson, C.A., Miller, B.R., Mathews, J.H., Mitchell, C.J., Roehrig, J.T., Wood, D.J., Taffs, F., Sil, B.K., Whitby, S.N., Whitby, J.E., Monath, T.P., Minor, P.D., Sanders, P.G. and Barrett, A.D.T. Analysis of a yellow fever virus isolated from a fatal case of vaccine-associated human encephalitis. J. Infect. Dis. 1994; 169:512–518.

Johnson, B.W., Chambers, T.V., Crabtree, M.B., Filippis, A.M., Vilarinhos, P.T., Resende, M.C., Macoris Mde, L. and Miller, B.R. Vector competence of Brazilian *Aedes aegypti* and *Ae. albopictus* for a Brazilian yellow fever virus isolate. Trans. R. Soc. Trop. Med. Hyg. 2002; 96:611–613.

Jones, C.T., Patkar, C.G. and Kuhn, R.J. Construction and applications of yellow fever virus replicons. Virology 2005; 331:247–259.

Julander, J.G., Furuta, Y., Shafer, K. and Sidwell, R.W. Activity of T-1106 in a hamster model of yellow fever virus infection. Antimicrob. Agents Chemother. 2007a; 51:1962–1966.

Julander, J.G., Morrey, J.D., Blatt, L.M., Shafer, K. and Sidwell, R.W. Comparison of the inhibitory effects of interferon alfacon-1 and ribavirin on yellow fever virus infection in a hamster model. Antiviral Res. 2007b; 73:140–146.

Kaplan, J.E., Nelson, D.B., Schonberger, L.B., Hatch, M.H., Monath, T.P., Lazuick, J.S., Calisher, C.H. and Rosa, F.W. The effect of immune globulin on the response to trivalent oral poliovirus and yellow fever vaccinations. Bull. WHO 1984; 62:585–590.

Kawai, T. and Akira, S. Innate immune recognition of viral infection. Nat. Immunol. 2006; 7:131–137.

Kelso, J.M., Mootrey, G.T. and Tsai, T.F. Anaphylaxis from yellow fever vaccine. J. Allergy Clin. Immunol. 1999; 103:698–701.

Kengsakul, K., Sathirapongsasuti, K. and Punyagupta, S. Fatal myeloencephalitis following yellow fever vaccination in a case with HIV infection. J. Med. Assoc. Thailand 2002; 85:131–134.

Kerr, J.A. The clinical aspects and diagnosis of yellow fever. In: Yellow Fever (G.K. Strode, Ed.), pp. 629–640. New York: McGraw-Hill, 1951.

Khromava, A.Y., Eidex, R.B., Weld, L.H., Kohl, K.S., Bradshaw, R.D., Chen, R.T. and Cetron, M.S.The Yellow Fever Vaccine Safety Working Group Yellow fever vaccine: an updated assessment of advanced age as a risk factor for serious adverse events. Vaccine 2005; 23:3256–3263.

Kitchener, S. Viscerotropic and neurotropic disease following vaccination with the 17D yellow fever vaccine, ARILVAX®. Vaccine 2004; 22:2103–2105.

Komar, O., Robbins, M.B., Klenk, K., Blitvich, B.J., Marlenee, N.L., Burkhalter, K.L., Gubler, D.J., Gonzalvez, G., Pena, C.J., Peterson, A.T. and Komar, N. West Nile virus transmission in resident birds, Dominican Republic. Emerg. Infect. Dis. 2003; 9:1299–1302.

Kroger, A.T., Atkinson, W.L., Marcuse, E.K. and Pickering, L.K. General Recommendations on Immunization. Recommendations of the Advisory Committee on Immunization Practices (ACIP). MMWR Recomm. Rep. 2006; 55:1–48.

Kuhn, R.J., Zhang, W., Rossmann, M.G., Pletnev, S.V., Corver, J., Lenches, E., Jones, C.T., Mukhopadhayay, S., Chipman, P.R., Strauss, E.G., Baker, T.S. and Strauss, J.H. Structure of dengue virus: implications for flavivirus organization, maturation, and fusion. Cell 2002; 108:717–725.

Kurane, I., Brinton, M.A., Samson, A.L. and Ennis, F.A. Dengue virus-specific, human CD4+ CD8- cytotoxic T-cell clones: multiple patterns of virus cross-reactivity recognized by NS3-specific T-cell clones. J. Virol. 1991; 65:1823–1828.

Lanciotti, R.S., Roehrig, J.T., Deubel, V., Smith, J., Parker, M., Steele, K., Crise, B., Volpe, K.E., Crabtree, M.B., Scherret, J.H., Hall, R.A., MacKenzie, J.S., Cropp, C.B., Panigrahy, B., Ostlund, E., Schmitt, B., Malkinson, M., Banet, C., Weissman, J., Komar, N., Savage, H.M., Stone, W., McNamara, T. and Gubler, D.J. Origin of the West Nile virus responsible for an outbreak of encephalitis in the northeastern United States. Science 1999; 286:2333–2337.

Lang, J., Zuckerman, J., Clarke, P., Barrett, P., Kirkpatrick, C. and Blondeau, C. Comparison of the immunogenicity and safety of two 17D yellow fever vaccines. Am J Trop Med Hyg Jun 1999; 60(6):1045–1050.

Lawrence, G.L., Burgess, M.A. and Kass, R.B. Age-related risk of adverse events following yellow fever vaccination in Australia. Commun. Dis. Intell. 2004; 28:244–248.

Ledger, T.N., Sil, B.K., Wills, M.R., Lewis, G., Kinney, R.M., Jennings, A.D., Stephenson, J.R. and Barrett, A.D. Variation in the biological function of envelope protein epitopes of yellow fever vaccine viruses detected with monoclonal antibodies. Biologicals 1992; 20:117–128.

Lepiniec, L., Dalgarno, L., Huong, V.T., Monath, T.P., Digoutte, J.P. and Deubel, V. Geographic distribution and evolution of yellow fever viruses based on direct sequencing of genomic cDNA fragments. J. Gen. Virol. 1994; 75:417–423.

Levenbrook, I.S., Pelleu, L.J. and Elisberg, B.L. The monkey safety test for neurovirulence of yellow fever vaccines: the utility of quantitative clinical evaluation and histological examination. J. Biol. Standardization 1987; 15:323–329.

Levin, A., Levin, C., Kristensen, D. and Matthias, D. An economic evaluation of thermostable vaccines in Cambodia, Ghana and Bangladesh. Vaccine 2007; 25:6945–6957.

Levy, S., Mullane, K., Miller, M., Siva, S., Barnes, D., Dhaliwal, P., Tiwari, S.C. and Julian, K.G. From the Centers for Disease Control and Prevention. Adverse events associated with 17D-derived yellow fever vaccination-United States, 2001–2002. J. Am. Med. Assoc. 2002; 288:2533–2535.

Lhuillier, M., Sarthou, J.L., Cordellier, R., Monteny, N., Gershy-Damet, G.M. and Bouchite, B. Endemic emergence of yellow fever in the Ivory Coast: the place of anti-yellow fever IgM detection in the strategy of surveillance. Bull. WHO 1986; 64:415–420.

Liu, T. and Chambers, T.J. Yellow fever virus encephalitis: Properties of the brain-associated T-cell response during virus clearance in normal and gamma interferon-deficient mice and requirement for CD4+ lymphocytes. J. Virol. 2001; 75:2107–2118.

Lobigs, M., Dalgarno, L., Schlesinger, J.J. and Weir, R.E. Location of a neutralization determinant in the E protein of yellow fever virus (17D vaccine strain). Virology 1987; 161:474–478.

Lourenco-de-Oliveira, R., Vazeille, M., de Filippis, A.M. and Failloux, A.B. *Aedes aegypti* in Brazil: genetically differentiated populations with high susceptibility to dengue and yellow fever viruses. Trans. R. Soc. Trop. Med. Hyg. 2004; 98:43–54.

Marchevsky, R.S., Mariano, J., Ferreira, V.S., Almeida, E., Cerqueira, M.J., Carvalho, R., Pissurno, J.W., Travassos Da Rosa, A.P.A., Simoes, M.C., Santos, C.N.D., Ferreira, I.I., Muylaert, I.R., Mann, G.F., Rice, C.M. and Galler, R. Phenotypic analysis of yellow fever virus derived from complementary DNA. Am. J. Trop. Med. Hyg. 1995; 52:75–80.

Marchous, E. and Simond, P.L. La transmission hereditaire du virus de la fievre jaune chez la Stegomyia fasciata. Comptes Rendus des Séances de la Société de Biologie et de ses Filiales 1905; 59:259.

Marfin, A.A., Barwick-Eidex, R. and Monath, T.P. Yellow fever. In: Tropical Infectious Diseases: Principles, Pathogens, and Practice (R.L. Guerrant, D.H. Walker and P.F. Weller, Eds.), 2nd ed., pp. 797–812. Philadelphia: Elsevier, 2006.

Marfin, A.A., Eidex, R.S., Kozarsky, P.E. and Cetron, M.S. Yellow fever and Japanese encephalitis vaccines: indications and complications. Infect. Dis. Clin. North America 2005; 19:151–168.

Marianneau, P., Steffan, A.M., Royer, C., Drouet, M.T., Kirn, A. and Deubel, V. Differing infection patterns of dengue and yellow fever viruses in a human hepatoma cell line. J. Infect. Dis. 1998; 178:1270–1278.

Martin, M., Tsai, T.F., Cropp, B., Chang, G.J., Holmes, D.A., Tseng, J., Shieh, W., Zaki, S.R., Al-Sanouri, I., Cutrona, A.F., Ray, G., Weld, L.H. and Cetron, M.S. Fever and multisystem organ failure associated with 17D-204 yellow fever vaccination: a report of four cases. Lancet 2001b; 358:98–104.

Martin, M., Weld, L.H., Tsai, T.F., Mootrey, G.T., Chen, R.T., Niu, M. and Cetron, M.S. Advanced age a risk factor for illness temporally associated with yellow fever vaccination. Emerg. Infect. Dis. 2001a; 7:945–951.

Martins, M.A., Silva, M.L., Marciano, A.P., Peruhype-Magalhaes, V., Eloi-Santos, S.M., Ribeiro, G.L., Correa-Oliveira, R., Homma, A., Kroon, E.G., Teixeira-Carvalho, A. and Martins-Filho, O.A. Activation/modulation of adaptive immunity emerges simultaneously after 17DD yellow fever first-time vaccination: is this the key to prevent severe adverse reactions following immunization? Clin. Exp. Immunol. 2007; 48:90–100.

Mason, P.W., Shustov, A.V. and Frolov, I. Production and characterization of vaccines based on flaviviruses defective in replication. Virology 2006; 351:432–443.

Massad, E., Coutinho, F.A., Burattini, M.N., Lopez, L.F. and Struchiner, C.J. Yellow fever vaccination: how much is enough? Vaccine 2005; 23:3908–3914.

Mateu, G.P., Marchevsky, R.S., Liprandi, F., Bonaldo, M.C., Coutinho, E.S., Dieudonné, M., Caride, E., Jabor, A.V., Freire, M.S. and Galler, R. Construction and biological properties of yellow fever 17D/dengue type 1 recombinant virus. Trans. R. Soc. Trop. Med. Hyg. 2007; 101:289–298.

Mathis, C., Sellards, A.W. and Laigret, J. Sensibilite du Macaca rhesus au virus de la fievre jaune. Comptes Rendus de l'Académie des Sciences (Paris) 1928; 188:604–606.

McAllister, A., Arbetman, A.E., Mandl, S., Pena-Rossi, C. and Andino, R. Recombinant yellow fever viruses are effective therapeutic vaccines for treatment of murine experimental solid tumors and pulmonary metastases. J. Virol. 2000; 74:9197–9205.

McArthur, M.A., Suderman, M.T., Mutebi, J.P., Xiao, S.Y. and Barrett, A.D. Molecular characterization of a hamster viscerotropic strain of yellow fever virus. J. Virol. 2003; 77:1462–1468.

McArthur, M.A., Xiao, S.Y. and Barrett, A.D. Phenotypic and molecular characterization of a non-lethal, hamster-viscerotropic strain of yellow fever virus. Virus Res. 2005; 110:65–71.

McBride, W.J. and Bielefeldt-Ohmann, H. Dengue viral infections; pathogenesis and epidemiology. Microbes Infect. 2000; 2:1041–1050.

McMahon, A.W., Eidex, R.B., Marfin, A.A., Russell, M., Sejvar, J.J., Markoff, L., Hayes, E.B., Chen, R.T., Ball, R., Braun, M.M. and Cetron, M.Yellow Fever Working Group Neurologic disease associated with 17D-204 yellow fever vaccination: a report of 15 cases. Vaccine 2007; 25:1727–1734.

Molenkamp, R., Kooi, E.A., Lucassen, M.A., Greve, S., Thijssen, J. C., Spaan, W.J. and Bredenbeek, P.J. Yellow fever virus replicons as an expression system for hepatitis C virus structural proteins. J. Virol. 2003; 77:1644–1648.

Monath, T.P. Yellow fever: a medically neglected disease. Report on a seminar. Rev. Infect. Dis. 1987; 9:165–175.

Monath, T.P. Dengue: the risk to developed and developing countries. Proc. Natl. Acad. Sci. U.S.A. 1994; 91:2395–2400.

Monath, T.P. Stability of yellow fever vaccine. Dev. Biol. Standardization 1996; 87:219–225.

Monath, T.P. Facing up to re-emergence of urban yellow fever. Lancet 1999; 353:1541.

Monath, T.P. Yellow fever: an update. Lancet Infect. Dis. 2001; 1:11–20.

Monath, T.P. Yellow fever vaccine. In: Vaccines (S.A. Plotkin, W.A. Orenstein and P.A. Offit, Eds.), 4th ed., pp. 1095–1176. New York: Saunders, 2004.

Monath, T.P. Yellow fever as an endemic/epidemic disease and priorities for vaccination. Bulletin de la Société de Pathologie Exotique 2006; 99:341–347.

Monath, T.P. and Barrett, A.D. Pathogenesis and pathophysiology of yellow fever. Adv Virus Res 2003; 60:343–395.

Monath, T.P., Ballinger, M., Miller, B.M. and Salaun, J. Detection of yellow fever viral RNA by nucleic acid hybridization and viral antigen by immunocytochemistry in fixed human liver tissues. Am. J. Trop. Med. Hyg. 1989; 40:663–668.

Monath, T.P., Brinker, K.R., Chandler, F.W., Kemp, G.E. and Cropp, C.B. Pathophysiologic correlations in a rhesus monkey model of yellow fever with special observations on the acute necrosis of B cell areas of lymphoid tissues. Am. J. Trop. Med. Hyg. 1981; 30:431–443.

Monath, T.P., Cetron, M.S., McCarthy, K., Nichols, R., Archambault, W.T., Weld, L. and Bedford, P. Yellow fever 17D vaccine safety and immunogenicity in the elderly. Hum. Vaccines 2005; 1:207–214.

Monath, T.P., Guirakhoo, F., Nichols, R., Yoksan, S., Schrader, R., Murphy, C., Blum, P., Woodward, S., McCarthy, K., Mathis, D., Johnson, C. and Bedford, P. Chimeric live, attenuated vaccine against Japanese encephalitis (ChimeriVax-JE): phase 2 clinical trials for safety and immunogenicity, effect of vaccine dose and schedule, and memory response to challenge with inactivated Japanese encephalitis antigen. J. Infect. Dis. 2003; 188:1213–1230.

Monath, T.P. and Cetron, M.S. Prevention of yellow fever in persons traveling to the tropics. Clin. Infect. Dis. 2002; 34:1369–1378.

Monath, T.P., Craven, R.B., Adjukiewicz, A., Germain, M., Francy, D. B., Ferrara, L., Samba, E.M., N'Jie, H., Cham, K., Fitzgerald, S.A., Crippen, P.H., Simpson, D.I., Bowen, E.T., Fabiyi, A. and Salaun, J.J. Yellow fever in the Gambia, 1978–1979: epidemiologic aspects with observations on the occurrence of orungo virus infections. Am J Trop Med Hyg Sep 1980; 29(5):912–928.

Monath, T.P. and Heinz, F.X. Flaviviruses. In: Fields Virology (B.N. Fields, D.M. Knipe, P.M. Howley, R.M. Chanock, T.P. Monath, J.L. Melnick, B. Roizman and S.E. Strauss, Eds.), Vol. 2, pp. 961–1024. Philadelphia: Raven Press, 1996.

Monath, T.P., Levenbrook, I., Soike, K., Zhang, Z.-X., Ratterree, M., Draper, K., Barrett, A.D.T., Nickols, R., Weltzin, R., Arroyo, J. and Guirakhoo, F. Chimeric yellow fever virus 17D-Japanese encephalitis vaccine: dose-response effectiveness and extended safety testing in Rhesus monkeys. J. Virol. 2000; 74:1742–1751.

Monath, T.P., Liu, J., Kanesa-Thasan, N., Myers, G.A., Nichols, R., Deary, A., McCarthy, K., Johnson, C., Ermak, T., Shin, S., Arroyo, J., Guirakhoo, F., Kennedy, J.S., Ennis, F.A., Green, S. and Bedford, P. A live, attenuated recombinant West Nile virus vaccine. Proc. Natl. Acad. Sci. U.S.A. 2006; 103:6694–6699. Erratum in: Proc. Natl. Acad. Sci. U.S.A. 103:10823.

Monath, T.P., McCarthy, K., Bedford, P., Johnson, C.T., Nichols, R., Yoksan, S., Marchesani, R., Knauber, M., Wells, K.H., Arroyo, J. and Guirakhoo, F. Clinical proof of principle for ChimeriVax: recombinant live, attenuated vaccines against flavivirus infections. Vaccine 2002b; 20:1004–1018.

Monath, T.P. and Nasidi, A. Should yellow fever vaccine be included in the expanded program of immunization in Africa? A cost-effectiveness analysis for Nigeria. Am. J. Trop. Med. Hyg. 1993; 48:274–299.

Monath, T.P., Nichols, R., Archambault, W.T., Moore, L., Marchesani, R., Tian, J., Shope, R.E., Thomas, N., Schrader, R., Furby, D. and Bedford, P. Comparative safety and immunogenicity of two yellow fever 17D vaccines (ARILVAX and YF-VAX) in a phase III multicenter, double-blind clinical trial. Am. J. Trop. Med. Hyg. 2002a; 66:533–541).

Mondet, B., da Rosa, A.P. and Vasconcelos, P.F. The risk of urban yellow fever outbreaks in Brazil by dengue vectors. *Aedes aegypti* and Aedes albopictus. Bulletin de la Société de Pathologie Exotique (1990) 1996; 89:107–113.

Mondet, B., Vasconcelos, P.F., Travassos da Rosa, A.P., Travassos da Rosa, E.S., Rodrigues, S.G., Travassos Rosa, J.F. and Bicout, D.J. Isolation of yellow fever virus from nulliparous *Haemagogus (Haemagogus) janthinomys* in eastern Amazonia. Vector Borne Zoonotic Dis. 2002; 2:47–50.

Moreau, J.P., Girault, G., Drame, I. and Perraut, R. Reemergence of yellow fever in West Africa: lessons from the past, advocacy for a control program. Bulletin de la Société de pathologie exotique (1990) 1999; 92:333–336.

Moss-Blundell, A.J, Bernstein S., Shepherd W.M., Langford, D.T., Ferris, R. and Kelly, A. A clinical study of stabilized 17D strain live attenuated yellow fever vaccine. J. Biol. Stand. Oct 1981; 9(4):445–452.

Mutebi, J.P. and Barrett, A.D. The epidemiology of yellow fever in Africa. Microbes Infect. 2002; 4:1459–1468.

Mutebi, J.P., Wang, H., Li, L., Bryant, J.E. and Barrett, A.D. Phylogenetic and evolutionary relationships among yellow fever virus isolates in Africa. J. Virol. 2001; 75:6999–7008.

Nasidi, A., Monath, T.P., Vandenberg, J., Tomori, O., Calisher, C.H., Hurtgen, X., Munube, G.R., Sorungbe, A.O., Okafor, G.C. and Wali, S. Yellow fever vaccination and pregnancy: a four-year prospective study. Trans. R. Soc. Trop. Med. Hyg. 1993; 87:337–339.

Niedrig, M., Lademann, M., Emmerich, P. and Lafrenz, M. Assessment of IgG antibodies against yellow fever virus after vaccination with 17D by different assays: neutralization test, haemagglutination inhibition test, immunofluorescence assay and ELISA. Trop. Med. Int. Health 1999; 4:867–871.

Nolla-Salas, J. and Sadalls-Radresa, J. Imported yellow fever in vaccinated tourist. Lancet 1989; 2:1275–1276.

Okell, C.C. Experiments with yellow fever vaccine in monkeys. Trans. R. Soc. Trop. Med. Hyg. 1930; 24:251–254.

Onyango, C.O., Grobbelaar, A.A., Gibson, G.V., Sang, R.C., Sow, A., Swaneopel, R. and Burt, F.J. Yellow fever outbreak, Southern Sudan, 2003. Emerg. Infect. Dis. 2004b; 10:1668–1670.

Onyango, C.O., Ofula, .O., Sang, R.C., Konongoi, S.L., Sow, A., De Cock, K.M., Tukei, P.M., Okoth, F.A., Swanepoel, R., Burt, F.J., Waters, N.C. and Coldren, R.L. Yellow fever outbreak, Imatong, Southern Sudan. Emerg. Infect. Dis. 2004a; 10:1063–1068.

Oudart, J.L. and Rey, M. Proteinuria, proteinaemia, and serum transaminase activity in 23 confirmed cases of yellow fever. Bull. WHO 1970; 42:95–102.

Palmer, D.R., Fernandez, S., Bisbing, J., Peachman, K.K., Rao, M., Barvir, D., Gunther, V., Burgess, T., Kohno, Y., Padmanabhan, R. and Sun, W. Restricted replication and lysosomal trafficking of yellow fever 17D vaccine virus in human dendritic cells. J. Gen. Virol. 2007; 88:148–156.

Patterson, K.D. Yellow fever epidemics and mortality in the United States, 1693–1905. Soc. Sci. Med. 1992; 34:855–865.

Penna, M.D. Production of 17D yellow fever vaccine. In: YF Vaccination (K.C. Smithburn, C. Duriex, R. Koerber, H.A. Penna, G.W.A. Dick, G. Courtois, C. de Sousa Manso, G. Stuart and P.H. Bonnel, Eds.), Monograph Series No. 30, pp. 67–68. Geneva: World Health Organization, 1965.

Pettit, A. Rapport sur la valeur immunisante des vaccins employés contre la fièvre jaune et la valeur thérapeutique du sérum antiamaril. Bull Acad Nat Méd 1931;105–522.

Pfister, M., Kürsteiner, O., Hilfiker, H., Favre, D., Durrer, P., Ennaji, A., L'Age-Stehr, J., Kaufhold, A. and Herzog, C. Immunogenicity and safety of BERNA-YF compared with two other 17D yellow fever vaccines in a phase 3 clinical trial. Am. J. Trop. Med. Hyg. 2005; 72(3):339–346.

Poland, J.D., Calisher, C.H., Monath, T.P., Downs, W.G. and Murphy, K. Persistence of neutralizing antibody 30–35 years after immunization with 17D yellow fever vaccine. Bull. WHO 1981; 59:895–900.

Porterfield, J.S. The haemagglutination-inhibition test in the diagnosis of yellow fever in man. Trans. R. Soc. Trop. Med. Hyg. 1954; 48:261–266.

Post, P.R., Santos, C.N., Carvalho, R., Cruz, A.C., Rice, C.M. and Galler, R. Heterogeneity in envelope protein sequence and N-linked glycosylation among yellow fever virus vaccine strains. Virology 1992; 188:160–167.

Powell, J.H. Bring Out Your Dead. Philadelphia: University of Pennsylvania Press, 1949.

Pugachev, K.V., Guirakhoo, F., Mitchell, F., Ocran, S.W., Parsons, M., Johnson, B.W., Kosoy, O.L., Lanciotti, R.S., Roehrig, J.T., Trent, D.W. and Monath, T.P. Construction of yellow fever/St. Louis encephalitis chimeric virus and the use of chimeras as a diagnostic tool. Am. J. Trop. Med. Hyg. 2004; 71:639–645.

Pugachev, K.V., Guirakhoo, F. and Monath, T.P. New developments in flavivirus vaccines with special attention to yellow fever. Curr. Opin. Infect. Dis. 2005; 18:387–394.

Putnak, J.R. and Schlesinger, J.J. Protection of mice against yellow fever virus encephalitis by immunization with a vaccinia virus recombinant encoding the yellow fever virus non-structural proteins, NS1, NS2a and NS2b. J. Gen. Virol. 1990; 71:1697–1702.

Querec, T., Bennouna, S., Alkan, S., Laouar, Y., Gorden, K., Flavell, R., Akira, S., Ahmed, R. and Pulendran, B. Yellow fever vaccine YF-17D activates multiple dendritic cell subsets via TLR2, 7, 8 and 9 to stimulate polyvalent immunity. J. Exp. Med. 2006; 203:413–424.

Querec, T.D. and Pulendran, B. Understanding the role of innate immunity in the mechanism of action of the live attenuated yellow fever vaccine 17D. Adv. Exp. Med. Biol. 2007; 590:43–53.

Reed, W., Carroll, J. and Agramonte, A. Experimental yellow fever. In: United States 61st Congress, 3rd Session. Senate document 822, pp. 110-130, 1901.

Reinhardt, B., Jaspert, R., Niedrig, M., Kostner, C. and L'age-Stehr, J. Development of viremia and humoral and cellular parameters of immune activation after vaccination with yellow fever virus strain 17D: a model of human flavivirus infection. J. Med. Virol. 1998; 56:159–167.

Rice, C.M., Grakoui, A., Galler, R. and Chambers, T.J. Transcription of infectious yellow fever RNA from full-length cDNA templates produced by in vitro ligation. New Biol. 1989; 1:285–296.

Rice, C.M., Lenches, E., Eddy, S.R., Shin, S.J., Sheets, R.L. and Strauss, J.H. Nucleotide sequence of yellow fever virus: implications for flavivirus gene expression and evolution. Science 1985; 229:726–733.

Rifakis, P.M., Benitez, J.A., De-la-Paz-Pineda, J. and Rodriguez-Morales, A.J. Epizootics of yellow fever in Venezuela (2004–2005): an emerging zoonotic disease. Ann. N. Y. Acad. Sci. 2006; 1081:57–60.

Robert, E., Vial, T., Schaefer, C., Arnon, J. and Reuvers, M. Exposure to yellow fever vaccine in early pregnancy. Vaccine 1999; 17:283–285.

Roberts, L. Infectious disease. Resurgence of yellow fever in Africa prompts a counter attack. Science 2007; 316:1109.

Robertson, S. Yellow Fever: The Immunological Basis for Immunization 8. Geneva: World Health Organization, 1993.

Robertson, S.E., Hull, B.P., Tomori, O., Bele, O., Le Duc, J.W. and Esteves, K. Yellow fever: a decade of reemergence. J. Am. Med. Assoc. 1996; 276:1157–1162.

Rosenzweig, E.C., Babione, R.W. and Wisseman, C.L., Jr. Immunological studies with group B arthropod-borne viruses. IV. Persistence of yellow fever antibodies following vaccination with 17D strain yellow fever vaccine. Am. J. Trop. Med. Hyg. 1963; 12:230–235.

Ruben, F.L., Smith, E.A., Foster, S.O., Casey, H.L., Pifer, J.M., Wallace, R.B., Atta, A.I., Jones, W.L., Arnold, R.B., Teller, B. E., Shaikh, Z.Q., Lourie, B., Eddins, D.L., Doko, S.M. and Foege, W.H. Simultaneous administration of smallpox, measles, yellow fever, and diphtheria-pertussis-tetanus antigens to Nigerian children. Bull. WHO 1973; 48:175–181.

Ryman, K.D., Ledger, T.N., Weir, R.C., Schlesinger, J.J. and Barrett, A.D. Yellow fever virus envelope protein has two discrete type-specific neutralizing epitopes. J. Gen. Virol. 1997; 78:1353–1356.

Ryman, K.D., Ledger, T.N., Campbell, G.A., Watowich, S.J. and Barrett, A.D. Mutation in a 17D-204 vaccine substrain-specific envelope protein epitope alters the pathogenesis of yellow fever virus in mice. Virology Apr 25 1998; 244(1):59–65.

Sabin, A.B. Research on dengue during World War II. Am. J. Trop. Med. Hyg. 1952; 1:30–50.

Santos, A.P., Bertho, A.L., Dias, D.C., Santos, J.R. and Marcovistz, R. Lymphocyte subset analyses in healthy adults vaccinated with yellow fever 17DD virus. Memórias do Instituto Oswaldo Cruz 2005; 100:331–337. Erratum in: Mem. Inst. Oswaldo Cruz 100:571.

Sareen Galbraith. Brain Infections Group, Divisions of Neurological Science and Medical Microbiology, University of Liverpool, Liverpool, UK

Sawyer, W.A. Persistence of yellow fever immunity. J. Prev. Med. 1931; 5:413–428.

Sbrana, E., Xiao, S.Y., Guzman, H., Ye, M., Travassos da Rosa, A.P. and Tesh, R.B. Efficacy of post-exposure treatment of yellow fever with ribavirin in a hamster model of the disease. Am. J. Trop. Med. Hyg. 2004; 71:306–312.

Sbrana, E., Xiao, S.Y., Popov, V.L., Newman, P.C. and Tesh, R.B. Experimental yellow fever virus infection in the golden hamster (Mesocricetus auratus) III. Clinical laboratory values. Am. J. Trop. Med. Hyg. 2006; 74:1084–1089.

Schlesinger, J.J., Brandriss, M.W., Cropp, C.B. and Monath, T.P. Protection against yellow fever in monkeys by immunization with yellow fever virus nonstructural protein NS1. J. Virol. 1986; 60:1153–1155.

Schlesinger, J.J., Brandriss, M.W. and Monath, T.P. Monoclonal antibodies distinguish between wild and vaccine strains of yellow fever virus by neutralization, hemagglutination inhibition and immune precipitation of the virus envelope protein. Virology 1983; 125:8–17.

Schlesinger, J.J., Brandriss, M.W., Putnak, J.R. and Walsh, E.E. Cell surface expression of yellow fever virus non-structural glycoprotein NS1: consequences of interaction with antibody. J. Gen. Virol. 1990; 71:593–599.

Schlesinger, J.J., Walsh, E.E. and Brandriss, M.W. Analysis of 17D yellow fever virus envelope protein epitopes using monoclonal antibodies. J. Gen. Virol. 1984; 65:1637–1644.

Seagroatt, V. and Magrath, D.I. A WHO collaborative study of in vitro and in vivo methods for the assay of yellow fever vaccines. J. Biol. Standardization 1983; 11:47–54.

Sellards, A.W. The behaviour of the virus of yellow fever in monkeys and mice. Proc. Natl. Acad. Sci. U.S.A. 1931; 17:339–343.

Sellards, A.W. and Bennett, B.L. Vaccination in yellow fever with non-infective virus. Ann. Trop. Med. Parasitol. 1937; 3:373.

Sibailly, T.S., Wiktor, S.Z., Tsai, T.F., Cropp, B.C., Ekpini, E.R., Adjorlolo-Johnson, G., Gnaore, E., DeCock, K.M. and Greenberg, A.E. Poor antibody response to yellow fever vaccination in children infected with human immunodeficiency virus type 1. Pediatr. Infect. Dis. J. 1997; 16:1177–1179.

Sil, B.K., Dunster, L.M., Ledger, T.N., Wills, M.R., Minor, P.D. and Barrett, A.D. Identification of envelope protein epitopes that are important in the attenuation process of wild-type yellow fever virus. J. Virol. 1992; 66:4265–4270.

Smith, H.H. Controlling yellow fever. In: Yellow Fever (G.K. Strode, J.C. Bugher, J. Austin-Kerr, H.H. Smith, K.C. Smithburn, R.M. Taylor, M. Theiler, A.J. Warren and L. Whitman, Eds.), pp. 539–628. New York: McGraw-Hill Book Company, Inc, 1951.

Smith, H.H., Penna, H.A. and Paoliello, A. Yellow fever vaccination with cultured virus (17D) without immune serum. Am J Trop Med Hyg 1938; 18:437–468.

Snijders, E.P., Postmus, S. and Schuffner, W. On the protective power of yellow fever sera and dengue sera against yellow fever virus. Am. J. Trop. Med. Hyg. 1934; 14:519–545.

Sood, D.K., Aggarwal, R.K., Kumar, S. and Sokhey, J. A rapid test for measuring the infectivity of yellow fever vaccine. Vaccine 1995; 13:427–428.

Stefano, I., Sato, H.K., Pannuti, C.S., Omoto, T.M., Mann, G., Freire, M.S., Yamamura, A.M., Vasconcelos, P.F., Oselka, G.W., Weckx, L.W., Salgado, M.F., Noale, L.F. and Souza, V.A. Recent immunization against measles does not interfere with the sero-response to yellow fever vaccine. Vaccine 1999; 17:1042–1046.

Strokes, A., Bauer, J.H. and Hudson, N.P. Experimental transmission of yellow fever to laboratory animals. Am. J. Trop. Med. Hyg. 1928; 8:103–164.

Struchiner, C.J., Luz, P.M., Dourado, I., Sato, H.K., Aguiar, S.G., Ribeiro, J.G., Soares, R.C. and Codeco, C.T. Risk of fatal adverse events associated with 17DD yellow fever vaccine. Epidemiol. Infect. 2004; 132:939–946.

Suzano, C.E., Amaral, E., Sato, H.K. and Papaiordanou, P.M. The effects of yellow fever immunization (17DD) inadvertently used in early pregnancy during a mass campaign in Brazil. Vaccine 2006; 24:1421–1426.

Sweet, B.H., Wisseman, C.J., Jr. and Kitaoka, M. Immunological studies with group B arthropod-borne viruses. II. Effect of prior infection with Japanese encephalitis virus on the viremia in human subject following administration of 17D yellow fever vaccine. Am. J. Trop. Med. Hyg. 1962; 11:562–570.

Takeda, K., Kaisho, T. and Akira, S. Toll-like receptors. Annu. Rev. Immunol. 2003; 21:335–376.

Tao, D., Barba-Spaeth, G., Rai, U., Nussenweig, V., Rice, C.M. and Nussenzweig, R.S. Yellow fever 17D as a vaccine vector for microbial CTL epitopes: protection in a rodent malaria model. J. Exp. Med. 2005; 201:201–209.

Tattevin, P., Depatureaux, A.G., Chapplain, J.M., Dupont, M., Souala, F., Arvieux, C., Poveda, J.D. and Michelet, C. Yellow fever vaccine is safe and effective in HIV-infected patients. AIDS 2004; 18:825–827.

Tauraso, N.M., Myers, M.G., Nau, E.V., O'Brien, T.C., Spindel, S. S. and Trimmer, R.W. Effect of interval between inoculation of live smallpox and yellow-fever vaccines on antigenicity in man. J. Infect. Dis. 1972; 126:362–371.

Taylor, R.M. Epidemiology. In: Yellow Fever (G.K. Strode, J.C. Bugher, J. Austin-Kerr, H.H. Smith, K.C. Smithburn, R.M. Taylor, M. Theiler, A.J. Warren and L. Whitman, Eds.), pp. 427–538. New York: McGraw-Hill Book Company, Inc., 1951.

Tesh, R.B., Guzman, H., da Rosa, A.P., Vasconcelos, P.F., Dias, L.B., Bunnell, J.E., Zhang, H. and Xiao, S.Y. Experimental yellow fever virus infection in the Golden Hamster (Mesocricetus auratus). I. Virologic, biochemical, and immunologic studies. J. Infect. Dis. 2001; 183:1431–1436.

Theiler, M. The virus. In: Yellow Fever (G.K. Strode, J.C. Bugher, J. Austin-Kerr, H.H. Smith, K.C. Smithburn, R.M. Taylor, M. Theiler, A.J. Warren and L. Whitman, Eds.), pp. 39–136. New York: McGraw-Hill Book Company, Inc., 1951.

Theiler, M. and Anderson, C.R. The relative resistance of dengue-immune monkeys to yellow fever virus. Am. J. Trop. Med. Hyg. 1975; 24:115–117.

Theiler, M. and Casals, J. The serological reactions in yellow fever. Am. J. Trop. Med. Hyg. 1958; 7:585–594.

Theiler, M. and Smith, H.H. Effect of prolonged cultivation in vivo upon pathogenicity of yellow fever virus. J. Exp. Med. 1937a; 65:767–786.

Theiler, M. and Smith, H.H. The use of yellow fever virus modified by in vitro cultivation for human immunization. J. Exp. Med. 1937b; 65:787–800.

Thonnon, J., Fontenille, D., Tall, A., Diallo, M., Renaudineau, Y., Baudez, B. and Raphenon, G. Re-emergence of yellow fever in Senegal in 1995. Am. J. Trop. Med. Hyg. 1998; 59:108–114.

Tigertt, W.D., Berge, T.O., Gochenour, W.S. et al. Experimental yellow fever. Trans. N. Y. Acad. Sci. 1960; 22:323–333.

Tomori, O. Yellow fever in Africa: public health impact and prospects for control in the 21st century. Biomédica: Revista del Instituto Nacional de Salud 2002; 22:178–210.

Tomori, O. Yellow fever: the recurring plague. Crit. Rev. Clin. Lab. Sci. 2004; 41:391–427.

Troillet, N. and Laurencet, F. Effects of yellow fever vaccination. Lancet 2001; 358:1908–1909.

Tsai, T.F., Bolin, R.A., Lazuick, J.S. and Miller, K.D. Chloroquine does not adversely affect the antibody response to yellow fever vaccine. J. Infect. Dis. 1986; 154:726–727.

Tsai, T.F., Chang, G.-J. and Yu, Y-X. Japanese Encephalitis Vaccines. Philadelphia: Saunders, 1999.

Tsai, T.F., Paul, R., Lynberg, M.C. and Letson, G.W. Congenital yellow fever virus infection after immunization in pregnancy. J. Infect. Dis. 1993; 168:1520–1523.

UN population division is part of the report given later in that sentence – United Nations, World Urbanization Prospects, the 1999 revision; Meeting the Urban Challenge, Population Reports Series M, number 16 United Nations, World Urbanization prospects, the 1999 revision.

Vainio, J. and Cutts, F. Yellow Fever: Global Programme for Vaccines and Immunization. Geneva: World Health Organization, 1998.

Van der Most, R.G., Harrington, L.E., Giuggio, V., Mahar, P.L. and Ahmed, R. Yellow fever virus 17D envelope and NS3 proteins are major targets of the antiviral T cell response in mice. Virology 2002; 296:117–124.

Van der Stuyft, P., Gianella, A., Pirard, M., Cespedes, J., Lora, J., Peredo, C., Pelegrino, J.L., Vorndam, V. and Boelaert, M. Urbanisation of yellow fever in Santa Cruz, Bolivia. Lancet 1999; 353:1558–1562.

Van Epps, HL. Broadening the horizons for yellow fever: new uses for an old vaccine. J. Exp. Med. 2005; 201:165–168.

Vasconcelos, P.F., Costa, Z.G., Travassos Da Rosa, E.S., Luna, E., Rodrigues, S.G., Barros, V.L., Dias, J.P., Monteiro, H.A., Oliva, O.F., Vasconcelos, H.B., Oliveira, R.C., Sousa, M.R., Barbosa Da Silva, J., Cruz, A.C., Martins, E.C. and Travassos Da Rosa, J.F. Epidemic of jungle yellow fever in Brazil, 2000: implications of climatic alterations in disease spread. J. Med. Virol. 2001c; 65:598–604.

Vasconcelos, P.F., Luna, E.J., Galler, R., Silva, E.J., Ciombra, T.L., Barros, V.L., Monath, T.P., Rodrigues, S.G., Laval, C., Costa, Z.G., Vilela, M.F., Santos, C.L., Papaiordanou, P.M., Alves, V.A., Andrade, L.D., Sato, H.H., Rosa, E.S., Froguas, G.B., Lacava, E., Almeida, L.M., Cruz, A.C., Rocco, I.M., Santos, R.T. and Oliva, O.F. Serious adverse events associated with yellow fever 17DD vaccine in Brazil: a report of two cases. Lancet 2001a; 358:91–97.

Vasconcelos, P.F., Rodrigues, S.G., Degallier, N., Moraes, M.A., da Rosa, J.F., da Rosa, E.S., Mondet, B., Barros, V.L. and da Rosa, A.P. An epidemic of sylvatic yellow fever in the southeast region of Maranhao State, Brazil, 1993–1994: epidemiologic and entomologic findings. Am. J. Trop. Med. Hyg. 1997; 57:132–137.

Vasconcelos, P.F., Rosa, A.P., Rodrigues, S.G., Rosa, E.S., Monteiro, H.A., Cruz, A.C., Barros, V.L., Souza, M.R. and Rosa, J.F. Yellow fever in Para State, Amazon region of Brazil, 1998-1999: entomologic and epidemiologic findings. Emerg. Infect. Dis. 2001b; 7:565–569.

Vasconcelos, P.F., Sperb, A.F., Monteiro, H.A., Torres, M.A., Sousa, M.R., Vasconcelos, H.B., Mardini, L.B. and Rodrigues, S.G. Isolations of yellow fever virus from *Haemagogus leucocelaenus* in Rio Grande do Sul State, Brazil. Trans. R. Soc. Trop. Med. Hyg. 2003; 97:60–62.

Vasconcelos, P.F.C. Yellow fever. Revista da Sociedade Brasileira de Medicina Tropical 2003; 36:275–293.

Vasconcelos, P.F.C., Bryant, J.E., Rosa, A.P., Tesh, R.B., Rodrigues, S.G. and Barrett, A.D. Genetic divergence and dispersal of yellow fever virus in Brazil: periodic expansions of the enzootic zone. Emerg. Infect. Dis. 2004; 10:1578–1584.

Vellozzi, C., Mitchell, T., Miller, E., Casey, C.G., Eidex, R.B. and Hayes, E.B. Yellow fever vaccine-associated viscerotropic disease (YEL-AVD) and corticosteroid therapy: eleven United States cases, 1996–2004. Am. J. Trop. Med. Hyg. 2006; 75:333–336.

Vicens, R., Robert, V., Pignon, D., Zeller, H., Ghipponi, P.M. and Digoutte, J.P. Yellow fever epidemic in the extreme North of Cameroon in 1990: first yellow fever virus isolation in Cameroon. Bull. WHO 1993; 71:173–176.

Waddell, D. Yellow fever in Europe in the early 19th century - Cadiz 1819. Report of Proceedings/The Scottish Society of the History of Medicine. 1990–1992; 20–34.

Wang, E., Ryman, K.D., Jennings, A.D., Wood, D.J., Taffs, J.F., Minor, P.D., Sanders, P.G. and Barrett, A.D.T. Comparison of the genomes of the wild-type French viscerotropic strain of yellow fever virus with its vaccine derivative French neurotropic vaccine. J. Gen. Virol. 1995; 76:2749–2755.

Wang, E., Weaver, S.C., Shope, R.E., Tesh, R.B., Watts, D.M. and Barrett, A.D. Genetic variation in yellow fever virus: duplication in the 3′ non coding region of strains from Africa. Virology 1996; 225:274–281.

Werfel, U. and Popp, W. Effects of yellow fever vaccination. Lancet 2001; 358:1909.

Wheelock, E.F. and Sibley, W.A. Circulating virus, interferon and antibody after vaccination with the 17-D strain of yellow-fever virus. N. Engl. J. Med. 1965; 273:194–198.

Wisseman, C.L., Jr., Sweet, B.H., Kitaoka, M. and Tamiya, T. Immunologocal studies with Group B arthropod-borne viruses. I. Broadened neutralizing antibody spectrum induced by strain 17D yellow fever vaccine in human subjects previously infected with Japanese encephalitis virus. Am. J. Trop. Med. Hyg. 1962; 11:550–561.

World Health Organization Standards for the manufacture and control of yellow fever vaccine. United Nations Relief and Rehabilitation Administration. WHO Epidemiol. Bull. 1945; 1:1–365.

World Health Organization Normes pour les substances biologiques. WHO Tech. Rep.Ser. 1959; 179:1–46.

World Health Organization Expert committee on yellow fever virus. Third report. WHO Tech. Rep. Ser. 1971; 479:1–56.

World Health Organization Expert committee on biological standardization: twenty seventh report. WHO Tech. Rep.Ser. 1976; 594:1–86.

World Health Organization Expert committee on biological standardization. WHO Tech. Rep. Ser. 1981; 658:1–325.

World Health Organization Yellow fever. Wkly. Epidemiol. Rec. 1986; 61:180.

World Health Organization The State of World Health: World Health Report. Geneva: World Health Organization, 1996.

World Health Organization Yellow fever, 1996–1997. Wkly. Epidemiol. Rec. 1998; 73:354–359.

World Health Organization Yellow fever, 1998–1999. Wkly. Epidemiol. Rec. 2000; 75:322–328.

World Health Organization Yellow Fever Fact Sheet Number 100 Revised December 2001. Geneva: World Health Organization, 2001.

World Health Organization Yellow fever vaccine. WHO position paper. Wkly. Epidemiol. Rec. 2003; 78:349–359.

World Health Organization/UNICEF. Yellow fever stockpile investment case. Submitted by Yellow Fever Task Force to the Global Alliance for Vaccines and Immunization. Global Alliance for Vaccines and Immunization, Geneva, 2005.

World Health Organization vaccine-preventable diseases: monitoring system 2006 global summary (http://www.who.int/vaccines-documents/).

World Health Organization Yellow fever situation in Africa and South America, 2005. Wkly. Epidemiol. Rec. 2006a; 81:317–324.

World Health Organization Outbreak news. Yellow fever, Cote d'Ivoire. Wkly. Epidemiol. Rec. 2006b; 81:410.

World Health Organization Outbreak news. Yellow fever, Togo—update. Wkly. Epidemiol. Rec. 2007a; 82:50.

World Health Organization Assessment of yellow fever epidemic risk-a decision-making tool for preventive immunization campaigns. Wkly. Epidemiol. Rec. 2007b; 82:153–604.

World Health Organization. The yellow fever initiative, 2007. Injection: one injection 10 years protection. Geneva: World

Health Organization, 2007c (http://www.who.int/csr/disease/yellowfev/introduction/en/index.html).

World Health Organization Standards for the manufacture and control of yellow fever vaccine. United Nations Relief and Rehabilitation Administration. World Health Organization Epidemiological Bulletin, 1945; 1:1–365.

Xiao, S.Y., Guzman, H., da Rosa, A.P., Zhu, H.B. and Tesh, R.B. Alteration of clinical outcome and histopathology of yellow fever virus infection in a hamster model by previous infection with heterologous flaviviruses. Am. J. Trop. Med. Hyg. 2003; 68:695–703.

Xiao, S.Y., Zhang, H., Guzman, H. and Tesh, R.B. Experimental yellow fever virus infection in the Golden hamster (Mesocricetus auratus). II. Pathology. J. Infect. Dis. 2001; 183:1437–1444.

Yvonnet, B., Coursaget, P., Deubel, V., Diop-Mar, I., Digoutte, J.P. and Chiron, J.P. Simultaneous administration of hepatitis B and yellow fever vaccines. J. Med. Virol. 1986; 19:307–311.

Zhou, Q., Liu, D.W., Shi, Y., Long, Y., Li, D.J., Liu, H.Z., Wang, H., Ni, X., Cai, W.S. and Wang, A.X. Adverse effects of attenuated yellow fever vaccine with multiple organ injury—a case report. Zhonghua Yi Xue Za Zhi 2005; 85:936.

SECTION IV

BACTERIAL VACCINES

CHAPTER

41

Anthrax

Jason E. Comer and Johnny W. Peterson
Department of Microbiology and Immunology, Sealy Center for Vaccine Development, University of Texas Medical Branch, Galveston, TX, USA

OUTLINE

ABSTRACT

The causative agent of anthrax is the spore-forming bacterium, *Bacillus anthracis*, which is considered a major bioterrorism threat because its spores are very stable, easily disseminated, and infectious by aerosol. The clinical form of the disease is dependent on the route of exposure and can manifest as cutaneous, gastrointestinal, or inhalational anthrax. Irrespective of the route acquired, septic anthrax is a devastating infection with high mortality. The virulence of *B. anthracis* is dependent on the production of an antiphagocytic capsule and two bipartite, proteinaceous toxins: lethal toxin and edema toxin. The toxins comprised three proteins, lethal factor (LF), edema factor (EF), and protective antigen (PA). PA, the binding moiety of each toxin, facilitates toxin entry into host cells. LF, the enzymatic portion of lethal toxin, is a zinc metalloprotease that cleaves the N-terminal region of mitogen-activated kinase kinases. This toxin interferes with the function of host immune cells and disrupts the endothelial barrier. EF, the enzymatic moiety of edema toxin, is a calmodulin-dependent adenylyl cyclase that converts ATP to cAMP, resulting in immune cell dysfunction and tissue edema. The pathology associated with an anthrax infection (i.e., hemorrhage, necrosis, and edema) is the result of the action of the toxins. The current FDA-licensed anthrax vaccine comprised filter-sterilized culture media, in which an attenuated (capsule-deficient) strain of *B. anthracis* has grown, combined with aluminum hydroxide adjuvant. While all secreted or released bacterial proteins are present in the vaccine formulation, the major immunogen is PA. The levels of antibodies to PA elicited by the vaccine directly correlate with survival in infected animal models. The need for a chemically defined vaccine has led to the development of a second-generation vaccine, consisting of recombinant PA and the aluminum hydroxide adjuvant. This new vaccine has proved efficacious in the nonhuman primate and rabbit models of inhalation anthrax, and clinical trials are currently evaluating the safety and immunogenicity in human volunteers. Apart from active immunization, passive transfer of antibodies to PA is efficacious in anthrax animal models. Compounds that inhibit the action of the toxins are also being investigated. These recent advances have increased the arsenal of therapeutics that can be used in the future to improve the prognosis of anthrax-infected individuals.

INTRODUCTION

A spore-forming bacterium, *Bacillus anthracis*, is the causative agent of anthrax. Outbreaks of this zoonotic disease date back to antiquity, and today the disease is endemic among animals worldwide. Herbivores are particularly susceptible and become infected while grazing by inhaling aerosolized spores or ingesting contaminated plant material. Once inside the animal, the spores germinate and the proliferating vegetative cells quickly kill the host. Close to or at the time of death, the animal hemorrhages and bacteria-laden blood is released into the environment from the mouth, nose, and rectum. Environmental parameters signal the bacteria to sporulate, and the spores remain viable long after the animal has decomposed. The spores can be disseminated by water, wind, or scavengers carrying away pieces of the contaminated carcass. The cycle repeats when another herbivore grazes in the contaminated area. Unlike most other pathogenic bacteria or viruses, *B. anthracis* must kill its host in order to infect another.

Humans can be infected when they come in close contact with *B. anthracis*-infected animals or contaminated animal products. We are also at risk of intentional infection by biowarfare or bioterrorism. Research on *B. anthracis* as a possible biological weapon began in the 1940s, and recent events have renewed interest in finding novel vaccines and therapies to combat the agent. The current anthrax vaccine has been licensed for human use since 1970. This chapter will focus on recent advances in the development of the vaccines to prevent anthrax in humans, as well as the characteristics of *B. anthracis* that make the bacterium a bioterrorism threat.

BRIEF HISTORY OF THE DISEASE

Anthrax has been described in antiquity and may have been responsible for the fifth and sixth plagues of Egypt described in the Bible. The Roman poet Virgil described the disease in both domestic and wild animals. In the 18th century, the sheep population of Europe was decimated by an anthrax outbreak, and the human form of the disease was termed wool sorters disease because of the rate of infection in individuals working with contaminated animal fibers (Sternbach, 2003).

Robert Koch was the first person to trace the complete life cycle of the bacterium and found that *B. anthracis* formed spores that remain viable even in adverse environments. Koch also cultured the organism in vitro and inoculated healthy animalsthat ultimately developed systemic infection characteristic of anthrax. *B. anthracis* became the prototypical organism of Koch's famous postulates for identifying the causative agents of infectious diseases. Louis Pasteur later created the first live vaccine against anthrax which comprised *B. anthracis* attenuated by passage at elevated incubation temperature (Sternbach, 2003).

ETIOLOGICAL AGENT

B. anthracis is a large ($1.0 - 1.2 \times 3 - 5\mu m$), Gram-positive bacillus capable of forming stable spores in unfavorable environments such as nutrient depletion (Fig. 41.1). The bacterium contains a ~5.3Mb chromosome and two virulence plasmids. The bacterial cells grow in long chains, and the large colonies have a ground glass-like appearance when cultured on solid agar. The organism has a biphasic life cycle; it can exist as a metabolically inactive endospore or as a rapidly proliferating vegetative cell. When bacteria are released from an animal by hemorrhage, environmental parameters such as temperature, pH, O_2 availability, sunlight, nutrient level, and certain cations signal the bacterium to sporulate. The spore comprises several layers with the internal core housing the genetic material and enzymes. The core is surrounded by an inner and outer membrane, which are constructed from the vegetative cell membrane during the sporulation process. The thick cortex is located between the two membranes

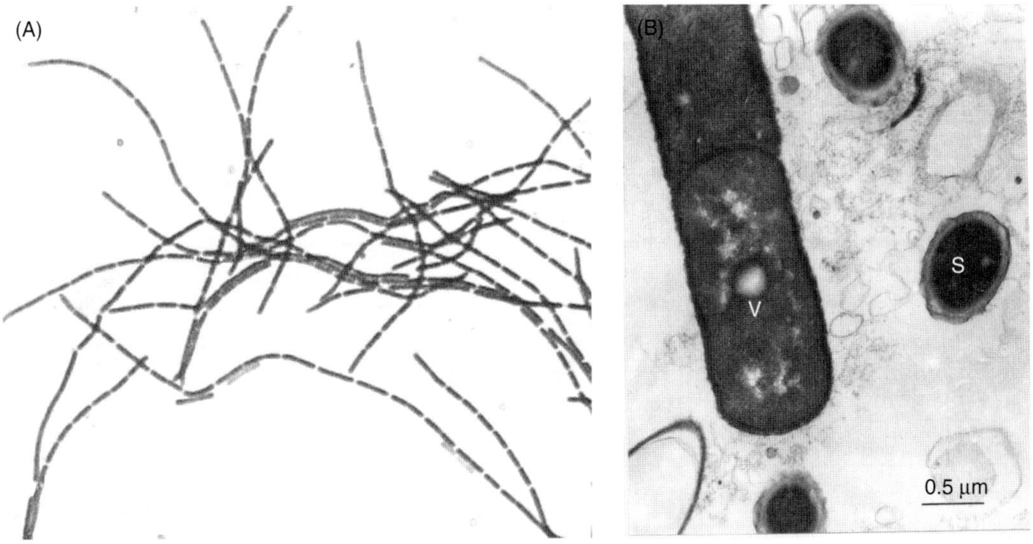

FIGURE 41.1 (A) Gram stain of *Bacillus anthracis* showing the bacilli growing in long chains. (B) Electron micrograph (EM) showing vegetative cells (v) and spores (s) at 47,000 ×. The bar denotes 0.5 μm. EM courtesy of Vsevolod Popov, Ph.D., Department of Pathology, University of Texas Medical Branch, Galveston, TX.

and is structurally similar to the bacterial cell wall. The cortex is heat resistant and is responsible for protecting the spore from dehydration. The outer membrane is encompassed by the thick spore coat, which is produced by the vegetative cell and is resistant to lytic enzymes. The outermost layer of the spore is the loose-fitting exosporium. Unlike other spore integuments, the exosporium is not composed of pre-existing membranes, but is produced at the time of sporulation. Further, the exosporium does not contain vegetative cellular antigens, except for PA, which has been reported on the surface of spores (Cote et al., 2005; Guidi-Rontani and Mock, 2002). The structure of the spore makes it resistant to most disinfectants; often 10–50% bleach, 10% formalin, or formaldehyde gas is used for decontamination.

The spores are 1–5 μm in diameter making them of ideal size for inhaling deep into the alveolar spaces. Once inside the host, *B. anthracis* has both intracellular and extracellular stages of growth. The spores are taken up by macrophages into an acidified endosome, where they germinate and escape the antimicrobial environment. Compounds such as L-alanine, L-tyrosine, and adenosine, which are abundant in a host, stimulate the germination process. The bacteria proliferate inside the macrophage eventually lysing the cell and continue to divide extracellularly.

ANTIGENS ENCODED BY AGENT

The virulence of *B. anthracis* is dependent on the genes encoded on two plasmids, pXO1 and pXO2. The genes for the synthesis of the poly-γ-D-glutamic acid capsule are located on pXO2. The electronegative charge of the capsule enables the bacteria to evade phagocytosis and killing and is required for dissemination to other tissues from the lungs in the mouse model of inhalation anthrax (Drysdale et al., 2005; Ezzell and Welkos, 1999).

Although pXO1 has many open-reading frames, three genes, *pag*, *lef*, and *cya*, encode protective antigen (PA), lethal factor (LF), and edema factor (EF), respectively (Mikesell et al., 1983). These three proteins comprise the two toxins, lethal toxin (LeTx) and edema toxin (EdTx) that belong to the family of bacterial binary A–B toxins. The A moieties (LF or EF) act within the cytosol of the host cell, and the B component (PA) binds the toxin complex to the target cell and facilitates the entry of the A subunit.

Transcription of toxin and capsular genes is affected by temperature (37°C) and bicarbonate/CO_2, two signals present in a mammalian host. The *atxA* gene located on pXO1 positively regulates transcription of toxin and capsule genes. Another virulence gene regulator encoded by pXO2, *acpA*, only induces expression of genes required for capsule synthesis. In addition, a chromosomal gene called *abrB* induces the toxin genes in a growth-phase-specific manner (Guidi-Rontani and Mock, 2002). This level of transcriptional control ensures that the bacterium only produces virulence factors when appropriate (i.e., when inside a host).

Two PA receptors on mammalian cells have been identified so far, namely Tumor Endothelial Marker 8 (TEM8) and Capillary Morphogenesis Gene-2 (CMG2), which are expressed by a variety of tissues and cells (Bradley et al., 2001; Scobie et al., 2003). The 83 kDa PA protein binds to the cell surface receptor and is cleaved at the N-terminal region by a host cell furin protease (Klimpel et al., 1992). The resulting 63 kDa protein (PA63) heptamerizes forming a ring structure with competitive binding sites for three molecules of LF and/or EF (Milne et al., 1994). Once the enzymatic moiety binds to the oligomerized PA63, the complex enters the cell via receptor-mediated endocytosis. The toxins enter an acidic endosomal compartment, and the low pH causes conformational changes in PA, such that it forms a membrane-spanning channel allowing for the translocation of EF and LF into the cytosol (Ascenzi et al., 2002; Fig. 41.2).

EF is a calmodulin-dependent adenylyl cyclase that forms 3′,5′-adenosine monophosphate (cAMP) from adenosine triphosphate (ATP) in many different types of cells (Leppla, 1982). The increase in cellular cAMP causes the efflux of water and electrolytes from the cell, resulting in the characteristic interstitial edema associated with *B. anthracis* infections. The toxin has also been shown to inhibit LPS-induced TNF-α production in monocytes and inhibit phagocytosis by neutrophils (Hoover et al., 1994; O'Brien et al., 1985). The increase in intracellular cAMP also inhibits the activation of murine CD4+ T cells (Paccani et al., 2005).

LF is a zinc metalloprotease with mitogen-activated kinase kinases (MAPKKs) 1, 2, 3, 4, 6, and 7 as the only known substrates (Duesbery et al., 1998; Pellizzari et al., 1999; Vitale et al., 2000). The toxin can enter and is proteolytically active in most cells but is specifically cytotoxic for macrophages from certain strains of inbred mice (Singh et al., 1989), suggesting some other substrate or function of the toxin (Chaudry et al., 2002). The susceptibility of murine macrophages is dependent on mutations in a single kinesin-like motor gene called *kif1C* (Watters et al., 2001). Laboratory rodents are killed by injection of purified PA and LF. LeTx readily kills laboratory mice regardless of macrophage sensitivity and evidence suggests that death is the result of TNF-α-independent hypoxia (Moayeri et al., 2003), although no direct measurements of pO_2 have

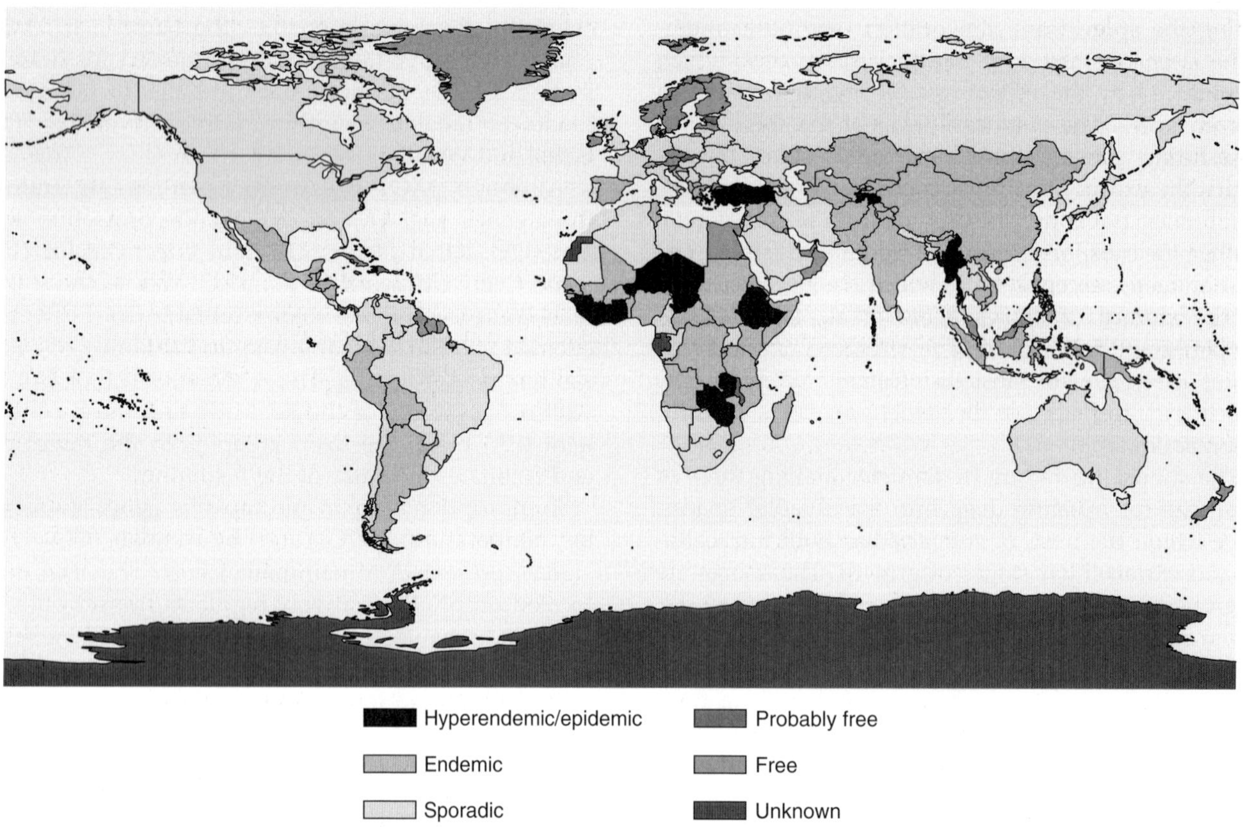

FIGURE 41.2 Mechanism of entry and action of anthrax toxins. Protective antigen (PA) binds to one of the two known anthrax toxin receptors, TEM8 or CMG2. PA is then activated by a host furin protease. The resulting 63 kDa PA molecules heptamerize, forming binding sites for three molecules of lethal factor (LF) or edema factor (EF). The holotoxin is taken into the cell by receptor-mediated endocytosis. Acidication of the endosome causes a conformational change in PA such that it forms a channel though the membrane, thus releasing LF and EF into the cytosol. LF is a zinc metalloprotease with mitogen-activated kinase kinases (MAPKKs) as its only know substrates. EF is a calmodulin-dependent adenylyl cyclase that converts ATP into cAMP (see color plate section).

confirmed this hypothesis. Infusion of the toxin into rats causes death by circulatory shock (Cui et al., 2004).

Human macrophages are not readily lysed by the toxin; however, Popov et al. (2002) showed that LeTx (1) inhibited the bactericidal activity of human monocytes, (2) suppressed production of proinflammatory cytokines, and (3) ultimately caused the cells to undergo apoptosis. The toxin also suppresses antigen presentation by murine dendritic cells, thereby inhibiting an adaptive immune response (Agrawal et al., 2003). The toxin further suppresses the adaptive immune response by inhibiting the activation of human CD4+ T cells (Fang et al., 2005) and B cells in mice (Fang et al., 2006). In addition to impairing immune cells, LeTx induces apoptosis in cultured endothelial cells (Kirby, 2004) and disrupts the endothelial barrier (Warfel et al., 2005), which may partly explain the hemorrhage observed in inhalational anthrax patients.

PROTECTIVE IMMUNE RESPONSE

There is little in the literature concerning the protective immune response to an anthrax infection in humans. Much of the data on protective immunity, namely, development of antibodies to PA, conferring protection against anthrax, are derived from studies with vaccines in animal models. In fact, as discussed later, PA is the dominant antigen in the current licensed anthrax vaccine adsorbed (AVA). Protection of rabbits vaccinated with recombinant PA (rPA) against aerosol challenge with virulent *B. anthracis* spores correlates directly with anti-PA titer (Little et al., 2004). The role in survival following infection played by anti-PA antibodies is not fully understood. Production of neutralizing antibodies, those that interfere with the anthrax toxin's ability to enter host cells, would inhibit the deleterious effects of the toxins. Further, antibodies to PA enhance phagocytosis of spores and killing of bacteria in vitro (Cote et al., 2005; Welkos et al., 2001). Although the humoral immune response is thought to be the defining factor determining immunity to anthrax, a cell-mediated response also develops. Shlyakhov et al. (1997) showed that healthy individuals vaccinated with live nonencapsulated spores respond to an anthrax-specific, tuberculin-like skin test. The relative importance of the humoral response and cell-mediated immunity in anthrax was demonstrated by passive transfer of serum and lymphocytes from PA-vaccinated mice to naïve animals. Mice receiving sera containing PA antibodies lived significantly longer than controls after challenge with a capsule-deficient strain of *B. anthracis*. In contrast, animals receiving lymphocytes died at the same rate as controls (Beedham et al., 2001).

EPIDEMIOLOGY

Anthrax is a worldwide disease affecting herbivores on most of the continents (Fig. 41.3). Anthrax is endemic in parts of South America, Asia, and Africa. North America, Europe, and Northern Asia have sporadic outbreaks of anthrax in animals and humans. Most recently, the Associated Press reported outbreaks of anthrax in cattle herds in North and South Dakota in 2005. A total of 500 unvaccinated head of cattle succumbed to the disease causing financial losses for the afflicted ranches.

Western Africa is heavily affected by anthrax, and an outbreak in Zimbabwe between October 1979 and March 1980 resulted in over 6000 cases of human anthrax, which were mostly the cutaneous form of the disease. Then between August and November 2004, a massive outbreak devastated the herds of herbivores in the Malilangwe Wildlife Reserve, Zimbabwe (Clegg et al., 2007). The slaughter and handling of raw meat from an infected cow resulted in 25 cases of cutaneous anthrax in Paraguay in 1987 (Cieslak and Eitzen, 1999). Between 1994 and 2004, 25 cases of cutaneous anthrax were reported in a clinic in the Elazig province of Turkey (Demirdag et al., 2003). Spain reported the largest number of cases in Europe, with 78–152 cases annually between 1991 and 1995, while neighboring France reported no cases between 1972 and 1994, until a fatal case of African-acquired anthrax was diagnosed. Then, in 1997 in two separate outbreaks in France, 95 cows were killed, and there were three fatal cases of anthrax in a farmer, a veterinarian, and young boy (Patra et al., 1998). Other cases of anthrax in Europe were caused by animal products imported from endemic regions (Schmid and Kaufmann, 2002).

Kazakhstan reported 73 human cases of anthrax between 1997 and 1998. A retrospective cohort analysis to determine risk factors found that all of these cases occurred following contact with infected animals, and persons butchering infected animals were at increased risk. In addition, the study reported that the risk may have been increased by the presence of abraded skin on the hands of the infected humans (Woods et al., 2004).

In recent years, the use of an attenuated veterinary vaccine in the United States and Europe has limited the number of anthrax outbreaks in livestock and thus decreased the incidence of human anthrax. In the United States, the number of human anthrax cases per

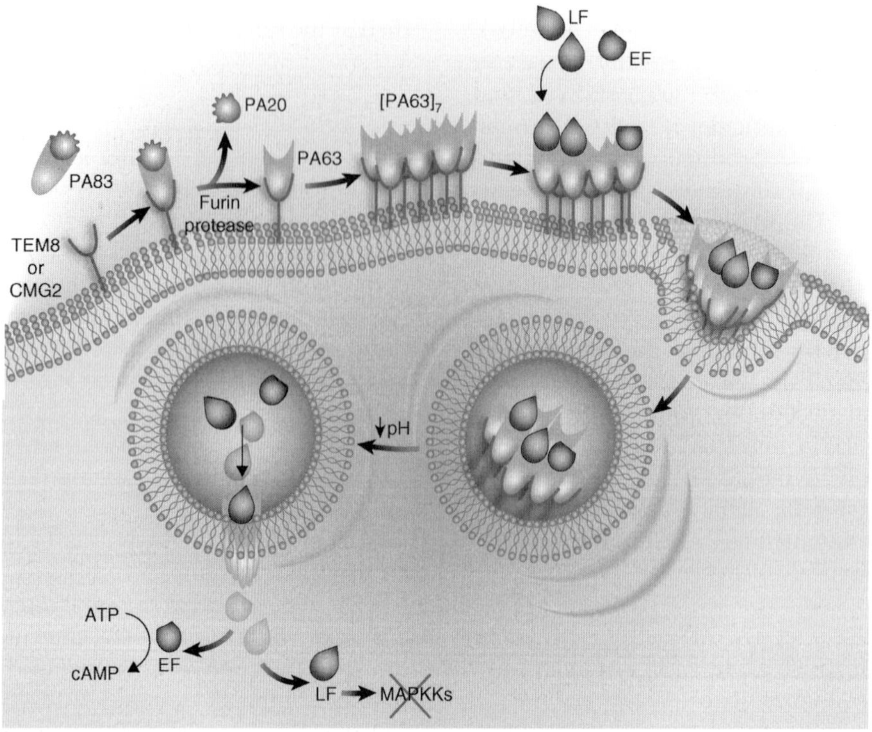

FIGURE 41.3 Worldwide occurrence of anthrax as of 2003. Map courtesy of Martin E. Hugh-Jones, Ph.D., WHO Collaborating Center for Remote Sensing and Geographic Information Systems for Public Health, Department of Epidemiology and Community Health, School of Veterinary Medicine, Louisiana State University, Baton Rouge, LA. (www.vetmed.lsu.edu/whocc/mp_world.htm (see color plate section).

year averaged about 35 in the 1950s and decreased to less than one per year since 1980. Between 1950 and 1999, 458 cases of human anthrax were reported nationally with 77% occurring before 1959. Forty-one field investigations conducted by The Centers for Disease Control (CDC) determined that the exposure sites comprised 24 agricultural settings, 11 textile mills, and 6 others. Of the 27 cases of anthrax in textile mills, 21 were cutaneous and 6 were inhalational with contaminated goat hair or sheep wool being the primary vehicles (Bales et al., 2002).

As demonstrated above, most human anthrax cases occur when individuals come in close contact with infected animals or animal products; however, some instances of human anthrax have more sinister causes. The largest outbreak of inhalation anthrax occurred in April and May of 1979 in Sverdlovsk in the former Soviet Union. Originally reported as gastrointestinal anthrax, further investigation determined that the outbreak was due to inhalation of anthrax spores. Tissue analysis of 11 victims identified four different strains of *B. anthracis* (Jackson et al., 1998). An epidemiological investigation determined that the likely source of the outbreak was the accidental release of spores from a military weapons factory, Compound 19, in southern Sverdlovsk. The incident infected 94 people and caused at least 64 deaths (Meselson et al., 1994). It was not until 1992 that Russian President Boris Yeltsin announced that the military installation was responsible for the outbreak.

Before the 2001 terrorist attack, the last reported case of inhalational anthrax in the United States occurred in California as a result of exposure to contaminated yarn imported from Pakistan (Suffin et al., 1978). In 2001, letters contaminated with anthrax spores, sent through the U.S. Postal Service, resulted in 22 cases of anthrax. In October 2001, the first case of bioterrorism-related anthrax occurred in Florida, and increased surveillance identified a total of 22 cases of anthrax between October 4 and November 20, 2001. Eleven of these cases were the cutaneous form of infection with the other 11 cases being inhalational anthrax. Five victims with inhalational anthrax died of the infection. Twenty of the 22 cases occurred in individuals who were mail handlers or exposed to work sites where the contaminated letters were processed or received (Jernigan et al., 2002). This event underscores the reason that *B. anthracis* is considered a significant threat as a biological warfare/bioterrorism agent.

Anthrax has a long history as an agent in the biological warfare arsenal. Germany investigated the possible use of anthrax against animals and humans in World War I and II. During this time, Japan also experimented with biological weapons including anthrax and used biological weapons on Chinese civilians. The allies including Great Britain and the United States also engaged in biological weapons research. The United States formed the War Research Service at Fort Detrick, Maryland, to investigate the offensive use of anthrax. The British tested anthrax bombs on a remote island northwest of Scotland. They also prepared *B. anthracis*-laden food pellets to infect German cattle. Although they actively engaged in research, the allies did not use bioweapons in military campaigns. In 1969, President Nixon renounced the use of biological weapons and ordered all offensive stockpiles destroyed, an action that prompted the International Biological Weapons Convention, which called for the complete elimination of all offensive biological weapons worldwide (Mobley, 1995).

The use of biological weapons in conventional warfare does not appear to be a major threat, but the possibility of an anthrax attack still exists from either rogue nations or terrorist groups, and would most probably target the civilian population. The World Health Organization estimated that the aerosolization of 50 kg of dried *B. anthracis* spores over a city with a population of 500,000 would incapacitate 125,000 people and kill 95,000. Such an event would overwhelm medical resources and disrupt the infrastructure of most cities (World Heath Organization, 1970). As part of a 1999 initiative from Congress to increase public health preparedness for response to biological terrorism, the CDC categorized biological agents as A, B, or C with category A agents having the greatest effect on public health. *B. anthracis* is a level A select agent and is ranked along with *Variola major*, the causative agent of smallpox, as the two greatest bioterrorism risks to public health. While there have been no reports on person-to-person transmission of inhalational anthrax, the ease of dissemination, infective stability of the spore, and virulence of the organism make it one of the highest potential biological threats (Rotz et al., 2002).

Although a biological attack of the enormity described by the WHO has not occurred, civilian populations have been the targets of anthrax attacks. In 1993, the Aum Shinrikyo cult released an aerosol cloud of anthrax spores in Kameido, Japan, but there were no reports of clinical anthrax. Molecular analysis of spores isolated from the site established that the released spores were from an attenuated stain of *B. anthracis* used to vaccinate animals in Japan (Keim et al., 2001). In comparison, the letters used to disseminate the spores in the U.S. attack were contaminated with fully virulent, weaponized *B. anthracis* Ames strain spores. This incident proved that terrorists could effectively distribute and disseminate anthrax spores without sophisticated equipment by exploiting the society's infrastructure.

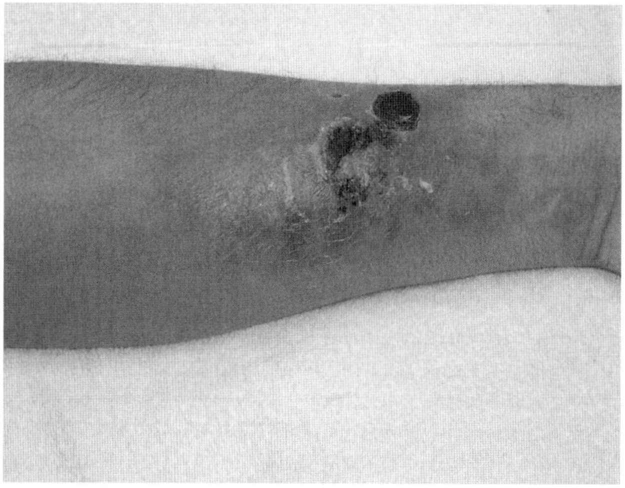

FIGURE 41.4 Case of cutaneous anthrax with characteristic eschar. A ranch hand in Edwards County Texas reportedly was infected after skinning a buffalo. The patient noticed a small pimple on his left arm 5 days after contact with the animal. The lesion progressed and the man was hospitalized for severe cellulitis, and treated with penicillin and ciprofloxacin. The lesion evolved into a single black eschar measuring 3 × 5 in. The individual was discharged after 9 days in the hospital. Picture courtesy of Army Major Margaret "Meg" Neidert during residency training at the Texas Department of State Health Services, Austin, Texas (www.tdh.state.tx.us./phpep/).

CLINICAL DISEASE

Anthrax occurs in three clinical forms: cutaneous, gastrointestinal, or inhalational. The type and severity of disease are dependent on the route of exposure. Ninety percent of all human cases of anthrax are cutaneous. Infection occurs when viable spores come in contact with abraded skin, and a small papule develops in 2–3 days. Within the next 24 h, a ring of vesicles forms around the papule, which starts to ulcerate and then dries resulting in the characteristic black eschar (Spencer, 2003; Fig. 41.4). The anthrax skin lesion is painless, which differentiates it from a similar painful lesion formed from the bite of the *Loxoceles reclusa* (brown recluse) spider. Even without treatment, the anthrax eschar usually resolves within 2–6 weeks. Twenty percent of untreated individuals become septic and die, but antibiotic intervention can reduce the mortality rate to <1% making this the most benign form of the disease.

Ingestion of viable spores from contaminated meat results in gastrointestinal anthrax and, depending on the site of inoculation, can take one of two forms, either oropharyngeal or abdominal with the lesion forming at the site of entry. Oropharyngeal anthrax occurs when spores enter the oropharynx, and symptoms include sore throat, dysphagia, fever, cervical lymphadenopathy, and edema. Infection of the terminal ileum and/or cecum with hemorrhage of the mesenteric lymph nodes results in abdominal anthrax, which also presents with nondescript symptoms including abdominal pain, hematemesis, and bloody diarrhea. If untreated, the bacteria can disseminate to the bloodstream, and once septic the patient rapidly succumbs to the infection. Although rare outside of Africa and Asia, gastrointestinal anthrax has a high mortality rate. The ambiguous symptoms characteristic of early infection make diagnosis difficult, and the delay in administration of appropriate antibiotics often leads to death (Spencer, 2003).

As discussed earlier in this chapter, inhalational anthrax was usually associated with occupational exposure to contaminated animal products. Until 2001, the last case of inhalational anthrax in the United States was reported in 1978 (Suffin et al., 1978). In this chapter, we will present clinical symptoms of the first 10 cases of the 2001 U.S. bioterrorism-related inhalational anthrax as reported by Jernigan et al. (2001). The median incubation time was 4 days from the time of exposure to symptom onset. As in the case of gastrointestinal anthrax the disease starts with nonspecific symptoms including fever or chills, sweats, fatigue with malaise, and minimal or nonproductive cough, dyspnea, and nausea. All patients had abnormal chest X-rays, characteristic of inhalational anthrax including infiltrates, pleural effusion, and widening of the mediastinum (Fig. 41.5). With multiple antibiotic treatment and supportive care, the mortality rate was 45%, substantially lower than the previously published estimate of 95%.

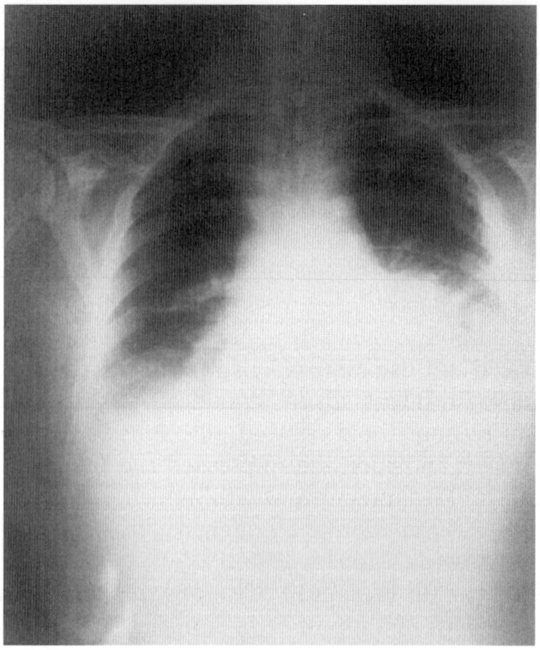

FIGURE 41.5 Chest radiograph of an inhalational anthrax patient in 1961. The radiograph was taken 22 h before death and demonstrates the mediastinal widening characteristic of this form of the disease. Radiograph was provided by Dr. Philip S. Brachman and provided by the Centers for Disease Control and Prevention (www.cdc.gov.).

TREATMENT

Successful treatment of inhalational anthrax requires early administration of antibiotics. The initial CDC-recommended treatment for inhalational anthrax is intravenous infusion of ciprofloxacin (400 mg bid) or doxycyline (100 mg bid) coupled with other antibiotics. Following this initial treatment, patients are switched to 500 mg of ciprofloxacin or 100 mg doxycycline orally bid for 60 days. The extended treatment period is required because of asynchronous germination of anthrax spores in the lungs of exposed individuals. The CDC also recommends postexposure vaccination, but the FDA has not approved postexposure use of the vaccine. Supportive care must also be given to combat the respiratory distress associated with the acute phase of the disease (e.g., draining of the massive pleural effusions).

PATHOGENESIS

Inhalation anthrax occurs when an individual inhales viable spores. The small size (1–3 μm) of the spores allows them to travel deep into the alveolar spaces of the lungs. Once there, they are taken up by alveolar macrophages and transported to the mediastinal lymph nodes, where the spores are thought to germinate, eventually lysing the macrophages. The vegetative cells continue to proliferate and spread throughout the body via the lymphatics and circulatory system (Spencer, 2003).

The unfortunate outbreak of inhalational anthrax in Sverdlovsk provided the largest set of clinical specimens to study the pathology of human anthrax. Necropsies of victims of the outbreak consistently showed pathologic characteristics of inhalational anthrax including necrotic hemorrhage of the thoracic lymph nodes, hemorrhagic mediastinitis, and pleural effusion. Fifty percent of the cases involved hemorrhagic meningitis, and 92% showed signs of gastrointestinal tract involvement (i.e., submucosal hemorrhagic lesions) (Abramova et al., 1993).

Quantitative microscopic findings showed that most of the severe pathologic lesions occurred in the mediastinum and mediastinal lymph nodes, the sites of initial replication of the bacterium. The investigators also observed peripheral transudate surrounding fibrin-rich edema, necrosis of veins and arteries, and apoptotic lymphocytes. The pathology (i.e., cellular death and edema) of the infection is associated with the two exotoxins produced by the bacterium (Grinberg et al., 2001). The five fatal cases of bioterrorism-related anthrax in the United States all demonstrated prominent pleural effusions and hemorrhagic mediastinitis (Guarner et al., 2003).

IMMUNE RESPONSE TO INFECTION

Anthrax survivors have been reported to develop antibodies to the bacterial capsule, PA, and, to a lesser extent, LF and EF. Anti-PA titers from the 11 cutaneous cases of the 2001 outbreak were detected as early as 12 days after symptoms developed. The six survivors of inhalational anthrax had significantly higher concentrations of anti-PA antibodies than the cutaneous anthrax patients, and concentrations remained higher more than 1 year after disease onset. Patients with detectable anti-PA titers 3–14 months after development of symptoms had detectable toxin-neutralizing antibody levels that peaked by 4–8 weeks after disease onset. Long-term immune memory from the anthrax patients was assessed by blood samples at 8–14 months after the onset of symptoms. All six inhalational anthrax survivors had PA-specific IgG memory B cells compared to only two of the seven cutaneous anthrax patients sampled (Quinn et al., 2004).

HISTORY OF ANTHRAX VACCINES

Louis Pasteur is usually credited with the discovery of the first anthrax vaccine in 1881; however, a literature review by Peter C.B. Turnbull determined that the first attenuated anthrax vaccine was developed by W.S. Greenfield at Brown Animal Sanatory Institution in London (Turnbull, 1991). The Pasteur vaccine became widely used in Europe and South America for the next 50 years. This vaccine consisted of two inoculations given2 weeks apart. The first dose comprised *B. anthracis* cells that were grown at 42–43°C for 15–20 days, while the cells for the second dose were grown at an elevated temperature for only 10–12 days. The cells grown for the first inoculation were only virulent in mice and young guinea pigs, and the second inoculum was less attenuated, because of the shorter incubation time at the higher temperature. Later analysis of the Pasteur strains determined that growth of the bacteria at 42–43°C cured the plasmid encoding the toxin genes, *pXO1*. This strain did not produce PA, an important antigen needed to induce protective immunity, and the cultures were probably made up of cured cells along with fully virulent cells capable of producing the toxin complexes (Turnbull, 1991).

The next notable advance in anthrax vaccine development occurred in 1937 when Sterne (1937) isolated a $pXO1^+$ $pXO2^-$ variant of *B. anthracis*. Sterne (1939) developed a live spore vaccine using nonencapsulated, toxigenic spores supplemented with 0.5% saponin in 50% glycine–saline. The original formulation has remained unchanged, except for an increase in the number of spores. The vaccine is considered effective and safe for use in livestock and is still used in many countries today (Turnbull, 1991). The vaccine strain still maintains some virulence for mice and guinea pigs (Turnbull, 1991). In fact, this strain is often used as a challenge strain in mice to test novel anthrax therapeutics. Also, some domestic animals such as goats and llamas are more susceptible to the vaccine strain than are other grazing livestock (Turnbull, 1991).

Shortly after Sterne's discovery, Russian scientists also isolated strains of nonencapsulated *B. anthracis* termed STI-1 and strain no. 3. STI-1 was partially protective in guinea pigs and rabbits and, to a greater extent, in sheep. This strain was subsequently used to vaccinate 38.4 million head of livestock in 1947 in Russia and increased to 100 million in 1960. In 1960, the incidence of anthrax in Russian farm animals was compared to that of 1947, and an almost 10-fold reduction in incidence of anthrax was observed. These attenuated strains were then used to prepare an attenuated human anthrax vaccine. The human vaccine consisted of spores of STI-1 and no. 3 strains administered by scarification with 1.3×10^8 spores or subcutaneous injection with 5×10^7 spores. In 1951–1952, a clinical trial tested the efficacy of the vaccine. Residents of rural districts in the Moldavian region were vaccinated against anthrax by either scarification or subcutaneous injection. An unvaccinated segment of the population served as controls. No significant adverse effects of vaccination were reported in this study, and follow-up surveillance determined that vaccination reduced the risk of anthrax infection by 5.4 fold within the 18-month postvaccination period. Several subsequent investigations confirmed the efficacy of the attenuated spore vaccine. Also, no adverse effects were reported in association with vaccination (Turnbull, 1991).

The Soviet Ministry of Health endorsed live anthrax vaccination by scarification in 1953 and by the subcutaneous route in 1959. Military microbiologists also experimented with a dry spore vaccine that could be aerosolized and used to safely and effectively vaccinate large numbers of individuals simultaneously; however, this route was not used for general vaccination and was reserved for emergency vaccinations (Turnbull, 1991).

The use of attenuated vaccines in western countries has not been as widely accepted for immunization of humans. Both the United States and the United Kingdom have licensed cell-free anthrax vaccines, consisting of filter-sterilized culture medium of an attenuated $pXO2^{-}$ strain of *B. anthracis* combined with aluminum hydroxide adjuvant. The U.S. vaccine, termed AVA, was licensed by the FDA (product license number 99, Bureau of Laboratories, Lansing, MI) in 1970 and consists of the culture filtrate adsorbed to aluminum hydroxide, the only vaccine adjuvant approved for use in humans in the United States. The discovery and formulation of AVA are discussed below (Turnbull, 1991).

Development of the AVA vaccine started in the 1950s when Wright et al. (1954) showed that PA, precipitated from bacterial culture with alum, could immunize and protect animals against anthrax. Isolation of a nonencapsulated, nonproteolytic derivative of *B. anthracis* strain V770 further advanced the development of an anthrax vaccine. This isolate produced greater PA levels and fewer enzymatic portions of the toxins, thereby increasing the amount of protective immunogen in the vaccine. In 1962, scientists discovered that growing the bacterium under anaerobic conditions increased the production of PA (Puziss and Wright, 1963; Wright et al., 1962). Later research by Puziss and Wright (1963) determined that PA adsorbed better to a preformed aluminum hydroxide gel, and the addition of a 1:40,000 dilution of benzethonium chloride and 0.009% formalin helped to stabilize the vaccine. This formulation provided protection from a virulent strain of *B. anthracis* in mice, guinea pigs, and rabbits. The researchers also tested human tolerance to the vaccine by injecting volunteers with 0.5 ml of the formulation at 0, 2, and 4 weeks subcutaneously into the deltoid region of the shoulder (Puziss and Wright, 1963). They reported no systemic reactions and only mild reactions at the inoculation site. The only scientific study to investigate the efficacy of the acellular anthrax vaccine was published by Brachman et al. (1962). The double-blind, controlled clinical trial was conducted between 1955 and 1959 in four goat hair mills in northeastern United States. Based on animal studies, the alum-adsorbed culture filtrate was given to 379 volunteers as a series of three initial 0.5 ml subcutaneous injections 2 weeks apart followed by three booster injections at 6-month intervals. The vaccinated individuals also received yearly boosters. In this study, 414 individuals received a placebo consisting of 0.1% alum. During the clinical evaluation of the vaccine, 21 cases of cutaneous anthrax and 5 cases of inhalational anthrax occurred in the control group, while only one case of cutaneous anthrax was reported in the vaccinated group (Brachman et al., 1962).

Today, AVA is marketed under the name BioThrax™ and is manufactured by the BioPort Corporation, Lansing, MI. The final vaccine consists of sterile bacterial culture filtrate containing PA, 1.2 mg/ml aluminum (added as aluminum hydroxide in 0.85% sodium chloride), benzethonium chloride (25 μg/ml), and formaldehyde (100 μg/ml) as preservatives. The vaccine is given as a 0.5 ml subcutaneous injection, and the recommended vaccination schedule is at 0, 2, and 4 weeks, with boosters at 6, 12, and 18 months. Yearly boosters are recommended to maintain immunity (BioThrax™ package insert).

The duration of immunity in humans is not known, and there are limited data on the antibody response to the vaccine. Two weeks after completion of the initial three injections, the vaccinated individuals had detectible antibody levels to PA; no anti-LF antibodies were detected (Turnbull et al., 1986). A cohort study of active duty military personnel receiving 1, 2, or 3 AVA injections showed that only 30% had detectible levels of antibodies to PA at 18–24 months after initial vaccination, but after subsequent injections, 99% produced detectable titers (Pittman et al., 2002a).

Because of the number of replicate injections needed in the current vaccination regimen, studies were conducted to investigate alternative vaccination strategies. Subjects given two AVA injections 4 weeks apart had comparable anti-PA antibody levels compared to individuals who received the standard regimen of three doses 2 weeks apart, suggesting that protective antibody titer may be reached with fewer doses of the vaccine (Pittman et al., 2002b). Additional studies would be needed to amend the current FDA-approved immunization schedule, but these studies will likely be performed on second-generation PA vaccines.

The data on the efficacy of AVA vaccine mostly are drawn from animal studies and the initial field trial of its alum-adsorbed precursor. Mice and guinea pigs cannot be consistently protected with AVA alone (Turnbull et al., 1986; Welkos and Friedlander, 1988). On the other hand, studies with rabbits and rhesus macaques showed 100% protection from aerosol challenge with virulent *B. anthracis* spores (Ivins et al., 1998; Pitt et al., 2001).

Vaccination with BioThrax™ is recommended for adults, who are likely to come in contact with potentially contaminated animals or animal products, such as laboratory workers, veterinarians, or others at risk of occupational exposure (Biothrax™ package insert). BioThrax™ is contraindicated for use in individuals who have developed an anaphylactic or anaphylactic-like reaction from previous inoculation with BioThrax™ or the vaccine components. The manufacturer also warns against the vaccination of pregnant women because of preliminary evidence of increased incidence of birth defects in children born to vaccinated women in the U.S. military. Individuals with a history of Guillain–Barre syndrome should not be vaccinated. The safety of the vaccine has not been established in individuals younger than 18 and older than 65.

Adverse effects of anthrax vaccination before licensure were obtained by evaluating 6985 individuals receiving the vaccine. Severe local reactions (edema or a reaction site area > 120 mm) occurred after only 1% of inoculations. Three percent of vaccinated individuals experienced moderate local reactions (edema and reaction site area of 120–130 mm), while 20% reported mild reactions (edema, reddening of inoculation site, and reaction area of <30 mm). Four of the vaccine recipients reported systemic reactions (fever, chills, body aches, or nausea) after inoculation. These adverse effects were deemed acceptable for licensure of the vaccine by the FDA.

Postlicensure data are available from the Vaccine Adverse Event Reporting System (VAERS). A total of 1,859,000 doses of anthrax vaccine were distributed in the United States from January 1, 1990 to August 31, 2000. From this, 1544 cases of adverse events were reported to the VAERS and, of these, 5% were considered serious, resulting in hospitalization, disability, or death. Most reported adverse events were reactions at the injection site and some systemic reactions included headache, joint pain, muscle weakness, and itching. Two deaths occurred in people receiving the vaccine, but the inoculations were determined to be unrelated to the deaths. Less than 10 cases were reported in which serious adverse events occurred, but again the events could not be clearly linked to administration of the vaccine (CDC, 2002).

Additional studies showed that the vaccine was not associated with serious side effects, and participants did not develop side effects 25 years following vaccination. Likewise, Persian Gulf syndrome, a malady occurring in veterans of the 1991 Gulf War, could not be correlated with vaccination against anthrax, although the issue is still being debated.

In 1998, in an effort to protect military personnel against biological warfare, the U.S. Department of Defense (DoD) initiated the Anthrax Immunization Program. As a result of a lawsuit against the Pentagon, charging that the FDA licensure had not met legal requirements, a federal judge ordered the DoD to halt compulsory anthrax vaccination (Dyer, 2004). The suspension was lifted after the FDA approved Bioport's renovated manufacturing facility, allowing voluntary immunization.

Although AVA has been established as a safe, effective vaccine, there are still reservations about its side effects. Also, because the vaccine contains all proteins secreted or released from the bacteria, a chemically defined alternative is needed. The numerous doses that are needed to successfully vaccinate military personnel and at-risk civilians heralds the need for a vaccine that is stable, highly immunogenic, rapidly prepared, and cost-effective. The major vaccine component is PA, and with the advent of molecular biology researchers are now able to develop newer vaccines containing a precise amount of purified PA. Still, the need for novel and effective adjuvants remains.

After the bioterrorist events of 2001, the U.S. government conceived the BioShield program to improve public defense against biological weapons (Check, 2004). To facilitate the rapid development of new therapeutics, the FDA amended its regulations pertaining to the requirement for human clinical trials in diseases in which it is not feasible or ethical, such as with anthrax infections (Food and Drug Administration, 2002). Currently, there are four stipulations that must be met before a product can be licensed for human use based upon animal studies (Food and Drug Administration, 2002):

1. There is a reasonably well-understood pathophysiological mechanism for the toxicity of the chemical, biological, radiological, or nuclear substance and its amelioration or prevention by the product;
2. The effect is demonstrated in more than one animal species expected to react with a response predictive for humans, unless the effect is demonstrated in a single animal species that represents a sufficiently well-characterized animal model (meaning that the model has been adequately evaluated for its responsiveness) for predicting the response in humans;
3. The animal study endpoint is clearly related to the desired benefit in humans, which is generally the enhancement of survival or prevention of major morbidity; and
4. The data or information on the pharmacokinetics and pharmacodynamics of the product or other relevant data or information in animals and

humans are sufficiently well understood to allow selection of an effective dose in humans, and it is therefore reasonable to expect the effectiveness of the product in animals to be a reliable indicator of its effectiveness in humans.

ANIMAL MODELS FOR EVALUATING VACCINE EFFICACY

Rabbits and nonhuman primates are often considered the best animal models to evaluate potential anthrax vaccine candidates because the pathophysiological response to disease most closely resembles human inhalational anthrax (Fritz et al., 1995; Zaucha et al., 1998). AVA consistently protects rabbits and nonhuman primates, but protection is variable in guinea pig and mouse models (Ivins et al., 1998; Pitt et al., 2001; Turnbull et al., 1986; Welkos and Friedlander, 1988). In addition, in the mouse model, anthrax toxins are not the predominant virulence factor, and the capsule has been shown to play a major role (Welkos et al., 1993). Hence, the murine model may not be the best for evaluating antitoxic-type vaccines against anthrax. Much vaccine research is still being performed with mice and guinea pigs, because of the lower cost, availability, and ease of housing infected animals. Each vaccine must comply with FDA guidelines and will likely be tested in rabbits and nonhuman primates.

RECOMBINANT PA VACCINES

The first cloning of PA was reported by Vodkin and Leppla (1983). rPA has been expressed in *B. subtilis*, attenuated strains of *B. anthracis*, and *E. coli* (Baillie et al., 1998; Farchaus et al., 1998; Laird et al., 2004). A recent report showed that multigram quantities (2.7 g/L) of highly purified, biologically active PA could be produced from *E. coli* expressing the *pag* gene (Gwinn et al., 2005).

PA alone does not elicit a strong protective immune response and must be combined with an appropriate adjuvant. Aluminum hydroxide (Alhydrogel) is the only FDA-approved human adjuvant. Several studies have been conducted to compare AVA to rPA + Alhydrogel. Mice vaccinated with rPA did not survive subcutaneous challenge with virulent spores. Likewise, AVA did not protect the mice after two doses. rPA coupled to aluminum hydroxide provided more protection in guinea pigs than in mice. At 6 weeks after vaccination, 60% survived intramuscular challenge with 4300 spores, and 89% survived challenge after 24 weeks. Protection dropped to 25% 8 weeks after challenge with 200,000 spores. AVA protected only 50% of the animals from the increased challenge dose (Ivins et al., 1992).

rPA provided far greater protection in the rabbit model. Rabbits immunized with 100 μg of rPA combined with Alhydrogel protected 93.3% of the animals against an aerosol challenge 4 weeks after a single inoculation. The investigators showed a dose-dependent relationship between the amount of rPA and protection, and established that anti-rPA titers and toxin-neutralizing titers were reliable predictors of survival (Little et al., 2004).

rPA also effectively protected rhesus macaques (Ivins et al., 1998). Animals were immunized once with 50 μg of rPA or AVA. Twenty percent of macaques immunized with AVA or PA + Alhydrogen were bacteremic 1 week after challenge; however, all survived infection (Ivins et al., 1998). These data demonstrated that PA coupled with the FDA-approved adjuvant provides protection equivalent to that of AVA.

rPA + Alhydrogel-based anthrax vaccines produced by Vaxgen and Dynport Vaccine Company have been approved for clinical trials sponsored by the National Institute of Allergy and Infectious Disease. Dynport's rPA anthrax vaccine is currently being evaluated in a phase 1 trial at the Walter Reed Army Institute of Research/Henry M. Jackson Foundation Vaccine Clinical Research Center in Rockville, MD. The rPA vaccine, produced in *E. coli*, will be administered by two intramuscular injections 4 weeks apart. The trial will compare different doses of rPA with and without adjuvant (Alhydrogel). Vaxgen has completed a multistate phase 1 clinical trial investigating the safety and immunogenicity of three doses of rPA + Alhydrogel compared to AVA. The company reported their rPA vaccine elicited immune responses comparable to those found with AVA, with no significant safety or reactogenicity issues. Varying doses of Vaxgen's rPA vaccine, rPA102, was given to healthy volunteers at weeks 0, 4, and 8. AVA was given at weeks 0 and 4. AVA and rPA102 elicited similar neutralizing antibody titers after two injections. The third rPA102 injection increased the anti-PA antibody response. Local reactogenicity (pain) was more common among subjects receiving AVA while systemic reactogenicity (headache) was higher in rPA102 vaccinees (Gorse et al., 2006). Vaxgen is currently involved in a phase-2 multicenter randomized dose-finding trial to access the safety and immunogenicity of two inoculations with rPA over two dose ranges and four adjuvant levels. Under a Project BioShield contract, VaxGen was awarded $879.2 million in November 2004 to produce 75 million doses of vaccine. The contract was terminated for default in December 2006 (Russell, 2007).

ADJUVANTS

As stated earlier, rPA must be administered with an adjuvant to be protective. rPA combined with microbial products such as monophosphoryl lipid A, trehalose dimycolate, and the delipidated, deproteinized cell wall of skeleton of *Mycobacterium* spp. improved the protection in the mouse model when compared to AVA or rPA + Alhydrogel. rPA combined with the microbial-derived products protected 100% of the guinea pigs challenged by intramuscular injection of virulent spores. rPA + Alhydrogel protection was increased with multiple inoculations, but did not provide the level of protection afforded by the microbial adjuvants (Ivins et al., 1992).

A second study investigated the efficacy of several adjuvants combined with PA in the guinea pig model after aerosol challenge of *B. anthracis* spores. Animals were vaccinated at 0 and 4 weeks, and then challenged at 10 weeks. PA + monophosphoryl lipid A in a squalene/lecithin/Tween 20 emulsion protected 50% of the animals while AVA and PA + Alhydrogel only protected 26% and 5%, respectively (Ivins et al., 1995).

The efficacy of rPA combined with different adjuvants was also compared to AVA in rhesus macaques (Ivins et al., 1998). Animals were immunized once with 50 μg of rPA and either saponin-QS21, Alhydrogel, or monophosphoryl lipid A (MPL) in squalene/lecithin/Tween 20 emulsion (STL). Another group was given one dose of AVA. All groups showed an IgM titer that peaked 2 weeks after immunization with IgG titers peaking at 4–5 weeks post inoculation. IgG titers were highest in the group receiving PA + Alhydrogel. All groups manifested antigen-specific lymphocyte proliferation and produced toxin-neutralizing antibodies. All vaccines conferred 100% protection except PA + MPL in STL, which was 90% effective (Ivins et al., 1998). These data suggested that Alhydrogel was an efficient adjuvant in the nonhuman primate model of anthrax and would probably be as effective in humans.

There is evidence that oligodeoxynucleotides containing unmethylated CpG improve the immunogenicity of AVA in multiple animal models (Gu et al., 2007; Klinman, 2006; Klinman et al., 2006). The effect of this adjuvant on second-generation rPA vaccines has yet to be elucidated.

Other laboratories are using animal models to investigate less invasive routes of immunization such as transdermal and intranasal routes (Gaur et al., 2002; Mikszta et al., 2005). The use of transcutaneous immunization with a needle-free patch and virus expression of PA is also being explored (Iacono-Connors et al., 1991; Kenney et al., 2004). DNA vaccines encoding the *pag* gene are also being developed (Galloway and Baillie, 2004).

CAPSULAR ANTIGEN VACCINES

While vaccines containing rPA will inhibit the action of the secreted toxins, they do not counteract the antiphagocytic capsule. The poly-γ-D-glutamic acid peptides comprising the capsule are poor immunogens and must be coupled to a carrier protein to evoke an anticapsule humoral response (Schneerson et al., 2003; Wang et al., 2004). The antibodies produced are oposonigenic and should aid the host in clearing the infection (Wang et al., 2004). Chabot et al. (2004) demonstrated that vaccination with capsular antigen protected mice from a lethal challenge with an encapsulated, toxin-deficient strain of *B. anthracis*. Conjugating the capsule to a carrier protein increased IgG titers, but abrogated the protection. Further, the capsule alone evoked a higher IgM titer and increased opsonic activity. Vaccination with capsular antigen or PA alone was not protective against a virulent strain of *B. anthracis*; however, protection improved when both antigens were used together (Chabot et al., 2004). These data suggested that vaccinating against both PA and capsule might increase the effectiveness of the anthrax vaccine.

PASSIVE IMMUNIZATION

Although active immunization has been shown to be an effective way to protect against anthrax, passive immunization with anti-PA antibodies is also being explored as a new therapeutic or prophylactic approach. Antibiotics must be administered early in infection in order to be effective, and there is a "point of no return" when antibiotic therapy fails to protect the patient, presumably because of the deleterious actions of the toxins (Dixon et al., 1999). Passive immunization with anti-PA antibodies that neutralize the toxins should protect the individual from their actions. Studies in guinea pigs determined that anti-PA serum given 24 h after an intranasal challenge with fully virulent *B. anthracis* spores protected 90% of the animals (Kobiler et al., 2002). Intravenous injection of 10 mg of a high-affinity monoclonal antibody to PA 30–45 min before aerosol exposure protected 90% of the challenged rabbits. Protection was dependent on the time of administration with an 80% survival rate of the animals administered 24 h after challenge. Only 50% treated with the antibody 36h post infection survived the study period (Mohamed et al., 2004). Several groups are studying the protective effects of passive immunization against anthrax, and one termed AbThrax produced by Human Genome Sciences, Inc. has completed phase 1 clinical trials and moved into phase 2 (Borio and Gronvall, 2005). While passive immunization against anthrax is promising as a prophylactic/therapeutic, it is likely that future recommendations for the individuals at risk of exposure will include active immunization and antibiotic treatment, as well as human monoclonal antibodies to PA. In fact, in 2006 a 44-year-old Pennsylvania man was diagnosed with naturally acquired inhalational anthrax and treated with antibiotics, supportive care, and human anthrax immunoglobulin, produced by Cangene Corp. Once the patient was diagnosed with anthrax, the CDC recommended the use of the immunoglobulin under the Emergency Investigational Drug use protocol. The patient survived the infection; however, the role of the immunoglobulin in recovery is difficult to determine and more studies on the passive immune therapy during anthrax infections are needed (Walsh et al., 2007). Along with active and passive immunity and antibiotics, laboratories are actively developing compounds that inhibit the action of LeTx and EdTx and other novel antimicrobial therapeutics.

ALTERNATIVE ANTHRAX THERAPIES

Investigation of antitoxic compounds centers around inhibiting the enzymatic activities of both LF and EF. The molecular structures of the PA, LF, and EF proteins have been derived by X-ray crystallography, allowing for the design of inhibitors that bind optimally and specifically to the active sites of these enzymes (Drum et al., 2002; Pannifer et al., 2001; Petosa et al., 1997). Several compounds that inhibit the proteolytic activity of LF have been identified (Dell'Aica et al., 2004; Lee et al., 2004; Numa et al., 2005; Panchal et al., 2004), but only a few have been tested against anthrax infection in animal models. One hydroxamate compound (LFI), produced by Merck Research Laboratories, inhibited the activity of LeTx in vitro and protected mice from challenge with purified toxin. The compound also doubled the mean time-to-death of rabbits challenged subcutaneously with virulent spores. Interestingly, when the hydroxamate compound was coupled with suboptimal doses of ciprofloxacin, 100% of the animals survived compared with 50% survival of animals receiving antibiotic alone (Shoop et al., 2005). A clinically approved drug, adefovir dipivoxil, used to treat chronic hepatitis B infections, also blocks the adenylyl cyclase activity of EF (Shen et al., 2004), but its effectiveness as an anthrax therapy has not been investigated.

Compounds that block entry into the cell are also being investigated. One target of great interest is the furin protease that activates PA (Kacprzak et al., 2004; Peinado et al., 2004). Another approach to disrupt toxin entry is with a dominant-negative mutant of PA. Singh et al. (2001) replaced the portion of PA responsible for membrane insertion with a similar segment of a toxin produced by *Clostridium perfringens*. The resulting protein was still activated by a host furin protease, formed oligomers, and bound LF. The mutant protein failed to translocate LF into the host cell cytosol. The mutant protein inhibited the action of wild-type PA and LF in vitro and protected laboratory rats from intoxication (Singh et al., 2001). Vaccination of mice with the mutant PA produced higher antibody titers than wild-type protein, and conjugating it to poly-γ-D-glutamate produced higher titers to PA and capsule than wild-type PA combined with poly-γ-D-glutamate. This study suggested that postexposure use of the dominant-negative PA might improve anti-PA titers while simultaneously inhibiting the action of the anthrax toxins (Aulinger et al., 2005). Finally, the use of naturally occurring antibacterial agents is being investigated as possible therapy against anthrax. The lytic enzyme of the *B. anthracis* bacteriophage, γ phage, has been isolated and shown to kill the bacteria.

These novel therapeutics are still in the early stages of development, but offer new and exciting approaches to combat anthrax.

KEY ISSUES

- Anthrax outbreaks, caused by the spore-forming bacterium, *B. anthracis*, date back to antiquity. The disease occurs in herbivores worldwide, and humans are at risk of infection when handling contaminated animal products.
- The anthrax can manifest in three forms of the disease depending on the route of exposure, including cutaneous, gastrointestinal, and inhalational anthrax.
- The spores are stable and easily disseminated, which make *B. anthracis* a major bioterrorism threat. Letters sent through the U.S. Postal Service in 2001 resulted in 22 cases of anthrax and 5 deaths. This incident prompted a renewed interest in *B. anthracis* and the development of novel vaccines and therapeutic agents.

- The virulence of *B. anthracis* is dependent on the synthesis of the poly-γ-D-glutamic acid capsule, which enables the bacteria to evade phagocytosis, and two toxins, lethal toxin and edema toxin. Both toxins share a binding moiety, called PA, which facilitates toxin entry into host cells. Lethal toxin is formed when PA combines with LF, a zinc metalloprotease that cleaves mitogen-activated kinase kinases. Edema toxin is formed when EF, a calmodulin-dependent adenylyl cyclase, binds to PA.
- Currently, the only treatment for anthrax is antibiotics, which must be given early in infection to be effective.
- The FDA-licensed anthrax vaccine, AVA, has been in use since 1970. AVA comprised filter-sterilized, attenuated *B. anthracis* growth medium adsorbed to aluminum hydroxide.
- The major immunogen in AVA is PA, and anti-PA antibody titers directly correlate with protection in animal models of anthrax. Antibodies to PA inhibit the action of the toxins and may help in the clearance of the infection.
- The second-generation anthrax vaccine is based on rPA. This vaccine provides protection in animal models comparable to AVA, and is chemically defined, allowing for the administration of known doses of PA. The rPA vaccine is currently undergoing safety and immunogenicity testing in human trials.
- Passive immunization, with antibodies to PA, is also being investigated as a possible anthrax therapy.
- Vaccines against the *B. anthracis* capsule and compounds that inhibit the action of the toxins are also being investigated.

References

Abramova, F.A., Grinberg, L.M., Yampolskaya, O.V. and Walker, D.H. Pathology of inhalational anthrax in 42 cases from the Sverdlovsk outbreak of 1979. Proc. Natl. Acad. Sci. USA 1993; 90:2291–2294.

Agrawal, A., Lingappa, J., Leppla, S.H., Agrawal, S., Jabbar, A., Quinn, C. and Pulendran, B. Impairment of dendritic cells and adaptive immunity by anthrax lethal toxin. Nature 2003; 424:329–334.

Ascenzi, P., Visca, P., Ippolito, G., Spallarossa, A., Bolognesi, M. and Montecucco, C. Anthrax toxin: a tripartite lethal combination. FEBS Lett. 2002; 531:384–388.

Aulinger, B.A., Roehrl, M.H., Mekalanos, J.J., Collier, R.J. and Wang, J.Y. Combining anthrax vaccine and therapy: a dominant-negative inhibitor of anthrax toxin is also a potent and safe immunogen for vaccines. Infect. Immun. 2005; 73:3408–3414.

Baillie, L., Moir, A. and Manchee, R. The expression of the protective antigen of *Bacillus anthracis* in *Bacillus subtilis*. J. Appl. Microbiol. 1998; 84:741–746.

Bales, M.E., Dannenberg, A.L., Brachman, P.S., Kaufmann, A.F., Klatsky, P.C. and Ashford, D.A. Epidemiologic response to anthrax outbreaks: field investigations, 1950–2001. Emerg. Infect. Dis. 2002; 8:1163–1174.

Beedham, R.J., Turnbull, P.C. and Williamson, E.D. Passive transfer of protection against *Bacillus anthracis* infection in a murine model. Vaccine 2001; 19:4409–4416.

Borio, L.L. and Gronvall, G.K. Anthrax countermeasures: current status and future needs. Biosecur. Bioterror. 2005; 3:102–112.

Brachman, P.S., Gold, H., Plotkin, S.A., Fekety, F.R., Werrin, M. and Ingraham, N.R. Field evaluation of a human anthrax vaccine. Am. J. Public Health 1962; 52:632–645.

Bradley, K.A., Mogridge, J., Mourez, M., Collier, R.J. and Young, J.A. Identification of the cellular receptor for anthrax toxin. Nature 2001; 414:225–229.

CDC Use of anthrax vaccine in response to terrorism: supplemental recommendations of the Advisory Committee on Immunization Practices. MMWR Morb. Mortal. Wkly. Rep. 2002; 51:1024–1026.

Chabot, D.J., Scorpio, A., Tobery, S.A., Little, S.F., Norris, S.L. and Friedlander, A.M. Anthrax capsule vaccine protects against experimental infection. Vaccine 2004; 23:43–47.

Chaudry, G.J., Moayeri, M., Liu, S. and Leppla, S.H. Quickening the pace of anthrax research: three advances point towards possible therapies. Trends Microbiol. 2002; 10:58–62.

Check, E. BioShield defence programme set to fund anthrax vaccine. Nature 2004; 429:4.

Cieslak, T.J. and Eitzen, E.M., Jr. Clinical and epidemiologic principles of anthrax. Emerg. Infect. Dis. 1999; 5:552–555.

Clegg, S.B., Turnbull, P.C., Foggin, C.M. and Lindeque, P.M. Massive outbreak of anthrax in wildlife in the Malilangwe Wildlife Reserve, Zimbabwe. Vet. Rec. 2007; 160:113–118.

Cote, C.K., Rossi, C.A., Kang, A.S., Morrow, P.R., Lee, J.S. and Welkos, S.L. The detection of protective antigen (PA) associated with spores of *Bacillus anthracis* and the effects of anti-PA antibodies on spore germination and macrophage interactions. Microb. Pathog. 2005; 38:209–225.

Cui, X., Moayeri, M., Li, Y., Li, X., Haley, M., Fitz, Y., Correa-Araujo, R., Banks, S.M., Leppla, S.H. and Eichacker, P.Q. Lethality during continuous anthrax lethal toxin infusion is associated with circulatory shock but not inflammatory cytokine or nitric oxide release in rats. Am. J. Physiol. Regul. Integr. Comp. Physiol. 2004; 286:R699–R709.

Dell'Aica, I., Dona, M., Tonello, F., Piris, A., Mock, M., Montecucco, C. and Garbisa, S. Potent inhibitors of anthrax lethal factor from green tea. EMBO Rep. 2004; 5:418–422.

Demirdag, K., Ozden, M., Saral, Y., Kalkan, A., Kilic, S.S. and Ozdarendeli, A. Cutaneous anthrax in adults: a review of 25 cases in the eastern Anatolian region of Turkey. Infection 2003; 31:327–330.

Dixon, T.C., Meselson, M., Guillemin, J. and Hanna, P.C. Anthrax. N. Engl. J. Med. 1999; 341:815–826.

Drum, C.L., Yan, S.Z., Bard, J., Shen, Y.Q., Lu, D., Soelaiman, S., Grabarek, Z., Bohm, A. and Tang, W.J. Structural basis for the activation of anthrax adenylyl cyclase exotoxin by calmodulin. Nature 2002; 415:396–402.

Drysdale, M., Heninger, S., Hutt, J., Chen, Y., Lyons, C.R. and Koehler, T.M. Capsule synthesis by *Bacillus anthracis* is required for dissemination in murine inhalation anthrax. Embo J. 2005; 24:221–227.

Duesbery, N.S., Webb, C.P., Leppla, S.H., Gordon, V.M., Klimpel, K.R., Copeland, T.D., Ahn, N.G., Oskarsson, M.K., Fukasawa, K., Paull, K.D. and Vande Woude, G.F. Proteolytic inactivation of MAP-kinase-kinase by anthrax lethal factor. Science 1998; 280:734–737.

Dyer, O. US judge halts compulsory anthrax vaccination for soldiers. Br. Med. J. 2004; 329:1062.

Ezzell, J.W. and Welkos, S.L. The capsule of *Bacillus anthracis*, a review. J. Appl. Microbiol. 1999; 87:250.

Fang, H., Cordoba-Rodriguez, R., Lankford, C.S. and Frucht, D.M. Anthrax lethal toxin blocks MAPK kinase-dependent IL-2 production in CD4+ T cells. J. Immunol. 2005; 174:4966–4971.

Fang, H., Xu, L., Chen, T.Y., Cyr, J.M. and Frucht, D.M. Anthrax lethal toxin has direct and potent inhibitory effects on B cell proliferation and immunoglobulin production. J. Immunol. 2006; 176:6155–6161.

Farchaus, J.W., Ribot, W.J., Jendrek, S. and Little, S.F. Fermentation, purification, and characterization of protective antigen from a recombinant, avirulent strain of *Bacillus anthracis*. Appl. Environ. Microbiol. 1998; 64:982–991.

Food and Drug Administration. New drug and biological drug products; evidence needed to demonstrate effectiveness of new drugs when human efficacy studies are not ethical or feasible. In "21 CFR 314 and 601". U.S. Food and Drug Administration, Department of Health and Human Services Rockville, MD, 2002.

Fritz, D.L., Jaax, N.K., Lawrence, W.B., Davis, K.J., Pitt, M.L., Ezzell, J.W. and Friedlander, A.M. Pathology of experimental inhalation anthrax in the rhesus monkey. Lab. Invest. 1995; 73:691–702.

Galloway, D.R. and Baillie, L. DNA vaccines against anthrax. Exp. Opin. Biol. Ther. 2004; 4:1661–1667.

Gaur, R., Gupta, P.K., Banerjea, A.C. and Singh, Y. Effect of nasal immunization with protective antigen of *Bacillus anthracis* on protective immune response against anthrax toxin. Vaccine 2002; 20:2836–2839.

Gorse, G.J., Keitel, W., Keyserling, H., Taylor, D.N., Lock, M., Alves, K., Kenner, J., Deans, L. and Gurwith, M. Immunogenicity and tolerance of ascending doses of a recombinant protective antigen (rPA102) anthrax vaccine: a randomized, double-blinded, controlled, multicenter trial. Vaccine 2006; 24:5950–5959.

Grinberg, L.M., Abramova, F.A., Yampolskaya, O.V., Walker, D.H. and Smith, J.H. Quantitative pathology of inhalational anthrax I: quantitative microscopic findings. Mod. Pathol. 2001; 14: 482–495.

Gu, M., Hine, P.M., James Jackson, W., Giri, L. and Nabors, G.S. Increased potency of BioThrax anthrax vaccine with the addition of the C-class CpG oligonucleotide adjuvant CPG 10109. Vaccine 2007; 25:526–534.

Guarner, J., Jernigan, J.A., Shieh, W.J., Tatti, K., Flannagan, L.M., Stephens, D.S., Popovic, T., Ashford, D.A., Perkins, B.A. and Zaki, S.R. Pathology and pathogenesis of bioterrorism-related inhalational anthrax. Am. J. Pathol. 2003; 163:701–709.

Guidi-Rontani, C. and Mock, M. Macrophage interactions. Curr. Top. Microbiol. Immunol. 2002; 271:115–141.

Gwinn, W., Zhang, M., Mon, S., Sampey, D., Zukauskas, D., Kassebaum, C., Zmuda, J.F., Tsai, A. and Laird, M.W. Scalable purification of *Bacillus anthracis* protective antigen from *Escherichia coli*. Protein Expr. Purif. 2005; 45:30–36.

Hoover, D.L., Friedlander, A.M., Rogers, L.C., Yoon, I.K., Warren, R.L. and Cross, A.S. Anthrax edema toxin differentially regulates lipopolysaccharide-induced monocyte production of tumor necrosis factor alpha and interleukin-6 by increasing intracellular cyclic AMP. Infect. Immun. 1994; 62:4432–4439.

Iacono-Connors, L.C., Welkos, S.L., Ivins, B.E. and Dalrymple, J.M. Protection against anthrax with recombinant virus-expressed protective antigen in experimental animals. Infect. Immun. 1991; 59:1961–1965.

Ivins, B., Fellows, P., Pitt, L., Estep, J., Farchaus, J., Friedlander, A. and Gibbs, P. Experimental anthrax vaccines: efficacy of adjuvants combined with protective antigen against an aerosol *Bacillus anthracis* spore challenge in guinea pigs. Vaccine 1995; 13:1779–1784.

Ivins, B.E., Pitt, M.L., Fellows, P.F., Farchaus, J.W., Benner, G.E., Waag, D.M., Little, S.F., Anderson, G.W., Jr., Gibbs, P.H. and Friedlander, A.M. Comparative efficacy of experimental anthrax vaccine candidates against inhalation anthrax in rhesus macaques. Vaccine 1998; 16:1141–1148.

Ivins, B.E., Welkos, S.L., Little, S.F., Crumrine, M.H. and Nelson, G.O. Immunization against anthrax with *Bacillus anthracis* protective antigen combined with adjuvants. Infect. Immun. 1992; 60:662–668.

Jackson, P.J., Hugh-Jones, M.E., Adair, D.M., Green, G., Hill, K.K., Kuske, C.R., Grinberg, L.M., Abramova, F.A. and Keim, P. PCR analysis of tissue samples from the 1979 Sverdlovsk anthrax victims: the presence of multiple *Bacillus anthracis* strains in different victims. Proc. Natl. Acad. Sci. USA 1998; 95:1224–1229.

Jernigan, D.B., Raghunathan, P.L., Bell, B.P., Brechner, R., Bresnitz, E.A., Butler, J.C., Cetron, M., Cohen, M., Doyle, T., Fischer, M., Greene, C., Griffith, K.S., Guarner, J., Hadler, J.L., Hayslett, J.A., Meyer, R., Petersen, L.R., Phillips, M., Pinner, R., Popovic, T., Quinn, C.P., Reefhuis, J., Reissman, D., Rosenstein, N., Schuchat, A., Shieh, W.J., Siegal, L., Swerdlow, D.L., Tenover, F.C., Traeger, M., Ward, J.W., Weisfuse, I., Wiersma, S., Yeskey, K., Zaki, S., Ashford, D.A., Perkins, B.A., Ostroff, S., Hughes, J., Fleming, D., Koplan, J.P. and Gerberding, J.L. Investigation of bioterrorism-related anthrax, United States, 2001: epidemiologic findings. Emerg. Infect. Dis. 2002; 8:1019–1028.

Jernigan, J.A., Stephens, D.S., Ashford, D.A., Omenaca, C., Topiel, M.S., Galbraith, M., Tapper, M., Fisk, T.L., Zaki, S., Popovic, T., Meyer, R.F., Quinn, C.P., Harper, S.A., Fridkin, S.K., Sejvar, J.J., Shepard, C.W., McConnell, M., Guarner, J., Shieh, W.J., Malecki, J.M., Gerberding, J.L., Hughes, J.M. and Perkins, B.A. Bioterrorism-related inhalational anthrax: the first 10 cases reported in the United States. Emerg. Infect. Dis. 2001; 7:933–944.

Kacprzak, M.M., Peinado, J.R., Than, M.E., Appel, J., Henrich, S., Lipkind, G., Houghten, R.A., Bode, W. and Lindberg, I. Inhibition of furin by polyarginine-containing peptides: nanomolar inhibition by nona-d-arginine. J. Biol. Chem. 2004; 279:36788–36794.

Keim, P., Smith, K.L., Keys, C., Takahashi, H., Kurata, T. and Kaufmann, A. Molecular investigation of the Aum Shinrikyo anthrax release in Kameido, Japan. J. Clin. Microbiol. 2001; 39:4566–4567.

Kenney, R.T., Yu, J., Guebre-Xabier, M., Frech, S.A., Lambert, A., Heller, B.A., Ellingsworth, L.R., Eyles, J.E., Williamson, E.D. and Glenn, G.M. Induction of protective immunity against lethal anthrax challenge with a patch. J. Infect. Dis. 2004; 190:774–782.

Kirby, J.E. Anthrax lethal toxin induces human endothelial cell apoptosis. Infect. Immun. 2004; 72:430–439.

Klimpel, K.R., Molloy, S.S., Thomas, G. and Leppla, S.H. Anthrax toxin protective antigen is activated by a cell surface protease with the sequence specificity and catalytic properties of furin. Proc. Natl. Acad. Sci. USA 1992; 89:10277–10281.

Klinman, D.M. CpG oligonucleotides accelerate and boost the immune response elicited by AVA, the licensed anthrax vaccine. Expert Rev. Vaccines 2006; 5:365–369.

Klinman, D.M., Xie, H. and Ivins, B.E. CpG oligonucleotides improve the protective immune response induced by the licensed anthrax vaccine. Ann. NY Acad. Sci. 2006; 1082:137–150.

Kobiler, D., Gozes, Y., Rosenberg, H., Marcus, D., Reuveny, S. and Altboum, Z. Efficiency of protection of guinea pigs against infection with *Bacillus anthracis* spores by passive immunization. Infect. Immun. 2002; 70:544–560.

Laird, M.W., Zukauskas, D., Johnson, K., Sampey, G.C., Olsen, H., Garcia, A., Karwoski, J.D., Cooksey, B.A., Choi, G.H., Askins, J., Tsai, A., Pierre, J. and Gwinn, W. Production and purification of *Bacillus anthracis* protective antigen from *Escherichia coli*. Protein Expr. Purif. 2004; 38:145–152.

Lee, L.V., Bower, K.E., Liang, F.S., Shi, J., Wu, D., Sucheck, S.J., Vogt, P.K. and Wong, C.H. Inhibition of the proteolytic activity of anthrax lethal factor by aminoglycosides. J. Am. Chem. Soc. 2004; 126:4774–4775.

Leppla, S.H. Anthrax toxin edema factor: a bacterial adenylate cyclase that increases cyclic AMP concentrations of eukaryotic cells. Proc. Natl. Acad. Sci. USA 1982; 79:3162–3166.

Little, S.F., Ivins, B.E., Fellows, P.F., Pitt, M.L., Norris, S.L. and Andrews, G.P. Defining a serological correlate of protection in rabbits for a recombinant anthrax vaccine. Vaccine 2004; 22:422–430.

Meselson, M., Guillemin, J., Hugh-Jones, M., Langmuir, A., Popova, I., Shelokov, A. and Yampolskaya, O. The Sverdlovsk anthrax outbreak of 1979. Science 1994; 266:1202–1208.

Mikesell, P., Ivins, B.E., Ristroph, J.D. and Dreier, T.M. Evidence for plasmid-mediated toxin production in *Bacillus anthracis*. Infect. Immun. 1983; 39:371–376.

Mikszta, J.A., Sullivan, V.J., Dean, C., Waterston, A.M., Alarcon, J.B., Dekker 3rd, J.P., Brittingham, J.M., Huang, J., Hwang, C.R., Ferriter, M., Jiang, G., Mar, K., Saikh, K.U., Stiles, B.G., Roy, C.J., Ulrich, R.G. and Harvey, N.G. Protective immunization against inhalational anthrax: a comparison of minimally invasive delivery platforms. J. Infect. Dis. 2005; 191:278–288.

Milne, J.C., Furlong, D., Hanna, P.C., Wall, J.S. and Collier, R.J. Anthrax protective antigen forms oligomers during intoxication of mammalian cells. J. Biol. Chem. 1994; 269:20607–20612.

Moayeri, M., Haines, D., Young, H.A. and Leppla, S.H. *Bacillus anthracis* lethal toxin induces TNF-alpha-independent hypoxia-mediated toxicity in mice. J. Clin. Invest. 2003; 112:670–682.

Mobley, J.A. Biological warfare in the twentieth century: lessons from the past, challenges for the future. Mil. Med. 1995; 160:547–553.

Mohamed, N., Li, J., Ferreira, C.S., Little, S.F., Friedlander, A.M., Spitalny, G.L. and Casey, L.S. Enhancement of anthrax lethal toxin cytotoxicity: a subset of monoclonal antibodies against protective antigen increases lethal toxin-mediated killing of murine macrophages. Infect. Immun. 2004; 72:3276–3283.

Numa, M.M., Lee, L.V., Hsu, C.C., Bower, K.E. and Wong, C.H. Identification of novel anthrax lethal factor inhibitors generated by combinatorial Pictet–Spengler reaction followed by screening in situ. Chembiochem 2005; 6:1002–1006.

O'Brien, J., Friedlander, A., Dreier, T., Ezzell, J. and Leppla, S. Effects of anthrax toxin components on human neutrophils. Infect. Immun. 1985; 47:306–310.

Paccani, S.R., Tonello, F., Ghittoni, R., Natale, M., Muraro, L., D'Elios, M.M., Tang, W.J., Montecucco, C. and Baldari, C.T. Anthrax toxins suppress T lymphocyte activation by disrupting antigen receptor signaling. J. Exp. Med. 2005; 201:325–331.

Panchal, R.G., Hermone, A.R., Nguyen, T.L., Wong, T.Y., Schwarzenbacher, R., Schmidt, J., Lane, D., McGrath, C., Turk, B.E., Burnett, J., Aman, M.J., Little, S., Sausville, E.A., Zaharevitz, D.W., Cantley, L.C., Liddington, R.C., Gussio, R. and Bavari, S. Identification of small molecule inhibitors of anthrax lethal factor. Nat. Struct. Mol. Biol. 2004; 11:67–72.

Pannifer, A.D., Wong, T.Y., Schwarzenbacher, R., Renatus, M., Petosa, C., Bienkowska, J., Lacy, D.B., Collier, R.J., Park, S., Leppla, S.H., Hanna, P. and Liddington, R.C. Crystal structure of the anthrax lethal factor. Nature 2001; 414:229–233.

Patra, G., Vaissaire, J., Weber-Levy, M., Le Doujet, C. and Mock, M. Molecular characterization of *Bacillus* strains involved in outbreaks of anthrax in France in 1997. J. Clin. Microbiol. 1998; 36:3412–3414.

Peinado, J.R., Kacprzak, M.M., Leppla, S.H. and Lindberg, I. Cross-inhibition between furin and lethal factor inhibitors. Biochem. Biophys. Res. Commun. 2004; 321:601–605.

Pellizzari, R., Guidi-Rontani, C., Vitale, G., Mock, M. and Montecucco, C. Anthrax lethal factor cleaves MKK3 in macrophages and inhibits the LPS/IFNgamma-induced release of NO and TNFalpha. FEBS Lett. 1999; 462:199–204.

Petosa, C., Collier, R.J., Klimpel, K.R., Leppla, S.H. and Liddington, R.C. Crystal structure of the anthrax toxin protective antigen. Nature 1997; 385:833–838.

Pitt, M.L., Little, S.F., Ivins, B.E., Fellows, P., Barth, J., Hewetson, J., Gibbs, P., Dertzbaugh, M. and Friedlander, A.M. In vitro correlate of immunity in a rabbit model of inhalational anthrax. Vaccine 2001; 19:4768–4773.

Pittman, P.R., Hack, D., Mangiafico, J., Gibbs, P., McKee, K.T., Jr., Friedlander, A.M. and Sjogren, M.H. Antibody response to a delayed booster dose of anthrax vaccine and botulinum toxoid. Vaccine 2002a; 20:2107–2115.

Pittman, P.R., Kim-Ahn, G., Pifat, D.Y., Coonan, K., Gibbs, P., Little, S., Pace-Templeton, J.G., Myers, R., Parker, G.W. and Friedlander, A.M. Anthrax vaccine: immunogenicity and safety of a dose-reduction, route-change comparison study in humans. Vaccine 2002b; 20:1412–1420.

Popov, S.G., Villasmil, R., Bernardi, J., Grene, E., Cardwell, J., Popova, T., Wu, A., Alibek, D., Bailey, C. and Alibek, K. Effect of *Bacillus anthracis* lethal toxin on human peripheral blood mononuclear cells. FEBS Lett. 2002; 527:211–215.

Puziss, M. and Wright, G.G. Studies on immunity in anthrax, X. Gel-adsorbed protective antigen for immunization of man. J. Bacteriol. 1963; 85:230–236.

Quinn, C.P., Dull, P.M., Semenova, V., Li, H., Crotty, S., Taylor, T.H., Steward-Clark, E., Stamey, K.L., Schmidt, D.S., Stinson, K.W., Freeman, A.E., Elie, C.M., Martin, S.K., Greene, C., Aubert, R.D., Glidewell, J., Perkins, B.A., Ahmed, R. and Stephens, D.S. Immune responses to *Bacillus anthracis* protective antigen in patients with bioterrorism-related cutaneous or inhalation anthrax. J. Infect. Dis. 2004; 190:1228–1236.

Rotz, L.D., Khan, A.S., Lillibridge, S.R., Ostroff, S.M. and Hughes, J.M. Public health assessment of potential biological terrorism agents. Emerg. Infect. Dis. 2002; 8:225–230.

Russell, P.K. Project BioShield: what it is, why it is needed, and its accomplishments so far. Clin. Infect. Dis. 2007; 45(Suppl 1):S68–S72.

Schmid, G. and Kaufmann, A. Anthrax in Europe: its epidemiology, clinical characteristics, and role in bioterrorism. Clin. Microbiol. Infect. 2002; 8:479–488.

Schneerson, R., Kubler-Kielb, J., Liu, T.Y., Dai, Z.D., Leppla, S.H., Yergey, A., Backlund, P., Shiloach, J., Majadly, F. and Robbins, J.B. Poly(gamma-d-glutamic acid) protein conjugates induce IgG antibodies in mice to the capsule of *Bacillus anthracis*: a potential addition to the anthrax vaccine. Proc. Natl. Acad. Sci. USA 2003; 100:8945–8950.

Scobie, H.M., Rainey, G.J., Bradley, K.A. and Young, J.A. Human capillary morphogenesis protein 2 functions as an anthrax toxin receptor. Proc. Natl. Acad. Sci. USA 2003; 100:5170–5174.

Shen, Y., Zhukovskaya, N.L., Zimmer, M.I., Soelaiman, S., Bergson, P., Wang, C.R., Gibbs, C.S. and Tang, W.J. Selective inhibition of anthrax edema factor by adefovir, a drug for chronic hepatitis B virus infection. Proc. Natl. Acad. Sci. USA 2004; 101:3242–3247.

Shlyakhov, E., Rubinstein, E. and Novikov, I. Anthrax post-vaccinal cell-mediated immunity in humans: kinetics pattern. Vaccine 1997; 15:631–636.

Shoop, W.L., Xiong, Y., Wiltsie, J., Woods, A., Guo, J., Pivnichny, J.V., Felcetto, T., Michael, B.F., Bansal, A., Cummings, R.T., Cunningham, B.R., Friedlander, A.M., Douglas, C.M., Patel, S.B., Wisniewski, D., Scapin, G., Salowe, S.P., Zaller, D.M., Chapman, K.T., Scolnick, E.M., Schmatz, D.M., Bartizal, K., MacCoss, M. and Hermes, J.D. Anthrax lethal factor inhibition. Proc. Natl. Acad. Sci. USA 2005; 102:7958–7963.

Singh, Y., Khanna, H., Chopra, A.P. and Mehra, V. A dominant negative mutant of *Bacillus anthracis* protective antigen inhibits anthrax toxin action in vivo. J. Biol. Chem. 2001; 276:22090–22094.

Singh, Y., Leppla, S.H., Bhatnagar, R. and Friedlander, A.M. Internalization and processing of *Bacillus anthracis* lethal toxin by toxin-sensitive and -resistant cells. J. Biol. Chem. 1989; 264:11099–11102.

Spencer, R.C. *Bacillus anthracis*. J. Clin. Pathol. 2003; 56:182–187.

Sternbach, G. The history of anthrax. J. Emerg. Med. 2003; 24:463–467.

Sterne, M. The effects of different carbon dioxide concentrations on the growth of virulent anthrax strains. Pathogenicity and immunity tests on guinea pigs and sheep with anthrax variants derived from virulent strains. Onderstepoort J. Vet. Sci. Anim. Ind. 1937; 12:279–302.

Sterne, M. The use of anthrax vaccines prepared from avirulent (unencapsulated) variants of *Bacillus anthracis*. Onderstepoort J. Vet. Sci. Anim. Ind. 1939; 13:307–312.

Suffin, S.C., Carnes, W.H. and Kaufmann, A.F. Inhalation anthrax in a home craftsman. Hum. Pathol. 1978; 9:594–597.

Turnbull, P.C. Anthrax vaccines: past, present and future. Vaccine 1991; 9:533–539.

Turnbull, P.C., Broster, M.G., Carman, J.A., Manchee, R.J. and Melling, J. Development of antibodies to protective antigen and lethal factor components of anthrax toxin in humans and guinea pigs and their relevance to protective immunity. Infect. Immun. 1986; 52:356–363.

Vitale, G., Bernardi, L., Napolitani, G., Mock, M. and Montecucco, C. Susceptibility of mitogen-activated protein kinase kinase family members to proteolysis by anthrax lethal factor. Biochem. J. 2000; 352(Pt 3):739–745.

Vodkin, M.H. and Leppla, S.H. Cloning of the protective antigen gene of *Bacillus anthracis*. Cell 1983; 34:693–697.

Walsh, J.J., Pesik, N., Quinn, C.P., Urdaneta, V., Dykewicz, C.A., Boyer, A.E., Guarner, J., Wilkins, P., Norville, K.J., Barr, J.R., Zaki, S.R., Patel, J.B., Reagan, S.P., Pirkle, J.L., Treadwell, T.A., Messonnier, N.R., Rotz, L.D., Meyer, R.F. and Stephens, D.S. A case of naturally acquired inhalation anthrax: clinical care and analyses of anti-protective antigen immunoglobulin G and lethal factor. Clin. Infect. Dis. 2007; 44:968–971.

Wang, T.T., Fellows, P.F., Leighton, T.J. and Lucas, A.H. Induction of opsonic antibodies to the gamma-d-glutamic acid capsule of *Bacillus anthracis* by immunization with a synthetic peptide–carrier protein conjugate. FEMS Immunol. Med. Microbiol. 2004; 40:231–237.

Warfel, J.M., Steele, A.D. and D'Agnillo, F. Anthrax lethal toxin induces endothelial barrier dysfunction. Am. J. Pathol. 2005; 166:1871–1881.

Watters, J.W., Dewar, K., Lehoczky, J., Boyartchuk, V. and Dietrich, W.F. Kif1C, a kinesin-like motor protein, mediates mouse macrophage resistance to anthrax lethal factor. Curr. Biol. 2001; 11:1503–1511.

Welkos, S., Little, S., Friedlander, A., Fritz, D. and Fellows, P. The role of antibodies to *Bacillus anthracis* and anthrax toxin components in inhibiting the early stages of infection by anthrax spores. Microbiology 2001; 147:1677–1685.

Welkos, S.L. and Friedlander, A.M. Comparative safety and efficacy against *Bacillus anthracis* of protective antigen and live vaccines in mice. Microb. Pathog. 1988; 5:127–139.

Welkos, S.L., Vietri, N.J. and Gibbs, P.H. Non-toxigenic derivatives of the Ames strain of *Bacillus anthracis* are fully virulent for mice: role of plasmid pX02 and chromosome in strain-dependent virulence. Microb. Pathog. 1993; 14:381–388.

Woods, C.W., Ospanov, K., Myrzabekov, A., Favorov, M., Plikaytis, B. and Ashford, D.A. Risk factors for human anthrax among contacts of anthrax-infected livestock in Kazakhstan. Am. J. Trop. Med. Hyg. 2004; 71:48–52.

World Heath Organization Health Aspects of Chemical and Biological Weapons. Geneva: WHO, 1970.

Wright, G.G., Green, T.W. and Kanode, R.G., Jr. Studies on immunity in anthrax. V. Immunizing activity of alum-precipitated protective antigen. J. Immunol. 1954; 73:387–391.

Wright, G.G., Puziss, M. and Neely, W.B. Studies on immunity in anthrax. IX. Effect of variations in cultural conditions on elaboration of protective antigen by strains of *Bacillus anthracis*. J. Bacteriol. 1962; 83:515–522.

www.clinicaltrials.gov.

www.tdh.state.tx.us./phpep/.

Zaucha, G.M., Pitt, L.M., Estep, J., Ivins, B.E. and Friedlander, A. M. The pathology of experimental anthrax in rabbits exposed by inhalation and subcutaneous inoculation. Arch. Pathol. Lab. Med. 1998; 122:982–992.

CHAPTER

42

Brucellosis

Thomas A. Ficht and L. Garry Adams
College of Veterinary Medicine, Texas A&M University, College Station, TX, USA

OUTLINE

ABSTRACT

Prevention of human brucellosis has relied specifically upon reduction of animal disease, which when coupled with the presence of antibiotic therapy, has been successful in virtually eliminating brucellosis from most developed nations. Consequently, little attention has been paid to the development of human vaccines that would offer the potential for direct intervention in the event of deliberate or natural disease. Furthermore, the animal vaccines employed today are fortuitous isolates attenuated in their ability to cause abortion due to reduced replication in reproductive tissues. Attenuation of these mutants does not extend to the reticuloendothelial disease observed in humans. Lack of genetic definition of these fortuitous isolates also limits their potential use in the development of improved vaccines, due to incomplete understanding of genetic stability, and warrants caution when applied to human use. Given the potential threat *Brucella* spp. pose to public health in both developed and underdeveloped nations, as well as the risk to front line defenders, development of human vaccine remains a critical necessity. The intracellular nature of this organism requires the stimulation of a cell-mediated immunity (Th1) favored by the use of attenuated live vectors capable of stimulating this arm of the immune system. Based on the success of "live" attenuated vaccine strains against brucellosis and other intracellular pathogens, one approach to human vaccine design is the identification of mutants of reduced persistence (survival) within macrophages, a primary target of infection and persistence. Better definition of the role of virulence determinants in survival of the organism and contribution to disease will help to define improved strains with enhanced safety and efficacy. Finally, increased knowledge of key immunogens and their role in survival will help to develop potential subunit vaccines that have provided little protection against infection with fully virulent smooth *Brucella* spp.

INTRODUCTION

Hippocrates is believed to have provided the first description of human brucellosis over two millennia ago (Eyre, 1908). Evidence preserved from the volcanic disruption near the city of Pompeii suggests that 17% of the human remains at this site had evidence of arthritic condition consistent with brucellosis (Capasso, 2002). Furthermore, the remnants of carbonized cheeses revealed forms consistent with both Lactobacillus and *Brucella* species. Luke, the physician, cites circa A.D. 63 in Acts 28:7–10 of the Bible that Paul, the apostle, healed the chief official Publius' father on the Isle of Malta who was suffering with a long-term fever consistent with the classical clinical manifestation of chronic brucellosis. Reference to a similar disease continues in works citing the post-crusades history of military and hospitaler orders of St. John, St. Lazarus, and the temple in Jerusalem. However, reliable accounts of the disease originate during the occupation of Malta by British troops circa 1860 (Marston, 1861). Bruce demonstrated that the etiologic agent of Malta Fever was a bacterium and fulfilled Koch's postulates by reproducing the disease in monkeys (Bruce, 1887, 1889). Ten years later Zammit led the group that was surprised to identify goat's milk as the source of infection rather than an insect vector (Vassallo, 1996).

Brucellosis is endemic in Mediterranean countries, and Central and South America, and is manifested as an undulant fever in humans that, if untreated, can develop into a chronic infection with symptoms persisting for several months or years (Bruce, 1888; Young, 1995). When cultured under laboratory conditions *Brucella* can form infectious aerosols, which have led to a significant number of laboratory-acquired illnesses (Miller et al., 1987; Fiori et al., 2000). In infected hosts, bacteria have been found to localize preferentially to organs that are rich in elements of the reticuloendothelial system (RES), such as liver, spleen, and lymph nodes.

The aim of research on infectious diseases is their prevention, and brucellosis is a classic example. Protection of susceptible hosts by vaccination still remains one of the best methods to accomplish this aim and early last century Bang (1906) reported that infection of cattle with virulent *Brucella abortus* often leads to immunity. Dr. Bang recommended that heifers should be vaccinated with a heavy suspension of cultured *B. abortus* sometime before first breeding. The protection afforded by this method of vaccination was determined by reduction in the rate of abortion rather than by the rate of protection against infection even though this "vaccination" procedure sometimes resulted in disease rather than prevention, but because of these problems, efforts were begun to identify strains of reduced virulence that could be used as live vaccines to provide safe protective immunity in cattle (Cunningham, 1968). Killed vaccines were also proposed as an alternative approach to the problems associated with live vaccines. Because killed *B. abortus* vaccine preparations were not effective in protecting cattle (Bang, 1906), several investigators used various combinations of live field strain isolates, doses, and inoculation routes. Among the numerous live strains developed and evaluated as vaccines are *B. abortus* Strain 19 (Buck, 1930), *B. abortus* strain 45/20 (McEwen and Roberts, 1936), *B. abortus* RB51 (Schurig et al., 1991a), mucoid variants of *Brucella suis* (Huddleston and Bennett, 1948), *Brucella melitensis* strain Rev 1 (Elberg and Faunce, 1957), and *Brucella suis* strain 2 (Bosseray and Plommet, 1990) of which only three, strain 19, RB51, and Rev 1, are currently used as live vaccines in ruminants but none are safe or efficacious for humans. Vaccination against most bacterial diseases, including brucellosis, has considerable room for improvement, since most vaccines have been used continuously for decades with little major improvements and which is the subject of this chapter. To provide ideal protection against brucellosis, the vaccines should:

1. prevent clinical disease manifestations,
2. prevent infection in both sexes at any age,
3. not produce disease in vaccinates,
4. provide long-term protection against clinical disease with a single vaccination,
5. possess Differentiating Vaccinated from Infected Animal (DIVA) or person properties,
6. not be transmitted to other individuals,
7. be biologically stable, with no revertants under in vitro and in vivo conditions, and
8. be readily grown in large-scale fermentation technology for low cost production.

ETIOLOGIC AGENT

Brucellae are small aerobes nonmotil Gram-negative coccobacilli that are facultative intracellular pathogen (Fig. 42.1). Over the past century, several species designations have been applied to isolates with apparent preference for different animal hosts. *B. melitensis* in sheep and goats, *B. abortus* in cattle and *B. suis* in swine are all associated with human disease (Bang, 1906). The close association of "nomen"-species with a particular host is thought to indicate specific adaptation or perhaps co-evolution (Chain et al., 2005). Two naturally occurring rough species *B. canis* and *B. ovis* cause abortion in dogs and epididimytis in rams (males of the species), respectively, but are rarely associated with human infection (Carmichael and Bruner, 1968; Wallach et al., 2004). *B. neotomae* was identified in the desert wood rat (*Neotoma lepida*) (Stoenner and Lackman, 1957), and possibly as many as three species are associated with marine mammals including dolphins, porpoises, and pinnepeds (Groussaud et al., 2007). Although the latter have not been widely observed, accidental exposures in humans and experimental exposures in animals have confirmed their pathogenic behavior (Sohn et al., 2003; McDonald et al., 2006).

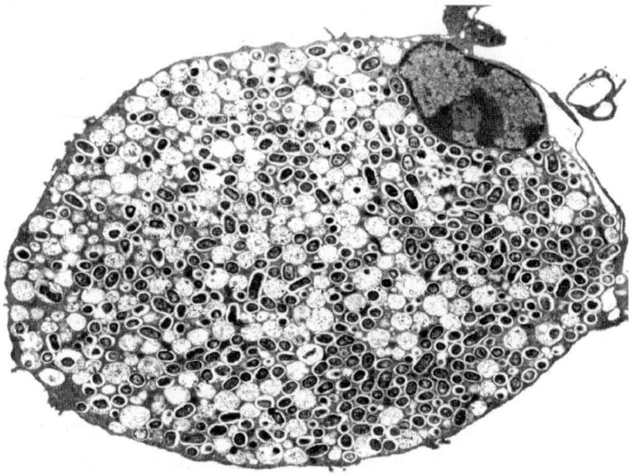

FIGURE 42.1 Transmission electron micrograph of intracellular *Brucella abortus* Strain 2308 undergoing massive replication within a primary cultured bovine macrophage.

Genus

Brucella spp. fall within the group 2 subdivision of the α-proteobactereaceae along with the plant pathogens rhizobia and agrobacteria. Features of *Brucella* that distinguish it from most genera within the order rhizobiales are the infection of mammalian cells, a feature *Brucella* share only with *Bartonella*. The second shared feature is a streamlined genome, at least in relation to the plant pathogens. However, despite these similarities, major differences exist between *Bartonella*, an obligate intracellular pathogen and *Brucella*, a facultative intracellular pathogen. First, the genome of *Brucella* spp. is 50–100% larger than *Bartonella* spp. genomes and has preserved more of the metabolic functions shared by the plant pathogens. Correspondence between genome size and host range is also observed when comparing the host range of *B. quintana* with *B. henselae* (Alsmark et al., 2004). Extended metabolic function is therefore suspected to be responsible for the broad host range of *Brucella*, but is also capable of enhancing survival in the environment.

Differences between host species may necessitate differences in cell surface structures (cell wall) and optimal growth conditions as well as specialized mechanisms for uptake and intracellular growth of mammalian pathogens. As a result, the ability to invade mammalian hosts is a feature that may have been acquired at least in part by both *Bartonella* and *Brucella* and may be expected to exhibit nucleotide composition that is distinct from genes conserved from progenitor organisms. Several candidates exist to fulfill this role including genes encoding biosynthesis of polysaccharides, secretion systems, adhesions, and invasions (DelVecchio et al., 2002; Paulsen et al., 2002; Halling et al., 2005). However, it is possible that genes involved in uptake or invasion of mammalian cells were also present in progenitor organisms, and lost from the plant pathogens. In this case the genes would not exhibit distinctive nucleotide compositions, and would require more direct approaches for identification.

Species

Speculation concerning the origin of *Brucella* species has focused on apparent adaptation to specific hosts. Co-evolution of *Brucella* species with their preferred hosts is an alternate possibility that is consistent with currently observed host preferences. Yet, this simple interpretation appears to ignore discrepancies in the overall genetic variation observed between host species and the limited variation observed between *Brucella* spp. Obviously, host and agent do not necessarily evolve at the same rate, but the overall similarities observed among host-adapted *Brucella* species either argues for restricted genetic flexibility or recent adaptation with only minor changes. Since in most cases the timing of separation of these different host species is not known with certainty, it is difficult to draw solid conclusions concerning the co-evolution of host and agent. However, the timing of some events is more widely accepted and may be used as reference points for the discussion presented. As a start, adaptation to individual hosts may be assumed to coincide with the timing of divergence from the main mammalian branch.

Host species primarily affected by *Brucella* belong to the order *Artiodactyla* in the ungulate family that exhibited its greatest divergence during the Miocene epoch. The origin of these lines is complex and still a matter of contention, but for this discussion the results of Blair Hedges and Kumar (2003) that takes into account multiple genes has been employed. Based on this analysis, the separation of cattle from sheep and goats occurred approximately 19.6 million years ago. In contrast, the separation of cattle from swine occurred 65 million years ago. The separation of artiodactyls from the carnivores including cats and dogs occurred 82 million years ago, and pinnepeds are derived from the carnivore line. Separation of rodents was not evaluated in this study due to wide variation in results, however a safe estimate places their separation at more than 100 million years ago while lagomorphs divergence is reported at 90 million years. Cetaceans were not included in the study by Blair Hedges and Kumar, but comparison of the gene encoding artiodactyl growth hormone suggests separation approximately 40–50 million years ago (Wallis et al., 2005). Interestingly, data now suggest a divergence of *Brucella* spp. affecting dolphin and porpoise that diverged from each other more than 11 million years ago.

The timeline described above makes certain simple predictions with regard to the potential co-evolution of *Brucella* organisms and their hosts. The separation between organisms co-evolving with a given species may be expected to reflect the distances suggested by the separation between host species. This may be an unsupported assumption given the differences in growth rates and generation times of different host species, and the assumption that genetic changes take place in each species at the same rate. Furthermore, it must be realized that there is yet no real agreement among scientists concerning the exact times of mammalian divergence. Genetic analysis and the study of host–pathogen interactions would be required to confirm or refute the following predictions.

The first prediction is that *B. abortus* and *B. melitensis* should exhibit fewer differences between each other than they do with *B. suis*. Similarly, *B. ovis* with specificity for ovines should be more closely related to *B. abortus* and *B. melitensis* than to *B. suis*. *B. neotomae* and *B. canis* may be expected to be the two most widely diverged species assuming adaptation was initiated at the time their hosts diverged from the artiodactyls, 100 and 82 million years, respectively. Additionally, the ability of *B. melitensis* to infect both bovine and ovine/caprine species may reflect the recently shared common ancestry of these hosts or a conservation of function that provides a broader specificity. Broad specificity is a property that is especially notable in *B. suis*, which infects a range of species including lagomorphs and artiodactyls. In contrast, *B. abortus* exhibits a much more narrow host range that may reflect specific adaptation and a concomitant loss of gene function. Finally, based on the earlier assumptions cetacean isolates should have more in common with *B. suis*/*B. abortus* or *B. melitensis*, while the isolates from the pinnepeds should have more in common with *B. canis*.

Based on overall sequence analysis of the genomes of *B. melitensis*, *B. abortus*, and *B. suis*, it is clear that *B. melitensis* and *B. abortus* are more closely related (Halling et al., 2005). A finding that is consistent with their hosts diverging a mere 20 million years ago. In support of this difference 7208 SNP mutations distinguish *B. abortus* from *B. suis* and 7844 distinguish *B. melitensis* from *B. suis*, while only 6242 distinguish *B. abortus* from *B. melitensis* (Halling et al., 2005). The fact that *B. abortus* shares 6 of 7 *B. melitensis* specific ORFs (nonphage related) while also sharing 12 of 14 *B. suis*-specific ORFs is surprising given that *B. abortus* and *B. melitensis* should have co-evolved for 25 million years after diverging from *B. suis*. However, it is possible that a specific loss of *B. suis*-specific segments from *B. melitensis* occurred following the divergence of sheep/goats and cattle over the last 20 million years. Such an outcome might be argued to predict broader host specificity for *B. abortus* than for *B. melitensis*, currently a point that is debatable. Final resolution of these issues can only be satisfied by analysis of the individual genes and their role in the survival of each species.

Of specific importance for this chapter is whether vaccination specific to each of these organisms will be necessary or more broadly based protection is possible. Although their close genetic similarity provides some promise that genus-wide protection may be readily obtained, adaptation to specific hosts could belie this simple assumption. Studies on cross-immunity with live virulent homologous and heterologous *Brucella* spp. and attenuated vaccines may also provide evidence to support approaches for developing improved live *Brucella* vaccines. Washko and Hutchings (1948) reported that cows fed *B. suis* failed to be protected against *B. abortus* challenge nor did Strain 19 protect against *B. suis* challenge in cattle or swine. *B. melitensis* rev. I protected cattle against *B. abortus* (van Drimmeln and Howell, 1964) with no evidence of rev. I shedding and goats were protected against *B. melitensis* challenge, but *B. abortus* Strain 19 was much less protective against *B. melitensis* challenge in goats. These data suggest that homologous *Brucella* spp. are usually more protective than heterologous *Brucella* spp. within a given livestock species, but in some cases heterologous *Brucella* spp. may induce protective immunity similar to the homologous *Brucella* spp. With the exception of swine, the general hierarchy of immunizing capacity of either vaccine or field strains of *Brucella* spp. for cattle, goats, and sheep is proposed to be *B. melitensis* ≥ *B. suis* > *B. abortus* (Adams, 1990). Yet, in this regard, vaccination with *B. abortus* and *B. melitensis* have been shown to provide significant cross protection despite significant differences in clearance (Kahl-McDonagh et al., 2007).

Antigens Encoded by the Agent

A recent study evaluating protein antigens recognized by the human humoral immune response included 42 immunogenic spots (Connolly et al., 2006). Of the 42, 23 were also recognized using bovine sera from an exposed animal. Although there was no attempt to distinguish between protected and unprotected or recovering patients, previous identification using two-dimensional approaches have met with limited success identifying candidate humoral or cell-mediated immunogens (Sowa et al., 1991, 1992). Among the proteins identified were GroEL, GroES, OMP25, Omp2b porin, OMP31 and the periplasmic immunogenic proteins. Several cytoplasmic enzymes (fumarate reductase flavoprotein subunit, F0F1-type ATP synthase α-subunit) were also identified, as were the ribosomal proteins (SSU S1P, L9, L13P).

Proteins encoded in the genome of *Brucella* and recognized as immunogens include subunit and killed vaccine candidates, which have proven to be effective in animals against natural rough strains *B. ovis* and *B. canis* (Bowden et al., 2000). However, subunit and killed vaccine approaches have shown little promise against "classic" smooth organisms. Despite the use of adjuvants, conjugation to carriers, immunization (+/−cytokines), or alternate routes of inoculation, attempts to develop subunit vaccines have met with limited success (Winter and Rowe, 1988; Bachrach et al., 1994; Golding et al., 1994; Vemulapalli et al., 1998, 2000a; Bae and Toth, 2000; Baloglu et al., 2000). The intracellular nature of this organism requires the stimulation of a cell-mediated immunity (T helper type 1, Th1) favored by the use of attenuated live vectors capable of stimulating this arm of the immune system. For this reason, *Brucella*, like *M. bovis* BCG, has been suggested as a vector for the delivery of immunogens of other intracellular agents (Vemulapalli et al., 2000b).

IMMUNE RESPONSE

The mouse model has been used to dissect the role of several components of innate and acquired immunity in control of *Brucella* infection, and to identify *Brucella* genes that contribute to the infection process. While many genes that are required for infection and persistence in the RES have been identified using this model, little is known about the *Brucella* gene products that participate in traversing mucosal surfaces.

Studies in mice and in macrophage cell lines have contributed to our current understanding of the host response to *Brucella* infection by identifying a few host gene expression patterns. The cytokines tumor necrosis factor alpha (TNF-α), interleukin-12 (IL-12), and interferon-gamma (IFN-γ) were found to be produced by murine cells during infection and to contribute to control of intracellular *B. abortus* growth (Jiang and Baldwin, 1993; Zhan et al., 1993b). In addition, mouse macrophages and BALB/c mice were shown to produce IL-1, IL-6, and TNF-α, whereas the human monocyte/macrophage line THP-1 expressed transcripts for IL-1, IL-6, and IL-8, but not TNF-α (Dornand et al., 2002). A recent study by Splitter and colleagues used mouse microarrays to study the response of RAW264.7 macrophages to *B. abortus* infection (Eskra et al., 2003). These investigators found that *B. abortus* infection induced expression of genes associated with inflammation, such as IL-1α, the chemokines MCP-1 and MIP-2, and TNF-α. Genes whose expression was reduced after infection, included genes related to apoptosis, vesicular trafficking, and the cell cycle (Eskra et al., 2003). While the responses of macrophage cell lines and mice to *Brucella* infection have been examined, no studies have focused on the response of mucosal surfaces to *Brucella* infection. This response is likely to be important, given that the natural route of infection of *Brucella* occurs through mucosal surfaces.

It is well established that *Brucella*-specific IgG levels become elevated during the early stages of infection in animals and humans, but there is no correlation between the appearance of these serologic responses and resolution of the infection. For *Brucella*, cellular immunity is understood to be critical for resolution of primary infection and for establishment of immunity to reinfection (Yingst and Hoover, 2003).

Considerable amounts of experimental evidence suggest that opsonization plays an important role in the early stages of *Brucella* infections (Gross et al., 1998; Eze et al., 2000). Although contact of *Brucella* with specific immunoglobulin G (IgG) enhances the brucellacidal activity of cultured macrophages, virulent strains of *B. abortus* can still resist killing by phagocytes and eventually demonstrate intracellular replication (Eze et al., 2000). Previous studies demonstrate that primary murine macrophages have limited ability to control infection with *B. melitensis*, even when activated by IFN-γ in the presence of highly opsonic concentrations of antibody and complement. The role of complement has been studied with *B. abortus* and *B. melitensis* strains, and it has been proposed that rough *Brucella* are less virulent due to their increased sensitivity to complement-mediated killing (Godfroid et al., 1998; Ugalde et al., 2000). This idea is supported by previous reports which demonstrate that disruption of the *wboA* gene in *B. abortus* and *B. melitensis* (which results in rough mutants) can bind more complement than their respective wild-type parental strains 2308 and 16M (Fernandez-Prada et al., 2001).

Neutrophils are probably the first immune cell to encounter *Brucella*. Rapid phagocytosis of *Brucella* strains by neutrophils occurs only after opsonization (Young et al., 1985). However, bacterial survival in neutrophils during early infection has been observed, suggesting that the transportation of *Brucella* to lymphoid tissues can be mediated by these cells. Finally, live virulent *Brucella* strains, but not cell wall fractions, inhibit the respiratory burst by neutrophils (Canning et al., 1985).

NK cells play a key role in host defense. They are implicated in an early immune response to a variety of pathogens. However, it has been shown that they do not control *Brucella* infection in mice (Fernandes et al., 1995). On the other hand, it was formally established that in the acute phase of the illness, the activity of the NK cells is impaired in humans developing brucellosis (Salmeron et al., 1992). A more recent study analyzed the behavior of *B. suis* in human macrophages infected in the presence of NK cells. The results suggest that NK cells are activated by *B. suis*-infected macrophages and that they inhibit the intracellular multiplication of the bacteria by lysing the infected cells, thus suggesting that NK cells could be one actor of the control of *Brucella* development in humans but not in mice (Dornand et al., 2004).

The role of dendritic cells in brucellosis is still uncertain. One report suggests that *B. abortus*, *B. melitensis*, and *B. suis* are capable of replicating in similar numbers inside human monocyte derived dendritic cells compared to macrophages. These data suggest that the bacteria may have a marked preference for DCs and that this property could support *Brucella* spreading (Billard et al., 2005).

T cells are the key players against *Brucella* infections. The major role of T cells in *Brucella* immunity can be divided into three principal functions: secretion of IFN-γ for the activation of bactericidal function in macrophages, cytotoxic T-lymphocyte activity, and contribution in IgG2a and IgG3 isotype switching (Ko and Splitter, 2003). It has been clearly demonstrated that a Th1 response is essential for resolution of the primary infection caused by *Brucella*, and the essential aspect of this response appears to be IFN-γ production (Mielke et al., 1988). The role of IFN-γ in the control of *Brucella* infections was initially demonstrated by supplementing BALB/c mice with recombinant IFN-γ. The treatment resulted in a 10-fold decrease in the number of bacteria 1 week after infection (Stevens et al., 1992). In accordance with these observations, depletion of IFN-γ by the administration of monoclonal antibodies which neutralize IFN-γ, resulted in the increase in the bacterial burden 1 week postinfection (Zhan and Cheers, 1993; Zhan et al., 1993a). The importance of IFN-γ in resolution of *Brucella* infection was recently supported by studies using BALB/c and C57BL/6 mice. C57BL/6 mice with gene deletions or disruptions in the IFN-γ, perforin or beta (2)-microglobulin genes had a decreased ability to control intracellular infections with *B. abortus* strain 2308 during the first week postinfection. However, only the IFN-γ knockout mice had a sustained inability to control the infection and this resulted in death of the mice at approximately 6 weeks postinfection (Murphy et al., 2001). In vitro studies have suggested that the mechanism by which IFN-γ enhances resistance is through activation of macrophages for anti-*Brucella* activity, largely by promoting production of reactive oxygen intermediates (Jiang et al., 1993).

The hallmarks of human brucellosis are measurable splenomegaly with increased lympho-histiocytic cells in the spleen, a reduced percentage of splenic $CD4^+$ and $CD8^+$ T cells, and increased percentage of splenic macrophages. Splenocytes in the infected host express more mRNA for IL-2, IFN-γ, and IL-10, and less mRNA for IL-4 than uninfected suggesting a Th1 response (Baldwin et al., 1993; Zaitseva et al., 1995; Liautard et al., 1996). However, increases in IL-10 are counterproductive and may explain the virulence of this organism and the failure of some vaccines to stimulate a protective immune response (Svetic et al., 1993; Hoover et al., 1999).

Reports to date have provided clear evidence that both humoral (Plommet and Plommet, 1983; Montaraz and Winter, 1986; Montaraz et al., 1986; Winter et al., 1989) and cell-mediated immune (CMI) responses (Pavlov et al., 1982; Montaraz and Winter, 1986; Plommet and Plommet, 1987; Bosseray and Plommet, 1988; Smith et al., 1990a, 1990b) participate in protection against brucellosis. The intracellular location of *Brucella* requires the stimulation of a cell-mediated immunity (Th1) for clearance favored by the use of attenuated live vectors potent in the stimulation of this arm of the immune system. Studies demonstrate that protective immunity against brucellosis in mice is due to the combined effects of antibodies and the CMI responses mediated by the CD4 and CD8 subsets of T cells (Araya et al., 1989). In comparative analyses of infections that were performed, the virulent strain 2308 was significantly more refractory to opsonization by antibodies than the attenuated Strain 19 (Araya et al., 1989). Splenic infection by both strains caused acute as well as chronic granulomatous responses, and a severe depletion of lymphoid tissue that was most prominent 3 weeks postinfection (Araya et al., 1989; Enright et al., 1990). Thereafter, a rapid decline in bacterial numbers was observed in mice infected with strain 19, along with a decrease in the numbers of granulomas in the liver and spleen (Enright et al., 1990). In contrast, infections performed with S2308 did not result in a decrease in bacterial counts, and granulomas persisted (Enright et al., 1990).

EPIDEMIOLOGY

Human infection results from the consumption of contaminated animal products or direct exposure to aborted tissue. As a result, the prevention of human brucellosis has focused upon the reduction of animal disease, and little attention has been paid to the development of human vaccines. In the chronic form of the disease observed in humans, organisms persist in macrophage and primarily affect the lymphoreticular system (Jones and Winter, 1992). Humans are normally dead-end hosts, although anecdotal evidence suggests sexual transmission may be possible (Stantic-Pavlinic et al., 1983). Intracellular persistence within the host is the basis for human disease and the cause of reduced efficacy of antibiotic treatment.

Acute brucellosis is only observed in pregnant ruminants and is characterized by fetal/placental colonization and abortion, resulting from organism replication to high density within reproductive tissues. Ruminants that have aborted frequently become chronic shedders of the organism in the milk due to colonization of the mammary gland, a primary source of transmission. Experimental results also suggest that approximately 12% of the animals that abort due to *Brucella* infection may abort a second and sometimes a third time without detection of the organism between pregnancies, thus encouraging immediate removal from the herd (ter Huurne et al., 1993). Over the past century, several species designations have been applied to isolates originating from different hosts (Bang, 1906). Three classical species, *B. melitensis*, *B. abortus*, and *B. suis* are associated with reproductive disease (abortion) in their primary hosts, sheep/goats, cattle, and pigs, respectively. Each presents an intact O-antigen (*n*-formyl perosamine homopolymer) on their outer surface. Rough derivatives of these organisms arise spontaneously, but are typically avirulent (Schurig et al., 1991b). However, two naturally occurring rough species, *B. canis* causes abortion in dogs, and *B. ovis*, unique among *Brucella*, causes epididimytis in rams (males of the species). Serological tests indicate *B. canis* exposure occurs in 10% of all dogs in the U.S. *B. neotomae*, another identified species, was isolated only once from the desert wood rat (*N. lepida*). Over the last 10 years, there have been numerous isolations of *Brucella* organisms from marine mammals (Ewalt et al., 1994; Garner et al., 1997; Miller et al., 1998). The atypical phenotype of these isolates led to their identification as new species. However, it must be emphasized that the brucellae are considered to be a monospecific genus (Verger et al., 1985). Although certain phenotypic properties have provided a useful system for epidemiological classification of isolates their long-term stability has been questioned (Moreira-Jacob, 1963; Banai et al., 1990; Banai et al., 1991).

Of course, the diversity among these organisms and the close association of "nomen"-species with a particular host may also suggest host-specific adaptation. Although *B. melitensis* primarily affects sheep and goats, cattle infection in endemic areas has been observed (Dafni et al., 1990). In many cases, however, such infections appear to be self-limiting, and are assumed exceptions rather than the rule. The overall similarities in the DNAs of this genus suggest that host-specific adaptations, if they exist, may only be defined by precise sequence comparison. Curiously, Bruce suggested that nothing similar to "Malta Fever" was observed in England (Bruce, 1887). Whether this was an oversight, the result of the introduction of pasteurization in 1864, or reflected differences in human virulence potential of *B. abortus* (cattle) and *B. melitensis* (sheep and goats) as reported for guinea pigs cannot be determined (Keppie et al., 1963). It is clear, however, that the organism was endemic in the United States by 1906 and is presumed to originate from cattle imported by settlers (Bang, 1906).

Significance as a Public Health Threat

An ancient disease, human brucellosis, may be the most common worldwide zoonotic bacterial infection (Fig. 42.2) (Young, 1995; Corbel, 1997; Pappas et al., 2006b), and even after over 100 years of research remains a serious challenge in the 21st century (Ariza et al., 2007). *B. melitensis* continues as a major cause of human disease worldwide, followed by *B. abortus* and *B. suis*, while rare persistent cases of *B. canis* human infection and disease by novel *Brucella* pathogens of marine mammals have also emerged. Brucellosis is re-emerging as a significant cause of travel-related disease (Memish and Balkhy, 2004).

B. melitensis represents a serious threat to worldwide animal production and public health in the Middle East, Latin America, and parts of Asia in which the organism is endemic in sheep- and/or goatherds. Humans are infected through consumption of contaminated animal products or direct exposure to aborted tissue, and are normally dead-end hosts.

The main species affecting the U.S. is *B. abortus* in wildlife reservoirs, including bison and elk in the Greater Yellowstone Area (GYA), where it has been estimated that up to 50% of the 4000 bison in the GYA and even higher numbers of elk have been exposed or are infected with *B. abortus*. Ruminant transmission of *Brucella* occurs as a result of abortion of the fetus and/or shedding in milk. Uncontrolled growth in the reproductive organs of pregnant animals is responsible and may be related to immune suppression in these tissues (Weinberg, 1987; Smith, 1994). Persistence within the host cells (including macrophages) results in repeated abortions, between periods of apparent latency (Ray et al., 1988; ter Huurne et al., 1993). Males and non pregnant females are less susceptible. The best approach to minimize animal disease is to prevent transmission via contact with aborted material or consumption of contaminated milk. Inhibiting replication in reproductive tissues prevents the release of massive numbers of organisms into the environment. For example, S19 does not replicate in the pregnant uterus due perhaps in part to erythritol present in elevated amounts in

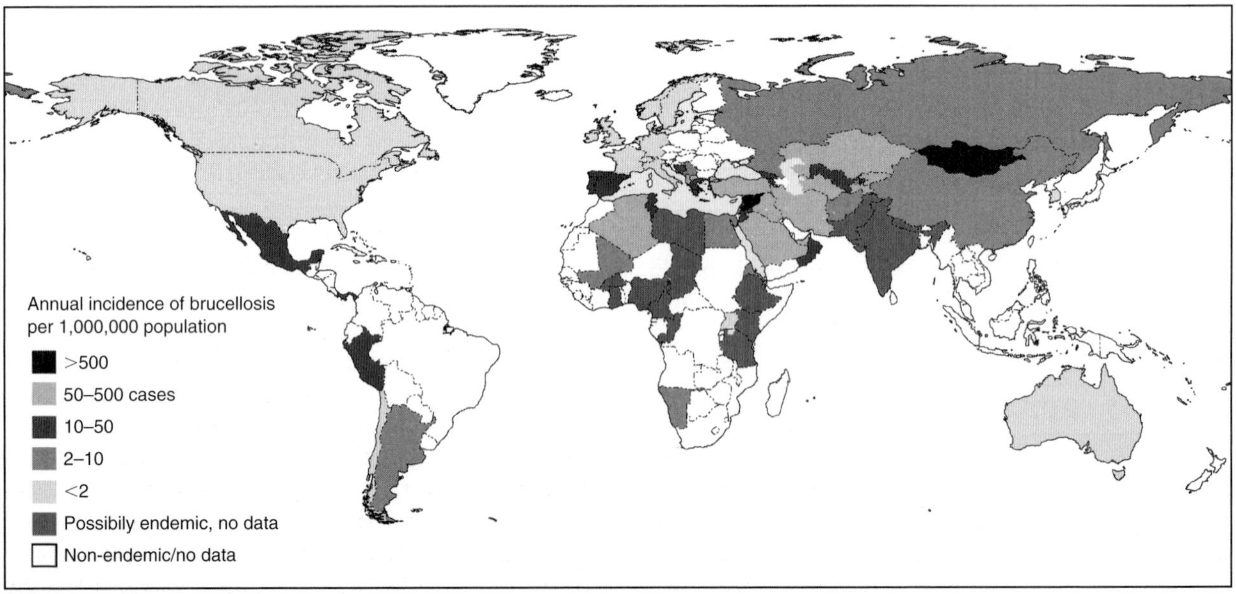

FIGURE 42.2 Worldwide distribution of human brucellosis. From Pappas et al. (2006b).

the pregnant uterus, although other defects may exist (Williams et al., 1962). Although a preferred carbon source in most *Brucella* sp., erythritol is toxic to S19 due to a gene deletion causing a toxic buildup of metabolic intermediate within the organism (Sangari et al., 1998). *B. melitensis* Rev1 a streptomycin-independent organism isolated following back-selection from a streptomycin-dependent organism is used to protect against *B. melitensis* infections worldwide (Elberg and Faunce, 1957). Both S19 and Rev1 provide protection in ruminants despite persistence in a percentage of vaccinates (Corner and Alton, 1981), but retain significant virulence in humans (Spink and Thompson, 1953; Bardenwerper, 1954; Spink et al., 1962). There are no reports of virulence in humans for the recently approved vaccine strain RB51, a rough derivative used with some success in ruminants (Schurig et al., 1991b), but with less success in other species (Jimenez de Bagues et al., 1995; Palmer and Cheville, 1997). The absence of O-antigen prevents development of cross-reacting immune responses that are indistinguishable from wild-type infection, but also increases the susceptibility of this strain to complement-mediated lysis, limiting survival and vaccine efficacy.

Potential as Biothreat Agent

Brucella, as Category B pathogens in the United States, have been the subject of offensive weaponization research in the recent past, because they can be readily aerosolized resulting in a high rate of morbidity with low mortality causing a prolonged incapacitating clinical disease requiring expensive long-term treatment (Pappas et al., 2006a). Very importantly, no human vaccine is currently available against this biothreat agent. If *Brucella* vaccines are developed and proven to be safe, and protect against aerosol and oral challenge in laboratory animals and nonhuman primates, these vaccines would be a deterrent to these biothreat agents. According to one study, the use of *Brucella* as a weapon was calculated to pose a substantial financial risk (less than anthrax or tularemia, but significant) (Kaufmann et al., 1997). *Brucella* organisms can be delivered via aerosol to infect humans (Kaufmann et al., 1980) with an infectious dose of fewer than 10–100 organisms (1000 cfu infectious dose assumed in economic impact study). Infection incapacitates human hosts with mostly flu-like symptoms, but will result in death if left untreated (Young, 1995).

The financial impact study did not attempt to determine the threshold at which financial risk may pose a risk to national security. Nor did the study outline scenarios in which the use of one organism might be favored over the use of others. The study did underscore the need to invest in research in all understudied organisms to prevent their use in this manner and suggested that decreased study of these organisms increase the potential consequences resulting from their use as weapons. Bioengineering poses the additional risk of introducing antibiotic resistance, rendering ineffective the most successful form of treatment.

CLINICAL DISEASE

Nonspecific, but characteristic symptoms of human brucellosis include pyrexia, diaphoresis, fatigue, loss of appetite, and muscle or joint pain. Depression, cephalalgia, and irritability also frequently occur. Infection of bones or joints occurs in about one in three patients, causing localized inflammation and edema. Some may also have cough, chest pain, and stomach upset. Although about one in four patients with brucellosis have respiratory symptoms, thoracic radiographs usually appear normal (Lubani et al., 1989; Madkour, 1989). Symptoms have also been described to recur years after the original infection. Chronic illness may result in infection of secondary tissues, including heart and brain, if the disease is left untreated. Brucellosis is most frequently acquired by man through contact with infected livestock and consumption of unpasteurized dairy products. Brucellosis in humans is often difficult to recognize and may present as an acute fever, or as a chronic or localized infection that is best diagnosed by culturing the bacterium from blood or other infected tissues. Due to the slow growth of *Brucella*, cultures may require several weeks for positive identification. Infection may also be diagnosed indirectly via detection of anti-*Brucella*-specific antibodies in patients' blood. Disease may begin abruptly or gradually from 3 days to several months after exposure. Less common sites of infection are the male reproductive system, brain, and heart valves. Symptoms often last for 3–6 months, but occasionally persist for a year or longer. Patients infected for a long time frequently experience weight loss, and many patients temporarily improve, then relapse (Ariza et al., 1986). Recurrence of human disease may be related to a latent form of survival demonstrated to occur at low frequency in at least one study (Ray et al., 1988; ter Huurne et al., 1993).

While brucellosis, like typhoid fever, is a nonenteric disease in which systemic illness is more common than localized disease, there have been reports from human cases of patchy mucosal hyperemia within the digestive tract, ulcerations of Peyer's patches, and acute ileitis (Petrella and Young, 1988; Sohn et al., 2003; McDonald et al., 2006). Furthermore, an association has been reported between ingestion of *Brucella* and specific digestive system complaints, including diarrhea and abdominal pain (Thapar and Young, 1986). Hence, whether it causes pathology in a particular case, *B. abortus* must pass the mucosal barrier of the gastrointestinal tract to cause systemic infection and target to the RES. Currently, very little is known about this process. A study in calves suggested that *B. abortus* is able to enter the lamina propria by transepithelial migration via lymphoepithelial cells (analogous to M cells) of the ileal Peyer's patches (Ackermann et al., 1988), a portal of entry shared by other pathogens, such as *Salmonella typhimurium* and *Mycobacterium avium* subsp. *paratuberculosis* (Santos et al., 2002). Although reports in the literature continue to suggest a link between brucellosis and abortion in women, direct isolation of the organism from human fetuses is seldom reported (Khan et al., 2001). Changes in hormone profiles are thought to contribute to these different outcomes whether through stress or suppressed immune function in the host or changes to the organism (Bienvenu and Young, 1962; Meyer, 1976; Ramirez-Romero, 1998).

In its natural bovine host, *B. abortus* infection is transmitted via the alimentary tract. *B. abortus* infection in the gravid cow causes late gestational abortion or perinatal death, and licking of aborted fetuses and placental tissue, or ingestion of organisms in the milk by calves result in contact of *B. abortus* with the mucosa of the gastrointestinal tract. The disease caused by this group of organisms may be divided into two types based on the course of disease and symptomology. Acute brucellosis observed in ruminants is characterized by fetal/placental colonization and abortion, resulting from replication to high density within reproductive tissues. Ruminants that have aborted frequently become chronic shedders of the organism in the milk due to colonization of the mammary gland. This form of the disease is not observed in humans. Chronic brucellosis is characterized by infection of the lymphatics and RES and is observed in nonpregnant ruminants, mice, and humans. Differences in the two forms of disease are attributable to the target cells infected. Trophoblasts are preferentially colonized in acute disease while the organism persists in macrophages during chronic disease (Jones and Winter, 1992).

TREATMENT

Treatment of human brucellosis remains complex, requiring protracted administration of more than one antibiotic which is mostly based on the principles applied half a century ago with few modifications being made even today despite the emergence of new antibiotic classes and different therapeutic approaches (Ariza et al., 2007). The Ioannina Recommendations panel suggested that the optimal treatment of uncomplicated brucellosis should be based on a 6-week regimen of doxycycline combined either with streptomycin for 2–3 weeks, or rifampicin for 6 weeks (Castillo et al., 1989; Ariza et al., 2007). Ruminants and mice will usually clear such infection, but humans require treatment with antibiotics to prevent sequelae that may include neurological and cardiac involvement (Young, 1995). Two recent studies have shown that the doxycycline–rifampin regimen is preferred by both clinicians and patients (Pappas et al., 2006c, 2007), even when they are aware of the relative superiority of the alternative regimen. The fact that doxycycline–rifampin is an all-oral regimen may allow for better implementation in clinical practice in areas with less well-developed health infrastructure, because it eliminates the need for parenteral administration of streptomycin as part of the doxycycline–streptomycin regimen. The cure is neither certain nor rapid, since the intracellular location reduces antibiotic efficacy and some affected individuals may require re-treatment if symptoms return.

PATHOGENESIS

Brucella spp. infect hosts principally by penetrating the natural mucosa from which it disseminates via bacteremia throughout the body particularly the RES and reproductive systems (Fig. 42.3). An experimental infection found that the oral cavity-pharynx and the intestine were the two main portals of entry for oral infection of *B. abortus* in calves (Carpenter, 1924). Conjunctivae, intranasal mucosa, and vagina were also permeable to *Brucella* infection (Plant et al., 1986; Meador and Deyoe, 1989; Mense et al., 2001). The evaluation of the invasion process of *B. abortus* through calf intestinal Peyer's patches revealed that lymphoepithelial, but not enteroabsorptive cells, engulfed individual brucellae that remain inside vesicles, phagolysosomes, and large vacuoles (Ackermann et al., 1988). After transepithelial migration, many brucellae were observed alive inside neutrophils, mononuclear phagocytes, or free in the interstitium and lymphatic vessels of lamina propria. The invasion process via conjunctiva was accompanied by a submucosal inflammatory reaction with the bacterium consistently cultured from the parotid lymph node 2–4 days postinoculation (Enright, 1990). These observations suggest that *Brucella* invade the host by

Attachment and Entry: Uterine exudates, fetal membranes, and fluids or semen containing massive numbers of *Brucella* spp. facilitate primary invasion of oronasopharynx and alimentary tract, genitalia, conjunctivae, and abraded or unbroken skin of susceptible hosts, also in utero exposure and infection of fetus

↓

Cellular Uptake: Intracellular invasion of macrophages, neutrophils, and/or Langerhan's and M cells to enter lymphatics draining into regional lymph nodes

↓

Lymphoreticular Colonization: Intracellular growth in reticuloendothelial cells with resulting regional lymphadenopathy

↓

Episodic Bacteremia: Migration into blood by infected neutrophils and macrophages to reach threshold levels for colonization other tissues or organs to establish a generalized systemic infection

↓

Target Tissue/Organ Colonization: Disseminated tropism for placental erythrophagocytic trophoblasts, chorionic trophoblasts, mammary macrophages, fetal lung and abomasum, macrophages of regional lymph nodes of mammary gland and reproductive tract, joints, and bursae for chronic intracellular survival and growth

↓

Pathologic Lesion Development: Regional lymphadenitis, necrotizing fibrinopurulent placentitis, lymphoplasmacytic endometritis, lymphoplasmacytic mastitis, pyogranulomatous epididymitis, necrotizing orchitis, seminal vesiculitis, ampullitis, bursitis

↓

Clinical Disease: Abortion, orchitis, epididymitis, mastitis, infertility, arthritis, fever

↓

Excretion: Uterine exudates, fetal membranes and fluids, milk, semen, urine

FIGURE 42.3 Composite schema of *Brucella* attachment, invasion, metastasis, dissemination, systemic disease, clinical disease, and excretion.

transepithelial migration, then escape into the submucosa and disseminate free or inside phagocytic cells to regional lymph nodes by lymphatic drainage.

In addition to in vivo studies required for full understanding of *Brucella* pathogenesis, in vitro assays performed on cell lines or primary cultures are informative for a more complete understanding of the host–agent relationship. The bacterial adhesion to cell membrane is an essential first step in the establishment of infection (Salyers and Whit, 2002). *Brucella* attach as a single organism to cultured epithelial cell lines via receptor molecules containing sialic acid or sulfated residues, and at 8h postinfection begin to form microcolonies which subsequently become larger with extended time (Castaneda-Roldan et al., 2004). It has been demonstrated that rough strains are more adherent and invasive to nonphagocytic cells than smooth ones (Detilleux et al., 1990), in part because the O-polysaccharide structure of the *Brucella* outer membrane protein obscures nonspecific ligands and consequently decreases the ability of the bacteria to bind the cell surface. Within a few minutes after binding nonprofessional phagocytic cells, *Brucella* are internalized by receptor-mediated phagocytosis (Detilleux et al., 1990, 1991). Detailed studies have shown that the *Brucella* uptake by epitheloid-like cells is not a passive but an active mechanism, where the bacteria induces its own internalization by activating small GTPases of the Rho subfamily (i.e., Rho, Rac, Cdc42) and modulate rearrangements of the host cell actin cytoskeleton and microtubules (Guzman-Verri et al., 2001). However, only a limited number of nonprofessional phagocytic cells in a monolayer are permissive to invasion (Detilleux et al., 1990; Pizarro-Cerda et al.,

1998a), suggesting heterogeneous susceptibility and/or invasiveness among cells and bacteria, respectively.

After in vivo invasion of nonprofessional phagocytic cells, such as trophoblasts, *Brucella* localize and replicate within the rough endoplasmic reticulum without restricting basic cellular functions. However, after extensive intracellular replication infected cells develop signs of degeneration and necrosis, leading to a release of free bacteria into the media and the infection of the other cells proceeds (Anderson et al., 1986). In vitro studies have found that during the first minutes after invasion, virulent *Brucella* transiently interact with an intracellular compartment related to the early endocytic network that is gradually transformed in a multimembranous autophagic vacuole. Later, *Brucella* are delivered into a rough endoplasmic reticulum–like compartment in the perinuclear area, where massive intracellular replication occurs (Detilleux et al., 1990; Pizarro-Cerda et al., 1998a). The benefits for the pathogen of being associated with the endoplasmic reticulum of nonprofessional phagocytic host cell, an event also observed in *Toxoplasma gondii* (Sinai and Joiner, 1997; Sinai et al., 1997), have not yet been identified, although the hypothesis of taking advantage of utilizing metabolites synthesized or translocated to this compartment is an attractive concept (Gorvel and Moreno, 2002). The Type IV-Secretion System (T4SS) encoded by *virB* operon in virulent *Brucella* controlling the intracellular trafficking, bypassing late-endosomal compartments and avoids its degradation by inhibiting phagosome–lysosome fusion in nonprofessional phagocytic cells (Pizarro-Cerda et al., 1998b; Comerci et al., 2001).

Once infected, natural hosts and humans can remain infected for life. After translocating the epithelium layer, *Brucella* are endocytosed by local macrophages and transported to the regional lymph nodes. Failure to destroy the bacteria at this stage results in dissemination to organs that are rich in elements of the RES, such as spleen, bone marrow, and lymph nodes (Enright et al., 1990). The chronic nature of the disease lies in the ability of virulent Brucellae to survive and replicate inside macrophages (Roop et al., 2004). On the contrary, as observed in nonprofessional phagocytic cells, differences in mechanisms of binding and entering of each individual bacterium into macrophages determines its fate within intracellular compartments and the outcome of the infection (Campbell et al., 1994; Rittig et al., 2001; Watarai et al., 2002). Opsonized *Brucella* bind professional phagocytic cells through complement and Fc receptors, whereas nonopsonized organisms seem to bind via lectin, fibronectin, and/or other unknown receptors (Campbell et al., 1994; Rittig et al., 2001). More recently, other papers have addressed the possibility that lipid rafts could serve as docking sites for nonopsonized *Brucella* to stabilize the bacteria–macrophage interaction previous to internalization (Naroeni and Porte, 2002; Watarai et al., 2002).

As soon as the bacteria contact the macrophage membrane, F-actin and annexin I–associated structures transiently accumulate beneath or around the pathogen, regardless of whether the bacteria were previously opsonized or not (Kusumawati et al., 2000). Phagocytosis occurs within a few minutes after contact, when opsonized *Brucella*, but not nonopsonized, induce major morphological changes in the macrophages cell membrane consisting on membrane ruffles that surround and capture the bacteria, in a zipper-like interaction between receptors and ligands (Kusumawati et al., 2000; Rittig et al., 2001). The initial formation of two types of *Brucella*-containing phagosomes, one tightly fitting, representing the survival-permitting compartment, and other spacious, representing the killing compartment, was observed after *Brucella* engulfment (Arenas et al., 2000; Rittig et al., 2001). The delayed phagocytosis observed in wild-type *Brucella* compared with *virB* mutant was interpreted as a necessary time for the fully virulent pathogen to alter the plasma membrane in order to create a specialized organelle permissive for survival and replication (Watarai et al., 2002; Bellaire et al., 2005). These *Brucella*-containing phagosomes briefly interact with early-endosomal compartments, and at 1 h postinfection colocalize with late-endosomal markers in cell lines (Arenas et al., 2000; Bellaire et al., 2005) but not in primary macrophages (Celli et al., 2003). Simultaneously, the delivery of protons by the ATPase proton pump to the phagosome membrane causes a rapid acidification inside the vacuole, a condition necessary during the early phase of infection for survival and replication of *Brucella* in macrophages (Porte et al., 1999; Arenas et al., 2000; Rittig et al., 2001; Bellaire et al., 2005). Paradoxically, the number of *Brucella* inside macrophages decreases in the first hours postinfection and only few phagocytosed organisms are capable of surviving, redirecting their intracellular trafficking, and replicating inside phagosomes, a process highly dependent on a functional T4SS (Arenas et al., 2000; Rittig et al., 2001; Celli et al., 2003; Bellaire et al., 2005; He et al., 2006). This initial killing is probably due to the capability of macrophages to generate toxic oxygen-dependent mechanisms during phagocytosis (Harmon et al., 1989).

After the adaptation phase, *Brucella* replication is initiated, however the *Brucella*-replicative niche in macrophages is controversial. It has been observed that *Brucella* replicates inside compartments with phagolysosomal characteristics (Arenas et al., 2000), endoplasmic reticulum-like compartments (Celli et al., 2003; Bellaire et al., 2005), and nonacidic phagosomes (Bellaire et al., 2005). Among all the differences reported, replication inside phagolysosome vacuoles is the most curious, because experimental evidence seems to support the proposal that *Brucella* intracellular survival in macrophages results from an inhibition of *Brucella*-containing phagosome and lysosome membrane fusion (Porte et al., 1999; Naroeni et al., 2001; Celli et al., 2003). However, the reported differences in *Brucella* intracellular trafficking within macrophages may in part be due to different cells (primary cultures vs. cell lines, human vs. mouse origin), *Brucella* strains, and conditions used (opsonized vs. nonopsonized bacteria).

In addition to macrophages and cells of the RES, *Brucella* have a major predilection for the gravid uterus ultimately causing abortion with huge numbers of organisms being expelled in the fetus and fetal fluids. In the ruminant placenta, *Brucella* replicate initially in the phagosomes of erythrophagocytic trophoblasts of the placentome and later in the rough endoplasmic reticulum of the periplacentomal chorioallantoic trophoblasts (Anderson et al., 1986). The subsequent invasion of the fetus and replication in placental cotyledons leads to extensive placentitis and abortion. Despite the important role that trophoblasts play in brucellosis, there is no detailed description of how *Brucella* invade and intracellularly traffic.

VACCINES

Prevention of human brucellosis has always relied upon reduction of animal disease. Consequently, little attention has been paid to the development of human vaccines. The animal vaccine strains employed today are fortuitous isolates attenuated in ability to cause abortion due to reduced replication in reproductive tissues. The attenuation of these mutants does not extend to reticuloendothelial disease observed in mice and in humans. Furthermore, the lack of genetic definition of fortuitous isolates limits the usefulness of vaccine strains, preventing complete description of their stability (Schurig et al., 1991b; Sangari et al., 1998).

Development of human vaccine strains must rely on different approaches. The best approach to human vaccine design has not been experimentally identified, however, reduced persistence (survival) within macrophages represents a better approach given the limited success observed using subunit and killed vaccines (Schurig et al., 2002). To date, functions targeted for inactivation are based on observations with other bacterial pathogens that typically cause acute infections. The chronic nature of brucellosis in humans suggests other approaches are warranted. Animal vaccine strains defective in important "virulence" determinants (O-antigen, SOD, heat shock proteins) often attenuate the survival of the organism either too severely or not severely enough (Latimer et al., 1992; Elzer et al., 1996; Robertson et al., 1996; Bellaire et al., 1999). This may result from sensitivity to extracellular killing mechanisms (O-antigen mutants) or sensitivity to macrophage killing activities prominent in early infection (SOD, stress response proteins). Better definition of the role of virulence determinants in brucellosis is required.

The potential reluctance of the general population to use live vaccines is based on safety issues and such thinking must not be used to deter the development of products based on otherwise sound scientific principles. The use of such vaccines in humans can be expected under extreme circumstances, such as protection against biological terrorism or biological warfare. Starting with the work of Louis Pasteur, live vaccines have offered the best possible solution for immune protection. Use in humans requires that safety be determined beyond a shadow of doubt. This is one of the reasons that we have proposed the construction of double-knockout mutants. Questions concerning the safety of such products warrant future testing in primate models or in *scid* mice expressing a reconstituted human immune system.

For this reason, numerous studies have explored gene functions necessary for persistence within macrophages using random mutagenesis (Allen et al., 1998; Hong et al., 2000; Lestrate et al., 2000; Ficht, 2002; Lestrate et al., 2003; Delrue et al., 2004; Wu et al., 2006a). In this approach, genes required for survival have been classified according to the duration of survival of the resulting mutant. Gene functions required early in the process reduce survival at both 2 and 8 weeks postinfection; although some may recover these are excluded from additional analysis (Hong et al., 2000). Gene functions required for long-term survival appear normal at 2 weeks, but are greatly reduced after 8 weeks. Examination of the vaccine potential of each group of organisms in the mouse model have revealed a direct correlation between survival and immune protection that will be explored in later sections of this chapter.

Of the currently available vaccine strains, only *B. abortus* S19 and *B. melitensis* Rev 1 have been tested in humans (Spink and Thompson, 1953; Spink et al., 1962). Rev 1 was found to be unsuitable for human use with 2/3 of the "volunteers" exhibiting symptoms of disease and colonization by the organism (Spink et al., 1962). In contrast, a subculture of S19 referred to as 19-BA provided results that may be more encouraging. Two volunteers (12%) exhibited symptoms of disease and the organism was only isolated from one. Strain 19-BA was originally used to vaccinate at least three million people in the Soviet Union (Vershilova, 1961), and investigators concluded that there were more problems due to hypersensitivity than to persistence of the organism. Eight percent complained of headache and malaise, and 2% showed signs of febrile illness. As a result of this vaccination program, a 60% decline (1952–1958) in human brucellosis was reported during a period when animal brucellosis exhibited no decline, and the investigators recommended the use of strain 19-BA in endemic areas. Although efficacious, vaccines with these side effects cannot be tolerated given the current knowledge of bacterial pathogenesis and the ability to manipulate microbial genomes. Reports of adverse reactions in humans also cast doubt on the recently approved vaccine strain RB51, a rough derivative used with some success in ruminants (Ashford et al., 2004). The absence of O-antigen prevents development of cross-reacting immune responses that are indistinguishable from wild-type infection, but also increases the susceptibility of this strain to host killing mechanisms limiting survival and vaccine efficacy (Jimenez de Bagues et al., 1995; Palmer and Cheville, 1997).

The pursuit for an ideal vaccine for eradicating brucellosis has kept scientists busy for nearly a century now, as the criteria for an ideal *Brucella* vaccine are very difficult to achieve. Before achieving clinical reality, the hurdles an ideal *Brucella* vaccine needs to overcome are to: (a) be harmless to the recipient, even on accidental inoculation; (b) be able to induce long-term immunity after one administration; (c) have minimal influence on diagnostic methods such as serologic tests; (d) prevent infection in both sexes, prevent abortion; and (e) be inexpensive, easily produced, stable, and of consistent quality.

In recent years, a stable rough variant of virulent *B. abortus* S2308 designated RB51 (Schurig et al., 1991b) was developed that has one advantage over other vaccines in that it does not stimulate the formation of serum-specific antibodies (Schurig et al., 1991b; Cheville et al., 1992; Tobias et al., 1992). However, the level of protection it affords against smooth virulent strains is lower than that conferred by the current live attenuated smooth vaccine strains (Roop et al., 1991; Samartino and Enright, 1992; Tobias et al., 1992), and inoculation of high doses (10^{10} organisms) leads to placental and fetal colonization of pregnant cattle. Unpublished data also suggests that in several developing countries where the challenge dose of *B. abortus* is very high, RB51 does not protect animals from brucellosis as effectively as strain 19. Besides causing abortion in reindeer and elk, it sometimes also localizes in the uterine trophoblasts of pregnant cows, and is shed in the milk of infected animals.

The use of available vaccines, Strain 19 and RB51, has been disappointing in that they are not sufficiently efficacious in bison, elk, or other species (Cook et al., 2002; Davis and Elzer, 2002; Elzer et al., 2002; Kreeger et al., 2002; Elzer et al., 2003; Olsen et al., 2003). Due to the shear numbers of elk (>30,000), it is not feasible to evaluate a parenteral delivery system in this population in which neither S19 nor RB51 are adequately efficacious (Cook et al., 2002; Kreeger et al., 2002).

Vaccines in Development

Given the potential threat this organism poses and the demonstrated potential for immune protection, the development of human vaccine strains appears to be a viable approach. However, better definition of the role of virulence determinants in survival of the organism and contribution to disease is required. For this reason, over the past 10 years we have characterized gene functions necessary for persistence within macrophages using transposon mutagenesis. In this approach, genes required for survival have been classified according to survival characteristics of deficient mutants (Hong et al., 2000; Wu et al., 2006b). Mutants are generally characterized as having early (rapid) and late (delayed) clearance. Examination of the vaccine potential of these mutants in the mouse model reveals improved immune protection associated with prolonged survival (Kahl-McDonagh et al., 2006; Kahl-McDonagh and Ficht, 2006). The use of live attenuated vaccines will require demonstration of safety as well as efficacy as addressed below.

Prospects for Future Vaccine Candidates

Two factors have been explored as measures of vaccine potential. The first is the safety of the candidate measured by reduced colonization and the absence of side effects. However, a direct relationship between survival and immune protection has been demonstrated in the mouse model (Kahl-McDonagh and Ficht, 2006). Yet, persistence enhances the possibility that the vaccine will colonize additional tissues. The solution is to enhance persistence of highly attenuated (safe) vaccines, and a promising approach has been the use of microencapsulation (A. Arenas-Gamboa, unpublished) in comparison with the standard *B. melitensis* vaccine strain Rev1 (Alton and Elberg, 1967; Elberg, 1981). Efficacy trials will evaluate vaccine candidates that have demonstrated significant immune protection in the mouse model. The second factor is the ability of the organism to induce a protective immune response.

Signature-tagged mutants have been identified using a survival-based strategy in the mouse model of infection (Hong et al., 2000; Wu et al., 2006b). Attenuated survival is subsequently confirmed based on clearance from the spleens of mice more rapidly than parental wild-type using either mixed infections (mutant:wild-type) or individual strains in the mouse model. Knockout mutants are then constructed to simplify potential field release with minimal environmental impact. The immune potential of vaccine candidates with different survival characteristics; rapid (1–2 weeks) versus delayed (>6 weeks) is evaluated in mice protection against colonization and persistence as a measure of efficacy. Mutants providing protection equivalent or superior to currently available vaccines are then selected for evaluation in the goat model (Kahl-McDonagh et al., 2006; Kahl-McDonagh and Ficht, 2006). Using this model, evaluation may be made between protection against clinical signs, infection (colonization), and abortion or premature birth of the kid. Current candidates of interest include Δ*luxR*, Δ*mucR*, and Δ*asp24* deletion mutants (Table 42.1). The choice of these mutants is based on preliminary results in which Δ*asp24* mutants persisted in the host, but provided elevated levels of protective immunity in mice and goats (Kahl-McDonagh et al., 2006; Kahl-McDonagh and Ficht, 2006). Examination of immune correlates in these animals will help to describe immune protection against brucellosis. Δ*luxR* mutants are rapidly cleared from the host, but still provide significant levels of protection. However, the Δ*luxR* mutants have also been shown to provide enhanced protection when delivered in microencapsulated form and have the added advantage of not stimulating (or preventing) splenomegaly when delivered in either format. These features of immune protection and safety make Δ*luxR* mutants ideal candidates for testing in nonhuman primates. Recently, DNA vaccines have been developed that are beginning to demonstrate a level of protection which may prove useful (Cassataro et al., 2005; Gonzalez-Smith et al., 2006; Luo et al., 2006).

Formulation and Microencapsulation

Oral or intranasal vaccine is relatively easy to distribute, low cost compared to parenteral immunization, simplifies multiple dosing, and has the potential for stimulation of mucosal immunity. Microencapsulation of *Brucella* antigens has been used successfully to provide immune protection in the mouse model of brucellosis (Murillo et al., 2001, 2002a, 2002b, 2002c). Microencapsulation provides advantages over delivery of vegetative or lyophilized *Brucella* vaccines. The microcapsules protect the vaccine, enhance the immune response via delivery to preferential sites, and allow exposure modulation (i.e., controlled release). Capsules of 5 μm diameter, delivered orally, typically enter Peyer's patches of the intestines, travel to local lymph nodes and to the spleen. Capsules of larger diameter typically reside in M cells of Peyer's patches and elicit a local immune response. The technique is also expected to increase the shelf life of vaccines. Microencapsulation takes advantage of the properties of early mutants, i.e., rapid clearance with minimal side effects, and is expected to enhance immunity through sustained release. Double-mutants represent additional potential candidates for microencapsulation, due to their elevated attenuation there typically is little induction of a protective immune response. Additional features that may be explored include pulse-release in response to specific

TABLE 42.1 Candidate vaccine *Brucella* mutants

Vaccine candidate	Secondary mutation	Phenotypic difference	Outcome/status
B. melitensis Δ*luxR*	None	Mutant highly attenuated; controls T4SS expression	IP and oral vaccination enhanced by encapsulation
	Δ*virB12*	Provides diagnostic marker	No effect on virulence; no change
	Δ*virB*	No additional effect on virulence; primary and secondary mutations affect same virulence genes	Reversion to virulence reduced; no change
	Δ*manBA*	Distinct virulence mechanism; aggregate defects	Attenuation enhanced; reversion to virulence reduced
	Δ*mucR*	Distinct virulence regulon; aggregate defects	Attenuation enhanced; reversion to virulence reduced

environments (low pH, reducing environment, etc.) or targeting to specific cell types.

A number of processes based on established technologies have recently been adapted for use as encapsulants for living cells to permit transplantation (Young et al., 2002). Hydrogel formation has been employed to encapsulate living cells in bioremediation applications (Russell et al., 2003), for live cell entrapment (Young et al., 2002; Russell et al., 2003), or for protein drug delivery (Qiu et al., 2003). PEG gels provide a protective environment for the cell mechanically blocking attack by cells of the immune system in cases of transplantation (Young et al., 2002). The pore size and therefore the release rate from the capsule will depend on the length of the PEG polymer utilized for the matrix. PEG-based spheres will be formed from a solution of PEG-diacrylate-575, a comonomer (allylamine, acrylic acid, or acrylamide) and living cells (Mellott et al., 2001). Pentaerythritol triacrylate (PETA) will serve as a cross-linking agent and 2,2′-dimethoxy-2-phenyl-acetophenone (DMPA) as photo-initiator for production of methyl radicals for cross-linking. This method has proven to be safe and efficient for the encapsulation of proteins and living cells (Mellott et al., 2001). An alternate hydrogel method utilizes thiol-containing PEG-copolymer injected as a liquid containing the payload (viable cells) and forms a polymer after injection due to the formation of disulfide cross-links in the reducing tissue environment. This depot slowly releases (2–4 weeks) the cellular payload as hydrolysis of the matrix occurs (Qiu et al., 2003).

Rationale for Second and Third New Generation Vaccines

Adams (1990) reviewed the future development of live, attenuated vaccines, and proposed that optimal immunity against *Brucella* spp. will require: (1) antibodies for immune clearance, and (2) sensitized T-helper-cells and delayed hypersensitivity T cells of the macrophage activating type are required for containment, killing, and elimination of *Brucella* spp., thus identification of B-cell and T-cell epitopes and/or modified virulence factors for effective delivery to the host was the next logical step. Furthermore, because the predominant natural exposure in brucellosis is per os, oral delivery of the immunogen to the gut-associated lymphoid tissue (GALT) that stimulates generalized secretory, humoral, and cellular immune responses is the optimal route for vaccination (Adams, 1990). A newer approach consists of identifying diagnostic immunodominant protein antigens and deleting the corresponding genes in the current live attenuated smooth *Brucella* strains. Mutants constructed thus should not induce antibody responses against these target proteins that will be immunodominant in infection and therefore good candidates for conventional serological tests. The introduction of stable and genetic or phenotypic markers used to replace the genes of interest should allow the newly constructed vaccine strain from being differentiated from field isolates, the so-called DIVA vaccines. *B. abortus* S19 mutants deleted for copper–zinc superoxide dismutase (CuZnSOD) or a 31 kDa protein have been constructed and have provided protective immunity against *B. abortus* in cattle similar to the response evoked by the parental strain (Cheville et al., 1993). However, the proteins chosen were not immunodominant in infected cattle and therefore inappropriate as diagnostic antigens.

The caprine model is proposed to be the next optimal model to evaluate gene deletion mutants previously characterized in the mouse model. Strains selected for high levels of protection induced in mice will then be evaluated for safety and efficacy in nonpregnant goats. Subsequent experiments should evaluate protective immunity induced against challenge with virulent organisms. The caprine model offers a natural ruminant host that is inexpensive, accessible, and can provide early-stage evaluation of vaccines with disease symptoms similar to those experienced in humans. However, the nonhuman primate model is the optimal model to evaluate protection against clinical signs that occur in human brucellosis as well as protection against infection. The caprine experiments provide optimal data to select the most appropriate vaccine candidate for nonhuman primate studies as well as reducing the number of nonhuman primates required for potency testing. Evaluations should include overall health, body temperature, pulse, respiration, leukocytosis, bacteremia, and colonization of selected tissues. The best candidates evaluated in both hosts are then used to provide support under the "two animal" U.S. Food and Drug Administration rule for projected Phase I testing.

Nonhuman Primates

Other species were examined for use in this protocol as a substitute for nonhuman primates. The first is the use of the goat model described above. The symptoms experienced in the goat model are similar to those observed in humans. A large amount of work performed previously was conducted in mice; however, these animals do not clinically duplicate brucellosis as seen in humans. In order for this project to test the safety and efficacy of a vaccine for human use, a nonhuman primate model must be used. There are no suitable nonhuman primate animal models, computer models, nor are there any in vitro tests, which can replace the rhesus monkey (*Macaca mullatta*) to evaluate the efficacy of *Brucella* vaccines. Rhesus monkeys, rather than cynomolgus monkeys, are preferable based on earlier studies and the availability of human monoclonal antibodies that cross-react against rhesus monkey proteins. This permits phenotyping immunological cell populations and measurement of cytokines in the rhesus monkeys and direct correlation with humans. These monoclonal antibodies, designed for human use, have been demonstrated to work in this species.

Early studies demonstrated that rhesus monkeys could be infected orally using whole agar slants or milk from infected animals (Huddleson and Hallman, 1929). In a recent report of USAMRIID studies (Mense et al., 2004), rhesus monkeys were shown to be susceptible to aerosol infection by *B. melitensis* strain 16M, and that the cellular and humoral immune responses correlate with clinical observations. Exposed animals became bacteremic and febrile following aerosol challenge, and the time from challenge to bacteremia appeared inversely related to the challenge dose. Histopathologic abnormalities were directly related to dose using morphometric techniques. Furthermore, spleen weights increased with increasing aerosol exposure dose. This is consistent with human disease where the lymphoreticular system is also a known target with *Brucella* infection (Spink and Thompson, 1953; Spink et al., 1962). Data from USAMRIID studies established that nonhuman primates could also be infected with virulent *B. melitensis* strain 16M by the intragastric route, and this infection may involve organs other than the digestive tract proper. Intragastric administration of 10^{12} CFU of the virulent 16M strain led to profound bacteremia and extensive tissue infection. It is anticipated that candidate vaccine strains will be initially delivered by intranasal inoculation and immune responses monitored both before and after challenge (Chen and Elberg, 1976). Microencapsulation may be expected to enhance protection via controlled release and boosting of the protective immune response and may permit the use of reduced vaccine doses.

SUMMARY OF VACCINE RESEARCH AIMS

Current genetic tools permit the development of novel *Brucella* vaccine candidates derived through deletion of virulence factors and subsequent removal of antibiotic resistance marker genes. The defined nature of targeted mutations is expected to simplify interpretation of safety and efficacy compared with the use of fortuitously derived spontaneously attenuated organisms. Evaluation in multiple hosts is expected to provide support for safety and efficacy in nonhuman primates. The identification of vaccine candidates should be based on the predictive capabilities of mouse and guinea pig models supported in another study prior to testing in goats and nonhuman primates to estimate safety and efficacy for future human testing. Combining multiple genetic mutations in a single organism is expected to enhance safety while maintaining a significant level of protection and preventing reversion to virulence. Immune parameters will be characterized with the intent of determining the correlation between mutation in specific genes, survival, and immune protection. Finally, timed release of viable vaccine strains through microencapsulation may enhance vaccine safety and efficacy by combining two features of *Brucella* vaccines that are usually incompatible, i.e., extreme attenuation and persistence within the host.

KEY ISSUES

- Brucellosis is a serious disease threat to human and animal populations in underdeveloped countries.
- The progress made in the control of animal brucellosis over the last 50 years, has prevented the development of improved strategies to directly control human disease.
- The continued public health threat posed by brucellosis is a direct result of the cycle of human poverty that results in self-reliance on small, local animal herds.
- The high infectivity, potential for aerosol delivery, and lack of human intervention strategies are characteristics of a biothreat agent.
- Development of improved vaccines may be expected to improve public health and economic standing of underdeveloped countries.
- Development of improved vaccines may be expected to provide protection for at-risk personnel in the war against terror.
- Development of significant levels of protection has only been associated with live attenuated vaccines; characterization of the genes necessary for intracellular survival will aid in the development of safe and efficacious vaccine strains.

- Understanding the mechanism(s) of interaction between host and agent is essential to the development of improved intervention strategies.
- An optimally attenuated *Brucella* vaccine may be used as a vector to deliver genes derived from other pathogens.

ACKNOWLEDGMENTS

The authors thank Dr. Angela Arenas, Dr. Melissa M. Kahl-McDonagh, Dr. Carlos A. Rossetti, and Dr. Allison Rice-Ficht for their thoughtful discussions, development of conceptual approaches, and their contributions to this manuscript.

References

Ackermann, M.R., Cheville, N.F. and Deyoe, B.L. Bovine ileal dome lymphoepithelial cells: endocytosis and transport of *Brucella abortus* Strain 19. Vet. Pathol. 1988; 25:28–35.

Adams, L.G. Development of live *Brucella* vaccines. In: Advances in Brucellosis Research (L.G. Adams, Ed.), pp. 101–128. College Station, TX: Texas A&M University Press, 1990.

Allen, C.A., Adams, L.G. and Ficht, T.A. Transposon-derived *Brucella abortus* rough mutants are attenuated and exhibit reduced intracellular survival. Infect. Immun. 1998; 66(3):1008–1016.

Alsmark, C.M., Frank, A.C., Karlberg, E.O., Legault, B.A., Ardell, D.H., Canback, B., Eriksson, A.S., Naslund, A.K., Handley, S.A., Huvet, M., La Scola, B., Holmberg, M. and Andersson, S.G. The louse-borne human pathogen Bartonella quintana is a genomic derivative of the zoonotic agent Bartonella henselae. Proc. Natl. Acad. Sci. U.S.A. 2004; 101(26):9716–9721.

Alton, G.G. and Elberg, S.S. Rev. 1 *Brucella melitensis* vaccine. A review of ten years of study. Vet. Bull. 1967; 37:793–800.

Anderson, T.D., Cheville, N.F. and Meador, V.P. Pathogenesis of placentitis in the goat inoculated with *Brucella abortus*. II. Ultrastructural studies. Vet. Pathol. 1986; 23(3):227–239.

Araya, L.N., Elzer, P.H., Rowe, G.E., Enright, F.M. and Winter, A.J. Temporal development of protective cell-mediated and humoral immunity in BALB/c mice infected with *Brucella abortus*. J. Immunol. 1989; 143:3330–3337.

Arenas, G.N., Staskevich, A.S., Aballay, A. and Mayorga, L.S. Intracellular trafficking of *Brucella abortus* in J774 macrophages. Infect. Immun. 2000; 68(7):4255–4263.

Ariza, J., Bosch, J., Gudiol, F., Linares, J., Fernandez Viladrich, P. and Martin, R. Relevance of in vitro antimicrobial susceptibility of *Brucella melitensis* to relapse rate in human brucellosis. Antimicrob. Agents Chemother. 1986; 30:958–960.

Ariza, J., Bosilkovski, M., Cascio, A., Colmenero, J.D., Corbel, M.J., Falagas, M.E., Memish, Z.A., Roushan, M.R., Rubinstein, E., Sipsas, N.V., Solera, J., Young, E.J. and Pappas, G. Perspectives for the treatment of brucellosis in the 21st century: the Ioannina recommendations. PLoS Med. 2007; 4(12):e317.

Ashford, D.A., di Pietra, J., Lingappa, J., Woods, C., Noll, H., Neville, B., Weyant, R., Bragg, S.L., Spiegel, R.A., Tappero, J. and Perkins, B.A. Adverse events in humans associated with accidental exposure to the livestock brucellosis vaccine RB51. Vaccine 2004; 22(25–26):3435–3439.

Bachrach, G., Banai, M., Bardenstein, S., Hoida, G., Genizi, A. and Bercovier, H. Brucella ribosomal protein L7/L12 is a major component in the antigenicity of brucellin INRA for DTH in brucella sensitized guinea pigs. Infect. Immun. 1994; 62:5361–5366.

Bae, J.E. and Toth, T.E. Cloning and kinetics of expression of *Brucella abortus* heat shock proteins by baculovirus recombinants. Vet. Microbiol. 2000; 75(2):199–204.

Baldwin, C.L., Jiang, X. and Fernandes, D.M. Macrophage control of *Brucella abortus*: influence of cytokines and iron. Trends Microbiol. 1993; 1:99–104.

Baloglu, S., Toth, T.E., Schurig, G.G., Sriranganathan, N. and Boyle, S.M. Humoral immune response of BALB/c mice to a vaccinia virus recombinant expressing *Brucella abortus* GroEL does not correlate with protection against a *B. abortus* challenge [in process citation]. Vet. Microbiol. 2000; 76(2):193–199.

Banai, M., Hoida, G., Rapoport, E., Zamir, O., Haimovich, M., Katz, E., Meyer, I., Cohen, A. and Davidson, M. 1991. Is *Brucella melitensis* undergoing evolutionary development or a change in the vaccinal Rev.1 strain?, p.152. In: Tumbay, A., Hilmi, S. and Ang, O. (ed.), Brucella and brucellosis in man and animals. Turkish Microbiological Society publication no.16. Ege University Press, Izmir, Turkey.

Banai, M., Mayer, I. and Cohen, A. Isolation, identification, and characterization in Israel of *Brucella melitensis* biovar 1 atypical strains susceptible to dyes and penicillin, indicating the evolution of a new variant. J. Clin. Microbiol. 1990; 28:1057–1059.

Bang, B. Infectious abortion in cattle. J. Comp. Pathol. 1906; 19:191–202.

Bardenwerper, H.W. Human sickness caused by *Brucella abortus*, Strain 19. JAMA 1954; 155:970–971.

Bellaire, B.H., Elzer, P.H., Baldwin, C.L. and Roop, R.M., 2nd. The siderophore 2,3-dihydroxybenzoic acid is not required for virulence of *Brucella abortus* in BALB/c mice. Infect. Immun. 1999; 67(5):2615–2618.

Bellaire, B.H., Roop, R.M., 2nd and Cardelli, J.A. Opsonized virulent *Brucella abortus* replicates within nonacidic, endoplasmic reticulum-negative, LAMP-1-positive phagosomes in human monocytes. Infect. Immun. 2005; 73(6):3702–3713.

Bienvenu, R.J. and Young, V.F. The relationship of pregnancy hormones to brucellacidal activity on bovine serum. Am. J. Vet. Res. 1962; 23:1027–1030.

Billard, E., Cazevieille, C., Dornand, J. and Gross, A. High susceptibility of human dendritic cells to invasion by the intracellular pathogens *Brucella suis*, *B. abortus*, and *B. melitensis*. Infect. Immun. 2005; 73(12):8418–8424.

Blair Hedges, S. and Kumar, S. Genomic clocks and evolutionary timescales. Trends Genet. 2003; 19(4):200–206.

Bosseray, N. and Plommet, M. Serum- and cell-mediated immune protection of mouse placenta and fetus against a *Brucella abortus* challenge: expression of barrier effect of placenta. Placenta 1988; 9:65–79.

Bosseray, N. and Plommet, M. *Brucella suis* S2, *Brucella melitensis* Rev. 1 and *Brucella abortus* S19 living vaccines: residual virulence and immunity induced against three Brucella species challenge strains in mice. Vaccine 1990; 8(5):462–468.

Bowden, R.A., Estein, S.M., Zygmunt, M.S., Dubray, G. and Cloeckaert, A. Identification of protective outer membrane antigens of *Brucella ovis* by passive immunization of mice with monoclonal antibodies. Microbes Infect. 2000; 2(5):481–488.

Bruce, D. Note on the discovery of a micro-organism in Malta fever. Practitioner 1887; 39:161.

Bruce, D. The micrococcus of Malta fever. Practitioner 1888; XL(4):241–249.

Bruce, D. Malta fever. Br. Med. J. 1889; I:1101–1107.

Buck, J.M. Studies of vaccination during calfhood to prevent bovine infectious abortion. J. Agric. Res. 1930; 22:667–689.

Campbell, G.A., Adams, L.G. and Sowa, B.A. Mechanisms of binding of *Brucella abortus* to mononuclear phagocytes from cows naturally resistant or susceptible to brucellosis. Vet. Immunol. Immunopathol. 1994; 41(3–4):295–306.

Canning, P.C., Roth, J.A., Tabatabai, L.B. and Deyoe, B.L. Isolation of components of *Brucella abortus* responsible for inhibition of function in bovine neutrophils. J. Infect. Dis. 1985; 152: 913–921.

Capasso, L. Bacteria in two-millennia-old cheese, and related epizoonoses in Roman populations. J. Infect. 2002; 45(2):122–127.

Carmichael, L.E. and Bruner, D.W. Characteristics of a newly-recognized species of Brucella responsible for infectious canine abortions. Cornell Vet. 1968; 58:579–592.

Carpenter, C.M. *Bacterium abortum* invasion of the tissues of calves from the ingestion of infected milk. Cornell Vet. 1924; 14: 16–31.

Cassataro, J., Velikovsky, C.A., de la Barrera, S., Estein, S.M., Bruno, L., Bowden, R., Pasquevich, K.A., Fossati, C.A. and Giambartolomei, G.H. A DNA vaccine coding for the Brucella outer membrane protein 31 confers protection against *B. melitensis* and *B. ovis* infection by eliciting a specific cytotoxic response. Infect. Immun. 2005; 73(10):6537–6546.

Castaneda-Roldan, E.I., Avelino-Flores, F., Dall'Agnol, M., Freer, E., Cedillo, L., Dornand, J. and Giron, J.A. Adherence of Brucella to human epithelial cells and macrophages is mediated by sialic acid residues. Cell. Microbiol. 2004; 6(5):435–445.

Castillo, J.D.C., Marquez, S.H., Iglesias, J.M.R., Franquela, F.C., Diaz, F.R. and Alonso, A. Comparative trial of doxycycline plus streptomycin versus doxycycline plus rifampin for the therapy of human brucellosis. Chemotherapy 1989; 35:146–152.

Celli, J., de Chastellier, C., Franchini, D.M., Pizarro-Cerda, J., Moreno, E. and Gorvel, J.P. Brucella evades macrophage killing via VirB-dependent sustained interactions with the endoplasmic reticulum. J. Exp. Med. 2003; 198(4):545–556.

Chain, P.S., Comerci, D.J., Tolmasky, M.E., Larimer, F.W., Malfatti, S.A., Vergez, L.M., Aguero, F., Land, M.L., Ugalde, R.A. and Garcia, E. Whole-genome analyses of speciation events in pathogenic Brucellae. Infect. Immun. 2005; 73(12):8353–8361.

Chen, T.H. and Elberg, S.S. Priming of *Macaca cynomolgus* philippinensis with purified antigen of *Brucella melitensis* before injection of Rev. I vaccine. J. Infect. Dis. 1976; 134(3):294–296.

Cheville, N.F., Halling, S.M., Tatum, F.M., Morfitt, D.C., Hennager, S.G., Frerichs, W.M. and Schurig, G. Bacterial survival, lymph node changes, and immunologic responses of cattle vaccinated with standard and mutant strains of *Brucella abortus*. Am. J. Vet. Res. 1992; 53:1881–1888.

Cheville, N.F., Stevens, M.G., Jensen, A.E., Tatum, F.M. and Halling, S.M. Immune responses and protection against infection and abortion in cattle experimentally vaccinated with mutant strains of *Brucella abortus*. Am. J. Vet. Res. 1993; 54:1591–1597.

Comerci, D.J., Martinez-Lorenzo, M.J., Sieira, R., Gorvel, J.P. and Ugalde, R.A. Essential role of the VirB machinery in the maturation of the *Brucella abortus*-containing vacuole. Cell. Microbiol. 2001; 3(3):159–168.

Connolly, J.P., Comerci, D., Alefantis, T.G., Walz, A., Quan, M., Chafin, R., Grewal, P., Mujer, C.V., Ugalde, R.A. and DelVecchio, V.G. Proteomic analysis of *Brucella abortus* cell envelope and identification of immunogenic candidate proteins for vaccine development. Proteomics 2006; 6(13):3767–3780.

Cook, W.E., Williams, E.S., Thorne, E.T., Kreeger, T.J., Stout, G., Bardsley, K., Edwards, H., Schurig, G., Colby, L.A., Enright, F. and Elzer, P.H. *Brucella abortus* strain RB51 vaccination in elk. I. Efficacy of reduced dosage. J. Wildl. Dis. 2002; 38(1):18–26.

Corbel, M.J. Brucellosis: an overview. Emerg. Infect. Dis. 1997; 3(2):213–221.

Corner, L.A. and Alton, G.G. Persistence of *Brucella abortus* Strain 19 infection in adult cattle vaccinated with reduced doses. Res.Vet. Sci. 1981; 31:342–344.

Cunningham, B. Brucellosis symposium. Part III. Vaccines and brucellosis. Irish Vet. J. 1968; 22:128–131.

Dafni, I., Hoida, G., Feinhaken, D. and Banai, M. Observation on *Brucella melitensis* infection in Israeli cattle herds. Israel J. Vet. Med. 1990; 46:13–19.

Davis, D.S. and Elzer, P.H. Brucella vaccines in wildlife. Vet. Microbiol. 2002; 90(1–4):533–544.

Delrue, R.M., Lestrate, P., Tibor, A., Letesson, J.J. and De Bolle, X. Brucella pathogenesis, genes identified from random large-scale screens. FEMS Microbiol. Lett. 2004; 231(1):1–12.

DelVecchio, V.G., Kapatral, V., Redkar, R.J., Patra, G., Mujer, C., Los, T., Ivanova, N., Anderson, I., Bhattacharyya, A., Lykidis, A., Reznik, G., Jablonski, L., Larsen, N., D'Souza, M., Bernal, A., Mazur, M., Goltsman, E., Selkov, E., Elzer, P.H., Hagius, S., O'Callaghan, D., Letesson, J.J., Haselkorn, R., Kyrpides, N. and Overbeek, R. The genome sequence of the facultative intracellular pathogen *Brucella melitensis*. Proc. Natl. Acad. Sci. U.S.A. 2002; 99(1):443–448.

Detilleux, P.G., Deyoe, B.L. and Cheville, N.F. Entry and intracellular localization of *Brucella* spp. in Vero cells: fluorescence and electron microscopy. Vet. Pathol. 1990; 27(5):317–328.

Detilleux, P.G., Deyoe, B.L. and Cheville, N.F. Effect of endocytic and metabolic inhibitors on the internalization and intracellular growth of *Brucella abortus* in Vero cells. Am. J. Vet. Res. 1991; 52(10):1658–1664.

Dornand, J., Gross, A., Lafont, V., Liautard, J., Oliaro, J. and Liautard, J.P. The innate immune response against Brucella in humans. Vet. Microbiol. 2002; 90(1–4):383–394.

Dornand, J., Lafont, V., Oliaro, J., Terraza, A., Castaneda-Roldan, E. and Liautard, J.P. Impairment of intramacrophagic *Brucella suis* multiplication by human natural killer cells through a contact-dependent mechanism. Infect. Immun. 2004; 72(4): 2303–2311.

Elberg, S.S. Rev 1 *Brucella melitensis* vaccine. Part II 1968–1980. Vet. Bull. 1981; 51:67–73.

Elberg, S.S. and Faunce, K., Jr. Immunization against Brucella infection. VI. Immunity conferred on goats by a non-dependent mutant from a streptomycin-dependent mutant strain of *Brucella melitensis*. J. Bacteriol. 1957; 73:211–217.

Elzer, P., Davis, D.S., Evans, D., Martin, B., Olsen, S.C., Rhyan, J. and Schurig, G. 107th Annual Meeting of the USAHA, San Diego, CA. 2003.

Elzer, P., Hagius, S., Roffe, T.J., Holland, S. and Davis, D.S. 106th Annual Meeting of USAHA, St. Louis, MO., 2002.

Elzer, P.H., Hagius, S.D., Robertson, G.T., Phillips, R.W., Walker, J.V., Fatemi, M.B., Enright, F.M. and Roop, R.M., 2nd. Behaviour of a high-temperature-requirement A (HtrA) deletion mutant of *Brucella abortus* in goats. Res. Vet. Sci. 1996; 60(1):48–50.

Enright, F.M. The pathogenesis and pathobiology of Brucella infection in domestic animals. In: Animal Brucellosis (K.A. Nielsen and J.R. Duncan, Eds.), pp. 301–320. Boca Raton, FL: CRC Press, 1990.

Enright, F.M., Araya, L.N., Elzer, P.H., Rowe, G.E. and Winter, A.J. Comparative histopathology in BALB/c mice infected with virulent and attenuated strains of *Brucella abortus*. Vet. Immunol. Immunopathol. 1990; 26:171–182.

Eskra, L., Mathison, A. and Splitter, G. Microarray analysis of mRNA levels from RAW264.7 macrophages infected with *Brucella abortus*. Infect. Immun. 2003; 71(3):1125–1133.

Ewalt, D.R., Payeur, J.B., Martin, B.M., Cummins, D.R. and Miller, W.G. Characteristics of a Brucella species from a bottlenose dolphin (*Tursiops truncatus*). J. Vet. Diagn. Invest. 1994; 6:448–452.

Eyre, J.W.H. *Melitensis Septicemia* (Malta or Mediterranean fever). Lancet 1908; 1:1677–1682.

Eze, M.O., Yuan, L., Crawford, R.M., Paranavitana, C.M., Hadfield, T.L., Bhattacharjee, A.K., Warren, R.L. and Hoover, D.L. Effects of opsonization and gamma interferon on growth of *Brucella melitensis* 16M in mouse peritoneal macrophages in vitro. Infect. Immun. 2000; 68(1):257–263.

Fernandes, D.M., Benson, R. and Baldwin, C.L. Lack of a role for natural killer cells in early control of *Brucella abortus* 2308 infections in mice. Infect. Immun. 1995; 63(10):4029–4033.

Fernandez-Prada, C.M., Nikolich, M., Vemulapalli, R., Sriranganathan, N., Boyle, S.M., Schurig, G.G., Hadfield, T.L. and Hoover, D.L. Deletion of wboA enhances activation of the lectin pathway of complement in *Brucella abortus* and *Brucella melitensis*. Infect. Immun. 2001; 69(7):4407–4416.

Ficht, T.A. Discovery of Brucella virulence mechanisms using mutational analysis. Vet. Microbiol. 2002; 90(1–4):311–315.

Fiori, P.L., Mastrandrea, S., Rappelli, P. and Cappuccinelli, P. *Brucella abortus* infection acquired in microbiology laboratories. J. Clin. Microbiol. 2000; 38(5):2005–2006.

Garner, M.M., Lambourn, D.M., Jeffries, S.J., Hall, P.B., Rhyan, J.C., Polzin, L.M. and Cheville, N.F. Evidence of Brucella infection in parafilaroides lungworms in a Pacific harbor seal (*Phoca vitulina richardsi*). J. Vet. Diagn. Invest. 1997; 9:298–303.

Godfroid, F., Taminiau, B., Danese, I., Denoel, P., Tibor, A., Weynants, V., Cloeckaert, A., Godfroid, J. and Letesson, J.J. Identification of the perosamine synthetase gene of *Brucella melitensis* 16M and involvement of lipopolysaccharide O side chain in Brucella survival in mice and in macrophages. Infect. Immun. 1998; 66(11):5485–5493.

Golding, B., Zaitseva, M. and Golding, H. The potential for recruiting immune responses toward type 1 or type 2 T cell help. Am. J. Trop. Med. Hyg. 1994; 50(4 Suppl):33–40.

Gonzalez-Smith, A., Vemulapalli, R., Andrews, E. and Onate, A. Evaluation of *Brucella abortus* DNA vaccine by expression of Cu-Zn superoxide dismutase antigen fused to IL-2. Immunobiology 2006; 211(1–2):65–74.

Gorvel, J.P. and Moreno, E. Brucella intracellular life: from invasion to intracellular replication. Vet. Microbiol. 2002; 90(1–4):281–297.

Gross, A., Spiesser, S., Terraza, A., Rouot, B., Caron, E. and Dornand, J. Expression and bactericidal activity of nitric oxide synthase in *Brucella suis*-infected murine macrophages. Infect. Immun. 1998; 66(4):1309–1316.

Groussaud, P., Shankster, S.J., Koylass, M.S. and Whatmore, A.M. Molecular typing divides marine mammal strains of Brucella into at least three groups with distinct host preferences. J. Med. Microbiol. 2007; 56(Pt 11):1512–1518.

Guzman-Verri, C., Chaves-Olarte, E., von Eichel-Streiber, C., Lopez-Goni, I., Thelestam, M., Arvidson, S., Gorvel, J.P. and Moreno, E. GTPases of the Rho subfamily are required for *Brucella abortus* internalization in nonprofessional phagocytes: direct activation of Cdc42. J. Biol. Chem. 2001; 276(48):44435–44443.

Halling, S.M., Peterson-Burch, B.D., Bricker, B.J., Zuerner, R.L., Qing, Z., Li, L.L., Kapur, V., Alt, D.P. and Olsen, S.C. Completion of the genome sequence of *Brucella abortus* and comparison to the highly similar genomes of *Brucella melitensis* and *Brucella suis*. J. Bacteriol. 2005; 187(8):2715–2726.

Harmon, B.G., Adams, L.G., Templeton, J.W. and Smith, R., 3rd. Macrophage function in mammary glands of *Brucella abortus*-infected cows and cows that resisted infection after inoculation of *Brucella abortus*. Am. J. Vet. Res. 1989; 50(4):459–465.

He, Y., Reichow, S., Ramamoorthy, S., Ding, X., Lathigra, R., Craig, J.C., Sobral, B.W., Schurig, G.G., Sriranganathan, N. and Boyle, S.M. *Brucella melitensis* triggers time-dependent modulation of apoptosis and down-regulation of mitochondrion-associated gene expression in mouse macrophages. Infect. Immun. 2006; 74(9):5035–5046.

Hong, P.C., Tsolis, R.M. and Ficht, T.A. Identification of genes required for chronic persistence of *Brucella abortus* in mice. Infect. Immun. 2000; 68(7):4102–4107.

Hoover, D.L., Crawford, R.M., Van De Verg, L.L., Izadjoo, M.J., Bhattacharjee, A.K., Paranavitana, C.M., Warren, R.L., Nikolich, M.P. and Hadfield, T.L. Protection of mice against brucellosis by vaccination with *Brucella melitensis* WR201(16MDeltapurEK). Infect. Immun. 1999; 67(11):5877–5884.

Huddleson, F. and Hallman, E.T. The pathogenicity of the species of the genus *Brucella* for monkeys. J. Infect. Dis. 1929; 45: 293–303.

Huddleston, I.F. and Bennett, G.R. The vaccinal immunizing value of a mucoid-growth phase of *Brucella suis* against brucellosis in cattle. Mich. State Coll. Agric. Exp. Stn. Bull. 1948; 31:139–156.

Jiang, X. and Baldwin, C.L. Effects of cytokines on intracellular growth of *Brucella abortus*. Infect. Immun. 1993; 61:124–134.

Jiang, X., Leonard, B., Benson, R. and Baldwin, C.L. Macrophage control of *Brucella abortus*: role of reactive oxygen intermediates and nitric oxide. Cell. Immunol. 1993; 151:309–319.

Jimenez de Bagues, M.P., Barberan, M., Marin, C.M. and Blasco, J.M. The *Brucella abortus* RB51 vaccine does not confer protection against *Brucella ovis* in rams. Vaccine 1995; 13:301–304.

Jones, S.M. and Winter, A.J. Survival of virulent and attenuated strains of *Brucella abortus* in normal and gamma interferon-activated murine peritoneal macrophages. Infect. Immun. 1992; 60:3011–3014.

Kahl-McDonagh, M., Pei, J., Ding, X. and Ficht, T.A. Aerosol infection of BALB/c mice with *Brucella melitensis* and *Brucella abortus* and protective efficacy against aerosol challenge. Infect. Immun. 2007; 75:4923–4932.

Kahl-McDonagh, M.M., Elzer, P.H., Hagius, S.D., Walker, J.V., Perry, Q.L., Seabury, C.M., den Hartigh, A.B., Tsolis, R.M., Adams, L.G., Davis, D.S. and Ficht, T.A. Evaluation of novel *Brucella melitensis* unmarked deletion mutants for safety and efficacy in the goat model of brucellosis. Vaccine 2006; 24(24):5169–5177.

Kahl-McDonagh, M.M. and Ficht, T.A. Evaluation of protection afforded by *Brucella abortus* and *Brucella melitensis* unmarked deletion mutants exhibiting different rates of clearance in BALB/c mice. Infect. Immun. 2006; 74(7):4048–4057.

Kaufmann, A.F., Fox, M.D., Boyce, J.M., Anderson, D.C., Potter, M.E., Martone, W.J. and Patton, C.M. Airborne spread of brucellosis. Ann. N.Y. Acad. Sci. 1980; 353:105–114.

Kaufmann, A.F., Meltzer, M.I. and Schmid, G.P. The economic impact of a bioterrorist attack: are prevention and postattack intervention programs justifiable? Emerg. Infect. Dis. 1997; 3:83–94.

Keppie, J., Witt, K. and Smith, H. Cross immunization of guinea-pigs with products of *Brucellae abortus*, *melitensis* and *suis*. Br. J. Exp. Pathol. 1963; 44:84–87.

Khan, M.Y., Mah, M.W. and Memish, Z.A. Brucellosis in pregnant women. Clin. Infect. Dis. 2001; 32(8):1172–1177.

Ko, J. and Splitter, G.A. Molecular host-pathogen interaction in brucellosis: current understanding and future approaches to vaccine development for mice and humans. Clin. Microbiol. Rev. 2003; 16(1):65–78.

Kreeger, T.J., Cook, W.E., Edwards, W.H., Elzer, P.H. and Olsen, S.C. *Brucella abortus* strain RB51 vaccination in elk. II. Failure of high dosage to prevent abortion. J. Wildl. Dis. 2002; 38(1):27–31.

Kusumawati, A., Cazevieille, C., Porte, F., Bettache, S., Liautard, J.P. and Sri Widada, J. Early events and implication of F-actin and annexin I associated structures in the phagocytic uptake of *Brucella suis* by the J-774A.1 murine cell line and human monocytes. Microb. Pathog. 2000; 28(6):343–352.

Latimer, E., Simmers, J., Sriranganathan, N., Roop, R.M., 2nd, Schurig, G.G. and Boyle, S.M. *Brucella abortus* deficient in copper/zinc superoxide dismutase is virulent in BALB/c mice. Microb. Pathog. 1992; 12:105–113.

Lestrate, P., Delrue, R.M., Danese, I., Didembourg, C., Taminiau, B., Mertens, P., De Bolle, X., Tibor, A., Tang, C.M. and Letesson, J.J. Identification and characterization of in vivo attenuated mutants of *Brucella melitensis*. Mol. Microbiol. 2000; 38(3):543–551.

Lestrate, P., Dricot, A., Delrue, R.M., Lambert, C., Martinelli, V., De Bolle, X., Letesson, J.J. and Tibor, A. Attenuated signature-tagged mutagenesis mutants of *Brucella melitensis* identified during the acute phase of infection in mice. Infect. Immun. 2003; 71(12): 7053–7060.

Liautard, J.P., Gross, A., Dornand, J. and Kohler, S. Interactions between professional phagocytes and *Brucella* spp. Microbiologia 1996; 12(2):197–206.

Lubani, M.M., Lulu, A.R., Araj, G.F., Khateeb, M.I., Qurtom, M.A. and Dudin, K.I. Pulmonary brucellosis. Q. J. Med. 1989; 71(264):319–324.

Luo, D., Ni, B., Li, P., Shi, W., Zhang, S., Han, Y., Mao, L., He, Y., Wu, Y. and Wang, X. Protective immunity elicited by a divalent DNA vaccine encoding both the L7/L12 and Omp16 genes of *Brucella abortus* in BALB/c mice. Infect. Immun. 2006; 74(5):2734–2741.

Madkour, M.M. Respiratory manifestations. In: Brucellosis (M.M. Madkour, Ed.), Ist ed., pp. 131–139. London, UK: Butterworths, 1989.

Marston, J.A. Report on fever (Malta). Great Britain Army Med. Dep. Rep. 1861; 3:486–521.

McDonald, W.L., Jamaludin, R., Mackereth, G., Hansen, M., Humphrey, S., Short, P., Taylor, T., Swingler, J., Dawson, C.E., Whatmore, A.M., Stubberfield, E., Perrett, L.L. and Simmons, G. Characterization of a Brucella sp. strain as a marine-mammal type despite isolation from a patient with spinal osteomyelitis in New Zealand. J. Clin. Microbiol. 2006; 44(12):4363–4370.

McEwen, A.D. and Roberts, R.S. Bovine contagious abortion. The use of guinea-pigs in immunisation studies. J. Comp. Pathol. Ther. 1936; 49:97–117.

Meador, V.P. and Deyoe, B.L. Intracellular localization of *Brucella abortus* in bovine placenta. Vet. Pathol. 1989; 26(6):513–515.

Mellott, M.B., Searcy, K. and Pishko, M.V. Release of protein from highly cross-linked hydrogels of poly(ethylene glycol) diacrylate fabricated by UV polymerization. Biomaterials 2001; 22(9):929–941.

Memish, Z.A. and Balkhy, H.H. Brucellosis and international travel. J. Travel Med. 2004; 11(1):49–55.

Mense, M.G., Borschel, R.H., Wilhelmsen, C.L., Pitt, M.L. and Hoover, D.L. Pathologic changes associated with brucellosis experimentally induced by aerosol exposure in rhesus macaques (*Macaca mulatta*). Am. J. Vet. Res. 2004; 65(5):644–652.

Mense, M.G., Van De Verg, L.L., Bhattacharjee, A.K., Garrett, J.L., Hart, J.A., Lindler, L.E., Hadfield, T.L. and Hoover, D.L. Bacteriologic and histologic features in mice after intranasal inoculation of *Brucella melitensis*. Am. J. Vet. Res. 2001; 62(3):398–405.

Meyer, M.E. Evolution and taxonomy in the genus *Brucella*: steroid hormone induction of filterable forms with altered characteristics after reversion. Am. J. Vet. Res. 1976; 37:207–210.

Mielke, M.E., Ehlers, S. and Hahn, H. T-cell subsets in delayed-type hypersensitivity, protection, and granuloma formation in primary and secondary *Listeria* infection in mice: superior role of Lyt-2^{+} cells in acquired immunity. Infect. Immun. 1988; 56:1920–1925.

Miller, C.D., Songer, J.R. and Sullivan, J.F. A twenty-five year review of laboratory-acquired human infections at the National Animal Disease Center. Am. Ind. Hyg. Assoc. J. 1987; 48:271–275.

Miller, W.G., Adams, L.G., Ficht, T.A., Cheville, N., Payeur, J., Harley, D. and Ridgeway, S. First case report of *Brucella* abortion in a bottlenose dolphin *Tursiops truncatus*. J. Zool. Wildl. Med., 1998; 30:100–110.

Montaraz, J.A. and Winter, A.J. Comparison of living and nonliving vaccines for *Brucella abortus* in BALB/c mice. Infect. Immun. 1986; 53:245–251.

Montaraz, J.A., Winter, A.J., Hunter, D.M., Sowa, B.A., Wu, A.M. and Adams, L.G. Protection against *Brucella abortus* in mice with O-polysaccharide-specific monoclonal antibodies. Infect. Immun. 1986; 51:961–963.

Moreira-Jacob, M. In vitro species (or type) transformation among strains of Brucella. Nature 1963; 197:400–406.

Murillo, M., Gamazo, C., Goni, M., Irache, J. and Blanco-Prieto, M. Development of microparticles prepared by spray-drying as a vaccine delivery system against brucellosis. Int. J. Pharm. 2002a; 242(1–2):341–344.

Murillo, M., Gamazo, C., Irache, J.M. and Goni, M.M. Polyester microparticles as a vaccine delivery system for brucellosis: influence of the polymer on release, phagocytosis and toxicity. J. Drug Target 2002b; 10(3):211–219.

Murillo, M., Goni, M.M., Irache, J.M., Arangoa, M.A., Blasco, J.M. and Gamazo, C. Modulation of the cellular immune response after oral or subcutaneous immunization with microparticles containing *Brucella ovis* antigens. J. Control. Release 2002c; 85(1–3):237–246.

Murillo, M., Grillo, M.J., Rene, J., Marin, C.M., Barberan, M., Goni, M.M., Blasco, J.M., Irache, J.M. and Gamazo, C. A *Brucella ovis* antigenic complex bearing poly-epsilon-caprolactone microparticles confer protection against experimental brucellosis in mice. Vaccine 2001; 19(30):4099–4106.

Murphy, E.A., Sathiyaseelan, J., Parent, M.A., Zou, B. and Baldwin, C.L. Interferon-gamma is crucial for surviving a *Brucella abortus* infection in both resistant C57BL/6 and susceptible BALB/c mice. Immunology 2001; 103(4):511–518.

Naroeni, A., Jouy, N., Ouahrani-Bettache, S., Liautard, J.P. and Porte, F. *Brucella suis*-impaired specific recognition of phagosomes by lysosomes due to phagosomal membrane modifications. Infect. Immun. 2001; 69(1):486–493.

Naroeni, A. and Porte, F. Role of cholesterol and the ganglioside GM(1) in entry and short-term survival of *Brucella suis* in murine macrophages. Infect. Immun. 2002; 70(3):1640–1644.

Olsen, S.C., Jensen, A.E., Stoffregen, W.C. and Palmer, M.V. Efficacy of calfhood vaccination with *Brucella abortus* strain RB51 in protecting bison against brucellosis. Res. Vet. Sci. 2003; 74(1):17–22.

Palmer, M.V. and Cheville, N.F. Effects of oral or intravenous inoculation with *Brucella abortus* strain RB51 vaccine in beagles. Am. J. Vet. Res. 1997; 58(8):851–856.

Pappas, G., Panagopoulou, P., Christou, L. and Akritidis, N. Brucella as a biological weapon. Cell. Mol. Life. Sci. 2006a; 63(19–20):2229–2236.

Pappas, G., Papadimitriou, P., Akritidis, N., Christou, L. and Tsianos, E.V. The new global map of human brucellosis. Lancet Infect. Dis. 2006b; 6(2):91–99.

Pappas, G., Papadimitriou, P., Christou, L. and Akritidis, N. Future trends in human brucellosis treatment. Expert Opin. Investig. Drugs 2006c; 15(10):1141–1149.

Pappas, G., Siozopoulou, V., Akritidis, N. and Falagas, M.E. Doxycycline-rifampicin: physicians' inferior choice in brucellosis or how convenience reigns over science. J. Infect. 2007; 54(5):459–462.

Paulsen, I.T., Seshadri, R., Nelson, K.E., Eisen, J.A., Heidelberg, J.F., Read, T.D., Dodson, R.J., Umayam, L., Brinkac, L.M., Beanan, M.J., Daugherty, S.C., Deboy, R.T., Durkin, A.S., Kolonay, J.F., Madupu, R., Nelson, W.C., Ayodeji, B., Kraul, M., Shetty, J., Malek, J., Van Aken, S.E., Riedmuller, S., Tettelin, H.,

Gill, S.R., White, O., Salzberg, S.L., Hoover, D.L., Lindler, L.E., Halling, S.M., Boyle, S.M. and Fraser, C.M. The *Brucella suis* genome reveals fundamental similarities between animal and plant pathogens and symbionts. Proc. Natl. Acad. Sci. U.S.A. 2002; 99(20):13148–13153.

Pavlov, H., Hogarth, M., McKenzie, I.F.C. and Cheers, C. In vivo and in vitro effects of monoclonal antibody to Ly antigens on immunity to infection. Cell. Immunol. 1982; 71:127–138.

Petrella, R. and Young, E.J. Acute Brucella ileitis. Am. J. Gastroenterol. 1988; 83:80–82.

Pizarro-Cerda, J., Meresse, S., Parton, R.G., van der Goot, G., Sola-Landa, A., Lopez-Goni, I., Moreno, E. and Gorvel, J.P. *Brucella abortus* transits through the autophagic pathway and replicates in the endoplasmic reticulum of nonprofessional phagocytes. Infect. Immun. 1998a; 66(12):5711–5724.

Pizarro-Cerda, J., Moreno, E., Sanguedolce, V., Mege, J.L. and Gorvel, J.P. Virulent *Brucella abortus* prevents lysosome fusion and is distributed within autophagosome-like compartments. Infect. Immun. 1998b; 66(5):2387–2392.

Plant, J.W., Eamens, G.J. and Seaman, J.T. Serological, bacteriological and pathological changes in rams following different routes of exposure to *Brucella ovis*. Aust. Vet. J. 1986; 63(12):409–412.

Plommet, M. and Plommet, A.M. Immune serum-mediated effects on Brucellosis evolution in mice. Infect. Immun. 1983; 41:97–105.

Plommet, M. and Plommet, A.M. Anti-Brucella cell-mediated immunity in mice vaccinated with a cell-wall fraction. Ann. Rech. Vet. 1987; 18(4):429–437.

Porte, F., Liautard, J.P. and Kohler, S. Early acidification of phagosomes containing *Brucella suis* is essential for intracellular survival in murine macrophages. Infect. Immun. 1999; 67(8):4041–4047.

Qiu, B., Stefanos, S., Ma, J., Lalloo, A., Perry, B.A., Leibowitz, M.J., Sinko, P.J. and Stein, S. A hydrogel prepared by in situ cross-linking of a thiol-containing poly(ethylene glycol)-based copolymer: a new biomaterial for protein drug delivery. Biomaterials 2003; 24(1):11–18.

Ramirez-Romero, R. Is *Brucella abortus* a facultative intracellular pathogen with mitochondria-like activity? Med. Hypotheses 1998; 51(1):41–45.

Ray, W.C., Brown, R.R., Stringfellow, D.A., Schnurrenberger, P.R., Scanlan, C.M. and Swann, A.I. Bovine brucellosis: an investigation of latency in progeny of culture-positive cows. J. Am. Vet. Med. Assoc. 1988; 192:182–186.

Rittig, M.G., Alvarez-Martinez, M.T., Porte, F., Liautard, J.P. and Rouot, B. Intracellular survival of *Brucella* spp. in human monocytes involves conventional uptake but special phagosomes. Infect. Immun. 2001; 69(6):3995–4006.

Robertson, G.T., Elzer, P.H. and Roop, R.M., 2nd. In vitro and in vivo phenotypes resulting from deletion of the high temperature requirement A (htrA) gene from the bovine vaccine strain *Brucella abortus* S19. Vet. Microbiol. 1996; 49(3–4):197–207.

Roop, R.M., 2nd, Bellaire, B.H., Valderas, M.W. and Cardelli, J.A. Adaptation of the Brucellae to their intracellular niche. Mol. Microbiol. 2004; 52(3):621–630.

Roop, R.M., 2nd, Jeffers, G., Bagchi, T., Walker, J., Enright, F.M. and Schurig, G.G. Experimental infection of goat fetuses in utero with a stable rough mutant of *Brucella abortus*. Res.Vet. Sci. 1991; 51:123–127.

Russell, A.J., Berberich, J.A., Drevon, G.F. and Koepsel, R.R. Biomaterials for mediation of chemical and biological warfare agents. Annu. Rev. Biomed. Eng. 2003; 5:1–27.

Salmeron, I., Rodriguez-Zapata, M., Salmeron, O., Manzano, L., Vaquer, S. and Alvarez-Mon, M. Impaired activity of natural killer cells in patients with acute brucellosis. Clin. Infect. Dis. 1992; 15(5):764–770.

Salyers, A.A. and Whit, D.D. Bacterial Pathogenesis: A Molecular Approach. Washington, DC: ASM Press, 2002.

Samartino, L.E. and Enright, F.M. Interaction of bovine chorioallantoic membrane explants with three strains of *Brucella abortus*. Am. J. Vet. Res. 1992; 53:359–363.

Sangari, F.J., Grillo, M.J., Jimenez De Bagues, M.P., Gonzalez-Carrero, M.I., Garcia-Lobo, J.M., Blasco, J.M. and Aguero, J. The defect in the metabolism of erythritol of the *Brucella abortus* B19 vaccine strain is unrelated with its attenuated virulence in mice. Vaccine 1998; 16(17):1640–1645.

Santos, R.L., Zhang, S., Tsolis, R.M., Baumler, A.J. and Adams, L.G. Morphologic and molecular characterization of *Salmonella typhimurium* infection in neonatal calves. Vet. Pathol. 2002; 39(2):200–215.

Schurig, G.G., Roop, R.M., 2nd, Bagchi, T., Boyle, S., Buhrman, D. and Sriranganathan, N. Biological properties of RB51; a stable rough strain of *Brucella abortus*. Vet. Microbiol. 1991a; 28(2):171–188.

Schurig, G.G., Roop, R.M., 2nd, Bagchi, T., Boyle, S., Buhrman, D. and Sriranganathan, N. Biological properties of RB51; a stable rough strain of *Brucella abortus*. Vet. Microbiol. 1991b; 28:171–188.

Schurig, G.G., Sriranganathan, N. and Corbel, M.J. Brucellosis vaccines: past, present and future. Vet. Microbiol. 2002; 90(1–4):479–496.

Sinai, A.P. and Joiner, K.A. Safe haven: the cell biology of nonfusogenic pathogen vacuoles. Annu. Rev. Microbiol. 1997; 51:415–462.

Sinai, A.P., Webster, P. and Joiner, K.A. Association of host cell endoplasmic reticulum and mitochondria with the *Toxoplasma gondii* parasitophorous vacuole membrane: a high affinity interaction. J. Cell Sci. 1997; 110(Pt 17):2117–2128.

Smith, C.A. Immune suppression challenges bovine practitioners. J. Am. Vet. Med. Assoc. 1994; 204(3):341–346.

Smith, R., 3rd, Kapatsa, J.C., Rosenbaum, B.A. and Adams, L.G. Bovine T-lymphocyte lines reactive with *Brucella abortus*. Am. J. Vet. Res. 1990a; 51(4):512–517.

Smith, R., 3rd, Kapatsa, J.C., Sherwood, S.J., 3rd, Ficht, T.A., Templeton, J.W. and Adams, L.G. Differential reactivity of bovine lymphocytes to species of Brucella. Am. J. Vet. Res. 1990b; 51(4):518–523.

Sohn, A.H., Probert, W.S., Glaser, C.A., Gupta, N., Bollen, A.W., Wong, J.D., Grace, E.M. and McDonald, W.C. Human neurobrucellosis with intracerebral granuloma caused by a marine mammal *Brucella* spp. Emerg. Infect. Dis. 2003; 9(4):485–488.

Sowa, B.A., Kelly, K.A., Ficht, T.A. and Adams, L.G. Virulence associated proteins of *Brucella abortus* identified by paired two-dimensional gel electrophoretic comparisons of virulent, vaccine and LPS deficient strains. Appl. Theor. Electrophor. 1992; 3(1):33–40.

Sowa, B.A., Kelly, K.A., Ficht, T.A., Frey, M. and Adams, L.G. SDS-soluble and peptidoglycan-bound proteins in the outer membrane-peptidoglycan complex of *Brucella abortus*. Vet. Microbiol. 1991; 27(3–4):351–369.

Spink, W.W., Hall, J.W., Finstad, J. and Mallet, E. Immunization with live Brucella organisms: results of safety test in humans. Bull. World Health Organ. 1962; 26:409–420.

Spink, W.W. and Thompson, H. Human brucellosis caused by *Brucella abortus*, Strain 19. JAMA 1953; 153:1162–1165.

Stantic-Pavlinic, M., Cec, V. and Mehle, J. Brucellosis in spouses and the possibility of interhuman infection. Infection 1983; 11(6):313–314.

Stevens, M.G., Pugh, G.W.J. and Tabatabai, L.B. Effects of gamma interferon and indomethacin in preventing *Brucella abortus* infections in mice. Infect. Immun. 1992; 60:4407–4409.

Stoenner, H.G. and Lackman, D.B. A new species of Brucella isolated from the desert wood rat, *Neotoma lepida* Thomas. Am. J. Vet. Res. 1957; 18:947–951.

Svetic, A., Jian, Y.C., Lu, P., Finkelman, F.D. and Gause, W.C. *Brucella abortus* induces a novel cytokine gene expression pattern characterized by elevated IL-10 and IFN in CD4 T cells. Int. Immunol. 1993; 5:877–887.

ter Huurne, A.A., Meijer, M. and Dijkerman, N.A. [Latency of *Brucella abortus* causes problems in oriented control: a review]. Tijdschr. Diergeneeskd. 1993; 118(21):679–683.

Thapar, M.K. and Young, E.J. Urban outbreak of goat cheese brucellosis. Pediatr. Inf. Dis. 1986; 5:640–643.

Tobias, L., Schurig, G.G. and Cordes, D.O. Comparative behaviour of *Brucella abortus* strains 19 and RB51 in the pregnant mouse. Res. Vet. Sci. 1992; 53:179–183.

Ugalde, J.E., Czibener, C., Feldman, M.F. and Ugalde, R.A. Identification and characterization of the *Brucella abortus* phosphoglucomutase gene: role of lipopolysaccharide in virulence and intracellular multiplication. Infect. Immun. 2000; 68(10):5716–5723.

van Drimmeln, G.C. and Howell, F.D. Preliminary findings with the use of *Brucella melitensis* rev 1 as a vaccine against brucellosis in cattle. Bull. Off. Int. Epizoot. 1964; 62:987–995.

Vassallo, D.J. The saga of brucellosis: controversy over credit for linking Malta fever with goats' milk. Lancet 1996; 348(9030):804–808.

Vemulapalli, R., Cravero, S., Calvert, C.L., Toth, T.E., Sriranganathan, N., Boyle, S.M., Rossetti, O.L. and Schurig, G.G. Characterization of specific immune responses of mice inoculated with recombinant vaccinia virus expressing an 18-kilodalton outer membrane protein of *Brucella abortus*. Clin. Diagn. Lab. Immunol. 2000a; 7(1):114–118.

Vemulapalli, R., Duncan, A.J., Boyle, S.M., Sriranganathan, N., Toth, T.E. and Schurig, G.G. Cloning and sequencing of yajC and secD homologs of *Brucella abortus* and demonstration of immune responses to YajC in mice vaccinated with *B. abortus* RB51. Infect. Immun. 1998; 66(12):5684–5691.

Vemulapalli, R., He, Y., Boyle, S.M., Sriranganathan, N. and Schurig, G.G. *Brucella abortus* strain RB51 as a vector for heterologous protein expression and induction of specific Th1 type immune responses. Infect. Immun. 2000b; 68(6):3290–3296.

Verger, J.M., Grimont, F., Grimont, P.A.D. and Crayon, M. Brucella, a monospecific genus as shown by deoxyribonucleic acid hybridization. Int. J. System. Bacteriol. 1985; 35:292–295.

Vershilova, P.A. The use of live vaccine for vaccination of human beings against in the USSR. Bull. World Health Organ. 1961; 24:85–89.

Wallach, J.C., Giambartolomei, G.H., Baldi, P.C. and Fossati, C.A. Human infection with M-strain of *Brucella canis*. Emerg. Infect. Dis. 2004; 10(1):146–148.

Wallis, O.C., Maniou, Z. and Wallis, M. Cloning and characterization of the gene encoding growth hormone in finback whale (*Balaenoptera physalus*). Gen. Comp. Endocrinol. 2005; 143(1):92–97.

Washko, F.W. and Hutchings, L.M. Susceptibility of cattle to *Brucella suis* following vaccination with *Brucella abortus* Strain 19. Am. J. Vet. Res. 1948; 8:343–349.

Watarai, M., Makino, S., Fujii, Y., Okamoto, K. and Shirahata, T. Modulation of Brucella-induced macropinocytosis by lipid rafts mediates intracellular replication. Cell. Microbiol. 2002; 4(6):341–355.

Weinberg, E.D. Pregnancy-associated immune suppression: risks and mechanisms. Microb. Pathog. 1987; 3:393–397.

Williams, A.E., Keppie, J. and Smith, H. The chemical basis of the virulence of *Brucella abortus*. III. Foetal erythritol a cause of the localisation of *Brucella abortus* in pregnant cows. Br. J. Exp. Pathol. 1962; 43:530–537.

Winter, A.J., Duncan, J.R., Santisteban, C.G., Douglas, J.T. and Adams, L.G. Capacity of passively administered antibody to prevent establishment of *Brucella abortus* infection in mice. Infect. Immun. 1989; 57(11):3438–3444.

Winter, A.J. and Rowe, G.E. Comparative immune responses to native cell envelope antigens and the hot sodium dodecyl sulfate insoluble fraction (PG) of *Brucella abortus* in cattle and mice. Vet. Immunol. Immunopathol. 1988; 18:149–163.

Wu, Q., Pei, J., Turse, C. and Ficht, T.A. Mariner mutagenesis of *Brucella melitensis* reveals genes with previously uncharacterized roles in virulence and survival. BMC Microbiol. 2006a; 6:102–116.

Wu, Q., Pei, J., Turse, C. and Ficht, T.A. Mariner mutagenesis of *Brucella melitensis* reveals genes with previously uncharacterized roles in virulence and survival. BMC Microbiol. 2006b; 6(1):102.

Yingst, S. and Hoover, D.L. T cell immunity to brucellosis. Crit. Rev. Microbiol. 2003; 29(4):313–331.

Young, E.J. An overview of human brucellosis. Clin. Infect. Dis. 1995a; 21(2):283–289, quiz 290.

Young, E.J., Borchert, M., Kretzer, F.L. and Musher, D.M. Phagocytosis and killing of Brucella by human polymorphonuclear leukocytes. J. Infect. Dis. 1985; 151:682–690.

Young, T.H., Chuang, W.Y., Hsieh, M.Y., Chen, L.W. and Hsu, J.P. Assessment and modeling of poly(vinyl alcohol) bioartificial pancreas in vivo. Biomaterials 2002; 23(16):3495–3501.

Zaitseva, M.B., Golding, H., Betts, M., Yamauchi, A., Bloom, E.T., Butler, L.E., Stevan, L. and Golding, B. Human peripheral blood CD4+ and CD8+ T cells express Th1-like cytokine mRNA and proteins following in vitro stimulation with heat-inactivated *Brucella abortus*. Infect. Immun. 1995; 63(7):2720–2728.

Zhan, Y. and Cheers, C. Endogenous gamma interferon mediates resistance to *Brucella abortus* infection. Infect. Immun. 1993; 61(11):4899–4901.

Zhan, Y., Kelso, A. and Cheers, C. Cytokine production in the murine response to brucella infection or immunization with antigenic extracts. Immunology 1993a; 80(3):458–464.

Zhan, Y., Yang, J. and Cheers, C. Cytokine response of T-cell subsets from *Brucella abortus* infected mice to soluble Brucella proteins. Infect. Immun. 1993b; 61:2841–2847.

CHAPTER

43

Burkholderia *mallei* and Burkholderia *pseudomallei*

Mitali Sarkar-Tyson[1] *and Richard W. Titball*[2]

[1]Defence Science and Technology Laboratory, Porton Down, Salisbury, Wiltshire, UK

[2]School of Biosciences, Geoffrey Pope Building, University of Exeter, Exeter, Devon, UK

OUTLINE

ABSTRACT

Burkholderia mallei and *Burkholderia pseudomallei* are gram-negative bacteria that cause two distinct diseases, glanders and melioidosis, respectively. Based on DNA sequence and 16S rRNA sequence comparisons, these organisms are closely related, and Multi-Locus Sequence Typing (MLST) suggests that *B. mallei* has evolved as a clonal derivative of *B. pseudomallei*. *B. mallei* is an obligate parasite of equines and was relatively common throughout the world until the early 20th century and the development of motorized transportation. Human glanders is usually a consequence of contact with infected animals. Acute glanders is usually fatal without medical intervention. The chronic form of the disease may present clinically with swollen lymph nodes, ulcerating nodules, and subcutaneous abscesses. *B. pseudomallei* is widely distributed throughout tropical countries. Unlike

B. mallei, the bacterium can survive outside of a mammalian host and be found in soil and water. Melioidosis affects animals and humans, has a diverse range of symptoms similar to those of glanders, and can be fatal in the acute septicemic form. In endemic areas, *B. pseudomallei* infection is a major cause of hospitalization, and the popularity of worldwide travel has led to increased numbers of cases being reported in non-endemic areas such as Europe and North America. There has been some effort to develop a vaccine against *B. pseudomallei* and *B. mallei*. The close genetic relatedness of the two species suggests that one vaccine may be effective against both diseases. The availability of the genome sequence data should help in the search for a new vaccine by facilitating the identification of potentially protective subunits.

INTRODUCTION

The aim of this chapter is to provide an insight into the biology of two closely related potential biowarfare/bioterrorism agents, *Burkholderia mallei* and *Burkholderia pseudomallei*. Recent genome and epidemiological analyses of these organisms suggest that *B. mallei* has diverged from *B. pseudomallei*. Both organisms cause diseases that affect humans and animals, cause similar clinical symptoms, and are inherently very resistant to a wide range of antibiotics. Despite the similarities, there are some differences. *B. mallei* is believed to be unable to survive outside of a mammalian host, whereas *B. pseudomallei* can be isolated from the environment in endemic areas. There are also differences in the presence of genes associated with virulence, such as type three secretion systems (TTSS), and a family of proteins related to hemagglutinins and invasins (Tiyawisutsri et al., 2007; Rainbow et al., 2002). Our understanding of the molecular basis of disease caused by both agents is rapidly developing, and there is an increasing understanding of virulence mechanisms. Genome sequencing and microarray technology has also identified organism-specific genes potentially involved in pathogenesis. In this chapter, we will consider both glanders and melioidosis, review the recent advances in our knowledge of the organisms, and discuss the potential to produce a vaccine.

HISTORY OF GLANDERS

Glanders is one of the oldest diseases known to mankind and was first described by Aristotle (Wilkinson, 1981). The disease terminology comes from the Greek word "melis" and its Latin derivative "malleus," meaning "severe disease." *B. mallei* was first isolated by Loeffler and Schutz in Germany in 1882, closely followed by Bouchard, Charrin, and Capitan in the same year (Steele, 1979). The organism is a gram-negative aerobic bacillus and is believed to be an obligate mammalian pathogen with solipeds as the natural reservoirs of infection (Pitt, 1990; Sanford, 1996).

The lethal nature of glanders renders *B. mallei* an ideal biological warfare/bioterrorism agent, and there are several reports of its use in this way. It is generally believed that glanders was deliberately spread during World War I by the Germans to infect a large number of Russian horses and mules on the Eastern front (Wheelis, 1998). This had a detrimental effect on the supply of convoys and troops and artillery movement, which were dependent on the horses and the mules. Possibly as a consequence of this dispersal of *B. mallei*, cases of human glanders in Russia increased after World War I. During World War II, the Japanese were believed to have infected horses, civilians, and prisoners of war with glanders at the Pinfang Institute in China. *B. mallei* was also studied by the United States as a biological warfare agent during the 1940s but not apparently weaponized. The former Soviet Union reportedly weaponized *B. mallei* and deployed it in Afghanistan during the 1980s (Alibek, 1999). Glanders has mostly been eradicated from the West since aggressive control measures were enforced in the early 20th century (Steele, 1979; Derbyshire, 2002). However, cases of glanders still periodically occur in certain parts of Asia, the Middle East, northern Africa, and various Mediterranean regions (Benenson, 1995; Kovalev, 1971).

HISTORY OF MELIOIDOSIS

In Rangoon, in 1911, Whitmore and Krishnaswami recorded several cases of a fatal septicemic disease in morphine addicts and neglected inhabitants of the town. After conducting a postmortem on the first victim, they observed a caseous consolidation of the lungs with the presence of abscesses in other organs including spleen, liver, and kidney. They initially described it as a glanders-like disease but were unsatisfied that the bacterium they isolated from the patient was actually glanders as the victim did not have close contact with horses. Subsequently, they conducted a series of experiments, including culturing on peptone agar and potato slopes, and performed virulence studies in guinea pigs. Although the bacterium was similar to the glanders bacillus, there were several differences. The organism isolated was motile, did not induce the characteristic Strauss reaction, and grew rapidly on agar. They concluded that they had identified a new bacterium that was closely related to *B. mallei* (Whitmore and Krishnaswami, 1912; Whitmore, 1913). Several more cases of the disease were described in Rangoon in the following years, and postmortem studies conducted on the victims showed the characteristic symptoms described above (Krishnaswami, 1917). Shortly after this, cases of melioidosis were reported in Malaya in animals at the Institute for Medical Research in Kuala Lumpur (Stanton and Fletcher, 1932). The disease was named melioidosis, derived from the Greek words "melis," which means distemper of asses, and "eidos," which means resemblance (Stanton and Fletcher, 1921).

B. pseudomallei is regarded as a potential biowarfare/bioterrorism agent and appears on the category B list of critical agents published by the U.S. Centers for Disease Control and Prevention. Cases of melioidosis were reported during and after World War II and in soldiers fighting in Vietnam (White, 2003). Melioidosis cases from the Vietnam war often presented as a chronic infection, and reactivation of the disease occurred many years after exposure (Rubin et al., 1963; Sanford and Moore, 1971). *B. pseudomallei* is also a major veterinary pathogen with several outbreaks in zoos. The most infamous case "Affaire du Jardin des Plantes" involved an outbreak in large mammals of the Paris zoological gardens. The infection was disseminated by a panda donated by Mao Ze Dong to President Pompidou (Dance and White, 1996).

Until recently, there were only a few reports of melioidosis cases in South and East Asia probably because of insufficient diagnostic facilities. The first reported case in Thailand was in 1955. Three more cases were reported by 1966 and, 20 years later, over 800 documented cases (White, 2003). Melioidosis is now recognized as an important human infection in Southeast Asia and northern Australia (Currie et al., 2000; Puthucheary et al., 1992). The number of cases of melioidosis that occurs globally is thought to be underestimated with many cases not being diagnosed (White, 2003).

ETIOLOGICAL AGENTS

B. mallei is a gram-negative, nonmotile, aerobic, rod-shaped bacterium, 0.3–0.8 μm in width and 2–5 μm in length (Timoney et al., 1988). The bacteria often stain irregularly and do not appear to have capsules or form spores. Electron microscopy of the bacteria has revealed a capsule-like structure covering the bacteria. This capsule has been shown to be a major virulence factor of *B. mallei* and is thought to protect against harsh environmental factors (Deshazer et al., 2001). The organism grows aerobically in ordinary media and anaerobically only in the presence of nitrate. The organism is slow to grow on glycerol agar, typically requiring 48 h at 37°C.

B. pseudomallei is a gram-negative, motile, non-spore-forming bacillus, 0.4–0.6 μm in width and 2–5 μm in length. The bacteria exhibit a characteristic bipolar staining and are seen singly, in pairs, or very occasionally in chains (Pitt, 1990). The organism grows aerobically in many simple media and will grow under strictly anaerobic conditions in a complex medium containing nitrate. The range of organic compounds that may be used as sole sources of carbon and energy is exceptionally wide, making *B. pseudomallei* one of the most nutritionally versatile members of the genus *Burkholderia*. The form and color of growth on solid media is highly variable, and colonies can range in structure from extreme rough to mucoid and in color from cream to bright orange (Redfearn et al., 1966; Chantratita et al., 2007). Figure 43.1 illustrates the differing colony morphologies on Ashdown agar. Chantratita et al. (2007) suggest that a particular colonial type (type II) is adapted to persistence, resulting in a low-virulence state. It has been shown that virulence is dependent on the presence of the capsular antigen (Atkins et al., 2002; Reckseidler et al., 2001). *B. pseudomallei* has been shown to secrete an abundant slime layer made up of polysaccharides. This layer has been termed the "glycocalyx," and it is thought that bacterial cells may live and divide within

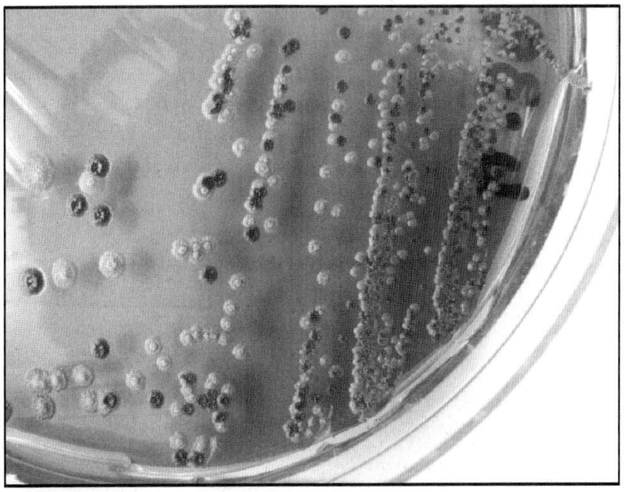

FIGURE 43.1 *B. pseudomallei* colonies on Ashdown media, kindly donated by Narisara Chantratita and Sharon Peacock, Mahidol University, Thailand.

it, protected from unfavorable environmental factors (Kanai and Kondo, 1994).

CLASSIFICATION

B. mallei and *B. pseudomallei* have previously been assigned to the genus *Bacillus*, *Acinetobacter*, *Loefferella*, *Actinobacillus*, *Malleomyces*, or *Pfeifferella* (NCBI Taxonomy). Most recently, based on their 16S ribosomal nucleic acid sequences, DNA homology, cellular lipid and fatty acid composition, and phenotypic characteristics, they were known as *Pseudomonas mallei* and *Pseudomonas pseudomallei* (Leelarasamee and Bovornkitti, 1989). Currently, these species are classified in the genus *Burkholderia*, which is named after U.S. microbiologist Walter Burkholder, who first described *Burkholderia cepecia*, formerly known as *Pseudomonas cepecia* (Yabuuchi et al., 1992).

The genome sequence of *B. pseudomallei* strain K96243 reveals that it is one of the largest prokaryotic genomes comprising two chromosomes of 4.07 and 3.17 Mb (Holden et al., 2004). The larger chromosome carries genes associated with housekeeping functions involved in processes such as cell growth and metabolism, and the smaller chromosome is thought to encode genes required for adaptation and survival in different environments. This proposal is also supported by the finding that genes important during the early phase growth of *B. pseudomallei* in vitro are preferentially located on chromosome 1, whereas genes involved in stationary-phase growth are biased toward chromosome 2 (Rodrigues et al., 2006). There is significant intraspecies diversity that is attributed to either DNA acquisition or loss (Ou et al., 2005). An unusual feature of the *B. pseudomallei* genome is the presence of 16 genomic islands, which make up 6.1% of the entire genome, and is thought to be acquired by horizontal gene transfer. These genome islands are absent in the *B. mallei* genome, which is smaller than that of *B. pseudomallei*, consisting to two chromosomes of 3.51 and 2.32 Mb (Nierman et al., 2004). The presence of a number of insertion sequence elements is thought to have mediated the extensive deletions and rearrangements of the genome relative to that of *B. pseudomallei* (Nierman et al., 2004). The genome downsizing supports previous Multi-Locus Sequence Typing (MLST)-derived conclusions that *B. mallei* has evolved from *B. pseudomallei* (Godoy et al., 2003). Subtractive hybridization between *B. mallei* and *B. pseudomallei* has identified several DNA fragments specific to the latter, which may provide functional detection tools (Monastyrskaya et al., 2004).

PROTECTIVE IMMUNE RESPONSE

The immunological basis of resistance against *B. pseudomallei* infection is poorly understood. However, limited studies in healthy humans exposed to *B. pseudomallei* in endemic areas or patients with overt disease indicate activation of both humoral and cell-mediated arms of the immune system. Individuals naturally exposed to *B. pseudomallei* develop antibodies that increase in titer during acute disease. Indeed, serodiagnostic tests for melioidosis depend on the development of antibodies to *B. pseudomallei* (Ashdown, 1987; Dance, 1991). The levels of IgA, IgM, and IgG antibodies to *B. pseudomallei* correlate with the severity of the disease (Wiersinga et al., 2006). Charuchaimontri et al. (1999) have reported that high levels of antibody to lipopolysaccharide II (LPS II) (now considered to be LPS) in humans suffering from melioidosis correlate with survival.

Experimental studies in mice also suggest that antibodies may play important roles in protection from disease. The immunization of mice with capsular polysaccharide or LPS can protect mice against challenge with *B. pseudomallei* (Nelson et al., 2004), suggesting a role for antibodies in protection. Polyclonal antibodies to LPS can protect diabetic rats against infection caused by *B. pseudomallei* (Brett and Woods, 1996), and in a previously reported study, four protective monoclonal antibodies (mabs) were identified (Jones et al., 2002), which were directed against capsular polysaccharide (antibodies 3VEI5 and 4VA5), LPS (antibody CC6), or an unidentified protein (antibody 4VH7). These antibodies were able to protect mice against an i.p. challenge with 10^4 cfu (approximately 250 LD50 doses) of *B. pseudomallei* strain NCTC4845 (Jones et al., 2002). Similar protection has been reported by Bottex et al. (2005) with a mab directed against the capsular polysaccharide and evaluated in an outbred strain of mice. Both capsular polysaccharide and LPS have been shown to be key virulence factors of *B. pseudomallei*. An antibody against these polysaccharides is reported to be opsonizing and to promote killing by polymorphonuclear leukocytes (PMNs) (Ho et al., 1997). Combinations of protective mabs provided protection, evidenced as an increased time to death, against an i.p. challenge with 10^6 cfu of *B. pseudomallei* (Jones et al., 2002).

Patients with severe melioidosis have high levels of pro-inflammatory cytokines (such as IL-12, IL-18, TNF, IFNγ, IL-15, and IL-6), immunoregulatory cytokines (such as IL-10), as well as granzymes A and B [components of natural killer (NK) cell and cytotoxic-T cell lytic granules], indicating extensive activation of the cellular immune compartment. Reportedly asymptomatic but seropositive individuals tend to show stronger cellular adaptive responses compared to individuals with clinical melioidosis, suggesting a key role for cellular immunity in the control of infection (Barnes et al., 2004). These observations are supported by experimental models of melioidosis in mice that have begun to identify the mechanisms of immunity (Haque et al., 2006; Holden et al., 2004; Santanirand et al., 1999). Nevertheless, detailed immunological knowledge of human infection is poor compared to many other infections, and there has been no previous attempt at a broader analysis of immune changes during acute infection. $CD8^+$ T cells have been shown in *B. pseudomallei* infections to be an important source of early IFNγ via a bystander activation pathway (Lertmemongkolchai et al., 2001). In the murine model of disease, $CD4^+$ T cells, rather than $CD8^+$ T cells, are strongly implicated in the control of infection (Haque et al., 2006). However, there is apparently no association between HIV infection and susceptibility to melioidosis, which questions the role of $CD4^+$ T cells in resistance to melioidosis in humans (Wiersinga et al., 2006).

The immune responses to infection with *B. mallei* are even less well understood. In solipeds, it is clear that antibody responses develop during infection since serodiagnostic tests are used to monitor possible cases of disease (Neubauer et al., 2005). In the murine model of disease, passive administration of monoclonal antibodies against LPS provides protection against an aerosol challenge with *B. mallei*, although sterile immunity was not achieved (Trevino et al., 2006). However, the ability of *B. mallei* to replicate inside host cells suggests that cellular responses might be evoked and linked to a protective response.

EPIDEMIOLOGY

Although glanders was once widespread throughout the world, it has been successfully eradicated by test and slaughter programs (van Zijderveld and Bongers, 2000). Sporadic cases of glanders occur in parts of Asia, Africa, countries in the former Soviet Union, Mexico, South America, and the Middle East (Fig. 43.2) (Arun et al., 1999; Scholz et al., 2006).

Melioidosis, on the other hand, is widespread (Fig. 43.3), and an excellent review of the epidemiology of the disease has been published (Cheng and Currie, 2005). Melioidosis in humans is reported

FIGURE 43.2 Worldwide distribution of *B. mallei*.

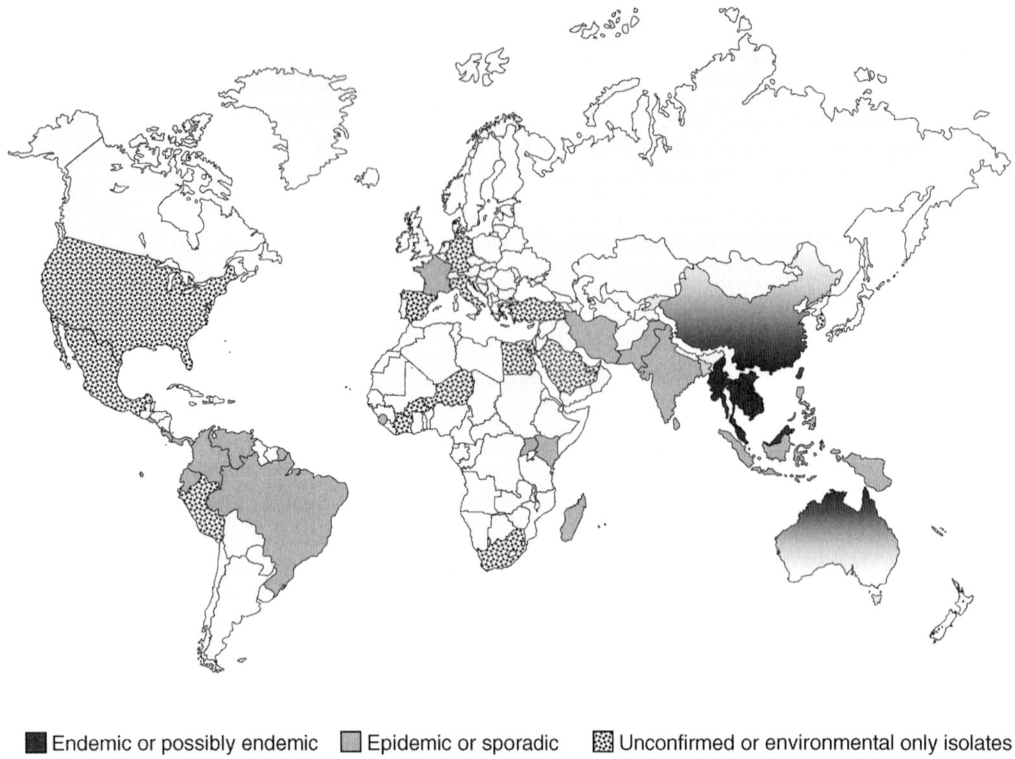

FIGURE 43.3 Worldwide distribution of *B. pseudomallei*.

mainly in northeastern Thailand, Singapore, and northern Australia, with sporadic cases occurring in surrounding areas. Cases reported from other parts of the world are usually associated with travelers returning from endemic countries (Dance et al., 1999; Nieminen and Vaara, 2005). More recently, there has been an increase in cases reported in South America probably due to the growing awareness of melioidosis and improved diagnostic methods (Inglis et al., 2006; Aardema et al., 2005).

CLINICAL DISEASE

Glanders

The incubation period of glanders ranges from 10 to 14 days, and there is some evidence that the disease may remain latent (Fritz et al., 2000). The disease can manifest itself in four forms dependent on the route of exposure: local cutaneous, pulmonary, septicemic, and chronic forms (Waag and Deshazer, 2004). Initial symptoms include fever, malaise, muscle aches, and headache. The cutaneous form is a localized infection with pus-forming ulcerations on the skin, which may spread over the body. Infections associated with mucous membranes of the nose, trachea, pharynx, and lungs results in purulent ulceration. The pulmonary form is associated with pneumonia, pulmonary abscesses, and pleural effusions. Chest X-rays reveal localized infections in the lobes of the lungs. The septicemic form is fatal, resulting in death within a few days if untreated. The chronic form, also known as farcy, is associated with multiple abscess formation in muscles of the arms and legs, or in the spleen, liver, and joints. Even with treatment at this stage, there is a high fatality rate with frequent relapses of infection.

Melioidosis

The incubation period of melioidosis is normally 2–3 days, but there are reports of the disease developing up to 60 years after exposure (Arakawa, 1990; Ngauy et al., 2005). The clinical spectrum of melioidosis is extremely diverse, consisting of four forms of disease: acute fulminant septicemia, subacute illness, chronic infection, and subclinical disease. The disease may be localized or disseminated, and virtually any organ may be affected. Depending on whether the disease is acute or chronic, melioidosis can mimic other bacterial infections, including typhoid, malaria, tuberculosis, and septicemia from other common bacterial organisms (Ip et al., 1995; Everett and Nelson, 1975).

The importance of early diagnosis and treatment has been stressed in the management of acute septicemic melioidosis. However, symptoms are so varied and the demise of the patient so rapid that, unless a clinician is alert to the possibility of a *B. pseudomallei* infection, the disease may be wrongly diagnosed (Arakawa, 1990). The acute septicemic form is usually associated with an underlying disease or condition such as diabetes, alcoholism, or immunodeficiency and, despite current antibiotic regimes, carries a high mortality rate (50–75%) (Majid, 1990).

Lesions apparent during the progression of melioidosis were shown to contain a large number of infiltrating cells, mainly macrophages, many of which contained phagocytosed bacilli (Wong et al., 1995). Most of the bacillary cells were surrounded by an electron-dense substance and had a clear halo between the substance and the cell wall (Narita et al., 1982). Following uptake, ingested *B. pseudomallei* escape from the vacuoles, possibly as a result of bacterium-mediated vacuolar membrane damage (Harley et al., 1994).

Each clinical case of melioidosis may represent the outcome of one of three possible processes: primary infection, reinfection, or reactivation of latent disease. Primary infection appears almost inevitable in children in northeast Thailand, where approximately 80% have developed antibodies to *B. pseudomallei* by the age of 4 years. Nearly all these infections appear to be asymptomatic or subclinical. In most cases, adult melioidosis results from reinfection or reactivation of latent infections (Maharjan et al., 2005; Suputtamongkol et al., 1994). Serological studies in Malaysia also found a high incidence of antibodies to *B. pseudomallei*, presumably due to mild, unrecognized, or inapparent infections (Strauss et al., 1969). *B. pseudomallei* infections are also prevalent in animals, and cases have been reported in captive dolphins, sheep, pigs, and goats (Currie et al., 1994; Dance, 1991; Vedros et al., 1988).

TREATMENT

Early positive laboratory diagnosis and appropriate chemotherapy are essential for the successful management of glanders and melioidosis. If these prerequisites are met, then the prognosis of these diseases are good and fatalities are limited to those patients who present with overwhelming, fulminating septicemia, which cannot be halted by current medical treatment (Ashdown et al., 1980). The antibiotic susceptibility profile of *B. mallei* and *B. pseudomallei* are similar, and both pathogens are susceptible in vitro to ceftazidime, imipenem, meropenem, and doxycycline (Kenny et al., 1999).

The small number of human cases of glanders means that there is limited information about antibiotic treatment, although sulfadiazine has been found to be effective in experimental animals and humans (Batmanov, 1993). An accidental exposure of a microbiologist to *B. mallei* provided an insight into the treatment and management of glanders (Srinivasan et al., 2001; Waag and Deshazer, 2004). The infection was not controlled by cephalosporin or clarithromycin, and symptoms returned after completion of the course of antibiotics. Intravenous tobramycin and doxycycline were more effective in controlling the infection, but lesions formed in the liver contained *B. mallei*. Treatment with imipenem and doxycycline, followed by oral doxycycline and azithromycin, resulted in resolution of the lesions in the spleen and the liver of the patient. There was no reported spread of the disease to whom the patient came in contact (Waag and Deshazer, 2004).

The conventional treatment regimen for melioidosis infections in Thailand was a combination of chloramphenicol, doxycycline, and cotrimoxazole (Dance et al., 1989). Fluoroquinolone antibiotics were found to be inappropriate for the treatment of melioidosis (Winton et al., 1988). Trials carried out in the 1980s recommended ceftazidime as the antibiotic of choice to replace the use of high doses of chloramphenicol, doxycycline, trimethoprim–sulfamethoxazole, and sometimes kanamycin (Chau et al., 1986). Ceftazidime, widely used as a single drug treatment against *Pseudomonas aeruginosa* septicemia, is safe, well tolerated, and highly active against *B. pseudomallei* in vitro, and its use in the treatment of melioidosis has halved the mortality of severe cases (White et al., 1989). The case fatality rate is still high (≈40%), with an increasing number of reports of resistance to currently used drug treatments (Dance et al., 1991; Godfrey et al., 1991). The treatment of choice is high-dose intravenous ceftazidime, meropenem, or imipenem/cilastatin administered for 10–14 days (or longer as clinically indicated), followed by oral trimethoprim–sulfamethoxazole–doxycyline for 12–20 weeks (Wuthiekanun and Peacock, 2006). Treatment of melioidosis patients who presented with septic shock in Australia with meropenem and granulocyte colony-stimulating factor (G-CSF) resulted in a large reduction in mortality from 95% to 10% (Cheng et al., 2004). However, this may be associated with improved intensive-care management during the study period. To address this issue, a randomized, placebo-controlled trial of G-CSF treatment has been conducted in Thailand (Peacock, 2006).

PATHOGENESIS

The pathogenic mechanisms of *B. mallei* and *B. pseudomallei* remain unclear, reflecting the various clinical presentations of both diseases. Current knowledge suggests that a complex interaction of a number of factors influence the course and nature of diseases. These factors include the size of the infectious dose, the route of infection, host factors, and bacterial virulence determinants. Although some virulence determinants in both *B. pseudomallei* and *B. mallei* have been identified, they are poorly characterized.

B. pseudomallei is reported to produce several toxic factors. Culture filtrates were found to contain both toxic and proteolytic activity and a heat-stable molecule with classical endotoxin activity (Colling et al., 1958; Heckly and Nigg, 1958). Further characterization showed the presence of a heat-labile exotoxin that is thought to suppress cellular immune functions (Heckly, 1964; Yahya, 1995). The production of hemolysin appears to be linked to population density. Well-isolated colonies of *B. pseudomallei* appear completely nonhemolytic; however, smaller colonies in a crowded environment are hemolytic (Ashdown, 1990; Liu, 1957).

Other putative extracellular virulence determinants include a 36-kDa protease (Sexton et al., 1994). Metalloproteases have been identified in many bacterial species and often play roles in the pathogenesis of disease (Kooi and Sokol, 1996). Cell-associated virulence determinants include pili and the extracellular polysaccharide (Brett et al., 1994). The flagella of *B. pseudomallei* have been suggested to play roles in virulence. However, nonmotile mutants are reported to be fully virulent in the diabetic rat model of disease but attenuated by the intranasal BALB/c mouse model and in the nematode model of infection (Essex-Lopresti et al., 2005; Deshazer et al., 1997). Therefore, the role of flagella in disease in humans is not clear.

TTSS play an important role in intracellular invasion and survival of gram-negative facultative intracellular pathogens. A type III secretion apparatus has been identified in *B. pseudomallei*, which shows strong homology to the *Salmonella enterica* pathogenicity island 1 (SPI-1) (Rainbow et al., 2002; Stevens et al., 2002). It has been demonstrated that mutations in *B. pseudomallei* genes involved in TTSS3 affect host cell invasion, resulting in bacteria being confined to endocytic vesicles (Stevens et al., 2002, 2003). Interestingly, strains of *B. pseudomallei* with mutations in TTSS3 demonstrated varying degrees of attenuation in animal models of infections (Stevens et al., 2004). TTSS are also thought to be important in facilitating cell-to-cell spread of *B. mallei* (Ribot and Ulrich, 2006). More recently, in *B. mallei*, a type VI secretion system has been identified and shown to be important for virulence in the hamster model of glanders infection (Schell et al., 2007). Furthermore, a protein secreted by the type VI secretion system showing homology to the hemolysin-coregulated protein (Hcp) family was recognized by glanders antisera, indicating that this protein is expressed in vivo (Schell et al., 2007).

Other key virulence determinants of both *B. mallei* and *B. pseudomallei* are surface polysaccharides, two of which have been well characterized. Capsular polysaccharide and LPS are also known as type I *O*-PS and type II *O*-PS, respectively (Deshazer et al., 1998; Reckseidler et al., 2001; Perry et al., 1995). The capsular polysaccharide is an unbranched homopolymer of -3)-2-*O*-acetyl-6-deoxy-β-d-manno-heptopyranose-(1-. The loss of capsule expression results in marked attenuation of *B. pseudomallei* and *B. mallei* (Atkins et al., 2002; Deshazer et al., 2001; Reckseidler et al., 2001). The significance of this polysaccharide in virulence is also reflected by the absence of the encoding gene cluster in *Burkholderia thailandensis*, a nonpathogenic bacterium that is closely related to *B. pseudomallei* (Ong et al., 2004; Kespichayawattana et al., 2004; Brett et al., 1998; Yu et al., 2006). The function of the capsular polysaccharide is not fully understood, but it may provide resistance to phagocytosis by reducing levels of complement C3 deposition (Reckseidler-Zenteno et al., 2005).

LPS consists of an unbranched heteropolymer with repeating d-glucose and l-talose units, with the structure -3)-β-d-glucopyranose-(-3)-6-deoxy-α-l-talopyranose-(1-. Approximately one-third of the l-talose residues bear 2-*O*-methyl and 4-*O*-acetyl substituents, and the remainders have 2-*O*-acetyl modifications (Knirel et al., 1992; Perry et al., 1995). The *B. pseudomallei O*-antigen appears to have properties typical of other bacterial *O*-antigens, for example, by providing resistance to serum killing (Deshazer et al., 1998). However, the LPS also appears to be less toxic and pyrogenic, compared to other enterobacterial LPS and is less active in stimulating murine macrophages, although it is more mitogenic (Matsuura et al., 1996). Antigenic studies suggest that the LPS is present in the avirulent *B. thailandensis* strain as well as *B. pseudomallei* and *B. mallei*, questioning the contribution of the LPS to infection (Anuntagool et al., 1998). However, previous studies have shown that LPS mutants of *B. pseudomallei* are markedly attenuated in animal models of disease (Deshazer et al., 1998).

Immunological studies with mabs demonstrate the expression of an exopolysaccharide by strains of *B. pseudomallei* and *B. mallei*. There was no cross-reactivity of the mab with other species of *Burkholderia* or *Pseudomonas* (Steinmetz et al., 1995). Two simultaneous, independent studies elucidated the structure of this polysaccharide to consist of a linear tetrasaccharide repeating unit comprising of three galactose residues, one bearing a 2-linked *O*-acetyl group, and a 3-deoxy-d-manno-2-octulosonic acid (KDO) residue (Masoud et al., 1997; Nimtz et al., 1997). The encoding gene cluster for the exopolysaccharide is yet to be determined. However, in addition to the gene clusters encoding capsular polysaccharide and LPS, two other gene clusters have been reported (Holden et al., 2004). The functions of these clusters are unknown, but mice challenged with *B. pseudomallei* strains with mutations in each of these clusters showed a delayed time to death, compared to wild-type strains (Sarkar-Tyson et al., 2007), suggesting that they play roles in virulence.

PROSPECTS FOR A VACCINE

There is no experimental or licensed vaccine against glanders or melioidosis, and the development of effective vaccines or alternative therapies is an essential requirement. Infection with both pathogens clearly involves the interaction of a number of virulence factors and is dependent on the host's susceptibility to infection. Because of the genetic and biochemical similarities between *B. pseudomallei* and *B. mallei* and the similar mechanisms of virulence, it is conceivable that one vaccine would provide protection against both glanders and melioidosis. Previous studies have shown that antibodies to the flagella, LPS, and capsular polysaccharides are protective against *B. pseudomallei* and *B. mallei* (Brett and Woods, 1996; Brett et al., 1994; Charuchaimontri et al., 1999; Ho et al., 1997; Jones et al., 2002; Nelson et al., 2004; Trevino et al., 2006). The possibility of an *O*-polysaccharide-antigen-flagellin protein conjugate was investigated and shown to elicit high titers of IgG, which was capable of protecting diabetic rats from *B. pseudomallei* challenge (Brett and Woods, 1996). However, this vaccine may not be effective against *B. mallei*, which does not express flagella. It has been reported that an acellular vaccine has been used successfully in captive marine mammals to protect against *B. pseudomallei* infection. The vaccine contains protein and polysaccharide (1:3) and induces high levels of specific antibodies to *B. pseudomallei* (Vedros et al., 1988). The use of this vaccine has significantly reduced mortality in marine mammals. Alternative vaccine strategies, such as DNA vaccine strategies, have been investigated using the flagellin-encoding gene of *B. pseudomallei*, *fliC*, and immunization resulted in both a humoral and cellular immune response (Chen et al., 2006b). Subsequent challenge with wild-type *B. pseudomallei* resulted in reduced bacteremia in the liver and spleens of vaccinated animals (Chen et al., 2006b). Introduction of CpG oligodeoxynucleotide to the DNA plasmid/*fliC* construct further improved the levels of resistance to *B. pseudomallei* infection, possibly by biasing a Th-1-type immune response (Chen et al., 2006a). There have been attempts to identify potential protective subunits, including immunoscreening of a lambda Zap II-expressed genomic library of *B. pseudomallei* genes and a proteomic-based method to specifically select for immunogenic, surface-located antigens (Harding et al., 2007; Jitsurong et al., 2003). Several immunogenic proteins have been identified and are currently being investigated for their protective efficacy against melioidosis and glanders (Harding et al., 2007).

Inactivated whole-cell vaccines also show potential as the basis for vaccines. Mice immunized with heat-inactivated *B. pseudomallei* were protected after lethal challenge (Razak et al., 1986). In our laboratory, we have shown that mice immunized with heat-inactivated *B. pseudomallei* are protected, compared to unvaccinated mice against a subsequent i.n. challenge with either *B. pseudomallei* or *B. mallei* (Sarkar-Tyson, unpublished data). These results indicate that a vaccine might be devised that protects against both diseases. The use of immunostimulatory CpG 1826 in conjunction with dendritic cells, pulsed with heat-killed *B. pseudomallei*, proved to be effective in providing protection against virulent wild-type strains in the BALB/c animal model of infection (Elvin et al., 2006). These studies demonstrate that inactivated strains may be effective as vaccines against melioidosis and glanders and merit further investigation.

Avirulent strains and auxotrophs of *B. pseudomallei* and *B. mallei* have been used to vaccinate animals prior to respiratory challenge with mixed results (Dannenberg and Scott, 1958; Levine and Maurer, 1958; Ulrich et al., 2005; Atkins et al., 2002). *B. pseudomallei* mutants with auxotrophies for purines, branched-chain amino acids, or aromatic amino acids have been shown to be attenuated and protective in murine models of disease. For example, immunization with a branched-chain amino acid auxotroph provided protection against at least 100 MLD doses of wild-type *B. pseudomallei* (Atkins et al., 2002). The abilities of these *B. pseudomallei* auxotrophs to induce protection against inhalation disease are not known. A *B. mallei* mutant with auxotrophy for branched-chain amino acids has also been constructed and tested as a vaccine (Ulrich et al., 2005). Aerogenic immunization with this mutant provided limited protection against a subsequent inhalation challenge with *B. mallei* (Ulrich et al., 2005). While live-attenuated strains of

B. pseudomallei and *B. mallei* show potential as vaccines, there are some significant concerns associated with their use. In the study reported by Ulrich et al. (2005), immunity was not sterile, and although "protected" animals did not develop acute diseases, abscesses were apparent in spleens and livers of immunized mice. Even if sterile immunity could be induced, it is not clear whether attenuated mutants would be licensable as vaccines because of the possibility of long-term persistence of the vaccine.

KEY ISSUES

- *B. mallei* and *B. pseudomallei* elicit humoral and cell-mediated immune responses to infection. Further research into these responses should provide essential information for vaccine development.
- Vaccination against disease caused by *B. mallei* or *B. pseudomallei* is possible. Both inactivated whole-cell and live-attenuated vaccines are effective, although it may be difficult to license live-attenuated vaccines.
- Subunit vaccines based on polysaccharides or proteins alone may not provide sufficient protection against melioidosis or glanders, and multicomponent or polysaccharide–protein conjugates may elicit more effective responses.

References

Aardema, H., Luijnenburg, E.M., Salm, E.F., Bijlmer, H.A., Visser, C.E. and Van't Wout, J.W. Changing epidemiology of melioidosis? A case of acute pulmonary melioidosis with fatal outcome imported from Brazil. Epidemiol. Infect. 2005; 133:871–875.

Alibek, K. Biohazard. Random House, New York, NY, 1999.

Anuntagool, N., Intachota, P., Wuthiekanun, V., White, N. and Sirisinha, S. Lipopolysaccharide from nonvirulent Ara^+ *Burkholderia pseudomallei* isolates is immunologically indistinguishable from lipopolysaccharide from virulent Ara^- clinical isolates. Clin. Diagn. Lab. Immunol. 1998; 5:225–229.

Arakawa, M. Infection with *Pseudomonas pseudomallei*. Rinsho Byori 1990; 11:1226–1231.

Arun, S., Neubauer, H., Gurel, A., Ayyildiz, G., Kuscu, B., Yesildere, T., Meyer, H. and Hermanns, W. Equine glanders in Turkey. Vet. Rec. 1999:255–258.

Ashdown, L.K.J. Production of hemolysin and other extracellular enzymes by clinical isolates of *Pseudomonas pseudomallei*. J. Clin. Microbiol. 1990; 28:2331–2334.

Ashdown, L., Duffy, V. and Douglas, R.A. Melioidosis. Med. J. Aust. 1980; 1:314–316.

Ashdown, L.R. Indirect hemagglutination test for melioidosis. Med. J. Aust. 1987; 147:364–365.

Atkins, T.P., Prior, R.G., Mack, K. et al. Characterisation of an acapsular mutant of *Burkholderia pseudomallei* identified by signature tagged mutagenesis. J. Med. Microbiol. 2002; 51:539–547.

Barnes, J.L., Warner, J., Melrose, W., Durrheim, D., Speare, R., Reeder, J.C. and Ketheesan, N. Adaptive immunity in melioidosis: a possible role for T cells in determining outcome of infection with *Burkholderia pseudomallei*. Clin. Immunol. 2004; 113:22–28.

Batmanov, V.P. Treatment of experimental glanders with combinations of sulfazine or sulfamonomethoxine with trimethoprim. Antibiot. Khimioter. 1993; 38:18–22.

Bottex, C., Gauthier, Y.P., Hagen, R.M., Finke, E.J., Splettstosser, W.D., Thibault, F.M., Neubauer, H. and Vidal, D.R. Attempted passive prophylaxis with a monoclonal anti-*Burkholderia pseudomallei* exopolysaccharide antibody in a murine model of melioidosis. Immunopharmacol. Immunotoxicol. 2005; 27:565–583.

Benenson, A.S. Control of communicable diseases manual, 6th edn, Washington, DC: American Public Health Association, 1995; 38:335–338.

Brett, P., Deshazer, D. and Woods, D. *Burkholderia thailandensis* sp. nov., a *Burkholderia pseudomallei*-like species. Int. J. Syst. Bacteriol. 1998; 48:317–320.

Brett, P. and Woods, D. Structural and immunological characterization of *Burkholderia pseudomallei* O-polysaccharide-flagellin protein conjugates. Infect. Immun. 1996; 64:2824–2828.

Brett, P.J., Mah, D.C. and Woods, D.E. Isolation and characterization of *Pseudomonas pseudomallei* flagellin proteins. Infect. Immun. 1994; 62:1914–1919.

Chantratita, N., Wuthiekanun, V., Boonbumrung, K. et al. Biological relevance of colony morphology and phenotypic switching by *Burkholderia pseudomallei*. J. Bacteriol. 2007; 189:807–817.

Charuchaimontri, C., Suputtamongkol, Y., Nilakul, C., Chaowagul, W., Chetchotisakd, P., Lertpatanasuwan, N., Intaranongpai, S., Brett, P. and Woods, D. Antilipopolysaccharide II: an antibody protective against fatal melioidosis. Clin. Infect. Dis. 1999; 29:813–818.

Chau, P.Y., Ng, W.S., Leung, Y.K. and Lolekha, S. In vitro susceptibility of strains of *Pseudomonas pseudomallei* isolated in Thailand and Hong Kong to some newer beta-lactam antibiotics and quinolone derivatives. J. Infect. Dis. 1986; 153:167–170.

Chen, Y.S., Hsiao, Y.S., Lin, H.H., Liu, Y. and Chen, Y.L. CpG-modified plasmid DNA encoding flagellin improves immunogenicity and provides protection against *Burkholderia pseudomallei* infection in BALB/c mice. Infect. Immun. 2006a; 74:1699–1705.

Chen, Y.S., Hsiao, Y.S., Lin, H.H., Yen, C.M., Chen, S.C. and Chen, Y.L. Immunogenicity and anti-*Burkholderia pseudomallei* activity in Balb/c mice immunized with plasmid DNA encoding flagellin. Vaccine 2006b; 24:750–758.

Cheng, A.C. and Currie, B.J. Melioidosis: epidemiology, pathophysiology, and management. Clin. Microbiol. Rev. 2005; 18:383–416.

Cheng, A.C., Stephens, D.P., Anstey, N.M. and Currie, B.J. Adjunctive granulocyte colony-stimulating factor for treatment of septic shock due to melioidosis. Clin. Infect. Dis. 2004; 38:32–37.

Colling, M., Nigg, C. and Heckly, R. Toxins of *Pseudomonas pseudomallei*. I. Production in vitro. J. Bacteriol. 1958; 76:422–426.

Currie, B., Smith Vaughan, H., Golledge, C., Buller, N., Sriprakash, K.S. and Kemp, D.J. *Pseudomonas pseudomallei* isolates collected over 25 years from a non-tropical endemic focus show clonality on the basis of ribotyping. Epidemiol. Infect. 1994; 113:307–312.

Currie, B.J., Fisher, D.A. and Howard, D.M. Endemic melioidosis in tropical northern Australia: a 10-year prospective study and review of the literature. Clin. Infect. Dis. 2000; 31:981–986.

Dance, D.A.B. Melioidosis: the tip of the iceberg? Clin. Microbiol. Rev. 1991; 4:52–60.

Dance, D.A.B., Smith, M.D., Aucken, H.M. and Pitt, T.L. Imported melioidosis in England and Wales. Lancet 1999; 353:208.

Dance, D.A.B. and White, N.J. Melioidosis. In: The Wellcome Trust Illustrated History of Tropical Diseases (F.E.G. Cox, Ed.), pp. 72–81. London: The Wellcome Trust, 1996.

Dance, D.A.B., Wuthiekanun, V., Chaowagul, W., Suputtamongkol, Y. and White, N.J. Development of resistance to ceftazidime and co-amoxiclav in *Pseudomonas pseudomallei*. J. Antimicrob. Chemother. 1991; 28:321–324.

Dance, D.A.B., Wuthiekanun, V., Chaowagul, W. and White, N.J. Interactions in vitro between agents used to treat melioidosis. J. Antimicrob. Chemother. 1989; 24:311–316.

Dannenberg, A.M. and Scott, E.M. Melioidosis: pathogenesis and immunity in mice and hamsters. II. Studies with avirulent strains of *Malleomyces pseudomallei*. Am. J. Pathol. 1958; 34:1099–1121.

Derbyshire, J.P. The eradication of glanders in Canada. Can. Vet. J. 2002; 43:722–726.

Deshazer, D., Brett, P., Carlyon, R. and Woods, D. Mutagenesis of *Burkholderia pseudomallei* with Tn5-OT182: isolation of motility mutants and molecular characterization of the flagellin structural gene. J. Bacteriol. 1997; 179:2116–2125.

Deshazer, D., Brett, P. and Woods, D. The type II O-antigenic polysaccharide moiety of *Burkholderia pseudomallei* lipopolysaccharide is required for serum resistance and virulence. Mol. Microbiol. 1998; 30:1081–1100.

Deshazer, D., Wagg, D., Fritz, D. and Woods, D. Identification of a *Burkholderia mallei* polysaccharide gene cluster by subtractive hybridisation and demonstration that the encoded capsule is an essential virulence determinant. Microb. Pathog. 2001; 30:253–269.

Elvin, S.J., Healey, G.D., Westwood, A., Knight, S.C., Eyles, J.E. and Williamson, E.D. Protection against heterologous *Burkholderia pseudomallei* strains by dendritic cell immunization. Infect. Immun. 2006; 74:1706–1711.

Essex-Lopresti, A.E., Boddey, J.A., Thomas, R. et al. A type IV pilin, PilA, contributes to adherence of *Burkholderia pseudomallei* and virulence in vivo. Infect. Immun. 2005; 73:1260–1264.

Everett, E. and Nelson, R. Pulmonary melioidosis. Am. Rev. Respir. Dis. 1975; 112:331–341.

Fritz, D.L., Vogel, P., Brown, D.R., Deshazer, D. and Waag, D.M. Mouse Model of sublethal and lethal intraperitoneal glanders (*Burkholderia mallei*). Vet. Pathol. 2000; 37:626–636.

Godfrey, A.J., Wong, S., Dance, D.A.B., Chaowagul, W. and Bryan, L.E. *Pseudomonas pseudomallei* resistance to beta-lactam antibiotics due to alterations in the chromosomally encoded beta-lactamase. Antimicrob. Agents Chemother. 1991; 35:1635–1640.

Godoy, D., Randle, G., Simpson, A.J., Aanensen, D.M., Pitt, T.L., Kinoshita, R. and Spratt, B.G. Multilocus sequence typing and evolutionary relationships among the causative agents of melioidosis and glanders, *Burkholderia pseudomallei* and *Burkholderia mallei*. J. Clin. Microbiol. 2003; 41:2068–2079.

Haque, A., Easton, A., Smith, D., O'Garra, A., Van Rooijen, N., Lertmemongkolchai, G., Titball, R.W. and Bancroft, G.J. Role of T cells in innate and adaptive immunity against murine *Burkholderia pseudomallei* infection. J. Infect. Dis. 2006; 193:370–379.

Harding, S.V., Sarkar-Tyson, M., Smither, S.J., Atkins, T.P., Oyston, P.C.F., Brown, K.A., Liu, Y.C., Wait, R. and Titball, R.W. The identification of surface proteins of *Burkholderia pseudomallei*. Vaccine 2007; 25:2664–2672.

Harley, V.S., Dance, D.A., Tovey, G. and Drasar, B.S. Interaction of *Pseudomonas pseudomallei* with macrophages. Biochem. Soc. Trans. 1994; 22:88S.

Heckly, R. Differentiation of exotoxin and other biologically active substances in *Pseudomonas pseudomallei* filtrates. J. Bacteriol. 1964; 88:1730–1736.

Heckly, R. and Nigg, C. Toxins of *Pseudomonas pseudomallei*. J. Bacteriol. 1958; 76:427–436.

Ho, M., Schollaardt, T., Smith, M., Perry, M., Brett, P., Chaowagul, W. and Bryan, L. Specificity and functional activity of anti-*Burkholderia pseudomallei* polysaccharide antibodies. Infect. Immun. 1997; 65:3648–3653.

Holden, M.T.G., Titball, R.W., Peacock, S.J. et al. Genomic plasticity of the causative agent of melioidosis, *Burkholderia pseudomallei*. Proc. Natl. Acad. Sci. U.S.A. 2004; 101:14240–14245.

Inglis, T.J.J., Rolim, D.B. and Sousa, A.D. Melioidosis in the Americas. Am. J. Trop. Med. Hyg. 2006; 75:947–954.

Ip, M., Osterberg, L.G., Chau, P.Y. and Raffin, T.A. Pulmonary melioidosis. Chest 1995; 108:1420–1424.

Jitsurong, S., Chirathaworn, C., Brown, N.F., Beacham, I.R. and Tumwasorn, S. Searching for virulence *Burkholderia pseudomallei* genes by immunoscreening the lambda Zap II expressed genomic library. Southeast Asian J. Trop. Med. Public Health 2003; 34:810–821.

Jones, S.M., Ellis, J.F., Russell, P., Griffin, K.F. and Oyston, P.C.F. Passive protection against *Burkholderia pseudomallei* infection in mice by monoclonal antibodies against capsular polysaccharide, lipopolysaccharide or proteins. J. Med. Microbiol. 2002; 51:1055–1062.

Kanai, K. and Kondo, E. Recent advances in biomedical sciences of *Burkholderia pseudomallei* (basonym: *Pseudomonas pseudomallei*). Jpn. J. Med. Sci. Biol. 1994; 47:1–45.

Kenny, D.J., Russell, P., Rogers, D., Eley, S.M. and Titball, R.W. In vitro susceptibility of *Burkholderia mallei* in comparison to those of other pathogenic *Burkholderia* spp. Antimicrob. Agents Chemother. 1999; 43:2773–2775.

Kespichayawattana, W., Intachote, P., Utaisincharoen, P. and Sirisinha, S. Virulent *Burkholderia pseudomallei* is more efficient than avirulent *Burkholderia thailandensis* in invasion of and adherence to cultured human epithelial cells. Microbial. Pathog. 2004; 36:287–292.

Knirel, Y., Paramonov, N.A., Shashkov, A.S., Kochetkov, N.K., Yassin, R.M., Farber, S.M. and Efremenko, V.I. Structure of the polysaccharide chains of *Pseudomonas pseudomallei* lipopolysaccharides. Carbohydr. Res. 1992; 233:185–193.

Kooi, C. and Sokol, P.A. Differentiation of thermolysins and serralysins by monoclonal antibodies. J. Med. Microbiol. 1971; 48:63–70.

Kovalev, G.K. Glanders. 2h. Mikrobiol. Epidemiol. Immunobiol. 1971; 48:63–70.

Krishnaswami, C.S. Morphia injectors' septicaemia. Ind. Med. Gaz. 1917; 52:296–299.

Leelarasamee, A. and Bovornkitti, S. Melioidosis: a review and update. Rev. Infect. Dis. 1989; 11:413–425.

Lertmemongkolchai, G., Cai, G.F., Hunter, C.A. and Bancroft, G.J. Bystander activation of CD8(+) T cells contributes to the rapid production of IFN-gamma in response to bacterial pathogens. J. Immunol. 2001; 166:1097–1105.

Levine, H. and Maurer, R. Immunization with an induced avirulent auxotrophic mutant of *Pseudomonas pseudomallei*. J. Immunol. 1958; 81:433–438.

Liu, P. Survey of hemolysin production among species of pseudomonads. J. Bacteriol. 1957; 74:718–727.

Maharjan, B., Chantratita, N., Vesaratchavest, M., Cheng, A., Wuthiekanun, V., Chierakul, W., Chaowagul, W., Day, N.P.J. and Peacock, S.J. Recurrent melioidosis in patients in northeast Thailand is frequently due to reinfection rather than relapse. J. Clin. Microbiol. 2005; 43:6032–6034.

Majid, A.A. Successful surgical management of a case of pulmonary and pericardial melioidosis. Aust. N.Z. J. Surg. 1990; 60:139–141.

Masoud, H., Ho, M., Schollaardt, T. and Perry, M. Characterization of the capsular polysaccharide of *Burkholderia (Pseudomonas) pseudomallei* 304b. J. Bacteriol. 1997; 179:5663–5669.

Matsuura, M., Kawahara, K., Ezaki, T. and Nakano, M. Biological activities of lipopolysaccharide of *Burkholderia (Pseudomonas) pseudomallei*. FEMS Microbiol. Lett. 1996; 137:79–83.

Monastyrskaya, G., Fushan, A., Abaev, I., Filyukova, O., Kostina, M., Pecherskih, E. and Sverdlov, E. Genome-wide comparison reveals great inter- and intraspecies variability in *B. pseudomallei* and *B. mallei* pathogens. Res. Microbiol. 2004; 155:781–793.

Narita, M., Loganathan, P., Hussein, A., Jamaluddin, A. and Joseph, P. Pathological changes in goats experimentally inoculated with *Pseudomonas pseudomallei*. Natl. Inst. Anim. Health Q. (Tokyo) 1982; 22:170–179.

Nelson, M., Prior, J.L., Lever, M.S., Jones, H.E., Atkins, T.P. and Titball, R.W. Evaluation of lipopolysaccharide and capsular polysaccharide as subunit vaccines against experimental melioidosis. J. Med. Microbiol. 2004; 53:1177–1182.

Neubauer, H., Sprague, D., Zachariah, R., Tomaso, H., Al Dahouk, S., Wernery, R., Wernery, U. and Scholz, H.C. Serodiagnosis of Burkholderia mallei infections in horses: state-of-the-art and perspectives. J. Vet. Med. B Infect. Dis. Vet. Public Health 2005; 52(5):201–205.

Ngauy, V., Lemeshev, Y., Sadkowski, L. and Crawford, G. Cutaneous melioidosis in a man who was taken as a prisoner of war by the Japanese during World War II. J. Clin. Microbiol. 2005; 43:970–972.

Nieminen, T. and Vaara, M. *Burkholderia pseudomallei* infections in Finnish tourists injured by the December 2004 tsunami in Thailand. Euro Surveill. 2005; 10.

Nierman, W.C., Deshazer, D., Kim, H.S. et al. Structural flexibility in the *Burkholderia mallei* genome. Proc. Natl. Acad. Sci. U.S.A. 2004; 101:14246–14251.

Nimtz, M., Wray, V., Domke, T., Brenneke, B., Haussler, S. and Steinmezt, I. Structure of an acidic exopolysaccharide of *Burkholderia pseudomallei*. Eur. J. Biochem. 1997; 250:608–616.

Ong, C., Ooi, C.H., Wang, D.L., Chong, H.L., Ng, K.C., Rodrigues, F., Lee, M.A. and Tan, P. Patterns of large-scale genomic variation in virulent and avirulent *Burkholderia* species. Genome Res. 2004; 14:2295–2307.

Ou, K.L., Ong, C., Koh, S.Y. et al. Integrative genomic, transcriptional, and proteomic diversity in natural isolates of the human pathogen *Burkholderia pseudomallei*. J. Bacteriol. 2005; 187:4276–4285.

Peacock, S.J. Melioidosis. Curr. Opin. Infect. Dis. 2006; 19:421–428.

Perry, M.B., MacLean, L.L., Schollaardt, T., Bryan, L.E. and Ho, M. Structural characterisation of the lipopolysaccharide O antigens of *Burkholderia pseudomallei*. Infect. Immun. 1995; 63:3348–3352.

Pitt, T. *Pseudomonas mallei* and *P. pseudomallei*. In: Topley & Wilson's Principles of Bacteriology, Virology and Immunity (M.T. Parker and L.H. Collier, Eds.), 8th ed., pp. 265–268. BC Decker Inc., Philadelphia, Pl. 1990.

Puthucheary, S., Parasakthi, N. and Lee, M.K. Septicaemic melioidosis: a review of 50 cases from Malaysia. Trans. R. Soc. Trop. Med. Hyg. 1992; 86:683–685.

Rainbow, L., Hart, C.A. and Winstanley, G. Distribution of type III secretion gene clusters in *Burkholderia pseudomallei, B. thailandensis* and *B. mallei*. J. Med. Microbiol. 2002; 51:374–384.

Razak, C.E., Ismail, G., Embi, N. and Omar, O. Protection studies using whole cells and partially purified toxic material (PPTM) of *Pseudomonas pseudomallei*. Malays. Appl. Biol. 1986; 15:105–111.

Reckseidler, S., Deshazer, D., Sokol, P. and Woods, D. Detection of bacterial virulence genes by subtractive hybridisation: identification of capsular polysaccharide of *Burkholderia pseudomallei* as a major virulence determinant. Infect. Immun. 2001; 69:34–44.

Reckseidler-Zenteno, S.L., DeVinney, R. and Woods, D.E. The capsular polysaccharide of *Burkholderia pseudomallei* contributes to survival in serum by reducing complement factor C3b deposition. Infect. Immun. 2005; 73:1106–1115.

Redfearn, M., Palleroni, N. and Stanier, R. A comparative study of *Pseudomonas pseudomallei* and *Bacillus mallei*. J. Gen. Microbiol. 1966; 43:293–313.

Ribot, W.J. and Ulrich, R.L. The animal pathogen-like type III secretion system is required for the intracellular survival of *Burkholderia mallei* within J774.2 macrophages. Infect. Immun. 2006; 74:4349–4353.

Rodrigues, F., Sarkar-Tyson, M., Harding, S.V. et al. Global map of growth-regulated gene expression in *Burkholderia pseudomallei*, the causative agent of melioidosis. J. Bacteriol. 2006; 188:8178–8188.

Rubin, H.L., Alexander, A.D. and Yager, R.H. Melioidosis—a military medical problem? Mil. Med. 1963; 128:538–542.

Sanford, J.P. *Pseudomonas* species (including melioidosis and glanders). In: Infectious Diseases and Their Etiologic Agents. (G.L. Mandell, J.A. Bennett, R. Dolin, Eds.), 4th ed., New York, NY: Churchill Livingstone 1995; pp. 2003–2009.

Sanford, J.P. and Moore, W.L. Recrudescent melioidosis: a Southeast Asian legacy? Am. Rev. Respir. Dis. 1971; 104:452–453.

Santanirand, P., Harley, V., Dance, D., Drasar, B. and Bancroft, G. Obligatory role of gamma interferon for host survival in a murine model of infection with *Burkholderia pseudomallei*. Infect. Immun. 1999; 69:3593–3600.

Sarkar-Tyson, M., Thwaite, J.E., Harding, S.V., Smither, S.J., Oytson, P.C.F., Atkins, T.P. and Titball, R.W. Polysaccharides and virulence of *Burkholderia pseudomallei*. J. Med. Microbiol. 2007; 56(pt 8): 1005–1010.

Schell, M.A., Ulrich, R.L., Ribot, W.J., Brueggemann, E.E., Hines, H.B., Chen, D., Lipscomb, L., Kim, H.S., Mrázek, J., Nierman, W.C. and Deshazer, D. Type VI secretion is a major virulence determinant in *Burkholderia mallei*. Mol. Microbiol. 2007; 64:1466–1485.

Scholz, H.C., Joseph, M., Tomaso, H. et al. Detection of the reemerging agent *Burkholderia mallei* in a recent outbreak of glanders in the United Arab Emirates by a newly developed fliP-based polymerase chain reaction assay. Diagn. Microbiol. Infect. Dis. 2006; 54:241–247.

Sexton, M.M., Jones, A.L., Chaowagul, W. and Woods, D.E. Purification and characterization of a protease from *Pseudomonas pseudomallei*. Can. J. Microbiol. 1994; 40:903–910.

Srinivasan, A., Kraus, C.N., Deshazer, D., Becker, P.M., Dick, J.D., Spacek, L., Bartlett, J.G., Byrne, W.R. and Thomas, D.L. Glanders in a military research microbiologist. N. Engl. J. Med. 2001; 345:256–258.

Stanton, A.T. and Fletcher, W. Melioidosis, a new disease of the tropics. Trans. Fourth Congr. Far East Assoc. Trop. Med. 1921; 2:196–198.

Stanton, A.T. and Fletcher, W. Melioidosis: Studies from the Institute of Medical Research, Federated Malay States. London: John Bale & Sons and Danielson, 1932.

Steele, J.H. Glanders. In: CRC Handbook Series in Zoonoses (J.H. Steele, Ed.), pp. 339–362. Boca Raton, FL: CRC Press, 1979.

Steinmetz, I., Rohde, M. and Brenneke, B. Purification and characterization of an exopolysaccharide of *Burkholderia (Pseudomonas) pseudomallei*. Infect. Immun. 1995; 63:3959–3965.

Stevens, M., Wood, M.W., Taylor, L.A., Monaghan, P., Hawes, P., Jones, P.W., Wallis, T.S. and Galyov, E.E. An Inv/Mxi-Spa-like type III protein secretion system in *Burkholderia pseudomallei* modulates intracellular behaviour of the pathogen. Mol. Microbiol. 2002; 46:649–659.

Stevens, M.P., Friebel, A., Taylor, L.A., Wood, M.W., Brown, P.J., Hardt, W.D. and Galyov, E.E. A *Burkholderia pseudomallei* type III secreted protein, BopE, facilitates bacterial invasion of epithelial cells and exhibits guanine nucleotide exchange factor activity. J. Bacteriol. 2003; 185:4992–4996.

Stevens, M.P., Haque, A., Atkins, T. et al. Attenuated virulence and protective efficacy of a *Burkholderia pseudomallei* bsa type III secretion mutant in murine models of melioidosis. Microbiology 2004; 150:2669–2676.

Strauss, J.M., Alexander, A.D., Rapmund, G. and Dorsey, A.E. Melioidosis in Malaysia. III. Antibodies to *Pseudomonas pseudomallei* in the human population. Am. J. Trop. Med. Hyg. 1969; 18:703–707.

Suputtamongkol, Y., Hall, A.J., Dance, D.A., Chaowagul, W., Rajchanuwong, A., Smith, M.D. and White, N.J. The epidemiology of melioidosis in Ubon Ratchatani, northeast Thailand. Int. J. Epidemiol. 1994; 23:1082–1090.

Timoney, J.F., Gillespie, J.H., Scott, F.W. and Barough, J.E. *Pseudomonas mallei*. In: Hagan and Bruner's Microbiology and Infectious Diseases of Domestic Animals: with Reference to Etiology, Epizootiology, Pathogenesis, Immunity, Diagnosis, and Antimicrobial Susceptibility, pp. 40–44. Ithaca, New York: Cornell University Press, 1988.

Tiyawisutsri, R., Holden, M.T.G., Tumapa, S., Rengpipat, S., Clarke, S.R., Foster, S.J., Nierman, W.C., Day, N.P.J. and Peacock, S.J. *Burkholderia* Hep_Hap autotransporter (BuHA) proteins elicit a strong antibody response during experimental glanders but not human melioidosis. BMC Microbiol. 2007; 7:19.

Trevino, S.R., Permenter, A.R., England, M.J., Parthasarathy, N., Gibbs, P.H., Waag, D.M. and Chanh, T.C. Monoclonal antibodies passively protect BALB/c mice against *Burkholderia mallei* aerosol challenge. Infect. Immun. 2006; 74:1958–1961.

Ulrich, R.L., Amemiya, K., Waag, D.M., Roy, C.J. and Deshazer, D. Aerogenic vaccination with a *Burkholderia mallei* auxotroph protects against aerosol-initiated glanders in mice. Vaccine 2005; 23:1986–1992.

van Zijderveld, F.G. and Bongers, J. Glanders. In: Manual of Standards for Diagnostic Tests and Vaccines, pp. 576–581. Paris: Office International des Epizooties, 2000.

Vedros, N., Chow, D. and Liong, E. Experimental vaccine against *Pseudomonas pseudomallei* infections in captive cetaceans. Dis. Aquat. Org. 1988; 5:157–161.

Waag, D. and Deshazer, D. Glanders: new insights into an old disease. In: Biological Weapons Defense: Infectious Diseases and Counterbioterrorism (L.E. Linder, F.J. Lebeda and G. Korch, Eds.), pp. 209–237. Totowa, NJ: Humana Press Inc., 2004.

Wheelis, M. First shots fired in biological warfare. Nature 1998; 395:213.

White, N.J. Melioidosis. Lancet 2003; 361:1715–1722.

White, N.J., Chaowagul, W., Wuthiekanun, V., Dance, D.A.B., Wattanagoon, Y. and Pitakwatchara, N. Halving of mortality of severe melioidosis by ceftazidime. Lancet 1989; 2:697–701.

Whitmore, A. An account of a glanders-like disease occurring in Rangoon. J. Hyg. (Lond.) 1913; 13:1–34.

Whitmore, A. and Krishnaswami, C.S. An account of the discovery of a hitherto undescribed infective disease occurring among the population of Rangoon. Ind. Med. Gaz. 1912; 47:262–267.

Wiersinga, W.J., van der Poll, T., White, N.J., Day, N.P. and Peacock, S.J. Melioidosis: insights into the pathogenicity of *Burkholderia pseudomallei*. Nat. Rev. Microbiol. 2006; 4:272–282.

Wilkinson, L. Glanders: medicine and veterinary medicine in common pursuit of a contagious disease. Med. Hist. 1981; 25:363–384.

Winton, M.D., Everett, E.D. and Dolan, S.A. Activities of 5 new fluoroquinolones against *Pseudomonas pseudomallei*. Antimicrob. Agents Chemother. 1988; 32:928–929.

Wong, K., Puthucheary, S. and Vadivelu, J. The histopathology of human melioidosis. Histopathology 1995; 26:51–55.

Wuthiekanun, V. and Peacock, S.J. Management of melioidosis. Expert Rev. Anti Infect. Ther. 2006; 4:445–455.

Yabuuchi, E., Kosako, Y., Oyaizu, H., Yano, I., Hotta, H., Hashimoto, Y., Ezaki, T. and Arakawa, M. Proposal of *Burkholderia* gen. nov. and transfer of seven species of the genus *Pseudomonas* homology group II to the new genus, with the type species *Burkholderia cepacia* (Palleroni and Holmes 1981) comb. nov. Microbiol. Immunol. 1992; 36:1251–1275.

Yahya, D.L.S. Suppression of delayed type hypersensitivity in mice by *Burkholderia pseudomallei*. Malays. Appl. Biol. 1995; 24:105–107.

Yu, Y.T., Kim, H.S., Chua, H.H. et al. Genomic patterns of pathogen evolution revealed by comparison of *Burkholderia pseudomallei*, the causative agent of melioidosis, to avirulent *Burkholderia thailandensis*. BMC Microbiol. 2006; 6:46.

CHAPTER

44

Chlamydia

Roger G. Rank
Department of Microbiology and Immunology, University of Arkansas for Medical Sciences and Arkansas Children's Hospital Research Institute, Little Rock, AR, USA

OUTLINE

ABSTRACT

Chlamydiae are responsible for a variety of infections in humans including genital infections in men and women, ocular infections in the form of inclusion conjunctivitis and trachoma, community-acquired pneumonia, and atypical pneumonia. Chlamydial genital infections have a worldwide distribution and are the major bacterial cause of sexually transmitted disease that may ascend the genital tract causing salpingitis with potential sequelae of involuntary infertility and ectopic pregnancy in women. Trachoma is the most common cause of preventable blindness in the world but is concentrated in underdeveloped countries. Both genital and ocular diseases are primarily caused by the acute inflammatory response elicited by the infection of the host cell and the T cell response to the organism. However, infections resolve through both antibody- and Th1-mediated mechanisms. Currently, there are no vaccines for either disease in humans although intensive efforts to develop a trachoma vaccine including human trials occurred in the 1960s. While the whole organism vaccines employed were successful in developing some immunity over the short term, no long-term protection could be produced and trials were halted because of some evidence of a higher conversion rate in individuals given the vaccine in contrast to control subjects. The trials also made clear that immunity was serovar specific. No clinical trials have since been performed. Since the cessation of those trials, there has been little effort on a vaccine for trachoma, and the emphasis has shifted to a vaccine against genital infections. Current vaccine efforts are concentrating on specific bacterial proteins rather than whole organisms because of the cross-reactivity between human and chlamydial hsp60. The major outer membrane protein of chlamydiae that is surface-exposed and highly immunogenic can elicit a protective response and has been the subject of much work related to vaccine. It has the disadvantage in that it is serovar specific and the protective epitopes are conformational, making production of a recombinant molecule problematic. Other molecules with genus specificity such as the chlamydial protease-like activity factor and polymorphic membrane protein D have been described but not as thoroughly explored as potential vaccine candidates. It is important that the vaccine does not exacerbate pathologic response or increase the incidence of clinical disease. A vaccine must be able to elicit long-term immunity, something which did not occur with the original trachoma trials and which does not occur with natural infection. A strategy of preventing ascending infection may in part facilitate longer term protection. Because there are multiple serovars that cause both genital and ocular infections, it is critical that a vaccine contain either multiple serovars and/or genus-specific antigens. Persistence of chlamydial infection, although still undefined and uncharacterized in humans and animals, is a major concern. A vaccine must be evaluated to determine whether it can prevent persistence or whether the organism is still present but prevented from causing disease. Before a vaccine can go to clinical trials, appropriate surrogates for infertility and ectopic pregnancy must be determined and evaluated in animal models.

INTRODUCTION

Chlamydiae are highly adapted, primarily mucosal pathogens, infecting virtually all mucosal sites, depending on the species of *Chlamydia* and the species of the host. Chlamydial diseases have had and continue to have a profound impact on the health of individuals throughout the world. *Chlamydia trachomatis* is not only the leading cause of preventable blindness in the world through the production of scarring disease in the conjunctiva, but also the most common bacterial cause of sexually transmitted infections in the world, often leading to tubal obstruction and involuntary infertility in women. Moreover, *Chlamydia pneumoniae* infects 20–50% of the population in the United States producing community-acquired pneumonia and has been associated with potential exacerbation of coronary artery disease. Finally, *Chlamydia psittaci*, an avian chlamydia, can be zoonotic for humans, eliciting a respiratory infection with a high rate of mortality in untreated individuals. The latter, because of its high virulence and respiratory route of transmission, has been considered a possible bioterrorism agent. Thus, there are four different diseases caused by chlamydiae which either affect a significant number of people or have the potential, as in the case of *C. psittaci*, to affect a large number if employed as a bioterrorism agent.

For the purposes of this chapter, however, we will emphasize the pursuit of a vaccine(s) for genital tract infections and trachoma, because the widespread nature of these diseases and the vast impact they have on the human population worldwide make the development of a vaccine the most effective means of preventing the diseases and contributing significantly to the health of a large portion of the world's population. Because respiratory infections caused by *C. pneumoniae* are generally not severe and the effects of intervention on the development of coronary artery disease not clear (Grayston et al., 2005), there has been less enthusiasm for the pursuit of a vaccine against *C. pneumoniae* in recent years. Similarly, natural infections in humans caused by *C. psittaci* are uncommon and the employment of this agent in bioterrorism is less likely compared to other agents; hence, there is little interest in developing a human vaccine against *C. psittaci*. Nevertheless, since chlamydiae are remarkably similar, especially with regard to protective antigens, it is quite feasible that a vaccine developed for either genital tract infection or trachoma could employ comparable antigens and be utilized to prevent *C. pneumoniae* and *C. psittaci* infections as well.

SHORT HISTORY OF THE DISEASES

Trachoma is an ancient disease and one of the first diseases recognized as a distinct clinical entity with the first references to treatment for it in China in the 27th century BC. The exudative aspects of the disease and treatment with copper salts were described in the Ebers Papyrus around 1500 BC in Egypt (Thygeson, 1962). It was widespread in ancient Greece and Rome, and several historical figures, including Paul of Tarsus, Cicero, Horace, and Pliny the Younger, were thought to have had trachoma. The term "trachoma" was first coined by a Sicilian physician, Pedanius Dioscorides, in 60 AD, and the four stages of the disease were later described by Galen. The disease was spread from the Middle East to Europe by the knights of the Crusades and later in the late 18th century by Napoleon's troops as a result of his campaign in Egypt.

The inclusion bodies associated with the organism were first described in 1907 by Halberstaedter and von Prowazek (1907) when they observed them in eye scrapings from experimental trachoma in nonhuman primates and then later in scrapings from humans. They named the organism "chlamydozoa" or "mantle bodies" because of their mantle-like structure on Giemsa stain. The similarity between the organism from trachoma and the etiologic agents of psittacosis and lymphogranuloma venereum was eventually recognized and Bedson (1953) accepted the agents of trachoma and inclusion conjunctivitis into the psittacosis-LGV group.

In contrast to trachoma, the history of genital infections caused by chlamydiae is more recent. It was observed at the end of the 19th century that conjunctivitis could be found in neonates who were not infected with gonorrhea (Kroner, 1884). Halberstaedter and von Prowazek (1909) found inclusions in infants with inclusion conjunctivitis and, later, Heymann (1910) observed inclusions in cervical smears and the urethra of parents of a child with inclusion conjunctivitis. The first actual recovery and isolation of chlamydiae from the cervix was demonstrated by Jones et al. (1959) in 1959. Shortly thereafter, organisms were isolated from the male urethra (Jones, 1964). While isolation of chlamydiae from the lower genital tract from males and females became quite common, it was not until 1977, that Mardh et al. (1977) documented the association of *C. trachomatis* with salpingitis and were able to isolate organisms from the fallopian tubes. With those findings and the wide distribution of chlamydial genital infections, the significance and necessity for a vaccine became apparent.

ETIOLOGIC AGENT(S)

Classification

Chlamydiae are obligate intracellular bacteria that are unique for their biphasic developmental cycle, the elementary body or the non-replicating infectious phase, and the reticulate body or the noninfectious replicating stage. While generally host species specific, chlamydiae can be found in nature infecting a wide range of mammals, including humans, mice, guinea pigs, sheep, cattle, cats, hamsters, swine, and koala bears, and birds including budgerigar, parakeets, parrots, cockatiel, pigeons, doves, and turkeys (Tanner et al., 1999).

Taxonomically, chlamydiae have their own order, Chlamydiales, which includes the family, Chlamydiaeceae (Moulder et al., 1984). Over the last several years, there has been dispute as to whether there are one or two genera; however, it has now been agreed upon by the research community, that there is a single genus, *Chlamydia*, with species related to the host species (Bavoil and Wyrick, 2006). Therefore, *C. trachomatis* is the primary human pathogen, which is the cause of trachoma and genital tract disease. *C. pneumoniae* is also a primary human pathogen and *C. psittaci* can be zoonotic for humans. The primary animal models for chlamydial infections include the mouse and guinea pig with their natural parasites, *Chlamydia muridarum* and *Chlamydia caviae*, respectively. *C. trachomatis* is subdivided into 18 serovars based on serovar-specific antigens in the major outer membrane protein (MOMP) (Wang and Grayston, 1971b). Serovars A, B, Ba, and C are primarily associated with trachoma while serovars D, Da, E, F, G, H, I, Ia, J, and K are associated with oculo-genital disease. Finally, L1, L2, L2a, and L3 are the causative agents of lymphogranuloma venereum.

Morphologically, the elementary body is approximately 0.3 μm in diameter and somewhat spherical in shape while the reticulate body is larger at 1 μm in diameter. A unique morphologic component of the elementary body is the presence of hexagonally arrayed projections on the surface which are now hypothesized to be the type III secretion apparatus originally described by Hsia and Bavoil (Hsia et al., 1997). Neither stage of the organism stains well with the gram stain, but they can be identified in their classic inclusion by aniline-based stains such as the Giemsa stain.

Chlamydiae have a small genome with only about 1×10^6 base pairs. One of the major stumbling blocks to chlamydial research, however, has been the total lack of the ability to manipulate the genome so that genes can be deleted or inserted, an important prerequisite to establish the function or role of many of the genes and open reading frames. Also, chlamydiae appear to lack de novo or salvage pathways for nucleotide biosynthesis and are, thus, forced into a life style of intracellular parasitism.

Antigens Encoded by Agent

Structurally, chlamydiae resemble gram-negative bacteria in that they have a trilaminar outer membrane. The main structural component of the outer membrane is the MOMP, a 38–42 kD protein with multiple cysteine residues providing disulfide cross-linking which gives the structural integrity to the outer membrane (Caldwell et al., 1981). In addition to being the major outer membrane structural protein, it also has porin activity and has been implicated in the attachment of chlamydiae to the host cell (Bavoil et al., 1984). MOMP is composed of essentially four loops, each having surface-exposed portions. There is greater amino acid diversity in the surface-exposed or variable domains, probably accounting for the antigenic variation among the *C. trachomatis* serovars. There are also genus- and species-specific epitopes associated with MOMP (Stephens et al., 1988). MOMP is an immunodominant, highly immunogenic protein in all chlamydial strains except for *C. pneumoniae* and, as such, has been identified as a major target antigen for the protective immune response. Other structural components of the outer membrane include a 60 kD cysteine-rich molecule, OmcB or Omp2, and a lipopolysaccharide (LPS) moiety. The LPS has mild endotoxic activity and resembles the "deep rough" (Re) mutant LPS of enteric bacteria (Nurminen et al., 1984). Another set of proteins, the polymorphic outer membrane proteins (PMPs) are associated with the outer membrane but appear to be variable in their surface exposure and immunogenicity.

PROTECTIVE IMMUNE RESPONSE

Genital Disease

It is clear that there is a substantial immune response to infection and that protective immunity to reinfection does indeed develop in humans following a chlamydial genital infection (Alani et al., 1977; Schachter et al., 1983; Jones and Batteiger, 1986), but that immunity is neither complete nor long-lasting, at least initially. Katz et al. (1987) observed that contacts of isolation-positive individuals were likely to be negative for chlamydiae if they had had a prior chlamydial infection less than 6 months previously, but if their prior chlamydial infection was greater than 6 months before, they were more likely to be positive. There is some evidence that solid immunity to reinfection can develop as older individuals appear more resistant to infection than adolescents, even when controlled for a number of exposures (Arno et al., 1994).

The primary animal models for chlamydial genital infection, including the nonhuman primate, mouse, and guinea pig all support the above observations. When animals are challenged shortly after resolution of a primary infection, immunity is complete. However, if challenged at later times, reinfection occurs, albeit at lower levels and of shorter duration (Rank et al., 1988; Ramsey et al., 1989; Wolner-Hanssen et al., 1991). Generally, the shorter interval of time between challenge infections corresponds to a high degree of immunity with respect to length and peak level of infection. That reinfection can occur following a natural infection is of concern because it suggests that it may be difficult for a vaccine to elicit a sterile immunity.

Antibody

While there is a very strong antibody response both locally in the genital tract and systemically in humans and animal models, the protective role of antibody remains somewhat controversial. Brunham et al. (1983) observed in patients that the number of chlamydiae isolated from a cervical swab was inversely proportional to the level of IgA in cervical secretions. They also reported that women who developed postabortal salpingitis had significantly lower serum antibody titers than women who did not develop the disease (Brunham et al., 1987). Using guinea pigs infected intravaginally with *C. caviae*, Rank et al. (1979) reported that animals deficient in antibody were unable to resolve the infection. Similarly, animals deficient in antibody but cured of the infection with tetracycline were susceptible to challenge infection (Rank and Barron, 1983b). Thus, in the guinea pig model, antibody was essential for both resolution of a primary infection and resistance to reinfection. Of interest was their observation that passive administration of serum antibody from immune animals to naïve animals was able to reduce the level of infection, suggesting that serum IgG has transudated into the genital tract and was able to provide a level of protection (Rank and Batteiger, 1989). It should be noted that IgG is the dominant antibody in the female genital tract (Masson et al., 1969). Based on in vitro experiments, it is likely that the antibody inhibits chlamydial infection by neutralization (Caldwell and Perry, 1982), i.e., the prevention of the attachment of elementary bodies to the cell and/or the opsonization with subsequent phagocytosis of the organism (Wyrick et al., 1978).

In contrast to the guinea pig, antibody does not appear to be required for resolution of infection in the mouse model of genital tract infection. When B cell deficient mice are infected with *C. muridarum*, the course of the infection is the same as in control, immunologically intact mice (Ramsey et al., 1988; Su et al., 1997). Recently, however, Morrison has shown some evidence that antibody may be involved to some extent in resistance to infection in the mouse, although the mechanism is not yet clear (Morrison and Morrison, 2005b).

Therefore, while one cannot state specifically that antibody is absolutely required for immunity to chlamydial genital infection in the human, the development of a strong antibody response in the genital tract should be the goal of a vaccine. Complete prevention of infection could only be accomplished by the presence of neutralizing or opsonizing antibodies in genital secretions.

Cell-Mediated Immunity

While the case for antibody remains to be confirmed, it is apparent that cell-mediated immunity (CMI) is an essential factor in the host protective response. Chlamydial genital infections in both guinea pigs and mice deficient in T cells fail to resolve (Rank and Barron, 1983a; Rank et al., 1985). When the relative contribution of both CD4 and CD8 was determined in mice, it was found that protection was dependent on the CD4 subpopulation. If CD4 cells were deficient, the mice failed to resolve their infections, while mice deficient in CD8 cells resolved their infections analogous to immunologically normal controls (Morrison et al., 1995). Moreover, the adoptive transfer of either CD4 cells or T cell lines enriched for CD4 cells into recipient mice were more efficient at resolving chlamydial genital infections in contrast to the adoptive transfer of CD8 cells or CD8-enriched T cell lines which had no effect on the course of the infection (Su and Caldwell, 1995; Ramsey and Rank, 1990; Morrison et al., 2000). Similarly, the adoptive transfer of a CD4 clone into nude mice resolved the infections in 80% of the mice (Igietseme et al., 1993). Of two CD8 clones tested, only one was able to resolve infections in nude mice and then in less than half of the recipients (Igietseme et al., 1994). A role for CD4 cells has also been implicated in humans. Kimani et al. (1996) reported that sex workers who were HIV positive had a higher incidence of chlamydial genital infections than did individuals who were HIV negative, suggesting that CD4 cells had a protective role.

It was further determined that Th1 cells were responsible for the protective response. When T cells from either the genital tract or draining lymph nodes of infected mice were assessed for IFN-γ or IL-4 by ELISPOT, the dominant cytokine was IFN-γ (Cain and Rank, 1995). Moreover, the adoptive transfer of a Th1 clone was able to resolve genital infections in nude mice while a Th2 clone was totally ineffective (Igietseme et al., 1993; Hawkins et al., 2002). The data would also suggest that the primary effector mechanism is an IFN-γ-mediated killing of intracellular chlamydiae (Cotter et al., 1997; Perry et al., 1997), although a recent study by Nelson et al. (2005) implicates a mechanism employing GTPases. Thus, not surprisingly, with respect to other bacterial infections, it is apparent that a Th1 CD4 T cell response is essential for the resolution of chlamydial genital infection and resistance to reinfection.

Trachoma

Just as with chlamydial genital infections, immunity to conjunctival infection also develops. With repeated infections, it becomes increasingly more difficult to isolate the organism. Ultimately, in the majority of the subjects, the infection resolves and the individual is resistant to reinfection (Taylor, 1985). Using a nonhuman primate model of trachoma, Taylor et al. (1981, 1982) demonstrated that ocular challenge of previously infected animals resulted in much abbreviated infections which were also substantially lower in the number of organisms isolated. With repeated infections, it became more difficult to isolate the organism, just as in human disease. Using the *C. caviae* guinea pig model of conjunctival infection, Monnickendam found similar results. Repeated infections demonstrated a marked degree of immunity with regard to the ability to isolate chlamydiae from the conjunctiva (Monnickendam et al., 1980a, 1980b). However, just as in genital infections, both nonhuman primates and guinea pigs could be reinfected and the ability to reinfect was related to the time interval between the prior and current challenge. For example, when guinea pigs were inoculated 35 days after a primary infection, there were few organisms isolated in contrast to a challenge 60 days after the primary infection when there were significantly more organisms detected for a longer period of time (Monnickendam et al., 1980a). It should be noted that, in general, specific immune mechanisms in trachoma are less well known than genital infections because of the lack of a mouse model in which one has a multitude of reagents and transgenic mice. Also, the major research emphasis with respect to chlamydial infections in the last 20 years has been on genital infections.

Antibody

In humans and animal models of trachoma, there is a strong local antibody response with the production of IgA and IgG as well as a strong systemic response. However, while there are ample studies to demonstrate that incubation of antibody with chlamydiae can prevent or diminish infection both in vivo and in vitro (Nichols et al., 1973; Barenfanger and MacDonald, 1974), there is little evidence in vivo that antibody is essential for the resolution of ocular infection or resistance to reinfection. The only study that suggests a role for antibody in resolution of the infection was performed by Modabber et al. (1976) in which they depressed the B cell response by treatment with cyclophosphamide and observed that the ocular infection with *C. caviae* was prolonged. There are several studies that show a correlation with the presence of both IgG and particularly sIgA antibodies and resistance to reinfection, but none are definitive in demonstrating a protective role for antibody (Pearce et al., 1981; Malaty et al., 1981).

Cell-Mediated Immunity

The lack of a mouse model and a significant amount of basic immunologic research done in animal models of trachoma has resulted in only minimal information on the role of CMI in protective immunity in the conjunctiva. Just as with the antibody response, there is a strong local and systemic CMI response to chlamydial conjunctival infection, and there are numerous studies indicating a prime role for CMI in the pathologic response (see below). Both CD4 and CD8 T cells have been found in conjunctival tissue in the monkey model (Whittum-Hudson et al., 1986) as well as Th1 cytokine expression (Holland et al., 1996); however, neither depletion nor adoptive transfer studies have been performed to confirm a protective role for CMI. Nevertheless, because Th1 T cells are required for immunity in genital tract infections, it would be highly surprising if a similar response were not required for the resolution of ocular infection as well.

EPIDEMIOLOGY

Genital Disease

Chlamydia genital infection in the United States is clearly of epidemic proportion, and is the most commonly reported disease in the country. The CDC estimates that 2.8 million people are infected each year in the United States alone, and from 1986 through 2005, the rates of reported infection have risen from 35.2 to 332.6 cases per 100,000 population. This increase may be the result in part of increased screening and more sensitive diagnostic techniques. It is virtually impossible to know the total infection rate worldwide but it clearly is distributed throughout the world and probably at similar rates in the United States.

The rates of chlamydial infection are highest in adolescents and particularly in adolescent women. In one prospective study of a cohort of individuals between the ages of 18 and 26 years in the United States, 4.2% of the individuals were infected with *C. trachomatis* (Miller et al., 2004). In addition to the infection being most frequent in adolescents and young adults, other risk factors for acquisition of chlamydial disease include the following: (1) multiple sex partners or a partner with other partners during the last 3 months or a recent new sex partner; (2) inconsistent use of barrier contraceptives; (3) unmarried status; (4) history of prior sexually transmitted disease; and (5) lower socioeconomic class or education not beyond high school.

From a public health perspective, there is a tremendous cost to the health care system in the treatment of chlamydial genital infections. Of even greater significance is the incalculable psychological cost to couples who are unable to have children as a result of tubal obstruction in women as well as the financial costs of in vitro fertilization for those couples.

Trachoma

Trachoma is caused by specific serovars of *C. trachomatis* and is the leading cause of preventable blindness in the world with roughly 10% of the world's population infected. The disease is concentrated in tropical to subtropical areas of the world in underdeveloped countries (Figure 44.1). The disease is associated with poor public health conditions where the organism can be easily spread by mechanical means via flies and personal contact. In some areas, over 90% of children are infected. Women are three times more likely to be infected because of their close contact with children.

Unfortunately, trachoma is a disease of underdeveloped countries that have no economic means to deal with the disease. Moreover, because blindness strikes adults in their prime working years, it can remove them from the workforce and hinder their ability to care for themselves and their families. When blindness strikes women, they can no longer perform their duties as caretakers of the home and often an older daughter must leave school and take over her mother's role, taking her away from a formal education. Thus, the disease can prevent families and societies from rising above poverty. It is estimated that the disease results in $2.9 billion of lost productivity each year for the affected countries.

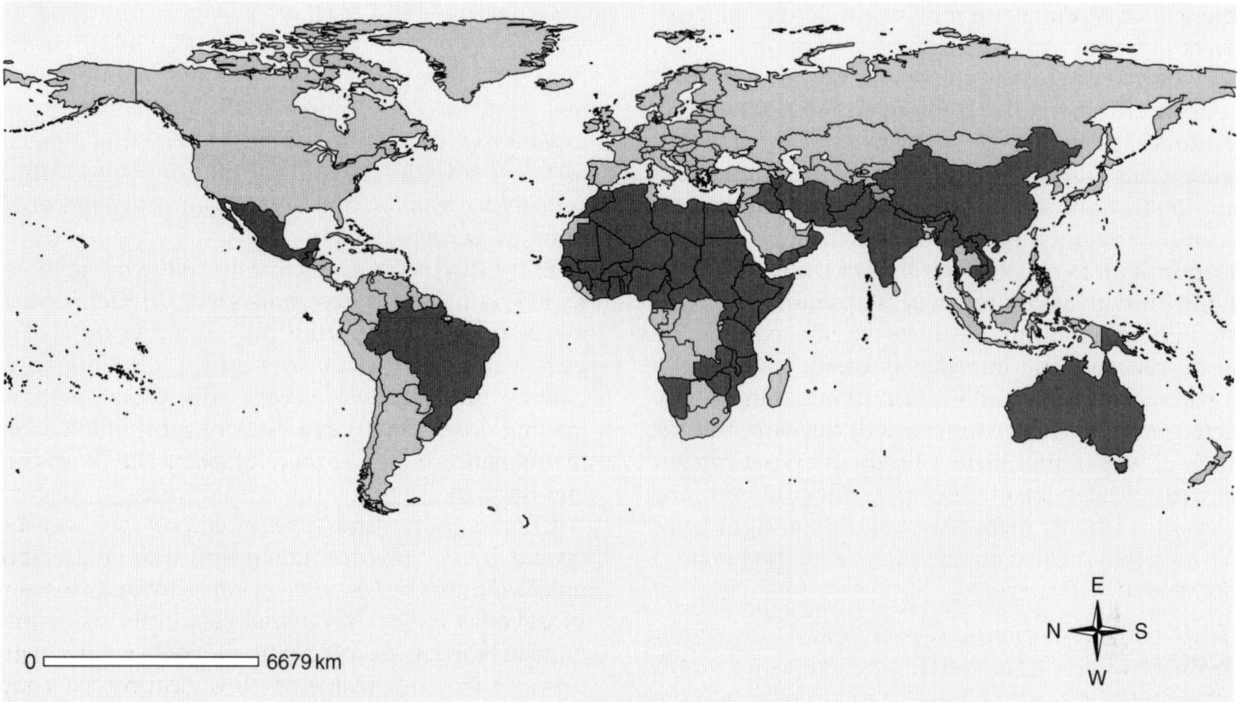

FIGURE 44.1 Distribution of trachoma.

Potential as Biothreat Agent

C. psittaci was originally included on the United States Select Agent list but is no longer on the list. While the organism can be transmitted by the respiratory route and has a high mortality rate in untreated individuals, it is a relatively labile organism and would not lend itself readily to use in biological warfare.

CLINICAL DISEASE

Genital Disease

C. trachomatis targets the endocervix in women, producing a range of symptomatology, from a mild to severe cervicitis, although about 70% of the infections in women are asymptomatic. Symptoms, if present, are generally nonspecific, but may include purulent or mucopurulent discharge from the endocervix, intermenstrual or postcoital bleeding, dysuria, urinary frequency, dyspareunia, and vulvovaginal irritation. Upon physical examination, again, a range of findings are observed, with purulent or mucopurulent discharge on the surface and/or exuding from the canal being a key finding. Cervical friability and cervical edema may also be present.

About 20–40% of women positive for chlamydiae in the lower genital tract will develop pelvic inflammatory disease (PID) that can lead to fibrosis and tubal obstruction of the fallopian tube and potentially involuntary infertility. Ectopic pregnancy has also been associated with PID caused by *C. trachomatis*. The primary indicator of PID is an onset of lower abdominal pain with diffuse tenderness in the lower quadrants which may be unilateral or bilateral. Rebound tenderness and decreased bowel sounds are common. About 50% of the patients may have fever. Pelvic exam may reveal purulent endocervical discharge and cervical motion tenderness or uterine/adnexal tenderness. It is also possible that PID may be subclinical. Among women who are confirmed with PID, there is a 20% chance that scarring and infertility will develop. Because of the high incidence of subclinical disease, women do not seek medical care, thereby increasing the risk of infertility and ectopic pregnancy.

In males, presentation is typically in the form of urethritis with a mucoid or watery discharge along with dysuria. However, about 40–50% of individuals may have asymptomatic infections. The incubation period is variable but is typically 5–10 days after exposure. Epididymitis may develop but is uncommon. It should be noted that it is not uncommon in men or women to have infections with more than one sexually transmitted agent. Moreover, it is possible for both men and women to develop inclusion conjunctivitis as a result of contaminating organisms from the genital tract.

Newborns may also acquire the infection at birth by passage through the birth canal. The most common clinical manifestation in the newborn is inclusion conjunctivitis within 5–14 days of delivery (Weinstock et al., 1994). Clinical findings range from mild swelling with a watery eye discharge, which becomes mucopurulent, to marked swelling of the eyelids with red and thickened conjunctivae. A pseudomembrane from adherence of the exudate to the conjunctiva may be present. The infection is easily treated with no residual sequeleae. In 5–30% of the infants born to infected mothers, a pneumonia will develop, but not until 4–12 weeks after birth. Cough and nasal congestion without significant discharge are common, and the infant is usually afebrile or has only a slight fever. Rales are often present on auscultation of the lungs.

Trachoma

Trachoma is a chronic disease process which can extend over many years and even decades and is the result of continual exposure to the infectious agent. The initial infection results in an inclusion conjunctivitis, involving both the bulbar and palpebral conjunctiva and consisting of congestion and edema. This gives way to the development of organized lymphoid follicles. Corneal involvement also occurs with the production of punctuate erosions of the epithelium and cellular infiltrates of the central corneal epithelium. The characteristic sign of trachoma is a neovascularization (pannus) of the superior limbus. Ultimately, the most significant damage occurs when the follicles of the palpebral conjunctiva scar and produce distortion of the eyelids so that the eyelashes continually scrape the cornea (trichiasis), causing further damage and scarring that results in blindness.

TREATMENT

Genital Disease

Fortunately, *C. trachomatis* is sensitive to several antibiotics and no resistance to any of the antibiotics has been confirmed. For treatment of urethritis, cervicitis, and conjunctivitis, azithromycin (1 g PO as a single dose) and doxycycline (100 mg PO BID for 7 days) are the two recommended regimens with both being equally efficacious (97% versus 98% cure rates) (Lau and Qureshi, 2002; Workowski and Berman, 2006). Alternative regimens include 7 days of erythromycin base (500 mg PO QID), erythromycin ethylsuccinate (800 mg PO), ofloxacin (300 mg PO BID), or levofloxacin (500 mg PO QD). In addition to antimicrobial treatment, it is important that partners be treated as well and that the patient and partner be evaluated for other sexually transmitted infections and safer-sex counseling be provided along with provision of condoms and contraception.

In the case of PID, either of the following parenteral regimens has been recommended: (1) Cefoxitin (2 g intravenously every 6 h) plus doxycycline (100 mg intravenously or orally every 12 h); (2) Clindamycin (900 mg intravenously every 8 h) plus gentamicin loading dose (2 mg/kg of body weight) followed by a maintenance dose (1.5 mg/kg) every 8 h (Workowski and Berman, 2006).

Conjunctivitis and pneumonia in the newborn caused by chlamydiae are treated with oral erythromycin (50 mg/kg/day PO in four divided doses) for 14 days for either chlamydial conjunctivitis or pneumonia (Workowski and Berman, 2006).

Trachoma

Trachoma can be readily treated with antibiotics but people are continually exposed to the infection; therefore, treatment becomes problematic, particularly in the parts of the world where trachoma occurs. Trichiasis can also be surgically treated, but, again, the logistics and lack of resources make this difficult. Currently, the International Trachoma Initiative has employed the SAFE strategy to eliminate trachoma. In this strategy, efforts are being made to correct advanced stages of trachoma with *surgery* while entire villages are being treated with *antibiotics* (Zithromax donated by Pfizer). People are being educated to do *face-washing* to reduce the chance for acquisition of trachoma. Finally, work is being done to change the *environment* by improving access to water and improving public hygiene.

PATHOGENESIS

The pathogenesis of chlamydial disease is directly associated with both the innate and adaptive host response to infection and is likely similar regardless of the site of infection. Since most chlamydiae enter the body via a mucosal tissue, the first target cell is an epithelial cell. In the process of infecting the cell, chlamydiae come into contact with toll-like receptors (TLR) and perhaps other yet-undefined receptors which initiate a signal transduction pathway resulting in the production of proinflammatory chemokines and cytokines, particularly IL-8, IL-1β, and TNF-α. Rasmussen et al. (1997) reported that infection of cervical and colonic epithelial cells with either *C. trachomatis* or *C. psittaci* up-regulated mRNA expression and secretion of the proinflammatory cytokines, IL-8, GROα, GM-CSF, and IL-6. IL-8 expression was seen as early as 2h after infection and NFκB translocation was detected as early as 30min after infection (Rasmussen et al., 1998). That these responses only occurred after infection with viable organisms and not by incubation of cells with inactivated organisms indicated that the process was initiated by a specific host/organism interaction.

In vivo in the mouse genital track model, TNF-α production can be observed as early as 6h after infection, suggesting the bacterial components or early steps in the cell infection process already elicit the response (Tseng and Rank, 1998). As a result the production of proinflammatory chemokines and cytokines, an acute inflammatory response is initiated. There may be other bacterial products that participate in the induction of the inflammatory response, such as via the type III secretion system but these are yet to be determined. Darville et al. (2003) reported that the toll pathway, especially via stimulation of TLR 2, was important in the pathologic response. They observed that TLR 2 knockout mice did not have upper tract pathology following genital infection with *C. muridarum* even though the infection course was not changed. Interestingly, mice lacking TLR 4 developed typical upper tract pathology. This is of particular interest since LPS and hsp60 activate TLR 4 and had been thought to be involved in eliciting the pathologic response.

We have noted that neutrophils can be observed entering the conjunctival epithelium 6h after *C. caviae* infection of the guinea pig conjunctiva with heavy inflammation by 24h, before a single developmental cycle has been completed (Rank, R.G., unpublished data). In all animal models, in the initial days of the infection, the histopathologic response is characterized by the acute inflammatory response, and it is generally accepted that the products of the neutrophil response cause tissue damage. In fact, Ramsey et al. (2005) have demonstrated a key role for matrix metalloprotease 9 in the induction of fibrosis in the mouse oviduct.

As the infection progresses, the adaptive immune response becomes activated and one finds an influx of mononuclear cells, particularly T cells, B cells, and macrophages, to the local site. In the MoPn infection of the mouse, this is mainly CD4 cells but in the guinea pig and subhuman primate, there are equal numbers of CD4 and CD8 cells (Van Voorhis et al., 1996; Whittum-Hudson et al., 1986; Kelly et al., 1996; Rank et al., 2003). However, the T cell response is clearly a Th1 response. These cells and the cytokines they secrete are critical for the resolution of infection, but it cannot be ruled out that they also contribute to the induction of pathology. It is apparent that upon reinfection, the mononuclear response is more intense, producing disease of increased severity (Wolner-Hanssen et al., 1986; Tuffrey et al., 1990; Rank et al., 1995).

It is likely that the severity of the pathologic outcome is dependent on two major factors: the infection dose and the number of infection episodes. Because the organism itself is responsible for eliciting proinflammatory cytokines from the host cells, it stands to reason that the more organisms that are introduced, the greater is the level of resultant chemokines and cytokines and, consequently, inflammatory infiltrate. The incubation period and level of peak infection are clearly associated with the number of chlamydiae inoculated (Rank et al., 2003), and the severity of disease is related to the number of organisms as well (Rank et al., 1979).

Upon repeated infections, the organisms will once again induce an inflammatory response, but in this case, there will be an anamnestic cell-mediated immune response, and it is quite likely that the mediators produced by CD4 and CD8 cells may also elicit tissue damage. Trachoma is clearly associated with the specific host adaptive response as it is difficult to isolate chlamydiae from individuals with chronic trachoma. Thus, the pathologic response is most likely a summation of the effects of both the innate inflammatory response and the T cell response.

VACCINES

History

When it became clear that the causative agent of trachoma was a bacterium and that it could be easily propagated in embryonated eggs (T'ang et al., 1957), a major effort was made in the 1960s and 1970s to develop a vaccine. During this time, four groups were working simultaneously in four separate areas of the world to evaluate their own versions of a trachoma vaccine. The investigative teams were optimistic about chances for success because many infected individuals did eventually resolve their infections and developed immunity to reinfection. As with most vaccines in this time period, either live or formalin-inactivated purified elementary bodies were used in all of the trials, with or without adjuvants, using a variety of regimens.

However, despite a massive effort by the four different groups, no universally effective vaccine was forthcoming. Nevertheless, a tremendous amount of information about the disease and immune mechanisms was accrued that can be used today in vaccine research. The work of these teams evaluating potential vaccines in actual human trials as well as in nonhuman primates is reviewed in great detail in two books (Bietti and Werner, 1967; Schachter and Dawson, 1978) but will be revisited in this chapter to give the reader a sense of the information gained from these trials. It should be noted that all of these trials were designed to develop a vaccine for trachoma. There have been no human trials of a vaccine for chlamydial genital infections.

By far the most extensive and detailed studies in both nonhuman primates and in clinical field trials were performed by the University of Washington group led by Thomas Grayston in collaboration with the United States Naval Medical Research Unit No. 2 (NAMRU-2) in Taiwan. In the first field trial of a potential vaccine from 1959 to 1960, they immunized uninfected younger siblings of first-grade children with active disease with two different formalized vaccines suspended in aluminum hydroxide (Grayston et al., 1963). One vaccine consisted of a locally (Taiwan) isolated serovar C and the second was a mixture of serovars B and C. Over a period of 2½ years, 405 children where included in the study and given 4 doses of the vaccine. Overall, 18% of the placebo group converted to active trachoma while only 8% of the vaccine group converted, demonstrating that the vaccine was effective. In 1965, a long-term follow-up was performed on these initial subjects (Woolridge et al., 1967b). It was observed that 1 year after the administration of the full immunization regimen, which was 2 ½ years after the initial injection, the protective effect began to break down, and by 6 years, there was no difference between the immunized and placebo groups. Thus, it was apparent that the immunity was limited in the length of its effectiveness.

Extensive studies were also performed in nonhuman primates to refine the immunization regimen and evaluate adjuvants (Wang et al., 1967). In these studies, they utilized yolk sac–grown organisms either in crude suspensions or purified by different methods. Serovars B and C were tested as well as the aggressive Bour strain (serovar E) alone or mixed with Freund's complete or incomplete adjuvant. Several observations were derived from these experiments. Protection was clearly related to the number of chlamydial particles included in the vaccine, and both Freund's complete and incomplete adjuvants appear to have comparable results. Of importance was the observation that immunity was serovar specific. Challenge of immunized monkeys with heterologous serovars did not result in a protective response.

Based on the results employing mineral oil (incomplete Freund's adjuvant) in monkeys, field trials were initiated in Taiwan in which either monovalent or bivalent vaccines of trachoma strains isolated from Taiwan (serovars B and C) were produced from yolk sac–grown material, formalin-inactivated and emulsified with mineral oil (Woolridge et al., 1967a). Second- and third-grade children were inoculated intramuscularly followed, 6 months later, by a booster inoculation. The bivalent vaccine demonstrated a protective response with only 5.1% of the children converting to trachoma over a period of 3 years versus 8.8% for the placebo group. Interestingly, the opposite response was seen in children immunized with the monovalent vaccine, 8% converting to trachoma in the vaccinated group in comparison to 3.3% with the placebo. The conversion rate in the placebo group was unusually low; hence, this may have affected the overall results. There was concern expressed that the majority of the conversions to positivity occurred prior to the booster immunization, similar to what was seen in monkeys challenged with a strain of chlamydiae heterologous to the vaccine type, possibly suggesting a hypersensitivity response (Wang et al., 1967).

Not only because trachoma rates were relatively low in Taiwan as a result of the improving public health conditions but also because of the encouraging results, the Seattle group initiated field trials with collaborators in India in an area where the trachoma incidence was 85% (Dhir et al., 1967). In these trials, two vaccines were prepared by either gradient purification or ether extraction from elementary bodies grown in yolk sacs and were then formalin-inactivated. Each vaccine contained serovars B and C at concentrations of particles 10^9/ml as it was previously found that a successful vaccine was dependent on having large numbers of elementary bodies. Four hundred fifty-one children with no evidence of trachoma between 3 months and 5 years of age were vaccinated and given a follow-up booster 3 months later. When examined 1 year after immunization, 36.8% of the placebo-inoculated subjects had converted to trachoma while only 10 and 18.5% of the two vaccine groups converted, suggesting a definite protective effect. The strongest protective effect was in the first 3 months following immunization and gradually declined thereafter, supporting previous observations that immunity induced by the vaccine was short lived. It is important to note that in this study, no difference in severity of disease was noted between immunized and control groups.

Follow-up studies with these vaccines were performed in monkeys (Grayston et al., 1971). Similar results were seen in that immunity did develop when high doses of organisms were contained in the vaccine; but, as in the human trials, immunity was short lived and serovar specific. They also noted that immunization with lower doses of organisms in the oil-based vaccine-containing serovars A, B, and C resulted in more severe disease upon challenge with each of the serovars.

Early human trials of a chlamydial vaccine were performed in Ethiopia by a group from Italy. For these trials, a formalized vaccine prepared from yolk sac–grown elementary bodies of a *C. trachomatis* isolate from Ethiopia was used. The vaccine was administered intramuscularly with either an aluminum hydroxide or water-in-oil adjuvant followed with an aqueous booster at 40 days and sometimes another aqueous booster at 6 months. The vaccines were standardized by the counting of chlamydial particles, the determination of DNA and nitrogen content, and the antigenic concentration of the batch with a standard rabbit antibody by the complement fixation assay. As in the majority of vaccine trials in this period, success was judged by the absence or lessening of clinical disease and sometimes in the reduction in the number of organisms detected on Giemsa-stained conjunctival smears. Initial trials in nonhuman primates and then in human volunteers demonstrated a high degree of protection against conjunctivitis upon challenge with lower doses of organisms although the intensity of the disease in volunteers becoming infected was not diminished. Nevertheless, these results suggested that a protective response could be elicited.

This group conducted field trials in Ethiopia from 1960 through 1962, immunizing children from 2 to 4 years of age who were either free of trachoma or had evidence of disease (Bietti and Werner, 1967). Thus, they were able to evaluate both the therapeutic and prophylactic effects of the vaccine. A summary of the trials over this period showed that 40% of the healthy patients developed disease versus 72% of healthy patients receiving no vaccine. Similarly, of those patients with severe follicular trachoma, vaccination reduced the level of disease in 63% of the cases versus only 23% of controls. If two booster doses were given, 83% of the immunized subjects had a positive outcome. If the disease was more advanced, disease reduction was 79% in vaccinated individuals versus 59% for controls. Therefore, it appeared that immunization was able to both prevent disease and to reduce disease in those already confirmed with trachoma.

Based on the success of these field trials, more expansive trials were conducted in the 1964–1965 with 10,000 subjects included. In two separate trials, 2500 subjects were given the vaccine and 2500 a placebo. In the trial for which data were published (Bietti and Werner, 1967), 17% of vaccinated individuals without previous disease developed clinical trachoma in contrast to 33% of subjects given placebo. For those with existing trachoma, 59% of the vaccinated people improved as compared to only 43% of the placebo-inoculated people. The data in this larger trial suggested that the vaccine still had positive prophylactic and therapeutic effects, although they were not as strong as in the initial field trials. This may have been the result of the trial being conducted in different region which had a higher standard of living.

The group from the Medical Research Council in London, headed by Collier and Blyth (1966a) initiated trials first in baboons. For immunogens, they used several different isolates, two of which were clearly genital isolates. The primary strain they used was the so-called "fast" strain which was later found to be LGV (L2) (Wang and Grayston, 1971a); therefore, for many of the studies, they were not using a typical trachoma strain, but live, potentially invasive LGV. The vaccines were grown predominantly in yolk sacs and were administered parenterally as live vaccines. The "fast" strain was also grown in HeLa cells because it was easy to grow in cell culture and was also administered as a live vaccine. In the course of their studies, they observed that they had good protection against challenge with the homologous strain but less effective protection using challenge with a heterologous strain. They also experimented with different methods of inactivation, including inactivation by heat, formalin, or exposure to UV light and reported that all reduced the immunogenicity of the vaccine (Collier and Blyth, 1967). Overall, their studies in baboons showed that the lack of cross-reactivity among strains was a problem, the immunity to reinfection was relatively short lived, and live vaccines were more effective. Moreover, the use of adjuvants may have led to increased disease via a hypersensitivity mechanism (Collier and Blyth, 1966b), but again, it should be noted that they were using a live strain of an aggressive chlamydiae, not at all typical of the human trachoma strains.

Based on their results in the baboon trials, the group moved on to field trials in The Gambia using live vaccines administered either subcutaneously or intramuscularly (Sowa et al., 1969). Results were mixed, with an initial trial producing short-term improvement in clinical disease; however, subsequent trials were inconsequential with either no evidence of protection or, in the case of the third trial in which the LGV-like "fast" strain was used, a possibility of exacerbated disease, although this may have been an artifact of the clinical scoring system used. The trials were halted when they subsequently found that the chlamydial strain they were using in the vaccine could become systemic in baboons (Collier and Smith, 1967; Collier and Mogg, 1969). This strain was later found to be an LGV contaminant, hence its "fast" or aggressive growth and systemic distribution in baboons.

Concomitant with the ongoing trials by the above groups, the group led by Nichols et al. (1966, 1969) from the Harvard School of Public Health was conducting field trials in Saudi Arabia followed later by experiments in New World monkeys. In these trials, children 3 years of age or younger, were grouped according to whether they had active trachoma, nontrachomatous conjunctivitis, or no disease, so that they could determine the effectiveness of the vaccine in both prophylactic and therapeutic modalities. They were immunized with vaccines containing equal portions of serovars B and C derived from yolk sacs and inactivated with formalin. Two preparations were in aqueous form at two different concentrations while the remaining vaccine was administered along with thimersol as adjuvant. A booster dose was administered in all three groups 6 months after the primary immunization.

In contrast to other groups, in addition to the clinical evaluation of the subjects, they also quantified the number of chlamydiae on conjunctival swabs. Both aqueous vaccines delayed the conversion of subjects to active trachoma while there was no such effect with the adjuvant vaccine; however, the overall microbiological conversion rate was not significantly affected by any of the vaccine regimens. Nevertheless, there was a reduction in numbers of inclusions detected in immunized subjects converting from negative to positive within 6 months after the primary immunization with either the low dose aqueous or adjuvant vaccines. In children who were positive at the outset, there was only evidence of reduction in inclusion number in those receiving the higher vaccine dose. Of interest was the observation that of the children who were clinically normal at the beginning of the study, those that received the higher dose aqueous vaccine developed disease three times more readily than control subjects, although their clinical disease was no worse than controls and there may have been a reduction in the number of organisms. Just as in other field trials, the effectiveness, even if marginal, was only apparent for a short period of time following immunization.

After approximately a decade of human vaccine testing, all studies were halted and to the current day, there have been no further trials in humans with potential vaccines against either trachoma or genital tract infections. So after all of these trials, what have we learned and what is the significance with regard to the current pursuit of a vaccine? In many of the studies, there was clear evidence for immunity. Generally, immunity was associated with the administration of high doses of formalin-treated organisms and was preventive if administered to previously uninfected individuals. The vaccine was also able to either prevent infection or reduce the infection so that if acquired, it was likely subclinical or otherwise undetectable. In support of this, the Harvard group was able to demonstrate a reduction in the number of organisms (Nichols et al., 1969). Evidence for a therapeutic effect in individuals who were already infected was also observed. Nevertheless, the consistent finding in all of the studies was that the immunity induced by the vaccine was very short lived, even with booster inoculations and natural exposure to the organism. That immunity lasted on the average only less than 1 year is clearly problematic when considering immunization of large populations, particularly in underdeveloped countries.

Another key observation that was supported by studies in animal models and in vitro was that the protective effect when seen was serovar specific. Especially in non human primate experiments, it was observed that immunization with one serovar did not protect against other serovars. In the case of trachoma or genital tract infections, this certainly indicates that a vaccine will have to be composed of either a single genus- or species-specific epitope or multiple serovar-specific epitopes.

Perhaps the most significant consequence of the human trials and from concurrent studies in monkeys which continues to impact on the development of a vaccine today was derived from the observations that in certain situations, there actually seemed to be either exacerbated disease or an increased rate of conversion to active trachoma in subjects receiving the vaccine (Grayston and Wang, 1975). In one of the early volunteer experiments, medical students in Taiwan were immunized with trachoma strains and then challenged in the eye (Grayston et al., 1961). Interestingly, the immunized students had an earlier onset of disease, and there was a correlation between the complement fixation antibody titer and the severity of the disease, suggesting some form of a hypersensitivity response. There is still much debate about how significant and representative the findings suggesting a hypersensitivity response were and whether the potential for a hypersensitivity response is a realistic concern.

In general, it appeared that the adverse responses were observed in situations where there was an inadequate amount of antigen in the vaccine or where there was a single serovar. Woolridge et al. (1967a) saw a slightly increased incidence of conversion in individuals vaccinated with a monovalent vaccine while Nichols et al. (1966) reported increased incidence of conversion to active trachoma in subjects administered a vaccine with a lower number of organisms. While the authors cautiously suggested that these results may have been associated with sensitization by the vaccine followed by a "hypersensitivity" response, the data are not all clear that there was a true hypersensitivity response. Exacerbated disease was not observed even in those vaccinated subjects who converted to active disease, although one can argue that sensitization by a vaccine would create a more rapid follicular response upon contact with the infectious agent in contrast to naïve individuals.

While the human trials only showed marginal data with regard to a hypersensitivity response, immunization of monkeys produced more definitive results. In early monkey experiments, Wang and Grayston (1967) observed that pannus occurred more frequently in vaccinated monkeys than in control monkeys and that pannus was also more frequent in monkeys immunized with the Bour strain. The Bour strain was actually a serovar E organism, albeit it was isolated from a trachoma case in the San Francisco area, and was considered to be an aggressive strain. This strain was originally used in the monkey model because it was able to produce a more characteristic disease than the serovar B and C strains. These data were simislar to those published by Collier and Blyth (1966b) in which they reported more severe disease in challenge of animals immunized with the 'fast' (likely LGV) strain and challenged with a non-trachoma isolate. Since these studies were done with exceptionally aggressive strains and in some cases, had heterologous challenges, one must question whether they are truly predictive for human vaccination. However, additional studies in monkeys using the same trivalent vaccine (serovars A, B, and C) as used in human trials did indeed show enhanced disease in monkeys immunized with a low-dose vaccine when challenged with a homologous strain of each serovar (Grayston et al., 1971).

Because the pathogenesis of trachoma is clearly associated with the cell-mediated immune response, the adverse reactions observed in humans and monkeys in these studies have raised major concerns among pharmaceutical companies that a vaccine, particularly a whole organism vaccine, might have potential deleterious effects. This concern was heightened several years ago in studies that suggested that hsp60 was an antigen which was responsible for eliciting the pathologic cell-mediated response (Morrison, 1991). Further studies showed that although hsp60 was highly immunogenic and an antibody response to it correlated with severe disease, it was not responsible for the hypersensitivity response (Rank and Sanders, 1992; Rank et al., 1995; Stephens, 2003). However, during the study of this antigen, it was noted that there was significant sequence homology between chlamydial and human hsp60 (Morrison et al., 1989). While there has been absolutely no evidence for an autoimmune response resulting from chlamydial infection, there is still major concern in the development of a vaccine that no component with cross-reaction to human antigens should be incorporated into a vaccine. Thus, it is highly unlikely that any future vaccine trials will utilize whole organisms, even if attenuated or killed, as was done in the original trachoma trials. The risk of litigation for vaccine companies is just too great.

Current Licensed Vaccines

There are currently no licensed vaccines for use in humans for either chlamydial genital infections or trachoma. Nevertheless, it should be mentioned that there are currently two licensed vaccines used in veterinary practice. Fevaxyn Pentofel™ from Fort Dodge, a division of Wyeth, is an inactivated veterinary vaccine for several diseases in cats, including chlamydial conjunctivitis caused by *Chlamydophila felis*. However, its protective value against conjunctivitis appears limited, reducing the length of clinical disease by only a few days. Moreover, it is not able to prevent the establishment of chronic infection, characteristic of the feline disease. Live vaccines have been used that also provide marginal protection but may result in actual conjunctival infection if they get inadvertently in contact with the eye.

A vaccine is also commercially available in Europe for the prevention of abortion in sheep caused by *Chlamydophila abortis*. Enzovax is a freeze-dried vaccine containing $10^{5.0}$–$10^{6.9}$ IFU of live, attenuated strain of *Chlamydophila abortus* per dose and is produced by Intervet in the United Kingdom. The attenuated strain is a temperature-sensitive mutant originally derived by Rodolakis which is infectious for sheep but produces a mild infection (Rodolakis and Souriau, 1983; Rodolakis and Bernard, 1984). As in the vaccine for *C. felis*, it is difficult to measure the outcome, but the vaccine itself does not exacerbate the disease and may have some mild protective effects. Not surprisingly, frequent boosters are required.

These vaccines are mentioned here because it is interesting that both of the vaccines are whole organism vaccines with the ovine vaccine being live attenuated; but they do not appear to elicit any untoward effects analogous to what was seen in the monkey experiments with *C. trachomatis*. Of course, clinical trials in the veterinary world are not as rigorously controlled and monitored as in human studies. Also, as has been seen over and over again in human and monkey trials, as well as in other animal models, immunity is not long-lasting and booster inoculations are essential.

VACCINES IN DEVELOPMENT

Despite the limited success and lengthy interval since the trachoma clinical trials 40 years ago, there has been a continued emphasis on research directed to the development of a vaccine. However, the target of vaccine efforts has changed from trachoma to genital tract infections caused by chlamydiae, with specific emphasis on the prevention of upper genital tract disease and involuntary infertility in women. Trachoma is no less important, but genital tract infections are epidemic in the developed countries, and there is a distinct economic advantage for vaccine companies to pursue a vaccine for chlamydial genital infections. The vast majority of research in the intervening years since the trachoma trials has addressed the immune mechanisms operative in natural immunity to chlamydial genital infection and the mechanisms of pathogenesis. Moreover, since the use of a whole organism is not feasible, there has also been continued interest in determining the potential protective antigens that could be incorporated into a vaccine. At this point, virtually all of the vaccine development is still in the discovery and preclinical development stages.

Discovery/Basic Science

The basic science work has centered on several important questions such as the following: (1) what arm(s) of the immune response is/are critical for protection; (2) can a long-lasting protective response be produced; (3) should a strategy of preventing infection or preventing disease be adopted; (4) what is/are the protective antigen(s); (5) will the candidate vaccine prime the individual for, actually elicit, or exacerbate a pathologic response? Each one of these questions is critical in the ultimate development of an effective vaccine.

It is evident from the review of the adaptive immune response that both antibody and cell-mediated immune responses are elicited during a natural infection. There is virtually universal agreement among researchers that a cell-mediated response is essential for a protective response, especially in a natural infection. Studies in mouse and guinea pig models of genital infection have definitely demonstrated that a CD4 Th1 cell-mediated immune response is essential for a protective immune response (Rank and Barron, 1983a; Rank et al., 1989; Morrison et al., 2000). Studies in humans have also supported the protective role of the CD4 cell (Kimani et al., 1996).

While it is the consensus that any vaccine will need to induce CD4 Th1 response, there are caveats associated with this response. Firstly, it is highly unlikely that a Th1 response will be able to prevent infection, especially over the long term. Since there is minimal resident lymphoid tissue in the genital tract, T cells must be recruited to the local site upon infection. The production of the appropriate chemokines will only be initiated upon infection and will occur only during the first developmental cycle through the stimulation of toll receptors and perhaps other yet-undefined mechanisms. Thus, it is unlikely that T cells will gain access to the site prior to the end of the first developmental cycle of the organism. During that time, the acute inflammatory response will already be under way. In the guinea pig conjunctival model, we have observed PMNs in the local tissue as early as 6 h after infection (Rank, R.G. and Wyrick, P.M., unpublished data). Even in an anamnestic response, it is unlikely that T cells will enter the site within the first 24–48 h of infection. Kelly and Rank (1997) observed a large influx of CD4 cells by 7 days after a challenge infection in the mouse, and although they did not look earlier, the high number suggested that the CD4 cells were present before 7 days.

The lack of resident lymphoid tissue in the genital tract may also explain why immunity to reinfection is short lived as seen in humans and all of the animal models. One can hypothesize that T cells are recruited to the local site and then bring about resolution of infection; however, when the organisms have been eliminated, there is no longer any stimulus to recruit and retain cells at the site. Upon a challenge infection, the cells must once again be recruited to the local site, a process which can take some time. Evidence for this mechanism was reported by Igietseme and Rank (1991) who observed that resolution of genital infection in the guinea pig was concomitant with an increase in the number of chlamydia-specific T cells in the genital tract. The animals were immune to reinfection as long as the T cells were still present, but they became susceptible to reinfection when T cells were absent. This is a difficult phenomenon to overcome with a vaccine, since it would obviously neither be practical nor desirable to maintain a constant T cell response in the genital tract. This may not be as critical an issue in the eye which has resident lymphoid tissue and after infection, follicles, which contain both B cells and T cells (Whittum-Hudson et al., 1986) and might explain in part why in trachoma, one has difficulty isolating organisms in the latter stages of the disease.

A vaccine must also be able to elicit a T cell response which can home to the genital tract. Because the genital tract is a mucosal site, T cells induced by a systemic inoculation may not necessarily have the appropriate homing receptors to allow them to readily move to the local genital tissue. The homing receptor $\alpha 4\beta 7$ has been found to be an important receptor for the homing of T cell clones to the genital tract in mice (Kelly and Rank, 1997).

While the role for CMI is clear, there has been substantial debate about the requirement for an antibody response. The data in the guinea pig model strongly indicate a role for antibody in resolution of a primary infection and resistance to reinfection (Rank et al., 1979; Rank and Barron, 1983b), but the role in the mouse model is not as clear; although recently Morrison and Morrison (2005a) have presented evidence for a role of antibody in resistance to reinfection in the mouse. There have been many reports in vivo and in vitro demonstrating that antibody can neutralize chlamydial infection (Rank, 1999), but it has been difficult to prove a correlation between the neutralizing antibody titer and a protective response (Jones and Van der Pol, 1994). Antibody would be the most obvious effector arm to either prevent infection or limit the course of infection in a time-sensitive manner. Moreover, especially with respect to infection in the female genital tract, IgG can be transudated into the genital tract from serum and is actually the dominant antibody in the female genital tract (Masson et al., 1969; Rank and Batteiger, 1989). Immunization by intramuscular or subcutaneous routes can induce a strong systemic antibody response that has the potential to be long-lasting. The continual presence of IgG in the genital tract of female guinea pigs was observed up to 825 days after a primary infection, long after IgA levels in secretions had diminished about day 50 (Rank et al., 1988). Thus, an ideal vaccine would elicit both an antibody- and cell-mediated immune response with the former neutralizing and decreasing the number of organisms while the latter would be necessary to clear the infection.

Since it will be very difficult to develop a vaccine that universally prevents infection, a vaccine for chlamydial genital infection could be considered to be successful if it is able to prevent ascending infection in women, so that salpingitis with its sequelae do not occur. It would be hoped that the antibody response could reduce the number of organisms present and delay or block the organisms from moving up the genital tract until antigen-specific T cells are recruited to the site to eliminate the infection. It has been demonstrated in the guinea pig that the incubation period, i.e., the time until organisms can be isolated from the lower genital tract following inoculation, increases as fewer organisms are used in the inoculum (Rank et al., 2003). Thus, the more effective the antibody response in the lower genital tract, the more likely is the possibility that the number of organisms will be diminished and ascending infection prevented. In addition, while adolescent females will be the target population for a vaccine, it would also be practical to immunize males. Ideally, this would not only diminish disease in males, but if there were a low level of infection and/or abbreviated infection in males, the inoculating dose for females would be lower and perhaps more readily controlled by the host response in an immunized female.

Vaccine Candidates

As one might expect, there has been a great deal of effort made in defining the protective antigens, and therefore, potential vaccine candidates. As described above, whole organism vaccines have been evaluated in trachoma clinical trials and two whole organism vaccines are in use in veterinary practice. However, even though all antigens are present, protection against challenge was variable and clearly short lived. Moreover, because of hypersensitivity and cross-reactive antigen issues, the use of whole organisms, even if attenuated, is not practical or safe for human use. Thus, researchers have been targeting individual proteins as candidates. To be a successful vaccine candidate, the protein should ideally have epitopes that bridge several serovars of *C. trachomatis*, have a structure such that one can construct a recombinant protein analogous to the native protein, and not elicit an exacerbated response upon subsequent challenge.

The vast majority of the research investigating potential vaccine candidates has centered on the MOMP, also referred to as the outer membrane protein A (ompA) (Table 44.1). The literature on MOMP and immunization studies with MOMP have been reviewed in detail previously (Rank, 1999). Caldwell first reported that antibodies to MOMP could neutralize infectivity of elementary bodies in cell culture (Caldwell and Perry, 1982), and Williams et al. (1984) later observed that incubation of *C. muridarum* elementary bodies with antibody to MOMP could neutralize infectivity in vivo in the mouse lung model. In studies on human sera as well as in infection studies in animals, MOMP was found to be highly immunogenic, appearing as a major band in immunoblot analyses, and correlations were observed with the resolution of infection. However, initial immunization

TABLE 44.1 Current vaccine candidates

Vaccine candidate	Location/function	In vivo protection
MOMP (ompA)	Outer membrane; porin	Yes
CPAF	Cytosol; protease	Yes
PmpD	Outer membrane	Not tested
PorB	Outer membrane; porin	Not tested

studies with MOMP or epitopes of MOMP were not highly successful even though antibodies generated to the protein and/or epitopes could neutralize infection in vivo. Subsequently, Zhang et al. (1989) observed that only those monoclonal antibodies which were able to recognize surface determinants were successful in neutralizing infectivity. It became apparent that conformational epitopes on MOMP were critical for the induction of protective immunity when *C. caviae* genital infection in guinea pigs could be significantly reduced in intensity by immunization with MOMP purified from whole elementary bodies with nonreducing methods but not by MOMP purified with sodium dodecyl sulfate which breaks disulfide bonds (Batteiger et al., 1993).

There is no doubt that MOMP is immunogenic and in immunization protocols with adjuvants approved for human use can markedly reduce the level of subsequent challenge infection and even prevent the development of infertility in mice (Pal et al., 2005); however, the major problem that needs to be surmounted is that a recombinant protein with the intact conformational epitopes must be produced. Without a recombinant protein, it will be logistically and economically not feasible to derive sufficient MOMP from chlamydiae for vaccine production. Moreover, MOMP clearly is serovar-specific although a vaccine could conceivably contain multiple MOMP molecules representing the dominant genital or ocular serovars. That a recombinant MOMP can be produced with the ability to elicit a protective response was reported by Eko et al. (2004). They transformed *Vibrio cholerae 01 strain H1* with recombinant MOMP and omp2 from serovar D and then produced outer membrane ghosts which they used for immunization of mice. The multivalent vaccine elicited a strong Th1 response as well as IgG2a and IgA antibody responses in genital secretions. That the antibodies were detected against whole elementary bodies in an ELISA assay suggested that they were recognizing surface determinants. When T cells from mice immunized with the multivalent vaccine were adoptively transferred into naïve mice and those mice challenged, a protective response was seen. Whether the recombinant MOMP and recombinant omp2 alone were able to induce a similar response was not assessed.

Another outer membrane protein of interest is PorB, an outer membrane protein with porin function which is conserved across *C. trachomatis* serovars (Kawa and Stephens, 2002). While natural infection in humans and intraperitoneal infection of mice with *C. trachomatis* results in minimal antibody responses to this protein, immunization of mice with PorB produces a substantial serum antibody response (Kawa et al., 2004). Moreover, antibodies to PorB neutralize in vitro, making it a possible vaccine candidate, although animal protection studies have not yet been performed.

Recently, Caldwell et al. (1975) have described and characterized the polymorphic membrane protein D (pmpD) (Crane et al., 2006). This is a high molecular weight molecule (~155 kD) which is recognized by convalescent sera from individuals with cervicitis, urethritis, LGV, and trachoma. The exciting aspect of this antigen is that it is a surface-exposed epitope that is conserved among *C. trachomatis* serotypes and that antibodies to pmpD can neutralize infectivity in cell culture. Thus, in addition to MOMP, it is the only other molecule that has been found to elicit neutralizing antibodies. A potential problem is that antibodies to both MOMP and LPS are able to block the neutralizing effect of the anti-pmpD antibodies, which could be a problem for individuals who have already been exposed to *C. trachomatis* prior to receiving a vaccine. Furthermore, if the protection generated by a pmpD vaccine was not sufficient to prevent infection, antibodies generated to MOMP and LPS in a low-level infection may abrogate the effects of the vaccine. Nevertheless, these findings are exciting, and pmpD holds promise as a possible vaccine candidate.

Recently, data have been published by Zhong et al. (2001) that an intracellular molecule secreted into the cytosol, chlamydial protease-like activity factor (CPAF) is highly immunogenic in humans. At least in vitro, CPAF degrades major histocompatibility complex transcription factors RFX-5 and USF-1 as well as keratin-8; however, its exact role in chlamydial infection of the host cell is not clear. Of interest is the observation that it is functionally conserved among several chlamydial species (Dong et al., 2005). Intranasal immunization of mice with recombinant CPAF, derived from the sequence of serovar L2, plus IL-12, was able to reduce the length of genital infection as well as significantly reduce the pathologic response in the oviduct (Murthy et al., 2007). These data suggest that this may be a viable vaccine candidate, especially if there is the potential for cross-serovar protection, although this needs to be evaluated further.

While the above molecules have been described in the literature as possible vaccine candidates, they are by no means the only candidates. With the small size of the chlamydial genome, it has not been difficult for those companies interested in vaccines to express virtually every open reading frame as either a protein or DNA vaccine and screen each for a possible protective role in relevant animal models. Most likely, some promising candidates are under evaluation behind closed doors and none have yet come to clinical trials.

Vaccine Candidates and Pathologic Response

Because of the experiences in the trachoma vaccine trials and subsequent studies on hsp60, there is obviously a great deal of concern that a vaccine may prime the host response so that upon infection, the individual may have more severe disease. It is important to point out that the trachoma vaccines for the most part were able to elicit a protective response, albeit short lived. Even in monkeys, there was a high degree of success in inducing a protective immune response. The reasons that some individuals had a higher rate of conversion to disease are not clear but could reasonably be associated with prior sensitization by the vaccine, since the pathologic response in chlamydial infections is indeed host mediated. Nevertheless, it is also important to note that in trachoma, the organism is inoculated in nature directly into the target tissue site that is also a site with resident lymphoid tissue. Any host cellular response will have the potential to have immediate pathologic consequences.

In contrast, the site of infection in the genital tract that results in morbidity is the upper genital tract of women. Based on animal model studies, it appears that the organism requires several days to reach the upper genital tract (Rank and Sanders, 1992). Therefore, it is quite conceivable that even if there is a substantial host response to the organism, analogous to the trachoma studies, the pathology may be restricted to the lower genital tract which is capable of repair and where any residual fibrosis would be of less long-term consequence. The infection could easily be resolved prior to the organism reaching the fallopian tubes. In fact, none of the immunization studies in animal models of genital tract infection have reported any evidence for enhanced upper tract disease upon challenge (Pal et al., 1997; Su et al., 1998; O'Connell et al., 2007; Murthy et al., 2007). Thus, a vaccine designed to protect against genital infection has an anatomical advantage on its side.

Clinical Trials

At the time of this writing, no vaccine candidates have advanced to clinical trials. Even if a vaccine candidate were ready today for testing, there are several important issues which must be determined for the proper evaluation of a vaccine in clinical trials. The most important issue is what population will be immunized. The obvious population to be immunized is the adolescent female population since the infection is epidemic in this population and the prevention of infertility in women is the major reason for producing a vaccine. Ideally, one should immunize adolescent females prior to the onset of sexual activity. Currently, with the HPV, meninogococcal and acellular pertussis vaccines being used in the adolescent population, there is clear precedent for an adolescent panel that has not been present before. Whether a vaccine will be effective if the subject has already had a chlamydial infection is also an important question. If the vaccine is successful in adolescent females, should the population be expanded to include adolescent males? Obviously, chlamydial infections are also epidemic in males but infertility in males following chlamydial infection is a rare event; so is it justifiable to immunize males in order to decrease the incidence of infection in females?

When the HPV vaccine was tested, investigators could not wait the length of time necessary to determine whether the vaccine would prevent cancer, so a surrogate endpoint was adopted. There is a similar issue in immunizing against chlamydial genital infections. It is quite apparent that it will not be practical to follow women to determine if infertility develops because of the long time interval that may occur before the onset or clinical detection; so an alternative endpoint must be determined. For the reasons given in earlier sections, it may be difficult to actually prevent infection with a vaccine, and given the already high incidence of subclinical infections, it may be problematic to determine whether the vaccine is reducing the infection to a subclinical state or whether the vaccine has had no effect, and the infection is naturally subclinical. Thus, clinical outcome in the short term may not be practical. One can potentially make use of the observation that the titer of antibody to hsp60 has been correlated with the presence of upper tract disease (Brunham et al., 1985; Wagar et al., 1990). While it does not appear that hsp60 itself is involved in inducing the pathologic response, it may well be a marker of either intensity and/or frequency of infection. In support of this, when antibody titers to hsp60 were measured in guinea pigs following one or more intravaginal infections with *C. caviae*, a strong correlation was observed between the titer and the number of times the animals were infected (Rank and Sanders, 1992). It was also found that upper genital tract pathology was more likely to develop with repeated infections, just as seen in humans (Rank et al., 1995). Thus, if immunized individuals have either a lower incidence of hsp60 antibody conversion or lower titers than unimmunized individuals, this may be considered predictive for the prevention of upper tract disease.

The longevity of a protective effect will also be an important factor to evaluate, since all of the human trials and animal studies showed that solid immunity is short lived. In the trachoma trials, protective immunity began to break down with 1–2 years of immunization; this may be a reasonable time over which to monitor the subjects in a clinical trial.

There is also the issue of persistence. While a great deal of effort has been expended on understanding persistence mechanisms in vitro, the in vitro models of persistence have no proven parallel in humans and are likely not predictive at all of the persistence in the human host. While there is no question that chlamydiae can persist in animals, no one has yet detected or demonstrated a uniquely "persistent" form in any animal model of chlamydial infection. It is more likely, based on precedents with other organisms such as in carrier states of various bacteria and long-term low-level infections of organisms such as *Plasmodia,* that persistent infection is simply a low-level chronic infection, in which the organism undergoes normal replication but is maintained at a low level by the host response. Nevertheless, clinical trials will have to address the potential for such an infection, even in the presence of a protective host response generated by the vaccine.

PROSPECTS FOR THE FUTURE

The prospects for a vaccine against chlamydial genital infection at this time are still somewhat daunting. There are several chlamydial antigens that have been identified as potential vaccine candidates—the most well-described is the MOMP that is surface exposed and highly immunogenic. However, it is clear that conformational epitopes are the key protective targets, and it has been difficult to produce recombinants that maintain the native conformation. Moreover, the protective epitopes of MOMP are serovar specific, although a vaccine could combine MOMPs from several of the most prevalent serotypes. There are other immunogens that have genus-specific epitopes, but these are still in the early stages of experimental evaluation in animal models. It is quite likely that a vaccine will need to contain multiple antigens to be most effective and to be effective against multiple serovars. A successful vaccine against a local genital tract infection, especially with the goal of eliciting an effective Th1 response may require specific adjuvants to produce that response locally.

Importantly, the trachoma vaccine trials revealed that a vaccine can indeed provide a certain level of protection, but there are clearly important issues that must be resolved. Among these, there must be certainty that a vaccine will not elicit more pathology or increase the risk of disease as seen in some of the trachoma vaccine trials. This issue alone and the observation that chlamydial hsp60 cross-reacts with human hsp60 have driven the vaccine research toward using a protein or subunit vaccine rather than the whole organism vaccines used in the original clinical trials. In addition, the trials told us that it is extremely difficult to elicit a protective response which is effective for more than 1–2 years, even with a whole organism or live vaccine. However, those studies were addressing immunity to trachoma, in which the organism is being directly deposited to the target tissue. With respect to a genital tract vaccine in women, the site of pathology is the upper tract, so there is a window of time before the organisms ascend the genital tract for existing effectors such as antibody to neutralize organisms and for an anamnestic T cell response to be generated. This anatomical difference in target tissues may facilitate a longer period of protection against upper genital tract disease. Studies in guinea pigs have shown that IgG which is derived from the serum can be detected in the genital tract for well over 2 years and that an exuberant anamnestic response occurs within a few days of challenge (Rank et al., 1988).

There are still some practical issues which must be solved before a vaccine can go to clinical trials. The primary issue is what surrogates for infertility and ectopic pregnancy can be used to determine vaccine effectiveness. The association of antibody to hsp60 and upper tract disease may be one parameter that can be monitored to determine vaccine effectiveness as well as simply the reduction in chlamydial infection and clinical disease in a prospective group.

Social issues regarding a vaccine against a sexually transmitted infection have always been of concern; however, the climate is now ripe for a vaccine since the HPV vaccine has been successfully marketed, and there is now an adolescent vaccine panel, including vaccines against the meningitis and pertussis, something which was not present only a few years ago. Thus, it is entirely conceivable that a vaccine against chlamydial infection could be easily attached to that panel.

The development of a vaccine against trachoma is a more difficult task because one must consider that the vaccine will have to be both prophylactic and therapeutic. It is reasonable to assume that the same vaccine candidates being investigated for genital tract vaccines could be potentially used, taking into consideration the different serovars associated with trachoma. However, a key factor in the success of a vaccine is that there is no time interval between the exposure to the organism and the arrival of the organism at the target site, so that the pathologic process, even in a reinfection situation, can begin immediately before an anamnestic response can be initiated. Moreover, because of practical economic reasons, the geographic distribution of trachoma, and the relative success of the SAFE program, there is less incentive for companies to invest a great deal of resources in a vaccine for trachoma. With these issues in mind, the prospects for a trachoma vaccine in the near future are somewhat bleak.

KEY ISSUES

- It is important that a vaccine does not exacerbate pathologic response or increase the incidence of clinical disease. There is precedent for this occurring in the trachoma trials, and the pathology produced by chlamydial infection is host mediated.
- A vaccine must be able to elicit long-term immunity, something which did not occur with the original trachoma trials and which does not occur with natural infection. A strategy of preventing ascending infection may in part facilitate long-term protection.
- Because there are multiple serovars that cause both genital and ocular infections, it is critical that a vaccine contain either multiple serovars or genus-specific antigens. It is likely that a vaccine will have multiple antigens.
- Persistence of chlamydial infection, although still undefined and uncharacterized in humans and animals, is still a major concern. A vaccine must be evaluated to determine whether it can prevent persistence or whether the organism is still present but prevented from causing disease.
- Before a vaccine can go to clinical trials, appropriate surrogates for infertility and ectopic pregnancy must be determined and evaluated in animal models.
- The success of the HPV vaccine and other vaccines administered to adolescents provides a golden opportunity for the addition of other vaccines against sexually transmitted infections to the adolescent panel.
- A vaccine for genital infections should target adolescent females before the beginning of sexual activity. Considering the lack of significant sequelae, it needs to be determined whether males should be targeted as well, although it would be reasonable so that infection is either prevented or reduced. This would also reduce the risk for women.
- Developing a vaccine against trachoma while feasible based on early clinical trials, may not have economic incentives for industry. Because of the emphasis on genital infections in recent years, there needs to be renewed efforts to understand the ocular immune response to chlamydial infections in animal models.

ACKNOWLEDGMENTS

I would like to thank Dr. Cho-chou Kuo and Dr. Bernhard Kaltenböck for the valuable information they provided me in the writing of this chapter. I also would like to acknowledge Dr. H. Marie Lacy for her helpful feedback in reviewing this manuscript.

References

Alani, M.D., Darougar, S., Burns, D.C., Thin, R.N. and Dunn, H. Isolation of *Chlamydia trachomatis* from the male urethra. Br. J. Vener. Dis. 1977; 53:88–92.

Arno, J.N., Katz, B.P., McBride, R., Carty, G.A., Batteiger, B.E., Caine, V.A. and Jones, R.B. Age and clinical immunity to infections with *Chlamydia trachomatis*. Sex. Transm. Dis. 1994; 21:47–52.

Barenfanger, J. and MacDonald, A.B. The role of immunoglobulin in the neutralization of trachoma infectivity. J. Immunol. 1974; 113:1607–1617.

Batteiger, B.E., Rank, R.G., Bavoil, P.M. and Soderberg, L.S.F. Partial protection against reinfection by immunization of guinea pigs with isolated outer-membrane proteins of the chlamydial agent of guinea pig inclusion conjunctivitis. J. Gen. Microbiol. 1993; 139:2965–2972.

Bavoil, P., Ohlin, A. and Schachter, J. Role of disulfide bonding in outer membrane structure and permeability in *Chlamydia trachomatis*. Infect. Immun. 1984; 44:479–485.

Bavoil, P.B. and Wyrick, P.B. Note from the editors. In: Chlamydia: Genomics and Pathogenesis (P.M. Bavoil and P.B. Wyrick, Eds.). Norfolk, UK: Horizon Bioscience, 2006.

Bedson, S.P. The psittacosis-lymphogranuloma venereum group of viruses. Br. Med. Bull. 1953; 9:226.

Bietti, G. and Werner, G.H. Trachoma: Prevention and Treatment. Springfield, IL: Charles C. Thomas, 1967.

Brunham, R.C., Kuo, C.-C., Cles, L. and Holmes, K.K. Correlation of host immune response with quantitative recovery of *Chlamydia trachomatis* from the human endocervix. Infect. Immun. 1983; 39:1491–1494.

Brunham, R.C., Maclean, I.W., Binns, B. and Peeling, R.W. Chlamydia trachomatis: its role in tubal infertility. J. Infect. Dis. 1985; 152:1275–1282.

Brunham, R.C., Peeling, R., Maclean, I., McDowell, J., Persson, K. and Osser, S. Postabortal *Chlamydia trachomatis* salpingitis: correlating risk with antigen-specific serological responses and with neutralization. J. Infect. Dis. 1987; 155:749–755.

Cain, T.K. and Rank, R.G. Local Th1-like responses are induced by intravaginal infection of mice with the mouse pneumonitis biovar of *Chlamydia trachomatis*. Infect. Immun. 1995; 63(5):1784–1789.

Caldwell, H.D., Kromhout, J. and Schachter, J. Purification and partial characterization of the major outer membrane protein of *Chlamydia trachomatis*. Infect. Immun. 1981; 31:1161–1176.

Caldwell, H.D., Kuo, C.C. and Kenny, G.E. Antigenic analysis of Chlamydiae by two dimensional immunoelectrophoresis. I. Antigenic heterogeneity between *C. trachomatis* and *C. psittaci*. J. Immunol. 1975; 115:963–968.

Caldwell, H.D. and Perry, L.J. Neutralization of *Chlamydia trachomatis* infectivity with antibodies to the major outer membrane protein. Infect. Immun. 1982; 38:745–754.

Collier, L.H. and Blyth, W.A. Immunogenicity of experimental trachoma vaccines in baboons. I. Experimental methods, and preliminary tests with vaccines prepared in chick embryos and in HeLa cells. J. Hyg. (Camb.) 1966a; 64:513–528.

Collier, L.H. and Blyth, W.A. Immunogenicity of experimental trachoma vaccines in baboons. II. Experiments with adjuvants, and tests of cross-protection. J. Hyg. (Camb.) 1966b; 64:529–544.

Collier, L.H. and Blyth, W.A. Immunogenicity of experimental trachoma vaccines in baboons. III. Experiments with inactivated vaccines. J. Hyg. (Camb.) 1967; 65:97–107.

Collier, L.H. and Mogg, A.E.N.S. Dissemination and immunogenicity of live TRIC agent in baboons after parenteral injection. II. Experiments with a 'slow-killing' strain. J. Hyg. 1969; 67:449–455.

Collier, L.H. and Smith, A. Dissemination and immunogenicity of live TRIC agent in baboons after parenteral injection. Am. J. Ophthalmol. 1967; 63:1589–1602.

Cotter, T.W., Ramsey, K.H., Miranpuri, G.S., Poulsen, C.E. and Byrne, G.I. Dissemination of *Chlamydia trachomatis* chronic genital tract infection in gamma interferon gene knockout mice. Infect. Immun. 1997; 65(6):2145–2152.

Crane, D.D., Carlson, J.H., Fischer, E.R., Bavoil, P., Hsia, R.C., Tan, C., Kuo, C.C. and Caldwell, H.D. *Chlamydia trachomatis* polymorphic membrane protein D is a species-common pan-neutralizing antigen. Proc. Natl. Acad. Sci. U.S.A. 2006; 103(6):1894–1899.

Darville, T., O'Neill, J.M., Andrews, C.W., Jr., Nagarajan, U.M., Stahl, L. and Ojcius, D.M. Toll-Like receptor-2, but not toll-like receptor-4, is essential for development of oviduct pathology in chlamydial genital tract infection. J. Immunol. 2003; 171(11):6187–6197.

Dhir, S.P., Agarwal, L.P., Detels, R., Wang, S.-P. and Grayston, J.T. Field trial of two bivalent trachoma vaccines in children of Punjab Indian villages. Am. J. Ophthalmol. 1967; 63:1639–1644.

Dong, F., Zhong, Y., Arulanandam, B. and Zhong, G. Production of a proteolytically active protein, chlamydial protease/proteasome-like activity factor, by five different Chlamydia species. Infect. Immun. 2005; 73(3):1868–1872.

Eko, F.O., He, Q., Brown, T., McMillan, L., Ifere, G.O., Ananaba, G.A., Lyn, D., Lubitz, W., Kellar, K.L., Black, C.M. and Igietseme, J.U. A novel recombinant multisubunit vaccine against Chlamydia. J. Immunol. 2004; 173:3375–3382.

Grayston, J.T., Kim, K.S.W., Alexander, E.R. and Wang, S.-P. Protective studies in monkeys with trivalent and monovalent trachoma vaccines. In: Trachoma and Related Disorders Caused by Chlamydial Agents (R.L. Nichols, Ed.), pp. 377–385. Amsterdam, The Netherlands: Excerpta Medica, 1971.

Grayston, J.T., Kronmal, R.A., Jackson, L.A., Parisi, A.F., Muhlestein, J.B., Cohen, J.D., Rogers, W.J., Crouse, J.R., Borrowdale, S.L., Schron, E. and Knirsch, C.ACES Investigators Azithromycin for the secondary prevention of coronary events. N. Engl. J. Med. 2005; 352(16):1637–1645.

Grayston, J.T. and Wang, S.-P. New knowledge of chlamydiae and the diseases they cause. J. Infect. Dis. 1975; 132:87–105.

Grayston, J.T., Wang, S.-P., Lin, H.M., Tai, F.H., Khaw, O.K. and Woolridge, R.L. Trachoma vaccine studies in volunteer students of the National Defense Medical Center. Chin. Med. J. (Engl.) 1961; 8:312–318.

Grayston, J.T., Woolridge, R.L., Wang, S., Yen, C., Yang, C., Cheng, K. and Chang, I. Field studies of protection from infection by experimental trachoma virus vaccine in preschool-aged children on Taiwan. Proc. Soc. Exp. Biol. Med. 1963; 112:589–595.

Halberstaedter, L. and von Prowazek, S. Ueber Zelleinschluesse parasitaerer Nature bein Trachom. Arb. Gesundhetisa. 1907; 26:44–47.

Halberstaedter, L. and von Prowazek, S. Über Chlamydozoenbefunde bei Blennorrhoea neonatorum non gonorrhoica. Klin. Wochenschr. 1909; 46:1839–1840.

Hawkins, R.A., Rank, R.G. and Kelly, K.A. A *Chlamydia trachomatis*-specific Th2 clone does not provide protection against a genital infection and displays reduced trafficking to the infected genital mucosa. Infect. Immun. 2002; 70(9):5132–5139.

Heymann, B. Ueber die Fundorte der Prowazek'schen Koerperchen. Berl. Klin. Wochenschr. 1910; 47:663–666.

Holland, M.J., Bailey, R.L., Conway, D.J., Culley, F., Miranpuri, G., Byrne, G.I., Whittle, H.C. and Mabey, D.C.W. T helper type-1 (Th1)/Th2 profiles of peripheral blood mononuclear cells (PBMC); Responses to antigens of *Chlamydia trachomatis* in subjects with severe trachomatous scarring. Clin. Exp. Immunol. 1996; 105(3):429–435.

Hsia, R.-C., Pannekoek, Y., Ingerowshi, E. and Bavoil, P.M. Type III secretion genes identify a putative virulence locus of *Chlamydia*. Mol. Microbiol. 1997; 25(2):351–359.

Igietseme, J.U., Magee, D.M., Williams, D.M. and Rank, R.G. Role for $CD8^+$ T cells in antichlamydial immunity defined by chlamydia-specific T-lymphocyte clones. Infect. Immun. 1994; 62:5195–5197.

Igietseme, J.U., Ramsey, K.H., Magee, D.M., Williams, D.M., Kincy, T.J. and Rank, R.G. Resolution of murine chlamydial genital infection by the adoptive transfer of a biovar-specific TH1 lymphocyte clone. Reg. Immunol. 1993; 5:317–324.

Igietseme, J.U. and Rank, R.G. Susceptibility to reinfection after a primary chlamydial genital infection is associated with a decrease of antigen-specific T cells in the genital tract. Infect. Immun. 1991; 59:1346–1351.

Jones, B.R. Ocular syndromes of TRIC virus infection and their possible genital significance. Br. J. Vener. Dis. 1964; 40:3–18.

Jones, B.R., Collier, L.H. and Smith, C.H. Isolation of virus from inclusion blennorrhea. Lancet 1959; 1:902–905.

Jones, R.B. and Batteiger, B.E. Human immune response to *Chlamydia trachomatis* infections. In: Chlamydial Infections (J. Oriel et al., Eds.), pp. 423–432. London, UK: Cambridge University Press, 1986.

Jones, R.B. and Van der Pol, B. Lack of correlation between acquisition of infection and ability of serum to neutralize chlamydial infectivity *in vitro*. In: Chlamydial Infections (J. Orfila et al., Eds.), pp. 95–98. Bologna, Italy: Societa Editrice Esculapio, 1994.

Katz, B.P., Batteiger, B.E. and Jones, R.B. Effect of prior sexually transmitted disease on the isolation of *Chlamydia trachomatis*. Sex. Transm. Dis. 1987; 14:160–164.

Kawa, D.E., Schachter, J. and Stephens, R.S. Immune response to the Chlamydia trachomatis outer membrane protein PorB. Vaccine 2004; 22(31–32):4282–4286.

Kawa, D.E. and Stephens, R.S. Antigenic topology of Chlamydial PorB protein and identification of targets for immune neutralization of infectivity. J. Immunol. 2002; 168(10):5184–5191.

Kelly, K.A. and Rank, R.G. Identification of homing receptors that mediate the recruitment of CD4 T cells to the genital tract following intravaginal infection with *Chlamydia trachomatis*. Infect. Immun. 1997; 65:5198–5208.

Kelly, K.A., Robinson, E.A. and Rank, R.G. Identification of cell adhesion molecules that mediate CD4 lymphocyte homing to the genital tract mucosa during infection with mouse pneumonitis biovar of Chlamydia trachomatis. In: Chlamydia Research (A. Stary, Ed.), Vol. 3, p. 76. Bologna, Italy: Societa Editrice Esculapio, 1996.

Kimani, J., Maclean, I.W., Bwayo, J.J., MacDonald, K., Oyugi, J., Maitha, G.M., Peeling, R.W., Cheang, M., Nagelkerke, N.J., Plummer, F.A. and Brunham, R.C. Risk factors for *Chlamydia trachomatis* pelvic inflammatory disease among sex workers in Nairobi, Kenya. J. Infect. Dis. 1996; 173(6):1437–1444.

Kroner, T. Zur Ätiologie der Ophthalmoblennorrhoea neonatorum. Zentralbl. Gynaekol. 1884; 8:643–645.

Lau, C.Y. and Qureshi, A.K. Azithromycin versus doxycycline for genital chlamydial infections: a meta-analysis of randomized clinical trials. Sex. Transm. Dis. 2002; 29(9):497–502.

Malaty, R., Dawson, C.R., Wong, I., Lyon, C. and Schachter, J. Serum and tear antibodies to Chlamydia after reinfection with guinea pig inclusion conjunctivitis agent. Invest. Ophthalmol. Vis. Sci. 1981; 21:833–841.

Mardh, P.A., Ripa, T., Svensson, L. and Westrom, L. *Chlamydia trachomatis* infection in patients with acute salpingitis. N. Engl. J. Med. 1977; 296:1377–1379.

Masson, P.L., Heremans, J.F. and Ferin, J. Clinical importance of the biochemical changes in the female genital tract. I. Studies on the proteins of cervical mucus. Int. J. Fertil. 1969; 14:1–7.

Miller, W.C., Ford, C.A., Morris, M., Handcock, M.S., Schmitz, J.L., Hobbs, M.M., Cohen, M.S., Harris, K.M. and Udry, J.R. Prevalence of chlamydial and gonococcal infections among young adults in the United States. JAMA 2004; 291(18): 2229–2236.

Modabber, F., Bear, S.E. and Cerny, J. The effect of cyclophosphamide on the recovery from a local chlamydial infection. Guinea pig inclusion conjunctivitis. Immunol. 1976; 30:929–933.

Monnickendam, M.A., Darougar, S., Treharne, J.D. and Tilbury, A.M. Development of chronic conjunctivitis with scarring and pannus, resembling trachoma, in guinea pigs. Br. J. Ophthalmol. 1980a; 64:284–290.

Monnickendam, M.A., Darougar, S., Treharne, J.D. and Tilbury, A.M. Guinea pig inclusion conjunctivitis as a model for the study of trachoma: clinical, microbiological, serological, and cytological studies of primary infection. Br. J. Ophthalmol. 1980b; 64:279–283.

Morrison, R.P. Chlamydial hsp60 and the immunopathogenesis of chlamydial disease. Semin. Immunol. 1991; 3:25–33.

Morrison, R.P., Belland, R.J., Lyng, K. and Caldwell, H.D. Chlamydial disease pathogenesis. The 57-kD chlamydial hypersensitivity antigen is a stress response protein. J. Exp. Med. 1989; 170:1271–1283.

Morrison, R.P., Feilzer, K. and Tumas, D.B. Gene knockout mice establish a primary protective role for major histocompatibility complex class II-restricted responses in *Chlamydia trachomatis* genital tract infection. Infect. Immun. 1995; 63(12):4661–4668.

Morrison, S.G. and Morrison, R.P. A predominant role for antibody in acquired immunity to chlamydial genital tract reinfection. J. Immunol. 2005a; 175(11):7536–7542.

Morrison, S.G. and Morrison, R.P. A predominant role for antibody in acquired immunity to chlamydial genital tract reinfection. J. Immunol. 2005b; 175(11):7536–7542.

Morrison, S.G., Su, H., Caldwell, H.D. and Morrison, R.P. Immunity to murine *Chlamydia trachomatis* genital tract reinfection involves B cells and CD4(+) T cells but not CD8(+) T cells. Infect. Immun. 2000; 68(12):6979–6987.

Moulder, J.W., Hatch, T.P., Kuo, C.-C., Schachter, J. and Storz, J. Genus Chlamydia. In: Bergey's Manual of Systematic Bacteriology (N.R. Krieg, Ed.), pp. 729–739. Baltimore, MD: Williams & Wilkins, 1984.

Murthy, A.K., Chambers, J.P., Meier, P.A., Zhong, G. and Arulanandam, B.P. Intranasal vaccination with a secreted chlamydial protein enhances resolution of genital *Chlamydia muridarum* infection, protects against oviduct pathology, and is highly dependent upon endogenous gamma interferon production. Infect. Immun. 2007; 75(2):666–676.

Nelson, D.E., Virok, D.P., Wood, H., Roshick, C., Johnson, R.M., Whitmire, W.M., Crane, D.D., Steele-Mortimer, O., Kari, L., McClarty, G. and Caldwell, H.D. Chlamydial IFN-{gamma} immune evasion is linked to host infection tropism. Proc. Natl Acad. Sci. 2005; 102(30):10658–10663.

Nichols, R.L., Bell, S.D., Jr., Haddad, N.A. and Bobb, A.A. Studies on trachoma VI. Microbiological observations in a field trial in Saudi Arabia of bivalent trachoma vaccine at three dosage levels. Am. J. Trop. Med. Hyg. 1969; 18:723–730.

Nichols, R.L., Bell, S.D., Jr., Murray, E.S., Haddad, N.A. and Bobb, A.A. Studies on trachoma. V. Clinical observations in a field trial of bivalent trachoma vaccine at three dosage levels in Saudi Arabia. Am. J. Trop. Med. Hyg. 1966; 15:639–647.

Nichols, R.L., Oertley, R.E., Fraser, C.E.O., MacDonald, A.B. and McComb, D.E. Immunity to chlamydial infections of the eye. VI. Homologous neutralization of trachoma infectivity for the owl monkey conjunctivae by eye secretions from humans with trachoma. J. Infect. Dis. 1973; 127:429–432.

Nurminen, M., Wahlstrom, E., Kleemola, M., Leinonen, M., Saikku, P. and Makela, P.H. Immunologically related ketodeoxyoctonate containing structures in Chlamydia trachomatis, Re:mutants of Salmonella species, and Acinetobacter calcoaceticus var. anitratus. Infect. Immun. 1984; 44:609–613.

O'connell, C.M., Ingalls, R.R., Andrews, C.W., Jr., Scurlock, A.M. and Darville, T. Plasmid-deficient *Chlamydia muridarum* fail to induce immune pathology and protect against oviduct disease. J. Immunol. 2007; 179(6):4027–4034.

Pal, S., Peterson, E.M., Rappuoli, R., Ratti, G. and de la Maza, L.M. Immunization with the *Chlamydia trachomatis* major outer membrane protein, using adjuvants developed for human vaccines, can induce partial protection in a mouse model against a genital challenge. Vaccine 2006; 24(6):766–775.

Pal, S., Theodor, I., Peterson, E.M. and de la Maza, L.M. Immunization with an acellular vaccine consisting of the outer membrane complex of *Chlamydia trachomatis* induces protection against a genital challenge. Infect. Immun. 1997; 65(8):3361–3369.

Pearce, J.H., Allan, I. and Ainsworth, S. Interaction of chlamydiae with host cells and mucous surfaces. Ciba Found. Symp. 1981; 80:234–251.

Perry, L.L., Feilzer, K. and Caldwell, H.D. Immunity to *Chlamydia trachomatis* is mediated by T helper 1 cells through IFN-gamma-dependent and -independent pathways. J. Immunol. 1997; 158:3344–3352.

Ramsey, K.H., Newhall, W.J. and Rank, R.G. Humoral immune response to chlamydial genital infection of mice with the agent of mouse pneumonitis. Infect. Immun. 1989; 57:2441–2446.

Ramsey, K.H. and Rank, R.G. The role of T cell subpopulations in resolution of chlamydial genital infection of mice. In: Chlamydial

Infections (W.R. Bowie et al., Eds.), pp. 241–244. New York, NY: Cambridge University Press, 1990.

Ramsey, K.H., Sigar, I.M., Schripsema, J.H., Shaba, N. and Cohoon, K.P. Expression of matrix metalloproteinases subsequent to urogenital *Chlamydia muridarum* infection of mice. Infect. Immun. 2005; 73(10):6962–6973.

Ramsey, K.H., Soderberg, L.S.F. and Rank, R.G. Resolution of chlamydial genital infection in B-cell-deficient mice and immunity to reinfection. Infect. Immun. 1988; 56:1320–1325.

Rank, R.G. Models of immunity. In: Chlamydia: Intracellular Biology, Pathogenesis, and Immunity (R.S. Stephens, Ed.), pp. 239–295. Washington, D.C.: American Society for Microbiology, 1999.

Rank, R.G. and Barron, A.L. Effect of antithymocyte serum on the course of chlamydial genital infection in female guinea pigs. Infect. Immun. 1983a; 41:876–879.

Rank, R.G. and Barron, A.L. Humoral immune response in acquired immunity to chlamydial genital infection of female guinea pigs. Infect. Immun. 1983b; 39:463–465.

Rank, R.G. and Batteiger, B.E. Protective role of serum antibody in immunity to chlamydial genital infection. Infect. Immun. 1989; 57:299–301.

Rank, R.G., Batteiger, B.E. and Soderberg, L.S.F. Susceptibility to reinfection after a primary chlamydial genital infection. Infect. Immun. 1988; 56:2243–2249.

Rank, R.G., Bowlin, A.K., Reed, R.L. and Darville, T. Characterization of chlamydial genital infection resulting from sexual transmission from male to female guinea pigs and determination of infectious dose. Infect. Immun. 2003; 71(11):6148–6154.

Rank, R.G., Dascher, C., Bowlin, A.K. and Bavoil, P.M. Systemic immunization with HSP60 alters the development of Chlamydial ocular disease. Invest. Ophthalmol. Vis. Sci. 1995; 36(7):1344–1351.

Rank, R.G. and Sanders, M.M. Pathogenesis of endometritis and salpingitis in a guinea pig model of chlamydial genital infection. Am. J. Pathol. 1992; 140:927–936.

Rank, R.G., Sanders, M.M. and Patton, D.L. Increased incidence of oviduct pathology in the guinea pig after repeat vaginal inoculation with the chlamydial agent of guinea pig inclusion conjunctivitis. J. Sex. Transm. Dis. 1995; 22(1):48–54.

Rank, R.G., Soderberg, L.S.F. and Barron, A.L. Chronic chlamydial genital infection in congenitally athymic nude mice. Infect. Immun. 1985; 48:847–849.

Rank, R.G., Soderberg, L.S.F., Sanders, M.M. and Batteiger, B.E. Role of cell-mediated immunity in the resolution of secondary chlamydial genital infection in guinea pigs infected with the agent of guinea pig inclusion conjunctivitis. Infect. Immun. 1989; 57:706–710.

Rank, R.G., White, H.J. and Barron, A.L. Humoral immunity in the resolution of genital infection in female guinea pigs infected with the agent of guinea pig inclusion conjunctivitis. Infect. Immun. 1979; 26:573–579.

Rasmussen, S.J., Eckmann, L., Quayle, A.J., Shen, L., Zhang, Y.X., Anderson, D.J., Fierer, J., Stephens, R.S. and Kagnoff, M.F. Secretion of proinflammatory cytokines by epithelial cells in response to *Chlamydia* infection suggests a central role for epithelial cells in chlamydial pathogenesis. J. Clin. Invest. 1997; 99(1):77–87.

Rasmussen, S.J., Sen, C.K. and Stephens, R.S. Temporal expression of interleukin-8 in *Chlamydia*-infected epithelial cells. In: Proceedings of the Ninth International Symposium on Human Chlamydial Infection (R.S. Stephens et al., Eds.), pp. 411–414. San Francisco: International Chlamydia Symposium, 1998.

Rodolakis, A. and Bernard, F. Vaccination with temperature sensitive mutant of *Chlamydia psittaci* against enzootic abortion of ewes. Vet. Rec. 1984; 114:193–194.

Rodolakis, A. and Souriau, A. Response of ewes to temperature sensitive mutants of *Chlamydia psittaci* (var.ovis) obtained by NTG mutagenesis. Ann. Rech. Vet. 1983; 14:155–161.

Schachter, J., Cles, L.D., Ray, R.M. and Hesse, F.E. Is there immunity to chlamydial infections of the human genital tract? Sex. Transm. Dis. 1983; 10:123–125.

Schachter, J. and Dawson, C.R. Human Chlamydial Infections. Littleton, MA: PSG Publishing Company, 1978.

Sowa, S., Sowa, J., Collier, L.H. and Blyth, W.A. Trachoma field trials in The Gambia. J. Hyg. 1969; 67:699–717.

Stephens, R.S. The cellular paradigm of chlamydial pathogenesis. Trends Microbiol. 2003; 11:44–51.

Stephens, R.S., Wagar, E.A. and Schoolnik, G.K. High-resolution mapping of serovar-specific and common antigenic determinants of the major outer membrane protein of *Chlamydia trachomatis*. J. Exp. Med. 1988; 167:817–831.

Su, H. and Caldwell, H.D. $CD4^+$ T cells play a significant role in adoptive immunity to *Chlamydia trachomatis* infection of the mouse genital tract. Infect. Immun. 1995; 63(9):3302–3308.

Su, H., Feilzer, K., Caldwell, H.D. and Morrison, R.P. *Chlamydia trachomatis* genital tract infection of antibody-deficient gene knockout mice. Infect. Immun. 1997; 65(6):1993–1999.

Su, H., Messer, R., Whitmore, W., Fischer, E., Portis, J.C. and Caldwell, H.D. Vaccination against chlamydial genital tract infection after immunization with dendritic cells pulsed ex vivo with nonviable chlamydiae. J. Exp. Med. 1998; 188(5):809–818.

T'ang, F.F., Chang, H.L., Huang, Y.T. and Wang, K.C. Studies on the etiology of trachoma with special reference to isolation of the virus in chick embryo. Chin. Med. J. (Peking) 1957; 75:429–446.

Tanner, M.A., Harris, J.K. and Pace, N.R. Molecular phylogeny of *Chlamydia* and relatives. In: Chlamydia: Intracellular Biology, Pathogenesis, and Immunity (R.S. Stephens, Ed.), pp. 1–8. Washington, D.C.: American Society for Microbiology, 1999.

Taylor, H.R. Ocular models of chlamydial infection. Rev. Infect. Dis. 1985; 7:737–740.

Taylor, H.R., Johnson, S.L., Prendergast, R.A., Schachter, J., Dawson, C.R. and Silverstein, A.M. An animal model of trachoma II. The importance of repeated reinfection. Invest. Ophthalmol. Vis. Sci. 1982; 23:507–515.

Taylor, H.R., Prendergast, R.A., Dawson, C.R., Schachter, J. and Silverstein, A.M. An animal model for cicatrizing trachoma. Invest. Ophthalmol. Vis. Sci. 1981; 21:422–433.

Thygeson, P. Trachoma virus: historical background and review of isolates. Ann. N. Y. Acad. Sci. 1962; 98:6–13.

Tseng, C.K. and Rank, R.G. Role of NK cells in the early host response to chlamydial genital infection. Infect. Immun. 1998; 66(12):5867–5875.

Tuffrey, M., Alexander, F. and Taylor-Robinson, D. Severity of salpingitis in mice after primary and repeated inoculation with a human strain of *Chlamydia trachomatis*. J. Exp. Pathol. 1990; 71:403–410.

Van Voorhis, W.C., Barrett, L.K., Sweeney, Y.T.C., Kuo, C.C. and Patton, D.L. Analysis of lymphocyte phenotype and cytokine activity in the inflammatory infiltrates of the upper genital tract of female macaques infected with *Chlamydia trachomatis*. J. Infect. Dis. 1996; 174(3):647–650.

Wagar, E.A., Schachter, J., Bavoil, P. and Stephens, R.S. Differential human serologic response to two 60,000 molecular weight *Chlamydia trachomatis* antigens. J. Infect. Dis. 1990; 162:922–927.

Wang, S.-P. and Grayston, J.T. Pannus with experimental trachoma and inclusion conjunctivitis agent infection of Taiwan monkeys. Am. J. Ophthalmol. 1967; 63:1133–1145.

Wang, S.-P. and Grayston, J.T. Studies on the identity of the 'fast' egg-killing chlamydia strains. In: Trachoma and Related Disorders Caused by Chlamydial Agents (R.L. Nichols, Ed.), pp. 322–336. Amsterdam, The Netherlands: Excerpta Medica, 1971a.

Wang, S.-P. and Grayston, J.T. Classification of TRIC and related strains with microimmunofluorescence. In: Trachoma and Related Disorders Caused by Chlamydial Agents (R.L. Nichols, Ed.), pp. 305–321. Excerpta Medica, 1971b.

Wang, S.P., Grayston, J.T. and Alexander, E.R. Trachoma vaccine studies in monkeys. Am. J. Ophthalmol. 1967; 63:1615–1630.

Weinstock, H., Dean, D. and Bolan, G. Chlamydia trachomatis infections. Infect. Dis. Clin. North Am. 1994; 4:797–819.

Whittum-Hudson, J.A., Taylor, H.R., Farazdaghi, M. and Prendergast, R.A. Immunohistochemical study of the local inflammatory response to chlamydial ocular infection. Invest. Ophthalmol. Vis. Sci. 1986; 27:64–69.

Williams, D.M., Schachter, J., Weiner, M.H. and Grubbs, B. Antibody in host defense against mouse pneumonitis agent (murine *Chlamydia trachomatis*). Infect. Immun. 1984; 45:674–678.

Wolner-Hanssen, P., Patton, D.L. and Holmes, K.K. Protective immunity in pig-tailed macaques after cervical infection with *Chlamydia trachomatis*. Sex. Transm. Dis. 1991; 18:21–25.

Wolner-Hanssen, P., Patton, D.L., Stamm, W.E. and Holmes, K.K. Severe salpingitis in pig-tailed macaques after repeated cervical infections followed by a single tubal inoculation with *Chlamydia trachomatis*. In: Chlamydia Infections (D. Oriel et al., Eds.), pp. 371–374. New York, NY: Cambridge University Press, 1986.

Woolridge, R.L., Grayston, J.T., Chang, I.H., Yang, C.Y. and Cheng, K.H. Field trial of a monovalent and of a bivalent mineral oil adjuvant trachoma vaccine in Taiwan school children. Am. J. Ophthalmol. 1967a; 63:1645–1650.

Woolridge, R.L., Grayston, J.T., Chang, I.H., Yang, C.Y. and Cheng, K.H. Long-term follow-up of the initial (1959–1960) trachoma vaccine field trial on Taiwan. Am. J. Ophthalmol. 1967b; 63:1650–1653.

Workowski, K.A. and Berman, S.M. Sexually transmitted diseases treatment guidelines, 2006. MMWR Recomm. Rep. 2006; 55:1–94

Wyrick, P.B., Brownridge, E.A. and Ivins, B.E. Interaction of *Chlamydia psittaci* with mouse peritoneal macrophages. Infect. Immun. 1978; 19:1061–1067.

Zhang, Y.-X., Stewart, S.J. and Caldwell, H.D. Protective monoclonal antibodies to *Chlamydia trachomatis* serovar- and serogroup-specific major outer membrane protein determinants. Infect. Immun. 1989; 57:636–638.

Zhong, G., Fan, P., Ji, H., Dong, F. and Huang, Y. Identification of a chlamydial protease-like activity factor responsible for the degradation of host transcription factors. J. Exp. Med. 2001; 193(8):935–942.

CHAPTER

45

Cholera

Chandrabali Ghose[1,2], *Stephen B. Calderwood*[1,4] *and Edward T. Ryan*[1,3,4,5]

[1]Division of Infectious Diseases, Massachusetts General Hospital, Boston, MA, USA

[2]Aaron Diamond AIDS Research Center, New York, NY, USA

[3]Tropical and Geographic Medicine Center, Massachusetts General Hospital, Boston, MA, USA

[4]Harvard Medical School, Boston, MA, USA

[5]Department of Immunology and Infectious Diseases, Harvard School of Public Health, Boston, MA, USA

OUTLINE

ABSTRACT

Cholera is a life-threatening diarrheal disease with epidemic potential caused by the Gram negative bacterium *Vibrio cholerae*, serogroups O1 and O139. Since most cases of cholera are either not detected or not reported, the global burden of cholera is not known with certainty; however, cholera causes significant morbidity and mortality worldwide, with an estimated 5–7 million cases occurring each year, resulting in greater than 100,000 deaths per year. At present, the world is experiencing its seventh recorded pandemic, and cholera is endemic in many countries in Asia, Africa, and Latin America. Large outbreaks, especially among impoverished or displaced persons, occur on a regular basis, and cholera may spread through travel or migration of infected individuals, including even short-term travelers or visitors. Natural infection with *V. cholerae* O1 or O139 induces long-lasting protective immunity against subsequent disease. Protection against cholera correlates best with the vibriocidal antibodies thought to be largely directed against *V. cholerae* lipopolysaccharide (LPS), although anti-LPS serum IgG levels have not been shown to correlate with protection against cholera. It is hypothesized that the vibriocidal response may be a surrogate marker for as yet poorly understood mucosal immune responses, since it is unclear why a serum complement-binding-based assay would predict protection against a noninvasive intestinal pathogen such as *V. cholerae*. In an effort to ameliorate the morbidity and mortality associated with cholera, several parenteral and oral cholera vaccines have been developed since the 1880s. Commercial production of the killed parenteral vaccine most widely available in the 20th century was recently discontinued. Currently, an oral killed cholera vaccine (rBS-WC) is the only international commercially available cholera vaccine, and is only available in some countries. No cholera vaccine is currently commercially available in the United States. Following repetitive administration in large field trials, rBS-WC induces approximately 85% protective immunity for at least 6 months, although protection wans over the subsequent 2–3 years following vaccination. An oral killed whole vaccine that targets both *V. cholerae* O1 and O139 is also available and used programmatically in Vietnam. A number of live attenuated cholera vaccines have also been developed, including classical O1 derivative CVD 103-HgR, and El Tor O1 derivative Peru-15. These live attenuated vaccines have been shown to be safe and immunogenic in volunteers, although their protective efficacy has not yet been demonstrated in large randomized field trials. Subunit vaccines and additional anti-O139 cholera vaccines are presently in earlier stages of development. To better understand immunity protective against cholera, high throughput genomic, proteomic, and immunomic technologies are being applied to *V. cholerae.*

INTRODUCTION

Vibrio cholerae serogroups O1 and O139 are the causative agents of cholera, a severe, dehydrating diarrheal illness of humans with epidemic potential, and *V. cholerae* is a category B agent. At present, the world is experiencing its seventh recorded cholera pandemic, and cholera is endemic in many countries in Asia, Africa, and Latin America. Cholera remains a disease of poverty, social disruption, and poor sanitation, and large and explosive outbreaks occur on a regular basis.

HISTORY OF DISEASE

Although scattered written historic reports of epidemics of severe watery diarrhea exist in Sanskrit, Chinese, and Arabic, perhaps one of the first written reports that may with some degree of certainty be ascribed to a cholera outbreak was written by the Portuguese explorer Gaspar Correira in 1543 in his work "Lendas da India," in which he described an outbreak of diarrhea among military troops in Calicut and Goa that left thousands dead. Correira wrote of a "Disease… (with) vomiting, with drought of water accompanying it, as if the stomach were parched up, and cramps that fixed the sinews of the joints…. Disease sudden-like, which struck with pain in the belly, so that a man did not last out eight hours of time." Although other large outbreaks were reported in the 1600s and 1700s in Asia, the global spread of cholera mirrored the growth of the global economy. By convention, since 1817, there have been seven global cholera pandemics, all of which have begun in Asia. The world is currently experiencing its seventh pandemic, which began in modern day Indonesia in 1961, spread to Africa and Europe in the 1970s, and then to the New World in the 1990s. The first six pandemics are historically ascribed to *V. cholerae* O1 classical biotype. The current pandemic is caused by *V. cholerae* O1 El Tor biotype (named for the El Tor medical field station in the Sinai Peninsula where this biotype was first isolated among religious pilgrims in the early 1900s). *V. cholerae* O139 first appeared in 1992 in South Asia, and since then has accounted for a minority of disease in 11 Asian countries.

It is hard to overstate the role that cholera has had on human societal, scientific, and biomedical development. Along with plague and smallpox, cholera was one of the pathogens of the highest historical global import. In large measure, attempts to control cholera drove the sanitary revolution of the 1800s, including the establishment of governmental and international boards of health, laws pertaining to hygiene, water treatment, and international travel, and construction of large public works including municipal sewer and water systems. Cholera was the first human illness ascribed to transmission by water, as suggested by John Snow in "On the Mode and Transmission of Cholera," and it was John Snow's epidemiological analysis of the London cholera outbreak surrounding the Broad Street pump of 1854 that is credited with initiating the field of evidence-based epidemiological intervention.

V. cholerae was also the first human pathogen identified microscopically when Filippo Pacini observed a curved bacillus in the stool of cholera victims in Florence during an 1854 outbreak. Pacini named the organism *Vibrio cholerae* since it appeared to vibrate under the microscope. Robert Koch then "rediscovered" the organism in the 1880s, and unaware of Pacini's work, referred to the organism as *Komma bacillus* (since it was comma shaped). Importantly, Koch was able not only to view the organism under the microscope, but also to develop the techniques required to grow the organism in pure culture in the laboratory. This permitted characterizing the cholera bacillus (and typhoid bacillus) by reactivity with specific antisera, leading to the agglutination assay still used to serogroup bacteria today. Koch also postulated production of a toxin by the cholera organism. Attempts to isolate this toxin led Pfeiffer to the identification of heat stable lipopolysaccharide (LPS) (i.e., "endotoxin"). Pfeiffer further observed that *V. cholerae* injected into the peritoneum of immunized guinea pigs became immobilized and lyzed, an observation which then led to the discovery of the complement system, which itself led to the development of the first specific laboratory-based serological assay for a human pathogen, the serum vibriocidal assay. In 1959, cholera toxin (CT) was also the first enterotoxin specifically identified.

CT causes a severe secretory diarrhea and cholera played a central role in advancing the understanding of fluid and salt physiology in the human body, was the indication for the first therapeutic use of intravenous fluid replacement therapy in humans (when Jaehnichen injected water into a cholera patient in 1830, and Latta injected water and salt into cholera patients in 1832), and played the central role in the development of oral replacement therapy (ORS), a simple and inexpensive intervention that has saved millions of lives from all causes of diarrhea in the last 40 years. Cholera has also played a central role in the history of vaccinology (see below).

ETIOLOGIC AGENT

Classification

V. cholerae is a Gram negative curved bacillus. The organism is flagellated and motile with a characteristic "shooting star" appearance in the watery diarrheal stools of cholera patients. *V. cholerae* are differentiated by the LPS component of the organism's outer membrane, and *V. cholerae* is currently classified into approximately 200 serogroups. *V. cholerae* strains belonging to serogroups O1 and O139 are capable of causing epidemic cholera. Non-O1, non-O139 *V. cholerae* strains have been associated with isolated cases or small outbreaks of gastroenteritis, cellulitis, or bacteremia, the latter two manifestations usually reported in individuals with chronic hepatic dysfunction. *V. cholerae* O1 strains can also be divided into two biotypes, classical and El Tor, based on phenotypic differences in hemolytic assays, phage characteristics, hemaglutination, and polymyxin B sensitivity. Based on O antigen differences, each O1 biotype can be further subdivided into three serotypes: Ogawa, Inaba, and Hikojima. During outbreaks or sustained transmission, *V. cholerae* O1 may switch between Ogawa and Inaba serotypes. The Hikojima serotype is rare and thought to be unstable. *V. cholerae* O1 EL Tor type strain N16961 is currently annotated to contain 3885 open reading frames on two circular chromosomes: a large chromosome containing approximately 2.96 mbp and a smaller chromosome containing approximately 1.07 mbp (Heidelberg et al., 2000). Although *V. cholerae* may contain plasmids that alter susceptibility to antimicrobial agents, full virulence of *V. cholerae* does not require virulence plasmids.

Antigens

Perhaps the two best studied *V. cholerae* antigens are LPS and CT. LPS is comprised of a lipid A domain associated with the *V. cholerae* outer membrane, a core antigen, and repetitive specific polysaccharides. Ogawa and Inaba specific polysaccharides differ by the replacement of a hydroxyl group with a methyl group on the terminal sugar (Villeneuve et al., 2000). The LPS of *V. cholerae* O1 and O139 are distinct and non cross protective, and *V. cholerae* O139 is surrounded by a polysaccharide capsule (Comstock et al., 1995).

CT is a heterodimeric exotoxin that contains an enzymatically active A subunit attached to a pentamer of nonenzymatically active B subunits. The B subunit pentamer mediates binding of CT to monosialoganglioside GM_1 on brush border intestinal epithelial cells.

The A subunit is an ADP-ribosylating enzyme that elevates intracellular cAMP, eventually resulting in a secretory diarrhea in infected humans. CtxA and CtxB are expressed as an operon, and are encoded on a filamentous phage CTXφ (Waldor and Mekalanos, 1996). CT of *V. cholerae* O1 El Tor and O139 are identical.

A number of other important antigens have also been identified in *V. cholerae*. Toxin coregulated pilus (TCP) is a type IV pilus colonization factor essential for colonization and virulence (Taylor et al., 1987; Herrington et al., 1988). The major structural unit of TCP is TcpA, and TcpA from El Tor and classical strains are approximately 80% homologous (Sun et al., 1990; Iredell and Manning 1994). TcpA from El Tor and O139 strains are identical. In addition to playing a role in intestinal colonization by *V. cholerae*, TCP functions as the *V. cholerae* receptor for the filamentous phage CTXφ (Waldor and Mekalanos, 1996). The *V. cholerae* genome encodes 11 other type IV pili, many of which have either not been studied in detail or have not been shown to be essential to intestinal colonization, and other less-well defined antigens possibly associated with virulence (Hang et al., 2003).

PROTECTIVE IMMUNE RESPONSE

Previous infection with *V. cholerae* induces long-lasting protective immunity (Cash et al., 1974a, 1974b; Levine et al., 1979). For at least 3 years after infection of North American volunteers, infection with *V. cholerae* O1 classical strain organisms provides 100% protection against rechallenge with a homologous strain, and El Tor infection provides 90% protection (Levine et al., 1981). Previous infection with either O1 biotype provides slightly less protection against challenge with the heterologous strain O1 biotype organism, and immunity is serogroup-specific (previous infection with *V. cholerae* O1 does not provide protection against O139 disease). Mirroring volunteer studies, infection with *V. cholerae* O1 El Tor in Bangladesh, an area of the world endemic for cholera, decreases the likelihood of symptomatic cholera over the next several years by approximately 90% (Glass et al., 1985). Unfortunately, the nature of immune responses protective against cholera are not currently well understood.

Antibody-Mediated Immune Responses

Serum Responses

The serum vibriocidal antibody response, measured in an assay that detects complement-mediated bactericidal activity, is currently the best identified predictor of protection against cholera. In Bangladesh, vibriocidal antibodies are detectable in 40–80% of individuals by adolescence, and in one study, every two-fold rise in vibriocidal titer was associated with an approximately 40% decrease in the risk of developing symptomatic cholera (Glass et al., 1985; Mosley et al., 1968a, 1968b). Despite this, symptomatic cholera has been reported in individuals with high baseline vibriocidal titers on exposure (greater than or equal 320), and no vibriocidal titer has been identified above which *V. cholerae* O1 infection is completely prevented (Saha et al., 2004). Adsorbing sera with *V. cholerae* LPS removes a majority of vibriocidal antibodies (Holmgren and Svennerholm, 1977), although serum IgG anti-LPS antibody levels do not correlate with protection from cholera. Interestingly, recent data suggest that serum IgA anti-LPS responses do correlate with protection against cholera (Harris et al., 2007). These and other data suggest that the serum vibriocidal response may be a surrogate marker of mucosal immune responses that are themselves truly protective against cholera.

Although serum IgG antibody levels to CT increase after cholera, serum anti-CT IgG responses have not been shown to correlate with protection against cholera (Glass et al., 1985). Serum IgA anti-CT responses have, however, recently been found to correlate with protection against *V. cholerae* infection, perhaps further supporting the importance of mucosal immune responses in mediating protection (Harris et al., 2007); however, previous infection with El Tor *V. cholerae* O1 does not confer protection from infection with *V. cholerae* O139, despite the fact that *V. cholerae* O1 El Tor and O139 express essentially identical CTs (Albert et al., 1993; Ramamurthy et al., 1993).

Anti-TcpA responses are also common following cholera in an endemic setting, occurring in serum, stool, or circulating mucosal lymphocytes of over 90% of Bangladeshis recovering from cholera caused by *V. cholerae* O1 El Tor (Asaduzzaman et al., 2004), and serum IgA anti-TcpA antibody levels have recently been shown to correlate with protection against cholera in Bangladesh (Harris et al., 2007). Immunization of mice with El Tor TcpA also reduces mortality following *V. cholerae* challenge by 50% (Rollenhagen et al., 2006); however, in humans, infection with El Tor *V. cholerae* O1 has not been associated with protection from infection with *V. cholerae* O139, even though both serogroups of organisms produce identical TcpA molecules.

Mucosal Immunity

Acquired protection against noninvasive mucosal pathogens is in large part mediated by secretory IgA (sIgA) produced locally at mucosal surfaces. Directly and reliably measuring local production of anti-*V. cholerae* sIgA at the intestinal surface can be problematic. As such, immune responses that are surrogate markers of intestinal immunity are often measured. In the case of *V. cholerae* infection, the serum vibriocidal and serum IgA antibody responses may be such markers. Following intestinal presentation of antigen, mucosal lymphocytes transiently migrate in the systemic circulation before re-homing to the intestinal mucosa. This transient migration peaks 7 days following intestinal exposure, and is measurable in an ELISPOT-format in an antibody-secreting cell (ASC) assay (Czerkinsky et al., 1988; Qadri et al., 1997). Following infection of North American volunteers with *V. cholerae*, anti-CT and anti-LPS IgA ASC responses were detectable in 50 and 83% of recipients, respectively (Losonsky et al., 1996), although protection from disease was 100% if rechallenge was done within 6 months of primary infection, suggesting that although anti-LPS and anti-CT IgA ASC responses correlate with protection from cholera, that protection against disease could occur even in the absence of detectable ASC responses to CT and LPS. Among Bangladeshi adults recovering from cholera, ASC responses against TcpA were also detectable in the majority, although no longitudinal study has been performed to see if such immunity correlates with protection from subsequent disease (Asaduzzaman et al., 2004).

Cell-Mediated Immune Responses

V. cholerae is a noninvasive mucosal pathogen, and although not investigated in detail, there are no current data to suggest that cell-mediated immune responses correlate with protection against cholera. Despite this, immune responses to CT, which is a potent immunoadjuvant in addition to being a potent enterotoxin, have been shown to be T cell dependent and MHC class II restricted (Cong et al., 1996).

EPIDEMIOLOGY

Significance as a Public Health Problem

V. cholerae is a human pathogen. There is no zoonotic infection of importance. *V. cholerae* exist in the environment in brackish water, especially favoring the mixed salinity of estuarine waters. *V. cholerae* has been shown to bind to chitin of crustacea, and outbreaks of *V. cholerae* are associated with zooplanktonic blooms (Schwartz et al., 2006; Huq et al., 2005) *V. cholerae* multiplication is inhibited by bacterial lysis mediated by strain-specific phages (Jensen et al., 2006). Reflecting these modifiers, *V. cholerae* outbreaks are often seasonal, and inversely correlated with the presence of environmental phages specific to *V. cholerae* (Jensen et al., 2006).

Humans usually acquire cholera through the ingestion of contaminated water or food. Infection usually involves the ingestion of a large inoculum of organisms (often 10^8–10^{11} organisms in individuals with normal gastric acidity), although lower inocula may be infectious in hypochlorhydric individuals. Infected humans excrete up to 10^8 organisms per milliliter of stool, and may expell 10–20 L of stool a day. Recent data suggest that passage of *V. cholerae* through an infected host may induce a "hyperinfectious" phenotype that may contribute to ongoing epidemic potential (Hartley et al., 2006; Alam et al., 2005; Merrell et al., 2002). Rapid spread of cholera is facilitated by lack of access to clean water and optimal sanitation.

Cholera is a reportable disease that causes significant morbidity and mortality worldwide, but the true global burden of cholera is not known since a specific diagnosis is often not made in areas of the world where cholera imposes its largest burden, and since many countries with cases do not report cholera to the World Health Organization (out of a fear of travel or trade-restrictions). Despite this, in 2005, 52 countries reported 131,943 cholera cases and 2272 cholera-related deaths to the WHO (Fig. 45.1) (WHO, 2006), although cholera may affect 5–7 million individuals each year, resulting in greater than 100,000 deaths per year. At present, the world is experiencing its seventh recorded pandemic, and cholera is endemic in many countries in Asia, Africa, and Latin America. Although the historic home of cholera is Asia, the majority of cholera cases currently reported to the WHO are occurring in Africa (WHO, 2006). Large outbreaks, especially among impoverished or displaced persons, occur on a regular basis, and cholera may spread through travel or migration of infected individuals, including even short-term travelers or visitors (Goma Epidemiology Group, 1995). Once introduced into an area, *V. cholerae* may become endemic, and the organism may persist in the local water supply and contribute to yearly outbreaks, especially following floods (Schwartz et al., 2006). Although previously unexposed individuals of all ages are affected during new outbreaks, cholera is often a disease of children and previously unexposed adults in endemic areas. Since its explosive appearance in 1992, cholera caused by *V. cholerae* O139 has been reported in 11 Asian countries, where it currently accounts for a minority of disease (Albert et al., 1993; Ramamurthy et al., 1993).

FIGURE 45.1 Countries reporting cholera cases to World Health Organization in 2005. Shaded countries reported locally acquired cases. Triangles represent countries reporting imported cases. Note that many countries in Asia, Africa, and Latin America do not report cholera to the WHO. Adapted from Map 1, p301, WHO (2006).

Use as Biothreat

V. cholerae is not transmitted by the respiratory route, usually requires the ingestion of a fairly large inoculum, does not exist in a spore form, is readily killed in appropriately treated water supplies, and is unlikely to result in secondary spread in areas of the world with adequate hygienic facilities. In addition, previously unexposed individuals with cholera are often symptomatic, can be readily identified by standard laboratory assays, and can be readily treated with fluid therapy. Despite this, *V. cholerae* does have potential as a biothreat agent. Introduction of *V. cholerae* in areas of the world with poor infrastructure can cause chaos, mass migration, and collapse of health and societal infrastructures, and can lead to social, political, and economic instability with global ramifications. Also, due to its historical connotations, even a small-scale intentional introduction of cholera among a resource-rich population would lead to public fear, bioterror, and the diverting of critical resources and personnel.

CLINICAL DISEASE

Depending on inoculum size, gastric pH, and absence of preexisting immunity, the incubation period for symptomatic cholera is usually 1–3 days (Cash et al., 1974a). Although asymptomatic intestinal colonization or mild-to-moderate diarrhea may occur in individuals with preexisting immunity, the hallmark clinical presentation of a nonimmune individual with cholera is explosive onset of a profuse watery diarrhea. Affected individuals may lose 10–20L/day, and may progress to unconsciousness and shock within a few hours of initial onset of diarrhea (cholera gravis) (Fig. 45.2). As intestinal contents are flushed out, diarrheal stools become increasingly watery with flecks of white mucus ("rice water stool") (Fig. 45.3). Vomiting is common. Since *V. cholerae* is a noninvasive organism, intestinal contents do not characteristically contain blood or fecal leukocytes. Despite this classic presentation, the severity of cholera actually reflects not only inoculum size and preexisting immunity, but also the identity of the infecting

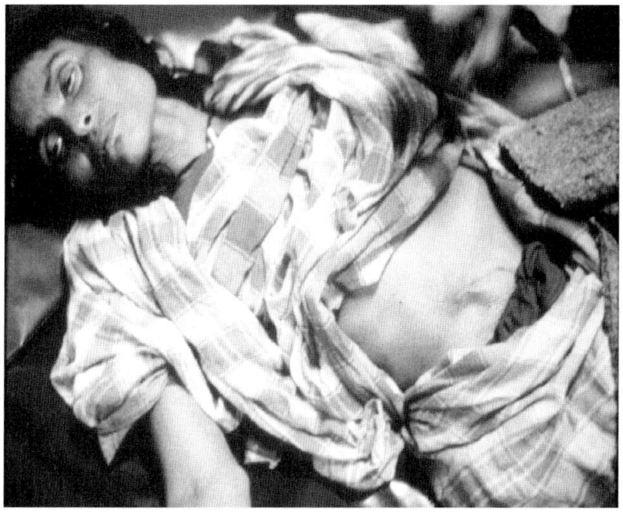

FIGURE 45.2 Adult with severe dehydration from cholera in Bangladesh. Patient is unresponsive and hypotensive. Note sunken eyes and tenting of skin on abdomen.

FIGURE 45.3 "Rice water" cholera stool. Following evacuation of bowel contents, the diarrhea of cholera patients may become progressively more watery, containing flecks of mucus.

biotype (classical biotype organisms cause more severe disease than El Tor organisms), and the blood type of the infected human (individuals with blood group O are less likely to become infected with *V. cholerae* following exposure, but are more likely to develop severe disease once infected) (Harris et al., 2005).

Individuals with cholera should be classified by level of dehydration as having mild, moderate, or severe disease. Individuals with mild dehydration may report thirst, but will have no or mild signs of dehydration, and will have lost ≤5% of their total body weight from diarrhea, with a fluid deficit of less than 50 ml/kg. Individuals with moderate dehydration will report thirst, may be restless, irritable, tachypneic and tachycardic, but will have preserved blood pressure and mentation. Mucus membranes will be dry, skin may retract slowly when pulled, eyes may be partially sunken, and urine output may be decreased. Individuals with moderate dehydration will have a fluid deficit of 60–100 ml/kg, representing 7–12% of total body water, and 6–10% of total body weight. Individuals with severe dehydration will have a nonpalpable or weak pulse, sunken eyes and dry mucus membranes, their skin will slowly retract when pulled, and they will produce no or minimal urine. Affected individuals will be tachypneic, often with labored breathing, and may be lethargic or comatose. Individuals with severe dehydration will have a fluid deficit of >100 ml/kg, representing >15% of their total body water, and >10% of their total body weight.

Symptomatic cholera is usually most severe within the first 2 days of onset, and usually lasts for 2–6 days. Hypoglycemia (especially in children) is common, as is acidosis (contributed to in part by the loss of bicarbonate in stool). Electrolyte abnormalities are common, and may contribute to persistent ileus. Rapid infusion of bicarbonate-containing intravenous fluid may lead to calcium shifts and tetany, which, with electrolyte abnormalities, most commonly manifest as muscle spasms. Pre-renal azotemia and acute tubular necrosis are common, although usually reversible upon rehydration. Among elderly individuals, the hypotension associated with severe dehydration may precipitate strokes and other ischemic events.

Although enterotoxigenic *E. coli* (ETEC) and rotavirus may cause similar diarrheal illnesses that can be quite severe in individual patients or among cohorts of young children, ETEC (see Chapter 51) and rotavirus (see Chapter 35) are unlikely to cause explosive epidemics that affect individuals of all ages, as does cholera. Definitive diagnosis of cholera rests on detection of *V. cholerae* O1 or O139 in the stool or vomit of a symptomatic human. Stool may be cultured on selective media to facilitate recovery of *V. cholerae*, and organisms may then be serogrouped and serotyped using slide agglutination assays. A rapid and preliminary diagnosis may also be made using darkfield microscopy to detect "shooting star" organisms in the stool of potentially infected individuals, or by using antigen detection strip assays. Retrospectively, a diagnosis of cholera can be confirmed serologically by detecting a fourfold change in the serum vibriocidal assay (when comparing convalescent to acute stage samples).

TREATMENT

Death from untreated cholera usually results from dehydration. Adequate, appropriate, and rapid fluid replacement is the cornerstone of effective treatment. Although fluid may be administered orally or intravenously, every effort should be made to replace fluid orally. Intravenous replacement should be reserved for initial treatment of individuals with severe dehydration or shock, or individuals progressing to those states despite initiation of oral rehydration therapy. Frequent reassessment of fluid status and diarrheal output is mandatory. Fluid replacement should involve solutions containing sodium, potassium, chloride, bicarbonate (or citrate), and sugar. Oral rehydration therapy rests upon the fact that despite CT's action causing the movement of chloride (and therefore sodium and water) out of affected intestinal epithelial cells (resulting in secretory diarrhea), that another transport system exists permitting simultaneous uptake of sodium (and therefore chloride and water) with glucose in intestinal epithelial cells, irrespective of the action of CT. The fundamental approach to effective treatment of individuals with cholera is, therefore, to first replace any fluid and electrolyte deficits, and then to match all ongoing losses.

Antibiotics play a secondary role in the treatment of individuals with cholera. The secretory diarrhea of cholera is due to the cellular action of CT on intestinal epithelial cells. Although antibiotics do not reverse the effects of CT once a cell is affected, administration of antibiotics decreases the intestinal burden of *V. cholerae*, and can thus decrease additional production of CT. Administration of antibiotics thus can decrease the duration and total volume of diarrhea, allowing critical resources to be used for more individuals in resource-poor areas of the world. Administration of antibiotics also decreases the likelihood of secondary spread and ongoing transmission of cholera. Unfortunately, *V. cholerae* are increasingly resistant to many antimicrobial agents (Saha et al., 2006; Guerrant, 2006).

Without treatment, the death rate from cholera during outbreaks may exceed 10–20%, with death rates of 70–100% of individuals with cholera gravis (Goma Epidemiology Group, 1995). Although death may occur in hours, it usually occurs 1–3 days after onset of diarrhea. With appropriate fluid replacement therapy, the death rate from even severe cholera should be lessened to <1% (Ryan et al., 2000).

PATHOGENESIS

Disease Process

Following ingestion of *V. cholerae*, the majority of organisms are killed by gastric acidity. Organisms that survive may multiply in the lumen of the small intestine. *V. cholerae* are flagellated and motile, and organisms that penetrate the intestinal layer of mucin may then colonize the intestinal epithelial surface. A recently proposed model suggests that TcpF and TcpA (members of the toxin coregulated pilus operon) and other factors are involved in this colonization process (Kirn and Taylor, 2005). Following colonization of the intestinal surface of the small bowel, additional environmental signals may be recognized, leading to the additional induction of the primary virulence factor of *V. cholerae*, CT. As mentioned above, CT is a secreted protein exotoxin that consists of a single, enzymatically active A subunit noncovalently associated with five B subunits. The pentamer of B subunits binds CT to ganglioside on the small bowel intestinal epithelial cell, and the A subunit is nicked, reduced, and translocated intracellularly, where it elevates cAMP, affecting the cystic fibrosis transmembrane conductance regulator (CFTR) chloride channel, leading to secretory diarrhea (Mekalanos et al., 1979; Cassel and Pfeuffer, 1978; Gill and King, 1975; Gill and Meren, 1978). Expression of the genes in the TCP operon, as well as expression of CT, are coordinately regulated by two pairs of membrane-localized transcription factors, ToxR/ToxS and TcpP/TcpH (Miller and Mekalanos, 1984; Miller et al., 1987, 1989; DiRita and Mekalanos, 1991; Carroll et al., 1997; Hase and Mekalanos, 1998). These pairs of transcription factors act via ToxT to activate transcription of both *ctxAB* and *tcpA* (DiRita et al., 1991).

VACCINES

History of Cholera Vaccine Development

Within 1 year of Koch's development of the techniques required to culture *V. cholerae* in the laboratory, a broth-cultured "attenuated" strain was incorporated into a parenteral vaccine by Ferran and administered to over 30,000 individuals during a cholera outbreak in Spain in 1884. Despite uncertain efficacy of this vaccine, this was the first large-scale roll out of a laboratory-derived vaccine strain of the modern microbiological era. By 1891, Haffkine and researchers at the Pasteur Institute developed a two-stage parenteral vaccine comprised of injection of an attenuated *V. cholerae* strain followed 6 days later with injection of a "hypervirulent" strain (serially passaged in guinea pigs). This vaccine was administered to over 40,000 individuals in India in 1893–1894. During this vaccine campaign, researchers administered the vaccine to subgroups on tea plantations and prisons, and analyzed results by vaccine status, representing the first attempt at determining vaccine efficacy in a large controlled field trial. Although the vaccine was thought to provide some efficacy, its production was hard to standardize, leading to the development of a killed parenteral vaccine by Kolle in 1896. This vaccine was used extensively in the early 1900s in Japan.

Cholera was also the target of the first oral vaccine when in 1893 Sawtschenko and colleagues ingested killed whole *V. cholerae* organisms in an attempt to induce immunity. By the 1920s, Bezredka and colleagues had developed an oral killed cholera vaccine mixed with bile, setting the stage for one of the first large comparative vaccine trials which occurred in the 1920s in India. In this trial, killed parenteral cholera vaccine was compared to the killed oral cholera vaccine (both vaccines were found to be approximately 80% effective at preventing cholera in the short-term). Since the bile salts caused diarrhea in some individuals, and since vaccine production was hard to standardize, the oral route was then abandoned for more than 50 years. By the 1960s, the parenteral whole cell vaccine was being mass-produced, and the World Health Assembly made cholera vaccination (as well as smallpox and yellow fever vaccination) required for certain international travel.

Current Licensed Vaccines

At present, there are a number of cholera vaccines licensed in various countries, as well as a number of candidate vaccines in various stages of development, evaluation, or licensure (Tables 45.1 and 45.2). At present, no cholera vaccine is commercially available and licensed in the United States.

Killed Parenteral Vaccine

A number of parenteral cholera vaccines have been developed, including a purified LPS vaccine (Oseasohn et al., 1965), a polysaccharide–CT conjugate vaccine (Gupta et al., 1998), killed whole cell vaccine with various adjuvants (Pal et al., 1980; Saroso et al., 1978), and killed whole cell vaccine (Mosley et al., 1972, 1973; Azurin et al., 1967; Ryan and Calderwood, 2000, 2001). Use of these vaccines has been associated with induction of relatively short-term immunity. Until approximately 1999/2000, the only FDA-approved cholera vaccine available in the United States was the whole cell killed vaccine, and was comprised of 4×10^9/ml of phenol-killed *V. cholerae* O1 Ogawa and 4×10^9/ml phenol-killed *V. cholerae* O1 Inaba organisms (usually classical strains NIH 41 and NIH 35A3). The organisms were grown on agar, suspended in sodium chloride, and killed with 0.5% phenol.

The vaccine was usually administered as a primary series of two injections separated by 7–28 days, and could be administered intramuscularly, subcutaneously, or intradermally (the latter route was approved for use in individuals 5 years of age or older). For children 6 months–4 years of age, the dose was 0.2 ml per inoculation. For children 5–10 years of age, the dose was 0.3 ml per inoculation. For individuals older than 10 years, the dose was 0.5 ml per inoculation. The vaccine was not administered to children younger than 6 months of age or pregnant women. The parenteral phenol-killed vaccine was associated with a high rate of moderate-to-severe local reactions (50%) including local induration, erythema, and pain, as well as a high rate (10–30%) of headache, malaise, and fever in vaccine recipients, especially following intramuscular or subcutaneous inoculation (Kaper et al., 1995; Tacket et al., 1992a). Intradermal inoculation was often better tolerated; however, there were minimal data on vaccine efficacy in immunologically naive individuals following intradermal administration (Ryan and Calderwood, 2001). Simultaneous administration of the killed parenteral cholera vaccine with yellow fever vaccine was associated with induction of decreased antibody levels to both vaccines, although there was no clinical evidence to suggest that protection against either disease was significantly altered. Although the vaccines were often administered simultaneously, if possible, an interval of 3 weeks between immunization against yellow fever and cholera was suggested.

Approximately 90% of individuals immunized with the killed parenteral whole cell vaccine developed a vibriocidal response following vaccination, with the highest response rates being observed in older children and adults in cholera-endemic areas (Mosley et al., 1969). Vibriocidal titers would, unfortunately, fall significantly by 6 months following vaccination. For approximately 6 months following vaccination, the vaccine provided approximately 50% protection against cholera caused by *V. cholerae* (Mosley et al., 1972, 1973; Azurin et al., 1967; Kaper et al., 1995; Levine and Pierce, 1992). Protection was thought to be present 6 days following the second injection. The vaccine was most thoroughly evaluated for efficacy in areas of the world endemic for cholera, confounding the ability to predict protective efficacy among immunologically naive populations. The vaccine was not very effective during cholera outbreaks, required more than one immunization, did not interrupt transmission of *V. cholerae*, and could divert critical resources during a cholera outbreak (Bart et al., 1970; Clemens et al., 1994a; Sommer and Mosley, 1973).

TABLE 45.1 Currently licensed, commercially-available, or programmatically used cholera vaccines

Vaccine	Composition	Route	Schedule	Age	Protective efficacy	Time to protection	Booster	Comments
Oral killed whole cell-recombinant B subunit vaccine (rBS-WC) (Dukoral, SBL Vaccin AB, Stockholm, Sweden/ Crucell Holland BV, Leiden, Netherlands))	Heat and formalin killed *Vibrio cholerae* O1, classical and El Tor strains, Inaba and Ogawa, with recombinant CtxB	Oral	>6 years of age: two doses separated by 7–42 days; 2–6 years of age: three doses, each separated by 7–42 days	≥2 years	Approximately 80% for 6 months, 50% at 3 years	Seven days after final dose	>6 years of age: 2 years; 2–6 years of age: 6 months	Well studied and well tolerated; licensed and commercially available in Europe; held effect increase population – Level protection
Oral live attenuated *V. cholerae* strain CVD 103-HgR (Orochol, Orochol E, Mutacol; Berna Swiss Serum Vaccine Institute, Bern, Switzerland/ Crucell Holland BV, Leiden, Netherlands)	Live attenuated *V. cholerae* O1 classical strain 569B derivative	Oral	Single dose	⩾2 years	62–100% in volunteer studies for 3 months; 14–80% in field studies	Within 8 days of dose	Six months	Well tolerated; licensed in Canada, Latin America, Europe, Asia; not currently manufactured/ available
Parenteral phenol-inactivated vaccine (Cholera vaccine, USP; Wyeth Laboratories, Inc., Marietta, PA)	Phenol-killed *V. cholerae* O1, Ogawa and Inaba	Intramuscular, subcutaneous, or intradermal (intradermal only approved for >5 years of age)	2 doses: >1–4 weeks apart	⩾6 months	Up to 50% for 3–6 months	Six days after second dose	Six months	High adverse event profile; only cholera vaccine licensed in United States; no longer commercially manufactured/ available
Bivalent oral killed whole cell vaccine (biv-WC) (Company for Vaccine and Biological Production No. 1 (VABIOTECH), Hanoi, Vietnam)	Heat and formalin killed *V. cholerae* O1 and O139	Oral	Two doses: 14 days apart	⩾2 years	50% against *V. cholerae* O1 3–5 years after vaccination	Probably within 7–14 days after second dose	Unknown	Only available in Vietnam where it is used programmatically; recently reformulated

If ongoing protection was indicated, booster immunizations were also required every 6 months. For these and other reasons, the World Health Assembly removed cholera vaccination from the list of required vaccines for certain international travel in 1973 (WHO, 1974). Although this vaccine is the only cholera vaccine licensed in the United States, the previous manufacturer of this vaccine (Wyeth Laboratories, Inc.) stopped commercial production of the vaccine in 1999/2000.

Oral Cholera Vaccines

Due to the poor immunogenicity, requirement for frequent booster vaccinations, and high adverse event profile of parenteral cholera vaccines, and due to the desire to induce immunity directly at the mucosal surface where *V. cholerae* pathogenic events unfold, researchers in the 1980s turned back to the development and evaluation of oral cholera vaccines (Ryan et al., 2006). At present, a number of oral cholera

TABLE 45.2 Candidate oral cholera vaccines thus far evaluated in humans[a]

Vaccine	Composition	Status
Vaccines targeting *V. cholerae* O1		
Peru-15	Live attenuated *V. cholerae* O1 El Tor Inaba C6709 derivative	Evaluated in volunteer studies in North America and Bangladesh, and Phase 1, 2 field studies in adults and children in Bangladesh
V. cholerae vaccine 638	Live attenuated *V. cholerae* O1 El Tor Ogawa C7258 derivative	Evaluated in volunteer studies in Cuba
Vaccines targeting *V. cholerae* O139		
B-O1/O139 WC	Killed *V. cholerae* O1 and O139 strains with CtxB	Evaluated in volunteer studies in Europe and Bangladesh
Biv-WC[b]	Killed *V. cholerae* O1 and O139 strains	Evaluated in volunteer studies in Vietnam
Bengal-15	Live attenuated *V. cholerae* O139 MO10 derivative	Evaluated in volunteer studies in the United States
CVD 112	Live attenuated *V. cholerae* O139 AI1837 derivative	Evaluated in volunteer studies in the United States

[a] A number of parenteral killed and subunit vaccines have been evaluated in small human volunteer studies since the 1980s.
[b] Biv-WC was recently reformulated and evaluated in a volunteer study.

vaccines are licensed in some countries (none in the United States).

Oral Whole Cell-B Subunit Vaccine (BS-WC and rBS-WC)

The oral whole cell killed vaccine contains 10^{11} killed *V. cholerae* O1 organisms (including classical and El Tor, Inaba, and Ogawa organisms of three strains, either formalin killed [to preserve protein antigens] or heat killed [to display LPS]), and 1 mg of nontoxic CT B subunit. Initial versions of the vaccine used nonrecombinant B subunit (BS); current formulations use recombinant B subunit (rBS). rBS-WC is manufactured by SBL Vaccin AB, Stockholm, Sweden/Crucell Holland BV, Leiden, Netherlands, and this vaccine is marketed in Europe as Dukoral/oral cholera vaccine, and is available in a number of countries (but not in the United States). To protect against cholera, the manufacturer has recommended the vaccine be administered in two doses (in individuals over 6 years of age) or three doses (in children 2–6 years of age), with each dose separated by 7–42 days (Wiedermann and Jong, 1997; Quiding et al., 1991; Svennerholm et al., 1984; Black et al., 1987). If more than 6 weeks separate the two doses, it has been recommended to restart the series. The vaccine comes in a 3 ml suspension in a glass vial, and is taken reconstituted in a glass of water with an alkaline buffer on an empty stomach (total volume 75 ml for children 2–6 years of age; 150 ml for individuals older than 6 years of age). The buffer is supplied in a sachet, and contains sodium carbonate, sodium citrate, sodium hydrogen carbonate, citric acid, saccharin, and flavoring. The manufacturer recommends avoiding food and drink 2 h prior and 1 h after ingestion of vaccine. If risk of cholera is ongoing, booster immunizations are recommended after 2 years for adults, and possibly as early as after 6 months for children 2–6 years of age. The vaccine should be stored at 2–8°C. Side effects are rare, with the most common being mild gastrointestinal upset, which occurs at a frequency similar to that among individuals receiving buffer alone. Although the vaccine has not been specifically studied during pregnancy, the manufacturer states that the vaccine may be administered during pregnancy and lactation. At present, there are no known vaccine interactions associated with rBS-WC.

The BS-WC and rBS-WC vaccines have been evaluated in a number of large and well-designed field studies. In Bangladesh, a study involving almost 90,000 individuals who took all of three scheduled vaccine doses showed that individuals receiving three doses of BS-WC at 6-week intervals developed excellent vibriocidal responses, and that immunization had 85% protective efficacy against cholera for 6 months, and 62% protective efficacy over 3 years of follow-up (Clemens et al., 1986). The vaccine protected equally well against severe and mild cholera. At 12 months, the protective efficacy of the vaccine was 62%. At 36 months, the vaccine was 50% protective. Within 6 months of immunization, the vaccine was equally protective among children, adolescents, and adults. However, protective efficacy was most short-lived among young children, falling to 26% at 36 months among children 2–5 years of age, compared with 63% among individuals older than 5 years of age (Clemens et al., 1990). The vaccine trial was carried out when both classical and El Tor *V. cholerae* O1 were circulating in Bangladesh (and before emergence of *V. cholerae* O139). Within 6 months of immunization, the vaccine protected equally well against cholera caused by either O1 biotype; however, by 36 months, protective efficacy was higher against classical biotype cholera (58%) than El Tor biotype cholera (39%) (Clemens et al., 1990). Individuals with blood group O were less protected than individuals with other blood groups. *V. cholerae* and ETEC express related toxins (CT and LT, respectively), which induce cross reactive antibodies, and individuals vaccinated with BS-WC were also protected in the short-term against ETEC (67% at 3 months, 21% at 12 months) (Clemens et al., 1988a). Initial analysis of the 3-year study of BS-WC disclosed that vaccination with BS-WC led to a 25% reduction in admission for all-cause diarrhea, a 50% reduction in hospital admission for life-threatening diarrhea, and a 45% reduction in mortality in women during a cholera epidemic (Holmgren et al., 1994; Clemens et al., 1988b).

Following replacement of the B subunit in BS-WC with recombinant B subunit, equivalency studies showed that two doses of rBS-WC conferred similar protective efficacy (86%) as did three doses of BS-WC in individuals greater than 5 years of age (Levine and Svennerholm, 1997). Evaluation of rBS-WC in Peru (soon after re-introduction of cholera to the New World in the early 1990s after a 100 year absence, meaning that cholera had not yet become endemic, and previous exposure to *V. cholerae* among vaccine recipients was unlikely) showed that the vaccine was perhaps less immunogenic and protective than in Bangladesh. In Peru, two doses of rBS-WC induced a twofold vibriocidal increase in 50% of individuals, and vibriocidal levels decreased to baseline within 1 year, leading to the recommendation that administration of yearly booster immunizations be considered for previously immunological naive individuals with ongoing risk of cholera (Begue et al., 1995a, 1995b). In a volunteer study in Peru, three doses of rBS-WC over 10 months produced a protective efficacy of 61% over a 2-year follow-up period, and during a mean follow-up period of 18 weeks, the vaccine provided 86% protection against *V. cholerae* O1 El Tor cholera (Sanchez et al., 1994).

In a recent field trial in Beira, Mozambique, a location with annual cholera outbreaks, receipt of one or more doses of rBS-WC was 78% protective against cholera within 6 months of vaccination, and importantly, the vaccine was equally effective in children younger than 5 years of age and in older individuals (Lucas et al., 2005). Of note, although not specifically assessed, this evaluation occurred in a location with a high prevalence of infection (20–30%) with human immunodeficiency virus, suggesting that rBS-WC could be effective even among populations with a high burden of immunocompromised individuals. In a small study that specifically measured virological parameters, administration of rBS-WC was associated with transient increases in HIV viral load, although the significance of this finding is currently uncertain (Ortigao-de-Sampaio et al., 1998; Lewis et al., 1994; Eriksson et al., 1993).

Overall, therefore, one to three doses of the killed oral whole cell cholera vaccine provides over 80% protective efficacy for at least 6 months against cholera caused by *V. cholerae* O1 (in young children as well as adults), and approximately 50–60% protective efficacy over 2–3 years, with the duration of immunity best correlating with increasing age. Protection is present within 7 days of the second immunization in previously nonimmune individuals. Recent reanalysis of the 1980s vaccine study in Bangladesh also suggests that BS-WC conferred herd immunity protection against cholera to neighboring nonvaccinated individuals, extending the efficacy of the vaccine from 62% to over 90% at the population level (Ali et al., 2005), and further strengthening the potential public health utility of the oral whole cell killed vaccine (Glass and Steele, 2005). For these and other reasons, the World Health Organization now suggests that vaccination with rBS-WC (along with other health interventions, including supplying potable water and sanitary facilities) be considered for populations at risk for cholera, including refugees (Lucas et al., 2005; Legros et al., 1999; Dorlencourt et al., 1999; Ivanoff, 2001).

WC and biv-WC Killed Oral Cholera Vaccine

Two killed oral whole cell cholera vaccines have also been developed and licensed for use in Vietnam: a killed oral whole cell vaccine (WC) targeting *V. cholerae* O1, and a bivalent killed oral cholera vaccine (biv-WC) targeting *V. cholerae* O1 and *V. cholerae* O139. The bivalent vaccine replaced the monovalent after introduction of *V. cholerae* O139 to Vietnam in the 1990s. Neither of these vaccines include exogenously added CtxB (as does rBS-WC), and as such, these vaccines are thought to be more acid resistant and do not require administration with buffer (Naficy et al., 2001). In an open, controlled field trial, a two dose regimen of the monovalent vaccine was found to induce a protective efficacy of 66% during a *V. cholerae* O1 El Tor Ogawa outbreak 8–10 months after immunization, and protection was similar among adults and children (Trach et al., 1997). Following programmatic use of the bivalent vaccine in Vietnam, a case-control evaluation showed an overall vaccine effectiveness of 50% (9–63%) against *V. cholerae* O1, 3–5 years after vaccination (Thiem et al., 2006). For programmatic use, 1.5 ml of a vaccine preparation containing a mixture of 2.5×10^{10} heat and formalin-killed *V. cholerae* O1 and O139 organisms are squirted into the mouth of vaccine recipients, who then drink water. A second dose is administered 2 weeks later. The vaccine is administered to nonpregnant individuals at least 2 years of age (Thiem et al., 2006).

Although promising for possible production and use in cholera-endemic areas outside of Vietnam, the biv-WC and WC vaccines are not currently recommended by the WHO for use outside of Vietnam. In order to decrease residual CT in the vaccine, the bivalent vaccine was recently reformulated, and in its current new preparation contains formalin-killed *V. cholerae* O1 Inaba El Tor strain Phil 6973, O1 Ogawa classical strain Cairo 50, and O139 strain 4260b, as well as heat-killed O1 Ogawa classical strain Cairo 50 and O1 Inaba classical strain Cairo 48; and dosage is based on ELISA units of LPS of each component strain (Anh et al., 2007). In a recent randomized placebo-controlled, safety and immunogenicity study among 143 Vietnamese adults, the reformulated vaccine was safe, well tolerated, and immunogenic (as judged by induction of anti-*V. cholerae* O1 vibriocidal responses) following two oral vaccinations separated by 2 weeks. Further evaluation of the vaccine is ongoing. The vaccine is produced by the Company for Vaccine and Biological Production No. 1 (VABIOTECH), Hanoi, Vietnam.

Since killed oral cholera vaccines often require repetitive administration, and since protection against cholera following vaccination with killed whole cell vaccines is often less prominent and of shorter duration than that following wild-type disease, investigators have also developed live, oral, attenuated *V. cholerae* vaccines.

CVD 103-HgR

Live oral cholera vaccines have the theoretical advantage that intestinal colonization and immune stimulation that mimic those that occur following infection with wild-type *V. cholerae* could follow a single oral administration. Over the last few decades, a number of live oral attenuated cholera vaccines have been developed (Ryan and Calderwood, 2000, 2001; Ryan et al., 2006). Although many of these vaccine strains have been immunogenic, many have been too reactogenic, resulting in abdominal cramps, diarrhea, fever, nausea, or anorexia in many recipients (Levine et al., 1988a). Eventually, a derivative of *V. cholerae* O1 classical strain 569B was developed, which was immunogenic without being too reactogenic. This vaccine strain is deficient in CtxA (and as such does not produce cholera holotoxin), and has a mercury resistance gene (HgR) inserted into the *hlyA* hemolysin gene (to permit recovery and identification of the vaccine strain without introducing antibiotic resistance) (Levine et al., 1988b). The vaccine was developed at the Center for Vaccine Development (CVD), University of Maryland, and is referred to as CVD 103-HgR.

CVD 103-HgR is approved for use in a number of countries, including countries in Asia, Latin America, Europe, and in Canada (and has been licensed under various names, including Orochol, Orochol-E, and Mutachol), although the vaccine is not licensed in the United States. In 2004, the manufacturer of CVD 103-HgR, Berna, Swiss Serum Vaccine Institute, Bern, Switzerland/Crucell Holland BV, Leiden, Netherlands, suspended production of CVD 103-HgR, and the vaccine is not currently commercially available anywhere in the world. The vaccine requires cold storage and administration with buffer. When produced, the vaccine was supplied in a sachet separately containing 2 g of lyophilized bacteria (approximately $2–10 \times 10^8$ cfu), sucrose, lactose, ascorbic acid, aspartame, and casein hydrolysate in one chamber, and buffer salts sodium bicarbonate and ascorbic acid in the other chamber. Vaccinees would mix the contents of the two chambers in 100 ml of clean cool or lukewarm water. The vaccine would be ingested 1 h before a meal. The need for booster immunization had not been established in detail; however, the manufacturer recommended that if the risk for cholera were ongoing, that boosters could be administered every 6 months. It was recommended not to administer the vaccine concurrently with antibiotics since it was a live bacterial vaccine. Limited data suggested that concomitant administration of chloroquine (but not mefloquine) decreased immunogenicity of the vaccine. The manufacturer recommended that administration of live oral cholera vaccine and live oral typhoid vaccine be separated by at least 8 h, due to effects of ingestion of the two buffers.

The most common side effects reported following ingestion of CVD 103-HgR were gastrointestinal, with 1–10% of recipients reporting diarrhea, abdominal cramps, or loss of appetite. The safety of the vaccine during pregnancy had not been specifically studied. The vaccine was approved for use in individuals 2 years of age and older.

CVD 103-HgR was both safe and immunogenic in North American volunteers (Levine et al., 1988b), and had not been recovered from the environment in field studies following oral administration. The vaccine strain was grown in fermenters in CF medium, and recovered bacteria were lyophilized, milled, and stored frozen. In North American volunteers, a single dose of 5×10^8 cfu resulted in a significant increase in vibriocidal antibodies in over 90% of recipients (Levine et al., 1988b). In volunteers, the vaccine also provided 94–100% protection against severe and moderate diarrhea caused by wild-type *V. cholerae* O1 (Levine et al., 1988b; Levine and Kaper, 1993; Tacket et al., 1999). Protection was 82–100% against challenge with classical biotype organisms (irrespective of serotype), and 62–80% protective against El Tor challenge (irrespective of serotype) (Kaper et al., 1995; Levine and Kaper, 1993; Tacket et al., 1999; Levine and Tacket, 1994). Protection was present within 8 days of vaccination, and lasted for at least 6 months in volunteers (Losonsky et al., 1996; Levine and Kaper, 1993; Tacket et al., 1999, 1992b; Losonsky et al., 1993).

Initial field evaluation of CVD 103-HgR showed that 5×10^9 cfu were required to induce vibriocidal seroconversion rates of 75–85% in resource-poor areas of the world (Su-Arehawaratana et al., 1992; Suharyono et al., 1992; Simanjuntak et al., 1993). In a large randomized field trial involving over 50,000 individuals in Indonesia, CVD 103-HgR was well tolerated, safe, and immunogenic, resulting in vibriocidal responses in 64–70% of vaccinees (Richie et al., 2000); however, the incidence of cholera was lower than expected during the study period, especially in the 6 months following vaccination, and only 93 evaluable cholera cases caused by *V. cholerae* O1 were detected in total (50 in placebo recipients, 43 in vaccine recipients, and only 7 within 6 months of vaccination), impeding the ability to judge the short-term protective efficacy of the vaccine. In this study, CVD 103-HgR had a protective efficacy of 14% (Richie et al., 2000). More recently, CVD 103-HgR was administered as part as a response plan to interrupt a cholera outbreak in Micronesia, and contributed to approximately 80% protection against cholera in this focused outbreak (Calain et al., 2004).

CVD 103-HgR has also been evaluated in a limited number of immunocompromised individuals. In a randomized, crossover, double-blind, placebo-controlled trial involving 38 individuals infected with human immunodeficiency virus but without clinical AIDS and 387 HIV-seronegative individuals in Mali, CVD 103-HgR was not recoverable from the stool of any of the vaccine recipients, and adverse events were comparable in the HIV seropositive and seronegative cohorts (Perry et al., 1998). Although CVD 103-HgR was immunogenic, the frequency and magnitude of the immune responses among individuals seropositive for HIV were less than those among individuals seronegative for HIV. The effects of vaccination with CVD 103-HgR on HIV viral load and CD4 cell counts were not measured in this study.

Vaccines in Development

A number of additional cholera vaccines are in various stages of development (Table 45.2).

Candidate Oral Live Attenuated Cholera Vaccines

Peru-15

Peru-15 is a live attenuated nonmotile derivative of *V. cholerae* O1 El Tor C6709 in which the entire CT genetic element and *attRS1* sites (used by CTXφ to insert into the *V. cholerae* chromosome) have been deleted, and the gene encoding CtxB has been inserted under the control of the *htpG* heat shock promoter into *recA* (lessening the likelihood of unwanted recombination events) (Kenner et al., 1995). The vaccine is produced by Avant Immunotherapeutics, Inc., Needham, MA.

In volunteer studies, Peru-15 was safe and immunogenic, with 90–100% vibriocidal seroconversion following single oral vaccination of freshly harvested 2×10^8 cfu (Kenner et al., 1995; Sack et al., 1997). In a pilot study, four of five controls, and two of five vaccinees developed diarrhea when challenged with wild-type *V. cholerae* 1 month following vaccination (Kenner et al., 1995). In a second small study, 12 volunteers received 10^7 or 10^8 cfu of freshly harvested Peru-15, and 50 individuals received either placebo or freeze-dried vaccine (10^8 or 10^9 cfu), in a double-masked three celled study (Sack et al., 1997). The vaccine was well tolerated and immunogenic, with the majority of individuals receiving Peru-15 developing vibriocidal responses. The vaccine strain was recoverable in the feces of approximately 60% of recipients (Sack et al., 1997). Peru-15 was then evaluated in a placebo-controlled, randomized, double-blind challenge study in which 59 individuals received either buffer alone or 2×10^8 cfu of lyophilized Peru-15 reconstituted in buffer. The vaccine was safe, well tolerated, immunogenic (98% of recipients developed vibiocidal responses), and protective (93% protective efficacy against diarrhea following challenge with wild-type *V. cholerae* O1 El Tor 3 months after vaccination) (Cohen et al., 2002).

Peru-15 was recently studied in staged Phase 1 and 2, randomized, double-blind, placebo-controlled studies in Bangladesh. In these studies, Peru-15 was first evaluated in individuals 18–45 years of age, then in children 2–5 years of age, then in infants 9–23 months of age (Qadri et al., 2005a, 2007). Adults received either 2×10^8 cfu of reconstituted lyophilized Peru-15 in buffer, or buffer alone. Peru-15 was safe and immunogenic. Vibriocidal responses were seen in 30 of 40 (75%) vaccine recipients, and 3 of 30 (10%) placebo recipients (Qadri et al., 2005a). Anti-LPS IgA responses were detected in the serum of 88% of vaccine recipients, and anti-LPS IgA and IgM ASC responses were detected in the blood of 78–88% of adult recipients. The vaccine strain was recovered from the stool of one adult recipient, and was identical to the inoculating strain.

Peru-15 was also safe and immunogenic in the studies involving 240 children. Two hundred and forty children 6 months–5 years of age were randomized to receive either buffer alone or 2×10^7 or 2×10^8 cfu of Peru-15 reconstituted in buffer. Among children, the vaccine was safe and immunogenic, with vibriocidal seroconversion rates of 84% (42/50) in children 2–5 years of age, and 70% (35/50) in children 9–23 months of age (Qadri et al., 2007). Overall, among children 5 years of age or younger receiving 2×10^8 cfu, Peru-15 induced vibriocidal seroconversion rates of 77%. Anti-LPS IgA antibody responses occurred in 60% of children 2–5 years of age, and 34% of children 6 months–2 years of age receiving 2×10^8 cfu. Approximately 40% of children developed serum anti-CT IgA responses (Qadri et al., 2007).

Other Live Oral Cholera Vaccines

A number of other live oral cholera vaccines are in earlier stages of development and evaluation. *V. cholerae* vaccine 638 is a live oral attenuated *V. cholerae* O1 Ogawa C7258 derivative from which the entire CTX genetic element has been deleted, and into which the *Clostridium thermocellum* endonuclease A gene has been inserted into the *V. cholerae* hemagglutinin/protease *hapA* gene (permitting identification of the vaccine strain on indicator media) (Ryan et al., 2006; Benitez et al., 1999). Among healthy Cuban adult volunteers, 4 of 42 vaccine recipients and 1 of 14 controls developed diarrhea or loose stools following ingestion of vaccine or placebo, respectively, and the vaccine strain was recoverable from the stool of 37 of 42 recipients. Volunteers received a range of doses (4×10^7–2×10^9 cfu). Serum vibriocidal responses occurred in 71–82%, and anti-Ogawa LPS IgA ASC responses occurred in 85–100% of vaccine recipients (Benitez et al., 1999). In a separate study involving 24 vaccinees, 96% developed vibriocidal responses, and 50% developed serum anti-LPS IgA responses in serum (Garcia et al., 2005). When challenged with 7×10^5 cfu of wild-type *V. cholerae* 3008 one month following vaccination, none of 12 vaccinees, and 7 of 9 placebo recipients developed diarrhea (Garcia et al., 2005).

Other live oral cholera vaccines include candidate vaccines VA1.3 in India, and IEM108 in China (Thungapathra et al., 1999; Liang et al., 2003). Although these strains have been immunogenic in animals, they have not yet been evaluated in humans.

V. cholerae O139

After its explosive emergence in 1992, *V. cholerae* O139 has been reported in 11 Asian countries, although *V. cholerae* O139 now accounts for the minority of disease even in affected areas. The reasons for the appearance, waning, but persistence of *V. cholerae* O139 are not well understood; however, *V. cholerae* O139 can cause explosive epidemics of cholera, even among individuals previously immune to *V. cholerae* O1, and as such, *V. cholerae* O139 has the potential to be a significant pathogen of global importance and a potential pandemic strain. *V. cholerae* O139 is surrounded by a capsule (*V. cholerae* O1 is not), and the presence of capsule complicates the ability to perform and standardize vibriocidal assays (important when judging immune responses following wild-type disease or vaccination) (Qadri et al., 2005b).

The only currently licensed vaccine that targets *V. cholerae* O139 is the biv-WC vaccine which is produced and used programmatically in Vietnam. Since the 1990s, millions of doses of biv-WC have been used in Vietnam, which remains the only cholera-endemic country with programmatic use of a cholera vaccine. As described above, biv-WC was recently reformulated and evaluated in a safety and immunogenicity volunteer study involving adults in Vietnam (Anh et al., 2007). Further evaluation is ongoing.

A bivalent oral whole cell cholera vaccine containing formalin-killed *V. cholerae* O139 and *V. cholerae* O1 organisms (B-O1/O139 WC) and CtxB has also been found to be both safe and immunogenic among Swedish volunteers (Jertborn et al., 1996). Following vaccination with B-O1/O139 WC, 10 of 12 vaccine recipients (83%) developed anti-O1 vibriocidal responses, and 8 (67%) developed anti-O139 vibriocidal responses. The magnitude and frequency of anti-CtxB and anti-O1 LPS responses were similar to those induced following vaccination with monovalent rBS-WC vaccine. Bivalent B-O1/O139 WC was also safe and immunogenic among 30 Bangladeshi volunteers, inducing vibriocidal responses in 50% (Qadri et al., 2005b).

A number of live oral attenuated anti-O139 vaccines have also been developed, including CVD 112, a derivative of *V. cholerae* O139 strain AI1837 made deficient in components of the CT genetic element (Waldor and Mekalanos, 1996; Tacket et al., 1998). Twelve volunteers have received either 10^6 or 10^8 cfu of CVD 112, with three of six receiving the higher dose developing mild diarrhea (Tacket et al., 1995, 1998). One of 8 vaccinees (13%) and 12 of 15 controls (80%) developed diarrhea when challenged with 10^6 cfu of wild-type *V. cholerae* O139 AI1837 five weeks after vaccination (protective efficacy, 84%) (Tacket et al., 1995, 1998). In other evaluations, 83–92% of 24 vaccine recipients developed anti-O139 vibriocidal responses, depending on which O139 strain was used in the vibriocidal assay (Losonsky et al., 1997).

Another live attenuated anti-O139 cholera vaccine candidate is Bengal-15, a nonmotile derivative of *V. cholerae* O139 MO10 from which the CT genetic element has been deleted and a copy of *ctxB* inserted into *recA* (Waldor and Mekalanos, 1994). In a small study involving four vaccine recipients, Bengal-15 produced fewer symptoms and was almost as immunogenic as MO10 (Coster et al., 1995). Among 10 vaccinees receiving 10^8 cfu of Bengal-15, no recipient developed diarrhea; 1 of 7 vaccine recipients and 5 of 6 controls developed diarrhea when challenged with 5×10^6 cfu of *V. cholerae* O139 one month following vaccination (protective efficacy, 83%) (Coster et al., 1995).

POSTEXPOSURE IMMUNOPROPHYLAXIS

V. cholerae is a noninvasive intestinal luminal pathogen. No effective postexposure immunoprophylaxis has been developed.

PROSPECTS FOR THE FUTURE

There are a number of potential uses for cholera vaccines. A prime indication for cholera vaccination would be to assist in preventing or controlling cholera outbreaks and epidemics. The WHO currently recommends that cholera vaccination (especially preemptively administered), along with efforts to supply safe water, improve sanitation, and identify and appropriately manage cases, be considered in emergency situations to control outbreaks among populations at high risk for cholera (Legros et al., 1999; Dorlencourt et al., 1999; Ivanoff, 2001; WHO, 1999, 2004). The effectiveness of killed oral whole cell vaccine is best substantiated in this regard (Lucas et al., 2005; Legros et al., 1999; Dorlencourt et al., 1999; Ivanoff, 2001; WHO, 2001, 2004), although live attenuated cholera vaccines could also assist in preventing or limiting outbreaks, and have the theoretical advantage of requiring a single immunization, and perhaps inducing more prominent early immune responses in immunologically naive individuals (WHO, 2001, 2004). As such, live oral cholera vaccines may be most beneficial if an outbreak has already started (WHO, 2001, 2004).

Although international travelers and workers can and do develop cholera (Taylor et al., 1996; Wittlinger et al., 1995; Sanchez and Taylor, 1997; Centers for Disease Control and Prevention, 1993), the risk is low (0.01–0.001% per month of stay in a developing country) (Ryan and Calderwood, 2001; Sanchez and Taylor, 1997; Steffen et al., 1987). There are fewer than 20–30 reported infections with *V. cholerae* each year in the United States, and the majority of these cases are often locally acquired in states bordering the Gulf of Mexico where *V. cholerae* is endemic. International travelers and workers who would be at highest risk for cholera are those who would be unable to secure safe water, especially individuals visiting or working within resource-poor areas during cholera outbreaks. Individuals who are hypochlorhydric may become infected following ingestion of even relatively small inocula. Although rBS-WC has been noted to provide protective immunity against ETEC, a very common cause of travelers' diarrhea, the relatively low protective immunity afforded by rBS-WC against "all causes of travelers' diarrhea" has limited use of this vaccine for this indication.

At present, the largest burden of cholera occurs in countries where *V. cholerae* is endemic. In these locations, cholera often exacts a large toll on children, much of the population may be undernourished or deficient in critical micronutrients, may be coinfected with other pathogens that may modify or misdirect immune responses, or be frankly immunocompromised. The majority of cases of cholera currently reported to the WHO are being reported from African countries where infection with human immunodeficiency virus is prevalent. A cholera vaccine with the greatest global benefit would, therefore, be safe and effective in children and immunocompromised individuals, inexpensive, and logistically simple to administer. Although a number of effective cholera vaccines are currently available in at least some countries, at present we do not fundamentally understand what mediates protective immunity against cholera, and why vaccines that appear promising in volunteer studies are less promising when evaluated in resource-poor areas. We also do not know why vaccine efficacy is often relatively short-lived, nor do we currently know how to improve the duration of vaccine efficacy. Although cholera can be prevented if safe water and proper sanitation are available, in reality, with 40% of the world's 6 billion human inhabitants currently subsisting on less than $2 a day, the growth of major mega-cities without proper infrastructure in many developing countries, and the potential impact of global warming on water supplies and flooding, it would perhaps be most prudent to assume that cholera will continue to be a plague of humans for the foreseeable future, and that the need for safe, inexpensive, and effective cholera vaccines may only continue to grow.

KEY ISSUES

- Cholera continues to be a pathogen of global importance, with the world currently experiencing its seventh pandemic, and an estimated 5–7 million cases per year, resulting in approximately 100,000 deaths.
- Cholera remains a disease of poverty, social disruption, and poor sanitation, and large and explosive outbreaks occur on a regular basis.
- Parenteral whole cell cholera vaccines require repetitive dosing and boosting, are associated with frequent adverse events, and are no longer commercially available.
- Oral cholera vaccines have been developed that are safe and protectively immunogenic.
- Only one oral cholera vaccine is currently commercially available and licensed in more than one country: the oral whole cell killed vaccine (rBS-WC, Dukoral, SBL Vaccine AB, Stockholm, Sweden/Crucell Holland BV, Leiden, Netherlands); this vaccine is not licensed or commercially available in the United States.
- Although rBS-WC is well studied, well tolerated, safe, and effective, vaccination with rBS-WC may require more than one immunization among immunologically naive populations, induces protective immunity of relatively short duration compared to that which occurs following wild-type disease, and is not protective against cholera caused by *V. cholerae* O139.
- *V. cholerae* O139, a cause of cholera that emerged in the 1990s, continues to be a cause of cholera in Asia, may have pandemic potential, and is not protected against by any presently internationally licensed, commercially available cholera vaccine.
- A killed oral cholera vaccine that targets both *V. cholerae* O1 and O139 is currently being locally produced and programmatically used in Vietnam. Efforts are underway to meet international standards and to evaluate effectiveness of this product.
- Oral live cholera vaccines, as well as other oral killed and parenteral subunit vaccines are also in various stages of development and evaluation.
- The role of cholera vaccines to control outbreaks needs to be more fully addressed.
- Public health strategies to optimally use cholera vaccines among populations in resource-poor areas need to be developed, evaluated, and implemented.
- The need of cholera vaccination among children in the developing world needs to be more fully addressed.
- The duration of protective immunity following vaccination against cholera needs to be lengthened.
- The safety and efficacy of cholera vaccination among immunocompromised individuals, including individuals infected with human immunodeficiency virus, needs to be more fully addressed.

ACKNOWLEDGMENTS

This work was supported in part by AI077883, AI40725, AI067342, and CI000514, all to Edward T. Ryan and AI58935 to Stephen B Calderwood. We are grateful to Jason B. Harris and Regina C. LaRocque for assistance with figures.

References

Alam, A., Larocque, R.C., Harris, J.B., Vanderspurt, C., Ryan, E.T., Qadri, F. and Calderwood, S.B. Hyperinfectivity of human-passaged *Vibrio cholerae* can be modeled by growth in the infant mouse. Infect. Immun. 2005; 73:6674–6679.

Albert, M.J., Siddique, A.K., Islam, M.S., Faruque, A.S., Ansaruzzaman, M., Faruque, S.M. and Sack, R.B. Large outbreak of clinical cholera due to *Vibrio cholerae* non-O1 in Bangladesh. Lancet 1993; 341:704.

Ali, M., Emch, M., von, S.L., Yunus, M., Sack, D.A., Rao, M., Holmgren, J. and Clemens, J.D. Herd immunity conferred by killed oral cholera vaccines in Bangladesh: a reanalysis. Lancet 2005; 366:44–49.

Anh, D.D., Canh, d.G., Lopez, A.L., Thiem, V.D., Long, P.T., Son, N.H., Deen, J., von, S.L., Carbis, R., Han, S.H., Shin, S.H., Attridge, S., Holmgren, J. and Clemens, J. Safety and immunogenicity of a reformulated Vietnamese bivalent killed, whole-cell, oral cholera vaccine in adults. Vaccine 2007; 25:1149–1155.

Asaduzzaman, M., Ryan, E.T., John, M., Hang, L., Khan, A.I., Faruque, A.S., Taylor, R.K., Calderwood, S.B. and Qadri, F. The major subunit of the toxin-coregulated pilus TcpA induces mucosal and systemic immunoglobulin A immune responses in patients with cholera caused by *Vibrio cholerae* O1 and O139. Infect. Immun. 2004; 72:4448–4454.

Azurin, J.C., Cruz, A., Pesigan, T.P., Alvero, M., Camena, T., Suplido, R., Ledesma, L. and Gomez, C.Z. A controlled field trial of the effectiveness of cholera and cholera El Tor vaccines in the Philippines. Bull. World Health Organ. 1967; 37:703–727.

Bart, K. J., Huq, Z., Khan, M. and Mosley, W.H. Seroepidemiologic studies during a simultaneous epidemic of infection with El Tor Ogawa and classical Inaba *Vibrio cholerae*. J. Infect. Dis. 1970; 121(Suppl.): 17.

Begue, R.E., Castellares, G., Cabezas, C., Sanchez, J.L., Meza, R., Watts, D.M. and Taylor, D.N. Immunogenicity in Peruvian volunteers of a booster dose of oral cholera vaccine consisting of whole cells plus recombinant B subunit. Infect. Immun. 1995a; 63:3726–3728.

Begue, R.E., Castellares, G., Ruiz, R., Hayashi, K.E., Sanchez, J.L., Gotuzzo, E., Oberst, R.B., Taylor, D.N. and Svennerholm, A.M. Community-based assessment of safety and immunogenicity of the whole cell plus recombinant B subunit (WC/rBS) oral cholera vaccine in Peru. Vaccine 1995b; 13:691–694.

Benitez, J.A., Garcia, L., Silva, A., Garcia, H., Fando, R., Cedre, B., Perez, A., Campos, J., Rodriguez, B.L., Perez, J.L., Valmaseda, T., Perez, O., Perez, A., Ramirez, M., Ledon, T., Jidy, M.D., Lastre, M., Bravo, L. and Sierra, G. Preliminary assessment of the safety and immunogenicity of a new CTXPhi-negative, hemagglutinin/protease-defective El Tor strain as a cholera vaccine candidate. Infect. Immun. 1999; 67:539–545.

Black, R.E., Levine, M.M., Clements, M.L., Young, C.R., Svennerholm, A.M. and Holmgren, J. Protective efficacy in humans of killed whole-vibrio oral cholera vaccine with and without the B subunit of cholera toxin. Infect. Immun. 1987; 55:1116–1120.

Calain, P., Chaine, J.P., Johnson, E., Hawley, M.L., O'Leary, M.J., Oshitani, H. and Chaignat, C.L. Can oral cholera vaccination play a role in controlling a cholera outbreak? Vaccine 2004; 22:2444–2451.

Carroll, P.A., Tashima, K.T., Rogers, M.B., DiRita, V.J. and Calderwood, S.B. Phase variation in tcpH modulates expression of the ToxR regulon in *Vibrio cholerae*. Mol. Microbiol. 1997; 25:1099–1111.

Cash, R.A., Music, S.I., Libonati, J.P., Craig, J.P., Pierce, N.F. and Hornick, R.B. Response of man to infection with *Vibrio cholerae*. II. Protection from illness afforded by previous disease and vaccine. J. Infect. Dis. 1974b; 130:325–333.

Cash, R.A., Music, S.I., Libonati, J.P., Snyder, M.J., Wenzel, R.P. and Hornick, R.B. Response of man to infection with *Vibrio cholerae*. I. Clinical, serologic, and bacteriologic responses to a known inoculum. J. Infect. Dis. 1974a; 129:45–52.

Cassel, D. and Pfeuffer, T. Mechanism of cholera toxin action: covalent modification of the guanyl nucleotide-binding protein of the adenylate cyclase system. Proc. Natl. Acad. Sci. U.S.A. 1978; 75:2669–2673.

Centers for Disease Control and Prevention. Cholera Western Hemisphere, 1992. MMWR Morb. Mortal. Wkly. Rep. 1993; 42:89–91.

Clemens, J., Spriggs, D. and Sack, D. Public health considerations for the use of cholera vaccines in cholera control programs. In: *Vibrio cholerae* and Cholera: Molecular to Global Perspectives (I.K. Wachsmuth, P.A. Blake and O. Olsvik, Eds.), pp. 425–442. Washington, DC: American Society for Microbiology Press, 1994.

Clemens, J.D., Sack, D.A., Harris, J.R., Chakraborty, J., Khan, M.R., Stanton, B.F., Ali, M., Ahmed, F., Yunus, M., Kay, B.A. et al. Impact of B subunit killed whole-cell and killed whole-cell-only oral vaccines against cholera upon treated diarrhoeal illness and mortality in an area endemic for cholera. Lancet 1988b; 1:1375–1379.

Clemens, J.D., Sack, D.A., Harris, J.R., Chakraborty, J., Khan, M.R., Stanton, B.F., Kay, B.A., Khan, M.U., Yunus, M., Atkinson, W. et al. Field trial of oral cholera vaccines in Bangladesh. Lancet 1986; 2:124–127.

Clemens, J.D., Sack, D.A., Harris, J.R., Chakraborty, J., Neogy, P.K., Stanton, B., Huda, N., Khan, M.U., Kay, B.A., Khan, M.R. et al. Cross-protection by B subunit-whole cell cholera vaccine against diarrhea associated with heat-labile toxin-producing enterotoxigenic *Escherichia coli*: results of a large-scale field trial. J. Infect. Dis. 1988a; 158:372–377.

Clemens, J.D., Sack, D.A., Harris, J.R., van, L.F., Chakraborty, J., Ahmed, F., Rao, M.R., Khan, M.R., Yunus, M., Huda, N. et al. Field trial of oral cholera vaccines in Bangladesh: results from three-year follow-up. Lancet 1990; 335:270–273.

Cohen, M.B., Giannella, R.A., Bean, J., Taylor, D.N., Parker, S., Hoeper, A., Wowk, S., Hawkins, J., Kochi, S.K., Schiff, G. and Killeen, K.P. Randomized, controlled human challenge study of the safety, immunogenicity, and protective efficacy of a single dose of Peru-15, a live attenuated oral cholera vaccine. Infect. Immun. 2002; 70:1965–1970.

Comstock, L.E., Maneval, D., Jr., Panigrahi, P., Joseph, A., Levine, M.M., Kaper, J.B., Morris, J.G., Jr. and Johnson, J.A. The capsule and O antigen in *Vibrio cholerae* O139 Bengal are associated with a genetic region not present in *Vibrio cholerae* O1. Infect. Immun. 1995; 63:317–323.

Cong, Y., Bowdon, H.R. and Elson, C.O. Identification of an immunodominant T cell epitope on cholera toxin. Eur. J. Immunol. 1996; 26:2587–2594.

Coster, T.S., Killeen, K.P., Waldor, M.K., Beattie, D.T., Spriggs, D.R., Kenner, J.R., Trofa, A., Sadoff, J.C., Mekalanos, J.J. and Taylor, D.N. Safety, immunogenicity, and efficacy of live attenuated *Vibrio cholerae* O139 vaccine prototype. Lancet 1995; 345:949–952.

Czerkinsky, C., Moldoveanu, Z., Mestecky, J., Nilsson, L.A. and Ouchterlony, O. A novel two colour ELISPOT assay. I. Simultaneous detection of distinct types of antibody-secreting cells. J. Immunol. Methods 1988; 115:31–37.

DiRita, V.J. and Mekalanos, J.J. Periplasmic interaction between two membrane regulatory proteins, ToxR and ToxS, results in signal transduction and transcriptional activation. Cell 1991; 64:29–37.

DiRita, V.J., Parsot, C., Jander, G. and Mekalanos, J.J. Regulatory cascade controls virulence in *Vibrio cholerae*. Proc. Natl. Acad. Sci. U.S.A. 1991; 88:5403–5407.

Dorlencourt, F., Legros, D., Paquet, C., Neira, M., Ivanoff, B. and Le, S.E. Effectiveness of mass vaccination with WC/rBS cholera vaccine during an epidemic in Adjumani district, Uganda. Bull. World Health Organ. 1999; 77:949–950.

Eriksson, K., Kilander, A., Hagberg, L., Norkrans, G., Holmgren, J. and Czerkinsky, C. Intestinal antibody responses to oral vaccination in HIV-infected individuals. AIDS 1993; 7:1087–1091.

Garcia, L., Jidy, M.D., Garcia, H., Rodriguez, B.L., Fernandez, R., Ano, G., Cedre, B., Valmaseda, T., Suzarte, E., Ramirez, M., Pino, Y., Campos, J., Menendez, J., Valera, R., Gonzalez, D., Gonzalez, I., Perez, O., Serrano, T., Lastre, M., Miralles, F., Del, C.J., Maestre, J.L., Perez, J.L., Talavera, A., Perez, A., Marrero, K., Ledon, T. and Fando, R. The vaccine candidate *Vibrio cholerae* 638 is protective against cholera in healthy volunteers. Infect. Immun. 2005; 73:3018–3024.

Gill, D.M. and King, C.A. The mechanism of action of cholera toxin in pigeon erythrocyte lysates. J. Biol. Chem. 1975; 250:6424–6432.

Gill, D.M. and Meren, R. ADP-ribosylation of membrane proteins catalyzed by cholera toxin: basis of the activation of adenylate cyclase. Proc. Natl. Acad. Sci. U.S.A. 1978; 75:3050–3054.

Glass, R.I. and Steele, A.D. The value of cholera vaccines reassessed. Lancet 2005; 366:7–9.

Glass, R.I., Svennerholm, A.M., Khan, M.R., Huda, S., Huq, M.I. and Holmgren, J. Seroepidemiological studies of El Tor cholera in Bangladesh: association of serum antibody levels with protection. J. Infect. Dis. 1985; 151:236–242.

Goma Epidemiology Group Public health impact of Rwandan refugee crisis: what happened in Goma, Zaire, in July, 1994? Lancet 1995; 345:339–344.

Guerrant, R.L. Cholera—still teaching hard lessons. N. Engl. J. Med. 2006; 354:2500–2502.

Gupta, R.K., Taylor, D.N., Bryla, D.A., Robbins, J.B. and Szu, S.C. Phase 1 evaluation of *Vibrio cholerae* O1, serotype Inaba, polysaccharide-cholera toxin conjugates in adult volunteers. Infect. Immun. 1998; 66:3095–3099.

Hang, L., John, M., Asaduzzaman, M., Bridges, E.A., Vanderspurt, C., Kirn, T.J., Taylor, R.K., Hillman, J.D., Progulske-Fox, A., Handfield, M., Ryan, E.T. and Calderwood, S.B. Use of in vivo-induced antigen technology (IVIAT) to identify genes uniquely expressed during human infection with *Vibrio cholerae*. Proc. Natl. Acad. Sci. U.S.A. 2003; 100:8508–8513.

Harris, J.B., Khan, A.I., Larocque, R.C., Dorer, D.J., Chowdhury, F., Faruque, A.S., Sack, D.A., Ryan, E.T., Qadri, F. and Calderwood, S.B.

Blood group, immunity, and risk of infection with *Vibrio cholerae* in an area of endemicity. Infect. Immun. 2005; 73:7422–7427.

Harris, J.B., Larocque, R.C., Chowdhury, F., Ashraful, I., Khan, A.I., Logvinenko, T., Faruque, A.S.G., Ryan, E.T., Qadri, F. and Calderwood, S.B. Susceptibility to *Vibrio cholerae* infection in a cohort of household contacts of patients with cholera in Bangladesh. PLoS Negl Trop Dis 2(4):e221, 2008.

Hartley, D.M., Morris, J.G. and Smith, D.L. Hyperinfectivity: a critical element in the ability of *V. cholerae* to cause epidemics? PLoS Med. 2006; 3:e7.

Hase, C.C. and Mekalanos, J.J. TcpP protein is a positive regulator of virulence gene expression in *Vibrio cholerae*. Proc. Natl. Acad. Sci. U.S.A. 1998; 95:730–734.

Heidelberg, J.F., Eisen, J.A., Nelson, W.C., Clayton, R.A., Gwinn, M.L., Dodson, R.J., Haft, D.H., Hickey, E.K., Peterson, J.D., Umayam, L., Gill, S.R., Nelson, K.E., Read, T.D., Tettelin, H., Richardson, D., Ermolaeva, M.D., Vamathevan, J., Bass, S., Qin, H., Dragoi, I., Sellers, P., McDonald, L., Utterback, T., Fleishmann, R.D., Nierman, W.C., White, O., Salzberg, S.L., Smith, H.O., Colwell, R.R., Mekalanos, J.J., Venter, J.C. and Fraser, C.M. DNA sequence of both chromosomes of the cholera pathogen *Vibrio cholerae*. Nature 2000; 406:477–483.

Herrington, D.A., Hall, R.H., Losonsky, G., Mekalanos, J.J., Taylor, R.K. and Levine, M.M. Toxin, toxin-coregulated pili, and the *toxR* regulon are essential for *Vibrio cholerae* pathogenesis in humans. J. Exp. Med. 1988; 168:1487–1492.

Holmgren, J., Osek, J. and Svennerholm, A.M. Protective oral cholera vaccine based on a combination of cholera toxin B subunit and inactivated cholera vibrios. In: *Vibrio cholerae* and Cholera: Molecular to Global Perspectives (I.K. Wachsmuth, P.A. Blake and O. Olsvik, Eds.), pp. 425–442. Washington, DC: American Society for Microbiology Press, 1994.

Holmgren, J. and Svennerholm, A.M. Mechanisms of disease and immunity in cholera: a review. J. Infect. Dis. 1977; 136(Suppl.):S105–S112.

Huq, A., Sack, R.B., Nizam, A., Longini, I.M., Nair, G.B., Ali, A., Morris, J.G., Jr., Khan, M.N., Siddique, A.K., Yunus, M., Albert, M.J., Sack, D.A. and Colwell, R.R. Critical factors influencing the occurrence of *Vibrio cholerae* in the environment of Bangladesh. Appl. Environ. Microbiol. 2005; 71:4645–4654.

Iredell, J.R. and Manning, P.A. Biotype-specific tcpA genes in *Vibrio cholerae*. FEMS Microbiol. Lett. 1994; 121:47–54.

Ivanoff, B. Cholera: vaccines present and future. Med. Trop. (Mars.) 2001; 61:245–248.

Jensen, M.A., Faruque, S.M., Mekalanos, J.J. and Levin, B.R. Modeling the role of bacteriophage in the control of cholera outbreaks. Proc. Natl. Acad. Sci. U.S.A. 2006; 103:4652–4657.

Jertborn, M., Svennerholm, A.M. and Holmgren, J. Intestinal and systemic immune responses in humans after oral immunization with a bivalent B subunit-O1/O139 whole cell cholera vaccine. Vaccine 1996; 14:1459–1465.

Kaper, J.B., Morris, J.G., Jr. and Levine, M.M. Cholera. Clin. Microbiol. Rev. 1995; 8:48–86.

Kenner, J.R., Coster, T.S., Taylor, D.N., Trofa, A.F., Barrera-Oro, M., Hyman, T., Adams, J.M., Beattie, D.T., Killeen, K.P., Spriggs, D.R. et al. Peru-15, an improved live attenuated oral vaccine candidate for *Vibrio cholerae* O1. J. Infect. Dis. 1995; 172:1126–1129.

Kirn, T.J. and Taylor, R.K. TcpF is a soluble colonization factor and protective antigen secreted by El Tor and classical O1 and O139 *Vibrio cholerae* serogroups. Infect. Immun. 2005; 73:4461–4470.

Legros, D., Paquet, C., Perea, W., Marty, I., Mugisha, N.K., Royer, H., Neira, M. and Ivanoff, B. Mass vaccination with a two-dose oral cholera vaccine in a refugee camp. Bull. World Health Organ. 1999; 77:837–842.

Levine, M.M., Black, R.E., Clements, M.L., Cisneros, L., Nalin, D.R. and Young, C.R. Duration of infection-derived immunity to cholera. J. Infect. Dis. 1981; 143:818–820.

Levine, M.M. and Kaper, J.B. Live oral vaccines against cholera: an update. Vaccine 1993; 11:207–212.

Levine, M.M., Kaper, J.B., Herrington, D., Ketley, J., Losonsky, G., Tacket, C.O., Tall, B. and Cryz, S. Safety, immunogenicity, and efficacy of recombinant live oral cholera vaccines, CVD 103 and CVD 103-HgR. Lancet 1988b; 2:467–470.

Levine, M.M., Kaper, J.B., Herrington, D., Losonsky, G., Morris, J.G., Clements, M.L., Black, R.E., Tall, B. and Hall, R. Volunteer studies of deletion mutants of *Vibrio cholerae* O1 prepared by recombinant techniques. Infect. Immun. 1988a; 56:161–167.

Levine, M.M., Nalin, D.R., Rennels, M.B., Hornick, R.B., Sotman, S., Van, B.G., Hughes, T.P., O'Donnell, S. and Barua, D. Genetic susceptibility to cholera. Ann. Hum. Biol. 1979; 6:369–374.

Levine, M.M. and Pierce, N.F. Immunity and vaccine development. In: Cholera (C. Barua and W.B. Greenough III., Eds.), pp. 285–327. New York: Plenum Medical Book Company, 1992.

Levine, M.M. and Svennerholm, A.M. Future enteric vaccines. In: Textbook of Travel Medicine and Health (H.L. DuPont and R. Steffen, Eds.), pp. 169–177. Hamilton, Ont.: B.C. Decker, Inc., 1997.

Levine, M.M. and Tacket, C.O. Recombinant live cholera vaccines. In: *Vibrio cholerae* and Cholera: Molecular to Global Perspectives (I.K. Wachsmuth, P.A. Blake and O. Olsvik, Eds.), pp. 395–414. Washington, DC: American Society for Microbiology Press, 1994.

Lewis, D.J., Gilks, C.F., Ojoo, S., Castello-Branco, L.R., Dougan, G., Evans, M.R., McDermott, S. and Griffin, G.E. Immune response following oral administration of cholera toxin B subunit to HIV-1-infected UK and Kenyan subjects. AIDS 1994; 8:779–785.

Liang, W., Wang, S., Yu, F., Zhang, L., Qi, G., Liu, Y., Gao, S. and Kan, B. Construction and evaluation of a safe, live, oral *Vibrio cholerae* vaccine candidate, IEM108. Infect. Immun. 2003; 71:5498–5504.

Losonsky, G.A., Lim, Y., Motamedi, P., Comstock, L.E., Johnson, J.A., Morris, J.G., Jr., Tacket, C.O., Kaper, J.B. and Levine, M.M. Vibriocidal antibody responses in North American volunteers exposed to wild-type or vaccine *Vibrio cholerae* O139: specificity and relevance to immunity. Clin. Diagn. Lab. Immunol. 1997; 4:264–269.

Losonsky, G.A., Tacket, C.O., Wasserman, S.S., Kaper, J.B. and Levine, M.M. Secondary *Vibrio cholerae*-specific cellular antibody responses following wild-type homologous challenge in people vaccinated with CVD 103-HgR live oral cholera vaccine: changes with time and lack of correlation with protection. Infect. Immun. 1993; 61:729–733.

Losonsky, G.A., Yunyongying, J., Lim, V., Reymann, M., Lim, Y.L., Wasserman, S.S. and Levine, M.M. Factors influencing secondary vibriocidal immune responses: relevance for understanding immunity to cholera. Infect. Immun. 1996; 64:10–15.

Lucas, M.E., Deen, J.L., von, S.L., Wang, X.Y., Ampuero, J., Puri, M., Ali, M., Ansaruzzaman, M., Amos, J., Macuamule, A., Cavailler, P., Guerin, P.J., Mahoudeau, C., Kahozi-Sangwa, P., Chaignat, C.L., Barreto, A., Songane, F.F. and Clemens, J.D. Effectiveness of mass oral cholera vaccination in Beira, Mozambique. N. Engl. J. Med. 2005; 352:757–767.

Mekalanos, J.J., Collier, R.J. and Romig, W.R. Enzymic activity of cholera toxin. II. Relationships to proteolytic processing, disulfide bond reduction, and subunit composition. J. Biol. Chem. 1979; 254:5855–5861.

Miller, V.L., DiRita, V.J. and Mekalanos, J.J. Identification of *toxS*, a regulatory gene whose product enhances *toxR*-mediated activation of the cholera toxin promoter. J. Bacteriol. 1989; 171:1288–1293.

Miller, V.L. and Mekalanos, J.J. Synthesis of cholera toxin is positively regulated at the transcriptional level by *toxR*. Proc. Natl. Acad. Sci. U.S.A. 1984; 81:3471–3475.

Miller, V.L., Taylor, R.K. and Mekalanos, J.J. Cholera toxin transcriptional activator *toxR* is a transmembrane DNA binding protein. Cell 1987; 48:271–279.

Mosley, W.H., Ahmad, S., Benenson, A.S. and Ahmed, A. The relationship of vibriocidal antibody titre to susceptibility to cholera in family contacts of cholera patients. Bull. World Health Organ. 1968a; 38:777–785.

Mosley, W.H., Aziz, K.M., Mizanur Rahman, A.S., Auddin Chowdhury, A.K., Ahmed, A. and Fahimuddin, M. Report of the 1966–67 cholera vaccine trial in rural East Pakistan. Bull. World Health Organ. 1972; 47:229–238.

Mosley, W.H., Aziz, K.M., Rahman, A.S., Chowdhury, A.K. and Ahmed, A. Field trials of monovalent Ogawa and Inaba cholera vaccines in rural Bangladesh: three years of observation. Bull. World Health Organ. 1973; 49:381–387.

Mosley, W.H., Benenson, A.S. and Barui, R. A serological survey for cholera antibodies in rural east Pakistan. 1. The distribution of antibody in the control population of a cholera-vaccine field-trial area and the relation of antibody titre to the pattern of endemic cholera. Bull. World Health Organ. 1968b; 38:327–334.

Mosley, W.H., McCormack, W.M., Ahmed, A., Chowdhury, A.K. and Barui, R.K. Report of the 1966–67 cholera vaccine field trial in rural East Pakistan. 2. Results of the serological surveys in the study population—the relationship of case rate to antibody titre and an estimate of the inapparent infection rate with *Vibrio cholerae*. Bull. World Health Organ. 1969; 40:187–197.

Naficy, A.B., Trach, D.D., Ke, N.T., Chuc, N.T., Sorkin, A., Rao, M.R., Sy, T.H., Thiem, V.D., Canh, D.G., Mahoney, R.T., Holmgren, J., Ivanoff, B. and Clemens, J.D. Cost of immunization with a locally produced, oral cholera vaccine in Vietnam. Vaccine 2001; 19:3720–3725.

Ortigao-de-Sampaio, M.B., Shattock, R.J., Hayes, P., Griffin, G.E., Linhares-de-Carvalho, M.I., Ponce de, L.A., Lewis, D.J. and Castello-Branco, L.R. Increase in plasma viral load after oral cholera immunization of HIV-infected subjects. AIDS 1998; 12:F145–F150.

Oseasohn, R.O., Benenson, A.S. and Fahimuddin, M. Field trial of cholera vaccine in rural East Pakistan. First year of observation. Lancet 1965; 1:450–452.

Pal, S.C., Deb, B.C., Sen Gupta, P.G., De, S.P., Sircar, B.K., Sen, D. and Sikdar, S.N. A controlled field trial of an aluminum phosphate-adsorbed cholera vaccine in Calcutta. Bull. World Health Organ. 1980; 58:741–745.

Perry, R.T., Plowe, C.V., Koumare, B., Bougoudogo, F., Kotloff, K.L., Losonsky, G.A., Wasserman, S.S. and Levine, M.M. A single dose of live oral cholera vaccine CVD 103-HgR is safe and immunogenic in HIV-infected and HIV-noninfected adults in Mali. Bull. World Health Organ. 1998; 76:63–71.

Qadri, F., Chowdhury, M.I., Faruque, S.M., Salam, M.A., Ahmed, T., Begum, Y.A., Saha, A., Al, T.A., Seidlein, L.V., Park, E., Killeen, K.P., Mekalanos, J.J., Clemens, J.D. and Sack, D.A. Peru-15, a live attenuated oral cholera vaccine, is safe and immunogenic in Bangladeshi toddlers and infants. Vaccine 2007; 25:231–238.

Qadri, F., Chowdhury, M.I., Faruque, S.M., Salam, M.A., Ahmed, T., Begum, Y.A., Saha, A., Alam, M.S., Zaman, K., Seidlein, L.V., Park, E., Killeen, K.P., Mekalanos, J.J., Clemens, J.D. and Sack, D.A. Randomized, controlled study of the safety and immunogenicity of Peru-15, a live attenuated oral vaccine candidate for cholera, in adult volunteers in Bangladesh. J. Infect. Dis. 2005a; 192:573–579.

Qadri, F., Jonson, G., Begum, Y.A., Wenneras, C., Albert, M.J., Salam, M. A. and Svennerholm, A.M. Immune response to the mannose-sensitive hemagglutinin in patients with cholera due to *Vibrio cholerae* O1 and O139. Clin. Diagn. Lab. Immunol. 1997; 4:429–434.

Qadri, F., Svennerholm, A.M., Shamsuzzaman, S., Bhuiyan, T.R., Harris, J.B., Ghosh, A.N., Nair, G.B., Weintraub, A., Faruque, S.M., Ryan, E.T., Sack, D.A. and Calderwood, S.B. Reduction in capsular content and enhanced bacterial susceptibility to serum killing of *Vibrio cholerae* O139 associated with the 2002 cholera epidemic in Bangladesh. Infect. Immun. 2005b; 73:6577–6583.

Quiding, M., Nordstrom, I., Kilander, A., Andersson, G., Hanson, L.A., Holmgren, J. and Czerkinsky, C. Intestinal immune responses in humans. Oral cholera vaccination induces strong intestinal antibody responses and interferon-gamma production and evokes local immunological memory. J. Clin. Invest. 1991; 88:143–148.

Ramamurthy, T., Garg, S., Sharma, R., Bhattacharya, S.K., Nair, G.B., Shimada, T., Takeda, T., Karasawa, T., Kurazano, H., Pal, A. et al. Emergence of novel strain of *Vibrio cholerae* with epidemic potential in southern and eastern India. Lancet 1993; 341:703–704.

Richie, E.E., Punjabi, N.H., Sidharta, Y.Y., Peetosutan, K.K., Sukandar, M.M., Wasserman, S.S., Lesmana, M.M., Wangsasaputra, F.F., Pandam, S.S., Levine, M.M., O'Hanley, P.P., Cryz, S.J. and Simanjuntak, C.H. Efficacy trial of single-dose live oral cholera vaccine CVD 103-HgR in North Jakarta, Indonesia, a cholera-endemic area. Vaccine 2000; 18:2399–2410.

Rollenhagen, J.E., Kalsy, A., Cerda, F., John, M., Harris, J.B., Larocque, R.C., Qadri, F., Calderwood, S.B., Taylor, R.K. and Ryan, E.T. Transcutaneous immunization with toxin-coregulated Pilin A induces protective immunity against *Vibrio cholerae* O1 El Tor challenge in mice. Infect. Immun. 2006; 74:5834–5839.

Ryan, E.T. and Calderwood, S.B. Cholera vaccines. Clin. Infect. Dis. 2000; 31:561–565.

Ryan, E.T. and Calderwood, S.B. Cholera vaccines. J. Travel Med. 2001; 8:82–91.

Ryan, E.T., Calderwood, S.B. and Qadri, F. Live attenuated oral cholera vaccines. Expert. Rev. Vaccines 2006; 5:483–494.

Ryan, E.T., Dhar, U., Khan, W.A., Salam, M.A., Faruque, A.S., Fuchs, G.J., Calderwood, S.B. and Bennish, M.L. Mortality, morbidity, and microbiology of endemic cholera among hospitalized patients in Dhaka, Bangladesh. Am. J. Trop. Med. Hyg. 2000; 63:12–20.

Sack, D.A., Sack, R.B., Shimko, J., Gomes, G., O'Sullivan, D., Metcalfe, K. and Spriggs, D. Evaluation of Peru-15, a new live oral vaccine for cholera, in volunteers. J. Infect. Dis. 1997; 176:201–205.

Saha, D., Karim, M.M., Khan, W.A., Ahmed, S., Salam, M.A. and Bennish, M.L. Single-dose azithromycin for the treatment of cholera in adults. N. Engl. J. Med. 2006; 354:2452–2462.

Saha, D., Larocque, R.C., Khan, A.I., Harris, J.B., Begum, Y.A., Akramuzzaman, S.M., Faruque, A.S., Ryan, E.T., Qadri, F. and Calderwood, S.B. Incomplete correlation of serum vibriocidal antibody titer with protection from *Vibrio cholerae* infection in urban Bangladesh. J. Infect. Dis. 2004; 189:2318–2322.

Sanchez, J.L. and Taylor, D.N. Cholera. Lancet 1997; 349:1825–1830.

Sanchez, J.L., Vasquez, B., Begue, R.E., Meza, R., Castellares, G., Cabezas, C., Watts, D.M., Svennerholm, A.M., Sadoff, J.C. and Taylor, D.N. Protective efficacy of oral whole-cell/recombinant-B-subunit cholera vaccine in Peruvian military recruits. Lancet 1994; 344:1273–1276.

Saroso, J.S., Bahrawi, W., Witjaksono, H., Budiarso, R.L., Brotowasisto, B.Z., Dewitt, W.E. and Gomez, C.Z. A controlled field trial of plain and aluminium hydroxide-adsorbed cholera vaccines in Surabaya, Indonesia, during 1973–1975. Bull. World Health Organ. 1978; 56:619–627.

Schwartz, B.S., Harris, J.B., Khan, A.I., Larocque, R.C., Sack, D.A., Malek, M.A., Faruque, A.S., Qadri, F., Calderwood, S.B., Luby, S.P.

and Ryan, E.T. Diarrheal epidemics in Dhaka, Bangladesh, during three consecutive floods: 1988, 1998, and 2004. Am. J. Trop. Med. Hyg. 2006; 74:1067–1073.

Simanjuntak, C.H., O'Hanley, P., Punjabi, N.H., Noriega, F., Pazzaglia, G., Dykstra, P., Kay, B., Suharyono, B.A., Rifai, A.R. et al. Safety, immunogenicity, and transmissibility of single-dose live oral cholera vaccine strain CVD 103-HgR in 24- to 59-month-old Indonesian children. J. Infect. Dis. 1993; 168:1169–1176.

Sommer, A. and Mosley, W.H. Ineffectiveness of cholera vaccination as an epidemic control measure. Lancet 1973; 1:1232–1235.

Steffen, R., Rickenbach, M., Wilhelm, U., Helminger, A. and Schar, M. Health problems after travel to developing countries. J. Infect. Dis. 1987; 156:84–91.

Su-Arehawaratana, P., Singharaj, P., Taylor, D.N., Hoge, C., Trofa, A., Kuvanont, K., Migasena, S., Pitisuttitham, P., Lim, Y.L., Losonsky, G. et al. Safety and immunogenicity of different immunization regimens of CVD 103-HgR live oral cholera vaccine in soldiers and civilians in Thailand. J. Infect. Dis. 1992; 165:1042–1048.

Suharyono, S.C., Witham, N., Punjabi, N., Heppner, D.G., Losonsky, G., Totosudirjo, H., Rifai, A.R., Clemens, J., Lim, Y. L. et al. Safety and immunogenicity of single-dose live oral cholera vaccine CVD 103-HgR in 5-9-year-old Indonesian children. Lancet 1992; 340:689–694.

Sun, D.X., Mekalanos, J.J. and Taylor, R.K. Antibodies directed against the toxin-coregulated pilus isolated from *Vibrio cholerae* provide protection in the infant mouse experimental cholera model. J. Infect. Dis. 1990; 161:1231–1236.

Svennerholm, A.M., Jertborn, M., Gothefors, L., Karim, A.M., Sack, D.A. and Holmgren, J. Mucosal antitoxic and antibacterial immunity after cholera disease and after immunization with a combined B subunit-whole cell vaccine. J. Infect. Dis. 1984; 149:884–893.

Tacket, C.O., Clemens, J. and Kaper, J.B. Cholera vaccines. Biotechnology 1992a; 20:53–68.

Tacket, C.O., Cohen, M.B., Wasserman, S.S., Losonsky, G., Livio, S., Kotloff, K., Edelman, R., Kaper, J.B., Cryz, S.J., Giannella, R.A., Schiff, G. and Levine, M.M. Randomized, double-blind, placebo-controlled, multicentered trial of the efficacy of a single dose of live oral cholera vaccine CVD 103-HgR in preventing cholera following challenge with *Vibrio cholerae* O1 El Tor Inaba three months after vaccination. Infect. Immun. 1999; 67:6341–6345.

Tacket, C.O., Losonsky, G., Nataro, J.P., Comstock, L., Michalski, J., Edelman, R., Kaper, J.B. and Levine, M.M. Initial clinical studies of CVD 112 *Vibrio cholerae* O139 live oral vaccine: safety and efficacy against experimental challenge. J. Infect. Dis. 1995; 172:883–886.

Tacket, C.O., Losonsky, G., Nataro, J.P., Cryz, S.J., Edelman, R., Kaper, J.B. and Levine, M.M. Onset and duration of protective immunity in challenged volunteers after vaccination with live oral cholera vaccine CVD 103-HgR. J. Infect. Dis. 1992b; 166:837–841.

Tacket, C.O., Taylor, R.K., Losonsky, G., Lim, Y., Nataro, J.P., Kaper, J.B. and Levine, M.M. Investigation of the roles of toxin-coregulated pili and mannose-sensitive hemagglutinin pili in the pathogenesis of *Vibrio cholerae* O139 infection. Infect. Immun. 1998; 66:692–695.

Taylor, R.K., Miller, V.L., Furlong, D.B. and Mekalanos, J.J. Use of *phoA* gene fusions to identify a pilus colonization factor coordinately regulated with cholera toxin. Proc. Natl. Acad. Sci. U.S.A. 1987; 84:2833–2837.

Taylor, D.N., Rizzo, J., Meza, R., Perez, J. and Watts, D. Cholera among Americans living in Peru. Clin. Infect. Dis. 1996; 22:1108–1109.

Thiem, V.D., Deen, J.L., von Seidlein, L., Canh Do, G., Anh, D.D., Park, J.K., Ali, M., Danovaro-Holliday, M.C., Son, N.D., Hoa, N.T., Holmgren, J. and Clemens, J.D. Long-term effectiveness against cholera of oral killed whole-cell vaccine produced in Vietnam. Vaccine 2006; 24:4297–4303.

Thungapathra, M., Sharma, C., Gupta, N., Ghosh, R.K., Mukhopadhyay, A., Koley, H., Nair, G.B. and Ghosh, A. Construction of a recombinant live oral vaccine from a non-toxigenic strain of *Vibrio cholerae* O1 serotype Inaba biotype E1 Tor and assessment of its reactogenicity and immunogenicity in the rabbit model. Immunol. Lett. 1999; 68:219–227.

Trach, D.D., Clemens, J.D., Ke, N.T., Thuy, H.T., Son, N.D., Canh, D.G., Hang, P.V. and Rao, M.R. Field trial of a locally produced, killed, oral cholera vaccine in Vietnam. Lancet 1997; 349:231–235.

Villeneuve, S., Souchon, H., Riottot, M.M., Mazie, J.C., Lei, P., Glaudemans, C.P., Kovac, P., Fournier, J.M. and Alzari, P.M. Crystal structure of an anti-carbohydrate antibody directed against *Vibrio cholerae* O1 in complex with antigen: molecular basis for serotype specificity. Proc. Natl. Acad. Sci. U.S.A. 2000; 97:8433–8438.

Waldor, M.K. and Mekalanos, J.J. Emergence of a new cholera pandemic: molecular analysis of virulence determinants in *Vibrio cholerae* O139 and development of a live vaccine prototype. J. Infect. Dis. 1994; 170:278–283.

Waldor, M.K. and Mekalanos, J.J. Lysogenic conversion by a filamentous phage encoding cholera toxin. Science 1996; 272:1910–1914.

Wiedermann, G. and Jong, E. Vaccine-preventable diseases. In: Textbook of Travel Medicine and Health (H.L. DuPont and R. Steffen, Eds.), pp. 143–154. Hamilton, Ont.: B.C. Decker, Inc., 1997.

Wittlinger, F., Steffen, R., Watanabe, H. and Handszuh, H. Risk of cholera among Western and Japanese travelers. J. Travel Med. 1995; 2:154–158.

World Health Organization International health regulations. Second annotated edition. Geneva: World Health Organization, 1974.

World Health Organization. Potential use of oral cholera vaccines in emergency situations. Report of a meeting, Geneva, 12–13 May, 1999, Geneva, World Health Organization. WHO/CDS/CSR/EDC/99. 4, 1999.

World Health Organization. Cholera vaccines: WHO position paper. Wly. Epidemiol. Rec. 2001; 76:117–124.

World Health Organization. Cholera vaccines: a new public health tool? WHO/CDS/CPE/ZFK/2004. 5, 2004.

World Health Organization. Cholera 2005. Wkly. Epidemiol Rec. 2006; 81:297–307.

CHAPTER

46

Botulinum Toxin

Lance L. Simpson
Center for Research on Bioterrorism and Biodefense,
Environmental Health Sciences, Jefferson Medical College,
Philadelphia, PA, USA

OUTLINE

ABSTRACT

Botulinum toxin is a remarkably potent substance that acts preferentially on cholinergic nerve endings to block transmitter release. Botulism, which is the disease caused by the toxin, can occur in several variations, but it most commonly presents as a form of food poisoning. Patients ingest food that is contaminated with the toxin, or they ingest organisms that can manufacture the toxin in the gut. Absorption occurs when the toxin binds to epithelial cells, then undergoes receptor-mediated endocytosis, transcytosis, and eventual release into the general circulation. Absorbed toxin is distributed throughout the periphery, but mainly to the cholinergic nerve endings. The toxin acts intracellularly to cleave polypeptides that are essential for transmitter release. Blockade of exocytosis produces muscle weakness and autonomic dysfunction. In extreme cases, patients suffer complete paralysis, including paralysis of the muscles of respiration.

In addition to being the etiologic agent in naturally occurring disease, botulinum toxin figures prominently in two other clinical settings. The toxin is recognized as a potential agent of bioterrorism and biological warfare, but it is also a valuable medicinal agent for the treatment of dystonia and related disorders. The fact that the toxin is a potential bioweapon has prompted intense efforts to develop a vaccine that can protect vulnerable populations. However, there is a keen awareness that administration of a vaccine would be detrimental to patients who receive the toxin as a medicinal agent. The existence of a vaccine that can be both indicated and contraindicated, depending on patient populations, is highly unusual in the practice of medicine. This unusual state of affairs means that rational development of a vaccine must address two broad areas of concern: *conventional issues*, such as choosing antigen, adjuvant, and route of administration; and *unconventional issues*, such as potential adverse consequences of vaccine administration.

There are no licensed, polyvalent vaccines against botulism. There are toxoids available under Investigational New Drug status, but these are likely to be replaced by recombinant or synthetic vaccines. The three major candidates for vaccines are: (1) holotoxin that has been rendered inactive by site-directed mutations, (2) nontoxic polypeptide fragments that represent functional domains of the holotoxin (i.e., tissue binding domain), and (3) short peptides that represent epitopes.

Preclinical studies indicate that recombinant vaccines can be highly effective against challenge doses of toxin. Unfortunately, this means that they would also be highly effective in negating therapeutic actions of the toxin. To date, no federal agency has advocated universal administration of a vaccine. However, the counterposing roles of any vaccine in protecting against botulism while simultaneously abolishing therapeutic benefits of the toxin raises questions about the ideal duration of vaccine action. This is a daunting question that requires serious attention.

INTRODUCTION

Botulinum toxin is one of the more extraordinary molecules encountered in medicine and science. There are many factors that contribute to this special status, but several are particularly well known. To begin with, botulinum toxin is generally considered the most potent of all biological toxins (Gill, 1982; Lamanna, 1959). In addition, this remarkably potent molecule must go through an unusually long and complex sequence of events to reach and poison its target organ, which is the peripheral cholinergic nerve ending (Simpson, 2004). However, the characteristic of the toxin that is most troubling in the context of vaccine development is the number and variety of clinical settings in which it is encountered. On the one hand, botulinum toxin is the etiologic agent that causes the disease botulism (Hatheway, 1995; Horowitz, 2005; Sobel, 2005). This disease can be naturally occurring, but it can also be the product of malice (i.e., bioterrorism or biological warfare). On the other hand, botulinum toxin is an approved medication for the treatment of a variety of disorders, such as dystonia (Comella et al., 2005). In the absence of botulinum toxin, patients would not have access to equally efficacious alternatives.

These contrasting clinical settings have fundamentally different relationships to the field of vaccine development. For patients who require active immunization, resistance would be the desired outcome; but for patients who receive toxin for its clinical benefit, resistance would be a therapeutic obstacle. The existence of a vaccine that can be either an enormous benefit or a serious detriment, depending on patient population, is rarely encountered in the practice of medicine.

This unusual conflict between benefit and risk would not be an issue if the real or potential incidence of botulism were limited to the naturally occurring disease. The number of outbreaks, and the number of cases per outbreak, is far too low to justify either vaccine development for, or vaccine administration to, the general public. Instead, the conflict between benefit and risk arises from the fact that botulinum toxin can be used as a weapon in acts of bioterrorism or biological warfare. The perceived threat posed by the toxin has greatly intensified the effort to find a vaccine.

For those investigators who are seeking to develop a vaccine, there is one observation upon which everyone can agree. There already exist techniques that can be used to evoke resistance to botulinum toxin. In fact, there are several and somewhat varied techniques that could be employed to evoke immunity. This means that not only is there urgency to finding a vaccine, but there is also urgency to identifying the strategy that will yield the best-possible product.

One of the factors that will facilitate the search for an ideal vaccine candidate is the substantial body of information that exists on the botulinum toxin molecule and on the epidemiology and clinical presentation of the disease botulism. For example, a great deal is known about the cellular, subcellular, and even molecular aspects of the many steps in the long progression of events that leads to poisoning. This means that it should be possible to engage in rational vaccine design that is tailored to the mechanism of toxin action (Levine and Sztein, 2004). Similarly, there is much known about the populations that are most at risk of contracting botulism, and about the signs and symptoms and the clinical course of the disease in patients who succumb to poisoning. This too is a fertile area that can provide clues on rational vaccine design.

BOTULINUM TOXIN

Structure of the Toxin

Botulinum toxin is produced by the organisms *Clostridium botulinum*, *Clostridium beratii*, and *Clostridium butyricum* (Smith and Sugiyama, 1988). The toxin is manufactured in seven different serotypes designated A, B, C, D, E, F, and G (Sakaguchi, 1983; Simpson, 1989). Each of these serotypes has been shown to occur, or is presumed to occur, in multiple subtypes (Arndt et al., 2006; Smith et al., 2005).

The complete primary structure for at least one representative of each serotype has been determined (Minton, 1995; Popoff and Marvaud, 1999). Sequence identity across all serotypes occurs only with low frequency, which in part explains why the different proteins are serologically distinct. Much less is known about the tertiary structures of the various serotypes. To date, complete three-dimensional structure data are available for serotypes A and B (Lacy and Stevens, 1999; Swaminathan and Eswaramoorthy, 2000), and partial structural data are available for serotypes D, E, F, and G (Agarwal et al., 2004, 2005; Arndt et al., 2005, 2006).

Early work on the protein structure of botulinum toxin indicated that it is composed of a light chain (LC) (N-terminus of the molecule; ca. 50 kDa) and a heavy chain (C-terminus of the molecule; ca. 100 kDa) linked by a single disulfide bond (DasGupta, 1989). The heavy chain in turn is composed of two domains of about 50 kDa each. The structural arrangement of the molecule, which suggests three somewhat independent lobes, is illustrated in Fig. 46.1. The three lobes in the tertiary structure represent the LC, the amino-terminal half of the heavy chain, and the carboxy-terminal half of the heavy chain. This structural organization of the molecule is key to understanding both mechanism of action and immune response (see below).

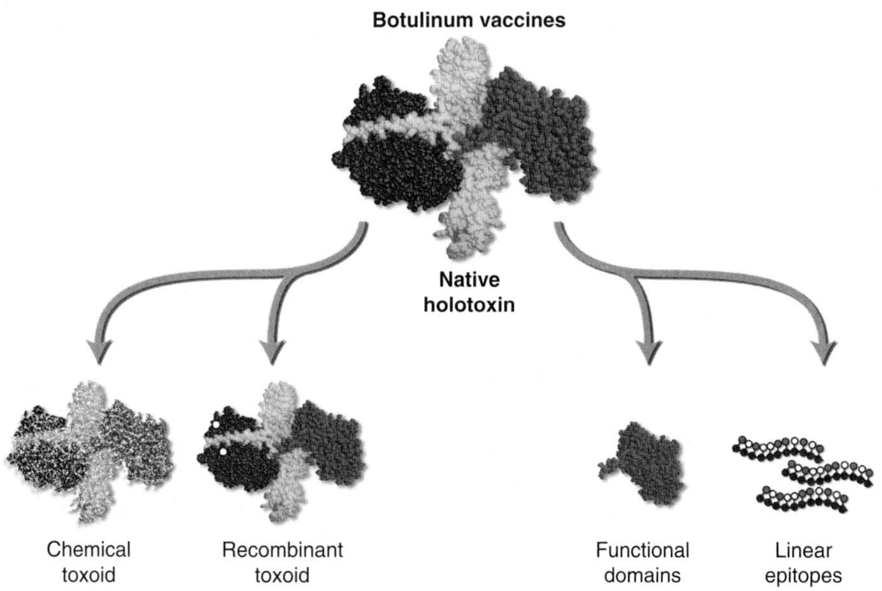

FIGURE 46.1 Botulinum vaccines. Botulinum toxin is a 150 kDa protein that is composed of three distinct domains (lobes). These represent the 50 kDa light chain (LC) (black tone), the 50 kDa amino-terminal half of the heavy chain (grey tone), and the 50 kDa carboxy-terminal half of the heavy chain (intermediate tone). Initial attempts to develop a vaccine relied upon chemical inactivation of impure preparations of the holotoxin (e.g., formalin treatment). More recent attempts have focused largely on homogeneous preparations of recombinant polypeptides. This approach has led to evaluation of three types of preparations: (a) a holotoxin that has been rendered nontoxic by point mutations in the histidine motif of the LC (white dots), (b) isolated functional domains, such as the carboxy-terminal half of the heavy chain (e.g., binding domain), and (c) relatively short linear epitopes. Each of these three can be administered either as a polypeptide vaccine or as a nucleic acid vaccine.

Botulinum toxin does not typically occur in bacterial cultures as an isolated neurotoxin. Instead, it is ordinarily part of a noncovalent complex that includes hemagglutinins of various sizes and a single nontoxin, nonhemagglutinin (Sakaguchi, 1983). These auxiliary proteins are not known to play any role in the signs and symptoms that are characteristic of the disease botulism.

As a matter of historical observation, one must note that initial efforts to develop a vaccine against botulinum toxin predated almost all work to define the structures of, and the interrelations among, the neurotoxin and the other molecules in the noncovalent complex (Byrne and Smith, 2000; Siegel, 1988). Hence, it is not surprising that early vaccine candidates are not considered equivalent to those currently under investigation. This is a matter of some importance because the only vaccine candidates that are presently available for patient administration are those that use the technologies that predate knowledge of toxin structure.

Mechanism of Action

Stated simply, botulinum toxin acts at vulnerable nerve endings in the periphery to block transmitter release. However, the sequence of events that culminates in blockade of exocytosis is rather complex. Fortunately, tremendous progress has been made in elucidating these events (Schiavo et al., 2000; Simpson, 2000, 2004; Simpson et al., 2001).

Most cases of botulism are oral in nature, although the disease can also be caused by inhalation exposure (Holzer, 1962). This means that the toxin must have an efficient mechanism to escape the lumen of the gut or the airway. In the case of the gut, absorption of botulinum toxin involves two distinct phenomena (Maksymowych and Simpson, 1998). Firstly, the toxin escapes metabolic degradation in the stomach by virtue of being intertwined with auxiliary proteins. This noncovalent complex is remarkably resistant to acid conditions and endoprotease exposure (Chen et al., 1998). When the toxin is stripped of auxiliary proteins, it is much less potent as an oral poison (Maksymowych et al., 1999). Secondly, the toxin traverses epithelial barriers in the small intestine to reach the general circulation (Fig. 46.2). It does this by binding exploitatively to receptors on the apical surface of epithelial cells. This is followed by receptor-mediated endocytosis, transcytosis, and eventual release of unmodified toxin on the basolateral surface of cells (Maksymowych and Simpson, 1998; Maksymowych and Simpson, 2004).

The airway does not contain the same harsh metabolic conditions as the gut, and thus there is little need for auxiliary proteins to protect the toxin. Nevertheless, the toxin does need to escape the airway to reach the general circulation, and it achieves this by active penetration of epithelial barriers. Similar to the gut, there is exploitative binding, endocytosis, transcytosis, and release (Park and Simpson, 2003).

There has been a limited amount of work to identify the epithelial cell population that transports the toxin, as well as to visualize the pathway used by the toxin as it crosses individual cells in this population (Ahsan et al., 2005). Interestingly, absorptive enterocytes are the key transport cells involved in poisoning. The M cell–Peyer's Patch complex is capable of transporting toxin, but it is not essential. Thus, knockout mice that do not develop M cell–Peyer's Patch

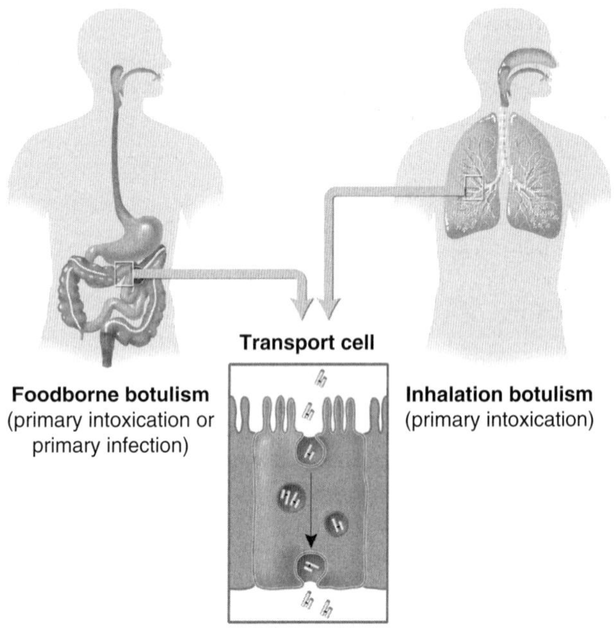

FIGURE 46.2 Absorption of botulinum toxin. The primary motive for vaccine development is to protect vulnerable populations against potential incidents of bioterrorism and biological warfare. Such incidents would likely be due to exposure by the oral route (which is also the route for most cases of naturally occurring botulism) or by the inhalation route. Oral poisoning leads to two variants of food-borne botulism: primary intoxication, or primary infection with secondary intoxication (see Box 46.1 for descriptions of clinical etiology). Inhalation poisoning leads to only one variant of the disease: primary intoxication. Regardless of route of exposure, the toxin escapes the gut or airway by interacting with epithelial cells that transport the toxin into the general circulation. The transport process involves binding to receptors on the apical surface of cells, followed by receptor-mediated endocytosis, transcytosis, and eventual release of the toxin on the basolateral surface of cells. Toxin that enters the general circulation is distributed to peripheral cholinergic nerve endings, where it acts to block transmitter release.

complexes have the same oral and inhalation sensitivity to the toxin as the wild-type strain.

Botulinum toxin that is delivered to the general circulation is distributed throughout the periphery. The toxin has little ability to penetrate the blood–brain barrier, and therefore the central nervous system is largely spared from its effects. Although the toxin is potentially available to all tissues and organs in the periphery, the molecule demonstrates high-affinity binding only at presynaptic nerve terminals of cholinergic fibers. The most exquisitely sensitive site is the cholinergic neuromuscular junction (Simpson, 1986, 2000). By virtue of acting on cholinergic nerve endings that govern motor activity, the toxin can produce flaccid paralysis. By acting on cholinergic fibers that innervate intercostal muscles and the diaphragm, the toxin can paralyze respiration.

The sequence of events that underlies poisoning at all cholinergic nerve endings is the same (Humeau et al., 2000; Schiavo et al., 2000; Simpson, 2004). The carboxy-terminal portion of the heavy chain binds to receptors on the surface, and this in turn leads to receptor-mediated endocytosis. There is evidence that binding actually involves two sites: a low-affinity receptor that brings the toxin into the plane of the membrane, and a high-affinity receptor that is presumably linked to endocytosis (Montecucco, 1986). For two of the toxin serotypes the dual receptor may have been identified [serotype A: polysialoganglioside and synaptic vesicle protein 2C (Dong et al., 2006; Mahrhold et al., 2006); serotype B: polysialoganglioside and synaptotagmin (Bullens et al., 2002; Dong et al., 2003; Kozaki et al., 1998; Nishiki et al., 1994, 1996; Yowler et al., 2002)]. In the case of serotype B, investigators have recently reported the three-dimensional structure of the toxin in association with its receptors (Chai et al., 2006; Jin et al., 2006).

The toxin that enters the cell by endocytosis is briefly trapped. However, as the proton pump in the membrane acidifies the lumen of the endosome, the toxin molecule undergoes a dramatic change in conformation. The amino-terminal half of the heavy chain exposes a previously occult hydrophobic domain, and this interacts with the membrane and with the LC to facilitate translocation of the latter (Hoch et al., 1985; Koriazova and Montal, 2003). At some point in this process, the interchain disulfide bond is reduced and the free LC reaches the cytosol (Simpson et al., 2004).

The LCs of all serotypes of botulinum neurotoxin are zinc-dependent endoproteases that are remarkably specific with respect to substrates and scissile bonds (see Table 46.1). The substrates for the toxins are three polypeptides that act cooperatively in the process of transmitter release (Humeau et al., 2000; Schiavo et al., 2000; Simpson, 2004). When these polypeptides are cleaved, exocytosis is blocked. It is this blockade of exocytosis that accounts for toxin-induced paralysis of movement and paralysis of respiration.

TABLE 46.1 Botulinum toxin and its substrates

Toxin	Histidine motif	Substrate
Serotype A	H · E · L · I · H	SNAP-25
Serotype B	H · E · L · I · H	VAMP
Serotype C	H · E · L · N · H	Syntaxin and SNAP-25
Serotype D	H · E · L · T · H	VAMP
Serotype E	H · E · L · I · H	SNAP-25
Serotype F	H · E · L · I · H	VAMP
Serotype G	H · E · L · I · H	VAMP

Note: Botulinum toxin exists in seven serotypes (A–G), each of which can occur in multiple subtypes. The mechanism of action of the various serotypes was deduced from primary sequence analysis. The light chain of each serotype has a "histidine motif" (H·E·X·X·H), which is a signature sequence for metalloendoprotease activity. The substrates for the various serotypes are SNAP-25 (synaptosomal protein of 25 kDa), VAMP (vesicle-associated membrane protein, also known as synaptobrevin), and syntaxin.

Implications for Vaccine Development

When one examines the information on structure and function of the toxin molecule, there are several points that emerge that bear on efforts to develop a vaccine. The most important of these are:

(1) Botulinum toxin can enter the body by absorption across the gut and airway epithelium. This suggests that it would be desirable to create a vaccine that can evoke a local IgA response. A mucosal immune response could serve as a barrier to block toxin absorption.
(2) Regardless of route of toxin absorption, the general circulation serves as a conduit to transport toxin to sensitive nerve endings. This suggests that a circulating IgG response could contribute to neutralization of toxin. The underlying mechanism for IgG-mediated neutralization could be enhanced clearance from the circulation secondary to enhanced accumulation in nontarget tissues (e.g., liver and spleen).
(3) The fact that the toxin progresses through multiple steps to produce neuromuscular blockade suggests that antibodies could act at one or more of these steps to diminish toxicity. Most obviously, an immune response could evoke antibodies that block toxin binding to, or entry into, cholinergic nerve endings.

It is worth noting that these several approaches to vaccine development are not mutually exclusive. Indeed, there are rational ways to design vaccine candidates that could evoke protection at all three levels.

HISTORY OF THE DISEASE

Evolutionary Timescale

The nucleotide sequences of the seven botulinum serotypes, as well those of numerous subtypes, are known (Minton, 1995; Popoff and Marvaud, 1999; Smith et al., 2005). Work has also been done to sequence the messages that encode some of the auxiliary proteins found in the botulinum toxin complex (East et al., 1996; Ohyama et al., 1995). These sequences have been used to construct genealogical trees that establish: (a) relatedness between and among the serotypes, and (b) likely points when precursors divided to give rise to descendants. These analyses provide insight into the evolution of botulinum toxin.

There are two other kinds of analyses that would also be helpful, and work in these areas is currently underway, but no results have actually been published. Several groups have sought to determine whether there is structural relatedness between botulinum toxin and any other proteins, with a particular emphasis on identifying the parent molecule from which the toxin might have evolved. To date, no precursor has been found. Another line of investigation would be to utilize a molecular clock, in combination with genealogical trees, to determine how ancient any supposed precursor might be. Nothing has emerged in this realm either.

In the absence of a molecular approach to describing and quantifying evolution, one can tentatively turn to a "victim clock." The target organ for botulinum toxin action is the cholinergic nerve ending. If there were no cholinergic nerves, there would be no obvious survival value that could select for emergence and persistence of the toxin. Interestingly, botulinum toxin is not a disease that is confined to mammals, nor even to vertebrates. There are cholinergic nerve endings in invertebrates, and some of these are susceptible to botulinum toxin action. This could mean that the toxin has an origin that reaches far back in time.

Ancient Origins and Innate Resistance

For any biological toxin that has a long evolutionary history, there is the possibility that organisms have emerged that are resistant. The study of innate resistance can help to clarify the mechanism of pathogen action, and it can also be helpful in analyzing phenomena such as host–pathogen interactions. Perhaps most importantly, the study of innate resistance can provide clues on ways to evoke resistance.

The identification of organisms that are resistant to botulinum toxin, and the use of these organisms to explore the nature of innate resistance, is a nascent field of research. Nevertheless, there are several observations that should be mentioned. The best-known example of resistance to botulinum toxin was reported in the middle of the last century. Burgen et al. (1949) found that rats were unusually resistant to serotype B, and this characteristic was not shared by mice. The underlying basis for this observation remained unknown for decades, until the subcellular actions of botulinum toxin were described. Serotype B acts in the cytosol to cleave a substrate (e.g., VAMP; Table 46.1) that is needed for exocytosis (Schiavo et al., 2000). Montecucco and his colleagues (Patarnello et al., 1993) found that VAMP in the peripheral nervous system of rats has a single modification (Gln → Val) at the cleavage site for serotype B. This same modification was found in chickens, which are also resistant to serotype B.

There are other examples of organisms that are resistant to botulinum toxin, but certainly the most provocative example pertains to human beings. Naturally occurring type C and type D botulism are so rare in the human population as to raise questions about resistance. This observation is all the more dramatic because of the widespread and devastating impact these serotypes have on other animals, such as birds. It can be stated with certainty that the human neuromuscular junction is not resistant to poisoning (Coffield et al., 1997). When studied on excised and viable human preparations, serotype C is as active as any other serotype in blocking transmission. In addition, the human substrate for serotype C (e.g., syntaxin) does not have any mutations at the scissile bond (Zhang et al., 1995). This means that any real or apparent resistance must relate to: (a) the relative absence in the human diet of foods contaminated with *C. botulinum* types C and D, or (b) the failure of these serotypes to be absorbed and distributed to the neuromuscular junction.

There is another and quite different area of investigation in which the question of human resistance has been raised. As discussed more fully below, botulinum toxin is an approved medication for the treatment of a variety of disorders in which efferent activity in cholinergic nerves must be managed. In the course of using the toxin as a medicinal agent, clinicians have encountered two types of resistance: (a) there are reportedly a small number of patients who do not respond when botulinum toxin therapy is initiated, and (b) there are patients who respond initially but who become refractory over time.

The underlying basis for the first phenomenon has not been explained, though most investigators agree that it is not an antibody-based mechanism (viz., prior unknown exposure to toxin or to a structurally similar antigen). If it should eventually prove true that some in the human population are inherently resistant to one or more serotypes, the underlying bases may be similar to those shown for other animals (e.g., alterations in receptors or substrates). The story that has emerged for patients who lose responsiveness over time is quite different. The underlying basis for most if not all cases of lost responsiveness is evoked immunity. When botulinum toxin is administered in sufficient amounts and frequencies, the therapeutic goal may be modulation of efferent cholinergic activity, but the occasional and unintended outcome is vaccination. It is appropriate to note here—and the matter will be explored more fully in a later section—that vaccination against botulinum toxin in certain patients should be described as an adverse event.

The conclusions that can be drawn from this narrow field of investigation are limited, although there is one notable point that does emerge. For most proteins that enter the body by the oral route, a powerful form of innate resistance is the combination of proteolysis and epithelial barriers. It is fascinating that botulinum toxin has evolved mechanisms to evade both. As explained earlier, the toxin is not ordinarily found as an isolated neurotoxin in bacterial culture. It is more customary to find the toxin as part of a noncovalent complex with auxiliary proteins, and the intertwining of these proteins confers substantial resistance to gut proteolysis (Chen et al., 1998). In addition, the toxin molecule possesses structural determinants that allow it to bind exploitatively to receptors on epithelial cells that mediate endocytosis and transcytosis. Thus, part of the extraordinary potency of the toxin molecule can be attributed to its ability to evade obstacles that for other pathogens would be the essence of innate resistance.

There is the possibility that not all serotypes have efficiently overcome the obstacles. For example, the low incidence of type C botulism in human beings could in part be due to the relative inability of this particular toxin to breach epithelial barriers (Maksymowych and Simpson, 1998). For those serotypes that do penetrate epithelial membranes, there are ways to recreate barriers and thus confer resistance. The most obvious of these is to induce an adaptive mucosal immune response that leads to clinically significant titers of luminal IgA that will block toxin absorption.

CLINICAL PRESENTATION OF THE DISEASE

Etiologic Factors

Almost all cases of human botulism are due to serotypes A, B, and E (Lindstrom and Korkeala, 2006). The signs and symptoms of types A, B, and E botulism can be attributed to the inherent pharmacologic properties of the toxins, as described above, and to the adverse effects of microbial products other than the toxins. It is intuitively clear that the former will always be present. The likelihood of the latter depends upon the circumstances of the exposure as well as the route of exposure.

Botulism can occur either as a primary intoxication or as a primary infection with secondary intoxication (Box 46.1). Primary intoxication occurs when clostridial organisms produce botulinum toxin outside the body, and this toxin subsequently enters the body to produce blockade of cholinergic transmission. Primary infection occurs when organisms enter the body and produce the toxin in situ. As before, the toxin will ultimately produce blockade of cholinergic transmission.

Primary intoxication and primary infection with secondary intoxication can occur in several reasonably well-defined variations (Box 46.1). As suggested above, all of these variations are associated with peripheral cholinergic blockade. Those variations that involve oral exposure may also present with a variety of gastrointestinal problems, including nausea, vomiting, and diarrhea. Some of these problems may be caused by botulinum toxin, but others are likely due to microbial products that are consumed as part of the disease process.

Signs and Symptoms

Botulinum toxin acts preferentially on peripheral cholinergic nerve endings to inhibit acetylcholine release. The two principal populations of vulnerable neurons are those at the neuromuscular junction and those at various autonomic sites.

Depending on severity, neuromuscular blockade can vary from slight weakness to frank paralysis (Hatheway, 1995; Horowitz, 2005; Sobel, 2005). Generally speaking, blockade tends to be both descending and bilateral. A classic patient might present with bilateral weakness of head and facial musculature, followed by weakness in the upper limbs, trunk, and lower limbs. The fact that weakness or paralysis is bilateral is understandable. The toxin is fully capable of poisoning transmission on both sides of the body. The fact that neuromuscular involvement is descending has not been so easy to explain.

BOX 46.1

CLINICAL PRESENTATION OF BOTULISM

There are two broad categories of botulism: (1) primary intoxication, and (2) primary infection with secondary intoxication. Each of these broad categories can be further subdivided, as follows:

1. Primary intoxication
   a. *Foodborne botulism*, which is caused by consumption of foods that are contaminated with preformed botulinum toxin. This form of the disease occurs when food that harbors *Clostridia* is prepared and/or stored in a way that allows bacterial growth. Toxin is released during microbial lysis, and thus it is available to cause disease when ingested.
   b. *Inhalation botulism*, which is due to inhalation of airborne particles or droplets contaminated with toxin. There is only one documented outbreak of human inhalation poisoning, which occurred in a laboratory that was handling the toxin.
   c. *Iatrogenic botulism*, which can be due to administration of the toxin for therapeutic or aesthetic purposes. If the administered dose is inappropriately high, or if a patient is unusually sensitive, disease can occur.
2. Primary infection with secondary intoxication
   a. *Foodborne botulism*, which is due to consumption of food that is contaminated with clostridial spores. These organisms colonize the gut and produce the toxin in situ. This form of the disease can present in two variants. *Infant botulism*, which is more prevalent, occurs in infants who have not yet acquired a normal gut flora. The relative absence of competing organisms allows *Clostridia* to establish a niche. *Adult botulism* occurs in people who have endogenous or exogenous factors that diminish or eliminate natural gut flora (e.g., antimicrobial agents). This creates a situation that is conducive to subsequent colonization by *Clostridia*.
   b. *Wound botulism*, which is due to introduction of clostridial spores, with subsequent colonization and production of toxin in situ. This form of the disease is usually secondary to puncture wounds associated with injection of illicit drugs (e.g., heroin). However, it can also be secondary to surgery.

The need for medical intervention is greatest when there is involvement of the muscles of respiration (diaphragm; intercostal muscles). It is likely that throughout most of human history severe poisoning resulted in death due to respiratory failure. The introduction of sophisticated intensive care practices and facilities, combined with heightened awareness and early diagnosis of the disease, has dramatically reduced morbidity and mortality in modern times.

Botulinum toxin can also produce a constellation of adverse autonomic effects. Among the more well-documented of these are fixed and unresponsive pupils; decreased salivation, with dryness of the mouth and throat and difficulty swallowing; orthostatic hypotension; paralytic ileus; and bladder and urinary dysfunction. Autonomic problems would not typically be a cause of death.

In contrast to peripheral cholinergic systems, other parts of the nervous system are generally preserved in terms of function. Sensory responses are normally intact, and there is little evidence to support any direct action of the toxin on the central nervous system. The toxin is a large protein and would not be expected to penetrate the blood–brain barrier. It has often been noted that the absence of overt effects of the toxin on the brain means that patients, including those who are fully paralyzed, are aware of their difficult circumstances.

Duration of Action

Botulinum toxin is generally viewed as the most potent biological substance known. However, this is only one of its remarkable features. Yet another is its ability to exert an unusually sustained effect.

In most respects the signs and symptoms associated with the various serotypes of botulism are similar. One striking exception is duration of illness. It appears that the signs and symptoms associated with serotype A poisoning are the longest, with a potential duration of months. However, the incidence of naturally occurring botulism is very low, and therefore efforts to quantify accurately the true duration of action are problematic. Ironically, most of the definitive work on duration of effect arises not from the study of botulinum toxin as a disease, but rather from work related to use of the toxin as a medicinal agent. Botulinum toxin is administered to patients by local injection to reduce the adverse and sometimes debilitating effects of uncontrolled efferent traffic in cholinergic fibers. The major drawback to this form of intervention is that the therapeutic benefits of the toxin wane with time, thus requiring readministration of toxin. The length of the intervals between administrations is an indicator of the duration of toxin action. Both clinical observation and research related to clinical use have verified that serotype A has the longest duration of action (Adler et al., 2001; Eleopra et al., 1998; Foran et al., 2003; Keller, 2006). This work has also shown that there is great variability in duration of effect, with recovery at certain sites occurring in 2–3 months, but recovery at other sites perhaps requiring more than 1 year. This is a stunning duration of effect for an exogenous protein.

Therapeutic Intervention

Treatment of patients can be somewhat arbitrarily divided into three categories: methods to reverse poisoning; methods to halt the progression of poisoning; and methods to provide supportive care while patients recover from poisoning. Sadly, there is nothing to report that falls into the first—and conceptually the most desirable—category. No clinically acceptable agent or procedure has been identified that can reverse the enzymatic actions of the toxin that account for blockade of exocytosis. When a toxin substrate has been cleaved, the only known intervention is to wait until the poisoned nerve synthesizes new substrate and reconstitutes the exocytotic process.

There are two possible ways to halt the progression of disease, which are passive immunization and pharmacologic antagonists (Arnon et al., 2001). There is an equine antitoxin, and administration of the material can lead to neutralization of circulating toxin. There is also a human immune globulin preparation that is intended for infants who ingest clostridial spores that manufacture toxin in situ (Arnon et al., 2006; Arnon, 2007). In either case, treating physicians must face this sobering reality: if signs and symptoms have begun to emerge, the therapeutic value of immune serum or neutralizing antibodies is much diminished. This loss of antibody effect is related to the ability of the toxin to enter nerve endings and block transmitter release. Neutralizing antibodies cannot chase the toxin because they have no mechanism for productive internalization by cells.

Many investigators have tried to identify drugs that can halt the progression of the disease. For example, there are several research groups trying to develop drugs that can block the enzymatic actions of the LC. There are reports of success in laboratory test systems, but there is nothing that is on the verge of clinical testing. One challenge to clinical progress is drug delivery. Toxin antagonists tend to be polar, and as such they may not be good candidates for mucosal delivery. Having to administer an antagonist by injection could be a difficult proposition when dealing with a toxin whose duration of action could be many weeks to months. A second challenge relates to membrane barriers. To antagonize botulinum toxin, a drug should be found that can cross neuronal membranes to reach toxin but not cross other membranes to enter non-neuronal cells. There are as yet no clear strategies for achieving selective membrane penetration, with one potential exception. There is the possibility that nontoxic variants of the toxin could be used as carriers to transport antagonists or even antibodies into cholinergic cells.

The third category of intervention is supportive care, and this is the one area in which there has been authentic success. The early literature on botulism indicated that fatal outcomes often reached 50% or more. With the advent of modern intensive care procedures, this fatality rate has fallen dramatically. This means that even those patients who become fully paralyzed, and who may need respiratory support for extended periods of time, have a good prospect of recovery.

Implications for Vaccine Development

There is an emerging body of information on the history of the disease botulism, as well as on the clinical presentation and treatment of the disease. This body of information supports several conclusions regarding vaccine development, as follows:

(1) Unless and until experimental or licensed therapeutic antibodies and/or drug antagonists become available for human administration, there will be a disproportionate reliance on current experimental vaccines. This reliance is heightened by the facts that: (a) the equine antitoxin that is currently available is not desirable for widespread use, either preincident or postincident; and (b) postincident use of antibody preparations, whether equine or human, has limited value in patients who have begun to manifest signs and symptoms. There is the additional limitation that therapeutic antibodies and pharmacologic antagonists may have durations of action that are much shorter than that of the toxin. This could, for example, create the necessity to administer

a pharmacologic antagonist repeatedly over an extended period of time.

(2) Botulinum toxin exists in seven serotypes, but only three of these are typically associated with human illness (A, B, and E). This suggests that the most immediate need for a product would be a polyvalent vaccine that evokes simultaneous resistance to serotypes A, B, and E.

(3) Botulinum toxin is released from bacteria as part of a noncovalent complex that includes auxiliary proteins. However, it has not been established when this complex dissociates to release free toxin, and this uncertainty applies both to primary intoxication and to primary infection with secondary intoxication. If it were to be found that toxin in the gut is always associated with auxiliary protein, this would raise the interesting possibility that a local IgA response against an auxiliary protein could block absorption and diminish toxicity (Kobayashi et al., 2005; Mahmut et al., 2002).

(4) In a similar vein, the botulinum toxin complex is able to defeat two of the most prominent mechanisms that contribute to innate immunity against toxic proteins. Firstly, the heteromolecular complex can withstand the conditions that ordinarily lead to gastric metabolism. Secondly, the toxin can bind exploitatively to receptors on epithelial cells, leading to absorption of the molecule rather than elimination. This serves to re-affirm the premise presented above (#3). Whether directed against the neurotoxin, the auxiliary proteins, or both, a secretory IgA response would create a barrier to toxin absorption that could substitute for circumvented mechanisms of innate resistance.

(5) For patients who are at risk of primary infection, there is the theoretical possibility that an attenuated vaccine could combat colonization by clostridial spores in the gut. However, there are powerful factors that weigh against this as a realistic option. The number of patients who develop primary infection as a natural disease is too low to warrant development of an attenuated vaccine. In addition, primary infection would be relatively difficult to implement as an act of bioterrorism, and it has essentially no relevance as an act of biological warfare.

(6) Botulinum toxin is not known to exert direct effects on the central nervous system. This means that an immune response must attack and neutralize serotypes A, B, and E before they enter peripheral nerve endings, but not central nerve endings.

EPIDEMIOLOGY

Naturally Occurring Disease

Reports have appeared at irregular intervals describing the incidence of botulism in the US (e.g., Sobel et al., 2004; Gupta et al., 2005). Over an extended period of time, both the number of outbreaks per year and the number of cases per outbreak has been low. An exact insight into the frequency of the disease can be gained from the epidemiologic report by Shapiro et al. (1998). Reviewing the data that were available for approximately a quarter of a century (1973–1996), these authors found that the median number of cases reported annually to the Centers for Disease Control and Prevention was ca. 100.

Few equivalent reports exist that quantify the worldwide incidence of botulism over time (but see Todd, 1997). However, given the way in which the disease presents when it is naturally occurring (Table 46.1), and given that the disease is not transmitted from patient to patient, one can deduce that the worldwide incidence is likely to be relatively low.

Bioterrorism and Biological Warfare

Botulinum toxin is considered a prime candidate as a potential bioweapon (Arnon et al., 2001). Indeed, there is strong evidence that the toxin, or cultures containing the toxin, have been isolated for the purpose of intentionally poisoning civilian (viz., bioterrorism) and military (viz., biological warfare) populations (Caya et al., 2004; Christopher et al., 1997; Klietmann and Ruoff, 2001; Zilinskas, 1997). For example, during the first half of the 20th century, cultures of *C. botulinum* were fed to Manchurian victims by their Japanese occupiers. Shortly thereafter, during World War II, a number of Western nations began accumulating clostridial cultures and semipurified toxin. This was often described as being for defensive purposes (i.e., toxoid production), but it is likely that offensive purposes were part of the primary motive. During the second half of the 20th century, certain nations that had signed the international convention banning offensive use of biological toxins accumulated frightening amounts of botulinum toxin (e.g., Iraq, the former Soviet Union), and there is no doubt that this was intended for use as a weapon. In addition, the Japanese cult Aum Shinriko not only accumulated the toxin, but also used it in aerosol attacks against civilian and military targets. Thankfully, these attempts at intentional poisoning have largely failed.

This unsettling history has caused botulinum toxin to become one of the agents that is on Select List A (NIAID, 2002). Each of the agents on this list possesses properties that contribute to a high level of risk. For botulinum toxin, the principal characteristics are:

- The extreme potency of the toxin.
- The relative ease with which the toxin can be isolated and then used with malice.
- The severity of the disease caused by the toxin, coupled with the prolonged and intensive medical care needed to treat serious poisoning.

Concerns about the potential use of the toxin as a bioweapon are heightened by the absence of adequate and targeted medical countermeasures. The three most obvious forms of intervention would be vaccination, therapeutic antibodies, and pharmacologic antagonists. At present there is no licensed polyvalent vaccine against botulinum toxin. Until the relatively recent past the only therapeutic antibody available was equine in origin. However, a significant advance occurred with the introduction of a licensed botulism immune globulin (BIG) (Arnon et al., 2006). This product is intended for use in infants with foodborne botulism (primary infection; see Table 46.1) due to serotypes A and B. In addition, substantial progress has been made in preclinical studies aimed at generating a new class of therapeutic antibodies (Nowakowski et al., 2002). It is likely that these antibodies will soon enter clinical trials.

Concerns about the potential use of botulinum toxin against civilian or military populations have dramatically intensified efforts to identify pharmacologic antagonists. Most investigators have focused on agents that either block toxin binding to neuronal membranes or that block intracellular expression of toxin catalytic activity (Dickerson and Janda, 2006). Unfortunately, none of these agents has entered clinical trials.

THE SEARCH FOR VACCINES

It would be gratifying to report that the field of vaccine development has been governed by the various principles that apply to rational design, but this would not be entirely true. Perhaps a truer assessment would be that the field has been governed by widely separated episodes of urgency, and two of these episodes have figured prominently in product development. The first began early in the last century, was intensified in the wake of World War II, and culminated in a chemical (formalin-inactivated) toxoid (Anderson and Lewis, 1981; Smith, 1977). The second began slightly before the turn of the present century, was intensified in the wake of the terrorist attacks on the World Trade Center and the bioterrorist attacks with anthrax, and culminated in a recombinant toxoid (Byrne and Smith, 2000; Smith, 1998). Both efforts should be viewed as milestones, and both resulted in Investigational New Drug status. However, neither toxoid has been licensed.

It is commendable that progress has been made toward developing a vaccine against botulinum toxin, but at the same time one must note that there are obvious and important ways to improve on the existing technology. Therefore, the sections that follow will survey the current field of vaccine development. The goal is to lay a framework for gauging the distance between current technologies and an "ideal" technology based on rational design.

Chemical Toxoids

Efforts to develop a vaccine against botulinum toxin predated knowledge about structure and structure–activity relationships of the molecule. For that matter, vaccine research began before the techniques for isolation and purification of toxin were fully developed. This meant that investigators who pioneered vaccine research were effectively limited to only one strategy: chemical inactivation of a bacterial culture fraction that was enriched in toxin (Fig. 46.1).

The initial work to develop a vaccine began almost 60 years ago, in response to concerns that the toxin would be used in biological warfare. The effort to generate a pentavalent product (A–E) required an extended period of time (Byrne and Smith, 2000). The process involved growing individual cultures that were acid-precipitated, processed to remove unwanted bacterial debris, and then precipitated again. The final product, which was still highly impure, was treated with formalin to denature, cross-link, and inactivate the toxin that was present. The individual formalin preparations were then combined to create a pentavalent toxoid. In subsequent work, a monovalent toxoid was generated against serotype F (Hatheway, 1995; Montgomery et al., 2000). In this case, a greater effort was made to fractionate material, and thus there was greater enrichment of toxin in the mixture treated with formalin. Both the pentavalent and the monovalent preparations were adsorbed to aluminum hydroxide, and both formulations were designed for parenteral administration.

Neither the pentavalent nor the monovalent toxoid has progressed through the entire regulatory process needed to secure United States Food and Drug Administration approval. However, both have received Investigational New Drug status, and limited quantities of the pentavalent toxoid are distributed by the Centers for Disease Control to persons for whom immunity is appropriate (i.e., laboratory workers who handle toxin).

Relatively little work is now being done on traditional toxoid preparations (for example, Montgomery et al., 2000; Mahmut et al., 2002; Torii et al., 2002; Kobayashi et al., 2005). This diminished interest in chemical toxoids can be attributed entirely to advances in alternative technologies. Recombinant techniques can now be used to express a holotoxin that is devoid of the ability to poison cholinergic transmission (e.g., point mutations that abolish LC catalytic activity; Kiyatkin et al., 1997). In addition, polypeptide domains within the toxin have been identified that are capable of evoking a robust immune response (Byrne et al., 1998). This means that a vaccine can be prepared without ever having had active toxin as a precursor or as an ingredient.

Recombinant Toxoids

There are three lines of investigation that have been pursued in an attempt to identify a recombinant vaccine that will evoke a neutralizing antibody response: (1) expression of holotoxin in an inactive form, (2) expression of large polypeptides that represent functional domains within the toxin, and (3) synthesis of small polypeptides that could represent linear epitopes. In all cases, the principal goal has been to elicit a high circulating titer of neutralizing IgG antibody.

Inactive Holotoxin

There is an obvious advantage of using the entire toxin molecule as an antigen: every epitope in the native toxin will be present. Furthermore, when compared to a formalin-inactivated toxoid, a recombinant holotoxin can be generated that has little or no change in structure. Yet another advantage is that the holotoxin, unlike specific functional domains (e.g., LC) or small polypeptides, retains the ability to bind and penetrate epithelial barriers. Therefore, the modified holotoxin could be evaluated as an oral or inhalation vaccine.

There are several possible ways in which a holotoxin can be expressed as an inactive variant. The most well-documented of these is to perform site-directed mutagenesis to alter the histidine motif in the LC that is essential for catalytic activity. This was the technique that was used to generate and test the first recombinant holotoxin as a parenteral and nonparenteral vaccine. Kiyatkin et al. (1997) expressed a full-length serotype C holotoxin that had three amino acid mutations (H229 → G; E230 → T; H233 → N) The isolated and mutated LC was also expressed. When tested in vivo (mouse toxicity assay) and in vitro (murine phrenic nerve-hemidiaphragm bioassay), the mutated holotoxin was devoid of neuromuscular blocking activity. When tested in a broken cell assay (isolated and ruptured synaptosomes) neither the mutated holotoxin nor the mutated LC possessed catalytic activity.

When mice were immunized with the two recombinant polypeptides, the results were largely in keeping with expectations (Fig. 46.3). Intramuscular administration of

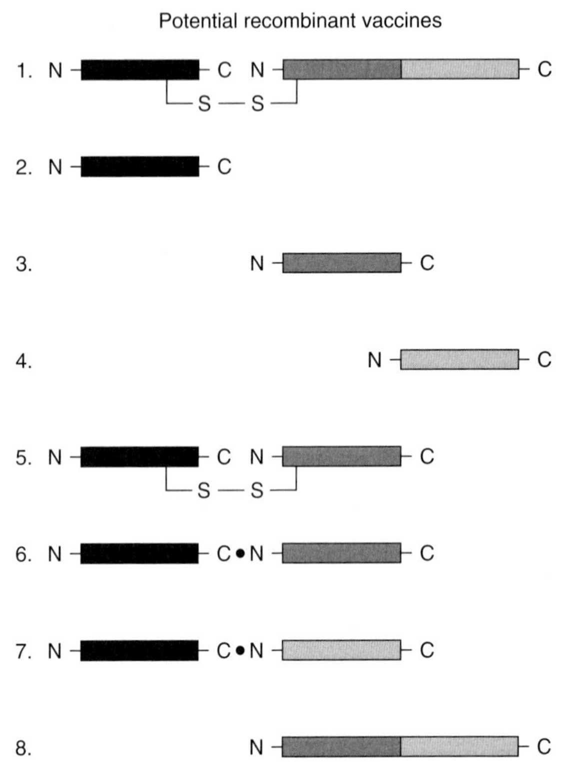

FIGURE 46.3 Potential recombinant vaccines. There are several different recombinant functional domains that are candidates as vaccines against botulinum toxin. These candidates represent the LC (black tone; 50,000 Da), the amino-terminal half of the heavy chain (HCN; grey tone; 50,000 Da), and the carboxy-terminal half of the heavy chain (HCC; intermediate tone; 50,000 Da), either individually or in combination, as follows: (1) holotoxin with point mutations to render it nontoxic, (2) the isolated LC, (3) the isolated HCN, (4) the isolated HCC, (5) a dipeptide containing LC linked to HCN, with a disulfide bridge that mimics that in the holotoxin (1), (6) a single-chain molecule in which LC and HCN are linked by an amide bond, (7) a single-chain molecule in which LC and HCC are linked by an amide bond, and (8) HC. The strategies of testing inactive holotoxin, individual functional domains, and combinations of functional domains have all been tested in laboratory animal experiments. To date, the only candidate that has reached human clinical trials is HCC (4).

either the holotoxin or the LC evoked a circulating titer of neutralizing IgG antibody and resistance to native toxin. Oral administration of the holotoxin similarly evoked an immune response, but oral administration of the LC did not evoke protection. The latter result could be attributed to the inability of the LC to bind and penetrate epithelial barriers.

Results such as these are promising, but there are issues that must be addressed, particularly regarding route of vaccine administration. If the goal is simply to create an injectable vaccine, a nontoxic holotoxin would be acceptable. If the goal is to create an oral vaccine—which is clearly preferable—there is a hurdle that must be overcome. When botulinum toxin is synthesized and released from clostridia, it is part of a noncovalent complex that contains other proteins, including hemagglutinins and a so-called nontoxin, nonhemagglutinin. These proteins intertwine in a way that produces remarkable resistance to the harsh conditions of low pH and proteolytic enzymes in the gut. When the toxin is stripped of this protective complex, it is less potent as an oral poison. It is, nonetheless, active by the oral route, and it is even more active when injected directly into the upper small intestine to bypass gastric metabolism (Maksymowych et al., 1999).

The concept of artificially wrapping a recombinant holotoxin with recombinant proteins may sound reasonable, but in practical terms it is not feasible. Fortunately, there is a relatively simple and straightforward alternative. The science of drug delivery has identified several techniques for protecting oral medications from metabolic degradation in the gut. Perhaps the best-known strategy involves encapsulation combined with enteric coating. This would appear to be an ideal approach for a botulinum vaccine. An appropriately designed delivery system could protect the molecule during its transit through the stomach, and the molecule itself could then bind and penetrate epithelial barriers in the intestine to reach the general circulation. This is an entirely feasible strategy that deserves serious consideration.

Functional Domains

The tactic of using recombinant functional domains as vaccines shows great promise, and it appears to be one that is en route to winning clinical approval. This approach could involve any one of six potential candidates: the isolated LC, the amino-terminal half of the heavy chain (HC_N), the carboxy-terminal half of the heavy chain (HC_C), a dimer of LC and HC_N, a dimer of LC and HC_C, or a dimer of HC_N and HC_C (i.e., intact heavy chain). Several of these candidates are being explored (Byrne et al., 1998, 2000; Byrne and Smith, 2000; Jensen et al., 2003; Kiyatkin et al., 1997; Park and Simpson, 2003; Yang et al., 2004), but greatest progress has been achieved with HC_C (Baldwin et al., 2005; Byrne et al., 1998, 2000; Byrne and Smith, 2000; Clayton et al., 1995; LaPenotiere et al., 1995). Vaccination of animals with this domain of serotypes A, B, C, D, E, and F has been shown to evoke protective antibody formation (circulating IgG) and resistance to the respective parent toxins.

The concept of using HC_C as a vaccine against botulinum toxin can be traced to earlier work on tetanus toxin. Botulinum toxin and tetanus toxin are both clostridial in origin; they have similar macrostructures (viz., functional domains); and they both act on nerve terminals to block transmitter release. Early work on the protein structure of tetanus toxin revealed that this molecule could be cleaved by trypsin to yield two main fragments: the LC still associated with the amino-terminal half of the heavy chain, and the remaining carboxy-terminal half of the heavy chain. The latter, which was called "fragment C", was shown to be an excellent immunogen (Fairweather et al., 1987; Helting, 1984; Makoff et al., 1989). These observations on cleavage products of tetanus toxin were the forerunners in efforts to use HC_C as a vaccine against botulinum toxin.

Work to demonstrate that the HC_C domain can be used as an injectable vaccine has been pioneered by investigators at the U.S. Army Medical Research Institute for Infectious Diseases (Byrne and Smith, 2000). In its initial iteration, this work focused on expression of the serotype A polypeptide in a traditional system (e.g., *Escherichia coli*). This approach worked, but it had notable limitations. Codon usage in clostridia is different from that in *E. coli*, and therefore expression efficiency of the HC_C polypeptide was only modest. Also, *E. coli* is generally not an adequate expression system for generation of the amounts of protein needed for eventual clinical testing and use. The first problem was addressed by creating a synthetic gene that was designed for optimization of HC_C expression in *E. coli* (Clayton et al., 1995). The second problem was solved by moving to an alternate and large-scale expression system (*Pichia pastoris*), while retaining the concept of using a synthetic gene that optimizes expression (Bouvier et al., 2003; Byrne et al., 1998; Potter et al., 2000).

The work on expression of the HC_C domains of serotypes A and B has progressed very nicely (Potter et al., 2000). Procedures for optimization of expression under GMP conditions have been achieved, and both an absence of toxicity and the presence of efficacy in preclinical models have been demonstrated. The U.S. Food and Drug Administration has granted

Investigational New Drug status to these products, and they are currently in clinical trials (Byrne and Smith, 2000; Smith et al., 2004).

Work similar to that with serotypes A and B is underway with other serotypes, including C and D (Arimitsu et al., 2004; Lee et al., 2007; Webb et al., 2007; Woodward et al., 2003), E (Smith et al., 2004), and F (Byrne et al., 2000; Holley et al., 2001; Johnson et al., 2003). There is reason to be optimistic that this approach could ultimately generate a polyvalent vaccine that combines HC_C domains of the various serotypes to produce multivalent resistance.

The discussion so far has focused on generation of an injectable vaccine against botulism. Interestingly, these same polypeptides can also be used as oral and inhalation vaccines. In a series of structure–function studies intended to localize the minimum essential domain that is needed for binding and penetration of human epithelial cells, Maksymowych and Simpson found that only a fraction of the holotoxin was necessary (Maksymowych and Simpson, 2004). Neither the LC nor the amino-terminal portion of the heavy chain was required. The isolated HC_C polypeptide possessed the same ability to penetrate human gut and airway epithelial cells as the parent holotoxin. The fact that HC_C retains the ability to cross epithelial barriers, combined with the fact that it can evoke a robust immune response, suggests that HC_C would be an excellent candidate for a mucosally administered vaccine (Lamm, 1997; Ravichandran et al., 2007).

Specific Epitopes

Ideally, the exact number, nature, and location of every epitope on each serotype of botulinum toxin would be mapped. This would be invaluable both to investigators trying to develop vaccines and to those trying to generate therapeutic antibodies. Sadly, this ideal is far from being realized. Of the seven serotypes, only one can be characterized as having been the subject of intense research (e.g., serotype A); and even for this serotype a complete epitope map is not yet available.

It is not clear whether the total number of epitopes in the intact botulinum toxin type A molecule is known. In the most detailed study quantifying the number of monoclonal (single chain variable fragment) binding sites, Chen et al. (1997) reported a total of 40 sites that were exposed in the intact molecule. An additional four sites were occult. Epitope mapping indicated that 22 of the exposed sites were in the carboxy-terminal portion of the heavy chain (receptor binding domain; HC_C), 3 were in the amino-terminal portion of the heavy chain (translocation domain; HC_N), and 15 were in the LC (catalytic domain).

Efforts to localize epitopes as potential vaccines have focused overwhelmingly on the heavy chain, and on the carboxy-terminal end of this chain in particular. This is simply a reflection of the fact that this portion of the molecule has long been known to be an efficient immunogen. Given that this is the only functional domain that has been advanced to the stage of initial clinical trials, attention will likely remain focused on this region of the molecule.

Two general strategies have been pursued in the quest to find clinically relevant epitopes (also known as putative vaccines). One approach has been to use the heavy chain or its isolated carboxy-terminal as an immunogen to raise monoclonal antibodies. Various techniques are then used to deduce or to determine the region or even the specific amino acid sequences that are antibody-binding sites. The second approach has been to synthesize relatively small (and sometimes overlapping) peptides on the premise that toxin epitopes are represented somewhere in this population of molecules. Each of the amino acid sequences is then tested to determine which will elicit antibodies, and preferably neutralizing antibodies. As far as linear epitopes are concerned, these two strategies should ultimately generate compatible outcomes. As far as conformational epitopes are concerned, the synthetic polypeptide approach will have limitations. This will certainly be true when conformational components arise from amino acids that are well separated in the primary sequence. The impact of this limitation cannot be stated with certainty because the relative number of linear and conformational epitopes is not known.

There are more than two decades of research in which investigators have generated monoclonal antibodies against inactive holotoxin or polypeptide fragments. Virtually none of the early work had vaccine development as a primary motive; instead, the typical goal was to use the antibodies as tools to map the structure and function of the toxin molecule (for a representative early paper, see Kozaki et al., 1986). More recently, several laboratories have sought to identify monoclonal antibodies that have significant neutralizing activity. This work has utilized toxoid preparations of holotoxins or specific functional domains, such as the HC_C domain, as antigens (for representative papers from different laboratories, see Amersdorfer et al., 1997; Brown et al., 1997; Kubota et al., 1997; Rosenberg et al., 1997; Wu et al., 2001; Yang et al., 2004). One encouraging point that emerges from all these and related studies is that it is possible to isolate antibodies that have appreciable neutralizing activity.

Unfortunately, the effort to translate monoclonal antibody binding sites into putative vaccine candidates has not met with great success. Thus, as an illustration, Wu et al. (2001) have identified monoclonal antibodies that have substantial neutralizing activity. Using phage display, they were able to identify a specific amino acid sequence which, when expressed, could bind to the antibodies. For technical reasons, it was not possible to use a small polypeptide representing the putative antibody-binding domain to evoke a meaningful immune response.

Bavari et al. (1998) used a somewhat different approach. Rather than generating an entire phage display library of polypeptides, they instead focused on two 25-mers that had been predicted to be antibody-binding sites based on conformational modeling studies that suggested the location of solvent-exposed loops. These synthetic polypeptides were indeed binding proteins for neutralizing antibodies, but when tested as immunogens only one of the peptides was active, and the level of induced resistance was modest.

This line of investigation sets the stage for the complementary approach, in which putative epitopes are the starting point. However, there is a cautionary note that must be inserted, especially in the context of using small polypeptides from HC_C as vaccines. Tavallaie et al. (2004) performed a comparison study in which they examined HC_C as well as each of the two ends of this polypeptide as vaccine candidates. They made the interesting discovery that neither half of the HC_C domain, nor the two used in combination, was equivalent to the intact HC_C as a vaccine. These results suggested that the interfacial region between the two halves of HC_C is critical to evoking a maximal protective response.

The effort to localize epitopes to specific and small linear sequences has been almost wholly the providence of one laboratory and its collaborators (Atassi and Oshima, 1999). The essence of this work has been to synthesize small, overlapping peptide sequences that reflect the linear sequence of the toxin. Most of this work has focused on HC_C (Atassi et al., 1996; Oshima et al., 1997, 1998), although HC_N has also been examined (Atassi and Dolimbek, 2004; Atassi et al., 2005). This work has contributed considerably to an understanding of the antibody-binding domains that are linear in nature. However, the investigators have set a high standard that, with respect to botulinum toxin, has not been entirely met. In earlier work on α-bungartoxin, Dolimbek and Atassi (1996) made the observation that a conjugate containing three synthetic peptides mimicking three domains within α-bungartoxin was a better vaccine than the holotoxin itself. A similar accomplishment has not been achieved with botulinum toxin, but if it were to be achieved, it would deservedly draw considerable attention.

Superantigens

It is inherent in the concept of serotypes that the different botulinum toxins are relatively or completely immunologically distinct. Generally speaking, investigators have accepted the premise that a vaccine must be polyvalent to afford protection against multiple serotypes. This could mean having to develop and subsequently mix seven monovalent preparations to achieve full spectrum protection. The issue becomes even more problematic when one considers subtypes. There is an emerging literature showing that individual serotypes can have multiple subtypes, these subtypes can have notable differences in amino acid sequences, and antibodies that efficiently neutralize one subtype may not be efficient in neutralizing others (Smith et al., 2005). The implications this has for vaccine development—and even more so for therapeutic antibody development—are obvious.

The existence of so many variants of botulinum toxin encourages the effort to re-examine and re-think the issue of distinctness. At a minimum, it would be desirable to identify shared epitopes across any given family of subtypes (e.g., serotype B subtypes); even better, it would be desirable to find common epitopes across all serotypes (e.g., A–G). This search is given added impetus by the two major lines of research summarized above. Both those who use antibodies to map epitopes and those who use putative epitopes to map antibody-binding sites agree on one fundamental premise. The evoking of antibodies, whether for active immunization or for passive immunization, is optimized with oligoclonal or polyclonal preparations. This simple and fundamental premise takes on considerable weight when discussing the possible existence of superantigens. It is not essential that each of the clonal members of a group be highly active. Even relatively weak antibodies can combine to produce a potent neutralizing outcome (Nowakowski et al., 2002). Thus, a superantigen does not have to evoke neutralization of all toxin variants, or for that matter even one variant, to be useful. If a superantigen can combine with a small number of other antigens to produce a highly protective response, it will have served an important purpose.

The effort to identify and localize superantigens for botulinum toxin cannot be described as a hotbed of current research! One might even say that the most important findings to date are relics from the past. For example, it has been known for decades that there is substantial cross-reactivity between serotypes C and D (Oguma et al., 1980). As another example, one often quoted paper that describes an antibody recognizing serotypes B, C, D, and E was published almost 20 years ago (Tsuzuki et al., 1988).

The tools for localizing putative superantigens are much more powerful now than they were in the recent past. In addition, immunologists are skilled at evoking immune responses to peptides and other agents that are ordinarily poor antigens. In the face of the growing multiplicity of botulinum toxin variants, the need to explore superantigens is urgent.

Mechanism of Vaccine Action

The key extracellular step in botulinum toxin action is association with peripheral cholinergic nerve endings. The carboxy-terminal portion of the heavy chain binds to membrane receptors, and this initiates the sequence of events that culminates in blockade of exocytosis. This is the same portion of the toxin molecule that is being vigorously examined as a vaccine candidate. Many laboratories have reported that this portion of the molecule possesses numerous epitopes. It has also been reported that antiserum raised against this domain can neutralize toxin action when tested in vivo, and it can delay onset of neuromuscular blockade when tested in vitro. This would seem to be a compelling argument that a vaccine possessing the HC_C domain leads to neutralization of toxin action by occluding the receptor-binding domain (Kozaki et al., 1989). Toxin that cannot bind to its target organ is left stranded until it is eliminated from the body.

The concept that an antibody can associate with the HC_C domain in a way that prevents subsequent toxin binding to vulnerable cells is absolutely sound. Furthermore, if the affinity constant governing the antibody–antigen reaction is respectable, and if the antibody titer following immunization is respectable, then it is a certainty that occlusion of the binding domain can delay or even prevent poisoning. However, these sound concepts should not be extrapolated to mean that occlusion of receptor binding is the only mechanism that underlies vaccine action. Indeed, it may not even be the most important action.

A recent series of studies on the systemic pharmacokinetics of botulinum toxin has revealed that there is another mechanism for antibody-induced protection, and this mechanism precedes that of protection at the neuromuscular junction. It has been reported that the circulating half-life for botulinum toxin in mouse and rat blood is several hours (Ravichandran et al., 2006). The majority of toxin in the circulation is in the free form, and as such it is available for distribution to other parts of the body, including the neuromuscular junction. The fate of toxin in the presence of a neutralizing titer of antitoxin is dramatically different. Within approximately 5 min, the majority of the toxin is cleared from the circulation. Enhanced clearance was due mainly to the accumulation of toxin in liver and spleen (Dodd et al., 2005; Ravichandran et al., 2007).

There are two implications that stem from these findings. Firstly, a botulinum toxin vaccine can provide multiple layers of protection. It initially removes a large fraction of the toxin from the circulation and thereby prevents its delivery to the neuromuscular junction. Of the fraction that does reach peripheral nerves, toxicity is antagonized by antibody occlusion of the binding domain. Secondly, it is the enhanced clearance that is quantitatively more important. When exposed to a sufficient titer of antitoxin antibodies, most of the toxin is removed from the circulation and thus never reaches its target organ. It is reassuring to know that a vaccine can evoke resistance by more than one mechanism, each of which by itself is quite effective (Ravichandran et al., 2007).

VARIATIONS ON A THEME

The preponderance of current vaccine research deals with one mode of administration (injection), one type of antigen (recombinant polypeptide), and one type of response (circulating IgG). Although this is an excellent starting point, it would not be appropriate to claim that this is the best approach that can be envisioned. There are a number of ways to improve an injectable vaccine that evokes a circulating IgG response. Furthermore, many of these improvements are within technical reach. Some of the more promising of these potential improvements are considered below.

Route of Administration

All agencies that have a major stake in vaccine development, such as the National Institutes of Health, the World Health Organization, and the Bill and Melinda Gates Foundation, have encouraged efforts to develop "needle-free" vaccines. The benefits that arise from oral, inhalation and other "needle-free" methods of administration are numerous, well known, and compelling (Mitragotri, 2005; Neutra and Kozlowski, 2006). These benefits would apply to a botulism vaccine as much as they would apply to any other vaccine.

One might argue that there is a biological imperative to developing an oral or inhalation vaccine against botulinum toxin. This statement is based on the fact that the toxin molecule itself has the inherent ability to bind and penetrate gut and airway epithelial barriers. Thus, it should not be surprising that a variant of the toxin that was engineered to lack catalytic activity (and therefore neurotoxicity) has been shown to be an oral vaccine (Kiyatkin et al., 1997). More recently, the native heavy chain (Park and Simpson, 2003) and the recombinant carboxy-terminal fragment (HC_C) have also been shown to be active when administered by mucosal routes (Ravichandran et al., 2007; Simpson et al., 2003).

The inhalation route might seem particularly desirable because the airway does not have the harsh conditions of low pH and proteolytic enzymes found in the stomach. This should mean that the auxiliary proteins that ordinarily protect the toxin from metabolism in the gut, or surrogates for these proteins such as enteric coatings, would be unnecessary. Even so, the advantages of not needing an enteric coating may be outweighed by other factors. It is well established that there are mechanisms by which xenobiotics delivered by the inhalation route can gain direct access to the brain. It is also well known that there are many common and not-so-common disorders of respiratory function that could mitigate attempts to deliver a vaccine by the inhalation route.

Perhaps the most prudent strategy at this point is to encourage research that would foster any viable "needle-free" vaccine. The development of an oral formulation that utilizes an enteric coating is not a particularly daunting challenge. Similarly, the design of experiments that could determine whether an inhalation formulation delivers foreign products to the brain is not difficult. A focused and well-conceived research plan could put a "needle-free" vaccine within reach rather quickly.

Source of Antigen

The ultimate product that a vaccine delivers to the immune system will be a polypeptide that mimics the parent toxin, but this does not mean that the polypeptide itself must be in the vaccine formulation. An alternative is to deliver the message that encodes the polypeptide. All three of the common approaches to delivering message have been reported. DNA vaccines have been tested for serotypes A (Clayton and Middlebrook, 2000; Shyu et al., 2000) and F (Bennett et al., 2003; Jathoul et al., 2004); a viral vector system has been tried with serotype A (Lee et al., 2001); and a bacterial expression system has been tested with serotype F (Foynes et al., 2003). In every instance, the goal was to elicit an immune response to the HC_C domain.

The initial efforts to create a DNA vaccine against serotype A were only partially successful (Clayton and Middlebrook, 2000; Shyu et al., 2000). Although a measurable titer of circulating IgG could be detected, the magnitude of the protective effect was relatively small. Subsequent efforts with a serotype F vaccine were more successful, and thus immunized animals were protected against challenge doses in the range of 10^4 MLD (Jathoul et al., 2004).

A propagation-deficient Venezuelan equine encephalitis virus vector was used to induce immunity to serotype A (Lee et al., 2001). This approach evoked a protective response against challenge doses of 10^5 MLD, and the protective immune response was still substantial 6 months and 1 year after immunization. Favorable results were also obtained with an attenuated *Salmonella enterica* var *typhimurium* expression system (Foynes et al., 2003). Induced immunity to serotype F afforded protection against challenge doses in the range of 10^4 MLD. This study did not monitor the duration of immunity.

The demonstration that high levels of protective immunity can be induced by viral and bacterial systems will no doubt fuel discussion about the relative merits of polypeptide vaccines versus nucleic acid vaccines. That discussion is beyond the scope of this review. On the other hand, it is worth noting that the two types of vaccines do have one feature in common. Depending on the formulation used for polypeptide vaccines, and depending on the virus or bacterium used for nucleic acid vaccines, both could be administered by the oral or inhalation routes.

Nature of the Immune Response

The concept of administering a polypeptide or a nucleic acid vector by a mucosal route has implications beyond those of eliminating the adverse consequences of using, storing, and disposing of needles and syringes. The most important of these is the prospect of evoking a local or universal secretory IgA response. At a minimum, IgA-mediated immunity at the mucosal level would provide an added layer of protection against botulinum toxin. Most cases of botulism are due to ingestion of toxin, or to ingestion of organisms that manufacture the toxin in the gut (see Box 46.1). Similarly, there is the anticipation that most cases of bioterrorism or biological warfare, should such incidents occur, would be due to oral or inhalation poisoning. These observations highlight the importance of an IgA-mediated protective barrier in the gut and airway.

In view of the potential importance of evoking a mucosal immune response, it is surprising that so little work has been published in this field (Fujihashi et al., 2007). One exception is a recent study that utilized toxoid rather than recombinant polypeptide, and added a mutant form of cholera toxin as an adjuvant (Kobayashi et al., 2005). This cocktail is not likely destined for clinical approval. Nevertheless, the work did demonstrate an awareness of evoking a mucosal immune response. Another example is a recent study in which recombinant HC_C was used as an antigen. This study describes a trivalent vaccine against botulinum toxin (serotypes A, B, and E), and it utilized adjuvants that have already been approved for use in human patients (Ravichandran et al., 2007).

Monovalent versus Polyvalent

There is a consensus among investigators and public health officials that there must be a vaccine that protects against serotypes A, B, and E. At some point, it may be desirable to have a vaccine against all seven serotypes. Regardless of the number of serotypes involved, the expectation is that a vaccine against any particular serotype will provide protection against all subtypes within that serotype.

All previous attempts to develop polyvalent vaccines have involved mixing of monovalent vaccines. There is nothing in the literature to suggest that current workers are approaching matters any differently. This is unfortunate because there may be a viable approach to creating a polyvalent vaccine within a single polypeptide (or nucleic acid encoding the peptide).

As discussed earlier, the botulinum toxin molecule binds and penetrates epithelial cells to enter the general circulation. Studies on the structure–function relationships that govern penetration of epithelial barriers have shown that the carboxy-terminal portion of the heavy chain is fully competent to achieve this task (Maksymowych and Simpson, 2004). Neither the amino-terminal portion of the heavy chain, which is needed for internalization by nerve cells, nor the LC, which acts enzymatically to block exocytosis, is needed. The fact that the HC_C domain can penetrate gut and airway barriers, combined with the fact that it can evoke a robust immune response, is the basis for its being evaluated as an oral or inhalation vaccine (Ravichandran et al., 2007).

There is yet another way to look at these data, and the examination will reveal a creative way for making polyvalent vaccines. The structure–function studies on transcytosis support a new way of envisioning the toxin molecule. The HC_C domain can be viewed as a "carrier," and its role is to transport the remainder of the molecule, which can be viewed as "cargo," across epithelial barriers (Simpson et al., 2003). This means that the 50 kDa HC_C polypeptide can act as a carrier to transport a 100 kDa polypeptide (the rest of the toxin). This view of the molecule prompts a logical question. If the naturally occurring cargo were to be removed from the carrier, would it be possible to substitute novel cargo? To put the question into terms more relevant to this discussion, can the carrier be used to transport heterologous antigens into the body?

There is already a substantial body of evidence to support the carrier–cargo concept. A host of in vitro and in vivo experiments have been conducted that confirm the wide breath of utility of the carrier polypeptide in transporting heterologous molecules across gut and airway epithelial barriers (Simpson et al., 2003). There is no reason why this concept cannot be applied to a polyvalent botulinum toxin vaccine. The carrier domain of one serotype could be used to transport the antigenic domain of another serotype (or two? or three?) across epithelial barriers. One obvious embodiment would be to link the HC_C domain of serotype A to the HC_C domain of serotype B to the HC_C domain of serotype E. The carriers at either end of this trivalent vaccine might even be oriented in a way to expose their respective binding domains. This would mean that a single polypeptide would have two epithelial binding domains working to transport a trivalent vaccine into the body. The author and his colleagues have already made a number of these constructs, which are undergoing biological testing.

A PROVOCATIVE QUESTION

From the Perspective of a Disease

Discussions about vaccine development seldom give serious consideration to the issue of an ideal duration of immunity. When these discussions deal with microbial pathogens, the unspoken premise is usually that an ideal duration would be "eternity." If a patient were to be vaccinated only once, and if this procedure were to evoke lifelong immunity, this would represent the Holy Grail.

It is certainly true that in many instances a sustained immune response would be desirable. Consider for example the tetanus vaccine, which is the most widely administered vaccine in the world. Current medical guidelines propose that boosters be given at 10-year intervals, and that these boosters be administered throughout life. A tetanus vaccine that could evoke lifelong immunity without the need for boosters might be described as ideal.

Tetanus toxin is only one of the pathogens for which a sustained immune response would be desirable. However, there are other pathogens for which this would not be true, and botulinum toxin happens to be one of them. If anything, a vaccine against botulinum toxin might be seen as a counterpoint to that against tetanus. Indeed, one might pose this provocative question: Would an ideal vaccine against botulinum toxin have a short duration rather than a long duration of action?

The answer to this question may emerge from a close examination of the role of the toxin as an agent that causes disease. There are three ways to view the disease, and each is related to a somewhat different population. To begin with, botulism can be a naturally occurring disease, and as such it can affect the entire population. It can also be an unnatural disease—a product of malice—in incidents of bioterrorism and biological warfare. Bioterrorism, by definition, applies to the civilian segment of the total population, whereas biological warfare applies to the military segment of this population.

Naturally occurring botulism has a very low incidence, especially in countries with a well-developed food processing industry. The incidence is so low that there would never be a justification for universal vaccination. Only a tiny fraction of the total population will actually contract the disease, or be at significant risk. It is noteworthy that among those who actually contract the disease, the largest proportion is infants who are only several months to about one year in age (Arnon, 1980; Ferrari and Weisse, 1995; Wigginton and Thill, 1993). It would be meaningless to describe this as a candidate population for lifelong immunization. Among those who are at significant risk, one could cite laboratory workers who handle the toxin. Conceivably there are investigators who could benefit from lifelong immunity, but their numbers are truly small.

Another way to view the disease is as an unnatural outcome that results from bioterrorism. This view is certainly warranted, but at the same time one must note that there has yet to be an incident of bioterrorism for which there was significant morbidity and mortality due to botulinum toxin. This has led many to conclude that, until there is a tragic event, there will be neither policy advocacy for, nor public acceptance of, the concept of universal vaccination. Even if there were to be an incident, it would likely have to be one of substantial proportions before the idea of universal vaccination would receive serious consideration. And even if the level of concern and alarm were to rise to high levels, a vaccine of limited duration of action would be more palatable to the general population than one of lengthy duration of action (see below).

With bioterrorism, as with naturally occurring botulism, the proportion of the population that one could identify as reasonable candidates for lifelong immunity is minuscule. For example, one might consider vaccinating those healthcare workers who would be called upon to handle environmental or patient specimens contaminated with toxin. The needs of this population are unquestionably legitimate, but their numbers are exceedingly small.

With biological warfare there similarly has never been an incident for which there was significant morbidity and mortality due to botulinum toxin. Nevertheless, a respectable proportion of the military population could be at risk in times of conflict, especially if that conflict is in an area where there is evidence that hostile forces possess the toxin. In instances like this, the proportion of the military that is deployed to areas of risk may be a candidate for vaccination. Current events in the Middle East can be taken as a case in point, and it is one that illustrates some key facts. In times of conflict, and even when biological warfare may be a concern, there has never been universal vaccination within the military. For those forces that have been deployed and may have been candidates for vaccination, the average length of deployment is less than 36 months. This duration of deployment is not a strong argument for long-term or lifelong immunity.

A reasonable assessment of the risk of disease, the likely duration of risk, and the likely numbers of persons at risk, does not justify the need for a vaccine that has a lengthy duration of action. A more thoughtful assessment is that the true need is for a vaccine that: (a) evokes a substantial immune response, (b) has a limited duration of action, and (c) can be easily administered, or re-administered, for episodic use.

From the Perspective of a Therapeutic Agent

The absence of justification for a long-acting vaccine is not the same thing as the presence of justification for a short-acting vaccine. Therefore, it is appropriate to ask whether an argument can be mounted that favors actively seeking a vaccine with a limited duration of action. Among those in the botulinum toxin field, such an argument is both well known and widely accepted.

In addition to being an agent that can cause disease, botulinum toxin is also an agent that can relieve disease (Comella et al., 2005). The toxin was first introduced for neuroophthalmologic disorders in which there is excessive and involuntary activity in efferent nerves ("dystonia"). The toxin has since been tested in a truly remarkable array of disorders, ranging from achalasia to migraine headaches to bladder dysfunction. For many patients who have these disorders, botulinum toxin is the only agent that provides a reasonable measure of relief.

For most of the disorders in which it is used as a medication, the toxin has an underlying mechanism of action that is similar to that described earlier. The toxin acts on target nerve endings to block exocytosis. The therapeutic goal of blocking transmitter release is achieved by injecting toxin locally into the vicinity of overactive nerves. Ideally, the toxin binds and enters nerve endings rather than diffusing away. However, there are patients who receive relatively large doses and/or frequent injections, and a fraction of this toxin does diffuse away, sometimes to evoke a systemic immune response. In essence, this is an unintended form of vaccination.

Clinical experience with patients receiving the toxin for its therapeutic benefit makes clear that an immune response can mitigate a therapeutic response. By extension, one can surmise that patients who are intentionally vaccinated would no longer enjoy the medicinal value of the toxin. Therein lies a quandary. Vaccines would be helpful in the context of bioterrorism or biological warfare, but there has never been an incident that has produced significant morbidity and mortality among any segment of the population. By contrast, vaccines would be detrimental to patients who need the toxin as a therapeutic agent, and the number of such patients among the civilian and military population is steadily growing. Furthermore, the disorders for which the toxin can be beneficial tend to be lengthy or lifelong in duration.

One can easily deduce how an imprudent vaccine policy could have unfortunate consequences. If a significant segment of the population were to receive a long-acting vaccine, sooner or later some fraction of this population would develop disorders for which botulinum toxin would ordinarily be administered therapeutically. These patients would be resistant to the toxin, and therefore unable to obtain satisfactory medical treatment. This creates a troubling risk-to-benefit analysis. One must gauge whether the potential benefit of vaccinating any given population against a biothreat can outweigh the authentic risk of depriving some in this population of a proven medicinal agent.

There is no way to know in advance which members of the population may face incidents of bioterrorism and biological warfare, or to know which will develop dystonia or related disorders. This means that any approach to dealing with the quandary will be less than satisfactory. Nevertheless, there are ways in which one could approach this matter thoughtfully. Vaccines that are intended as medical countermeasures against bioterrorism or biological warfare with botulinum toxin should possess two properties. First, the antigen, the formulation, and the route of administration should be designed in a way that is intended to achieve a robust magnitude but limited duration of action. A duration of immunity of 10 years, as is seen with the tetanus vaccine, is too long. Something closer to one-quarter or one-third of this may be more rational. Secondly, a booster should be designed that is so easy to dispense that re-establishing immunity for those who require long-term resistance could be easily achieved. An oral formulation might be exactly the right thing for this purpose.

LINKING THEORY AND REALITY

The goal of this review is to extract from the basic science and clinical science literature those findings that could be most helpful in terms of designing an ideal vaccination paradigm. The expectation is not that an ideal paradigm has already been determined. To the contrary, the expectation is that a careful survey of the literature will help to clarify the distance between current vaccine procedures and a putative ideal procedure.

Perhaps the most instructive way to proceed is to propose three broad conclusions that reflect the current state of the literature. These conclusions relate to selection of an antigen, selection of a route of administration, and determination of an acceptable duration of action. In addition, these conclusions embody, either directly or indirectly, all of the implications for vaccine development that were enumerated earlier in this review.

1. The vaccine candidate that has drawn greatest attention as a possible antigen is the HC_C domain of botulinum toxin. This polypeptide may not be the best possible candidate one can envision. However, efforts to improve upon it may approach the realm of diminishing returns.
2. The routes of administration that have been the focus of most research are intramuscular and subcutaneous injection. Clearly, this is not ideal. Other routes of administration are superior, both in practical terms and in clinical terms.
3. The duration of vaccine action that has been sought, at least implicitly, is one that is lengthy and may approximate that of the tetanus vaccine. Once again, this is not ideal. Although the concept of seeking a vaccine with a limited duration of action may seem counter-intuitive, there are compelling reasons to pursue this course of action.

An examination of the literature on the structure of the botulinum toxin molecule, as well as the history and clinical presentation of the disease botulism, provides ample evidence to support these three conclusions. This examination also suggests that there are rational ways in which investigators can translate the first and second conclusions into authentic vaccine candidates. Additional information will be needed to translate the third conclusion into practice.

Selection of Antigen

There are relatively sound bases for identifying vaccine candidates that would be desirable, those that would be of questionable value, and those that are not worth pursuing. Beginning with the latter, there is no obvious reason to pursue an attenuated vaccine. Although it is true that most cases of naturally occurring botulism are due to primary infection, the actual number of cases is so small that there is no impetus to develop a vaccine against the organism. In addition, primary intoxication is far more likely than primary infection to be the mode of any acts of bioterrorism or biological warfare, so again there is little impetus for seeking an attenuated vaccine.

In the category of vaccines that are of questionable value, perhaps the most obvious example would be the one aimed at auxiliary proteins in the progenitor toxin complex. Vaccination with these polypeptides has not yielded an efficacy equivalent to that which is customary with vaccines directed against the neurotoxin or its subunits (Mahmut et al., 2002). This is unfortunate because there appears to be conceptual merit to this approach. If the progenitor complex remains intact in the stomach and intestine, then one would expect a secretory IgA response to produce at least one antibody clone that could attach to the complex and block its absorption. A phenomenon such as this would produce the most desirable form of resistance—blockade of toxin entry into the general circulation—and it would have a substantial secondary benefit. An immune response directed against auxiliary proteins would not abolish the therapeutic potential of botulinum toxin. The importance of this secondary benefit suggests that a renewed effort to assess mucosal immunity against auxiliary proteins may be in order.

In the category of vaccines that are desirable, the neurotoxin itself is the prime candidate. The only question is whether to use the holotoxin (viz., mutated holotoxin that does not express catalytic activity) or a polypeptide domain (viz., truncation mutant). The principal argument in favor of the holotoxin is that it possesses all the epitopes. However, it should be noted that the toxin has a large number of epitopes, and most of these are in one domain (HC_C). The number of epitopes in this polypeptide (ca. 20) is so large that a robust, polyclonal antiserum is virtually assured. It would, of course, be possible to use a larger polypeptide, but it is not clear that the presence of additional epitopes would lead to an increase in resistance that could be measured in clinical terms. This conclusion is based on the observation that antibodies directed against the HC_C domain provide all three of the major categories of resistance (i.e., blockade of absorption, enhanced clearance from the circulation, and inhibition of neuronal binding; Ravichandran et al., 2007).

For any given HC_C domain, there is a requirement that it elicit immunity against all subtypes relevant to that domain. For example, the serotype A HC_C should evoke immunity against all known subtypes of A. Little work has been done to assess cross-neutralization within subtypes, so it is too early to know whether this expectation will always be realistic (Baldwin et al., 2005; Smith et al., 2005).

Another issue relates to multivalency. An ideal vaccine candidate would evoke immunity to the seven serotypes, but a more pressing need is for a vaccine that will confer resistance against the three serotypes most commonly implicated in human illness (A, B, and E). A trivalent vaccine could arise from the admixture of three monovalent vaccines. Alternatively, it may be desirable to explore a single chimeric antigen that is the fusion product of multiple HC_C domains.

One final matter relates to the actual nature of the vaccine. Stated as a question, should the preferred embodiment be a polypeptide vaccine or a nucleic acid vaccine? The literature on botulinum toxin vaccines is not sufficient to support strong conclusions, although some investigators may be tempted to extrapolate from the literature on other vaccines. This sort of extrapolation could be misleading or even wrong. One must recall that there is a relatively unique feature of a rational botulinum toxin vaccine. The ideal candidate should evoke strong immunity, but of a limited duration. Unless and until polypeptide vaccines and nucleic acid vaccines against botulinum toxin are compared in all meaningful ways, including duration of action, the effort to anoint one or the other is premature.

Choosing a Route of Administration

The preponderance of research on botulinum toxin vaccines has focused on intramuscular or subcutaneous administration. In other words, research in this field has been much the same as that for most other vaccines.

However, there are sound reasons to abandon the convenience of the status quo. Three of these arguments are especially compelling.

The Modern Era of Vaccine Development

Major healthcare entities that have a stake in vaccine development are virtually unanimous in their advocacy that most injection vaccines should be abandoned in favor of noninjection ("needle-free") vaccines. The reasons cited for this are numerous, with the more prominent being that needle-free vaccines: (a) reduce the need for, and thus the cost of, healthcare personnel; (b) eliminate the possibility of accidental needle sticks, and thus the prospect for accidental transmission of blood-borne diseases; (c) reduce the need to store, monitor, and dispose off needle-related medical waste; and (d) eliminate the possibility of needle-related environmental contamination. There is no reason why a botulinum toxin vaccine should be exempt from these considerations.

Rational Vaccine Design

If a bioterrorism or biowarfare incident were to occur, it would likely involve oral or inhalation exposure. One of the first steps in botulinum toxin action is association with receptors on the apical surface of gut and airway epithelial cells. This is an essential prelude to toxin absorption.

The fact that the toxin must bind and be transported across epithelial barriers makes clear that a needle-free (viz., mucosal) vaccine would be superior to an injection vaccine. The former has the potential to evoke a secretory IgA response that could greatly diminish or even abolish absorption of toxin. If an oral or inhalation formulation could evoke not only a local IgA response but also a circulating IgG response, then the superiority of the needle-free vaccine over the injection vaccine would be beyond dispute.

Logistic Responses to a Biothreat Agent

If the need should ever arise to vaccinate large numbers of the civilian and/or military populations, the difficulty of having patients return for one or more booster injections can pose a logistic challenge. It would be far easier to dispense oral or inhalation formulations in the proper number of doses (viz., blister pack) to achieve the intended outcome.

Duration of Action

The reasons for seeking a vaccine with a limited duration of action have been presented in detail. In that earlier discussion it was not possible to cite a body of work on botulism vaccines that addresses the issue of duration because no such body of work exists.

When one considers the issue of duration, there are two broad categories of concern: the immune system and the antigen. The goal here is to focus on antigen, in three senses: (a) as an agent to evoke initial immunization; (b) as an agent to extend effective immunization; and (c) as an agent that can evoke or extend unwanted immunization.

When evoking an immune response to botulinum toxin, the magnitude of that response and the duration of that response are influenced by the amount and frequency of antigen (and adjuvant) administration. To the extent that there is literature to support these relationships, it does not arise from the field of immunology. Instead, it is a by-product of observations on patients who have been treated with the toxin as a therapeutic agent. This patient literature also contains the finding that, within a finite amount of time, immunity to the toxin wanes.

There is now a need for investigators who are interested in vaccine development to revisit the concept of "optimization." Rather than seeking formulations and dosing regimens that maximize the duration of an initial immunization, the aim should be to produce a robust response of limited duration. This aim is tightly linked to the concept of extending immunization. A vaccination regimen that evokes a limited response will be more acceptable, especially to those who need long-term protection, if there is a simple approach to extending duration. If, for example, there were an oral booster, this would carry two significant benefits. In mechanistic terms, it could evoke mucosal immunity that would serve as a barrier to toxin absorption. In patient compliance terms, it would afford a simple and painless way to extend immunity.

Finally, there is the issue of immunization as an iatrogenic problem. The unintended evoking of immunity as a byproduct of toxin therapy is well documented. The unintended boosting of an earlier and intentional vaccination has not been described, but this may simply be a reflection of the relatively small number of persons in whom such a problem could emerge. If this population were to be significantly larger—which could occur if there were an authentic bioterrorist incident—the prospects for iatrogenic boosting would increase.

To conclude on a positive note, the potential for iatrogenic boosting of an immune response may be diminishing at this very point in time. Botulinum toxin that is given as a therapeutic agent has historically been an impure preparation. The toxin has been administered with auxiliary proteins that are presumed to have an adjuvant-like effect, and there are other proteins that are also present. In the recent past, clinical trials have been initiated with neurotoxin preparations that have fewer impurities, and initial indications are that this material is less likely to evoke unwanted immunity (Jankovic et al., 2003). Thus, local administration of small amounts of pure toxin for therapeutic purposes may have diminished ability to evoke immunity or to boost prior immunity.

KEY ISSUES

- Botulinum toxin, which is produced by the organisms *C. botulinum*, *C. beratii*, and *C. butyricum*, is the agent that causes the disease botulism.
- Naturally occurring botulism is relatively rare, and it does not warrant widespread administration of a vaccine.
- Botulism that results from acts of bioterrorism or biological warfare could affect significant numbers of people. Therefore, a vaccine may be needed to protect vulnerable populations.
- Botulinum toxin is also a therapeutic agent that is used in a wide array of clinical disorders, some of which can be debilitating.
- A vaccine against botulism would protect patients against bioterrorism and biological warfare caused by the toxin, but it would also deprive patients of the therapeutic benefits provided by the toxin.
- There currently exists an experimental vaccine that was generated by formalin inactivation of a bacterial preparation enriched in toxin.
- Efforts are underway to develop a new generation of vaccines that utilize nontoxic recombinant polypeptides derived from the toxin.
- An ideal recombinant vaccine would be one that can be administered by the oral or inhalation route.
- A mucosal vaccine should evoke antibodies that provide at least three layers of protection: (a) in the lumen of the airway or gut to block toxin absorption, (b) in the general circulation to enhance clearance, and (c) at nerve endings to block toxin binding.
- Various recombinant polypeptides are being evaluated as potential vaccines, but none has yet been approved by the Food and Drug Administration.

ACKNOWLEDGMENTS

This work was supported in part by the National Institutes of Health contract N01-AI30028, National Institutes of Health grants GM57342 and AI065967, and by a sponsored research agreement with DOR Biopharma.

References

Adler, M., Keller, J.E., Sheridan, R.E. and Deshpande, S.S. Persistence of botulinum neurotoxin A demonstrated by sequential administration of serotypes A and E in rat EDL muscle. Toxicon 2001; 39:233–243.

Agarwal, R., Binz, T. and Swaminathan, S. Structural analysis of botulinum neurotoxin serotype F light chain: implications on substrate binding and inhibitor design. Biochemistry 2005; 44:11758–11765.

Agarwal, R., Eswaramoorthy, S., Kumaran, D., Binz, T. and Swaminathan, S. Structural analysis of botulinum neurotoxin type E catalytic domain and its mutant Glu212→Gln reveals the pivotal role of the Glu212 carboxylate in the catalytic pathway. Biochemistry 2004; 43:6637–6644.

Ahsan, C.R., Hajnoczky, G., Maksymowych, A.B. and Simpson, L.L. Visualization of binding and transcytosis of botulinum toxin by human intestinal epithelial cells. J. Pharmacol. Exp. Ther. 2005; 315:1–8.

Amersdorfer, P., Wong, C., Chen, S., Smith, T., Deshpande, S., Sheridan, R. and Marks, J.D. Molecular characterization of murine humoral immune response to botulinum neurotoxin type A binding domain as assessed by using phage antibody libraries. Infect. Immun. 1997; 65:3743–3752.

Anderson, J.H. and Lewis, G.E. Clinical evaluation of botulinum toxoids. In: Biomedical Aspects of Botulism (G.E. Lewis, Jr. and P.S. Angel, Eds.). New York: Academic Press, 1981.

Arimitsu, H., Lee, J.C., Sakaguchi, Y., Hayakawa, Y., Hayashi, M., Nakaura, M., Takai, H., Lin, S.N., Mukamoto, M., Murphy, T. and Oguma, K. Vaccination with recombinant whole heavy chain fragments of *Clostridium botulinum* type C and D neurotoxins. Clin. Diag. Lab. Immunol. 2004; 11:496–502.

Arndt, J.W., Jacobson, M.J., Abola, E.E., Forsyth, C.M., Tepp, W. H., Marks, J.D., Johnson, E.A. and Stevens, R.C. A structural perspective of the sequence variability within botulinum neurotoxin subtypes a1–a4. J. Mol. Biol. 2006; 362:733–742.

Arndt, J.W., Yu, W., Bi, F. and Stevens, R.C. Crystal structure of botulinum neurotoxin type G light chain: serotype divergence in substrate recognition. Biochemistry 2005; 44:9574–9580.

Arnon, S.S. Infant botulism. Ann. Rev. Med. 1980; 31:541–560.

Arnon, S.S. Creation and development of the public service orphan drug human botulism immune globulin. Pediatrics 2007; 119:785–789.

Arnon, S.S., Schechter, R. and Inglesby, T.V. Botulinum toxin as a biological weapon: medical and public health management. JAMA 2001; 285:1059–1070.

Arnon, S.S., Schechter, R., Maslanka, S.E., Jewell, N.P. and Hatheway, C.L. Human botulism immune globulin for the treatment of infant botulism. N. Engl. J. Med. 2006; 354:462–471.

Atassi, M.Z. and Dolimbek, B.Z. Mapping of the antibody-binding regions on the H[N] domain (residues 449–859) of botulinum neurotoxin A with antitoxin antibodies from four host species: full profile of the continuous antigenic regions of the H-chain of botulinum neurotoxin A. Protein J. 2004; 23:39–52.

Atassi, M.Z., Dolimbek, B.Z., Hayakari, M., Middlebrook, J.L., Whitney, B. and Oshima, M. Mapping of the antibody-binding regions on botulinum neurotoxin H-chain domain 855–1296 with antitoxin antibodies from three host species. J. Protein Chem. 1996; 15:691–700.

Atassi, M.Z., Dolimbek, G.S., Deitiker, P.R., Aoki, K.R. and Dolimbek, B.Z. Submolecular recognitions profiles in two mouse strains of non-protective antibodies against botulinum neurotoxin A. Mol. Immunol. 2005; 42:1509–1520.

Atassi, M.Z. and Oshima, M. Structure, activity, and immune (T and B cell) recognition of botulinum neurotoxins. Crit. Rev. Microbiol. 1999; 19:219–260.

Baldwin, M.R., Tepp, W.H., Pier, C.L., Bradshaw, M., Ho, M., Wilson, B.A., Fritz, R.B., Johnson, E.A. and Barbieri, J.T. Characterization of the antibody response to the receptor binding domain of botulinum neurotoxin serotypes A and E. Infect. Immun. 2005; 73:6998–7005.

Bavari, S.B., Pless, D.D., Torres, E.R., Lebeda, F.J. and Olson, M.A. Identifying the principal protective antigenic determinants of type A botulinum neurotoxin. Vaccine 1998; 16:1850–1856.

Bennett, A.M., Perkins, S.D. and Holley, J.L. DNA vaccination protects against botulinum neurotoxin type F. Vaccine 2003; 21:3110–3117.

Bouvier, A., Chapline, J., Boerner, R., Jeyarajah, S., Cook, S., Acharya, P.S., Henderson, I., Schrimsher, J.L. and Shepard, S.R. Identifying and modulating disulfide formation in the biopharmaceutical production of a recombinant protein vaccine candidate. J. Biotechnol. 2003; 103:257–271.

Brown, D.R., Lloyd, D.P. and Schmidt, J.J. Identification and characterization of a neutralizing monoclonal antibody against botulinum neurotoxin serotype F, following vaccination with active toxin. Hybridoma 1997; 16:447–456.

Bullens, R.W.M., O'Hanlon, G.M., Wagner, E., Molenaar, P.C. and Furukawa, K. Complex gangliosides at the neuromuscular junction are membrane receptors for autoantibodies and botulinum neurotoxin but redundant for normal synaptic function. J. Neurosci. 2002; 22:6876–6884.

Burgen, A.S.V., Dickens, J. and Zatman, L.J. The action of botulinum toxin on the neuro-muscular junction. J. Physiol. 1949; 109:10–24.

Byrne, M.P. and Smith, L.A. Development of vaccines for prevention of botulism. Biochimie 2000; 82:955–966.

Byrne, M.P., Smith, T.J., Montgomery, V.A. and Smith, L.A. Purification, potency, and efficacy of the botulinum neurotoxin type A binding domain from *Pichia pastoris* as a recombinant vaccine candidate. Infect. Immun. 1998; 66:4817–4822.

Byrne, M.P., Titball, R.W., Holley, J. and Smith, L.A. Fermentation, purification, and efficacy of a recombinant vaccine candidate against botulinum neurotoxin type F from *Pichia pastoris*. Protein Expr. Purif. 2000; 18:327–337.

Caya, J.C., Agni, R. and Miller, J.E. *Clostridium botulinum* and the clinical laboratorian: a detailed review of botulism, including biological warfare ramifications of botulinum toxin. Arch. Pathol. Lab. Med. 2004; 128:653–662.

Chai, Q., Arndt, J.W., Dong, M., Tepp, W.H., Johnson, E.A., Chapman, E.R. and Stevens, R.C. Structural basis of cell surface receptor recognition by botulinum neurotoxin B. Nature 2006; 444:1096–1100.

Chen, F., Kuziemko, G.M., Amersdorfer, P., Wong, C., Marks, J.D. and Stevens, R.C. Antibody mapping to domains of botulinum neurotoxin serotype A in the complexed and uncomplexed forms. Infect. Immun. 1997; 65:1626–1630.

Chen, F., Kuziemko, G.M. and Stevens, R.C. Biophysical characterization of the stability of the 150-kilodalton botulinum toxin, the nontoxic component, and the 900-kilodalton botulinum toxin complex species. Infect. Immun. 1998; 66:2420–2425.

Christopher, G.W., Cieslak, T.J., Pavlin, J.A. and Eitzen, E.M. Biological warfare: a historical perspective. JAMA 1997; 278:412–417.

Clayton, J. and Middlebrook, J.L. Vaccination of mice with DNA encoding a large fragment of botulinum neurotoxin serotype A. Vaccine 2000; 18:1855–1862.

Clayton, M.A., Clayton, J.M., Brown, D.R. and Middlebrook, J.L. Protective vaccination with a recombinant fragment of *Clostridium botulinum* neurotoxin serotype A expressed from a synthetic gene in *Escherichia coli*. Infect. Immun. 1995; 63:2738–2742.

Coffield, J.A., Bakry, N.M., Zhang, R.D., Carlson, J., Gomella, L.G. and Simpson, L.L. In vitro characterization of botulinum toxin types A, C, and D action on human tissues: combined electrophysiologic, pharmacologic and molecular biologic approaches. J. Pharmacol. Exp. Ther. 1997; 280:1489–1498.

Comella, C.L., Simpson, L.L. and Jankovic, J. Botulinum toxins: transformation of a toxin into a treatment. In: From Neuroscience to Neurology (S. Waxman, Ed.). Amsterdam: Elsevier Academic Press, 2005.

DasGupta, B.R. The structure of botulinum neurotoxin. In: L.L. Simpson (Ed.), Botulinum Neurotoxin and Tetanus Toxin. San Diego: Academic Press, 1989.

Dickerson, T.J. and Janda, K.D. The use of small molecules to investigate molecular mechanisms and therapeutic targets for treatment of botulinum neurotoxin A intoxication. ACS Chem. Biol. 2006; 1:359–369.

Dodd, S.L., Selsby, J., Payne, A., Judge, A. and Dott, C. Botulinum neurotoxin type A causes shifts in myosin heavy chain composition in muscle. Toxicon 2005; 46:196–203.

Dolimbek, B.Z. and Atassi, M.Z. Protection against α-bungartoxin poisoning by immunization with synthetic toxin peptides. Mol. Immunol. 1996; 33:681–689.

Dong, M., Richards, D.A., Goodnough, M.C., Tepp, W.H., Johnson, E.A. and Chapman, E.R. Synaptotagmins I and II mediate entry of botulinum neurotoxin B into cells. J. Cell Biol. 2003; 162:1293–1303.

Dong, M., Yeh, F., Tepp, W.H., Dean, C., Johnson, E.A., Janz, R. and Chapman, E.R. SV2 is the protein receptor for botulinum neurotoxin A. Science 2006; 312:592–596.

East, A.K., Bhandari, M., Stacey, J.M., Campbell, K.D. and Collins, M.D. Organization and phylogenetic interrelationships of genes encoding components of the botulinum toxin complex in proteolytic *Clostridium botulinum* types A, B and F: evidence of chimeric sequences in the gene encoding the nontoxic nonhemagglutinin component. Int. J. Syst. Bacteriol. 1996; 46:1105–1112.

Eleopra, R., Tugnoli, V., Rossetto, O., De Grandis, D. and Montecucco, C. Different time courses of recovery after poisoning with botulinum neurotoxin serotypes A and E in humans. Neurosci. Lett. 1998; 256:135–138.

Fairweather, N.F., Lyness, V.A. and Maskell, D.J. Immunization of mice against tetanus with fragments of tetanus toxin synthesized in *Escherichia coli*. Infect. Immun. 1987; 55:2541–2545.

Ferrari, N.D. and Weisse, M.E. Botulism. Adv. Pediatr. Infect. Dis. 1995; 10:81–91.

Foran, P.G., Mohammed, N., Lisk, G.O., Nagwaney, S., Lawrence, G.W., Johnson, E., Smith, L., Aoki, K.R. and Dolly, J.O. Evaluation of the therapeutic usefulness of botulinum neurotoxin B, C1, E, and F compared with the long lasting type A. J. Biol. Chem. 2003; 278:1363–1371.

Foynes, S., Holley, J.L., Garmory, H.S., Titball, R.W. and Fairweather, N.F. Vaccination against type F botulinum toxin using attenuated salmonella enterica var typhimurium strains expressing the BoNT/F H(C) fragment. Vaccine 2003; 21:1052–1059.

Fujihashi, K., Staats, H.F., Kozaki, S. and Pascual, D.W. Mucosal vaccine development for botulinum intoxication. Expert Rev. Vacc. 2007; 6:35–45.

Gill, D.M. Bacterial toxins: a table of lethal amounts. Microbiol. Rev. 1982; 46:86–94.

Gupta, A., Sumner, C.J., Castor, M., Maslanka, S. and Sobel, J. Adult botulism type F in the United States, 1981–2002. Neurology 2005; 65:1694–1700.

Hatheway, C.L. Botulism: the present status of the disease. Curr. Top. Microbiol. Immunol. 1995; 195:55–75.

Helting, T.B.N. and Nau, H.H. Analysis of the immune response to papain digestion products of tetanus toxin. Acta Path. Microbiol. Immunol. Scand. 1984; 92:59–63.

Hoch, D.H., Romero-Mira, M., Ehrlich, B.E., Finkelstein, A., DasGupta, B.R. and Simpson, L.L. Channels formed by botulinum, tetanus, and diphtheria toxins in planar lipid bilayers: relevance to translocation of proteins across membranes. Proc. Natl. Acad. Sci. 1985; 82:1692–1696.

Holley, J.L., Elmore, M., Mauchline, M., Minton, N. and Titball, R. W. Cloning, expression and evaluation of a recombinant sub-unit vaccine against *Clostridium botulinum* type F toxin. Vaccine 2001; 19:288–297.

Holzer, V.E. Botulismus durch inhalation. Med. Klin. 1962; 57:1735–1738.

Horowitz, B.Z. Botulinum toxin. Crit. Care Clin. 2005; 21:825–839.

Humeau, Y., Doussau, F., Grant, N.J. and Poulain, B. How botulinum and tetanus neurotoxins block neurotransmitter release. Biochimie 2000; 82:427–446.

Jankovic, J., Vuong, K.D. and Ahsan, J. Comparison of efficacy and immunogenicity of original versus current botulinum toxin in cervical dystonia. Neurology 2003; 60:1186–1188.

Jathoul, A.P., Holley, J.L. and Garmory, H.S. Efficacy of DNA vaccines expressing the type F botulinum toxin hc fragment using different promoters. Vaccine 2004; 22:3942–3946.

Jensen, M.J., Smith, T.J., Ahmed, S.A. and Smith, L.A. Expression, purification, and efficacy of the type A botulinum neurotoxin catalytic domain fused to two translocation domain variants. Toxicon 2003; 41:691–701.

Jin, R., Rummel, A., Binz, T. and Brunger, A.T. Botulinum neurotoxin B recognizes its protein receptor with high affinity and specificity. Nature 2006; 444:1092–1095.

Johnson, S.K., Zhang, W., Smith, L.A., Hywood-Potter, K.J., Todd Swanson, S., Schlegel, V.L. and Meagher, M.M. Scale-up of the fermentation and purification of the recombinant heavy chain fragment C of botulinum neurotoxin serotype F, expressed in *Pichia pastoris*. Protein Expr. Purif. 2003; 32:1–9.

Keller, J.E. Recovery from botulinum neurotoxin poisoning in vivo. Neuroscience 2006; 139:629–637.

Kiyatkin, N., Maksymowych, A.B. and Simpson, L.L. Induction of immune response by oral administration of recombinant botulinum toxin. Infect. Immun. 1997; 65:4586–4591.

Klietmann, W.F. and Ruoff, X. Bioterrorism: implications for the clinical microbiologist. Clin. Microbiol. Rev. 2001; 14:364–381.

Kobayashi, R., Kohda, T., Kataoka, K., Ihara, H., Kozaki, S., Pascual, D.W., Staats, H.F., Kiyono, H., McGhee, J.R. and Fujihashi, K. A novel neurotoxoid vaccine prevents mucosal botulism. J. Immunol. 2005; 174:2190–2195.

Koriazova, L.K. and Montal, M. Translocation of botulinum neurotoxin light chain protease through the heavy chain channel. Nat. Struct. Biol. 2003; 10:13–18.

Kozaki, S., Kamata, Y., Nagai, T., Ogasawara, J. and Sakaguchi, G. The use of monoclonal antibodies to analyze the structure of *Clostridium botulinum* type E derivative toxin. Infect. Immun. 1986; 52:786–791.

Kozaki, S., Kamata, Y., Watarai, S., Nishiki, T. and Mochida, S. Ganglioside GT1b as a complementary receptor component for *Clostridium botulinum* neurotoxins. Microb. Pathog. 1998; 25:91–99.

Kozaki, S., Miki, A., Kamata, Y., Ogasawara, J. and Sakaguchi, G. Immunological characterization of papain-induced fragments of *Clostridium botulinum* type A neurotoxin and interaction of the fragments with brain synaptosomes. Infect. Immun. 1989; 57:2634–2639.

Kubota, T., Watanabe, T., Yokosawa, N., Tsuzuki, K., Indoh, T., Moriishi, K., Sanda, K., Maki, Y., Inoue, K. and Fuji, N. Epitope regions in the heavy chain of *Clostridium botulinum* type E neurotoxin recognized by monoclonal antibodies. Appl. Environ. Microbiol. 1997; 63:1214–1218.

Lacy, D.B. and Stevens, R.C. Sequence homology and structural analysis of the clostridial neurotoxins. J. Mol. Biol. 1999; 291:1091–1104.

Lamanna, C. The most poisonous poison. Science 1959; 130:763–772.

Lamm, M.E. Interaction of antigens and antibodies at mucosal surfaces. Ann. Rev. Microbiol. 1997; 51:311–340.

LaPenotiere, H.F., Clayton, M.A. and Middlebrook, J.L. Expression of a large, nontoxic fragment of botulinum neurotoxin serotype A and its use as an immunogen. Toxicon 1995; 35:1383–1386.

Lee, J.C., Hwang, H.J., Sakaguchi, Y., Yamamoto, Y., Arimitsu, H., Tsuji, T., Watanabe, T., Ohyama, T., Tsuchiya, T. and Oguma, K. C-terminal half fragment (50 kDa) of heavy chain components of *Clostridium botulinum* type C and D neurotoxins can be used as an effective vaccine. Microbiol. Immunol. 2007; 51:445–455.

Lee, J.S., Pushko, P., Parker, M.D., Dertzbaugh, M.T., Smith, L.A. and Smith, J.F. Candidate vaccine against botulinum neurotoxin serotype A derived from a Venezuelan equine encephalitis virus vector system. Infect. Immun. 2001; 69:5709–5715.

Levine, M.M. and Sztein, M.B. Vaccine development strategies for improving immunization: the role of modern immunology. Nat. Immunol. 2004; 5:460–464.

Lindstrom, M. and Korkeala, H. Laboratory diagnostics of botulism. Clin. Microbiol. Rev. 2006; 19:298–314.

Mahmut, N., Inoue, K., Fujinaga, Y., Arimitsu, H., Sakaguchi, Y., Hughes, L., Hirst, R., Murphy, T., Tsuji, T., Watanabe, T., Ohayama, T., Karasawa, T., Nakamura, S., Yokata, K. and Oguma, K. Mucosal immunization with *Clostridium botulinum* type C 16 S toxoid and its non-toxic component. J. Med. Microbiol. 2002; 51:813–820.

Mahrhold, S., Rummel, A., Bigalke, H., Davletov, B. and Binz, T. The synaptic vesicle protein 2C mediates the uptake of botulinum neurotoxin A into phrenic nerves. FEBS Lett. 2006; 580:2011–2014.

Makoff, A.J., Ballantine, S.P., Smallwood, A.E. and Fairweather, N.F. Expression of tetanus toxin fragment C in *E. coli*: its purification and potential use as a vaccine. Bio/Technol. 1989; 7:1043–1046.

Maksymowych, A.B., Reinhard, M., Malizio, C.J., Goodnough, M.C., Johnson, E.A. and Simpson, L.L. Pure botulinum toxin is absorbed from the stomach and small intestine and produces peripheral neuromuscular blockade. Infect. Immun. 1999; 67:4708–4712.

Maksymowych, A.B. and Simpson, L.L. Binding and transcytosis of botulinum neurotoxin by polarized human colon carcinoma cells. J. Biol. Chem. 1998; 273:21950–21957.

Maksymowych, A.B. and Simpson, L.L. Structural features of the botulinum neurotoxin molecule that govern binding and transcytosis across polarized human intestinal epithelial cells. J. Pharmacol. Exp. Ther. 2004; 310:633–641.

Minton, N.P. Molecular genetics of clostridial neurotoxins. Curr. Top. Microbiol. Immunol. 1995; 195:161–194.

Mitragotri, S. Immunization without needles. Nat. Rev. Immunol. 2005; 5:905–916.

Montecucco, C. How do tetanus and botulinum toxins bind to neuronal membranes? Trends Biochem. Sci. 1986; 11:315–317.

Montgomery, V.A., Makuch, R.S., Brown, J.E. and Hack, D.C. The immunogenicity in humans of a botulinum type F vaccine. Vaccine 2000; 18:728–735.

Neutra, M.R. and Kozlowski, P.A. Mucosal vaccines: the promise and the challenge. Nat. Rev. Immunol. 2006; 6:148–158.

NIAID. NIAID biodefense research agenda for CC category A agents. 2002.

Nishiki, T., Kamata, Y., Nemoto, Y., Omori, A., Ito, T., Takahashi, M. and Kozaki, S. Identification of protein receptor for *Clostridium botulinum* type B neurotoxin in rat brain synaptosomes. J. Biol. Chem. 1994; 269:10498–10503.

Nishiki, T., Tokuyama, Y., Kamata, Y., Nemoto, Y., Yoshida, A., Sato, K., Sekiguchi, M., Takahashi, M. and Kozaki, S. The high-affinity binding of *Clostridium botulinum* type B neurotoxin to synaptotagmin II associates with gangliosides GT1b and GD1a. FEBS Lett. 1996; 378:253–257.

Nowakowski, A., Wang, C., Powers, D.B., Amersdorfer, P., Smith, T.J., Montgomery, V.A., Sheridan, R., Blake, R., Smith, L.A. and Marks, J.D. Potent neutralization of botulinum neurotoxin by recombinant oligoclonal antibody. Proc. Natl. Acad. Sci. USA 2002; 99:11346–11350.

Oguma, K., Syuto, B., Iida, H. and Kubo, S. Antigenic similarity of toxins produced by *Clostridium botulinum* type C and D strains. Infect. Immun. 1980; 30:656–660.

Ohyama, T., Watanabe, T., Fujinaga, Y., Inoue, K., Sunagawa, H., Fujii, N., Inoue, K. and Oguma, K. Characterization of nontoxic-nonhemagglutinin component of the two types of progenitor toxin (M and L) produced by *Clostridium botulinum* type D CB-16. Microbiol. Immunol. 1995; 39:457–465.

Oshima, M., Hayakari, M., Middlebrook, J.L. and Atassi, M.Z. Immune recognition of botulinum neurotoxin type A: regions recognized by T cells and antibodies against the protective H(C) fragment (residues 855–1296) of the toxin. Mol. Immunol. 1997; 34:1031–1040.

Oshima, M., Middlebrook, J.L. and Atassi, M.Z. Antibodies and T cells against synthetic peptides of the C-terminal domain (hc) of botulinum neurotoxin type A and their cross-reaction with hc. Immunol. Lett. 1998; 60:7–12.

Park, J.B. and Simpson, L.L. Inhalational poisoning by botulinum toxin and inhalation vaccination with its heavy-chain component. Infect. Immun. 2003; 71:1147–1154.

Patarnello, T., Bargelloni, L., Rossetto, O., Schiavo, G. and Montecucco, C. Neurotransmission and secretion. Nature 1993; 364:581–582.

Popoff, M.R. and Marvaud, J.C. Structural and genomic features of clostridial neurotoxins. In: The Comprehensive Sourcebook of Bacterial Protein Toxins (J.E. Alouf and J.H. Freer, Eds.). London: Academic Press, 1999.

Potter, K.J., Zhang, W., Smith, L.A. and Meagher, M.M. Production and purification of the heavy chain fragment C of botulinum neurotoxin, serotype A, expressed in the methylotropic yeast *Pichia pastoris*. Protein Expr. Purif. 2000; 19:393–402.

Ravichandran, E., Al-Saleem, F.H., Ancharski, D.M., Singh, A., Elias, M., Gong, Y. and Simpson, L.L. A trivalent vaccine against botulinum toxin (serotypes A,B, and E) that can be administered by the mucosal route. Infect. Immun. 2007; 75:3043–3054.

Ravichandran, E., Gong, Y., Al-Saleem, F.H., Ancharski, D.M., Joshi, S.G. and Simpson, L.L. An initial assessment of the systemic pharmacokinetics of botulinum toxin. J. Pharmacol. Exp. Ther. 2006; 318:1343–1351.

Rosenberg, J.S., Middlebrook, J.L. and Atassi, M.Z. Localization of the regions on the c-terminal domain of the heavy chain of botulinum toxin A recognized by T-lymphocytes and by antibodies after immunization of mice with pentavalent toxoid. Immunol. Invest. 1997; 26:491–504.

Sakaguchi, G. *Clostridium botulinum* toxins. Pharmacol. Ther. 1983; 19:165–194.

Schiavo, G., Matteoli, M. and Montecucco, C. Neurotoxins affecting neuroexocytosis. Physiol. Rev. 2000; 80:717–766.

Shapiro, R.L., Hatheway, C. and Swerdlow, D.L. Botulism in the United States: a clinical and epidemiological review. Ann. Intern. Med. 1998; 129:221–228.

Shyu, R.H., Shaio, M.F., Tang, S.S., Shyu, H.F., Tsai, M.H., Smith, J.E., Huang, H.H., Wey, J.J., Huang, J.L. and Chang, H.H. DNA vaccination using the fragment C of botulinum neurotoxin type A provided protective immunity in mice. J. Biomed. Sci. 2000; 7:51–57.

Siegel, L.S. Human immune response to botulinum pentavalent (ABCDE) toxoid determined by a neutralization test and by an enzyme-linked immunosorbent assay. J. Clin. Microbiol. 1988; 26:2351–2356.

Simpson, L.L. Molecular pharmacology of botulinum toxin and tetanus toxin. Ann. Rev. Pharmacol. Toxicol. 1986; 26:427–453.

Simpson, L.L. Botulinum Neurotoxin and Tetanus Toxin. San Diego: Academic Press, 1989.

Simpson, L.L. Identification of the characteristics that underlie botulinum toxin potency: implications for designing novel drugs. Biochimie 2000; 82:943–953.

Simpson, L.L. Identification of the major steps in botulinum toxin action. Ann. Rev. Pharmacol. Toxicol. 2004; 44:167–193.

Simpson, L.L., Maksymowych, A.B. and Hao, S. The role of zinc binding in the biological activity of botulinum toxin. J. Biol. Chem. 2001; 276:27034–27041.

Simpson, L.L., Maksymowych, A.B. and Park, J.B. Compositions and methods for transepithelial molecular transport. 2003; 60/384,949.

Simpson, L.L., Maksymowych, A.B., Park, J.B. and Bora, R.S. The role of the interchain disulfide bond in governing the pharmacological actions of botulinum toxin. J. Pharmacol. Exp. Ther. 2004; 308:857–864.

Smith, L.A. Development of recombinant vaccines for botulinum neurotoxin. Toxicon 1998; 36:1539–1548.

Smith, L.A., Jensen, M.J., Montgomery, V.A., Brown, D.R., Ahmed, S.A. and Smith, T.J. Roads from vaccines to therapies. Mov. Disord. 2004; 19:S48–S52.

Smith, L.D. Botulism: The Organism, its Toxins, the Disease. Springfield, IL, USA: Charles C. Thomas, 1977.

Smith, L.D.S. and Sugiyama, H. Botulism: The Organism, its Toxins, the Disease. Springfield, IL: Charles C. Thomas, 1988.

Smith, T.J., Lou, J., Geren, I.N., Forsyth, C.M., Tsai, R., LaPorte, S.L., Tepp, W.H., Bradshaw, M., Johnson, E.A., Smith, L.A. and Marks, J.D. Sequence variation with botulinum neurotoxin serotypes impacts antibody binding and neutralization. Infect. Immun. 2005; 73:5450–5457.

Sobel, J. Botulism. Clin. Infect. Dis. 2005:1167–1173.

Sobel, J., Tucker, N., Sulka, A., McLaughlin, J. and Maslanka, S. Foodborne Botulism in the United States, 1990–2000. Emerg. Infect. Dis. 2004; 10:1606–1611.

Swaminathan, S. and Eswaramoorthy, S. Structural analysis of the catalytic and binding sites of *Clostridium botulinum* neurotoxin B. Nat. Struct. Biol. 2000; 7:693–699.

Tavallaie, M., Chenal, A., Gillet, D., Pereira, Y., Manich, M. and Gibert, M. Interaction between the two subdomains of the C-terminal part of the botulinum neurotoxin A is essential for the generation of protective antibodies. FEBS Lett. 2004; 572:299–306.

Todd, E.C.D. Epidemiology of foodborne diseases: a worldwide review. World Health Stat. Q. 1997; 50:30–50.

Torii, Y., Tokumaru, Y., Kawaguchi, S., Izumi, N., Maruyama, S., Mukamoto, M., Kozaki, S. and Takahashi, M. Production and immunogenic efficacy of botulinum tetravalent (A, B, E, F) toxoid. Vaccine 2002; 20:2256–2561.

Tsuzuki, K., Yokosawa, N., Syuto, B., Ohishi, I., Fujii, N., Kimura, K. and Oguma, K. Establishment of a monoclonal antibody recognizing an antigenic site common to *Clostridium botulinum* type B, C1, D, and E toxins and tetanus toxin. Infect. Immun. 1988; 56:898–902.

Webb, R.P., Smith, T.J., Wright, P.M., Montgomery, V.A., Meagher, M.M. and Smith, L.A. Protection with recombinant *Clostridium botulinum* C1 and D binding domain subunit (Hc) vaccines against C and D neurotoxins. Vaccine 2007; 25(21):4273–4282.

Wigginton, J.M. and Thill, P. Infant botulism: a review of the literature. Clin. Pediatr. 1993; 32:669–674.

Woodward, L.A., Arimitsu, H., Hirst, R. and Oguma, K. Expression of Hc Subunits from *Clostridium botulinum* types C and D and their evaluation as candidate vaccine antigens in mice. Infect. Immun. 2003; 71:2941–2944.

Wu, H.C., Yeh, C.T., Huang, Y.L., Tarn, L.J. and Lung, C.C. Characterization of neutralizing antibodies and identification of neutralizing. Appl. Environ. Microbiol. 2001; 67:3201–3207.

Yang, G.H., Kim, K.S., Kim, H.W., Jeong, S.T., Huh, G.H., Kim, J.C. and Jung, H.H. Isolation and characterization of a neutralizing antibody specific to internalization domain of *Clostridium botulinum* neurotoxin type B. Toxicon 2004:19–25.

Yowler, B.C., Kensinger, R.D. and Schengrund, C.L. Botulinum neurotoxin A activity is dependent upon the presence of specific gangliosides in neuroblastoma cells expressing synaptotagmin I. J. Biol. Chem. 2002; 227:32815–32819.

Zhang, R., Maksymowych, A.B. and Simpson, L.L. Cloning and sequence analysis of a cDNA encoding human syntaxin 1A, a polypeptide essential for exocytosis. Gene 1995; 159:293–294.

Zilinskas, R.A. Iraq's biological weapons: past as future? JAMA 1997; 278:418–424.

CHAPTER

47

Ehrlichia

Jere W. McBride

Departments of Pathology and Microbiology and Immunology, Institute for Human Infections and Immunity, Center for Biodefense and Emerging Infectious Diseases, Sealy Center for Vaccine Development, University of Texas Medical Branch, Galveston, TX, USA

OUTLINE

ABSTRACT

Ehrlichiae are small, obligately intracellular, gram-negative, tick-transmitted bacteria responsible for life-threatening emerging human zoonoses and diseases of veterinary importance, and one species is recognized as a significant agricultural biothreat. Organisms in the genus *Ehrlichia* were first described in the early 20th century, and the contemporary genus *Ehrlichia* consists of six named species that have regional or global distribution. The emergence of *Ehrlichia chaffeensis* and *Ehrlichia ewingii* as human pathogens is attributed to a combination of ecologic and demographic factors that have converged to establish these new human zoonoses in North America. Conversely,

Ehrlichia ruminantium, an established pathogen of agricultural importance, is confined to Africa and a few eastern Caribbean islands, causes high mortality in ruminants, and can be efficiently transmitted by *Ambloymma* vectors native to North America. Thus, *E. ruminantium* is considered a potential agricultural biothreat to the United States agricultural industry.

Ehrlichiae have small genomes but have evolved complex mechanisms that allow intracellular survival in arthropod hosts and persistent infections in vertebrate hosts, by evading the innate and adaptive immune responses. Ehrlichiae enter the host cell by receptor-mediated endocytosis that requires cellular signaling events and calcium influx. They reside in cytoplasmic vacuoles that do not fuse with lysosomes, and they alter transcription of host cell genes involved in apoptosis, cell cyclins, membrane trafficking, and various cytokines, by inhibiting MAK phosphorylation and potentially by translocation of ehrlichial proteins to the host cell nucleus that modulate gene expression. Ehrlichiae have a small group of major immunoreactive proteins that include multigene families of paralogous proteins potentially involved in immune evasion and serine/threonine-rich proteins with tandem repeats that elicit vigorous antibody responses. Ehrlichiae elicit an innate immune response dominated by inflammatory chemokines, but the pattern recognition receptors and pathogen-associated molecular patterns involved in innate immune recognition are unknown. Natural killer cells play a role in innate immune recognition of ehrlichiae, and antibodies and cell-mediated immune responses both play a role in ehrlichial clearance. The pathophysiology of the disease appears to involve overproduction of cytokines associated with toxic shock by cytotoxic T cells.

Live, attenuated, killed, nucleic acid, and recombinant vaccines have been developed against *Ehrlichia* spp. However, development of new-generation vaccines that are effective and practical will be facilitated by the availability of the genome sequence of several ehrlichial pathogens, recent characterization of many of the major immunoreactive proteins and pathogenic mechanisms, and a more comprehensive understanding of protective and pathologic immune mechanisms.

INTRODUCTION

Ehrlichioses were once considered diseases only of veterinary importance, but the first recognized human infection with the obligately intracellular bacterium, *Ehrlichia chaffeensis*, in 1986 signaled the emergence of *Ehrlichia* species as zoonotic pathogens (Maeda et al., 1987). Subsequently in 1999, *Ehrlichia ewingii*, a veterinary pathogen associated with granulocytic ehrlichiosis in dogs was molecularly identified in four patients from Missouri presenting with fever, headache, and thrombocytopenia (Buller et al., 1999). By the end of the 20th century, human monocytotropic ehrlichiosis (HME) and ehrlichiosis ewingii were established as tick-transmitted zoonoses caused by *E. chaffeensis* and *E. ewingii*, respectively. Demographic and ecologic factors have contributed to the emergence of human ehrlichioses, and these factors in addition to increased surveillance and diagnostic capability are likely to result in increasing incidence of HME in the future. *E. chaffeensis* is now recognized as a prototypical emerging pathogen, and HME is the most prevalent life-threatening tick-transmitted infectious disease in North America (Paddock and Childs, 2003).

The veterinary pathogen *Ehrlichia ruminantium* (previously known as *Cowdria ruminantium*) causes severe production losses to livestock in Africa and is considered an agricultural biothreat for the United States mainland, and *Ehrlichia canis*, the first ehrichial agent identified in North America, is the primary etiologic agent of canine monocytic ehrlichiosis (CME) and the only globally distributed ehrlichial agent. Progress towards the molecular characterization of these difficult-to-cultivate agents, development of animal models, and understanding of protective and pathologic host immune responses is facilitating the development of vaccines for emerging human ehrlichioses and established veterinary ehrlichioses.

HISTORY OF EHRLICHIAL DISEASES

Ehrlichiae were initially recognized in the 1920s and 1930s in Africa as agents of veterinary importance, when Cowdry (1925) identified *Rickettsia ruminantium* (now known as *E. ruminantium*) as the etiologic agent of heartwater, a highly fatal disease affecting ruminants (Cowdry, 1925). Heartwater was recognized in South Africa as early as 1860, which coincided with the appearance of the vector *Amblyomma hebraeum* in the region (Cowdry, 1925). Subsequently, Donatien and Lestoquard (1935) described a rickettsia-like agent *Rickettsia canis* (now known as *E. canis*) in the monocytes of Algerian dogs. Ehrlichial diseases remained in relative obscurity until the early 1960s when *E. canis* was first described in the United States by the academic veterinarian Sidney Ewing (1963), and was later attributed to epizootic outbreaks of canine tropical pancytopenia (CTP) (Wilkins et al., 1967; Huxsoll et al., 1969, 1972) afflicting British and American military dogs (sentry and scout dogs) in southeast Asia by veterinary pathologist, David Huxsoll. CTP ultimately caused a substantial number of deaths

(19%) in United States military canine units between 1963 and 1970 (Wilkins et al., 1967; Kelch, 1984). CME is now recognized as a prevalent and the only globally distributed ehrlichial disease.

In April 1986, a critically ill 51-year-old man with history of a tick bite was admitted to a Detroit hospital with acute symptoms that included fever, malaise, myalgia, and headache following a trip to Arkansas 3 weeks earlier (Maeda et al., 1987). The agent responsible for his illness was presumptively identified as *E. canis* based on morphologic study of the patient's leukocytes and serologic tests, and the patient recovered after doxycycline treatment and 12 weeks of hospitalization. A few years later, in 1990, an ehrlichial agent was isolated from an ill 21-year-old army reservist at Fort Chaffee, Arkansas. Molecular characterization of this agent revealed a new *Ehrlichia* sp., and was subsequently named *E. chaffeensis* in 1991 (Anderson et al., 1991; Dawson et al., 1991). *E. chaffeensis* is now recognized as an emerging human pathogen of public health importance, and HME is considered one of the most prevalent tick-transmitted diseases in the United States. HME, however, was not the only ehrlichiosis to emerge in humans. In 1999, 12 years after the first recognized case of HME was reported, four patients from Missouri were confirmed by molecular methods to have infections with *E. ewingii*, a pathogen previously considered to be only of veterinary importance as the cause of canine granulocytic ehrlichiosis (CGE) (Buller et al., 1999). This emerging human ehrlichiosis, referred to as ehrlichiosis ewingii, manifests primarily in immunocompromised individuals, and thus, appears to be an opportunistic infection in humans.

ETIOLOGIC AGENTS

Recently, the members of the genus *Ehrlichia* were reclassified using contemporary molecular methods, which have determined that previous criteria used for taxonomy (primarily morphology and intracellular existence) had resulted in incorrect genetic relationships of numerous organisms in several genera containing obligately intracellular bacteria. Hence, a new molecularly based taxonomy of the genus *Ehrlichia* using two highly conserved genes, *rrs* (16S ribosomal RNA genes) and *groESL* (Dumler et al., 2001), has resulted in an amended and now smaller *Ehrlichia* genus consisting of five members (*E. canis*, *E. chaffeensis*, *E. muris*, *E. ruminantium*, and *E. ewingii*) following the reassignment of six previously recognized members to the genera, *Anaplasma* (*E. phagocytophila*, *E. equi*, *E. platys*, and *E. bovis*) and *Neorickettsia* (*E. sennetsu* and *E. risticii*), and acquisition of one new member from the genus *Cowdria* (*C. ruminantium*). *E. chaffeensis* and *E. ewingii* are recognized as human pathogens as well as pathogens of veterinary importance that include *E. canis* and *E. ruminantium* (Table 47.1). The genus *Ehrlichia* is now part of a newly created family *Anaplasmataceae*, which also includes the genera *Anaplasma*, *Wolbachia*, and *Neorickettsia*, but remains in the order *Rickettsiales*.

Ehrlichiae are coccoid and coccobacillary and exhibit two morphologic cell types, reticulate (0.4–0.6 μm by 0.7–1.9 μm) and dense-cored (0.4–0.6 μm in diameter) (Fig. 47.1A–C), and typically reside as microcolonies of bacteria in cytoplasmic vacuoles derived from early endosomes in immature and mature hemopoietic cells (monocytes/macrophages and neutrophils) and endothelial cells (Barnewall et al., 1997). Morulae appear as dark blue to purple intracytoplasmic inclusions by light microscopy using Romanovsky-type stains (Rikihisa, 1991) (Figs 47.1 and 47.2). Examination of infected cells by electron microscopy demonstrates that in each cell numerous (1 to >400) morulae (1.0–6.0 μm) are usually present and contain as few as one, but more often numerous ehrichiae (>40). Both forms have a gram-negative cell wall structure,

TABLE 47.1 Summary of medically important ehrlichiae, vectors, natural hosts, and distribution

Agent	Vector	Natural host(s)	Disease	Distribution
E. chaffeensis	*A. americanum*, *D. variabilus*	White-tailed deer, canines	Human monocytotropic ehrlichiosis (HME), canine monocytic ehrlichiosis (CME)	Southeastern, south-central United States
E. ewingii	*A. americanum* *D. variabilus*	Canines	Ehrlichiosis ewingii, canine granulocytic ehrlichiosis (CGE)	Southeastern, south-central United States
E. canis	*R. sanguineus*	Canines	CME	Worldwide
E. ruminantium	*A. haebreum*	Ruminants	Heartwater	Sub-Saharan Africa; Caribbean

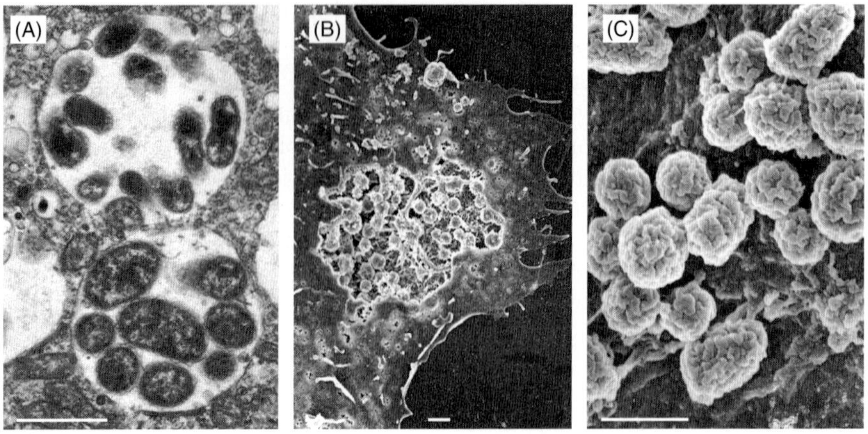

FIGURE 47.1 (Panel A) *Ehrlichia chaffeensis* dense-cored (top) and reticulate cell (bottom) morphologic types; (Panel B) scanning electron micrograph (SEM) of *E. chaffeensis* infected macrophage showing ehrlichial organisms within the cell; and (Panel C) SEM of *E. chaffeensis* attached to the surface of the macrophage prior to entry (Bars = 1 μm; reproduced with permission of V. L. Popov, University of Texas Medical Branch).

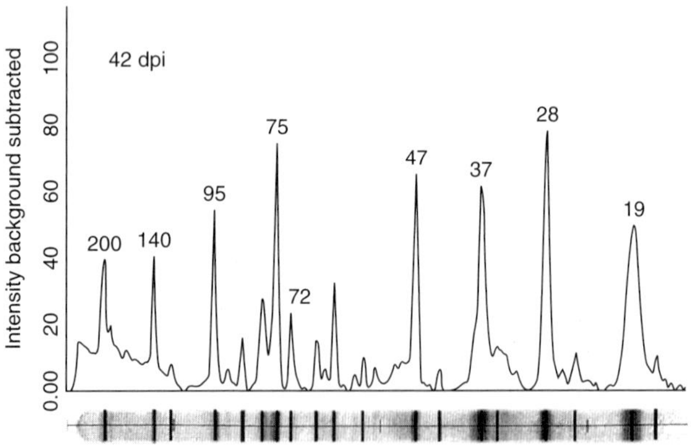

FIGURE 47.2 *Ehrlichia canis* major immunoreactive proteins as determined by Western immunoblotting of convalescent serum from an infected dog. Peak intensity and average relative quantity were used to identify major immunoreactive proteins. Molecular masses of proteins represented are labeled above each peak (Copyright permission ASM Press, McBride et al., 2003).

characterized by a cytoplasmic membrane and outer membrane separated by a periplasmic space, but do not appear to have peptidoglycan. Reticulate cells are pleomorphic and have uniformly dispersed nucleoid filaments and ribosomes, and dense-cored cells are typically coccoid and have centrally condensed nucleoid filaments and ribosomes (Popov et al., 1995, 1998). Small and large morulae containing both reticulate and dense-cored cells or exclusively containing dense-cored cells usually in loosely packed clusters can be observed within a single infected cell (Popov et al., 1995, 1998). The intramorular space in some morulae contains a fibrillar matrix of ehrlichial origin (Popov et al., 1995).

This chapter focuses on the agents of two emerging human ehrlichioses, *E. chaffeensis* and *E. ewingii*, a veterinary pathogen, *E. ruminantium* (heartwater), considered to be a potential agricultural biothreat to the United States, and *E. canis*, a globally distributed ehrlichial agent of dogs. Vaccine development for the ehrlichioses is being fueled by the commercial interest in vaccines for CME (*E. canis*) and heartwater (*E. ruminantium*).

E. chaffeensis

E. chaffeensis is the etiologic agent of HME (Anderson et al., 1992), which is the most severe of the human ehrlichioses, and can also cause mild-to-severe disease in dogs (Dawson and Ewing, 1992; Breitschwerdt et al., 1998). *E. chaffeensis* exhibits tropism for monocytes/macrophages. Many (40–60%) *E. chaffeensis* infections of humans require hospitalization (Eng et al., 1990; Fishbein et al., 1994), and there is case fatality rate of 3% due to the difficulty in making an accurate diagnosis. *E. chaffeensis* is maintained in nature in a zoonotic cycle potentially involving many vertebrate species. However, the white-tailed deer appears to be the primary reservoir for *E. chaffeensis*, but dogs may also be a significant natural reservoir (Paddock and Childs, 2003). The primary vector is the lone star tick, *Amblyomma americanum*, which is distributed from west central Texas throughout the southeastern, south-central, and mid-Atlantic states (Paddock and Childs, 2003; Childs and Paddock, 2003). Larval ticks become infected with *E. chaffeensis* after feeding on an infected vertebrate hosts and maintain a transstadial infection. The emergence of *E. chaffeensis* appears to coincide with changes in demographic and ecologic factors including increases in vector and mammalian host populations and human contact with natural foci, immunocompromised and aging human populations, and improved diagnosis and reporting (Paddock and Childs, 2003). *E. chaffeensis* can be cultivated in vitro in various mammalian and tick cell lines (Dawson et al., 1991), but causes only transient subclinical infection in immunocompetent mice without any pathology (Winslow et al., 1998).

E. ewingii

Human granulocytic ehrlichiosis (ehrlichiosis ewingii) is caused by *E. ewingii* and is the most recently described emerging human ehrlichiosis (Buller et al., 1999). *E. ewingii* is also a veterinary pathogen first described in 1971 (Ewing et al., 1971), and causes infections with two distinct clinical syndromes, anemia and polyarthritis in dogs (Goldman et al., 1998). *E. ewingii* exhibits host cell tropism for granulocytes (neutrophils), and is transmitted by the lone star tick, *A. americanum* (Anziani et al., 1990). Dogs are likely to be the main reservoir for *E. ewingii*, with many of the documented human cases reporting contact with dogs before onset of symptoms (Buller et al., 1999). Most cases of ehrlichiosis ewingii are manifested in immunocompromised patients, and thus, *E. ewingii* appears to be an opportunistic pathogen (Buller et al., 1999; Paddock et al., 2001). There is some serologic cross reactivity between *E. chaffeensis* and *E. ewingii*, but this is directed at higher molecular weight antigens and not the major outer membrane protein (p28) (Buller et al., 1999). *E. ewingii* has not been cultivated in vitro; thus molecular studies of this agent have been limited.

E. ruminantium

E. ruminantium is strictly a veterinary pathogen that is found on the continent of Africa (sub-Saharan) and on a few eastern Caribbean islands after importation in the 1800s (Uilenberg, 1983). *E. ruminantium* infects the endothelium and causes severe acute infection in wild and domestic ruminants known as heartwater, which can result in high mortality (50–90%) (Uilenberg, 1983). *E. ruminantium* is responsible for major production losses to the African livestock industry, and is also considered to pose a potential danger to the southeastern United States where native competent tick vectors are present (Uilenberg, 1982; Barre et al., 1987a; Mahan et al., 2000) and climate is appropriate for introduction of exotic tick species capable of transmitting the disease (Burridge et al., 2000, 2002a, 2002b). In Africa, *E. ruminantium* is transmitted by *A. variegatum* and *A. hebraeum*, but other *Amblyomma* spp. ticks including two North American species, *A. maculatum* and *A. cajennense*, appear to have similar transmission efficiency (Mahan et al., 2000). *E. ruminantium* can be cultivated in vitro in bovine endothelial cell lines and tick cells (Bezuidenhout et al., 1985; Bell-Sakyi et al., 2000), and various strains of mice can be infected with *E. ruminantium* (du Plessis et al., 1991; Byrom et al., 2000b).

E. canis

E. canis is the type species for the genus *Ehrlichia* and is the primary etiologic agent of CME, a serious and sometimes fatal, globally distributed disease of dogs (Keefe et al., 1982). *E. canis* is transmitted by the brown dog tick, *Rhipicephalus sanguineus* (Groves et al., 1975), and infects monocytes/macrophages in dogs. In the United States, *E. canis* was initially described in dogs in 1963 (Ewing, 1963), but received more attention after *E. canis* was identified as the agent responsible for outbreaks of an idiopathic hemorrhagic disease called tropical canine pancyopenia in American and British military dog units on duty in southeast Asia (Wilkins et al., 1967; Huxsoll et al., 1969; Seamer and Snape, 1970). Human infections with *E. canis* have been reported in Venezuela (Perez et al., 1996, 2006). The clinical manifestations of acute infection with *E. canis* are similar to those observed in humans infected with *E. chaffeensis*. *E. canis* can be propagated in numerous mammalian cell lines in vitro and acute *E. canis* infections in dogs are a useful comparative model for HME.

Genetic, Antigenic, and Phenotypic Characteristics

The genomes of *E. chaffeensis*, *E. canis*, and *E. ruminantium* are relatively small (1.2, 1.3, and 1.5 Mbp, respectively) and contain approximately 900 protein-encoding genes with an average G + C content of 27–29% (Collins et al., 2005; Mavromatis et al., 2006). These genomes have a substantially lower than average ratio of coding to noncoding sequence (62–72% coding) and contain pseudogenes (17–32) (Collins et al., 2005; Mavromatis et al., 2006). The genomes of these ehrlichiae reveal a high degree of genomic synteny, gene duplications, tandem repeats, proteins with ankyrin domains, a large group of secreted proteins with transmembrane helicies suggestive of a complex membrane structure and protein families that may be involved in immune evasion (Collins et al., 2005; Mavromatis et al., 2006). Comparison of three sequenced *E. ruminantium* genomes suggests active expansion and contraction of these genomes through addition or deletion of tandem repeats in noncoding regions (Frutos et al., 2006).

Ehrlichia are unable to utilize glucose as a carbon or energy source because they lack essential genes for the glycolytic pathway, ATP/ADP translocases are absent, and they lack genes for the biosynthesis of lipopolysaccharide and peptidoglycan, but have complete pathways for the synthesis of purines and pyrimidines, genes for enzymes involved in aerobic respiration, possess a limited set of genes involved in synthesis of lipids and phospholipids and have tRNAs for all amino acids (Collins et al., 2005; Mavromatis et al., 2006). *E. ewingii* has not been cultivated in vitro, and very little genomic and molecular characterization of this agent has been accomplished.

Prior to genome sequence information, many genes were molecularly characterized in *E. chaffeensis*, *E. canis*, and *E. ruminantium*, including functionally conserved genes, *rrs*, *rrl*, *rrf* (16S, 23S, and 5S rRNA, respectively) (Anderson et al., 1991; Allsopp et al., 1997; Massung et al., 2002), *groEL* and *groESL* (Sumner et al., 1993, 1997; Lally et al., 1995), thio-disulfide oxidoreductase (*dsb*) (Barbet et al., 2001; McBride et al., 2002), ferric-ion binding protein (*fbp*) (Yu et al., 1999a; Doyle et al., 2005), quinolate synthetase A (*nadA*) (Yu and Walker, 1997), citrate synthase (*gltA*) (Inokuma et al., 2001), and a *virB* operon (type IV secretion machinery) (Ohashi et al., 2002). A small group of major immunoreactive proteins of *E. chaffeensis* and *E. canis* has been identified on the basis of immunoblot reactivity, but the functions of these proteins remain largely unknown (Fig. 47.2). Major immunoreactive *E. chaffeensis* proteins are 200, 120, 88, 55, 47, 40, 28, and 23 kDa (Rikihisa et al., 1994; Chen et al., 1994); *E. canis*, 200, 140, 95, 75, 47, 36, 28, and 19 kDa (McBride et al., 2003a, 2003b); and *E. ruminantium*, 160, 85, 58, 46, 40, 32, and 21 kDa (Mahan et al., 1994a). *E. chaffeensis* immunoreactive proteins (gp200, gp120, gp47, p32 [VLPT], p28s [22 genes], and MAP2) have been molecularly characterized as well as the corresponding orthologs in *E. canis* (gp200, gp140, gp36, gp19 [VLPT], p28s [25 genes], and MAP2, respectively). Fewer of these orthologs have been molecularly identified and characterized in *E. ruminantium* (MAP1 [16 genes], MAP2, and mucin-like protein [clone hw26; gp36/47 ortholog]) (Jongejan and Thielemans, 1989; Mahan et al., 1994a; Sulsona et al., 1999), but additional immunoreactive genes have been identified in *E. ruminantium* (Barbet et al., 2001) that have not been described in *E. chaffeensis* or *E. canis*. Immunoreactivity of the *E. chaffeensis* and *E. canis* MAP2 is primarily dependent on a major conformational epitope and does not react by Western immunoblot (Alleman et al., 2000; Knowles et al., 2003), while *E. ruminantium* MAP2 appears to have a linear B cell epitope (Mahan et al., 1994a).

Many of the major immunoreactive *E. chaffeensis* proteins (gp200, gp120, gp47, gp28) that have been identified and characterized, exhibit larger than predicted molecular mass and two (gp120 and gp47) have serine-rich tandem repeats (Yu et al., 1997; McBride et al., 2003; Doyle et al., 2006b). The VLPT protein of *E. chaffeensis* also exhibits a mass twice the predicted size and has characteristics (tandem repeats and serine-rich composition) found in other ehrlichial proteins (Sumner et al., 1999). Strain-dependent sequence polymorphisms have been demonstrated in the number (2–5) of serine/threonine-rich tandem repeats of gp120 (Yu et al., 1997; Chen et al., 1997b; Yabsley et al., 2003) and VLPT (3–6) (Sumner et al., 1999), and in genes encoding the p28 multigene family (Yu et al., 1999b; Reddy and Streck 2000). The observed molecular masses of these proteins vary depending on the number of repeat units (gp120, 80–130 kDa; VLPT, 35–55 kDa) (Chen et al., 1997a, 1997b; Sumner et al., 1999), and variation in sizes of other homologous proteins has been observed on Western immunoblots reacted with monoclonal antibodies (MAb, 7C1-C, and 7C1-B) (Chen et al., 1997b). Antigenic heterogeneity has also been demonstrated with MAb 1A9 specific for p28 proteins, which does not react with all *E. chaffeensis* isolates (Chen et al., 1997b).

The *E. chaffeensis* gp120 is a surface protein differentially expressed on dense-cored cells and is also a component of the fibrillar matrix found in some morulae. The gp120 of *E. chaffeensis* type strain (Arkansas) has four 240-bp repeats. Antibodies in sera from patients with HME recognize the gp120 regardless of the number of repeats (Yu et al., 1997, 1999a). The gp120 is highly reactive with antibodies in convalescent sera from patients with HME and may be useful as a diagnostic antigen (Yu et al., 1996, 1999a). Similarly, the *E. canis* gp140 ortholog is a highly immunoreactive glycoprotein that contains 14 identical 108-bp repeats that have a high serine content appear to contain a major epitope (Yu et al., 2000a). Unlike *E. chaffeensis* gp120, the *E. canis* gp140 is highly conserved in geographically dispersed North American isolates each having the same number of repeat units (Yu et al., 2000a).

The gp200 orthologs are the largest major immunoreactive protein identified in *E. chaffeensis* (1438 amino acids) and *E. canis* (1406 amino acids) by Western immunoblot and have predicted molecular mass of 156 and 153 kDa, respectively, but exhibit larger masses (~200 kDa) by gel electrophoresis (McBride et al., 2003; Nethery et al., 2007). Five major species-specific antibody epitopes have been identified in terminal acidic domains within the *E. canis* gp200

(Nethery et al., 2007). The *E. chaffeensis* and *E. canis* gp200s have eukaryote-like ankyrin repeats (21) that may mediate protein–protein interactions, exhibit chromosomal synteny, but are substantially divergent at the molecular level (nucleic and amino acid). Ankyrin repeats are also present in numerous eukaryotic and prokaryotic proteins including *A. phagocytophilum* AnkA (Caturegli et al., 2000). The gp200s are present primarily in the cytoplasm of ehrlichiae and in the nucleus (on condensed chromatin) of infected host cells where they may be involved in modulation of host cell gene expression (Nethery et al., 2005).

The *E. chaffeensis* gp47 and *E. canis* gp36 are protein orthologs that are differentially expressed on the surface of dense-cored cells and contain variable numbers of serine-rich repeats depending on the isolate (Doyle et al., 2006b) (Fig. 47.3). The tandem repeats (nine amino acids) of *E. canis* gp36 have a conserved amino acid sequence but can vary in number. In contrast, two completely divergent *E. chaffeensis* gp47 repeat units (19 and 33 amino acids) have been identified in different isolates and vary in number (Doyle

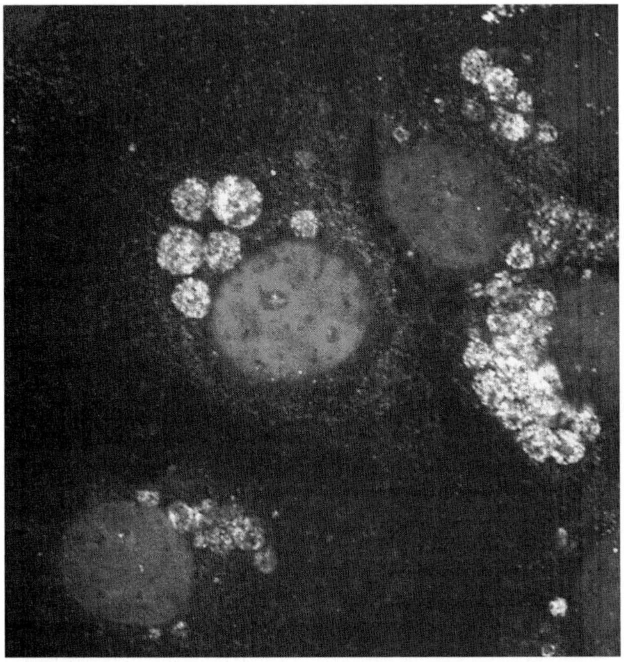

FIGURE 47.3 *E. chaffeensis* infected macrophages (DH82 cells) exhibiting differential expression of gp47 as visualized by three-color scanning laser confocal fluorescent microscopy. *E. chaffeensis* infected cells were dually stained with rabbit anti-*Ehrlichia* disulfide bond formation (Dsb) (green: Alexa Fluor 488), and mouse anti-*E. chaffeensis* gp47 (red: Alexa Fluor 568). Host cell nuclei were counterstained with 4′, 6′-diamidino-2-phenylindole, dihydrochloride (DAPI) and images merged. Infected macrophage (center) exhibits single (green: Dsb or red: gp47) and dual-labeled (yellow) ehrlichiae in individual morulae (reproduced with permission of J.W. McBride, University of Texas Medical Branch) (see color plate section).

et al., 2006a; Zhang et al., 2008). Carbohydrate has been detected on recombinant forms of these proteins, and the major epitope of the *E. canis* gp36 has been characterized in a two-repeat unit (18 amino acids) and in a single repeat (19 amino acids) in the *E. chaffeensis* gp47 (Doyle et al., 2006b). The gp47 and gp36 are secreted and observed in the fibrillar intramorular matrix and are released into the supernatant of infected cell cultures (Doyle et al., 2006b).

The *E. chaffeensis* VLPT is an immunoreactive protein that has 3–6 imperfect 30-amino acid tandem repeats (Sumner et al., 1999; Luo et al., 2008) and has been used to determine genetic diversity of *E. chaffeensis* isolates based on the number of repeat units. The *E. chaffeensis* VLPT gene amplified from blood of infected patients most often contains five repeat units (Standaert et al., 2000). A nonrepeat containing 19 kDa ortholog of VLPT in *E. canis* (gp19) has been identified (McBride et al., 2006). The *E. canis* gp19 elicits an early antibody response, which is directed against a single major serine-rich glycopeptide epitope (25 amino acids) located in the N-terminal region of the protein. The role of this protein in ehrlichial pathobiology remains unknown.

Ehrlichia species have a homologous multigene family that encodes major outer membrane proteins. The *E. chaffeensis*, *E. canis*, and *E. ruminantium* p28 (MAP1 for *E. ruminantium*) multigene families consist of 22, 25, and 16 nonidentical paralogous genes, respectively, that encode mature surface-exposed major outer membrane proteins of 26–32 kDa with 20–80% amino acid sequence identity (Yu et al., 2000b; Ohashi et al., 2001; Collins et al., 2005). All are located in a single chromosomal locus, except in *E. canis* which has a primary locus with 22 genes and an additional three duplicated genes in a separate chromosomal locus. These proteins contain three hypervariable regions, and a major B cell epitope has been identified in the first hypervariable region within *E. chaffeensis* p28 proteins (Li et al., 2002). The homology of p28 proteins among *Ehrlichia* spp. accounts for some antigenic cross reactivity among these agents. The potential for antigenic diversity by differential expression of p28 genes exists, but simultaneous transcription of all p28 genes does occur in vitro and in *E. chaffeensis*-infected dogs after needle inoculation (Long et al., 2002; Unver et al., 2002; Zhang et al., 2004a). The expression of only one paralog (*omp-1B*) has been observed in ticks (Unver et al., 2002), but the p28 genes expressed in the mammalian host after tick inoculation are still unknown. The function of MAP2 is not known, but the protein is conserved (~83% amino acid identity) among *E. chaffeensis*, *E. canis*, and *E. ruminantium*, and sequence analysis of *E. ruminantium* isolates indicates that MAP2 is highly conserved in this species (Bowie et al., 1999).

IMMUNITY

Understanding the adaptive immune mechanisms involved in immunity to *Ehrlichia* has been slowed by resistance of inbred mice to infection with *E. chaffeensis* (Winslow et al., 1998). Conversely, *E. ruminantium*, *E. muris*, and *Ixodes ovatus* ehrlichia (IOE) infect a variety of inbred mouse strains, and important insights to protective innate and adaptive immune mechanisms to ehrlichiae have been provided using these agents in both natural infections and experimental animal models (du Plessis 1970; du Plessis and Kumm 1971; du Plessis et al., 1991; Mwangi et al., 1998a, 1998b, 2002; Byrom et al., 2000a, 2000b). It is becoming evident that both humoral and cellular immunity plays a role in host defense against ehrlichiae. The role of specific immune system elements in immunity to *E. chaffeensis* has been investigated with inbred mice deficient in various immune system components including major histocompatibility complex (MHC), toll-like receptors (TLRs), class II, FcγRI, and B cells, and mice with severe combined immunodeficiency (SCID). Immune mechanisms involved in immunity to *E. ruminantium* have been investigated with immunocompetent mice and in natural hosts (du Plessis, 1970; du Plessis et al., 1991; Byrom et al., 1993, 2000a, 2000b; Totte et al., 1997; Mwangi et al., 1998a, 1998b; Esteves et al., 2004b). All of these investigations utilizing natural infections, established murine models, and newly developed murine models have provided new insight that suggests roles for innate, cellular, and humoral immune mechanisms in the elimination of *Ehrlichia*.

Innate Immunity

Ehrlichial replication appears to be partially controlled by several innate antimicrobial mechanisms. Induction of nitric oxide production by *E. ruminantium* infected endothelial cells via inducible nitric oxide synthase, resulted in a reduction of *E. ruminantium* (Mutunga et al., 1998). Others have reported that mice depleted of complement or deficient for complement receptors (1 and 2) or phox 91 subunit of NADPH oxidase are also susceptible to infection with IOE (Yager et al., 2005), demonstrating the importance of these innate immune components. However, the innate immune response to *Ehrlichia* spp. appears to be driven by atypical pathogen-associated molecular patterns (PAMPs) and a newly recognized innate defense mechanism. *Ehrlichia* lack the genes required for synthesis of two common gram-negative bacterial cellular components, lipopolysaccharide (LPS) and peptidoglycan (PGN) (Lin and Rikihisa 2003a), well-characterized PAMPs recognized by the innate immune response through TLRs. The pattern recognition receptors involved in recognition of *E. chaffeensis* are not known, but *tlr4*-deficient mice have depressed nitric oxide and interleukin-6 production and have more prolonged infections with *E. chaffeensis* than wild-type mice (Ganta et al., 2002), suggesting a minimal role for TLR4 recognition. Others have reported downregulation or no induction of major pattern recognition receptor gene expression by THP-1 cells in response to *E. chaffeensis* (Lin and Rikihisa, 2004; Zhang et al., 2004b), suggesting potential evasion of conventional TLR-mediated innate immune response by ehrlichiae. However, monocytes incubated with live and killed *E. chaffeensis* respond with strong production of IL-8 and minimal production of IL-1β and IL-10, while expression of TNF-α and IL-6 are absent (Lee and Rikihisa, 1996), which suggest that other uncharacterized ehrlichial PAMPs interact with an undefined PRRs present on macrophages. More extensive analysis of cytokine/chemokine response to *E. chaffeensis* revealed a predominant chemokine response to viable and killed *E. chaffeensis* characterized by production of inflammatory chemokines including IL-8, MCP-1, MCP-2, MCP-4, MIP-1β, and RANTES, while in contrast the production of proinflammatory cytokines, including TNF-α, IL-1β, or IL-10, were not observed (McBride, unpublished). Induction of this innate chemokine response is linked to an *E. chaffeensis* carbohydrate component and is abrogated by periodate treatment (Lee and Rikihisa, 1996). The role of this innate immune response by macrophages in pathogenesis or immunity to *E. chaffeensis* is not clear. Other less-characterized innate responses appear to be involved in the clearance of ehrlichiae. Mattner et al. (2005) demonstrated that *Ehrlichia* activate innate-like natural killer T cells (NKT) that are known to recognize glycolipids presented by CD1d, a MHC-like molecule. Futhermore, NKT cell-deficient mice are unable to clear ehrlichial infection (Mattner et al., 2005), suggesting that NKT cells are important for innate immunity to *Ehrlichia* spp. and for generation of the adaptive immune response.

Cell-Mediated Immunity

Numerous studies with *E. ruminantium* suggest that IFN-γ is an essential mediator in protection (Totte et al., 1993, 1996; Mahan et al., 1994b, 1996; Mutunga et al., 1998). IFN-γ production by $CD4^+$ and $CD8^+$ T cells in a protective response to *E. ruminantium* has been demonstrated (Esteves et al., 2004a, 2004b). In addition, transfer of the combination of IOE-specific polyclonal antibody and IFN-γ producing *Ehrlichia*-specific $CD4^+$ and $CD8^+$ type 1 cells protects naïve mice against lethal challenge (Ismail et al., 2004). A second study also concluded that resistance to sublethal IOE infection was dependent on $CD4^+$ T lymphocytes and required IFN-γ and TNF-α, but not IL-4 (Bitsaktsis et al., 2004). Taken together, these studies provide convincing evidence that classical cell-mediated immune mechanisms involving CD4 cells and type 1 cytokines are responsible for macrophage activation and elimination of *Ehrlichia*. Notably, similar conclusions regarding the importance of MHC class I, $CD4^+$, and $CD8^+$ T cells, the synergistic roles of IFN-γ and TNF-α, and the role of antibody have been reported in mice infected with *E. muris* (Feng and Walker, 2004). A more important role for $CD4^+$ T cells in immunity to *E. ruminanitum* has been suggested in studies using mice and cattle (Totte et al., 1997; Byrom et al., 2000a), and others have reported that γδ T cells contribute substantially to the IFN-γ response in cattle (Mwangi et al., 1998a). Similarly, mice lacking functional MHC class II genes are unable to clear *E. chaffeensis* following infection, suggesting that $CD4^+$ T cells are essential for ehrlichial clearance (Ganta et al., 2002). The roles of specific T cell subsets involved in cellular immunity to *E. chaffeensis* are not well defined.

Humoral Immunity

Humoral immunity appears to play a significant role in clearance of *E. chaffeensis*. Infection of SCID mice (B and T cell deficient) with *E. chaffeensis* results in an overwhelming infection, but passive transfer of anti-*E. chaffeensis* immune serum or MAbs directed against the first hypervariable domain of one (p28-19) of the p28 major outer membrane proteins before or during infection protects SCID mice from disease, but does not eliminate the organism (Winslow et al., 2000; Li et al., 2001, 2002). Futhermore, mice lacking B cells or FcγRI are unable to resolve a sublethal infection by IOE and passive transfer of antibodies in these mice resulted in significant reduction in bacterial load (Yager et al., 2005). Similarly, passive transfer of anti-*E. muris* antibodies, but not Fab fragments, protected mice against lethal infection (Feng and Walker, 2004). The specific anti-ehrlichial antibody-mediated mechanism is not fully understood, but appears to involve binding of antibody to the Fc receptor (Lee and Rikihisa, 1997; Yager et al., 2005) and subsequent generation of a proinflammatory cytokine response (Lee and Rikihisa, 1997) and exposure to oxidative defenses (Yager et al., 2005). Studies with *E. ruminantium* have suggested little role for antibody in immunity to *E. ruminantium*. In fact, the majority of studies suggest that antibodies provide no protection (du Plessis, 1970, 1984); however, one study has demonstrated the ability of antibodies to neutralize *E. ruminantium* in vitro (Byrom et al., 1993). Further studies with *E. ruminantium* are needed address inconsistencies in the role of antibodies in protection compared to that reported with other *Ehrlichia* spp.

PATHOGENESIS

Advances in understanding the molecular pathogenesis of ehrlichial diseases has been obtained primarily through studies with *E. chaffeensis*, *E. canis*, and closely related *E. muris* and IOE. Ehrlichial pathogenesis begins with injection of ehrlichiae by a feeding tick, hematogenous spread, and attachment to the target cell (monocyte/macrophage/neutrophil/endothelial cell), entry, and intracellular survival. Host cell receptors utilized by *E. chaffeensis* include, but are not limited to, E- and L-selectin, and binding of these receptors results in unconventional clathrin-independent receptor-mediated endocytosis (Barnewall et al., 1997; Zhang et al., 2003) involving caveolae and glycosylphosphatidylinositol-anchored proteins (Lin and Rikihisa, 2003b). Endocytosis of *E. chaffeensis* requires signaling events including protein crosslinking by transglutaminase, protein tyrosine phosphorylation, phospholipase C (PLC)-γ2 activation, and calcium influx (Lin et al., 2002). Following endocytosis, tyrosine phosphorylated proteins and PLC-γ2 colocalize with *E. chaffeensis* inclusions (Lin et al., 2002), suggesting their role in bacterial survival and proliferation in the macrophage. The unique endosomes containing *E. chaffeensis* express several early endosomal markers including an early endosomal antigen 1 (EEA1) and rab5, and do not fuse with lysosomes to form phagolysosomes (Barnewall et al., 1997; Mott et al., 1999), but colocalize with vesicle-associated membrane protein 2 (VAMP2), MHC class I and MHC class II molecules, β_2-microglobulin, and transferrin receptor (Barnewall et al., 1997, 1999; Mott et al., 1999). After entry, *E. chaffeensis* modulates the expression of various cytokines, TLRs, transcription factors, apoptosis inhibitors, cell cyclins, membrane trafficking proteins, and transferrin receptor gene expression in the macrophage (Barnewall et al., 1999; Lin and Rikihisa 2004; Zhang et al., 2004b). The inhibition of p38 MAPK by *E. chaffeensis* has been linked to the downregulation of transcription factor PU.1, which can also regulate the expression of TLRs and other genes (Lin and Rikihisa, 2004). Within 24h of infection *E. chaffeensis* blocks tyrosine phosphorylation of Janus kinase (Jak) and signal transducer and activator of transcription (Stat) signaling, inhibiting the anti-ehrlichial effect of IFN-γ (Lee and Rikihisa, 1998). *E. chaffeensis* has eight genes encoding type IV secretion machinery that may be associated with intracellular survival and replication by delivering effector macromolecules to the host cell and are transcribed in the blood of acutely ill patients (Ohashi et al., 2002). Some of these genes (*virB3*, *virB4*, and *virB6*) are downstream of superoxide dismutase B (*sodB*) and are cotranscribed with *sodB* through a *sodB* promoter (Ohashi et al., 2002). Iron plays a critical role in the survival of *E. chaffeensis* in the host cell, and iron acquisition mechanisms in *E. chaffeensis* appear to be evolutionarily divergent from those of other gram-negative bacteria (Doyle et al., 2005). A structurally and functionally conserved ferric-ion binding protein (Fbp) has been identified in *E. chaffeensis* and *E. canis* that binds Fe(III), but not Fe(II), and Fbp is preferentially expressed on the surface of dense-cored ehrlichiae and is secreted extracellularly (Doyle et al., 2005). The lack of iron availability prevents the survival of *E. chaffeensis*, which has been demonstrated with the intracellular iron chelator deferoxamine (Barnewall and Rikihisa, 1994). IFN-γ also appears to mediate killing of *E. chaffeensis* by downregulation of transferrin receptor, thus reducing the available labile cytoplasmic iron in human monocytes (Barnewall and Rikihisa, 1994). This anti-ehrlichial effect can be overcome by addition of iron-saturated transferrin, demonstrating the direct role for iron in ehrlichial survival. The role of the p28 multigene family in the pathogenesis of HME is unknown, but differential expression of p28 genes appears to occur in mammalian and tick hosts (Unver et al., 2002). *E. chaffeensis* demonstrates expression of 16 *p28* genes concurrently in vitro and in vivo (Long et al., 2002; Unver et al., 2002). These findings suggest that the p28s may play a role in ehrlichial adaptation in tick and mammalian hosts and are less likely to be involved in immune evasion by antigenic variation.

The relatively low bacterial burden in the blood and tissues in nonimmunocompromised patients with HME suggests that the pathophysiology of ehrlichiosis may involve immunopathologic responses that are manifest as a toxic shock-like syndrome (Maeda et al., 1987; Fichtenbaum et al., 1993). The first murine model of fatal human ehrlichiosis (Sotomayor et al., 2001) has been instrumental in understanding the mechanisms behind the toxic shock-like syndrome of severe and fatal human ehrlichiosis. This model has recently been used to investigate the immunologic mechanisms involved in the development of severe monocytotropic ehrlichiosis. Mice inoculated with an ehrlichia (IOE) closely related to *E. chaffeensis* develop histopathology resembling that observed in HME patients, and a similar disease course is observed in the IOE murine model. Lethal infections with IOE are accompanied by extremely high levels of serum TNF-α, a high frequency of TNF-α producing $CD8^+$ splenic T cells, decreased *Ehrlichia*-specific $CD4^+$ T lymphocyte proliferation, low IL-12 levels in the spleen, and a 40-fold decrease in the number of antigen-specific IFN-γ producing $CD4^+$ Th1 cells (Ismail et al., 2004). Furthermore, in mice lacking TNF receptor I/II were more resistant to IOE induced liver injury, but exhibited higher bacterial burdens (Ismail et al., 2006).

EPIDEMIOLOGY AND PUBLIC HEALTH IMPORTANCE

Many factors have contributed to the emergence of *E. chaffeensis* as a zoonotic pathogen in the United States including increased density of *A. americanum* ticks (Ginsberg et al., 1991), expanded vector geographic distribution (Means and White, 1997), and vertebrate host populations of the tick vector (Means and White, 1997), increase in reservoir host populations for *E. chaffeensis* (McCabe and McCabe, 1997), increased human contact with natural foci of infection through recreational and occupational activities (Tal and Shannahan, 1995; Standaert et al., 1995), increased size, longevity, and immunocompromised status of the human population (Palella et al., 1998; Paddock et al., 2001), and the availability of diagnostic reagents, and improved surveillance (McQuiston et al., 1999; Childs et al., 1999). The most important factor in the emergence of *E. ewingii* appears to be increased immunocompromised populations, because the infection has been observed primarily in human immunodeficiency virus (HIV)-infected individuals and patients on immunosuppressive therapies (Buller et al., 1999; Paddock et al., 2001).

HME is a seasonal disease with most reported cases occurring in the spring and summer months (peak incidence during May–July) (Fishbein et al., 1994; Standaert et al., 1995) and occurring primarily within the geographic range of the tick vector *A. americanum* that extends from west central Texas and east along the Gulf Coast, north through Oklahoma and Missouri, eastward to the Atlantic Coast, and proceeds northeast through New Jersey, encompassing all the south-central, southeastern, and mid-Atlantic states. Passive surveillance of HME underestimates the true incidence of disease due to inadequate clinical and laboratory diagnosis and reporting (McQuiston et al., 1999). Nevertheless, ehrlichiosis has similar infection rates as those reported for Rocky Mountain spotted fever (Treadwell et al., 2000), with the highest incidence reported in Arkansas, North Carolina, Missouri, Oklahoma, and New Jersey (McQuiston et al., 1999). Active surveillance in HME-endemic locations including southeastern Georgia and southeastern Missouri indicates an incidence that could be 10–100 times higher than those reported by individual states by passive surveillance (Fishbein et al., 1989; Olano et al., 2003). HME is more often diagnosed in male (>2:1) patients >40 years of age, the majority (>80%) report a tick bite (Fishbein et al., 1994; Olano et al., 2003), and HME outbreaks are associated with recreational or occupational activities (Petersen et al., 1989; Standaert et al., 1995). *E. chaffeensis* has been found in 5–15% of *A. americanum* ticks collected from endemic areas in the eastern US (Whitlock et al., 2000; Ijdo et al., 2000; Stromdahl et al., 2001) and has been detected in ticks collected from at least 15 states. *E. chaffeensis* has not been isolated outside the United States, and the only proven tick vector is restricted to North America, but *E. chaffeensis* or a closely related *Ehrlichia* sp. has been detected in other tick species located in other regions of the world, including Russia, Korea, Thailand, and China (Ravyn et al., 1999; Cao et al., 2000; Alekseev et al., 2001; Kim et al., 2003; Parola et al., 2003). Human infections with *E. chaffeensis* or antigenically related ehrlichiae have been reported in Europe (Nuti et al., 1998), Asia (Heppner et al., 1997), South America (Ripoll et al., 1999), and Africa (Uhaa et al., 1992). *A. americanum* readily feed on humans and white-tailed deer, which are considered to be the primary natural hosts for maintaining the *E. chaffeensis* transmission cycle (Lockhart et al., 1997). Antibodies reactive with *E. chaffeensis* have been detected in white-tailed deer (Lockhart et al., 1995), and laboratory-raised *A. americanum* ticks can acquire *E. chaffeensis* infection from deer (Ewing et al., 1995). Other potentially important reservoirs that are naturally infected with *E. chaffeensis* include goats, domestic dogs, and coyotes (Breitschwerdt et al., 1998; Kocan et al., 2000; Dugan et al., 2000).

CLINICAL SPECTRUM

HME and ehrlichiosis ewingii manifest as undifferentiated febrile illnesses 1–3 weeks after the bite of an infected tick, and infected individuals usually seek medical care within 4 days after onset of illness (Eng et al., 1990; Fishbein et al., 1994). For HME, the most frequent clinical findings reported anytime during acute illness are fever, malaise, headache, dizziness, chills, and myalgias (Fishbein et al., 1989, 1994; Eng et al., 1990; Olano et al., 2003). Other signs including respiratory or central nervous system involvement, lymphadenopathy, and rash are less frequent (Eng et al., 1990; Fishbein et al., 1994; Olano et al., 2003). Patients with ehrlichiosis ewingii present with a milder disease and few complications (Buller et al., 1999). A majority of these patients are immunocompromised, further suggesting that *E. ewingii* is less pathogenic. A large portion of HME patients (60–70%), including those that are immunocompromised, develop more serious manifestations or multisystem involvement including renal failure, disseminated intravascular coagulation, cardiomegaly, acute respiratory distress syndrome, spontaneous hemorrhage, and neurological manifestations requiring hospitalization (Eng et al., 1990; Fishbein et al., 1994; Paddock et al., 2001). Prevention of severe manifestations correlates well with diagnosis and treatment with tetracyclines within the first week of illness (Eng et al., 1990). Hematologic and biochemical abnormalities usually include leucopenia, thrombocytopenia, anemia, mildly elevated serum hepatic transaminase activities, and hyponatremia (Fishbein et al., 1994; Paddock et al., 2001; Olano et al., 2003). Lymphocytosis characterized by a predominance of $\gamma\delta$ T cells is often seen in patients during recovery (Caldwell et al., 1995). A high proportion of immunocompetent (41–62%) and immunocompromised patients (86%) require hospitalization (Fishbein et al., 1994; Paddock et al., 2001; Olano et al., 2003). The case fatality rate is estimated to be 3% (McQuiston et al., 1999), and fatal disease is most often described in older patients, and patients debilitated by underlying disease or immunodeficiency (Eng et al., 1990; Paddock et al., 2001). Immunocompromised patients (HIV-infected persons, transplant recipients, corticosteroid-treated patients) have a high risk of fatal infection associated with overwhelming infection not typically observed in immunocompetent patients (Paddock et al., 2001). No deaths have been reported as a result of infection with *E. ewingii* (Buller et al., 1999; Paddock et al., 2001).

TREATMENT AND PREVENTION

Most patients with HME seek medical attention 4 days after onset of symptoms, which is usually before antibodies diagnostic of infection can be detected. Therefore, patients suspected to have ehrlichiosis should be treated empirically with doxycycline or tetracycline because therapeutic delay can result in the development of severe clinical manifestations and fatal outcome (Fishbein et al., 1994; Olano et al., 2003). Doxycycline and rifampin are highly active against *E. chaffeensis* in vitro (Brouqui and Raoult 1992;

Branger et al., 2004). Therefore, tetracyclines are considered to be the drug of choice for treatment of human ehrlichioses, but rifampin may be preferred in pregnant women, and cases have been reported of successful treatment of closely related *Anaplasma phagocytophilum* infections with this antibiotic (Buitrago et al., 1998; Krause et al., 2003). *E. chaffeensis* is resistant to chloramphenicol in vitro (Brouqui and Raoult 1992), but both treatment failures (Fichtenbaum et al., 1993) and successes with this antibiotic have been reported (Eng et al., 1990; Fishbein et al., 1994); thus it should not be considered a primary therapeutic option.

Disease prevention begins with reducing tick exposure through the use of tick repellents containing *n,n*-diethyl-*m*-toluamide (DEET) and barrier clothing when recreational or occupational activities will result in potential tick exposure. Early removal of any attached ticks will lower the risk of transmission, as it may require 24 or more hours for ehrlichiae to be inoculated.

VACCINES IN DEVELOPMENT

Conventional Vaccines

The emergence of human ehrlichioses in the last decade and the risk to public health has elicited interest in the development of vaccines for HME. The use of vaccines to prevent HME may be especially useful for military personnel and other persons who are active outdoors and are at an increased risk level for acquiring the disease. Ehrlichioses, such as heartwater and CME, are important veterinary diseases. Thus, considerable effort has been made to develop vaccines for *E. ruminantium*, which causes large economic losses to the livestock industry in third world countries, and creates limitations on livestock export. Because *E. ruminantium* is present on a few eastern Caribbean islands (Perreau et al., 1980; Birnie et al., 1985), it poses a threat to the United States livestock industry (Barre et al., 1987b; Burridge et al., 2002a). Consequently, much of the vaccine development for ehrlichial diseases has been focused on prevention of heartwater, where various vaccines have been developed and tested, including live, attenuated, nucleic acid, and recombinant compositions.

Animals that survive infection with *E. ruminantium* develop immunity and are solidly protected against challenge. Such observations led to the development of an effective "infection and treatment" strategy introduced by Neitz and Alexander (1941), which has been the only commercially available vaccine strategy for the last 50 years (Zweygarth et al., 2005). This method of vaccination is cumbersome, requires a cold chain delivery system (liquid nitrogen or dry ice) that makes it inconvenient and impractical for most farmers and private practitioners. In addition, this method has vaccine-associated deaths, risk of transmission of other blood-borne pathogens, lack of standardization and heterologous protection, and is time consuming and expensive (Van der, 1987; Mahan et al., 1999). Consequently, the development and testing of attenuated, inactivated, and recombinant vaccines has been ongoing in an effort to replace the current heartwater "vaccine." Spontaneous attenuated strains of *E. ruminantium* (Senegalese and Welgevonden) were obtained by in vitro cell culture passage (Jongejan, 1991; Zweygarth et al., 2005). These attenuated strains do not produce disease when inoculated into animals, but elicit protective immunity against homologous and heterologous strains (Jongejan, 1991; Zweygarth et al., 2005). The ability of attenuated *E. ruminantium* vaccines to induce protection has been established, but attenuated vaccines also require a cold storage delivery system, and they cannot be used in areas where the vaccine strain is not endemic due to the possibility of reversion to wild-type. In order to develop a vaccine absent of the constraints associated with live attenuated vaccines, inactivated *E. ruminantium* vaccines have been produced and tested. Substantial overall protection has been demonstrated in goats (65%), sheep (79%), and cattle (100%) of vaccinated animals to homologous and heterologous experimental and field challenges as determined by numerous studies (Martinez et al., 1994, 1996; Mahan et al., 1995, 1998; Totte et al., 1997). Inactivated vaccines also have the advantage of being useful in regions where the disease is not endemic, but where the vectors and the potential exist for infections to occur.

Recombinant and Nucleic Acid Vaccines

Development of recombinant subunit and nucleic acid vaccines that has been limited primarily to those utilizing a major surface protein ortholog of *Ehrlichia* spp. (designated MAP1 in *E. ruminantium*, p28 in *E. chaffeensis*, and p28/p30 in *E. canis*), which is a member of a paralogous nonidentical multigene family of outer membrane protein genes (16–25 genes, see section "Genetic, Antigenic, and Phenotypic Characteristics") in each respective *Ehrlichia* spp. Partial protection using a recombinant version of the *E. chaffeensis* P28 protein was demonstrated in mice after homologous challenge (Ohashi et al., 1998). More extensive protection studies with this vaccine candidate have been conducted with *E. ruminantium* demonstrating similar results. Mice vaccinated with a DNA vaccine containing *map1* generated MAP1-specific antibody, had lymphocytes that proliferated in response *E. ruminantium*, secreted IFN-γ, and increased survival of immunized mice against homologous challenge was noted (Nyika et al., 1998). A follow-up study demonstrated significantly increased protection against homologous challenge using MAP1 DNA vaccination and recombinant protein boost in a mouse model (Nyika et al., 2002). Unfortunately there is substantial divergence in *map1/p28* genes among different isolates of *E. ruminantium* and *E. chaffeensis*, and therefore this diversity may complicate development and implementation. Conversely, the *p28/p30* genes of *E. canis* appear to be highly conserved among geographically dispersed strains, and thus, may facilitate more rapid development of effective vaccines utilizing this antigen. Several newly identified protein orthologs (gp120/140, gp36/47, and gp200s) from *E. chaffeensis* and *E. canis*, respectively, are also immunodominant and are consistently recognized by antibodies in convalescent antisera. The ability of these proteins to protect against homologous challenge has not been determined. Additional antigens of *E. ruminantium* have been tested as DNA vaccines with limited, inconsistent, or no protection (Brayton et al., 1998a, 1998b; Pretorius et al., 2002; Louw et al., 2002; Collins et al., 2003).

PROSPECTS FOR THE FUTURE

Many questions remain with regard to our understanding of ehrlichial pathogenesis and mechanisms of protective immunity. The emergence of human ehrlichioses in the late 20th century has focused new resources and efforts to improve diagnosis and treatment, and to understand pathogenic and protective immune mechanisms that will facilitate vaccine development. The completion of several genome sequences from *Ehrlichia* spp. has provided new insight into how these organisms have evolved and clues to the unique strategies that they utilize to survive in both invertebrate and vertebrate hosts and their interaction with and dependence on the host cell for survival. Molecular identification and characterization of the majority of the major immunoreactive proteins has been accomplished. These new prospects coupled with a more complete understanding of ehrlichial pathobiology and interaction with the innate and adaptive host immune responses will undoubtedly stimulate the development of new and more effective nucleic acid or subunit vaccines for human and veterinary use in the future.

KEY ISSUES

- HME is an emerging life-threatening tick-borne zoonosis caused by the obligately intracellular pathogen, *Ehrlichia chaffeensis*.
- *Ehrlichia ruminantium* is considered to be an agricultural biothreat.
- Commercially, vaccines for human and veterinary ehrlichioses are not available.
- Ehrlichiae evade innate and adaptive host immune responses and establish persistent infections in natural hosts.
- Cellular and humoral immune mechanisms are involved in immunity to *Ehrlichia*.
- *Ehrlichia* major immunoreactive proteins that may be protective vaccine candidates have been identified, but immunoprotective antigens of *Ehrilichia* spp. remain relatively undefined.
- Genetic and antigenic variability exists among strains of *E. chaffeensis* and *E. ruminantium*.
- Natural infection solidly protects against challenge (*E. ruminantium*).
- Infection and treatment technique is the only effective immunization strategy for *E. ruminantium*.
- Partial protection has been achieved with inactivated and attenuated *E. ruminantium* vaccines.
- Little is known about the antigenic composition of *Ehrlichia* spp. in the invertebrate host.

References

Alekseev, A.N., Dubinina, H.V., Semenov, A.V. and Bolshakov, C.V. Evidence of ehrlichiosis agents found in ticks (Acari: Ixodidae) collected from migratory birds. J. Med. Entomol. 2001; 38:471–474.

Alleman, A.R., Barbet, A.F., Bowie, M.V., Sorenson, H.L., Wong, S.J. and Belanger, M. Expression of a gene encoding the major antigenic protein 2 homolog of *Ehrlichia chaffeensis* and potential application for serodiagnosis. J. Clin. Microbiol. 2000; 38:3705–3709.

Allsopp, M., Visser, E.S., du Plessis, J.L., Vogel, S.W. and Allsopp, B.A. Different organisms associated with heartwater as shown by analysis of 16S ribosomal RNA gene sequences. Vet. Parasitol. 1997; 71:283–300.

Anderson, B.E., Dawson, J.E., Jones, D.C. and Wilson, K.H. *Ehrlichia chaffeensis*, a new species associated with human ehrlichiosis. J. Clin. Microbiol. 1991; 29:2838–2842.

Anderson, B.E., Sumner, J.W., Dawson, J.E., Tzianabos, T., Greene, C.R., Olson, J.G., Fishbein, D.B., Olsen-Rasmussen, M., Holloway, B.P. and George, E.H. Detection of the etiologic agent of human ehrlichiosis by polymerase chain reaction. J. Clin. Microbiol. 1992; 30:775–780.

Anziani, O.S., Ewing, S.A. and Barker, R.W. Experimental transmission of a granulocytic form of the tribe Ehrlichieae by *Dermacentor variabilis* and *Amblyomma americanum* to dogs. Am. J. Vet. Res. 1990; 51:929–931.

Barbet, A.F., Whitmire, W.M., Kamper, S.M., Simbi, B.H., Ganta, R.R., Moreland, A.L., Mwangi, D.M., McGuire, T.C. and Mahan, S.M. A subset of *Cowdria ruminantium* genes important for immune recognition and protection. Gene 2001; 275:287–298.

Barnewall, R.E., Ohashi, N. and Rikihisa, Y. *Ehrlichia chaffeensis* and *E. sennetsu*, but not the human granulocytic erhlichiosis agent, colocalize with transferrin receptor and up-regulate transferrin receptor mRNA by activating iron-responsive protein 1. Infect. Immun. 1999; 67:2258–2265.

Barnewall, R.E. and Rikihisa, Y. Abrogation of gamma interferon- induced inhibition of *Ehrlichia chaffeensis* infection in human monocytes with iron-transferrin. Infect. Immun. 1994; 62: 4804–4810.

Barnewall, R.E., Rikihisa, Y. and Lee, E.H. *Ehrlichia chaffeensis* inclusions are early endosomes which selectively accumulate transferrin receptor. Infect. Immun. 1997; 65:1455–1461.

Barre, N., Uilenberg, G., Morel, P.C. and Camus, E. Danger of introducing heartwater onto the American mainland: potential role of indigenous and exotic *Amblyomma* ticks. Onderstepoort J. Vet. Res. 1987a; 54:405–417.

Barre, N., Uilenberg, G., Morel, P.C. and Camus, E. Danger of introducing heartwater onto the American mainland: potential role of indigenous and exotic *Amblyomma* ticks. Onderstepoort J. Vet. Res. 1987b; 54:405–417.

Bell-Sakyi, L., Paxton, E.A., Munderloh, U.G. and Sumption, K.J. Growth of *Cowdria ruminantium*, the causative agent of heartwater, in a tick cell line. J. Clin. Microbiol. 2000; 38:1238–1240.

Bezuidenhout, J.D., Paterson, C.L. and Barnard, B.J. In vitro cultivation of *Cowdria ruminantium*. Onderstepoort J. Vet. Res. 1985; 52:113–120.

Birnie, E.F., Burridge, M.J., Camus, E. and Barre, N. Heartwater in the Caribbean: isolation of *Cowdria ruminantium* from Antigua. Vet. Rec. 1985; 116:121–123.

Bitsaktsis, C., Huntington, J. and Winslow, G. Production of IFN-gamma by CD4 T cells is essential for resolving ehrlichia infection. J. Immunol. 2004; 172:6894–6901.

Bowie, M.V., Reddy, G.R., Semu, S.M., Mahan, S.M. and Barbet, A.F. Potential value of major antigenic protein 2 for serological diagnosis of heartwater and related ehrlichial infections. Clin. Diagn. Lab. Immunol. 1999; 6:209–215.

Branger, S., Rolain, J.M. and Raoult, D. Evaluation of antibiotic susceptibilities of *Ehrlichia canis*, *Ehrlichia chaffeensis*, and *Anaplasma phagocytophilum* by real-time PCR. Antimicrob. Agents Chemother. 2004; 48:4822–4828.

Brayton, K.A., van der, W.M., Vogel, S.W. and Allsopp, B.A. A partially protective clone from *Cowdria ruminantium* identified by using a *Salmonella* vaccine delivery system. Ann. N.Y. Acad. Sci. 1998a; 849: 247–252.

Brayton, K.A., Vogel, S.W. and Allsopp, B.A. Expression library immunization to identify protective antigens from *Cowdria ruminantium*. Ann. N.Y. Acad. Sci. 1998b; 849:369–371.

Breitschwerdt, E.B., Hegarty, B.C. and Hancock, S.I. Sequential evaluation of dogs naturally infected with *Ehrlichia canis*, *Ehrlichia chaffeensis*, *Ehrlichia equi*, *Ehrlichia ewingii*, or *Bartonella vinsonii*. J. Clin. Microbiol. 1998; 36:2645–2651.

Brouqui, P. and Raoult, D. In vitro antibiotic susceptibility of the newly recognized agent of ehrlichiosis in humans, *Ehrlichia chaffeensis*. Antimicrob. Agents Chemother. 1992; 36:2799–2803.

Buitrago, M.I., Ijdo, J.W., Rinaudo, P., Simon, H., Copel, J., Gadbaw, J., Heimer, R., Fikrig, E. and Bia, F.J. Human granulocytic ehrlichiosis during pregnancy treated successfully with rifampin. Clin. Infect. Dis. 1998; 27:213–215.

Buller, R.S., Arens, M., Hmiel, S.P., Paddock, C.D., Sumner, J.W., Rikhisa, Y., Unver, A., Gaudreault-Keener, M., Manian, F.A., Liddell, A.M., Schmulewitz, N. and Storch, G.A. *Ehrlichia ewingii*, a newly recognized agent of human ehrlichiosis. N. Engl. J. Med. 1999; 341:148–155.

Burridge, M.J., Simmons, L.A., Peter, T.F. and Mahan, S.M. Increasing risks of introduction of heartwater onto the American mainland associated with animal movements. Ann. N.Y. Acad. Sci. 2002a; 969:269–274.

Burridge, M.J., Simmons, L.A., Simbi, B.H., Mahan, S.M., Fournier, P.E. and Raoult, D. Introduction of the exotic tick *Amblyomma hebraeum* into Florida on a human host. J. Parasitol. 2002b; 88:800–801.

Burridge, M.J., Simmons, L.A., Simbi, B.H., Peter, T.F. and Mahan, S.M. Evidence of *Cowdria ruminantium* infection (heartwater) in *Amblyomma sparsum* ticks found on tortoises imported into Florida. J. Parasitol. 2000; 86:1135–1136.

Byrom, B., Barbet, A.F., Obwolo, M. and Mahan, S.M. CD8(+) T cell knockout mice are less susceptible to *Cowdria ruminantium* infection than athymic, CD4(+) T cell knockout, and normal C57BL/6 mice. Vet. Parasitol. 2000a; 93:159–172.

Byrom, B., Mahan, S.M. and Barbet, A.F. The development of antibody to *Cowdria ruminantium* in mice and its role in heartwater disease. Rev. Elev. Med. Vet. Pays Trop. 1993; 46:197–201.

Byrom, B., Obwolo, M., Barbet, A.F. and Mahan, S.M. A polarized Th1 type immune response to *Cowdria ruminantium* infection is detected in immune DBA/2 mice. J. Parasitol. 2000b; 86:983–992.

Caldwell, C.W., Everett, E.D., McDonald, G., Yesus, Y.W. and Roland, W.E. Lymphocytosis of gamma/delta T cells in human ehrlichiosis. Am. J. Clin. Pathol. 1995; 103:761–766.

Cao, W.C., Gao, Y.M., Zhang, P.H., Zhang, X.T., Dai, Q.H., Dumler, J.S., Fang, L.Q. and Yang, H. Identification of *Ehrlichia chaffeensis* by nested PCR in ticks from Southern China. J. Clin. Microbiol. 2000; 38:2778–2780.

Caturegli, P., Asanovich, K.M., Walls, J.J., Bakken, J.S., Madigan, J.E., Popov, V.L. and Dumler, J.S. *ankA*: an *Ehrlichia phagocytophila* group gene encoding a cytoplasmic protein antigen with ankyrin repeats. Infect. Immun. 2000; 68:5277–5283.

Chen, S.M., Cullman, L.C. and Walker, D.H. Western immunoblotting analysis of the antibody responses of patients with human monocytotropic ehrlichiosis to different strains of *Ehrlichia chaffeensis* and *Ehrlichia canis*. Clin. Diag. Lab. Immunol. 1997a; 4:731–735.

Chen, S.M., Dumler, J.S., Feng, H.M. and Walker, D.H. Identification of the antigenic constituents of *Ehrlichia chaffeensis*. Am. J. Trop. Med. Hyg. 1994; 50:52–58.

Chen, S.M., Yu, X.J., Popov, V.L., Westerman, E.L., Hamilton, F.G. and Walker, D.H. Genetic and antigenic diversity of *Ehrlichia chaffeensis*: comparative analysis of a novel human strain from Oklahoma and previously isolated strains. J. Infect. Dis. 1997b; 175:856–863.

Childs, J.E. and Paddock, C.D. The ascendancy of *Amblyomma americanum* as a vector of pathogens affecting humans in the United States. Annu. Rev. Entomol. 2003; 48:307–337.

Childs, J.E., Sumner, J.W., Nicholson, W.L., Massung, R.F., Standaert, S.M. and Paddock, C.D. Outcome of diagnostic tests using samples from patients with culture-proven human monocytic ehrlichiosis: implications for surveillance. J. Clin. Microbiol. 1999; 37:2997–3000.

Collins, N.E., Liebenberg, J., de Villiers, E.P., Brayton, K.A., Louw, E., Pretorius, A., Faber, F.E., van, H.H., Josemans, A., van, K.M., Steyn, H.C., van Strijp, M.F., Zweygarth, E., Jongejan, F., Maillard, J.C., Berthier, D., Botha, M., Joubert, F., Corton, C.H., Thomson, N.R., Allsopp, M.T. and Allsopp, B.A. The genome of the heartwater agent *Ehrlichia ruminantium* contains multiple tandem repeats of actively variable copy number. Proc. Natl. Acad. Sci. U.S.A. 2005; 102:838–843.

Collins, N.E., Pretorius, A., van, K.M., Brayton, K.A., Allsopp, M.T., Zweygarth, E. and Allsopp, B.A. Development of improved attenuated and nucleic acid vaccines for heartwater. Dev. Biol. (Basel) 2003; 114:121–136.

Cowdry, E.V. Studies on the etiology of heartwater. J. Exp. Med. 1925; 42.

Dawson, J.E., Anderson, B.E., Fishbein, D.B., Sanchez, J.L., Goldsmith, C.S., Wilson, K.H. and Duntley, C.W. Isolation and characterization of an *Ehrlichia* sp. from a patient diagnosed with human ehrlichiosis. J. Clin. Microbiol. 1991; 29:2741–2745.

Dawson, J.E. and Ewing, S.A. Susceptibility of dogs to infection with *Ehrlichia chaffeensis*, causative agent of human ehrlichiosis. Am. J. Vet. Res. 1992; 53:1322–1327.

Donatien, A. and Lestoquard, F. Existence en Algerie d'uen rickettsia du chien. Bull. Soc. Path. Exot. 1935; 28:418–419.

Doyle, C.K., Cardenas, A.M., Aguiar, D.M., Labruna, M.B., Ndip, L.M., Yu, X.J. and McBride, J.W. Molecular characterization of *E. canis* gp36 and *E. chaffeensis* gp47 tandem repeats among different geographic locations. Ann. N.Y. Acad. Sci. 2006a; 1063.

Doyle, C.K., Nethery, K.A., Popov, V.L. and McBride, J.W. Differentially expressed and secreted major immunoreactive protein orthologs of *Ehrlichia canis* and *E. chaffeensis* elicit early antibody responses to epitopes on glycosylated tandem repeats. Infect. Immun. 2006b; 74:711–720.

Doyle, C.K., Zhang, X., Popov, V.L. and McBride, J.W. An immunoreactive 38-kilodalton protein of *Ehrlichia canis* shares structural homology and iron-binding capacity with the ferric ion-binding protein family. Infect. Immun. 2005; 73:62–69.

du Plessis, J.L. Immunity in heartwater. I. A preliminary note on the role of serum antibodies. Onderstepoort J. Vet. Res. 1970; 37:147–149.

du Plessis, J.L. Colostrum-derived antibodies to *Cowdria ruminantium* in the serum of calves and lambs. Onderstepoort J. Vet. Res. 1984; 51:275–276.

du Plessis, J.L., Berche, P. and Van Gas, L. T cell-mediated immunity to *Cowdria ruminantium* in mice: the protective role of Lyt-2+ T cells. Onderstepoort J. Vet. Res. 1991; 58:171–179.

du Plessis, J.L. and Kumm, N.A. The passage of *Cowdria ruminantium* in mice. J. S. Afr. Vet. Med. Assoc. 1971; 42:217–221.

Dugan, V.G., Little, S.E., Stallknecht, D.E. and Beall, A.D. Natural infection of domestic goats with *Ehrlichia chaffeensis*. J. Clin. Microbiol. 2000; 38:448–449.

Dumler, J.S., Barbet, A.F., Bekker, C.P., Dasch, G.A., Palmer, G.H., Ray, S.C., Rikihisa, Y. and Rurangirwa, F.R. Reorganization of genera in the families *Rickettsiaceae* and *Anaplasmataceae* in the order *Rickettsiales*: unification of some species of *Ehrlichia* with *Anaplasma*, *Cowdria* with *Ehrlichia* and *Ehrlichia* with *Neorickettsia*, descriptions of six new species combinations and designation of *Ehrlichia equi* and 'HGE agent' as subjective synonyms of *Ehrlichia phagocytophila*. Int. J. Syst. Evol. Microbiol. 2001; 51:2145–2165.

Eng, T.R., Harkess, J.R., Fishbein, D.B., Dawson, J.E., Greene, C.N., Redus, M.A. and Satalowich, F.T. Epidemiologic, clinical, and laboratory findings of human ehrlichiosis in the United States, 1988. J. Am. Med. Assoc. 1990; 264:2251–2258.

Esteves, I., Vachiery, N., Martinez, D. and Totte, P. Analysis of *Ehrlichia ruminantium*-specific T1/T2 responses during vaccination with a protective killed vaccine and challenge of goats. Parasite Immunol. 2004a; 26:95–103.

Esteves, I., Walravens, K., Vachiery, N., Martinez, D., Letesson, J.J. and Totte, P. Protective killed *Ehrlichia ruminantium* vaccine elicits IFN-gamma responses by CD4+ and CD8+ T lymphocytes in goats. Vet. Immunol. Immunopathol. 2004b; 98:49–57.

Ewing, S.A. Canine ehrlichiosis. J. Am. Med. Assoc. 1963; 143:503–506.

Ewing, S.A., Dawson, J.E., Kocan, A.A., Barker, R.W., Warner, C.K., Panciera, R.J., Fox, J.C., Kocan, K.M. and Blouin, E.F. Experimental transmission of *Ehrlichia chaffeensis* (Rickettsiales: Ehrlichieae) among white-tailed deer by *Amblyomma americanum* (Acari: Ixodidae). J. Med. Entomol. 1995; 32:368–374.

Ewing, S.A., Roberson, W.R., Buckner, R.G. and Hayat, C.S. A new strain of *Ehrlichia canis*. J. Am. Vet. Med. Assoc. 1971; 159:1771–1774.

Feng, H.M. and Walker, D.H. Mechanisms of immunity to *Ehrlichia muris*: a model of monocytotropic ehrlichiosis. Infect. Immun. 2004; 72:966–971.

Fichtenbaum, C.J., Peterson, L.R. and Weil, G.J. Ehrlichiosis presenting as a life-threatening illness with features of the toxic shock syndrome. Am. J. Med. 1993; 95:351–357.

Fishbein, D.B., Dawson, J.E. and Robinson, L.E. Human ehrlichiosis in the United States, 1985 to 1990. Ann. Intern. Med. 1994; 120:736–743.

Fishbein, D.B., Kemp, A., Dawson, J.E., Greene, N.R., Redus, M.A. and Fields, D.H. Human ehrlichiosis: prospective active surveillance in febrile hospitalized patients. J. Infect. Dis. 1989; 160:803–809.

Frutos, R., Viari, A., Ferraz, C., Morgat, A., Eychenie, S., Kandassamy, Y., Chantal, I., Bensaid, A., Coissac, E., Vachiery, N., Demaille, J. and Martinez, D. Comparative genomic analysis of three strains of *Ehrlichia ruminantium* reveals an active process of genome size plasticity. J. Bacteriol. 2006; 188:2533–2542.

Ganta, R.R., Wilkerson, M.J., Cheng, C., Rokey, A.M. and Chapes, S.K. Persistent *Ehrlichia chaffeensis* infection occurs in the absence of functional major histocompatibility complex class II genes. Infect. Immun. 2002; 70:380–388.

Ginsberg, H.S., Ewing, C.P., O'Connell, A.F., Jr., Bosler, E.M., Daley, J.G. and Sayre, M.W. Increased population densities of *Amblyomma americanum* (Acari: Ixodidae) on Long Island, New York. J. Parasitol. 1991; 77:493–495.

Goldman, E.E., Breitschwerdt, E.B., Grindem, C.B., Hegarty, B.C., Walls, J.J. and Dumler, J.S. Granulocytic ehrlichiosis in dogs from North Carolina and Virginia. J. Vet. Intern. Med. 1998; 12:61–70.

Groves, M.G., Dennis, G.L., Amyx, H.L. and Huxsoll, D.L. Transmission of *Ehrlichia canis* to dogs by ticks (*Rhipicephalus sanguineus*). Am. J. Vet. Res. 1975; 36:937–940.

Heppner, D.G., Wongsrichanalai, C., Walsh, D.S., McDaniel, P., Eamsila, C., Hanson, B. and Paxton, H. Human ehrlichiosis in Thailand. Lancet 1997; 350:785–786.

Huxsoll, D.L., Amyx, H.L., Hemelt, I.E., Hildebrandt, P.K., Nims, R.M. and Gochenour, W.S., Jr. Laboratory studies of tropical canine pancytopenia. Exp. Parasitol. 1972; 31:53–59.

Huxsoll, D.L., Hildebrandt, P.K., Nims, R.M., Ferguson, J.A. and Walker, J.S. *Ehrlichia canis*—the causative agent of a haemorrhagic disease of dogs? Vet. Rec. 1969; 85:587.

Ijdo, J.W., Wu, C., Magnarelli, L.A., Stafford, K.C., III, Anderson, J.F. and Fikrig, E. Detection of *Ehrlichia chaffeensis* DNA in *Amblyomma americanum* ticks in Connecticut and Rhode Island. J. Clin. Microbiol. 2000; 38:4655–4656.

Inokuma, H., Brouqui, P., Drancourt, M. and Raoult, D. Citrate synthase gene sequence: a new tool for phylogenetic analysis and identification of *Ehrlichia*. J. Clin. Microbiol. 2001; 39:3031–3039.

Ismail, N., Soong, L., McBride, J.W., Valbuena, G., Olano, J.P., Feng, H.M. and Walker, D.H. Overproduction of TNF-alpha by CD8+ type 1 cells and down-regulation of IFN-gamma production by CD4+ Th1 cells contribute to toxic shock-like syndrome in an animal model of fatal monocytotropic ehrlichiosis. J. Immunol. 2004; 172:1786–1800.

Ismail, N., Stevenson, H.L. and Walker, D.H. Role of tumor necrosis factor alpha (TNF-alpha) and interleukin-10 in the pathogenesis of severe murine monocytotropic ehrlichiosis: increased resistance of TNF receptor p55- and p75-deficient mice to fatal ehrlichial infection. Infect. Immun. 2006; 74:1846–1856.

Jongejan, F. Protective immunity to heartwater (*Cowdria ruminantium* infection) is acquired after vaccination with in vitro-attenuated rickettsiae. Infect. Immun. 1991; 59:729–731.

Jongejan, F. and Thielemans, M.J. Identification of an immunodominant antigenically conserved 32-kilodalton protein from *Cowdria ruminantium*. Infect. Immun. 1989; 57:3243–3246.

Keefe, T.J., Holland, C.J., Salyer, P.E. and Ristic, M. Distribution of *Ehrlichia canis* among military working dogs in the world and selected civilian dogs in the United States. J. Am. Vet. Med. Assoc. 1982; 181:236–238.

Kelch, W.J. The canine ehrlichiosis (tropical canine pancytopenia) epizootic in Vietnam and its implications for the veterinary care of military working dogs. Mil. Med. 1984; 149:327–331.

Kim, C.M., Kim, M.S., Park, M.S., Park, J.H. and Chae, J.S. Identification of *Ehrlichia chaffeensis*, *Anaplasma phagocytophilum*, and *A. bovis* in *Haemaphysalis longicornis* and *Ixodes persulcatus* ticks from Korea. Vector Borne Zoonotic Dis. 2003; 3:17–26.

Knowles, T.T., Alleman, A.R., Sorenson, H.L., Marciano, D.C., Breitschwerdt, E.B., Harrus, S., Barbet, A.F. and Belanger, M. Characterization of the major antigenic protein 2 of *Ehrlichia canis* and *Ehrlichia chaffeensis* and its application for serodiagnosis of ehrlichiosis. Clin. Diagn. Lab. Immunol. 2003; 10:520–524.

Kocan, A.A., Levesque, G.C., Whitworth, L.C., Murphy, G.L., Ewing, S.A. and Barker, R.W. Naturally occurring *Ehrlichia chaffeensis* infection in coyotes from Oklahoma. Emerg. Infect. Dis. 2000; 6:477–480.

Krause, P.J., Corrow, C.L. and Bakken, J.S. Successful treatment of human granulocytic ehrlichiosis in children using rifampin. Pediatrics 2003; 112:e252–e253.

Lally, N.C., Nicoll, S., Paxton, E.A., Cary, C.M. and Sumption, K.J. The *Cowdria ruminantium* groE operon. Microbiology 1995; 141:2091–2100.

Lee, E.H. and Rikihisa, Y. Absence of tumor necrosis factor alpha, interleukin-6 (IL-6), and granulocyte-macrophage colony-stimulating factor expression but presence of IL-1beta, IL-8, and IL-10 expression in human monocytes exposed to viable or killed *Ehrlichia chaffeensis*. Infect. Immun. 1996; 64:4211–4219.

Lee, E.H. and Rikihisa, Y. Anti-*Ehrlichia chaffeensis* antibody complexed with *E. chaffeensis* induces potent proinflammatory cytokine mRNA expression in human monocytes through sustained reduction of IkappaB-alpha and activation of NF-kappaB. Infect. Immun. 1997; 65:2890–2897.

Lee, E.H. and Rikihisa, Y. Protein kinase A-mediated inhibition of gamma interferon-induced tyrosine phosphorylation of Janus kinases and latent cytoplasmic transcription factors in human monocytes by *Ehrlichia chaffeensis*. Infect. Immun. 1998; 66:2514–2520.

Li, J.S., Chu, F., Reilly, A. and Winslow, G.M. Antibodies highly effective in SCID mice during infection by the intracellular bacterium *Ehrlichia chaffeensis* are of picomolar affinity and exhibit preferential epitope and isotype utilization. J. Immunol. 2002; 169:1419–1425.

Li, J.S., Yager, E., Reilly, M., Freeman, C., Reddy, G.R., Reilly, A.A., Chu, F.K. and Winslow, G.M. Outer membrane protein-specific monoclonal antibodies protect SCID mice from fatal infection by the obligate intracellular bacterial pathogen *Ehrlichia chaffeensis*. J. Immunol. 2001; 166:1855–1862.

Lin, M. and Rikihisa, Y. *Ehrlichia chaffeensis* and *Anaplasma phagocytophilum* lack genes for lipid A biosynthesis and incorporate cholesterol for their survival. Infect. Immun. 2003a; 71:5324–5331.

Lin, M. and Rikihisa, Y. Obligatory intracellular parasitism by *Ehrlichia chaffeensis* and *Anaplasma phagocytophilum* involves caveolae and glycosylphosphatidylinositol-anchored proteins. Cell. Microbiol. 2003b; 5:809–820.

Lin, M. and Rikihisa, Y. *Ehrlichia chaffeensis* downregulates surface toll-like receptors 2/4, CD14 and transcription factors PU.1 and inhibits lipopolysaccharide activation of NF-kappa B, ERK 1/2 and p38 MAPK in host monocytes. Cell. Microbiol. 2004; 6:175–186.

Lin, M., Zhu, M.X. and Rikihisa, Y. Rapid activation of protein tyrosine kinase and phospholipase C-gamma2 and increase in cytosolic free calcium are required by *Ehrlichia chaffeensis* for internalization and growth in THP-1 cells. Infect. Immun. 2002; 70:889–898.

Lockhart, J.M., Davidson, W.R., Dawson, J.E. and Stallknecht, D.E. Temporal association of *Amblyomma americanum* with the presence of *Ehrlichia chaffeensis* reactive antibodies in white-tailed deer. J. Wildl. Dis. 1995; 31:119–124.

Lockhart, J.M., Davidson, W.R., Stallknecht, D.E., Dawson, J.E. and Howerth, E.W. Isolation of *Ehrlichia chaffeensis* from wild white-tailed deer (*Odocoileus virginianus*) confirms their role as natural reservoir hosts. J. Clin. Microbiol. 1997; 35:1681–1686.

Long, S.W., Zhang, X.F., Qi, H., Standaert, S., Walker, D.H. and Yu, X.J. Antigenic variation of *Ehrlichia chaffeensis* resulting from differential expression of the 28-kilodalton protein gene family. Infect. Immun. 2002; 70:1824–1831.

Luo, T., Zhang, X.F., Wakeel, A., Popov, V. and McBride, J.W. A variable-length PCR target protein of *Ehrlichia chaffeensis* contains major species-specific epitopes in acidic serine-rich tandem repeats. Infect. Immun. 2008; 76:1572–1580.

Louw, E., Brayton, K.A., Collins, N.E., Pretorius, A., Van, S.F. and Allsopp, B.A. Sequencing of a 15-kb *Ehrlichia ruminantium* clone and evaluation of the cpg1 open reading frame for protection against heartwater. Ann. N.Y. Acad. Sci. 2002; 969:147–150.

Maeda, K., Markowitz, N., Hawley, R.C., Ristic, M., Cox, D. and McDade, J.E. Human infection with *Ehrlichia canis*, a leukocytic rickettsia. N. Engl. J. Med. 1987; 316:853–856.

Mahan, S.M., Allsopp, B., Kocan, K.M., Palmer, G.H. and Jongejan, F. Vaccine strategies for *Cowdria ruminantium* infections and their application to other ehrlichial infections. Parasitol. Today 1999; 15:290–294.

Mahan, S.M., Andrew, H.R., Tebele, N., Burridge, M.J. and Barbet, A.F. Immunisation of sheep against heartwater with inactivated *Cowdria ruminantium*. Res. Vet. Sci. 1995; 58:46–49.

Mahan, S.M., Kumbula, D., Burridge, M.J. and Barbet, A.F. The inactivated *Cowdria ruminantium* vaccine for heartwater protects against heterologous strains and against laboratory and field tick challenge. Vaccine 1998; 16:1203–1211.

Mahan, S.M., McGuire, T.C., Semu, S.M., Bowie, M.V., Jongejan, F., Rurangirwa, F.R. and Barbet, A.F. Molecular cloning of a gene encoding the immunogenic 21 kDa protein of *Cowdria ruminantium*. Microbiology 1994a; 140(Pt 8):2135–2142.

Mahan, S.M., Peter, T.F., Simbi, B.H., Kocan, K., Camus, E., Barbet, A.F. and Burridge, M.J. Comparison of efficacy of American and African *Amblyomma* ticks as vectors of heartwater (*Cowdria ruminantium*) infection by molecular analyses and transmission trials. J. Parasitol. 2000; 86:44–49.

Mahan, S.M., Sileghem, M., Smith, G.E. and Byrom, B. Neutralization of bovine concanavalin-A T cell supernatant-mediated anti-*Cowdria ruminantium* activity with antibodies specific to interferon gamma but not to tumor necrosis factor. Parasite Immunol. 1996; 18:317–324.

Mahan, S.M., Smith, G.E. and Byrom, B. Conconavalin A-stimulated bovine T-cell supernatants inhibit growth of *Cowdria ruminantium* in bovine endothelial cells in vitro. Infect. Immun. 1994b; 62:747–750.

Martinez, D., Maillard, J.C., Coisne, S., Sheikboudou, C. and Bensaid, A. Protection of goats against heartwater acquired by immunisation with inactivated elementary bodies of *Cowdria ruminantium*. Vet. Immunol. Immunopathol. 1994; 41:153–163.

Martinez, D., Perez, J.M., Sheikboudou, C., Debus, A. and Bensaid, A. Comparative efficacy of Freund's and Montanide ISA50 adjuvants for the immunisation of goats against heartwater with inactivated *Cowdria ruminantium*. Vet. Parasitol. 1996; 67:175–184.

Massung, R.F., Lee, K., Mauel, M. and Gusa, A. Characterization of the rRNA genes of *Ehrlichia chaffeensis* and *Anaplasma phagocytophila*. DNA Cell Biol. 2002; 21:587–596.

Mattner, J., Debord, K.L., Ismail, N., Goff, R.D., Cantu, C., III, Zhou, D., Saint-Mezard, P., Wang, V., Gao, Y., Yin, N., Hoebe, K., Schneewind, O., Walker, D., Beutler, B., Teyton, L., Savage, P.B. and Bendelac, A. Exogenous and endogenous glycolipid antigens activate NKT cells during microbial infections. Nature 2005; 434:525–529.

Mavromatis, K., Doyle, C.K., Lykidis, A., Ivanova, N., Chain, P., Shin, M., Malfatti, S., Larimer, F., Copeland, A., Detter, J.C., Land, M., Richardson, P.M., Yu, X.J., Walker, D.H., McBride, J.W. and Kyripides, N.C. The genome of obligately intracellular *Ehrlichia canis* reveals themes of complex membrane structure and immune evasion strategies. J. Bacteriol. 2006; 188:4015–4023.

McBride, J.W., Comer, J.E. and Walker, D.H. Novel immunoreactive glycoprotein orthologs of *Ehrlichia* spp. Ann. N.Y. Acad. Sci. 2003b; 990:678–684.

McBride, J.W., Corstvet, R.E., Gaunt, S.D., Boudreaux, C., Guedry, T. and Walker, D.H. Kinetics of antibody response to *Ehrlichia canis* immunoreactive proteins. Infect. Immun. 2003a; 71:2516–2524.

McBride, J.W., Doyle, C.K., Zhang, X., Cardenas, A.M., Popov, V.L., Nethery, K.A. and Woods, M.E. Identification of a glycosylated *Ehrlichia canis* 19-kDa major immunoreactive protein with a species-specific serine-rich glycopeptide epitope. Infect. Immun. 2006; 75:74–82.

McBride, J.W., Ndip, L.M., Popov, V.L. and Walker, D.H. Identification and functional analysis of an immunoreactive DsbA-like thio-disulfide oxidoreductase of *Ehrlichia* spp. Infect. Immun. 2002; 70:2700–2703.

McCabe, T.R. and McCabe, R.E. Recounting whitetails past. In: The Science of Over Abundance: Deer Ecology and Population Management (W.J. McShea, H.B. Underwood and J.H. Rappole, Eds.), pp. 11–26. Washington, DC: Smithsonian Institution Press, 1997.

McQuiston, J.H., Paddock, C.D., Holman, R.C. and Childs, J.E. Human ehrlichioses in the United States. Emerg. Infect. Dis. 1999; 5:635–642.

Means, R.G. and White, D.J. New distribution records of *Amblyomma americanum* (Acari: Ixodidae) in New York State. J. Vect. Ecol. 1997; 22:133–145.

Mott, J., Barnewall, R.E. and Rikihisa, Y. Human granulocytic ehrlichiosis agent and *Ehrlichia chaffeensis* reside in different cytoplasmic compartments in HL-60 cells. Infect. Immun. 1999; 67:1368–1378.

Mutunga, M., Preston, P.M. and Sumption, K.J. Nitric oxide is produced by *Cowdria ruminantium*-infected bovine pulmonary endothelial cells in vitro and is stimulated by gamma interferon. Infect. Immun. 1998; 66:2115–2121.

Mwangi, D.M., Mahan, S.M., Nyanjui, J.K., Taracha, E.L. and McKeever, D.J. Immunization of cattle by infection with *Cowdria ruminantium* elicits T lymphocytes that recognize autologous, infected endothelial cells and monocytes. Infect. Immun. 1998a; 66:1855–1860.

Mwangi, D.M., McKeever, D.J. and Mahan, S.M. Cellular immune responses of cattle to *Cowdria ruminantium*. Dev. Biol. Stand. 1998b; 92:309–315.

Mwangi, D.M., McKeever, D.J., Nyanjui, J.K., Barbet, A.F. and Mahan, S.M. Immunisation of cattle against heartwater by infection with *Cowdria ruminantium* elicits T lymphocytes that recognise major antigenic proteins 1 and 2 of the agent. Vet. Immunol. Immunopathol. 2002; 85:23–32.

Neitz, W. and Alexander, R.A. The immunization of calves against heartwater. J. S. Afr. Vet. Assoc. 1941; 12:103–111.

Nethery, K.A., Doyle, C.K., Elsom, B.L., Herzog, N.K., Popov, V.L. and McBride, J.W. Ankyrin repeat containing immunoreactive 200 kD glycoprotein (gp200) orthologs of *Ehrlichia chaffeensis* and *Ehrlichia canis* are translocated to the nuclei of infected monocytes. In: 4th International Conference on Rickettsiae and Rickettsial Diseases, Longrono, Spain, 2005; p. O-60.

Nethery, K.A., Doyle, C.K., Zhang, X. and McBride, J.W. *Ehrlichia canis* gp200 contains dominant species-specific antibody epitopes in terminal acidic domains. Infect. Immun. 2007; 75:4900–4908.

Nuti, M., Serafini, D.A., Bassetti, D., Ghionni, A., Russino, F., Rombola, P., Macri, G. and Lillini, E. *Ehrlichia* infection in Italy. Emerg. Infect. Dis. 1998; 4:663–665.

Nyika, A., Barbet, A.F., Burridge, M.J. and Mahan, S.M. DNA vaccination with map1 gene followed by protein boost augments protection against challenge with *Cowdria ruminantium*, the agent of heartwater. Vaccine 2002; 20:1215–1225.

Nyika, A., Mahan, S.M., Burridge, M.J., McGuire, T.C., Rurangirwa, F. and Barbet, A.F. A DNA vaccine protects mice against the rickettsial agent *Cowdria ruminantium*. Parasite Immunol. 1998; 20:111–119.

Ohashi, N., Rikihisa, Y. and Unver, A. Analysis of transcriptionally active gene clusters of major outer membrane protein multigene family in *Ehrlichia canis* and *E. chaffeensis*. Infect. Immun. 2001; 69:2083–2091.

Ohashi, N., Zhi, N., Lin, Q. and Rikihisa, Y. Characterization and transcriptional analysis of gene clusters for a type IV secretion machinery in human granulocytic and monocytic ehrlichiosis agents. Infect. Immun. 2002; 70:2128–2138.

Ohashi, N., Zhi, N., Zhang, Y. and Rikihisa, Y. Immunodominant major outer membrane proteins of *Ehrlichia chaffeensis* are encoded by a polymorphic multigene family. Infect. Immun. 1998; 66:132–139.

Olano, J.P., Masters, E., Hogrefe, W. and Walker, D.H. Human monocytotropic ehrlichiosis, Missouri. Emerg. Infect. Dis. 2003; 9:1579–1586.

Paddock, C.D. and Childs, J.E. *Ehrlichia chaffeensis*: a prototypical emerging pathogen. Clin. Microbiol. Rev. 2003; 16:37–64.

Paddock, C.D., Folk, S.M., Shore, G.M., Machado, L.J., Huycke, M.M., Slater, L.N., Liddell, A.M., Buller, R.S., Storch, G.A., Monson, T.P., Rimland, D., Sumner, J.W., Singleton, J., Bloch, K.C., Tang, Y.W., Standaert, S.M. and Childs, J.E. Infections with *Ehrlichia chaffeensis* and *Ehrlichia ewingii* in persons coinfected with human immunodeficiency virus. Clin. Infect. Dis. 2001; 33:1586–1594.

Palella, F.J., Jr., Delaney, K.M., Moorman, A.C., Loveless, M.O., Fuhrer, J., Satten, G.A., Aschman, D.J. and Holmberg, S.D. Declining morbidity and mortality among patients with advanced human immunodeficiency virus infection. HIV Outpatient Study Investigators. N. Engl. J. Med. 1998; 338:853–860.

Parola, P., Cornet, J.P., Sanogo, Y.O., Miller, R.S., Thien, H.V., Gonzalez, J.P., Raoult, D., Telford, I.I.I., Sr. and Wongsrichanalai, C. Detection of *Ehrlichia* spp, *Anaplasma* spp., *Rickettsia* spp., and other eubacteria in ticks from the Thai-Myanmar border and Vietnam. J. Clin. Microbiol. 2003; 41:1600–1608.

Perez, M., Bodor, M., Zhang, C., Xiong, Q. and Rikihisa, Y. Human infection with *Ehrlichia canis* accompanied by clinical signs in Venezuela. Ann. N.Y. Acad. Sci. 2006; 1078:110–117.

Perez, M., Rikihisa, Y. and Wen, B. *Ehrlichia canis*-like agent isolated from a man in Venezuela: antigenic and genetic characterization. J. Clin. Microbiol. 1996; 34:2133–2139.

Perreau, P., Morel, P.C., Barre, N. and Durand, P. Cowdriosis (Heartwater) by *Cowdria ruminantium* in ruminants of French Indies (Guadeloupe) and Mascarene Islands (La Reunion and Mauritius). Rev. Elev. Med. Vet. Pays Trop. 1980; 33:21–22.

Petersen, L.R., Sawyer, L.A., Fishbein, D.B., Kelley, P.W., Thomas, R.J., Magnarelli, L.A., Redus, M. and Dawson, J.E. An outbreak of ehrlichiosis in members of an Army Reserve unit exposed to ticks. J. Infect. Dis. 1989; 159:562–568.

Popov, V.L., Chen, S.M., Feng, H.M. and Walker, D.H. Ultrastructural variation of cultured *Ehrlichia chaffeensis*. J. Med. Microbiol. 1995; 43:411–421.

Popov, V.L., Han, V.C., Chen, S.M., Dumler, J.S., Feng, H.M., Andreadis, T.G., Tesh, R.B. and Walker, D.H. Ultrastructural differentiation of the genogroups in the genus *Ehrlichia*. J. Med. Microbiol. 1998; 47:235–251.

Pretorius, A., Van, S.F., Brayton, K.A., Collins, N.E. and Allsopp, B.A. Genetic immunization with *Ehrlichia ruminantium* GroEL and GroES homologues. Ann. N.Y. Acad. Sci. 2002; 969:151–154.

Ravyn, M.D., Korenberg, E.I., Oeding, J.A., Kovalevskii, Y.V. and Johnson, R.C. Monocytic *Ehrlichia* in *Ixodes persulcatus* ticks from Perm, Russia [letter]. Lancet 1999; 353(9154):722–723.

Reddy, G.R. and Streck, C.P. Variability in the 28-kDa surface antigen protein multigene locus of isolates of the emerging disease agent *Ehrlichia chaffeensis* suggests that it plays a role in immune evasion [published erratum appears in Mol. Cell Biol. Res. Commun. 2000 Jan; 3(1):66]. Mol. Cell. Biol. Res. Commun. 2000; 1:167–175.

Rikihisa, Y. The tribe Ehrlichieae and ehrlichial diseases. Clin. Microbiol. Rev. 1991; 4:286–308.

Rikihisa, Y., Ewing, S.A. and Fox, J.C. Western immunoblot analysis of *Ehrlichia chaffeensis*, *E. canis*, or *E. ewingii* infections in dogs and humans. J. Clin. Microbiol. 1994; 32:2107–2112.

Ripoll, C.M., Remondegui, C.E., Ordonez, G., Arazamendi, R., Fusaro, H., Hyman, M.J., Paddock, C.D., Zaki, S.R., Olson, J.G. and Santos-Buch, C.A. Evidence of rickettsial spotted fever and ehrlichial infections in a subtropical territory of Jujuy, Argentina. Am. J. Trop. Med. Hyg. 1999; 61:350–354.

Seamer, J. and Snape, T. Tropical canine pancytopaenia and *Ehrlichia canis* infection. Vet. Rec. 1970; 86:375.

Sotomayor, E., Popov, V.L., Feng, H.M., Walker, D.H. and Olano, J.P. Animal model of fatal human monocytotropic ehrlichiosis. Am. J. Path. 2001; 158:757–769.

Standaert, S.M., Dawson, J.E., Schaffner, W., Childs, J.E., Biggie, K.L., Singleton, J.J., Gerhardt, R.R., Knight, M.L. and Hutcheson, R.H. Ehrlichiosis in a golf-oriented retirement community. N. Engl. J. Med. 1995; 333:420–425.

Standaert, S.M., Yu, T., Scott, M.A., Childs, J.E., Paddock, C.D., Nicholson, W.L., Singleton, J., Jr. and Blaser, M.J. Primary isolation of *Ehrlichia chaffeensis* from patients with febrile illnesses: clinical and molecular characteristics. J. Infect. Dis. 2000; 181:1082–1088.

Stromdahl, E.Y., Evans, S.R., O'Brien, J.J. and Gutierrez, A.G. Prevalence of infection in ticks submitted to the human tick test kit program of the U.S. Army Center for Health Promotion and Preventive Medicine. J. Med. Entomol. 2001; 38:67–74.

Sulsona, C.R., Mahan, S.M. and Barbet, A.F. The *map1* gene of *Cowdria ruminantium* is a member of a multigene family containing both conserved and variable genes. Biochem. Biophys. Res. Commun. 1999; 257:300–305.

Sumner, J.W., Childs, J.E. and Paddock, C.D. Molecular cloning and characterization of the *Ehrlichia chaffeensis* variable-length PCR target: an antigen-expressing gene that exhibits interstrain variation. J. Clin. Microbiol. 1999; 37:1447–1453.

Sumner, J.W., Nicholson, W.L. and Massung, R.F. PCR amplification and comparison of nucleotide sequences from the groESL heat shock operon of *Ehrlichia* species. J. Clin. Microbiol. 1997; 35:2087–2092.

Sumner, J.W., Sims, K.G., Jones, D.C. and Anderson, B.E. *Ehrlichia chaffeensis* expresses an immunoreactive protein homologous to the *Escherichia coli* GroEL protein. Infect. Immun. 1993; 61:3536–3539.

Tal, A. and Shannahan, D. Ehrlichiosis presenting as a life-threatening illness [letter; comment]. Am. J. Med. 1995; 98: 318–319.

Totte, P., Blankaert, D., Zilimwabagabo, P. and Werenne, J. Inhibition of *Cowdria ruminantium* infectious yield by interferons alpha and gamma in endothelial cells. Rev. Elev. Med. Vet. Pays Trop. 1993; 46:189–194.

Totte, P., McKeever, D., Martinez, D. and Bensaid, A. Analysis of T-cell responses in cattle immunized against heartwater by vaccination with killed elementary bodies of *Cowdria ruminantium*. Infect. Immun. 1997; 65:236–241.

Totte, P., Vachiery, N., Martinez, D., Trap, I., Ballingall, K.T., MacHugh, N.D., Bensaid, A. and Werenne, J. Recombinant bovine interferon gamma inhibits the growth of *Cowdria ruminantium* but fails to induce major histocompatibility complex class II following infection of endothelial cells. Vet. Immunol. Immunopathol. 1996; 53:61–71.

Treadwell, T.A., Holman, R.C., Clarke, M.J., Krebs, J.W., Paddock, C.D. and Childs, J.E. Rocky Mountain spotted fever in the United States, 1993–1996. Am. J. Trop. Med. Hyg. 2000; 63:21–26.

Uhaa, I.J., MacLean, J.D., Greene, C.R. and Fishbein, D.B. A case of human ehrlichiosis acquired in Mali: clinical and laboratory findings. Am. J. Trop. Med. Hyg. 1992; 46:161–164.

Uilenberg, G. Experimental transmission of *Cowdria ruminantium* by the Gulf coast tick *Amblyomma maculatum*: danger of introducing heartwater and benign African theileriasis onto the American mainland. Am. J. Vet. Res. 1982; 43:1279–1282.

Uilenberg, G. Heartwater (*Cowdria ruminantium* infection): current status. Adv. Vet. Sci. Comp. Med. 1983; 27:427–480.

Unver, A., Rikihisa, Y., Stich, R.W., Ohashi, N. and Felek, S. The omp-1 major outer membrane multigene family of *Ehrlichia chaffeensis* is differentially expressed in canine and tick hosts. Infect. Immun. 2002; 70:4701–4704.

Van der, M.L. The infection and treatment method of vaccination against heartwater. Onderstepoort J. Vet. Res. 1987; 54:489–491.

Whitlock, J.E., Fang, Q.Q., Durden, L.A. and Oliver, J.H., Jr. Prevalence of *Ehrlichia chaffeensis* (Rickettsiales: Rickettsiaceae) in *Amblyomma americanum* (Acari: Ixodidae) from the Georgia coast and barrier islands. J. Med. Entomol. 2000; 37:276–280.

Wilkins, J.H., Bowden, R.S. and Wilkinson, G.T. A new canine disease syndrome. Vet. Rec. 1967; 81:57–58.

Winslow, G.M., Yager, E., Shilo, K., Collins, D.N. and Chu, F.K. Infection of the laboratory mouse with the intracellular pathogen *Ehrlichia chaffeensis*. Infect. Immun. 1998; 66:3892–3899.

Winslow, G.M., Yager, E., Shilo, K., Volk, E., Reilly, A. and Chu, F. K. Antibody-mediated elimination of the obligate intracellular bacterial pathogen *Ehrlichia chaffeensis* during active infection. Infect. Immun. 2000; 68:2187–2195.

Yabsley, M.J., Little, S.E., Sims, E.J., Dugan, V.G., Stallknecht, D.E. and Davidson, W.R. Molecular variation in the variable-length PCR target and 120-kilodalton antigen genes of *Ehrlichia chaffeensis* from white-tailed deer (*Odocoileus virginianus*). J. Clin. Microbiol. 2003; 41:5202–5206.

Yager, E., Bitsaktsis, C., Nandi, B., McBride, J.W. and Winslow, G. Essential role for humoral immunity during *Ehrlichia* infection in immunocompetent mice. Infect. Immun. 2005; 73:8009–8016.

Yu, X.J., Crocquet-Valdes, P., Cullman, L.C. and Walker, D.H. The recombinant 120-kilodalton protein of *Ehrlichia chaffeensis*, a potential diagnostic tool. J. Clin. Microbiol. 1996; 34:2853–2855.

Yu, X.J., Crocquet-Valdes, P. and Walker, D.H. Cloning and sequencing of the gene for a 120-kDa immunodominant protein of *Ehrlichia chaffeensis*. Gene 1997; 184:149–154.

Yu, X.J., Crocquet-Valdes, P.A., Cullman, L.C., Popov, V.L. and Walker, D.H. Comparison of *Ehrlichia chaffeensis* recombinant proteins for serologic diagnosis of human monocytotropic ehrlichiosis. J. Clin. Microbiol. 1999a; 37:2568–2575.

Yu, X.J., McBride, J.W., Diaz, C.M. and Walker, D.H. Molecular cloning and characterization of the 120-kilodalton protein gene of *Ehrlichia canis* and application of the recombinant 120-kilodalton protein for serodiagnosis of canine ehrlichiosis. J. Clin. Microbiol. 2000a; 38:369–374.

Yu, X.J., McBride, J.W. and Walker, D.H. Genetic diversity of the 28-kilodalton outer membrane protein gene in human isolates of *Ehrlichia chaffeensis*. J. Clin. Microbiol. 1999b; 37:1137–1143.

Yu, X.J., McBride, J.W., Zhang, X.F. and Walker, D.H. Characterization of the complete transcriptionally active *Ehrlichia chaffeensis* 28 kDa outer membrane protein multigene family. Gene 2000b; 248:59–68.

Yu, X.J. and Walker, D.H. Sequence and characterization of an *Ehrlichia chaffeensis* gene encoding 314 amino acids highly homologous to the NAD A enzyme. FEMS Microbiol. Lett. 1997; 154:53–58.

Zhang, J.Z., Guo, H., Winslow, G.M. and Yu, X.J. Expression of members of the 28-kilodalton major outer membrane protein family of *Ehrlichia chaffeensis* during persistent infection. Infect. Immun. 2004a; 72:4336–4343.

Zhang, J.Z., McBride, J.W. and Yu, X.J. L-selectin and E-selectin expressed on monocytes mediating *Ehrlichia chaffeensis* attachment onto host cells. FEMS Microbiol. Lett. 2003; 227: 303–309.

Zhang, J.Z., Sinha, M., Luxon, B.A. and Yu, X.J. Survival strategy of obligately intracellular *Ehrlichia chaffeensis*: novel modulation of immune response and host cell cycles. Infect. Immun. 2004b; 72:498–507.

Zhang, X.F., Luo, T., Keysary, A., Baneth, G., Miyashiro, S., Strenger, C., Waner, T. and McBride, J.W. Genetic and antigenic diversities of major immunoreactive proteins in globally distributed *Ehrlichia canis* strains. Clin. Vaccine Immunol. 2008; 1080–1088.

Zweygarth, E., Josemans, A.I., van Strijp, M.F., Lopez-Rebollar, L., van, K.M. and Allsopp, B.A. An attenuated *Ehrlichia ruminantium* (Welgevonden stock) vaccine protects small ruminants against virulent heartwater challenge. Vaccine 2005; 23:1 695–1702.

CHAPTER

48

Extraintestinal Pathogenic *Escherichia coli*

Thomas A. Russo[1] *and James R. Johnson*[2]

[1]VA Medical Center, Department of Medicine, Department of Microbiology, The Center for Microbial Pathogenesis, and The Center for Excellence in Bioinformatics and Life Sciences, University at Buffalo, Buffalo, NY, USA

[2]Medical Service, VA Medical Center, and Department of Medicine, University of Minnesota, Minneapolis, MN, USA

OUTLINE

ABSTRACT

Extaintestinal pathogenic *Escherichia coli* (ExPEC) cause a wide variety of infections in humans and domestic animals. This chapter will focus on ExPEC strains as they relate to humans. ExPEC are responsible for significant morbidity, mortality, and costs to our healthcare system. The development of an efficacious ExPEC vaccine will minimize disease and may be cost-effective in women with recurrent urinary tract infection (UTI) and probably other groups at risk for ExPEC infection such as the elderly. Surface polysaccharides, such as capsule, have been traditional targets for vaccine development. Because the surface polysaccharides present in various ExPEC strains exhibit significant antigenic heterogeneity, their use as vaccine candidates will be challenging. ExPEC adhesins (e.g., P fimbriae and the type 1 pilus), iron-regulated outer membrane proteins (e.g., IroN), and lipid A/core saccharides of lipopolysaccharide (LPS) have been identified as potential vaccine candidates and have been shown to confer protection in various animal infection models. Human phase 1 studies have been completed for the adhesin of the type 1 pilus (FimH and its chaperone protein FimC) and for detoxified J5 LPS (lipid A/core saccharides) complexed to Neisseria meningitidis group B outer membrane proteins, but no phase 2/3 data have been published. A variety of killed whole-cell vaccines, administered via the vaginal or oral route for the prevention of recurrent UTI, have been assessed in animal models and human clinical trials in Europe. These hold some promise, but require further evaluation. We envision that one option for a successful ExPEC vaccine will be the development of a polyvalent subunit vaccine comprising some combination of adhesins, outer membrane proteins, and perhaps detoxified lipid A/core saccharides of LPS. An alternative approach will be the development of a killed whole-cell vaccine that contains multiple strains of genetically engineered derivatives of pathogenic wild-type ExPEC designed to generate an optimal immune response. Development of either a polyvalent subunit vaccine or a genetically engineered killed whole-cell vaccine will be challenging. However, achieving this goal is important because of the medical–economic burden attributable to infections due to ExPEC.

INTRODUCTION

A wide variety of infections continue to be responsible for significant morbidity, mortality, and costs to our healthcare system. The development of new vaccines directed against the responsible pathogens will minimize disease and could be highly cost-effective. *Escherichia coli* cause a wide variety of infections in both humans and domestic animals, including intestinal and extraintestinal disease. Strains of *E. coli* responsible for extraintestinal infections are genetically distinct from those that cause intestinal disease and have been termed extraintestinal pathogenic *E. coli* (ExPEC) (Russo and Johnson, 2000). This chapter will focus on ExPEC as they relate to humans. ExPEC are the most common enteric Gram-negative organisms to cause extraintestinal infection in the ambulatory, long-term-care, and hospital settings (McBean and Rajamani, 2001; Gransden et al., 1990; Roberts et al., 1991; Vastag, 2001; Siegman-Igra et al., 2002). Typical extraintestinal infections due to *E. coli* include urinary tract infection (UTI), diverse intraabdominal infections, pneumonia, surgical site infection, meningitis, intravascular device infection, osteomyelitis, and soft tissue infections, any of which can be accompanied by bacteremia. *E. coli* is a leading cause of severe sepsis (Fluit et al., 2001; Bernard et al., 2001; McBean and Rajamani, 2001; Siegman-Igra et al., 2002), with sepsis ranked as the tenth overall cause of death in the United States (Minino and Smith, 2001). In this era in which health news often sensationalizes uncommon infection syndromes or pathogens, the strains of ExPEC that cause extraintestinal infection are an increasingly important endemic problem and under-appreciated "killers." Billions of health-care dollars, millions of workdays, and hundreds of thousands of lives are lost each year to extraintestinal infections due to *E. coli*. Therefore, given the large number and diversity of infections due to ExPEC, an efficacious vaccine has the potential to have a significant clinical impact on a wide variety of medically important syndromes.

CLINICAL CLASSIFICATION OF *E. COLI*

From a genetic and clinical perspective, *E. coli* strains of biological significance to humans can be broadly categorized as: (1) commensal strains, (2) intestinal pathogenic (i.e., enteric or diarrheagenic) strains, and (3) ExPEC (Russo and Johnson, 2000).

Commensal *E. coli* variants, which constitute the major portion of the normal facultative intestinal flora in most humans and other mammals, for the most part confer benefits to their hosts such as colonization resistance. These strains generally lack the specialized virulence traits that enable intestinal and ExPEC to cause disease within or outside of the gastrointestinal tract, respectively. However, commensal *E. coli* can participate in extraintestinal infections when an aggravating factor is present, such as a foreign body (e.g., urinary catheter), host compromise (e.g., local anatomical or functional abnormalities such as urinary or biliary tract obstruction, or immunocompromise), or a high or a mixed bacterial species inoculum (e.g., with fecal contamination of the peritoneal cavity).

Intestinal pathogenic strains of *E. coli* have evolved a special ability to cause gastrointestinal disease, including enteritis, enterocolitis, and colitis. These strains are largely incapable of causing disease outside the intestinal tract. At least five distinct pathotypes of intestinal pathogenic *E. coli* exist: (1) Shiga-toxin-producing *E. coli*/enterohemorrhagic *E. coli*, (2) enterotoxigenic *E. coli*, (3) enteropathogenic *E. coli*, (4) enteroinvasive *E. coli*, and (5) enteroaggregative *E. coli* (Kaper et al., 2004). Diffusely adherent *E. coli* (DAEC) and cytodetaching *E. coli* are additional putative pathotypes (Le Bouguenec and Servin, 2006). Disease due to this group of pathogens occurs primarily in developing countries, with the exception of the enterohemorragic or Shiga-toxin-producing strains of *E. coli* and perhaps enteroaggregative *E. coli* (Nataro et al., 2006). Intestinal pathogenic strains of *E. coli* are rarely encountered in the fecal flora of healthy hosts, at least in the industrialized world, and instead appear to be essentially obligate pathogens. The effort to develop an efficacious vaccine directed against intestinal pathogenic *E. coli* is ongoing and has important medical–economic implications, and is discussed in Chapter 51.

The majority of *E. coli* isolates from symptomatic infection of the urinary tract, bloodstream, cerebral spinal fluid (Johnson and Stell, 2000; Picard et al., 1999), respiratory tract (Johnson et al., 2002a), and peritoneum (spontaneous bacterial peritonitis) (Soriano et al., 1995), as indicated by their virulence factor profiles and phylogenetic background, are distinct from commensal and intestinal pathogenic *E. coli*. It has been recently proposed that these strains of *E. coli* be termed ExPEC, rather than uropathogenic *E. coli* (UPEC), sepsis associated *E. coli* (SEPEC), or neonatal meningitis associated *E. coli* (NEMEC), to reflect their ability to cause disease at multiple anatomical sites (Johnson and Russo, 2002b; Russo and Johnson, 2000). Evaluation of a limited number of strains has established that ExPEC are also capable of causing surgical wound infection, osteomyelitis, and myositis, but the number of cases evaluated to date is too small to reliably assess proportions (Johnson and Russo, 2002a). Like commensal *E. coli* (but in contrast to intestinal pathogenic *E. coli*), ExPEC strains are often found within the normal intestinal flora and do not cause gastroenteritis in humans.

The fact that ExPEC are the major cause of most (and perhaps all) types of extraintestinal infection due to *E. coli* is an important concept. It implies that if preventive measures (e.g., vaccination) or therapies could be developed to specifically target the ExPEC fraction of the *E. coli* population, this could have a major beneficial impact on extraintestinal infections due to *E. coli*. For example, a vaccine strategy that leads to the development of bactericidal or interfering antibodies directed against ExPEC-specific virulence factors has the potential to prevent infections due to ExPEC without perturbing the commensal strains of *E. coli* that make up an important component of the normal intestinal flora.

EPIDEMIOLOGY

In healthy humans, commensal *E. coli* and ExPEC are the predominant species of Gram-negative bacilli in the colonic flora. *E. coli* and ExPEC only transiently colonize the oropharynx and skin of healthy individuals. In contrast, in long-term care and hospital settings, a variety of Gram-negative bacilli, including ExPEC, emerge as the dominant flora of both mucosal and skin surfaces, particularly in association with antimicrobial use, severe illness, and extended lengths of stay; and this colonization may lead to subsequent infection (e.g., oro-pharygeal colonization and pneumonia). Entry of ExPEC into an organ or anatomic location outside of these colonization sites is the limiting step for infection. As a result, ExPEC is a global pathogen and not a potential biothreat agent. All age groups, all types of hosts, and nearly all organs and anatomical sites are susceptible to infection by ExPEC. Previously healthy hosts infected with ExPEC can become severely ill or die; however, adverse outcomes are more likely in the presence of comorbid illnesses and compromised host defenses. *E. coli* is the most common enteric Gram-negative species to cause extraintestinal infection in ambulatory, long-term care, and hospital settings.

PATHOGENESIS

Since ExPEC strains are commonly part of the normal bowel flora, acquisition is not the rate-limiting step for infection. Instead, entry of an ExPEC strain from its site of colonization (e.g., the colon, vagina, or oropharynx) into a normally sterile extraintestinal site (e.g., the urinary tract, peritoneal cavity, or lungs) is the critical step in pathogenesis. Innate host defense systems (urine flow and mucociliary clearance, complement, antimicrobial peptides, professional phagocytes) and humoral immunity are the principal host-defense components. As a result, both susceptibility to and severity of infection are increased with dysfunction or deficiencies of these components (e.g., neutrophils). Multiple bacterial virulence factors are required for the pathogenesis of infections caused by ExPEC. Many putative virulence factors were identified epidemiologically as being more prevalent among clinical isolates as compared to fecal isolates from uninfected hosts (Arthur et al., 1989; Bingen et al., 1997; Maslow et al., 1993; Mitsumori et al., 1999; Johanson et al., 1993). However, such factors should be considered as "associated with," not necessarily "contributing to," virulence, since epidemiological associations do not guarantee causality. Animal models studies are a more direct test for whether a given factor contributes to pathogenesis and thereby warrants the designation of being a virulence factor (Table 48.1) (Johnson and Russo, 2005). Possession of specialized virulence genes is what defines ExPEC (e.g., versus commensal *E. coli*) and enables them to infect the host efficiently.

It is becoming clear that hosts and their cognate pathogens (e.g., humans and ExPEC) have been coadapting throughout evolutionary history. During this host–pathogen "chess match" over time, various and redundant strategies have emerged in both the pathogens and their hosts that enable these partners to maintain their coexistence. These virulence factors can be grouped by functional category, e.g., adhesins (Connell et al., 1996; Marklund et al., 1992; Mitsumori et al., 1998; Nowicki et al., 1990), toxins (Guyer et al., 2000; Rippere-Lampe et al., 2001b; Scott and Kaper, 1994; Welch et al., 1981), nutrient acquisition systems (iron, etc.) (Russo et al., 1996b, 2001, 2002; Opal et al., 1990; Schubert et al., 1998; Torres et al., 2001), protectins (e.g., surface polysaccharides) (Burns and Hull, 1999; Russo et al., 1994), and traits of miscellaneous or unknown function (Johnson, 2002a, 2003b; Kurazono et al., 2000) (Table 48.1). Clinical isolates often contain multiple virulence factors from a particular functional category. Conversely, many seemingly virulent strains lack known representatives of one or more of these functional categories (Johnson et al., 2001e, 2002c). The biological implications of these apparent redundancies and deficits are poorly understood. However, a given ExPEC strain usually possesses multiple adhesins for binding to a variety of host cells (e.g., in *E. coli*: type 1 fimbriae, sfa/foc fimbriae, P pili), many genes for nutrient acquisition (e.g., iron via siderophores) that are necessary but not sufficient for pathogenesis, and the ability to resist the bactericidal activity of complement and phagocytes in the absence of antibody (e.g., as conferred by capsule or O-antigen of lipopolysaccharide (LPS)), which is one of the defining traits of an extracellular pathogen (Russo et al., 1996a, 1993, 2003a). Host cell damage (e.g., as mediated by hemolysin in the case of ExPEC) may facilitate spread (e.g., intestinal translocation), release nutrients, and/or impair the function of phagocytes (Bhakdi et al., 1994; Russo et al., 2005, 2007b; Troeger et al., 2007). Many important ExPEC virulence genes doubtlessly await identification. ExPEC also possess the ability to induce septic shock. The lipid A moiety of LPS (via interaction with host Toll-like receptor 4) and probably also other bacterial factors stimulate a proinflammatory host response which, if overexuberant, results in shock.

ExPEC virulence factors that are surface exposed and fulfill a number of additional criteria, as will be discussed below, are potential vaccine candidates. Although is it unclear whether vaccination with an antigen shared by commensal *E. coli* will deleteriously affect our normal colonic flora, virulence factors unique to ExPEC are appealing since that consideration becomes a nonissue.

CLINICAL DISEASE

ExPEC are capable of causing infection in nearly every organ and site outside of the intestinal tract (Johnson and Russo, 2002a, 2002b; Johnson, et al., 2003a; Russo and Johnson, 2003; Smith et al., 2007). Several of the major ExPEC-associated infection syndromes are discussed below.

Urinary Tract Infection

The urinary tract is the most frequent extraintestinal site infected by *E. coli* (Johnson, 2003). UTIs are best considered by clinical syndrome (e.g., uncomplicated cystitis, pyelonephritis, catheter-associated, recurrent cystitis, recurrent pyelonephritis, asymptomatic bacteriuria, complicated UTI, and prostatitis) and within the context of specific hosts (e.g., premenopausal women, pregnant women, postmenopausal women, males, compromised host). Uncomplicated cystitis, the most common acute UTI syndrome, is characterized by dysuria, frequency, and suprapubic pain. Furthermore, 20% of women with an initial cystitis episode develop frequent recurrences (0.3 to >20 per year). Fever and/or back pain suggests progression to pyelonephritis. Asymptomatic bacteriuria warrants treatment in pregnant women, prior to urologic procedures, and perhaps in selected compromised hosts (Nicolle, 2006).

Abdominal Infection

The abdomen/pelvis is the second most frequent extraintestinal site for infections due to *E. coli*. A large

TABLE 48.1 Virulence-associated traits of extraintestinal pathogenic *Escherichia coli* (ExPEC) by functional category

Category	Gene(s) or operon	Comment	Evidence for trait as virulence factor: Epidemiological[a]	Evidence for trait as virulence factor: Experimental (in vivo)[b]
Adhesins[c]	*pap*	Pilus associated with pyelonephritis (P fimbriae)	Yes (Arthur et al., 1989; O'Hanley et al., 1985)	Yes (Roberts et al., 1994); no (Mobley et al., 1993)
	sfa/foc	S and F1C fimbriae	Yes (Schönian et al., 1992)	Yes (Marre and Hacker, 1987)
	sfa	S fimbriae (sialic acid-specific)	Yes (Johnson et al., 2002c)	Yes (Marre and Hacker, 1987)
	foc	F1C fimbriae	Yes (Pere et al., 1985)	No
	afa/dra	Dr antigen-specific adhesin operons (AFA I-III, Dr, F1845)	Yes (Nowicki et al., 1989; Zhang et al., 1997)	Yes (Goluszko et al., 1997)
	afaE-8	Afimbrial adhesin VIII	Yes (Le Bougenec et al., 2001)	No
	iha	Iron-regulated-gene-homologue adhesin (from O157:H7 and CFT073)	Yes (Johnson et al., 2000b; Kanamura et al., 2003)	Yes (Johnson et al., 2005a)
	bmaE	Blood group M-specific adhesin	No	No
	gafD	N-acetyl-D-glucosamine-specific (G, F17c) fimbriae adhesin	Yes (Bertin et al., 1996)	No
	clpG	CS31A adhesin (K88-related)	Yes (Bertin et al., 1996)	No
	nfa	NFA-1, -2, -3, -4 (nonfimbrial adhesins)	No (Goldhar et al., 1987; Grünberg et al., 1988; Hoschützky et al., 1989)	No
	csgA	Curli	Yes (Bian et al., 2000)	Yes (Bian et al., 2001)
	fim	D-mannose-specific adhesin, type 1 fimbriae	Yes (Mobley et al., 1987; Sokurenko et al., 1995)	Yes (Connell et al., 1996; Langermann et al., 1997)
	Antigen 43	Self-recognizing adhesin	Yes (Restieri et al., 2007)	Yes (Ulett et al., 2007)
Toxinsb	*hly*	a-Hemolysin	Yes (Brooks et al., 1980)	Yes (O'Hanley et al., 1991; Welch et al., 1981)
	cnf1	Cytotoxic necrotizing factor 1	Yes (Caprioli et al., 1987)	Yes (Khan et al., 2002; Rippere-Lampe et al., 2001a, 2001b)
	cdtB	Cytolethal distending toxin	Yes (Johnson et al., 2002c)	No
	sat	Secreted autotransporter toxin (serine protease autotransporter)	Yes (Guyer et al., 2000)	Yes (Guyer et al., 2002)
	pic, tsh	PIC and TSH, serine protease autotransporters	Yes (Heimer et al., 2004)	Yes (Dozois et al., 2000; Heimer et al., 2004)
	astA	EAST1 (heat-stable cytotoxin associated with enteroaggregative *E. coli*)	Yes (Janben et al., 2001)	No (Savarino et al., 1993)
Nutrition	*irp, fyuA*	Yersiniabactin (siderophore) synthesis and receptor	Yes (Clermont et al., 2001; Schubert et al., 1998)	Yes (Schubert et al., 2002)
	iuc, iutA	Aerobactin (siderophore) synthesis and receptor	Yes (Clermont et al., 2001; Fernandez-Beros et al., 1990)	Yes (Torres et al., 2001)
	iroN	Catecholate siderophore receptor	Yes (Bauer et al., 2002b; Johnson et al., 2000b)	Yes (Negre et al., 2004; Russo et al., 2002)
	ireA	Iron-regulated element (catecholate siderophore receptor)	Yes (Russo et al., 2001)	Yes (Russo et al., 2001)

(Continued)

TABLE 48.1 (Continued)

Category	Gene(s) or operon	Comment	Evidence for trait as virulence factor: Epidemiological[a]	Experimental (in vivo)[b]
	chuA	Heme receptor	Yes (Clermont et al., 2001)	Yes (Torres et al., 2001)
	entF	Enterobactin synthesis	No	Yes (Negre et al., 2004)
	tonB	Siderophore expression	N/a	Yes (Negre et al., 2004)
	guaA	Guanine synthesis	N/a	Yes (Russo et al., 1996b)
	argC	Arginine synthesis	N/a	Yes (Russo et al., 1996b)
	scrA,B,R,Y	Sucrose-specific phosphotransferase	Yes (Sorsa et al., 2007)	N/a
Protectins	*kpsMT* II	Group II capsular polysaccharide synthesis (e.g., K1, K5, K12)	Yes (Cross et al., 1984; Johnson and Stell, 2000; Orskov et al., 1982)	Yes (Kim et al., 1992)
	kpsMT III	Group III capsular polysaccharide synthesis (e.g., K3, K10, K54)	Yes (Cross et al., 1984; Johnson and Stell, 2000; Orskov et al., 1982)	Yes (Russo et al., 1994, 2003a)
	rfc	O4 lipopolysaccharide (LPS) synthesis	Yes (Korhonen et al., 1985; Opal et al., 1988)	Yes (Russo et al., 1996a, 2003a)
	cvaC	Microcin ColV; on plasmids with *traT*, *iss*, antibiotic resistance genes, *iuc/iut*	Yes (Fernandez-Beros et al., 1990)	Yes (Aguero et al., 1989)
	iss	Increased serum survival (outer membrane protein)	Yes (Delicato et al., 2003, Fernandez-Beros et al., 1990)	Yes (Binns et al., 1979)
	traT	Surface exclusion, serum survival-associated (outer membrane protein)	Yes (Fernandez-Beros et al., 1990)	No (Montenegro et al., 1985)
	proP	Osmoprotection; proline permease	N/a	Yes (Culham et al., 1998)
	oxyR	Global oxygen stress regulator	N/a	Yes (Johnson et al., 2006)
Misc.	*ompT*	Outer membrane protein T (protease)	Yes (Lundrigan and Webb, 1991)	No
	ompA	Outer membrane protein A (cellular invasion)	N/a	Yes (Kim, 2001)
	ibeA-C	Invasion of brain endothelium IbeA (Ibe10), B, and C	Yes (IbeA) (Johnson et al., 2002b)	Yes (Kim, 2001)
	aslA	Cellular invasion (arylsulfatase-like gene)	N/a	Yes (Kim, 2001)
	traJ	Cellular invasion (F-like plasmid transfer region homologue)	No	Yes (Kim, 2001)
	usp	Uropathogenic-specific protein (bacteriocin)9	Yes (Kurazono et al., 2000)	Yes (Yamamoto et al., 2001)
	malX	Pathogenicity-associated island marker (from strain CFT073)	Yes (Johnson et al., 2001a, 2003b)	No (Johnson et al., 2002a)
	fli	Flagellin	N/a	(Lane et al., 2005)
	ukoE-K	Unknown	Yes (Sorsa et al., 2007)	N/a

Note: The list is not comprehensive even for recognized markers, and many as-yet-unknown markers remain to be characterized. Conversely, not all of the listed traits necessarily contribute to virulence; some are only epidemiologically linked with virulence, and/or confer phenotypes that are suspected of promoting virulence. Additionally, some of these traits also are prominent among intestinal pathogenic *E. coli*, e.g., cytolethal distending toxin, certain Dr-binding adhesins, and EAST1. Misc., miscellaneous.
Source: Modified from Johnson and Russo (2005).
[a]Statistically associated with clinical isolates, or with specific host characteristics; or highly prevalent in a specific extraintestinal infection syndrome.
[b]Based on animal model infection studies, not necessarily using isogenic strains or complemented mutants.
[c]Certain adhesins and toxins also function as invasins, e.g., type 1 fimbriae, certain Dr-binding adhesins, and CNF1 (Khan et al., 2002; Mulvey et al., 2000; Selvarangan et al., 2001).

variety of clinical syndromes occur in this location, including acute peritonitis secondary to fecal contamination, spontaneous bacterial peritonitis, dialysis-associated peritonitis, diverticulitis, appendicitis, intraperitoneal or visceral abscesses (hepatic, pancreatic, splenic), infected pancreatic pseudo-cysts, septic cholangitis and/or cholecystitis, and pelvic inflammatory disease. *E. coli* can be isolated either alone or, as often occurs, with other facultative and/or anaerobic members of the intestinal flora.

Pulmonary Infection

E. coli is not usually regarded as a significant cause of pneumonia. Indeed, enteric Gram-negative bacilli account for only 2–5% of community-acquired pneumonia episodes, in part because in healthy hosts they colonize the oropharynx only transiently and in only a minority of individuals. In contrast, oral colonization with *E. coli* and other Gram-negative bacilli increases with severity of illness and antibiotic use. As a result, Gram-negative bacilli are a substantially more common cause of pneumonia among residents of long-term-care institutions (Muder, 1998) and are the most frequent cause (60–70%) of hospital-acquired pneumonia, particularly in postoperative patients and in the intensive care unit (Craven et al., 1990; Emori and Gaynes, 1993; Fluit et al., 2001; Center for Disease Control NNIS System, 1996). Although significant institutional variation occurs, *E. coli* is generally the third or fourth most commonly isolated enteric Gram-negative bacillus in nosocomial pneumonia, accounting for 5–8% of episodes in both U.S. and European-based studies (Cook et al., 1998; Fluit et al., 2001; Horan et al., 1988; Center for Disease Control NNIS System, 1996).

Surgical Site Infection

E. coli is the fourth most common organism implicated in surgical site infections. It accounts for 8% of the total 300,000 to 800,000 surgical site infections that occur annually in the United States (Center for Disease Control NNIS System, 1996).

Meningitis

E. coli is one of the two leading causes of neonatal meningitis (along with group B *Streptococcus*), causing 20–40% of an estimated 400 cases annually in the United States (De Louvois, 1994; Murphy, 2000). Other than in neonates, *E. coli* meningitis occurs predominantly in the setting of disruption of the meninges due to surgery or trauma, or in the presence of cirrhosis where presumably the meninges are seeded from poorly cleared portal-source episodes of bacteremia (Pauwels et al., 1997).

Additional Infections

E. coli is capable of causing infection in nearly every organ and anatomical site. Other extraintestinal infections that can be caused by *E. coli* include: infections of ulcers due to pressure or ischemia (particularly in patients with diabetes or other causes of neurovascular compromise), cellulitis, burn wound infections, osteomyelitis (via contiguous spread or hematogenously; particularly of vertebral bodies), orthopedic device-associated infection, and myositis/fasciitis (Elliott et al., 2000; Grayson et al., 1994; Revathi et al., 1998; Saccente, 1998; Sapico, 1999; Johnson and Russo, 2002a). *E. coli* occasionally causes complicated sinusitis and rarely causes endocarditis, vascular graft infection, endophthalmitis, or brain abscesses.

Bacteremia-Sepsis

E. coli bacteremia can result secondarily from any of the above primary sites of infection, from percutaneous intravascular devices, or from the increased intestinal mucosal permeability that occurs in neonates and accompanies neutropenia, chemotherapy-induced mucositis, trauma, and burns. In studies from all continents, *E. coli* accounts for 17–37% of clinically significant blood isolates, with isolates originating in the community versus hospital being approximately equal (Bernard et al., 2001; Elhanan et al., 1995; Fluit et al., 2001; Gosbell et al., 2000; Gransden et al., 1990; Ismail et al., 1997; McBean and Rajamani, 2001; McGregor and Colligan, 1993; Olesen et al., 1998; Rayner and Willcox, 1988; Sader et al., 2002; Siegman-Igra et al., 2002; Uslan et al., 2007). Isolation of *E. coli* from a blood culture is almost always clinically significant, and is typically accompanied by sepsis syndrome, severe sepsis (sepsis-induced dysfunction of at least one organ or system), or septic shock (American College of Chest Physicians, 1992).

SIGNIFICANCE AS PUBLIC HEALTH PROBLEM

The medical and economic impacts of extraintestinal *E. coli* infections are evident from a review of the following specific syndromes (Table 48.2). It is important to note, however, that data reported

TABLE 48.2 Estimated frequency and associated annual costs in the treatment of selected extraintestinal infection syndromes due to ExPEC in United States

Infection syndrome	Estimated frequency of cases	Estimated direct costs (U.S. dollars)
Uncomplicated cystitis in premenopausal women	6–8 million	1 billion
Uncomplicated pyelonephritis	250,000	175 million
Catheter-associated UTI	130,000–525,000	47–350 million
Pneumonia	14,100–23,400	80–133 million
Surgical site infection	24,000–64,000	94–252 million
Sepsis	127,500	1.1–5.9 billion

Source: Modified from Russo and Johnson (2003).

below have been generated primarily from studies based in the United States and Europe. The extent to which these data can be extrapolated to other countries is unknown. Clearly, costs will be dependent on health-care system. Additional data, in particular from developing countries, would be desirable.

Urinary Tract Infection

UTI is a common infection in ambulatory patients, accounting for 1% of ambulatory care visits in the United States, and second only to lower respiratory tract infection among the infections responsible for hospitalization (Hooton and Stamm, 1997; Stamm and Hooton, 1993). From both clinical and economic points of view, UTI is best considered by clinical syndrome within the context of specific hosts. *E. coli* is the single most prevalent pathogen for all UTI syndrome–host group combinations. *E. coli* causes 85–95% of episodes of uncomplicated cystitis in premenopausal women (Hooton and Stamm, 1997; Stamm and Hooton, 1993). The estimated total number of cases of uncomplicated cystitis is 6–8 million/year in the United States and 130–175 million globally. In the United States, uncomplicated cystitis is responsible for an estimated $1 billion of direct health-care costs annually (Harding and Ronald, 1994; Patton et al., 1991; Schappert, 1994). Furthermore, each episode results in a mean of two restricted activity days and six symptom days (Foxman and Frerichs, 1985). Because of this, since women now constitute nearly 50% of the work force, cystitis results in considerable additional indirect costs. *E. coli* are the cause of over 90% cases of uncomplicated pyelonephritis in premenopausal women (Talan et al., 2000). An estimated 250,000 cases of pyelonephritis occur each year in the United States, 100,000 of which require hospitalization (McCarthy, 1980; Stamm et al., 1989). Based on population ratios (U.S. population 301 million, global population 6.6 billion), this would translate into an estimated 5.5 million cases worldwide per annum. In the United States, the mean direct cost per cure is approximately $700 (Talan et al., 2000). Therefore, in the United States alone, the estimated *E. coli*-associated costs for pyelonephritis are $175 million/year. An estimated 1–1.5 million episodes of catheter-associated UTI (Fluit et al., 2001) occur annually in the United States, with an associated treatment cost of $363–676 per episode (Saint, 2000; Stamm et al., 1991; Tambyah et al., 2002). *E. coli* accounts for 13–35% of the episodes. Thus, the estimated direct health-care cost in the United States associated with catheter-associated UTI due to *E. coli* is $47–350 million/year.

Abdominal Infection

Specific data are not available for intraabdominal and pelvic infections regarding overall infection rates, the proportion of episodes in which *E. coli* is isolated, or associated costs. However, considering the frequency of these infections, the common participation of *E. coli*, and the multiple interventions and prolonged hospital stays often necessary for successful management, the cost of intraabdominal/pelvic infections due to *E. coli* is undoubtedly substantial.

Pulmonary Infection

Nosocomial pneumonia prolongs hospitalization by an estimated mean of 5.9 days and adds an estimated $5000 in costs per episode (Jarvis, 1996; Wenzel, 1989). Therefore, the total annual associated cost of nosocomial pneumonia in the United States is estimated to be $1.5 billion/year, with estimated *E. coli*-associated costs of $75–120 million/year.

Surgical Site Infection

Approximately 34 million surgical procedures are performed annually in the United States. As a result of these procedures, 300,000–800,000 surgical site infections occur every year (Vital and Health Statistics, 1998). Surgical site infection is associated with a 2–2.5-fold increase in the 6-month mortality rate (Kirkland et al., 1999; Poulsen et al., 1995). Surgical site infection results in a mean increase of 10.2 days of hospitalization and an added cost of $4000–5000 per patient (Kirkland et al., 1999; Zoutman et al., 1998). Therefore, the total estimated associated costs of surgical site infection in the United States is $1.2–4.0 billion/year, with the estimated *E. coli*-associated portion being $94–252 million/year.

Meningitis

The number of cases of *E. coli* meningitis may be increasing due to the increased use of ampicillin designed to reduce the vertical transmission of group B *Streptococcus*. Although this strategy has been effective in decreasing the prevalence of group B streptococcal infections, a concomitant increase in infections due to *E. coli* (85% were ampicillin resistant) has been observed (Stoll et al., 2002). Case-fatality rates range from 25–40%, and 33–50% of survivors develop neurological sequelae (De Louvois, 1994; Franco et al., 1992; Unhanand et al., 1993). No cost data are available for these infections, but the prolonged hospitalizations that they commonly entail and the extensive supportive and rehabilitative care that often is needed even after the infection is cured doubtlessly results in substantial direct costs despite the modest number of cases.

Bacteremia-Sepsis

Septicemia was the tenth ranked cause of mortality in the United States in 2000 and was responsible for 1.3% of deaths overall (Minino and Smith, 2001). A study, based on U.S. data, estimated that 751,000 cases of severe sepsis occurred in 1995. With the observed 28.6% associated mortality rate (Angus et al., 2001), this translates into an estimated 215,000 deaths. Among the 15 leading causes of death in the United States since 1950, the largest increase in death rate has been due to septicemia, increasing 38-fold (i.e., from 0.3 in 1950,to 11.5 in 2000, per 100,000 U.S. standard population) (Minino and Smith, 2001; Murphy, 2000). Estimated hospitalizations rates for septicemia in Medicare beneficiaries have also increased dramatically, from 90,965 in 1986 to 219,350 in 1997 (a 241% increase) (McBean and Rajamani, 2001). The incidence of severe sepsis and its associated mortality has been projected to increase in the coming years by 1.5% per annum (Angus et al., 2001). Although associated mortality overestimates true attributable mortality, in the context of sepsis this difference may be less than one might predict. The associated mortality for severe sepsis in patients with and without underlying comorbidity was 31.8% and 25%, respectively (Angus et al., 2001). Since the latter value derives from hosts without comorbid conditions that could confound the association of sepsis with mortality, it probably approximates the true attributable mortality due to sepsis. These data thus strongly suggests that most of the mortality associated with severe sepsis is actually due to sepsis itself and not to comorbidities.

In 1997, the approximately 219,350 hospitalizations for septicemia were estimated to cost Medicare $1.8 billion in hospitalization costs alone (McBean and Rajamani, 2001). The true cost is acknowledged to be significantly higher, since neither prehospital and posthospital costs, nor costs for physician services in hospital, were included. In 1997, a total of 773,530 cases of severe sepsis were estimated to occur nationwide (Angus et al., 2001). Therefore, extrapolation of estimated Medicare costs for septicemia would yield a conservative estimate of $6.5 billion/year. Considerably higher estimates have been provided with the average cost per case of severe sepsis calculated to be $22,100 (1995 dollars) (Angus et al., 2001) or $44,600 (2001 dollars) (Weycker et al., 2003). These figures project to an annual United States cost for all-cause severe sepsis of $16.7–34.5 billion.

By using the conservative estimate that *E. coli* causes 17% of the cases of severe sepsis (Bernard et al., 2001; Fluit et al., 2001; McBean and Rajamani, 2001), severe sepsis due to *E. coli* would be associated with an estimated 40,000 deaths in 2001, and an estimated *E. coli*-associated direct health-care costs of $1.1–2.8 billion/year in the United States alone. From a global perspective, the World Health Organization's report regarding the leading infectious causes of death worldwide in 1999 is informative (World health Organization, 2000). Although two of the infections listed (i.e., acute respiratory tract infection and acquired immunodeficiency syndrome) in some instances cause sepsis, sepsis per se did not appear on the list. Differentiating sepsis-associated mortality from mortality caused by an infection at a defined anatomical location may be difficult, if not impossible. Nonetheless, it is important to acknowledge sepsis as a distinct, infectious clinical entity. A sepsis-specific therapy has been approved in the United States (Bernard et al., 2001), and additional sepsis-specific therapies are in development. Therefore, regardless of whether sepsis is primary or secondary, its prevalence, associated mortality, and economic consequences warrant individual recognition. If the U.S. data on sepsis-associated mortality are extrapolated worldwide, sepsis-associated mortality would supplant acute lower respiratory infections (3,963,000 estimated deaths per annum) as the number one infectious cause of death (4,950,000 estimated deaths per annum). Sepsis-associated mortality due to *E. coli* (using the conservative estimate of *E. coli* causing 17% of sepsis cases (Bernard et al., 2001; Fluit et al., 2001; McBean and Rajamani, 2001)) would be ranked seventh overall (868,000 estimated deaths per annum). This calculation assumes that sepsis-associated mortality is uniform among all pathogens. This may not be the case, since some studies have suggested a higher sepsis-associated mortality with *Pseudomonas* and/or *Candida* than with other pathogens. Nonetheless, these figures leave little doubt that the global burden of *E. coli* sepsis-associated deaths is substantial.

SHORT HISTORY OF THE DISEASE

ExPEC, or their evolutionary precursor, have likely been part of the colonic flora of humans since the dawn of humankind. Likewise, infections due to ExPEC almost certainly have an equally long history. In the past, ExPEC have typically been highly susceptible to antibiotics, hence readily eradicated with antibiotic therapy. Unfortunately, this situation has changed as of late. Antimicrobial resistance among *E. coli* strains has increased (see "Treatment"). Furthermore, the incidence of serious extraintestinal infection due to *E. coli* increases with age (Angus et al., 2001; Jackson et al., 2005; McBean and Rajamani, 2001). As the proportion of elderly patients increases in the United States and other developed countries, so likely will the number of extraintestinal *E. coli* infections. The combination of increasing numbers and increasing antimicrobial resistance predictably will make the future management of extraintestinal *E. coli* infections more challenging and costlier than ever.

TREATMENT

In general, the frequency of resistance precludes empirical use of ampicillin, even in community-acquired infections. The prevalence of resistance to first-generation cephalosporins and trimethoprim-sulfamethoxazole is increasing among community-acquired strains in the United States (currently 10–40%) and is even higher outside North America. Until recently, trimethoprim-sulfamethoxazole was the drug of choice for the treatment of uncomplicated cystitis in many locales. Although continued empirical use of trimethoprim-sulfamethoxazole will predictably result in ever-diminishing cure rates, a wholesale switch to alternative agents (e.g., fluoroquinolones) will just as predictably accelerate the widespread emergence of resistance to these antimicrobial classes, as is already occurring in many areas. Resistance to fluoroquinolones has been steadily increasing over the last decade, with 2002–2005 prevalence values ranging from 5–20% in North America and even higher in other regions. The prevalence of resistance is higher in settings with extensive use of fluoroquinolone prophylaxis (e.g., in patients with leukemia, transplants, and cirrhosis), and among isolates from long-term-care facilities and hospitals. Among fluoroquinolone-resistant strains, a significant prevalence of coresistance (30–40%) to amoxicillin/clavulanic acid and piperacillin has been reported. The prevalence of coresistance to more advanced cephalosporins (second-, third-, and fourth-generation), monobactams (e.g., aztreonam), piperacillin-tazobactam, and the non-amikacin aminoglycosides is increasing but still generally <10%. Carbapenems (e.g., imipenem) and amikacin are the most predictably active agents, but recently resistance has been described against these agents as well (Bogaerts et al., 2007; Lartigue et al., 2007). Although clinical experience is limited, tigecycline and polymyxin B (an agent of last resort due to potential toxicities) are highly active in vitro (Stein and Craig, 2006).

PROTECTIVE IMMUNE RESPONSE

ExPEC are typical extracellular bacterial pathogens. These strains are inherently resistant to innate host defense factors such as complement, cationic antimicrobial peptides, and phagocytosis in the absence of opsonization. Given the extracellular lifestyle of ExPEC, the development of bactericidal antibodies should lead to protective immunity (Ahmed et al., 2002). However, despite the peaceful coexistence of ExPEC strains with humans (and other mammals and birds) on the intestinal (+/− the vaginal and oropharyngeal) mucosal surfaces, the host is unable to develop a protective immune response as a result of colonization. In fact, the host is susceptible not only to an initial infection at an extraintestinal site, but also to recurrent infections from both homologous and heterologous strains (Russo et al., 1995; Olesen et al., 1998; Maslow et al., 1994; Brauner et al., 1994). This suggests that the development of a protective immune response is not always a consequence of natural ExPEC infection. We hypothesize that ExPEC have evolved mechanisms to avoid and/or subvert an acquired protective immune response from the host. To successfully develop a vaccine to protect against ExPEC, an understanding is required of (1) the mechanisms by which these strains escape the host's acquired immune response, and (2) optimal immunization approaches.

Despite the host's apparent inability to develop a protective immune response to natural infection, we hypothesize that a successful immunization strategy can be developed that will confer protection against ExPEC. An immunologic correlate for protection against ExPEC infection (e.g., threshold antibody titer) has not been established. Further, individuals with deficiencies in humoral or cell-mediated immunity do not seem to have an increased incidence of ExPEC-mediated infections such as UTI. However, abundant experimental evidence supports that humoral immunity is important in conferring protection against ExPEC infection. In animal models, passive or active immunization against capsule, O-specific antigen, and iron-regulated outer membrane proteins (OMPs) have afforded protection against systemic infection (Kaijser and Ahlstedt, 1977; Salles et al., 1989; Bolin and Jensen, 1987), and immunization with capsule, O-antigen, and P and type 1 fimbriae, and the siderophore receptor IroN are protective against UTI from ExPEC strains expressing these virulence factors (Kaijser et al., 1978, 1983; Langermann et al., 1997, 2000; O'Hanley et al., 1991; Pecha et al., 1989; Russo et al., 2003b). Further, in cynomologus monkeys, more rapid clearance of *E. coli* from the urinary tract correlated with anti-*E. coli* serum IgM and urine secretory IgA and IgG (Hopkins and Uehling, 1995). Taken together, these data support that generation of an appropriate systemic and perhaps a mucosal (e.g., against cystitis) acquired immune response will prevent infections due to ExPEC.

ExPEC VACCINES

At present, no human ExPEC vaccines are licensed in the United States. A number of whole-cell vaccines that contain ExPEC and a variety of other pathogens that cause UTI have been studied and are available for use in Europe for the prevention of UTI. These include Urovac (SolcoBasel, Basel, Switzerland), Uro-Vaxom (OM Pharma, Myerlin, Switzerland), Urvakol (Institute of Microbiology, Czech Republic), and Urostim (National Center for Infectious and Parasitic Diseases, Bulgaria). However, the data that support the use of these vaccines are limited and/or far from convincing, and many studies lack scientific rigor. Therefore, we will discuss whole-cell vaccines under "ExPEC Vaccine Candidates" (Russo and Johnson, 2006).

ExPEC VACCINE CANDIDATES

Surface Polysaccharides

Development of a successful vaccine designed to protect against infections due to ExPEC presents several challenges (Russo and Johnson, 2006). First, although surface polysaccharides such as capsule have formed the basis for several successful vaccines directed against extracellular pathogens such as *Haemophilus influenzae* type b, *Streptococcus pneumoniae*, and certain serotypes of *Neisseria meningitidis* (Smit et al., 1977; Lepow et al., 1986; Centers for Disease Control and Prevention., 1991), this approach is less practical for ExPEC. More than 80 capsular serotypes exist for *E. coli* strains (Jann and Jann, 1990; Orskov et al., 1977). Further, for the O-antigen moiety of LPS, >150 major variants, some with multiple subtypes, have been described in *E. coli* (Whitfield and Valvano, 1993). Although only a fraction of these capsular and O-antigen variants are encountered among ExPEC, surface polysaccharides from ExPEC isolates nonetheless exhibit considerable antigenic diversity. Further, certain capsular polysaccharides (e.g., K1, K5) are poorly immunogenic, which may be due to their antigenic similarities to host tissues (Roberts, 1995). These considerations would be predicted to make the development of an ExPEC vaccine based on surface polysaccharides extremely challenging (Whitfield and Valvano, 1993). Despite this, the development of a polyvalent conjugated vaccine based on a combination of the most common capsular and O-antigen serotypes is still a consideration.

Adhesins

Assessment of surface-exposed protein moieties as putative vaccine candidates has also been pursued. Adhesins are a logical choice because of their surface exposure and potential role in pathogenesis. Antibodies directed against these elements have the promise to enhance bactericidal activity mediated by complement or by professional phagocytes via opsonization, and/or to inhibit bacterial binding to host structures, a potentially critical step in the pathogenesis of infection. P fimbriae, which via their PapG adhesin molecule recognize glycolipids that contain a Galα(1-4)Gal moiety, are associated with the 70–90% of ExPEC isolates that cause pyelonephritis and in 30–50% of those that cause cystitis. Systemic immunization with purified P fimbriae (Pecha et al., 1989; O'Hanley et al., 1991) and synthetic peptides corresponding to the protective epitope of the P fimbria major structural subunit (Schmidt et al., 1988) conferred protection in a murine pyelonephritis model. Likewise, immunization with purified P fimbriae or with purified PapDG protein complex conferred protection in a nonhuman primate model (Roberts et al., 1984, 2004), and passive immunization with anti-PapG conferred protection to chickens in an air sac infection model (Kariyawasam et al., 2004b). Human studies to assess the efficacy of a vaccine against P fimbriae or its components thereof have not been performed. FimH, the adhesin for the mannose binding type 1 pilus that has been shown to mediate binding to uroepithelial cells via uroplakin receptors, is a critical determinant in the pathogenesis of cystitis in mice, and is expressed in >90% of *E. coli* strains. Systemic immunization with recombinant FimH and its chaperone FimC was protective in a murine cystitis model (Langermann et al., 1997) and prevented cystitis in three of the four cynomolgus monkeys (Langermann et al., 2000). A human phase 2 trial has been undertaken, but published data are not available. If the FimH-FimC vaccine is successful in preventing cystitis, it has the potential to prevent bacteremia originating from this site. However, most cases of ExPEC bacteremia from a urinary tract origin occur in the setting of concomitant renal infection. The impact of a FimH-FimC vaccine on pyelonephritis is unclear. Further, the utility of this vaccine will probably not extend to the many other types of infection caused by extraintestinal *E. coli* outside of the urinary tract (Vandemaele et al., 2005; Kariyawasam et al., 2004b), since its protective effect appears to result from inhibition of bladder-specific adherence mediated by type 1 pilus (Langermann et al., 1997).

Immunization with purified Dr fimbriae reduced mortality after challenge with a homologous strain in a chronic UTI model, using C3H/HeJ mice, but bacterial titers in the urinary tract were not affected (Wroblewska-Seniuk et al., 2005). Although Dr fimbriae and related adhesins are associated with ExPEC isolates that cause UTI in children and pregnant women, overall their prevalence in surveys of ExPEC isolates is generally <10% (Gordon et al., 2005; Johnson et al., 2005c; Sannes et al., 2004), making them a less than ideal vaccine candidate, particularly as a "stand alone" immunogen. S and F1C fimbriae are more prevalent among ExPEC isolates, but have not been assessed as vaccine candidates.

Outer Membrane Proteins

OMPs are also potential vaccines candidates. Several iron-regulated OMPs have been assessed to date. This group of proteins is also logical given that (1) many are surface exposed; (2) iron acquisition is a requisite for pathogenesis, since free iron is largely unavailable in animal hosts; (3) a variety of iron-regulated and non-iron-regulated OMPs have been shown to exhibit increased expression in vivo in mice (Snyder et al., 2004); (4) unlike adhesins, which display considerable antigenic variability (Weissman et al., 2006), OMPs assessed as putative vaccine candidates demonstrate a high degree of conservation (Durant et al., 2007).

One of the most extensively studied iron-regulated OMPs has been the siderophore receptor IroN. Expression of IroN is increased in human blood, ascites, and urine (Russo et al., 1999; Hagan and Mobley, 2007), contributes to invasion of uroepithelial cells in vitro (Feldmann et al., 2007), and contributes to virulence in a mouse UTI model (Russo et al., 2002), in the avian air sac infection model (Dozois et al., 2003), and in the rat neonatal meningitis model (Negre et al., 2004). Systemic immunization with denatured IroN conferred protection against renal but not bladder infection in a mouse UTI model (Russo et al., 2002). A recent study demonstrated that both active and passive systemic immunization with IroN conferred protection against ExPEC challenge in a lethal sepsis model (Durant et al., 2007). The prevalence of IroN in human ExPEC isolates ranged from 39–83% (mean 50%, 461/919), depending on the site of infection (Johnson et al., 2001c, 2000a, 2002a, 2001d; Bauer et al., 2002b; Russo et al., 1999; Hagan and Mobley, 2007).

Limited published studies have pursued other OMPs as potential vaccine candidates. Several conserved OMPs have been shown to be expressed during UTI, leading the authors to conclude that such antigens could form the basis of an ExPEC vaccine (Hagan and Mobley, 2007). Through the use of a "reverse vaccinology" approach, FyuA and the antigens C0393, C4424, and C3389, but not ChuA, were shown to confer protection in a lethal sepsis model (Durant et al., 2007). Passive immunization with rabbit antisera, raised against iron-regulated OMPs in *E. coli* (O78/K80/H9), protected turkeys from bacteremia and death in an air sac inoculation model (Bolin and Jensen, 1987; Kariyawasam et al., 2004b). Active nasal immunization with purified aerobactin receptor IutA or passive immunization with anti-IgY raised against IutA protected chickens after challenge with the homologous strain and both the homologous and heterologous strains, respectively, in an air sac infection model (Kariyawasam et al., 2002, 2004b). A number of ExPEC iron-regulated OMPs have been demonstrated to be urovirulence factors (e.g., IreA, Iha) but have not yet been assessed as potential vaccine candidates (Johnson et al., 2005b; Russo et al., 2001).

OmpT, a non-iron-regulated OMP, is highly prevalent among ExPEC isolates, but has not yet been assessed experimentally as either a urovirulence factor or a potential vaccine candidate (Gordon et al., 2005; Johnson et al., 2005c; Sannes et al., 2004). No human studies have been performed assessing OMPs as vaccine candidates. However, OMPs possess their own challenges, which include difficulties in purification and in maintaining the proper conformational structure. Further, in ExPEC strains, although some OMPs are widely prevalent, others (e.g., IreA) are less so (Russo et al., 2001). Moreover, nearly all ExPEC strains possess capsule and O-antigen, which may impede antibody binding to OMP antigens. Lastly, given the variable prevalence of many OMPs in ExPEC, it is unlikely that a single OMP will be able to confer broad protection, thus necessitating the need for a polyvalent vaccine.

Lipid A/Core Saccharides of LPS

Because of the relative antigenic conservation of lipid A and the core saccharides of LPS, not only among the different *E. coli* strains but also among *Enterobacteriaceae* in general, there has been significant interest in the role of antibodies directed against these moieties in protecting against a variety of infections. The demonstration that passive administration of polyclonal antisera developed against an O-antigen-deficient *E. coli* strain (J5) could significantly decrease mortality against Gram-negative bacteremia (35/91 treated patients had *E. coli* bacteremia) generated tremendous excitement (Ziegler et al., 1982). It was hypothesized that the protective effect of J5 antisera was due to antibodies directed against core saccharides or lipid A, which neutralized the biologic activity of lipid A or increased bacterial clearance. Disappointingly, however, subsequent studies, which treated patients with Gram-negative sepsis with human monoclonal antibodies directed against lipid A, did not support the initial findings (Greenman et al., 1991; Ziegler et al., 1991). It has been subsequently demonstrated that J5 antiserum contains antibodies directed against ≥3 OMPs (e.g., OmpA, anti-murein lipoprotein, and peptidoglycan-associated lipoprotein), which suggested that the protective effect initially seen with polyclonal J5 antiserum (Ziegler et al., 1982) may actually have been due to antibodies directed against these alternative targets, rather than LPS per se (Hellman et al., 1997). Other investigators are continuing to pursue the development of a subunit vaccine composed of detoxified J5 LPS complexed to *N. meningitidis* group B OMPs (Cross et al., 2003).

Toxins

Alpha-hemolysin (HlyA) has been shown to be highly conserved (O'Hanley et al., 1993). In a mouse model of pyelonephritis, systemic immunization with purified HlyA was associated with decreased renal damage, but did not affect the clearance of *E. coli* (O'Hanley et al., 1991), and in another study both active and passive immunization with antisera raised against HlyA protected mice (Linggood and Ingram, 1982). Other toxins that are variably present in ExPEC isolates (Gordon et al., 2005; Johnson et al., 2005c; Sannes et al., 2004), such as cytotoxic necrotizing factor-1 (Rippere-Lampe et al., 2001b) and secreted autotransporter toxin (Guyer et al., 2002), have been shown to contribute to uropathogenesis, but have not been assessed as potential vaccine candidates (Guyer et al., 2002).

Whole-Cell Vaccines

Four standardized whole-cell vaccine formulations have been tested for their efficacy in preventing UTIs.

1. Urovac (SolcoBasel, Basel, Switzerland) consists of 10 heat-killed uropathogens (*E. coli*, $N = 6$; *P. mirabilis*, $N = 1$; *M. morganii*, $N = 1$; *K. pneumoniae*, $N = 1$; *E. faecalis*, $N = 1$: 1×10^8 cfu of each strain per dose). Urovac conferred protection in mouse, rat, and cynomolgus monkey UTI models when administered intramuscularly (IM) or via the intravaginal route (Kruze et al., 1992; Uehling et al., 1991, 1994). In an open comparative study of 400 women who had a UTI, there was a significant decrease in recurrent UTI in women treated with Urovac IM. However 47% had reactions that were attributed to endotoxin (Grischke and Ruttgers, 1987). As a result, subsequent studies explored intravaginal administration to minimize adverse reactions and maximize a mucosal immune response. A double-blind, placebo controlled, phase 2 clinical trial in 54 women demonstrated that six intravaginal immunizations with Urovac (0, 1, 2, 6, 10, 14 weeks), but not three intravaginal immunizations (0, 1, 2), increased the time to reinfection compared to placebo (Uehling et al., 2003). This protective effect was not associated with an increase in antibody levels against the uropathogens included in Urovac. Therefore, it is unclear whether the protection observed was due to an acquired immune response or an immunostimulatory effect on the host's innate defense response that may require regular booster

doses. A small (n = 24–26 per group), recent study demonstrated that women with recurrent UTI that underwent "primary" intravaginal immunization (0, 1, 2 weeks) and "primary" immunization with boosters (0, 1, 2, 6, 10, 14 weeks) were significantly more likely to remain free of recurrent *E. coli* infection, but not UTIs by any bacteria, compared to placebo (Gautam and Khera, 2007). This protective effect was not associated with differences in *E. coli* IgA and IgG antibodies in urine and vaginal fluid samples.

2. OM-89 or Uro-Vaxom (OM Pharma, Myerlin, Switzerland) is a lyophilized extract from 18 *E. coli* strains and is administered via the oral route. The proposed mechanism of OM-89 is induction of antibody response to antigens, both present and not present in the vaccine (Huber et al., 2000). A meta-analysis of five placebo-controlled double blind studies (1 capsule/day for 90 days with a 6-month study duration) supported that oral therapy with OM-89 was associated with a decrease in recurrent UTI compared to placebo (Bauer et al., 2002a). The same conclusion was reached in a recent multicenter, double-blind study conducted in 453 women over 12 months (1 capsule/day for months 1–3, no treatment in months 4–6, 1 capsule/day for the first 10 days of months 7–9) (Bauer et al., 2005).
3. Urvakol (Institute of Microbiology, Czech Republic) (Koukalova et al., 1999a, 1999b), which contains inactivated whole cells of *E. coli*, *P. mirabilis*, *P. aeruginosa*, and *E. faecalis*, and 4.Urostim (National Center for Infectious and Parasitic Diseases, Bulgaria) (Marinova et al., 2005, Nenkov et al., 1995), which contains lyophilized lysates of *E. coli*, *P. mirabilis*, *K. pneumoniae*, and *E. faecalis*, are also administered orally. These vaccine formulations have been studied both in animals and in human clinical trials. Published data are limited but suggest that these formulations have a nonspecific immunostimulatory effect on both innate and acquired immunity. It is unclear how the immunostimulatory properties of OM-89, Urvakol, and Urostim are more effective than those bestowed by the daily bacterial load from ingested food or from the endogenous colonic flora, which contains the uropathogens destined to cause UTI.

Other nonstandardized whole-cell formulations have been assessed in a variety of animal infection models. Vaginal immunization with a formalin-killed *E. coli* strain (O6, P fimbriae, type 1 pilus positive) conferred protection against cystitis to cynomolgus monkeys after challenge with the homologous strain (Uehling et al., 1987). Systemic immunization with formalin killed *E. coli* (O4, P fimbriae, type 1 pilus, hemolysin positive) afforded some protection against renal damage, but not bacterial clearance (Roberts et al., 1995). Immunization of chickens, via spray droplets, with attenuated *E. coli* constructs with deletions in either *galE*, *purA*, or *aroA* resulted in an appropriate immune response and conferred protection against pulmonary challenge with the wild-type homologous parent (O78), but not against a heterologous *E. coli* strain (O2) (Kariyawasam et al., 2004a). Thus, although the available data regarding whole-cell *E. coli* vaccines to date is far from convincing and many studies lack scientific rigor, based on successes with other pathogens (e.g., *B. pertussis*) the concept should not be discarded.

Genetically-Engineered Whole-Cell Vaccines

Our laboratory has generated data, which lay the foundation for the logical development of a killed, whole-cell *E. coli* vaccine (Russo et al., 2007a). We demonstrated that nasal immunization with a formalin-killed ExPEC derivative deficient in capsule and O-antigen results in (1) a significantly greater overall humoral response compared to its wild-type derivative (which demonstrates that capsule and/or the O-antigen impede the development of an optimal humoral immune response) and (2) a significantly greater immune response against noncapsular and O-antigen epitopes. We also demonstrated that a similar amount of antibodies that recognize surface epitopes on CP923 (capsule and O-antigen minus) is generated after immunization with live and formalin-killed CP923, suggesting that antigen presentation, surface architecture, and conformational epitopes are relatively conserved with formalin treatment. These antibodies also bound to a subset of heterologous ExPEC strains and enhanced neutrophil-mediated bactericidal activity against both the homologous and a heterologous strain. Taken together, these studies support the concept that formalin-killed genetically engineered ExPEC derivatives are whole-cell vaccine candidates to prevent infections due to ExPEC. We envision the potential for additional improvement of the humoral immune response. First, since the genetics of capsule and O-antigen production in ExPEC are well-described (Russo, 2002), construction of additional capsule and O-antigen-deficient derivatives should be straightforward. Our data support that a molecular epidemiologic approach, possibly incorporating virulence gene profiling and multilocus sequence typing, has the potential to identify prototypic strains from relatedness groups for use in a polyvalent vaccine containing multiple formalin-killed capsule and O-antigen-deficient constructs. The goal for such a vaccine would be to generate a polyclonal antibody response against surface-exposed epitopes in 90–100% of ExPEC isolates. Second, we hypothesize that ExPEC factors other than capsule and O-antigen may adversely modulate the host's acquired immune response or that a dominant antigen may narrow the immune response. If/when identified, disruption of the corresponding genes may further improve the host's antibody response. Third, adjuvants have the potential to enhance the immune response. Fourth, although lipidA toxicity should be less of an issue with mucosal immunization, if necessary, constructs can be generated in which the bioactive properties of lipidA have been minimized (Somerville et al., 1996). Lastly, if certain epitopes/proteins are shown in future studies to be critical in conferring protection against ExPEC infection, they can be overexpressed in the genetically engineered background that results in the optimal antibody response.

PROSPECTS FOR THE FUTURE

Although the development of an effective vaccine against ExPEC will be challenging, continued efforts in this area will likely lead to success (Russo and Johnson, 2006). Pursuit of this goal will be driven in part by the need for a human vaccine, but also by vaccine development against *E. coli* strains that cause infections in domesticated animals, which are responsible for a variety of infectious syndromes that result in a significant economic burden. Animal and human isolates of *E. coli* share common virulence factors, global genomic similarities, and in some cases are indistinguishable (Johnson et al., 2001b, 2005d, 2007; Moulin-Schouleur et al., 2007). In humans, the first target population will probably be women who suffer from recurrent UTIs. However, given the overall number of ExPEC infections and the fact that ExPEC is a leading cause of severe sepsis, with sepsis ranked as the tenth overall cause of death in the United States, we envision a broader population benefiting from an efficacious ExPEC vaccine. The high rate of *E. coli* bacteremia (3× higher than *S. pneumoniae*) in the elderly identifies them as one such population (Jackson et al., 2005). Two major approaches are likely to be pursued.

The first will be the continued identification and evaluation of ExPEC vaccine candidates (e.g., adhesins, iron-regulated and constitutively expressed OMPs, and perhaps lipid A/core saccharides of LPS). Because surface polysaccharides (e.g., capsule, O-antigen) exhibit antigenic heterogeneity, limited immunogenicity, and poor immunological memory, if not protein conjugated, at present they seem to be less viable candidates. Likewise, because adhesins possess significant antigenic variability, development of a broadly efficacious ExPEC vaccine based solely on adhesin antigens will also be challenging. At present, a number of studies support the use of OMPs as potential ExPEC vaccine candidates. The OMPs assessed to date are relatively conserved and have been shown to confer protection in several animal infection models. A potential problem may be that nearly all ExPEC strains possess capsule and O-antigen, and these surface polysaccharides may impede antibody binding to OMP antigens. Due to the variable prevalence of virulence factors in ExPEC strains, and the diversity of virulence factor profiles observed in different clonal groups of ExPEC, we predict that a polyvalent vaccine will be necessary to confer broad protection. Besides eliciting bactericidal antibodies (i.e., those that enhance complement-mediated bactericidal activity and/or bactericidal activity mediated by professional phagocytes via opsonization) against a variety of antigenic targets, a polyvalent vaccine also has the potential to inhibit the function of multiple virulence factors (e.g., adhesins, iron-acquisition proteins) necessary in the pathogenesis of ExPEC infection, thereby maximizing its efficacy. A potential limitation of this approach is the effort/money needed to identify and assess multiple vaccine candidates. Further, purification of proteins, especially OMPs, in an antigenically appropriate form may be challenging.

An alternative and more novel paradigm would be the continued development of a whole-cell vaccine. Immunization with whole *E. coli* has certain potential advantages. First, given that commensal *E. coli* and ExPEC are part of the normal human flora, mucosal immunization with whole organisms will likely be safe. Second, utilization of whole organisms has the potential for the development of bactericidal antibodies to multiple antigenic targets. Third, whole organisms may possess natural adjuvants (Jeannin et al., 2002, 2003). Lastly, immunization with whole organisms has the potential for development of bactericidal antibodies directed against conformational and linear epitopes. However, this approach also has potential limitations. Nasal immunization with killed, whole ExPEC strains is likely to generate antibodies directed against antigens present in commensal *E. coli* strains that constitute part of our normal intestinal flora. Whether this has the potential to modify the carriage of commensal *E. coli*, and if so, whether this might adversely affect the host, is unknown. Although nasal immunization can generate both a mucosal and a systemic response, an adjuvant may be needed to generate an immune response of sufficient magnitude to confer protection (Couch, 2004). There is strong evidence implicating the use of heat-labile *E. coli* enterotoxin as an adjuvant (for nasal immunization with an inactivated influenza vaccine) in the development of Bell's palsy (Mutsch et al., 2004). Therefore, prior to consideration of nasal immunization with an ExPEC vaccine formulation in humans, it will be necessary to develop a safe mucosal adjuvant. Lastly, given the antigenic complexity of a killed ExPEC vaccine and the possibility of biologically active components, there will undoubtedly be increased scrutiny by regulatory bodies regarding safety issues.

In summary, development of a polyvalent subunit vaccine or a genetically engineered killed whole-cell vaccine will be challenging. However, achieving this goal is important because of the medical–economic burden attributable to infections due to ExPEC.

KEY ISSUES

- An efficacious ExPEC vaccine would prevent disease and may be cost effective in women with recurrent UTIs and probably other high-risk groups for ExPEC infection, such as the elderly.
- ExPEC surface polysaccharides possess significant antigenic heterogeneity, making their use as vaccine targets challenging.
- ExPEC adhesins (e.g., P fimbriae and type 1 pilus), iron-regulated OMPs (e.g., IroN), and lipid A/core saccharides of LPS are immunogenic, and when used as immunogens confer protection in various animal infection models.
- Human phase 1 studies have been completed for vaccines based on the type 1 pilus adhesin (FimH and its chaperone protein FimC) and on detoxified J5 LPS (lipid A/core saccharides) complexed to *N. meningitidis* group B OMPs; however, no phase 2/3 data have been published.
- A variety of whole-cell vaccine formulations, administered via the vaginal or oral route for the prevention of recurrent UTI, have been assessed in animal models and human clinical trials in Europe. These hold some promise but require further evaluation.
- One option for a successful ExPEC vaccine will be the development of a polyvalent subunit vaccine composed of some combination of adhesins, OMPs, and perhaps detoxified lipid A/core saccharides of LPS.
- An alternative approach will be the construction of a whole-cell vaccine that contains multiple strains of killed, genetically engineered derivatives of pathogenic wild-type ExPEC in which factors that impede an optimal host immune response (e.g., capsule, O-antigen) have been inactivated and critical antigenic determinants are overexpressed.
- Development of either a polyvalent subunit vaccine or a genetically engineered killed whole-cell vaccine will be challenging. However, achieving this goal is important because of the medical–economic burden attributable to infections due to ExPEC.

ACKNOWLEDGMENTS

This material is based upon the work supported by Office of Research and Development, Medical Research Service, Department of Veterans Affairs (TAR and JRJ).

References

Aguero, M.E., De La Fuente, G., Vivaldi, E. and Cabello, F. ColV increases the virulence of *Escherichia coli* K1 strains in animal models of neonatal meningitis and urinary infection. Med. Microbiol. Immunol. 1989; 178:211–216.

Ahmed, R., Lanier, J. and Pamer, E. Immunological memory and infection. In: Immunology of Infectious Diseases (S. Kaufmann, A. Sher and R. Ahmed, Eds.). Washington DC: ASM Press, 2002.

American College of Chest Physicians—Society of Critical Care Medicine Consensus Conference Definitions for sepsis and organ failure and guidelines for the use of innovative therapies in sepsis. Crit. Care Med. 1992; 20:864–875.

Angus, D., Linde-Zwirble, W., Lidlicker, J., Clermont, G., Carcillo, J. and Pinsky, M. Epidemiology of severe sepsis in the United States: analysis of incidence, outcome, and associated costs of care. Crit. Care Med. 2001; 29:1303–1310.

Arthur, M., Johnson, C.E., Rubin, R.H., Arbeit, R.D., Campanelli, C., Kim, C., Steinbach, S., Agarwal, M., Wilkinson, R. and Goldstein, R. Molecular epidemiology of adhesin and hemolysin virulence factors among uropathogenic *Escherichia coli*. Infect. Immun. 1989; 57:303–313.

Bauer, H.W., Alloussi, S., Egger, G., Blumlein, H.M., Cozma, G. and Schulman, C.C. A long-term, multicenter, double-blind study of an *Escherichia coli* extract (OM-89) in female patients with recurrent urinary tract infections. Eur. Urol. 2005; 47:542–548, discussion 548. Epub Jan. 21, 2005.

Bauer, H.W., Rahlfs, V.W., Lauener, P.A. and Blessmann, G.S. Prevention of recurrent urinary tract infections with immuno-active *E. coli* fractions: a meta-analysis of five placebo-controlled double-blind studies. Int. J. Antimicrob. Agents 2002a; 19:451–456.

Bauer, R., Zhang, L., Foxman, B., Siitonen, A., Jantunen, M., Saxen, H. and Marrs, C. Molecular epidemiology of 3 putative virulence genes for *Escherichia coli* urinary tract infection—*usp*, *iha*, and *iroN* (*E. coli*). J. Infect. Dis. 2002b; 185:1521–1524.

Bernard, G., Vincent, J., Laterre, P., Larosa, S., Dhainaut, J., Lopez-Rodriguez, A., Steingrub, J., Garber, G., Helterbrand, J., Ely, E. and Fisher, C.J. Recombinant human protein C Worldwide Evaluation in Severe Sepsis (PROWESS) study group Efficacy and safety of recombinant human activated protein C for severe sepsis. N. Engl. J. Med. 2001; 344:699–709.

Bertin, Y., Girardeau, J.-P., Darfeuille-Michaud, A. and Contrepois, M. Characterization of 20K fimbria, a new adhesin of septicemic and diarrhea-associated *Escherichia coli* strains, that belongs to a family of adhesins with *N*-acetyl-D-glucosamine recognition. Infect. Immun. 1996; 64:332–342.

Bhakdi, S., Grimminger, F., Suttorp, N., Walmrath, D. and Seeger, W. Proteinaceous bacterial toxins and pathogenesis of sepsis syndrome and septic shock: the unknown connection. Med. Microbiol. Immunol. 1994; 183:119–144.

Bian, Z., Brauner, A., Li, Y. and Normark, S. Expression of and cytokine activation by Escherichia coli curli fibers in human sepsis. J. Infect. Dis. 2000; 181:602–612.

Bian, Z., Yan, Z.-Q., Hansson, G.K., Thoren, P. and Normark, S. Activation of inducible nitric oxide synthast/nitric oxide by curli fibers leads to a fall in blood pressure during systemic *Escherichia coli* infection in mice. J. Infect. Dis. 2001; 183:612–619.

Bingen, E., Bonacorsi, S., Brahimi, N., Denamur, E. and Elion, J. Virulence patterns of *Escherichia coli* K1 strains associated with neonatal meningitis. J. Clin. Microbiol. 1997; 35:2981–2982.

Binns, M.M., Davies, D.L. and Hardy, K.G. Cloned fragments of the plasmid ColV, I-K94 specifying virulence and serum resistance. Nature 1979; 279:778–781.

Bogaerts, P., Galimand, M., Bauraing, C., Deplano, A., Vanhoof, R., De Mendonca, R., Rodriguez-Villalobos, H., Struelens, M. and Glupczynski, Y. Emergence of *ArmA* and *RmtB* aminoglycoside resistance 16S rRNA methylases in Belgium. J. Antimicrob. Chemother. 2007; 59:459–464, Epub Jan. 15, 2007.

Bolin, C. and Jensen, A. Passive immunization with antibodies against iron-regulated outer membrane proteins protects turkeys from *Escherichia coli* septicemia. Infect. Immun. 1987; 55:1239–1242.

Brauner, A., Kaijser, B. and Kuhn, I. Recurrent *Escherichia coli* bacteraemia—clinical characteristics and bacterial properties. J. Infect. 1994; 28:49–57.

Brooks, H.J.L., O'Grady, F., Mcsherry, M.A. and Cattell, W.R. Uropathogenic properties of *Escherichia coli* in recurrent urinary-tract infection. J. Med. Microbiol. 1980; 13:57–68.

Burns, S.M. and Hull, S.I. Loss of resistance to ingestion and phagocytic killing by O(-) and K(-) mutants of a uropathogenic *Escherichia coli* O75:K5 strain. Infect. Immun. 1999; 67:3757–3762.

Caprioli, A., Falbo, V., Ruggeri, F.M., Baldassarri, L., Bisicchia, R., Ippolito, G., Romol, E. and Donelli, G. Cytotoxic necrotizing factor production by hemolytic strains of *Escherichia coli* causing extraintestinal infections. J. Clin. Microbiol. 1987; 25:146–149.

Centers for Disease Control and Prevention. Recommendation of the Immunization Practices Advisory Committee (ACIP) *Haemophilus* b conjugate vaccines for prevention of *Haemophilus influenzae* type b disease among infants and children two months of age and older. MMWR Morb. Mortal. Wkly Rep. 1991; 40:1–17.

Center for Disease Control NNIS System National nosocomial infections surveillance (NNIS) report, data summary from October 1986–April 1996, issued May 1996. Am. J. Infect. Control 1996; 24:380–388.

Clermont, O., Bonacorsi, S. and Bingen, E. The *Yersinia* high-pathogenicity island is highly predominant in virulence-associated phylogenetic groups of *Escherichia coli*. FEMS Microbiol. Lett. 2001; 196:153–157.

Connell, I., Agace, W., Klemm, P., Schembri, M., Marild, S. and Svanborg, C. Type 1 fimbrial expression enhances *Escherichia coli* virulence for the urinary tract. Proc. Natl. Acad. Sci. U.S.A. 1996; 93:9827–9832.

Cook, D., Guyatt, G., Marshall, J., Leasa, D., Fuller, H., Hall, R., Peters, S., Rutledge, F., Griffith, L., Mclellan, A., Wood, G. and Kirby, A. A comparison of sucralfate and ranitidine for the prevention of upper gastrointestinal bleeding in patients requiring mechanical ventilation. New Engl. J. Med. 1998; 338: 791–797.

Couch, R. Nasal vaccination, *Escherichia coli* enterotoxin, and Bell's palsy. New Engl. J. Med. 2004; 350:860–861.

Craven, D., Barber, T., Steger, K. and Montecalvo, M. Nosocomial pneumonia in the 1990s: update of epidemiology and risk factors. Semin. Respir. Infect. 1990; 5:157–172.

Cross, A.S., Gemski, P., Sadoff, J.C., Orskov, F. and Orskov, I. The importance of the K1 capsule in invasive infections caused by *Escherichia coli*. J. Infect. Dis. 1984; 149:184–193.

Cross, A.S., Opal, S.M., Palardy, J.E., Drabick, J.J., Warren, H.S., Huber, C., Cook, P. and Bhattacharjee, A.K. Phase I study of detoxified *Escherichia coli* J5 lipopolysaccharide (J5dLPS)/group B meningococcal outer membrane protein (OMP) complex vaccine in human subjects. Vaccine 2003; 21:4576–4587.

Culham, D.E., Dalgado, C., Gyles, C.L., Mamelak, D., Maclellan, S. and Wood, J.M. Osmoregulatory transporter ProP influences colonization of the urinary tract by *Escherichia coli*. Microbiology 1998; 144:91–102.

Delicato, E.R., De Brito, B.G., Gaziri, L.C. and Vidotto, M.C. Virulence-associated genes in *Escherichia coli* isolates from poultry with colibacillosis. Vet. Microbiol. 2003; 94:97–103.

De Louvois, J. Acute bacterial meningitis in the newborn. J. Antimicrob. Chemother. 1994; 34(Suppl. A):61–73.

Dozois, C., Daigle, F. and Curtiss, R.I. Identification of pathogen-specific and conserved genes expressed in vivo by an avian pathogenic *Escherichia coli* strain. Proc. Natl. Acad. Sci. U.S.A. 2003; 100:247–252.

Dozois, C.M., Dho-Moulin, M., Bree, A., Fairbrother, J.M., Desautels, C. and Curtiss 3rd., R. Relationship between the Tsh autotransporter and pathogenicity of avian *Escherichia coli* and localization and analysis of the Tsh genetic region. Infect. Immun. 2000; 68:4145–4154.

Durant, L., Metais, A., Soulama-Mouze, C., Genevard, J.M., Nassif, X. and Escaich, S. Identification of candidates for a subunit vaccine against extraintestinal pathogenic *Escherichia coli*. Infect. Immun. 2007; 75:1916–1925, Epub Dec. 4, 2006.

Elhanan, G., Raz, R., Pitlik, S., Sharir, R., Konisberger, H., Samra, Z., Kennes, Y., Drucker, M. and Leibovici, L. Bacteraemia in a community and a university hospital. J. Antimicrob. Chemother. 1995; 36:681–695.

Elliott, D., Kufera, J. and Myers, R. The microbiology of necrotizing soft tissue infections. Am. J. Surg. 2000; 179:361–366.

Emori, T. and Gaynes, R. An overview of nosocomial infections, including the role of the microbiology laboratory. Clin. Microbiol. Rev. 1993; 6:428–442.

Feldmann, F., Sorsa, L.J., Hildinger, K. and Schubert, S. The salmochelin siderophore receptor IroN contributes to invasion of urothelial cells by extraintestinal pathogenic *Escherichia coli* in vitro. Infect. Immun. 2007; 75:3183–3187, Epub Mar. 12, 2007.

Fernandez-Beros, M.E., Kissel, V., Lior, H. and Cabello, F.C. Virulence-related genes in ColV plasmids of *Escherichia coli* isolated from human blood and intestines. J. Clin. Microbiol. 1990; 28:742–746.

Fluit, A., Schmitz, F., Verhoef, J. and Group., E.S.P. Frequency of isolation of pathogens from bloodstream, nosocomial pneumonia, skin and soft tissue, and urinary tract infections occurring in European patients. Eur. J. Clin. Microbiol. Infect. Dis. 2001; 20:188–191.

Foxman, B. and Frerichs, R. Epidemiology of urinary tract infection. 1. Diaphragm use and sexual intercourse. Am. J. Public Health 1985; 75:1308–1313.

Franco, S., Cornelius, V. and Andrews, B. Long-term outcome of neonatal meningitis. Am. J. Dis. Child 1992; 146:567–571.

Gautam, G. and Khera, R. Vaginal mucosal vaccine for recurrent urinary tract infections. Indian J. Urol. 2007; 23:335–336.

Goldhar, J., Perry, R., Golecki, J.R., Hoschutzky, H., Jann, B. and Jann, K. Nonfimbrial, mannose-resistant adhesins from uropathogenic *Escherichia coli* 083:K1:H4 and 014:K?:H11. Infect. Immun. 1987; 55:1837–1842.

Goluszko, P., Moseley, S.L., Truong, S.L., Kaul, A., Williford, J.R., Selvarangan, R. and Nowicki, S. Development of experimental model of chronic pyelonephritis with *Escherichi coli* O75:K5: H- bearing Dr fimbriae. J. Clin. Invest. 1997; 99:1661–1672.

Gordon, D.M., Stern, S.E. and Collignon, P.J. Influence of the age and sex of human hosts on the distribution of Escherichia coli ECOR groups and virulence traits. Microbiology 2005; 151:15–23.

Gosbell, I., Newton, P. and Sullivan, E. Survey of blood cultures from five community hospitals in south-western Sydney, Australia, 1993–1994. Aust. N.Z. J. Med. 2000; 30:684–692.

Gransden, W., Eykyn, S., Phillips, I. and Rowe, B. Bacteremia due to *Escherichia coli*: a study of 861 episodes. Rev. Infect. Dis. 1990; 12:1008–1018.

Grayson, M., Gibbons, G., Habershaw, G., Freeman, D., Pomposelli, F., Rosenblum, B., Levin, E. and Karchmer, A. Use of ampicillin/sulbactam versus imipenem/cilastatin in the treatment of limb-threatening foot infections in diabetic patients. Clin. Infect. Dis. 1994; 18:683–693.

Greenman, R.L., Schein, R.M., Martin, M.A., Wenzel, R.P., Macintyre, N.R., Emmanuel, G., Chmel, H., Kohler, R.B., Mccarthy, M., Plouffe, J. et al. A controlled clinical trial of E5 murine monoclonal IgM antibody to endotoxin in the treatment of Gram-negative sepsis. The XOMA Sepsis Study Group. JAMA 1991; 266:1097–1102.

Grischke, E.M. and Ruttgers, H. Treatment of bacterial infections of the female urinary tract by immunization of the patients. Urol. Int. 1987; 42:338–341.

Grünberg, J., Perry, R., Hoschützky, H., Jann, B., Jann, K. and Goldhar, J. Nonfimbrial blood group N-specific adhesin (NFA-3) from *Escherichia coli* O2O:KX104:H-, causing systemic infection. FEMS Microbiol. Lett. 1988; 56:241–246.

Guyer, D.M., Henderson, I.R., Nataro, J.P. and Mobley, H.L. Identification of Sat, an autotransporter toxin produced by uropathogenic *Escherichia coli*. Mol. Microbiol. 2000; 38:53–66.

Guyer, D.M., Radulovic, S., Jones, F.E. and Mobley, H.L. Sat, the secreted autotransporter toxin of uropathogenic *Escherichia coli*, is a vacuolating cytotoxin for bladder and kidney epithelial cells. Infect. Immun. 2002; 70:4539–4546.

Hagan, E.C. and Mobley, H.L. Uropathogenic *Escherichia coli* outer membrane antigens expressed during urinary tract infection. Infect. Immun. 2007; 75:3941–3949, Epub May 21, 2007.

Harding, G. and Ronald, A. The management of urinary tract infections: what we have learned in the last decade. Int. J. Antimicrob. Agents 1994; 4:83–88.

Heimer, S.R., Rasko, D.A., Lockatell, C.V., Johnson, D.E. and Mobley, H.L. Autotransporter genes *pic* and *tsh* are associated with *Escherichia coli* strains that cause acute pyelonephritis and are expressed during urinary tract infection. Infect. Immun. 2004; 72:593–597.

Hellman, J., Zanzot, E.M., Loiselle, P.M., Amato, S.F., Black, K.M., Ge, Y., Kurnick, J.T. and Warren, H.S. Antiserum against *Escherichia coli* J5 contains antibodies reactive with outer membrane proteins of heterologous Gram-negative bacteria. J. Infect. Dis. 1997; 176:1260–1268.

Hooton, T. and Stamm, W. Diagnosis and treatment of uncomplicated urinary tract infection. Infect. Dis. Clin. North Am. 1997; 11:551–581.

Hopkins, W.J. and Uehling, D.T. Resolution time of *Escherichia coli* cystitis is correlated with levels of preinfection antibody to the infecting *Escherichia coli* strain. Urology 1995; 45:42–46.

Horan, T., Culver, D. and Jarvis, W. Pathogens causing nosocomial infections. Antimicrob. Newslett. 1988; 5:65–67.

Hoschützky, H., Nimmich, W., Lottspeich, F. and Jann, K. Isolation and characterization of the non-fimbrial adhesin NFA-4 from uropathogenic *Escherichia coli* O7:K98:H6. Microb. Pathog. 1989; 6:351–359.

Huber, M., Krauter, K., Winkelmann, G., Bauer, H.W., Rahlfs, V.W., Lauener, P.A., Blessmann, G.S. and Bessler, W.G. Immunostimulation by bacterial components: II. Efficacy studies and meta-analysis of the bacterial extract OM-89. Int. J. Immunopharmacol. 2000; 22:1103–1111.

Ismail, N., Lieu, P., Lien, C. and Ling, M. Bacteremia in the elderly. Ann. Acad. Med. Singapore 1997; 26:593–598.

Jackson, L.A., Benson, P., Neuzil, K.M., Grandjean, M. and Marino, J.L. Burden of community-onset *Escherichia coli* bacteremia in seniors. J. Infect. Dis. 2005; 191:1523–1529, Epub Mar. 22, 2005.

Janben, T., Schwarz, C., Preikschat, P., Voss, M., Philipp, H.C. and Wieler, L.H. Virulence-associated genes in avian pathogenic *Escherichia coli* (APEC) isolated from internal organs of poultry having died from colibacillosis. Int. J. Med. Microbiol. 2001; 291:371–378.

Jann, B. and Jann, K. Structure and biosynthesis of the capsular antigens of *Escherichia coli*. Curr. Top. Microbiol. Immunol. 1990; 150:19–42.

Jarvis, W. Selected aspects of the socioeconomic impact of nosocomial infections: morbidity, mortality, cost, and prevention. Infect. Control Hosp. Epidemiol. 1996; 17:552–557.

Jeannin, P., Magistrelli, G., Goetsch, L., Haeuw, J., Thieblemont, N., Bonnefoy, J. and Delneste, Y. Outer membrane protein A (OmpA): a new pathogen-associated molecular pattern that interacts with antigen presenting cells-impact on vaccine strategies. Vaccine 2002; 20:A23–A27.

Jeannin, P., Magistrelli, G., Herbault, N., Goetsch, L., Godefroy, S., Charbonnier, P., Gonzalez, A. and Delneste, Y. Outer membrane protein A renders dendritic cells and macrophages responsive to CCL21 and triggers dendritic cell migration to secondary lymphoid organs. Eur. J. Immunol. 2003; 33:326–333.

Johanson, I.M., Plos, K., Marklund, B.I. and Svanborg, C. Pap, papG and prsG DNA sequences in *Escherichia coli* from the fecal flora and the urinary tract. Microb. Pathog. 1993; 15:121–129.

Johnson, J., Delavari, P., Kuskowski, M. and Stell, A. Phylogenetic distribution of extraintestinal virulence-associated traits in *Escherichia coli*. J. Infect. Dis. 2001a; 183:78–88.

Johnson, J., Delavari, P., Stell, A., Whittam, T., Carlino, U. and Russo, T. Molecular comparison of extraintestinal *Escherichia coli* isolates of the same electrophoretic lineages from humans and domestic animals. J. Infect. Dis. 2001b; 183:154–159.

Johnson, J., Kuskowski, M., O'Bryan, T. and Maslow, J. Epidemiological correlates of virulence genotype and phylogenetic background among *Escherichia coli* blood isolates from adults with diverse-source bacteremia. J. Infect. Dis. 2002a; 185:1439–1447.

Johnson, J., O'Bryan, T., Delavari, P., Kuskowski, M., Stapleton, A., Carlino, U. and Russo, T. Clonal relationships and extended virulence genotypes among *Escherichia coli* isolates from women with a first or recurrent episode of cystitis. J. Infect. Dis. 2001c; 183:1508–1517.

Johnson, J., Oswald, E., O'Bryan, T., Kuskowski, M. and Spanjaard, L. Phylogenetic distribution of virulence-associated genes among *Escherichia coli* isolates associated with neonatal bacterial meningitis in the Netherlands. J. Infect. Dis. 2002b; 185:774–784.

Johnson, J. and Russo, T. "Uropathogenic" *Escherichia coli* as agents of diverse non-urinary-tract extraintestinal infections. J. Infect. Dis. 2002a; 186:859–864.

Johnson, J., Russo, T., Tarr, P., Carlino, U., Bilge, S., Vary, J.J. and Stell, A. Molecular epidemiological and phylogenetic associations of two novel putative virulence genes, *iha* and *iroN* (*E. coli*), among *Escherichia coli* isolates from patients with urosepsis. Infect. Immun. 2000a; 68:3040–3047.

Johnson, J. and Stell, A. Extended virulence genotypes of *Escherichia coli* strains from patients with urosepsis in relation to phylogeny and host compromise. J. Infect. Dis. 2000; 181:261–272.

Johnson, J., Stell, A., Delavari, P., Murray, A., Kuskowski, M. and Gaastra, W. Phylogenetic and pathotypic similarites between *Escherichia coli* isolates from human urinary tract infections in dogs and extraintestinal infections in humans. J. Infect. Dis. 2001d; 183:897–906.

Johnson, J.R. Microbial virulence determinants and the pathogenesis of urinary tract infection. Infect. Dis. Clin. North Am. 2003; 17(June):261–278.

Johnson, J.R., Clabots, C. and Rosen, H. Effect of inactivation of the global oxidative stress regulator oxyR on the colonization ability of *Escherichia coli* O1:K1:H7 in a mouse model of ascending urinary tract infection. Infect. Immun. 2006; 74:461–468.

Johnson, J.R., Gajewski, A., Lesse, A.J. and Russo, T.A. Extraintestinal pathogenic *Escherichia coli* as a cause of invasive nonurinary infections. J. Clin. Microbiol. 2003a; 41:5798–5802.

Johnson, J.R., Jelacic, S., Schoening, L.M., Clabots, C., Mobley, H.L.T. and Tarr, P.I. The IrgA homologue adhesin (Iha) is an *Escherichia coli* virulence factor in murine urinary tract infection. Infect. Immun. 2005a; 73:965–971.

Johnson, J.R., Jelacic, S., Schoening, L.M., Clabots, C., Shaikh, N., Mobley, H.L. and Tarr, P.I. The IrgA homologue adhesin Iha is an *Escherichia coli* virulence factor in murine urinary tract infection. Infect. Immun. 2005b; 73:965–971.

Johnson, J.R., Kaster, N., Kuskowski, M.A. and Ling, G.V. Identification of urovirulence traits in *Escherichia coli* by comparison of urinary and rectal *E. coli* isolates from dogs with urinary tract infection. J. Clin. Microbiol. 2003b; 41:337–345.

Johnson, J.R., Kuskowski, M.A., Gajewski, A., Soto, S., Horcajada, J.P., Jimenez de Anta, M.T. and Vila, J. Extended virulence genotypes and phylogenetic background of *Escherichia coli* isolates from patients with cystitis, pyelonephritis, or prostatitis. J. Infect. Dis. 2005c; 191:46–50, Epub Nov. 30, 2004.

Johnson, J.R., Kuskowski, M.A., Smith, K., O'Bryan, T.T. and Tatini, S. Antimicrobial-resistant and extraintestinal pathogenic *Escherichia coli* in retail foods. J. Infect. Dis. 2005d; 191:1040–1049, Epub Mar. 1, 2005.

Johnson, J.R., O'Bryan, T.T., Kuskowski, M. and Maslow, J.N. Ongoing horizontal and vertical transmission of virulence genes and papA alleles among *Escherichia coli* blood isolates from patients with diverse-source bacteremia. Infect. Immun. 2001e; 69:5363–5374.

Johnson, J.R., Oswald, E., O'Bryan, T.T., Kuskowski, M.A. and Spanjaard, L. Phylogenetic distribution of virulence-associated genes among *Escherichia coli* isolates associated with neonatal bacterial meningitis in the Netherlands. J. Infect. Dis. 2002c; 185:774–784, Epub Feb. 14, 2002.

Johnson, J.R. and Russo, T.A. Extraintestinal pathogenic *Escherichia coli*: "the other bad E. coli". J. Lab. Clin. Med. 2002b; 139:155–162.

Johnson, J.R. and Russo, T.A. Molecular epidemiology of extraintestinal pathogenic (uropathogenic) *Escherichia coli*. Int. J. Med. Microbiol. 2005; 295:383–404.

Johnson, J.R., Russo, T.A., Tarr, P.I., Carlino, U., Bilge, S.S., Vary, J.C.J. and Stell, A.L. Molecular epidemiological and phylogenetic associations of two novel putative virulence genes, *iha* and $iroN_{E.\ coli}$, among *Escherichia coli* isolates from patients with urosepsis. Infect. Immun. 2000b; 68:3040–3047.

Johnson, T.J., Kariyawasam, S., Wannemuehler, Y., Mangiamele, P., Johnson, S.J., Doetkott, C., Skyberg, J.A., Lynne, A.M., Johnson, J.R. and Nolan, L.K. The genome sequence of avian pathogenic *Escherichia coli* strain O1:K1:H7 shares strong similarities with human extraintestinal pathogenic *E. coli* genomes. J. Bacteriol. 2007; 189:3228–3236, Epub Feb. 9, 2007.

Kaijser, B. and Ahlstedt, S. Protective capacity of antibodies against *Escherichia coli* O and K antigens. Infect. Immun. 1977; 17:286–289.

Kaijser, B., Larsson, P., Nimmich, W. and Soderstrom, T. Antibodies to *Escherichia coli* K and O antigens in protection against acute pyelonephritis. Prog. Allergy 1983; 33:275–288.

Kaijser, B., Larsson, P. and Olling, S. Protection against ascending *Escherichia coli* pyelonephritis in rats and significance of local immunity. Infect. Immun. 1978; 20:78–81.

Kanamura, S., Kurazono, H., Ishitoya, S., Terai, A., Habuchi, T., Nakano, M., Ogawa, O. and Yamamoto, S. Distribution and genetic association of putative uropathogenic virulence factors *iroN*, *iha*, *kpsMT*, *ompT* and *usp* in *Escherichia coli* isolated from urinary tract infections in Japan. J. Urol. 2003; 170:2490–2493.

Kaper, J.B., Nataro, J.P. and Mobley, H.L. Pathogenic *Escherichia coli*. Nat. Rev. Microbiol. 2004; 2:123–140.

Kariyawasam, S., Wilkie, B.N. and Gyles, C.L. Construction, characterization, and evaluation of the vaccine potential of three genetically defined mutants of avian pathogenic *Escherichia coli*. Avian Dis. 2004a; 48:287–299.

Kariyawasam, S., Wilkie, B.N. and Gyles, C.L. Resistance of broiler chickens to *Escherichia coli* respiratory tract infection induced by passively transferred egg-yolk antibodies. Vet. Microbiol. 2004b; 98:273–284.

Kariyawasam, S., Wilkie, B.N., Hunter, D.B. and Gyles, C.L. Systemic and mucosal antibody responses to selected cell surface antigens of avian pathogenic *Escherichia coli* in experimentally infected chickens. Avian Dis. 2002; 46:668–678.

Khan, N.A., Wang, Y., Kim, K.J., Chung, J.W., Wass, C.A. and Kim, K.S. Cytotoxic necrotizing factor-1 contributes to *Escherichia coli* K1 invasion of the central nervous system. J. Bacteriol. 2002; 277:15607–15612.

Kim, K.S. *Escherichia coli* translocation at the blood–brain barrier. Infect. Immun. 2001; 69:5217–5222.

Kim, K.S., Itabashi, H., Gemski, P., Sadoff, J., Warren, R.L. and Cross, A.S. The K1 capsule is the critical determinant in the development of *Escherichia coli* meningitis in the rat. J. Clin. Invest. 1992; 90:897–905.

Kirkland, K., Briggs, J., Trivette, S., Wilkinson, W. and Sexton, D. The impact of surgical-site infections in the 1990s: attributable mortality, excess length of hospitalization, and extra costs. Infect. Control Hosp. Epidemiol. 1999; 20:725–730.

Korhonen, T.K., Valtonen, M.V., Parkkinen, J., Vaisanen-Rhen, V., Fine, J., Orskov, F., Orskov, I., Svenson, S.B. and Makela, P.H. Serotypes, hemolysin production, and receptor recognition of *Escherichia coli* strains associated with neonatal sepsis and meningitis. Infect. Immun. 1985; 48:486–491.

Koukalova, D., Krocova, Z., Vitek, P., Macela, A. and Hajek, V. Immunostimulatory activity of the vaccine used in the treatment of recurrent urinary infections. II. Bratisl. Lek. Listy 1999a; 100:215–217.

Koukalova, D., Reif, R., Hajek, V., Mucha, Z., Vesely, J., Viktorinova, M. and Lochman, I. [Immunomodulation of recurrent urinary tract infections with Urvakol vaccine]. Bratisl. Lek. Listy 1999b; 100:246–251.

Kruze, D., Biro, K., Holzbecher, K., Andrial, M. and Bossart, W. Protection by a polyvalent vaccine against challenge infection and pyelonephritis. Urol. Res. 1992; 20:177–181.

Kurazono, H., Yamamoto, S., Nakano, M., Nair, G.B., Terai, A., Chaicumpa, W. and Hayashi, H. Characterization of a putative virulence island in the chromosome of uropathogenic *Escherichia coli* possessing a gene encoding a uropathogenic-specific protein. Microb. Pathog. 2000; 28:183–189.

Lane, M.C., Lockatell, V., Monterosso, G., Lamphier, D., Weinert, J., Hebel, J.R., Johnson, D.E. and Mobley, H.L. Role of motility in the colonization of uropathogenic *Escherichia coli* in the urinary tract. Infect. Immun. 2005; 73:7644–7656.

Langermann, S., Mollby, R., Burlein, J., Palaszynski, S., Auguste, C., De Fusco, A. and Al., E. Vaccination with FimH adhesin protects cynomologus monkeys from colonization and infection by uropathogenic *Escherichia coli*. J. Infect. Dis. 2000; 181:774–778.

Langermann, S., Palaszynski, S., Barnhart, M., Auguste, G., Pinkner, J., Burlein, J. and Al., E. Prevention of mucosal *Escherichia coli* infection by Fim-H-adhesin-based systemic vaccination. Science 1997; 276:607–611.

Lartigue, M.F., Poirel, L., Poyart, C., Reglier-Poupet, H. and Nordmann, P. Ertapenem resistance of *Escherichia coli*. Emerg. Infect. Dis. 2007; 13:315–317.

Le Bougenec, C., Lalioui, L., Du Merle, L., Jouve, M., Courcoux, P., Bouzari, S., Selvarangan, R., Nowicki, B.J., Y, G., Andremont, A., Gounon, P. and Garcia, M.-I. Characterization of AfaE adhesins produced by extraintestinal and intestinal human *Escherichia coli* isolates: PCR assays for detection fo Afa adhesins that do or do not recognize Dr blood group antigens. J. Clin. Microbiol. 2001; 39:1738–1745.

Le Bouguenec, C. and Servin, A.L. Diffusely adherent *Escherichia coli* strains expressing Afa/Dr adhesins (Afa/Dr DAEC): hitherto unrecognized pathogens. FEMS Microbiol. Lett. 2006; 256:185–194.

Lepow, M.L., Beeler, J., Randolph, M., Samuelson, J.S. and Hankins, W.A. Reactogenicity and immunogenicity of a quadrivalent combined meningococcal polysaccharide vaccine in children. J. Infect. Dis. 1986; 154:1033–1036.

Linggood, M.A. and Ingram, P.L. The role of alpha haemolysin in the virulence of *Escherichia coli* for mice. J. Med. Microbiol. 1982; 15:23–30.

Lundrigan, M.D. and Webb, R.M. Prevalence of *ompT* among *Escherichia coli* isolates of human origin. FEMS Microbiol. Lett. 1991; 97:51–56.

Marinova, S., Nenkov, P., Markova, R., Nikolaeva, S., Kostadinova, R., Mitova, I. and Vretenarska, M. Cellular and humoral systemic and mucosal immune responses stimulated by an oral polybacterial immunomodulator in patients with chronic urinary tract infections. Int. J. Immunopathol. Pharmacol. 2005; 18:357–473.

Marklund, B.I., Tennent, J.M., Garcia, E., Hamers, A., Baga, M., Lindberg, F., Gaastra, W. and Normark, S. Horizontal gene transfer of the *Escherichia coli* pap and prs pili operons as a mechanism for the development of tissue-specific adhesive properties. Mol. Microbiol. 1992; 6:2225–2242.

Marre, R. and Hacker, J. Role of S- and common-type 1-fimbriae of *Escherichia coli* in experimental upper and lower urinary tract infection. Microb. Pathog. 1987; 2:223–226.

Maslow, J.N., Mulligan, M.E., Adams, K.S., Justis, J.C. and Arbeit, R.D. Bacterial adhesins and host factors: role in the development and outcome of *Escherichia coli* bacteremia. Clin. Infect. Dis. 1993; 17:89–97.

Maslow, J.N., Mulligan, M.E. and Arbeit, R.D. Recurrent *Escherichia coli* bacteremia. J. Clin. Microbiol. 1994; 32:710–714.

Mcbean, M. and Rajamani, S. Increasing rates of hospitalization due to septicemia in the US elderly population, 1986–1997. J. Infect. Dis. 2001; 183:596–603.

Mccarthy, E. Inpatient utilization of short-stay hospitals, by diagnosis: United States 1980. Hyattsville, MD:National Center for Health Statisitcs, 1983. DHHS publication no. (PHS) 83-1735.

Mcgregor, A. and Colligan, P. Bacteraemia and fungaemia in an Australian general hospital—associations and outcomes. Med. J. Aust. 1993; 158:671–674.

Minino, A. and Smith, B. Deaths: preliminary data for 2000. Natl. Vital Stat. Rep. 2001; 49:1–40.

Mitsumori, K., Terai, A., Yamamoto, S., Ishitoya, S. and Yoshida, O. Virulence characteristics of *Escherichia coli* in acute bacterial prostatitis. J. Infect. Dis. 1999; 180:1378–1381.

Mitsumori, K., Terai, A., Yamamoto, S. and Yoshida, O. Identification of S, F1C and three PapG fimbrial adhesins in uropathogenic *Escherichia coli* by polymerase chain reaction. FEMS Immunol. Med. Microbiol. 1998; 21:261–268.

Mobley, H.L.T., Chippendale, G.R., Tenney, J.H., Hull, R.A. and Warren, J.W. Expression of type 1 fimbriae may be required for persistence of *Escherichia coli* in the catheterized urinary tract. J. Clin. Microbiol. 1987; 25:2253–2257.

Mobley, H.L.T., Jarvis, K.G., Elwood, J.P., Whittle, D.I., Lockatell, C.V., Russell, R.G., Johnson, D.E., Donnenberg, M.S. and Warren, J.W. Isogenic P-fimbrial deletion mutants of pyelonephritogenic *Escherichia coli*: the role of aGal(1-4)-bGal binding in virulence of a wild-type strain. Mol. Microbiol. 1993; 10:143–155.

Montenegro, M.A., Bitter-Suermann, D., Timmis, J.K., Aguero, M.E., Cabello, F.C., Sanyal, S.C. and Timmis, K.N. *traT* gene sequences, serum resistance and pathogenicity-related factors in clinical isolates of *Escherichia coli* and other Gram-negative bacteria. J. Gen. Microbiol. 1985; 131:1511–1521.

Moulin-Schouleur, M., Reperant, M., Laurent, S., Bree, A., Mignon-Grasteau, S., Germon, P., Rasschaert, D. and Schouler, C. Extra-intestinal pathogenic *Escherichia coli* of avian and human origin: link between phylogenetic relationships and common virulence patterns. J. Clin. Microbiol. 2007; 45:3366–3376.

Muder, R. Pneumonia in residents of long-term care facilities: epidemiology, etiology, management, and prevention. Am. J. Med. 1998; 105:319–330.

Mulvey, M.A., Schilling, J.D., Martinez, J.J. and Hultgren, S.J. Bad bugs and beleaguered bladders: Interplay between uropathogenic *Escherichia coli* and innate host defenses. Proc. Natl. Acad. Sci. U.S.A. 2000; 97:8829–8835.

Murphy, S. Deaths: Final Data for 1998. National Vital Statistics reports, 2000; 48.

Mutsch, M., Zhou, W., Rhodes, P., Bopp, M., Chen, R.T., Linder, T., Spyr, C. and Steffen, R. Use of the inactivated intranasal influenza vaccine and the risk of Bell's palsy in Switzerland. N. Engl. J. Med. 2004; 350:896–903.

Nataro, J.P., Mai, V., Johnson, J., Blackwelder, W.C., Heimer, R., Tirrell, S., Edberg, S.C., Braden, C.R., Glenn Morris, J., Jr. and Hirshon, J.M. Diarrheagenic *Escherichia coli* infection in Baltimore, Maryland, and New Haven, Connecticut. Clin. Infect. Dis. 2006; 43:402–407, Epub Jul. 11, 2006.

Negre, V.L., Bonacorsi, S., Schubert, S., Bidet, P., Nassif, X. and Bingen, E. The siderophore receptor IroN, but not the high-pathogenicity island or the hemin receptor ChuA, contributes to the bacteremic step of *Escherichia coli* neonatal meningitis. Infect. Immun. 2004; 72:1216–1220.

Nenkov, P., Mitov, I., Marinova, S. and Markova, R. Experimental investigations on the immunomodulating activity of polybacterial preparation for peroral immunotherapy and immunoprophylaxis of uroinfections. Int. J. Immunopharmacol. 1995; 17:489–496.

Nicolle, L.E. Asymptomatic bacteriuria: review and discussion of the IDSA guidelines. Int. J. Antimicrob. Agents. 2006; 28:S42–S48, Epub Jul. 10, 2006.

Nowicki, B., Labigne, A., Moseley, S., Hull, R., Hull, S. and Moulds, J. The Dr hemagglutinin, afimbrial adhesins AFA-I and AFA-III, and F1845 fimbriae of uropathogenic and diarrhea-associated *Escherichia coli* belong to a family of hemagglutinins with Dr receptor recognition. Infect. Immun. 1990; 58:279–281.

Nowicki, B., Svanborg Eden, C., Hull, R. and Hull, S. Molecular analysis and epidemiology of the Dr hemagglutinin of uropathogenic *Escherichia coli*. Infect. Immun. 1989; 57:446–451.

O'Hanley, P., Lalonde, G. and Ji, G. Alpha-hemolysin contributes to the pathogenicity of piliated digalactoside-binding *Escherichia coli* in the kidney: efficacy of an alpha-hemolysin vaccine in preventing renal injury in the Balb/c mouse model of pyelonephritis. Infect. Immun. 1991; 59:1153–1161.

O'Hanley, P., Low, D., Romero, I., Lark, D., Vosti, K., Falkow, S. and Schoolnik, G. Gal-Gal binding and hemolysin phenotypes and genotypes associated with uropathogenic *Escherichia coli*. N. Engl. J. Med. 1985; 7:414–420.

O'Hanley, P., Marcus, R., Baek, K.H., Denich, K. and Ji, G.E. Genetic conservation of hlyA determinants and serological conservation of HlyA: basis for developing a broadly cross-reactive subunit *Escherichia coli* alpha-hemolysin vaccine. Infect. Immun. 1993; 61:1091–1097.

Olesen, B., Kolmos, H.J., Orskov, F. and Orskov, I. *Escherichia coli* bacteraemia in patients with and without haematological malignancies: a study of strain characters and recurrent episodes. J. Infect. 1998; 36:93–100.

Opal, S.M., Cross, A., Gemski, P. and Lyhte, L.W. Survey of purported virulence factors of *Escherichia coli* isolated from blood, urine and stool. Eur. J. Clin. Microbiol. Infect. Dis. 1988; 7:425–427.

Opal, S.M., Cross, A.S., Gemski, P. and Lyhte, L.W. Aerobactin and alpha-hemolysin as virulence determinants in *Escherichia coli* isolated from human blood, urine, and stool. J. Infect. Dis. 1990; 161:794–796.

Orskov, F., Orskov, I., Jann, B. and Jann, K. Serology, chemistry, and genetics of O and K antigens of *Escherichia coli*. Bacteriol. Rev. 1977; 41:667–710.

Orskov, I., Orskov, F., Birch-Andersen, A., Kanamori, M. and Svanborg Eden, C. O, K, H and fimbrial antigens in *Escherichia coli* serotypes associated with pyelonephritis and cystitis. Scand. J. Infect. Dis. 1982; 33:18–25.

Patton, J., Nash, D. and Abrutyn, E. Urinary tract infection: economic considerations. Med. Clin. North Am. 1991; 75:495–513.

Pauwels, A., Pines, E., Abboura, M., Chiche, I. and Levy, V.G. Bacterial meningitis in cirrhosis: review of 16 cases. J. Hepatol. 1997; 27: 830–834.

Pecha, B., Low, D. and O'Hanley, P. Gal-Gal pili vaccines prevent pyelonephritis by piliated *Escherichia coli* in a murine model: single component Gal-Gal pili vaccines prevent pyelonephritis by homologous and heterologous piliated *E. coli* strains. J. Clin. Invest. 1989; 83:2102–2108.

Pere, A., Leinonen, M., Vaisanen-Rhen, V., Rhen, M. and Korhonen, T.K. Occurrence of type-1C fimbriae on *Escherichia coli* strains isolates from human extraintestinal infections. J. Gen. Microbiol. 1985; 131:1705–1711.

Picard, B., Sevali-Garcia, J., Gouriou, S. and Al., E. The link between phylogeny and virulence in *Escherichia coli* extraintestinal infection. Infect. Immun. 1999; 67:546–553.

Poulsen, K., Wachmann, C., Bremmelgaard, A. and Al., E. Survival of patients with surgical wound infection: a case-control study of common surgical interventions. Br. J. Surg. 1995; 82:208–209.

Rayner, B. and Willcox, P. Community-acquired bacteraemia; a prospective survey of 239 cases. Q. J. Med. 1988; 69:907–919.

Restieri, C., Garriss, G., Locas, M.C. and Dozois, C.M. Autotransporter-encoding sequences are phylogenetically distributed among *Escherichia coli* clinical isolates and reference strains. Appl. Environ. Microbiol. 2007; 73:1553–1562, Epub Jan. 12, 2007.

Revathi, G., Puri, J. and Jain, B. Bacteriology of burns. Burns 1998; 24:347–349.

Rippere-Lampe, K.E., Lang, M., Ceri, H., Olson, M., Lockman, H.A. and O'Brien, A. Cytotoxic necrotizing factor type 1-positive *Escherichia coli* causes increased inflammation and tissue damage to the prostate in a rat prostatitis model. Infect. Immun. 2001a; 69:5615–6519.

Rippere-Lampe, K.E., O'Brien, A.D., Conran, R. and Lockman, H.A. Mutation of the gene encoding cytotoxic necrotizing factor type 1 (cnf(1)) attenuates the virulence of uropathogenic *Escherichia coli*. Infect. Immun. 2001b; 69:3954–3964.

Roberts, F., Geere, I. and Coldman, A. A three-year study of positive blood cultures, with emphasis on prognosis. Rev. Infect. Dis. 1991; 13:34–46.

Roberts, I.S. Bacterial polysaccharides in sickness and in health. The 1995 Fleming lecture. Microbiology 1995; 141:2023–2031.

Roberts, J.A., Hardaway, K., Kaack, B., Fussell, E.N. and Baskin, G. Prevention of pyelonephritis by immunization with P-fimbriae. J. Urol. 1984; 131:602–607.

Roberts, J.A., Kaack, M.B., Baskin, G., Chapman, M.R., Hunstad, D.A., Pinkner, J.S. and Hultgren, S.J. Antibody responses and protection from pyelonephritis following vaccination with purified *Escherichia coli* PapDG protein. J. Urol. 2004; 171:1682–1685.

Roberts, J.A., Kaack, M.B., Baskin, G. and Svenson, S.B. Vaccination with a formalin-killed P-fimbriated *E. coli* whole-cell vaccine prevents renal scarring from pyelonephritis in the non-human primate. Vaccine 1995; 13:11–16.

Roberts, J.A., Marklund, B.-I., Ilver, D., Haslam, D., Kaack, M.B., Baskin, G., Louis, M., Möllby, R., Winberg, J. and Normark, S. The Gal(a1-4)Gal-specific tip adhesin of *Escherichia coli* P-fimbriae is needed for pyelonephritis to occur in the normal urinary tract. Proc. Natl. Acad. Sci. U.S.A. 1994; 91:11889–11893.

Russo, T. Capsule and lipopolysaccharide. In: *Escherichia coli*: Virulence Mechanisms of a Versatile Pathogen (M. Donnenberg, Ed.). London, UK: Academic Press, 2002.

Russo, T., Brown, J., Jodush, S. and Johnson, J. The O4 specific antigen moiety of lipopolysaccharide but not the K54 group 2 capsule is important for urovirulence in an extraintestinal isolate of *Escherichia coli*. Infect. Immun. 1996a; 64:2343–2348.

Russo, T., Carlino, U. and Johnson, J. Identification of *ireA*, a new iron regulated virulence gene in an extraintestinal pathogenic isolate of *Escherichia coli*. Infect. Immun. 2001; 69:6209–6216.

Russo, T., Carlino, U., Mong, A. and Jodush, S. Identification of genes in an extraintestinal isolate of *Escherichia coli* with increased expression after exposure to human urine. Infect. Immun. 1999; 67:5306–5314.

Russo, T., Jodush, S., Brown, J. and Johnson, J. Identification of two previously unrecognized genes (*guaA, argC*) important for uropathogenesis. Mol. Microbiol. 1996b; 22:217–229.

Russo, T. and Johnson, J. A proposal for a new inclusive designation for extraintestinal pathogenic isolates of *Escherichia coli*: ExPEC. J. Infect. Dis. 2000; 181:1753–1754.

Russo, T. and Johnson, J. Medical and economic impact of extraintestinal infections due to *Escherichia coli*: focus on an increasingly important endemic problem. Microbes Infect. 2003; 5:449–456.

Russo, T., Liang, Y. and Cross, A. The presence of K54 capsular polysaccharide increases the pathogenicity of *Escherichia coli* in vivo. J. Infect. Dis. 1994; 169:112–118.

Russo, T., Mcfadden, C., Carlino, U., Beanan, J., Barnard, T. and Johnson, J. IroN functions as a siderophore receptor and is a urovirulence factor in an extraintestinal pathogenic isolate of *Escherichia coli*. Infect. Immun. 2002; 70:7156–7160.

Russo, T., Moffitt, M., Hammer, C. and Frank, M. TnphoA mediated disruption of K54 capsular polysaccharide genes in *Escherichia coli* confers serum sensitivity. Infect. Immun. 1993; 61:3578–3582.

Russo, T., Stapleton, A., Wenderoth, S., Hooton, T. and Stamm, W. Chromosomal restriction fragment length polymorphism analysis of *Escherichia coli* strains causing recurrent urinary tract infection. J. Infect. Dis. 1995; 172:440–445.

Russo, T.A., Beanan, J.M., Olson, R., Genagon, S.A., Macdonald, U., Cope, J.J., Davidson, B.A., Johnston, B. and Johnson, J.R. A killed, genetically engineered derivative of a wild-type extraintestinal pathogenic *E. coli* strain is a vaccine candidate. Vaccine 2007a; 25:3859–3870, Epub Feb. 5, 2007.

Russo, T.A., Davidson, B.A., Carlino-Macdonald, U.B., Helinski, J.D., Priore, R.L. and Knight 3rd., P.R. The effects of *Escherichia coli* capsule, O-antigen, host neutrophils, and complement in a rat model of Gram-negative pneumonia. FEMS Microbiol. Lett. 2003a; 226:355–361.

Russo, T.A., Davidson, B.A., Genagon, S.A., Warholic, N.M., Macdonald, U., Pawlicki, P.D., Beanan, J.M., Olson, R., Holm, B.A. and Knight 3rd., P.R. *E. coli* virulence factor hemolysin induces neutrophil apoptosis and necrosis/lysis in vitro and necrosis/lysis and lung injury in a rat pneumonia model. Am. J. Physiol. Lung Cell Mol. Physiol. 2005; 289:L207–L216, Epub Apr. 1, 2005.

Russo, T.A. and Johnson, J.R. Extraintestinal isolates of *Escherichia coli*: identification and prospects for vaccine development. Exp. Rev. Vaccines 2006; 5:45–54.

Russo, T.A., Mcfadden, C.D., Carlino-Macdonald, U.B., Beanan, J.M., Olson, R. and Wilding, G.E. The siderophore receptor IroN of extraintestinal pathogenic *Escherichia coli* is a potential vaccine candidate. Infect. Immun. 2003b; 71:7164–7169.

Russo, T.A., Wang, Z., Davidson, B.A., Genagon, S.A., Beanan, J.M., Olson, R., Holm, B.A., Knight 3rd., P.R., Chess, P.R. and Notter, R.H. Surfactant dysfunction and lung injury due to the E. coli virulence factor hemolysin in a rat pneumonia model. Am. J. Physiol. Lung Cell Mol. Physiol. 2007b; 292:L632–L643.

Saccente, M. Periprosthetic joint infections: a review for clinicians. Infect. Dis. Clin. Pract. 1998; 7:431–441.

Sader, H., Jones, R., Andrade-Baiocchi, S. and Biedenbach, D. Four-year evaluation of frequency of occurrence and antimicrobial susceptibility patterns of bacteria from bloodstream infections in Latin American medical centers. Diagn. Microbiol. Infect. Dis. 2002; 44:273–280.

Saint, S. Clinical and economic consequences of nosocomial catheter-related bacteriuria. Am. J. Infect. Control 2000; 28:68–75.

Salles, M., Mandine, E., Zalisz, R., Guenounou, M. and Smets, P. Protective effects of murine monoclonal antibodies in experimental septicemia: *E. coli* antibodies protect against different serotypes of *E. coli*. J. Infect. Dis. 1989; 159:641–647.

Sannes, M.R., Kuskowski, M.A., Owens, K., Gajewski, A. and Johnson, J.R. Virulence factor profiles and phylogenetic background of *Escherichia coli* isolates from veterans with bacteremia and uninfected control subjects. J. Infect. Dis. 2004; 190:2121–2128, Epub Nov. 15, 2004.

Sapico, F. Pyogenic hematogenous spinal infections. Infect. Dis. Clin. Pract. 1999; 8:267–269.

Savarino, S.J., Fasano, A., Watson, J., Martin, B.M., Levine, M.M., Guandalini, S. and Guerry, P. Enteroaggregative *Escherichia coli* heat-stable enterotoxin 1 represents another subfamily of *E. coli* heat-stable toxin. Proc. Natl. Acad. Sci. U.S.A. 1993; 90:3093–3097.

Schappert, S. National ambulatory medical care survey: 1992 survey. Hyattsville, MD. DHHS publication no. (PHS) 94-1250 (Advanced data for vital and health statistics no. 253), 1994.

Schmidt, M.A., O'Hanley, P., Lark, D. and Schoolnik, G.K. Synthetic peptides corresponding to protective epitopes of *Escherichia coli* digalactoside-binding pilin prevent infection in a murine pyelonephritis model. Proc. Natl. Acad. Sci. U.S.A. 1988; 85:1247–1251.

Schönian, G., Sokolowkska-Köhler, W., Rollmann, R., Schubert, A., Gräser, Y. and Presber, W. Determination of S fimbriae among *Escherichia coli* strains from extraintestinal infections by colony hybridization and dot enzyme immunoassay. Zentralbl. Bakteriol. 1992; 276:273–279.

Schubert, S., Picard, B., Gouriou, S., Heesemann, J. and Denamur, E. *Yersinia* high-pathogenicity isoland contributes to virulence in *Escherichia coli* causing extraintestinal infections. Infect. Immun. 2002; 70:5335–5337.

Schubert, S., Rakin, A., Karch, H., Carniel, E. and Heesemann, J. Prevalence of the "high-pathogenicity island" of Yersinia species among *Escherichia coli* strains that are pathogenic to humans. Infect. Immun. 1998; 66:480–485.

Scott, D.A. and Kaper, J.B. Cloning and sequencing of the genes encoding *Escherichia coli* cytolethal distending toxin. Infect. Immun. 1994; 62:244–251.

Selvarangan, R., Goluszko, P., Popov, V., Singhal, J., Pham, T., Lublin, D.M., Nowicki, S. and Nowicki, B. Role of decay-accelerating factor domains and anchorage in internalization of Dr-fimbriated *Escherichia coli*. Infect. Immun. 2001; 68:1391–1399.

Siegman-Igra, Y., Fourer, B., Orni-Wasserlauf, R., Golan, Y., Noy, A., Schwartz, D. and Giladi, M. Reappraisal of community-acquired bacteremia: a proposal of a new classification for the spectrum of acquisition of bacteremia. Clin. Infect. Dis. 2002; 34:1431–1439.

Smit, P., Oberholzer, D., Hayden-Smith, S., Koornhof, H.J. and Hilleman, M.R. Protective efficacy of pneumococcal polysaccharide vaccines. JAMA 1977; 238:2613–2616.

Smith, J.L., Fratamico, P.M. and Gunther, N.W. Extraintestinal pathogenic *Escherichia coli*. Foodborne Pathog. Dis. 2007; 4:134–163.

Snyder, J.A., Haugen, B.J., Buckles, E.L., Lockatell, C.V., johnson, D.E., Donnenberg, M.S., Welch, R.A. and Mobley, H.L. Transcriptome of uropathogenic *Escherichia coli* during urinary tract infection. Infect. Immun. 2004; 72:6373–6381.

Sokurenko, E.V., Courtney, H.S., Maslow, J., Siitonen, A. and Hasty, D.L. Quantitative differences in adhesiveness of type 1 fimbriated *Escherichia coli* due to structural differences in *fimH* genes. J. Bacteriol. 1995; 177:3680–3686.

Somerville, J.J., Cassiano, L., Bainbridge, B., Cunningham, M. and Darveau, R. A novel *Escherichia coli* lipid A mutant that produces an antiinflammaory lipopolysaccharide. J. Clin. Invest. 1996; 97:359–365.

Soriano, G., Coll, P., Guarner, C. and Al., E. *Escherichia coli* capsular polysaccharide and spontaneous bacterial peritionitis. Hepatology 1995; 21:668–673.

Sorsa, L.J., Feldmann, F., Hildinger, K., Dufke, S. and Schubert, S. Characterization of four novel genomic regions of uropathogenic *Escherichia coli* highly associated with the extraintestinal virulent phenotype: a jigsaw puzzle of genetic modules. Int. J. Med. Microbiol. 2007; 297:83–95, Epub Feb. 5, 2007.

Stamm, W. and Hooton, T. Management of urinary tract infection in adults. N. Engl. J. Med. 1993; 329:1328–1334.

Stamm, W., Hooton, T., Johnson, J. and Al., E. Urinary tract infections. J. Infect. Dis. 1989; 15:400–406.

Stamm, W., Mckevitt, M., Roberts, P. and White, N. Natural history of recurrent urinary tract infections in women. Rev. Infect. Dis. 1991; 13:77–84.

Stein, G.E. and Craig, W.A. Tigecycline: a critical analysis. Clin. Infect. Dis. 2006; 43:518–524, Epub Jun. 26, 2006.

Stoll, B., Hansen, N., Fanaroff, A. and Al., E. Changes in pathogens causing early-onset sepsis in very-low-birth-weight infants. New Engl. J. Med. 2002; 347:240–247.

Talan, D., Stamm, W., Hooton, T., Moran, G., Burke, T., Iravani, A., Reuning-Scherer, J. and Church, D. Comparison of ciprofloxacin (7 days) and trimethoprim-sulfamethoxazole (14 days) for acute uncomplicated pyelonephritis in women: a randomized trial. JAMA 2000; 283:1583–1590.

Tambyah, P.A., Knasinski, V. and Maki, D.G. The direct costs of nosocomial catheter-associated urinary tract infection in the era of managed care. Infect. Control Hosp. Epidemiol. 2002; 23:27–31.

Torres, A.G., Redford, P., Welch, R.A. and Payne, S.M. TonB-dependent systems of uropathogenic *Escherichia coli*: aerobactin and heme transport and TonB are required for virulence in the mouse. Infect. Immun. 2001; 69:6179–6185.

Troeger, H., Richter, J.F., Beutin, L., Gunzel, D., Dobrindt, U., Epple, H.J., Gitter, A.H., Zeitz, M., Fromm, M. and Schulzke, J.D. *Escherichia coli* alpha-haemolysin induces focal leaks in colonic epithelium: a novel mechanism of bacterial translocation. Cell Microbiol. 2007; 9:2530–2540.

Uehling, D.T., Hopkins, W.J., Elkahwaji, J.E., Schmidt, D.M. and Leverson, G.E. Phase 2 clinical trial of a vaginal mucosal vaccine for urinary tract infections. J. Urol. 2003; 170:867–869.

Uehling, D.T., Hopkins, W.J., James, L.J. and Balish, E. Vaginal immunization of monkeys against urinary tract infection with a multi-strain vaccine. J. Urol. 1994; 151:214–216.

Uehling, D.T., Hopkins, W.J., Jensen, J. and Balish, E. Vaginal immunization against induced cystitis in monkeys. J. Urol. 1987; 137:327–329.

Uehling, D.T., James, L.J., Hopkins, W.J. and Balish, E. Immunization against urinary tract infection with a multi-valent vaginal vaccine. J. Urol. 1991; 146:223–226.

Ulett, G.C., Valle, J., Beloin, C., Sherlock, O., Ghigo, J.M. and Schembri, M.A. Functional analysis of antigen 43 in uropathogenic *Escherichia coli* reveals a role in long-term persistence in the urinary tract. Infect. Immun. 2007; 75:3233–3244, Epub Apr. 9, 2007.

Unhanand, M., Mustafa, M., Mccracken, G. and Nelson, J. Gram-negative enteric bacillary meningitis: a twenty-year experience. J. Pediatr. 1993; 122:15–21.

Uslan, D.Z., Crane, S.J., Steckelberg, J.M., Cockerill 3rd., F.R., St Sauver, J.L., Wilson, W.R. and Baddour, L.M. Age- and sex-associated trends in bloodstream infection: a population-based study in Olmsted County, Minnesota. Arch. Intern. Med. 2007; 167:834–839.

Vandemaele, F., Ververken, C., Bleyen, N., Geys, J., D'Hulst, C., Addwebi, T., Van Empel, P. and Goddeeris, B.M. Immunization with the binding domain of FimH, the adhesin of type 1 fimbriae, does not protect chickens against avian pathogenic *Escherichia coli*. Avian Pathol. 2005; 34:264–272.

Vastag, B. New vaccine decreases rate of nosocomial infections. JAMA 2001; 285:1565.

Vital and Health Statistics. Ambulatory and in-patient procedures in the United States, 1995. Hyattsville, MD: Department of Health and Human Services, Centers for Disease Control and Prevention, National Center for Health Statistics. Series 13: data from the national health care survey, no. 135, DHHS publication no. (PHS) 98-1796, 1998.

Weissman, S.J., Chattopadhyay, S., Aprikian, P., Obata-Yasuoka, M., Yarova-Yarovaya, Y., Stapleton, A., Ba-Thein, W., Dykhuizen, D., Johnson, J.R. and Sokurenko, E.V. Clonal analysis reveals high rate of structural mutations in fimbrial adhesins of extraintestinal pathogenic *Escherichia coli*. Mol. Microbiol. 2006; 59:975–988.

Welch, R.A., Dellinger, E.P., Minshew, B. and Falkow, S. Haemolysin contributes to virulence of extra-intestinal *E. coli* infections. Nature 1981; 294:665–667.

Wenzel, R. Hospital-acquired pneumonia: overview of the current state of art for prevention and control. Eur. J. Clin Microbiol. 1989; 8:56–60.

Weycker, D., Akhras, K.S., Edelsberg, J., Angus, D.C. and Oster, G. Long-term mortality and medical care charges in patients with severe sepsis. Crit. Care Med. 2003; 31:2316–2323.

Whitfield, C. and Valvano, M. Biosynthesis and expression of cell-surface polysaccharides in Gram-negative bacteria. Adv. Micro. Physiol. 1993; 35:135–246.

World Health Organization. The World Health Report 2000—Health Systems: Improving Performance. Geneva: World Health Organization, 2000.

Wroblewska-Seniuk, K., Selvarangan, R., Hart, A., Pladzyk, R., Goluszko, P., Jafari, A., Du Merle, L., Nowicki, S., Yallampalli, C., Le Bouguenec, C. and Nowicki, B. Dra/AfaE adhesin of uropathogenic Dr/Afa + *Escherichia coli* mediates mortality in pregnant rats. Infect. Immun. 2005; 73:7597–7601.

Yamamoto, S., Nakano, M., Terai, A., Yuri, K., Nakata, K., Nair, G.B., Kurazono, H. and Ogawa, O. The presence of the virulence island containing the *usp* gene in uropathogenic *Escherichia coli* is associated with urinary tract infection in an experimental mouse model. J. Urol. 2001; 165:1347–1351.

Zhang, L., Foxman, B., Tallman, P., Cladera, E., Le Bougenec, C. and Marrs, C.F. Distribution of *drb* genes coding for Dr binding adhesins among uropathogenic and fecal *Escherichia coli* isolates and identification of new subtypes. Infect. Immun. 1997; 65:2011–2018.

Ziegler, E.J., Fisher, C.J., Jr., Sprung, C.L., Straube, R.C., Sadoff, J.C., Foulke, G.E., Wortel, C.H., Fink, M.P., Dellinger, R.P., Teng, N.N. et al. Treatment of Gram-negative bacteremia and septic shock with HA-1 A human monoclonal antibody against endotoxin. A randomized, double-blind, placebo-controlled trial. The HA-1 A Sepsis Study Group. N. Engl. J. Med. 1991; 324:429–436.

Ziegler, E.J., Mccutchan, J.A., Fierer, J., Glauser, M.P., Sadoff, J.C., Douglas, H. and Braude, A.I. Treatment of Gram-negative bacteremia and shock with human antiserum to a mutant *Escherichia coli*. N. Engl. J. Med. 1982; 307:1225–1230.

Zoutman, D., Mcdonald, S. and Vethanayagan, D. Total and attributable costs of surgical-wound infections at a Canadian tertiary-care center. Infect. Control Hosp. Epidemiol. 1998; 19:254–259.

CHAPTER

49

Gonorrhea

Michael W. Russell[1] *and Edward W. Hook III*[2]

[1]Department of Microbiology and Immunology, Witebsky Center for Microbial Pathogenesis and Immunology, University at Buffalo, Buffalo, NY, USA

[2]Department of Medicine, Division of Infectious Diseases, University of Alabama at Birmingham, Birmingham, AL, USA

OUTLINE

ABSTRACT

Along with other sexually transmitted diseases, gonorrhoeae has an enormous impact on global human health, as the World Health Organization estimates that 62 million new gonococcal infections occur annually. Syndromes range from uncomplicated cervicitis in women and urethritis in men to upper tract infection and pelvic inflammatory disease in women with consequent reproductive tract damage, and rarer cases of invasive disease; asymptomatic infections are also common in women. Ocular infections that threaten eyesight can occur in infants born to infected women. Untreated gonorrhea also facilitates the spread of HIV-AIDS. Although the causative agent, *Neisseria gonorrhoeae*, is usually sensitive to modern antibiotic treatments, the spread of antibiotic resistance is causing concern that these options may become limited. Most major gonococcal surface antigens are hypervariable and their expression is subject to phase variation, but considerable advances have been made in elucidating the molecular mechanisms that govern these, and in comprehending gonococcal pathogenesis. Conserved antigens of current interest include NspA and the transferrin-binding proteins, but comparison with *Neisseria meningitidis* suggests that others might exist. The systemic and genital tract immune response to uncomplicated gonorrhea is generally weak and poorly understood, and repeated infections are common. While *N. gonorrhoeae* deploys multiple mechanisms to evade destruction by host immune defenses, the concept is emerging that it also avoids inducing adaptive immunity and manipulates host responses to its own advantage. A previous attempt to develop vaccines based on pilus antigen proved ineffective, and efforts to create a porin-based vaccine have stalled. Experimental studies are hampered by the lack of good animal models as gonorrhea is an exclusively human infection. Concerted, new, and sustained effort is required to identify suitable gonococcal antigenic targets, define the parameters of immune protection against gonococcal infections in vivo, comprehend the mechanisms of both male and female genital tract immune defense against infection, and devise novel immunization strategies for eliciting immune protection in the male and female genital tracts. Sociopolitical barriers to the deployment of the resources required to accomplish these goals need to be overcome.

INTRODUCTION

Gonorrhea is an ubiquitous, worldwide sexually transmitted disease (STD) that is most common in developing nations and it disproportionately afflicts underprivileged and urban populations. It is also widely acknowledged that gonorrhea is underreported, and its true incidence is probably much greater than the official numbers. While about 90% of men typically display symptoms within a few days of infection, in women symptoms are more nonspecific and at least 30–50% (possibly more) of infected women are asymptomatic or go unrecognized (Hook and Handsfield, 1999). Moreover, women bear the brunt of morbidity as untreated cervicitis can progress to upper tract infection and pelvic inflammatory disease (PID), leading to subsequent tubal scarring, infertility, and risk for ectopic pregnancy. Eye infections can be acquired by babies born to infected women, or in adults by the direct transfer of infected material, in both cases leading to corneal scarring and blindness. Though an infrequent complication of genital infection, disseminated gonococcal infection is a systemic bacteremic condition most often involving the skin or joints, leading to septic dermatitis, arthritis, or other manifestations. Untreated gonorrhea is a risk factor for the acquisition of HIV, although the mechanisms underlying this are incompletely understood.

Since the beginning of the antibiotic era, the gonococcus (*Neisseria gonorrhoeae*) has regularly developed resistance to antibiotics used for treatment, including penicillin, tetracycline, and most recently the fluoroquinolone, ciprofloxacin. Although anecdotal evidence suggests that some cases of uncomplicated gonorrhea may be self-limiting and clear spontaneously after a few months, it can also persist for years. Furthermore, it does not leave a state of effective specific immunity, but can be acquired repeatedly with no apparent diminution in probability, severity, or duration. This has led to the speculation that *N. gonorrhoeae* interacts with the human immune system in a manner that is little understood, but that results in failure to generate an adaptive immune response.

Gonorrhea therefore represents a challenging problem in infectious disease, both from the public health and clinical standpoint, and scientifically. Great strides have been made in comprehending the molecular biology of *N. gonorrhoeae* (its genome has been completely sequenced), and in elucidating its pathogenic mechanisms and colonization of the human male and female genital tracts (Edwards and Apicella, 2004). Comprehension of the immune response against the gonococcus and how this can be manipulated to generate protective immunity to gonorrhea are, however, limited, and efforts over many years to develop an effective vaccine have not yet been successful.

HISTORICAL ASPECTS OF GONORRHEA

While gonorrhea is likely an ancient human affliction and apparent references to it can be seen in the Bible and other ancient writings, it was not until the 18th century that it was recognized as a distinct disease from syphilis; credit for this is given to Benjamin Bell (1749–1806), a surgeon in Edinburgh, Scotland (Benedek, 2005). The causative agent, now known as *N. gonorrhoeae*, was first recognized microscopically in specimens of pus by Albert Neisser (1855–1916) at the University of Breslau in 1879, and confirmed by Arpad Bokai (1856–1919) at the University of

Budapest in 1880. Thus, the gonococcus was among the earliest bacteria to be associated with human disease. However, controversy persisted for several years because it proved impossible to infect animals such as rabbits or dogs, and brave efforts on the part of investigators to infect themselves or other volunteers with cultures were initially unsuccessful. Indeed the difficulty of culturing the gonococcus and of distinguishing it morphologically from other micrococci undoubtedly contributed to this until the gram stain was developed (1884) and better culture media were devised.

The British lawyer James Boswell (1740–1795), better known as the biographer of Samuel Johnson, left copious personal papers describing 12–19 episodes of gonorrhea over the course of some 30 years (Ober, 1970). These reveal, besides his profligate lifestyle, that the infections subsided (presumably cleared) after several months or more, but left no evident state of immunity against repeated infection—or recrudescence. The advent of sulfonamides in 1936, followed by penicillin in 1943, provided effective and rapid therapy, but antibiotic resistance soon emerged. In the United States, gonorrhea rates peaked in 1975 at >460 reported cases per 100,000 population, and thereafter declined substantially. However, in recent years there has been concern that the trend has leveled out or may even be increasing again in both the United States and Western Europe. The worldwide incidence, estimated at 62 million cases per year (WHO, 2001), however, reveals that gonorrhea represents a major international health issue for which there is a huge unmet need to provide treatment and effective preventative measures.

ETIOLOGIC AGENT: *NEISSERIA GONORRHOEAE*

Classification

N. gonorrhoeae is one of two pathogenic species of the genus *Neisseria*, the other being *Neisseria meningitidis* (Sparling, 1999). *Neisseria* is classified in the family Neisseriaceae within the phylum Proteobacteria (β subgroup) of the Eubacteria (Janda and Gaydos, 2007). *N. gonorrhoeae* typically appears as an aerobic gram-negative diplococcus that lacks a polysaccharide capsule (Fig. 49.1). It is nutritionally demanding, requiring enriched media such as chocolate agar or modified Thayer-Martin agar for growth. In the past, diagnosis has usually depended on microscopy and culture (Janda and Gaydos, 2007), but in developed nations nucleic acid amplification tests have been found somewhat more sensitive and easier to use (Peeling et al., 2006), and have become the preferred method for diagnosis of urogenital infection.

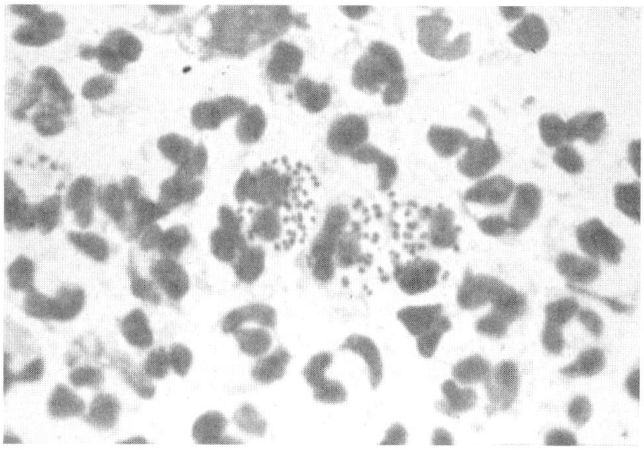

FIGURE 49.1 *N. gonorrhoeae* in a gram-stained smear of exudate from a case of gonorrhea. Note the presence of gram-negative diplococci within neutrophil leukocytes.

Antigens Encoded by N. *gonorrhoeae*

The antigens of *N. gonorrhoeae* are complex and highly variable, but unlike the closely related *N. meningitidis* (see Chapter 53) it does not possess a polysaccharide capsule. Most of the major surface components are subject to phase variation in which their expression can be up- or down-regulated spontaneously or modified in response to environmental signals. These factors have complicated both serological classification of *N. gonorrhoeae* (and hence its utility for epidemiological tracing) and analysis of the immune response to infection. *N. gonorrhoeae* is also naturally transformable, thus horizontal transfer of genes and the proteins, including antigens, encoded by them can readily occur. The major surface antigens are listed in Table 49.1, and are discussed below.

Lipooligosaccharide (LOS)

N. gonorrhoeae possesses a lipooligosaccharide (LOS) anchored in the outer membrane through lipid A, with short, triple-branched, and variable glycan chains (Apicella et al., 1987; Burch et al., 1997). These depend on glycosyltransferases some of which are subject to off–on expression through phase variation determined by a slip–strand mispairing mechanism resulting from poly-G tracts in the coding sequences of the genes (Yang and Gotschlich, 1996). However, some of the oligosaccharides mimic host glycolipid epitopes (Mandrell et al., 1988). In addition, LOS can be sialylated by means of a gonococcal sialyltransferase

TABLE 49.1 Major antigens of *Neisseria gonorrhoeae*

Antigen	Functions and properties[a]	Comments[a]
LOS	Outer membrane integrity; adhesin; host cell interactions	Abs bactericidal (if LOS not sialylated); T-independent Ag; glycan chains variable
Porin	Major OMP; anion exchange channel; host cell interactions	Constitutive; 2 main serovars plus several subtypes; Abs bactericidal if LOS not sialylated and in absence of Ab to Rmp
Opa proteins	Host cell interactions; immunosuppressive	Expression phase variable; >50 types
Rmp	Protects porin	Highly immunogenic, induces blocking Ab
Pili	Adhesin; host cell interactions	Expression phase variable; highly heterogeneous; type-specific Abs inhibit homologous pilus adherence
H.8	Conserved lipoprotein epitope	Present in *Neisseria meningitidis* and other neisseriae
NspA	Conserved small OMP; function unknown	Abs conformation dependent, bactericidal; present in *N. meningitidis*
Tbp	Iron uptake (binds human transferrin)	Abs inhibit iron uptake and adherence, bactericidal
IgA1 protease	Cleaves human IgA1 and lysosomal LAMP1; inhibits S-IgA function; promotes intracellular survival	Constitutive; few antigenic variants; Abs inhibit enzyme activity

[a] See text for explanations and further discussion.

for which the host supplies the substrate, cytidine monophosphate-*N*-acetylneuraminic acid (Mandrell et al., 1990; Smith et al., 1995). Some, though not all, of the glycan subunits are immunogenic, and antibodies to them can induce complement-dependent bacteriolysis. However, sialic acid residues on LOS interfere with this bactericidal activity by facilitating the binding of the complement regulatory factor H (Ram et al., 1998). Other LOS glycan structures enhance resistance to complement by binding the regulatory protein, C4b-binding protein (Ram et al., 2007).

Gonococcal LOS is highly endotoxic, probably through interaction with Toll-like receptor (TLR) 4 and CD14, and contributes significantly to the inflammatory response induced by gonococcal infection and the resulting damage to tissues such as the fallopian tube (Gregg et al., 1981). However, LOS is not considered a prime candidate for a vaccine antigen unless its toxicity can be reduced or eliminated. In addition, carbohydrate antigens are usually T-independent unless coupled to proteins. Moreover, sialylation of LOS inhibits the bactericidal activity of complement-fixing antibodies against it (Estabrook et al., 1997).

Porin

Porin (P.I) is the major outer membrane protein (OMP) constituting ~30% of all gonococcal surface protein. Only one porin is present, designated PorB (in contrast to *N. meningitidis* which has two porins, PorA and PorB). Gonococcal porin is the basis of a serovar classification determined by means of a panel of monoclonal antibodies (Sarafian and Knapp, 1989). Two major porin types exist, P.I-A and P.I-B, each having numerous subtypes. Porin is a homotrimeric molecule, which forms a pore in the outer membrane, and probably functions as an ion channel. As a major surface protein that is constitutively expressed in all strains and one of the least variable, it has been proposed as a vaccine antigen (see below). This is supported by findings that antibodies against it mediate in vitro complement-dependent bacteriolytic and opsonic activities (Virji et al., 1986, 1987). Within the gonococcal outer membrane, porin is closely associated with both LOS and another protein designated Rmp (reduction-modifiable protein) or P.III, and conventional biochemical purification usually results in preparations of porin that are contaminated with Rmp. When these are used to immunize animals, antibodies induced by Rmp may block complement-dependent bacteriolysis mediated by antibodies to porin (Rice et al., 1986). Thus efforts at developing vaccines based on porin have required procedures that eliminate Rmp yet preserve the native conformation of porin (Wetzler et al., 1992). Sialylation of adjacent LOS molecules also interferes with the bactericidal activity of antibodies to porin (Elkins et al., 1992).

Porin has been found to stimulate human T and B lymphocytes (Wetzler et al., 1996; Mackinnon et al., 1999) and interact with TLR2 (Massari et al., 2002; Singleton et al., 2005). Thus, it is likely to be of significance for the interactions of gonococci with the human immune system (Massari et al., 2003).

Opacity (Opa) Proteins

The Opa proteins (P.II) represent a large family of polymorphic OMPs found in *N. gonorrhoeae*; they were named because their expression governs the appearance (opacity) of the colonies on agar plates. Approximately

50–60 proteins and genes have been described, and each isolate appears to possess about 10–12 *opa* genes. Opa protein expression is controlled by a slip–strand mispairing mechanism arising from the presence of a variable number of a pentanucleotide (CTCTT) repeats in the signal sequences of the *opa* genes. An individual gonococcus can simultaneously express up to 3 or 4 Opa proteins, which can be switched off or on again spontaneously. Opa switching has been observed in vivo, both in humans and in colonized mice (Jerse et al., 1994; Jerse, 1999), but it does not appear to be driven by selective pressure arising from the host's specific immune responses.

Whereas a few gonococcal Opa proteins have been found to bind to heparan sulfate proteoglycans, the majority bind in a species-specific manner to certain human cell surface receptors of the carcinoembryonic antigen-related cell adhesion molecule (CEACAM) family (Popp et al., 1999; Virji, 2000). This interaction probably facilitates gonococcal colonization and invasion of epithelial cells, which express CEACAM1, CEACAM5, and CEACAM6 on their apical surfaces. Recognition of CEACAM1, the only member of the family that is expressed on T cells, by Opa-expressing gonococci results in the suppression of T-helper cell activity by signaling through an ITIM (immunoreceptor tyrosine-based inhibition motif) present on the intracellular tail of CEACAM1 (Boulton and Gray-Owen, 2002). Similar effects have been reported in human B cells (Pantelic et al., 2005). This is postulated to represent one mechanism by which *N. gonorrhoeae* suppresses the development of specific immunity in the host, and it has therefore been suggested that, for example, outer membrane vesicles used as a vaccine candidate should be depleted of CEACAM1-binding Opa proteins (Lee et al., 2007). However, CEACAM3, an activating receptor having an intracellular ITAM (immunoreceptor tyrosine-based activating motif), has been proposed as an adaptation that allows human neutrophils to phagocytose and kill Opa-expressing gonococci (Sarantis and Gray-Owen, 2007).

Pili

Gonococcal type IV pili are surface filamentous structures that mediate initial host cell contact, and are considered to be important virulence factors essential for human infection (Swanson et al., 1987). Pilus expression is subject to phase variation. Structurally, pili consist of numerous pilin subunits, which are generated when "silent" non-expressed genes in several *pilS* loci are recombined into the *pilE* expression locus (Segal et al., 1985; Hamrick et al., 2001; Criss et al., 2005). This results in an extremely large number of highly variable pilin sequences, yet the conserved regions of pilin are shrouded in the core of the assembled pilus (Parge et al., 1995), and there is little or no antibody response to the conserved residues (Forest et al., 1996; Hansen et al., 2007). Phosphorylation and glycosylation add to the structural heterogeneity (Hegge et al., 2004). PilC has been identified as the adhesin moiety at the pilus tip (Rudel et al., 1995) that interacts with CD46 on epithelial cells (Källström et al., 1997), although the identification of CD46 as the pilus receptor has been questioned (Tobiason and Seifert, 2001). CD46 is a complement regulatory protein (membrane cofactor protein) that serves as a receptor for several other human pathogens, including measles virus and *Streptococcus pyogenes*, and its ligation results in immunosuppressive effects (Riley-Vargas et al., 2004). Ligation of CD46 on human $CD4^+$ T cells leads to the development of T-regulatory cells and the production of the immunosuppressive cytokine, IL-10 (Kemper et al., 2003). Indeed piliated gonococci have been shown to induce IL-10 secretion by human T cells (Plant and Jonsson, 2006).

Several studies have shown that antibodies to gonococcal pili can block adherence and mediate opsonization (Rothbard et al., 1985; Virji and Heckels, 1985), although the protective epitopes are highly variable. However, as the pilus type of infecting gonococci changes rapidly, well before any immune response can be mounted (Swanson et al., 1987; Seifert et al., 1994) it is unlikely that an immune response could be effective against such a moving target.

H.8 Lipoprotein

H.8 is a conserved antigenic determinant found on *N. gonorrhoeae* and *N. meningitidis* and some commensal neisseriae, and defined by a monoclonal antibody (Cannon et al., 1984). Further investigation revealed that at least two proteins contain the repeated epitope, AAEAP, one being a lipoprotein with an azurin-like sequence (Kawula et al., 1987). Its role in gonococcal pathogenesis is unknown, although antibodies against it have been found in the sera of convalescent patients (Black et al., 1985).

NspA

A search for conserved proteins in meningococci using a panel of monoclonal antibodies against outer membrane preparations identified one such protein that was present in all strains of *N. meningitidis*, designated NspA (Martin et al., 1997). Interest was stimulated by the finding that the defining monoclonal antibody exerted complement-dependent bacteriolysis and conferred passive protection against challenge. NspA was later identified also in *N. gonorrhoeae* with few amino acid variations (Plante et al., 1999), and antibodies against it were found in sera of infected patients (Hedges and Russell, unpublished observations). However, NspA is a small OMP (~16 kDa) and its crystallographic structure reveals it to be mostly inserted in the membrane with only three small surface-exposed loops (Vandeputte-Rutten et al., 2003). While bactericidal antibodies recognize the L3 loop, this is a conformational epitope whose structural integrity depends on the association of NspA with the lipid membrane and detergent-solubilized NspA is not recognized (Cadieux et al., 1999; Hou et al., 2003). Thus, it has proven difficult to generate protective antibodies that recognize native NspA on the gonococcal surface, although efforts to exploit it as a potential vaccine antigen against meningococci have continued (Martin et al., 2000). The biological function of NspA remains unknown. NspA deletion mutants of *N. meningitidis* grow satisfactorily in culture media and are virulent in an infant rat bacteremia model (Moe et al., 2001), but as it is constitutively expressed in all clinical isolates presumably it has an important role for the growth of the bacteria in the natural human host.

Transferrin-Binding Protein

Like most other pathogenic bacteria, *N. gonorrhoeae* has evolved several mechanisms to acquire iron from the host in which it occurs only in protein-bound form (Cornelissen and Sparling, 1994; Schryvers and Stojiljkovic, 1999). Indeed, pathogenic neisseriae utilize a ferric iron-uptake regulator (Fur) mechanism to sense low environmental iron concentrations and regulate the expression of many genes in response to such conditions in vivo (Grifantini et al., 2003). Among the proteins that are up-regulated in low iron conditions are the transferrin-binding proteins, TbpA and TbpB (Cornelissen et al., 1997, 2000; Ronpirin et al., 2001), which have been found to be essential for virulence in a human challenge model (Cornelissen et al., 1998). Moreover, Tbps are constitutively expressed, not subject to phase variation, and are relatively well conserved (Cornelissen et al., 2000; Agarwal et al., 2005). Surface-exposed loops of TbpA have been defined (Yost-Daljev and Cornelissen, 2004). The specificity of gonococcal Tbps for human transferrin might contribute to the unique host specificity of *N. gonorrhoeae* for humans. However, as with most other gonococcal antigens, natural uncomplicated infection does not lead to the development of potent antibody responses against Tbp (Price et al., 2004). These considerations make gonococcal Tbps potential vaccine candidates. Intranasal immunization of mice with gonococcal TbpA and TbpB conjugated to cholera toxin (CT) B subunit induces circulating and vaginal antibodies that exert bactericidal activity against heterologous strains (Price et al., 2005). In a further development of this approach, functional domains from TbpA and TbpB were genetically fused to the A2 subunit of CT and co-expressed with CT B subunit to create chimeric immunogens. Intranasal immunization of mice with these constructs also induced serum bactericidal and vaginal growth-inhibiting antibodies (Price et al., 2007).

N. gonorrhoeae also produces analogous lactoferrin-binding proteins (LbpA and LbpB) that are able to sequester iron from human lactoferrin, the iron-binding protein found in secretions (Blanton et al., 1990). However, only about half of all clinical isolates express Lbp, and it therefore does not appear to be an essential virulence factor, although it can partly substitute for Tbp in the male challenge model (Anderson et al., 2003).

IgA1 Protease

Both *N. gonorrhoeae* and *N. meningitidis* produce an IgA1 protease that cleaves human IgA1 (not IgA2) at a Pro–Ser or Pro–Thr bond in the hinge region (Kilian and Russell, 2005). This enzyme is constitutively expressed in all clinical isolates, and is believed to serve as a virulence factor. However, definitive proof of this assertion has been difficult to obtain, because of the unique specificity of IgA1 proteases for human (and other anthropoid ape) IgA1, thereby making conventional animal models inapplicable. Because an IgA1 protease-deletion mutant was found to colonize isolated fallopian tube explants to the same extent as wild-type gonococci it was originally suggested that IgA1 protease is not an essential virulence factor (Cooper et al., 1984). However, this conclusion is undermined by the absence of the relevant enzyme substrate, IgA1 anti-gonococcal antibodies. Vaginal washes from infected women contain gonococcal IgA1 protease that can cleave IgA1 in vitro (Blake et al., 1979), but we were unable to identify the characteristic cleavage fragments of IgA1 in genital secretions from infected women (Hedges et al., 1998a), possibly due to the presence of IgA1 protease-inhibiting antibodies (Lomholt et al., 1995a). Gonococcal IgA1 protease also cleaves the lysosomal protein LAMP-1, thereby enhancing intracellular survival (Hauck and Meyer, 1997; Lin et al., 1997), although degradation of LAMP-1 alone appears to be insufficient to promote gonococcal survival within phagosomes (Binker et al., 2007). In contrast to other bacterial pathogens, there is relatively little antigenic variation in neisserial IgA1 proteases

(Lomholt et al., 1995b), thus they too remain a potential target antigen for inclusion in a gonococcal vaccine.

AniA

AniA (originally called Pan1) is the major anaerobically induced OMP of *N. gonorrhoeae* that possesses nitrite reductase activity (Clark *et al.*, 1987). Antibodies to AniA have been detected in patients with both uncomplicated and complicated gonorrhea as well as disseminated gonococcal infection (Clark *et al.*, 1988), indicating that it is expressed in vivo. AniA expression has been linked to enhanced resistance to complement-mediated killing of gonococci (Cardinale and Clark, 2000), suggesting that it might be a vaccine candidate, but AniA appears to be antigenically heterogeneous and no specific studies of its potential as a vaccine antigen have been reported.

OmpA-Like Protein

A 23 kDa gonococcal OMP homologous to OmpA of *Escherichia coli* has been described and shown to have a significant role in adherence to and invasion of human cervical and endometrial carcinoma cells (Serino *et al.*, 2007). This protein is also required for intracellular survival of *N. gonorrhoeae* within macrophages following its uptake, and a deletion mutant shows diminished recovery in the murine vaginal infection model. Thus gonococcal OmpA appears to be a virulence factor, but its potential as a vaccine antigen has not been reported.

Other Potential Antigens: Lessons from Meningococcal Vaccine Development

Considerable effort has been devoted in recent years to the development of new and improved vaccines against *N. meningitidis*, because capsular polysaccharide vaccines are of limited value or unavailable in the case of serogroup B. Given the relatedness of the two species, the search for meningococcal protein-based vaccine candidates may also suggest useful avenues of exploration for gonococcal vaccine development. Neisserial outer membrane vesicles containing surface protein antigens and depleted of reactogenic components such as LOS, or incorporating mutant LOS of diminished reactogenicity, show considerable promise (Bakke et al., 2001; Fisseha et al., 2005). Genome mining has identified several meningococcal genome-derived neisserial antigens (GNA), including GNA33 (Adu-Bobie et al., 2004), GNA1870, GNA1946, GNA2132 (Jacobsson et al., 2006), and also NadA (Comanducci et al., 2004). A universal meningococcal vaccine has been developed from a cocktail of such antigens (Giuliani et al., 2006). As the genome of at least one strain of *N. gonorrhoeae* has been sequenced, genome mining should be a feasible approach in this case also.

PROTECTIVE IMMUNE RESPONSE

It is a well-known clinical observation that gonorrhea can be contracted repeatedly, and despite multiple exposures no state of effective immunity is conferred after recovery (Noble et al., 1977). A study in commercial sex workers in Nairobi, Kenya, suggested that multiple exposures diminished the probability of reinfection with the homologous serovar of *N. gonorrhoeae*, implying the development of partial serovar-specific immunity (Plummer et al., 1989), but this finding was not repeated elsewhere (Fox et al., 1999) although few persons are at comparable risk for infection as those who took part in the former study. Numerous studies have assessed antibody responses in both serum and the local genital secretions. However, these efforts have been complicated by several factors, including the extraordinary antigenic variability of the gonococcus, which makes it difficult to select target antigenic preparations against which to measure antibody responses. In addition, longitudinal and prospective studies that would allow the measurement of antibody levels before, during, and after an infection have been difficult to conduct among persons at highest risk for infection, such as the clientele of STD clinics. The sensitive nature of the disease has also obviated accurate record keeping, such that previous episodes of infection cannot always be reliably documented. Moreover, as cervical infection in women may frequently be asymptomatic, cases often go undetected for prolonged periods and it is not always possible to determine when an infection was acquired in a positive-testing individual. All of the above factors have contributed to difficulty in defining correlates of protective immunity to gonorrhea. Indeed, the prevailing paradigm holds that whereas the host might generate an adaptive immune response to *N. gonorrhoeae*, its extraordinary capacity to vary its surface antigens and to resist complement-mediated lysis enables the gonococcus to evade the impact of the host's specific immune responses against it. However, an emerging view is that *N. gonorrhoeae* also avoids inducing effective immune responses in the first place. Most studies on the immune response to gonorrhea have examined subjects with a spectrum of disease conditions, adding to the complexity of interpretation, or have focused on uncomplicated lower tract infections, i.e., gonococcal cervicitis and urethritis. Rigorous, comparative, and quantitative studies of immune responses to different states of gonococcal infection, i.e., asymptomatic, uncomplicated, salpingitis or PID in women or epididymitis in men, and disseminated infection, have not been conducted.

Antibody

A large number of studies have documented the presence of antibodies to *N. gonorrhoeae* in the serum and local genital secretions of both men and women infected with gonorrhea (Kearns et al., 1973a, 1973b; O'Reilly et al., 1976; Tapchaisri and Sirisinha, 1976; Tramont, 1977; McMillan et al., 1979a, 1979c, 1979b; Tramont et al., 1980; Ison et al., 1986; Tramont, 1992; Price et al., 2004). However, some anti-gonococcal antibodies probably induced by cross-reactive antigens of *N. meningitidis* or commensal neisseriae can also be detected in individuals who have never been exposed to *N. gonorrhoeae*. We applied quantitative methods to the analysis of IgM, IgG, and IgA antibodies serum and secretions of subjects at enrollment and treatment and at recall approximately 2 and 4 weeks later, in comparison with age- and gender-matched control subjects presenting at the same clinic but proven microbiologically not to be currently infected with *N. gonorrhoeae* (Hedges et al., 1998b, 1998a, 1999). To reduce the impact of antigenic variability, antibodies were assayed against a standard laboratory strain and against each patient's homologous isolate. Although modest levels of antibodies were detected in serum and secretions, and more to the homologous isolate, in general antibodies were not much higher in infected than in currently uninfected subjects. Antibody levels mostly declined at recall visits, and no correlation was found with known history of prior infection. We concluded that despite the typical generation of an acute inflammatory reaction in symptomatic patients, the adaptive humoral immune response, both local and systemic, to uncomplicated gonorrhea is minimal.

As discussed above, several studies have shown that serum antibodies (predominantly IgG) to various gonococcal antigens, including LOS, porin, and pili, can induce complement- or neutrophil-mediated killing of gonococci in vitro. Whether these conditions are always met in vivo is questionable, because *N. gonorrhoeae* at least in uncomplicated disease remains largely a superficial mucosal infection that colonizes and invades the genital mucosal epithelium. It has been proposed that the ratio: [(anti-porin IgG + anti-LOS IgG)/anti-Rmp IgG] predicts the bactericidal activity of human serum and hence immunity to gonococci (Rice et al., 1994). However, this hypothesis has not been rigorously tested, and while it is plausible that complement-mediated bacteriolysis plays an important role in invasive gonococcal infection (as it does in meningococcal disease), the extent to which it is essential for immune defense within the lumen and mucosa of the genital tract is open to question. Gonococci are able to survive partially within neutrophils (Shafer and Rest, 1989; Jerse and Rest, 1997; Simons et al., 2005), and as noted above possess several mechanisms for avoiding destruction by complement. Furthermore, the immune defense mechanisms in the genital tract, especially that of the male, are not well understood (Russell and Mestecky, 2002). Although secretory IgA (S-IgA) is found in both male and female secretions and has been shown to be locally produced and transported especially in the cervix of the female tract (Kutteh, 1999) or in the penile epithelial glands of Littré in the male (Anderson and Pudney, 1999), IgG, probably largely of systemic origin, is the predominant Ig isotype in both male and female tracts. Although antibody-mediated inhibition of gonococcal attachment to epithelium has been demonstrated (Tramont, 1977), and is a known function for S-IgA antibodies in general (Russell and Kilian, 2005), the function of IgG antibodies in the genital tract is less clear. Whether all complement components are present in genital secretions at sufficiently high concentration to form a functional bacteriolytic system is unclear. Furthermore, IgA antibodies or their $F(ab')_2$ fragments generated by the action of IgA1 proteases have been shown to interfere with IgG antibody-mediated complement activation (Griffiss and Goroff, 1983; Russell et al., 1989; Jarvis and Griffiss, 1991; Hamadeh et al., 1995). Thus, although complement-mediated bacteriolysis is considered to be a major mechanism of defense against systemic meningococcal disease and possibly also in disseminated gonococcal infection, it is debatable whether it holds the same significance for local mucosal gonococcal infections.

Cell-Mediated Immunity

Cellular immune responses in gonorrhea have been even less well investigated, probably because the clinical picture of the disease does not suggest the involvement of significant cell-mediated immunity (CMI) or delayed-type hypersensitivity reactions to *N. gonorrhoeae*. Lymphocyte proliferation and leukocyte migration-inhibition assays have revealed that patients with gonorrhea display CMI to gonococcal extracts, usually to a greater degree than control subjects (Kraus et al., 1970; Esquanazi and Streitfeld, 1973; Grimble and McIllmurray, 1973; Kearns et al., 1973b; Wyle et al., 1977; Landolfo et al., 1981). Peripheral blood T cells from patients with gonorrhea proliferate in response to gonococcal porin, and a significant proportion of these cells are Th2 cells that express α4β7 integrin, the receptor for the mucosal addressin MAdCAM-1 (Simpson et al., 1999). Neisserial porins enhance the expression of the costimulatory ligand B7-2 (CD86) on murine B cells, thereby enhancing T cell stimulation, and they also stimulate responses in neutrophils (Bjerknes et al., 1995; Wetzler et al., 1996; Mackinnon et al., 1999; Simpson et al., 1999). Further studies are necessary to investigate the cellular and regulatory aspects of the immune response in gonorrhea. As antibody responses require the participation of T cells and antigen-presenting cells, it is possible that the paucity of both antibody and cytokine responses in gonorrhea might be due to inadequate activation of these cell types, or their modulation by gonococcal factors.

Inflammatory cytokines IL-1β, IL-6, IL-8, and TNF-α have been found in the urine and plasma of men experimentally infected with *N. gonorrhoeae* (Ramsey et al., 1995), and exposure of epithelial cells in culture to *N. gonorrhoeae* induces the synthesis of IL-1β and TNF-α, probably through the NF-κB pathway (Naumann et al., 1997). Some women with gonorrhea, especially those with concomitant infections due to *Chlamydia trachomatis* or *Trichomonas vaginalis*, showed high circulating levels of inflammatory cytokines but not in the genital secretions (Hedges et al., 1998b). There is little published information on the cytokine response to naturally acquired gonorrhea in men; we found that the cytokine response in men with gonococcal urethritis was weak, although IL-8, which is normally present in semen (Pate et al., 2001), was elevated (Hedges and Russell, unpublished observations).

EPIDEMIOLOGY

Significance as a Public Health Problem

Gonorrhea has worldwide distribution, and WHO estimates that 62 million new infections occur annually, most of these occurring in sub-Saharan Africa and South and Southeast Asia (Fig. 49.2). Although by the 1990s it had been almost eliminated from Western

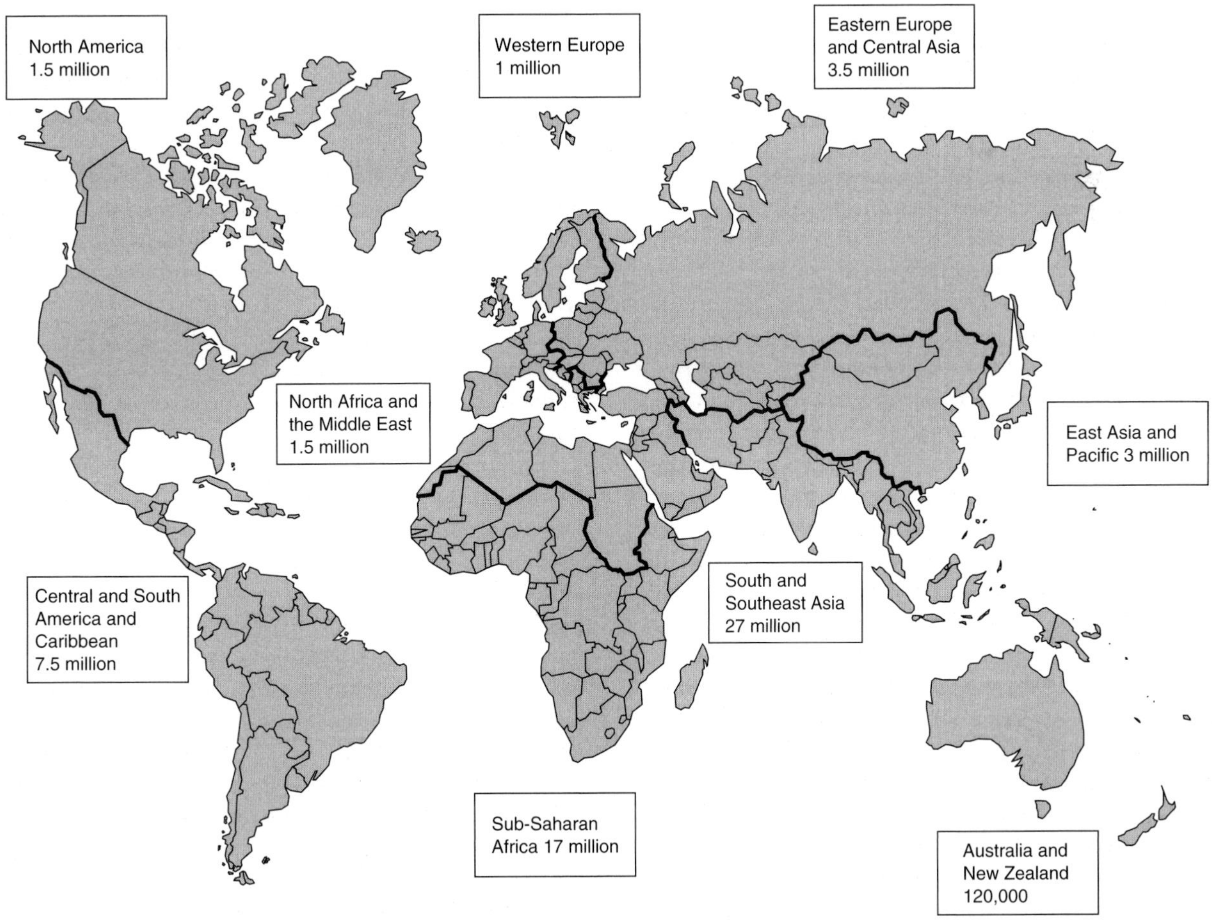

FIGURE 49.2 Worldwide distribution of gonorrhea according to WHO estimates for 1999 (WHO, 2001).

Europe (<20 cases per 100,000 population, even fewer in Sweden at 2.4 cases per 100,000 in 1996) (WHO, 2001), recent increases are thought to have resulted from increased migration from the former Soviet Union and Eastern European countries where rates remained higher. Among Western nations, the United States has the highest rate at >327,000 reported cases in 2006 (>100 cases per 100,000 population) making it the second-most-frequent, notifiable infectious disease (CDC, 2007), but the majority of cases occur in the Southeastern states and among African-Americans. In all communities, gonorrhea occurs overwhelmingly among the sexually active age-groups, i.e., 15–50 years of age. With an estimated incidence of >60 million new infections annually, it is clear that gonorrhea represents a major international health issue that is inadequately addressed.

Potential as a Biothreat Agent

N. gonorrhoeae is not considered to constitute any threat as a potential agent for biowarfare or terrorism. It is far too sensitive to environmental conditions other than enriched growth media, including dehydration and non-physiological temperatures, to survive for longer than a few minutes outside of the human body.

CLINICAL DISEASE

The clinical spectrum of gonococcal infections comprises asymptomatic mucosal infection of any site potentially exposed in the course of sexual contact, through symptomatic local infections, and upper genital tract and systemic complications (Hook and Handsfield, 1999). Thus, sexually exposed persons may acquire gonococcal infection not only at urogenital sites but also at the pharynx or rectum where infections are more often asymptomatic than not. Asymptomatic and unidentified infections at all mucosal sites are associated with increased risk for complications of infection (see below) and they contribute substantially to the persistence of *N. gonorrhoeae* in communities. Persons who are unaware of their infections lack the cues provided by signs and symptoms to seek health care and may thus unknowingly transmit the organism to others.

In men, the most common clinical manifestation of gonorrhea is symptomatic urethritis that typically presents with urethral discharge and dysuria about 3–5 days following exposure to an infected sex partner. The discharge of classical gonococcal urethritis is most often grossly purulent and on gram-stained smears large numbers of polymorphonuclear leukocytes are present, some of which contain ingested organisms seen as gram-negative diplococci (Fig. 49.1). In a small proportion of infected males, the spread of urethral infection to the epididymis and testis leads to acute gonococcal epididymitis.

The description of the clinical spectrum of disease in women is more complex. Gonococcal infection is one of many potential causes of vaginal discharge and the most common symptoms, vaginal discharge or dysuria, are present in about 50% of infected persons. On speculum-guided pelvic examination, some women with gonococcal cervicitis may have purulent endocervical discharge. In addition to the uterine cervix, other common sites of gonococcal infection in women include the urethra and Bartholin's glands. In 15–40% of infected women, untreated infection may ascend to cause complicated upper genital tract infection involving the endometrium and fallopian tubes, even giving rise to peritonitis. This clinical syndrome is often referred to as PID and the scarring resulting from upper genital tract infections in turn substantially increases the risk of infertility and ectopic pregnancy, as well as chronic pelvic pain and increased risk of repeat episodes of PID.

In addition to local genital infections, a small proportion (approximately 1%) of untreated infections may spread hematogenously to cause disseminated gonococcal infection, a bacteremic illness which is most often manifest as a septic dermatitis or asymmetric septic arthritis. Rarely, gonococcal bacteremia may infect heart valves to cause endocarditis or even, very rarely, meningitis.

TREATMENT

Uncomplicated gonococcal infections can be readily cured by the administration of single-dose antibiotics, while persons with complicated infections are treated longer, typically for 10–14 days. In the United States, gonorrhea treatment is guided by the Centers for Disease Control and Prevention's STD Treatment Guidelines (CDC, 2006), an evidence-based document which is periodically revised to represent current best practices for treatment of most STDs other than HIV-AIDS.

Since the availability of sulfa drugs heralded the beginning of the modern antibiotic age, however, *N. gonorrhoeae* has regularly become resistant to the drugs used to treat it. Gonococci develop antibiotic resistance either through the selection of chromosomal mutations conferring resistance to drugs, as has occurred for penicillins, tetracyclines, macrolides, spectinomycin, and fluoroquinolone antibiotics, or through acquisition of plasmids containing antibiotic-resistant genes, as has occurred for both penicillins and tetracyclines to date. Until recently the regular availability of new antibiotics has allowed the recommendation of alternative therapies as gonococcal antibiotic resistance has emerged. In recent years, however, increasingly widespread resistance to fluoroquinolone antibiotics such as ciprofloxacin has left third-generation cephalosporin antibiotics (e.g., ceftriaxone or cefixime) as the primary drugs recommended for treatment of gonorrhea with few alternative therapies. It is reasonable to expect that *N. gonorrhoeae* will eventually become resistant to cephalosporin and related beta-lactam drugs, and the question is "when" rather than "whether" this will happen. This in turn emphasizes the importance of efforts to develop vaccines to prevent gonorrhea.

PATHOGENESIS

The pathogenic process of gonococcal infections has been extensively reviewed by Edwards and Apicella (2004) who discuss at length the differences between infections of the male and female tracts. These authors proposed a model of the infectious process incorporating the following features. In males, following exposure to an infected sex partner, initial attachment of gonococci to urethral epithelial cells depends on pili, and is succeeded by endocytosis involving Opa proteins, LOS, and porin. A quiescent phase lasts for up to 40 h after infection during which gonococci cannot be cultivated, probably because they are residing within epithelial cells. As the organism grows, LOS is largely responsible for eliciting the acute inflammatory response that ensues after a few days, including the production of cytokines such as TNFα, IL-1β, IL-6, and IL-8 and the recruitment of neutrophils, giving rise to the symptoms of infection. Opa proteins are also involved in the interaction of gonococci with neutrophils and their non-opsonic phagocytic uptake by these cells. Indeed, it has been proposed that Opa-CEACAM3 interactions are responsible for neutrophil activation and subsequent intracellular killing of gonococci. However, at least some gonococci are able to survive within neutrophils. Endocytosis by urethral epithelial cells appears to result from the interactions between LOS and the asialoglycoprotein receptor. However, sialylation of LOS impairs both invasion of epithelial cells and phagocytic uptake by neutrophils.

Although gonococci transmitted from men to women are mostly sialylated and sialylation can occur within the female tact, infection of cervical epithelial cells does not appear to be affected by sialylation. Indeed it is speculated that desialylation by host neuraminidases facilitates the transmittal from women to men (Edwards and Apicella, 2004). Although cell lines derived from female tract epithelium can release inflammatory cytokines when cultured with gonococci in vitro, this is not necessarily what happens in vivo, and female infections are frequently asymptomatic, i.e., they display no evidence of an inflammatory response. However, production of complement components by cervical epithelium results in deposition of C3b on the gonococcal surface by alternative pathway activation and the rapid inactivation of this to C3bi, which allows invasion of the epithelial cells through their expression of complement receptor 3. Pili and porin are also involved, but probably not Opa proteins. Progression to upper tract infection, which can occur in 45% of women with untreated gonococcal cervicitis, is facilitated by hormonal changes and menstruation is known to enhance the onset of PID or disseminated infection. Gonococci adhere to non-ciliated cells in the fallopian tubes, and this leads to loss of ciliated cells possibly through the production of inflammatory cytokines. Disseminated infections have been associated particularly with gonococci that express porin type P.IA, which binds complement factor H leading to inhibition of alternative pathway activity and enhancing resistance to complement-mediated lysis.

The immune response to infection has been discussed above.

VACCINES

History

In the early 1900s crude whole-cell vaccines were often applied as a treatment for gonorrhea that was believed to work by enhancing opsonization (Eyre and Stewart, 1909)—then much in vogue following its demonstration by Wright and Douglas (1903). Subsequent efforts to develop prophylactic vaccines have, however, been frustrated by the difficulty of creating suitable animal models of the disease. Small animal models (rabbits, mice) were attempted in the 1930s and 1940s [reviewed in Arko (1989)]; later more extensive work was done using chimpanzees (Lucas et al., 1971) and subcutaneous chamber models in various laboratory animals (Arko, 1972). While skin chamber models, particularly in guinea pigs, allowed characterization of some aspects of pathology and the definition of various "immunotypes" (Chandler et al., 1976; Wong et al., 1979), they, like most other species, have proved to be inadequate models of human genital infection. Furthermore, susceptibility to guinea pig complement did not accord with resistance to human complement (Arko, 1989). The establishment of gonococcal urethritis in male chimpanzees and transfer of the infection to females afforded a model more closely resembling the human disease (Brown et al., 1972), but several practical, economic, and now ethical issues have impeded the widespread use of primate models to study gonococcal pathogenesis and immunity. Interestingly, it was found that parenteral immunization with formaldehyde-fixed gonococci induced substantially greater protective immunity than prior experimental infection (Arko et al., 1976). However, a vaccine made from lysed gonococci proved to be ineffective when tested in humans (Greenberg et al., 1974).

Several groups have developed human challenge models to study various aspects of virulence and immune response in humans (Swanson et al., 1987; Cohen and Cannon, 1999; Schmidt et al., 2001). However, for ethical reasons (i.e., the far higher risk for upper genital tract complications and sequelae in women) these are limited to males only. Moreover, as infections are terminated by antibiotic treatment on development of symptoms or after 1 week, only short-term effects can be studied. Nevertheless, such models may provide a preliminary test of vaccine efficacy. Meanwhile, a mouse vaginal challenge model has been successfully resurrected and shown to be reproducible (Jerse, 1999), although it inevitably suffers the shortcomings of all rodent models in not replicating all facets of the human disease, it can be used to test initial vaccine potential (Plante et al., 2000).

Because of demonstrations in vitro that antibodies to pili could block gonococcal attachment to epithelial cells, which were supported by findings in subcutaneous chamber models, a major effort was mounted to develop a gonococcal pilus vaccine (Tramont et al., 1981; Siegel et al., 1982). This vaccine was successful in eliciting genital antibody responses to homologous pili (McChesney et al., 1982), but in a large-scale field trial it proved completely ineffective (Boslego et al., 1991). Given what is now known about gonococcal pilus structure and the molecular basis of its antigenic hypervariability, this outcome viewed in hindsight is not surprising and there is little current interest in pursuing pilus-based gonococcal vaccines.

The finding that some monoclonal antibodies were bactericidal against a range of gonococci of different porin serovar spurred the development of a porin-based vaccine (Heckels et al., 1989; Elkins et al., 1992). Because of the association of porin with LOS and Rmp, recombinant porin was renatured and incorporated into liposomes that induced antibodies capable of binding to the gonococcal surface and causing bacteriolysis (Elkins et al., 1994). An alternative formulation was prepared from gonococci engineered to delete expression of Rmp (Wetzler et al., 1992). Such preparations are highly immunogenic, probably on account of the adjuvant effect of porin (Lowell et al., 1988). A gonococcal porin vaccine was prepared by Wyeth-Lederle Vaccines (Matsuka et al., 1998), but it does not appear to have been tested as yet in clinical trials.

Current Licensed Vaccines

There are no currently licensed vaccines against gonorrhea.

Vaccines in Development

Although a few candidate vaccine antigens continue to be investigated in laboratories (Table 49.2), none since the gonococcal pilus vaccine has progressed to clinical trial, and pursuit of a porin-based vaccine appears to have stalled. The current sociopolitical and funding scenario in the United States is not conducive to the development of gonococcal vaccines. No dedicated charitable foundations exist to support research into STDs (other than HIV-AIDS), and major pharmaceutical manufacturers seem averse to embarking on new vaccine programs. Studies on gonococcal pathogenesis and immunity, the discovery of gonococcal antigens, and efforts to develop vaccines have been discussed above.

TABLE 49.2 Gonococcal antigens under investigation as potential vaccine candidates

Antigen	Status[a]
Porin	Liposomal formulations developed; immunogenicity and protective potential shown in animals and in vitro; clinical trials pending?
NspA	Immunogenicity and protective potential shown in animals and in vitro (meningococcal vaccines)
Tbp	Immunogenicity shown in animals; protective potential of Ab shown in vitro
LOS	Not known; potential for coupling constitutive oligosaccharides to protein carriers, or using anti-idiotypic or peptide mimics

[a] See text for discussion.

POSTEXPOSURE IMMUNOPROPHYLAXIS

Postexposure immunoprophylaxis, either passive or active, has not been considered for gonorrhea as treatment with appropriate antibiotics is usually effective. Moreover, the short incubation period before the development of symptoms (at least in males), would probably not allow time for the development of an active immune response *de novo*, and inadequate understanding of the correlates of immune protection does not afford a basis for passive administration of preformed antibodies. Early efforts to use therapeutic immunization to treat gonorrhea (see above), although apparently partially successful, did not include adequate controls, and they were supplanted by the advent of chemotherapy.

PROSPECTS FOR THE FUTURE

Efforts at developing vaccines against gonorrhea have fluctuated over several decades as enthusiasm has grown and receded, according to measures of success in comprehending gonococcal pathogenesis and immunity or, conversely, the perceived difficulty, as the complexity of these issues has emerged. Undoubtedly, an important factor is the sociopolitical priority accorded to a disease that is all too often regarded as a matter of morality with strong religious overtones. At the same time it is clear that *N. gonorrhoeae* presents significant scientific challenges because it is an extremely well-adapted and successful human pathogen that exploits weaknesses in not only the human reproductive tract and immune system but also in our behavior. The unfortunate outcome of the pilus vaccine initiative probably had a knock-on effect in undermining the effort to develop a porin-based vaccine, which has not yet proceeded to even a Phase I clinical trial. Whether or not it might be ultimately successful in progressing toward a workable vaccine, a thoughtfully planned trial could yield an abundance of valuable information about human genital tract immune responses.

The hypervariability of most of the major surface antigens of *N. gonorrhoeae* probably means that LOS, pilus, and Opa proteins are not viable candidates for vaccine antigens. Even a "cocktail" vaccine consisting of multiple components selected to represent the spectrum of commonly encountered gonococcal antigens seems an unlikely prospect at present, and might make the cost of such a vaccine too high for use in underdeveloped nations where it is most needed. Progress has stalled over the identification of novel, conserved gonococcal antigens, yet the genome-mining approaches recently pursued for the identification of novel antigens in *N. meningitidis* could surely also be applied to *N. gonorrhoeae*. In addition, major efforts to develop new meningococcal vaccines based on conserved protein antigens or native outer membrane vesicles containing the non-capsular surface components could usefully suggest similar approaches to a gonococcal vaccine.

An important aspect is an improved understanding of immunity to infection in the male and female genital tracts (Russell and Mestecky, 2002). The focus of reproductive immunology has hitherto been on the remarkable mechanisms involved in permitting the engraftment of a semi-allogeneic fetus in the uterine wall without resulting in immunological rejection. The mechanisms of immune response against and delivery of immune effectors to infectious agents have been much less explored, even in the female tract, and almost not at all in the male tract. Yet, recent successes in developing vaccines against human papillomaviruses and herpes simplex virus type II, both of which are genital infections, indicate that it should be possible to develop vaccines against other genital tract pathogens. It has been known for some time that the genital tract (both male and female) is an effector site of the so-called "common mucosal immune system" (Russell and Mestecky, 2002), in the sense that mucosal immune responses induced at remote inductive sites elsewhere can result in the development of S-IgA antibody responses in the genital secretions. Neither the male nor female tracts, however, contain such inductive tissues as typified by intestinal Peyer's patches or Waldeyer's ring in the upper respiratory tract. In these mucosa-associated lymphoid tissues (MALT) organized lymphoid follicles are overlain by a specialized follicle-associated epithelium which samples the luminal contents and passes antigenic materials to the underlying immune-responsive cells. The MALT structures appear to be necessary to disseminate the responding B and T cells to remote mucosal effector sites. For reasons that have been incompletely elucidated, but probably involve the selective expression of chemokines as well as mucosal addressins and homing receptors on the cells involved, mucosal IgA responses induced in the upper respiratory MALT (tonsils in humans, nasal lymphoid tissue in rodents) have a strong propensity to be expressed in the genital tract (Wu et al., 2000; Gallichan et al., 2001). In addition, antigen-presenting dendritic cells have been identified within and underneath the genital tract epithelium, where they likely function to induce local immune responses (Hussain and Lehner,

1995). Investigation and exploitation of these immune response mechanisms for the protection of the genital tract against pathogens need to be vigorously pursued.

A major new advance in immunology in the past few years has been the discovery of a third subset of helper T cells designated "Th17" on account of their ability to produce IL-17 (Weaver et al., 2007). Th17 cells span the divide between innate and adaptive immunity, and by producing IL-17 as well as other downstream factors, elicit inflammatory responses including the mobilization of neutrophils and defensins. While on one hand this results in inflammatory and autoimmune damage—which has been the main focus of interest in Th17 responses so far—it is becoming clear that the Th17 axis of immunity plays an important role in defense against bacterial pathogens (Happel et al., 2003; Yu et al., 2007), including those of mucosal surfaces, and especially those involving an acute inflammatory response. Yet, investigators in microbial pathogenesis and infectious diseases have been slow to pursue this development. Whether *N. gonorrhoeae* is among the numerous bacterial pathogens that elicit Th17 responses remains to be determined, but initial evidence suggests that it might do so (Feinen et al., 2008). If that is correct, then new paradigms in gonococcal pathogenesis and immunity might emerge.

KEY ISSUES

- Sustained effort is required to identify novel, conserved gonococcal antigens as targets for protective immunity.
- The mechanisms and surrogate measures of protective immunity to gonococcal infection in the genital tract should be defined.
- An improved understanding is needed of immune defense mechanisms, their induction and delivery, within both male and female genital tracts.
- Immunization strategies should be devised for the generation of protective immune responses within the genital tract.
- Efforts should be directed to generate and maintain stable, long-term political and financial commitment to the development of vaccines against STDs (other than HIV-AIDS).

ACKNOWLEDGMENTS

Studies in the authors' laboratories have been supported by U.S. National Institutes of Health grant AI46561 (Mucosal Vaccines against Gonorrhea; to MWR) and contract HHSN266200400073C (Sexually Transmitted Infections Clinical Trials Group; to EWH) from the National Institute of Allergy and Infectious Diseases.

References

Adu-Bobie, J., Lupetti, P., Brunelli, B., Granoff, D., Norais, N., Ferrari, G., Grandi, G., Rappuoli, R. and Pizza, M. GNA33 of *Neisseria meningitidis* is a lipoprotein required for cell separation, membrane architecture, and virulence. Infect. Immun. 2004; 72:1914–1919.

Agarwal, S., King, C.A., Klein, E.K., Soper, D.E., Rice, P.A., Wetzler, L.M. and Genco, C.A. The gonococcal Fur-regulated tbpA and tbpB genes are expressed during natural mucosal gonococcal infection. Infect. Immun. 2005; 73:4281–4287.

Anderson, D.J. and Pudney, J. Human male genital tract immunity and experimental models. In: Mucosal Immunology (P.L. Ogra, J. Mestecky, M.E. Lamm, W. Strober, J. Bienenstock and J.R. McGhee, Eds.), pp. 1411–1422. San Diego: Academic Press, 1999.

Anderson, J.E., Hobbs, M.M., Biswas, G.D. and Sparling, P.F. Opposing selective forces for expression of the gonococcal lactoferrin receptor. Mol. Microbiol. 2003; 48:1325–1337.

Apicella, M.A., Shero, M., Jarvis, G.A., Griffiss, J.M., Mandrell, R.E. and Schneider, H. Phenotypic variation in epitope expression of the *Neisseria gonorrhoeae* lipooligosaccharide. Infect. Immun. 1987; 55:1755–1761.

Arko, R.J. *Neisseria gonorrhoeae*: experimental infection of laboratory animals. Science 1972; 177:1200–1201.

Arko, R.J. Animal models for pathogenic *Neisseria* species. Clin. Microbiol. Rev. 1989; 2(Suppl.):S56–S59.

Arko, R.J., Duncan, W.P., Brown, W.J., Peacock, W.L. and Tomizawa, T. Immunity in infection with *Neisseria gonorrhoeae*: duration and serological response in the chimpanzee. J. Infect. Dis. 1976; 133:441–447.

Bakke, H., Lie, K., Haugen, I.L., Korsvold, G.E., Hoiby, E.A., Nass, L.M.r., Holst, J., Aaberge, I.S., Oftung, F. and Haneberg, B. Meningococcal outer membrane vesicle vaccine given intranasally can induce immunological memory and booster responses without evidence of tolerance. Infect. Immun. 2001; 69:5010–5015.

Benedek, T.G. Gonorrhea and the beginnings of clinical research ethics. Perspect. Biol. Med. 2005; 48:54–73.

Binker, M.G., Cosen-Binker, L.I., Terebiznik, M.R., Mallo, G.V., McCaw, S.E., Eskelinen, E.-L., Willenborg, M., Brumell, J.H., Saftig, P., Grinstein, S. and Gray-Owen, S.D. Arrested maturation of *Neisseria*-containing phagosomes in the absence of the lysosome-associated membrane proteins, LAMP-1 and LAMP-2. Cell. Microbiol. 2007; 9:2153–2166.

Bjerknes, R., Guttormsen, H.-K., Solberg, C.O. and Wetzler, L.M. Neisserial porins inhibit human neutrophil actin polymerization, degranulation, opsonin receptor expression, and phagocytosis but prime the neutrophils to increase their oxidative burst. Infect. Immun. 1995; 63:160–167.

Black, J.R., Black, W.J. and Cannon, J.G. Neisserial antigen H.8 is immunogenic in patients with disseminated gonococcal and meningococcal infections. J. Infect. Dis. 1985; 151:650–657.

Blake, M., Holmes, K.K. and Swanson, J. Studies on gonococcus infection. XVII. IgA1-cleaving protease in vaginal washings from women with gonorrhea. J. Infect. Dis. 1979; 139:89–92.

Blanton, K.J., Biswas, G.D., Tsai, J., Adams, J., Dyer, D.W., Davis, S.M., Koch, G.G., Sen, P.K. and Sparling, P.F. Genetic evidence that *Neisseria gonorrhoeae* produces specific receptors for transferrin and lactoferrin. J. Bacteriol. 1990; 172: 5225–5235.

Boslego, J.W., Tramont, E.C., Chung, R.C., McChesney, D.G., Ciak, J., Sadoff, J.C., Piziak, M.V., Brown, J.D., Brinton, C.C., Wood, S.W. and Bryan, J.R. Efficacy trial of a parenteral gonococcal pilus vaccine in men. Vaccine 1991; 9:154–162.

Boulton, I.C. and Gray-Owen, S.D. Neisserial binding to CEACAM1 arrests the activation and proliferation of $CD4^+$ T lymphocytes. Nat. Immunol. 2002; 3:229–236.

Brown, W.J., Lucas, C.T. and Kuhn, U.S. Gonorrhoea in the chimpanzee. Infection with laboratory-passed gonococci and by natural transmission. Br. J. Vener. Dis. 1972; 48:177–178.

Burch, C.L., Danaher, R.J. and Stein, D.C. Antigenic variation in *Neisseria gonorrhoeae*: Production of multiple lipooligosaccharides. J. Bacteriol. 1997; 179:982–986.

Cadieux, N., Plante, M., Rioux, C.R., Hamel, J., Brodeur, B.R. and Martin, D. Bactericidal and cross-protective activities of a monoclonal antibody directed against *Neisseria meningitidis* NspA outer membrane protein. Infect. Immun. 1999; 67: 4955–4959.

Cannon, J.G., Black, W.J., Nachamkin, I. and Stewart, P.W. Monoclonal antibody that recognizes an outer membrane antigen common to the pathogenic *Neisseria* species but not to most nonpathogenic *Neisseria* species. Infect. Immun. 1984; 43:994–999.

Cardinale, J.A. and Clark, V.L. Expression of AniA, the major anaerobically induced outer membrane protein of *Neisseria gonorrhoeae*, provides protection against killing by normal human sera. Infect. Immun. 2000; 68:4368–4369.

CDC Sexually Transmitted Diseases Treatment Guidelines, 2006. MMWR 2006; 55:1–93.

CDC Notifiable Diseases/Deaths in Selected Cities Weekly Information. MMWR 2007; 55:1396–1407.

Chandler, F.W., Kraus, S.J. and Watts, J.C. Pathological features of experimental gonococcal infection in mice and guinea pigs. Infect. Immun. 1976; 13:909–914.

Clark, V.L., Campbell, L.A., Palermo, D.A., Evans, T.M. and Klimpel, K.W. Induction and repression of outer membrane proteins by anaerobic growth of *Neisseria gonorrhoeae*. Infect. Immun. 1987; 55:1359–1364.

Clark, V.L., Knapp, J.S., Thompson, S. and Klimpel, K.W. Presence of antibodies to the major anaerobically induced gonococcal outer membrane protein in sera from patients with gonococcal infections. Microb. Pathog. 1988; 5:381–390.

Cohen, M.S. and Cannon, J.G. Human experimentation with *Neisseria gonorrhoeae*: progress and goals. J. Infect. Dis. 1999; 179(Suppl. 2):S375–S379.

Comanducci, M., Bambini, S., Caugant, D.A., Mora, M., Brunelli, B., Capecchi, B., Ciucchi, L., Rappuoli, R. and Pizza, M. NadA diversity and carriage in *Neisseria meningitidis*. Infect. Immun. 2004; 72:4217–4223.

Cooper, M.D., McGee, Z.A., Mulks, M.H., Koomey, J.M. and Hindman, T.L. Attachment to and invasion of human fallopian tube mucosa by an IgA1 protease-deficient mutant of *Neisseria gonorrhoeae* and its wild-type parent. J. Infect. Dis. 1984; 150:737–744.

Cornelissen, C.N., Anderson, J.E., Boulton, I.C. and Sparling, P.F. Antigenic and sequence diversity in gonococcal transferrin-binding protein A. Infect. Immun. 2000; 68:4725–4735.

Cornelissen, C.N., Anderson, J.E. and Sparling, P.F. Characterization of the diversity and the transferrin-binding domain of gonococcal transferrin-binding protein 2. Infect. Immun. 1997; 65:822–828.

Cornelissen, C.N., Kelley, M., Hobbs, M.M., Anderson, J.E., Cannon, J.G., Cohen, M.S. and Sparling, P.F. The transferrin receptor expressed by gonococcal strain FA1090 is required for the experimental infection of human male volunteers. Mol. Microbiol. 1998; 27:611–616.

Cornelissen, C.N. and Sparling, P.F. Iron piracy: acquisition of transferrin-bound iron by bacterial pathogens. Mol. Microbiol. 1994; 14:843–850.

Criss, A.K., Kline, K.A. and Seifert, H.S. The frequency and rate of pilin antigenic variation in *Neisseria gonorrhoeae*. Mol. Microbiol. 2005; 58:510–519.

Edwards, J.L. and Apicella, M.A. The molecular mechanisms used by *Neisseria gonorrhoeae* to initiate infection differ between men and women. Clin. Microbiol. Rev. 2004; 17:965–981.

Elkins, C., Barkley, K.B., Carbonetti, N.H., Coimbre, A.J. and Sparling, P.F. Immunobiology of purified recombinant outer membrane porin protein I of *Neisseria gonorrhoeae*. Mol. Microbiol. 1994; 14:1059–1075.

Elkins, C., Carbonetti, N.H., Varela, V.A., Stirewalt, D., Klapper, D.G. and Sparling, P.F. Antibodies to N-terminal peptides of gonococcal porin are bactericidal when gonococcal lipopolysaccharide is not sialylated. Mol. Microbiol. 1992; 6:2617–2628.

Esquanazi, V. and Streitfeld, M.M. Transformation of lymphocytes in gonorrhea before and after therapy. Infect. Immun. 1973; 8:503–509.

Estabrook, M.M., Griffiss, J.M. and Jarvis, G.A. Sialylation of *Neisseria meningitidis* lipooligosaccharide inhibits serum bactericidal activity by masking lacto-N-neotetraose. Infect. Immun. 1997; 65:4436–4444.

Eyre, J.W.H. and Stewart, B.H. The treatment of gonococcus infections by vaccines. Lancet 1909; 174:76–81.

Feinen, B., Jerse, A.E., Gaffen, S.L. and Russell, M.W. *Neisseria gonorrhoeae* induces a Th17 response. ASM abstract, 2008.

Fisseha, M., Chen, P., Brandt, B., Kijek, T., Moran, E. and Zollinger, W. Characterization of native outer membrane vesicles from lpxL mutant strains of *Neisseria meningitidis* for use in parenteral vaccination. Infect. Immun. 2005; 73:4070–4080.

Forest, K.T., Bernstein, S.L., Getzoff, E.D., So, M., Tribbick, G., Geysen, H.M., Deal, C.D. and Tainer, J.A. Assembly and antigenicity of the *Neisseria gonorrhoeae* pilus mapped with antibodies. Infect. Immun. 1996; 64:644–652.

Fox, K.K., Thomas, J.C., Weiner, D.H., Davis, R.H., Sparling, P.F. and Cohen, M.S. Longitudinal evaluation of serovar-specific immunity to *Neisseria gonorrhoeae*. Am. J. Epidemiol. 1999; 149:353–358.

Gallichan, W.S., Woolstencroft, R.N., Guarasci, T., McCluskie, M.J., Davis, H.L. and Rosenthal, K.L. Intranasal immunization with CpG oligodeoxynucleotides as an adjuvant dramatically increases IgA and protection against herpes simplex virus-2 in the genital tract. J. Immunol. 2001; 166:3451–3457.

Giuliani, M.M., Adu-Bobie, J., Comanducci, M., Arico, B., Savino, S., Santini, L., Brunelli, B., Bambini, S., Biolchi, A., Capecchi, B., Cartocci, E., Ciucchi, L., Di Marcello, F., Ferlicca, F., Galli, B., Luzzi, E., Masignani, V., Serruto, D., Veggi, D., Contorni, M., Morandi, M., Bartalesi, A., Cinotti, V., Mannucci, D., Titta, F., Ovidi, E., Welsch, J.A., Granoff, D., Rappuoli, R. and Pizza, M. A universal vaccine for serogroup B meningococcus. Proc. Natl. Acad. Sci. U.S.A. 2006; 103:10834–10839.

Greenberg, L., Diena, B.B., Ashton, F.A., Wallace, R., Kenny, C.P., Znamirowski, R., Ferrari, H. and Atkinson, J. Gonococcal vaccine studies in Inuvik. Can. J. Public Health 1974; 65:29–33.

Gregg, C.R., Melly, M.A., Hellerqvist, C.G., Coniglio, J.G. and McGee, Z.A. Toxic activity of purified lipopolysaccharide of *Neisseria gonorrhoeae* for human fallopian tube mucosa. J. Infect. Dis. 1981; 143:432–439.

Grifantini, R., Sebastian, S., Frigimelica, E., Draghi, M., Bartolini, E., Muzzi, A., Rappuoli, R., Grandi, G. and Genco, C.A.

Identification of iron-activated and -repressed Fur-dependent genes by transcriptome analysis of *Neisseria meningitidis* group B. Proc. Natl. Acad. Sci. U.S.A. 2003; 100:9542–9547.

Griffiss, J.M. and Goroff, D.K. IgA blocks IgM and IgG-initiated immune lysis by separate molecular mechanisms. J. Immunol. 1983; 130:2882–2885.

Grimble, A.S. and McIllmurray, M.B. Cell-mediated immune response in gonorrhoea. Br. J. Vener. Dis. 1973; 49:446–449.

Hamadeh, R.M., Estabrook, M.M., Zhou, P., Jarvis, G.A. and Griffiss, J.M. Anti-gal binds to pili of *Neisseria meningitidis*: The immunoglobulin A isotype blocks complement-mediated killing. Infect. Immun. 1995; 63:4900–4906.

Hamrick, T.S., Dempsey, J.A., Cohen, M.S. and Cannon, J.G. Antigenic variation of gonococcal pilin expression in vivo: analysis of the strain FA1090 pilin repertoire and identification of the pilS gene copies recombining with pilE during experimental human infection. Microbiology 2001; 147:839–849.

Hansen, J.K., Demick, K.P., Mansfield, J.M. and Forest, K.T. Conserved regions from *Neisseria gonorrhoeae* pilin are immunosilent and not immunosuppressive. Infect. Immun. 2007; 75:4138–4147.

Happel, K.I., Zheng, M., Young, E., Quinton, L.J., Lockhart, E., Ramsay, A.J., Shellito, J.E., Schurr, J.R., Bagby, G.J., Nelson, S. and Kolls, J.K. Cutting edge: roles of Toll-like receptor 4 and IL-23 in IL-17 expression in response to *Klebsiella pneumoniae* infection. J. Immunol. 2003; 170:4432–4436.

Hauck, C.R. and Meyer, T.F. The lysosomal/phagosomal membrane protein h-lamp-1 is a target of the IgA1 protease of *Neisseria gonorrhoeae*. FEBS Lett. 1997; 405:86–90.

Heckels, J.E., Fletcher, J.N. and Virji, M. The potential protective effect of immunization with outer-membrane protein I from *Neisseria gonorrhoeae*. J. Gen. Microbiol. 1989; 135:2269–2276.

Hedges, S.R., Mayo, M.S., Kallman, L., Mestecky, J., Hook, E.W. and Russell, M.W. Evaluation of immunoglobulin A1 (IgA1) protease and IgA1 protease-inhibitory activity in human female genital infection with *Neisseria gonorrhoeae*. Infect. Immun. 1998a; 66:5826–5832.

Hedges, S.R., Mayo, M.S., Mestecky, J., Hook, E.W. and Russell, M.W. Limited local and systemic antibody responses to *Neisseria gonorrhoeae* during uncomplicated genital infections. Infect. Immun. 1999; 67:3937–3946.

Hedges, S.R., Sibley, D.A., Mayo, M.S., Hook, E.W., III and Russell, M.W. Cytokine and antibody responses in women infected with *Neisseria gonorrhoeae*: Effect of concomitant infections. J. Infect. Dis. 1998b; 178:742–751.

Hegge, F.T., Hitchen, P.G., Aas, F.E., Kristiansen, H., Lovold, C., Egge-Jacobsen, W., Panico, M., Leong, W.Y., Bull, V., Virji, M., Morris, H.R., Dell, A. and Koomey, M. Unique modifications with phosphocholine and phosphoethanolamine define alternate antigenic forms of *Neisseria gonorrhoeae* type IV pili. Proc. Natl. Acad. Sci. U.S.A. 2004; 101:10798–10803.

Hook, E.W. and Handsfield, H.H. Gonococcal infections in the adult. In: Sexually Transmitted Diseases (K.K. Holmes, P.F. Sparling, P.-A. Mårdh, S.M. Lemon, W.E. Stamm, P. Piot and J.N. Wasserheit, Eds.), 3rd Edn., pp. 451–466. New York: McGraw-Hill, 1999.

Hou, V.C., Moe, G.R., Raad, Z., Wuorimaa, T. and Granoff, D.M. Conformational epitopes recognized by protective anti-neisserial surface protein A antibodies. Infect. Immun. 2003; 71:6844–6849.

Hussain, L. and Lehner, T. Comparative investigation of Langerhans' cells and potential receptors for HIV in oral, genital and rectal epithelia. Immunology 1995; 85:475–484.

Ison, C.A., Hadfield, S.G., Bellinger, C.M., Dawson, S.G. and Glynn, A.A. The specificity of serum and local antibodies in female gonorrhoea. Clin. Exp. Immunol. 1986; 65:198–205.

Jacobsson, S., Thulin, S., Molling, P., Unemo, M., Comanducci, M., Rappuoli, R. and Olcen, P. Sequence constancies and variations in genes encoding three new meningococcal vaccine candidate antigens. Vaccine 2006; 24:2161–2168.

Janda, W.M. and Gaydos, C.A. Neisseria. In: Manual of Clinical Microbiology (P.R. Murray, E.J. Baron, J.H. Jorgensen, M.L. Landry and M.A. Pfaller, Eds.), 9th Edn., pp. 601–620. Washington, DC: ASM Press, 2007.

Jarvis, G.A. and Griffiss, J.M. Human IgA1 blockade of IgG-initiated lysis of *Neisseria meningitidis* is a function of antigen-binding fragment binding to the polysaccharide capsule. J. Immunol. 1991; 147:1962–1967.

Jerse, A.E. Experimental gonococcal genital tract infection and opacity protein expression in estradiol-treated mice. Infect. Immun. 1999; 67:5699–5708.

Jerse, A.E., Cohen, M.S., Drown, P.M., Whicker, L.G., Isbey, S.F., Seifert, S. and Cannon, J.G. Multiple gonococcal opacity proteins are expressed during experimental urethral infection in the male. J. Exp. Med. 1994; 179:911–920.

Jerse, A.E. and Rest, R.F. Adhesion and invasion by the pathogenic *Neisseria*. Trends Microbiol. 1997; 5:217–221.

Källström, H., Liszewski, M.K., Atkinson, J.P. and Jonsson, A.B. Membrane cofactor protein (MCP or CD46) is a cellular pilus receptor for pathogenic *Neisseria*. Mol. Microbiol. 1997; 25:639–647.

Kawula, T.H., Spinola, S.M., Klapper, D.G. and Cannon, J.G. Localization of a conserved epitope and an azurin-like domain in the H.8 protein of pathogenic *Neisseria*. Mol. Microbiol. 1987; 1:179–185.

Kearns, D.H., O'Reilly, R.J., Lee, L. and Welch, B.G. Secretory IgA antibodies in the urethral exudate of men with uncomplicated urethritis due to *Neisseria gonorrhoeae*. J. Infect. Dis. 1973a; 127:99–101.

Kearns, D.H., Seibert, G.B., O'Reilly, R., Lee, L. and Logan, L. Paradox of the immune response to uncomplicated gonococcal urethritis. N. Engl. J. Med. 1973b; 289:1170–1174.

Kemper, C., Chan, A.C., Green, J.M., Brett, K.A., Murphy, K.M. and Atkinson, J.P. Activation of human $CD4^+$ cells with CD3 and CD46 induces a T-regulatory cell 1 phenotype. Nature 2003; 421:388–392.

Kilian, M. and Russell, M.W. Microbial evasion of IgA functions. In: Mucosal Immunology (J. Mestecky, J. Bienenstock, M.E. Lamm, L. Mayer, W. Strober and J.R. McGhee, Eds.), pp. 291–303. Amsterdam: Elsevier/Academic Press, 2005.

Kraus, S.J., Perkins, G.H. and Geller, R.C. Lymphocyte transformation in repeated gonococcal urethritis. Infect. Immun. 1970; 2:655–658.

Kutteh, W.H. Mucosal immunity in the human female reproductive tract. In: Mucosal Immunology (P.L. Ogra, J. Mestecky, M.E. Lamm, W. Strober, J. Bienenstock and J.R. McGhee, Eds.), pp. 1423–1434. San Diego: Academic Press, 1999.

Landolfo, P.J., Marrie, T.J., Nelson, N.A. and Ronald, A.R. Cell-mediated immune response in gonococcal infection. Can. J. Microbiol. 1981; 27:76–80.

Lee, H.S., Boulton, I.C., Reddin, K., Wong, H., Halliwell, D., Mandelboim, O., Gorringe, A.R. and Gray-Owen, S.D. Neisserial outer membrane vesicles bind the coinhibitory receptor carcinoembryonic antigen-related cellular adhesion molecule 1 and suppress $CD4^+$ T lymphocyte function. Infect. Immun. 2007; 75:4449–4455.

Lin, L., Ayala, P., Larson, J., Mulks, M., Fukuda, M., Carlsson, S.R., Enns, C. and So, M. The *Neisseria* type 2 IgA1 protease cleaves LAMP1 and promotes survival of bacteria within epithelial cells. Mol. Microbiol. 1997; 24:1083–1094.

Lomholt, H., Lind, I. and Kilian, M. Neisseria gonorrhoeae IgA1 proteases share epitopes recognized by neutralizing antibodies. Vaccine 1995a; 13:1213–1219.

Lomholt, H., Poulsen, K. and Kilian, M. Antigenic and genetic heterogeneity among *Haemophilus* and *Neisseria* IgA1 proteases. Adv. Exp. Med. Biol. 1995b; 371A:599–603.

Lowell, G.H., Ballou, W.R., Smith, L.F., Wirtz, R.A., Zollinger, W.D. and Hockmeyer, W.T. Proteosome-lipopeptide vaccines: enhancement of immunogenicity for malaria CS peptides. Science 1988; 240:800–802.

Lucas, C.T., Chandler, F., Martin, J.E. and Schmale, J.D. Transfer of gonococcal urethritis from man to chimpanzee. An animal model for gonorrhea. JAMA 1971; 216:1612–1614.

Mackinnon, F.G., Ho, Y., Blake, M.S., Michon, F., Chandraker, A., Sayegh, M.H. and Wetzler, L.M. The role of B/T costimulatory signals in the immunopotentiating activity of neisserial porin. J. Infect. Dis. 1999; 180:755–761.

Mandrell, R.E., Griffiss, J.M. and Macher, B.A. Lipooligosaccharides (LOS) of *Neisseria gonorrhoeae* and *Neisseria meningitidis* have components that are immunochemically similar to precursors of human blood group antigens. J. Exp. Med. 1988; 168:107–126.

Mandrell, R.E., Lesse, A.J., Sugai, J.V., Shero, M., Griffiss, J.M., Cole, J.A., Parsons, N.J., Smith, H., Morse, S.A. and Apicella, M.A. In vitro and in vivo modification of *Neisseria gonorrhoeae* lipooligosaccharide epitope structure by sialylation. J. Exp. Med. 1990; 171:1649–1664.

Martin, D., Brodeur, B.R., Hamel, J., Couture, F., De Alwis, U., Lian, Z.R., Martin, S., Andrews, D. and Ellis, R.W. Candidate *Neisseria meningitidis* NspA vaccine. J. Biotechnol. 2000; 83:27–31.

Martin, D., Cadieux, N., Hamel, J. and Brodeur, B.R. Highly conserved *Neisseria meningitidis* surface protein confers protection against experimental infection. J. Exp. Med. 1997; 185:1173–1183.

Massari, P., Henneke, P., Ho, Y., Latz, E., Golenbock, D.T. and Wetzler, L.M. Cutting edge: immune stimulation by neisserial porins is toll-like receptor 2 and MyD88 dependent. J. Immunol. 2002; 168:1533–1537.

Massari, P., Ram, S., Macleod, H. and Wetzler, L.M. The role of porins in neisserial pathogenesis and immunity. Trends Microbiol. 2003; 11:87–93.

Matsuka, Y.V., Dilts, D.A., Hoiseth, S. and Arumugham, R. Characterization of a subunit structure and stability of the recombinant porin from *Neisseria gonorrhoeae*. J. Protein Chem. 1998; 17:719–728.

McChesney, D., Tramont, E.C., Boslego, J.W., Ciak, J., Sadoff, J. and Brinton, C.C. Genital antibody response to a parenteral gonococcal pilus vaccine. Infect. Immun. 1982; 36:1006–1012.

McMillan, A., McNeillage, G. and Young, H. Antibodies to *Neisseria gonorrhoeae*: a study of the urethral exudates of 232 men. J. Infect. Dis. 1979a; 140:89–95.

McMillan, A., McNeillage, G., Young, H. and Bain, S.S.R. Secretory antibody response of the cervix to infection with *Neisseria gonorrhoeae*. Br. J. Vener. Dis. 1979b; 55:265–270.

McMillan, A., McNeillage, G., Young, H. and Bain, S.R. Serum immunoglobulin response in uncomplicated gonorrhoea. Br. J. Vener. Dis. 1979c; 55:5–9.

Moe, G.R., Zuno-Mitchell, P., Lee, S.S., Lucas, A.H. and Granoff, D.M. Functional activity of anti-neisserial surface protein A monoclonal antibodies against strains of *Neisseria meningitidis* serogroup B. Infect. Immun. 2001; 69:3762–3771.

Naumann, M., Wessler, S., Bartsch, C., Wieland, B. and Meyer, T.F. *Neisseria gonorrhoeae* epithelial cell interaction leads to the activation of the transcription factors nuclear factor κB and activator protein 1 and the induction of inflammatory cytokines. J. Exp. Med. 1997; 186:247–258.

Noble, R.C., Kirk, N.M., Slagel, W.A., Vance, B.J. and Somes, G.W. Recidivism among patients with gonococcal infection presenting to a venereal disease clinic. Sex. Transm. Dis. 1977; 4:39–43.

Ober, W.B. Boswell's clap. JAMA 1970; 212:91–95.

O'Reilly, R.J., Lee, L. and Welch, B.G. Secretory IgA antibody responses to *Neisseria gonorrhoeae* in the genital secretions of infected females. J. Infect. Dis. 1976; 133:113–125.

Pantelic, M., Kim, Y.-J., Bolland, S., Chen, I., Shively, J. and Chen, T. *Neisseria gonorrhoeae* kills carcinoembryonic antigen-related cellular adhesion molecule 1 (CD66a)-expressing human B cells and inhibits antibody production. Infect. Immun. 2005; 73:4171–4179.

Parge, H.E., Forest, K.T., Hickey, M.J., Christensen, D.A., Getzoff, E.D. and Tainer, J.A. Structure of the fibre-forming protein pilin at 2.6 A resolution. Nature 1995; 378:32–38.

Pate, M.S., Hedges, S.R., Sibley, D.A., Russell, M.W., Hook, E.W. and Mestecky, J. Urethral cytokine and immune responses in *Chlamydia trachomatis*-infected males. Infect. Immun. 2001; 69:7178–7181.

Peeling, R.W., Mabey, D., Herring, A. and Hook, E.W. Why do we need quality-assured diagnostic tests for sexually transmitted infections? Nat. Rev. Microbiol. 2006; 4:909–921.

Plant, L.J. and Jonsson, A.-B. Type IV pili of *Neisseria gonorrhoeae* influence the activation of human $CD4^+$ T cells. Infect. Immun. 2006; 74:442–448.

Plante, M., Cadieux, N., Rioux, C.R., Hamel, J., Brodeur, B.R. and Martin, D. Antigenic and molecular conservation of the gonococcal NspA protein. Infect. Immun. 1999; 67:2855–2861.

Plante, M., Jerse, A., Hamel, J., Couture, F., Rioux, C.R., Brodeur, B.R. and Martin, D. Intranasal immunization with gonococcal outer membrane preparations reduces the duration of vaginal colonization of mice by *Neisseria gonorrhoeae*. J. Infect. Dis. 2000; 182:848–855.

Plummer, F.A., Simonsen, J.N., Chubb, H., Slaney, L., Kimata, J., Bosire, M., Ndinya-Achola, J.O. and Ngugi, E.N. Epidemiologic evidence for the development of serovar-specific immunity after gonococcal infection. J. Clin. Invest. 1989; 83:1472–1476.

Popp, A., Dehio, C., Grunert, F., Meyer, T.F. and Gray-Owen, S.D. Molecular analysis of neisserial Opa protein interactions with the CEA family of receptors: identification of determinants contributing to the differential specificities of binding. Cell. Microbiol. 1999; 1:169–181.

Price, G.A., Hobbs, M.M. and Cornelissen, C.N. Immunogenicity of gonococcal transferrin binding proteins during natural infections. Infect. Immun. 2004; 72:277–283.

Price, G.A., Masri, H.P., Russell, M.W. and Cornelissen, C.N. Gonococcal transferrin binding protein chimeras induce bactericidal and growth inhibitory antibodies in mice. Vaccine 2007; 25:7247–7260.

Price, G.A., Russell, M.W. and Cornelissen, C.N. Intranasal administration of recombinant *Neisseria gonorrhoeae* transferrin binding proteins A and B conjugated to the cholera toxin B subunit induces systemic and vaginal antibodies in mice. Infect. Immun. 2005; 73:3945–3953.

Ram, S., Ngampasutadol, J., Cox, A.D., Blom, A.M., Lewis, L.A., St. Michael, F., Stupak, J., Gulati, S. and Rice, P.A. Heptose I glycan substitutions on *Neisseria gonorrhoeae* lipooligosaccharide influence C4b-binding protein binding and serum resistance. Infect. Immun. 2007; 75:4071–4081.

Ram, S., Sharma, A.K., Simpson, S.D., Gulati, S., McQuillen, D.P., Pangburn, M.K. and Rice, P.A. A novel sialic acid binding site on factor H mediates serum resistance of sialylated *Neisseria gonorrhoeae*. J. Exp. Med. 1998; 187:743–752.

Ramsey, K.H., Schneider, H., Cross, A.S., Boslego, J.W., Hoover, D.L., Staley, T.L., Kuschner, R.A. and Deal, C.D. Inflammatory cytokines produced in response to experimental human gonorrhea. J. Infect. Dis. 1995; 172:186–191.

Rice, P.A., McQuillen, D.P., Gulati, S., Jani, D.B., Wetzler, L.M., Blake, M.S. and Gotschlich, E.C. Serum resistance of *Neisseria gonorrhoeae*. Does it thwart the inflammatory response and facilitate the transmission of infection? Ann. N. Y. Acad. Sci. 1994; 730:7–14.

Rice, P.A., Vayo, H.E., Tam, M.R. and Blake, M.S. Immunoglobulin G antibodies directed against protein III block killing of serum resistant *Neisseria gonorrhoeae* by immune serum. J. Exp. Med. 1986; 164:1735–1748.

Riley-Vargas, R.C., Gill, D.B., Kemper, C., Liszewski, M.K. and Atkinson, J.P. CD46: expanding beyond complement regulation. Trends Immunol. 2004; 25:496–503.

Ronpirin, C., Jerse, A.E. and Cornelissen, C.N. Gonococcal genes encoding transferrin-binding proteins A and B are arranged in a bicistronic operon but are subject to differential expression. Infect. Immun. 2001; 69:6336–6347.

Rothbard, J.B., Fernandez, R., Wang, L., Teng, N.N. and Schoolnik, G.K. Antibodies to peptides corresponding to a conserved sequence of gonococcal pilins block bacterial adhesion. Proc. Natl. Acad. Sci. U.S.A. 1985; 82:915–919.

Rudel, T., Scheuerpflug, I. and Meyer, T.F. *Neisseria* PilC protein identified as type-4 pilus tip-located adhesin. Nature 1995; 373:357–359.

Russell, M.W. and Kilian, M. Biological activities of IgA. In: Mucosal Immunology (J. Mestecky, J. Bienenstock, M.E. Lamm, L. Mayer, W. Strober and J.R. McGhee, Eds.), pp. 267–289. San Diego: Academic Press, 2005.

Russell, M.W. and Mestecky, J. Humoral immune responses to microbial infections in the genital tract. Microbes Infect. 2002; 4:667–677.

Russell, M.W., Reinholdt, J. and Kilian, M. Anti-inflammatory activity of human IgA antibodies and their Fabα fragments: inhibition of IgG-mediated complement activation. Eur. J. Immunol. 1989; 19:2243–2249.

Sarafian, S.K. and Knapp, J.S. Molecular epidemiology of gonorrhea. Clin. Microbiol. Rev. 1989; 2(Suppl.):S49–S55.

Sarantis, H. and Gray-Owen, S.D. The specific innate immune receptor CEACAM3 triggers neutrophil bactericidal activities via a Syk kinase-dependent pathway. Cell. Microbiol. 2007; 9:2167–2180.

Schmidt, K.A., Schneider, H., Lindstrom, J.A., Boslego, J.W., Warren, R.A., Van de Verg, L., Deal, C.D., McClain, J.B. and Griffiss, J.M. Experimental gonococcal urethritis and reinfection with homologous gonococci in male volunteers. Sex. Transm. Dis. 2001; 28:555–564.

Schryvers, A.B. and Stojiljkovic, I. Iron acquisition systems in the pathogenic *Neisseria*. Mol. Microbiol. 1999; 32:1117–1123.

Segal, E., Billyard, E., So, M., Storzbach, S. and Meyer, T.F. Role of chromosomal rearrangement in *N. gonorrhoeae* pilus phase variation. Cell 1985; 40:293–300.

Seifert, H.S., Wright, C.J., Jerse, A.E., Cohen, M.S. and Cannon, J.G. Multiple gonococcal pilin antigenic variants are produced during experimental human infections. J. Clin. Invest. 1994; 93:2744–2749.

Serino, L., Nesta, B., Leuzzi, R., Fontana, M.R., Monaci, E., Mocca, B.T., Cartocci, E., Masignani, V., Jerse, A.E., Rappuoli, R. and Pizza, M. Identification of a new OmpA-like protein in *Neisseria gonorrhoeae* involved in the binding to human epithelial cells and in vivo colonization. Mol. Microbiol. 2007; 64:1391–1403.

Shafer, W.M. and Rest, R.F. Interactions of gonococci with phagocytic cells. Annu. Rev. Microbiol. 1989; 43:121–145.

Siegel, M., Olsen, D., Critchlow, C. and Buchanan, T.M. Gonococcal pili: safety and immunogenicity in humans and antibody function in vitro. J. Infect. Dis. 1982; 145:300–310.

Simons, M.P., Nauseef, W.M. and Apicella, M.A. Interactions of *Neisseria gonorrhoeae* with adherent polymorphonuclear leukocytes. Infect. Immun. 2005; 73:1971–1977.

Simpson, S.D., Ho, Y., Rice, P.A. and Wetzler, L.M. T lymphocyte response to *Neisseria gonorrhoeae* porin in individuals with mucosal gonococcal infections. J. Infect. Dis. 1999; 180:762–773.

Singleton, T.E., Massari, P. and Wetzler, L.M. Neisserial porin-induced dendritic cell activation is MyD88 and TLR2 dependent. J. Immunol. 2005; 174:3545–3550.

Smith, H., Parsons, N.J. and Cole, J.A. Sialylation of neisserial lipopolysaccharide: a major influence on pathogenicity. Microb. Pathog. 1995; 19:365–377.

Sparling, P.F. Biology of *Neisseria gonorrhoeae*. In: Sexually Transmitted Diseases (K.K. Holmes, P.F. Sparling, P.-A. Mårdh, S.M. Lemon, W.E. Stamm, P. Piot and J.N. Wasserheit, Eds.), 3rd Edn., pp. 433–449. New York: McGraw-Hill, 1999.

Swanson, J., Robbins, K., Barrera, O., Corwin, D., Boslego, J., Ciak, J., Blake, M. and Koomey, J.M. Gonococcal pilin variants in experimental gonorrhea. J. Exp. Med. 1987; 165:1344–1357.

Tapchaisri, P. and Sirisinha, S. Serum and secretory antibody responses to *Neisseria gonorrhoeae* in patients with gonococcal infections. Br. J. Vener. Dis. 1976; 52:374–380.

Tobiason, D.M. and Seifert, H.S. Inverse relationship between pilus-mediated gonococcal adherence and surface expression of the pilus receptor, CD46. Microbiology 2001; 147:2333–2340.

Tramont, E.C. Inhibition of adherence of *Neisseria gonorrhoeae* by human genital secretions. J. Clin. Invest. 1977; 59:117–124.

Tramont, E.C. Local gonococcal antibodies. Vaccine Res. 1992; 1:287–293.

Tramont, E.C., Ciak, J., Boslego, J., McChesney, D.C., Brinton, C.C. and Zollinger, W. Antigenic specificity of antibodies in vaginal secretions during infection with *Neisseria gonorrhoeae*. J. Infect. Dis. 1980; 142:23–31.

Tramont, E.C., Sadoff, J.C., Boslego, J.W., Ciak, J., McChesney, D., Brinton, C.C., Wood, S. and Takafuji, E. Gonococcal pilus vaccine: studies of antigenicity and inhibition of attachment. J. Clin. Invest. 1981; 68:881–888.

Vandeputte-Rutten, L., Bos, M.P., Tommassen, J. and Gros, P. Crystal structure of neisserial surface protein A (NspA), a conserved outer membrane protein with vaccine potential. J. Biol. Chem. 2003; 278:24825–24830.

Virji, M. The structural basis of CEACAM-receptor targeting by neisserial opa proteins: response. Trends Microbiol. 2000; 8:260–261.

Virji, M., Fletcher, J.N., Zak, K. and Heckels, J.E. The potential effect of monoclonal antibodies to gonococcal outer membrane protein IA. J. Gen. Microbiol. 1987; 133:2639–2646.

Virji, M. and Heckels, J.E. Role of anti-pilus antibodies in host defense against gonococcal infection studied with monoclonal anti-pilus antibodies. Infect. Immun. 1985; 49:621–628.

Virji, M., Zak, K. and Heckels, J.E. Monoclonal antibodies to gonococcal outer membrane protein IB: use in investigation of the potential protective effect of antibodies directed against conserved and type-specific epitopes. J. Gen. Microbiol. 1986; 132:1621–1629.

Weaver, C.T., Hatton, R.D., Mangan, P.R. and Harrington, L.E. IL-17 family cytokines and the expanding diversity of effector T cell lineages. Annu. Rev. Immunol. 2007; 25:821–852.

Wetzler, L.M., Blake, M.S., Barry, K. and Gotschlich, E.C. Gonococcal porin vaccine evaluation: comparison of Por proteosomes, liposomes, and blebs isolated from *rmp* deletion mutants. J. Infect. Dis. 1992; 166:551–555.

Wetzler, L.W., Ho, Y. and Reiser, H. Neisserial porins induce B lymphocytes to express costimulatory B7-2 molecules and to proliferate. J. Exp. Med. 1996; 183:1151–1159.

WHO Global Prevalence and Incidence of Selected Curable Sexually Transmitted Infections: Overview and Estimates. Geneva: World Health Organization, 2001.

Wong, K.H., Arko, R.J., Schalla, W.O. and Steurer, F.J. Immunological and serological diversity of *Neisseria gonorrhoeae*: identification of new immunotypes and highly protective strains. Infect. Immun. 1979; 23:717–722.

Wright, A.E. and Douglas, S.R. An experimental investigation of the role of the blood fluids in connection with phagocytosis. Proc. R. Soc. 1903; 72:357–370.

Wu, H.Y., Abdu, S., Stinson, D. and Russell, M.W. Generation of female genital tract antibody responses by local or central (common) mucosal immunization. Infect. Immun. 2000; 68:5539–5545.

Wyle, F.A., Rowlett, C. and Blumenthal, T. Cell-mediated immune response in gonococcal infections. Br. J. Vener. Dis. 1977; 53:353–359.

Yang, Q.L. and Gotschlich, E.C. Variation of gonococcal lipooligosaccharide structure is due to alterations in poly-G tracts in lgt genes encoding glycosyl transferases. J. Exp. Med. 1996; 183:323–327.

Yost-Daljev, M.K. and Cornelissen, C.N. Determination of surface-exposed, functional domains of gonococcal transferrin-binding protein A. Infect. Immun. 2004; 72:1775–1785.

Yu, J.J., Ruddy, M.J., Wong, G.C., Sfintescu, C., Baker, P.J., Smith, J.B., Evans, R.T. and Gaffen, S.L. An essential role for IL-17 in preventing pathogen-initiated bone destruction: recruitment of neutrophils to inflamed bone requires IL-17 receptor-dependent signals. Blood 2007; 109:3794–3802.

CHAPTER

50

Helicobacter pylori

Victor E. Reyes[1,2], *Giovanni Suárez*[2], *Johanna C. Sierra*[1] *and Ellen J. Beswick*[1]

[1]Department of Pediatrics, University of Texas Medical Branch, Galveston, TX, USA

[2]Department of Microbiology and Immunology, Sealy Center for Vaccine Development, University of Texas Medical Branch, Galveston, TX, USA

OUTLINE

ABSTRACT

Helicobacter pylori (*H. pylori*) is possibly the most widespread human pathogen worldwide. Since it was initially suggested in 1983 by Marshall and Warren to be implicated in gastritis and peptic ulcer disease, *H. pylori* has also been implicated in gastric carcinoma and was classified as a class I carcinogen. During the last two decades, a significant body of research has revealed the multiple processes that this Gram-negative bacterium activates to cause gastroduodenal disease in humans. Most infections are acquired early in life and may persist for the entire life of the individual. While infected individuals mount an inflammatory response, which becomes chronic, along with a detectable adaptive immune response, these responses are ineffective in clearing the infection. *H. pylori* has unique features that allow it to reside within the harsh conditions of the gastric environment and also to evade the host immune response. Although only a fraction of infected individuals develop the significant clinical diseases associated with this infection, it is important to eradicate *H. pylori* from humans. Current therapies based on proton pump inhibitors and antimicrobials are effective but fail at times due to poor compliance and the development of antibiotic resistance by *H. pylori*. Therefore, an effective vaccine is clearly needed. This chapter reviews the association of *H. pylori* with human disease, the best-characterized virulence factors, the host response, and vaccine development efforts to date. Although major advances have been made in our understanding of how *H. pylori* interacts with humans, this bacterium keeps on surprising investigators and there are still many unanswered questions that need to be addressed in order to develop an effective vaccine.

INTRODUCTION

Helicobacter pylori (*H. pylori*) is perhaps the most common human pathogen, and has the unique ability to colonize the human gastric mucosa. The clinical importance of this bacterium has become recognized relatively recently. Before the discovery of *H. pylori*, the stomach was not generally considered a niche that pathogens could infect because it is an inhospitable environment where the gastric juice and low pH are able to digest most things and clearly possesses a strong bactericidal action.

SHORT HISTORY OF THE DISEASE

In the 19th century, there were some initial reports of spiral-shaped bacteria in the stomach of dogs and humans, but their characterization and relevance to disease were not determined until almost 100 years later. There is evidence that *H. pylori* has been infecting humans for thousands of years, since pre-Columbian mummies and mummified human feces at least 3000 years old have been found to harbor *H. pylori* (Allison et al., 1999). Also, a study by Blaser and colleagues suggested that human infection by *H. pylori* may have occurred at least 11,000 years ago based on the finding of Asian strains of *H. pylori* in Amerindian populations in remote regions of the Amazon (Ghose et al., 2002). They interpreted their findings to suggest that the ancestors of present-day Amerindians were infected with *H. pylori* when they migrated from Asia and crossed the Bering Strait 11,000 years ago to colonize the Americas.

These gastric bacteria were isolated and cultured by two investigators in Perth, Western Australia, Barry Marshall and J. Warren, who detected the curved bacilli in the stomachs of patients with chronic gastritis and peptic ulceration (Marshall et al., 1984; Marshall and Warren, 1984). Isolation and growth of the bacteria was not a straightforward undertaking since the bacteria were slow to grow. When these investigators initially suggested that gastritis and ulcers might be associated with infection with these bacteria, the scientific community met their proposal with significant skepticism. To address this skepticism, Marshall and another volunteer acted as guinea pigs to try to fulfill Koch's postulates. They self-administered broth cultures of the bacteria. Both volunteers developed gastritis, underwent endoscopy, and provided biopsies from which the bacteria were reisolated. Following the reports of this team of Australians, who initially referred to this bacteria as *Campylobacter pyloridis*, others were also able to culture *H. pylori* from gastric biopsies of patients with gastritis. These studies triggered substantial interest in determining the involvement of this bacterium in other gastric diseases. The association of the infection with this bacterium and ulcers was deduced later from epidemiological studies and was strengthened by observations that antimicrobial therapy had a curative effect on ulcers.

A decade after the seminal studies of Warren and Marshall, the World Health Organization International Agency for Research in Cancer classified *H. pylori* as a class I carcinogen, after epidemiological and statistical studies linked the infection with a higher risk of gastric malignancy. In 2005, Warren and Marshall became the recipients of the Nobel Prize in Physiology or Medicine for their discovery of *H. pylori* and its role in gastritis and peptic ulcer disease.

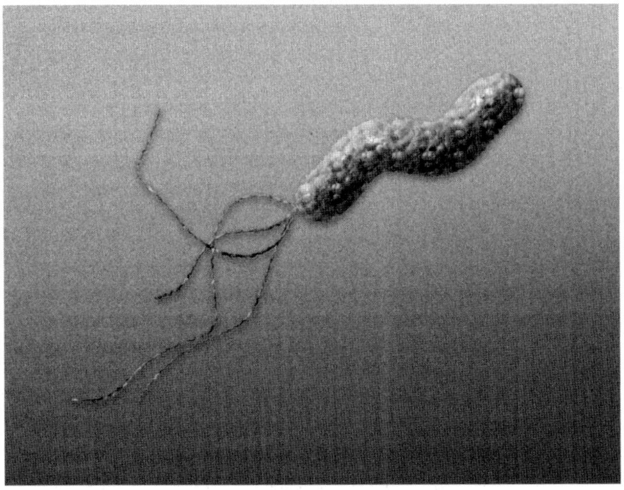

FIGURE 50.1 The flagellated, spiral-shaped bacteria, *Helicobacter pylori.*

Since the studies of Marshall and Warren, significant advances have been made in our understanding of how this human pathogen interacts with its host. This chapter will review historical aspects of the diseases associated with *H. pylori* infection, characteristics of the bacterium, the host response, virulence mechanisms, how the disease is treated, and what preventive measures are in development.

ETIOLOGIC AGENTS

The *Helicobacter* genus was introduced in 1989 with *pylori* as the first species classified. Members of the genus *Helicobacter* are Gram-negative, spiral-shaped, microaerophilic bacteria that are often characterized by sheathed flagella (Fig. 50.1). There are at least 20 characterized species of the genus *Helicobacter*, with several species capable of colonizing the stomach of humans and other animals (Table 50.1). The best recognized and characterized of these species is *H. pylori*. There is also one species, *Helicobacter hepaticus,* which infects the human liver. The shape of this microorganism and its motility, made possible by flagella, are crucial characteristics that facilitate penetration and colonization of the mucosal environment. The mucous covering the epithelial lining of the stomach provides a protective niche for *Helicobacter* from hydrochloric acid in the lumen of the stomach, while the epithelium protects the bacteria from immune responses. Upon making its way through the mucosal layer, *H. pylori* utilize some of their surface proteins as adhesins to attach to membrane-associated receptors on gastric epithelial cells.

TABLE 50.1 Species of the *Helicobacter* genus known to infect humans or animals and the tissue infected

Species	Host	Tissue colonized
Helicobacter pylori	Human	Stomach
Helicobacter acinonychis	Feline	Stomach
Helicobacter hellmanni	Human, feline, canine	Stomach
Helicobacter felis	Feline, canine	Stomach
Helicobacter bizzozeronii	Feline, canine	Stomach
Helicobacter salomonis	Feline, canine	Stomach
Helicobacter pullorum	Poultry	Intestine, liver
Helicobacter rodentium	Rodent	Liver, intestine, gallbladder
Helicobacter fennelliae	Canine	Intestine
Helicobacter trogontum	Rodent	Intestine
Helicobacter hepaticus	Rodent	Liver
Helicobacter muridarum	Rodent	Intestine
Helicobacter canis	Canine, feline	Intestine, liver
Helicobacter bilis	Rodent	Liver
Helicobacter cinaedi	Human, primate	Intestine
Helicobacter pametensis	Poultry	Stomach
Helicobacter mastomyrinus	Rodent	Intestine, liver
Helicobacter cholecystus	Rodent	Gallbladder, liver
Helicobacter mustelae	Ferret	Stomach

VIRULENCE FACTORS

Adhesins

In order to colonize the unwelcoming niche of the stomach and to induce pathogenesis, *H. pylori* expresses a variety of virulence factors. Adherence factors are considered virulence factors because these adhesins are necessary for colonization and maintenance of infection (Evans, Jr. and Evans, 2000). Multiple adhesins have been identified for *H. pylori*, but the ones responsible for pathogenic events are still being investigated. Several well-known adhesins are BabA, SabA, AlpAB, HopZ, HpaA, and urease. BabA and SabA bind to fucosylated and sialylated ABO blood group antigens, respectively (Fig. 50.2). There are clearly multiple adhesins and receptors for *H. pylori* because only half of the strains in the US have detectable BabA (Hennig et al., 2004). While the attachment of *H. pylori* using BabA as an adhesin does not appear to induce signaling or immune responses from host cells, SabA appears to be required for activation of neutrophils and the resulting oxidative burst by binding to sialylated neutrophil receptors (Unemo et al., 2005). Although the AlpAB receptor is unknown,

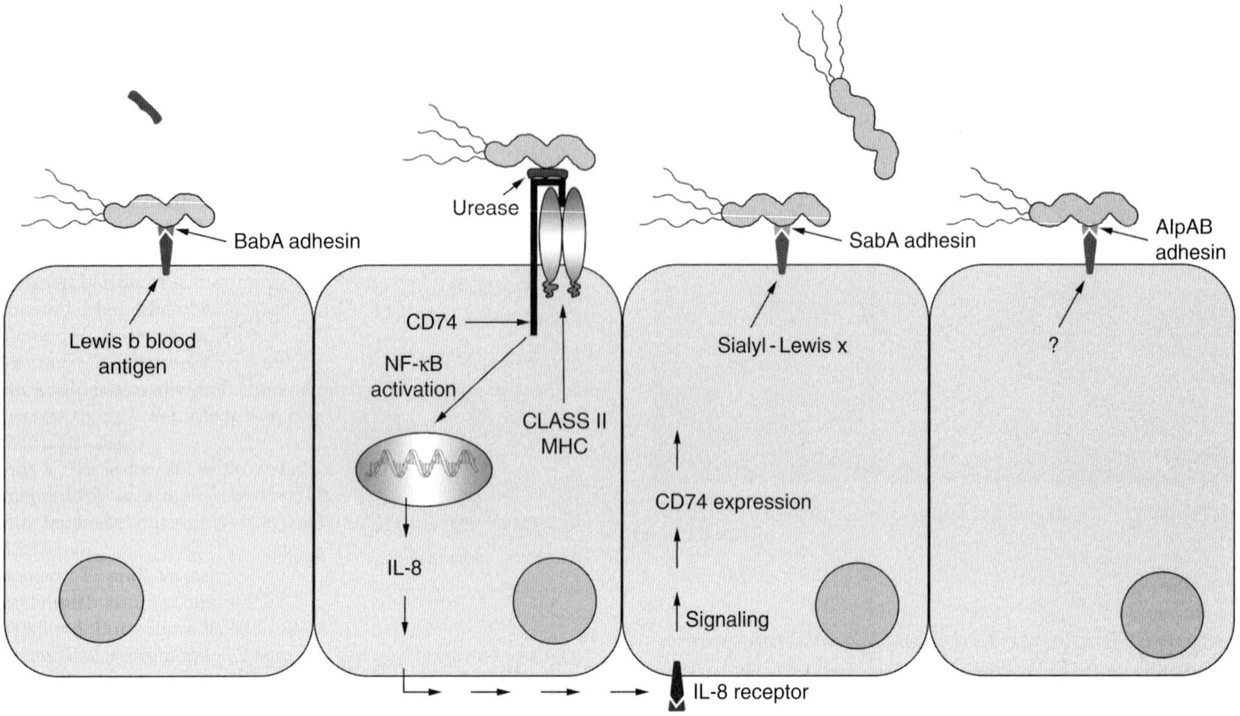

FIGURE 50.2 *H. pylori* adhesins, BabA, SabA, AlpAB, and urease binding to their respective receptors, Lewis b blood antigen, sialyl Lewis x, unknown, and CD74 and class II MHC. Engaging CD74 results in NF-κB signaling and IL-8 production, which further increases CD74 expression.

it may be even more important as an adhesin because studies with knockout strains dramatically reduced adherence of the bacteria to some cells (Odenbreit et al., 2002). In addition, $AlpAB^{+}$ strains activate signaling processes in epithelial cells that influence inflammatory cytokine production (Lu et al., 2007). For instance, deletion of AlpAB reduced interleukin (IL)-6 induction in gastric epithelial cells, whereas IL-8 induction was reduced by deletion of AlpAB in East Asian but not in Western strains of *H. pylori*. These observations suggested that AlpAB may induce gastric injury by mediating adherence to gastric epithelial cells and by modulating proinflammatory intracellular signaling cascades. HopZ, another adhesin being investigated, also showed decreased adherence when a knockout strain was utilized (Peck et al., 1999), but not as dramatically as the AlpAB knockout strain. The HpaA protein is another outer membrane protein (MP) that was shown to be an important adhesin in a mouse model (Carlsohn et al., 2006). *H. pylori* urease can also act as an adhesin (Fan et al., 2000). Urease present on the bacterial surface due to bacterial lysis or release (Bode et al., 1993; Rokita et al., 2000) binds to class II MHC molecules on host cells, and may induce their apoptosis (Fan et al., 2000). Since there have been multiple *H. pylori* adhesins described, bacterial adhesion is clearly a complex mechanism with multiple outcomes depending on the host cell receptor engaged.

cag Pathogenicity Island

One of the best-recognized *H. pylori* virulence factors is the *cag* pathogenicity island. This pathogenicity island is made up of 31 genes, which encode for a type IV secretion system. The system is composed of a needle-like organelle that pierces the host cell membrane and injects bacterial factors into the host cells. The most-recognized protein translocated into host cells by the type IV secretion system is the cytotoxin associated gen A (CagA) virulence factor, which is injected into the cytoplasm of host gastric epithelial cells after attachment to these cells (Odenbreit et al., 2000). CagA does not appear to have homologs in other *Helicobacter* or other bacterial species. While several genes within the island are known to be antigenic, CagA is the most virulent, and is known to undergo tyrosine phosphorylation by the Src family kinases once inside the host cell. CagA has differing numbers of tyrosine phosphorylation motifs (EPIYA motifs), which determine the virulence of the *H. pylori* strain and host cell response to it. The amount of EPIYA motifs is directly

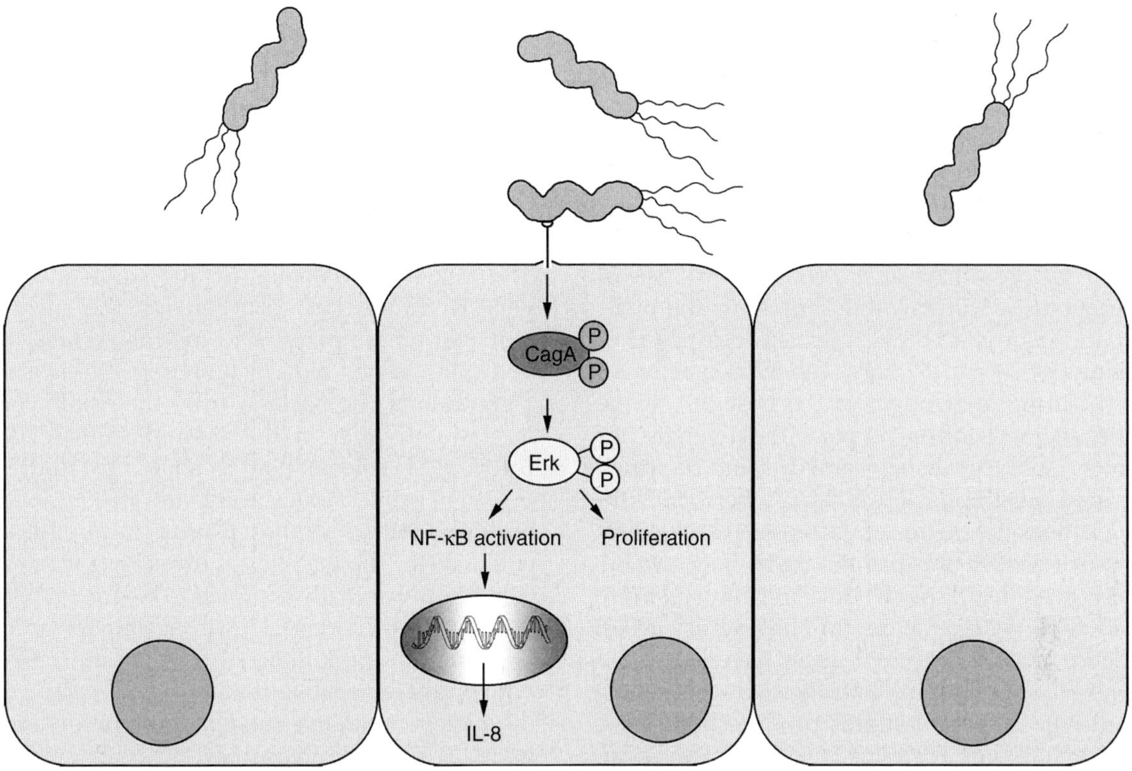

FIGURE 50.3 *H. pylori* infects CagA into a host gastric epithelial cell. After phosphorylation, ERK 1/2 signaling is induced, which leads to NF-κB activation and proinflammatory cytokine production, or changes in cell-cycle events.

related to the levels of phosphorylation and cytoskeletal rearrangement seen in epithelial cells (Figueiredo et al., 2005). Phosphorylated CagA interacts with a tyrosine phosphatase, SHP-2. Upon tyrosine phosphorylation, host cell signaling is activated, such as ERK1/2 signaling, which is responsible for changes in epithelial cell morphology and cell cycle events (Higashi et al., 2004). Through ERK1/2 signaling, NF-κB may also be activated, and upregulation of proinflammatory cytokines may also ensue (Brandt et al., 2005; Fig. 50.3). The activation of the Ras/MEK/ERK pathway occurs following the interaction of CagA with Grb2. Because CagA interacts with important signaling mediators in the host cells, it is considered responsible for changes in cell morphology, adhesion, and turnover. The CagA virulence factor has also been suggested to be associated with the induction of *H. pylori* gastric cancer (Al Marhoon et al., 2004; Chiba, 2004). Interestingly, in Asian and South American countries where 90% of strains are CagA positive, gastric cancer is more prevalent. Western countries have a higher incidence of *cag*$^-$ strains, which are more likely to lead to ulcers or asymptomatic outcomes of infection. CagA found in Eastern strains has also been proposed to be more virulent and undergo more tyrosine phosphorylation than CagA from Western strains (Azuma, 2004). A recent study showed that CagA physically interacts with E-cadherin independently of CagA tyrosine phosphorylation and induces aberrant expression of an intestinal-differentiation marker (Murata-Kamiya et al., 2007). These studies suggest that CagA may play a role in the transdifferentiation of gastric mucosa to an intestinal type, a premalignant transdifferentiation that gives rise to intestinal-type gastric adenocarcinoma.

Other bacterial components proposed to be injected into host cells through the type IV secretion system are bacterial cell wall components such as peptidoglycan. One study indicates that an intracellular sensor for peptidoglycan, Nod1, which initiates NF-κB signaling and upregulation of proinflammatory immune responses (Viala et al., 2004) recognizes *H. pylori* peptidoglycan injected via the type IV secretion system. This is one of several mechanisms whereby NF-κB signaling is initiated by *H. pylori*, and the response seen during infection is likely due to several events since other bacterial factors mentioned below also induce NF-κB activation.

VacA

The vacuolating toxin (VacA), is expressed by about half of all *H. pylori* strains. Like CagA, VacA appears to be unique to *H. pylori* since no other species have a homolog. There is substantial allelic diversity for VacA between strains of *H. pylori*, which contributes to the degree of virulence it projects. VacA has two possible signal regions, s1 or s2. It also may have one of two midregions, m1 or m2, that represents the binding domain. VacA is an 88 kDa protein composed of two subunits of 33 and 55 kDa. The 55 kDa segment is utilized during attachment to the host cell, while the smaller segment forms the pore. The diversity has led to many subtypes being named, some of which can be traced geographically. Diversity between subtypes determines the degree of vacuolation and ability to form pores (Cover and Blanke, 2005). VacA s1/m1 strains are considered more virulent and are more associated with gastric cancer (Miehlke et al., 2000). The presence of *vacA* and *cagA* genes have led to the division of *H. pylori* strains into two broad families, type I and type II. Type I strains produce both VacA and CagA, while type II strains lack their expression. Type I strains are regarded as being more pathogenic and with the greater potential to cause disease.

VacA has effects on many cell types, including gastric epithelial cells, antigen-presenting cells, mast cells, and lymphocytes, which makes it an important virulence factor. This toxin is secreted by *H. pylori*, and it binds to the plasma membrane of host cells where it forms anion-selective channels in the plasma membrane of the cells. This process results in the release of nutrients from the cell, and the bacteria may use these for survival (Szabo et al., 1999). As the anion concentration becomes higher inside the cell through these pores, proton pumping also increases, as does an influx of weak bases. The weak bases are protonated and trapped inside, causing osmotic swelling and the formation of a vacuole (Cover and Blanke, 2005). VacA can also disrupt mitochondrial membrane potential and affect cellular ATP concentrations, which disrupt the cell cycle progression and lead to apoptosis (Cover et al., 2003; Kimura et al., 1999). Another significant way in which VacA contributes to pathogenesis is by inhibiting T cell activation and proliferation. When mixed with T cell lines, VacA inhibited IL-2 production and downregulated surface expression of their IL-2 receptors, which are required for T cell proliferation and viability (Cover and Blanke, 2005). Multiple signaling pathways of T cell activation are also affected by VacA exposure, which is one mechanism *H. pylori* may use to evade immune responses.

Urease

Another major virulence factor of *H. pylori* is urease, which is expressed by all strains. Urease is composed of two subunits, α, which is approximately 24 kDa, and β, which is approximately 68 kDa. *H. pylori* produces a large amount of urease, representing 5–10% of the total protein content of the bacterium. This enzyme is essential for the survival and pathogenesis of the bacteria. Perhaps the most important role urease plays is to hydrolyze urea into CO_2 and NH_3, which aids in buffering the area immediately around the bacteria from the acidic conditions it may encounter in the stomach. Urease is crucial for *H. pylori* colonization, as shown by studies where urease-negative strains were not able to colonize in multiple animal studies (Karita et al., 1995; Tsuda et al., 1994). The inability of urease-negative strains to colonize was initially assumed to be due to their inability to buffer their niche. However, similar studies under hypochloridic conditions also led to the same results where urease-negative mutants of *H. pylori* could not colonize in an animal model. These observations suggested a role for urease beyond its enzymatic function. Although much urease is located intracellularly, there is some present on the bacterial surface (Phadnis et al., 1996; Rokita et al., 2000). *H. pylori* surface-associated urease can act as an adhesin for the bacteria, which induces the production of inflammatory cytokines from both gastric epithelial cells and macrophages (Harris et al., 1998; Tanahashi et al., 2000), along with their apoptosis (Fan et al., 2000). While the mechanism of action associated with these responses elevated by urease is not entirely clear, the induction of apoptosis may result as a consequence of binding to class II MHC (Fan et al., 1998). Other studies from this group also suggest that the urease B subunit binds to CD74, or the class II MHC-associated invariant chain (Ii), and induces IL-8 production by gastric epithelial cells (Beswick et al., 2006). Both of these responses are important in the overall pathogenesis seen during *H. pylori* infection.

In addition to urease A and B, there are several other proteins involved in the urease system that are crucial to the success of the buffering system ensuring bacterial survival. The UreI protein of *H. pylori* is an integral MP that functions as a proton-gated urea channel (Weeks et al., 2000). It regulates cytoplasmic urease activity by controlling permeability to urea, which is essential for maintaining a pH where the bacteria can survive (Rektorschek et al., 2000). The importance of UreI in maintaining pH makes it crucial for survival of *H. pylori*, and its necessity has been shown in studies where UreI knockout strains failed to colonize in animal models (Mollenhauer-Rektorschek et al., 2002; Skouloubris et al., 1998). The other urease system proteins, UreE–H, are accessory proteins required for assembly of active urease enzyme.

OipA

Another important disease-associated virulence factor of *H. pylori* is the outer inflammatory protein, or OipA. OipA is part of a family of 32 outer-MPs characterized as part of the *H. pylori* genome. This protein has been suggested to induce proinflammatory responses from gastric epithelial cell lines. In one study with *H. pylori* clinical isolates, isolates expressing OipA, but not the *cag* pathogenicity island, were able to induce IL-8 production from gastric epithelial cell lines at three times the level of strains that did not express either (Yamaoka et al., 2000). Isolates from Japan all expressed OipA, while isolates from the United States did not, and thus it is thought that the presence of OipA may make Japanese strains more virulent. When the signaling induced by cagPAI was compared to OipA, OipA was found to induce phosphorylation of Stat1, while the cagPAI induced NF-κB activation (Yamaoka et al., 2004). Both of these signaling pathways contribute to induction of IL-8 production, but act in conjunction with one another to fully activate the IL-8 promoter.

Other Virulence Factors

Lipopolysaccharide expressed by *H. pylori* is a very weak immunogen compared to that of other Gram-negative bacteria, although it has been shown to induce proinflammatory cytokines (Bliss et al., 1998). Other bacterial factors that may play a role in the induction of immune responses are neutrophil-activating protein (NAP) and heat shock protein 60 (Hsp 60) (Bai et al., 2003; Yamaguchi et al., 1998). Hsp 60 has been shown to induce proinflammatory cytokines by macrophages and gastric epithelial cells (Gobert et al., 2004; Takenaka et al., 2004), which appears to be mediated by Toll-like receptors (TLRs).

THE IMMUNE RESPONSE TO *H. PYLORI*

As reviewed in detail below, epidemiological studies have shown that about 50% of the world's population is infected by *H. pylori* and most infections are asymptomatic, making the infection lifelong without an effective bacterial elimination (McColl, 1996; Rhen et al., 2003; Svennerholm and Quiding-Jarbrink, 2003). Since the bacteria induce an immune response, the persistence of the infection suggests that the response is not effective in elimination of the infection. Furthermore, multiple lines of evidence suggest that the immune response contributes to the pathogenesis associated with the infection. As a result, the immune response induced by *Helicobacter pylori* is a theme of continuous study that has stimulated multiple questions.

The inability of the host response to clear *H. pylori* infection could reflect downregulatory mechanisms that are acting to ensure immune responses without simultaneous harmful inflammation as a mechanism to protect the host. As a result, the chronic immune response induced may not be sufficiently robust or not directed against protective antigens (Svennerholm and Quiding-Jarbrink, 2003) and could thus provide a colonization advantage for the bacteria by providing enhanced availability of adhesion places. Another mechanism where *H. pylori* can downregulate the immune response is through the VacA protein, as described in the previous section. This cytotoxin can interfere with the processing and presentation of antigens by antigen-presenting cells (APCs; Molinari et al., 1998), and also inhibiting T cell activation through interference of the calcineurin-associated IL-2 signaling pathway (Gebert et al., 2003).

Multiple studies have provided data supporting the role of some bacterial components in triggering cellular as well as humoral immune responses, while suppressing other responses. These observations have led to models that help explain how the bacteria can persist in the gastric environment generating a noneffective immune response that, together with the host factors, can determine the severity of the disease.

The Humoral Response

The noninfected human stomach is relatively free of inflammatory cells, whereas *H. pylori* infection results in dense infiltration of mononuclear cells triggering an immune response. Most individuals infected with the bacteria develop specific antibodies, which are found in serum and in gastric aspirates or extracts of stomach. Thus, high titers of IgG and IgA antibodies against MP, flagellin, urease, LPS, and HpaA have been reported in patients infected with *H. pylori* (Mattsson et al., 1998a). However, those titers do not differ between asymptomatic patients and patients with duodenal ulcer (Mattsson et al., 1998b). Studies performed with biopsies from the antral region of the stomach of noninfected and *H. pylori*-infected patients of IgM- and IgA-producing cells showed frequencies of 40- to 50-fold higher in infected subjects.

However, IgG-producing cell numbers are the same for noninfected and infected *H. pylori* subjects. Those results suggest that the infection induces a large recruitment of immune cells into the gastric mucosa, particularly IgA-producing cells.

The T-Cell Response

H. pylori infection also induces the recruitment of $CD4^+$ and $CD8^+$ T cells into the gastric mucosa, but there appears to be preferential activation of $CD4^+$ cells rather than $CD8^+$ cells (Lundgren et al., 2003). Circulating $CD8^+$ cells produce IFN-γ when they are stimulated with *H. pylori* antigens. Thus, the presence of this subset of T cells in the infected mucosa suggests that cytotoxic T cell activity can be elicited by *H. pylori* (Fan et al., 1994; Quiding-Jarbrink et al., 2001). Other subsets of T cells whose frequencies have also been reported to be increased in the antral lamina propria of subjects infected with *H. pylori* are $CD45RO^+$ memory T cells as well as activated $CD69^+$ and $CD25^+$ T cells (Stromberg et al., 2003a, 2003b).

Several studies have noted that the T helper cell response to *H. pylori* is polarized, since $CD4^+$ T cells in the gastric mucosa of infected individuals produce IL-12 and IFN-γ, whereas IL-4 production by these T cells is absent (Bamford et al., 1998; Haeberle et al., 1997). While there is a demonstrated infiltration of T cells in the gastric mucosa and most of those are $CD4^+$ T cells with markers of activation, various studies have tried to address the inefficiency of the host response in clearing the infection. Different studies have demonstrated that *H. pylori* infection can decrease T cell response as well as induce T cell anergy (Lundgren et al., 2003; Stromberg et al., 2003a). For example, memory T cells isolated from peripheral blood from infected people responded less to stimulation with *H. pylori* antigens than cells isolated from noninfected subjects (Karttunen et al., 1990; Lundgren et al., 2003; Quiding-Jarbrink et al., 2001). These results suggested the presence of regulatory T (T-reg) cells $CD4^+$ $CD25^+$ in the peripheral blood of *H. pylori* infected individuals. This notion was supported by observations that a higher responsiveness was obtained after depletion of *H. pylori*-T-reg specific cells (Baecher-Allan et al., 2001; Lundgren et al., 2003). Hence, these studies may help explain the inability of the host response to eliminate the infection due to the activation of T-reg cells. Such cells may simultaneously reduce mucosal damage mediated by T cells as well as reduce specific T cell responses that can be effective in protection against the infection.

The Gastric Epithelium as an Active Player in the Mucosal Response

In terms of providing protection, the gastric epithelium has typically been regarded as a physical barrier; however, multiple studies have provided evidence to suggest that the gastric epithelium plays a key role in the inflammatory and immune responses induced by *H. pylori*. The epithelium is the only cell phenotype in the gastric mucosa that is in direct contact with the pathogen. This feature places the epithelium in a strategic situation to interact with *H. pylori* and with the immune elements in the lamina propria. There is strong evidence to suggest that the gastric epithelium is an active player in the response while performing functions associated with antigen-presenting cells (Barrera et al., 2001; Ye et al., 1997), in addition to its well-documented ability to produce cytokines that trigger the recruitment of inflammatory cells into the gastric lamina propria (Beswick et al., 2005a; Fig. 50.4).

The production of IL-8 in response to *H. pylori* infection is one of the first epithelial responses. This chemokine acts as a chemoattractant inducing the migration of immunological components into the gastric mucosa from the periphery, particularly polymorphonuclear cells that contribute to epithelial damage (Yoshikawa and Naito, 2001) as well as macrophages. Macrophages produce nitric oxide in response to *H. pylori* urease leading to induction of additional inflammatory mediators (Gobert et al., 2002). As reviewed above, one of the major mechanisms of IL-8 induction by epithelial cells is through the injection of CagA by the type IV secretion system (Fischer et al., 2001; Henderson and Nataro, 2001; Sansonetti, 2002). This system releases CagA into the epithelial cell cytosol inducing cell proliferation and IL-8 production (Mimuro et al., 2002). Our group has recently described the interaction of *H. pylori* with CD74 on gastric epithelial cells leading to the production of IL-8, via NF-κB activation (Beswick, et al., 2005a). Interestingly, IL-8 induced by *H. pylori*, in addition to its effect in the recruitment of inflammatory cells, also acts in an autocrine manner, since gastric epithelial cells express the receptors for IL-8, and induces further expression of CD74 (Beswick et al., 2005b). Some of the interactions of the epithelium with *H. pylori* can be detrimental to the integrity of the epithelium. For instance, it has been shown that *H. pylori* use class II MHC as receptors on gastric epithelial cells and this interaction leads to apoptosis (Fan et al., 2000). This interaction is mediated via *H. pylori* urease. It has also been reported that cag genes may upregulate Fas ligand (FasL) expression leading T cells to undergo apoptosis (Wang et al., 2000).

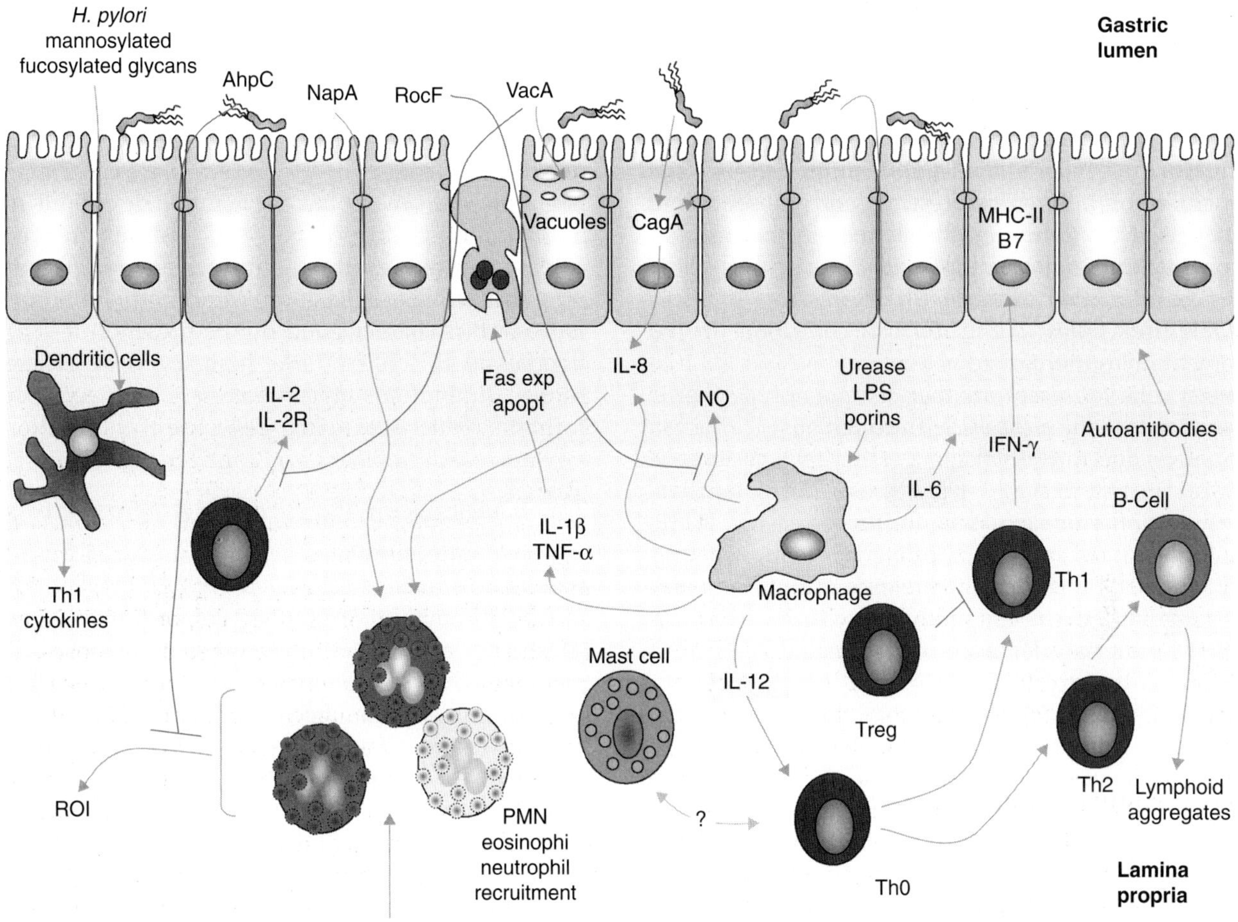

FIGURE 50.4 Immune response induced by *H. pylori*.

The Innate Response to *H. pylori*

Other potential interactions that lead to production of proinflammatory cytokines are the interactions of *H. pylori* with TLR expressed by epithelial cells. It has been reported that gastric epithelial cells express TLR4, TLR5, and TLR9 (Ishihara et al., 2004; Schmausser et al., 2004; Schmausser et al., 2005; Smith et al., 2003) that interact with LPS, flagellin, and CpG motifs, correspondingly. The expression of those receptors by epithelial cells is of importance in innate immunity against *H. pylori*.

The infected gastric mucosa contains a significant macrophage population that produce nitric oxide and IL-12 that help drive a T helper 1 response responsible for the production of IFN-γ, and little or no IL-4 and IL-5 (D'Elios et al., 1997). As alluded to earlier, the predominant Th1 response induced by *H. pylori* is not effective in controlling the infection and may have adverse effects on the host (D'Elios et al., 1997; Fan et al., 1994; Lindholm et al., 1998). Macrophages also produce other cytokines in response to *H. pylori* and these include IL-1β, IL-6, monocyte chemotactic protein 1 (MCP-1), IL-8, and TNF-α (Bliss et al., 1998). In addition to macrophages, another innate cell population that has been shown to respond to *H. pylori* with the production of cytokines includes dendritic cells that produce IL-6, IL-8, IL-10, and IL-12, and have increased expression of CD80, CD83, CD86, and HLA-DR as a result of their stimulation with *H. pylori* (Galgani et al., 2004). Thus, together with the well-documented cytokine response by the epithelium, it is not surprising that immunochemical reports show that the staining levels for most cytokines (IL-8, IL-6, TNF-α, TGF-β) are substantially increased in asymptomatic or gastric ulcer *H. pylori* positive patients when compared to basal levels expressed by noninfected individuals (Lindholm et al., 1998).

As may be expected due to the nature of the cytokines produced in response to *H. pylori*, this innate response to *H. pylori* may have a direct effect in the subsequent adaptive response through the enhancement of processing and presentation of antigen by host cells, mainly gastric epithelial cells, and in influencing the polarity of T helper cells. The expression of class II MHC as well as other molecules, such as cathepsins, CD74 and CD86, important in the regulation of antigen processing and presentation, is increased by gastric epithelial cells in the infected mucosa (Barrera et al., 2001, 2005;Ye et al., 1997). Paradoxically, this increase in expression of class II MHC molecules and CD74 can contribute to the colonization as mentioned above.

Mast cells represent another cell phenotype that is present within the *Helicobacter*-infected gastric mucosa of humans and mice (Nakajima et al., 1997). Mast cells are also present in the gastric mucosa during nonsteroidal anti-inflammatory drug induced gastritis. While these cells represent an innate defense component and may kill bacteria through the release of proteases and other mediators, a recent study showed that they can mediate bacterial clearance in vaccinated mice, and were suggested to do so via a cross talk with $CD4^+$ T cells (Velin et al., 2005).

Autoimmunity Associated with *H. pylori* Infection

Due to the fact that almost every individual has autoreactive T cells, it has been suggested that pathogens can induce activation and proliferation of those autoreactive clones, thereby triggering autoimmune diseases. This predisposition, together with the molecular similarity between some bacterial antigens and autoantigen epitopes, provides the basis of molecular mimicry (Albert and Inman, 1999; Murali-Krishna et al., 1998; Scherer et al., 1993). Molecular mimicry may also play an important role in upregulating the immune response to *H. pylori*.

Autoimmune Gastritis

There are several known *H. pylori* antigens that have cross-reactivity with host components. Perhaps the best known is the expression of Lewis antigens by *H. pylori* LPS. This antigen is similar to that expressed by human cells, including gastric epithelial cells (Appelmelk et al., 1996). Thus, in experiments using *H. pylori* infected mice it was observed that they developed antibodies against LPS, Lewis X, and Lewis Y antigen. Interestingly, these mice also developed antibodies to H^+, K^+-ATPase, which is a major target autoantigen in chronic *H. pylori* gastritis with corpus atrophy (Appelmelk et al., 1998; Claeys et al., 1998). In humans, serum antibodies recognize native and recombinant H^+, K^+-ATPase lacking Lewis antigens (Claeys et al., 1998); hence, human anti-H^+, K^+-ATPase antibodies associated with *H. pylori* do not involve Lewis antigen mimicry as it is the case in infected mice (Alderuccio et al., 2002). Due to this cross-reactivity between bacterial antigens and some mucosal components, *H. pylori* may play a role in the development of autoimmune gastritis (AIG). AIG is an organ-specific disease that can result in pernicious anemia. This type of gastritis involves the corpus and fundus, but not the autrum. It is characterized by the presence of autoantibodies to H^+, K^+-ATPase, high levels of gastrin in serum, diminutions in acid secretion, and pepsinogen I/II ratio (Faller et al., 1997, 1998; Toh et al., 1992).

Thrombocytopenic Purpura and Atherosclerosis

Other autoimmune diseases outside of digestive tract that have been associated with *H. pylori* infection include thrombocytopenic purpura and atherosclerosis (Gasbarrini et al., 2003). In immune thrombocytopenic purpura (ITP) cross mimicry between *H. pylori* and platelet antigens has been suggested (Michel et al., 2002). Platelet-associated immunoglobulin G from patients with ITP has been shown to cross-react with CagA, but the titers of these antibodies decreased after *H. pylori* eradication (Takahashi et al., 2004). Others studies have also demonstrated a positive correlation between the eradication of *H. pylori* and platelet counts (Hashino et al., 2003; Hino et al., 2003; Michel et al., 2002; Veneri et al., 2002). Epidemiological and clinical reports have identified an association between chronic infections with *H. pylori* and coronary heart disease (Aceti et al., 1996; Mendall et al., 1994; Ossei-Gerning et al., 1997; Patel et al., 1995) promoting acute cardiovascular events by provoking atherosclerotic plaque instability or rupture (Aceti et al., 1996). The bacteria may provoke a systemic inflammatory response, with an increase in plasma fibrinogen and other markers of inflammation such as C reactive protein, white cell counts, serum albumin, and plasma viscosity, which are implicated in the plaque instability or rupture (Mehta and Li, 1999; Mendall et al., 1996; Patel et al., 1994). In addition, it has been suggested that *H. pylori* HSP60-like subunits and antibodies against these antigens can mediate cytotoxicity on stressed endothelial cells that could trigger atherogenesis and plaque instability (Birnie et al., 1998; Mayr et al., 2000). More recently, the presence of *H. pylori* DNA has been demonstrated in a considerable number of atherosclerotic plaques, which supports the hypothesis that the bacteria can have a direct role in the development of atherosclerosis and the plaque instability or rupture (Ameriso et al., 2001; Farsak et al., 2000). Other reports have shown that anti-CagA antibodies cross-react with vascular wall antigens, providing another hypothetical pathogenic link between *H. pylori* and atherosclerosis (Franceschi et al., 2002).

Evasion of the Host Immune Response by *H. pylori*

In order to maintain persistent infection, *H. pylori* has developed strategies to evade the host response. Some of these mechanisms involve innate immunity while others involve adaptive immunity.

The immune response to *H. pylori* is composed of Th1 and Th2-cells types, but the cytokines profiles are polarized to a Th1 response. Some studies suggest that the Th1 response is dysfunctional and may play a role in the pathogenesis of infection. One way *H. pylori* may evade immune responses is by utilizing mannosylated and fucosylated glycans on the bacterial surface to bind to DC-SING on dendritic cells. This binding induces the release of Th1-type cytokines by dendritic cells. Thus, the interaction between *H. pylori* and dendritic cells may influence antigen presentation and cytokines secretion by these cells. The bacteria use this interaction to trigger ineffective immune responses that permit its persistence and the development of gastric inflammation and injury (Appelmelk et al., 2003).

Another mechanism used by *H. pylori* to elude the immune response is by limiting the effect of proinflammatory molecules. Thus, *H. pylori* has a gene, *rocF*, that encodes a functional arginase to compete with host enzyme iNOS, which limits the production of nitric oxide (Gobert et al., 2002). There are also other enzymes encoded by *H. pylori* that contribute to immunological evasion such as alkylhydroperoxide reductase (AhpC), which catalyzes the reduction of organic peroxides to alcohols, and neutrophil-activating protein, NapA, which encodes an iron-binding neotrophil-activating protein. These two proteins have an essential role in controlling hydroperoxides and reactive oxygen, which thereby decreases essential immune responses (Ding et al., 2004).

Host defense can also be avoided by *H. pylori* during the innate immunity mediated by TLRs. It has developed strategies to evade activation through TLR; thus, its LPS has low immunogenic activity due to lipid A core modification. Through this modification in the LPS it allows the bacteria to avoid signaling via TLR4 (Ogawa et al., 2003). Another modification is with the flagellin protein. Whereas flagellins secreted by Gram-negative pathogens can activate TLR5-mediated proinflammatory responses, *H. pylori* flagellin is not secreted and consequently is noninflammatory (Gewirtz et al., 2004). Another TLR that *H. pylori* can circumvent is TLR9 because the *H. pylori* DNA is highly methylated (Takata et al., 2002); recognition and response by TLR9 is limited . Yet another way *H. pylori* avoids response by a component of the innate immune response is by producing polyamines, which induce macrophages apoptosis (Chaturvedi et al., 2004).

H. pylori can further avoid elements of the immune response by utilizing a mechanism called antigen variation. This involves the gene *cagY*, which is a surface protein of *H. pylori* that contains a number of amino acid repeat patterns in the amino terminus and the midregion. This pattern can be explained by the high number of DNA repeats that facilitate sequence variation through intragenomic recombination. This system provides a mechanism for rapid changes in amino acid composition when the epitopes of CagY is exposed. Studies have shown there is both strain-specific protein variation, as well as absent or minimal host response to CagY proteins. Several other *H. pylori* surface proteins, such as BabA and BabB, contain repetitive DNA sequences as well, suggesting that CagY immune evasion may be a model for a common mechanism in *H. pylori* immune response evasion (Aras et al., 2003b). A mechanism similar to this is described for the gene *cagA*. The bacteria can regulate the number of tyrosine phosphorylation motifs present with CagA in order to modify the degree of disruption in host cells. This can be in response to selective pressures, and thereby the bacteria may regulate the inflammatory immune response and persist for long term (Aras et al., 2003b, 2003a).

Not only does *H. pylori* make use of various immune avoidance strategies, but it is also able to subvert elements of the host immune response for its own advantage. The use of immunologically relevant molecules such as class II MHC and CD74 as receptors to facilitate its binding and to promote biological responses by gastric epithelial cells are clear examples of subversion.

PD-L1 (also known as B7-H1 or CD279) is one of the two ligands that bind to the programmed cell death receptor, PD-1, on activated T cells and it functions to control the induction and maintenance of peripheral T cell tolerance. The other ligand is known as PD-L2 or B7-DC. PD-L1 is constitutively expressed on cells of hematopoietic origin such as T cells, B cells, DCs, macrophages, mesenchymal stem cells. PD-L1 expression on nonhematopoietic cell types suggests that PD-L1 protects tissues from autoimmune attack. A series of recent studies suggest that microbes that cause chronic infection take advantage of the interaction between PD-1 and its ligands to evade host immune effector mechanisms. We recently showed that *H. pylori* infection leads to upregulated expression of PD-L1 on human gastric epithelial cell lines (Das et al., 2006). In those studies anti-PD-L1 antibody enhanced T cell proliferation and IL-2 production. PD-L1 is upregulated in gastric biopsies from people infected with *H. pylori* compared with samples from uninfected people, suggesting that *H. pylori* enhances the expression of PD-L1 on the epithelium to facilitate its persistence in the host.

EPIDEMIOLOGY

Initial Infection

The acquisition of *H. pylori* infection usually occurs in childhood, and could persist for decades in some infected subjects. The route of transmission appears to be person to person, and there is also evidence of waterborne transmission. Animals do not seem to be a source of infection, since *H. pylori* is strictly a human pathogen. Evidence of intrafamilial transmission has been reported (Bamford et al., 1993; Kivi et al., 2003) where parents, specially the mother, may play an important role in the transmission of the infection. Poor hygiene is a predisposing factor to infection, with oral–oral and fecal–oral transmission as the most probable modes of transmission. The fecal–oral transmission could be a result of direct contact with the infected person or through contaminated water or food. The oral–oral route has been supported by the PCR detection of *H. pylori* DNA in saliva and dental plaque. However, the presence of DNA does not indicate viability of the bacteria. Waterborne transmission has been reported in Colombia, Peru, China, and Japan (Goodman et al., 1996; Karita et al., 2003; Klein et al., 1991; Zhang et al., 1996). Waterborne factors identified in Colombian children included swimming streams and pools, and drinking water from streams. The study in Peru concluded that the water supply was exposed to bacterial contamination. The study conducted by Karita et al. in Japan demonstrated that there is a strong relationship between serological prevalence of *H. pylori* and a history of drinking well water.

Disease Association

As alluded to earlier in this chapter, *H. pylori* has been identified as the causal agent for chronic gastritis, and duodenal and gastric ulcers, and is associated with the development of gastric cancer, but little is known about what determines one disease or the other. The outcome of the infection may involve a combination of the bacterial infection, as well as environmental factors. A high percentage of the duodenal and gastric ulcers are reported to be attributable to *H. pylori*, around 95% and 70%, respectively. Multiple virulence genes have been studied for disease association, and these genes include the *cag* pathogenicity island, BabA, OipA, and VacA. Although these genes have been associated with a higher risk of developing gastric pathologies, none of them can be specifically associated with a particular disease. However, a recent report suggests that the presence of *jhp0917-0918* gene from *H. pylori* is correlated to an increased risk of duodenal ulcer. This gene is present in 42% of duodenal ulcers compared with 21% of gastritis. Its presence was also associated with increased IL-8 production and was a marker of protection for gastric cancer (Lu et al., 2005).

The association between gastric cancer and *H. pylori* has been demonstrated by many different studies (Forman et al., 1991; Parsonnet et al., 1991; Watanabe et al., 1998). Some of the evidence comes from studies of serum samples from patients with gastric cancer and control subjects free of cancer, where a significant association was found between the infection by *H. pylori* and gastric cancer (Talley et al., 1991). Further evidence suggests that 70% of distal gastric cancers are attributable to *H. pylori* (Ekstrom et al., 2001) when IgG anti-*H. pylori* and immunoblot to detect antibodies against CagA are combined. Since gastric carcinogenesis is a complex, multistep, and multifactorial process, it is important to remember that the association between *H. pylori* and gastric cancer is modified by environmental factors such as tobacco smoking, alcohol drinking, and deficiency of dietary antioxidants (Lunet and Barros, 2003).

The prevalence of the infection varies between countries, but there are two patterns for developed and developing countries. For developed countries, the prevalence of infection increases with age where the seropositivity detected in the Eurogast study was between 35% in the 25–34 years age group and 62% in the 55–64 years age group (1993). In addition, a study from Germany showed an overall prevalence of 40% in participants between 18 and 79 years (Rothenbacher and Brenner, 2003). However, the frequency of the infection is decreasing in many developed countries. For example, a report from the Netherlands in children and young adolescents showed that from 1978 to 1993 *H. pylori* prevalence declined from 19% to 9% in children up to 8 years of age and from 23% to 11% in adolescents (Roosendaal et al., 1997).

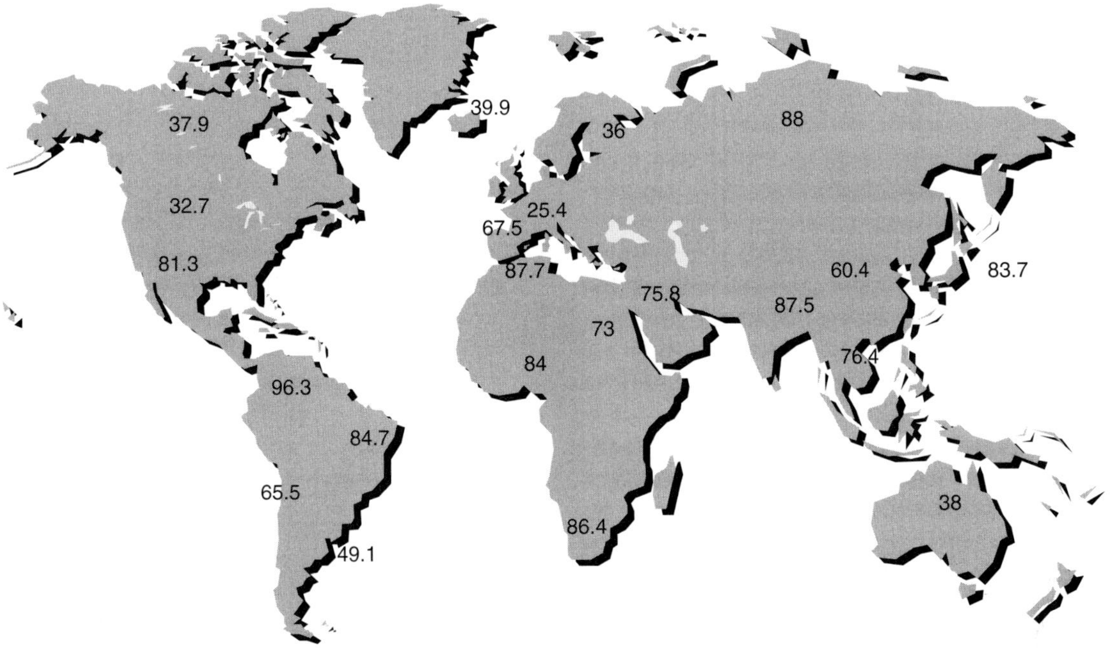

FIGURE 50.5 Prevalence of *Helicobacter pylori* infection around the world per 100,000 population (Lunet and Barros, 2003).

In contrast, the data in developing countries suggests that increasing age is a predisposing factor for infection since the prevalence of infection for the developing countries is more than 50% by 5 years and by 20 years the infection rate exceeds 90% (Bardhan, 1997; Fig. 50.5). Possibly, the age when *H. pylori* is acquired could be considered a risk factor in the serious diseases associated with the infection, since that will determine the length of the chronic inflammation. The prevalence of *H. pylori* varies between the different countries. In a study from Brazil the prevalence in adults was 82% compared with 65% in adults from Guatemala (Dowsett et al., 1999; Oliveira et al., 1994).

Pathogen versus Commensal

Lately, the prevalence of the infection by *H. pylori* has been decreasing in developed countries as a consequence of the improvement in the hygienic conditions and also due to the widespread use of antibiotic therapy for the eradication of *H. pylori* (Parsonnet, 1995; Rothenbacher and Brenner, 2003). The reduction in the prevalence of the infection is accompanied by a diminished rate of the gastric pathologies associated with *H. pylori* including gastritis and gastric cancer. However, the incidence of esophageal adenocarcinoma has been increasing in the United States, Canada, and South Australia and in six European countries (Scotland, Denmark, Iceland, Finland, Sweden, and Norway) (Vizcaino et al., 1998). The absence of *H. pylori* infection has been associated with an increased risk of development of esophageal adenocarcinoma in different case-controlled studies (de Martel et al., 2005; Ye et al., 2004). A possible explanation, provided by Blaser and colleagues, to the apparent protection conferred by *H. pylori* infection in esophageal adenocarcinoma may reside within the delicate balance between the microbe and the host (Blaser, 2005). Blaser proposes a cross talk between the microbe and the host where they could be sending signals to each other. For example, during the infection the acidity of the stomach can increase and *H. pylori*, in turn, signals the host to reduce the stress on the bacteria, using CagA as a signaling molecule, that induces an inflammatory immune response affecting the acid-producing cells and in consequence lowering the pH (Blaser, 2005). In that way, the bacterium downregulates some process in the host preventing the damage of the gastric mucosa and thus keeping their habitat. On the other hand, it is well known that CagA-positive strains increase the risk of gastric cancer, but these strains also modulate the acid production in the stomach, and this balance could be really important in the pathogenesis of the esophageal diseases (Blaser, 2005). Avoiding anti-*H. pylori* treatment is still not an option besides the possible deleterious effects in the esophagus because it is well known that the eradication reduces the risk of gastric cancer significantly.

CLINICAL DISEASE

Diseases Associated with *H. pylori* Infection

Gastritis

The symptoms associated with acute infection with *H. pylori* include nausea and abdominal pain that may persist for several days. These are the same symptoms reported by the volunteers in Marshall et al. (1985) study. In this study, the gastric mucosa was found to be inflamed within 1 week after exposure. A hallmark feature of infection with *H. pylori* is a pronounced inflammatory response, and the inability of the host to clear the infection results in a persistent infection with associated chronic inflammation. Chronic gastritis is elicited by pathogenic products of *H. pylori*, as will be reviewed below, together with the host response to the infection. In addition to a panel of proinflammatory chemokines elicited by the bacteria, the host mounts a polarized T helper cell response. Studies of infected antral biopsies have shown IL-12, IFN-γ, and TNF-α, but not IL-4 mRNA, or protein expression (Bamford et al., 1998; Haeberle et al., 1997). Various independent studies suggest that this polarization in the T cell response contributes to the severity of the associated diseases. As reviewed earlier, chronic gastritis is also thought to include an autoimmune component because several studies have revealed that patients with AIG were infected with *H. pylori*. This chronic inflammation contributes to an increased risk for ulceration and cancer. In fact, infection with *H. pylori* is recognized as a major contributor to chronic gastritis and peptic ulcer formation, and is strongly associated with gastric carcinoma and mucosa-associated lymphoid tissue (MALT) lymphoma (Correa, 1995; Solnick and Tompkins, 1992; Talley et al., 1991). While all individuals infected with *H. pylori* have histological gastritis, only a fraction of the infected persons develop gastric ulcers, duodenal ulcers, gastric cancer, or MALT lymphoma.

Peptic and Duodenal Ulcers

Before *H. pylori* was implicated in peptic ulceration, most ulcers were attributed to nonsteroidal anti-inflammatory drugs (NSAIDs), alcohol consumption, or were labeled as of unknown causes. Among the lay population, ulcers were generally attributed to spicy foodstuffs or stress. Today, the two most recognized etiologic causes responsible for peptic ulcers are chronic infection with *H. pylori* and NSAIDs. Infection with *H. pylori* is found in 90% of patients with duodenal ulcers and between 70% and 90% of patients with gastric ulcers (Cohen, 2000). Patients in the United States who are infected with *H. pylori* have a 3.5 times increased risk of developing peptic ulcer disease than uninfected persons (Nomura et al., 1994). While peptic ulcer was at one time a relatively common disease in the United States, its incidence has been decreasing during the last 30 years, and is currently estimated to be about 10% or 3.7 million new cases each year. According to the National Institutes of Diabetes and Digestive and Kidney Diseases (NIDDK, http://www.niddk.nih.gov/), the prevalence of peptic ulcers in the United States is 5 million people with 6500 fatalities on an annual basis. This decrease in peptic ulcer incidence has paralleled a decrease in the prevalence of *H. pylori* infection in the US population. The etiologic role of *H. pylori* in peptic ulcer has been confirmed by studies where antimicrobial treatment of infected individuals promoted ulcer healing and a decrease in the rate of ulcer recurrence (Graham et al., 1992).

Gastric Cancer

Gastric carcinoma remains among the most common forms of cancer worldwide. Interestingly, gastric cancer was the leading cause of cancer mortality in the United States until the late 1950s, but like the incidence of peptic ulcers, the prevalence of gastric cancer has also been declining in parallel with a decline in the infection with *H. pylori*. The incidence of gastric cancer in the general US population by 2001 was 7.1 per 100,000; it was about twice as frequent in minority populations (www.CDC.gov). In 2003, there were 22,400 new cases of gastric cancer diagnosed in the United States and 12,100 related deaths (Jemal et al., 2003). Until the early 1980s, gastric cancer was the leading cause of cancer-related deaths worldwide and now is second only to lung cancer. Given the strength of the evidence supporting an association between adenocarcinomas of the gastric mucosa and *H. pylori* infection, *H. pylori* became classified as a class I carcinogen by the International Agency for Research on Cancer in affiliation with the World Health Organization (1993). Thus, it became the first bacterial agent to be classified as a carcinogen. While animal models to study the linkage of infection with *H. pylori* and the associated diseases were initially difficult to establish, a study in Mongolian gerbils showed that *H. pylori* infection could result in gastric adenocarcinoma (Watanabe et al., 1998). A recent study examining molecular mechanisms to explain the linkage between *H. pylori* infection and cancer showed, using Mongolian gerbils, that the CagA virulence factor from *H. pylori* induced dysregulation of β-catenin-dependent pathways (Franco et al., 2005). As pointed out by the authors of the study, activation of β-catenin leads to transcriptional upregulation of genes implicated in cancer development.

The prevalence rates of *H. pylori* seropositivity and the incidence of gastric cancer are highly associated within several populations from various countries (1993). For instance, seropositivity may be as high as 80–100% in some age groups in some countries or in minorities with lower incomes in the United States (Blaser and Parsonnet, 1994; Correa, 1995). These groups have the highest risk of developing gastric cancer and/or gastric ulcer.

Since only a fraction (<1%) of the infected individuals eventually develop gastric carcinoma, other factors are thought to contribute to carcinoma development. In addition to virulence of *H. pylori*, other components include the host response, genetic susceptibility, duration of chronic inflammation, and environmental factors. The process whereby infection with *H. pylori* is associated with gastric carcinoma is believed to be a multistep process initiated by chronic gastritis, followed by chronic atrophic gastritis, intestinal metaplasia, and dysplasia (Correa, 1995).

Gastric MALT Lymphoma

H. pylori has also been implicated in the onset of MALT lymphoma. This association has been supported by multiple studies. For instance, while the normal stomach lacks organized lymphoid tissue, one study showed that most patients with gastric MALT lymphoma were also infected with *H. pylori* (Wotherspoon et al., 1991). Studies using animal models have also linked the infection with other *Helicobacter* species with the development of MALT lymphoma. Infection of ferrets with *H. mustelae* or mice with *H. felis* reproduced MALT development, which could be treated with antimicrobials. In addition, several studies in humans have demonstrated that clearance with antimicrobials also resulted in regression of gastric MALT lymphoma in patients who remained in remission (Isaacson et al., 1999; Wotherspoon et al., 1993). MALT lymphoma is found in a small fraction of patients infected with *H. pylori* with an estimated incidence of 1 per 80,000. Chronic antigenic stimulation with *H. pylori* may be responsible for the development of MALT lymphoma associated with this infection. The cells in MALT lymphoma are largely B cells whose surface immunoglobulin recognize autoantigens rather than *H. pylori* antigens (Hussell et al., 1993), which suggests that they originate from autoreactive B cells.

Thus, *H. pylori*-associated pathogenesis encompasses a range of diseases that affect people worldwide. There are significant global and domestic problems resulting from these diseases, which lead to considerable morbidity, mortality, and societal costs.

Treatment

The most common treatment of *H. pylori* symptomatic infection involves three drugs that include two antimicrobials such as amoxicillin and clarithromycin. The other drug is a proton pump inhibitor (PPI), such as omeprazole (Losec) and ezomeprazole (Nexium). This represents the first-line therapy that is administered twice a day and ranges from 7 days outside the United States to 10 or 14 days within the United States (Bytzer and O'morain, 2005; McLoughlin et al., 2005). This triple therapy has had a success rate above 80%. The eradication of *H. pylori* is recommended as the most cost-effective treatment of gastric and duodenal ulcers. Failures in the treatment have been attributed to inadequate patient compliance and the development of antibiotic resistance. Because of the increase in antibiotic resistance, there have been efforts to study the efficacy of quadruple therapy that includes bismuth in addition to the two antibiotics and a PPI. This is considered a second-line treatment (McLoughlin et al., 2005). However, these regimens are associated with side effects such as abdominal pain or nausea, poor compliance due to the high number of tablets to be taken daily, and high cost. The emergence of antibiotic resistant strains makes eradication difficult to achieve. Furthermore, reinfection is frequent in areas with high level of transmission, making antibiotic therapy ineffective in the global control of the infection.

Due to the increase in antibiotic resistance and the associated treatment failure, some studies have considered alternative agents such as phytomedicines, probiotics, and antioxidants to treat *H. pylori* infection. While the use of these agents in vitro or in animal models has provided promising results, the few reported in vivo human trials have shown these agents to be largely ineffective.

PATHOGENESIS

The evidence accumulated through multiple studies suggests that pathogenesis of *H. pylori* infection is the result of both bacterial factors and the host immune response. An added factor into the equation is the contribution of environmental factors to the pathogenesis associated with the infection. Different combinations of these factors determine disease outcome of the infection. *H. pylori* vary in the expression of virulence factors, and the variation affects the intensity of the host immune responses. Chronic infection is common, as the bacteria evade some host responses, and the host fails to respond effectively to eradicate the bacteria.

There are two main phenotypes exhibited with patients who develop disease due to *H. pylori* infection. When high acid secretion and chronic inflammation are seen, more commonly in the antrum of the stomach, gastric or duodenal ulcers normally result. In the case of gastritis of the corpus, decreased acid secretion, and gastric atrophy, gastric cancer is more likely (El Omar et al., 1997). Another difference often seen between these two phenotypes is the density of bacteria. In ulcer patients, high amounts of colonized bacteria have often been observed, whereas in the case of hyposecretion of gastric acid, much lower bacterial density is noted.

Inflammation

Perhaps the most important response for both phenotypes is inflammation. Both antigen specific and nonspecific responses contribute to inflammation during infection. These responses contribute to fighting infection, but are also responsible for mucosal damage. Adhesion of *H. pylori* to the host epithelium, or bacterial factors such as urease or the *cag* pathogenicity island, induce signaling that upregulates proinflammatory cytokines and chemokines such as IL-8 and GRO-α. Although the *cag* pathogenicity island is the most recognized factor inducing inflammatory responses, there are several other interactions known to upregulate these responses. We have recently discovered that through *H. pylori* attachment to CD74, or cross-linking CD74 on gastric epithelial cells, NF-κB activation occurs, leading to IL-8 production (Beswick et al., 2005a). Blocking this interaction with monoclonal antibodies resulted in a substantial decrease in the amount of IL-8 produced in response to *H. pylori*. Other bacterial factors that induce inflammatory responses are HSP60, which was shown to induce IL-8 through TLR pathways (Takenaka et al., 2004), and urease, which induced responses from both gastric epithelial cells and peripheral blood mononuclear cells (Tanahashi et al., 2000).

The cytokine responses, in turn, recruit other immune cells to the site of infection. IL-8 is one of the initial responses that recruit neutrophils to the site of infection. Neutrophils are then activated by *H. pylori* or its soluble products, and proceed to release reactive oxygen species (ROS) and more IL-8 (Shimoyama et al., 2003). Recently, *H. pylori* was also shown to induce production of macrophage migration inhibitory factor (MIF) from gastric epithelial cells, T cells, and macrophages (He et al., 2005; Xia et al., 2004). MIF is a versatile cytokine that mediates both innate and adaptive immunity and plays a role in chronic inflammation. It has been shown to upregulate NF-κB, Erk1/2, activator protein-1 (AP-1), and protein kinase C pathways leading to increased expression of IL-1β and IL-8 mRNA (Onodera et al., 2004; Ren et al., 2004). MIF may be an important early response to infection that upregulates proinflammatory pathways. Macrophages respond to *H. pylori* bacterial factors with the production of proinflammatory cytokines such as IL-1β, IL-6, and TNF-α, and nitric oxide, which are important in epithelial cell injury (Gobert et al., 2001). IL-1β is important in signaling leading to the production of other cytokines. It has also been implicated in inhibiting gastric acid production, which may be an important factor in the development of gastric carcinoma (El Omar, 2001). IL-6 has also been correlated with gastric cancer and severity of disease because of its antiapoptotic effect (Lin et al., 2001). It also induces T cells to produce IFN-γ that contributes to the overall response by upregulating expression of class II MHC on various cells and reducing proliferative rates of cells.

Ulcers

The ulcer phenotype is usually the result of upregulation of the inflammatory response by adhesion to the gastric epithelium. The inflammatory cytokines induced by the bacteria from the host cell lead to the recruitment of T cells and neutrophils to the site of infection. Upon activation, these cells produce proinflammatory cytokines that recruit B cells, mast cells, and macrophages, which damage gastric structure and function. Since *H. pylori* is persistent in infection and not easily cleared by antibiotics, chronic infection is common. Persistent infection leading to chronic inflammation causes tissue damage and ulceration (Fig. 50.6).

Carcinogenesis

The induction of carcinogenesis associated with *H. pylori* is multifaceted. During this process, the cell cycle must be modified so that cell proliferation rates overtake cell death rates. In addition, genetic mutation must occur and be maintained, leading to uncontrolled cell proliferation. The inflammatory response upregulated during infection also plays a role in carcinogenesis. IL-1β and TNF-α are both known to decrease gastric acid production, which may decrease tissue damage seen during ulceration and increase gastric atrophy. MIF, produced in response to *H. pylori*, may lead to

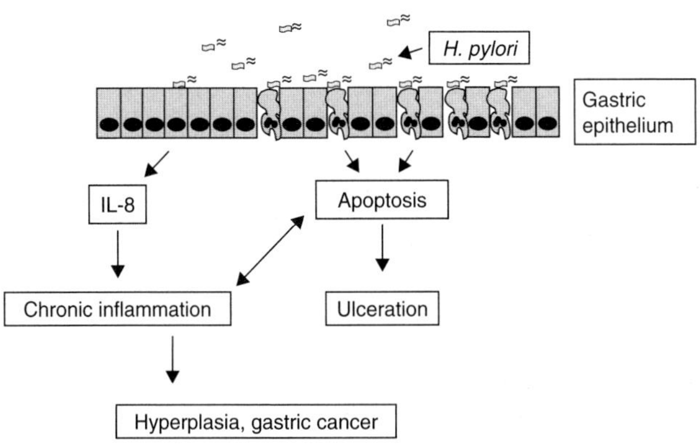

FIGURE 50.6 Inflammatory responses and altered cell turnover by the gastric epithelium are initial events leading to clinically significant diseases associated with *H. pylori* infection.

an increase in gastric epithelial cell proliferation (Xia et al., 2004). Inflammatory cytokines are also known to induce Cox-2 expression, which is upregulated during *H. pylori* infection (Romano et al., 1998; Zarrilli et al., 1999). Cox-2 catalyzes the conversion of arachidonic acid to prostagladins that play a role in inflammation and inhibit apoptosis. Cox-2 is found at elevated levels in many cancer types, and may be important in gastric cancer as well (Fig. 50.6).

During *H. pylori* infection, ROS are produced by the gastric mucosa (Jung et al., 2001). Neutrophils are one of the major sources of ROS, but gastric epithelial cells are also thought to produce it (Bagchi et al., 1996; Shimoyama et al., 2003). Epidemiologic studies have suggested that ROS may play a crucial role in the development of gastric cancer. Gastric epithelial cells may undergo DNA damage in the presence of high amounts of ROS, which could contribute to carcinogenesis.

Various studies have focused on the gastric epithelial cells affected by *H. pylori* and that become transformed. A study took advantage of laser microdissection to isolate the three main epithelial cell lineages from stomach tissue of mice infected with the *H. pylori* strain SS1 to show by gene expression profiling that the mucus-producing cell was the only epithelial cell type displaying a profound transcriptional response to the infection (Mueller et al., 2004). Neither parietal cells nor chief cells showed differential regulation of genes. In a model of gastric cancer, chronic inflammation was shown to result in the recruitment of bone marrow-derived epithelial cells (Houghton et al., 2004). In these studies on C57BL/6 mice transplanted with labeled bone marrow cells and infected with *H. felis*, most metaplasia, dysplasia, and cancer were found to be of donor origin. The group that made these observations suggested that persistent inflammation leads to tissue injury and eventual failure of stem cells in the tissue. The inflammatory process then leads to the recruitment and engraftment of bone marrow-derived circulating cells into the tissue stem cell niche where they replace the tissue stem cells (Houghton and Wang, 2005).

An interesting occurrence to note about the *H. pylori* disease process is that while expression of CagPAI and VacA virulence factors, along with the inflammatory responses induced during infection, is associated with both ulcerogenesis and carcinogenesis, these two outcomes are mutually exclusive of one another. This observation supports the notion that host factors or genetic susceptibility may play a role in disease outcome associated with *H. pylori* infections.

THE USE OF PROTEOMICS TO UNDERSTAND THE *H. PYLORI*-ASSOCIATED PATHOGENESIS

Proteomic analysis is a powerful technology that permits comprehensive inspection of protein expression in cells, tissues, and bodily fluids. Comparison of protein expression between control tissue and disease tissue, or tissue exposed to an insult, permits the detection of proteins altered in their expression. Thus, proteins whose expression is altered during disease could potentially be used as biomarkers for diagnosis and monitoring of infection and disease outcome. This technology has also been applied to analyze the full complement of proteins expressed by pathogens and *H. pylori* is one pathogen whose proteome has been extensively analyzed using two-dimensional electrophoresis (2-DE) and LC-MS (Govorun et al., 2003; Jungblut et al., 2000). However, this technology has not been exploited to understand how *H. pylori* induces disease. Three relevant recent studies have examined resected or biopsy tissue from gastric cancer patients by a similar approach that included 2-DE analysis followed by MALDI-TOF-MS (Baek et al., 2004; Ebert et al., 2005; Ryu et al., 2003). As controls, they used nonaffected tissue from the same patients. There was little congruence in the array of proteins whose expression was altered in the three studies, and there was no confirmation by an independent method such as Western blotting, with the exception of one protein in one of the studies (Ebert et al., 2005). The differences observed may be due to the study of different populations, differences in infecting strains, or both. A limitation of the studies to date is that they examined tissue that has been infected for decades or is already transformed, and the samples contained multiple cell phenotypes, many of which are normal. Das et al. (2005) examined gastric epithelial cells that are the target of infection, and also cells that become transformed to give rise to gastric carcinoma. Using SELDI-TOF-MS combined with retentate chromatography, where proteins were selectively retained based on their chemical properties, they detected a panel of proteins whose expression differed ($p \leq 0.05$) in cells exposed to an inflammatory cytokine (IFN-γ), *H. pylori*, or both (Das et al., 2005). A total of 15 biomarkers in the N87 total cell lysate with a statistically significant $p \leq 0.05$ with at least twofold changes in their level of expression were discovered. We ensured that each of the peaks at the specified condition where it is induced had a signal to noise ratio ≥ 5.0. Untreated gastric epithelial cells were used as a control to get the ratio from the epithelial cells treated with *H. pylori*, IFN-γ, and IFN-γ plus *H. pylori*. It was observed in these studies that the more differentially expressed biomarkers were obtained from IFN-pretreated cells exposed to *H. pylori* samples compared to cells treated only with *H. pylori*. Peaks were selected as induced if the ratio was $\geq$ 2.0, and repressed if the ratio was ≤ 0.5. This might reflect the effect of IFN-γ on the expression of receptors used by *H. pylori*, which in turn helps the interaction of *H. pylori* with GEC (Fan et al., 1998). Annexin II was one biomarker identified in *H. pylori* infected GEC. Annexin II is a multifunctional protein that has been suggested to be involved in the pathogenesis of carcinoma.

VACCINES

The failures associated with the therapy include poor compliance, emergence of antibiotic resistant strains, and the potential for reinfection. These problems could be overcome with the use of an efficacious vaccine. Substantial interest has been shown in developing both therapeutic and prophylactic vaccines. The development of a prophylactic vaccine may not be cost-effective in the United States and Japan. However, regions with high infection rates and prevalent cancerous outcomes may benefit from such a vaccine.

Initial studies in the development of a vaccine require animal models to examine initial preparations. Since *H. pylori* is a strictly human pathogen, an animal model was one of the initial barriers in vaccine development efforts. After an animal model of *Helicobacter* infection was finally presented, Chen et al. (1992) were able to demonstrate that prophylactic immunization could protect against the challenge with *Helicobacter* species. A couple of years after this study was reported, mice infected with *H. felis* were used to show that therapeutic immunization could be effective (Doidge et al., 1994). Another important advance in vaccine development was the discovery of strains (e.g., Sidney strain) isolated from human gastric biopsies that would colonize mice. These adapted strains were crucial in the establishment of a mouse model of *H. pylori* infection. The guidelines for this model require that this strain of *H. pylori* consistently colonize multiple strains of mice to a high level, and establishes infection that may persist for many months (Lee and Chen, 1994) Both prophylactic and therapeutic immunizations have been found to be effective in the control of *H. pylori* colonization in mice (Ghiara et al., 1997; Kleanthous et al., 1998; Sanchez et al., 2001). However, these results in mice have yet to be reproduced in humans. There are several considerations that must be kept in mind in trying to extrapolate the results from studies in mice to humans. The mouse model of *Hp* infection using the Sydney strain (SS1) has a number of limitations, such as the lack of pathologic gastritis seen in humans. The SS1 strain lacks a functional cag pathogenicity island, and does not induce vacuole formation in HeLa cells or IL-8 production in human gastric epithelial cells. Another mouse model of infection uses the *H. felis* strain. However, the virulence mechanisms of this model may be radically different from those of *H. pylori* (no cytotoxin, no cag pathogenicity island, no adherence of the organism to epithelial cells). In the mouse model of *Hp* infection, TNF-α does not seem to play a role in the development of the inflammatory response, in contrast to humans where TNF-α is consistently found to be increased in the infected gastric mucosa of *H. pylori* infected individuals (Yamamoto et al., 2004). Furthermore, studies of *H. pylori* infection in outbred mice, which more closely resemble the outbred human population, showed that these mice mounted a Th2 response, while infections of humans with *H. pylori* induce a Th1 cell dominated response (Robinson et al., 2005). Differences in the immune response between humans and mice may reflect differences in the regulation of the gastric immune response that in mice do not lead to neutrophil recruitment, while in humans this is a hallmark of *H. pylori* infection.

Over the last few years, different groups have been working on developing vaccines against *H. pylori*, following a variety of vaccination strategies in humans and in animal models. The methods that have been attempted to develop a vaccine include whole-cell, single-antigen, and multicomponent preparations. Others have also incorporated the use of different vectors, such as those from *Salmonella* species.

Whole-Cell Vaccines

The fist attempt at vaccination using a whole-cell lysate employed oral immunization in mice and ferrets. This study resulted in the induction of IgG and IgA in gastrointestinal secretions and sera of these animals. Then, in studies where antigen preparation was supplemented with cholera toxin (CT), the IgG and IgA response was significantly increased (Czinn and Nedrud, 1991). However, at that time there was no acceptable animal model for *H. pylori* infection; hence, it was not possible to assess the protective effect of that vaccine preparation. The first animal model of the infection was based on *H. felis*, a gastric pathogen present in cats and dogs. In order to prove the protective effect of this vaccine candidate, mice were immunized with *H. felis* sonicates and CT, and subsequently challenged with *H. felis*. This study resulted in protection that reached a level of 96% against the acute infection (Lee and Chen, 1994). Later, various groups tested these vaccination schemes in a wide range of animals, including ferrets, gnotobiotic piglets, and monkeys, and also achieved high success rates (Eaton et al., 1998).

Other studies using *H. pylori* lysates also met some success in animal models. Using both oral and intramuscular immunization in mice, one group demonstrated the importance of LPS in the development of a vaccine. In a mouse model, *H. pylori* lysates containing LPS induced strong innate and antigen-specific Th1 responses, while mice receiving vaccine devoid of LPS induced a Th2 response (Taylor et al., 2006).

Although the protection achieved by vaccination with whole-cell extracts was very high, such an approach could encounter quality control and safety problems due to the variability of lot preparations and the presence of antigens potentially dangerous for humans. In addition to these problems, the use of CT as an adjuvant in humans is not acceptable. Nevertheless, the safety and immunogenicity of a formalin-inactivated, oral *H. pylori* whole-cell vaccine was tested in healthy and infected volunteers (Kotloff et al., 2001). In this study, the whole-cell vaccine was administered with or without a modified form of *Escherichia coli* heat-labile toxin (LT) as a mucosal adjuvant. After the first immunization, diarrhea, nausea and vomiting were observed. Also, a significant increase in fecal and salivary anti-*H. pylori* whole-cell IgA was observed in infected and uninfected subjects. This vaccination approach elicited significant increases in lymphocyte proliferation and production of IFN-γ in noninfected volunteers. However, the effectiveness of the vaccine in the reduction of the bacterial load was not reported.

Outer Membrane Vesicles

Like other Gram-negative bacteria, *H. pylori* sheds part of its outer membrane as vesicles. These outer membrane vesicles (OMV) contain porins, LPS, VacA, and the Lpp20 lipoprotein among other protein components. Thus, these components were incorporated into vaccine studies. In one study, mice immunized with OMV are protected from the challenge with *H. pylori*. The protection measured correlated with an IgG antibody response to an OMV component of 18 kDa, which is commonly expressed by *H. pylori* strains (Keenan et al., 1998). This OMV component was identified as lipoprotein 20 (Lpp20) using a monoclonal antibody. Further examination indicated that hybridoma backpacks secreting IgG antibodies against Lpp20 considerably reduced the bacterial load in infected mice (Keenan et al., 1998). Additionally, previous evidence indicates that this kind of antigen is a good vaccine candidate. Lpp20 was identified as a protective antigen in a mouse model, since administration of Lpp20 before the challenge resulted in significantly reduced bacterial numbers (Keenan et al., 2000). In order to further investigate this vaccine candidate, the efficacy of the immunization with OMV was compared between mouse and guinea pig models. These animals received intranasal immunization of recombinant Lpp20 (rLpp20) or OMV preparations from *H. pylori*, using CT as a mucosal adjuvant (Keenan et al., 2000). After immunization with rLpp20, the animals that were challenged with *H. pylori* SS1 strain induced a specific antibody response and a reduced level of *H. pylori* colonization in mice, but not in guinea pigs. On the other hand, immunization with OMV in both models elicited systemic and intestinal local immune response, characterized by specific IgG in serum and gastric samples in addition to IgA in bile and the gastric mucosa.

Recombinant Antigen-Based Vaccines

Studies by various groups showed that protection can be achieved by immunization with single antigens such as urease, VacA, catalase, NapA, heat-shock protein Hsp60, and the outer MP HpA. Most of those antigens were selected as potential vaccine candidates because they are abundant, surface exposed, well-conserved *H. pylori* proteins, and/or because they are major virulence factors.

Urease

The urease antigen is one of the most studied of the *H. pylori* major virulence factors, and thus has received the attention of researchers for vaccine studies. Immunization with urease has had much success in mice (Ermak et al., 1998; Kleanthous et al., 1998; Sanchez et al., 2001). Studies suggested that protection of mice against *H. pylori* infection after immunization is dependent on MHC class II antigen presentation; hence, antiurease antibodies may not be necessary for protection (Ermak et al., 1998). Although it is a good vaccine candidate, the experience of *H. pylori* vaccine studies using recombinant urease in humans has been limited. A study conducted to evaluate the safety and immunogenicity of recombinant urease administered with *E. coli* heat-labile enterotoxin (LT) in healthy *H. pylori*-infected volunteers resulted in an increase in the titers of antiurease serum IgA and also in circulating antiurease IgA producing cells (Michetti et al., 1999). Although this effect did not result in complete removal of the bacteria, there was a significant reduction in gastric *H. pylori* density. Despite these promising results, side effects due to the administration of *E. coli* LT were evident in 66% of the volunteers. In order to try to circumvent the side effects due to the oral administration of LT, a subsequent study used a different route of administration. Based on experiments in mice showing that rectal and nasal immunization routes are as effective as the oral route, the subsequent study tested the safety and immunogenicity of *H. pylori* urease vaccine administered rectally using LT as adjuvant in infected and uninfected volunteers (Sougioultzis et al., 2002). Although LT and urease delivered rectally were well tolerated, the immune response to LT was more vigorous compared with the response to urease, since just 16.7% of the subjects had antiurease IgG antibodies and 8.3% developed antiurease IgA antibodies. In another study to determine the adjuvant properties of low doses of LT, a range of doses were evaluated in human volunteers against *H. pylori* using recombinant urease (Banerjee et al., 2002). The urease was administered enterically using acid-resistant capsules, and was also administered orally using soluble urease. Both methods were determined to be safe, without significant adverse effects. After the immunization, 50% of the volunteers who received the higher doses developed mild diarrhea. The proportion of volunteers developing mild diarrhea with the lower doses of LT did not differ from the baseline or from those receiving no LT. Seroconversion (defined as fourfold increase in titers of IgA and IgG or as more than 15 urease-specific IgG antibody secreting cells) was observed in 67%, 17%, 33%, and 25% of the subjects who received 2.5μg, 0.5μg, 0.1μg, and no LT, respectively. Significant cellular immune responses were observed only in the subjects receiving the highest dose of LT. These studies also concluded that the percentage of $\alpha2\beta7$ $CD69^+$ circulating lymphocytes, and $\alpha2\beta7$ $CD4^+$ and $\alpha2\beta7$ $CD45RO^+$ lymphocytes was significantly increased. This study showed that a dose of 2.5μg LT is required, but this dose also induced mild side effects in half of the volunteers.

VacA, NAP, and CagA

The rationale for using virulence factors vaccines is that the immunity induced against them confers protection. Among the other representative virulence factors of *H. pylori* are the CagA, the VacA, and the NAP.

One of the first attempts to use purified VacA as a prospective vaccine candidate showed that immunization with this antigen protected mice from the infection with an *H. pylori* strain expressing both CagA and VacA (Marchetti et al., 1995). A posterior study using recombinant VacA in a mouse model of infection showed that a chronic infection by *H. pylori* Type I strain ($CagA^+$ and $VacA^+$) can be successfully eradicated by intragastric immunization with VacA in combination with a nontoxic mucosal adjuvant in the form of a detoxified mutant of *E. coli* LT. Additionally, these investigators demonstrated that these protected mice were resistant to reinfection (Ghiara et al., 1997). Recently, the Beagle dog model was used to evaluate the safety, immunogenicity, and efficacy of therapeutic vaccine preparations (Rossi et al., 2004). The dogs were immunized with recombinant VacA, CagA, and NAP with alum intramuscularly. A good dose-dependent IgG response was generated without obvious adverse effects. The bacterial colonization and gastritis diminished after the immunization. This successful study suggests that antigen vaccines are a feasible approach to developing an *H. pylori* vaccine.

An important parameter usually considered in immunization protocols is the route of administration. Different administration routes have been combined to determine whether there are differences between them as well as to establish which combination is better. Using CagA and NAP to immunize BALB/c mice Vajdi and colleagues compared the mucosal route with parental routes of immunization. At the same time they wanted to establish whether the combination of mucosal or systemic priming followed by systemic or mucosal boosting would enhance humoral immunity (Vajdy et al., 2003). Their studies showed that mucosal and systemic antibody responses were enhanced with mucosal followed by parenteral immunization. Any single route or combination of immunization routes with CagA and NAP induced antigen-specific splenic IL-4 secreting cells. Interestingly, serum IgG1 was induced in response to intranasal immunization alone or in combination with intramuscular immunization.

In order to further these studies, the safety and immunogenicity of a multicomponent vaccine administered with aluminum hydroxide as adjuvant was evaluated in human volunteers (Ruggiero et al., 2003). The vaccine consisted of CagA, VacA, and NAP plus aluminum hydroxide. This combination of antigens was administered intramuscularly to *H. pylori* uninfected individuals three times. The results of this study indicated that the vaccine was safe, with only some mild and temporary effects observed at the site of the injection. The results also showed that the vaccine was immunogenic and capable of inducing antibody responses to all three antigens.

A disadvantage of immunization with recombinant CagA and VacA is that not all strains express these proteins. Therefore, infection with such negative strains could not be prevented with these vaccines. However, these two virulence factors remain important targets since their presence is associated with the most virulent strains, as explained earlier in this chapter.

Salmonella-Vectored Vaccine

Attenuated *Salmonella* strains have been shown to be efficient live vaccine vectors, and able to induce mucosal, cellular, and humoral immune responses (Emoto et al., 1992; Tacket et al., 1997). For these vaccine studies, defined mutations have been incorporated into the *Salmonella* chromosome to obtain nonvirulent strains able to induce a strong immune response after oral administration. One of the first attempts made at this type of vaccine used plasmid-encoded urease A and B from *H. pylori* expressed in the attenuated *Salmonella typhimurium* strain SL3261 (Gomez-Duarte et al., 1998). Oral immunization of BALB/c mice with a ureA and ureB expressing *Salmonella* vaccine protected the mice from colonization with a mouse-adapted *H. pylori* strain and also induced serum and mucosal specific antiurease antibody responses (Gomez-Duarte et al., 1998). Similar results were reported by Corthesy-Theulaz et al. (1998) using live *S. typhimurium* phoPc expressing urease A and B from *H. pylori*. The combination of mucosal priming with *S. typhi* expressing *H. pylori* urease and parenteral boost with urease plus alum effectively enhanced the immune response and protected the mice against *H. pylori* infection (Londono-Arcila et al., 2002).

The first human immunization study of this type used a *S. typhi* strain where the *phoP/phoQ* virulence region was deleted. This strain was modified to express *H. pylori* urease subunits A and B, and was administered orally to volunteers (DiPetrillo et al., 1999). The immunized volunteers developed strong immune responses against *S. typhi* flagella and lipopolysaccharide antigens, but, intriguingly, there were no detectable mucosal or systemic antibody responses to urease. A later study, using *S. typhimurium* serovar expressing urease from *H. pylori* to immunize volunteers, resulted in a better response compared with the previous report (Angelakopoulos and Hohmann, 2000). Three of the six volunteers had urease-specific immunoglobulin production by blood mononuclear cells, indicating that attenuated *S. typhimurium* appears to be more effective than *S. typhi*. A summary of clinical trials to date is presented in Table 50.2.

Other Vectored Vaccines

Recently, a new potential *H. pylori* vaccine based on a poliovirus vector has been reported. In this vector, the poliovirus capsid genes were substituted with the

TABLE 50.2 Vaccine candidate trials in humans using different approaches

H. pylori status	Antigen	Route	Result	Reference
Uninfected	Urease and LT	Oral	Significant immune response (IgA and IgG)	Banerjee et al. (2002)
Infected	Urease and LT	Oral	No eradication of the infection, decrease in *H. pylori* density	Michetti et al. (1999)
Infected and Uninfected	Whole-cell (formalin-inactivated) and mutant LT (R192G)	Oral	Mucosal and systemic immune response, no eradication	Kotloff et al. (2001)
Uninfected	*Salmonella typhi* Δ*phoP/phoQ* expressing Urease A and B	Oral	No humoral or mucosal immune response	DiPetrillo et al. (1999)
Uninfected	*Salmonella typhimurium* Δ*phoP/phoQ* expressing Urease A and B	Oral	Antiurease specific antibodies	Angelakopoulos and Hohmann (2000)

gene for *H. pylori* urease B (UreB replicon) (Smythies et al., 2005). Immunization with UreB replicons induced humoral and cellular mediated immune responses in C57BL/6/DAB mice transgenic for the human poliovirus receptor (Novak et al., 1999). In testing the vaccine for the prophylactic and protective capacity, it was discovered that only 20% of the mice vaccinated before the challenge became infected compared with 80% percent of mice infected in the control group. Interestingly, 73% of the mice with previous *H. pylori* infection and then therapeutically vaccinated cleared the infection, indicating that therapeutic and prophylactic protection may be achieved through the use of this vaccine candidate (Smythies et al., 2005).

DNA-Based Vaccines

An alternative approach to the above-mentioned vaccines are the administration of DNA-based vaccines. Studies have shown that DNA vaccines are effective at eliciting humoral and cellular immune responses. Additionally, DNA vaccines are easy to produce and purify. One group of investigators showed that intracutaneous immunization of mice with DNA vaccines encoding *H. pylori* heat shock protein A and B induced specific antibodies against *H. pylori* and significantly suppressed bacterial colonization of the stomach as well as reduced the degree of gastritis in vaccinated mice (Todoroki et al., 2000). In a similar study, C57/BL6 mice were immunized intranasally or intracutaneously with a DNA vaccine encoding *H. pylori* catalase. This study showed that vaccinated mice were significantly protected from colonization, and had a lower degree of gastritis than controls (Miyashita et al., 2002). Furthermore, the innate and adaptative immune responses induced by an *H. pylori* ureB DNA construct administered intramuscularly and subcutaneously were analyzed in BALB/c mice (Hatzifoti et al., 2004). Immunized mice showed a sustained increase in the expression of IL-10 mRNA, as well as in the mRNA expression of b1-defensin. The antibody response against ureB was significantly increased after intramuscular immunization, whereas mice subcutaneously immunized did not increase the antibody response.

PROSPECTS FOR THE FUTURE

H. pylori is widely accepted as a very common human pathogen responsible for significant morbidity and mortality. This pathogen has developed unique features that have allowed it to chronically infect its host. Those properties were likely developed over the thousands of years that it has been in close association with humans. Work over that past decade has provided valuable insights to explain how it interacts with the host and causes disease. However, there are still many questions that need to be addressed to prevent infection and disease. Effective prophylactic and therapeutic vaccines are clearly needed. While trials have focused on key virulence factors of *H. pylori* as target antigens in immunization attempts, main targets that have been widely ignored are the adhesins responsible for initial binding to the host epithelium.

As stated earlier, several receptors have been described on the host cells, but those that are shown to deliver intracellular signals after *H. pylori* binds to them may be of major importance in novel therapeutic strategies. Likewise, the *H. pylori* adhesins that are shown to bind to these host cell receptors could become key targets of immunization.

With the advances that are being made in this field and the adaptation of novel technologies, such as proteomics, to characterize critical interactions between *H. pylori* and its host, the eradication of this human pathogen is probably not too distant.

KEY ISSUES

- Why only a fraction of infected individuals develops disease whereas most remain asymptomatic?
- Why are ulcerogenesis and carcinogenesis among infected individuals mutually exclusive processes?
- Critical in the development of vaccines is the understanding of how to elicit a protective immune response while avoiding responses that may become immunopathogenic.
- It will be critical to define what elements of the innate and/or adaptive immune response contribute to peptic and duodenal ulcer development as well as gastric carcinoma.
- It may be critical to examine whether the cytokines induced by the bacterium to promote inflammation and clearance could be central to gastric carcinoma development. It will be necessary to determine the most effective target antigens.
- In studies utilizing urease to immunize humans, observations made in mice have not been reproduced in humans.
- Urease is a logical target for immunization given its overall abundance, its location in the cell, and its functions.
- Human T cells, unlike mouse T cells, express class II MHC after they become activated and this

could be yet another mechanism whereby *H. pylori* circumvents the host response. It is not known at what stage *H. pylori* eradication may protect from cancer development. It is unclear whether individuals with advanced premalignant lesions can be helped by antibiotic clearance of *H. pylori*.

ACKNOWLEDGMENT

This work was supported by the National Institutes of Health Grants DK50669 and DK56338, which funds the Texas Gulf Coast Digestive Diseases Center. Ellen J. Beswick was a recipient of a fellowship under National Institutes of Health T32 AI007536-06 Training Grant. The authors are grateful to Kurt F. Beswick for his assistance with the illustrations included in this manuscript.

References

An international association between *Helicobacter pylori* infection and gastric cancer. The EUROGAST Study Group published erratum appears in Lancet 1993 Jun 26; 341(8861):1668: see comments. Lancet 1993; 341:1359–1362.

Aceti, A., Mazzacurati, G., Amendolea, M., Pennica, A., Zechini, B., Trappolini, M. and Puletti, M. Relation of C reactive protein to cardiovascular risk factors. *H. pylori* and *C. pneumoniae* infections may account for most acute coronary syndromes. Br. Med. J. 1996; 313(7054):428–429.

Albert, L.J. and Inman, R.D. Molecular mimicry and autoimmunity. N. Engl. J. Med. 1999; 341(27):2068–2074.

Alderuccio, F., Sentry, J.W., Marshall, A.C., Biondo, M. and Toh, B.H. Animal models of human disease: experimental autoimmune gastritis—a model for autoimmune gastritis and pernicious anemia. Clin. Immunol. 2002; 102(1):48–58.

Allison, M.J., Bergman, T. and Gerszten, E. Further studies on fecal parasites in antiquity. Am. J. Clin. Pathol. 1999; 112(5):605–609.

Al Marhoon, M.S., Nunn, S. and Soames, R.W. The association between $CagA^+$ *H. pylori* infection and distal gastric cancer: a proposed model. Dig. Dis. Sci. 2004; 49(7–8):1116–1122.

Ameriso, S.F., Fridman, E.A., Leiguarda, R.C. and Sevlever, G.E. Detection of *Helicobacter pylori* in human carotid atherosclerotic plaques. Stroke 2001; 32(2):385–391.

Angelakopoulos, H. and Hohmann, E.L. Pilot study of phoP/phoQ-deleted *Salmonella enterica* serovar *typhimurium* expressing *Helicobacter pylori* urease in adult volunteers. Infect. Immun. 2000; 68(4):2135–2141.

Appelmelk, B.J., Faller, G., Claeys, D., Kirchner, T. and Vandenbroucke-Grauls, C.M. Bugs on trial: the case of *Helicobacter pylori* and autoimmunity. Immunol. Today 1998; 19(7):296–299.

Appelmelk, B.J., Simoons-Smit, I., Negrini, R., Moran, A.P., Aspinall, G.O., Forte, J.G., De Vries, T., Quan, H., Verboom, T., Maaskant, J.J., Ghiara, P., Kuipers, E.J., Bloemena, E., Tadema, T.M., Townsend, R.R., Tyagarajan, K., Crothers, J.M., Jr., Monteiro, M.A., Savio, A. and De Graaff, J. Potential role of molecular mimicry between *Helicobacter pylori* lipopolysaccharide and host Lewis blood group antigens in autoimmunity. Infect. Immun. 1996; 64(6):2031–2040.

Appelmelk, B.J., van, D.I., van Vliet, S.J., Vandenbroucke-Grauls, C.M., Geijtenbeek, T.B. and van Kooyk, Y. Cutting edge: carbohydrate profiling identifies new pathogens that interact with dendritic cell-specific ICAM-3-grabbing nonintegrin on dendritic cells. J. Immunol. 2003; 170(4):1635–1639.

Aras, R.A., Fischer, W., Perez-Perez, G.I., Crosatti, M., Ando, T., Haas, R. and Blaser, M.J. Plasticity of repetitive DNA sequences within a bacterial (Type IV) secretion system component. J. Exp. Med. 2003a; 198(9):1349–1360.

Aras, R.A., Kang, J., Tschumi, A.I., Harasaki, Y. and Blaser, M.J. Extensive repetitive DNA facilitates prokaryotic genome plasticity. Proc. Natl. Acad. Sci. USA 2003b; 100(23):13579–13584.

Azuma, T. *Helicobacter pylori* CagA protein variation associated with gastric cancer in Asia. J. Gastroenterol. 2004; 39(2):97–103.

Baecher-Allan, C., Brown, J.A., Freeman, G.J. and Hafler, D.A. CD4+CD25 high regulatory cells in human peripheral blood. J. Immunol. 2001; 167(3):1245–1253.

Baek, H.Y., Lim, J.W., Kim, H., Kim, J.M., Kim, J.S., Jung, H.C. and Kim, K.H. Oxidative-stress-related proteome changes in *Helicobacter pylori*-infected human gastric mucosa. Biochem. J. 2004; 379(Pt 2):291–299.

Bagchi, D., Bhattacharya, G. and Stohs, S.J. Production of reactive oxygen species by gastric cells in association with *Helicobacter pylori*. Free Radic. Res. 1996; 24(6):439–450.

Bai, Y., Li, L.R., Wang, J.D., Chen, Y., Jin, J.F., Zhang, Z.S., Zhou, D.Y. and Zhang, Y.L. Expression of *Helicobacter pylori* Hsp60 protein and its immunogenicity. World J. Gastroenterol. 2003; 9(12):2711–2714.

Bamford, K.B., Bickley, J., Collins, J.S., Johnston, B.T., Potts, S., Boston, V., Owen, R.J. and Sloan, J.M. *Helicobacter pylori*: comparison of DNA fingerprints provides evidence for intrafamilial infection. Gut 1993; 34:1348–1350.

Bamford, K.B., Fan, X.J., Crowe, S.E., Leary, J.F., Gourley, W.K., Luthra, G., Brooks, E.G., Graham, D.Y., Reyes, V.E. and Ernst, P.B. Lymphocytes during infection with *Helicobacter pylori* have a helper 1 (Th1) phenotype. Gastro 1998; 114:1–12.

Banerjee, S., Medina-Fatimi, A., Nichols, R., Tendler, D., Michetti, M., Simon, J., Kelly, C.P., Monath, T.P. and Michetti, P. Safety and efficacy of low dose *Escherichia coli* enterotoxin adjuvant for urease based oral immunisation against *Helicobacter pylori* in healthy volunteers. Gut 2002; 51(5):634–640.

Bardhan, P.K. Epidemiological features of *Helicobacter pylori* infection in developing countries. Clin. Infect. Dis. 1997; 25(5):973–978.

Barrera, C., Ye, G., Espejo, R., Gunasena, S., Almanza, R., Leary, J., Crowe, S., Ernst, P. and Reyes, V.E. Expression of cathepsins B, L, S, and D by gastric epithelial cells implicates them as antigen presenting cells in local immune responses. Hum. Immunol. 2001; 62(10):1081–1091.

Barrera, C.A., Beswick, E.J., Sierra, J.C., Bland, D., Espejo, R., Mifflin, R., Adegboyega, P., Crowe, S.E., Ernst, P.B. and Reyes, V.E. Polarized expression of CD74 by gastric epithelial cells. J. Histochem. Cytochem. 2005; 53:1481–1489.

Beswick, E.J., Bland, D.A., Suarez, G., Barrera, C.A., Fan, X. and Reyes, V.E. *Helicobacter pylori* binds to CD74 on gastric epithelial cells and stimulates interleukin-8 production. Infect. Immun. 2005a; 73(5):2736–2743.

Beswick, E.J., Das, S., Pinchuk, I.V., Adegboyega, P., Suarez, G., Yamaoka, Y. and Reyes, V.E. *Helicobacter pylori*-induced IL-8 production by gastric epithelial cells up-regulates CD74 expression. J. Immunol. 2005b; 175(1):171–176.

Beswick, E.J., Pinchuk, I.V., Minch, K., Suarez, G., Sierra, J.C., Yamaoka, Y. and Reyes, V.E. The *Helicobacter pylori* Urease B subunit binds to CD74 on gastric epithelial cells and induces NF-?B activation and interleukin-8 production. Infect. Immun. 2006; 74:1148–1155.

Birnie, D.H., Holme, E.R., McKay, I.C., Hood, S., McColl, K.E. and Hillis, W.S. Association between antibodies to heat shock protein 65 and coronary atherosclerosis. Possible mechanism of action of *Helicobacter pylori* and other bacterial infections in increasing cardiovascular risk. Eur. Heart J. 1998; 19(3):387–394.

Blaser, M.J. An endangered species in the stomach. Sci. Am. 2005; 292(2):38–45.

Blaser, M.J. and Parsonnet, J. Parasitism by the "slow" bacterium *Helicobacter pylori* leads to altered gastric homeostasis and neoplasia. J. Clin. Invest. 1994; 94:4–8.

Bliss, C.M., Jr., Golenbock, D.T., Keates, S., Linevsky, J.K. and Kelly, C.P. *Helicobacter pylori* lipopolysaccharide binds to CD14 and stimulates release of interleukin-8, epithelial neutrophil-activating peptide 78, and monocyte chemotactic protein 1 by human monocytes. Infect. Immun. 1998; 66(11):5357–5363.

Bode, G., Malfertheiner, P., Lehnhardt, G., Nilius, M. and Ditschuneit, H. Ultrastructural localization of urease of *Helicobacter pylori*. Med. Microbiol. Immunol. (Berl.) 1993; 182(5): 233–242.

Brandt, S., Kwok, T., Hartig, R., Konig, W. and Backert, S. NF-kappaB activation and potentiation of proinflammatory responses by the *Helicobacter pylori* CagA protein. Proc. Natl. Acad. Sci. USA 2005; 102(26):9300–9305.

Bytzer, P. and O'morain, C. Treatment of *Helicobacter pylori*. Helicobacter 2005; 10(Suppl. 1):40–46.

Carlsohn, E., Nystrom, J., Bolin, I., Nilsson, C.L. and Svennerholm, A. M. HpaA is essential for *Helicobacter pylori* colonization in mice. Infect. Immun. 2006; 74(2):920–926.

Chaturvedi, R., Cheng, Y., Asim, M., Bussiere, F.I., Xu, H., Gobert, A.P., Hacker, A., Casero, R.A., Jr. and Wilson, K.T. Induction of polyamine oxidase 1 by *Helicobacter pylori* causes macrophage apoptosis by hydrogen peroxide release and mitochondrial membrane depolarization. J. Biol. Chem. 2004; 279(38):40161–40173.

Chen, M., Lee, A. and Hazell, S. Immunisation against gastric *Helicobacter* infection in a mouse/*Helicobacter felis* model. Lancet 1992; 339(8801):1120–1121.

Chiba, N. CagA-seropositive strains of *Helicobacter pylori* increase the risk for gastric cancer more than the presence of *H. pylori* alone. Can. J. Gastroenterol. 2004; 18(5):341–343.

Claeys, D., Faller, G., Appelmelk, B.J., Negrini, R. and Kirchner, T. The gastric H^+, K^+-ATPase is a major autoantigen in chronic *Helicobacter pylori* gastritis with body mucosa atrophy. Gastroenterology 1998; 115(2):340–347.

Cohen, H. Peptic ulcer and *Helicobacter pylori*. Gastroenterol. Clin. N. Am. 2000; 29(4):775–789.

Correa, P. *Helicobacter pylori* and gastric carcinogenesis. Am. J. Surg. Pathol. 1995; 19:S37–S43.

Corthesy-Theulaz, I.E., Hopkins, S., Bachmann, D., Saldinger, P.F., Porta, N., Haas, R., Zheng-Xin, Y., Meyer, T., Bouzourene, H., Blum, A.L. and Kraehenbuhl, J.P. Mice are protected from *Helicobacter pylori* infection by nasal immunization with attenuated *Salmonella typhimurium* phoPc expressing urease A and B subunits. Infect. Immun. 1998; 66(2):581–586.

Cover, T.L. and Blanke, S.R. *Helicobacter pylori* VacA, a paradigm for toxin multifunctionality. Nat. Rev. Microbiol. 2005; 3(4):320–332.

Cover, T.L., Krishna, U.S., Israel, D.A. and Peek, R.M., Jr. Induction of gastric epithelial cell apoptosis by *Helicobacter pylori* vacuolating cytotoxin. Cancer Res. 2003; 63(5):951–957.

Czinn, S.J. and Nedrud, J.G. Oral immunization against *Helicobacter pylori*. Infect. Immun. 1991; 59(7):2359–2363.

Das, S., Sierra, J.C., Soman, K.V., Suarez, G., Mohammad, A.A., Dang, T.A.T., Luxon, B.A. and Reyes, V.E. Differential protein expression profiles of gastric epithelial cells following *Helicobacter pylori* infection using ProteinChips. J. Proteome Res. 2005; 4:000.

Das, S., Suarez, G., Beswick, E.J., Sierra, J.C., Graham, D.Y. and Reyes, V.E. Expression of B7-H1 on gastric epithelial cells: its potential role in regulating T cells during *Helicobacter pylori* infection. J. Immunol. 2006; 176(5):3000–3009.

D'Elios, M.M., Manghetti, M., De Carli, M., Costa, F., Baldari, C.T., Burroni, D., Telford, J.L., Romagnani, S. and Del Prete, G. T helper 1 effector cells specific for *Helicobacter pylori* in the gastric antrum of patients with peptic ulcer disease. J. Immunol. 1997; 158(2):962–967.

de Martel, C., Llosa, A.E., Farr, S.M., Friedman, G.D., Vogelman, J.H., Orentreich, N., Corley, D.A. and Parsonnet, J. *Helicobacter pylori* infection and the risk of development of esophageal adenocarcinoma. J. Infect. Dis. 2005; 191(5):761–767.

Ding, S.Z., O'Hara, A.M., Denning, T.L., Dirden-Kramer, B., Mifflin, R.C., Reyes, V.E., Ryan, K.A., Elliott, S.N., Izumi, T., Boldogh, I., Mitra, S., Ernst, P.B. and Crowe, S.E. *Helicobacter pylori* and H_2O_2 increase AP endonuclease-1/redox factor-1 expression in human gastric epithelial cells. Gastroenterology 2004; 127(3):845–858.

DiPetrillo, M.D., Tibbetts, T., Kleanthous, H., Killeen, K.P. and Hohmann, E.L. Safety and immunogenicity of phoP/phoQ-deleted *Salmonella typhi* expressing *Helicobacter pylori* urease in adult volunteers. Vaccine 1999; 18(5–6):449–459.

Doidge, C., Crust, I., Lee, A., Buck, F., Hazell, S. and Manne, U. Therapeutic immunisation against *Helicobacter* infection. Lancet 1994; 343(8902):914–915.

Dowsett, S.A., Archila, L., Segreto, V.A., Gonzalez, C.R., Silva, A., Vastola, K.A., Bartizek, R.D. and Kowolik, M.J. *Helicobacter pylori* infection in indigenous families of Central America: serostatus and oral and fingernail carriage. J. Clin. Microbiol. 1999; 37(8):2456–2460.

Eaton, K.A., Ringler, S.S. and Krakowka, S. Vaccination of gnotobiotic piglets against *Helicobacter pylori*. J. Infect. Dis. 1998; 178(5):1399–1405.

Ebert, M.P., Kruger, S., Fogeron, M.L., Lamer, S., Chen, J., Pross, M., Schulz, H.U., Lage, H., Heim, S., Roessner, A., Malfertheiner, P. and Rocken, C. Overexpression of cathepsin B in gastric cancer identified by proteome analysis. Proteomics 2005; 5(6):1693–1704.

Ekstrom, A.M., Held, M., Hansson, L.E., Engstrand, L. and Nyren, O. *Helicobacter pylori* in gastric cancer established by CagA immunoblot as a marker of past infection. Gastroenterology 2001; 121(4):784–791.

El Omar, E.M. The importance of interleukin 1beta in *Helicobacter pylori* associated disease. Gut 2001; 48(6):743–747.

El Omar, E.M., Oien, K., El Nujumi, A., Gillen, D., Wirz, A., Dahill, S., Williams, C., Ardill, J.E. and McColl, K.E. *Helicobacter pylori* infection and chronic gastric acid hyposecretion. Gastroenterology 1997; 113(1):15–24.

Emoto, M., Danbara, H. and Yoshikai, Y. Induction of gamma/delta T cells in murine salmonellosis by an avirulent but not by a virulent strain of *Salmonella choleraesuis*. J. Exp. Med. 1992; 176(2):363–372.

Ermak, T.H., Giannasca, P.J., Nichols, R., Myers, G.A., Nedrud, J., Weltzin, R., Lee, C.K., Kleanthous, H. and Monath, T.P. Immunization of mice with urease vaccine affords protection against *Helicobacter pylori* infection in the absence of antibodies and is mediated by MHC class II-restricted responses. J. Exp. Med. 1998; 188(12):2277–2288.

Evans, D.J., Jr. and Evans, D.G. *Helicobacter pylori* adhesins: review and perspectives. Helicobacter 2000; 5(4):183–195.

Faller, G., Steininger, H., Kranzlein, J., Maul, H., Kerkau, T., Hensen, J., Hahn, E.G. and Kirchner, T. Antigastric autoantibodies in *Helicobacter pylori* infection: implications of histological and clinical parameters of gastritis. Gut 1997; 41(5):619–623.

Faller, G., Winter, M., Steininger, H., Konturek, P., Konturek, S.J. and Kirchner, T. Antigastric autoantibodies and gastric secretory function in *Helicobacter pylori*-infected patients with duodenal ulcer and non-ulcer dyspepsia. Scand. J. Gastroenterol. 1998; 33(3):276–282.

Fan, X., Crowe, S.E., Behar, S., Gunasena, H., Ye, G., Haeberle, H., Van Houten, N., Gourley, W.K., Ernst, P.B. and Reyes, V.E. The effect of class II major histocompatibility complex expression on adherence of *Helicobacter pylori* and induction of apoptosis in gastric epithelial cells: a mechanism for T helper cell type 1-mediated damage. J. Exp. Med. 1998; 187(10): 1659–1669.

Fan, X., Gunasena, H., Cheng, Z., Espejo, R., Crowe, S.E., Ernst, P.B. and Reyes, V.E. *Helicobacter pylori* urease binds to class II MHC on gastric epithelial cells and induces their apoptosis. J. Immunol. 2000; 165(4):1918–1924.

Fan, X.J., Chua, A., Shahi, C.N., McDevitt, J., Keeling, P.W. and Kelleher, D. Gastric T lymphocyte responses to *Helicobacter pylori* in patients with *H. pylori* colonisation. Gut 1994; 35(10):1379–1384.

Farsak, B., Yildirir, A., Akyon, Y., Pinar, A., Oc, M., Boke, E., Kes, S. and Tokgozoglu, L. Detection of *Chlamydia pneumoniae* and *Helicobacter pylori* DNA in human atherosclerotic plaques by PCR. J. Clin. Microbiol. 2000; 38(12):4408–4411.

Figueiredo, C., Machado, J.C. and Yamaoka, Y. Pathogenesis of *Helicobacter pylori* infection. Helicobacter 2005; 10(Suppl. 1): 14–20.

Fischer, W., Puls, J., Buhrdorf, R., Gebert, B., Odenbreit, S. and Haas, R. Systematic mutagenesis of the *Helicobacter pylori* cag pathogenicity island: essential genes for CagA translocation in host cells and induction of interleukin-8. Mol. Microbiol. 2001; 42(5):1337–1348.

Forman, D., Newell, D.G., Fullerton, F., Yarnell, J.W., Stacey, A.R., Wald, N. and Sitas, F. Association between infection with *Helicobacter pylori* and risk of gastric cancer: evidence from a prospective investigation. Br. Med. J. 1991; 302(6788):1302–1305.

Franceschi, F., Sepulveda, A.R., Gasbarrini, A., Pola, P., Silveri, N.G., Gasbarrini, G., Graham, D.Y. and Genta, R.M. Cross-reactivity of anti-CagA antibodies with vascular wall antigens: possible pathogenic link between *Helicobacter pylori* infection and atherosclerosis. Circulation 2002; 106(4):430–434.

Franco, A.T., Israel, D.A., Washington, M.K., Krishna, U., Fox, J.G., Rogers, A.B., Neish, A.S., Collier-Hyams, L., Perez-Perez, G.I., Hatakeyama, M., Whitehead, R., Gaus, K., O'Brien, D.P., Romero-Gallo, J. and Peek, R.M., Jr. Activation of beta-catenin by carcinogenic *Helicobacter pylori*. Proc. Natl. Acad. Sci. USA 2005; 102(30):10646–10651.

Galgani, M., Busiello, I., Censini, S., Zappacosta, S., Racioppi, L. and Zarrilli, R. *Helicobacter pylori* induces apoptosis of human monocytes but not monocyte-derived dendritic cells: role of the cag pathogenicity island. Infect. Immun. 2004; 72(8):4480–4485.

Gasbarrini, A., Carloni, E., Gasbarrini, G. and Menard, A. *Helicobacter pylori* and extragastric diseases—other helicobacters. Helicobacter 2003; 8(Suppl. 1):68–76.

Gebert, B., Fischer, W., Weiss, E., Hoffmann, R. and Haas, R. *Helicobacter pylori* vacuolating cytotoxin inhibits T lymphocyte activation. Science 2003; 301(5636):1099–1102.

Gewirtz, A.T., Yu, Y., Krishna, U.S., Israel, D.A., Lyons, S.L. and Peek, R.M., Jr. *Helicobacter pylori* flagellin evades toll-like receptor 5-mediated innate immunity. J. Infect. Dis. 2004; 189(10):1914–1920.

Ghiara, P., Rossi, M., Marchetti, M., Di Tommaso, A., Vindigni, C., Ciampolini, F., Covacci, A., Telford, J.L., De Magistris, M.T., Pizza, M., Rappuoli, R. and Del Giudice, G. Therapeutic intragastric vaccination against *Helicobacter pylori* in mice eradicates an otherwise chronic infection and confers protection against reinfection. Infect. Immun. 1997; 65(12):4996–5002.

Ghose, C., Perez-Perez, G.I., Dominguez-Bello, M.G., Pride, D.T., Bravi, C.M. and Blaser, M.J. East Asian genotypes of *Helicobacter pylori* strains in Amerindians provide evidence for its ancient human carriage. Proc. Natl. Acad. Sci. USA 2002; 99(23):15107–15111.

Gobert, A.P., Bambou, J.C., Werts, C., Balloy, V., Chignard, M., Moran, A.P. and Ferrero, R.L. *Helicobacter pylori* heat shock protein 60 mediates interleukin-6 production by macrophages via a toll-like receptor (TLR)-2-, TLR-4-, and myeloid differentiation factor 88-independent mechanism. J. Biol. Chem. 2004; 279(1):245–250.

Gobert, A.P., Cheng, Y., Wang, J.Y., Boucher, J.L., Iyer, R.K., Cederbaum, S.D., Casero, R.A., Jr., Newton, J.C. and Wilson, K.T. *Helicobacter pylori* induces macrophage apoptosis by activation of arginase II. J. Immunol. 2002; 168(9):4692–4700.

Gobert, A.P., McGee, D.J., Akhtar, M., Mendz, G.L., Newton, J. C., Cheng, Y., Mobley, H.L. and Wilson, K.T. *Helicobacter pylori* arginase inhibits nitric oxide production by eukaryotic cells: a strategy for bacterial survival. Proc. Natl. Acad. Sci. USA 2001; 98(24):13844–13849.

Gomez-Duarte, O.G., Lucas, B., Yan, Z.X., Panthel, K., Haas, R. and Meyer, T.F. Protection of mice against gastric colonization by *Helicobacter pylori* by single oral dose immunization with attenuated *Salmonella typhimurium* producing urease subunits A and B. Vaccine 1998; 16(5):460–471.

Goodman, K.J., Correa, P., Tengana Aux, H.J., Ramirez, H., DeLany, J.P., Guerrero, P.O., Lopez, Q.M. and Collazos, P.T. *Helicobacter pylori* infection in the Colombian Andes: a population-based study of transmission pathways. Am. J. Epidemiol. 1996; 144(3):290–299.

Govorun, V.M., Moshkovskii, S.A., Tikhonova, O.V., Goufman, E.I., Serebryakova, M.V., Momynaliev, K.T., Lokhov, P.G., Khryapova, E.V., Kudryavtseva, L.V., Smirnova, O.V., Toropyguine, I.Y., Maksimov, B.I. and Archakov, A.I. Comparative analysis of proteome maps of *Helicobacter pylori* clinical isolates. Biochemistry (Mosc.) 2003; 68(1):42–49.

Graham, D.Y., Lew, G.M., Klein, P.D., Evans, D.G., Evans, D.J., Jr., Saeed, Z.A. and Malaty, H.M. Effect of treatment of *Helicobacter pylori* infection on the long-term recurrence of gastric or duodenal ulcer. A randomized, controlled study. Ann. Intern. Med. 1992; 116(9):705–708.

Haeberle, H.A., Kubin, M., Bamford, K.B., Garofalo, R., Graham, D.Y., El Zaatari, F., Karttunen, R., Crowe, S.E., Reyes, V.E. and Ernst, P.B. Differential stimulation of interleukin-12 (IL-12) and IL-10 by live and killed *Helicobacter pylori* in vitro and association of IL-12 production with gamma interferon-producing T cells in the human gastric mucosa. Infect. Immun. 1997; 65(10): 4229–4235.

Harris, P.R., Ernst, P.B., Kawabata, S., Kiyono, H., Graham, M.F. and Smith, P.D. Recombinant *Helicobacter pylori* urease activates primary mucosal macrophages. J. Infect. Dis. 1998; 178(5):1516–1520.

Hashino, S., Mori, A., Suzuki, S., Izumiyama, K., Kahata, K., Yonezumi, M., Chiba, K., Kondo, T., Ota, S., Toyashima, N., Kato, N., Tanaka, J., Imamura, M. and Asaka, M. Platelet recovery in patients with idiopathic thrombocytopenic purpura after eradication of *Helicobacter pylori*. Int. J. Hematol. 2003; 77(2):188–191.

Hatzifoti, C., Bajaj-Elliott, M., Dorrell, N., Anyim, M., Prentice, M.B., Nye, K.E., Wren, B. and Morrow, W.J. A plasmid immunization construct encoding urease B of *Helicobacter pylori* induces an antigen-specific antibody response and upregulates the expression of beta-defensins and IL-10 in the stomachs of immunized mice. Vaccine 2004; 22(20):2651–2659.

He, X.X., Yang, J., Zheng, X.L., Ding, Y.W., Shen, Q.Y., Liu, W. and Zhao, Y.H. The effect of *Helicobacter pylori* infection on expression of macrophage migration inhibitory factor by T cells and macrophages in gastric mucosa. Chin Med. J. (Engl.) 2005; 118(14):1201–1205.

Henderson, I.R. and Nataro, J.P. Virulence functions of autotransporter proteins. Infect. Immun. 2001; 69(3):1231–1243.

Hennig, E.E., Mernaugh, R., Edl, J., Cao, P. and Cover, T.L. Heterogeneity among *Helicobacter pylori* strains in expression of the outer membrane protein BabA. Infect. Immun. 2004; 72(6):3429–3435.

Higashi, H., Nakaya, A., Tsutsumi, R., Yokoyama, K., Fujii, Y., Ishikawa, S., Higuchi, M., Takahashi, A., Kurashima, Y., Teishikata, Y., Tanaka, S., Azuma, T. and Hatakeyama, M. *Helicobacter pylori* CagA induces Ras-independent morphogenetic response through SHP-2 recruitment and activation. J. Biol. Chem. 2004; 279(17):17205–17216.

Hino, M., Yamane, T., Park, K., Takubo, T., Ohta, K., Kitagawa, S., Higuchi, K. and Arakawa, T. Platelet recovery after eradication of *Helicobacter pylori* in patients with idiopathic thrombocytopenic purpura. Ann. Hematol. 2003; 82(1):30–32.

Houghton, J., Stoicov, C., Nomura, S., Rogers, A.B., Carlson, J., Li, H., Cai, X., Fox, J.G., Goldenring, J.R. and Wang, T.C. Gastric cancer originating from bone marrow-derived cells. Science 2004; 306(5701):1568–1571.

Houghton, J. and Wang, T.C. *Helicobacter pylori* and gastric cancer: a new paradigm for inflammation-associated epithelial cancers. Gastroenterology 2005; 128(6):1567–1578.

Hussell, T., Isaacson, P.G., Crabtree, J.E. and Spencer, J. The response of cells from low-grade B-cell gastric lymphomas of mucosa-associated lymphoid tissue to *Helicobacter pylori* [see comments]. Lancet 1993; 342(8871):571–574.

Isaacson, P.G., Diss, T.C., Wotherspoon, A.C., Barbazza, R., de Boni, M. and Doglioni, C. Long-term follow-up of gastric MALT lymphoma treated by eradication of *H. pylori* with antibodies. Gastroenterology 1999; 117(3):750–751.

Ishihara, S., Rumi, M.A., Kadowaki, Y., Ortega-Cava, C.F., Yuki, T., Yoshino, N., Miyaoka, Y., Kazumori, H., Ishimura, N., Amano, Y. and Kinoshita, Y. Essential role of MD-2 in TLR4-dependent signaling during *Helicobacter pylori*-associated gastritis. J. Immunol. 2004; 173(2):1406–1416.

Jemal, A., Murray, T., Samuels, A., Ghafoor, A., Ward, E. and Thun, M.J. Cancer statistics, 2003. CA Cancer J. Clin. 2003; 53(1):5–26.

Jung, H.K., Lee, K.E., Chu, S.H. and Yi, S.Y. Reactive oxygen species activity, mucosal lipoperoxidation and glutathione in *Helicobacter pylori*-infected gastric mucosa. J. Gastroenterol. Hepatol. 2001; 16(12):1336–1340.

Jungblut, P.R., Bumann, D., Haas, G., Zimny-Arndt, U., Holland, P., Lamer, S., Siejak, F., Aebischer, A. and Meyer, T.F. Comparative proteome analysis of *Helicobacter pylori*. Mol. Microbiol. 2000; 36(3):710–725.

Karita, M., Teramukai, S. and Matsumoto, S. Risk of *Helicobacter pylori* transmission from drinking well water is higher than that from infected intrafamilial members in Japan. Dig. Dis. Sci. 2003; 48(6):1062–1067.

Karita, M., Tsuda, M. and Nakazawa, T. Essential role of urease in vitro and in vivo *Helicobacter pylori* colonization study using a wild-type and isogenic urease mutant strain. J. Clin. Gastroenterol. 1995; 21(Suppl. 1):S160–S163.

Karttunen, R., Andersson, G., Poikonen, K., Kosunen, T.U., Karttunen, T., Juutinen, K. and Niemela, S. *Helicobacter pylori* induces lymphocyte activation in peripheral blood cultures. Clin. Exp. Immunol. 1990; 82(3):485–488.

Keenan, J., Oliaro, J., Domigan, N., Potter, H., Aitken, G., Allardyce, R. and Roake, J. Immune response to an 18-kilodalton outer membrane antigen identifies lipoprotein 20 as a *Helicobacter pylori* vaccine candidate. Infect. Immun. 2000; 68(6):3337–3343.

Keenan, J.I., Allardyce, R.A. and Bagshaw, P.F. Lack of protection following immunisation with *H. pylori* outer membrane vesicles highlights antigenic differences between *H. felis* and *H. pylori*. FEMS Microbiol. Lett. 1998; 161(1):21–27.

Kimura, M., Goto, S., Wada, A., Yahiro, K., Niidome, T., Hatakeyama, T., Aoyagi, H., Hirayama, T. and Kondo, T. Vacuolating cytotoxin purified from *Helicobacter pylori* causes mitochondrial damage in human gastric cells. Microb. Pathog. 1999; 26(1):45–52.

Kivi, M., Tindberg, Y., Sorberg, M., Casswall, T.H., Befrits, R., Hellstrom, P.M., Bengtsson, C., Engstrand, L. and Granstrom, M. Concordance of *Helicobacter pylori* strains within families. J. Clin. Microbiol. 2003; 41(12):5604–5608.

Kleanthous, H., Myers, G.A., Georgakopoulos, K.M., Tibbitts, T.J., Ingrassia, J.W., Gray, H.L., Ding, R., Zhang, Z.Z., Lei, W., Nichols, R., Lee, C.K., Ermak, T.H. and Monath, T.P. Rectal and intranasal immunizations with recombinant urease induce distinct local and serum immune responses in mice and protect against *Helicobacter pylori* infection. Infect. Immun. 1998; 66(6):2879–2886.

Klein, P.D., Graham, D.Y., Gaillour, A., Opekun, A.R. and Smith, E.O. Water source as risk factor for *Helicobacter pylori* infection in Peruvian children. Gastrointestinal Physiology Working Group. Lancet 1991; 337(8756):1503–1506.

Kotloff, K.L., Sztein, M.B., Wasserman, S.S., Losonsky, G.A., DiLorenzo, S.C. and Walker, R.I. Safety and immunogenicity of oral inactivated whole-cell *Helicobacter pylori* vaccine with adjuvant among volunteers with or without subclinical infection. Infect. Immun. 2001; 69(6):3581–3590.

Lee, A. and Chen, M. Successful immunization against gastric infection with *Helicobacter* species: use of a cholera toxin B-subunit-whole-cell vaccine. Infect. Immun. 1994; 62(8):3594–3597.

Lin, M.T., Juan, C.Y., Chang, K.J., Chen, W.J. and Kuo, M.L. IL-6 inhibits apoptosis and retains oxidative DNA lesions in human gastric cancer AGS cells through up-regulation of anti-apoptotic gene mcl-1. Carcinogenesis 2001; 22(12):1947–1953.

Lindholm, C., Quiding-Jarbrink, M., Lonroth, H., Hamlet, A. and Svennerholm, A.M. Local cytokine response in *Helicobacter pylori*-infected subjects. Infect. Immun. 1998; 66(12):5964–5971.

Londono-Arcila, P., Freeman, D., Kleanthous, H., O'Dowd, A.M., Lewis, S., Turner, A.K., Rees, E.L., Tibbitts, T.J., Greenwood, J., Monath, T.P. and Darsley, M.J. Attenuated *Salmonella enterica serovar typhi* expressing urease effectively immunizes mice against *Helicobacter pylori* challenge as part of a heterologous mucosal priming–parenteral boosting vaccination regimen. Infect. Immun. 2002; 70(9):5096–5106.

Lu, H., Hsu, P.I., Graham, D.Y. and Yamaoka, Y. Duodenal ulcer promoting gene of *Helicobacter pylori*. Gastroenterology 2005; 128(4):833–848.

Lu, H., Wu, J.Y., Beswick, E.J., Ohno, T., Odenbreit, S., Haas, R., Reyes, V.E., Kita, M., Graham, D.Y. and Yamaoka, Y. Functional and intracellular signaling differences associated with the *Helicobacter pylori* AlpAB adhesin from Western and East Asian strains. J. Biol. Chem. 2007; 282(9):6242–6254.

Lundgren, A., Suri-Payer, E., Enarsson, K., Svennerholm, A.M. and Lundin, B.S. *Helicobacter pylori*-specific $CD4^+$ CD25 high regulatory T cells suppress memory T-cell responses to *H. pylori* in infected individuals. Infect. Immun. 2003; 71(4):1755–1762.

Lunet, N. and Barros, H. *Helicobacter pylori* infection and gastric cancer: facing the enigmas. Int. J. Cancer 2003; 106(6):953–960.

Marchetti, M., Arico, B., Burroni, D., Figura, N., Rappuoli, R. and Ghiara, P. Development of a mouse model of *Helicobacter pylori* infection that mimics human disease. Science 1995; 267(5204): 1655–1658.

Marshall, B., Warren, R., Royce, H., Armstrong, J., Sanderson, C. and Waters, T. *Campylobacter pyloridis*—a new species of bacteria isolated from the stomach of patients with gastritis and peptic-ulceration. Aust. N. Z. J. Med. 1984; 14(6):906.

Marshall, B.J., Armstrong, J.A., McGechie, D.B. and Glancy, R.J. Attempt to fulfil Koch's postulates for pyloric *Campylobacter*. Med. J. Aust. 1985; 142(8):436–439.

Marshall, B.J. and Warren, J.R. Unidentified curved bacilli in the stomach of patients with gastritis and peptic ulceration. Lancet 1984; 1(8390):1311–1315.

Mattsson, A., Quiding-Jarbrink, M., Lonroth, H., Hamlet, A., Ahlstedt, I. and Svennerholm, A. Antibody-secreting cells in the stomachs of symptomatic and asymptomatic *Helicobacter pylori*-infected subjects. Infect. Immun. 1998a; 66(6):2705–2712.

Mattsson, A., Tinnert, A., Hamlet, A., Lonroth, H., Bolin, I. and Svennerholm, A.M. Specific antibodies in sera and gastric aspirates of symptomatic and asymptomatic *Helicobacter pylori*-infected subjects. Clin. Diagn. Lab. Immunol. 1998b; 5(3):288–293.

Mayr, M., Kiechl, S., Willeit, J., Wick, G. and Xu, Q. Infections, immunity, and atherosclerosis: associations of antibodies to *Chlamydia pneumoniae*, *Helicobacter pylori*, and cytomegalovirus with immune reactions to heat-shock protein 60 and carotid or femoral atherosclerosis. Circulation 2000; 102(8):833–839.

McColl, K.E. *Helicobacter pylori* infection and its role in human disease—an overview. Pharm. World Sci. 1996; 18(2):49–55.

McLoughlin, R.M., O'Morain, C.A. and O'Connor, H.J. Eradication of *Helicobacter pylori*: recent advances in treatment. Fundam. Clin. Pharmacol. 2005; 19(4):421–427.

Mehta, J.L. and Li, D.Y. Inflammation in ischemic heart disease: response to tissue injury or a pathogenetic villain? Cardiovasc. Res. 1999; 43(2):291–299.

Mendall, M.A., Goggin, P.M., Molineaux, N., Levy, J., Toosy, T., Strachan, D., Camm, A.J. and Northfield, T.C. Relation of *Helicobacter pylori* infection and coronary heart disease. Br. Heart J. 1994; 71(5):437–439.

Mendall, M.A., Patel, P., Ballam, L., Strachan, D. and Northfield, T.C. C reactive protein and its relation to cardiovascular risk factors: a population based cross sectional study. Br. Med. J. 1996; 312(7038):1061–1065.

Michel, M., Khellaf, M., Desforges, L., Lee, K., Schaeffer, A., Godeau, B. and Bierling, P. Autoimmune thrombocytopenic purpura and *Helicobacter pylori* infection. Arch. Intern. Med. 2002; 162(9):1033–1036.

Michetti, P., Kreiss, C., Kotloff, K.L., Porta, N., Blanco, J.L., Bachmann, D., Herranz, M., Saldinger, P.F., Corthesy-Theulaz, I., Losonsky, G., Nichols, R., Simon, J., Stolte, M., Ackerman, S., Monath, T.P. and Blum, A.L. Oral immunization with urease and *Escherichia coli* heat-labile enterotoxin is safe and immunogenic in *Helicobacter pylori*-infected adults. Gastroenterology 1999; 116(4):804–812.

Miehlke, S., Kirsch, C., Agha-Amiri, K., Gunther, T., Lehn, N., Malfertheiner, P., Stolte, M., Ehninger, G. and Bayerdorffer, E. The *Helicobacter pylori* vacA s1, m1 genotype and CagA is associated with gastric carcinoma in Germany. Int. J. Cancer 2000; 87(3):322–327.

Mimuro, H., Suzuki, T., Tanaka, J., Asahi, M., Haas, R. and Sasakawa, C. Grb2 is a key mediator of *Helicobacter pylori* CagA protein activities. Mol. Cell 2002; 10(4):745–755.

Miyashita, M., Joh, T., Watanabe, K., Todoroki, I., Seno, K., Ohara, H., Nomura, T., Miyata, M., Kasugai, K., Tochikubo, K., Itoh, M. and Nitta, M. Immune responses in mice to intranasal and intracutaneous administration of a DNA vaccine encoding *Helicobacter pylori*-catalase. Vaccine 2002; 20(17–18): 2336–2342.

Molinari, M., Salio, M., Galli, C., Norais, N., Rappuoli, R., Lanzavecchia, A. and Montecucco, C. Selective inhibition of Ii-dependent antigen presentation by *Helicobacter pylori* toxin VacA. J. Exp. Med. 1998; 187(1):135–140.

Mollenhauer-Rektorschek, M., Hanauer, G., Sachs, G. and Melchers, K. Expression of UreI is required for intragastric transit and colonization of gerbil gastric mucosa by *Helicobacter pylori*. Res. Microbiol. 2002; 153(10):659–666.

Mueller, A., Merrell, D.S., Grimm, J. and Falkow, S. Profiling of microdissected gastric epithelial cells reveals a cell type-specific response to *Helicobacter pylori* infection. Gastroenterology 2004; 127(5):1446–1462.

Murali-Krishna, K., Altman, J.D., Suresh, M., Sourdive, D.J., Zajac, A.J., Miller, J.D., Slansky, J. and Ahmed, R. Counting antigen-specific CD8 T cells: a reevaluation of bystander activation during viral infection. Immunity 1998; 8(2):177–187.

Murata-Kamiya, N., Kurashima, Y., Teishikata, Y., Yamahashi, Y., Saito, Y., Higashi, H., Aburatani, H., Akiyama, T., Peek, R.M., Jr., Azuma, T. and Hatakeyama, M. *Helicobacter pylori* CagA interacts with E-cadherin and deregulates the beta-catenin signal that promotes intestinal transdifferentiation in gastric epithelial cells. Oncogene 2007; 26:4617–4626.

Nakajima, N., Kuwayama, H., Ito, Y., Iwasaki, A. and Arakawa, Y. *Helicobacter pylori*, neutrophils, interleukins, and gastric epithelial proliferation. J. Clin. Gastroenterol. 1997; 25(Suppl. 1):S198–S202.

Nomura, A., Stemmermann, G.N., Chyou, P.H., Perez-Perez, G.I. and Blaser, M.J. *Helicobacter pylori* infection and the risk for duodenal and gastric ulceration. Ann. Intern. Med. 1994; 120(12):977–981.

Novak, M.J., Smythies, L.E., McPherson, S.A., Smith, P.D. and Morrow, C.D. Poliovirus replicons encoding the B subunit of *Helicobacter pylori* urease elicit a Th1 associated immune response. Vaccine 1999; 17(19):2384–2391.

Odenbreit, S., Faller, G. and Haas, R. Role of the alpAB proteins and lipopolysaccharide in adhesion of *Helicobacter pylori* to human gastric tissue. Int. J. Med. Microbiol. 2002; 292(3–4):247–256.

Odenbreit, S., Puls, J., Sedlmaier, B., Gerland, E., Fischer, W. and Haas, R. Translocation of *Helicobacter pylori* CagA into gastric epithelial cells by type IV secretion. Science 2000; 287(5457):1497–1500.

Ogawa, T., Asai, Y., Sakai, Y., Oikawa, M., Fukase, K., Suda, Y., Kusumoto, S. and Tamura, T. Endotoxic and immunobiological activities of a chemically synthesized lipid A of *Helicobacter pylori* strain 206-1. FEMS Immunol. Med. Microbiol. 2003; 36(1–2):1–7.

Oliveira, A.M., Queiroz, D.M., Rocha, G.A. and Mendes, E.N. Seroprevalence of *Helicobacter pylori* infection in children of low socioeconomic level in Belo Horizonte, Brazil. Am. J. Gastroenterol. 1994; 89(12):2201–2204.

Onodera, S., Nishihira, J., Koyama, Y., Majima, T., Aoki, Y., Ichiyama, H., Ishibashi, T. and Minami, A. Macrophage migration inhibitory factor up-regulates the expression of interleukin-8 messenger RNA in synovial fibroblasts of rheumatoid arthritis patients: common transcriptional regulatory mechanism between interleukin-8 and interleukin-1beta. Arthritis Rheum. 2004; 50(5):1437–1447.

Ossei-Gerning, N., Moayyedi, P., Smith, S., Braunholtz, D., Wilson, J.I., Axon, A.T. and Grant, P.J. *Helicobacter pylori* infection is related to atheroma in patients undergoing coronary angiography. Cardiovasc. Res. 1997; 35(1):120–124.

Parsonnet, J. The incidence of *Helicobacter pylori* infection. Aliment. Pharmacol. Ther. 1995; 9(Suppl. 2):45–51.

Parsonnet, J., Friedman, G.D., Vandersteen, D.P., Chang, Y., Vogelman, J.H., Orentreich, N. and Sibley, R.K. *Helicobacter pylori* infection and the risk of gastric carcinoma. N. Engl. J. Med. 1991; 325(16):1127–1131.

Patel, P., Carrington, D., Strachan, D.P., Leatham, E., Goggin, P., Northfield, T.C. and Mendall, M.A. Fibrinogen: a link between chronic infection and coronary heart disease. Lancet 1994; 343(8913):1634–1635.

Patel, P., Mendall, M.A., Carrington, D., Strachan, D.P., Leatham, E., Molineaux, N., Levy, J., Blakeston, C., Seymour, C.A. and Camm, A.J. Association of *Helicobacter pylori* and *Chlamydia pneumoniae* infections with coronary heart disease and cardiovascular risk factors. Br. Med. J. 1995; 311(7007):711–714.

Peck, B., Ortkamp, M., Diehl, K.D., Hundt, E. and Knapp, B. Conservation, localization and expression of HopZ, a protein involved in adhesion of *Helicobacter pylori*. Nucl. Acids Res. 1999; 27(16):3325–3333.

Phadnis, S.H., Parlow, M.H., Levy, M., Ilver, D., Caulkins, C.M., Connors, J.B. and Dunn, B.E. Surface localization of *Helicobacter pylori* urease and a heat shock protein homolog requires bacterial autolysis. Infect. Immun. 1996; 64(3):905–912.

Quiding-Jarbrink, M., Lundin, B.S., Lonroth, H. and Svennerholm, A.M. $CD4^+$ and $CD8^+$ T cell responses in *Helicobacter pylori*-infected individuals. Clin. Exp. Immunol. 2001; 123(1):81–87.

Rektorschek, M., Buhmann, A., Weeks, D., Schwan, D., Bensch, K.W., Eskandari, S., Scott, D., Sachs, G. and Melchers, K. Acid resistance of *Helicobacter pylori* depends on the UreI membrane protein and an inner membrane proton barrier. Mol. Microbiol. 2000; 36(1):141–152.

Ren, Y., Chan, H.M., Li, Z., Lin, C., Nicholls, J., Chen, C.F., Lee, P.Y., Lui, V., Bacher, M. and Tam, P.K. Upregulation of macrophage migration inhibitory factor contributes to induced N-Myc expression by the activation of ERK signaling pathway and increased expression of interleukin-8 and VEGF in neuroblastoma. Oncogene 2004; 23(23):4146–4154.

Rhen, M., Eriksson, S., Clements, M., Bergstrom, S. and Normark, S.J. The basis of persistent bacterial infections. Trends Microbiol. 2003; 11(2):80–86.

Robinson, K., Loughlin, M.F., Potter, R. and Jenks, P.J. Host adaptation and immune modulation are mediated by homologous recombination in *Helicobacter pylori*. J. Infect. Dis. 2005; 191(4):579–587.

Rokita, E., Makristathis, A., Hirschl, A.M. and Rotter, M.L. Purification of surface-associated urease from *Helicobacter pylori*. J. Chromatogr. B Biomed. Sci. Appl. 2000; 737(1–2):203–212.

Romano, M., Ricci, V., Memoli, A., Tuccillo, C., Di Popolo, A., Sommi, P., Acquaviva, A.M., Del Vecchio, B.C., Bruni, C.B. and Zarrilli, R. *Helicobacter pylori* up-regulates cyclooxygenase-2 mRNA expression and prostaglandin E2 synthesis in MKN 28 gastric mucosal cells in vitro. J. Biol. Chem. 1998; 273(44):28560–28563.

Roosendaal, R., Kuipers, E.J., Buitenwerf, J., van Uffelen, C., Meuwissen, S.G., van Kamp, G.J. and Vandenbroucke-Grauls, C.M. *Helicobacter pylori* and the birth cohort effect: evidence of a continuous decrease of infection rates in childhood. Am. J. Gastroenterol. 1997; 92(9):1480–1482.

Rossi, G., Ruggiero, P., Peppoloni, S., Pancotto, L., Fortuna, D., Lauretti, L., Volpini, G., Mancianti, S., Corazza, M., Taccini, E., Di Pisa, F., Rappuoli, R. and Del Giudice, G. Therapeutic vaccination against *Helicobacter pylori* in the beagle dog experimental model: safety, immunogenicity, and efficacy. Infect. Immun. 2004; 72(6):3252–3259.

Rothenbacher, D. and Brenner, H. Burden of *Helicobacter pylori* and *H. pylori*-related diseases in developed countries: recent developments and future implications. Microbes Infect. 2003; 5(8):693–703.

Ruggiero, P., Peppoloni, S., Rappuoli, R. and Del Giudice, G. The quest for a vaccine against *Helicobacter pylori*: how to move from mouse to man? Microbes Infect. 2003; 5(8):749–756.

Ryu, J.W., Kim, H.J., Lee, Y.S., Myong, N.H., Hwang, C.H., Lee, G.S. and Yom, H.C. The proteomics approach to find biomarkers in gastric cancer. J. Korean Med. Sci. 2003; 18(4):505–509.

Sanchez, V., Gimenez, S., Haensler, J., Geoffroy, C., Rokbi, B., Seguin, D., Lissolo, L., Harris, B., Rizvi, F., Kleanthous, H., Monath, T., Cadoz, M. and Guy, B. Formulations of single or multiple *H. pylori* antigens with DC Chol adjuvant induce protection by the systemic route in mice. Optimal prophylactic combinations are different from therapeutic ones. FEMS Immunol. Med. Microbiol. 2001; 30(2):157–165.

Sansonetti, P. Host–pathogen interactions: the seduction of molecular cross talk. Gut 2002; 50(Suppl. 3):III2–III8.

Scherer, M.T., Ignatowicz, L., Winslow, G.M., Kappler, J.W. and Marrack, P. Superantigens: bacterial and viral proteins that manipulate the immune system. Annu. Rev. Cell Biol. 1993; 9:101–128.

Schmausser, B., Andrulis, M., Endrich, S., Lee, S.K., Josenhans, C., Muller-Hermelink, H.K. and Eck, M. Expression and subcellular distribution of toll-like receptors TLR4, TLR5 and TLR9 on the gastric epithelium in *Helicobacter pylori* infection. Clin. Exp. Immunol. 2004; 136(3):521–526.

Schmausser, B., Andrulis, M., Endrich, S., Muller-Hermelink, H.K. and Eck, M. Toll-like receptors TLR4, TLR5 and TLR9 on gastric carcinoma cells: an implication for interaction with *Helicobacter pylori*. Int. J. Med. Microbiol. 2005; 295(3):179–185.

Shimoyama, T., Fukuda, S., Liu, Q., Nakaji, S., Fukuda, Y. and Sugawara, K. *Helicobacter pylori* water soluble surface proteins prime human neutrophils for enhanced production of reactive oxygen species and stimulate chemokine production. J. Clin. Pathol. 2003; 56(5):348–351.

Skouloubris, S., Thiberge, J.M., Labigne, A. and De Reuse, H. The *Helicobacter pylori* UreI protein is not involved in urease activity but is essential for bacterial survival in vivo. Infect. Immun. 1998; 66(9):4517–4521.

Smith, M.F., Jr., Mitchell, A., Li, G., Ding, S., Fitzmaurice, A.M., Ryan, K., Crowe, S. and Goldberg, J.B. Toll-like receptor (TLR) 2 and TLR5, but not TLR4, are required for *Helicobacter pylori*-induced NF-kappa B activation and chemokine expression by epithelial cells. J. Biol. Chem. 2003; 278(35): 32552–32560.

Smythies, L.E., Novak, M.J., Waites, K.B., Lindsey, J.R., Morrow, C.D. and Smith, P.D. Poliovirus replicons encoding the B subunit of *Helicobacter pylori* urease protect mice against *H. pylori* infection. Vaccine 2005; 23(7):901–909.

Solnick, J.V. and Tompkins, L.S. *Helicobacter pylori* and gastroduodenal disease: pathogenesis and host–parasite interaction. Infect. Agents Dis. 1992; 1:294–309.

Sougioultzis, S., Lee, C.K., Alsahli, M., Banerjee, S., Cadoz, M., Schrader, R., Guy, B., Bedford, P., Monath, T.P., Kelly, C.P. and Michetti, P. Safety and efficacy of *E. coli* enterotoxin adjuvant for urease-based rectal immunization against *Helicobacter pylori*. Vaccine 2002; 21(3–4):194–201.

Stromberg, E., Edebo, A., Svennerholm, A.M. and Lindholm, C. Decreased epithelial cytokine responses in the duodenal mucosa of *Helicobacter pylori*-infected duodenal ulcer patients. Clin. Diagn. Lab. Immunol. 2003a; 10(1):116–124.

Stromberg, E., Lundgren, A., Edebo, A., Lundin, S., Svennerholm, A.M. and Lindholm, C. Increased frequency of activated T-cells in the *Helicobacter pylori*-infected antrum and duodenum. FEMS Immunol. Med. Microbiol. 2003b; 36(3):159–168.

Svennerholm, A.M. and Quiding-Jarbrink, M. Priming and expression of immune responses in the gastric mucosa. Microbes Infect. 2003; 5(8):731–739.

Szabo, I., Brutsche, S., Tombola, F., Moschioni, M., Satin, B., Telford, J.L., Rappuoli, R., Montecucco, C., Papini, E. and

Zoratti, M. Formation of anion-selective channels in the cell plasma membrane by the toxin VacA of *Helicobacter pylori* is required for its biological activity. EMBO J. 1999; 18(20): 5517–5527.

Tacket, C.O., Sztein, M.B., Losonsky, G.A., Wasserman, S.S., Nataro, J.P., Edelman, R., Pickard, D., Dougan, G., Chatfield, S.N. and Levine, M.M. Safety of live oral *Salmonella typhi* vaccine strains with deletions in htrA and aroC aroD and immune response in humans. Infect. Immun. 1997; 65(2):452–456.

Takahashi, T., Yujiri, T., Shinohara, K., Inoue, Y., Sato, Y., Fujii, Y., Okubo, M., Zaitsu, Y., Ariyoshi, K., Nakamura, Y., Nawata, R., Oka, Y., Shirai, M. and Tanizawa, Y. Molecular mimicry by *Helicobacter pylori* CagA protein may be involved in the pathogenesis of *H. pylori*-associated chronic idiopathic thrombocytopenic purpura. Br. J. Haematol. 2004; 124(1):91–96.

Takata, T., Aras, R., Tavakoli, D., Ando, T., Olivares, A.Z. and Blaser, M.J. Phenotypic and genotypic variation in methylases involved in type II restriction-modification systems in *Helicobacter pylori*. Nucl. Acids Res. 2002; 30(11):2444–2452.

Takenaka, R., Yokota, K., Ayada, K., Mizuno, M., Zhao, Y., Fujinami, Y., Lin, S.N., Toyokawa, T., Okada, H., Shiratori, Y. and Oguma, K. *Helicobacter pylori* heat-shock protein 60 induces inflammatory responses through the Toll-like receptor-triggered pathway in cultured human gastric epithelial cells. Microbiology 2004; 150(Pt 12):3913–3922.

Talley, N.J., Zinsmeister, A.R., Weaver, A., DiMagno, E.P., Carpenter, H.A., Perez-Perez, G.I. and Blaser, M. Gaastric adenocarcinoma and *Helicobacter pylori* infection. J. Natl. Cancer Inst. 1991; 23:1734–1739.

Tanahashi, T., Kita, M., Kodama, T., Yamaoka, Y., Sawai, N., Ohno, T., Mitsufuji, S., Wei, Y.P., Kashima, K. and Imanishi, J. Cytokine expression and production by purified *Helicobacter pylori* urease in human gastric epithelial cells. Infect. Immun. 2000; 68(2):664–671.

Taylor, J.M., Ziman, M.E., Huff, J.L., Moroski, N.M., Vajdy, M. and Solnick, J.V. *Helicobacter pylori* lipopolysaccharide promotes a Th1 type immune response in immunized mice. Vaccine 2006; 24(23):4987–4994.

Todoroki, I., Joh, T., Watanabe, K., Miyashita, M., Seno, K., Nomura, T., Ohara, H., Yokoyama, Y., Tochikubo, K. and Itoh, M. Suppressive effects of DNA vaccines encoding heat shock protein on *Helicobacter pylori*-induced gastritis in mice. Biochem. Biophys. Res. Commun. 2000; 277(1):159–163.

Toh, B.H., Van Driel, I.R. and Gleeson, P.A. Autoimmune gastritis: tolerance and autoimmunity to the gastric H^+/K^+ ATPase (proton pump). Autoimmunity 1992; 13(2):165–172.

Tsuda, M., Karita, M., Mizote, T., Morshed, M.G., Okita, K. and Nakazawa, T. Essential role of *Helicobacter pylori* urease in gastric colonization: definite proof using a urease-negative mutant constructed by gene replacement. Eur. J. Gastroenterol. Hepatol. 1994; 6(Suppl. 1):S49–S52.

Unemo, M., Aspholm-Hurtig, M., Ilver, D., Bergstrom, J., Boren, T., Danielsson, D. and Teneberg, S. The sialic acid binding SabA adhesin of *Helicobacter pylori* is essential for nonopsonic activation of human neutrophils. J. Biol. Chem. 2005; 280(15):15390–15397.

Vajdy, M., Singh, M., Ugozzoli, M., Briones, M., Soenawan, E., Cuadra, L., Kazzaz, J., Ruggiero, P., Peppoloni, S., Norelli, F., Del Giudice, G. and O'Hagan, D. Enhanced mucosal and systemic immune responses to *Helicobacter pylori* antigens through mucosal priming followed by systemic boosting immunizations. Immunology 2003; 110(1):86–94.

Velin, D., Bachmann, D., Bouzourene, H. and Michetti, P. Mast cells are critical mediators of vaccine-induced *Helicobacter* clearance in the mouse model. Gastroenterology 2005; 129(1):142–155.

Veneri, D., Franchini, M., Gottardi, M., D'Adda, M., Ambrosetti, A., Krampera, M., Zanetti, F. and Pizzolo, G. Efficacy of *Helicobacter pylori* eradication in raising platelet count in adult patients with idiopathic thrombocytopenic purpura. Haematologica 2002; 87(11):1177–1179.

Viala, J., Chaput, C., Boneca, I.G., Cardona, A., Girardin, S.E., Moran, A.P., Athman, R., Memet, S., Huerre, M.R., Coyle, A.J., DiStefano, P.S., Sansonetti, P.J., Labigne, A., Bertin, J., Philpott, D.J. and Ferrero, R.L. Nod1 responds to peptidoglycan delivered by the *Helicobacter pylori* cag pathogenicity island. Nat. Immunol. 2004; 5(11):1166–1174.

Vizcaino, A.P., Moreno, V., Bosch, F.X., Munoz, N., Barros-Dios, X.M. and Parkin, D.M. International trends in the incidence of cervical cancer: I. Adenocarcinoma and adenosquamous cell carcinomas. Int. J. Cancer 1998; 75(4):536–545.

Wang, J., Fan, X., Lindholm, C., Bennett, M., O'Connoll, J., Shanahan, F., Brooks, E.G., Reyes, V.E. and Ernst, P.B. *Helicobacter pylori* modulates lymphoepithelial cell interactions leading to epithelial cell damage through Fas/Fas ligand interactions. Infect. Immun. 2000; 68(7):4303–4311.

Watanabe, T., Tada, M., Nagai, H., Sasaki, S. and Nakao, M. *Helicobacter pylori* infection induces gastric cancer in Mongolian gerbils. Gastroenterology 1998; 115(3):642–648.

Weeks, D.L., Eskandari, S., Scott, D.R. and Sachs, G. A H^+-gated urea channel: the link between *Helicobacter pylori* urease and gastric colonization. Science 2000; 287(5452):482–485.

Wotherspoon, A.C., Doglioni, C., Diss, T.C., Pan, L., Moschini, A., de Boni, M. and Isaacson, P.G. Regression of primary low-grade B-cell gastric lymphoma of mucosa-associated lymphoid tissue type after eradication of *Helicobacter pylori* [see comments]. Lancet 1993; 342(8871):575–577.

Wotherspoon, A.C., Ortiz-Hidalgo, C., Falzon, M.R. and Isaacson, P.G. *Helicobacter pylori*-associated gastritis and primary B-cell gastric lymphoma [see comments]. Lancet 1991; 338(8776):1175–1176.

Xia, H.H., Lam, S.K., Huang, X.R., Wong, W.M., Leung, S.Y., Yuen, S.T., Lan, H.Y. and Wong, B.C. *Helicobacter pylori* infection is associated with increased expression of macrophage migratory inhibitory factor—by epithelial cells, T cells, and macrophages—in gastric mucosa. J. Infect. Dis. 2004; 190(2):293–302.

Yamaguchi, H., Osaki, T., Taguchi, H., Hanawa, T., Yamamoto, T. and Kamiya, S. Relationship between expression of HSP60, urease activity, production of vacuolating toxin, and adherence activity of *Helicobacter pylori*. J. Gastroenterol. 1998; 33(Suppl. 10):6–9.

Yamamoto, T., Kita, M., Ohno, T., Iwakura, Y., Sekikawa, K. and Imanishi, J. Role of tumor necrosis factor-alpha and interferon-gamma in *Helicobacter pylori* infection. Microbiol. Immunol. 2004; 48(9):647–654.

Yamaoka, Y., Kudo, T., Lu, H., Casola, A., Brasier, A.R. and Graham, D.Y. Role of interferon-stimulated responsive element-like element in interleukin-8 promoter in *Helicobacter pylori* infection. Gastroenterology 2004; 126(4):1030–1043.

Yamaoka, Y., Kwon, D.H. and Graham, D.Y. A M(r) 34,000 proinflammatory outer membrane protein (oipA) of Helicobacter pylori. Proc. Natl. Acad. Sci. USA 2000; 97(13):7533–7538.

Ye, G., Barrera, C., Fan, X., Gourley, W.K., Crowe, S.E., Ernst, P.B. and Reyes, V.E. Expression of B7-1 and B7-2 costimulatory molecules by human gastric epithelial cells: potential role in $CD4^+$ T cell activation during *Helicobacter pylori* infection. J. Clin. Invest. 1997; 99(7):1628–1636.

Ye, W., Held, M., Lagergren, J., Engstrand, L., Blot, W.J., McLaughlin, J.K. and Nyren, O. *Helicobacter pylori* infection and gastric atrophy: risk of adenocarcinoma and squamous-cell carcinoma of the esophagus and adenocarcinoma of the gastric cardia. J. Natl. Cancer Inst. 2004; 96(5):388–396.

Yoshikawa, T. and Naito, Y. The role of neutrophils and inflammation in gastric mucosal injury. Free Radic. Res. 2001; 33(6):785–794.

Zarrilli, R., Tuccillo, C., Santangelo, M., Nardone, G. and Romano, M. Increased COX-2, but not COX-1, mRNA expression in *Helicobacter pylori* gastritis. Am. J. Gastroenterol. 1999; 94(11): 3376–3378.

Zhang, L., Blot, W.J., You, W.C., Chang, Y.S., Kneller, R.W., Jin, M.L., Li, J.Y., Zhao, L., Liu, W.D., Zhang, J.S., Ma, J.L., Samloff, I.M., Correa, P., Blaser, M.J., Xu, G.W. and Fraumeni, J.F., Jr. *Helicobacter pylori* antibodies in relation to precancerous gastric lesions in a high-risk Chinese population. Cancer Epidemiol. Biomarkers Prev. 1996; 5(8):627–630.

CHAPTER

51

Intestinal Pathogenic *Escherichia coli*

Alfredo G. Torres

Departments of Pathology and Microbiology and Immunology, Sealy Center for Vaccine Development, University of Texas Medical Branch, Galveston, TX, USA

OUTLINE

ABSTRACT

Intestinal pathogenic *Escherichia coli* strains that cause diarrhea can be divided into at least six different categories, and each type is associated with a distinct pathogenic mechanism. Taken together, these organisms represent the most common causes of pediatric diarrhea worldwide. In recent years, a large amount of information has been generated regarding the virulence mechanisms used by the different categories of diarrheagenic *E. coli* to produce several distinct clinical syndromes. This review describes how their pathogenic schemes underlie the clinical manifestations, treatment approach, and epidemiologic investigation of these important pathogens and examines in detail the past and current progress in the development of vaccines against these categories of diarrheagenic *E. coli*.

INTRODUCTION

Diarrhea is a common problem that usually lasts a day or two and gradually disappears without any special treatment. However, prolonged diarrhea can be a sign of other problems. Diarrhea can cause dehydration, which is particularly dangerous in children and the elderly, and must be treated promptly to avoid serious health problems. Diarrhea caused by a variety of *E. coli* strains is a major problem in developing countries, although traveler's diarrhea and hemorrhagic colitis are also a problem for industrialized countries. It has been estimated that *E. coli* is responsible for an estimated 780–900 million cases of diarrhea worldwide and at least 300,000–500,000 deaths annually. Taken together, these organisms represent the most common cause of pediatric diarrhea worldwide (Bern et al., 1992). There are numerous categories of diarrheagenic *E. coli* (Kaper et al., 2004), but only a few serotype combinations are associated with diarrheal disease (Nataro and Kaper, 1998). This chapter provides a brief description of the different categories of *E. coli*, as well as their pathogenic mechanisms, clinical relevance, and treatment. Further, this chapter also presents past and present progress made in the area of vaccinology against diarrheagenic *E. coli* and discusses future lines of research being pursued to improve current vaccines.

ETIOLOGIC AGENT(S)

E. coli can cause diarrhea by at least six different mechanisms, and each type is associated with a different pathotype and different virulence determinants (Table 51.1; Kaper et al., 2004; Torres et al., 2005). Enterotoxigenic *E. coli* (ETEC) is the major cause of traveler's diarrhea. Enteropathogenic *E. coli* (EPEC) causes watery diarrhea in infants and children. Enterohemorrhagic *E. coli* (EHEC) and enteroinvasive *E. coli* (EIEC) strains both produce bloody diarrhea, but while EIEC causes a more *Shigella*-like dysentery, EHEC infections more frequently cause kidney failure (hemolytic uremic syndrome [HUS]). Of these first four categories, ETEC, EPEC, and EIEC are responsible for the vast majority of cases in the developing world. Enteroaggregative *E. coli* (EAEC) is primarily associated with persistent diarrhea in children in developing countries and in AIDS-associated diarrhea, while diffusely adhering *E. coli* (DAEC) is a cause of childhood diarrhea and also traveler's diarrhea in Mexico and North Africa. ETEC, EPEC, EAEC, and DAEC preferentially colonize the small bowel, and EIEC and EHEC colonize the large bowel prior to causing disease. In general, the pathogenic process of diarrheagenic *E. coli* strains include: (i) ingestion of the organism; (ii) bacterial attachment via pili (fimbriae), fibrils, or outer membrane proteins; (iii) colonization of the intestinal mucosa; (iv) expression of bacterial mechanisms to manipulate the host cell cytoskeleton or to evade host defenses; (v) multiplication of the bacteria, and (vi) host damage and production of disease. The description of the mechanisms associated with disease, epidemiology, clinical manifestations, and treatment of the individual categories of diarrheagenic *E. coli* are discussed below.

TABLE 51.1 Summary of clinical syndromes and pathogenic mechanisms of diarrheagenic *Escherichia coli*

Classification (category)	Site affected	Clinical syndrome	Proposed pathogenic mechanism
Enterotoxigenic *E. coli* (ETEC)	Small intestine	Traveler's diarrhea. Watery stool, low fever, cramps, nausea	Hypersecretion of fluids and electrolytes stimulated by plasmid-mediated enterotoxins
Enteropathogenic *E. coli* (EPEC)	Small intestine	Infantile diarrhea with fever, nausea, vomiting	A/E[a] lesion formation and BFP-mediated microcolony stabilization
Enterohemorrhagic *E. coli* (EHEC)	Large intestine	Hemorrhagic colitis. Severe abdominal pain, watery diarrhea, possible leading to hemolytic uremic syndrome	Inhibition of protein synthesis by cytotoxins and A/E[a] lesion formation
Enteroinvasive *E. coli* (EIEC)	Large intestine	*Shigella*-like diarrhea. Fever, cramps, watery diarrhea followed by bloody stool	Tissue invasion and destruction of epithelial cells (plasmid-mediated)
Enteroaggregative *E. coli* (EAEC)	Small intestine	Infant diarrhea in developing countries and diarrhea in AIDS patients	Inhibition of fluid absorption by plasmid-mediated aggregative adherence (biofilm formation)
Diffusely adhering *E. coli* (DAEC)	Various sites	Diarrhea	Little is known about mechanisms causing diarrhea

[a]A/E, Attaching and effacing lesion formation causes destruction of enterocytes; BFP, bundle-forming pilus.

Enterotoxigenic *E. coli*

ETEC causes a self-limited illness, usually lasting fewer than 5 days, and is considered a major cause of childhood diarrheal disease in developing countries as well as a frequent cause of enteric illness among travelers and in soldiers deployed to those countries (Nataro and Kaper, 1998; Qadri et al., 2005). ETEC are acquired by ingestion of contaminated food and water and produce a range of symptoms, usually of moderate severity, including nonbloody, watery diarrhea (cholera-like symptoms), and abdominal cramps.

Pathogenesis

ETEC colonize the proximal small intestine and adhere to enterocytes by means of the plasmid-encoded fimbrial factors (Fig. 51.1). Colonization is mediated by one or more proteinaceous fimbrial or fibrillar colonization factors (CFs) that are designated as colonization factor antigens (CFAs, e.g. CFA I and CFA II), coli surface antigen (CS), or putative colonization factors (PCFs) in ETEC infections of humans, or K-88 and K-99 in the case of animals (Gaastra and Svennerholm, 1996; Torres et al., 2005). After colonization, ETEC strains produce two classes of plasmid-encoded enterotoxins. The first class comprises the high-molecular-weight, heat-labile enterotoxins (LTI and LTII), which are structurally and functionally related to cholera toxin, and a second class, the low-molecular-weight, heat-stable enterotoxins (ST_A and ST_B) (Nataro and Kaper, 1998; Sears and Kaper, 1996). ETEC strains may express either one or both enterotoxins, and both classes are responsible for net secretion of fluid into the intestine and diarrhea. LTs produce diarrhea by activating the main chloride channel, and by stimulating prostaglandin synthesis and the enteric nervous system, both of which can stimulate secretion and inhibit absorption. STs are small peptides that resemble a mammalian peptide hormone (guanylin) and bind to the guanylin receptor. This binding results in elevation of cyclic AMP, which, through a series of steps, leads to secretion of chloride (Nataro and Kaper, 1998). In addition, ETEC strains have been reported to invade cultured human epithelial cells, although there are no clinical data to suggest that ETEC invades cells in vivo (Elsinghorst and Kopecko, 1992; Elsinghorst and Weitz, 1994). The ETEC factors associated with this invasion phenotype have been identified with the characterization of the Tia pathogenicity island (Tia-PAI) in some clinical ETEC isolates (Elsinghorst and Kopecko, 1992; Elsinghorst and Weitz, 1994).

Epidemiology

ETEC is usually a frequent cause of diarrhea in infants younger than 2 years of age, with the incidence of ETEC infections in developing countries decreasing between the ages of 5 and 15 years (Girard et al., 2006; Qadri et al., 2005). ETEC also is a major cause of traveler's diarrhea and accounts for approximately 280 million diarrhea episodes and more than 400,000 deaths annually (Steffen et al., 2005). ETEC is responsible for about 25% of persistent diarrheas and 26%

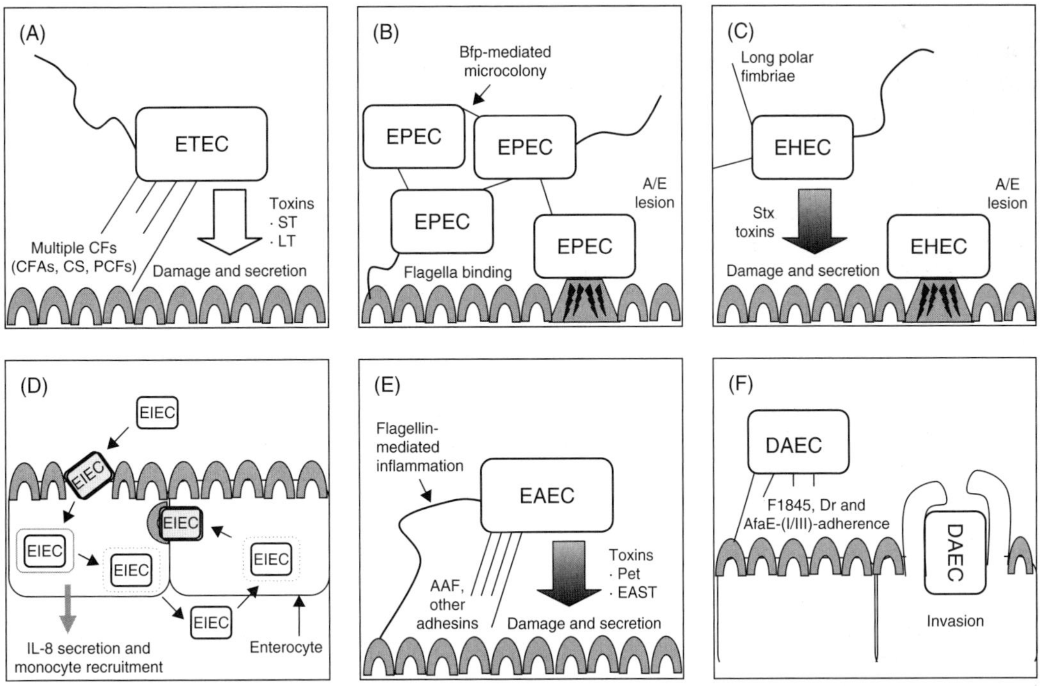

FIGURE 51.1 Diarrheagenic *E. coli* interactions with epithelial cells. The six categories of diarrheagenic *E. coli* each have unique features in their interaction with intestinal epithelial cells. Here, the interactions (e.g. mediated by fimbriae or other adhesins) of each category with a target cell, followed by damage (e.g. toxin-mediated) and the subsequent fluid secretion, are schematically represented. See text for further details.

of severe diarrheas requiring hospitalization and perfusions worldwide (Fig. 51.2; Qadri et al., 2005). Despite the overall low incidence, outbreaks of ETEC are being increasingly recognized in the United States, and just between 1996 and 2003, 16 ETEC outbreaks were identified; three occurred on cruise ships and the rest within US territory (Beatty et al., 2004).

Clinical Disease

Infection with ETEC can cause profuse watery diarrhea and abdominal cramping. Fever, nausea (with or without vomiting), chills, appetite loss, headache, muscle aches, and bloating can also occur, but are less common. Illness develops 1–3 days after exposure and usually lasts 3–4 days. Some infections may take a week or longer to resolve. Symptoms rarely last more than 3 weeks. Most patients recover with supportive measures alone and do not require hospitalization or antibiotics (Qadri et al., 2005).

Treatment

Most infected persons will recover within a few days without requiring any specific treatment. In the case of infants and children with diarrhea, clear liquids or premixed oral rehydration solutions are recommended to prevent dehydration and loss of electrolytes (CDC, 2005). Bismuth subsalicylate also may be used as treatment (one fluid ounce or two 262-mg tablets every 30 min for up to eight doses in a 24-h period). ETEC is frequently resistant to common antibiotics, including trimethoprim-sulfamethoxazole and ampicillin. Because resistance to antibiotics is increasing worldwide, the decision to use an antibiotic should be carefully weighed against the severity of illness and the risk of adverse reactions. Currently, fluoroquinolones are the drugs of choice. Commonly prescribed regimens are 500 mg of ciprofloxacin twice a day or 400 mg of norfloxacin twice a day for 3–5 days (CDC, 2005).

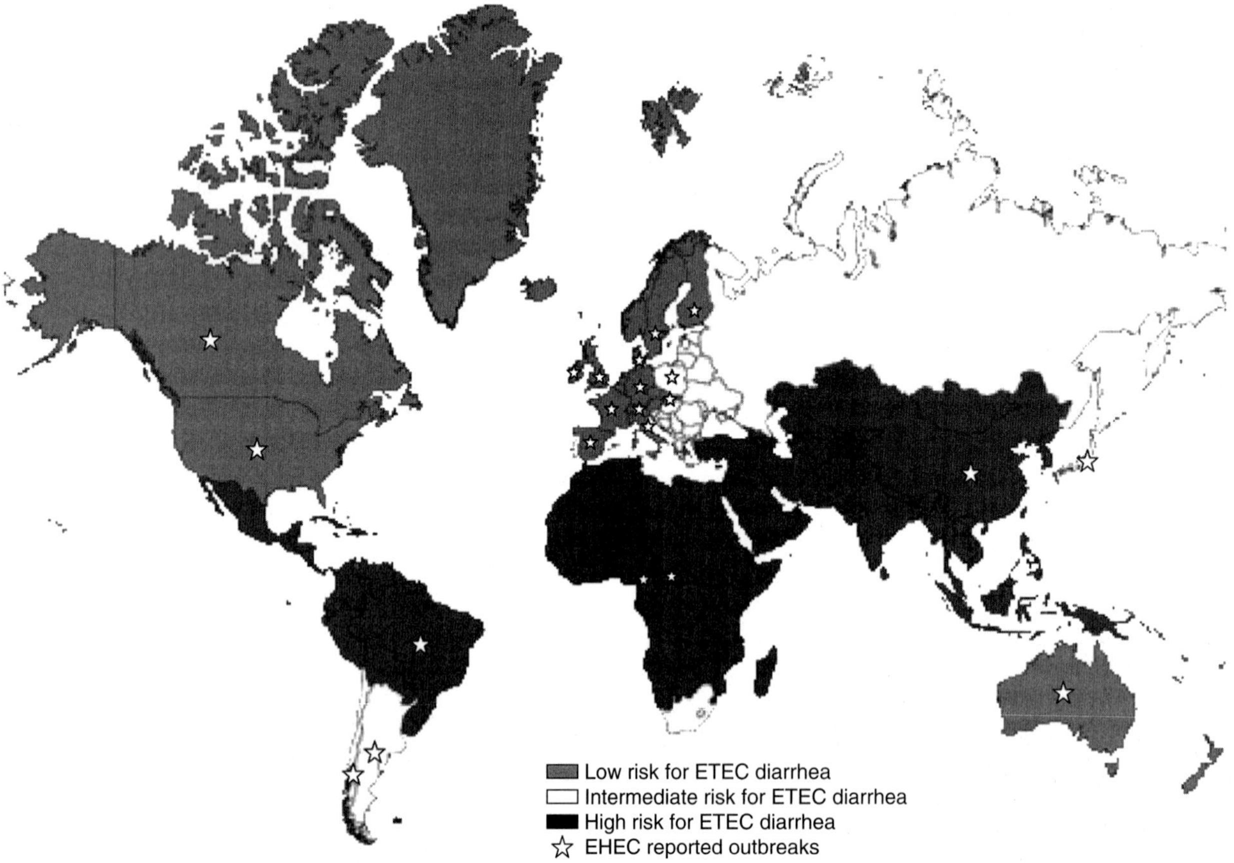

FIGURE 51.2 Areas of risk for traveler's diarrhea caused by enterotoxigenic *E. coli* and distribution of reported outbreaks due to enterohemorrhagic *E. coli* (EHEC) worldwide. The world is divided into three risk levels (low, intermediate, and high) of traveler's diarrhea for travelers from industrialized countries. The star symbols represent countries reporting outbreaks of enterohemorrhagic *E. coli*, O157:H7 being one of the predominant serotypes.

Enteropathogenic *E. coli*

EPEC is an important category of diarrheagenic *E. coli* that has been linked to infant diarrhea in the developing world. However, in industrialized countries, the frequency of these organisms has decreased, but they continue to be an important cause of diarrhea (Nataro and Kaper, 1998). EPEC strains can be classified as typical and atypical isolates. In general terms, EPEC produces a characteristic histopathological lesion known as attaching and effacing (A/E) on intestinal cells and do not produce Shiga toxins (Stx). Typical EPEC of human origin possesses a virulence plasmid known as the EAF (EPEC adherence factor) plasmid, while an atypical EPEC does not posses this plasmid; this is the basic difference between the two groups of organisms.

Pathogenesis

The central mechanism of EPEC pathogenesis is the adhesion to small bowel enterocytes and destruction of the normal microvillar architecture, initiating the formation of A/E lesions, which are characterized by microvilli destruction, intimate adherence of bacteria to the intestinal epithelium, pedestal formation, and aggregation of polarized actin and other elements of the cytoskeleton at sites of bacterial attachment. Cytoskeleton derangements are accompanied by an inflammatory response and diarrhea. The majority of genetic determinants for the production of A/E lesions are located on the pathogenicity island known as locus of enterocyte effacement (LEE) (Fig. 51.1) (Kaper et al., 2004; Torres et al., 2005). The LEE is present in all EPEC strains as well as in other A/E positive strains, including strains of EHEC and other enteric pathogens (Torres and Kaper, 2001). Many EPEC strains produce a characteristic adherence pattern, called localized adherence, in tissue-cultured cells (Scaletsky et al., 1984). This phenomenon is associated with the presence of the large EAF plasmid, which carries the so-called EAF sequence and also encodes type IV fimbriae, known as bundle-forming pili (BFP), which interconnect bacteria within microcolonies and thus promote their stabilization (Nataro and Kaper, 1998). The EAF plasmid is not essential for the formation of A/E lesions, although its presence enhances the efficiency of lesion formation due to the presence of the regulatory region known as Per (Mellies et al., 1999), as well as BFP, whose role in cell adhesion would similarly increase the efficiency of A/E lesion formation (Torres et al., 2005).

Epidemiology

EPEC serotypes are strongly associated with diarrhea in children <1 year of age. In this age group, these serotypes have been found to be the main cause of endemic diarrhea in several, well-controlled studies carried out in Brazil (Gomes et al., 1991; Toledo et al., 1983). The frequency of typical EPEC serotypes in children >1 year of age is lower and similar to the frequency in controls (2–4%). Adult infections are rare and usually associated with other conditions (Nataro and Kaper, 1998). A remarkable epidemiologic difference between typical and atypical EPEC serotypes is their geographic distribution. Typical EPEC serotypes have traditionally been associated with outbreaks of infantile diarrhea, and, in fact, the first EPEC strains isolated in different countries were of serotypes O55: H6 and O111:H2 (Trabulsi et al., 2002). In the past, these epidemic serotypes were frequently identified in industrialized countries as a cause of outbreaks and sporadic cases of diarrhea, but at present they are very rare (Nataro and Kaper, 1998). In these countries today, serotypes without the EAF plasmid predominate (Bokete et al., 1997; Scotland et al., 1996). The situation in developing countries is not well defined, but several studies in Brazil in the 1980s and early 1990s showed a high frequency of typical serotypes (Trabulsi et al., 2002). In the case of atypical strains, these isolates have been associated with prolonged diarrhea in children (Nguyen et al., 2006).

Clinical Disease

EPEC primarily causes acute diarrhea, although many cases of protracted EPEC diarrhea have also been reported (Donnenberg, 1995; Levine and Edelman, 1984). The infection can often be quite severe, and many clinical reports emphasize the severity of the disease (Rothbaum et al., 1982). In outbreaks from developing countries, a mortality rate of up to 30% has been reported (Senerwa et al., 1989), and in developed countries with the full range of modern treatment available, mortality is much lower, although deaths can still result (Rothbaum et al., 1982). In addition to profuse watery diarrhea, vomiting and low-grade fever are common symptoms of EPEC infection. The presence of the A/E lesion is associated with disarrangement of the digestive–absorptive enzyme system, leading to malabsorption of nutrients (Nataro and Kaper, 1998).

Treatment

As with other diarrheal pathogens, the primary goal of treatment of EPEC diarrhea is to prevent dehydration by correcting fluid and electrolyte imbalances. Oral rehydration may be sufficient for milder cases, but more severe cases require parenteral rehydration. A variety of antibiotics have been used to treat EPEC and have proved useful in many cases (Donnenberg, 1995), but multiple antibiotic resistance is common for EPEC (Donnenberg, 1995).

Enterohemorrhagic *E. coli*

EHEC were first recognized as a cause of human disease in 1982 (Riley et al., 1983) and are associated with diarrhea and hemorrhagic colitis (Karmali, 1989). EHEC is the only pathotype commonly responsible for diarrhea in children in the United States (Frenzen et al., 2005) and *E. coli* O157:H7, in particular, has emerged as the most virulent member of this pathotype, responsible both for outbreaks and sporadic cases of diarrhea in North America (Kaper et al., 2004). EHEC are defined by their ability to produce one or more Shiga toxins (Stx) (Tesh and O'Brien, 1991), which mediate the systemic complications of EHEC infections, and the induction of the characteristic attaching and effacing lesions on intestinal epithelia (Fig. 51.1) (Kaper et al., 2004; Torres and Kaper, 2001). In some people, particularly children under 5 years of age and the elderly, the infection can also cause a complication called HUS. In the United States, HUS is the principal cause of acute kidney failure in children, and most cases of HUS are caused by *E. coli* O157:H7.

Pathogenesis

Pathogenesis involves establishment of the organism in the gut, where it adheres to colon epithelial cells, probably mediated by the long polar fimbriae, and induces the attaching and effacing lesions produced by LEE pathogenicity island-encoded proteins as well as other factors found in O157-specific islands elsewhere in the genome (Torres et al., 2005). EHEC produce one or more Shiga toxins (also known as verocytotoxins) whose systemic absorption leads to potentially life-threatening complications (Fig. 51.1; Robinson et al., 2006). Not all Stx-producing *E. coli* strains contain the LEE and the terms Shiga toxin-producing *E. coli* (STEC) or verotoxin-producing *E. coli* (VTEC) are used to describe any *E. coli* strain that produces Stx, whereas EHEC is used to denote only the subset of LEE-positive, Stx-producing *E. coli*. The ability to produce one or more phage-encoded Stxs is the most important virulence characteristic of EHEC strains (Robinson et al., 2006). Toxin production mediates both local and systemic disease. Local intestinal effects cause the development of bloody diarrhea as Stx is internalized by the cells of the gut where it blocks cellular protein synthesis and may lead to apoptosis. HUS results from microvascular disease when the toxins enter the blood stream and bind to receptors on endothelial cells abundant in kidneys and brain (Nataro and Kaper, 1998).

Epidemiology

EHEC/STEC strains are the most frequently identified diarrheagenic *E. coli* serotypes in North America and Europe, with *E. coli* O157:H7 as the most common STEC serotype isolated in these countries (Frenzen et al., 2005). The CDC estimates indicate that STEC O157 infections cause 73,000 illnesses annually, resulting in more than 2000 hospitalizations and 60 deaths (Fig. 51.2) (Frenzen et al., 2005). From 1998 to 2002, the U. S. officials in 49 states reported 350 outbreaks representing 8598 cases of *E. coli* O157:H7; 1493 hospitalizations (17%); 354 cases of HUS (4%); and 40 deaths (0.5%) (Rangel et al., 2005). Recent data collected by the CDC in 2006 revealed that from 17,252 laboratory-confirmed cases of foodborne infections, 590 cases were attributed to *E. coli* O157 infections and 209 to non-STEC O157 infections, which demonstrate that the incidence of infections caused by STEC O157 have not decreased significantly as compared with previous years, and suggesting that further measures are needed to prevent foodborne illness in the United States (CDC, 2007).

Clinical Disease

The clinical manifestations of EHEC, particularly *E. coli* O157:H7, typically include the sudden onset of abdominal cramps followed by watery diarrhea, which becomes bloody (Nataro and Kaper, 1998). People generally become ill by 2–8 days (average of 3–4) after being exposed to *E. coli* O157:H7. Diarrhea usually lasts 2–14 days and sometimes the infection leads to nonbloody diarrhea. Nausea, vomiting, and usually little or no fever are typical. Progression to HUS, characterized by microangiopathic hemolytic anemia, thrombocytopenia, and renal failure, complicates 2–7% of *E. coli* O157:H7 cases (Griffin and Tauxe, 1991).

Treatment

Treatment is usually supportive and antibiotics are not usually recommended in the case of EHEC/STEC infections (AMA et al., 2004). Some studies have suggested that children who suffered hemorrhagic colitis are at greater risk of HUS if they are treated with antibiotics (Wong et al., 2000). For this reason, most clinicians do not recommend antibiotics for treatment of EHEC/STEC infections. Further, children suffering from bloody diarrhea should not be given antimotility agents.

Enteroinvasive *E. coli*

EIEC may produce an illness known as bacillary dysentery. The EIEC strains responsible for this disease are closely related to *Shigella* spp., and therefore EIEC infections are often associated with fever. Following the ingestion of EIEC, the organisms invade the epithelial cells of the intestine, where they induce a local inflammatory response. Diarrhea can be bloody, but is usually watery and without blood or mucus.

Pathogenesis

The characteristic virulence determinants of EIEC are indistinguishable from those of *Shigella* spp. and include the presence of a virulence plasmid of ~220 kbp and the pathogens' ability to induce their entry into epithelial cells and disseminate from cell to cell (Fig. 51.1; Parsot, 2005). The pathogenicity factors found in the virulence plasmid include a type three secretion apparatus (TTSA) that spans the bacterial envelope, translocators that travel through the TTSA and insert into the host cell membrane to form a pore (translocon), and effectors that travel through the TTSA and the translocon and are injected within the eukaryotic cell. Once in the colonic mucosa, EIEC (similar to *Shigella* spp.) are proposed to cross the epithelial layer by invading M cells to be delivered to resident macrophages. Subsequently, they induce apoptosis and reach the basolateral pole of epithelial cells, in which they induce their entry. Movement of intracellular EIEC leads to the formation of protrusions and dissemination of bacteria within the epithelium. Release of cytokines and chemokines, including IL-1 by apoptotic macrophages and IL-8 by infected enterocytes, promotes recruitment of monocytes that migrate through the epithelial barrier, facilitating the entry of luminal bacteria into epithelial cells and increasing invasion of the epithelium (Prats and Llovet, 1995).

Epidemiology

EIEC is less common than ETEC or EPEC in the developing world and is associated with only a few characteristic serotypes (Nataro and Kaper, 1998). Though it has been responsible for outbreaks in both adults and children, it is much less common than ETEC, EPEC or EHEC. The incidence of EIEC in developed countries is believed to be low, but occasional foodborne outbreaks have been identified.

Clinical Disease

The clinical presentation closely resembles that of shigellosis with fever, abdominal cramps, rectal urgency, tenesmus, and watery or bloody diarrhea, lasting 5–7 days. The disease can appear within 24 h of infection with EIEC. In some persons, especially young children and the elderly, the diarrhea can be so severe that the patient needs to be hospitalized.

Treatment

Treatment recommendations are identical to that proposed for shigellosis. The greatest concern in treating EIEC disease is to replace the fluids and electrolytes (sodium, potassium, etc.) lost in the diarrhea. No controlled trials have examined the role of antimicrobial therapy though use of trimethoprim and sulfamethoxazole helps to eradicate susceptible strains of EIEC (DuPont, 1997). Further, ampicillin, nalidixic acid, or ciprofloxacin are also recommended; however, persons with mild infections will usually recover quickly without antibiotic treatment.

Enteroaggregative *E. coli*

EAEC is a diarrheal enteropathogen defined by aggregative adherence (AA) to cultured epithelial cells. EAEC has been implicated as an emerging cause of acute and chronic diarrhea in both developing and developed countries (Kahali et al., 2004; Nataro et al., 1998). Several epidemiological investigations of diarrhea suggest that EAEC is an important etiologic agent (Nataro et al., 1995; Nataro and Kaper, 1998) and, recently, it been shown that EAEC is an unrecognized cause of community-acquired diarrhea in infants in the United States (DuPont, 2005).

Pathogenesis

EAEC is defined by its ability to adhere to cultured cells in a "stacked brick" appearance (Okeke and Nataro, 2001). This phenotype, termed AA, is associated with specific fimbriae (AAFs) encoded by plasmids (pAAs) (Fig. 51.1; Harrington et al., 2006; Nataro et al., 1998). Other factors are associated with this adherence pattern, indicating its multifactorial nature (Torres et al., 2005). Several toxins expressed by EAEC strains have already been identified. Among the well-known toxins are EAEC heat-stable toxin (EAST1) and plasmid-encoded toxin (Pet), both of which are encoded by pAAs (Nataro et al., 1998). Furthermore, a serine protease involved in the colonization process (Pic) is encoded by the chromosome from prototype strain 042 (Henderson et al., 1999). Inflammation is a hallmark in EAEC-infected patients, and recent data indicated that the flagellin of EAEC strains induced the release of IL-8 from Caco-2 cells in culture (Steiner et al., 2000).

Epidemiology

EAEC strains were first recognized in 1987 and are most often associated with illnesses in different settings, including endemic diarrhea of infants in both industrialized and developing countries (Harrington et al., 2006; Okeke and Nataro, 2001), chronic diarrhea among human immunodeficiency virus/acquired immunodeficiency syndrome patients, and traveler's diarrhea. Asymptomatic infection can cause subclinical inflammatory enteritis and growth disturbances. Several outbreaks of EAEC diarrhea have also been described. Prospective studies suggest that in well-nourished populations in industrialized countries, EAEC may be a cause of endemic sporadic diarrhea. For example, a large prospective surveillance study in the UK implicated EAEC among the major pathogens at all ages (Wilson et al., 2001).

Clinical Disease

EAEC infection comprises watery diarrhea, occasionally with blood and mucus. Several studies have suggested that patients infected with EAEC manifest intestinal inflammation, marked by the presence of fecal lactoferrin and proinflammatory cytokines, notably interleukin (IL)-8 (Greenberg et al., 2002; Steiner et al., 2000). Further data have suggested that even asymptomatic carriage of EAEC strains can result in evidence of low-level enteritis (Steiner et al., 1998).

Treatment

Patients who are dehydrated should receive oral administration of solutions with electrolytes, and antibiotics are not usually recommended because EAEC can be resistant to treatment with antibiotics. However, a clinical trial of ciprofloxacin showed that diarrhea improved in AIDS patients infected with EAEC (Wanke et al., 1998).

Diffusely Adhering *E. coli*

DAEC strains are a heterogeneous group of isolates, all of which exhibit diffuse adherence (DA) to epithelial cells in the classical laboratory assay of adherence to HEp-2 or HeLa cells (Cravioto et al., 1991). The association of DAEC with diarrheal disease is not as strong as the other *E. coli* categories (Jallat et al., 1993, 1994), and little is known about the mechanism of pathogenesis. However, it is clear that DAEC isolates are associated with watery diarrhea that can become persistent in young children (Kaper et al., 2004; Le Bouguenec, 1999).

Pathogenesis

The DA pattern of DAEC isolates is due to the production of adhesins. About 75% of DAEC strains produce a fimbrial adhesin (designated F1845) encoded by the *daa* operon or express adhesins encoded by a family of *afa/dra/daa*-related operons (Fig. 51.1; Le Bouguenec and Servin, 2006; Torres et al., 2005). Unlike the virulence factors from pathovars associated with severe acute diarrhea (EHEC, ETEC, and EIEC), and the virulence factors only found in extraintestinal *E. coli*, the *afa/dra/daa* operons are genes that are expressed in a variety of genetic backgrounds (Escobar-Paramo et al., 2004). The *afa/dra/daa* operons are also interesting because they reportedly code for both afimbrial (such as AfaE-I and AfaE-III) and fimbrial (such as F1845 and Dr) adhesive structures on the bacterial surface and constitute the first known example of a single operon promoting both adhesion to epithelial cells and invasion into these cells (Le Bouguenec and Servin, 2006).

Epidemiology

DAEC infections are not well studied but have been recognized as a cause of diarrhea in the developing world, particularly among children. DAEC are infrequently identified in the United States. In a number of epidemiological studies, DAEC strains have been associated with diarrheal disease in different geographic areas (Lopes et al., 2005). However, DAEC virulence markers associated with diarrhea are yet to be clarified.

Clinical Disease

The clinical features of DAEC isolates are poorly described, and the diagnosis available is based on tissue culture assays looking for DA. However, a retrospective case-control study performed with a small group of children revealed that, among clinical signs, vomiting, but not diarrhea, was significantly associated with the presence of DAEC in children suffering from gastroenteritis (Poitrineau et al., 1995). Further prospective studies of children in France and Brazil showed a significant correlation between DAEC infection and diarrhea (Forestier et al., 1996; Scaletsky et al., 2002).

Treatment

As with many other diarrheagenic *E. coli* pathovars, oral rehydration therapy can help replace fluids and salts lost due to *E. coli*-induced illness. No information is currently available about antibiotic-resistant DAEC clinical isolates.

VACCINES

The development of vaccines against diarrheagenic *E. coli* strains, like any other enteric pathogen, represents a serious challenge because of the large number of serotypes involved and the requirement to induce immunity that is effective in the gut (Nataro and Kaper, 1998). Ideally, a vaccine against diarrheagenic *E. coli* should be administered by the mucosal route in order to elicit mucosal protection. Vaccines using live, attenuated bacteria have been developed to determine whether or not they can protect via the mucosal route against heterologous diarrheagenic *E. coli* infections (Girard et al., 2006). New, efficient mucosal adjuvants for human use have been designed based on, among others, *E. coli* toxins, with the aim of improving the induction of mucosal Thl and Th2 responses (Eriksson and Holmgren, 2002). Further, transgenic plants offer a new strategy for the development of safe, oral subunit vaccines against diarrheal diseases such as ETEC infection (Tacket, 2004). Even with all the recent progress, there are no vaccines currently approved to prevent diarrhea caused by diarrheagenic *E. coli* (with the possible exception of ETEC infection). The following section summarizes the state of the art in vaccine research and development to combat diarrheagenic *E. coli* infections, with a special emphasis on three pathotypes of *E. coli* (ETEC, EHEC, and EPEC), where most of the progress in vaccine development has accumulated in recent years.

Vaccines against ETEC

Several attempts have been made to develop a vaccine that will be effective against pathogenic *E. coli*, specifically ETEC (Svennerholm and Steele, 2004). ETEC comprises many different O:H serotypes, multiple antigenically distinct CFs types, and three different toxin phenotypes (LT only, ST only, or LT/ST). To confer broad-spectrum protection, vaccines must immunize against this heterogeneous array of ETEC pathogens (Nataro and Kaper, 1998; Steffen et al., 2005). Natural history studies of ETEC infections in children in developing countries, epidemiological evidence in children and adults, and results from experimental challenge studies in volunteers, suggest that ETEC infections are immunizing, as reflected by declining rates of ETEC diarrhea and lower ratios of symptomatic-to-asymptomatic ETEC infections with increasing age, and that multiple infections with antigenically diverse ETEC strains lead to broad-spectrum protection against ETEC diarrhea (Levine and Svennerholm, 2001). It has been suggested that protective immunity appears to be mediated by secretory IgA (SIgA) antibodies directed against fimbriae, other surface antigens, and LT. Interestingly, the ST toxin does not elicit neutralizing antibodies following natural infection (Svennerholm and Savarino, 2004). Overall, cumulative evidence suggests that immunization against ETEC early in life may be an effective preventive strategy. Further, travelers going from industrialized to developing countries constitute another important potential target population for vaccination against ETEC. Therefore, different strategies have been used to deliver ETEC fimbriae and toxin antigens in the form of inactivated or live candidate vaccines to elicit protective human immune responses (Levine and Svennerholm, 2001).

Inactivated and Subunit Vaccines

One of the best-studied ETEC vaccines has been developed by investigators at the University of Goteborg (Sweden), and is an oral, whole-cell vaccine consisting of five strains expressing the CFs: CFA/I, CFA/IV (CS4 and CS6), CFA/IV (CS5 and CS6), CFA/II (CS2 and CS3), and recombinant cholera toxin B subunit (CTB). This composition is designed to induce protection against the most common CFs expressed by ETEC strains. Phase II studies of a two-dose regimen of this vaccine have been conducted in Israel, Nicaragua, the United States, and in immunologically naive populations of Sweden (Jertborn et al., 1998; Svennerholm and Steele, 2004). In these studies, the vaccine was found to be safe and immunogenic, as manifested by induction of mucosal antibody responses to CTB and to the CFA components of the vaccine. These results have been supported by further studies in Egyptian adults and children (Hall et al., 2001) and in Bangladeshi children (Qadri et al., 2003). A pilot efficacy trial of this vaccine in Austrian tourists traveling to developing country destinations found the vaccine to confer about 80% protection against ST–ETEC diarrhea (the only toxin phenotype detected in this study) (Wiedermann et al., 2000). The vaccine was well tolerated in all studies with only mild side effects. However, a preliminary report on the efficacy of this vaccine in Egyptian infants and children indicates that the vaccine was relatively inefficient (conferring less than 20% protection) against ETEC infections in the very young (Boedeker, 2005; Savarino et al., 2002).

One vaccine for which protective efficacy against ETEC has been demonstrated in phase III studies is a whole-cell/recombinant B subunit (WC/rBS) oral cholera vaccine (Dukoral™), which also provides cross protection against ETEC (Girard et al., 2006). The WC/rBS vaccine was designed to stimulate antitoxic as well as antibacterial immunity against cholera and consists of two main vaccine components. The first is the nontoxic CTB, which is known to induce protective antitoxic IgA immunity. The second is heat- and formalin-killed, whole bacterial cells of *Vibrio cholerae*. The WC/rBS vaccine has also been shown to have up to 60% protective efficacy against LT–ETEC-mediated traveler's diarrhea (Clemens et al., 1988; Haberberger et al., 1991; Scerpella et al., 1995). Thus, the protection against ETEC diarrhea after vaccination with the WC/rBS vaccine is mediated specifically through antitoxic immunity rather than by antibodies against the bacterial cells. The phase III efficacy field trial with this vaccine was initiated in 1985 in rural Bangladesh to study the protection afforded by the WC/rBS vaccine against LT–ETEC (Clemens et al., 1988). The protective efficacy against LT-producing ETEC observed in this trial was 67%. However, it was not possible to demonstrate protection for more than 3 months after vaccination, but since the study was done in an ETEC-endemic population, the data were sufficient to prove efficacy against traveler's diarrhea. The efficacy of the WC/rBS vaccine has also been investigated in a study involving tourists who traveled to Morocco from Finland (Peltola et al., 1991). The oral WC/rBS vaccine was found to prevent 23% of all diarrhea episodes and 52% of episodes due to ETEC, particularly LT-producing ETEC, but this protection reportedly did not last longer than a few months. A third efficacy study was conducted in 1992 among US students going to Mexico (Scerpella et al., 1995). In this study, the WC/rBS vaccine was given after arrival in Mexico, with protection expected to be conferred within 7 days of administration of the second dose. This study confirmed the findings of the Finnish study, with a 50% protective efficacy conferred.

Other vaccines currently in development include one using transcutaneous immunization (TCI) with a mixture of fimbrial antigen CS6 and LT administered on a 0-, 1-, and 3-month schedule via skin patches. This trial regimen was found to induce serum antibodies and antibody-secreting cells in a majority of adult volunteers. Delayed-type hypersensitivity, suggesting T-cell responses, was seen in 14 of the 19 volunteers receiving LT and CS6. This study has demonstrated that protein antigens delivered by a simple patch could induce systemic immune responses but only in the presence of an adjuvant such as LT (Guerena-Burgueno et al., 2002). Whether TCI also elicits protective mucosal immunity at the intestinal level still remains to be demonstrated. Lastly, feeding volunteers with transgenic corn or potatoes expressing the LTB resulted in significant serum and gut antibody responses (Tacket, 2004).

Live Vaccines

Evidence indicates that anti-CFA and anti-LT immune responses play important roles in mediating protection against ETEC disease (Svennerholm et al., 1997; Tacket and Levine, 1997). The idea of using live, toxin-free mutants as effective oral vaccines came from previous findings, indicating that a spontaneous mutant, which had lost its ST and LT toxins but continued to express CFs (e.g. CFA/II), provided 75% protection against challenge with an LT^+, ST^+, CFA/II^+ strain (Levine, 1986). Therefore, further attenuation was needed to produce a prototype vaccine for ETEC and different strategies have been attempted thereafter.

Considerable success has been achieved utilizing attenuated ETEC strains as vectors of key protective antigens. Two nontoxigenic ETEC strains (variant of a CFs [CS1 plus CS3]-positive ETEC) were attenuated by mutagenesis of the *aroC* and *ompR* genes or the *aroC*, *ompC*, and *ompF* genes, respectively, to be used as candidate live, oral attenuated vaccines. The mutagenized strains were found to be well tolerated and immunogenic when fed to human volunteers (Turner et al., 2001). Based on these results, the development of a vaccine that protects against a diverse range of ETEC strains was the next step. Therefore, an attenuated CFA/I-expressing ETEC vaccine candidate was constructed by deleting the ST and EAST1 (enteroaggregative *E. coli* heat-stable toxin) genes and introducing defined deletion mutations into the *aroC*, *ompC*, and *ompF* chromosomal genes. The vaccine candidate, administered in two formulations, fresh culture and frozen suspension, was tested in a phase I clinical evaluation of the safety and immunogenicity, and results showed no evidence of significant adverse events related to vaccination (Turner et al., 2006). The vaccine was well tolerated and immunogenic in healthy adult volunteers, and this finding represents a significant milestone in the development of a live attenuated ETEC vaccine.

The live ETEC vaccine approach is also being pursued by investigators at the Center for Vaccine Development at the University of Maryland. Their strategy utilizes live, attenuated, *Shigella*-based multivalent *Shigella*/ETEC hybrid vaccines for expression of ETEC fimbrial and LT antigens (Altboum et al., 2001). In their studies, the genes encoding the production of CFs (CS2 and CS3 fimbriae) were isolated and expressed in an attenuated *Shigella flexneri* 2a strain CVD 1204, which carried a mutation in the *guaBA* genes. Animals immunized with *Shigella* expressing CS2 and/or CS3 developed serum antibodies that agglutinated *Shigella*, as well as ETEC strains bearing the homologous fimbriae, findings that supported the feasibility of a multivalent vaccine against shigellosis and ETEC diarrhea. CVD 1204 has also been used to express individually CFA/I, CS2, CS3, and CS4, as well as a detoxified version of the human ETEC variant of a heat-labile toxin (LThK63) (Barry et al., 2003). Following mucosal immunization of guinea pigs with a mixed formulation of two or more *Shigella* strains expressing the different CS antigens and modified LT, immune responses were observed against each ETEC antigen plus the *Shigella* vector. When each strain expresses multiple ETEC antigens, combinations of attenuated *Shigella* strains allow development of a mixed formulation containing relevant immunogens from both pathogens. Accordingly, new prototype attenuated strains have been constructed, and each contains expression plasmids encoding one or two ETEC antigens (Barry et al., 2006). Therefore, live attenuated strains of *S. flexneri* 2a, *S. sonnei*, and *S. dysenteriae* 1 containing deletions in *guaBA* genes as well as in genes encoding enterotoxins have been constructed and subsequently used as a live vector for the expression of one or two critical ETEC antigens. The resulting *Shigella* derivative strains were able to elicit serum and mucosal antibody responses against the live vector as well as the ETEC antigens, further supporting the use of mixtures of live *Shigella* derivatives expressing ETEC antigens as immunogenic multivalent vaccines.

Current and Future Status of ETEC Vaccine Development

As described above, several attempts have been made to develop a vaccine that will be effective against ETEC infections. To date, only the WC/rBS ETEC vaccine has been evaluated for protective efficacy in a field trial in children in endemic areas. Unfortunately, this vaccine did not induce significant protection in this important target group (Svennerholm and Savarino, 2004). Further studies may potentially provide information that will contribute to the development of a more effective vaccine because to date, no prospective alternative ETEC vaccine is available for licensing within the next 3–5 years.

Vaccines against EHEC

Enterohemorrhagic *Escherichia coli* O157:H7, for which cattle are an important reservoir, is an important pathogen of humans and the main source of illness, either through direct contact (petting zoos) or by consumption of contaminated meat products, or indirectly, through consumption of fruits and vegetables that have been watered with contaminated water. Therefore, scientific efforts have focused on the development of vaccines for people and cattle to prevent or treat food contaminated by this particular *E. coli* serotype. So far, no vaccines have been approved, but some vaccine candidates, which are discussed below, have advanced to clinical trials.

Subunit Vaccines

It has been documented that high serum titers of antibodies to the lipopolysaccharide (LPS) of *E. coli* O157:H7 are detectable after asymptomatic infection (Bell et al., 1997; Chart et al., 1991). The protective role that these antibodies play is suggested in populations having frequent contact with cattle, such as dairy farm workers, who have higher serum titers of antibodies to *E. coli* O157:H7 and a lower risk of symptomatic infection than does the general population (Haack et al., 2003; Reymond et al., 1996). Based on these premises, a research group at the U.S. National Institutes of Health produced a polysaccharide conjugate vaccine. This vaccine consisted of an *E. coli* O157:H7 O-specific polysaccharide covalently linked to recombinant exoprotein A of *Pseudomonas aeruginosa* (O157-rEPA). In a phase I trial, the investigators showed that the conjugate vaccine administered to healthy adults was safe and elicited high serum IgG LPS antibody titers with bactericidal activity (Konadu et al., 1998). The same vaccine has been tested in a phase II clinical trial in 2- to 5-year-old children and the results demonstrated that children who received one or two doses had increased titers of serum IgG LPS antibodies (Ahmed et al., 2006). Further, 1 week after the first dose was administered, most patients (81%) responded with a >fourfold increase in serum IgG LPS antibodies, and, at 6 weeks, all children responded with an >eight-fold increase; a second dose did not elicit a booster response. Serum samples of patients at 26 weeks after the first dose was administered had high titers of bactericidal activity that correlated roughly with serum IgG LPS antibody titers (Ahmed et al., 2006). Because the O157-rEPA vaccine was shown to be safe and immunogenic in young children, a phase III trial concurrent with routine infant immunization is planned.

Other compelling targets for vaccine development are the cytotoxins (Stx1 and Stx2), because expression of these Shiga toxins is universal in antigenically diverse EHEC strains and serotypes. In addition, it has been shown that they play a central role in severe pathogenesis and there is a good correlation between EHEC isolates expressing prototypic Stx2 and a more severe course of illness (Marcato et al., 2005). Following this rationale, an acellular vaccine consisting of the nontoxic B subunit from prototypic Stx2 was constructed and tested in animals. Data indicated that rabbits immunized with a recombinant preparation of the prototypic Stx2 B subunit were protected from a subsequent challenge with a lethal dose of Stx2 holotoxin (Marcato et al., 2001). However, effective vaccination in this study was found to be unpredictable unless LPS was included with the antigen. Since the presence of endotoxin would be unacceptable in a human vaccine, a second study was designed to investigate ways to safely augment the immunogenicity of the recombinant Stx2 B subunit containing <1 endotoxin unit per ml. The study revealed that sera from mice immunized with such a preparation, conjugated to keyhole limpet hemocyanin and administered with the Ribi adjuvant system, displayed the highest Stx2 B-subunit-specific IgG titers, and, when animals were challenged with a lethal dose of Stx2 holotoxin, 100% of the vaccinated mice were protected, supporting further evaluation of a Stx2 B-subunit-based human EHEC vaccine (Marcato et al., 2005).

Another promising approach being pursued in the development of EHEC candidate vaccines is a plant-based oral vaccine to either reduce colonization of *E. coli* O157:H7 or protect against systemic intoxication by Stx2. Transgenic plants offer the flexibility to function as low-cost, efficient, and practical oral delivery systems for vaccine antigens to stimulate mucosal immunity or to boost and shift initial immunity to a mucosal antibody response (Lauterslager et al., 2001; Tacket, 2004). The first EHEC antigen to be tested was the LEE-encoded adhesin intimin. The selection of this adhesin as a candidate vaccine was based on previous studies showing that antibodies against the carboxy-terminal third of intimin blocked adherence of wild-type *E. coli* O157:H7 to HEp-2 cells (Gansheroff et al., 1999), and that colostrum from pigs immunized with *E. coli* O157:H7 intimin contains anti-intimin antibodies that can protect suckling piglets from colonization with *E. coli* O157:H7 (Gansheroff et al., 1999). Since the data suggest that antibodies specific to intimin play an important role in blocking adherence of the bacterium to host cells and can protect the host from *E. coli* O157:H7-mediated disease, the *Nicotiana tabacum* cell line NT-1 was modified by means of *Agrobacterium*-mediated transformation to express the C-terminal binding domain of intimin (Judge et al., 2004). Mice immunized intraperitoneally with intimin expressed from the plant cells, fed transgenic plant cells, or both, generated an intimin-specific mucosal immune response when primed parenterally and then boosted orally and also exhibited a reduced duration of *E. coli* O157:H7 fecal shedding after challenge (Judge et al., 2004). Because the results with the plant-based vaccine against intimin were so encouraging, this group used the same strategy to express a genetically inactivated Stx2 toxoid to stimulate an immune response in the gut against Stx2 and, more importantly, to protect against systemic Stx2 intoxication (Wen et al., 2006). In this study, the immunized mice produced Stx2-specific mucosal IgA and Stx2-neutralizing serum IgG. It was also shown that oral immunization fully protected mice from the challenge, demonstrating that a plant-based oral vaccine can confer protection against lethal systemic intoxication (Wen et al., 2006). The EHEC oral plant-based vaccine shows promise for protection against the severe, sometimes life-threatening, consequences of STEC infection, against which no other suitable treatments or therapies are currently available; however, further work is required to determine whether or not the use of this expression system is the best approach in developing an EHEC vaccine.

Current and Future Status of EHEC Vaccine Development

A number of early studies strongly suggested that decreased shedding of *E. coli* O157:H7, as well as a reduction in the number of cattle that excrete this serotype, could cause a significant reduction in the prevalence of the bacteria in cattle and the farm environment, which could potentially lead to a decline in the incidence of human *E. coli* O157:H7-related disease (Hancock et al., 1998; Jordan et al., 1999; Zhao et al., 1998). Based on these observations, investigators at the University of British Columbia have designed and initiated field studies of *E. coli* O157:H7 secreted products (effectors from the LEE-encoded type III secretion system) as a subcutaneously administered bovine vaccine (Potter et al., 2004). The *E. coli* O157:H7 vaccine was tested in cattle in an initial trial and showed reduction in the duration, frequency, and quantity of *E. coli* O157:H7 shed in cattle following experimental challenge. In addition, the experimental vaccine was shown to significantly reduce shedding of the organism under conditions of natural exposure (Potter et al., 2004). A subsequent field trial was designed to determine the effectiveness of an *E. coli* O157:H7 vaccine in reducing fecal shedding of *E. coli* O157:H7 in pens of feedlot cattle (Van Donkersgoed et al., 2005). This study demonstrated that the *E. coli* O157:H7 vaccine did not significantly reduce the proportion of cattle shedding *E. coli* O157:H7 in their feces. Failure to see any vaccine efficacy was attributed to a poor immunological response to the vaccine either because of alterations of protein conformation, and hence epitopic changes due to formalin fixation, or due to application of a vaccination strategy different from the one used in the previous study (Van Donkersgoed et al., 2005). Despite this drawback and some technical issues, which remain to be resolved, the use of vaccination to decrease food-borne pathogens in reservoir animals seems to hold significant theoretical promise, and has been demonstrated by the purchase of the rights to produce and commercialize the vaccine Bioniche Life Sciences Inc., a research-based, technology-driven Canadian biopharmaceutical company. Recent controlled challenge studies demonstrated that vaccination with Bioniche's *E. coli* O157:H7 Type III secreted protein vaccine (three doses, 3 weeks apart) resulted in a 36.5% reduction in the number of days the bacterium was shed in the feces, and a 2.28 log_{10} reduction in the number of bacteria being shed in the feces; resulting in a vaccine efficacy of more than 99% (Rogan, 2007). Further, a large-scale feedlot study was also recently conducted in 2004 across 19 commercial feedlots to evaluate the efficacy of only two vaccinations, upon entry and again at reimplantation. This study demonstrated that, as compared to unvaccinated cattle on a pen basis, vaccinated cattle had a 75% lower probability of being colonized by *E. coli* O157:H7, indicating that the vaccine reduced environmental exposure of cattle to *E. coli* O157:H7 (Rogan, 2007). Based on these results, the company is scaling up production of the vaccine in anticipation of its commercial launch in Canada because Bioniche Life Sciences has received authorization from the Canadian Food Inspection Agency to make the vaccine available to veterinarians. However, the approval is conditional, and the company has to provide substantial evidence that the vaccine works.

Vaccines against EPEC

No vaccines against EPEC infections have been approved, but some vaccine candidates have been developed. Some of these candidate vaccines are derived from observations indicating that experimental infection of human volunteers with EPEC-produced IgG antibodies against the protein intimin (Levine et al., 1985). It was observed in another study that human volunteers rechallenged with homologous or heterologous EPEC strains developed an IgG response mainly against LPS O antigen and intimin, but not to bundle-forming pili major subunit (BfpA) (Donnenberg et al., 1998). Further, antibody responses against BfpA, intimin, and the type III secretion proteins EspA and EspB have been investigated in Brazilian children naturally infected with EPEC, and the results showed that IgG antibodies against BfpA and EspB were the most commonly found, followed by anti-EspA and intimin antibodies (Martinez et al., 1999). Overall, these data suggest that Bfp and LEE-encoded products are produced in vivo during natural EPEC infections and elicit an immune response against heterologous EPEC virulence determinants, which serve as the basis for the development of candidate vaccines. Therefore, an EPEC recombinant BfpA and EspB candidate vaccine has been generated. Next, the frequencies of naturally occurring IgA antibody activities to the EPEC recombinant vaccine were compared in feces of breast-fed and artificially fed Brazilian infants hospitalized for acute watery diarrhea (de Souza Campos Fernandes et al., 2003). The study found that the frequencies of anti-EPEC IgA copro-antibodies in either breast or artificially fed children <6 months of age were comparable, and would allow a correlation to be established between the presence of anti-EPEC copro-antibodies and protection to clinical infection during infancy (de Souza Campos Fernandes et al., 2003). The high prevalence of anti-BfpA fecal antibodies demonstrates that this virulence factor is highly immunogenic in vivo; however, further studies with human volunteers are required to determine whether the EPEC recombinant vaccine is protective against natural EPEC infections.

Another approach being used in the development of a candidate vaccine is the construction of an attenuated EPEC strain. The *ler* (LEE-encoded regulator) gene product is a central regulator for the genes encoded on the LEE-pathogenicity island of A/E pathogens, including human EPEC and EHEC as well as animal isolates (Torres and Kaper, 2001). Although an in vivo role for Ler in bacterial virulence has not been documented, it has been hypothesized that a *ler* deletion mutant should be attenuated for virulence, and may retain immunogenicity (Zhu et al., 2006). To test this hypothesis, investigators at the University of Maryland constructed a *ler* mutant of a rabbit EPEC (rEPEC) strain (rEPEC constitutes a member of A/E-producing *E. coli* family and isolates of different serotypes have been shown to be causative agents of rabbit enteritis) to examine the effect of *ler* on in vivo virulence, and to determine if intragastric inoculation of an attenuated rEPEC *ler* mutant was immunogenic and could protect rabbits against subsequent challenge with the wild-type strain. The preliminary results demonstrated that the Ler protein of rEPEC is critical for both in vitro pathogenic effects and in vivo virulence, and that immunization with this isogeneic *ler* mutant protected rabbits from fatal challenge with the virulent parent strain (Zhu et al., 2006). The same investigators have now assessed the degree of homologous (serotype-specific) and heterologous (cross-serotype) protection induced by immunization with either rEPEC *ler* mutant strains of different serotypes, or with a prototype strain which expresses a full array of *ler* upregulated proteins (Zhu et al., 2007). The data demonstrated serotype-specific protection induced by the isogenic *ler* mutants. However, cross-serotype protection was not induced by the *ler* mutant, but only by the prototypical rEPEC strain expressing a full set of LEE-encoded proteins. These results are consistent with the hypothesis that cross-serotype protection requires the expression and induction of antibodies to LEE-encoded virulence factors, including intimin and secreted proteins, whereas serotype-specific protection can be induced even when the expression of these virulence factors is markedly downregulated (Zhu et al., 2007). This serotype-specific protection is a drawback in the development of an effective EPEC (and potentially an EHEC) vaccine because although serotype-specific vaccines may be useful for some serotypes (e.g. EHEC O157:H7) that are highly prevalent in certain areas, ideal vaccines should cross protect against all of the major serotypes likely to be encountered.

PROSPECTS FOR THE FUTURE

The genomic plasticity observed in intestinal pathogenic *E. coli* strains resulted in a highly versatile species that is capable of colonizing, and multiplying in and damaging diverse environments. The increased understanding of the mechanisms by which pathogenic *E. coli* can cause diarrhea has dramatically changed the current perspective of this species. However, the ability of various *E. coli* virulence factors to affect such a wide range of cellular functions has produced an unexpected high level of complexity in order to develop an effective vaccine.

KEY ISSUES

- Although significant advances have been made in the field of intestinal pathogenic *E. coli* vaccine research and development, there still remain several outstanding issues and challenges, which need to be addressed.
- The immunogenicity and efficacy of licensed vaccines, such as WC/rBS oral cholera/ETEC vaccine, remain to be elucidated in infants, and they have yet to be fully implemented in areas where the disease is endemic.
- Vaccine candidates against ETEC and other diarrheagenic *E. coli* strains also face serious challenges.
- For those candidate vaccines already in clinical trials, there are initial indications that the immunogenicity might be lower in infants in less-developed countries than in those from industrialized countries, and, therefore, the clinical efficacy of these vaccines needs to be evaluated in infant populations in developing countries.
- The diarrheagenic *E. coli* candidate vaccines also have to cope with the multiplicity of protective antigens that need to be included in the vaccine and with the geographical diversity of circulating serotypes of the bacteria.
- The immune correlates of protection are not clearly understood for any of these intestinal pathogenic *E. coli* vaccines.
- Although many of these challenges remain to be overcome, the involvement of the public health sector and the high priority given to vaccine research and development in this field bode well for the future.
- A combined public health effort coupled with the extensive and active investigations being conducted by the academic and research community should finally bring hope for the control of diarrheal diseases in infants and young children around the globe.

ACKNOWLEDGMENTS

We thank Mardelle Susman and Christopher Allen for critical reading of the manuscript and Jennifer Junemann for assistance with figure designs.

References

Ahmed, A., Li, J., Shiloach, Y., Robbins, J.B. and Szu, S.C. Safety and immunogenicity of *Escherichia coli* O157 O-specific polysaccharide conjugate vaccine in 2–5-year-old children. J. Infect. Dis. 2006; 193:515–521.

Altboum, Z., Barry, E.M., G., L., Galen, J.E. and Levine, M.M. Attenuated *Shigella flexneri* 2a Delta *guaBA* strain CVD 1204 expressing enterotoxigenic *Escherichia coli* (ETEC) CS2 and CS3 fimbriae as a live mucosal vaccine against *Shigella* and ETEC infection. Infect. Immun. 2001; 69:3150–3158.

AMA, ANA-ANF, CDC, CFSAN-FDA, and FSIS-USDA Diagnosis and management of foodborne illnesses: a primer for physicians and other health care professionals. MMWR Recomm. Rep. 2004; 53:1–33.

Barry, E.M., Altboum, Z., Losonsky, G. and Levine, M.M. Immune responses elicited against multiple enterotoxigenic *Escherichia coli* fimbriae and mutant LT expressed in attenuated *Shigella* vaccine strains. Vaccine 2003; 21:333–340.

Barry, E.M., Wang, J., Wu, T., Davis, T. and Levine, M.M. Immunogenicity of multivalent *Shigella*-ETEC candidate vaccine strains in a guinea pig model. Vaccine 2006; 24:3727–3734.

Beatty, M.E., Bopp, C.A., Wells, J.G., Greene, K.D., Puhr, N.D. and Mintz, E.D. Enterotoxin-producing *Escherichia coli* O169:H41, United States. Emerg. Infect. Dis. 2004; 10:518–521.

Bell, B.P., Griffin, P.M., Lozano, P., Christie, D.L., Kobayashi, J.M. and Tarr, P.I. Predictors of hemolytic uremic syndrome in children during a large outbreak of *Escherichia coli* O157:H7 infections. Pediatrics 1997; 100:E12.

Bern, C., Martines, J., de Zoysa, I. and Glass, R.I. The magnitude of the global problem of diarrhoeal disease: a ten-year update. Bull. World Health Organ. 1992; 70:705–714.

Boedeker, E.C. Vaccines for enterotoxigenic *Escherichia coli*: current status. Curr. Opin. Gastroenterol. 2005; 21:15–19.

Bokete, T.N., Whittam, T.S., Wilson, R.A., Clausen, C.R., O'Callahan, C.M., Mosely, S.L., Fritsche, T.R. and Tarr, P. I. Genetic and phenotypic analysis of *Escherichia coli* with enteropathogenic characteristics isolated from Seattle children. J. Infect. Dis. 1997; 175:1382–1389.

CDC. Travelers' diarrhea. Frequently asked questions. 2005.

CDC Preliminary FoodNet data on the incidence of infection with pathogens transmitted commonly through food—10 states, 2006. MMWR Morb. Mortal. Wkly. Rep. 2007; 56:336–339.

Chart, H., Smith, H.R., Scotland, S.M., Rowe, B., Milford, D.V. and Taylor, C.M. Serological identification of *Escherichia coli* O157:H7 infection in haemolytic uraemic syndrome. Lancet 1991; 337: 138–140.

Clemens, J.D., Sack, D.A., Harris, J.R., Chakraborty, J., Neogy, P.K., Stanton, B., Huda, N., Khan, M.U., Kay, B.A., Khan, M.R. et al. Cross-protection by B subunit-whole cell cholera vaccine against diarrhea associated with heat-labile toxin-producing enterotoxigenic *Escherichia coli* results of a large-scale field trial. J. Infect. Dis. 1988; 158:372–377.

Cravioto, A., Tello, A., Navarro, A., Ruiz, J., Villafan, H., Uribe, F. and Eslava, C. Association of *Escherichia coli* HEp-2 adherence patterns with type and duration of diarrhoea. Lancet 1991; 337:262–264.

de Souza Campos Fernandes, R.C., Quintana Flores, V.M., Sousa de Macedo, Z. and Medina-Acosta, E. Coproantibodies to the enteropathogenic *Escherichia coli* vaccine candidates BfpA and EspB in breastfed and artificially fed children. Vaccine 2003; 21:1725–1731.

Donnenberg, M.S. Enteropathogenic *Escherichia coli*. In: Infections of the Gastrointestinal Tract (M.J. Blaser, P.D. Smith, J.I. Ravdin, H.B. Greenberg and R.L. Guerrant, Eds.), pp. 709–726. New York, NY: Raven Press, 1995.

Donnenberg, M.S., Tacket, C.O., Losonsky, G., Frankel, G., Nataro, J.P., Dougan, G. and Levine, M.M. Effect of prior experimental human enteropathogenic *Escherichia coli* infection on illness following homologous and heterologous rechallenge. Infect. Immun. 1998; 66:52–58.

DuPont, H.L. Guidelines on acute infectious diarrhea in adults. The Practice Parameters Committee of the American College of Gastroenterology. Am. J. Gastroenterol. 1997; 92:1962–1975.

DuPont, H.L. What's new in enteric infectious diseases at home and abroad. Curr. Opin. Infect. Dis. 2005; 18:407–412.

Elsinghorst, E.A. and Kopecko, D.J. Molecular cloning of epithelial cell invasion determinants from enterotoxigenic *Escherichia coli*. Infect. Immun. 1992; 60:2409–2417.

Elsinghorst, E.A. and Weitz, J.A. Epithelial cell invasion and adherence directed by the enterotoxigenic *Escherichia coli tib* locus is associated with a 104-kilodalton outer membrane protein. Infect. Immun. 1994; 62:3463–3471.

Eriksson, K. and Holmgren, J. Recent advances in mucosal vaccines and adjuvants. Curr. Opin. Immunol. 2002; 14:666–672.

Escobar-Paramo, P., Clermont, O., Blanc-Potard, A.B., Bui, H., Le Bouguenec, C. and Denamur, E. A specific genetic background is required for acquisition and expression of virulence factors in *Escherichia coli*. Mol. Biol. Evol. 2004; 21:1085–1094.

Forestier, C., Meyer, M., Favre-Bonte, S., Rich, C., Malpuech, G., Le Bouguenec, C., Sirot, J., Joly, B. and De Champs, C. Enteroadherent *Escherichia coli* and diarrhea in children: a prospective case-control study. J. Clin. Microbiol. 1996; 34:2897–2903.

Frenzen, P.D., Drake, A. and Angulo, F.J. Group E.I.P.F.W. Economic cost of illness due to *Escherichia coli* O157 infections in the United States. J. Food Prot. 2005; 68:2623–2630.

Gaastra, W. and Svennerholm, A.M. Colonization factors of human enterotoxigenic *Escherichia coli* (ETEC). TRENDS Microbiol. 1996; 4:444–452.

Gansheroff, L.J., Wachtel, M.R. and O'Brien, A.D. Decreased adherence of enterohemorrhagic *Escherichia coli* to HEp-2 cells in the presence of antibodies that recognize the C-terminal region of intimin. Infect. Immun. 1999; 67:6409–6417.

Girard, M.P., Steele, D., Chaignat, C.L. and Kieny, M.P. A review of vaccine research and development: human enteric infections. Vaccine 2006; 24:2732–2750.

Gomes, T.A.T., Rassi, V., MacDonald, K.L., Ramos, S.R., Trabulsi, L.R., Vieira, M.A., Guth, B.E., Candeias, J.A., Ivey, C., Toledo, M.R. et al. Enteropathogens associated with acute diarrheal disease in urban infants in Sao Paulo, Brazil. J. Infect. Dis. 1991; 164:331–337.

Greenberg, D.E., Jiang, Z., Steffen, R., Verenker, M.P. and DuPont, H.L. Markers of inflammation in bacterial diarrhea among travelers, with a focus on enteroaggregative *Escherichia coli* pathogenicity. J. Infect. Dis. 2002; 185:944–949.

Griffin, P.M. and Tauxe, R.V. The epidemiology of infections caused by *Escherichia coli* O157:H7, other enterohemorrhagic *E. coli*, and the associated hemolytic uremic syndrome. Epidemiol. Rev. 1991; 13:60–98.

Guerena-Burgueno, F., Hall, E.R., Taylor, D.N., Cassels, F.J., Scott, D.A., Wolf, M.K., Roberts, Z.J., Nesterova, G.V., Alving, C.R. and Glenn, G.M. Safety and immunogenicity of a prototype enterotoxigenic *Escherichia coli* vaccine administered transcutaneously. Infect. Immun. 2002; 70:1874–1880.

Haack, J.P., Jelacic, S., Besser, T.E., Weinberger, E., Kirk, D.J., McKee, G.L., Harrison, S.M., Musgrave, K.J., Miller, G., Price, T.H. and Tarr, P.I. *Escherichia coli* O157 exposure in Wyoming and Seattle: serologic evidence of rural risk. Emerg. Infect. Dis. 2003; 9:1226–1231.

Haberberger, R.L., Mikhail, I.A., Burans, J.P., Hyams, K.C., Glenn, J.C., Diniega, B.M., Sorgen, S., Mansour, N., Blacklow, N.R. and Woody, J.N. Travelers' diarrhea among United States military personnel during joint American–Egyptian armed forces exercises in Cairo, Egypt. Mil. Med. 1991; 156:27–30.

Hall, E.R., Wierzba, T.F., Ahren, C., Rao, M.R., Bassily, S., Francis, W., Girgis, F.Y., Safwat, M., Lee, Y.J., Svennerholm, A.M., Clemens, J.D. and Savarino, S.J. Induction of systemic antifimbria and antitoxin antibody responses in Egyptian children and adults by an oral, killed enterotoxigenic *Escherichia coli* plus cholera toxin B subunit vaccine. Infect. Immun. 2001; 69:2853–2857.

Hancock, D.D., Besser, T.E., Rice, D.H. and Tarr, P.I. Ecology of *Escherichia coli* O157:H7 in cattle and impact of management practices. In: *Escherichia coli* O157:H7 and other Shiga Toxin-Producing *E. coli* Strains (J.B. Kaper and A.D. O'Brien, Eds.), pp. 85–91. Washington, DC: ASM Press, 1998.

Harrington, S.M., Dudley, E.G. and Nataro, J.P. Pathogenesis of enteroaggregative *Escherichia coli* infection. FEMS Microbiol. Lett. 2006; 254:12–18.

Henderson, I.R., Czeczulin, J., Eslava, C., Noriega, F. and Nataro, J.P. Characterization of pic, a secreted protease of *Shigella flexneri* and enteroaggregative *Escherichia coli*. Infect. Immun. 1999; 67:5587–5596.

Jallat, C., Darfeuille-Michaud, A., Rich, C. and Joly, B. Survey of clinical isolates of diarrhoeogenic *Escherichia coli*: diffusely adhering *E. coli* strains with multiple adhesive factors. Res. Microbiol. 1994; 145:621–632.

Jallat, C., Livrelli, V., Darfeuille-Michaud, A., Rich, C. and Joly, B. *Escherichia coli* strains involved in diarrhea in France: high prevalence and heterogeneity of diffusely adhering strains. J. Clin. Microbiol. 1993; 31:2031–2037.

Jertborn, M., Ahren, C., Holmgren, J. and Svennerholm, A.M. Safety and immunogenicity of an oral inactivated enterotoxigenic *Escherichia coli* vaccine. Vaccine 1998; 16:225–260.

Jordan, D., McEwen, S.A., Lammerding, A.M., McNab, W.B. and Wilson, J.B. Pre-slaughter control of *Escherichia coli* O157 in beef cattle: a simulation study. Prev. Vet. Med. 1999; 41:55–74.

Judge, N.A., Mason, H.S. and O'Brien, A.D. Plant cell-based intimin vaccine given orally to mice primed with intimin reduces time of *Escherichia coli* O157:H7 shedding in feces. Infect. Immun. 2004; 72:168–175.

Kahali, S., Sarkar, B., Rajendran, K., Khanam, J., Yamasaki, S., Nandy, R.K., Bhattacharya, S.K. and Ramamurthy, T. Virulence characteristics and molecular epidemiology of enteroaggregative *Escherichia coli* isolates from hospitalized diarrheal patients in Kolkata, India. J. Clin. Microbiol. 2004; 42:4111–4120.

Kaper, J.B., Nataro, J.P. and Mobley, H.L. Pathogenic *Escherichia coli*. Nat. Rev. Microbiol. 2004; 2:123–140.

Karmali, M.A. Infection by verocytotoxin-producing *Escherichia coli*. Clin. Microbiol. Rev. 1989; 2:15–38.

Konadu, E.Y., Parke, J.C.J., Tran, H.T., Bryla, D.A., Robbins, J.B. and Szu, S.C. Investigational vaccine for *Escherichia coli* O157: phase 1 study of O157 O-specific polysaccharide–*Pseudomonas aeruginosa* recombinant exoprotein A conjugates in adults. J. Infect. Dis. 1998; 177:383–387.

Lauterslager, T.G., Florack, D.E., van der Wal, T.J., Molthoff, J.W., Langeveld, J.P., Bosch, D., Boersma, W.J. and Hilgers, L.A. Oral immunisation of naive and primed animals with transgenic potato tubers expressing LT-B. Vaccine 2001; 19:2749–2755.

Le Bouguenec, C. Diarrhea-associated diffusely adherent *Escherichia coli*. Clin. Microbiol. Rev. 1999; 12:180–181.

Le Bouguenec, C. and Servin, A.L. Diffusely adherent *Escherichia coli* strains expressing Afa/Dr adhesins (Afa/Dr DAEC): hitherto unrecognized pathogens. FEMS Microbiol. Lett. 2006; 256:185–194.

Levine, M.M. Fimbriae (pili) as vaccines. In: Protein–Carbohydrate Interactions in Biological Systems: The Molecular Biology of Microbial Pathogenicity (D.L. Lark, Ed.), pp. 143–145. London, United Kingdom: Academic Press, Ltd., 1986.

Levine, M.M. and Edelman, R. Enteropathogenic *Escherichia coli* of classic serotypes associated with infant diarrhea: epidemiology and pathogenesis. Epidemiol. Rev. 1984; 6:31–51.

Levine, M.M., Nataro, J.P., Karch, H., Baldini, M.M., Kaper, J.B., Black, R.E., Clements, M.L. and O'Brien, A.D. The diarrheal response of humans to some serotypes of enteropathogenic *Escherichia coli* is dependent on a plasmid encoding an enteroadhesiveness factor. J. Infect. Dis. 1985; 152:550–559.

Levine, M.M. and Svennerholm, A.-M. Enteric vaccines: present and future. In: Textbook of Travel Medicine and Health (H.L. DuPont and H. Steffen, Eds.), pp. 252–263. New York: BC Decker, 2001.

Lopes, L.M., Fabbricotti, S.H., Ferreira, A.J., Kato, M.A., Michalski, J. and Scaletsky, I.C. Heterogeneity among strains of diffusely adherent *Escherichia coli* isolated in Brazil. J. Clin. Microbiol. 2005; 43:1968–1972.

Marcato, P., Griener, T.P., Mulvey, G.L. and Armstrong, G.D. Recombinant Shiga toxin B-subunit-keyhole limpet hemocyanin conjugate vaccine protects mice from Shigatoxemia. Infect. Immun. 2005; 73:6523–6529.

Marcato, P., Mulvey, G., Read, R.J., Vander Helm, K., Nation, P.N. and Armstrong, G.D. Immunoprophylactic potential of cloned Shiga toxin 2 B subunit. J. Infect. Dis. 2001; 183:435–443.

Martinez, M.B., Taddei, C.R., Ruiz-Tagle, A., Trabulsi, L.R. and Giron, J.A. Antibody response of children with enteropathogenic *Escherichia coli* infection to the bundle-forming pilus and locus of enterocyte effacement-encoded virulence determinants. J. Infect. Dis. 1999; 179:269–274.

Mellies, J.L., Elliott, S.J., Sperandio, V., Donnenberg, M.S. and Kaper, J.B. The Per regulon of enteropathogenic *Escherichia coli*: identification of a regulatory cascade and a novel transcriptional activator, the locus of enterocyte effacement (LEE)-encoded regulator (Ler). Mol. Microbiol. 1999; 33:296–306.

Nataro, J.P. and Kaper, J.B. Diarrheagenic *Escherichia coli*. Clin. Microbiol. Rev. 1998; 11:142–201.

Nataro, J.P., Steiner, T. and Guerrant, R.L. Enteroaggregative *Escherichia coli*. Emerg. Infect. Dis. 1998; 4:251–261.

Nataro, J.P., Yikang, D., Cookson, S., Cravioto, A., Savarino, S.J., Guers, L.D., Levine, M.M. and Tacket, C.O. Heterogeneity of enteroaggregative *Escherichia coli* virulence demonstrated in volunteers. J. Infect. Dis. 1995; 171:465–468.

Nguyen, R.N., Taylor, L.S., Tauschek, M. and Robins-Browne, R.M. Atypical enteropathogenic *Escherichia coli* infection and prolonged diarrhea in children. Emerg. Infect. Dis. 2006; 12:597–603.

Okeke, I.N. and Nataro, J.P. Enteroaggregative *Escherichia coli*. Lancet Infect. Dis. 2001; 1:304–313.

Parsot, C. *Shigella* spp. and enteroinvasive *Escherichia coli* pathogenicity factors. FEMS Microbiol. Lett. 2005; 252:11–18.

Peltola, H., Siitonen, A., Kyronseppa, H., Simula, I., Mattila, L., Oksanen, P., Kataja, M.J. and Cadoz, M. Prevention of travelers' diarrhea by oral B-subunit/whole-cell cholera vaccine. Lancet 1991; 338:1285–1289.

Poitrineau, P., Forestier, C., Meyer, M., Jallat, C., Rich, C., Malpuech, G. and De Champs, C. Retrospective case-control study of diffusely adhering *Escherichia coli* and clinical features in children with diarrhea. J. Clin. Microbiol. 1995; 33:1961–1962.

Potter, A.A., Klashinsky, S., Li, Y., Frey, E., Townsend, H., Rogan, D., Erickson, G., Hinkley, S., Klopfenstein, T., Moxley, R.A., Smith, D.R. and Finlay, B.B. Decreased shedding of *Escherichia coli* O157:H7 by cattle following vaccination with type III secreted proteins. Vaccine 2004; 22:362–369.

Prats, G. and Llovet, T. Enteroinvasive *Escherichia coli*. Pathogenesis and epidemiology. Microbiologia 1995; 11:91–96.

Qadri, F., Svennerholm, A.M., Faruque, A.S. and Sack, R.B. Enterotoxigenic *Escherichia coli* in developing countries: epidemiology, microbiology, clinical features, treatment, and prevention. Clin. Microbiol. Rev. 2005; 18:465–483.

Qadri, F.A., Ahmed, T., Ahmed, F., Bradley Sack, R., Sack, D.A. and Svennerholm, A.M. Safety and immunogenicity of an oral, inactivated enterotoxigenic *Escherichia coli* plus cholera toxin B subunit vaccine in Bangladeshi children 18–36 months of age. Vaccine 2003; 21:2394–2403.

Rangel, J.M., Sparling, P.H., Crowe, C., Griffin, P.M. and Swerdlow, D.L. Epidemiology of *Escherichia coli* O157:H7 outbreaks, United States, 1982–2002. Emerg. Infect. Dis. 2005; 11:603–609.

Reymond, D., Johnson, R.D., Karmali, M.A., Petric, M., Winkler, M., Johnson, S., Rahn, K., Renwick, S., Wilson, J., Clarke, R.C. and Spika, J. Neutralizing antibodies to *Escherichia coli* Vero cytotoxin 1 and antibodies to O157 lipopolysaccharide in healthy farm family members and urban residents. J. Clin. Microbiol. 1996; 34:2053–2057.

Riley, L.W., Remis, R.S., Helgerson, S.D., McGee, H.B., Wells, J.G., Davis, B.R., Hebert, R.J., Olcott, E.S., Johnson, L.M., Hargrett, N.T., Blake, P.A. and Cohen, M.L. Hemorrhagic colitis associated with a rare *Escherichia coli* serotype. N. Engl. J. Med. 1983; 308:681–685.

Robinson, C.M., Sinclair, J.F., Smith, M.J. and O'Brien, A.D. Shiga toxin of enterohemorrhagic *Escherichia coli* type O157:H7 promotes intestinal colonization. Proc. Natl. Acad. Sci. USA 2006; 103:9667–9672.

Rogan, D. Vaccination reduces shedding and colonization of *E. coli* 0157:H7 in cattle. In 4th International Conference on Vaccines for Enteric Diseases Lisbon, Portugal., 2007.

Rothbaum, R., McAdams, A.J., Giannella, R. and Partin, J.C. A clinicopathological study of enterocyte-adherent *Escherichia coli*: a cause of protracted diarrhea in infants. Gastroenterology 1982; 83:441–454.

Savarino, S.J., Hall, E.R., Bassily, S., Wierzba, T.F., Youssef, F.G., Peruski, L.F.J., Abu-Elyazeed, R., Rao, M., Francis, W.M., El Mohamady, H., Safwat, M., Naficy, A.B., Svennerholm, A.M., Jertborn, M., Lee, Y.J., Clemens, J.D. and Group, P.S. Introductory evaluation of an oral, killed whole cell enterotoxigenic *Escherichia coli* plus cholera toxin B subunit vaccine in Egyptian infants. Pediatr. Infect. Dis. J. 2002; 21:322–330.

Scaletsky, I.C., Fabbricotti, S.H., Carvalho, R.L., Nunes, C.R., Maranhao, H.S., Morais, M.B. and Fagundes-Neto, U. Diffusely adherent *Escherichia coli* as a cause of acute diarrhea in young children in Northeast Brazil: a case-control study. J. Clin. Microbiol. 2002; 40:645–648.

Scaletsky, I.C., Silva, M.L. and Trabulsi, L.R. Distinctive patterns of adherence of enteropathogenic *Escherichia coli*. Infect. Immun. 1984; 45:534–536.

Scerpella, E.G., Sanchez, J.L., Mathewson III, J.J., Torres-Cordero, J. V., Sadoff, J.C., Svennerholm, A.M., DuPont, H.L., Taylor, D.N. and Ericsson, C.D. Safety, immunogenicity, and protective efficacy of the wholecell/recombinant B subunit (WC/rBS) oral cholera vaccine against travelers' diarrhea. J. Travel Med. 1995; 2:22–27.

Scotland, S.M., Smith, H.R., Cheasty, T., Said, B., Willshaw, G.A., Stokes, N. and Rowe, B. Use of gene probes and adhesion tests to characterize *Escherichia coli* belonging to enteropathogenic serogroups isolated in the United Kingdom. J. Med. Microbiol. 1996; 44:438–443.

Sears, C.L. and Kaper, J.B. Enteric bacterial toxins: mechanisms of action and linkage to intestinal secretion. Microbiol. Rev. 1996; 60:167–215.

Senerwa, D., Olsvik, O., Mutanda, L.N., Lindqvist, K.J., Gathuma, J.M., Fossum, K. and Wachsmuth, K. Enteropathogenic *Escherichia coli* serotype O111:HNT isolated from preterm neonates in Nairobi, Kenya. J. Clin. Microbiol. 1989; 27:1307–1311.

Steffen, R., Castelli, F., Dieter Nothdurft, H., Rombo, L. and Jane Zuckerman, N. Vaccination against enterotoxigenic *Escherichia coli*, a cause of travelers' diarrhea. J. Travel Med. 2005; 12:102–107.

Steiner, T.S., Lima, A.A.M., Nataro, J.P. and Guerrant, R.L. Enteroaggregative *Escherichia coli* produce intestinal inflammation and growth impairment and cause interleukin-8 release from intestinal epithelial cells. J. Infect. Dis. 1998; 177:88–96.

Steiner, T.S., Nataro, J.P., Poteet-Smith, C.E., Smith, J.A. and Guerrant, R.L. Enteroaggregative *Escherichia coli* expresses a novel flagellin that causes IL-8 release from intestinal epithelial cells. J. Clin. Invest. 2000; 105:1769–1777.

Svennerholm, A.M., Ahren, C. and Jertborn, M. Vaccines against enterotoxigenic *Escherichia coli* infections. I. Oral inactivated vaccines against enterotoxigenic *Escherichia coli*. In: New Generation Vaccines (M.M. Levine, G.C. Woodrow, J.B. Kaper and G.S. Cobon, Eds.), pp. 865–873. New York, NY: Marcel Dekker, Inc., 1997.

Svennerholm, A.-M. and Savarino, S.J. Oral inactivated whole cell B subunit combination vaccine against enterotoxigenic *Escherichia coli*. In: New Generation Vaccines (M.M. Levine, J.B. Kaper, R. Rappuoli, M.A. Liu, M.F. Good Eds.). New York, NY: Marcel Dekker, 2004.

Svennerholm, A.M. and Steele, D. Microbial–gut interactions in health and disease. Progress in enteric vaccine development. Best Pract. Res. Clin. Gastroenterol. 2004; 18:421–445.

Tacket, C.O. Plant-derived vaccines against diarrhoeal diseases. Exp. Opin. Biol. Ther. 2004; 4:719–728.

Tacket, C.O. and Levine, M.M. Vaccines against enterotoxigenic *Escherichia coli* infections. II. Live oral vaccines and subunit (purified fimbriae and toxin subunit vaccines). In: New Generation Vaccines (M.M. Levine, G.C. Woodrow, J.B. Kaper and G.S. Cobon, Eds.), pp. 875–883. New York, NY: Marcel Dekker, Inc., 1997.

Tesh, V.L. and O'Brien, A.D. The pathogenic mechanisms of Shiga toxin and the Shiga-like toxins. Mol. Microbiol. 1991; 5:1817–1822.

Toledo, M.R.F., Alvariza, M.C.B., Murahovschi, J., Ramos, S.R.T.S. and Trabulsi, L.R. Enteropathogenic *Escherichia coli* serotypes and endemic diarrhea in infants. Infect. Immun. 1983; 39:586–589.

Torres, A.G. and Kaper, J.B. PAIs of intestinal *E. coli*. In: Pathogenicity Islands (PAIs) and the Evolution of Pathogenic Microbes (J. Hacker and J.B. Kaper, Eds.), Vol. 1, pp. 31–48. Berlin: Springer-Verlag, 2001.

Torres, A.G., Zhou, X. and Kaper, J.B. Adherence of diarrheagenic *Escherichia coli* strains to epithelial cells. Infect. Immun. 2005; 73:18–29.

Trabulsi, L.R., Keller, R. and Tardelli Gomes, T.A. Typical and atypical enteropathogenic *Escherichia coli*. Emerg. Infect. Dis. 2002; 8:508–513.

Turner, A.K., Beavis, J.C., Stephens, J.C., Greenwood, J., Gewert, C., Thomas, N., Deary, A., Casula, G., Daley, A., Kelly, P., Randall, R. and Darsley, M.J. Construction and phase I clinical evaluation of the safety and immunogenicity of a candidate enterotoxigenic *Escherichia coli* vaccine strain expressing colonization factor antigen CFA/I. Infect. Immun. 2006; 74:1062–1071.

Turner, A.K., Terry, T.D., Sack, D.A., Londono-Arcila, P. and Darsley, M.J. Construction and characterization of genetically defined *aro omp* mutants of enterotoxigenic *Escherichia coli* and preliminary studies of safety and immunogenicity in humans. Infect. Immun. 2001; 69:4969–4979.

Van Donkersgoed, J., Hancock, D., Rogan, D. and Potter, A.A. *Escherichia coli* O157:H7 vaccine field trial in 9 feedlots in Alberta and Saskatchewan. Can. Vet. J. 2005; 46:724–728.

Wanke, C.A., Gerrior, J., Blais, V., Mayer, H. and Acheson, D. Successful treatment of diarrheal disease associated with enteroaggregative *Escherichia coli* in adults infected with human immunodeficiency virus. J. Infect. Dis. 1998; 178:1369–1372.

Wen, S.X., Teel, L.D., Judge, N.A. and O'Brien, A.D. A plant-based oral vaccine to protect against systemic intoxication by Shiga toxin type 2. Proc. Natl. Acad. Sci. USA 2006; 103:7082–7087.

Wiedermann, G., Kollaritsch, H., Kundi, M., Svennerholm, A.M. and Bjare, U. Double-blind, randomized, placebo controlled pilot study evaluating efficacy and reactogenicity of an oral ETEC B-subunit-inactivated whole cell vaccine against travelers' diarrhea (preliminary report). J. Travel. Med. 2000; 7:27–29.

Wilson, A., Evans, J., Chart, H., Cheasty, T., Wheeler, J.G., Tompkins, D. and Smith, H.R. Characterization of strains of enteroaggregative *Escherichia coli* isolated during the infectious intestinal disease study in England. Eur. J. Epidemiol. 2001; 17:1125–1130.

Wong, C.A., Jelacic, S., Habeeb, R.L., Watkins, S.L. and Tarr, P.I. The risk of the hemolytic–uremic syndrome after antibiotic treatment of *Escherichia coli* O157:H7 infections. N. Engl. J. Med. 2000; 342:1930–1936.

Zhao, T., Doyle, M.P., Harmon, B.G., Brown, C.A., Mueller, P.O. and Parks, A.H. Reduction of carriage of enterohemorrhagic *Escherichia coli* O157:H7 in cattle by inoculation with probiotic bacteria. J. Clin. Microbiol. 1998; 36:641–647.

Zhu, C., Feng, S., Thate, T.E., Kaper, J.B. and Boedeker, E.C. Towards a vaccine for attaching/effacing *Escherichia coli*: a LEE encoded regulator (ler) mutant of rabbit enteropathogenic *Escherichia coli* is attenuated, immunogenic, and protects rabbits from lethal challenge with the wild-type virulent strain. Vaccine 2006; 24:3845–3855.

Zhu, C., Feng, S., Yang, Z., Davis, K., Rios, H., Kaper, J.B. and Boedeker, E.C. LEE-encoded regulator (Ler) mutants elicit serotype-specific protection, but not cross protection, against attaching and effacing *E. coli* strains. Vaccine 2007; 25:1884–1892.

CHAPTER

52

Lyme Disease

Christopher G. Earnhart[1] *and Richard T. Marconi*[1,2]

[1]Department of Microbiology and Immunology, Medical College of Virginia at Virginia Commonwealth University, Richmond, VA, USA

[2]Center for the Study of Biological Complexity, Medical College of Virginia at Virginia Commonwealth University, Richmond, VA, USA

OUTLINE

ABSTRACT

Lyme disease is the most prevalent vector-borne disease in North America and Europe. It is an important emerging disease, with increasing incidence and expanding endemic regions. Lyme disease was first described as a clinical entity in North America in the late 1970s and its spirochetal etiology identified in 1982. Clinical manifestations consistent with Lyme disease were described in Europe in the early 1900's. The infection is caused by a subset of species of the *Borrelia burgdorferi* sensu lato complex. In nature, the *Borrelia* are obligate host-associated spirochetes that are maintained in an enzootic cycle involving *Ixodes* ticks, and small mammals, birds, and lizards. Humans are incidental hosts, and are of no significance in the enzootic cycle. Clinical manifestations of early-stage disease in humans, with the exception of the pathognomonic erythema migrans rash, are generally nondescript. However, long-term infection can lead to serious dermatological, arthritic, cardiac, and neurological sequelae that can be debilitating. The potentially serious and costly nature of Lyme disease dictates the need for improved preventive strategies. There is currently no vaccine that is commercially available for use in humans. This chapter reviews the history, current status, and future directions of Lyme disease vaccine development.

INTRODUCTION

Lyme disease is an emerging zoonotic infection caused by spirochete species of the *Borrelia burgdorferi* sensu lato complex (Burgdorfer et al., 1982; Benach et al., 1983). This complex consists of 10 species, with *B. garinii*, *B. afzelii*, and *B. burgdorferi* being most frequently associated with disease in humans (Barbour and Hayes, 1986; Baranton et al., 1992; Marconi and Garon, 1992b; Welsh et al., 1992; Fukunaga et al., 1993, 1996; Kawabata et al., 1993; Marconi et al., 1995; Postic et al., 1998). While all three species occur in Europe and Asia, only *B. burgdorferi* has been demonstrated in North America. Additional species that may be associated to a lesser extent with human disease include *B. valaisiana* (Rijpkema et al., 1997; Diza et al., 2004) and *B. lusitaniae* (Collares-Pereira et al., 2004) which have been identified in Europe, and *B. lonestari* in the southern United States. *B. lonestari* is thought to be the causative agent of a Lyme-like disease referred to as southern tick-associated rash illness (STARI) or Masters disease (Masters et al., 1998; Varela et al., 2004). The true incidence of STARI remains to be determined and has been the subject of considerable debate.

The Lyme disease spirochetes are maintained in an enzootic cycle involving *Ixodes* ticks and a wide variety of animals and reptiles (Brown and Lane, 1992; Keirans et al., 1996; Gern et al., 1998). *Ixodes scapularis* (formerly *I. dammini* (Oliver et al., 1993)) is the primary arthropod vector in the eastern and midwestern regions of North America, and *I. pacificus* in the western United States (Lane et al., 1991). In Europe, *I. ricinus* is the dominant vector, while *I. persulcatus* and *I. nipponensis* are important in Asia (Hengge et al., 2003). *Borrelia lonestari* is vectored by the lone star tick, *Amblyomma americanum*, and is the only putative Lyme disease spirochete that is carried by a tick species that is not part of the *Ixodes* complex.

The Lyme disease spirochetes are strictly dependent on animal reservoirs for population maintenance, as transovarial transmission of the spirochetes in ticks does not occur at a biologically significant level (Barbour and Hayes, 1986). Lyme spirochetes are maintained in a 2-year enzootic cycle (Fig. 52.1). In spring, uninfected larvae hatch from eggs and take a blood meal from a small mammal or bird in late summer. If the larvae feed to repletion on an infected mammal they have the potential to become infected. After completion of the blood meal, the larvae molt to the nymph stage and become dormant as they overwinter. When the infected nymphs emerge during the late spring/early summer of the following year they feed on, and potentially infect, a wide range of small animals that are competent to serve as reservoirs for the Lyme disease spirochetes (Keirans et al., 1996; Gern et al., 1998; Steere et al., 2004). In North America, the white-footed mouse, *Peromyscus leucopus*, is a major, though not the only, mammalian reservoir. After nymphs complete the blood meal, they molt to become sexually differentiated adults, and feed on larger mammals. Deer and other large mammals are important in the enzootic cycle primarily due to the

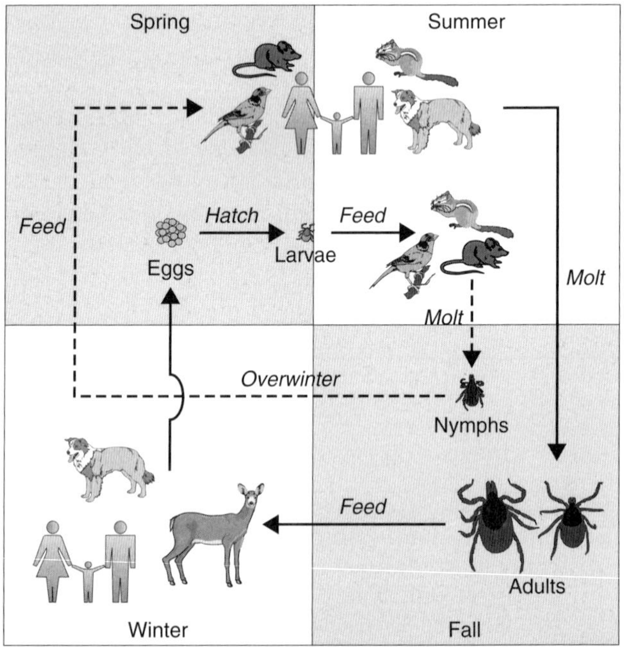

FIGURE 52.1 Life cycle of the *Ixodes* vector ticks and enzootic transmission of the Lyme disease spirochetes. The schematic depicts the general life cycle of *Ixodes* ticks and the enzootic cycle of the Lyme disease spirochetes. The dashed arrows demonstrate the portion of the life cycle during which transmission from infected to uninfected mammals may occur.

fact that they can support large numbers of adult ticks and can serve to distribute ticks over a wider geographic area. Humans are incidental hosts and are not significant in the enzootic cycle.

HISTORY OF LYME DISEASE

Lyme disease, originally termed Lyme arthritis, was first recognized as a clinical entity in North America in 1977 (Steere et al., 1977a, 1977b, 1978). The occurrence of an unusually large number of cases of arthritis in children in the three contiguous communities of Lyme, Old Lyme, and East Haddam, Connecticut prompted epidemiological studies that culminated in 1982 with the identification of a previously unrecognized tick-borne spirochete as the causative agent (Burgdorfer et al., 1982; Benach et al., 1983; Steere et al., 1983). This agent was determined to belong to the genus *Borrelia* and was subsequently designated *B. burgdorferi* (Johnson et al., 1984). Retrospective PCR analyses of preserved ticks and tissue samples collected from *Peromyscus* mice demonstrated the presence of *B. burgdorferi* in North America well before it was recognized as a distinct clinical entity. *B. burgdorferi* was demonstrated in some samples dating back to 1894 (Persing et al., 1990; Marshall et al., 1994). Early 20th century reports of an arthritic condition common on Long Island, New York, and locally referred to as "Montauk knee," appear to represent the first clinical descriptions of Lyme disease in North America. In Europe, clinical manifestations consistent with Lyme disease had been separately described in the early 20th century, though the individual symptoms were not recognized under the umbrella of a single clinical entity (Herxheimer and Hartmann, 1902; Afzelius, 1921; Bannwarth, 1944; Hellerstrom, 1951; Hollstrom, 1951). The description of Lyme disease and its spirochetal etiology in America was followed by its recognition as a cause of disease in Europe (Ackermann, 1983; Ackermann et al., 1984).

ETIOLOGIC AGENT

The spirochetes share a common morphology and ultrastructure; however, at the genetic level they comprise a highly diverse monophyletic phylum, the *Spirochaetes* (Paster et al., 1991). The order *Spirochaetales* is divided into three families, the *Leptospiraceae*, *Brachyspiraceae* (alternatively *Serpulinaceae*), and *Spirochaetaceae*. The family *Spirochetaceae* contains six major genera: *Treponema*, *Brevinema*, *Spirochaeta*, *Cristispira*, *Spironema*, and *Borrelia* (Paster et al., 1991; Paster and Dewhirst, 2001). *B. burgdorferi* was the first *Borrelia* species to be demonstrated as a causative agent of Lyme disease. Subsequent molecular phylogenetic analyses have demonstrated the existence of 10 closely related species that form the *B. burgdorferi* sensu lato complex (Schmid et al., 1984; Marconi and Garon, 1992a, 1992b; Marconi et al., 1992; Wang et al., 1999b). These species include *B. burgdorferi*, *B. garinii* (Baranton et al., 1992), *B. afzelii* (Busch et al., 1996), *B. valaisiana* (Wang et al., 1999b), *B. lusitaniae* (Le Fleche et al., 1997), *B. andersonii* (Marconi et al., 1995), *B. bissettii* (Postic et al., 1998), *B. lonestari* (Barbour et al., 1996b), *B. japonica* (Kawabata et al., 1993; Marconi et al., 1995), *B. tanukii* (Fukunaga et al., 1996), *B. turdae* (Fukunaga et al., 1996), and *B. sinica* (Masuzawa et al., 2001). As described above, only a subset of these species has been definitively demonstrated as causative agents of disease in humans.

The *Borrelia* are microaerophilic, Gram-negative bacteria that have a helical or flat-wave morphology that can be readily visualized using dark-field microscopy (Fig. 52.2). They range from 0.18 to 0.25 μm in diameter and from 4 to 30 μm in length. Cultivation can be achieved in a complex culture medium (e.g., BSK-H, BSK II) supplemented with rabbit serum (6%) (Barbour, 1984; Ruzic-Sabljic et al., 2006). The optimal temperature range for growth is from 33 to 37°C (Barbour, 1984; Ruzic-Sabljic et al., 2006). The *Borrelia* have a two membrane system similar to that of Gram-negative bacteria, but lack lipopolysaccharide

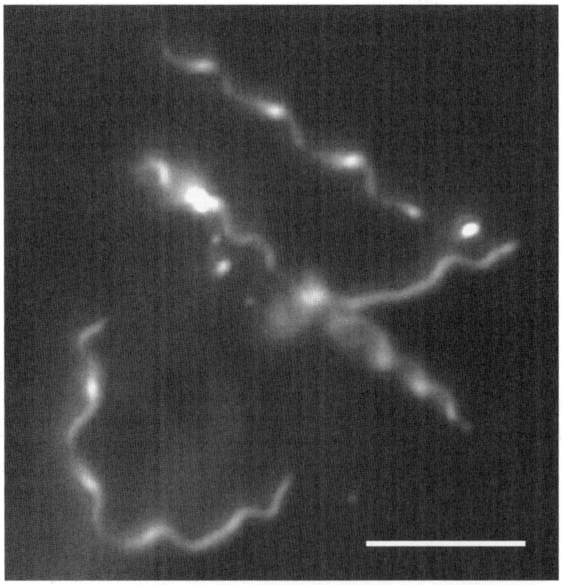

FIGURE 52.2 Darkfield photomicrograph of *Borrelia burgdorferi*. B31MI strain grown at 33°C in BSK-H medium, 100× original magnification, scale bar is 10 μm. Note the characteristic spiral morphology.

(Takayama et al., 1987). The spirochetes are unique among bacteria in that they possess two sets of internal flagella that anchor within the inner membrane but reside primarily within the periplasmic space. The individual sets of flagella anchor at opposite ends of the cell and extend approximately three-fourths of the length of the cell, overlapping at the center. Mutants lacking flagella are nonmotile and lack the characteristic spiral morphology (Sadziene et al., 1991). The "corkscrew like" motility of the spirochetes is a critical feature that allows for their efficient movement through viscous media such as the dermal extracellular matrix (Charon et al., 1992; Goldstein et al., 1994, 1996; Li et al., 2000; Wolgemuth et al., 2006).

At the genetic level, the *Borrelia* possess a unique segmented genome that consists of linear and circular DNA (Barbour and Garon, 1987, 1988). Complete or partial genome sequences have been determined for several Lyme disease spirochete strains and for the related relapsing fever spirochetes, *Borrelia turicatae* and *B. hermsii* (Dunn et al., 1994; Barbour et al., 1996a; Fraser et al., 1997; Caimano et al., 2000; Glockner et al., 2004, 2006). The genome of the Lyme disease spirochete, *B. burgdorferi* strain B31MI, was among the first bacterial genomes to be completely determined (Fraser et al., 1997). The B31MI genome has 1516 open reading frames encoded on a linear chromosome (910 kbp), 12 linear plasmids (lp; 5–54 kbp), and 9 circular plasmids (cp; 9–32 kbp), some of which appear to be prophage (Fraser et al., 1997; Eggers and Samuels, 1999; Casjens et al., 2000; Eggers et al., 2000; Zhang and Marconi, 2005). The telomeric structure of the chromosome and linear plasmids are similar to that of the poxviruses, with covalently closed terminal hairpins (Barbour and Garon, 1987; Hinnebusch et al., 1990; Casjens et al., 1997). Plasmid composition varies considerably among isolates (Iyer et al., 2003; Terekhova et al., 2006), and can change during in vitro culture or in vivo passage (Barbour, 1988; Schwan et al., 1988; Persing et al., 1994; McDowell et al., 2001a). Some plasmids are universal among isolates and appear to be critical (e.g., cp26), while others are not required for survival but correlate with a high-infectivity phenotype (e.g., lp28-1, lp25) (Marconi et al., 1993; Tilly et al., 1997; Purser and Norris, 2000; Labandeira-Rey and Skare, 2001; Labandeira-Rey et al., 2003; Byram et al., 2004; Grimm et al., 2004). It is likely that infectivity and virulence are mediated by combined individual or synergistic effects of genes on multiple plasmids (McDowell et al., 2001a). Recent studies suggest that specific "ribosomal spacer types" may be an indicator of the invasiveness and dissemination properties of a strain (Iyer et al., 2001; Ojaimi et al., 2005). Consistent with the fastidious growth requirements of the *Borrelia* and their host dependence in vivo, the genome contains only a limited number of genes that encode proteins involved in biosynthetic pathways. The *Borrelia* are largely incapable of de novo amino acid, lipid, nucleotide, or enzyme cofactor synthesis (Fraser et al., 1997).

EPIDEMIOLOGY

Lyme disease is the most prevalent vector-borne disease in Europe and North America. A standardized case definition was developed in 1991 and revised in 1996 (Wharton et al., 1990; Anonymous, 1997). Lyme disease is a reportable disease in the United States, with an average of 21,600 cases reported per year (2002–2006) (Fig. 52.3). However, the actual number is certainly much greater, as physician compliance with reporting ranges from 7 to 36% (Matteson et al., 1992; Coyle et al., 1996; Meek et al., 1996; Campbell et al., 1998; Young, 1998; Orloski et al., 2000; Naleway et al., 2002). In the United States, Lyme disease is concentrated in the northeast, with the highest per capita incidences occurring in Delaware, Connecticut, New Jersey, Massachusetts, Pennsylvania, and New York (McNabb et al., 2007) (Fig. 52.4). In Europe, a standard case definition is not in place and there is limited data regarding the incidence of Lyme disease (Anonymous, 1995). At least 85,000 cases are documented each year; however, as in the United States this is thought to be a significant underestimate of the actual incidence (Lindgren and Jaenson, 2006). Recent studies estimate that the incidence in Europe may be approximately 200 cases per 100,000 persons per year (Derdakova and Lencakova, 2005; Smith and Takkinen, 2006) and steadily rising (Smith and Takkinen, 2006). The highest reported incidences occur in Slovenia, Austria, the Netherlands, Germany, and Sweden. Lyme disease incidence correlates with the population dynamics of its mammalian reservoirs and tick vectors. Changes in these variables are due at least in part to climatic and environmental changes that alter the natural vegetation and other ecological factors. The historical trend involving the transformation of farmland into suburban woodlots, an environmental transition that is favorable for deer and *Ixodes* ticks, appears to have been a significant contributor to the emergence of Lyme disease. It has been suggested that recent climate changes have altered the distribution of vector ticks in Europe (reviewed in Lindgren and Jaenson, 2006). Ongoing climatic changes in North America are predicted to result in the continuing expansion of the northern range of *I. scapularis* (Brownstein et al., 2003; Ogden et al., 2006a, 2006b).

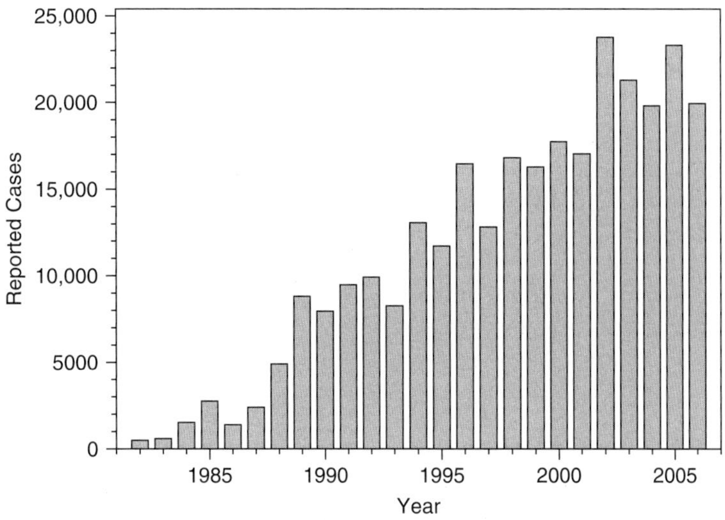

FIGURE 52.3 Reported cases of Lyme disease per year, 1982–2006. The data represent the total cases per year for the United States, as reported to the U.S. Centers for Disease Control and Prevention. A standard case definition was introduced in 1991, and revised in 1996. Lyme disease remains a reportable disease, though there is evidence that a significant proportion of cases go unreported.

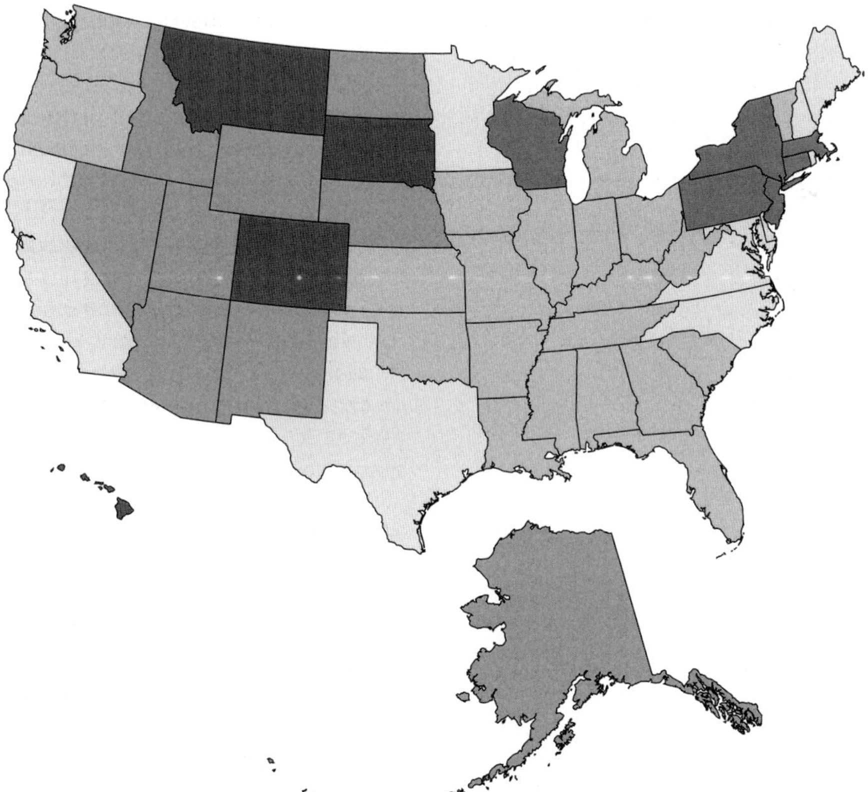

FIGURE 52.4 Reported cases of Lyme disease by state. The data represent the aggregate number of cases reported to the U.S. Centers for Disease Control and Prevention between 1992 and 2006. Each change in color represents a logarithmic increase in the total number of cases, shown as purple (0–10 cases), blue (10–99), green (100–999), yellow (1000–9999), and orange (>10,000). The greatest majority of Lyme disease cases occur in the northeast states and in Wisconsin, likely due to environmental conditions suitable for the tick vector (see color plate section).

Public Health Significance

With the increasing incidence of Lyme disease, expansion of its endemic areas, and its recognition in areas previously thought to be free of this disease, it is clear that this tick-transmitted infection remains a significant public health threat. At the present time, preventive strategies focus primarily on tick avoidance and promotion of public awareness of the signs and symptoms of disease. Demographic profiles of persons with Lyme disease indicate primarily a suburban and rural risk. The disease has a significant socioeconomic impact and has altered recreational and behavioral patterns out of concern over tick exposure. Fortunately, treatment for early Lyme disease is generally successful. If, however, the infection is not diagnosed early or remains untreated, it can become a debilitating disorder with significant physical, emotional, and economic impacts on infected individuals. Numerous studies have been undertaken to assess the cost of Lyme disease treatment, and the point at which vaccination becomes cost-effective. A recent study assessing the economic impacts of Lyme disease found that the direct medical cost of early Lyme disease treatment was $464, and late disease was $1380 (Zhang et al., 2006). When direct medical costs, indirect costs, nonmedical costs, and productivity losses were included, the average Lyme disease case costs approximately $8200 (Zhang et al., 2006). Calculations based on the number of cases reported indicate that over $200 million is spent nationwide. Another study estimated the costs for Lyme disease treatment in the United States at approximately $500 million per year (based on an estimated national incidence of 4.73 per 100,000 persons and 1996 dollars) (Maes et al., 1998). Both of these figures are likely significant underestimates, since the large majority of cases are not reported (Matteson et al., 1992; Coyle et al., 1996; Meek et al., 1996; Campbell et al., 1998; Young, 1998; Orloski et al., 2000; Naleway et al., 2002). Although more difficult to quantify, as discussed above concerns over contracting Lyme disease have also led to behavioral modifications, such as decreased outdoor activity that have hidden socioeconomic costs. Because of its significant health and economic impacts, reducing the number of annual Lyme disease cases has been made a national priority by the Healthy People 2010 program of the United States Department of Health and Human Services (Anonymous, 2000; Poland, 2001).

CLINICAL DISEASE

Lyme disease is transmitted to humans primarily by infected nymphal stage ticks. Successful transmission requires a feeding period of approximately 48 hours (Piesman et al., 1991; Piesman, 1993). Prompt and careful removal of ticks can dramatically decrease the potential for spirochete transmission and resulting disease. With initiation of a blood meal, the spirochetes transit from the tick midgut into the salivary glands and then into the skin of the mammal. The presence and amplification of the spirochete population in the skin, along with their lateral dissemination through the dermis results in the pathognomonic, expanding, annular or macular skin rash, erythema migrans (EM) (Afzelius, 1921). In recognition of the central clearing that may occur with the expansion of the EM, the rash is commonly described as having a "bull's eye" appearance. EM develops in only 60–85% of infected individuals and can be highly variable in appearance with no central clearing. In the absence of central clearing, EM can often mimic rashes or local reactions induced by other arthropods. In some cases secondary EM lesions occur at sites distant from the tick bite. Secondary EM lesions are caused by spirochetes that disseminate from the initial site of infection and establish new foci of infection. This phenomenon tends to occur at a higher frequency with *B. burgdorferi* infections. The EM rash may be accompanied by local mild itching, paresthesia, or in rare cases, pain (Nadelman and Wormser, 1995; Steere, 2001; Hengge et al., 2003; Stanek and Strle, 2003). Constitutional symptoms of early infection include headache, fever, arthralgia, regional lymphadenopathy, and myalgia (Steere et al., 1977a; Nadelman and Wormser, 1995). If not diagnosed early and treated promptly, the spirochetes can colonize distal tissues and persist. Late-stage clinical manifestations may include serious cardiac, dermatological, arthritic, and/or neurological sequelae (Nadelman et al., 1990; Goodman et al., 1995; Wormser et al., 2005). Some late-stage clinical manifestations appear to be species-specific. Severe dermatological manifestations (acrodermatitis chronica atrophicans, ACA), arthritis, or neurological complications correlate with infection by *B. afzelii*, *B. burgdorferi*, and *B. garinii*, respectively (van Dam et al., 1993; Lebech et al., 1994; Balmelli and Piffaretti, 1995; Busch et al., 1996; Rijpkema et al., 1997; Lunemann et al., 2001).

The development of arthritis is a significant concern in patients with late-stage disease, particularly in United States where *B. burgdorferi* is the dominant etiologic agent of Lyme disease. Nearly 60% of patients will experience intermittent arthritic attacks that are typically unilateral and commonly affect large joints (Steere et al., 1977b, 1987; Eiffert et al., 1998; Steere and Glickstein, 2004; Drouin et al., 2008). Though viable spirochetes are difficult to demonstrate in synovial fluid, the presence of *Borrelia* DNA has been demonstrated by PCR (Karch and Huppertz, 1993; Nocton et al., 1994; Jaulhac et al., 1996). Lyme arthritis appears to result from strong innate and adaptive immune responses to the spirochetes (Hardin et al., 1979; Steere et al., 1979; Yssel et al., 1991; Yin et al., 1997; Gross et al., 1998b; Glatzel et al., 2002; Brown et al., 2003; Steere and Glickstein, 2004; Guerau-de-Arellano et al., 2005). In some cases, the arthritis may progress to a chronic and highly debilitating joint inflammation that does not resolve with antibiotic treatment. A large proportion of patients with chronic Lyme arthritis harbor HLA-DR4 alleles of MHC II, particularly DRB1*0401. It has been proposed that the basis for the development of Lyme arthritis is autoimmunity induced by one or more cross-reactive T-cell epitopes of *Borrelia* outer surface protein A (OspA) (Kalish et al., 1993; Kamradt et al., 1996; Gross et al., 1998a; Chen et al., 1999; Gross and Huber, 2000; Meyer et al., 2000; Steere et al., 2001; Trollmo et al., 2001; Steere and Glickstein, 2004; Drouin et al., 2007). In a hamster model of Lyme arthritis, vaccination with r-OspA has been demonstrated to exacerbate Lyme arthritis (Croke et al., 2000).

Cardiac involvement in Lyme disease patients in both Europe and the United States has been well documented. Carditis, marked by macrophage and $CD4^+$ lymphocyte infiltration, may develop in up to 10% of Lyme disease patients during early infection (Barthold et al., 1991; Asch et al., 1994; Ruderman et al., 1995; Nagi et al., 1996; Guerau-de-Arellano et al., 2005). Cardiac manifestations may include conduction disorders (Steere et al., 1980; van der Linde et al., 1990, 1991; Sigal, 1995; Nagi et al., 1996), ventricular arrhythmias (Cox and Krajden, 1991; Bartunek et al., 2001), and ventricular dysfunction (Steere et al., 1980; Karadag et al., 2004). Several studies have documented an association of cardiomyopathy with the presence of Lyme spirochetes in the myocardium (de Koning et al., 1989; Stanek et al., 1990; van der Linde et al., 1990; Lardieri et al., 1993), and with Lyme disease seropositivity (Klein et al., 1991; Stanek et al., 1991; Seinost et al., 1998; Bartunek et al., 2001).

In addition to EM, other dermatological manifestations of Lyme disease include ACA and lymphocytoma. These manifestations are common during late-stage disease, particularly in European patients. ACA is a chronic and progressive atrophic skin disorder that occurs primarily on extensor surfaces of the hands and feet (Asbrink et al., 1984; Asbrink and Hovmark, 1990; Asbrink, 1991; Hengge et al., 2003). It begins with skin edema and discoloration progressing very slowly to atrophy, with thin, wrinkled, discolored, sometimes sclerotic skin (Asbrink, 1991; Stanek and Strle, 2003). In approximately 60% of patients with ACA, peripheral neuropathy, muscle weakness, cramping, and pain occur (Kindstrand et al., 1997, 2000, 2002; Mullegger, 2004). Less frequently, lesions of the bones and joints may be found in the general area of the involved skin (Mullegger, 2004). Borrelial lymphocytoma is a less severe manifestation characterized by local infiltration of B-lymphocytes at a defined site, typically in the earlobe, areola, or scrotum (Hovmark et al., 1986; Asbrink and Hovmark, 1990; Strle et al., 1992, 1996; Berger, 1997; Maraspin et al., 2002; Hengge et al., 2003; Mullegger, 2004). Lyme lymphocytoma presents as a sharply demarcated bluish red nodule or plaque that is not typically associated with other constitutional symptoms. Lymphocytomas generally resolves within 1–12 weeks following antibiotic treatment (Asbrink, 1991; Strle et al., 1992, 1996; Wormser et al., 2006).

Neurological complications of Lyme disease (neuroborreliosis) may develop within weeks of infection in approximately 10% of untreated patients (Ackermann et al., 1988; Logigian et al., 1990; Gerber et al., 1996; Oschmann et al., 1998). Neuroborreliosis most commonly manifests itself as radiculoneuritis involving the abdomen and chest, or by cranial neuritis, often affecting the facial nerve and resulting in palsy (i.e., Bell's palsy) (Halperin et al., 1988, 1989, 1990a; Oschmann et al., 1998). Other neurological complications include peripheral neuropathy with paresthesia (Logigian et al., 1990), and aseptic meningitis with intermittent mild-to-severe headaches (Oschmann et al., 1998; Coyle and Schutzer, 2002; Nachman and Pontrelli, 2003). Mild encephalopathies have also been reported, especially in association with *B. burgdorferi* infection (Halperin et al., 1990b, 1991; Logigian et al., 1990).

Treatment

Approaches to the treatment of Lyme disease have been the subject of some controversy. Specifically, there are divergent opinions on the duration of antibiotic treatment and the route of administration (Reid et al., 1998; Auwaerter, 2007; Feder et al., 2007; Stricker, 2007). The origins and arguments for and against differing treatment approaches are beyond the scope of this chapter; however, the data clearly indicate that when Lyme disease is diagnosed early, most cases are responsive to antibiotic therapy. A 2–4 week course of doxycycline has proven effective (Wormser et al., 2006). Amoxicillin is an alternative in those patient groups where doxycycline is contraindicated. Intravenous ceftriaxone (2–4 week course) is generally reserved for patients with neuroborreliosis, cardiac involvement, or late Lyme disease. All of the above treatment regimes are appropriate for individuals with Lyme arthritis. The possibly autoimmune-mediated antibiotic-refractory form of Lyme arthritis is treated with NSAIDs or synovectomy (Schoen et al., 1991; Steere and Angelis, 2006; Wormser et al., 2006).

Immune Response to Infection

Infection with the Lyme disease spirochetes typically evokes a slowly developing but strong humoral immune response. Several immunodominant antigens have been identified, characterized, and exploited in diagnostics or explored for their potential as vaccine candidates. Recently, the utility of a peptide in the C-terminal domain of the VlsE antigenic variation protein (Zhang and Norris, 1998b) as a diagnostic antigen for Lyme disease has been demonstrated (Liang and Philipp, 1999). The C6 peptide ELISA is now considered to be a key tool in Lyme disease diagnostics. While several antigens have been demonstrated to elicit bactericidal antibody, natural infection does not culminate in immune clearance or lead to the development of protective immunity. In untreated naturally or experimentally infected mammals, the Lyme spirochetes can persist long-term, and perhaps indefinitely.

The ability to persist despite a strong humoral response is essential for completion of the enzootic cycle and for population maintenance in nature. The spirochetes must survive long enough within their mammalian hosts to allow transmission into feeding, uninfected ticks which can then pass the infection to naïve animals. Several divergent and potentially synergistic mechanisms that contribute to immune evasion and persistence have been described. Differential expression of the expansive repertoire of surface protein encoding genes over the course of infection may allow the spirochetes to alter their antigenic profile (Revel et al., 2002; Ojaimi et al., 2005; Caimano et al., 2007). This process may also aid in the establishment of infection in immunologically primed hosts.

Antigenic variation may also be important in evasion of antibody responses and persistence (Zhang and Norris, 1998a; Liang and Philipp, 1999; McDowell et al., 2002; Norris, 2006). In the Lyme disease spirochetes, the underlying mechanism for antigenic variation is unidirectional segmental gene conversion (Zhang and Norris, 1998a). Partial *vlsE* gene cassettes undergo recombination with the single *vlsE* expression site in the *Borrelia* genome resulting in the generation of new antigenic variants of this dominant surface antigen. Successful immune evasion and long-term colonization of mammals also requires mechanisms that allow for complement evasion. Serum-resistant strains of *Borrelia* have been demonstrated to evade complement through the binding of negative regulators of the complement cascade such as factor H, and factor H–like protein 1 (Hellwage et al., 2001; Kraiczy et al., 2001, 2004a, 2004b; Alitalo et al., 2002; McDowell et al., 2003; Brooks et al., 2005; Rogers and Marconi, 2007). The binding of these regulatory proteins increases serum resistance.

VACCINES

First-Generation OspA Subunit Vaccines

Following the discovery of the etiologic agent of Lyme disease, significant research effort was expended on vaccine development. OspA, a protein that is highly expressed by the Lyme disease spirochetes during in vitro cultivation was the focus of most of the early vaccine development strategies. Preclinical trials of OspA vaccination demonstrated protection in animal models of Lyme disease, either as a recombinant protein or lipoprotein, or expressed on the surface of *E. coli, Salmonella typhimurium*, or *Mycobacterium bovis* (BCG) (Fikrig et al., 1990, 1991; Schaible et al., 1990; Simon et al., 1991; Stover et al., 1993; Aydintug et al., 1994; Langermann et al., 1994; Probert and LeFebvre, 1994; Chang et al., 1995; Dunne et al., 1995; Lovrich et al., 1995; Luke et al., 1997b; Philipp et al., 1997; Gomes-Solecki et al., 2006). Two OspA-based vaccine candidates, ImuLyme (Pasteur Merieux-Connaught) and LYMErix (SmithKlineBeecham), were subsequently developed and tested. The ImuLyme vaccine consists of a recombinant OspA lipoprotein purified following expression in *E. coli* (Sigal et al., 1998). In the LYMErix formulation, a lipid moiety was added to recombinant OspA after protein purification, and the vaccine was adsorbed to alum adjuvant (Erdile et al., 1993; Steere et al., 1998). The presence of the lipid motif enhances vaccine immunogenicity by interaction with TLR2 (Yoder et al., 2003). Since there is significant sequence diversity in *ospA*, the sequence chosen for vaccine development was one that represented the most prevalent OspA phyletic type in North American Lyme disease spirochete isolates (Wilske et al., 1996). Initial clinical testing demonstrated safety and efficacy in human subjects (Keller et al., 1994; Wallich et al., 1994; Schoen et al., 1995; Van Hoecke et al., 1996; Sigal et al., 1998; Steere et al., 1998). In phase III studies, an enrollment of 11,000 was achieved for each of the trials. The vaccines were administered as two 30 μg doses, administered 1 month apart, with a booster vaccination at 12 months. The vaccine efficacy following the first two doses was 68% and 49% for ImuLyme and LYMErix, respectively. In the second year, following the 12 month booster vaccination, efficacy rose to 92% for ImuLyme and 76% for LYMErix. Overall, OspA-based vaccines appeared to be safe and well tolerated during phase III studies. Side effects were not remarkable and consisted of those typically associated with vaccination (e.g., minor soreness, redness, swelling at the injection site, low grade fever, chills). FDA approval was given for the LYMErix vaccine at the end of 1998; no application for ImuLyme approval was submitted (Nigrovic and Thompson, 2007). The LYMErix vaccine was subsequently recommended by the CDC Advisory Committee on Immunization Practices for use in populations at high risk for contracting Lyme disease (Anonymous, 1999).

OspA is a linear-plasmid-encoded 28 kDa lipoprotein (Howe et al., 1986; Barbour and Garon, 1988) that structurally consists of two globular domains composed almost entirely of beta-sheets that are connected by a unique single-layer beta-sheet (Fig. 52.5) (Li et al., 1997; Makabe et al., 2006). Crystallographic structures have indicated that there is a possible ligand-binding site formed in the C-terminal globular domain (Li et al., 1997; Makabe et al., 2006). OspA is expressed in the tick vector (Fingerle et al., 1995; Leuba-Garcia et al., 1998), and may be responsible for adhesion of spirochetes to the wall of the tick midgut (Pal et al., 2004a). During transit from the midgut to the tick salivary glands, OspA expression is down-regulated; thus, spirochetes entering the mammalian host do not express the OspA protein (Schwan et al., 1995). Vaccination with OspA induces an anti-OspA antibody response that blocks transmission of spirochetes from the tick to the mammalian host by killing or immobilizing the spirochetes within the tick midgut (de Silva et al., 1996, 1999). Since the spirochetes are, in large part, killed prior to entry into the mammalian host, this mechanism of action largely precludes the elicitation of anamnestic responses. The effectiveness of the vaccine is thus dependent on the presence of sufficient levels of circulating anti-OspA antibody. The requirement for high circulating antibody titer likely explains the need for yearly booster immunizations to maintain efficacy. During the second year of the ImuLyme trial, for example, there was no difference in protection between unvaccinated patients and vaccinated patients that had received the two initial vaccinations, but not the 12 month booster (Sigal et al., 1998). The incomplete (49% in year one; 76% in year two) protection provided by the LYMErix vaccine, the lack of safety and efficacy data for children, and apparent requirement for frequent booster immunizations to maintain protection are all factors that suggest that the development of new and improved Lyme disease vaccines are needed.

Within a year of its introduction, LYMErix was being anecdotally associated with development of arthritis. The possible association between the vaccine and these adverse events was well publicized, particularly by Lyme disease advocacy groups. Lawsuits, including a class-action, alleging vaccine-related harm

FIGURE 52.5 Crystal structure of OspA. The ribbon diagram is derived from the OspA crystal structure 2G8C (Makabe et al., 2006). The conformationally determined epitope recognized by a protective mAb (LA-2) is noted in blue (Golde et al., 1997; Ding et al., 2000). The peptide that may be associated with autoimmune-inducing cross-reactivity with the LFA-1 protein is highlighted in red (see text). The N-terminus is at the upper right of the figure (see color plate section).

began to be filed in 1999. OspA is expressed to some degree during natural mammalian infection (Coyle et al., 1995; Schutzer et al., 1997; Akin et al., 1999), and it was recognized early in vaccine development and confirmed in later studies that the immune response to OspA may be associated with development of antibiotic-refractory arthritis. In particular, there was reported to be an association between arthritis development and particular MHC II alleles (HLA-DR4) (Kalish et al., 1993; Kamradt et al., 1996; Gross et al., 1998a; Chen et al., 1999; Gross and Huber, 2000; Meyer et al., 2000; Steere et al., 2001; Trollmo et al., 2001; Steere and Glickstein, 2004; Drouin et al., 2008). During investigations into possible mechanisms, a predicted OspA T-cell epitope was found to be similar to one predicted for lymphocyte function antigen 1 (LFA-1), prompting speculation that this epitope could be the basis of an autoimmune arthritis (Gross et al., 1998a; Trollmo et al., 2001). It is still not clear, however, that OspA-induced LFA-1 cross-reactivity is, in fact, the cause of autoimmune arthritis (Steere et al., 1990, 2001; Nocton et al., 1994; Gross et al., 1998a; Chen et al., 1999; Hemmer et al., 1999; Meyer et al., 2000; Rose et al., 2001; Trollmo et al., 2001; Drouin et al., 2008).

While concerns, real or perceived, continued to mount, post-approval phase IV studies of LYMErix safety were undertaken by the manufacturer. These studies were originally designed to monitor 25,000 vaccine recipients, and 75,000 age- and sex-matched unvaccinated controls; however, by November of 2000, only 10% of the planned enrollment had been met. The results of the phase IV study to date were presented to an FDA advisory panel in January, 2001. The data did not demonstrate an association between vaccination and development of autoimmunity (Anonymous, 2001a, 2001b). Also presented were data from the Vaccine Adverse Event Reporting System (VAERS), in which the CDC and FDA analyzed adverse events reported between approval in December, 1998, and July, 2000. With 1.4 million distributed doses, there were 905 reported adverse events, of which 102 were coded as arthritis, arthrosis, or rheumatoid arthritis and 12 as facial paralysis, both of which had been postulated to be autoimmune sequelae of Lyme disease (Lathrop et al., 2002). There was no apparent temporal association between the development of arthritic problems and vaccine administration, though reports of arthritis were more often reported after the second and third immunizations. The reported rate of both arthritis and facial paralysis was below what was expected as background (Lathrop et al., 2002). The conclusion of the VAERS study was that there were no unexpected or unusual patterns of adverse events associated with LYMErix vaccination. The 2001 FDA panel found no reason to remove the vaccine from the market or modify its labeling. Continued media coverage of the potential vaccine-associated adverse events in combination with the above described limitations of the vaccine led to decreased acceptance of LYMErix, and sales continued to decline. Citing poor sales performance, GSK removed the LYMErix vaccine from the market in 2002.

Licensed Veterinary Vaccines

In endemic areas of North America and Europe, a large proportion of companion pets are exposed to *Ixodes* ticks and the Lyme disease spirochetes, a subset of which develop arthritic disease (reviewed in (Skotarczak, 2002). Several veterinary Lyme disease vaccines are currently available. Three are formulations of *B. burgdorferi* bacterin (e.g., Merilym, Merial, Germany; LymeVax, Fort Dodge, USA; Galaxy Lyme, Schering Plough, USA) (Chu et al., 1992; Wiedemann and Milward, 1999; Levy, 2002). A European bacterin vaccine uses a combination of *B. garinii* and *B. afzelii* cells (Biocan B, Bioveta, Czech Republic) (Rapuntean et al., 2003; Nepereny et al., 2007). Recombinant OspA-based vaccines, administered either with (ProLyme, Intervet, USA) or without (Recombitek Lyme, Merial, USA) adjuvant (Levy et al., 1993; Conlon et al., 2000; Wikle et al., 2006) are also available. All of the veterinary vaccines employ a two-dose immunization protocol with boosters recommended yearly. Very limited data are available regarding adverse events or efficacy. A comparison of the humoral immune response to veterinary vaccine preparations demonstrated that while all elicit an antibody response that is reactive with the vaccine strain from which the bacterin or OspA was derived, there was only minor reactivity with heterologous *Borrelia* species (Topfer and Straubinger, 2007), a point that has also been demonstrated in several other studies (Johnson et al., 1988; Ma et al., 1995; Gern et al., 1997; Shang et al., 2001). This suggests that the currently available vaccines will not convey broad protective coverage.

VACCINES IN DEVELOPMENT

Rationale for Second-Generation Vaccines

Pharmacoeconomic studies indicate that the benefit of vaccination against Lyme disease will be greatest when based on individual risk assessments, with cost effectiveness occurring when used for persons with an annual risk of contracting Lyme disease that exceeds 1% (Meltzer et al., 1999; Shadick et al., 2001). Rates of infection between 1% and 3% occur in certain focal areas of the United States and Europe (Hanrahan et al., 1984; Fahrer et al., 1991; Orloski et al., 2000). In the U.S., the reported incidence in 2005 ranged from 0.019% to 0.077% in the 10 states with the highest reported levels of Lyme disease (McNabb et al., 2007). As discussed above, the number of reported cases underestimates the actual incidence, which in these states may exceed 1% (Matteson et al., 1992; Coyle et al., 1996; Meek et al., 1996; Campbell et al., 1998; Young, 1998; Orloski et al., 2000; Naleway et al., 2002). The calculations of the disease incidence at which vaccination becomes cost-effective have been based on assumptions that may be invalid for a second-generation vaccine. Specifically, models have assumed a requirement for three successive vaccinations, with yearly boosters, or projected a cost for vaccination repeated on a yearly basis. While these assumptions were reasonable for the first-generation vaccines (discussed above), future vaccines may not have the same requirements. Lyme vaccination could also have other, less tangible, benefits that are not included in typical cost/benefit analyses, including increased outdoor recreation and reduced use of prophylactic antibiotic treatment secondary to uncomplicated tick bite.

Discovery and Basic Science

Of critical importance in Lyme disease vaccine development is an understanding of the subset of proteins or antigens that are expressed during each phase of the enzootic cycle and specifically during mammalian infection. Several studies have documented environmentally mediated changes in global gene expression profiles (Liang et al., 2002a; Revel et al., 2002; Narasimhan et al., 2003; Ojaimi et al., 2005; Caimano et al., 2007). *Borrelia* gene regulation is not completely understood and the mechanisms that alter gene expression during the infectious cycle are complex and multifactorial. In an effort to develop growth conditions that more closely mimic those encountered by spirochetes in the mammalian environment, Akins et al. (1998) developed a rat dialysis membrane chamber implant model. In this model, cultures of *Borrelia* are placed in dialysis membranes and implanted into the peritoneal cavity of host mammals (Akins et al., 1998). After explanting the chambers several days later, the spirochetes can be harvested and the transcriptome and proteome assessed. Bacteria cultivated under these conditions are referred to as "host-adapted." While this model has its limitations, it appears that the expression patterns in host-adapted bacteria are similar to that seen in infected mice. For example, OspA production, which has been demonstrated to be downregulated with the onset of tick feeding, is also downregulated in the host-adapted spirochete model. As mentioned above, OspA is expressed by spirochetes residing in the tick midgut, but it is rapidly downregulated when the spirochetes enter the mammalian host (Schwan et al., 1995). The production patterns of outer surface protein C (OspC) in host-adapted bacteria, a potential vaccine candidate discussed in detail below, also mimics that seen in infected mammals. Spirochetes in unfed ticks do not produce OspC. With the initiation of the blood meal, OspC expression is upregulated and maintained through the early stages of mammalian infection. OspC is a dominant surface antigen and its expression is critical for the establishment of mammalian infection (Tilly et al., 1997, 2006, 2007; Grimm et al., 2004; Pal et al., 2004b; Stewart et al., 2006; Fingerle et al., 2007; Gilbert et al., 2007). Transcriptional control of both OspA and OspC is dependent, at least in part, on a complex regulatory network involving RpoS (Hubner et al., 2001; Alverson et al., 2003; Caimano et al., 2004, 2005; Yang et al., 2005; Lybecker and Samuels, 2007). The ideal Lyme disease vaccine candidate(s) will be either constitutively expressed, or upregulated by tick feeding and expressed during infection in mammals. Antibody to vaccine candidates with these properties have the potential to block transmission of spirochetes from the tick vector and/or kill spirochetes that have been transmitted into vaccinated mammals. In addition, the vaccine candidate should display sufficient conservation, or at least manageable diversity, within naturally exposed epitopes that are presented by intact spirochetes. It should be genetically stable and be encoded by an essential plasmid or be chromosomally encoded. Of those genes upregulated in the mammal, only a small subset will exhibit these properties and thus be of utility in vaccine development. Some of the proteins that have been assessed as potential vaccine candidates include OspC, DbpA, OspE (p21), Elp (paralogous gene family 163), OspF (bbk2.10, pG), Mlp proteins, p35 (BbK32), and p37 (BbK50) (Akins et al., 1995, 1998, 1999; Schwan et al., 1995; Stevenson et al., 1995; Marconi et al., 1996; Fikrig et al., 1997; Yang et al., 2000, 2003a, 2003b; Hefty et al., 2001, 2002; McDowell et al., 2001b; Liang et al., 2002a, 2002b; Brooks et al., 2003; Ojaimi et al., 2003). OspA, OspB, OspC, OspF, DbpA, BBK32, and VraA have been shown to provide some protection against challenge in mice (Telford et al., 1993; Nguyen et al., 1994; Probert and LeFebvre, 1994; Wallich et al., 1994; de Silva et al., 1996; Fikrig et al., 1997; Feng et al., 1998; Hanson et al., 1998; Liang et al., 2001; Labandeira-Rey et al., 2001). Immunogenic, upregulated proteins that have not been shown to be protective include OspD, OspE, BmpA, and VlsE (Nguyen et al., 1994; Probert and LeFebvre, 1994; Gilmore et al., 1996; Liang et al., 2001). In spite of the large number of potential vaccine candidates that have been investigated, OspA and OspC remain the primary foci of subunit vaccine development.

Nonprotein targets are also being assessed for potential utility in Lyme disease vaccine development. The *Borrelia* do not produce lipopolysaccharide, but do produce glycolipids on their outer surface that may be unique to the *Borrelia*. During infection, a strong immune response is mounted against some of these nonprotein antigens (Wheeler et al., 1993; Hossain et al., 2001). One glycolipid in particular, acylated cholesteryl galactoside (ACG), has been the focus of investigation as a vaccinogen because of its immunogenicity during infection (Ben-Menachem et al., 2003; Schroder et al., 2003). ACG induces an IgM antibody response when administered to mice or rabbits in micellar form in PBS, or in DMSO, squalene, or Freund's adjuvants (Ben-Menachem et al., 2003). A protein conjugate of ACG has been synthesized, and is currently being tested for use as a Lyme disease vaccine (Pozsgay and Kubler-Kielb, 2007).

Critical to the testing of Lyme disease vaccine candidates is the recognition that the protein production profiles of the spirochetes can be significantly altered by cultivation and differ from that of spirochetes present in fed ticks. Hence, the route of infection can influence the outcome of challenge experiments. The majority of challenge experiments have been carried out using cultivated spirochetes that are needle inoculated either subcutaneously or intradermally. Decorin-binding protein A (DbpA), a candidate vaccinogen that is expressed in the mammalian host, has been shown to be protective against infection by needle inoculation, but not against infection by tick bite (Cassatt et al., 1998; Feng et al., 1998; Hagman et al., 1998, 2000; Hanson et al., 2000). In contrast, mice immunized with OspA are protected when challenge is administered by either the tick or needle inoculation route (Fikrig et al., 1992; Telford et al., 1993). The OspC protein has also been shown to be protective against infection by either route (Gilmore et al., 1996; Probert et al., 1997). In challenge experiments, consideration must also be given to potential immunological changes at the site of infection that are mediated by biologically active components of tick saliva (Anguita et al., 2002; Brossard and Wikel, 2004; Ramamoorthi et al., 2005; Titus et al., 2006). These findings highlight the complexity of both bacterial protein regulation and the environment surrounding the bacterium during transit from the tick and early infection, and the need to consider this complexity in designing and interpreting vaccine challenge experiments.

Second-Generation OspA Subunit Vaccines

Efforts are now underway to alter the composition of the original OspA-based vaccines to improve theoretical coverage and reduce the potential for adverse events. As described above, there was concern that the LYMErix vaccine formulation may induce autoimmune responses in patients having an HLA-DR4 MHC II allele. Within the OspA protein, there is a predicted T-cell epitope that displays some homology to a sequence in the human LFA-1 protein (Fig. 52.5). It has been speculated that this specific domain may be the basis of putative autoimmune reactions (Gross et al., 1998a; Trollmo et al., 2001). A modified recombinant *B. burgdorferi* OspA in which this potentially cross-reactive epitope was altered has been generated and tested in mice. This modified OspA protein elicited protective antibodies in mice at a level similar to that of the wild-type OspA protein, but did not stimulate HLA-DR4-positive T-cells specific for the wild-type OspA molecule (Willett et al., 2004). The elimination of reactivity with the potentially autoimmune-inducing peptide may alleviate some concern about the safety of OspA-based Lyme disease vaccines; however, there remains significant doubt about the mechanism of OspA-induced autoimmune arthritis and the identity of potential autoreactive proteins (Drouin et al., 2008).

Studies with monoclonal antibodies suggest that the protective epitopes of OspA reside in the C-terminal domain of the protein and are conformationally determined (e.g., the LA-2 epitope) (Fig. 52.5) (Sears et al., 1991; Huang et al., 1998; Ding et al., 2000). Several studies have attempted to use OspA truncations lacking the N-terminal domain of the protein to both reduce vaccine size and eliminate peptides potentially associated with autoimmune reactions. Recombinant OspA truncations expressed with N-terminal polyhistidine fusions (but not with glutathione S-transferase N-terminal tags) elicited protective antibody responses (Bockenstedt et al., 1996). Immune responses to OspA fragments could also be enhanced by targeted replacement of charged amino acids in a C-terminal domain cavity with hydrophobic amino acids. The replacements increased the domain stability, and the resultant protein had antigenic properties similar to wild-type OspA (Koide et al., 2005).

Future efforts to exploit OspA in the development of a more broadly protective vaccine must address issues pertaining to OspA sequence variation.

There are at least seven serotypes of OspA. Serotype 1 is produced by *B. burgdorferi*, serotype 2 by *B. afzelii*, and serotypes 3–7 by *B. garinii* (Wilske et al., 1993). Multivalent-chimeric OspA proteins have been developed that incorporate the protective epitopes from three or more OspA serotypes (Gern et al., 1997). A trivalent vaccine that included OspA from each Lyme disease genospecies provided broad protection for mice challenged by tick infection (Gern et al., 1997). Baxter Vaccines is currently planning clinical trials of an OspA-based chimeric vaccine that incorporates the protective regions of six predominant OspA serotypes on three recombinant proteins (Anonymous, 2007).

OspC Subunit Vaccines

OspC Structure, Regulation, and Function

OspC has received considerable attention in the effort to develop a broadly protective Lyme disease vaccine. OspC possesses several features that render it a desirable vaccine candidate. It is a 22 kDa immunodominant lipoprotein that is tethered to the spirochete outer membrane by an N-terminal tripalmitoyl-S-glyceryl-cysteine (Brooks et al., 2006). It is encoded on a ubiquitous and highly stable 26 kbp circular plasmid (cp26) (Marconi et al., 1993) and the sequence of OspC does not undergo mutation during infection (Stevenson and Barthold, 1994; Hodzic et al., 2000). Crystal structures have been obtained for several OspC variants. These analyses demonstrated OspC to be largely helical in structure, with five alpha helices connected by loops (Eicken et al., 2001; Kumaran et al., 2001). Based on size exclusion, cross-linking, immunological, and structural analyses, OspC appears to form homodimers, and based on the extensive buried interface of the dimer, it has been postulated that it is the dimeric form of the protein that is biologically active (Bunikis and Barbour, 1999; Huang et al., 1999; Eicken et al., 2001; Kumaran et al., 2001; Zuckert et al., 2001; Earnhart et al., 2005).

OspC is an essential virulence factor that is critical for the establishment of early infection in mammals (Margolis and Rosa, 1993; Tilly et al., 1997, 2006, 2007; Byram et al., 2004; Grimm et al., 2004; Pal et al., 2004b; Rosa, 2005; Stewart et al., 2006). *ospC* deletion mutants in several *Borrelia* strains are not infectious in mammals, either by subcutaneous injection or tick transmission (Grimm et al., 2004; Pal et al., 2004b). Although its precise role in the transmission process has not yet been definitively demonstrated it has also been postulated to be required for the passage of spirochetes from the midgut to the salivary glands in the tick (Probert and LeFebvre, 1994; Gilmore et al., 1996; Pal et al., 2004b; Fingerle et al., 2007). Additionally, OspC may bind and transport an immunomodulatory tick salivary gland protein (Salp15) that interferes with CD4+ T-lymphocyte function (Anguita et al., 2002; Ramamoorthi et al., 2005). These analyses collectively suggest that anti-OspC antibodies have the potential to interfere with some of the critical early events in transmission and in the establishment of infection in mammals.

OspC Phylogenetics

A complicating factor in the exploitation of OspC as a vaccine candidate is that it is highly variable in sequence. However, recent analyses have shed important light on the nature of OspC variation. Molecular phylogenetic analyses have revealed that OspC sequences form at least 38 distinct OspC phyletic clusters or OspC types (Seinost et al., 1999a; Wang et al., 1999c; Earnhart and Marconi, 2007c). Within OspC phyletic clusters, amino acid sequences are >95% identical, whereas between types, identity ranges from approximately 55–80% (Wang et al., 1999c; Attie et al., 2007; Earnhart and Marconi, 2007c). Sequence variation within OspC occurs in distinct domains that are surface-exposed and antigenic (Earnhart et al., 2005, 2007). While an individual strain expresses only a single OspC type, multiple OspC types have been demonstrated in spirochetes recovered form ticks, mammals, or Lyme disease patients from small geographic areas (Wang et al., 1999c; Lin et al., 2002; Alghaferi et al., 2005; Earnhart et al., 2005; Anderson and Norris, 2006). The maintenance in nature of different OspC types has been postulated to be driven by balancing selection, where the maintenance of multiple relatively stable alleles in a population is more advantageous than the directional selection of a single, most fit allele (Wang et al., 1999c; Rannala et al., 2000; Kreitman and Di Rienzo, 2004). The maintenance of multiple alleles in the population may aid in circumventing the immune response in naturally immunologically primed reservoir animals. Early studies hypothesized strains producing OspC types A, B, I, and K were primarily associated with invasive infections in humans (Seinost et al., 1999a; Lagal et al., 2003); however, based on more recent studies it is now apparent that strains producing several additional OspC types can also cause invasive infection (Alghaferi et al., 2005; Earnhart et al., 2005). OspC variation arises primarily through genetic exchange and recombination and not by hypermutation with concomitant immune selection (Livey et al., 1995; Marti Ras et al., 1997; Wang et al., 1999a; Dykhuizen and Baranton, 2001; Bunikis et al., 2004; Attie et al., 2007; Earnhart and Marconi, 2007c).

This is of importance in vaccine design, as it implies potential natural restrictions on the extent to which the OspC sequence can change. With the recent significant advances in our understanding of the sequence diversity of OspC, it is now possible to devise strategies to exploit the variation within OspC in vaccine development.

Immune Response to OspC

While OspC is highly antigenic and can elicit protective antibody responses, the protective range is narrow (Probert and LeFebvre, 1994; Gilmore et al., 1996; Bockenstedt et al., 1997; Probert et al., 1997; Mathiesen et al., 1998; Rousselle et al., 1998; Gilmore and Mbow, 1999; Mbow et al., 1999; Seinost et al., 1999b; Wallich et al., 2001; Jobe et al., 2003; Scheiblhofer et al., 2003; Brown et al., 2005). The basis for the limited protective range of OspC has been determined to be due to the above described sequence variation. In a recent study, clonal populations of *B. burgdorferi* producing different OspC types were used to infect mice, serum was harvested and then used to screen an immunoblot consisting of a panel of OspC proteins of different phyletic types (Earnhart et al., 2005). These analyses demonstrated that the antibody response elicited during early infection is OspC type-specific suggesting that it is the type-specific domains of the protein that are presented at the bacterial cell surface. As infection progresses, a broader array of OspC epitopes in widely separated domains of the protein become recognized by the antibody response (Earnhart et al., 2005, 2007). This is most likely due to the antibody-mediated lysis of cells, release of OspC, and the recognition of epitopes that are normally shielded in viable intact spirochetes. Further analyses have demonstrated that antibody elicited to the variable domains displays OspC type-specific bactericidal activity (Buckles et al., 2006; Earnhart et al., 2007; Earnhart and Marconi, 2007b). The identification of epitopes that are presented by native OspC during the early stages of infection and that elicit bactericidal antibody is an important step towards the design of an effective and broadly protective OspC-based vaccine.

In an effort to develop a polyvalent OspC vaccine, Baxter Vaccines tested an OspC recombinant protein cocktail consisting of up to 14 full-length OspC proteins of different types. However, these formulations did not elicit sufficient antibody titers against the essential type-specific epitopes of each OspC variant included in the vaccine. In addition, unacceptable reactigenicity was reported (Hanson and Edelman, 2004). The inclusion of numerous full-length proteins in the vaccine formulation may have resulted in the misdirection of the immune response by presenting immunodominant but nonprotective epitopes. A potential approach to solving this problem is to generate a recombinant chimeric protein that consists entirely of protective epitopes from those OspC types associated with disease in humans.

OspC Epitope Mapping

The successful development of a polyvalent chimeric vaccine requires the systematic identification of protective epitopes from diverse OspC sequences. Surprisingly few studies have sought to identify the naturally presented epitopes in different OspC type proteins. Initial studies of type A OspC suggested at least one protective epitope may be conformationally defined (Gilmore et al., 1996; Gilmore and Mbow, 1999; Mbow et al., 1999). While conformationally defined epitopes can be important in eliciting protective immunity, linear epitopes are more easily incorporated into chimeric constructs. The C-terminus of OspC harbors a linear epitope that is highly conserved among OspC types and which has been shown to induce antibodies in infected mammals and in humans (Lovrich et al., 1991, 2005, 2007; Mathiesen et al., 1998; Jobe et al., 2003). This highly conserved C-terminal sequence has not, however, been demonstrated to elicit a broadly protective response and hence it is not in and of itself sufficient to elicit broad protection against strains expressing different OspC types. OspC type-specific linear epitopes that may be of utility in the construction of a chimeric, polyvalent vaccine have been identified by screening series of OspC truncations and subfragments derived from multiple OspC types with sera derived from mice experimentally infected with clonal *B. burgdorferi* isolates (Earnhart et al., 2005, 2007). Two linear epitopes have been identified: the "loop 5" and the "helix 5" epitopes. The loop 5 epitope is comprised of the C-terminal portion of alpha helix 3 and loop 5 (Eicken et al., 2001; Kumaran et al., 2001), while the helix 5 epitope encompasses the fifth alpha helix and a small region C-terminal to this helix (Fig. 52.6) (Earnhart et al., 2005, 2007; Yang et al., 2006). These epitopes are recognized by antibody that develops during experimental infection in mice and natural infection in humans (Earnhart et al., 2005, 2007; Buckles et al., 2006). Both epitopes map to surface-exposed domains of OspC that are highly variable between OspC types but conserved within each type. A recombinant protein containing the type A OspC loop 5 epitope elicited antibodies in mice that demonstrate type-specific, complement-dependent antibody-mediated killing in vitro (Buckles et al., 2006). The ability to generate

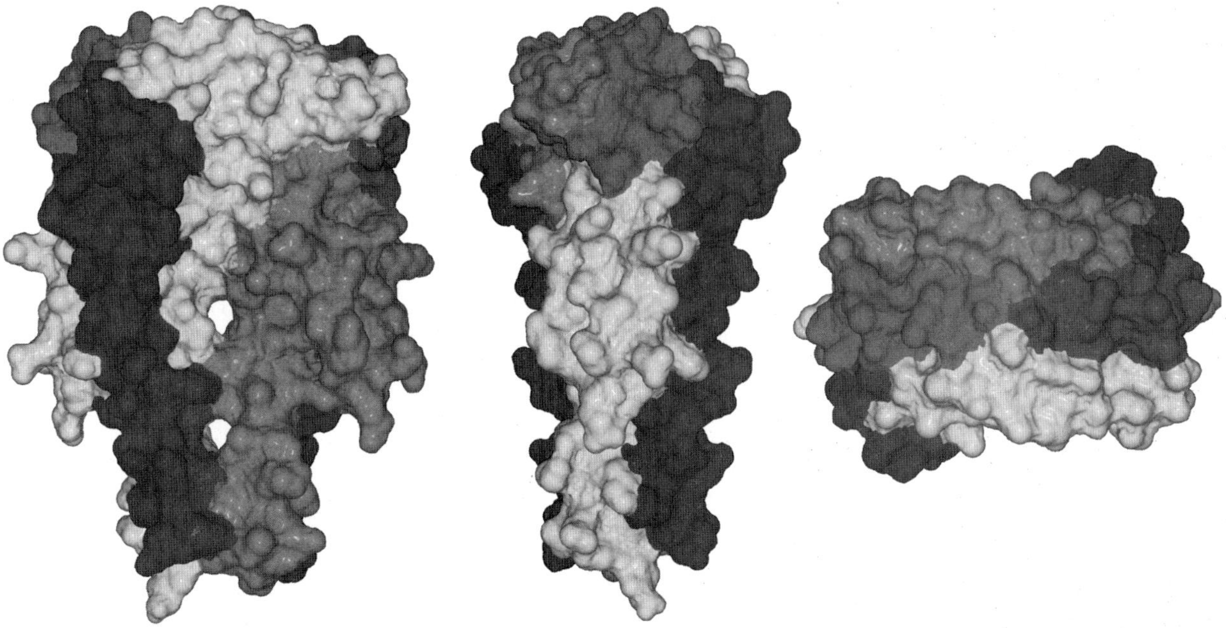

FIGURE 52.6 Crystal structure of outer surface protein C. The diagram is derived from the type A OspC crystal structure 1GGQ (Kumaran et al., 2001). A surface representation of the OspC molecule is shown in three rotations. The locations of the two regions in which epitopes have been mapped in multiple OspC types are denoted in red (loop 5 region) and blue (helix 5 region). The molecule is a homodimer; the component monomers are differentiated using darker and lighter color shades. A portion of both the N- and C-termini is omitted from the structure (see color plate section).

a protective response using isolated linear epitope-containing sequences further indicates the potential of a polyvalent, epitope-based chimeric OspC vaccine.

Development of a Polyvalent OspC Vaccine Based on Mapped Epitopes

Chimeric vaccines consisting of a series of immunodominant epitopes have been explored in the development of vaccines against malaria (Hanson and Edelman, 2004; Caro-Aguilar et al., 2005), group A streptococci (Dale, 1999; Hu et al., 2002; Kotloff et al., 2004; Kotloff and Dale, 2004; Dale et al., 2005; McNeil et al., 2005), and several viruses (Wang et al., 1999d; Bouche et al., 2005; Fan et al., 2005; Apt et al., 2006). Data suggest that a broadly protective OspC vaccine will require the inclusion of epitopes from approximately 28 OspC types (Earnhart and Marconi, 2007c). Such a construct is predicted to provide protection against all major Lyme disease spirochete species associated with human disease, and to be effective in both Europe and North America. Possible cross-protection elicited by some epitopes may reduce the total number of epitopes required to achieve this goal.

A prototype tetravalent chimeric recombinant OspC-based vaccine has been produced that incorporates epitope-containing regions from types A, B, K, and D. This "ABKD" vaccine elicited antibodies in mice that bind OspC as presented on the surface of intact and viable spirochetes and mediate bactericidal activity by a complement-dependent mechanism (Buckles et al., 2006; Earnhart et al., 2007). It is noteworthy that a decrease in epitope-specific titer was observed for epitopes progressing from the N- to the C-terminus of the chimeric protein. The antibody titer to the type D epitope was 1.7 logs lower than that observed for the N-terminally located type A epitope (Earnhart et al., 2007). This effect did not appear to be due to C-terminal degradation of the construct since the addition of C-terminal tags that have been reported to stabilize recombinant proteins did not improve antibody titer (Earnhart and Marconi, 2007a). This observation suggests that the tertiary structure of the chimeric protein influences antibody titer to specific epitopes. The antibody titer to C-terminal epitopes of the ABKD construct was improved by epitope reiteration and by increasing the solubility of the construct (Earnhart and Marconi, 2007a). A similar observation of decreasing antibody titer to C-terminal epitopes had been reported for the streptococcal M protein vaccine (Dale et al., 1993, 1996).

The ABKD OspC construct has recently been expanded to include four additional OspC types (E, N, I, and C) that were predicted, based on phylogenetic analysis, to provide broad protection (Earnhart and Marconi, 2007c). This octavalent vaccine (OspC-A8.1) induced type-specific antibodies against all incorporated epitope-containing regions, though there was decreased titer to the C-terminal epitope similar to the tetravalent vaccine (Earnhart and Marconi, 2007b). The OspC-A8.1 vaccine induced IgG2b and IgG2a antibody isotypes, which have the ability to fix complement (Earnhart and Marconi, 2007b). In addition, the OspC-A8.1 hyperimmune serum efficiently labeled intact *B. burgdorferi* strains bearing each of the component OspC types (Earnhart and Marconi, 2007b). These initial studies with OspC-based chimeric proteins represent the first attempts in the Lyme disease field to construct a vaccine consisting exclusively of epitopes that are recognized during early infection, and that may protect against diverse strains from both Europe and North America. Vaccines that provide global coverage are essential in this era of frequent and rapid world travel.

A theoretical concern with vaccination against a subset of epitopes is the potential to drive selection in nature toward OspC types not included in the vaccine. This would tend to increase the proportion of the population bearing formerly rare OspC types. However, as humans are incidental hosts, it is unlikely that vaccination of specific high-risk human populations would significantly alter the population distribution of strains expressing specific OspC types in nature. Studies are currently underway to extend the breadth of potential protection using additional constructs that incorporate OspC types from the three Lyme disease *Borrelia* species.

Combination Subunit Vaccines

The combination of two or more different recombinant immunogens in a Lyme disease vaccine formulation has been investigated in a limited number of studies. Studies that assessed the efficacy of a cocktail of recombinant OspC proteins derived from different Lyme disease spirochete isolates are described above. Attempts have also been made to enhance protective efficacy using combinations of unrelated proteins. In one study, possible synergy in immune responses were assessed using a combination of OspA and DbpA (Hanson et al., 2000). Like OspA, DbpA is an established surface-exposed lipoprotein (Hagman et al., 1998) that elicits some degree of protection as assessed using the mouse infection model (Feng et al., 1998; Hanson et al., 1998). When compared with the protection elicited by each individual protein, the OspA/DbpA combination protected against a 100-fold greater challenge dose and also conveyed a higher degree of protection against challenge with heterologous strains. Tick challenge was not assessed in this study and subsequent analyses suggest that immunization with DbpA does not protect against tick challenge (Hagman et al., 2000). In a different analysis, a combination of three different antigens were tested: OspC, DbpA, and fibronectin-binding protein (BBK32) (Brown et al., 2005) were generated as recombinant proteins and administered as a cocktail. BBK32 is a surface-exposed lipoprotein (Guo et al., 1995; Probert and Johnson, 1998) that is thought to contribute to adherence by the Lyme disease spirochetes (Seshu et al., 2006). The triple combination vaccine was shown to be more effective than single or paired vaccination. In these analyses, only the needle challenge model was assessed.

DNA-Based Vaccines

The use of DNA vaccines encoding *Borrelia* outer surface proteins has also been explored. Mice immunized with plasmid or vaccinia virus expressing *Borrelia* OspA or OspC exhibited varying degrees of protection against different *B. burgdorferi* strains (Simon et al., 1996; Zhong et al., 1996; Luke et al., 1997a; Scheiblhofer et al., 2003; Scheckelhoff et al., 2006). A DNA vaccine encoding an OspA–OspC fusion that was administered by intramuscular injection also elicited a protective immune response (Wallich et al., 2001). However, in the case of the OspA–OspC fusion it was determined that protection was due solely to the antibody response to OspA. In contrast, in a separate study, *ospC* fused to the human tissue plasminogen activator leader sequence in a pCMV vector elicited protective responses. This construct was introduced into mice by two routes: intradermal needle inoculation (Th1 response) and gene gun (Th2 response). Both approaches were equally effective in eliciting protective responses, suggesting that a successful Lyme disease vaccine is not restricted to induction of a certain Th type of immune response (Scheiblhofer et al., 2003). It is important to note however that no studies employing DNA-based vaccines that have been published to date have demonstrated broad protective capability.

Wildlife Vaccine Strategies

Additional approaches being considered in Lyme disease prevention and control include reducing the carriage rate of the Lyme disease spirochetes in ticks through the vaccination of mammalian reservoirs or the development of anti-tick vaccines that can be used to control tick populations. Focusing first on pathogen-specific control approaches, in one study approximately 1000 wild *Peromyscus* mice in southern Connecticut were immunized intraperitoneally with up to three doses of OspA or with a control protein in Freund's adjuvants. The reduction in Lyme disease transmission efficiency was reported to be 48%, 92%, and 99% for 1, 2, or 3 doses of OspA, compared with control mice (Tsao et al., 2004). The mean decrease in *B. burgdorferi* infection prevalence in the test area was determined to be 16%. While the outcome was significant, the data also served to highlight the importance of other mammalian species in the Lyme disease enzootic cycle. The ability to effectively reduce the carriage rate of the Lyme disease spirochetes in endemic areas will require the development of vaccine delivery strategies that can be applied to a diverse range of mammals (Tsao et al., 2004).

Parenteral vaccination wildlife strategies are, for numerous reasons, not a practical approach for Lyme disease control. Oral delivery is clearly the most viable approach to wildlife vaccination, and could theoretically be designed to reach a broad range of reservoir hosts. OspA has been the focus of most efforts to develop an oral bait vaccine (Gomes-Solecki et al., 2006; Scheckelhoff et al., 2006). Gomes-Solecki et al. (2006) demonstrated that mice immunized by oral gavage with *E. coli* expressing OspA developed a IgG2a/2b systemic antibody response that led to 89% protection against challenge by the tick route. In addition, an eightfold reduction in spirochetal numbers in ticks fed on vaccinated mice was observed. Protective responses have also been demonstrated using other bacterial delivery platforms (Fikrig et al., 1990, 1991; Dunne et al., 1995). Oral administration of purified lipidated OspA has been demonstrated to elicit a protective antibody response in mice (Luke et al., 1997b). An orally delivered recombinant vaccinia virus expressing *ospA* has also been demonstrated to evoke protective responses (Scheckelhoff et al., 2006). The overall feasibility of oral delivery strategies in wildlife remains unclear, and is complicated by the diversity of animals that can serve as reservoirs and the variations in their diets and feeding patterns. It remains to be determined if a baiting strategy can be developed that will prove all-encompassing. It is possible that species-specific baiting strategies will allow for the maintenance of Lyme disease spirochete populations in a subset of animal species and that once protective immunity wanes (or bait vaccine is no longer distributed) Lyme disease spirochete populations will become rapidly re-established. In addition, will it be economically feasible to distribute bait vaccines targeting multiple reservoir species over the large and expanding endemic regions for Lyme disease?

As discussed above, an alternative approach to prevent transmission of Lyme disease and other tick-borne disease in humans is to control tick populations. Several studies have sought to identify tick proteins that can be used to as vaccines. Two broad groups of vaccine candidates have been defined: the "concealed" and "exposed" antigens. Concealed antigens are those to which the mammalian immune cells are not typically exposed, such as proteins on the surface of the tick gut. Exposed antigens, such as tick salivary proteins, are those with the potential to elicit responses within mammals during the feeding process (Nuttall et al., 2006). The *Boophilus microplus* Bm86 gut epithelium glycoprotein serves as one example of a concealed tick antigen. *B. microplus* is a tick species associated with debilitating infestations of cattle (Rand et al., 1989; Willadsen, 2004, 2006). Vaccination with Bm86 elicits antibody that can enter the tick and inhibit feeding, reduce tick burden and lower tick reproductive efficiency (Jonsson et al., 2000; Trimnell et al., 2005). Two examples of concealed antigens that may be of utility in ultimately reducing *Ixodes* tick populations (and possibly other ticks as well) are subolesin (4D8) and 5′ nucleotidase (4F8). Vaccination with these proteins has been demonstrated to significantly reduce tick infestation on sheep (Almazan et al., 2003, 2005a, 2005b). Exposed tick antigens that are being investigated for reducing tick populations include a series of *Ixodes* salivary proteins. Ten potential candidate proteins of *Ixodes scapularis* have been described (Xu et al., 2005). Some of these, such as the tick-produced serine protease inhibitors, have known functional roles (Mulenga et al., 2001, 2003; Sugino et al., 2003; Imamura et al., 2005, 2006). Immunization with the recombinant salivary serine protease inhibitor, Iris, partially protects against adult female and nymphal ticks (Prevot et al., 2007). Other exposed antigens of interest in vaccine development include Salp14, Salp15, TROSPA, and 64P (Anguita et al., 2002; Pal et al., 2004a; Ramamoorthi et al., 2005; Soares et al., 2005; de la Fuente and Kocan, 2006; Pedra et al., 2006).

POSTEXPOSURE IMMUNOPROPHYLAXIS

Postexposure passive immunization as a treatment modality for Lyme disease has been assessed in animal models. Passive immunization with serum from mice that were experimentally infected with *B. burgdorferi* does provide some degree of protection but only when administered prior to challenge (Barthold and Bockenstedt, 1993). Once an active infection is established, passive immunization with "infection-derived serum" reduces disease manifestations and decreases spirochetal burden but does not eliminate infection

(Barthold and Bockenstedt, 1993; Shih et al., 1995; Barthold et al., 1996, 2006). Passive immunization with hyperimmune serum generated against several recombinant *Borrelia* proteins including arthritis-related protein and OspA has been reported to reduce specific clinical manifestations and overall disease severity but not to clear infection (Johnson et al., 1986; Schaible et al., 1990; Fikrig et al., 1993; Zhong et al., 1997; Feng et al., 2000; Barthold et al., 2006). Passive immunization with hyperimmune serum to DbpA has also been investigated (Barthold et al., 2006); however, while it is clear that there is some improvement in disease there are conflicting reports regarding infection resolution (Feng et al., 1998; Hanson et al., 1998; Barthold et al., 2006). In contrast to that observed with OspA, DbpA, and Arp, passive immunization with anti-OspC immunization is protective (Mbow et al., 1999) and can resolve established infection in mice (Zhong et al., 1997, 1999). The efficacy of passive immunization has not been assessed in humans as a means of immunoprophylaxis. The role of immunoprophylaxis in the treatment of Lyme disease is likely to be limited, particularly in light of the effectiveness of antibiotic therapy.

PROSPECTS FOR THE FUTURE

It has been nearly 30 years since the recognition of Lyme disease as a distinct clinical entity. This disease, caused by a unique and highly complex group of bacteria, remains a significant human health concern. As the incidence of Lyme disease continues to increase and its endemic regions spread, there is a pressing need for the development of new preventive strategies. After years of intense research effort in which there were major steps forward in our understanding of the pathogenesis of the Lyme disease spirochetes, it is somewhat ironic that some of the first proteins identified in these bacteria, OspA and OspC, remain at the forefront of subunit vaccine development. When the LYMErix vaccine was voluntarily pulled from the market in 2002, the general consensus appeared to be that the need for, and possibly the commercial viability of, a Lyme disease vaccine had passed. However, in the past few years it has been increasingly clear that the need remains. Efforts are underway to revive OspA-based vaccines or to develop new polyvalent OspC-based chimeric recombinant proteins. In addition, vaccination strategies that seek to interfere with the maintenance of the Lyme disease spirochetes in nature, either by decreasing their carriage by mammalian reservoirs or by reducing tick populations, are also being actively pursued. With the dramatic steps forward that have occurred in our understanding of the enzootic cycle of Lyme disease and the immunological responses to the pathogen, there is reason to be confident that new highly effective vaccines that are commercially viable in North America and Europe will be developed.

KEY ISSUES

- Lyme disease is the leading vector-borne disease in North America and Europe and, despite underreporting and incomplete data, it is clear that its incidence is increasing.
- In view of the serious sequelae of Lyme disease, and the possibility that early disease will go unrecognized, immunoprophylaxis is highly desirable, particularly in regions of high disease endemicity.
- Though the OspA-based LYMErix vaccine was never conclusively associated with adverse vaccine reactions, the negative publicity surrounding its voluntary removal from the market has created a potential barrier to the adoption of future Lyme disease vaccines.
- The use of OspA vaccines reformulated to eliminate potentially autoimmune-inducing T-cell epitopes may have poor acceptance because of the limited duration of protection and requirement for annual boosters.
- Lyme disease vaccines based on OspA or OspC must take into account protein sequence heterogeneity among and within *Borrelia* species.
- Using host-adaptation techniques and microarray and immunoproteome analyses, new potential vaccinogens have been identified that are expressed and immunogenic in the mammalian host.
- OspC is a leading second-generation vaccine candidate, but its development has been hampered by significant sequence heterogeneity in the population of Lyme disease *Borrelia*.
- Epitope mapping studies and phylogenetic analyses have enabled the creation of polyvalent, epitope-based OspC vaccines with the potential for broad protection, and the ability to rely on an anamnestic response during infection.
- Vaccination of wildlife involved in the enzootic cycle is an attractive means by which the incidence of Lyme disease could be reduced; however, due to the complex array of potential carriers, the results to date have not been encouraging.

- Vaccination against tick vector proteins may be an alternative approach to reduce the incidence of Lyme disease by targeting the vector, particularly if the anti-tick vaccine is efficacious at reducing or eliminating pathogen transmission.

References

Ackermann, R. The spirochetal etiology of erythema chronicum migrans and Garin-Bujadoux-Bannwarth meningo-polyneuritis. Fortschr. Med. 1983; 101:1167–1170.

Ackermann, R., Kabatzki, J., Boisten, H.P., Steere, A.C., Grodzicki, R.L., Hartung, S. and Runne, U. Spirochete etiology of erythema chronicum migrans disease. Dtsch. Med. Wochenschr. 1984; 109:92–97.

Ackermann, R., Rehse-Kupper, B., Gollmer, E. and Schmidt, R. Chronic neurologic manifestations of erythema migrans borreliosis. Ann. N. Y. Acad. Sci. 1988; 539:16–23.

Afzelius, A. Erythema chronicum migrans. Acta Derm. Venereol. 1921; 2:120–125.

Akin, E., McHugh, G.L., Flavell, R.A., Fikrig, E. and Steere, A.C. The immunoglobulin (IgG) antibody response to OspA and OspB correlates with severe and prolonged Lyme arthritis and the IgG response to P35 correlates with mild and brief arthritis. Infect. Immun. 1999; 67:173–181.

Akins, D.R., Bourell, K.W., Caimano, M.J., Norgard, M.V. and Radolf, J.D. A new animal model for studying Lyme disease spirochetes in a mammalian host-adapted state. J. Clin. Invest. 1998; 101:2240–2250.

Akins, D.R., Porcella, S.F., Popova, T.G., Shevchenko, D., Baker, S.I., Li, M., Norgard, M.V. and Radolf, J.D. Evidence for in vivo but not in vitro expression of a *Borrelia burgdorferi* outer surface protein F (OspF) homologue. Mol. Microbiol. 1995; 18:507–520.

Alghaferi, M.Y., Anderson, J.M., Park, J., Auwaerter, P.G., Aucott, J.N., Norris, D.E. and Dumler, J.S. *Borrelia burgdorferi ospC* heterogeneity among human and murine isolates from a defined region of northern Maryland and southern Pennsylvania: lack of correlation with invasive and noninvasive genotypes. J. Clin. Microbiol. 2005; 43:1879–1884.

Alitalo, A., Meri, T., Lankinen, H., Seppala, I., Lahdenne, P., Hefty, P.S., Akins, D. and Meri, S. Complement inhibitor factor H binding to Lyme disease spirochetes is mediated by inducible expression of multiple plasmid-encoded outer surface protein E paralogs. J. Immunol. 2002; 169:3847–3853.

Almazan, C., Blas-Machado, U., Kocan, K.M., Yoshioka, J.H., Blouin, E.F., Mangold, A.J. and de la Fuente, J. Characterization of three *Ixodes scapularis* cDNAs protective against tick infestations. Vaccine 2005a; 23:4403–4416.

Almazan, C., Kocan, K.M., Bergman, D.K., Garcia-Garcia, J.C., Blouin, E.F. and de la Fuente, J. Identification of protective antigens for the control of *Ixodes scapularis* infestations using cDNA expression library immunization. Vaccine 2003; 21:1492–1501.

Almazan, C., Kocan, K.M., Blouin, E.F. and de la Fuente, J. Vaccination with recombinant tick antigens for the control of *Ixodes scapularis* adult infestations. Vaccine 2005b; 23:5294–5298.

Alverson, J., Bundle, S.F., Sohaskey, C.D., Lybecker, M.C. and Samuels, D.S. Transcriptional regulation of the *ospAB* and *ospC* promoters from *Borrelia burgdorferi*. Mol. Microbiol. 2003; 48:1665–1677.

Anderson, J.M. and Norris, D.E. Genetic diversity of *Borrelia burgdorferi* sensu stricto in *Peromyscus leucopus*, the primary reservoir of Lyme disease in a region of endemicity in southern Maryland. Appl. Environ. Microbiol. 2006; 72:5331–5341.

Anguita, J., Ramamoorthi, N., Hovius, J.W., Das, S., Thomas, V., Persinski, R., Conze, D., Askenase, P.W., Rincon, M., Kantor, F.S. and Fikrig, E. Salp15, an *Ixodes scapularis* salivary protein, inhibits CD4(+) T cell activation. Immunity 2002; 16:849–859.

Anonymous. WHO workshop on Lyme borreliosis diagnosis and surveillance. Warsaw, Poland, 1995.

Anonymous. Case definitions for infectious conditions under public health surveillance. Centers for disease control and prevention. Morb. Mortal. Wkly. Rep. 1997; 46:1–55.

Anonymous. Recommendations for the use of Lyme disease vaccine. Recommendations of the Advisory Committee on Immunization Practices (ACIP). Morb. Mortal. Wkly. Rep. 1999; 48:1–17, 21–15.

Anonymous. Section 14.8 Lyme Disease. In: Healthy People 2010: Understanding and improving health and objectives for improving health (2nd ed.). pp. 14.11–14.49. Washington, DC: U.S. Government Printing Office, 2000.

Anonymous. Synopsis of LYMRrix Phase IV Observational Study. Bethesda, MD: United States Food and Drug Administration Center for Biologics Evaluation and Research, 2001a.

Anonymous. Vaccines and Related Biological Products Advisory Committee Meeting, January 31, 2001. Bethesda, MD: United States Food and Drug Administration Center for Biologics Evaluation and Research, 2001b.

Anonymous. Baxter vaccines: Lyme borreliosis vaccine. Baxter Vaccines Product Pipeline, 2007.

Apt, D., Raviprakash, K., Brinkman, A., Semyonov, A., Yang, S., Skinner, C., Diehl, L., Lyons, R., Porter, K. and Punnonen, J. Tetravalent neutralizing antibody response against four dengue serotypes by a single chimeric dengue envelope antigen. Vaccine 2006; 24:335–344.

Asbrink, E. Cutaneous manifestations of Lyme borreliosis. Clinical definitions and differential diagnoses. Scand. J. Infect. Dis. Suppl. 1991; 77:44–50.

Asbrink, E. and Hovmark, A. Lyme borreliosis: aspects of tick-borne *Borrelia burgdorferi* infection from a dermatologic viewpoint. Semin. Dermatol. 1990; 9:277–291.

Asbrink, E., Hovmark, A. and Hederstedt, B. The spirochetal etiology of acrodermatitis chronica atrophicans Herxheimer. Acta Derm. Venereol. 1984; 64:506–512.

Asch, E.S., Bujak, D.I., Weiss, M., Peterson, M.G. and Weinstein, A. Lyme disease: an infectious and postinfectious syndrome. J. Rheumatol. 1994; 21:454–461.

Attie, O., Bruno, J.F., Xu, Y., Qiu, D., Luft, B.J. and Qiu, W.G. Coevolution of the outer surface protein C gene (*ospC*) and intraspecific lineages of *Borrelia burgdorferi* sensu stricto in the northeastern United States. Infect. Genet. Evol. 2007; 7:1–12.

Auwaerter, P.G. Point: antibiotic therapy is not the answer for patients with persisting symptoms attributable to Lyme disease. Clin. Infect. Dis. 2007; 45:143–148.

Aydintug, M.K., Gu, Y. and Philipp, M.T. *Borrelia burgdorferi* antigens that are targeted by antibody-dependent, complement-mediated killing in the rhesus monkey. Infect. Immun. 1994; 62:4929–4937.

Balmelli, T. and Piffaretti, J.C. Association between different clinical manifestations of Lyme disease and different species of *Borrelia burgdorferi* sensu lato. Res. Microbiol. 1995; 146:329–340.

Bannwarth, A. Zur klinik und pathogenese der "chronischen lymphocytaren meningitis". Arch. Psychiatr. Nervenkr. 1944; 117:161–185.

Baranton, G., Postic, D., Saint Girons, I., Boerlin, P., Piffaretti, J.C., Assous, M. and Grimont, P.A. Delineation of *Borrelia burgdorferi* sensu stricto, *Borrelia garinii* sp. nov., and group VS461 associated with Lyme borreliosis. Int. J. Syst. Bacteriol. 1992; 42:378–383.

Barbour, A.G. Isolation and cultivation of Lyme disease spirochetes. Yale J. Biol. Med. 1984; 57:521–525.

Barbour, A.G. Plasmid analysis of *Borrelia burgdorferi*, the Lyme disease agent. J. Clin. Microbiol. 1988; 26:475–478.

Barbour, A.G., Carter, C.J., Bundoc, V. and Hinnebusch, J. The nucleotide sequence of a linear plasmid of *Borrelia burgdorferi* reveals similarities to those of circular plasmids of other prokaryotes. J. Bacteriol. 1996a; 178:6635–6639.

Barbour, A.G. and Garon, C.F. Linear plasmids of the bacterium *Borrelia burgdorferi* have covalently closed ends. Science 1987; 237:409–411.

Barbour, A.G. and Garon, C.F. The genes encoding major surface proteins of *Borrelia burgdorferi* are located on a plasmid. Ann. N. Y. Acad. Sci. 1988; 539:144–153.

Barbour, A.G. and Hayes, S.F. Biology of *Borrelia* species. Microbiol. Rev. 1986; 50:381–400.

Barbour, A.G., Maupin, G.O., Teltow, G.J., Carter, C.J. and Piesman, J. Identification of an uncultivable *Borrelia* species in the hard tick *Amblyomma americanum*: possible agent of a Lyme disease-like illness. J. Infect. Dis. 1996b; 173:403–409.

Barthold, S.W. and Bockenstedt, L.K. Passive immunizing activity of sera from mice infected with *Borrelia burgdorferi*. Infect. Immun. 1993; 61:4696–4702.

Barthold, S.W., deSouza, M. and Feng, S. Serum-mediated resolution of Lyme arthritis in mice. Lab. Invest. 1996; 74:57–67.

Barthold, S.W., Hodzic, E., Tunev, S. and Feng, S. Antibody-mediated disease remission in the mouse model of Lyme borreliosis. Infect. Immun. 2006; 74:4817–4825.

Barthold, S.W., Persing, D.H., Armstrong, A.L. and Peeples, R.A. Kinetics of *Borrelia burgdorferi* dissemination and evolution of disease after intradermal inoculation of mice. Am. J. Pathol. 1991; 139:263–273.

Bartunek, P., Mrazek, V., Gorican, K., Bina, R., Listvanova, S. and Zapletalova, J. *Borrelia* infection as a cause of carditis (a long-term study). Wien. Klin. Wochenschr. 2001; 113:38–44.

Benach, J.L., Bosler, E.M., Hanrahan, J.P., Coleman, J.L., Habicht, G.S., Bast, T.F., Cameron, D.J., Ziegler, J.L., Barbour, A.G., Burgdorfer, W., Edelman, R. and Kaslow, R.A. Spirochetes isolated from the blood of two patients with Lyme disease. N. Engl. J. Med. 1983; 308:740–742.

Ben-Menachem, G., Kubler-Kielb, J., Coxon, B., Yergey, A. and Schneerson, R. A newly discovered cholesteryl galactoside from *Borrelia burgdorferi*. Proc. Natl. Acad. Sci. U. S. A. 2003; 100:7913–7918.

Berger, B.W. Current aspects of Lyme disease and other *Borrelia burgdorferi* infections. Dermatol. Clin. 1997; 15:247–255.

Bockenstedt, L.K., Fikrig, E., Barthold, S.W., Flavell, R.A. and Kantor, F.S. Identification of a *Borrelia burgdorferi* OspA T cell epitope that promotes anti-OspA IgG in mice. J. Immunol. 1996; 157:5496–5502.

Bockenstedt, L.K., Hodzic, E., Feng, S., Bourrel, K.W., de Silva, A., Montgomery, R.R., Fikrig, E., Radolf, J.D. and Barthold, S.W. *Borrelia burgdorferi* strain-specific OspC-mediated immunity in mice. Infect. Immun. 1997; 65:4661–4667.

Bouche, F.B., Steinmetz, A., Yanagi, Y. and Muller, C.P. Induction of broadly neutralizing antibodies against measles virus mutants using a polyepitope vaccine strategy. Vaccine 2005; 23:2074–2077.

Brooks, C.S., Hefty, P.S., Jolliff, S.E. and Akins, D.R. Global analysis of *Borrelia burgdorferi* genes regulated by mammalian host-specific signals. Infect. Immun. 2003; 71:3371–3383.

Brooks, C.S., Vuppala, S.R., Jett, A.M. and Akins, D.R. Identification of *Borrelia burgdorferi* outer surface proteins. Infect. Immun. 2006; 74:296–304.

Brooks, C.S., Vuppala, S.R., Jett, A.M., Alitalo, A., Meri, S. and Akins, D.R. Complement regulator-acquiring surface protein 1 imparts resistance to human serum in *Borrelia burgdorferi*. J. Immunol. 2005; 175:3299–3308.

Brossard, M. and Wikel, S.K. Tick immunobiology. Parasitology 2004; 129(Suppl):S161–S176.

Brown, C.R., Blaho, V.A. and Loiacono, C.M. Susceptibility to experimental Lyme arthritis correlates with KC and monocyte chemoattractant protein-1 production in joints and requires neutrophil recruitment via CXCR2. J. Immunol. 2003; 171:893–901.

Brown, E.L., Kim, J.H., Reisenbichler, E.S. and Hook, M. Multicomponent Lyme vaccine: three is not a crowd. Vaccine 2005; 23:3687–3696.

Brown, R.N. and Lane, R.S. Lyme disease in California: A novel enzootic transmission cycle of *Borrelia burgdorferi*. Science 1992; 256:1439–1442.

Brownstein, J.S., Holford, T.R. and Fish, D. A climate-based model predicts the spatial distribution of the Lyme disease vector *Ixodes scapularis* in the United States. Environ. Health Perspect. 2003; 111:1152–1157.

Buckles, E.L., Earnhart, C.G. and Marconi, R.T. Analysis of antibody response in humans to the type A OspC loop 5 domain and assessment of the potential utility of the loop 5 epitope in Lyme disease vaccine development. Clin. Vaccine Immunol. 2006; 13:1162–1165.

Bunikis, J. and Barbour, A.G. Access of antibody or trypsin to an integral outer membrane protein (P66) of *Borrelia burgdorferi* is hindered by Osp lipoproteins. Infect. Immun. 1999; 67:2874–2883.

Bunikis, J., Garpmo, U., Tsao, J., Berglund, J., Fish, D. and Barbour, A.G. Sequence typing reveals extensive strain diversity of the Lyme borreliosis agents *Borrelia burgdorferi* in North America and *Borrelia afzelii* in Europe. Microbiology 2004; 150:1741–1755.

Burgdorfer, W., Barbour, A.G., Hayes, S.F., Benach, J.L., Grunwaldt, E. and Davis, J.P. Lyme disease—a tick-borne spirochetosis? Science 1982; 216:1317–1319.

Busch, U., Hizo-Teufel, C., Boehmer, R., Fingerle, V., Nitschko, H., Wilske, B. and Preac-Mursic, V. Three species of *Borrelia burgdorferi* sensu lato (*B. burgdorferi* sensu stricto, *B. afzelii*, and *B. garinii*) identified from cerebrospinal fluid isolates by pulsed-field gel electrophoresis and PCR. J. Clin. Microbiol. 1996; 34:1072–1078.

Byram, R., Stewart, P.E. and Rosa, P. The essential nature of the ubiquitous 26-kilobase circular replicon of *Borrelia burgdorferi*. J. Bacteriol. 2004; 186:3561–3569.

Caimano, M.J., Eggers, C.H., Gonzalez, C.A. and Radolf, J.D. Alternate sigma factor RpoS is required for the in vivo-specific repression of *Borrelia burgdorferi* plasmid lp54-borne *ospA* and *lp6.6* genes. J. Bacteriol. 2005; 187:7845–7852.

Caimano, M.J., Eggers, C.H., Hazlett, K.R. and Radolf, J.D. RpoS is not central to the general stress response in *Borrelia burgdorferi* but does control expression of one or more essential virulence determinants. Infect. Immun. 2004; 72:6433–6445.

Caimano, M.J., Iyer, R., Eggers, C.H., Gonzalez, C., Morton, E.A., Gilbert, M.A., Schwartz, I. and Radolf, J.D. Analysis of the RpoS regulon in *Borrelia burgdorferi* in response to mammalian host signals provides insight into RpoS function during the enzootic cycle. Mol. Microbiol. 2007; 65:1193–1217.

Caimano, M.J., Yang, X., Popova, T.G., Clawson, M.L., Akins, D.R., Norgard, M.V. and Radolf, J.D. Molecular and evolutionary characterization of the cp32/18 family of supercoiled plasmids in *Borrelia burgdorferi* 297. Infect. Immun. 2000; 68:1574–1586.

Campbell, G.L., Fritz, C.L., Fish, D., Nowakowski, J., Nadelman, R.B. and Wormser, G.P. Estimation of the incidence of Lyme disease. Am. J. Epidemiol. 1998; 148:1018–1026.

Caro-Aguilar, I., Lapp, S., Pohl, J., Galinski, M.R. and Moreno, A. Chimeric epitopes delivered by polymeric synthetic linear peptides induce protective immunity to malaria. Microbes Infect. 2005; 7:1324–1337.

Casjens, S., Murphy, M., DeLange, M., Sampson, L., van Vugt, R. and Huang, W.M. Telomeres of the linear chromosomes of Lyme disease spirochaetes: nucleotide sequence and possible exchange with linear plasmid telomeres. Mol. Microbiol. 1997; 26:581–596.

Casjens, S., Palmer, N., van Vugt, R., Huang, W.M., Stevenson, B., Rosa, P., Lathigra, R., Sutton, G., Peterson, J., Dodson, R.J., Haft, D., Hickey, E., Gwinn, M., White, O. and Fraser, C.M. A bacterial genome in flux: the twelve linear and nine circular extrachromosomal DNAs in an infectious isolate of the Lyme disease spirochete *Borrelia burgdorferi*. Mol. Microbiol. 2000; 35:490–516.

Cassatt, D.R., Patel, N.K., Ulbrandt, N.D. and Hanson, M.S. DbpA, but not OspA, is expressed by *Borrelia burgdorferi* during spirochetemia and is a target for protective antibodies. Infect. Immun. 1998; 66:5379–5387.

Chang, Y.F., Appel, M.J., Jacobson, R.H., Shin, S.J., Harpending, P., Straubinger, R., Patrican, L.A., Mohammed, H. and Summers, B.A. Recombinant OspA protects dogs against infection and disease caused by *Borrelia burgdorferi*. Infect. Immun. 1995; 63:3543–3549.

Charon, N.W., Greenberg, E.P., Koopman, M.B. and Limberger, R.J. Spirochete chemotaxis, motility, and the structure of the spirochetal periplasmic flagella. Res. Microbiol. 1992; 143:597–603.

Chen, J., Field, J.A., Glickstein, L., Molloy, P.J., Huber, B.T. and Steere, A.C. Association of antibiotic treatment-resistant Lyme arthritis with T cell responses to dominant epitopes of outer surface protein A of *Borrelia burgdorferi*. Arthritis Rheum. 1999; 42:1813–1822.

Chu, H.J., Chavez, L.G., Jr., Blumer, B.M., Sebring, R.W., Wasmoen, T.L. and Acree, W.M. Immunogenicity and efficacy study of a commercial *Borrelia burgdorferi* bacterin. J. Am. Vet. Med. Assoc. 1992; 201:403–411.

Collares-Pereira, M., Couceiro, S., Franca, I., Kurtenbach, K., Schafer, S.M., Vitorino, L., Goncalves, L., Baptista, S., Vieira, M.L. and Cunha, C. First isolation of *Borrelia lusitaniae* from a human patient. J. Clin. Microbiol. 2004; 42:1316–1318.

Conlon, J.A., Mather, T.N., Tanner, P. and Gallo, G. Efficacy of a nonadjuvanted, outer surface protein A, recombinant vaccine in dogs after challenge by ticks naturally infected with *Borrelia burgdorferi*. Vet. Ther. 2000; 1:96–107.

Cox, J. and Krajden, M. Cardiovascular manifestations of Lyme disease. Am. Heart J. 1991; 122:1449–1455.

Coyle, B.S., Strickland, G.T., Liang, Y.Y., Pena, C., McCarter, R. and Israel, E. The public health impact of Lyme disease in Maryland. J. Infect. Dis. 1996; 173:1260–1262.

Coyle, P.K. and Schutzer, S.E. Neurologic aspects of Lyme disease. Med. Clin. North Am. 2002; 86:261–284.

Coyle, P.K., Schutzer, S.E., Deng, Z., Krupp, L.B., Belman, A.L., Benach, J.L. and Luft, B.J. Detection of *Borrelia burgdorferi*-specific antigen in antibody-negative cerebrospinal fluid in neurologic Lyme disease. Neurology 1995; 45:2010–2015.

Croke, C.L., Munson, E.L., Lovrich, S.D., Christopherson, J.A., Remington, M.C., England, D.M., Callister, S.M. and Schell, R.F. Occurrence of severe destructive Lyme arthritis in hamsters vaccinated with outer surface protein A and challenged with *Borrelia burgdorferi*. Infect. Immun. 2000; 68:658–663.

Dale, J.B. Multivalent group A streptococcal vaccine designed to optimize the immunogenicity of six tandem M protein fragments. Vaccine 1999; 17:193–200.

Dale, J.B., Chiang, E.Y. and Lederer, J.W. Recombinant tetravalent group A streptococcal M protein vaccine. J. Immunol. 1993; 151:2188–2194.

Dale, J.B., Penfound, T., Chiang, E.Y., Long, V., Shulman, S.T. and Beall, B. Multivalent group A streptococcal vaccine elicits bactericidal antibodies against variant M subtypes. Clin. Diagn. Lab. Immunol. 2005; 12:833–836.

Dale, J.B., Simmons, M., Chiang, E.C. and Chiang, E.Y. Recombinant, octavalent group A streptococcal M protein vaccine. Vaccine 1996; 14:944–948.

de Koning, J., Hoogkamp-Korstanje, J.A., van der Linde, M.R. and Crijns, H.J. Demonstration of spirochetes in cardiac biopsies of patients with Lyme disease. J. Infect. Dis. 1989; 160:150–153.

de la Fuente, J. and Kocan, K.M. Strategies for development of vaccines for control of Ixodid tick species. Parasite Immunol. 2006; 28:275–283.

Derdakova, M. and Lencakova, D. Association of genetic variability within the *Borrelia burgdorferi* sensu lato with the ecology, epidemiology of Lyme borreliosis in Europe. Ann. Agric. Environ. Med. 2005; 12:165–172.

de Silva, A.M., Telford, S.R., 3rd, Brunet, L.R., Barthold, S.W. and Fikrig, E. *Borrelia burgdorferi* OspA is an arthropod-specific transmission-blocking Lyme disease vaccine. J. Exp. Med. 1996; 183:271–275.

de Silva, A.M., Zeidner, N.S., Zhang, Y., Dolan, M.C., Piesman, J. and Fikrig, E. Influence of outer surface protein A antibody on *Borrelia burgdorferi* within feeding ticks. Infect. Immun. 1999; 67:30–35.

Ding, W., Huang, X., Yang, X., Dunn, J.J., Luft, B.J., Koide, S. and Lawson, C.L. Structural identification of a key protective B-cell epitope in Lyme disease antigen OspA. J. Mol. Biol. 2000; 302:1153–1164.

Diza, E., Papa, A., Vezyri, E., Tsounis, S., Milonas, I. and Antoniadis, A. *Borrelia valaisiana* in cerebrospinal fluid. Emerg. Infect. Dis. 2004; 10:1692–1693.

Drouin, E.E., Glickstein, L., Kwok, W.W., Nepom, G.T. and Steere, A.C. Human homologues of a *Borrelia* T cell epitope associated with antibiotic-refractory Lyme arthritis. Mol. Immunol. 2008; 45:180–189.

Dunn, J.J., Buchstein, S.R., Butler, L.L., Fisenne, S., Polin, D.S., Lade, B.N. and Luft, B.J. Complete nucleotide sequence of a circular plasmid from the Lyme disease spirochete, *Borrelia burgdorferi*. J. Bacteriol. 1994; 176:2706–2717.

Dunne, M., al-Ramadi, B.K., Barthold, S.W., Flavell, R.A. and Fikrig, E. Oral vaccination with an attenuated *Salmonella typhimurium* strain expressing *Borrelia burgdorferi* OspA prevents murine Lyme borreliosis. Infect. Immun. 1995; 63:1611–1614.

Dykhuizen, D.E. and Baranton, G. The implications of a low rate of horizontal transfer in *Borrelia*. Trends Microbiol. 2001; 9:344–350.

Earnhart, C.G., Buckles, E.L., Dumler, J.S. and Marconi, R.T. Demonstration of OspC type diversity in invasive human Lyme disease isolates and identification of previously uncharacterized epitopes that define the specificity of the OspC murine antibody response. Infect. Immun. 2005; 73:7869–7877.

Earnhart, C.G., Buckles, E.L. and Marconi, R.T. Development of an OspC-based tetravalent, recombinant, chimeric vaccinogen that elicits bactericidal antibody against diverse Lyme disease spirochete strains. Vaccine 2007; 25:466–480.

Earnhart, C.G. and Marconi, R.T. Construction and analysis of variants of a polyvalent Lyme disease vaccine: approaches for improving the immune response to chimeric vaccinogens. Vaccine 2007a; 25:3419–3427.

Earnhart, C.G. and Marconi, R.T. An octavalent Lyme disease vaccine induces antibodies that recognize all incorporated OspC type-specific sequences. Hum. Vaccines 2007b; 3(6):281–289.

Earnhart, C.G. and Marconi, R.T. OspC phylogenetic analyses support the feasibility of a broadly protective polyvalent chimeric Lyme disease vaccine. Clin. Vaccine Immunol. 2007c; 14:628–634.

Eggers, C.H., Casjens, S., Hayes, S.F., Garon, C.F., Damman, C.J., Oliver, D.B. and Samuels, D.S. Bacteriophages of spirochetes. J. Mol. Microbiol. Biotechnol. 2000; 2:365–373.

Eggers, C.H. and Samuels, D.S. Molecular evidence for a new bacteriophage of *Borrelia burgdorferi*. J. Bacteriol. 1999; 181:7308–7313.

Eicken, C., Sharma, V., Klabunde, T., Owens, R.T., Pikas, D.S., Hook, M. and Sacchettini, J.C. Crystal structure of Lyme disease antigen outer surface protein C from *Borrelia burgdorferi*. J. Biol. Chem. 2001; 276:10010–10015.

Eiffert, H., Karsten, A., Thomssen, R. and Christen, H.J. Characterization of *Borrelia burgdorferi* strains in Lyme arthritis. Scand. J. Infect. Dis. 1998; 30:265–268.

Erdile, L.F., Brandt, M.A., Warakomski, D.J., Westrack, G.J., Sadziene, A., Barbour, A.G. and Mays, J.P. Role of attached lipid in immunogenicity of *Borrelia burgdorferi* OspA. Infect. Immun. 1993; 61:81–90.

Fahrer, H., van der Linden, S.M., Sauvain, M.J., Gern, L., Zhioua, E. and Aeschlimann, A. The prevalence and incidence of clinical and asymptomatic Lyme borreliosis in a population at risk. J. Infect. Dis. 1991; 163:305–310.

Fan, C.F., Zeng, R.H., Sun, C.Y., Mei, X.G., Wang, Y.F. and Liu, Y. Fusion of DsbA to the N-terminus of CTL chimeric epitope, F/M2:81-95, of respiratory syncytial virus prolongs protein- and virus-specific CTL responses in Balb/c mice. Vaccine 2005; 23:2869–2875.

Feder, H.M., Jr., Johnson, B.J.B., O'Connell, S., Shapiro, E.D., Steere, A.C. and Wormser, G.P. A critical appraisal of "chronic Lyme disease". N. Engl. J. Med. 2007; 357:1422–1430.

Feng, S., Hodzic, E. and Barthold, S.W. Lyme arthritis resolution with antiserum to a 37-kilodalton *Borrelia burgdorferi* protein. Infect. Immun. 2000; 68:4169–4173.

Feng, S., Hodzic, E., Stevenson, B. and Barthold, S.W. Humoral immunity to *Borrelia burgdorferi* N40 decorin binding proteins during infection of laboratory mice. Infect. Immun. 1998; 66: 2827–2835.

Fikrig, E., Barthold, S.W. and Flavell, R.A. OspA vaccination of mice with established *Borrelia burgdorferi* infection alters disease but not infection. Infect. Immun. 1993; 61:2553–2557.

Fikrig, E., Barthold, S.W., Kantor, F.S. and Flavell, R.A. Protection of mice against the Lyme disease agent by immunizing with recombinant OspA. Science 1990; 250:553–556.

Fikrig, E., Barthold, S.W., Kantor, F.S. and Flavell, R.A. Protection of mice from Lyme borreliosis by oral vaccination with *Escherichia coli* expressing OspA. J. Infect. Dis. 1991; 164:1224–1227.

Fikrig, E., Barthold, S.W., Marcantonio, N., Deponte, K., Kantor, F.S. and Flavell, R.A. Roles of OspA, OspB, and flagellin in protective immunity to Lyme borreliosis in laboratory mice. Infect. Immun. 1992; 60:657–661.

Fikrig, E., Barthold, S.W., Sun, W., Feng, W., Telford, S.R., 3rd and Flavell, R.A. *Borrelia burgdorferi* P35 and P37 proteins, expressed in vivo, elicit protective immunity. Immunity 1997; 6:531–539.

Fingerle, V., Goettner, G., Gern, L., Wilske, B. and Schulte-Spechtel, U. Complementation of a *Borrelia afzelii* OspC mutant highlights the crucial role of OspC for dissemination of *Borrelia afzelii* in *Ixodes ricinus*. Int. J. Med. Microbiol. 2007; 297:97–107.

Fingerle, V., Hauser, U., Liegl, G., Petko, B., Preac-Mursic, V. and Wilske, B. Expression of outer surface proteins A and C of *Borrelia burgdorferi* in *Ixodes ricinus*. J. Clin. Microbiol. 1995; 33:1867–1869.

Fraser, C.M., Casjens, S., Huang, W.M., Sutton, G.G., Clayton, R., Lathigra, R., White, O., Ketchum, K.A., Dodson, R., Hickey, E.K., Gwinn, M., Dougherty, B., Tomb, J.F., Fleischmann, R.D., Richardson, D., Peterson, J., Kerlavage, A.R., Quackenbush, J., Salzberg, S., Hanson, M., van Vugt, R., Palmer, N., Adams, M.D., Gocayne, J., Weidman, J., Utterback, T., Watthey, L., McDonald, L., Artiach, P., Bowman, C., Garland, S., Fujii, C., Cotton, M.D., Horst, K., Roberts, K., Hatch, B., Smith, H.O. and Venter, J.C. Genomic sequence of a Lyme disease spirochaete, *Borrelia burgdorferi*. Nature 1997; 390:580–586.

Fukunaga, M., Hamase, A., Okada, K. and Nakao, M. *Borrelia tanukii* sp. nov. and *Borrelia turdae* sp. nov. found from Ixodid ticks in Japan: rapid species identification by 16S rRNA gene-targeted PCR analysis. Microbiol. Immunol. 1996; 40:877–881.

Fukunaga, M., Sohnaka, M. and Yanagihara, Y. Analysis of *Borrelia* species associated with Lyme disease by rRNA gene restriction fragment length polymorphism. J. Gen. Microbiol. 1993; 139: 1141–1146.

Gerber, M.A., Shapiro, E.D., Burke, G.S., Parcells, V.J. and Bell, G.L. Lyme disease in children in southeastern Connecticut. Pediatric Lyme Disease Study Group. N. Engl. J. Med. 1996; 335:1270–1274.

Gern, L., Estrada-Pena, A., Frandsen, F., Gray, J.S., Jaenson, T.G., Jongejan, F., Kahl, O., Korenberg, E., Mehl, R. and Nuttall, P.A. European reservoir hosts of *Borrelia burgdorferi* sensu lato. Zentralbl. Bakteriol 1998; 287:196–204.

Gern, L., Hu, C.M., Voet, P., Hauser, P. and Lobet, Y. Immunization with a polyvalent OspA vaccine protects mice against *Ixodes ricinus* tick bites infected by *Borrelia burgdorferi* ss, *Borrelia garinii* and *Borrelia afzelii*. Vaccine 1997; 15:1551–1557.

Gilbert, M.A., Morton, E.A., Bundle, S.F. and Samuels, D.S. Artificial regulation of *ospC* expression in *Borrelia burgdorferi*. Mol. Microbiol. 2007; 63:1259–1273.

Gilmore, R.D., Jr., Kappel, K.J., Dolan, M.C., Burkot, T.R. and Johnson, B.J. Outer surface protein C (OspC), but not P39, is a protective immunogen against a tick-transmitted *Borrelia burgdorferi* challenge: evidence for a conformational protective epitope in OspC. Infect. Immun. 1996; 64:2234–2239.

Gilmore, R.D., Jr. and Mbow, M.L. Conformational nature of the *Borrelia burgdorferi* B31 outer surface protein C protective epitope. Infect. Immun. 1999; 67:5463–5469.

Glatzel, A., Entschladen, F., Zollner, T.M., Kraiczy, P., Brade, V., Kaufmann, R., Janssen, O., Lengl-Janssen, B., Wesch, D. and Kabelitz, D. The responsiveness of human V delta 1 gamma delta T cells to *Borrelia burgdorferi* is largely restricted to synovial-fluid cells from patients with Lyme arthritis. J. Infect. Dis. 2002; 186:1043–1046.

Glockner, G., Lehmann, R., Romualdi, A., Pradella, S., Schulte-Spechtel, U., Schilhabel, M., Wilske, B., Suhnel, J. and Platzer, M. Comparative analysis of the *Borrelia garinii* genome. Nucleic Acids Res. 2004; 32:6038–6046.

Glockner, G., Schulte-Spechtel, U., Schilhabel, M., Felder, M., Suhnel, J., Wilske, B. and Platzer, M. Comparative genome analysis: selection pressure on the *Borrelia vls* cassettes is essential for infectivity. BMC Genomics 2006; 7:211.

Golde, W.T., Piesman, J., Dolan, M.C., Kramer, M., Hauser, P., Lobet, Y., Capiau, C., Desmons, P., Voet, P., Dearwester, D. and Frantz, J.C. Reactivity with a specific epitope of outer surface protein A predicts protection from infection with the Lyme disease spirochete, *Borrelia burgdorferi*. Infect. Immun. 1997; 65:882–889.

Goldstein, S.F., Buttle, K.F. and Charon, N.W. Structural analysis of the *Leptospiraceae* and *Borrelia burgdorferi* by high-voltage electron microscopy. J. Bacteriol. 1996; 178:6539–6545.

Goldstein, S.F., Charon, N.W. and Kreiling, J.A. *Borrelia burgdorferi* swims with a planar waveform similar to that of eukaryotic flagella. Proc. Natl. Acad. Sci. U. S. A. 1994; 91:3433–3437.

Gomes-Solecki, M.J., Brisson, D.R. and Dattwyler, R.J. Oral vaccine that breaks the transmission cycle of the Lyme disease spirochete can be delivered via bait. Vaccine 2006; 24:4440–4449.

Goodman, J.L., Bradley, J.F., Ross, A.E., Goellner, P., Lagus, A., Vitale, B., Berger, B.W., Luger, S. and Johnson, R.C. Bloodstream invasion in early Lyme disease: results from a prospective, controlled, blinded study using the polymerase chain reaction. Am. J. Med. 1995; 99:6–12.

Grimm, D., Tilly, K., Byram, R., Stewart, P.E., Krum, J.G., Bueschel, D.M., Schwan, T.G., Policastro, P.F., Elias, A.F. and Rosa, P.A. Outer-surface protein C of the Lyme disease spirochete: a protein induced in ticks for infection of mammals. Proc. Natl. Acad. Sci. U. S. A. 2004; 101:3142–3147.

Gross, D.M., Forsthuber, T., Tary-Lehmann, M., Etling, C., Ito, K., Nagy, Z.A., Field, J.A., Steere, A.C. and Huber, B.T. Identification of LFA-1 as a candidate autoantigen in treatment-resistant Lyme arthritis. Science 1998a; 281:703–706.

Gross, D.M. and Huber, B.T. Cellular and molecular aspects of Lyme arthritis. Cell. Mol. Life Sci. 2000; 57:1562–1569.

Gross, D.M., Steere, A.C. and Huber, B.T. T helper 1 response is dominant and localized to the synovial fluid in patients with Lyme arthritis. J. Immunol. 1998b; 160:1022–1028.

Guerau-de-Arellano, M., Alroy, J. and Huber, B.T. Beta2 integrins control the severity of murine Lyme carditis. Infect. Immun. 2005; 73:3242–3250.

Guo, B.P., Norris, S.J., Rosenberg, L.C. and Hook, M. Adherence of *Borrelia burgdorferi* to the proteoglycan decorin. Infect. Immun. 1995; 63:3467–3472.

Hagman, K.E., Lahdenne, P., Popova, T.G., Porcella, S.F., Akins, D.R., Radolf, J.D. and Norgard, M.V. Decorin-binding protein of *Borrelia burgdorferi* is encoded within a two-gene operon and is protective in the murine model of Lyme borreliosis. Infect. Immun. 1998; 66:2674–2683.

Hagman, K.E., Yang, X., Wikel, S.K., Schoeler, G.B., Caimano, M.J., Radolf, J.D. and Norgard, M.V. Decorin-binding protein A (DbpA) of *Borrelia burgdorferi* is not protective when immunized mice are challenged via tick infestation and correlates with the lack of DbpA expression by *B. burgdorferi* in ticks. Infect. Immun. 2000; 68:4759–4764.

Halperin, J., Luft, B.J., Volkman, D.J. and Dattwyler, R.J. Lyme neuroborreliosis. Peripheral nervous system manifestations. Brain 1990a; 113(Pt 4):1207–1221.

Halperin, J.J., Krupp, L.B., Golightly, M.G. and Volkman, D.J. Lyme borreliosis-associated encephalopathy. Neurology 1990b; 40:1340–1343.

Halperin, J.J., Luft, B.J., Anand, A.K., Roque, C.T., Alvarez, O., Volkman, D.J. and Dattwyler, R.J. Lyme neuroborreliosis: central nervous system manifestations. Neurology 1989; 39:753–759.

Halperin, J.J., Pass, H.L., Anand, A.K., Luft, B.J., Volkman, D.J. and Dattwyler, R.J. Nervous system abnormalities in Lyme disease. Ann. N. Y. Acad. Sci. 1988; 539:24–34.

Halperin, J.J., Volkman, D.J. and Wu, P. Central nervous system abnormalities in Lyme neuroborreliosis. Neurology 1991; 41: 1571–1582.

Hanrahan, J.P., Benach, J.L., Coleman, J.L., Bosler, E.M., Morse, D.L., Cameron, D.J., Edelman, R. and Kaslow, R.A. Incidence and cumulative frequency of endemic Lyme disease in a community. J. Infect. Dis. 1984; 150:489–496.

Hanson, M.S., Cassatt, D.R., Guo, B.P., Patel, N.K., McCarthy, M.P., Dorward, D.W. and Hook, M. Active and passive immunity against *Borrelia burgdorferi* decorin binding protein A (DbpA) protects against infection. Infect. Immun. 1998; 66:2143–2153.

Hanson, M.S. and Edelman, R. Vaccines against Lyme disease. In: New Generation Vaccines (M. Levine, J.B. Kaper, R. Rappuoli, M.A. Liu and M.F. Good, Eds.), Vol. 3, pp. 487–498. New York: Marcel Dekker AG, 2004.

Hanson, M.S., Patel, N.K., Cassatt, D.R. and Ulbrandt, N.D. Evidence for vaccine synergy between *Borrelia burgdorferi* decorin binding protein A and outer surface protein A in the mouse model of Lyme borreliosis. Infect. Immun. 2000; 68:6457–6460.

Hardin, J.A., Steere, A.C. and Malawista, S.E. Immune complexes and the evolution of Lyme arthritis. Dissemination and localization of abnormal C1q binding activity. N. Engl. J. Med. 1979; 301:1358–1363.

Hefty, P.S., Jolliff, S.E., Caimano, M.J., Wikel, S.K. and Akins, D.R. Changes in temporal and spatial patterns of outer surface lipoprotein expression generate population heterogeneity and antigenic diversity in the Lyme disease spirochete, *Borrelia burgdorferi*. Infect. Immun. 2002; 70:3468–3478.

Hefty, P.S., Jolliff, S.E., Caimano, M.J., Wikel, S.K., Radolf, J.D. and Akins, D.R. Regulation of OspE-related, OspF-related, and Elp lipoproteins of *Borrelia burgdorferi* strain 297 by mammalian host-specific signals. Infect. Immun. 2001; 69:3618–3627.

Hellerstrom, S. Erythema chronicum migrans Afzelius with meningitis. Acta Derm. Venereol. 1951; 31:227–234.

Hellwage, J., Meri, T., Heikkila, T., Alitalo, A., Panelius, J., Lahdenne, P., Seppala, I.J. and Meri, S. The complement regulator factor H binds to the surface protein OspE of *Borrelia burgdorferi*. J. Biol. Chem. 2001; 276:8427–8435.

Hemmer, B., Gran, B., Zhao, Y., Marques, A., Pascal, J., Tzou, A., Kondo, T., Cortese, I., Bielekova, B., Straus, S.E., McFarland, H.F., Houghten, R., Simon, R., Pinilla, C. and Martin, R. Identification of candidate T-cell epitopes and molecular mimics in chronic Lyme disease. Nat. Med. 1999; 5:1375–1382.

Hengge, U.R., Tannapfel, A., Tyring, S.K., Erbel, R., Arendt, G. and Ruzicka, T. Lyme borreliosis. Lancet Infect. Dis. 2003; 3:489–500.

Herxheimer, K. and Hartmann, K. Ueber acrodermatitis chronica atrophicans. Arch. Dermatol. (Berl.) 1902; 61:57–76.

Hinnebusch, J., Bergstrom, S. and Barbour, A.G. Cloning and sequence analysis of linear plasmid telomeres of the bacterium *Borrelia burgdorferi*. Mol. Microbiol. 1990; 4:811–820.

Hodzic, E., Feng, S. and Barthold, S.W. Stability of *Borrelia burgdorferi* outer surface protein C under immune selection pressure. J. Infect. Dis. 2000; 181:750–753.

Hollstrom, E. Successful treatment of erythema migrans Afzelius. Acta Derm. Venereol. 1951; 31:235–243.

Hossain, H., Wellensiek, H.J., Geyer, R. and Lochnit, G. Structural analysis of glycolipids from *Borrelia burgdorferi*. Biochimie 2001; 83:683–692.

Hovmark, A., Asbrink, E. and Olsson, I. The spirochetal etiology of lymphadenosis benigna cutis solitaria. Acta Derm. Venereol. 1986; 66:479–484.

Howe, T.R., LaQuier, F.W. and Barbour, A.G. Organization of genes encoding two outer membrane proteins of the Lyme disease agent *Borrelia burgdorferi* within a single transcriptional unit. Infect. Immun. 1986; 54:207–212.

Hu, M.C., Walls, M.A., Stroop, S.D., Reddish, M.A., Beall, B. and Dale, J.B. Immunogenicity of a 26-valent group A streptococcal vaccine. Infect. Immun. 2002; 70:2171–2177.

Huang, X., Link, K., Koide, A., Dunn, J.J., Luft, B.J. and Koide, S. ^{1}H, ^{13}C, and ^{15}N NMR backbone assignments of 37 kDa surface antigen OspC from *Borrelia burgdorferi*. J. Biomol. NMR 1999; 14:283–284.

Huang, X., Yang, X., Luft, B.J. and Koide, S. NMR identification of epitopes of Lyme disease antigen OspA to monoclonal antibodies. J. Mol. Biol. 1998; 281:61–67.

Hubner, A., Yang, X., Nolen, D.M., Popova, T.G., Cabello, F.C. and Norgard, M.V. Expression of *Borrelia burgdorferi* OspC and DbpA is controlled by a RpoN-RpoS regulatory pathway. Proc. Natl. Acad. Sci. U. S. A. 2001; 98:12724–12729.

Imamura, S., da Silva Vaz Junior, I., Sugino, M., Ohashi, K. and Onuma, M. A serine protease inhibitor (serpin) from *Haemaphysalis longicornis* as an anti-tick vaccine. Vaccine 2005; 23:1301–1311.

Imamura, S., Namangala, B., Tajima, T., Tembo, M.E., Yasuda, J., Ohashi, K. and Onuma, M. Two serine protease inhibitors (serpins) that induce a bovine protective immune response against *Rhipicephalus appendiculatus* ticks. Vaccine 2006; 24:2230–2237.

Iyer, R., Kalu, O., Purser, J., Norris, S., Stevenson, B. and Schwartz, I. Linear and circular plasmid content in *Borrelia burgdorferi* clinical isolates. Infect. Immun. 2003; 71:3699–3706.

Iyer, R., Liveris, D., Adams, A., Nowakowski, J., McKenna, D., Bittker, S., Cooper, D., Wormser, G.P. and Schwartz, I. Characterization of *Borrelia burgdorferi* isolated from erythema migrans lesions: interrelationship of three molecular typing methods. J. Clin. Microbiol. 2001; 39:2954–2957.

Jaulhac, B., Chary-Valckenaere, I., Sibilia, J., Javier, R.M., Piemont, Y., Kuntz, J.L., Monteil, H. and Pourel, J. Detection of *Borrelia burgdorferi* by DNA amplification in synovial tissue samples from patients with Lyme arthritis. Arthritis Rheum. 1996; 39:736–745.

Jobe, D.A., Lovrich, S.D., Schell, R.F. and Callister, S.M. C-terminal region of outer surface protein C binds borreliacidal antibodies in sera from patients with Lyme disease. Clin. Diagn. Lab. Immunol. 2003; 10:573–578.

Johnson, R.C., Kodner, C. and Russell, M. Passive immunization of hamsters against experimental infection with the Lyme disease spirochete. Infect. Immun. 1986; 53:713–714.

Johnson, R.C., Kodner, C., Russell, M. and Duray, P.H. Experimental infection of the hamster with *Borrelia burgdorferi*. Ann. N. Y. Acad. Sci. 1988; 539:258–263.

Johnson, R.C., Schmid, G.P., Hyde, F.W., Stiegerwalt, A.G. and Brenner, D.J. *Borrelia burgdorferi* sp. nov.: etiologic agent of Lyme disease. Int. J. Syst. Bacteriol. 1984; 34:496–497.

Jonsson, N.N., Matschoss, A.L., Pepper, P., Green, P.E., Albrecht, M.S., Hungerford, J. and Ansell, J. Evaluation of tickGARD(PLUS), a novel vaccine against *Boophilus microplus*, in lactating Holstein-Friesian cows. Vet. Parasitol. 2000; 88:275–285.

Kalish, R.A., Leong, J.M. and Steere, A.C. Association of treatment-resistant chronic Lyme arthritis with HLA-DR4 and antibody reactivity to OspA and OspB of *Borrelia burgdorferi*. Infect. Immun. 1993; 61:2774–2779.

Kamradt, T., Lengl-Janssen, B., Strauss, A.F., Bansal, G. and Steere, A.C. Dominant recognition of a *Borrelia burgdorferi* outer surface protein A peptide by T helper cells in patients with treatment-resistant Lyme arthritis. Infect. Immun. 1996; 64:1284–1289.

Karadag, B., Spieker, L.E., Schwitter, J., Ruschitzka, F., Luscher, T.F., Noll, G. and Corti, R. Lyme carditis: *restitutio ad integrum* documented by cardiac magnetic resonance imaging. Cardiol. Rev. 2004; 12:185–187.

Karch, H. and Huppertz, H.I. Repeated detection of *Borrelia burgdorferi* DNA in synovial fluid of a child with Lyme arthritis. Rheumatol. Int. 1993; 12:227–229.

Kawabata, H., Masuzawa, T. and Yanagihara, Y. Genomic analysis of *Borrelia japonica* sp. nov. isolated from *Ixodes ovatus* in Japan. Microbiol. Immunol. 1993; 37:843–848.

Keirans, J.E., Hutcheson, H.J., Durden, L.A. and Klompen, J.S. *Ixodes scapularis* (Acari:Ixodidae): redescription of all active stages, distribution, hosts, geographical variation, and medical and veterinary importance. J. Med. Entomol. 1996; 33:297–318.

Keller, D., Koster, F.T., Marks, D.H., Hosbach, P., Erdile, L.F. and Mays, J.P. Safety and immunogenicity of a recombinant outer surface protein A Lyme vaccine. JAMA 1994; 271:1764–1768.

Kindstrand, E., Nilsson, B.Y., Hovmark, A., Nennesmo, I., Pirskanen, R., Solders, G. and Asbrink, E. Polyneuropathy in late Lyme borreliosis—a clinical, neurophysiological and morphological description. Acta Neurol. Scand. 2000; 101:47–52.

Kindstrand, E., Nilsson, B.Y., Hovmark, A., Pirskanen, R. and Asbrink, E. Peripheral neuropathy in acrodermatitis chronica atrophicans—a late *Borrelia* manifestation. Acta Neurol. Scand. 1997; 95:338–345.

Kindstrand, E., Nilsson, B.Y., Hovmark, A., Pirskanen, R. and Asbrink, E. Peripheral neuropathy in acrodermatitis chronica atrophicans—effect of treatment. Acta Neurol. Scand. 2002; 106:253–257.

Klein, J., Stanek, G., Bittner, R., Horvat, R., Holzinger, C. and Glogar, D. Lyme borreliosis as a cause of myocarditis and heart muscle disease. Eur. Heart J. 1991; 12(Suppl D):73–75.

Koide, S., Yang, X., Huang, X., Dunn, J.J. and Luft, B.J. Structure-based design of a second-generation Lyme disease vaccine based on a C-terminal fragment of *Borrelia burgdorferi* OspA. J. Mol. Biol. 2005; 350:290–299.

Kotloff, K.L., Corretti, M., Palmer, K., Campbell, J.D., Reddish, M.A., Hu, M.C., Wasserman, S.S. and Dale, J.B. Safety and immunogenicity of a recombinant multivalent group a streptococcal vaccine in healthy adults: phase 1 trial. JAMA 2004; 292:709–715.

Kotloff, K.L. and Dale, J.B. Progress in group A streptococcal vaccine development. Pediatr. Infect. Dis. J. 2004; 23:765–766.

Kraiczy, P., Hartmann, K., Hellwage, J., Skerka, C., Kirschfink, M., Brade, V., Zipfel, P.F., Wallich, R. and Stevenson, B. Immunological characterization of the complement regulator factor H-binding CRASP and Erp proteins of *Borrelia burgdorferi*. Int. J. Med. Microbiol. 2004a; 293(Suppl 37):152–157.

Kraiczy, P., Hellwage, J., Skerka, C., Becker, H., Kirschfink, M., Simon, M.M., Brade, V., Zipfel, P.F. and Wallich, R. Complement resistance of *Borrelia burgdorferi* correlates with the expression of BbCRASP-1, a novel linear plasmid-encoded surface protein that interacts with human factor H and FHL-1 and is unrelated to Erp proteins. J. Biol. Chem. 2004b; 279:2421–2429.

Kraiczy, P., Skerka, C. and Zipfel, P.F. Further characterization of complement regulator-acquiring surface proteins of *Borrelia burgdorferi*. Infect. Immun. 2001; 69:7800–7809.

Kreitman, M. and Di Rienzo, A. Balancing claims for balancing selection. Trends Genet. 2004; 20:300–304.

Kumaran, D., Eswaramoorthy, S., Luft, B.J., Koide, S., Dunn, J.J., Lawson, C.L. and Swaminathan, S. Crystal structure of outer surface protein C (OspC) from the Lyme disease spirochete, *Borrelia burgdorferi*. EMBO J. 2001; 20:971–978.

Labandeira-Rey, M., Seshu, J. and Skare, J.T. The absence of linear plasmid 25 or 28-1 of *Borrelia burgdorferi* dramatically alters the kinetics of experimental infection via distinct mechanisms. Infect. Immun. 2003; 71:4608–4613.

Labandeira-Rey, M. and Skare, J.T. Decreased infectivity in *Borrelia burglorferi* strain B31 is associated with loss of linear plasmid 25 or 28–1. Infect. Immun. 2001; 69:446–455.

Labandeira-Rey, M., Baker, E.A. and Skare, J.T. VraA (BBI16) protein of *Borrelia burgdorferi* is a surface-exposed antigen with a repetitive motif that confers partial protection against experimental Lyme borreliosis. Infect. Immun. 2001; 69:1409–1419, #2108.

Lagal, V., Postic, D., Ruzic-Sabljic, E. and Baranton, G. Genetic diversity among *Borrelia* strains determined by single-strand conformation polymorphism analysis of the *ospC* gene and its association with invasiveness. J. Clin. Microbiol. 2003; 41:5059–5065.

Lane, R.S., Piesman, J. and Burgdorfer, W. Lyme borreliosis: relation of its causative agent to its vectors and hosts in North America and Europe. Annu. Rev. Entomol. 1991; 36:587–609.

Langermann, S., Palaszynski, S., Sadziene, A., Stover, C.K. and Koenig, S. Systemic and mucosal immunity induced by BCG vector expressing outer-surface protein A of *Borrelia burgdorferi*. Nature 1994; 372:552–555.

Lardieri, G., Salvi, A., Camerini, F., Cinco, M. and Trevisan, G. Isolation of *Borrelia burgdorferi* from myocardium. Lancet 1993; 342:490.

Lathrop, S.L., Ball, R., Haber, P., Mootrey, G.T., Braun, M.M., Shadomy, S.V., Ellenberg, S.S., Chen, R.T. and Hayes, E.B. Adverse event reports following vaccination for Lyme disease: December 1998–July 2000. Vaccine 2002; 20:1603–1608.

Lebech, A.M., Hansen, K., Wilske, B. and Theisen, M. Taxonomic classification of 29 *Borrelia burgdorferi* strains isolated from patients with Lyme borreliosis: a comparison of five different phenotypic and genotypic typing schemes. Med. Microbiol. Immunol. 1994; 183:325–341.

Le Fleche, A., Postic, D., Girardet, K., Peter, O. and Baranton, G. Characterization of *Borrelia lusitaniae* sp. nov. by 16S ribosomal DNA sequence analysis. Int. J. Syst. Bacteriol. 1997; 47:921–925.

Leuba-Garcia, S., Martinez, R. and Gern, L. Expression of outer surface proteins A and C of *Borrelia afzelii* in *Ixodes ricinus* ticks and in the skin of mice. Zentralbl. Bakteriol. 1998; 287:475–484.

Levy, S.A. Use of a C6 ELISA test to evaluate the efficacy of a whole-cell bacterin for the prevention of naturally transmitted canine *Borrelia burgdorferi* infection. Vet. Ther. 2002; 3:420–424.

Levy, S.A., Lissman, B.A. and Ficke, C.M. Performance of a *Borrelia burgdorferi* bacterin in borreliosis-endemic areas. J. Am. Vet. Med. Assoc. 1993; 202:1834–1838.

Li, C., Motaleb, A., Sal, M., Goldstein, S.F. and Charon, N.W. Spirochete periplasmic flagella and motility. J. Mol. Microbiol. Biotechnol. 2000; 2:345–354.

Li, H., Dunn, J.J., Luft, B.J. and Lawson, C.L. Crystal structure of Lyme disease antigen outer surface protein A complexed with an Fab. Proc. Natl. Acad. Sci. U. S. A. 1997; 94:3584–3589.

Liang, F.T., Jacobs, M.B. and Philipp, M.T. C-terminal invariable domain of VlsE may not serve as target for protective immune response against *Borrelia burgdorferi*. Infect. Immun. 2001; 69:1337–1343.

Liang, F.T., Nelson, F.K. and Fikrig, E. DNA microarray assessment of putative *Borrelia burgdorferi* lipoprotein genes. Infect. Immun. 2002a; 70:3300–3303.

Liang, F.T., Nelson, F.K. and Fikrig, E. Molecular adaptation of *Borrelia burgdorferi* in the murine host. J. Exp. Med. 2002b; 196:275–280.

Liang, F.T. and Philipp, M.T. Analysis of antibody response to invariable regions of VlsE, the variable surface antigen of *Borrelia burgdorferi*. Infect. Immun. 1999; 67:6702–6706.

Lin, T., Oliver, J.H., Jr. and Gao, L. Genetic diversity of the outer surface protein C gene of southern *Borrelia* isolates and its possible epidemiological, clinical, and pathogenetic implications. J. Clin. Microbiol. 2002; 40:2572–2583.

Lindgren, E. and Jaenson, T.G. Lyme Borreliosis in Europe: Influences of Climate and Climate Change, Epidemiology, Ecology and Adaptation Measures. World Health Organization, Copenhagen, Denmark, 2006.

Livey, I., Gibbs, C.P., Schuster, R. and Dorner, F. Evidence for lateral transfer and recombination in OspC variation in Lyme disease *Borrelia*. Mol. Microbiol. 1995; 18:257–269.

Logigian, E.L., Kaplan, R.F. and Steere, A.C. Chronic neurologic manifestations of Lyme disease. N. Engl. J. Med. 1990; 323:1438–1444.

Lovrich, S.D., Callister, S.M., DuChateau, B.K., Lim, L.C., Winfrey, J., Day, S.P. and Schell, R.F. Abilities of OspA proteins from different seroprotective groups of *Borrelia burgdorferi* to protect hamsters from infection. Infect. Immun. 1995; 63:2113–2119.

Lovrich, S.D., Callister, S.M., Schmitz, J.L., Alder, J.D. and Schell, R.F. Borreliacidal activity of sera from hamsters infected with the Lyme disease spirochete. Infect. Immun. 1991; 59:2522–2528.

Lovrich, S.D., Jobe, D.A., Schell, R.F. and Callister, S.M. Borreliacidal OspC antibodies specific for a highly conserved epitope are immunodominant in human Lyme disease and do not occur in mice or hamsters. Clin. Diagn. Lab. Immunol. 2005; 12:746–751.

Lovrich, S.D., La Fleur, R.L., Jobe, D.A., Johnson, J.C., Asp, K.E., Schell, R.F. and Callister, S.M. Borreliacidal OspC antibody response of canines with Lyme disease differs significantly from that of humans with Lyme disease. Clin. Vaccine Immunol. 2007; 14:635–637.

Luke, C.J., Carner, K., Liang, X. and Barbour, A.G. An OspA-based DNA vaccine protects mice against infection with *Borrelia burgdorferi*. J. Infect. Dis. 1997a; 175:91–97.

Luke, C.J., Huebner, R.C., Kasmiersky, V. and Barbour, A.G. Oral delivery of purified lipoprotein OspA protects mice from systemic infection with *Borrelia burgdorferi*. Vaccine 1997b; 15:739–746.

Lunemann, J.D., Zarmas, S., Priem, S., Franz, J., Zschenderlein, R., Aberer, E., Klein, R., Schouls, L., Burmester, G.R. and Krause, A. Rapid typing of *Borrelia burgdorferi* sensu lato species in specimens from patients with different manifestations of Lyme borreliosis. J. Clin. Microbiol. 2001; 39:1130–1133.

Lybecker, M.C. and Samuels, D.S. Temperature-induced regulation of RpoS by a small RNA in *Borrelia burgdorferi*. Mol. Microbiol. 2007; 64:1075–1089.

Ma, J., Bulger, P.A., Dante, S., Davis, D.R., Perilli-Palmer, B. and Coughlin, R.T. Characterization of canine humoral immune responses to outer surface protein subunit vaccines and to natural infection by Lyme disease spirochetes. J. Infect. Dis. 1995; 171:909–915.

Maes, E., Lecomte, P. and Ray, N. A cost-of-illness study of Lyme disease in the United States. Clin. Ther. 1998; 20:993–1008, discussion 1992.

Makabe, K., Tereshko, V., Gawlak, G., Yan, S. and Koide, S. Atomic-resolution crystal structure of *Borrelia burgdorferi* outer surface protein A via surface engineering. Protein Sci. 2006; 15:1907–1914.

Maraspin, V., Cimperman, J., Lotric-Furlan, S., Ruzic-Sabljic, E., Jurca, T., Picken, R.N. and Strle, F. Solitary borrelial lymphocytoma in adult patients. Wien. Klin. Wochenschr. 2002; 114:515–523.

Marconi, R.T. and Garon, C.F. Identification of a third genomic group of *Borrelia burgdorferi* through signature nucleotide analysis and 16S rRNA sequence determination. J. Gen. Microbiol. 1992a; 138:533–536.

Marconi, R.T. and Garon, C.F. Phylogenetic analysis of the genus *Borrelia*: a comparison of North American and European isolates of *Borrelia burgdorferi*. J. Bacteriol. 1992b; 174:241–244.

Marconi, R.T., Liveris, D. and Schwartz, I. Identification of novel insertion elements, restriction fragment length polymorphism patterns, and discontinuous 23S rRNA in Lyme disease spirochetes: phylogenetic analyses of rRNA genes and their intergenic spacers in *Borrelia japonica* sp. nov. and genomic group 21038 (*Borrelia andersonii* sp. nov.) isolates. J. Clin. Microbiol. 1995; 33:2427–2434.

Marconi, R.T., Lubke, L., Hauglum, W. and Garon, C.F. Species-specific identification of and distinction between *Borrelia burgdorferi* genomic groups by using 16S rRNA-directed oligonucleotide probes. J. Clin. Microbiol. 1992; 30:628–632.

Marconi, R.T., Samuels, D.S. and Garon, C.F. Transcriptional analyses and mapping of the *ospC* gene in Lyme disease spirochetes. J. Bacteriol. 1993; 175:926–932.

Marconi, R.T., Sung, S.Y., Hughes, C.A. and Carlyon, J.A. Molecular and evolutionary analyses of a variable series of genes in *Borrelia burgdorferi* that are related to *ospE* and *ospF* constitute a gene family, and share a common upstream homology box. J. Bacteriol. 1996; 178:5615–5626.

Margolis, N. and Rosa, P.A. Regulation of expression of major outer surface proteins in *Borrelia burgdorferi*. Infect. Immun. 1993; 61:2207–2210.

Marshall, W.F., 3rd, Telford, S.R., 3rd, Rys, P.N., Rutledge, B.J., Mathiesen, D., Malawista, S.E., Spielman, A. and Persing, D.H. Detection of *Borrelia burgdorferi* DNA in museum specimens of *Peromyscus leucopus*. J. Infect. Dis. 1994; 170:1027–1032.

Marti Ras, N., Postic, D., Foretz, M. and Baranton, G. *Borrelia burgdorferi* sensu stricto, a bacterial species "made in the U.S.A."? Int. J. Syst. Bacteriol. 1997; 47:1112–1117.

Masters, E., Granter, S., Duray, P. and Cordes, P. Physician-diagnosed erythema migrans and erythema migrans-like rashes following lone star tick bites. Arch. Dermatol. 1998; 134:955–960.

Masuzawa, T., Takada, N., Kudeken, M., Fukui, T., Yano, Y., Ishiguro, F., Kawamura, Y., Imai, Y. and Ezaki, T. Borrelia sinica sp. nov., a Lyme disease-related Borrelia species isolated in China. Int. J. Syst. Evol. Microbiol. 2001; 51:1817–1824.

Mathiesen, M.J., Holm, A., Christiansen, M., Blom, J., Hansen, K., Ostergaard, S. and Theisen, M. The dominant epitope of *Borrelia garinii* outer surface protein C recognized by sera from patients with neuroborreliosis has a surface-exposed conserved structural motif. Infect. Immun. 1998; 66:4073–4079.

Matteson, E.L., Beckett, V.L., O'Fallon, W.M., Melton, L.J., 3rd and Duffy, J. Epidemiology of Lyme disease in Olmsted County, MN, 1975–1990. J. Rheumatol. 1992; 19:1743–1745.

Mbow, M.L., Gilmore, R.D., Jr. and Titus, R.G. An OspC-specific monoclonal antibody passively protects mice from tick-transmitted infection by *Borrelia burgdorferi* B31. Infect. Immun. 1999; 67:5470–5472.

McDowell, J.V., Sung, S.Y., Hu, L.T. and Marconi, R.T. Evidence that the variable regions of the central domain of VlsE are antigenic during infection with Lyme disease spirochetes. Infect. Immun. 2002; 70:4196–4203.

McDowell, J.V., Sung, S.Y., Labandeira-Rey, M., Skare, J.T. and Marconi, R.T. Analysis of mechanisms associated with loss of infectivity of clonal populations of *Borrelia burgdorferi* B31MI. Infect. Immun. 2001a; 69:3670–3677.

McDowell, J.V., Sung, S.Y., Price, G. and Marconi, R.T. Demonstration of the genetic stability and temporal expression of select members of the Lyme disease spirochete OspF protein family during infection in mice. Infect. Immun. 2001b; 69:4831–4838.

McDowell, J.V., Wolfgang, J., Tran, E., Metts, M.S., Hamilton, D. and Marconi, R.T. Comprehensive analysis of the factor H binding capabilities of *Borrelia* species associated with Lyme disease: delineation of two distinct classes of factor H binding proteins. Infect. Immun. 2003; 71:3597–3602.

McNabb, S.J., Jajosky, R.A., Hall-Baker, P.A., Adams, D.A., Sharp, P., Anderson, W.J., Javier, A.J., Jones, G.J., Nitschke, D.A., Worshams, C.A. and Richard, R.A., Jr. Summary of notifiable diseases—United States, 2005. Morb. Mortal. Wkly. Rep. 2007; 54:1–92.

McNeil, S.A., Halperin, S.A., Langley, J.M., Smith, B., Warren, A., Sharratt, G.P., Baxendale, D.M., Reddish, M.A., Hu, M.C., Stroop, S.D., Linden, J., Fries, L.F., Vink, P.E. and Dale, J.B. Safety and immunogenicity of 26-valent group A streptococcus vaccine in healthy adult volunteers. Clin. Infect. Dis. 2005; 41:1114–1122.

Meek, J.I., Roberts, C.L., Smith, E.V., Jr. and Cartter, M.L. Underreporting of Lyme disease by Connecticut physicians, 1992. J. Public Health Manag. Pract. 1996; 2:61–65.

Meltzer, M.I., Dennis, D.T. and Orloski, K.A. The cost effectiveness of vaccinating against Lyme disease. Emerg. Infect. Dis. 1999; 5:321–328.

Meyer, A.L., Trollmo, C., Crawford, F., Marrack, P., Steere, A.C., Huber, B.T., Kappler, J. and Hafler, D.A. Direct enumeration of *Borrelia*-reactive CD4 T cells ex vivo by using MHC class II tetramers. Proc. Natl. Acad. Sci. U. S. A. 2000; 97:11433–11438.

Mulenga, A., Sugino, M., Nakajim, M., Sugimoto, C. and Onuma, M. Tick-encoded serine proteinase inhibitors (serpins): potential target antigens for tick vaccine development. J. Vet. Med. Sci. 2001; 63:1063–1069.

Mulenga, A., Tsuda, A., Onuma, M. and Sugimoto, C. Four serine proteinase inhibitors (serpin) from the brown ear tick, *Rhiphicephalus appendiculatus*; cDNA cloning and preliminary characterization. Insect Biochem. Mol. Biol. 2003; 33:267–276.

Mullegger, R.R. Dermatological manifestations of Lyme borreliosis. Eur. J. Dermatol. 2004; 14:296–309.

Nachman, S.A. and Pontrelli, L. Central nervous system Lyme disease. Semin. Pediatr. Infect. Dis. 2003; 14:123–130.

Nadelman, R.B., Pavia, C.S., Magnarelli, L.A. and Wormser, G.P. Isolation of *Borrelia burgdorferi* from the blood of seven patients with Lyme disease. Am. J. Med. 1990; 88:21–26.

Nadelman, R.B. and Wormser, G.P. Erythema migrans and early Lyme disease. Am. J. Med. 1995; 98:15S–23S.

Nagi, K.S., Joshi, R. and Thakur, R.K. Cardiac manifestations of Lyme disease: a review. Can. J. Cardiol. 1996; 12:503–506.

Naleway, A.L., Belongia, E.A., Kazmierczak, J.J., Greenlee, R.T. and Davis, J.P. Lyme disease incidence in Wisconsin: a comparison of state-reported rates and rates from a population-based cohort. Am. J. Epidemiol. 2002; 155:1120–1127.

Narasimhan, S., Caimano, M.J., Liang, F.T., Santiago, F., Laskowski, M., Philipp, M.T., Pachner, A.R., Radolf, J.D. and Fikrig, E. *Borrelia burgdorferi* transcriptome in the central nervous system of non-human primates. Proc. Natl. Acad. Sci. U. S. A. 2003; 100:15953–15958.

Nepereny, J., Chumela, J., Tuhackova, J. and Vrzal, V. Verification of vaccine effectivity against Lyme borreliosis in dogs by challenge test. Veterinarstvi 2007; 57:213–217.

Nguyen, T.P., Lam, T.T., Barthold, S.W., Telford, S.R., 3rd, Flavell, R.A. and Fikrig, E. Partial destruction of *Borrelia burgdorferi* within ticks that engorged on OspE- or OspF-immunized mice. Infect. Immun. 1994; 62:2079–2084.

Nigrovic, L.E. and Thompson, K.M. The Lyme vaccine: a cautionary tale. Epidemiol. Infect. 2007; 135:1–8.

Nocton, J.J., Dressler, F., Rutledge, B.J., Rys, P.N., Persing, D.H. and Steere, A.C. Detection of *Borrelia burgdorferi* DNA by polymerase chain reaction in synovial fluid from patients with Lyme arthritis. N. Engl. J. Med. 1994; 330:229–234.

Norris, S.J. Antigenic variation with a twist—the Borrelia story. Mol. Microbiol. 2006; 60:1319–1322.

Nuttall, P.A., Trimnell, A.R., Kazimirova, M. and Labuda, M. Exposed and concealed antigens as vaccine targets for controlling ticks and tick-borne diseases. Parasite Immunol. 2006; 28:155–163.

Ogden, N.H., Barker, I.K., Beauchamp, G., Brazeau, S., Charron, D.F., Maarouf, A., Morshed, M.G., O'Callaghan, C.J., Thompson, R.A., Waltner-Toews, D., Waltner-Toews, M. and Lindsay, L.R. Investigation of ground level and remote-sensed data for habitat classification and prediction of survival of *Ixodes scapularis* in habitats of southeastern Canada. J. Med. Entomol. 2006a; 43:403–414.

Ogden, N.H., Maarouf, A., Barker, I.K., Bigras-Poulin, M., Lindsay, L.R., Morshed, M.G., O'Callaghan C, J., Ramay, F., Waltner-Toews, D. and Charron, D.F. Climate change and the potential for range expansion of the Lyme disease vector *Ixodes scapularis* in Canada. Int. J. Parasitol. 2006b; 36:63–70.

Ojaimi, C., Brooks, C., Casjens, S., Rosa, P., Elias, A., Barbour, A., Jasinskas, A., Benach, J., Katona, L., Radolf, J., Caimano, M., Skare, J., Swingle, K., Akins, D. and Schwartz, I. Profiling of temperature-induced changes in *Borrelia burgdorferi* gene expression by using whole genome arrays. Infect. Immun. 2003; 71:1689–1705.

Ojaimi, C., Mulay, V., Liveris, D., Iyer, R. and Schwartz, I. Comparative transcriptional profiling of *Borrelia burgdorferi* clinical isolates differing in capacities for hematogenous dissemination. Infect. Immun. 2005; 73:6791–6802.

Oliver, J.H., Jr., Owsley, M.R., Hutcheson, H.J., James, A.M., Chen, C., Irby, W.S., Dotson, E.M. and McLain, D.K. Conspecificity of the ticks *Ixodes scapularis* and *I. dammini* (Acari: Ixodidae). J. Med. Entomol. 1993; 30:54–63.

Orloski, K.A., Hayes, E.B., Campbell, G.L. and Dennis, D.T. Surveillance for Lyme disease—United States, 1992–1998. Morb. Mortal. Wkly. Rep. 2000; 49:1–11.

Oschmann, P., Dorndorf, W., Hornig, C., Schafer, C., Wellensiek, H.J. and Pflughaupt, K.W. Stages and syndromes of neuroborreliosis. J. Neurol. 1998; 245:262–272.

Pal, U., Li, X., Wang, T., Montgomery, R.R., Ramamoorthi, N., Desilva, A.M., Bao, F., Yang, X., Pypaert, M., Pradhan, D., Kantor, F.S., Telford, S., Anderson, J.F. and Fikrig, E. TROSPA, an *Ixodes scapularis* receptor for *Borrelia burgdorferi*. Cell 2004a; 119:457–468.

Pal, U., Yang, X., Chen, M., Bockenstedt, L.K., Anderson, J.F., Flavell, R.A., Norgard, M.V. and Fikrig, E. OspC facilitates

Borrelia burgdorferi invasion of *Ixodes scapularis* salivary glands. J. Clin. Invest. 2004b; 113:220–230.

Paster, B.J. and Dewhirst, F.E. Phylogenetic foundation of spirochetes. In: The Spirochetes: Molecular and Cellular Biology (M.H. Saier and J. Garcia-Lara, Eds.), pp. 5–9. Wymondham, UK: Horizon Scientific Press, 2001.

Paster, B.J., Dewhirst, F.E., Weisburg, W.G., Tordoff, L.A., Fraser, G.J., Hespell, R.B., Stanton, T.B., Zablen, L., Mandelco, L. and Woese, C.R. Phylogenetic analysis of the spirochetes. J. Bacteriol. 1991; 173:6101–6109.

Pedra, J.H., Narasimhan, S., Deponte, K., Marcantonio, N., Kantor, F.S. and Fikrig, E. Disruption of the salivary protein 14 in *Ixodes scapularis* nymphs and impact on pathogen acquisition. Am. J. Trop. Med. Hyg. 2006; 75:677–682.

Persing, D.H., Mathiesen, D., Podzorski, D. and Barthold, S.W. Genetic stability of *Borrelia burgdorferi* recovered from chronically infected immunocompetent mice. Infect. Immun. 1994; 62:3521–3527.

Persing, D.H., Telford, S.R., 3rd, Rys, P.N., Dodge, D.E., White, T.J., Malawista, S.E. and Spielman, A. Detection of *Borrelia burgdorferi* DNA in museum specimens of *Ixodes dammini* ticks. Science 1990; 249:1420–1423.

Philipp, M.T., Lobet, Y., Bohm, R.P., Jr., Roberts, E.D., Dennis, V.A., Gu, Y., Lowrie, R.C., Jr., Desmons, P., Duray, P.H., England, J.D., Hauser, P., Piesman, J. and Xu, K. The outer surface protein A (OspA) vaccine against Lyme disease: efficacy in the rhesus monkey. Vaccine 1997; 15:1872–1887.

Piesman, J. Dynamics of *Borrelia burgdorferi* transmission by nymphal *Ixodes dammini* ticks. J. Infect. Dis. 1993; 167:1082–1085.

Piesman, J., Maupin, G.O., Campos, E.G. and Happ, C.M. Duration of adult female *Ixodes dammini* attachment and transmission of *Borrelia burgdorferi*, with description of a needle aspiration isolation method. J. Infect. Dis. 1991; 163:895–897.

Poland, G.A. Prevention of Lyme disease: a review of the evidence. Mayo Clin. Proc. 2001; 76:713–724.

Postic, D., Ras, N.M., Lane, R.S., Hendson, M. and Baranton, G. Expanded diversity among Californian *Borrelia* isolates and description of *Borrelia bissettii* sp. nov. (formerly *Borrelia* group DN127). J. Clin. Microbiol. 1998; 36:3497–3504.

Pozsgay, V. and Kubler-Kielb, J. Synthesis of an experimental glycolipoprotein vaccine against Lyme disease. Carbohydr. Res. 2007; 342:621–626.

Prevot, P.P., Couvreur, B., Denis, V., Brossard, M., Vanhamme, L. and Godfroid, E. Protective immunity against *Ixodes ricinus* induced by a salivary serpin. Vaccine 2007; 25:3284–3292.

Probert, W.S., Crawford, M., Cadiz, R.B. and LeFebvre, R.B. Immunization with outer surface protein (Osp) A, but not OspC, provides cross-protection of mice challenged with North American isolates of *Borrelia burgdorferi*. J. Infect. Dis. 1997; 175:400–405.

Probert, W.S. and Johnson, B.J. Identification of a 47 kDa fibronectin-binding protein expressed by *Borrelia burgdorferi* isolate B31. Mol. Microbiol. 1998; 30:1003–1015.

Probert, W.S. and LeFebvre, R.B. Protection of C3H/HeN mice from challenge with *Borrelia burgdorferi* through active immunization with OspA, OspB, or OspC, but not with OspD or the 83-kilodalton antigen. Infect. Immun. 1994; 62:1920–1926.

Purser, J.E. and Norris, S.J. Correlation between plasmid content and infectivity in *Borrelia burgdorferi*. Proc. Natl. Acad. Sci. U. S. A. 2000; 97:13865–13870.

Ramamoorthi, N., Narasimhan, S., Pal, U., Bao, F., Yang, X.F., Fish, D., Anguita, J., Norgard, M.V., Kantor, F.S., Anderson, J.F., Koski, R.A. and Fikrig, E. The Lyme disease agent exploits a tick protein to infect the mammalian host. Nature 2005; 436:573–577.

Rand, K.N., Moore, T., Sriskantha, A., Spring, K., Tellam, R., Willadsen, P. and Cobon, G.S. Cloning and expression of a protective antigen from the cattle tick *Boophilus microplus*. Proc. Natl. Acad. Sci. U. S. A. 1989; 86:9657–9661.

Rannala, B., Qiu, W.G. and Dykhuizen, D.E. Methods for estimating gene frequencies and detecting selection in bacterial populations. Genetics 2000; 155:499–508.

Rapuntean, G., Boldizsar, E., Brudasca, E., Fartan, S. and Popescu, R. Consideration on immunogenic capacity of Biocan B vaccine against borreliosis. Buletinul Universitatii de Stiinte Agricole si Medicina Veterinara Cluj-Napoca Seria Medicina Veterinara 2003; 60:304–306.

Reid, M.C., Schoen, R.T., Evans, J., Rosenberg, J.C. and Horwitz, R.I. The consequences of overdiagnosis and overtreatment of Lyme disease: an observational study. Ann. Intern. Med. 1998; 128:354–362.

Revel, A.T., Talaat, A.M. and Norgard, M.V. DNA microarray analysis of differential gene expression in *Borrelia burgdorferi*, the Lyme disease spirochete. Proc. Natl. Acad. Sci. U. S. A. 2002; 99:1562–1567.

Rijpkema, S.G., Tazelaar, D.J., Molkenboer, M.J., Noordhoek, G.T., Plantinga, G., Schouls, L.M. and Schellekens, J.F. Detection of *Borrelia afzelii*, *Borrelia burgdorferi* sensu stricto, *Borrelia garinii* and group VS116 by PCR in skin biopsies of patients with erythema migrans and acrodermatitis chronica atrophicans. Clin. Microbiol. Infect. 1997; 3:109–116.

Rogers, E.A. and Marconi, R.T. Delineation of species-specific binding properties of the CspZ protein (BBH06) of the Lyme disease spirochetes: evidence for new contributions to the pathogenesis of *Borrelia* spp. Infect. Immun. 2007; 75(11):5272–5281.

Rosa, P. Lyme disease agent borrows a practical coat. Nat. Med. 2005; 11:831–832.

Rose, C.D., Fawcett, P.T. and Gibney, K.M. Arthritis following recombinant outer surface protein A vaccination for Lyme disease. J. Rheumatol. 2001; 28:2555–2557.

Rousselle, J.C., Callister, S.M., Schell, R.F., Lovrich, S.D., Jobe, D.A., Marks, J.A. and Wieneke, C.A. Borreliacidal antibody production against outer surface protein C of *Borrelia burgdorferi*. J. Infect. Dis. 1998; 178:733–741.

Ruderman, E.M., Kerr, J.S., Telford, S.R., 3rd, Spielman, A., Glimcher, L.H. and Gravallese, E.M. Early murine Lyme carditis has a macrophage predominance and is independent of major histocompatibility complex class II-CD4+ T cell interactions. J. Infect. Dis. 1995; 171:362–370.

Ruzic-Sabljic, E., Lotric-Furlan, S., Maraspin, V., Cimperman, J., Logar, M., Jurca, T. and Strle, F. Comparison of isolation rate of *Borrelia burgdorferi* sensu lato in MKP and BSK-II medium. Int. J. Med. Microbiol. 2006; 296:267–273.

Sadziene, A., Thomas, D.D., Bundoc, V.G., Holt, S.C. and Barbour, A.G. A flagella-less mutant of *Borrelia burgdorferi*. Structural, molecular, and in vitro functional characterization. J. Clin. Invest. 1991; 88:82–92.

Schaible, U.E., Kramer, M.D., Eichmann, K., Modolell, M., Museteanu, C. and Simon, M.M. Monoclonal antibodies specific for the outer surface protein A (OspA) of *Borrelia burgdorferi* prevent Lyme borreliosis in severe combined immunodeficiency (scid) mice. Proc. Natl. Acad. Sci. U. S. A. 1990; 87:3768–3772.

Scheckelhoff, M.R., Telford, S.R. and Hu, L.T. Protective efficacy of an oral vaccine to reduce carriage of *Borrelia burgdorferi* (strain N40) in mouse and tick reservoirs. Vaccine 2006; 24:1949–1957.

Scheiblhofer, S., Weiss, R., Durnberger, H., Mostbock, S., Breitenbach, M., Livey, I. and Thalhamer, J. A DNA vaccine encoding the outer surface protein C from *Borrelia burgdorferi* is able to induce protective immune responses. Microbes Infect. 2003; 5:939–946.

Schmid, G.P., Steigerwalt, A.G., Johnson, S.E., Barbour, A.G., Steere, A.C., Robinson, I.M. and Brenner, D.J. DNA characterization of the spirochete that causes Lyme disease. J. Clin. Microbiol. 1984; 20:155–158.

Schoen, R.T., Aversa, J.M., Rahn, D.W. and Steere, A.C. Treatment of refractory chronic Lyme arthritis with arthroscopic synovectomy. Arthritis Rheum. 1991; 34:1056–1060.

Schoen, R.T., Meurice, F., Brunet, C.M., Cretella, S., Krause, D.S., Craft, J.E. and Fikrig, E. Safety and immunogenicity of an outer surface protein A vaccine in subjects with previous Lyme disease. J. Infect. Dis. 1995; 172:1324–1329.

Schroder, N.W., Schombel, U., Heine, H., Gobel, U.B., Zahringer, U. and Schumann, R.R. Acylated cholesteryl galactoside as a novel immunogenic motif in *Borrelia burgdorferi* sensu stricto. J. Biol. Chem. 2003; 278:33645–33653.

Schutzer, S.E., Coyle, P.K., Krupp, L.B., Deng, Z., Belman, A.L., Dattwyler, R. and Luft, B.J. Simultaneous expression of *Borrelia* OspA and OspC and IgM response in cerebrospinal fluid in early neurologic Lyme disease. J. Clin. Invest. 1997; 100:763–767.

Schwan, T.G., Burgdorfer, W. and Garon, C.F. Changes in infectivity and plasmid profile of the Lyme disease spirochete, *Borrelia burgdorferi*, as a result of in vitro cultivation. Infect. Immun. 1988; 56:1831–1836.

Schwan, T.G., Piesman, J., Golde, W.T., Dolan, M.C. and Rosa, P.A. Induction of an outer surface protein on *Borrelia burgdorferi* during tick feeding. Proc. Natl. Acad. Sci. U. S. A. 1995; 92:2909–2913.

Sears, J.E., Fikrig, E., Nakagawa, T.Y., Deponte, K., Marcantonio, N., Kantor, F.S. and Flavell, R.A. Molecular mapping of Osp-A mediated immunity against *Borrelia burgdorferi*, the agent of Lyme disease. J. Immunol. 1991; 147:1995–2000.

Seinost, G., Dykhuizen, D.E., Dattwyler, R.J., Golde, W.T., Dunn, J.J., Wang, I.N., Wormser, G.P., Schriefer, M.E. and Luft, B.J. Four clones of *Borrelia burgdorferi* sensu stricto cause invasive infection in humans. Infect. Immun. 1999a; 67:3518–3524.

Seinost, G., Gasser, R., Reisinger, E., Rigler, M.Y., Fischer, L., Keplinger, A., Dattwyler, R.J., Dunn, J.J. and Klein, W. Cardiac manifestations of Lyme borreliosis with special reference to contractile dysfunction. Acta Med. Austriaca 1998; 25:44–50.

Seinost, G., Golde, W.T., Berger, B.W., Dunn, J.J., Qiu, D., Dunkin, D.S., Dykhuizen, D.E., Luft, B.J. and Dattwyler, R.J. Infection with multiple strains of *Borrelia burgdorferi* sensu stricto in patients with Lyme disease. Arch. Dermatol. 1999b; 135:1329–1333.

Seshu, J., Esteve-Gassent, M.D., Labandeira-Rey, M., Kim, J.H., Trzeciakowski, J.P., Hook, M. and Skare, J.T. Inactivation of the fibronectin-binding adhesin gene bbk32 significantly attenuates the infectivity potential of *Borrelia burgdorferi*. Mol. Microbiol. 2006; 59:1591–1601.

Shadick, N.A., Liang, M.H., Phillips, C.B., Fossel, K. and Kuntz, K.M. The cost-effectiveness of vaccination against Lyme disease. Arch. Intern. Med. 2001; 161:554–561.

Shang, E.S., Wu, X.Y., Lovett, M.A., Miller, J.N. and Blanco, D.R. Homologous and heterologous *Borrelia burgdorferi* challenge of infection-derived immune rabbits using host-adapted organisms. Infect. Immun. 2001; 69:593–598.

Shih, C.M., Spielman, A. and Telford, S.R., 3rd Short report: mode of action of protective immunity to Lyme disease spirochetes. Am. J. Trop. Med. Hyg. 1995; 52:72–74.

Sigal, L.H. Early disseminated Lyme disease: cardiac manifestations. Am. J. Med. 1995; 98:25S–28S, discussion 28S–29S.

Sigal, L.H., Zahradnik, J.M., Lavin, P., Patella, S.J., Bryant, G., Haselby, R., Hilton, E., Kunkel, M., Adler-Klein, D., Doherty, T., Evans, J., Molloy, P.J., Seidner, A.L., Sabetta, J.R., Simon, H.J., Klempner, M.S., Mays, J., Marks, D. and Malawista, S.E. A vaccine consisting of recombinant *Borrelia burgdorferi* outer-surface protein A to prevent Lyme disease. Recombinant outer-surface Protein A Lyme disease vaccine study consortium. N. Engl. J. Med. 1998; 339:216–222.

Simon, M.M., Gern, L., Hauser, P., Zhong, W., Nielsen, P.J., Kramer, M.D., Brenner, C. and Wallich, R. Protective immunization with plasmid DNA containing the outer surface lipoprotein A gene of *Borrelia burgdorferi* is independent of an eukaryotic promoter. Eur. J. Immunol. 1996; 26:2831–2840.

Simon, M.M., Schaible, U.E., Wallich, R. and Kramer, M.D. A mouse model for *Borrelia burgdorferi* infection: approach to a vaccine against Lyme disease. Immunol. Today 1991; 12:11–16.

Skotarczak, B. Canine borreliosis—epidemiology and diagnostics. Ann. Agric. Environ. Med. 2002; 9:137–140.

Smith, R. and Takkinen, J. Lyme borreliosis: Europe-wide coordinated surveillance and action needed? Euro Surveill. 2006; 11:E060622–E060621.

Soares, C.A., Lima, C.M., Dolan, M.C., Piesman, J., Beard, C.B. and Zeidner, N.S. Capillary feeding of specific dsRNA induces silencing of the *isac* gene in nymphal *Ixodes scapularis* ticks. Insect Mol. Biol. 2005; 14:443–452.

Stanek, G., Klein, J., Bittner, R. and Glogar, D. Isolation of *Borrelia burgdorferi* from the myocardium of a patient with longstanding cardiomyopathy. N. Engl. J. Med. 1990; 322:249–252.

Stanek, G., Klein, J., Bittner, R. and Glogar, D. *Borrelia burgdorferi* as an etiologic agent in chronic heart failure? Scand. J. Infect. Dis. Suppl. 1991; 77:85–87.

Stanek, G. and Strle, F. Lyme borreliosis. Lancet 2003; 362: 1639–1647.

Steere, A.C. Lyme disease. N. Engl. J. Med. 2001; 345:115–125.

Steere, A.C. and Angelis, S.M. Therapy for Lyme arthritis: strategies for the treatment of antibiotic-refractory arthritis. Arthritis Rheum. 2006; 54:3079–3086.

Steere, A.C., Batsford, W.P., Weinberg, M., Alexander, J., Berger, H.J., Wolfson, S. and Malawista, S.E. Lyme carditis: cardiac abnormalities of Lyme disease. Ann. Intern. Med. 1980; 93:8–16.

Steere, A.C., Broderick, T.F. and Malawista, S.E. Erythema chronicum migrans and Lyme arthritis: epidemiologic evidence for a tick vector. Am. J. Epidemiol. 1978; 108:312–321.

Steere, A.C., Coburn, J. and Glickstein, L. The emergence of Lyme disease. J. Clin. Invest. 2004; 113:1093–1101.

Steere, A.C., Dwyer, E. and Winchester, R. Association of chronic Lyme arthritis with HLA-DR4 and HLA-DR2 alleles. N. Engl. J. Med. 1990; 323:219–223.

Steere, A.C. and Glickstein, L. Elucidation of Lyme arthritis. Nat. Rev. Immunol. 2004; 4:143–152.

Steere, A.C., Grodzicki, R.L., Kornblatt, A.N., Craft, J.E., Barbour, A.G., Burgdorfer, W., Schmid, G.P., Johnson, E. and Malawista, S.E. The spirochetal etiology of Lyme disease. N. Engl. J. Med. 1983; 308:733–740.

Steere, A.C., Gross, D., Meyer, A.L. and Huber, B.T. Autoimmune mechanisms in antibiotic treatment-resistant Lyme arthritis. J. Autoimmun. 2001; 16:263–268.

Steere, A.C., Hardin, J.A., Ruddy, S., Mummaw, J.G. and Malawista, S.E. Lyme arthritis: correlation of serum and cryoglobulin IgM with activity, and serum IgG with remission. Arthritis Rheum. 1979; 22:471–483.

Steere, A.C., Malawista, S.E., Hardin, J.A., Ruddy, S., Askenase, W. and Andiman, W.A. Erythema chronicum migrans and Lyme arthritis. The enlarging clinical spectrum. Ann. Intern. Med. 1977a; 86:685–698.

Steere, A.C., Malawista, S.E., Snydman, D.R., Shope, R.E., Andiman, W.A., Ross, M.R. and Steele, F.M. Lyme arthritis: an epidemic of oligoarticular arthritis in children and adults in three Connecticut communities. Arthritis Rheum. 1977b; 20:7–17.

Steere, A.C., Schoen, R.T. and Taylor, E. The clinical evolution of Lyme arthritis. Ann. Intern. Med. 1987; 107:725–731.

Steere, A.C., Sikand, V.K., Meurice, F., Parenti, D.L., Fikrig, E., Schoen, R.T., Nowakowski, J., Schmid, C.H., Laukamp, S.,

Buscarino, C. and Krause, D.S. Vaccination against Lyme disease with recombinant *Borrelia burgdorferi* outer-surface lipoprotein A with adjuvant. Lyme Disease Vaccine Study Group. N. Engl. J. Med. 1998; 339:209–215.

Stevenson, B. and Barthold, S.W. Expression and sequence of outer surface protein C among North American isolates of *Borrelia burgdorferi*. FEMS Microbiol. Lett. 1994; 124:367–372.

Stevenson, B., Schwan, T.G. and Rosa, P.A. Temperature-related differential expression of antigens in the Lyme disease spirochete, *Borrelia burgdorferi*. Infect. Immun. 1995; 63:4535–4539.

Stewart, P.E., Wang, X., Bueschel, D.M., Clifton, D.R., Grimm, D., Tilly, K., Carroll, J.A., Weis, J.J. and Rosa, P.A. Delineating the requirement for the *Borrelia burgdorferi* virulence factor OspC in the mammalian host. Infect. Immun. 2006; 74:3547–3553.

Stover, C.K., Bansal, G.P., Hanson, M.S., Burlein, J.E., Palaszynski, S.R., Young, J.F., Koenig, S., Young, D.B., Sadziene, A. and Barbour, A.G. Protective immunity elicited by recombinant bacille Calmette-Guerin (BCG) expressing outer surface protein A (OspA) lipoprotein: a candidate Lyme disease vaccine. J. Exp. Med. 1993; 178:197–209.

Stricker, R.B. Counterpoint: long-term antibiotic therapy improves persistent symptoms associated with Lyme disease. Clin. Infect. Dis. 2007; 45:149–157.

Strle, F., Maraspin, V., Pleterski-Rigler, D., Lotric-Furlan, S., Ruzic-Sabljic, E., Jurca, T. and Cimperman, J. Treatment of borrelial lymphocytoma. Infection 1996; 24:80–84.

Strle, F., Pleterski-Rigler, D., Stanek, G., Pejovnik-Pustinek, A., Ruzic, E. and Cimperman, J. Solitary borrelial lymphocytoma: report of 36 cases. Infection 1992; 20:201–206.

Sugino, M., Imamura, S., Mulenga, A., Nakajima, M., Tsuda, A., Ohashi, K. and Onuma, M. A serine proteinase inhibitor (serpin) from Ixodid tick *Haemaphysalis longicornis*; cloning and preliminary assessment of its suitability as a candidate for a tick vaccine. Vaccine 2003; 21:2844–2851.

Takayama, K., Rothenberg, R.J. and Barbour, A.G. Absence of lipopolysaccharide in the Lyme disease spirochete, *Borrelia burgdorferi*. Infect. Immun. 1987; 55:2311–2313.

Telford, S.R., 3rd, Fikrig, E., Barthold, S.W., Brunet, L.R., Spielman, A. and Flavell, R.A. Protection against antigenically variable *Borrelia burgdorferi* conferred by recombinant vaccines. J. Exp. Med. 1993; 178:755–758.

Terekhova, D., Iyer, R., Wormser, G.P. and Schwartz, I. Comparative genome hybridization reveals substantial variation among clinical isolates of *Borrelia burgdorferi* sensu stricto with different pathogenic properties. J. Bacteriol. 2006; 188:6124–6134.

Tilly, K., Bestor, A., Jewett, M.W. and Rosa, P. Rapid clearance of Lyme disease spirochetes lacking OspC from skin. Infect. Immun. 2007; 75:1517–1519.

Tilly, K., Casjens, S., Stevenson, B., Bono, J.L., Samuels, D.S., Hogan, D. and Rosa, P. The *Borrelia burgdorferi* circular plasmid cp26: conservation of plasmid structure and targeted inactivation of the *ospC* gene. Mol. Microbiol. 1997; 25:361–373.

Tilly, K., Krum, J.G., Bestor, A., Jewett, M.W., Grimm, D., Bueschel, D., Byram, R., Dorward, D., Vanraden, M.J., Stewart, P. and Rosa, P. *Borrelia burgdorferi* OspC protein required exclusively in a crucial early stage of mammalian infection. Infect. Immun. 2006; 74:3554–3564.

Titus, R.G., Bishop, J.V. and Mejia, J.S. The immunomodulatory factors of arthropod saliva and the potential for these factors to serve as vaccine targets to prevent pathogen transmission. Parasite Immunol. 2006; 28:131–141.

Topfer, K.H. and Straubinger, R.K. Characterization of the humoral immune response in dogs after vaccination against the Lyme borreliosis agent: a study with five commercial vaccines using two different vaccination schedules. Vaccine 2007; 25:314–326.

Trimnell, A.R., Davies, G.M., Lissina, O., Hails, R.S. and Nuttall, P.A. A cross-reactive tick cement antigen is a candidate broad-spectrum tick vaccine. Vaccine 2005; 23:4329–4341.

Trollmo, C., Meyer, A.L., Steere, A.C., Hafler, D.A. and Huber, B.T. Molecular mimicry in Lyme arthritis demonstrated at the single cell level: LFA-1 alpha L is a partial agonist for outer surface protein A-reactive T cells. J. Immunol. 2001; 166:5286–5291.

Tsao, J.I., Wootton, J.T., Bunikis, J., Luna, M.G., Fish, D. and Barbour, A.G. An ecological approach to preventing human infection: vaccinating wild mouse reservoirs intervenes in the Lyme disease cycle. Proc. Natl. Acad. Sci. U. S. A. 2004; 101:18159–18164.

van Dam, A.P., Kuiper, H., Vos, K., Widjojokusumo, A., de Jongh, B.M., Spanjaard, L., Ramselaar, A.C., Kramer, M.D. and Dankert, J. Different genospecies of *Borrelia burgdorferi* are associated with distinct clinical manifestations of Lyme borreliosis. Clin. Infect. Dis. 1993; 17:708–717.

van der Linde, M.R. Lyme carditis: clinical characteristics of 105 cases. Scand. J. Infect. Dis. Suppl. 1991; 77:81–84.

van der Linde, M.R., Crijns, H.J., de Koning, J., Hoogkamp-Korstanje, J.A., de Graaf, J.J., Piers, D.A., van der Galien, A. and Lie, K.I. Range of atrioventricular conduction disturbances in Lyme borreliosis: a report of four cases and review of other published reports. Br. Heart J. 1990; 63:162–168.

Van Hoecke, C., Comberbach, M., De Grave, D., Desmons, P., Fu, D., Hauser, P., Lebacq, E., Lobet, Y. and Voet, P. Evaluation of the safety, reactogenicity and immunogenicity of three recombinant outer surface protein (OspA) Lyme vaccines in healthy adults. Vaccine 1996; 14:1620–1626.

Varela, A.S., Luttrell, M.P., Howerth, E.W., Moore, V.A., Davidson, W.R., Stallknecht, D.E. and Little, S.E. First culture isolation of *Borrelia lonestari*, putative agent of southern tick-associated rash illness. J. Clin. Microbiol. 2004; 42:1163–1169.

Wallich, R., Kramer, M.D. and Simon, M.M. A recombinant vaccine for Lyme disease. Behring Inst. Mitt. 1994:106–108.

Wallich, R., Siebers, A., Jahraus, O., Brenner, C., Stehle, T. and Simon, M.M. DNA vaccines expressing a fusion product of outer surface proteins A and C from *Borrelia burgdorferi* induce protective antibodies suitable for prophylaxis but not for resolution of Lyme disease. Infect. Immun. 2001; 69:2130–2136.

Wang, G., van Dam, A.P. and Dankert, J. Evidence for frequent OspC gene transfer between *Borrelia valaisiana* sp. nov. and other Lyme disease spirochetes. FEMS Microbiol. Lett. 1999a; 177:289–296.

Wang, G., van Dam, A.P., Schwartz, I. and Dankert, J. Molecular typing of *Borrelia burgdorferi* sensu lato: taxonomic, epidemiological, and clinical implications. Clin. Microbiol. Rev. 1999b; 12:633–653.

Wang, I.N., Dykhuizen, D.E., Qiu, W., Dunn, J.J., Bosler, E.M. and Luft, B.J. Genetic diversity of *ospC* in a local population of *Borrelia burgdorferi* sensu stricto. Genetics 1999c; 151:15–30.

Wang, L., Huang, J.A., Nagesha, H.S., Smith, S.C., Phelps, A., Holmes, I., Martyn, J.C., Coloe, P.J. and Reeves, P.R. Bacterial expression of the major antigenic regions of porcine rotavirus VP7 induces a neutralizing immune response in mice. Vaccine 1999d; 17:2636–2645.

Welsh, J., Pretzman, C., Postic, D., Saint Girons, I., Baranton, G. and McClelland, M. Genomic fingerprinting by arbitrarily primed polymerase chain reaction resolves *Borrelia burgdorferi* into three distinct phyletic groups. Int. J. Syst. Bacteriol. 1992; 42:370–377.

Wharton, M., Chorba, T.L., Vogt, R.L., Morse, D.L. and Buehler, J.W. Case definitions for public health surveillance. Morb. Mortal. Wkly. Rep. 1990; 39:1–43.

Wheeler, C.M., Garcia Monco, J.C., Benach, J.L., Golightly, M.G., Habicht, G.S. and Steere, A.C. Nonprotein antigens of *Borrelia burgdorferi*. J. Infect. Dis. 1993; 167:665–674.

Wiedemann, C. and Milward, F. Efficacy and safety studies with a new vaccine against Lyme borreliosis in dogs (Merilyme(R)). Tierarztl. Umsch. 1999; 54:242–249.

Wikle, R.E., Fretwell, B., Jarecki, M. and Jarecki-Black, J.C. Canine Lyme disease: one-year duration of immunity elicited with a canine OspA monovalent Lyme disease vaccine. Int. J. Appl. Res. Vet. Med. 2006; 4:23–28.

Willadsen, P. Anti-tick vaccines. Parasitology 2004; 129(Suppl): S367–S387.

Willadsen, P. Vaccination against ectoparasites. Parasitology 2006; 133(Suppl):S9–S25.

Willett, T.A., Meyer, A.L., Brown, E.L. and Huber, B.T. An effective second-generation outer surface protein A-derived Lyme vaccine that eliminates a potentially autoreactive T cell epitope. Proc. Natl. Acad. Sci. U. S. A. 2004; 101:1303–1308.

Wilske, B., Busch, U., Fingerle, V., Jauris-Heipke, S., Preac Mursic, V., Rossler, D. and Will, G. Immunological and molecular variability of OspA and OspC. Implications for *Borrelia* vaccine development. Infection 1996; 24:208–212.

Wilske, B., Preac-Mursic, V., Gobel, U.B., Graf, B., Jauris, S., Soutschek, E., Schwab, E. and Zumstein, G. An OspA serotyping system for *Borrelia burgdorferi* based on reactivity with monoclonal antibodies and OspA sequence analysis. J. Clin. Microbiol. 1993; 31:340–350.

Wolgemuth, C.W., Charon, N.W., Goldstein, S.F. and Goldstein, R.E. The flagellar cytoskeleton of the spirochetes. J. Mol. Microbiol. Biotechnol. 2006; 11:221–227.

Wormser, G.P., Dattwyler, R.J., Shapiro, E.D., Halperin, J.J., Steere, A.C., Klempner, M.S., Krause, P.J., Bakken, J.S., Strle, F., Stanek, G., Bockenstedt, L., Fish, D., Dumler, J.S. and Nadelman, R.B. The clinical assessment, treatment, and prevention of Lyme disease, human granulocytic anaplasmosis, and babesiosis: clinical practice guidelines by the Infectious Diseases Society of America. Clin. Infect. Dis. 2006; 43:1089–1134.

Wormser, G.P., McKenna, D., Carlin, J., Nadelman, R.B., Cavaliere, L.F., Holmgren, D., Byrne, D.W. and Nowakowski, J. Brief communication: hematogenous dissemination in early Lyme disease. Ann. Intern. Med. 2005; 142:751–755.

Xu, Y., Bruno, J.F. and Luft, B.J. Identification of novel tick salivary gland proteins for vaccine development. Biochem. Biophys. Res. Commun. 2005; 326:901–904.

Yang, X., Goldberg, M.S., Popova, T.G., Schoeler, G.B., Wikel, S.K., Hagman, K.E. and Norgard, M.V. Interdependence of environmental factors influencing reciprocal patterns of gene expression in virulent *Borrelia burgdorferi*. Mol. Microbiol. 2000; 37:1470–1479.

Yang, X., Li, Y., Dunn, J.J. and Luft, B.J. Characterization of a unique borreliacidal epitope on the outer surface protein C of *Borrelia burgdorferi*. FEMS Immunol. Med. Microbiol. 2006; 48:64–74.

Yang, X.F., Alani, S.M. and Norgard, M.V. The response regulator Rrp2 is essential for the expression of major membrane lipoproteins in *Borrelia burgdorferi*. Proc. Natl. Acad. Sci. U. S. A. 2003a; 100:11001–11006.

Yang, X.F., Hubner, A., Popova, T.G., Hagman, K.E. and Norgard, M.V. Regulation of expression of the paralogous Mlp family in *Borrelia burgdorferi*. Infect. Immun. 2003b; 71:5012–5020.

Yang, X.F., Lybecker, M.C., Pal, U., Alani, S.M., Blevins, J., Revel, A.T., Samuels, D.S. and Norgard, M.V. Analysis of the *ospC* regulatory element controlled by the RpoN-RpoS regulatory pathway in *Borrelia burgdorferi*. J. Bacteriol. 2005; 187:4822–4829.

Yin, Z., Braun, J., Neure, L., Wu, P., Eggens, U., Krause, A., Kamradt, T. and Sieper, J. T cell cytokine pattern in the joints of patients with Lyme arthritis and its regulation by cytokines and anticytokines. Arthritis Rheum. 1997; 40:69–79.

Yoder, A., Wang, X., Ma, Y., Philipp, M.T., Heilbrun, M., Weis, J.H., Kirschning, C.J., Wooten, R.M. and Weis, J.J. Tripalmitoyl-S-glyceryl-cysteine-dependent OspA vaccination of toll-like receptor 2-deficient mice results in effective protection from *Borrelia burgdorferi* challenge. Infect. Immun. 2003; 71:3894–3900.

Young, J.D. Underreporting of Lyme disease. N. Engl. J. Med. 1998; 338:1629.

Yssel, H., Shanafelt, M.C., Soderberg, C., Schneider, P.V., Anzola, J. and Peltz, G. *Borrelia burgdorferi* activates a T helper type 1-like T cell subset in Lyme arthritis. J. Exp. Med. 1991; 174:593–601.

Zhang, H. and Marconi, R.T. Demonstration of cotranscription and 1-methyl-3-nitroso-nitroguanidine induction of a 30-gene operon of *Borrelia burgdorferi*: evidence that the 32-kilobase circular plasmids are prophages. J. Bacteriol. 2005; 187:7985–7995.

Zhang, J.R. and Norris, S.J. Genetic variation of the *Borrelia burgdorferi* gene *vlsE* involves cassette-specific, segmental gene conversion. Infect. Immun. 1998a; 66:3698–3704.

Zhang, J.R. and Norris, S.J. Kinetics and in vivo induction of genetic variation of *vlsE* in *Borrelia burgdorferi*. Infect. Immun. 1998b; 66:3689–3697.

Zhang, X., Meltzer, M.I., Pena, C.A., Hopkins, A.B., Wroth, L. and Fix, A.D. Economic impact of Lyme disease. Emerg. Infect. Dis. 2006; 12:653–660.

Zhong, W., Gern, L., Stehle, T., Museteanu, C., Kramer, M., Wallich, R. and Simon, M.M. Resolution of experimental and tick-borne *Borrelia burgdorferi* infection in mice by passive, but not active immunization using recombinant OspC. Eur. J. Immunol. 1999; 29:946–957.

Zhong, W., Stehle, T., Museteanu, C., Siebers, A., Gern, L., Kramer, M., Wallich, R. and Simon, M.M. Therapeutic passive vaccination against chronic Lyme disease in mice. Proc. Natl. Acad. Sci. U. S. A. 1997; 94:12533–12538.

Zhong, W., Wiesmuller, K.H., Kramer, M.D., Wallich, R. and Simon, M.M. Plasmid DNA and protein vaccination of mice to the outer surface protein A of *Borrelia burgdorferi* leads to induction of T helper cells with specificity for a major epitope and augmentation of protective IgG antibodies in vivo. Eur. J. Immunol. 1996; 26:2749–2757.

Zuckert, W.R., Kerentseva, T.A., Lawson, C.L. and Barbour, A.G. Structural conservation of neurotropism-associated VspA within the variable *Borrelia* Vsp-OspC lipoprotein family. J. Biol. Chem. 2001; 276:457–463.

CHAPTER

53

Neisseria meningitidis

Nadine G. Rouphael[1], *Shanta M. Zimmer*[1] *and David S. Stephens*[1,2]
[1]Division of Infectious Diseases, Department of Medicine, Emory University School of Medicine, Atlanta, GA, USA
[2]Laboratories of Microbial Pathogenesis, Atlanta VA Medical Center, Decatur, GA, USA

OUTLINE

ABSTRACT

Neisseria meningitidis causes epidemic and sporadic meningitis, and/or septicemia, resulting in significant morbidity or death. Invasive meningococcal disease is the result of microbial factors influencing the virulence of the organism, environmental conditions facilitating exposure and acquisition, and host susceptibility factors favoring bacterial colonization, invasion, and survival. *N. meningitidis* colonizes 8–25% of human upper respiratory tracts with the highest rates in adolescents. Fulminant meningococcemia is a dramatic presentation. Within hours, otherwise healthy subjects may become severely ill, develop hypotension and organ failure, and despite effective antibiotics and aggressive supportive care, 10–15% of patients die and up to 20% of survivors have lifelong sequelae. Meningococci are classified according to their capsular polysaccharide into 13 serogroups of which A, B, C, Y, X, and W-135 are responsible for most endemic and epidemic disease. *N. meningitidis* causes an estimated 1.2 million cases per year with an estimated death toll of 135,000. Disease rates, in particular meningitis rates, are highest in the African "meningitis belt," where serogroup A outbreaks are prevalent. Recently, serogroup W-135 has caused outbreaks associated with the Muslim hajj pilgrimage, and serogroup Y emerged as a cause of endemic disease in the United States, Israel, and South Africa. Chemoprophylaxis is effective in known contacts and limited outbreak settings. Rapid emergence of resistance (e.g., rifampin) and access and availability of drugs limit this approach in large epidemic outbreaks. Prevention of meningococcal disease by vaccination remains the best control strategy. Polysaccharide vaccines against serogroups A, C, Y, and W-135, in use since the 1970s, are safe and effective; however, they do not produce long-lasting immunity and are of limited effectiveness in children under 2 years of age, and repeated doses may result in hyporesponsiveness. New meningococcal polysaccharide–protein conjugate vaccines are now impacting incidence. In late 1999, the United Kingdom successfully introduced conjugate serogroup C meningococcal vaccines, later introduced by other European countries and Canada. In 2005, a quadrivalent (A, C, Y, and W-135) meningococcal conjugate vaccine was licensed in the United States. Even though the duration of protection is unknown, these conjugate vaccines induce herd immunity, memory responses, antibody affinity maturation, and better immunogenicity in infants, young children, and adults. However, certain infant strategies are associated with rapidly waning immunity, and memory responses alone do not appear to protect because of the short incubation of the disease. A major hurdle in achieving control of meningococcal disease is the development of an effective vaccine strategy for sub-Saharan Africa. Another challenge is the development of safe and immunogenic serogroup B meningococcal vaccines due to the poor immunogenicity of the serogroup B polysaccharide and the theoretical risk of inducing autoimmunity with serogroup B polysaccharide-containing vaccines. Vaccines against serogroup B meningococcal disease have been developed based on outer-membrane vesicles. These vaccines are effective against epidemic strains but usually do not confer cross-protection against heterologous strains. Through the recent availability of the meningococcal genome sequences, new protein-based vaccine candidates have been identified for serogroup B.

INTRODUCTION

Neisseria meningitidis is a leading worldwide cause of sepsis and meningitis, with the highest increase in infants, young children, and adolescents resulting in significant morbidity and mortality. The pathogen has a unique ability among the different etiologic agents of bacterial meningitis to cause major epidemics, case clusters, and hyperendemic as well as endemic disease. Meningococcal epidemics in sub-Saharan Africa have occurred for 100 years (Greenwood, 1999) as large outbreaks at a frequency of 5–10 years or smaller yearly outbreaks in a region of over 18 countries described as the meningitis belt (Lepeyssonnie, 1963). Other regions of Africa also have high rates of meningococcal disease. Thousands of cases occur in the United States, Canada, Europe, parts of South America, Australia, New Zealand, and China. Up to 70–85% of cases were fatal in the pre-serum therapy and pre-antibiotic eras; today, the overall mortality rate in meningococcemia remains high, between 10% and 15% (Sharip et al., 2006). Vaccination against meningococci is the most effective means of disease prevention. Significant advances have improved our understanding of meningococcal biology, pathogenesis, epidemiology, transmission and carriage, clinical presentation, diagnosis, and treatment and led to progress in the development of the next generation of meningococcal vaccines.

HISTORY

Weichselbaum (1887) was the first to identify the meningococcus from the cerebrospinal fluid (CSF) of a patient with meningitis. Epidemics of meningococcal meningitis were first described in 1805 in Geneva, Switzerland, by Vieusseux (1805), and in 1806 in Medfield, Massachusetts, by Danielson and Mann (1806). By the beginning of the 20th century, the recognition of meningococcal carriage in healthy individuals was made (Kiefer, 1896), especially in crowded situations such as those seen in military recruit camps (Glover, 1918). Treatment for meningococcal disease included serum therapy introduced in 1913 (Flexner, 1913) and sulfonamides first introduced in 1937 (Schwentker et al., 1937). Sulfonamides were used not only for treatment but also for chemoprophylaxis (Khuns et al., 1943). Emergence of resistance to sulfonamides (Schoenback and Phair, 1948) in the 1960s prompted the development of the first vaccines against meningococci (Artenstein et al., 1970).

ETIOLOGIC AGENT

N. meningitidis is a fastidious, encapsulated, aerobic gram-negative diplococcus and an exclusive pathogen

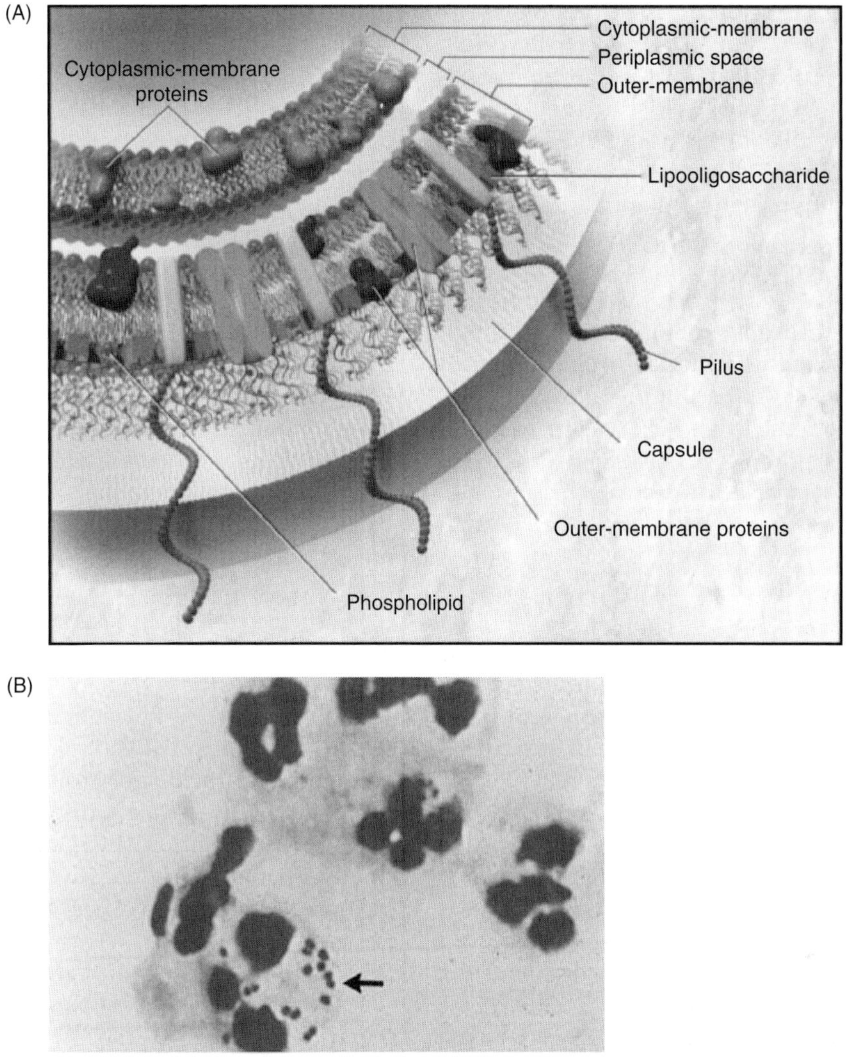

FIGURE 53.1 (A) Cross-sectional view of the meningococcal cell membrane showing the cytoplasmic membrane, outer membrane, and capsule. Phospholipid, lipooligosaccharide, outer-membrane proteins, and pili extending through the capsule are all part of the outer membrane. [Adapted from Rosenstein et al. (2001) with permission from Massachusetts Medical Society.] (B) *Neisseria meningitidis* are intracellular gram-negative diplococci as seen in the CSF of patient with meningitis. [Adapted from Rosenstein et al. (2001) with permission from Massachusetts Medical Society.]

of humans (Fig. 53.1). Meningococci are traditionally classified according to serologic typing (Frasch et al., 1985). This classification is based on structural differences in capsule (serogroup), major outer-membrane protein (OMP) porins (serotype), other OMPs (serosubtype), and lipooligosaccharides (LOS) (immunotype). At least 13 serologically distinct meningococcal groups have been defined on the basis of their capsule (Branham, 1953). Meningococci are highly transformable, and the transformation can lead to a genetic change in capsule structure and therefore in serogroup (Swartley et al., 1997; Vogel et al., 2000). Molecular and genetic techniques have been developed to identify related meningococcal strains. The molecular techniques used (Caugant et al., 1986) include pulsed-field gel electrophoresis (PFGE) (Bevanger et al., 1998), multilocus enzyme electrophoresis (MLEE) (Weis and Lind, 1998), multilocus sequence typing (MLST) (Maiden et al., 1998), and polymerase chain reaction (PCR) (Mothershed et al., 2004).

PROTECTIVE IMMUNE RESPONSE

Bactericidal antibodies and complement are considered major mechanisms in the protection against meningococcal disease. Opsonization and phagocyte killing also appear important.

The meningococcal carrier state is an immunizing process with appearance of antibodies in serum 2 weeks after nasopharyngeal colonization of pathogenic and nonpathogenic meningococci. Goldschneider et al. (1969a, 1969b) found that age-specific incidence of meningococcal disease is inversely proportional to the prevalence of serum bactericidal antibodies (SBA) to the meningococcus. Up to 60–80% of young adults will have detectable serum meningococcal bactericidal antibodies. Further, sera from susceptible hosts lack antibodies to the causative meningococcal strains during an epidemic. Finally, individuals with congenital or acquired deficiencies of immunoglobulins or complement are at increased risk for invasive meningococcal disease (Castagliuolo et al., 1986). Although specific immunity is generally protective, this immunity is not absolute; meningococcal disease can occur in a rare number of individuals with preexisting antibody titers that are considered protective (Greenwood et al., 1987; Kayhty et al., 1981).

Complement is required for both bactericidal activity and opsonophagocytosis (Nicholson and Lepow, 1979). Individuals deficient in both the early and late components of the complement system are at increased risk for the disease (Densen, 1991; Nielsen et al., 1989; Alper et al., 1970; Ellison et al., 1983). Compared to the general population, these patients usually experience less severe but more recurrent disease at an older age with uncommon serogroups (Fijen et al., 1989, 1999). Up to 10–20% of invasive meningococcal disease in adults is associated with a complement defect. The mannose-binding lectin (MBL) pathway of complement activation can be genetically variable and has been associated with differences in susceptibility to meningococcal disease in one-third of all cases (Hibberd et al., 1999; Bax et al., 1999). Other genetic polymorphisms affecting the risk of acquiring meningococcal disease have also been described (including polymorphisms in TNF, FcγRIIA, FcγRIII, PAI-1, ACE-1, IL-1Ra, IL-1β, and TLR4) (Emonts et al., 2003).

PATHOGENESIS

Humans are the only known host for the meningococcus. Nasopharyngeal colonization is present in 8–25% of adults (Greenfield et al., 1971; Stephens, 1999; Caugant et al., 1994), and the duration of carriage can vary from days to many months (Fig. 53.2). Transmission among humans occurs through respiratory secretions, but the inoculum size needed for transmission is unknown. Acquisition of *N. meningitidis* in the upper respiratory tract may be asymptomatic, result in colonization, infrequently cause local inflammation or disease (e.g., pneumonia), produce fulminant sepsis after access to the bloodstream, may produce fulminant sepsis or seed secondary sites such as CSF, joints or pericardium. Disease usually occurs 1–14 days after acquisition of the pathogen.

Invasive disease is dependent on environmental factors, meningococcal virulence, and host factors (refer to Protective Immune Response). Environmental factors that impair the integrity of the human nasopharyngeal mucosa such as tobacco (Fischer et al., 1997), exposure to low humidity, dust, and coinfections (Moore et al., 1990) increase the incidence of colonization, transmission, mucosal invasion, and invasive meningococcal disease.

Major contributors to the virulence of *N. meningitidis* include capsular polysaccharide, other surface structures [pili, OMPs (e.g., PorA, PorB, Opa, and Opc), LOS], and virulence factors associated with the genotype. The capsule, composed of sialic acid derivatives (with the exception of serogroup A), provides the organism with antiphagocytic, antibactericidal, and antiadherent properties enhancing its survival in the bloodstream or in the central nervous system (CNS). Meningococci that do not express capsular polysaccharides rarely cause invasive disease. LOS is in the family of lipopolysaccharide (LPS) or endotoxin expressed by many other gram-negative bacteria but lacks the repeating *O*-side chain of LPS of enteric gram-negative

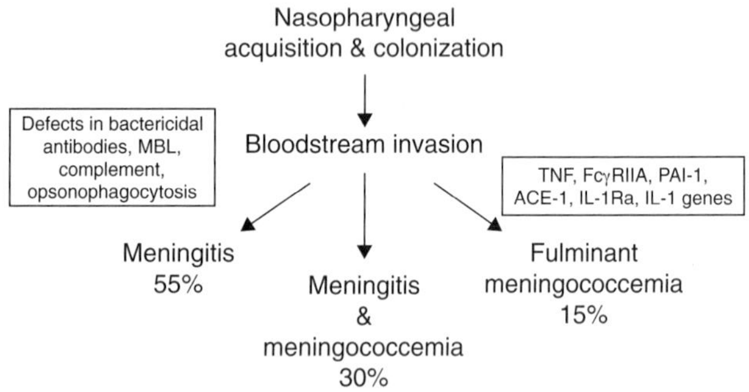

FIGURE 53.2 Pathogenesis of meningococcal disease. Acquisition of *N. meningitidis* in the upper respiratory tract may be asymptomatic, result in colonization, infrequently cause local inflammation or local disease (e.g., pneumonia), produce fulminant sepsis after access to the bloodstream, or seed secondary sites, particularly the CSF. In individuals who lack bactericidal antibodies, terminal or alternative pathway complement-deficiency states, and other genetic polymorphisms may influence the severity of the ensuing host response and the clinical presentation. Legend: MBL, mannose-binding lectin; TNF, tumor necrosis factor; FcγRIIA, FcγRIIA R131 allele; PAI-1, plasminogen activator inhibitor 1; ACE-1, angiotension-converting enzyme 1; IL, interleukin. [Adapted from Kasper et al. (2005) with permission of the McGraw-Hill Companies.]

bacilli. It plays a role in the adherence of the meningococcus (Kahler and Stephens, 1998) and activation of the innate immune system. The severity of meningococcal sepsis has been correlated with circulating levels of meningococcal LOS (Brandtzaeg et al., 1992). Other important virulence factors are pili (Stephens et al., 1985) and other OMPs (Virji et al., 1993) facilitating the adherence of the meningococcus to epithelial and endothelial surfaces. Importantly, meningococcal virulence is influenced by genetic mechanisms [transformation, slip-strand mispairing, insertion sequence (IS) element movement] that produce phase and antigenic variation and allow for genomic plasticity and adaptation (Stephens and Zimmer, 2002). The pathogen has also developed sophisticated pathways for iron acquisition, essential for colonization and infection, including an array of iron transport mechanisms (Schryvers and Stojiljkovic, 1999).

EPIDEMIOLOGY

Meningococcal infection is a global but nonuniform problem occurring as sporadic, hypersporadic, and epidemic disease. Disease patterns vary widely among regions, age groups, and serogroups, and in developing as well as developed countries (Fig. 53.3). In the last 60 years, large serogroup A epidemics have not occurred in industrialized countries for reasons not understood, but serogroup C and prolonged serogroup B outbreaks do occur. In the United States, the attack rate is less than 1 case per 100,000 per year while it is higher in Europe (Cartwright et al., 2001); in the United Kingdom, the attack rate was 5 cases per 100,000 per year prior to the introduction of the meningococcal C (MenC) conjugate vaccines. In the meningitis belt, rates vary between 20 and 600 and up to 1000 per 100,000 depending on the year and the presence or absence of epidemics. Meningococcal disease peak coincides with respiratory viral illnesses (Artenstein et al., 1967; Young et al., 1972) during winter months; in Africa, meningococcal disease occurs in the dry season (Greenwood et al., 1985).

Since the remarkable decline in disease due to the *Haemophilus influenzae* and pneumococcal polysaccharide–protein conjugate vaccines, the meningococcus is now the most common cause of bacterial meningitis in children and young adults in the United States (Harrison et al., 1999), mostly affecting children less than 2 years of age (Kaplan et al., 2006; Rosenstein et al., 1999). However, even though peak incidence occurs among infants and adolescents; one-third to one-half of the cases are seen in adults older than 18 years.

The epidemiology of meningococcal disease differs depending on the serogroup (Tzeng and Stephens, 2000). Six serogroups (A, B, C, Y, W-135, and recently X) (Boisier et al., 2007; Gagneux et al., 2002) constitute most of the cases of meningococcal disease worldwide. Serogroup A was the cause of most meningococcal disease in the first part of the 20th century and is now responsible for recurring epidemics in developing countries, particularly in sub-Saharan Africa (Hart and Cuevas, 1997). In the largest meningococcal epidemic outbreak recorded, an estimated 300,000 cases and 30,000 deaths occurred in sub-Saharan Africa during 1996–1997 due to serogroup A *N. meningitidis*. From 2000 to 2002, epidemics of serogroup W-135 occurred, related to spread during the hajj pilgrimage, affecting the health of these travelers and their contacts in countries throughout the world (Lingappa et al., 2003; Wilder-Smith et al., 2003; Hahne et al., 2002). In 2002, a large outbreak of W-135 occurred in Burkina Faso (Koumare et al., 2007; Traore et al.,

FIGURE 53.3 Serogroup distribution of invasive meningococcal disease worldwide. [Adapted from Stephens et al. (2007) with permission from Elsevier and from Caugant (1998) with permission from Wiley-Blackwell Publishing.]

2006; Decosas and Koama, 2002; Ouedraogo-Traore et al., 2002). Serogroup B is the most important cause of endemic disease in developed countries (30–40% and 80% of disease in the United States and European countries, respectively), and serogroup C has variable rates of endemic disease (currently around 30%) in industrialized countries (Jackson et al., 1995). Serogroup Y has emerged in the last decade in the United States and caused one-third of the disease related to meningococci in the country (Rosenstein et al., 1999), usually affecting older age groups. Interestingly, meningococcal pneumonia is usually due to serogroups Y and W-135.

CLINICAL DISEASE

The clinical spectrum of meningococcal disease can vary from transient bacteremia (Sullivan and LaScolea, 1987) to fulminant disease (Rosenstein et al., 2001). The predominant clinical syndrome is one of acute bacterial meningitis (Stephens et al., 1995; Thompson et al., 2006). Patients present with fever, headache, photophobia, neck stiffness, occasional seizures, altered mental status, nausea, vomiting, myalgia, and petechial or purpuric rash (28–77%) (Rosenstein et al., 2001). However, the rash may be absent or difficult to detect, and signs of sepsis (hypotension, organ failure, confusion) may be the presenting features. Approximately 30% of patients with meningococcal disease present with both meningitis and signs of meningococcemia. Up to 10–20% of patients will present with fulminant meningococcemia with severe sepsis leading to organ failure, disseminated intravascular coagulation (DIC), and death in up to 50% of these cases. The Waterhouse-Friderichsen syndrome, including adrenal hemorrhage in the setting of meningococcemia, is a dramatic example of microthrombosis from DIC, bleeding, and tissue injury. Up to 10–20% of patients with meningitis and meningococcemia will develop permanent sequelae (Erickson et al., 2001; Edwards and Baker, 1981). Complications are seen in meningitis in the form of different neurological deficits, especially hearing loss. In fulminant meningococcemia, loss of limb or digits from ischemia and necrosis can be observed.

Fifteen percent of patients present with pneumonia, frequently at an older age and more commonly in association with serogroup Y (Winstead et al., 2000). Other manifestations of meningococcal disease include conjunctivitis (Moraga et al., 1990; Newton and Wilson, 1977), pharyngitis (Mattila and Carlson, 1998; Pether et al., 1994; Wiesner et al., 1973), pericarditis (El Bashir et al., 2004; Morgan et al., 2002; Baevsky, 1999; Blaser et al., 1984; Chow et al., 1975), endocarditis (Benes et al., 2003; Joao et al., 2001), myocarditis (Garcia et al., 1999; Hardman and Earle, 1969), arthritis (Schaad, 1980; Bilavsky et al., 2006), urethritis (D'Antuono et al., 1999; Karolus et al., 1980; Miller et al., 1979), or cervicitis (Hagman et al., 1991; Odegaard and Gundersen, 1977). Rarely, chronic meningococcemia syndrome is reported. It is characterized by recurrent fever, skin rash, and headache (Nielsen, 1970) and may have a rheumatological presentation of weeks to months. Diagnosis is made by isolation of the meningococcus in sterile sites such as the blood or the CSF. Antigen detection assays can be used when cultures are negative and in developing settings where cultures are not available. PCR has rapidly gained acceptance in the diagnosis of meningococcal infection since it is capable of serogroup and strain typing, and its sensitivity is not affected by prior antibiotics (Ni et al., 1992; Newcombe et al., 1996; Bryant et al., 2004).

TREATMENT

Respiratory isolation of the patient is recommended for the first 24 h of hospital stay. For treatment, a third-generation cephalosporin (e.g., ceftriaxone) is recommended. Penicillin can also be given if susceptibility is confirmed. In patients with severe allergic reaction to beta-lactam drugs, fluoroquinolones or chloramphenicols are alternatives (McCrumb et al., 1951; Blondeau and Yaschuk, 1995). Several prognostic studies (Barquet et al., 1997; Niklasson et al., 1971) indicated that early administration of antibiotics results in decrease in mortality and morbidity (Cartwright et al., 1992), and prehospital antibiotic treatment is advocated in many countries. In meningococcal meningitis, steroid use is controversial (Weisfelt et al., 2006; Casella et al., 2004; Gupta and Tuladhar, 2004), but some experts recommend that they be given prior to, or simultaneously with, the first dose of antibiotics (van de Beek et al., 2004). For fulminant meningococcemia, often seen with serogroup C *N. meningitidis,* intensive care is required. The role of activated protein C for patients with meningococcemia is not clearly defined, with some studies showing promising results in subgroups (White et al., 2000; de Kleijn et al., 2003) while others did not show benefit (Nadel et al., 2007) and may suggest increase risks for side effects in patients with meningococcal disease (Vincent et al., 2005). Overall mortality from invasive meningococcal disease ranges from 10% to 15%; however, if purpura fulminans is present, mortality is increased to up to 50%. In the United Kingdom, the case-fatality rate of meningococcal infections was unchanged from 1963 to 1998 (Goldacre et al., 2003), but use of aggressive supportive therapy has reduced mortality. In the United States, between 1994 and 2002, the case-fatality rate was increased during outbreaks compared to sporadic disease (Brooks et al., 2006).

POSTEXPOSURE IMMUNOPROPHYLAXIS

Chemoprophylaxis eradicates nasopharyngeal carriage and prevents secondary cases in close contacts. Chemoprophylaxis consists of rifampin twice daily for a total of four doses or a single dose of ciprofloxacin. Ceftriaxone as a single intramuscular dose can also be offered. Azithromycin has also shown efficacy (Girgis et al., 1998) among children but is not routinely recommended. Knowing that the attack rate among household or other close contacts of meningococcal cases is more than 400-fold greater than the general population (De Wals et al., 1981), chemoprophylaxis for individuals at risk should be offered (Fraser et al., 2006; Bilukha and Rosenstein, 2005). Close contacts include those living in the same household, day-care center contacts, and anyone directly exposed to oral secretions of the patient with meningococcal disease. Prophylaxis is not indicated if the exposure to the index case was brief (Gilmore et al., 2000). Chemoprophylaxis should be administered as soon as the index case is identified and up to 7 days following exposure; after 14 days, chemoprophylaxis is ineffective. Since treatment with penicillin alone is not reliable in eradicating nasopharyngeal carriage (Abramson and Spika, 1985), terminal prophylaxis should be given to index patients at hospital discharge. Vaccination rather than chemoprophylaxis is usually offered during outbreaks in large populations because of cost of chemoprophylaxis and the increased likelihood of antimicrobial resistance.

VACCINES

History

Early in the 20th century, attempts were made to develop meningococcal vaccines using killed whole cells (Sophian and Black, 1912; Gates, 1918). Those attempts failed due to uncertainty in the ability of the vaccines to confer protective immunity and high rates of observed side effects. In the 1930s, "inactivated toxin" was used as potential vaccine candidate (Ferry and Steele, 1935; Khuns et al., 1938). Though immunogenic, those vaccines carried the risk of side effects with other components of the meningococcal outer membrane. The successful introduction of sulfonamides in 1937 in both the treatment and the prevention of meningococcal disease somewhat dampened the enthusiasm for vaccine development. However, the emergence of antimicrobial resistance to sulfonamides in the 1960s led to renewed efforts in the development of a vaccine to prevent meningococcal disease. In 1969–1971, the purification of meningococcal capsular polysaccharides by Gotschlich et al. (1969a, 1969b) at the Walter Reed Army Institute of Research led to the discovery of meningococcal A and C polysaccharide vaccines. These vaccines were used to control serogroup C outbreaks in the U.S. military and serogroup A epidemics in Africa. In 1978, the quadrivalent (A, C, Y, and W-135) polysaccharide vaccine was approved. The main limitations of polysaccharide vaccines are their inability to induce a T-cell-dependent response limiting effectiveness in children and failure to induce long-term immunologic memory. The chemical conjugation to protein carriers converts the polysaccharide to a T-cell-mediated response. The successful introduction of the conjugate *H. influenzae* serotype b vaccine (Heath, 1998) preceded the first human trial with a MenA–MenC conjugate vaccine developed by Salvo Vaccines (Sienna) in 1991 (Costantino et al., 1992). In late 1999, the United Kingdom was the first country to introduce the conjugate serogroup C vaccines. In 2005, a tetravalent conjugate vaccine including serogroups A, C, Y, and W-135 was licensed in the United States.

Since 1980, researchers have been working on a vaccine for serogroup B, and to date, none of the vaccines has been shown to be broadly protective. Since the serogroup B polysaccharide is poorly immunogenic due to its identity with human glycoproteins [e.g., neural cell adhesion molecule (NCAM)] and carries a theoretical risk of inducing autoimmunity through molecular mimicry, other antigens are used as targets in vaccine development for serogroup B. The currently available serogroup B outer-membrane vesicle (OMV) vaccines protect against particular strains and have been used with success in outbreak settings. Since the sequencing of the serogroup B genome in 2000, recent advances in reverse vaccinology have revealed new molecules as potential serogroup B vaccine candidates.

Polysaccharide Vaccines

Polysaccharide vaccines are either monovalent or multivalent, containing one to four of the serogroup

TABLE 53.1 Meningococcal vaccines for serogroups A, C, Y, W-135, and B

Meningococcal polysaccharide vaccines	
Menomune® (Sanofi Pasteur)	A, C, Y, and W-135
ACWY Vax (GlaxoSmithKline)	A, C, Y, and W-135
Trivalent ACW vaccine (GlaxoSmithKline)	A, C, and W-135
AC Vax® (GlaxoSmithKline)	A and C
Mengiva A+C® (Sanofi Pasteur)	A and C
Meningococcal polysaccharide–protein conjugate vaccines	
Meningitec® (Wyeth)	C with CRM_{197}
Menjugate® (Novartis)	C with CRM_{197}
NeisVac-C® (Baxter)	C with tetanus toxoid
Menactra® (Sanofi Pasteur)	A, C, Y, and W-135 with CRM_{197}
Serogroup B meningococcal vaccines (in development)	
MonoMen®, HexaMen®, NonaMen® (Netherlands Vaccine Institute)	
Finlay Institute vaccine/VA-MENGOC-BC (Havana, Cuba)	
National Institute of Public Health vaccine (Oslo, Norway)	
B+OMV (Walter Reed Army Institute of Research, Washington, D.C., USA)	
MeNZB® (New Zealand)	

A, C, Y, and W-135 capsular polysaccharides (Table 53.1). A single subcutaneous injection of quadrivalent meningococcal polysaccharide vaccine containing 50 μg of purified bacterial polysaccharide for serogroups A, C, Y, and W-135 [meningococcal polysaccharide vaccine (MPSV-4), Menomune®, Sanofi Pasteur, PA, USA] has been licensed in the United States since 1981. The vaccine is both safe and effective, inducing serum bactericidal immunity in 80–95% of adults (Bilukha and Rosenstein, 2005). Side effects are usually mild, with local reactions (4–46%) lasting 1–2 days. Severe side effects [systemic allergic reactions (0–0.1/100,000), infrequent neurological reactions] are rare (Lepow et al., 1986). The immunologic response to each of the four polysaccharides contained in the vaccine is serogroup specific and independent. In particular, the serogroup C capsular polysaccharide is poorly immunogenic in children less than 2 years of age (Gold et al., 1979), and SBA titers in adults decline markedly within 2 years of immunization (Zangwill et al., 1994). Serogroup A polysaccharide is capable of offering protection in children as young as 3 months (Gold et al., 1979); however, this protection is of short duration, and additional doses are required for longer protection. The duration of protection of the quadrivalent polysaccharide vaccine is therefore less than 5 years, and repeated doses of the vaccine may result in hyporesponsiveness, especially in the C component, among children and young adults (MacLennan et al., 1999). Polysaccharide vaccines do not confer long-term herd immunity. In a study from Gambia, immunization with the bivalent serogroup A/C vaccine had little effect on nasopharyngeal carriage of serogroup A meningococcus 6 months after vaccine administration (Hassan-King et al., 1988). Polysaccharide meningococcal vaccine is not recommended for routine use in the general population. It can be used in outbreak settings (Masterton et al., 1988) and high-risk populations. Since the introduction of meningococcal capsular polysaccharide vaccination for all recruits in the United States in 1972, epidemics in the military have not occurred (Brundage et al., 2002).

Conjugate Vaccines

The limitations of the polysaccharide vaccines, including lack of herd immunity and poor immunogenicity in young children, were the driving forces behind the development of protein–polysaccharide conjugate meningococcal vaccines (Zimmer and Stephens, 2004). The polysaccharide is covalently linked to a carrier protein and therefore converted to a thymus-dependent antigen enhancing memory B cells (Kelly et al., 2006) and IgG anticapsular antibody production in infants. In 1999, in an attempt to control hyperendemic serogroup C meningococcal disease, a campaign to vaccinate adolescents, children, and infants was launched in the United Kingdom, using serogroup C meningococcal polysaccharide–protein conjugate vaccines (MenC). The vaccination campaign was begun on the basis of safety and immunogenicity data, without prior studies demonstrating clinical efficacy. The vaccine program was a success, with dramatic reduction in invasive meningococcal serogroup C disease, and was replicated in other countries (Snape and Pollard, 2005). Major advantages to this vaccine over polysaccharide vaccines are the induction of herd immunity, memory responses, and better immunogenicity in infants and young children.

The vaccines used are conjugated either to CRM_{197} (a nontoxic mutant of diphtheria toxin) or to tetanus toxoid. The vaccines were developed from isolates containing *O*-acetyl group in their polysaccharides (OAC+ve). However, 12% (Borrow et al., 2000) of meningococcal serogroup C disease is caused by strains that lack the *O*-acetyl group (OAC−ve). Therefore, an alternative version of the monovalent MenC vaccine was developed conjugating an OAC−ve polysaccharide to tetanus toxoid (MenC-TT).

Currently, three vaccines are licensed: two OAC+ve–CRM_{197} conjugate vaccines (Meningitec® Wyeth, NJ, USA, and Menjugate® Novartis, Siena, Italy) and one OAC−ve–tetanus toxoid conjugate (NeisVac-C® Baxter Bioscience, MD, USA). All three vaccines are safe and immunogenic. Local (0.15%) and systemic reactions (seizure 1/60,000, anaphylaxis 1/500,000) are rare (2000). The efficacy of the vaccines is around 90% among all age groups (Snape and Pollard, 2005). In the United Kingdom, incidence of serogroup C disease declined by 81–87% among vaccinees (Balmer et al., 2002; Miller et al., 2001). In addition to the reduction in meningococcal disease, significant differences were seen in meningococcal carriage with a 66% reduction in serogroup C nasopharyngeal carriage in students 15–17 years of age (Maiden and Stuart, 2002). A herd immunity effect was noted with a 67% reduction in meningococcal disease reported for unvaccinated children (Ramsay et al., 2003). The use of the MenC conjugate vaccine was also effective in an outbreak setting (De Wals et al., 2004). Surveillance data after vaccine introduction have not demonstrated (Alonso et al., 2007), as of yet, significant capsule switching (Alcala et al., 2002; Diggle and Clarke, 2005) or serogroup replacement (Trotter et al., 2006). Although the duration of immunity is unknown, SBA titers wane rapidly after primary immunization (Trotter et al., 2004) but memory response persists (Richmond et al., 1999). However, in confirmed cases of vaccine failure (Auckland et al., 2006), this memory response alone did not provide protection against invasive disease. For this reason, a booster dose at 12–14 months should be considered after routine infant vaccination, and this change was made in the U.K. schedule. Other countries followed the successful U.K. example—Ireland (Dell et al., 2001) (2000), Spain (Salleras et al., 2003) (2000), Canada (De Wals et al., 2004) (Quebec) (2001), the Netherlands (Welte et al., 2004) (2002), Belgium (De Schrijver and Maes, 2003) (2002), Iceland (Bergthornorsson et al., 2004) (2002), Australia (Booy et al., 2007) (2003), Portugal (De Queiros et al., 2004) (2006), and Germany (Wiese-Posselt et al., 2006) (2006)—and a 54–96% reduction in invasive serogroup C meningococcal disease was noted (Snape and Pollard, 2005). Meningococcal conjugate vaccines are given as part of a routine childhood immunization program, but scheduling, dosing, and catch-up campaigns have varied among the countries (Trotter and Ramsay, 2007). The vaccine has been given as a two- or three-dose series in infancy or as one dose for toddlers at 12–14 months. Recent data suggest that while there is no clear difference if the immunization was given during infancy or in toddlers (Vu et al., 2006), a booster dose after 12 months is needed. In the United Kingdom, for instance, a booster dose is now given at 12 months after the two doses given in infancy. In addition to the childhood immunization program, all persons in the United Kingdom under 24 years of age were included in the initial catch-up campaign.

A new meningococcal quadrivalent conjugate vaccine (A, C, Y, and W-135 conjugated with diphtheria toxoid) [meningococcal conjugate vaccine (MCV-4), Menactra®, Aventis Pasteur, PA, USA] has been licensed in the United States since 2005 (Maiden and Spratt, 1999). Data comparing the immunogenicity and overall safety of Menactra to Menomune showed noninferiority (Bilukha and Rosenstein, 2005). Local reactions were more common in Menactra, likely due to the diphtheria toxoid component of the vaccine. Post licensure, a slight increase in cases of Guillain-Barré syndrome (GBS) was reported among Menactra recipients (2006), especially those over 15 years old. However, compared to the number of doses given and the impact on meningococcal disease incidence, those reports have not resulted in any change in vaccine recommendations. Duration of immunity to the vaccine is still unknown. A single intramuscular dose of vaccine is recommended as part of routine immunization for children aged 11–12 years, other adolescents, and groups at increased risk (2005). Vaccination is recommended in high-risk groups such as military recruits, microbiologists routinely exposed to meningococci (Sejvar et al., 2005), college freshmen living in dormitories, pilgrims to the hajj, travelers to sub-Saharan Africa during the dry months or to countries with known epidemics, and persons with increased susceptibility to meningococcal disease (terminal complement deficiency or anatomic/functional asplenia). MCV-4 has been recently licensed for children 2–11 years of age with risk factors (e.g., complement or antibody deficiency) for meningococcal disease. Recent data comparing MCV-4 to MPSV-4 in 2- to 10-year-olds showed a similar safety profile and elicited significantly higher and more persistent SBA responses (Pichichero et al., 2005). MCV-4 is contraindicated in persons with a history of GBS or an allergy to any component of the vaccine.

Serogroup B Vaccines

Development of a vaccine that can elicit broad protection against diverse serogroup B meningococcal strains presents several challenges (Zimmer and Stephens, 2006). The poor immunogenicity of the serogroup B polysaccharide is attributable to the fact that this polysaccharide resembles host antigens (Finne et al., 1983). The capsule is of identical structure to polysialic structures expressed in fetal neural tissue. The development of a polysaccharide vaccine for serogroup B has been limited by the theoretical risk of inducing autoimmunity (Stein et al., 2006). Even though IgM against serogroup B polysaccharide can contribute to serum bactericidal activity, IgG antibodies to serogroup B polysaccharide are nonspecific and poorly protective in animal models (Granoff et al., 1995; Mandrell et al., 1995) as well as humans (Bruge et al., 2004). Recent data suggest that the meningococcal serogroup B capsular polysaccharide may be engineered to express unique bactericidal epitopes suitable as vaccine targets (Moe et al., 2006; Granoff et al., 1998).

Other noncapsular antigens such as OMVs, containing OMPs and LOS, are therefore important vaccine candidates for serogroup B. The challenge of developing an effective vaccine against meningococcal serogroup B disease is of regional importance. In the late 1980s and in the 1990s, several trials of serogroup B OMV vaccines were undertaken in Cuba, Brazil, Chile, and Norway, which showed efficacy between 50% and 80%. Similar to the polysaccharide vaccines, serogroup B OMV vaccines have limited efficacy in infants and very young children, and the single-strain OMV vaccines may not provide cross-protection against a variety of heterologous meningococcal isolates. Usually, the protection is of short duration, and nasopharyngeal carriage is not affected.

Norway has had high rates of meningococcal disease, up to 80%, due to serogroup B. Between 1988 and 1991, the Norwegian Institute of Public Health (NIPH) enrolled 171,800 students in a double-blind, placebo-controlled, efficacy trial to evaluate the effect of OMV vaccine against serogroup B meningococci (MenBvac) (Bjune et al., 1991). Efficacy was around 50%. The Finlay Institute in collaboration with GlaxoSmithKline produced a purified OMP vaccine (VA-MENGOC-BC™) that was tested in Cuba (106,000 vaccinees) with great success (Sierra et al., 1991). The vaccine has been licensed in many Latin American countries. However, the high efficacy observed in the original Cuban trial was not replicated in case control studies in Brazil including 4 million vaccinees from Sao Paolo and Rio de Janeiro (Noronha et al., 1995; de Moraes et al., 1992). Two other OMV vaccines developed by the Walter Reed Army Institute of Research (Boslego et al., 1995) and the NIPH/Chiron (Tappero et al., 1999) (VA-MENGOC-BC) were tested in Chile. Because of the geographical and temporal diversity of OMPs, the OMP-based approach may be best suited for the development of vaccines for specific outbreaks and is best illustrated by the New Zealand experience. In New Zealand, in 1997, the rate of meningococcal disease rose to 16.9 cases per 100,000 with more than 90% of the disease due to serogroup B (Baker et al., 2001). As this is typical of serogroup B outbreaks, this epidemic has persisted for a decade. The stability of OMVs from serogroup B isolates during this meningococcal outbreak supported the use of a strain-specific OMV vaccine to control the epidemic (Dyet and Martin, 2005; Devoy et al., 2005). MeNZB™ was produced as a result of collaboration between the New Zealand government, the University of Auckland, and Chiron (now Novartis) (O'Hallahan et al., 2004). A three-dose regimen of the vaccine containing PorA, PorB, and LOS was safe and immunogenic in 75 healthy adults (Thornton et al., 2006) and children (Wong et al., 2007) (16–24 months). New Zealand's regulatory authority granted licensure to deliver MeNZB™, and 1 million individuals under age 20 received the vaccine (Sexton et al., 2004), which has diminished the outbreak.

An estimation of potential coverage by current OMP vaccines to European serogroup B strains was performed (Trotter and Ramsay, 2007). The Cuban, Norwegian, and New Zealand vaccines cover less than one-third of the strains circulating in Europe between 1999 and 2004, and multivalent OMP vaccines such as those used in the Netherlands can cover more than 70% of the strains. Both a hexavalent OMV vaccine (HexaMen™) containing two OMVs with three PorA subtypes each and a monovalent OMV vaccine (MonoMen™) have been studied in infants, toddlers, and children. These vaccines showed promise but limited cross-protection to other serosubtypes not contained in the vaccine (de Kleijn et al., 2000a, 2000b). Three additional vesicles have been added to a new nonavalent formulation (NonaMen™), and trials are underway. In the United States, it has been estimated that 20 different PorA proteins need to be incorporated in the serogroup B vaccine to cover 80% of circulating strains (Tondella et al., 2000).

Vaccines in Development

Other Serogroup B Vaccines

The lack of broad efficacy provided by PorA-based serogroup B vaccines suggests that other vaccine strategies should be further investigated. The challenge is to identify proteins that are exposed at the surface, conserved across various strains, and consistently expressed. Those proteins should be immunogenic and capable of inducing bactericidal antibodies and long-lasting immunity (Ruggeberg and Pollard, 2004).

Proteins involved in the meningococcal iron metabolism have been evaluated as potential vaccine candidates (Johnson et al., 1999; Gomez et al., 1998; Thompson et al., 2003; West et al., 2001; Lehmann et al., 1999). Among these proteins are the transferrin-binding proteins. Opsonins are recovered in humans during the convalescence (Lehmann et al., 1999) period, and there is protection in animal models (West et al., 2001). However, phase I human trials led to disappointing results (Ruggeberg and Pollard, 2004). Neisserial surface protein A (NspA) is highly conserved among meningococcal strains, although the function of the protein is unknown. Active and passive immunizations confer protection in mice (Martin et al., 1997). However, focusing on meningococcal serogroup B vaccines, other researchers were not able to fully replicate these results (Moe et al., 2001). A vaccine based on purified NspA has been developed by Shire Pharmaceuticals and evaluated in phase I trials. Though protective in the mouse model and safe in humans, the vaccine did not induce SBA in a phase I trial (Halperin et al., 2007). Other OMVs have also been considered as vaccine candidates (e.g., Opa, Opc, App, pilin, Rmp, PorB, and NadA) (Stephens et al., 1985; Wright et al., 2002; Hadi et al., 2001; de Jonge et al., 2003, 2004; Rosenqvist et al., 1999; Comanducci et al., 2002; Bowe et al., 2004) (Table 53.2).

The range of novel antigens is not restricted to proteins. Lipoprotein LP2086, displayed on the surface of 91% meningococcal strains, has been shown to induce serum bactericidal activity against a broad variety of meningococcal strains and prevent colonization in animal models (Pillai et al., 2005; Zhu et al., 2005, 2006). LPS is a key player in pathogenesis and may be useful as vaccine adjuvant. Even though it is highly conserved among meningococci, complete lack of protection among certain strains has been shown (Plested et al., 2003).

Other approaches such as use of commensal neisserial species or a genomic approach are being considered in the quest for a broader vaccine against serogroup B meningococcal disease. *Neisseria lactamica* is a commensal of the upper respiratory tract, is often carried by infants and young children, and shares many common antigens with *N. meningitidis*. Epidemiologic data suggest that colonization with *N. lactamica* confers some protection against invasive meningococcal disease (Olsen et al., 1991). Vaccines based of the OMVs of *N. lactamica* have been tested in animals and were found to induce cross-protection (Gorringe, 2005; Li et al., 2006).

Finally, in recent years, with the completion of meningococcal genomes, the use of reverse vaccinology has led to the discovery of new meningococcal

TABLE 53.2 Selected outer-membrane protein meningococcal vaccine candidates

OMP	References	Comments
Transferrin-binding protein (TbpA/TbpB)	Dyet et al., 2005	TbpB is better exposed and more immunogenic, but TbpA is more conserved, providing better protection in an animal model
Ferric enterobactin receptor (FetA)	Baker et al., 2001	Protein diversity and poor expression of cross-reactive epitopes limit the use as vaccine candidate
Lactoferrin-binding protein (LbpA/LbpB)	Boslego et al., 1995	Poorly immunogenic
Ferric-binding protein (FbpA)	Tappero et al., 1999	Well-conserved and -exposed protein but induces low bactericidal activity
Adhesion penetration protein (App)	De Kleijn et al., 2000	Conserved protein with presence of antibodies during human convalescence as well as bactericidal rabbit anti-App
Opacity-associated protein (Opa/Opc)	De Kleijn et al., 2000 Tondella et al., 2000	Anti-Opa interference with colonization and intranasal vaccine in mice induce IgA and IgG responses
Pilin	Kahler et al., 1998	High degree of variability but potential for antibodies to block pathogen adhesion
Porin B (PorB)	Sexton et al., 2004	Antibody to PorB mediates serotype-specific complement-mediated killing
Neisserial surface protein (NspA)	O'Hallahon et al., 2004 Wong et al., 2007	Highly conserved protein. Passive and active immunization conferring protection in mice. Vaccines in phase I trial show safety and immunogenicity but absence of bactericidal antibodies
NadA	Thompson et al., 2003	Intranasal vaccine provides protection in rats
Reduction-modifiable protein (Rmp)	Ruggeberg et al., 2004	No bactericidal or opsonic activity

Source: Adapted from Ruggeberg and Pollard (2004) with permission from the publisher.

serogroup B vaccine candidates (Pizza et al., 2000). Five antigens with aluminum hydroxide adjuvants induced SBA in mice against 78% of a large panel of meningococcal serogroup B isolates representative of global strains (Giuliani et al., 2006). Coverage could reach 90% if additional antigens or different adjuvants are used. Another promising vaccine candidate based on genome-derived neisserial antigen 1870 (recently renamed factor H-binding protein, FHBP) provides protection in rats and induces SBA in mice (Beernink et al., 2006).

Vaccines for Africa

The current approach in controlling meningococcal epidemics in sub-Saharan Africa is based on early detection of disease and mass vaccination with meningococcal polysaccharide vaccine once a weekly incidence threshold has been crossed (Lewis et al., 2001). A conjugate A/C vaccine has been studied in the African "meningitis belt" in the Gambia and Niger, and the safety and the immunogenicity of the vaccine were demonstrated (Campagne et al., 2000; Leach et al., 1997; Twumasi et al., 1995). However, the response of the serogroup A conjugate component was not as robust as the response to the serogroup C conjugate component. Currently, efforts are underway to develop conjugate vaccines against serogroup A since it is the most prevalent serogroup in the African meningitis belt.

In 2001, a partnership between the Program for Appropriate Technology in Health (PATH) and the World Health Organization (WHO) resulted in the creation of the Meningitis Vaccine Project (MVP) (Laforce et al., 2007) with the goal of eliminating meningococcal epidemics in Africa, through the development, licensure, introduction, and widespread use of an affordable (US$0.40 per dose) MenA conjugate vaccine. The MVP has identified a European manufacturer (SynCo Bio Partners, Amsterdam, Netherlands) for polysaccharide serogroup A, a developing-country manufacturer of high-quality tetanus toxoid used as a carrier (Serum Institute of India), and a U.S. public biological research center (CBER/FDA) to provide expertise in the design of conjugate vaccines. A phase I study of the vaccine in India has shown that the vaccine is both safe and immunogenic. Phase II studies have begun in Gambia and Mali, and a large demonstration study of the conjugate vaccine is envisioned for 2008–2009. After consultations with African public health officials, a vaccine introduction plan will aim at including the MenA conjugate vaccine into standard immunization schedules. Emphasis should be put on vaccination of 1- to 29-year-olds to induce herd immunity, a strategy shown to be highly effective when the MenC conjugate vaccine was introduced in several European countries.

Correlates of Protection

The standardization of the laboratory evaluation of meningococcal vaccines remains a challenge. The definition and harmonization of surrogate endpoints of efficacy, the reference strains, and the reagents to be utilized in assays and the applicability of such methodologies across all classes of vaccines and meningococcal subtypes responsible for widespread epidemic outbreaks must be established.

The currently accepted serological correlate of protection is SBA using either a rabbit or human as source of complement (Borrow et al., 2005). SBA titers greater than four using human complement and greater than eight using rabbit complement are current standards of protection. Also, data suggest that the absence of serum bactericidal activity is not necessarily associated with susceptibility to meningococcal disease. A different approach consists of assessment of a greater than fourfold rise in antibody titer after immunization. Other correlates of protection, such as a whole-blood assay that may be more sensitive than the SBA but is difficult to standardize, can be used. The serological correlates of long-term protection are not well defined. Standardization of assays to assess immunologic memory is also needed. Memory alone may be insufficient for protection since it takes at least 4–7 days for a vaccinated primed adult to mount a memory anticapsular response, and the time between acquisition of the meningococcus and development of invasive disease can be less. Memory can be assessed by the immune response to polysaccharide challenge or an increase in antibody avidity over time (Snape and Pollard, 2005). The important herd immunity features of conjugate vaccines are not easily predicted or quantified.

PROSPECTS FOR THE FUTURE

Global elimination of meningococcal disease could be an achievable target when potent and affordable vaccines against meningococcal A, C, Y, and W-135 conjugate vaccines and a broadly effective serogroup B vaccine are available. The introduction of conjugate vaccines for encapsulated organisms such as *H. influenzae, Streptococcus pneumoniae,* and *N. meningitidis* offer protection to the pediatric populations and herd immunity to the overall population, drastically decreasing the number of bacterial meningitis cases worldwide. Because of the rapid increase in the number of vaccines given during infancy, there is a need to develop combination vaccines that will protect against meningococcal disease without creating immunological interference (Southern et al., 2006; Tejedor et al., 2007; Buttery et al., 2005). Standardization of correlates of protection is important for the comparison among the different future vaccine candidates.

KEY ISSUES

- Meningococci cause meningitis/septicemia with worldwide epidemic and endemic disease, leading to high morbidity and morbidity.
- Vaccination remains the mainstay for disease prevention.
- Polysaccharide vaccines to serogroups A, C, Y, and W-135 are safe and effective; however, they do not produce herd immunity, memory responses, or immunogenicity in infants and young children. Now, available conjugate vaccines are able to overcome these limitations.
- No serogroup B polysaccharide vaccine is available due to the lack of immunogenicity and the theoretical risk of inducing autoantibodies to the polysaccharide. Serogroup B vaccines based on OMV can be used in epidemic settings but do not provide cross-protection in some age groups.
- Control of meningococcal disease relies on the development of an affordable conjugate vaccine for Africa and a broadly effective vaccine against serogroup B, using new vaccine candidates.

References

Meningococcal disease. Secondary attack rate and chemoprophylaxis in the United States, 1974. JAMA 1976; 235(3):261–265.

Prevention and control of meningococcal disease: recommendations for use of meningococcal vaccines in pediatric patients. Pediatrics 2005; 116(2):496–505.

Safety of meningococcal group C vaccines. Curr. Prob. Pharmacovigil. 2000; 26(14).

Centers for Disease Control and Prevention (CDC): Guillain-Barré syndrome among recipients of Menactra meningococcal conjugate vaccine—United States, June 2005–September 2006. MMWR Morb. Mortal. Wkly. Rep. 2006; 55(41):1120–1124.

Abramson, J.S. and Spika, J.S. Persistence of *Neisseria meningitidis* in the upper respiratory tract after intravenous antibiotic therapy for systemic meningococcal disease. J. Infect. Dis. 1985; 151(2):370–371.

Alcala, B., Arreaza, L., Salcedo, C., Uria, M.J., De La Fuente, L. and Vazquez, J.A. Capsule switching among C:2b:P1.2,5 meningococcal epidemic strains after mass immunization campaign, Spain. Emerg. Infect. Dis. 2002; 8(12):1512–1514.

Alonso, J.M., Gilmet, G., Rouzic, E.M. et al. Workshop on vaccine pressure and *Neisseria meningitidis*, Annecy, France, 9–11 March 2005. Vaccine 2007; 25(21):4125–4129.

Alper, C.A., Abramson, N., Johnston, R.B., Jr., Jandl, J.H. and Rosen, F.S. Increased susceptibility to infection associated with abnormalities of complement-mediated functions and of the third component of complement (C3). N. Engl. J. Med. 1970; 282(7):350–354.

Artenstein, M.S., Gold, R., Zimmerly, J.G., Wyle, F.A., Schneider, H. and Harkins, C. Prevention of meningococcal disease by group C polysaccharide vaccine. N. Engl. J. Med. 1970; 282(8): 417–420.

Artenstein, M.S., Rust, J.H., Jr., Hunter, D.H., Lamson, T.H. and Buescher, E.L. Acute respiratory disease and meningococcal infection in army recruits. JAMA 1967; 201(13):1004–1007.

Auckland, C., Gray, S., Borrow, R. et al. Clinical and immunologic risk factors for meningococcal C conjugate vaccine failure in the United Kingdom. J. Infect. Dis. 2006; 194(12):1745–1752.

Baevsky, R.H. Primary meningococcal pericarditis. Clin. Infect. Dis. 1999; 29(1):213–215.

Baker, M.G., Martin, D.R., Kieft, C.E. and Lennon, D. A 10-year serogroup B meningococcal disease epidemic in New Zealand: descriptive epidemiology, 1991–2000. J. Paediatr. Child Health 2001; 37(5):S13–S19.

Balmer, P., Borrow, R. and Miller, E. Impact of meningococcal C conjugate vaccine in the UK. J. Med. Microbiol. 2002; 51(9):717–722.

Barquet, N., Domingo, P., Cayla, J.A. et al. Prognostic factors in meningococcal disease. Development of a bedside predictive model and scoring system. Barcelona Meningococcal Disease Surveillance Group. JAMA 1997; 278(6):491–496.

Bax, W.A., Cluysenaer, O.J., Bartelink, A.K., Aerts, P.C., Ezekowitz, R.A. and van Dijk, H. Association of familial deficiency of mannose-binding lectin and meningococcal disease. Lancet 1999; 354(9184):1094–1095.

Beernink, P.T., Leipus, A. and Granoff, D.M. Rapid genetic grouping of factor h-binding protein (genome-derived neisserial antigen 1870), a promising group B meningococcal vaccine candidate. Clin. Vaccine Immunol. 2006; 13(7):758–763.

Benes, J., Dzupova, O., Kabelkova, M., Krizova, P. and Gabrielova, A. Infective endocarditis due to *Neisseria meningitidis*: two case reports. Clin. Microbiol. Infect. 2003; 9(10):1062–1064.

Bergthornorsson, G.I., Matthiasson, T., Guethnason, T. and Briem, H. Cost-effectiveness analysis on meningococca serogroup C vaccination. Laeknabladid 2004; 90(5):379–383.

Bevanger, L., Bergh, K., Gisnas, G., Caugant, D.A. and Froholm, L.O. Identification of nasopharyngeal carriage of an outbreak strain of *Neisseria meningitidis* by pulsed-field gel electrophoresis versus phenotypic methods. J. Med. Microbiol. 1998; 47(11):993–998.

Bilavsky, E., Yarden-Bilavsky, H., Zevit, N. and Amir, J. Primary meningococcal arthritis in a child: case report and literature review. Scand. J. Infect. Dis. 2006; 38(5):396–399.

Bilukha, O.O. and Rosenstein, N. Prevention and control of meningococcal disease. Recommendations of the Advisory Committee on Immunization Practices (ACIP). MMWR Recomm. Rep. 2005; 54(RR-7):1–21.

Bjune, G., Hoiby, E.A., Gronnesby, J.K. et al. Effect of outer membrane vesicle vaccine against group B meningococcal disease in Norway. Lancet 1991; 338(8775):1093–1096.

Blaser, M.J., Reingold, A.L., Alsever, R.N. and Hightower, A. Primary meningococcal pericarditis: a disease of adults associated with serogroup C *Neisseria meningitidis*. Rev. Infect. Dis. 1984; 6(5):625–632.

Blondeau, J.M. and Yaschuk, Y. In vitro activities of ciprofloxacin, cefotaxime, ceftriaxone, chloramphenicol, and rifampin against fully susceptible and moderately penicillin-resistant *Neisseria meningitidis*. Antimicrob. Agents Chemother. 1995; 39(11):2577–2579.

Boisier, P., Nicolas, P., Djibo, S. et al. Meningococcal meningitis: unprecedented incidence of serogroup X-related cases in 2006 in Niger. Clin. Infect. Dis. 2007; 44(5):657–663.

Booy, R., Jelfs, J., El Bashir, H. and Nissen, M.D. Impact of meningococcal C conjugate vaccine use in Australia. Med. J. Aust. 2007; 186(3):108–109.

Borrow, R., Balmer, P. and Miller, E. Meningococcal surrogates of protection—serum bactericidal antibody activity. Vaccine 2005; 23(17–18):2222–2227.

Borrow, R., Longworth, E., Gray, S.J. and Kaczmarski, E.B. Prevalence of de-O-acetylated serogroup C meningococci before the introduction of meningococcal serogroup C conjugate vaccines in the United Kingdom. FEMS Immunol. Med. Microbiol. 2000; 28(3):189–191.

Boslego, J., Garcia, J., Cruz, C. et al. Efficacy, safety, and immunogenicity of a meningococcal group B (15:P1.3) outer membrane protein vaccine in Iquique, Chile. Chilean National Committee for Meningococcal Disease. Vaccine 1995; 13(9):821–829.

Bowe, F., Lavelle, E.C., McNeela, E.A. et al. Mucosal vaccination against serogroup B meningococci: induction of bactericidal antibodies and cellular immunity following intranasal immunization with NadA of *Neisseria meningitidis* and mutants of *Escherichia coli* heat-labile enterotoxin. Infect. Immun. 2004; 72(7):4052–4060.

Brandtzaeg, P., Bryn, K., Kierulf, P. et al. Meningococcal endotoxin in lethal septic shock plasma studied by gas chromatography, mass-spectrometry, ultracentrifugation, and electron microscopy. J. Clin. Invest. 1992; 89(3):816–823.

Branham, S. Serological relationships among meningococci. Bacteriol. Rev. 1953; 17:175–188.

Brooks, R., Woods, C.W., Benjamin, D.K., Jr. and Rosenstein, N.E. Increased case-fatality rate associated with outbreaks of *Neisseria meningitidis* infection, compared with sporadic meningococcal disease, in the United States, 1994–2002. Clin. Infect. Dis. 2006; 43(1):49–54.

Bruge, J., Bouveret-Le Cam, N., Danve, B., Rougon, G. and Schulz, D. Clinical evaluation of a group B meningococcal N-propionylated polysaccharide conjugate vaccine in adult, male volunteers. Vaccine 2004; 22(9–10):1087–1096.

Brundage, J.F., Ryan, M.A., Feighner, B.H. and Erdtmann, F.J. Meningococcal disease among United States military service members in relation to routine uses of vaccines with different serogroup-specific components, 1964–1998. Clin. Infect. Dis. 2002; 35(11):1376–1381.

Bryant, P.A., Li, H.Y., Zaia, A. et al. Prospective study of a real-time PCR that is highly sensitive, specific, and clinically useful for diagnosis of meningococcal disease in children. J. Clin. Microbiol. 2004; 42(7):2919–2925.

Buttery, J.P., Riddell, A., McVernon, J. et al. Immunogenicity and safety of a combination pneumococcal-meningococcal vaccine in infants: a randomized controlled trial. JAMA 2005; 293(14):1751–1758.

Campagne, G., Garba, A., Fabre, P. et al. Safety and immunogenicity of three doses of a *Neisseria meningitidis* A + C diphtheria conjugate vaccine in infants from Niger. Pediatr. Infect. Dis. J. 2000; 19(2):144–150.

Cartwright, K., Noah, N. and Peltola, H. Meningococcal disease in Europe: epidemiology, mortality, and prevention with conjugate vaccines. Report of a European advisory board meeting Vienna, Austria, 6–8 October, 2000. Vaccine 2001; 19(31):4347–4356.

Cartwright, K., Reilly, S., White, D. and Stuart, J. Early treatment with parenteral penicillin in meningococcal disease. BMJ 1992; 305(6846):143–147.

Casella, E.B., Cypel, S., Osmo, A.A. et al. Sequelae from meningococcal meningitis in children: a critical analysis of dexamethasone therapy. Arq. Neuropsiquiatr. 2004; 62(2B):421–428.

Castagliuolo, P.P., Nisini, R., Quinti, I., Fattorossi, A. and D'Amelio, R. Immunoglobulin deficiencies and meningococcal disease. Ann. Allergy 1986; 57(1):68–70.

Caugant, D.A. Population genetics and molecular epidemiology of *Neisseria meningitidis*. APMIS 1998; 106(5):505–525.

Caugant, D.A., Froholm, L.O., Bovre, K. et al. Intercontinental spread of a genetically distinctive complex of clones of *Neisseria meningitidis* causing epidemic disease. Proc. Natl. Acad. Sci. U.S.A. 1986; 83(13):4927–4931.

Caugant, D.A., Hoiby, E.A., Magnus, P. et al. Asymptomatic carriage of *Neisseria meningitidis* in a randomly sampled population. J. Clin. Microbiol. 1994; 32(2):323–330.

Chow, A.W., Culver, B., Yoshikawa, T.T. and Guze, L.B. Primary meningococcal pericarditis presenting with tamponade. Chest 1975; 67(5):611–612.

Comanducci, M., Bambini, S., Brunelli, B. et al. NadA, a novel vaccine candidate of *Neisseria meningitidis*. J. Exp. Med. 2002; 195(11):1445–1454.

Costantino, P., Viti, S., Podda, A., Velmonte, M.A., Nencioni, L. and Rappuoli, R. Development and phase 1 clinical testing of a conjugate vaccine against meningococcus A and C. Vaccine 1992; 10(10):691–698.

Danielson, L. and Mann, E. A history of a singular and very noted disease, which lately made its appearance in Medfield. Med. Agric. Reg. 1806; 1:65–69.

D'Antuono, A., Andalo, F. and Varotti, C. Acute urethritis due to *Neisseria meningitidis*. Sex. Transm. Infect. 1999; 75(5):362.

Decosas, J. and Koama, J.B. Chronicle of an outbreak foretold: meningococcal meningitis W135 in Burkina Faso. Lancet Infect. Dis. 2002; 2(12):763–765.

de Jonge, M.I., Hamstra, H.J., Jiskoot, W. et al. Intranasal immunisation of mice with liposomes containing recombinant meningococcal OpaB and OpaJ proteins. Vaccine 2004; 22(29–30):4021–4028.

de Jonge, M.I., Vidarsson, G., van Dijken, H.H. et al. Functional activity of antibodies against the recombinant OpaJ protein from *Neisseria meningitidis*. Infect. Immun. 2003; 71(5):2331–2340.

de Kleijn, E.D., de Groot, R., Hack, C.E. et al. Activation of protein C following infusion of protein C concentrate in children with severe meningococcal sepsis and purpura fulminans: a randomized, double-blinded, placebo-controlled, dose-finding study. Crit. Care Med. 2003; 31(6):1839–1847.

de Kleijn, E.D., de Groot, R., Labadie, J. et al. Immunogenicity and safety of a hexavalent meningococcal outer-membrane-vesicle vaccine in children of 2–3 and 7–8 years of age. Vaccine 2000a; 18(15):1456–1466.

de Kleijn, E.D., de Groot, R., Lafeber, A.B. et al. Immunogenicity and safety of monovalent p1.7(h),4 meningococcal outer membrane vesicle vaccine in toddlers: comparison of two vaccination schedules and two vaccine formulations. Vaccine 2000b; 19(9–10):1141–1148.

Dell, E., Bedford, D., McMahon, B. and Nicholson, A. Meningococcal disease—management of serogroup C clusters in a hyperendemic area. Ir. Med. J. 2001; 94(6):166, Comment: 168–169.

de Moraes, J.C., Perkins, B.A., Camargo, M.C. et al. Protective efficacy of a serogroup B meningococcal vaccine in Sao Paulo, Brazil. Lancet 1992; 340(8827):1074–1078.

Densen, P. Complement deficiencies and meningococcal disease. Clin. Exp. Immunol. 1991; 86(Suppl. 1):57–62.

De Queiros, L., Castro, L., Ferreira, M.C. and Goncalves, G. [Acceptance of the new conjugate vaccines. Meningococcal and pneumococcal vaccines, in the cohort born in 1999, in the North Region of Portugal]. Acta Med. Port. 2004; 17(1):49–53.

De Schrijver, K. and Maes, I. An outbreak of serogroup C meningococcal disease in the province of Antwerp (Belgium) in 2001–2002. Eur. J. Epidemiol. 2003; 18(11):1073–1077.

Devoy, A.F., Dyet, K.H. and Martin, D.R. Stability of PorA during a meningococcal disease epidemic. J. Clin. Microbiol. 2005; 43(2):832–837.

De Wals, P., Deceuninck, G., Boulianne, N. and De Serres, G. Effectiveness of a mass immunization campaign using serogroup C meningococcal conjugate vaccine. JAMA 2004; 292(20):2491–2494.

De Wals, P., Hertoghe, L., Borlee-Grimee, I. et al. Meningococcal disease in Belgium. Secondary attack rate among household, daycare nursery and pre-elementary school contacts. J. Infect. 1981; 3(Suppl. 1):53–61.

Diggle, M.A. and Clarke, S.C. Increased genetic diversity of *Neisseria meningitidis* isolates after the introduction of meningococcal serogroup C polysaccharide conjugate vaccines. J. Clin. Microbiol. 2005; 43(9):4649–4653.

Dyet, K.H. and Martin, D.R. Sequence variation in the porB gene from B:P1.4 meningococci causing New Zealand's epidemic. J. Clin. Microbiol. 2005; 43(2):838–842.

Edwards, M.S. and Baker, C.J. Complications and sequelae of meningococcal infections in children. J. Pediatr. 1981; 99(4):540–545.

El Bashir, H., Klaber, R., El Baki, A. and Booy, R. W135 meningococcal pericarditis: report of two cases and review of the literature. Pediatr. Infect. Dis. J. 2004; 23(10):969–970.

Ellison, R.T., III, Kohler, P.F., Curd, J.G., Judson, F.N. and Reller, L.B. Prevalence of congenital or acquired complement deficiency in patients with sporadic meningococcal disease. N. Engl. J. Med. 1983; 308(16):913–916.

Emonts, M., Hazelzet, J.A., de Groot, R. and Hermans, P.W. Host genetic determinants of *Neisseria meningitidis* infections. Lancet Infect. Dis. 2003; 3(9):565–577.

Erickson, L.J., De Wals, P., McMahon, J. and Heim, S. Complications of meningococcal disease in college students. Clin. Infect. Dis. 2001; 33(5):737–739.

Ferry, N. and Steele, A. Active immunization with meningococcus toxin. JAMA 1935; 104:983–984.

Fijen, C.A., Kuijper, E.J., Hannema, A.J., Sjoholm, A.G. and van Putten, J.P. Complement deficiencies in patients over ten years old with meningococcal disease due to uncommon serogroups. Lancet 1989; 2(8663):585–588.

Fijen, C.A., Kuijper, E.J., te Bulte, M.T., Daha, M.R. and Dankert, J. Assessment of complement deficiency in patients with meningococcal disease in The Netherlands. Clin. Infect. Dis. 1999; 28(1):98–105.

Finne, J., Leinonen, M. and Makela, P.H. Antigenic similarities between brain components and bacteria causing meningitis. Implications for vaccine development and pathogenesis. Lancet 1983; 2(8346):355–357.

Fischer, M., Hedberg, K., Cardosi, P. et al. Tobacco smoke as a risk factor for meningococcal disease. Pediatr. Infect. Dis. J. 1997; 16(10):979–983.

Flexner, S. The results if the serum treatment in thirteen hundred cases of epidemic meningitis. J. Exp. Med. 1913; 17:553.

Frasch, C.E., Zollinger, W.D. and Poolman, J.T. Serotype antigens of *Neisseria meningitidis* and a proposed scheme for designation of serotypes. Rev. Infect. Dis. 1985; 7(4):504–510.

Fraser, A., Gafter-Gvili, A., Paul, M. and Leibovici, L. Antibiotics for preventing meningococcal infections. Cochrane Database Syst. Rev. 2006; 4:CD004785.

Gagneux, S.P., Hodgson, A., Smith, T.A. et al. Prospective study of a serogroup X *Neisseria meningitidis* outbreak in northern Ghana. J. Infect. Dis. 2002; 185(5):618–626.

Garcia, N.S., Castelo, J.S., Ramos, V., Rezende, G.S. and Pereira, F.E. Frequency of myocarditis in cases of fatal meningococcal infection in children: observations on 31 cases studied at autopsy. Rev. Soc. Bras. Med. Trop. 1999; 32(5):517–522.

Gates, F. A report on antimeningitis vaccination and observations on agglutinins in the blood of chronic meningococcus carriers. J. Exp. Med. 1918; 28:449–474.

Gilmore, A., Stuart, J. and Andrews, N. Risk of secondary meningococcal disease in health-care workers. Lancet 2000; 356(9242):1654–1655.

Girgis, N., Sultan, Y., Frenck, R.W., Jr., El-Gendy, A., Farid, Z. and Mateczun, A. Azithromycin compared with rifampin for eradication of nasopharyngeal colonization by *Neisseria meningitidis*. Pediatr. Infect. Dis. J. 1998; 17(9):816–819.

Giuliani, M.M., Adu-Bobie, J., Comanducci, M. et al. A universal vaccine for serogroup B meningococcus. Proc. Natl. Acad. Sci. U.S.A. 2006; 103(29):10834–10839.

Glover, J. The cerebrospinal fever epidemic of 1917 at "X" depot. J. R. Army Med. Corps 1918; 30:23.

Gold, R., Lepow, M.L., Goldschneider, I., Draper, T.F. and Gotshlich, E.C. Kinetics of antibody production to group A and group C meningococcal polysaccharide vaccines administered during the first six years of life: prospects for routine immunization of infants and children. J. Infect. Dis. 1979; 140(5):690–697.

Goldacre, M.J., Roberts, S.E. and Yeates, D. Case fatality rates for meningococcal disease in an English population, 1963–98: database study. BMJ. 2003; 327(7415):596–597.

Goldschneider, I., Gotschlich, E.C. and Artenstein, M.S. Human immunity to the meningococcus. I. The role of humoral antibodies. J. Exp. Med. 1969a; 129(6):1307–1326.

Goldschneider, I., Gotschlich, E.C. and Artenstein, M.S. Human immunity to the meningococcus. II. Development of natural immunity. J. Exp. Med. 1969b; 129(6):1327–1348.

Gomez, J.A., Criado, M.T. and Ferreiros, C.M. Bactericidal activity of antibodies elicited against the *Neisseria meningitidis* 37-kDa ferric binding protein (FbpA) with different adjuvants. FEMS Immunol. Med. Microbiol. 1998; 20(1):79–86.

Gorringe, A.R. Can *Neisseria lactamica* antigens provide an effective vaccine to prevent meningococcal disease? Expert Rev. Vaccines 2005; 4(3):373–379.

Gotschlich, E.C., Goldschneider, I. and Artenstein, M.S. Human immunity to the meningococcus. IV. Immunogenicity of group A and group C meningococcal polysaccharides in human volunteers. J. Exp. Med. 1969a; 129(6):1367–1384.

Gotschlich, E.C., Liu, T.Y. and Artenstein, M.S. Human immunity to the meningococcus. 3. Preparation and immunochemical properties of the group A, group B, and group C meningococcal polysaccharides. J. Exp. Med. 1969b; 129(6):1349–1365.

Granoff, D.M., Bartoloni, A., Ricci, S. et al. Bactericidal monoclonal antibodies that define unique meningococcal B polysaccharide epitopes that do not cross-react with human polysialic acid. J. Immunol. 1998; 160(10):5028–5036.

Granoff, D.M., Kelsey, S.K., Bijlmer, H.A. et al. Antibody responses to the capsular polysaccharide of *Neisseria meningitidis* serogroup B in patients with meningococcal disease. Clin. Diagn. Lab. Immunol. 1995; 2(5):574–582.

Greenfield, S., Sheehe, P.R. and Feldman, H.A. Meningococcal carriage in a population of "normal" families. J. Infect. Dis. 1971; 123(1):67–73.

Greenwood, B. Manson Lecture. Meningococcal meningitis in Africa. Trans. R. Soc. Trop. Med. Hyg. 1999; 93(4):341–353.

Greenwood, B.M., Bradley, A.K. and Wall, R.A. Meningococcal disease and season in sub-Saharan Africa. Lancet 1985; 2(8459):829–830.

Greenwood, B.M., Greenwood, A.M., Bradley, A.K. et al. Factors influencing susceptibility to meningococcal disease during an epidemic in The Gambia, West Africa. J. Infect. 1987; 14(2): 167–184.

Gupta, S. and Tuladhar, A.B. Does early administration of dexamethasone improve neurological outcome in children with meningococcal meningitis? Arch. Dis. Child. 2004; 89(1):82–83.

Hadi, H.A., Wooldridge, K.G., Robinson, K. and Ala'Aldeen, D.A. Identification and characterization of App: an immunogenic autotransporter protein of *Neisseria meningitidis*. Mol. Microbiol. 2001; 41(3):611–623.

Hagman, M., Forslin, L., Moi, H. and Danielsson, D. *Neisseria meningitidis* in specimens from urogenital sites. Is increased awareness necessary? Sex. Transm. Dis. 1991; 18(4):228–232.

Hahne, S.J., Gray, S.J., Jean, F. et al. W135 meningococcal disease in England and Wales associated with Hajj 2000 and 2001. Lancet 2002; 359(9306):582–583.

Halperin, S.A., Langley, J.M., Smith, B. et al. Phase 1 first-in-human studies of the reactogenicity and immunogenicity of a recombinant meningococcal NspA vaccine in healthy adults. Vaccine 2007; 25(3):450–457.

Hardman, J.M. and Earle, K.M. Myocarditis in 200 fatal meningococcal infections. Arch. Pathol. 1969; 87(3):318–325.

Harrison, L.H., Dwyer, D.M., Maples, C.T. and Billmann, L. Risk of meningococcal infection in college students. JAMA 1999; 281(20):1906–1910.

Hart, C.A. and Cuevas, L.E. Meningococcal disease in Africa. Ann. Trop. Med. Parasitol. 1997; 91(7):777–785.

Hassan-King, M.K., Wall, R.A. and Greenwood, B.M. Meningococcal carriage, meningococcal disease and vaccination. J. Infect. 1988; 16(1):55–59.

Heath, P.T. *Haemophilus influenzae* type b conjugate vaccines: a review of efficacy data. Pediatr. Infect. Dis. J. 1998; 17(Suppl. 9):S117–S122.

Hibberd, M.L., Sumiya, M., Summerfield, J.A., Booy, R. and Levin, M. Association of variants of the gene for mannose-binding lectin with susceptibility to meningococcal disease. Meningococcal Research Group. Lancet 1999; 353(9158):1049–1053.

Jackson, L.A., Schuchat, A., Reeves, M.W. and Wenger, J.D. Serogroup C meningococcal outbreaks in the United States. An emerging threat. JAMA 1995; 273(5):383–389.

Joao, I., Marques, P., Namora, J. et al. *Neisseria meningitidis* native valve endocarditis. A case report. Rev. Port. Cardiol. 2001; 20(9): 877–880.

Johnson, A.S., Gorringe, A.R., Mackinnon, F.G., Fox, A.J., Borrow, R. and Robinson, A. Analysis of the human Ig isotype response to lactoferrin binding protein A from *Neisseria meningitidis*. FEMS Immunol. Med. Microbiol. 1999; 25(4):349–354.

Kahler, C.M. and Stephens, D.S. Genetic basis for biosynthesis, structure, and function of meningococcal lipooligosaccharide (endotoxin). Crit. Rev. Microbiol. 1998; 24(4):281–334.

Kasper, D.L., Fauci, A.S., Longo, D.L., Braunwald, E., Hauser, S.L., Jameson, J.L. Harrison's Principles of Internal Medicine, 16th ed. McGraw-Hill Companies, 2005.

Kaplan, S.L., Schutze, G.E., Leake, J.A. et al. Multicenter surveillance of invasive meningococcal infections in children. Pediatrics 2006; 118(4):e979–e984.

Karolus, J.J., Gandelman, A.L. and Nolan, B.A. Urethritis caused by *Neisseria meningitidis*. J. Clin. Microbiol. 1980; 12(2):284–285.

Kayhty, H., Jousimies-Somer, H., Peltola, H. and Maketa, P.H. Antibody response to capsular polysaccharides of groups A and C *Neisseria meningitidis* and *Haemophilus influenzae* type b during bacteremic disease. J. Infect. Dis. 1981; 143(1):32–41.

Kelly, D.F., Snape, M.D., Clutterbuck, E.A. et al. CRM197-conjugated serogroup C meningococcal capsular polysaccharide, but not the native polysaccharide, induces persistent antigen-specific memory B cells. Blood 2006; 108(8):2642–2647.

Khuns, D., Kisner, P., Williams, M. and Moorman, P. The control of meningococcic meningitis epidemics by active immunization with meningococcus soluble toxin: further studies. JAMA 1938; 110:484–487.

Khuns, D., Nelson, C., Feldman H.A., et al. The prophylactic value of sulfadiazine in the control of meningococcic meningitis. JAMA 1943; 123:335–339.

Kiefer, F. Zur differential Diagnose des Erregers der epidemischen Cerebrospinalmeningitis und der Gonorrhoea. Berl. Klin. Wochenschr. 1896; 33:628.

Koumare, B., Ouedraogo-Traore, R., Sanou, I., et al. The first large epidemic of meningococcal disease caused by serogroup W135, Burkina Faso, 2002. Vaccine 2007; 25(Suppl 1):A37–A41.

Laforce, F.M., Konde, K., Viviani, S. and Preziosi, M.P. The Meningitis Vaccine Project. Vaccine 2007; 25(Suppl 1):A97–A100.

Leach, A., Twumasi, P.A., Kumah, S. et al. Induction of immunologic memory in Gambian children by vaccination in infancy with a group A plus group C meningococcal polysaccharide-protein conjugate vaccine. J. Infect. Dis. 1997; 175(1):200–204.

Lehmann, A.K., Gorringe, A.R., Reddin, K.M., West, K., Smith, I. and Halstensen, A. Human opsonins induced during meningococcal disease recognize transferrin binding protein complexes. Infect. Immun. 1999; 67(12):6526–6532.

Lepeyssonnie, L. La meningite cerebrospinale en Afrique. Bull. World Health Organ. 1963; 28(Suppl.):53–114.

Lepow, M.L., Beeler, J., Randolph, M., Samuelson, J.S. and Hankins, W.A. Reactogenicity and immunogenicity of a quadrivalent combined meningococcal polysaccharide vaccine in children. J. Infect. Dis. 1986; 154(6):1033–1036.

Lewis, R., Nathan, N., Diarra, L., Belanger, F. and Paquet, C. Timely detection of meningococcal meningitis epidemics in Africa. Lancet 2001; 358(9278):287–293.

Li, Y., Zhang, Q., Winterbotham, M., Mowe, E., Gorringe, A. and Tang, C.M. Immunization with live *Neisseria lactamica* protects mice against meningococcal challenge and can elicit serum bactericidal antibodies. Infect. Immun. 2006; 74(11):6348–6355.

Lingappa, J.R., Al-Rabeah, A.M., Hajjeh, R. et al. Serogroup W-135 meningococcal disease during the Hajj, 2000. Emerg. Infect. Dis. 2003(6):665–671.

MacLennan, J., Obaro, S., Deeks, J. et al. Immune response to revaccination with meningococcal A and C polysaccharides in Gambian children following repeated immunisation during early childhood. Vaccine 1999; 17(23–24):3086–3093.

Maiden, M.C., Bygraves, J.A., Feil, E. et al. Multilocus sequence typing: a portable approach to the identification of clones within populations of pathogenic microorganisms. Proc. Natl. Acad. Sci. U.S.A. 1998; 95(6):3140–3145.

Maiden, M.C. and Spratt, B.G. Meningococcal conjugate vaccines: new opportunities and new challenges. Lancet 1999; 354(9179):615–616.

Maiden, M.C. and Stuart, J.M. Carriage of serogroup C meningococci 1 year after meningococcal C conjugate polysaccharide vaccination. Lancet 2002; 359(9320):1829–1831.

Mandrell, R.E., Azmi, F.H. and Granoff, D.M. Complement-mediated bactericidal activity of human antibodies to poly alpha 2→8 N-acetylneuraminic acid, the capsular polysaccharide of *Neisseria meningitidis* serogroup B. J. Infect. Dis. 1995; 172(5):1279–1289.

Martin, D., Cadieux, N., Hamel, J. and Brodeur, B.R. Highly conserved *Neisseria meningitidis* surface protein confers protection against experimental infection. J. Exp. Med. 1997; 185(7):1173–1183.

Masterton, R.G., Youngs, E.R., Wardle, J.C., Croft, K.F. and Jones, D.M. Control of an outbreak of group C meningococcal meningitis with a polysaccharide vaccine. J. Infect. 1988; 17(2):177–182.

Mattila, P.S. and Carlson, P. Pharyngolaryngitis caused by *Neisseria meningitidis*. Scand. J. Infect. Dis. 1998; 30(2):198–200.

McCrumb, F., Hall, H. and Merideth, A. Chloramphenicol in the treatment of meningococcal meningitis. Am. J. Med. 1951; 10:696.

Miller, E., Salisbury, D. and Ramsay, M. Planning, registration, and implementation of an immunisation campaign against meningococcal serogroup C disease in the UK: a success story. Vaccine 2001; 20(Suppl. 1):S58–S67.

Miller, M.A., Millikin, P., Griffin, P.S., Sexton, R.A. and Yousuf, M. *Neisseria meningitidis* urethritis. A case report. JAMA 1979; 242(15):1656–1657.

Moe, G.R., Dave, A. and Granoff, D.M. Molecular analysis of anti-N-propionyl *Neisseria meningitidis* group B polysaccharide monoclonal antibodies. Mol. Immunol. 2006; 43(9):1424–1431.

Moe, G.R., Zuno-Mitchell, P., Lee, S.S., Lucas, A.H. and Granoff, D.M. Functional activity of anti-Neisserial surface protein A monoclonal antibodies against strains of *Neisseria meningitidis* serogroup B. Infect. Immun. 2001; 69(6):3762–3771.

Moore, P.S., Hierholzer, J., DeWitt, W. et al. Respiratory viruses and mycoplasma as cofactors for epidemic group A meningococcal meningitis. JAMA 1990; 264(10):1271–1275.

Moraga, F.A., Domingo, P., Barquet, N., Gasser, I. and Gallart, A. Invasive meningococcal conjunctivitis. JAMA 1990; 264(3): 333–334.

Morgan, D.R., Spence, M., Crowe, M. and O'Keeffe, D.B. Primary (isolated) meningococcal pericarditis. Clin. Cardiol. 2002; 25(6):305–307.

Mothershed, E.A., Sacchi, C.T., Whitney, A.M. et al. Use of real-time PCR to resolve slide agglutination discrepancies in serogroup identification of *Neisseria meningitidis*. J. Clin. Microbiol. 2004; 42(1):320–328.

Nadel, S., Goldstein, B., Williams, M.D. et al. Drotrecogin alfa (activated) in children with severe sepsis: a multicentre phase III randomised controlled trial. Lancet 2007; 369(9564):836–843.

Newcombe, J., Cartwright, K., Palmer, W.H. and McFadden, J. PCR of peripheral blood for diagnosis of meningococcal disease. J. Clin. Microbiol. 1996; 34(7):1637–1640.

Newton, D.A. and Wilson, W.G. Primary meningococcal conjunctivitis. Pediatrics 1977; 60(1):104–106.

Ni, H., Knight, A.I., Cartwright, K., Palmer, W.H. and McFadden, J. Polymerase chain reaction for diagnosis of meningococcal meningitis. Lancet 1992; 340(8833):1432–1434.

Nicholson, A. and Lepow, I.H. Host defense against *Neisseria meningitidis* requires a complement-dependent bactericidal activity. Science 1979; 205(4403):298–299.

Nielsen, H.E., Koch, C., Magnussen, P. and Lind, I. Complement deficiencies in selected groups of patients with meningococcal disease. Scand. J. Infect. Dis. 1989; 21(4):389–396.

Nielsen, L.T. Chronic meningococcemia. Arch. Dermatol. 1970; 102(1):97–101.

Niklasson, P.M., Lundbergh, P. and Strandell, T. Prognostic factors in meningococcal disease. Scand. J. Infect. Dis. 1971; 3(1):17–25.

Noronha, C.P., Struchiner, C.J. and Halloran, M.E. Assessment of the direct effectiveness of BC meningococcal vaccine in Rio de Janeiro, Brazil: a case-control study. Int. J. Epidemiol. 1995; 24(5):1050–1057.

Odegaard, K. and Gundersen, T. Isolation of *Neisseria meningitidis* in urogenital/rectal infections. Acta Derm. Venereol. 1977; 57(2):173–176.

O'Hallahan, J., Lennon, D. and Oster, P. The strategy to control New Zealand's epidemic of group B meningococcal disease. Pediatr. Infect. Dis. J. 2004; 23(Suppl. 12):S293–S298.

Olsen, S.F., Djurhuus, B., Rasmussen, K. et al. Pharyngeal carriage of *Neisseria meningitidis* and *Neisseria lactamica* in households with infants within areas with high and low incidences of meningococcal disease. Epidemiol. Infect. 1991; 106(3):445–457.

Ouedraogo-Traore, R., Hoiby, E.A., Sanou, I. et al. Molecular characteristics of *Neisseria meningitidis* strains isolated in Burkina Faso in 2001. Scand. J. Infect. Dis. 2002; 34(11):804–807.

Pether, J.V., Scott, R.J. and Hancock, P. Do meningococci cause sore throats? Lancet 1994; 344(8937):1636.

Pichichero, M., Casey, J., Blatter, M. et al. Comparative trial of the safety and immunogenicity of quadrivalent (A, C, Y, W-135) meningococcal polysaccharide-diphtheria conjugate vaccine versus quadrivalent polysaccharide vaccine in two- to ten-year-old children. Pediatr. Infect. Dis. J. 2005; 24(1):57–62.

Pillai, S., Howell, A., Alexander, K. et al. Outer membrane protein (OMP) based vaccine for *Neisseria meningitidis* serogroup B. Vaccine 2005; 23(17–18):2206–2209.

Pizza, M., Scarlato, V., Masignani, V. et al. Identification of vaccine candidates against serogroup B meningococcus by whole-genome sequencing. Science 2000; 287(5459):1816–1820.

Plested, J.S., Harris, S.L., Wright, J.C. et al. Highly conserved *Neisseria meningitidis* inner-core lipopolysaccharide epitope confers protection against experimental meningococcal bacteremia. J. Infect. Dis. 2003; 187(8):1223–1234.

Ramsay, M.E., Andrews, N.J., Trotter, C.L., Kaczmarski, E.B. and Miller, E. Herd immunity from meningococcal serogroup C conjugate vaccination in England: database analysis. BMJ 2003; 326(7385):365–366.

Richmond, P., Borrow, R., Miller, E. et al. Meningococcal serogroup C conjugate vaccine is immunogenic in infancy and primes for memory. J. Infect. Dis. 1999; 179(6):1569–1572.

Rosenqvist, E., Musacchio, A., Aase, A. et al. Functional activities and epitope specificity of human and murine antibodies against the class 4 outer membrane protein (Rmp) of *Neisseria meningitidis*. Infect. Immun. 1999; 67(3):1267–1276.

Rosenstein, N.E., Perkins, B.A., Stephens, D.S., Popovic, T. and Hughes, J.M. Meningococcal disease. N. Engl. J. Med. 2001; 344(18):1378–1388.

Rosenstein, N.E., Perkins, B.A., Stephens, D.S. et al. The changing epidemiology of meningococcal disease in the United States, 1992–1996. J. Infect. Dis. 1999; 180(6):1894–1901.

Ruggeberg, J.U. and Pollard, A.J. Meningococcal vaccines. Paediatr. Drugs 2004; 6(4):251–266.

Salleras, L., Dominguez, A. and Cardenosa, N. Impact of mass vaccination with polysaccharide conjugate vaccine against serogroup C meningococcal disease in Spain. Vaccine 2003; 21(7–8):725–728.

Schaad, U.B. Arthritis in disease due to *Neisseria meningitidis*. Rev. Infect. Dis. 1980; 2(6):880–888.

Schoenback, E. and Phair, J. The sensitivity of meningococci to sulfadiazine. Am. J. Hyg. 1948; 47:177–186.

Schryvers, A.B. and Stojiljkovic, I. Iron acquisition systems in the pathogenic *Neisseria*. Mol. Microbiol. 1999; 32(6):1117–1123.

Schwentker, F., Gelman, S. and Long, P. The treatment of meningococcic meningitis with sulfonamide. Preliminary report. JAMA 1937; 108:1407.

Sejvar, J.J., Johnson, D., Popovic, T. et al. Assessing the risk of laboratory-acquired meningococcal disease. J. Clin. Microbiol. 2005; 43(9):4811–4814.

Sexton, K., Lennon, D., Oster, P. et al. Proceedings of the meningococcal vaccine strategy World Health Organization satellite meeting, 10 March 2004, Auckland, New Zealand. N.Z. Med. J. 2004; 117(1200), 1 p. preceding U1027.

Sharip, A., Sorvillo, F., Redelings, M.D., Mascola, L., Wise, M. and Nguyen, D.M. Population-based analysis of meningococcal disease mortality in the United States: 1990–2002. Pediatr. Infect. Dis. J. 2006; 25(3):191–194.

Sierra, G.V., Campa, H.C., Varcacel, N.M. et al. Vaccine against group B *Neisseria meningitidis*: protection trial and mass vaccination results in Cuba. NIPH Ann. 1991; 14(2):195–207, discussion 208–210.

Snape, M.D. and Pollard, A.J. Meningococcal polysaccharide-protein conjugate vaccines. Lancet Infect. Dis. 2005; 5(1):21–30.

Sophian, A. and Black, J. Prophylactic vaccination against epidemic meningitis. JAMA 1912; 59:527–532.

Southern, J., Crowley-Luke, A., Borrow, R., Andrews, N. and Miller, E. Immunogenicity of one, two or three doses of a meningococcal C conjugate vaccine conjugated to tetanus toxoid, given as a three-dose primary vaccination course in UK infants at

2, 3 and 4 months of age with acellular pertussis-containing DTP/Hib vaccine. Vaccine 2006; 24(2):215–219.

Stein, D.M., Robbins, J., Miller, M.A., Lin, F.Y. and Schneerson, R. Are antibodies to the capsular polysaccharide of *Neisseria meningitidis* group B and *Escherichia coli* K1 associated with immunopathology? Vaccine 2006; 24(3):221–228.

Stephens, D.S. Uncloaking the meningococcus: dynamics of carriage and disease. Lancet 1999; 353(9157):941–942.

Stephens, D.S., Greenwood, B. and Brandtzaeg, P. Epidemic meningitis, meningococcaemia, and *Neisseria meningitidis*. Lancet 2007; 369(9580):2196–2210.

Stephens, D.S., Hajjeh, R.A., Baughman, W.S., Harvey, R.C., Wenger, J.D. and Farley, M.M. Sporadic meningococcal disease in adults: results of a 5-year population-based study. Ann. Intern. Med. 1995; 123(12):937–940.

Stephens, D.S., Whitney, A.M., Rothbard, J. and Schoolnik, G.K. Pili of *Neisseria meningitidis*. Analysis of structure and investigation of structural and antigenic relationships to gonococcal pili. J. Exp. Med. 1985; 161(6):1539–1553.

Stephens, D.S. and Zimmer, S.M. Pathogenesis, therapy, and prevention of meningococcal sepsis. Curr. Infect. Dis. Rep. 2002; 4(5):377–386.

Sullivan, T.D. and LaScolea, L.J., Jr. *Neisseria meningitidis* bacteremia in children: quantitation of bacteremia and spontaneous clinical recovery without antibiotic therapy. Pediatrics 1987; 80(1): 63–67.

Swartley, J.S., Marfin, A.A., Edupuganti, S. et al. Capsule switching of *Neisseria meningitidis*. Proc. Natl. Acad. Sci. U.S.A. 1997; 94(1):271–276.

Tappero, J.W., Lagos, R., Ballesteros, A.M. et al. Immunogenicity of 2 serogroup B outer-membrane protein meningococcal vaccines: a randomized controlled trial in Chile. JAMA 1999; 281(16):1520–1527.

Tejedor, J.C., Moro, M., Ruiz-Contreras, J. et al. Immunogenicity and reactogenicity of primary immunization with a novel combined *Haemophilus influenzae* type b and *Neisseria meningitidis* serogroup C-tetanus toxoid conjugate vaccine coadministered with a diphtheria-tetanus-acellular pertussis-hepatitis B-inactivated poliovirus vaccine at 2, 4 and 6 months. Pediatr. Infect. Dis. J. 2007; 26(1):1–7.

Thompson, E.A., Feavers, I.M. and Maiden, M.C. Antigenic diversity of meningococcal enterobactin receptor FetA, a vaccine component. Microbiology 2003; 149(Pt. 7):1849–1858.

Thompson, M.J., Ninis, N., Perera, R. et al. Clinical recognition of meningococcal disease in children and adolescents. Lancet 2006; 367(9508):397–403.

Thornton, V., Lennon, D., Rasanathan, K. et al. Safety and immunogenicity of New Zealand strain meningococcal serogroup B OMV vaccine in healthy adults: beginning of epidemic control. Vaccine 2006; 24(9):1395–1400.

Tondella, M.L., Popovic, T., Rosenstein, N.E. et al. Distribution of *Neisseria meningitidis* serogroup B serosubtypes and serotypes circulating in the United States. The Active Bacterial Core Surveillance Team. J. Clin. Microbiol. 2000; 38(9):3323–3328.

Traore, Y., Njanpop-Lafourcade, B.M., Adjogble, K.L. et al. The rise and fall of epidemic *Neisseria meningitidis* serogroup W135 meningitis in Burkina Faso, 2002–2005. Clin. Infect. Dis. 2006; 43(7):817–822.

Trotter, C.L., Andrews, N.J., Kaczmarski, E.B., Miller, E. and Ramsay, M.E. Effectiveness of meningococcal serogroup C conjugate vaccine 4 years after introduction. Lancet 2004; 364(9431):365–367.

Trotter, C.L. and Ramsay, M.E. Vaccination against meningococcal disease in Europe: review and recommendations for the use of conjugate vaccines. FEMS Microbiol. Rev. 2007; 31(1):101–107.

Trotter, C.L., Ramsay, M.E., Gray, S., Fox, A. and Kaczmarski, E. No evidence for capsule replacement following mass immunisation with meningococcal serogroup C conjugate vaccines in England and Wales. Lancet Infect. Dis. 2006; 6(10):616–617, author reply 617–618.

Twumasi, P.A., Jr., Kumah, S., Leach, A. et al. A trial of a group A plus group C meningococcal polysaccharide-protein conjugate vaccine in African infants. J. Infect. Dis. 1995; 171(3):632–638.

Tzeng, Y.L. and Stephens, D.S. Epidemiology and pathogenesis of *Neisseria meningitidis*. Microbes Infect. 2000; 2(6):687–700.

van de Beek, D., de Gans, J., McIntyre, P. and Prasad, K. Steroids in adults with acute bacterial meningitis: a systematic review. Lancet Infect. Dis. 2004; 4(3):139–143.

Vieusseux, M. Memoire sur la maladie qui a regne a Geneva au printemps de 1805. J. Med. Chir. Pharm. 1805; 11:163–182.

Vincent, J.L., Nadel, S., Kutsogiannis, D.J. et al. Drotrecogin alfa (activated) in patients with severe sepsis presenting with purpura fulminans, meningitis, or meningococcal disease: a retrospective analysis of patients enrolled in recent clinical studies. Crit. Care 2005; 9(4):R331–R343.

Virji, M., Makepeace, K., Ferguson, D.J., Achtman, M. and Moxon, E.R. Meningococcal Opa and Opc proteins: their role in colonization and invasion of human epithelial and endothelial cells. Mol. Microbiol. 1993; 10(3):499–510.

Vogel, U., Claus, H. and Frosch, M. Rapid serogroup switching in *Neisseria meningitidis*. N. Engl. J. Med. 2000; 342(3):219–220.

Vu, D.M., Kelly, D., Heath, P.T., McCarthy, N.D., Pollard, A.J. and Granoff, D.M. Effectiveness analyses may underestimate protection of infants after group C meningococcal immunization. J. Infect. Dis. 2006; 194(2):231–237.

Weichselbaum, A. Ueber die Aetiologie der akuten meningitis cerebrospinalis. Fortschr. Med. 1887; 5:573.

Weis, N. and Lind, I. Epidemiological markers in *Neisseria meningitidis*: an estimate of the performance of genotyping vs phenotyping. Scand. J. Infect. Dis. 1998; 30(1):69–75.

Weisfelt, M., Hoogman, M., van de Beek, D., de Gans, J., Dreschler, W.A. and Schmand, B.A. Dexamethasone and long-term outcome in adults with bacterial meningitis. Ann. Neurol. 2006; 60(4):456–468.

Welte, R., van den Dobbelsteen, G., Bos, J.M. et al. Economic evaluation of meningococcal serogroup C conjugate vaccination programmes in The Netherlands and its impact on decision-making. Vaccine 2004; 23(4):470–479.

West, D., Reddin, K., Matheson, M. et al. Recombinant *Neisseria meningitidis* transferrin binding protein A protects against experimental meningococcal infection. Infect. Immun. 2001; 69(3):1561–1567.

White, B., Livingstone, W., Murphy, C., Hodgson, A., Rafferty, M. and Smith, O.P. An open-label study of the role of adjuvant hemostatic support with protein C replacement therapy in purpura fulminans-associated meningococcemia. Blood 2000; 96(12):3719–3724.

Wiese-Posselt, M., Hellenbrand, W., Siedler, A. and Mayer, C. Universal childhood immunisation with pneumococcal vaccine and meningococcal serogroup C vaccine introduced in Germany. Euro. Surveill. 2006; 11(9):E060907 4.

Wiesner, P.J., Tronca, E., Bonin, P., Pedersen, A.H. and Holmes, K.K. Clinical spectrum of pharyngeal gonococcal infection. N. Engl. J. Med. 1973; 288(4):181–185.

Wilder-Smith, A., Goh, K.T., Barkham, T. and Paton, N.I. Hajj-associated outbreak strain of *Neisseria meningitidis* serogroup W135: estimates of the attack rate in a defined population and the risk of invasive disease developing in carriers. Clin. Infect. Dis. 2003; 36(6):679–683.

Winstead, J.M., McKinsey, D.S., Tasker, S., De Groote, M.A. and Baddour, L.M. Meningococcal pneumonia: characterization and review of cases seen over the past 25 years. Clin. Infect. Dis. 2000; 30(1):87–94.

Wong, S., Lennon, D., Jackson, C. et al. New Zealand epidemic strain meningococcal B outer membrane vesicle vaccine in children aged 16–24 months. Pediatr. Infect. Dis. J. 2007; 26(4):345–350.

Wright, J.C., Williams, J.N., Christodoulides, M. and Heckels, J.E. Immunization with the recombinant PorB outer membrane protein induces a bactericidal immune response against *Neisseria meningitidis*. Infect. Immun. 2002; 70(8):4028–4034.

Young, L.S., LaForce, F.M., Head, J.J., Feeley, J.C. and Bennett, J.V. A simultaneous outbreak of meningococcal and influenza infections. N. Engl. J. Med. 1972; 287(1):5–9.

Zangwill, K.M., Stout, R.W., Carlone, G.M. et al. Duration of antibody response after meningococcal polysaccharide vaccination in US Air Force personnel. J. Infect. Dis. 1994; 169(4):847–852.

Zhu, D., Barniak, V., Zhang, Y., Green, B. and Zlotnick, G. Intranasal immunization of mice with recombinant lipidated P2086 protein reduces nasal colonization of group B *Neisseria meningitidis*. Vaccine 2006; 24(26):5420–5425.

Zhu, D., Zhang, Y., Barniak, V., Bernfield, L., Howell, A. and Zlotnick, G. Evaluation of recombinant lipidated P2086 protein as a vaccine candidate for group B *Neisseria meningitidis* in a murine nasal challenge model. Infect. Immun. 2005; 73(10):6838–6845.

Zimmer, S.M. and Stephens, D.S. Meningococcal conjugate vaccines. Expert Opin. Pharmacother. 2004; 5(4):855–863.

Zimmer, S.M. and Stephens, D.S. Serogroup B meningococcal vaccines. Curr. Opin. Investig. Drugs 2006; 7(8):733–739.

CHAPTER

54

Plague

E. Diane Williamson
DSTL Porton Down, Salisbury, Wiltshire, UK

OUTLINE

ABSTRACT

Killed whole cell vaccines for plague were first produced as long ago as the late 1890s and modified versions of these are still used, with evidence that they are efficacious against bubonic plague. Renewed efforts with modern technology have yielded new candidate vaccines that are less reactogenic, can be produced in a conventional pharmaceutical manufacturing plant, and are protective against the life-threatening pneumonic form of the disease. This chapter reviews the threat still posed by plague in the world today, the rationale for the research and development of new vaccine formulations and assesses the likely impact of a prophylactic vaccine for pneumonic plague.

INTRODUCTION

Plague is an ancient disease which has not been eradicated from the modern world. During the last two millennia, the causative bacterium *Yersinia pestis* has been responsible for social and economic devastation on a scale unmatched by other infectious diseases or armed conflicts. In the modern era, bubonic plague is primarily a disease of small rodents and mammals that is spread by fleas in endemic areas, establishing enzootic foci which occasionally erupt as an epizootic outbreak. Humans can be infected either by flea bite or by inhalational exposure through a secondary host, e.g., wild rabbit or prairie dog, or domestic cat. Endemic areas for the disease include China, southern Africa, vast areas of Asia and South America and the southwest portion of the United States (Perry and Fetherston, 1997) while an antibiotic-resistant strain of plague emerged in Madagascar in 1995 (Guiyoule et al., 1997, 2001).

Each year there are several thousand reported cases of the disease worldwide (Dennis et al., 1999) with a fatality rate between 5 and 15% (Anon, 1996, 2000), and plague has been classified as a re-emerging disease by the World Health Organization. In the United States, an average of 18 cases is reported each year with a death rate of 1 in 7. In 2003, two individuals infected in New Mexico, traveled to New York where they developed feverish symptoms, later diagnosed as plague (reported in the *Sunday Telegraph*, London, UK, 11 May 2003). Antibiotic therapy was started in time to save their lives, but this case history indicates the insidious threat that the infection causes to individuals and to the wider population. Additional threats to modern human life posed by plague include the emergence of strains resistant to multiple drugs (Gallimand et al., 1997) and the potential use of plague as an agent of biological warfare.

The consequences of infection with plague in humans are serious and the infection needs to be detected and treated promptly to prevent serious morbidity leading to death. Transmission to humans by feeding fleas leads to the swelling of the draining lymph nodes, followed by a septicemic illness, characteristic of bubonic plague (Perry and Fetherston, 1997). However, animal-to-human or human-to-human transmission can occur via droplet nuclei spread by the coughing of individuals with bubonic or septicemic plague who have developed pulmonary lesions (Meyer, 1961; Poland and Barnes, 1979). Protection against pneumonic plague is the paramount requirement to prevent epidemic spread.

In the long term, comprehensive vaccination against this disease holds the brightest prospect for prevention and control. Killed whole cell (KWC) vaccines for plague have been used in humans since the late 1890s and although evidence from the 20th century indicates that these formulations have reduced the incidence of bubonic plague (Cavanaugh et al., 1974), there is no evidence to suggest that they are effective against pneumonic plague (Anderson et al., 1996; Andrews et al., 1996; Reisman, 1970; Williamson et al., 1995, 2001). In the late 1990s renewed efforts led to recombinant vaccine candidates with potential to protect against the virulent pneumonic form of plague and progress toward the development and licensing of these formulations for human use is reported below.

SHORT HISTORY OF THE DISEASE

The first major epidemic of plague probably occurred in China around 224 BC. Credible sources suggest that the Justinian plague (AD 541–750) killed an estimated 100 million people. This originated in central Africa and rapidly spread along the trading routes from Egypt's main port, Alexandria, to Central and South Asia, Arabia, North Africa, and much of Southern Europe. By the end of 544, it is estimated that 20–25% of the population of Europe south of the Alps had been killed by the plague (Sodeinde et al., 1992).

The most well-known epidemic of the plague, the Black Death, occurred during the 14th century (Table 54.1). This epidemic probably began in China about AD 1338 and 1351, killing an estimated 60% of the population (Dols, 1977) and may have been associated with natural disasters such as floods and earthquakes which destroyed the habitats of the plague-bearing rodents, forcing them into contact with other rodent

TABLE 54.1 Transmission of the Black Death along trading routes during the second plague pandemic

Major trading region	Year of first arrival
Central Asia	1338
Volga River	1345
Anatolia	1347
Lower Egypt	
Southern Italy	
Palestine	1348
Arabia	
Tunisia	
Northern Italy	
Iberia	
France	
England	1348
Northern Germany	

Source: Reprinted from www.american.edu/TED/bubonic.html.

populations and thus spreading their fleas (Gottfried, 1983). From China, the plague spread along the trade route known as the Silk Road into Central Asia and throughout the Caucasus. The northern and eastern shores of the Black Sea became ridden with plague, from where infected rats and fleas were picked up in merchant ships on the spice-trading sea routes and carried south (Gottfried, 1983). Thus, the plague eventually reached the Mediterranean and Southern Europe. By 1350, the plague had run its course through the Mediterranean Basin, killing 35–40% of the population in the region. However, following the sea lanes and rivers, the disease had moved north, into France, reaching Paris in 1348, where the population was decimated (Gottfried, 1983). The plague continued northward, killing an estimated 45–55% of the Scandinavian population, and along the trade routes to Iceland, where it killed 60% of the population, and then onto Greenland, reaching there probably in the winter of 1350 and annihilating the small population (Gottfried, 1983).

The Black Death is best documented in Britain, where it is believed to have arrived on Gascon ships to the port of Melcombe Regis, in 1348. Other outbreaks followed across southwest England. It reached London in September 1348 and Ireland the following year. The plague ravaged London until the spring of 1350, killing an estimated 20,000 people. There is evidence that the port of Bristol and city of Winchester were also severely affected (Gottfried, 1983).

Plague persisted on the European continental land mass for several centuries, and culminated in an epidemic of bubonic plague, known as the Great Plague of London. This started in London in the summer of 1665 killing more than 30,000 people in the city and spread to many parts of England, and was curtailed by the killing of the infected fleas with the arrival of winter.

The third plague pandemic started in China in the middle of the 19th century, and spread east and west, causing 10 million deaths in India alone. In the 1890s, plague finally made its way to North America. Rats infected with the disease arrived in San Francisco in 1900 aboard ships which had sailed from Asia. In the 20th century, outbreaks of plague in Asia, prevalent during the 1960s and 1970s, were attributed to the ecological disruption caused by the Vietnam War and consequently increased contact between the animal reservoir of the disease and humans. Credible estimates indicate that in total almost 200 million deaths can be attributed to plague, which swept across Europe in the three major epidemics (Duplaix, 1998; Perry and Fetherston, 1997).

ETIOLOGIC AGENT

Classification

The etiological agent of plague is *Y. pestis*, a gram-negative bacterium which is a member of the Enterobacteriacae family. *Y. pestis* is closely related to the other human pathogenic Yersiniae. However, unlike *Yersinia enterocolitica* and *Yersinia pseudotuberculosis*, *Y. pestis* does not infect the host by the enteric route. Some of the genes that are required for invasion by this route in *Y. pseudotuberculosis* and *Y. enterocolitica*, such as the *inv* and *yadA*, are present in *Y. pestis* but are not expressed as a consequence of mutations (Brubaker, 1991; Perry and Fetherston, 1997). Another major difference is that *Y. pestis* is unable to survive outside an animal host, whereas *Y. enterocolitica* and *Y. pseudotuberculosis* can survive in the environment. These findings suggest that *Y. pestis* might have evolved from the other human pathogenic Yersiniae. This suggestion is supported by a comparison of the genetic diversity of several housekeeping genes in *Y. pestis* and *Y. pseudotuberculosis* which indicates that *Y. pestis* likely evolved from *Y. pseudotuberculosis* 1500–20,000 years ago (Achtman et al., 1999).

The recent genome sequencing of the CO92 strain of *Y. pestis* (Parkhill et al., 2001) has allowed comparison with the genome sequence of *Y. pseudotuberculosis* and has revealed numerous genetic differences reflecting changes in virulence and niche (Wren, 2003). In the transition from an enteropathogen to an arthropod-vectored systemic pathogen, *Y. pestis* has acquired, inactivated, and rearranged genes (Parkhill et al., 2001;

Wren, 2003). In addition to the 70 kb virulence plasmid (pYV/pCD1) common to all the pathogenic yersiniae, *Y. pestis* possesses two unique plasmids, a 100–110 kb plasmid denoted pFra/pMT1 and a small pPST/pPCP1 plasmid of 9.5 kb. These plasmids encode a number of virulence determinants, described below. A total of 149 pseudogenes has been identified in the genome of the fully virulent *Y. pestis* strain CO92 (Parkhill et al., 2001) indicating the genetic evolution undertaken by the organism to adapt from being an enteropathogen to a flea-vectored pathogen (Oyston and Isherwood, 2005), with genome reduction.

Because of its evolution to a flea-vectored lifecycle, environmental temperature would be expected to be influential in the regulation of virulence in *Y. pestis*. In the flea where the temperature does not exceed 28°C, the *caf1* and the pCD1 Type three secretion TTS operons are repressed so that the bacterium does not have a capsule or secrete yersinia outer proteins (YOPs). The evidence that the two-component response regulatory system, PhoPQ, may have a role in the survival of *Y. pestis* in the flea, comes from a study where inactivation of the *phoP* gene allowed the expression of a number of Pho-regulated proteins, usually at 28°C, to occur at 37°C (Oyston et al., 2000). The *PhoP*$^{-}$ *Y. pestis* was more susceptible to macrophage killing and less virulent in the mouse model (Oyston et al., 2000). However, the mutation of a high temperature response gene, *HtrA*, did not significantly attenuate *Y. pestis* in the mouse model, although the mutant could not grow *in vitro* at 39°C (Williams et al., 2000).

ANTIGENS

The three plasmids borne by *Y. pestis* encode an array of virulence factors which are active at different stages in the life cycle of the bacterium. These virulence factors are also potential antigens for isolation, toxoiding if necessary and inclusion in a potential new vaccine for plague. To have protective potential, the antigens must be secreted from the bacterium in a niche accessible to the host immune system which could then be primed by vaccination to neutralize them. In this context, therefore, it is helpful to review the virulence factors associated with the stages in the life cycle of the plague bacterium.

Y. pestis, present in a blood meal ingested by the flea from an infected mammal, is protected by an intracellular phospholipase D, encoded by pFra/pMT1 (Hinnebusch et al., 2000), from lysis by plasma components. The ability of the bacteria to aggregate in the flea and block the midgut is now known to be due to their ability to form a biofilm which in turn is dependent on the *hms* locus (Hinnebusch et al, 1996). *Y. pestis* uses additional complex regulatory mechanisms to survive within the flea, which are influenced by environmental temperature and which include changes to the surface structure of the bacteria. The small pPST/pPCP1 plasmid encodes plasminogen activator (Pla), a surface protease, which is required for dissemination of the bacteria from the initial site of infection by promoting the cleavage of fibrin clots (Sodeinde et al., 1992) through the activation of plasminogen to plasmin and by facilitating adhesion to extracellular matrices and the invasion of mammalian endothelial cells (Lahteenmaki et al., 1998, 2001).

Once in the mammalian host, the bacteria are phagocytosed in macrophages and likely transported to a draining lymph node. There is controversial evidence that phagocytosis may be facilitated by the chemokine receptor CCR5 (Elvin et al., 2004; Mecsas et al., 2004; Stephens et al., 1998), prevalent on macrophages and dendritic cells, but not on neutrophils. *Y. pestis* can survive and multiply in the phagolysosome (Straley and Harmon, 1984a, 1984b), eventually killing the macrophage (Straley and Harmon, 1984b). While intracellular and at 37°C, additional temperature and calcium-regulated virulence factors are expressed. One of these is the Fraction 1 antigen (F1) which becomes surface expressed in the capsule surrounding the bacterium, preventing further phagocytosis and aiding the eventual establishment of an extracellular infection (Titball et al., 2003). Also important in this process is the pCD1/pYV-encoded type III secretion system, the function of which is to secrete an array of YOPs with cytotoxic or anti-phagocytic activity into the host cell in a highly regulated manner (Cornelis, 2002). A key regulatory antigen in this system is the low calcium response virulence or lcrV antigen (Burrows, 1963) which has been visualized on the bacterial cell surface (Petterson et al., 1999) and is important in establishing host cell contact and initiating TTS. In addition, the *psa* operon of *Y. pestis* encodes the pH6 antigen adhesin which appears as surface fimbriae on the bacteria on release from the macrophage and which may be involved in adhesion (Payne et al., 1998) to secondary targets *in vivo*. Although, *Y. pestis* has a rough LPS phenotype, with lipid A and core components, it has no O antigen (Prior et al., 2001a, 2001b; Skurnik et al., 2000) to confer resistance to complement. It may be that surface proteases such as Pla are able to confer serum resistance by cleaving complement components (Sodeinde et al., 1992).

Many of the *Y. pestis* virulence factors have been isolated in native or recombinant forms and screened as immunogens, including LPS (Prior et al., 2001a),

TABLE 54.2 Immunogenicity and protective efficacy of subunit antigens of *Yersinia pestis* in the mouse model

Subunit	Description	Immunogenic	Protective (references)
Pla	Surface plasminogen activator protease	Y	Not tested
PH6 antigen	Putative surface adhesin	Y	Bubonic—not protective (Unpublished data, Payne et al., 1998)
LPS	Lipopolysaccharide	Y	Bubonic—not protective (Prior et al., 2001a)
F1 antigen	Surface capsule	Y	Bubonic & pneumonic—protective (Andrews et al., 1996; Lawton et al., 1963; Miller et al., 1998; Simpson et al., 1990; Williamson et al., 1995)
YopD	Type III system—translocation Yop	Y	Bubonic—partially protective (Andrews et al., 1999; Leary et al., 1999)
YopH	Type III system—PTPase effector Yop	Y	Bubonic—minimally protective (Andrews et al., 1999)
YopE	Type III system—cytotoxin effector Yop	Y	Bubonic—not protective (Leary et al., 1999)
YopN	Type III system—regulates Yop release?	Y	Bubonic—not protective (Andrews et al., 1999; Leary et al., 1999)
YopK	Type III system—regulates Yop release?	Y	Bubonic—not protective (Andrews et al., 1999; Leary et al., 1999)
YopM	Type III system—effector Yop	Y	Bubonic—not protective (Andrews et al., 1999; Leary et al., 1999)
Ypk A	Type III system—Ser/Thr kinase effector Yop	Y	Bubonic—delayed time to death (Andrews et al., 1999)
YscF	Type III system—part of injection needle structure	Y	Bubonic—protective (Matson et al., 2005)
V antigen	Type III system—part of the injectosome	Y	Bubonic & pneumonic—protective (Anderson et al., 1996; Lawton et al., 1963; Leary et al., 1995; Williamson et al., 1995)

YOPs (Andrews et al., 1999; Leary et al., 1999; Matson et al., 2005), pH6 antigen (unpublished data), Pla (unpublished data) and F1 and V (Heath et al., 1998; Lawton et al., 1963; Williamson et al., 1995), (Table 54.2). Only the F1 and V proteins have been demonstrated to have significant protective efficacy in animal models as either individual proteins, or when coadministered or fused (Heath et al., 1998; Lawton et al., 1963; Simpson et al., 1990; Williamson et al., 1995).

PROTECTIVE IMMUNE RESPONSE

Much effort has been made in understanding the mechanism of protective immunity in plague, with early work identifying the F1 and V (VW) proteins as protective antigens (Burrows, 1963; Lawton et al., 1963). A significant amount of work has been carried out in the mouse (Williamson, 2001), guinea pig (Jones et al., 2003), and in nonhuman primate models (Williamson, 2004) from which immune correlates have been proposed. Although a titer of neutralizing antibody provides protection against exposure, it is clear that the development of cell-mediated immunity (CMI) is also critical to full protection and eventual clearance of the bacterial challenge from the host (Elvin et al., 2004).

ANTIBODY

Mice immunized with antigens derived from *Y. pestis* or in recombinant form have been observed to develop specific antibody titers to the YOPs (Leary et al., 1999; Matson et al., 2005) and to recombinant F1 (rF1) and rV (Heath et al., 1998; Williamson et al., 1995). The major research effort in recent years has focused on the rF1 and rV antigens and these have been found to be immunogenic individually (Anderson et al., 1996; Andrews et al., 1996; Lawton et al., 1963; Leary et al., 1995; Miller et al., 1998; Simpson et al., 1990; Williamson et al., 1995) in different strains of mice (Elvin and Williamson, 2000, 2004; Heath et al., 1998; Williamson et al., 2001) and when tested in different laboratories (Heath et al., 1998; Lawton et al., 1963; Williamson et al., 1995). A systematic study in which mice of four different haplotypes and each gender was used, showed that all responded to immunization with the rF1 and rV antigens with high antibody titers (Jones et al., 2001). The response to either antigen is affected by the conformation of the protein (Williamson et al., 1999) and an optimum combination of rF1 to rV was found to be when rF1 was in a twofold molar excess to rV. However, mice can also mount an antibody response to these antigens when the C-terminal of the F1 antigen is genetically fused to the N-terminus of the V antigen (Heath et al., 1998). In immunized Balb/c mice, the titer of IgG1 directed toward the combined antigens correlated significantly ($P < 0.005$) with protection against challenge (Williamson et al., 1999). Antibody titer to rF1 + rV in guinea pigs also correlated with protection and passive transfer of specific antibody (IgG) from the guinea pig into naïve mice, protecting the latter against challenge (Jones et al., 2003). Cynomolgus macaques immunized with rF1 and rV antigens produced high titers of specific antibody which also passively protected naïve mice against challenge, when used as a reference serum (Williamson et al., 2005). A murine monoclonal antibody (Mab7.3) specific for the V antigen, has been identified to passively protect naïve mice against live organism challenge (Hill et al., 1997) and also to compete with mouse, macaque, and human serum, derived from rF1 + rV immunized subjects, for binding to V antigen *in vitro* (Williamson et al., 2005) indicating the existence of a B-cell epitope in the V antigen which is highly conserved and protective. In addition, a monoclonal antibody to F1 is synergistic in effect with Mab 7.3 in conferring protection by passive transfer in the mouse against plague challenge (Hill et al., 2003).

However, the protection conferred by passive transfer is not infinite and as the mass of delivered antibody decays with time, breakthrough can occur in protection (Green et al., 1999), indicating the importance of inducing a CMI response by immunization.

CMI

CMI has traditionally been much more difficult than antibody to evaluate satisfactorily. However, the induction of CMI has been detected in mice immunized with a combination of rF1 and rV antigens by measurement of the recall response of lymphocytes isolated from the vaccinees to the antigens *in vitro* (Williamson et al., 1996, 1999). Moreover, mice immunized with the rF1 + rV antigens by the nasal or transdermal routes have been found to have splenic recall responses, when splenocytes have been isolated and restimulated *in vitro* (Eyles et al., 2004). Attempts to measure changes in cellular activation by flow cytometry following immunization of individuals in a Phase I clinical trial of the rF1 + rV vaccine, failed to discern measurable trends (Williamson et al., 2005), possibly due to the inherent variation present in a genetically diverse target population. A simplified *in vitro* assay in which peripheral blood mononuclear cells (PBMCs) from clinical trial vaccinees are restimulated with rF1 and rV has also not been sufficiently sensitive to detect a significant activation signal for these cells over background noise.

However, CMI is clearly of importance in the survival of immunized animals following live organism challenge. This conclusion has been reached by the testing of the vaccine in mice with targeted gene deletions which impair either the antibody or cell-mediated axes of the immune response. IL4 was found not to be essential for protective efficacy, in that IL4 knockout mice were able to respond to the vaccine with measurable antibody and were protected against challenge (Elvin and Williamson, 2000). However, a targeted gene deletion in the Stat4 signaling pathway (preventing the ligation of the Stat4 receptor with IL12 signaling) and abrogating a Th1 response, resulted in reduced protection, whereas mice with a targeted gene deletion in the Stat 6 pathway and no Th2 activity, were fully protected against challenge (Elvin and Williamson, 2004). In general terms, antibody is highly protective against this predominantly extracellular infection. Supplementation of the specific antibody responses with a balanced Th1/Th2 activity (Elvin and Williamson, 2004) appears to provide an optimum strategy for protection.

EPIDEMIOLOGY

In the modern world, plague is maintained in endemic areas by enzootic foci which occasionally erupt in an epizootic outbreak (Fig. 54.1). In these areas, *Y. pestis* cycles through populations of rodents and is transmitted to and between mammals by the bite of infected fleas. Infected fleas can survive for long periods in close association with infected animals (Engelthaler and Gage, 2000; Kir'iakova, 1973) and periodically there is an epizootic outbreak, sometimes associated with major environmental change, e.g., an earthquake, but often unattributable (Oyston and Isherwood, 2005). In this complex life cycle maintained by the bacterium, human is an accidental host, either through close contact with infected rodents and their fleas or following exposure to an infected wild

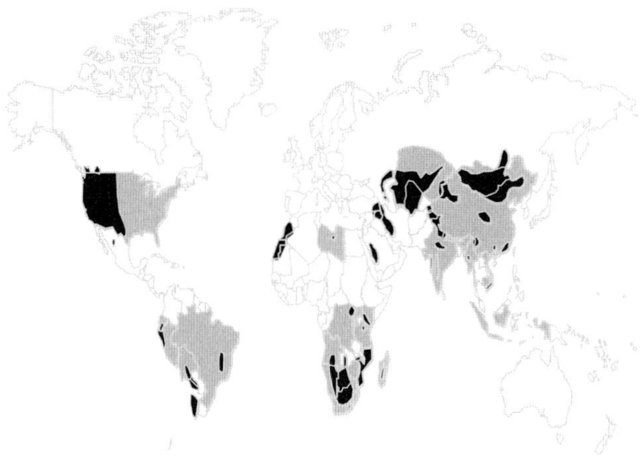

FIGURE 54.1 Occurrence of plague in the modern world. The map depicts the probable foci of plague in red, with countries reporting plague (1975–1995) in ocher.

animal or to an aerosol of bacteria from an infected domestic cat (Doll et al., 1994). In these instances, subsequent spread of plague bacilli to the lungs leads to the development of the pneumonic form of the disease; person-to-person transmission by respiratory droplets (Poland and Barnes, 1979) can result in rapid epidemic spread of the disease (Meyer, 1961). It is the pneumonic form of the disease that is most feared and associated with a mortality rate approaching 100% when untreated (Christie and Corbel, 1990; Parkhill et al., 2001; Poland and Barnes, 1979).

SIGNIFICANCE AS PUBLIC HEALTH PROBLEM

Plague continues to affect people all over the world. Plague returned to Madagascar in 1997, having been absent from the country since the 1930s. The recurrence has been attributed mainly to a reduction in rat control measures and control over the disease is still trying to be regained, complicated by the identification of a multi-drug-resistant isolate (Gallimand et al., 1997; Guiyoule et al., 1997, 2001). An outbreak of plague in the Surat region of India in 1994 caused 876 presumptive cases of plague and 54 fatalities (Anon, 1995). In 2005, an outbreak of pneumonic plague at a diamond mine in Northern Congo in January–February 2005, caused 114 confirmed infections among the miners, 54 of whom died. The outbreak was limited only by the dispersal of miners fleeing in panic from the area (WHO, 2005).

POTENTIAL AS BIOTHREAT AGENT

Due to its propensity to cause epidemic disease, plague is a potential biothreat agent. The development of vaccines with comprehensive efficacy against both the bubonic and pneumonic forms of plague provides a robust defense against the risk of the deliberate use of this serious pathogen as a biothreat agent.

CLINICAL DISEASE

There are three recognized forms of plague in human. Bubonic plague is the most common form of disease and arises following a bite from a flea that has previously fed on an infected animal (Poland and Barnes, 1979). The bacteria are disseminated from the initial site of infection to the draining lymph nodes, which become swollen and tender, forming a bubo. The bubo can reach the size of a hen's egg and is the classical feature of bubonic plague. A bacteremia may develop with blood culture counts in the range $<10–4 \times 10^7$ cfu/ml (Perry and Fetherston, 1997). Symptoms usually begin within a week of a person's exposure to the bacteria. The typical presentation involves fever of sudden onset, chills, and acutely swollen and extremely tender lymph nodes, or buboes, usually in the groin or axilla. As the disease progresses, sepsis may develop along with disseminated intravascular coagulation (DIC) and purpuric skin lesions, eventually producing necrosis of the digits or nose (hence the name Black Death).

In a small percentage of patients, a septicemic form of plague will develop, in which the buboes are not noticeable but which is characterized by an elevated temperature, chills, headache, malaise, and gastrointestinal disturbances (Poland and Barnes, 1979). Septicemic plague can be primary or secondary to pneumonic or bubonic plague and can result in DIC, purpura, meningitis, and pneumonia. Thrombosis of acral blood vessels can result in gangrene of the fingers and nose. Survival rates are good if plague is diagnosed and treatment is started promptly but because of the generalized nature of these symptoms a diagnosis is often delayed, and if this is the case, even with medical intervention, 50% of patients may die, probably as a result of the induction of the systemic inflammatory response syndrome (Perry and Fetherston, 1997). Some patients may also have secondary pneumonia caused by hematogenous spreading of bacteria to the lungs (Poland and Barnes, 1979). Pneumonic plague results in the production of highly infectious bloody sputum. Coughing results in the

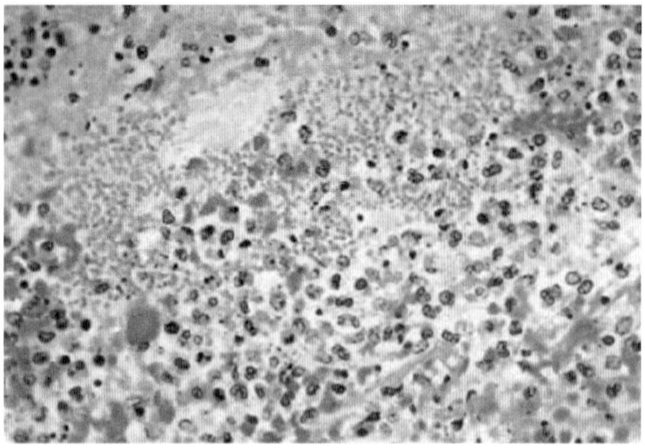

FIGURE 54.2 Pneumonic plague. This Giemsa stain reveals pneumonia and many *Yersinia pestis* bacteria in the lung. Image (PHIL ID NO. 741) courtesy of Dr. Marshall Fox/CDC; available at http://phil.cdc.gov/phil/ (see color plate section).

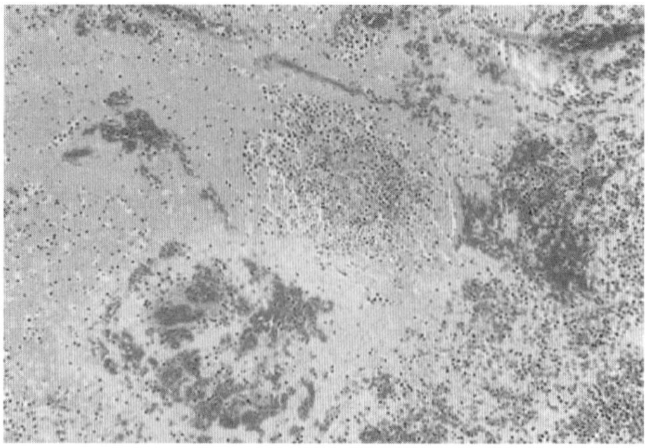

FIGURE 54.3 Evidence of plague in the lymph nodes. In this image, medullary necrosis with fluid and *Yersinia pestis* bacteria are evident. Image (PHIL ID NO. 731) courtesy of Dr. Marshall Fox/CDC; available at http://phil.cdc.gov/phil/ (see color plate section).

production of airborne droplets containing bacteria, which can be inhaled by susceptible individuals, causing a primary pneumonic plague (Meyer, 1961; Poland and Barnes, 1979). The infectious dose by inhalation is estimated to be 100–500 organisms (Franz et al., 1997). The pneumonic form of the disease is feared because of the rapidity with which the disease develops (1–3 days), infected persons having fever, cough, and dyspnea. Respiratory symptoms include pleuritic chest pain, dyspnea, cyanosis, and a productive cough (hemoptysis). Gastrointestinal symptoms (nausea, vomiting, abdominal pain, and diarrhea) may be present. Meningitis, septicemia, and DIC can also occur. There is a high mortality rate in infected individuals (approximately 100%) and potential for rapid spread of disease from human to human (Poland and Barnes, 1979). The clinical course follows that of a severe and rapidly progressive pneumonia. Marked intra-alveolar edema and congestion of the lungs are common (Centers for Disease Control; Bioterrorism preparedness website) (Fig. 54.2). Pulmonary lesions include areas of central exudate with peripheral congestion. This pattern initially is lobular, but usually progresses to lobar consolidation (Dennis and Meier, 1997). Distinguishing primary pneumonic plague from secondary hematogenous spread to the lungs can be difficult. Features that occur more commonly with primary pneumonic plague include the following: tracheal and bronchial mucosal hemorrhages, fibrinous pleuritis and subpleural hemorrhages overlying areas of exudative pneumonia, less inflammation and necrosis and more exudation in lobular foci of the parenchyma, foci of pneumonia along medium and large bronchi, and more involvement of hilar lymph nodes (Dennis and Meier, 1997) (Fig. 54.3). There is less evidence of severe disease in organs other than the lungs, if such evidence is present. Symptoms may progress rapidly to respiratory failure and full-blown sepsis with DIC. Chest films commonly show bilateral infiltrates or consolidation. The results of laboratory studies of seriously ill patients usually reveal what would be expected to accompany sepsis and end organ failure—leukocytosis, coagulation abnormalities, elevated transaminase levels, and azotemia.

DIAGNOSIS

Plague would probably be detected through the usual microbiologic investigations performed for patients with severe pneumonia, and DIC. A Gram's stain of blood, sputum, or node aspirate may reveal gram-negative coccobacilli. A Wright, Giemsa, or Wayson stain may show a characteristic bipolar pattern. Cultures of sputum, blood, or node aspirate will show growth in about 24–48 h. Where available, specialist assays may be deployed to identify plague antigens by ELISA for IgM, or fluorescent antibody staining of sputum or blood specimens, or polymerase chain reaction. Chest radiograph findings in pneumonic plague are nonspecific and include infiltrates and consolidation.

TREATMENT

Although study data are very limited regarding the success of treatment targeted against plague, *Y. pestis* is susceptible to a number of antibiotics. Historically, streptomycin has been the preferred treatment, although it is not commonly used now and its availability is limited. Other options include gentamicin, doxycycline, ciprofloxacin, levofloxacin, and chloramphenicol (Centers for Disease Control; Bioterrorism preparedness website). The same antibiotics are administered to adults and children alike, with dosage scaled down for body weight for children. Intravenous therapy is usually recommended for seriously ill patients. In the case of an epidemic, during which resources may be limited, oral therapy with ciprofloxacin or doxycycline could be effective. Once exposure to plague has been established, prompt antibiotic therapy should be administered to any patient who presents with fever or cough, because to delay such treatment before confirmatory test results are obtained would threaten a patient's chance of survival.

While the mortality of untreated bubonic plague is estimated to be 60%, antibiotic therapy should reduce this to less than 5%. In pneumonic plague, the disease progresses rapidly to shock and death if not treated with antibiotics within 24h of onset. In septicemic plague the overall case-fatality rate is 30–50% but approaches 100% without therapy.

PROPHYLACTIC VACCINATION

Prophylactic vaccination should hold the best prospect for disease prevention. The availability of vaccines is reviewed below. Primarily, these have been suspensions of killed whole organisms and there is evidence that their use, e.g., during the Vietnam War, did reduce the incidence of bubonic plague in military personnel in this endemic region of the world. However, there is also evidence that these formulations are ineffective against the life-threatening pneumonic plague, a fact that has accelerated the search in the last 20 years for improved vaccine(s).

VACCINES

History

KWC vaccines for plague have been used in humans since 1897, when Waldemar Haffkine introduced a KWC vaccine, following an outbreak of plague in Bombay in 1896. Subsequently a number of iterations of the Haffkine vaccine were produced including the Army Vaccine in 1946, the Cutter USP vaccine in 1994, and the Greer vaccine (1994–1999). These KWC vaccine formulations were sterile suspensions of killed *Y. pestis* (1.8–2.2×10^9 organisms per ml), suspended in aqueous medium incorporating a preservative such as phenol. The plague bacilli were killed by formaldehyde exposure. These vaccines were generally rather reactogenic, inducing local and systemic side effects, which although transient, could be incapacitating. Despite these limitations, there is evidence to suggest that the KWC vaccines did provide some protection against bubonic plague, principally from KWC-vaccinated U.S. servicemen in Vietnam in whom the incidence of disease was 1 case/10^6 people whereas the incidence in unvaccinated Vietnamese was 333 cases/10^6 people (Anon, 1996; Cavanaugh et al., 1974). There is, however, experimental evidence that protection against the pneumonic form of the disease was negligible (Williamson et al., 2001).

A series of live attenuated vaccines, the EV series, has also been used, mainly in the former Soviet Union and French colonies, e.g., Madagascar. EV76 (in use since 1908), EV Saigon, and EV Madagascar have questionable efficacy in evoking an immune response in humans and some severe side effects (Meyer et al., 1974; Naumov et al., 1992). It was reported that 12/12 human volunteers administered EV76 developed severe systemic reactions including febrile responses with headache, weakness, and malaise of sufficient severity to warrant the hospitalization of some vaccinees (Meyer, 1970). Moreover, there is experimental evidence that EV76 protects mice against bubonic and pneumonic plague (Russell et al., 1995). However, the EV vaccines are not licensed for clinical use in Europe/United States, possibly for reasons of reactogenicity and instability. EV76 is a pigmentation mutant of *Y. pestis* in which the *pgm* locus, affecting not only pigmentation but also iron aquisition and siderophore production, was spontaneously deleted but in which there is also evidence for the insertional activation of some genes.

Current Vaccine

The only currently available vaccine for plague is a KWC formulation, made by the Commonwealth Serum Laboratories (CSL) in Australia and comprises heat-killed agar-grown *Y. pestis* organisms (3×10^9 organisms per ml) in a saline suspension containing 0.5% w/v phenol. The strains used in the preparation of the vaccine were derived from the Haffkine Institute, India. This vaccine is a prescription medicine. The primary immunization schedule comprises an initial subcutaneous 0.5ml dose, followed by a 0.5ml booster dose at 1–4 weeks and every 6 months thereafter. The recommended dosage for children is scaled down from the adult dose, depending on age (Table 54.3).

TABLE 54.3 Recommended dosage (ml) for CSL plague vaccine

Age	1st dose	2nd dose	3rd dose	Booster dose
6 months–2 years	0.1	0.1	0.1	0.1
3–6 years	0.2	0.2	0.2	0.2
7–11 years	0.3	0.3	0.3	0.3
Over 12 years	0.5	0.5	–	0.5

Source: Derived from Data sheet issued by CSL Ltd, Parkville 3052, Victoria, Australia.

TABLE 54.4 Recommendations for the use of plague vaccines

Situation
Vaccination recommended
1. Following natural disaster and/or at times when regular sanitary practices are disrupted
2. All laboratory and field personnel who are working with *Y. pestis* organisms resistant to antibiotics
3. Persons engaged in aerosol experiments with *Y. pestis*
4. Persons engaged in field operations in plague enzootic areas where prevention of exposure is not possible (e.g., some disaster areas)
Selective vaccination recommended
1. Laboratory personnel regularly working with *Y. pestis* or plague-infected rodents
2. Workers (e.g., Peace Corps volunteers or agricultural advisers) who reside in plague enzootic areas or plague epidemic rural areas where avoidance of rodents or fleas is impossible
3. Persons whose vocation brings them into regular contact with wild rodents or rabbits in plague enzootic areas

Booster doses every 6 months are recommended for persons living in areas where plague is prevalent. If booster doses are required for people who have had reactions to this vaccine previously, it is recommended to give a reduced dose (0.1 ml) by intradermal injection.

Indications for Vaccination and Target Populations

The indication for the current licensed vaccine is to protect against risk of exposure to plague (Table 54.4). Currently such a risk could arise from environmental exposure of individuals in high-risk areas in countries in which the disease is endemic, including the rural mountains or upland areas of South America, Asia, and Africa. Routine vaccination is not necessary however for persons residing in plague enzootic areas (such as those living in the Western United States) nor for travelers in countries where cases have been reported, particularly if travel is limited to urban areas. It is generally believed that the use of plague vaccine greatly increases the chances of recovery in those vaccinated individuals who may develop the bubonic form of infection. The degree of protection afforded against the pneumonic form is unknown and vaccinated persons exposed to the pneumonic form should be given daily, adequate doses of a suitable antibiotic over a 7–10-day period (Centers for Disease Control; Bioterrorism preparedness website).

Contraindications of Vaccination

There are no particular contraindications to the use of the vaccine. It is advisable not to give any injections during upper respiratory tract infections, because more severe reactions may ensue. In general, inactivated vaccines can be administered simultaneously at different sites. It should be noted however that when vaccines commonly associated with local or systemic side effects (such as cholera, typhoid, and plague vaccines) are given simultaneously, it is possible that the side effects may be accentuated. When practical, therefore, it is recommended that these vaccines should be given on separate occasions.

Duration of Immunity

The duration of protection against infection following administration of the primary series of injections of plague vaccine is thought to be brief (i.e., 6–12 months) and booster doses in approximately 6 monthly intervals are required for continued protection.

Adverse Events

Local and general reactions to the plague vaccine are usually mild and infrequent. They will generally become more frequent with increasing numbers of booster injections. The adverse events reported include injection site reaction, rigors, fever, increased sweating, neuropathy, paresthesia, paresis, dizziness, syncope, abnormal coordination, speech disorder, muscle atrophy, and anorexia. As with other injectable vaccines, appropriate medical treatment and supervision should always be available in case of anaphylactic reactions. Adrenalin should always be readily available whenever the injection is given.

IMPROVED VACCINES

Rationale for Second-Generation Vaccines

The limitations of the live attenuated and KWC vaccines detailed above have prompted significant efforts in recent years to devise improved vaccines against plague.

Evidence that the existing KWC vaccines provided some protection against bubonic plague and the predominant protective immunogen in these vaccines was the F1-antigen (Williams et al., 1986) provided a rationale for inclusion of this antigen in next-generation vaccines. In addition, KWC-vaccinated animals had serum antibody to F1-antigen which appeared to correlate with the level of protection against a subcutaneous challenge with *Y. pestis* (Meyer, 1970). Although the EV series of live attenuated vaccines have some disadvantages, they have been observed to be protective in animal models and this was attributed partly to their content of both F1 and V antigens (Williamson et al., 1995). Subsequently, it was observed that immunization of mice with either the F1-capsular antigen or the LcrV (V-antigen) component of the TTS alone provides protection against plague (Anderson et al., 1996; Leary et al., 1995; Miller et al., 1998). Combining these subunits together significantly enhanced the protective efficacy of the vaccine (Williamson et al., 1995). The use of the F1 antigen alone in a subunit vaccine is not desirable since, as noted above, virulent but F1-negative strains of *Y. pestis* have been reported (Friedlander et al., 1995).

The production of the F1 and V antigens has been greatly facilitated by the development of recombinant systems for the over-expression of the encoding genes in *Escherichia coli* (Leary et al., 1995; Titball et al., 1997). Formulations incorporating the individual subunits or a recombinant F1/V-antigen fusion protein have been reported (Heath et al., 1998); the use of individual subunits allows the optimal molar ratio of rF1: rV of 2:1 to be achieved in the final vaccine formulation (Williamson et al., 1999). In addition, the use of the rF1-antigen subunit may allow oligomerization which appears to enhance the immunogenicity of the protein (Miller et al., 1998). A vaccine based on rF1 and rV antigens, and formulated with an alhydrogel adjuvant, has been shown to be efficacious in a range of animal models (Anderson et al., 1996; Andrews et al., 1996; Jones et al., 2001, 2003). Importantly, unlike the KWC vaccines, in mice this vaccine provides good protection against an airborne challenge with *Y. pestis* (Heath et al., 1998; Williamson et al., 2001) suggesting that it can be developed as a new prophylactic for pneumonic as well as bubonic plague in humans. Although a single dose of this vaccine has been shown to induce good protective responses in animals (Williamson et al., 2001), the immunization schedule trial in humans for safety and immunogenicity entailed two doses of the rF1 + rV vaccine given on days 1 and 21 (Williamson et al., 2005).

The basis of protection afforded by the rF1 + rV vaccine is not fully elucidated. However, the passive transfer of immune sera into naive mice provides protection against a subsequent challenge (Green et al., 1999), and it has been shown that IgG1 subclass antibody correlates with protection (Williamson et al., 1999). Therefore, it seems likely that the antibody plays a significant role in protection. However, CMI is also important in the development of protective immunity and evidence for induction of a CMI response has been derived from animal studies (Williamson et al., 1995). The concerted immune response to the rF1 + rV vaccine has been shown to be protective in animal species tested and has led to the clearance of *Y. pestis* from the spleens, livers, and lungs of immunized animals 14 days after challenge (Jones et al., 2001).

In addition, the observation that Mab7.3, previously shown to protect mice by passive immunization against plague (Hill et al., 1997) competes with rV-immune macaque and human serum for binding to rV *in vitro*, indicates that this B cell epitope in rV is conserved across the species (Williamson et al., 2005). Work is also in progress to map murine T-cell epitopes in rF1 and rV in order to estimate the degree of conservation of CMI responses between animal species and man (Parent et al., 2005; Williamson, 2004).

ALTERNATIVE VACCINE APPROACHES: RATIONALLY ATTENUATED MUTANTS AND DNA VACCINES

There are several reports that immunization with naked DNA vaccines encoding either the F1 or V antigens is able to provide protection against plague (Garmory et al., 2004; Grosfield et al., 2003; Williamson et al., 2002). Multiple vaccinations with naked DNA are required and, at least in the case of a naked DNA construct expressing rF1, protection appears to be antibody mediated (Grosfield et al., 2003). In some studies a prime-boost strategy, where naked DNA-immunized mice were boosted with purified protein, was necessary to provide good protection (Garmory et al., 2004; Williamson et al., 2002), while gene gun–mediated vaccination was significantly more efficient than manual immunization with a DNA vaccine (Bennett et al., 2000).

Rationally attenuated mutants of *Y. pestis* have also been derived and tested as candidate vaccines. The findings that immunization with *Y. pestis* mutants, such as strain EV76, results in protection against subcutaneous and inhalation challenges indicate the feasibility of a live attenuated vaccine. The main limitations of the EV76 vaccine are the lack of detailed knowledge on the basis of attenuation and a partial retention of virulence. In other bacterial pathogens, such as *Salmonella enterica*, rationally attenuated mutants have proven to be highly effective vaccines. Therefore, some workers have applied the technologies developed with *S. enterica* to *Y. pestis*. In general, the results of this work have been disappointing. Mutations in the *aroA*, *phoP*, and *htrA* genes, which markedly attenuate other pathogens, have only a limited effect on the virulence of *Y. pestis* (Oyston et al., 1996, 2000; Williams et al., 2000). However, the exposure of mice to sublethal doses of these mutants can result in the induction of protective immunity (Oyston et al., 1996, 2000).

An alternative approach is to use a live attenuated salmonella as a vaccine vector from which to express plague antigens. Both the F1 and V antigens have been expressed from *S. enterica* serovar *typhimurium* (Garmory et al., 2005; Morton et al., 2004; Titball et al., 1997) which has then been used to dose mice orally or intranasally. The salmonella F1 construct, derived by cloning the entire *caf* operon from *Y. pestis* in plasmid form into *Salmonella typhimurium*, induced full protective immunity in Balb/c mice after two oral doses (Titball et al., 1997) while a similar construct conferred some protection against aerosol challenge with plague after only a single oral dose (Garmory et al., 2005). Unusually for an attenuated live vaccine, the recombinant plasmid in this *S. typhimurium* did not incorporate any antibiotic resistance markers (Garmory et al., 2005). Stable maintenance of the recombinant plasmid in the salmonella vector was achieved alternatively by means of Ortho-repressor titration (Cranenburgh et al., 2001) in which a gene product from the plasmid was essential for growth and survival of the attenuated salmonella vector (Garmory et al., 2005).

VACCINES IN DEVELOPMENT

A fully recombinant subunit vaccine for plague has been identified (Williamson et al., 1995), which in animal models of both bubonic and pneumonic infections, provides enhanced protection and which represents a new candidate human vaccine (Williamson et al., 2005).

Much effort is being expended in the development of the combination of rF1 and rV as a second-generation plague vaccine (Williamson, 2001).

DISCOVERY/BASIC SCIENCE

Preclinical Development, Including Relevant Animal Models

Mouse

The immunogenicity and efficacy of the rF1 and rV immunogens have been extensively studied in the mouse in which the synergistic effect of these proteins was first reported (Heath et al., 1998; Williamson et al., 1995). Subsequently, regions of the V antigen associated with protection were identified in the mouse model and Mab 7.3, derived against V135-275, was found to passively protect mice against live plague organism challenge (Hill et al., 1997). Recombinant F1 antigen protected Balb/c mice optimally when allowed to naturally aggregate, although attempts to denature this protein reduced, but did not abolish, efficacy (Miller et al., 1998). The development of an IgG1 titer to the rF1 + rV antigens in the Balb/c mouse correlated with protection against challenge (Williamson et al., 1999), although evidence for the existence of CMI responses has also been found (Williamson et al., 1995, 1996). Mouse haplotype and gender did not significantly influence the response to the F1 + V vaccine so that CBA(H-2k), C57Bl6 (H-2b), Balb/c (H-2d), and the CB57Bl6 x Balb/c strain (CB6F1) all responded and were protected to a similar high level against either subcutaneous or aerosol challenge with plague (Jones et al., 2001). Similar levels of immunogenicity and efficacy have been observed in outbred mouse strains immunized with the rF1 and rV proteins, either in combination (Williamson et al., 2001) or as a genetic fusion (Heath et al., 1998). Passive transfer of antiserum to rF1 + rV from the Balb/c mouse into the servere combined immunodeficient with Beige mutation (Scid/Bge) mouse fully protected the latter against plague challenge, again indicating the importance of high titer-specific antibody in protection (Green et al., 1999).

Guinea Pig

Although the guinea pig is susceptible to plague, it has been much less used to evaluate subunit vaccines incorporating the rF1 and rV antigens. However, in a single study Jones et al. (2003) reported that guinea pigs responded with specific antibody to each antigen, although the response to rF1 was more variable than the response to rV. Actively immunized guinea pigs were also fully protected against subcutaneous challenge with 10^5 colony-forming units (cfu) of *Y. pestis* (equivalent to at least 10 lethal doses). Moreover, antiserum raised in the guinea pig and specific for rF1 + rV passively protected naive Balb/c mice from challenge with plague, indicating a conservation of B cell epitopes in F1 and V between the two species.

Macaque

Cynomolgus macaques, immunized with the rF1 + rV vaccine in alhydrogel have responded with high titer antibody to each subunit antigen (Williamson, 2004). The positive reference macaque antiserum raised (Williamson et al., 2005) has been demonstrated to protect mice in a dose-related manner against challenge by passive transfer and to compete *in vitro* with murine Mab7.3 for binding to rV antigen on a solid phase, in a competitive ELISA (Williamson et al., 2005). These data indicate that cynomolgus macaques immunized with rF1 + rV in alhydrogel, responded with protective antibody and again indicates the conservation of B-cell epitopes between the macaque and the mouse.

CLINICAL TRIALS

Phase I

The human immune response to the recombinant plague vaccine, comprising rF1 and rV antigens in alhydrogel, has been assessed during a Phase 1 safety and immunogenicity trial in 32 healthy male volunteers in Europe in 2001 (Williamson et al., 2005). The vaccine was found to be safe with no vaccine-related serious adverse events reported. All the subjects produced specific IgG in serum after the priming dose, which peaked in value after the booster dose (day 21), with the exception of one individual in the lowest dose-level group, who responded to rF1 only. Three subjects, found to have an anti-rV titer at screening, were excluded from the overall analysis. Human antibody functionality was assessed by quantification of antibody competing for binding to rV *in vitro* and also by the transfer of protective immunity in human serum into the naïve mouse. Human and macaque IgG competed for binding to rV *in vitro* with Mab7.3, previously shown to protect mice against challenge with plague, suggesting that this protective B-cell epitope on rV is conserved between these three species. Total IgG to rV in individuals and the titer of IgG competing for binding to rV, correlated significantly at days 21 ($r = 0.72$; $p < 0.001$) and 28 ($r = 0.82$; $p < 0.001$). Passive transfer of protective immunity into mice also correlated significantly with total IgG titer to rF1 plus rV at days 21 ($r^2 = 98.6\%$; $p < 0.001$) and 28 ($r^2 = 76.8\%$; $p < 0.03$). However, no significant vaccination-related change in the activation of PBMC was detected at any point of time. Potential serological immune correlates of protection were identified, but no trends specific to vaccination could be detected in cellular markers.

Phase II

Expanded clinical trials have been conducted to further address issues of safety and immunogenicity in male and female volunteers together with vaccine dose ranging studies. The ultimate aim of these studies is to arrive at a final formulation and regimen for licensure of the vaccine.

Phase III

In Europe, Phase III post-marketing safety studies may be conducted after marketing authorization has been received for the vaccine, in a sufficient number of individuals to provide a statistically sound evaluation of vaccine safety. In North America, Phase III studies to evaluate vaccine safety in an expanded population are normally required prior to the application for Biologic Licensing Authorization (BLA).

Postexposure Immunoprophylaxis

It is possible that a recombinant subunit vaccine with enhanced immunogenicity could provide some benefit in conjunction with antibiotic, in a postexposure scenario. In mouse studies, the earliest protective response against aerosol challenge, induced to a single high dose vaccination with rF1 + rV was observed at approximately day 6 (Fig. 54.4) and so by extrapolation, postexposure vaccination in man with an enhanced formulation could be beneficial.

Third-Generation Vaccines

Third-generation vaccines comprising rF1 + rV delivered in association with different adjuvants and/or by a non-parenteral route, or alternative physical formulations or rF1 + rV and rF1-V are also being researched. For this first option, rF1 + rV simply coadministered with a mucosal adjuvant such as cholera toxin B subunit (CTB), induced a high level of

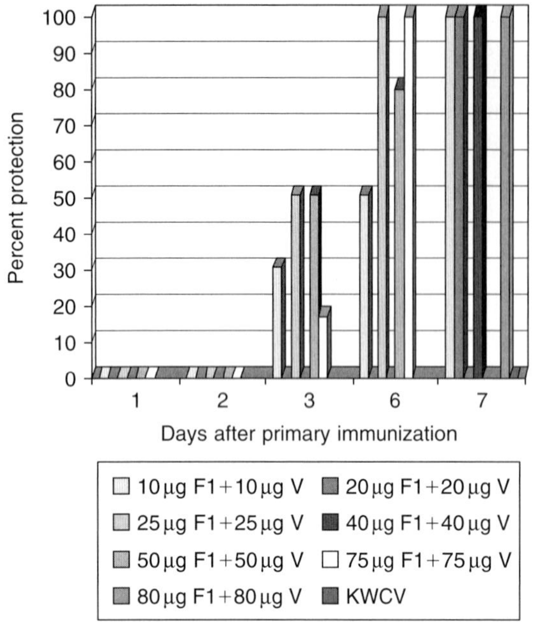

FIGURE 54.4 Time to immunity against pneumonic plague in the Balb/c mouse model; groups of five mice received a single immunizing dose of the rF1 + rV vaccine on day zero at escalating dose-levels of each subunit, followed by inhalational challenge with >300 MLD (3×10^5 cfu) *Y. pestis* strain GB on days 1–7. The figure shows percent protection at 14 days post-challenge (see color plate section).

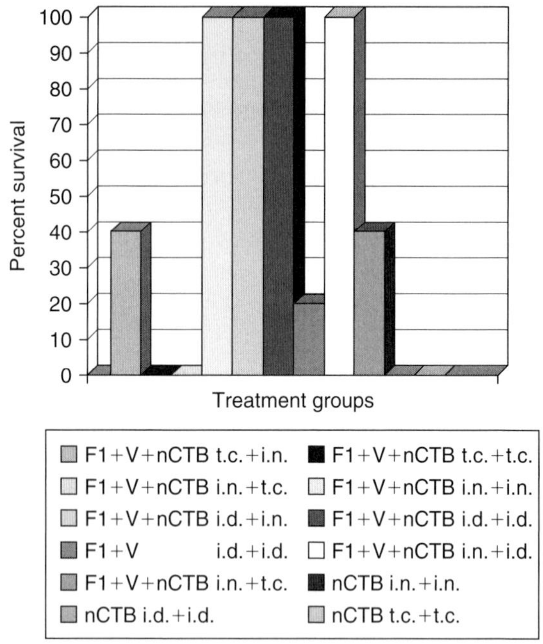

FIGURE 54.5 The protective efficacy of rF1 + rV coadministered with CTB by the transcutaneous, intradermal, or intranasal routes, or combinations of these. The figure shows the percentage survival for groups of 5 Balb/c mice immunized by the routes indicated (1 bar per group) on two occasions (days 1 and 21) with 10 µg rF1 + 40 µg rV, with or without 10 µg CTB (Sigma, Poole, United Kingdom) in 50 µl phosphate-buffered saline per mouse. All groups were challenged subcutaneously at day 90 with 5 × 106 MLD *Y. pestis* (GB strain) (Somavarapu et al., 2001) (see color plate section).

protective immunity in Balb/c mice when delivered intranasally (i.n.) or intradermally (i.d.) (Fig. 54.5). Interestingly, the primary immune response induced by this formulation given i.d. could be boosted by i.n. dosing and the primary response induced to i.n. dosing, could be boosted i.d., either of these combinations leading to full protective immunity, and indicating that there is significant cross talk between the nasal-associated lymphoid tissue and the skin. Moreover, transcutaneous (t.c.) priming and boosting with this formulation did not induce protective immunity although 40% protection was achieved when t.c.-primed mice were boosted i.n. However, the substitution of cholera holotoxin (CT) for CTB in the formulation and the use of three t.c. immunizations, did lead to protective immunity in Balb/c mice against 10^2 median lethal doses of *Y. pestis* and t.c. priming with this formulation could be effectively boosted i.n. or i.d (Eyles et al., 2004). T.c. immunization with this formulation induced CMI responses in the spleens of immunized mice.

Significant progress has also been made using formulations in which the rF1 and rV proteins are encapsulated in polymeric microspheres which further enhances their immunogenicity and facilitates their delivery by mucosal routes such as nasal (Eyles et al., 1998), or inhalational (Eyles et al., 2000). Some of these formulations are in early stage of development and represent potential vaccine formulations for the future. These formulation types have the advantage that they are generally more stable than liquid formulations, do not require cold chain storage, and *in vivo* act as effective depot release systems, so that they are expected to need less boosting. These are particularly ideal qualities for a biodefense vaccine which may be required for use in the field or on an intermittent basis, depending on an assessment of risk. Microencapsulated formulations of rF1 + rV have been extensively studied (Williamson et al., 1996) and optimized (Eyles et al., 2003) so that in the mouse model a single dose provides slow release sufficient to prime and boost the immune system (Elvin and Williamson, 2006), to induce substantial antibody titers and CMI which confers protection against virulent plague challenge (Eyles et al., 2000; Williamson et al., 2001).

PROSPECTS FOR THE FUTURE

For a number of reasons it is clear that the existing KWC or live attenuated vaccines are not suitable for the immunization of populations to protect against pneumonic plague. However, until improved vaccines become available the KWC vaccine has a role to play in the immunization of individuals who might contract bubonic plague.

A number of approaches to the development of an improved vaccine have been reported. Live attenuated vaccines offer some advantages but a suitable candidate has yet to be identified and the licensing of such a vaccine could be difficult. Naked DNA vaccines do not appear to offer any significant benefits over subunit vaccines, especially since there is no evidence of the need for cytotoxic T cell responses to control infection. The most likely candidate vaccine, which could provide protection against bubonic and pneumonic plague, is an injected subunit vaccine. Most effort to date has focused on recombinant *Y. pestis* proteins, which are produced in *E. coli*, and this seems to be the simplest strategy for the development of an improved vaccine. A work has already shown that a vaccine based on rF1 and rV is effective and this vaccine is currently undergoing clinical trials in humans. Such a vaccine could therefore be licensed for use within the next few years. However, the identification of additional protective subunits could be used to provide alternative vaccine formulations for the future.

In the long term, the licensing of a recombinant subunit vaccine would support ongoing studies to devise plague vaccine formulations which could be given noninvasively. Such vaccines would be ideally suited to the large-scale immunization of populations. Licensed vaccines based on attenuated *S. enterica* (or other vaccine vectors), which can be given orally and in which the maintenance of recombinant plasmids *in vivo* does not require antibiotic resistance, or microencapsulated subunit protein vaccines which may be delivered by a mucosal or parenteral route, appear to be a realizable prospect within the next decade.

KEY ISSUES

- Plague is an ancient life-threatening disease which still exists in some parts of the modern world, with the potential for epidemic spread.
- The only vaccine available for human use in the Western world is a KWC formulation which provides some protection against the bubonic form of the disease, but probably provides little or no protection against the more serious pneumonic plague syndrome.
- A live attenuated mutant strain of plague, EV76, has been used in human but can cause serious side effects.
- A number of alternative vaccines have been researched to include live rationally attenuated strains, a live vaccine vector expressing plague subunit proteins, DNA vaccines, and a fully recombinant subunit vaccine.
- The subunit proteins F1 and V, which occur naturally in the causative organism, are protective immunogens which are synergistic in effect and protective against pneumonic plague in an experimental model.
- Fully recombinant subunit vaccines, comprising the F1 and V antigens in an alhydrogel vehicle, are in development and have been shown to be immunogenic in a range of species as well as in human.
- A Phase I clinical trial has shown that the vaccine is safe and well tolerated and Phase II trials are ongoing.
- The future application of the rF1 + rV vaccine is expected to provide the capability to protect people against pneumonic plague.

References

Achtman, M., Zurth, K., Moerlli, G., Torrea, G., Guiyoule, A. and Carniel, E. *Yersinia pestis*, the cause of plague, is a recently emerged clone of *Yersinia pseudotuberculosis*. Proc. Natl. Acad. Sci. U.S.A. 1999; 96:14043–14048.

Andrews, G.P., Heath, D.G., Anderson, G.W.J., Welkos, S.L. and Friedlander, A.M. Fraction 1 capsular antigen (F1) purification from *Yersinia pestis* CO92 and from an *Escherichia coli* recombinant strain and efficacy against lethal challenge. Infect. Immun. 1996; 64:2180–2187.

Anderson, G.W., Leary, S.E.C., Williamson, E.D., Titball, R.W., Welkos, S.L., Worsham, P.L. and Friedlander, A.M. Recombinant V antigen protects mice against pneumonic and bubonic plague caused by F1-capsule-positive and negative strains of *Y. pestis*. Infect. Immun. 1996; 64:4580–4585.

Andrews, G.P., Strachan, S.T., Benner, G.E., Sample, A.K., Anderson, G.W., Adamovicz, J.J., Welkos, S.L., Pullen, J.K. and Friedlander, A.M. Protective efficacy of recombinant *Yersinia* outer proteins against bubonic plague caused by encapsulated and nonencapsulated *Yersinia pestis*. Infect. Immun. 1999; 67:1533–1537.

Anon Plague. World Health Organ. Wkly. Epidemiol. Rec. 1995; 5:35.

Anon Human plague in 1994. World Health Organ. Wkly. Epidemiol. Rec. 1996; 71:165–172.

Anon Human plague in 1998 and 1999. World Health Organ. Wkly. Epidemiol. Rec. 2000; 75:337–344.

Bennett, A.M., Phillpotts, R.J., Perkins, S.D., Jacobs, S.C. and Williamson, E.D. Gene gun mediated vaccination is superior to manual delivery for immunisation with DNA vaccines expressing protective antigens from *Yersinia pestis* or Venezuelan Equine Encephalitis Virus. Vaccine 2000; 18:588–596.

Brubaker, R.R. Factors promoting acute and chronic diseases caused by yersiniae. Clin. Microbiol. Rev. 1991; 4:309–324.

Burrows, T.W. Virulence of *Pasteurella pestis* and immunity to plague. In: Ergebnisse der mikrobiologi immunitatsforshing und experimentellen therapie. Band 37 (W. Henle, W. Kikuth,

K.F. Meyer, E.G. Nauck and J. Tomcsik, Eds.), pp. 59–113. Berlin: Springer-Verlag, 1963.

Cavanaugh, D.C., Elisberg, B.L., Llewellyn, C.H., Marshall, J.D., Rust, J.H., Williams, J.E. and Meyer, K.F. Plague immunization. V. Indirect evidence for the efficacy of plague vaccine. J. Infect. Dis. 1974; 129:S37–S40.

Centers for Disease Control; Bioterrorism preparedness website.

Christie, A.B. and Corbel, M.J. Plague and other yersinial diseases. In: Topley and Wilson's Principles of Bacteriology, Virology and Immunity (G.R. Smith, Ed.), pp. 399–410. London: Edward Arnold, 1990.

Cornelis, G.R. The yersinia YSC-YOP 'type III' weaponry. Nat. Rev. Mol. Cell. Biol. 2002; 3:742–752.

Cranenburgh, R.M., Hanak, J.A.J., Williams, S.G. and Sherratt, D.J. *Escherichia coli* strains that allow antibiotic free plasmid selection and maintenance by repressor titration. Nucleic Acids Res. 2001; 29:e26.

Dennis, D. and Meier, F. Plague. In: Pathology of Emerging Infections (C.R. Horsburgh and A.M. Nelson, Eds.), pp. 21–47. Washington, D.C.: ASM Press, 1997.

Dennis, D.T., Gage, K.L., Grantz, N., Poland, J.D. and Tikhomirov, E. Plague Manual: Epidemiology, Distribution, Surveillance and Control. Geneva: World Health Organization, 1999.

Doll, J.M., Zeitz, P.S., Ettestad, P., Bucholtz, A., Davis, T. and Gage, K. Cat-transmitted fatal pneumonic plague in a person who travelled from Colorado to Arizona. Am. J. Trop. Med. Hygiene. 1994; 51:109–114.

Dols, M. The Black Death in the Middle East. New Jersey: Princeton University Press, 1977.

Duplaix, N. The lethal lepers. National Geographic. 1988; 173:672–694.

Elvin, S.J. and Williamson, E.D. The F1 and V subunit vaccine protects against plague in the absence of IL-4 driven immune responses. Microb. Pathogen. 2000; 29:223–230.

Elvin, S.J. and WIlliamson, E.D. Stat 4 but not Stat 6 mediated immune mechanisms are essential in protection against plague. Microb. Pathogen. 2004; 37:177–184.

Elvin, S.J. and WIlliamson, E.D. Protection against bubonic and pneumonic plague with a single dose microencapsulated sub-unit vaccine. Vaccine 2006; 24:4433–4439.

Elvin, S.J., Williamson, E.D., Scott, J.C., Smith, J.N., De Lema, G.P., Chilla, S., Clapham, P., Schlondorff, D. and Luckow, B. Ambiguous role of CCR5 in *Yersinia pestis* infection. Nature. 22 July 2004; 430 (xiii).

Engelthaler, D.M. and Gage, K.L. Quantities of *Yersinia pestis* in fleas (Siphonaptera: Pulicidae, Ceratophyllidae, and Hystrichopsyllidae) collected from areas of known or suspected plague activity. J. Med. Entomol. 2000; 37:422–426.

Eyles, J., Sharpe, G.E.J., Williamson, E.D., Spiers, I.D. and Alpar, H.O. Intra-nasal administration of poly lactic acid microsphere co-encapsulated *Yersinia pestis* sub-units confers protection from pneumonic plague in the mouse. Vaccine 1998; 16:698–707.

Eyles, J.E., Carpenter, Z.C., Alpar, H.O. and Williamson, E.D. Immunolgical aspects of polymer microsphere vaccine delivery systems. J. Drug Target. 2003; 11:509–514.

Eyles, J.E., Elvin, S.J., Westwood, A., LeButt, C.S., Alpar, H.O., Somavarapu, S. and Williamson, E.D. Immunisation against plague by transcutaneous and intradermal application of sub-unit antigens. Vaccine 2004; 22:4365–4373.

Eyles, J.E., Williamson, E.D., Spiers, I.D. and Alpar, H.O. Protection studies following bronchopulmonary and intramuscular immunsiation with *Yersinia pestis* F1 and V sub-unit vaccines coencapsulated in biodegradable microspheres: a comparison of efficacy. Vaccine 2000; 18:3266–3271.

Franz, D.R., Jahrling, P.B., Friedlander, A.M. et al. Clinical recognition and management of patients exposed to biological warfare agents. JAMA 1997; 278(5):399–411.

Friedlander, A.M., Welkos, S.L., Worsham, P.L. et al. Relationship between virulence and immunity as revealed in recent studies of the F1 capsule of *Yersinia pestis*. Clin. Infect. Dis. 1995; 21:S178–S181.

Gallimand, M., Guiyoule, A., Gerbaud, G., Rasoamanana, B., Chanteau, S., Carniel, E. and Courvalin, P. Multidrug resistance in *Yersinia pestis* mediated by a transferable plasmid. N. Engl. J. Med. 1997; 337:677–680.

Garmory, H.S., Freeman, D., Brown, K.A. and Titball, R.W. Protection against plague afforded by immunisation with DNA vaccines optimised for expression of the *Yersinia pestis* V antigen. Vaccine 2004; 22:947–957.

Garmory, H.S., Leckenby, M.W., Griffin, K.F., Elvin, S.J., Taylor, R.R., Hartley, M.G., Hanak, J.A., Williamson, E.D. and Cranenburgh, R.M. Antibiotic- free plasmid stabilisation by Operator-Repressor Titration for vaccine delivery by using live *Salmonella enterica* serovar *typhimurium*. Infect. Immun. 2005; 73:2005–2011.

Gottfried, R. The Black Death: Natural and Human Disaster in Medieval Europe. New York: Free Press, 1983.

Green, M., Rogers, D., Russell, P., Stagg, A.J., Bell, D.L., Eley, S.M., Titball, R.W. and Williamson, E.D. The SCID/Beige mouse as a model to investigate protection against *Yersinia pestis*. FEMS Immunol. Med. Microbiol. 1999; 23:107–113.

Grosfield, H., Cohen, S., Bino, T. et al. Effective protective immunity to *Yersinia pestis* infection conferred by DNA vaccine coding for derivatives of the F1 capsular antigen. Infect. Immun. 2003; 71:374–383.

Guiyoule, A., Gerbaud, G., Buchrieser, C., Galimand, M., Rahalison, L., Chanteau, S., Courvalin, P. and Carniel, E. Transferable plasmid-mediated resistance to streptomycin in a clinical isolate of *Yersinia pestis*. Emerg. Infect. Dis. 2001; 7:43–48.

Guiyoule, A., Rasoamanana, B., Buchrieser, C., Michel, P., Chanteau, S. and Carniel, E. Recent emergence of new variants of *Yersinia pestis* in Madagascar. J. Clin. Microbiol. 1997; 35:2826–2833.

Heath, D.G., Anderson, G.W.J., Mauro, J.M. et al. Protection against experimental bubonic and pneumonic plague by a recombinant capsular F1-V fusion protein vaccine. Vaccine 1998; 16:1131–1137.

Hill, J., Copse, C., Leary, S., Stagg, A.J., Williamson, E.D. and Titball, R.W. Synergistic protection of mice against plague with monoclonal antibodies specific for the F1 and V antigens of *Yersinia pestis*. Infect. Immun. 2003; 71:2234–2238.

Hill, J., Leary, S.E.C., Griffin, K.F., Williamson, E.D. and Titball, R.W. Regions of *Yersinia pestis* V antigen that contribute to protection against plague identified by passive and active immunisation. Infect. Immun. 1997; 65:4476–4482.

Hinnebusch, B.J., Perry, R.D. and Schwan, T.G. Role of the *Yersinia pestis* hemin storage (hms) locus in the transmission of plague by fleas. Science 1996; 273:367–370.

Hinnebusch, J., Cherepanov, P., Du, Y., Rudolph, A., Dixon, J.D., Schwan, T. and Forsberg, A. Murine toxin of *Yersinia pestis* shows phospholipase D activity but is not required for virulence in mice. Int. J. Med. Microbiol. 2000; 290:483–487.

Jones, S.M., Day, F., Stagg, A.J. and Williamson, E.D. Protection conferred by a fully-recombinant sub-unit vaccine against *Yersinia pestis* in male and female mice of four inbred strains. Vaccine 2001; 19:358–366.

Jones, S.M., Griffin, K.F., Hodgson, I. and Williamson, E.D. Protective efficacy of a fully recombinant plague vaccine in the guinea pig. Vaccine 2003; 21:3912–3918.

Kir'iakova, A.N. Lifespan of fleas in burrows. Parazitologiya 1973; 7:261–263.

Lahteenmaki, K., Kukkonen, M. and Korhonen, T.K. The Pla surface protease/adhesin of *Yersinia pestis* mediates bacterial invasion into human endothelial cells. FEBS Lett. 2001; 504:69–72.

Lahteenmaki, K., Virkola, R., Saren, A., Emody, L. and Kohonen, T.K. Expression of plasminogen activator Pla of *Yersinia pestis* enhances bacterial attachment to the mammalian extracellular matrix. Infect. Immun. 1998; 66:5755–5762.

Lawton, W.D., Erdman, R.L. and Surgalla, M.J. Biosynthesis and purification of V and W antigen in *Pasteurella pestis*. J. Immun. 1963; 91:179–184.

Leary, S.E.C., Griffin, K.F., Galyov, E.E., Hewer, J., Williamson, E.D., Holmstrom, A., Forsberg, A. and Titball, R.W. Yersinia outer proteins (YOPS) E, K and N are antigenic but non-protective compared to V antigen, in a murine model of bubonic plague. Microb. Pathog. 1999; 26:159–169.

Leary, S.E.C., Williamson, E.D., Griffin, K.F., Russell, P., Eley, S.M. and Titball, R.W. Active immunisation with recombinant V antigen from *Yersinia pestis* protects mice against plague. Infect. Immun. 1995; 63:2854–2858.

Matson, J.S., Durick, K.A., Bradley, D.S. and Nilles, M.L. Immunization of mice with YscF provides protection from *Yersinia pestis* infections. BMC Microbiol. 2005; 5:38–55.

Mecsas, J., Franklin, G., Kuziel, W.A., Brubaker, R.R., Falkow, S. and Mosier, D.E. CCR5 mutation and plague protection. Nature 2004; 427:606.

Meyer, K.F. Pneumonic plague. Bacteriol. Rev. 1961; 35:249–261.

Meyer, KF. Effectiveness of live or killed plague vaccines in man. Bull. World Health Organ. 1970; 42:653–666.

Meyer, K.F., Cavanaugh, D.C., Bartelloni, P.J. and Marshall, J.D. Plague immunization. I. Past and present trends. J. Infect. Dis. 1974; 129:S13–S18.

Miller, J., Williamson, E.D., Lakey, J.H., Pearce, M.J., Jones, S.M. and Titball, R.W. Macromolecular organisation of recombinant *Yersinia pestis* F1 antigen and the effect of structure on immunogenicity. FEMS Immunol. Med. Microbiol. 1998; 21:213–223.

Morton, M., Garmory, H.S., Perkins, S.D., O'Dowd, A.M., Griffin, K.F., Turner, A.K., Bennett, A.M. and Titball, R.W. A Salmonella enterica serovar typhi vaccine expressing F1 antigen on its surface provides protection against plague. Vaccine 2004; 22:2524–2532.

Naumov, A.V., Ledvanov, M.Y. and Drozdov, I.G. Plague Immunology. Saratov, Russia: Russian Research Anti-Plague Institute 'Microbe', 1992.

Oyston, P.C.F., Dorrell, N., Williams, K., Li, S.R., Green, M., Titball, R.W. and Wren, B.W. The response regulator PhoP is important for survival under conditions of macrophage-induced stress and virulence in *Yersinia pestis*. Infect. Immun. 2000; 68:3419–3425.

Oyston, P.C.F. and Isherwood, K.E. The many and varied niches occupied by *Yersinia pestis* as an arthropod-vectored zoonotic pathogen. Antonie Van Leeuwenhoek 2005; 87:171–177.

Oyston, P.C.F., Russell, P., Williamson, E.D. and Titball, R.W. An Aro A mutant of *Yersinia pestis* is attenuated in guinea pigs, but virulent in mice. Microbiology 1996; 142:1847–1853.

Parent, M.A., Berggren, K.N., Mullarky, I.K., Szaba, F.M., Kummer, L.W., Adamovicz, J.J. and Smiley, S.T. *Yersinia pestis* V protein epitopes recognised by CD4 T cells. Infect. Immun. 2005; 73:2197–2204.

Parkhill, J., Wren, B.W., Thomson, N.R., Titball, R.W., Holden, M.T.G., Prentice, M.B. et al. Genome sequence of *Yersinia pestis*, the causative agent of plague. Nature 2001; 413:523–527.

Payne, D., Tatham, D., Williamson, E.D. and Titball, R.W. The pH6 antigen of *Yersinia pestis* binds to b1-linked galactosyl residues in glycosphingolipids. Infect. Immun. 1998; 66:4545–4548.

Perry, R.D. and Fetherston, J.D. *Yersinia pestis*—etiologic agent of plague. Clin. Microbiol. Rev. 1997; 10:35–66.

Petterson, J., Holmstrom, A., Hill, J., Leary, S., Frithz-Lindsten, E., von Euler-Matell, A., Carlsson, E., Titball, R., Forsberg, A. and Wolf-Watz, H. The V antigen of Yersinia is surface-exposed before target cell contact and involved in virulence protein translocation. Mol. Microbiol. 1999; 32:961–976.

Poland, J.D. and Barnes, A.M. Plague. In: CRC Handbook Series in Zoonoses (J.H. Steele, H. Stoenner, W. Kaplan and M. Torten, Eds.), pp. 515–519. Boca Raton, FL: CRC Press, 1979.

Prior, J.L., Hitchen, P.G., Williamson, E.D., Reason, A.J., Morris, H.R., Dell, A., Wren, B.W. and Titball, R.W. Characterisation of the LPS of *Y. pestis*. Microb. Pathog. 2001a; 30:49–57.

Prior, J.L., Parkhill, J., Hitchen, P.G. et al. The failure of different strains of *Yersinia pestis* to produce lipopolysaccharide O-antigen under different growth conditions is due to mutations in the O-antigen gene cluster. FEMS Microbiol. Lett. 2001b; 197:229–233.

Reisman, R.E. Allergic reactions due to plague vaccine. J. Allergy 1970; 46:49–56.

Russell, P., Eley, S.M., Hibbs, S.E., Manchee, R.J., Stagg, A.J. and Titball, R.W. A comparison of plague USP and EV76 vaccine induced protection against *Yersinia pestis* in a murine model. Vaccine 1995; 13:1551–1556.

Simpson, W.J., Thomas, R.E. and Schwan, T.H. Recombinant capsular antigen (Fraction 1) from *Yersinia pestis* induces a protective antibody response in Balb/c mice. Am. J. Trop. Med. Hyg. 1990; 43:389–396.

Skurnik, M., Peippo, A. and Ervela, E. Characterization of the O-antigen gene clusters of *Yersinia pseudotuberculosis* and the cryptic O-antigen gene cluster of *Yersinia pestis* shows that the plague bacillus is most closely related to and has evolved from *Y. pseudotuberculosis* serotype O:1b. Mol. Microbiol. 2000; 37:316–330.

Sodeinde, O.A., Subrahmanyam, Y.V.B.K., Stark, K., Quan, T., Bao, Y. and Goguen, J.D. A surface protease and the invasive character of plague. Science 1992; 258:1004–1007.

Somavarapu, S., Eyles, J.E., Flick-Smith, H.C., Jones, O., Alpar, H.O. and Williamson, E.D. Cross-talk between trans-cutaneous, intradermal and nasal routes of immunization. Proceedings of the BSI, Harrogate, 2001.

Stephens, J.C., Reich, D.E., Goldstein, D.B. et al. Dating the origin of the CCR5-Delta 32 AIDS-resistance allele by the coalescence of haplotypes. Am. J. Hum. Gen. 1998; 62:1507–1515.

Straley, S.C. and Harmon, P.A. Growth in mouse peritoneal macrophages of *Yersinia pestis* lacking established virulence determinants. Infect. Immun. 1984a; 45:649–654.

Straley, S.C. and Harmon, P.A. *Yersinia pestis* grows within phagolysosomes in mouse peritoneal macrophages. Infect. Immun. 1984b; 45:655–659.

Titball, R.W., Hill, J., Lawton, D.G. and Brown, K.A. *Yersinia pestis* and plague. Biochem. Soc. Trans. 2003; 31:104–107.

Titball, R.W., Howells, A.M., Oyston, P.C.F. and Williamson, E.D. Expression of the *Yersinia pestis* capsular antigen (F1 antigen) on the surface of an *Aro A* mutant of *Salmonella typhimurium* induces high level protection against plague. Infect. Immun 1997; 65:1926–1930.

WHO Plague, Democratic Republic of the Congo. Wkly. Epidemiol. Rec. 2005; 80:86.

Williams, J.E., Arntzen, L., Tyndal, G.L. and Isaacson, M. Application of enzyme immunoassays for the confirmation of clinically suspect plagued in Namibia, 1982. Bull. World Health Organ. 1986; 64:745–752.

Williams, K., Oyston, P.C.F., Dorrell, N., Li, S.-R., Titball, R.W. and Wren, B.W. Investigation into the role of the serine protease

HtrA in *Yersinia pestis* pathogenesis. FEMS Microbiol. Lett. 2000; 186:281–286.

Williamson, E.D. Plague vaccine research and development. J. Appl. Microbiol. 2001; 91:606–608.

Williamson, E.D. Oral presentation on 'The role of antibodies and cell-mediated immunity in conferring protection against plague'. Proceedings of the Public Workshop on Animal Models and correlates of protection for plague vaccines. Sponsored by DHHS/FDA/CBER, October 2004. Washington, D.C., 2004.

Williamson, E.D., Bennett, A.M., Perkins, S.D., Beedham, R.J., Miller, J. and Baillie, L.W.J. Co-immunisation with a plasmid DNA cocktail effectively primes mice against anthrax and plague. Vaccine 2002; 20:2933–2941.

Williamson, E.D., Eley, S.M., Griffin, K.F., Green, M., Russell, P., Leary, S.E.C., Oyston, P.C.F., Easterbrook, T., Reddin, K.M., Robinson, A. and Titball, R.W. A new improved sub-unit vaccine for plague: the basis of protection. FEMS Immunol. Med. Microbiol. 1995; 12:223–230.

Williamson, E.D., Eley, S.M., Stagg, A.J., Green, M., Russell, P. and Titball, R.W. A single dose sub-unit vaccine protects against pneumonic plague. Vaccine 2001; 19:566–571.

Williamson, E.D., Flick-smith, H.C., Lebutt, C.S., Rowland, C.A., Jones, S.M., Waters, E.L., Gwyther, R.J., Miller, J., Packer, P.J. and Irving, M. Human immune response to a plague vaccine comprising recombinant F1 and rV antigens. Infect. Immun. 2005; 73(6):3598–3608.

Williamson, E.D., Sharpe, G.E.J., Eley, S.M., Vesey, P.M., Pepper, T., Titball, R.W. and Alpar, H.O. Local and systemic immune response to a microencapsulated sub-unit vaccine for plague. Vaccine 1996; 14:1613–1619.

Williamson, E.D., Vesey, P.M., Gillhespy, K.J., Eley, S.M., Green, M. and Titball, R.W. An IgG1 titre to the F1 and V antigens correlates with protection against plague in the mouse model. Clin. Exp. Immun. 1999; 116:107–114.

Wren, B.W. The yersiniae-A model genus to study the rapid evolution of bacterial pathogens. Nat. Rev. Microbiol. 2003; 1:55–64.

CHAPTER

55

Ricin: A Type II Ribosome-Inactivating Protein

Robert N. Brey[1] *and Nicholas J. Mantis*[2]

[1]DOR BioPharma, Inc., Ewing, NJ, USA

[2]Division of Infectious Disease, Wadsworth Center, New York State Department of Health, Albany, NY, USA

OUTLINE

ABSTRACT

Ricin is a plant cytotoxin that poses a significant risk as a biothreat, primarily as a toxin that could be delivered by aerosolization in acts of bioterrorism and biological warfare. Ricin is a type II ribotoxin that is closely related in the mechanism of action to other plant toxins and several bacterial toxins, including the family of Shiga toxins, a cause of diarrhea-associated hemolytic uremia syndrome. There has been a huge amount of research directed at understanding the molecular basis of toxicity of ricin and related toxins and the use of those toxins in targeted immunotoxins for therapeutics. Lately, because of the interest in developing countermeasures and vaccines for category A and B biothreats, this research has culminated in the development of subunit vaccines for ricin that are now in advanced preclinical and clinical studies. Depending on dose and route, ricin exposure results in three separately identifiable pathways of toxicity: direct cytotoxicity through the inhibition of protein synthesis resulting in necrosis, yet-unresolved signaling of programmed cell death (apoptosis), and upregulation of inflammatory cascades regulated by a variety of mitogen-activated protein kinases (MAPK) and transcription factors. Both tissue damage and organ damage are a result of direct cytotoxicity, highest in those mucosal regions directly exposed to toxin, and inflammation. Antibodies to ricin holotoxin have historically been shown to confer protection to naive animals by passive transfer. Antibodies to ricin can rescue exposed animals, but only if given shortly after exposure. Several recombinant vaccine immunogens have been developed in the past several years, the most promising ones being derived by mutation or deletion from the ricin toxin A (RTA) chain. The RTA mutant vaccines induce protection from systemic toxin exposure in animal models and neutralizing antibodies that are correlated to protection. RTA vaccines also induce protection from oral and aerosol toxin exposure and largely protect against mucosal damage. The role of local and systemic antibodies in animal models is being studied as part of the program to correlate human immune responses to protective efficacy in animal models.

INTRODUCTION

Protein toxins are principal determinants of pathogenicity for a large number of bacterial diseases of animals and humans. In many cases, toxins are logical targets for vaccine development since inactivation of the toxins by specific circulating antibodies that participate in toxin neutralization or clearance prevents the toxins from reaching intracellular targets. For several bacteria that manifest toxins as their principal virulence determinants, passive immunization with antitoxins had been a means to treat or prevent human disease before the availability of vaccines. It is well known that several of the most successful vaccines used, since the first quarter of the 20th century, as public health tools are based on the chemical inactivation of bacterial toxins secreted into culture supernatants and the fact that such toxoids induce antibodies that recognize and neutralize the native toxins. The most notable examples of antitoxin vaccines are diphtheria and tetanus toxoids that are usually given as a pediatric series in combination with whole-cell pertussis (DTP) or acellular pertussis components (DTaP), and the principal component of anthrax vaccines. Although antitoxic immunity would not be expected to inhibit initial infections, induction of antitoxic antibodies by vaccination can control morbidity and mortality due to downstream elaboration of toxins from these infections. Just as the development of effective and safe toxoids had been successful for controlling diphtheria and tetanus epidemics, new vaccines can be generated from toxin targets of bacterial pathogens that are emerging as public health threats and toxins of plant and fungal origin that could be used in biowarfare or bioterrorism.

Over the last several decades, elucidation of the molecular mechanisms of toxin action across the spectrum of potential pathogens has led to the greater understanding of how toxins interact with animal and human hosts. This fundamental knowledge has been crucial to the development of new vaccine candidates and therapies for emerging pathogens that express toxins that determine pathogenicity.

This chapter focuses on the progress of vaccine development against ricin toxin as a paradigm for vaccines against other ribosome-inactivating toxins. Although this subject has been recently reviewed (Marsden et al., 2005; Mantis, 2005), an update is warranted as there has been significant progress over the past 3 years in both the advancement of several vaccine candidates and knowledge of toxin pathogenesis. In this chapter, we will highlight the recent advances in our understanding of ricin toxicity, pathology, and immunity and relate these to vaccines. Ordinarily, the subject of ricin intoxication and vaccine development would not be the subject of discussion in the context of emerging infectious diseases. However, ricin bears striking similarities to bacterial toxins, notably the Shiga family of toxins (Stx1 and Stx2, formerly called verotoxins), which cause human and animal diseases of public health significance. It is becoming increasingly apparent that the lessons learned with respect to ricin will have direct implication for countermeasures against these related bacterial toxins.

Considering that ricin intoxication is relatively rare [e.g., there have been less than 1000 cases of known ricin intoxication documented during the 1900s (Rauber and Heard, 1985], one is tempted to ask whether efforts to develop vaccines against ricin are truly necessary. Historically, efforts to understand immunity to ricin have been associated with the potential use of ricin, or more specifically the ricin enzymatic subunit, to create engineered immunotoxins for targeted therapy of cancers and select infectious diseases (Olsnes et al., 1989; Vitetta and Thorpe, 1991; Thrush et al., 1996). While these efforts are ongoing, the resurgence in ricin-related research has been driven by the proven potential of the toxin to be used as an agent of bioterrorism.

RICIN AS A CATEGORY B BIOTHREAT

Ricin toxin is a natural product of the castor bean plant, *Ricinis communis*, which is cultivated on industrial levels around the world for the production of castor oil. Recent economic surveys have shown that over 1 million tons of castor beans are produced in the world annually, with approximately 800,000 tons originating from India and around 100,000 tons from Brazil. Ricin toxin can constitute up to 5% of the total protein of castor bean and can be extracted from the mash produced as a by-product of castor oil production by several simple enrichment steps (Franz and Jaax, 1997). Because of the copious amounts of processed castor beans and the high content of ricin, a rough calculation indicates that, in any single year, it may be possible to purify over 50,000 tons of pure ricin.

In addition to its relative ease of preparation, ricin in semi-purified or purified form is extremely toxic. Only botulinum toxin is more potent than ricin (Arnon et al., 2001; Schechter and Arnon, 2000). Whereas botulinum toxin is an endoprotease that selectively targets peripheral neurons, resulting in descending paralysis of peripheral nerve function, ricin's potency is due to its promiscuity (it can intoxicate all known cell types) and extremely efficient enzymatic activity. Ricin was developed, but never used, as a weapon by the United States during World War II, under the code name of "Compound W." Since World War II, there has been little need to produce castor oil in the United States. Before the First Gulf War in 1990, the Iraqi military had attempted to devise ways to disseminate ricin as an explosive bomb (Shea and Groton, 2004; Stone, 2002), attempts that were forestalled by the war and ensuing events. Ricin has also been detected in a powder form sent in a letter addressed to Senator Frist in 2004 and several other similar but less publicized incidents.

For the reasons mentioned above, ricin has been classified by the Centers for Disease Control and Prevention (CDC) as a category B select agent. The CDC classification system categorizes biological agents based upon their risks for causing mass casualties in the event of a bioterrorist attack (Khan et al., 2000; Rotz et al., 2002). The highest priority agents (category A) are organisms or toxins that can be easily disseminated or transmitted person to person, cause high mortality with potential for major public health impact, might cause public panic and social disruption, and require special action for public health preparedness. Category A agents include variola major (smallpox), *Bacillus anthracis*, *Yersinia pestis*, *Clostridium botulinum* toxin, *Francisella tularensis*, Ebola virus, Lassa fever virus, and several other related viruses. Category B includes biothreats that are relatively easy to disseminate and cause moderate morbidity and low mortality. In addition to ricin, the category B list includes *Coxiella burnetti* (Q fever), *Brucella* species (brucellosis), *Burkholderia mallei* (glanders), alphaviruses, Venezuelan equine encephalomyelitis (VEE), eastern and western equine encephalomyelitis, epsilon toxin of *Clostridium perfringens*, *Staphylococcus* enterotoxin B, and select foodborne or waterborne pathogens, including *Salmonella* species and *Vibrio cholerae*. Although Shiga toxin itself is not considered a biothreat agent, Shiga toxin-producing bacteria notably *Shigella dysenteriae* and *Escherichia coli* O157:H7, are classified as category B agents and represent a significant threat to public health. Finally, category C agents include emerging pathogens that could be engineered for mass dissemination in the future because of availability, ease of production and dissemination, and potential for high morbidity and mortality and major health impact. They include Nipah virus, hantaviruses, tick-borne hemorrhagic fever viruses, tick-borne encephalitis viruses, yellow fever, and multidrug-resistant tuberculosis.

Thus, the toxicity and availability of purified and semi-purified ricin has been the greatest impetus for development of vaccines and therapeutics. Although it would be technically difficult to achieve mass dissemination of ricin, it is clear that localized dissemination via aerosol or adulteration of food chains could disrupt public health and the perceived safety of the United States from bioterrorists. The development of vaccines and therapeutics will absolutely depend on the correct interpretation of the correlates of protection, which can be obtained only through evaluation of efficacy in relevant animal models. Additionally, a vaccine must be designed to protect against the circulating toxin as well as the local effects on mucosal tissues in the lung or the gastrointestinal (GI) tract. Further, a postexposure therapeutic may have to address the temporal nature of the rapidity of toxin action and the downstream effects of the toxins that cannot be addressed by neutralizing antibodies or molecules that inactivate the toxins per se. It is becoming increasingly apparent that the host's response to intoxication (i.e., inflammation) may actually amplify the damage elicited by the toxin and be more detrimental than the toxin itself.

SIMILARITIES BETWEEN RICIN AND SHIGA TOXINS

Ricin and Shiga toxins are similar in a number of respects, including enzymatic activity, intracellular trafficking, and activation of cellular stress pathways.

The family of Shiga toxins consists of several antigenically related toxins (e.g., Stx1 and Stx2) that are expressed by certain strains of *Shigella* and *E. coli*, most notably *E. coli* O157:H7, that are increasingly associated with foodborne outbreaks of hemorrhagic colitis and hemolytic uremia syndrome (HUS). HUS is the major cause of acute renal failure in infants and young children. Enzymatically, ricin and Shiga toxins are identical. These toxins are RNA *N*-glycosidases, whose sole substrate is a highly conserved adenine residue within ribosomal RNA (Endo and Tsurugi, 1987; Endo et al., 1987). Both ricin and Shiga toxins are members of the A–B family of toxins in which the A subunit confers enzymatic (i.e., cytotoxic) activity, whereas the B subunit mediates toxin attachment to cell surfaces. Shiga toxin is an A_1:B_5 toxin, meaning that five B subunits assemble as a donut-shaped pentamer in which a single A subunit sits within its center. In the case of ricin, the A–B subunits associate in a 1:1 stoichiometry and are linked by a single disulfide bond. By definition, ricin and Shiga toxins are categorized as type II ribosome-inactivating proteins (RIPs) (i.e., an A and B subunit). Type I RIPs, on the other hand, consist solely of a single catalytic subunit analogous to the A chain and are invariably derived from plant sources. Although ricin and Shiga toxins bind to different cellular receptors (see below), both toxins reach their ribosomal targets by retrograde transport through the Golgi apparatus and the endoplasmic reticulum (ER). Finally, both ricin and Shiga toxins activate the so-called ribotoxic stress response, which triggers the initiation of events that lead to apoptotic cell death, as opposed to necrosis, and the activation of multiple pathways that lead to secretion of proinflammatory cytokines and chemokines.

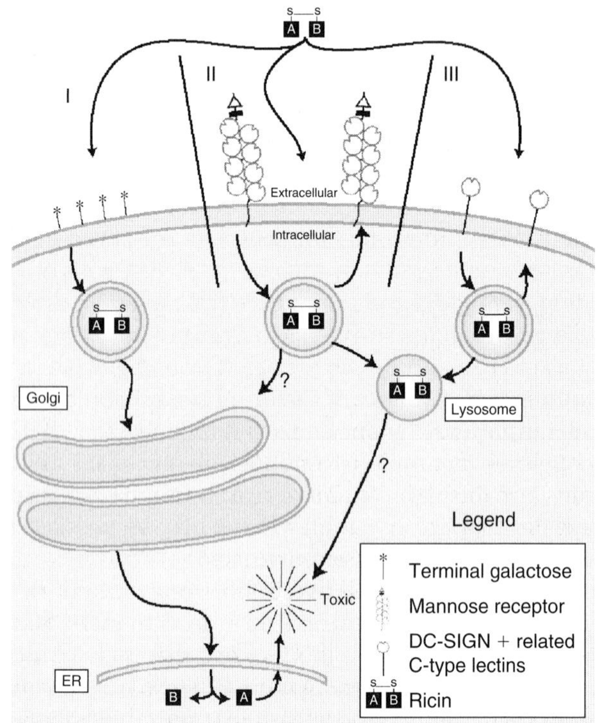

FIGURE 55.1 Pathways of ricin uptake into host cells. There are two known (and one postulated) pathways (I–III) by which ricin is internalized into host cells. In *Pathway I*, RTB binds to galactose-containing glycoproteins and glycolipids on host cell surfaces. The toxin is internalized by all known endocytic processes. A fraction of the internalized toxin is transported retrograde to the Golgi apparatus, and eventually to the ER. The toxin subunits dissociate, and RTA is translocated into the cytoplasm via a Sec61-dependent process. RTA exerts its cytotoxic properties once within the cytoplasm. In *Pathway II*, ricin is recognized by the mannose receptor (MR), which is expressed on macrophages and hepatic endothelial cells. Uptake by the MR is approximately 100 times more efficient than the Pathway I, although virtually nothing is known about the details of how this occurs. It is postulated that ricin initially resides within an acidic endosome, but it is unclear whether the RTA gains access to the cytoplasm by merging with Pathway I (i.e., via retrograde transport) or is translocated across the endocytic membrane (see question marks). *Pathway III* remains speculative but proposes that ricin may utilize other mannose-specific lectins (e.g., DC-SIGN) to gain access to the inside of host cells.

RICIN TOXIN STRUCTURE AND FUNCTION

Ricin Toxin A (RTA) and B (RTB) Chain Function

As mentioned above, ricin is a prototypic A–B toxin. It exists as a 64-kilodalton (kDa) globular glycoprotein heterodimer consisting of a 32-kDa RTA chain linked via a single disulfide bond to the 32-kDa RTB chain (Montfort et al., 1987). The RTA chain is enzymatically active and is responsible for the induction of the toxicity (Olsnes and Pihl, 1973). RTA enters the cytosol by at least two distinct pathways, as will be discussed below (Fig. 55.1). Once in the cytoplasm, RTA depurinates a single adenine residue (A4324) near the 3′-end of 28S RNA in ribosomes, the identical site for Shiga toxins (Yamasaki et al., 2004), blocking protein synthesis (Barbieri et al., 1993). The X-ray crystal structure of RTA has been determined, and the residues responsible for depurination of 28S RNA have been identified (Ready et al., 1991; Lamb et al., 1985; Montfort et al., 1987; Mlsna et al., 1993; Weston et al., 1994). Residues Y80, Y123, E177, R180, N209, and W211 constitute RTA's enzymatically active site (Lebeda and Olson, 1999; Olson, 1997). Mutations in some of these amino acid residues have yielded RTA mutants with negligible toxicity as determined by the inhibition of protein synthesis (Kim and Robertus, 1992). The latest vaccine candidates have been developed with the purpose of reducing the enzymatic activity of the A chain to levels below detection of activity.

RTB is a bivalent lectin specific for β1,3-linked galactose and *N*-acetylgalactosamine and is capable of binding to both glycolipids and glycoproteins on host cell membranes (Baenziger and Fiete, 1979; Wu et al., 2006). The primary function of RTB is to mediate attachment of ricin to cell surfaces (Fig. 55.1). RTB consists of two globular domains with identical folding topologies (Montfort et al., 1987). Each of the two domains (1 and 2) themselves comprise three homologous sub-domains (α, β, γ) that probably arose by gene duplication from a "primordial" carbohydrate recognition domain (CRD) (Rutenber et al., 1987). Only domains 1α and 2β retain functional carbohydrate binding activity, as inferred by the crystal structure of ricin in the presence of lactose (Rutenber and Robertus, 1991), as well as binding studies with RTB mutants (Swimmer et al., 1992). The two galactoside pockets are separated by 70 Å, suggesting that each pocket binds a different galactoside-containing side chain when attached to the surface of a cell (Rutenber and Robertus, 1991).

It is important to note that both RTA and RTB carry *N*-linked mannose side chains (Foxwell et al., 1985c; Kimura et al., 1988). RTA has a single side chain composed of $GlcNAc_2Man_4$ at position N10, while RTB has two side chains: $GlcNAc_2Man_6$ at position N95 and $GlcNAc_2Man_7$ at position N135 (Kimura et al., 1988). The significance of the mannose side chains is revealed by the fact that deglycosylated ricin (dgRicin) is ~90% less toxic than unmodified holotoxin in standard in vitro cytotoxicity assays and in mouse models of systemic intoxication (Foxwell et al., 1987; Simeral et al., 1980; Simmons et al., 1986; Thorpe et al., 1985). Discerning the basis of this attenuation has relied on the use of hybrid toxins in which only one of the two subunits is deglycosylated. Removal of RTA mannose side chains has no effect on the RTA's ability to inactivate ribosomes or the capacity of the holotoxin to kill host cells (Simmons et al., 1986; Thorpe et al., 1985). In contrast, holotoxins consisting of deglycosylated RTB are highly attenuated (Simeral et al., 1980; Simmons et al., 1986). Because deglycosylation had no effect on RTB's lectin activity, it has been postulated that the mannose side chains function to either interact with mannose-binding proteins during intracellular transport or stabilize RTB intracellularly (Simeral et al., 1980; Simmons et al., 1986). It has recently been shown that ricin interacts with a α-mannosidase during retrotranslocation, although it is unclear if this enzyme interacts with the mannose side chains on RTB (Slominska-Wojewodzka et al., 2006).

Ricin Uptake into Host Cells via RTB and Retrograde Transport: Pathway I

Although there are specific differences between Shiga and ricin toxins, both toxins enter cells and traffick intracellularly through common pathways. The intracellular trafficking pathways followed by both toxins have been well reviewed recently (Sandvig and van Deurs, 2005; Sandvig et al., 2004). The salient features of toxin trafficking that relate to the pathogenesis of these toxins are the steps in which the toxins bind to surface receptors and are taken up through clathrin and clathrin-independent endocytosis, with possible release of some toxin directly from the endosomes into the cytosol, fusion of endosomes with the trans-Golgi network (TGN), translocation into the ER, and retrotranslocation into the cytosol. Currently, there is little controversy that ricin and Shiga toxins enter the cytosol through retrograde transport through the ER, although the precise molecular steps are not well defined. As the first part in the uptake process, receptor-bound toxins are slowly endocytosed (Sandvig et al., 1976) and a portion of the internalized ricin is transported back to the cell surface; another portion is delivered to lysosomes and degraded, and a part is transported to the TGN (van Deurs et al., 1986). Subsequently, RTA is translocated to the cytosol from the ER (Wales et al., 1992, 1993; Tagge et al., 1996). The first convincing evidence of the involvement of the ER in this was demonstrated in experiments that documented the *N*-glycosylation of nonglycosylated recombinant RTA that was reconstituted with RTB, since glycosylation occurred only within the ER (Llorente et al., 2003; Rapak et al., 1997). Intimate association of RTA with ER-located Sec61 by cross-linking (Wesche et al., 1999) demonstrated the presence of RTA in the ER. Although ricin itself has not been visualized in the ER by electron microscopy, Shiga toxin has been directly detected, suggesting that perhaps a greater portion of Shiga toxin actually enters the ER pathway than ricin toxin (Sandvig et al., 1993). It is possible that only a small portion of endocytosed ricin (~5%) productively reaches the TGN (van Deurs et al., 1988). Hence, the precise fate of all of endocytosed ricin is still unclear, and there is also evidence that ricin can completely avoid Golgi processing (Llorente et al., 2003). The TGN and Golgi steps involved are partly clarified. For example, ricin can intoxicate cells inhibited in both the classical constitutive photomorphogenesis 1 (COP1)-dependent and the less-well-defined COP1-independent pathways (Chen et al., 2003). Nonetheless, intracellular routing of ricin and Shiga toxins differs somewhat from that of several other bacterial toxins (e.g., cholera toxin) that have C-terminal Lys-Asp-Glu-Leu (KDEL) sequences that are thought to act as retrieval signals (Chaudhary et al., 1990). Ricin does not contain such a sequence, but there is some experimental evidence that RTB interacts with the chaperone calreticulin, which possesses a KDEL retrieval signal (Day et al., 2001). The fact that calreticulin-deficient cells are sensitive to ricin suggests that ricin exploits additional pathways to enter the ER. Addition of the KDEL signal to RTA by genetic fusion enhanced the toxicity of both a reconstituted AB holotoxin and the A chain alone (Wesche et al., 1999; Wales et al., 1992, 1993).

Ricin travels from the Golgi complex to the ER, where it likely subverts the ER-associated degradation (ERAD) pathway and unfolds and refolds through the Sec61 protein, a translocating conserved heterotrimeric membrane protein (Sandvig et al., 2002, 2004; Rapak et al., 1997; Lord and Roberts, 1998; Wesche et al., 1999). The Sec61 pathway is normally used to redirect misfolded and incorrectly assembled proteins to the cytosol for ubiquitination and proteasomal degradation (Plemper and Wolf, 1999). This is thought to hold true for the A chains not only of ricin (Rapak et al., 1997) but also of *Pseudomonas* exotoxin (Kreitman and Pastan, 1995), cholera toxin (Jackson et al., 1999), and Shiga toxin (Sandvig et al., 2004). For RTA to act, the disulfide bonds between RTA and RTB must be reduced to expose the active site. Protein disulfide isomerase (PDI) is thought to be involved with the reduction of RTB (Bellisola et al., 2004) and other toxins (Di Cola et al., 2001). RTA appears to be unfolded for retrotranslocation (Beaumelle et al., 1997) and may be refolded in the cytoplasm in contact with substrate ribosomes (Argent et al., 2000). RTA is low in available lysine sites for ubiquitination (Deeks et al., 2002) and avoids substantial proteasome-mediated degradation.

Ricin Uptake into Host Cells via the Mannose Receptor: Pathway II

It has been recognized for more than two decades that ricin exploits the mannose receptor (MR) as a second pathway (independent of RTB's galactose-binding activity) by which to deliver RTA into the cytoplasm of host cells (Simeral et al., 1980; Simmons et al., 1986; Thorpe et al., 1985; Frankel et al., 1997). The MR (CD206) is a 175-kDa transmembrane endocytic receptor that recognizes complex oligosaccharides terminating in mannose, fucose, or *N*-acetylglucosamine (East and Isacke, 2002; Taylor et al., 2005). It was first identified on alveolar macrophages (Largent et al., 1984; Shepherd et al., 1981) and later discovered to be expressed on a variety of cell types, including hepatic sinusoidal endothelial cells (HSECs) and Kupffer cells. Skilleter et al. (1981) noted that ^{125}I-labelled ricin accumulated in rat liver non-parenchymal cells to a much greater extent than parenchymal cells and showed that this accumulation could be inhibited by D-mannose, fucose, or ovalbumin. Magnusson and colleagues in a series of papers demonstrated ex vivo that ricin uptake by Kupffer cells and HSECs was in fact MR mediated in the sense that the process was inhibited by mannan (Magnusson and Berg, 1993; Magnusson et al., 1991, 1993). Intravenous injection of plant-derived RTA could selectively deplete mouse Kupffer cells associated with uptake through the MR (Zenilman et al., 1988). Macrophage cell cultures that expressed the MR were shown to be sensitive to mannosylated plant-derived RTA but not to *E. coli*-derived RTA, and the toxicity of plant-derived RTA could be blocked by mannan. These lines of evidence strongly suggest that there are two pathways operable in vivo for the uptake of ricin, some of it occurring through the RTB galactosyl pathway and a portion not entirely dependent upon the RTB subunit and mediated by mannosyl residues on RTB and RTA.

Despite the fact that MR-mediated uptake is an important route by which ricin is internalized into cells in vivo, virtually nothing is known about the underlying mechanism by which this occurs. Simmons et al. (1986), using rat bone marrow-derived macrophages, demonstrated that RTB-mediated and MR-mediated uptake pathways are biochemically distinct from each other. For example, MR-mediated uptake results in ricin passing through an acidic vesicle prior to translocation of RTA into the cytoplasm, whereas RTB-mediated uptake does not (see above). This is consistent with the MR's role as an endocytic receptor that continuously cycles between the plasma membrane and the endosomal compartment (East and Isacke, 2002). However, it is completely unknown how RTA escapes acidic endosomes and translocates into the cytosol. The fact that RTA itself (without RTB) is toxic to cells following uptake by the MR indicates that the mechanism of RTA translocation is distinct from the well-studied retrograde transport/ER translocation that occurs following RTB-mediated uptake (Rapak et al., 1997; Simmons et al., 1986) (see below). However, the relative contributions of uptake through galactosyl-mediated RTB receptors and the MR may be dependent on the cell type. The realization over the past decade that the MR is only one member of a large family of mannose-specific lectins expressed on lymphoid and non-lymphoid cells suggests that much remains to be explored with respect to how RTA gains access to the cytosol of host cells (van Kooyk and Geijtenbeek, 2003).

Vascular Leak Syndrome

An additional site and residues involved in the binding of RTA to endothelial cells have also been identified, which occur extracellularly and do not require toxin entry into host cells (Baluna et al., 2000; Baluna and Vitetta, 1999; Smallshaw et al., 2003). The endothelial binding site on RTA is implicated in the damage to isolated human umbilical vein endothelial cells (HUVECs) and to the induction of vascular leak syndrome (VLS), which has been determined to be a dose-limiting toxicity in the use of RTA-containing immunotoxins (Smallshaw et al., 2003). VLS has been a side effect of the intravenous administration of monoclonal antibody conjugates constructed using enzymatically deglycosylated RTA (dgRTA) during Phase I and II trials in cancer (Sauseville and Vitetta, 1997). In VLA, fluids leak from blood vessels leading to hypoalbinemia, weight gain, and pulmonary edema. The portion of RTA involved in both pulmonary vascular leak, vascular leak in human skin xenografts in severe combined immunodeficient (SCID) mice, and HUVECs appears to involve amino acid residues L74, D75, and V76 (Baluna et al., 1999). The residues involved in vascular leak were predicted from comparison of motifs in other toxins and interleukin-2 (IL-2). In one of the vaccine candidates under development (RiVax™), residue V76 has been modified by mutagenesis (i.e., V76M) to abrogate the VLS activity and minimize any possible endothelial cell damage.

TOXICITY OF RICIN

Little human data are available for the toxicity of ricin, although it is predicted from animal models that pure ricin will be toxic to humans by inhalation, oral, and systemic exposure. Essentially all of the data on the toxicity comes from studies in animals, with the exception of the single well-known human case of Georgi Markov, who was assassinated from an injection of sub-milligram quantities of ricin (Crompton and Gall, 1980). Although ricin toxin acts quickly on cells in vitro, clinical signs after ingestion, injection, or inhalation appear within 24h of exposure. In the Markov incident, symptoms over the 4 days preceding his death were well characterized. In essence, intramuscular (i.m.) or subcutaneous (s.c.) administration induced local lymphoid necrosis, hemorrhages in the GI tract, liver necrosis, nephritis, pulmonary failure, and cardiac arrest. Data concerning oral exposure to ricin come from 750 reported cases of castor bean or ricin ingestion and a single study of oral administration of a crude castor bean extract in animals (Balint, 1974).

The toxicity of ricin in animals depends on the route of exposure (Franz and Jaax, 1997). Inhalation induces death in 60h with an LD_{50} of 3–5μg/kg (Franz and Jaax, 1997; Bradberry et al., 2003). By intraveneous (i.v.) injection, the LD_{50} is 5μg/kg and death occurs in 90h. Following s.c. injection, the LD_{50} is 24μg/kg and death occurs at 100h. Because of the rapid internalization of ricin by endocytosis and the catalytic amounts of ricin needed to inhibit protein synthesis, the toxic action of ricin is irreversible (Lord et al., 1994). Ricin is toxic in many animal species, and there is no evidence that there is a species-specific toxicity (Table 55.1). Thus, for vaccine development, there is every reason that no specific animal model is preferred as a model of pathogenesis and immune response. That is, protection data generated in mice can be generally translated to what can be expected in humans to the extent that if antibodies generated in a mouse protect a mouse, then it is expected that antibodies generated in a human will protect a human.

TABLE 55.1 Published lethal doses for ricin

Species/route	Toxic dose	References
BALB/c mouse/ inhalation[a]	12μg/kg LD_{50}	Roy et al. (2003)
CD-1 mice/ inhalation	<3μg/kg 3μg/kg, 80% death 10μg/kg, 100% lethality	Doebler et al. (1995)
Rhesus monkey/ inhalation	21–42μg/kg LD_{100} in 48h	Wilhelmsen and Pitt (1996)
Rat/oral	30mg/kg LD_{100} in 36h	Ishiguro et al. (1992d)
Mouse/intragastric (oral gavage)	20mg/kg	Franz and Jaax (1997)
Swiss Webster[d] mouse/oral	10 μg/kg	Smallshav et al. (2007)
Swiss albino mouse/ i.p.[b]	25μg/kg	Kumar et al. (2003)
CD-1 mouse/i.p.	20–60μg/kg[c]	Pincus et al. (2002)
B6D2 mice/i.v.	2μg/kg	Fodstad et al. (1976)
Mice/i.v.	5μg/kg	Franz and Jaax (1997)
Rats/i.v.	0.35–5μg/kg	Fodstad et al. (1976)
Rabbits/i.v.	LD_{50} 0.54μg/kg; minimum lethal dose of 0.44μg/kg	Christiansen et al. (1994)

Note: Published LD_{50} data from parenteral, aerosol, and gastrointestinal exposure in several animal species.

[a] 1-μm MMAD particle. MMAD = mass median aerosol diameter.

[b] Mice sacrificed at 24h, therefore an estimate of LD_{50}. Significant increases in serum enzymes indicative of liver and kidney injury were observed at this dose.

[c] Significant decreases in blood glucose occurred at 60μg/kg, but not 20μg/kg, suggesting an LD_{50} between the two doses.

[d] Mice fasted 24hr prior to oral ricin garage.

Respiratory Toxicity

The effect of aerosolized ricin on the respiratory tract has been studied in rodents (Brown and White, 1997; DaSilva et al., 2003; Doebler et al., 1995; Griffiths et al., 1999) and rhesus monkeys (Wilhelmsen and Pitt, 1996). In laboratory animals, ricin inhalation results in respiratory distress and airway and pulmonary lesions. Inhalation of ricin in rats leads to rapid apoptotic changes in alveolar macrophages within 6h after exposure, finally culminating in interalveolar edema at 12 and 15h after exposure, mixed inflammatory cell infiltrates, and alveolar flooding followed by tissue necrosis (Brown and White, 1997). Similar pathological and histological changes have been seen in nonhuman primate studies, where death from ricin inhalation occurs 48h after the challenge in a dose-dependent fashion. Aerosol administration to primates results in the sudden onset of congestion, itchiness of the eyes, tightness of the chest, wheezing, and asthma (Franz and Jaax, 1997). This rapidly progresses to diffuse necrotizing pneumonia, interstitial and alveolar inflammation, edema, and alveolar flooding (Franz and Jaax, 1997).

When administered as an aerosol to mice, ^{125}I-labled ricin accumulated in both the respiratory and GI tracts, with a small fraction detected in other organs (Doebler et al., 1995). Monkeys exposed to lethal doses of ricin (~20–40μg/kg) died (or were sacrificed upon onset of respiratory distress) roughly 36–48h postexposure (Wilhelmsen and Pitt, 1996). Death was preceded by widespread necrosis in the airways and alveoli, peribronchovascular edema, and mixed inflammatory cell infiltrates. Similar results were observed in rats (Brown and White, 1997). Similarly, rodents exposed to lethal doses of aerosolized ricin develop necrotizing airway lesions, interstitial edema, and pulmonary inflammation, including massive neutrophil influx, and die 2–4 days later (Griffiths et al., 1997, 1999; Hewetson et al., 1993; Poli et al., 1996). In both monkeys and rodents, death was attributed to massive pulmonary alveolar flooding. There are no case reports of inhalation ricin exposure in humans.

Gastrointestinal Toxicity

Compared to aerosol exposure, ricin administered orally is considerably less toxic. The literature indicates that the oral LD_{50} is 20mg/kg (Table 55.1), approximately 1000-fold higher than the aerosol or systemic LD_{50} values. Ingestion of whole castor beans results in severe abdominal pain, vomiting, diarrhea, and (depending on the number of beans and degree of mastication) death (Audi et al., 2005; Bradberry et al., 2003; Mantis, 2005; Olsnes, 2004). As there are no known reported cases of intentional ricin intoxication by the oral route in humans, our understanding of both the pathogenesis and mechanisms of intestinal immunity to ricin relies exclusively on experimental animal models, namely, rodents (Sekine et al., 1986; Smallshaw et al., 2007; Yoder et al., 2007) and, in some instances, rabbits (Hazen, 1927). Ishiguro and colleagues reported, in a series of short and highly descriptive papers, the histologic and physiologic changes that occur in the rat jejunum following ricin intoxication (Ishiguro et al., 1983, 1984, 1992a, 1992b, 1992c, 1992d; Sekine et al., 1986). Rats challenged with ricin by gavage (1–30mg/kg) developed dose-dependent lesions in the stomach and the proximal small intestine (Sekine et al., 1986). In the latter tissue, there was widespread villus atrophy, crypt elongation, sloughing of the epithelium, and infiltration of inflammatory cells, including eosinophils and neutrophils. Similar histopathologic changes have been observed in mice challenged intragastrically with ricin (Yoder et al., 2007). It is postulated that following intestinal exposure, ricin ultimately disseminates from the mucosa into circulation (Ishiguro et al., 1992d). Indeed, ricin has been shown to cross polarized epithelial cell monolayers in vitro, although how much toxin crosses the intestinal barrier in vivo remains unknown (Mantis et al., 2006; van Deurs et al., 1990).

The inherent resistance of the GI tract to ricin is likely due to a number of factors that impede toxin absorption, including intestinal proteases, digestive enzymes, mucus, and secretory IgA, whose galactose-rich oligosaccharide can competitively inhibit ricin attachment to the apical surfaces of intestinal epithelial cells (Mantis et al., 2004). It should be mentioned that Smallshaw et al. (2007) have recently challenged the accepted LD_{50} based on results from a mouse model in which animals are fasted 20h prior to gavage to ensure that food particles in the stomach do not interfere with toxin absorption. Under these circumstances, doses of ricin equal to or exceeding 100μg/kg were consistently lethal. Although intestinal histology appeared largely normal in animals exposed to this fasting/challenging regimen, it should be kept in mind that food deprivation of this duration is known to have dramatic effects on the intestinal physiology of mice (Lenaerts et al., 2006).

Cellular Basis for Downstream Effects of Ricin and Shiga Toxins

While tissue damage is a hallmark of intestinal and pulmonary ricin intoxication, it remains unclear how much of the damage is a result of the direct effects of the toxin (i.e., protein synthesis arrest and apoptosis) and how much is a consequence inflammation. There are now many studies documenting that epithelial cells, for example, exposed to Shiga or ricin toxin elicit a number of proinflammatory chemokines and cytokines, including IL-8, growth-regulated oncogene-alpha (GRO-alpha), GRO-beta, GRO-gamma, epithelial cell-derived neutrophil-activating protein-78 (ENA-78), and monocyte chemotactic protein 1 (MCP-1), also known as CCL2 (O'Loughlin and Robins-Browne, 2001; Thorpe et al., 1999, 2001; Yamasaki et al., 2004). Some of these observations have been performed in animal models of disease. These studies have essentially been iterated in a number of cell lines of human and murine origin, including endothelial, epithelial, and macrophagic cell lines (Table 55.2). Generally, the effects of ricin and Shiga toxins constitute the activation of inflammatory pathways and pathways acting through nuclear factor kappa B (NF-κB) and mitogen-activated protein (MAP) kinase p38 (p38 MAPK), extracellular regulated kinase (ERK), and c-jun *N*-terminal kinase (JNK) (Fig. 55.2). This results in the secretion of numerous proinflammatory cytokines and chemokines. Chemokines of the CXC subfamily, such as IL-8, can recruit neutrophils into areas of infection (Baggiolini et al., 1995; Baggiolini and Loetscher, 2000; Baggiolini, 2001). IL-8 can also stimulate neutrophil granule content release and the respiratory burst. The release of granule contents, such as proteases, toxic oxygen products, and other proinflammatory substances, can subsequently damage the intestinal mucosa (Grisham and Granger, 1988).

To determine whether ricin intoxication is accompanied by a local increase in proinflammatory cytokines in vivo, Yoder et al. (2007) challenged mice intragastrically (i.g.) with ricin at a range of doses (1–10 mg/kg) for various time periods (0, 5, 12, and 24 h) and then measured cytokine/chemokine levels in intestinal homogenates. There was no detectable increase in local IFN-γ, IL-1, IL-6, IL-12p70, or TNF-α levels following ricin challenge. In contrast, there was a dose- and time-dependent increase in MCP-1. MCP-1 has been previously implicated in mediating intestinal inflammation (Reinecker et al., 1995) and could possibly be involved in ricin-mediated tissue damage. MCP-1 is a chemoattractant for lymphocytes, monocytes, and macrophages, and can stimulate macrophages to undergo respiratory burst resulting in release of oxidative intermediates (Carr et al., 1994;

TABLE 55.2 Mediators of inflammation and apoptosis induced by ricin or Shiga toxin in cell cultures

Toxin	Cell type(s)	Pathway	Mediators/factors upregulated	Results	References
Ricin	Primary mouse bone marrow macrophages alveolar macrophages	p38 MAPK, JNK	IL-1β, TNF-α, CXCL1/ GRO-α	Activation of p38, JNK, TNF-α secretion	Korcheva et al. (2007)
	Human primary airway cell cultures	NF-κB, p38 MAPK, JNK	CXCL1, CCL2, IL-8, IL-1β, IL-6, TNF-α	Increased chemokines, transcription factors, cytokines, chemokines, inflammatory gene families by microarray	Wong et al. (2007b)
	Human monocytes/ macrophages 28SC, U937	p38 MAPK	IL-8	p38 inhibitor SB202190 suppressed IL-8	Gonzalez et al. (2006)
	Human PMN	Activation	TNF-α, IL-1β	Induction of cytokines by ricin	Licastro et al. (1993)
	U937	Apoptosis	Caspase-3	Induction of apoptosis/ correspondence of caspase-3	Kageyama et al. (2002)
Stx2	HMEC-1 human endothelial cells	Platelet stimulation/ thrombosis	Rantes (CCL5), IL-8, SDF-1α (CXCL12)	Platelet aggregation by Stx/ LPS-stimulated HMEC cells	Guessous et al. (2005)
Stx1/Stx2	Hct-8 human colonic carcinoma cells	c-jun	IL-8, CXCL1/ GRO-α	Induction of chemokines, p38 inhibitor diminished IL-8	Thorpe et al. (1999, 2001)
Stx1/Stx2	Primary murine peritoneal macrophages; THP-1 monocytes	Macrophage activation; apoptosis	TNF-α, IL-6, E1fFe transcription factor	Stimulations by LPS+Stx, inhibition of IL-8/IIL-1β by p38, ERK, and JNK inhibitors	Tesh et al. (1994), Cherla et al. (2006), Harrison et al. (2004)

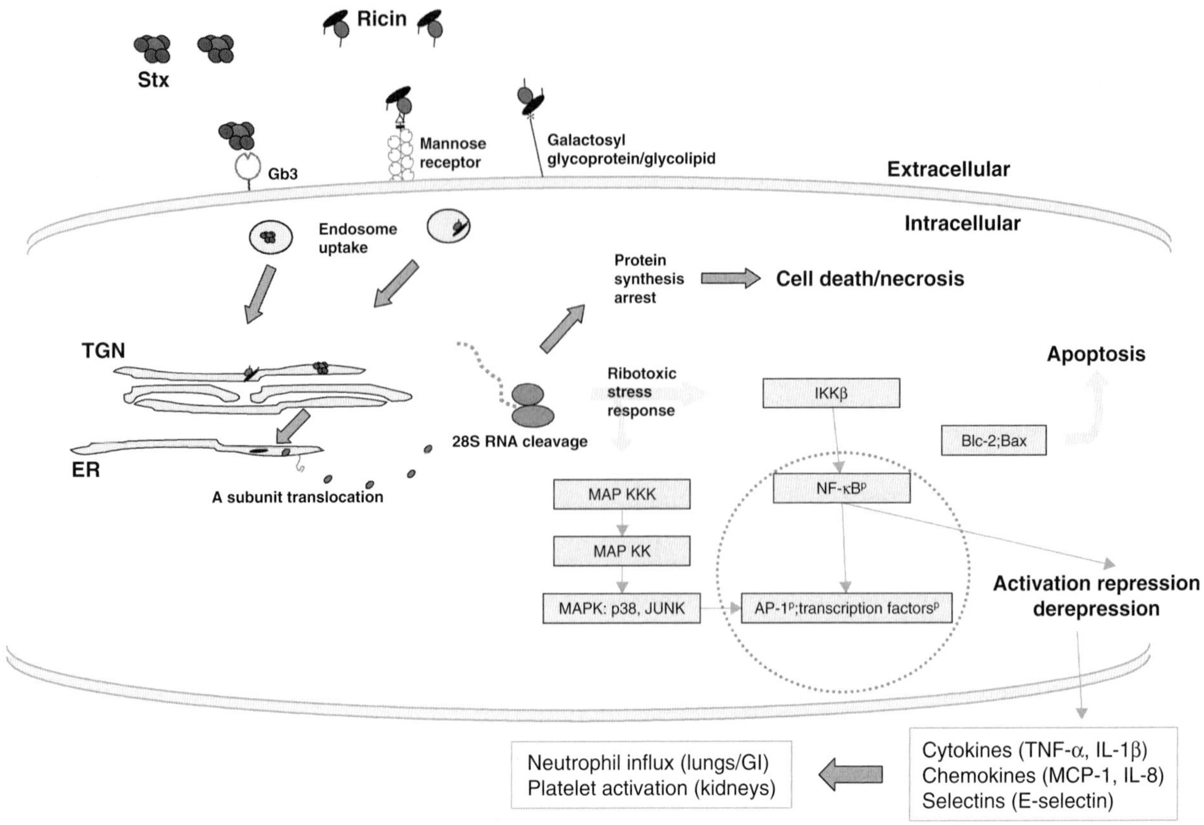

FIGURE 55.2 Pathways of intoxication implicated in the cellular actions of ricin and Shiga toxins. Ricin is a 64-kDa dimer of RTA and RTB. Ricin binds, via specific recognition sites, on RTB to galactosyl moieties on glycoproteins and glycolipids present virtually on all mammalian cell types and alternatively to the MR through *N*-linked mannosyl residues on either RTA or RTB. Shiga toxin is an A-B_5 toxin of 64 kDa that binds exclusively to globotriaosylceramide receptors that are primarily located in kidney and microglial cells. Through clatharin-dependent and -independent endocytosis, both Shiga and ricin toxins are taken up by endosomes and traffick through the TGN to the ER, where reductive cleavage of the subunit occurs, and the A subunits of each toxin translocate to the cytoplasm by subversion of the ERAD pathway. Once in the cytosol, the A subunits cleave specific sites in 28S rRNA, resulting in cessation of protein chain elongation. Through a series of unknown signaling steps, p38 MAPKs, ERK, and JNK are activated, resulting in the upregulation of cytokines, chemokines, and other gene families. The events culminate in recruitment of neutrophils to the sites, and separately to apoptosis.

Bischoff et al., 1992; Connor et al., 2004; Lu et al., 1998; Rollins et al., 1991). Moreover, MCP-1 is released by intestinal epithelial cell lines following bacterial infection (Hu and Hickey, 2005; Jung et al., 1995) and exposure to microbial toxins, including *Bacteroides fragilis* enterotoxin (Kim et al., 2005) and *Clostridium difficile* A toxin (Kim et al., 2002). MCP-1 expression is regulated in part by the p38 MAPK (Waterhouse et al., 2001). The availability of MCP-1 knockout mice permits one to examine the contribution of the chemokine-driven inflammation in toxin-induced tissue damage and the potential of p38 MAPK inhibitors to serve as postexposure therapeutics.

The inflammation associated with Shiga toxin-induced HUS is also indicated by the release of cytokines and chemokines, and experiments conducted in isolated cell cultures and animals also indicate the expression of proinflammatory mediators. The levels of IL-8 and MCP-1 are significantly increased in urine samples collected from HUS patients, suggesting the local action of cytokine on glomerular endothelium (van Setten et al., 1998). As with ricin, treatment of intestinal epithelial cell types with Stx1 or Stx2 leads to increased expression of CXC chemokines (Thorpe et al., 2001). The expression of inflammatory cytokines, such as IL-6 and IL-8, in macrophages in response to Shiga toxin and lipopolysaccharides (LPS) was reported more than 10 years ago (Tesh et al., 1994) and in other monocytic cells more recently (THP-1 monocytic cells) (Harrison et al., 2005; Table 55.2). These studies have also been extended to a variety of endothelial and epithelial cell lines. Stx2-treated HUVECs have been shown to release IL-8 and MCP-1, which stimulated adhesion and transmigration of leukocytes (Zoja et al., 2002). Development of HUS may also depend on other proinflammatory stimuli, such as LPS, TNF-α, or IL-1β (van de Kar et al., 1992), that are associated with increased Gb3 levels in human brain and kidney endothelial cells (Stricklett et al., 2002), which increase sensitivity to the toxins by upregulating Gb3 receptor (van de Kar et al., 1992). Consistent with this, LPS and TNF-α increase the direct cytotoxic effect of Shiga toxin on HUVECs in vitro (Louise and Obrig, 1991, 1992).

Shiga toxins gain access to the systemic vasculature from the gut lumen, possibly by direct transcytosis or intracellular replication of *E. coli* cells, and circulate to target organs and cause damage to microvascular endothelial cells. There are no comparable data that ricin is also capable of gaining access to target organs escaping from the GI lumen, but recently acquired indirect data have indicated that lethal amounts of ricin given to mice by oral gavage act in the absence of obvious lesions, although inflammatory responses were not measured (Smallshaw et al., 2007). These results suggest that some ricin enters through undetected lesions or there is a minor mechanism for uptake of ricin directly through transcytosis. By the same token, some detailed studies exploring the intratracheal instillation of ricin at high doses, twofold above the LD_{50}, indicated induction of inflammatory responses not only in the lungs but also in kidney tissues (Wong et al., 2007a). This was accompanied by deposition of fibrin/fibrinogen in lung tissue of mice after 48 h, plus influx of neutrophils, focal areas of hemorrhage, necrosis and apoptosis of the airway epithelium and alveolae, increased expression of proinflammatory transcripts (TNFα, IL-1β, IL-6, CXCL1 and CXCL2, and transcription factors c-jun and c-fos), and activation of ERK and JNK. Interestingly, at lethal doses, elevated transcripts were seen in kidney tissue and other organ tissues, which was associated with lesions in 28S rRNA in the tissue samples examined, indicating presence of ricin in those tissues. At sublethal doses (0.2 LD_{50}), JNK and ERK activation was accompanied by increased expression of NK-κB and neutrophil infiltrates, and differed from the lethal dose as there was no evidence of hemorrhage or fibrin/fibrinogen deposition in lung or airway tissue. There was no evidence of distal organ damage. This suggests a situation in which local necrosis of tissue may allow the penetration of ricin, as suggested for Shiga toxin, and the effects may be dose related. These results are important for consideration of the types of endpoints that may be examined in vaccine studies since protection against high- and low-dose exposures may be indicative of significantly different processes.

IMMUNITY TO RICIN

For more than a century, ricin has served as a powerful model to understand the fundamental interactions between toxins and the adaptive arm of the immune system. In the late 1800s, for example, Paul Ehrlich demonstrated that feeding mice or rabbits castor bean "cakes" induced immunity in the animals to subsequent challenges with normally lethal doses of castor bean extract, demonstrating for the first time the possibility of adaptive immunization against a biological toxin. We now appreciate that immunity to ricin is in fact mediated by antibodies, although the underlying mechanisms by which this occurs remain largely unknown (Chanh and Hewetson, 1993; Foxwell et al., 1985a; Godal et al., 1983; Griffiths et al., 1995; Hewetson et al., 1995; Houston, 1982; Lemley et al., 1994; Smallshaw et al., 2002). All of the published vaccine protection data are obtained from challenge studies and passive transfer studies performed in rodents. There are, however, unpublished studies performed by the U.S. Army that indicate nonhuman primates are protected against ricin challenge in an antibody-mediated fashion (Chad Roy and Charles Millard, personal communication). Mice immunized with sublethal doses of ricin (Godal et al., 1983; Hewetson et al., 1993) or formalin-inactivated ricin toxoid (Yan et al., 1995, 1996; Kende et al., 2002; Foxwell et al., 1985b) produce high-titer anti-ricin IgG antibodies in serum and are protected from lethal doses of ricin given i.v., s.c., and intraperitoneally. Protection can also be imparted to mice by passive transfer of polyclonal or certain monoclonal anti-ricin antibodies (Chanh and Hewetson, 1993; Colombatti et al., 1987; Foxwell et al., 1985a; Houston, 1982; Lemley et al., 1994). Polyclonal antibodies against either RTA or RTB are protective (Foxwell et al., 1985b). However, protection does not necessarily correlate with total antitoxin serum antibody titers, suggesting that a specific subset of antitoxin antibodies is responsible for neutralizing ricin in vivo (Smallshaw et al., 2002). It should be noted that the utility of postexposure passive antibody therapy to rescue an animal from the toxic effects of ricin is questionable, as the therapeutic window is likely to be very narrow. For example, rabbit antitoxin polyclonal antisera against ricin toxoid, RTA, or RTB could only rescue mice from ricin intoxication, if given intravenously within 20–40 min of toxin exposure (Foxwell et al., 1985a).

Similarly, results were obtained using a neutralizing murine monoclonal antibody (Chanh et al., 1993).

Monoclonal Antibodies

A number of investigators have produced anti-ricin monoclonal antibodies (MAbs) with the goal of identifying the subpopulation of antibodies involved in neutralization of ricin in vivo (Colombatti et al., 1986; Dertzbaugh et al., 2005; Lebeda and Olson, 1999; Maddaloni et al., 2004; Mantis et al., 2006; McGuinness and Mantis, 2006) (Table 55.3). The most potent MAb identified to date is UNIVAX 70/138, also referred to as R70, which is a mouse IgG_1 directed against RTA. R70 was produced from mice immunized with ricin toxoid and identified as being capable of protecting mouse leukemia cells from the cytotoxic effects of ricin in vitro. The in vitro neutralization activity of R70 proved to be consistent with its in vivo activity. When administered intravenously to mice before or concurrently with a lethal dose of ricin (18μg/kg), R70 fully protected animals from death. It was estimated that protection was achieved when antibody-to-antigen ratios were greater than 80 (w/w) (Lemley et al., 1994). In a subsequent study, Lebeda and Olson (1999) suggest that R70 recognizes an epitope within residues Y91-F108, a solvent alpha helix that borders the toxin's active site. Using peptide array, the epitope specificity has been further confined to residues N97-F108 (C. McGuinness and Mantis, unpublished data) was determined. R70's equilibrium dissociation constant (K_D) to be 3.2×10^{-9}M (McGuinness and Mantis, 2006).

Two other anti-RTA-neutralizing IgG MAbs, referred to as RAC17 and RAC18, have been described by Maddaloni et al. (2004). RAC17 is postulated to recognize amino acids H65–L68 within "domain 1" of the protein (Katzin et al., 1991), whereas RAC18 is proposed to bind a discontinuous epitope consisting of residues A178, Q177, and W211 (Maddaloni et al., 2004). Residue A178 is immediately adjacent to two residues that are part of the toxin's active site, suggesting RAC18 interferes with RTA's enzymatic activity. Indeed, this is supported by in vitro protein synthesis inhibition assays in which RAC18 prevented ricin from ribosome inactivation (Maddaloni et al., 2004). Additional anti-RTA IgG MAbs with complete or partial neutralization activity have been described (Chanh and Hewetson, 1993; Colombatti et al., 1986; Gao et al., 2002; Lemley et al., 1994), but the epitopes recognized by these antibodies have not been identified, nor have they been characterized in vivo. It should also be noted that two reports have identified anti-RTA MAbs that enhance ricin cytotoxicity in vitro and in vivo (Colombatti et al., 1986; Maddaloni et al., 2004).

It remains largely unknown how antibodies against RTA neutralize ricin. R70, for example, reduces ricin attachment to cell surfaces but not to an extent that can account for its potent neutralizing activity (McGuinness and Mantis, 2006). RAC18 interferes with RTA's capacity to arrest protein synthesis in vitro, yet it has not been investigated if toxin–MAb immunocomplexes can be internalized, trafficked, and translocated across the ER membrane into the host cell cytoplasm without disassociating. It is more likely that anti-RTA MAbs interfere with steps involved in intracellular trafficking or retrotranslocation, possibly

TABLE 55.3 Characterized anti-ricin MAbs

Specificity	MAb	Isotype and subclass	Epitope	Neutralizing in vitro	Neutralizing in vivo	References
Ricin	BG11-G2	IgG_1	n.d.	+++	+++	Chanh et al. (1993)
RTA	UNIVAX 70/138 (R70)	IgG_1	N97-F108	+++	+++	Lemley et al., (1994), Lebeda and Olson, (1999)
	RAC17	IgG_1	H66-L69	+++	+++	Maddaloni et al. (2004)
	RAC18	IgG_{2a}	Discontinuous	+++	+++	Maddaloni et al. (2004)
	RAC23	IgG_1	GTXS	+++	–	Maddaloni et al. (2004)
	23D7	IgA	n.d.	+++	n.d.	Mantis et al. (2006)
	25A4	IgA	N97-F108	+++	n.d.	Mantis et al. (2006)
RTB	75/3B12	IgG_{2a}	n.d.	+++	+++	(Colombatti et al. (1987)
	24B11	IgG_1	P38-T43	+++	n.d.	McGuinness and Mantis (2006)
	TFTB-1	IgG_1	P249-H251	–	–	Fulton et al. (1986)
	33G2	IgA	n.d.	+++	n.d.	Mantis et al. (2006)
	35H6	IgA	R235-S241	+++	n.d.	Mantis et al. (2006)

Note: The properties and epitope specificities of monoclonal antibodies to RTA and RTB.

by blocking ricin's association with host molecules involved in these processes (Slominska-Wojewodzka et al., 2006; Utskarpen et al., 2006). In support of this model, there are some preliminary results using Vero cells that toxin–MAb immunocomplexes formed extracellularly are internalized and accumulate within the Golgi apparatus (C. McGuinness and Mantis, unpublished data).

Numerous MAbs against RTB have been described, although only two have been characterized in detail (Colombatti et al., 1986; Maddaloni et al., 2004; Mantis et al., 2006; McGuinness and Mantis, 2006). The MAb 75/3B12 is a murine IgG_1 that blocks ricin attachment to cell surfaces in vitro (Colombatti et al., 1987). Antibody recognition of RTB was competitively inhibited by the addition of lactose, but not sucrose or glucose, suggesting 75/3B12 recognizes one or both of RTB's galactose-binding domains. 75/3B12 protected mice against a lethal injection with ricin (270 ng of toxin/per animal given i.v.), although it was not as protective as R70 (Colombatti and Dosio, 2001; Lemley et al., 1994). The second well-characterized anti-RTB MAb is 24B11 (McGuinness and Mantis, 2006). In vitro, 24B11 is approximately twice as effective at neutralizing ricin as R70. The equilibrium dissociation constants of 24B11 ($K_D = 4.2 \times 10^{-9}$ M) is virtually identical to that of R70, indicating the relative differences in neutralizing activity are not due to differences in affinity for ricin. Like 75/3B12, 24B11 blocks ricin attachment to galactoside receptors on primary mouse splenocytes and on the apical surfaces of human mucosal epithelial cell monolayers. Unlike 75/3B12, however, the binding of 24B11 to RTB is not affected by galactose or galactose-containing glycoproteins. Using a phage-displayed peptide library, it was determined that 24B11 binds an epitope on RTB adjacent to, but not within, one of the two galactose-binding domains, which has led us to propose that 24B11 neutralizes ricin by steric hindrance (i.e., physically obstructing RTB from attaching to galactose residues on cell surfaces). It should be noted that the association of any MAb with RTB is not sufficient to interfere with toxin attachment. This is exemplified by TFTB-1, a murine IgG_1 MAb directed against RTB, which binds ricin as well as 24B11 but shows no detectable neutralizing activity (McGuinness and Mantis, 2006).

Mucosal Immunity to Ricin

The mucosal epithelia of the respiratory and GI tracts are vulnerable to ricin intoxication, and protecting these tissues is an important consideration when developing a ricin vaccine (Mantis et al., 2006). There is a general consensus that protection against lung injury requires antibodies in respiratory secretions to effectively intercept ricin before it can adversely affect the respiratory epithelium (Griffiths et al., 1997, 1998, 1999; Poli et al., 1996; Yan et al., 1996). For example, animals immunized parenterally (i.e., subcutaneously) with ricin toxoid were protected against a lethal aerosol challenge but suffered significant lung damage (Hewetson et al., 1995; Poli et al., 1996). Similar experiments conducted in rats immunized with ricin toxoid revealed little lung damage in immunized animals 28 days after challenge, although rats developed some symptoms during the first week after challenge (Griffiths et al., 1995). Protection against lethal challenge was associated with high titers of serum anti-ricin antibodies. Protection against lung lesions was accomplished by passively pretreating animals with an aerosolized goat IgG cocktail enriched with anti-ricin antibodies (Poli et al., 1996). Active vaccination of rats intratracheally with liposomal formulations of ricin toxoid or plant-derived RTA was associated with increased antitoxin secretory immunoglobulin A (SIgA) in lungs and protection against intratracheal challenge with ricin that was correlated with decrees in neutrophil influx into the lungs (Griffiths et al., 1997, 1998, 1999). Similar studies with ricin toxoid encapsulated in poly-lactide co-glycolide microspheres in mice vaccinated orally or intranasally have also suggested an association of more effective protection against aerosol challenge by stimulation of local SIgA (Yan et al., 1996; Kende et al., 2002).

The concept that antitoxin serum antibodies protect against death but not damage to the respiratory mucosa is reinforced by a recent paper by Smallshaw et al. (2007) in which mice were immunized intramuscularly with an attenuated RTA vaccine (RiVax) and then challenged with $10LD_{50}$s of ricin by aerosol. All animals in groups receiving the two highest doses of vaccine (10 or 3 μg three times at 4-week intervals) survived challenge, whereas other groups suffered 20–80% death rate. However, even animals immunized with the highest dose of vaccine suffered some lung damage, as assessed by plethysmography (a measure of lung function) and histological examination. Survival and lung function were correlated to total serum anti-RTA antibodies. Lung function in mice vaccinated with high-dose vaccines and having the highest anti-RTA titers returned to baseline values 3 days following aerosol exposure, although there was evidence, 21 days following exposure of mild epithelial hyperplasia, suggestive of ongoing lung repair. The actual contribution of SIgA in protecting the respiratory epithelium will not be known until similar vaccination studies are performed in available IgA- and polymeric immunoglobulin receptor (pIgR)-deficient mice (Mbawuike et al., 1999).

In the intestinal tract, mucosal immunity to ricin in mice correlates with antitoxin IgG and IgA levels in serum as well as elevated antitoxin IgA antibodies in intestinal secretions (Yoder et al., 2007). Intragastric immunization with ricin toxoid elicits IgA antibodies against RTA and RTB that are capable of protecting polarized epithelial cell monolayers from ricin in vitro. In tissue section overlay assays, IgA MAbs against RTB blocked ricin attachment to the luminal surfaces of human duodenum, suggesting that if these antibodies were present in mucosal secretions, they would be capable of interfering with the earliest steps in the intoxication process (Mantis et al., 2006). IgA MAbs directed against RTA are equally effective at neutralizing ricin in vitro, although they have little effect on toxin attachment to epithelial cell surfaces. The availability of anti-ricin IgG and IgA MAbs (Mantis et al., 2006), in combination with a well-characterized mouse model (Yoder et al., 2007), now permits us to identify the isotypes and epitope specificities of antibodies involved in intestinal immunity to ricin in vivo.

VACCINE IMMUNOGEN CANDIDATES

The molecular immunogens proposed and studied in animal models as vaccine candidates over the last decade have progressed from plant-derived toxoid, ricin holotoxin, to isolated plant-derived subunits, such as chemically deglycosylated RTA, to recombinant derivatives of RTA and RTB. A variety of formulation approaches have been advocated with the intent of delivering as either a conventional injected vaccine with a conventional approach to immunological adjuvants, or more exploratory delivery approaches that are intended to alter the route of administration, principally for the stimulation of mucosal immunity in the lungs and the GI tract or to generate more long-lasting and robust immunity in fewer doses. In addition, the formulation approach to subunit vaccine development is aimed at providing chemical and conformational stability to the candidate immunogens. To date, the immunogens that are candidates for more advanced development have centered on the biophysical, biochemical, and immunological properties of RTA derivatives that lack either enzymatic or VLS activity (Vitetta et al., 2006; Carra et al., 2007). The focus on RTA derivatives is supported by a number of studies that indicate that protection, associated with the induction of serum neutralizing antibodies and perhaps neutralizing antibodies in mucosal secretions, can be generated by active vaccination with the A chain and supported by some of the studies described above with the high neutralizing activity of several MAbs to RTA. There are few published studies that directly compare the protection induced by RTA and RTB, but one recent study (Maddaloni et al., 2004) demonstrates a more robust protective immune response to RTA in comparison to RTB. It is intuitive that antibodies to RTB may be more efficient at the inhibition of ricin binding to galactosyl cellular receptors. However, it is surprising that RTA antibodies protect animals and neutralize in vitro unless one considers that uptake through the MR is mediated through RTA and takes into account that antibodies to the A chain may modulate the intracellular trafficking of RTA.

Ricin Toxoid and dgRTA

Treatment of ricin toxin with formalin results in a toxoid that induces protective immunity in rodents and primates (Hewetson, 1996; Franz and Jaax, 1997). Although active immunization with the toxoid vaccine protects against lethal inhalation of ricin toxin in mice, the vaccine may only partially protect against lung damage (Hewetson et al., 1995; Poli et al., 1996). Data regarding the toxoid vaccine made by the U.S. Army Medical Research Institute of Infectious Diseases (USAMRIID) were submitted to the Food and Drug Administration (FDA) in 1995. However, clinical trials were not done with this vaccine due to concern over residual toxicity. The vaccine may not be completely inactivated, and it was difficult to obtain reproducibly inactivated batches of ricin. Published data have suggested that long-term treatment with formaldehyde may be necessary to substantially inactivate ricin (Griffiths et al., 1997). Plant-derived ricin holotoxin is extensively glycosylated, and it is well established that chemically deglycosylated native RTA (dgRTA) is effective at protecting mice against systemically administered and aerosolized ricin, although surviving mice expressed some lung damage as evidenced by elevations of albumin in bronchial fluids. Immunized mice had high levels of anti-RTA antibody in their sera (Hewetson, 1996). Bronchial fluids were not evaluated for antibody levels. dgRTA is stable as a lyophilized formulation for 22 months and after reconstitution is stable for 24 h at 4°C. Safety testing in rodents showed that female mice lost more weight than male mice, but that weight loss was not dose related and resulted from decreased food and water intakes. A single i.m. injection was associated with a local inflammatory response and degenerative changes in the muscle that continued beyond 15 days. These changes could be related to direct muscle damage since similar symptoms have been observed in patients treated with dgRTA-containing immunotoxins (Sauseville and Vitetta, 1997). There was mild but reversible neutropenia and limited single cell necrosis in the intestinal crypt epithelium that resolved in 15 days. Hence, although this vaccine had efficacy, it also had some toxicity. In addition, it is expensive and dangerous to make, and commercial sources of dgRTA are no longer available.

Recombinant Vaccines

There are several recombinant vaccines that are in various stages of development by different groups. Three of these candidates have been recently described in the literature, and all of them are derived from RTA (McHugh et al., 2004; Smallshaw et al., 2002; Marsden et al., 2004) The isolated A and B subunits of ricin can be produced in *E. coli* and other recombinant hosts. RTB is also a candidate for inclusion in a vaccine, especially when protection is focused on mucosal compartments (Maddaloni et al., 2004). Although RTA in the absence of RTB is at least 1000-fold less toxic than the native ricin, it still retains residual enzymatic activity that may result in toxicity when used as a vaccine (Thorpe et al., 1985).

One approach to making an RTA vaccine has involved creating a double mutant of RTA containing both a single mutation in its enzymatic site (Y80A) and mutations in the vascular leak site (V76M). Several variants have been initially characterized, of which one has been studied in great detail and produced for clinical trials (Smallshaw et al., 2002). This vaccine has been given the trade name RiVax, has progressed through preliminary mouse protection studies and rabbit good laboratory practice (GLP) toxicology, and has been evaluated in a Phase I safety and immunogenicity trial in a "trial" formulation.

Another RTA-based vaccine candidate has been produced by inserting a peptide into a surface-exposed loop of RTA distal from the active site. This mutant has substantially reduced catalytic activity, resistance to proteolysis, and virtually no cytotoxicity. It induced immunity to systemic ricin challenge in mice (Marsden et al., 2004).

A third approach to an RTA vaccine has involved a structural approach to eliminate parts of the A chain that are implicated in low thermostability of RTA (McHugh et al., 2004). This particular candidate contains an internal deletion that improves solubility and an extensive deletion in the carboxy-terminal region, resulting in stability in aqueous solutions under a variety of conditions. This vaccine candidate is more stable to thermal denaturation than the parental native RTA and protects mice against ricin lethality (Olson et al., 2004).

Each of these candidates protects mice from ricin exposure and appears to contain the epitopes that have been characterized from the study of the human response of cancer patients given multiple doses of RTA-antibody conjugate immunotoxin to which antibodies to the A chain had developed. These studies have determined that there is a single, major linear B cell determinant in the A chain that is associated with ricin neutralization in cell-based assays (L161-E185) (Tommasi et al., 2001; Castelletti et al., 2004). This epitope is contained in a single helical loop that impinges upon the active site. A second epitope described by the peptides that bound to a neutralizing MAb, located in an exposed loop (residues 95–110) (Lebeda and Olson, 1999), is also present in each of the vaccine candidates. These two epitopes have been characterized as being involved in ricin neutralization. The neutralizing antibody that has been described by Pincus (MAb 18) (Maddaloni et al., 2004) and discussed above is thought to overlap the active site but has not been mapped to a linear peptide. It may be a conformationally dependent neutralizing epitope. There may be additional conformationally dependent epitopes or other linear B and T cell epitopes that have not yet been characterized in animal or human studies. The introduction of mutations in or near the active site has the possibility of disrupting nearby B cell epitopes, which raises the concern that vaccine immunogenicity can be affected in some of the candidates (Marsden et al., 2005).

PRECLINICAL AND CLINICAL STUDIES

There are several key factors in manufacture and formulation that impinge upon the development of any of the ricin vaccine candidates mentioned above. The success of vaccine studies in the research stage can reveal that a vaccine has the appropriate characteristics to induce protective immunity in a relevant animal model and later in development considerations of manufacture and formulation with adjuvants, and delivery systems become extremely important as the vaccine progresses toward Phase I and II human testing (and pivotal animal trials). RTA is known to be extremely thermally unstable in aqueous buffers, but conditions that allow for retention of maximal conformational stability can be identified (Peek et al., 2007) for molecules closely related in structure to RTA, such as RiVax immunogen (Smallshaw et al., 2005). This vaccine candidate induces rabbit and human antibodies that are neutralizing in an in vitro cell-based assay, which indicates the total functionality of the antibodies (Vitetta et al., 2006; Smallshaw et al., 2005) and is likely to predict the capacity of a serum to protect an animal. Furthermore, those antibodies, purified and passively transferred to naive mice, protect against parenteral challenge.

A high-yield current good manufacturing practice (cGMP) fermentation and purification process for RiVax has been developed and implemented at a scale anticipated to be in the range for commercial production (Brey et al., unpublished data). It is now known, based on crystal structure data, that this protein is identical in structure to native RTA (Charles Millard, unpublished data), indicating that the point mutations contained in the molecule do not disrupt any potential tertiary structure (or potential conformationally dependent epitopes). At the same time, this molecule is known to induce antibodies that compete with known neutralizing MAbs, suggesting that the important complement of structural epitopes is still contained within the molecule. The aqueous stability of RiVax can be improved by the addition of excipients and the optimization of pH and salt conditions, which among other things can modify water structure, and the drug substance (the protein per se) can be maintained in solution in the presence of glycerol for up to 2 years or longer without loss of conformation and physical properties. The physical properties of the protein are associated with the ability of the protein to induce protective immunity in animal models and protective antibodies in humans. The presence of glycerol results in a 10°C increase in the melting temperature of RTA, the point at which the protein begins to unfold and aggregate irreversibly. The relevance of stability in solution is thought to relate to long-term stability that would be required if the vaccine were produced for a stockpile that would be used only under emergency conditions or in very limited target populations. The real stability issue concerns the final formulation of the vaccine rather than the solution stability per se, and the stability concerns the endpoints that are not necessarily totally related to conformation, that is, the ability to induce protective immunity or correlates of protective immunity. The ideal vaccine would have high thermal stability that would be predictive of long-term storage under ambient conditions or under refrigeration, but those conditions must be evaluated with the final clinical or marketable formulations.

The RTA1-33/44-198 vaccine candidate has elements of built-in solution stability resulting from the deletion of regions and residues that otherwise are obscured by RTB in the holotoxin (McHugh et al., 2004; Carra et al., 2007). This vaccine has also been formulated with aluminum salt adjuvants, and recent evidence suggests that it contains long-term stability while bound to aluminum adjuvants in terms of potency in a mouse model of protection against ricin challenge.

A recent study of RiVax was conducted with an investigational formulation that did not contain an immunological adjuvant and was stabilized against gross aggregation by the inclusion of sucrose and Tween detergent (Vitetta et al., 2006). A safety and immunogenicity trial was performed with this formulation with the motive of addressing the potential toxicity of the subunit in the absence of any potential interference from an adjuvant. The vaccine was safely tolerated at doses up to 100 μg given three times at monthly intervals, with little evidence of local side effects attributable to the immunogen. More importantly, the vaccine stimulated seroconversion in a dose-dependent manner, and antibodies purified from seroconverters mixed with ricin toxin were protective upon transfer to mice. The fact that human serum also contained antibodies that neutralized ricin in a cell-based assay (Daudi cells, a human B cell line) indicates that humans, rabbits, and mice are capable of generating protective and functional antibodies elicited by a totally inactive RTA. More recent unpublished studies have indicated that RiVax, similarly to RTA1-33/44-198, maintains potency in a dose-dependent manner for long periods of time when adsorbed to aluminum hydroxide (Alhydrogel®), suggesting that adsorption of a relatively unstable protein immunogen to the crystalline surface may aid in the long-term stability of the vaccine. This aluminum adjuvant version of this vaccine is currently being evaluated in nonhuman primate trials to examine the potential of this vaccine to protect against lethality and lung damage after parenteral vaccination. Concurrently, human trials are underway to establish an optimized dosage regimen and dose level that can be correlated to protection in the most relevant animal model.

The most relevant animal model is a nonhuman primate model, and Rhesus macaque and African green monkey models have been developed (Chad Roy, Tulane University, personal communication), but other mammalian species, such as the rabbit, develop neutralizing antibodies in response to RTA vaccines (Smallshaw et al., 2005) and are susceptible to ricin. In developing the correlates of immunity and pathology during vaccine development, there is no way to know the actual dose of ricin to which a human might become exposed, but there is a way of examining extreme situations of high-dose exposure in animal models and beginning to correlate the outcomes of those extremes with immune responses in humans. For example, if a certain vaccine dosage regimen is shown to protect against a high-dose aerosol challenge that induces significant pathology in unvaccinated animals, the levels and duration of antibodies in serum and secretions can be determined and used in functional assays of antibody activity, including avidity, neutralization, and passive transfer to naive animals. However, the actual measure of pathology, such as inflammatory markers, local cytokines, chemokines, influx of neutrophils, and microarray analysis of upregulated genes, can be performed only in a challenged animal.

Further development of these vaccine candidates in human clinical trials is dependent on establishing the relevant correlates of immunity in animal models that will include endpoints of lung pathology, inflammation markers, as well as the determination of functional antibodies in serum and mucosal secretions.

PROSPECTS FOR THE FUTURE

Recombinant subunit vaccines for ricin based on RTA have been shown to induce antibodies that are protective in rodent models and correlated with protection against ricin exposure by aerosol and oral exposure in rabbits and humans. That is, the antibodies present in the serum are neutralizing in vitro and can passively convey protection to naive animals. The role of secretory antibodies and possibly other effector mechanisms may implicate compartmentalization of responses and suggest that these antibodies can function to prevent binding of toxin to luminal cell surfaces. Thus, the role of secretory antibodies in preventive vaccination may be important in protecting epithelial surfaces against toxin exposure. Equally, the possible transudation of antitoxin IgG across luminal surfaces may aid to inactivate toxin prior to binding. Therefore, it may be possible to induce protective immunity to mucosal tissue damage, primarily through stimulation of humoral antibodies. This remains to be thoroughly tested in animal models. The current data indicate that RTA subunit vaccines, in common with prior vaccine candidates, can induce protection against lethal toxin exposure but may not completely prevent all morbidity; this, of course, is related to toxin dose exposure.

A vaccine for ricin toxin will have utility solely as a biodefense countermeasure, as there is no burden of human or animal disease. The existence of vaccines in an advanced state of development and readiness for production is key to several key objectives: first, to provide a deterrent to further development of the toxin weapon and, second, to provide a vaccine that can actually be used in emergency situations prior to marketing authorization by the FDA and other regulatory authorities while the vaccine is proceeding through pivotal animal and human safety/immunogenicity trials. A vaccine for ricin may provide the best approach since postexposure therapy must be instituted within a very narrow window after exposure, and once the toxin has stimulated an inflammatory cascade, it cannot be counteracted by antitoxin antibodies. The similarities of the action of ricin and Shiga toxins and the connections to the induction of inflammatory cascades provide additional leads into areas of therapeutics development for ricin poisoning per se and for HUS.

KEY ISSUES

- Long-term stability of vaccine formulations. A stockpile of vaccine that is not intended to be used in routine immunization must have long-term stability, possibly at storage temperatures that avoid the cold chain.
- Determination of the correlates of immunity in appropriate animal models. The U.S. FDA has mandated development of well-understood animal models that are to be used as correlates of protection in human vaccine studies. The correlates of immunity involve most likely antibodies that can be evaluated for functionality in vitro and in vivo by passive transfer. A vaccine for ricin would probably have an indication that involves the description of a respiratory exposure endpoint.
- The major question in human studies is the magnitude as duration of protective antibodies and how to induce them most effectively
- A parenterally administered vaccine, for instance, by i.m. injection, is expected to induce primarily serum antibodies to the subunit and little secretory IgA. The role of antibodies present at the mucosal surface arising from parenteral vaccination is not clear and may be critical to the degree of protection to aerosolized toxin.

References

Argent, R.H., Parrott, A.M., Day, P.J., Roberts, L.M., Stockley, P.G., Lord, J.M. and Radford, S.E. Ribosome-mediated folding of partially unfolded ricin A-chain. J. Biol. Chem. 2000; 275: 9263–9269.

Arnon, S.S., Schechter, R., Inglesby, T.V., Henderson, D.A., Bartlett, J.G., Ascher, M.S., Eitzen, E., Fine, A.D., Hauer, J., Layton, M., Lillibridge, S., Osterholm, M.T., O'Toole, T., Parker, G., Perl, T.M., Russell, P.K., Swerdlow, D.L. and Tonat, K. Botulinum toxin as a biological weapon: medical and public health management. JAMA 2001; 285:1059–1070.

Audi, J., Belson, M., Patel, M., Schier, J. and Osterloh, J. Ricin poisoning: a comprehensive review. JAMA 2005; 294:2342–2351.

Baenziger, J.U. and Fiete, D. Structural determinants of *Ricinus communis* agglutinin and toxin specificity for oligosaccharides. J. Biol. Chem. 1979; 254:9795–9799.

Baggiolini, M. Chemokines in pathology and medicine. J. Intern. Med. 2001; 250:91–104.

Baggiolini, M. and Loetscher, P. Chemokines in inflammation and immunity. Immunol. Today 2000; 21:418–420.

Baggiolini, M., Loetscher, P. and Moser, B. Interleukin-8 and the chemokine family. Int. J. Immunopharmacol. 1995; 17:103–108.

Balint, G.A. Ricin: the toxic protein of castor oil seeds. Toxicology 1974; 2:77–102.

Baluna, R., Coleman, E., Jones, C., Ghetie, V. and Vitetta, E.S. The effect of a monoclonal antibody coupled to ricin A chain-derived peptides on endothelial cells in vitro: insights into toxin-mediated vascular damage. Exp. Cell Res. 2000; 258:417–424.

Baluna, R., Rizo, J., Gordon, B.E., Ghetie, V. and Vitetta, E.S. Evidence for a structural motif in toxins and interleukin-2 that may be responsible for binding to endothelial cells and initiating vascular leak syndrome. Proc. Natl. Acad. Sci. U.S.A. 1999; 96:3957–3962.

Baluna, R. and Vitetta, E.S. An in vivo model to study immunotoxin-induced vascular leak in human tissue. J. Immunother. 1999; 22:41–47.

Barbieri, L., Battelli, M.G. and Stirpe, F. Ribosome-inactivating proteins from plants. Biochim. Biophys. Acta 1993; 1154:237–282.

Beaumelle, B., Taupiac, M.P., Lord, J.M. and Roberts, L.M. Ricin A chain can transport unfolded dihydrofolate reductase into the cytosol. J. Biol. Chem. 1997; 272:22097–22102.

Bellisola, G., Fracasso, G., Ippoliti, R., Menestrina, G., Rosen, A., Solda, S., Udali, S., Tomazzolli, R., Tridente, G. and Colombatti, M. Reductive activation of ricin and ricin A-chain immunotoxins by protein disulfide isomerase and thioredoxin reductase. Biochem. Pharmacol. 2004; 67:1721–1731.

Bischoff, S.C., Krieger, M., Brunner, T. and Dahinden, C.A. Monocyte chemotactic protein 1 is a potent activator of human basophils. J. Exp. Med. 1992; 175:1271–1275.

Bradberry, S.M., Dickers, K.J., Rice, P., Griffiths, G.D. and Vale, J.A. Contrib Title: Ricin poisoning. Toxicol. Rev. 2003; 22:65–70.

Brown, R.F. and White, D.E. Ultrastructure of rat lung following inhalation of ricin aerosol. Int. J. Exp. Pathol. 1997; 78:267–276.

Carr, M.W., Roth, S.J., Luther, E., Rose, S.S. and Springer, T.A. Monocyte chemoattractant protein 1 acts as a T-lymphocyte chemoattractant. Proc. Natl. Acad. Sci. U.S.A. 1994; 91:3652–3656.

Carra, J.H., Wannemacher, R.W., Tammariello, R.F., Lindsey, C.Y., Dinterman, R.E., Schokman, R.D. and Smith, L.A. Improved formulation of a recombinant ricin A-chain vaccine increases its stability and effective antigenicity. Vaccine 2007; 25:4149–4158.

Castelletti, D., Fracasso, G., Righetti, S., Tridente, G., Schnell, R., Engert, A. and Colombatti, M. A dominant linear B-cell epitope of ricin A-chain is the target of a neutralizing antibody response in Hodgkin's lymphoma patients treated with an anti-CD25 immunotoxin. Clin. Exp. Immunol. 2004; 136:365–372.

Chanh, T.C. and Hewetson, J.F. Polyclonal anti-idiotypes induce antibody responses protective against ricin cytotoxicity. Immunology 1993; 79:681–684.

Chanh, T.C., Romanowski, M.J. and Hewetson, J.F. Monoclonal antibody prophylaxis against the in vivo toxicity of ricin in mice. Immunol. Invest. 1993; 22:63–72.

Chaudhary, V.K., Jinno, Y., FitzGerald, D. and Pastan, I. Pseudomonas exotoxin contains a specific sequence at the carboxyl terminus that is required for cytotoxicity. Proc. Natl. Acad. Sci. U.S.A. 1990; 87:308–312.

Chen, A., AbuJarour, R.J. and Draper, R.K. Evidence that the transport of ricin to the cytoplasm is independent of both Rab6A and COPI. J. Cell Sci. 2003; 116:3503–3510.

Cherla, R.P., Lee, S.Y., Mees, P.L. and Tesh, V.L. Shiga toxin 1-induced cytokine production is mediated by MAP kinase pathways and translation initiation factor eIF4E in the macrophage-like THP-1 cell line. J. Leukoc. Biol. 2006; 79:397–407.

Christiansen, V.J., Hsu, C.H., Dormer, K.J. and Robinson, C.P. The cardiovascular effects of ricin in rabbits. Pharmacol. Toxicol. 1994; 74:148–152.

Colombatti, M. and Dosio, F. Synthesis of monensin derivatives and their effect on the activity of ricin A-chain immunotoxins. Methods Mol. Biol. 2001; 166:55–70.

Colombatti, M., Johnson, V.G., Skopicki, H.A., Fendley, B., Lewis, M.S. and Youle, R.J. Identification and characterization of a monoclonal antibody recognizing a galactose-binding domain of the toxin ricin. J. Immunol. 1987; 138:3339–3344.

Colombatti, M., Pezzini, A. and Colombatti, A. Monoclonal antibodies against ricin: effects on toxin function. Hybridoma 1986; 5:9–19.

Connor, S.J., Paraskevopoulos, N., Newman, R., Cuan, N., Hampartzoumian, T., Lloyd, A.R. and Grimm, M.C. CCR2 expressing CD4+ T lymphocytes are preferentially recruited to the ileum in Crohn's disease. Gut 2004; 53:1287–1294.

Crompton, R. and Gall, D. Georgi Markov—death in a pellet. Med. Leg. J. 1980; 48:51–62.

DaSilva, L., Cote, D., Roy, C., Martinez, M., Duniho, S., Pitt, M.L., Downey, T. and Dertzbaugh, M. Pulmonary gene expression profiling of inhaled ricin. Toxicon 2003; 41:813–822.

Day, P.J., Owens, S.R., Wesche, J., Olsnes, S., Roberts, L.M. and Lord, J.M. An interaction between ricin and calreticulin that may have implications for toxin trafficking. J. Biol. Chem. 2001; 276:7202–7208.

Deeks, E.D., Cook, J.P., Day, P.J., Smith, D.C., Roberts, L.M. and Lord, J.M. The low lysine content of ricin A chain reduces the risk of proteolytic degradation after translocation from the endoplasmic reticulum to the cytosol. Biochemistry (Mosc.) 2002; 41:3405–3413.

Dertzbaugh, M.T., Rossi, C.A., Paddle, B.M., Hale, M., Poretski, M. and Alderton, M.R. Monoclonal antibodies to ricin: in vitro inhibition of toxicity and utility as diagnostic reagents. Hybridoma 2005; 24:236–243.

Di Cola, A., Frigerio, L., Lord, J.M., Ceriotti, A. and Roberts, L.M. Ricin A chain without its partner B chain is degraded after retrotranslocation from the endoplasmic reticulum to the cytosol in plant cells. Proc. Natl. Acad. Sci. U.S.A. 2001; 98:14726–14731.

Doebler, J.A., Wiltshire, N.D., Mayer, T.W., Estep, J.E., Moeller, R.B., Traub, R.K., Broomfield, C.A., Calamaio, C.A., Thompson, W.L. and Pitt, M.L. The distribution of [^{125}I]ricin in mice following aerosol inhalation exposure. Toxicology 1995; 98:137–149.

East, L. and Isacke, C.M. The mannose receptor family. Biochim. Biophys. Acta 2002; 1572:364–386.

Endo, Y., Mitsui, K., Motizuki, M. and Tsurugi, K. The mechanism of action of ricin and related toxic lectins on eukaryotic ribosomes. The site and the characteristics of the modification in 28 S ribosomal RNA caused by the toxins. ssue Series Title: J. Biol. Chem. 1987; 262:5908–5912.

Endo, Y. and Tsurugi, K. RNA N-glycosidase activity of ricin A-chain. Mechanism of action of the toxic lectin ricin on eukaryotic ribosomes. J. Biol. Chem. 1987; 262:8128–8130.

Fodstad, O., Olsnes, S. and Pihl, A. Toxicity, distribution and elimination of the cancerostatic lectins abrin and ricin after parenteral injection into mice. Br. J. Cancer 1976; 34:418–425.

Foxwell, B.M., Blakey, D.C., Brown, A.N., Donovan, T.A. and Thorpe, P.E. The preparation of deglycosylated ricin by recombination of glycosidase-treated A- and B-chains: effects of deglycosylation on toxicity and in vivo distribution. Biochim. Biophys. Acta 1987; 923:59–65.

Foxwell, B.M., Detre, S.I., Donovan, T.A. and Thorpe, P.E. The use of anti-ricin antibodies to protect mice intoxicated with ricin. Toxicology 1985a; 34:79–88.

Foxwell, B.M., Detre, S.I., Donovan, T.A. and Thorpe, P.E. The use of anti-ricin antibodies to protect mice intoxicated with ricin. Toxicology 1985b; 34:79–88.

Foxwell, B.M., Donovan, T.A., Thorpe, P.E. and Wilson, G. The removal of carbohydrates from ricin with endoglycosidases H, F and D and alpha-mannosidase. Biochim. Biophys. Acta 1985c; 840:193–203.

Frankel, A.E., Fu, T., Burbage, C., Tagge, E., Harris, B., Vesely, J. and Willingham, M.C. Lectin-deficient ricin toxin intoxicates cells bearing the D-mannose receptor. Carbohydr. Res. 1997; 300:251–258.

Franz, D.R. and Jaax, N.K. In: Textbook of Military Medicine (F.R. Sidell, E.T. Takafuji and D.R. Franz, Eds.), pp. 631–642. Washington, DC: Office of the Surgeon General, Department of the Army, 1997.

Fulton, R.J., Blakey, D.C., Knowles, P.P., Uhr, J.W., Thorpe, P.E. and Vitetta, E.S. Purification of ricin A1, A2, and B chains and characterization of their toxicity. J. Biol. Chem. 1986; 261:5314–5319.

Gao, C., Mao, S., Kaufmann, G., Wirsching, P., Lerner, R.A. and Janda, K.D. A method for the generation of combinatorial antibody libraries using pIX phage display. Proc. Natl. Acad. Sci. U.S.A. 2002; 99:12612–12616.

Godal, A., Fodstad, O. and Pihl, A. Antibody formation against the cytotoxic proteins abrin and ricin in humans and mice. Int. J. Cancer 1983; 32:515–521.

Gonzalez, T.V., Farrant, S.A. and Mantis, N.J. Ricin induces IL-8 secretion from human monocyte/macrophages by activating the p38 MAP kinase pathway. Mol. Immunol. 2006; 43:1920–1923.

Griffiths, G.D., Bailey, S.C., Hambrook, J.L. and Keyte, M.P. Local and systemic responses against ricin toxin promoted by toxoid or peptide vaccines alone or in liposomal formulations. Vaccine 1998; 16:530–535.

Griffiths, G.D., Bailey, S.C., Hambrook, J.L., Keyte, M., Jayasekera, P., Miles, J. and Williamson, E. Liposomally-encapsulated ricin toxoid vaccine delivered intratracheally elicits a good immune response and protects against a lethal pulmonary dose of ricin toxin. Vaccine 1997; 15:1933–1939.

Griffiths, G.D., Lindsay, C.D., Allenby, A.C., Bailey, S.C., Scawin, J.W., Rice, P. and Upshall, D.G. Protection against inhalation toxicity of ricin and abrin by immunisation. Hum. Exp. Toxicol. 1995; 14:155–164.

Griffiths, G.D., Phillips, G.J. and Bailey, S.C. Comparison of the quality of protection elicited by toxoid and peptide liposomal vaccine formulations against ricin as assessed by markers of inflammation. Vaccine 1999; 17:2562–2568.

Grisham, M.B. and Granger, D.N. Neutrophil-mediated mucosal injury. Role of reactive oxygen metabolites. Dig. Dis. Sci. 1988; 33:6S–15S.

Guessous, F., Marcinkiewicz, M., Polanowska-Grabowska, R., Kongkhum, S., Heatherly, D., Obrig, T. and Gear, A.R. Shiga toxin 2 and lipopolysaccharide induce human microvascular endothelial cells to release chemokines and factors that stimulate platelet function. Infect. Immun. 2005; 73:8306–8316.

Harrison, L.M., van den Hoogen, C., van Haaften, W.C. and Tesh, V.L. Chemokine expression in the monocytic cell line THP-1 in response to purified shiga toxin 1 and/or lipopolysaccharides. Infect. Immun. 2005; 73:403–412.

Harrison, L.M., van Haaften, W.C. and Tesh, V.L. Regulation of proinflammatory cytokine expression by Shiga toxin 1 and/or lipopolysaccharides in the human monocytic cell line THP-1. Infect. Immun. 2004; 72:2618–2627.

Hazen, E.L. General and local immunity to ricin. J. Immunol. 1927; 13:171–218.

Hewetson, J. Immune response and protection of rats and mice from ricin aerosol exposure after vacccination with deglyglosylated ricin A chain. Fundam. Appl. Toxicol. 1996; 30S:59.

Hewetson, J.F., Rivera, V.R., Creasia, D.A., Lemley, P.V., Rippy, M.K. and Poli, M.A. Protection of mice from inhaled ricin by vaccination with ricin or by passive treatment with heterologous antibody. Vaccine 1993; 11:743–746.

Hewetson, J.F., Rivera, V.R., Lemley, P.V., Pitt, M.L., Creasia, D.A. and Thompson, W.L. A formalinized toxoid for protection of mice from inhaled ricin. Vaccine Res. 1995; 4:179–187.

Houston, L.L. Protection of mice from ricin poisoning by treatment with antibodies directed against ricin. J. Toxicol. Clin. Toxicol. 1982; 19:385–389.

Hu, L. and Hickey, T.E. Campylobacter jejuni induces secretion of proinflammatory chemokines from human intestinal epithelial cells. Infect. Immun. 2005; 73:4437–4440.

Ishiguro, K., Ho, P.T. and Sartorelli, A.C. Characterization of the defect in a variant of HL-60 promyelocytic leukemia cells with reduced transferrin receptor expression. Somat. Cell Mol. Genet. 1992a; 18:45–63.

Ishiguro, M., Harada, H., Ichiki, O., Sekine, I., Nishimori, I. and Kikutani, M. Effects of ricin, a protein toxin, on glucose absorption by rat small intestine (biochemical studies on oral toxicity of ricin. II). Chem. Pharm. Bull. (Tokyo) 1984; 32:3141–3147.

Ishiguro, M., Matori, Y., Tanabe, S., Kawase, Y., Sekine, I. and Sakakibara, R. Biochemical studies on oral toxicity of ricin. V. The role of lectin activity in the intestinal absorption of ricin. Chem. Pharm. Bull. (Tokyo) 1992b; 40:1216–1220.

Ishiguro, M., Mitarai, M., Harada, H., Sekine, I., Nishimori, I. and Kikutani, M. Biochemical studies on oral toxicity of ricin. I. Ricin administered orally can impair sugar absorption by rat small intestine. Chem. Pharm. Bull. (Tokyo) 1983; 31:3222–3227.

Ishiguro, M., Nakashima, H., Tanabe, S. and Sakakibara, R. Interaction of toxic lectin ricin with epithelial cells of rat small intestine in vitro. Chem. Pharm. Bull. (Tokyo) 1992c; 40:441–445.

Ishiguro, M., Tanabe, S., Matori, Y. and Sakakibara, R. Biochemical studies on oral toxicity of ricin. IV. A fate of orally administered ricin in rats. J. Pharmacobio-dyn. 1992d; 15:147–156.

Jackson, M.E., Simpson, J.C., Girod, A., Pepperkok, R., Roberts, L.M. and Lord, J.M. The KDEL retrieval system is exploited by *Pseudomonas* exotoxin A, but not by Shiga-like toxin-1, during retrograde transport from the Golgi complex to the endoplasmic reticulum. J. Cell Sci. 1999; 112:467–475.

Jung, H.C., Eckmann, L., Yang, S.K., Panja, A., Fierer, J., Morzycka-Wroblewska, E. and Kagnoff, M.F. A distinct array of proinflammatory cytokines is expressed in human colon epithelial cells in response to bacterial invasion. J. Clin. Invest. 1995; 95:55–65.

Kageyama, A., Kusano, I., Tamura, T., Oda, T. and Muramatsu, T. Comparison of the apoptosis-inducing abilities of various protein synthesis inhibitors in U937 cells. Biosci. Biotechnol. Biochem. 2002; 66:835–839.

Katzin, B.J., Collins, E.J. and Robertus, J.D. Structure of ricin A-chain at 2.5 A. Proteins 1991; 10:251–259.

Kende, M., Yan, C., Hewetson, J., Frick, M.A., Rill, W.L. and Tammariello, R. Oral immunization of mice with ricin toxoid vaccine encapsulated in polymeric microspheres against aerosol challenge. Vaccine 2002; 20:1681–1691.

Khan, A.S., Levitt, A.M. and Sage, M.J. Biological and chemical terrorism: strategic plan for preparedness and response. MMWR 2000; 49(RR-4):1–14.

Kim, J.M., Jung, H.Y., Lee, J.Y., Youn, J., Lee, C.H. and Kim, K.H. Mitogen-activated protein kinase and activator protein-1 dependent signals are essential for *Bacteroides fragilis* enterotoxin-induced enteritis. Eur. J. Immunol. 2005; 35:2648–2657.

Kim, J.M., Kim, J.S., Jun, H.C., Oh, Y.K., Song, I.S. and Kim, C.Y. Differential expression and polarized secretion of CXC and CC chemokines by human intestinal epithelial cancer cell lines in response to *Clostridium difficile* toxin A. Microbiol. Immunol. 2002; 46:333–342.

Kim, Y. and Robertus, J.D. Analysis of several key active site residues of ricin A chain by mutagenesis and X-ray crystallography. Protein Eng. 1992; 5:775–779.

Kimura, Y., Hase, S., Kobayashi, Y., Kyogoku, Y., Ikenaka, T. and Funatsu, G. Structures of sugar chains of ricin D. J. Biochem. (Tokyo) 1988; 103:944–949.

Korcheva, V., Wong, J., Lindauer, M., Jacoby, D.B., Iordanov, M.S. and Magun, B. Role of apoptotic signaling pathways in regulation of inflammatory responses to ricin in primary murine macrophages. Mol. Immunol. 2007; 44:2761–2771.

Kreitman, R.J. and Pastan, I. Importance of the glutamate residue of KDEL in increasing the cytotoxicity of *Pseudomonas* exotoxin derivatives and for increased binding to the KDEL receptor. Biochem. J. 1995; 307:29–37.

Kumar, O., Sugendran, K. and Vijayaraghavan, R. Oxidative stress associated hepatic and renal toxicity induced by ricin in mice. Toxicon 2003; 41:333–338.

Lamb, F.I., Roberts, L.M. and Lord, J.M. Nucleotide sequence of cloned cDNA coding for preproricin. Eur. J. Biochem. 1985; 148:265–270.

Largent, B.L., Walton, K.M., Hoppe, C.A., Lee, Y.C. and Schnaar, R.L. Carbohydrate-specific adhesion of alveolar macrophages to mannose-derivatized surfaces. J. Biol. Chem. 1984; 259:1764–1769.

Lebeda, F.J. and Olson, M.A. Prediction of a conserved, neutralizing epitope in ribosome-inactivating proteins. Int. J. Biol. Macromol. 1999; 24:19–26.

Lemley, P.V., Amanatides, P. and Wright, D.C. Identification and characterization of a monoclonal antibody that neutralizes ricin toxicity in vitro and in vivo. Hybridoma 1994; 13:417–421.

Lenaerts, K., Sokolovic, M., Bouwman, F.G., Lamers, W.H., Mariman, E.C. and Renes, J. Starvation induces phase-specific changes in the proteome of mouse small intestine. J. Proteome Res. 2006; 5:2113–2122.

Licastro, F., Morini, M.C., Bolognesi, A. and Stirpe, F. Ricin induces the production of tumour necrosis factor-alpha and interleukin-1 beta by human peripheral-blood mononuclear cells. Biochem. J. 1993; 294:517–520.

Llorente, A., Lauvrak, S.U., van Deurs, B. and Sandvig, K. Induction of direct endosome to endoplasmic reticulum transport in Chinese hamster ovary (CHO) cells (LdlF) with a temperature-sensitive defect in epsilon-coatomer protein (epsilon-COP). J. Biol. Chem. 2003; 278:35850–35855.

Lord, J.M. and Roberts, L.M. Toxin entry: retrograde transport through the secretory pathway. J. Cell Biol. 1998; 140:733–736.

Lord, J.M., Roberts, L.M. and Robertus, J.D. Ricin: structure, mode of action, and some current applications. FASEB J. 1994; 8:201–208.

Louise, C.B. and Obrig, T.G. Shiga toxin-associated hemolytic-uremic syndrome: combined cytotoxic effects of Shiga toxin, interleukin-1 beta, and tumor necrosis factor alpha on human vascular endothelial cells in vitro. Infect. Immun. 1991; 59: 4173–4179.

Louise, C.B. and Obrig, T.G. Shiga toxin-associated hemolytic uremic syndrome: combined cytotoxic effects of shiga toxin and lipopolysaccharide (endotoxin) on human vascular endothelial cells in vitro. Infect. Immun. 1992; 60:1536–1543.

Lu, B., Rutledge, B.J., Gu, L., Fiorillo, J., Lukacs, N.W., Kunkel, S.L., North, R., Gerard, C. and Rollins, B.J. Abnormalities in monocyte recruitment and cytokine expression in monocyte chemoattractant protein 1-deficient mice. J. Exp. Med. 1998; 187:601–608.

Maddaloni, M., Cooke, C., Wilkinson, R., Stout, A.V., Eng, L. and Pincus, S.H. Immunological characteristics associated with the protective efficacy of antibodies to ricin. J. Immunol. 2004; 172:6221–6228.

Magnusson, S. and Berg, T. Endocytosis of ricin by rat liver cells in vivo and in vitro is mainly mediated by mannose receptors on sinusoidal endothelial cells. Biochem. J. 1993; 291:749–755.

Magnusson, S., Berg, T., Turpin, E. and Frenoy, J.P. Interactions of ricin with sinusoidal endothelial rat liver cells. Different involvement of two distinct carbohydrate-specific mechanisms in surface binding and internalization. Biochem. J. 1991; 277:855–861.

Magnusson, S., Kjeken, R. and Berg, T. Characterization of two distinct pathways of endocytosis of ricin by rat liver endothelial cells. Exp. Cell Res. 1993; 205:118–125.

Mantis, N.J. Vaccines against the category B toxins: staphylococcal enterotoxin B, epsilon toxin and ricin. Adv. Drug Deliver. Rev. 2005; 57:1424–1439.

Mantis, N.J., Farrant, S.A. and Mehta, S. Oligosaccharide side chains on human secretory IgA serve as receptors for ricin. J. Immunol. 2004; 172:6838–6845.

Mantis, N.J., McGuinness, C.R., Sonuyi, O., Edwards, G. and Farrant, S.A. Immunoglobulin A antibodies against ricin A and B subunits protect epithelial cells from ricin intoxication. Infect. Immun. 2006; 74:3455–3462.

Marsden, C.J., Knight, S., Smith, D.C., Day, P.J., Roberts, L.M., Phillips, G.J. and Lord, J.M. Insertional mutagenesis of ricin A chain: a novel route to an anti-ricin vaccine. Vaccine 2004; 22:2800–2805.

Marsden, C.J., Smith, D.C., Roberts, L.M. and Lord, J.M. Ricin: current understanding and prospects for an antiricin vaccine. Expert Rev. Vaccines 2005; 4:229–237.

Mbawuike, I.N., Pacheco, S., Acuna, C.L., Switzer, K.C., Zhang, Y. and Harriman, G.R. Mucosal immunity to influenza without IgA: an IgA knockout mouse model. J. Immunol. 1999; 162:2530–2537.

McGuinness, C.R. and Mantis, N.J. Characterization of a novel high-affinity monoclonal immunoglobulin G antibody against the ricin B subunit. Infect. Immun. 2006; 74:3463–3470.

McHugh, C.A., Tammariello, R.F., Millard, C.B. and Carra, J.H. Improved stability of a protein vaccine through elimination of a partially unfolded state. Protein Sci. 2004; 13:2736–2743.

Mlsna, D., Monzingo, A.F., Katzin, B.J., Ernst, S. and Robertus, J.D. Structure of recombinant ricin A chain at 2.3 A. Protein Sci. 1993; 2:429–435.

Montfort, W., Villafranca, J.E., Monzingo, A.F., Ernst, S.R., Katzin, B., Rutenber, E., Xuong, N.H., Hamlin, R. and Robertus, J.D. The three-dimensional structure of ricin at 2.8 A. J. Biol. Chem. 1987; 262:5398–5403.

O'Loughlin, E.V. and Robins-Browne, R.M. Effect of Shiga toxin and Shiga-like toxins on eukaryotic cells. Microbe 2001; 3:493–507.

Olsnes, S. The history of ricin, abrin and related toxins. Toxicon 2004; 44:361–370.

Olsnes, S. and Pihl, A. Different biological properties of the two constituent peptide chains of ricin, a toxic protein inhibiting protein synthesis. Biochemistry (Mosc.) 1973; 12:3121–3126.

Olsnes, S., Sandvig, K., Petersen, O.W. and van Deurs, B. Immunotoxins—entry into cells and mechanisms of action. Immunol. Today 1989; 10:291–295.

Olson, M.A. Ricin A-chain structural determinant for binding substrate analogues: a molecular dynamics simulation analysis. Proteins 1997; 27:80–95.

Olson, M.A., Carra, J.H., Roxas-Duncan, V., Wannemacher, R.W., Smith, L.A. and Millard, C.B. Finding a new vaccine in the ricin protein fold. Protein Eng. Des. Sel. 2004; 17:391–397.

Peek, L.J., Brey, R.N. and Middaugh, C.R. A rapid, three-step process for the preformulation of a recombinant ricin toxin A-chain vaccine. J. Pharm. Sci. 2007; 96:44–60.

Pincus, S.H., Eng, L., Cooke, C.L. and Maddaloni, M. Identification of hypoglycemia in mice as a surrogate marker of ricin toxicosis. Comp. Med. 2002; 52:530–533.

Plemper, R.K. and Wolf, D.H. Retrograde protein translocation: eradication of secretory proteins in health and disease. Trends Biochem. Sci. 1999; 24:266–270.

Poli, M.A., Rivera, V.R., Pitt, M.L. and Vogel, P. Aerosolized specific antibody protects mice from lung injury associated with aerosolized ricin exposure. Toxicon 1996; 34:1037–1044.

Rapak, A., Falnes, P.O. and Olsnes, S. Retrograde transport of mutant ricin to the endoplasmic reticulum with subsequent translocation to cytosol. Proc. Natl. Acad. Sci. U.S.A. 1997; 94:3783–3788.

Rauber, A. and Heard, J. Castor bean toxicity re-examined: a new perspective. Vet. Hum. Toxicol. 1985; 27:498–502.

Ready, M.P., Kim, Y. and Robertus, J.D. Site-directed mutagenesis of ricin A-chain and implications for the mechanism of action. Proteins 1991; 10:270–278.

Reinecker, H.C., Loh, E.Y., Ringler, D.J., Mehta, A., Rombeau, J.L. and MacDermott, R.P. Monocyte-chemoattractant protein 1 gene expression in intestinal epithelial cells and inflammatory bowel disease mucosa. Gastroenterology 1995; 108:40–50.

Rollins, B.J., Walz, A. and Baggiolini, M. Recombinant human MCP-1/JE induces chemotaxis, calcium flux, and the respiratory burst in human monocytes. Blood 1991; 78:1112–1116.

Rotz, L.D., Khan, A.S., Lillibridge, S.R., Ostroff, S.M. and Hughes, J.M. Public health assessment of potential biological terrorism agents. Emerg. Infect. Dis. 2002; 8:225–230.

Roy, C.J., Hale, M., Hartings, J.M., Pitt, L. and Duniho, S. Impact of inhalation exposure modality and particle size on the respiratory deposition of ricin in BALB/c mice. Inhal. Toxicol. 2003; 15:619–638.

Rutenber, E., Ready, M. and Robertus, J.D. Structure and evolution of ricin B chain. Nature 1987; 326:624–626.

Rutenber, E. and Robertus, J.D. Structure of ricin B-chain at 2.5Å resolution. Proteins 1991; 10:260–269.

Sandvig, K., Dubinina, E., Garred, O., Prydz, K., Kozlov, J.V., Hansen, S.H. and Van Deurs, B. Entry of Shiga toxin into cells. Zentralbl. Bakteriol. [Orig.] 1993; 278:296–305.

Sandvig, K., Grimmer, S., Lauvrak, S.U., Torgersen, M.L., Skretting, G., van Deurs, B. and Iversen, T.G. Pathways followed by ricin and Shiga toxin into cells. Histochem. Cell Biol. 2002; 117:131–141.

Sandvig, K., Olsnes, S. and Pihl, A. Kinetics of binding of the toxic lectins abrin and ricin to surface receptors of human cells. J. Biol. Chem. 1976; 251:3977–3984.

Sandvig, K., Spilsberg, B., Lauvrak, S.U., Torgersen, M.L., Iversen, T.G. and van Deurs, B. Pathways followed by protein toxins into cells. Int. J. Med. Microbiol. 2004; 293:483–490.

Sandvig, K. and van Deurs, B. Delivery into cells: lessons learned from plant and bacterial toxins. Gene Ther. 2005; 12:865–872.

Sauseville, E.A. and Vitetta, E.S. In: Monoclonal Antibody-Based Therapy of Cancer (M.L. Grossbard, Ed.). Boston: Marcel Dekker, 1997.

Schechter, R. and Arnon, S.S. Extreme potency of botulinum toxin. Lancet 2000; 355:237–238.

Sekine, I., Kawase, Y., Nishimori, I., Mitarai, M., Harada, H., Ishiguro, M. and Kikutani, M. Pathological study on mucosal changes in small intestine of rat by oral administration of ricin. I. Microscopical observation. Acta Pathol. Jpn. 1986; 36:1205–1212.

Shea, D. and Groton, F. Ricin: Technical Background and Potential Role in Terrorism, 2004. CRS Report for Congress RS21383 at http://www.fas.org/irp/crs/RS21383.pdf.

Shepherd, V.L., Lee, Y.C., Schlesinger, P.H. and Stahl, P.D. L-fucose-terminated glycoconjugates are recognized by pinocytosis receptors on macrophages. Proc. Natl. Acad. Sci. U.S.A. 1981; 78: 1019–1022.

Simeral, L.S., Kapmeyer, W., MacConnell, W.P. and Kaplan, N.O. On the role of the covalent carbohydrate in the action of ricin. J. Biol. Chem. 1980; 255:11098–11101.

Simmons, B.M., Stahl, P.D. and Russell, J.H. Mannose receptor-mediated uptake of ricin toxin and ricin A chain by macrophages. Multiple intracellular pathways for a chain translocation. J. Biol. Chem. 1986; 261:7912–7920.

Skilleter, D.N., Paine, A.J. and Stirpe, F. A comparison of the accumulation of ricin by hepatic parenchymal and non-parenchymal cells and its inhibition of protein synthesis. Biochim. Biophys. Acta 1981; 677:495–500.

Slominska-Wojewodzka, M., Gregers, T.F., Walchli, S. and Sandvig, K. EDEM is involved in retrotranslocation of ricin from the endoplasmic reticulum to the cytosol. Mol. Biol. Cell 2006; 17:1664–1675.

Smallshaw, J.E., Firan, A., Fulmer, J.R., Ruback, S.L., Ghetie, V. and Vitetta, E.S. A novel recombinant vaccine which protects mice against ricin intoxication. Vaccine 2002; 20:3422–3427.

Smallshaw, J.E., Ghetie, V., Rizo, J., Fulmer, J.R., Trahan, L.L., Ghetie, M.A. and Vitetta, E.S. Genetic engineering of an immunotoxin to eliminate pulmonary vascular leak in mice. Nat. Biotechnol. 2003; 21:387–391.

Smallshaw, J.E., Richardson, J.A., Pincus, S., Schindler, J. and Vitetta, E.S. Preclinical toxicity and efficacy testing of RiVax, a recombinant protein vaccine against ricin. Vaccine 2005; 23:4775–4784.

Smallshaw, J.E., Richardson, J.A. and Vitetta, E.S. RiVax, a recombinant ricin subunit vaccine, protects mice against ricin delivered by gavage or aerosol. Vaccine 2007; 25:7459–7469.

Stone, R. Biodefense. Peering into the shadows: Iraq's bioweapons program. Science 2002; 297:1110–1112.

Stricklett, P.K., Hughes, A.K., Ergonul, Z. and Kohan, D.E. Molecular basis for up-regulation by inflammatory cytokines of Shiga toxin 1 cytotoxicity and globotriaosylceramide expression. J. Infect. Dis. 2002; 186:976–982.

Swimmer, C., Lehar, S.M., McCafferty, J., Chiswell, D.J., Blattler, W.A. and Guild, B.C. Phage display of ricin B chain and its single binding domains: system for screening galactose-binding mutants. Proc. Natl. Acad. Sci. U.S.A. 1992; 89:3756–3760.

Tagge, E., Chandler, J., Tang, B.L., Hong, W., Willingham, M.C. and Frankel, A. Cytotoxicity of KDEL-terminated ricin toxins correlates with distribution of the KDEL receptor in the Golgi. J. Histochem. Cytochem. 1996; 44:159–165.

Taylor, P.R., Gordon, S. and Martinez-Pomares, L. The mannose receptor: linking homeostasis and immunity through sugar recognition. Trends Immunol. 2005; 26:104–110.

Tesh, V.L., Ramegowda, B. and Samuel, J.E. Purified Shiga-like toxins induce expression of proinflammatory cytokines from murine peritoneal macrophages. Infect. Immun. 1994; 62: 5085–5094.

Thorpe, C.M., Hurley, B.P., Lincicome, L.L., Jacewicz, M.S., Keusch, G.T. and Acheson, D.W. Shiga toxins stimulate secretion of interleukin-8 from intestinal epithelial cells. Infect. Immun. 1999; 67:5985–5993.

Thorpe, C.M., Smith, W.E., Hurley, B.P. and Acheson, D.W. Shiga toxins induce, superinduce, and stabilize a variety of C-X-C chemokine mRNAs in intestinal epithelial cells, resulting in increased chemokine expression. Infect. Immun. 2001; 69: 6140–6147.

Thorpe, P.E., Detre, S.I., Foxwell, B.M., Brown, A.N., Skilleter, D.N., Wilson, G., Forrester, J.A. and Stirpe, F. Modification of the carbohydrate in ricin with metaperiodate-cyanoborohydride mixtures. Effects on toxicity and in vivo distribution. Eur. J. Biochem. 1985; 147:197–206.

Thrush, G.R., Lark, L.R., Clinchy, B.C. and Vitetta, E.S. Immunotoxins: an update. Annu. Rev. Immunol. 1996; 14:49–71.

Tommasi, M., Castelletti, D., Pasti, M., Fracasso, G., Lorenzetti, I., Sartoris, S., Pera, C., Ferrara, G.B., Tridente, G. and Colombatti, M. Identification of ricin A-chain HLA class II-restricted epitopes by human T-cell clones. Clin. Exp. Immunol. 2001; 125:391–400.

Utskarpen, A., Slagsvold, H.H., Iversen, T.G., Walchli, S. and Sandvig, K. Transport of ricin from endosomes to the Golgi apparatus is regulated by Rab6A and Rab6A'. Traffic 2006; 7:663–672.

van de Kar, N.C., Monnens, L.A., Karmali, M.A. and van Hinsbergh, V.W. Tumor necrosis factor and interleukin-1 induce expression of the verocytotoxin receptor globotriaosylceramide on human endothelial cells: implications for the pathogenesis of the hemolytic uremic syndrome. Blood 1992; 80:2755–2764.

van Deurs, B., Hansen, S.H., Petersen, O.W., Melby, E.L. and Sandvig, K. Endocytosis, intracellular transport and transcytosis of the toxic protein ricin by a polarized epithelium. Eur. J. Cell Biol. 1990; 51:96–109.

van Deurs, B., Sandvig, K., Petersen, O.W., Olsnes, S., Simons, K. and Griffiths, G. Estimation of the amount of internalized ricin that reaches the trans-Golgi network. J. Cell Biol. 1988; 106:253–267.

van Deurs, B., Tonnessen, T.I., Petersen, O.W., Sandvig, K. and Olsnes, S. Routing of internalized ricin and ricin conjugates to the Golgi complex. J. Cell Biol. 1986; 102:37–47.

van Kooyk, Y. and Geijtenbeek, T.B. DC-SIGN: escape mechanism for pathogens. Nat. Rev. Immunol. 2003; 3:697–709.

van Setten, P.A., van Hinsbergh, V.W., van den Heuvel, L.P., Preyers, F., Dijkman, H.B., Assmann, K.J., van der Velden, T.J. and Monnens, L.A. Monocyte chemoattractant protein-1 and interleukin-8 levels in urine and serum of patents with hemolytic uremic syndrome. Pediatr. Res. 1998; 43:759–767.

Vitetta, E.S., Smallshaw, J.E., Coleman, E., Jafri, H., Foster, C., Munford, R. and Schindler, J. A pilot clinical trial of a recombinant ricin vaccine in normal humans. Proc. Natl. Acad. Sci. U.S.A. 2006; 103:2268–2273.

Vitetta, E.S. and Thorpe, P.E. Immunotoxins containing ricin or its A chain. Semin. Cell Biol. 1991; 2:47–58.

Wales, R., Chaddock, J.A., Roberts, L.M. and Lord, J.M. Addition of an ER retention signal to the ricin A chain increases the cytotoxicity of the holotoxin. Exp. Cell Res. 1992; 203:1–4.

Wales, R., Roberts, L.M. and Lord, J.M. Addition of an endoplasmic reticulum retrieval sequence to ricin A chain significantly increases its cytotoxicity to mammalian cells. J. Biol. Chem. 1993; 268:23986–23990.

Waterhouse, C.C., Joseph, R.R., Winsor, G.L., Lacombe, T.A. and Stadnyk, A.W. Monocyte chemoattractant protein-1 production by intestinal epithelial cells in vitro: a role for p38 in epithelial chemokine expression. J. Interferon Cytokine Res. 2001; 21:223–230.

Wesche, J., Rapak, A. and Olsnes, S. Dependence of ricin toxicity on translocation of the toxin A-chain from the endoplasmic reticulum to the cytosol. J. Biol. Chem. 1999; 274:34443–34449.

Weston, S.A., Tucker, A.D., Thatcher, D.R., Derbyshire, D.J. and Pauptit, R.A. X-ray structure of recombinant ricin A-chain at 1.8 Å resolution. J. Mol. Biol. 1994; 244:410–422.

Wilhelmsen, C.L. and Pitt, M.L. Lesions of acute inhaled lethal ricin intoxication in rhesus monkeys. Vet. Pathol. 1996; 33:296–302.

Wong, J., Korcheva, V., Jacoby, D.B. and Magun, B. Intrapulmonary delivery of ricin at high dosage triggers a systemic inflammatory response and glomerular damage. Am. J. Pathol. 2007a; 170:1497–1510.

Wong, J., Korcheva, V., Jacoby, D.B. and Magun, B.E. Proinflammatory responses of human airway cells to ricin involve stress-activated protein kinases and NF-{kappa}B. Am. J. Physiol. Lung Cell. Mol. 293:1385–1394.

Wu, J.H., Singh, T., Herp, A. and Wu, A.M. Carbohydrate recognition factors of the lectin domains present in the *Ricinus communis* toxic protein (ricin). Biochimie 2006; 88:201–217.

Yamasaki, C., Nishikawa, K., Zeng, X.T., Katayama, Y., Natori, Y., Komatsu, N. and Oda, T. Induction of cytokines by toxins that have an identical RNA N-glycosidase activity: Shiga toxin, ricin, and modeccin. Biochim. Biophys. Acta 2004; 1671:44–50.

Yan, C., Rill, W.L., Malli, R., Hewetson, J., Naseem, H., Tammariello, R. and Kende, M. Intranasal stimulation of long-lasting immunity against aerosol ricin challenge with ricin toxoid vaccine encapsulated in polymeric microspheres. Vaccine 1996; 14:1031–1038.

Yan, C., Rill, W.L., Malli, R., Hewetson, J., Tammariello, R. and Kende, M. Dependence of ricin toxoid vaccine efficacy on the structure of poly(lactide-co-glycolide) microparticle carriers. Vaccine 1995; 13:645–651.

Yoder, J.M., Aslam, R.U. and Mantis, N.J. Evidence for widespread epithelial damage and coincident production of monocyte chemotactic protein 1 in a murine model of intestinal ricin intoxication. Infect. Immun. 2007; 75:1745–1750.

Zenilman, M.E., Fiani, M., Stahl, P., Brunt, E. and Flye, M.W. Use of ricin A-chain to selectively deplete Kupffer cells. J. Surg. Res. 1988; 45:82–89.

Zoja, C., Angioletti, S., Donadelli, R., Zanchi, C., Tomasoni, S., Binda, E., Imberti, B., te Loo, M., Monnens, L., Remuzzi, G. and Morigi, M. Shiga toxin-2 triggers endothelial leukocyte adhesion and transmigration via NF-kappaB dependent up-regulation of IL-8 and MCP-1. Kidney Int. 2002; 62:846–856.

CHAPTER

56

Rickettsia

Xuejie Yu, Nahed Ismail and David H. Walker

Department of Pathology, Sealy Center for Vaccine Development, Center for Biodefense and Emerging infectious Diseases, University of Texas Medical Branch, Galveston, TX, USA

OUTLINE

ABSTRACT

Among the 17 named species of the Rickettsiaceae that have been documented to cause human illness, 5 are significant threats of a fatal outcome, namely *Rickettsia rickettsii*, *Rickettsia prowazekii*, *Orientia tsutsugamushi*, *Rickettsia typhi*, and *Rickettsia conorii*. These *Rickettsia* can be obtained in nature, propagated by available methods, and prepared in a stable state of infectivity by a low dose. Thus, they could be used by terrorists. After a series of attacks, the public would seek to receive a protective vaccine. Rocky Mountain spotted fever, louse-borne typhus, and scrub typhus are currently reemerging infectious diseases with increasing incidence. Epidemic louse-borne typhus requires a vaccine to control large epidemics that occur in association with natural and manmade disasters including wars. The vast majority of emerging rickettsial diseases have existed for eons, are seldom life threatening, and were only discovered recently: African tick bite fever (*Rickettsia africae*), Flinders Island spotted fever (*Rickettsia honei*), Japanese spotted fever (*Rickettsia japonica*), tick-borne lymphadenopathy (*Rickettsia slovaca*), maculatum spotted fever (*Rickettsia parkeri*), flea-borne spotted fever (*Rickettsia felis*), the unnamed eschar-associated infections (*Rickettsia aeschlimannii* and *Rickettsia massiliae*), and the unnamed infection caused by *Rickettsia monacensis*. Other rickettsioses (North Asian tick typhus, *Rickettsia sibirica*; Queensland tick typhus, *Rickettsia australis*; rickettsialpox, *Rickettsia akari*) are rarely severe illnesses. A vaccine should be developed against *R. prowazekii* and would likely provide good crossprotection against *R. typhi*. Because *R. rickettsii* is the most virulent *Rickettsia* and genetic engineering to complete antibiotic resistance is highly feasible, development of a vaccine against Rocky Mountain spotted fever would be a very wise biodefense strategy. Such a vaccine would very likely provide crossprotection against other spotted fever group rickettsiae. The preparation of a vaccine that protects against all *Rickettsia* species is a genuine possibility. Such a vaccine would have a commercial value for use in travelers, especially those going on safari in southern Africa, where more hunters and tourists suffer from African tick bite fever than malaria.

Development of a vaccine that is protective against scrub typhus poses tremendous challenges owing to transient active immunity and vast antigenic diversity of the agent.

HISTORY OF RICKETTSIAL DISEASES

Records of the recognition of diseases that are now known to be caused by rickettsiae were clearly described by Ge Hong (scrub typhus) in 313, Hieronymus Fracastorius (epidemic typhus) in 1546, E.E. Maxey (Rocky Mountain spotted fever) in 1899, and Conor (boutonneuse fever) in 1910. Two pathologists, Louis Wilson of the Minnesota State Board of Health and William Chowning of the University of Minnesota, described the geographic and seasonal distribution, the clinical features of the severe form of Rocky Mountain spotted fever, age- and gender-specific incidence and case fatality rate, and pathologic lesions. They determined the absence of a conventional bacterial agent and predicted an infectious etiology and tick vector. Their erroneous identification of intraerythrocytic protozoa as the cause led to the investigations of Howard Ricketts, who between 1906 and 1910 proved that Rocky Mountain spotted fever was an infectious disease transmitted by tick bite that resulted in solid protective immunity. Experimental retention of infectivity by Berkefeld filters led to his prediction that the agent would be visible microscopically, which was supported by his visualization of small diplobacilli in blood and ticks. That the agent was in fact a novel type of obligately intracellular bacterium was established by its histochemical detection in endothelial cells of infected humans and in ticks by S. Burt Wolbach between 1916 and 1919, propagation in embryonated chicken eggs by Herald Cox in 1938, recognition of the therapeutic efficacy of tetracycline and chloramphenicol in the late 1940s and early 1950s, visualization in living cell culture by Mosielo Schaechter in 1957, and electron microscopic characterization as gram-negative cytosolic organisms by Anderson in 1965.

The clinical similarities of typhus and Rocky Mountain spotted fever were recognized by physicians prior to the discovery of the causative agents. A patient diagnosed as having typhus by the world's foremost physician William Osler died and was examined postmortem by the father of American academic pathology, William Welch, at Johns Hopkins University in 1903. Later reexamination of the tissues by immunohistochemistry revealed that the patient had actually died of Rocky Mountain spotted fever, which was not recognized to occur in the eastern US until more than 25 years later.

The early discoveries relied upon experimental transmission of infectivity from human blood or vector arthropods to guinea pigs and nonhuman primates. The vector-association of these diseases was also discovered in early studies such as that of Charles Nicolle, who demonstrated experimentally in 1909 that typhus was transmitted by human body lice. The recognition that *Rickettsia prowazekii* is the etiology of typhus relied upon observations of Ricketts and von Prowazek, who both died of typhus, da Rocha-Lima, and Wolbach, who visualized the bacteria in humans, experimental lesions and lice, which were used as a means of cultivation by xenodiagnosis.

Recrudescent typhus fever was described clinically by Nathan Brill in 1896 and 1910, and was shown by Hans Zinsser in 1934 to be caused by reactivation of latent *R. prowazekii* infection. A similar illness, now known as murine typhus, was differentiated clinically and epidemiologically from other typhus fevers by Maxey in 1926, and Dyer isolated the agent, *Rickettsia typhi*, from rats and fleas, the natural zoonotic cycle, in 1931.

Other spotted fevers and their agents continue to be discovered around the globe. An illustrative example is African tick bite fever, which was described clinically by MacNaughton in southern Africa in 1911, and was distinguished from boutonneuse fever by Pijper in 1936. Not until 1996 was the agent rediscovered and designated taxonomically as *Rickettsia africae*. Other spotted fever group (SFG) rickettsioses and rickettsiae were identified by Mill (North Asian tick typhus) in 1936 and Krontovskaya (*Rickettsia sibirica*) in 1938;

rickettsialpox by Shankman and *Rickettsia akari* by Huebner in 1945; Queensland tick typhus and *Rickettsia australis* by Andrew reported in 1946; Japanese spotted fever by Mahara in 1984 and its causative agent, *Rickettsia japonica*, by Uchida in 1992; Flinders Island spotted fever by Stewart in 1991 and its etiologic agent, *Rickettsia honei*, by Stenos in 1998; *Rickettsia slovaca* in 1976 and the disease that it causes, tick-borne lymphadenopathy, by Lakos in 1997; *Rickettsia parkeri* by Lackman in 1949 and the first human infection by Paddock in 2004; *Rickettsia felis* and its human infection first reported by Azad in 1994.

Pioneering studies of these organisms included investigations of metabolic activities of the bacteria separated from their host cells by Bovarnick in 1949, of rickettsial antigens by Shepard in 1945, of rickettsial proteins by Anacker in 1983, and the genome of *R. prowazekii* by Andersson in 1998.

Orientia tsutsugamushi were visualized as intracellular organisms in smears from animals by Nagayo in 1924. In 1929 Ogata passaged the organism 80 times in rabbit testis and also observed them in stained macrophages. Three years later he developed the method for isolation of *O. tsutsugamushi* by intraperitoneal inoculation of mice. In 1995 Tamura separated *Orientia* into its own genus on the basis of significant genetic divergence and phenotypic differences from *Rickettsia*.

ETIOLOGIC AGENTS

Classification

The family Rickettsiaceae includes the genera *Rickettsia* and *Orientia*, which are small gram-negative, obligately intracellular bacteria that reside in the cytosol of host cells. *Rickettsia* (0.3–0.5 × 0.8–1.0 μm) are smaller than *Orientia* (0.5–0.8 × 1.2–3.0 μm). *Rickettsia* organisms have undergone genome reduction resulting in a smaller genome (approximately 1 Mb) than *Orienta* (2.4–2.7 Mb), and has lost genes encoding enzymes for sugar metabolism, lipid biosynthesis, nucleotide synthesis, and amino acid synthesis. Although *Rickettsia* can synthesize adenosine triphosphate (ATP), *Rickettsia* are also energy parasites. They transport ATP into the rickettsiae from the host cell cytoplasm using the ATP/ADP translocase, which is unique to *Rickettsia* and another intracellular bacterium *Chlamydia* (Andersson et al., 1998; Plano and Winkler 1991).

Rickettsia

Organisms in the genus *Rickettsia* are divided into two antigenically distinct groups based on their lipopolysaccharide (LPS): the typhus group (TG) and the SFG as well as an ancestral group and transitional group.

TG rickettsiae include two species: *R. prowazekii*, the agent of epidemic typhus, and *R. typhi*, the agent of murine typhus. SFG rickettsiae include 16 validated species and numerous unvalidated rickettsial species. However, most of these named SFG rickettsiae are so closely related that they could be considered strains of a single species. There is less than 0.5% divergence of the 16S rRNA gene among the core species of SFG rickettsiae including *Rickettsia rickettsii*, *Rickettsia conorii*, *R. sibirica*, *R. africae*, *R. parkeri*, *R. slovaca*, and many others. The tick-borne SFG rickettsiae range from avirulent to highly virulent. Studies of rickettsial phylogeny have led to a proposal to consider *R. akari*, *R. australis*, and *R. felis* as the transitional group branching between the TG and SFG and an ancestral group that includes *Rickettsia bellii* and *Rickettsia canadensis* (Vitorino et al., 2007).

Multiplication of *R. rickettsii* occurs primarily in the cytoplasm, but intranuclear growth is sufficiently prominent to have stimulated early investigators to use it as a criterion for the classification of SFG. Because of cell-to-cell spread via actin-based mobility and greater cytopathogenicity, plaques on monolayers occur earlier and are larger than in the case of TG rickettsiae.

Orientia

Genus *Orientia* contains a single species, *O. tsutsugamushi*, which was formerly called *Rickettsia tsutsugamushi*. *Orientia* possess an unusual gram-negative cell wall that lacks LPS and peptidoglycan (Amano, et al., 1987). The outer leaflet of the cell wall is considerably thicker than the inner leaflet, while the opposite is true of *Rickettsia* (Silverman and Wisseman, 1978). The diversity among strains of *O. tsutsugamushi* is greater than the diversity among the species of *Rickettsia* as determined by the 16S rRNA gene. Phylogenetic analyses based on homologies of 56-kDa type-specific antigen genes classify *Orientia* isolates from China, Korea, Japan, and Southeast Asia into seven genotypes. All isolates originating in southeast Asia, including the prototype Gilliam and Karp strains isolated in Burma and New Guinea, respectively, are distantly located in the phylogenetic tree from those isolates in Japan, Korea, and China, indicating that strains of *O. tsutsugamushi* distributed in northeastern and southeastern Asia have diverged evolutionarily (Enatsu et al., 1999).

Rickettsial Antigens

The major antigens of *Rickettsia* that are recognized by the humoral immune response are LPS, lipoprotein, outer membrane proteins (OMPs), and heat shock proteins. The Weil–Felix reaction was used as a presumptive diagnostic test for rickettsial diseases. It is based on the crossreaction of antibodies to rickettsial antigens from primary rickettsial infections, apparently owing to IgM, with the antigens of two bacteria: *Proteus vulgaris* (OX19 or OX2) and *Proteus mirabilis* (OXK). They do not elicit an anamnestic antibody response in recrudescent typhus. The antibodies of Japanese spotted fever patients are reactive with the LPS of OX2 as well as OX19. The antibodies of scrub typhus patients recognize OXK LPS only (Amano et al., 1996). Thus, the crossreactive antigens between rickettsiae and *Proteus* are most likely present in the LPS. The *O*-polysaccharides of typhus rickettsial LPS are composed of glucose, glucosamine, quinovosamine, and phosphorylated hexosamine, which are also found in the *O*-polysaccharide of the LPS from *P. vulgaris* OX19 used in the Weil–Felix test, suggesting that they may represent the antigens common to LPS from TG rickettsiae and *P. vulgaris* OX19 (Amano et al., 1998). *O. tsutsugamushi* lacks LPS.

A 17-kDa-protein is a genus-common protein of rickettsiae and has been identified in all *Rickettsia* species examined (Anderson, 1990). The sequences of the 17-kDa-protein gene are conserved among rickettsial species, indicating the importance of it to the survival of the rickettsiae. The protein is predicted to be a lipoprotein, and part of the protein is surface exposed. Thus, it has been speculated that the 17-kDa protein may play a scaffolding and protective role in the rickettsiae (Anderson, 1990).

All rickettsiae have a 135-kDa outer membrane protein B (OmpB) that has been identified as an S-layer protein of rickettsiae (Ching et al., 1990). OmpB is the most abundant rickettsial surface protein, and it contains species-, group-, and genus-specific epitopes (Anacker et al., 1987a, 1987b). OmpB consists of 10–15% of the total cellular protein and is released as a soluble fraction, when the cells are suspended in hypotonic solution lacking Mg^{2+} and are incubated at 45 °C for 20 min (Dasch, 1981). The amino acid sequence of OmpB is very conserved among *Rickettsia* species.

Spotted fever and transitional group rickettsiae have an additional outer membrane protein, OmpA, which contains a hydrophilic region of tandem repeat units, that comprises over 40% of the amino acid sequences (Anderson, 1990). The repeat units are not identical and thus are divided into three types with type I of 75 amino acids, type II of 72 amino acids, and type III of 85 amino acids. The type II repeats are less conserved and are further divided into two subtypes (IIa and IIb) (Anderson, 1990). The *ompA* sequences upstream and downstream of the repeat region are conserved among SFG rickettsial species. Although the sequences of the repeat units are conserved among *R. rickettsii*, *R. conorii*, and *R. akari*, the number and the order of arrangement of the repeat units are varied among SFG rickettsiae (Gilmore, 1993); however, the type III of repeat unit that is found in *R. australis* differs greatly from other SFG rickettsial repeat units. *R. australis* contains only type III repeat units, which have only 21% identity to the type I repeat of *R. rickettsii* (Stenos and Walker, 2000). The antigenic diversity of SFG rickettsiae is determined in large part by the number, order, and type of repeat units.

OmpA of *R. felis* is truncated by a premature stop codon at one-third of the sequence from the N-terminus of the protein. *Rickettsia peacockii*, a tick symbiont, does not express OmpA, and the *ompA* of *R. peacockii* contains a deletion of G at 403 bp and insertions of GT and A following 4872 and 5828 bp, respectively, resulting in premature stop codons (Baldridge et al., 2004). The *ompA* of avirulent *R. rickettsii* Iowa strain was also interrupted by a nucleotide deletion that introduces a stop codon truncating *ompA* at 660 bp from the start codon (Ellison et al., 2008).

O. tsutsugamushi Antigens

SDS-polyacrimide gel analysis reveals as many as 30 proteins of *O. tsutsugamushi*, and it is difficult to correlate the molecular sizes of the proteins in different studies (Tamura et al., 1985; Hanson, 1985; Stover, et al., 1990). The most abundant antigen of *O. tsutsugamushi* is the 56 kDa surface protein (Tamura et al., 1985). It comprises 10–15% of the total bacterial cellular protein content, and it is the most immunodominant protein (Hanson, 1985; Ohashi, et al., 1989, 1988). The 56-kDa protein varies among geographic isolates of *O. tsutsugamushi* and thus is designated the type-specific protein (Ohashi et al., 1992). The 56-kDa protein has the feature of transmembrane proteins with alternating hydrophobic and hydrophilic regions. Analysis of the 56-kDa protein of *Orientia* variants revealed four variable domains with spans of 16–40 amino acids. The variable domains are located in the hydrophilic regions of the molecule that are likely surface exposed and have different amino acid sequences among the strains (Ohashi et al., 1992).

Another major immunodominant protein is the 60-kDa protein, which is the homolog of the GroEL (heat shock) protein family (Ohashi et al., 1988) (Stover et al., 1990). Genetic analysis of the DNA sequences of *groESL* of several geographic foci revealed a limited number of variants within each focus and diversity of strains from focus to focus.

CLINICAL DISEASE

Rickettsial diseases are characterized by systemic symptoms at onset: fever, severe headache, and muscle aches. The characteristic maculopapular rash typically appears 3–5 days later. In severe disease, petechiae subsequently appear in the center of the maculopapules. The rickettsioses that are usually less severe such as African tick bite fever and murine typhus have a lower incidence of rash. However, a small portion of patients with fatal Rocky Mountain spotted fever have no rash but suffer lethal damage to vital organs such as the brain and lungs. Erythematous maculopapules are difficult to see in darkly pigmented skin. The rash in Rocky Mountain spotted fever is described as centripetal, beginning on the wrists and ankles and later appearing on the trunk and vice versa for epidemic typhus, a centrifugal rash. However, this generalization is not a constant observation. Great emphasis is placed on the involvement of the palms and soles in SFG rickettsioses, but a significant fraction do not have this manifestation, and in others it appears late in the course. Early in the course of Rocky Mountain spotted fever, nausea, vomiting, and abdominal pain occur more often than rash. Focal skin necrosis with a dark scab (an eschar) at the site of tick or mite feeding is a common feature of boutonneuse fever, African tick bite fever, rickettsialpox, North Asian tick typhus, Queensland tick typhus, Japanese spotted fever, Flinders Island spotted fever, and tick-borne lymphadenopathy, but is rare in Rocky Mountain spotted fever, and nonexistent in epidemic and murine typhus.

Systemic vascular infection in Rocky Mountain spotted fever results in encephalitis, leading to stupor (25%), coma (10%), and seizures (8%), as well as interstitial pneumonia and noncardiogenic pulmonary edema causing cough (33%), pneumonitis (15%), and even adult respiratory distress syndrome. Hypovolemia and hypotensive shock lead to acute renal failure. In the preantibiotic era the case fatality rates were 23% for Rocky Mountain spotted fever and 15% for epidemic louse-borne typhus. Currently case fatality rates are still significant: approximately 4% for Rocky Mountain spotted fever, boutonneuse fever, and epidemic typhus and 1% for murine typhus, largely owing to misdiagnosis or late diagnosis leading to treatment with ineffective antibiotics or delayed effective antirickettsial treatment.

Scrub typhus frequently manifests an eschar at the site of chigger (larval mite) feeding during the incubation period. Regional lymphadenopathy, fever, headache, and myalgias begin abruptly. Hearing loss is present in a third of cases. Conjunctival redness and generalized lymphadenopathy are frequently present. The occurrence of rash is variable. Cough, tachypnea, and radiologic pulmonary infiltrates are frequent presenting findings as are confusion and mild personality changes. Severe cases may manifest adult respiratory distress syndrome, acute renal failure, and hypotensive shock. In the preantibiotic era, case fatality rates varied drastically from less than 1 to 30%.

Animal Models

Guinea Pig

Guinea pigs are the standard animal model for Rocky Mountain spotted fever and epidemic typhus. After infection with rickettsiae, guinea pigs develop fever and scrotal swelling. A fatal outcome may occur in *R. rickettsii* infection. Guinea pigs also develop humoral and cellular immune responses. Rickettsiae can be isolated from blood, liver, spleen, and testes. Development of illness depends on the strain of guinea pigs, route of inoculation, and dose. Rocky Mountain spotted fever can be induced experimentally in Hartley strain guinea pigs by the subcutaneous, intraperitoneal, intranasal, aerosol, or conjunctival route. Aerosol, intranasal, and subcutaneous routes are the most sensitive, and guinea pigs can be infected by as low a dose as one organism via these routes. The conjunctival route is less sensitive; 10^6 organisms are required to infect guinea pigs (Kenyon, et al., 1979).

Nonhuman Primates

A rhesus monkey (*Macaca mulatta*) model has been used for both Rocky Mountain spotted fever and epidemic typhus (Saslaw et al., 1966). Due to the limited availability of the rhesus monkey, a cynomolgus (*Macaca fascicularis*) monkey model was developed for further vaccine testing for both Rocky Mountain spotted fever and epidemic typhus. Both monkey species develop clinical signs similar to epidemic typhus and Rocky Mountain spotted fever in humans.

For Rocky Mountain spotted fever, the severity of the disease in the monkey depends on the route and dose of the rickettsial inoculation. When 100 organisms are inoculated into monkeys via the intravenous route, 100% of monkeys develop fever, and 85% of them die. Subcutaneous inoculation requires a higher dose than intravenous inoculation and yields inconsistent results. Aerosol inoculation is the most sensitive means of transmission, causing 100% of monkeys to develop fever and 75% die with as small a dose as one organism. The incubation period is 3–9 days, and is not dependent on the inoculation route or the dose. The most frequently observed microscopic lesions are vasculitis and thrombosis of the capillaries, arterioles, and venules of in numerous tissues. The lesions are characterized primarily by perivascular infiltration of lymphocytes and plasma cells. Coagulopathy occurs in some monkeys. Blood chemistry changes include mildly increased leukocyte counts due to an absolute neutrophilia during the febrile period. Skin rashes are less frequently observed in cynomolgus than in rhesus monkeys due to the darker cutaneous pigmentation of the cynomolgus monkey (Gonder, et al., 1979).

Monkeys are susceptible to *R. prowazekii* infection and develop clinical signs of illness and pathological changes characteristic of epidemic typhus infection in humans. Monkeys can be infected with 10^3 organisms when rickettsiae are injected intravenously. However, intravenous inoculation of 10^7 plaque-forming units of *R. prowazekii* are required for the monkeys to develop clinical signs of illness and pathological changes characteristic of epidemic typhus infection in humans. The incubation period is 1–9 days, and is longer for the low dose inoculum. Typical typhus nodules are observed in the brain (Gonder et al., 1980).

Monkeys are not a good disease model for scrub typhus although they have been used for scrub typhus studies since 1918. The early studies used wild silver leaf monkeys, and difficulty in maintaining the silver leaf monkeys in captivity prompted investigators to use cynomolgus monkeys. The only detectable signs of infection in monkeys are an eschar at the site of intradermal inoculation of *O. tsutsugamushi* and rickettsemia. Rickettsemia lasts 1–2 weeks (Chattopadhyay et al., 2005).

Mice

The C3H/HeN mouse inoculated intravenously with *R. conorii* provides an excellent model for boutonneuse fever, and intravenous inoculation of *R. typhi* is an excellent model of epidemic typhus or murine typhus. These intravenous infection mouse models faithfully represent the human infections in the aspects of disseminated endothelial infection and injury, cellular immune response, histopathology of the vascular lesions, critical target organs with characteristic life-threatening lesions such as meningoencephalitis and interstitial pneumonia (Feng et al., 1994; Walker et al., 1994, 2000). Some mouse strains intravenously inoculated with other SFG rickettsiae also develop rickettsial disease associated with disseminated endothelial infection by rickettsiae, e.g., *R. australis* in Balb/c mice and *R. sibirica* in C3H/HeN, Balb/c, or Swiss-Webster mice.

PROTECTIVE IMMUNE RESPONSE

Although most natural infections with *Rickettsia* enter via the skin, most animal models characterizing immune responses to this pathogen employ intravenous inoculation to initiate infection. Thus, most studies of infection with SFG or TG rickettsiae focus on immune responses to systemic *Rickettsia* infection. In contrast, dermal immune responses to infection with *Rickettsia* are less well characterized but are increasingly the focus of investigation. Most of our understanding of the immune response against *Rickettsia* is derived from in vitro studies as well as the murine models of rickettsioses.

Innate Response

Following infection with *Rickettsia*, innate immune responses are rapidly triggered and contribute to host survival, as indicated by the impact of infection on mice that lack various molecules with important innate immune functions such as IFN-γ, tumor necrosis factor-alpha (TNF-α), and NK cells. IFN-γ and TNF-α are essential for primary defense against infection with *Rickettsia*, and mice that lack these cytokines rapidly succumb to infection (Walker et al., 2000; Billings et al., 2001; Feng et al., 1994; Feng and Walker, 1993). IFN-γ and TNF-α mediated resistance is partly attributed to the production of microbicidal effector molecules such as nitric oxide (NO) (Feng et al., 1994; Feng and Walker, 1993; Walker et al., 1993; Feng and Walker, 2000). Mice that lack these cytokines have decreased NO production and an overwhelming rickettsial infection (Feng et al., 1994). Similarly immunocompetent mice treated with NO synthesis inhibitor N^G-monomethyl-l-arginine (N^GMMLA) fail to control rickettsial growth (Feng et al., 1994). Mouse endothelial cells activated in vitro with IFN-γ and TNF-α trigger the expression of inducible NO synthase (iNOS), an enzyme involved in NO production, and control rickettsial infection (Feng and Walker, 1993, 2000; Walker et al., 1993).

IFN-γ is essential for resistance to *Rickettsia* infection, and multiple cell types have the ability to rapidly secrete IFN-γ (Yeaman et al., 1998; Ohteki et al., 1999; Munder et al., 1998; Fultz et al., 1993). NK cells are a major source of IFN-γ early in rickettsial infection (Billings et al., 2001). The activity of NK cells is increased early during the course of murine infection with *R. typhi* and *R. conorii*. Depletion of NK cells results in increased rickettsial infectivity titers in the spleen of infected mice (2). NK cells secrete IFN-γ in response to IL-12 (Pillarisetty et al., 2005; Gerosa et al., 2005). IL-12 is a cytokine produced mainly by activated monocytes, macrophages, and B cells and has immunoregulatory effects on NK cells and T cells (Gately et al., 1998). IL-12 and IFN-γ are both important components of the Th1 immune response that effectively controls intracellular pathogens including *Rickettsia* (Walker et al., 2000; Biron and Gazzinelli, 1995).

Several lines of evidence point to the protective role of proinflammatory cytokines such as IL-1 and TNF-α

in host defense against *Rickettsia* (Sporn and Marder, 1996; Vitale et al., 1999). In vitro activated macrophages and endothelial cells infected by *R. conorii* or *R. rickettsii* synthesize proinflammatory cytokines such as TNF-α and IL-6 (Manor and Sarov, 1990; Kaplanski et al., 1995). Production of bactericidal NO by LPS-activated endothelial cells is dependent on endogeneous production of IL-1α and TNF-α. In an animal model of SFG rickettsiosis employing a sublethal dose of *R. conorii*, depletion of TNF-α results in fatal infection, with high rickettsial burden in endothelial cells, macrophages, and hepatocytes (Feng et al., 1994; Feng and Walker, 1993), suggesting a protective role of TNF-α in host response against *Rickettsia*. The biologic activities of TNF are mediated by two structurally related, but functionally distinct, tumor necrosis factor receptors (TNFR), p55 and p75 (Locksley et al., 2001; Mauri et al., 1998; Monastra et al., 1996). TNFRp55 is the primary signaling receptor on most cell types through which the majority of inflammatory responses and antimicrobial host defenses classically attributed to TNF occur. High plasma levels of soluble TNF-receptors, mainly TNFRp55, are present in patients with severe Mediterranean spotted fever (MSF) (Vitale et al., 2001). These data suggest that soluble TNFR released into the circulation of patients with severe rickettsiosis binds to the secreted or membrane form of TNF-α, thus preventing its binding to membrane TNFRp55, and results in impaired TNF-α mediated activation of intracellular bactericidal effector killing mechanisms.

Consistent with the above studies, we found that eschar lesions from patients with mild-to-moderate MSF express high mRNA levels of TNF, IFN-γ, and IL-10. Interestingly, very high levels of intralesional IL-10 were inversely correlated with lower levels of IFN-γ and TNF. This mixed proinflammatory and anti-inflammatory response in skin lesions from patients with MSF was also associated with significantly increased mRNA of RANTES, indoleamine-2,3-dioxygenase (IDO; an enzyme involved in limiting rickettsial growth by tryptophan degradation), and (iNOS: a source of microbicidal nitric oxide). Thus, humans with rickettsial infections express many of the effector mechanisms that have been identified in experimental animals and in vitro. Skin lesions from patients with severe MSF have significantly higher RANTES expression than those with mild or moderate disease. RANTES is a chemokine that mediates recruitment of T cells to the site of infection. In addition, RANTES activates *Rickettsia*-infected target cells, which results in intracellular killing of rickettsiae by an NO-mediated mechanism. However, very high concentration of RANTES in vivo can lead to T cell apoptosis, and subsequent disease progression. Further studies are needed to determine whether these cytokines and chemokines responses in patients with MSF are simply a correlate of mild and severe disease or contribute to antirickettsial immunity and pathogenesis.

At the cellular level, human hepatocytes and human umbilical vein endothelial cells (HUVECs) stimulated in vitro by IFN-γ, TNF-α, IL-1β, and RANTES (regulated by activation, normal T cell-expressed and secreted chemokine; CCL5) kill intracellular rickettsiae by a NO-dependent mechanism (Feng and Walker, 2000). Neither primary human monocytes nor the monocytic cell line THP-1 produces NO with the same stimuli; however, an antirickettsial effect is elicited in these cells by the combination of cytokines (without CCL5) (Feng and Walker, 2000). This activity is mediated by production of hydrogen peroxide and tryptophan starvation of intracellular bacteria due to degradation of tryptophan by the enzyme IDO. Hydrogen peroxide-dependent intracellular killing of *Rickettsia* also occurs in HUVECs, but not in the human hepatocytes. Thus, *Rickettsia*-infected nonprofessional phagocytic host cells, such as endothelial cells and hepatocytes, possess intracellular microbicidal effector mechanisms similar to those mounted by professional phagocytic cells such as macrophages.

Since the primary target cell for *Rickettsia* is the endothelium, examination of factors that control the interaction between endothelial cells and T lymphocytes is important. Among these factors are chemokines, particularly inflammatory chemokines that specifically target activated T cells through the CXCR3 receptor such as CXCL9 (Mig) and CXCL10 (IP-10). CXCL9 and CXCL10 are inflammatory chemokines that are expressed in vascular lesions in the brain of patients with fatal Rocky Mountain spotted fever (Valbuena and Walker, 2004; Valbuena et al., 2003). In animal models of SFG rickettsiosis, CXCL9 and CXCL10 (CXCR3 ligands) are expressed early in the liver and lungs of mice infected with a sublethal dose of *R. conorii* (Valbuena et al., 2003; Valbuena and Walker, 2005). The mouse endothelial cell line SVEC4-10 expresses CXCL9 and CXCL10 when stimulated with IL-1, IFN-γ, and TNF-α (Kaplanski et al., 1995; Sporn and Marder, 1996; Dignat-George et al., 1997). In infections with other intracellular pathogens, CXCL10 is expressed in inflamed tissues in which the Th1-type cytokine IFN-γ is upregulated (Qin et al., 1998; Kim et al., 2001; Sallusto et al., 1998; Bonecchi et al., 1998). Similarly, CXCR3 is expressed predominantly on Th1 clones and on a significant fraction of circulating memory T cells and is required for migration of in vivo activated T cells, particularly effector CD8+ T cells to dermal sites of inflammation (Nishimura et al., 2002; Ng et al., 1985). Therefore, it is possible that the early rise in the expression of CXCL9 and CXCL10 may participate in the Th1 polarization and effector cell generation that are so important in the clearance of rickettsiae from infected hosts. However, antibody-mediated neutralization of CXCL9 and CXCL10 and absence of CXCR3 expression in knockout mice did not affect survival or bacterial loads of these mice infected with rickettsiae (Valbuena and Walker, 2004), suggesting a possible compensatory or alternative mechanism that circumvents the need for these chemokines in triggering antirickettsial effector mechanisms.

Many studies have examined the interaction between *Rickettsia* and their main target, endothelial cells. However, endothelial cells are not professional antigen presenting cells and lack the required costimulatory molecules necessary for the induction of T cell responses. In addition, the fact that rickettsiae are transmitted via the dermis suggests that resident dendritic cells could play a more important role than endothelial cells in innate immunity as well as in induction of specific T cell responses in vivo. Analysis of the in vitro interaction of rickettsiae with bone marrow-derived DCs (BMDCs) from mice that are resistant (C57BL/6) and susceptible (C3H/HeN) to lethal *R. conorii* infection suggests that dendritic cells not only contribute to the initial clearance of rickettsiae during the innate response but also determine the type of acquired immune response against *Rickettsia*. Compared to the immediate and exclusive cytoplasmic localization within endothelial cells, rickettsiae efficiently enter and localize in both phagosomes and cytosol of BMDCs from both mouse strains (Fang et al., 2007). This dual intracellular localization of rickettsiae within dendritic cells would enable efficient priming of antigen-specific $CD4^+$ and $CD8^+$ T cells via major histocompatibility complex (MHC) class II and I pathways, respectively. *R. conorii* infection induces maturation of BMDCs from both mouse strains as evidenced by increased expression of MHC and costimulatory molecules. However, compared to susceptible mice, BMDCs from resistant mice exhibit higher expression levels of MHC class II and Th-1 promoting cytokines such as interleukin-12p40 (IL-12p40) upon rickettsial infection. Importantly, BMDC from resistant C57BL/6 background are very efficient in activation of naive $CD4^+$ T cells, and promotion of their differentiation into Th1 cells producing IFN-γ production is indispensable for activation of intracellular microbicidal functions of infected target cells and, thus, killing of intracellular rickettsiae. In contrast, in vitro coculture of naive $CD4^+$ T cells from susceptible C3H mice with *Rickettsia*-infected syngeneic BMDC reveals delayed $CD4^+$ T cell activation and suppressed expansion of $CD4^+$ Th1 and Th2 cells compared to unstimulated naive T cells (Fang et al., 2007). The suppressive $CD4^+$ T cell phenotypes or responses detected in vitro upon coculture of T cells and BMDC from susceptible C3H mice are associated with expansion of $Foxp3^+$ natural T regulatory cells. Analysis of T cell priming functions of DC compared to macrophages indicates that DC are more efficient inducers of T cell activation, as characterized by the production of significant quantities of the prototypical Th1 cytokines, IFN-γ and IL-2 (Jordan et al., 2007). In addition, *R. conorii*-infected DCs are able to activate naive $CD8^+$ T lymphocytes in vitro in the absence of T cell help (Jordan et al., 2007). However, further investigation of the $CD8^+$ T cell priming function of DC is necessary to determine whether a similar phenomenon occurs in vivo. Interestingly, transfer of *R. conorii*-infected DCs protect C3H/HeN mice against lethal rickettsial infection. This protection is associated with significant expansion of splenic lymphocytes and NK cells in vivo. Together, these data suggest that differential interactions of rickettsiae with BMDCs might contribute greatly to different host susceptibilities to rickettsial diseases. IL-12p40-producing mature DC could drive the differentiation of $CD4^+$ T cells toward a protective Th1 phenotype in the resistant host, while decreased MHC class II expression and lack of IL-12p40 production by DC in the susceptible host stimulate T regulatory cells and suppress effector Th1 or Th2 cells. Whether suppression of effector CD4+ T cell responses by DC from the susceptible host is mediated indirectly by T regulatory cells or directly by *Rickettsia*-stimulated DC remains to be investigated.

The immune response to *O. tsutsugamushi* has been studied in laboratory mice. This bacterium infects a variety of host cells in vitro and in vivo, including macrophages and endothelial cells, where it replicates in the cytoplasm without being surrounded by a phagosomal membrane (Ng et al., 1985; Chi et al., 1997). Analysis of early immune responses to *O. tsutsugamushi* infection in mice showed that macrophage-mediated cellular immunity is essential for resolution of this infection (Nacy and Groves, 1981). The pathogenesis of scrub typhus and protection of *O. tsutsugamushi*-infected animals are influenced by at least three factors: (1) route of inoculation, (2) *Orientia* strain, and (3) natural resistance of the host. Natural resistance of mice to lethal infections of *O. tsutsugamushi* is controlled by a single, autosomal, dominant gene, which is designated *Ric*, with "*r*" and "*s*" representing the resistant and susceptible alleles, respectively (Nacy and Groves, 1981; Groves et al., 1980; Jerrells and Osterman, 1981). The expression of the early T lymphocyte activation (*Eta-1*)/Osteopontin (*op*) gene, which maps to the natural resistance gene

(*Ric*), correlates with resistance to *O. tsutsugamushi* infections in mice (Patarca et al., 1989). The expression of Eta-1a allele represents an essential early step in the pathway that leads to Th1 immunity, an effective arm of immune response against *Orientia*, by balancing IL-12/IL-10 production. In addition to its immunomodulatory effect, Eta-1a/op enhances resistance to rickettsial infection by affecting the ability of macrophages to migrate to the site of infection and/or to exert bactericidal activity (Singh et al., 1990). Replication of *O. tsutsugamushi* is controlled by a cellular immune response dominated by type 1 cytokines, in particular IFN-γ. In a mouse model of scrub typhus, IFN-γ production was closely associated with the development of immunity against *O. tsutsugamushi* (Koh et al., 2004; Yun et al., 2005), and elevated levels of circulating IFN-γ have been observed in a limited number of patients with scrub typhus (Chierakul et al., 2004; de Fost et al., 2005; Watt and Parola, 2003). In addition, the plasma concentrations of the IFN-γ-inducing cytokines IL-12p40, IL-18, IL-15, and TNF-α are also elevated in patients with scrub typhus (Yun et al., 2005). Although data obtained from patients with scrub typhus must be evaluated carefully since cytokine levels vary on a daily basis and the levels of these cytokines/chemokines can be influenced by several confounding factors in humans, these data suggest that a cell-mediated immune response is critical in host defense against *O. tsutsugamushi*.

Adaptive Immunity

The generation of protective T cell responses to infectious agents is a complex process in which cytokines and costimulatory molecules provide signals that direct the development of adaptive immunity. Understanding the events that influence the initial generation and expansion of effector cells and the maintenance of memory is critical for the rational design of vaccines. Similar to other intracellular pathogens, cell-mediated immunity characterized by substantial generation of IFN-γ producing Th1 and CD8 type-1 cells plays a critical role in host defenses against rickettsial infections (Feng et al., 1997; Walker et al., 2000, 2001). $CD8^+$ T cells provide a greater contribution to long-term protective immunity against rickettsiae than $CD4^+$ T cells. MHC-class I knockout mice that lack CD8+ T cells are highly susceptible to *R. australis* rickettsial infection (Walker et al., 2001). All MHC-I gene knockout mice succumb to infection with *R. australis* and develop severe subcutaneous edema and pleural and peritoneal effusions. Furthermore an ordinarily sublethal dose of *R. conorii* is lethal or causes persistent infection in mice depleted of CD8+ T lymphocytes (Feng et al., 1997; Walker et al., 2000, 2001). In contrast, mice depleted of CD4+ T cells clear the infection in a similar time frame as normal mice (Feng et al., 1997). Adoptive transfer of immune CD4+ or CD8+ T lymphocytes before infection with *R. conorii* protects immunocompetent naive mice (Feng et al., 1997). However, the immunity provided by the transferred cells is dose-dependent, as neither CD4 nor CD8+ T cells protect mice against a higher dose of rickettsiae. $CD4^+$ T cell-mediated protective immunity requires T cell production of IFN-γ, whereas $CD8^+$ T cells mediate protection via IFN-γ production and cytotoxic killing (Feng et al., 1997; Walker et al., 2001). The importance of IFN-γ in immunity to rickettsiae is emphasized by the observation that the IFN-γ gene knockout mice with the C57BL/6 background are 100 times more susceptible to a lethal outcome of *R. australis* infection than wild-type C57BL/6 mice (Walker et al., 2001). Despite the increased susceptibility of IFN-γ knockout mice, these mice survived lower doses of rickettsiae infection. This observation together with the finding that mice do not survive the lowest dose of rickettsial infection without CD8+ T cells suggests that an IFN-γ-independent CD8 T cell-mediated cytotoxic effector function plays a major role in host defense against *Rickettsia*. MHC-class-I-restricted $CD8^+$ T cell responses to *Rickettsia* determined by CTL activity against *Rickettsia*-infected, MHC-matched target cells reach peak frequencies approximately 10 days after intravenous infection. Several cytotoxic cell death-mediating effector molecules including Fas/Fas ligand, TNF/TNFR, perforin, and granulysin may contribute to function of CD8+ T cells (Shresta et al., 1998; Doherty et al., 1997). Perforin gene knockout mice are much more susceptible than wild-type mice indicating that perforin-mediated activity accounts for a large component, but not all, of the CTL-mediated antirickettsial effect (Walker et al., 2001). A recent study in Thai patients with scrub typhus caused by *O. tsutsugamushi* displayed elevated plasma concentrations of granzymes A and B together with IFNγ and IFN-γ-inducible protein 10 (IP-10) (de Fost et al., 2005). These data suggest that activation of cytotoxic lymphocytes is part of the early host response to scrub typhus.

Taken together, these data strongly suggest that there are two important effector components of acquired immune response against *Rickettsia*, namely IFN-γ production by $CD4^+$ and CD8+ type-1 cells, which activates intracellular bactericidal mechanisms of endothelial cells and macrophages, and the generation of *Rickettsia*-specific cytotoxic CD8+ T cells that lyse infected target cells via pathways involving perforin and/or granzymes.

Humoral Response

The concept that only cell-mediated immunity plays a role in resistance to intracellular pathogens in large part stems from work initially begun in the 1960s by George Mackaness examining *Listeria* infection in mice (Mackaness, 1962, 1969). These experiments demonstrated that immune splenocytes could transfer protection via the activation of macrophages. Earlier studies of correlates of protection of patients with SFG rickettsiosis (Ricketts, 1908; Topping, 1940, 1943) or in guinea pigs (Murphy et al., 1980; Sumner et al., 1995) passively immunized with immune sera against *Rickettsia* also provided little evidence of a role for antibodies in primary host defense against *Rickettsia.* On the other hand, pretreatment of rickettsiae with immune serum before infection prevents lethal disease. Because the latter situation mimics a secondary rather than primary immune response against *Rickettsia*, investigations have focused on cellular immunity as the main mechanism of protection against primary infection with these intracellular pathogens. However, it should be noted that some monoclonal antibodies (MAbs) of a defined specificity and isotype can provide protection. Natural infection does not result in the production of protective antibodies prior to clearance of rickettsiae. These observations indicate that humoral immunity may be more important in preventing reinfection than in clearance of primary infection. In the past few recent years, provocative studies have shown the role of antibody in protection against certain intracellular pathogens (Edelson et al., 1999; Edelson and Unanue, 2000a, 2000b).

In the animal model of SFG rickettsiosis, C3H/SCID mice that receive MAbs against *Rickettsia* OmpA or OmpB or polyclonal antisera specific to *Rickettsia* 48h prior to intravenous infection with 10 LD_{50} of *R. conorii* are completely protected, while anti-LPS antibodies or Fab fragment of the polyclonal antibodies are not protective (Feng et al., 2004a). Polyclonal antibodies or MAbs against OmpA or OmpB given after the establishment of infection are effective in killing rickettsiae, presumably interacting with rickettsiae that exist extracellularly as they spread from one cell to another. Kinetic analysis of antibody development against *R. conorii* reveals that antibodies to LPS appear on day 6 during the illness, while antibodies to OmpA and OmpB appear on day 12 after recovery from the infection. The T cell response against *Rickettsia* reaches a peak at 9–12 days postinfection (Feng et al., 1997; Walker et al., 2001), which suggests that antibodies to OmpA and OmpB are T cell dependent, while antibodies against LPS are T cell independent. Since antibodies to OmpA and OmpB were detected on day 12, after recovery from the illness had occurred, it seems more likely that an antibody response would play an important role in protective immunity against reinfection than in clearance of the primary infection in an immunocompetent host. Recently, the potential mechanism(s) of antibody-mediated protection against *Rickettsia* has been examined (Feng et al., 2004b). The observation that Fab fragments of polyclonal antiserum do not provide protection in vivo indicates that antibodies mediate protection most likely via enhancement of opsonization and phagocytosis of organisms. In vitro cultured mouse endothelial and macrophage-like cell lines incubated with *R. conorii*-coated with different types of antibodies reveal that the Fc portion of the antibodies enhances adherence of opsonized rickettsiae and subsequent internalization. Binding of polyclonal antibodies or MAbs against OmpA or OmpB to *Rickettsia* enhances eventual intracellular killing. In addition, polyclonal antibodies and anti-OmpB MAb inhibit the escape of *R. conorii* from the phagosome, resulting in intraphagolysosomal rickettsial death mediated by NO, reactive oxygen intermediates, and L-tryptophan starvation (Feng et al., 2004b).

Persistence of Rickettsial Infection, Latent Infection, and Reactivation

Although rickettsiosis has been identified as an acute infectious disease, several lines of evidence suggest that *Rickettsia* can cause chronic infection. *R. rickettsii* organisms were isolated from lymph nodes of patients 1 year after recovery from RMSF (Parker et al., 1954). Persistence of *Rickettsia*-specific antibodies in patients recovered from RMSF suggests a latent infection. Similarly, in naturally infected dogs, antibody titers can be detected at least 12 months after *R. rickettsii* infection, suggesting a persistent infection (Tesouro et al., 1998). *O. tsutsugamushi* also persist in tissues of patients recovered from scrub typhus (Smadel et al., 1952). *R. prowazekii*, the etiologic agent of epidemic typhus and recrudescent typhus (Brill–Zinsser disease), causes latent infection (Zinsser and Castaneda, 1933; Walker et al., 2003).

The long-term persistence of pathogens in a host that is also able to maintain strong resistance to reinfection is a characteristic of certain infectious diseases. The ability of certain intracellular pathogens to establish latency in immune individuals is often associated with a suppressed immune response or antigenic variation (Salgame, 2005; McCulloch, 2004). To the best of our knowledge, antigenic variation does not occur in *Rickettsia.* Sterile immunity in resistant *R. conorii*-infected mice maintained by $CD4^+$ T cells, IL-12, IFN-γ, and inducible iNOS suggests that impairment of these responses may result in rickettsial persistence and possible reactivation. The explanation as to why these control mechanisms fail to completely eliminate the *Rickettsia* is not known. Transient immunosuppression occurs in vivo during chronic infection of mice with *O. tsutsugamushi* administered subcutaneously (Jerrells, 1985). Spleen cells from subcutaneously infected mice do not proliferate in vitro in response to a polyclonal T cell stimulator such as ConA. Additionally, mice infected subcutaneously with *O. tsutsugamushi* develop chronic infection and are unable to mount an antibody response to either T cell dependent antigens such as SRBC or T cell independent antigens such as TNP-ficoll (Jerrells, 1985). However the suppressed antibody response to the T-independent antigen TNP-ficoll was not as severely affected as antigens requiring T cell help to produce antibodies, suggesting that the transient immune suppression following *O. tsutsugamushi* infection is due to suppression of T cells. Nevertheless, the mechanism(s) of the observed immunosuppression is not completely understood, and it is controversial as to whether or not T suppressor cells play a role in this mechanism. More importantly, if such transient immunosuppression occurs following natural rickettsial infection and whether this could be a mechanism of persistence of rickettsiae remain to be determined. In support of the concept of immunosuppression is the observation that *O. tsutsugamushi* infection suppresses the production of proinflammatory cytokines (e.g., TNF-α) in murine macrophages cocultured in vitro with the rickettsiae or conditioned medium from *O. tsutsugamushi*-infected macrophages (Kim and Kang, 2001). Further study revealed that *Orientia*-mediated suppression of macrophage activation is due to production of immunosuppressive cytokine IL-10 (Kim et al., 2005). In humans, elevated serum levels of IL-10 are also present in patients with boutonneuse fever (Cillari et al., 1996). Nevertheless, the source of IL-10 in murine or human rickettsiosis has not been determined. Consistent with the hypothesis of immunosuppression in severe rickettsial disease, infection of C3H/HeN mice with *R. conorii* that closely mimics severe human rickettsioses with similar disseminated vascular injury is associated with inability to control bacterial replication resulting in an overwhelming infection (Walker et al., 1994). Splenocytes from these lethally infected mice exhibit suppressed proliferation to in vitro polyclonal stimulation (Feng et al., 1994). *Rickettsia*-infected DC stimulate significant expansion of Foxp3T regulatory cells, which correlates with suppressive adaptive immune responses (Fang et al., 2007). A similar immunosuppressive mechanism appears to occur in vivo following lethal *R. conorii* infection in susceptible C3H/HeN mice. Higher bacterial burden and fatal disease in mice infected with lethal dose of *R. conorii* are associated with unresponsiveness of CD4+ T cells as evidenced by suppressed nonantigen-specific proliferation and IL-2 production, which are strongly associated with significant expansion of inducible antigen-specific T regulatory cells producing IL-10 and IFN-γ.

Another mechanism of persistence is derived from studies with virulent and attenuated strains of *R. prowazekii*. Resistance to IFN-γ-mediated intracellular killing has been observed in vitro following culture of avirulent Madrid E strain of *R. prowazeki* in IFN-γ-treated fibroblast L929 cells (Turco and Winkler, 1989). Since the passage of rickettsiae in culture selects IFN-γ resistant strains, it is possible that this avirulent phenotype could emerge in vivo after recovery from primary infection, and acquire the ability to persist due to their IFN-γ resistance (Turco and Winkler, 1989, 1990; Turco et al., 1989). The ability of avirulent Madrid E strain to revert to the virulent phenotype may explain, to some extent, reactivation of *R. prowazekii* infection and development of recrudescent typhus.

EPIDEMIOLOGY

Natural rickettsial infections are all transmitted by arthropods. The four *Rickettsia* species that are threats for use by bioterrorists are all highly infectious by inhalation of a low dose of rickettsiae by aerosol. TG rickettsiae have an insect vector, the human body louse (*Pediculus humanus corporis*) for *R. prowazekii* and fleas (particularly *Xenopsylla cheopis*, the Oriental rat flea, worldwide and currently *Ctenocephalides felis* cat flea in the US).

TG rickettsiae are transmitted mainly in their vectors' feces, which contains a stable infectious form of the bacterium that is scratched into the skin, rubbed into mucous membranes such as the conjunctiva, or inhaled. *R. prowazekii* is highly lethal for its louse host; 100% die. Thus, the louse is not a reservoir host of *R. prowazekii*. Humans who survive acute louse-borne typhus maintain a latent infection and serve as reservoir host of the rickettsiae. Years later some latently infected persons fail to maintain effective immune control of the rickettsiae, allowing for reactivation of rickettsial growth, recrudescent typhus fever, and bacteremia that infects 5% of lice that are feeding on them. Human body lice live in the clothing of the person, where the temperature is suitable for them, and they leave a febrile person because of their aversion to the higher temperature. This departure seeking another host accomplishes transfer of the rickettsiae to a susceptible person. After 5–7 days the growth of *R. prowazekii* in louse gut epithelium results in rickettsial shedding in the feces onto the skin of the new host. Successive cycles of these events in nonimmune, louse-infested persons under crowded conditions result in an outbreak. Travel of persons during the incubation period (e.g., displaced by warfare) spreads the infection leading to a widespread epidemic such as occurred among 100,000 persons during the civil war that occurred in the mid-1990s in Burundi.

R. prowazekii is also maintained in a zoonotic cycle involving flying squirrels (*Glaucomys volans*) and their specific flea and louse. Humans become infected from this cycle, presumably when contacting the infected fleas or infected feces from the lice and fleas. *R. prowazekii* also resides in some ticks.

The first biologic weapon developed by the Soviets was *R. prowazekii*. The history of Japanese biologic weapon development during World War II also includes typhus.

Murine typhus occurs in persons who contact the feces of the infected fleas and less often by flea bite. Globally murine typhus has a very high incidence although typically neither diagnosed nor treated effectively. Analyses such as disability adjusted days of life lost are blind to the importance of murine typhus and other unreported rickettsioses. The most common natural reservoir of *R. typhi* is a zoonotic cycle involving the Oriental rat flea and *Rattus* species. Infected fleas maintain *R. typhi* throughout a normal lifespan, and via transovarian transmission to a small portion of their eggs. Rats are easily infected, develop a sufficient concentration of *R. typhi* in their blood to infect fleas that feed on their blood, and do not become ill.

SFG and transitional group rickettsiae are maintained in nature largely via transovarian transmission in ticks, mites (*R. akari* only), or fleas (*R. felis* only). Some *Rickettsia* species are pathogenic for the vector tick (e.g., *R. rickettsii* causes reduced survival of *Dermacentor* ticks). Thus, infection of new ticks is required to establish new lines of transovarian rickettsial maintenance (e.g., *Dermacentor variabilis* ticks acquire *R. rickettsii* while feeding on rickettsemic cotton rats). The pathologic effects of *R. rickettsii* on ticks apparently account for their extremely low (less than 0.1%) prevalence in their tick hosts. The vectors of Rocky Mountain spotted fever are *D. variabilis* (American dog tick) in the eastern two-thirds of the US and regions of the Pacific coast states, *Dermacentor andersoni* (wood tick) in the Rocky Mountain states, *Rhipicephalus sanguineus* (brown dog tick) in the southwestern US and northern Mexico, and *Amblyomma cajennense* and *Amblyomma aureolatum* in South America.

Each *Rickettsia* has its particular vector(s), e.g., *R. conorii* (*R. sanguineus* tick), *R. africae* (*Amblyomma variegatum* and *Amblyomma hebraeum* ticks), *R. felis* (*C. felis* flea), *and R. akari* (*Liponyssoides sanguineus* mite). In some cases, there is a known amplifying vertebrate host, e.g., the domestic mouse (*Mus musculus*) for *R. akari* during outbreaks of rickettsialpox.

Some rickettsiae are apparently not significantly pathogenic for their vectors in which their prevalence is high (e.g., *R. africae* in *A. variegatum* ticks). The most prevalent rickettsiae are of no known pathogenicity and are apparently maintained purely by vertical transovarian transmission (e.g., *R. peacockii*).

The low dose infectivity of rickettsiae is well illustrated by *R. rickettsii*, which is 1000-fold more infectious by aerosol than the spores of *Bacillus anthracis*, an indication of their danger as a potential biologic weapon.

SFG rickettsioses such as Rocky Mountain spotted fever, boutonneuse fever (MSF), and African tick bite fever are transmitted by ticks during their feeding. Ticks attach to the skin and inject saliva that contains pharmacologically active substances that prevent pain, modulate host defenses, and inhibit coagulation. Sharp tick mouth parts cut dermal blood vessels creating a blood-filled cavity from which blood is ingested and into which *Rickettsia*-infected saliva is injected as soon as 6 h after attachment and later during feeding that typically lasts for a week or longer.

The seasonal and geographic distribution of each rickettsiosis reflects the months of activity of the vector and its contact with humans. Rocky Mountain spotted fever occurs between May and September in the southeastern, mid-Atlantic, and south-central US.

O. tsutsugamushi is maintained entirely by transovarial transmission in trombiculid mites. Humans are thus accidental hosts who become infected by feeding of the larval stage of mite for 2–10 days on the skin. The other life stages of the mites are free living, feed in the soil, and therefore do not transmit orientiae. The seasonal occurrence is determined by the feeding activity of the vector, e.g., *Leptotrombidium pallidum* in autumn-to-winter and late spring-to-early summer; *Leptotrombidium scutellare* in autumn and winter; and *Leptotrombidium deliense* year-round in tropical areas.

PATHOGENESIS

Rickettsia interact with their eukaryote hosts by way of parasitism or mutualism. Rickettsiae are arthropod-borne intracellular bacteria that in vertebrates primarily infect endothelial cells and cause acute diseases. Natural infection with *Rickettsia* involves intradermal inoculation of the organisms together with saliva during tick feeding or with the feces of a louse or flea. Therefore, analysis of the interactions among *Rickettsia*, their initial target cells including potentially dermal microvessel endothelial cells, dendritic cells, and macrophages and infiltrating T cells in the skin is of paramount importance in understanding the pathogenesis of the disease and design of a rational vaccination strategy.

Because of experimental limitations imposed by the obligate intracellular nature of rickettsiae, little is known about specific virulence determinants utilized by these organisms that mediate initial rickettsiae–host interaction. SFG *Rickettsia* attach via its OmpA and OmpB, the latter of which binds to a transmembrane protein Ku70 in the endothelial cell membrane (Li and Walker, 1998; Uchiyama, 1999; Martinez et al., 2005). Binding of *Rickettsia* to Ku70 induces focal host cell cytoskeletal actin rearrangements at the site of attachment, resulting in induced phagocytosis (Uchiyama, 1999; Li and Walker, 1992; Martinez, 2004). Interestingly, *Rickettsia* bind to Ku70 via OmpB, and this binding results in activation of Ku70. Activation of Ku70 is postulated to lead to the activation of a cascade of signaling events, including the small GTPase, Cdc42, phosphoinositidyl-3-kinase, src-family tyrosine kinases, and the tyrosine phosphorylation of focal adhesion kinase (Martinez and Cossart, 2004). These signaling events are known to be strongly associated with activation of β1-integrin (Parsons, 2003) and bacterial entry (Martinez and Cossart, 2004). *R. conorii* infection also stimulates the ubiquitination of Ku70 that is mediated by recruitment of the ubiquitin ligase c-Cbl to *R. conorii* entry foci. The role of Ku70 ubiquitination in rickettsial entry is evident by data showing that down regulation of endogenous c-Cbl blocks Ku70 ubiquitination and bacterial entry. However, further studies are essential to dissect whether Ku70 is utilized directly by *R. conorii* or that indirectly through other host cell receptors or adhesion molecules that facilitate bacterial entry into host cells.

Internalized rickettsiae are initially bound within a phagosome (Teysseire et al., 1995). An increase in rickettsial membranolytic activity (phospholipase D and tlyC) occurs concomitantly with rickettsial entry, and facilitates rickettsial escape from the phagosome into the cytoplasm prior to phagolysosomal fusion, thus avoiding exposure to the lysosomal enzymes (Whitworth, et al., 2005; Walker et al., 2001; Renesto et al., 2003; Martinez et al., 2004). Once in the cytosol, rickettsiae are free to exploit the nutrient-rich environment and interact with host structural components (Austin and Winkler, 1988; Winkler and Turco, 1988). In the cytosol, they replicate by binary fission. The spread of TG rickettsiae occurs via bursting of the massively infected host cell. In contrast, spotted fever rickettsiae are released from the infected cells via long thin cell projections either extracellularly or into the adjacent cell. These projections from which they are released are caused by propulsion by the host cell's actin filaments (Heinzen et al., 1993; Nath, 2004; Teysseire et al., 1992). The nucleation of actin polymerization is activated by Arp2/3, which induces the continuous formation of a network of short and highly branched filaments of actin that appears as tails that propel the rickettsiae (Gouin et al., 2004). In *Rickettsia*-infected cells, Arp2/3 is detected on the rickettsial surface but not in actin tails (Nath, 2004; Gouin et al., 2004). Arp2/3 is normally inactive and has to be activated by proteins of the WASP/N-WASP/Scar/Wave-family proteins in eukaryotic cells (Pollard and Borisy, 2003; Frischknecht and Way, 2001; Gouin et al., 1999). Comparison of genome sequences of *R. prowazekii* and *R. conorii* has identified a 2-kb region that is present in the *R. conorii* genome and is absent from that of *R. prowazekii* (Gouin et al., 2004). This locus encodes a 517-amino acid protein, RickA, which recruits Arp2/3, activates it, and induces actin polymerization. Thus, RickA is a rickettsial actin nucleator that is most closely related to the WASP/N-WASP-family proteins, which are involved in actin-based motility (Gouin et al., 1999; Welch et al., 1998; Machesky et al., 1999; Boujemaa-Paterski et al., 2001). RickA is expressed on the surface of most SFG rickettsiae in infected cells (Gouin et al., 1999, 2004). However, it is unknown how RickA is expressed on the bacterial surface and whether the type IV secretion system predicted by the genome sequence is involved in targeting RickA to the surface. It is possible that posttranslational modification of RickA such as phosphorylation might control its activity. Whether another *Rickettsia* protein also participates in the actin-based motility has to be addressed.

Nevertheless, the mechanism of SFG rickettsial cell-to-cell spread in the microcirculation results in disseminated infection of contiguous networks of endothelial cells, which coincides with lesion development (e.g., maculopapular rash). *R. rickettsii* is more invasive than other rickettsiae, routinely spreading to infect adjacent vascular smooth muscle cells. Because SFG rickettsiae are adapted for both intracellular survival and direct cell-to-cell spread, they are protected, to a large degree, against antibody-mediated complement activation that plays a major role in elimination of extracellular bacteria.

Pathophysiology of Rickettsial Diseases

The fundamental pathology of both spotted fever and TG rickettsioses is vasculitis caused by proliferation of organisms in the endothelial lining of small arteries, veins, and capillaries. This *Rickettsia*-induced vascular injury accounts for most or all of the clinicopathologic abnormalities including the presence of an eschar at the site of the tick bite followed by a hemorrhagic rash, which extends over the entire body, including the palms and soles of the feet. The vascular lesions that underlie the rash lead to increased vascular permeability, with resulting edema due to leakage of fluid from the bloodstream, and its accumulation in the surrounding tissues (Davidson et al., 1990). The interstitial edema in certain organs such as lung and brain can pose a life-threatening situation. Generalized leakage of fluid from the circulation results in hypovolemia and ischemia. In severe cases, rickettsiae cause noncardiogenic pulmonary edema, interstitial pneumonia, and adult respiratory distress syndrome associated with hypoxemia, interstitial myocarditis, perivascular glial nodules of the central nervous system, and similar vascular lesions in the skin, gastrointestinal tract, pancreas, liver, skeletal muscles, and kidneys (Walker et al., 1980; Walker and Mattern, 1980; Randall and Walker, 1984; Adams and Walker, 1981). However, focal lesions or injury in some organs such as the kidneys are not sufficiently extensive to result in organ failure. Rocky Mountain spotted fever and boutonneuse fever induce a procoagulant state including release of procoagulant components, activation of the coagulation cascade with thrombin generation, platelet activation, increased antifibrinolytic factors, consumption of natural anticoagulants, and secretion of coagulation-promoting cytokines (George et al., 1993; Sporn et al., 1991; Shi et al., 1996, 2000; Elghetany and Walker, 1999). It has been speculated that thrombus-mediated vascular occlusion, including disseminated intravascular coagulation (DIC), might be an important pathogenic mechanism in severe rickettsioses. However, DIC rarely occurs, and postmortem studies of fatal cases reveal that thrombi are few and do not constitute an important general pathogenic mechanism. Increased adherence of platelets to infected endothelial cells has been demonstrated in vitro. Consumption of platelets results in thrombocytopenia in 32–52% of patients with Rocky Mountain spotted fever. Analysis of the sequential changes in many hematologic and biochemical parameters in 108 patients with travel-associated African tick bite fever during the first 2 weeks of illness and prior to the institution of antirickettsial therapy revealed several changes that, although less pronounced, mimic those of potentially more severe SFG rickettsioses (Jensenius et al., 2003). For example, there are significant decreases in mean total leukocyte count (leucopenia), mean absolute lymphocyte count (lymphopenia), and mean platelet count (thrombocytopenia), and significant increases in mean serum levels of C-reactive protein (CRP) and serum level of liver enzymes (ALT and AST) (Jensenius et al., 2003). Interestingly, an increased serum level of CRP and lymphopenia are the two most consistent findings. These observations are consistent with sharp decreases in serum IFN-γ in Sicilian patients with acute boutonneuse fever in whom IFN-γ levels dropped sharply during the second week after infection (Cillari et al., 1996). The fall in IFN-γ in boutonneuse fever patients coincides with reduction in the levels of CD4+ T cells in peripheral blood and decreased serum levels of IL-6, IL-10, and TNF (Cillari et al., 1996), indications of a return toward homeostasis.

In scrub typhus, infections with *O. tsutsugamushi* cause local inflammation accompanying eschars at the site of infection, and then spread systemically. Inflammation is initiated by infected macrophages and endothelial cells in the dermis. In the murine model of mild and severe rickettsiosis caused by infection with low and high virulence strains of *O. tsutsugamushi*, an early strong PMN response, followed by mononuclear cell influx is associated with enhanced susceptibility of C3H/HeN mice to *O. tsutsugamushi* infection and coincides with a high mortality rate (Cho et al., 2000; Rikihisa and Ito, 1979). The prominent cellular influx in susceptible mice correlates with delayed but sustained increase in the production of chemokines including MIP-1α, MIP-2, and MCP-1, which are important chemokines for activating neutrophils and monocytes, respectively, recruiting these circulating immune cells to the sites of inflammation. These data suggest that increased PMNs and macrophages infiltration in susceptible hosts may be an indicator of more severe local tissue injury or even mediate tissue damage via secretion of oxidative molecules, rather than providing antirickettsial immunity.

Mechanisms of Cell Injury by *Rickettsia*

Several pathophysiologic mechanisms have been proposed to explain the tissue and organ injury in severe rickettsiosis. These include: (i) reduced delivery of nutrients and oxygen to tissues owing to hypoperfusion and hypoxemia due to decreased blood volume, pulmonary edema, and in severe focal lesions, thrombus-mediated vascular occlusion and (ii) increased vascular permeability associated with intra-endothelial infection (Valbuena et al., 2002). *Rickettsia*-induced increase in vascular permeability is considered to be the result of interplay between host and microbial factors, particularly in the heavily parasitized vasculature. A critical host factor that plays a dominant role in influencing vascular permeability in rickettsial infections is TNF-α. TNF-α increases endothelial permeability with subsequent vascular leakage contributing to severe organ dysfunction. Significant increases in plasma levels of TNF-α in patients with severe MSF suggest that TNF-α could be a causal contributor to pathogenesis (Kern et al., 1996; Oristrell et al., 1994). Another pathogenic mechanism is production of high levels of reactive oxygen species (ROS) by *Rickettsia*-infected endothelial and phagocytic host cells, causing tissue damage via lipid perioxidation of host cell membranes (Silverman and Santucci, 1988, 1990; Santucci et al., 1992; Eremeeva and Silverman, 1998a; Eremeeva et al., 2001). The oxidative stress-mediated injury of cultured endothelial cells is associated with depletion of defenses against oxidative stress. These components include enzymes such as glucose-6-phosphate dehydrogenase, glutathione reductase, heme oxygenase, and catalase (Silverman and Santucci, 1990; Santucci et al., 1992; Eremeeva and Silverman, 1998a; Eremeeva et al., 2001; Hong et al., 1998; Rydkina et al., 2002). The ameliorating effect of antioxidant molecules such as α-lipoic acid in an in vitro infected endothelial cell model supports a significant role of *Rickettsia*-induced oxidative stress as a pathogenic mechanism of cell injury (Eremeeva and Silverman, 1998a, 1998b; Eremeeva et al., 2001). However, the role and importance of ROS-mediated endothelial cell injury has not been established in vivo although there is evidence of oxidative stress in *R. conorii*-infected mice. It is important to bear in mind that ROS together with NO exert microbicidal functions that contribute to the control of intracellular pathogens (Hong et al., 1998; Rydkina et al., 2002, 2004; Eremeeva and Silverman, 1998b). The paradox of ROS acting as a host defense and a pathologic mechanism is unresolved, but presumably is due to localization in different cellular compartments and an imbalance between the formation of ROS and their removal by endogenous antioxidant scavenging compounds (Eremeeva et al., 2001; Hong et al., 1998; Rydkina et al., 2002, 2004; Eremeeva and Silverman, 1998b). Oxidative stress and endothelial cell damage occur when there is an excessive generation of ROS. The latter could be due to failure of the innate immune response to control intracellular bacterial replication. Rickettsiae may induce persistent and substantial endothelial cell activation, resulting in excess generation of proinflammatory cytokines and oxidative molecules, both of which could play a role in tissue injury by inducing the activation of transcription factors of the NF-κB family. NF-κB regulates the expression of cytokines, chemokines, adhesion molecules, and other early response genes. These transcription factors remain in the cytoplasm bound to the inhibitor IκB, which upon phosphorylation is targeted to the proteasome for degradation, resulting in exposure of the nuclear translocation signals of NF-κB. *R. rickettsii* directly activates NF-κB, causing upregulation of antiapoptotic factors that protect against host cell death and support cell cycle progression (Joshi et al., 2003, 2004; Clifton et al., 2005a). When *R. rickettsii*-induced activation of NF-κB is inhibited, apoptosis of infected but not uninfected endothelial cells rapidly ensues. Therefore, the antiapoptotic effect of *Rickettsia* via NF-κB activation is considered as an immune evasion mechanism that potentially allows the host cell to remain as a site of infection, thus supporting rickettsial survival and replication. Thus, continuous activation of host cells by persistently replicating rickettsiae could contribute to pathology and tissue damage in rickettsial infection. Finally, rickettsial phospholipase(s) or protease(s) may also play a role in cell injury (Walker et al., 1983, 1984; Silverman et al., 1992). It is most likely that tissue injury by *Rickettsia* is a multifactorial process, in which proinflammatory cytokines, oxidative molecules, procoagulant factors, rickettsial phospholipase(s) and proteases act in damaging host cells and tissues.

Role of Chemokines in Pathogenesis of Rickettsiosis

Studies of two major pathogenic species of SFG rickettsiae, namely *R. rickettsii* and *R. conorii*, have established the occurrence of significant changes in the expression of important adhesion molecules and regulatory chemokines during in vitro and in vivo infection of host endothelial cells (Eremeeva et al., 2000; Valbuena et al., 2003; Clifton et al., 2005b; Valbuena and Walker, 2005). In vitro *R. conorii*-infected endothelial cells express higher levels of E-selectin, intercellular adhesion molecule-1 (ICAM-1), and vascular cell adhesion molecule-1 (VCAM-1) than uninfected cells (Dignat-George et al., 1997). The activation mechanism involves autocrine/paracrine stimulation with IL-1α. Patients with severe boutonneuse fever have significantly higher plasma levels of soluble E-selectin and soluble VCAM-1 than patients with mild disease, who also have increased levels of soluble ICAM-1 (Dignat-George et al., 1999). Thus the soluble forms of adhesion molecules may play a pathogenic role in rickettsiosis as they enhance endothelial adhesiveness and leukocyte accumulation. Another study in Sicilian patients with boutonneuse fever suggested that soluble forms of adhesion molecules may produce an immunomodulatory, anti-inflammatory effect by blocking adhesion between leukocytes and endothelial cells (Vitale et al., 1999).

TREATMENT

Early treatment of the spotted fever rickettsiosis is the most important factor in speeding convalescence and reducing mortality. Antibiotic therapy that is begun early in the course results in rapid resolution of clinical abnormalities. Empiric treatment of rickettsioses should be started before laboratory confirmation of the diagnosis.

The preferred treatment for RMSF, boutonneuse fever, Japanese spotted fever, African tick bite fever, and other spotted fever rickettsioses is doxycycline (Vibramycin), 100 mg intravenously or orally every 12 h for adults and children who weigh more than 45 kg (Holman et al., 2001; Masters et al., 2003; Purvis and Edwards, 2000). The dose for children who weigh less than 45 kg and for children less than 8 years of age is 3 mg/kg body weight in two divided doses (maximum dose 200 mg/day). Adjunctive measures such as oxygen therapy, blood transfusion, anticonvulsants, mechanical ventilation, and hemodialysis are sometimes needed in severe cases. Although there are no published studies on the optimal duration of therapy for Rocky Mountain spotted fever, doxycycline therapy is usually continued for 5–10 days or until the patient has been afebrile for at least 48 h. Doxycycline is also the treatment of choice for children with Rocky Mountain spotted fever as the risk of dental staining is minimal when short doses of therapy are given. In patients with severe hypersensitivity to tetracyclines, 50–75 mg/kg of body weight/day of chloramphenicol can be considered as an alternative therapy, but it is less effective and its use is limited by side effects. Doxycycline should not be used in pregnant women (who should be treated with chloramphenicol [Chloromycetin] 500 mg intravenously every 6 h). Although both doxycycline and chloramphenicol are effective, neither drug is rickettsicidal. The antibiotics inhibit rickettsial growth until an adequate immune response by the patient eradicates the infection. Several quinolones have also been documented to be effective for treatment of patients with boutonneuse fever. Newer macrolides, such as clarithromycin and azithromycin have been proposed as alternatives to doxycycline and chloramphenicol in treatment of children and pregnant women with boutonneuse fever (Segura and Anton, 2002; Cascio et al., 2002, 2001). Azithromycin seems to be a better treatment option as it is administered once a day and has proven effective after a shorter duration of therapy (3 days) compared to clarithromycin (7 days). Because boutonneuse fever is a less severe disease and the macrolides were used to treat only children with mild illness, neither fluoroquinolones nor macrolides are recommended for treating severe rickettsioses such as Rocky Mountain spotted fever. Many classes of broad-spectrum antibiotics, including penicillins, cephalosporins, and aminoglycosides, are ineffective as therapy for rickettsial diseases (Rolain et al., 1998). Sulfa-containing antimicrobials are not only ineffective but also exacerbate rickettsioses (Ruiz Beltran and Herrero Herrero, 1992).

The treatment of choice for murine and epidemic typhus is oral doxycycline; even a single 200 mg dose usually leads to defervescence within 48–72 h.

The treatment of choice for scrub typhus is doxycycline. Scrub typhus cases from northern Thailand with a diminished response to doxycycline and chloramphenicol have been described (Watt et al., 2000), but neither the geographic distribution of resistance nor its mechanism have been defined. Limited data suggest that rifampicin is effective in areas where scrub typhus responds poorly to standard antirickettsial drugs (Watt et al., 2000). Roxithromycin, a macrolid antibiotic, was as effective as doxycycline and chloramphenicol in a trial of 39 Korean children (Lee et al., 2003), and azithromycin and clarithromycin have been used in pregnancy and where doxycycline and chloramphenicol resistance is a problem.

VACCINES

History

The history of development of vaccines against rickettsial diseases contains numerous failures and limited success in preventing or ameliorating disease. Technologic approaches to the preparation of vaccine materials have reflected advances in knowledge and methods. The three vaccines that have been developed to prevent Rocky Mountain spotted fever illustrate these advances. When it had been established that *R. rickettsii* not only survived but also replicated in ticks, a killed rickettsial preparation from infected ticks was developed by Spencer and Parker in 1924. Because preparation of this vaccine was cumbersome and dangerous to workers dealing with live infected ticks, the method of propagating *R. rickettsii* in yolk sac of embryonated chicken eggs was adapted soon after its development by Cox in 1938. A third killed rickettsial vaccine that was free of yolk sac antigens was prepared from cell culture-propagated organisms by the US Army in the 1970s.

The original tick rickettsial vaccine produced severe local inoculation site reactions. Evaluation of its protective effect in field use was impressive by the standards of the day. The fatality rate from Rocky Mountain spotted fever among vaccine recipients was reduced dramatically. Annual booster vaccinations were recommended. Vaccinated persons continued to get infected, but generally appeared to suffer less severe disease.

The killed yolk sac rickettsial vaccine was never field tested. Like recipients of the tick vaccine, some vaccine recipients became ill with Rocky Mountain spotted fever. A challenge trial in human volunteers was conducted in 1973. Neither the yolk sac vaccine nor the tick vaccine prevented the illness, which, of course, was treated to prevent severe illness or death. The yolk sac vaccine was withdrawn from the market in 1978. Subsequent challenge trial of the cell culture killed-*R. rickettsii* vaccine yielded protection of 25% of the volunteers who received it. Evaluation of the recipients' immune responses revealed failure to stimulate sustained cellular immunity.

Attempts to develop a vaccine against epidemic typhus fever used living organisms as well as killed rickettsiae. A low dose of *R. prowazekii* from human blood or guinea pig brain or crossreactive less pathogenic *R. typhi* suspended in ox bile or in egg yolk and olive oil was reported to have protected some persons and to have caused severe illness, even death, in other vaccinees.

Propagation of *R. prowazekii* for preparation of killed rickettsial vaccines employed intrarectal inoculation of lice, infected louse nymphs, tunica vaginalis of intraperitoneally inoculated animals, and lungs of intranasally inoculated animals. Weigl's louse-derived vaccine required intrarectal infection of 30 lice, dissection of the lice's intestines, and phenol-inactivation for a single person's vaccination. Cox's chicken embryo yolk sac-propagated, killed-*R. prowazekii* vaccine was massively produced during World War II to protect US military personnel. Epidemic typhus caused many deaths in civilians, Axis military personnel, and concentration camp prisoners in North Africa, western Asia, and Europe, particularly on the Eastern Front during this period. Not a single death was caused by louse-borne typhus in US soldiers in North Africa and the Middle East between 1942 and 1945. Use of this vaccine was continued until 1968 when great variability in the potency of different vaccine batches was discovered. The FDA vaccine potency testing method was the level of stimulation of antibodies that neutralized the mouse toxicity phenomenon. This surrogate test depends on the presence of antibodies to conformational epitopes on OmpA and OmpB for spotted fever rickettsiae and to OmpB for TG rickettsiae. Antibodies to these rickettsial antigens may indeed play a role in prevention of infection or clearance of rickettsiae early in the infection before onset of illness.

Development of a vaccine against scrub typhus has proven extremely difficult. The approaches that have been attempted include vaccine containing killed organisms, live putatively attenuated *O. tsutsugamushi*, viable gamma irradiated orientiae unable to replicate, and a regimen of induced infection and prophylactic anti, rickettsial treatment. Immunity to neither homologous nor heterologous strains of *O. tsutsugamushi* has been achieved. Attenuation was not successful.

Rickettsial Antigens and Protective T Cell Epitopes

Crossprotection between SFG and TG rickettsiae, which are genetically related but antigenically different, in intravenously infected animal models is mediated by T cells, which recognize crossreacting antigens between the two rickettsial groups. Human T lymphocytes also are stimulated by shared antigens. The SFG rickettsiae differ from the TG rickettsiae not only in their capacity to stimulate host cell actin polymerization for cell-to-cell movement, but also in their larger genome size, the presence of species-specific LPS antigen, and rickettsial OmpA. The nature of the rickettsial antigens triggering acquired immune responses has been the subject of much research directed toward the development of a vaccine that may offer solid protection comparable to that acquired by natural infection. However, the immunodominant rickettsial antigens responsible for CD4+ and CD8+ T cell stimulation are only partially known. OmpA and OmpB are two major OMPs that were the first candidates identified based on their reactivity to immune sera, protection of guinea pigs by immunization with recombinant *R. rickettsii* or *R. conorii* OmpA, and protection of mice against lethal *R. typhi* challenge by immunization with native *R. typhi* OmpB. These proteins possess not only B cell epitopes but also T cell epitopes.

Attempts to identify the immunodominant epitopes of these OMPs was addressed by subcloning small fragments and assaying their stimulatory and protective effect by DNA immunization with different fragments from OmpA and OmpB ($ompA_{2176-3933}$, $ompA_{4999-6710}$, $ompB_{1550-2738}$, and $ompB_{2459-4123}$) and subsequent booster immunization with the homologous recombinant protein. These immunizations stimulate T lymphocyte proliferation and IFN-γ secretion in mice. All immunized mice are protected from lethal challenge by immunization with a combination of the four fragments. Immunization of mice with recombinant *Mycobacterium vaccae* expressing $ompA_{3006-3960}$ and subsequent boosting with the homologous protein produces partial protection against lethal challenge with *R. conorii*, as does DNA priming with $ompA_{2331-3976}$ with booster immunizations with the homologous protein. Despite the protection mediated by immunization with the OmpA and OmpB fragments, all mice exhibit signs of disease compared to no illness in rechallenged mice that have recovered from sublethal infection with *R. conorii*.

Similar observations have been made with the Karp strain of *O. tsutsugamushi*, where 56-kDa Omp and other proteins of this strain employed as vaccine candidates for prevention of scrub typhus are not as efficient as infection with live organisms in preventing reinfection, despite the ability to induce strong humoral and cellular immune responses. These observations suggest that memory T cell responses to other unidentified rickettsial antigens or subdominant epitopes that are expressed upon infection with the whole organism are critical for controlling rickettsial replication and long-term protection.

The number of epitopes involved in the effector CD8+ CTL response to microbial infection is small relative to the total number of epitopes potentially available because effector CTL responses to dominant epitopes tend to suppress responses to subdominant epitopes on the same protein from which immunodominant epitopes are derived. This phenomenon is referred to as immunodominance. However, CTL responses to subdominant epitopes readily compensate for the loss of an effector CTL response to the dominant epitope. Thus, the failure of rickettsial fragment derived from OMPs to establish solid immunity and complete protection compared to natural infection could be due to absence of memory CD8 CTL precursors that are specific to subdominant epitopes on OmpA or OmpB, which could compensate for defective or inadequate generation of effector CTL to immunodominant epitopes. A genome-wide approach is thus essential to identify both immunodominant and subdominant epitopes.

VACCINES IN DEVELOPMENT

Rationale for New Generation Vaccine

Vaccines are required for two rickettsial diseases: epidemic typhus and Rocky Mountain spotted fever. The requirement of vaccines for these diseases is determined by the fact that both diseases can be fatal even with the availability of antibiotic treatment that is effective when given sufficiently early in the course, and antibiotic resistant strains have been obtained by natural selection in the laboratory for *R. prowazekii* and can be genetically engineered for both *R. prowazekii* and *R. rickettsii*. These agents are genuine threats for bioterrorism. The previously developed vaccines are either ineffective (e.g., killed *R. rickettsii*) or easily revert to the virulent state (e.g., attenuated *R. prowazekii* E strain).

Vaccine Candidates

Vaccine candidates for rickettsiae include an attenuated strain of *R. prowazekii* and subunit vaccines for both spotted fever and typhus rickettsiae.

Attenuated E Strain Vaccine

E strain was evaluated as a vaccine on a large scale in South America in the 1950s and in Burundi in the 1960s; 94% (170/181) of immunized persons were protected from natural infection by epidemic typhus compared to the unvaccinated controls in a 14-month period after vaccination (Kenyon et al., 1979). In another study, Fox and colleagues evaluated the duration of immunity stimulated by E strain vaccine in volunteers (Fox et al., 1957). Ninety-six percent (Gonder et al., 1979, 1980) of volunteers who were vaccinated with E strain and challenged with Breinl strain at intervals from 2 months to 36 months remained healthy following challenge, and 83% (5/6) of the volunteers who were challenged at 48–66 months were protected. However, the E strain vaccine caused a late reaction in up to 14% of vaccinated persons 9–14 days after inoculation. The late reaction varied from simple malaise and mild headache to modified typhus characterized by fever, headache, malaise, and occasionally a rash in a small proportion of subjects (Kenyon et al., 1979). One possible explanation for this reaction in some persons immunized with E strain is that E strain might have reverted to virulence in these vaccinated persons after initiating stimulation of immunity. In the early 1970s, Russian scientists reported that one or a few passages of E strain in guinea pigs resulted in the stable enhancement of virulence, and they isolated the virulent revertant Evir strain from animals (Balayeva and Nikolskaya, 1971).

Since the avirulent E strain reverts to virulence when passaged in animals, the mutation that affects the virulence of E strain must be reversible. Comparison of the genomes of *R. prowazekii* E strain with the genomes of *R. conorii*, *R. typhi*, *R. rickettsii*, and *R. sibirica*, identified one gene (Rp028/Rp027) that is inactivated by a frameshift mutation in E strain, but the mutation is not present in virulent revertant Evir strain and wild-type virulent Breinl strain (Zhang et al., 2007). This situation suggests that the mutation reverted to wild-type in the virulent revertant. Rp028/Rp027 shares homology with the regulatory domain of an S-adenosylmethionine (SAM)-dependent methyltransferase. The patterns of methylation of lysine in the surface antigens of E, Evir, and Breinl strains are different. The surface antigen of the virulent Breinl and Evir strains contains more N^{ε}-Me3-lysine and less N^{ε}-Me-lysine than to the avirulent E strain. OmpB, the major OMP of rickettsiae from the virulent strains, is heavily methylated while the OmpB from the attenuated strain is hypomethylated. Western blot analysis of partially digested OmpB revealed that one of the reactive fragments was located at the N-terminus (aa 33-272), in which the lysine residues were either mono-, di-, or tri-methylated (Ching et al., 1992, 1993). However, neither the significance of the difference in methylation between the avirulent E strain and virulent strains of *R. prowazekii* nor the mechanism of the difference in methylation has been determined. The difference in OmpB methylation is most likely not caused by amino acid differences because there is only one amino acid difference in OmpB between the avirulent E strain and virulent Breinl strain. Thus, it seems that the deficiency in lysine methylation in E strain is caused by a mutation in the lysine methyltransferase rather than the mutation in OmpB. The correlation between the SAM-dependent methyltransferase mutation and the virulence of *R. prowazekii* needs to be investigated by genetic complementation of the avirulent E strain or gene knockout in virulent revertant Evir strain of *R. prowazekii*. With the progress in transforming rickettsiae that has been made in recent years, the *R. prowazekii* methyltransferase gene may be inactivated in the future to evaluate whether it is the determinant that attenuates E strain. A substantial knockout in the mutated gene of a virulence factor may prevent the reversion of E strain to the virulent state, thus creating an irreversibly attenuated vaccine against *R. prowazekii*.

Subunit Vaccine for Rickettsia

Two surface protein antigens of *R. rickettsii*, OmpA and OmpB, have been identified as major protective antigens and are candidates for use as subunit vaccines. The first evidence that OmpA and OmpB contain protective epitopes came from the studies of MAbs to heat sensitive epitopes of OmpA and OmpB, which neutralized *R. rickettsii* toxicity in mice and infection in guinea pigs (Anacker et al., 1987b; Li et al., 1988). The *E. coli*-expressed OmpA N-terminal fragment that includes the domain II repetitive peptide and a shorter nonrepetitive region after the repetitive region partially protects guinea pigs against a lethal challenge dose of *R. rickettsii* (McDonald et al., 1987, 1988). A fragment from the N-terminus of *R. conorii* OmpA protects guinea pigs against experimental infection with *R. conorii* and partially protects guinea pigs from challenge with the heterologous *R. rickettsii* (Vishwanath et al., 1990). The full length recombinant OmpA using the baculovirus-eukaryotic cell expression system confers complete protection in guinea pigs against homologous *R. rickettsii* challenge (Sumner et al., 1995). Fragments of the *ompA* and *ompB* genes have been tested as DNA vaccines. In a regime of DNA immunization followed by boosters of the corresponding peptide, mice immunized with one of the following *R. rickettsii ompA* or *ompB* fragments: $ompA_{2176-3933}$, $ompA_{4999-6710}$, $ompB_{1550-2739}$, or $ompB_{2459-4123}$ are partially protected against a lethal challenge with heterologous *R. conorii* (Crocquet-Valdes et al., 2002; Diaz-Montero et al., 2001). The recombinant protein alone from each DNA fragment above also provides partial protection to mice against heterologous *R. conorii* challenge, albeit less protection. However, DNA vaccination alone failed to stimulate protection to mice. It is not known whether the incomplete protection of OmpA and OmpB to the heterologous *Rickettsia* species challenge in these experiments is caused by the antigenic differences between the rickettsial species, the immunization regime, or the antigen composition. Immunization with a combination of all four fragments protected completely against death, but not against illness.

Broad Range Vaccine for Rickettsial Diseases

C3H/HeN mice infected with a sublethal dose of *R. conorii* are completely protected from lethal dose challenge of heterologous *R. australis* or vice versa. In contrast, heterologous immune sera did not protect mice against a lethal dose (2 LD_{50}) of rickettsiae in the mouse toxicity neutralization assay, the standard method for evaluation of rickettsial vaccine potency. These observations suggest that development of a broadly protective vaccine against SFG rickettsiae is feasible, and the results indicate that mouse toxicity neutralization is an inappropriate method for evaluation of candidate vaccines (Feng and Walker, 2002). T cell-mediated crossprotection of mice against challenge with a lethal dose of *R. conorii* or *R. typhi* suggests that a broadly crossprotective vaccine is feasible (Valbuena and Walker, 2004).

Subunit vaccine for O. tsutsugamushi

The antigenic variability of *Orientia* poses a challenge for vaccine design. Immunization with *Orientia* confers relatively strong protection that lasts only 1–3 years against challenge by the homologous strain, whereas protection is very weak and short-lived against heterologous strains (Seong et al., 2001; Choi et al., 2006). Despite the antigenic variation, most studies on subunit vaccine are focused on the 56-kDa protein of *O. tsutsugamushi*. Immunization with the recombinant 56-kDa protein results in potent lymphocyte proliferation and IFN-γ production, but the immunized mice are only partially protected from lethal dose challenge with *O. tsutsugamushi* (Seong et al., 1997). Immunization of monkeys with truncated 56-kDa protein of Karp strain of *O. tsutsugamushi* results in T cell proliferation and IFN-γ production, but does not provide sterile immunity to the monkeys (Chattopadhyay et al., 2005). Subsequent studies with conserved antigens of *O. tsutsugamushi* have demonstrated either partial or no protection. Elucidation of the immune mechanisms of protective immunity, mechanisms of immune avoidance, fate of anti-*O. tsutsugamushi* memory T lymphocytes, and discovery of the antigens that stimulate protective immunity may be required before an effective scrub typhus vaccine is developed.

CONCLUSIONS/FUTURE PROSPECTIVE

The protective immune response to infection with *Rickettsia* spp. is highly dynamic and involves both innate and adaptive immune responses. In SFG rickettsiosis, infection of mice with *R. conorii* has shown that $CD8^+$ T cells, and $CD4^+$ T cells, producing IFN-γ, as well as antibodies and B cells, are required to clear infection and to prevent reinfection.

The search for a vaccine against infection with *Rickettsia* continues to be a complex task. Nevertheless, progress has been achieved in the past few years and has led to the identification of various protective rickettsial antigens as potential vaccine candidates. Although immunization regimens involving priming with DNA vaccines and boosting with protein-based vaccines have been found to be protective, to some extent, in mice, their practical application in humans remains unclear. Given that multisubunit protein vaccines seem to be more effective than vaccines based on single antigens, in the future, rickettsial vaccine candidates are likely to include several antigens.

Rickettsial vaccine research will continue to focus on the identification of additional rickettsial antigens that induce protective T cell responses and on the mechanisms that promote protective immunity against these obligate intracellular pathogens that are transmitted mainly via skin. The gaps in our knowledge include the role of dendritic cells in vivo, and the role of proinflammatory cytokines in influencing the Th1/Th2 response bias. Further data are required to understand the mechanisms that allow evasion or down regulation of the immune responses in cases of latent infection with *R. prowazekii*, which would result in persistence of rickettsiae with possible reactivation. Potential immunomodulatory mechanisms in rickettsial infections could include the effects of tick saliva-containing immunosuppressive factors, anti-inflammatory cytokines as IL-10 and TGF-β, and the possible regulatory effect of particular T cell populations. Finally, a better definition of human immune response correlates with protective immunity against different rickettsial species and disease pathogenesis needs to remain an important research priority if we are to develop a protective vaccine against rickettsial infection.

KEY ISSUES

- *R. prowazekii*, a category US B bioterror threat, and *R. rickettsii*, a US category C organism, are Select Agents that because of potentially engineered antibiotic resistance and high fatality rate after aerosol infection need vaccines for strategic deterrence.
- Subunit vaccines that stimulate antibodies to OmpB of TG rickettsiae and outer membrane proteins A and B of SFG rickettsiae confer protection, presumably by opsonization, and enhancing intracellular killing.
- Recovery from rickettsial infection is mediated by cytokine-activated endothelial cell killing of rickettsiae and CD8 cytotoxic T lymphocyte elimination of infected cells.
- An effective live attenuated vaccine against typhus, *R. prowazekii* Madrid E strain, which reverts to virulence because it depends on a single point mutation, is an excellent candidate for further development by a permanently inactivating mutation of the methyltransferase gene.
- The existence of broadly crossprotective T lymphocyte immunity suggests that development of a pan-*Rickettsia* vaccine is feasible.
- Knowledge of the mechanisms of protective immunity in scrub typhus, reasons for its transience, and poor crossprotection among strains of *O. tsutsugamushi* are severely lacking and very likely hold the key to future progress in vaccine development.

References

Adams, J.S. and Walker, D.H. The liver in Rocky Mountain spotted fever. Am. J. Clin. Pathol. 1981; 75:156–161.

Amano, K., Cedzynski, M., Swierzko, A.S. et al. Comparison of serological reactions of Rickettsiae-infected patients and rabbit anti-*Proteus* OX antibodies with *Proteus* OX2, OX19 and OXK lipopolysaccharides. Arch. Immunol. Ther. Exp. (Warsz.) 1996; 44:235–240.

Amano, K., Tamura, A., Ohashi, N., Urakami, H., Kaya, S. and Fukushi, K. Deficiency of peptidoglycan and lipopolysaccharide components in *Rickettsia tsutsugamushi*. Infect. Immun. 1987; 55:3392–5590.

Amano, K.I., Williams, J.C. and Dasch, G.A. Structural properties of lipopolysaccharides from *Rickettsia typhi* and *Rickettsia prowazekii* and their chemical similarity to the lipopolysaccharide from *Proteus vulgaris* OX19 used in the Weil–Felix test. Infect. Immun. 1998; 66:923–926.

Anacker, R.L., Mann, R.E. and Gonzales, C. Reactivity of monoclonal antibodies to *Rickettsia rickettsii* with spotted fever and typhus group rickettsiae. J. Clin. Microbiol. 1987a; 25:167–171.

Anacker, R.L., McDonald, G.A., List, R.H. et al. Neutralizing activity of monoclonal antibodies to heat-sensitive and heat-resistant epitopes of *Rickettsia rickettsii* surface proteins. Infect. Immun. 1987b; 55:825–827.

Anderson, B.E. The 17-kilodalton protein antigens of spotted fever and typhus group rickettsiae. Ann. N.Y. Acad. Sci. 1990; 590:326–333.

Andersson, S.G., Zomorodipour, A., Andersson, J.O. et al. The genome sequence of *Rickettsia prowazekii* and the origin of mitochondria. Nature 1998; 396:133–140.

Austin, F.E. and Winkler, H.H. Relationship of rickettsial physiology and composition to the rickettsia–host cell interaction. In: Biology of Rickettsial Diseases (D.H. Walker, Ed.), Vol. 1, pp. 29–49. Boca Raton: CRC Press, 1988.

Balayeva, N.M. and Nikolskaya, V.N. Enhanced virulence of the vaccine strain E of Rickettsia prowazekii on passaging in white mice and guinea pigs. Acta Virol. 1971; 16:80–82.

Baldridge, G.D., Burkhardt, N.Y., Simser, J.A., Kurtti, T.J. and Munderloh, U.G. Sequence and expression analysis of the ompA gene of *Rickettsia peacockii*, an endosymbiont of the Rocky Mountain wood tick, *Dermacentor andersoni*. Appl. Environ. Microbiol. 2004; 70:6628–6636.

Billings, A.N., Feng, H.M., Olano, J.P. et al. Rickettsial infection in murine models activates an early anti-rickettsial effect mediated by NK cells and associated with production of gamma interferon. Am. J. Trop. Med. Hyg. 2001; 65:52–56.

Biron, C.A. and Brossay, L. NK cells and NKT cells in innate defense against viral infections. Curr. Opin. Immunol. 2001; 13:458–464.

Biron, C.A. and Gazzinelli, R.T. Effects of IL-12 on immune responses to microbial infections: a key mediator in regulating disease outcome. Curr. Opin. Immunol. 1995; 7:485–496.

Bonecchi, R., Bianchi, G., Bordignon, P.P. et al. Differential expression of chemokine receptors and chemotactic responsiveness of type 1T helper cells (Th1s) and Th2s. J. Exp. Med. 1998; 187:129–134.

Boujemaa-Paterski, R., Gouin, E., Hansen, G., Samarin, S., Le, C., Didry, D., Dehoux, P., Cossart, P., Kocks, C., Carlier, M.F. and Pantaloni, D. *Listeria* protein ActA mimics WASp family proteins: it activates filament barbed end branching by Arp2/3 complex. Biochemistry 2001; 40:11390–11404.

Cascio, A., Colomba, C., Antinori, S. et al. Clarithromycin versus azithromycin in the treatment of Mediterranean spotted fever in children: a randomized controlled trial. Clin. Infect. Dis. 2002; 34:154–158.

Cascio, A., Colomba, ., Di Rosa, D. et al. Efficacy and safety of clarithromycin as treatment for Mediterranean spotted fever in children: a randomized controlled trial. Clin. Infect. Dis. 2001; 33:409–411.

Chattopadhyay, S., Jiang, J., Chan, T.C. et al. Scrub typhus vaccine candidate Kp r56 induces humoral and cellular immune responses in cynomolgus monkeys. Infect. Immun. 2005; 73:5039–5047.

Chi, W.C., Huang, J.J., Sung, J.M. et al. Scrub typhus associated with multiorgan failure: a case report. Scand. J. Infect. Dis. 1997; 29:634–635.

Chierakul, W., de Fost, M., Suputtamongkol, Y. et al. Differential expression of interferon-gamma and interferon-gamma-inducing cytokines in Thai patients with scrub typhus or leptospirosis. Clin. Immunol. 2004; 113:140–144.

Ching, W.M., Carl, M. and Dasch, G.A. Mapping of monoclonal antibody binding sites on CNBr fragments of the S-layer protein antigens of Rickettsia typhi and Rickettsia prowazekii. Mol Immunol. 1992; 29:95–105.

Ching, W.M., Wang, H., Davis, J. and Dasch G.A. Amino acid analysis and multiple methylation of lysine residues in the surface protein antigen of *Rickettsia prowazekii*, p. 307–314. In: Techniques in Protein Chemistry (R.H. Angeletti, Ed.), Vol. IV. Academic Press, San Diego.

Ching, W.M., Dasch, G.A., Carl, M. et al. Structural analyses of the 120-kDa serotype protein antigens of typhus group rickettsiae. Comparison with other S-layer proteins. Ann. N.Y. Acad. Sci. 1990; 590:334–351.

Cho, N.H., Seong, S.Y., Huh, M.S. et al. Expression of chemokine genes in murine macrophages infected with *Orientia tsutsugamushi*. Infect. Immun. 2000; 68:594–602.

Choi, Y., Kim, K.-S., Kim, T.-Y. and Ahn, B-Y. Long-term egg-yolk adaptation of the *Orientia tsutsugamushi* for preparation of a formalinized immunogen. Vaccine 2006; 24:1438–1445.

Cillari, E., Milano, S., D'Agostino, P. et al. Depression of CD4T cell subsets and alteration in cytokine profile in boutonneuse fever. J. Infect. Dis. 1996; 174:1051–1057.

Clifton, D.R., Rydkina, E., Freeman, RS. et al. NF-kappaB activation during *Rickettsia rickettsii* infection of endothelial cells involves the activation of catalytic IkappaB kinases IKKalpha and IKKbeta and phosphorylation-proteolysis of the inhibitor protein IkappaBalpha. Infect. Immun. 2005a; 73:155–165.

Clifton, D.R., Rydkina, E., Huyck, H. et al. Expression and secretion of chemotactic cytokines IL-8 and MCP-1 by human endothelial cells after *Rickettsia rickettsii* infection: regulation by nuclear transcription factor NF-κB. Int. J. Med. Microbiol. 2005b; 295(4):267–278.

Crocquet-Valdes, P.A., Diaz-Montero, C.M., Feng, H.M. et al. Immunization with a portion of rickettsial outer membrane protein A stimulates protective immunity against spotted fever rickettsiosis. Vaccine 2002; 20:979–988.

Dasch, G.A. Isolation of species-specific protein antigens of *Rickettsia typhi* and *Rickettsia prowazekii* for immunodiagnosis and immunoprophylaxis. J. Clin. Microbiol. 1981; 14:333–341.

Davidson, M.G., Breitschwerdt, E.B., Walker, D.H. et al. Vascular permeability and coagulation during *Rickettsia rickettsii* infection in dogs. Am. J. Vet. Res. 1990; 51:165–170.

de Fost, M., Chierakul, W., Pimda, K. et al. Activation of cytotoxic lymphocytes in patients with scrub typhus. Am. J. Trop. Med. Hyg. 2005; 72:465–467.

Diaz-Montero, C.M., Feng, H.-M., Crocquet-Valdes, P.A. et al. Identification of protective components of two major outer membrane proteins of spotted fever group rickettsiae. Am. J. Trop. Med. Hyg. 2001; 65:371–378.

Dignat-George, N., Teysseire, M., Mutin, N. et al. *Rickettsia conorii* infection enhances vascular cell adhesion molecule-1- and intercellular adhesion molecule-1-dependent mononuclear cell adherence to endothelial cells. J. Infect. Dis. 1997; 175:1142–1152.

Dignat-George, F., Tissot-Dupont, H., Grau, G.E. et al. Differences in levels of soluble E-selectin and VCAM-1 in malignant versus non malignant Mediterranean spotted fever. Thromb. Haemost. 1999; 82:1610–1613.

Doherty, P.C., Topham, D.J., Tripp, R.A. et al. Effector CD4+ and CD8+ T-cell mechanisms in the control of respiratory virus infections. Immunol. Rev. 1997; 159:105–117.

Edelson, BT., Cossart, P. and Unanue, ER. Cutting edge: paradigm revisited: antibody provides resistance to *Listeria* infection. J. Immunol. 1999; 163:4087–4090.

Edelson, B.T. and Unanue, E.R. Immunity to *Listeria* infection. Curr. Opin. Immunol. 2000a; 12:425–431.

Edelson, B.T. and Unanue, E.R. Intracellular antibody neutralizes *Listeria* growth. Immunity 2000b; 14:503–512.

Elghetany, T.M. and Walker, D.H. Hemostatic changes in Rocky Mountain spotted fever and Mediterranean spotted fever. Am. J. Clin. Pathol. 1999; 112:159–168.

Ellison, D.W., Clark, T.R., Sturdevant, D.E., Virtaneva, K., Porcella, S.F. and Hackstadt, T. Genomic comparison of virulent *Rickettsia rickettsii* Sheila Smith and avirulent *Rickettsia rickettsii* Iowa. Infect. Immun. 2008; 76:542–550.

Enatsu, T., Urakami, H. and Tamura, A. Phylogenetic analysis of *Orientia tsutsugamushi* strains based on the sequence homologies of 56-kDa type-specific antigen genes. FEMS Microbiol. Lett. 1999; 180:163–169.

Eremeeva, M.E., Dasch, G.A. and Silverman, D.J. Interaction of rickettsiae with eukaryotic cells. Adhesion, entry, intracellular growth, and host cell responses. Subcell. Biochem. 2000; 33:479–516.

Eremeeva, M.E., Dasch, G.A. and Silverman, D.J. Quantitative analyses of variations in the injury of endothelial cells elicited by 11 isolates of *Rickettsia rickettsii*. Clin. Diagn. Lab. Immunol. 2001; 8:788–795.

Eremeeva, M.E. and Silverman, D.J. *Rickettsia rickettsii* infection of the EA. hy926 endothelial cell line: morphological response to infection and evidence for oxidative injury. Microbiology 1998a; 144:2037–2048.

Eremeeva, M.E. and Silverman, D.J. Effects of the antioxidant α-lipoic acid on human umbilical vein endothelial cells infected with *Rickettsia rickettsii*. Infect. Immun. 1998b; 66:2290–2299.

Fang, R., Ismail, N., Soong, L., Popov, V.L., Whitworth, T., Bouyer, D.H. and Walker, D.H. Differential interaction of dendritic cells with *Rickettsia*: impact on host susceptibility to murine spotted fever rickettsiosis. Infect. Immun. 2007; 75:3112–3123.

Feng, H., Popov, V.L., Yuoh, G. and Walker, D.H. Role of T lymphocyte subsets in immunity to spotted fever group Rickettsiae. J. Immunol. 1997; 158:5314–5320.

Feng, H.M., Popov, V.L. and Walker, D.H. Depletion of gamma interferon and tumor necrosis factor alpha in mice with *Rickettsia conorii*-infected endothelium: impairment of rickettsicidal nitric oxide production resulting in fatal, overwhelming rickettsial disease. Infect. Immun. 1994; 62:1952–1960.

Feng, H.M. and Walker, D.H. Interferon-gamma and tumor necrosis factor-alpha exert their antirickettsial effect via induction of synthesis of nitric oxide. Am. J. Pathol. 1993; 143:1016–1023.

Feng, H.M. and Walker, D.H. Mechanisms of intracellular killing of *Rickettsia conorii* in infected human endothelial cells, hepatocytes, and macrophages. Infect. Immun. 2000; 68:6729–6736.

Feng, H.-M. and Walker, D.H. Cross-protection between distantly related spotted fever group rickettsiae. Vaccine 2002; 21:3901–3905.

Feng, H.-M., Whitworth, T., Olano, J.P., Popov, V.L. and Walker, D. H. Fc-dependent polyclonal antibodies and antibodies to outer membrane proteins A and B, but not to lipopolysaccharide, protect SCID mice against fatal *Rickettsia conorii* infection. Infect. Immun. 2004a; 72:2222–2228.

Feng, H.M., Whitworth, T., Popov, V. and Walker, D.H. Effect of antibody on the rickettsia–host cell interaction. Infect. Immun. 2004b; 72:3524–3530.

Fox, J.P., Jordan, M.E. and Gelfand, H.E. Immunization of man against epidemic typhus by infection with avirulent *Rickettsia prowazekii* strain E. J Immunol. 1957; 79:348–354.

Frischknecht, F. and Way, M. Surfing pathogens and the lessons learned for actin polymerization. Trends Cell Biol. 2001; 11:30–38.

Fultz, M., Barber, S., Dieffenbach, C. and Vogel, S. Induction of IFN-gamma in macrophages by lipopolysaccharide. Int. Immunol. 1993; 5:1383–1392.

Gately, M.K., Renzetti, L.M., Magram, J., Stern, A.S., Adorini, L., Gubler, U. and Presky, D.H. The interleukin-12/interleukin-12-receptor system: role in normal and pathologic immune responses. Ann. Rev. Immunol. 1998; 16:495–521.

George, F., Brouqui, P., Boffa, M.-C. et al. Demonstration of *Rickettsia conorii*-induced endothelial injury *in vivo* by measuring circulating endothelial cells, thrombomodulin, and von Willebrand factor in patients with Mediterranean spotted fever. Blood 1993; 82:2109–2116.

Gerosa, F., Gobbi, A., Zorzi, P., Burg, S., Briere, F., Carra, G. and Trinchieri, G. The reciprocal interaction of NK cells with plasmacytoid or myeloid dendritic cells profoundly affects innate resistance functions. J. Immunol. 2005; 174:727–734.

Gilmore, R.D., Jr. Comparison of the rompA gene repeat regions of *Rickettsiae* reveals species-specific arrangements of individual repeating units. Gene 1993; 125:97–102.

Gonder, J.C., Kenyon, R.H. and Pedersen, C.E., Jr. Evaluation of a killed Rocky Mountain spotted fever vaccine in cynomolgus monkeys. J. Clin. Microbiol. 1979; 10:719–723.

Gonder, J.C., Kenyon, R.H. and Pedersen, C.E., Jr. Epidemic typhus infection in cynomolgus monkeys (Macaca fascicularis). Infect. Immun. 1980; 30:219–223.

Gouin, E., Egile, C., Dehoux, P., Villiers, V., Adams, J., Gertler, F., Li, R. and Cossart, P. The RickA protein of *Rickettsia conorii* activates the Arp2/3 complex. Nature 2004; 427:457–461.

Gouin, E., Gantelet, H., Egile, C., Lasa, I., Ohayon, H., Villiers, V., Gounon, P., Sansonetti, P.J. and Cossart, P. A comparative study of the actin-based motilities of the pathogenic bacteria *Listeria monocytogenes*, *Shigella flexneri* and *Rickettsia conorii*. J. Cell Sci. 1999; 112:1697–1708.

Groves, M.G., Rosenstreich, D.L., Taylor, B.A. and Osterman, J.V. Host defenses in experimental scrub typhus: mapping the gene that controls natural resistance in mice. J. Immunol. 1980; 125:1395–1399.

Hanson, B. Identification and partial characterization of *Rickettsia tsutsugamushi* major protein immunogens. Infect. Immun. 1985; 50:603–609.

Heinzen, R.A., Hayes, S.F., Peacock, M.G. and Hackstadt, T. Directional actin polymerization associated with spotted fever group *Rickettsia* infection of Vero cells. Infect. Immun. 1993; 61:1926–1935.

Holman, R.C., Paddock, C.D., Curns, A.T., Krebs, J.W., McQuiston, J.H. and Childs, J.E. Analysis of risk factors for fatal Rocky Mountain spotted fever: evidence for superiority of tetracyclines for therapy. J. Infect. Dis. 2001; 184:1437–1444.

Hong, J.E., Santucci, L.A., Tian, X. et al. Superoxide dismutase-dependent, catalase-sensitive peroxides in human endothelial cells infected by *Rickettsia rickettsii*. Infect. Immun. 1998; 66:1293–1298.

Jensenius, M., Fournier, P.E., Hellum, K.B., Wesslen, L., Caruso, G., Prio, T., Lohne, K., Vene, S., Raoult, D. and Myrvang, B. Sequential changes in hematologic and biochemical parameters in African tick bite fever. Clin. Microbiol. Infect. 2003; 9:678–683.

Jerrells, T.R. Immunosuppression associated with the development of chronic infections with *Rickettsia tsutsugamushi*: adherent suppressor cell activity and macrophage activation. Infect. Immun. 1985; 50:175–182.

Jerrells, T.R. and Osterman, J.V. Host defenses in experimental scrub typhus: inflammatory response of congenic C3H mice differing at the *Ric* gene. Infect. Immun. 1981; 31:1014–1022.

Jordan, J.M., Woods, M.E., Feng, H.-M., Soong, L. and Walker, D.H. Rickettsiae-stimulated dendritic cells mediate protection against lethal rickettsial challenge in an animal model of spotted fever rickettsiosis. J. Infect. Dis. 2007; 196:629–638.

Joshi, S.G., Francis, C.W., Silverman, D.J. and Sahni, S.K. Nuclear factor kappa B protects against host cell apoptosis during *Rickettsia rickettsii* infection by inhibiting activation of apical and effector caspases and maintaining mitochondrial integrity. Infect. Immun. 2003; 71:4127–4136.

Joshi, S.G., Francis, C.W., Silverman, D.J. and Sahni, S.K. NF-kappaB activation suppresses host cell apoptosis during *Rickettsia rickettsii* infection via regulatory effects on intracellular localization or levels of apoptogenic and anti-apoptotic proteins. FEMS Microbiol. Lett. 2004; 34:333–341.

Kim, M.J., Kim, M.K. and Kang, J.S. *Orientia tsutsugamushi* inhibits tumor necrosis factor alpha production by inducing interleukin 10 secretion in murine macrophages. Microb. Pathog. 2005; 32:1–7.

Kim, M.K. and Kang, J.S. *Orientia tsutsugamushi* suppresses the production of inflammatory cytokines induced by its own heat-stable component in murine macrophages. Microb. Pathog. 2001; 31:145–150.

Kaplanski, G., Teysseire, N., Farnarier, C., Kaplanski, S., Lissitzky, J.C., Durand, J.M., Soubeyrand, J., Dinarello, C.A. and Bongrand, P. IL-6 and IL-8 production from cultured human endothelial cells stimulated by infection with *Rickettsia conorii* via a cell-associated IL-1 alpha-dependent pathway. J. Clin. Investig. 1995; 96:2839–2844.

Kenyon, R.H., Kishimoto, R.A. and Hall, W.C. Exposure of guinea pigs to *Rickettsia rickettsii* by aerosol, nasal, conjunctival, gastric, and subcutaneous routes and protection afforded by an experimental vaccine. Infect. Immun. 1979; 25:580–582.

Kern, W.V., Oristrell, J., Segura-Porta, F. and Kern, P. Release of soluble tumor necrosis factor receptors in Mediterranean spotted fever rickettsiosis. Clin. Diagn. Lab. Immunol. 1996; 3:233–235.

Kim, C.H., Rott, L., Kunkel, E.J., Genovese, M.C., Andrew, D.P., Wu, L. and Butcher, E.C. Rules of chemokine receptor association with T cell polarization in vivo. J. Clin. Investig. 2001; 108:1331–1339.

Koh, Y.S., Yun, J.H., Seong, S.Y., Choi, M.S. and Kim, I.S. Chemokine and cytokine production during *Orientia tsutsugamushi* infection in mice. Microb. Pathog. 2004; 36:51–57.

Lee, K.Y., Lee, H.S., Hong, J.H. et al. Roxithromycin treatment of scrub typhus (tsutsugamushi disease) in children. Pediatr. Infect. Dis. J. 2003; 22:130–133.

Li, H., Lenz, B. and Walker, D.H. Protective monoclonal antibodies recognize heat-labile epitopes on surface proteins of spotted fever group rickettsiae. Infect. Immun. 1988; 56:2587–2593.

Li, H. and Walker, D.H. Characterization of rickettsial attachment to host cells by flow cytometry. Infect. Immun. 1992; 60:2030–2035.

Li, H. and Walker, D.H. rOmpA is a critical protein for the adhesion of *Rickettsia rickettsii* to host cells. Microb. Pathog. 1998; 24:289–298.

Locksley, R.M., Killeen, N. and Lenardo, M.J. The TNF and TNF receptor superfamilies: integrating mammalian biology. Cell 2001; 104:487.

Mackaness, G.B. Cellular resistance to infection. J. Exp. Med. 1962; 116:381–406.

Mackaness, G.B. The influence of immunologically committed lymphoid cells on macrophage activity in vivo. J. Exp. Med. 1969; 129:973–992.

Machesky, L.M., Mullins, R.D., Pantaloni, D., Higgs, H.N., Kaiser, D.A., Blanchoin, L., May, R.C., Hall, M.E. and Scar, a WASp-related protein, activates nucleation of actin filaments by the Arp2/3 complex. Proc. Natl. Acad. Sci. U. S. A. 1999; 96:3739–3744.

Manor, E. and Sarov, I. Inhibition of *Rickettsia conorii* growth by recombinant tumor necrosis factor alpha: enhancement of inhibition by gamma interferon. Infect. Immun. 1990; 58:1886–1890.

Martinez, J.J. and Cossart, P. Early signaling events involved in the entry of *Rickettsia conorii* into mammalian cells. J. Cell Sci. 2004; 117:5097–5106.

Martinez, J.J., Seveau, S., Veiga, E., Matsuyama, S. and Cossart, P. Ku70, a component of DNA-dependent protein kinase, is a mammalian receptor for *Rickettsia conorii*. Cell 2005; 123:1013–1023.

Masters, E.J., Olson, G.S., Weiner, S.J. and Paddock, C.D. Rocky Mountain spotted fever—a clinician's dilemma. Arch. Intern. Med. 2003; 163:769–774.

Mauri, D.N., Ebner, R., Montgomery, R.I., Kochel, K.D., Cheung, T.C., Yu, G.L., Ruben, S., Murphy, M., Eisenberg, R.J. and Cohen, G.H. LIGHT, a new member of the TNF superfamily, and lymphotoxin α are ligands for herpesvirus entry mediator. Immunity 1998; 8:21–23.

McCulloch, R. Antigenic variation in African trypanosomes: monitoring progress. Trends Parasitol. 2004; 20:117–121.

McDonald, G.A., Anacker, R.L. and Garjian, K. Cloned gene of *Rickettsia rickettsii* surface antigen: candidate vaccine for Rocky Mountain spotted fever. Science 1987; 235:83–85.

McDonald, G.A., Anacker, R.L., Mann, R.E. and Milch, L.J. Protection of guinea pigs from experimental Rocky Mountain spotted fever with a cloned antigen of *Rickettsia rickettsii*. J. Infect. Dis. 1988; 158:228–231.

Monastra, G., Cabrelle, A., Zambon, A., Rosato, A., Macino, B., Collavo, D. and Zanovello, P. Membrane form of TNFα induces both cell lysis and apoptosis in susceptible target cells. Cell Immunol. 1996; 171:102–105.

Munder, M., Mallo, M., Eichmann, K. and Modolell, M. Murine macrophages secrete interferon gamma upon combined stimulation with interleukin (IL)-12 and IL-18: a novel pathway of autocrine macrophage activation. J. Exp. Med. 1998; 187:2103–2108.

Murphy, J.R., Wisseman, C.L., Jr. and Fiset, P. Mechanisms of immunity in typhus infection: analysis of immunity to *Rickettsia mooseri* infection of guinea pigs. Infect. Immun. 1980; 27:730–738.

Nacy, C.A. and Groves, M.G. Macrophages in resistance to rickettsial infections: early host defense mechanisms in experimental scrub typhus. Infect. Immun. 1981; 31:1239–1250.

Nath, D. Cell biology: pathogen propulsion. Nature 2004; 427:407–409.

Ng, F.K.P., Oaks, S.C. Jr., Lee, M., Groves, M.G. and Lewis, G. E. Jr. A scanning and transmission electron microscopic examination of *Rickettsia tsutsugamushi*-infected human endothelial, MRC-5, and L-929 cells. Jpn. J. Med. Sci. Biol. 1985; 38:125–139.

Nishimura, M., Umehara, H., Nakayama, T., Yoneda, O., Hieshima, K., Kakizaki, M., Dohmae, N., Yoshie, O. and Imai, T. Dual functions of fractalkine/CX3C ligand 1 in trafficking of perforin+/granzyme B+ cytotoxic effector lymphocytes that are defined by CX3CR1 expression. J. Immunol. 2002; 168:6173–6180.

Ohashi, N., Nashimoto, H., Ikeda, H. and Tamura, A. Diversity of immunodominant 56-kDa type-specific antigen (TSA) of *Rickettsia tsutsugamushi*. Sequence and comparative analyses of the genes encoding TSA homologues from four antigenic variants. J. Biol. Chem. 1992; 267:12728–12735.

Ohashi, N., Tamura, A., Ohta, M. and Hayashi, K. Purification and partial characterization of a type-specific antigen of *Rickettsia tsutsugamushi*. Infect. Immun. 1989; 57:1427–1431.

Ohashi, N., Tamura, A. and Suto, T. Immunoblotting analysis of anti-rickettsial antibodies produced in patients of Tsutsugamushi disease. Microbiol. Immunol. 1988; 32:1085–1092.

Ohteki, T., Fukao, T., Suzue, K., Maki, C., Ito, M., Nakamura, M. and Koyasu, S. Interleukin 12-dependent interferon gamma production by CD8alpha^{+} lymphoid dendritic cells. J. Exp. Med. 1999; 189:1981–1986.

Oristrell, J., Amengual, M.J., Font-Creus, B., Casanovas, A. and Segura-Porta, F. Plasma levels of tumor necrosis factor alpha in patients with Mediterranean spotted fever: clinical and analytical correlations. Clin. Infect. Dis. 1994; 19:1141–1143.

Parker, H.T., Menon, P.G. and Merideth, A.M. Persistence of *Rickettsia rickettsii* in a patient recovered from Rocky Mountain spotted fever. J. Immunol. 1954; 73:383–386.

Parsons, J.T. Focal adhesion kinase: the first ten years. J. Cell Sci. 2003; 116:1409–1416.

Patarca, R., Freeman, G.J., Singh, R.P., Wei, F.Y., Durfee, T., Blattner, F., Regnier, D.C., Kozak, C.A., Mock, B.A., Morse, H.C., III et al. Structural and functional studies of the early T lymphocyte activation 1 (Eta-1) gene. Definition of a novel T cell-dependent response associated with genetic resistance to bacterial infection. J. Exp. Med. 1989; 170:145–161.

Pillarisetty, V.G., Katz, S.C., Bleier, J.I., Shah, A.B. and Dematteo, R.P. Natural killer dendritic cells have both antigen presenting and lytic function and in response to CpG produce IFN-gamma via autocrine IL-12. J. Immunol. 2005; 174:2612–2618.

Plano, G.V. and Winkler, H.H. Identification and initial topological analysis of the *Rickettsia prowazekii* ATP/ADP translocase. J. Bacteriol. 1991; 173:3389–3396.

Pollard, T.D. and Borisy, G.G. Cellular motility driven by assembly and disassembly of actin filaments. Cell 2003; 112:453–465.

Purvis, J.J. and Edwards, M.S. Doxycycline use for rickettsial disease in pediatric patients. Pediatr. Infect. Dis. J. 2000; 19:871–874.

Qin, S., Rottman, J.B., Myers, P., Kassam, N., Weinblatt, M., Loetscher, M., Koch, A.E., Moser, B. and Mackay, C.R. The chemokine receptors CXCR3 and CCR5 mark subsets of T cells associated with certain inflammatory reactions. J. Clin. Invest. 1998; 101:746–754.

Randall, M.B. and Walker, D.H. Rocky Mountain spotted fever: gastrointestinal and pancreatic lesions and rickettsial infection. Arch. Pathol. Lab. Med. 1984; 108:963–967.

Renesto, P., Dehoux, P., Gouin, E., Touqui, L., Cossart, P. and Raoult, D. Identification and characterization of a phospholipase D-superfamily gene in rickettsiae. J. Infect. Dis. 2003; 188:1276–1283.

Ricketts, H.T. and Gomez, L. Studies on immunity in Rocky Mountain spotted fever. First communication. J. Infect. Dis. 1908; 5:221–244.

Rikihisa, Y. and Ito, S. Intracellular localization of *Rickettsia tsutsugamushi* in polymorphonuclear leukocytes. J. Exp. Med. 1979; 150:703–708.

Rolain, J.M., Maurin, M., Vestris, G. and Raoult, D. In vitro susceptibilities of 27 rickettsiae to 13 antimicrobials. Antimicrob. Agents Chemother. 1998; 42:1537–1541.

Ruiz Beltran, R. and Herrero Herrero, J.I. Deleterious effect of trimethoprim-sulfamethoxazole in Mediterranean spotted fever. Antimicrob. Agents Chemother. 1992; 36:1342–1344.

Rydkina, E., Sahni, A., Silverman, D.J. and Sahni, S.K. *Rickettsia rickettsii* infection of cultured human endothelial cells induces heme oxygenase 1 expression. Infect. Immun. 2002; 70:4045–4052.

Rydkina, E., Sahni, S.K., Santucci, L.A., Turpin, L.C., Baggs, R.B. and Silverman, D.J. Selective modulation of antioxidant enzyme activities in host tissues during *Rickettsia conorii* infection. Microb. Pathog. 2004; 36:293–301.

Salgame, P. Host innate and Th1 responses and the bacterial factors that control *Mycobacterium tuberculosis* infection. Curr. Opin. Immunol. 2005; 17:374–380.

Sallusto, F., Lenig, D., Mackay, C.R. and Lanzavecchia, A. Flexible programs of chemokine receptor expression on human polarized T helper 1 and 2 lymphocytes. J. Exp. Med. 1998; 187:875–883.

Santucci, L.A., Gutierrez, P.L. and Silverman, D.J. *Rickettsia rickettsii* induces superoxide radical and superoxide dismutase in human endothelial cells. Infect. Immun. 1992; 60:5113–5118.

Segura, F. and Anton, E. Clarithromycin for the treatment of Mediterranean spotted fever. Clin. Infect. Dis. 2002; 34:560.

Seong, S.Y., Choi, M.S. and Kim, I.S. *Orientia tsutsugamushi* infection: overview and immune responses. Microb. Infect. 2001; 3:11–21.

Seong, S.Y., Huh, M.S., Jang, W.J., Park, S.G., Kim, J.G., Woo, S.G., Choi, M.S., Kim, I.S. and Chang, W.H. Induction of homologous immune response to *Rickettsia tsutsugamushi* Boryong with a partial 56-kilodalton recombinant antigen fused with the maltose-binding protein MBP-Bor56. Infect. Immun. 1997; 65:1541–1545.

Shi, R.-J., Simpson-Haidaris, P.J., Marder, V.J. et al. Increased expression of plasminogen activator inhibitor-1 in *R. rickettsii*-infected endothelial cells. Thromb. Haemost. 1996; 75:600–606.

Shi, R-J., Simpson-Haidaris, P.J., Marder, V.J. et al. Post-transcriptional regulation of endothelial cell plasminogen activator inhibitor-1 expression during *R. rickettsii* infection. Microb. Pathog. 2000; 28:127–133.

Shresta, S., Pham, C.T., Thomas, D.A., Graubert, T.A. and Ley, T.J. How do cytotoxic lymphocytes kill their targets? Curr. Opin. Immunol. 1998; 10:581–597.

Silverman, D.J. and Santucci, L.A. Potential for free radical-induced lipid peroxidation as a cause of endothelial cell injury in Rocky Mountain spotted fever. Infect. Immun. 1988; 56:3110–3115.

Silverman, D.J. and Santucci, L.A. A potential protective role for thiols against cell injury caused by *Rickettsia rickettsii*. Ann. N.Y. Acad. Sci. 1990; 590:111–117.

Silverman, D.J., Santucci, L.A., Meyers, N. et al. Penetration of host cells by *Rickettsia rickettsii* appears to be mediated by a phospholipase of rickettsial origin. Infect. Immun. 1992; 60:2733–2740.

Silverman, D.J. and Wisseman, C.L. Comparative ultrastructural study on the cell envelopes of *Rickettsia prowazekii*, *Rickettsia rickettsii*, and *Rickettsia tsutsugamushi*. Infect. Immun. 1978; 21:1020–1023.

Singh, R.P., Patarca, R., Schwartz, J., Singh, P. and Cantor, H. Definition of a specific interaction between the early T lymphocyte activation 1 (Eta-1) protein and murine macrophages in vitro and its effect upon macrophages in vivo. J. Exp. Med. 1990; 171:1931–1942.

Smadel, J.E., Ley, H.L., Diercks, R.H. and Cameron, J.A.. Persistence of *Rickettsia tsutsugamushi* in tissues of patients recovered from scrub typhus. Am. J. Hyg. 1952; 56:294–302.

Sousa, R., Ismail, N., Nobrega, S.D., Franca, A., Amaro, M., Anes, M., Pocas, J., Coelho, R., Torgai, J., Bacellar, F. and Walker, D.H. Intralesional IFN-γ, TNF-α, IL-10, NOS2, indoleamine-2, 3-dioxygenase, and RANTES are major immune effectors in Mediterranean spotted fever. J. Infect. Dis. 2007; 196:770–781.

Sporn, L.A. and Marder, V.J. Interleukin-1 alpha production during *Rickettsia rickettsii* infection of cultured endothelial cells: potential role in autocrine cell stimulation. Infect. Immun. 1996; 64:1609–1613.

Sporn, L.A., Shi, R-J., Lawrence, S.O. et al. *Rickettsia rickettsii* infection of cultured endothelial cells induces release of large von Willebrand factor multimers from Weibel-Palade bodies. Blood 1991; 78:2595–2602.

Stenos, J. and Walker, D.H. The rickettsial outer-membrane protein A and B genes of *Rickettsia australis*, the most divergent rickettsia of the spotted fever group. Int. J. Syst. Evol. Microbiol. 2000; 50:1775–1779.

Stover, C.K., Marana, D.P., Dasch, G.A. and Oaks, E.V. Molecular cloning and sequence analysis of the Sta58 major antigen gene of *Rickettsia tsutsugamushi*: sequence homology and antigenic comparison of Sta58 to the 60-kilodalton family of stress proteins. Infect. Immun. 1990; 58:1360–1368.

Sumner, J.W., Sims, K.G., Jones, D.C. and Anderson, B.E.. Protection of guinea-pigs from experimental Rocky Mountain spotted fever by immunization with baculovirus-expressed *Rickettsia rickettsii* rOmpA protein. Vaccine 1995; 13:29–35.

Tamura, A., Ohashi, N., Urakami, H., Takahashi, K. and Oyanagi, M. Analysis of polypeptide composition and antigenic components of *Rickettsia tsutsugamushi* by polyacrylamide gel electrophoresis and immunoblotting. Infect. Immun. 1985; 48:671–675.

Tesouro, M.A., Bacellar, F., Sainz, A. and Filips, A. Persistence of antibodies to *Rickettsia conorii* in dogs. Ann. N.Y. Acad. Sci. 1998; 849:441–443.

Teysseire, N., Boudier, J.A. and Raoult, D. *Rickettsia conorii* entry into Vero cells. Infect. Immun. 1995; 63:366–374.

Teysseire, N., Chiche-Portiche, C. and Raoult, D. Intracellular movements of *Rickettsia conorii* and R. typhi based on actin polymerization. Res. Microbiol. 1992; 143:821–829.

Topping, N.H. Rocky Mountain spotted fever. Treatment of infected laboratory animals with immune rabbit serum. Public Health Rep. 1940; 55:41–46.

Topping, N.H. Rocky Mountain spotted fever: further experience in the therapeutic use of immune rabbit serum. Public Health Rep. 1943; 58:757–775.

Turco, J., Keysary, A. and Winkler, H.H. Interferon-gamma- and rickettsia-induced killing of macrophage-like cells is inhibited by anti-rickettsial antibodies and does not require the respiratory burst. J. Interferon Res. 1989a; 9:615–629.

Turco, J. and Winkler, H.H. Isolation of *Rickettsia prowazekii* with reduced sensitivity to gamma interferon. Infect. Immun. 1989; 57:1765–1772.

Turco, J. and Winkler, H.H. Isolation of *Rickettsia prowazekii* with reduced sensitivity to gamma interferon. Infect. Immun. 1989b; 57:1765–1772.

Turco, J. and Winkler, H.H. Selection of alpha/beta interferon- and gamma interferon-resistant rickettsiae by passage of *Rickettsia prowazekii* in L929 cells. Infect. Immun. 1990; 58:3279–3285.

Turco, J. and Winkler, H.H. Cytokine sensitivity and methylation of lysine in Rickettsia prowazekii EVir and interferon-resistant R. prowazekii strains. Infect. Immun. 1994; 62:3172–3177.

Uchiyama, T. Role of major surface antigens of *Rickettsia japonica* in the attachment to host cells. In: Rickettsiae and Rickettsial Diseases (J. Kazar and D. Raoult, Eds.), pp. 182–188. Bratislava: Publishing House of the Slovak Academy of Sciences, 1999.

Valbuena, G., Bradford, W. and Walker, D.H. Expression analysis of the T-cell-targeting chemokines CXCL9 and CXCL10 in mice and humans with endothelial infections caused by rickettsiae of the spotted fever group. Am. J. Pathol. 2003; 163:1357–1369.

Valbuena, G., Feng, H.M. and Walker, D.H. Mechanisms of immunity against rickettsiae. New perspectives and opportunities offered by unusual intracellular parasites. Microb. Infect. 2002; 4:625–633.

Valbuena, G., Jordan, J.M. and Walker, D.H. T cells mediate cross-protective immunity between spotted fever group rickettsiae and typhus group rickettsiae. J. Infect. Dis. 2004; 190:1221–1227.

Valbuena, G. and Walker, D.H. Effect of blocking the CXCL9/10-CXCR3 chemokine system in the outcome of endothelial-target rickettsial infections. Am. J. Trop. Med. Hyg. 2004; 71:393–399.

Valbuena, G. and Walker, D.H. Expression of CX3CL1 (fractalkine) in mice with endothelial-target rickettsial infection of the spotted-fever group. Virchows Arch. 2005; 446:21–27.

Vishwanath, S., McDonald, G.A. and Watkins, N.G. A recombinant *Rickettsia conorii* vaccine protects guinea pigs from experimental boutonneuse fever and Rocky Mountain spotted fever. Infect. Immun. 1990; 58:646–653.

Vitale, S., Mansueto, G., Gambino, C., Mocciaro, C., La Russa, P., Mansueto, M.A., Zambito, V., Ferlazzo, C., Barbera, M., La Rosa, S. and Milano and Cillari, E. Differential up-regulation of circulating soluble selectins and endothelial adhesion molecules in Sicilian patients with boutonneuse fever. Clin. Exp. Immunol. 1999; 117:304–308.

Vitale, G., Mansueto, S., Gambino, G., Mocciaro, C., Spinelli, A., Rini, G.B., Affronti, M., Chifari, N., La Russa, C., Di Rosa, S., Colletti, P., Barbera, C., La Rosa, M., Di Bella, G., Ferlazzo, V., Milano, S., D'Agostino, P. and Cillari, E. The acute phase response in Sicilian patients with boutonneuse fever admitted to hospitals in Palermo, 1992–1997. J. Infect. Dis. 2001; 42:33–39.

Vitorino, L., Chelo, I.M., Bacellar, F. and Ze-Ze, L. Rickettsiae phylogeny: a multigenic approach. Microbiology 2007; 153: 160–168.

Walker, D.H., Crawford, C.G., Cain, B.G. et al. Rickettsial infection of the pulmonary microcirculation: the basis for interstitial pneumonitis in Rocky Mountain spotted fever. Hum. Pathol. 1980; 11:263–272.

Walker, D.H., Feng, H.M. and Popov, V.L. Rickettsial phospholipase A_2 as a pathogenic mechanism in a model of cell injury by typhus and spotted fever group rickettsiae. Am. J. Trop. Med. Hyg. 2001; 65:936–942.

Walker, D.H., Firth, W.T., Ballard, J.G. et al. Role of phospholipase-associated penetration mechanism in cell injury by *Rickettsia rickettsii*. Infect. Immun. 1983; 40:840–842.

Walker, D.H. and Mattern, W.D. Renal failure in Rocky Mountain spotted fever. Arch. Intern. Med. 1980; 140:867.

Walker, D.H. Rocky Mountain spotted fever and other rickettsioses. In: Mechanisms of Microbial Disease (M. Schaechter, G. Medoff, and B. I. Eisenstein, Ed.), Williams & Wilkins, Baltimore. 1993; p. 358–367.

Walker, D.H., Olano, J.P. and Feng, H.M. Critical role of cytotoxic T lymphocytes in immune clearance of rickettsial infection. Infect. Immun. 2001; 69:1841–1846.

Walker, D.H., Popov, V.L. and Feng, H.M. Establishment of a novel endothelial target mouse model of a typhus group rickettsiosis: evidence for critical roles for gamma interferon and CD8 T lymphocytes. Lab. Invest. 2000; 80:1361–1372.

Walker, D.H., Popov, V.L., Wen, J. and Feng, H.M. *Rickettsia conorii* infection of C3H/HeN mice A model of endothelial-target rickettsiosis. Lab. Invest. 1994; 70:358–368.

Walker, D.H., Tidwell, R.R., Rector, T.M. et al. Effect of synthetic protease inhibitors of the amidine type on cell injury by *Rickettsia rickettsii*. Antimicrob. Chemother. 1984; 25:582–585.

Walker, D.H., Valbuena, G.A. and Olano, J.P. Pathogenic mechanisms of diseases caused by Rickettsia. Ann. N.Y. Acad. Sci. 2003; 990:1–11.

Watt, G., Kantipong, P., Jongsakul, K., Watcharapichat, P., Phulsuksombati, D. and Strickman, D. Doxycycline and rifampicin for mild scrub-typhus infections in northern Thailand: a randomized trial. Lancet 2000; 356:1057–1061.

Watt, G. and Parola, P. Scrub typhus and tropical rickettsioses. Curr. Opin. Infect. Dis. 2003; 16:429–436.

Welch, M.D., Rosenblatt, J., Skoble, J., Portnoy, D.A. and Mitchison, T.J. Interaction of human Arp2/3 complex and the *Listeria monocytogenes* ActA protein in actin filament nucleation. Science 1998; 281:105–108.

Whitworth, T., Popov, V.L., Yu, X.J., Walker, D.H. and Bouyer, D.H. Expression of *Rickettsia prowazekii pld* or *thlyC* gene in *Salmonella enterica* serovar typhimurium mediates phagosomal escape. Infect. Immun. 2005; 73:6668–6673.

Winkler, H.H. and Turco, J. *Rickettsia prowazekii* and the host cell: entry, growth and control of the parasite. Curr. Top. Microbiol. Immunol. 1988; 138:81–107.

Yeaman, G.R., Collins, J.E., Currie, J.K., Guyre, P.M., Wira, C.R. and Fanger, M.W. IFN-gamma is produced by polymorphonuclear neutrophils in human uterine endometrium and by cultured peripheral blood polymorphonuclear neutrophils. J. Immunol. 1998; 160:5145–5153.

Yun, J.H., Koh, Y.S., Lee, K.H., Hyun, J.W., Choi, Y.J., Jang, W.J., Park, K.H., Cho, N.H., Seong, S.Y., Choi, M.S. and Kim, I.S. Chemokine and cytokine production in susceptible C3H/HeN mice and resistant BALB/c mice during *Orientia tsutsugamushi* infection. Microbiol. Immunol. 2005; 49:551–557.

Zinsser, H. and Castaneda, M.R. On the isolation from a case of Brill's disease of a typhus strain resembling the European type. N. Engl. J. Med. 1933; 209:815–819.

CHAPTER

57

Typhoid Fever

Robert A. Kingsley and Gordon Dougan

The Wellcome Trust Sanger Institute, The Wellcome Trust Genome Campus, Hinxton, Cambridge, UK

OUTLINE

ABSTRACT

Typhoid fever, caused by *Salmonella enterica* serotype Typhi, is an infection of global importance with an estimated 21.5 million infections and 200,000 deaths per year (estimated in the year 2000). *S.* Typhi is a potential bioterrorist agent that could be disseminated in untreated water supplies and food, resulting in moderate morbidity and low mortality. Resistance to chloramphenicol, ampicillin and trimethoprim/sulfamethoxazole is widespread and resistance to the fluoroquinolones is currently spreading throughout Asia. Typhoid has been a target for vaccine development for decades and whole cell, live oral and subunit vaccines have been developed and proved both safe and efficacious. Two currently licensed vaccines are available but are not commonly used in endemic areas and the need for extensive vaccination programmes are of critical importance.

INTRODUCTION

Typhoid fever is the clinical syndrome caused by the enteric bacterial pathogen *Salmonella enterica* serotype Typhi (*S.* Typhi). An English physician Thomas Willis (1621–1675) made perhaps the first clinical descriptions of the syndrome in 1659. A clinically similar syndrome called paratyphoid is caused by antigenically and phylogenetically distinct serotypes of *S. enterica*, namely *S.* Paratyphi A, *S.* Paratyphi B, and *S.* Paratyphi C. Initially, discrimination between typhoid and typhus fever was not routinely made and this continued to be the case until 1837 when William Wood Gerhard defined these as distinct syndromes. Identification of the typhoid bacillus in 1880 by Carl Joseph Eberth and development of the Widal agglutination test by Georges Widal in 1896 resulted in considerable progress in the diagnosis of typhoid fever. Typhoid was endemic in most countries of Europe and North America in the late 19th and early 20th centuries at least in part due to urbanization of populations during the industrial revolution and poorly planned public water systems. At that time the disease burden was estimated to be 200–800 cases per 100,000 population and an associated 10–20% mortality (JAMA, 1920), rates that are similar to the current estimates for disease burden in some regions of Asia where typhoid remains endemic. Introduction of sanitation in the form of water chlorination and sand filtration drastically reduced the incidence of typhoid fever in the first half of the 20th century in the Western world.

Typhoid fever is still endemic in many areas of Asia, Africa, and South America, where sanitation of water supply and waste treatment is inadequate. In developed countries the risk of contracting typhoid fever is limited to travel in areas with endemic typhoid and for laboratory workers, where vaccination and regular boosting or revaccination is recommended. The severity of typhoid fever is underlined by the fact that up to 32% mortality is reported in some regions of the world (Hoffman et al., 1984).

The potential use of *S.* Typhi in bioterrorist attacks is generally considered to be minimal. Since *S.* Typhi is normally transmitted through water supplies and sewerage, large-scale attacks in the Western world would likely be limited due to water treatment and sanitation procedures that are in place in these countries. Smaller scale attacks on the other hand are possible and this has been underlined by events that took place in The Dalles, OR, USA in 1984. In this case a local commune, called the Rajneeshee, attacked diners at "Shakey's Pizza" by sprinkling a nontyphoidal *S. enterica* serotype into the salad bar. Although no deaths were recorded, the resulting outbreak was the largest in Oregon history with 751 confirmed cases of *Salmonella* gastroenteritis. The attack could have been considerably worse if *S.* Typhi had been used as the bioterrorist agent in which case several deaths would have been expected. Disturbingly, Federal investigators found an invoice at the commune's ranch headquarters indicating that the Rajneeshee had ordered and received several pathogenic bacterial strains from the American-type culture collection (ATCC) including *S.* Typhi. Similar attacks in the future cannot be discounted.

ETIOLOGIC AGENT

The genus *Salmonella* contains a diverse group of Gram-negative rod-shaped bacilli in the same proteobacterial group as *Escherichia coli*. Variants within the genus, called serovars, are traditionally distinguished on the basis of their major antigens; lipopolysaccharide (LPS), flagella, and in the case of *S.* Typhi, the polysaccharide capsule. *S.* Typhi most commonly have the antigenic formula 9,12, (Vi):d:-. That is, they express the immune-dominant somatic antigen O9, and also the O12 somatic antigen, associated with LPS and are monophasic with respect to flagella, expressing the d variant. *S.* Typhi also characteristically express the Vi polysaccharide capsule that is genetically unstable during passage in the laboratory. Serovar variants are further distinguished by phage typing, a powerful epidemiological tool based on the differential susceptibility of isolates to a panel of lytic bacteriophage.

EPIDEMIOLOGY

The true incidence of typhoid fever is difficult to ascertain since active surveillance and microbiological facilities are expensive and not widely available in areas where the disease is endemic. The best sources of data are large-scale field trials of vaccine candidates in which blood samples from placebo control groups are analyzed definitively. The most recent estimate indicates a global burden of 21.6 million cases of typhoid resulting in 216,500 deaths in the year 2000 (Crump et al., 2004). The incidence is greatest in south-central Asia, south-east Asia, and Southern Africa (>100 per 100,000), with fewer cases in the rest of Asia, Africa, Latin America, and Oceania (10–100 per 100,000). The highest incidence is in India in the age-group 0–40 years in which 980 cases per 100,000 population was estimated (Sinha et al., 1999), and similarly high incidence was reported in Southern Africa and Indonesia (Klugman et al., 1987). Typhoid is normally transmitted by consumption of contaminated food or water and specific risk factors include drinking contaminated water supply, ice cream, iced drinks, and fruit and vegetables grown in fields fertilized with sewage. Furthermore, a history of certain clinical conditions, including infection with *Helicobacter pylori* is a risk factor probably due to the reduced gastric acidity associated with *H. pylori* infection (Bhan et al., 2002; Black et al., 1985; Levine et al., 1982; Luby et al., 1998; Mermin et al., 1999).

CLINICAL DISEASE

Following ingestion of *S.* Typhi an asymptomatic period is observed, usually lasting 7–14 days, but ranging from 3 to 60 days. Patients frequently present to the hospital with fever, flu-like symptoms, dull frontal headache, malaise, anorexia, abdominal discomfort, a dry cough, and myalgia (Bhan et al., 2005; Parry et al., 2002; WHO, 2003). The fever frequently rises in a stepwise fashion becoming persistent and high-grade by the second week and, without treatment, lasts for up to 4 weeks. Following this there is a return to normal temperature but malaise and lethargy may remain for several weeks. Other physical features that may be present include relative bradycardia, coated tongue, hepatomegaly, splenomegaly, and the patient may experience altered bowel movements that may manifest as diarrhea or constipation. Characteristic erythematous maculopapular lesions, also referred to as rose spots, are sometimes observed on the back of arms, legs, and the back during the first week of illness. Liver involvement is common and is accompanied by elevated levels of serum bilirubin and alanine transferase.

Up to 10–15% of untreated patients develop severe disease that is typified by a range of complications including gastrointestinal hemorrhage, gastrointestinal perforation, and a variety of neuropsychiatric manifestations. Gastrointestinal hemorrhage occurs during the latter stages of infection and is heralded by a rapid drop in body temperature and then subsequent increase at the onset of peritonitis. Hemorrhage is most common in the terminal ileum where erosion of Peyer's patch results in rupture of blood vessels. Colonoscopy reveals variable sized punched-out ulcers with slightly raised margins (Lee et al., 2004). Blood loss is most often limited to occult blood in stool or malaena, although breach of large blood vessels occurs occasionally and is rapidly fatal if untreated. Gastrointestinal perforation occurs in 1–3% of hospitalized cases and has an overall mortality rate of 40%. This complication most frequently affects the terminal ileum and requires rapid surgical intervention (van Basten and Stockenbrugger, 1994). Some typhoid fever patients exhibit neuropsychiatric manifestations, which have been described as muttering delirium or coma vigil but curiously these symptoms appear subject to geographical variation (Parry et al., 2002). Less frequent complications include disseminated intravascular coagulation or hemorrhages, severe pneumonia, granulomas (hepatic, splenic, or bone marrow) (Mert et al., 2004), abscesses (hepatic or splenic) (Chaudhry et al., 2003), multiple organ dysfunction syndrome (Snyder et al., 2004), hemolytic uremic syndrome (Albaqali et al., 2003), endocarditis (Wani et al., 2004), and pericarditis (Baysal et al., 1998).

Relapse occurs in 5–10% of cases, generally within a month of resolution of fever. The symptoms are generally milder than the original infection and in cases where molecular typing of *S.* Typhi has been applied the relapse is generally due to the same strain as the original isolate. Reinfection with a distinct strain may also occur (Wain et al., 1999).

One of the distinctive features of typhoid fever is the carrier state that has been popularized due in part to the "Typhoid Mary" episode. Mary Mallor, a resident of New York City and a domestic servant in a number of households during the early 20th century, was responsible for a number of typhoid fever outbreaks. Her unwillingness to accept that an apparently healthy person can spread disease contributed to a series of outbreaks that followed Mary from job to job. Following detective work by local health officials Mary was found to be an asymptomatic carrier of the bacillus and she was eventually removed form the city to a secluded location. Approximately 10% of convalescent patients excrete *S.* Typhi in their stool for at least 3 months and 1–5% go on to become chronic carriers (defined as isolation from stool for more than 1 year). Detection and treatment of chronic carriers is especially problematic since they are generally asymptomatic and one in four have no history of typhoid fever. It has been speculated that carrier status is a major factor for persistence of *S.* Typhi in human populations during interepidemic periods since unlike nontyphoidal Salmonellosis, *S.* Typhi has no known animal reservoir (Kingsley and Bäumler, 2000).

TREATMENT

The most important contributing factor to a poor outcome for typhoid fever patients is a delay in initiating effective antibiotic treatment. It is important to assure eradication of the *S.* Typhi from the patient in order to reduce the likelihood of relapse or carrier status. The more severe cases that present to hospital with abdominal distension, persistent vomiting, or severe diarrhea require admission to hospital and parenteral antibiotic treatment. However, the majority of cases (up to 90%) are managed at home with oral antibiotic therapy and close medical follow-up (Bhan et al., 2005).

Introduction of chloramphenicol treatment of typhoid fever transformed what was frequently fatal to a readily treatable disease, often without need for admission to hospital. Currently the most effective group of antimicrobial drugs for the treatment of typhoid fever are the fluoroquinolones (Akalin, 1999). In randomized, controlled trials these drugs were more rapidly effective and associated with lower levels of carriage than the previous first-line drugs chloramphenicol and trimethriprim-sulfamoxazole. Cases involving quinolone-resistant *S.* Typhi that are generally also multidrug resistant can be treated with third generation cephalosporins (ceftriaxone, cefixime, cefotaxime, and cefoperazone) that achieve only marginally poorer results than the fluoroquinolones. The major drawback of these drugs is there relative expense and lack of availability. Chloramphenicol, amoxicillin, and trimethoprim remain excellent treatments if the infecting strain is fully susceptible. Although widely available and relatively inexpensive these drugs have significant drawbacks. In addition to the widespread resistance to chloramphenicol, amoxicillin, and trimethoprim, these drugs also require extended treatment for 2–3 weeks with four times a day oral regimen for full effectiveness. Consequently, noncompliance is a significant problem resulting in increased relapse and carrier rates.

ANTIBIOTIC RESISTANCE

S. Typhi isolates with resistance to chloramphenicol were first reported in a typhoid outbreak in Mexico in 1972 (Olarte and Galindo, 1973). A similar outbreak with a chloramphenicol resistant strain occurred in the late 1970s in Lima, Peru (Goldstein et al., 1986). In each case, following an initial large increase in incidence of resistant strains, within a few years resistance was markedly reduced. This may have been influenced by the shift to an alternative therapy, using amoxicillin and trimethoprim, thereby removing the selective pressure for maintenance of the plasmid encoded chloramphenicol resistance. Chloramphenicol resistance associated with an R factor also emerged in Vietnam in the early 1970s (Butler et al., 1973). On the whole, the incidence of chloramphenicol resistant strains was low during the 1980s and even when sporadic outbreaks occurred these strains were generally fully sensitive to amoxicillin and trimethoprim. However, in the early 1990s isolates which were resistant to amoxicillin and trimethoprim in addition to chloramphenicol were reported and termed multidrug resistance (MDR) (Anand et al., 1990; Gupta, 1994). MDR is associated with incH1 plasmids which are related to the R27 plasmid first reported in *S. enterica* in the 1960s (Parry et al., 2002; Taylor et al., 1985). MDR strains require therapy with quinolones or third generation cephalosporins, and perhaps due to the use of these alternative drugs, a trend toward a lower incidence of MDR strains has since been observed. For example, in one study the incidence of MDR in blood isolates in Bangladesh indicated a peak of 3.2% in 1994 that reduced to 1.0% by 1996 (Rahman et al., 2002) and similarly fluctuations in incidence of MDR *S.* Typhi strains were reported in Vietnam (Parry, 2004). The current emerging problem in treating typhoid fever is the emergence of partial or full resistance to fluoroquinolones in several areas of Asia (Mehta et al., 2001; Mermin et al., 1999; Wain et al., 1997). Low-level resistance to ciprofloxacin, which is invariably linked to complete resistance to nalidixic acid, is frequently not picked up in clinical microbiology labs but is of clinical significance.

PATHOGENESIS

Since *S.* Typhi is host restricted to man and higher primates, investigations of pathogenesis of *S.* Typhi infections are limited to a handful of studies with chimpanzees, biopsies of typhoid fever patients, volunteer trials, and autopsy reports. The first primate experiments were carried out as early as 1904 (Grenbaum, 1904) with further more comprehensive studies in the middle of the 20th century (Edsall et al., 1960; Gaines et al., 1968; Hornick et al., 1970a, 1970b). Monetary expense and ethical considerations has since limited the use of this model for studies of pathogenesis and vaccine development. Commonly, a surrogate murine model is used as an alternative. In 1892 Loeffler described the causative agent of murine typhoid, now called *S. enterica* serotype Typhimurium (*S.* Typhimurium) using current nomenclature. The pathology of internal organs and the distribution of the *S.* Typhimurium in the host superficially resembled that of *S.* Typhi in typhoid fever patients and experimentally infected primates. These early reports thereby established *S.* Typhimurium infections of mice as a surrogate model for typhoid fever (Santos et al., 2001). Consequently much of our understanding of typhoid fever pathogenesis results from extrapolating data from experimental infections of *S.* Typhimurium infections of mice.

Typhoid fever is initiated following ingestion of between 10^3 and 10^6 organisms depending on many factors, but likely dominated by the buffering effects of certain foods (Hornick et al., 1970a). A significant barrier to the progression of infection is the low pH of the stomach, hence the association of achlorhydria with increased susceptibility (Bhan et al., 2005). Following passage through the stomach, bacteria adhere to and invade mucosal cells most likely at the Peyer's patch—lymphoid aggregates associated with the intestinal wall. M-cells, which are specialized epithelial cells involved in internalizing lumen contents for antigen presentation to the immune system, likely represent a significant portal of entry. Binding to the cystic fibrosis transmembrane receptor (CFTR) mediated by the *S.* Typhi type IV pilus may be important in the process that culminates in colonization of the gut-associated lymphoid tissue (Pier et al., 1998). Significantly, the early interactions of *S.* Typhi with the intestinal mucosa does not normally result in a polymorphonuclear infiltrate, a characteristic of infections by nontyphoidal serotypes that cause gastroenteritis. Instead, *S.* Typhi induces a predominantly mononuclear response that is not associated with diarrheagenesis. Evidence is now beginning to emerge that Vi polysaccharide capsule expressed by *S.* Typhi may be involved in modulation of the immune response that would otherwise lead to PMN influx during early stages of infection (Raffatellu et al., 2005; Sharma and Qadri, 2004).

Following invasion of the intestinal mucosa *S.* Typhi next translocate to mesenteric lymph nodes and cells of the reticuloendothelial cells of the liver and spleen during a period of transient bacteremia. Part of this dissemination may involve trafficking of infected dendritic cells (Vazquez-Torres et al., 1999). *S.* Typhi are known to survive and multiply in mononuclear phagocytic cells in vitro and likely do so at this stage within similar cells of the lymphoid follicles, liver, and spleen. After a period of intracellular replication, viable *S.* Typhi become detectable in the blood stream initiating fever in the patient, a progression in disease for which the stimuli are not currently known. Following this second episode of bacteremia, the organism is widely disseminated to secondary sites of colonization including the liver, spleen, gallbladder, bone marrow, and Peyer's patches of the terminal ileum. It has been suggested that necrosis of the Peyer's patches observed in typhoid fever patients in the third week of infection may result from the interaction of host immunological mediators and bacterial factors by a mechanism similar to the Shwartzman and Koch reactions during secondary exposure of these lymphoid tissue to high concentrations of pathogen antigens (Everest et al., 2001).

ANTIGENS ENCODED BY S. TYPHI

The major antigenic determinants of intact *S.* Typhi are LPS, the Vi polysaccharide capsule, and the flagella. The two dominant antigenic components of *S.* Typhi LPS are the repeating sugar backbone composed of 2-α-mannose-1,4-α-Rhamnose-1, 3-α-galactose-1 (O12 somatic antigen) and the immune-dominant antigen tyvelose, a sugar side chain (O9 somatic antigen) (Reeves, 1993). Vi is a polymer of α-1,4-galacturonic acid with an *N*-acetyl at position C-2 and variable *O*-acetylation at C-3 (Mai et al., 2003). The presence of this polysaccharide capsule correlates with virulence since just 1 of 2222 clinical *S.* Typhi isolates tested, was negative for Vi as indicated by serology or multiplex PCR to assay for the presence of the biosynthetic genes in a survey in Pakistan (Wain et al., 2005). *S.* Typhi lacking Vi had a 10,000-fold greater lethal dose 50% in the mouse mucin model of infection following intraperitoneal inoculation (Hone et al., 1988). Furthermore, human volunteers inoculated with Vi expressing *S.* Typhi isolates had a higher incidence of typhoid fever than those inoculated with isolates not expressing Vi (Hornick et al., 1970a, 1970b). *S.* Typhi are monophasic with respect to flagella and generally express the H:d or H:j antigen. However, some isolates from Indonesia express an alternative flagellin termed the H:z66 antigen (Huang et al., 2004; Tamura et al., 1988). *S.* Typhi also expresses iron-regulated outer-membrane proteins (IROMPs) and porins that are targets for the immune system during infection (Blanco et al., 1993; Sood et al., 2005).

S. Typhi encodes a number of pili (fimbriae), proteinaceous hair-like appendages commonly encoded by enteric bacteria and involved in interactions with the host or self association. A type IV pilus, encoded by the *pil* locus on the *S.* Typhi *Salmonella* pathogenicityisland 7 (SPI-7), may be important during initial phases of infection in the intestine where it mediates interaction with the CFTR receptor on epithelial cells (Zhang et al., 2000). *S.* Typhi also encodes an ortholog of the *agf* fimbriae of the curli family, and 12 putative fimbriae of the usher-chaperone-family (Townsend et al., 2001). Six of the usher-chaperone-family fimbrial operons contain pseudogenes resulting from the presence of frame shift and/or a stop codon in the coding sequence and therefore may not be expressed. The genomic degradation of fimbrial operons may in part be responsible for the reduced intestinal involvement in typhoid fever compared to nontyphoidal serotypes.

PROTECTIVE IMMUNE RESPONSE

Experimental infections of mice with attenuated *S.* Typhimurium or *S. enteritidis* results in antibody responses to a diverse set of antigens including LPS, Vi polysaccharide, porins, outer membrane proteins, lipoproteins, heat shock proteins, flagella, and fimbriae (Brown and Hormaeche, 1989; Harrison et al., 1997; Kuusi et al., 1979; Matsui and Arai, 1989; Ogunniyi et al., 1994). Antibody response to somatic LPS, flagella, and Vi antigen are generally monitored in human volunteer studies with live attenuated vaccine candidate *S.* Typhi strains. Vaccinees inoculated with live oral vaccine exhibit a humoral immune response to LPS or flagella with the immune response varying from low to relatively high titer, depending on the nature of attenuation of the vaccine strain and design of the trial (Dilts et al., 2000; Hindle et al., 2002; Hohmann et al., 1996; Levine et al., 1987a; Tacket et al., 2000). Killed vaccine preparations and subunit vaccines are considered by some experts to be potentially inferior to live oral vaccines because dead vaccines predominantly generate a humoral immune response. In contrast some live vaccines generate a potent cell-mediated immune (CMI) response in addition to a humoral immune response. Although killed vaccines of *S.* Typhimurium are able to generate a protective immunity against challenge with virulent live *S.* Typhimurium in genetically resistant mice, these vaccines provide poor protection in genetically susceptible mice (Eisenstein et al., 1984; Hormaeche et al., 1995). Live vaccines generate protective immunity in both genetically resistant and susceptible strains of mice. Furthermore, live vaccines are administered orally, and are therefore capable of generating a mucosal immune response that may be important in immunity during the early stages of infection (Hormaeche et al., 1995). However, the success of the Vi subunit vaccine shows that dead vaccines can offer considerable efficacy.

Robust CMI responses to attenuated *S.* Typhimurium and *S.* Typhi following immunization of mice or volunteers have been described. The responses are predominantly of the Th1 type, as indicated by delayed type hypersensitivity (DTH) and by production of IL-2 and IFN-γ cytokines on stimulation of sensitized T-cells in vitro (Harrison et al., 1997; Sztein et al., 1994). The antigens involved in cell-mediated responses in experimental infections of mice include unspecified proteinaceous antigens, porins, flagellin, and pilin (Cao et al., 1992; Cookson and Bevan, 1997). Cell-mediated response has been monitored by various assays including tracking proliferation of sensitized lymphocytes in vitro, detection of cytokine production by in vitro stimulated T-cells, cytolytic assays, ELISPOT assays, and monitoring the DTH response following stimulation.

The protective antigens of live vaccine strains are not well defined currently. Attempts to answer this question are frustrated by the complexity of live oral *S.* Typhi vaccines in which multiple antigens are expressed in the context of a sublethal infection. It seems likely that a combination of antibody and cell-mediated responses to immune-dominant LPS antigens as well as protein antigens are required for high levels of protection against virulent *S.* Typhi following a single dose of live oral vaccine (Hormaeche et al., 1991, 1996; Kuusi et al., 1979; Matsui and Arai, 1989).

VACCINES

History

There is a long history of typhoid vaccine development, production, and marketing. The first typhoid vaccine, a parentally administered heat-killed preparation of *S.* Typhi, was introduced in 1896 and has been variably attributed to the British pathologist Almroth Write and the German bacteriologist Richard Pfeiffer (Groschel and Hornick, 1981). It was first demonstrated that specific antibody against *S.* Typhi was able to protect guinea pigs against test challenge with virulent *S.* Typhi. Write tested his vaccine in two officers in the Indian Army Corps, demonstrating general safety. A larger scale test of efficacy in 2835 army volunteers then revealed significant side effects that were nevertheless deemed acceptable in the light of the encouraging level of protection against typhoid. This vaccine was subsequently employed on a large scale in soldiers during the Boer War in South Africa (1899) and for decades to follow. However, due to a high frequency of systemic and localized reactions associated with the heat-killed parental vaccine new vaccines have subsequently been developed and are now licensed for use in many countries in recent years. A summary of currently licensed vaccines is presented in Table 57.1.

Current Licensed Vaccines

Heat-Killed Whole Cell Vaccine

Although rarely used today a whole cell killed vaccine, essentially the same as that developed by

TABLE 57.1 S. Typhi currently licensed vaccines

Vaccine	Primary course	Booster/revaccination	Adverse reactions	Protective efficacy[a]
Killed whole cell	Adults: 0.5 ml sc or im Children (1–10 years): 0.25 ml sc or im Children <1 year: not recommended	Adults and children every 3 years	Local reactions; persistent redness, swelling, pain and tenderness Systemic; malaise, nausea, headache, pyrexia	51–67%
Vi poly-saccharide	Adults and children >18 months: 0.5 ml im or deep sc Children <18 months: not recommended	Adults and children every 3 years	Mild and transient local reactions and less frequent systemic reactions than the whole cell vaccine	55–75%
Live attenuated Ty21a	Adults and children >6 years: 1 capsule on alternate days × 3 doses Children <6 years: not recommended	Adults and children every 1–3 years	Infrequent transient mild nausea, vomiting, abdominal cramps, diarrhea and urticarian rash	62–79%

[a] Reported by WHO (http://www.who.int/en/).

Alroth Write in 1896, is available and is relatively inexpensive. Various methods of preparation have been described including treatment with heat, alcohol, and acetone, and protective efficacy rates ranging from 40 to 89% in field trials have been reported (Engels et al., 1998). The major drawbacks are that these formulations are associated with systemic reactions in 9–34% of recipients and variability in production quality can affect efficacy.

Vi Polysaccharide Capsular Vaccine

The current subunit vaccine is a preparation of Vi polysaccharide from *S.* Typhi currently produced under the commercial names Typhim® and Typherix®. Vaccine is administered deep subcutaneously or intramuscularly as a single dose of 25 μg in 0.5 ml of preparation. Protection against typhoid fever begins days after injection and revaccination by a single dose is recommended every 3 years to maintain protection. In a trial in Nepal with individuals aged 5–44, a level of 75% protection during a 20-month period of surveillance was reported (Acharya et al., 1987). In a second study in South Africa involving children aged 5–16 years, a 55% efficacy was reported 3 years after vaccination and the serological correlate of protection was estimated to be 1 μg/ml of Vi-specific antibody (Klugman et al., 1996). The vaccine is not effective in children less than 2 years of age and due to poor immunological responses may not be cost effective in children 2–5 years of age. The Vi polysaccharide vaccine can be given with other vaccines suggested for travelers and has no contraindications other than previous severe reaction. Adverse reactions are less frequent than with the killed whole cell vaccine but may include fever (<1% recipients), headache (1.5–3% of recipients), and erythema or induration at the site of injection (7% or recipients). In common with other polysaccharide vaccines there is no booster effect suggesting a lack of immune memory. New Vi–protein conjugate vaccines that are currently under development may overcome this drawback of the Vi vaccine (Canh do et al., 2004; Kossaczka et al., 1999; Szu et al., 1994).

Attenuated Live Oral Vaccine Strain Ty21a

S. Typhi strain Ty21a that is currently manufactured under the commercial name Vivotif®, was derived from the wild-type strain Ty2 by two rounds of nitrosoguanidine (NTG) chemical mutagenesis. Ty21a is normally administered orally as an enteric-coated capsule but liquid formulations have also been tested. When three doses are administered 2 days apart, protection may be established within days following the last dose and lasts for several years. However, a booster is recommended every 1–3 years depending on whether the recipient is visiting or living in an endemic area. A protective efficacy of 67% over 7 years has been demonstrated for the standard enteric-coated capsules. In a second trial involving 36,000 vaccinees aged 5–19 years, using a liquid formulation, a protective efficacy of 79% was reported after 5 years. Generally Ty21a is extremely well tolerated but may cause transient mild nausea, vomiting, abdominal cramps, diarrhea, and urticarian rash. Ty21a may be taken together with other vaccines, including live vaccines. As Ty21a is a live vaccine it should not therefore be used if the recipient is taking antimicrobial agents, or if mefloquine is being taken as malaria prophylaxis. The vaccine is safe for use by HIV+ individuals with T-cell count (CD4) above 200/mm^3.

The mechanism of attenuation for Ty21a is not known, and this is a major drawback of this vaccine. However, it is known that Ty21a does not express the Vi polysaccharide capsule and carries a null mutation in the *galE* gene. The *galE* gene encodes galactose epimerase that catalyses the synthesis of diphosphate–galactose from uridine diphosphate–galactose. Galactose is an important component of LPS core, the major component of the outer membrane. In the absence of exogenous galactose this LPS deficiency gives rise to a rough colony-morphology that in *S*. Typhimurium, is associated with attenuation with respect to murine pathogenesis. Indeed, *S*. Typhimurium *galE* mutants are attenuated and effective as live oral vaccines in mice (Hone et al., 1987). However, a defined *galE* mutant of *S*. Typhi Ty2, that also lacked the Vi polysaccharide, retained virulence in human volunteers (Hone et al., 1988) indicating that these mutations alone do not account for the full attenuation of Ty21a.

Vaccines in Development

Currently available typhoid vaccines have a number of drawbacks. Neither the Vi polysaccharide capsular vaccine nor the Ty21a live oral vaccine confer complete protective efficacy, although this is a goal not readily attainable by any vaccine. Although the Vi vaccine is effective following a single parenteral inoculum, it is relatively expensive as it requires skilled personnel to administer the vaccine using a syringe and needle. The available live oral vaccine is only weakly immunogenic and requires three doses to provide maximal protective efficacy. Two approaches are being used to develop improved vaccines. First, Vi–protein conjugates are being developed that induce a memory component to the immune response to the subunit. Second, a series of live oral vaccines have been constructed (summarized in Table 57.2) with the aim to develop a vaccine that is safe, immunogenic, and efficacious following a single dose.

Subunit Vaccines in Development

To improve the immunogenicity of Vi polysaccharide various proteins have been conjugated including the B-subunit of the heat-labile toxin of *E. coli* or the recombinant exoprotein A of *Pseudomonas aeruginosa* (Vi-rEPA). In guinea pigs and mice the antibody response to Vi was four- to fivefold greater in conjugates compared to unconjugated Vi 26 weeks after vaccination. Furthermore, a second immunization with the conjugates resulted in a booster response not seen in the Vi vaccine (Szu et al., 1994). In human volunteer trials of Vi-rEPA the vaccine conjugate elicited no adverse reactions and was impressively immunogenic (Kossaczka et al., 1999; Lin et al., 2001). The vaccine was safe and had a protective efficacy of 89% in a double-blind, randomized, and placebo-controlled field trial with 2–5 year old children in Vietnam (Mai et al., 2003). This compares favorably with the 55% protective efficacy of the currently licensed vaccine (Klugman et al., 1996). A second Vi–porin conjugate vaccine has also been developed that in mice, induced a systemic and mucosal immune response against Vi and porins and was sixfold more protective than Vi alone (Singh et al., 1999).

Live Attenuated Vaccine Strains in Development

A number of approaches to generate attenuated *S*. Typhi strains suitable for development of live typhoid vaccines have been described. Early studies described the attenuating properties of auxotrophic mutant derivatives of *Salmonella* including those requiring purine, aspartic acid, and *p*-aminobenzoate (pABA) (Bacon et al., 1950a, 1950b, 1951). *Salmonella* require pABA as a precursor of folates, aromatic amino acids, and the siderophore enterochelin. Bacteria synthesize pABA from chorismate, the terminal product of the aromatic biosynthesis pathway and indeed the precursor of all aromatic metabolites including enterochelin, tyrosine, phenylalanine, and tryptophan. Loss of enzyme-function in the aromatic pathway due to mutations in the *aro* genes creates a requirement for endogenous sources of the three aromatic amino acids and pABA. Additionally *aro* strains are unable to biosynthesize enterochelin, a low molecular weight iron chelating molecule involved in iron scavenging. However, enterochelin-mediated iron supply to *Salmonella* is at least partially dispensable during infection by some strains (Benjamin et al., 1985) and therefore mutation of this gene does not appear to be the mechanism for attenuation of *aro* strains. While *Salmonella* can transport and use endogenous aromatic amino acids, endogenous folate cannot be scavenged, perhaps due to its bulky structure that precludes diffusion through the porins in the outer membrane. It is, therefore, most likely that a lack of available folate or pABA is the main reason for the observed attenuation (Stocker, 2000). The first report to suggest the potential utility of aromatic auxotrophs as vaccine strains was about a quarter of a century ago. An *aroA* strain of *S*.

TABLE 57.2 *S.* Typhi attenuated live oral vaccines in development

Strain	Mutation	Basis of attenuation	Immune response	Tolerance	References
CVD908	*aroC, aroD*	Aromatic metabolite auxotrophy	IgG seroconversion to *S.* Typhi O-antigen in 83% of vaccinees and stimulated specific IgA-secreting gut-derived lymphocytes in 100% of vaccinees	Well tolerated, no febrile or other adverse response; bacteremia in a small proportion of vaccinees	Hone et al. (1992)
541Ty and 543Ty	*aroA, purA*	Aromatic metabolite auxotrophy, purine biosynthesis	Meager humoral antibody response to O, H, and Vi; 69% of vaccines responded to *S.* Typhi particulate or O-antigen	Well tolerated, no febrile or other adverse response	Levine et al. (1987b)
CVD908-*htrA*	*aroC, aroD htrA*	Aromatic metabolite auxotrophy, oxidative stress response	Serum immune response to *S.* Typhi O-antigen in 75–100% of vaccinees and lymphoproliferative responses to flagella and particulate antigens with high dose	Generally well tolerated, but 3 of 36 volunteers had mild diarrhea and 1 had mild fever; in phase II trial no febrile responses were reported	Tacket et al. (1997b, 2000)
χ3927	*crp, cya*	Regulation of transport and breakdown of catabolites and biogenesis of flagella	53% developed IgG response to *S.* typhi O-antigen	2 of 12 vaccinees had vaccine bacteremia without symptoms; 1 of 12 had fever	Tacket et al. (1992)
Ty800	*phoP, phoQ*	Regulation of antimicrobial peptide resistance genes and SPI-2 genes	10 of 11 volunteers developed a mucosal immune response (IgA secreting cells in blood); 9 of 11 had humoral immune response	Generally well tolerated; two volunteers had self-limiting side effects but no vaccine bacteremia was reported	Hohmann et al. (1996)
Ty445	*aroA, phoP, phoQ*	As Ty800 but additional aromatic metabolite auxotrophy	2 of 14 volunteers seroconverted to whole *S.* typhi or O-antigen	Well tolerated, no febrile or other adverse response	Hohmann et al. (1996)

Typhimurium was found to protect mice even when genetically susceptible mice were challenged by the intraperitoneal route at a 1000-fold higher dose than the lethal dose 50% (Hoiseth and Stocker, 1981). In order to reduce the likelihood of the emergence of *aro*+ revertants by back mutation or complementation during passage or preparation of the vaccine, a *S.* Typhi vaccine candidate (CVD908) was constructed that harbored paired attenuating mutations, in the *aroC* and *aroD* genes that widely separated on the genome. In early clinical studies CVD908 appeared safe and immunogenic following a single dose. However, the presence of the vaccine in the blood of some volunteers was considered an unfavorable characteristic and this strain was not further developed (Hone et al., 1992).

Several attempts to further attenuate the *S.* Typhi aromatic biosynthesis mutant strains have since been made. An approach involving the introduction of purine auxotrophy by mutation of the *purA* gene failed, as this strain was over attenuated and poorly immunogenic. Only a fraction of vaccinees responded to *S.* Typhi antigens in a lymphocyte proliferation assay and antibody responses were very low (Levine et al., 1987b). In contrast, addition of *htrA* or *ssaV* mutations to the aromatic biosynthesis (*aro*) background has been extremely productive and such *S.* Typhi vaccine candidates are well advanced in clinical trials. The *Salmonella htrA* gene is homologous to a heat stress protein of *E. coli.* However, *S.* Typhimurium *htrA* mutants are not temperature sensitive but rather exhibit increased sensitivity to oxidative stress in vitro (Pallen and Wren, 1997). It is likely that this mutation results in an increased susceptibility to oxidative killing by polymorphonucleocytes (PMN) and macrophages in vivo, and that this may be the basis of the observed attenuation of *htrA* mutants (Johnson et al., 1991). An *aroC*, *aroD*, and *htrA S.* Typhi triple mutant (CVD908 *htrA*) has been demonstrated to be safe in humans and induce humoral immune response to O-antigen and flagella. Furthermore, cellular immune response to flagella and particulate antigens, as indicated by lymphoproliferation and production of interferon-γ, was also reported (Tacket et al., 1997b, 2000). Another candidate vaccine called *S.* Typhi strain M01ZH09, is based on an *aroC* mutation combined with a mutation in the *ssaV* gene encoded on the *Salmonella* pathogenicity island 2 (SPI-2). SPI-2 is a horizontally acquired genetic island present only in *S. enterica*. The bulk of the island encodes the apparatus for a type III secretion system (TTSS-2), a macromolecular "syringe" involved in the transfer of effector proteins into the host cell cytoplasm. The TTSS encoded on SPI-2 is upregulated upon internalization of *Salmonella* into macrophages and subsequently several injected effectors modify the phagosome compartment such that the bacteria can survive and replicate (Meresse et al., 2001; Vazquez-Torres et al., 2000). ssaV is an integral protein of the secretion apparatus and in its absence no effector proteins are secreted into target cells. In clinical trials, the *S.* Typhi strain M01ZH09 was well tolerated, although a small fraction of volunteers in the medium and high dose groups had slightly elevated temperatures. However, no viable M01ZH09 were recovered from the blood and no adverse reactions were reported. All subjects tested in the highest dose group had a significant antibody-secreting cell (ASC) response and 75% had an increase in antibody titer for O or H-antigen (Hindle et al., 2002; Kirkpatrick et al., 2005a, 2005b).

Additional *S.* Typhi candidate vaccines have been developed that are based on mutations in regulator genes. One candidate harbors mutations in adenylate cyclase (*cya*) and the cyclic AMP receptor (*crp*). These genes regulate the transcription of genes involved in the transport and utilization of carbohydrates and amino acids (Alper and Ames, 1978), and in the expression of outer-membrane proteins, flagella, and fimbriae (Komeda et al., 1976; Movva et al., 1981; Saier et al., 1978). A *S.* Typhi strain designated χ3927 with defined mutations in *cya* and *crp* was found to be highly immunogenic in a volunteer study, but produced fever in 1 of the 12 volunteers and vaccine bacteremia in 2 of the volunteers (Tacket et al., 1992). A second candidate vaccine strain was engineered with a mutation in the *phoP* and *phoQ* regulatory genes. These genes form a two-component regulator comprising the membrane bound sensor-kinase (PhoQ) and a cytoplasmic transcriptional regulator (PhoP) (Groisman et al., 1989; Miller et al., 1989). PhoP/PhoQ is involved in the activation or repression of a number of unlinked genes, many of which have been associated with virulence. For example, the PhoP/PhoQ regulon includes the TTSS-2 that is essential for macrophage survival (Bijlsma and Groisman, 2005), genes required for resistance to antimicrobial peptides (Gunn and Miller, 1996; Shi et al., 2004), and genes associated with resistance to bile (van Velkinburgh and Gunn, 1999). A *S.* Typhi vaccine candidate called Ty800 was constructed that contained defined deletion of the *phoP/phoQ* locus. In a clinical trial Ty800 was more immunogenic in volunteers following a single oral dose compared to control vaccinees that were given four doses of the conventional Ty21a (Hohmann et al., 1996). A potential drawback of the *phoP/phoQ* vaccine is that it is attenuated due to deletion of a single locus, therefore raising the possibility of complementation back to full virulence by a single event. A second *S.* Typhi vaccine candidate called Ty445 was constructed that contained defined deletion of the *aroA* gene in addition to the *phoP/phoQ* locus. In a clinical trial in which escalating doses of Ty445 *S.* Typhi strain was compared with Ty21a, the candidate vaccine was found to be safe and nonreactogenic but was considerably less immunogenic even after administration of two inoculations at the highest dose compared to the conventional vaccine strain (Hohmann et al., 1996).

S. Typhi Vaccines for Delivery of Heterologous Antigens

In the last two decades considerable effort has been directed to the development of bivalent *S.* Typhi vaccines engineered to express antigens from heterologous pathogens. The aim is to develop a live oral *S.* Typhi strain capable of protecting against multiple pathogens such as bacterial and viral pathogens and even eukaryotic parasites. This technology has been developed mainly using *S.* Typhimurium as an experimental vector since this can be genetically manipulated and the immunogenicity and protective efficacy easily evaluated in the murine model of infection. Many antigens have been expressed in *Salmonella* and the effect of expression levels, choice of promoter, and context of expression on immunogenicity assessed. For example, the effect of gene dosage on immmunogenicity of *E. coli* heat-labile toxin (LTK63) expressed in a *S.* Typhimurium *cya/crp* vaccine strain was evaluated (Covone et al., 1998). They found that high expression levels especially at the early time points of infection, were of particular importance. The choice of promoter was evaluated by driving expression of tetanus toxin fragment C (TetC) from a number of promoters. It was reported that the *nirB* promoter, which is activated in response to low oxygen tension, was particularly good for stable expression of TetC from multicopy vectors during infection (Chatfield et al., 1992).

The context of expression has been found to be of considerable importance. That is, antigen expressed on the surface or secreted into the milieu, generally results in an improved immune response compared to expression in the cytosol of the vector strain. Various strategies have been used to express heterologous antigens on the surface of *S.* Typhimurium. For example, epitopes of colonization factor antigen I (CFA/I) and cholera toxin have been fused to flagellin and expressed in *S.* Typhimurium vaccine strains (Chauhan et al., 2005; das Gracas Luna et al., 2000). In each case an immune response to the epitope in the chimeric flagellin was detectable following immunization of mice. A second strategy involved the presentation of antigens as part of engineered autotransporter proteins. The autotransporter-family of proteins share a common mechanism of secretion in which the C-terminus forms a β-barrel structure in the outer membrane through which the N-terminal passenger domain is secreted (Henderson et al., 1998). For example, antigen 43 and AIDA of *E. coli* have been used to express foreign antigens on the surface of *S.* Typhimurium vaccine strains (Kjaergaard et al., 2002; Kramer et al., 2003). Furthermore, autotransporter-based delivery systems can be engineered to release the "passenger antigen" into the external milieu where it may be more immunogenic. For example, the major B-cell epitope of the *Plasmodium falciparum* circumsporozoitic protein (NANP) has been fused with the *S.* Typhimurium MisL C-terminal transporter domain including protease (OmpT) recognition sequence (Ruiz-Olvera et al., 2003). NANP was detectable both associated with the surface of a *S.* Typhimurium vaccine strain and free in the milieu.

Despite the relative success in expressing heterologous antigens in *S.* Typhimurium vaccine vectors that result in the generation of impressive immune responses in animal trials, development of *S.* Typhi vaccines expressing heterologous antigens has been generally problematic. One of the main issues associated with further development is that there are no animal models available to optimize expression and presentation of the antigen in *S.* Typhi. One model that has been explored is the intranasal inoculation of *S.* Typhi vaccine candidates using the murine model. In this system, a vaccine candidate *S.* Typhi CVD908 (*aroC, aroD*) expressing TetC was shown to induce high titers of anti-TetC antibody and confer protection against a tetanus toxin challenge. A handful of bivalent *S.* Typhi vaccines have been evaluated in clinical trials with generally disappointing results. Two volunteer studies of the *S.* Typhi χ4073 (*cyr/crp*) vaccine candidate expressing hepatitis B antigen resulted in poor immune response to the heterologous antigen (Nardelli-Haefliger et al., 1996; Tacket et al., 1997a). Similarly vaccine candidate *S.* Typhi Ty800 expressing *H. pylori* urease failed to induce an immune response in volunteers even following oral boost with purified urease A/B with *E. coli* heat-labile toxin as adjuvant (DiPetrillo et al., 1999). At least one study has provided some promising observations. In a volunteer study in which candidate vaccine *S.* Typhi strain CVD908-htrA expressed TetC from a plasmid, three of the nine volunteers receiving medium to high doses generated immune response to TetC (Tacket et al., 2000). Furthermore, this study established that it was possible to generate protective levels of serum antitoxin.

Clearly more development is required before the possibility of efficacious bivalent or even multivalent *S.* Typhi vaccines becomes commercial reality. An important consideration that remains to be fully investigated is the impact of prior immunity to *S.* Typhi. One study reported that exposure to *S.* Dublin 2, 6, or 9 weeks previously did not decrease the immune response to *E. coli* heat-labile toxin expressed by *S.* Dublin in a second inoculation. Indeed there was evidence in these experiments that previous exposure may even potentiate the subsequent response. However, contradictory evidence have since been reported (Attridge et al., 1997; Roberts et al., 1999). Preexisting immunity to *S.* Typhimurium and to a lesser extent *S.* Dublin, depressed the humoral immune response to heterologous antigen TetC that was expressed in a *S.* Typhimurium vaccine strain administered 44 days later (Roberts et al., 1999). It is possible that differences in these reports are serotype or strain-specific. These considerations are important if such vaccines are to be used in areas where typhoid is endemic and the target populations are likely to have preexisting exposure to the vector species.

PROSPECTS FOR THE FUTURE

There is reason for considerable optimism for future developments in the prophylaxis of typhoid fever. Improved subunit vaccines based on protein–Vi conjugates have been demonstrated to be safe and highly efficacious in phase II volunteer studies. At least three candidates CVD908-htrA (*aroC, aroD, htrA*), M01ZH09 (*aroC, htrA*), and Ty800 (*crp, cya*), are promising live oral vaccine strains and candidates to replace Ty21. At least one of these, *S.* Typhi strain M01ZH09, is soon to be evaluated in bridging trials in preparation for phase III field trials. Some of the candidate live oral vaccines have already been tested as carriers for heterologous antigens, although further development is required to optimize stability and the elicitation of immune response.

KEY ISSUES

- In the face of a rise in incidence of paratyphoid fever caused by *S.* Paratyphi A, a vaccine effective against this serotype is needed particularly for use in the Asian sub-continent.
- The development of a typhoid conjugant vaccine or live oral vacine safe and efficacious in children under two years of age.
- An understanding of the effect of vaccine programs on bacterial populations targeted by the vaccine and populations of related pathogens. For example, what are the effects of *S.* Typhi vaccination on the population structure of *S.* Typhi and *S.* Paratyphi A.
- The development of a vaccine to combat the rising problem of non-typhoidal *Salmonella* (NTS) in sub-saharan Africa.

References

Acharya, I.L., Lowe, C.U., Thapa, R., Gurubacharya, V.L., Shrestha, M.B., Cadoz, M., Schulz, D., Armand, J., Bryla, D.A., Trollfors, B. et al. Prevention of typhoid fever in Nepal with the Vi capsular polysaccharide of *Salmonella typhi*. A preliminary report. N. Engl. J. Med. 1987; 317:1101–1104.

Akalin, H.E. Quinolones in the treatment of typhoid fever. Drugs 1999; 58(Suppl. 2):52–54.

Albaqali, A., Ghuloom, A., Al Arrayed, A., Al Ajami, A., Shome, D.K., Jamsheer, A., Al Mahroos, H., Jelacic, S., Tarr, P.I., Kaplan, B.S. and Dhiman, R.K. Hemolytic uremic syndrome in association with typhoid fever. Am. J. Kidney Dis. 2003; 41:709–713.

Alper, M.D. and Ames, B.N. Transport of antibiotics and metabolite analogs by systems under cyclic AMP control: positive selection of *Salmonella typhimurium* cya and crp mutants. J. Bacteriol. 1978; 133:149–157.

Anand, A.C., Kataria, V.K., Singh, W. and Chatterjee, S.K. Epidemic multiresistant enteric fever in eastern India. Lancet 1990; 335:352.

Attridge, S.R., Davies, R. and LaBrooy, J.T. Oral delivery of foreign antigens by attenuated Salmonella: consequences of prior exposure to the vector strain. Vaccine 1997; 15:155–162.

Bacon, G.A., Burrows, T.W. and Yates, M. The effects of biochemical mutation on the virulence of *Bacterium typhosum*: the virulence of mutants. Br. J. Exp. Pathol. 1950a; 31:714–724.

Bacon, G.A., Burrows, T.W. and Yates, M. The effects of biochemical mutation on the virulence on *Bacterium typhosum*: the induction and isolation of mutants. Br. J. Exp. Pathol. 1950b; 31:703–713.

Bacon, G.A., Burrows, T.W. and Yates, M. The effects of biochemical mutation on the virulence of *Bacterium typhosum*; the loss of virulence of certain mutants. Br. J. Exp. Pathol. 1951; 32:85–96.

Baysal, K., Sancak, R., Ozturk, F., Uysal, S. and Gurses, N. Cardiac involvement due to *Salmonella typhi* infections in children. Ann. Trop. Paediatr. 1998; 18:23–25.

Benjamin, W.H., Turnbough, C.L., Posey, B.S. and Briles, D.E. The ability of *Salmonella typhimurium* to produce the siderophore enterobactin is not a virulence factor in mouse typhoid. Infect. Immun. 1985; 50:392–397.

Bhan, M.K., Bahl, R. and Bhatnagar, S. Typhoid and paratyphoid fever. Lancet 2005; 366:749–762.

Bhan, M.K., Bahl, R., Sazawal, S., Sinha, A., Kumar, R., Mahalanabis, D. and Clemens, J.D. Association between *Helicobacter pylori* infection and increased risk of typhoid fever. J. Infect. Dis. 2002; 186:1857–1860.

Bijlsma, J.J. and Groisman, E.A. The PhoP/PhoQ system controls the intramacrophage type three secretion system of *Salmonella enterica*. Mol. Microbiol. 2005; 57:85–96.

Black, R.E., Cisneros, L., Levine, M.M., Banfi, A., Lobos, H. and Rodriguez, H. Case-control study to identify risk factors for paediatric endemic typhoid fever in Santiago, Chile. Bull. World Health Organ. 1985; 63:899–904.

Blanco, F., Isibasi, A., Raul Gonzalez, C., Ortiz, V., Paniagua, J., Arreguin, C. and Kumate, J. Human cell mediated immunity to porins from *Salmonella typhi*. Scand. J. Infect. Dis. 1993; 25:73–80.

Brown, A. and Hormaeche, C.E. The antibody response to salmonellae in mice and humans studied by immunoblots and ELISA. Microb. Pathog. 1989; 6:445–454.

Butler, T., Linh, N.N., Arnold, K. and Pollack, M. Chloramphenicol-resistant typhoid fever in Vietnam associated with R factor. Lancet 1973; 2:983–985.

Canh do, G., Lin, F.Y., Thiem, V.D., Trach, D.D., Trong, N.D., Mao, N.D., Hunt, S., Schneerson, R., Robbins, J.B., Chu, C., Shiloach, J., Bryla, D.A., Bonnet, M.C., Schulz, D. and Szu, S.C. Effect of dosage on immunogenicity of a Vi conjugate vaccine injected twice into 2- to 5-year-old Vietnamese children. Infect. Immun. 2004; 72:6586–6588.

Cao, Y., Wen, Z. and Lu, D. Construction of a recombinant oral vaccine against *Salmonella typhi* and *Salmonella typhimurium*. Infect. Immun. 1992; 60:2823–2827.

Chatfield, S.N., Charles, I.G., Makoff, A.J., Oxer, M.D., Dougan, G., Pickard, D., Slater, D. and Fairweather, N.F. Use of the nirB promoter to direct the stable expression of heterologous antigens in Salmonella oral vaccine strains: development of a single-dose oral tetanus vaccine. Biotechnology (N.Y.) 1992; 10:888–892.

Chaudhry, R., Mahajan, R.K., Diwan, A., Khan, S., Singhal, R., Chandel, D.S. and Hans, C. Unusual presentation of enteric fever: three cases of splenic and liver abscesses due to *Salmonella typhi* and *Salmonella paratyphi* A. Trop. Gastroenterol. 2003; 24:198–199.

Chauhan, N., Kumar, R., Badhai, J., Preet, A. and Yadava, P.K. Immunogenicity of cholera toxin B epitope inserted in *Salmonella flagellin* expressed on bacteria and administered as DNA vaccine. Mol. Cell. Biochem. 2005; 276:1–6.

Cookson, B.T. and Bevan, M.J. Identification of a natural T cell epitope presented by Salmonella-infected macrophages and recognized by T cells from orally immunized mice. J. Immunol. 1997; 158:4310–4319.

Covone, M.G., Brocchi, M., Palla, E., Dias da Silveira, W., Rappuoli, R. and Galeotti, C.L. Levels of expression and immunogenicity of attenuated *Salmonella enterica* serovar typhimurium strains expressing *Escherichia coli* mutant heat-labile enterotoxin. Infect. Immun. 1998; 66:224–231.

Crump, J.A., Luby, S.P. and Mintz, E.D. The global burden of typhoid fever. Bull. World Health Organ. 2004; 82:346–353.

das Gracas Luna, M., Sardella, F.F. and Ferreira, L.C. *Salmonella flagellin* fused with a linear epitope of colonization factor antigen

I (CFA/I) can prime antibody responses against homologous and heterologous fimbriae of enterotoxigenic *Escherichia coli*. Res. Microbiol. 2000; 151:575–582.

Dilts, D.A., Riesenfeld-Orn, I., Fulginiti, J.P., Ekwall, E., Granert, C., Nonenmacher, J., Brey, R.N., Cryz, S.J., Karlsson, K., Bergman, K., Thompson, T., Hu, B., Bruckner, A.H. and Lindberg, A.A. Phase I clinical trials of aroA aroD and aroA aroD htrA attenuated *S. typhi* vaccines; effect of formulation on safety and immunogenicity. Vaccine 2000; 18:1473–1484.

DiPetrillo, M.D., Tibbetts, T., Kleanthous, H., Killeen, K.P. and Hohmann, E.L. Safety and immunogenicity of phoP/phoQ-deleted *Salmonella typhi* expressing *Helicobacter pylori* urease in adult volunteers. Vaccine 1999; 18:449–459.

Edsall, G., Gaines, S., Landy, M., Tigertt, W.D., Sprinz, H., Trapani, R.J., Mandel, A.D. and Benenson, A.S. Studies on infection and immunity in experimental typhoid fever. I. Typhoid fever in chimpanzees orally infected with *Salmonella typhosa*. J. Exp. Med. 1960; 112:143–166.

Eisenstein, T.K., Killar, L.M. and Sultzer, B.M. Immunity to infection with *Salmonella typhimurium*: mouse-strain differences in vaccine- and serum-mediated protection. J. Infect. Dis. 1984; 150:425–435.

Engels, E.A., Falagas, M.E., Lau, J. and Bennish, M.L. Typhoid fever vaccines: a meta-analysis of studies on efficacy and toxicity. BMJ 1998; 316:110–116.

Everest, P., Wain, J., Roberts, M., Rook, G. and Dougan, G. The molecular mechanisms of severe typhoid fever. Trends Microbiol. 2001; 9:316–320.

Gaines, S., Sprinz, H., Tully, J.G. and Tigertt, W.D. Studies on infection and immunity in experimental typhoid fever. VII. The distribution of *Salmonella typhi* in chimpanzee tissue following oral challenge, and the relationship between the numbers of bacilli and morphologic lesions. J. Infect. Dis. 1968; 118:293–306.

Goldstein, F.W., Chumpitaz, J.C., Guevara, J.M., Papadopoulou, B., Acar, J.F. and Vieu, J.F. Plasmid-mediated resistance to multiple antibiotics in *Salmonella typhi*. J. Infect. Dis. 1986; 153:261–266.

Grenbaum, A. Some experiments on enterica, scarlet fever and measles in chimpanzee. Br. Med. J. 1904; 1:817.

Groisman, E.A., Chiao, E., Lipps, C.J. and Heffron, F. *Salmonella typhimurium* phoP virulence gene is a transcriptional regulator. Proc. Natl. Acad. Sci. U.S.A. 1989; 86:7077–7081.

Groschel, D.H. and Hornick, R.B. Who introduced typhoid vaccination: Almroth Write or Richard Pfeiffer? Rev. Infect. Dis. 1981; 3:1251–1254.

Gunn, J.S. and Miller, S.I. PhoP-PhoQ activates transcription of pmrAB, encoding a two-component regulatory system involved in *Salmonella typhimurium* antimicrobial peptide resistance. J. Bacteriol. 1996; 178:6857–6864.

Gupta, A. Multidrug-resistant typhoid fever in children: epidemiology and therapeutic approach. Pediatr. Infect. Dis. J. 1994; 13:134–140.

Harrison, J.A., Villarreal-Ramos, B., Mastroeni, P., Demarco de Hormaeche, R. and Hormaeche, C.E. Correlates of protection induced by live Aro *Salmonella typhimurium* vaccines in the murine typhoid model. Immunology 1997; 90:618–625.

Henderson, I.R., Navarro-Garcia, F. and Nataro, J.P. The great escape: structure and function of the autotransporter proteins. Trends Microbiol. 1998; 6:370–378.

Hindle, Z., Chatfield, S.N., Phillimore, J., Bentley, M., Johnson, J., Cosgrove, C.A., Ghaem-Maghami, M., Sexton, A., Khan, M., Brennan, F.R., Everest, P., Wu, T., Pickard, D., Holden, D.W., Dougan, G., Griffin, G.E., House, D., Santangelo, J.D., Khan, S.A., Shea, J.E., Feldman, R.G. and Lewis, D.J. Characterization of *Salmonella enterica* derivatives harboring defined aroC and Salmonella pathogenicity island 2 type III secretion system (ssaV) mutations by immunization of healthy volunteers. Infect. Immun. 2002; 70:3457–3467.

Hoffman, S.L., Punjabi, N.H., Kumala, S., Moechtar, M.A., Pulungsih, S.P., Rivai, A.R., Rockhill, R.C., Woodward, T.E. and Loedin, A.A. Reduction of mortality in chloramphenicol-treated severe typhoid fever by high-dose dexamethasone. N. Engl. J. Med. 1984; 310:82–88.

Hohmann, E.L., Oletta, C.A., Killeen, K.P. and Miller, S.I. phoP/phoQ-deleted *Salmonella typhi* (Ty800) is a safe and immunogenic single-dose typhoid fever vaccine in volunteers. J. Infect. Dis. 1996; 173:1408–1414.

Hohmann, E.L., Oletta, C.A. and Miller, S.I. Evaluation of a phoP/phoQ-deleted, aroA-deleted live oral *Salmonella typhi* vaccine strain in human volunteers. Vaccine 1996; 14:19–24.

Hoiseth, S.K. and Stocker, B.A.D. Aromatic-dependent *Salmonella typhimurium* are non-virulent and effective as live oral vaccines. Nature 1981; 291:238–239.

Hone, D., Morona, R., Attridge, S. and Hackett, J. Construction of defined galE mutants of Salmonella for use as vaccines. J. Infect. Dis. 1987; 156:167–174.

Hone, D.M., Attridge, S.R., Forrest, B., Morona, R., Daniels, D., LaBrooy, J.T., Bartholomeusz, R.C., Shearman, D.J. and Hackett, J. A *galE via* (Vi antigen-negative) mutant of *Salmonella typhi* Ty2 retains virulence in humans. Infect. Immun. 1988; 56:1326–1333.

Hone, D.M., Tacket, C.O., Harris, A.M., Kay, B., Losonsky, G. and Levine, M.M. Evaluation in volunteers of a candidate live oral attenuated *Salmonella typhi* vector vaccine. J. Clin. Invest. 1992; 90:412–420.

Hormaeche, C.E., Anjam Khan, C.M., Mastroeni, P., Villarreal, B., Dougan, G., Roberts, M. and Chatfield, S. Salmonella Vaccines: Mechanisms of Immunity and their use as Carriers of Recombinant Antigens. Wiley, London, 1995.

Hormaeche, C.E., Joysey, H.S., Desilva, L., Izhar, M. and Stocker, B.A. Immunity conferred by Aro-Salmonella live vaccines. Microb. Pathog. 1991; 10:149–158.

Hormaeche, C.E., Mastroeni, P., Harrison, J.A., Demarco de Hormaeche, R., Svenson, S. and Stocker, B.A. Protection against oral challenge three months after i.v. immunization of BALB/c mice with live Aro *Salmonella typhimurium* and *Salmonella enteritidis* vaccines is serotype (species)-dependent and only partially determined by the main LPS O antigen. Vaccine 1996; 14:251–259.

Hornick, R.B., Greisman, S.E., Woodward, T.E., DuPont, H.L., Dawkins, A.T. and Snyder, M.J. Typhoid fever: pathogenesis and immunologic control. N. Engl. J. Med. 1970a; 283:686–691.

Hornick, R.B., Greisman, S.E., Woodward, T.E., DuPont, H.L., Dawkins, A.T. and Snyder, M.J. Typhoid fever: pathogenesis and immunologic control. 2. N. Engl. J. Med. 1970b; 283:739–746.

Huang, X., Phung le, V., Dejsirilert, S., Tishyadhigama, P., Li, Y., Liu, H., Hirose, K., Kawamura, Y. and Ezaki, T. Cloning and characterization of the gene encoding the z66 antigen of *Salmonella enterica* serovar Typhi. FEMS Microbiol. Lett. 2004; 234:239–246.

JAMA Typhoid in the large cities of the United States in 1919. JAMA 1920:672–675.

Johnson, K., Charles, I., Dougan, G., Pickard, D., O'Gaora, P., Costa, G., Ali, T., Miller, I. and Hormaeche, C. The role of a stress-response protein in *Salmonella typhimurium* virulence. Mol. Microbiol. 1991; 5:401–407.

Kingsley, R.A. and Bäumler, A.J. Host adaptation and the emergence of infectious disease: the salmonella paradigm. Mol. Microbiol. 2000; 36:1006–1014.

Kirkpatrick, B.D., McKenzie, R., O'Neill, J.P., Larsson, C.J., Bourgeois, A.L., Shimko, J., Bentley, M., Makin, J., Chatfield, S., Hindle, Z., Fidler, C., Robinson, B.E., Ventrone, C.H., Bansal, N.,

Carpenter, C.M., Kutzko, D., Hamlet, S., Lapointe, C. and Taylor, D.N. Evaluation of *Salmonella enterica* serovar Typhi (Ty2 aroC-ssaV-) M01ZH09, with a defined mutation in the Salmonella pathogenicity island 2, as a live, oral typhoid vaccine in human volunteers. Vaccine 2005a; 24(2):116–123.

Kirkpatrick, B.D., Tenney, K.M., Larsson, C.J., O'Neill, J.P., Ventrone, C., Bentley, M., Upton, A., Hindle, Z., Fidler, C., Kutzko, D., Holdridge, R., Lapointe, C., Hamlet, S. and Chatfield, S.N. The novel oral typhoid vaccine M01ZH09 is well tolerated and highly immunogenic in 2 vaccine presentations. J. Infect. Dis. 2005b; 192:360–366.

Kjaergaard, K., Hasman, H., Schembri, M.A. and Klemm, P. Antigen 43-mediated autotransporter display, a versatile bacterial cell surface presentation system. J. Bacteriol. 2002; 184:4197–4204.

Klugman, K.P., Gilbertson, I.T., Koornhof, H.J., Robbins, J.B., Schneerson, R., Schulz, D., Cadoz, M. and Armand, J. Protective activity of Vi capsular polysaccharide vaccine against typhoid fever. Lancet 1987; 2:1165–1169.

Klugman, K.P., Koornhof, H.J., Robbins, J.B. and Le Cam, N.N. Immunogenicity, efficacy and serological correlate of protection of *Salmonella typhi* Vi capsular polysaccharide vaccine three years after immunization. Vaccine 1996; 14:435–438.

Komeda, Y., Suzuki, H., Ishidsu, J.I. and Iino, T. The role of cAMP in flagellation of *Salmonella typhimurium*. Mol. Gen. Genet. 1976; 142:289–298.

Kossaczka, Z., Lin, F.Y., Ho, V.A., Thuy, N.T., Van Bay, P., Thanh, T.C., Khiem, H.B., Trach, D.D., Karpas, A., Hunt, S., Bryla, D.A., Schneerson, R., Robbins, J.B. and Szu, S.C. Safety and immunogenicity of Vi conjugate vaccines for typhoid fever in adults, teenagers, and 2- to 4-year-old children in Vietnam. Infect. Immun. 1999; 67:5806–5810.

Kramer, U., Rizos, K., Apfel, H., Autenrieth, I.B. and Lattemann, C.T. Autodisplay: development of an efficacious system for surface display of antigenic determinants in Salmonella vaccine strains. Infect. Immun. 2003; 71:1944–1952.

Kuusi, N., Nurminen, M., Saxen, H., Valtonen, M. and Makela, P.H. Immunization with major outer membrane proteins in experimental salmonellosis of mice. Infect. Immun. 1979; 25:857–862.

Lee, J.H., Kim, J.J., Jung, J.H., Lee, S.Y., Bae, M.H., Kim, Y.H., Son, H.J., Rhee, P.L. and Rhee, J.C. Colonoscopic manifestations of typhoid fever with lower gastrointestinal bleeding. Dig. Liver Dis. 2004; 36:141–146.

Levine, M.M., Black, R.E. and Lanata, C. Precise estimation of the numbers of chronic carriers of *Salmonella typhi* in Santiago, Chile, an endemic area. J. Infect. Dis. 1982; 146:724–726.

Levine, M.M., Ferreccio, C., Black, R.E. and Germanier, R. Large-scale field trial of Ty21a live oral typhoid vaccine in enteric-coated capsule formulation. Lancet 1987a; 1:1049–1052.

Levine, M.M., Herrington, D., Murphy, J.R., Morris, J.G., Losonsky, G., Tall, B., Lindberg, A.A., Svenson, S., Baqar, S., Edwards, M.F. et al. Safety, infectivity, immunogenicity, and in vivo stability of two attenuated auxotrophic mutant strains of *Salmonella typhi*, 541Ty and 543Ty, as live oral vaccines in humans. J. Clin. Invest. 1987b; 79:888–902.

Lin, F.Y., Ho, V.A., Khiem, H.B., Trach, D.D., Bay, P.V., Thanh, T.C., Kossaczka, Z., Bryla, D.A., Shiloach, J., Robbins, J.B., Schneerson, R. and Szu, S.C. The efficacy of a *Salmonella typhi* Vi conjugate vaccine in two-to-five-year-old children. N. Engl. J. Med. 2001; 344:1263–1269.

Luby, S.P., Faizan, M.K., Fisher-Hoch, S.P., Syed, A., Mintz, E.D., Bhutta, Z.A. and McCormick, J.B. Risk factors for typhoid fever in an endemic setting, Karachi, Pakistan. Epidemiol. Infect. 1998; 120:129–138.

Mai, N.L., Phan, V.B., Vo, A.H., Tran, C.T., Lin, F.Y., Bryla, D.A., Chu, C., Schiloach, J., Robbins, J.B., Schneerson, R. and Szu, S.C. Persistent efficacy of Vi conjugate vaccine against typhoid fever in young children. N. Engl. J. Med. 2003; 349:1390–1391.

Matsui, K. and Arai, T. Protective immunity induced by porin in experimental mouse salmonellosis. Microbiol. Immunol. 1989; 33:699–708.

Mehta, G., Randhawa, V.S. and Mohapatra, N.P. Intermediate susceptibility to ciprofloxacin in *Salmonella typhi* strains in India. Eur. J. Clin. Microbiol. Infect. Dis. 2001; 20:760–761.

Meresse, S., Unsworth, K.E., Habermann, A., Griffiths, G., Fang, F., Martinez-Lorenzo, M.J., Waterman, S.R., Gorvel, J.P. and Holden, D.W. Remodelling of the actin cytoskeleton is essential for replication of intravacuolar Salmonella. Cell. Microbiol. 2001; 3:567–577.

Mermin, J.H., Villar, R., Carpenter, J., Roberts, L., Samaridden, A., Gasanova, L., Lomakina, S., Bopp, C., Hutwagner, L., Mead, P., Ross, B. and Mintz, E.D. A massive epidemic of multidrug-resistant typhoid fever in Tajikistan associated with consumption of municipal water. J. Infect. Dis. 1999; 179:1416–1422.

Mert, A., Tabak, F., Ozaras, R., Ozturk, R., Aki, H. and Aktuglu, Y. Typhoid fever as a rare cause of hepatic, splenic, and bone marrow granulomas. Intern. Med. 2004; 43:436–439.

Miller, S.I., Kukral, A.M. and Mekalanos, J.J. A two-component regulatory system (*phoP phoQ*) controls *Salmonella typhimurium* virulence. Proc. Natl. Acad. U.S.A. 1989; 86:5054–5058.

Movva, R.N., Green, P., Nakamura, K. and Inouye, M. Interaction of cAMP receptor protein with the ompA gene, a gene for a major outer membrane protein of *Escherichia coli*. FEBS Lett. 1981; 128:186–190.

Nardelli-Haefliger, D., Kraehenbuhl, J.P., Curtiss, R., III, Schodel, F., Potts, A., Kelly, S. and De Grandi, P. Oral and rectal immunization of adult female volunteers with a recombinant attenuated *Salmonella typhi* vaccine strain. Infect. Immun. 1996; 64:5219–5224.

Ogunniyi, A.D., Manning, P.A. and Kotlarski, I. A *Salmonella enteritidis* 11RX pilin induces strong T-lymphocyte responses. Infect. Immun. 1994; 62:5376–5383.

Olarte, J. and Galindo, E. *Salmonella typhi* resistant to chloramphenicol, ampicillin, and other antimicrobial agents: strains isolated during an extensive typhoid fever epidemic in Mexico. Antimicrob. Agents Chemother. 1973; 4:597–601.

Pallen, M.J. and Wren, B.W. The HtrA family of serine proteases. Mol. Microbiol. 1997; 26:209–221.

Parry, C.M. The treatment of multidrug-resistant and nalidixic acid-resistant typhoid fever in Viet Nam. Trans. R. Soc. Trop. Med. Hyg. 2004; 98:413–422.

Parry, C.M., Hien, T.T., Dougan, G., White, N.J. and Farrar, J.J. Typhoid fever. N. Engl. J. Med. 2002; 347:1770–1782.

Pier, G.B., Grout, M., Zaidi, T., Meluleni, G., Mueschenborn, S.S., Banting, G., Ratcliff, R., Evans, M.J. and Colledge, W.H. *Salmonella typhi* uses CFTR to enter intestinal epithelial cells. Nature 1998; 393:79–82.

Raffatellu, M., Chessa, D., Wilson, R.P., Dusold, R., Rubino, S. and Baumler, A.J. The Vi capsular antigen of *Salmonella enterica* serotype *Typhi* reduces Toll-like receptor-dependent interleukin-8 expression in the intestinal mucosa. Infect. Immun. 2005; 73:3367–3374.

Rahman, M., Ahmad, A. and Shoma, S. Decline in epidemic of multidrug resistant *Salmonella typhi* is not associated with increased incidence of antibiotic-susceptible strain in Bangladesh. Epidemiol. Infect. 2002; 129:29–34.

Reeves, P. Evolution of Salmonella O antigen variation by interspecific gene transfer on a large scale. Trends Genet. 1993; 9:17–22.

Roberts, M., Bacon, A., Li, J. and Chatfield, S. Prior immunity to homologous and heterologous Salmonella serotypes suppresses local and systemic anti-fragment C antibody responses and protection from tetanus toxin in mice immunized with

Salmonella strains expressing fragment C. Infect. Immun. 1999; 67:3810–3815.

Ruiz-Olvera, P., Ruiz-Perez, F., Sepulveda, N.V., Santiago-Machuca, A., Maldonado-Rodriguez, R., Garcia-Elorriaga, G. and Gonzalez-Bonilla, C. Display and release of the *Plasmodium falciparum* circumsporozoite protein using the autotransporter MisL of *Salmonella enterica*. Plasmid 2003; 50:12–27.

Saier, M.H., Jr., Schmidt, M.R. and Leibowitz, M. Cyclic AMP-dependent synthesis of fimbriae in *Salmonella typhimurium*: effects of cya and pts mutations. J. Bacteriol. 1978; 134:356–358.

Santos, R.L., Zhang, S., Tsolis, R.M., Kingsley, R.A., Adams, L.G. and Baumler, A.J. Animal models of Salmonella infections: enteritis versus typhoid fever. Microbes Infect. 2001; 3:1335–1344.

Sharma, A. and Qadri, A. Vi polysaccharide of *Salmonella typhi* targets the prohibitin family of molecules in intestinal epithelial cells and suppresses early inflammatory responses. Proc. Natl. Acad. Sci. U.S.A. 2004; 101:17492–17497.

Shi, Y., Cromie, M.J., Hsu, F.F., Turk, J. and Groisman, E.A. PhoP-regulated Salmonella resistance to the antimicrobial peptides magainin 2 and polymyxin B. Mol. Microbiol. 2004; 53:229–241.

Singh, M., Ganguly, N.K., Kumar, L. and Vohra, H. Protective efficacy and immunogenicity of Vi-porin conjugate against *Salmonella typhi*. Microbiol. Immunol. 1999; 43:535–542.

Sinha, A., Sazawal, S., Kumar, R., Sood, S., Reddaiah, V.P., Singh, B., Rao, M., Naficy, A., Clemens, J.D. and Bhan, M.K. Typhoid fever in children aged less than 5 years. Lancet 1999; 354:734–737.

Snyder, G.E., Shaps, H.J. and Nelson, M. Multiple organ dysfunction syndrome associated with *Salmonella typhi* infection. Am. J. Emerg. Med. 2004; 22:138–139.

Sood, S., Rishi, P., Vohra, H., Sharma, S. and Ganguly, N.K. Cellular immune response induced by *Salmonella enterica* serotype *Typhi* iron-regulated outer-membrane proteins at peripheral and mucosal levels. J. Med. Microbiol. 2005; 54:815–821.

Stocker, B.A. Aromatic-dependent salmonella as anti-bacterial vaccines and as presenters of heterologous antigens or of DNA encoding them. J. Biotechnol. 2000; 83:45–50.

Sztein, M.B., Wasserman, S.S., Tacket, C.O., Edelman, R., Hone, D., Lindberg, A.A. and Levine, M.M. Cytokine production patterns and lymphoproliferative responses in volunteers orally immunized with attenuated vaccine strains of *Salmonella typhi*. J. Infect. Dis. 1994; 170:1508–1517.

Szu, S.C., Taylor, D.N., Trofa, A.C., Clements, J.D., Shiloach, J., Sadoff, J.C., Bryla, D.A. and Robbins, J.B. Laboratory and preliminary clinical characterization of Vi capsular polysaccharide-protein conjugate vaccines. Infect. Immun. 1994; 62:4440–4444.

Tacket, C.O., Galen, J., Sztein, M.B., Losonsky, G., Wyant, T.L., Nataro, J., Wasserman, S.S., Edelman, R., Chatfield, S., Dougan, G. and Levine, M.M. Safety and immune responses to attenuated *Salmonella enterica* serovar typhi oral live vector vaccines expressing tetanus toxin fragment C. Clin. Immunol. 2000; 97:146–153.

Tacket, C.O., Hone, D.M., Curtiss, R.D., Kelly, S.M., Losonsky, G., Guers, L., Harris, A.M., Edelman, R. and Levine, M.M. Comparison of the safety and immunogenicity of delta aroC delta aroD and delta cya delta crp *Salmonella typhi* strains in adult volunteers. Infect. Immun. 1992; 60:536–541.

Tacket, C.O., Kelly, S.M., Schodel, F., Losonsky, G., Nataro, J.P., Edelman, R., Levine, M.M. and Curtiss, R., III. Safety and immunogenicity in humans of an attenuated *Salmonella typhi* vaccine vector strain expressing plasmid-encoded hepatitis B antigens stabilized by the Asd-balanced lethal vector system. Infect. Immun. 1997a; 65:3381–3385.

Tacket, C.O., Sztein, M.B., Losonsky, G.A., Wasserman, S.S., Nataro, J.P., Edelman, R., Pickard, D., Dougan, G., Chatfield, S.N. and Levine, M.M. Safety of live oral *Salmonella typhi* vaccine strains with deletions in htrA and aroC aroD and immune response in humans. Infect. Immun. 1997b; 65:452–456.

Tacket, C.O., Sztein, M.B., Wasserman, S.S., Losonsky, G., Kotloff, K.L., Wyant, T.L., Nataro, J.P., Edelman, R., Perry, J., Bedford, P., Brown, D., Chatfield, S., Dougan, G. and Levine, M.M. Phase 2 clinical trial of attenuated *Salmonella enterica* serovar typhi oral live vector vaccine CVD 908-htrA in U.S. volunteers. Infect. Immun. 2000; 68:1196–1201.

Tamura, K., Sakazaki, R., Kuramochi, S. and Nakamura, A. Occurrence of H-antigen Z66 of R phase in cultures of *Salmonella* serovar typhi originated from Indonesia. Epidemiol. Infect. 1988; 101:311–314.

Taylor, D.E., Chumpitaz, J.C. and Goldstein, F. Variability of IncHI1 plasmids from *Salmonella typhi* with special reference to Peruvian plasmids encoding resistance to trimethoprim and other antibiotics. Antimicrob. Agents Chemother. 1985; 28:452–455.

Townsend, S. M., Kramer, N.E., Edwards, R., Baker, S., Hamlin, N., Simmonds, M., Stevens, K., Maloy, S., Parkhill, J., Dougan, G. and Bäumler, A.J. *Salmonella enterica* serotype *Typhi* possesses a unique repertoire of fimbrial gene sequences. Infect. Immun. 2001; 69(5):2894–2901.

van Basten, J.P. and Stockenbrugger, R. Typhoid perforation. A review of the literature since 1960. Trop. Geogr. Med. 1994; 46:336–339.

van Velkinburgh, J.C. and Gunn, J.S. PhoP-PhoQ-regulated loci are required for enhanced bile resistance in Salmonella spp. Infect. Immun. 1999; 67:1614–1622.

Vazquez-Torres, A., Jones-Carson, J., Baumler, A.J., Falkow, S., Valdivia, R., Brown, W., Le, M., Berggren, R., Parks, W.T. and Fang, F.C. Extraintestinal dissemination of Salmonella by CD18-expressing phagocytes. Nature 1999; 401:804–808.

Vazquez-Torres, A., Xu, Y., Jones-Carson, J., Holden, D.W., Lucia, S.M., Dinauer, M.C., Mastroeni, P. and Fang, F.C. Salmonella pathogenicity island 2-dependent evasion of the phagocyte NADPH oxidase. Science 2000; 287:1655–1658.

Wain, J., Hien, T.T., Connerton, P., Ali, T., Parry, C.M., Chinh, N.T., Vinh, H., Phuong, C.X., Ho, V.A., Diep, T.S., Farrar, J.J., White, N.J. and Dougan, G. Molecular typing of multiple-antibiotic-resistant *Salmonella enterica* serovar Typhi from Vietnam: application to acute and relapse cases of typhoid fever. J. Clin. Microbiol. 1999; 37:2466–2472.

Wain, J., Hoa, N.T., Chinh, N.T., Vinh, H., Everett, M.J., Diep, T.S., Day, N.P., Solomon, T., White, N.J., Piddock, L.J. and Parry, C.M. Quinolone-resistant *Salmonella typhi* in Viet Nam: molecular basis of resistance and clinical response to treatment. Clin. Infect. Dis. 1997; 25:1404–1410.

Wain, J., House, D., Zafar, A., Baker, S., Nair, S., Kidgell, C., Bhutta, Z., Dougan, G. and Hasan, R. Vi antigen expression in *Salmonella enterica* serovar Typhi clinical isolates from Pakistan. J. Clin. Microbiol. 2005; 43:1158–1165.

Wani, T., Kakru, D.K., Shaheen, R., Nazir, A., Lone, R., Shakeel, S. and Shah, A. Infective endocarditis due to *Salmonella typhi*: a case report. Indian J. Pathol. Microbiol. 2004; 47:76–77.

WHO. Posting date. Background Document: The Diagnosis, Treatment and Prevention of Typhoid Fever. WHO/V&B?03.07. [Online] 2003.

Zhang, X.L., Tsui, I.S., Yip, C.M., Fung, A.W., Wong, D.K., Dai, X., Yang, Y., Hackett, J. and Morris, C. *Salmonella enterica* serovar typhi uses type IVB pili to enter human intestinal epithelial cells. Infect. Immun. 2000; 68:3067–3073.

CHAPTER

58

Shigellosis

Malabi Venkatesan, Robert W. Kaminski and Ryan T. Ranallo
Department of Pediatrics, Division of Bacterial and Rickettsial Diseases,
Walter Reed Army Institute of Research, Silver Spring, MD, USA

OUTLINE

ABSTRACT

Shigellosis is an acute, inflammatory gastrointestinal disease of humans and primates, characterized by diarrhea, fever, nausea, tenesmus, and often dysentery, which is excretion of bloody, small volume stool with mucus. It is caused by a gram-negative, nonmotile bacterium that is orally ingested in contaminated food and water. There are four major serogroups each with one or more serotypes that are differentiated by the structure of the LPS-linked O antigen. *Shigella* is a Category B pathogen and the ricin-like Shiga toxin produced by *Shigella dysenteriae* 1 strains can cause significant toxicity and sometimes hemolytic uremic syndrome which can be fatal. All virulent strains of *Shigella* carry a large nonconjugative ~220 kb plasmid, which encodes determinants that induce a phagocytic uptake within the colonic epithelial layer. Invasion is followed by intercellular spread within the tissue, resulting in necrosis and release of proinflammatory cytokines and chemokines. Eventually this leads to an infiltration of neutrophils at the site of inflammation and resolution of infection. *Shigella* infections lead to systemic and mucosal responses that are primarily against the LPS O-antigen polysaccharide. Hence, immunity is serotype-specific although some responses are also seen against the major plasmid-encoded invasion plasmid antigens. Over the last 50 years subunit vaccines based on presenting LPS to the mucosal system as well as live-attenuated noninvasive and invasive vaccines have been tested in Phase I and field trials. Many lessons have been learnt from these trials and today, a more complete understanding of the molecular details of pathogenesis has helped to design specific mutated *Shigella* strains that have shown promise as live oral vaccine candidates. At the same time, novel ways of introducing the LPS antigen to the host strains are being evaluated as subunit vaccine candidates with significant success. Currently, only a single live, noninvasive bivalent *Shigella* vaccine exists in China which provides ~60% efficacy in adults. In the next few years a mixture of the current crop of vaccine candidates, both live and subunit, as well as combinations of the two types of vaccines will undergo testing in Phase I, II, and possibly Phase III trials. Data from these studies is expected to provide for a safe and effective vaccine development strategy that will protect against the predominant circulating serotypes of *Shigella* seen worldwide.

INTRODUCTION

Shigella is an enteroinvasive bacterium, evolved from *Escherichia coli*, and is the etiologic agent of bacillary dysentery and diarrhea. The bacterium is ingested orally through contaminated food and water, and fecal–oral and person-to-person transmission predominate as the primary sources of infection. The symptoms of shigellosis includes a sudden onset of severe abdominal cramping, high-grade fever, vomiting, anorexia, and watery diarrhea, often followed by tenesmus, fecal incontinence, and dysentery which is diarrhea with frank blood (Box 58.1). *S. dysenteriae* 1 strains are capable of epidemic diarrhea, hemorrhagic colitis, and in ~5% of patients, potentially life-threatening hemolytic uremic syndrome (HUS) due to Shiga toxin 1 production. Although the challenge dose in volunteer trials is 500–1000 virulent bacteria, as few as 10 bacteria have been shown to cause dysentery in some volunteers. Since *Shigella* is a food-borne pathogen, it has the potential to be deliberately introduced into the food supply with the intent to harm civilian populations. Thus, *Shigella* has been categorized as a Category B pathogen by the U.S. Center for Disease Control. While antibiotics are used to treat infections the ever-increasing problem of antibiotic resistance necessitates the development of effective vaccines and therapeutics.

BOX 58.1

CLINICAL PRESENTATION OF SHIGELLOSIS

Dehydration	Diarrhea
Fever	Tenesmus
Anorexia	Dysentery
Nausea	Colitis
Vomiting	

SHORT HISTORY OF THE DISEASE

An excellent history of diarrheal disease can be obtained in Henock Blaise and Dovie (2007). In the mid-1800s with the isolation of *Vibrio cholerae* as the etiologic agent for cholera, it became recognized that diarrheal diseases had an infectious disease component (Henock Blaise and Dovie, 2007). In 1896, during a severe epidemic of dysentery in Japan, Kiyoshi Shiga (1871–1957) isolated the organism from feces and intestinal walls of 36 dysentery patients. At the time, Kiyoshi Shiga was working as a research assistant in the laboratory of Dr. Shibasaburo Kitasato (discoverer of the tetanus bacillius) at the Institute of Infectious Diseases in Tokyo. The isolated bacillus (originally named *Bacillus dysentericus*, now known as *Shigella dysenteriae*) was shown to be negative by gram-staining, fermented dextrose, was negative in the indole reaction, and did not form acid from mannitol. Dr. Shiga was able to demonstrate the production of Shiga toxin and the agglutination of the organism when exposed to the serum from an infected patient. In the years immediately following Shiga's discovery of the dysentery bacillus, three other groups of similar organisms were isolated and named in honor of the lead scientists investigating their discovery. In 1899, Simon Flexner, a professor of pathological anatomy at the Johns Hopkins University, located in Baltimore, MD, isolated *Bacillus dysenteriae* (now *Shigella flexneri*), during a stay in Manila in the Philippine islands. Mark

Frederick Boyd, an American bacteriologist and epidemiologist isolated *Shigella boydii* in 1900 and *Shigella sonnei* was identified in 1915 by Danish microbiologist Carl Olaf Sonne. The genus was first termed *Shigella* in the 1930 edition of Bergey's Manual of Determinative Bacteriology and in the 1940s, four species of the new genus were recognized, *S. dysenteriae*, *S. flexneri*, *S. boydii*, and *S. sonnei*, also known as *Shigella* subgroups A, B, C, and D, respectively (Ewing, 1949). Since the 1960s, epidemiological surveys instituted by the Centers for Disease Control (CDC in Atlanta, GA) and the World Health Organization (WHO) have contributed significantly toward understanding the prevalence of *Shigella* species in different areas, the age groups most affected, and the socioeconomic factors that increase the risk of infection with *Shigella*.

ETIOLOGIC AGENT(S)

Shigella are gram-negative rods, nonmotile, and facultatively anaerobic. The lack of motility is due to an incomplete set of fimbrial, curli, and flagellar genes found after sequence completion of several *Shigella* genomes (Jin et al., 2002; Wei et al., 2003; Yang et al., 2005). There are very few biochemical reactions with which to distinguish *Shigella* from enteroinvasive *E. coli* (EIEC), which also causes bacillary dysentery. *Shigella* strains typically ferment glucose, cannot ferment sucrose or lactose (mutations and deletions in the *lacZYA* genes), are oxidase negative (lack cytochrome c), urease negative, do not decarboxylate lysine (lack *cadA*), and fail to produce hydrogen sulfide in triple-sugar-iron agar (TSI agar). Unlike the other three serogroups, *S. dysenteriae* (Group A) does not ferment mannitol (lacks *mtlA*, *mtlD*) while unlike *S. sonnei* (group D), the other three serogroups do not give a positive biochemical reaction with ornithine (Melito et al., 2005). *S. sonnei* and *S. dysenteriae* type 1 strains are *O*-nitrophenyl-b-d galactopyranoside (ONPG)-positive (Melito et al., 2005).

Selective media are often used to discriminate between *Shigella* and other enteric bacteria from stool samples. These include *Salmonella–Shigella* agar (SS agar), xylose lysine deoxycholate agar (XLD agar), Hektoen-enteric agar (HE) and MacConkey agar (MAC agar) all of which contain a variety of sugars, dyes, and bile salts to inhibit gram-positive bacteria and differentiate between different organisms based on their metabolic activities. For example, MAC agar contains lactose on which lactose-positive colonies such as most *E. coli* strains are red to pink in color, while lactose-negative strains like *Shigella* are usually transparent. On SS agar, *Shigella*, like Salmonella forms colorless colonies but without black centers (lack of hydrogen sulfide production by *Shigella*). On XLD agar, *Shigella* remains red because it does not ferment xylose or decarboxylate lysine while Salmonella ferments xylose, decarboxylates lysine, and forms hydrogen sulfide from thiosulfate-forming red colonies with black centers. Confirmation of biochemical identity is done serologically with *Shigella* grouping and typing antisera (DuPont et al., 1989).

CLASSIFICATION

Traditionally, the genus *Shigella* has been classified in the Family Enterobacteriaceae, with four major serogroups, each with one or more serotypes and subtypes. *S. dysenteriae* constitutes group A (with 15 serotypes), *S. flexneri* group B (16 serotypes), *S. boydii* is group D (20 serotypes), and *S. sonnei* is group C (one serotype) (Box 58.2). Subdivision into serotypes is based on the O antigen only, because *Shigella* strains lack the flagella H antigen and capsular K antigen also used in typing other *E. coli*. Some O antigens associated with EIEC are identical to those found in *Shigella* spp.

The O-antigen structure is part of the bacterial lipopolysaccharide (LPS), which is a major structural element in the outer membrane of gram-negative

BOX 58.2

CLASSIFICATION OF SHIGELLA

Family: *Enterobacteriaceae*
Genus: *Escherichia*
Species: *Escherichia coli*
Serovar: *Shigella*
Serogroups
- A. *S. dysenteriae*
- B. *S. flexneri*
- C. *S. boydii*
- D. *S. sonnei*

Serotypes and subtypes
- A. Type 1–15
- B. Type 1–6, X, Y (subtypes a, b, c, etc.)
- C. Type 1–20
- D. Type 1

bacteria. Besides the O-antigen, the LPS is composed of the core oligosaccharide and lipid A anchor. The core consists of a single sequence of heptoses and hexoses while the O-antigen is a polysaccharide unit composed of a repetitive sequence of hexoses, further modified by the addition of glucosyl and acetyl groups. The *S. flexneri* serotypes 1 through 5 (subtypes include *S. flexneri* 1a and 1b, 2a and 2b, etc.) are closely related and have a common basic structure encoded by a common O-antigen gene cluster (Cheah et al., 1991; Simmons and Romanowska, 1987). The differences between these serotypes are conferred by modification of the basic structure of glucosyl and/or O-acetyl residues generating the diversity of group (group antigens 3, 4, 6, 7, 8) and type antigens (type I–VI) (Huan et al., 1997). *S. flexneri* serotypes 6 and 6a are not genetically similar to the other *S. flexneri* forms, but crossreact with them serologically because of similarity in part of the O-antigen structure (Cheah et al., 1991). The O-antigen repeat of *S. boydii* and *S. dysenteriae* are unique among *Shigella*, although many *S. boydii* and *S. dysenteriae* O antigens are either identical to, or related to, a conventional *E. coli* O antigen (Jonsson et al., 2006; Perepelov et al., 2007; Valvano and Marolda, 1991; Wang et al., 2001). The *S. sonnei* and *S. flexneri* O antigens are not found in *E. coli* (Pupo et al., 2000). So far, approximately 50 serotypes, including subtypes, of *Shigella* have been recognized and the number is expected to grow. Monoclonal antibodies (mAbs) have been generated to help in typing *Shigella* (Carlin et al., 1986, 1989; Carlin and Lindberg, 1987).

Although *Shigella* and *E. coli* have always been known to be very similar, *Shigella* has historically been treated as a different species that is distinct from *E. coli*. This is based primarily on its pathogenic traits and a few metabolic defects that were used to differentiate *Shigella* from *E. coli*. Population genetic studies using multilocus enzyme electrophoresis (MLEE) showed that *Shigella* strains fall within *E. coli*, confirming previous evidence from DNA hybridization data (Ochman et al., 1983; Riley and Anilionis, 1980). More recently, sequencing of eight housekeeping genes in four regions of the chromosome of ~50 *Shigella* strains belonging to the four traditional *Shigella* species indicated that, all but five of the *Shigella* strains (*S. sonnei*, *S. dysenteriae* 1, and *S. flexneri* VIa are three of these five strains) fall into one of three major clusters within *E. coli* species, and these the same strains are in each cluster for all four regions of the chromosome (Pupo et al., 2000). While the strains within each cluster are very similar, there is little in common among the three clusters suggesting that *Shigella* strains, like the other pathogenic forms of *E. coli*, do not have a single evolutionary origin but have multiple origins. By a process of convergent evolution of phenotypic properties, including gene gain (capture of the virulence plasmid) and gene loss or "black holes" (e.g., losses of the chromosomal *cadB* and the *nadB* gene), *Shigella* has uniquely adapted its pathogenic lifestyle within the host (Maurelli, 2007; Pupo et al., 2000).

EPIDEMIOLOGY

Shigellosis affects an estimated 164 million people worldwide and although it is considered a disease of developing countries, over 14,000 laboratory-confirmed cases are reported to occur in the US annually (Kotloff et al., 1999). *Shigella* species are not known to be zoonotic pathogens, and natural infections have been documented only for humans and primates. While the infectious dose for humans is in the range of 100–1000 bacteria, doses of 10^9–10^{10} cfu are required to observe dysentery in rhesus monkeys (DuPont et al., 1989; Formal et al., 1965). Epidemiological and antibiotic resistance data on *Shigella* is typically monitored in travelers returning from developing countries or by in-country, population-based national surveillance systems, such as those found at local hospitals and clinics. The isolates are often typed, subtyped, and characterized using specific antisera, biochemical profile, ribotyping, plasmid analysis, PCR-based epidemiological tools, and antibiotic resistance patterns (Korpela et al., 1995; Materu et al., 1997; Na-Ubol et al., 2006; Navia et al., 1999; Penatti et al., 2007; Taneja et al., 2005; Zafar et al., 2005). A global view of the incidence of shigellosis was estimated from current literature (Fig. 58.1) and indicates that approximately 99% of all shigellosis occurs in developing countries with the highest burden of *Shigella* infection in northern Africa (Chompook et al., 2005; Lee et al., 2005; Ram et al., 2007; von Seidlein et al., 2006). The lowest incidence of shigellosis occurs in the US, Europe, and Australia (Kotloff et al., 1999). As noted by many epidemiological studies, shigellosis incidence data are almost nonexistent in sub-Saharan Africa and represents a significant challenge when estimating the global epidemiology. *S. flexneri* is the most prevalent serotype worldwide and is found most often in the developing countries of Africa and South and East Asia while *S. sonnei* predominates in the US and Europe (Box 58.3). New isolates of each major serogroups are being isolated such as *S. flexneri* 1c, *S. flexneri* 4X, *S. boydii* 20, and *S. dysenterie* type strain KIVI 162 (Grimont et al., 2007; Talukder et al., 2002, 2003, 2006, 2007).

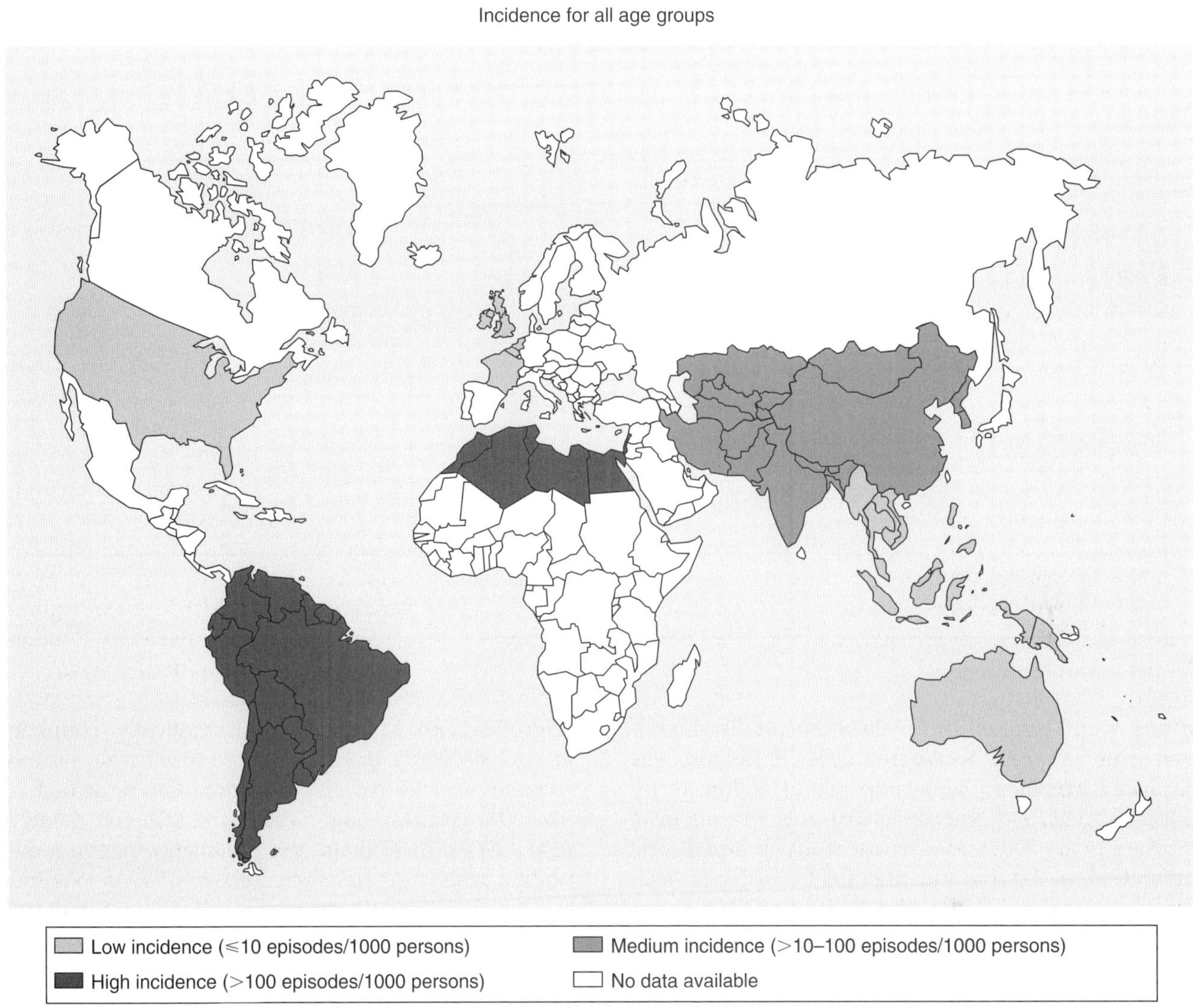

FIGURE 58.1 World incidence of shigellosis.

In developing countries where substandard hygiene and poor water quality persist children under 5 years of age are most at risk (>6 months and <2 years are at highest risk) for developing shigellosis (Elamreen et al., 2007; Faruque et al., 1998; Nguyen et al., 1998). Worldwide, approximately 600,000 deaths occur annually; 60% of such deaths are seen in children under 5 years of age (Sur et al., 2004). The bacteria are primarily transmitted through direct contact with an infected person or by ingestion of contaminated food or water. Several factors increase the risk for shigellosis such as age and socioeconomic status including, sanitation, and malnutrition (Elamreen et al., 2007; Sakdisiwasdi et al., 1982). Flies can serve as mechanical vectors for the transmission of *Shigella* under particularly poor environmental conditions (Bidawid et al., 1978; Chompook et al., 2006; Dupont, 2005).

In the US, *Shigella* infections constitute the third most common cause of gastroenteritis, after *Campylobacter* and *Salmonella* infections. Particularly susceptible are children in day-care centers, migrant workers, travelers to developing countries, and homosexual men (CDC, 2002; Garrett et al., 2006; Monteville et al., 2006; Morgan et al., 2006; Riddle et al., 2006). An association of *Helicobacter pylori* with *Shigella* gastroenteritis in children has also been reported (Shmuely et al., 2004). From 1989 to 2003, a total of 208,368 cases of laboratory-confirmed *Shigella* infections were reported to the Centers for Disease Control and Prevention wherein, *S. sonnei* accounted for the majority of the cases (72%), followed by *S. flexneri* (18%), *S. boydii* (2%), and *S. dysenteriae* (~1%) (Gupta et al., 2004). To compare a different part of the world, in Kolkata, India, 193 *Shigella*

BOX 58.3

COMPARATIVE EPIDEMIOLOGY OF *SHIGELLA*

Aspect	Developing	Industrialized
Annual cases/year	160 million	~100,000
Deaths	~1.1 million	<10,000
Likely means	Overcrowding, lack of sanitation	Camps, daycare centers, cruise ships, airlines
Most susceptible	Young (<5 years), old, malnourished, immunocompromised	Non-immune
Species prevalence	*S. flexneri* (60%)[a] *S. sonnei* (23%) *S. dysenteriae* (10%) *S. boydii* (6%)	*S. flexneri* (18%)[b] *S. sonnei* (72%) *S. dysenteriae* (2%) *S. boydii* (1%)

[a]*Source*: Pazhani et al. (2005).
[b]*Source*: Gupta et al. (2004).

strains were isolated from 2489 hospitalized children from Jan 2001 to August 2004. *S. flexneri* was the most prevalent serogroup (60%) followed by *S. sonnei* (23%), *S. dysenteriae* (10%), and *S. boydii* (6%) (Pazhani et al., 2005). In a recent study in a pediatric population in Egypt, the prevalent serotypes seen were *S. flexneri* 2a (37.1%), 1b (18.6%), 1c (17.5%), and 6 (15.5%), which together comprised over 88.7% of the total isolates. These data also show that *S. flexneri* 1c has emerged as a dominant *S. flexneri* serotype in Egypt (Ahmed et al., 2006).

ANTIBIOTIC RESISTANCE

Wide spread resistance to first line antimicrobials has complicated the selection of drugs for the treatment of shigellosis. A survey of the literature suggests that a majority of *Shigella* isolates from around the world, including the US, demonstrate some resistance to commonly used antibiotics such as ampicillin, chloramphenicol, TMP-SMX, and tetracycline (Ahmed et al., 2006; Al-Moyed et al., 2006; Pan et al., 2006; Replogle et al., 2000; Sivapalasingam et al., 2006; Toro et al., 2005). In fact, multidrug resistance to these antibiotics has made them no longer appropriate for empirical treatment in the US. Resistance to these antimicrobials appears to be even higher in developing countries, like India and China, where resistance to both traditional antibiotics as well as fluoroquinolones (ciprofloxacin) is increasing dramatically (Talukder et al., 2003, 2007). In parts of these countries resistance to nalidixic acid in *S. flexneri* strains can be as high as 92–100% (Haukka and Siitonen, 2007; von Seidlein et al., 2006). In Kolkata, the prevalence of fluoroquinolone resistance has increased to 25% in endemic *S. flexneri* isolates (Pazhani et al., 2005). In a pediatric population in Egypt, marked resistance to ampicillin (87.6%), tetracycline (84.5%), and trimethoprim-sulfamethoxazole (63.9%) was seen among the strains isolated (Ahmed et al., 2006). Increasing resistance to antimicrobials in developing countries highlights the perils of over-the-counter sale and indiscriminate use of antibiotics for treating infectious diseases (Ahmed et al., 2006; Al-Moyed et al., 2006; Pan et al., 2006; Toro et al., 2005).

SIGNIFICANCE AS PUBLIC HEALTH PROBLEM

The low infectious dose, the fecal–oral route of transmission, and the emergence of resistance to multiple antibiotics among *Shigella* isolates poses a major public health problem throughout the developing world. Areas of political upheaval and natural disasters are particularly susceptible to high rates of morbidity and mortality due to diarrheal diseases, in particular, because of the lack of clean water, proper sanitary conditions, overcrowding in refugee camps and villages. *S. dysenteriae* 1 has the highest mortality rate (as high as 5–15%) of any *Shigella* species due in part to the expression of the ricin-like Shiga toxin which can cause renal complications leading to HUS. Since the late 1960s, pandemics of dysentery, caused by *S. dysenteriae* 1, have occurred in Central America, sub-Saharan Africa, Middle East, and Southeast Asia (Cahill et al., 1966; Gangarosa et al., 1970; Rahaman et al., 1975; Simchen et al., 1991). During a regional epidemic of *Shigella*-based dysentery that swept through Guatemala and neighboring countries in Central America during 1969–1971, an estimated 20,000 cases of dysentery-induced deaths were reported. One particularly disturbing feature was the resistance of the bacteria to the most commonly used antibacterial drugs (Gangarosa et al., 1970). Dysentery spared infants under 6 months of age, the attack rate being highest in the age group 7–23 months and thereafter was more or less equally distributed between the ages and the sexes (Gangarosa et al., 1970). During the civil war in Rwanda that started in mid-1994, among the estimated 500,000–800,000 refugees who crossed over into Goma in neighboring Zaire, an unprecedented number of deaths, ~50,000, were reported during the early months of the emergency situation, between the months of July–August, 1994. Eighty-five to 90% of the mortality rates were ascribed to diarrheal diseases, more specifically due to successive cholera and *S. dysenteriae* 1-induced outbreaks of bloody diarrhea (Legros et al., 2001). International intervention with provision of clean water contributed thereafter to significant reduction in the mortality rates (Legros et al., 2001). *S. dysenteriae* epidemics occur in roughly 10 to 14-year cycles with the most recent outbreak documented in northern India and Bangladesh (Hens et al., 2005; Pazhani et al., 2005; Talukder et al., 2003; Taneja et al., 2005). The most troubling finding from these outbreaks is the emergence of fluoroquinolone-resistant strains capable of clonal spread (Talukder et al., 2006; Taneja et al., 2005).

POTENTIAL AS A BIOTHREAT AGENT

In October 1996, a disgruntled lab worker in Dallas, TX, deliberately infected pastries to be consumed by colleagues with *S. dysenteriae* 2, causing illness in 12 people and sending 4 to the hospital (Kolavic et al., 1997). Such an intentional and malevolent use of a multidrug resistant Shiga toxin-producing *S. dysenteriae* as an agent of food terrorism on a larger scale can pose a huge burden on the public health system and has the potential to disrupt the social, economic, and political stability of the region. "Early detection of disease resulting from covert food terrorism depends on sensitive surveillance systems for communicable disease at the local and national levels, with close cooperation and communication among clinicians, laboratories and public health officials" (World Health Organization, 2002).

CLINICAL DISEASE

During the Guatemalan epidemic in early 1969, the onset of *S. dysenteriae* 1-induced shigellosis in patients was described as follows and typifies the clinical symptoms of the disease: "in patients with Shiga dysentery, diarrhea started abruptly, usually without blood. Three fourths of the patients had fever, generally below 102°F. Twelve to 72 h after onset, blood and pus appeared in the stools. Many patients passed only mucus and blood during the acute phase of the disease. Abdominal pain and cramps were commonly observed. Vomiting was recorded in 47% of cases. The most severely ill had 10–40 evacuations a day and intense tenesmus and rectal pain at the height of illness. The duration of symptoms for most patients was 5–7 days. A few were ill for less than 48 h and others for more than a month" (Mata et al., 1970). Pathology reports of autopsy cases indicated a diffusely inflamed mucosa of the sigmoid colon and shallow ulcers no deeper than the muscularis mucosa. The cause of death was dehydration and exhaustion in most cases.

In endemic areas shigellosis is rare in infants less than 6 months of age, presumably, because antibodies in their mother's milk protect them from infection. Shigellosis in the neonatal period results from mother-to-infant fecal–oral transmission during labor and delivery, usually from asymptomatic carriage. Bloody diarrhea is uncommon, however, intestinal perforation and mortality are more common in this group than in older children. In a postmortem examination of the intestine of 37 Bangladeshi children, ~age 2 years, who died of shigellosis, the regional pattern of lesions ranged from involvement of the rectosigmoid only (3 cases) to the entire colon plus the ileum (20 cases) (Box 58.4) (Butler et al., 1989). Children who died of infections due to *S. dysenteriae* 1 and 2, *S. flexneri*, and *S. boydii*, were also compromised by severe malnutrition, hypoproteinemia, and sometimes hypoglycemia. Pathology reports indicated that the common lesion was erythema and edema denoting congestion and inflammation. Histopathology indicated increased numbers of leukocytes, predominantly mononuclear,

BOX 58.4

CLINICAL PRESENTATION OF SHIGELLOSIS

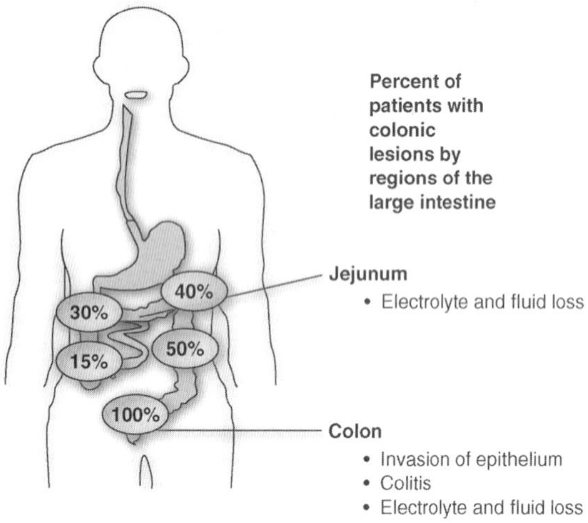

BOX 58.5

POTENTIAL COMPLICATIONS FROM SHIGELLOSIS

Necrotizing enterocolitis
Intestinal obstruction or perforation
Haemolytic uremic syndrome (HUS)
Toxic megacolon
Rectal prolapse
Bacteraemia and sepsis
Neurological disorders (seizures)
Reactive arthritis (HLA-B27-linked)
Chronic malnutrition

in the lamina propria and sometimes in the submucosa (Butler et al., 1989).

In approximately 5% of *S. dysenteriae* 1 infections, HUS (bloody urine), microangiopathic hemolytic anemia, thrombocytopenia, and renal failure have been reported because of vasculopathy mediated by Shiga toxin or Stx (Tesh and O'Brien, 1991). Shiga toxin belongs to a family of toxins characterized by a receptor binding B subunit (7.7 kDa), which forms a pentameric ring and a single 32 kDa enzymatic A subunit. The B subunit binds to the glycolipid receptor Gb_3 found in the walls of the vascular endothelial cells, allowing entry of the toxin A subunit which is an *N*-glycosidase that cleaves the 28S ribosomal RNA component of ribosomes halting protein synthesis (Tesh and O'Brien, 1991). The ensuing endothelial cell death destroys the lining of the vascular walls leading to hemorrhage. The principal organ affected in shiga toxin-mediated HUS is the kidney where there is abundant baseline expression and high inducibility of the receptor Gb_3 in the glomerular microcirculation. Direct Stx-mediated injury to vascular endothelial cells leads to tissue ischemia and dysfunction and Stx-mediated release of cytokines and chemokines, including IL-6, CXCL8 (IL-8), TNF-α which leads to a systemic inflammatory response. Extraintestinal symptoms have been recorded and include severe headache, lethargy, meningismus, delirium, and seizures, especially with *S. dysenteriae* (Chapel et al., 2005; Islam et al., 1994). Neurologic manifestations during shigellosis have been described and include rarely severe toxic encephalopathy. The pathogenesis of neurologic manifestations during shigellosis is unclear and various bacterial (Stx, LPS) and host factors (TNF-α, IL-1β, corticotrophin releasing hormone, CRH) have been implicated from studies in mice (Yuhas et al., 2002, 2004). Reiter syndrome (arthritis, urethritis, conjunctivitis), a postinfectious sequalae, is commonly observed in adults carrying human leukocyte antigen (HLA)-B27 histocompatibility antigen (Finch et al., 1986) (Box 58.5).

TREATMENT

While most diarrheal episodes should be treated with oral rehydration therapy (ORT), antibiotics are the main drugs for treating shigellosis (Alam and Ashraf, 2003; Khan et al., 1997; King et al., 2003; Sur and Bhattacharya, 2006). An assessment of the antibiotic resistance profiles of circulating strains is important when choosing which antibiotic to use in a particular region (Box 58.6). Fluoroquinolones such as ciprofloxacin, levofloxacin, and norfloxacin introduced in the late 1980s inhibit DNA replication by inhibiting DNA gyrase. These antibiotics seem to be the antibiotics of choice in US adults, however, they are not approved for use in children in the US for whom azithromycin, an azalide antibiotic is the prescribed drug (ciprofloxacin dose 500 mg given bid for 3–5 days) (Bennish et al., 1992; Ehrenpreis and Ehrenpreis, 2001; Murphy et al., 1993). Other antibiotics include trimethoprim-sulfamethoxazole (Bactrim or Septra, adult dose 800 mg

BOX 58.6

TREATMENT OF SHIGELLOSIS

Current treatments for shigellosis
Antimicrobial therapies
Supportive care
Oral rehydration therapy
Anti-pyretic for fever
Analgesic for pain

Treatments in development
Zinc supplementation
Short-chain fatty acids

Antidotes for Shiga toxins (in development)
Monoclonal antibodies
Receptor mimics
Genetic toxoids
Active and passive immunization strategies

SMX/160 mg of TMP, given orally bid for 5 days) which inhibits bacterial growth by inhibiting synthesis of dihydrofolic acid and third generation cephalosporins such as cefixime and ceftriaxone, which bind to penicillin-binding proteins and disrupt or inhibit bacterial cell wall synthesis (Ehrenpreis and Ehrenpreis, 2001).

Among new developments in this area, Rifaximin, a nonabsorbed oral antibiotic given at 200 mg three times a day for 3 days, protected volunteers from shigellosis after being challenged with virulent *S. flexneri* 2a in a double-blind placebo-controlled study (Taylor et al., 2006a, 2006b). Oral butyrate represents an exciting new treatment for shigellosis. This short-chain fatty acid induces the endogenous cathelicidin CAP-18 in the colonic epithelium, stimulating the release of the active antimicrobial peptide CAP-18, which inhibits *Shigella* (Raqib et al., 2006). Promising effects of zinc, given as a micronutrient to pediatric populations, has also been reported (Roy et al., 2007).

There are currently no FDA-approved therapies or vaccines in the United States to combat or prevent illness from Shiga toxin, which is not only produced by *S. dysenteriae*, but also Shiga toxin-producing *E. coli* (STEC). STEC infections represent a significant problem in the US. Incidentally, STEC strains produce two Shiga toxins, Stx1 (identical to Shiga toxin) and Stx2 (O'Brien et al., 1992). Several different strategies to inhibit Shiga toxin and Shiga toxin-like activity include receptor mimics, active and passive immunization strategies, Stx1 and Stx2 genetic toxoids, a plant-based Stx2 toxoid, and a chimeric StxA2/StxB1 toxoid (Karmali, 2004; Paton et al., 2001; Smith et al., 2006; Tzipori et al., 2004; Wen et al., 2006a, 2006b). Passive immunization strategies utilizing humanized or human monoclonal antibodies that neutralize one or more Stx's are also being evaluated (Sheoran et al., 2005; Tzipori et al., 2004).

PATHOGENESIS

The use of cell culture and different animal models of infection have proven useful for elucidating the primary steps of infection with *Shigella*. Data from each model system has been combined to form the current pathogenic paradigm of shigellosis (Fig. 58.2, A–D) (Parsot, 2005; Sansonetti, 2006). After being orally ingested the bacteria transit through the stomach and small intestine where it survives the gastric acidity and antimicrobial peptides using mechanisms that are not completely understood. At the distal colon and rectal mucosa the bacteria transit through the specialized M cells of the follicle-associated epithelium (FAE) to the subepthelial dome where they are taken up into phagocytic vacuoles by resident phagocytes (macrophages and dendritic cells). *Shigella* rapidly escape from the vacuole and induce a proinflammatory cell death during which the loss of membrane integrity coincides with the release of numerous proinflammatory cytokines (TNF-α, IL-1β, IL-18) (Fig. 58.3) (Haimovic and Venkatesan, 2005; Suzuki et al., 2007).

After escaping the phagocytic cells, *Shigella* invades the adjacent enterocytes at the basolateral surface and are taken up within an endocytic vacuole through the coordinated activities of numerous proteins, which include the products of the *ipa* genes (IpaA, B, C, and D) and the *mxi-spa* genes that form a needle-like complex called the type three secretion system (TTSS) (Buchrieser et al., 2000; Sansonetti, 2006). The *ipa-mxi-spa-loci* comprises a 31 kb coding region located on a large >200 kb virulence plasmid present in all pathogenic isolates of *Shigella*. The IpaB and IpaC proteins are stored in the bacterial cytoplasm complexed with chaperones, such as IpgC, while others such as IpaD form a complex at the tip of the TTSS thereby controlling its secretion activity (Demers et al., 1998; Espina et al., 2006; Parsot et al., 2005; Picking et al., 2005). The invasion plasmid antigen (Ipa) proteins constitute key antigens to which antibodies are seen in serum from infected humans and monkeys

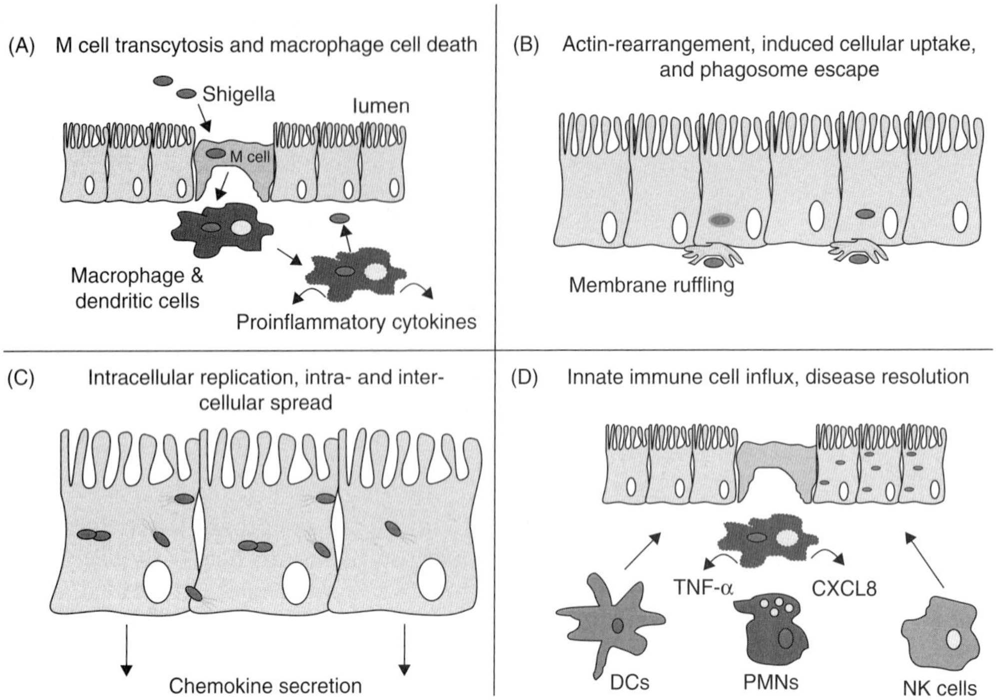

FIGURE 58.2 Major steps of *Shigella* pathogenesis (see color plate section).

(Oaks et al., 1986) (Box 58.7). Upon contact with the surface of epithelial cells, the IpaB and IpaC proteins exit through the TTSS and are believed to form a pore through which a set of type III effector proteins are secreted within the host cell. Secretion of these effectors (IpaA, IpaB, IpaC, IpgB1, IpgD, and VirA) into host cells causes major cytoskeletal rearrangements involving both F-actin and microtubules thus allowing for engulfment of the bacteria within an endocytic vacuole (Sansonetti, 2006; Tran Van Nhieu et al., 2000).

The bacteria lyse the endocytic vacuole, presumably with the help of some of the Ipa proteins, and begin to replicate using host cell metabolites to fulfill energy requirements (Fig. 58.3). Intracellular movement of the bacteria is facilitated by the polar expression of the VirG(IcsA), a critical virulence protein which belongs to the family of type V autotransporters, and mediates localized actin polymerization on one end of the bacteria (Ogawa and Sasakawa, 2006; Wing et al., 2005). VirG(IcsA) mediates intracellular movement by interacting with and activating N-WASP which in turn leads to recruitment and activation of the Arp2/3 complex (Ally et al., 2004; Suzuki and Sasakawa, 2001). VirG(IcsA) deletion mutants remain invasive in epithelial cells but are severely attenuated in plaque formation in cultured cells and in animal models of infection (Fig. 58.4). Therefore the *virG(icsA)* mutation has become a principle attenuating feature of multiple live attenuated vaccine strains (see below).

Intracellular movement is followed by intercellular spread through the enterocytes lining the intestinal mucosa. Replication and cell-to-cell movement within enterocytes results in the secretion of cytokines and chemokines, such as IL-8, a potent chemoattractant that promotes the recruitment of polymorphonuclear leucocytes (PMNLs) and neutrophils to the site of infection (Francois et al., 2000). Secretion of CXCL8 by enterocytes is induced by bacterial products such as LPS and peptidoglycan, which are recognized by host cell pattern-recognition receptors such as Toll-like receptors (TLRs) and nucleotide-oligomerization domain (NOD)-like receptors (NLRs) (Girardin et al., 2001; Viala et al., 2004). The lipid A component of extracellular LPS is recognized by TLR4 through a series of interactions involving LPS binding protein (LBP), soluble and membrane bound forms of CD14, and myeloid differentiation protein 2 (MD-2). The TLRs and NLR's activate central signaling cascades involving mitogen activated protein kinases (MAPKs) pathways and transcription factors, such as NF-kB, leading to upregulation of key cytokine synthesis such

BOX 58.7

KEY ANTIGENS RECOGNIZED AFTER NATURAL INFECTION WITH *SHIGELLA*

Antigens	Size (kDal)	Proposed role(s)
IpaA	78	Binds vinculin to induce F-actin polymerization
IpaB	62	Involved with invasion Cross-reactive with myosin Required for contact hemolytic activity Required for escape from the phagosome in infected cells
IpaC	43	Involved with invasion Required for contact hemolytic activity Required for escape from the phagosome in infected cells Polymerizes actin
IpaD	38	Control flux of proteins through secretion
VirG	120	Required for actin polymerization and intracellular motility
LPS	Variable	Protects against phagocytosis Barrier function

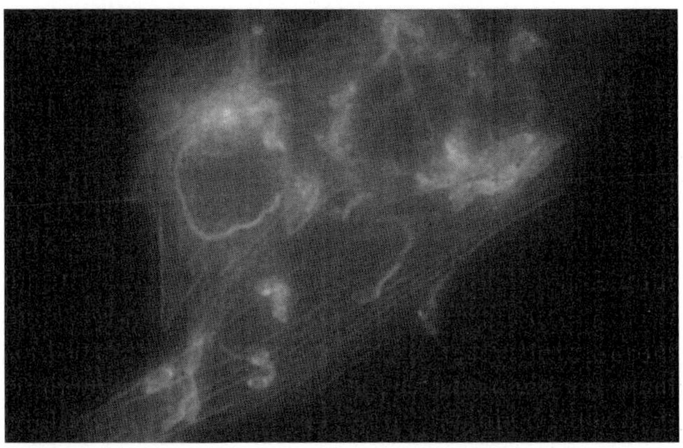

FIGURE 58.3 *Shigella*-induced cellular actin polymerization used for intra- and intercellular spread of the bacteria.

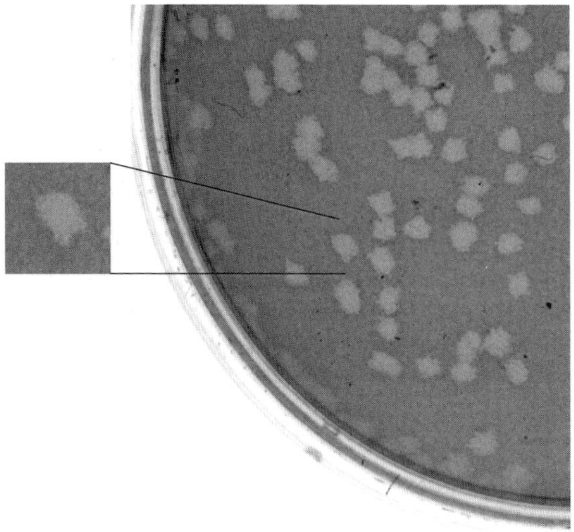

FIGURE 58.4 *Shigella*-mediated cellular plaque formation.

as IL-8 and activation of the innate immune system (Philpott et al., 2000; Viala et al., 2004).

However, recent evidence suggests that invasive bacteria like *Shigella* have evolved counter defenses capable of evading or suppressing, at least temporarily, the innate immune response. For *Shigella* these counter defenses involve the secretion of additional TTSS substrates once the bacterium has entered the host cells. For example, the *Shigella* OspG protein targets host ubiquitin-transfer proteins, thereby blocking IkB degradation and subsequent activation of NF-kB (Kim et al., 2005). OspF has been shown to block NF-kB mediated transcription by targeting MAPKs such as Erk1/2 and p38 by a phosphatase or a phosphothreonine lyase activity, which removes phosphate groups from the phosphothreonine residue in the MAPK activation loop (Arbibe et al., 2007; Li et al., 2007). According to one report, this results in reduced histone H3 phosphorylation, and chromatin remodeling, leading to inaccessibility at the NF-kB regulated IL-8 promoter (Arbibe et al., 2007). Finally, a member of the IpaH family ($IpaH_{9.8}$) has been shown to interact with the mRNA splicing factor $U2AF^{35}$. This interaction appears to affect the expression of both chemokine and cytokine genes (CXCL8, RANTES, IL-β) resulting in the suppression of neutrophil migration (Okuda et al., 2005). In a mouse pulmonary model of infection, $IpaH_{9.8}$ mutants produced more inflammation than wild-type strains (Okuda et al., 2005). As more counter defense mechanisms are elucidated, strategies for utilizing these proteins to improve current vaccine efforts will be forthcoming. These elaborate mechanisms, however, are clearly inadequate to stop the secretion of IL-8 and other chemokines by the host tissues which results in a massive influx of inflammatory cells (neutrophils) to the site of infection causing increased destruction of the colonic epithelium including loosening of the cell tight junctions, a process that seems to hasten intra- and intercellular spread. The presence of activated neutrophils is an important step in resolving the infection because neutrophils have been shown to efficiently kill *Shigella* {(Francois et al., 2000, Fig. 58.2, panel D)}.

The potency of the lipid A region of the bacterial LPS as an immune stimulant is dependent on structural features such as the number, length, and degree of acyl chain saturation. It has been demonstrated that underacylated lipid A has reduced ability to stimulate endotoxic responses (Clementz et al., 1997; Somerville et al., 1999). In *Shigella* several enzymes (encoded by the *htrB* and *msbB* genes) are responsible for secondary acylation of lipid A (Clementz et al., 1997). *Shigella* has two *msbB* genes, both of which encode late acyltransferases that are required for producing full acyl-oxyacylated lipid A under standard laboratory growth conditions (D'Hauteville et al., 2002). Furthermore *msbB* deletion mutants were found to be attenuated for virulence using animal models making them attractive targets for reducing the endotoxicity of live attenuated vaccines strains (D'Hauteville et al., 2002).

Although advanced shigellosis is usually characterized by low volume stools with blood and mucus, watery or large-volume stools often precede this stage of infection and sometimes can be the only clinical manifestation in mild infections. This stage is often overlooked as a pathogenic step, however, it occurs frequently in patients infected with *Shigella*. A detailed investigation into the cause of watery diarrhea revealed the presence of at least two enterotoxins termed *Shigella* enterotoxin 1 and 2 (ShET1 and ShET2). Both proteins were shown to contribute to enterotoxic activity using ligated rabbit ileal loops and Ussing Chambers (Fasano et al., 1995, 1997; Nataro et al., 1995). ShET1 is uniquely situated in a pathogenicity island (SHI-1) present in all *S. flexneri* 2a isolates, but rarely in other serotypes (Nataro et al., 1995). The two genes that encode ShET1 make up an operon (*setAB*) that is embedded in the opposite DNA strand of a secreted mucinase termed Pic (Behrens et al., 2002). The proteins encoded in these genes most likely form a classic A_1-B_5 oligomer and induce secretion, however, a detailed mechanism remains to be established. The gene that encodes ShET2 (*senA* or *ospD3*) is found on the large invasion plasmid and is present in all *Shigella* serogroups. The protein does not resemble any known toxin and its mechanism of action is completely unknown. Interestingly, at least one other paralog of ShET2 (encoded by *senB* or *ospD2* gene) is also present on the *Shigella* invasion plasmid (Jiang et al., 2005; Venkatesan et al., 2001). The presence of this gene could explain residual enterotoxic activity seen in the absence of both ShET1 and ShET2 (Nataro et al., 1995). The enterotoxic activity of ShET1 and ShET2 was confirmed by results from a recent clinical trial, which compared two different attenuated strains of *S. flexneri* 2a (Kotloff et al., 2004). These data show that deletion of both *senA* and *setAB* dramatically reduces the incidence of watery diarrhea in experimentally infected volunteers, a finding that has significantly advanced our ability to construct safe efficacious live attenuated *Shigella*-based vaccine strains.

Replication within epithelial cells requires multiple systems capable of scavenging limited oxygen and other nutrients found within host cells. For example, *Shigella* has multiple iron acquisition systems, which allow for growth in the iron restricted intracellular environment (Boulette and Payne, 2007; Runyen-Janecky et al., 2003). One of the most well-characterized systems is the aerobactin *iucABCD* operon, which encodes an iron siderophore capable of transporting ferric iron (Fe^{+3}) to the bacteria while in the host cell cytoplasm (Boulette and Payne, 2007). Aerobactin mutants are less virulent when compared to wild-type strains presumably because of the lack of adequate iron levels, which limits intracellular growth (Runyen-Janecky et al., 2003). Indeed deletion of *iuc* has been used as a secondary mutation in the live attenuated *Shigella* vaccine strain SC602 (discussed below). Additional attenuation strategies that take advantage of the limited biosynthetic precursors found inside host cells, include the use of *aroA*, *aroD*, and *guaBA* mutations. The most extensively tested example of such attenuation strategies can be found with the live attenuated *Shigella* vaccines from the University of Maryland Center for Vaccine Development (CVD) in Baltimore, MD. The principle attenuating feature of these strains is a deletion of the *guaBA* operon rendering the cells auxotrophic for purine biosynthesis and attenuated for both invasion and intracellular growth.

IMMUNE RESPONSE AFTER INFECTION

Innate Immune Response

Natural infection with *Shigella* results in a pronounced inflammatory response due to engagement of the innate immune system (Fernandez and Sansonetti,

BOX 58.8

SHIGELLA-SPECIFIC IMMUNE RESPONSES POSTINFECTION

Immune responses to infection with *Shigella*

Innate immune response

The initial inflammatory response induced by the release of IL-1β, IL-6, TNF-α, and IFN-γ from infiltrating PMNs, necrotic macrophages, and other lymphocytes.

Cellular immune response

Although IFN-γ and IL-10 are specifically elevated in patients 21 days after experimental infection T cell responses still remain ill defined.

Humoral immune response

Natural infections induce serum (IgA and IgG) and mucosal (sIgA) immune responses to O-antigen polysaccharide (LPS) and outer membrane proteins like the TTSS-associated invasion plasmid antigen (Ipa) proteins (IpaA, B, C, D) and VirG(IcsA).

2003). Soon after transcytosis across the gut epithelium via M cells (Box 58.8), *Shigella* induce inflammatory cytokine release after interaction with phagocytes (Fernandez-Prada et al., 1997; Viala et al., 2004). In addition, during both acute and convalescent stages of shigellosis, significant levels of IL-1α, IL-1β, TNF-α, IL-6, IL-4, IL-10, and IFN-γ can be found in rectal biopsies (Raqib et al., 1995a, 1995b, 1995c). The release of inflammatory cytokines and chemokines initiates the influx of inflammatory cells, primarily neutrophils and monocytes to the site of infection. Additional release of inflammatory cytokines likely occurs from cellular activation of epithelial cells by both intracellular and extracellular bacterial components, such as LPS. In addition to neutrophils and macrophages, dendritic cells are also recruited to the site of infection. The influx of innate cells to the site of infection largely controls the infection and is responsible for the self-limited disease.

Adaptive Immune Response

Induction of adaptive immunity also occurs after infection with *Shigella*. During acute infection, there is an increase in CD8+ T cells in the surface epithelium, an increase in the number of CD4+ and CD8+ T cells in the lamina propria, and an induction of HLA-DR antigens on the epithelial surface (Raqib et al., 1994). As an intracellular pathogen, *Shigella* induces cell-mediated immunity, both cellular proliferation and secretion of immunomodulatory cytokines, IFN-γ, and IL-10 after infection (Samandari et al., 2000; Way et al., 1998). The humoral response to infection has been well-characterized. Patients infected with *Shigella* respond with robust serum antibody responses to LPS, the Ipa proteins, and VirG (Barzu et al., 1993; Coster et al., 1999; Munoz et al., 1995; Oberhelman et al., 1991; Venkatesan and Ranallo, 2006). Mucosal antibodies, directed to both LPS and the Ipa proteins, have also been demonstrated after natural infection or immunization with live-attenuated vaccine strains in fecal samples, urine, and intestinal secretions (Cohen et al., 1996; Kotloff et al., 2004, 2002; Oberhelman et al., 1991; Van de Verg et al., 1990; Venkatesan and Ranallo, 2006). Subtle changes in the bacterial strain or species prevents full protection from disease in subsequent exposures leading to the notion that immunity is serotype-specific and conferred by LPS structure (Black et al., 1987; Cohen et al., 1997). However, the specific immune mechanisms that mediate resistance to infection have not been clearly defined and are currently being debated. For example, in contrast to anti-LPS antibodies, anti-Ipa antibodies may limit the spread and severity of infection (Oberhelman et al., 1991). Humans and rhesus macaques infected with *Shigella* mount a vigorous serum antibody response to the Ipa proteins (IpaA, IpaB, IpaC, and IpaD) and to the VirG protein (Hayani et al., 1992; Oaks et al., 1986, 1996), which is exposed on the bacterial surface. The strongest responses are generally directed to IpaB and the weakest to IpaD (Li et al., 1994a, 1994b).

VACCINES TO PREVENT SHIGELLOSIS

Early History

Early attempts to make vaccines against *Shigella* used both whole bacterial cells and whole-cell extracts (Tamura, 1967). Whole bacterial cells given parenterally were toxic and caused severe local and systemic reactions. Extracts made with various solvents were also extremely toxic. Saline extracts of acetone dried *Shigella* were highly antigenic but also exhibited considerable toxicity in mice and humans (Tully and Tamura, 1954). Since extracts from stable, colonially smooth *Shigella* was critical to obtain immunogenic extracts, it was suggested that immunity was due to specific somatic antigens, which are present only in smooth colonial types (Box 58.9).

Studies in human volunteers demonstrated that parenteral administration of *Shigella* endotoxin and other products as well as oral administration of whole-cell inactivated vaccines did not afford consistent and significant protection. This led to the concept that live vaccines, that survived in the gastrointestinal tract longer, and prolonged the presentation of the antigen to the intestinal lymphoid tissue would be more protective (Formal et al., 1965). Several noninvasive strains were developed whose principal goal was to deliver the primary immunogen, the O antigen of LPS.

BOX 58.9

HISTORICAL AND CURRENT *SHIGELLA* VACCINE DEVELOPMENT STRATEGIES

Historical *Shigella* vaccines	*Shigella* vaccines in development
Whole bacterial cells	Whole, killed bacterial cells
Whole-cell extracts	Live, attenuated
Live, non-invasive	Δ*virG*-based
Sm-D	Δ*guaBA*-based
T-32 Istrati	
Live hybrid vaccines	Live hybrid vaccines
Live *E. coli–Shigella*	Live *Salmonella–Shigella*
Live *Salmonella–Shigella*	Subunit
Live *Shigella*-based	*Shigella* Invaplex
Pur-Rif	LPS-conjugate
ΔaroD-based	Ribosomal
ΔaroA-based	LPS-proteosome

Live, Noninvasive, Shigella-Based Vaccines

Sm-D Vaccines

In the mid-1960s, streptomycin-dependent (Sm-D) strains of *S. flexneri* and *S. sonnei* were used as live, oral vaccine candidates and tested in controlled field trials, in soldiers and children, living in hyperendemic areas in Yugoslavia (Mel et al., 1965a, 1965b). The vaccines were usually administered orally four to five times at high doses ranging from 3×10^{10} to 5×10^{10} cfu, preceded by a bicarbonate solution to neutralize gastric acidity. Following numerous field trials using the Sm-D strains several important observations were made. These included demonstration of (a) serotype-specific protection against shigellosis; (b) a positive correlation between vaccine efficacy and the number of organisms administered; (c) lack of antagonistic effects when two serotypes were combined into a single dose; (d) the safety and efficacy of lyophilized vaccines; (e) the utility of bicarbonate ingestion before vaccination to increase stomach pH and thus intestinal colonization; (f) a direct correlation between reactogenic symptoms and vaccine dose; (g) increasing reactions in children with decreasing age; (h) the requirement of booster doses after primary vaccination to reinforce protection (Mel et al., 1968, 1971). In the United States, a strep-dependent *S. sonnei* live oral vaccine was tested in a double-blind field trial in institutionalized children where *S. sonnei* was endemic (Levine et al., 1976). A causal relationship was observed between vaccination and protection against disease. However, the efficacy was compromised by inadvertent child-to-child vaccine transmission (Levine et al., 1976).

T-32 Istrati (Vadizen)

The live, T-32 Istrati strain is a spontaneous Sereny negative avirulent *S. flexneri* 2a strain that was tested as a live oral vaccine candidate extensively in Romania in several field trials in over 60,000 individuals and in 8000 individuals in China from 1976 to 1980 (Bingrui, 1984; Meitert et al., 1978, 1984a, 1984b). The avirulent status of the vaccine strain was later shown to be due to the loss of the ipa-mxi-spa region on the invasion plasmid (Venkatesan et al., 1991). The vaccine was initially manufactured at Bucharest as a liquid refrigerated formulation (trade name Vadizen) and administered with buffer in Romanian children and adult collectives. Individuals received five increasing doses of the vaccine, from 5×10^{10} to 2×10^{11} cfu at 3-day intervals. Using historical data for comparison, an efficacy of 81% was calculated against *S. flexneri* 2a and 89% protection against heterologous *Shigella* species, including *S. sonnei* (Meitert et al., 1984b). The vaccine was also manufactured by Lanzhou and Beijing Institute of Biological Products in China as an enteric-coated pill. In seven placebo-controlled field trials, 11,000 individuals were tested with the vaccine or placebo (Bingrui, 1984). The vaccine was safe and demonstrated considerable protection against both homologous and heterologous serotypes (Bingrui, 1984). More recently, a deleted version of the *S. sonnei* plasmid expressing its O-antigen, has been transferred into T-32 Istrati and is used as a licensed bivalent *Shigella* vaccine in China (Levine et al., 2007).

Live, E. coli–Shigella *and* Salmonella–Shigella *Hybrid Vaccines*

Several live invasive, hybrid vaccine candidates were constructed and tested in the 1970s–1980s, with the aim of delivering the primary *Shigella* LPS antigen more effectively to the mucosal immune system. None of these strains are being pursued today; data from these trials are of historical interest.

PGA142-1-15

The first hybrid *Shigella* vaccines were constructed by the conjugal transfer of genetic markers between *E. coli* K-12 and *Shigella*. These hybrid strains were invasive in cultured cells but Sereny negative. One such strain PGA145-1-15 was tested as one or two doses at 3×10^{10} cfu per dose and showed significant colonization of the intestinal lumen. However, immunogenicity was poor and no protection was seen after challenge with virulent *S. flexneri* 2a strain 2457T (Levine et al., 1977). This suggested that colonization alone was not sufficient for protection.

EC104

The identification of the virulence plasmid and its role in promoting invasiveness in cultured cells, led to the design of EC104 which contained the Tn5-tagged virulence plasmid (pWR110) of *S. flexneri* 5 in addition to chromosomal markers from *S. flexneri* 2a in an *E. coli* background (Formal et al., 1984; Sansonetti et al., 1981, 1982). EC104 was invasive in cultured cells but negative in the Sereny reaction, which was later shown to be linked to the *kcpA* locus in *E. coli*. Safety trials in human volunteers indicated that doses that were safe and showed colonization of the intestine were not protective (Hale, 1990).

EcSf2a-1 and EcSf2a-2

Two additional derivatives (EcSf2a-1 and EcSf2a-2) were constructed based on conjugal transfer of the *Shigella* virulence plasmid and chromosomal markers into *E. coli*. Like EC104, both EcSf2a-1 and EcSf2a-2 were invasive in cultured cells but negative in the Sereny reaction. EcSf2a-1 was tested in three doses at 10^6–10^9 cfu per dose. The vaccine was shed for 7–12 days and at the highest dose, several volunteers had fever, diarrhea, or dysentery. Although 30–40% of the volunteers indicated significant responses to LPS and Ipa proteins, at the lower dose where the vaccine was safe, no protection was observed between immunized and unimmunized volunteers. Introduction of an *aroD* mutation into EcSf2a-1 resulted in EcSf2a-2 (Kotloff et al., 1992). The *aroD* mutation made EcSf2a-2 dependent upon exogenous PABA for growth, and limited multiplication within epithelial cells. Three doses of EcSf2a-2 at 2×10^9 cfu per dose was fairly safe while two of the four volunteers had dysentery at 10^{10} cfu. In two subsequent efficacy studies the vaccine showed variable efficacy. No correlation was observed between protection and fourfold rises in antibody titers to LPS or the Ipa proteins. An outpatient phase II trial at Fort Ord, CA employing a four-dose regimen of 5×10^8 cfu showed EcSf2a-2 to be safe and immunogenic but in a subsequent trial, vaccine efficacy with the four doses of EcSf2a-2 was only 27% (Kotloff et al., 1995b). Since EcSf2a-1 and EcSf2a-2 occasionally formed tiny plaques, which upon isolation proved to be positive in the Sereny reaction, deletions of the *virG(icsA)* gene was introduced into both strains generating EcSf2a-3 and EcSf2a-5 respectively (Alexander et al., 1996). However, these did not enter clinical trials due to instability of these strains to retain invasiveness in cultured cells (M. Venkatesan, unpublished) and the move away from designing live, *E. coli–Shigella* hybrid vaccines towards constructing *Shigella*-based live vaccines (see below).

Salmonella typhi–S. sonnei *Bivalent Vaccine 5076-1C*

The oral typhoid vaccine Ty21a was also used for the transfer of the *S. sonnei* invasion plasmid generating the transconjugate 5076-1C. This strain agglutinated with antisera to both *S. typhi* and *S. sonnei*, and chemical analysis indicated that the *S. sonnei* O-antigen repeat unit is transported to the surface of the hybrid strain as a polymer unlinked to the *S. typhi* lipid A core (Formal et al., 1981; Seid et al., 1984; Tramont et al., 1984). The vaccine was initially tested in military personnel as either one or three doses ranging from 10^6 to 10^9 cfu per dose. The vaccine showed good colonization rates but poor immunogenicity (Tramont et al., 1984). Subsequently, several lots of a lyophilized vaccine were tested as three doses at 10^9 cfu per dose. Upon challenge with *S. sonnei*, two of the three vaccine lots tested showed 53% protection against diarrhea and 71% protection against bloody diarrhea. The rates of serum IgG or IgA responses to the protective vaccine lots were similar as the lot that did not protect (Black et al., 1987). To reduce the lot-to-lot variation of the vaccine, a lyophilized seed lot was prepared and used for the production of another three lots of vaccine. These new lots were also not protective (Black et al., 1987). Comparison of protective lots versus nonprotective lot under the EM indicated that the nonprotective lot of 5076-1C strain lacked intact flagella. By systematic changes in the manufacturing process a new vaccine lot was prepared (lot 87-5-1), which resembled the original protective lots biochemically, serologically, and by EM visualization. Vaccination studies showed that although the strain was excreted and showed good immune responses, there was no evidence of protection against challenge with *S. sonnei* (Herrington et al., 1990). 5076-1C was later shown to have deletions in the invasion plasmid. Recently, this hybrid strain has been reengineered, wherein the Ty21a strain has been transformed with a plasmid that encodes the *S. sonnei* O-antigen operon. This new hybrid strain has shown protection in mice and guinea pigs (Xu de et al., 2007; Xu et al., 2002).

Live, *Shigella*-Based Vaccines

Linde's pur-/Rif^r Strains

Purine auxotrophs (*pur*) carrying RNA polymerase mutants (Rifampicin resistance) in *S. flexneri* 2a, strain Pur-/Rif vc77 and *S. sonnei* strain Pur-/Rif vc3359 were tested in adult volunteers and children in Sofia, Bulgaria (Linde et al., 1990). These strains retained invasiveness, displayed attenuated Sereny reaction, and showed no fluid production in rabbit ileal loops. The volunteers received one, two, or three doses of increasing amounts of the vaccines from 3×10^8 to 3×10^{10} cfu, spaced 4–5 days apart. None of the vaccinees had dysentery or any major reactions. Of four vc77-immunized volunteers, two volunteers with fecal antibody titers of >1:40, did not contract the disease after challenge with the *S. flexneri* 2a virulent strain suggesting a correlation between fecal antibody titers and immunity. In general, the vaccine was shed for 2–3 days after doses of 3×10^{10} cfu and no revertants were found. The vaccines appeared safe in a larger group of children aged 2–17 years, who were given a total dose of $3–5 \times 10^8$ cfu of either vc77 or vc3359, but no excretion of the vaccine was observed (Dentchev et al., 1990).

aroD *and* aroA*-based Vaccines*

Two *aroD*-based live attenuated vaccines, SFL124 and SFL1070 were made in the late 1980s by investigators at the Karolinska Institute in Sweden (Karnell et al., 1992). The *aroD* gene codes for a 3-dehydroquinase enzyme that catalyzes the conversion of dehydroquinate to dehydroshikimate, a metabolite on the pathway to PABA biosynthesis. *aroD* mutants are therefore, auxotrophic for PABA, a precursor of folic acid, which is a cofactor in multiple essential biochemical reactions including the de novo biosynthesis pathway of purines. Since eukaryotic cells do not contain PABA, a *Shigella aroD* mutant is expected to invade epithelial cells but undergo limited multiplication. SFL124, was derived from a *S. flexneri* serotype Y strain background which contains the common repeating tetrasaccharide unit of the O-antigen of the *S. flexneri* serotypes (except *S. flexneri* 6) (Karnell et al., 1992). The *S. flexneri* Y strain also expresses the group 3:4 antigen and recently has been shown to have originated from a *S. flexneri* 2a strain (Roberts et al., 2005). Besides providing cross protection against other *S. flexneri* serotypes, the *S. flexneri* Y strain can be modified by glucosylating and acetylating phages to form other serotypes (Verma et al., 1991). SFL1070 has the same *aroD* mutation, however, a more virulent wild-type *S. flexneri* 2a isolate was used as the parent strain (Karnell et al., 1995). SFL124 was initially given orally to Swedish volunteers in one or three doses at 2×10^9 cfu per dose and proved to be safe and immunogenic (Li et al., 1992). In general, more volunteers responded with ASCs and local sIgA production against both LPS and Ipa than with serum antibodies. Compared to Swedish volunteers, SFL124 was both less reactogenic and less immunogenic in Vietnamese adult volunteers, probably because most of these volunteers had higher prevaccination serum anti-LPS and anti-Ipa titers (Li et al., 1993). Nonetheless, there was evidence of strong LPS-specific ASC response after primary vaccination. In addition, there were increases in sIgA to LPS and to the Ipa antigens after booster doses as well as stimulation of mucosal memory functions that may have contributed to lower excretion rates after the booster dose. In three groups of Vietnamese children, aged 9–14 years, SFL124 was given in a single dose at 10^7–10^9 cfu (Li et al., 1994a). A dose-dependent LPS-specific ASC response as well as sIgA response was seen in 50–90% of the vaccinated children. Since the children had high preexisting antibody titers to *S. flexneri* Y LPS and water extract, modest, insignificant serum antibody responses were noted. SFL1070 was tested in adult Swedish volunteers at 10^5, 10^7, 10^8, and 10^9 cfu (Karnell et al., 1995). At 10^8 cfu the vaccine was safe but at 10^9 cfu the vaccine was quite reactogenic. A vigorous LPS-specific and water extract-specific IgA ASC response, intestinal sIgA responses, and serum IgA, IgG, and IgM responses was seen at the two highest doses. The requirement for multiple high doses of these vaccines made further development unfeasible. However, based on the safety and immunogenicity of the *aroD* vaccine strains, investigators at the CVD, introduced an *aroA* and a *virG(icsA)* deletion in the virulent *S. flexneri* 2a strain 2457T (Kotloff et al., 1996; Noriega et al., 1996). Like SFL124 and SFL1070, CVD 1203 was auxotrophic for PABA. CVD1203 was tested at a single dose of 10^6 cfu and at two doses each at 10^8 and 10^9 cfu. The strain colonized 80–95% of the volunteers at the two highest doses, however, reactogenicity was also evident (Kotloff et al., 1996). Like SFL124 and SFL1070, more volunteers responded with LPS-specific sIgA and ASCs than serum antibodies and 90–100% of the volunteers had elevated levels of serum TNF-α. In more current CVD constructs the *aroA* mutant background has been replaced with deletions in the *guaBA* genes (see below).

VACCINES IN DEVELOPMENT

Challenge Dose for Efficacy Trials of *Shigella* Vaccines

In earlier studies in naive US volunteers, doses of 100–200 cfu of 2457T, produced illness in 25–40% of the volunteers while doses ranging from 5×10^3 to 5×10^8 cfu produced illness in ~60% of the volunteers with no clear relationship between cfu given and illness (DuPont et al., 1989). *S. dysenteriae* 1, strain M-131 produced illness in 75% of the volunteers when given doses from 2×10^3 to 1×10^4 cfu, although a dose of 10 cfu caused 1 out of 10 volunteers to get sick (DuPont et al., 1989). *S. sonnei* strain 53G given at 5×10^3 cfu in this study produced illness in 35% of the volunteers (DuPont et al., 1989). In order to obtain a more consistent attack rate, volunteers were given 120 ml of bicarbonate buffer to neutralize gastric acidity prior to ingesting the 2457T challenge strain (Kotloff et al., 1995a). This time the attack rate for a 1×10^3 dose of 2457T was 92% in naive controls. In ongoing vaccine trials, administration of the vaccine is usually preceded by ingestion of a bicarbonate solution to neutralize gastric acidity.

Several subunit and live vaccines are in different stages of development (Table 58.1) and are described below.

Parenteral Vaccines

Ribosomal Shigella *Vaccines*

Crude endotoxin from *S. flexneri* was used extensively as a parenteral vaccine in the Russian army

TABLE 58.1 Developmental stage of current *Shigella vaccines*

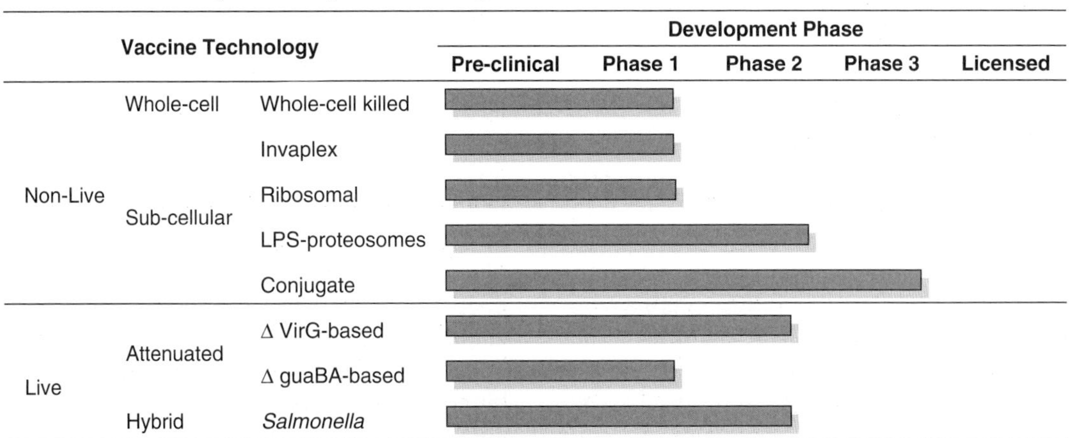

during second world war but was subsequently shown to be nonprotective (mentioned in Levenson et al., 1991). Based on published reports of protection with ribosomal preparations from *M. tuberculosis*, *Salmonella*, *Streptococcus mutans*, and *Candida albicans*, several studies using a ribosomal vaccine preparation from *Shigella* were described between the years 1978 and 1988, including a clinical trial in human volunteers in 1985 in Russia (Gregory, 1986). In the US, ribosomal *Shigella* vaccine preparations (SRV) were isolated from fermenter cultures of avirulent strains of Form I *S. sonnei* and *S. flexneri* 2b. After sonication the ribosomes were collected by ultracentrifugation or precipitation. After further processing, the SRV was shown to consist of RNA (~55%) and protein (37%) with 0.2% by weight of LPS based on O-antigen neutralizing capacity. Since the SRV O-specific component was devoid of lipid A and KDO, that are characteristic of classical LPS, it is believed to be composed of polymerized O-side chains that constitute cytoplasmic precursors of LPS, that are noncovalently linked to the ribosomal vector in SRV preparations (Levenson and Egorova, 1990). One to two injections of 100–1000 μg of SRV given subcutaneously (s.c.) to mice, guinea pigs, or monkeys resulted in high levels of serum O-antibodies. Immunized animals were protected against intraperitoneal (mice on day 7), ocular (guinea pig on day 14), or oral challenge (monkeys at 4–20 weeks) with live virulent strain that caused disease in unimmunized animals. No correlation was seen between individual O-antibody titers and the resistance to challenge in guinea pigs (Levenson et al., 1991).

More recently, a ribosomal *Shigella* vaccine has been manufactured by the International Vaccine Institute (IVI) at Seoul, Korea for field testing in developing countries (Shim et al., 2007). Mice were immunized with a ribosomal preparation containing 2.5 μg dose of O-antigen, on days 0, 1, either intranasally (i.n.) or s.c. and compared with SC602 (a virG(icsA) and iuc mutant of *S. flexneri* 2a, see below) given i.n. Both routes of administration yielded high levels of antigen-specific serum IgG. Intranasal administration also showed high levels of IgA in mucosal secretions such as nasal washes, saliva, vaginal washes, and fecal extracts and the values were similar to that seen with SC602. Also, i.n. administration elicited significantly higher numbers of IgA ASCs in the mononuclear cells isolated from nasal passage, lung, and submandibular gland tissues than s.c. administration or SC602 given i.n. The serum and ASCs levels were maintained up to 28 weeks past immunization. Mice immunized i.n. were protected to a greater extent than those immunized s.c. after challenge with virulent strains 1 week after immunization (Shim et al., 2007). Mice with knockouts in MyD88, TLR2, TLR4, and TLR5 and immunized i.n. with SRV resulted in induction of significant levels of serum IgG and IgA as well as IgA antibodies in nasal washes similar to wild-type mice. In contrast, levels of serum IgA and IgA ASCs were significantly lower in the lung, saliva, and fecal extracts of MyD88 and TLR2 knockouts (Shim et al., 2007). Using polymeric immunoglobulin receptor knockout mice (pIgR−/−) immunized i.n. with SRV, significant levels of IgG and IgA antibodies were detected in serum and lung lavage samples, but not in saliva and fecal extracts and such mice were poorly protected after i.n. challenge with virulent strains. This indicates that sIgA induced after i.n. administration is critical for protection against *Shigella* infection (Shim et al., 2007).

Conjugate **Shigella** *Vaccines*

These vaccine candidates were designed based on the role of the O-antigen as the major protective antigen during natural infections of *Shigella* or during immunization with attenuated strains (Formal et al., 1989). Furthermore, prospective studies in recruits in Israel Defense Forces (IDF) showed that LPS-specific serum IgG correlated with resistance to shigellosis (Cohen et al., 1991; Robbins et al., 1992). In initial studies, LPS from *S. dysenteriae* 1 strain 1617, *S. flexneri* 2a strain 2457T, and *S. sonnei* strain 53G were extracted using hot phenol extraction and ethanol precipitation and then treated with 1% boiling glacial acetic acid to extract O-antigen-specific polysaccharides (O-SP). These O-SP products were derivatized using adipic acid hydrazide as a linker and initially conjugated to tetanus toxoid (TT) (Chu et al., 1991). Since *S. sonnei* does not stably express the O-SP, later versions of *S. sonnei* conjugate vaccines used the LPS from *Pleisomonas shigelloides*, which has an identical O-specific structure as *S. sonnei*. Injection of 25 μg of *S. dysenteriae* 1, *S. flexneri* 2a, and *P. shigelloides* conjugates into adult volunteers elicited mild local reactions only. Each conjugate induced a significant rise of the geometric mean LPS-specific serum IgG, IgM, and IgA levels (Taylor et al., 1993). A second injection 6 weeks later did not elicit booster responses, and adsorption of the conjugates onto alum did not enhance their immunogenicity. Conjugate-induced levels of IgA, but not IgG or IgM, declined to preimmunization levels at day 56. The levels of postimmunization antibodies of the three immunoglobulin classes were similar to or higher than those of recruits in the IDF following shigellosis caused by *S. flexneri* type 2a or *S. sonnei*. In later versions the O-SP was conjugated to recombinant mutant exoprotein A of *Pseudomonas aeruginosa* (rEPA) (Cohen et al., 1997). A *S. sonnei*-rEPA was given to recruits of the IDF who were at increased risk of shigellosis under field conditions (Cohen et al., 1997). The vaccine was given as a single 0.5 ml dose intramuscularly (i.m.) (contains 25 μg of polysaccharide, 75 μg protein in saline with 0.01% thimerosal). In this study, the control group comprised volunteers who received either four oral doses of EcSf2a-2 ($5–9 \times 10^8$ cfu, given on days 0, 3, 14, 17) or a parenteral tetravalent meningococcal vaccine. Overall 559 recruits received the *S. sonnei*-rEPA conjugate and 839 recruits served as controls. The overall attack rate of *S. sonnei* shigellosis was 2.2% in recipients of *S. sonnei*-rEPA, compared with 8.6% in controls indicating an efficacy of 74%. *S. sonnei*-rEPA recipients who developed culture positive shigellosis had lower geometric mean titers of serum IgG and IgA to *S. sonnei* LPS and a lower proportion of such recruits showed a fourfold or higher increase in antibody titers (Cohen et al., 1997). In 4 to 7-year old children, *S. sonnei* and *S. flexneri* 2a conjugate vaccines elicited significant LPS-specific serum IgG titers against the homologous serotypes, which were boosted after a second dose at 6 weeks and remained high for up to 6 months after vaccination (Ashkenazi et al., 1999). Succinylation of the carrier protein was shown to improve the immunogenicity of the conjugates in mice and therefore a clinical trial of O-SP conjugates of *S. sonnei* and *S. flexneri* 2a bound to succinylated rEPA (rEPAsucc) or native or succinylated *Corynebacterium diphtheriae* toxin mutant (CRM9 or CRM9succ) was conducted in healthy adults in Israel (Passwell et al., 2001). All three conjugates of *S. sonnei* elicited significant rises of GMT of IgG anti-LPS within 1 week of injection that remained high up to 26 weeks while the *S. flexneri* 2a-rEPAsucc elicited significantly higher titers to LPS than those elicited by *S. flexneri* 2a-rCRM-9succ. Both conjugates elicited statistically significant rises of serum antibodies to the injected carrier protein. A similar profile of immunogenicity was also observed in healthy 1- to 4-year-old children given two 0.5-ml i.m. injections 6 weeks apart of either *S. sonnei*-CRM9 or *S. flexneri* 2a-rEPAsucc (Passwell et al., 2003). Five to 10% of the children had low transient fever lasting 24–48 h. *S. sonnei*-CRM9 elicited a >fourfold rise in IgG anti-LPS in 92% of the immunized children and *S. flexneri* 2a-rEPA(succ) in 85% of the children after the second injection; both conjugates elicited type-specific booster responses (Passwell et al., 2003). At 2 years the geometric mean concentrations of IgG anti-LPS were significantly higher than preimmunization levels. A >4-fold rise of IgG anti-diphtheria (66%) and IgG anti-EPA (78%) was also observed. A study of these newer derivatives of conjugate vaccines in infants is ongoing (Passwell et al., 2003).

Intranasal *Shigella* Vaccines

LPS-proteosomes

This product was developed by combining purified preparations of outer membrane proteins (OMPs) from Group B *Neisseria meningitides* that form multimolecular vesicular structures with purified antigen. In the case of *Shigella* the antigen is purified LPS and the antigen is noncovalently bound to the OMP complex in approximately equimolar proportion. In an initial clinical trial the *S. flexneri* 2a proteosome preparation was given in two i.n. doses ranging from 0.1 to 1.5 mg, with a spray device. The primary symptoms seen in most volunteers were mild-to-moderate rhinorrhea and headache, which did not appear to interfere with daily activity. All dose groups demonstrated a significant ASC response in all antibody isotypes (50% of the volunteers had LPS-specific IgA ASCs of 20–25 per 10^6 PBMCs at the 1 and 1.5 mg doses). *S. flexneri* 2a LPS-specific serum IgG and IgA rose sharply between days 0 and 14 with 40% of the volunteers showing ~5-fold increases in IgG titers at the 1 mg dose. A >twofold increase in IgG geometric mean titers to LPS in stools were seen in 50–80% of the volunteers at all doses. A subsequent challenge study failed to show adequate protection (TL Hale, unpublished communication).

Invaplex Subunit Vaccine

The *Shigella* invasin complex or Invaplex is a subcellular, macromolecular complex isolated from water extracts of virulent, intact, viable *Shigella* using step gradients of NaCl on an FPLC ion-exchange column (Turbyfill et al., 2008). Fractions eluting from the column using either 0.24 or 0.5 M NaCl and containing LPS, IpaB, and IpaC are pooled to form Invaplex 24 or Invaplex 50, respectively. Both Invaplex 24 and Invaplex 50 contain LPS, IpaB, IpaC, and IpaD, whereas only Invaplex 50 contains a truncated form of VirG and IpaA. Invaplex has been isolated from each of the four species of *Shigella* and is void of major OMPs (Turbyfill et al., 2000). Although Invaplex contains several unknown proteins, many of which are nonimmunogenic, recent work has isolated and defined the constituents responsible for the immunogenicity and protective efficacy to be LPS, IpaB, and IpaC. Significant work is currently underway to generate Invaplex from the individual components, potentially resulting in a more defined product that can be customized.

Invaplex is delivered i.n. to mice, guinea pigs, and humans and induces both systemic and mucosal antibody responses. Invaplex is not delivered with an adjuvant. In fact, Invaplex functions as a mucosal adjuvant with both protein-based and DNA-based vaccines, highlighting its potential use with other enteric vaccine antigens for use in a combination vaccine (Kaminski et al., 2006). In mice, the mucosal immune response is induced at sites both proximal (lungs) and distal (intestines) to the site of immunization. The serum antibody response is directed primarily to LPS, IpaB, and IpaC. *Shigella*-specific cell-mediated immune responses are also induced, which have been phenotyped as primarily Th2, with high levels of antigen-specific IgG1 and the secretion of IL-4, IL-5, and IL-10 after *in vitro* restimulation of immune cells. There also is a Th1 component to the immune response, indicated with high levels of IFN-gamma secretion. Serotype-specific, homologous protective efficacy is achieved in both the mouse pulmonary and guinea pig keratoconjunctitis models (Turbyfill et al., 2000). The subunit Invaplex approach is highly adaptable to creating a multivalent vaccine, by simply combining the necessary monovalent vaccines prior to immunization (Oaks and Turbyfill, 2006).

Shigella flexneri 2a Invaplex 50 has been evaluated in Phase 1 safety and immunogenicity trials using a dose-escalation study design. Human volunteers were administered three i.n. vaccinations, separated by 2 weeks, of either 10, 50, 240, or 480 μg. No severe adverse reactions were noted in any of the dose groups. Immunization with ≥50 μg was immunogenic, with ≥240 μg induced mucosal and systemic immune responses with increased frequency and magnitude (Tribble, 2007). Positive immunogenicity results combined with an excellent safety record from the first clinical trial has warranted further evaluation in humans, which is currently underway.

Oral Vaccines

Inactivated Whole-Cell Vaccine

An inactivated *S. sonnei* whole-cell vaccine candidate SsWC, was recently tested in seven volunteers with three placebo recipients, in a Phase 1 clinical trial (McKenzie et al., 2006). To increase the expression of Ipa proteins, the strain was grown in BHI broth supplemented with 0.1% sodium deoxycholate. In vitro studies previously indicated that growth in the presence of bile salts increased invasiveness of *Shigella* (Pope et al., 1995). The cell pellet obtained after harvesting the fermentation culture was inactivated by 1% formaldehyde. The vaccine was formulated at a concentration of 2×10^{10} bacterial cells/ml in phosphate buffered saline, which also constituted a single dose. Volunteers were given either a three does (on days 0, 14, 28) or a five-dose regimen (0, 2, 4, 6, 28) with one group serving as placebo controls. One volunteer in the three-dose group had moderate abdominal pain, mild diarrhea, and one episode of vomiting after the first, second, and third dose respectively, while one volunteer in the five-dose group had moderate nausea after the fourth dose. No significant differences were seen between the two dosing groups in the mean antibody levels and each of the seven volunteers developed a fourfold or greater rise in serum IgG or IgA to at least two of the three antigens: six had significant response to SsWC, five to IpaC, and four to LPS. Five vaccines also had significant rises in fecal IgA to SsG and three volunteers to LPS and IpaC. One placebo recipient seroconverted and also had 7.5-fold rise in fecal IgA to LPS.

guaBA-*Based Vaccines Developed at CVD*

A series of live attenuated *Shigella* vaccines are being developed at the CVD based on alterations of genes associated with the biosynthesis of guanine nucleotides (*guaBA*) with additional deletions in *virG(icsA)*, enterotoxin genes *sen* and *set*. CVD 1204 contains a specific 918 bp deletion in the *guaBA* operon. The *guaBA* operon is part of the purine metabolic pathway (Noriega et al., 1994). *guaB* encodes IMP dehydrogenase and converts inosinic acid (IMP) to XMP. *guaA* encodes GMP synthetase and converts XMP to GMP. In HeLa cells CVD 1204 was less invasive than the parent strain 2457T. Additional loss of the *virG(icsA)* gene generated CVD 1205. Further suicide-vector-based deletions in the chromosomal gene *setAB*, encoding *Shigella* enterotoxin 1 (ShET1, 459 bp deletion in *set*A encoding the subunit with putative enzymatic activity) and the plasmid gene *senA*, encoding *Shigella* enterotoxin 2 (ShET2, 397 bp deletion in *senA*) resulted in CVD 1207 (Kotloff et al., 2000). Immunization with CVD 1207 was protective against challenge in guinea pigs by the ocular route even though it was significantly less invasive in HeLa cells and Sereny reaction negative. Groups of three to seven volunteers were given a single oral dose of CVD 1207 starting at 1×10^6 cfu and increasing the dose, in 10-fold increments, to 10^{10} cfu (Kotloff et al., 2000). No subject experienced fever or dysentery, although at the two highest doses a small percentage of volunteers experienced diarrhea and emesis. All of the volunteers receiving 10^8–10^{10} cfu excreted the vaccine strain for an average of 1–3 days, and 65–100% of these volunteers also demonstrated a positive ASC response to LPS. The magnitude of the LPS-specific IgA ASC responses (35 per 10^6 PBMCs) was modest compared to the response following virulent infection (239 per 10^6 PBMCs). Serum antibody responses to LPS and the

Ipa proteins were modest. In a subset of volunteers tested, increases in IFN-γ, IL-10, and TGF-β indicated a Th1 response and proliferative responses to IpaC and IpaD were also observed in a few volunteers.

More recently, CVD 1204 (*guaBA*) was compared with CVD 1208 (*guaBA, set, sen*) in groups of 16–18 volunteers who received sequentially a single oral dose of 10^7, 10^8, or 10^9 cfu of either of the two vaccine strains or placebo (Kotloff et al., 2004). While 8 of 23 recipients ingesting CVD 1204 had an adverse clinical reaction, only 1 of 21 recipients receiving CVD 1208 had mild fever, and only at the highest dose. Both vaccines elicited significant LPS-specific ASC responses while responses to the Ipa proteins were lower in magnitude and less frequent. Significant serum antibody responses to LPS and to the IpaB protein were measured with both vaccine strains, and >4-fold increases in fecal IgA to LPS were seen in 100 and 86% of volunteers at the highest dose. These results established the association of *senA* and *setAB* enterotoxins with diarrheal symptoms previously observed with live vaccines (Kotloff et al., 2004). It is anticipated that further studies with CVD 1208 will confirm its safety and immunogenicity profile.

virG(icsA)-*Based Vaccines being Developed at Walter Reed Army Institute for Research (WRAIR)*

A series of vaccine candidates based on loss of VirG(IcsA) have been tested by investigators at WRAIR and shown to be both immunogenic and in one case, efficacious against dysentery. SC602 was derived from *S. flexneri* 2a, strain 454, and constructed at the Institut Pasteur, Paris, France. In addition to the loss of the *virG(icsA)* gene, SC602 also contained a deletion in the chromosomal *iuc* gene encoding aerobactin. By itself, an *iuc* mutant gives a slightly attenuated Sereny reaction as well as a compromised reaction in rabbit ileal loops. SC602 was manufactured as a lyophilized product and orally administered in a single dose to volunteers after ingestion of a bicarbonate solution. Initially, groups of three volunteers (and three placebo controls) were vaccinated sequentially, with increasing doses from 10^2 to 10^8 cfu. There were no cases of dysentery at any dose (Coster et al., 1999). At the highest dose, two of three volunteers got diarrhea, fever, and severe constitutional symptoms. A follow-up expanded study was carried out in 15 volunteers at the 10^6 cfu dose. This time 47% of volunteers had diarrhea, 33% had fever, 40% had severe intestinal problems, and 40% had severe constitutional symptoms. An additional cohort of 12 volunteers was vaccinated with 10^4 cfu. Only one volunteer's symptoms met the criteria of diarrhea while 10–30% of the volunteers reported mild-to-moderate intestinal and constitutional symptoms. Robust and prolonged colonization was seen, as measured by excretion of the SC602, which was proportional to the dose given (Coster et al., 1999). For example 100 and 90% of the volunteers fed 10^6 and 10^4 cfu respectively, excreted the vaccine for 7 days. Seven of 12 volunteers given 10^4 cfu had significant peak IgA ASCs. Four of the seven volunteers had fourfold or greater rises in serum and urine IgA titer to LPS. Fewer volunteers showed response to the Ipa proteins. An efficacy study was performed using seven SC602-vaccinated volunteers from the 10^4 cfu dose study, and challenged with 2457T two months after immunization along with seven unvaccinated controls (Coster et al., 1999). Four of the vaccinated volunteers showed no symptoms of the disease while three volunteers met the clinical definition of diarrhea that lasted for 24h. Three of the four volunteers who were completely protected had demonstrated significant LPS-specific IgA ASCs and had greater than fourfold increase in serum and urine LPS-specific IgA titers. Of the three vaccinated volunteers who had diarrhea after challenge, two lacked an LPS-specific IgA ASC response as well as an adequate increase in serum or urine IgA titer after vaccination. These results indicate that the immune correlates of protection must include a combination of LPS-specific ASC responses, along with three to fourfold increases in LPS-specific serum antibody titers. Fecal IgA responses were not measured in this study. In contrast to SC602-vaccinated volunteers, six of seven control volunteers developed shigellosis after 2457T challenge that included diarrhea, fever, severe gastrointestinal and constitutional symptoms, and dysentery. Thus, vaccination with SC602 was efficacious against disease. A community-based evaluation of SC602 was carried out in 12 inpatient and 34 outpatient volunteers in the 10^3–10^4 cfu dose range. Approximately 15% of the volunteers had transient fever and diarrhea while headache and abdominal cramps appeared to be the main constitutional symptoms reported. Approximately 50% of the volunteers had a threefold serum IgA ELISA response to LPS, and approximately 75% had an ASC response to the same antigen confirming earlier trial data with this vaccine candidate (Katz et al., 2004).

WRSS1 was derived from a stable *S. sonnei* strain, Moseley, and contains a 212bp deletion in the *virG(icsA)* gene (Hartman and Venkatesan, 1998). WRSS1 was manufactured as a lyophilized product and the protocol for administering it to volunteers was similar to that followed with SC602. WRSS1 was tested in a placebo-controlled dose escalating study in 27 volunteers who received a single oral dose of either

10^3, 10^4 10^5, or 10^6 cfu (Kotloff et al., 2002). No volunteer receiving WRSS1 had dysentery or a fever >102°F. Approximately 30% of the volunteers in the two highest doses developed diarrhea and low-grade fever. At the highest dose tested, the vaccine was excreted for at least a week showing robust colonization, which elicited strong LPS-specific IgA ASCs, serum antibodies, and fecal IgA antibodies. An increase in IFN-γ production was seen in response to purified Ipa proteins, IpaB, IpaC, and IpaD at the two highest doses in some of the volunteers. In general, WRSS1 was considered safe given as a single oral dose in the range of 10^3–10^4 cfu. A community-based study at 10^3, 10^4, and 10^5 cfu dose of WRSS1 was carried out in Israeli volunteers (Orr et al., 2005). At the 10^5 cfu dose, approximately 30% of the volunteers had mild fever and/or diarrhea whereas the 10^3–10^4 cfu dose was well tolerated, with no reported fever and one report of moderate diarrhea for 1 day. The vaccine was excreted for an average of 5 days by 80% of the volunteers ingesting the 10^3–10^4 dose, and induced robust LPS-specific IgA ASCs and increases in serum IgA titers. Most importantly, in spite of the strong colonization seen with WRSS1, the vaccine was not transmitted to household contacts. The low but consistent rate of fever and diarrhea that was seen with SC602 and WRSS1 at the 10^4 cfu dose suggests that, further attenuation is needed to reduce these symptoms.

In a recent Phase 1 trial, a *virG(icsA)*-based *S. dysenteriae* 1 vaccine strain WRSd1 was fed to five groups of eight subjects who were given escalating doses from 10^3 to 10^7 cfu at the Johns Hopkins Medical Center in Baltimore, MD. The parent strain, 1617, was obtained from a 1960s epidemic dysentery outbreak in Guatemala. WRSd1 contained a 10 kb deletion on the plasmid that included the loss of the *virG(icsA)* gene, and an additional ~20 kb deletion on the chromosome, that eliminated not only the *stxAB* genes but also an important transcriptional regulator of anaerobic gene expression, the *fnr* gene, among others (Venkatesan et al., 2002). WRSd1 is invasive in Hela cells but Sereny negative, and it was less immunogenic than SC602 and WRSS1 in guinea pigs. In the Phase I trial no subject developed fever or abdominal pain. Fecal shedding of the vaccine strain was poor and identified in 9 of 40 subjects (23%) and 8 of these 9 had anti-LPS IgA ASC responses (McKenzie, 2007). About one-third of the volunteers showed a serum anti-LPS IgA response. Eight of the 40 subjects (20%) had diarrhea, and 2 subjects in the 10^4 CFU group passed a single loose, stool containing blood without any other constitutional symptoms. The rate and severity of diarrhea were not related to the dose of the vaccine. Poor colonization may have resulted from the lack of the *fnr* gene since it has been implicated to play a critical role in the gastrointestinal colonization of *E. coli* in a rat model.

Based on these trials, second generation *virG (icsA)* vaccine strains have been constructed with retention of the *fnr* gene and deletion of additional genes with a goal of improving colonization, tolerability, and immunogenicity. The presence of the *setAB* gene (encoding ShET1) on the chromosome of SC602, the *senA* gene (encoding ShET2-1), and *senB* gene (encoding ShET2-2) on the invasion plasmid of SC602, WRSS1, and WRSd1 indicate that deletions of enterotoxin genes may alleviate the reactogenic symptoms seen with these vaccine candidates (Fasano et al., 1995, 1997; Nataro et al., 1995). Additionally, loss of the *msbB* gene on the plasmid (*msbB2*) has been incorporated into the design of the second generation *virG(icsA)*-based vaccines to reduce the febrile reactions seen with the earlier vaccines. A *S. flexneri* 2a strain WRSf2G11 with deletion of the *virG(icsA)*, *setAB* and *senA* genes has been shown to be immunogenic and protective in a guinea pig model. Further deletion of the *senB* gene (WRSf2G12) as well as the *msbB2* gene (WRSf2G15) have been constructed and are being tested in animal models (Ranallo et al., 2007). Similar derivatives have been designed and constructed in *S. sonnei* and *S. dysenteriae* 1 background and have been shown to be immunogenic and protective in animal models (Ranallo, 2007; Ranallo et al., 2006; Venkatesan and Ranallo, 2006). These newer derivatives await manufacture and testing in Phase 1 trials.

Current Licensed Vaccines

FS Bivalent Vaccine

Currently this is the only licensed oral, live noninvasive *Shigella* vaccine, and is used in China (Levine et al., 2007). It was constructed by mobilizing a spontaneously modified deleted form of the *S. sonnei* invasion plasmid, tagged with Tn5, into the *S. flexneri* 2a vaccine strain T32-ISTRATI. The FS vaccine strain is a bivalent vaccine designed to provide protection against both *S. sonnei* and *S. flexneri* 2a. Initial testing in humans with freshly harvested bacteria provided safety colonization and immunogenicity data at three doses of 2–5 × 10^{10} cfu. The vaccine was excreted for 1–3 days by 65–75% of the volunteers. A lyophilized form of the vaccine was then tested in field studies in 17,500 adults and children and similar numbers of placebo controls in Changge City, China. Passive surveillance for the next 5–6 months indicated 61–65% protective efficacy (PE) against *S. flexneri*, 50–72% against *S. sonnei*, and 48–52% against other *Shigella* species. Similar data is currently unavailable in infants and toddlers.

PROSPECTS FOR THE FUTURE

Several live attenuated vaccines are being developed. At the same time several new approaches to subunit vaccines are also being evaluated. A combination of live and subunit vaccines should be tested in animal models for immunogenicity and efficacy. Phase 1 trials will be needed to determine the value of individual vaccine candidates. Such trials should also incorporate bivalent and multivalent mixtures of vaccines. Challenge studies can take the place of large-scale field trials to show efficacy. Dose–response studies in endemic populations will determine the range to be tested for effectiveness in these groups. Information about the geographical distribution of *Shigella* serotypes will be needed to determine the vaccine serotypes to be tested in a particular region of the world.

KEY ISSUES

- Protection is serotype-specific, therefore an effective vaccine, whether live or subunit, will have to incorporate components of several serotypes.
- Development and application of a *Shigella* vaccine should consider the diversity of *Shigella* serotypes within a geographical region.
- The vaccine must be easy to manufacture so that it is cost effective, especially if it is to be used in developing countries.
- Delivery of the vaccine should be needle free, if possible.
- The technology for making and manufacturing the vaccine must be user friendly.
- Reliance on a cold chain for storage of the vaccine must be minimized.
- The vaccine should be immunogenic in endemic populations.
- Engagement of local health care policy makers, scientists, and technicians must be carried out prior to testing vaccines in endemic regions of the world.

References

Ahmed, S.F., Riddle, M.S., Wierzba, T.F., Messih, I.A., Monteville, M.R., Sanders, J.W. and Klena, J.D. Epidemiology and genetic characterization of *Shigella flexneri* strains isolated from three paediatric populations in Egypt (2000–2004). Epidemiol. Infect. 2006; 134:1237–1248.

Alam, N.H. and Ashraf, H. Treatment of infectious diarrhea in children. Paediatr. Drugs 2003; 5:151–165.

Alexander, W.A., Hartman, A.B., Oaks, E.V. and Venkatesan, M.M. Construction and characterization of virG (icsA)-deleted *Escherichia coli* K12-*Shigella flexneri* hybrid vaccine strains. Vaccine 1996; 14:1053–1061.

Ally, S., Sauer, N.J., Loureiro, J.J., Snapper, S.B., Gertler, F.B. and Goldberg, M.B. Shigella interactions with the actin cytoskeleton in the absence of Ena/VASP family proteins. Cell. Microbiol. 2004; 6:355–366.

Al-Moyed, K.A., Harmal, N.S., Al-Harasy, A.H. and Al-Shamahy, H.A. Increasing single and multi-antibiotic resistance in Shigella species isolated from shigellosis patients in Sana'a, Yemen. Saudi Med. J. 2006; 27:1157–1160.

Arbibe, L., Kim, D.W., Batsche, E., Pedron, T., Mateescu, B., Muchardt, C., Parsot, C. and Sansonetti, P.J. An injected bacterial effector targets chromatin access for transcription factor NF-kappaB to alter transcription of host genes involved in immune responses. Nat. Immunol. 2007; 8:47–56.

Ashkenazi, S., Passwell, J.H., Harlev, E., Miron, D., Dagan, R., Farzan, N., Ramon, R., Majadly, F., Bryla, D.A., Karpas, A B., Robbins, J.B. and Schneerson, R. Safety and immunogenicity of *Shigella sonnei* and *Shigella flexneri* 2a O-specific polysaccharide conjugates in children. J. Infect. Dis. 1999; 179: 1565–1568.

Barzu, S., Nato, F., Rouyre, S., Mazie, J.C., Sansonetti, P. and Phalipon, A. Characterization of B-cell epitopes on IpaB, an invasion-associated antigen of *Shigella flexneri*: identification of an immunodominant domain recognized during natural infection. Infect. Immun. 1993; 61:3825–3831.

Behrens, M., Sheikh, J. and Nataro, J.P. Regulation of the overlapping pic/set locus in *Shigella flexneri* and enteroaggregative *Escherichia coli*. Infect. Immun. 2002; 70:2915–2925.

Bennish, M.L., Salam, M.A., Khan, W.A. and Khan, A.M. Treatment of shigellosis: III. Comparison of one- or two-dose ciprofloxacin with standard 5-day therapy. A randomized, blinded trial. Ann. Intern. Med. 1992; 117:727–734.

Bidawid, S.P., Edeson, J.F., Ibrahim, J. and Matossian, R.M. The role of non-biting flies in the transmission of enteric pathogens (Salmonella species and Shigella species) in Beirut, Lebanon. Ann. Trop. Med. Parasitol. 1978; 72:117–121.

Bingrui, W. Study on the effect of oral immunization of T32-Istrati strain against bacillary dysentery in field trials. Arch. Roum. Pathol. Exp. Microbiol. 1984; 43:285–289.

Black, R.E., Levine, M.M., Clements, M.L., Losonsky, G., Herrington, D., Berman, S. and Formal, S.B. Prevention of shigellosis by a *Salmonella typhi–Shigella sonnei* bivalent vaccine. J. Infect. Dis. 1987; 155:1260–1265.

Boulette, M.L. and Payne, S.M. Anaerobic regulation of *Shigella flexneri* virulence: ArcA regulates Fur and iron acquisition genes. J. Bacteriol. 2007; 189:6957–6967.

Buchrieser, C., Glaser, P., Rusniok, C., Nedjari, H., D'Hauteville, H., Kunst, F., Sansonetti, P. and Parsot, C. The virulence plasmid pWR100 and the repertoire of proteins secreted by the type III secretion apparatus of *Shigella flexneri*. Mol. Microbiol. 2000; 38:760–771.

Butler, T., Dunn, D., Dahms, B. and Islam, M. Causes of death and the histopathologic findings in fatal shigellosis. Pediatr. Infect. Dis. J. 1989; 8:767–772.

Cahill, K.M., Davies, J.A. and Johnson, R. Report on an epidemic due to *Shigella dysenteriae*, type 1, in the Somali interior. Am. J. Trop. Med. Hyg. 1966; 15:52–56.

Carlin, N.I., Gidney, M.A., Lindberg, A.A. and Bundle, D.R. Characterization of *Shigella flexneri*-specific murine monoclonal antibodies by chemically defined glycoconjugates. J. Immunol. 1986; 137:2361–2366.

Carlin, N.I. and Lindberg, A.A. Monoclonal antibodies specific for *Shigella flexneri* lipopolysaccharides: clones binding to type IV, V, and VI antigens, group 3,4 antigen, and an epitope common to all *Shigella flexneri* and *Shigella dysenteriae* type 1 stains. Infect. Immun. 1987; 55:1412–1420.

Carlin, N.I., Rahman, M., Sack, D.A., Zaman, A., Kay, B. and Lindberg, A.A. Use of monoclonal antibodies to type *Shigella flexneri* in Bangladesh. J. Clin. Microbiol. 1989; 27:1163–1166.

CDC *Shigella sonnei* outbreak among men who have sex with men—San Francisco, California, 2000–2001. JAMA 2002; 287:37–38.

Chapel, H., Puel, A., von Bernuth, H., Picard, C. and Casanova, J.L. *Shigella sonnei* meningitis due to interleukin-1 receptor-associated kinase-4 deficiency: first association with a primary immune deficiency. Clin. Infect. Dis. 2005; 40:1227–1231.

Cheah, K.C., Beger, D.W. and Manning, P.A. Molecular cloning and genetic analysis of the rfb region from *Shigella flexneri* type 6 in *Escherichia coli* K-12. FEMS Microbiol. Lett. 1991; 67:213–218.

Chompook, P., Samosornsuk, S., von Seidlein, L., Jitsanguansuk, S., Sirima, N., Sudjai, S., Mangjit, P., Kim, D.R., Wheeler, J.G., Todd, J., Lee, H., Ali, M., Clemens, J., Tapchaisri, P. and Chaicumpa, W. Estimating the burden of shigellosis in Thailand: 36-month population-based surveillance study. Bull. World Health Organ. 2005; 83:739–746.

Chompook, P., Todd, J., Wheeler, J.G., von Seidlein, L., Clemens, J. and Chaicumpa, W. Risk factors for shigellosis in Thailand. Int. J. Infect. Dis. 2006; 10:425–433.

Chu, C.Y., Liu, B.K., Watson, D., Szu, S.S., Bryla, D., Shiloach, J., Schneerson, R. and Robbins, J.B. Preparation, characterization, and immunogenicity of conjugates composed of the O-specific polysaccharide of *Shigella dysenteriae* type 1 (Shiga's bacillus) bound to tetanus toxoid. Infect. Immun. 1991; 59:4450–4458.

Clementz, T., Zhou, Z. and Raetz, C.R. Function of the *Escherichia coli* msbB gene, a multicopy suppressor of htrB knockouts, in the acylation of lipid A. Acylation by MsbB follows laurate incorporation by HtrB. J. Biol. Chem. 1997; 272:10353–10360.

Cohen, D., Ashkenazi, S., Green, M.S., Gdalevich, M., Robin, G., Slepon, R., Yavzori, M., Orr, N., Block, C., Ashkenazi, I., Shemer, J., Taylor, D.N., Hale, T.L., Sadoff, J.C., Pavliakova, D., Schneerson, R. and Robbins, J.B. Double-blind vaccine-controlled randomised efficacy trial of an investigational *Shigella sonnei* conjugate vaccine in young adults. Lancet 1997; 349:155–159.

Cohen, D., Green, M.S., Block, C., Slepon, R. and Ofek, I. Prospective study of the association between serum antibodies to lipopolysaccharide O antigen and the attack rate of shigellosis. J. Clin. Microbiol. 1991; 29:386–389.

Cohen, D., Orr, N., Robin, G., Slepon, R., Ashkenazi, S., Ashkenazi, I. and Shemer, J. Detection of antibodies to Shigella lipopolysaccharide in urine after natural Shigella infection or vaccination. Clin. Diagn. Lab. Immunol. 1996; 3:451–455.

Coster, T.S., Hoge, C.W., VanDeVerg, L.L., Hartman, A.B., Oaks, E.V., Venkatesan, M.M., Cohen, D., Robin, G., Fontaine-Thompson, A., Sansonetti, P.J. and Hale, T.L. Vaccination against shigellosis with attenuated *Shigella flexneri* 2a strain SC602. Infect. Immun. 1999; 67:3437–3443.

Demers, B., Sansonetti, P.J. and Parsot, C. Induction of type III secretion in *Shigella flexneri* is associated with differential control of transcription of genes encoding secreted proteins. EMBO J. 1998; 17:2894–2903.

Dentchev, V., Marinova, S., Vassilev, T., Bratoyeva, M. and Linde, K. Live *Shigella flexneri* 2a and *Shigella sonnei* I vaccine candidate strains with two attenuating markers. II. Preliminary results of vaccination of adult volunteers and children aged 2–17 years. Vaccine 1990; 8:30–34.

D'Hauteville, H., Khan, S., Maskell, D.J., Kussak, A., Weintraub, A., Mathison, J., Ulevitch, R.J., Wuscher, N., Parsot, C. and Sansonetti, P.J. Two msbB genes encoding maximal acylation of lipid A are required for invasive *Shigella flexneri* to mediate inflammatory rupture and destruction of the intestinal epithelium. J. Immunol. 2002; 168:5240–5251.

Dupont, H.L. Shigella species (bacillary dysentery). In: Principles and Practice of Infectious Diseases, 2005 (G.L. Mandell, J.E. Bennett and R. dolin, Eds.), pp. 2655–2661. Philadelphia, PA: Elsevier, 2005.

DuPont, H.L., Levine, M.M., Hornick, R.B. and Formal, S.B. Inoculum size in shigellosis and implications for expected mode of transmission. J. Infect. Dis. 1989; 159:1126–1128.

Ehrenpreis, S. and Ehrenpreis, E. Clinician's Handbook of Prescription Drugs. New York: McGraw Hill, 2001.

Elamreen, F.H., Sharif, F.A. and Deeb, J.E. Isolation and antibiotic susceptibility of Salmonella and Shigella strains isolated from children in Gaza, Palestine from 1999 to 2006. J. Gastroenterol. Hepatol. 2008; 23:e330–3.

Espina, M., Olive, A.J., Kenjale, R., Moore, D.S., Ausar, S.F., Kaminski, R.W., Oaks, E.V., Middaugh, C.R., Picking, W.D. and Picking, W.L. IpaD localizes to the tip of the type III secretion system needle of *Shigella flexneri*. Infect. Immun. 2006; 74:4391–4400.

Ewing, W.H. Shigella nomenclature. J. Bacteriol. 1949; 57:633–638.

Faruque, A.S., Teka, T. and Fuchs, G.J. Shigellosis in children: a clinico-epidemiological comparison between *Shigella dysenteriae* type I and *Shigella flexneri*. Ann. Trop. Paediatr. 1998; 18:197–201.

Fasano, A., Noriega, F.R., Liao, F.M., Wang, W. and Levine, M.M. Effect of shigella enterotoxin 1 (ShET1) on rabbit intestine in vitro and in vivo. Gut 1997; 40:505–511.

Fasano, A., Noriega, F.R., Maneval, D.R., Jr., Chanasongcram, S., Russell, R., Guandalini, S. and Levine, M.M. Shigella enterotoxin 1: an enterotoxin of *Shigella flexneri* 2a active in rabbit small intestine in vivo and in vitro. J. Clin. Invest. 1995; 95: 2853–2861.

Fernandez, M.I. and Sansonetti, P.J. Shigella interaction with intestinal epithelial cells determines the innate immune response in shigellosis. Int. J. Med. Microbiol. 2003; 293:55–67.

Fernandez-Prada, C.M., Hoover, D.L., Tall, B.D. and Venkatesan, M.M. Human monocyte-derived macrophages infected with virulent *Shigella flexneri* in vitro undergo a rapid cytolytic event similar to oncosis but not apoptosis. Infect. Immun. 1997; 65:1486–1496.

Finch, M., Rodey, G., Lawrence, D. and Blake, P. Epidemic Reiter's syndrome following an outbreak of shigellosis. Eur. J. Epidemiol. 1986; 2:26–30.

Formal, S.B., Baron, L.S., Kopecko, D.J., Washington, O., Powell, C. and Life, C.A. Construction of a potential bivalent vaccine strain: introduction of *Shigella sonnei* form I antigen genes into the galE *Salmonella typhi* Ty21a typhoid vaccine strain. Infect. Immun. 1981; 34:746–750.

Formal, S.B., Hale, T.L. and Kapfer, C. Shigella vaccines. Rev. Infect. Dis. 1989; 11(Suppl. 3):S547–S551.

Formal, S.B., Hale, T.L., Kapfer, C., Cogan, J.P., Snoy, P.J., Chung, R., Wingfield, M.E., Elisberg, B.L. and Baron, L.S. Oral vaccination of monkeys with an invasive *Escherichia coli* K-12 hybrid expressing *Shigella flexneri* 2a somatic antigen. Infect. Immun. 1984; 46:465–469.

Formal, S.B., Labrec, E.H., Palmer, A. and Falkow, S. Protection of monkeys against experimental Shigellosis with attenuated vaccines. J. Bacteriol. 1965; 90:63–68.

Francois, M., Le Cabec, V., Dupont, M.A., Sansonetti, P.J. and Maridonneau-Parini, I. Induction of necrosis in human neutrophils by *Shigella flexneri* requires type III secretion, IpaB and

IpaC invasins, and actin polymerization. Infect. Immun. 2000; 68:1289–1296.

Gangarosa, E.J., Perera, D.R., Mata, L.J., Mendizabal-Morris, C., Guzman, G. and Reller, L.B. Epidemic Shiga bacillus dysentery in Central America. II. Epidemiologic studies in 1969. J. Infect. Dis. 1970; 122:181–190.

Garrett, V., Bornschlegel, K., Lange, D., Reddy, V., Kornstein, L., Kornblum, J., Agasan, A., Hoekstra, M., Layton, M. and Sobel, J. A recurring outbreak of *Shigella sonnei* among traditionally observant Jewish children in New York City: the risks of daycare and household transmission. Epidemiol. Infect. 2006; 134:1231–1236.

Girardin, S.E., Tournebize, R., Mavris, M., Page, A.L., Li, X., Stark, G.R., Bertin, J., DiStefano, P.S., Yaniv, M., Sansonetti, P.J. and Philpott, D.J. CARD4/Nod1 mediates NF-kappaB and JNK activation by invasive *Shigella flexneri*. EMBO Rep. 2001; 2:736–742.

Gregory, R.L. Microbial ribosomal vaccines. Rev. Infect. Dis. 1986; 8:208–217.

Grimont, F., Lejay-Collin, M., Talukder, K.A., Carle, I., Issenhuth, S., Le Roux, K. and Grimont, P.A. Identification of a group of shigella-like isolates as *Shigella boydii* 20. J. Med. Microbiol. 2007; 56:749–754.

Gupta, A., Polyak, C.S., Bishop, R.D., Sobel, J. and Mintz, E.D. Laboratory-confirmed shigellosis in the United States, 1989–2002: epidemiologic trends and patterns. Clin. Infect. Dis. 2004; 38:1372–1377.

Haimovic, B. and Venkatesan, M.M. Shigella and Salmonella: death as a means of survival. Microbes Infect. 2006; 8:568–577.

Hale, T.L. Hybrid vaccines using *Escherichia coli* as an antigen carrier. Res. Microbiol. 1990; 141:913–919.

Hartman, A.B. and Venkatesan, M.M. Construction of a stable attenuated *Shigella sonnei* DeltavirG vaccine strain, WRSS1, and protective efficacy and immunogenicity in the guinea pig keratoconjunctivitis model. Infect. Immun. 1998; 66:4572–4576.

Haukka, K. and Siitonen, A. Emerging resistance to newer antimicrobial agents among Shigella isolated from Finnish foreign travellers. Epidemiol. Infect. 2008; 136:476–482.

Hayani, K.C., Guerrero, M.L., Morrow, A.L., Gomez, H.F., Winsor, D.K., Ruiz-Palacios, G.M. and Cleary, T.G. Concentration of milk secretory immunoglobulin A against Shigella virulence plasmid-associated antigens as a predictor of symptom status in Shigella-infected breast-fed infants. J. Pediatr. 1992; 121:852–856.

Henock Blaise, N.Y. and Dovie, D.B. Diarrheal diseases in the history of public health. Arch. Med. Res. 2007; 38:159–163.

Hens, D.K., Niyogi, S.K. and Kumar, R. Epidemic strain *Shigella dysenteriae* Type 1 Dt66 encodes several drug resistances by chromosome. Arch. Med. Res. 2005; 36:399–403.

Herrington, D.A., Van de Verg, L., Formal, S.B., Hale, T.L., Tall, B.D., Cryz, S.J., Tramont, E.C. and Levine, M.M. Studies in volunteers to evaluate candidate Shigella vaccines: further experience with a bivalent *Salmonella typhi–Shigella sonnei* vaccine and protection conferred by previous *Shigella sonnei* disease. Vaccine 1990; 8:353–357.

Huan, P.T., Bastin, D.A., Whittle, B.L., Lindberg, A.A. and Verma, N.K. Molecular characterization of the genes involved in O-antigen modification, attachment, integration and excision in *Shigella flexneri* bacteriophage SfV. Gene 1997; 195:217–227.

Islam, M.M., Azad, A.K., Bardhan, P.K., Raqib, R. and Islam, D. Pathology of shigellosis and its complications. Histopathology 1994; 24:65–71.

Jiang, Y., Yang, F., Zhang, X., Yang, J., Chen, L., Yan, Y., Nie, H., Xiong, Z., Wang, J., Dong, J., Xue, Y., Xu, X., Zhu, Y., Chen, S. and Jin, Q. The complete sequence and analysis of the large virulence plasmid pSS of *Shigella sonnei*. Plasmid 2005; 54:149–159.

Jin, Q., Yuan, Z., Xu, J., Wang, Y., Shen, Y., Lu, W., Wang, J., Liu, H., Yang, J., Yang, F., Zhang, X., Zhang, J., Yang, G., Wu, H., Qu, D., Dong, J., Sun, L., Xue, Y., Zhao, A., Gao, Y., Zhu, J., Kan, B., Ding, K., Chen, S., Cheng, H., Yao, Z., He, B., Chen, R., Ma, D., Qiang, B., Wen, Y., Hou, Y. and Yu, J. Genome sequence of *Shigella flexneri* 2a: insights into pathogenicity through comparison with genomes of *Escherichia coli* K12 and O157. Nucleic Acids Res. 2002; 30:4432–4441.

Jonsson, K.H., Weintraub, A. and Widmalm, G. Structural studies of the O-antigenic polysaccharides from *Shigella dysenteriae* type 3 and *Escherichia coli* O124, a reinvestigation. Carbohydr. Res. 2006; 341:2986–2989.

Kaminski, R.W., Turbyfill, K.R. and Oaks, E.V. Mucosal adjuvant properties of the Shigella invasin complex. Infect. Immun. 2006; 74:2856–2866.

Karmali, M.A. Prospects for preventing serious systemic toxemic complications of Shiga toxin-producing *Escherichia coli* infections using Shiga toxin receptor analogues. J. Infect. Dis. 2004; 189:355–359.

Karnell, A., Li, A., Zhao, C.R., Karlsson, K., Nguyen, B.M. and Lindberg, A.A. Safety and immunogenicity study of the auxotrophic *Shigella flexneri* 2a vaccine SFL1070 with a deleted aroD gene in adult Swedish volunteers. Vaccine 1995; 13:88–99.

Karnell, A., Stocker, B.A., Katakura, S., Reinholt, F.P. and Lindberg, A.A. Live oral auxotrophic *Shigella flexneri* SFL124 vaccine with a deleted aroD gene: characterization and monkey protection studies. Vaccine 1992; 10:389–394.

Katz, D.E., Coster, T.S., Wolf, M.K., Trespalacios, F.C., Cohen, D., Robins, G., Hartman, A.B., Venkatesan, M.M., Taylor, D.N. and Hale, T.L. Two studies evaluating the safety and immunogenicity of a live, attenuated *Shigella flexneri* 2a vaccine (SC602) and excretion of vaccine organisms in North American volunteers. Infect. Immun. 2004; 72:923–930.

Khan, W.A., Seas, C., Dhar, U., Salam, M.A. and Bennish, M.L. Treatment of shigellosis: V. Comparison of azithromycin and ciprofloxacin. A double-blind, randomized, controlled trial. Ann. Intern. Med. 1997; 126:697–703.

Kim, D.W., Lenzen, G., Page, A.L., Legrain, P., Sansonetti, P.J. and Parsot, C. The *Shigella flexneri* effector OspG interferes with innate immune responses by targeting ubiquitin-conjugating enzymes. Proc. Natl. Acad. Sci. U.S.A. 2005; 102:14046–14051.

King, C.K., Glass, R., Bresee, J.S. and Duggan, C. Managing acute gastroenteritis among children: oral rehydration, maintenance, and nutritional therapy. MMWR Recomm. Rep. 2003; 52:1–16.

Kolavic, S.A., Kimura, A., Simons, S.L., Slutsker, L., Barth, S. and Haley, C.E. An outbreak of *Shigella dysenteriae* type 2 among laboratory workers due to intentional food contamination. JAMA 1997; 278:396–398.

Korpela, J., Karpanoja, P., Taipalinen, R. and Siitonen, A. Subtyping of *Shigella sonnei* for tracing nosocomial transmission. J. Hosp. Infect. 1995; 30:261–266.

Kotloff, K.L., Herrington, D.A., Hale, T.L., Newland, J.W., Van De Verg, L., Cogan, J.P., Snoy, P.J., Sadoff, J.C., Formal, S.B. and Levine, M.M. Safety, immunogenicity, and efficacy in monkeys and humans of invasive *Escherichia coli* K-12 hybrid vaccine candidates expressing *Shigella flexneri* 2a somatic antigen. Infect. Immun. 1992; 60:2218–2224.

Kotloff, K.L., Losonsky, G.A., Nataro, J.P., Wasserman, S.S., Hale, T.L., Taylor, D.N., Newland, J.W., Sadoff, J.C., Formal, S.B. and Levine, M.M. Evaluation of the safety, immunogenicity, and efficacy in healthy adults of four doses of live oral hybrid *Escherichia coli-Shigella flexneri* 2a vaccine strain EcSf2a-2. Vaccine 1995b; 13:495–502.

Kotloff, K.L., Nataro, J.P., Losonsky, G.A., Wasserman, S.S., Hale, T.L., Taylor, D.N., Sadoff, J.C. and Levine, M.M. A modified Shigella volunteer challenge model in which the inoculum is administered with bicarbonate buffer: clinical experience and implications for Shigella infectivity. Vaccine 1995a; 13:1488–1494.

Kotloff, K.L., Noriega, F., Losonsky, G.A., Sztein, M.B., Wasserman, S.S., Nataro, J.P. and Levine, M.M. Safety, immunogenicity, and transmissibility in humans of CVD 1203, a live oral *Shigella flexneri* 2a vaccine candidate attenuated by deletions in aroA and virG. Infect. Immun. 1996; 64:4542–4548.

Kotloff, K.L., Noriega, F.R., Samandari, T., Sztein, M.B., Losonsky, G.A., Nataro, J.P., Picking, W.D., Barry, E.M. and Levine, M.M. *Shigella flexneri* 2a strain CVD 1207, with specific deletions in virG, sen, set, and guaBA, is highly attenuated in humans. Infect. Immun. 2000; 68:1034–1039.

Kotloff, K.L., Pasetti, M.F., Barry, E.M., Nataro, J.P., Wasserman, S.S., Sztein, M.B., Picking, W.D. and Levine, M.M. Deletion in the Shigella enterotoxin genes further attenuates *Shigella flexneri* 2a bearing guanine auxotrophy in a phase 1 trial of CVD 1204 and CVD 1208. J. Infect. Dis. 2004; 190:1745–1754.

Kotloff, K.L., Taylor, D.N., Sztein, M.B., Wasserman, S.S., Losonsky, G.A., Nataro, J.P., Venkatesan, M., Hartman, A., Picking, W.D., Katz, D.E., Campbell, J.D., Levine, M.M. and Hale, T.L. Phase I evaluation of delta virG *Shigella sonnei* live, attenuated, oral vaccine strain WRSS1 in healthy adults. Infect. Immun. 2002; 70:2016–2021.

Kotloff, K.L., Winickoff, J.P., Ivanoff, B., Clemens, J.D., Swerdlow, D.L., Sansonetti, P.J., Adak, G.K. and Levine, M.M. Global burden of Shigella infections: implications for vaccine development and implementation of control strategies. Bull. World Health Organ. 1999; 77:651–666.

Lee, H., Kotloff, K., Chukaserm, P., Samosornsuk, S., Chompook, P., Deen, J.L., von Seidlein, L., Clemens, J.D. and Wanpen, C. Shigellosis remains an important problem in children less than 5 years of age in Thailand. Epidemiol. Infect. 2005; 133:469–474.

Legros, D., Paquet, C. and Nabeth, P. Forced Migration and Mortality: The Evolution of Mortality among Rwandan Refugees in Zaire between 1994 and 1997. Washington, DC: National Academies Press, 2001.

Lerman, Y., Yavzori, M., Ambar, R., Sechter, I., Wiener, M. and Cohen, D. Epidemic spread of *Shigella sonnei* shigellosis and evidence for development of immunity among children attending day-care centers in a communal settlement (Kibbutz). J. Clin. Microbiol. 1994; 32:1092–1094.

Levenson, V.I. and Egorova, T.P. Polysaccharide nature of O antigen in protective ribosomal preparations from Shigella: experimental evidence and implications for the ribosomal vaccine concept. Res. Microbiol. 1990; 141:707–720.

Levenson, V.I., Egorova, T.P., Belkin, Z.P., Fedosova, V.G., Subbotina, J.L., Rukhadze, E.Z., Dzhikidze, E.K. and Stassilevich, Z.K. Protective ribosomal preparation from *Shigella sonnei* as a parenteral candidate vaccine. Infect. Immun. 1991; 59:3610–3618.

Levine, M.M., Gangarosa, E.J., Barrow, W.B. and Weiss, C.F. Shigellosis in custodial institutions. V. Effect of intervention with streptomycin-dependent *Shigella sonnei* vaccine in an institution with endemic disease. Am. J. Epidemiol. 1976; 104:88–92.

Levine, M.M., Kotloff, K.L., Barry, E.M., Pasetti, M.F. and Sztein, M.B. Clinical trials of Shigella vaccines: two steps forward and one step back on a long, hard road. Nat. Rev. Microbiol. 2007; 5:540–553.

Levine, M.M., Woodward, W.E., Formal, S.B., Gemski, P., Jr., DuPont, H.L., Hornick, R.B. and Snyder, M.J. Studies with a new generation of oral attenuated shigella vaccine: *Escherichia coli* bearing surface antigens of *Shigella flexneri*. J. Infect. Dis. 1977; 136:577–582.

Li, A., Cam, P.D., Islam, D., Minh, N.B., Huan, P.T., Rong, Z.C., Karlsson, K., Lindberg, G. and Lindberg, A.A. Immune responses in Vietnamese children after a single dose of the auxotrophic, live *Shigella flexneri* Y vaccine strain SFL124. J. Infect. 1994; 28:11–23.

Li, A., Karnell, A., Huan, P.T., Cam, P.D., Minh, N.B., Tram, L.N., Quy, N.P., Trach, D.D., Karlsson, K., Lindberg, G. et al. Safety and immunogenicity of the live oral auxotrophic *Shigella flexneri* SFL124 in adult Vietnamese volunteers. Vaccine 1993; 11:180–189.

Li, A., Pal, T., Forsum, U. and Lindberg, A.A. Safety and immunogenicity of the live oral auxotrophic *Shigella flexneri* SFL124 in volunteers. Vaccine 1992; 10:395–404.

Li, A., Zhao, C.R., Ekwall, E. and Lindberg, A.A. Serum IgG antibody responses to Shigella invasion plasmid-coded antigens detected by immunoblot. Scand. J. Infect. Dis. 1994b; 26:435–445.

Li, H., Xu, H., Zhou, Y., Zhang, J., Long, C., Li, S., Chen, S., Zhou, J.M. and Shao, F. The phosphothreonine lyase activity of a bacterial type III effector family. Science 2007; 315:1000–1003.

Linde, K., Dentchev, V., Bondarenko, V., Marinova, S., Randhagen, B., Bratoyeva, M., Tsvetanov, Y. and Romanova, Y. Live *Shigella flexneri* 2a and *Shigella sonnei* I vaccine candidate strains with two attenuating markers. I. Construction of vaccine candidate strains with retained invasiveness but reduced intracellular multiplication. Vaccine 1990; 8:25–29.

Mata, L.J., Gangarosa, E.J., Caceres, A., Perera, D.R. and Mejicanos, M.L. Epidemic Shiga bacillus dysentery in Central America. I. Etiologic investigations in Guatemala, 1969. J. Infect. Dis. 1970; 122:170–180.

Materu, S.F., Lema, O.E., Mukunza, H.M., Adhiambo, C.G. and Carter, J.Y. Antibiotic resistance pattern of Vibrio cholerae and Shigella causing diarrhoea outbreaks in the eastern Africa region: 1994–1996. East Afr. Med. J. 1997; 74:193–197.

Maurelli, A.T. Black holes, antivirulence genes, and gene inactivation in the evolution of bacterial pathogens. FEMS Microbiol. Lett. 2007; 267:1–8.

McKenzie, R. Presented at the Vaccines for Enteric Diseases, Lisbon, Portugal, 2007.

McKenzie, R., Walker, R.I., Nabors, G.S., Van De Verg, L.L., Carpenter, C., Gomes, G., Forbes, E., Tian, J.H., Yang, H.H., Pace, J.L., Jackson, W.J. and Bourgeois, A.L. Safety and immunogenicity of an oral, inactivated, whole-cell vaccine for *Shigella sonnei*: preclinical studies and a Phase I trial. Vaccine 2006; 24:3735–3745.

Meitert, T., Ciudin, L., Pencu, E., Tonciu, M. and Gheorghe, G. Improvement of production and control methods of live dysentery vaccine Vadizen (*Shigella flexneri* 2a T32-Istrati). Arch. Roum. Pathol. Exp. Microbiol. 1984a; 43:357–367.

Meitert, T., Pencu, E., Ciudin, L. and Tonciu, M. Vaccine strain Sh. flexneri T32-Istrati. Studies in animals and in volunteers. Antidysentery immunoprophylaxis and immunotherapy by live vaccine Vadizen (Sh. flexneri T32-Istrati). Arch. Roum. Pathol. Exp. Microbiol. 1984b; 43:251–278.

Meitert, T., Radu, I., Ciudin, L., Pencu, E. and Tonciu, M. The cellular and humoral immune response after administration to the guinea pig of *Shigella flexneri* 2a T32-Istrati vaccinal strain by different routes. Arch. Roum. Pathol. Exp. Microbiol. 1978; 37:267–276.

Mel, D.M., Arsic, B.L., Nikolic, B.D. and Radovanic, M.L. Studies on vaccination against bacillary dysentery. 4. Oral immunization with live monotypic and combined vaccines. Bull. World Health Organ. 1968; 39:375–380.

Mel, D., Gangarosa, E.J., Radovanovic, M.L., Arsic, B.L. and Litvinjenko, S. Studies on vaccination against bacillary dysentery. 6. Protection of children by oral immunization with streptomycin-dependent Shigella strains. Bull. World Health Organ. 1971; 45:457–464.

Mel, D.M., Papo, R.G., Terzin, A.L. and Vuksic, L. Studies on vaccination against bacillary dysentery. 2. Safety tests and reactogenicity studies on a live dysentery vaccine intended for use in field trials. Bull. World Health Organ. 1965a; 32:637–645.

Mel, D.M., Terzin, A.L. and Vuksic, L. Studies on vaccination against bacillary dysentery. 3. Effective oral immunization against *Shigella flexneri* 2a in a field trial. Bull. World Health Organ. 1965b; 32:647–655.

Melito, P.L., Woodward, D.L., Munro, J., Walsh, J., Foster, R., Tilley, P., Paccagnella, A., Isaac-Renton, J., Ismail, J. and Ng, L.K. A novel *Shigella dysenteriae* serovar isolated in Canada. J. Clin. Microbiol. 2005; 43:740–744.

Monteville, M.R., Riddle, M.S., Baht, U., Putnam, S.D., Frenck, R.W., Brooks, K., Moustafa, M., Bland, J. and Sanders, J.W. Incidence, etiology, and impact of diarrhea among deployed US military personnel in support of Operation Iraqi Freedom and Operation Enduring Freedom. Am. J. Trop. Med. Hyg. 2006; 75:762–767.

Morgan, O., Crook, P., Cheasty, T., Jiggle, B., Giraudon, I., Hughes, H. and Jones, S.M. *Shigella sonnei* outbreak among homosexual men, London. Emerg. Infect. Dis. 2006; 12:1458–1460.

Munoz, C., Baqar, S., van de Verg, L., Thupari, J., Goldblum, S., Olson, J.G., Taylor, D.N., Heresi, G.P. and Murphy, J.R. Characteristics of *Shigella sonnei* infection of volunteers: signs, symptoms, immune responses, changes in selected cytokines and acute-phase substances. Am. J. Trop. Med. Hyg. 1995; 53:47–54.

Murphy, G.S., Bodhidatta, L., Echeverria, P., Tansuphaswadikul, S., Hoge, C.W., Imlarp, S. and Tamura, K. Ciprofloxacin and loperamide in the treatment of bacillary dysentery. Ann. Intern. Med. 1993; 118:582–586.

Nataro, J.P., Seriwatana, J., Fasano, A., Maneval, D.R., Guers, L.D., Noriega, F., Dubovsky, F., Levine, M.M. and Morris, J.G., Jr. Identification and cloning of a novel plasmid-encoded enterotoxin of enteroinvasive *Escherichia coli* and Shigella strains. Infect. Immun. 1995; 63:4721–4728.

Na-Ubol, M., Samosornsuk, S., Von Seidlein, L., Tapchaisri, P., Ali, M., Clemens, J.D. and Chaicumpa, W. Molecular characteristics of Shigella spp. isolated from patients with diarrhoea in a new industrialized area of Thailand. Epidemiol. Infect. 2006; 134:997–1003.

Navia, M.M., Capitano, L., Ruiz, J., Vargas, M., Urassa, H., Schellemberg, D., Gascon, J. and Vila, J. Typing and characterization of mechanisms of resistance of Shigella spp. isolated from feces of children under 5 years of age from Ifakara, Tanzania. J. Clin. Microbiol. 1999; 37:3113–3117.

Nguyen, B.M., Lanata, C.F., Black, R.E., Gil, A.I., Karnell, A. and Wretlind, B. Age-related prevalence of Shigella and Salmonella antibodies and their association with diarrhoeal diseases in Peruvian children. Scand. J. Infect. Dis. 1998; 30:159–164.

Noriega, F.R., Losonsky, G., Wang, J.Y., Formal, S.B. and Levine, M.M. Further characterization of delta aroA delta virG *Shigella flexneri* 2a strain CVD 1203 as a mucosal Shigella vaccine and as a live-vector vaccine for delivering antigens of enterotoxigenic *Escherichia coli*. Infect. Immun. 1996; 64:23–27.

Noriega, F.R., Wang, J.Y., Losonsky, G., Maneval, D.R., Hone, D.M. and Levine, M.M. Construction and characterization of attenuated delta aroA delta virG *Shigella flexneri* 2a strain CVD 1203, a prototype live oral vaccine. Infect. Immun. 1994; 62:5168–5172.

Oaks, E.V., Hale, T.L. and Formal, S.B. Serum immune response to Shigella protein antigens in rhesus monkeys and humans infected with Shigella spp. Infect. Immun. 1986; 53:57–63.

Oaks, E.V., Picking, W.D. and Picking, W.L. Antibody response of monkeys to invasion plasmid antigen D after infection with Shigella spp. Clin. Diagn. Lab. Immunol. 1996; 3:242–245.

Oaks, E.V. and Turbyfill, K.R. Development and evaluation of a *Shigella flexneri* 2a and *S. sonnei* bivalent invasin complex (Invaplex) vaccine. Vaccine 2006; 24:2290–2301.

Oberhelman, R.A., Kopecko, D.J., Salazar-Lindo, E., Gotuzzo, E., Buysse, J.M., Venkatesan, M.M., Yi, A., Fernandez-Prada, C., Guzman, M., Leon-Barua, R. et al. Prospective study of systemic and mucosal immune responses in dysenteric patients to specific Shigella invasion plasmid antigens and lipopolysaccharides. Infect. Immun. 1991; 59:2341–2350.

O'Brien, A.D., Tesh, V.L., Donohue-Rolfe, A., Jackson, M.P., Olsnes, S., Sandvig, K., Lindberg, A.A. and Keusch, G.T. Shiga toxin: biochemistry, genetics, mode of action, and role in pathogenesis. Curr. Top. Microbiol. Immunol. 1992; 180:65–94.

Ochman, H., Whittam, T.S., Caugant, D.A. and Selander, R.K. Enzyme polymorphism and genetic population structure in *Escherichia coli* and Shigella. J. Gen. Microbiol. 1983; 129:2715–2726.

Ogawa, M. and Sasakawa, C. Intracellular survival of Shigella. Cell. Microbiol. 2006; 8:177–184.

Okuda, J., Toyotome, T., Kataoka, N., Ohno, M., Abe, H., Shimura, Y., Seyedarabi, A., Pickersgill, R. and Sasakawa, C. Shigella effector IpaH9.8 binds to a splicing factor U2AF(35) to modulate host immune responses. Biochem. Biophys. Res. Commun. 2005; 333:531–539.

Orr, N., Katz, D.E., Atsmon, J., Radu, P., Yavzori, M., Halperin, T., Sela, T., Kayouf, R., Klein, Z., Ambar, R., Cohen, D., Wolf, M.K., Venkatesan, M.M. and Hale, T.L. Community-based safety, immunogenicity, and transmissibility study of the *Shigella sonnei* WRSS1 vaccine in Israeli volunteers. Infect. Immun. 2005; 73: 8027–8032.

Pan, J.C., Ye, R., Meng, D.M., Zhang, W., Wang, H.Q. and Liu, K.Z. Molecular characteristics of class 1 and class 2 integrons and their relationships to antibiotic resistance in clinical isolates of *Shigella sonnei* and *Shigella flexneri*. J. Antimicrob. Chemother. 2006; 58:288–296.

Parsot, C. Shigella spp. and enteroinvasive *Escherichia coli* pathogenicity factors. FEMS Microbiol. Lett. 2005; 252:11–18.

Parsot, C., Ageron, E., Penno, C., Mavris, M., Jamoussi, K., d'Hauteville, H., Sansonetti, P. and Demers, B. A secreted antiactivator, OspD1, and its chaperone, Spa15, are involved in the control of transcription by the type III secretion apparatus activity in *Shigella flexneri*. Mol. Microbiol. 2005; 56:1627–1635.

Passwell, J.H., Ashkenazi, S., Harlev, E., Miron, D., Ramon, R., Farzam, N., Lerner-Geva, L., Levi, Y., Chu, C., Shiloach, J., Robbins, J.B. and Schneerson, R. Safety and immunogenicity of *Shigella sonnei*-CRM9 and *Shigella flexneri* type 2a-rEPAsucc conjugate vaccines in one- to four-year-old children. Pediatr. Infect. Dis. J. 2003; 22:701–706.

Passwell, J.H., Harlev, E., Ashkenazi, S., Chu, C., Miron, D., Ramon, R., Farzan, N., Shiloach, J., Bryla, D.A., Majadly, F., Roberson, R., Robbins, J.B. and Schneerson, R. Safety and immunogenicity of improved Shigella O-specific polysaccharide-protein conjugate vaccines in adults in Israel. Infect. Immun. 2001; 69: 1351–1357.

Paton, J.C., Rogers, T.J., Morona, R. and Paton, A.W. Oral administration of formaldehyde-killed recombinant bacteria expressing a mimic of the Shiga toxin receptor protects mice from fatal challenge with Shiga-toxigenic *Escherichia coli*. Infect. Immun. 2001; 69:1389–1393.

Pazhani, G.P., Ramamurthy, T., Mitra, U., Bhattacharya, S.K. and Niyogi, S.K. Species diversity and antimicrobial resistance of Shigella spp. isolated between 2001 and 2004 from hospitalized children with diarrhoea in Kolkata (Calcutta), India. Epidemiol. Infect. 2005; 133:1089–1095.

Penatti, M.P., Hollanda, L.M., Nakazato, G., Campos, T.A., Lancellotti, M., Angellini, M., Brocchi, M., Rocha, M.M. and Dias da Silveira, W. Epidemiological characterization of resistance and PCR typing of *Shigella flexneri* and *Shigella sonnei* strains isolated from bacillary dysentery cases in Southeast Brazil. Braz. J. Med. Biol. Res. 2007; 40:249–258.

Perepelov, A.V., Liu, B., Senchenkova, S., N., Shashkov, A.S., Feng, L., Knirel, Y.A. and Wang, L. Close relation of the O-polysaccharide structure of *Escherichia coli* O168 and revised structure of the O-polysaccharide of *Shigella dysenteriae* type 4. Carbohydr. Res. 2007; 342:2676–2681.

Philpott, D.J., Yamaoka, S., Israel, A. and Sansonetti, P.J. Invasive *Shigella flexneri* activates NF-kappa B through a lipopolysaccharide-dependent innate intracellular response and leads to IL-8 expression in epithelial cells. J. Immunol. 2000; 165:903–914.

Picking, W.L., Nishioka, H., Hearn, P.D., Baxter, M.A., Harrington, A.T., Blocker, A. and Picking, W.D. IpaD of *Shigella flexneri* is independently required for regulation of Ipa protein secretion and efficient insertion of IpaB and IpaC into host membranes. Infect. Immun. 2005; 73:1432–1440.

Pope, L.M., Reed, K.E. and Payne, S.M. Increased protein secretion and adherence to HeLa cells by Shigella spp. following growth in the presence of bile salts. Infect. Immun. 1995; 63:3642–3648.

Pupo, G.M., Lan, R. and Reeves, P.R. Multiple independent origins of Shigella clones of *Escherichia coli* and convergent evolution of many of their characteristics. Proc. Natl. Acad. Sci. U.S.A. 2000; 97:10567–10572.

Rahaman, M.M., Khan, M.M., Aziz, K.M., Islam, M.S. and Kibriya, A.K. An outbreak of dysentery caused by *Shigella dysenteriae* type 1 on a coral island in the Bay of Bengal. J. Infect. Dis. 1975; 132:15–19.

Ram, P.K., Crump, J.A., Gupta, S.K., Miller, M.A. and Mintz, E.D. Part II. Analysis of data gaps pertaining to Shigella infections in low and medium human development index countries, 1984–2005. Epidemiol. Infect. 2008; 136:577–603.

Ranallo, R. Presented at the Vaccines for Enteric Diseases, Lisbon, Portugal, 2007.

Ranallo, R.T., Barnoy, S., Thakkar, S., Urick, T. and Venkatesan, M.M. Developing live Shigella vaccines using lambda Red recombineering. FEMS Immunol. Med. Microbiol. 2006; 47:462–469.

Ranallo, R.T., Thakkar, S., Chen, Q. and Venkatesan, M.M. Immunogenicity and characterization of WRSF2G11: a second generation live attenuated *Shigella flexneri* 2a vaccine strain. Vaccine 2007; 25:2269–2278.

Raqib, R., Lindberg, A.A., Bjork, L., Bardhan, P.K., Wretlind, B., Andersson, U. and Andersson, J. Down-regulation of gamma interferon, tumor necrosis factor type I, interleukin 1 (IL-1) type I, IL-3, IL-4, and transforming growth factor beta type I receptors at the local site during the acute phase of Shigella infection. Infect. Immun. 1995a; 63:3079–3087.

Raqib, R., Lindberg, A.A., Wretlind, B., Bardhan, P.K., Andersson, U. and Andersson, J. Persistence of local cytokine production in shigellosis in acute and convalescent stages. Infect. Immun. 1995b; 63:289–296.

Raqib, R., Reinholt, F.P., Bardhan, P.K., Karnell, A. and Lindberg, A.A. Immunopathological patterns in the rectal mucosa of patients with shigellosis: expression of HLA-DR antigens and T-lymphocyte subsets. APMIS 1994; 102:371–380.

Raqib, R., Sarker, P., Bergman, P., Ara, G., Lindh, M., Sack, D.A., Nasirul Islam, K.M., Gudmundsson, G.H., Andersson, J. and Agerberth, B. Improved outcome in shigellosis associated with butyrate induction of an endogenous peptide antibiotic. Proc. Natl. Acad. Sci. U.S.A. 2006; 103:9178–9183.

Raqib, R., Wretlind, B., Andersson, J. and Lindberg, A.A. Cytokine secretion in acute shigellosis is correlated to disease activity and directed more to stool than to plasma. J. Infect. Dis. 1995c; 171:376–384.

Replogle, M.L., Fleming, D.W. and Cieslak, P.R. Emergence of antimicrobial-resistant shigellosis in Oregon. Clin. Infect. Dis. 2000; 30:515–519.

Riddle, M.S., Sanders, J.W., Putnam, S.D. and Tribble, D.R. Incidence, etiology, and impact of diarrhea among long-term travelers (US military and similar populations): a systematic review. Am. J. Trop. Med. Hyg. 2006; 74:891–900.

Riley, M. and Anilionis, A. Conservation and variation of nucleotide sequences within related bacterial genomes: enterobacteria. J. Bacteriol. 1980; 143:366–376.

Robbins, J.B., Chu, C. and Schneerson, R. Hypothesis for vaccine development: protective immunity to enteric diseases caused by nontyphoidal salmonellae and shigellae may be conferred by serum IgG antibodies to the O-specific polysaccharide of their lipopolysaccharides. Clin. Infect. Dis. 1992; 15:346–361.

Roberts, F., Jennison, A.V. and Verma, N.K. The *Shigella flexneri* serotype Y vaccine candidate SFL124 originated from a serotype 2a background. FEMS Immunol. Med. Microbiol. 2005; 45:285–289.

Roy, S.K., Raqib, R., Khatun, W., Azim, T., Chowdhury, R., Fuchs, G.J. and Sack, D.A. Zinc supplementation in the management of shigellosis in malnourished children in Bangladesh. Eur. J. Clin. Nutr. 2008; 62:849–855.

Runyen-Janecky, L.J., Reeves, S.A., Gonzales, E.G. and Payne, S.M. Contribution of the *Shigella flexneri* Sit, Iuc, and Feo iron acquisition systems to iron acquisition in vitro and in cultured cells. Infect. Immun. 2003; 71:1919–1928.

Sakdisiwasdi, O., Achananuparp, S., Limsuwan, A., Nanna, P. and Barnyen, L. Salmonella and Shigella carrier rates and environmental sanitation in a rural district, Central Thailand. Southeast Asian J. Trop. Med. Public Health 1982; 13:380–384.

Samandari, T., Kotloff, K.L., Losonsky, G.A., Picking, W.D., Sansonetti, P.J., Levine, M.M. and Sztein, M.B. Production of IFN-gamma and IL-10 to Shigella invasins by mononuclear cells from volunteers orally inoculated with a Shiga toxin-deleted *Shigella dysenteriae* type 1 strain. J. Immunol. 2000; 164:2221–2232.

Sansonetti, P.J. The bacterial weaponry: lessons from Shigella. Ann. N.Y. Acad. Sci. 2006; 1072:307–312.

Sansonetti, P.J., Kopecko, D.J. and Formal, S.B. *Shigella sonnei* plasmids: evidence that a large plasmid is necessary for virulence. Infect. Immun. 1981; 34:75–83.

Sansonetti, P.J., Kopecko, D.J. and Formal, S.B. Involvement of a plasmid in the invasive ability of *Shigella flexneri*. Infect. Immun. 1982; 35:852–860.

Seid, R.C., Jr., Kopecko, D.J., Sadoff, J.C., Schneider, H., Baron, L.S. and Formal, S.B. Unusual lipopolysaccharide antigens of a *Salmonella typhi* oral vaccine strain expressing the *Shigella sonnei* form I antigen. J. Biol. Chem. 1984; 259:9028–9034.

Sheoran, A.S., Chapman-Bonofiglio, S., Harvey, B.R., Mukherjee, J., Georgiou, G., Donohue-Rolfe, A. and Tzipori, S. Human antibody against shiga toxin 2 administered to piglets after the onset of diarrhea due to *Escherichia coli* O157:H7 prevents fatal systemic complications. Infect. Immun. 2005; 73:4607–4613.

Shim, D.H., Chang, S.Y., Park, S.M., Jang, H., Carbis, R., Czerkinsky, C., Uematsu, S., Akira, S. and Kweon, M.N. Immunogenicity and protective efficacy offered by a ribosomal-based vaccine from *Shigella flexneri* 2a. Vaccine 2007; 25:4828–4836.

Shmuely, H., Samra, Z., Ashkenazi, S., Dinari, G., Chodick, G. and Yahav, J. Association of Helicobacter pylori infection with Shigella gastroenteritis in young children. Am. J. Gastroenterol. 2004; 99:2041–2045.

Simchen, E., Jeeraphat, S., Shihab, S. and Fattal, B. An epidemic of waterborne Shigella gastroenteritis in Kibbutzim of western Galilee in Israel. Int. J. Epidemiol. 1991; 20:1081–1087.

Simmons, D.A. and Romanowska, E. Structure and biology of *Shigella flexneri* O antigens. J. Med. Microbiol. 1987; 23: 289–302.

Sivapalasingam, S., Nelson, J.M., Joyce, K., Hoekstra, M., Angulo, F.J. and Mintz, E.D. High prevalence of antimicrobial resistance among Shigella isolates in the United States tested by the National Antimicrobial Resistance Monitoring System from 1999 to 2002. Antimicrob. Agents Chemother. 2006; 50:49–54.

Smith, M.J., Carvalho, H.M., Melton-Celsa, A.R. and O'Brien, A.D. The 13C4 monoclonal antibody that neutralizes Shiga toxin Type 1 (Stx1) recognizes three regions on the Stx1 B subunit and prevents Stx1 from binding to its eukaryotic receptor globotriaosylceramide. Infect. Immun. 2006; 74:6992–6998.

Somerville, J.E., Jr., Cassiano, L. and Darveau, R.P. *Escherichia coli* msbB gene as a virulence factor and a therapeutic target. Infect. Immun. 1999; 67:6583–6590.

Sur, D. and Bhattacharya, S.K. Acute diarrhoeal diseases—an approach to management. J. Indian Med. Assoc. 2006; 104:220–223.

Sur, D., Ramamurthy, T., Deen, J. and Bhattacharya, S.K. Shigellosis: challenges and management issues. Indian J. Med. Res. 2004; 120:454–462.

Suzuki, T., Franchi, L., Toma, C., Ashida, H., Ogawa, M., Yoshikawa, Y., Mimuro, H., Inohara, N., Sasakawa, C. and Nunez, G. Differential regulation of caspase-1 activation, pyroptosis, and autophagy via Ipaf and ASC in Shigella-infected macrophages. PLoS Pathog. 2007; 3:e111.

Suzuki, T. and Sasakawa, C. Molecular basis of the intracellular spreading of Shigella. Infect. Immun. 2001; 69:5959–5966.

Talukder, K.A., Islam, M.A., Dutta, D.K., Hassan, F., Safa, A., Nair, G.B. and Sack, D.A. Phenotypic and genotypic characterization of serologically atypical strains of *Shigella flexneri* type 4 isolated in Dhaka, Bangladesh. J. Clin. Microbiol. 2002; 40:2490–2497.

Talukder, K.A., Islam, M.A., Khajanchi, B.K., Dutta, D.K., Islam, Z., Safa, A., Alam, K., Hossain, A., Nair, G.B. and Sack, D.A. Temporal shifts in the dominance of serotypes of *Shigella dysenteriae* from 1999 to 2002 in Dhaka, Bangladesh. J. Clin. Microbiol. 2003; 41:5053–5058.

Talukder, K.A., Khajanchi, B.K., Islam, M.A., Dutta, D.K., Islam, Z., Khan, S.I., Nair, G.B. and Sack, D.A. The emerging strains of *Shigella dysenteriae* type 2 in Bangladesh are clonal. Epidemiol. Infect. 2006; 134:1249–1256.

Talukder, K.A., Mondol, A.S., Islam, M.A., Islam, Z., Dutta, D.K., Khajanchi, B.K., Azmi, I.J., Hossain, M.A., Rahman, M., Cheasty, T., Cravioto, A., Nair, G.B. and Sack, D.A. A novel serovar of *Shigella dysenteriae* from patients with diarrhoea in Bangladesh. J. Med. Microbiol. 2007; 56:654–658.

Tamura, J.T. A cell-free polyvalent Shigella vaccine. J. Infect. Dis. 1967; 117:353–359.

Taneja, N., Lyngdoh, V., Vermani, A., Mohan, B., Rao, P., Singh, M., Dogra, A., Singh, M.P. and Sharma, M. Re-emergence of multidrug resistant *Shigella dysenteriae* with added resistance to ciprofloxacin in north India and their plasmid profiles. Indian J. Med. Res. 2005; 122:348–354.

Taylor, D.N., Bourgeois, A.L., Ericsson, C.D., Steffen, R., Jiang, Z.D., Halpern, J., Haake, R. and Dupont, H.L. A randomized, double-blind, multicenter study of rifaximin compared with placebo and with ciprofloxacin in the treatment of travelers' diarrhea. Am. J. Trop. Med. Hyg. 2006a; 74:1060–1066.

Taylor, D.N., McKenzie, R., Durbin, A., Carpenter, C., Atzinger, C.B., Haake, R. and Bourgeois, A.L. Rifaximin, a nonabsorbed oral antibiotic, prevents shigellosis after experimental challenge. Clin. Infect. Dis. 2006b; 42:1283–1288.

Taylor, D.N., Trofa, A.C., Sadoff, J., Chu, C., Bryla, D., Shiloach, J., Cohen, D., Ashkenazi, S., Lerman, Y., Egan, W. et al. Synthesis, characterization, and clinical evaluation of conjugate vaccines composed of the O-specific polysaccharides of *Shigella dysenteriae* type 1, *Shigella flexneri* type 2a, and *Shigella sonnei* (*Plesiomonas shigelloides*) bound to bacterial toxoids. Infect. Immun. 1993; 61:3678–3687.

Tesh, V.L. and O'Brien, A.D. The pathogenic mechanisms of Shiga toxin and the Shiga-like toxins. Mol. Microbiol. 1991; 5:1817–1822.

Toro, C.S., Farfan, M., Contreras, I., Flores, O., Navarro, N., Mora, G.C. and Prado, V. Genetic analysis of antibiotic-resistance determinants in multidrug-resistant Shigella strains isolated from Chilean children. Epidemiol. Infect. 2005; 133:81–86.

Tramont, E.C., Chung, R., Berman, S., Keren, D., Kapfer, C. and Formal, S.B. Safety and antigenicity of typhoid-*Shigella sonnei* vaccine (strain 5076-1C). J. Infect. Dis. 1984; 149:133–136.

Tran Van Nhieu, G., Bourdet-Sicard, R., Dumenil, G., Blocker, A. and Sansonetti, P.J. Bacterial signals and cell responses during Shigella entry into epithelial cells. Cell. Microbiol. 2000; 2:187–193.

Tribble, D. Presented at the Vaccines for Enteric Diseases, Lisbon, Portugal, 2007.

Tully, J.G. and Tamura, J.T. Preparation, characterization, and antigenicity of a saline soluble fraction from Shigella types. Proc. Soc. Exp. Biol. Med. 1954; 86:612–616.

Turbyfill, K.R., Hartman, A.B. and Oaks, E.V. Isolation and characterization of a *Shigella flexneri* invasin complex subunit vaccine. Infect. Immun. 2000; 68:6624–6632.

Turbyfill, K.R., Kaminski, R.W. and Oaks, E.V. Immunogenicity and efficacy of highly purified invasion complex vaccine from *Shigella flexneri* 2a. Vaccine 2008; 26:1353–1364.

Tzipori, S., Sheoran, A., Akiyoshi, D., Donohue-Rolfe, A. and Trachtman, H. Antibody therapy in the management of shiga toxin-induced hemolytic uremic syndrome. Clin. Microbiol. Rev. 2004; 17:926–941, table of contents.

Valvano, M.A. and Marolda, C.L. Relatedness of O-specific lipopolysaccharide side chain genes from strains of *Shigella boydii* type 12 belonging to two clonal groups and from *Escherichia coli* O7: K1. Infect. Immun. 1991; 59:3917–3923.

Van de Verg, L., Herrington, D.A., Murphy, J.R., Wasserman, S.S., Formal, S.B. and Levine, M.M. Specific immunoglobulin A-secreting cells in peripheral blood of humans following oral immunization with a bivalent *Salmonella typhi–Shigella sonnei* vaccine or infection by pathogenic *S. sonnei*. Infect. Immun. 1990; 58:2002–2004.

Venkatesan, M., Fernandez-Prada, C., Buysse, J.M., Formal, S.B. and Hale, T.L. Virulence phenotype and genetic characteristics of the T32-ISTRATI *Shigella flexneri* 2a vaccine strain. Vaccine 1991; 9:358–363.

Venkatesan, M.M., Goldberg, M.B., Rose, D.J., Grotbeck, E.J., Burland, V. and Blattner, F.R. Complete DNA sequence and analysis of the large virulence plasmid of *Shigella flexneri*. Infect. Immun. 2001; 69:3271–3285.

Venkatesan, M.M., Hartman, A.B., Newland, J.W., Ivanova, V.S., Hale, T.L., McDonough, M. and Butterton, J. Construction, characterization, and animal testing of WRSd1, a *Shigella dysenteriae* 1 vaccine. Infect. Immun. 2002; 70:2950–2958.

Venkatesan, M.M. and Ranallo, R.T. Live-attenuated Shigella vaccines. Expert Rev. Vaccines 2006; 5:669–686.

Verma, N.K., Brandt, J.M., Verma, D.J. and Lindberg, A.A. Molecular characterization of the O-acetyl transferase gene of converting bacteriophage SF6 that adds group antigen 6 to *Shigella flexneri*. Mol. Microbiol. 1991; 5:71–75.

Viala, J., Sansonetti, P. and Philpott, D.J. Nods and 'intracellular' innate immunity. C. R. Biol. 2004; 327:551–555.

von Seidlein, L., Kim, D.R., Ali, M., Lee, H., Wang, X., Thiem, V.D., Canh do, G., Chaicumpa, W., Agtini, M.D., Hossain, A., Bhutta, Z.A., Mason, C., Sethabutr, O., Talukder, K., Nair, G. B., Deen, J.L., Kotloff, K. and Clemens, J. A multicentre study of Shigella diarrhoea in six Asian countries: disease burden, clinical manifestations, and microbiology. PLoS Med. 2006; 3:e353.

Wang, L., Qu, W. and Reeves, P.R. Sequence analysis of four *Shigella boydii* O-antigen loci: implication for *Escherichia coli* and Shigella relationships. Infect. Immun. 2001; 69:6923–6930.

Way, S.S., Borczuk, A.C., Dominitz, R. and Goldberg, M.B. An essential role for gamma interferon in innate resistance to *Shigella flexneri* infection. Infect. Immun. 1998; 66:1342–1348.

Wei, J., Goldberg, M.B., Burland, V., Venkatesan, M.M., Deng, W., Fournier, G., Mayhew, G.F., Plunkett, G., III, Rose, D.J., Darling, A., Mau, B., Perna, N.T., Payne, S.M., Runyen-Janecky, L.J., Zhou, S., Schwartz, D.C. and Blattner, F.R. Complete genome sequence and comparative genomics of *Shigella flexneri* serotype 2a strain 2457T. Infect. Immun. 2003; 71:2775–2786.

Wen, S.X., Teel, L.D., Judge, N.A. and O'Brien, A.D. A plant-based oral vaccine to protect against systemic intoxication by Shiga toxin type 2. Proc. Natl. Acad. Sci. U.S.A. 2006a; 103:7082–7087.

Wen, S.X., Teel, L.D., Judge, N.A. and O'Brien, A.D. Genetic toxoids of Shiga toxin types 1 and 2 protect mice against homologous but not heterologous toxin challenge. Vaccine 2006b; 24:1142–1148.

WHO Terrorist Threats To Food: Guidance for Establishing and Strengthening Prevention and Response Systems. Geneva: World Health Organization, 2002.

Wing, H.J., Goldman, S.R., Ally, S. and Goldberg, M.B. Modulation of an outer membrane protease contributes to the virulence defect of *Shigella flexneri* strains carrying a mutation in the virK locus. Infect. Immun. 2005; 73:1217–1220.

Xu, D.Q., Cisar, J.O., Ambulos, N., Jr., Burr, D.H. and Kopecko, D.J. Molecular cloning and characterization of genes for *Shigella sonnei* form I O polysaccharide: proposed biosynthetic pathway and stable expression in a live Salmonella vaccine vector. Infect. Immun. 2002; 70:4414–4423.

Xu de, Q., Cisar, J.O., Osorio, M., Wai, T.T. and Kopecko, D.J. Core-linked LPS expression of *Shigella dysenteriae* serotype 1 O-antigen in live Salmonella Typhi vaccine vector Ty21a: preclinical evidence of immunogenicity and protection. Vaccine 2007; 25:6167–6175.

Yang, F., Yang, J., Zhang, X., Chen, L., Jiang, Y., Yan, Y., Tang, X., Wang, J., Xiong, Z., Dong, J., Xue, Y., Zhu, Y., Xu, X., Sun, L., Chen, S., Nie, H., Peng, J., Xu, J., Wang, Y., Yuan, Z., Wen, Y., Yao, Z., Shen, Y., Qiang, B., Hou, Y., Yu, J. and Jin, Q. Genome dynamics and diversity of Shigella species, the etiologic agents of bacillary dysentery. Nucleic Acids Res. 2005; 33:6445–6458.

Yuhas, Y., Nofech-Mozes, Y., Weizman, A. and Ashkenazi, S. Enhancement of pentylenetetrazole-induced seizures by *Shigella dysenteriae* in LPS-resistant C3H/HeJ mice: role of the host response. Med. Microbiol. Immunol. 2002; 190:173–178.

Yuhas, Y., Weizman, A., Chrousos, G.P., Ovadia, H. and Ashkenazi, S. Involvement of the neuropeptide corticotropin-releasing hormone in an animal model of Shigella-related seizures. J. Neuroimmunol. 2004; 153:36–39.

Zafar, A., Sabir, N. and Bhutta, Z.A. Frequency of isolation of Shigella serogroups/serotypes and their antimicrobial susceptibility pattern in children from slum areas in Karachi. J. Pak. Med. Assoc. 2005; 55:184–188.

CHAPTER

59

Syphilis

Lorenzo Giacani, Sheila Lukehart and Arturo Centurion-Lara
Department of Medicine, School of Medicine, University of Washington, Seattle, WA, USA

OUTLINE

ABSTRACT

Syphilis continues to be a common and serious disease, affecting at least 25 million persons worldwide. It is a recognized cofactor in the transmission and acquisition of human immunodeficiency virus, and is a major cause of stillbirth and perinatal morbidity in the developing world. Despite the availability of safe, effective, and inexpensive diagnostic tools and treatment, syphilis is still uncontrolled or has reemerged in many areas of the world, thus demonstrating the failure of conventional efforts to control transmission and emphasizing the need for an effective and practical vaccine. The peculiar biology of *Treponema pallidum*, along with the inability to continuously grow this pathogen in vitro, has hindered progress in this field. The public health importance of syphilis prompted numerous vaccination attempts with whole cell vaccines in the mid-1900s. Only one protective vaccine has been described which, though completely protective in the rabbit model, cannot be translated for human use. The availability of new tools such as genomics, proteomics, and recombinant DNA technology permits today's investigators to focus on the identification of surface-exposed antigens and other virulence factors to target their vaccine efforts. The combination of modern in silico analysis tools and biological assays has led to the identification of several promising candidate antigens, including the members of the Tpr family, newly identified adhesin molecules, lipoproteins, and other putative outer membrane proteins. While early subunit vaccine studies have not yielded complete protection, a number of antigens have induced sufficient immunity to significantly alter disease progression following infectious challenge. These encouraging results serve as the basis for a renewed interest in syphilis vaccine development.

Syphilis has long been a topic of controversy, consternation, and fascination by historians who argue about its origin, physicians who marvel and despair at its protean clinical manifestations, and scientists who struggle to unravel the secrets of its causative agent *Treponema pallidum* subsp. *pallidum* (*T. pallidum*). Although effective therapy with penicillin has significantly decreased the burden of syphilis in developed countries, this infection still results in significant morbidity and mortality, especially in the developing world. The seeming dichotomy of a fragile and metabolically crippled bacterium that is able to cause lifelong and potentially destructive infection has yet to be resolved. This chapter reviews our current understanding of *T. pallidum* and the disease process, then focuses on efforts to develop a practical and effective vaccine.

HISTORY

The first definitive clinical descriptions of syphilis resulted from the outbreak of an apparently new disease in Italy following the Siege of Naples by the French in 1495. The sexually transmissible nature of the infection was recognized early, and the infection was dubbed "the French Disease" and later the "Great Pox" (to distinguish it from smallpox). Because of the coincidence of this outbreak with the return of Columbus and his sailors from the New World, it has been speculated that syphilis was brought to Europe by these men. Others believe that syphilis had existed in Europe and Asia before 1492, and that the sudden widespread recognition of the disease was the result of increased movement of armies throughout Europe and the greater dissemination of information at the end of the Middle Ages.

Regardless of its origin, syphilis became a common disease, with an estimated ~10% of the population of Europe and the United States being infected in the early 20th century. Before the discovery of penicillin, fear of the destructive late manifestations of syphilis, including insanity, paralysis, blindness, and death, led to the use of a number of dangerous treatments, including the use of hotboxes to raise the body temperature, fever induced by intentional malaria infection, and mercury salves and vapors.

ETIOLOGIC AGENT

Treponema pallidum is a spiral organism with a structure similar to gram-negative bacteria. The organism has both an outer membrane (OM) and a cytoplasmic membrane, with a periplasmic space containing the periplasmic (or endo-) flagella (Fig. 59.1). The lipid-rich OM of *T. pallidum* has an unusually small number of integral membrane proteins (Radolf et al., 1989b; Walker et al., 1989) and is very fragile, complicating physical and biochemical analyses of surface structures. At this writing, there is no consensus about the identity of any surface-exposed protein on the organism. The genome sequence of *T. pallidum* (Fraser et al., 1998) revealed an organism with significant metabolic limitations, including the inability to synthesize lipids, nucleotides, and most amino acids de novo; a lack of Kreb's cycle enzymes; and no evidence for oxidative phosphorylation. This renders the organism highly dependent on the host for nutrients, and efforts to propagate *T. pallidum* in vitro in continuous culture have been unsuccessful.

CLASSIFICATION

T. pallidum is one of four recognized human pathogens within the genus *Treponema*: *T. pallidum* subsp.

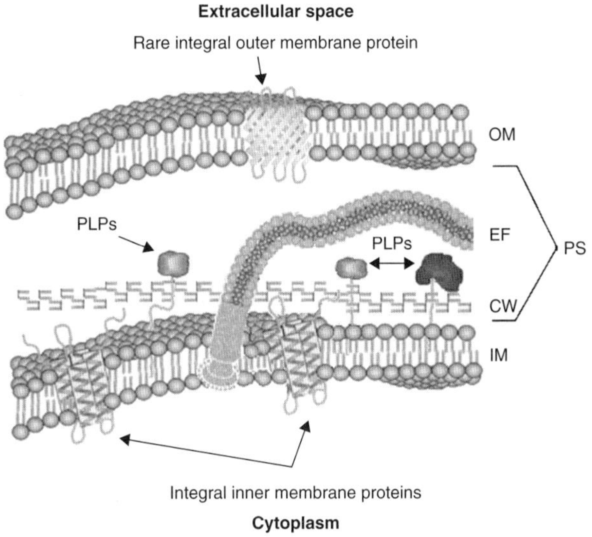

FIGURE 59.1 A schematic representation of the *T. pallidum* cell envelope. Similar to conventional Gram-negative bacteria, the outer membrane (OM) separates the internal cellular compartments from the extracellular space and, along with the inner membrane (IM), defines the periplasmic space (PS). Unlike conventional Gram-negative bacteria, *T. pallidum* OM does not contain LPS, and has only rare integral membrane proteins compared to the IM. Furthermore, in *T. pallidum*, the peptidoglycan layer (cell wall, CW) overlies the cytoplasmic membrane rather than being associated with the inner leaflet of the outer membrane. The most antigenic *T. pallidum* proteins, such as the components of the endoflagella (EF) and the periplasmic lipoproteins (PLPs, anchored to the IM by a lipidic moiety) are found within the PS. Adapted with permission from Mary S. Lipton (www.PNL.gov/biology/programs/MSD/characterization.stm).

pallidum (syphilis), *T. pallidum* subsp. *pertenue* (yaws), *T. pallidum* subsp. *endemicum* (bejel), and *T. carateum* (pinta). The infections caused by these organisms are similar in that they appear in distinct clinical stages, are chronic, and can cause serious destructive late sequelae. The *Treponema* also include oral treponemes that are associated with periodontal disease, *Treponema paraluiscuniculi* which causes a venereal disease in rabbits, a group of organisms (including *Treponema brennaborense*) associated with papillomatous digital dermatitis and ulcerative mammary dermatitis in cattle and sheep (Stamm et al., 2002), and symbionts of the termite hindgut (Lilburn et al., 2001).

ANTIGENS

A number of antigens have been identified in *T. pallidum* using biochemical and genetic tools. Of the 1039 predicted proteins in the *T. pallidum* genome, 882 were expressed and tested for reactivity with antisera from rabbits infected with syphilis; of these, 106 antigens were recognized by the antisera. (McKevitt et al., 2005). Antisera from humans with syphilis showed significant reactivity to 34 of 908 peptides tested (Brinkman et al., 2006). These genome-wide screens undoubtedly underestimate the actual number of *T. pallidum* antigens, possibly because of the inability to express some *T. pallidum* antigens in *Escherichia coli*. Major classes of *T. pallidum* antigens are introduced briefly below, and are then discussed in more detail as they relate to vaccine studies.

Lipoproteins

Three of the major antigens of *T. pallidum* with molecular masses of approximately 47, 17, and 15 kDa are now recognized as lipoproteins (Chamberlain et al., 1989a; Purcell et al., 1990), as are TmpA, B, and C (Schouls et al., 1989b, 1991). The genome predicts 22 lipoproteins, including roughly one-third of those antigens that were highly reactive with rabbit and human syphilis sera above. Lipoproteins are not only highly immunogenic, but the lipid moiety has been demonstrated to be recognized by toll-like receptor (TLR) 2, thus implicating these in induction of the innate immune response (Aliprantis et al., 1999; Lien et al., 1999; Salazar et al., 2002) and in early induction of adaptive immunity.

Endoflagella

The major structural components of the endoflagella (37, 34.5, 33, and 30 kDa) and associated flagellar proteins are readily recognized by immune serum from rabbits and humans. The sheath protein (37 kDa) is unique to the pathogenic treponemes while the core proteins (34.5, 33, and 30 kDa) share cross-reactive epitopes with the cultivable treponemes (Lukehart et al., 1982, 1985).

Tpr Family

Most of the members of the 12-member Tpr family (TprA–TprL) have been shown to induce the production of antibodies during infection in rabbits (Leader et al., 2003; Sun et al., 2004). The most uniformly immunogenic are TprK, TprC, TprD, TprF, and TprI. These have predicted cleavable signal peptides and may therefore be OM associated (Centurion-Lara et al., 1999); these *tpr* genes are highly transcribed (Giacani et al., 2007b), consistent with their expression during infection. Mechanisms of transcriptional regulation of several of the *tpr* genes have recently been identified (Giacani et al., 2007a).

Lipids

Lipids also induce immune responses during syphilis infection. The VDRL and RPR serological tests are widely used in syphilis diagnosis, and these tests detect antibodies to cardiolipin, phosphatidylcholine, and cholesterol complexes. The presence of these lipids on the surface of *T. pallidum* is suggested by the demonstration that antibodies raised against VDRL antigen can effectively opsonize *T. pallidum* for phagocytosis (Baker-Zander et al., 1993b).

PROTECTIVE IMMUNE RESPONSES

Immunity to *T. pallidum* does develop during infection, but the level of protection against reinfection depends on the length of the initial infection. Physicians have long recognized that persons treated for early syphilis can be repeatedly reinfected. Numerous reinfection studies have been conducted in humans (reviewed in Magnuson et al., 1956) but the most definitive work was conducted at Sing Sing prison in the 1950s (Magnuson et al., 1956). In this study, persons with various stages of treated and untreated syphilis (and uninfected controls) were inoculated on the forearm with an infectious dose of *T. pallidum* Nichols strain. Infection was defined as development of a lesion at the challenge site or by an increase in antibody titer. After a 4-month observation period (or earlier if the subjects developed lesions), all subjects were treated with the then-recommended dose of penicillin. Only those persons with late latent syphilis prior to challenge showed significant levels of protection against reinfection: none of 5 with untreated, and 10 of 26 with treated, late latent syphilis showed evidence of reinfection, while the majority of subjects with earlier stages of treated syphilis were reinfected upon challenge. This study clearly demonstrates that prolonged antigen exposure is required for protective immunity.

The rabbit provides an excellent model for investigation of protective immunity. As in humans, infection immunity develops only after lengthy infections (Turner and Hollander, 1957). This model also provides the opportunity to examine the development of immunity to heterologous strains of *T. pallidum*; intensive studies conducted by Turner and others, reviewed in Turner and Hollander (1957), demonstrated that, while homologous immunity was uniformly seen, immunity to symptomatic infection with other strains is variable. These data suggest that the "protective antigen(s)" of *T. pallidum* may not be conserved or may be differentially expressed across strains.

The specific mechanisms that contribute to the protective immune response have not been defined, although it is widely accepted that organisms are cleared from the early lesions of syphilis via phagocytosis and killing of opsonized *T. pallidum* by interferon-γ (IFNγ)- activated macrophages (Lukehart et al., 1980a; Baker-Zander and Lukehart, 1992; Baker-Zander et al., 1993a). This process requires the production of antibodies that react with surface-exposed antigens (opsonization), as well as secretion of IFNγ by specifically sensitized CD4+ and CD8+ T lymphocytes.

Humoral Immunity

Specific anti-treponemal antibodies first become demonstrable during the primary stage in human infection (Hanff et al., 1982; Baker-Zander et al., 1985), and within 1–2 weeks of infection in the rabbit model of syphilis (Lukehart et al., 1986). Titers generally peak during the secondary stage in humans and during healing of the primary lesion in rabbits. Both IgG and IgM are generally detectable through untreated infection, even in infections of several years duration. Despite the existence of high titers of antibodies during primary and secondary syphilis, some organisms escape and persist for decades within the untreated host.

Serological testing is a major tool in diagnosis of syphilis and in screening settings. Antibodies can be measured using the lipoidal tests (e.g., VDRL, RPR) or treponemal tests such as the TPPA, FTA-ABS, and recombinant-based immunochromatographic and ELISA tests. The identified functions of anti-*T. pallidum* antibodies include complement-mediated immobilization (Nelson and Mayer, 1949) or neutralization (Bishop and Miller, 1976b), opsonization (Lukehart and Miller, 1978; Baker-Zander et al., 1993a), and inhibition of attachment to host cells (Fitzgerald et al., 1984). The molecular targets of these immune functions have not been definitively identified, although opsonic and anti-attachment antibodies are likely to be directed against surface antigens. Blanco et al. (2005) recently demonstrated that a monoclonal antibody with specificity for a phosphorylcholine lipid moiety has immobilizing activity against *T. pallidum* in vitro in the presence of complement. The antibody response alone, however, does not protect against infection per se (Bishop and Miller, 1976a) suggesting that a cellular component is also necessary for protective immunity in vivo.

Cell-Mediated Immunity

Primary and secondary lesions in humans (Tosca et al., 1988; Van Voorhis et al., 1996b) and rabbits

(Lukehart et al., 1980a; Leader et al., 2007) show an impressive T lymphocyte infiltration (comprised of both CD4+ and CD8+ lymphocytes) with variable numbers of infiltrating B lymphocytes. RT-PCR analysis of human (Van Voorhis et al., 1996a) and rabbit (Godornes et al., 2007; Leader et al., 2007) lesions demonstrates a Th1 cytokine milieu, with significant IFN-γ production and very little detectable IL-4.

Specifically sensitized T lymphocytes can be demonstrated in infected rabbits within a few days of infection (Lukehart et al., 1980b); these recognize a variety of *T. pallidum* antigens (Baker-Zander et al., 1988; Giacani et al., 2007b). Stimulation of lymphocytes in vitro with *T. pallidum* antigens results in production of IFN-γ (Arroll et al., 1999). Macrophages appear just prior to the clearance of treponemes from the lesions, and treponemal antigens can be seen within macrophages in healing lesions (Lukehart et al., 1980a; Sell et al., 1982). These data suggest that activated macrophages are the primary effector cell in bacterial clearance, and that the phagocytosis and killing of *T. pallidum* that is demonstrable in vitro mimics the activities of macrophages in vivo.

EPIDEMIOLOGY

Syphilis is a leading cause of genital ulcers worldwide and is a cofactor for acquisition of HIV (CDC, 1998). In 1998, the World Health Organization estimated 12 million new cases per year, with an overall prevalence of at least 25 million, mostly in developing countries (Gerbase et al., 1998). The highest prevalence is found in developing nations and among people with limited or no access to health care; globally, sub-Saharan Africa, Southeast Asia, and Latin America have the highest incidence of syphilis (Fig. 59.2). Naturally, prevalence rates vary among countries and populations. During the past 25 years, prevalence rates in regions of sub-Saharan Africa have ranged from 2.5 to 12.5% in antenatal women (Ratnam et al., 1982; Creek et al., 2005; Pham et al., 2005; Potter et al., 2006). In South America, seropositivity rates in female commercial sex workers were 3.2, 13, and 28% in Peru, urban Guyana, and Buenos Aires, respectively (Zapiola et al., 1996; Persaud et al., 1999; Trujillo et al., 1999). After the dissolution of the Soviet Union, there was a dramatic increase in syphilis and other sexually transmitted infections in the newly independent states (Renton et al., 1998), and emigration carried syphilis to many eastern European countries that previously had very low rates of syphilis. In all high-prevalence areas of the world, congenital syphilis is a major cause of preventable miscarriage, perinatal morbidity and mortality (Watson-Jones et al., 2002).

In the United States and Europe, the past decade has shown a significant increase in new cases, affecting primarily men who have sex with men (MSM). The incidence of primary and secondary syphilis in U.S. males doubled from 2000 to 2005 (CDC, 2006b), and provisional data suggest a continuing increase of 5–8% per year since that time. Similar outbreaks of syphilis have been reported in MSM in Western Europe (Cowan, 2004; Marcus et al., 2004; Righarts et al., 2004; Sasse et al., 2004). In the United States, racial differences in syphilis prevalence still persist, but an important decline of syphilis in African Americans has been observed in the last decade, due, at least in part, to the efforts of the Centers for Disease Control (CDC)-sponsored Syphilis Elimination Program (Hook, 1998; CDC, 1999) which focused on areas of the southeastern United States where rates in African Americans were very high. The last decade has also brought a dramatic increase in syphilis in China, with rates increasing from 0.2 to 5.7 cases per 100,000 between 1993 and 2005 (Chen et al., 2007). Recent studies report very high prevalence among incarcerated Chinese female sex workers (15.7%) (Ruan et al., 2006), drug users (7–15.6%) (Liu et al., 2006), and MSM (13.5%) (Choi et al., 2007). These recent examples of new outbreaks of this readily curable infection demonstrate the failure of efforts for controlling transmission and highlight the urgent need for the development of a syphilis vaccine.

NATURAL HISTORY AND PATHOGENESIS

Syphilis is transmitted primarily during sexual contact with an infected individual when treponemes breach the mucosal surfaces or skin of an uninfected host. The infection can also be transmitted transplacentally to the fetus (congenital syphilis), or rarely through blood transfusion or organ transplantation. *T. pallidum* is a highly invasive organism that can reach virtually any tissue in the human body, resulting in a multitude of possible clinical presentations. If untreated, syphilis can result in serious cardiovascular and neurological complications or even death.

Clinically, syphilis is characterized by episodes of active disease interrupted by periods of asymptomatic, latent infection. *T. pallidum* multiplies at the site of contact (entry) and an ulcerative lesion (chancre) develops, containing millions of spirochetes per gram

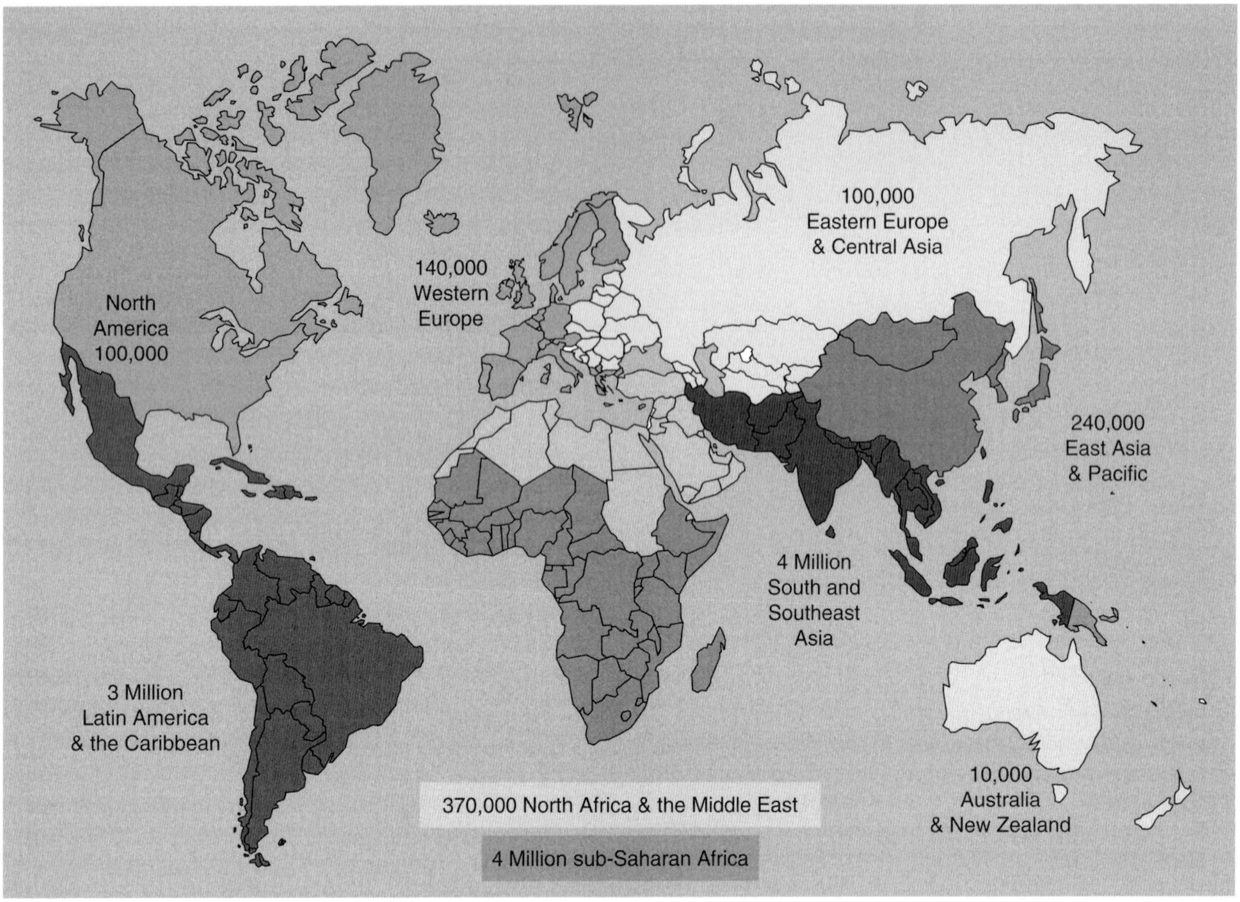

FIGURE 59.2 The incidence of syphilis in different regions of the world. The World Health Organization estimated in 1998 that there are 12 million new cases of venereal syphilis annually worldwide. Over 90% of these new cases are in developing countries in Latin America and sub-Saharan Africa, although increasing numbers of cases were reported in Eastern Europe, Central Asia, and East Asia. Outbreaks have been reported in several cities in Western Europe and North America among men who have sex with men. In the United States, a program for the elimination of syphilis was proposed in the late 1990s, but the number of reported cases has increased in the past 5 years.

of tissue. Concurrently, treponemes gain access to the circulation and rapidly disseminate throughout the body seeding many tissues, including the central nervous system. After several weeks, the chancre heals spontaneously and infection is next manifest by the secondary stage, characterized by disseminated skin lesions; these also resolve spontaneously over a period of several weeks or months. During primary and secondary syphilis, a strong host humoral and cellular immune response eliminates the great majority of treponemes, but a small population of organisms escapes to cause persistent infection. The untreated host then enters the latent stage, which can persist for decades or a lifetime; few or no treponemes can be detected during the latent stage of infection. A small minority of untreated persons will develop the serious manifestations of tertiary syphilis, including tabes dorsalis, general paresis, or aortic aneurysm.

The role of the inflammatory and immune responses in pathogenesis of syphilis is now better understood. Salazar et al. (2002) has proposed that inflammation driven by *T. pallidum* lipoproteins leads to the unique histopathological changes observed in syphilis. *T. pallidum* lipoproteins act as potent proinflammatory agonists (Radolf et al., 1991, 1995a; Salazar et al., 2002) and they stimulate dendritic and other cells via interactions with CD14, TLR1, and TLR2 expressed by monocytes/macrophages and dendritic cells (Lien et al., 1999). These cells then serve to present antigen to CD4+ lymphocytes which orchestrate the production of cytokines and signaling factors that result in the production of specific antibodies and the activation of both CD4+ and CD8+ lymphocytes and macrophages. The role of IFN-γ, activated macrophages, and opsonic antibodies in clearance of *T. pallidum* from early lesions is described above.

Primary Syphilis

T. pallidum establishes infection when it penetrates skin through intact mucosa or microabrasions at the site of contact, likely by attaching first to host cells via surface antigens (Fitzgerald et al., 1977a, 1977b). The treponeme has also been shown to bind to serum components, cell membranes, and several extracellular matrix (ECM) components, including fibronectin (Peterson et al., 1983; Thomas et al., 1985b), laminin, collagen I, and hyaluronic acid (Fitzgerald et al., 1984). Recently, Cameron et al. (2004) have identified *T. pallidum* proteins that bind fibronectin and laminin (Cameron, 2003); binding to ECM may serve as a bridge for *T. pallidum* to attach to host cells prior to invasion and dissemination to different tissues. The bacteria multiply at the site of contact and trigger the innate and acquired immune responses, resulting in lymphocyte and macrophage infiltration of the local site to form a chancre. The histopathology of the lesions shows proliferation of endothelial cells in small vessels, with infiltrating CD4+ and CD8+ lymphocytes and plasma cells. Macrophages infiltrate just prior to bacterial clearance, and phagocytosis of opsonized *T. pallidum* by activated macrophages is thought to be the main mechanism of bacterial clearance (LaFond and Lukehart, 2006).

During the early stages of infection, *T. pallidum* disseminates rapidly through the body. It is still unclear which strategies *T. pallidum* uses to survive without being eliminated by the host; however, the chronicity of this infection suggests intricate mechanisms for adaptation and immune evasion. Several mechanisms have been postulated including localization in immunologically privileged tissues (Medici, 1972), intracellular location in non-phagocytic cells (Azar et al., 1970; Sykes and Miller, 1971; Sykes et al., 1974), the presence of a protective outer coat of host origin masking the bacterial surface (Christiansen, 1963; Alderete and Baseman, 1979), a poorly antigenic OM (Radolf et al., 1989b; Radolf, 1994), and antigenic variation (Centurion-Lara et al., 2004; LaFond et al., 2006b).

Secondary Syphilis

Hematogenous dissemination of *T. pallidum* during primary infection leads to development of the systemic secondary stage, which is accompanied by malaise, generalized lymphadenopathy, and a disseminated macular or papular mucocutaneous rash (Chapel, 1980; Baughn and Musher, 2005). Signs and symptoms usually appear 4–10 weeks after the primary lesion has disappeared. The clinical presentation of secondary syphilis varies from inconspicuous lesions to lues maligna with massive necrosis (Baughn and Musher, 2005). Other manifestations include condylomata lata, mucous patches, gastric and renal involvement, and hepatitis (Chapel, 1980).

It is thought that the distribution of skin lesions on palms and soles reflects the preference of *T. pallidum* for lower temperatures. Although mucocutaneous lesions are the hallmark of secondary syphilis, this stage involves virtually any organ or tissue, and bacterial burden is thought to be high. The relatively mild systemic response to the presence of viable organisms is presumably due to the lack of lipopolysaccharide (LPS) (Hardy and Levin, 1983; Fraser et al., 1998) in *T. pallidum*; however, following antibiotic treatment for primary and secondary syphilis, the release of treponemal lipoproteins likely triggers the relatively common Jarisch-Herxheimer reaction. The histopathology of secondary and primary syphilis is very similar. As in primary syphilis, the immune response of the host is again able to clear the great majority of treponemes from lesions by macrophage-mediated phagocytosis. Measurement of cytokine mRNA in skin samples from infected individuals shows a Th1 cytokine pattern (Van Voorhis et al., 1996a), consistent with this immune mechanism.

Latent and Tertiary Syphilis

After resolution of secondary syphilis, the infected individual enters the latent stage, in which infection is subclinical and may last for the lifetime of the patient. In the pre-antibiotic era, approximately 30% of untreated individuals with latent infection developed tertiary syphilis after years or decades of infection (Gjestland, 1955). The manifestations of tertiary syphilis include gummas, cardiovascular syphilis, and late neurosyphilis, but these are rarely seen today.

Gummas are granulomatous lesions with a central area of necrosis affecting skin, bones, liver, heart, brain, stomach, or upper respiratory tract. In a few cases, spirochetes have been demonstrated in these lesions (Handsfield et al., 1983). In cardiovascular syphilis, the ascending aorta is typically affected. Recently, DNA amplification demonstrated the presence of *T. pallidum* in an aortic aneurysm (O'Regan et al., 2002), indicating that an active infection may result in tissue damage. Preexisting atheromatous conditions of the aorta might facilitate adhesion of treponemes to exposed ECM components (Cameron, 2003; Cameron et al., 2004).

Neurosyphilis is classified as asymptomatic, meningeal, meningovascular, or parenchymatous. In the absence of treatment, asymptomatic neurosyphilis either resolves spontaneously or progresses to symptomatic disease (Moore and Hopkins, 1930). Vasculitis of the small vessels and proliferative endarteritis are the hallmarks of meningeal and meningovascular syndromes. The parenchymatous syndromes (generalized paresis and tabes dorsalis) involve neuronal loss, demyelination, and astrocyte proliferation with gliosis. The mechanisms that give rise to these lesions are unknown to date.

Treatment

Penicillin G is the preferred drug for the treatment of all stages of syphilis (CDC, 2006a), and no penicillin resistance has been documented. The long-acting forms, such as benzathine penicillin G, are widely used for all forms except neurosyphilis, which is treated with aqueous crystalline or aqueous procaine penicillin G to ensure sufficient penicillin levels in the cerebrospinal fluid. Other classes of antibiotics with efficacy against *T. pallidum* include the tetracyclines and macrolides. Azythromycin, a macrolide compound that can be administered orally, has been shown to be effective for treatment of early syphilis and for sexual contacts of infected patients (Mashkilleyson et al., 1996; Hook et al., 1999, 2002). Unfortunately, macrolide resistance associated with a point mutation in the 23S rRNA genes (Stamm and Bergen, 2000) has been documented in the United States, Canada, and Ireland (Lukehart et al., 2004; Morshed and Jones, 2006). Although azythromycin may be a safe and effective alternative to benzathine penicillin in areas without resistant strains of *T. pallidum* and where proper clinical evaluation and follow-up can be guaranteed, penicillin remains the cornerstone of syphilis treatment.

VACCINE STUDIES TO DATE

Although remarkable efforts have been made in the last two decades to unravel the complex pathogenesis and immunology of syphilis, allowing the identification of a wide variety of treponemal antigens, a practical and effective vaccine is still unavailable. Furthermore, the recently discovered ability of *T. pallidum* to use antigenic variation to evade the host immune response complicates vaccine development. In the remaining part of this chapter, the rationale for the choice of past and current syphilis vaccine candidates will be critically reviewed in light of our current understanding of the peculiar *T. pallidum* ultrastructure and the difficulty in definitive identification of surface-exposed antigens. We will also describe how the most recent advances in vaccinology could benefit and improve our ability to control this too often neglected disease.

The Rationale for a Syphilis Vaccine

Syphilis is a disease whose characteristics make it amenable to control by vaccination. Importantly, like smallpox, there is no known animal reservoir for syphilis. Although it is a chronic infection, syphilis is transmissible only during the early symptomatic stages, so the period of transmissibility is limited. Readily available and simple serological tests make it possible to determine the prevalence of infection in specific populations or geographical regions, so that interventions can be focused appropriately. Moreover, the relatively low incidence of syphilis with respect to other curable sexually transmitted infections (Gerbase et al., 1998), the ease and low cost of diagnosis and treatment (Larsen et al., 1995; CDC, 2002), and the concentration of this infection in somewhat marginalized populations, likely reduce the focus on syphilis vaccine development in the United States. Syphilis is still a major health problem, however, in many countries and a vaccine could provide the means for a global syphilis eradication effort similar to the U.S. Syphilis Elimination Program initiated in 1999 (CDC, 1999). Globally, mass or focused treatment could be combined with immunization in the areas where syphilis incidence is high while, in regions where the disease is mainly restricted to defined social categories (e.g., MSM and intravenous drug users in North America and Europe), vaccination could selectively target individuals at higher risk.

Impediments to the Identification of Putative Protective Surface Antigens

Several major limitations hamper the ability of scientists to apply standard biochemical and immunological approaches to identification of putative candidate vaccine antigens in *T. pallidum*. First, the syphilis spirochete cannot be propagated in vitro, and the bacterial cells rapidly die outside of a susceptible host. The New Zealand white rabbit (*Oryctolagus cuniculus*) is used for in vivo propagation of *T. pallidum* and has become the preferred animal model for syphilis in that, following experimental infection, the clinical and pathological manifestations of the disease and immune response to the pathogen closely resemble those seen during natural infection in humans (Turner and Hollander, 1957; Sell and Norris, 1983). The slow growth rate of *T. pallidum* (generation time is ~30–33 h) (Magnuson et al.,

1948; Cumberland and Turner, 1949), along with the rapid onset of both innate and adaptive host immune response in the rabbit, make it possible to recover only limited numbers of organisms from an infected animal, prohibiting any experimental approach that requires large amounts of bacterial cells. The failure to maintain *T. pallidum* viability in culture (Fieldsteel et al., 1982; Riley and Cox, 1988; Cox et al., 1990) also precludes any kind of genetic manipulation, limiting the direct analysis of potential virulence factors. The ability to functionally inactivate, delete, and complement genes would in fact provide important information on which determinants *T. pallidum* requires to survive and persist in its host; such an approach would likely be particularly useful in the setting of the very small *T. pallidum* genome (Fraser et al., 1998), where the low number of genes potentially allows only limited functional redundancy.

The ongoing quest for vaccine candidate antigens finds its major obstacle in the biochemical composition and structural organization of the *T. pallidum* membrane systems (Fig. 59.1). Unlike more conventional gram-negative bacteria, the *T. pallidum* OM does not contain LPS (Hardy and Levin, 1983; Penn et al., 1985; Radolf and Norgard, 1988; Belisle et al., 1994; Fraser et al., 1998; Weinstock et al., 1998), and there is close chemical similarity between the lipid composition of the outer and inner membranes (Radolf et al., 1995b). Furthermore, while in most gram-negative bacteria, the peptidoglycan layer is physically associated with the inner leaflet of the OM, this layer overlies the cytoplasmic membrane in *T. pallidum* (Hovind-Hougen, 1983; Radolf et al., 1989a). These characteristics result in the extremely fragile nature of the *T. pallidum* OM, and difficulty in applying common immunological methods (i.e., surface immunofluorescence staining or immunoelectron microscopy) without compromising the OM integrity and exposing internal antigens to the immunological probes. Further, the substantial chemical homogeneity between the outer and inner membranes (Radolf et al., 1995b) discourages the use of separation techniques (i.e., sucrose density centrifugation) to isolate OM from the protoplasmic cylinder, in that such methods repeatedly result in contamination of the OM fractions with periplasmic antigens (Chamberlain et al., 1989a).

Criteria for Measuring Post-Challenge Protection against Syphilis Infection

The rabbit model provides a straightforward and relevant method for assessing post-immunization protection against infection. After the last immunization, test and control rabbits are challenged by intradermal injection of infectious *T. pallidum* on their shaved backs; multiple sites can be injected per animal. Lesion appearance, development, and progression (to ulceration) in control rabbits closely resemble that of primary chancres in humans, and the degree of alteration of this course in immunized rabbits is proportional to the level of immunity, with complete immunity resulting in no clinical or serological evidence of infection post-challenge. Partial immunity may be manifest as development of atypical lesions, such as delayed lesion appearance, flat nonprogressive lesions, a lower proportion of challenge sites progressing to ulceration (Fig. 59.3), and fewer challenge sites containing *T. pallidum* cells demonstrable by darkfield microscopy. Partial protection, however, does not prevent the establishment of the infection or seroconversion of the animals. In addition to monitoring lesion development, rabbit infectivity tests (RITs) can be performed for detection of inapparent infection of lymph nodes or other tissues in asymptomatic rabbits (Turner et al., 1969), particularly when the employment of whole-cell vaccines does not allow monitoring of seroconversion in the challenged animals. In this practice, a tissue sample from the test animal is injected into a naïve rabbit, and the eventual seroconversion of the recipient confirms infection in the tissue donor. Overall, vaccination with a partially protective antigen attenuates the symptoms of the disease, and

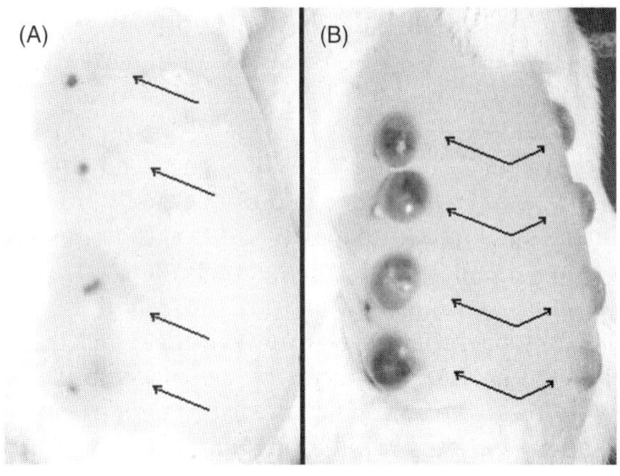

FIGURE 59.3 An example of altered lesion development following intradermal infectious challenge in rabbits immunized with a chimeric Tpr protein containing the NH_2-terminus of Tpr F/I and the TprG central region (Panel A) compared to unimmunized animals (Panel B). In immunized rabbits, small flat and non-ulcerative lesions (indicated by arrows in Panel A), containing few or no detectable treponemes, were observed compared to typical large ulcerative lesions in control animals at the same time post-challenge. In both animals, black ink dots mark the left of the injection site to facilitate lesion identification.

vaccinees may not develop ulcerative lesions, even though becoming infected. It is unclear whether individuals lacking ulcerative lesions could transmit the infection to others.

First-Generation Syphilis Vaccines: Whole Cell Approaches

The only successful vaccination ever reported to have induced durable sterile immunity against syphilis was described by Miller (1973). In this experiment, Miller immunized Dutch Belt rabbits intravenously with 3.7×10^9 γ-irradiated *T. pallidum* (Nichols strain) cells over a period of 37 weeks, with injections of weekly or semiweekly frequency. Ten days after the last immunization, 24 rabbits were challenged intradermally with 10^3 or 10^6 homologous *T. pallidum* cells, without developing any visible lesions in the following 3-month observation period. Asymptomatic infection was excluded by the negative outcome of the RITs performed using testicular tissue and lymph nodes from the challenged animals. A second group of 11 rabbits was challenged with 800 virulent *T. pallidum* cells one year after the immunization procedure, showing the durability of sterile immunity in these animals (Miller, 1973). Although this study was never independently confirmed due to its laborious nature, it clearly provides proof of concept for successful development of protective immunity against syphilis following immunization.

Although Miller's experiment represents the only success over a very long series of failed attempts to protect rabbits and other animals using killed or attenuated *T. pallidum,* reviewed previously by Lukehart (1985), Miller himself recognized that such a demanding procedure could not be easily translated from animals into humans. In addition to the exhausting immunization protocol, the inability to adequately separate *T. pallidum* cells from the rabbit tissue in which the bacteria were grown, and the fact that the serologic reactivity induced by the vaccination protocol would be difficult to distinguish from that caused by active syphilis infection, are major obstacles. Nonetheless, Miller's success provides important fundamental concepts that can guide those involved today in the search for a syphilis vaccine. First, a key aspect in the antigen preparation was the preservation of the native structure of the antigens inducing sterile immunity. His use of a physical (γ-irradiation) rather than chemical method to inactivate *T. pallidum* targeted the microorganism's genetic material, rather than its antigens, thus preserving molecules that could stimulate protective immunity in the host. In addition, the series of studies preceding Miller's successful vaccination suggest that the preservation of the fragile OM of *T. pallidum* (e.g., no centrifugation, etc.) was essential to success. It is still unclear whether the protective antigen(s) are protein, carbohydrate, or lipid, but it seems apparent that they are readily removed from the intact organism. The use of a high dose of treponemes (3.7×10^9 treponemes), in combination with a prolonged immunization cycle ($\sim$60 injections per rabbit), strongly suggests that the antigens able to elicit protective immunity are minimally expressed in *T. pallidum* or are weakly immunogenic. In previous attempts using lower doses of γ-irradiated treponemes and shorter immunization periods, either absent or partial protection was observed (Miller, 1967). In a subsequent attempt to reproduce the original results with a less demanding immunization schedule, Miller (1976) increased the immunizing dose to 12×10^9 *T. pallidum* cells administrated over 9 weeks vs. 37 weeks, and obtained only partial protection. Lastly, the durable sterile immunity demonstrated by Miller was not dependent on the development of immobilizing antibody. No other antibody functions (e.g., opsonization, neutralization) or cellular mechanisms of immunity were measured in the immunized animals, so the immune correlates of protection remain elusive.

The most recent vaccination experiments using whole-cell preparations were attempted in the early 1990s by Fitzgerald (1992), who endorsed the now-discounted theory that *T. pallidum*'s ability to escape clearance during early lesion resolution could be facilitated by a premature down-regulation of the initial Th1-mediated response in favor of Th2. Fitzgerald's hypothesized association between a premature Th1/Th2 switch and *T. pallidum*'s ability to establish a chronic infection was supported by the perceived parallels between the immune response to early syphilis and other chronic infections such as leishmaniasis (Boom et al., 1990; Scott, 1991), and trypanosomiasis (Reed, 1988; Silva et al., 1992). Fitzgerald immunized rabbits with a single injection of heat-inactivated *T. pallidum* cells and supported the Th1-type response with the administration of cyclophosphamide, monophosphoryl lipid A, and indomethacin (Fitzgerald, 1991; Fitzgerald et al., 1991). Despite the significant increase in the number of atypical lesions in the challenged animals, this approach did not prove effective in inducing sterile immunity, and in fact increased dissemination of organisms to distant sites (Fitzgerald, 1991; Fitzgerald et al., 1991).

Syphilis vaccinology has also known a remarkable number of attempts to immunize with different species of the genus *Treponema*, based on their antigenic relatedness to *T. pallidum* and their ability to be grown in vitro. All immunizations with *Treponema phagedenis* (Reiter strain) (Gelperin, 1951; Al-Samarrai and Henderson, 1976), *Treponema refringens* (Izzat et al., 1970; Al-Samarrai and Henderson, 1976), *Treponema minutum*, *Treponema ambiguum*, *Treponema microdentium* (Al-Samarrai and Henderson, 1976), *T. paraluiscuniculi* (Graves, 1981), and *Spirochaeta aurantia* (Graves et al., 1984) resulted, however, in absent, partial, or questionable protection after challenge with *T. pallidum* (reviewed by Lukehart, 1985).

Second-Generation Syphilis Vaccines: Subunit Immunogens

The increasing understanding of antigenic relatedness among different treponemal species and the application of DNA engineering techniques quickly fostered the use of individual subcellular fractions and recombinant proteins as experimental vaccines (Table 59.1) instead of whole-cell preparations. For example, in their work, Hindersson et al. (1985) immunized rabbits with purified endoflagella from *T. phagedenis* in the attempt to induce sterile immunity against syphilis. The outcome of the experiment showed no alteration of lesion development following challenge with virulent *T. pallidum*; nonetheless, the rationale for selecting such antigen was justified by the presence of shared epitopes between the *T. phagedenis* and *T. pallidum* homologous peptides (Blanco et al., 1986). To increase the specificity of the immune response, a similar experiment was attempted by Champion et al. (1990) with purified *T. pallidum* endoflagella; in this case, infection occurred although lesions were atypical in the immunized animals. These two examples are noteworthy in that both groups tried to induce sterile immunity against syphilis with an antigen already known to be located in the *T. pallidum* periplasm and therefore lacking any surface exposure. Based on the disappointing results of these studies, contrasted with the success of Miller's studies described above, the general concept emerged that antigens exposed on the *T. pallidum* surface would be the most likely targets to induce sterile immunity; this launched an active effort in many laboratories to identify surface-exposed antigens of *T. pallidum*.

In the 1980s, supported by the introduction of recombinant DNA technology in syphilis research, the identification of putative *T. pallidum* integral membrane proteins seemed to be easily achieved. Libraries were constructed with *T. pallidum* genomic DNA and propagated in *E. coli* strains suitable for the expression of exogenous antigens, and *T. pallidum*'s major immunogens began to be identified based on the reactivity of *E. coli* colonies with immune sera from experimentally infected rabbits or syphilitic patients (Walfield et al., 1982; Norgard and Miller, 1983; van Embden et al., 1983). The *Treponema* membrane protein A (TmpA, or TpN44.5), B (TmpB) (Hansen et al., 1985), and C (TmpC, TpN35) (Schouls et al., 1991) were among the antigens first isolated with this newly introduced approach, and their membrane location was supported by the evidence that these antigens were synthesized in *E. coli* as larger precursors, that the amino terminal sequence was predicted to contain a signal peptide, and finally that fractionation studies of *E. coli* cells expressing the Tmp proteins showed the tendency of these antigens to localize in the hydrophobic phase containing the cell membranes. Surprisingly, however, only partial or absent protection were reported following immunization with recombinant antigens based on the Tmp protein sequences (Wicher et al., 1989; Wicher et al., 1991) (Table 59.1); later studies of radiolabeled fatty acid incorporation into *T. pallidum* membranes lead to the conclusion that these antigens are lipoproteins whose characteristics of hydrophobicity, previously attributed to membrane-spanning domains, are due to protein acylation with lipid moieties, responsible for the anchoring of the antigens to the periplasmic side of the cellular membrane (Chamberlain et al., 1989a; Schouls et al., 1989a). The 47 kDa antigen of *T. pallidum* was also cloned early and, similarly, is a lipoprotein (Norgard et al., 1986a; Chamberlain et al., 1989a, 1989b).

The list of *T. pallidum* lipoproteins initially mistaken for OM antigens is quite extensive (Fehniger et al., 1986a, 1986b; Norgard et al., 1986a, 1986b; Radolf et al., 1986; Cunningham et al., 1988; Chamberlain et al., 1989a, 1989b; Schouls et al., 1989a; Swancutt et al., 1990). Although these proteolipids represent major *T. pallidum* immunogens, all have uniformly failed to induce sterile immunity against syphilis when used in immunization/challenge experiments (Table 59.1) (Borenstein et al., 1988; Centurion-Lara et al., 1997), similar to the previously attempted immunization with preparations of periplasmic endoflagella (Champion et al., 1990). Unpublished studies from our laboratories have failed to demonstrate any protective immunity following immunization with the 47 kDa antigen.

THE *T. PALLIDUM* OUTER MEMBRANE AND THE QUEST FOR SURFACE-EXPOSED ANTIGENS

In the last two decades, the desire to deepen the study of the *T. pallidum* OM and unravel some of its

TABLE 59.1 Summary of recent studies on the development of a vaccine against *Treponema pallidum*[a]

Immunogen	Route[b]	Adjuvant[c]	Duration (weeks)	Total dose[d] (μg)	Protection[e]	References
Subcellular fractions						
Treponema phagedenis (Reiter) endoflagella	SC	IFA	3	150	None	Hindersson et al. (1985)
T. pallidum endoflagella	IM + SC	CFA/IFA	32	450	Partial	Champion et al. (1990)
Recombinant proteins: rabbit model of infection						
TpN24	SC	±CFA	16	6–18	None	Hsu et al. (1988)
TpN19 (4D, TpF1)	IM	CFA/IFA	19	1250	None	Borenstein et al. (1988)
	IM + IV	CFA	14	2000	Partial	Borenstein et al. (1988)
	iv	Ribi	14	600	Partial	Borenstein et al. (1988)
Gpd (GlpQ)	IM/SC/ID/IP	Ribi	9	300	Partial	Cameron et al. (1998)
GlpQ (Gpd)	SC	CFA/IFA	8	200	None	Shevchenko et al. (1999)
TpN15	SC	IFA	10	500	None	Centurion-Lara et al. (1997)
	IM/SC/ID/IP	Ribi	10	500	None	Centurion-Lara et al. (1997)
TprK						
Partial-length protein	IM/SC/ID/IP	Ribi	9	450	Partial	Centurion-Lara et al. (1999)
Partial and full length	IM/SC/ID/IP	Ribi	2	250	None	Hazlett et al. (2001)
Amino terminal region	IM/SC/ID/IP	Ribi	16	750	Partial	Morgan et al. (2002a)
TprF						
Full-length protein	IM/SC/ID/IP	Ribi	9	625	Partial	Sun et al. (2004)
Amino terminal region					Partial	Sun et al. (2004)
TprI full-length protein	IM/SC/ID/IP	Ribi	9	450	Partial	Giacani et al. (2005b)
TprC/D or TprJ central regions	IM/SC/ID/IP	Ribi	9	450	Partial	Unpublished results from the authors' laboratory
TprE/G/J carboxyl terminal region	IM/SC/ID/IP	Ribi	9	450	Partial	Unpublished results from the authors' laboratory
TprE, I, or G central region	IM/SC/ID/IP	Ribi	9	450	None	Unpublished results from the authors' laboratory
TprA, B, or H, full-length peptides	IM/SC/ID/IP	Ribi	9	450	None	Unpublished results from the authors' laboratory
Tp0326 (Tp92)	IM/SC/ID/IP	Ribi	11, 13, 16	375–625	Partial	Cameron et al. (2000)
Recombinant proteins: guinea pig model of infection						
TpN 44.5 (TmpA)	SC + IP	Ribi	Not reported	600	None	Wicher et al. (1991)
TpN 36 (TmpB)	SC + IP	Ribi	Not reported	600	Partial	Wicher et al. (1991)
TpN 35 (TmpC)	SC + IP	Ribi	Not reported	600	None	Wicher et al. (1991)
Tmp A + B + C	SC + IP	Ribi	Not reported	200 each	None	Wicher et al. (1991)

[a] A summary of previous published studies (1948–1984) involving the use of whole treponemal cells as immunizing preparations is not included here, as it is already available in the literature (Lukehart, 1985).
[b] Route of immunization. ID, intradermal; IP, intraperitoneal; IM, intramuscular; SC, subcutaneous.
[c] Adjuvant. CFA, Complete Freund's adjuvant (a water-in-oil emulsion formulated with non-metabolizable mineral oil, the detergent mannide monoleate, and heat-killed *Mycobacterium tuberculosis*, *Mycobacterium butyricum* or their extracts); IFA, Incomplete Freund`s adjuvant (same formulation as CFA but lacking Mycobacterium components); Ribi = Ribi adjuvants contain a metabolizable oil (squalene) and Tween-80, with refined Mycobacterial cellular components (trehalose dimycolate, cell wall skeleton) and bacterial monophosphoryl lipid A as immunomodulating agents.
[d] Total dose of immunogen administered.
[e] Protection is defined as resistance to challenge with *T. pallidum*. Partial, significantly altered lesion progression or asymptomatic infection; None, lesion development equivalent to that in unimmunized control animals.

peculiar characteristics prompted a series of pivotal investigations that demonstrated the OM to be poorly antigenic and almost devoid of integral membrane proteins. The perceived poor immunologic reactivity of the outer surface of *T. pallidum* is based on the evidence that the binding activity of syphilitic sera to undamaged *T. pallidum* cells is modest, while binding is highly enhanced by any intervention which damages the OM integrity (Penn and Rhodes, 1982; Robertson et al., 1982; Fehniger et al., 1986b; Radolf et al., 1986; Stamm et al., 1987; Cox et al., 1992). Furthermore, complement-mediated killing (immobilization and neutralization) and phagocytosis of opsonized *T. pallidum* cells have been shown to occur only upon use of high concentrations of antisera and extended incubation times (Nelson and Mayer, 1949; Bishop and Miller, 1976b; Lukehart and Miller, 1978; Alder et al., 1989, 1990). One long-standing hypothesis to explain the limited reactivity of intact *T. pallidum* with antibodies has been the ability of *T. pallidum* to gather proteins from the host to form a self-like protective shield surrounding the OM; although this hypothesis has not been completely disregarded, it became a less-plausible explanation when freeze fracture electron microscopy studies revealed that the *T. pallidum* OM possesses an extremely low density of integral membrane proteins (Radolf et al., 1989b; Walker et al., 1989, 1991). More specifically, the outer and inner leaflets of the OM were calculated to harbor 70 and 100 membrane particles per μm^2 (Walker et al., 1989), approximately 1% of the number found in the *E. coli* inner and outer leaflets of the OM, respectively. These fascinating discoveries provided the first plausible explanation to the fact that conventional methods to identify putative surface exposed epitopes (e.g., immunofluorescence, immunoelectron microscopy, or surface protein radiolabeling) find very limited application in *T. pallidum*. Equally interestingly, *T. pallidum*'s surface proteins were shown to have a nonuniform but ordinate distribution throughout the bacterial surface, and to be capable of only limited mobility within the lipid bilayer (Radolf et al., 1989b; Blanco et al., 1990; Bourell et al., 1994), which might correlate with the limited ability of immune sera to cross-link these antigens and the very slow rate at which complement-mediated killing of *T. pallidum* occurs (Nelson and Mayer, 1949; Bishop and Miller, 1976b).

The modest immunological reactivity and the protein-poor content of the *T. pallidum* OM have not discouraged investigators from attempting alternative experimental routes to identify surface-exposed antigens. These methods can be generally classified into two groups, according to the use of a direct or indirect identification approach. Direct approaches encompass all of the biochemical methodologies aimed to selectively isolate the *T. pallidum* OM after detaching it from the underlying cellular structures (endoflagella, peptidoglycan, and the cytoplasmic membrane with its array of proteolipids) which could constitute a source of contaminating proteins. A generic hypertonic plasmolysis of the OM, using a sucrose gradient followed by isopycnic centrifugation to recover the OM fractions (Radolf et al., 1995b; Shevchenko et al., 1997, 1999), and subsequent analysis of the protein constituents by two-dimensional electrophoresis and immunoblotting, identified at least two new plausible candidates, p30.5 (Tp0453) and GPD/GlpQ (Shevchenko et al., 1997). In addition, at least three known periplasmic lipoproteins were also represented on the immunoblot, underlining how direct isolation of OM fragments is fraught with contamination by proteins from other cellular compartments.

It is by indirect analytical methods, however, that the majority of the *T. pallidum* putative surface antigens have been identified, mostly using in silico analysis of the Nichols strain genome. The initial search for homology between *T. pallidum* predicted proteins and previously characterized OM proteins of related bacteria, has allowed the identification of a first group of candidates. These include the twelve genes encoding for the Tpr proteins, which are homologous to the major sheath protein of the oral spirochete *Treponema denticola* (Weinstock et al., 1998; Centurion-Lara et al., 1999), and Tp0326 (Tp92), which is homologous to the Omp85/D15 antigens described in many pathogenic bacteria (Cameron et al., 2000). The combination of bioinformatics tools for prediction of signal peptides, membrane spanning domains, and protein cellular location has also led to the identification of three proteins which, in a series of intensive experiments, were shown to encode for laminin-binding (Tp0751) (Cameron, 2003) and fibronectin-binding (Tp0483 and Tp0155) proteins (Cameron et al., 2004). These proteins are likely to be surface exposed and involved in the attachment of *T. pallidum* to the host cells and ECM.

Many of the hypothetical *T. pallidum* OM proteins identified on the basis of the methods described above have been the subject of intensive investigation to confirm or refute the original hypotheses suggested by the in silico analysis of their sequences. Although none of those antigens has been unequivocally shown to reside within the OM, and immunoprotection assays have thus far failed to induce sterile immunity, very promising immunization studies have been performed. The ample knowledge accumulated on these proteins has strongly renewed the interest in the study of this elusive pathogen and has provided new analytical instruments to further investigate the pathogenesis of syphilis. An increasing body of information is available today on these proteins regarding their immunogenic capabilities during experimental or natural infection, their functions, their structures, the conservation of their genes among *T. pallidum* strains, subspecies and treponemal species, their gene organization into transcription units, their transcriptional patterns, and mechanisms of transcriptional regulation. This knowledge will eventually help in the characterization of these and other antigens that could serve as more effective immunogens for inducing protective immunity.

Tpr Antigens

The recognition of the gene-family coding for the *T. pallidum* repeat (Tpr) antigens is one of the major findings of the *T. pallidum* genome project (Fraser et al., 1998), in that these hypothetical proteins were immediately identified as potential virulence factors and vaccine candidates (Weinstock et al., 1998; Centurion-Lara et al., 1999). Several of these genes had been identified earlier using other methods (Hardham, 1995; Centurion-Lara et al., 1999), but the existence of a 12-member family (named *tprA-tprL*) was not anticipated until the genome sequence was analyzed. The encoded antigens share homology with the major sheath protein (Msp) of *T. denticola* (Weinstock et al., 1998), a virulence factor reported to be surface exposed and to have both porin and adhesin proprieties (Fenno et al., 1996; Mathers et al., 1996). These *tpr* genes account for a high percentage (~2%) of the very sparse *T. pallidum* genome (Centurion-Lara et al., 1999), suggesting an important role for these antigens in the survival of an organism that has lost many biosynthetic capabilities during evolution (Fraser et al., 1998). The Tpr antigens are divided into three subfamilies by their predicted amino acid homology: Subfamily I (TprC, D, F, and I), Subfamily II (TprE, G, and J), and Subfamily III (TprA, B, H, K, and L) (Centurion-Lara et al., 1999). The Subfamily I members, along with Tpr J and K, are predicted to have cleavable signal sequences and possible OM localization (Centurion-Lara et al., 1999; Giacani et al., 2007b), although definitive localization studies have not yet been performed. Within the Subfamily I and Subfamily II members, the proteins have conserved NH_2- and COOH-terminal regions, with central domains that are variable in sequence and length, allowing the differentiation of the individual proteins (Centurion-Lara et al., 1999). In the Nichols strain, where the Tprs were first characterized, Subfamily I TprC and D are identical in sequence but their genes reside in two distinct genomic loci, and TprF is predicted to lack the conserved COOH-terminal region and part of the central variable domain due to a deletion and frame-shift mutation in the coding sequence (Fraser et al., 1998; Centurion-Lara et al., 1999).

A comparison of the *tpr* genes among different *T. pallidum* isolates has revealed strain-to-strain sequence variability in several members of this gene family. Because the Tpr proteins stimulate humoral and cellular immune responses during infection (Morgan et al., 2002b; Leader et al., 2003; Sun et al., 2004; Giacani et al., 2007b), variations in the gene sequence has potential ramifications for immune function and cross-immunity, which is relevant to the development of a broadly effective syphilis vaccine. Among Subfamily I members, an alternative allele, named *tprD2*, can be found within the *tprD* locus in about 50% of syphilis isolates tested (Centurion-Lara et al., 2000b). Diversity was described also for the *tprC* gene locus, in that all of the strains analyzed so far with the *tprD* sequence in the D locus also harbor an identical sequence in the C locus (as in the Nichols strain); in contrast, strains in which a *tprD2* allele is found in the D locus, contain a *tprC* allele containing a variable number of base changes resulting in new predicted amino acids at those sites (Sun et al., 2004). Sequences for *tprF*, *tprI*, and *tprE* have not yet been found to vary among syphilis strains (Sun et al., 2004). In the Sea 81-4 strain, the central region of Subfamily II *tprJ* has been replaced by the corresponding region of *tprG*, resulting into a hybrid *tprG-J* gene in the J locus (Giacani et al., 2005a). In this same isolate, a frameshift in the G locus predicts a premature termination of TprG (Giacani et al., 2005a). Among Subfamily III members, the Nichols strain *tprA* contains a frameshift mutation resulting in a premature stop codon which is absent in the Sea 81-4 strain and in the rabbit treponeme *T. paraluiscuniculi* (Giacani et al., 200, 2007b). Interestingly, in the Nichols strain, the *tprA* frame-shift is caused by a deletion of a dinucleotide (CT) located in a region containing three dinucleotide repeats; conversely, in the Sea 81-4 strain, the presence of four CT repeats at this location reverts the frame shift and generates a full length *tprA* open reading frame (ORF) (Giacani et al., 2007b).

Among Subfamily III members, however, the greatest degree of sequence heterogeneity has been reported for the *tprK* gene which, among human isolates, undergoes extensive sequence diversification shown to accumulate during the course of infection (Centurion-Lara et al., 2000a; LaFond et al., 2003, 2006a; Morgan et al., 2003). Among all of the Tpr proteins, TprK is the most highly expressed (Giacani et al., 2007b), induces the earliest and most vigorous immune response in the infected host (Leader et al.,

2003; Giacani et al., 2007b), and has been reported to be a target for opsonic antibody (Centurion-Lara et al., 1999), suggesting possible surface exposure of this antigen. This sequence variation in TprK is the first antigenic variation mechanism for *T. pallidum*. Although the function and location of TprK have not been finally resolved (Hazlett et al., 2001), this variation is hypothesized to provide an efficient way for the organism to evade the host immune response, leading to persistence and chronic infection despite a vigorous immune response.

In more detail, *tprK* heterogeneity is restricted to seven discrete variable regions (named V1–V7) (Centurion-Lara et al., 2000a), mostly located in the COOH-terminal half of the gene; these regions are flanked by conserved sequences, and variability is generated by a nonreciprocal gene conversion-like mechanism which mediates the recombination of a DNA template (called a donor cassette) into the expression site. Over 50 identified donor cassettes are located in two regions distant from the expression site on the chromosome (Centurion-Lara et al., 2004). Morgan et al. (2002b) demonstrated that antibodies arising in infected rabbits are directed against the V regions of TprK, while cellular immune responses are focused on the conserved regions. This segregation of epitopes is consistent with a major role for antibody in the immune response to *T. pallidum*. Morgan also demonstrated a correlation between the level of protection following immunization and the specificity of the induced V region antibodies (Morgan et al., 2003). In an intensive series of experiments, LaFond et al. (2006b) demonstrated that new variant V region sequences arising during infection are not recognized by antibodies developed against V regions of the infecting strain. These results strongly support not only the involvement of this gene in the generation of antigenic diversity as mechanism of immune escape, but also the TprK OM location. It is difficult to imagine why *T. pallidum* has evolved such a complex mechanism to generate so many TprK antigenic variants if portions of the protein are not exposed to the host immune system.

The surface exposure of TprK and other Tprs, however, is still the object of some debate: surface exposure for TprK and several other Tpr antigens is supported by experimental evidence, PSORTb (http://www.psort.org/psortb) analysis (Giacani et al., 2007b), and predicted cleavable signal peptides in TprF, I and K. Antibodies elicited against TprK have been shown to exert opsonic activity (Centurion-Lara et al., 1999) suggesting that at least some TprK epitopes are surface exposed; this result, however, has not been reproduced by others (Hazlett et al., 2001). TprI's cellular location has been investigated using electron microscopy after colloidal gold labeling using a polyclonal antiserum raised against the recombinant full-length antigen (Giacani et al., 2005b); although gold particles were shown to label the *T. pallidum* OM, the periplasm was also targeted.

Rabbits inoculated with different *T. pallidum* strains develop antibody (Leader et al., 2003) and CD4+ reactivity (Giacani et al., 2007b) in varying magnitudes to different repertoires of the Tpr antigens. This is likely to be a consequence of the differential expression of the *tpr*s during infection. Recent data on transcriptional analysis of the *tpr* genes using quantitative real time RT-PCR, confirmed that the *tpr* genes are differentially expressed within and among *T. pallidum* isolates (Giacani et al., 2007b), suggesting the existence of transcriptional regulatory mechanisms. Subfamily I and II *tpr* expression is regulated, at least in part, by the variable length of homopolymeric G-repeat sequences immediately upstream of the transcriptional start site (Giacani et al., 2007a); this resembles the phase variation mechanism described for *Neisseria meningitidis opc* genes (van der Ende et al., 1995; van der Woude and Baumler, 2004; Cullen and Cameron, 2006). Tight control of transcription initiation is a recognized feature of virulence factors, and the hypothesis that Subfamily I and II *tpr*s could, in addition to TprK, contribute to the antigenic diversity of the *T. pallidum* surface cannot be excluded. Both the sequence variation and the differential expression of individual Tpr among strains of *T. pallidum* have important ramifications for the design of a broadly protective syphilis vaccine.

Tpr Proteins as Vaccine Candidates

Given the impressive and steadily growing body of information concerning the Tpr antigens, these were naturally selected for immunization trials and studies were independently performed in different laboratories (Table 59.1). In all of these studies, adult male rabbits were immunized with purified *E. coli* expressed recombinant peptides, often representing only a portion of the predicted protein antigen; these rabbits were then challenged intradermally with virulent *T. pallidum* (differing doses) and observed for lesion development. Although none of these immunizations yielded sterile immunity, several studies demonstrated significant alterations in the appearance, progression (to ulceration), and content of detectable *T. pallidum* (i.e., presence or absence of *T. pallidum* in darkfield microscopic examination of lesion aspirates) following infectious challenge with the Nichols strain. Antigens examined include peptides from

TprK (Centurion-Lara et al., 1999; Hazlett et al., 2001; Morgan et al., 2002a), TprI (Giacani et al., 2005b) and F full-length peptides (Sun et al., 2004), the TprC/D and J central unique regions (unpublished data), the NH_2 conserved region of Subfamily I Tprs, and the COOH-terminal domain of Subfamily I members (Sun et al., 2004). Further studies identified the NH_2-terminal portion of the TprK antigen as the fragment inducing the highest level of protection, compared to the central and COOH-terminal regions (Morgan et al., 2002a). One independent study reported lack of protective activity (Hazlett et al., 2001) upon immunization with a mixture of full-length TprK plus a peptide analogous to that reported by Centurion-Lara et al. (1999). No alteration of disease progression was seen upon immunization with TprA, B, and H full-length peptides, and the TprI, E, and G central regions, while significant alteration of lesion development was seen upon immunization with the Subfamily II COOH-terminal domain (unpublished data from the authors' laboratory) (Table 59.1). TprL could not be expressed in our laboratory as a recombinant peptide, and no results are available in the literature on the protective capability of this antigen.

There is reason to be enthusiastic about Tpr antigens as vaccine candidates, and studies are in progress to refine the antigens and the immunization protocols. The preliminary studies described above strongly suggest that significant levels of immunity are induced by the NH_2-terminal regions of TprK and the Subfamily I, and the central region of Tpr C/D. Ongoing work in the authors' laboratories is focused on the development of chimeric peptides, including selected regions of Tpr antigens in a single larger recombinant peptide. Given the expected limited exposure of epitopes on the surface of *T. pallidum*, it is hoped that the chimeric immunogens, inducing antibodies to carefully selected regions of multiple antigens, may yield increased levels of protection.

Adhesins

For *T. pallidum*, as for many bacterial pathogens, attachment to host cells or tissue components is a crucial first step in the pathogenic process that will eventually lead to the establishment of infection in the host. *T. pallidum*'s ability to attach to various cultured mammalian cell types was first reported in the 1970s (Fitzgerald et al., 1975, 1977a; Quist et al., 1983), and subsequent studies noted the ability of *T. pallidum* to bind to a variety of components of the ECM, including fibronectin, laminin, and collagen (Baseman and Hayes, 1980; Peterson et al., 1983; Fitzgerald et al., 1984; Fitzgerald and Repesh, 1985; Thomas et al., 1985a, 1985b). Only recently, however, has such interaction been attributed to specific, receptor-mediated attachment through the identification of three ORFs coding for *T. pallidum* ECM-binding proteins: Tp0751 was found to bind laminin (Cameron, 2003) and Tp0155 and Tp0483 were shown to specifically recognize fibronectin (Cameron, 2003; Cameron et al., 2004). Studies by Cameron et al. demonstrated the attachment of viable *T. pallidum* to fibronectin- or laminin-coated surfaces, with specific inhibition of that activity by the respective recombinant proteins. These functional studies argue strongly for surface exposure of the proteins and, correspondingly, for the plausibility of these antigens to induce a protective immune response. Immunization experiments involving the ECM-binding proteins are currently in progress (Cullen and Cameron, 2006).

Lipoproteins

Past attempts to induce protective immunity using major *T. pallidum* lipoproteins, discussed above, have been ineffective. In more recent years, however, the identification of a lipoprotein likely to be associated with *T. pallidum* OM has stimulated a reexploration of this category of antigens in immunoprotection assays. Among these, the proteolipid encoded by the Tp0453 ORF is currently being examined (Cullen and Cameron, 2006). Tp0453 encodes for a 30.5 kDa protein previously identified in *T. pallidum* OM fractions (Shevchenko et al., 1997), but only recently characterized at the genetic, structural, and functional level by Hazlett et al. (2005). Structural and functional analysis showed that, even in absence of its lipid anchor, Tp0453 retains its amphiphilicity and possesses an intriguing ability to destabilize artificial membranes. Such evidence induced the authors to hypothesize that Tp0453, although physically located in the periplasmic space, is able to reversibly increase the permeability of the OM, perhaps functioning as a hidden porin that would allow the pathogen to acquire nutrients from the external environment in a nonselective way without exposing any potential epitope to the host immune system.

Similarly to Tp0453, the *T. pallidum* lipoprotein molecule glycerophosphodiester phosphodiesterase (also called Gpd or GlpQ) may also be associated with the OM. This antigen was initially identified both by screening a *T. pallidum* expression library with opsonizing antibodies (Stebeck et al., 1997; Cameron et al., 1998), and also by reverse genomics applied to protein constituents purified after OM isolation

(Shevchenko et al., 1997); subsequently, its proteolipid nature was shown by its resistance to N-terminal Edman degradation and by protein labeling studies using radioactive fatty acids (Shevchenko et al., 1997). The evidence that Gpd binds the Fc portion of some classes of human immunoglobulins, that anti-Gpd immune sera show decreased reactivity with lysates *T. pallidum* lacking the OM, and the high similarity between the *T. pallidum* Gpd sequence (Tp0257) and its surface-exposed homologs in *Borrelia hermsii* (Gpd) and *Hemophilus influenzae* (Hpd, or Protein D) (Ruan et al., 1990; Janson et al., 1992; Shang et al., 1997), clearly support the hypothesis that Gpd is associated with the *T. pallidum* OM (Cameron et al., 1998). Conversely, however, attempts to identify Gpd in *T. pallidum* using surface immunofluorescence and opsonophagocytosis assays were unsuccessful, thus questioning the association between this antigen and the OM (Shevchenko et al., 1999). Immunoprotection assays using recombinant *T. pallidum* Gpd independently performed in the two laboratories in which the antigen was characterized (Cameron et al., 1998; Shevchenko et al., 1999) were also discordant, one showing significant alteration in lesions following challenge, and the other showing no effect of immunization.

Tp92

As with Gpd, Tp92 (encoded by the Tp0326 ORF) was initially identified both by differential antibody screening of a *T. pallidum* genomic library (using opsonizing and non-opsonizing rabbit immune sera) and by in silico analysis of the Nichols strain genome (Cameron et al., 2000). Tp92 is predicted by PSORT analysis to contain a cleavable signal peptide and to have a significant likelihood of localizing in the bacterial OM. Experimental evidence supporting OM location of Tp92 was provided by the absence of immunoblot reactivity of Tp92-specific antiserum against washed treponemal cells (in which the OM had likely been sheared off), and by the opsonizing capability of antisera from Tp92-immunized rabbits (Cameron et al., 2000). The function of Tp92 has not been determined in *T. pallidum*, but homologous antigens are widely present among gram-negative bacteria. The Tp92 homolog Omp85, described in both *N. meningitidis* and *Neisseria gonorrheae* (Manning et al., 1998), has been shown to be surface exposed, and functional characterization studies performed on *N. meningitidis* Omp85 (Voulhoux et al., 2003; Bos and Tommassen, 2004; Voulhoux and Tommassen, 2004) strongly suggest that this protein is part of a multi-subunit complex involved in correct OM protein sorting and assembly. Furthermore, immunization with the Tp92-homologs of *H. influenzae* and *Pasteurella multocida* (called D15 and Oma87, respectively) were shown to be protective against infection in animal models (Flack et al., 1995; Ruffolo and Adler, 1996). Protection experiments with *T. pallidum* Tp92, however, induced partial protection in immunized rabbits which developed atypical lesions but did show seroconversion for syphilis after infectious challenge (Cameron et al., 2000).

GENOMICS AND RECOMBINANT DNA TECHNOLOGY: GREAT HELP WITH SOME LIMITATIONS

The new sciences of genomics, proteomics, and recombinant DNA technology have revolutionized modern microbiology and are important allies in the challenging path to vaccine development. Given the great potential of these techniques, it might seem surprising that since their introduction, only three vaccines against bacterial pathogens (*Bordetella pertussis*, *Vibrio cholerae*, and *Borrelia burgdorferi*) (Grandi, 2001) have been developed using recombinant DNA technologies. Although DNA engineering provides a virtually unlimited availability of recombinant antigens, there are significant limitations of the application of these methods for production of membrane proteins. Because of their chemical nature, membrane proteins most often are produced in the insoluble (sometimes membrane-associated) fraction of the recombinant host, and the harsh chemical procedures (often 6M urea) needed to solubilize the material for purification of the single recombinant protein destroys native conformation. Reestablishment of native conformation prior to use in immunization requires tedious application of multiple refolding protocols, most often in a setting where the native conformation is completely unknown. As proposed by Cullen and Cameron (2006), experimental methods that might be utilized include limiting the level of expression of a recombinant antigen to favor production of soluble recombinant protein, and promotion of antigen transport to the *E. coli* OM by replacing the *T. pallidum* signal sequences with those of *E. coli* OM proteins. These approaches may offer improvements but may not be the full answer. Antibody-mediated protective immunity in many cases is directed at conformational epitopes formed by the proteins as they loop in and out of the OM, and purified proteins may or may not form such epitopes in the absence of an environment mimicking the OM.

ANIMAL MODELS AND ADJUVANT SELECTION

The rabbit is the most widely used animal model to study experimental syphilis (Turner and Hollander, 1957), and has advantages over guinea pigs, hamsters, mice, and nonhuman primates because of the many parallels between the experimental infection in rabbits and natural human infection. Similar to humans, rabbits are highly susceptible to *T. pallidum*, with an equivalent ID_{50} (Magnuson et al., 1948, 1956), and develop skin lesions following infection that are clinically and histologically very similar to human primary chancres (Turner and Hollander, 1957; Sell and Norris, 1983). In rabbit and humans, similarities also include the repertoire of antigens recognized by humoral immune responses, the IFNγ-predominant cytokine milieu at the sites of infection, and the ultimate development of immunity to reinfection (Turner and Hollander, 1957; Sell and Norris, 1983). As in humans, rabbits remain chronically infected for a lifetime unless adequately treated (Turner and Hollander, 1957), and infectious organisms can be detected in lymph nodes and other tissues (via transfer of tissue to a susceptible rabbit) even after years of latent infection. Secondary lesions are sometimes seen in infected rabbits if the fur is kept clipped to reduce the temperature of the skin surface, but manifestations of tertiary syphilis are not observed in syphilitic rabbits (Turner and Hollander, 1957). In rabbits, *T. pallidum* disseminates via the bloodstream and lymphatic circulation to distant organs, including the central nervous system, similar to human infection. The disadvantages of the rabbit model include the high cost of the animals, the lack of inbred strains, and the lack of species-specific immunological reagents. The rabbit genome sequence is being completed (http://www.ncbi.nlm.nih.gov/projects/genome/guide/rabbit/) which will facilitate development of reagents. With regard to rabbit strains, it may be worth noting that the successful immunization experiment by Miller (1973) was conducted in Dutch Belt rabbits, while current investigators use the New Zealand White strain; the possible effect of this change on modern immunization studies is unexplored.

Hamsters and guinea pigs have been used as alternative models (Schell et al., 1980; Schell, 1983; Wicher and Wicher, 1989) because inbred strains are available; however, hamsters have reduced susceptibility to the syphilis spirochete compared to the other *T. pallidum* subspecies (subsp. *endemicum* and *pertenue*, the causative agents of endemic syphilis and yaws, respectively), and guinea pigs require inoculation of extraordinarily high numbers of treponemes to induce the development of ulcerative lesions (Wicher and Wicher, 1989). The documented immunological differences between guinea pigs and humans (i.e., absence of production of Th1 cytokines during experimental infection) and the different histopathology of skin lesions also disfavor this animal model with respect to rabbit. Although infection can be experimentally achieved in mice, these animals do not develop clinical syphilis, and therefore the model has not been deemed adequate (Schell et al., 1975; Folds et al., 1983).

Although the vast majority of immunoprotection experiments have been performed in rabbits (except for the TmpA, B, and C immunizations performed in guinea pigs) (Wicher et al., 1989; Wicher et al., 1991), there has been wide variability in the immunization protocols used (Table 59.1), particularly with regard to duration of immunization, dose of antigen, and route of immunization. These factors are critical to successful immunization studies, yet optimal protocols are determined almost completely empirically. There is no effective way to determine the correct combination of these parameters without actually conducting the lengthy and labor-intensive experiments. Adjuvant selection is another critical experimental variable and, in the absence of knowledge of known markers of protective immunity, little guidance can be offered. Many recent immunization studies performed in the authors' laboratories have used the Ribi adjuvant, which contains monophosphoryl lipid A (a TLR4 agonist), synthetic trehalose dicorynomycolate, and cell wall (CW) skeleton. This adjuvant, which enhances the development of cellular immunity in addition to antibody, has not been commercially available for several years, but is soon to become available again. Other than complete or incomplete Freund's adjuvant, other more modern adjuvants have not been explored with *T. pallidum* vaccine candidate antigens.

ALTERNATIVE IMMUNIZATION APPROACHES

DNA vaccination and live attenuated bacterial (LAB) vaccines constitute two very promising branches of modern vaccinology (Ulmer et al., 2006) whose application has not been reported yet for syphilis. Unpublished studies from our laboratories, conducted in collaboration with Cameron, using a DNA vaccination approach for Gpd immunization, failed to induce significant protection. Although at the present time DNA vaccination has no practical use in humans (so far no clinical trial for DNA vaccines have gone beyond Phase I/II) (Liu and Ulmer, 2005; Laddy and Weiner,

2006), their efficacy in many animal models has been widely documented (Ulmer et al., 2006). It is encouraging that DNA vaccination stimulates both branches of the immune response, and that the initial limitations due to poor delivery of the antigen-encoding DNA are gradually being overcome by many innovative DNA delivery systems (including the employment of microparticles targeting antigen-presenting cells, electroporation, and intradermal DNA "tattooing") (O'Hagan et al., 2006; Ulmer et al., 2006). Furthermore, the employment of DNA-vaccine adjuvant (i.e., CpG motifs in the vector backbone to activate intracellular TLR and accessory genes coding for cytokines and chemokines) (Coban et al., 2005) has greatly improved the effective immunogenicity of these vaccines. Particularly interesting is also the evidence seen in recent studies that DNA priming followed by protein boosting can increase the level of protection above that obtained with the protein alone (Hall et al., 2002); this may be an interesting approach to enhance the immune response against many *T. pallidum* antigens already shown to induce partial protection.

The use of LAB vaccines is another interesting strategy (Daudel et al., 2007), though not yet reported for syphilis; the administration of *T. pallidum* antigens using a living carrier, rather than a coding sequence as in DNA vaccination, could provide many advantages, including the mimicry of a natural infection, the natural adjuvant proprieties of bacterial components, and potentially better immunogen conformation. Live bacterial vaccines might also be administered orally or nasally through aerosolization, with the advantage of stimulating mucosal immunity, which is not achieved by the more traditional immunization practices discussed so far. Although very appealing, this technique also has potential hazards that will need to be addressed carefully before the application of LAB vaccines to humans, including the possibility of systemic infection, reversion of the carrier bacterial strain from attenuated to virulent, generic perturbation of the natural flora of the gastrointestinal or respiratory tracts, and accidental transfer of genetically modified organisms to unimmunized individuals (Detmer and Glenting, 2006).

CONCLUSIONS

It is clear today that syphilis pathogenesis involves a very complex strategy that *T. pallidum* uses to evade the immune response of the host. This strategy exploits the limited surface immunogenicity due to a remarkable paucity of integral OM proteins, the rapid generation of antigenic variants of TprK, and probably additional yet-unidentified mechanisms. The recognition of these multi-faceted strategies has strongly underlined the necessity to identify those rare surface-exposed antigens in the attempt to develop more effective vaccines. While the extreme fragility of the *T. pallidum* OM and the absence of a host-free cultivation system impede that search, significant progress has been made in the past decade in our understanding of *T. pallidum* and its antigenic structure. Even with cheap and effective therapy, syphilis is uncontrolled throughout much of the world; a vaccine is the best hope for disease control and eradication. The improvement of existing molecular and immunological methodologies, the promise of even more sophisticated advances in vaccinology, and the public health need for a syphilis vaccine continually serve to bolster the enthusiasm and resolve of the investigators who tackle this difficult organism.

KEY ISSUES

- Syphilis is a major public health problem globally, with the highest burden in developing countries. In addition to adult infection, syphilis is responsible for a high proportion of stillbirth, spontaneous abortion, and neonatal morbidity and mortality in developing countries.
- Cheap and effective diagnosis and treatment of syphilis are available, yet the disease burden remains high. Lack of access to medical care in many countries and failure of traditional disease control methods to eliminate syphilis underscore the need for a vaccine.
- The frequent and prolonged administration of *T. pallidum* attenuated by gamma-irradiation has been shown to provide complete and long-lasting protection against infectious challenge in the rabbit model. This approach, however, cannot be adapted for human studies.
- Because *T. pallidum* cannot be genetically manipulated or grown in culture, and its OM is extraordinarily fragile, its protective antigen(s) have not been identified. The availability of the *T. pallidum* genome sequence, however, has facilitated the identification of key putative surface -exposed proteins involved in attachment and antigenic variation. Many of these have been tested for their ability to induce protective immunity.
- Immunization studies in rabbits using recombinant antigens have demonstrated significant attenuation of lesion development and progression following

infectious challenge, but have not provided complete protection against infection.

- The TprK antigen, which undergoes antigenic variation, and likely contributes to *T. pallidum*'s ability to evade the host immune response, is among the most promising targets for the development of a vaccine against syphilis infection. Production of a broadly protective TprK-based vaccine, however, will likely require immunization with cocktails of peptides representing the variable regions.
- New approaches to immunization under investigation include DNA prime/protein boost strategies and development of chimeric recombinant proteins.

ACKNOWLEDGMENTS

The authors are grateful to Elizabeth Robertson for manuscript preparation. We thank Christa Castro, Barbara Molini, Charmie Godornes and Lynn Barrett for technical assistance with previously unpublished studies. This work was supported by grants AI 34616, AI 42143, and AI 63940 from the National Institutes of Health.

References

Al-Samarrai, H.T. and Henderson, W.G. Immunity in syphilis. Studies in active immunity. Br. J. Vener. Dis. 1976; 52:300–308.

Alder, J.D., Daugherty, N., Harris, O.N., Liu, H., Steiner, B.M. and Schell, R.F. Phagocytosis of *Treponema pallidum pertenue* by hamster macrophages on membrane filters. J. Infect. Dis. 1989; 160:289–297.

Alder, J.D., Friess, L., Tengowski, M. and Schell, R.F. Phagocytosis of opsonized *Treponema pallidum* subsp. pallidum proceeds slowly. Infect. Immun. 1990; 58:1167–1173.

Alderete, J.F. and Baseman, J.B. Surface-associated host proteins on virulent *Treponema pallidum*. Infect. Immun. 1979; 26:1048–1056.

Aliprantis, A.O., Yang, R.B., Mark, M.R., Suggett, S., Devaux, B., Radolf, J.D., Klimpel, G.R., Godowski, P. and Zychlinsky, A. Cell activation and apoptosis by bacterial lipoproteins through toll-like receptor-2. Science 1999; 285:736–739.

Arroll, T.W., Centurion-Lara, A., Lukehart, S.A. and Van Voorhis, W.C. T-cell responses to *Treponema pallidum* subsp. *pallidum* antigens during the course of experimental syphilis infection. Infect. Immun. 1999; 67:4757–4763.

Azar, H.A., Pham, T.D. and Kurban, A.K. An electron microscopic study of a syphilitic chancre. Engulfment of *Treponema pallidum* by plasma cells. Arch. Pathol. 1970; 90:143–150.

Baker-Zander, S.A., Fohn, M.J. and Lukehart, S.A. Development of cellular immunity to individual soluble antigens of *Treponema pallidum* during experimental syphilis. J. Immunol. 1988; 141:4363–4369.

Baker-Zander, S.A., Hook, E.W., III, Bonin, P., Handsfield, H.H. and Lukehart, S.A. Antigens of *Treponema pallidum* recognized by IgG and IgM antibodies during syphilis in humans. J. Infect. Dis. 1985; 151:264–272.

Baker-Zander, S.A. and Lukehart, S.A. Macrophage-mediated killing of opsonized *Treponema pallidum*. J. Infect. Dis. 1992; 165:69–74.

Baker-Zander, S.A., Shaffer, J.M. and Lukehart, S.A. Characterization of the serum requirement for macrophage-mediated killing of *Treponema pallidum* ssp. *pallidum*: relationship to the development of opsonizing antibodies. FEMS Immunol. Med. Microbiol. 1993a; 6:273–279.

Baker-Zander, S.A., Shaffer, J.M. and Lukehart, S.A. VDRL antibodies enhance phagocytosis of *Treponema pallidum* by macrophages. J. Infect. Dis. 1993b; 167:1100–1105.

Baseman, J.B. and Hayes, E.C. Molecular characterization of receptor binding proteins and immunogens of virulent *Treponema pallidum*. J. Exp. Med. 1980; 151:573–586.

Baughn, R.E. and Musher, D.M. Secondary syphilitic lesions. Clin. Microbiol. Rev. 2005; 18:205–216.

Belisle, J.T., Brandt, M.E., Radolf, J.D. and Norgard, M.V. Fatty acids of *Treponema pallidum* and *Borrelia burgdorferi* lipoproteins. J. Bacteriol. 1994; 176:2151–2157.

Bishop, N.H. and Miller, J.N. Humoral immunity in experimental syphilis. I. The demonstration of resistance conferred by passive immunization. J. Immunol. 1976a; 117:191–196.

Bishop, N.H. and Miller, J.N. Humoral immunity in experimental syphilis. II. The relationship of neutralizing factors in immune serum to acquired resistance. J. Immunol. 1976b; 117:197–207.

Blanco, D.R., Champion, C.I., Dooley, A., Cox, D.L., Whitelegge, J.P., Faull, K. and Lovett, M.A. A monoclonal antibody that conveys in vitro killing and partial protection in experimental syphilis binds a phosphorylcholine surface epitope of *Treponema pallidum*. Infect. Immun. 2005; 73:3083–3095.

Blanco, D.R., Radolf, J.D., Lovett, M.A. and Miller, J.N. The antigenic interrelationship between the endoflagella of *Treponema phagedenis* biotype Reiter and *Treponema pallidum* Nichols strain. I. Treponemicidal activity of cross-reactive endoflagellar antibodies against *T. pallidum*. J. Immunol. 1986; 137:2973–2979.

Blanco, D.R., Walker, E.M., Haake, D.A., Champion, C.I., Miller, J.N. and Lovett, M.A. Complement activation limits the rate of in vitro treponemicidal activity and correlates with antibody-mediated aggregation of *Treponema pallidum* rare outer membrane protein. J. Immunol. 1990; 144:1914–1921.

Boom, W.H., Liebster, L., Abbas, A.K. and Titus, R.G. Patterns of cytokine secretion in murine leishmaniasis: correlation with disease progression or resolution. Infect. Immun. 1990; 58:3863–3870.

Borenstein, L.A., Radolf, J.D., Fehniger, T.E., Blanco, D.R., Miller, J.N. and Lovett, M.A. Immunization of rabbits with recombinant *Treponema pallidum* surface antigen 4D alters the course of experimental syphilis. J. Immunol. 1988; 140:2415–2421.

Bos, M.P. and Tommassen, J. Biogenesis of the Gram-negative bacterial outer membrane. Curr. Opin. Microbiol. 2004; 7: 610–616.

Bourell, K.W., Schulz, W., Norgard, M.V. and Radolf, J.D. *Treponema pallidum* rare outer membrane proteins: analysis of mobility by freeze-fracture electron microscopy. J. Bacteriol. 1994; 176:1598–1608.

Brinkman, M.B., McKevitt, M., McLoughlin, M., Perez, C., Howell, J., Weinstock, G.M., Norris, S.J. and Palzkill, T. Reactivity of antibodies from syphilis patients to a protein array representing the *Treponema pallidum* proteome. J. Clin. Microbiol. 2006; 44:888–891.

Cameron, C.E. Identification of a *Treponema pallidum* laminin-binding protein. Infect. Immun. 2003; 71:2525–2533.

Cameron, C.E., Castro, C., Lukehart, S.A. and Van Voorhis, W.C. Function and protective capacity of *Treponema pallidum* subsp. *pallidum* glycerophosphodiester phosphodiesterase. Infect. Immun. 1998; 66:5763–5770.

Cameron, C.E., Brown, E.L., Kuroiwa, J.M., Schnapp, L.M. and Brouwer, N.L. *Treponema pallidum* fibronectin-binding proteins. J. Bacteriol. 2004; 186:7019–7022.

Cameron, C.E., Lukehart, S.A., Castro, C., Molini, B., Godornes, C. and Van Voorhis, W.C. Opsonic potential, protective capacity, and sequence conservation of the *Treponema pallidum* subspecies *pallidum* Tp92. J. Infect. Dis. 2000; 181:1401–1413.

CDC. *Haemophilus* b conjugate vaccines for prevention of *Haemophilus influenzae* type b disease among infants and children two months of age and older. Recommendations of the immunization practices advisory committee (ACIP). MMWR Recomm. Rep. Atlanta, GA: U.S. Department of Health and Human Services, Centers for Disease Control and Prevention, 1991.

CDC. HIV prevention through early detection and treatment of other sexually transmitted diseases—United States. MMWR Recomm. Rep. Atlanta, GA: U.S. Department of Health and Human Services, Centers for Disease Control and Prevention, 1998.

CDC. The national plan to eliminate syphilis in the United States. Atlanta, GA: U.S. Department of Health and Human Services, Centers for Disease Control and Prevention, 1999.

CDC. Sexually Transmitted Diseases Treatment Guidelines 2002. MMWR. Atlanta, GA: U.S. Department of Health and Human Services, Centers for Disease Control and Prevention, 2002.

CDC. Sexually Transmitted Diseases Treatment Guidelines, 2006. MMWR Recomm. Rep. Atlanta, GA: U.S. Department of Health and Human Services, Centers for Disease Control and Prevention, 2006a.

CDC. Sexually Transmitted Disease Surveillance, 2005 Supplement. Syphilis Surveillance Report. Atlanta, GA: U.S. Department of Health and Human Services, Centers for Disease Control and Prevention, 2006b.

Centurion-Lara, A., Arroll, T., Castillo, R., Shaffer, J.M., Castro, C., Van Voorhis, W.C. and Lukehart, S.A. Conservation of the 15-kilodalton lipoprotein among *Treponema pallidum* subspecies and strains and other pathogenic treponemes: genetic and antigenic analyses. Infect. Immun. 1997; 65:1440–1444.

Centurion-Lara, A., Castro, C., Barrett, L., Cameron, C., Mostowfi, M., Van Voorhis, W.C. and Lukehart, S.A. *Treponema pallidum* major sheath protein homologue Tpr K is a target of opsonic antibody and the protective immune response. J. Exp. Med. 1999; 189:647–656.

Centurion-Lara, A., Godornes, C., Castro, C., Van Voorhis, W.C. and Lukehart, S.A. The *tprK* gene is heterogeneous among *Treponema pallidum* strains and has multiple alleles. Infect. Immun. 2000a; 68:824–831.

Centurion-Lara, A., LaFond, R.E., Hevner, K., Godornes, C., Molini, B.J., Van Voorhis, W.C. and Lukehart, S.A. Gene conversion: a mechanism for generation of heterogeneity in the *tprK* gene of *Treponema pallidum* during infection. Mol. Microbiol. 2004; 52:1579–1596.

Centurion-Lara, A., Sun, E.S., Barrett, L.K., Castro, C., Lukehart, S.A. and Van Voorhis, W.C. Multiple alleles of *Treponema pallidum* repeat gene D in *Treponema pallidum* isolates. J. Bacteriol. 2000b; 182:2332–2335.

Chamberlain, N.R., Brandt, M.E., Erwin, A.L., Radolf, J.D. and Norgard, M.V. Major integral membrane protein immunogens of *Treponema pallidum* are proteolipids. Infect. Immun. 1989a; 57:2872–2877.

Chamberlain, N.R., DeOgny, L., Slaughter, C., Radolf, J.D. and Norgard, M.V. Acylation of the 47-kilodalton major membrane immunogen of *Treponema pallidum* determines its hydrophobicity. Infect. Immun. 1989b; 57:2878–2885.

Champion, C.I., Miller, J.N., Borenstein, L.A., Lovett, M.A. and Blanco, D.R. Immunization with *Treponema pallidum* endoflagella alters the course of experimental rabbit syphilis. Infect. Immun. 1990; 58:3158–3161.

Chapel, T.A. The signs and symptoms of secondary syphilis. Sex. Transm. Dis. 1980; 7:161–164.

Chen, Z.Q., Zhang, G.C., Gong, X.D., Lin, C., Gao, X., Liang, G.J., Yue, X.L., Chen, X.S. and Cohen, M.S. Syphilis in China: results of a national surveillance programme. Lancet 2007; 369: 132–138.

Choi, K.H., Ning, Z., Gregorich, S.E. and Pan, Q.C. The influence of social and sexual networks in the spread of HIV and syphilis among men who have sex with men in Shanghai, China. J. Acquir. Immune. Defic. Syndr. 2007; 45:77–84.

Christiansen, S. Protective layer covering pathogenic treponematosis. Lancet 1963; 1:423–425.

Coban, C., Ishii, K.J., Gursel, M., Klinman, D.M. and Kumar, N. Effect of plasmid backbone modification by different human CpG motifs on the immunogenicity of DNA vaccine vectors. J. Leukoc. Biol. 2005; 78:647–655.

Cowan, S. Syphilis in Denmark—outbreak among MSM in Copenhagen, 2003-2004. Euro. Surveill. 2004; 9:25–27.

Cox, D.L., Chang, P., McDowall, A.W. and Radolf, J.D. The outer membrane, not a coat of host proteins, limits antigenicity of virulent *Treponema pallidum*. Infect. Immun. 1992; 60:1076–1083.

Cox, D.L., Riley, B., Chang, P., Sayahtaheri, S., Tassell, S. and Hevelone, J. Effects of molecular oxygen, oxidation-reduction potential, and antioxidants upon in vitro replication of *Treponema pallidum* subsp. *pallidum*. Appl. Environ. Microbiol. 1990; 56:3063–3072.

Creek, T.L., Thuku, H., Kolou, B., Rahman, M. and Kilmarx, P.H. Declining syphilis prevalence among pregnant women in northern Botswana: an encouraging sign for the HIV epidemic? Sex. Transm. Infect. 2005; 81:453–455.

Cullen, P.A. and Cameron, C.E. Progress towards an effective syphilis vaccine: the past, present and future. Expert. Rev. Vaccines 2006; 5:67–80.

Cumberland, M.C. and Turner, T.B. The rate of multiplication of *Treponema pallidum* in normal and immune rabbits. Am. J. Syph. Gonorrhea Vener. Dis. 1949; 33:201–212.

Cunningham, T.M., Walker, E.M., Miller, J.N. and Lovett, M.A. Selective release of the *Treponema pallidum* outer membrane and associated polypeptides with Triton X-114. J. Bacteriol. 1988; 170:5789–5796.

Daudel, D., Weidinger, G. and Spreng, S. Use of attenuated bacteria as delivery vectors for DNA vaccines. Expert. Rev. Vaccines 2007; 6:97–110.

Detmer, A. and Glenting, J. Live bacterial vaccines-a review and identification of potential hazards. Microb. Cell Fact. 2006; 5:23.

Fehniger, T.E., Radolf, J.D. and Lovett, M.A. Properties of an ordered ring structure formed by recombinant *Treponema pallidum* surface antigen 4D. J. Bacteriol. 1986a; 165:732–739.

Fehniger, T.E., Radolf, J.D., Walfield, A.M., Cunningham, T.M., Miller, J.N. and Lovett, M.A. Native surface association of a recombinant 38-kilodalton *Treponema pallidum* antigen isolated from the *Escherichia coli* outer membrane. Infect. Immun. 1986b; 52:586–593.

Fenno, J.C., Muller, K.H. and McBride, B.C. Sequence analysis, expression, and binding activity of recombinant major outer sheath protein (Msp) of *Treponema denticola*. J. Bacteriol. 1996; 178:2489–2497.

Fieldsteel, A.H., Cox, D.L. and Moeckli, R.A. Further studies on replication of virulent *Treponema pallidum* in tissue cultures of Sf1Ep cells. Infect. Immun. 1982; 35:449–455.

Fitzgerald, T.J. Syphilis vaccine: up-regulation of immunogenicity by cyclophosphamide, Ribi adjuvant, and indomethacin confers significant protection against challenge infection in rabbits. Vaccine 1991; 9:266–272.

Fitzgerald, T.J. The Th1/Th2-like switch in syphilitic infection: is it detrimental? Infect. Immun. 1992; 60:3475–3479.

Fitzgerald, T.J., Johnson, R.C., Miller, J.N. and Sykes, J.A. Characterization of the attachment of *Treponema pallidum* (Nichols strain) to cultured mammalian cells and the potential relationship of attachment to pathogenicity. Infect. Immun. 1977a; 18:467–478.

Fitzgerald, T.J., Johnson, R.C., Sykes, J.A. and Miller, J.N. Interaction of *Treponema pallidum* (Nichols strain) with cultured mammalian cells: effects of oxygen, reducing agents, serum supplements, and different cell types. Infect. Immun. 1977b; 15:444–452.

Fitzgerald, T.J., Miller, J.N. and Sykes, J.A. *Treponema pallidum* (Nichols strain) in tissue cultures: cellular attachment, entry, and survival. Infect. Immun. 1975; 11:1133–1140.

Fitzgerald, T.J. and Repesh, L.A. Interactions of fibronectin with *Treponema pallidum*. Genitourin. Med. 1985; 61:147–155.

Fitzgerald, T.J., Repesh, L.A., Blanco, D.R. and Miller, J.N. Attachment of *Treponema pallidum* to fibronectin, laminin, collagen IV, and collagen I, and blockage of attachment by immune rabbit IgG. Br. J. Vener. Dis. 1984; 60:357–363.

Fitzgerald, T.J., Tomai, M.A., Trachte, G.J. and Rice, T. Prostaglandins in experimental syphilis: treponemes stimulate adherent spleen cells to secrete prostaglandin E2, and indomethacin upregulates immune functions. Infect. Immun. 1991; 59:143–149.

Flack, F.S., Loosmore, S., Chong, P. and Thomas, W.R. The sequencing of the 80-kDa D15 protective surface antigen of *Haemophilus influenzae*. Gene 1995; 156:97–99.

Folds, J.D., Rauchbach, A.S., Shores, E. and Saunders, J.M. Evaluation of the inbred mouse as a model for experimental *Treponema pallidum* infection. Scand. J. Immunol. 1983; 18:201–206.

Fraser, C.M., Norris, S.J., Weinstock, G.M., White, O., Sutton, G.G., Dodson, R., Gwinn, M., Hickey, E.K., Clayton, R., Ketchum, K.A., Sodergren, E., Hardham, J.M., McLeod, M.P., Salzberg, S., Peterson, J., Khalak, H., Richardson, D., Howell, J.K., Chidambaram, M., Utterback, T., McDonald, L., Artiach, P., Bowman, C., Cotton, M.D. and Venter, J.C. Complete genome sequence of *Treponema pallidum*, the syphilis spirochete. Science 1998; 281:375–388.

Gelperin, A. Immunochemical studies of the Reiter spirochete. Am. J. Syph. 1951; 35:1–13.

Gerbase, A.C., Rowley, J.T., Heymann, D.H., Berkely, S.F. and Piot, P. Global prevalence and incidence estimates of selected curable STDs. Sex. Transm. Infect. 1998; 74:S12–S16.

Giacani, L., Hevner, K. and Centurion-Lara, A. Gene organization and transcriptional analysis of the *tprJ*, *tprI*, *tprG* and *tprF* loci in the Nichols and Sea 81-4 *Treponema pallidum* isolates. J. Bacteriol. 2005a; 187:6084–6093.

Giacani, L., Lukehart, S. and Centurion-Lara, A. Length of guanosine homopolymeric repeats modulates promoter activity of Subfamily II *tpr* genes of *Treponema pallidum* ssp. *pallidum*. FEMS Immunol. Med. Microbiol. 2007a; 51:289–301.

Giacani, L., Molini, B., Godornes, C., Barrett, L., Van Voorhis, W.C., Centurion-Lara, A. and Lukehart, S.A. Quantitative analysis of *tpr* gene expression in *Treponema pallidum* isolates: differences among isolates and correlation with T-cell responsiveness in experimental syphilis. Infect. Immun. 2007b; 75:104–112.

Giacani, L., Sambri, V., Marangoni, A., Cavrini, F., Storni, E., Donati, M., Corona, S., Lanzarini, P. and Cevenini, R. Immunological evaluation and cellular location analysis of the TprI antigen of *Treponema pallidum* subsp. *pallidum*. Infect. Immun. 2005b; 73:3817–3822.

Giacani, L., Sun, E.S., Hevner, K., Molini, B.J., Van Voorhis, W.C., Lukehart, S.A. and Centurion-Lara, A. Tpr homologs in *Treponema paraluiscuniculi* Cuniculi A strain. Infect. Immun. 2004; 72:6561–6576.

Gjestland, T. The Oslo study of untreated syphilis. Acta Derm. Venereol. 1955; 35:11–368.

Godornes, C., Leader, B.T., Molini, B.J., Centurion-Lara, A. and Lukehart, S.A. Quantitation of rabbit cytokine mRNA by real-time RT-PCR. Cytokine 2007; 38:1–7.

Grandi, G. Antibacterial vaccine design using genomics and proteomics. Trends Biotechnol. 2001; 19:181–188.

Graves, S. Sequential changes in susceptibility to *Treponema pallidum* of rabbits previously infected with *Treponema paraluis-cuniculi*. Br. J. Vener. Dis. 1981; 57:11–14.

Graves, S., Drummond, L. and Strugnell, R. Immunization of rabbits with *Spirochaeta aurantia* does not induce resistance to *Treponema pallidum*. Sex. Transm. Dis. 1984; 11:1–5.

Hall, R., Adamson, R. and Boulter, N. Prime-boost: the way forward for recombinant vaccines against apicomplexan parasites. A Theileria perspective. Parasitol. Res. 2002; 88:S45–S47.

Handsfield, H.H., Lukehart, S.A., Sell, S., Norris, S.J. and Holmes, K.K. Demonstration of *Treponema pallidum* in a cutaneous gumma by indirect immunofluorescence. Arch. Dermatol. 1983; 119: 677–680.

Hanff, P.A., Fehniger, T.E., Miller, J.N. and Lovett, M.A. Humoral immune response in human syphilis to polypeptides of *Treponema pallidum*. J. Immunol. 1982; 129:1287–1291.

Hansen, E.B., Pedersen, P.E., Schouls, L.M., Severin, E. and van Embden, J.D. Genetic characterization and partial sequence determination of a *Treponema pallidum* operon expressing two immunogenic membrane proteins in *Escherichia coli*. J. Bacteriol. 1985; 162:1227–1237.

Hardham, J.M. Identification and Characterization of *Treponema pallidum* Genes Encoding Exported and Membrane Proteins (Dissertation), University of North Carolina, Chapel Hill, NC, 1995.

Hardy, P.H., Jr. and Levin, J. Lack of endotoxin in *Borrelia hispanica* and *Treponema pallidum*. Proc. Soc. Exp. Biol. Med. 1983; 174:47–52.

Hazlett, K.R., Cox, D.L., Decaffmeyer, M., Bennett, M.P., Desrosiers, D.C., La Vake, C.J., La Vake, M.E., Bourell, K.W., Robinson, E.J., Brasseur, R. and Radolf, J.D. TP0453, a concealed outer membrane protein of *Treponema pallidum*, enhances membrane permeability. J. Bacteriol. 2005; 187:6499–6508.

Hazlett, K.R.O., Sellati, T.J., Nguyen, T.T., Cox, D.L., Clawson, M.L., Caimano, M.J. and Radolf, J.D. The TprK protein of *Treponema pallidum* is periplasmic and is not a target of opsonic antibody or protective immunity. J. Exp. Med. 2001; 193:1015–1026.

Hindersson, P., Petersen, C.S. and Axelsen, N.H. Purified flagella from *Treponema phagedenis* biotype Reiter does not induce protective immunity against experimental syphilis in rabbits. Sex. Transm. Dis. 1985; 12:124–127.

Hook, E.W., III. Is elimination of endemic syphilis transmission a realistic goal for the USA? Lancet 1998; 351(Suppl. 3):19–21.

Hook, E.W., III, Martin, D.H., Stephens, J., Smith, B.S. and Smith, K. A randomized, comparative pilot study of azithromycin versus benzathine penicillin G for treatment of early syphilis. Sex. Transm. Dis. 2002; 29:486–490.

Hook, E.W., III, Stephens, J. and Ennis, D.M. Azithromycin compared with penicillin G benzathine for treatment of incubating syphilis. Ann. Intern. Med. 1999; 131:434–437.

Hovind-Hougen, K. Morpholgy. In: Pathogenesis and Immunology of Treponemal Infection (R.F. Schell and D.M. Musher, (Eds.)). New York, NY: Marcel Dekker, Inc, 1983.

Izzat, N.N., Dacres, W.G., Knox, J.M. and Wende, R. Attempts at immunization against syphilis with a virulent *Treponema pallidum*. Br. J. Vener Dis. 1970; 46:451–453.

Janson, H., Heden, L.O. and Forsgren, A. Protein D, the immunoglobulin D-binding protein of *Haemophilus influenzae,* is a lipoprotein. Infect. Immun. 1992; 60:1336–1342.

Laddy, D.J. and Weiner, D.B. From plasmids to protection: a review of DNA vaccines against infectious diseases. Int. Rev. Immunol. 2006; 25:99–123.

LaFond, R.E., Centurion-Lara, A., Godornes, C., Rompalo, A.M., Van Voorhis, W.C. and Lukehart, S.A. Sequence diversity of *Treponema pallidum* subsp. *pallidum tprK* in human syphilis lesions and rabbit-propagated isolates. J. Bacteriol. 2003; 185:6262–6268.

LaFond, R.E., Centurion-Lara, A., Godornes, C., Van Voorhis, W.C. and Lukehart, S.A. TprK sequence diversity accumulates during infection of rabbits with *Treponema pallidum* subsp. *pallidum* Nichols strain. Infect. Immun. 2006a; 74:1896–1906.

LaFond, R.E. and Lukehart, S.A. Biological basis for syphilis. Clin. Microbiol. Rev. 2006; 19:29–49.

LaFond, R.E., Molini, B.J., Van Voorhis, W.C. and Lukehart, S.A. Antigenic variation of TprK V regions abrogates specific antibody binding in syphilis. Infect. Immun. 2006b; 74: 6244–6251.

Larsen, S.A., Steiner, B.M. and Rudolph, A.H. Laboratory diagnosis and interpretation of tests for syphilis. Clin. Microb. Rev. 1995; 8:1–21.

Leader, B.T., Godornes, C., Vanvoorhis, W.C. and Lukehart, S.A. CD4+ Lymphocytes and gamma interferon predominate in local immune responses in early experimental syphilis. Infect. Immun. 2007; 75:3021–3026.

Leader, B.T., Hevner, K., Molini, B.J., Barrett, L.K., Van Voorhis, W.C. and Lukehart, S.A. Antibody responses elicited against the *Treponema pallidum* repeat proteins differ during infection with different isolates of *Treponema pallidum* subsp. *pallidum*. Infect. Immun. 2003; 71:6054–6057.

Lien, E., Sellati, T.J., Yoshimura, A., Flo, T.H., Rawadi, G., Finberg, R.W., Carroll, J.D., Espevik, T., Ingalls, R.R., Radolf, J.D. and Golenbock, D.T. Toll-like receptor 2 functions as a pattern recognition receptor for diverse bacterial products. J. Biol. Chem. 1999; 274:33419–33425.

Lilburn, T.G., Kim, K.S., Ostrom, N.E., Byzek, K.R., Leadbetter, J.R. and Breznak, J.A. Nitrogen fixation by symbiotic and free-living spirochetes. Science 2001; 292:2495–2498.

Liu, H., Grusky, O., Li, X. and Ma, E. Drug users: a potentially important bridge population in the transmission of sexually transmitted diseases, including AIDS, in China. Sex. Transm. Dis. 2006; 33:111–117.

Liu, M.A. and Ulmer, J.B. Human clinical trials of plasmid DNA vaccines. Adv. Genet. 2005; 55:25–40.

Lukehart, S.A. Prospects for development of a treponemal vaccine. Rev. Infect. Dis. 1985; 7(Suppl. 2):S305–S313.

Lukehart, S.A., Baker-Zander, S.A. and Gubish, E.R., Jr. Identification of *Treponema pallidum* antigens: comparison with a nonpathogenic treponeme. J. Immunol. 1982; 129:833–838.

Lukehart, S.A., Baker-Zander, S.A., Lloyd, R.M. and Sell, S. Characterization of lymphocyte responsiveness in early experimental syphilis. II. Nature of cellular infiltration and *Treponema pallidum* distribution in testicular lesions. J. Immunol. 1980a; 124:461–467.

Lukehart, S.A., Baker-Zander, S.A. and Sell, S. Characterization of lymphocyte responsiveness in early experimental syphilis. I. In vitro response to mitogens and *Treponema pallidum* antigens. J. Immunol. 1980b; 124:454–460.

Lukehart, S.A., Baker-Zander, S.A. and Sell, S. Characterization of the humoral immune response of the rabbit to antigens of *Treponema pallidum* after experimental infection and therapy. Sex. Transm. Dis. 1986; 13:9–15.

Lukehart, S.A., Godornes, C., Molini, B.J., Sonnett, P., Hopkins, S., Mulcahy, F., Engelman, J., Mitchell, S.J., Rompalo, A.M., Marra, C.M. and Klausner, J.D. Macrolide resistance in *Treponema pallidum* in the United States and Ireland. N. Engl. J. Med. 2004; 351:154–158.

Lukehart, S.A. and Miller, J.N. Demonstration of the in vitro phagocytosis of *Treponema pallidum* by rabbit peritoneal macrophages. J. Immunol. 1978; 121:2014–2024.

Lukehart, S.A., Tam, M.R., Hom, J., Baker-Zander, S.A., Holmes, K.K. and Nowinski, R.C. Characterization of monoclonal antibodies to *Treponema pallidum*. J. Immunol. 1985; 134:585–592.

Magnuson, H.J., Eagle, H. and Fleischman, R. The minimal infectious inoculum of *Spirochaeta pallida* (Nichols Strain), and a consideration of its rate of multiplication in vivo. Am. J. Syph. 1948; 32:1–18.

Magnuson, H.J., Thomas, E.W., Olansky, S., Kaplan, B.I., DeMello, L. and Cutler, J.C. Inoculation syphilis in human volunteers. Medicine 1956; 35:33–82.

Manning, D.S., Reschke, D.K. and Judd, R.C. Omp85 proteins of *Neisseria gonorrhoeae* and *Neisseria meningitidis* are similar to *Haemophilus influenzae* D-15-Ag and *Pasteurella multocida* Oma87. Microb. Pathog. 1998; 25:11–21.

Marcus, U., Bremer, V. and Hamouda, O. Syphilis surveillance and trends of the syphilis epidemic in Germany since the mid-90s. Euro. Surveill. 2004; 9:11–14.

Mashkilleyson, A.L., Gomberg, M.A., Mashkilleyson, N. and Kutin, S.A. Treatment of syphilis with azithromycin. Int. J. STD AIDS 1996; 7(Suppl. 1):13–15.

Mathers, D.A., Leung, W.K., Fenno, J.C., Hong, Y. and McBride, B.C. The major surface protein complex of *Treponema denticola* depolarizes and induces ion channels in HeLa cell membranes. Infect. Immun. 1996; 64:2904–2910.

McKevitt, M., Brinkman, M.B., McLoughlin, M., Perez, C., Howell, J.K., Weinstock, G.M., Norris, S.J. and Palzkill, T. Genome scale identification of *Treponema pallidum* antigens. Infect. Immun. 2005; 73:4445–4450.

Medici, M.A. The immunoprotective niche—a new pathogenic mechanism for syphilis, the systemic mycoses and other infectious diseases. J. Theor. Biol. 1972; 36:617–625.

Miller, J.N. Immunity in experimental syphilis. V. The immunogenicity of *Treponema pallidum* attenuated by g-irradiation. J. Immunol. 1967; 99:1012–1016.

Miller, J.N. Immunity in experimental syphilis. VI. Successful vaccination of rabbits with *Treponema pallidum*, Nichols strain, attenuated by g-irradiation. J. Immunol. 1973; 110:1206–1215.

Miller, J.N. Potential for vaccines for venereal diseases. Bull. N.Y. Acad. Med. 1976; 52:986–1003.

Moore, J.E. and Hopkins, H.H. Asymptomatic neurosyphilis. VI. The prognosis of early and late asymptomatic neurosyphilis. JAMA 1930; 95:1637–1641.

Morgan, C.A., Lukehart, S.A. and Van Voorhis, W.C. Immunization with the N-terminal portion of *Treponema pallidum* repeat protein K attenuates syphilitic lesion development in the rabbit model. Infect. Immun. 2002a; 70:6811–6816.

Morgan, C.A., Lukehart, S.A. and Van Voorhis, W.C. Protection against syphilis correlates with specificity of antibodies to the variable regions of *Treponema pallidum* repeat protein K. Infect. Immun. 2003; 71:5605–5612.

Morgan, C.A., Molini, B.J., Lukehart, S.A. and Van Voorhis, W.C. Segregation of B and T cell epitopes of *Treponema pallidum* repeat protein K to variable and conserved regions during experimental syphilis infection. J. Immunol. 2002b; 169: 952–957.

Morshed, M.G. and Jones, H.D. *Treponema pallidum* macrolide resistance in BC. CMAJ 2006; 174:349.

Nelson, R.A. and Mayer, M.M. Immobilization of *Treponema pallidum* in vitro by antibody produced in syphilis infection. J. Exp. Med. 1949; 89:369–393.

Norgard, M.V., Chamberlain, N.R., Swancutt, M.A. and Goldberg, M.S. Cloning and expression of the major 47-kilodalton surface immunogen of *Treponema pallidum* in *Escherichia coli*. Infect. Immun. 1986a; 54:500–506.

Norgard, M.V., Marchitto, K.S. and Cox, D.L. Phenotypic expression of the major 47 kDa surface immunogen of *Treponema pallidum* in virulent, tissue-cultured treponemes. J. Gen. Microbiol. 1986b; 132(Pt.6):1775–1778.

Norgard, M.V. and Miller, J.N. Cloning and expression of *Treponema pallidum* (Nichols) antigen genes in *Escherichia coli*. Infect. Immun. 1983; 42:435–445.

O'Hagan, D.T., Singh, M. and Ulmer, J.B. Microparticle-based technologies for vaccines. Methods 2006; 40:10–19.

O'Regan, A.W., Castro, C., Lukehart, S.A., Kasznica, J.M., Rice, P.A. and Joyce-Brady, M.F. Barking up the wrong tree? Use of polymerase chain reaction to diagnose syphilitic aortitis. Thorax 2002; 57:917–918.

Penn, C.W., Cockayne, A. and Bailey, M.J. The outer membrane of *Treponema pallidum*: biological significance and biochemical properties. J. Gen. Microbiol. 1985; 131:2349–2357.

Penn, C.W. and Rhodes, J.G. Surface-associated antigens of *Treponema pallidum* concealed by an inert outer layer. Immunology 1982; 46:9–16.

Persaud, N.E., Klaskala, W., Tewari, T., Shultz, J. and Baum, M. Drug use and syphilis. Co-factors for HIV transmission among commercial sex workers in Guyana. West Indian Med. J. 1999; 48:52–56.

Peterson, K.M., Baseman, J.B. and Alderete, J.F. *Treponema pallidum* receptor binding proteins interact with fibronectin. J. Exp. Med. 1983; 157:1958–1970.

Pham, L., Woelk, G.B., Ning, Y., Madzime, S., Mudzamiri, S., Mahomed, K. and Williams, M.A. Seroprevalence and risk factors of syphilis infection in pregnant women delivering at Harare Maternity Hospital, Zimbabwe. Cent. Afr. J. Med. 2005; 51:24–30.

Potter, D., Goldenberg, R.L., Read, J.S., Wang, J., Hoffman, I.F., Saathoff, E., Kafulafula, G., Aboud, S., Martinson, F.E., Dahab, M. and Vermund, S.H. Correlates of syphilis seroreactivity among pregnant women: the HIVNET 024 Trial in Malawi, Tanzania, and Zambia. Sex. Transm. Dis. 2006; 33:604–609.

Purcell, B.K., Swancutt, M.A. and Radolf, J.D. Lipid modification of the 15 kiloDalton major membrane immunogen of *Treponema pallidum*. Mol. Microbiol. 1990; 4:1371–1379.

Quist, E.E., Repesh, L.A., Zeleznikar, R. and Fitzgerald, T.J. Interaction of *Treponema pallidum* with isolated rabbit capillary tissues. Br. J. Vener. Dis. 1983; 59:11–20.

Radolf, J.D. Role of outer membrane architecture in immune evasion by *Treponema pallidum* and *Borrelia burgdorferi*. Trends Microbiol. 1994; 2:307–311.

Radolf, J.D., Arndt, L.L., Akins, D.R., Curetty, L.L., Levi, M.E., Shen, Y., Davis, L.S. and Norgard, M.V. *Treponema pallidum* and *Borrelia burgdorferi* lipoproteins and synthetic lipopeptides activate monocytes/macrophages. J. Immunol. 1995a; 154:2866–2877.

Radolf, J.D., Fehniger, T.E., Silverblatt, F.J., Miller, J.N. and Lovett, M.A. The surface of virulent *Treponema pallidum*: resistance to antibody binding in the absence of complement and surface association of recombinant antigen 4D. Infect. Immun. 1986; 52:579–585.

Radolf, J.D., Moomaw, C., Slaughter, C.A. and Norgard, M.V. Penicillin-binding proteins and peptidoglycan of *Treponema pallidum* subsp. pallidum. Infect. Immun. 1989a; 57:1248–1254.

Radolf, J.D. and Norgard, M.V. Pathogen specificity of *Treponema pallidum* subsp. *pallidum* integral membrane proteins identified by phase partitioning with Triton X-114. Infect. Immun. 1988; 56:1825–1828.

Radolf, J.D., Norgard, M.V., Brandt, M.E., Isaacs, R.D., Thompson, P.A. and Beutler, B. Lipoproteins of *Borrelia burgdorferi* and *Treponema pallidum* activate cachectin/tumor necrosis factor synthesis. Analysis using a CAT reporter construct. J. Immunol. 1991; 147:1968–1974.

Radolf, J.D., Norgard, M.V. and Schulz, W.W. Outer membrane ultrastructure explains the limited antigenicity of virulent *Treponema pallidum*. Proc. Natl. Acad. Sci. U.S.A. 1989b; 86:2051–2055.

Radolf, J.D., Robinson, E.J., Bourell, K.W., Akins, D.R., Porcella, S.F., Weigel, L.M., Jones, J.D. and Norgard, M.V. Characterization of outer membranes isolated from *Treponema pallidum*, the syphilis spirochete. Infect. Immun. 1995b; 63:4244–4252.

Ratnam, A.V., Din, S.N., Hira, S.K., Bhat, G.J., Wacha, D.S., Rukmini, A. and Mulenga, R.C. Syphilis in pregnant women in Zambia. Br. J. Vener Dis. 1982; 58:355–358.

Reed, S.G. In vivo administration of recombinant IFN-gamma induces macrophage activation, and prevents acute disease, immune suppression, and death in experimental *Trypanosoma cruzi* infections. J. Immunol. 1988; 140:4342–4347.

Renton, A.M., Borisenko, K.K., Meheus, A. and Gromyko, A. Epidemics of syphilis in the newly independent states of the former Soviet Union. Sex. Transm. Infect. 1998; 74:165–166.

Righarts, A.A., Simms, I., Wallace, L., Solomou, M. and Fenton, K.A. Syphilis surveillance and epidemiology in the United Kingdom. Euro Surveill. 2004; 9:21–25.

Riley, B.S. and Cox, D.L. Cultivation of cottontail rabbit epidermal (Sf1Ep) cells on microcarrier beads and their use for suspension cultivation of *Treponema pallidum* subsp. *pallidum*. Appl. Environ. Microbiol. 1988; 54:2862–2865.

Robertson, S.M., Kettman, J.R., Miller, J.N. and Norgard, M.V. Murine monoclonal antibodies specific for virulent *Treponema pallidum* (Nichols). Infect. Immun. 1982; 36:1076–1085.

Ruan, M.R., Akkoyunlu, M., Grubb, A. and Forsgren, A. Protein D of Haemophilus influenzae. A novel bacterial surface protein with affinity for human IgD. J. Immunol. 1990; 145:3379–3384.

Ruan, Y., Cao, X., Qian, H.Z., Zhang, L., Qin, G., Jiang, Z., Song, B., Hu, W., Liang, S., Chen, K., Yang, Y., Li, X., Wang, J., Chen, X., Hao, C., Song, Y., Xing, H., Wang, N. and Shao, Y. Syphilis among female sex workers in southwestern China: potential for HIV transmission. Sex. Transm. Dis. 2006; 33:719–723.

Ruffolo, C.G. and Adler, B. Cloning, sequencing, expression, and protective capacity of the *oma*87 gene encoding the *Pasteurella multocida* 87-kilodalton outer membrane antigen. Infect. Immun. 1996; 64:3161–3167.

Salazar, J.C., Hazlett, K.R. and Radolf, J.D. The immune response to infection with *Treponema pallidum*, the stealth pathogen. Microb. Infect. 2002; 4:1133–1140.

Sasse, A., Defraye, A. and Ducoffre, G. Recent syphilis trends in Belgium and enhancement of STI surveillance systems. Euro. Surveill. 2004; 9:6–8.

Schell, R.F. Rabbit and hamster models of treponemal infection. In: Pathogenesis and Immunology of Treponemal Infection (R.F. Schell and D.M. Musher, Eds.). New York: Marcel Dekker, Inc., 1983.

Schell, R.F., LeFrock, J.L., Chan, J.K. and Bagasra, O. LSH hamster model of syphilitic infection. Infect. Immun. 1980; 28:909–913.

Schell, R.F., Musher, D.M., Jacobson, K. and Schwethelm, P. New evidence for the non-infectivity of *Treponema pallidum* for mice. Br. J. Vener. Dis. 1975; 51:19–21.

Schouls, L.M., Ijsselmuiden, O.E., Weel, J. and van Embden, J.D. Overproduction and purification of *Treponema pallidum* recombinant-DNA-derived proteins TmpA and TmpB and their potential use in serodiagnosis of syphilis. Infect. Immun. 1989a; 57:2612–2623.

Schouls, L.M., Mout, R., Dekker, J. and van Embden, J.D. Characterization of lipid-modified immunogenic proteins of *Treponema pallidum* expressed in *Escherichia coli*. Microb. Pathog. 1989b; 7:175–188.

Schouls, L.M., van der Heide, H.G. and van Embden, J.D. Characterization of the 35-kilodalton *Treponema pallidum* subsp. *pallidum* recombinant lipoprotein TmpC and antibody response to lipidated and nonlipidated *T. pallidum* antigens. Infect. Immun. 1991; 59:3536–3546.

Scott, P. IFN-gamma modulates the early development of Th1 and Th2 responses in a murine model of cutaneous leishmaniasis. J. Immunol. 1991; 147:3149–3155.

Sell, S., Baker-Zander, S. and Powell, H.C. Experimental syphilitic orchitis in rabbits: ultrastructural appearance of *Treponema pallidum* during phagocytosis and dissolution by macrophages in vivo. Lab. Invest. 1982; 46:355–364.

Sell, S. and Norris, S.J. The biology, pathology, and immunology of syphilis. Int. Rev. Exp. Pathol. 1983; 24:203–276.

Shang, E.S., Skare, J.T., Erdjument-Bromage, H., Blanco, D.R., Tempst, P., Miller, J.N. and Lovett, M.A. Sequence analysis and characterization of a 40-kilodalton *Borrelia hermsii* glycerophosphodiester phosphodiesterase homolog. J. Bacteriol. 1997; 179:2238–2246.

Shevchenko, D.V., Akins, D.R., Robinson, E.J., Li, M., Shevchenko, O.V. and Radolf, J.D. Identification of homologs for thioredoxin, peptidyl prolyl cis-trans isomerase, and glycerophosphodiester phosphodiesterase in outer membrane fractions from *Treponema pallidum*, the syphilis spirochete. Infect. Immun. 1997; 65:4179–4189.

Shevchenko, D.V., Sellati, T.J., Cox, D.L., Shevchenko, O.V., Robinson, E.J. and Radolf, J.D. Membrane topology and cellular location of the *Treponema pallidum* glycerophosphodiester phosphodiesterase (GlpQ) ortholog. Infect. Immun. 1999; 67:2266–2276.

Silva, J.S., Morrissey, P.J., Grabstein, K.H., Mohler, K.M., Anderson, D. and Reed, S.G. Interleukin 10 and interferon gamma regulation of experimental *Trypanosoma cruzi* infection. J. Exp. Med. 1992; 175:169–174.

Stamm, L.V. and Bergen, H.L. A point mutation associated with bacterial macrolide resistance is present in both 23S rRNA genes of an erythromycin-resistant *Treponema pallidum* clinical isolate. Antimicrob. Agents Chemother. 2000; 44:806–807.

Stamm, L.V., Bergen, H.L. and Walker, R.L. Molecular typing of papillomatous digital dermatitis-associated *Treponema* isolates based on analysis of 16S-23S ribosomal DNA intergenic spacer regions. J. Clin. Microbiol. 2002; 40:3463–3469.

Stamm, L.V., Hodinka, R.L., Wyrick, P.B. and Bassford, P.J., Jr. Changes in the cell surface properties of *Treponema pallidum* that occur during in vitro incubation of freshly extracted organisms. Infect. Immun. 1987; 55:2255–2261.

Stebeck, C.E., Shaffer, J.M., Arroll, T.W., Lukehart, S.A. and Van Voorhis, W.C. Identification of the *Treponema pallidum* subsp. *pallidum* glycerophosphodiester phosphodiesterase homologue. FEMS Microbiol. Lett. 1997; 154:303–310.

Sun, E.S., Molini, B.J., Barrett, L.K., Centurion-Lara, A., Lukehart, S.A. and Van Voorhis, W.C. Subfamily I *Treponema pallidum* repeat protein family: sequence variation and immunity. Microb. Infect. 2004; 6:725–737.

Swancutt, M.A., Radolf, J.D. and Norgard, M.V. The 34-kilodalton membrane immunogen of *Treponema pallidum* is a lipoprotein. Infect. Immun. 1990; 58:384–392.

Sykes, J.A. and Miller, J.N. Intracellular location of *Treponema pallidum* (Nichols strain) in the rabbit testis. Infect. Immun. 1971; 4:307–314.

Sykes, J.A., Miller, J.N. and Kalan, A.J. *Treponema pallidum* within cells of a primary chancre from a human female. Br. J. Vener. Dis. 1974; 50:40–44.

Thomas, D.D., Baseman, J.B. and Alderete, J.F. Fibronectin mediates *Treponema pallidum* cytadherence through recognition of fibronectin cell-binding domain. J. Exp. Med. 1985a; 161:514–525.

Thomas, D.D., Baseman, J.B. and Alderete, J.F. Fibronectin tetrapeptide is target for syphilis spirochete cytadherence. J. Exp. Med. 1985b; 162:1715–1719.

Tosca, A., Lehou, J., Hatjivasiliou, M., Varelzidis, A. and Stratigos, J.D. Infiltrate of syphilitic lesions before and after treatment. Genitourin. Med. 1988; 64:289–293.

Trujillo, L., Munoz, D., Gotuzzo, E., Yi, A. and Watts, D.M. Sexual practices and prevalence of HIV, HTLV-I/II, and *Treponema pallidum* among clandestine female sex workers in Lima, Peru. Sex. Transm. Dis. 1999; 26:115–118.

Turner, T.B., Hardy, P.H. and Newman, B. Infectivity tests in syphilis. Br. J. Vener. Dis. 1969; 45:183–196.

Turner, T.B. and Hollander, D.H. Biology of the Treponematoses. Geneva: World Health Organization, 1957.

Ulmer, J.B., Wahren, B. and Liu, M.A. Gene-based vaccines: recent technical and clinical advances. Trends Mol. Med. 2006; 12:216–222.

van der Ende, A., Hopman, C.T., Zaat, S., Essink, B.B., Berkhout, B. and Dankert, J. Variable expression of class I outer membrane protein in *Neisseria meningitidis* is caused by variation in the spacing between the -10 and -35 regions of the promoter. J. Bacteriol. 1995; 177:2475–2480.

van der Woude, M.W. and Baumler, A.J. Phase and antigenic variation in bacteria. Clin. Microbiol. Rev. 2004; 17:581–611.

van Embden, J.D., van der Donk, H.J., van Eijk, R.V., van der Heide, H.G., de Jong, J.A., van Olderen, M.F., Osterhaus, A.B. and Schouls, L.M. Molecular cloning and expression of *Treponema pallidum* DNA in *Escherichia coli* K-12. Infect. Immun. 1983; 42:187–196.

Van Voorhis, W.C., Barrett, L.K., Koelle, D.M., Nasio, J.M., Plummer, F.A. and Lukehart, S.A. Primary and secondary syphilis lesions contain mRNA for Th1 cytokines. J. Infect. Dis. 1996a; 173:491–495.

Van Voorhis, W.C., Barrett, L.K., Nasio, J.M., Plummer, F.A. and Lukehart, S.A. Lesions of primary and secondary syphilis contain activated cytolytic T cells. Infect. Immun. 1996b; 64:1048–1050.

Voulhoux, R., Bos, M.P., Geurtsen, J., Mols, M. and Tommassen, J. Role of a highly conserved bacterial protein in outer membrane protein assembly. Science 2003; 299:262–265.

Voulhoux, R. and Tommassen, J. Omp85, an evolutionarily conserved bacterial protein involved in outer-membrane-protein assembly. Res. Microbiol. 2004; 155:129–135.

Walfield, A.M., Hanff, P.A. and Lovett, M.A. Expression of Treponema pallidum antigens in *Escherichia coli*. Science 1982; 216:522–523.

Walker, E.M., Borenstein, L.A., Blanco, D.R., Miller, J.N. and Lovett, M.A. Analysis of outer membrane ultrastructure of pathogenic *Treponema* and *Borrelia* species by freeze-fracture electron microscopy. J. Bacteriol. 1991; 173:5585–5588.

Walker, E.M., Zampighi, G.A., Blanco, D.R., Miller, J.N. and Lovett, M.A. Demonstration of rare protein in the outer membrane of *Treponema pallidum* subsp. *pallidum* by freeze-fracture analysis. J. Bacteriol. 1989; 171:5005–5011.

Watson-Jones, D., Changalucha, J., Gumodoka, B., Weiss, H., Rusizoka, M., Ndeki, L., Whitehouse, A., Balira, R., Todd, J., Ngeleja, D., Ross, D., Buve, A., Hayes, R. and Mabey, D. Syphilis in pregnancy in Tanzania. I. Impact of maternal syphilis on outcome of pregnancy. J. Infect. Dis. 2002; 186:940–947.

Weinstock, G.M., Hardham, J.M., McLeod, M.P., Sodergren, E.J. and Norris, S.J. The genome of *Treponema pallidum*: new light on the agent of syphilis. FEMS Microb. Rev. 1998; 22:323–332.

Wicher, K., van Embden, J.D., Schouls, L.M., Zabek, J., Jakubowski, A. and Wicher, V. Immunogenicity of three recombinant *Treponema pallidum* antigens examined in guinea pigs. Int. Arch. Allergy Appl. Immunol. 1989; 89:128–135.

Wicher, K., Schouls, L.M., Wicher, V., Van Embden, J.D. and Nakeeb, S.S. Immunization of guinea pigs with recombinant TmpB antigen induces protection against challenge infection with *Treponema pallidum* Nichols. Infect. Immun. 1991; 59: 4343–4348.

Wicher, K. and Wicher, V. Experimental syphilis in guinea pig. Crit. Rev. Microbiol. 1989; 16:181–234.

Zapiola, I., Salomone, S., Alvarez, A., Scolastico, M.C., Koessel, R.A., Lemus, J., Wainstein, C. and Muchinik, G. HIV-1, HIV-2, HTLV-I/II and STD among female prostitutes in Buenos Aires, Argentina. Eur. J. Epidemiol. 1996; 12:27–31.

CHAPTER

60

Mycobacterium tuberculosis

Leonard Moise[1,2], *Julie McMurry*[2] *and Anne S. De Groot*[1,2]
[1]Collaborative for Vaccine Research and Development, Institute for Immunology and Informatics, University of Rhode Island, RI, USA
[2]EpiVax, Inc., Providence, RI, USA

OUTLINE

ABSTRACT

Tuberculosis (TB) is one of the most prevalent infectious diseases worldwide and accounts for a large portion of all preventable deaths. Latent TB infection is also extremely common, affecting as many as one-third of humans alive today. Fortunately, only about 10% of TB infection leads to active TB disease. TB is curable with proper treatment, but treatment programs are labor intensive and increasingly threatened by drug resistance. Furthermore, the long treatment regimen poses compliance problems, and lack of access to TB care is common. As a consequence of these combined factors, TB continues to claim 2 million lives per year. An effective vaccine reliably preventing TB disease in adults would significantly reduce the number of deaths due to TB; however, no such vaccine is available. A live attenuated strain of *Mycobacterium bovis*, bacille Calmette–Guérin (BCG), is used with variable efficacy to vaccinate children against TB in many countries throughout the world. Because BCG protects against meningeal TB in children and against leprosy in adults, it will be difficult to justify replacing BCG with a new TB vaccine unless that vaccine is an improved version of BCG, which may take decades to develop and test. Effective vaccines that could be administered against a background of BCG immunization and/or preexisting *Mycobacterium tuberculosis* (Mtb) infection ("postinfection" vaccines) stand to make the greatest and most immediate impact on TB control. This chapter highlights the most recent history of TB vaccine development and describes TB vaccines that are in the research and clinical pipelines.

INTRODUCTION

About one in every three persons alive carries the tuberculosis (TB) infection and is at risk for TB disease (9 million new cases of TB per year) and death (2 million deaths per year) (WHO, 2007a, 2007b). Most cases of active TB occur in regions of Africa and Asia, where TB case finding is difficult or nonexistent and treatment monitoring is unavailable (Fig. 60.1); thus the actual impact of TB in those regions is underestimated in the World Health Organization (WHO) global TB program reports (WHO, 2007). TB is a treatable disease; however, it remains one of the most prevalent diseases because of the critical disconnect between the need for care and access to it. Sub-Saharan Africa, which has the highest rate of TB, reports the lowest treatment completion rate in the world: of those cases identified and placed in treatment programs, less than 75% complete treatment.

One of the reasons that very few individuals complete treatment is that TB treatment takes at least 6 months, and strict adherence to treatment is essential to achieve a cure. Completion rates differ dramatically between TB control programs in the developed and developing world. Both prompt diagnosis and effective treatment require trained and motivated staff, fully functioning laboratories, and reliable drug supplies. Despite some improvements since the late 1990s, the WHO reported in 2007 that every national treatment program in the world experienced ruptures in TB drug supply, too few laboratories, weak quality control, and limited facilities to carry out culture and drug-susceptibility testing. In most countries, even where TB treatment is free or low cost, physicians on the front line lack the most basic means of testing for active TB infection [the ability to perform a sputum smear for acid-fast bacilli (AFB)].

At the level of patient care, difficulties associated with adherence to a course of therapy over 6 months result in the development of drug resistance. Drug-resistant TB has become more common: multidrug-resistant (MDR) TB accounted for 1 million of the 8 million new cases of TB identified in 2004, the latest date for which the international statistics were compiled, and extensively drug-resistant TB (XDR TB) has emerged. Providing patient monitoring and continuous access to TB medications is a significant challenge for TB programs that globally face a funding shortfall of $1 billion for the next few years (Feuer, 2007).

Despite these figures, some gains have been made in recent years. Indeed, a "leveling off" of global TB incidence was reported (WHO, 2007). However, it is unlikely that widespread access to TB treatment alone will solve the problem of TB. The only possible intervention for TB that does not require extended treatment over 6–9 months, trained clinical personnel, careful monitoring of adherence, and uninterrupted supply chains is a TB vaccine.

Currently, there is no effective vaccine for reliably preventing TB in adults (Colditz et al., 1994; Sepkowitz, 2001). A live attenuated strain of *Mycobacterium bovis*, bacille Calmette–Guérin (BCG), is used with variable efficacy to vaccinate children against TB in many countries (Fine, 1995). Because BCG protects against meningeal TB in children and leprosy in adults and is very widely used in children, it will be difficult to perform studies that justify replacing BCG with a new TB vaccine unless that vaccine is an improved version of BCG. Because of the chronology of TB infection and disease (10–20 years, or more), a replacement vaccine for BCG will take

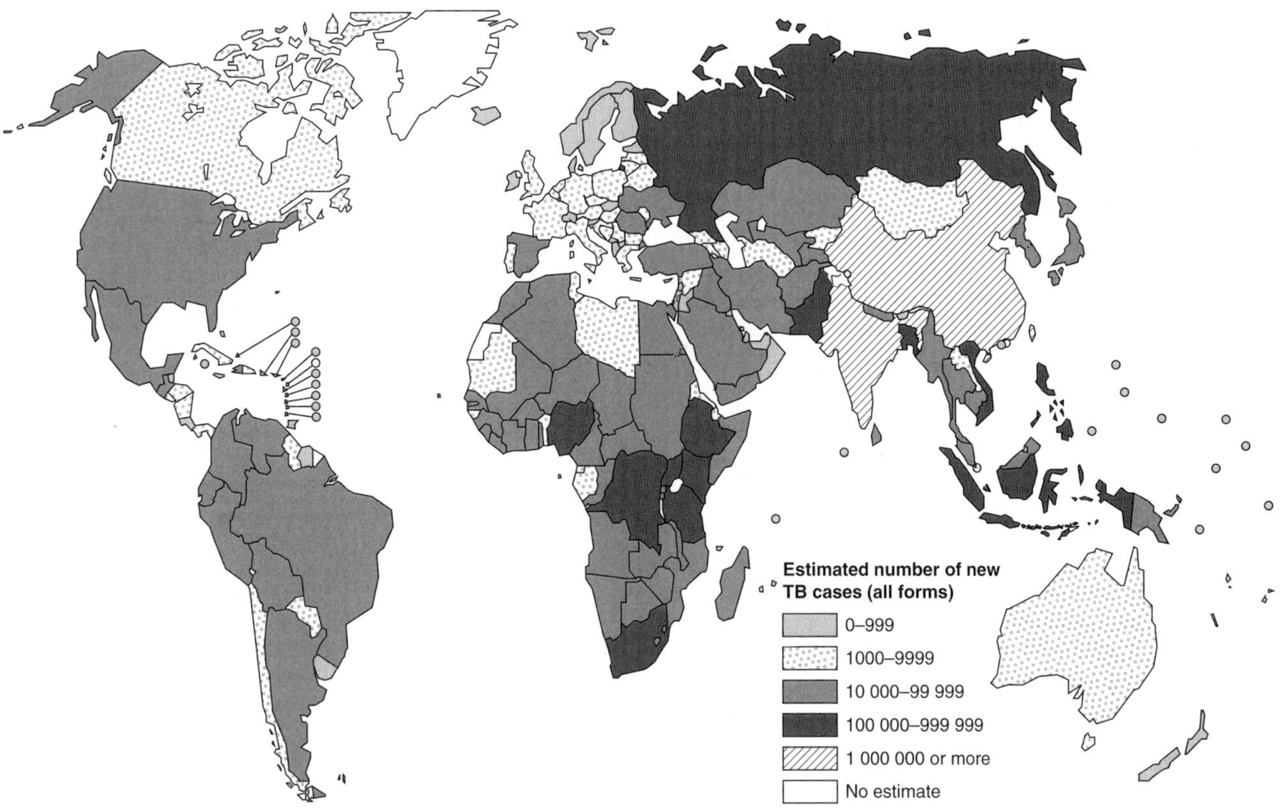

FIGURE 60.1 Estimated TB incidence rates, 2005. Reproduced with permission from the 2007 WHO report, Global Tuberculosis Control: Surveillance, Planning, Financing.

decades to develop and test. As illustrated in this chapter, TB is an extremely successful and robust pathogen for which several vaccine attempts have failed. However, if an effective vaccine can be developed, it could have a substantial impact on TB control, even if the efficacy is limited to delaying disease outbreak and transmission. In particular, vaccines that could be administered against a background of BCG immunization and/or preexisting *Mycobacterium tuberculosis* (Mtb) infection ("postinfection" vaccines) stand to have the greatest and most immediate impact (Nacy and Sacksteder, 2002; Young and Stewart, 2002).

SHORT HISTORY OF THE DISEASE

Paleomicrobiologists date the earliest evidence of TB back to the Pleistocene epoch, according to a study that detected the pathogen in 17,000-year-old bison remains (Rothschild et al., 2001). How and when TB became a human disease remains unclear, although evidence from Egyptian mummies from BC 2400 provides an estimate of the latest point it reached humans (Zink et al., 2003). Only in 1882 was the causative agent, Mtb, isolated by Robert Koch with the advice of Fanny Hesse, who recommended that he cultured bacteria in agar instead of a gelatin-based medium. Isolation of Mtb heralded the beginning of the fight against TB, which still continues today. Vaccine and drug approaches, intensely pursued over the years, have yielded products for disease prevention or treatment. Most notably, the BCG vaccine, developed in the early 1900s, has provided billions of people worldwide with protection from childhood TB, meningeal TB, and leprosy. Because BCG does not adequately protect against adult pulmonary TB, efforts to develop a better vaccine continue. The emergence of MDR TB and the increased risk of TB in persons coinfected with HIV have raised awareness of the importance of TB vaccine development.

ETIOLOGIC AGENT

Classification

Mtb is the causative agent of TB. It is a facultative intracellular, nonmotile, non-sporulating, and rod-shaped

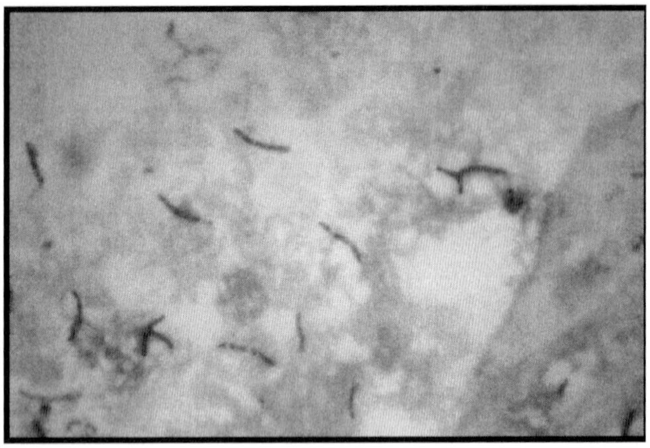

FIGURE 60.2 This photomicrograph reveals Mtb bacteria using acid-fast Ziehl–Neelsen stain; magnified 1000×. The acid-fast stains depend on the ability of mycobacteria to retain dye when treated with mineral acid or an acid–alcohol solution, such as the Ziehl–Neelsen or the Kinyoun stains that are carbolfuchsin methods specific for Mtb. *Source*: phil.cdc.gov CDC-PHIL ID #5789; Date: created 1979; *Author*: CDC/Dr. George P. Kubica. This image is a work of the Centers for Disease Control and Prevention, part of the United States Department of Health and Human Services, taken or made during the course of an employee's official duties. As a work of the U.S. federal government, the image is in the public domain.

bacterium (Fig. 60.2). Interestingly, this obligate aerobe is neither classified as gram-positive nor gram-negative as its cell wall structure, being rich in mycolic acids and lipids, is atypical of either class (Prescott et al., 1996). *Mycobacterium* is a member of the Mycobacteriaceae family in the Actinobacteria phylum. It is classified into several major groups for diagnosis and treatment. Members of the Mtb complex that can cause TB include *M. tuberculosis*, *M. bovis*, *Mycobacterium africanum*, and *Mycobacterium microti*. Nontuberculous mycobacteria (NTM) can cause pulmonary disease resembling TB, lymphadenitis, skin disease, or disseminated disease. *Mycobacterium leprae* causes Hansen's disease or leprosy. *Mycobacterium avium* complex is a group of mycobacterial species that are a significant cause of death in patients who are immunocompromised. Species in this complex include *M. avium silvaticum*, *M. avium hominissuis*, and *M. avium*. *M. avium paratuberculosis* has been implicated in Crohn's disease in humans and Johne's disease in sheep.

Antigens

Sequencing and annotation is complete for Mtb H37Rv and Mtb CDC1551 as well as for closely related mycobacteria, including *M. bovis*, *M. leprae*, and *M. avium*. Mtb H37Rv encodes 3995 proteins of which 2058 (52%) have functional annotations based on sequence homology and 376 putative proteins that may be unique to Mtb (Camus et al., 2002). The CDC1551 genome contains 4189 protein-coding genes of which 2235 (53%) have a matching database entry (Fleischmann et al., 2002). A complete genome comparison between H37Rv and CDC1551 has identified polymorphic sequences related to pathogenesis and immunity (Fleischmann et al., 2002).

Antigens that are secreted by Mtb have been the focus of several TB vaccine development efforts since these antigens have been shown to induce potent immune responses in humans (Andersen, 1994, 1997; Boesen et al., 1995; Horwitz et al., 2000). Additionally, some studies have shown that responses to secreted TB antigens have been greater than to those antigens of cytosolic origin (Mustafa et al., 1998). Secreted antigens have been shown to elicit significant protection (in animal models), equal in magnitude to BCG, while also priming responses to a wider range of the antigens than BCG does (Andersen, 1994; Boesen et al., 1995; Demissie et al., 1999; Sonnenberg and Belisle 1997). Despite the evidence that these proteins are involved in protection, no single protein or group of proteins has yet been shown to protect against the development of active TB infection. When considering that there are more than 3000 open reading frames in the average mycobacterial genome, it becomes clear that much remains to be done to discover proteins that might contribute to vaccine development. Many of the proteins being considered for vaccine development are summarized in Table 60.1.

Young et al. (1985) have successfully made expression libraries of TB proteins and identified antigens that are reactogenic with serum and/or peripheral blood mononuclear cells (PBMC) from animals and humans exposed to TB. Several of the antigens have been epitope mapped, using overlapping peptides, and while many of these antigen-derived peptides have been confirmed as epitopes, some such epitopes have been confirmed as being immunodominant (Mustafa, 1994, 1995).

Yet another approach to antigen discovery for TB vaccines has been to emphasize Mtb-unique proteins that are upregulated during the growth of Mtb under stress conditions. Mtb thrives in the most oxygenated parts of the body and, when reactivated, preferentially establishes infection in the upper lobes of the lungs (Adler and Rose, 1996). During infection, Mtb bacilli are inhaled and then phagocytosed by alveolar macrophages. Before the host macrophages become immunologically active, Mtb blocks the development of a phagosome into a phagolysome and establishes long-term residence (Russell, 2001). In order to survive under the hypoxic conditions of

TABLE 60.1 Selected antigens under consideration for TB vaccine development

	Description	Relevant literature	Vaccines in development	Developers
Ag85A and 85B	These *Mycobacterium tuberculosis* proteins were among the first to be defined and elicit responses in healthy individuals who are exposed to *M. tuberculosis*. However, the BCG vaccine already expresses these proteins, as do all environmental mycobacteria, so the relevance of these proteins for boosting responses in developing countries is under evaluation.	Goonetilleke et al., 2003; McShane et al., 2004; Horwitz and Harth, 2003	Modified vaccinia virus Ankara that expresses antigen 85: this might be given to individuals who have already been vaccinated with BCG in infancy to boost both CD4+ and cytotoxic CD8+ T cell responses. Recombinant BCG that over-expresses antigen 85B: this aims to boost responses to antigen 85 to a greater extent than BCG usually does.	85A: University of Oxford with funding from the Wellcome Trust and the European Commission 85B: Marcus Horwitz and Aeras Vaccine Foundation
ESAT-6	ESAT-6 is one of the low-molecular-weight proteins that is unique to Mtb. For that reason, it appears to be a critical antigen for protective immune response, but the use in vaccines is controversial as the protein is now used in special Mtb-specific diagnostic tests.	Skjot et al., 2000; Okkels et al., 2003	Fusion protein of early secretory antigenic target 6 (ESAT-6) and antigen 85B: this aims to prime or boost the interferon response to these proteins, which are immunogenic antigens that are secreted by *M. tuberculosis*.	Peter Anderson, Statens Serum Institute, Intercell and Aeras
TB10.4	ESAT-6 homologue, low-molecular-weight protein of Mtb	Dietrich et al., 2005; Skjot et al., 2002	Included in perfringen vaccine as described by Sadoff (2005)	Aeras
Mtb32 and 39 and recombinant fusion protein Mtb72F	These are immunogenic proteins from *M. tuberculosis* that are recognized by T cells from infants who have been vaccinated with BCG and from healthy individuals who live in regions in which tuberculosis (TB) is endemic.	Skeiky et al., 2004 Alderson et al., 2000	Mtb72F is a 72-kDa polyprotein that consists of Mtb32 and Mtb39.	Steve Reed, GlaxoSmithKline and Aeras
Mtb19	The 19-kDa antigen is a cell wall-associated lipoprotein present in *M. tuberculosis* and in BCG vaccine strains. The 19-kDa protein has a detrimental effect on the efficacy of vaccination with live mycobacteria.	Yeremeev et al., 2000	Targeted elimination of the 19-kDa gene (by RNAi) is one consideration for live Mtb vaccines.	No known vaccines in development
CFP-10	10-kDa culture filtrate protein	Mustafa et al., 2002	Focus has been placed on CFP-10 for diagnostic purposes.	No known vaccines in development
HSP 65	HSPs are important target antigens and have poorly understood immunoregulatory properties.	Johansen et al., 2006	DNA vaccine that encodes HSP65: this induces CD4+ and CD8+ T-cell responses, and it downregulates Th2 cell responses in animal models of TB78.	Doug Lowrie
16-kDa alpha crystallin	Sherman et al., 2001	Agrewala and Wilkinson, 1998		Ken Huygen
Mtb39a-e	These antigens were found primarily in the cell lysate with low levels detected in the culture filtrate.	Dillon et al., 1999.		No known vaccines in development
Mtb9.9				
LppX	LppX is a lipoprotein required for the translocation of phthiocerol dimycocerosates to the surface of *M. tuberculosis*.	Lefevre et al., 2000	Sheep immunized with the recombinant LppX of paratuberculosis analog mounted cellular and humoral responses.	No known human vaccines in development
MPT64	The MPT64 protein and its homologues form a highly conserved family of secreted proteins with unknown function that are found within the pathogenic Mtb genus.	Bhigjee et al., 2007; Cai et al., 2005	Focus has been more recently placed on MPT64 for diagnostic purposes, although it is still being investigated as a vaccine in animal models.	No known human vaccines in development

the macrophage phagosome, Mtb alters its metabolism and gene expression to reflect the conditions of the organelle (Schnappinger et al., 2003). Culturing Mtb under hypoxic conditions is one way to mimic the metabolic state of latent TB infection (LTBI) such that RNA can be isolated and used to generate cDNA probes for microarray analysis (Sherman et al., 2001). Proteins that upregulated under these conditions may be important for effective containment of the LTBI.

PROTECTIVE IMMUNE RESPONSE

Antibody

The role of antibody in the protection against TB is still under study. A recent study by Chambers et al. (2004) showed that the monoclonal antibody MBS43 (IgG2b isotype) to MPB83 protected mice against both high- and low-dose challenge with *M. bovis*. Other studies have shown that passive transfer of antibodies in animal models does not confer protection against Mtb (Dunlap and Briles, 1993). There is a correlation between specific antibody profiles and TB state in humans. In one study, antibody levels to six Mtb antigens were evaluated in serum derived from human subjects grouped into four diagnostic categories: active disease, inactive (past) TB, latent infection without radiographic chest abnormalities, and infection free (Davidow et al., 2005). Statistical data analyses showed that the latter two groups were serologically indistinguishable and active TB and inactive TB were characterized by different antibody profiles. Antibodies to the 38-kDa antigen, alanine dehydrogenase, and Rv2626c were associated with active TB, while antibodies to the 16-kDa antigen, ferredoxin A, and ESAT-6 were associated with inactive TB. This finding has been disputed more recently in studies performed in populations living in TB-endemic regions of the world (Hoff et al., 2007).

Some of the disparities among studies may be due to chronological differences in the expression of the protein antigens. Bacterial transcription profiling has been used to explore the correlation between antigen production and infection stage in human macrophages (Cappelli et al., 2006). Whether these fluctuations in antigen presentation might influence immune response, and thereby need to be considered in vaccine development, is unknown.

CMI

In contrast with antibody response, adoptive transfer of T cells does confer protection from TB in the mouse model (Lefford, 1975). In humans, induction of broad T-cell-mediated immunity to Mtb requires the involvement of type 1 cytokines: interleukin (IL)-2, interferon (IFN)-gamma, and tumor necrosis factor (TNF)-alpha (Kaufmann and Hess, 1999). In addition, it is clear that CD4+ T cells are involved: The rate of TB reactivation in HIV-infected individuals is inversely correlated with the number of circulating CD4+ T cells (Shen et al., 2001). It is believed that this is due to the gradual depletion of T cell clones specific for the coinfecting pathogen, which in turn is due to HIV-mediated destruction of activated T cell clones (Hirsch et al., 2001; Kovacs et al., 2001; Lawn et al., 2001; Moriuchi et al., 2000). By extension, activation of immune response increases susceptibility to HIV (Garrait et al., 1997). Based on the level of immune compromise at which TB begins to develop relatively early in HIV infection, CD4+ T cells are likely to be the critical component of cell-mediated immunity to LTBI. Importantly, T-helper-1 (Th1)-mediated immune responses have been demonstrated in mouse protection studies as well (Cooper et al., 1993; Flynn et al., 1993; Scanga et al., 2000). Some studies have shown that high levels of antigen-specific IFN-gamma production are found in subjects with active minimal disease as compared with lower levels in subjects with advanced disease (Boesen et al., 1995; Sodhi et al., 1997). Similarly, as compared with active TB patients, household contacts of these patients were shown to have greater reactivity to the antigen 85 (Ag85) complex (Torres et al., 1998). However, these studies do not establish a causal link between IFN-gamma production levels and progression to active disease. Levels of TB-specific IFN-gamma production vary considerably between those in the periphery and those at the site of disease (Barnes et al., 1993).

IFN-gamma plays a role in activating specific macrophages and increasing natural killer (NK) cell activity, both of which are host protective. Th1-biased T cell responses to Mtb epitopes not contained in BCG may help control immune response in latently infected individuals. While Th1 cytokines are essential for protection, their levels of production do not alone explain the immunity–susceptibility dichotomy. The spatial and temporal complexity of the TB granuloma has a significant bearing on the nature and effectiveness of the cytokine response. The dynamics of granuloma formation and maintenance are reviewed in Russell (2007).

CD8+ T cells also play a significant role in TB immunity (Serbina and Flynn 2001; Serbina et al., 2001). Importantly, studies of human CD8+ -restricted responses to Mtb antigens have revealed that these cells play an important role in the control of Mtb replication in the alveolar macrophage (Canaday et al., 2001;

Lewinsohn et al., 2007). Indeed, CD8+ T cells specific for Mtb antigens were derived from peripheral blood samples obtained from Mtb-immune donors (Cho et al., 2000). CD8+ T cells may play an important role in the clearance of Mtb as alveolar macrophages have been shown to be effective targets of CD8+ T cells (Tan et al., 1997). Ag85A-specific CD8+ T cell responses have also been identified in patients who have active pulmonary TB (Weichold et al., 2007). Taken together, these data suggest that stimulation of Th1-biased and CD8+ T cell responses to Mtb epitopes may control immune responses in latently infected individuals.

Tregs

Recently, several authors have proposed that the immune responses to latent and active TB also involve regulatory T cells (Guerkov et al., 2003; Boussiotis et al., 2000). T regulatory cells (Tregs) express both CD4+ and CD25+ on their cell surfaces, express the regulatory T-cell-associated transcription factor forkhead box P3 (FOXP3), and produce Th1 suppressive cytokines, such as IL-10 and TGF-beta (Bluestone and Abbas, 2003). $CD4^{+}CD25^{high}FOXP3^{+}$ T cells have been shown to be elevated in peripheral blood of individuals diagnosed with active TB disease in comparison with uninfected control subjects or LTBI individuals (Guyot-Revol et al., 2006; Hougardy et al., 2007). Ex vivo depletion of Tregs from peripheral blood of active TB patients, but not LTBI subjects, increases Mtb antigen-specific IFN-gamma responses (Guyot-Revol et al., 2006; Hougardy et al., 2007; Ribeiro-Rodrigues et al., 2006). This regulatory phenotype is also expressed on the cytokine level as the Th1 suppressive cytokines IL-10 and TGF-beta suppress T cell responses in active TB patients (Boussiotis et al., 2000; Hirsch et al., 1996). A large population-based study in Ethiopia found that TB patients have decreased levels of Th1 cytokines and increased levels of IL-10, when compared with healthy infected and noninfected controls (Demissie et al., 2004). While active TB and LTBI exhibit different regulatory T cell responses, it has not yet been determined if this difference is the cause, or result, of disease.

Regulatory T cells may also explain anergy in patients with TB who respond negatively to a skin test with tuberculin (an Mtb-derived antigen) and whose T cells secrete the regulatory cytokine IL-10 but release little IL-2 or IFN-gamma in response to Mtb (Delgado et al., 2002). Some T cells that secrete IFN-gamma in the lungs of patients with TB also secrete IL-10 (Gerosa et al., 1999), which indicates that they may be regulatory T cells, instead of true Th1 cells.

These data corroborate similar observations made in murine studies. Transgenic mice that overexpress IL-10 show evidence of reactivation of chronic TB with a significant increase in bacterial numbers in the lungs (Turner et al., 2002). Studies using FOXP3-GFP (green fluorescent protein) mice illustrate that Tregs suppress immune responses and contribute to higher bacterial burden upon aerosol Mtb challenge (Scott-Browne et al., 2007). These findings underline the need to improve the Th1, effector T cell bias of TB vaccines and suggest that LTBI vaccine candidates will need to overcome a Th2 and regulatory T cell bias that may occur by coinfection with other pathogenic organisms, such as roundworms (Rook and Zumla, 2001), which are factors common in areas where a TB vaccine is most needed. Complicating the situation, however, is the role that Tregs play in limiting immunopathological damage. Host suppressor mechanisms that maintain immune homeostasis may qualify Tregs as friend, not only foe (Mills, 2004). For example, in early stages of Mtb infection, IL-10 also enhances the function of NK cells, leading to increased antigen availability through pathogen destruction (Mocellin et al., 2003).

EPIDEMIOLOGY

Significance as Public Health Problem

During the 1980s and the 1990s, TB reemerged in the developed world as a result of TB/HIV coinfection (Daley et al., 1992), but recent efforts to address this problem have decreased the level of TB transmission in the Americas and Europe (WHO, 2007). Despite the decline in TB case rates in the United States, recommendations to test all TB patients for HIV infection are followed in less than 30% of cases (CDC, 2007a, 2007b, 2007c). For similar reasons, TB remains a significant burden and major contributor to morbidity in developing countries located in sub-Saharan Africa and the former Soviet Union.

TB is the 10th leading contributor to the global burden of disease, measured as a function of disability-adjusted life years (DALYs) (Lopez et al., 2006). TB accounts for the loss of 34.5 million DALYs, or 2.5% of total losses in low- and middle-income countries. In sub-Saharan Africa, which is highly dependent on manual labor, the impact is even greater—it is the eighth leading cause of DALY losses, representing 10 million or 2.7% of its total. As a result of the structural inequalities between developed and developing nations, TB remains one of the leading causes of preventable death throughout the underdeveloped world.

Directly observed treatment, short-course (DOTS) was invented to address the problem of adherence to therapy. Due to delayed diagnosis of active TB cases, relapse rates as high as 5% after DOTS, and the inherent difficulty in compliance with the long DOTS regimen, DOTS alone is not expected to eradicate TB (Dye et al., 1999; WHO, 2000).

The task of TB control is further complicated by the HIV epidemic and the emergence of drug-resistant TB strains (Wells et al., 2007). HIV coinfection accelerates the rate of conversion to active TB: LTBI individuals who are also HIV-infected have an estimated 7–10% chance of developing TB within 1 year (Selwyn et al., 1989), compared to within a lifetime for counterparts not infected with HIV. Thus TB is currently the leading cause of death from AIDS (Raviglione et al., 1997; Alban and Andersen, 2007). As the global epidemic of HIV continues to expand into countries with high rates of TB, more active TB cases can be expected (Sepkowitz, 2001).

Both suboptimal TB care and poor patient compliance with the treatment regimen can result in the emergence of MDR TB. The WHO is now reporting that resistance to first-line anti-TB medications (including those in the DOTS regimen) was already as high as 40% in some countries in the late 1990s (Pablos-Mendez et al., 1998). In fact, resistance to isoniazid (INH), the first-line drug against Mtb, is nearly 100% in parts of Russia and Eastern Europe (Andersen, 2001). Unfortunately, due to the emergence of MDR TB, it is likely that many of the individuals who have latent Mtb infection in those regions of the world have MDR LTBI. As a consequence of the continuing expansion of the MDR TB and the HIV pandemics, the development of an effective TB vaccine has become a matter of increasing urgency.

XDR TB

The worldwide emergence of XDR TB and a provisional definition for this form of TB were first reported in 2005. In October 2006, the WHO convened an Emergency Global Task Force on XDR TB, which revised the case definition to specify resistance to at least INH and rifampin among first-line anti-TB drugs, any fluoroquinolone, and at least a second-line injectable drug (amikacin, capreomycin, or kanamycin). During 1993–2006, a total of 49 cases in the United States (3% of evaluative MDR TB cases) met the revised case definition for XDR TB. Seventeen (35%) were reported during 2000–2006, and a trend toward more frequent reporting of XDR TB in foreign-born individuals was also noted (CDC, 2007b). XDR TB presents a challenge to TB control activities worldwide. To prevent the spread of XDR TB, renewed vigilance is needed in drug-susceptibility testing, case reporting, specialized care, infection control, and expanded capacity for outbreak detection and response.

Potential as Biothreat Agent

MDR TB is classified by the National Institute of Allergy and Infectious Diseases as a Category C priority pathogen (NIH, 2007). As a third-tier biothreat, it is recognized as an emerging pathogen that may be widely disseminated because of its accessibility, ease of production, and dispersal and likelihood of high morbidity and mortality rates (CDC, 2007). Fortunately, the bacillus is sensitive to radiation (such as that contained in sunlight) and thus not likely to survive a dissemination attempt by large-scale aerosolization. Nonetheless, XDR TB and MDR TB both have greater potential to have a negative impact on the populations of developing and Eastern European countries; therefore, TB detection, treatment, and vaccine development should remain high priorities for the defense of public health.

CLINICAL DISEASE

Diagnosis

Active TB

Chronic cough, hemoptysis, and wasting are all symptoms of active pulmonary TB. A positive diagnosis is made using sputum smear and culture for AFB. Extrapulmonary TB is rare (less than 10% of TB cases) but most common in children and immunocompromised individuals. The presence of AFB can also be detected in urine, pleural fluid, and cerebrospinal fluid.

LTBI

In the absence of active, ongoing disease, the diagnosis of LTBI is usually based on a positive skin test and the absence of ongoing TB disease. The tuberculin skin test (TST) remains the cornerstone of LTBI diagnosis—especially in countries where the test result is not complicated by prior BCG vaccination. New laboratory diagnostics, such as the enzyme-linked immunosorbent spot (ELISpot), have been developed and may be better at discriminating between prior BCG vaccination and TB infection as they include antigens that are present in Mtb but not BCG (Johnson et al., 1999). Individuals who have received BCG vaccine are still considered to have LTBI if their TST is positive (>10mm). If the TST or ELISpot is positive, a chest radiograph is performed to evaluate the presence of pulmonary lesions. If lesions or other suspicious aspects are detected, the clinician should perform a sputum examination for AFB and culture the sputum for Mtb.

Treatment of TB Disease

The American Thoracic Society, Infectious Diseases Society of America, and Centers for Disease Control and Prevention issued updated guidelines for the treatment of TB (CDC, 2003). The most recent TB treatment guidelines provide a complete discussion of the drugs currently available to treat TB, including dosing, dose adjustments needed for renal or hepatic dysfunction, toxicities, management of common adverse effects, and information about interactions between anti-TB drugs and other drugs. Since one of the key anti-TB agents, rifamycin (rifampin, rifampicin (RIF), has the potential for drug–drug interactions, including some antiretroviral medications, special attention must be given to the treatment of the patient who has coinfection with HIV and Mtb.

DOTS

DOTS programs focus on increasing patient adherence, the primary determinant in treatment success rates. DOTS is currently the WHO-recommended strategy for TB control. DOTS involves treatment with a four-drug regimen [INH, RIF, pyrazinamide (PZA), and ethambutol (EMB)] for 6–9 months. DOTS must be used throughout the entire course of therapy for best cure rates. The core intervention involves a health worker who observes the patient when s/he is swallowing each dose of anti-TB medication. DOTS also involves the regular collection of sputum specimens until two consecutive samples test negative for AFB. Although DOTS is labor intensive, the standardized first-line regimen costs only around $10 per patient and is associated with a 90% cure rate in the absence of drug-resistant strains. Up to 89% of countries, both developed and developing, had implemented DOTS by the end of 2006. However, since case detection rate is as low as 50% in some countries, and completion rates for those cases of TB that are detected also remain low, at a rate of approximately 75% (2004), the implementation of DOTS remains at best in progress in most developing countries.

Treatment of LTBI

The preferred treatment for LTBI is 9 months of INH taken either daily or biweekly. Without DOTS, the cure rate of the biweekly regimen is worse than that of the daily regimen since a missed dose in intermittent treatment constitutes a greater percentage of total doses. Because the levels of primary INH resistance are on the rise, and because of a 0.3% rate of INH side effects, not all countries recommend treating LTBI in this way, if at all. In the United States, LTBI affects 10–15 million individuals, each with a 10% chance of progressing to TB disease without treatment; thus, in the 1950s, the U.S. Public Health Service (USPHS) Commissioned Corps began to recommend and financially support the treatment of LTBI with INH. Whether for TB disease or LTBI, cure rates are highest when treatment is administered by those experienced in the management of TB, or in consultation with someone with such experience. This is especially true for drug-resistant cases.

PATHOGENESIS

TB Disease Process

TB vaccine development has been hampered by our lack of understanding of the details of TB pathogenesis. Protective immunity against Mtb occurs since only 8–10% of Mtb-infected individuals go on to develop TB, following exposure. However, the set of host immune responses that determine effective protection against TB disease remain an enigma.

Mtb is spread in tiny aerosol droplets expelled by one person and taken up into the lung of another. Mtb finds its home in alveolar macrophages, where it prevents endosome maturation by inhibiting acidification through expression of UreC. Infected macrophages participate with other cells (see next section) in the formation of a granulomatous lesion that contains the bacteria together with host immune cells (Fig. 60.3). Surrounded by a fibrotic wall, the granuloma is a low-oxygen environment that forces Mtb to enter a reversible state of dormancy (Boshoff and Barry, 2005). Mtb respiration and metabolism are significantly diminished and growth stops. This state of dormancy, or latency from the perspective of the host, is the basis of a truce between host and pathogen where infection ensues without manifestation of clinical disease. Upon disruption of this fine balance, oxygen enters the lesion, leading to resuscitation, a molecular process of increasing metabolic and replicative activity and production of resuscitation

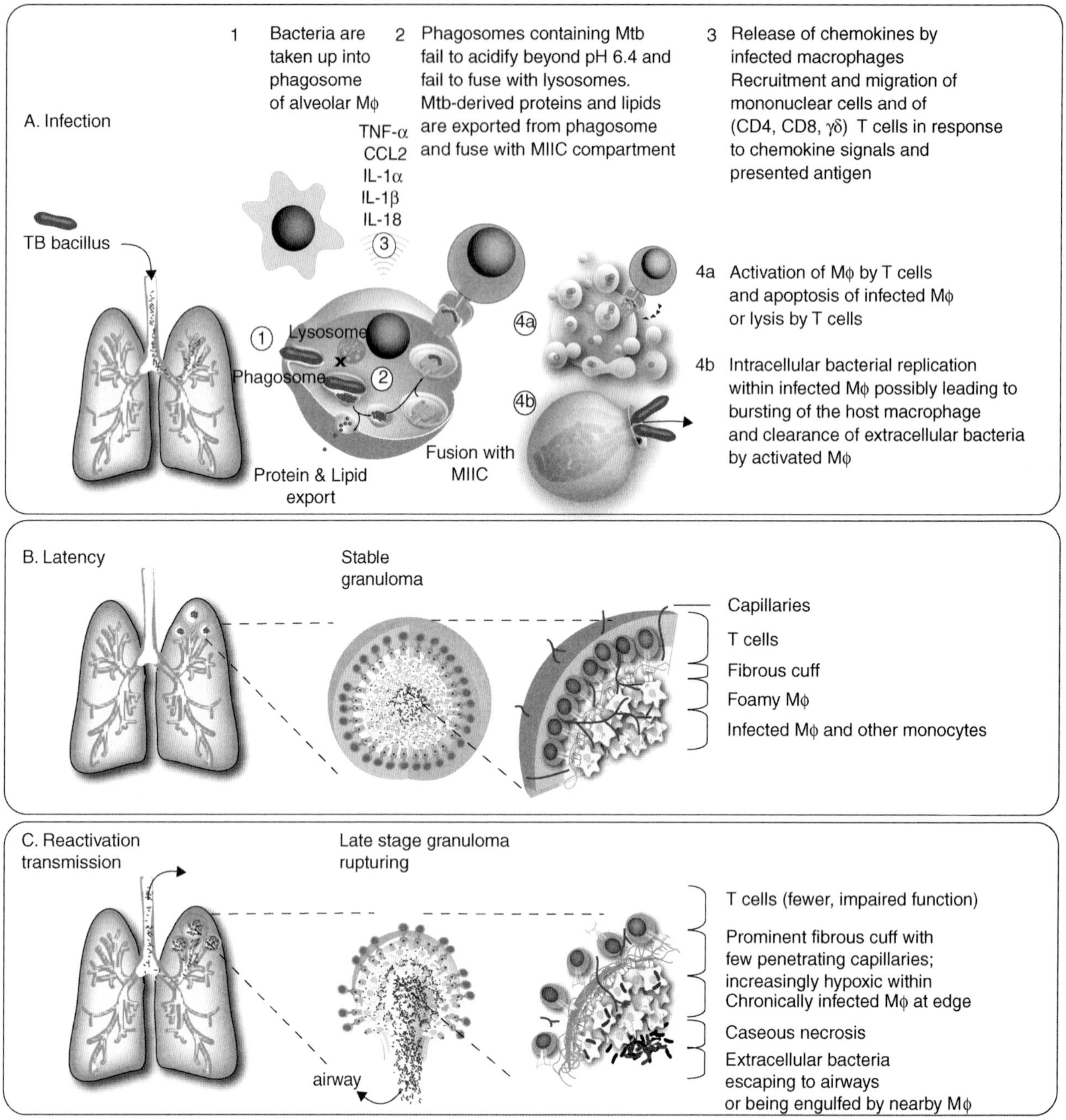

FIGURE 60.3 Pathogenesis of pulmonary tuberculosis. Infectious Mtb bacilli are inhaled into the lung where they are phagocytosed by alveolar macrophages. Mtb-containing phagosome fails to acidify beyond 6.4. Infected macrophages produce a localized proinflammatory response that leads to the recruitment of macrophages and other mononuclear phagocytes and T cells. The recruited T cells can cause infected macrophages to apoptose or lyse. Alternatively, bacterial replication inside the macrophage can lead to its bursting open. Extracellular bacteria are then engulfed by surrounding macrophages; together the recruited cells form a granuloma at the center of which are infected macrophages and foamy macrophages surrounded by a fibrous cuff and a layer of T cells. The stable granuloma walls off the bacteria, preventing them from causing outward manifestations of disease and from being transmitted to other hosts. In later stages, the granuloma develops a collagenous fibrous sheath through which few capillaries penetrate and consequently the outer boundary of these granulomas hypoxic. When the immune status of the host is impaired due to age, malnutrition or coinfection with HIV, the number or function of the patrolling T cells is compromised, resulting in the structural decay of the granuloma center into a caseous mass of cell debris. The failure of the granuloma leads to the development of a cough that propels the bacteria out through the airways.

promoting factors (Rpfs) (Downing et al., 2004, 2005). Rpfs send signals to neighboring dormant bacteria, enabling them to recover. It is thought that resuscitation may lead to reactivation where high bacterial burden can develop in lesions causing caseation and cavitation of the lung. Active disease ensues with lung tissue deterioration leading to weakened lung function and bacterial dissemination to other organs.

Immune Response to Infection

Mtb-infected cells migrate to nearby draining lymph nodes, where the host immune response is initiated. TB bacilli residing in the arrested phagosome shed antigens that are processed via the major histocompatibility complex (MHC) Class II antigen presentation pathway (Russell, 2007), leading to a predominantly CD4+ Th1-type response. IFN-gamma and TNF-alpha are cytokines associated with this response that leads to formation of the granuloma to contain Mtb. Granuloma formation is a complex process involving recruitment of CD4+ T cells, CD8+ T cells, and CD1-restricted $\alpha\beta$ T cells. The CD8+ T cell response arises from cross-priming on MHC Class I molecules via Mtb antigens that are released from apoptosed infected macrophages and taken up by dendritic cells.

As described above, both effector and memory Th1 cells develop in response to antigens expressed early in infection. Memory cells against early antigens may persist during latency and prevent reactivation but may not be able to contain dormant Mtb, which expresses a differential set of antigens that are controlled by *DosR* genes (Voskuil et al., 2003; Sherman et al., 2001; Schnappinger et al., 2003). Indeed, non-Th1 (Th2 and regulatory) cytokine levels are elevated in patients in advanced stages of disease (van Rie et al., 1999; Rook et al., 2004,), which suggests that the T cell response may be refocused from Th1 to Th2 or suppressed by Tregs in the course of reactivation. This idea is supported by the finding that IL-10-overexpressing mice have a higher bacterial burden in later stages of infection (Turner et al., 2002). Much remains to be learned about the host immune response to Mtb infection.

VACCINES

History

Prophylactic Vaccines

BCG, the live attenuated vaccine for TB developed by Albert Calmette and Camille Guérin between 1908 and 1921 remains the most widely distributed vaccine in use in the world today. It is estimated that 76% of children born in 2002 received (BCG) vaccination, recommended by the WHO because it confers protection against childhood TB (Trunz et al., 2006) and leprosy (Sterne et al., 1996). Despite this wide coverage, BCG's efficacy is highly variable, ranging from 0% to 80% against adult pulmonary TB (Colditz et al., 1994). The suboptimal efficacy of BCG occurs for a number of possible reasons, including waning efficacy over time, differences between BCG sub-strains (Wu et al., 2007), deletion of protective antigens found in Mtb, and failure of BCG to stimulate adequate balanced CD4+ and CD8+ T cell responses.

A number of different approaches to developing novel TB vaccines are being actively pursued. In the next section, we will review prophylactic vaccination (live vaccines such as augmented BCG, live attenuated Mtb, and subunit) and then focus on "therapeutic vaccination" approaches.

Currently Licensed Vaccines

The only currently licensed TB vaccine is BCG. There are a number of BCG strains (BCG "Pasteur," BCG "Copenhagen," etc.), each of which was attenuated from an original BCG strain through serial passage. The strains used in BCG are well known to be divergent from the original strain developed at the Pasteur Institute. As a result of years of passage in geographically distant laboratories, the following vaccine strains are now recognized: BCG Russia/Moreau/Japan; BCG Sweden/Birkhaug; BCG Glaxo/Danish (passage 1331)/Merieux (passage 1077)/Prague; BCG Pasteur (passage 1173)/Tice/Frappier/Connaught/Phipps (Brosch et al., 2007) (Fig. 60.4).

Vaccine Recommendations, Including Potential to Administer with Other Vaccines

BCG is given as a single intradermal injection in the deltoid. It is rarely given to adults, and in fact, BCG is not used in the United States, except for therapy of certain cancers (bladder cancer, for example). In the developing world, BCG is given to newborns; vaccination coverage is approximately 80%. In Europe, 10 countries recommend BCG for children under 12 months, five countries give the vaccine to older children, and 10 countries give the vaccine to children considered to be at increased risk (due to the country of origin, close contact, or travel). Seven countries do not recommend the use of BCG as a routine vaccination. Revaccination is given to children in four

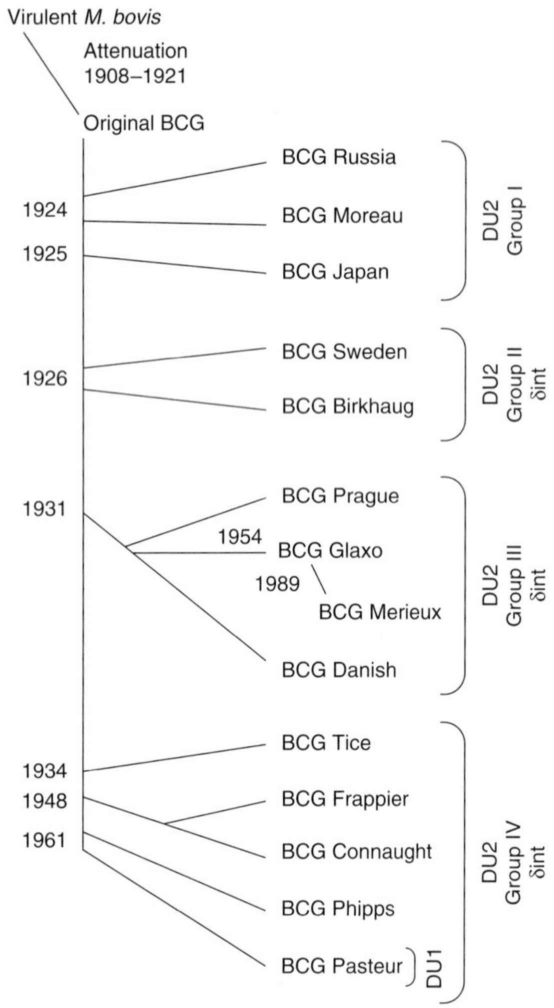

FIGURE 60.4 Genealogy of BCG vaccines. Repeated passages of virulent *M. bovis* gave rise to the live attenuated BCG vaccine in the early 1920s. BCG was circulated to multiple laboratories where daughter strains with differing genomes were produced by further passage over the years. BCG vaccines are distributed into four major groups according to their major genomic polymorphisms, tandem duplications (DU), and internal deletions (Dint). Adapted from Brosch et al. (2007).

countries. Where BCG is provided as part of universal vaccination, BCG coverage is very high (83–99.8%). Despite vaccination, TB has been observed in vaccinated children, and due to the lack of efficacy and the potential for side effects (see below), nearly half the countries in Europe were considering revisions of their vaccine policies in 2005 (Infuso and Falzon, 2006).

Countries where Vaccine is Licensed, Including Manufacturers

BCG has near universal licensure and is produced by 40 or more manufacturers around the world. The major commercial producers include Pasteur Merieux Connaught, Evans-Medeva, and Japan BCG Laboratory (WHO, 1999). As noted above, not all BCG vaccine strains are identical.

Indications for Vaccination/Target Populations

Despite the fact that BCG is one the most widely implemented vaccines, the mechanism of action of this vaccine is not well understood, and different strains of BCG have demonstrated wide-ranging degrees of efficacy in studies carried out in different regions of the world (from 0% protective efficacy to greater than 80%). Some of the countries in which protective efficacy is low or lacking choose not to administer BCG at all. Where BCG is administered, age at vaccination varies from at birth to 15 years of age, depending on the country.

Duration of Immunity

Duration of immunity afforded by BCG is not well characterized. Studies that show a protective effect of BCG indicate that immunity lasts 10–20 years (Sterne et al., 1998). Interestingly, a retrospective study of BCG vaccines found a sex difference in waning vaccine efficacy with a decline for men but not women (Aronson et al., 2004). While this finding corroborates data from a herpes simplex virus type 2 vaccine study (Stanberry et al., 2002), it remains to be confirmed in a future, prospective BCG study that is designed to address this hypothesis.

Contraindications of Vaccination, Including Special Risk Groups

BCG is contraindicated for immunocompromised individuals since the vaccine can cause disseminated infection in these individuals (WHO, 2007b). Despite widespread knowledge about the potential effects of BCG in immunocompromised individuals, this recommendation was only recently extended to HIV-infected children who have received the bulk of BCG vaccination since the beginning of the HIV epidemic. The prevalence of BCG disease that develops in HIV-infected infants who have received BCG is yet to be determined, although recent reports are alarming: One study concluded that the risk of disseminated BCG disease is increased several hundred-fold in HIV-infected infants compared to the documented risk in HIV-uninfected infants (Hesseling et al., 2007). Population- and hospital-based surveillance will be required to accurately estimate morbidity and mortality associated with BCG vaccination in HIV-exposed and infected infants.

Adverse Events

BCG has an excellent safety record when properly administered to healthy persons, although BCG can cause pain and scarring at the site of injection in healthy individuals. If BCG is delivered subcutaneously, it can cause skin infection that can spread to lymph nodes. It should also be noted that BCG should not be given to purified protein derivative (PPD) skin test-positive (LTBI) individuals. Skin testing is recommended prior to BCG vaccination because BCG can cause significant local inflammation that contraindicates its use for persons who are already skin test positive.

The Case for Further TB Vaccine Research and Development

INH treatment for LTBI provides an excellent yardstick against which the potential commercial success of an effective TB vaccine can be measured. In the United States alone, between 1954 and 1997, the use of anti-TB treatment reduced the number of newly diagnosed cases of active TB by 32%, the number of mortalities by 81%, the number of life years lost by 87%, and the cost of medical treatment by 76%. The total financial burden of illness over this time period (including the value of lost life years) was reduced from $894 billion to $128 billion in 1997 (Javitz and Ward, 2002). Thus, a TB vaccine that would prevent the progression of LTBI to active disease would reduce medical expenditures in the United States by approximately $250 million per year.

As already described, vaccines against TB can be divided into two types: vaccines to prevent primary TB infection and vaccines to prevent reactivation of latent TB disease. Vaccination to prevent primary TB infection would eliminate mycobacteria before they are able to take up residence in host macrophages, thereby eliminating the potential for the development of latency and eliminating the potential for later development of full-blown TB disease. Vaccination during LTBI would contain infection, preventing the multiplication of Mtb bacilli and the development of full-blown TB disease.

The ideal vaccine would be an effective vaccine to *prevent* TB infection. Such a vaccine would help reduce the global burden of TB. However, replacing BCG with a new TB vaccine is likely to be very difficult, for many of the reasons already stated—it has been shown to be protective in infants. Demonstrating superiority of a new vaccine over BCG will also be difficult since a Phase III prevention study comparing BCG to a new vaccine requires a time frame of 10–20 years.

On the other hand, an effective vaccine targeting LTBI would significantly reduce the global burden of TB and is predicted to be cost-effective (Dye, 2003; Aeras, 2007). Furthermore, it could also be used in the context of BCG pre-immunization since the goal of such a vaccine would be to "boost" preexisting immune responses. Performing Phase III trials with a dual-purpose vaccine in LTBI- or BCG-immunized adults would be simpler and pose less risk than performing large, prospective Phase III comparisons of BCG with a new preventative TB vaccine in children.

Which approach should be used when building a vaccine for TB? Clearly, the most relevant approach would be to stimulate immune responses that correlate with protective immune response in natural TB infection. Fortunately, a protective immune response to TB can indeed be generated (unlike the situation with HIV) since only 8–10% of Mtb-infected individuals go on to develop TB. Unfortunately, it is not perfectly clear what the critical components, or correlates, of immunity to TB might be.

Discovery/Basic Science

A variety of vaccine approaches are currently under exploration. These include attenuated live vaccines, subunit vaccines, vectored vaccines, and epitope-driven vaccines; each will be described in the following sections.

Attenuated Live Mtb Vaccines Attenuated live Mtb vaccines, produced by targeted inactivation of regulatory proteins, contain antigens that are otherwise missing in BCG without being virulent themselves. For example, inactivation of PhoP diminished Mtb virulence and elicited immune responses similar to parental Mtb (Soto et al., 2004). Auxotrophic Mtb mutants, such as the one produced by targeted deletion of the *panC*, *panD*, and *lysA* genes, are similarly live vaccines that are infective but replication deficient. The *panCpanDlysA* vaccine produced similar responses to the BCG vaccine (Sambandamurthy et al., 2005). A similar *panCpanD* mutant with deletion of the RD1 region, which is thought to be partly responsible for attenuation of BCG, also produced responses comparable to BCG (Sambandamurthy et al., 2006). A major drawback of the live TB vaccine approach is safety concerns about reverting mutations that may restore virulence.

Moreover, the delicate balance between sufficient attenuation and protective efficacy is difficult to strike.

Recombinant BCG Vaccines rBCGΔureC:Hly is a recombinant BCG that expresses listeriolysin and carries a urease deletion mutation. It is being developed by Vakzine Projekt Management GmbH under license from the Max Planck Society. This vaccine secretes listeriolysin to form pores in the endosome and enable escape of BCG into the cytoplasm of infected cells to induce CD4+ and CD8+ T cell responses. Urease is deleted to provide the optimal pH conditions for listeriolysin function. In preclinical studies, this vaccine induced increased efficacy compared with the parental BCG strain (Grode et al., 2005). The vaccine is currently under good manufacturing practice (GMP) production, and Phase I clinical trials are expected to begin in early 2008. A similar recombinant BCG vaccine that combines antigen overexpression and endosome escape approaches is under development by the Aeras Global TB Vaccine Foundation. Aeras-X03 overexpresses Mtb antigens Ag85A, Ag85B, Rv3407, and TB10.4. To enable endosomal escape of Mtb antigens for induction of CD4+ and CD8+ T cells, this modified BCG expresses perfringolysin (similar to listeriolysin) and bears a urease deletion mutation (Skeiky and Sadoff, 2006). It is in late-stage preclinical development and expected to enter Phase I clinical trials in 2008.

Bacterial Vectors Aeras-X05 is a bacteria-vectored oral vaccine under development by Aeras. Mtb antigens Ag85A, Ag85B, and Rv3407, encoded in *Shigella*-delivered double-stranded RNA nucleocapsids, induce high levels of CD8+ T cells. It is in late-stage preclinical development and scheduled for Phase I clinical trials in 2008.

Clinical Trials

Some newer approaches are currently in clinical trials. These include modified BCG vaccines, vectored vaccines, and subunit vaccines; they are described according to the phase of clinical trail in the next section.

Phase I

Viral-Vectored Vaccines Aeras and Crucell Holland B.V. developed Aeras-402, a viral-vectored TB vaccine that entered Phase I safety trials in the United States in 2006. The replication-incompetent adenovirus 35 vector carries genes for the expression of Mtb antigens Ag85A, Ag85B, and TB10.4 to induce IFN-gamma-producing CD4+ and CD8+ T cells. In preclinical studies, the vaccine exhibited protective T cell responses in two mouse strains with distinct MHC backgrounds. This result suggests that the Mtb antigens may encode sufficient numbers of protective T cell epitopes to provide broad coverage of heterogeneous MHC haplotypes in humans (Havenga et al., 2006; Radosevic et al., 2007).

Subunit Vaccines Two protein subunit vaccines are currently in Phase I clinical trials in the United States and Europe. In 2006, GlaxoSmithKline (GSK) and Aeras joined forces to begin a clinical trial of M72, a recombinant fusion of Mtb antigens Rv1196 and Rv0125. The vaccine is delivered with GSK proprietary adjuvant AS02. Like M72, HyVac4 is an adjuvant subunit vaccine. It was developed on the heels of Hybrid-1, a fusion molecule of the Ag85B and ESAT-6 antigens (Agger et al., 2006), which was abandoned because ESAT-6 had shown utility in diagnostic tests. HyVac4 is a recombinantly expressed fusion of Mtb antigens Ag85B and TB10.4 that is delivered in IC31, an Intercell proprietary adjuvant based on a cationic peptide $KLKL_5KLK$ and immunostimulatory oligodeoxynucleotide ODN1a. It was developed by the Statens Serum Institute, Intercell, and Aeras (Aeras-404) and entered Phase I clinical trials in 2007.

Phase II

The most advanced new TB vaccine in clinical trials is MVA85A, a viral-vectored vaccine developed by Adrian Hill and Helen McShane at the University of Oxford with funding from the Wellcome Trust and the European Commission. MVA85A, which is intended to be used as a BCG booster, is a modified vaccinia Ankara (MVA) virus that expresses Mtb antigen Ag85A (Chambers et al., 2004). MVA is an attenuated strain of vaccinia virus with a safety record established in humans during the smallpox eradication campaign. MVA85A completed Phase I in Gambia, South Africa, and the United Kingdom in BCG-naive and -vaccinated healthy adults. The vaccine was safe and generated high levels of antigen-specific IFN-gamma-secreting T cells in both patient groups (Dunlap and Briles, 1993). In Phase II, which began in July 2007 in Gambia, 471 HIV-negative 4-month-old babies will be evaluated for MVA85A-induced immune responses at various dosing levels as well as for interactions with other vaccinations, including those for diphtheria, pertussis, tetanus, and *Haemophilus influenzae* type b. In South Africa, the vaccine is being tested in HIV-infected adults.

Phase III

Although there are currently no new TB vaccines in Phase III clinical trials, the Working Group on New TB Vaccines under the Stop TB Partnership has set the target for licensure of a safe, effective vaccine available at reasonable cost by 2015.

New TB Vaccines in Development

A number of vaccines to prevent TB are under development although few of these vaccines are directed at the problem of LTBI. As no adequate animal model of TB exists, most TB vaccine research is performed in mice (who, unlike humans, do not develop cavitary TB lesions in their lungs) and in guinea pigs (who develop TB lesions that resemble human disease). Since animal models are poor predictors of TB vaccine efficacy, trial sites are being prepared in locations where TB is highly prevalent around the world.

Vaccines to Prevent Mtb Disease in the LTBI Host

Revaccinating subjects who have LTBI with BCG has never been considered to be a very effective means for preventing TB disease—indeed, the first trial of BCG for this indication failed more than 80 years ago. One reason that this approach may have failed is that exposure to Mtb antigens could exacerbate latent disease. This effect was recently demonstrated in the context of live or heat-treated BCG, with or without lipid adjuvants, on the recurrence of Mtb in latently infected mice (Moreira et al., 2002). In these experiments, increased antigen-specific T cell proliferation was observed, but there was no reduction in bacterial load in the lungs of BCG-vaccinated mice rechallenged with Mtb and, perhaps more worrisome, larger lung granulomas were discovered. This effect seemed to be due to increased expression of TNF-alpha and IL-6. The studies appeared to demonstrate that providing Mtb antigens in the context of prior Mtb infection leads to immune activation, exacerbating lung pathology. The exacerbation may be due to TNF-alpha-induced inflammation. It is however possible that the effect of immunization with whole BCG, including bacterial cell membranes, lipids, and other components, shifted the immune response to Th2-type responses, which would be associated with TNF secretion.

In contrast, immunization of mice with pure (recombinant) antigens, as in the experiments performed with DNA by Repique et al. (2002), did not appear to exacerbate LTBI or increase disease pathology following rechallenge. A number of other researchers have also failed to demonstrate that postexposure vaccination prevents relapse in mice with latent TB. For example, Turner et al. (2000) showed that a DNA vaccine that was effective against a primary infection did not protect when administered immunotherapeutically in a postexposure model. In addition, immunization with either BCG or a heat-killed *Mycobacterium vaccae* vaccine did not alter the relapse rate in a murine model of dormant TB (Collins and Mackaness, 1970).

A DNA vaccine expressing the heat shock protein 65 (hsp 65) vaccine has been the subject of a great deal of controversy. This vaccine was first shown to be successful in the prevention of TB recurrence in the Cornell model of MDR TB by Lowrie and associates (Tascon et al., 1996). Only a few years later, another group of researchers claimed that the hsp 65 vaccine might exacerbate TB in mice (Orme et al., 2001). In latter studies, mice appeared to develop discrete regions of cellular necrosis throughout the lung. Similar reactions were seen in mice given a vaccine with DNA coding for the Ag85 antigen of Mtb (Taylor et al., 2003). The relevance of this study to DNA immunization in humans is yet to be determined.

Thus, at least four recent attempts at developing LTBI vaccines have failed in murine models or had equivocal results. Morris and others have postulated that this failure may be due to the effect of LTBI itself on immune cells. While competent immune response against latent TB appears to be maintained in most hosts because the bacteria are contained inside the granulomas, the type of response that is required to clear infection may be qualitatively different. To reduce or eliminate LTBI, a vaccine may need to promote greater macrophage (or dendritic cell) activation. For example, Britton and coworkers have shown that DNA immunization in conjunction with plasmids expressing IL-12 are effective means of inducing TB immunity (Palendira et al., 2002). IL-12 can also be given as a recombinant protein, which avoids the problem of prolonged IL-12 exposure due to DNA delivery. Alternatively, coadministration with adjuvants that increase Th1 response, such as CpG motif-containing plasmids, may overcome the refractory nature of local immune response.

New Approach to TB Vaccine Development: Epitope-Driven Vaccines

Taking the Lowrie, Morris, and Kaplan studies in context, "whole" TB vaccines (based on BCG, or attenuated Mtb) for LTBI may have an adverse effect on the immune response due to the priming of the immune system for Th2-type responses. As neither Lowrie nor Morris observed exacerbation in LTBI mice vaccinated with DNA vaccines, removing the protein antigens from the context of the whole or killed bacteria appears to reduce the Th2 effect associated with mycobacterial cell wall components, which may in turn be responsible for the increased pathology. This suggests that it is worth pursuing a vaccine based on the delivery of multiple antigens—or epitopes derived from these antigens—in an immunostimulatory adjuvant or perhaps in conjunction with the TB vaccine developed by Lowrie. Thus, as an alternative to replacing BCG, it may be possible to develop such a genome-derived multi-epitope/antigen vaccine as would afford additional protection from TB disease in individuals already either vaccinated with BCG or latently infected with TB.

A new paradigm of TB vaccine development is therefore emerging, based on the concept that an *ensemble of epitopes* could stimulate protective Th1 immune responses, while the whole antigen and/or the whole organism might suppress immune responses. This new paradigm is based on the following observations. Following exposure to a pathogen, epitope-specific memory T cell clones are established. These clones respond rapidly and efficiently upon any subsequent infection, elaborating cytokines, killing infected host cells, and marshaling other cellular defenses against the pathogen. Broad T cell response may resemble multidrug TB therapy in that multiple T cell clones may be able to contain the bacteria or significantly impair bacterial replication, leading to clearance of the bacilli.

PROSPECTS FOR THE FUTURE

An effective TB vaccine will have a significant economic benefit in terms of global health. Experts estimate that such a vaccine would be worth at least $25 billion in terms of avoided medical spending worldwide. More specifically, the 1 billion people who are currently at risk of developing TB following exposure could expect to save $25 in medical costs if they had received a 75%-effective vaccine for 10 years (Weis et al., 1999; Bishai and Mercer, 2001). An effective LTBI vaccine could easily be priced in the same range as the hepatitis vaccine ($50 per dose or $150 per three-dose regimen) in the developed world. Pricing in that range would generate more than $500 million in revenue in the U.S. market and as much or more in other developed countries where the LTBI rate is as high as 50 times the rate in the United States. In the developing world, the current intervention for active TB costs $5 per DALY. An effective vaccine priced at $1 per dose would be equivalent to the cost per DALY of a directly observed short-course therapy for LTBI. According to the WHO, an acceptable price for an effective TB vaccine might be as high as $10 per dose in developing countries, given the urgency of the TB epidemic.

Significant progress is being made in terms of new vaccines in the pipeline, due to the infusion of funding from philanthropic organizations in the past decade. The expansion of the number of vaccines should not diminish the sense of urgency for a new TB vaccine but, rather, should serve to stimulate additional funding to support the development of vaccines from bench to clinic, so that we do not fall short of this important goal.

KEY ISSUES

- Mtb transmission is a worldwide public health threat.
- Upon infection, the host mounts an immune response that contains but does not eradicate Mtb.
- Infected individuals have a 10% lifetime risk of Mtb reactivation that leads to lung disease and dissemination to other organs.
- HIV-co-infected individuals have a 7–10% risk of reactivation in 1 year. TB has emerged as the leading cause of death from AIDS.
- The TB pandemic has continued unabated despite the use of directly observed chemotherapy programs. Moreover, multidrug-resistant and extensively drug-resistant TB have emerged and require control to prevent spread.
- Vaccination is a means to bring TB under control by preventing transmission or providing therapy.
- BCG, the only licensed TB vaccine available today, is administered to millions of people every year. While BCG reduces the incidence of TB in children, meningeal TB, and leprosy, it does not reliably prevent adult pulmonary TB, the most common form of the disease.
- New vaccines to replace or boost BCG are under development from the discovery stages to preclinical development to Phase II clinical trials today.
- Vaccine development approaches include live attenuated Mtb, recombinant BCG, viral- and bacterial-vectored, and subunit- and epitope-driven vaccines.

- TB research and development underfunding threatens to prevent achievement of the goals for new diagnostics, drugs, and vaccines by 2015 as set out by the Stop TB Partnership.

References

Adler, J.J. and Rose, D.N. Tuberculosis. Boston: Little, Brown, 1996, pp. 129–140.

Aeras. 2007. http://www.aeras.org/tb/epidemiology/index.html

Agger, E.M., Rosenkrands, I., Olsen, A.W., Hatch, G., Williams, A., Kritsch, C., Lingnau, K., von Gabain, A., Andersen, C.S., Korsholm, K.S. and Andersen, P. Protective immunity to tuberculosis with Ag85B-ESAT-6 in a synthetic cationic adjuvant system IC31. Vaccine 2006; 24:5452–5460.

Agrewala, J.N. and Wilkinson, R.J. Differential regulation of TH1 and TH2 cells by p91–110 and p21–40 peptides of the 16-kD "-crystallin antigen of Mycobacterium tuberculosis. Clin. Exp. Immunol. 1998; 114:392–397.

Alban, A. and Andersen, N.B. HIV/AIDS' spread and impact on other health-related millennium development goals—secondary publication. Dan. Med. Bull. 2007; 54(2):163–166.

Alderson, M.R., Bement, T., Day, C.H., Zhu, L., Molesh, D., Skeiky, Y.A., Coler, R., Lewinsohn, D.M., Reed, S.G. and Dillon, D. C. Expression cloning of an immunodominant family of Mycobacterium tuberculosis antigens using human CD4(+) T cells. J. Exp. Med. 2000; 191:551–560.

Andersen, P. Effective vaccination of mice against *Mycobacterium tuberculosis* infection with a soluble mixture of secreted mycobacterial proteins. Infect. Immun. 1994; 62:2536–2544.

Andersen, P. Host responses and antigens involved in protective immunity to *Mycobacterium tuberculosis*. Scand. J. Immunol. 1997; 45:115–1131.

Andersen, P. TB vaccines, progress and problems. Trends Immunol. 2001; 22:160–168.

Aronson, N.E., Santosham, M., Comstock, G.W., Howard, R.S., Moulton, L.H., Rhoades, E.R. and Harrison, L.H. Long-term efficacy of BCG vaccine in American Indians and Alaska natives: a 60-year follow-up study. JAMA 2004; 291:2086–2091.

Barnes, P.F., Lu, S., Abrams, J.S., Wang, E., Yamamura, M. and Modlin, R.L. Cytokine production at the site of disease in human tuberculosis. Infect. Immun. 1993; 61:3482–3489.

Bhigjee, A.I., Padayachee, R., Paruk, H., Hallwirth-Pillay, K.D., Marais, S. and Connoly, C. Diagnosis of tuberculous meningitis: clinical and laboratory parameters. Int. J. Infect. Dis. 2007; 11:348–354.

Bishai, D.M. and Mercer, D. Modeling the economic benefits of better TB vaccines. Int J Tuberc Lung Dis. 2001; 5(11):984–993.

Boesen, H., Jensen, B.N., Wilcke, T. and Andersen, P. Human T-cell responses to secreted antigen fractions of *Mycobacterium tuberculosis*. Infect. Immun. 1995; 63:1491–1497.

Bluestone, J.A. and Abbas, A.K. Natural versus adaptive regulatory T cells. Nat. Rev. Immunol. 2003; 3:253–257.

Boshoff, H.I. and Barry, C.E. Tuberculosis-metabolism and respiration in the absence of growth. Nat. Rev. Microbiol. 2005; 3:70–80.

Boussiotis, V.A., Tsai, E.Y., Yunis, E.J., Thim, S., Delgado, J.C., Dascher, C.C., Berezovskaya, A., Rousset, D., Reynes, J.M. and Goldfeld, A.E. IL-10-producing T cells suppress immune responses in anergic tuberculosis patients. J. Clin. Invest. 2000; 105:1317–1325.

Brosch, R., Gordon, S.V., Garnier, T., Eiglmeier, K., Frigui, W., Valenti, P., Dos Santos, S., Duthoy, S., Lacroix, C., Garcia-Pelayo, C., Inwald, J.K., Golby, P., Garcia, J.N., Hewinson, R.G., Behr, M.A., Quail, M. A., Churcher, C., Barrell, B.G., Parkhill, J. and Cole, S.T. Genome plasticity of BCG and impact on vaccine efficacy. Proc. Natl. Acad. Sci. U.S.A. 2007; 104(13):5596–5601.

Cai, H., Yu, D.H., Tian, X. and Zhu, Y.X. Coadministration of interleukin 2 plasmid DNA with combined DNA vaccines significantly enhances the protective efficacy against Mycobacterium tuberculosis. DNA Cell Biol. 2005; 24:605–613.

Camus, J.C., Pryor, M.J., Medigue, C. and Cole, S.T. Re-annotation of the genome sequence of *Mycobacterium tuberculosis* H37Rv. Microbiology 2002; 148:2967–2973.

Canaday, D.H., Wilkinson, R.J., Li, Q., Harding, C.V., Silver, R.F. and Boom, W.H. CD4(+) and CD8(+) T cells kill intracellular *Mycobacterium tuberculosis* by a perforin and Fas/Fas ligand-independent mechanism. J. Immunol. 2001; 167:2734–2742.

Cappelli, G., Volpe, E., Grassi, M., Liseo, B., Colizzi, V. and Mariani, F. Profiling of *Mycobacterium tuberculosis* gene expression during human macrophage infection: upregulation of the alternative sigma factor G, a group of transcriptional regulators, and proteins with unknown function. Res. Microbiol. 2006; 157(5):445–455.

CDC. Treatment of tuberculosis. MMWR Morb. Mortal. Wkly. Rep. 2003; 52 (RR11):1–77.

CDC. Reported HIV status of tuberculosis patients—United States, 1993–2005. MMWR Morb. Mortal. Wkly. Rep. Oct. 26, 2007a; 56(42):1103–1106.

CDC. Extensively drug-resistant tuberculosis—United States, 1993–2006. MMWR Morb. Mortal. Wkly. Rep. 2007b; 56:250–253.

CDC. 2007c. http://emergency.cdc.gov/agent/agentlist-categoryasp#3

Chambers, M.A., Gavier-Widen, D. and Hewinson, R.G. Antibody bound to the surface antigen MPB83 of *Mycobacterium bovis* enhances survival against high dose and low dose challenge. FEMS Immunol. Med. Microbiol. 2004; 41:93–100.

Cho, S., Mehra, V., Thoma-Uszynski, S., Stenger, S., Serbina, N., Mazzaccaro, R.J., Flynn, J.L., Barnes, P.F., Southwood, S., Celis, E., Bloom, B.R., Modlin, R.L. and Sette, A. Antimicrobial activity of MHC Class I-restricted CD8+ T cells in human tuberculosis. Proc. Natl. Acad. Sci. U.S.A. 2000; 97:12210–12215.

Colditz, G.A., Brewer, T.F., Berkey, C.S., Wilson, M.E., Burdick, E., Fineberg, H.V. and Mosteller, F. Efficacy of BCG vaccine in the prevention of tuberculosis. Meta-analysis of the published literature. J. Am. Med. Assoc. 1994; 271:698–702.

Collins, F.M. and Mackaness, G.B. The relationship of delayed hypersensitivity to acquired antituberculous immunity. I. Tuberculin sensitivity and resistance to reinfection in BCG-vaccinated mice. Cell. Immunol. 1970; 1:253–265.

Cooper, A.M., Dalton, D.K., Stewart, T.A., Griffin, J.P., Russell, D.G. and Orme, I.M. Disseminated tuberculosis in interferon gamma gene-disrupted mice. J. Exp. Med. 1993; 178:2243–2247.

Daley, C.L., Small, P.M., Schecter, G.F., Schoolnik, G.K., McAdam, R.A., Jacobs, W.R., Jr. and Hopewell, P.C. An outbreak of tuberculosis with accelerated progression among person infected with the human immunodeficiency virus. An analysis using restriction fragment length polymorphisms. N. Engl. J. Med. 1992; 326:231–325.

Davidow, A., Kanaujiam, G.V., Shi, L., Kaviar, J., Guo, X., Sung, N., Kaplan, G., Menzies, D. and Gennaro, M.L. Antibody profiles characteristic of *Mycobacterium tuberculosis* infection state. Infect. Immun. 2005; 73:6846–6851.

Delgado, J.C., Tsai, E.Y., Thim, S., Baena, A., Boussiotis, V.A., Reynes, J.M., Sath, S., Grosjean, P., Yunis, E.J. and Goldfeld, A.E. Antigen-specific and persistent tuberculin anergy in a cohort of pulmonary tuberculosis patients from rural Cambodia. Proc. Natl. Acad. Sci. U.S.A. 2002; 99:7576–7581.

Demissie, A., Abebe, M., Aseffa, A., Rook, G., Fletcher, H., Zumla, A., Weldingh, K., Brock, I., Andersen, P. and Doherty, T.M. VACSEL Study Group. Healthy individuals that control a latent infection with *Mycobacterium tuberculosis* express high levels of

TH1 cytokines and the IL-4 antagonist IL-4delta2. J. Immunol. 2004; 172:6938–6943.

Demissie, A., Ravn, P., Olobo, J., Doherty, TM., Eguale, T., Geletu, M., Hailu, W., Andersen, P. and Britton, S. T-cell recognition of *Mycobacterium tuberculosis* culture filtrate fractions in tuberculosis patients and their household contacts. Infect. Immun. 1999; 67:5967–5971.

Dietrich, J., Aagaard, C., Leah, R. et al. Exchanging ESAT6 with TB10.4 in an Ag85B fusion molecule-based tuberculosis subunit vaccine: efficient protection and ESAT6-based sensitive monitoring of vaccine efficacy. J. Immunol. 2005; 174:6332–6339.

Dillon, D.C., Alderson, M.R., Day, C.H., Lewinsohn, D.M., Coler, R., Bement, T., Campos-Neto, A., Skeiky, Y.A., Orme, I.M., Roberts, A., Steen, S., Dalemans, W., Badaro, R. and Reed, S.G. Molecular characterization and human T-cell responses to a member of a novel Mycobacterium tuberculosis mtb39 gene family. Infect. Immun. 1999; 67:2941–2950.

Downing, K.J., Betts, J.C., Young, D.I., McAdam, R.A., Kelly, F., Young, M. and Mizrahi, V. Global expression profiling of strains harbouring null mutations reveals that the five rpf-like genes of *Mycobacterium tuberculosis* show functional redundancy. Tuberculosis (Edinb.) 2004; 84:167–179.

Downing, K.J., Mischenko, V.V., Shleeva, M.O., Young, D.I., Young, M., Kaprelyants, A.S., Apt, A.S. and Mizrahi, V. Mutants of *Mycobacterium tuberculosis* lacking three of the five rpf-like genes are defective for growth in vivo and for resuscitation in vitro. Infect. Immun. 2005; 73:3038–3043.

Dunlap, N.E. and Briles, D.E. Immunology of tuberculosis. Med. Clin. North Am. 1993; 77:1235–1251.

Dye, C. Epidemiology in Clinical Tuberculosis, 3rd ed. London: Arnold, 2003.

Dye, C., Scheele, S., Dolin, P., Pathania, V. and Raviglione, M.C. Global burden of tuberculosis, estimated incidence, prevalence, and mortality by country. J. Am. Med. Assoc. 1999; 282: 677–686.

Feuer, C. Tuberculosis Research and Development: A Critical Analysis of Funding Trends, 2005–2006 (J. Syed, M. Harrington, B. Huff, Eds.). New York: Treatment Action Group, 2007; pp. 1–43.

Fine, P.E. Variation in protection by BCG, implications of and for heterologous immunity. Lancet 1995; 346:1339–1345.

Fleischmann, R.D., Alland, D., Eisen, J.A., Carpenter, L., White, O., Peterson, J., DeBoy, R., Dodson, R., Gwinn, M., Haft, D., Hickey, E., Kolonay, J.F., Nelson, W.C., Umayam, L.A., Ermolaeva, M., Salzberg, S.L., Delcher, A., Utterback, T., Weidman, J., Khouri, H., Gill, J., Mikula, A., Bishai, W., Jacobs, W.R., Jr., Venter, J.C. and Fraser, C.M. Whole-genome comparison of *Mycobacterium tuberculosis* clinical and laboratory strains. J. Bacteriol. 2002; 184:5479–5490.

Flynn, J.L., Chan, J., Triebold, K.J., Dalton, D.K., Stewart, T.A. and Bloom, B.R. An essential role for interferon gamma in resistance to *Mycobacterium tuberculosis* infection. J. Exp. Med. 1993; 178:2249–2254.

Garrait, V., Cadranel, J., Esvant, H., Herry, I., Morinet, P., Mayaud, C. and Israel-Biet, D. Tuberculosis generates a microenvironment enhancing the productive infection of local lymphocytes by HIV. J. Immunol. 1997; 159:2824–2830.

Gerosa, F., Nisii, C., Righetti, S., Micciolo, R., Marchesini, M., Cazzadori, A. and Trinchieri, G. CD4+ T cell clones producing both interferon-gamma and interleukin-10 predominate in bronchoalveolar lavages of active pulmonary tuberculosis patients. Clin. Immunol. 1999; 92:224–234.

Goonetilleke, N.P., McShane, H., Hannan, C.M., Anderson, R.J., Brookes, R.H. and Hill, A.V. Enhanced immunogenicity and protective efficacy against Mycobacterium tuberculosis of bacille Calmette-Guerin vaccine using mucosal administration and boosting with a recombinant modified vaccinia virus Ankara. J. Immunol. 2003; 171:1602–1609.

Grode, L., Seiler, P., Baumann, S., Hess, J., Brinkmann, V., Nasser Eddine, A., Mann, P., Goosmann, C., Bandermann, S., Smith, D., Bancroft, G.J., Reyrat, J.M., van Soolingen, D., Raupach, B. and Kaufmann, S.H. Increased vaccine efficacy against tuberculosis of recombinant *Mycobacterium bovis* bacille Calmette-Guérin mutants that secrete listeriolysin. J. Clin. Invest. 2005; 115: 2472–2479.

Guerkov, R.E., Targoni, O.S., Kreher, C.R., Boehm, B.O., Herrera, M.T., Tary-Lehmann, M., Lehmann, P.V. and Schwander, S.K. Detection of low-frequency antigen-specific IL-10 producing CD4(+). T cells via ELISPOT in PBMC, cognate vs. nonspecific production of the cytokine. J. Immunol. Methods 2003; 279:111–121.

Guyot-Revol, V., Innes, J.A., Hackforth, S., Hinks, T. and Lalvani, A. Regulatory T cells are expanded in blood and disease sites in patients with tuberculosis. Am. J. Respir. Crit. Care Med. 2006; 173:803–810.

Havenga, M., Vogels, R., Zuijdgeest, D., Radosevic, K., Mueller, S., Sieuwerts, M., Weichold, F., Damen, I., Kaspers, J., Lemckert, A., van Meerendonk, M., van der Vlugt, R., Holterman, L., Hone, D., Skeiky, Y., Mintardjo, R., Gillissen, G., Barouch, D., Sadoff, J. and Goudsmit, J. Novel replication-incompetent adenoviral B-group vectors, high vector stability and yield in PER.C6 cells. J. Gen. Virol. 2006; 87:2135–2143.

Hesseling, A.C., Marais, B.J., Gie, R.P., Schaafa, H.S., Fine, P.E.M., Godfrey-Faussett, P. and Beyers, N. The risk of disseminated Bacille Calmette-Guerin (BCG) disease in HIV-infected children. Vaccine 2007; 25(1):14–18.

Hirsch, C.S., Hussain, R., Toossi, Z., Dawood, G., Shahid, F. and Ellner, J.J. Cross-modulation by transforming growth factor beta in human tuberculosis: suppression of antigen-driven blastogenesis and interferon gamma production. Proc. Natl. Acad. Sci. U.S.A. 1996; 93:3193–3198.

Hirsch, C.S., Toossi, Z., Johnson, J.L., Luzze, H., Ntambi, L., Peters, P., McHugh, M., Okwera, A., Joloba, M., Mugyenyi, P., Mugerwa, R.D., Terebuh, P. and Ellner, J.J. Augmentation of apoptosis and interferon-gamma production at sites of active Mtb infection in human tuberculosis. J. Infect. Dis. 2001; 183:779–788.

Hoff, S.T., Abebe, M., Ravn, P., Range, N., Malenganisho, W., Rodriques, D.S., Kallas, E.G., Soborg, C., Mark Doherty, T., Andersen, P. and Weldingh, K. Evaluation of *Mycobacterium tuberculosis*—specific antibody responses in populations with different levels of exposure from Tanzania, Ethiopia, Brazil, and Denmark. Clin. Infect. Dis. 2007; 45:575–582.

Horwitz, M.A., Harth, G., Dillon, B.J. and Maslesa-Galic, S. Recombinant bacillus Calmette-Guerin (BCG) vaccines expressing the *Mycobacterium tuberculosis* 30-kDa major secretory protein induce greater protective immunity against tuberculosis than conventional BCG vaccines in a highly susceptible animal model. Proc. Natl. Acad. Sci. U.S.A. 2000; 97:13853–13858.

Horwitz, M.A. and Harth, G. A new vaccine against tuberculosis affords greater survival after challenge than the current vaccine in the guinea pig model of pulmonary tuberculosis. Infect. Immun. 2003; 71:1672–1679.

Hougardy, J.M., Place, S., Hildebrand, M., Drowart, A., Debrie, A.S., Locht, C. and Mascart, F. Regulatory T cells depress immune responses to protective antigens in active tuberculosis. Am. J. Respir. Crit. Care Med. 2007; 176:409–416.

Infuso, A. and Falzon, D. European survey of BCG vaccination policies and surveillance in children, 2005. Euro. Surveill. 2006; 11(3):6–11.

Javitz, H.S. and Ward, M. Value of antimicrobials in the prevention and treatment of tuberculosis in the United States. Int. J. Tuberc. Lung Dis. 2002; 6:275–288.

Johnson, P.D., Stuart, R.L., Grayson, M.L., Olden, D., Clancy, A., Ravn, P., Andersen, P., Britton, W.J. and Rothel, J.S. Tuberculin-purified protein derivative-, MPT-64-, and ESAT-6-stimulated gamma interferon responses in medical students before and after *Mycobacterium bovis* BCG vaccination and in patients with tuberculosis. Clin. Diagn. Lab. Immunol. 1999; 6:934–937.

Johansen, P., Raynaud, C., Yang, M., Colston, M.J., Tascon, R.E. and Lowrie, D.B. Anti-mycobacterial immunity induced by a single injection of M. leprae Hsp65-encoding plasmid DNA in biodegradable microparticles. Immunol. Lett. 2003; 90:81–85.

Kaufmann, S.H. and Hess, J. Impact of intracellular location of and antigen display by intracellular bacteria, implications for vaccine development. Immunol. Lett. 1999; 65:81–84.

Kovacs, J.A., Lempicki, R.A., Sidorov, I.A., Adelsberger, J.W., Herpin, B., Metcalf, J.A., Sereti, I., Polis, M.A., Davey, R.T., Tavel, J., Falloon, J., Stevens, R., Lambert, L., Dewar, R., Schwartzentruber, D.J., Anver, M.R., Baseler, M.W., Masur, H., Dimitrov, D.S. and Lane, H.C. Identification of dynamically distinct subpopulations of T lymphocytes that are differentially affected by HIV. J. Exp. Med. 2001; 194:1731–1741.

Lawn, S.D., Butera, S.T. and Folks, T.M. Contribution of immune activation to the pathogenesis and transmission of human immunodeficiency virus type 1 infection. Clin. Microbiol. Rev. 2001; 14:753–777.

Lefford, M.J. Transfer of adoptive immunity to tuberculosis in mice. Infect. Immun. 1975; 11:1174–1181.

Lefevre, P., Denis, O., De Wit, L., Tanghe, A., Vandenbussche, P., Content, J. and Huygen, K. Cloning of the gene encoding a 22-kilodalton cell surface antigen of Mycobacterium bovis BCG and analysis of its potential for DNA vaccination against tuberculosis. Infect. Immun. 2000; 68:1040–1047.

Lewinsohn, D.A., Winata, E., Swarbrick, G.M., Tanner, K.E., Cook, M.S., Null, M.D., Cansler, M.E., Sette, A., Sidney, J. and Lewinsohn, D.M. Immunodominant tuberculosis CD8 antigens preferentially restricted by HLA-B. PLoS Pathog. 2007; 3(9):1240–1249.

Lopez, A.D., Mathers, C.D., Ezzati, M., Jamison, D.T. and Murray, C.J. Global and regional burden of disease and risk factors, 2001: systematic analysis of population health data. Lancet 2006; 367(9524):1747–1757.

McShane, H., Pathan, A.A., Sander, C.R., Keating, S.M., Gilbert, S.C., Huygen, K., Fletcher, H.A. and Hill, A.V. Recombinant modified vaccinia virus Ankara expressing antigen 85A boosts BCG-primed and naturally acquired antimycobacterial immunity in humans. Nat. Med. 2004; 10:1240–1244.

Mills, K.H. Regulatory T cells: friend or foe in immunity to infection? Nat. Rev. Immunol. 2004; 4:841–855.

Mocellin, S., Panelli, M.C., Wang, E., Nagorsen, D. and Marincola, F.M. The dual role of IL-10. Trends Immunol. 2003; 24:36–43.

Moreira, A.L., Tsenova, L., Aman, M.H., Bekker, L.G., Freeman, S., Mangaliso, B., Schröder, U., Jagirdar, J., Rom, W.N., Tovey, M. G., Freedman, V.H. and Kaplan, G. Mycobacterial antigens exacerbate disease manifestations in Mycobacterium tuberculosis-infected mice. Infect Immun 2002; 70(4):2100–2107.

Moriuchi, H., Moriuchi, M., Mizell, S.B., Ehler, L.A. and Fauci, A.S. In vitro reactivation of human immunodeficiency virus 1 from latently infected, resting CD4+ T cells after bacterial stimulation. J. Infect. Dis. 2000; 181:2041–2044.

Mustafa, A.S. Isolation and characterization of the genes of pathogenic mycobacteria that express antigens for T-cell reactivity. Nutrition 1995a; 11(Suppl. 5):653–656.

Mustafa, A.S., Amoudy, H.A., Wiker, H.G., Abal, A.T., Ravn, P., Oftung, F. and Andersen, P. Comparison of antigen-specific T-cell responses of tuberculosis patients using complex or single antigens of *Mycobacterium tuberculosis*. Scand. J. Immunol. 1998; 48:535–543.

Mustafa, A.S. Development of new vaccines and diagnostic reagents against tuberculosis. Mol. Immunol. 2002; 39:113–119.

Nacy, C.A. and Sacksteder, K.A. New tuberculosis vaccine development. Expert Opin. Biol. Ther. 2002; 2:741–749.

NIH. 2007. http://www3.niaid.nih.gov/topics/BiodefenseRelated/Biodefense/PDF/Cat.pdf

Okkels, L.M., Doherty, T.M. and Andersen, P. Selecting the components for a safe and efficient tuberculosis subunit vaccine: recent progress and post-genomic insights. Curr. Pharma. Biotech. 2003; 4:69–83.

Pablos-Mendez, A., Raviglione, M.C., Laszlo, A., Binkin, N., Rieder, H.L., Bustreo, F., Cohn, D.L., Lambregts-van Weezenbeek, C.S., Kim, S.J., Chaulet, P. and Nunn, P. Global surveillance for antituberculosis-drug resistance, 1994–1997. World Health Organization-International Union against Tuberculosis and Lung Disease Working Group on Anti-Tuberculosis Drug Resistance Surveillance. N. Engl. J. Med. 1998; 338:1641–1649.

Palendira, U., Kamath, A.T., Feng, C.G., Martin, E., Chaplin, P.J., Triccas, J.A. and Britton, W.J. Coexpression of interleukin-12 chains by a self-splicing vector increases the protective cellular immune response of DNA and *Mycobacterium bovis* BCG vaccines against *Mycobacterium tuberculosis*. Infect. Immun. 2002; 70:1949–1956.

Prescott, L.M., Harley, J.P. and Klein, D.A. Microbiology, 3rd ed. Dubuque, IA: William C. Brown Publishers, 1996, 415–476, 416–437, 491–502.

Radosevic, K., Wieland, C.W., Rodriguez, A., Weverling, G.J., Mintardjo, R., Gillissen, G., Vogels, R., Skeiky, Y.A., Hone, D.M., Sadoff, J.C., van der Poll, T., Havenga, M. and Goudsmit, J. Protective immune responses to a recombinant adenovirus type 35 tuberculosis vaccine in two mouse strains, CD4 and CD8 T-cell epitope mapping and role of gamma interferon. Infect. Immun. 2007; 75:4105–4115.

Raviglione, M.C., Dye, C., Schmidt, S. and Kochi, A. Assessment of worldwide TB control. Lancet 1997; 350:624–627.

Repique, C.J., Li, A., Collins, F.M. and Morris, S.L. DNA immunization in a mouse model of latent tuberculosis: effect of DNA vaccination on reactivation of disease and on reinfection with a secondary challenge. Infect. Immun. 2002; 70:3318–3323.

Ribeiro-Rodrigues, R., Resende Co, T., Rojas, R., Toossi, Z., Dietze, R., Boom, W.H., Maciel, E. and Hirsch, C.S. A role for CD4+ CD25+ T cells in regulation of the immune response during human tuberculosis. Clin. Exp. Immunol. 2006; 144:25–34.

Rigden, R.C., Jandhyala, D.M., Dupont, C., Crosbie-Caird, D., Lopez-Villalobos, N., Maeda, N., Gicquel, B. and Murray, A. Humoral and cellular immune responses in sheep immunized with a 22 kilodalton exported protein of Mycobacterium avium subspecies paratuberculosis. J. Med. Microbiol. 2006; 55:1735–1740.

Rook, G.A., Hernandez-Pando, R., Dheda, K. and Teng Seah, G. IL-4 in tuberculosis, implications for vaccine design. Trends Immunol. 2004; 25:483–488.

Rook, G.A. and Zumla, A. Advances in the immunopathogenesis of pulmonary tuberculosis. Curr. Opin. Pulm. Med. 2001; 7:116–123.

Rothschild, B.M., Martin, L.D., Lev, G., Bercovier, H., Bar-Gal, G.K., Greenblatt, C., Donoghue, H., Spigelman, M. and Brittain, D. *Mycobacterium tuberculosis* complex DNA from an extinct bison dated 17,000 years before the present. Clin. Infect. Dis. 2001; 33:305–311.

Russell, D.G. *Mycobacterium tuberculosis*: here today, and here tomorrow. Nat. Rev. Mol. Cell Biol. 2001; 2:569–577.

Russell, D.G. Who puts the tubercle in tuberculosis? Nat. Rev. Microbiol. 2007; 5:39–47.

Sambandamurthy, V.K., Derrick, S.C., Hsu, T., Chen, B., Larsen, M.H., Jalapathy, K.V., Chen, M., Kim, J., Porcelli, S.A., Chan, J.,

Morris, S.L. and Jacobs, W.R., Jr. *Mycobacterium tuberculosis* DeltaRD1 DeltapanCD: a safe and limited replicating mutant strain that protects immunocompetent and immunocompromised mice against experimental tuberculosis. Vaccine 2006; 24(37–39):6309–6320.

Sadoff, J. Public private partnership approach to vaccine development Presented at New Approaches to Vaccine Development Sept 8-10, Berlin, Germany (2005).

Sambandamurthy, V.K., Derrick, S.C., Jalapathy, K.V., Chen, B., Russell, R.G., Morris, S.L. and Jacobs, W.R., Jr. Long-term protection against tuberculosis following vaccination with a severely attenuated double lysine and pantothenate auxotroph of *Mycobacterium tuberculosis*. Infect. Immun. 2005; 73(2):1196–1203.

Scanga, C.A., Mohan, V.P., Yu, K., Joseph, H., Tanaka, K., Chan, J. and Flynn, J.L. Depletion of CD4(+)T cells causes reactivation of murine persistent tuberculosis despite continued expression of interferon gamma and nitric oxide synthase 2. J. Exp. Med. 2000; 192:347–358.

Schnappinger, D., Ehrt, S., Voskuil, M.I., Liu, Y., Mangan, J.A., Monahan, I.M., Dolganov, G., Efron, B., Butcher, P.D., Nathan, C. and Schoolnik, G.K. Transcriptional adaptation of *Mycobacterium tuberculosis* within macrophages, insights into the phagosomal environment. J. Exp. Med. 2003; 198:693–704.

Scott-Browne, J.P., Shafiani, S., Tucker-Heard, G., Ishida-Tsubota, K., Fontenot, J.D., Rudensky, A.Y., Bevan, M.J. and Urdahl, K.B. Expansion and function of Foxp3-expressing T regulatory cells during tuberculosis. J. Exp. Med. 2007; 204(9):2159–2169.

Selwyn, P.A., Hartel, D., Lewis, V.A., Schoenbaum, E.E., Vermund, S.H., Klein, R.S., Walker, A.T. and Friedland, G.H. A prospective study of the risk of tuberculosis among intravenous drug users with human immunodeficiency virus infection. N. Engl. J. Med. 1989; 320:545–550.

Sepkowitz, K.A. Tuberculosis control in the 21st century. Emerg. Infect. Dis. 2001; 7:259–262.

Serbina, N.V. and Flynn, J.L. CD8(+) T cells participate in the memory immune response to *Mycobacterium tuberculosis*. Infect. Immun. 2001; 69:4320–4328.

Serbina, N.V., Lazarevic, V. and Flynn, J.L. CD4(+) T cells are required for the development of cytotoxic CD8(+) T cells during *Mycobacterium tuberculosis* infection. J. Immunol. 2001; 167:6991–7000.

Shen, Y., Shen, L., Sehgal, P., Zhou, D., Simon, M., Miller, M., Enimi, E.A., Henckler, B., Chalifoux, L., Sehgal, N., Gastron, M., Letvin, N.L. and Chen, Z.W. Antiretroviral agents restore mycobacterium-specific T-cell immune responses and facilitate controlling a fatal tuberculosis-like disease in macaques coinfected with Simian immunodeficiency virus and *Mycobacterium bovis* BCG. J. Virol. 2001; 75:8690–8696.

Sherman, D.R., Voskuil, M., Schnappinger, D., Liao, R., Harrell, M.I. and Schoolnik, G.K. Regulation of the *Mycobacterium tuberculosis* hypoxic response gene encoding alpha-crystallin. Proc. Natl. Acad. Sci. U.S.A. 2001; 98:7534–7539.

Skeiky, Y.A., Alderson, M.R., Ovendale, P.J., Guderian, J.A., Brandt, L., Dillon, D.C., Campos-Neto, A., Lobet, Y., Dalemans, W., Orme, I. M. and Reed, S.G. Differential immune responses and protective efficacy induced by components of a tuberculosis polyprotein vaccine, Mtb72F, delivered as naked DNA or recombinant protein. J. Immunol. 2004; 172:7618–7628.

Skjot, R.L., Oettinger, T., Rosenkrands, I., Ravn, P., Brock, I., Jacobsen, S. and Andersen, P. Comparative evaluation of low-molecular-mass proteins from Mycobacterium tuberculosis identifies members of the ESAT-6 family as immunodominant T-cell antigens. Infect. Immun. 2000; 68:214–220.

Skjot, R.L., Brock, I., Arend, S.M., Munk, M.E., Theisen, M., Ottenhoff, T.H. and Andersen, P. Epitope mapping of the immunodominant antigen TB10.4 and the two homologous proteins TB10.3 and TB12.9, which constitute a subfamily of the esat-6 gene family. Infect. Immun. 2002; 70:5446–5453.

Skeiky, Y.A. and Sadoff, J.C. Advances in tuberculosis vaccine strategies. Nat. Rev. Microbiol. 2006; 4:469–476.

Sodhi, A., Gong, J., Silva, C., Qian, D. and Barnes, P.F. Clinical correlates of interferon gamma production with tuberculosis. Clin. Infect. Dis. 1997; 25:617–620.

Sonnenberg, M.G. and Belisle, J.T. Definition of *Mycobacterium tuberculosis* culture filtrate proteins by two-dimensional polyacylamide gel electrophoresis, *N*-terminal amino acid sequencing, and electrospray mass spectrometry. Infect. Immun. 1997; 65:4515–4524.

Soto, C.Y., Menendez, M.C., Perez, E., Samper, S., Gomez, A.B., Garcia, M.J. and Martin, C. IS6110 mediates increased transcription of the phoP virulence gene in a multidrug-resistant clinical isolate responsible for tuberculosis outbreaks. J. Clin. Microbiol. 2004; 42(1):212–219.

Stanberry, L.R., Spruance, S.L., Cunningham, A.L., Bernstein, D.I., Mindel, A., Sacks, S., Tyring, S., Aoki, F.Y., Slaoui, M., Denis, M., Vandepapeliere, P. and Dubin, G. GlaxoSmithKline Herpes Vaccine Efficacy Study Group Glycoprotein-D-adjuvant vaccine to prevent genital herpes. N. Engl. J. Med. 2002; 347:1652–1661.

Sterne, J.A., Fine, P.E., Ponnighaus, J.M., Sibanda, F., Munthali, M. and Glynn, J.R. Does bacille Calmette-Guerin scar size have implications for protection against tuberculosis or leprosy? Tuber. Lung Dis. 1996; 77:117–123.

Sterne, J.A., Rodrigues, L.C. and Guedes, I.N. Does the efficacy of BCG decline with time since vaccination? Int. J. Tuberc. Lung Dis. 1998; 2:200–207.

Tan, J.S., Canaday, D.H., Boom, W.H., Balaji, K.N., Schwander, S.K. and Rich, E.A. Human alveolar T lymphocyte responses to *Mycobacterium tuberculosis* antigens, role for CD4+ and CD8+ cytotoxic T cells and relative resistance of alveolar macrophages to lysis. J. Immunol. 1997; 159:290–297.

Tascon, R.E., Colston, M.J., Ragno, S., Stavropoulos, E., Gregory, D. and Lowrie, D.B. Vaccination against tuberculosis by DNA injection. Nat. Med. 1996; 2:888–892.

Taylor, J.L., Turner, O.C., Basaraba, R.J., Belisle, J.T., Huygen, K. and Orme, I.M. Pulmonary necrosis resulting from DNA vaccination against tuberculosis. Infect. Immun. 2003; 71:2192–2198.

Taylor, J.L., Ordway, D.J., Troudt, J., Gonzalez-Juarrero, M., Basaraba, R.J. and Orme, I.M. Factors associated with severe granulomatous pneumonia in Mycobacterium tuberculosis-infected mice vaccinated therapeutically with hsp65 DNA. Infect. Immun. 2005; 73(8):5189–5193.

Torres, M., Herrera, T., Villareal, H., Rich, E.A. and Sada, E. Cytokine profiles for peripheral blood lymphocytes from patients with active pulmonary tuberculosis and healthy household contacts in response to the 30-kilodalton antigen of *Mycobacterium tuberculosis*. Infect. Immun. 1998; 66:176–180.

Trunz, B.B., Fine, P. and Dye, C. Effect of BCG vaccination on childhood tuberculous meningitis and miliary tuberculosis worldwide: a meta-analysis and assessment of cost-effectiveness. Lancet 2006; 367:1173–1180.

Turner, J., Gonzalez-Juarrero, M., Ellis, D.L., Basaraba, R.J., Kipnis, A., Orme, I.M. and Cooper, A.M. In vivo IL-10 production reactivates chronic pulmonary tuberculosis in C57BL/6 mice. J. Immunol. 2002; 169:6343–6351.

Turner, J., Rhoades, E.R., Keen, M., Belisle, J.T., Frank, A.A. and Orme, I.M. Effective preexposure tuberculosis vaccines fail to protect when they are given in an immunotherapeutic mode. Infect. Immun. 2000; 68:1706–1709.

van Rie, A., Warren, R., Richardson, M., Victor, T.C., Gie, R.P., Enarson, D.A., Beyers, N. and van Helden, P.D. Exogenous reinfection as a cause of recurrent tuberculosis after curative treatment. N. Engl. J. Med. 1999; 341:1174–1179.

Voskuil, M.I., Schnappinger, D., Visconti, K.C., Harrell, M.I., Dolganov, G.M., Sherman, D.R. and Schoolnik, G.K. Inhibition of respiration by nitric oxide induces a *Mycobacterium tuberculosis* dormancy program. J. Exp. Med. 2003; 198:705–713.

Weichold, F.F., Mueller, S., Kortsik, C., Hitzler, W.E., Wulf, M.J., Hone, D.M., Sadoff, J.C. and Maeurer, M.J. Impact of MHC class I alleles on the *M. tuberculosis* antigen-specific CD8+ T-cell response in patients with pulmonary tuberculosis. Genes Immun. 2007; 8(4):334–343.

Weis, S.E., Foresman, B., Matty, K.J., Brown, A., Blais, F.X., Burgess, G., King, B., Cook, P.E. and Slocum, P.C. Treatment costs of directly observed therapy and traditional therapy for Mycobacterium tuberculosis: a comparative analysis. Int J Tuberc Lung Dis. 1999; 3(11):976–984.

Wells, C.D., Cegielski, J.P., Nelson, L.J., Laserson, K.F., Holtz, T.H., Finlay, A., Castro, K.G. and Weyer, K. HIV infection and multidrug-resistant tuberculosis: the perfect storm. J. Infect. Dis. 2007; 196(Suppl. 1):S86–S107.

WHO. 1999. http://www.who.int/vaccines-documents/DocsPDF99/www9943.pdf

WHO. Recommendations of the Scientific Working Group on Tuberculosis, 2000. Geneva, Feb. 9–11, 2000.

WHO. Global Tuberculosis Control (2007), Key Findings. 2007a. Accessed at http://www.who.int/tb/publications/global_report/2007/key_findings/en/index.html on 9/1/07.

WHO. 2007b. http://www.who.int/vaccine_safety/topics/bcg/immunocompromised/Dec_2006/en/index.html

Wu, B., Huang, C., Garcia, L., Ponce de Leon, A., Osornio, J.S., Bobadilla-del-Valle, M., Ferreira, L., Canizales, S., Small, P., Kato-Maeda, M., Krensky, A.M. and Clayberger, C. Unique gene expression profiles in infants vaccinated with different strains of *Mycobacterium bovis* bacille Calmette-Guerin. Infect. Immun. 2007; 75(7):3658–3664.

Yeremeev, V.V., Lyadova, I.V., Nikonenko, B.V., Apt, A.S., Abou-Zeid, C., Inwald, J. and Young, D.B. The 19-kD antigen and protective immunity in a murine model of tuberculosis. Clin. Exp. Immunol. 2000; 120:274–279.

Young, D.B. and Stewart, G.R. Tuberculosis vaccines. Br. Med. Bull. 2002; 62:73–86.

Young, R.A., Bloom, B.R., Grosskinsky, C.M., Ivanyi, J., Thomas, D. and Davis, R.W. Dissection of *Mycobacterium tuberculosis* antigens using recombinant DNA. Proc. Natl. Acad. Sci. U.S.A. 1985; 82:2583–2587.

Zink, A.R., Sola, C., Reischl, U., Grabner, W., Rastogi, N., Wolf, H. and Nerlich, A.G. Characterization of *Mycobacterium tuberculosis* complex DNAs from Egyptian mummies by spoligotyping. J. Clin. Microbiol. 2003; 41:359–367.

CHAPTER

61

Francisella tularensis

Richard W. Titball[1] and Petra C.F. Oyston[2]

[1]School of Biosciences, Geoffrey Pope Building, University of Exeter, Exeter, Devon, UK

[2]Defence Science and Technology Laboratory, Salisbury, Wiltshire, UK

OUTLINE

ABSTRACT

Francisella tularensis has four subspecies: *tularensis, holarctica, mediasiatica,* and *novicida*. It is a zoonotic pathogen, capable of infecting a wide range of hosts, from amoebae to rodents and humans. Of the subspecies, *F. tularensis* subspecies *tularensis* is the most virulent for humans. There are diverse presentations of tularemia, depending on the route of infection. The most common is ulceroglandular tularemia, which is acquired by the bite of an infected insect vector, or through infection of skin abrasions. The most acute form of disease is respiratory tularemia, arising from inhalation of the bacterium, for example, through contaminated hay dusts. The pathogen has a very low infectious dose for humans by the inhalational route. Infection responds well to antibiotic therapy, but if left untreated mortality rates can be high. For these reasons, the pathogen has long been of concern as a biothreat agent. An attenuated live vaccine strain, strain LVS, was awarded investigational new drug status and shown to be effective in preventing tularemia. However, this vaccine has never been fully licensed, and its investigational new drug status has now been revoked. Thus, there is renewed effort underway to develop an effective licensable vaccine against *F. tularensis* infection. The availability of genome sequence data should help in the search for a new vaccine, by facilitating the identification of attenuation targets or potentially protective subunits. However, the immunity required for protection against high-virulence strains of *F. tularensis* requires both humoral and cell-mediated responses to be induced.

INTRODUCTION

Francisella tularensis is one of the most infectious pathogens known and is able to cause the disease tularemia after entering the body via a number of routes. Respiratory tularemia is of greatest concern in humans because of the acute nature of the disease. So infectious is this pathogen by the airborne route that Overholt et al. (1961) stated that "The hazard of infection with *F. tularensis* is well recognized; few persons escape illness if they continue to work with the organism." It is probably the high infectivity by the airborne route and the ease of culture that in the past attracted some to develop this bacterium as a bioweapon. During the middle of the twentieth century, there was a significant effort to devise a vaccine against tularemia. A live vaccine was shown to be effective in laboratory workers, but this vaccine is currently not licensed. Against the background of an increased threat from biological weapons and increased research into the biology of this organism, there is an urgent need for an effective licensed vaccine against tularemia. This chapter reviews past work to devise vaccines against tularemia and assesses the prospect for improved vaccines.

SHORT HISTORY OF THE DISEASE

The first reported isolation of *F. tularensis* was in 1911 in Tulare County, California and followed an outbreak of a plague-like disease in rodents (McCoy and Chapin, 1912). The bacterium was originally named *Bacterium tularensis*, then subsequently placed in the *Pasturella* genus followed by a move to the *Brucella* until finally being assigned to a new genus entitled *Francisella* in 1947 (Dorofe'ev, 1947). Although the bacterium was originally isolated from rodents, naturally occurring tularemia has since been reported in mammals, birds, amphibians, fish, and even in invertebrates (Morner, 1992). Until recently, the only reported cases of tularemia in humans or in animals were in the Northern Hemisphere. However, there has recently been a report from Australia of an infection in humans caused by a strain of *F. tularensis* subspecies *novicida* (Whipp et al., 2003). At this stage it is not clear whether this is an exceptional case of disease, or the bacterium is more widely distributed in the Southern Hemisphere than was previously thought.

ETIOLOGIC AGENT

Tularemia is caused by the etiological agent *F. tularensis*. The bacteria stain weakly Gram-negative. The cells are small, 0.2 to 0.5 μm by 0.7 to 1.0 μm, single coccobacilli. The organism is fastidious, requiring enriched media for growth, and has a requirement for cysteine. The organism will grow on general microbiological agars used routinely in US clinical laboratories, such as Thayer-Martin, sheep blood, and chocolate agars. Traditionally, the organism has been cultured on cysteine glucose blood agar, but cysteine heart agar supplemented with heated sheep red blood cells to give a chocolate agar has also been used (Anonymous, 2001). *F. tularensis* is slow-growing on solid media, requiring 2–4 days incubation to produce individual colonies at 37°C. On blood agar, a small zone of alpha-hemolysis develops. *F. tularensis* does not grow well in liquid media and requires a heavy inoculum for growth. The organism will not grow in standard laboratory liquid media used by US clinical laboratories, such as brain heart infusion or trypticase soy broth. Growth in aerobic enriched media can take up to a week to produce visible turbidity; static cultures will take a minimum of 10 days. A synthetic, defined medium for the growth of *Francisella* is available (Chamberlain, 1965).

The identification of *F. tularensis* and differentiation of its subspecies has traditionally been accomplished by observing growth characteristics and biochemical analysis. Besides being time consuming and tedious, cultivation results in a high risk of laboratory-acquired infection (Burke, 1977). In view of these problems, the diagnosis of tularemia has relied mainly on clinical picture backed by serology. However, antibodies are usually not detected before the second week of disease (Koskela and Salminen, 1985). Moreover, due to its fastidious growth requirements and slow growth rate, isolation of the bacterium directly from the environment is hampered since other bacteria can grow more rapidly on most media used to culture *F. tularensis*.

CLASSIFICATION

F. tularensis is a member of the gamma subdivision of the *Proteobacteria*. There are two species in the *Francisella* based on 16S rDNA sequencing and fatty acid composition; *F. tularensis* and *Francisella philomiragia* (Hollis et al., 1989; Forsman et al., 1994). There are currently four recognized subspecies of *F. tularensis*. *F. tularensis* subspecies *tularensis* is the most virulent of the four subspecies and is found primarily in North America (Fig. 61.1), whereas *F. tularensis* subspecies *holarctica* is less virulent and is found mainly in Europe and Asia. *F. tularensis* subspecies *mediasiatica* is isolated in central Asia and *F. tularensis* subspecies *novicida* primarily in North America. Subspecies *mediasiatica* and *novicida* rarely cause disease in humans. The highly virulent *F. tularensis* subspecies *tularensis* strains have infectious doses (IDs) for humans of less than 10 CFU, making this one of the most highly infectious bacterial pathogens known.

Francisella-specific 16S rRNA sequence signatures have allowed the development of PCR-based methods for the identification of *Francisella* strains at the genus level and differentiation at the species level. PCR amplification of the rRNA gene cluster combined with endonuclease digestion has been used for typing of *Francisella* strains, allowing differentiation at the species level but not at the subspecies level. Repetitive extragenic palindromic sequence (REP)-PCR has been applied to identify strains of subspecies *novicida*, but patterns from subspecies *holarctica* and subspecies *tularensis* strains were found to be similar. Recent studies have evaluated the use of PCR based on the use of various arbitrary primers as well as of primers specific to REP and enterobacterial repetitive intragenic consensus sequences (ERIC). It was concluded that the methods were useful for rapid and a technically simple strategy for discrimination of subspecies but not of individual strains. A 30-bp sequence heterogeneity among the genomes of various *F. tularensis* strains was found by the use of PCR and arbitrary priming; by targeting this genomic region a PCR was

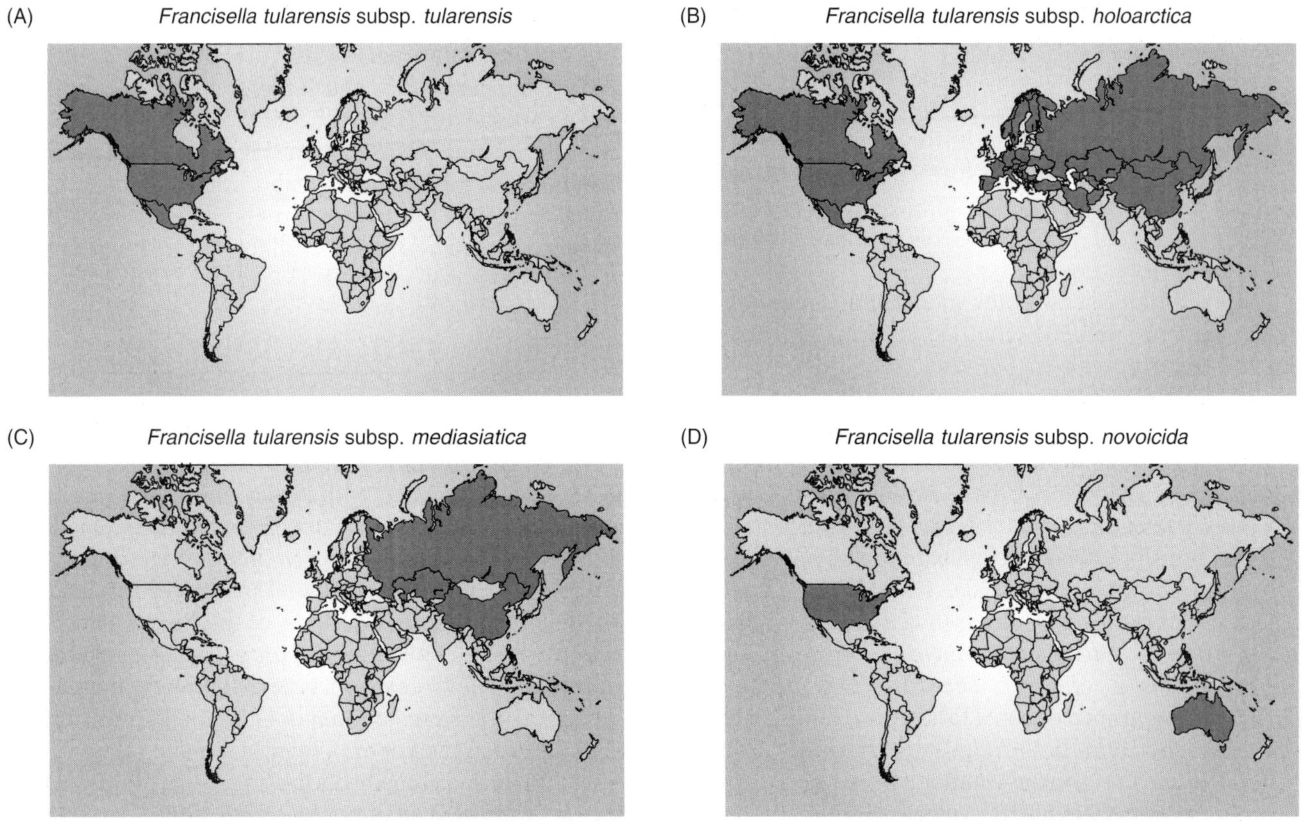

FIGURE 61.1 Worldwide distribution of *F. tularensis* subspecies. Reproduced from Oyston et al. (2004).

developed that distinguished the two clinically most important subspecies, *tularensis* and *holarctica*.

Work to determine the complete genome sequence of a virulent subspecies *tularensis* strain, SCHU S4, also allowed the production of an ordered clone microarray. Hybridization patterns of DNA isolated from a range of strains representing the various subspecies revealed a hyperplastic region different in each subspecies. This allowed the development of a PCR whereby the four subspecies could be differentiated (Broekhuijsen et al., 2003). However, it must be borne in mind that the ordered clone array was developed prior to complete assembly and analysis of the genome sequence data, and as such there may be gaps in the coverage provided by this array.

ANTIGENS EXPRESSED BY THE ORGANISM

The genus *Francisella* is antigenically coherent and the subspecies are almost indistinguishable by serological methods, although the LPS of *F. tularensis* subspecies *novicida* discriminates this subspecies from the others (Prior et al., 2003; Vinogradov et al., 2004). Serological tests for confirmation of tularemia rely on the detection of antibodies produced against LPS.

Several proteins have been identified that are able to induce the proliferation of T cells isolated from LVS vaccinees or convalescent individuals (Table 61.1), two of which were putative surface proteins of 32 and 17 kDa (Sandstrom et al., 1987; Sjostedt et al., 1990a, 1990b; Surcel et al., 1991). The role of these proteins in infection is not known.

Western blotting of outer membrane preparations, with sera from patients who had contracted tularemia during an outbreak in Sweden, showed patient-to-patient variation in immune response (Bevanger et al., 1988). However, one protein of 43 kDa termed FopA appeared to be immunodominant as all 12 patients tested produced antibodies against this antigen. Recently, a proteomic analysis was undertaken to identify the immunologic proteins expressed by strain LVS, recognized by C3H/HeN and C3H/HeJ mice, and naturally infected humans (Havlasová et al., 2005). Strong murine antibody responses were detected to 36 *F. tularensis* proteins, 27 of which were identified for the first time as being immunogenic (Table 61.1). Eight proteins were recognized by both mouse and human: ClpB, an outer membrane protein, malonyl-Co-A-acyl carrier protein transacylase, GrpE, oxidoreductase, a hypothetical protein, 10 kDa chaperonin, and a probable sigma modulation protein.

TULAREMIA AS A PUBLIC HEALTH PROBLEM

Tularemia has been isolated from several parts of the Northern Hemisphere but has only rarely been found in the Southern Hemisphere (Fig. 61.1). The majority of cases of human disease are reported from Scandinavia, Northern America, Japan, and Russia (Boyce, 1975; Ohara et al., 1996; Stewart, 1996; Tarnvik et al., 1996b; Berdal et al., 2000). Although outbreaks of tularemia are rare, they can involve a high number of cases. For example, in a recent outbreak in Bulgaria, 262 cases were confirmed, mainly presenting with oropharyngeal tularemia (Christova et al., 2004). The outbreak was caused by *F. tularensis* subspecies *holarctica*, probably contracted from food and water that had been contaminated by infected rodents. In an outbreak in Sweden, almost 700 cases of respiratory tularemia were reported, and in another more than 500 cases of ulceroglandular tularemia reported. The source of pathogen was most likely either carcasses or secretions of infected animals.

TULAREMIA AS A BIOTHREAT AGENT

F. tularensis subspecies *tularensis* has been an organism of concern since the large state-funded biological weapons programs of the 1950s, when the USA and the former Soviet Union (fSU) first evaluated the organism as a biological weapon. It is a highly infectious organism inducing a potentially fatal disease, but no licensed vaccine is available for prophylaxis.

CLINICAL DISEASE

The typical symptoms of tularemia occur after an incubation period of 3–5 days and the onset of disease is often abrupt with fever, chills, malaise, sore throat, and headache (Evans, 1985; Evans et al., 1985). However, other symptoms of disease are dependent largely on the route of entry of the bacterium into the body. The most common form of the disease, and representing 90% of all cases in Europe, is ulceroglandular tularemia. The disease is acquired following direct contact with an infected animal or is vector-borne (Tarnvik et al., 1996a; Ohara et al., 1998). At the site of the vector bite, a primary ulcer develops (Fig. 61.2a) followed by the development of a pustule surrounded by inflammation (Evans, 1985; Evans et al., 1985). In some cases, the ulcer is inconspicuous and heals within a week

TABLE 61.1 Immunogenic proteins identified from *F. tularensis*

Antigen	Recognition by convalescent antisera?	Recognition by convalescent T cells?
43 kDa outer membrane protein (FopA)	Yes (Bevanger et al., 1988)	NR
Outer membrane protein	Yes (Havlasová et al., 2005)	NR
Chaperone DnaK	Yes (Havlasova et al., 2002)	Yes (Ericsson et al., 2001)
Omp-1-like proptein	Yes (Havlasová et al., 2005)	NR
17 kDa lipoprotein (Tul4)	Yes (Havlasova et al., 2002)	Yes (Sjostedt et al., 1990a, 1990b)
Cationic 19 kDa OMP precursor	Yes (Havlasová et al., 2005)	NR
Hsp60	Yes (Havlasova et al., 2002)	Yes (Ericsson et al., 2001)
Hsp10	Yes (Havlasova et al., 2002)	Yes (Ericsson et al., 2001)
GrpE protein	Yes (Havlasová et al., 2005)	NR
HtpG heat shock protein	Yes (Havlasová et al., 2005)	NR
Elongation factor TU	Yes (Havlasova et al., 2002)	NR
Elongation factor Ts	Yes (Havlasová et al., 2005)	NR
Glycine cleavage system T1 protein	Yes (Havlasova et al., 2002)	NR
Hypothetical protein	Yes (Havlasova et al., 2002)	NR
Hypothetical protein	Yes (Havlasová et al., 2005)	NR
Putative periplasmic protein	Yes (Havlasová et al., 2005)	NR
Malonyl CoA-acyl carrier protein	Yes (Havlasová et al., 2005)	NR
Acetyl-coenzyme A carboxylase transferase subunit	Yes (Havlasová et al., 2005)	NR
Biotin carboxylase subunit of acetyl CoA carboxylase	Yes (Havlasová et al., 2005)	NR
Oxidoreductase	Yes (Havlasova et al., 2002)	NR
Succinyl-CoA-synthetase subunit	Yes (Havlasová et al., 2005)	NR
Biotin carrier protein	Yes (Havlasova et al., 2002)	NR
club	Yes (Havlasová et al., 2005)	NR
Pyrrolidone-carboxylate peptidase	Yes (Havlasová et al., 2005)	NR
Hfq	Yes (Havlasová et al., 2005)	NR
Chitinase	Yes (Havlasová et al., 2005)	NR
Isocitrate dehydrogenase	Yes (Havlasová et al., 2005)	NR
Peroxidase catalase	Yes (Havlasová et al., 2005)	NR
Malate dehydrogenase	Yes (Havlasová et al., 2005)	NR
Transketolase	Yes (Havlasová et al., 2005)	NR
Citrate synthase	Yes (Havlasová et al., 2005)	NR
Nucleoside diphosphate kinase	Yes (Havlasová et al., 2005)	NR
Glyceraldehyde-3-phosphate dehydrogenase	Yes (Havlasová et al., 2005)	NR
50S ribosomal protein	Yes (Havlasova et al., 2002)	NR
Probable bacterioferritin	Yes (Havlasova et al., 2002)	NR
3-dehydroquinase	Yes (Havlasova et al., 2002)	NR
Histone-like protein	Yes (Havlasova et al., 2002)	NR
Probable signal modulation protein	Yes (Havlasová et al., 2005)	NR
Two-component response regulator	Yes (Havlasová et al., 2005)	NR
Single-strand binding protein	Yes (Havlasová et al., 2005)	NR

and may be mistaken for an insect bite. Subsequently the most obvious sign of disease is enlargement of the draining lymph nodes (Evans, 1985; Evans et al., 1985). If appropriate antibiotic therapy is not instituted within 7–10 days, the enlargement may become even more prominent and in 30–40% of cases suppuration eventually results (Helvaci et al., 2000). Recovery from ulceroglandular tularemia can be protracted but the

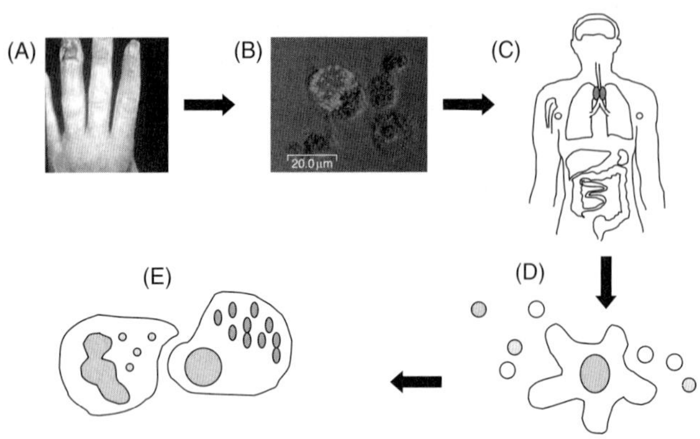

FIGURE 61.2 Pathogenesis of *F. tularensis* infection. (a) In typical human ulceroglandular tularemia, a skin lesion first appears at the site of infection after 3–5 days. (b) The bacteria (shown here labeled with green fluorescent protein) are taken up by macrophages, where they escape killing and multiply to high levels within the cytoplasm of the cell. (c) A transient bacteremia occurs seeding the organism throughout reticuloendothelial tissues of the body. (d) A wide range of cytokines are expressed in the reticuloendothelial tissues; for example, infected hepatocytes produce TNF-α, IL-10, IL12, and IFN-γ. (e) As infection progresses, T cells play a more significant role in protection. Neutrophils also appear to play a role in defence by ingesting and killing microorganisms, lysing infected hepatocytes, and acting as a source of cytokines.

mortality rate associated with this form of disease is typically less than 3% (Evans et al., 1985).

Respiratory tularemia is the consequence of the inhalation of bacteria, and naturally occurring cases of disease are usually the result of farming activities when dusts contaminated with the bacterium are generated (Tarnvik et al., 1996b). Often exposure occurs during the clearing of barns or other storage areas that infected rodents have colonized (Syrjala et al., 1985; Stewart 1996). The symptoms of respiratory tularemia are quite variable and the disease commonly presents as a systemic illness with fever but without obvious signs of respiratory disease (Dahlstrand et al., 1971; Evans et al., 1985). Respiratory infection caused by *F. tularensis* subspecies *holarctica* is often a relatively mild infection. Other common symptoms of disease included headache, myalgia, dry cough, and arthralgia (Syrjala et al., 1985). Respiratory tularemia caused by *F. tularensis* subspecies *tularensis* is a much more acute disease characterized by high fever, chills, malaise, dyspnea, and cough (Evans, 1985; Evans et al., 1985). Pulmonary symptoms, including cough, substernal pain, and tachypnea are more common than in the respiratory disease caused by *F. tularensis* subspecies *holarctica*. Without antibiotic therapy the case fatality rate is typically up to 30–60% but with appropriate therapy this can be reduced to less than 2% (Dennis et al., 2001). The respiratory from of the disease caused by *F. tularensis* subspecies *tularensis* is the most likely consequence of the deliberate release of the bacterium.

Other forms of tularemia are much less common. Typhoidal tularemia is a relatively ill-defined form of the disease in which septicemia develops without lymphadenopathy or the appearance of an ulcer. Oculoglandular tularemia can occur following the contamination of the eye with bacteria (Steinemann et al., 1999). The disease is marked by the development of ulcers and nodules on the conjunctiva, and the infection can spread to the local lymph nodes. Oropharyngeal or gastrointestinal tularemia can develop following the ingestion of bacteria in water of foodstuffs (Stewart, 1996; Tarnvik et al., 1996b). Oropharyngeal disease is characterized by a sore throat with enlargement of the tonsils and the formation of a yellow-white pseudomembrane (Ellis et al., 2002). The severity of gastrointestinal tularemia can range from a mild and persistent diarrhea to an acute fatal disease with extensive ulceration of the bowel (Ellis et al., 2002).

TREATMENT

Following the possible exposure of an individual to *F. tularensis*, chemotherapy should commence immediately and continue for at least 14 days. The isolation of the individual is not necessary since human-to-human transmission of the disease has not been reported. *F. tularensis* is susceptible to a range of antibiotics in vitro, and historically the aminoglycoside streptomycin has been the drug of choice for the treatment of tularemia (Enderlin et al., 1994). Nowadays streptomycin is not the preferred drug of choice because of the associated ototoxicity, but it is an option for the treatment of tularemia meningitis. Gentamicin provides an acceptable alternative. However, because of the need for parenteral administration and the continuous monitoring of serum levels, aminoglycosides are only the drugs of choice for serious forms of tularemia and in cases where there are no alternative efficacious regimens available, such as treatment of children and pregnant women (Dennis et al., 2001).

Tetracyclines have also been used extensively to treat tularemia, and due to its excellent pharmacokinetics, doxycycline is the preferred choice (Tärnvik and Berglund 2003). It is a second-line treatment for uncomplicated forms of ulceroglandular tularemia. To avoid the possibility of relapse, tetracycline therapy should continue for at least 2 weeks.

Other antibiotics might also be of value for the treatment of tularemia, although clinical experience using these drugs for this disease is more limited. In vitro data indicate that *F. tularensis* is susceptible to carbapenems, ceftriazone, ceftazidime, rifampin, and certain macrolides, but there is a lack of clinical data to recommend any of them for clinical use (Tärnvik and Berglund, 2003). Additionally, there is widespread natural resistance to erythromycin among strains of subspecies *holarctica*. The quinolones have excellent in vitro efficacy against *F. tularensis* (MIC 0.01–0.1 mg/L (Johansson et al., 2002)) and several reports have demonstrated the utility of ciprofloxacin for the treatment of tularemia (Enderlin et al., 1994; Johansson et al., 2000, 2001; Perez-Castrillon et al., 2001). Based on published clinical evidence and microbiological data, ciprofloxacin is becoming the drug of choice for treatment of uncomplicated tularemia and as a second line of treatment for severe infections (Dennis et al., 2001). It may also be of value for treatment of tularemia in pregnant women. Excellent clinical efficacy for treatment of children has been reported (Johansson et al., 2000). In a mass casualty setting, it has been suggested that the oral administration of ciprofloxacin is the drug regime of choice (Dennis et al., 2001).

PATHOGENESIS

F. tularensis is able to infect a wide range of hosts, with a typical ID in humans of only ten bacteria (McCrumb, 1961). Very few classical virulence factors have been identified for this pathogen. It does not apparently produce secreted exotoxins, and the LPS of *F. tularensis* does not exhibit the properties of a classical endotoxin (Sandstrom et al., 1992).

In typical human ulceroglandular tularemia, a skin lesion first appears at the site of infection 3–5 days after infective exposure (Fig. 61.2). During this initial phase, T cells appear to play little role in combating infection, but the cytokines tumor necrosis factor (TNF) and interferon gamma (IFN-γ) play a key role in response to infection (Elkins et al., 1993). A transient bacteremia occurs, during which the pathogen must resist lysis by complement. The capsule is reported to be essential for serum resistance, but not required for survival following phagocytosis (Sandstrom et al., 1988). The bacteremic phase allows the organism to be seeded throughout the body, infecting all reticuloendothelial tissues. As infection progresses, T cells appear to play a more significant role in protection (Conlan et al., 1994; Yee et al., 1996; Fulop et al., 2001). Neutrophils also appear to play a role in defence by ingesting and killing microorganisms, lysing infected hepatocytes, and acting as a source of cytokines (Conlan et al., 1994; Sjostedt et al., 1994).

The pathogenicity of intracellular bacteria depends upon their ability to survive within macrophages, although other cell types such as hepatocytes may play a role (Conlan and North, 1992). The intracellular lifestyle of *F. tularensis* is unique compared to other intracellular bacterial pathogens. The phagosomes containing *F. tularensis* show limited maturation in the endocytic pathway (Golovliov et al., 2003; Clemens et al., 2004). The phagosome matures to a late endosome-like stage, but there is no lysosome fusion. The bacterium then escapes into the cytoplasm by disruption of the phagosomal membrane and multiplies to high numbers.

Little is known about the molecular mechanisms of *Francisella* pathogenesis. The recently published genome sequence of a highly virulent strain revealed a paucity of potential virulence factors (Larsson et al., 2005). However, a genomic island required for intracellular survival and growth has been reported, the *iglC* gene contained therein being responsible for the inhibition of lysosomal fusion (Santic et al., 2005).

IMMUNE RESPONSES TO INFECTION

Following a natural infection with *F. tularensis*, most of the antibody that develops is directed against the LPS. Additionally, several studies (Bevanger et al., 1988; Sjostedt et al., 1990a, 1990b; Ericsson et al., 2001; Havlasova et al., 2002, 2005) have identified at least some of the protein antigens which are recognized by either immune sera or T cells (Table 61.1).

Recovery from natural infection generally provides long-lasting protection against reinfection with *F. tularensis*. However, there have been relatively few studies on the nature of protective immunity to tularemia. In a review published in 1989 (Tarnvik, 1989), and based largely on studies carried out during the period 1940–1960, both humoral and cellular responses were identified as being important for protection, but the relative contributions of these arms of the immune system to protection were different. During the 1930s and 1940s, there were several studies to evaluate immune sera for the treatment of tularemia (Foshay 1934; Foshay 1946). These studies yielded contradictory results. For example, Foshay, (1934, 1946) claimed that antisera was effective in the treatment of tularemia, whereas Francis and Felton (1942) claimed that antisera was of little value in treating disease. Overall, the current consensus is that antibody to *F. tularensis* can play a role in protection or recovery from tularemia caused by low-virulence strains. However, the role of antibody in protection or recovery from disease caused by high-virulence strains is at best minimal (Tarnvik, 1989).

Conversely, the evidence points to cellular responses playing a major role in protection from tularemia caused by all strains of *F. tularensis* (Tarnvik, 1989). The transfer of mononuclear leukocytes from immune to naive mice or from immune to naive rats provides protection against a subsequent challenge with a virulent strain of *F. tularensis* (Allen, 1962; Thorpe and Marcus, 1965). However, these studies do not clarify the roles of activated phagocytes and committed leukocytes. During the early stages of the disease, T cells appear to play a minor role in the control of the infection (Elkins et al., 1993). However, cytokines such as IFN-γ and TNF-α are important for the resolution of disease (Elkins et al., 1993; Fortier et al., 1994), possibly because they play roles in the activation of phagocytic cells such as macrophages. Later during infection, both CD4+ and CD8+ T cells play a role in protection. Mice devoid of either subset of T cells were capable of controlling but not eliminating infection (Conlan et al., 1994; Yee et al., 1996; Fulop et al., 2001).

Circulating $\gamma\delta$ T cells might also play an important role in protection or recovery from tularemia and the expansion of this subset of T cell has been reported in individuals recovering from tularemia (Poquet et al., 1998; Kroca et al., 2000). Phosphoantigens that are expressed in vivo are strongly implicated in the development of this response (Poquet et al., 1998). It is known that $\gamma\delta$ T cells play a key role in the control of some intracellular infections (Hiromatsu et al., 1992), but the role of these cells in the control of tularemia has yet to be determined.

VACCINES

History

Live Attenuated Vaccines

During the early part of the twentieth century research to devise a vaccine against tularemia was carried out both in the USA and in the fSU. In the fSU, work appeared to be focused on the development of a live attenuated vaccine, and in 1934 El'bert demonstrated proof of principle that a live attenuated vaccine was a realistic prospect (Ellis et al., 2002). Subsequently, a live attenuated vaccine (strain Moscow) was reportedly tested in several thousand human volunteers, before being lost (Tigertt, 1962). Subsequent work resulted in the generation of strain 15, which had reduced virulence in guinea pigs. This strain, given subcutaneously, was tested extensively in humans and it was concluded that this vaccine was safe (Tigertt, 1962). The strain 15 vaccine was used in a number of mass-vaccination campaigns in the fSU, in areas where outbreaks of tularemia occurred naturally. In total, as many as 60 million individuals were immunized. Subsequently, an additional live vaccine termed strain 155 was developed and produced along with strain 15 at the Gamleya Institute. Both of these strains were transferred to the USA in 1956. These strains were propagated and subcultured to provide the so-called "Live Vaccine strain" (LVS), though it is uncertain whether LVS was derived from strain 15 or strain 155.

All of the evidence points towards these live attenuated vaccines as being effective for the prevention of tularemia in humans. In one study in the USA, both individuals immunized with a live attenuated vaccine and naive individuals were exposed to airborne *F. tularensis* (McCrumb, 1961). This study showed that the degree of protection afforded by immunization depended on the challenge dose (Table 61.2). Good protection against a challenge of 10 IDs was reported but when challenged with 1000 IDs the degree of protection was marginal (McCrumb, 1961). In another study, the incidence of laboratory-acquired tularemia at Fort Detrick in Maryland was shown to reduce markedly following the introduction of the LVS vaccine (Burke, 1977). Interestingly, this study also shows that the use of microbiological safety cabinets alone was insufficient to reduce the incidence of disease. Collectively, these findings indicate that a live attenuated vaccine is effective for the control of tularemia.

Although live attenuated vaccines were generally given by the intradermal/subcutaneous route, there is a significant body of evidence that oral, aerogenic, and intranasal immunization also provides protection against tularemia (Eigelsbach and Tulis, 1961; Eigelsbach et al., 1962). Some of these studies suggested that the aerogenic rather than intradermal route of immunization provided a higher level of protection against a subsequent respiratory challenge (Hornick and Eigelsbach, 1966). This finding might be in accordance with more recent studies which have shown that mice immunized with the LVS strain by the intranasal route are better protected against a subsequent challenge than mice immunized by the intradermal/subcutaneous route (Conlan et al., 2005; Wu et al., 2005).

During the selection of the LVS strain in the USA, the phenomenon of blue and gray variant colonies was first highlighted. When grown on certain solid media and viewed under obliquely transmitted light, colonies appeared to have either a gray or a blue tint (Eigelsbach and Downs, 1961). Subsequent studies in mice and in guinea pigs showed that only immunization with blue variant colonies resulted in protection against a subsequent challenge with *F. tularensis* (Eigelsbach and Downs, 1961). During the subsequent manufacture of

TABLE 61.2 Protection afforded by immunization with a live attenuated strain of *F. tularensis*

Group	Challenge[a]	No illness	Transient illness	Modified disease	Unmodified disease[b]
Control	10 IDc	None	None	None	2/2
	100 ID	None	None	None	2/2
	1000 ID	None	None	None	2/2
LVS vaccinated	10 ID	5/6	1/6	None	None
	100 ID	3/5	1/5	1/5	None
	1000 ID	None	None	2/3	1/3

[a] By the airborne route.
[b] These individuals were treated with streptomycin.
[c] ID, human infectious dose; 1 ID = 200 cfu by the airborne route.

batches of the vaccine by the National Drug Company (USA), the control of the proportion of gray variant colonies was an area of significant concern. In addition to the potential for the LVS strain to segregate into different colony variants, there is evidence that marginal differences in cultural conditions can affect the immunogenicity of the vaccine (Hambleton et al., 1974; Cherwonogrodzky et al., 1994). The potential for significant variation in batches of vaccine is highlighted by comparing the properties of different batches. For example, the NDBR101 lot of vaccine contained 1% gray colony variants whereas one of the last vaccine lots produced contained 20% gray cells (Sandstrom, 1994).

Whilst in the past this vaccine has been used in humans as an investigational new drug, currently the LVS vaccine is not available. The potential for batch-to-batch variation of vaccine lots, along with the lack of information on the molecular or biochemical basis of attenuation, has so far precluded the full licensing of this vaccine for use in humans. It has been suggested (Sandstrom, 1994; Ellis et al., 2002) that any future work to license this vaccine should especially focus on characterizing the LVS strain in the following ways. It would be essential to establish a reliable seedstock and to devise a method for the reliable quantification of blue and gray variant colonies in batches of vaccine. This may then allow methods for the culture of the LVS strain to be established which ensure that the proportion of blue variant colonies is reproducible. Additional work is required to establish the genetic or biochemical basis of attenuation of the LVS strain and identify mechanism of protection.

Killed Whole-Cell Vaccines

Much of the work on killed whole-cell vaccines took place in the USA during the 1930s and 1940s, and was driven by an early observation that immune sera could modify the course of tularemia in humans (Foshay, 1934, 1946). However, the significance of this finding is still debated. As highlighted above, antibody can place a role in protection against low-to-medium virulence strain of *F. tularensis*. The individuals who were involved in these trials of antisera were naturally infected and it is not clear whether they had been infected with strains of *F. tularensis* subspecies *holarctica* or *F. tularensis* subspecies *tularensis*. Without this information it is difficult to make a meaningful judgment on the value of antisera to treat disease.

Notwithstanding this concern, there were a number of reports on the preparation and efficacy of killed whole-cell vaccines during the 1940s, 50s, and 60s. Various methods for killing the bacteria cells and extracting immunogenic components were reported including acetone, heat, and phenol treatments (Foshay et al., 1942; Foshay, 1946, 1950). Some of these vaccines reportedly caused significant side effects in human volunteers, including the development of lesions at the site of immunization (Foshay, 1950). The available evidence does not conclusively indicate that these vaccines are of value in preventing tularemia. In animal models of disease, such as the mouse, guinea pig, or rabbit, killed vaccines failed to increase the level of resistance to infection or to modify the severity of the disease caused by *F. tularensis* subspecies *tularensis* given intraperitoneally, intracerebrally, or subcutaneously (Eigelsbach and Downs, 1961). In monkeys immunized with a killed vaccine, the susceptibility to infection by the subcutaneous or intradermal route was not modified. However, the severity of disease after challenge with 10 organisms (but not after challenge with 750 or 2600 organisms) was reduced (Eigelsbach and Downs, 1961). The phenolized and acetone-extracted vaccines (or the Foshay vaccines) underwent some limited trials in humans. The immunization of laboratory workers or hunters with the Foshay vaccine provided only slight protection against disease (Eigelsbach and Downs, 1961). Volunteer studies indicated that the vaccine reduced the severity of disease caused after intracutaneous challenge with 10 organisms of *F. tularensis* subspecies *tularensis* (strain SCHU S4), but did not provide any protection against disease caused by the inhalation of 50 organisms (Eigelsbach and Tulis, 1961). Overall, it can be concluded that killed whole-cell vaccines, at least those prepared in the ways reported above, are not tularemia vaccines.

Subunit Vaccines

There have been a limited number of studies to identify protective polysaccharides or proteins from *F. tularensis*, which might be exploited as components of a subunit vaccine. These studies have generally met with limited success. The immunization of mice with purified LPS induced a humoral response that was able to protect mice against low-virulence strains of *F. tularensis* subspecies *holarctica* (Fulop et al., 1995, 2001). However, this immune response served only to extend the time to death following challenge with high-virulence subspecies *tularensis* strains (Fulop et al., 2001; Conlan et al., 2002). This difference in protective efficacy does not appear to reflect differences in the structure of LPS. Other workers have evaluated antigens such as the 17 kDa TUL4 protein (Sjostedt et al., 1992; Golovliov et al., 1995), the FopA protein (Fulop et al., 1995), or heat shock protein 60 (Hartley et al., 2004) as protective antigens, without success.

Current Licensed Vaccines

There is no current licensed vaccine for use in Europe or in the USA.

Vaccines in Development

The history of immunization of humans with live attenuated vaccine strains has shown that an attenuated mutant can induce a protective immune response against virulent strains of *F. tularensis*. However, because these vaccines were all derived from strains of *F. tularensis* subspecies *holarctica*, a much debated question is whether or not an attenuated mutant of *F. tularensis* subspecies *tularensis* would provide improved protection against disease. The main problems associated with developing such defined attenuated strains have been the lack of genetic tools to manipulate *F. tularensis* and the difficulty of transforming subspecies *tularensis* strains. However, methods for producing defined isogenic allelic replacement mutants has been published recently (Lauriano et al., 2003); (Santic et al., 2005) opening the way to the production of attenuated vaccines.

Defined allelic replacement mutagenesis has been used in a variety of bacterial species to produce vaccine strains capable of limited replication in the host. In many cases, the genes targeted are in biosynthetic pathways, such as purine or aromatic amino acid biosynthetic pathways. The genes for these pathways are all present in *F. tularensis*, and the organism can grow in media lacking purines or tyrosine, respectively, showing the pathways to be fully functional (Karlsson et al., 2000). The availability of the *F. tularensis* subspecies *tularensis* SCHU S4 genome sequence (Larsson et al., 2005) will aid the identification of target genes for inactivation, especially when compared to the strain LVS genome once this is completed (http://bbrp.llnl.gov/bbrp/html/microbe.html).

The genome sequence data, supported by proteomic analysis (Lenco et al., 2005; Pavkova et al., 2005), will also help in the search for suitable protective subunits. However, it may be that in order to stimulate a suitable immune response, a protein antigen would have to be delivered with an appropriate adjuvant or using a vector system that would allow induction of a cellular immune response, for example, using a DNA vaccine.

POSTEXPOSURE IMMUNOPROPHYLAXIS

There is no therapy available in the form of vaccination or antibody therapy available for use post exposure. Antibiotics are effective for the treatment of infection (reviewed above).

PROSPECTS FOR THE FUTURE

Efforts are underway to develop an effective licensed vaccine to prevent tularemia. Enhanced funding, especially by US agencies such as the National Institute of Allergy and Infectious Diseases, has resulted in an increased research effort on this pathogen. However, a downside to this is the increased risk of laboratory-acquired infection due to a lack of experience in handling such a highly infectious pathogen. For example, in a recent incident at Boston University three researchers were infected with a highly virulent strain (Dalton, 2005).

KEY ISSUES

- *F. tularensis* is one of the most infectious pathogens known.
- Respiratory tularemia is an acute disease.
- *F. tularensis* has long been of concern as a biothreat agent.
- Antibiotic therapy is effective.
- There is no licensed vaccine available.
- A live attenuated vaccine (LVS) has been shown to be effective in humans, but is probably not licensable.
- Protection appears to be dependent on cellular immunity.
- To date no protective subunits have been identified.

References

Allen, W. Immunity against tularemia: passive protection of mice by transfer of immune tissues. J. Exp. Med. 1962; 115: 411–420.

Anonymous. Basic laboratory protocols for the presumptive identification of *Francisella tularensis*. Atlanta, GA, Centers for Disease Control and Prevention. 2001.

Berdal, B.P., Mehl, R. Haaheim, H., Loksa, M., Grunow, R. (2000) Field detection of *Francisella tularensis*. Scandinavian Journal of Infectious Disease 2000; 32:287–291.

Bevanger, L., Maeland, J.A. et al. Agglutinins and antibodies to *Francisella tularensis* outer membrane antigens in the early diagnosis of disease during an outbreak of Tularemia. J. Clin. Microbiol. 1988; 26(3):433–437.

Boyce, J.M. Recent trends in the epidemiology of tularemia in the United States. J. Infect. Dis. 1975; 131:197–199.

Broekhuijsen, M., Larsson, N. et al. Genome-wide DNA microarray analysis of *Francisella tularensis* strains demonstrates extensive genetic conservation within the species but identifies regions that are unique to the highly virulent *F. tularensis* subsp *tularensis*. J. Clin. Microbiol. 2003; 41(7):2924–2931.

Burke, D.S. Immunization against tularemia: analysis of the effectiveness of live *Francisella tularensis* vaccine in prevention of laboratory acquired tularemia. J. Infect. Dis. 1977; 135(1):55–60.

Chamberlain, R.E. Evaluation of live tularemia vaccine prepared in a chemically defined medium. Appl. Microbiol. 1965; 13(2):232–235.

Cherwonogrodzky, J.W., Knodel, M.H. et al. Increased encapsulation and virulence of *Francisella tularensis* live vaccine strain (LVS) by subculturing on synthetic medium. Vaccine 1994; 12(9):773–775.

Christova, I., Velinov, T. et al. Tularaemia outbreak in Bulgaria. Scand. J. Infect. Dis. 2004; 36(11–12):785–789.

Clemens, D.L., Lee, B.-Y. et al. Virulent and avirulent strains of *Francisella tularensis* prevent acidification and maturation of their phagosomes and escape into the cytoplasm in human macrophages. Infect. Immun. 2004; 72:3204–3217.

Conlan, J.W. and North, R.J. Early pathogenesis of infection in the liver with the facultative intracellular bacteria *Listeria monocytogenes*, *Francisella tularensis*, and *Salmonella typhimurium* involves lysis of infected hepatocytes by leukocytes. Infect. Immun. 1992; 60(12):5164–5171.

Conlan, J.W., Shen, H. et al. Mice vaccinated with the O-antigen of *Francisella tularensis* LVS lipopolysaccharide conjugated to bovine serum albumin develop varying degrees of protective immunity against systemic or aerosol challenge with virulent type A and type B strains of the pathogen. Vaccine 2002; 20:3465–3471.

Conlan, J.W., Shen, H. et al. Aerosol-, but not intradermal-immunization with the live vaccine strain of *Francisella tularensis* protects mice against subsequent aerosol challenge with a highly virulent type A strain of the pathogen by an alphabeta T cell- and interferon gamma-dependent mechanism. Vaccine 2005; 23:2477–2485.

Conlan, J.W., Sjostedt, A. et al. CD4+ and CD8+ T-cell-dependent and -independent host defence mechanisms can operate to control and resolve primary and secondary *Francisella tularensis* LVS infection in mice. Infect. Immun. 1994; 62(12):5603–5607.

Dahlstrand, S., Ringertz, O. et al. Airborne tularemia in Sweden. Scand. J. Infect. Dis. 1971; 7:7–16.

Dalton, R. Infection scare inflames fight against biodefence network. Nature 2005; 433:344.

Dennis, D.T., Inglesby, T.V. et al. Tularemia as a biological weapon—medical and public health management. J. Am. Med. Assoc. 2001; 285(21):2763–2773.

Dorofe'ev, K.A. Classification of the causative agent of tularemia. Symp. Res. Works Inst. Epidemiol. Mikrobiol. Chita. 1947; 1:170–180.

Eigelsbach, H.T. and Downs, C.M. Prophylactic effectiveness of live and killed tularemia vaccines. I. Production of vaccine and evaluation in the white mouse and guinea pig. J. Immunol. 1961; 87:415–425.

Eigelsbach, H.T. and Tulis, J.J. Aerogenic immunization of the monkey and the guinea pig with live tularemia vaccine. Proc. Soc. Exp. Biol. Med. 1961; 108:732–734.

Eigelsbach, H.T., Tulis, J.J. et al. Live tularemia vaccine. I. Host–parasite relationship in monkeys vaccinated intracutaneously or aerogenically. J. Bacteriol. 1962; 84(1020–1027).

Elkins, K.L., Rhinehart-Jones, T. et al. T-cell-independent resistance to infection and generation of immunity to *Francisella tularensis*. Infect. Immun. 1993; 61:823–829.

Ellis, J., Oyston, P.C.F. et al. Tularemia. Clin. Microbiol. Rev. 2002; 15(4):631–646.

Enderlin, G., Morales, L. et al. Streptomycin and alternative agents for the treatment of Tularemia: review of the literature. Clin. Infect. Dis. 1994; 19:42–47.

Ericsson, M., Kroca, M. et al. Long-lasting recall response of CD4(+) and CD8(+) alpha beta T cells, but not gamma delta T cells, to heat shock proteins of *Francisella tularensis*. Scand. J. Infect. Dis. 2001; 33(2):145–152.

Evans, M.E. *Francisella tularensis*. Infect. Control 1985; 6(9):381–383.

Evans, M.E., Gregory, D.W. et al. Tularemia: a 30 year experience with 88 cases. Medicine (Baltimore) 1985; 64(4):251–269.

Forsman, M., Sandstrom, G. et al. Analysis of 16S ribosomal DNA sequences of *Francisella* strains and utilization for determination of the phylogeny of the genus and for identification of strains by PCR. Int. J. Syst. Bacteriol. 1994; 44(1):38–46.

Fortier, A.H., Green, S.J. et al. Life and death of an intracellular pathogen: *Francisella tularensis* and the macrophage. Immunol. Ser. 1994:349–361.

Foshay, L. Tularemia treated by a new specific antiserum. Am. J. Med. Sci. 1934; 187:235–245.

Foshay, L. A comparative study of the treatment of tularemia with immune serum, hyperimmune serum and streptomycin. Am. J. Med. 1946; 1:180–188.

Foshay, L. Tularemia. Ann. Rev. Microbiol. 1950; 4:313–330.

Foshay, L., Hesselbrock, W.H. et al. Vaccine prophylaxis against tularemia in man. Am. J. Pub. Health 1942; 32:1131–1145.

Francis, E. and Felton, L.D. Antitularemic serum. Public Health Rep. 1942; 57:44–55.

Fulop, M., Manchee, R. et al. Role of lipopolysaccharide and a major outer-membrane protein from *Francisella tularensis* in the induction of immunity against tularemia. Vaccine 1995; 13(13):1220–1225.

Fulop, M., Mastroeni, P. et al. Role of antibody to lipopolysaccharide in protection against low- and high-virulence strains of *Francisella tularensis*. Vaccine 2001; 19(31):4465–4472.

Golovliov, I., Baranov, V. et al. An attenuated strain of the facultative intracellular bacterium *Francisella tularensis* can escape the phagosome of monocytic cells. Infect. Immun. 2003; 71:5940–5950.

Golovliov, I., Ericsson, M. et al. Adjuvanticity of Iscoms incorporating a T-cell-reactive lipoprotein of the facultative intracellular pathogen *Francisella tularensis*. Vaccine 1995; 13(3):261–267.

Hambleton, P., Evans, C.G.T. et al. Vaccine potencies of the live vaccine strain of *Francisella tularensis* and isolated bacterial components. Br. J. Exp. Pathol. 1974; 55:363–373.

Hartley, M.G., Green, M. et al. Protection afforded by heat shock protein 60 from *Francisella tularensis* is due to copurified lipopolysaccharide. Infect. Immun. 2004; 72:4109–4113.

Havlasova, J., Hernychova, L. et al. Mapping of immunoreactive antigens of *Francisella tularensis* live vaccine strain. Proteomics 2002; 2:857–867.

Havlasová, J., Hernychová, L. et al. Proteomic analysis of anti-*Francisella tularensis* LVS antibody response in murine model of tularemia. Proteomics 2005; 5:2090–2103.

Helvaci, S., Gedikoglu, S. et al. Tularemia in Bursa, Turkey: 205 cases in ten years. Eur. J. Epidemiol. 2000; 16(3):271–276.

Hiromatsu, K., Yoshikai, Y. et al. A protective role of gamma delta T-cells in primary infection with *Listeria monocytogenes* in mice. J. Exp. Med. 1992; 175:49–56.

Hollis, D.G., Weaver, R.E. et al. *Francisella philomiragia* comb. nov. (formerly *Yersinia philomiragia*) and *Francisella tularensis* biogroup novicida (formerly *Francisella novicida*) associated with human disease. J. Clin. Microbiol. 1989; 27(7):1601–1608.

Hornick, R.B. and Eigelsbach, H.T. Aerogenic immunization of man with live tularemia vaccine. Bacteriol. Rev. 1966; 30:532–538.

Johansson, A., Berglund, L. et al. Ciprofloxacin for treatment of tularemia in children. Pediatr. Infect. Dis. J. 2000; 19:449–453.

Johansson, A., Berglund, L. et al. Ciprofloxacin for treatment of tularemia. Clin. Infect. Dis. 2001; 33(2):267–268.

Johansson, A., Urich, S.K. et al. In vitro susceptibility to quinolones of *Francisella tularensis* subspecies tularensis. Scand. J. Infect. Dis. 2002; 34(5):327–330.

Karlsson, J., Prior, R.G. et al. Sequencing of the *Francisella tularensis* strain Schu 4 genome reveals the shikimate and purine metabolic pathways, targets for the construction of a rationally attenuated auxotrophic vaccine. Microb. Comp. Genom. 2000; 5(1):25–39.

Koskela, P. and Salminen, A. Humoral immunity against *Francisella tularensis* after natural infection. J. Clin. Microbiol. 1985; 22:973–979.

Kroca, M., Tarnvik, A. et al. The proportion of circulating gamma delta T cells increases after the first week of onset of tularaemia and remains elevated for more than a year. Clin. Exp. Immunol. 2000; 120(2):280–284.

Larsson, P., Oyston, P.C.F. et al. The complete genome sequence of *Francisella tularensis*, the causative agent of tularemia. Nat. Genet. 2005; 37(2):153–159.

Lauriano, C.M., Barker, J.R. et al. Allelic exchange in *Francisella tularensis* using PCR products. FEMS Microbiol. Lett. 2003; 229(2):195–202.

Lenco, J., Pavkova, I. et al. Insights into the oxidative stress response in *Francisella tularensis* LVS and its mutant Delta iglC1+2 by proteomics analysis. FEMS Microbiol. Lett. 2005; 246(1):47–54.

McCoy, G.W. and Chapin, C.W. Further observations on a plague-like disease of rodents with a preliminary note on the causative agent, *Bacterium tularense*. J. Infect. Dis. 1912; 10:61–72.

McCrumb, F.R. Aerosol infection of man with *Pasteurella tularensis*. Bacteriol. Rev. 1961; 25:262–267.

Morner, T. The ecology of tularemia. Rev. Sci. Tech. Off. Int. Epiz. 1992; 11(4):1123–1130.

Ohara, Y., Sato, T. et al. Epidemiological analysis of tularemia in Japan (*yato-byo*). FEMS Immunol. Med. Microbiol. 1996; 13:185–189.

Ohara, Y., Sato, T. et al. Arthropod-borne tularemia in Japan: clinical analysis of 1,374 cases observed between 1924 and 1996. J. Med. Entomol. 1998; 35(4):471–473.

Overholt, E.L., Tigertt, W.D. et al. An analysis of forty-two cases of laboratory-acquired tularemia. Treatment with broad spectrum antibiotics. Am. J. Med. 1961; 30:785–806.

Oyston, P.C.F., Sjostedt, A. et al. Tularemia: bioterrorism defence renews interest in *Francisella tularensis*. Nat. Rev. Microbiol. 2004; 2:967–979.

Pavkova, I., Hubalek, M. et al. *Francisella tularensis* live vaccine strain: proteomic analysis of membrane proteins enriched fraction. Proteomics 2005; 5(9):2460–2467.

Perez-Castrillon, J.L., Bachiller-Luque, P. et al. Tularemia epidemic in northwestern Spain: clinical description and therapeutic response. Clin. Infect. Dis. 2001; 33:573–576.

Poquet, Y., Kroca, M. et al. Expansion of V gamma 9 V delta 2 T cells is triggered by *Francisella tularensis*-derived phosphoantigens in tularemia but not after tularemia vaccination. Infect. Immun. 1998; 66(5):2107–2114.

Prior, J.L., Prior, R.G. et al. Characterisation of the O-antigen gene cluster and structural analysis of the O-antigen of a type A strain of *Francisella tularensis*. J. Infect. Dis. 2003; 52:845–851.

Sandstrom, G. The tularemia vaccine. J. Chem. Technol. Biotechnol. 1994; 59:315–320.

Sandstrom, G., Lofgren, S. et al. A capsule-deficient mutant of *Francisella tularensis* LVS exhibits enhanced sensitivity to killing by serum but diminished sensitivity to killing by polymorphonuclear leukocytes. Infect. Immun. 1988; 56(5):1194–1202.

Sandstrom, G., Sjostedt, A. et al. Immunogenicity and toxicity of lipopolysaccharide from *Francisella tularensis* LVS. FEMS Microbiol. Immunol. 1992; 5:201–210.

Sandstrom, G., Tarnvik, A. et al. Immunospecific T-lymphocyte stimulation by membrane proteins from *Francisella tularensis*. J. Clin. Microbiol. 1987; 25(4):641–644.

Santic, M., Molmeret, M. et al. The *Francisella tularensis* pathogenicity island protein IglC and its regulator MglA are essential for modulating phagosome biogenesis and subsequent escape into the cytoplasm. Cell. Microbiol. 2005; 7:969–979.

Sjostedt, A., Conlan, J.W. et al. Neutrophils are critical for host defence against primary infection with the facultative intracellular bacterium *Francisella tularensis* in mice and participate in defence against reinfection. Infect. Immun. 1994; 62:2779–2783.

Sjostedt, A., Sandstrom, G. et al. Several membrane polypeptides of the live vaccine strain *Francisella tularensis* LVS stimulate T cells from naturally infected individuals. J. Clin. Microbiol. 1990a; 28(1):43–48.

Sjostedt, A., Sandstrom, G. et al. Nucleotide sequence and T cell epitopes of a membrane protein of *Francisella tularensis*. J. Immunol. 1990b; 145(1):311–317.

Sjostedt, A., Sandstrom, G. et al. Humoral and cell-mediated-immunity in mice to a 17-kiloDalton lipoprotein of *Francisella tularensis* expressed by *Salmonella typhimurium*. Infect. Immun. 1992; 60(7):2855–2862.

Steinemann, T.L., Sheikholeslami, M.R. et al. Oculoglandular tularemia. Arch. Ophthalmol. 1999; 117(1):132–133.

Stewart, S.J. Tularemia: association with hunting and farming. FEMS Immunol. Med. Microbiol. 1996; 13:197–199.

Surcel, H.-M., Syrjala, H. et al. Development of *Francisella tularensis* antigen responses measured as T-lymphocyte proliferation and cytokine production (Tumor necrosis factor alpha, Gamma interferon, and Interleukin-2 and -4) during human tularemia. Infect. Immun. 1991; 59.

Syrjala, H., Kujala, P. et al. Airborne transmission of tularemia in farmers. Scand. J. Infect. Dis. 1985; 17(4):371–375.

Tarnvik, A. Nature of protective immunity to *Francisella tularensis*. Rev. Infect. Dis. 1989; 11(3):440–451.

Tärnvik, A. and Berglund, L. Tularaemia. Eur. Respir. J. 2003; 21:361–373.

Tarnvik, A., Ericsson, M. et al. Orchestration of the protective immune response to intracellular bacteria: *Francisella tularensis* as a model organism. FEMS Immunol. Med. Microbiol. 1996; 13:221–225.

Tarnvik, A., Sandstrom, G. et al. Epidemiological analysis of tularemia in Sweden 1931–1993. FEMS Immunol. Med. Microbiol. 1996; 13:201–204.

Thorpe, B.D. and Marcus, S. Phagocytosis and intracellular fate of *Pasteurella tularensis*. III. In vivo studies with passively transferred cells and sera. J. Immunol. 1965; 94(4):578–585.

Tigertt, W.D. Soviet viable *Pasteurella tularensis* vaccines: a review of selected articles. Bacteriol. Rev. 1962; 26:354–373.

Vinogradov, E., Conlan, W.J. et al. Characterization of the lipopolysaccharide O-antigen of *Francisella novicida* (U112). Carbohydrate Res. 2004; 339(3):649–654.

Whipp, M.J., Davis, J.M. et al. Characterization of a novicida-like subspecies of *Francisella tularensis* isolated in Australia. J. Med. Microbiol. 2003; 52:839–842.

Wu, T.H., Hutt, J.A. et al. Intranasal vaccination induces protective immunity against intranasal infection with virulent *Francisella tularensis* biovar A. Infect. Immun. 2005; 73:2644–2654.

Yee, D., Rhinehart-Jones, T.R. et al. Loss of either CD4+ or CD8+ T cells does not affect the magnitude of protective immunity to an intracellular pathogen, *Francisella tularensis* strain LVS. J. Immunol. 1996; 157:5042–5048.

SECTION V

PARASITIC VACCINES

CHAPTER

62

African Trypanosomiasis

Andrew J. Nok
Department of Biochemistry, Ahmadu Bello University, Zaria, Nigeria

OUTLINE

ABSTRACT

African trypanosomiasis, otherwise known as sleeping sickness in humans and Nagana in cattle, is resurgent in Africa. Research suggests that the development of a vaccine is still far away and that existing drugs are becoming ineffective on account of the emergence of drug-resistant trypanosomes. These infections contribute to heavy economic losses and sociopolitical crises in the continent thus underscoring the pressure to intensify research for inexpensive less toxic and affordable trypanocides and effective vaccines. African trypanosomes are transmitted from reservoir hosts to livestock by several species of tsetse flies and other biting insects. In each host the parasite undergoes many life-cycle stages involving forms with discrete morphologies, patterns of gene expression, and proliferation. In each case there developmental changes are precisely programmed.

Infected animals develop fever, weight loss, and progressively become weak and unproductive; breeding animals, if left untreated, may abort or become infertile. Most animals die of anemia, heart failure, or intercurrent bacterial infections as a result of the animal's weakened resistance. Highly productive exotic breeds tend to be susceptible. In human trypanosomiasis, a similar course of events take place with parasites ultimately invading the brain, resulting in the disease syndrome known as "sleeping sickness."

In Africa, trypanosomiasis occurs in 37 countries, extending over 10 million square kilometers which is about a third of the continent. It is recorded in this region that an estimated 50 million cattle, about 30% of Africa's total cattle population, are exposed to the risk of infection. Africa produces about 70 times less animal protein per hectare than Europe, and this is on account of the widespread incidence of trypanosomiasis.

Trypanosomiasis control is focused on the use of insecticide spraying to control the vector tsetse populations and on regular treatment of livestock at risk with trypanocidal drugs. The high cost of regular drug and insecticidal treatment, the limited effectiveness of insecticide application in high-rainfall areas and concerns on environmental pollution, the emergence of parasites resistant to available drugs make tsetse and trypanosomiasis control difficult and expensive.

The development of a vaccine against trypanosomiasis by conventional approaches has been unsuccessful. All African trypanosomes are covered by a thick glycoprotein coat made up of variant surface glycoproteins (VSG), which act as shield for underlying invariant immunogenic proteins. VSG, the predominant surface antigen in African trypanosomes, prevents trypanosome lysis by alternative complement pathway, and enables the organism to avoid adaptive immune responses via the phenomenon of antigenic variation.

In order to pursue chemotherapeutic control of trypanosomiasis, intensive research into several aspects of the basic biology of the parasites, the role of tsetse flies in transmitting disease, and the responses to infection of different breeds and species of livestock are mandatory. This chapter will review developments in trypanosomiasis research and the current progress in vaccine research.

INTRODUCTION

Sleeping sickness, or human African trypanosomiasis (HAT), is an endemic parasitic disease exclusively located in intertropical Africa (Fig. 62.1) where it is transmitted by its unique vector, the tsetse fly or *Glossina* (Fig. 62.2a) (Vickerman, 1985). This vector is exclusive to the African continent. East African trypanosomiasis is caused by the parasite *Trypanosoma brucei rhodesiense* and West African trypanosomiasis by *Trypanosoma brucei gambiense*. East African trypanosomiasis occurs in parts of Eastern and Central Africa, including Uganda, Kenya, Tanzania, Malawi, Ethiopia, Zaire, Zimbabwe, and Botswana. West African trypanosomiasis is found in parts of Western and Central Africa (Fig. 62.2b). The inoculation of trypanosomes into mammalian hosts triggers a series

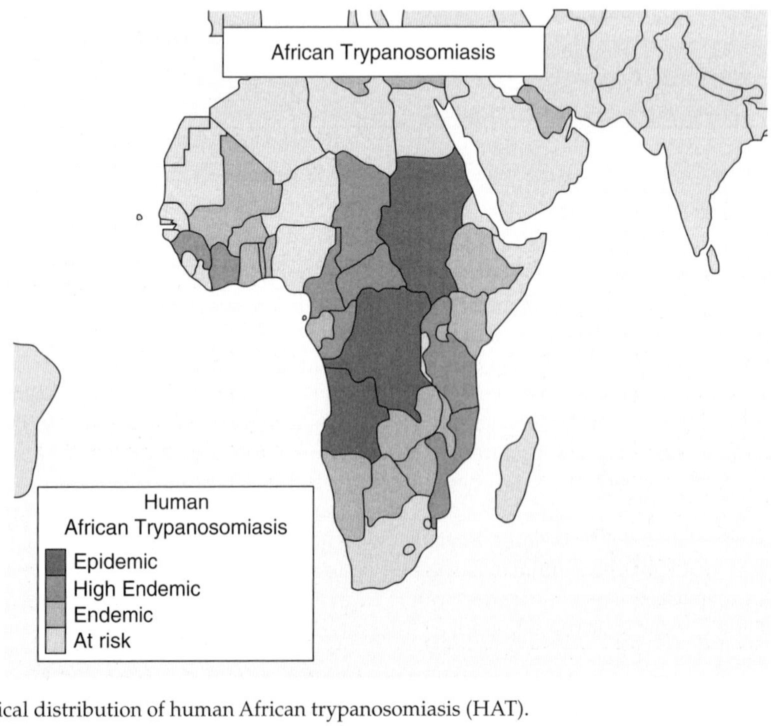

FIGURE 62.1 Geographical distribution of human African trypanosomiasis (HAT).

of events involving, innate and adaptive immunity. The latter requires an efficient presentation of parasitic antigens, activation of T and B cells implying specific antigen receptor recognition, and the development of effector cells and molecules. These mechanisms are highly regulated by multiple signals delivered through a large number of receptors transduced across the plasma membrane and processed.

During co-evolution with their hosts, trypanosomes have learnt to cope with host immune systems, by penetrating, diverting, and altering the numerous steps leading to the generation of an effective immune response. Major modifications of immune systems have been observed in trypanosomiasis: lymphadenopathy, splenomegaly (up to 30 times the normal size) with destruction of lymphatic tissue architecture and hypergammaglobulinemia (Maurice and Pearce, 1987).

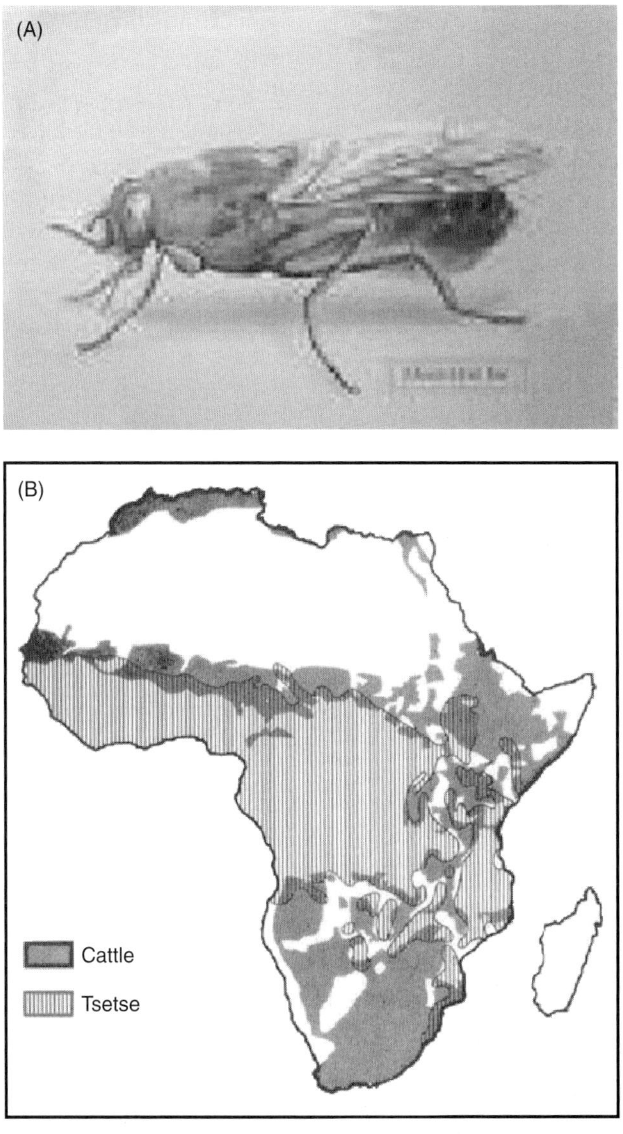

FIGURE 62.2 (a) The tsetse fly, *Glossina morsitans*; (b) geographical distribution of tsetse flies and cattle in the African continent.

Animal trypanosomiasis occurs throughout the tropical regions of Africa and in large areas of Asia and South America. It affects cattle, sheep, goats, pigs, horses, camels, and man. Also, wild animals can be infected with the parasites; however, they seldom suffer from disease. They serve as the reservoir of infection for domestic animals. The most important trypanosome species affecting domestic livestock in Africa are *Trypanosoma congolense*, *T. vivax*, and *T. b. brucei* in cattle, sheep, and goats; *T. simiae* in pigs; and *T. evansi* in camels. *T. vivax* also has a significant impact on cattle production in South America, while *T. evansi* affects camels in Asia and horses, cattle, and domestic buffalo in South America, India, and Southeast Asia.

African trypanosomiasis is transmitted between mammalian host by a variety of *Glossina* species including *Glossina morsitans* and *Glossina* palpalis usually termed tsetse flies (Fig. 62.2). The vector-mediated mode of transmission is relatively inefficient and requires the development of a chronic rather than acute infection to give the parasite sufficient opportunity for reuptake by the vector in order to complete the cycle and sustain the population (Maurice and Pearce, 1987). Therefore, trypanosomes must survive and grow for prolonged periods in their mammalian hosts. This requirement necessitates the existence of a mechanism that regulates parasite growth and ensures the survival of host at least in the short to medium term. Central to this mechanism is the continuous decrease of parasite number.

East African trypanosomiasis is usually confined to woodland and savannah areas away from places where people live. Tourists, hunters, game wardens, fishermen, and other persons who work in or visit game parks in East and Central Africa are at greatest risk for illness. The tsetse flies that spread West African trypanosomiasis can be found in Western and Central African forests and in areas of thick shrubbery and trees by rivers and waterholes. The risk of infection increases with the number of times a person is bitten by a tsetse fly. Trypanosomes use two distinct pathways for respiration in the different environments of the tsetse and the mammalian bloodstream. In the bloodstream, respiration is entirely dependent upon the trypanosome alternative oxidase (TAO), a mitochondrial glycerophosphate oxidase, which is unique to trypanosomes and might provide a prime target for trypanocidal drugs (Chaudhuri et al., 1995). Procyclic trypanosomes utilize a mitochondrial cytochrome electron transport system for 70–80% of respiration and the TAO for only 20–30%. Metacyclic parasites also use both systems, but the relative contribution of each system is reversed with 20–30% derived from mitochondrial cytochrome electron transport and 70–80% from the TAO pathway; this suggests that the metacyclic forms are an intermediate differentiation stage between procyclic and bloodstream parasites (Fig. 62.3).

The current options for trypanosomiasis treatment are poor, making the search for improved drugs essential. Moreover, because those in the lower socioeconomic class are disproportionately affected by the disease, there is a great need for the development of not only less toxic drugs but also less expensive agents.

Another challenging aspect of trypanosomiasis is the parasite's dense glycoprotein coat. The parasite's genome has 1000 different genes encoding antigenically distinct versions of the coat, thus affording protection from adaptive immune responses through antigenic variation (Cross, 1978). This strategy makes the prospect of a vaccine problematic (Pays and Nolan, 1998). The moment African trypanosomes are injected into the bloodstream of the mammalian host by means of the tsetse fly bite, they are exposed to host blood and host immune system. Yet these unicellular eukaryotes live freely in blood dividing by binary fission every 4–6h.

Most of the data concerning African trypanosomiasis have been obtained from naturally infected animals or experimental animal models. Few studies have examined host immune responses to the parasite. Genetic analysis of resistance and susceptibility to infection in inbred and congenic animal strains form the basis for research into equivalent human genes. Introduction of double-stranded RNA (ds RNA) into parasites induces potent and specific gene silencing a phenomenon called RNA interference (RNAi) and is a valuable tool to investigate trypanosome gene functions. The recent sequencing of the entire genome of *T. brucei* is an important breakthrough to investigate immunology and immunopathology (Beverley, 2003).

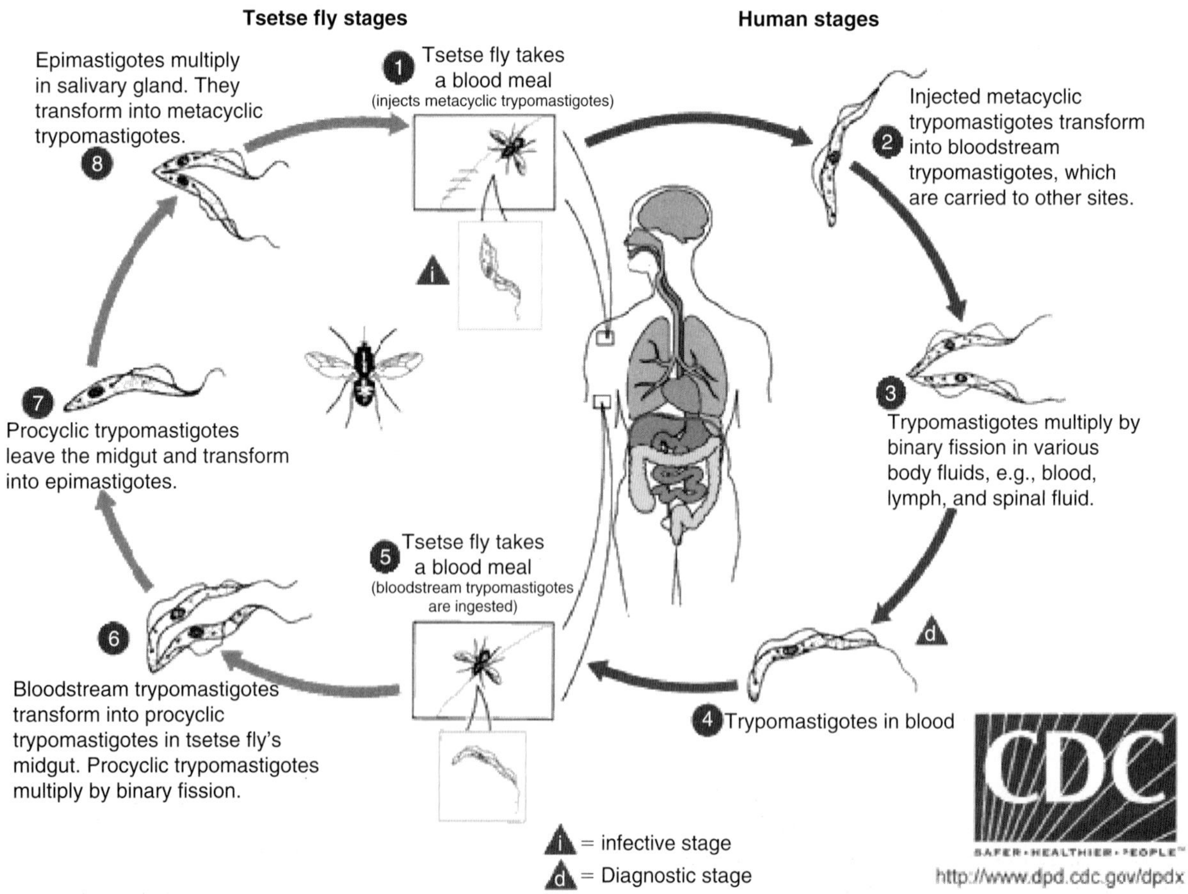

FIGURE 62.3 The developmental stages of trypanosome infection between the insect vector and human host.

CLASSIFICATION

Trypanosoma are of the class kinetoplastida, a monophyletic group of unicellular parasitic protozoa. The name is derived from the Greek *trypaô* (boring) and *soma* (body) because of their corkscrew-like motion. Trypanosomes infect a variety of hosts and cause the fatal disease, sleeping sickness, in humans. Trypanosoma undergo a complex life cycle which includes several different morphological forms (Fig. 62.3). For example, *T. brucei* is transmitted between mammalian hosts through a tsetse fly vector and undergoes a series of morphological and metabolic changes to adapt to these very different environments. In contrast, *T. equiperdum* are sexually transmitted between horses and have no insect part of the life cycle.

Characteristic of this order is the mitochondrial genome, known as the kinetoplast. Structurally it appears as a highly complex series of catenatated circles and minicircles; a set of proteins are required for organization during cell division.

Species of *Trypanosoma* include the following:

- *T. avium*, which causes trypanosomiasis in birds
- *T. boissoni*, in elasmobranch
- *T. brucei*, which causes sleeping sickness in humans and *nagana* in cattle
- *T. carassii*, in freshwater *teleosts*
- *T. cruzi*, which causes *Chagas disease* in humans
- *T. congolense*, which causes *nagana* in cattle, horses, and camels
- *Trypanosoma equinum*, in horses infected by Tabanidae, South America
- *T. equiperdum*, which causes dourine or covering sickness in horses and other *Equidae*
- *T. evansi*, which causes one form of the disease *surra* in certain animals
- *Trypanosoma levisi*, in rats
- *Trypanosoma melophagium*, in sheep infected by *Melophagus ovinus*
- *Trypanosoma percae*, in fish: *Perca fluviatilis*
- *Trypanosoma rangeli*, believed to be nonpathogenic to humans
- *T. rotatorium*, in amphibian
- *T. simiae*, which causes *nagana* in animals
- *T. suis*, which causes a different form of surra
- *T. theileri*, a large trypanosome infecting ruminants
- *T. triglae*, in marine teleosts
- *T. vivax*, which causes the disease *nagana*

The new taxonomy tools used in African trypanosomes (isoenzyme characterization, DNA analysis) have permitted scientists to separate the *T. brucei* class in several subspecies. Two are infective for humans: *T. b. gambiense*, and *T. b. rhodesiense*.

The *brucei* group is made up of *T. b. brucei*, *T. b. gambiense*, *T. b. rhodesiense*, *T. evansi*, *T. simiae*, *T. lewisi*, and *T. congolense*.

Antigens Encoded by Agent

The entire trypanosome surface is covered with a homogenous and dense glycoprotein coat made of approximately five million dimers of single antigen termed variant surface glycoprotein (VSG) that is repeatedly changed in a fraction of population (Fig. 62.4). This allows the trypanosomes to escape antibody-mediated killing and repopulate the host thereby resulting in the development of long-lasting chronic infection (Barry and McCulloch, 2001; Vanhamme et al., 1995; Pays et al., 2004). The VSG constitutes an important molecular interface between trypanosomes and the host immune system. It prevents trypanosome lysis by complement alternative pathway and enables them to avoid the specific immune response via antigenic variation (Fig. 62.5). It also has several effects on immune elements such as induction of autoantibodies and cytokines, in particular tumor necrosis factor (TNF) (Tachado and Schofield, 1994; Okomo-Assoumou et al., 1995a, Magez et al., 2002) (Fig. 62.6).

VSG are immunogenic and are eventually recognized by the host immune system but prior to immune recognition a minority of parasites will switch to an immunologically distinct coat to facilitate the next wave of infection. Despite successful evasion of antibody-mediated clearance by antigenic variation, the parasitemia is attenuated by immune responses

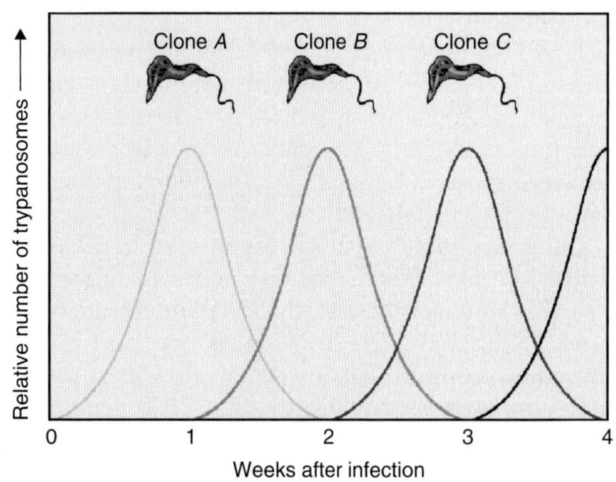

FIGURE 62.4 Schematic representation of the switch to different forms of parasites with varied surface antigens.

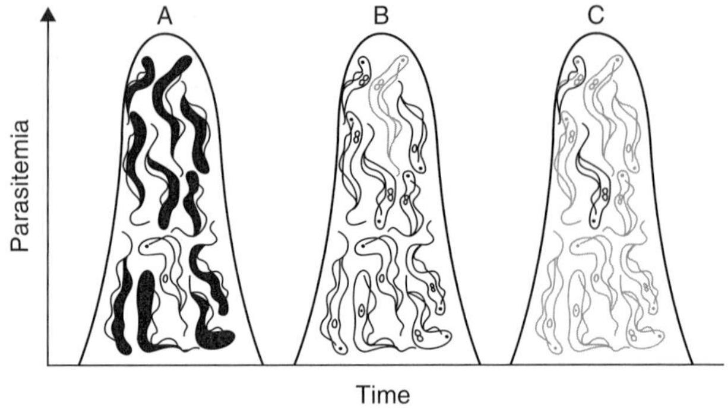

FIGURE 62.5 Schematic representation of the different forms of the trypanosome parasite at the various bouts of parasitemia.

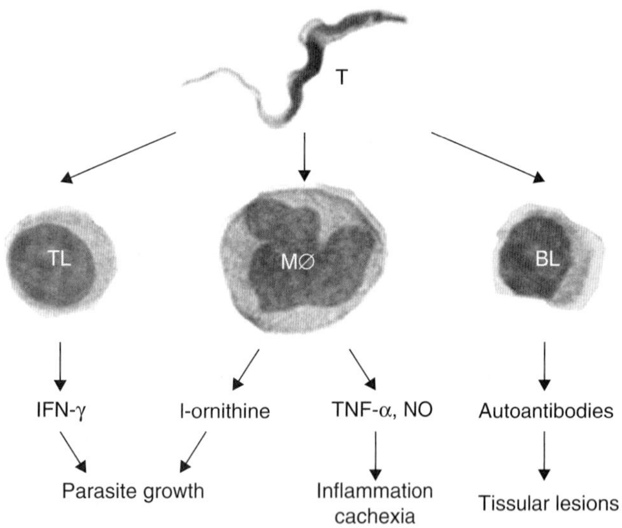

FIGURE 62.6 Trypanosome-induced secretion of various components from immune cells. T, trypanosomes; MØ, macrophage; TL, T lymphocyte; BL, B lymphocyte.

such as TNF (Magez et al., 1997, 1999). The continual release of TNF-α from activated macrophages causes the hallmark chronic wasting disease (cachexia) associated with trypanosomiasis. The parasites invade the central nervous system (CNS) resulting in neurological diseases, coma, and eventual death.

The genes encoding VSGs are located in telomeric expression sites. There are two types of expression sites (ES1) metacyclic ES1 (M-ES1) and bloodstream forms ES2 (B-ES1). The B-ES1 takes over and is used throughout chronic infection (Borst and Ulbert, 2001).

Trypanosomes can change the VSG gene transcribed in a B-ES by replacing the VSG gene in an active expression site by a different one. It is the most important source of VSG antigenic variation and requires DNA rearrangement. Also they can switch expression sites by turning off the active site and activating an earlier silent site. As there are only 20 B-ES1s per trypansosome nucleus, the switching of ES makes a very modest contribution to antigenic variation of the VSG coat. It has been postulated that multiple B-ESs exist to allow expression of different sets of the ESAG1 (Bitter et al., 1998).

Control of Metacyclic Expression

Prior to entry into the mammalian host, trypanosomes that have reached the salivary gland of the tsetse fly acquire a VSG coat. These metacyclic parasites lack functional ESAGs and the VSG gene in this phase are rarely replaced and antigenic variation in metacyclic phase of the life cycle is only accomplished by differential control of the M-ES.

As the wild animals in tropical Africa are chronically infected with trypanosomes, antigenic diversity of the inoculated trypanosome greatly increases incoming trypanosomes gain a foothold within its new host. Up to 27 different M-VATs have been detected in the metacyclic trypanosome population (Turner et al., 1988), hence at least 27 M-ES. Each trypanosome expresses a single VSG gene and continues to express a single VSG gene as long as the metacyclic stage continues. Hence each trypanosome has to decode which M-ES should be activated and how to keep the other 26 silent. The M-ES characterized thus far are all located at the end of large chromosomes and their activation is not associated with DNA rearrangement.

The exact start site for transcription is controversial. Promoter screens using M-ES fragments expressed in procyclic trypanosomes led to the identification of a set of promoters that had some semblance to B-ES promoter (Nagoshi et al., 1995; Vanhamme et al., 1995). Other trypanosome components and soluble factors, such as a trypanosome-released triggering factor (TLTF) which triggers interferon (IFN) production by T cells, are also involved in modulation of the immune system by acting on the synthesis of immune elements (Olsson et al., 1991; Vaidya et al., 1997). Furthermore, increased levels of circulating endotoxins are a feature of human and experimental trypanosomiasis. These endotoxins, potent immunomodulatory molecules, participate to the immune disorders observed in trypanosomiasis. Elaboration of escape mechanisms to host immune defenses and induction of parasite growth factor production are well developed by trypanosomes. In a recently discovered escape mechanism, host arginase induction by trypanosomes decrease immune response efficiency and increase the production of L-ornithine, an essential growth factor (Gobert et al., 2000; Vincendeau et al., 2003).

Flagellar Pocket Antigens

The flagellar pocket of the trypanosome contains receptors that are involved in specific host macromolecule uptake. One example of a flagellar pocket protein is trypanosomal transferrin-binding protein (TFBP), whose gene is present in multiple versions, and to which antibodies are produced in chronic infection. We showed that azanthraquinone coupled to transferrin enhanced the killing effect of the compound on trypanosomes (Nok and Nock, 2002). In a study with organotin palmitate, a known trypanocidal compound, we showed that the flagellar pocket is intrinsically affected by the compound and the topoisomerase of the parasite are equally strongly inhibited (Shuaibu et al., 2004). It has been conjectured that flagellar pocket proteins are highly conserved among flagellated protozoans. For this reason, a vaccine against the flagellar proteins of *T. brucei* may also confer protection against other parasites. In one study a flagellar pocket antigen from *T. b. rhodesiense*, with bovine serum albumin as a carrier and alum as the adjuvant, was used to inoculate 90 cattle in Kenya (Mkunza et al., 1995). The rate of infection was reduced from 13% to 0.9%. This study is of particular note because it was carried out in an environment of natural exposure with a naturally susceptible species.

Congopain

A second approach in vaccine control has been to prevent disease (limit pathogenicity) associated with infection without preventing the infection itself. Cysteine proteases of microorganisms can degrade host proteins such as immunoglobulins and complement factors. They can also modulate cytokine activities, and are suspected of interfering with antigen presentation and processing. Some cattle species in Africa are more resistant to infection than others. Studies have shown that a *T. congolense* cysteine protease (congopain) may play a role in the different levels of tolerance (Lalmanach et al., 2002). The more resistant cattle generate a stronger IgG response to congopain than the less resistant cattle. Attempts are being made to use congopain antigens in a vaccine that would generate antibodies capable of neutralizing the enzyme's activity.

Intracellular Antigens

Members of the Trypanosomatidae family have subpellicular microtubules cross-linked to each other and to the plasma membrane by unique trypanosomal microtubule-associated proteins (MAPs). The trypanosomal MAP (p52) has been used in an antigenic preparation with the enzymes aldolase and GAPDH as a vaccine in mice and rats. The p52 was isolated from *T. b. brucei*, and when the animals were immunized three times over a period of 3 weeks 100% protection was achieved (Rasooly and Balaban, 2004). The serum from immunized animals also cross reacted with *T. b. rhodesiense* and *T. b. evansi*, which gives hope for cross protection at least among trypanome species, and possibly for the whole family.

PROTECTIVE IMMUNE RESPONSE

Host Antibody Production

In infected animals, antibodies directed against trypanosomal VSGs attach to the surface of the parasites and mark them for destruction by macrophages or by the complement system. Previous studies of *T. b. brucei* infection in mice suggested that host antibody responses do not play a role in stimulating trypanosome transformation in the bloodstream, but only becomes effective after most of the parasites have already switched to nondividing forms. In vitro experiments to examine the effects of anti-trypanosomal antibodies on rapidly dividing forms of the parasite show that the polyvalent and monoclonal antibodies raised against trypanosome VSG bound to the surface of rapidly dividing bloodstream forms of *T. b. brucei.* Parasites exposed to moderate levels of the antibody stop multiplying but when the antibody was removed they began multiplying again, indicating that they had not made an irreversible switch to nondividing forms. The antibodies added to the cultures were rapidly destroyed; most likely, they were endocytosed by the parasites (ILRAD, 1994). Mice that are highly susceptible to trypanosomiasis do not produce detectable antibodies against the parasites, although their B lymphocytes are extensively activated and transformed into plasma cells which contain antibody. In vitro analysis of spleen cells from infected susceptible mice show that their activated B cells were defective in their capacity both to synthesize and to secrete antibody. When resistant mice were infected with trypanosomes, they produced antibodies which controlled the first parasitemic wave, but showed similar defective plasma cell function during the second parasitemic wave. However, in resistant mice antibody synthesis and secretion eventually recovered and parasitemia was controlled. At the time of parasite remission, vast amounts of antibody were synthesized and secreted although the number of cells containing antibody did not increase.

Anemia

In most early cases of trypanosomiasis there is an acute onset of anemia corresponding closely with the detection of parasites in the bloodstream. Later, in chronic infections, levels of anemia are observed which appear unrelated to the rise and fall of parasite populations. The N'Dama breed of cattle is known to withstand the pathogenic effects of trypanosome infection better than susceptible breeds (Shugaba et al., 1994). The etiology of anemia in trypanosomiasis has been linked to the enzymes phospholipase A_2 (Nok et al., 1993) and sialidases (Esievo et al., 1990; Nok et al., 2003b, Nok and Balogun, 2003). The PLA_2 is secreted by dying trypanosomes leading to the hydrolysis of red blood cells (RBC) phospholipids, thus generating free fatty acid and lyso-lipid derivative. The latter is a powerful hemolysing agent that induces high rate of RBC hydrolysis leading to anemia. Studies on potential inhibitors of PLA_2-like organotin derivatives of different fatty acids have proved valuable in experimental chemotherapy of the disease (Nok et al., 1993; Shuaibu et al., 2004). Our work has also revealed that the enzyme sialidase is abundant in bloodstream form of the trypanosome parasite. It mediates its pathogenic property by hydrolysis of membrane-bound sialic acid. When the enzyme attacks the sialyl residue on RBC, asialo-epitopes are generated which are easily recognized by macrophages and cleared accordingly, thereby leading to anemia. Interestingly, the *T. evansi* sialidase has capacity to hydrolize gangliosides, providing a link to the pathology of the brain stage of the disease during the chronic phase. Anemia in trypanosome-infected animals maybe due in part to impaired bone marrow function. Undifferentiated stem cells in the bone marrow give rise to both erythrocytes and white blood cells of different types. Trypanosome infection in susceptible animals could have an inhibitory effect on cell production in the bone marrow and/or result in the premature destruction of developing cells.

Endocrine Dysfunction

In susceptible livestock, growth, fertility, and immune responses are all impaired during trypanosome infection. These important functions are influenced by hormones. Previous studies showed that *T. b. brucei* parasites, perhaps when dying, release enzymes into the bloodstream of infected mice which were capable of degrading host proteins. In a recent report, we showed that the modulation of semen α-fucosidase and the attendant sialic acid surged in experimental trypanosomiasis in rams, suggesting a possible role in the loss of reproductive capacity (Okubanjo et al., 2007). Further, histological examination of the testis of rams chronically infected with *T. congolense* showed severe degeneration (Fig. 62.7).

CLINICAL DISEASE

Hemolymphatic Stage

After a painful tsetse bite, a chancre, from several millimeters to centimeters in size, develops at the site of the bite. It is characterized by local erythema, edema, tenderness, and a lack of suppuration. Trypanosomes are present in the inflammatory tissues. The chancre disappears within 2 or 3 weeks. The disease progresses through two distinct successive phases, the hemolymphatic stage and the meningoencephalitic stage (Molyneux et al., 1996;

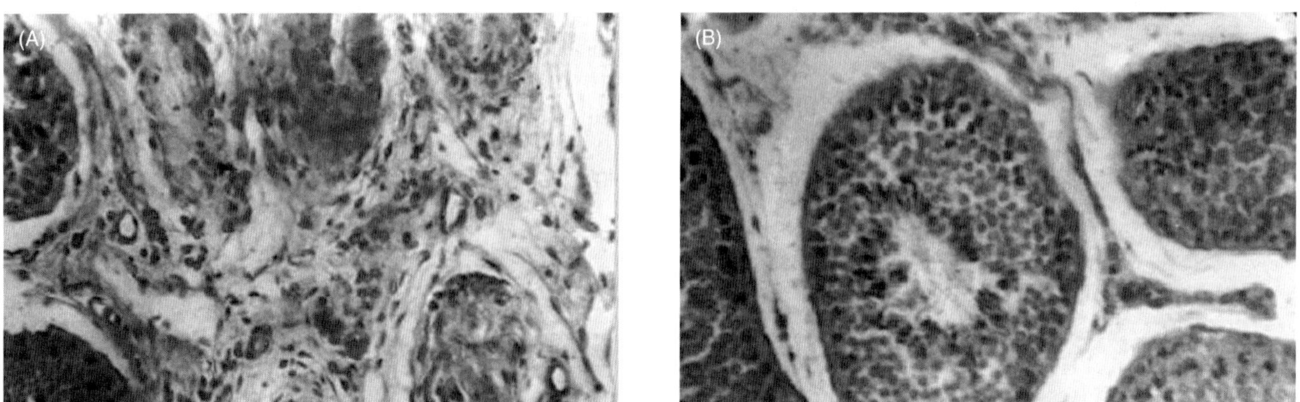

FIGURE 62.7 (a) Severely degenerated testis in *T. congolense*–infected ram (H&E, 400×); (b) normal testis in control ram (H&E, 400×).

Dumas and Bisser, 1999). Systemic signs appear very early. Intermittent fever develops in association with successive waves of parasitemia. Adenopathies, splenomegaly, or even hepatological signs mark the invasion of the reticuloendothelial system. Skin eruptions or trypanides are commonly observed. Severe pruritus can become unbearable for the patient. Cardiovascular alterations are less prominent, especially in the Gambian form. Irregular febrile episodes are accompanied by headaches, malaise, exhaustion, anorexia, extreme thirst, muscle and joint pains, pruritus, anemia, rash, and often deep hyperesthesia. The lymph nodes are generally rubbery and mobile, and painful at the beginning. Palpation of the subclavicular region (Winterbottom sign) is an important part of the diagnostic procedure in the Gambian form. Any adenopathy accompanied by fever should evoke the diagnosis of sleeping sickness in patients from endemic areas. Later, pruritus generalizes. Edema of the face and extremities also develops.

Minor neurological and endocrine findings maybe early indicators of dissemination. Daytime somnolence or nighttime insomnia may develop and electroencephalographic (EEG) tracings may reveal abnormalities. Psychiatric signs, with irritability, changes in personality or mood affecting the daily and professional life of the patients may be an early manifestation of the disease. Endocrine involvement can be marked by a permanent feeling of coldness, lack of appetite, or in contrast hyperphagia, polydipsia and impotence, amenorrhea, or infertility.

Meningoencephalitic Stage

This stage appears over a period of months or years, depending on the trypanosome. The clinical signs remain reversible for a long time with treatment attesting to the predominance of potentially reversible inflammatory lesions over irreversible demyelinating lesions. The general signs of the hemolymphatic stage do not completely disappear: spikes of fever (but sometimes hypothermia), adenopathies and splenomegaly, cardiovascular manifestations, endocrine disturbances, and typical pruritus. The development of the neurological symptoms is progressive. As neurological signs occur, the only means to confirm CNS invasion is based on cerebrospinal fluid (CSF) examination: more than 5 cells/μL and/or the presence of trypanosomes. A range of symptoms may be encountered including daytime somnolence and hence the name, "sleeping sickness." Patient's typically also experience nocturnal insomnia and the patients may be described as being "sleepy by day and restless by night." The sleep–wake cycle disturbances are accompanied by other symptoms including headaches, sensory disturbances with diffuse superficial or deep sensations (muscle and bone hyperesthesia, either spontaneous or provoked; hyperpathia), presence of primitive reflexes (palm-mental reflex, sucking reflex), exaggerated deep tendon reflexes, psychiatric disorders (confusion, mood swings, agitation, aggressive behavior, euphoria, absent gaze, mutism, indifference), and tremor (fine and diffuse without any myoclonic jerk at rest or during movement). Pyramidal alterations revealed by a Babinski sign can also be observed along with alterations in muscle tone, numbness, or sensory deficit.

An abnormal number of monocytic cells are observed in the CSF. The early neurological symptoms correlate with the widespread meningeal inflammation, which occurs in both forms of HAT. The selective CNS locations explain in part the principal clinical neurological signs. Sleep–wake disturbances may result from invasion of the median eminence by parasites, which also accounts for neuroendocrine dysfunctions with the involvement of the suprachiasmatic nuclei. Disorders of the sleep–wake cycle are accompanied by a state of apathy in the patient, the loss of muscle tone especially in the neck muscles and a drooping of the eyelids. Extrapyramidal symptoms signal the involvement of the striatum. Deep sensory disturbances with hyperpathia may result from the involvement of the thalamus and the early invasion of posterior spinal roots. Apart from the disruptions of the circadian rhythm of the sleep–wake cycle, other biological rhythms are disturbed, such as body temperature, cortisol and prolactin, and growth hormone secretion. The invasion of the subthalamic and hypophyseal regions account for the persistence of endocrine disturbances such as impotence, amenorrhea or infertility, and the development of disturbed sensations of hunger and thirst, with often hyperphagia and polydipsia in contrast to the poor general state of malnutrition of the patients. At the terminal phase of the disease, CNS demyelination and atrophy are accompanied with disturbances in consciousness and the development of dementia with incoherence, incontinence, and seizures. The patient dies in a state of cachexia and physiological misery.

TREATMENT

Some of the currently available compounds for the treatment of trypanosomiasis include arsenicals, pentamidines, suramin, isometadium eflornithine, homidium bromide, and nifurtimox. In treating patients who have progressed to the encephalitic stage, drugs must be used that have the ability to cross the blood–brain barrier and kill the *T. b. gambiense* and *T. b. rhodesiense* parasites that are present in the CSF. Melarsoprol was first introduced in 1949 for the treatment of late-stage trypanosomiasis and has remained the main drug of choice (Denise and Barret, 2001). The administration of melarsoprol to patients is complicated. In *T. b. rhodesiense* infection, it is usually given in three series of four intravenous injections with an interval of 10 days between each series (Denise and Barret, 2001). The treatment of late-stage African human trypanosomiasis with melarsoprol is accompanied by some adverse effects such as cutaneous reactions, polyneuropathy, headache, tremor, fever, abdominal pain, chest pain, skin rash, which affects about 19% of the patients (Barett, 2000). Treatment failures have been reported in late-stage sleeping sickness patients treated with melarsoprol (Burri and Keiser, 2001).

Pentamidine is an aromatic diamine 1,5-bis (4 amidi-) phenoxypentane. It is used for the chemotherapy of African trypanosomiasis, antimony-resistant Leishmaniasis, and *Pneumocytis carini pneumonia* (Sands et al., 1985). It is especially useful in the treatment of early-stage *T. b. gambiense* sleeping sickness. On account of hydrodynamic properties, the role of transporters is central in the mode of action of pentamidine in *Trypanosoma brucei.* Treatment failures and adverse effects such as nephrotoxicity and diabetes mellitus have been reported (Bitonti et al., 1986).

Suramin (Germin, Naganol, Morany Bayer 205) is a colorless polysulfonated symmetrical naphthalene derivative. It has six negative charges at physiological pH as such it cannot cross the blood–brain barrier (Hawking, 1940). Suramin is usually the drug of choice for early stages of African human trypanosomiasis especially *T. b. rhodesiense* infections.

Isometamidium chloride a conjugate of homidium and is used exclusively as a veterinary trypanocide. It is used for both prophylactic and therapeutic purposes. Isometamidium transport and resistance has largely been studied in *T. congolense,* which together with *T. brucei* and *T. vivax* is the main cause of trypanosomiasis in African livestock. Resistance to the drug is a severe problem in many parts of sub-Saharan Africa. Resistance by parasites with respect to reduced uptake has been reported (Sutherland et al., 1992).

INNATE IMMUNITY

African cattle trypanosomiasis, mainly due to *T. congolense* and *T. vivax,* causes anemia and weight loss, leading to death. Some African cattle breeds (N'Dama), however, are able to live and be productive in endemic areas and are considered to be trypanotolerant. Trypanotolerance is genetically controlled and is an innate character, but can be increased by repeated infections. Trypanotolerant cattle are not less susceptible to trypanosome infection but limit trypanosome proliferation and have lower parasite counts (*Bos taurus, B. indicus*). This resistance depends on the nutritional, physiological, and stress conditions of the animal. Studies have revealed that the ability to produce IL-4 plays a role in the susceptibility to *T. brucei* disease (Bakhiet et al., 1996). Recently, mouse strain susceptibility to trypanosome infection has been correlated with an increase in host arginase production (Duleu et al., 2004). Evidence of human trypanotolerance is suggested by case reports (Lapeyssonnie, 1960) and results from immunological screening (Lemesre et al., 1988; Authié et al., 1992). Subjects with a positive card agglutination trypanosomiasis test (CATT) have been reported to be asymptomatic despite having evidence of blood parasites. Moreover, although many Bantou people from Mbomo foci in the Congo were infected with *T. b. gambiense,* none of the pygmy population was infected. This effect, present before infection and unrelated to antibody production, is dependent on innate immunity factors. Moreover, in a recent study in the HAT focus of Sinfra (Côte d'Ivoire), single nucleotide polymorphisms within TNF-α and IL-10 promoters and genes were associated with susceptibility to HAT (Courtin et al., 2006).

Natural Immunity

Normal human sera injected into *T. b. brucei*–infected mice has been shown to reduce parasitemia. This phenomenon cannot be reproduced with the human trypanosome strains *T. b. gambiense* and *T. b. rhodesiense.* Trypanolytic factors (TLF) contained in normal human serum were identified as high-density lipoproteins. Recently, two TLFs have been characterized in human serum, the first one (TLF1) is inhibited by haptoglobin, while the second factor, TLF2, is not inhibited by haptoglobin (Raper et al., 1996). The main trypanolytic effect is thought to be due to TLF2. The trypanocidal effect of Cape buffalo serum has been attributed to xanthine oxidase (Muranjan et al., 1997). Recently a trypanosome lysosomal protein (SRA) was found to be associated with resistance to normal human serum. SRA is a truncated form of VSG and interacts with serum apolipoprotein L-I in the parasite lyzosome (Vanhamme and Pays, 2004).

Complement

Both in humans and animals, complement activation by two pathways occurs in HAT. The alternative pathway, independent of specific antibodies, was studied by the induction of trypanosome lysis (*T. congolense* and *T. b. brucei*) observed after the addition of fresh serum. Serum could induce trypanosome lysis only on uncoated VSG trypanosomes, as observed during the cycle of this parasite (procyclic forms). However, the appearance of VSG on parasites prevents trypanosome lysis by this alternative pathway (Ferrante and Allison, 1983). In another strain of *T. b. gambiense*, it was demonstrated that the alternative pathway was incompletely activated without generation of the terminal complex (C5–C9) able to induce membrane lysis (Devine et al., 1986). The classical pathway, mediated by specific antibodies against trypanosomes, has also described and may be involved in parasite clearance by antibody-mediated lysis and/or opsonization. The coated stages of *T. b. brucei* are lysed by antibodies with activation of complement by the classical pathway. Nevertheless, during these complement activations, the appearance of soluble fragments, including C3a and C5a anaphylatoxins and the C567 complex, could induce, on the one hand, the chemoattaction of neutrophils and monocytes and, on the other hand, the release of amines involved in vasoconstriction and an increase in vascular permeability participating in the initial inflammatory response in the chancre. Immune complexes can also activate complement. These immune complexes are constituted by antibodies specific to trypanosomes (e.g., anti-VSG antibody) leading to a rapid elimination of complement-fixing immune complexes (Russo et al., 1994) or by autoantibodies, such as rheumatoid factor or anti-nucleic acid antibodies. These immune complexes with complement activation are also involved in some adverse effects, especially in tissue damage mediated by immune complex deposits (Nielsen, 1985), such as thrombosis and glomerular involvement (Bruijn et al., 1988; van Velthuysen et al., 1994).

Natural Killer (NK) and T Cells

In *T. brucei*–infected mice, NK activity was not modified in the early stages of infection, but was severely reduced from day 9 onwards (Askonas and Bancroft, 1984). By contrast, NK cells were activated in mice infected with a natural extracellular trypanosome (*T. musculi*) and their critical role was demonstrated by the effects of their depletion by antiserum against asialo GM_1 or their activation by polycytidylic copolymer (Albright et al., 1997). Initial studies have evidenced alterations in T cell functions in trypanosomiasis, both in vivo and in vitro (Mansfield and Wallace, 1974). Histological examination show massive B cell expansion in the lymph nodes and spleen that replaced the thymus-dependent area in *T. b. brucei* TREU 667–infected mice. These changes persist from 7 days postinfection to at least 70 days. Trypanosome antigen-specific T cell response was difficult to identify. In several studies, a transient proliferative T cell response to trypanosome antigens was noted in the first days of the infection followed by loss of the response (Gasbarre et al., 1980). The kinetoplastid membrane protein 11 of African trypanosomes is a potent stimulator of T lymphocyte proliferation (Tolson et al., 1994). In *T. b. brucei*–infected mice, an increased proliferation of T cells was noted in the first days of infection in spleen and bone marrow, T blasts disappeared very rapidly. Th1 cells expressing a functional T cell response directed to VSG are generated in *T. b. rhodesiense*–infected mice. VSG-specific T cells were found predominantly in the peritoneum. These cells did not proliferate but made a substantial IFN-γ and IL-2 cytokine response (Schleifer et al., 1993). The cellular phenotype of VSG-responsive T cells ($CD4^+$ $CD3^+$) indicates that the VSG appear to preferentially stimulate a Th1 cell subset during infection. Analysis of lymphocyte subsets in regional lymph nodes of *T. congolense*–infected N'Dama (trypanotolerant) and Boran (trypanosusceptible) revealed a significant decrease in the percentage of $CD2^+$ and $CD4^+$ T associated with increase in the percentage of $CD8^+$ T cells, B cells, and $\gamma\delta$T cells. VSG and two invariant antigens (33 kDa cysteine protease and 66 kDa antigen homologous to immunoglobulin heavy chain binding protein hsp70/Bip) induced in vitro proliferation and synthesis of IL-2 and IFN-γ (Authié et al., 1992).

B Cells

In African trypanosomiasis, a dramatic increase in IgM is a main feature (Shoda et al., 2001). In *T. b. brucei*–infected mice, B lymphocytes display an aberrant activation phenotype (Sacco et al., 1994). Antibodies specific to trypanosomes are induced by several parasite antigens, including variant and invariant glycoprotein epitopes, as well as membrane, cytoplasmic, and nuclear antigens (Reinitz and Mansfield, 1990). Antibodies directed against trypanosome VSG components appear in sera and their binding to the surface coat of the trypanosomes induce a decrease in parasitemia, both in the blood and extravascular spaces, specifically by immune lysis of parasites and their destruction by the Kupffer cells in the liver. Only heterologous antigenic variants (<0.1%) remain to repopulate the blood and tissues. Parasites are eliminated due to VSG-specific IgM (appearing at high levels, 3–4 days after infection). During infection, B cell nonspecific stimulation is enhanced as T-independent B cell responses to the VSG successive parasitemias. In contrast, specific trypanosome B cell response, depending on T cell regulation, is depressed. Several factors may contribute to this immunosuppression. Macrophages may become unable to present antigens to T cells (by defects in antigen processing and association of epitopes with MHC Class II) and produce immunosuppressive factors as nitric oxide (NO), prostaglandins (PG), and cytokines. An increase in immunosuppressive cytokines, such as IFN-γ and transforming growth factor (TGF)-β, is observed during infection. However, TGF-β is known to inhibit the production of IL-4, IL-5, and IL-6, the major cytokines implied in B cell proliferation and differentiation (Fargeas et al., 1992). Autoantibodies directed against components of CNS myelin have also been reported. They are specific for the major glycosphingolipids of myelin, the galactocerebrosides, and are detected in the sera of experimentally infected animals (Jauberteau et al., 1991) and patients from the Ivory Coast (Amevigbe et al., 1992). Other autoantibodies directed against not yet characterized proteins have been described in HAT patients (Asonganyi et al., 1989) as well as antibodies directed at myelin basic protein in experimentally infected animals (Hunter et al., 1992). Other antibodies were raised against an epitope containing L-tryptophan, a precursor to the neurotransmetter serotonin (Okomo-Assoumou et al., 1995b), or recognized some neuronal components of the cytoskeleton, neurofilament proteins. In some cases, these autoantibodies (antigalactocerebrosides and antineurofilaments) are associated with the neurological stage of the disease and their detection in sera and CSF could contribute toward defining the neurological involvement of HAT (Courtioux et al., 2005). In vivo demyelinization has been produced by purified antibodies to galactocerebroside (Saida et al., 1979).

Macrophages

Macrophages may play an important role in protection against trypanosomes, particularly in the presence of homologous antiserum. The immunological clearance of [75Se]-methionine-labeled *T. brucei* in mice was studied to investigate the respective roles of antibodies, macrophage activation, and complement in the removal of circulating parasites. The clearance was largely accomplished by antibody-mediated hepatic phagocytosis. C3 is necessary for the full opsonic activity present in murine clearance in passively immunized mice (MacAskill et al., 1980). These in vivo studies extend previous studies on the in vitro phagocytic function of macrophages in the presence of immune serum (Takayanagi et al., 1992). Macrophages from trypanosome-infected mice also synthesize reactive nitrogen intermediates (RNI). Trypanosomes are highly sensitive to the cytostatic/cytotoxic effects of these compounds (Vincendeau and Daulouède, 1991; Vincendeau et al., 1992). It has been shown that *T. b. gambiense* are highly sensitive to S-nitroso compounds, which are new effector molecules synthesized by activated human macrophages in vitro (Mnaimneh et al., 1997). Nitrosylated compounds could represent new effector molecules with a potent effect on targets distant from macrophages. In a recent study, DNA from *T. b. brucei* have increased macrophage production of IL-12, TNF-α, and NO (Shoda et al., 2001). Macrophages are also active in secreting PGs which modulate lymphocyte and macrophage functions. During a *T. b. brucei* infection, the ratio of PGE2/PGF1a is reversed, with an overproduction of PGE2. Macrophages are involved in immunosuppressive mechanism, and VSG also inhibits macrophage functions (Schleifer et al., 1993). Macrophages respond to, and synthesize, a large number of cytokines and the production of IL-1 is increased in *T. b. brucei*–infected mice, but this increase may be due to release rather than synthesis. Human monocytes can also be induced by trypanosomes and secreted factors from trypanosomes to express TNF-α RNA transcripts and secrete TNF-α in culture supernatants (Daulouède et al., 2001). Classical and alternative states of macrophage activation are observed in trypanosomiasis. Classical activation precedes alternative activation in murine trypanosomiasis. However, both activation states are expressed in these mice. By inducing alternative macrophage activation, trypanosomes induce host arginase which both decrease trypanocidal nitrosylated compound synthesis and increases L-ornithine production (Gobert et al., 2000). L-ornithine is the first step of polyamine synthesis, essential for parasite growth and trypanothione synthesis (Vincendeau et al., 2003).

Cytokines and Chemokines

A dysregulation of the cytokine network occurs in trypanosomiasis. The first evidence of overproduction of TNF-α/cachectin was reported in *T. b. brucei*–infected rabbits (Rouzer and Cerami, 1980). TNF-α is known to induce fever, asthenia, cachexia, and hypertriglyceridemia. High levels of TNF-α are associated with potent inflammatory signs in the early phase of human trypanosomiasis and major neurologic signs in the late phase (Okomo-Assoumou et al., 1995). Initial control of parasitemia in *T. b. brucei*–infected mice was diminished by the injection of anti-TNF-α antibodies (Lucas et al., 1993). VSG can trigger TNF-α production by macrophages. Moreover, TNF-α production can be stimulated by IFN-γ and TGF-β and can be produced by CD8 T cells activated by TLTF released by *T. b. brucei* (Vaidya et al., 1997). Mice chronically infected with *T. b. brucei* develop inflammatory lesions of the CNS after treatment with subcurative doses of a trypanocidal agent (Hunter et al., 1991). Chemokines favor macrophage and lymphocyte recruitment to the CNS of *T. b. brucei*–infected animals. Such recruitment to selected areas of the CNS may be associated with sleep and endocrine disorders.

Nitric Oxide

NO induced by dysregulation of the cytokine network may lead to alteration of immune response and may also be involved in pathophysiological mechanisms. Nitrotyrosine, a marker of peroxynitrite formation, and iNOS are detected in the brains of *T. b. brucei*–infected mice. Nitrotyrosine-positive cells are associated with the development of neurological signs (Keita et al., 2000). In HAT, nitrite production is increased at first and detection of antibodies against nitrosylated antigens indicates the presence of NO-adducts (Semballa et al., 2004). However, in trypanosome-infected mice, a decrease in plasma L-arginine leads to a decreased NO production. By inducing arginase, trypanosomes bypass NO production and facilitate growth factor production. Arginase induction by parasites might be considered as a new strategy elaborated by parasites to escape host defense.

Immunoprotection

The resistance of mice to African trypanosomes can be nonspecifically increased by immunomodulators such as Calmette–Guérin bacilli and *Propionibacterium acnes* (Murray and Morrison, 1979; Black et al., 1989). These immunomodulators activate macrophages. *P. acnes*–treated macrophages inhibit *T. brucei* growth in vitro (Black et al., 1989). In a 10-year study, in Zaire, of adults previously diagnosed and treated for HAT, the risk of a second episode of HAT was greatly reduced compared to the risk of a first episode (Khonde et al., 1995).

VACCINE

Induction of protective immunity by vaccination is an important goal, but in the case of trypanosomiasis the vaccine will need to induce immunity to a large number of antigenic variants. An important feature of the trypanosome parasite is the dense glycoprotein coat covering the entire surface of the parasite. With 1000 different genes encoding antigenically distinct versions of the coat, the parasites have the capacity to engage in an immune escape through the process of antigenic variation. This has proven particularly problematic in the development of a vaccine against trypanosomiasis. However, previous studies have shown some promise in the use of other parasite surface proteins. Scott et al. (1978) reported protection in zebu cattle following double vaccination with 11 isolates of *T. congolense* given either as live or dead organisms and trypanocidal therapy when the cattle were challenged with nine of the original isolates. Neither the live or killed organism regimen prevented the animals from becoming parasitemic nor eventually succumbing to the infection, however, the mean survival times were greater in the live vaccine group compared to the killed vaccine and control groups. Whitelaw et al. (1979) reported that in mice immunized with louping-ill virus vaccine and challenged with *T. brucei* or *T. congolense*, the antibody response was completely suppressed, while in immunized Ethiopian cattle experimentally infected with *T. congolense* or *T. vivax* the antibody response to the vaccine was only 10% of that observed in uninfected animals. In contrast, the response of cattle infected with *T. brucei* was not significantly reduced, and this was attributed to their relatively light and transient parasitemias. Trypanocidal chemotherapy (diminazine aceturate) administered on the same day as vaccination largely restored the competence of the immune response of both mice and cattle infected with *T. congolense*.

Studies have shown that the immunization of animals and humans with trypanosome tubulin protein protects against trypanosomes. A vaccine consisting of a tubulin extract from *T. brucei* has been reported to protect animals and humans against heterologous strains of different species of *Trypanosoma*. Recent studies examining a vaccine containing the trypanosome microtubule-associated protein p15 (MAP p15) in mice found that p15 (native or recombinant) provided up to 100% protection from an otherwise lethal challenge of a heterologous strain of *T. brucei* (Rasooly and Balaban, 2004).

Vaccine Development

DNA vaccination is a relatively recent development. DNA vaccines are generally less costly to produce than peptide or protein vaccines, and are chemically stable under a variety of conditions. There has been some promising work with a related trypanosome family, *Leishmania*. Recently, we reported the protection of mice immunized with a DNA plasmid encoding parasite gamma-glutamylcysteine synthetase and experimentally challenged with *Leishmania donovani* (Carter et al., 2007). The DNA vaccine induced specific IgG1 and IgG2a antibodies and resulted in significantly lower liver parasite burdens compared to controls. Protection was also associated with a significant increase in cell-mediated immunity, demonstrated as an increase in nitrite production by ConA-stimulated splenocytes, an increase in the percentage of splenic $CD3^+$ $CD4^+$ cells, and enhanced granuloma maturation, compared to control values. It is hoped that similar DNA vaccine strategies can be developed for control of African trypanosomiasis. One strategy currently being explored is the development of a DNA vaccine encoding sialidase, an enzyme implicated in the etiology of anemia during trypanosomiasis (Esievo et al., 1990; Nok et al., 2003a; Nok and Balogun, 2003).

Based on the current state of the field it is clear that progress in developing an effective vaccine will likely require complex strategies that may include the use of trypanocidal drugs.

KEY ISSUES

- In Africa, trypanosomiasis occurs in 37 countries extending over 10 million square kilometers.
- An estimated 50 million cattle, about 30% of African total cattle population, are exposed to the risk of infection.
- Trypanosomiasis control is focused on the use of insecticide spraying to control vector tsetse population and on regular treatment with trypanocidal drugs.
- Toxicity of existing drugs, high rate of treatment failures, and the emergence of drug-resistant trypanosomes make the development of alternative control measures mandatory.
- Development of a vaccine by conventional methods has been unsuccessful due to antigenic variation by the parasite.
- The trypanosome parasite has at least 1000 different genes encoding antigenically distinct versions of the surface coat and has been a major obstacle toward the development of a vaccine.
- MAP p15 tested as a vaccine in mice generated 100% protection against lethal *T. brucei*.
- Current effort on the control is focused on the development of DNA vaccines.

References

Albright, J.W., Jiang, D. and Albright, J.F. Innate control of the early course of infection in mice inoculated with *Trypanosoma musculi*. Cell. Immunol. 1997; 176:146–152.

Amevigbe, M., Jauberteau-Marchan, M.O., Bouteille, B., Doua, F., Breton, J.C., Nicolas, J.A. and Dumas, M. Human African trypanosomiasis: presence of antibodies to galactocerebrosides. Am. J. Trop. Hyg. 1992; 45:652–662.

Askonas, B.A. and Bancroft, G.J. Interaction of African trypanosomes with the immune system. Philos. Trans. R. Soc. Lond., B, Biol. Sci. 1984; 307:41–50.

Asonganyi, T., Lando, G. and Ngu, J.L. Serum antibodies against human brain myelin proteins in Gambian trypanosomiasis. Ann. Soc. Belg. Trop. Med. 1989; 69:213–221.

Authié, E., Muteti, D.K., Mbawa, Z., Lonsdale-Eccles, J.D., Webtser, P. and Wells, C.W. Identification of a 33-kilodalton immunodominant antigen of *Trypanosoma congolense* as a cysteine protease. Mol. Biochem. Parasitol. 1992; 56:103–116.

Bakhiet, M., Jansson, L., Büscher, P., Holmdahl, R., Kristensson, K. and Olsson, T. Control of parasitemia and survival during *Trypanosoma brucei brucei* infection is related to strain-dependent ability to produce IL-4. J. Immunol. 1996; 157:3518–3526.

Barett, M. Problems for the chemotherapy of human African trypanosomiasis. Curr. Opin. Infect. Dis. 2000; 13:647.

Beverley, S.M. Protozomics: trypanosomatid parasite comes of age. Nat. Rev. 2003; 4:11–19.

Bitonti, A.J., Dumony, A.J. and McCann, P.P. Characterization of *T. b.brucei* S-adenosyl methionine decarboxylase and its inhibition by berenil, pentamidine, methylglyoxal(bis guynylhydrazone). Biochem. J. 1986; 237:685–689.

Bitter, W., Gerrits, H., Kieft, R. and Borst, P. How *Trypanosoma brucei* may cope with diversity of transferrins in its mammalian hosts. Nature 1998; 391:499–502.

Black, S.J., Murray, M., Shapiro, S.Z., Kaminsky, R., Borowy, N.K., Musanga, R. and Otieno-Omondi, F. Analysis of Propionibacterium acnes-induced non-specific immunity to *Trypanosoma brucei* in mice. Parasite Immunol. 1989; 11:371–383.

Borst, P. and Ulbert, S. Control of VSG gene expression sites. Mol. Biochem. Parasitol. 2001; 114:17–27.

Bruijn, J.A., Oemar, B.S., Ehrich, J.H. and Fleuren, G.J. Immune complex formation in the kidney: recent observations in experimental trypanosomiasis. Ann. Soc. Belg. Trop. Med. 1988; 68:11–14.

Burri, C. and Keiser, J. Pharmacokinetic investigations in patients from northern Angola refractory to melarsoprol treatment. Trop. Med. Int. Health 2001; 6:412–420.

Carter, K.C., Henriquez, F.L., Campbell, S.A., Roberts, C.W., Nok, A.J., Mullen, A.B. and McFarlane, F. DNA vaccination against the parasite enzyme gamma glutaniylcysteinesynthetase confers protection against *Leishtnania donovani* infection. Vaccine 2007; (In press).

Chaudhuri, M., Ajayi, W., Temple, S. and Hill, G.C. Identification and partial purification of a stage specific 33 kDa mitochondrial protein as alternative oxidase of *T. b. brucei* blood stream trypomastigotes. J. Eucaryot. Microbiol. 1995; 42:467–475.

Courtin, D., Argiro, L., Jamonneau, V., N'dri, L., N'guessan, P., Abel, L., Dessein, A., Cot, M., Laveissière, C. and Garcia, A. Interest of tumor necrosis factor-alpha -308 G/A and interleukin-10 -592 C/A polymorphisms in human African trypanosomiasis. Infect. Genet. Evol. 2006; 6:123–129.

Courtioux, B., Bisser, S., M'belesso, P., Ngoungou, E., Girard, M., Nangouma, A., Josenando, T., Jauberteau-Marchan, M. and Bouteille, B. Dot-ELISA for a more reliable human African trypanosomiasis staging. J. Clin. Microbiol. 2005; 43:4789–4795.

Cross, G. Antigenic variation in trypanosomes. Proc. R. Soc. Lond. B 1978; 202:55–75.

Daulouède, S., Bouteille, B., Moynet, D.E., Baetselier, P., Courtois, P., Lemesre, J.L., Buguet, A., Cespuglio, R. and Vincendeau, P. Human macrophage tumor necrosis factor(TNF)-alpha production induced by *Trypanosomabrucei gambiense* and the role of TNF-alpha in parasite control. J. Infect. Dis. 2001; 183:988–991.

Denise, H. and Barret, M.P. Uptake and mode of action of drugs used against sleeping sickness. Biochem. Pharmacol. 2001; 61:1–5.

Devine, D.V., Falk, R.J. and Balber, A.E. Restriction of the alternative pathway of human complement by intact *Trypanosoma brucei* subsp. *gambiense*. Infect. Immun. 1986; 52:223–229.

Duleu, S., Vincendeau, P., Courtois, P., Semballa, S., Lagroye, I., Daulouède, S., Boucher, J.L., Wilson, K.T., Veyret, B. and Gobert, P.A. Mouse strain susceptibility to trypanosome infection: an arginase-dependent effect. J. Immunol. 2004; 172:6298–6303.

Dumas, M. and Bisser, S. Clinical aspects of human African trypanosomiasis. In: Progress in Human African Trypanosomiasis, Sleeping Sickness (M. Dumas, B. Bouteille and A. Buguet, Eds.), pp. 215–233. Paris: Springer Verlag, 1999.

Esievo, K.A.N., Jaye, A., Nok, A.J., Saror, D., Ruth, E., Eduvie, E. and Njoku, C.O. Electrophoresis of bovine erythrocyte sialic acids: existence of additional band in trypanotolerant *Ndama* cattle. J. Comp. Pathol. 1990; 102:357–361.

Fargeas, C., Wu, C.Y., Nakajima, T., Cox, D., Nutman, T. and Delespesse, G. Differential effect of transforming growth factor beta on the synthesis of Th1- and Th2-like lymphokines by human T lymphocytes. Eur. J. Immunol. 1992; 22:2173–2176.

Ferrante, A. and Allison, A.C. Alternative pathway activation of complement by African trypanosomes lacking a glycoprotein coat. Parasite Immunol. 1983; 5:491–498.

Gasbarre, L.C., Hug, K. and Louis, J.A. Murine T lymphocyte specificity for African trypanosomes. I. Induction of a T lymphocyte dependent proliferative response to *Trypanosoma brucei*. Clin. Exp. Immunol. 1980; 41:97–106.

Gobert, A., Daulouède, S., Lepoivre, M., Boucher, J.L., Cespuglio, R., Buguet, A., Bouteille, B., Veyret, B. and Vincendeau, P. L-arginine availability modulates local NO production and parasite killing in experimental trypanosomiasis. Infect. Immun. 2000; 68:4653–4657.

Hawking, F. Concentration of Bayer 205 (Germanin) in human blood and cerebrospinal fluid after treatment. Trans. R. Soc. Trop. Med. Hyg. 1940; 34:37–52.

Hunter, C.A., Jennings, F.W., Kennedy, P.G.E. and Murray, M. Astrocyte activation correlates with cytokine production in central nervous system of *Trypanosoma brucei brucei*-infected mice. Lab. Invest. 1992; 67:635–642.

ILRAD. Trypanosomiasis. International Laboratory for Research on Animal Diseases. Reports. Nairobi: ILRAD, 1994, pp. 21–29.

Jauberteau, M.O., Ben Younes-Chenoufi, A., Amevigbe, M., Bouteille, B., Dumas, M., Breton, J.C. and Baumann, N. Galactocerebrosides are antigens for immunoglobulins in sera of an experimental model of trypanosomiasis in sheep. J. Neurol. Sci. 1991; 101:82–86.

Keita, M., Vincendeau, P., Buguet, A., Cespuglio, R., Vallat, J.M., Dumas, M. and Bouteille, B. Inducible nitric oxide synthase and nitrotyrosine in the central nervous system of mice chronically infected with *Trypanosoma brucei brucei*. Exp. Parasitol. 2000; 95:19–27.

Khonde, N., Pepin, J., Niyonsenga, T., Milord, F. and De Wals, P. Epidemiological evidence for immunity following *Trypanosoma brucei gambiense* sleeping sickness. Trans. R. Soc. Trop. Med. Hyg. 1995; 89:607–611.

Lalmanach, G., Boulangé, A., Serveau, C., Lecaille, F., Scharfstein, J., Gauthier, F. and Authié, E. Congopain from *Trypanosoma congolense:* drug target and vaccine candidate. Biol. Chem. 2002; 383:739–749.

Lapeyssonnie, L. Seconde note sur un cas exceptionnel de trypanosomiase. Présence de parasites pendant 21 ans, sans signes cliniques appréciables, chez une femme traitée sans succès pendant 10 ans. Bull. Soc. Pathol. 1960; 53:20–32.

Lemesre, J.L., Noireau, F., Makoundou, M.I., Louembet, M.T. and Frézil, J.L. Apport des techniques sérologiques dans l'analyse du liquide céphalo-rachidien de patients congolais atteints de la maladie du sommeil. Bull. Soc. Pathol. 1988; 81:506–510.

Lucas, R., Magez, S., Songa, B., Darji, A., Hamers, R. and De Baetselier, P. A role for TNF during African trypanosomiasis: involvement in parasite control, immunosuppression and pathology. Res. Immunol. 1993; 144:370–375.

Macaskill, J.A., Holmes, P.H., Whitelaw, D.D., McConnel, I, Jennings, F.W. and Urquhart, G.M. Immunological clearance of 75Se-labelled *Trypanosoma brucei* in mice. II. Mechanisms in immune animals. Immunology 1980; 40:629–635.

Magez, S., Geuskens, M., Beschin, A., Del Favero, H., Verschueren, H., Lucas, R., Pays, E.D.E. and Baetselier, P. Specific uptake of tumor necrosis factor is involved in growth control of *Trypanosoma brucei*. J. Cell Biol. 1997; 137:715–727.

Magez, S., Radwanska, M., Beschin, A., Sekikawa, K. and Baetselier, P. Tumor necrosis factor alpha is a key mediator in regulation of experimental *Trypanosoma brucei* infections. Infect. Immun. 1999; 67:3128–3132.

Magez, S., Stijlemans, B., Baral, T. and De Baetselier, P. DNA from protozoan parasites *Babesia bovis*, *Trypanosoma cruzi*, and *T. brucei* is mitogenic for B lymphocytes and stimulates macrophage expression of interleukin-12, tumor necrosis factor-alpha, and nitric oxide. Microbes Infect. 2002; 4:999–1006.

Mamani-Matsuda, M., Rambert, J., Lejoly-Boisseau, H., Daulouède, S., Malvy, D., Thiolat, D., Coves, S., Courtois, P., Vincendeau, P. and Mossalayi, M.D. Quercetin induces theapoptosis of *Trypanosoma brucei gambiense* and decreases the pro-inflammatory response of human macrophages. Antimicrob. Agents Chemother. 2004; 48:924–929.

Mansfield, J.M. and Wallace, J.H. Suppression of cell-mediated immunity in experimental African trypanosomiasis. Infect. Immun. 1974; 10:335–339.

Maurice, J.N.S. and Pearce, A.M. (Eds.). Tropical Disease Research. "A Global Partnership". S. A. Atar Geneva: World Health Organization, 1987, p. 75.

Mkunza, F., Olaho, W.M. and Powell, C.N. Partial protection against natural trypanosomiasis after vaccination with a flagellar pocket antigen from *Trypanosoma brucei rhodesiense*. Vaccine 1995; 13:151–154.

Mnaimneh, S., Geffard, M., Veyret, B. and Vincendeau, P. Albumin nitrosylated by activated macrophages possesses antiparasitic effectsneutralized by anti-NO-acetylated-cysteine antibodies. J. Immunol. 1997; 158:308–314.

Molyneux, D.H., Pentreath, V. and Doua, F. African trypanosomiasis in man. In: Manson's Tropical Diseases (G.C. Cook, Ed.), 20th ed, pp. 1183–1196. London, UK: WB Saunders Company Ltd., 1996.

Muranjan, M., Wang, Q., Li, Y.L., Hamilton, E., Otieno-Omondi, F.P., Wang, J., Van Praagh, A., Grootenhus, J.G. and Black, S.J. The trypanocidal cape buffalo serum protein is xanthineoxidase. Infect. Immun. 1997; 65:3806–3814.

Murray, M. and Morrison, W.I. Non-specific induction of increased resistance in mice to *Trypanosoma congolense* and *T. brucei* by immunostimulants. Parasitology 1979; 79:349–366.

Nagoshi, Y.L., Alarcon, C.M. and Donelson, J.E. The putative promoter for a metacyclic VSG gene in African trypanosomes. Mol. Biochem. Parasitol. 1995; 72:33–45.

Nielsen, K.H. Complement in trypanosomiasis. In: Immunology and Pathogenesis of Trypanosomiasis (I. Tizard, Ed.), pp. 133–144. Boca Raton, FL: CRC Press, 1985.

Nok, A.J. and Balogun, E. A blood stream *Trypanosoma congolence* Sialidase could be involved in anaemia during experimental trypanosomiasis. J. Biochem. 2003; 133:725–730.

Nok, A.J., Esievo, K.A.N., Ibrahim, S., Ukoha, A.I. and Ikediobi, C.O. Phospholipase A_2 from *Trypanosoma congolense*: characterization and hematological properties. Cell Biochem. Funct. 1993; 11:125–130.

Nok, A.J. and Nock, I.H. Transferrin coupled azanthraquinone enhances the killing effect on trypanosomes. The role of lysosomal amannosidase. Parasite 2002; 9:375–379.

Nok, A.J., Nock, I.H. and Bonire, J.J. The cholesterol pathway of *Trypanosoma congolence* could be target for triphenyltinsalicylate and triphenylsiliconsalicylate inhibition. Appl. Organomet. Chem. 2003a; 17:17–22.

Nok, A.J., Nzelibe, H. and Yako, S.K. *Trypanosoma evansi* sialidase: surface localization, properties and hydrolysis of ghost red blood cells and brain cells – implications in trypanosomiasis. Z. Naturforsch. 2003b; 58:594–601.

Okomo-Assoumou, M.C., Daulouède, S., Lemesre, J.L., N'zila-Mouanda, A. and Vincendeau, P. Correlation of high serum levels of tumor necrosis factor—a with disease severity in human African trypanosomiasis. Am. J. Trop. Med. Hyg. 1995a; 53:539–543.

Okomo-Assoumou, M.C., Geffard, M., Daulouède, S., Chaugie, R.C., Lemesre, J.L. and Vincendeau, P. Circulating antibodies directed against tryptophan-like epitopes in sera of patients with human African trypanosomiasis. Am. J. Trop. Med. Hyg. 1995b; 52:461–467.

Okubanjo, O., Inuwa, H.M., Ajanusi, J.A., Ogwu, D., Sekoni, V. and Nok, A.J. Semen sialic acid surge and modulation of α-L-fucosidase activity: possible link to loss in reproductive capacity during trypanosomiasis. Cell Biochem. Funct. 2007; 24:1–4.

Olsson, T., Bakhiet, M., Edlund, C., Hojeberg, B., Van Der Meide, P.H. and Kristensson, K. Bidirectional activating signals between *Trypanosoma brucei* and CD8+ T cells: a trypanosome released factor triggers interferon-g production that stimulates parasite growth. Eur. J. Immunol. 1991; 21:2447–2454.

Pays, E. and Nolan, D.P. Expression and function of surface proteins in *Trypanosoma brucei*. Mol. Biochem. Parasitol. 1998; 91:3–36.

Raper, J., Nussenzweig, V. and Tomlinson, S. The main lytic factor of *Trypanosoma brucei brucei* in normal human serum is not high density lipoprotein. J. Exp. Med. 1996; 183:1023–1029.

Rasooly, R. and Balaban, N. Trypanosome microtubule-associated protein p15 as a vaccine for the prevention of African sleeping sickness. Vaccine 2004; 22:1007–1015.

Reinitz, D.M. and Mansfield, J.M. T-cell-independent and T-cell-dependent B-cell responses to exposed variant surface glycoprotein epitopes in trypanosome-infected mice. Infect. Immun. 1990; 58:2337–2342.

Rouzer, C.A. and Cerami, I.A. Hypertriglyceridemia associated with *Trypanosoma brucei brucei* infection in rabbits: role of defective triglyceride removal. Mol. Biochem. Parasitol. 1980; 2:31–38.

Russo, D.C., Williams, D.J. and Grab, D.J. Mechanisms for the elimination of potentially lytic complement-fixing variable surface glycoprotein antibody-complexes in *Trypanosoma brucei*. Parasitol. Res. 1994; 80:487–492.

Sacco, R.E., Hagen, M., Donelson, J.E. and Lynch, R.G. B lymphocytes of mice display an aberrant activation phenotype and are cell cycle arrested in G0/G1A during acute infection with *Trypanosoma brucei*. J. Immunol. 1994; 153:1714–1723.

Saida, K., Saida, T., Alvingl, G.R., Brown, M.J., Silbergerg, D.H. and Asbur, A.K. In vivo demyelinisation produced by purified antibodies to galactocerebroside. J. Neuropathol. Exp. Neurol. 1979; 38:338–348.

Sands, M., Kron, M.A. and Brown, R.B. Pentamidine: a review. Rev. Infect. Dis. 1985; 7:625–634.

Schleifer, K.W., Filutowicz, L.R., Schopf, L.R. and Mansfield, J.M. Characterization of T helper cell responses to the trypanosome variant surface glycoprotein. J. Immunol. 1993; 150:2910–2919.

Scott, J.M., Holmes, P.H., Jennings, F.W. and Urquhart, G.M. Attempted protection of zebu cattle against trypanosomiasis using a multi-stabilate vaccine. Res. Vet. Sci. 1978; 25:115–117.

Semballa, S., Geffard, M., Daulouède, S., Veyret, B., Lemesre, J.L., Holzmuler, P., Malvy, D. and Vincendeau, P. Antibodies directed against nitrosylated neoepitopes in sera of patients with human African trypanosomiasis. Trop. Med. Int. Health Med. 2004; 9:1104–1110.

Shoda, L.K., Kegerreis, K.A., Suarez, C.E., Roditi, I., Corral, R.S., Bertot, G.M., Norimine, J. and Brow, N.W.C. DNA from protozoan parasites *Babesia bovis*, *Trypanosoma cruzi*, and *T. brucei* is mitogenic for B lymphocytes and stimulates macrophage expression of interleukin-12, tumor necrosis factor alpha, and nitric oxide. Infect. Immun. 2001; 69:2162–2171.

Shuaibu, M.N., Kanbara, H., Yanagi, T., Ichinose, A., Ameh, D., Bonire, J.J. and Nok, A.J. In vitro trypanocidal activity of dibutyltin chloride and its fatty acid derivatives. Parasitol. Res. 2003; 91:5–11.

Shuaibu, M.N., Kanbara, H., Yanagi, T., Ichinose, T., Ameh, D., Bonire, J.J. and Nok, A.J. Effect of dibutyltin (IV) on the ultrastructure of African Trypanosomes. Parasitol. Res. 2004; 92:65–73.

Shugaba, A., Omage, J., Ibrahim, N.D.G., Nok, A.J., Ukoha, A.I., Saror, D. and Esievo, K.A.N. Biochemical differences (*O*-acetyl and glycolyl groups) in erythrocyte surface sialic acids of trypanotolerant Ndama and trypanosusceptible Zebu cattle. J. Comp. Pathol. 1994; 110:91–95.

Sutherland, I.A., Mounsey, A. and Holmes, P.H. Reduced accumulation of sometamidium by drug resistant *T. congolense*. Parasitology 1992; 104:461–467.

Tachado, D.S. and Schofield, L. Glycosylphosphatidylinositol toxin of *Trypanosoma brucei* regulates Il-1a and TNF-α expression in macrophages by protein tyrosine kinase mediated signal transduction. Biochem. Biophys. Res. Commun. 1994; 205:984–991.

Takayanagi, T., Kawaguchi, H., Yabu, Y., Itoh, M. and Yano, K. Inhibition of IgM antibody-mediated aggregation of *Trypanosoma gambiense* in the presence of complement. Experientia 1992; 48:1002–1006.

Tolson, D.I., Jardim, A., Schnur, L.F., Stebeck, C., Tuckey, C., Beecroft, R.P., The, H.S., Olafson, R.W. and Pearson, T.W. The kinetoplastid membrane protein 11 of *Leishmania donovani* and African trypanosomes is a potent stimulator of T-lymphocyte proliferation. Infect. Immun. 1994; 62:4893–4899.

Turner, C.M.R., Barry, J.D., Maudlin, I. and Vickerman, K. An estimate of the size of metacyclic variable antigen repertoire of *T. b. rhodesiense*. Parasitology 1988; 97:269–276.

Vaidya, T., Bakhiet, M., Hill, K.L., Olsson, T., Kristensson, K. and Donelson, J.E. The gene for a lymphocyte triggering factor from African trypanosomes. J. Exp. Med. 1997; 186:433–438.

Vanhamme, L., Berberof, M., Le Ray, D. and Pays, E. Stimuli of differentiation regulate RNA elongation in the transcription units for the major stage-specific antigens of *Trypanosoma brucei*. Nucleic Acids Res. 1995; 23:1862–1869.

Vanhamme, L. and Pays, E. The trypanosome lytic factor of human serum and the molecular basis of sleeping sickness. Int. J. Parasitol. 2004; 34:887–898.

Van Velthuysen, M.L., Mayen, A.E., Prins, F.A., De Heer, E., Bruijn, J.A. and Fleuren, G.J. Phagocytosis by glomerular endothelial cells in infection-related glomerulopathy. Nephrol. Dial. Transplant. 1994; 9:1077–1083.

Vickerman, K. Developmental cycles and biology of pathogenic trypanosomes. Br. Med. Bull. 1985; 41:105–114.

Vincendeau, P. and Daulouède, S. Macrophage cytostatic effect on *Trypanosoma musculi*involves an l-arginine-dependent mechanism. J. Immunol. 1991; 146:4338–4343.

Vincendeau, P., Daulouède, S., Veyret, B., Dardé, M.L., Bouteille, B. and Lemesre, J.L. Nitric oxide-mediated cytostatic activity on *Trypanosoma brucei gambiense* and *Trypanosoma brucei brucei*. Exp. Parasitol. 1992; 75:353–360.

Vincendeau, P., Gobert, A.P., Daulouède, S., Moynet, D. and Mossalayi, M.D. Arginases in parasitic diseases. Trends Parasitol. 2003; 19:9–12.

Whitelaw, D.D., Scott, J.M., Reid, H.W., Holmes, P.H., Jennings, F.W. and Urquhart, G.M. Immunosuppression in bovine trypanosomiasis: studies with louping-ill vaccine. Res. Vet. Sci. 1979; 26:102–107.

CHAPTER

63

Leishmaniasis

Lynn Soong

Departments of Microbiology and Immunology and Pathology,
Center for Biodefense and Emerging Infections Diseases,
Sealy Center for Vaccine Development,
University of Texas Medical Branch,
Galveston, TX, USA

OUTLINE

ABSTRACT

Leishmania parasites are transmitted by phlebotomine sand flies and can cause diverse clinical manifestations known as leishmaniasis. Drug resistance/toxicity and *Leishmania*/HIV coinfection call for a safe, effective vaccine. Recent research focuses on generating vaccines targeted on parasite- or sand fly-derived components. Defining the optimal combinations of antigens and adjuvants that elicit durable cellular immunity is critical for the control of leishmaniasis.

INTRODUCTION

Leishmaniasis contains a group of (mainly) zoonotic diseases transmitted to humans and animals by the bite of phlebotomine sand flies (Fig. 63.1A). Leishmaniasis is currently considered the third most important vector-borne parasitic disease after malaria and lymphatic filariasis. It was estimated recently that leishmaniasis was responsible for 1.8 million disability-adjusted life years and 59,000 deaths annually. Leishmaniasis is widespread in 22 countries in the New World and 66 nations in the Old World (but is not found in Southeast Asia). Human infections are found in 16 countries in Europe, including France, Italy, Greece, Malta, Spain, and Portugal (Fig. 63.2) (http://www.who.int/tdr/diseases/leish/default.htm).

Over the past 10–20 years, there has been a dramatic increase in the number of reported human leishmaniasis cases. This trend is partially due to improved diagnostic methods and reporting systems and due to changes in transmission epizoology, failure of vector and reservoir control campaigns, emergence of *Leishmania* as an opportunistic infection in HIV-infected people, increase in drug-resistant parasite strains, and the massive relocation of susceptible individuals to leishmaniasis-endemic areas because of war or economic pressures (Reithinger and Davies, 2002). For example, during August 2002–February 2004, the Department of Defense staff identified 522 parasitologically confirmed cases of cutaneous leishmaniasis (CL) in military personnel who were deployed to three countries (Afghanistan, Iraq, and Kuwait). Following isoenzyme electrophoresis of cultured parasites, *Leishmania major* was found to be the etiologic agent for all 176 cases for which species data were available (http://www.cdc.gov/ncidod/dpd/parasites/leishmania/default.htm).

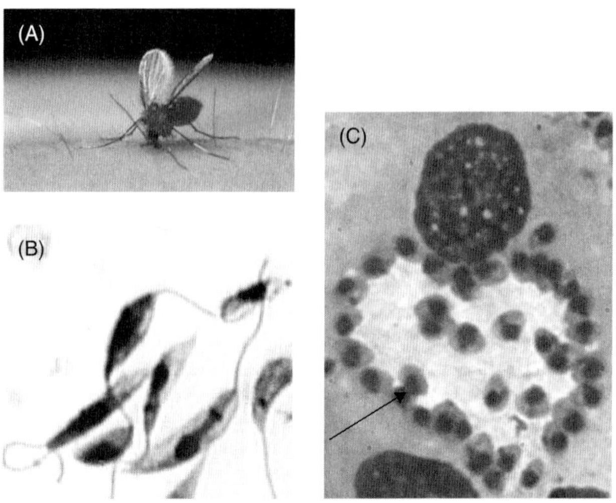

FIGURE 63.1 The vector and parasites that cause leishmaniaisis. (A) Phlebotomine sand flies transmit *Leishmania* parasites to the vertebrate hosts, while they take blood meals. (B) The mobile, elongated promastigotes replicate extracellularly in the midgut of the sand fly. Metacyclic promastigotes are injected together with sand fly saliva to the skin, where they enter target cells and establish infection in macrophages and dendritic cells. (C) Nonmobile amastigotes replicate within the phagolysosomal compartment of an infected macrophage. The arrow indicates amastigotes within the parasitophorous vacuole.

Given the public importance of *Leishmania* infection, extensive studies aimed at understanding the cellular and molecular basis of protective immunity and disease pathogenesis have been conducted over the past several decades. Despite the wealth of information gained from murine models of leishmaniasis, and much effort by the scientific community to develop an anti-leishmanial vaccine, there is yet no efficient vaccine for any of the forms of leishmaniasis. Current control measures rely on chemotherapy and vector control and the control of reservoir host populations. The chemotherapeutic agents used presently are inadequate, expensive, and often associated with side effects. This chapter will emphasize recent advances in vaccine development for the control of leishmaniasis and the particular difficulties encountered in generating efficient vaccines.

BIOLOGY OF *LEISHMANIA* PARASITES

Leishmania parasites are named after W.B. Leishman, who identified and generated one of the earliest strains of the *Leishmania* parasite in 1901. The genus *Leishmania* is taxonomically classified as belonging to the order of Kinetoplastida and the family Trypanosomatidae because of the presence of the kinetoplast-mitochondrion and other trypanosomal features. *Leishmania* parasites have a dimorphic life cycle characterized by a free-living promastigote form and an intracellular amastigote form. Metacyclic promastigotes (Fig. 63.1B) replicate extracellularly in the sand fly midgut and are transmitted to the mammalian host by the bite of infected female sand flies. Once inside the host, the promastigotes are phagocytosed by macrophages or dendritic cells (DCs), where they differentiate into nonmotile amastigotes and replicate within the amino acid-rich phagolysomal compartment (Fig. 63.1C) (McConville et al., 2007). The promastigote form can be maintained in vitro in cell-free cultures (at 23°C, pH 7.0), whereas the amastigote form can be isolated from mouse lesions or infected macrophages or, for some species, maintained in vitro by the differentiation of promastigotes using defined pH and temperature conditions (32°C, pH 5.0). The doubling time for *Leishmania* parasites in culture media is around 16 h. Given the technical difficulty in generating large quantities of amastigotes, the vast majority of immunological, biochemical, and vaccine studies have been conducted using the promastigote forms.

The genome of Old World *Leishmania* (*L. major* and *Leishmania donovani* groups) is spread over 36 chromosome pairs (0.28–2.8 Mb in size), yielding a total genome size of about 34 Mb. The genome sequence projects for *L. major* Friendlin (http://www.genedb.org/genedb/leish) and *Leishmania infantum* (http://www.genedb.org/genedb/linfantum) were completed in 2005, revealing 8272 protein-coding genes, only 36% of which can be ascribed a putative function (Ivens et al., 2005). Therefore, the function of the majority of the open-reading frames is still under investigation. In comparison, New World *Leishmania* species have only 34 or 35 chromosome pairs, with chromosomes 8 + 29 and 20 + 36 fused in the *Leishmania mexicana* group and 20 + 34 in the *Leishmania braziliensis* group, respectively (Stiles et al., 1999). Gene order and sequence are highly conserved among the ~30 *Leishmania* species. Little is known about the mechanism of transcription initiation and functional promoters in *Leishmania* and other related trypanosomatids (*Trypanosoma cruzi* and *Trypanosoma brucei*). Trypanosomatid mRNA processing is distinctive: Almost all mature mRNAs have a trans-spliced leader RNA added to their 5′-ends, and all protein-coding genes are transcribed polycistronically by RNA polymerase II in a divergent or convergent manner on DNA strands of the different chromosomes (Myler et al., 2000).

Developmental regulation of gene expression in *Leishmania* is mediated almost exclusively at the posttranscriptional level and often involves the 3′-untranslated region of the RNA, which determines RNA stability and translational efficiency (Rochette et al., 2005). In addition, infectivity of the parasites is also dependent on extensive posttranslational modifications (Ilgoutz and McConville, 2001). Often mRNA abundance does not reflect protein levels in *Leishmania*, and methodologies to globally analyze gene expression at the protein level are especially useful for this organism. Recently, global analyses

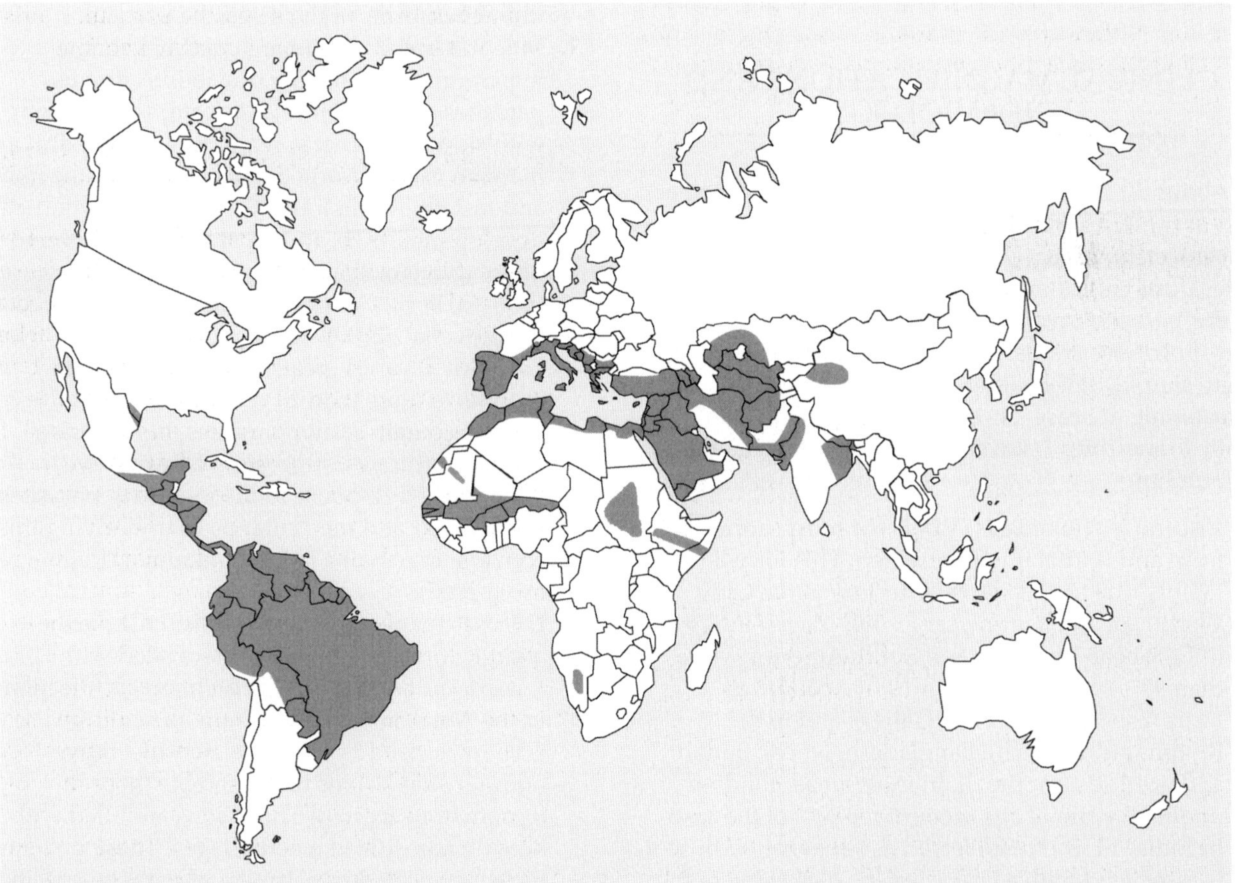

FIGURE 63.2 Map of geographic distribution of leishmaniasis. Major forms of leishmaniasis are endemic in 88 countries on four continents. According to the World Health Organization reports, more than 90% of cutaneous leishmaniasis cases occur in Iran, Afghanistan, Syria, Saudi Arabia, Brazil, and Peru, while more than 90% of visceral leishmaniasis cases occur in Bangladesh, Brazil, India, and Sudan.

of *Leishmania* gene expression indicate that the vast majority (>95%) of genes are constitutively expressed in all life stages, and among differentially expressed genes between amastigotes and promastigotes, most are species specific (Cohen-Freue et al., 2007). For example, among two dozen amastigote-specific genes identified each in *L. major* and *L. mexicana*, none are present in both species (Cohen-Freue et al., 2007). This limited amastigote-specific gene expression obviously posts additional challenges to identify the contribution of parasite molecules to diverse host immune responses, following infection with various *Leishmania* species. Although molecular technologies such as targeted gene disruption (see the vaccine section below) and RNA interference (Robinson and Beverley, 2003) have been successfully applied in the study of *Leishmania* biology and host–parasite interactions, they have not been widely used as in the study of bacteria and viruses. Because of its importance, *Leishmania* has been heavily studied and continues to be investigated in numerous areas, such as cell differentiation, host–parasite interactions, and new drugs and vaccine targets.

CLINICAL MAGNIFICATIONS AND EPIDEMIOLOGY

About 30 *Leishmania* species have been identified, and of these, approximately 20 species and subspecies can infect humans (Banuls et al., 2007). The diverse clinical magnifications are determined by the species of the parasite involved, as well as the host immune and nutrition status. Human leishmaniasis ranges from simple, self-healing skin ulcers to severe, life-threatening disease and can be classified into four main forms (http://www.who.int/tdr/diseases/leish/default.htm):

1. Visceral leishmaniasis (VL) is the most serious form and is fatal if left untreated. This form is mainly due to *L. donovani* infection in the Old World, and *L. infantum/Leishmania chagasi* infection in European countries and South America. More than 90% of VL cases occur in five countries—India (especially near the Ganges and in the Brahmaputra plains), Bangladesh, Nepal, Sudan, and northeastern Brazil. In most areas, children under 10 years of age account for 80% of the cases. In India, VL is often referred as kala-azar because it is associated with a characteristic darkening of the skin. VL is characterized by fever, splenomegaly, hepatomegaly, hypergammaglobulinemia, and pancytopenia, with a high mortality (up to 36%), especially in late-stage cases and individuals with severe malnutrition (Guerin et al., 2002). During the active stage of VL, patients display an antigen-specific suppression of cellular immune responses and nonresponsiveness to leishmanin and other antigen-stimulated skin tests.
2. CL is the most common form among the four clinical forms. While skin lesions in most individuals can heal spontaneously within a few months, they leave unsightly scars. More than 90% of CL cases occur in seven countries—Afghanistan, Iran, Syria, Algeria, Saudi Arabia, Brazil, and Peru. Parasites in the *Leishmania tropica* complex, including *L. major* and *Leishmania aethiopica*, are responsible for CL in the Old World, and lesions are mostly limited to the skin. Two *Leishmania* subgenera (*L. Leishmania mexicana* and *L. Viannia braziliensis*), with up to 11 species of parasites, are responsible for American (or New World) CL (Reithinger et al., 2007). While a majority of American CL patients can mount adequate cell-mediated immunity to parasites and heal the primary lesion(s) spontaneously, some progress into other forms of the disease.
3. Mucocutaneous leishmaniasis (MCL) is the hyperergic form, mainly due to infection by parasites in the *L. braziliensis* complex. It begins with skin ulcers that spread, causing dreadful and massive tissue destruction, especially of the nose and mouth. About 5% of CL cases will eventually develop into MCL, and most of these are found in Brazil (Silveira et al., 2004) and Andean countries (Bolivia, Peru, Colombia, Ecuador, and Venezuela) (Davies et al., 2000). The time of conversion from CL to MCL varies greatly, ranging from a few months to more than 20 years. The mucosal lesions usually contain scanty parasites, but massive cellular infiltrates, suggesting an excessive, uncontrolled response to the involving parasites. The cellular and molecular basis underlying this conversion remains largely unknown (Reithinger et al., 2007).
4. Diffuse cutaneous leishmaniasis (DCL) is the anergic form and is mainly associated with *L. mexicana* and *Leishmania amazonensis* infection in the Amazon basin. This rare form of the disease is characterized by multiple, non-ulcerative nodular skin lesions that resemble those of lepromatous leprosy. The lesions are filled with heavily parasitized macrophages. These patients have a profound impairment of parasite-specific responses and are often refractory to currently available treatment (Reithinger et al., 2007; Silveira et al., 2004).

Leishmania parasites have been recognized as an opportunistic pathogen in immunosuppressed individuals, including those infected with HIV-1. Leishmaniasis and AIDS overlap in several subtropical and tropical regions around the world, including the Mediterranean region, Europe, Brazil, India, and Africa (Desjeux and Alvar, 2003; Reithinger et al., 2007). Since *Leishmania* and HIV-1 can infect cells of myeloid or lymphoid origin (e.g., macrophages and DCs), probably through the binding of the same host receptor on the cell surface (e.g., DC-SIGN) (Garg et al., 2007), this coinfection is of special concern: (1) It changes the epidemiology of leishmaniasis (especially VL) because of an increased risk of human-to-human transmission of parasites, probably due to needle sharing (Molina et al., 2003); (2) it reduces accuracy of immune-based diagnosis because of a compounding immunosuppression associated with each infection (Deniau et al., 2003); (3) it reduces the efficiency of therapeutic treatment for leishmaniasis (Laguna, 2003) and poses a new challenge for those seeking to develop live, parasite-based vaccines.

A presumptive diagnosis of leishmaniasis can be made based on clinical signs, epidemiologic data, and laboratory tests; however, a definitive diagnosis requires demonstration of the parasite. Serological diagnosis for VL can be accomplished using a direct agglutination test, ELISA, or immunofluorescence. Testing with a commercially available immunochromatographic strip that uses recombinant leishmanial antigen K39 has proven highly specific and sensitive (98–100%) (Singh and Sivakumar, 2003). The leishmanin skin (Montenegro) test, which measures the cutaneous cellular responses to *Leishmania*-derived antigens, is still frequently used for clinical diagnosis and in epidemiological surveys on the prevalence of leishmanial infection. However, these tests fail to distinguish between current and past leishmanial infections. The detection of parasites by examination of smears from bone marrow, lymph nodes, or splenic aspirate, or detection of parasite kinetoplast DNA in biopsy materials, remains an essential part of any definitive diagnosis (Reithinger et al., 2007). In chronic cases of CL and MCL, a definitive diagnosis is sometimes difficult because of the paucity of parasites in lesions.

TREATMENT AND PREVENTION

Most cases of CL heal without treatment, leaving the person immune to further infection. Other forms of leishmaniasis are extremely difficult to treat, often requiring a long course (20- to 28-day regimen) of intravenous (i.v.) pentavalent antimony drugs, such as meglumine antimonate (e.g., Glucantime) or sodium stibogluconate (e.g., Pentostam) (Vanlerberghe et al., 2007). Recently, resistance to these drugs has been reported, requiring the use of more toxic drugs, such as amphotericin B. Most available drugs are costly, require long treatment regimens, and are becoming more and more ineffective, necessitating the discovery of new drugs. Several new drugs, including Miltefosine and Sitamaquine (both of which can be given orally), are in the pipeline; however, their efficacy needs to be evaluated further (Olliaro et al., 2005; Vanlerberghe et al., 2007). Leishmaniasis, as a complex of diseases, is as yet impossible to control with a single approach or tool, although a vaccine has potential to be a cost-effective exception. In cases in which the reservoir is infected humans, control depends on early case detection and drug treatment. Vector and reservoir host control measures are expensive, requiring good infrastructure and maintenance—often giving results that are short lived. Vector control, based on spraying with residual insecticides, can be effective where transmission occurs in and around the home. In areas known to be the site of transmission, individuals should use some form of protection, such as insect repellent or insecticide.

REGULATION OF HOST PROTECTIVE IMMUNITY AND PATHOGENESIS

Although the clinical manifestations of leishmaniasis are determined by the species of *Leishmania* causing the infection and the host immune responses, the results of most studies indicate that the outcome of *Leishmania* infection seems dependent on regulation of the balance between the parasite and reactive T cell populations. The murine models of *L. major* infection have been extremely helpful for understanding the genetic mechanisms of protection conferred by T cells (von Stebut, 2007). Most inbred mouse strains (e.g., C57BL/6, C3H, CBA/J, and B10D2) are genetically resistant to *L. major* infection, mainly due to the production of high levels of IL-12, IFN-γ, TNF-α, IL-1, and Th1-promoting chemokines (e.g., IP-10/CXCL-10) at early stages of the infection. In vitro studies have confirmed that these cytokines promote macrophage effector activities by stimulating the expression of inducible nitric oxide (NO) synthase (iNOS) and the production NO and, consequently, the killing of intracellular parasites (Fig. 63.3). Suppression/alteration of these Th1 responses by the targeted depletion of cytokine genes or their corresponding receptors,

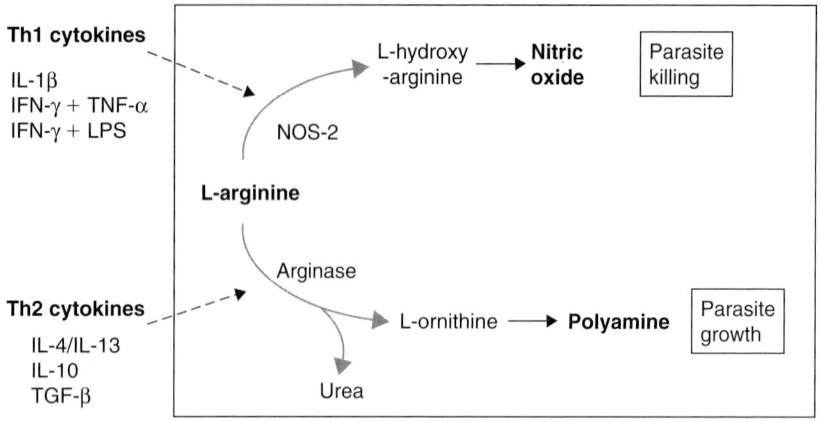

FIGURE 63.3 The arginine metabolism pathways in macrophage determine the fate of intracellular *Leishmania* parasites. Th1 cytokines preferentially stimulate the inducible nitric oxide synthase (iNOS), and the resulting nitric oxide (NO) mediates the killing of intracellular amastigotes. Th2 cytokines preferentially activate arginase, and the resulting polyamines are essential components for replication of amastigotes because parasites are incapable of making their own polyamines.

administration of neutralizing antibodies to block the action of these cytokines, or adoptive transfer of $CD4^+CD25^+$ regulatory T cells (Treg, derived from infected skin tissues or spleen of naive mice) can render these otherwise resistant mice susceptible to the infection (Belkaid et al., 2002).

In contrast, BALB/c mice are highly susceptible to *L. major* infection, developing extensive skin lesions and metastasized infections in the liver, spleen, and bone marrow, mainly due to the uncontrolled production of IL-4, IL-10, and TGF-β. At the cellular level, these Th2-type cytokines suppress macrophage activation by stimulating the arginase-expression-polyamine-production pathway and, consequently, the growth of intracellular parasites (Fig. 63.3). Suppression/alteration of these Th2 responses by targeted gene depletion, injection of neutralizing antibodies to block the action of these cytokines, or adoptive transfer of Treg cells can render these otherwise susceptible mice resistant to infection (Belkaid et al., 2002; Ji et al., 2005). Thus, a fine regulation/balance among Th1/Th2 effector T cells and Treg cells at the local site determines the outcome of the infection.

However, this Th1/Th2 paradigm learned from murine *L. major* infection cannot fully explain healing/nonhealing infection models associated with other *Leishmania* species (Alexander and Bryson, 2005; McMahon-Pratt and Alexander, 2004). First, mice infected with parasite species responsible for different forms of CL in South America (e.g., *L. amazonensis, L. mexicana, L. braziliensis*), or species responsible for VL (e.g., *L. donovani, L. chagasi*), do not show a clear Th1 or Th2 dominancy. Likewise, leishmaniasis patients usually mount a Th1/Th2-mixed response. Second, a given cytokine can display diverse functions, depending on the time and magnitude of its production, as well as the presence of other cytokines/chemokines in the microenvironment. For example, while Th2 responses can be induced in the absence of IL-4, under certain circumstances, IL-4 can prime for IL-12 production and a type-1 response. On the other hand, while IFN-γ and TNF-α synergistically activate macrophage's leishmanicidal activity, IFN-γ alone can actually enhance the growth of parasites within macrophages (Qi et al., 2004) by promoting the arginase pathway with no activation of the iNOS pathway (Wanasen et al., 2007).

Regardless of the mechanism underlying the pathogenesis associated with different models of *Leishmania* infection, what we have learned from animal studies is critical for designing an efficient vaccine for the control of leishmaniasis. First, since Th1 responses favor resistance to leishmaniasis, antigens that preferentially stimulate Th1 cells may not necessarily be good vaccine candidates. Second, immunization with antigens that preferentially stimulate Th2 cells would counterbalance Th2 responses and provide protection in mice (see LACK antigen below) (Campos-Neto, 2005). Recently, Zaph et al. have demonstrated distinct requirements for generating and maintaining different subpopulations of $CD4^+$ memory T cells after infection with *L. major* in mice: short-lived, pathogen-dependent effector cells and long-lived, pathogen-independent central memory cells. Therefore, central memory T cells should be the targets for non-live vaccines against infectious diseases requiring cell-mediated immunity (Scott et al., 2004; Zaph et al., 2004).

VACCINE DEVELOPMENT FOR THE CONTROL OF CL AND VL: TARGETING PARASITE-SPECIFIC COMPONENTS

In endemic areas, a significant number of individuals acquire subclinical infections associated with the development of parasite-specific T cell responses and resistance to leishmaniasis. The acquisition of protective immunity after resolved CL suggests that vaccination, if it induces an appropriate cellular immune response, is a feasible approach toward the control of leishmaniasis. Vaccination against leishmaniasis has passed through many developmental stages, beginning with the ancient practice of "leishmanization" (use of live parasites) as far back as in 1911, during the Iran–Iraq war. However, this practice has been largely discontinued due to the inherent danger of developing disease in a more exaggerated form following immunization with live parasites. Rather, the first-generation vaccines include killed or attenuated parasites whose virulence has been reduced by long-term culture, γ-irradiation, chemical mutagenesis, or gene inactivation/replacement. While injection with a small dose of attenuated organisms is sometimes more efficient in stimulating a protective Th1 response than in natural infection, the spread of HIV infections and use of immunosuppressive drugs in organ transplantation or cancer chemotherapy pose some concerns for the potential of sudden outburst of otherwise latent *Leishmania* infection in these immune-compromised individuals.

During the past few years, significant efforts have been directed toward developing second-generation vaccines that are composed of defined antigenic components and designed to induce potent cellular immunity against *Leishmania* infections (Coler and Reed, 2005; Kubar and Fragaki, 2005; Sukumaran and Madhubala, 2004). Since there are no suitable murine models for the study of MCL caused by *L. braziliensis* infection, or DCL caused by *L. amazonensis* infection, the majority of vaccine studies have been centered on CL caused by *L. major* infection or VL caused by *L. donovani* infection.

Vaccine Studies for CL in Mouse Models

Several approaches have recently been taken to identify vaccine candidates for the control of CL. The first one focused on live virulent *L. major* with a targeted deletion of the dihydrofolate-thymidylate synthase gene locus (*DHFR-TS*). Since these knockout parasites can invade target cells but are deficient for replication within infected macrophages, they are much safer than live parasite vaccines that are generated via long-term in vitro cultivation or chemical mutagenesis (Titus et al., 1995). Targeted gene deletion has been used to generate other mutant lines. The surface of the promastigotes is coated with a variety of interrelated glycoconjugates, including lipophosphoglycan (LPG) and glycosylphosphatidylinositol (GPI)-anchored proteins, such as gp63 and gp46 (Naderer et al., 2004). In the sand fly, stage-specific modifications of LPG are responsible for parasite attachment to, and release from, the sand fly midgut. In the mammalian host, LPG is considered a virulence factor, enhancing the parasite's resistance to complement-mediated lysis and oxidative stress but suppressing the macrophage's effector and antigen-presentation functions (Olivier et al., 2005). The Beverley and Turco groups have generated an *L. major* *lpg1*$^-$ mutant by deleting a gene involved in the biosynthesis of the LPG core glycan and found that this attenuated parasite still induced disease in mice (Spath et al., 2000). However, *L. major lpg2*$^-$ mutants, which are deficient in assembly of disaccharide–phosphate repeats of LPG and other phosphoglycan-containing molecules (Spath et al., 2003), are highly attenuated parasites. These *lpg2*$^-$ mutants persist in mice without causing any overt cutaneous lesions and provide protection in susceptible BALB/c mice against challenge with wild-type *L. major* (Uzonna et al., 2004). While deleting the genes involved in the synthesis of LPG results in attenuated *L. major* parasites, similar manipulation does not yield attenuated *L. mexicana* strains (Turco et al., 2001). Mice infected with parasites in the *L. mexicana* group do not show classical Th1/Th2 polarization, as seen in *L. major* infection (Ji et al., 2003), and evidence from biochemical and in vivo studies collectively indicates discernible differences in the biology of the parasites and their associated disease pathogenesis (Xin et al., 2007) and the need for additional vaccine studies for New World CL.

When live parasites are used as candidate vaccines for control of *L. major* infection, the route of vaccination can greatly influence the type, strength, and duration of the memory responses. In the C57BL/6 mouse model (which more closely resembles immune responses seen in *L. major*-infected humans than the BALB/c model), subcutaneous (s.c.) immunization with live parasites in the footpad provides superior protection to that seen with intradermal (i.d.) immunization in the ear (Tabbara et al., 2005). The high level of protection achieved by footpad vaccination was correlated with efficient expansion and recruitment of IFN-γ-producing CD4$^+$ T cells, but delayed recruitment of IL-10-producing Treg cells, to the infection site. This study indicates that a delicate balance between effector T cells and Treg cells at the site of injection can influence the efficacy of a live leishmanial vaccine.

Based on the Th1/Th2 paradigm defined in murine models of *L. major* infection, leishmanial antigens that preferentially activate type-1 responses (Th1 cells and $CD8^+$ cytotoxic T lymphocytes) are referred to as "protective antigens" and actively investigated as vaccine candidates, whereas antigens that fail to stimulate Th1 cells and/or preferentially activate Th2 responses are not considered favorably as vaccine candidates. The antigens investigated include subcellular (soluble or membrane-enriched) fractions, purified antigens, poly-subunits, mosaic subunits, as well as cDNA expression libraries that express leishmanial proteins. These antigens (or DNA) are usually co-injected with a Th1-promoting adjuvant (rIL-12 or BCG) or constructed into viral vectors (e.g., vaccinia viruses or adenovirus) for prolonged antigen release (Gabaglia et al., 2004).

Among the 15–20 defined *Leishmania* vaccine candidates that have been investigated in detail (Coler and Reed, 2005; Kubar and Fragaki, 2005; Sukumaran and Madhubala, 2004), the 36-kDa LACK antigen (*Leishmania* analogue of the receptors of activated C kinase) has attracted the greatest attention, due to its high conservation among *Leishmania* species and presence in both developmental stages of the parasites. The LACK antigen was initially identified via the use of a T cell expression cloning strategy (Mougneau et al., 1995). The rapid production of IL-4 by LACK-reactive Vβ4Vα8 $CD4^+$ T cells is known to be responsible for Th2 development and susceptibility of BALB/c mice to *L. major* infection (Launois et al., 1997). Mice made tolerant to LACK by transgenic expression of the antigen in the thymus exhibit both a diminished Th2 response and a healing phenotype (Julia et al., 1996). Vaccines composed of the LACK antigen mixed with an adjuvant or DNA encoding LACK antigen provided satisfactory protection in mice against *L. major* infection; however, these types of vaccines fail to show protection in New World CL caused by *L. mexicana* infection or VL caused by *L. donovani* infection (Dumonteil et al., 2003; Melby et al., 2001). Interestingly, mice intradermally primed with DNA encoding the LACK antigen and then intraperitoneally boosted with vaccinia viruses expressing the LACK antigen developed protective immunity against *L. infantum* infection (Dondji et al., 2005). Therefore, the heterologous prime-boost vaccination strategy, which involves the sequential administration of vaccines that contain different antigens, appears to be a promising approach toward the development and maintenance of a strong cellular immune response for the control of leishmaniasis. To date, there have been relatively limited vaccine studies conducted for New World CL caused by *L. amazonensis* (Armijos et al., 2004; Hernandez et al., 2004) and *L. mexicana* (Dumonteil et al., 2003; Rosado-Vallado et al., 2005).

Vaccine Studies for VL in Mouse and Hamster Models

In nature, VL in humans and animals is initiated via a parasite infection in the skin, and yet there is limited success in establishing VL following s.c./i.d. injection of BALB/c mice with *L. donovani* or *L. infantum/L. chagasi*, either alone or mixed with sand fly salivary gland extracts (Ahmed et al., 2003). Therefore, most murine VL studies are initiated by intravenous (i.v.) injection of several millions of parasites; however, such BALB/c mice develop foci of splenic and hepatic lesions, which self-heal several months later. In comparison, the Syrian golden hamster (*Mesocricetus auratus*) is considered a better small animal model that resembles some clinical features of VL (Melby et al., 2001); however, dogs are superior to rodents for vaccine studies.

In comparison to CL, much less is known regarding experimental vaccines for VL. It has been shown that s.c. inoculation of high doses (10^7) of live virulent *L. chagasi* causes a subclinical infection that protects against a subsequent i.v. challenge with *L. chagasi*; however, a similar immunization with attenuated *L. chagasi*, *L. donovani*, or *L. major* parasites that are deficient in the *DHFR-TS* gene fails to provide protection (Streit et al., 2001), suggesting the requirement for a subclinical infection for eliciting protective immunity. It has been found that i.v. immunization with 5×10^7 *L. donovani* deficient in the gene for biopterin transport (*BT1*) provides partial protection against i.v. challenge with virulent parasites (Papadopoulou et al., 2002), and it appears that, besides the dose and route of immunization, the in vivo replication of parasites is an essential factor to be considered for a VL vaccine. Given that alum-precipitated, autoclaved *L. donovani* mixed with *Macrobacterium bovis* Bacille Calmette–Guerin (BCG), but not *L. donovani*-derived LPG mixed with BCG, provides protective immunity in mice and hamsters against VL (Srivastava et al., 2003; Tonui et al., 2003), it appears that complex antigenic preparations (and antigenic epitopes) are required for adequate activation of multiple defense mechanisms for the control of VL.

In the search for novel, second-generation vaccine candidates for VL, two approaches have been taken: a systematic search via immunization of mice with fractions of an *L. donovani* cDNA expression library (Melby et al., 2000) and a targeted search via a screening of the expression library with $CD4^+$ T cell lines generated from human subjects (Arora et al., 2005). Among several identified vaccine candidates, the fucose–mannose ligand (FML) antigen complex of *L. donovani* holds high promise. When the FML protein was tested in combination with different adjuvants, the FML mixed with Quil A or QS21 saponin provided better protection in terms of reducing parasite load in the liver and stimulating delayed-type hypersensitivity (DTH) and IgG2a responses than did the FML mixed with BCG or IL-12 (Santos et al., 2002). The main component in the FML complex is a 36-kDa nucleoside hydrolase (NH36). When BALB/c mice were intramuscularly (i.m.) immunized with different regimens (purified FML or recombinant NH36 mixed with saponin as an adjuvant, or with NH36 expressed in the VR1012 DNA vaccine vector), the best protection was observed in mice immunized with the NH36 DNA (Aguilar-Be et al., 2005). This study indicates a higher efficacy of DNA-based vaccines over that of protein-based ones. In addition, a DNA-based vaccine study involving the kinetoplastid membrane protein-11 (KMP-11), a highly conserved surface membrane protein present in all members of the family Kinetoplastidae, gave complete protection against challenge with both pentavalent antimonial-sensitive and -resistant strains of *L. donovani* in the highly susceptible hamster model (Basu et al., 2005).

Dendritic cells (DC)-based Vaccine Studies in Mice

Examination of the utility of DC vaccination as cellular immunotherapy for cancers has yielded some encouraging results and prompted similar studies in animal models of leishmaniasis. Although DC-based vaccination may lack a conspicuous application in the context of such infections predominately in developing countries, it serves as a valuable tool to define DC-targeted control methods. Using *L. major* infection as a model, Moll and colleagues have provided solid evidence that a single i.v. injection of ex vivo antigen-pulsed Langerhans cells (LCs) (Berberich et al., 2003), myeloid DCs (mDCs) (Ramirez-Pineda et al., 2004), or plasmacytoid DCs (pDCs) (Remer et al., 2007) (5×10^4–5×10^5 cells per mouse) is sufficient to induce protective immunity against a single or repeated parasite challenges in susceptible BALB/c mice. Interestingly, antigen-pulsed LCs and pDCs can mediate protection independent of additional DC stimuli (CpG, TNF-α, LPS, or CD40 ligation); however, an overnight exposure to CpG motifs is required for antigen-loaded mDCs to induce specific Th1 responses (Ramirez-Pineda et al., 2004), a phenomenon also observed for mDCs pulsed with the *L. infantum* antigen (Carrion et al., 2007). In addition, while a polarized Th1 cytokine profile is linked to protection in LC- and CpG/mDC-immunized mice, no strong correlation (or specific roles for IL-12 and IFN-α) is observed for pDC-immunized mice (Remer et al., 2007). These findings reemphasize the functional diversity of DC subsets in the control of *L. major* infection and the importance of mounting adequate local immune responses for the control of the infection (Fajardo-Moser et al., 2008). However, direct lesional injection of *L. amazonensis* antigen-pulsed mDCs (10^6 cells per mouse), together with recombinant IL-12, failed to promote healing in mice chronically infected with *L. amazonensis* (Vanloubbeeck et al., 2004). Therefore, simply providing in vitro-activated DCs or primed Th1 cells is not sufficient to ensure healing in *L. amazonensis*-infected hosts, and additional treatments aimed at reducing intracellular parasite loads and/or blocking parasite's enzymatic activities are needed in order to overcome the parasites in their persistent state. Given the unique features for DC–amastigote interaction, DC-based vaccination studies should be extended to hosts with established infection. Targeted delivery of leishmanial antigens to endogenous DCs (Zubairi et al., 2004) and the use of genetically engineered parasites that express the DC activator (Field et al., 2007) are other attractive approaches.

Vaccine Studies for Canine VL

Foxes, wild canines and dogs, and domestic dogs are important reservoirs for maintaining the population of *Leishmania* parasites that can lead to VL in humans and considered major targets in vaccine control of VL (Lemesre et al., 2007; Reithinger and Davies, 2002). Serological studies have estimated that 2.5 million dogs (16.7%, from 15 million dogs tested) in European countries are infected with *Leishmania* parasites, and higher (33–36%) seroprevalences have been reported in South American countries. Dogs 8 years old and above usually show the highest seroprevalence (up to 40–50%). In 2002, it was reported that the *Leishmania* parasite has infected more than 1000 hunting dogs in 21 U.S. states and the Canadian Province of Ontario (Enserink, 2000); however, the basis for this infection is unknown. Given that only a quarter of the seropositive domestic dogs develop clinical signs, eliminating infected dogs may not be an accepted option for dog owners. Thus, developing vaccines for canine leishmaniasis is a viable alternative for reducing the number of parasites in infected dogs and, subsequently, transmission of the parasites.

To this end, a few phase I/II vaccine trials have been performed in dogs, with killed *Leishmania* promastigotes, purified leishmanial fractions, or recombinant DNA (Giunchetti et al., 2007; Kullberg et al., 2002). For example, an FML-enriched fraction of *L. donovani* was used in a phase III vaccine trial against symptomatic canine VL, with a nearly 80% clinical efficacy. The experimental vaccine conferred about 90% protection from disease progression when used for the immunotherapy of asymptomatic dogs (Borja-Cabrera et al., 2002, 2004). At present, Leishmune® vaccine, which is composed of the FML antigen isolated from the aqueous extract of stationary-phase promastigotes of *L. donovani* (the Sudan strain), is the first licensed vaccine against canine VL and scaled up for commercialization in Brazil (Patent/INPI number: PI1100173-9) (Nogueira et al., 2005). The potential of Leishmune in the interruption of the *natural transmission* of the disease was assessed in vaccinated ($n = 32$) and control ($n = 40$) dogs in a Brazilian epidemic area. The animals were injected on the flank with three s.c. doses of Leishmune with a 21-day interval between injections. Eleven months after vaccination, the vaccinated dogs showed a 100% seropositivity to the FML and a complete absence of clinical signs and parasites in skin, lymph node, and blood for leishmanial DNA, indicating the Leishmune-vaccinated dogs were non-infectious (Nogueira et al., 2005). In the Meshkin-Shahr district, Iran, a double-blind randomized efficacy field trial was conducted with aluminum hydroxide (alum)-precipitated *L. major* (alum-ALM) vaccine plus BCG against canine VL. The vaccine group ($n = 182$) and control dogs ($n = 165$) were injected intradermally with a single dose of vaccine (200 kg of protein in 0.1 ml) or saline mixed with BCG. During a 16-month follow-up period, the incidence of *Leishmania* infection was 3.7% and 12% in the vaccinated and control groups, respectively, and the efficacy of the vaccine was 69.3% (Mohebali et al., 2004).

In contrast, a phase III field trial of a tandemly linked, multi-subunit recombinant leishmanial polyprotein (also known as Leish-111f) formulated in monophosphoryl lipid A (MPL®) adjuvant was conducted in a VL endemic area of southern Italy. After three monthly injections before the transmission season, immunized dogs showed no protection against *L. infantum* infection and disease progression, suggesting a limited utility of this vaccine to prevent *Leishmania* infection in dogs under field conditions (Gradoni et al., 2005). While variations in vaccine efficacy observed in different field trials emphasize the need to optimize the vaccine formula for different endemic regions, the development of such vaccines to reduce parasite loads in domestic dogs is highly advantageous. Other control strategies, including insecticide-impregnated dog collars and immunizing dogs with a sand fly saliva-based vaccine, would also be useful in reducing the canine and human disease incidence.

Vaccine Trials in Human Volunteers

An ideal vaccine for human leishmaniasis should be safe, free of animal products, require a minimal number of immunizations, and provide long-term protective immunity. An effective vaccine against complex diseases, such as leishmaniasis, would also require a multivalent cocktail of several antigens containing a broad range of protective epitopes that would cover a wide range of MHC types in a heterogeneous population. Vaccines made of killed whole parasites can meet some of these requirements but have shown only limited protection in studies in Rhesus monkeys (Amaral et al., 2002) and humans (Velez et al., 2005). A recent phase III trial with three i.m. injections of merthiolate-killed *L. amazonensis* (Leishvacin®) in 2597 healthy volunteers (Columbian soldiers) confirmed that the vaccine was safe but offered no protection against CL caused by *Leishmania panamensis*, according to findings during a 1-year follow-up survey in the field setting (Velez et al., 2005). Due to issues related to the use of live parasites and killed whole parasites, more attention has been given to second-generation vaccines against leishmaniasis as well as suitable adjuvants for human use.

Recently, scientists in the United States have generated Leish-111f, formulated in MPL adjuvant, referred to as Leish-111f–MPL-SE (Coler et al., 2002; Coler and Reed, 2005; Skeiky et al., 2002). This single recombinant polyprotein (111 kDa) is composed of three immunodominant leishmanial antigens (TSA, LmSTI1, and LeIF), previously identified in the context of host response to infection in infected donors and BALB/c mice, as well as their ability to elicit partial protection against *L. major* infection in BALB/c mice. MPL, a detoxified form of lipid A derived from the LPS of *Salmonella minnesota* R595, was used as an immunostimulatory adjuvant because of its ongoing clinical trials for a range of vaccines, including malaria, hepatitis B, genital herpes, allergy desensitization, and human papillomavirus (Coler and Reed, 2005). The Leish-111f–MPL-SE was shown to protect BALB/c mice against *L. major* infection to a degree equal or superior to those seen with any of the individual components of the vaccine construct or SLA, a soluble *Leishmania* lysate (Coler et al., 2002; Skeiky et al., 2002). These studies demonstrate the feasibility of manufacturing a single recombinant vaccine comprising multiple, protective antigens and the potential use of MPL-SE as a substitute for IL-12. Thus, Leish-111f is the first anti-leishmanial human vaccine to enter phase I, double-blind, dose–response trials in healthy volunteers (Coler et al., 2002; Coler and Reed, 2005).

VACCINES FOR BLOCKING THE TRANSMISSION CYCLE OF LEISHMANIASIS: TARGETING SAND FLY-SPECIFIC COMPONENTS

About 30 different species of sand flies can carry and transmit *Leishmania* parasites. Old World *Leishmania* parasites are transmitted by sand flies of the genus *Phlebotomus*, while New World ones are transmitted by flies of the genus *Lutzomyia*. These tiny, sand-colored, blood-feeding flies (Fig. 63.1A) breed in forested areas, caves, or the burrows of small rodents. Sand flies become infected by ingesting amastigotes while taking blood from infected reservoir hosts or humans. When an infected sand fly takes blood meals, it injects a few hundred metacyclic promastigotes together with some of its saliva. The natural infection rates of sand flies with *Leishmania* parasites vary in different field studies but are generally low (0.05–2%). Thus, animals and humans in endemic areas are more often bitten by uninfected sand flies and, therefore, mount anti-saliva antibodies, as well as a cell-mediated, delayed-type hypersensitivity reaction at the bite site (Belkaid et al., 2000).

Sand fly saliva and its vasodilator peptide, maxadilan, have multiple immunomodulatory properties, in favor of establishing *Leishmania* infection at the initial stage of infection and exacerbating the infection in several murine models of CL (Andrade et al., 2007). Interestingly, mice can also be protected against *L. major* infection when pre-exposed to noninfected sand flies or pre-immunized with sand fly salivary gland extracts (SGE) or one of its components, such as *Lutzomyia longipalpis* maxadilan or a *Phlebotomus papatasi* SP-15 protein (Valenzuela et al., 2004). These results indicate that, at least in mice, a vaccine targeting sand fly-specific components would be efficient in changing local responses in the skin whenever a sand fly bites, making the microenvironment more difficult for the parasite to colonize. This type of vaccine has an advantage over vaccines targeting parasite-specific components because of repeated boosts during sand fly bites in nature. If proven effective in human trials, it would be a useful component of the strategy for the control of leishmaniasis and other vector-transmitted diseases as well. A major challenge for this type of vaccine is to define protective antigens that are shared by several sand fly species and the optimal immunization regimen for infection with *Leishmania* species other than *L. major* (Thiakaki et al., 2005).

MAJOR CHALLENGES IN VACCINE DEVELOPMENT FOR LEISHMANIASIS

In many respects, *Leishmania* parasites and *Mycobacterium tuberculosis* share some common features in their biology and relationship with mammalian hosts: slow growth in vitro, multiplying in an acidic environment within the phagolysosome, actively suppressing the macrophage's effector functions, causing chronic and persistent infection, and requiring IFN-γ- and NO-induced Th1 responses for the control of the disease. In comparison to other pathogens, *Leishmania* parasites have several unique features that can be utilized for vaccine development: (1) Unlike influenza viruses or *T. brucei*, which are notorious for antigenic switches or variations, leishmanial proteins have limited antigenic variation; (2) unlike pathogenic bacteria, *Leishmania* parasites do not secrete toxin for direct cytolysis; (3) there is limited cross-reactivity between leishmanial and host proteins (i.e., repeated injections of dead parasites or parasite antigens do not result in leishmaniasis); (4) a battery of mouse models has been developed for the study of CL (and to a lesser extent VL), and a hamster model is increasingly being used in vaccine studies for VL; (5) in most cases, parasites freshly isolated from leishmaniasis patients can be used to directly infect animals (mouse, hamster, dog) for the study of protective immunity or pathogenesis. Therefore, with the advent of effective culture conditions for the various stages of the life cycles, together with a wealth of information on the genome of *L. major* and protective immunity to this pathogen, it is tempting to speculate that *L. major* would be the first efficient vaccine for control of parasitic diseases. Why do we still not have a vaccine for CL?

KEY ISSUES

Developing an efficient vaccine for the control of leishmaniasis is not as easy as it is in the case of bacterial or viral infections (Modabber et al., 2007). The major challenges and critical issues include:

- *Leishmania* parasites have a very large diploid genome and complex, poorly understood regulation of stage-specific proteins, and molecular tools for genetic manipulation to delete or add a gene are not as widely used as they are for other microorganisms.

- The transmission cycle involves insects and small animal reservoirs, and there are complex interactions between the parasite and its sand fly vector and between the parasite and its mammalian hosts.
- Diverse parasite species are responsible for a broad spectrum of clinical manifestations—one parasite species can result in different clinical forms, and one clinical form can be caused by different parasite species.
- Disease pathogenesis is due to inadequate immune responses to parasite infection.
- Impairment in protective Th1 responses would be IL-4- and IL-10-dependent or -independent.
- Most reported vaccine studies reveal short-lived (or partial) protection in experimental models. Thus, the induction of protective immunity for leishmaniasis is complex, relying on the generation of cellular immune responses that are of considerable strength, longevity, and anatomical distribution to effectively fight the pathogen.

ACKNOWLEDGMENTS

The author wishes to thank Dr. Jiaren Sun and Ms. Mardelle Susman for critical reading and Ms. Belinda Iles for assisting in preparation of this chapter. This work is supported in part by National Institutes of Health Grant AI43003.

References

Aguilar-Be, I., da Silva Zardo, R., Paraguai de Souza, E., Borja-Cabrera, G.P., Rosado-Vallado, M., Mut-Martin, M., Garcia-Miss, M.d.R., Palatnik de Sousa, C.B. and Dumonteil, E. Cross-protective efficacy of a prophylactic *Leishmania donovani* DNA vaccine against visceral and cutaneous murine leishmaniasis. Infect. Immun. 2005; 73:812–819.

Ahmed, S., Colmenares, M., Soong, L., Goldsmith-Pestana, K., Munstermann, L., Molina, R. and McMahon-Pratt, D. Intradermal infection model for pathogenesis and vaccine studies of murine visceral leishmaniasis. Infect. Immun. 2003; 71:401–410.

Alexander, J. and Bryson, K. T helper (h)1/Th2 and *Leishmania*: paradox rather than paradigm. Immunol. Lett. 2005; 99:17–23.

Amaral, V.F., Teva, A., Oliveira-Neto, M.P., Silva, A.J., Pereira, M.S., Cupolillo, E., Porrozzi, R., Coutinho, S.G., Pirmez, C., Beverley, S.M. and Grimaldi, G., Jr. Study of the safety, immunogenicity and efficacy of attenuated and killed *Leishmania* (*Leishmania*) *major* vaccines in a rhesus monkey (*Macaca mulatta*) model of the human disease. Mem. Inst. Oswaldo Cruz 2002; 97:1041–1048.

Andrade, B.B., de Oliveira, C.I., Brodskyn, C.I., Barral, A. and Barral-Netto, M. Role of sand fly saliva in human and experimental leishmaniasis: current insights. Scand. J. Immunol. 2007; 66:122–127.

Armijos, R.X., Weigel, M.M., Calvopina, M., Mancheno, M. and Rodriguez, R. Comparison of the effectiveness of two topical paromomycin treatments versus meglumine antimoniate for New World cutaneous leishmaniasis. Acta Trop. 2004; 91:153–160.

Arora, S.K., Pal, N.S. and Mujtaba, S. *Leishmania donovani*: identification of novel vaccine candidates using human reactive sera and cell lines. Exp. Parasitol. 2005; 109:163–170.

Banuls, A.-L., Hide, M. and Prugnolle, F. *Leishmania* and the leishmaniases: a parasite genetic update and advances in taxonomy, epidemiology and pathogenicity in humans. Adv. Parasitol. 2007; 64:1–109.

Basu, R., Bhaumik, S., Basu, J.M., Naskar, K., De, T. and Roy, S. Kinetoplastid membrane protein-11 DNA vaccination induces complete protection against both pentavalent antimonial-sensitive and -resistant strains of *Leishmania donovani* that correlates with inducible nitric oxide synthase activity and IL-4 generation: evidence for mixed Th1- and Th2-like responses in visceral leishmaniasis. J. Immunol. 2005; 174:7160–7171.

Belkaid, Y., Piccirillo, C.A., Mendez, S., Shevach, E.M. and Sacks, D. L. $CD4^+CD25^+$ regulatory T cells control *Leishmania major* persistence and immunity. Nature 2002; 420:502–507.

Belkaid, Y., Valenzuela, J.G., Kamhawi, S., Rowton, E., Sacks, D.L. and Ribeiro, J.M. Delayed-type hypersensitivity to *Phlebotomus papatasi* sand fly bite: an adaptive response induced by the fly? Proc. Natl. Acad. Sci. U.S.A. 2000; 97:6704–6709.

Berberich, C., Ramirez-Pineda, J.R., Hambrecht, C., Alber, G., Skeiky, Y.A. and Moll, H. Dendritic Cell (DC)-based protection against an intracellular pathogen is dependent upon DC-derived IL-12 and can be induced by molecularly defined antigens. J. Immunol. 2003; 170:3171–3179.

Borja-Cabrera, G.P., Correia Pontes, N.N., da Silva, V.O., Paraguai de Souza, E., Santos, W.R., Gomes, E.M., Luz, K.G., Palatnik, M. and Palatnik de Sousa, C.B. Long lasting protection against canine kala-azar using the FML-QuilA saponin vaccine in an endemic area of Brazil (Sao Goncalo do Amarante, RN). Vaccine 2002; 20:3277–3284.

Borja-Cabrera, G.P., Cruz Mendes, A., Paraguai de Souza, E., Hashimoto Okada, L.Y., de, A.T.F.A., Kawasaki, J.K., Costa, A.C., Reis, A.B., Genaro, O., Batista, L.M., Palatnik, M. and Palatnik-de-Sousa, C.B. Effective immunotherapy against canine visceral leishmaniasis with the FML-vaccine. Vaccine 2004; 22:2234–2243.

Campos-Neto, A. What about Th1/Th2 in cutaneous leishmaniasis vaccine discovery? Braz. J. Med. Biol. Res. 2005; 38:979–984.

Carrion, J., Nieto, A., Soto, M. and Alonso, C. Adoptive transfer of dendritic cells pulsed with *Leishmania infantum* nucleosomal histones confers protection against cutaneous leishmaniosis in BALB/c mice. Microbes Infect. 2007; 9:735–743.

Cohen-Freue, G., Holzer, T.R., Forney, J.D. and McMaster, W.R. Global gene expression in *Leishmania*. Int. J. Parasitol. 2007; 37:1077–1086.

Coler, R.N. and Reed, S.G. Second-generation vaccines against leishmaniasis. Trends Parasitol. 2005; 21:244–249.

Coler, R.N., Skeiky, Y.A., Bernards, K., Greeson, K., Carter, D., Cornellison, C.D., Modabber, F., Campos-Neto, A. and Reed, S.G. Immunization with a polyprotein vaccine consisting of the T-cell antigens thiol-specific antioxidant, *Leishmania major* stress-inducible protein 1, and *Leishmania* elongation initiation factor protects against leishmaniasis. Infect. Immun. 2002; 70:4215–4225.

Davies, C.R., Reithinger, R., Campbell-Lendrum, D., Feliciangeli, D., Borges, R. and Rodriguez, N. The epidemiology and control of leishmaniasis in Andean countries. Cad. Saude Publica 2000; 16:925–950.

Deniau, M., Canavate, C., Faraut-Gambarelli, F. and Marty, P. The biological diagnosis of leishmaniasis in HIV-infected patients. Ann. Trop. Med. Parasitol. 2003; 97(Suppl. 1):115–133.

Desjeux, P. and Alvar, J. *Leishmania*/HIV co-infections: epidemiology in Europe. Ann. Trop. Med. Parasitol. 2003; 97(Suppl. 1):3–15.

Dondji, B., Perez-Jimenez, E., Goldsmith-Pestana, K., Esteban, M. and McMahon-Pratt, D. Heterologous prime-boost vaccination with the LACK antigen protects against murine visceral leishmaniasis. Infect. Immun. 2005; 73:5286–5289.

Dumonteil, E., Maria Jesus, R.S., Javier, E.O. and Maria del Rosario, G.M. DNA vaccines induce partial protection against *Leishmania mexicana*. Vaccine 2003; 21:2161–2168.

Enserink, M. Has leishmaniasis become endemic in the U.S.? Science 2000; 290:1881–1883.

Fajardo-Moser, M., Berzel, S. and Moll, H. Mechanisms of dendritic cell-based vaccination against infection. Int. J. Med. Microbiol. 2008; 298(1–2):11–20.

Field, A.E., Wagage, S., Conrad, S.M. and Mosser, D.M. Reduced pathology following infection with transgenic *Leishmania major* expressing murine CD40 ligand. Infect. Immun. 2007; 75:3140–3149.

Gabaglia, C.R., Sercarz, E.E., Diaz-De-Durana, Y., Hitt, M., Graham, F.L., Gauldie, J. and Braciak, T.A. Life-long systemic protection in mice vaccinated with *L. major* and adenovirus IL-12 vector requires active infection, macrophages and intact lymph nodes. Vaccine 2004; 23:247–257.

Garg, R., Trudel, N. and Tremblay, M.J. Consequences of the natural propensity of *Leishmania* and HIV-1 to target dendritic cells. Trends Parasitol. 2007; 23:317–324.

Giunchetti, R.C., Corrêa-Oliveira, R., Martins-Filho, O.A., Teixeira-Carvalho, A., Roatt, B.M., de Oliveira Aguiar-Soares, R.D., de Souza, J.V., das Dores Moreira, N., Malaquias, L.C., Mota, E., Castro, L.L., de Lana, M. and Reis, A.B. Immunogenicity of a killed *Leishmania* vaccine with saponin adjuvant in dogs. Vaccine 2007; 25:7674–7686.

Gradoni, L., Foglia Manzillo, V., Pagano, A., Piantedosi, D., De Luna, R., Gramiccia, M., Scalone, A., Di Muccio, T. and Oliva, G. Failure of a multi-subunit recombinant leishmanial vaccine (MML) to protect dogs from *Leishmania infantum* infection and to prevent disease progression in infected animals. Vaccine 2005; 23:5245–5251.

Guerin, P.J., Olliaro, P., Sundar, S., Boelaert, M., Croft, S.L., Desjeux, P., Wasunna, M.K. and Bryceson, A.D.M. Visceral leishmaniasis: current status of control, diagnosis, and treatment, and a proposed research and development agenda. Lancet Infect. Dis. 2002; 2:494–501.

Hernandez, R., Nombela, C., Diez-Orejas, R. and Gil, C. Two-dimensional reference map of *Candida albicans* hyphal forms. Proteomics 2004; 4:374–382.

Ilgoutz, S.C. and McConville, M.J. Function and assembly of the *Leishmania* surface coat. Int. J. Parasitol. 2001; 31:899–908.

Ivens, A.C., Peacock, C.S., Worthey, E.A., Murphy, L., Aggarwal, G., Berriman, M., Sisk, E., Rajandream, M.A., Adlem, E., Aert, R., Anupama, A., Apostolou, Z., Attipoe, P., Bason, N., Bauser, C., Beck, A., Beverley, S.M., Bianchettin, G., Borzym, K., Bothe, G., Bruschi, C.V., Collins, M., Cadag, E., Ciarloni, L., Clayton, C., Coulson, R.M., Cronin, A., Cruz, A.K., Davies, R.M., De Gaudenzi, J., Dobson, D.E., Duesterhoeft, A., Fazelina, G., Fosker, N., Frasch, A.C., Fraser, A., Fuchs, M., Gabel, C., Goble, A., Goffeau, A., Harris, D., Hertz-Fowler, C., Hilbert, H., Horn, D., Huang, Y., Klages, S., Knights, A., Kube, M., Larke, N., Litvin, L., Lord, A., Louie, T., Marra, M., Masuy, D., Matthews, K., Michaeli, S., Mottram, J.C., Muller-Auer, S., Munden, H., Nelson, S., Norbertczak, H., Oliver, K., O'Neil, S., Pentony, M., Pohl, T.M., Price, C., Purnelle, B., Quail, M.A., Rabbinowitsch, E., Reinhardt, R., Rieger, M., Rinta, J., Robben, J., Robertson, L., Ruiz, J.C., Rutter, S., Saunders, D., Schafer, M., Schein, J., Schwartz, D.C., Seeger, K., Seyler, A., Sharp, S., Shin, H., Sivam, D., Squares, R., Squares, S., Tosato, V., Vogt, C., Volckaert, G., Wambutt, R., Warren, T., Wedler, H., Woodward, J., Zhou, S., Zimmermann, W., Smith, D.F., Blackwell, J.M., Stuart, K.D., Barrell, B. and Myler, P.J. The genome of the kinetoplastid parasite, *Leishmania major*. Science 2005; 309:436–442.

Ji, J., Masterson, J., Sun, J. and Soong, L. $CD4^+CD25^+$ regulatory T cells restrain pathogenic responses during *Leishmania amazonensis* infection. J. Immunol. 2005; 174:7147–7153.

Ji, J., Sun, J. and Soong, L. Impaired expression of inflammatory cytokines and chemokines at early stages of infection with *Leishmania amazonensis*. Infect. Immun. 2003; 71:4278–4288.

Julia, V., Rassoulzadegan, M. and Glaichenhaus, N. Resistance to *Leishmania major* induced by tolerance to a single antigen. Science 1996; 274:421–423.

Kubar, J. and Fragaki, K. Recombinant DNA-derived *Leishmania* proteins: from the laboratory to the field. Lancet Infect. Dis. 2005; 5:107–114.

Kullberg, M.C., Jankovic, D., Gorelick, P.L., Caspar, P., Letterio, J.J., Cheever, A.W. and Sher, A. Bacteria-triggered $CD4^+$ T regulatory cells suppress *Helicobacter hepaticus*-induced colitis. J. Exp. Med. 2002; 196:505–515.

Laguna, F. Treatment of leishmaniasis in HIV-positive patients. Ann. Trop. Med. Parasitol. 2003; 97(Suppl. 1):135–142.

Launois, P., Maillard, I., Pingel, S., Swihart, K.G., Xenarios, I., Acha-Orbea, H., Diggelmann, H., Locksley, R.M., MacDonald, H.R. and Louis, J.A. IL-4 rapidly produced by V beta 4 V alpha 8 $CD4^+$ T cells instructs Th2 development and susceptibility to *Leishmania major* in BALB/c mice. Immunity 1997; 6:541–549.

Lemesre, J.-L., Holzmuller, P., Goncalves, R.B., Bourdoiseau, G., Hugnet, C., Cavaleyra, M. and Papierok, G. Long-lasting protection against canine visceral leishmaniasis using the LiESAp-MDP vaccine in endemic areas of France: double-blind randomised efficacy field trial. Vaccine 2007; 25:4223–4234.

McConville, M.J., de Souza, D., Saunders, E., Likic, V.A. and Naderer, T. Living in a phagolysosome; metabolism of *Leishmania* amastigotes. Trends Parasitol. 2007; 23:368–375.

McMahon-Pratt, D. and Alexander, J. Does the *Leishmania major* paradigm of pathogenesis and protection hold for New World cutaneous leishmaniases or the visceral disease? Immunol. Rev. 2004; 201:206–224.

Melby, P.C., Ogden, G.B., Flores, H.A., Zhao, W., Geldmacher, C., Biediger, N.M., Ahuja, S.K., Uranga, J. and Melendez, M. Identification of vaccine candidates for experimental visceral leishmaniasis by immunization with sequential fractions of a cDNA expression library. Infect. Immun. 2000; 68:5595–5602.

Melby, P.C., Yang, J., Zhao, W., Perez, L.E. and Cheng, J. *Leishmania donovani* p36(LACK) DNA vaccine is highly immunogenic but not protective against experimental visceral leishmaniasis. Infect. Immun. 2001; 69:4719–4725.

Modabber, F., Buffet, P.A., Torreele, E., Milon, G. and Croft, S.L. Consultative meeting to develop a strategy for treatment of cutaneous leishmaniasis. Institute Pasteur, Paris, 13–15 June, 2006. Kinetoplastid Biol. Dis. 2007; 6:3.

Mohebali, M., Khamesipour, A., Mobedi, I., Zarei, Z. and Hashemi-Fesharki, R. Double-blind randomized efficacy field trial of alum precipitated autoclaved *Leishmania major* vaccine mixed with BCG against canine visceral leishmaniasis in Meshkin-Shahr district, I.R. Iran. Vaccine 2004; 22:4097–4100.

Molina, R., Gradoni, L. and Alvar, J. HIV and the transmission of *Leishmania*. Ann. Trop. Med. Parasitol. 2003; 97(Suppl. 1):29–45.

Mougneau, E., Altare, F., Wakil, A.E., Zheng, S., Coppola, T., Wang, Z.-E., Waldmann, R., Locksley, R.M. and Glaichenhaus, N. Expression cloning of a protective *Leishmania* antigen. Science 1995; 268:563–566.

Myler, P.J., Sisk, E., McDonagh, P.D., Martinez-Calvillo, S., Schnaufer, A., Sunkin, S.M., Yan, S., Madhubala, R., Ivens, A. and Stuart, K. Genomic organization and gene function in *Leishmania*. Biochem. Soc. Trans. 2000; 28:527–531.

Naderer, T., Vince, J.E. and McConville, M.J. Surface determinants of *Leishmania* parasites and their role in infectivity in the mammalian host. Curr. Mol. Med. 2004; 4:649–665.

Nogueira, F.S., Moreira, M.A., Borja-Cabrera, G.P., Santos, F.N., Menz, I., Parra, L.E., Xu, Z., Chu, H.J., Palatnik-de-Sousa, C.B. and Luvizotto, M.C. Leishmune® vaccine blocks the transmission of canine visceral leishmaniasis. Absence of *Leishmania* parasites in blood, skin and lymph nodes of vaccinated exposed dogs. Vaccine 2005; 23:4805–4810.

Olivier, M., Gregory, D.J. and Forget, G. Subversion mechanisms by which *Leishmania* parasites can escape the host immune response: a signaling point of view. Clin. Microbiol. Rev. 2005; 18:293–305.

Olliaro, P.L., Guerin, P.J., Gerstl, S., Haaskjold, A.A., Rottingen, J.-A. and Sundar, S. Treatment options for visceral leishmaniasis: a systematic review of clinical studies done in India, 1980–2004. Lancet Infect. Dis. 2005; 5:763–774.

Papadopoulou, B., Roy, G., Breton, M., Kundig, C., Dumas, C., Fillion, I., Singh, A.K., Olivier, M. and Ouellette, M. Reduced infectivity of a *Leishmania donovani* biopterin transporter genetic mutant and its use as an attenuated strain for vaccination. Infect. Immun. 2002; 70:62–68.

Qi, H., Ji, J., Wanasen, N. and Soong, L. Enhanced replication of *Leishmania amazonensis* amastigotes in gamma interferon-stimulated murine macrophages: implications for the pathogenesis of cutaneous leishmaniasis. Infect. Immun. 2004; 72:988–995.

Ramirez-Pineda, J.R., Frohlich, A., Berberich, C. and Moll, H. Dendritic cells (DC) activated by CpG DNA *ex vivo* are potent inducers of host resistance to an intracellular pathogen that is independent of IL-12 derived from the immunizing DC. J. Immunol. 2004; 172:6281–6289.

Reithinger, R. and Davies, C.R. Canine leishmaniasis: novel strategies for control. Trends Parasitol. 2002; 18:289–290.

Reithinger, R., Dujardin, J.-C., Louzir, H., Pirmez, C., Alexander, B. and Brooker, S. Cutaneous leishmaniasis. Lancet Infect. Dis. 2007; 7:581–596.

Remer, K.A., Apetrei, C., Schwarz, T., Linden, C. and Moll, H. Vaccination with plasmacytoid dendritic cells induces protection against infection with *Leishmania major* in mice. Eur. J. Immunol. 2007; 37:2463–2473.

Robinson, K.A. and Beverley, S.M. Improvements in transfection efficiency and tests of RNA interference (RNAi) approaches in the protozoan parasite *Leishmania*. Mol. Biochem. Parasitol. 2003; 128:217–228.

Rochette, A., McNicoll, F., Girard, J., Breton, M., Leblanc, E., Bergeron, M.G. and Papadopoulou, B. Characterization and developmental gene regulation of a large gene family encoding amastin surface proteins in *Leishmania* spp. Mol. Biochem. Parasitol. 2005; 140:205–220.

Rosado-Vallado, M., Mut-Martin, M., Garcia-Miss, M.D. and Dumonteil, E. Aluminium phosphate potentiates the efficacy of DNA vaccines against *Leishmania mexicana*. Vaccine 2005; 23:5372–5379.

Santos, W.R., de Lima, V.M.F., de Souza, E.P., Bernardo, R.R., Palatnik, M. and Palatnik de Sousa, C.B. Saponins, IL-12 and BCG adjuvant in the FML-vaccine formulation against murine visceral leishmaniasis. Vaccine 2002; 21:30–43.

Scott, P., Artis, D., Uzonna, J. and Zaph, C. The development of effector and memory T cells in cutaneous leishmaniasis: the implications for vaccine development. Immunol. Rev. 2004; 201:318–338.

Silveira, F.T., Lainson, R. and Corbett, C.E. Clinical and immunopathological spectrum of American cutaneous leishmaniasis with special reference to the disease in Amazonian Brazil: a review. Mem. Inst. Oswaldo Cruz 2004; 99:239–251.

Singh, S. and Sivakumar, R. Recent advances in the diagnosis of leishmaniasis. J. Postgrad. Med. 2003; 49:55–60.

Skeiky, Y.A., Coler, R.N., Brannon, M., Stromberg, E., Greeson, K., Crane, R.T., Webb, J.R., Campos-Neto, A. and Reed, S.G. Protective efficacy of a tandemly linked, multi-subunit recombinant leishmanial vaccine (Leish-111f) formulated in MPL adjuvant. Vaccine 2002; 20:3292–3303.

Spath, G.F., Epstein, L., Leader, B., Singer, S.M., Avila, H.A., Turco, S.J. and Beverley, S.M. Lipophosphoglycan is a virulence factor distinct from related glycoconjugates in the protozoan parasite *Leishmania major*. Proc. Natl. Acad. Sci. U.S.A. 2000; 97:9258–9263.

Spath, G.F., Lye, L.F., Segawa, H., Sacks, D.L., Turco, S.J. and Beverley, S.M. Persistence without pathology in phosphoglycan-deficient *Leishmania major*. Science 2003; 301:1241–1243.

Srivastava, J.K., Misra, A., Sharma, P., Srivastava, B., Naik, S. and Dube, A. Prophylactic potential of autoclaved *Leishmania donovani* with BCG against experimental visceral leishmaniasis. Parasitology 2003; 127:107–114.

Stiles, J.K., Hicock, P.I., Shah, P.H. and Meade, J.C. Genomic organization, transcription, splicing and gene regulation in *Leishmania*. Ann. Trop. Med. Parasitol. 1999; 93:781–807.

Streit, J.A., Recker, T.J., Filho, F.G., Beverley, S.M. and Wilson, M.E. Protective immunity against the protozoan *Leishmania chagasi* is induced by subclinical cutaneous infection with virulent but not avirulent organisms. J. Immunol. 2001; 166:1921–1929.

Sukumaran, B. and Madhubala, R. Leishmaniasis: current status of vaccine development. Curr. Mol. Med. 2004; 4:667–679.

Tabbara, K.S., Peters, N.C., Afrin, F., Mendez, S., Bertholet, S., Belkaid, Y. and Sacks, D.L. Conditions influencing the efficacy of vaccination with live organisms against *Leishmania major* infection. Infect. Immun. 2005; 73:4714–4722.

Thiakaki, M., Rohousova, I., Volfova, V., Volf, P., Chang, K.P. and Soteriadou, K. Sand fly specificity of saliva-mediated protective immunity in *Leishmania amazonensis*-BALB/c mouse model. Microbes Infect. 2005; 7:760–766.

Titus, R.G., Gueiros-Filho, F.J., de Freitas, L.A.R. and Beverley, S.M. Development of a safe live *Leishmania* vaccine line by gene replacement. Proc. Natl. Acad. Sci. U.S.A. 1995; 92: 10267–10271.

Tonui, W.K., Mpoke, S.S., Orago, A.S., Turco, S.J., Mbati, P.A. and Mkoji, G.M. *Leishmania donovani*-derived lipophosphoglycan plus BCG induces a Th1 type immune response but does not protect Syrian golden hamsters (*Mesocricetus auratus*) and BALB/c mice against *Leishmania donovani*. Onderstepoort J. Vet. Res. 2003; 70:255–263.

Turco, S.J., Spath, G.F. and Beverley, S.M. Is lipophosphoglycan a virulence factor? A surprising diversity between *Leishmania* species. Trends Parasitol. 2001; 17:223–226.

Uzonna, J.E., Spath, G.F., Beverley, S.M. and Scott, P. Vaccination with phosphoglycan-deficient *Leishmania major* protects highly susceptible mice from virulent challenge without inducing a strong Th1 response. J. Immunol. 2004; 172:3793–3797.

Valenzuela, J.G., Garfield, M., Rowton, E.D. and Pham, V.M. Identification of the most abundant secreted proteins from the salivary glands of the sand fly *Lutzomyia longipalpis*, vector of *Leishmania chagasi*. J. Exp. Biol. 2004; 207:3717–3729.

Vanlerberghe, V., Diap, G., Guerin, P.J., Meheus, F., Gerstl, S., Van der Stuyft, P. and Boelaert, M. Drug policy for visceral leishmaniasis: a cost-effectiveness analysis. Trop. Med. Int. Health 2007; 12:274–283.

Vanloubbeeck, Y.F., Ramer, A.E., Jie, F. and Jones, D.E. $CD4^+$ Th1 cells induced by dendritic cell-based immunotherapy in mice chronically infected with *Leishmania amazonensis* do not promote healing. Infect. Immun. 2004; 72:4455–4463.

Velez, I.D., Gilchrist, K., Arbelaez, M.P., Rojas, C.A., Puerta, J.A., Antunes, C.M., Zicker, F. and Modabber, F. Failure of a killed *Leishmania amazonensis* vaccine against American cutaneous leishmaniasis in Colombia. Trans. R. Soc. Trop. Med. Hyg. 2005; 99:593–598.

von Stebut, E. Cutaneous *Leishmania* infection: progress in pathogenesis research and experimental therapy. Exp. Dermatol. 2007; 16:340–346.

Wanasen, N., MacLeod, C.L., Ellies, L.G. and Soong, L. l-arginine and cationic amino acid transporter 2B regulate growth and survival of *Leishmania amazonensis* amastigotes in macrophages. Infect. Immun. 2007; 75:2802–2810.

Xin, L., Li, Y. and Soong, L. Role of interleukin-1β in activating the $CD11c^{high}$ $CD45RB^-$ dendritic cell subset and priming *Leishmania amazonensis*-specific $CD4^+$ T cells in vitro and in vivo. Infect. Immun. 2007; 75:5018–5026.

Zaph, C., Uzonna, J., Beverley, S.M. and Scott, P. Central memory T cells mediate long-term immunity to *Leishmania major* in the absence of persistent parasites. Nat. Med. 2004; 10:1104–1110.

Zubairi, S., Sanos, S.L., Hill, S. and Kaye, P.M. Immunotherapy with OX40L-Fc or anti-CTLA-4 enhances local tissue responses and killing of *Leishmania donovani*. Eur. J. Immunol. 2004; 34:1433–1440.

CHAPTER

64

Leptospirosis

Nobuo Koizumi[1], *and Haruo Watanabe*[2]

[1]Department of Bacteriology, National Institute of Infectious Diseases, Toyama Shinjuku-ku, Toyama, Tokyo, Japan

[2]Deputy of Director General, National Institute of Infectious Diseases, Toyama, Shinjuku-ku, Tokyo, Japan

OUTLINE

ABSTRACT

Leptospirosis, caused by pathogenic *Leptospira*, is a globally important zoonotic disease that affects humans in both rural and urban settings. This disease may occur in both industrialized and developing countries; in particular, leptospirosis has become an important public health concern in Asia and Latin America. However, even in endemic areas, it remains under diagnosed due to nonspecific, broad spectrum of clinical manifestation, lacking clinical suspicion, and shortage of established early diagnostic methods. Leptospirosis in humans is transmitted by direct contact with infected animals or by exposure to water or soil contaminated by the urine of infected animals.

The spectrum of the disease is wide, ranging from asymptomatic infection to severe disease affecting multiple organs with high mortality. Much attention has recently been paid in the severe pulmonary form of leptospirosis, which is globally emerging as an important clinical form of the disease.

The efficacy of a vaccine for preventing leptospirosis has been recognized, but serovar-specific protection limits the effectiveness of currently available vaccines and a globally accepted vaccine has not been developed. Progress has been made in the areas of pathogenesis of *Leptospira* and immunity against leptospirosis; in particular, pathogenic mechanisms in kidney and lung, identification of adhesion proteins and role of cell-mediated immunity against leptospirosis. However, the molecular mechanism still remains inadequately understood due to lack of genetic tools to study pathogenic *Leptospira*. The whole-genome sequences of pathogenic *Leptospira* have been completed. More recently, random mutagenesis in pathogenic *Leptospira* has been successful using the *Himar1 mariner* transposon. Genome information and this technique will facilitate the identification of virulence factors as well as the understanding of the pathogenesis of leptospires, which is promising to develop a new effective vaccine for leptospirosis.

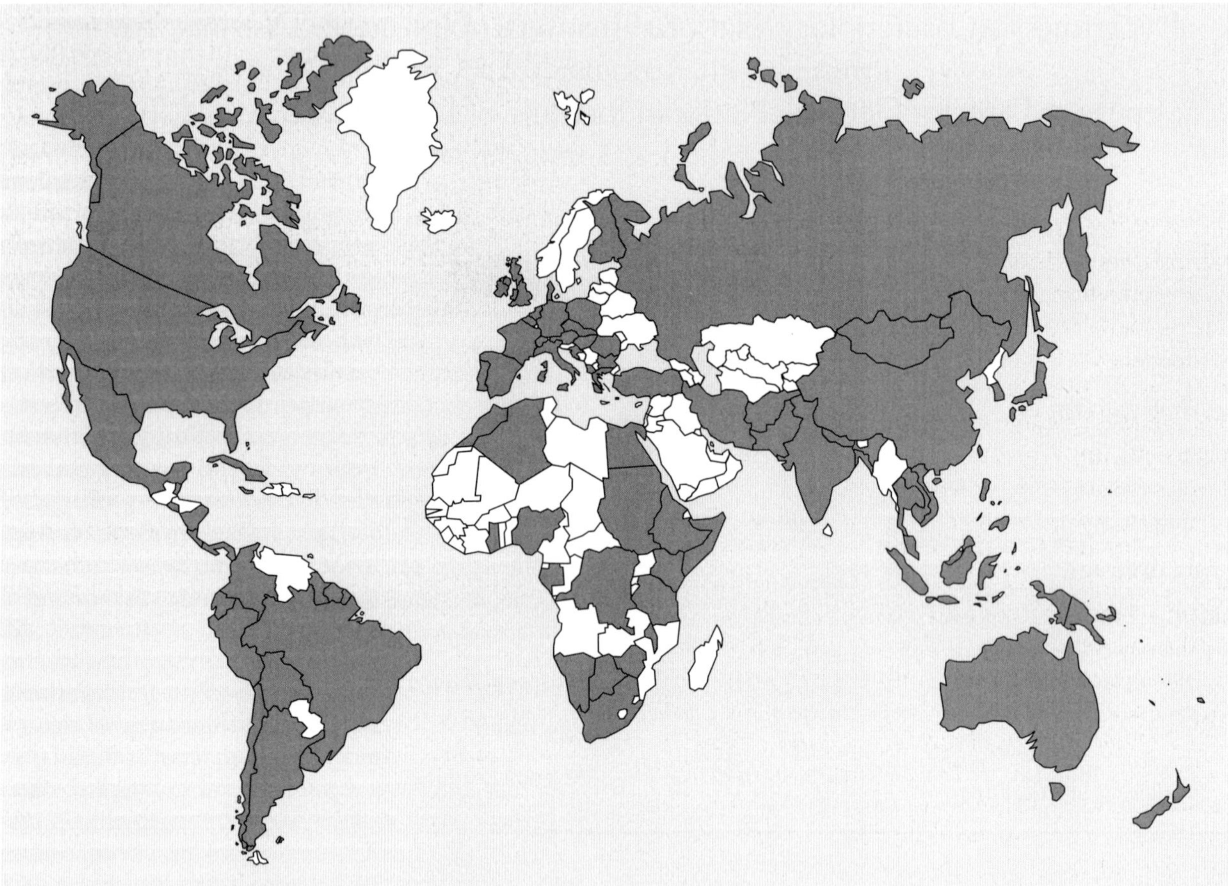

This map contains countries where not only incidences of human leptospirosis but also animal leptospirosis and results of seroepidemiology of human and animal leptospirosis were reported.

INTRODUCTION

Leptospirosis is a worldwide zoonotic disease affecting humans in both rural and urban settings, and in both industrialized and developing countries (Levett, 2001; Bharti et al., 2003; Cachay and Vinetz, 2005; McBride et al., 2005; Ricaldi and Vinetz, 2006). It is well known that leptospirosis vaccine prevents the disease (Koizumi and Watanabe, 2005). However, the development of a new vaccine is urgently needed because currently available vaccines have disadvantages, including short duration of immunity, adverse reactions, and serovar-specific effectiveness. In this chapter, we have summarized recent progress on basic research of leptospirosis, which may contribute to the development of new effective vaccines for leptospirosis.

HISTORY OF THE DISEASE

Leptospirosis is a zoonosis distributed worldwide, caused by infection with pathogenic *Leptospira*. The syndrome, an infectious disease (leptospirosis) with jaundice and nephritis, was first reported by Weil in 1886, whose name was given to severe leptospirosis (Weil's disease) (Weil, 1886). Although this is regarded as the first description of severe leptospirosis, an apparently identical syndrome and its endemicity was described before Weil's report (Levett, 2001; Faine et al., 1999).

The causative agent of Weil's disease was discovered by Inada and colleagues in Japan in 1914–1915. They could consistently succeed in transmitting infection to guinea pigs from patient's blood. They detected a spirochete in the liver tissue of a guinea pig injected with the blood of a patient suffering from Weil's disease. Besides the discovery of the causative organism of the disease, they clarified the source and route of infection, its pathology and morbid anatomy, distribution of the organisms in the tissues, excretion of the spirochetes, its division, filterability, and morphological characterization (Kobayashi, 2001). In October 1915, German groups reported independently the detection of spirochetes in the blood of guinea pigs inoculated with the blood of patients (Hübener and Reiter, 1915; Uhlenhuth and Fromme, 1915). After Inada's work was published in English, the occurrence of leptospirosis was confirmed rapidly in Europe (Faine et al., 1999).

Although the etiologic agent of Weil's disease was demonstrated in 1915, it is now recognized that *Leptospira* was first described by Stimson (1907). He observed spirochetes with hooked ends in the tubules of the kidney of a patient who was believed to have died of yellow fever, and called them *Spirochaeta interrogans* due to their resemblance to a question mark.

As mentioned above, Inada's group described the role of rats as a source of human infection soon after the discovery of *Leptospira*. Leptospirosis in domestic animals, such as dogs and livestock, was recognized during the 1920s and 1930s and the late in 1930s, respectively (Faine et al., 1999).

ETIOLOGIC AGENT

Classification

Leptospires are spirochetes that belong to the genus *Leptospira*, the family Leptospiraceae, in the order Spirochaetales (Fig. 64.1) (Faine et al., 1999). The family also contains other genera *Leptonema* and *Turneriella* (Faine et al., 1999; Levett et al., 2005). *Leptospira* consists of both pathogenic and saprophytic (nonpathogenic) species. *Leptospira* species are defined according to DNA relatedness and there are now 13 named species and 4 unnamed genomospecies (Table 64.1) (Brenner et al., 1999; Levett et al., 2006). Leptospires

FIGURE 64.1 *Leptospira interrogans*. Department of Bacteriology, National Institute of Infectious Diseases.

TABLE 64.1 *Leptospira* species and their pathogenicity

Species	Pathogenicity
L. alexanderi	+
L. biflexa	−
L. borgpetersenii	+
L. broomii	+
L. fainei	±
L. inadai	±
L. interrogans	+
L. kirschneri	+
L. meyeri	±
L. noguchii	+
L. santarosai	+
L. weilii	+
L. wolbachii	−

±, Pathogenic status not clear. Referred from Levett et al. (2006); World Health Organization (2003).

are divided into numerous serovars defined by a cross-agglutination–adsorption test. There are more than 230 serovars recognized among pathogenic *Leptospira* and over 60 serovars of nonpathogenic leptospires. Antigenically related serovars are grouped into serogroups (Faine et al., 1999).

Leptospires are helicoidal bacteria, usually 6–20 μm long and 0.1 μm diameter. One or both ends of the cells distinctively hooked (Faine et al., 1999). The leptospiral cells are composed of cytoplasmic membrane and peptidoglycan cell wall, which are surrounded by an outer membrane. Two flagella (axial filaments) are located at each end of the cell and in the periplasmic space. Leptospires are obligate aerobic spirochetes with optimal growth at pH 7.2–7.6 and temperature 28–30°C. The long chain fatty acids are sources of both carbon and energy. Leptospires are cultivated in artificial media containing either 8–10% rabbit serum or Tween 80/bovine serum albumin, and in low-protein or protein-free media (World Health Organization, 2003).

Antigens Encoded by Agent

Lipopolysaccharide (LPS)

Leptospires possess LPS, which is absent in the outer membrane other pathogenic spirochetes, *Borrelia* and *Treponema*. Variations in the carbohydrate composition of LPS reflect the antigenic diversity (serovars) among pathogenic leptospires (Faine et al., 1999). Leptospiral LPS has lower endotoxic activity than enterobacterial LPS (Nahori et al., 2005). It was recently found that the LPS O-antigen content of leptospires in liver and kidney of acutely infected guinea pigs was markedly reduced compared with those in the kidney of chronically infected rats (Nally et al., 2005a). This result suggests that regulation of O-antigen synthesis may be different between disseminated acute infection and chronic infection in animals.

PROTEIN ANTIGENS

The immunogenic proteins, especially the outer membrane surface proteins of pathogenic *Leptospira*, may be effective as immunogens and potential virulence determinants. Protein antigens, which are recognized by human patient sera and experimentally infected or immunized animals sera, have been described and some of them have been cloned and characterized (Haake et al., 1991, 1993, 1998, 1999, 2000; Shang et al., 1996; Park et al., 1999; Guerreiro et al., 2001; Branger et al., 2001; Nally et al., 2001, 2005b, 2007; Matsunaga et al., 2002, 2003, 2006; Cullen et al., 2002, 2003, 2004, 2005; Haake and Matsunaga, 2002; Palaniappan et al., 2002; Koizumi and Watanabe, 2003a, 2003b, 2004; Artiushin et al., 2004; Hsieh et al., 2005; Verma et al., 2005). Proteomic analysis combined with available genome sequence identified several novel outer membrane proteins (Nally et al., 2005b). Furthermore, a surfaceome strategy has been developed and potential surface-exposed proteins have been identified (Cullen et al., 2005). These analyses have been performed using in vitro cultivated leptospires. Recently, Nally et al. (2007) have developed a method of characterization of the proteome of hydrophobic membrane proteins in *Leptospira* expressed during acute infection relative to that of in vitro cultivated *Leptospira*.

PROTECTIVE IMMUNE RESPONSE

Antibody

The primary mechanism of immunity to *Leptospira* is believed to be solely humoral (Adler and Faine, 1977). Protective immunity was conferred by passive transfer of monoclonal antibody alone (Jost et al., 1986, 1989). These antibodies were directed against leptospiral LPS. Active immunization with LPS and epitopes derived from LPS protected experimental animals from lethal challenge (Schoone et al., 1989; Sonrier et al., 2000); very low levels of antibody were protective (Schoone et al., 1989). The protective immunity conferred by leptospiral LPS as an immunogen is generally serovar-specific (Sonrier et al., 2000). Interestingly, sera from patients with leptospirosis cross-reacted with antigens from nonpathogenic (saprophytic) *Leptospira biflexa* serovar Patoc (strain Patoc I) (Addamiano and Babudieri, 1968; Turner, 1968). Matsuo and colleagues extracted the antigenic components from *L. biflexa* Patoc I by a hot phenol–water method, which strongly reacted with not only anti-*L. biflexa* serum, but also with various antisera against pathogenic leptospires, regardless of their serovars or serogroups. Nuclear magnetic resonance studies demonstrated that the main structural part of the antigenic components of LPS had a repeating disaccharide of →3)-β-D-Man*p*-(1→4)-β-D-Man*p*-(1→ (Matsuo et al., 2000a). Furthermore, not only antisera against various pathogenic leptospires, but also sera from patients with leptospirosis, reacted with exocellular mannans from *Rhodotorula glutinis* that have the same structural backbone of →3)-β-D-Man*p*-(1→4)-β-D-Man*p*-(1→ (Matsuo et al., 2000b; Gorin et al., 1965). Administration of *L. biflexa* LPS preparation to hamsters was protective against a challenge with virulent *L. interrogans* serovar Manilae without any side effects (Matsuo et al., 2000c).

Since protein-conjugated polysaccharide vaccines against *Haemophilus influenzae* type b, pneumococci and meningococci have been successful; it is assumed similar conjugated polysaccharide vaccines for leptospirosis could be developed.

In addition to antibody against LPS, sera from patients with leptospirosis contain antibodies against several protein antigens (Guerreiro et al., 2001). Protein extracts prepared from a pathogenic *Leptospira* can induce protective immunity against challenge with a heterologous serovar strain in an experimental animal model (Sonrier et al., 2000). Some leptospiral protein antigens have been shown to elicit protective immunity in animal models. Immunization with recombinant proteins, such as OmpL1 and LipL41, and Lig proteins conferred protective immunity (Haake et al., 1999; Koizumi and Watanabe, 2004; Palaniappan et al., 2006). Vaccination using an adenovirus vector and plasmid vector encoding the *lipL32/hap-1* gene induced cross-protection in the gerbil model of leptospirosis (Branger et al., 2001, 2005) (see "2Vaccines" section).

Cell-Mediated Immunity

Although immunity against leptospiral infection was thought to be primarily humoral (Faine et al., 1999), recent studies point to a role for cell-mediated immunity in protection from leptospirosis. Peripheral blood mononuclear cells (PBMCs) from cattle immunized with a killed *Leptospira borgpetersenii* serovar Hardjo vaccine proliferated and produced IFN-γ after in vitro stimulation with leptospiral antigens (Naiman et al., 2001). $CD4^+$ $\alpha\beta$ T cells and $WC1.1^+$ $\gamma\delta$ T cells were the main source of IFN (Naiman et al., 2001, 2002; Rogers et al., 2005). $CD4^+$ T cells were shown to be required for $\gamma\delta$ T cells (Blumerman et al., 2007). The PBMCs from Hardjo-vaccine immunized cattle also responded to stimulation with serovar Grippotyphosa antigen preparation (Brown et al., 2003). PBMCs from nonvaccinated cattle also responded to leptospiral antigens, but the responses were lower than those of vaccinated cattle. The low cell-mediated immune response in unvaccinated cattle may be associated with lack of protection from chronic infection (Naiman et al., 2002). Thus, protective immunity against serovar Hardjo infection in cattle correlates with establishment of Th1 immunity. Pronounced expression of Th1 cytokines mRNAs was also detected in hamsters after inoculation with *Leptospira* (Vernel-Pauillac and Merien, 2006). An uncharacterized leptospiral glycolipoprotein elicited in vitro production of TNF-α and IL-10, and upregulated the expression of CD69 and HLA-DR on human PBMCs (Diament et al., 2002). This glycoprotein also induced IL-6 production and monocyte expansion in human PBMCs (Dorigatti et al., 2005). The heat-killed leptospires also induced the production of IFN-γ, IL-12 and TNF-α in human whole blood in vitro (de Fost et al., 2003). Naïve human PBMCs from healthy individuals proliferated and produced IFN-γ, IL-12, and TNF-α in vitro after stimulation with *L. interrogans* (Klimpel et al., 2003). High numbers of leptospires caused the expansion of $\gamma\delta$ T cells, while low numbers of leptospires induced $\alpha\beta$ T cells. In patients with acute leptospirosis the number of peripheral blood $\gamma\delta$ T cells increased significantly (Klimpel et al., 2003; Barry et al., 2006). The in vivo role of $\gamma\delta$ T cells in protection from leptospiral disease remains to be elucidated.

EPIDEMIOLOGY

Significance as Public Health Problem

Leptospirosis has worldwide distribution. Based on global data collected by International Leptospirosis Society surveys, 300,000–500,000 cases of leptospirosis are estimated to occur annually (World Health Organization, 1999; Hartskeerl, 2006). However, this number is probably an underestimate and the true extent of leptospirosis remains unknown due to lack of worldwide surveillance. The incidence of human infection is higher in tropical areas than in temperate regions and in developing countries than in industrialized countries. Leptospirosis is currently recognized as a globally important public health problem, especially in Asia and Latin America (Bharti et al., 2003; McBride et al., 2005; Ricaldi and Vinetz, 2006). In these tropic areas, large outbreaks of leptospirosis are most likely to occur after floods, hurricanes, or other disasters, which affect both rural and urban areas (Ko et al., 1999; Trevejo et al., 1998; Sanders et al., 1999; Easton, 1999; Mudur, 2005; Maskey et al., 2006). Occupational exposure, such as rice farming and other agricultural activities is still significant in rural areas of the tropics (Tangkanakul et al., 2000; Sharma et al., 2006). Leptospirosis has also become an urban problem in the tropics. Outbreaks occur in poor urban slum communities during the same seasonal period of heavy rainfall annually (Ko et al., 1999; Bharadwaj et al., 2002; Johnson et al., 2004; LaRocque et al., 2005). In urban slums, infrastructural deficiencies, such as open sewers, produce the ecological conditions for rodent-borne transmission and contribute the transmission of leptospirosis during epidemics (McBride et al., 2005). Risk for infection in urban dwellings is not limited to tropical areas as the importance of urban leptospirosis was recognized in a U.S. inner-city population (Vinetz et al., 1996). In metropolitan Tokyo, more than 20% of rats harbor leptospires and patients with Weil's disease are thought to be due to contact with environments contaminated with rat urine (our unpublished data).

Leptospires infect the proximal renal tubules of various mammals and are excreted in the urine of carrier animals. Probably every known species of mammals is a potential carrier and excretor of leptospires (Faine et al., 1999). Important animal reservoirs of leptospirosis for human transmission are rats and mice near human habitats, domestic animals such as cattle and swine, companion animals, especially dogs, and wild animals, especially rodents. Leptospirosis in humans is transmitted by direct contact with infected animals or by exposure to water or soil contaminated by the urine of infected animals (Faine et al., 1999).

Leptospirosis has been considered to be primarily an occupational disease associated with activities such as rice and other agricultural practices, mining, sewer maintenance, livestock farming, and fish farming (Levett, 2001). There is a significant risk associated with recreational exposures occurring during water sports such as swimming, canoeing, and rafting (Reisberg et al., 1997; Narita et al., 2005; Nakamura et al., 2006; Nardone et al., 2004), and in travel and adventure tourism (Reisberg et al., 1997; van Creel et al., 1994; Sejvar et al., 2003; Jansen et al., 2005). The large outbreaks of leptospirosis in recreational activities are associated with competitive events (Morgan et al., 2002), even in international adventure race (Sejvar et al., 2003). In endemic areas, humans contract leptospirosis from activities of daily life (Johnson et al., 2004).

Potential as Biothreat Agent

Leptospira is not considered as a potential biothreat agent. However, there is merit to the attention of leptospirosis in bioterrorism since its clinical manifestation can mimic those of viral hemorrhagic fevers (Bharti et al., 2003; Monsuez et al., 1997).

CLINICAL DISEASE

Leptospirosis is an acute febrile illness. The spectrum of symptoms in leptospirosis is extremely broad; from asymptomatic infection or mild manifestation to severe disease including jaundice, renal failure, hemorrhage, and death. Asymptomatic seroconversion is common in human infection occurring in up to 70% of all serologically identified cases (Ashford et al., 2000; Phraisuwan et al., 2002). Acute leptospirosis starts with the initial symptoms such as fever, chills, headache, myalgia, and conjunctival suffusion after an incubation period of 2–30 days (usual range is 5–14 days) (Levett, 2001; Faine et al., 1999). About 90% of recognized infections are mild and self-limited. Five to ten percent of the patients develop severe forms of disease, such as Weil's disease characterized by jaundice, renal failure, hemorrhage, and severe pulmonary hemorrhage syndrome with or without jaundice and renal failure. Mortality rate can be as high as 5–15% and 30–60% (Dolhnikoff et al., 2007), respectively.

The jaundice in leptospirosis does not seem to be associated with hepatocyte damage, but seems to be more related to the cholestasis of sepsis (Bharti et al., 2003). Serum bilirubin levels may be very high and transamilase levels are typically increased only for 2–3 times above normal level, suggesting leptospirosis rather than viral hepatitis (Faine et al., 1999).

Renal involvement is common in leptospirosis; the majority of the patients have abnormalities on urinalysis. Blood urea and creatinine elevate rapidly (Faine et al., 1999). Acute renal failure (ARF), which occurs in 10 to over 60% of cases (Visith and Kearkiat, 2005), frequently accompanies hypokalemia and nonoliguria, even polyuria. In the patients with ARF, oliguria is a significant predictor of death (Daher et al., 1999). Thrombocytopenia occurs in up to 50% of the patients with leptospirosis, which does not seem to result from disseminated intravascular coagulation (Edwards et al., 1982, 1986; Nicodemo et al., 1997).

Pulmonary involvement in leptospirosis has recently gained much more attention as several epidemics of severe pulmonary form of leptospirosis have been reported (Ko et al., 1999; Zaki and Shieh, 1996; Yersin et al., 2000; Seijo et al., 2002; Vijayachari et al., 2003; Segura et al., 2005). Occurrence of respiratory manifestation in leptospirosis varies from 20% to 70%. Symptoms can be mild and nonspecific, including chest pain, cough, and dyspnea but the presence of severe pulmonary hemorrhage with or without acute respiratory distress syndrome is associated with higher mortality. Respiratory symptoms usually appear between fourth and sixth day of disease and may lead to death in less than 72h (Dolhnikoff et al., 2007).

Aseptic meningitis may be found in up to 25% of cases. Cardiac involvement is probably more common than is recognized, including myocarditis, nonspecific electrocardiogram abnormalities, and atrial fibrillation (Plank and Dean, 2000). Leptospirosis involves ocular manifestation. Conjunctival suffusion is a common clinical finding in the acute stage of illness (Martins et al., 1998). Uveitis is a rare but an important sequel of leptospirosis (Rathinam, 2002), which is presumed to occur by an immune-mediated process; although isolation and/or DNA detection of *Leptospira* has been demonstrated in aqueous humor of patients (Mancel et al., 1999).

Leptospirosis has diverse clinical manifestations easily confused with many other diseases in tropics, such as dengue fever or dengue hemorrhagic fever, malaria, and scrub typhus (Sanders et al., 1999; LaRocque et al., 2005; Zaki and Shieh, 1996; Wongsrichanalai et al., 2003; Suputtamongkol et al., 2004; Karande et al., 2005). The differential diagnosis also include influenza, HIV seroconversion, and hantavirus infections (Bharti et al., 2003).

TREATMENT

Controversy remains about efficacy of antibiotics in treating leptospirosis (Pappas and Cascio, 2006). However, many experts recommend the use of antibiotic therapy for severe leptospirosis and clinically and epidemiologically suspected cases (McBride et al., 2005; Vinetz, 2003; Yang et al., 2005). Although optimal treatment of leptospirosis is not clearly defined (Pappas and Cascio, 2006; Griffith et al., 2006), doxycycline is recommended for mild disease and severe cases of leptospirosis should be treated with intravenous penicillin (World Health Organization, 2003; McClain et al., 1984; Watt et al., 1988). Recent clinical trials support the use of cephalosporins (Suputtamongkol et al., 2004; Panaphut et al., 2003). Intravenous ceftriaxone was shown to be as effective against severe leptospirosis as intravenous penicillin. The median duration of fever was same in both groups and there was no significant difference in the duration of organ dysfunction and the overall mortality rate between two groups (Panaphut et al., 2003). Another study revealed that cefotaxime and doxycycline were each as effective as penicillin for treatment of severe leptospirosis. The mortality rate, time to defervescence, and progression of dysfunction of vital organs was similar among the three groups (Suputtamongkol et al., 2004).

A wide range of antimicrobial agents has an activity against *Leptospira* in vitro and/or animal models, but their relevance in clinical outcomes remains unknown (Pappas and Cascio, 2006; Griffith et al., 2006).

The utility of doxycycline (200 mg/week) in preventing susceptible population from the disease has shown in a study using U.S. soldiers in Panama (Takafuji et al., 1984). Doxycycline prophylaxis in endemic areas or recreational activity was also shown to be protective; although these results were not statistically significant (Sejvar et al., 2003; Gonsalez et al., 1998; Sehgal et al., 2000).

In severe cases, supportive treatment for dehydration, hypotension, renal failure, and pulmonary involvement is also required. Nitric oxide inhalation has been anecdotally resulted in marked clinical improvement of pulmonary hemorrhage (Borer et al., 1999). Corticosteroids have been shown to reduce mortality and change the outcome significantly when they were used early in the management of pulmonary leptospirosis (Shenoy et al., 2006).

PATHOGENESIS

Pathogenic leptospires enter into bloodstream via small abrasions and breaches of skin, conjunctiva, and mucosa (Faine et al., 1999). Pathogenic *Leptospira* is able to translocate through polarized Madin-Darby canine kidney cell monolayers rapidly, suggesting that evasion from killing by host cells permits pathogenic leptospires to quickly reach the bloodstream and disseminate to multiple organs (Barocchi et al., 2002). Leptospires circulate and survive in the bloodstream in the initial phase of disease, indicating that they are able to escape from killing by the complement system, especially the alternative pathway. Pathogenic leptospires showed serum resistance whereas nonpathogenic *Leptospira* was sensitive (Meri et al., 2005; Verma et al., 2006). Serum resistance was found to correlate with the binding of the soluble complement regulator factor H on the pathogenic leptospiral cell surface. As a consequence, pathogenic leptospires deposited much less C5, C6, C8, and C5b-9 on their surfaces than the nonpathogenic *Leptospira*, resulting in serum resistance of pathogenic leptospires (Meri et al., 2005). Verma et al. (2006) have recently identified the human factor H-binding protein of *Leptospira*, LfhA. LfhA is a putative lipoprotein and located in the outer membrane of the cells. LfhA is expressed during natural infection, suggesting its relevance in the protection of leptospires from complement-mediated killing and immune evasion.

The primary lesions to small endothelial blood vessels caused by *Leptospira* lead to hemorrhage and localized ischemia in multiple organs. Several hemolysins have been identified in leptospires (Artiushin et al., 2004; Segers et al., 1990; Lee et al., 2000, 2002; Zhang et al., 2005; Hauk et al., 2005). These hemolysins are divided into two groups, sphingomyelinase or nonsphingomyelinase hemolysins. SphH is a pore-forming hemolysin which shows cytotoxic activity on mammalian cells (Lee et al., 2002). Lk73.5 showed cytotoxicity on equine pulmonary endothelial cells and is suggested to be expressed only in vivo (Artiushin et al., 2004), although its orthologous protein gene in the other serovar strain was transcribed in vitro (Zhang et al., 2005).

Pathogenic *Leptospira* disseminates hematogenously to various tissues, where leptospires colonize. Leptospires interact with various cultured cells, including fibroblasts, macrophages, microglial, endothelial, and epithelial cells (Barocchi et al., 2002; Vihn et al., 1984; Thomas and Higbie, 1990; Merien et al., 1997, 2000; Cinco et al., 2006). A 36 kDa fibronectin-binding protein identified only in virulent *Leptospira* functions as an adhesion protein (Merien et al., 2000). Lsa24 was recently shown to be a laminin-binding protein (Barbosa et al., 2006). This protein has been identified as LfhA (Verma et al., 2006) described above. Lig proteins are surface-exposed outer membrane proteins punctuated by tandem repeats of about 90 amino acids of bacterial immunoglobulin (Ig) like domains (Palaniappan et al., 2002, 2004; Matsunaga et al., 2003; Koizumi and Watanabe, 2004). LigA consists only of Ig-like domains, whereas LigB has an additional unique domain at the C terminus. Expression of Lig protein and *lig* mRNA was lost after attenuation of *Leptospira kirshneri* and *L. interrogans*, suggesting that Lig proteins are associated with virulence (Matsunaga et al., 2003). Expression of Lig proteins, surface exposure of LigB, and extracellular release of LigA are all enhanced by physiological osmolarity (Matsunaga et al., 2005). It has been recently demonstrated that Lig proteins bound to extracellular matrix (ECM) proteins such as fibronectin, collagen and laminin, and fibrinogen (Choy et al., 2007). Unique Ig-like domains of both LigA and LigB are responsible for the interaction with fibronectin.

In kidney, *Leptospira* colonizes and multiplies in proximal tubule epithelial cells resulting in ARF and tubulointerstitial nephritis (Yang et al., 2001). Yang et al. (2002) have shown that leptospiral outer membrane proteins (LOMPs) extracted from pathogenic *Leptospira*, but not from nonpathogenic *Leptospira* stimulated the production of a set of pro-inflammatory chemokine such as CCL2/MCP-1, CCL5/RANTES, TNF-α in cultured mouse proximal tubule cells. Toll-like receptor (TLR) 2 pathway mediated the induction of these chemokines, at least CCL2/MCP1, caused by LOMPs (Yang et al., 2006a). Recombinant LipL32, one of the most abundant outer membrane lipoproteins in *Leptospira* (Haake et al., 2000), also stimulates the TLR2-mediated induction of expression and secretion of CCL2/MCP1 (Yang et al., 2006a). They also showed that LOMPs also induced upregulation of chemokines CXCL2/MIP-2 and CXCL1/KC through the TLR2 pathway (Hung et al., 2006b, 2006a). CXC chemokines, including CXCL8/IL8, CXCL1/KC/GRO-α, and CXCL2/MIP-2, are molecules that are potentially responsible for the infiltration of leukocytes and the initiation of tubulointerstitial nephritis (Tang et al., 1997). Thus, the production of these chemokines and subsequent influx of leukocytes in kidneys results in the development of tubulointerstitial nephritis in *Leptospira* infection.

If untreated, tubulointerstitial nephritis caused by leptospiral infection leads consequently to interstitial fibrosis (Sterling and Thiermann, 1981; Penna et al., 1963). Sustained expression of TGF-β1 leads to the accumulation of ECM proteins through an increase of ECM synthesis and a decrease of ECM degradation, which result in tubulointerstitial fibrosis (Yamamoto et al., 1994; Baricos et al., 1999). It has been shown that LOMPs induced the increase of the production of type I and IV collagen, but not the enhancement of their degradation, through TGF-β1/Smad-dependent pathway in cultured human renal proximal tubular cells (Tian et al., 2006).

Leptospirosis causes a unique ARF with hypokalemia and nonoliguria. Hypokalemia is secondary to renal potassium wasting. Patients infected with *Leptospira santarosai* serovar Shermani were suspected to have a defect in the thick ascending limb (TAL) sodium reabsorption (Wu et al., 2004). LOMPs from *L. santarosai*, but not *L. interrogans*, inhibited the function of Na^+–K^+–Cl^- cotransporter NKCC2, a major transporter for apical sodium reabsorption by TAL, and the expression of *mNKCC2* mRNA in mouse TAL cells (Wu et al., 2004). On the other hand, Andrade et al. (2007) showed that expression of sodium/hydrogen exchanger isoform 3, which plays an important role in reabsorption of sodium and fluid by the proximal tubule, was significantly decreased in kidneys of hamsters infected with *L. interrogans*. These results suggest that *Leptospira* species specifically target different transporters. Alternatively, it has been shown that leptospiral glycolipoprotein endotoxin and its nonesterified unsaturated fatty acid components are a potent inhibitor of renal Na^+–K^+ ATPase (Burth et al., 1997, 2005). Leptospirosis patients showed an increased in serum concentrations of oleic and linoleic acid but a decrease in serum albumin concentrations. Correlation between disease severity and serum oleic acid:albumin or oleic plus linoleic acid:albumin ratio was revealed (Burth et al., 2005).

Severe pulmonary hemorrhage leptospirosis is now recognized as an important public health problem. Two mechanisms, a toxin-mediated and/or an immune-mediated mechanism, are suggested (Dolhnikoff et al., 2007). The amount of leptospiral antigens was not correlated with the severity of the lesion in lung autopsy (Nicodemo et al., 1997). Fewer leptospires were detected in infected hemorrhagic lung tissues than in liver and kidney, which are without evidence of hemorrhage, in guinea pigs (Nally et al., 2004). These observations would support a host immune-mediated mechanism. Disseminated intravascular coagulation has shown to have no relation to pulmonary hemorrhage (Nicodemo et al., 1997). Nally et al. (2004) have demonstrated that IgM, IgG, IgA, and C3 were deposited along the alveolar basement membrane in infected guinea pigs, suggesting that an autoimmune process may be the cause of severe pulmonary hemorrhage leptospirosis. Deposition of Ig in alveolar septum and alveolar space was observed in human autopsy, which showed diffused pulmonary hemorrhage although diagnosis of leptospirosis was made only by Warthin-Starry stain in the lung autopsy (Yang and Hsu, 2005). The etiology of pulmonary edema is also unclear. It has been documented that resolution of pulmonary edema occurs as a consequence of active sodium transport across the alveolar epithelium (Matthay et al., 2002). Andrade et al. (2007) also observed that leptospiral infection decreased the expression of the alpha subunit of the epithelium sodium channel (α-ENaC) and induced the expression of the Na^+–K^+–Cl^- cotransporter NKCC1 in lungs of hamster model of leptospirosis. This result suggests that leptospirosis profoundly influences the sodium transport capacity of alveolar epithelial cells, resulting in impaired pulmonary fluid clearance and subsequent disturbance of lung function and increased susceptibility of the lung to injury.

Ocular leptospirosis is widely considered to be an immune-mediated disease in both humans and horses (Rathinam, 2002), although isolation and/or DNA detection of *Leptospira* has been demonstrated in aqueous humor of both uveitic people and horses (Mancel et al., 1999; Faber et al., 2000). A leptospiral DNA fragment that encodes cross-reactive protein with equine cornea has been cloned (Lucchesi and Parma, 1999). Verma et al. (2005) have recently identified novel leptospiral lipoproteins, LruA and LruB. Eye fluids of uveitic horses contained significantly high titers of Ig specific to both lipoproteins. Further, antisera raised against these proteins reacted with equine ocular components, suggesting an immune-mediated process in leptospiral uveitis. The relevance of this finding in human ocular leptospirosis remains to be confirmed.

In addition to leptospiral virulence determinants and mechanisms, host factors are presumably related to pathogenesis of leptospirosis. Elevated plasma levels of tumor necrosis factor-α are associated with disease severity and mortality among patients with leptospirosis (Tajiki and Salomão, 1996). Recently, Lingappa et al. (2004) found the association between human leukocyte antigen (HLA)-DQ6 and the risk of infection due to swallowing of waters contaminated with leptospires in an outbreak of leptospirosis.

Immune Response to Infection

Innate immunity relies on signaling by TLRs to alert the immune system against bacterial infections (Gerold et al., 2007). TLR activation leads to the release of cytokines that allow for effective innate and adaptive immune responses. Werts et al. (2001) have shown that leptospiral LPS activates human macrophages through TLR2 instead of TLR4, which is a central component for Gram-negative bacteria. Lipid A is the active component of LPS responsible for its toxic activity and functions. The structure of leptospiral lipid A moiety has been recently deciphered (Que-Gewirth et al., 2004). It has been revealed that leptospiral lipid A possesses some peculiar characteristics compared with lipid A from Gram-negative bacteria. Species specificity of recognition of leptospiral lipid A/LPS has been reported (Nahori et al., 2005). Lipid A stimulated cells through murine TLR4/MD2 but not through human TLR4/MD2, nor murine and human TLR2. On the other hand, LPS activates human cells through TLR2/TLR1 whereas both TLR2 and TLR4 are involved in signaling against LPS in murine cells. Mice are resistant to *Leptospira* infection while humans are not. The species specificity of TLRs for recognition of LPS may be a part of an effective innate immune protection against leptospirosis in mice but not in humans.

In the mouse model of leptospirosis, it has been demonstrated that TLR4 plays a crucial role in protection from acute lethal infection and control of leptospiral burden in tissues during sublethal chronic infection (Viriyakosol et al., 2006). Heat-killed *Leptospira* did not stimulate cytokine induction in $TLR4^{-/-}$ macrophages whereas cytokines were induced in both $TLR2^{-/-}$ and $CD14^{-/-}$ macrophages, indicating that cytokine responses in macrophages correlated with leptospiral clearance in vivo. Induction of cytokines by leptospiral cells was abolished by polymixin B, suggesting that these TLR4-dependent but CD14 and TLR2-independent responses are likely mediated by a leptospiral ligand(s) other than LPS.

VACCINES

History

The efficacy of a vaccine for preventing leptospirosis was shown soon after *Leptospira* was proven to be the causative agent of Weil's disease in Japan. The heat-killed whole cell vaccine made from *Leptospira* cultures dramatically protected coal miners from Weil's disease in Kyushu where the disease was endemic (Wani, 1933). The development of killed whole cell vaccine against leptospirosis in the middle of the last century has been reviewed (Babudieri, 1959), and subsequent results of vaccine field trials were published (Torten et al., 1973; Fukumura, 1984; Chen, 1985; Ikoev et al., 1999; Sánchez et al., 2000; Yan et al.,2003; Martínez et al., 2004).

Current Licensed Vaccines

- Weil's disease and Akiyami combined vaccine (Denka Seiken, Japan) (Koizumi and Watanabe, 2005)
  Formalin-killed whole cell vaccine
  The concentrations and serovars are 2.5×10^8 cells/ml each of Australis, Autumnalis, and Hebdomadis and 5×10^8 cells/ml of Copenhageni.
  Initial vaccination: two subcutaneous injections of 1.0 ml of vaccine given at a 7-day interval.
  Booster injection: subcutaneous injection of 1.0 ml of vaccine given within 5 years after the second initial dose.
  For children (7–13 years): half dose of adult (mentioned above)

- SPIROLEPT (Thea Laboratories, France) (http://sante-az.aufeminin.com/w/sante/m3514716/medicaments/spirolept.html, 2007; Laurichesse et al., 2007)
  Formaldehyde-killed whole cell vaccine
  Supplied as a syringe with a 1-ml dose of vaccine 2×10^8 cells *Leptospira interrogans* serogroup Icterohaemorrhagiae
  Target population: people who, because of their profession or their activities of leisures, are in contact with contaminated waters or animals.
  Initial vaccination: two subcutaneous injections at a 15-day interval.
  Booster injection: 4–6 months after the second initial dose
  Ulterior booster: every 2 years
  No data is available for the children.

- Leptospirosis vaccine (Shanghai Institute of Biological Products, China) (Yan et al., 2003; http://www.siobp.com/CPJS/YFZP_13.htm, 2007)
  Polyvalent vaccine (strains/serovars not specified) composed of outer envelope fraction of leptospiral cells
  Target population: 7–60 years in epidemic area of leptospirosis. People who go to epidemic area and have not been vaccinated before.
  Initial vaccination: two subcutaneous injections (1st: 0.5 ml; 2nd: 1 ml) at a 7 to 10-day interval.
  Booster injection: subcutaneous injection of 1 ml of vaccine at 1 year after the second initial dose.
  Children (7–13 years): 1/2 dose of adult (mentioned above)
  Child (under 7) not exceed 1/4 dose of adult, if needed

- Finlay Institut, Cuba (Martínez et al., 2004; http://www.leptonet.net/html/vaccination.asp, 2007)
  Vax-SPIRAL
  Trivalent vaccine: *Leptospira interrogans* serovars Canicola, Copenhageni, and Mozak.
  Inactivated vaccine

The antibody titers (MAT titers) after vaccination were significantly lower than those developed after natural infection and seroconversion occurred with low frequency (about 20–60%), although a high seroconversion rate has been reported (90–100%) (Koizumi and Watanabe, 2005; Torten et al., 1973; Ikoev et al., 1999; Laurichesse et al., 2007; Martínez Sánchez et al., 1998; Sánchez et al., 2002). However, protection was reported to be high in such populations and efficacy rates of whole cell vaccines were about 60–100% (Fukumura, 1984; Sánchez et al., 2000; Martínez et al., 2004). Concerning the duration of immunity induced by killed vaccines, one study reported that the duration of immunity ranged between 6 months and 1 year at best while other studies reported a duration of at least 3 years (Babudieri, 1959; Fukumura, 1984; Chen, 1985). To maintain immunity, the French vaccine recommends a booster every 2 years and the Japanese vaccine at least every 5 years.

These vaccines (at least Japanese and French vaccine according to the product summary) must not be used during pregnancy.

Side effects of the whole cell vaccines have been reported, which included both systemic (such as fever, headaches, chill, malaise) and local reactions (such as redness, swelling, pain) at various frequencies.

Vaccines in Development

Although killed whole cell vaccines are used for prevention of human leptospirosis, there still remain several problems. In addition to issues on persistence of immunity and adverse events, the effectiveness of the killed vaccine is generally serovar-specific (Fukumura, 1984; Chen, 1985; Philip and Tennent, 1966). There are more than 230 serovars among the pathogenic leptospires. The local variability in serovars of endemic leptospiral strains complicates the development of a vaccine that can be used worldwide (Levett, 2001; Faine et al., 1999). Thus, new vaccine strategies are directed to the exploration for antigens that can generate cross-protection against various serovars. The immunogenic proteins, especially the outer membrane surface proteins, of pathogenic *Leptospira*, may be effective as immunogens. The identification of proteins, which are conserved among pathogenic leptospires and can elicit cross-protection against various serovars, has become a major focus of leptospirosis research (see "Protein Antigens" section). Subunit vaccines would also have fewer side effects than killed whole cell vaccine. Sera from patients with leptospirosis have antibodies against several protein antigens (Naiman et al., 2002). Protein extracts prepared from a pathogenic *Leptospira* can induce protective immunity against challenge with a heterologous serovar strain in an experimental animal model (Narita et al., 2005). These data point to the potential use of leptospiral protein(s) as candidates for a new vaccine that could induce good protection against diverse serovars.

OmpL1 is a transmembrane porin (Haake et al., 1993; Shang et al., 1995), and LipL41 is a surface-exposed outer membrane lipoprotein (Shang et al., 1996). OmpL1 and LipL41 act synergistically to induce immunoprotection in a hamster model of leptospirosis, although neither of the individual proteins induces protective immunity (Haake et al., 1999). Southern and western blot analyses revealed that *ompL1* gene product and LipL41 protein were present in various pathogenic leptospiral serovars, but not in nonpathogenic leptospires (Haake et al., 1993; Shang et al., 1996). Patients with leptospirosis have antibodies against these proteins in their sera (Haake et al., 2000; Flannery et al., 2001). Whether or not these proteins will confer cross-protection against a heterologous challenge remains to be determined.

LipL32 is one of the most abundant proteins in *Leptospira* (Haake et al., 2000). LipL32 is a surface-expressed outer membrane lipoprotein that is conserved, both genetically and immunologically, in the various pathogenic leptospires (Haake et al., 2000; Cullen et al., 2005). LipL32 antigen induces the production of antibody in patients with leptospirosis (Flannery et al., 2001). LipL32 is also called hemolysis-associated protein-1 (Hap-1) showing hemolytic activity alone or synergistically with another hemolysin (Lee et al., 2000; Hauk et al., 2005). Vaccination using an adenovirus vector and plasmid vector encoding the *lipL32/hap-1* gene induced cross-protection in a gerbil model of leptospirosis (Branger et al., 2001, 2005).

Lig proteins are surface-exposed outer membrane proteins and their expression is upregulated during infection of mammalian hosts (Palaniappan et al., 2002, 2004; Matsunaga et al., 2003; Koizumi and Watanabe, 2004). The *lig* genes are present among various pathogenic, but not nonpathogenic, leptospires (Matsunaga et al., 2003; Koizumi and Watanabe, 2004; Palaniappan et al., 2004). Sera from human patients infected with different leptospiral serogroup strains reacted with Lig proteins (Koizumi and Watanabe, 2004; Croda et al., 2007). Furthermore, in a mouse model of leptospirosis, the Lig proteins elicited protective immunity against challenge, not only with the homologous serovar Manilae infection (Koizumi and Watanabe, 2004), but also the heterologous serovar Icterohaemorrhagiae (our unpublished data). Lig proteins also conferred protection against leptospiral infection in hamster model (Palaniappan et al., 2006).

The ability to rapidly determine full-length genome sequences has opened a new approach to vaccine design that may be relevant for the treatment of bacterial infections (Plotkin, 2005). The strategy of "reverse vaccinology," in which the full-length genome is "mined" by various computer algorithms, for genes that encode proteins with desired characteristics, has been applied to some bacteria and novel vaccine candidate sequences have been identified (Rappuoli, 2001; Serruto et al., 2004). The full-length genome sequences of two strains of *L. interrogans*, serovar Lai and serovar Copenhageni, and two strains of *L. borgpetersenii* serovar Hardjo have been published (Ren et al., 2003; Nascimento et al., 2004a; Bulach et al., 2006). A full-length genome analysis of serovar Copenhageni strain has been used to identify candidate antigens for a leptospiral vaccine (Gamberini et al., 2005). A total of 206 genes had been selected and 150 of them were expressed in *Escherichia coli*, purified, and used for immunoblotting with leptospirosis patient sera. A total of 16 proteins reacted with convalescent patient sera in immunoblotting. Four of the 10 proteins tested were highly conserved among different pathogenic leptospires.

The other group identified 226 genes as potential vaccine candidates against *L. interrogans* using in silico analysis (identification of genes that encode surface-exposed proteins), comparative genome hybridization (genes conserved in the 10 pathogenic leptospiral serovar strains) and transcriptional analysis (genes have high transcriptional level in vitro) (Yang et al., 2006b). Further research that evaluates the protective activity of these candidates is needed.

The availability of genome sequences of *Leptospira* also enabled investigation of the transcriptome using microarray technology. Following excretion in the urine from reservoir animals, *L. interrogans* is able to survive in the environment for weeks (Koizumi and Watanabe, 2005), and then be transmitted to humans, which involves a shift from ambient temperatures to 37°C and more and an upshift in osmolarity during the early stage of infection. Recent microarray analyses identified the genes that are induced or repressed when *Leptospira* cultivated at 20 or 30°C is shifted to temperatures at 37 or 39°C or *Leptospira* is shifted to culture media supplemented with sodium chloride to physiologic osmolarity (Qin et al., 2006; Lo et al., 2006; Matsunaga et al., 2007). Several genes that were induced by physiological osmolarity were also upregulated by temperature shifts (Matsunaga et al., 2007). In contrast to *L. interrogans*, *L. borgpetersenii* cannot survive in nutrient poor environments and is transmitted by direct contact with *Leptospira* containing body fluids. Recent comparative genomic analysis revealed that *L. borgpetersenii* has undergone genome reduction relative to *L. interrogans* (Bulach et al., 2006). Many genes encoding environmental sensing, metabolite transport, and utilization functions are especially absent or pseudogenes in *L. borgpetersenii* but not in *L. interrogans.* Transcriptome analysis on osmolarity shift also demonstrated that genes that were repressed at physiological osmolarity were much more likely than other genes to be impaired in *L. borgpetersenii*, suggesting that physiological osmolarity is a key signal for the environment-to-host transmission (Matsunaga et al., 2007).

PROSPECTS FOR THE FUTURE

Both the mechanisms of *Leptospira* pathogenesis and nature of protective immunity against leptospiral infection are still poorly understood, in particular, due to the lack of efficient genetic tools. It has been reported that genetic transformation using a shuttle vector has been successfully undertaken, followed by gene knockouts by allelic exchange in the saprophytic *Leptospira* (Girons et al., 2000; Picardeau et al., 2001; Bauby et al., 2003). However, these techniques have not been successful for pathogenic leptospires. Although the availability of full-length genome sequences of *Leptospira* strains will be helpful for elucidating the pathogenesis of *Leptospira*, it has revealed that about 41% (serovar Lai), 31% (serovar Copenhageni), and 23% (serovar Hardjo) of the total genes in *Leptospira* encode unique hypothetical proteins that do not show homology with any other bacterial proteins (Ren et al., 2003; Nascimento et al., 2004a, 2004b; Bulach et al., 2006). To analyze the function of these *Leptospira* specific gene products, it is essential to develop genetic analysis tools for pathogenic leptospires. More recently, Bourhy et al. (2005) showed that the *Himar1 mariner* transposon permits random mutagenesis in pathogenic *Leptospira*. This technique will accelerate the identification of virulence factors as well as our understanding of the pathogenesis of leptospires. As new information evolves, insights will be gained into ways to improve existing, and develop effective new, vaccines for leptospirosis.

KEY ISSUES

- Leptospirosis is a globally important zoonotic disease that affects humans in both rural and urban settings, and in both industrialized and developing countries. Large outbreaks of leptospirosis occur especially in Asia and Latin America.
- The effectiveness of currently available vaccines is generally serovar-specific. Thus, new vaccine strategies are directed to the exploration for antigens that can generate cross-protection against various serovars.
- The availability of full-length genome sequences and genetic tools of pathogenic *Leptospira* will open new ways to elucidate the pathogenesis of *Leptospira* and immune mechanism(s) against leptospiral infection, as well as to develop a new effective vaccine for leptospirosis.

ACKNOWLEDGMENTS

We thank C.W. Yang for comment to this manuscript.

References

Addamiano, L. and Babudieri, B. Water strains of *Leptospira* in the serodiagnosis of human and animal leptospirosis. Bull. World Health Organ. 1968; 39:925–934.

Adler, B. and Faine, S. Host immunological mechanisms in the resistance of mice to leptospiral infections. Infect. Immun. 1977; 17:67–72.

Andrade, L., Rodrigues, A.C., Jr., Sanches, T.R.C., Souza, R.B. and Seguro, A.C. Leptospirosis leads to dysregulation of sodium transporters in the kidney and lung. Am. J. Physiol. Renal Physiol. 2007; 292:F586–F592.

Artiushin, S., Timoney, J.F., Nally, J. and Verma, A. Host-inducible immunogenic sphingomyelinase-like protein, Lk73.5, of *Leptospira interrogans*. Infect. Immun. 2004; 72:742–749.

Ashford, D.A., Kaiser, R.M., Spiegel, R.A. et al. Asymptomatic infection and risk factors for leptospirosis in Nicaragua. Am. J. Trop. Med. Hyg. 2000; 63:249–254.

Babudieri, B. Vaccine against Leptospirosis. Fifth International Meeting of Biological Standardization. Jerusalem: The Weizmann Science Press of Israel; 1959:313–350.

Barbosa, A., Abreu, P.A.E., Neves, F.O. et al. A newly identified leptospiral adhesin mediates attachment to laminin. Infect. Immun. 2006; 74:6356–6364.

Baricos, W.H., Cortez, S.L., Deboisblanc, M. and Xin, S. Transforming growth factor-β is a potent inhibitor of extracellular matrix degradation by cultured human mesangial cells. J. Am. Soc. Nephrol. 1999; 10:790–795.

Barocchi, M.A., Ko, A.I., Reis, M.G., McDonald, K.M. and Riley, L.W. Rapid translocation of polarized MDCK cell monolayers by *Leptospira interrogans*, an invasive but nonintracellular pathogen. Infect. Immun. 2002; 70:6926–6932.

Barry, M., Wisnewski, A.V., Matthias, M.A., Inoue, S.K. and Vinetz, J.M. Suburban leptospirosis: atypical lymphocytosis and $\gamma - \delta$ T cell response. Clin. Infect. Dis. 2006; 43:1304–1307.

Bauby, H., Saint Girons, I. and Picardeau, M. Construction and complementation of the first auxotrophic mutant in the spirochaete *Leptospira meyeri*. Microbiology 2003; 149:689–693.

Bharadwaj, R., Bal, A.M., Joshi, S.A. et al. An urban outbreak of leptospirosis in Mumbai, India. Jpn. J. Infect. Dis. 2002; 55:194–196.

Bharti, A.R., Nally, J.E., Ricaldi, J.N., Matthias, M.A., Diaz, M.M., Lovett, M.A. et al. Leptospirosis: a zoonotic disease of global importance. Lancet Infect. Dis. 2003; 3:757–771.

Blumerman, S.L., Herzig, C.T. and Baldwin, C.L. $WC1^+$ $\gamma\delta$ T cell memory population is induced by killed bacterial vaccine. Eur. J. Immunol. 2007; 37:1204–1216.

Borer, A., Metz, I., Gilad, J. et al. Massive pulmonary hemorrhage caused by leptospirosis successfully treated with nitric oxide inhalation and haemofiltration. J. Infect. 1999; 38:42–45.

Bourhy, P., Louvel, H., Saint Girons, I. and Pacardeau, M. Random insertional mutagenesis of *Leptospira interrogans*, the agent of leptospirosis, using a *mariner* transposon. J. Bacteriol. 2005; 187:3255–3258.

Branger, C., Chatrenet, B., Gauvrit, A., Aviat, F., Aubert, A., Bach, J.M. and Andrè-Fontaine, G. Protection against *Leptospira interrogans* sensu lato challenge by DNA immunization with the gene encoding hemolysin-associated protein 1. Infect. Immun. 2005; 73:4062–4069.

Branger, C., Sonrier, C., Chatrenet, B., Klonjkowski, B., Ruvoen-Clouet, N., Aubert, A. et al. Identification of the hemolysis-associated protein 1 as a cross-protective immunogen of *Leptospira interrogans* by adenovirus-mediated vaccination. Infect. Immun. 2001; 69:6831–6838.

Brenner, D.J., Kaufmann, A.F., Sulzer, K.R., Steigerwalt, A.G., Rogers, F.C. and Weyant, R.S. Further determination of DNA relatedness between serogroups and serovars in the family Leptospiraceae with a proposal for *Leptospira alexanderi* sp. nov. and four new *Leptospira* genomospecies. Int. J. Syst. Bacteriol. 1999; 49:839–858.

Brown, R.A., Blumerman, S., Gay, C., Bolin, C., Duby, R. and Baldwin, C.L. Comparison of three different leptospiral vaccines for induction of a type 1 immune response to *Leptospira borgpetersenii* serovar Hardjo. Vaccine 2003; 21:4448–4458.

Bulach, D.M., Zuerner, R.L., Wilson, P., Seemann, T., McGrath, A., Cullen, P.A. et al. Genome reduction in *Leptospira borgpetersenii* reflects limited transmission potential. Proc. Natl. Acad. Sci. U.S.A. 2006; 103:14560–14565.

Burth, P., Younes-Ibrahim, M., Gonçalez, F.H.F.S., Costa, E.R. and Castro Faria, M.V. Purification and characterization of a Na^+, K^+ ATPase inhibitor found in a endotoxin of *Leptospira interrogans*. Infect. Immun. 1997; 65:1557–1560.

Burth, P., Younes-Ibrahim, M., Santos, F.C.B., Casto-Faria Neto, H.C. and Castro Faria, M.V. Role of nonesterified unsaturated fatty acids in the pathophysiological processes of leptospiral infection. J. Infect. Dis. 2005; 191:51–57.

Cachay, E.R. and Vinetz, J.M. A global research agenda for leptospirosis. J. Postgrad. Med. 2005; 52:174–178.

Chen, T.Z. Development and present status of leptospiral vaccine and technology of vaccine production in China. Jpn. J. Bacteriol. 1985; 40:755–762.

Choy, H.A., Kelley, M.M., Chen, T.L., Møller, A.K., Matsunaga, J. and Haake, D.A. Physiological osmotic induction of *Leptospira interrogans* adhesin: LigA and LigB bind extracellular matrix proteins and fibrinogen. Infect. Immun. 2007; 75:2441–2450.

Cinco, M., Domenis, R., Perticarari, S., Presani, G., Marangoni, A. and Blasi, E. Interaction of leptospires with murine microglial cells. New Microbiol. 2006; 29:193–199.

Croda, J., Ramos, J.G., Matsunaga, J. et al. *Leptospira* immunoglobulin-like proteins as a serodiagnostic marker for acute leptospirosis. J. Clin. Microbiol. 2007; 45:1528–1534.

Cullen, P., Xu, X., Matsunaga, J., Sanchez, Y., Ko, A.I., Haake, D.A. and Adler, B. Surfaceome of *Leptospira* spp. Infect. Immun. 2005; 73:4853–4863.

Cullen, P.A., Cordwell, S.J., Bulach, D.M., Haake, D.A. and Adler, B. Global analysis of outer membrane proteins from *Leptospira interrogans* serovar Lai. Infect. Immun. 2002; 70:2311–2318.

Cullen, P.A., Haake, D.A. and Adler, B. Outer membrane proteins of pathogenic spirochetes. FEMS Microbiol. Rev. 2004; 28:291–318.

Cullen, P.A., Haake, D.A., Bulach, D.M., Zuerner, R.L. and Adler, B. LipL21 is a novel surface-exposed lipoprotein of pathogenic *Leptospira* species. Infect. Immun. 2003; 71:2414–2421.

Daher, E., Zanetta, D.M., Cavalcante, M.B. and Abdulkader, R.C. Risk factors for death and changing pattern in leptospirosis acute renal failure. Am. J. Trop. Med. Hyg. 1999; 61:630–634.

de Fost, M., Hartskeerl, R.A., Groenendijk, M.R. and van der Poll, T. Interleukin 12 in part regulates gamma interferon release in human whole blood stimulated with *Leptospira interrogans*. Clin. Diagn. Lab. Immunol. 2003; 10:332–335.

Diament, D., Brunialti, M.K., Romero, E.C., Kallas, E.G. and Salomao, R. Peripheral blood mononuclear cell activation induced by *Leptospira interrogans* glycolipoprotein. Infect. Immun. 2002; 70:1677–1683.

Dolhnikoff, M., Mauad, T., Bethlem, E.P. and Carvalho, C.R.R. Leptospiral pneumonias. Curr. Opin. Pulm. Med. 2007; 13: 230–235.

Dorigatti, F., Brunialti, M.K.C., Romero, E.C., Kallas, E.G. and Salomão, R. *Leptospira interrogans* activation of peripheral blood monocyte glycolipoprotein demonstrated in whole blood by the release of IL-6. Braz. J. Med. Biol. Res. 2005; 38:99–104.

Easton, A. Leptospirosis in Philippine floods. BMJ 1999; 319:212.

Edwards, C.N., Nicholson, G.D. and Everard, C.O.R. Thrombocytopenia in leptospirosis. Am. J. Trop. Med. Hyg. 1982; 31:827–829.

Edwards, C.N., Nicholson, G.D., Hassel, T.A., Everard, C.O.R. and Callender, J. Thrombocytopenia in leptospirosis: the absence of evidence for disseminated intravascular coagulation. Am. J. Trop. Med. Hyg. 1986; 35:352–354.

Faber, N.A., Crawford, M., LeFebvre, R.B., Buyukmihci, N.C., Madigan, J.E. and Willits, N.H. Detection of *Leptospira* spp. in the aqueous humor of horses with naturally acquired recurrent uveitis. J. Clin. Microbiol. 2000; 38:2731–2733.

Faine, S., Adler, B., Bolin, C. and Perolat, P. *Leptospira* and Leptospirosis, 2nd ed. Melbourne: MediSci, 1999.

Flannery, B., Costa, D., Carvalho, F.P., Guerreiro, H., Matsunaga, J., Da Silva, E.D. et al. Evaluation of recombinant *Leptospira* antigen-based enzyme-linked immunosorbent assays for the serodiagnosis of leptospirosis. J. Clin. Microbiol. 2001; 39:3303–3310.

Fukumura, K. Epidemiological studies on the leptospirosis in Okinawa Part 1. Prevalence of the leptospirosis in Izena Island and the prevention by vaccination. Yamaguchi Igaku 1984; 33:257–268.

Gamberini, M., Gomez, R.M., Atzingen, M.V., Martins, E.A., Vasconcellos, S.A., Romero, E.C. et al. Whole-genome analysis of *Leptospira interrogans* to identify potential vaccine candidates against leptospirosis. FEMS Microbiol. Lett. 2005; 244: 305–313.

Gerold, G., Zychlinsky, A. and de Diego, J.L. What is the role of Toll-like receptors in bacterial infections? Semin. Immunol. 2007; 19:41–47.

Girons, I.S., Bourhy, P., Ottone, C., Picardeau, M., Yelton, D., Hendrix, R.W. et al. The LE1 bacteriophage replicates as a plasmid within *Leptospira biflexa*: construction of an *L. biflexa–Escherichia coli* shuttle vector. J. Bacteriol. 2000; 182:5700–5705.

Gonsalez, C.R., Casseb, J., Monterio, F.G. et al. Use of doxycycline for leptospirosis after high-risk exposure in Sao Paulo, Brazil. Rev. Inst. Med. Trop. Sao Paulo 1998; 40:59–61.

Gorin, P.A., Horitsu, K. and Spencer, J.F.T. An exocellular mannan, alternately linked 1, 3-β and 1, 4- β from *Rhodotorula glutinis*. Can. J. Chem. 1965; 43:950–954.

Griffith, M.E., Hospenthal, D.R. and Murray, C.K. Antimicrobial therapy of leptospirosis. Curr. Opin. Infect. Dis. 2006; 19:533–537.

Guerreiro, H., Croda, J., Flannery, B., Mazel, M., Matsunaga, J., Galvão Reis, M. et al. Leptospiral proteins recognized during the humoral immune response to leptospirosis in humans. Infect. Immun. 2001; 69:4958–4968.

Haake, D.A., Champion, C.I., Martinich, C., Shang, E.S., Blanco, D.R., Miller, J.N. et al. Molecular cloning and sequence analysis of the gene encoding OmpL1, a transmembrane outer membrane protein of pathogenic *Leptospira* spp. J. Bacteriol. 1993; 175:4225–4234.

Haake, D.A., Chao, G., Zuerner, R.L., Barnett, J.K., Barnett, D., Mazel, M. et al. The leptospiral major outer membrane protein LipL32 is a lipoprotein expressed during mammalian infection. Infect. Immun. 2000; 68:2276–2285.

Haake, D.A., Martinich, C., Summers, T.A., Shang, E.S., Pruetz, J.D., McCoy, A.M. et al. Characterization of leptospiral outer membrane lipoprotein LipL36: downregulation associated with late-log-phase growth and mammalian infection. Infect. Immun. 1998; 66:1579–1587.

Haake, D.A. and Matsunaga, J. Characterization of the leptospiral outer membrane and description of three novel leptospiral membrane proteins. Infect. Immun. 2002; 70:4936–4945.

Haake, D.A., Mazel, M.K., McCoy, A.M., Milward, F., Chao, G., Matsunaga, J. et al. Leptospiral outer membrane proteins OmpL1 and LipL41 exhibit synergistic immunoprotection. Infect. Immun. 1999; 67:6572–6582.

Haake, D.A., Walker, E.M., Blanco, D.R., Bolin, C.A., Miller, M.N. and Lovett, M.A. Changes in the surface of *Leptospira interrogans* serovar grippotyphosa during in vitro cultivation. Infect. Immun. 1991; 59:1131–1140.

Hartskeerl, R.A. Leptospirosis: current status and future trends. Indian J. Med. Microbiol. 2006; 24:309.

Hauk, P., Negrotto, S., Romero, E.C. et al. Expression and characterization of HlyX hemolysin from *Leptospira interrogans* serovar Copenhageni: potentiation of hemolytic activity by LipL32. Biochem. Biophys. Res. Commun. 2005; 333:1341–1347.

Hsieh, W.J., Chang, Y.F., Chen, C.S. and Pan, M.J. Omp52 is a growth-phase-regulated outer membrane protein of *Leptospira santarosai* serovar Shermani. FEMS Microbiol. Lett. 2005; 243:339–345.

http://www.leptonet.net/html/vaccination.asp last accessed on 30 July, 2007.

http://sante-az.aufeminin.com/w/sante/m3514716/medicaments/spirolept.html last accessed on 30 July, 2007.

http://www.siobp.com/CPJS/YFZP_13.htm last accessed on 30 July, 2007.

Hübener, E.A. and Reiter, H. Beiträge zur Aetioligie der Weilschen Krankheit. Dtsch. Med. Wochenschr. 1915; 41:1275–1277.

Hung, C.C., Chang, C.T., Chen, K.H. et al. Upregulation of chemokine CXCL1/KC by leptospiral membrane protein preparation in renal tubule epithelial cells. Kidney Int. 2006a; 69:1814–1822.

Hung, C.C., Chang, C.T., Tian, Y.C. et al. Leptospiral membrane proteins stimulate pro-inflammatory chemokines secretion by renal tubule epithelial cells through toll-like receptor 2 and p38 mitogen activated protein kinase. Nephrol. Dial. Transplant. 2006b; 21:898–910.

Ikoev, V.N., Gorbunov, M.A., Vachaev, B.F., Iagovkin, E.A., Kondratenko, V.F., Anan'ina, IuV. et al. The evaluation of the reactogenicity and immunogenic activity of a new concentrated inactivated leptospirosis vaccine. Zh. Mikrobiol. Epidemiol. Immunobiol. 1999; (4):39–43.

Jansen, A., Schöneberg, I., Frank, C., Alpers, K., Schneider, T. and Stark, K. Leptospirosis in Germany, 1962–2003. Emerg. Infect. Dis. 2005; 11:1048–1054.

Johnson, M.A., Smith, H., Joseph, P. et al. Environmental exposure and leptospirosis, Peru. Emerg. Infect. Dis. 2004; 10:1016–1022.

Jost, B.H., Adler, B. and Faine, S. Experimental immunization of hamsters with lipopolysaccharide antigens of *Leptospira interrogans*. J. Med. Microbiol. 1989; 29:115–120.

Jost, B.H., Adler, B., Vihn, T. and Faine, S. A monoclonal antibody reacting with a determinant on leptospiral lipopolysacharide protects guinea pigs against leptospirosis. J. Med. Microbiol. 1986; 22:269–275.

Karande, S., Gandhi, D., Kulkarni, M., Bharadwaj, R., Pol, S., Thakare, J. and De, A. Concurrent outbreak of leptospirosis and dengue in Mumbai, India, 2002. J. Trop. Pediatr. 2005; 51:174–181.

Klimpel, G.R., Matthias, M.A. and Vinetz, J.M. *Leptospira interrogans* activation of human peripheral blood mononuclear cells: preferential expansion of TCR $\gamma\delta^+$ T cells vs. TCR $\alpha\beta^+$ T cells. J. Immunol. 2003; 171:1447–1455.

Ko, A.I., Reis, M.G., Dourado, C.M. et al. Urban epidemic of severe leptospirosis in Brazil. Lancet 1999; 354:820–825.

Kobayashi, Y. Discovery of the causative organism of Weil's disease: historical view. J. Infect. Chemother. 2001; 7:10–15.

Koizumi, N. and Watanabe, H. Identification of a novel antigen of pathogenic *Leptospira* spp. that reacted with the convalescent mice sera. J. Med. Microbiol. 2003a; 52:585–589.

Koizumi, N. and Watanabe, H. Leptospiral immunoglobulin-like proteins elicit protective immunity. Vaccine 2004; 22:1545–1552.

Koizumi, N. and Watanabe, H. Leptospirosis vaccine: past, present, and future. J. Postgrad. Med. 2005; 51:210–214.

Koizumi, N. and Watanabe, H. Molecular cloning and characterization of a novel leptospiral lipoprotein with OmpA domain. FEMS Microbiol. Lett. 2003b; 226:215–219.

LaRocque, R.C., Breiman, R.F., Ari, M.D. et al. Leptospirosis during Dengue outbreak, Bangladesh. Emerg. Infect. Dis. 2005; 11:766–769.

Laurichesse, H., Gourdon, F., Smits, H.L., Abdoe, T.H., Estavoyer, J.M., Rebika, H. et al. Safety and immunogenicity of subcutaneous or intramuscular administration of a monovalent inactivated vaccine against *Leptospira interrogans* serogroup Icterohaemorrhagiae in healthy volunteers. Clin. Microbiol. Infect. 2007; 13:395–403.

Lee, S.H., Kim, S., Park, S.C. and Kim, M.J. Cytotoxic activities of *Leptospira interrogans* hemolysin SphH as a pore-forming protein on mammalian cells. Infect. Immun. 2002; 70:315–322.

Lee, S.H., Kim, K.A., Park, Y.K., Seong, I.W., Kim, M.J. and Lee, Y.J. Identification and partial characterization of a novel hemolysin from *Leptospira interrogans* serovar *lai*. Gene 2000; 254:19–28.

Levett, P.N. Leptospirosis. Clin. Microbiol. Rev. 2001; 14:296–326.

Levett, P.N., Morey, R.E., Galloway, R., Steigerwalt, A.G. and Ellis, W.A. Reclassification of *Leptospira parva* Hovind-Hougen et al. 1982 as *Turneriella parva* gen. nov., comb. nov. Int. J. Syst. Evol. Microbiol. 2005; 55:1497–1499.

Levett, P.N., Morey, R.E., Galloway, R.L. and Steigerwalt, A.G. *Leptospira broomii* sp. nov., isolated from humans with leptospirosis. Int. J. Syst. Evol. Microbiol. 2006; 56:671–673.

Lingappa, J., Kuffner, T., Tappero, J., Whitworth, W., Mize, A., Kaiser, R. et al. HLA-DQ6 and ingestion of contaminated water: possible gene-environment interaction in an outbreak of leptospirosis. Genes Immun. 2004; 5:197–202.

Lo, M., Bulach, D.M., Powell, D.R., Haake, D.A., Matsunaga, J., Paustian, M.L., Zuerner, R.L. and Adler, B. Effects of temperature on gene expression patterns in *Leptospira interrogans* serovar Lai as assessed by whole-genome microarrays. Infect. Immun. 2006; 74:5848–5859.

Lucchesi, P.M.A. and Parma, A.E. A DNA fragment of *Leptospira interrogans* encodes a protein which shares epitopes with equine cornea. Vet. Immunol. Immunopathol. 1999; 71:173–179.

Mancel, E., Merien, F., Pesenti, L., Salino, D., Angibaud, G. and Perolat, P. Clinical aspects of ocular leptospirosis in New Caledonia (South Pacific). Aust. N.Z. J. Ophthalmol. 1999; 27:380–386.

Martínez Sánchez, R., Obregón Fuentes, A.M., Pérez Sierra, A., Baly Gil, A., Díaz González, M., Baró Suárez, M. et al. The reactogenicity and immunogenicity of the first Cuban vaccine against human leptospirosis. Rev. Cubana Med. Trop. 1998; 50:159–166.

Martínez, R., Pérez, A., Quiñones, Mdel.C., Cruz, R., Álvarez, Á., Armesto, M. et al. Efficacy and safety of a vaccine against human leptospirosis in Cuba. Rev. Panam. Salud. Pública 2004; 15:249–255.

Martins, M.G., Matos, K.T., da Silva, M.V. and de Abreu, M.T. Ocular manifestations in the acute phase of leptospirosis. Ocul. Immunol. Inflamm. 1998; 6:75–79.

Maskey, M., Shastri, J.S., Saraswathi, K., Surpam, R. and Vaidya, N. Leptospirosis in Mumbai: post-deluge outbreak 2005. Indian J. Med. Microbiol. 2006; 24:337–338.

Matsunaga, J., Barocchi, M.A., Croda, J., Young, T.A., Sanchez, Y., Siqueira, I. et al. Pathogenic *Leptospira* species express surface-exposed proteins belonging to the bacterial immunoglobulin superfamily. Mol. Microbiol. 2003; 49:929–945.

Matsunaga, J., Lo, M., Bulach, D.M., Zuerner, R.L., Adler, B. and Haake, D.A. Response of *Leptospira interrogans* to physiologic osmolarity: relevance in signaling the environment-to-host transition. Infect. Immun. 2007; 75:2864–2874.

Matsunaga, J., Sanchez, Y., Xu, X. and Haake, D.A. Osmolarity, a key environmental signal controlling expression of leptospiral proteins LigA and LigB and the extracellular release of LigA. Infect. Immun. 2005; 73:70–78.

Matsunaga, J., Werneid, K., Zuerner, R., Frank, A. and Haake, D.A. LipL46 is a novel surface-exposed lipoprotein expressed during leptospiral dissemination in the mammalian host. Microbiology 2006; 152:3777–3786.

Matsunaga, J., Young, T.A., Barnett, J.K., Barnett, D., Bolin, C.A. and Haake, D.A. Novel 45-kilodalton leptospiral protein that is processed to a 31-kilodalton growth-phase-regulated peripheral membrane protein. Infect. Immun. 2002; 70:323–334.

Matsuo, K., Isogai, E. and Araki, Y. Occurrence of [→3)-β-d-Man*p*-(1→4)- β-d-Man*p*-(1→]$_n$ units in the antigenic polysaccharides from *Leptospira biflexa* serovar patoc strain Patoc I. Carbohydr. Res. 2000a; 328:517–524.

Matsuo, K., Isogai, E. and Araki, Y. Utilization of exocellular mannan from *Rhodotorula glutinis* as an immunoreactive antigen in diagnosis of leptospirosis. J. Clin. Microbiol. 2000b; 38: 3750–3754.

Matsuo, K., Isogai, E. and Araki, Y. Control of immunologically crossreactive leptospiral infection by administration of lipopolysaccharides from a nonpathogenic strain of *Leptospira biflexa*. Microbiol. Immunol. 2000c; 44:887–890.

Matthay, M.A., Folkesson, H.G. and Clerici, C. Lung epithelial fluid transport and the resolution of pulmonary edema. Physiol. Rev. 2002; 82:569–600.

McBride, A.J.L., Athanazio, D.A., Reis, M.G. and Ko, A.I. Leptospirosis. Curr. Opin. Infect. Dis. 2005; 18:376–386.

McClain, J.B., Ballou, W.R., Harrison, S.M. and Steinweg, D.L. Doxycycline therapy for leptospirosis. Ann. Intern. Med. 1984; 100:696–698.

Meri, T., Murgia, R., Stefanel, P., Meri, S. and Cinco, M. Regulation of complement activation at C3-level by serum resistant leptospires. Microb. Pathog. 2005; 39:139–147.

Merien, F., Baranton, G. and Perolat, P. Invasion of Vero cells and induction of apoptosis in macrophages by pathogenic *Leptospira interrogans* are correlated with virulence. Infect. Immun. 1997; 65:729–738.

Merien, F., Toruccolo, J., Baranton, G. and Perolat, P. Identification of a 36-kDa fibronectin-binding protein expressed by a virulent variant of *Leptospira interrogans* serovar icterohaemorrhagiae. FEMS Microbiol. Lett. 2000; 185:17–22.

Monsuez, J.J., Kidouche, R., Le Gueno, B. and Postic, D. Leptospirosis presenting as haemorrhagic fever in visitor to Africa. Lancet 1997; 349:254–255.

Morgan, J., Bornstein, S.L., Karpati, A.M. et al. Outbreak of leptospirosis among triathlon participants and community residents in Springfield, Illinois, 1998. Clin. Infect. Dis. 2002; 34:1593–1599.

Mudur, G. Mumbai braces itself for leptospirosis and waterborne infections. BMJ 2005; 331:307.

Nahori, M.A., Fournie-Amazouz, E., Que-Gewirth, N.S., Balloy, V., Chignard, M., Raetz, C.R. et al. Differential TLR recognition of leptospiral lipid A and lipopolysaccharide in murine and human cells. J. Immunol. 2005; 175:6022–6031.

Naiman, B.M., Alt, D., Bolin, C.A., Zuerner, R. and Baldwin, C.L. Protective killed *Leptospira borgpetersenii* vaccine induces potent Th1 immunity comprising responses by CD4 and γδ T lymphocytes. Infect. Immun. 2001; 69:7550–7558.

Naiman, B.M., Blumerman, S., Alt, D., Bolin, C.A., Brown, R., Zuerner, R. and Baldwin, C.L. Evaluation of type 1 immune response in naïve and vaccinated animals following challenge with *Leptospira borgpetersenii* serovar Hardjo: involvement of WC1$^+$ γδ and CD4 T cells. Infect. Immun. 2002; 70:6147–6157.

Nakamura, M., Taira, K., Itokazu, K., Kudaka, J., Asato, R., Kise, T. and Koizumi, N. Sporadic cases and an outbreak of leptospirosis probably associated with recreational activities in rivers in the northern part of Okinawa Main Island. J. Vet. Med. Sci. 2006; 68:83–85.

Nally, J.E., Artiushin, S. and Timoney, J.F. Molecular characterization of thermoinduced immunogenic proteins Q1p42 and Hsp15 of *Leptospira interrogans*. Infect. Immun. 2001; 69:7616–7624.

Nally, J.E., Chantranuwat, C., Wu, X.Y. et al. Alveolar septal deposition of immunoglobulin and complement parallels pulmonary hemorrhage in a guinea pig model of severe pulmonary leptospirosis. Am. J. Pathol. 2004; 163:1115–1127.

Nally, J.E., Chow, E., Fishbein, M.C., Blanco, D.R. and Lovett, M.A. Changes in lipopolysaccharide O antigen distinguish acute versus chronic *Leptospira interrogans* infections. Infect. Immun. 2005a; 73:3251–3260.

Nally, J.E., Whitelegge, J.P., Aguilera, R., Pereira, M.M., Blanco, D.R. and Lovett, M.A. Purification and proteomic analysis of outer membrane vesicles from a clinical isolate of *Leptospira interrogans* serovar Copenhageni. Proteomics 2005b; 5:144–152.

Nally, J.E., Whitelegge, J.P., Bassilian, S., Blanco, D.R. and Lovett, M.A. Characterization of the outer membrane proteome of *Leptospira interrogans* expressed during acute lethal infection. Infect. Immun. 2007; 75:766–773.

Nardone, A., Capek, I., Baranton, G., Campèse, C., Postic, D., Vaillant, V. et al. Risk factors for leptospirosis in Metropolitan France: results of a national case-control study, 1999–2000. Clin. Infect. Dis. 2004; 39:751–753.

Narita, M., Fujitani, S., Haake, D.A. and Paterson, D.L. Leptospirosis after recreational exposure to water in the Yaeyama Islands, Japan. Am. J. Trop. Med. Hyg. 2005; 73:652–656.

Nascimento, A.L., Ko, A.I., Martins, E.A., Monteiro-Vitorello, C.B., Ho, P.L., Haake, D.A. et al. Comparative genomics of two *Leptospira interrogans* serovars reveals novel insights into physiology and pathogenesis. J. Bacteriol. 2004a; 186: 2164–2172.

Nascimento, A.L., Verjovski-Almeida, S., Van Sluys, M.A., Monteiro-Vitorello, C.B., Camargo, L.E., Digiampietri, L.A. et al. Genome features of *Leptospira interrogans* serovar Copenhageni. Braz. J. Med. Biol. Res. 2004b; 37:459–477.

Nicodemo, A.C., Duarte, M.I.S., Alves, D.V.A.F., Takakura, C.F.H., Santos, R.T.M. and Nicodemo, E.L. Lung lesions in human leptospirosis: microscopic, immunohistochemical, and ultrastructural features related to thrombocytopenia. Am. J. Trop. Med. Hyg. 1997; 56:181–187.

Palaniappan, R.U., Chang, Y.F., Hassan, F., McDonough, S.P., Pough, M., Barr, S.C. et al. Expression of leptospiral immunoglobulin-like protein by *Leptospira interrogans* and evaluation of its diagnostic potential in a kinetic ELISA. J. Med. Microbiol. 2004; 53:975–984.

Palaniappan, R.U., Chang, Y.F., Jusuf, S.S., Artiushin, S., Timoney, J.F., McDonough, S.P. et al. Cloning and molecular characterization of an immunogenic LigA protein of *Leptospira interrogans*. Infect. Immun. 2002; 70:5924–5930.

Palaniappan, R.U., McDonough, S.P., Divers, T.J. et al. Immunoprotection of recombinant leptospiral immunoglobulin-like protein A against *Leptospira interrogans* serovar Pomona infection. Infect. Immun. 2006; 74:1745–1750.

Panaphut, T., Domrongkitchaiporn, S., Vibhagool, A., Thinkamrop, B. and Susaengrat, W. Ceftriaxone compared with sodium penicillin G for treatment of severe leptospirosis. Clin. Infect. Dis. 2003; 36:1507–1513.

Pappas, G. and Cascio, A. Optimal treatment of leptospirosis: queries and projections. Int. J. Antimicrob. Agents 2006; 28:491–496.

Park, S.H., Ahn, B.Y. and Kim, M.J. Expression and immunologic characterization of recombinant heat shock protein 58 of *Leptospira* species: a major target antigen of the humoral immune response. DNA Cell. Biol. 1999; 18:903–910.

Penna, D., De Brito, T., Pupo, A.A., Machado, M.M., Ayroza, P.A. and De Almeida, S.S. Kidney biopsy in human leptospirosis. Am. J. Trop. Med. Hyg. 1963; 12:896–901.

Philip, N.A. and Tennent, R.B. Leptospirosis: a report from one practice on the use of a leptospiral vaccine for a period of three years. N.Z. Med. J. 1966; 65(Suppl. 15):13–19.

Phraisuwan, P., Spotts Whitney, E.A., Tharmaphornpilas, P. et al. Leptospirosis: skin wounds and control strategies, Thailand, 1999. Emerg. Infect. Dis. 2002; 8:1455–1459.

Picardeau, M., Brenot, A. and Saint Girons, I. First evidence for gene replacement in *Leptospira* spp. Inactivation of *L. biflexa flaB* results in non-motile mutants deficient in endoflagella. Mol. Microbiol. 2001; 40:189–199.

Plank, R. and Dean, D. Overview of the epidemiology, microbiology, and pathogenesis of *Leptospira* spp. in humans. Microbes Infect. 2000; 2:1265–1276.

Plotkin, S.A. Six revolutions in vaccinology. Pediatr. Infect. Dis. J. 2005; 24:1–9.

Qin, J.H., Sheng, Y.Y., Zhang, Z.M. et al. Genome-wide transcriptional analysis of temperature shift in *L. interrogans* serovar lai strain 56601. BMC Microbiol. 2006; 6:51.

Que-Gewirth, N.L.S., Ribeiro, A.A., kalb, S.R., Cotter, R.J., Bulach, D.M., Adler, B. et al. A methylated phosphate group and four amide-linked acyl chains in *Leptospira interrogans* lipid A. J. Biol. Chem. 2004; 279:25420–25429.

Rappuoli, R. Reverse vaccinology, a genome-based approach to vaccine development. Vaccine 2001; 19:2688–2691.

Rathinam, S.R. Ocular leptospirosis. Curr. Opin. Ophthalmol. 2002; 13:381–386.

Reisberg, B.E., Wurtz, R., Diaz, P., Francis, B., Zakowski, P., Fannin, S. et al. Outbreak of leptospirosis among white-water rafters—Costa Rica, 1996. Morb. Mortal. Wkly. Rep. 1997; 46:577–579.

Ren, S.X., Fu, G., Jiang, X.G., Zeng, R., Miao, Y.G., Xu, H. et al. Unique physiological and pathogenic features of *Leptospira interrogans* revealed by whole-genome sequencing. Nature 2003; 422:888–893.

Ricaldi, J.N. and Vinetz, J.M. Leptospirosis in the tropics and in travelers. Curr. Infect. Dis. Rep. 2006; 8:51–58.

Rogers, A.N., Vanburen, D.G., Hedblom, E.E., Tilahun, M.E., Telfer, J.C. and Baldwin, C.L. γδ T cell function varies with the expressed WC1 coreceptor. J. Immunol. 2005; 174:3386–3393.

Sánchez, R.M., Sierra, A.P., Obregón Fuentes, A.M., González, I.R., Gil, A.B., Suárez, M.B. et al. Reactogenicity and immunogenicity of Cuban trivalent inactivated vaccine against human leptospirosis in different vaccination schedules. Rev. Cubana Med. Trop. 2002; 54:37–43.

Sánchez, R.M., Sierra, A.P., Suárez, M.B., Álvarez, Á.M., Hernández, J.M., González, M.D. et al. Evaluation of the effectiveness of a new vaccine against human leptospirosis in groups at risk. Rev. Panam. Salud. Pública 2000; 8:385–392.

Sanders, E.J., Rigau-Pérez, J.G., Smith, H.L. et al. Increase of leptospirosis in dengue-negative patients after a hurricane in Puerto Rico in 1996 [correction of 1966]. Am. J. Trop. Med. Hyg. 1999; 61:399–404.

Schoone, G.J., Everard, C.O.R., Korver, H., Carrington, D.G., Inniss, V.A. and Terpstra, W.J. An immunoprotective monoclonal antibody directed against *Leptospira interrogans* serovar *copenhageni*. J. Gen. Microbiol. 1989; 135:73–78.

Segers, P.A.M., van der Drift, A., de Nijs, A., Corcione, P., ven der Zeijst, B.A.M. and Gaastra, W. Molecular analysis of a sphingomyelinase C gene from Leptospira interrogans serovar Hardjo. Infect. Immun. 1990; 58:2177–2185.

Segura, E.R., Gonaza, C., Campos, K. et al. Clinical spectrum of pulmonary involvement in leptospirosis in a region of endemicity, with quantification of leptospiral burden. Clin. Infect. Dis. 2005; 40:343–351.

Sehgal, S.C., Sugunan, A.P., Murhekar, M.V., Sharma, S. and Vijayachari, P. Randomized controlled trial doxycycline prophylaxis against leptospirosis in an endemic area. Int. J. Antimicrob. Agents 2000; 13:249–255.

Seijo, A., Coto, H., San Juan, J. et al. Lethal leptospirosis pulmonary hemorrhage: an emerging disease in Buenos Aires, Argentina. Emerg. Infect. Dis. 2002; 8:1004–1005.

Sejvar, J., Bancroft, E., Winthrop, K. et al. Leptospirosis in eco-challenge athletes, Malaysian Borneo, 2000. Emerg. Infect. Dis. 2003; 9:702–707.

Serruto, D., Adu-Bobie, J., Capecchi, B., Rappuoli, R., Pizza, M. and Masignani, V. Biotechnology and vaccines: application of functional genomics to *Neisseria meningitidis* and other bacterial pathogens. J. Biotechnol. 2004; 113:15–32.

Shang, E.S., Exner, M.M., Summers, T.A., Martinich, C., Champion, C.I., Hancock, R.E. et al. The rare outer membrane protein, OmpL1, of pathogenic *Leptospira* species is a heat-modifiable porin. Infect. Immun. 1995; 63:3174–3181.

Shang, E.S., Summers, T.A. and Haake, D.A. Molecular cloning and sequence analysis of the gene encoding LipL41, a surface-exposed lipoprotein of pathogenic *Leptospira* species. Infect. Immun. 1996; 64:2322–2330.

Sharma, S., Vijayachari, P., Sugunan, A.P., Natarajaseenivasan, K. and Sehgal, S.C. Seroprevalence of leptospirosis among high-risk population of Andaman Islands, India. Am. J. Trop. Med. Hyg. 2006; 74:278–283.

Shenoy, V.V., Nagar, V.S., Chowdhury, A.A., Bhalgat, P.S. and Juvale, N.I. Pulmonary leptospirosis: an excellent response to bolus methylprednisolone. Postgrad. Med. J. 2006; 82:602–606.

Sonrier, C., Branger, C., Michel, V., Ruvoën-Clouet, N., Ganière, J.P. and André-Fontaine, G. Evidence of cross-protection within *Leptospira interrogans* in an experimental model. Vaccine 2000; 19:86–94.

Sterling, C.R. and Thiermann, A.B. Urban rats as chronic carriers of leptospirosis: an ultrastructural investigation. Vet. Pathol. 1981; 18:628–637.

Stimson, A.M. Note on an organism found in yellow fever tissue. Public Health Rep. 1907; 22:541.

Suputtamongkol, Y., Niwattayakul, K., Suttinont, C. et al. An open, randomized, controlled trial of penicillin, doxycycline, and cefotaxime for patients with severe leptospirosis. Clin. Infect. Dis. 2004; 39:1417–1424.

Tajiki, H. and Salomão, R. Association of plasma levels of tumor necrosis factor alpha with severity of disease and mortality among patients with leptospirosis. Clin. Infect. Dis. 1996; 23:1177–1178.

Takafuji, E.T., Kirkpatrick, J.W., Miller, R.N. et al. An efficacy trial of doxycycline chemoprophylaxis against leptospirosis. N. Engl. J. Med. 1984; 310:497–500.

Tang, W.W., Qi, M., Warren, J.S. et al. Chemokine expression in experimental tubulointerstitial nephritis. J. Immunol. 1997; 159:870–876.

Tangkanakul, W., Tharmaphornpil, P., Plikaytis, B.D. et al. Risk factors associated with leptospirosis in northeastern Thailand, 1998. Am. J. Trop. Med. Hyg. 2000; 63:204–208.

Thomas, D.D. and Higbie, L.M. In vitro association of leptospires with host cells. Infect. Immun. 1990; 58(3):581–585.

Tian, Y.C., Chen, Y.C., Hung, C.C. et al. Leptospiral outer membrane protein induces extracellular matrix accumulation through a TGF-β1/Smad-dependent pathway. J. Am. Soc. Nephrol. 2006; 17:2792–2798.

Torten, M., Shenberg, E., Gerichter, C.B., Neuman, P. and Klingberg, M.A. A new leptospiral vaccine for use in man. II. Clinical and serologic evaluation of a field trial with volunteers. J. Infect. Dis. 1973; 128:647–651.

Trevejo, R.T., Rigau-Perez, J.G., Ashford, D.A. et al. Epidemic leptospirosis associated with pulmonary hemorrhage—Nicaragua, 1995. J. Infect. Dis. 1998; 178:1457–1463.

Turner, L.H. Leptospirosis. II. Serology. Trans. R. Soc. Trop. Med. Hyg. 1968; 62:880–899.

Uhlenhuth, P. and Fromme, W. Experimentelle Untersuchungen über die sogenannte Weilsche Krankheit (ansteckende Gelbsucht). Med. Klin. 1915; 44:1202–1203.

van Creel, R., Speelman, P., Gravekamp, C. and Terpstra, W.J. Leptospirosis in travelers. Clin. Infect. Dis. 1994; 19:132–134.

Verma, A., Artiushin, S., Matsunaga, J., Haake, D.A. and Timoney, J.F. LruA and LruB, novel lipoproteins of pathogenic *Leptospira interrogans* associated with equine recurrent uveitis. Infect. Immun. 2005; 73:7259–7266.

Verma, A., Hellwage, J., Artiushin, S. et al. LfhA, a novel factor H-binding protein of *Leptospira interrogans*. Infect. Immun. 2006; 74:2659–2666.

Vernel-Pauillac, F. and Merien, F. Proinflammatory and immunomodulatory cytokine mRNA time course profiles in hamsters infected with a virulent variant of *Leptospira interrogans*. Infect. Immun. 2006; 74:4172–4179.

Vihn, T., Faine, S. and Adler, B. Adhesion of leptospires to mouse fibroblast (L929) and its enhancement by specific antibody. J. Med. Microbiol. 1984; 18:73–85.

Vijayachari, P., Sehgal, S.C., Goris, M.G. et al. *Leptospira interrogans* serovar Valbuzzi: a cause of severe pulmonary hemorrhage in the Andaman Islands. J. Med. Microbiol. 2003; 52:913–918.

Vinetz, J.M. A mountain out of molehill: do we treat acute leptospirosis, and if so, with what? Clin. Infect. Dis. 2003; 36:1514–1515.

Vinetz, J.M., Glass, G.E., Flexner, C.E. et al. Sporadic urban leptospirosis. Ann. Intern. Med. 1996; 125:794–798.

Viriyakosol, S., Matthias, M.A., Swancutt, M.A., Kirkland, T.N. and Vinetz, J.M. Toll-like receptor 4 protects against lethal *Leptospira interrogans* serovar Icterohaemorrhagiae infection and contributes to in vivo control of leptospiral burden. Infect. Immun. 2006; 74:887–895.

Visith, S. and Kearkiat, P. Nephropathy in leptospirosis. J. Postgrad. Med. 2005; 51:184–188.

Wani, H. Ueber die prophylaxe der Spirochaetosis Icterohaemorrhagica Inada (Weilschen Krankheit) durch schutzimpfung. Zeitschrift Immunitätsforschung Experimentelle Therapie 1933; 79:1–26.

Watt, G., Padre, L.P., Tuazon, M.L. et al. Placebo-controlled trial of intravenous penicillin for severe and late leptospirosis. Lancet 1988; 1:433–435.

Weil, A. Über eine eigentümliche, mit Milztumor, Ikterus und nephritis einhergehende, acute Infectionskrankheit. Deutsch. Arch. Klin. Med. 1886; 39:209–232.

Werts, C., Tapping, R.I., Mathison, J.C., Chuang, T.H., Kravchenko, V., Saint Girons, I. et al. Leptospiral lipopolysaccharide activates cells through a TLR2-dependent mechanism. Nat. Immunol. 2001; 2:346–352.

Wongsrichanalai, C., Murray, C.K., Gray, M., Miller, R.S., McDaniel, P., Liao, W.J., Pickard, A.L. and Magill, A.J. Co-infection with malaria and leptospirosis. Am. J. Trop. Med. Hyg. 2003; 68:583–585.

World Health Organization. Leptospirosis worldwide, 1999. Wkly. Epidemiol. Rec. 1999; 74:237–242.

World Health Organization. Human leptospirosis: guidance for diagnosis, surveillance, and control. Geneva: World Health Organization, 2003.

Wu, M.S., Yang, C.W., Pan, M.J., Chang, C.T. and Chen, Y.C. Reduced renal Na^+–K^+–Cl^- co-transporter activity and inhibited *NKCC2* mRNA expression by *Leptospira shermani*: from bed-side to bench. Nephrol. Dial. Transplant. 2004; 19:2472–2479.

Yamamoto, T., Noble, N.A., Miller, D.E. and Border, W.A. Sustained expression of TGF-β1 underlies development of progressive kidney fibrosis. Kidney Int. 1994; 45:916–927.

Yan, Y., Chen, Y., Liou, W., Ding, J., Chen, J., Zhang, J. et al. An evaluation of the serological and epidemiological effects of the outer envelope vaccine to leptospira. J. Chin. Med. Assoc. 2003; 66:224–230.

Yang, C.W., Hung, C.C., Wu, M.S. et al. Toll-like receptor 2 mediates early inflammation by leptospiral outer membrane proteins in proximal tubule cells. Kidney Int. 2006a; 69:815–822.

Yang, C.W., Wu, M.S. and Pan, M.J. Leptospirosis renal disease. Nephrol. Dial. Transplant. 2001; 16:73–77.

Yang, C.W., Wu, M.S., Pan, M.J., Hsieh, W.J., Vandewalle, A. and Huang, C.C. The *Leptospira* outer membrane protein LipL32 induces tubulointerstitial nephritis-mediated gene expression in mouse proximal tubule cells. J. Am. Soc. Nephrol. 2002; 13:2037–2045.

Yang, G.G. and Hsu, Y.H. Nitric oxide production and immunoglobulin deposition in leptospiral hemorrhagic respiratory failure. J. Formos Med. Assoc. 2005; 104:759–763.

Yang, H.L., Zhu, Y.Z., Qin, J.H., He, P., Jiang, X.C., Zhao, G.P. and Guo, X.K. *In silico* and microarray-based genomic approaches to identifying potential vaccine candidates against *Leptospira interrogans*. BMC Genomics 2006b; 7:293.

Yang, H.Y., Hsu, P.Y., Pan, M.J., Wu, M.S., Lee, C.H., Yu, C.C., Hung, C.C. and Yang, C.W. Clinical distinction and evaluation of leptospirosis in Taiwan—a case-control study. J. Nephrol. 2005; 18:45–53.

Yersin, C., Bovet, P., Merien, F. et al. Pulmonary hemorrhage as a predominant cause of death in leptospirosis in Seychelles. Trans. R. Soc. Trop. Med. Hyg. 2000; 94:71–76.

Zaki, S.R. and Shieh, W. Leptospirosis associated with outbreak of acute febrile illness and pulmonary hemorrhage, Nicaragua, 1995. The Epidemic Working Group at Ministry of Health in Nicaragua. Lancet 1996; 347:535–536.

Zhang, Y.X., Geng, Y., Bi, B. et al. Identification and classification of all potential hemolysin encoding genes and their products from *Leptospira interrogans* serogroup Icterohaemorrhagiae serovar Lai. Acta Pharmacol. Sin. 2005; 26:453–461.

CHAPTER

65

Malaria

Thomas L. Richie and Falgunee K. Parekh
U.S. Military Malaria Vaccine Program, Naval Medical Research Center/Walter Reed Army Institute of Research, Silver Spring, MD, USA

OUTLINE

ABSTRACT

Malaria is the most important parasitic disease with its vast distribution across the tropics and subtropics, immense public health burden, and tremendous economic impact on affected populations. Efforts to control malaria are being conducted on multiple fronts including vector control through residual house spraying, distribution of insecticide-treated bednets and improved diagnosis and treatment, including the development of new drugs and drug combinations effective against resistant strains. Despite these efforts, malaria remains prevalent in at least 87 countries with approximately 40% of the world's population living at risk; the greatest burden of malaria falls on young children with one child dying from malaria every 30 s. Civil disturbance, inadequate health infrastructure, and poverty contribute to the lack of effective control. Experience with other infections such as smallpox, measles, and polio demonstrates that vaccines can be a highly effective and cost-efficient method of controlling an infectious agent with global impact.

The complexity of the malarial parasite has made the task of developing an efficacious vaccine difficult. Malaria is a vector-borne disease caused by a protozoan parasite of the genus *Plasmodium*, and transmitted by the bite of an *Anopheles* mosquito. The *Plasmodium* parasite has a complex life cycle with multiple stages, differential expression of antigens with each stage, and a high degree of antigenic variability. Additionally, unlike smallpox, measles, and many other infectious agents, protective immunity develops slowly, requires repeated infections, is never strong enough to prevent reinfection, and is quickly lost in the absence of exposure. Despite these challenges, there are two models of immunity that support the feasibility of developing a malaria vaccine: (1) immunization with radiation-attenuated sporozoites, which induces sterile protection against the sporozoite and liver stages of the parasite and (2) naturally acquired immunity, which is characterized by partial protection against the asexual blood stages. These models provide the basis for many of the malaria vaccine candidates currently under development, including vaccines designed to prevent infection (targeting sporozoite and liver stages) or to prevent disease (targeting blood stages). In addition, the sexual stages of the parasite can be targeted by vaccines designed to prevent transmission, an intriguing approach aimed at communitywide as opposed to individual benefit.

Currently, over 50 malaria vaccine candidates have reached various stages of clinical development. Most are subunit vaccines based on one or few antigens, and target one particular stage of the parasite life cycle. The subunit vaccines are constructed using a variety of platforms, including synthetic peptides or recombinant protein in adjuvant, viral vectors, DNA plasmids, and prime-boost combinations of these platforms. As it has become more apparent that a single antigen may not be sufficient to induce strong protective immunity against such a complex organism, a multiantigen, multistage approach is also being explored. More recently, there is focus on the attenuated whole parasite approach, which precludes the need to identify the antigens or immune mechanisms associated with protection.

This chapter will provide a summary of the history, biology, epidemiology, and clinical pathology of this important disease, and will then review progress in developing an effective malaria vaccine for widespread use.

INTRODUCTION

Malaria is the most important parasitic disease in the world, taking into account its global distribution, the number and severity of infections, and the appalling public burden. With an annual worldwide incidence exceeding 600 million clinical cases caused by the two most common malaria species, *Plasmodium falciparum* and *P. vivax* (Mendis et al., 2001; Snow et al., 2005), leading to 1.1–2.7 million deaths (attributable primarily to *P. falciparum*) (World Health Organization, 2005), malaria causes great suffering among people living in tropical and subtropical regions, with a disproportionate impact on infants, children, and pregnant women (Abdullah et al., 2007). Malaria is endemic in at least 87 countries, notably much of sub-Saharan Africa and large areas of Latin America, South Asia, Southeast Asia, and the Western Pacific, currently inhabited by at least 40% of the world's population, placing 2.5 billion people at risk (Guerra et al., 2006, 2007, 2008; Hay et al., 2004) (Fig. 65.1). Sub-Saharan Africa experiences the greatest impact, with adverse effects on health, economic development, and educational achievement (Fernando et al., 2003; Gallup and Sachs, 2001; Kihara et al., 2006; Sachs and Malaney, 2002; Worrall et al., 2007). In this region, malaria has been estimated to directly or indirectly decrease the gross domestic product by 1–5% (Gallup and Sachs, 2001; Sachs and Malaney, 2002). Malaria also poses a threat to travelers, often constituting the most significant health risk for those visiting malarious areas, whether via international or within-country travel (Richie, 2004). Development projects relying on the influx of personnel from nonmalarious areas can be compromised, contributing to the economic burden. As an added concern, coinfection with malaria can adversely influence the severity of other pathogens, such as HIV, and vice versa (Abu-Raddad et al., 2006).

Although the application of control measures such as insecticide-treated bednets and residual insecticide spraying, and the introduction of efficacious new drug regimens such as artemisinin combination therapies (ACTs) can significantly reduce the burden of malaria in a given community (Bhattarai et al., 2007; Bosman and Mendis, 2007; Rieckmann, 2006), historically such interventions have proven difficult to sustain. This is due to a variety of causes, including costs, poor public health infrastructure, civil disturbances, the development of insecticide resistance by the vector, and drug resistance by the parasite. As exacerbating forces, climate change or environmental disruptions can create more favorable habitats for mosquito breeding, leading to the expansion of vector populations into regions

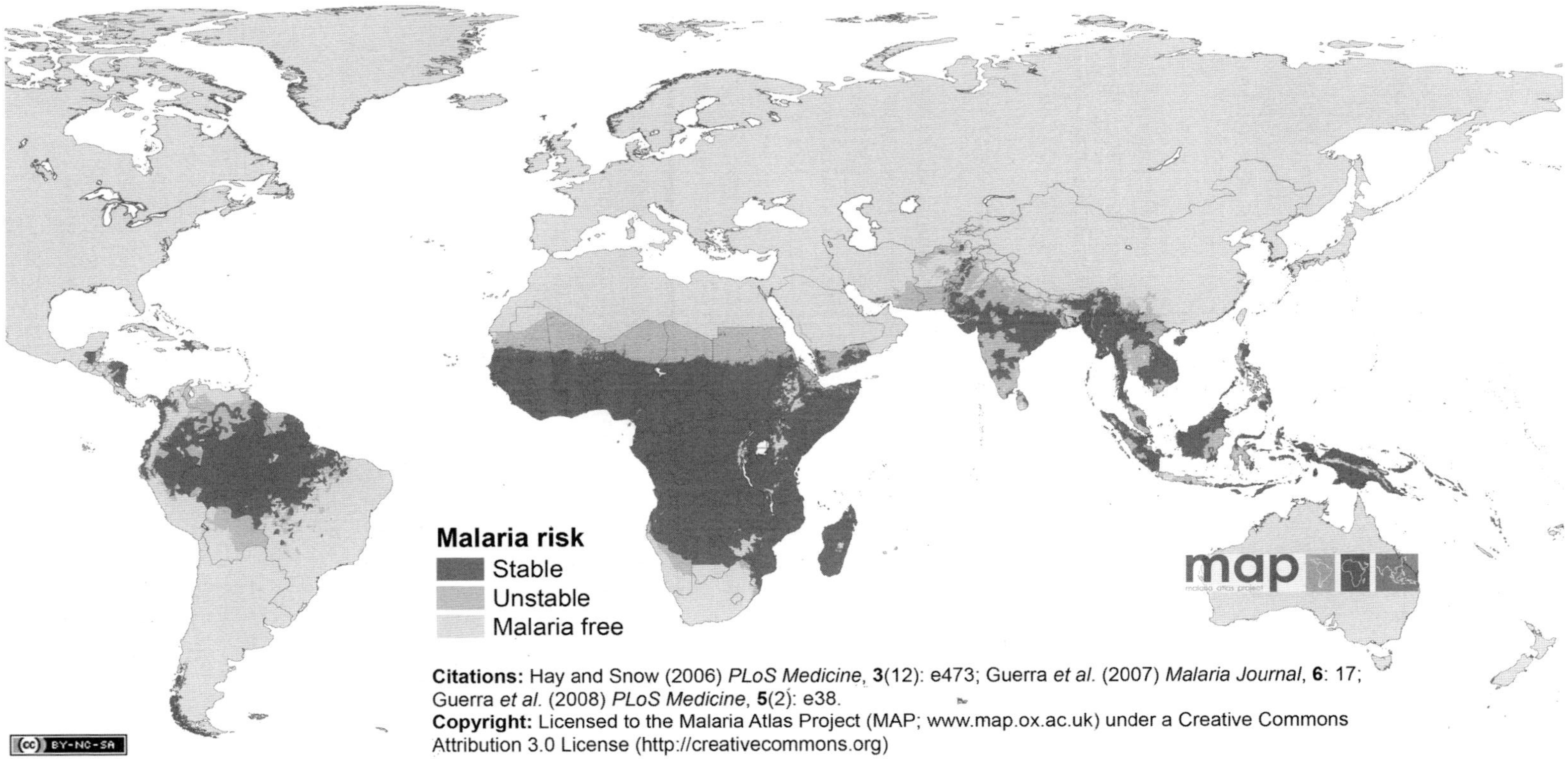

FIGURE 65.1 Global risk of *P. falciparum* malaria. The map is derived from nationally reported case-incidence data, medical intelligence, and temperature and aridity boundaries for the dominant *Anopheles* vectors, with risk expressed as the number of people infected per 1000 per annum (the annual parasite incidence, or API) (Guerra et al., 2008). Dark shading indicates stable malaria (API > 0.1 per 1000 per annum), intermediate shading indicates unstable malaria (API < 0.1 per 1000 per annum), and light shading indicates no risk. The distribution of *P. vivax* adds additional countries, particularly in subtropical regions, including areas well outside the tropics such as Korea (data not shown). Reprinted with permission from PLoS Medicine. (Guerra et al., 2007, 2008; Hay and Snow, 2006). Copyright licensed to the Malaria Atlas Project (MAP; http://www.map.ox.ac.uk) under a Creative Commons Attribution 3.0 License (http://creativewemones.org).

of irrigation or higher altitudes (Afrane et al., 2006; Githeko et al., 2006). For these reasons, a strongly protective, long-lasting malaria vaccine would provide an extremely cost-effective intervention, and is therefore considered an urgent international health priority. The ongoing efforts to create a vaccine constitute one of the great public health enterprises of our time.

Preventing malaria through immunization is a fascinating biomedical challenge, and has drawn scientists from many disciplines and spawned major achievements, including the sequencing of the *P. falciparum* genome (Gardner et al., 2002) as well as the genomes of several other *Plasmodium* species (Hall et al., 2005), the establishment of online databases containing extensive genomic, transcriptomic, and proteomic information (Di Girolamo et al., 2005; Le Roch et al., 2003; Winzeler, 2006; Zhou et al., 2008) and the development of large reagent repositories (Adams et al., 2000). Great progress has been made elucidating, at the molecular level, the biology and antigenic structure of this complex parasite (Conway, 2007; Cowman and Crabb, 2002). After decades of effort, however, a licensed vaccine is still not available and the leading candidate in development is only modestly effective. This chapter will provide a synopsis of the history, biology, epidemiology, and clinical expression of this important disease and then will review the progress being made toward developing an effective malaria vaccine for widespread use.

HISTORY

Malaria has most certainly been afflicting humans for centuries. Early medical writings from India, China, and Assyria described illnesses characterized by intermittent fevers (Cox, 2002). The early writings of the Sumerians, the first civilization that settled between the Tigris and Euphrates rivers, frequently described fevers typical of malaria (Desowitz, 1991). Around the same time, the Chinese medical volume dated from 2700 B.C., the Nei Ching, provided an account of tertian and quartan fevers, and accurately described the spleen enlargement that often follows. The Nei Ching purported that these fevers were caused by an imbalance of the Yin and Yang, and prescribed medicinal plants as cures (Cox, 2002; Desowitz, 1991). Around 1600 B.C., the Hindu epic religious texts, the Vedas, provided records of fevers that caused severe illness and death, suggesting that malaria was prevalent in India at that time (Desowitz, 1991). Hippocrates, in the fifth century B.C., accurately described clinical manifestations of malaria including the tertian and quartan fevers and the enlarged, stiff spleens (Cox, 2002; Desowitz, 1991). Hippocrates speculated that this illness was caused by an external agent, particularly the vapors and mists coming from the marshes. Indeed, the name of this disease, malaria (mal means bad; aria is air), which was adopted in the mid-eighteenth century, reflects this belief that malaria was caused by a miasma rising from the swamps in the environs of Rome. This observation was made when travelers from the city developed the telltale signs of the infection within a week or two of visiting the countryside. The theory that malaria was caused by evil air rising from marshy waters was further supported by repeated observations of a decline in malaria cases after draining of swamps (Desowitz, 1991).

Despite the assortment of historical texts describing this distinct illness over the centuries, it was not until 1880, when Alphonse Laveran first observed the crescent-shaped gametocyte of *P. falciparum*, the causative agent of falciparum, or malignant tertian malaria, that the true etiologic agent was discovered (Bruce-Chwatt, 1987; Desowitz, 1991). From the blood of a feverish soldier stationed in Algeria, Laveran discerned several different morphologies of the malarial parasite, including "crescent-shaped bodies" (the gametocytes of *P. falciparum*), "rounded bodies with dancing/lashing surface filaments" (the exflagellating male gametocyte), and "ring-like hyaline bodies" within red blood cells (ring stages) (Desowitz, 1991). Finally, in the 1890s, Ronald Ross and Battista Grassi proved that the gametocytes are taken up by the female *Anopheles* mosquito during a blood meal, thereby demonstrating the route of transmission of this deadly infection (Desowitz, 1991).

ETIOLOGIC AGENTS

Classification

Malaria is a vector-borne disease caused by a protozoan parasite of the genus *Plasmodium*, transmitted by the bite of a female *Anopheles* mosquito. Of the numerous species of *Anopheles* occurring worldwide, approximately 60 are involved in malaria transmission (Oaks et al., 1991). Because of the nocturnal habits of the *Anopheles* mosquito, malaria is transmitted between the hours of dusk and dawn. The *Plasmodium* parasites belong to the protozoan Order Apicomplexa, which includes other pathogens such as Babesia, Toxoplasma, and Cryptosporidium. Apicomplexa are morphologically distinguished by the presence of a specialized complex of apical organelles (micronemes, rhoptries, and dense granules) involved in host-cell invasion (Binder and Kim, 2004; Cowman and Crabb, 2006). There are scores of species of *Plasmodium* that infect a wide variety of hosts (e.g., reptiles, amphibians, birds, and a variety of mammals including non-human primates (NHPs)), but only four that primarily or exclusively infect humans: *P. falciparum, P. vivax, P. malariae*, and *P. ovale* (Fig. 65.2). A fifth species, *P. knowlesi*, resides in NHPs but can be transmitted to humans and has been recognized as an important zoonotic infection in parts of Malaysia (Cox-Singh et al., 2008; Vythilingam et al., 2006). The human malarias are thought to have originated in Southeast Asia from simian species, from where they spread to Africa and Europe (Cox, 2002). Malaria was probably introduced into the Americas in the sixteenth century, and was a significant public health problem in the USA through the first half of the twentieth century until its eradication in 1951.

Plasmodium Life Cycle

The life cycle of *Plasmodium* is shown in Figs 65.2 and 65.3. The infection begins when sporozoites (Fig. 65.2A) are injected into cutaneous or subcutaneous tissue or directly into microvessels of the blood or lymphatics with the saliva of the feeding mosquito (Jin et al., 2007). Sporozoites deposited into interstitial spaces disperse rapidly using a gliding motility and are able to identify and enter blood vessels. Movies of this process, made using intravital cinematography, can be seen online (Vanderberg and Frevert, 2004). The circulatory system carries the sporozoites to the liver, where they invade hepatocytes. Recent evidence suggests that sporozoites initially penetrate the cell membranes of several cells, first Kupffer cells then multiple hepatocytes, passing through the cytosol and exiting again, leaving behind components of their surface coat and sometimes destroying the hepatocyte on traversal (Ishino et al., 2004; Mota et al., 2001). Movies of this process are likewise available on the Internet (Frevert et al., 2005).

After the unexplained migration, which results in a local inflammatory response and the presentation of surface coat antigens to the immune system (Bongfen et al., 2007), sporozoites invade their host hepatocyte nondestructively, by inducing an invagination of the cell membrane, resulting in the formation of a parasitophorous vacuole within which the sheltered parasite can obtain nutrients from the host cell as well as from the interstitial fluids beyond. After increasing many-fold in size, the parasite undergoes asexual (clonal) replication, known as exoerythrocytic schizogony, whereby it divides into thousands to tens of thousands of merozoites, depending on the species (approximate range 2000–40,000) (Fig. 65.2B). When the nuclear and cytoplasmic division processes are complete, merozoites are released into the circulation. It has recently been shown in the murine malaria species, *P. berghei* and *P. yoelii*, that merozoites may be released in membrane-bound packets called merosomes or extrusomes, which potentially circumvent phagocytic cells in the liver as well as neutralizing antibodies (Baer et al., 2007; Sturm et al., 2006).

Liver stage development is generally completed in 5–14 days, with some variation by species. For *P. falciparum* and *P. malariae*, all parasites complete development, and merozoite progeny are directly released into the bloodstream. The liver stage parasites from *P. vivax* and *P. ovale*, however, can enter a state of dormancy (hypnozoite forms), reactivating weeks, months, or years later to complete development, leading to *relapse* of the blood stage infection. Remarkably, a relapse of *P. ovale* has been recorded 4 years after primary infection (Trager and Most, 1963). In some areas, such as Korea, the hypnozoite life form has enabled *P. vivax* to adapt to seasonal climates where cold winter eliminates the activity of the vector; the parasite remains dormant for several months, emerging the following spring when renewed vector activity permits transmission. These relapses are to be contrasted with *recrudescences*, which occur regularly with all species of human malaria and constitutes new waves of blood-stage parasites originating from subpatent populations of blood-stage parasites. Relapse and recrudescence both cause recurrent parasitemia in *P. vivax* and *P. ovale*, and can be difficult to distinguish, whereas only recrudescences cause recurrent parasitemia in *P. falciparum* and *P. malariae*.

Merozoites released into the bloodstream invade erythrocytes. Protein products of the apical organelles of the parasite (rhoptries, micronemes) engage receptors on the erythrocyte membrane propelling invagination and, as in the liver, formation of a parasitophorous vacuole (Charpian, 2008) (Fig. 65.4). Once inside the erythrocyte, parasites enter a second trophic period in which the parasite enlarges, similar in nature but not as long lasting as the trophic period in the liver. The early trophozoite is known as the ring form because of its unique ring-shaped morphology (Fig. 65.2C, D). The enlargement of the trophozoite is accompanied by the ingestion of erythrocyte cytoplasm and the proteolysis of hemoglobin. A by-product of this metabolism, hemozoin, accumulates as refractile, pigmented granules visible within the infected erythrocyte and within macrophages that have ingested erythrocyte debris (Egan, 2007; Goldberg, 1993; Hanscheid et al., 2007). The enlarging trophozoite undergoes clonal nuclear divisions,

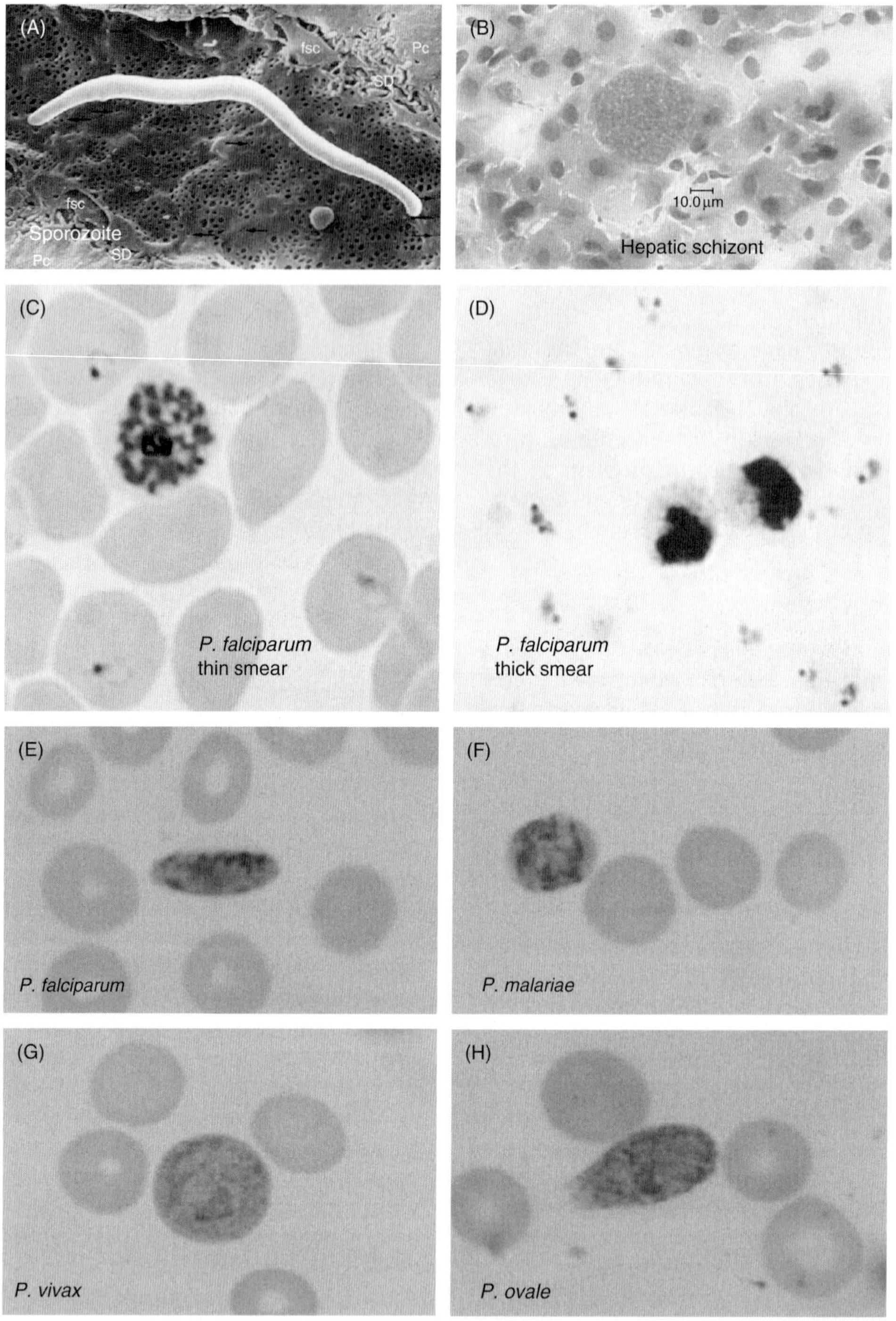

FIGURE 65.2 The malaria parasite. (A) Scanning electron micrograph of a *P. cynomolgi* sporozoite superimposed on SEM of liver sinusoidal sieve plates; parasite is approximately 10 μm in length (photograph composite by Ute Frevert; http://www.med.nyu.edu/parasitology/faculty/ufrevert.html) (Cochrane et al., 1976; Wisse et al., 1985). (B) Hematoxylin and eosin stain of 6-day old *P. falciparum* schizont in chimpanzee liver surrounded by normal-appearing parenchyma; 40 μm diameter (can range to 80 μm) (10 μm bar) (photograph by John Sacci). (C), (D) Giemsa stain of thin smear and thick smear of *P. falciparum* blood culture, showing ring-infected erythrocytes; erythrocytic stage schizont is seen on the left with individual merozoites ready for release; this form normally not seen in peripheral blood due to cytoadherence and sequestration (photographs by Ted Hall). (E)–(H) Giemsa stain of characteristic erythrocytic stages of *P. falciparum* (banana-shaped gametocyte), *P. malariae* (band-shaped trophozoite in a normocyte), *P. vivax* (large amoeboid trophozoite in a reticulocyte with Schuffner's dots in the erythrocyte cytoplasm), and *P. ovale* (fibrionated edge to the parasitized oval erythrocyte, also with Schuffner's dots), from a child with quadruple infection (photographs by Purnomo) (Purnomo et al., 1999).

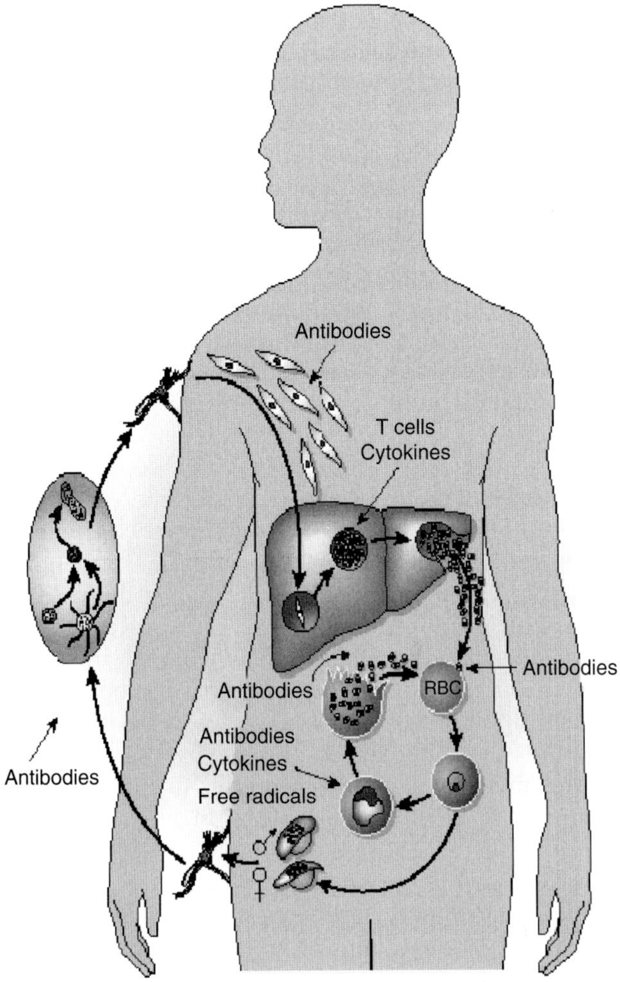

FIGURE 65.3 *Plasmodium* life cycle. Parasite morphology, mobility, host-cell specificities, patterns of protein expression, and resulting immunologic vulnerabilities vary from stage to stage. The induction of antibodies is particularly important for vaccines targeting sporozoites, the induction of cell-mediated immune responses for targeting liver stages, and the induction of both antibodies and antibody-dependent cytotoxic responses for targeting blood stages. The ideal vaccine will induce all these responses against multiple stages. The depiction of the life cycle in the insect vector is simplified; see text for details. Reprinted by permission from Macmillan Publishers Ltd (Richie and Saul, 2002).

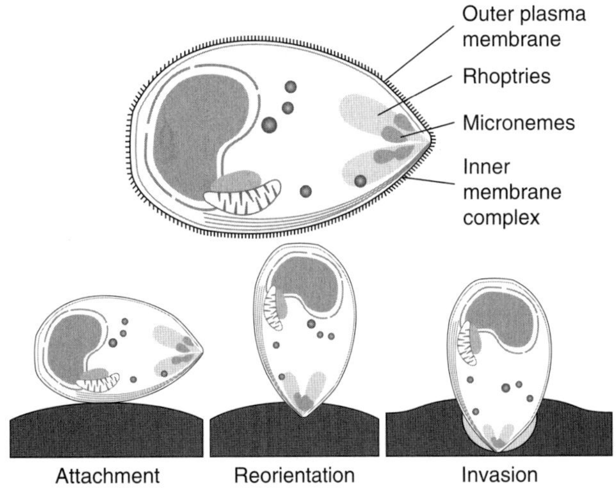

FIGURE 65.4 Merozoite invasion of erythrocyte. This figure illustrates a functionally critical element of host-parasite biology—the invasion of the erythrocyte. The interaction begins with low-affinity binding between the surfaces of the two cells, followed by apical reorientation powered by an actin–myosin motor. Once the apical end contacts the erythrocyte surface, a tight junction is formed utilizing a different set of parasite proteins secreted from the apical organelles. This junction migrates circumferentially along the merozoite, leading to invagination and shedding of the merozoite surface coat. Finally, the erythrocyte membrane reseals, leaving the parasite encased in the parasitophorous vacuole. The entire process takes about 30 s. The secretion of key parasite proteins *after* the apical end is juxtaposed may obstruct access by antibodies, making it difficult to block the process. At least 48 proteins are associated with the merozoite surface and apical organelles (Cowman and Crabb, 2006), including MSP1, AMA1, and EBA175. Reprinted with permission from Alan Cowman, Brendan Crabb, and Jake Baum (http://www.wehi.edu.au/facweb/indexresearch.php?id = 16).

generating a blood-stage schizont (Fig. 65.2C), which releases merozoites when the erythrocyte ruptures. *P. falciparum* or *P. vivax*, each produce approximately 5–40 merozoites per schizont (depending on the species), leading to logarithmic expansion of the parasite population.

Most of these parasite progeny will again undergo asexual growth and schizogony, the cycle lasting 48 h for *P. falciparum*, *P. vivax*, and *P. ovale*, 72 h for *P. malariae*, and 24 h for *P. knowlesi*. However, a proportion develops into the sexual forms of the parasite—micro- and macrogametocytes—which are targeted for ingestion by a female *Anopheles* mosquito. Conversion to gametocytes differs by species, *P. vivax* developing gametocytes soon after release from the liver stage, *P. falciparum* developing gametocytes after several days (Mota et al., 2001; Weatherall et al., 2002).

The ingestion of the gametocytes by the mosquito induces gametogenesis, which consists of exflagellation of the microgametocyte, further development of the macrogametocyte into the macrogamete, and fertilization of the macrogamete by the motile microgamete, all taking place in the mosquito midgut. The now diploid zygote develops into a motile ookinete that traverses the intestinal wall and transforms into an oocyst on the external gut surface. Within the oocyst, the parasite undergoes multiple rounds of asexual replication, including reductive divisions restoring the haploid state (Sinden, 1991). The oocyst releases thousands of sporozoites, which travel through the insect hemocoel to the salivary glands. After penetrating the salivary glands and entering the lumen, they are ready to be injected into another human host. The duration of the sporogonic cycle varies by species and is temperature-sensitive, but may typically take 14–18 days.

Antigens Encoded by Plasmodium

Proteins on the sporozoite surface, including the circumsporozoite protein (CSP) and the thrombospondin-related adhesion protein (TRAP) (Roussilhon et al., 2007) (also called the sporozoite surface protein 2, or SSP2) enable gliding motility, invasion of hepatocytes (binding to heparin sulfate proteoglycans), and additional functions (Garcia et al., 2006; Sibley, 2004; Sinnis and Coppi, 2007; Sinnis and Nardin, 2002). Many additional proteins are likely expressed on the sporozoite surface. Following hepatocyte invasion, additional proteins, such as liver stage antigen 1 (LSA1), UIS3 (UIS, upregulation in sporozoites) and UIS4 are newly expressed or expressed in greater quantity. These liver stage proteins play hundreds of roles in the complex interaction between parasite and host, permitting the breakdown of host-cell structure, the establishment of a new architecture, and the uptake of nutrients from outside the cell. These events occur through the interface of the parasitophorous vacuole membranes as well as through various channels and pores.

Rapid progress is now being made functionally characterizing preerythrocytic stage (sporozoite plus liver stage) proteins by using the technique of targeted genetic disruption (gene knockout) (D'Ombrain et al., 2007; Labaied et al., 2007; Mueller et al., 2005a). Absence of a particular protein may halt development at a certain stage, and the morphological and molecular characteristics of these arrested parasites provide information regarding the role of the missing molecule.

As hepatic stage schizonts mature, erythrocyte proteins are expressed. Merozoite surface molecules are involved in the attachment, apical reorientation and invasion of erythrocytes (Fig. 65.4), including several merozoite surface proteins (MSPs), the apical membrane antigen 1 (AMA1) and the erythrocyte-binding antigen 175 (EBA175), all leading vaccine candidates. Key invasion proteins may be secreted after completion of reorientation of the parasite and approximation of the apical organelles with the erythrocyte membrane. This last-minute release lessens the accessibility of plasma and thus the susceptibility to antibody attack. Invasion itself is an active process characterized by a ring of attachment that migrates up the oval cylinder of the merozoite, pulling/pushing it into the expanding parasitophorous vacuole. Once within the erythrocyte, additional proteins are expressed to mediate nutrition and growth, and a system of channels and pores develops, as within the hepatocyte (Merckx et al., 2008). Of particular interest are the antigens expressed on the surface of the infected erythrocyte, some of which are highly immunogenic and elicit antibody responses that limit parasite replication. The best studied erythrocyte surface protein is *P. falciparum* erythrocyte membrane protein 1 (PfEMP1), encoded by approximately 50–60 different genetic loci within the parasite genome (*var* gene family). By expressing alleles in sequence on knobby protruberances appearing on the surface of infected erythrocytes during the late trophozoite/schizont stages, PfEMP1 undergoes antigenic variation, and this is likely an important factor responsible for the successive, recrudescent waves of parasitemia that characterize malarial infection. PfEMP1 is also one of the molecules responsible for sequestration of mature blood stages in the microvasculature (see "Pathogenesis" section below).

The final stage of the parasite, the gametocyte, expresses its own set of proteins in preparation for ingestion by the mosquito and the subsequent sudden change of environment resulting from transition from human blood to the mosquito midgut (Kooij and Matuschewski, 2007; Matuschewski and Mueller, 2007). Mature gametocytes appear relatively silent metabolically and immunologically, circulating in the blood and possibly arresting in skin microvessels (Gautret et al., 1996) until ingested by the vector.

PROTECTIVE IMMUNE RESPONSES

SSPs are considered to be potential targets for antibody responses, and indeed, sporozoites coated with antibodies recognizing CSP are immobilized, leading to a process called the circumsporozoite reaction whereby the surface coat is shed (McCarthy and Clyde, 1977; Vanderberg et al., 1969). Videos of this process can be viewed on the Internet (Vanderberg and Frevert, 2004). Antibody-mediated attack is thus a potential mechanism for vaccine-induced immunity during the first minutes following inoculation by the mosquito, offering an opportunity to disrupt gliding motility and thus immobilize the sporozoites in the skin or block the ligand-receptor binding required for hepatocyte invasion. Once the sporozoite has successfully entered a host hepatocyte, the ability of antibodies to affect development is restricted, although there is evidence that antibodies to antigens such as LSA1 and EXP1 expressed by liver stage parasites may nevertheless be associated with protection in endemic settings (John et al., 2008; Meraldi et al., 2004).

In the liver, cell-mediated immunity emerges as the primary protective response, as demonstrated by the transfer of protection in conjunction with the adoptive transfer of splenocytes recognizing sporozoite and liver stage antigens (Khusmith et al., 1991, 1994, Rodrigues et al., 1991). Protection results from the recognition by lymphocytes of parasite molecules expressed on the surface of infected hepatocytes in the context of Class I MHC molecules. The parasite is killed by CD8+ cytotoxic T lymphocytes (CTL) or through the induction of nitric oxide triggered by interferon-gamma, with both CD8+ and CD4+ T cells playing a role (Doolan and Martinez-Alier, 2006). In addition, natural killer cells may contribute to protection (Hansen et al., 2007).

During liver stage development, parasites are relatively few in number (likely less than 100) and restricted to a single organ, whereas blood stage parasites occur in billions and circulate throughout the body. Thus, the liver stage provides a window of opportunity for completely aborting the infection. For this reason, many vaccine developers are focused on targeting liver stage antigens and inducing potent cell-mediated immunity (Hill, 2006b). Ideally, vaccines will induce both antibody- and cell-mediated responses thereby targeting both sporozoites and liver stages, to provide high-grade resistance to infection (Schofield et al., 1987).

After release from hepatocytes, the parasite enters the blood and is again susceptible to antibody attack. Merozoite surface proteins are considered prime candidates for inclusion in vaccines and in fact comprise the majority of vaccines in development. Antibodies to these proteins can prevent merozoite adherence, apical reorientation, and invasion of the erythrocyte, an effect measured in vitro through growth inhibition assays (Bergmann-Leitner et al., 2006; Miura et al., 2007b). Following erythrocyte invasion, parasite proteins expressed on the surface of the erythrocyte represent another target, and antibodies could potentially interrupt nutrient uptake or cytoadherence (see "Pathogenesis" section below). Parasitized erythrocytes have decreased deformability, and if prevented from sequestering in postcapillary venules through antibody-mediated reversal of cytoadherence are subject to removal from circulation by splenic macrophages and other immune cells of the reticuloendothelial system. Antibodies recognizing surface-expressed parasite proteins can also facilitate parasite killing through antibody-dependent cellular cytotoxicity (ADCC) (e.g., IgG1 or IgG3) (Bouharoun-Tayoun et al., 1990, 1995).

The effect of antibodies targeting blood-stage proteins has been demonstrated in vivo, through the transfusion of immunoglobulin purified from semi-immune human adults to children with acute malarial infection. The children experience clinical improvement and a reduction in the number of circulating parasites (Bouharoun-Tayoun et al., 1990; Sabchareon et al., 1991). It has been more difficult to demonstrate a dominant role for cell-mediated immunity in blood stage infection in vivo (Kumar and Miller, 1990; Weidanz et al., 1990). One supportive finding is the protection observed in humans following immunization with a whole parasite blood stage vaccine, which appeared to be associated with cell-mediated immune responses in the absence of detectable antibody (Pombo et al., 2002).

Another protective mechanism targeting blood stage infection is the potential neutralization of malarial toxins released from infected erythrocytes at the time of rupture. Parasite-derived molecules, such as the glycosylphosphatidylinositol (GPI) sequences that anchor many membrane-bound parasite proteins, have pyrogenic and other toxic properties. It is hypothesized that a significant component of naturally acquired immunity (NAI) is derived from the ability of antibodies to neutralize malarial toxins (Schofield et al., 1987, 2002). Although targeting malarial toxins released on schizont rupture might not limit parasite multiplication, it could lessen the clinical expression of the infection by curtailing cytokine release and other inflammatory events.

Finally, antibodies may potentially play a role in preventing sporogony in the mosquito gut. Antibodies induced by a vaccine containing gamete or gametocyte proteins normally not expressed or not exposed in the human host can be passively imbibed by a feeding mosquito, and then attack the parasite proteins newly expressed or newly exposed in the mosquito gut. This can be demonstrated in an in vitro/in vivo model in which human blood harboring gametocytes is used for membrane feeding by mosquito vectors. When the mosquito guts are examined several days later, oocyct numbers can be reduced if an antibody targeting various sexual stage proteins is added to the blood prior to its ingestion (Arevalo-Herrera et al., 2005; Miura et al., 2007a). This has given rise to the search for a transmission-blocking vaccine, which would benefit a given vaccine recipient not through protection against infection but by reducing transmission throughout the community (see "Transmission-Blocking Vaccines" section below).

EPIDEMIOLOGY

Significance as a Public Health Problem

Worldwide, *P. vivax* and *P. falciparum* are the most common of the four human malarias. Both are prevalent throughout much of the tropical regions of South and Southeast Asia, the Western Pacific and Latin America, with *P. vivax* often the predominant species. In sub-Saharan Africa, however, *P. falciparum* nearly always predominates. This is particularly true in West Africa where *P. vivax* may be virtually absent due to the high prevalence of the Duffy negative phenotype, which precludes erythrocyte invasion by *P. vivax* (see "*P. vivax* Vaccines" section below). *P. malariae* is also widespread across the tropics, often occurring in sympatry with other species and occasionally found to be the predominant species. *P. ovale* has the most limited distribution; occurring in sub-Saharan Africa, in the region of New Guinea and in parts of the Philippines and Southeast Asia (Carter and Mendis, 2002; Collins and Jeffery, 2005; Lysenko and Beljaev, 1969).

Malaria epidemiology can be characterized by several indices, two of the most important being the prevalence of malarial infection, determined as the number of positive blood smears relative to the number examined (often separately determined by age group), and the incidence of malarial infection, measured as the rate of acquiring new infections. The latter is formally measured by tallying new-onset parasitemia in cohorts of drug-cleared volunteers followed prospectively for a defined period, but as this is labor intensive, costly, and entails risks for the volunteers, incidence is usually approximated by the easier-to-measure, closely related entomologic inoculation rate (EIR), which is the number of infective bites received per person per year (Beier et al., 1994). This is calculated by capturing human-landing mosquitoes during nocturnal surveys and determining the proportion that harbor sporozoites.

The need to provide a general classification of malaria burden in a given geographic area has led to distinguishing different levels of endemicity, focusing on prevalence: hypoendemic areas have a prevalence less than 10% and lower EIRs/transmission rates; mesoendemic areas have a prevalence ranging from 11% to 50% and moderate EIRs/transmission rates, while hyperendemic areas show a prevalence of 50% or more and have higher EIRs/transmission rates. In some classification systems, spleen rates in young children have been incorporated into the determination of endemicity but this system is not used commonly at the present time.

The conduct of point prevalence surveys to measure prevalence, of prospective cleared-cohort studies to measure incidence or of nocturnal mosquito landing collections to measure EIR can be impractical for characterizing the malaria burden across a nation. As a result, countrywide data are usually assembled by determining the annual parasite incidence (Branch et al., 2005) defined as the annual number of parasite-specific local malarial infections by geographic unit (Guerra et al., 2006, 2007, 2008). API figures are based on available medical intelligence, usually febrile malaria cases reported from health clinics supplemented with the results of surveys where blood smears are taken from residents with fever. API emphasizes clinical malaria and provides less information on the potentially large number of asymptomatic infections that may be present. It is useful as a surveillance technique for monitoring increasing or decreasing trends in the burden of malaria.

Another approach to characterizing malaria epidemiology focuses on the continuity of transmission and resulting patterns of clinical immunity, defining three categories: (1) stable malaria (2) unstable malaria, and (3) epidemic malaria (Carter and Mendis, 2002). Superimposed on this is the seasonality of transmission, which usually reflects patterns of rainfall. In stable malaria, a population is exposed to a fairly constant rate of malarial infection throughout the year, with at least one clinical infection experienced by most individuals on an annual basis. In these regions, regular exposure to infection allows for the development of naturally acquired protective immunity (Hviid, 2005; Schofield and Mueller, 2006). The risk of clinical malaria in these regions is therefore age-dependent, with the greatest morbidity occurring in infants and young children (<5 years) who have had few exposures to infection and whose ability to mount an effective immune response may be limited, relative to adults, by age-dependent differences in the immune system (Reyburn et al., 2005; Schofield and Mueller, 2006). In contrast, teenagers and adults, although frequently infected, experience clinical malaria less frequently, and when they do, the symptoms are usually mild. This protected state in older children and adults, called naturally acquired immunity (NAI), is characteristic of the high transmission areas of sub-Saharan Africa. However, the development of an effective protective immunity has also been noted in other areas of the world with lower, but stable levels of transmission (Baird, 1995; Baird, et al., 1998; Branch et al., 2005). To add to the complexity, there may be significant differences in age-dependent clinical expression of disease within high transmission areas, with the highest transmission areas characterized by the development of severe malarial anemia in infants and young children and the somewhat lower transmission areas characterized by cerebral malaria in older children (Snow and Marsh, 2002).

In unstable malaria, there may be permanent malaria transmission; however, there are large fluctuations in the rate of exposure to infection among individuals. The irregularity of transmission may result in greater intervals between infection precluding the development or maintenance of effective protective immunity. In areas of unstable malaria (Fig. 65.1), the risk of clinical malaria is less age-dependent, and individuals tend to be susceptible to the severe complications of malaria throughout life. This phenomenon may partially explain the frustrating fact that as malaria control is successfully instituted in a given region, and the level of endemicity is reduced, the public health impact of the infection may remain unchanged or even worsen (Dicko et al., 2007; Mbogo et al., 1995; Snow and Marsh, 2002) (Fig. 65.5). This is important when planning for the fielding of a partially protective malaria vaccine, because the semieradication of the parasite may serve to increase the susceptibility of the general population to clinical malaria, and thus sustain or increase morbidity and mortality rates despite progress against the parasite.

In the third broad category, epidemic malaria, there is a sudden increase in vector populations, or alternatively malaria may be reintroduced to an area from which the parasite (but not the vector) has been eradicated. Devastating malaria epidemics occurred in the early 1930s in South Africa (Sharp and le Sueur, 1996), in 1958 in Ethiopia (Fontaine et al., 1961), and in 1986 in Madagascar (Mouchet et al., 1998), and have been a recurrent feature in many areas with periodic excessive or failing rainfall, illustrated by the history of epidemics in India (Bouma and van der Kaay, 1996). Epidemics often occur in situations where there has been limited previous exposure and thus limited opportunity to acquire protective immunity, and partly for this reason, epidemic malaria can cause extreme morbidity and mortality throughout a population, especially with *P. falciparum* infection, with mortality rates exceeding 10%. Factors precipitating epidemic malaria in a population can relate to importation of the disease by travelers, failure of public health surveillance, deforestation, and climate change. There is concern that global warming could abruptly expand the range of malaria transmission by increasing vector populations at higher altitudes and by shortening the duration of the sporogonic cycle in the mosquito, which is temperature-dependent (Munga et al., 2006; Patz and Olson, 2006). Epidemic malaria could also become a significant concern following the fielding of a successful malaria vaccine. If the vaccine eradicates malaria from areas with high vector capacity and then the parasite is reintroduced later, either due to waning vaccine coverage or to the evolution of resistance to the vaccine by the parasite, epidemic malaria could result.

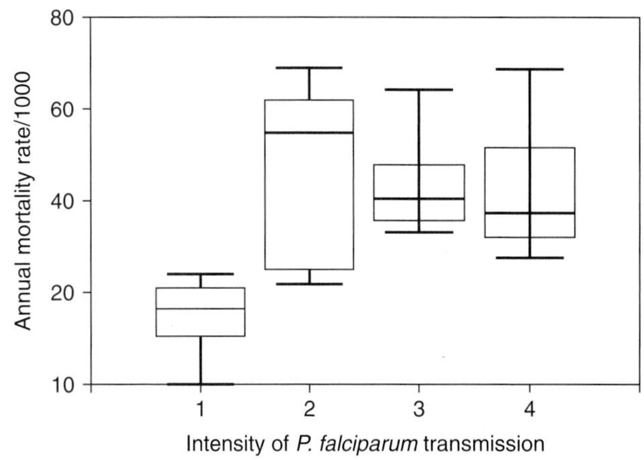

FIGURE 65.5 Relationship between all-cause mortality rate in early childhood and *P. falciparum* malaria transmission intensity in Africa. Box plot showing median , 25%, 75% quartile ranges around the median (box depth), and upper and lower limits of all-cause childhood mortality per 1000 children aged 0–4 years per annum at four transmission intensities. Malaria transmission above a certain threshold is associated with significant increases in mortality rates, as shown in this figure by the striking increase (approximately two-fold) in all-cause annual mortality rates comparing low-endemicity areas (category 1, childhood infection prevalence ranging from 0% to 24%) to low-to-moderate endemicity areas (category 2, prevalence 25–50%). As transmission intensity rises further, there appears to be a plateau effect, with mortality rates associated with moderate-to-high endemicity areas (category 3, prevalence 51–74%) and with high-endemicity areas (category 4, prevalence 75–100%) similar to those associated with low-to-moderate endemicity areas. These data suggest that a malaria control strategy such as vaccination may not significantly improve childhood mortality rates unless substantial reductions in prevalence of malaria are achieved. Note that in this analysis, a covariate such as differences in socioeconomic status among the 26 sites included in the data set could influence both rates of malarial infection and all-cause mortality. Reprinted from Advances in Parasitology, Vol 52, Snow and Marsh, 2002, with permission from Elsevier.

Across these categories, EIR may range widely, from less than 1 to greater than 1000 infective bites per year. In general, higher EIRs characterize stable malaria, but transmission can also be stable at lower EIRs, and both very low (Lindblade et al., 1999) as well as very high EIRs may characterize epidemic malaria.

Pregnancy-Associated Malaria (PAM)

At all levels of endemicity, including highly endemic areas characterized by the acquisition and maintenance of NAI, pregnant women are at greater risk for clinical malaria than nonpregnant women. The risk is especially great when experiencing a first pregnancy, a phenomenon originally described by Clark in 1915 during studies conducted during malaria epidemics (Clark, 1915). It is now well established that pregnancy is associated with an increased frequency (up to 12-fold in some studies) and density of parasitemia with *P. falciparum*, as well as increased severity of infection (Brabin, 1983, 1991; Brabin et al., 1988; Bruce-Chwatt, 1983; Menendez, 1995, 2006; Okoko et al., 2003). When it occurs during pregnancy, malaria increases the risk of maternal mortality twofold, and is the most common cause of maternal death in hyperendemic areas (Brabin, 1991). This is a major health challenge, given that each year approximately 24 million women living in malaria-endemic regions become pregnant.

Malarial infection during pregnancy is associated with adverse fetal and infant outcomes, including abortions, stillbirths, premature delivery, congenital malaria, intrauterine growth retardation, low birthweight (LBW), and infant death (Okoko et al., 2003; Uneke, 2007). Placental malaria is often a key mediator of these adverse fetal outcomes because the accumulation of parasites in the placenta and the associated maternal immune response can interfere with the efficient transfer of oxygen and nutrients to the fetus (Ismail et al., 2000; Menendez et al., 2000; Okoko et al., 2002). The most severe fetal consequences, such as abortions, stillbirths, and congenital malaria, are frequently encountered in regions of unstable or epidemic transmission (Menendez et al., 2000), while in regions of high, stable transmission, the primary adverse infant outcome is LBW (Guyatt and Snow, 2004). LBW in turn is associated with diminished survival in the first year of life (McMormick, 1985), mediated in part by an increased risk of acquiring other infections. In fact, it is estimated that 5.7% of infant deaths in sub-Saharan African are indirectly associated with maternal malarial infection (Guyatt and Snow, 2001), and approximately 200,000 infants die each year as a result of malaria during pregnancy (Steketee et al., 2001).

One reason proposed for the increased risk of malaria during pregnancy may be the pregnancy-associated alteration in the immune system to prevent immunological rejection of the fetus. Specifically, the proinflammatory Th1, cell-mediated response is downregulated, while the Th2 response is enhanced. This immunologic shift has been hypothesized to increase the susceptibility of pregnant women to many infections, particularly intracellular organisms requiring strong cell-mediated immunity, such as *Plasmodium*. In addition, the placenta provides a novel sequestration site for *P. falciparum*. It is believed that placental malaria results from a specific population of parasites that preferentially sequester and cytoadhere to chondroitin sulfate A in the placenta through the binding of PfEMP1 variants binding preferentially to this molecule (Fried and Duffy, 1996; Fried et al., 2004; Hviid, 2007; Rogerson et al., 1995, 2007). Natural protective immunity against PAM is acquired during the course of multiple pregnancies, with the development of anti-PfEMP1 antibodies recognizing placental infected erythrocytes. This proposed mechanism for explaining the severity of malaria in first pregnancies and for the reduction in risk in multiparous women has instigated research into developing a pregnancy-specific malaria vaccine with CSA-binding PfEMP1 as the primary antigenic target (see "Pregnancy-Specific Vaccines" section below).

It is worth noting that *P. vivax* also contributes to PAM and its complications, and in fact *P. vivax* is an important cause of congenital malaria (Lesko et al., 2007).

Potential as a Biothreat Agent

Malaria is generally considered to have low potential as a biothreat agent, as it can be transmitted only by the insect vector, by parenteral administration (e.g., blood transfusion) or transplacentally. It also has an incubation period of a week or more if transmitted via mosquito bite, so there is time to administer presumptive treatment in case of exposure to infectious mosquitoes. Therefore, it cannot be effectively weaponized. On the other hand, malaria is amenable to abuse by terrorists, e.g., if infected mosquitoes are released surreptitiously in crowded areas in a nonmalarious area.

CLINICAL DISEASE

In contrast to liver-stage infection, which is entirely without clinical signs, symptoms, or changes in routine hematological or chemical parameters, the multiplication of parasites within the blood results in an accelerating clinical syndrome, particularly in nonimmune individuals. The classic fevers of blood stage malaria coincide with the synchronous lysis of millions of infected erythrocytes and are probably mediated by pyrogenic parasite-derived products such as GPI, as mentioned previously. These malaria toxins precipitate the release of inflammatory cytokines such as tumor necrosis factor, contributing to headache, body aches, fever, chills, nausea, vomiting, diarrhea, general weakness, and prostration. These symptoms are similar to those of other infectious organisms, however; so the diagnosis of malaria requires the demonstration of parasites on a Giemsa-stained blood film. On the other hand, the characteristic malaria paroxysm, when observed in the right setting, is a strong indicator of the diagnosis.

A paroxysm typically begins with dramatic shaking chills (rigors) lasting 10–15 min or longer, which may rattle the bed and often completely incapacitate the sufferer. This period, during which the infected individual may experience a feeling of extreme cold, is followed by a hot stage, in which the patient appears agitated, restless, and disoriented. The body temperature may exceed 40°C, and the patient may complain of severe headache, backache, and pain in the limbs. This is followed by a defervescent phase, during which the patient sweats profusely and generally feels better, although left with weakness and fatigue, and often falling asleep. Body temperature may be normal upon awakening, and the patient may remain free of symptoms and able to go about daily activities until the onset of the next paroxysm 2–3 days later.

Paroxysms can occur in tertian cycles (48 h for each cycle, called tertian because fever occurs on the first and third days) or quartan cycles (72 h) depending on the species. *P. falciparum*, *P. vivax*, and *P. ovale* have a tertian cycle and *P. malariae* a quartan cycle, reflecting the underlying periodicity of blood stage schizogony. Quite often, however, the classic periodicity is not present because there are multiple asynchronous cohorts of parasites infecting the blood at the same time, so fevers come and go without a recognizable pattern.

The first wave of parasitemia, if controlled by the immune system or by subcurative drug therapy, will in nearly all cases be followed by recurrent waves of parasites (termed recrudescences, as previously described). In addition, patients with infections caused by *P. vivax* or *P. ovale* may experience malaria relapse weeks, months, or more than a year following the primary infection.

Complications

Uncomplicated malaria, as caused by all five species infecting humans, can be extremely debilitating, particularly in nonimmune individuals. As the infection progresses, white blood cell and platelet counts fall, and the red blood cells are gradually depleted as a result of direct lysis and diminished production from germinal centers. After several days, enlargement and tenderness of the liver and spleen may occur. The severity of infections with *P. vivax*, *P. ovale*, and *P. malariae* generally does not progress beyond this point, with serious complications generally limited to individuals with preexisting conditions. In contrast, *P. falciparum* frequently leads to progressive clinical deterioration, engendering a variety of severe complications. In sub-Saharan Africa, where the majority of falciparum-related deaths occur, approximately 1 in 100 pediatric cases of *P. falciparum* malaria will progress to severe disease (Greenwood et al., 1991).

Cerebral malaria is one of the most dreaded complications of *P. falciparum* infection. It is characterized by decreased level of consciousness, progressing from delirium to obtundation, stupor, coma, and death. Patients may present with a severe headache followed by drowsiness, increasing confusion and prostration. Seizures are a key feature of cerebral malaria and may often herald the onset of coma. Other serious complications of falciparum malaria which may occur with or without altered consciousness include hypoglycemia, hyperpyrexia, hyperparasitemia, pulmonary failure, renal failure, jaundice, bleeding disorders, sepsis, severe metabolic acidosis, and a variety of other conditions. The clinical manifestations of severe disease are thus varied, giving rise to malignant tertian malaria's reputation as a great imitator, mimicking meningitis, encephalitis, stroke, hepatitis, pulmonary edema, acute abdomen, acute renal failure, and many other conditions. All manifestations of severe malaria have, in common, the need for immediate diagnosis and institution of parenteral antimalarial therapy with maximum supportive care. The World Health Organization has published a list of criteria for severe malaria, aiming to promote appreciation for the importance of the diagnosis and the institution of appropriate treatment (World Health Organization, 2005).

Most patients surviving severe malaria will recover completely. However, some patients, particularly pediatric patients, may experience increased intracranial pressure, brain ischemia, and in some cases transtentorial herniation during the infection. Those that recover may be left with significant neurological sequellae, including cortical blindness, hemiparesis, generalized spasticity, cerebellar ataxia, hypotonia, or more subtle neurological deficits (Weatherall et al., 2002).

Another prominent syndrome of severe and complicated malaria is severe malarial anemia (SMA), which is particularly common in areas of sub-Saharan Africa with high transmission intensities (Snow and Marsh, 2002). In these settings, SMA is often the leading cause of malaria-related mortality in infants and young children (Obonyo et al., 2007).

The other species of human malaria may, on occasion, lead to complicated disease, although it is unusual. *P. vivax* can cause fatal splenic rupture (Mokashi et al., 1992; Ozsoy et al., 2004; Yagmur et al., 2000; Zingman and Viner, 1993), and there are now several reports that this species may cause other severe complications reminiscent of *P. falciparum* (Barcus et al., 2007; Kochar et al., 2005; Kumar et al., 2007; Lomar et al., 2005). Chronic *P. malariae* has classically been associated with nephrotic syndrome (Collins and Jeffery, 2007). The zoonotic parasite, *P. knowlesi*, has been reported to lead to fatal complications (Cox-Singh et al., 2008), possibly associated with its very rapid growth in the blood—the only species infecting humans with a 24h cycle.

TREATMENT AND PROPHYLAXIS

A large arsenal of drugs exists for the treatment of uncomplicated malaria, although there is often a problem with their availability in a given location (Lufesi et al., 2007). Unfortunately, there has been a recent series of distributions of fraudulent drugs that lack antimalarial activity in several countries (Newton et al., 2008). For standard treatment and prevention of malaria, the reader is referred to Griffith et al. (2007), Magill (2006), and Whitty et al. (2006). Malarone or mefloquine are highly effective, but additional regimens, such as ACTs, sulfadoxine/pyrimethamine, chloroquine (for sensitive strains of *P. falciparum* and for the nonfalciparum species), or quinine/doxycyline can be highly effective, with special considerations required for pregnant women and infants (Coll et al., 2008). In countries where artemisinin derivatives are available (compounds derived from the Chinese herbal plan *Artemesia diminuta*), there are often a variety of ACTs in the market, such as artemether-lumefantrine (Hatz et al., 2008). Artemisinin compounds cause a rapid knockdown of parasite populations in the blood, and when combined with a second agent to synergize in the killing of parasites and to prevent the recrudescence of resistant organisms, they are extremely effective antimalarials, and indeed are transforming the treatment of malaria in countries where they have been adopted as first-line therapy (Bosman and Mendis, 2007). These same compounds are now the treatment of choice for severe *P. falciparum* malaria (Day and Dondorp, 2007), and if available should be selected in favor of parenteral quinine or quinidine. The lack of artemisinin compounds in the USA is being addressed by a partnership between the Walter Reed Army Institute of Research and Sigma-Tau Pharmaceutical, the latter planning to seek licensure of artesunate for parenteral use during the latter half of 2008 (A. Magill, personal communication).

In addition to the immediate institution of parenteral antimalarials, patients with severe malaria require emergent and intensive supportive care to control seizures, hypoglycemia, brain edema, respiratory failure, acidosis, dehydration, sepsis, renal failure, and other complications. Many experimental compounds and procedures, such as administration of steroids and antitumor necrosis factor have been tested and found not to improve the success of treatment (Hoffman, 1988; Warrell, 1982; van Hensbroek, 1996; Enwere, 2005). Even with optimal support, treatment of severe malaria may be unsuccessful, and mortality rates can rise to 50% in remote areas where intensive care, dialysis, and mechanical respiratory support are not available, although they are generally much lower (Ranque et al., 2008; Winkler et al., 2008). A major improvement in the treatment of severe malaria has accompanied the development of rectal administration of artesunate, which can now be given in an outpatient setting while the patient is being transported to the hospital and can be lifesaving (Karunajeewa et al., 2007).

Ironically, the treatment of severe malaria can be poor in highly developed countries. This is due primarily to lack of experience by physicians working in parts of the world where malaria in not endemic. The travel history may not come to light, diagnostic procedures may be inadequate, and in some cases the required drugs may not be available (Magill, 2006). Travelers or their family members must insist that caregivers consider malaria as a cause of illness, and in some cases proactively help to identify clinicians qualified to care for this potentially devastating infection.

Many drugs that kill blood stage parasites do not kill hypnozoites, with the exception of 8-aminoquinolines such as primaquine (Crockett and Kain, 2007). If a patient has been diagnosed with *P. vivax* or *P. ovale*, a course of primaquine should be initiated as part of follow-up care in order to prevent relapse.

Many of the same drugs used for treating malaria are used for malaria prophylaxis and therefore prescribed for the traveler. Weekly chloroquine is the drug of choice only in those few areas of the world where chloroquine-resistance has never been reported, including parts of the Caribbean and parts of the Middle East. In most other areas, weekly mefloquine, daily Malarone, or daily doxycycline are recommended, with daily primaquine another possibility in those with normal glucose-6-phosphate dehydrogenase activity. Tafenoquine is under study as another potential prophylactic agent (Crockett and

Kain, 2007). Malarone, primaquine, and tafenoquine have the advantage of killing parasites during liver stage development, and thus their administration can be stopped a week after departure from a malaria-endemic area, whereas the blood schizonticides such as chloroquine or mefloquine need to be continued for a month to wait for all parasites to exit the liver and enter the blood, where they can be killed by these drugs, and are thus less convenient for short trips.

As a general rule, the prescription of a drug for malaria prophylaxis for a given individual should be performed by a travel medicine specialist, taking into account drug-resistance patterns in the area to be visited, the traveler's age and pregnancy/lactation status, and various key aspects of the traveler's medical history (seizures, heart disease, drug allergies) and preferences (weekly vs. daily administration, tolerance of side effects (Senn, 2007)). The complexity of prescribing the best antimalarial and the difficulties of side effects and compliance are major reasons why a "fire and forget" vaccine would be particularly helpful for travelers.

In addition to treatment and prophylaxis, antimalarial drugs are gaining increasing importance as tools of public health. It has now been shown that intermittent "presumptive treatment" (IPT) of malaria during key stages of life where the risk of clinical and severe malaria is greatest, namely infancy, early childhood, and pregnancy, can markedly reduce morbidity and mortality. Thus, IPT for infants (IPTi) (Sokhna et al., 2008) and pregnant women (IPTp) (Asa et al., 2008) has been instituted as national health policy in some countries although implementation has proven to be challenging (Hill and Kazembe, 2006).

The evolution of resistance to antimalarial drugs has historically been and currently remains an exceedingly important problem, especially for *P. falciparum* (Ekland and Fidock, 2007) but also for *P. vivax* (Baird, 2004) and occasionally *P. malariae* (Maguire et al., 2002). There must be continuous monitoring for this phenomenon in endemic areas (Laufer et al., 2007; Price et al., 2007; Vestergaard and Ringwald, 2007), and the institution of national treatment policy changes regarding first- and second-line drugs when significant resistance is detected. A component of retaining the upper hand against the parasite is the ongoing development and licensure of novel antimalarial drugs.

POSTEXPOSURE IMMUNOPROPHYLAXIS

It has long been known that treatment with immunoglobulin derived from semi-immune adults can lead to reductions in parasite density and clinical improvement (Bouharoun-Tayoun et al., 1990; Sabchareon et al., 1991) and that the presence of partial immunity enhances the curative effect of drug treatment, even permitting clearance of parasites that are resistant to the drug employed (Enevold et al., 2007). In addition, immune sera can assist in the identification of antigenic targets, aiding the development of effective vaccines (Lundquist et al., 2006; McIntosh et al., 2007; Nixon et al., 2005). However, due to the availability of excellent antimalarial drugs, there is no application for postexposure immunoprophylaxis.

PATHOGENESIS

Description of the Disease Process

The pathogenesis of malarial infection relates directly to the biological events associated with each stage of the life cycle, which differ markedly in the degree to which they disrupt normal host processes. Although sporozoites migrate destructively through the liver parenchyma, their numbers are few (500 are deposited in the skin during a 15 min mosquito bite, with typical numbers reaching the liver in nature likely far fewer) (Jin et al., 2007), damage is localized, and there is no detectable compromise in liver function. Tissue sections taken from infected mouse livers reveal an inflammatory reaction at the site of sporozoite migration (Mota and Rodriguez, 2004), but this is clinically silent. In marked contrast, blood stage infection by all five malarial parasites infecting humans precipitates a dramatic clinical syndrome, including extensive erythrocyte destruction, toxin release, febrile response, and a host of metabolic derangements. *P. falciparum* infection, however, has an additional feature arising out of its distinct pattern of sequestration in the deep venous microvasculature of organs throughout the body including the brain, lung, liver, kidney, dermis, bone marrow, and placenta (Schofield and Grau, 2005; Trampuz et al., 2003). This sequestration characterizes the late trophozoite and schizont stages of the parasite, explaining why *P. falciparum* schizonts are rarely seen in the peripheral blood. It follows the direct binding to endothelial cells of parasite antigens expressed on the surface of infected erythrocytes (with modified host erythrocyte surface proteins contributing). As mentioned previously, an important parasite protein responsible for binding is PfEMP1, the immunodominant antigen undergoing sequential expression of antigenically variant alleles enabling repeated immune escape and recrudescent parasitemia. Other molecules are involved as well, and a variety of receptors expressed on endothelial cells bind PfEMP1 and other ligands, including ICAM1 and VCAM1 (Sherman et al., 2003).

Although cytoadherence is a primary factor in sequestration and the associated disruption of normal flow, additional processes contribute, including rosetting (the binding of nonparasitized erythrocytes to parasitized erythrocytes), reduced red cell deformability, and the accumulation of unbound, uninfected erythrocytes as a result of locally poor vascular flow. This vascular "log jam" leads to the local release of inflammatory cytokines and an associated cascade of events that precipitate serious metabolic derangement including local tissue hypoxia and acidosis. As a general rule, it appears that the site of sequestration determines the degree of organ compromise. For example, sequestration in the cerebral vasculature results in cerebral malaria, sequestration in the kidney results in acute renal failure, etc. Sequestration in the placenta is a special case that explains the prevalence of severe disease in pregnant women, particularly primagravida. The awareness that sequestration is an important contributor to severe disease has led to the concept of developing vaccines or immunotherapies that prevent or reverse cytoadherence, thereby protecting the patient against the adverse manifestations of malaria.

Although sequestration is a major factor associated with pathogenesis, there are a host of additional pathogenic features of malarial infection, including the release of malaria toxins, the destruction of erythrocytes, and the enlargement of the spleen. In the latter case, chronic malaria contributes to the debilitating and potentially fatal tropical splenomegaly syndrome, also called hyperreactive malarial splenomegaly due to the associated increases in IgM and hyperplasia of the lymphoreticular system (Singh, 2007).

The consequences of malarial infection are so severe and the parasite so common that selection pressures have been exerted on the human population, making malaria, as expressed by one expert, "the strongest known force for evolutionary selection in the recent history of the human genome" (Kwiatkowski, 2005). These selection pressures have led to the genetically based adaptive mechanisms that in the presence of endemic malaria provide a selective advantage. Many of these have been described, including the changes in proteins and their structure underlying sickle cell anemia, the thalassemia, G6PD deficiency, ovalocytosis and many similar conditions including a role in shaping the ABO blood grouping. More recently, additional polymorphisms have been described affecting human immune system and inflammatory genes that likewise confer a selective advantage by offering protection against severe malaria. The reader is referred to excellent reviews for a detailed description of the field (Hill, 2006a; Kwiatkowski, 2005; Timmann et al., 2007).

IMMUNE RESPONSE TO INFECTION

While many protective immune mechanisms have been demonstrated, as described in "Protective Immune Responses" section above, the immune response to natural infection is complex and also fundamentally different from the persistent sterile immunity that follows many acute infectious diseases. The parasite is adept at living chronically in the host, either by evading the immune response, or by actively limiting its potency. NAI is never sufficient to prevent infection or to prevent the circulation of parasites at low levels or the generation of gametocytes that can infect a mosquito, even after a lifetime of exposure. The mechanisms whereby malaria evades or suppresses the immune response are diverse, including:

1. escape in time (the entry of merozoites into an erythrocyte is completed in less than 30 s),
2. escape within host cells (avoids exposure to plasma and thus to antibodies and complement),
3. sequestration of maturing blood stages in the microvasculature (escaping splenic filtration),
4. presentation of immunodominant repeat sequences to divert the immune responses (e.g., CSP repeat sequences),
5. inhibition of expression of host costimulatory molecules (e.g., CD80) needed for activation of lymphocytes (Sponaas et al., 2006; Steers et al., 2005),
6. inhibition of antigen processing and/or expression of MHC Class I or II molecules on the surface of antigen presenting cells (Steers et al., 2005),
7. stimulation of regulatory T cells (Tregs—CD4+, CD25+, FOXP3+) to downregulate T-cell responses against the liver and blood stage parasites (Hisaeda et al., 2008),
8. allelic/antigenic variation of key antigens (e.g., PfEMP1),
9. induction of blocking antibodies that limit the effectiveness of antibodies otherwise able to disrupt vital molecular pathways such as MSP1 processing and erythrocyte invasion (Uthaipibull et al., 2001),
10. mimicry of host sequences of commensal microorganisms in order to fall under the umbrella of tolerance (Good et al., 1993; Wide et al., 2006),

11. presentation of variant sequences to induce altered peptide ligand antagonism (Plebanski et al., 1997),
12. use of alternative pathways for key processes such as the invasion of erythrocytes (Baum et al., 2005; Stubbs et al., 2005), and
13. inhibition of the respiratory burst in Kupffer cells (Usynin et al., 2007) and many others.

The more closely parasite biology is studied, the more it appears that the parasite is capable of orchestrating the immune response to its advantage, enabling chronic infection and ongoing production and transmission of gametocytes. In sum, the natural immune response to acute infection can be considered ineffective if judged by the host's ability to eliminate the parasite, although, as described previously, a degree of clinical immunity gradually develops with repeated exposure. Countervailing this perversion of immune function in order to induce a highly effective immune response is a significant challenge for vaccine developers.

VACCINES

The development of a malaria vaccine represents one of the most important scientific and public health enterprises of our time (Epstein et al., 2007a). Attempts began early in the twentieth century (Desowitz, 1991), but despite tremendous advances in our understanding of the parasite and of the host immune response, the goal remains elusive. Providing high-level protection against the malarial parasite will entail overcoming multiple hurdles and likely require a complex vaccine design (Mahanty et al., 2003). Some of the principal impediments to progress are as follows.

1. Stage-specific antigen expression: *Plasmodium* expresses a different, complex array of antigens on its surface and on the surface of the infected host cell at each stage of its life cycle, reflecting different tissue environments and growth requirements. The predominant antigens expressed during the preerythrocytic stages are largely distinct from those expressed during the asexual or sexual blood stages. A vaccine targeting one stage of the parasite life cycle would have to be 100% effective in order to prevent progression to the next stage where growth would occur unchecked, or alternatively the vaccine would have to include antigens from multiple stages. This indicates that a simple vaccine based on one or two antigens is unlikely to be broadly protective.
2. Allelic variation: Key antigens demonstrate extensive allelic diversity (de Souza-Neiras et al., 2007; Ferreira et al., 2007; Mahajan et al., 2005), with the encoding genes harboring significantly greater variability than immediately adjacent flanking regions of DNA, indicating that the antigens have experienced strong immune selection favoring polymorphism (Amodu et al., 2007). It is not surprising that, given the history of selection, immune responses are generally allele-specific, with cross-protection developing only gradually following repeated exposure, if developing at all. Therefore, vaccines based on a single allele are unlikely to provide protection in the field where multiple alleles are circulating and where individuals may typically be infected by multiple strains (Fig. 65.6). The finding that a newly discovered antigen exhibits extensive polymorphism is usually interpreted as prima facie evidence of the importance of the antigen as an immunological target. However, this same information conveys the difficulty of employing the antigen effectively as a vaccine candidate. One way to circumvent allelic variation is to design chimeric immunogens or combine immunogens into mixtures or epitope strings that induce cross-protective responses or responses targeting multiple alleles. The evaluation of these vaccines will require careful assessment of allele-specific vs. cross-protective responses when tested in the field. To enable this detailed analysis, investigators are characterizing the genetic

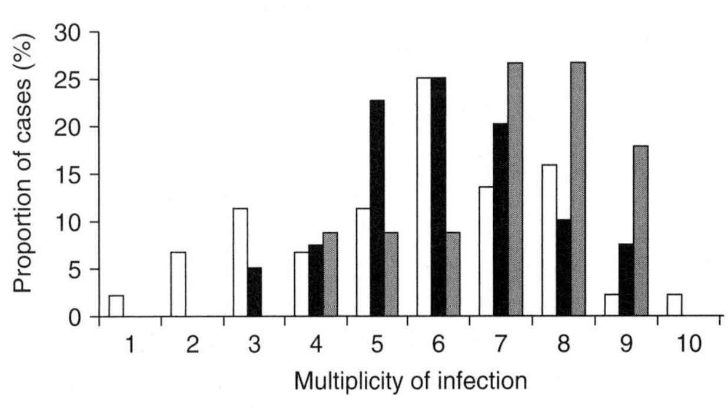

FIGURE 65.6 Multiplicity of infection in peripheral, placental, and cord blood in Senegalese women. The graph shows the number of parasite genotypes detected by PCR in peripheral (black bars), placental (white bars), and cord (hatched bars) blood at the time of delivery. These data are based on the diversity exhibited by only the MSP1 and MSP2 antigens, therefore likely underestimating the true multiplicity of infection in these samples. Journal of Clinical Microbiology, Vol 43, Jafari-Guemouri et al., 2005, with permission from American Society for Microbiology.

polymorphisms present in parasite populations at vaccine testing sites so that trials can be adequately powered to detect allele-specific protective effects (Takala et al., 2007).

3. Antigenic variation: Individual *Plasmodium* parasite clones have the ability to sequentially replace and thereby antigenically vary key protective proteins, exemplified by PfEMP1 expression on the surface of the infected erythrocyte (Kyes et al., 2007). A review of *var* and other variant gene families discovered in *P. falciparum* including *rif, stevor*, and *Pfmc-2TM*, is provided by Dzikowski et al. (2006). To date, is has not been possible to identify conserved and immunogenic sequences inducing strongly cross-protective responses targeting these antigens.
4. Genetic restriction of the host immune response: Key immune responses to malarial antigens, such as those targeting liver-stage parasites, are cell-mediated and thus genetically restricted. Individuals with different HLA haplotypes respond to different T-cell epitopes from a given antigen, or may not respond at all. Therefore, vaccines designed to act via the induction of cell-mediated immunity must contain a variety of CD4+ and CD8+ T-cell epitopes in order to provide broad population coverage, and this may be difficult to achieve if vaccines are based on one or two antigens (Doolan et al., 1997).

In the most general sense, barriers to developing a malaria vaccine emerge from the contradiction inherent in the parasite's ability to evade the host immune response and establish a chronic infection. One may wonder if the proposal to harness the immune response to reject the parasite is not an exercise in futility.

Evidence for the Feasibility of Developing a Malaria Vaccine

Two models of immunity help to allay the above concerns, indicating that developing an effective malaria vaccine is indeed feasible: (1) immunization with irradiated sporozoites and (2) repeated blood stage infection, through natural exposure or malarial therapy (the treatment of neurosyphilis with malarial fevers in the preantibiotic era).

Immunization with Irradiated Sporozoites: Model of Sterile Immunity

In 1967, Nussenzweig et al. (1967) first demonstrated that mice immunized with radiation-attenuated sporozoites were protected against challenge with fully infectious sporozoites. Similar studies in NHPs likewise demonstrated protection or partial protection (Collins and Contacos, 1972; Jahiel et al., 1970). These findings were rapidly extended to humans, with sterile protection (i.e., absence of blood stage infection) first reported in 1973 (Clyde et al., 1973a, 1973b; Hoffman et al., 2002).

In this model, malaria-infected mosquitoes are exposed to irradiation, which damages but does not kill the sporozoites the mosquitoes are harboring. The optimal dose of 12,000–15,000 cGy was empirically determined—higher radiation doses reduce protection, and lower doses lead to breakthrough blood stage infections. The mosquitoes are allowed to feed on hosts, or alternatively, crude preparations of irradiated sporozoites isolated from infected mosquitoes by dissection may be injected intravenously in the case of animal models. The weakened sporozoites travel to the liver and undergo partial development, expressing new antigens, such as LSA1, but then stop developing, without releasing merozoites into the bloodstream. A portion of the injected sporozoites migrate to the lymph nodes rather than to the liver, where presentation of malarial antigens in germinal centers contributes to the immune response (Chakravarty et al., 2007).

In humans, sterile protection has been induced in 13 of the 14 volunteers by this method, provided that the infectious mosquito bites total at least 900–1000. In the few volunteers studied, protection was induced following immunization with *P. vivax* as well, although protection was not cross-specific (Hoffman et al., 2002). Intriguingly, protection did appear to be cross-protective among strains, suggesting that the attenuated sporozoite approach could help to circumvent the problem of antigenic polymorphism that hinders the development of effective subunit vaccines.

This sterile protection induced in the irradiated sporozoite model is thought to be mediated by CD8+ T-cell responses recognizing preerythrocytic stage antigens expressed on the surface of infected hepatocytes, with CD4+ T-cell responses and antibody responses contributing (Jobe et al., 2007; Tarun et al., 2007; Vanderberg et al., 2007); however, the mechanisms of protection are not completely understood, nor have the antigenic targets been identified. Although responses targeting the CSP contribute to protection, they are not required and 100% protection can be achieved in the absence of these responses (Gruner et al., 2007; Kumar et al., 2006).

In summary, the irradiated sporozoite model demonstrates that it is possible to induce immune responses directed toward antigens expressed by preerythrocytic stage parasites that completely prevent blood stage infection in humans. If these antigens could be identified and the correct immune responses generated, it should be possible to mimic the irradiated sporozoite vaccine by using a subunit approach and thereby achieve high-level protection.

Repeated Exposure to Infection: Model of Partial Immunity

NAI through repeated exposure to blood stage parasites provides a second model of protection supporting the feasibility of developing malaria vaccines, in this case vaccines targeting blood stage antigens. There appear to be two components to this immunity, developing at different rates: (1) antidisease immunity and (2) antiparasite immunity. In endemic areas, the risk of fatal malaria is highest during the first 2–3 years of life and then considerably diminishes by age 5 with sufficient exposure. Antidisease immunity develops rapidly during this period, with a reduction in the risk of life-threatening disease after one or more infections (Gupta et al., 1999). Antiparasite immunity develops more gradually, manifest as low-grade parasitemias with spontaneous clearance (Branch et al., 2005). The development of both clinical and parasitological immunity has also been observed with malarial therapy for the treatment of neurosyphilis (Collins and Jeffery, 1999a, 1999b, 2005, 2007; Collins et al., 1989, 2004; Molineaux et al., 2002).

The protection underlying NAI is thought to be mediated by antibody responses targeting blood stage antigens, since immunoglobulin preparations from semi-immune African adults reduced symptoms and parasitemia when administered to children with acute malarial infections (Bouharoun-Tayoun et al., 1990; Sabchareon et al., 1991). Additionally, it has been shown that antibody responses to specific blood stage antigens, such as MSP1, are associated with clinical resistance to malaria in residents of endemic areas. In summary, NAI demonstrates the feasibility of developing a partially protective vaccine targeting blood stage parasites that limits parasite density and reduces the risk of clinical disease and death.

These two models broadly define approaches to malaria vaccine development: (1) preerythrocytic stage vaccines (antiinfection vaccines) aimed at mimicking the irradiated sporozoite model that entirely prevent blood stage infection by killing sporozoites and developing liver-stage parasites, and (2) erythrocytic stage vaccines (antidisease vaccines) that mimicking NAI, reducing parasite density, morbidity, and mortality in infected individuals.

Although this classification is supported by a strong rationale, the distinction is often blurred. For example, it has been demonstrated that a vaccine based on the CSP, a preerythrocytic stage protein not expressed by blood stage parasites, may reduce the frequency of clinical episodes in African children for a period of at least 18 months following immunization, with an even stronger effect on reducing severe disease (see discussion of the RTS,S vaccine below), thereby mimicking NAI. It is presumed that the mechanism is related to reduced liver-stage parasite burden and reduced frequency of merozoite release into the blood, which may limit genetic diversity of the circulating parasites and thus the likelihood of a pathogenic strain entering the blood (Enosse et al., 2006).

In addition to preerythrocytic- and erythrocytic-stage vaccines, there is a third category —transmission-blocking vaccines—which target the sexual stages (gametocytes) of the parasite and aim to reduce or prevent transmission, and these are generally placed in a group by themselves (see "Transmission-Blocking Vaccines" section below).

History of Malaria Vaccine Development

Protein-Based Vaccines

As just described, the first substantial advances in malaria vaccines were achieved in the 1960s and 1970s with the demonstration that irradiated sporozoites could provide sterile protection, but no further development of this approach was attempted due to the impracticality of a vaccine delivered by a mosquito bite. Consequently, most development efforts during the past 30 years have focused on the subunit approach, which involves formulating defined partial- or full-length antigens to induce potent immune responses targeting the antigens involved in key biological processes required for parasite development. The subunit approach was launched with great enthusiasm in the early 1980s following the identification and cloning of the first preerythrocytic-stage antigen, the CSP (Dame et al., 1984; Yoshida et al., 1980). It was anticipated by many that this discovery would rapidly lead to an effective vaccine.

Early studies in animal models using CSP or CSP epitopes as the vaccine antigen were promising, but when tested in humans, several formulations including peptide–protein conjugates and recombinant proteins failed to provide consistent protection against experimental sporozoite challenge, although in a few individuals sterile protection was achieved (Ballou et al., 1985; Brown et al., 1994; Fries et al., 1992; Heppner et al., 1996; Herrington et al., 1987; Rickman et al.,

1991; Sherwood et al., 1996). Early vaccines were limited by suboptimal epitope configuration and adjuvant formulation. It was not until the antigen was engineered as the virus-like particle RTS,S and adjuvanted with a proprietary oil-in-water emulsion that it induced protection (Gordon et al., 1995; Stoute et al., 1997). This vaccine, RTS,S/AS02a, is discussed in "RTS,S" section below as it remains a leading candidate vaccine. Other efforts to develop CSP or CSP epitopes as vaccine antigens, including formulation as multiantigen peptides (MAPs) or virus-like particles similar to RTS,S, but utilizing hepatitis B core antigen rather than hepatitis B surface antigen, have induced strong immune responses but failed to protect or were sidelined due to reactogenicity (Nardin, 2000, 2001, 2004; Edelman, 2002; Oliveira, 2005; Walther, 2007; Gregson, 2008).

Several blood stage antigens were also cloned and sequenced in the 1980s, identified from cDNA expression libraries screened with growth-inhibitory serum obtained from semi-immune humans living in malaria-endemic areas or from *Aotus* monkeys exposed to repeated *P. falciparum* infections (Ardeshir et al., 1985; Kemp et al., 1983). Many of the identified antigens remain leading vaccine candidates to this day, although, disappointingly, none has proven to protect against infection when tested in humans. There have been some encouraging findings, however: the Combo B vaccine, derived from three blood stage antigens (MSP1, MSP2, and RESA) formulated in Montanide ISA 720, was tested in 5–9-year-old children in Papua New Guinea and reduced the parasite density in the study group not pretreated with sulfadoxine-pyrimethamine, relative to the comparator group (Genton et al., 2002). This finding corresponded to a paucity of the 3D7 allele of MSP2 (the allele included in the vaccine) in the infections acquired by vaccine recipients during the observation period, whereas the nonvaccine FC27 allele showed no differences among study groups, implying that vaccination did protect against infection with parasite harboring the 3D7 allele. These finding have encouraged vaccine developers, and the search for an effective blood stage vaccine remains viable.

Another development enterprise that figured prominently in the field was led by Dr. Manuel Patarroyo of the National University of Colombia. This vaccine, called SPf66, combined three synthetic peptides from blood stage antigens, joined with epitopes from the CSP. SPf66 was tested extensively in South America, Africa, and Asia in the 1990s in a variety of settings and age groups. After the completion of these studies, a meta-analysis of 10 randomized, placebo-controlled trials involving nearly 10,000 study subjects was performed to assess the value of the SPf66 vaccine. This revealed moderate (28%) efficacy against *P. falciparum* infections in South America, but, disappointingly, no significant protection in trials conducted in Africa or Asia (Graves and Gelband, 2006). The vaccine has not been developed further.

DNA Vaccines

In the early 1990s, a novel vaccine platform—immunization with "naked" DNA—was first tested against malaria. A murine model was used to demonstrate that intramuscular administration of recombinant DNA plasmids encoding the CSP of *P. yoelii* permitted the uptake of the plasmids by host cells and subsequent expression of the malaria transgene, and induced an immune response affording protection against sporozoite challenge (Sedegah et al., 1994). Because of the intracellular synthesis and processing of the antigenic targets, DNA vaccines allow for HLA presentation of Class I T-cell epitopes via the endogenous presentation pathway, thereby inducing the cytotoxic T lymphocyte (CTL) responses believed to contribute to the destruction of developing liver stage parasites. DNA has the advantage that transgenes can be switched without changing manufacturing processes, codon usage can be optimized or harmonized (Hillier et al., 2005; Narum et al., 2000, 2001), key epitopes can be strung together (Doolan et al., 1997) or genes can be shuffled to increase antigen coverage and immunogenicity (Cai et al., 2007).

Based on these considerations and the demonstration of protection in mice, the U.S. Navy engaged in a program called MuStDO (multistage DNA vaccine operation) in the late 1990s to develop DNA vaccines for preventing malaria in humans. The first two clinical trials demonstrated safety (Epstein et al., 2004; Le et al., 2000) and induced CTL and interferon gamma responses targeting CSP (Wang et al., 1998, 2001). However, there were no measurable antibodies, a departure from the murine model in which both CTL and antibodies were routinely induced by DNA immunization. Assuming that multiple antigens would be required for protection, a mixture of five plasmids encoding preerythrocytic stage antigens (CSP, SSP2/TRAP, EXP1, LSA1, and LSA3) was tested with or without a sixth plasmid encoding human granulocyte macrophage-colony stimulating factor (hGM-CSF). As in the initial trials, cell-mediated immune responses were detected, but there were negligible antibody responses and no protection against sporozoite challenge (Wang et al., 2005). Based on the results of this trial, further testing of naked DNA as a stand-alone regimen was not pursued. Now, after a gap of several years, intensive efforts to optimize the platform have resulted in improved immunogenicity. Electroporation has dramatically increased tranfection efficiency (Cemazar and Sersa, 2007), and this, along with many additional improvements (Laddy and Weiner, 2006), may soon bring DNA back to the forefront of vaccine development.

Even as the testing of DNA as a stand-alone regimen was sidelined, the importance of DNA as a priming agent was recognized. Volunteers primed with three doses of DNA encoding CSP and boosted with RTS,S/AS02A (see "RTS,S" section below) more than a year later showed stronger cell-mediated immunity than volunteers receiving RTS,S/AS02A without priming (Wang et al., 2004). These data, supported by similar findings in animal models, established the value of naked DNA as a priming agent able to induce long-lasting memory T cells capable of improving the immunogenicity and protection of a booster vaccine given subsequent to the DNA prime (see "Prime-Boost Strategy" section below).

Virally Vectored Vaccines

In the late 1990s and early 2000s, efforts shifted to viral vectors as a possibly better way to deliver DNA encoding malarial antigens. Viral vectors retain a key attribute of DNA, namely the intracellular synthesis of the parasite protein by a eukaryotic host cell, leading to Class I as well as Class II presentation and thus the induction of both CD8+ and CD4+ T-cell responses. Eukaryotic host cell synthesis favors native protein folding, which is difficult to achieve when parasite proteins are synthesized by a prokaryote such as *E. coli*, although this problem with bacterial expression can be overcome with codon harmonization or special refolding steps during the manufacturing process (Hillier et al., 2005). However, unlike DNA, viral vectors may target particular host cells such as dendritic cells, and are intrinsically designed for efficient cell entry and transfection. Moreover, viral vectors provide a stimulatory danger signal that activates the innate immune system via Toll-like receptors (TLRs), enhancing immunogenicity. As an added attribute, viral vectors can be designed to express multiple transgenes placed into different expression cassettes. Although in some cases manufacturing and scale-up can be more difficult for viral vectors than for DNA, and there may be concerns regarding the adequacy of attenuation, these challenges can generally be addressed successfully.

Promising viral vectors include attenuated mammalian poxviruses such as Modified Vaccinia Ankara (MVA), avian poxviruses such as canary pox or fowl pox (FP), human adenoviruses attenuated via genetically-engineered gene deletions, simian adenoviruses (Sridhar et al., 2008), viral replicon particles (VRPs, derived from Venezuelan equine encephalitis virus), and the yellow fever vaccine vector (Tao et al., 2005). One of the first studies of malarial antigens delivered by a viral vector involved the recombination of seven different malaria genes into a NYVAC poxvirus platform, creating the NYVAC Pf7 vaccine (Tine et al., 1996), a joint effort by the U.S. Army and Virogenetics, Inc. When tested in the clinic, this vaccine was immunogenic and achieved a modest level of protection against experimental sporozoite challenge: one out of 35 volunteers was sterilely protected and the onset of parasitemia was delayed relative to controls in others (Ockenhouse et al., 1998). Many considered this to be a promising result meriting further study, but the vaccine was not pursued further. Other viral platforms, such as adenovectors, are now entering clinical testing and will be described below in "Vaccines Currently under Development" section (Li et al., 2007).

Prime-Boost Strategy

The most recent evolution in the history of subunit vaccines is sequential immunization with two different vaccine platforms, which has proven superior to immunization with either platform alone. Some platforms are more effective as priming agents (e.g., DNA), while others can serve both to prime and boost (adenovirus, poxvirus). The superiority of heterologous over homologous regimens may relate to the ability of heterologous regimens to reduce competition between the antigen expressed by the transgene and the multiple viral antigens expressed by the vector, since the transgene is the only protein expressed in common between the two immunizations, enabling an anamnestic response to the target antigen. The heterologous approach may also avoid neutralization by antivector responses induced by the prime, leading to higher host cell transfection rates and therefore enhanced transgene expression when the boost is administered.

In a remarkable series of clinical studies, the malaria vaccine development program at Oxford University systematically explored a variety of heterologous prime-boost regimens in humans, focusing on DNA or FP followed by MVA (Dunachie et al., 2006a, 2006b; Moore and Hill, 2004). In most of these studies, the immunogen was ME-TRAP, a multiepitope string coupled to the full-length preerythrocytic stage antigen TRAP (SSP2) (McConkey, 2003) spliced into a DNA plasmid or poxvector, although the CSP was also studied (Walters and Sperandio, 2006). These regimens achieved markedly improved cellular immune responses, and were associated with a calculated average 92% reduction in liver stage burden combining data from several human trials (Bejon et al., 2005). In a few studies, there were volunteers sterilely protected against sporozoite challenge, with delays in onset of parasitemia exhibited in others (Webster et al., 2005). These prime-boost regimens involving DNA and poxvectors were transitioned to testing in Africa, but, disappointingly, in two different Phase 2b trials, no protection was observed (Bejon et al., 2006, 2007; Moorthy et al., 2004). These vaccines, like many others, have been sidelined while efforts are underway to improve the viral platforms.

Many clinical trials testing heterologous prime-boost regimens can be anticipated in the coming years. Tempering the excitement, however, is an awareness of the difficulties associated with developing two different vaccines for use in one regimen, including the challenges of forming the business partnerships that are often required, the complex administration schedules and increased expense. These concerns currently motivate efforts to improve homologous regimens to the point where they may equal the immunogenicity of heterologous regimens.

Other Platforms

Bacteria have been evaluated as delivery systems for several decades. The transgene is expressed from plasmids residing in the host bacterium. Efforts to develop Salmonella as a delivery platform for malarial antigens continue, e.g., at the University of Maryland (Chinchilla et al., 2007) and the University of Newcastle (McKelvie et al., 2008). Bacillus Calmette-Guérin (BCG), a live and attenuated strain of *Mycobacterium bovis*, is also under development as a platform for malaria vaccines (Beveridge et al., 2007; Rapeah and Norazmi, 2006; Zheng et al., 2002). Protozoa, including Toxoplasmosis have also been used as delivery systems for malarial antigens (Charest et al., 2000).

Whole Organism Vaccines

Spurred by the many challenges facing subunit vaccines, the last 5 years have seen the reemergence of the whole organism approach. During more than 40 years, irradiated sporozoites provided a highly meritorious model but pointedly lacked viability as a licensable product (Luke and Hoffman, 2003). Rather than targeting an identified biological process, the whole organism approach seeks to identify the best immunogen, presenting to the immune system a complex array of antigens that are tested for protection (Pinzon-Charry and Good, 2008). In this approach, neither the antigens nor the protective immune mechanism needs to be identified. This empirical approach is supported by the fact that most licensed vaccines have been developed in this fashion, and consist of whole, attenuated organisms or material isolated from whole organisms such as a toxin or crude antigen preparations.

Most pathogens are less complex than *Plasmodium*, however, which has a genome comprising more than 5300 genes. Moreover, most of these pathogens can be easily cultured, facilitating attenuation and enabling scale-up for mass production. The only stage of any human malarial parasite that can be routinely and easily cultured is the blood stage of *P. falciparum*, and this requires human blood as the substrate—thus, there has never been a way to attenuate preerythrocytic stages via repeated passage nor to collect significant quantities of immunogen from the preerythrocytic stage of any malarial parasite or from the blood stage of any species other than *P. falciparum*. In terms of the irradiated sporozoite model, the only way to produce sporozoites is from mosquitoes.

There are multiple additional barriers to a whole sporozoite vaccine, including administration (using mosquitoes as the method for administration is not practical), suitability for parenteral injection (the mosquito "factory" for producing sporozoites originates in swamp-like open trays of water hardly amenable to good manufacturing practices), and storage (the sporozoites must be maintained in a metabolically active state in order to undergo partial development in the liver and induce protective responses). Finally, there are concerns about attenuation, since a single sporozoite completing liver-stage development could turn a protective vaccine into a lethal malarial infection.

Remarkably, many of the barriers have been overcome by creative bioengineering, enabling the revitalization of the attenuated sporozoite concept. There are now at least two approaches under development, both supported by the Malaria Vaccine Initiative at PATH, with funds provided by the Bill and Melinda Gates Foundation. One approach is based on sporozoites genetically attenuated via irradiation and the other is based on genetic attenuation via targeted gene deletion (knockout). There are intriguing data suggesting that a whole, attenuated blood stage vaccine may be feasible as well (Pombo et al., 2002). The attenuated whole organism approach is now on an accelerating development track, in parallel with subunit vaccines, and the malaria vaccine field is galvanized to observe the outcome of the first clinical trials.

Within this historical framework, the next section will review a selection of active vaccine candidates focusing on those in clinical development. There are many additional vaccines not described, most of which have not yet reached clinical testing. Exhaustive lists of candidates have recently been published, and are available for perusal on the Internet (World Health Organization Rainbow Table; Moran et al., 2007).

Vaccines Currently under Development

Preerythrocytic Stage Vaccines

Preerythrocytic stage malaria vaccines have induced complete protection against blood stage infection by targeting sporozoites and liver stage parasites. Partially effective preerythrocytic stage vaccines have reduced liver stage parasite burden and the frequency and severity of clinical episodes of malaria.

Vaccines Based on Circumsporozoite Protein

CSP forms the surface coat of the sporozoite and is involved in sporozoite invasion of hepatocytes, with a recently described role in sporozoite morphogenesis (Kappe, 2004). Humans protected with irradiated sporozoites generate antibodies and lymphocytes recognizing CSP (Egan et al., 1993; Krzych et al., 1995), and sera induced by CSP have been shown to inhibit sporozoite invasion (Hollingdale et al., 1987) and CTL to destroy infected hepatocytes (Franke et al., 2000). The antigen is protective in animal models (Bharadwaj et al., 1998; Charest et al., 2000; Doolan et al., 1996; Lanar et al., 1996; Sedegah et al., 1994; Wang et al., 1995), and CD8+ and CD4+ T-cell responses to specific epitopes within CSP have been described (Doolan and Good, 1992; Marussig et al., 1997; Zevering et al., 1998). Although there have been many vaccine compositions based on CSP, only the RTS,S recombinant protein vaccine, developed in 1987, has consistently induced protective responses in humans, and only when formulated in particular adjuvant systems (Ballou and Cahill, 2007; Heppner et al., 1996, 2001, 2005).

RTS,S (CSP) (Adjuvanted Hybrid CSP-HepB Particle) The world's leading malaria vaccine, RTS,S, is the product of a twenty year collaboration initiated in 1984 between the Walter Reed Army Institute of Research (WRAIR) and GlaxoSmithKline Biologicals (GSK) (Ballou and Cahill, 2007; Heppner et al., 1996, 2001; 2005). The CSP antigen in RTS,S is a hybrid protein containing part of the tandem **r**epeat region of CSP (R), a contiguous segment of the C-terminal flanking region harboring key **T**-cell epitopes (T) and the N-terminus of the hepatitis B **s**urface antigen (HBsAg) (S). This fusion protein (RTS) is expressed by the yeast *Saccharomyces* in 1:4 ratio with unfused HBsAg (S). The RTS and S proteins spontaneously assemble into virus like particles with CSP arrayed on the surface. RTS,S was first successfully adjuvanted with GSK's proprietary adjuvant AS02A (an oil-in-water emulsion containing the two immunostimulants QS21 and MPL) and more recently with the liposomal AS01B and with AS02D, two additional proprietary adjuvant systems developed by GSK (Mettens et al., 2008).

In a meta-analysis of Phase 2a trials, RTS,S/AS02A was shown to have induced sterile protection against experimental sporozoite challenge in 41% of malaria naïve adult volunteers (95% CI [22–56%], $p = 0.0006$) and partial protection in the majority of those not sterilely protected, as evidenced by a significant delay in onset of parasitemia (Kester et al., 2001). A subsequent Phase 2b trial conducted in semi-immune Gambian adults with lifetime exposure to malaria demonstrated a 34% reduction in the incidence of first parasitemia (95% CI [8–53%]) over a 16-week follow-up period, providing evidence for protection against naturally transmitted malaria (Bojang et al., 2001).

Following these promising results, the vaccine advanced to Phase 1 studies in children in the Gambia and Mozambique, where it was shown to be safe and immunogenic (Bojang, 2005). Subsequently, a Phase 2b, randomized, double-blind, controlled trial conducted in children 1 to 4 years of age in Mozambique demonstrated an overall reduction in time to the first clinical episode of malaria of 30% (95% CI [11–45%]) at 6 months, with a one year extension to the surveillance period showing sustained protection (35%, 95% CI [22–47%]) (Alonso et al., 2004, 2005) (Fig. 65.7). Furthermore, RTS,S/AS02A showed 58% (95% CI [16–81%]) and 49% (95% CI [12–71%]) protection against severe malaria during these same observation periods, at 6 and 18 months, respectively (Alonso et al., 2004, 2005). Most recently, RTS,S/AS02D reduced the rate of new onset parasitemia in infants over a six month observation period by 66% (95% CI [43–80]%) (Aponte et al., 2007). These results demonstrate the potential of RTS,S to reduce the burden of malaria in endemic areas, and have established it as a leading malaria vaccine candidate.

A large Phase 3 trial is scheduled to begin in late 2008 in seven countries across Africa under a partnership agreement between GSK and PATH MVI, a grantee of the Bill and Melinda Gates Foundation (Ballou and Cahill, 2007), with costs shared between GSK and MVI. This Phase 3 trial will determine if the encouraging findings in Mozambique extend to other geographic areas encompassing diverse populations and transmission conditions. If all goes well in the

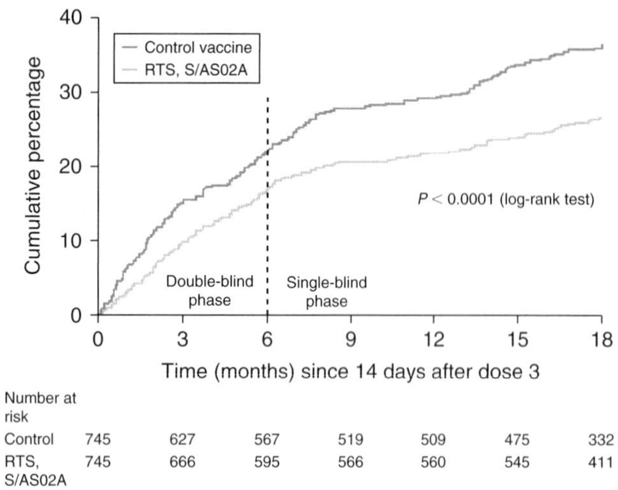

FIGURE 65.7 Kaplan-Meier curves for the cumulative proportion of children with at least one episode of clinical malaria. Protection against clinical malaria was sustained for 18 months in 1–4-year-old children living in Mozambique, following immunization with RTS,S/ASO2A. Reprinted from The Lancet, Vol. 366, Alonso et al., 2005, with permission from Elsevier.

Phase 3 trial, the sponsor, GSK, anticipates submitting its file to the European Medicines Agency for a positive opinion under the Article 58 procedure by 2011. In anticipation of a successful outcome to this Phase 3 trial, the processes required for host country approval and national public health policy decision-making are already being developed.

RTS,S Combinations There have been several efforts to improve RTS,S by (1) prime-boost combination with other vaccine platforms including DNA (Epstein et al., 2004) and poxvirus (Dunachie et al., 2006b) and (2) through combination with other preerythrocytic (TRAP/SSP2) and blood stage (MSP1) antigens. None has led to significant improvements, and in one case (combination with TRAP/SSP2) appeared to have compromised the protection afforded by RTS,S (Ballou and Cahill, 2007; Heppner et al., 2005). On the horizon are prime-boost regimens pairing RTS,S with adenovirus, a combination that strongly boosts cell-mediated immune responses without compromising the already excellent antibody responses (Hutchings et al., 2007; Rodriguez et al., 2008; Stewart et al., 2007). Although the clinical development pathway for an adenovirus-recombinant protein prime-boost regimen for CSP has not been defined, it is a high priority.

Ad35.CS (Gene Encoding CSP) (Adenovirus Serotype 35) The National Institute of Allergy and Infectious Diseases (NIAID) at the National Institutes of Health (NIH) has partnered with the biotechnology company, Crucell, to develop a CSP-based vaccine using an adenovirus serotype 35 vector. This vaccine, called Ad35.CS, is currently being tested in a Phase 1 trial by Vanderbilt and Stanford Universities, and was demonstrated to be safe and immunogenic (B.F. Hall, personal communication). Assuming that appropriate safety and immunogenicity criteria are met, NIAID currently plans further evaluation of this product in clinical trials in Africa and elsewhere.

Vaccines Based on TRAP/SSP2 TRAP/SSP2 is a highly conserved SSP associated with gliding motility and hepatocyte invasion (Muller et al., 1993). Immunization with TRAP/SSP2 has induced protection or partial protection against sporozoite challenge in mice (Khusmith et al., 1991; Schneider et al., 1998; Wang et al., 1996), monkeys (Bhardwaj et al., 2003), and humans (Dunachie et al., 2006a; Webster et al., 2005). CD8+ T-cell-dependent cytolytic activity mediated by HLA-restricted epitopes in TRAP/SSP2 has been identified in humans immunized with irradiated sporozoites (Wizel et al., 1995a, 1995b) as well as humans living in malaria-endemic areas (Aidoo et al., 1995).

TRAP/SSP2 1 Multiepitope String (Gene Encoding this Antigen) (MVA and FP) Adrian Hill at Oxford, whose pioneering clinical studies of prime-boost regimens were described earlier, is developing TRAP/SSP2 for delivery by a simian adenovirus-vector (Reyes-Sandoval et al., 2008). The antigen is expressed as a fusion protein linking the TRAP/SSP2 molecule to a multiepitope string of T- and B-cell epitopes (ME-TRAP). A clinical trial funded by the Medical Research Council and the Wellcome Trust is currently underway at the Jenner Institute to test this approach, in collaboration with the biopharmaceutical company, Okairòs. Future plans include using simian adenovectors in prime-boost approaches with MVA and building internal adjuvants into the vector to enhance immunogenicity, a project supported by the Foundation for NIH (Bill and Melinda Gates Foundation Grand Challenge in Global Health Program). Clinical trials of blood stage antigens delivered by simian adenovirus and MVA vectors are also planned.

Vaccines Based on Liver-Stage Antigen 1 LSA1 is a 230 kDa preerythrocytic stage protein containing a large central region consisting of over eighty 17 amino acid residue repeat units flanked by highly conserved C- and N-terminal regions (Fidock et al., 1994; Zhu and Hollingdale, 1991). LSA1 is expressed only by liver stage parasites, not by sporozoites; currently, its function is not known. Humans immunized with irradiated sporozoites demonstrate proliferative responses against LSA1 (Krzych et al., 1995), and people living in endemic regions demonstrate interferon gamma responses (John et al., 2004; Migot-Nabias et al., 2000) and CTL responses (Hill et al., 1992) against specific LSA1 epitopes. A study conducted in Kenya indicated that high levels of IgG targeting LSA1 are associated with protection from infection (John et al., 2008).

FMP011 (LSA1) (Recombinant Protein) Based on these data, WRAIR and GSK developed an LSA1-based recombinant protein vaccine, called falciparum malaria protein 011 (FMP011) (Hillier et al., 2005). After demonstrating immunogenicity in mice and NHPs (Brando et al., 2007; Pichyangkul et al., 2008), FMP011 was tested in malaria-naïve healthy adult volunteers using AS02A and AS01B as adjuvants. In a Phase 1/2a clinical trial, the vaccine was shown to be safe and to induce strong antibody responses and strong CD4+ T-cell responses. However, no protection or delay in parasitemia was observed following sporozoite challenge (J. Cummings, personal communication), indicating that a different quality of immune response may be required to achieve protection with LSA1. An adenovirus construct containing the same sequence used to make FMP011 was tested in a prime-boost regimen with FMP011, and this regimen induced high CD8 responses in mice (Rodriguez et al., 2008) suggesting that adenovirus protein prime-boost may prove more effective for the induction of a protective immune response in humans than FMP-11 alone.

Vaccines Based on Liver Stage Antigen 3 LSA3 is a relatively conserved preerythrocytic stage antigen selected by differential immune responses from irradiated sporozoite immunized volunteers, either protected or not upon challenge. Using lipopeptide formulations, DNA, adjuvated peptide, or recombinant protein, the antigen is protective in mice (Sauzet et al., 2001) and induces CD4 Th1-related protection against *P. falciparum* challenges in *Aotus* monkeys (Perlaza et al., 2003) and chimpanzees (BenMohamed et al., 2004; Daubersies et al., 2000).

PfLSA-3-rec (LSA3) (Recombinant Protein) A vaccine based on LSA-3 is being developed by the Institute Pasteur, as a recombinant protein and, under consideration, as a lipopeptide. The recombinant protein construct, PfLSA-3-rec, has been formulated in adjuvant and is being tested at Radboud University Nijmegen Medical Centre in the Netherlands with funding by the European Commission, for safety, immunogenicity, and protective efficacy against experimental sporozoite challenge in humans.

Vaccines Based on Multiple Preerythrocytic Stage Antigens

FP9-PP and MVA-PP (Gene Expressing Polyprotein Linking LSA3, STARP, Pfs16, Exp1, TRAP/SSP2, LSA1) (MVA and FP) Adrian Hill at Oxford has constructed a polyprotein 3240 amino acids in length called L3SEPTL that includes six preerythrocytic stage antigens: LSA3, sporozoite threonine and asparagine rich protein (STARP), exported protein-1 (Exp1), Pfs16 (an integral membrane protein expressed by sexual stage parasites as well as sporozoites), TRAP/SSP2, and LSA1 (Prieur et al., 2004). This linked sequence of genes has been cloned into DNA, MVA, and FP vaccine delivery platforms. A Phase 1/2a trial was conducted in 2006 at Oxford University using prime-boost combination of FP and MVA; although the vaccine was safe and moderately immunogenic, there was no protection against experimental sporozoite challenge (A. Hill, personal communication).

Vaccines Based on Sporozoites

PfSPZ (Metabolically Active, Radiation-Attenuated Sporozoites) PfSPZ is a metabolically active, nonreplicating (attenuated by radiation) sporozoite vaccine being developed by Stephen Hoffman at Sanaria, Inc, a biotechnology company aiming to induce complete protection against malarial infection by using sporozoites as immunogens. The company has overcome the previously discussed biological and manufacturing obstacles by designing a process meeting regulatory and commercial standards for safety, purity, potency, and stability (Fig. 65.8). With scale-up, this process should allow for the production of enough vaccine to fulfill global requirements. PfSPZ will be assessed for safety, tolerability, immunogenicity, and protective efficacy in healthy adult volunteers in 2009. The clinical trial will be conducted by the U.S. Naval Medical Research Center and the University of Maryland Center for Vaccine Development, with support provided by PATH MVI. The trial will explore whether or not the high levels of protection demonstrated in humans 30 years ago following the delivery of fresh, radiation-attenuated sporozoites via mosquito bite can be reproduced by a commercially viable product that is cryopreserved, thawed, and delivered via

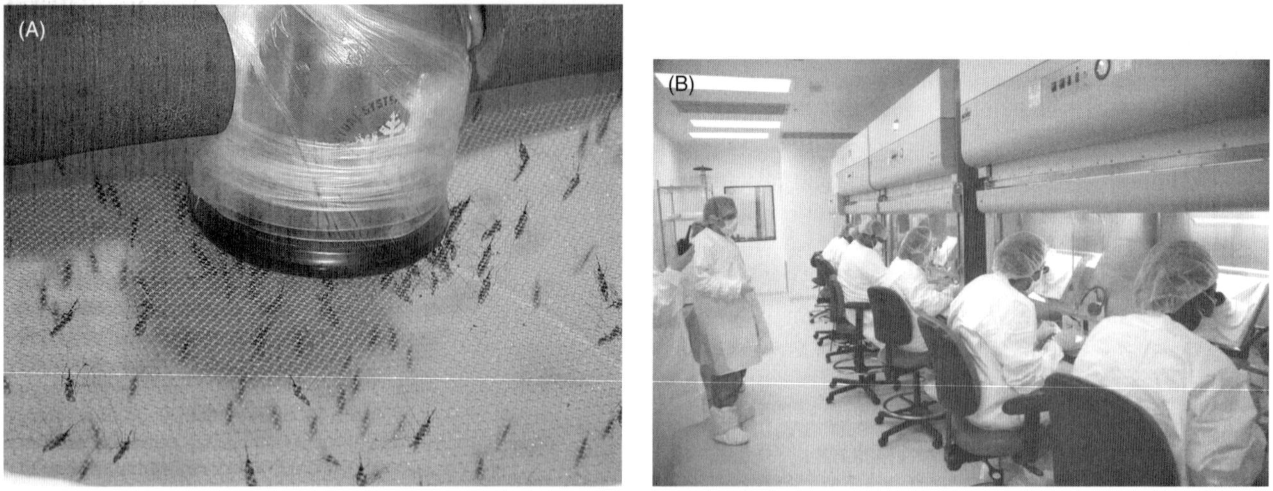

FIGURE 65.8 Manufacturing sporozoites. (A) Feeding sterile mosquitoes on sterile gametocytemic blood cultures is part of the manufacturing process required to produce sporozoites for injection. (B) Approximately 2 weeks later, the sporozoites are harvested from the salivary glands and then purified. Photographs by Robert C. Thompson, Sanaria, Inc.

needle and syringe. Sanaria had recently established two partnerships to develop and test sporozoites attenuated by a different approach, targeted gene deletion. One is with Columbia University and the second is with Top Institute Pharma, the Radboud University Nijmegen Medical Centre, and the Leiden University Medical Center.

Genetically Attenuated Sporozoites (Gene Knockout Sporozoites) The initial development of targeted gene deletions to create genetically attenuated sporozoites took place at the Seattle Biomedical Research Institute in collaboration with the University of Heidelburg (Mueller et al., 2005a, 2005b) and independently at Leiden University (van Dijk et al., 2005). The basic approach is to identify genes in a murine malaria model that, when deleted from the genome, lead to arrested development in the liver. These gene knockout parasite lines are tested for protective efficacy by immunizing mice with the knockout sporozoites and then challenging with intact sporozoites. Those attenuated parasites inducing protective responses with no occurrence of breakthrough blood stage infection are reproduced in *P. falciparum* by knocking out the orthologous genes, ideally creating a double knockout to ensure adequate attenuation (Jobe et al., 2007; Labaied et al., 2007). A major project based on this approach is led by Stefan Kappe at the Seattle Biological Research Institute in partnership with the Walter and Elisa Hall Institute in Melbourne and the Walter Reed Army Institute of Research (WRAIR), supported by the Foundation for NIH (Bill and Melinda Gates Foundation Grand Challenges in Global Health Program). Investigators have deleted preerythrocytic stage-essential genes from the *P. falciparum* genome, and will test these attenuated lines in a clinical trial conducted at WRAIR, delivering the sporozoites via mosquito bite. It is hoped that this vaccine will induce high levels of protection, and that the protection will prove long-lasting, since studies in the rodent models have indicated that 100% protection against sporozoite challenge can last up to 180 days after immunization with gene knockout sporozoites (Tarun et al., 2007). If successful in this proof-of-principle study, the vaccine will be engineered into a GMP production system such as that being developed for the PfSPZ vaccine.

Intact Sporozoites An interesting proof-of-principle trial has been conducted at Radboud University Nijmegen Medical Centre in the Netherlands to see if immunity to preerythrocytic stage malaria can be induced by intact sporozoites. In this clinical trial, led by Robert Sauerwein, volunteers are being exposed three times to the bites of infectious mosquitoes while protected against blood stage malaria with chloroquine. Following immunization, volunteers are challenged to look for protective immunity. While immunization with intact sporozoites under chloroquine protection is not likely a feasible strategy for a commercial vaccine, the trial will have significant bearing on research into whole sporozoite vaccines.

Asexual Erythrocytic Stage Vaccines

Asexual erythrocytic (blood) stage vaccines are designed to limit parasite density, or to counter the effects of malaria toxins, thereby reducing the clinical manifestations of malaria.

Vaccines Based on Merozoite Surface Protein 1

MSP1 is one of the best characterized blood stage antigens and is a leading vaccine candidate. It is a 195 kDa protein, the most abundant of the nine GPI-anchored proteins on the surface of the merozoite. It undergoes proteolytic cleavage to yield four fragments (as seen on SDS-PAGE analysis of merozoite extracts), including the C-terminal 42 kDa fragment, MSP1-42. This fragment undergoes another cleavage to yield a 19 kDa fragment (MSP1-19) and a 33 kDa fragment, which are involved in merozoite invasion, with the 19 kDa fragment internalized during invasion. Antibodies to the C-terminal region of PfMSP1 inhibit merozoite invasion in vitro (Long et al., 1994), confer passive immunity in rodents (Burns et al., 1989, 2004), and are associated with resistance to clinical malaria caused by *P. falciparum* in humans in some (Branch et al., 1998; Egan et al., 1996; Riley et al., 1992) but not all (Osier et al., 2008) studies. The presence of anti-MSP1 antibodies, particularly IgG, has been associated with a decrease in blood stage infections and febrile illness (Huaman et al., 2008). Antibodies against MSP1 appear to be an indicator of NAI, and may reduce severe complications of malaria including severe malarial anemia (Dobano et al., 2008). For these reasons, MSP1-based vaccines have been extensively investigated as potential antidisease vaccines for use in endemic regions. There have been more than 20 different MSP1 constructs in various stages of preclinical and clinical development (Genton and Reed, 2007). Two development projects will be described.

FMP1/AS02A (MSP1-42) (Recombinant Protein) This recombinant protein vaccine was developed by WRAIR and GSK with support from USAID, and has progressed farthest in clinical trials among the MSP1-42 vaccine candidates. The vaccine uses the MSP1-42 from the 3D7 clone of *P. falciparum* and is expressed in *E. coli* and formulated with the adjuvant AS02A. A Phase 1 trial conducted in malaria-naïve adults in the USA demonstrated that the vaccine was safe and immunogenic (Ockenhouse et al., 2006a); a second Phase 1/2a trial in which FMP1/AS02A was delivered alone and also in combination with RTS,S confirmed these findings but did not show any protection induced by FMP1 alone or any significant enhancement of the protection afforded by RTS,S when given in combination (J. Cummings, personal communication). FMP1/AS02A without RTS,S was transitioned to Africa, where it was demonstrated to be safe and immunogenic, initially in two Phase 1b trials conducted in malaria-exposed adults, one in Kenya conducted by the Kenya Medical Research Institute (KEMRI) and Walter Reed Project (Stoute et al., 1997, 2007) and the second in Mali conducted by the Malaria Research and Training Center (MRTC), University of Bamako, with support from the University of Maryland (Thera et al., 2006), and then in a Phase 1b trial conducted in children in Kenya by KEMRI/WRAIR (Withers et al., 2006). These studies were followed by a Phase 2b trial also in Kenyan children, also conducted by the KEMRI/WRAIR with additional support from PATH MVI; unfortunately, this trial did not demonstrate any protective effect in reducing parasitemia or clinical malaria (B. Ogutu, personal communication), although a subanalysis aimed at identifying any allele-specific effects is ongoing (C. Ockenhouse, personal communication). While the FMP01 vaccine, which is based on the 3D7 allele, has been sidelined from further testing due to relatively poor induction of growth inhibitory activity in vitro and the lack of cross-strain protection seen in the Phase 2b trial, the FVO version of the MSP1-42 vaccine (FMP010) remains under clinical development by WRAIR/GSK and will soon be tested in a Phase 1 study adjuvanted with AS01B.

MSP1-42-C1/Alhydrogel + CPG7909(MSP1-42)(Recombinant Protein) The Malaria Vaccine Development Branch (MVDB) at NIAID is developing a second MSP1-42-based vaccine, a recombinant protein containing a mixture of the 3D7 and FVO alleles formulated on Alhydrogel. The vaccine antigens are produced in *E. coli* and then refolded. Both the FVO and 3D7 forms of MSP1-42 when formulated independently on Alhydrogel proved modestly immunogenic in a first Phase 1 study (Huaman et al., 2008; Malkin et al., 2007). The antibodies generated in response to vaccination recognized both forms of the protein equally whereas the T cells predominantly recognized the immunizing allele. A second trial evaluated a vaccine composed of both the FVO and 3D7 alleles of MSP1-42 (MSP1-42-C1) formulated on Alhydrogel with or without the Toll-like receptor 9 immunostimulant CPG 7909; this trial was recently completed in the USA by the Johns Hopkins University. The vaccine combination was demonstrated to be safe, and the addition of CPG 7909 significantly enhanced the antibody response over Alhydrogel alone (L. Martin, personal communication). Despite the enhancement in specific antibody levels, parasite growth inhibition in vitro was only moderately enhanced by the immunostimulant. Thus, further development of the CPG 7909 formulation of MSP1-42-C1/Alhydrogel has recently been halted by MVDB.

Vaccines Based on Apical Membrane Antigen 1 AMA1 is an 83kDa membrane protein originating from micronemes in the apical complex, expressed on the surface of merozoites and involved in erythrocyte invasion (Mitchell et al., 2004). The AMA1 gene appears to have extensive cross-species homology; unlike many other malarial antigens it has no repeat residues. Antibodies to AMA1 have a strong inhibitory effect on parasite growth in vitro (Kocken et al., 2002), but this potential advantage as a vaccine candidate is balanced against the protein's high degree of genetic polymorphism (Bai et al., 2005; Escalante et al., 2001; Healer et al., 2004). This polymorphism probably indicates the importance of AMA1 as a target of protective immunity and therefore underlines the difficulties to be faced in designing a broadly protective formulation without including multiple alleles, although there is recent progress in developing chimeric constructs that induce cross-protective responses (Dutta et al., 2007). Antibodies against AMA1 have been associated with protective immunity in malaria-endemic areas (Polley, 2004; Gray et al., 2007; Osier et al., 2008), and immunization with AMA1 has afforded protection in animal models (Anders et al., 1998; Burns et al., 2004; Narum et al., 2000; Salazar et al., 2002). Recently, it has been discovered that AMA1 is present on the surface of sporozoites as well as merozoites and plays a role in hepatocyte invasion (Silvie et al., 2004), indicating that an AMA1 vaccine may be able to target both the preerythrocytic and erythrocytic stages of the parasite. This leading vaccine candidate (Remarque et al., 2008) is being developed in parallel as a recombinant protein by several different institutions. Three development efforts will be described.

AMA1-C1 (AMA1) (Recombinant Protein) The AMA1-C1 vaccine developed by MVDB at NIAID is the AMA1 construct farthest along the clinical testing pathway. It contains recombinant AMA1 proteins representing the ectodomains from both the FVO and 3D7 alleles of *P. falciparum*, expressed in *Pichia pastoris* and formulated with the adjuvant Alhydrogel. The inclusion of two alleles aims to address the antigenic heterogeneity of AMA1. Clinical trials conducted in malaria-naive adult volunteers in the USA demonstrated that the vaccine was safe and elicited functional antibody responses inhibiting the growth of asexual stage *P. falciparum* parasites in vitro (Malkin et al., 2005). Subsequent Phase 1 studies conducted in adults and children in Mali by the MRTC also showed the vaccine to be safe and immunogenic in endemic populations (Dicko et al., 2008). Based on these results, the vaccine was advanced to a Phase 2b trial in a 2–3-year-old pediatric population in Bancoumana, Mali, but did not show any protection against parasitemia or clinical disease during the first 5 months of follow-up (A. Dicko, personal communication). As with its MSP1-42 vaccine, MVDB is combining AMA1-C1/Alhydrogel with CPG-7909, and this new formulation is undergoing clinical testing in adults in the USA and in Doneguebougou, Mali (MRTC). This vaccine was found to be safe in both populations and to enhance the specific antibody response over the same formulation without the addition of CPG 7909, so it is advancing to Phase 1 evaluation in young children in Mali and will be considered for Phase 2b efficacy testing. A third formulation, AMA1-C1/ISA 720, is being tested in malaria-naïve adults in Brisbane, Australia. This trial is still ongoing, and no safety issues have been noted to date (L. Martin, personal communication).

FMP2.1/AS02A (AMA1) (Recombinant Protein) The FMP2.1 vaccine was developed by WRAIR and GSK with support provided by USAID. It consists of the ectodomain of the 3D7 allele, is produced in *E. coli*, and formulated with AS02A. Safety, tolerability, and immunogenicity were demonstrated in a Phase 1 study conducted in healthy adults in the USA (Polhemus et al., 2007), and the vaccine was then transitioned to Mali where safety and immunogenicity were confirmed in adults (Thera et al., 2008) and children (M. Thera, personal communication). There followed a Phase 2b efficacy trial, conducted in Bandiagara by the Malaria Research and Training Center (University of Bamako) in partnership with the University of Maryland; this trial was recently completed and protection (and other) results are pending. WRAIR has also conducted a Phase 1/2a challenge study in malaria-naïve adults in the USA comparing AMA1 formulated in the adjuvants AS01B and AS02A. Although the vaccine was immunogenic and strongly induced growth-inhibitory antibodies, there was no protection against sporozoite challenge (M. Spring, personal communication). However, it is difficult to rule out an antidisease effect using the experimental challenge model; information on antidisease effect should be forthcoming from the pediatric Phase 2b trial recently completed in Mali.

PfAMA1-FVO[25-545] (AMA1) (Recombinant Protein) In a third development effort, the AMA1 ectodomain from the FVO allele expressed in *P. pastoris* is being developed by a European/African consortium, including the Biomedical Primate Research Centre (BPRC) in Rijswijk, the European Malaria Vaccine Initiative (EMVI), the MRTC in Mali and, as the sponsor, the African Malaria Network Trust (AMANET). After Phase 1 testing at Radboud University Nijmegen Medical Centre in the Netherlands, comparing several adjuvants, the vaccine was advanced to Mali, where a clinical trial conducted in Bandiagara demonstrated that the vaccine, formulated in Alhydrogel, was safe and immunogenic in African adults (M. Thera, personal communication). The future of this vaccine is unclear, given the lack of protection seen with the MVDB product in Mali and the pending protection results from the WRAIR/GSK product.

Vaccines Based on Merozoite Surface Protein 3

MSP3 is another merozoite surface protein associated with erythrocyte invasion. MSP3 is supported as a vaccine candidate by epidemiological evidence from endemic areas, where antibody levels to both conserved and polymorphic regions may be positively associated with protection against clinical malaria (Polley, 2007; Osier, 2007; Roussilhon et al., 2007; Osier et al., 2008) and also by studies in NHP models where immunization led to protection against blood stage challenge (Carvalho et al., 2004, 2005; Hisaeda et al., 2002). One development effort will be described.

MSP3-LSP (MSP3) (Long Synthetic Peptide) The merozoite surface protein 3 long synthetic peptide (MSP3-LSP) vaccine, developed by Pierre Druilhe at the Institute Pasteur, comprises the conserved C terminal region of the molecule (amino acids 186–276) associated with the induction of cytophilic antibodies stimulating monocyte-mediated protection against blood stage malaria (Oeuvray et al., 1994; Singh et al., 2004). The vaccine, formulated on Montanide ISA 720 or on aluminum hydroxide, was studied in a Phase 1 trial at the University of Lausanne, and was found to be immunogenic for antibodies, proliferative responses, and interferon-gamma production (Audran et al., 2005). Immunoglobulin isolated from 60% of the vaccine recipients up to a year after immunization inhibited *P. falciparum* erythrocytic growth in a monocyte-dependent manner in vitro and in *P. falciparum*-infected humanized SCID mice in vivo (Druilhe et al., 2005). Due to unacceptable local reactogenicity associated with the Montanide preparation in the initial trial, subsequent studies have used the aluminum hydroxide adjuvant. In a follow-on Phase 1 trial in malaria-exposed adults in Burkina Faso, the vaccine was well tolerated and induced lymphoproliferative responses, but did not increase antibodies which were already at high levels in these malaria-exposed adults (Sirima et al., 2007). MSP3-LSP is currently being studied in two Phase 1 trials in children, one conducted by the National Institute for Medical Research (NIMR) in Tanzania and the second by the Centre National de Recherche et de Formation sur le Paludisme in Burkina Faso, under sponsorship from AMANET, and safety profiles appear favorable to date. The vaccine is progressing to Phase 2b efficacy testing in Sotuba, Mali.

Vaccines Based on Other Blood Stage Proteins

Vaccines based on other merozoite surface proteins are being developed, including the development of MSP2 at LaTrobe University in Melbourne, Australia, in association with Q Pharm, Brisbane, Australia, and MSP4 and MSP5 at Monash University also in Melbourne. Both projects are receiving support from PATH MVI. In addition, the extramural program at NIH is developing the erythrocyte binding antigen EBA-175, which has undergone Phase 1 clinical testing. Another blood stage candidate, the serine repeat antigen (SERA), is being developed by Osaka University and BIKEN Pharmaceuticals in Japan. This vaccine, called SE36, is a recombinant protein consisting of the amino terminal domain (amino acids 17–382) of *P. falciparum* SERA 5, produced in *E. coli* and formulated on alum. A Phase 1 clinical trial has been conducted in malaria-naive adults in Japan by Toshi Horii.

Vaccines Based on Multiple Blood Stage Proteins

PfCP-2.9 (MSP1-19 + AMA1-domain III) (Recombinant Fusion Protein) This vaccine is a fusion antigen consisting of the C-terminal region of AMA1 and the 19 kDa portion of MSP1 produced in *P. pastoris* and formulated on Montanide ISA 720 (Pan et al., 2004). It is being developed by Shanghai Wanxing Bio-Pharmaceuticals and the Second Military Medical University in Shanghai with support from WHO and PATH MVI. The safety and immunogenicity of this vaccine were demonstrated in a Phase 1 clinical trial conducted in malaria-naïve adults at the Shanghai Changhai Hospital (Hu et al., 2008), and a second trial is planned for the same site prior to a decision whether or not to test the vaccine in endemic areas.

GMZ2 (GLURP 1 MSP3) (Recombinant Fusion Protein) This recombinant protein hybrid of two blood stage antigens, glutamate rich protein (GLURP) and MSP3, produced in *Lactococcus lactis* and formulated in aluminum hydroxide, has been developed by Michael Theisen at Statens Serum Institut in Copenhagen. The vaccine was studied in a dose escalation Phase 1 trial in malaria-naive adults at Tübingen University in Germany, under sponsorship by EMVI. The vaccine is currently undergoing Phase 1 testing in adult Gabonese at the Medical Research Unit of Albert Schweitzer Hospital in Lambarene under sponsorship by AMANET. GLURP, formulated as a long synthetic peptide rather than as a recombinant protein, had previously been studied as a single antigen vaccine at Radboud University Nijmegen Medical Centre in the Netherlands in a Phase 1 trial sponsored by EMVI, where it was well tolerated and induced antibodies inhibiting the growth of *P. falciparum* in vitro (Hermsen et al., 2007). However, further development of GLURP as a single antigen, either as a long synthetic peptide (which proved difficult to manufacture) or as a recombinant protein (which has not yet been manufactured as a GMP product), has been sidelined pending results of the GMZ2 trials. GLURP is supported as a vaccine antigen by epidemiological data showing a strong association between antibody responses and protection against clinical malaria (Lusingu et al., 2005; Meraldi et al., 2004; Nebie et al., 2008; Soe et al., 2004) and by the induction of protective responses in NHPs (Carvalho et al., 2004, 2005).

Vaccines Based on Whole Blood Stage Parasites

Blood Stage Whole Organism Vaccine Whole blood stage parasites are being pursued as immunogens in a development effort similar to the development of PfSPZ and genetically attenuated sporozoites. Michael Good and colleagues at the Queensland Institute for Medical Research demonstrated that repeated low-dose infections (injecting only 30 *P. falciparum*-infected red blood cells) followed by drug-treatment-induced protection in malaria-naïve adults, as seen on challenge with a homologous strain of blood stage parasites (Pombo et al., 2002). The majority of volunteers (80%) developed immune responses, including T-cell responses (both CD4+ and CD8+ T cells) and interferon-gamma production, but no antibody responses. These results, supported by animal models also showing protection following blood stage immunization, including against heterologous strain challenge (Elliott et al., 2005), have led to plans for a dose-escalation Phase 1 trial. The candidate vaccine will consist of crude (killed) preparations of blood stage parasite delivered in adjuvant to potentiate cell-mediated immune responses, and will be tested in malaria-naïve adults (M. Good, personal communication).

Multistage Vaccines

The complexities of the malarial parasite, with antigenically distinct life cycle stages and extensive polymorphisms identified in leading candidate antigens, have led investigators to conclude that a multiantigen, multistage vaccine may be required to induce robust protective responses (Doolan and Hoffman, 2001). Multistage vaccines may, in particular, provide a safety net of protection should parasites break through the first (preerythrocytic) barrier of defense, thereby functioning as both antiinfection and antidisease vaccines (Fig. 65.9). Two such efforts are here described.

PEV302 and PEV301 (CSP + AMA1 Peptides) (Virosome) This vaccine is a two-component vaccine developed by the Swiss Tropical Institute and Pevion Biotech Ltd, consisting of peptides derived from CSP and AMA1 formulated in two influenza virosomes, called PEV302 and PEV301, respectively. The virosomes are composed of natural and synthetic phospholipids mixed with influenza surface glycoproteins, forming spherical vesicles that enhance the delivery and immunogenicity of peptide antigens. In a first clinical trial conducted by the Swiss Tropical Institute in malaria-naïve adults, the vaccine induced relatively strong, long-lived antibody responses that inhibited sporozoite migration and hepatocyte invasion in vitro (Genton et al., 2007; Okitsu et al., 2007). In a second clinical trial, three doses of the PEV301/302 mixture were tested alone and in combination with the Oxford ME-TRAP vaccine delivered as an MVA-MVA-FP prime-boost regimen. Encouragingly, there was evidence of protection in the virosome-alone group (Thompson et al., 2008a). Based on these results, the vaccine, which will contain 4–5 small synthetic peptides, is being transitioned to a Phase 1 trial in Tanzania, to be conducted by the Bagamoyo Research and Training Unit.

NMRC-M3V-Ad-PfCA (Genes Expressing CSP + AMA1) (Adenovirus Serotype 5) Developed by the Naval Medical Research Center in partnership with GenVec, Inc and with support provided by USAID and the U.S. Dept. of Defense, this vaccine is comprised of a mixture of two recombinant adenovirus vectors (serotype 5) encoding CSP and the AMA1 ectodomain, respectively. Currently, a Phase 1/2a clinical trial is underway at the U.S. National Naval Medical Center to test the safety, immunogenicity, and protective efficacy of this vaccine. A second adenovectored vaccine, also comprising two constructs, includes five antigens: CSP, LSA1, Ag2/CelTOS, MSP1-42 and AMA1. This vaccine will be tested in 2009 with support from

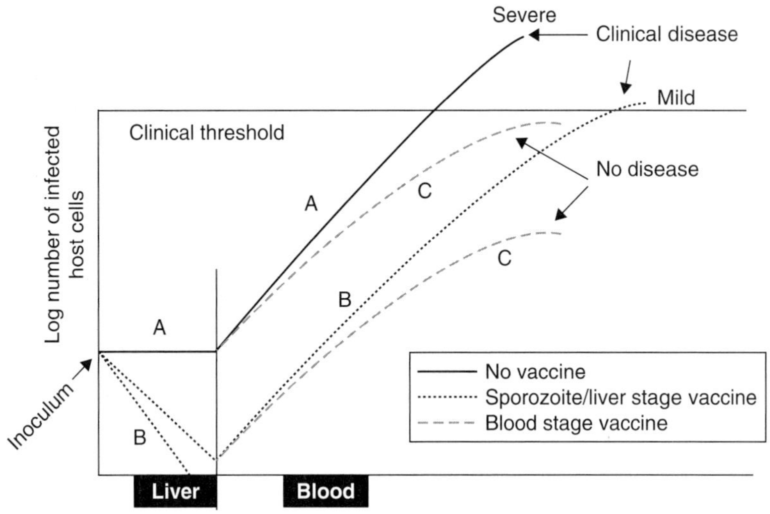

FIGURE 65.9 Two-tiered malaria vaccine. The solid line (A) portrays the normal development of *Plasmodium* in a malaria-naïve host. Following inoculation of sporozoites by mosquito bite, hepatocytes are infected (intersection of line with Y-axis), develop for five or more days and release merozoites into the blood. The parasite then multiplies many orders of magnitude, inducing clinical symptoms which in a naïve host can be severe. The incubation period in the liver provides a window of opportunity for a vaccine-induced immune response to abort the developing infection; this effect is portrayed by the dotted lines. (B) If all parasites are killed, there will be no blood stage infection (dotted line intersects X-axis). If a few parasites get through this first line of defense, the blood stage infection may be less clinically severe because the immune system has more time to respond, causing parasitemia to peak at a lower density, resulting in milder disease (dotted line B). The prediction that a preerythrocytic stage vaccine can reduce blood stage severity in this fashion is born out by the impact of RTS,S on reducing morbidity in children in Mozambique (Alonso et al., 2004) (Fig. 65.7). A blood stage vaccine, on the other hand, acts directly to limit the parasite multiplication in the blood (dashed lines, C), preventing disease even if not the infection. This prediction is born out by the impact of the Combo B vaccine on reducing parasite density (Genton et al., 2002). The two-tiered approach (preerythrocytic + erythrocytic) could be supplemented with a third component—a transmission blocking vaccine—to prevent the spread of vaccine-resistant mutants. Reprinted by permission from S. Karger, A.G. Basel (Kumar et al., 2002).

PATH MVI. Ag2 (antigen 2, or CelTOS) (Kariu et al., 2006) is a novel preerythrocytic stage antigen recognized by lymphocytes from human volunteers immunized with irradiated sporozoites and protected against sporozoite challenge (Doolan et al., 2003b).

Pregnancy-Specific Vaccines

Two genes from the 50 member *var* gene family encoding PfEMP1, var1CSA, and var2CSA are associated with the CSA-binding ability of *P. falciparum* parasites in the placenta that underlies PAM. Var2CSA, in particular, has been shown to be consistently upregulated in various CSA-binding *P. falciparum* genotypes, and is an important antigenic target under consideration for development of a pregnancy-specific vaccine (Gamain et al., 2007). Studies in mice showed that monoclonal antibodies against *P. falciparum* variant surface proteins expressed by CSA-binding parasites inhibit CSA binding (Avril et al., 2006) and that polyclonal antibodies raised in rabbits to defined VAR2CSA Duffy binding-like (DBL) domains from PfEMP1 or induced by DNA vaccines encoding VAR2CSA recognize certain polymorphic VAR2CSA epitopes held in common across geographically diverse CSA-binding parasite lines (Avril et al., 2008). These promising results support further research and investigation into the potential for a pregnancy-specific vaccine. Several such vaccine development projects are underway in various institutions, including the Seattle Biological Research Institute and the University of Copenhagen.

Transmission-Blocking Vaccines

Transmission-blocking vaccines are designed to protect the community by reducing transmission rates, even though not directly protecting individual vaccine recipients. Proponents of transmission-blocking vaccines reason that if a sufficient proportion of the population were to be immunized, malaria-related morbidity and mortality would be reduced, much as occurs through the use of bednets, which also reduce rates of transmission. Transmission-blocking vaccine would thus result in a special type of herd immunity in which none of the individuals would be immunologically protected against the infectious agent. With this type of immunity, a single infectious person entering the community could still spread the infection to others. The levels of immunity needed in a population for a given effect on malaria transmission have been modeled and suggest that current candidate vaccines should be sufficient to provide a significant reduction in transmission intensity (Saul, 2008).

Transmission blocking is achieved by targeting proteins expressed during the sexual stages of the parasite (Pradel, 2007). Antibodies are passively imbibed by the mosquito during feeding and attack newly revealed sexual stage antigens in the gut of the mosquito, preventing the formation of oocysts and thus sporozoites.

The MVDB of the U.S. NIH is developing several transmission-blocking vaccines against *P. falciparum* and *P. vivax*. The leading candidate antigens, Pfs25, Pfs28, Pvs25, and Pvs28, are expressed on the surface of the ookinete (Pf signifying the *P. falciparum* antigen, Pv the *P. vivax* antigen). Pfs25 was originally produced in *S. cerevisiae* as a recombinant protein TBV25H. TBV25H was formulated on Alhydrogel and tested in two clinical trials, but the product was not pursued further, due in part to concerns over hypersensitivity reactions (Kaslow, 2002; Y. Wu, personal communication). More recently, Pvs25, also expressed as a recombinant protein by *S. cerevisiae*, was formulated on Alhydrogel and assessed in a Phase 1 study. The vaccine was safe and immunogenic, and antibody from the volunteers inhibited parasite development in a membrane feeding assay, with antibody titers correlating with the degree of inhibition (Malkin et al., 2005, 2007), a correlation which has been confirmed by more recent work (Miura et al., 2007a). A second trial assessed Pfs25 and Pvs25, formulated, individually, with Montanide ISA 51. The trial was halted due to systemic reactogenicity including erythema nodosum and leukemoid reaction, although Pfs25/ISA51 induced transmission-blocking antibody responses (Y. Wu, personal communication). New formulations are now under development at MVDB. Other efforts to develop transmission-blocking vaccines are being pursued at the Johns Hopkins Bloomberg School of Public Health in Maryland (LeBlanc et al., 2008), Nijmegen Medical Center, The Netherlands (Outchkourov et al., 2008), and the Ehime University in Matsuyama, Japan (Arakawa et al., 2005).

An important role for transmission-blocking vaccines may be their combination with preerythrocytic or erythrocytic stage vaccines in order to prevent the transmission of vaccine escape mutants and thus stave off the degradation of vaccine efficacy over time through the gradual spread of vaccine-resistant parasite strains. A significant challenge for testing the efficacy of transmission-blocking vaccines, whether used alone or in combination with other malaria vaccines, will be developing field-based indices for measuring transmission intensity as an outcome variable. Field trials may require randomization of villages to treatment arms and then comparison of community-specific transmission rates or clinical disease incidence.

P. vivax Vaccines

The burden of *P. vivax* across the world is huge, with 2.6 billion people at risk and 50–70 million cases annually (Guerra et al., 2006, 2007, 2008; Mendis et al., 2001). In many areas outside of Africa, *P. vivax* may often be the dominant malaria species present. Moreover, the clinical manifestations of infection (Song et al., 2003) are equivalent in terms of morbidity to those of uncomplicated *P. falciparum* malaria. Clearly, a vaccine for preventing *P. vivax* is sorely needed.

However, less attention and resources have been applied to *P. vivax* than to *P. falciparum* (Herrera et al., 2007). Although the potential severity and sheer numbers of *P. falciparum* infections are the most important reasons for this difference, there are others. One relates to the ease of maintaining blood stage parasites in culture, which is routine for *P. falciparum* (Trager and Jensen, 1976). *P. vivax* requires reticulocytes as host cells for efficient invasion and growth, and this has hindered the establishment of routine culture, although there has been significant progress (Chotivanich et al., 2001; Golenda et al., 1997; Panichakul et al., 2007; Suwanabun et al., 2001; Udomsangpetch et al., 2007). In the case of *P. falciparum*, blood stage parasite material can be obtained from in vitro cultures, and with the development of techniques to induce gametocyte production (Campbell et al., 1982), sporozoites, and liver stage material as well. These sources of parasites have not been available for *P. vivax*, creating a major impediment for those studying the biology of this species.

The lack of in vitro culture has also hindered vaccine testing, as there has been no sporozoite challenge model available since the cessation of malarial therapy. Malarial therapy was accomplished by direct human to human blood passage, or in some instances sporozoites were obtained by feeding mosquitoes on gametocyte donors. Recently, this deficit has been partially overcome by Socrates Herrera at the Malaria Vaccine and Drug Development Center in Cali Colombia. Relying on the availability of *P. vivax* cases at local health clinics for providing the donations of gametocytemic blood required to infect the mosquitoes, two successful challenges have been conducted. This should significantly accelerate progress in developing and testing *P. vivax* vaccines.

Some of the most notable progress in developing *P. vivax* vaccines reflects distinct differences between the biology of *P. vivax* and *P. falciparum*. *P. vivax* invades erythrocytes by binding to the Duffy antigen/receptor for chemokines (DARC), a different entry mechanism than for *P. falciparum*. Remarkably, the invasion of erythrocytes appears to be totally dependent upon this one molecular interaction, since lack of DARC results in the erythrocyte being refractory to *P. vivax* invasion (Miller et al., 1976), with heterozygotes demonstrating reduced infection (Kasehagen et al., 2007). This dependence on the Duffy antigen is thought to explain why the parasite is uncommon or even absent from many parts of Africa, where the Duffy negative phenotype is common, particularly in West Africa.

The parasite-binding ligand, the Duffy-binding protein (DBP), is released from micronemes during the process of invasion. Studies in endemic areas have indicated a possible association between the presence of antibodies to the DBP and protection against clinical malaria (Ceravolo et al., 2008), and erythrocyte invasion can be inhibited in vitro by antibodies directed to DBP (Grimberg et al., 2007; Singh et al., 2002). Recently, the crystal structure and binding epitopes of DBP have been defined (Hans et al., 2005; McHenry and Adams, 2006; Singh et al., 2006; VanBuskirk et al., 2004). It appears that the receptor-binding site of this molecule is relatively conserved, and thus may be amenable to the development of a cross-protective vaccine.

With these supporting data, the DBP has emerged as a vaccine candidate. Vaccines based on the DBP have been tested for immunogenicity and protective efficacy in *Aotus* monkeys (Arevalo-Herrera et al., 2005), and a focused effort to develop this antigen as a vaccine for human use is underway at the International Centre for Genetic Engineering and Biotechnology in New Delhi, India, led by Chetan Chitnis and supported by the PATH MVI.

Other *P. vivax* antigens under development as vaccines include PvCSP, which has been tested in humans as a mixture of three synthetic peptides (Herrera et al., 2005), PvTRAP (Castellanos et al., 2007), PvAMA1 (Kocken, 1999), PvMSP1 (Kaushal et al., 2007), and several novel antigens (Gaur et al., 2007). An interesting PvCSP chimeric recombinant protein that incorporates important sequence variation in an effort to increase cross-strain protection has been developed and manufactured at WRAIR and is slated for clinical testing (Yadava et al., 2007).

DISCOVERY/BASIC SCIENCE

Success in making a vaccine against malaria is hardly assured, given that there are no licensed vaccines against any human parasitic infections. Fortunately, the scientific community is engaged in wide-ranging basic research that should improve the chances of success.

One of the most prominent contributions has been the sequencing of malarial genomes, including those of *P. falciparum* (Gardner et al., 2002), *P. vivax* (Carlton, 2003), and several murine and NHP malarias (Carlton et al., 2002; Hall et al., 2005). This work has led to the elucidation of malaria transcriptomes (Kaiser et al., 2004; Le Roch et al., 2003; Tarun et al., 2008; Watanabe et al., 2007; Zhou et al., 2008), proteomes (Hall et al., 2005; Tarun, 2007; Lasonder et al., 2002), interactomes (Date and Stoeckert, 2006; LaCount et al., 2005), and other derivatives. This has opened an entirely new field of endeavor, namely, malaria functional genomics or, more broadly, systems biology (Fraunholz, 2005; Winzeler, 2006).

Among many applications, these complex data sets enable the use of reverse genetics for designing vaccines. In this "genomes to vaccines" approach, screening procedures are developed for down-selecting antigen candidates, starting with the 5300+ open reading frames contained in the *P. falciparum* genome, which are available in public domain databases, such as PlasmoDB (Stoeckert et al., 2006), maintained by David Roos at the University of Pennsylvania (www.plasmodb.org), and Comparasite (http://comparasite.hgc.jp/), which includes full-length cDNA sequence data from seven subdatabases of apicomplexa protozoa (Watanabe et al., 2004). These screening algorithms have led to significant progress in the identification of new candidate antigens (Doolan et al., 2003a; Villard et al., 2007), although lack of reagents, particularly for liver stage parasites (Gruner et al., 2003), the difficulty with expression and poor immunogenicity of many malarial antigens (Aguiar et al., 2004), and the problem of developing screening tools revealing the protective potential of candidate antigens have hindered progress. With new interest in the identification of novel vaccine candidate antigens by funders such as the Bill and Melinda Gates Foundation, these efforts will surely bring payoffs for vaccine development. At least one novel antigen, currently known as antigen 2 (Ag2) (Doolan et al., 2003b), was identified using reagents from human volunteers immunized with irradiated sporozoites. Ag2 is slated for clinical testing in 2009 through a partnership between the Naval Medical Research Center, GenVec, Inc and PATH MVI. An independent group in Japan also discovered the same antigen, calling it CelTOS (Kariu et al., 2006).

Genomic data have impacted other related disciplines. For example, they have allowed the development of high-throughput processes for identifying the crystal structure of proteins (Vedadi et al., 2007) to support our understanding of parasite molecules and to assist in identifying new antimalarial drugs; the development of protein microarrays to permit the serological screening of hundreds of malaria proteins (Sundaresh et al., 2006) in order to reveal the complex patterns of humoral responses associated with protection (Gray et al., 2007); and the elucidation of the evolutionary history of malarial parasites (Wootton et al., 2002). An important aim of the genomics revolution is the creation of a full haplotype map of *P. falciparum*, which should be particularly helpful in confronting the problem of antigenic polymorphism that challenges subunit vaccines (Carlton, 2007).

Genomics data from the host side have also had a major impact on the research on malaria. For example, high-density oligonucleotide microarrays have allowed the examination of a full panel of host immune response genes for up- or downregulation during the course of malarial infection. Generating expression profiles from the peripheral blood mononuclear cell transcriptome in this fashion may help to define the functionally complex networks of gene and protein interactions that underlie protective as well as detrimental immune mechanisms (Ockenhouse et al., 2006b; Schaecher et al., 2005). A large project, funded by the Gates Grand Challenges in Global Health Program, aims to define the natural variations in the human genome that determine an individual's susceptibility to malarial infection (Ntoumi et al., 2007).

Another area of tremendous progress has been the development of genetically altered or transgenic models, both for the parasite and the host. From the parasite's side, targeted genetic disruption has been a very productive technique for elucidating the function of the missing protein, such as UIS4 (Mueller et al., 2005a), TRSP (Labaied et al., 2007), or PfEMP1 (D'Ombrain et al., 2007), to cite just three examples. In cases where absence of the antigen may preclude growth, replacement with an ortholog may be revealing. For example, replacing the native CSP of the *P. berghei* parasite with the CSP of *P. falciparum* allowed the demonstration that protective immunity could be induced by irradiated *P. berghei* sporozoites in the absence of PbCSP, indicating that other antigens besides CSP contribute to and are sufficient for protection (Gruner et al., 2007). As another example, the expression of the ookinete antigen P25 from *P. vivax* in *P. berghei* allowed the determination of transmission-blocking activity in sera from a clinical trial of Pvs25 (Ramjanee et al., 2007). A similar model has been generated for Pfs25 (Mlambo et al., 2008). In some cases, it has been useful to express hybrid molecules including segments of both the native and transgenic molecule together, such as the integration of PfCSP repeat residues into PbCSP expressed on the surface of the *P. berghei* sporozoite to permit testing the protective efficacy of PfCSP vaccines targeting the repeat region in a murine model (Persson et al., 2002).

Of particular benefit has been the transfection of malarial parasites with green fluorescent protein (Mikolajczak et al., 2008; Ono et al., 2007) or luciferase (Franke-Fayard et al., 2006), permitting the visualization of biological processes, including the injection of parasites into the skin by the mosquito (Vanderberg and Frevert, 2004), their migration through the liver (Frevert et al., 2005), intracellular trafficking of the enzyme falcipain 2 within the infected erythrocyte (Dasaradhi et al., 2007), and the dynamics of sequestration and its role in pathology (Franke-Fayard et al., 2006). The expression of GFP by parasites has also allowed the purification of parasitized erythrocytes (Janse et al., 2006) and hepatocytes (Tarun et al., 2008), greatly facilitating the generation of parasite material for transcriptomic and proteomic analyses.

From the host side, transgenic mice have permitted the partial development of *P. falciparum* in an otherwise nonpermissive host (Morosan et al., 2006) and have been used to study the pathogenesis of malaria (Hernandez-Valladares et al., 2007) and the immunological mechanisms of protection (Romero et al., 2007; Thompson et al., 2008b). A fully "humanized" mouse model permissive for the human malarial parasites is a long-term goal. RNAi-mediated gene silencing has been another useful technique for manipulating the host in order to better understand the functions of key molecules involved in host–parasite interactions (Brown and Catteruccia, 2006). For example, this technique has helped to elucidate antiplasmodial protective mechanisms taking place in the mosquito gut (Abraham et al., 2005; Dinglasan et al., 2007).

There are many other areas where basic research may greatly impact malaria vaccine development, ranging from the axenic culture of sporozoites from gametocytes (Hurd et al., 2003) to the development of in vitro platforms for assessing immunogenicity (Byers, 2007). These studies contribute not only to the field of malaria vaccines, but to our basic understanding of the requirements for developing vaccines against other difficult pathogens.

PRECLINICAL DEVELOPMENT

Although not widely recognized by the medical community, the genus *Plasmodium* is primarily a parasite of reptiles and birds, which harbor the majority of malaria species. *Plasmodium* is also found in a variety of mammals, including both Old and New World primates (Coatney, 1971) and a small number of African rodents (Peters et al., 1978). Several of these parasites have been adapted as laboratory models, with the murine species *P. yoelii* and *P. berghei* used most commonly. The natural host for the rodent malarias is the African thicket rat, but the laboratory mouse is susceptible to both sporozoite and blood stage challenge. Both outbred and inbred mice including immunodeficient and transgenic strains have been valuable for defining the mechanisms of protective immunity to malaria which have proven to be remarkably diverse (Doolan and Hoffman, 2000). Murine models have also been useful for testing and developing novel vaccine platforms (Gilbert et al., 2002; Hedstrom et al., 1994), and for assessing the protective potential of antigens (Khusmith et al., 1991).

The NHP malarias have played an important role by providing a model system closer to humans for evaluating protective efficacy. Some New World NHPs are directly susceptible to blood stage infection with human parasites, such as the owl monkey, *Aotus*, while others, like the rhesus macaque, require a simian parasite such as *P. knowlesi*. The original selection of antigens for the SPf66 vaccine was made in *Aotus* (Rodriguez et al., 1990), and this model has also been used effectively to demonstrate the protective potential of key antigens such as MSP1 (Siddiqui et al., 1987) and AMA1 (Stowers et al., 2002) and to explore novel technologies such as DNA vaccines and DNA adjuvants (Gramzinski et al., 1997; Jones et al., 1999) and prime-boost approaches (Jones et al., 2001). *P. knowlesi* in the rhesus monkey, a model for which a reliable sporozoite challenge is available (unlike *P. falciparum* in *Aotus*), has been very useful for demonstrating proof-of-principle for the protection afforded by prime-boost approaches (Rogers et al., 2002).

Rodents and larger animals can also be used to test immunogenicity and safety of *P. falciparum* or *P. vivax* vaccines directly if a protection read-out is not required, and thus can be used to effect improvement in technologies without having to construct vaccines using murine or simian malaria species. Examples include the assessment of antigen mixtures (Sedegah et al., 2004), novel pairing of antigens and adjuvants (Pichyangkul et al., 2008), and prime-boost immunization regimens (Stewart et al., 2007).

To effectively use these animal models, new antigens and technologies are typically assessed first in the mouse, such as the initial testing of DNA vaccines (Sedegah et al., 1994), and are then transitioned to a larger species such as NHP before proceeding to humans. The value of the NHP step, particularly the use of *Aotus* blood stage challenge, has been debated (Heppner et al., 2001; Stowers and Miller, 2001), but is clearly useful in many cases. In each model, if the goal is to demonstrate protection, vaccines need to be reformulated using the antigens of the *Plasmodium* species infecting the selected host, with the exception of blood stage challenge in *Aotus* and *Sciurius*. There is thus reliance on the assumption that orthologous antigens fulfill the same function and will demonstrate the same protective potential following transition to humans. Unfortunately, this is not always born out, and a "species gap" is often revealed. For example, in many cases, it has proven relatively straightforward to protect mice, harder to protect monkeys, and extremely difficult to protect humans. Moreover, novel technologies may work well in the mouse system, e.g., adjuvanting DNA plasmids with granulocyte macrophage-colony stimulating factor (Weiss et al., 1998) but not in the human system (Wang et al., 2005). The models are thus a mixed blessing, aiding progress but potentially misleading at the same time.

Counterbalancing this drawback is the existence of the human challenge model for *P. falciparum*, which allows an immediate assessment of vaccine efficacy, often in the initial first-in-humans trial. Disappointing results in the clinic prompt investigators to return to preclinical research, leading to an iterative vaccine development strategy, as exemplified by the development of CSP as a vaccine antigen which was formulated multiple times for human use, with intervening animal studies in most cases, before success was achieved (Ballou and Cahill, 2007). This iterative approach, illustrated in Fig. 65.10, has enabled malaria vaccine investigators to pioneer and test novel technologies, demonstrating their potential for human applications, and in some cases make improvements.

CLINICAL TRIALS

The clinical trials associated with the development malaria vaccines have a number of distinguishing attributes that collectively place them in a unique position in the field of vaccinology. There are no other vaccines against human pathogens that currently combine such an urgent need with the ability to (1) experimentally challenge volunteers in nonendemic areas and then (2) transition to the field where natural exposure provides attack rates high enough to permit the reliable assessment of efficacy using sample sizes in the hundreds and follow-up periods of 6 months or less. These favorable testing parameters allow for the efficient evaluation of the safety, tolerability,

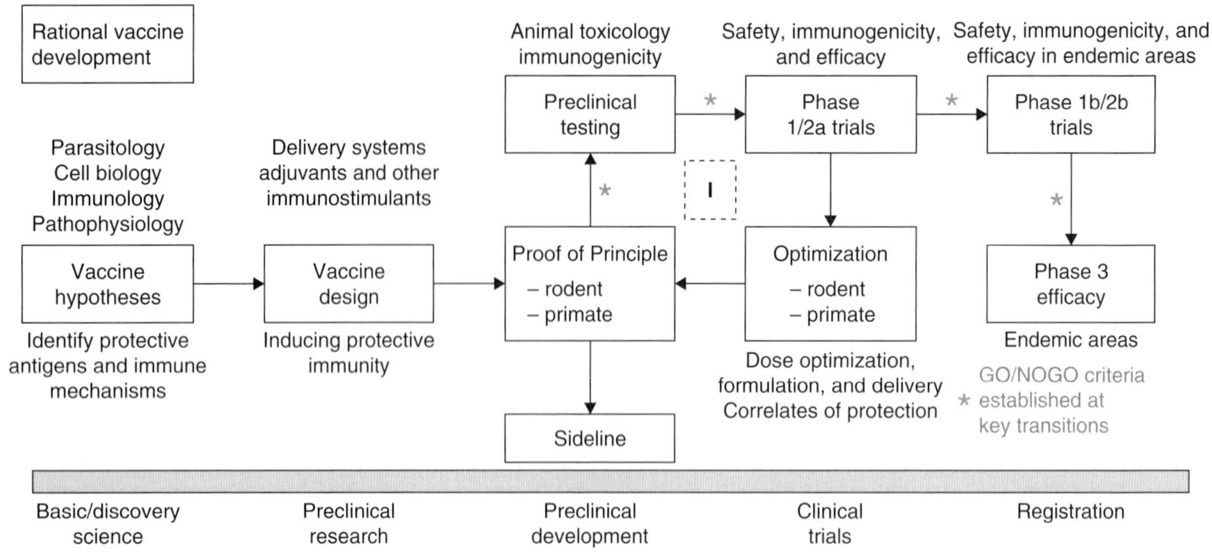

FIGURE 65.10 Iterative approach to malaria vaccine development. As there are no licensed vaccines against any parasitic infection to serve as a model, many developmental efforts have started with discovery research in an effort to understand the biological processes underlying host–parasite interactions at the molecular level. Protective antigens need to be identified, and protective immune responses revealed. This understanding is used to construct hypotheses regarding how a vaccine could be rationally designed. Prototype vaccines are then designed on the basis of these hypotheses, selecting delivery systems, including adjuvants and immunostimulants, to induce the identified protective immune responses against the identified target antigens. Reagent grade vaccines are produced (e.g., using *P. yoelii*, *P. berghei*, or *P. knowlesi* antigens) and tested to establish proof of principle in animal models. Typically, this is done with a challenge to allow testing for protection. Vaccines meeting preestablished go criteria are committed to the clinical testing pathway. At this point, a vaccine based on *P. falciparum* antigens must be made, processes worked out for manufacturing, and toxicology and immunogenicity studies performed prior to clinical testing. As described in the text, Phase 1 studies may be combined with a Phase 2a sporozoite challenge, allowing for rapid assessment of efficacy. If protection is not adequate, the vaccine is returned to the preclinical level for optimization and possible reentry into the clinical pipeline (circular pathway around the I [=iteration] in the diagram). If protection is achieved in the clinic, vaccines can move forward to testing in endemic areas. In the case of blood stage vaccines, for which experimental sporozoite challenge is less useful, vaccines can move forward to endemic areas on the basis of safety. Printed with permission from Gary Nabel, Vaccine Research Center, National Institutes of Health.

immunogenicity, and protective efficacy of vaccines, leading some investigators to conclude that malaria could be selected as a "focus disease" for testing and optimizing novel vaccine technologies that could then be applied to other infectious agents such as HIV where experimental challenge is impossible and field trials require thousands of individuals carefully selected for high risk and requiring years of follow-up (Gay et al., 2007).

Experimental malaria sporozoite challenge, described in the preceding section, is possibly the vaccine developer's most powerful tool. It is effected through the bites of five infected mosquitoes and has proven to be a safe procedure (Church et al., 1997; Epstein et al., 2007b; Hoffman, 1997; Verhage et al., 2005). It is generally well tolerated, although a proportion of volunteers will miss 1–2 days from work as a result of infection. Treatment is initiated when the blood smear becomes positive as detected by traditional microscopy, although PCR, which permits diagnosis 1–2 days earlier than microscopy, has opened the possibility of accelerating diagnosis and treatment and thereby further limiting clinical manifestations.

Sporozoite challenge is very useful for testing the efficacy of preerythrocytic stage vaccines designed to prevent infection, but is less useful for testing blood stage vaccines aiming to limit disease, due to the requirement to treat immediately upon diagnosis, which obviates the possibility of monitoring symptoms. It is theoretically possible to use PCR to measure the rate of multiplication in the blood during the roughly 2-day window between PCR diagnosis and microscopic diagnosis through repeated PCR measurements and calculation of a slope (Bejon et al., 2005; Hermsen et al., 2004), but the wide fluctuations in parasite density resulting from cyclic sequestration can be problematic for this approach. The consensus is that the efficacy of a blood stage vaccine designed to reduce disease is best tested in endemic areas with the incidence of clinical disease used as the outcome measure.

Based on these considerations, there is agreement regarding the terminology used for describing the phases of malaria vaccine testing. Phase 1 is divided into 1a and 1b according to whether the safety testing is performed in malaria-naïve or malaria-exposed individuals, the former typically enrolled at trial sites located in nonendemic areas and the latter in endemic areas. Similarly, Phase 2a refers to experimental sporozoite challenge and Phase 2b refers to natural challenge in the field. Phase 3 has the same meaning as used generally.

Experimental blood stage challenges have been employed in the past (Cheng et al., 1997; Pombo et al., 2002) and remain an acceptable approach, but need refinement and standardization. Another potential evolution in challenge procedures is being pursued by Sanaria, Inc, which is developing the capability for generating a cryostabillate of sterile, living sporozoites that could be injected by needle and syringe. This would permit a standardized challenge in the absence of an insectary, although it is unclear if needle injection can provide data as applicable to protection against mosquito transmission as the traditional mosquito bite challenge. Finally, there are efforts underway at Radboud University Nijmegen Medical Centre in the Netherlands to generate multiple parasite strains for sporozoite challenge in order to permit assessment of heterologous strain protection (the current challenge available in the USA, Great Britain, and the Netherlands employs the 3D7 clone or parent NF54 strain).

An important question for clinical trials research is the degree to which experimental challenge predicts protection in the field. RTS,S proves that there is indeed predictive value, but because efficacy has been calculated by different methods in Phase 2a and 2b trials, it is difficult to establish a conversion factor. The time-to-event analysis used for the RTS,S field trials yields a higher figure for efficacy than if a simple proportional reduction in events analysis had been used. As a second example with data available from both Phase 2a and 2b trials of the same vaccine, the ME-TRAP antigen delivered with heterologous prime-boost regimens that sometimes sterilely protected volunteers in Phase 2a trials at Oxford (Webster et al., 2005) provided no protection when tested in the Gambia (Moorthy et al., 2004) or Kenya (Bejon et al., 2006, 2007).

There are several areas of active debate when considering trials in endemic areas. Four are discussed:

1. Case definitions: The design of field trials is complicated by the difficulty of defining incident malaria when a large proportion of the population chronically harbors the parasite in their blood, as typically occurs in sub-Saharan Africa. Case definitions vary, and often involve a threshold parasitemia combined with clinical measures such as fever. The selection of the threshold parasitemia is critical, and can strongly influence subsequent efficacy measurements (Mwangi et al., 2005; Rogers et al., 2006).
2. Radical cure: A second question is whether or not to provide radical cure (elimination of all liver and blood stage parasites through drug treatment) during vaccination. Radical cure allows a clearer assessment of incident parasitemia, but may also provide an unrealistic estimate of efficacy since a licensed vaccine is likely be delivered to both uninfected and infected individuals, and in the latter case the presence of malarial parasites could suppress the immune response to immunization (Ocana-Morgner et al., 2003). In addition, radical cure might increase the risk of subsequent clinical malaria (Owusu-Agyei et al., 2002), and at least in one trial, appeared to adversely affect the protection afforded by the vaccine (Genton et al., 2002).
3. Outcome measures: A third question relates to identifying the most useful outcome measure, e.g., incident parasitemia, clinical disease, severe disease, or mortality. One solution to this question is to establish multiple cohorts, each designed to assess a particular outcome measure, so that data on several can be collected. For each outcome, there is a trade-off in costs associated with achieving an adequate sample size to measure accurately. These questions have been addressed in several reviews (Greenwood, 2005; Guinovart and Alonso, 2007). Moreover, there is controversy regarding the most appropriate statistical analyses, such as the debate over proportional vs. time-to-event calculations for efficacy. The World Health Organization sponsored a meeting in June 2008 to discuss this question.
4. Molecular epidemiology: A fourth issue involves the need to match the testing of subunit vaccines based on particular alleles to field sites where these alleles circulate in the parasite population. Although obviously a cross-protective vaccine is desired, initial proof-of-concept field trials could attain a measure of success if the vaccines demonstrate allele-specific protective effects, indicating that a multiallele vaccine might be broadly effective. These considerations illustrate the concept of molecular monitoring during vaccine trials (Felger et al., 2003) and have led to sophisticated analyses of parasite haplotypes at field sites where testing a particular vaccine candidate is contemplated (Takala et al., 2007).

The clinical testing of malaria vaccines would be greatly facilitated if there were a surrogate marker for efficacy. Currently, this does not exist, even for RTS,S, for which it has been a challenge to define immune correlates of protection. It can be anticipated that as more effective vaccines are developed, it should be possible to better define a surrogate marker to facilitate vaccine evaluation.

PROSPECTS FOR THE FUTURE

The development of an effective malaria vaccine for global use faces complex scientific challenges, as illustrated by the many failed vaccines and associated barriers described in this chapter. The scientific challenges are compounded by a variety of economic and socio-political obstacles. For example, the fact that the primary burden of malaria is shouldered by the world's poorest people, who often live in regions facing critical health, and social and economic issues, reduces the financial incentive for pharmaceutical companies which might otherwise prioritize malaria vaccine development. Moreover, the scientific enterprise involves testing novel platforms and delivery systems as well as addressing complex safety issues that are beyond the means of the malaria-endemic countries.

As a result of these disjunctions, malaria vaccine development has evolved into a "north-south" collaboration. Financial resources and research capacity from more affluent, malaria-free countries have joined with outstanding health institutes and initiatives present in less-prosperous malaria-endemic countries across Africa, Asia, and Latin America where the commitment to developing malaria vaccines is strongest and where there is willingness to establish the clinical trials capacity needed to field test candidates as they come through the product pipeline (Kilama et al., 2007). Donor agencies have stepped forward to assume research and development costs, and interested scientists, clinicians, and pubic health officials have joined in collegial association to provide the needed innovation and leadership. The importance of malaria prevention for international travelers (Richie, 2004) and existing market incentives (Group, 2005) encourage the involvement of developed countries and industry. In addition, military establishments, which have experienced the disastrous impact of malaria in conflicts played out across the tropics throughout history (Aronson et al., 2006; Beadle et al., 1994; Shanks and Karwacki, 1991), have a vested interest and are adding resources to the fight. Pharmaceutical firms and biotechnology companies are leading development efforts in many cases, and will shoulder increasing responsibilities as candidate vaccines pass initial safety and efficacy hurdles thereby establishing a more favorable economic equation. International organizations such as the World Health Organization and the World Bank serve to coordinate and fill existing gaps, such as assuring adequate clinical trials capacity (Moran et al., 2007). A recent product of this global partnership is the Malaria Vaccine Technology Roadmap, a document providing a framework for the coordinated development of malaria vaccines (Roadmap, 2005).

With the evolution of these public–private partnerships and the presence of promising vaccine candidates in the pipeline, the malaria vaccine enterprise appears to have a reasonable chance of success. If a potent malaria vaccine is combined with other control modalities available and in development, such as altering the genetic structure of vector populations to reduce vectorial capacity, we can contemplate, as a long-term strategy, the eradication of malaria.

KEY ISSUES

- Identification of the most protective antigens.
- Identification of protective immune mechanisms targeting these antigens.
- Developing vaccine delivery systems including adjuvants and immunostimulants that potently induce these protective responses.
- Developing formulations of antigens with sufficient valency to protect all vaccinees against all parasite strains despite (1) stage-specific expression of antigens, (2) allelic polymorphism, (3) antigenic variation, and (4) genetically restricted host responses.
- Developing immune surrogates and validated animal models.
- Developing standardized assays allowing comparison among vaccine candidates.
- Developing vaccines with adequate safety profiles.
- Developing vaccines targeting all five species of malaria infecting humans.
- Establishing cost-effective processes for manufacturing and scale-up.
- Developing workable marketing and licensure strategies enabling industry to shoulder the costs, risks, and responsibilities associated with the development, manufacture, sale, and distribution of malaria vaccines.
- Developing strategies to assure adequate distribution and administration of malaria vaccines

in malaria-endemic areas in coordination with existing vaccines and in spite of poverty and human displacement.

- Developing processes for detecting and eliminating vaccine-resistant parasite strains.

ACKNOWLEDGMENTS

The authors wish to gratefully acknowledge the comments and suggestions of the following individuals, who were contacted to assure the accuracy of particular sections of this chapter: Ripley Ballou (RTS,S), Ashley Birkett (CSP vaccine), Blaise Genton (PEV302 and PEV301), David Lanar (FMP011), Lee Hall (Ad35.CS), Adrian Hill (ME-TRAP, FP9-PP, and MVA-PP), Pierre Druilhe (PfLSA-3-rec, MSP3-LSP), Stephen Hoffman (PfSPZ), Stefan Kappe (genetically attenuated sporozoites), Evelina Angov (FMP1/AS02A, FMP010, FMP2.1/AS02A), Laura Martin (MSP142-C1/Alhydrogel + CPG7909, AMA1-C1), Robert Sauerwein (PfLSA-3-rec, PfAMA1-FVO[25-545], GMZ2), Michael Good (whole organism blood stage vaccine), Zarifah Reed (PfCP-2.9, SE36), Yimin Wu (transmission-blocking vaccines), and Alan Magill (licensure of IV artesunate in the USA).

The following are sincerely thanked for their assistance in providing photographs or figures: Ute Frevert (sporozoite), John Sacci (hepatic stage parasite), Purnomo (blood stage parasites), Ted Hall (*P. falciparum* ring stages), Stephen Hoffman and Robert Thompson (infecting mosquitoes, dissecting mosquitoes), and Gary Nabel (vaccine development algorithm).

Our appreciation is expressed to the following authors for allowing previously published material to be reproduced in this chapter: Carlos Guerra, Simon Hay, and Robert Snow (world map of *P. falciparum*), Alan Saul (life cycle), Alan Cowman, Brendan Crabb, and Jacob Baum (merozoite invasion), Robert Snow and Kevin Marsh (graph of mortality vs. transmission intensity), Sayeh Jafari-Guemouri and Phillipe Deloron (multiplicity of infection), Pedro Alonso (RTS,S Kaplan-Meier curves), and Sanjai Kumar and Judy Epstein (two-tiered malaria vaccine).

Finally, our thanks are offered to those who have allowed the referencing of unpublished data as a personal communication: Chris Ockenhouse (allele-specific protection with MSP1), Lee Hall (results of Ad35.CS Phase 1a trial), Jim Cummings (results of LSA1 Phase 1/2a trial and MSP1 + RTS,S Phase 1/2a trial), Bernhards Ogutu (results of MSP1 Phase 2b trial), Laura Martin (results of MSP142-C1/Alhydrogel + CPG7909 Phase 1a trial), Alassane Dicko (results of AMA1-C1 Phase 2b trial), Mahamadou Thera (results of FMP2.1/AS02A Phase 1b trial and PfAMA1-FVO Phase 1b trial), Michael Good (clinical testing plans for blood stage whole organism vaccine), and Yimin Wu (results of TBV25H Phase 1a trial and Pfs25/Pvs25 Phase 1a trial).

The studies reported herein that were conducted by the Naval Medical Research Center Malaria Program were conducted in accordance with U.S. Navy regulations governing the protection of human subjects in medical research. All protocols involving human subjects were reviewed and approved by the Naval Medical Research Center's Institutional Review Board in accordance with the U.S. Navy regulations (SECNAVINST 3900.39B) governing the use of human subjects in medical research. The opinions and assertions herein are the private ones of the authors and are not to be construed as official or as reflecting the views of the U.S. Navy or the Department of Defense. Work was supported by funds allocated to the Naval Medical Research Center by the U.S. Army Medical Research and Materiel Command.

References

Abdullah, S., Adazu, K., Masanja, H., Diallo, D., Hodgson, A., Ilboudo-Sanogo, E., Nhacolo, A., Owusu-Agyei, S., Thompson, R., Smith, T. and Binka, F.N. Patterns of age-specific mortality in children in endemic areas of sub-Saharan Africa. Am. J. Trop. Med. Hyg. 2007; 77:99–105.

Abraham, E.G., Pinto, S.B., Ghosh, A., Vanlandingham, D.L., Budd, A., Higgs, S., Kafatos, F.C., Jacobs-Lorena, M. and Michel, K. An immune-responsive serpin, SRPN6, mediates mosquito defense against malaria parasites. Proc. Natl. Acad. Sci. U.S.A. 2005; 102:16327–16332.

Abu-Raddad, L.J., Patnaik, P. and Kublin, J.G. Dual infection with HIV and malaria fuels the spread of both diseases in sub-Saharan Africa. Sciences 2006; 314:1603–1606.

Adams, J.H., Wu, Y. and Fairfield, A. Malaria Research and Reference Reagent Resource Center. Parasitol. Today (Personal ed) 2000; 16:89.

Afrane, Y.A., Zhou, G., Lawson, B.W., Githeko, A.K. and Yan, G. Effects of microclimatic changes caused by deforestation on the survivorship and reproductive fitness of Anopheles gambiae in western Kenya highlands. Am. J. Trop. Med. Hyg. 2006; 74:772–778.

Aguiar, J.C., LaBaer, J., Blair, P.L., Shamailova, V.Y., Koundinya, M., Russell, J.A., Huang, F., Mar, W., Anthony, R.M., Witney, A., Caruana, S.R., Brizuela, L., Sacci, J.B., Jr., Hoffman, S.L. and Carucci, D.J. High-throughput generation of *P. falciparum* functional molecules by recombinational cloning. Genome Res. 2004; 14:2076–2082.

Aidoo, M., Lalvani, A., Allsopp, C.E., Plebanski, M., Meisner, S.J., Krausa, P., Browning, M., Morris-Jones, S., Gotch, F., Fidock, D.A. et al. Identification of conserved antigenic components for a cytotoxic T lymphocyte-inducing vaccine against malaria. Lancet 1995; 345:1003–1007.

Alonso, P.L., Sacarlal, J., Aponte, J.J., Leach, A., Macete, E., Aide, P., Sigauque, B., Milman, J., Mandomando, I., Bassat, Q., Guinovart, C., Espasa, M., Corachan, S., Lievens, M., Navia, M.M., Dubois, M.C.,

Menendez, C., Dubovsky, F., Cohen, J., Thompson, R. and Ballou, W.R. Duration of protection with RTS,S/AS02A malaria vaccine in prevention of *Plasmodium falciparum* disease in Mozambican children: single-blind extended follow-up of a randomised controlled trial. Lancet 2005; 366: 2012–2018.

Alonso, P.L., Sacarlal, J., Aponte, J.J., Leach, A., Macete, E., Milman, J., Mandomando, I., Spiessens, B., Guinovart, C., Espasa, M., Bassat, Q., Aide, P., Ofori-Anyinam, O., Navia, M.M., Corachan, S., Ceuppens, M., Dubois, M.C., Demoitie, M.A., Dubovsky, F., Menendez, C., Tornieporth, N., Ballou, W.R., Thompson, R. and Cohen, J. Efficacy of the RTS,S/AS02A vaccine against *Plasmodium falciparum* infection and disease in young African children: randomised controlled trial. Lancet 2004; 364:1411–1420.

Amodu, O.K., Hartl, D.L. and Roy, S.W. Patterns of polymorphism in genomic regions flanking three highly polymorphic surface antigens in *Plasmodium falciparum*. Mol. Biochem. Parasitol. 2007; 159:1–6.

Anders, R.F., Crewther, P.E., Edwards, S., Margetts, M., Matthew, M.L., Pollock, B. and Pye, D. Immunisation with recombinant AMA-1 protects mice against infection with *Plasmodium chabaudi*. Vaccine 1998; 16:240–247.

Aponte, J.J., Aide, P., Renom, M., Mandomando, I., Bassat, Q., Sacarlal, J., Manaca, M.N., Lafuente, S., Barbosa, A., Leach, A., Lievens, M., Vekemans, J., Sigauque, B., Dubois, M.C., Demoitie, M.A., Sillman, M., Savarese, B., McNeil, J.G., Macete, E., Ballou, W.R., Cohen, J. and Alonso, P.L. Safety of the RTS,S/AS02D candidate malaria vaccine in infants living in a highly endemic area of Mozambique: a double blind randomised controlled phase I/IIb trial. Lancet 2007; 370:1543–1551.

Arakawa, T., Komesu, A., Otsuki, H., Sattabongkot, J., Udomsangpetch, R., Matsumoto, Y., Tsuji, N., Wu, Y., Torii, M. and Tsuboi, T. Nasal immunization with a malaria transmission-blocking vaccine candidate, Pfs25, induces complete protective immunity in mice against field isolates of *Plasmodium falciparum*. Infect. Immun. 2005; 73:7375–7380.

Ardeshir, F., Flint, J.E. and Reese, R.T. Expression of *Plasmodium falciparum* surface antigens in *Escherichia coli*. Proc. Natl. Acad. Sci. U.S.A. 1985; 82:2518–2522.

Arevalo-Herrera, M., Solarte, Y., Zamora, F., Mendez, F., Yasnot, M. F., Rocha, L., Long, C., Miller, L.H. and Herrera, S. *Plasmodium vivax*: transmission-blocking immunity in a malaria-endemic area of Colombia. Am. J. Trop. Med. Hyg. 2005; 73:38–43.

Aronson, N.E., Sanders, J.W. and Moran, K.A. In harm's way: infections in deployed American military forces. Clin. Infect. Dis. 2006; 43:1045–1051.

Asa, O.O., Onayade, A.A., Fatusi, A.O., Ijadunola, K.T. and Abiona, T.C. Efficacy of intermittent preventive treatment of malaria with sulphadoxine-pyrimethamine in preventing anaemia in pregnancy among Nigerian women. Matern. Child Health J. 2008.

Audran, R., Cachat, M., Lurati, F., Soe, S., Leroy, O., Corradin, G., Druilhe, P. and Spertini, F. Phase I malaria vaccine trial with a long synthetic peptide derived from the merozoite surface protein 3 antigen. Infect. Immun. 2005; 73:8017–8026.

Avril, M., Gamain, B., Lepolard, C., Viaud, N., Scherf, A. and Gysin, J. Characterization of anti-var2CSA-PfEMP1 cytoadhesion inhibitory mouse monoclonal antibodies. Microbes Infect./Institut Pasteur 2006; 8:2863–2871.

Avril, M., Kulasekara, B.R., Gose, S.O., Rowe, C., Dahlback, M., Duffy, P.E., Fried, M., Salanti, A., Misher, L., Narum, D.L. and Smith, J.D. Evidence for globally shared, cross-reacting polymorphic epitopes in the pregnancy-associated malaria vaccine candidate VAR2CSA. Infect. Immun. 2008; 76:1791–1800.

Baer, K., Klotz, C., Kappe, S.H., Schnieder, T. and Frevert, U. Release of hepatic *Plasmodium yoelii* merozoites into the pulmonary microvasculature. PLoS Pathog. 2007; 3:e171.

Bai, T., Becker, M., Gupta, A., Strike, P., Murphy, V.J., Anders, R.F. and Batchelor, A.H. Structure of AMA1 from *Plasmodium falciparum* reveals a clustering of polymorphisms that surround a conserved hydrophobic pocket. Proc. Natl. Acad. Sci. U.S.A. 2005; 102:12736–12741.

Baird, J.K. Host age as a determinant of naturally acquired immunity to *Plasmodium falciparum*. Parasitol. Today (Personal ed) 1995; 11:105–111.

Baird, J.K. Chloroquine resistance in *Plasmodium vivax*. Antimicrob. Agents Chemother. 2004; 48:4075–4083.

Baird, J.K., Masbar, S., Basri, H., Tirtokusumo, S., Subianto, B. and Hoffman, S.L. Age-dependent susceptibility to severe disease with primary exposure to *Plasmodium falciparum*. J. Infect. Dis. 1998; 178:592–595.

Ballou, W.R. and Cahill, C.P. Two decades of commitment to malaria vaccine development: GlaxoSmithKline Biologicals. Am. J. Trop. Med. Hyg. 2007; 77:289–295.

Ballou, W.R., Rothbard, J., Wirtz, R.A., Gordon, D.M., Williams, J.S., Gore, R.W., Schneider, I., Hollingdale, M.R., Beaudoin, R.L., Maloy, W.L. et al. Immunogenicity of synthetic peptides from circumsporozoite protein of *Plasmodium falciparum*. Science 1985; 228:996–999.

Barcus, M.J., Basri, H., Picarima, H., Manyakori, C., Sekartuti, Elyazar, I., Bangs, M.J., Maguire, J.D. and Baird, J.K. Demographic risk factors for severe and fatal vivax and falciparum malaria among hospital admissions in northeastern Indonesian papua. Am. J. Trop. Med. Hyg. 2007; 77:984–991.

Baum, J., Maier, A.G., Good, R.T., Simpson, K.M. and Cowman, A.F. Invasion by *P. falciparum* merozoites suggests a hierarchy of molecular interactions. PLoS Pathog. 2005; 1:e37.

Beadle, C., Long, G.W., Weiss, W.R., McElroy, P.D., Maret, S.M., Oloo, A.J. and Hoffman, S.L. Diagnosis of malaria by detection of *Plasmodium falciparum* HRP-2 antigen with a rapid dipstick antigen-capture assay. Lancet 1994; 343:564–568.

Beier, J.C., Oster, C.N., Onyango, F.K., Bales, J.D., Sherwood, J.A., Perkins, P.V., Chumo, D.K., Koech, D.V., Whitmire, R.E., Roberts, C.R. et al. *Plasmodium falciparum* incidence relative to entomologic inoculation rates at a site proposed for testing malaria vaccines in western Kenya. Am. J. Trop. Med. Hyg. 1994; 50:529–536.

Bejon, P., Andrews, L., Andersen, R.F., Dunachie, S., Webster, D., Walther, M., Gilbert, S.C., Peto, T. and Hill, A.V. Calculation of liver-to-blood inocula, parasite growth rates, and preerythrocytic vaccine efficacy, from serial quantitative polymerase chain reaction studies of volunteers challenged with malaria sporozoites. J. Infect. Dis. 2005; 191:619–626.

Bejon, P., Mwacharo, J., Kai, O., Mwangi, T., Milligan, P., Todryk, S., Keating, S., Lang, T., Lowe, B., Gikonyo, C., Molyneux, C., Fegan, G., Gilbert, S.C., Peshu, N., Marsh, K. and Hill, A.V. A phase 2b randomised trial of the candidate malaria vaccines FP9 ME-TRAP and MVA ME-TRAP among children in Kenya. PLoS Clin. Trials 2006; 1:e29.

Bejon, P., Ogada, E., Mwangi, T., Milligan, P., Lang, T., Fegan, G., Gilbert, S.C., Peshu, N., Marsh, K. and Hill, A.V. Extended follow-up following a phase 2b randomized trial of the candidate malaria vaccines FP9 ME-TRAP and MVA ME-TRAP among children in Kenya. PLoS ONE 2007; 2:e707.

BenMohamed, L., Thomas, A. and Druilhe, P. Long-term multiepitopic cytotoxic-T-lymphocyte responses induced in chimpanzees by combinations of *Plasmodium falciparum* liver-stage peptides and lipopeptides. Infect. Immun. 2004; 72: 4376–4384.

Bergmann-Leitner, E.S., Duncan, E.H., Mullen, G.E., Burge, J.R., Khan, F., Long, C.A., Angov, E. and Lyon, J.A. Critical evaluation of different methods for measuring the functional activity of antibodies against malaria blood stage antigens. Am. J. Trop. Med. Hyg. 2006; 75:437–442.

Beveridge, N.E., Price, D.A., Casazza, J.P., Pathan, A.A., Sander, C.R., Asher, T.E., Ambrozak, D.R., Precopio, M.L., Scheinberg, P., Alder, N.C., Roederer, M., Koup, R.A., Douek, D.C., Hill, A.V. and McShane, H. Immunisation with BCG and recombinant MVA85A induces long-lasting, polyfunctional *Mycobacterium* tuberculosis-specific CD4+ memory T lymphocyte populations. Eur. J. Immunol. 2007; 37:3089–3100.

Bharadwaj, A., Sharma, P., Joshi, S.K., Singh, B. and Chauhan, V.S. Induction of protective immune responses by immunization with linear multiepitope peptides based on conserved sequences from *Plasmodium falciparum* antigens. Infect. Immun. 1998; 66:3232–3241.

Bhardwaj, D., Kushwaha, A., Puri, S.K., Herrera, A., Singh, N. and Chauhan, V.S. DNA prime-protein boost immunization in monkeys: efficacy of a novel construct containing functional domains of *Plasmodium cynomolgi* CS and TRAP. FEMS Immunol. Med. Microbiol. 2003; 39:241–250.

Bhattarai, A., Ali, A.S., Kachur, S.P., Martensson, A., Abbas, A.K., Khatib, R., Al-Mafazy, A.W., Ramsan, M., Rotllant, G., Gerstenmaier, J.F., Molteni, F., Abdulla, S., Montgomery, S.M., Kaneko, A. and Bjorkman, A. Impact of artemisinin-based combination therapy and insecticide-treated nets on malaria burden in Zanzibar. PLoS Med. 2007; 4:e309.

Binder, E.M. and Kim, K. Location, location, location: trafficking and function of secreted proteases of Toxoplasma and Plasmodium. Traffic 2004; 5:914–924.

Bojang, K.A., Milligan, P.J., Pinder, M., Vigneron, L., Alloueche, A., Kester, K.E., Ballou, W.R., Conway, D.J., Reece, W.H., Gothard, P., Yamuah, L., Delchambre, M., Voss, G., Greenwood, B.M., Hill, A., McAdam, K.P., Tornieporth, N., Cohen, J.D. and Doherty, T. Efficacy of RTS,S/AS02 malaria vaccine against *Plasmodium falciparum* infection in semi-immune adult men in The Gambia: a randomised trial. Lancet 2001; 358:1927–1934.

Bojang, K.A., Olodude, F., Pinder, M., Ofori-Anyinam, O., Vigneron, L., Fitzpatrick, S., Njie, F., Kassanga, A., Leach, A., Milman, J., Rabinovich, R., McAdam, K.P., Kester, K.E., Heppner, D.G., Cohen, J.D., Tornieporth, N. and Milligan, P.J. Safety and immunogenicty of RTS,S/AS02A candidate malaria vaccine in Gambian children. Vaccine 2005; 23:4148–4157.

Bongfen, S.E., Torgler, R., Romero, J.F., Renia, L. and Corradin, G. Plasmodium berghei-infected primary hepatocytes process and present the circumsporozoite protein to specific CD8+ T cells in vitro. J. Immunol. 2007; 178:7054–7063.

Bosman, A. and Mendis, K.N. A major transition in malaria treatment: the adoption and deployment of artemisinin-based combination therapies. Am. J. Trop. Med. Hyg. 2007; 77:193–197.

Bouharoun-Tayoun, H., Attanath, P., Sabchareon, A., Chongsuphajaisiddhi, T. and Druilhe, P. Antibodies that protect humans against *Plasmodium falciparum* blood stages do not on their own inhibit parasite growth and invasion in vitro, but act in cooperation with monocytes. J. Exp. Med. 1990; 172:1633–1641.

Bouharoun-Tayoun, H., Oeuvray, C., Lunel, F. and Druilhe, P. Mechanisms underlying the monocyte-mediated antibody-dependent killing of *Plasmodium falciparum* asexual blood stages. J. Exp. Med. 1995; 182:409–418.

Bouma, M.J. and van der Kaay, H.J. The El Nino southern oscillation and the historic malaria epidemics on the Indian subcontinent and Sri Lanka: an early warning system for future epidemics? Trop. Med. Int. Health 1996; 1:86–96.

Brabin, B. An assessment of low birthweight risk in primiparae as an indicator of malaria control in pregnancy. Int. J. Epidemiol. 1991; 20:276–283.

Brabin, B.J. An analysis of malaria in pregnancy in Africa. Bull. World Health Organ. 1983; 61:1005–1016.

Brabin, B.J., Brabin, L.R., Sapau, J. and Alpers, M.P. A longitudinal study of splenomegaly in pregnancy in a malaria endemic area in Papua New Guinea. Trans. R. Soc. Trop. Med. Hyg. 1988; 82:677–681.

Branch, O., Casapia, W.M., Gamboa, D.V., Hernandez, J.N., Alava, F.F., Roncal, N., Alvarez, E., Perez, E.J. and Gotuzzo, E. Clustered local transmission and asymptomatic *Plasmodium falciparum* and *Plasmodium vivax* malaria infections in a recently emerged, hypoendemic Peruvian Amazon community. Malar. J. 2005; 4:27.

Branch, O.H., Udhayakumar, V., Hightower, A.W., Oloo, A.J., Hawley, W.A., Nahlen, B.L., Bloland, P.B., Kaslow, D.C. and Lal, A.A. A longitudinal investigation of IgG and IgM antibody responses to the merozoite surface protein-1 19-kilo-Dalton domain of *Plasmodium falciparum* in pregnant women and infants: associations with febrile illness, parasitemia, and anemia. Am. J. Trop. Med. Hyg. 1998; 58:211–219.

Brando, C., Ware, L.A., Freyberger, H., Kathcart, A., Barbosa, A., Cayphas, S., Demoitie, M.A., Mettens, P., Heppner, D.G. and Lanar, D.E. Murine immune responses to liver-stage antigen 1 protein FMP011, a malaria vaccine candidate, delivered with adjuvant AS01B or AS02A. Infect. Immun. 2007; 75:838–845.

Brown, A.E. and Catteruccia, F. Toward silencing the burden of malaria: progress and prospects for RNAi-based approaches. BioTechniques 2006(Suppl.):38–44.

Brown, A.E., Singharaj, P., Webster, H.K., Pipithkul, J., Gordon, D.M., Boslego, J.W., Krinchai, K., Su-archawaratana, P., Wongsrichanalai, C., Ballou, W.R. et al. Safety, immunogenicity and limited efficacy study of a recombinant *Plasmodium falciparum* circumsporozoite vaccine in Thai soldiers. Vaccine 1994; 12:102–108.

Bruce-Chwatt, L.J. Malaria and pregnancy. Br. Med. J. 1983; 286:1457–1458.

Bruce-Chwatt, L.J. From Laveran's discovery to DNA probes: new trends in diagnosis of malaria. Lancet 1987; 2:1509–1511.

Burns, J.M., Jr., Flaherty, P.R., Nanavati, P. and Weidanz, W.P. Protection against *Plasmodium chabaudi* malaria induced by immunization with apical membrane antigen 1 and merozoite surface protein 1 in the absence of gamma interferon or interleukin-4. Infect. Immun. 2004; 72:5605–5612.

Burns, J.M., Jr., Majarian, W.R., Young, J.F., Daly, T.M. and Long, C.A. A protective monoclonal antibody recognizes an epitope in the carboxyl-terminal cysteine-rich domain in the precursor of the major merozoite surface antigen of the rodent malarial parasite, *Plasmodium yoelii*. J. Immunol. 1989; 143:2670–2676.

Byers, A.M. Vaccine testing potential of an in vitro human tissue model. Keystone Symposium—Immunological Intervention in Human Disease, Big Sky Resort, Montana, 2007.

Cai, Q., Peng, G., Bu, L., Lin, Y., Zhang, L., Lustigmen, S. and Wang, H. Immunogenicity and in vitro protective efficacy of a polyepitope *Plasmodium falciparum* candidate vaccine constructed by epitope shuffling. Vaccine 2007; 25:5155–5165.

Campbell, C.C., Collins, W.E., Nguyen-Dinh, P., Barber, A. and Broderson, J.R. *Plasmodium falciparum* gametocytes from culture in vitro develop to sporozoites that are infectious to primates. Science 1982; 217:1048–1050.

Carlton, J. The *Plasmodium vivax* genome sequencing project. Trends Parasitol. 2003; 19:227–231.

Carlton, J.M. Toward a malaria haplotype map. Nat. Genet. 2007; 39:5–6.

Carlton, J.M., Angiuoli, S.V., Suh, B.B., Kooij, T.W., Pertea, M., Silva, J.C., Ermolaeva, M.D., Allen, J.E., Selengut, J.D., Koo, H.L., Peterson, J.D., Pop, M., Kosack, D.S., Shumway, M.F., Bidwell, S.L., Shallom, S.J., van Aken, S.E., Riedmuller, S.B., Feldblyum, T.V., Cho, J.K., Quackenbush, J., Sedegah, M., Shoaibi, A., Cummings, L.M., Florens, L., Yates, J.R., Raine, J.D., Sinden, R.E., Harris, M.A., Cunningham, D.A., Preiser, P.R., Bergman, L.W., Vaidya, A.B., van Lin, L.H., Janse, C.J., Waters, A.P., Smith, H.O., White, O.R., Salzberg, S.L., Venter, J.C., Fraser, C.M., Hoffman, S.L., Gardner, M.J. and Carucci, D.J. Genome sequence and comparative analysis of the model rodent malaria parasite *Plasmodium yoelii yoelii*. Nature 2002; 419:512–519.

Carter, R. and Mendis, K.N. Evolutionary and historical aspects of the burden of malaria. Clin. Microbiol. Rev. 2002; 15:564–594.

Carvalho, L.J., Alves, F.A., Bianco, C., Jr., Oliveira, S.G., Zanini, G.M., Soe, S., Druilhe, P., Theisen, M., Muniz, J.A. and Daniel-Ribeiro, C.T. Immunization of Saimiri sciureus monkeys with a recombinant hybrid protein derived from the *Plasmodium falciparum* antigen glutamate-rich protein and merozoite surface protein 3 can induce partial protection with Freund and Montanide ISA720 adjuvants. Clin. Diagn. Lab. Immunol. 2005; 12: 242–248.

Carvalho, L.J., Oliveira, S.G., Theisen, M., Alves, F.A., Andrade, M.C., Zanini, G.M., Brigido, M.C., Oeuvray, C., Povoa, M.M., Muniz, J.A., Druilhe, P. and Daniel-Ribeiro, C.T. Immunization of Saimiri sciureus monkeys with *Plasmodium falciparum* merozoite surface protein-3 and glutamate-rich protein suggests that protection is related to antibody levels. Scand. J. Immunol. 2004; 59:363–372.

Castellanos, A., Arevalo-Herrera, M., Restrepo, N., Gulloso, L., Corradin, G. and Herrera, S. *Plasmodium vivax* thrombospondin related adhesion protein: immunogenicity and protective efficacy in rodents and *Aotus* monkeys. Mem. Inst. Oswaldo Cruz 2007; 102:411–416.

Cemazar, M. and Sersa, G. Electrotransfer of therapeutic molecules into tissues. Curr. Opin. Mol. Ther. 2007; 9:554–562.

Ceravolo, I.P., Souza-Silva, F.A., Fontes, C.J., Braga, E.M., Madureira, A.P., Krettli, A.U., Souza, J.M., Brito, C.F., Adams, J.H. and Carvalho, L.H. Inhibitory properties of the antibody response to *Plasmodium vivax* Duffy binding protein in an area with unstable malaria transmission. Scand. J. Immunol. 2008; 67:270–278.

Chakravarty, S., Cockburn, I.A., Kuk, S., Overstreet, M.G., Sacci, J.B. and Zavala, F. CD8+ T lymphocytes protective against malaria liver stages are primed in skin-draining lymph nodes. Nat. Med. 2007; 13:1035–1041.

Charest, H., Sedegah, M., Yap, G.S., Gazzinelli, R.T., Caspar, P., Hoffman, S.L. and Sher, A. Recombinant attenuated *Toxoplasma gondii* expressing the *Plasmodium yoelii* circumsporozoite protein provides highly effective priming for CD8+ T cell-dependent protective immunity against malaria. J. Immunol. 2000; 165: 2084–2092.

Cheng, Q., Lawrence, G., Reed, C., Stowers, A., Ranford-Cartwright, L., Creasey, A., Carter, R. and Saul, A. Measurement of *Plasmodium falciparum* growth rates in vivo: a test of malaria vaccines. Am. J. Trop. Med. Hyg. 1997; 57:495–500.

Chinchilla, M., Pasetti, M.F., Medina-Moreno, S., Wang, J.Y., Gomez-Duarte, O.G., Stout, R., Levine, M.M. and Galen, J.E. Enhanced immunity to *Plasmodium falciparum* circumsporozoite protein (PfCSP) by using *Salmonella enterica* serovar Typhi expressing PfCSP and a PfCSP-encoding DNA vaccine in a heterologous prime-boost strategy. Infect. Immun. 2007; 75:3769–3779.

Chotivanich, K., Silamut, K., Udomsangpetch, R., Stepniewska, K.A., Pukrittayakamee, S., Looareesuwan, S. and White, N.J. Ex-vivo short-term culture and developmental assessment of *Plasmodium vivax*. Trans. R. Soc. Trop. Med. Hyg. 2001; 95:677–680.

Church, L.W., Le, T.P., Bryan, J.P., Gordon, D.M., Edelman, R., Fries, L., Davis, J.R., Herrington, D.A., Clyde, D.F., Shmuklarsky, M.J., Schneider, I., McGovern, T.W., Chulay, J.D., Ballou, W.R. and Hoffman, S.L. Clinical manifestations of *Plasmodium falciparum* malaria experimentally induced by mosquito challenge. J. Infect. Dis. 1997; 175:915–920.

Clark, H. The diagnostic value of the placental blood film in aestivo-autumnal malaria. J. Exp. Med. 1915; 22:427–444.

Clyde, D.F., McCarthy, V.C., Miller, R.M. and Hornick, R.B. Specificity of protection of man immunized against sporozoite-induced falciparum malaria. Am. J. Med. Sci. 1973a; 266: 398–403.

Clyde, D.F., Most, H., McCarthy, V.C. and Vanderberg, J.P. Immunization of man against sporozite-induced falciparum malaria. Am. J. Med. Sci. 1973b; 266:169–177.

Coatney, G.R. The simian malarias: zoonoses, anthroponoses, or both? Am. J. Trop. Med. Hyg. 1971; 20:795–803.

Cochrane, A.H., Aikawa, M., Jeng, M. and Nussenzweig, R.S. Antibody-induced ultrastructural changes of malarial sporozoites. J. Immunol. 1976; 116:859–867.

Coll, O., Menendez, C., Botet, F., Dayal, R., Xavier Carbonell-Estrany, A.T., Weisman, L.E., Anceschi, M.M., Greenough, A., Gibss, R.S. and Ville, Y. Treatment and prevention of malaria in pregnancy and newborn. J. Perinat. Med. 2008; 36:15–29.

Collins, W.E. and Contacos, P.G. Immunization of monkeys against *Plasmodium cynomolgi* by X-irradiated sporozoites. Nat.: New Biol. 1972; 236:176–177.

Collins, W.E. and Jeffery, G.M. A retrospective examination of sporozoite- and trophozoite-induced infections with *Plasmodium falciparum* in patients previously infected with heterologous species of Plasmodium: effect on development of parasitologic and clinical immunity. Am. J. Trop. Med. Hyg. 1999a; 61:36–43.

Collins, W.E. and Jeffery, G.M. A retrospective examination of the patterns of recrudescence in patients infected with *Plasmodium falciparum*. Am. J. Trop. Med. Hyg. 1999b; 61:44–48.

Collins, W.E. and Jeffery, G.M. Plasmodium ovale: parasite and disease. Clin. Microbiol. Rev. 2005; 18:570–581.

Collins, W.E. and Jeffery, G.M. *Plasmodium malariae*: parasite and disease. Clin. Microbiol. Rev. 2007; 20:579–592.

Collins, W.E., Jeffery, G.M. and Roberts, J.M. A retrospective examination of reinfection of humans with *Plasmodium vivax*. Am. J. Trop. Med. Hyg. 2004; 70:642–644.

Collins, W.E., Nussenzweig, R.S., Ballou, W.R., Ruebush, T.K., II, Nardin, E.H., Chulay, J.D., Majarian, W.R., Young, J.F., Wasserman, G.F., Bathurst, I. et al. Immunization of Saimiri sciureus boliviensis with recombinant vaccines based on the circumsporozoite protein of *Plasmodium vivax*. Am. J. Trop. Med. Hyg. 1989; 40:455–464.

Conway, D.J. Molecular epidemiology of malaria. Clin. Microbiol. Rev. 2007; 20:188–204.

Cowman, A.F. and Crabb, B.S. A parasite genome sheds light on an old enemy. Nat. Biotechnol. 2002; 20:1098–1099.

Cowman, A.F. and Crabb, B.S. Invasion of red blood cells by malaria parasites. Cell 2006; 124:755–766.

Cox, F.E. History of human parasitology. Clin. Microbiol. Rev. 2002; 15:595–612.

Cox-Singh, J., Davis, T.M., Lee, K.S., Shamsul, S.S., Matusop, A., Ratnam, S., Rahman, H.A., Conway, D.J. and Singh, B. Plasmodium knowlesi malaria in humans is widely distributed and potentially life threatening. Clin. Infect. Dis. 2008; 46:165–171.

Crockett, M. and Kain, K.C. Tafenoquine: a promising new antimalarial agent. Expert Opin. Investig. Drugs 2007; 16:705–715.

Dame, J.B., Williams, J.L., McCutchan, T.F., Weber, J.L., Wirtz, R.A., Hockmeyer, W.T., Maloy, W.L., Haynes, J.D., Schneider, I., Roberts, D. et al. Structure of the gene encoding the immunodominant surface antigen on the sporozoite of the human malaria parasite *Plasmodium falciparum*. Sciences 1984; 225:593–599.

D'Ombrain, M.C., Voss, T.S., Maier, A.G., Pearce, J.A., Hansen, D.S., Cowman, A.F. and Schofield, L. *Plasmodium falciparum* erythrocyte membrane protein-1 specifically suppresses early production of host interferon-gamma. Cell Host Microbe 2007; 2:130–138.

Dasaradhi, P.V., Korde, R., Thompson, J.K., Tanwar, C., Nag, T.C., Chauhan, V.S., Cowman, A.F., Mohmmed, A. and Malhotra, P. Food vacuole targeting and trafficking of falcipain-2, an important cysteine protease of human malaria parasite *Plasmodium falciparum*. Mol. Biochem. Parasitol. 2007; 156:12–23.

Date, S.V. and Stoeckert, C.J., Jr. Computational modeling of the *Plasmodium falciparum* interactome reveals protein function on a genome-wide scale. Genome Res. 2006; 16:542–549.

Daubersies, P., Thomas, A.W., Millet, P., Brahimi, K., Langermans, J.A., Ollomo, B., BenMohamed, L., Slierendregt, B., Eling, W., Van Belkum, A., Dubreuil, G., Meis, J.F., Guerin-Marchand, C., Cayphas, S., Cohen, J., Gras-Masse, H. and Druilhe, P. Protection against *Plasmodium falciparum* malaria in chimpanzees by immunization with the conserved pre-erythrocytic liver-stage antigen 3. Nat. Med. 2000; 6:1258–1263.

Day, N. and Dondorp, A.M. The management of patients with severe malaria. Am. J. Trop. Med. Hyg. 2007; 77:29–35.

de Souza-Neiras, W.C., de Melo, L.M. and Machado, R.L. The genetic diversity of *Plasmodium vivax*—a review. Mem. Inst. Oswaldo Cruz 2007; 102:245–254.

Desowitz, R.S. The Malaria Capers: Tales of Parasites and People. WW Norton and Company, 1991.

Dicko, A., Sagara, I., Diemert, D., Sogoba, M., Niambele, M.B., Dao, A., Dolo, G., Yalcouye, D., Diallo, D.A., Saul, A., Miller, L.H., Toure, Y.T., Klion, A.D. and Doumbo, O.K. Year-to-year variation in the age-specific incidence of clinical malaria in two potential vaccine testing sites in Mali with different levels of malaria transmission intensity. Am. J. Trop. Med. Hyg. 2007; 77:1028–1033.

Dicko, A., Sagara, I., Ellis, R.D., Miura, K., Guindo, O., Kamate, B., Sogoba, M., Niambele, M.B., Sissoko, M., Baby, M., Dolo, A., Mullen, G.E., Fay, M.P., Pierce, M., Diallo, D.A., Saul, A., Miller, L.H. and Doumbo, O.K. Phase 1 study of a combination AMA1 blood stage malaria vaccine in Malian children. PLoS ONE 2008; 3:e1563.

Di Girolamo, F., Raggi, C., Bultrini, E., Lanfrancotti, A., Silvestrini, F., Sargiacomo, M., Birago, C., Pizzi, E., Alano, P. and Ponzi, M. Functional genomics: new tools in malaria research. Ann. dell'Istit. Super. Sanita 2005; 41:469–477.

Dinglasan, R.R., Alaganan, A., Ghosh, A.K., Saito, A., van Kuppevelt, T.H. and Jacobs-Lorena, M. *Plasmodium falciparum* ookinetes require mosquito midgut chondroitin sulfate proteoglycans for cell invasion. Proc. Natl. Acad. Sci. U.S.A. 2007; 104:15882–15887.

Dobano, C., Rogerson, S.J., Mackinnon, M.J., Cavanagh, D.R., Taylor, T.E., Molyneux, M.E. and McBride, J.S. Differential antibody responses to *Plasmodium falciparum* merozoite proteins in Malawian children with severe malaria. J. Infect. Dis. 2008; 197:766–774.

Doolan, D.L., Aguiar, J.C., Weiss, W.R., Sette, A., Felgner, P.L., Regis, D.P., Quinones-Casas, P., Yates, J.R., 3rd, Blair, P.L., Richie, T.L., Hoffman, S.L. and Carucci, D.J. Utilization of genomic sequence information to develop malaria vaccines. J. Exp. Biol. 2003a; 206:3789–3802.

Doolan, D.L. and Good, M.F. *Plasmodium falciparum* CS protein—prime malaria vaccine candidate: definition of the human CTL domain and analysis of its variation. Mem. Inst. Oswaldo Cruz 1992; 87(Suppl. 3):241–247.

Doolan, D.L. and Hoffman, S.L. The complexity of protective immunity against liver-stage malaria. J. Immunol. 2000; 165:1453–1462.

Doolan, D.L. and Hoffman, S.L. DNA-based vaccines against malaria: status and promise of the multi-stage malaria DNA vaccine operation. Int. J. Parasitol. 2001; 31:753–762.

Doolan, D.L., Hoffman, S.L., Southwood, S., Wentworth, P.A., Sidney, J., Chesnut, R.W., Keogh, E., Appella, E., Nutman, T.B., Lal, A.A., Gordon, D.M., Oloo, A. and Sette, A. Degenerate cytotoxic T cell epitopes from *P. falciparum* restricted by multiple HLA-A and HLA-B supertype alleles. Immunity 1997; 7:97–112.

Doolan, D.L. and Martinez-Alier, N. Immune response to pre-erythrocytic stages of malaria parasites. Curr. Mol. Med. 2006; 6:169–185.

Doolan, D.L., Sedegah, M., Hedstrom, R.C., Hobart, P., Charoenvit, Y. and Hoffman, S.L. Circumventing genetic restriction of protection against malaria with multigene DNA immunization: CD8+ cell-, interferon gamma-, and nitric oxide-dependent immunity. J. Exp. Med. 1996; 183:1739–1746.

Doolan, D.L., Southwood, S., Freilich, D.A., Sidney, J., Graber, N.L., Shatney, L., Bebris, L., Florens, L., Dobano, C., Witney, A.A., Appella, E., Hoffman, S.L., Yates, J.R., III, Carucci, D.J. and Sette, A. Identification of *Plasmodium falciparum* antigens by antigenic analysis of genomic and proteomic data. Proc. Natl. Acad. Sci. U.S.A. 2003b; 100:9952–9957.

Druilhe, P., Spertini, F., Soesoe, D., Corradin, G., Mejia, P., Singh, S., Audran, R., Bouzidi, A., Oeuvray, C. and Roussilhon, C. A malaria vaccine that elicits in humans antibodies able to kill *Plasmodium falciparum*. PLoS Med. 2005; 2:e344.

Dunachie, S.J., Walther, M., Epstein, J.E., Keating, S., Berthoud, T., Andrews, L., Andersen, R.F., Bejon, P., Goonetilleke, N., Poulton, I., Webster, D.P., Butcher, G., Watkins, K., Sinden, R.E., Levine, G.L., Richie, T.L., Schneider, J., Kaslow, D., Gilbert, S.C., Carucci, D.J. and Hill, A.V. A DNA prime-modified vaccinia virus Ankara boost vaccine encoding thrombospondin-related adhesion protein but not circumsporozoite protein partially protects healthy malaria-naive adults against *Plasmodium falciparum* sporozoite challenge. Infect. Immun. 2006a; 74:5933–5942.

Dunachie, S.J., Walther, M., Vuola, J.M., Webster, D.P., Keating, S.M., Berthoud, T., Andrews, L., Bejon, P., Poulton, I., Butcher, G., Watkins, K., Sinden, R.E., Leach, A., Moris, P., Tornieporth, N., Schneider, J., Dubovsky, F., Tierney, E., Williams, J., Heppner, D.G., Jr., Gilbert, S.C., Cohen, J. and Hill, A.V. A clinical trial of prime-boost immunisation with the candidate malaria vaccines RTS,S/AS02A and MVA-CS. Vaccine 2006b; 24:2850–2859.

Dutta, S., Lee, S.Y., Batchelor, A.H. and Lanar, D.E. Structural basis of antigenic escape of a malaria vaccine candidate. Proc. Natl. Acad. Sci. U.S.A. 2007; 104:12488–12493.

Dzikowski, R., Templeton, T.J. and Deitsch, K. Variant antigen gene expression in malaria. Cell. Microbiol. 2006; 8:1371–1381.

Edelman, R., Wasserman, S.S., Kublin, J.G., Bodison, S.A., Nardin, E.H., Oliveira, G.A., Ansari, S., Diggs, C.L., Kashala, O.L., Schmeckpeper, B.J. and Hamilton, R.G. Immediate-type hypersensitivity and other clinical reactions in volunteers immunized with a synthetic multi-antigen peptide vaccine (PfCS-MAP1NYU) against Plasmodium falciparum sporozoites. Vaccine 2002 Dec 13; 21(3-4):269–280, Erratum in: Vaccine 2003; 21:2689.

Egan, A.F., Morris, J., Barnish, G., Allen, S., Greenwood, B.M., Kaslow, D.C., Holder, A.A. and Riley, E.M. Clinical immunity to *Plasmodium falciparum* malaria is associated with serum antibodies to the 19-kDa C-terminal fragment of the merozoite surface antigen, PfMSP-1. J. Infect. Dis. 1996; 173:765–769.

Egan, J.E., Hoffman, S.L., Haynes, J.D., Sadoff, J.C., Schneider, I., Grau, G.E., Hollingdale, M.R., Ballou, W.R. and Gordon, D.M. Humoral immune responses in volunteers immunized with irradiated *Plasmodium falciparum* sporozoites. Am. J. Trop. Med. Hyg. 1993; 49:166–173.

Egan, T.J. Recent advances in understanding the mechanism of hemozoin (malaria pigment) formation. J. Inorg. Biochem. 2007; 102:1288-1299.

Ekland, E.H. and Fidock, D.A. Advances in understanding the genetic basis of antimalarial drug resistance. Curr. Opin. Microbiol. 2007; 10:363–370.

Elliott, S.R., Kuns, R.D. and Good, M.F. Heterologous immunity in the absence of variant-specific antibodies after exposure to subpatent infection with blood-stage malaria. Infect. Immun. 2005; 73:2478–2485.

Enevold, A., Nkya, W.M., Theisen, M., Vestergaard, L.S., Jensen, A.T., Staalsoe, T., Theander, T.G., Bygbjerg, I.C. and Alifrangis, M. Potential impact of host immunity on malaria treatment outcome in Tanzanian children infected with *Plasmodium falciparum*. Malar. J. 2007; 6:153.

Enosse, S., Dobano, C., Quelhas, D., Aponte, J.J., Lievens, M., Leach, A., Sacarlal, J., Greenwood, B., Milman, J., Dubovsky, F., Cohen, J., Thompson, R., Ballou, W.R., Alonso, P.L., Conway, D.J. and Sutherland, C.J. RTS,S/AS02A malaria vaccine does not induce parasite CSP T cell epitope selection and reduces multiplicity of infection. PLoS Clin. Trials 2006; 1:e5.

Enwere, G. A review of the quality of randomized clinical trials of adjunctive therapy for the treatment of cerebral malaria. Trop. Med. Int. Health 2005; 10:1171–1175.

Epstein, J.E., Charoenvit, Y., Kester, K.E., Wang, R., Newcomer, R., Fitzpatrick, S., Richie, T.L., Tornieporth, N., Heppner, D.G., Ockenhouse, C., Majam, V., Holland, C., Abot, E., Ganeshan, H., Berzins, M., Jones, T., Freydberg, C.N., Ng, J., Norman, J., Carucci, D.J., Cohen, J. and Hoffman, S.L. Safety, tolerability, and antibody responses in humans after sequential immunization with a PfCSP DNA vaccine followed by the recombinant protein vaccine RTS,S/AS02A. Vaccine 2004; 22:1592–1603.

Epstein, J.E., Giersing, B., Mullen, G., Moorthy, V. and Richie, T.L. Malaria vaccines: are we getting closer? Curr. Opin. Mol. Ther. 2007a; 9:12–24.

Epstein, J.E., Rao, S., Williams, F., Freilich, D., Luke, T., Sedegah, M., de la Vega, P., Sacci, J., Richie, T.L. and Hoffman, S.L. Safety and clinical outcome of experimental challenge of human volunteers with *Plasmodium falciparum*-infected mosquitoes: an update. J. Infect. Dis. 2007b; 196:145–154.

Escalante, A.A., Grebert, H.M., Chaiyaroj, S.C., Magris, M., Biswas, S., Nahlen, B.L. and Lal, A.A. Polymorphism in the gene encoding the apical membrane antigen-1 (AMA-1) of *Plasmodium falciparum*. X. Asembo Bay Cohort Project. Mol. Biochem. Parasitol. 2001; 113:279–287.

Felger, I., Genton, B., Smith, T., Tanner, M. and Beck, H.P. Molecular monitoring in malaria vaccine trials. Trends Parasitol. 2003; 19: 60–63.

Fernando, S.D., Gunawardena, D.M., Bandara, M.R., De Silva, D., Carter, R., Mendis, K.N. and Wickremasinghe, A.R. The impact of repeated malaria attacks on the school performance of children. Am. J. Trop. Med. Hyg. 2003; 69:582–588.

Ferreira, M.U., Zilversmit, M. and Wunderlic, G. Origins and evolution of antigenic diversity in malaria parasites. Curr. Mol. Med. 2007; 7:588–602.

Fidock, D.A., Gras-Masse, H., Lepers, J.P., Brahimi, K., Benmohamed, L., Mellouk, S., Guerin-Marchand, C., Londono, A., Raharimalala, L., Meis, J.F. et al. *Plasmodium falciparum* liver stage antigen-1 is well conserved and contains potent B and T cell determinants. J. Immunol. 1994; 153:190–204.

Fontaine, R.E., Najjar, A.E. and Prince, J.S. The 1958 malaria epidemic in Ethiopia. Am. J. Trop. Med. Hyg. 1961; 10:795–803.

Franke, E.D., Sette, A., Sacci, J., Jr., Southwood, S., Corradin, G. and Hoffman, S.L. A subdominant CD8(+) cytotoxic T lymphocyte (CTL) epitope from the *Plasmodium yoelii* circumsporozoite protein induces CTLs that eliminate infected hepatocytes from culture. Infect. Immun. 2000; 68:3403–3411.

Franke-Fayard, B., Waters, A.P. and Janse, C.J. Real-time in vivo imaging of transgenic bioluminescent blood stages of rodent malaria parasites in mice. Nat. Protoc. 2006; 1:476–485.

Fraunholz, M.J. Systems biology in malaria research. Trends Parasitol. 2005; 21:393–395.

Frevert, U., Engelmann, S., Zougbede, S., Stange, J., Ng, B., Matuschewski, K., Liebes, L. and Yee, H. Intravital observation of *Plasmodium berghei* sporozoite infection of the liver. PLoS Biol. 2005; 3:e192.

Fried, M. and Duffy, P.E. Adherence of *Plasmodium falciparum* to chondroitin sulfate A in the human placenta. Sciences 1996; 272:1502–1504.

Fried, M., Wendler, J.P., Mutabingwa, T.K. and Duffy, P.E. Mass spectrometric analysis of *Plasmodium falciparum* erythrocyte membrane protein-1 variants expressed by placental malaria parasites. Proteomics 2004; 4:1086–1093.

Fries, L.F., Gordon, D.M., Richards, R.L., Egan, J.E., Hollingdale, M.R., Gross, M., Silverman, C. and Alving, C.R. Liposomal malaria vaccine in humans: a safe and potent adjuvant strategy. Proc. Natl. Acad. Sci. U.S.A. 1992; 89:358–362.

Gallup, J.L. and Sachs, J.D. The economic burden of malaria. Am. J. Trop. Med. Hyg. 2001; 64:85–96.

Gamain, B., Smith, J.D., Viebig, N.K., Gysin, J. and Scherf, A. Pregnancy-associated malaria: parasite binding, natural immunity and vaccine development. Int. J. Parasitol. 2007; 37:273–283.

Garcia, J.E., Puentes, A. and Patarroyo, M.E. Developmental biology of sporozoite-host interactions in *Plasmodium falciparum* malaria: implications for vaccine design. Clin. Microbiol. Rev. 2006; 19:686–707.

Gardner, M.J., Hall, N., Fung, E., White, O., Berriman, M., Hyman, R.W., Carlton, J.M., Pain, A., Nelson, K.E., Bowman, S., Paulsen, I.T., James, K., Eisen, J.A., Rutherford, K., Salzberg, S.L., Craig, A., Kyes, S., Chan, M.S., Nene, V., Shallom, S.J., Suh, B., Peterson, J., Angiuoli, S., Pertea, M., Allen, J., Selengut, J., Haft, D., Mather, M.W., Vaidya, A.B., Martin, D.M., Fairlamb, A.H., Fraunholz, M.J., Roos, D.S., Ralph, S.A., McFadden, G.I., Cummings, L.M., Subramanian, G.M., Mungall, C., Venter, J.C., Carucci, D.J., Hoffman, S.L., Newbold, C., Davis, R.W., Fraser, C.M. and Barrell, B. Genome sequence of the human malaria parasite *Plasmodium falciparum*. Nature 2002; 419:498–511.

Gaur, D., Singh, S., Singh, S., Jiang, L., Diouf, A. and Miller, L.H. Recombinant *Plasmodium falciparum* reticulocyte homology protein 4 binds to erythrocytes and blocks invasion. Proc. Natl. Acad. Sci. U.S.A. 2007; 104:17789–17794.

Gautret, P., Miltgen, F., Gantier, J.C., Chabaud, A.G. and Landau, I. Enhanced gametocyte formation by *Plasmodium chabaudi* in immature erythrocytes: pattern of production, sequestration, and infectivity to mosquitoes. J. Parasitol. 1996; 82:900–906.

Gay, C.G., Richie, T.L., Pastoret, P.P., Minguez-Tudela, I., de Baetselier, P., Gobel, T., Goddeeris, B., Kaiser, P., Morrison, I., Sanchez-Vizcaino, J.M., Anderson, K., Baillie, L.W., Brown, W.C., Estes, D.M., Herrera, E., Nara, P.L., Ockenhouse, C.F., Roth, J.A. and Sztein, M.B. Advances in immunology and vaccine discovery report of the United States–European Commission workshop. Vaccine 2007; 25:7007–7011.

Genton, B., Al-Yaman, F., Betuela, I., Anders, R.F., Saul, A., Baea, K., Mellombo, M., Taraika, J., Brown, G.V., Pye, D., Irving, D.O., Felger, I., Beck, H.P., Smith, T.A. and Alpers, M.P. Safety and immunogenicity of a three-component blood-stage malaria vaccine (MSP1, MSP2, RESA) against *Plasmodium falciparum* in Papua New Guinean children. Vaccine 2003; 22:30–41.

Genton, B., Betuela, I., Felger, I., Al-Yaman, F., Anders, R.F., Saul, A., Rare, L., Baisor, M., Lorry, K., Brown, G.V., Pye, D., Irving, D.O., Smith, T.A., Beck, H.P. and Alpers, M.P. A recombinant blood-stage malaria vaccine reduces *Plasmodium falciparum* density and exerts selective pressure on parasite populations in a phase 1-2b trial in Papua New Guinea. J. Infect. Dis. 2002; 185: 820–827.

Genton, B., Pluschke, G., Degen, L., Kammer, A.R., Westerfeld, N., Okitsu, S.L., Schroller, S., Vounatsou, P., Mueller, M.M., Tanner, M. and Zurbriggen, R. A randomized placebo-controlled phase ia malaria vaccine trial of two virosome-formulated synthetic peptides in healthy adult volunteers. PLoS ONE 2007; 2:e1018.

Genton, B. and Reed, Z.H. Asexual blood-stage malaria vaccine development: facing the challenges. Curr. Opin. Infect. Dis. 2007; 20:467–475.

Gilbert, S.C., Schneider, J., Hannan, C.M., Hu, J.T., Plebanski, M., Sinden, R. and Hill, A.V. Enhanced CD8T cell immunogenicity and protective efficacy in a mouse malaria model using a recombinant adenoviral vaccine in heterologous prime-boost immunisation regimes. Vaccine 2002; 20:1039–1045.

Githeko, A.K., Ayisi, J.M., Odada, P.K., Atieli, F.K., Ndenga, B.A., Githure, J.I. and Yan, G. Topography and malaria transmission heterogeneity in western Kenya highlands: prospects for focal vector control. Malar. J. 2006; 5:107.

Goldberg, D.E. Hemoglobin degradation in Plasmodium-infected red blood cells. Semin. Cell Biol. 1993; 4:355–361.

Golenda, C.F., Li, J. and Rosenberg, R. Continuous in vitro propagation of the malaria parasite *Plasmodium vivax*. Proc. Natl. Acad. Sci. U.S.A. 1997; 94:6786–6791.

Good, M.F., Zevering, Y., Currier, J. and Bilsborough, J. 'Original antigenic sin', T cell memory, and malaria sporozoite immunity: an hypothesis for immune evasion. Parasite Immunol. 1993; 15:187–193.

Gordon, D.M., McGovern, T.W., Krzych, U., Cohen, J.C., Schneider, I., LaChance, R., Heppner, D.G., Yuan, G., Hollingdale, M., Slaoui, M. et al. Safety, immunogenicity, and efficacy of a recombinantly produced *Plasmodium falciparum* circumsporozoite protein-hepatitis B surface antigen subunit vaccine. J. Infect. Dis. 1995; 171:1576–1585.

Gramzinski, R.A., Maris, D.C., Doolan, D., Charoenvit, Y., Obaldia, N., Rossan, R., Sedegah, M., Wang, R., Hobart, P., Margalith, M. and Hoffman, S. Malaria DNA vaccines in *Aotus* monkeys. Vaccine 1997; 15:913–915.

Graves, P. and Gelband, H. Vaccines for preventing malaria (SPf66). Cochrane Database Syst. Rev. 2006; (2):CD005966.

Gray, J.C., Corran, P.H., Mangia, E., Gaunt, M.W., Li, Q., Tetteh, K.K., Polley, S.D., Conway, D.J., Holder, A.A., Bacarese-Hamilton, T., Riley, E.M. and Crisanti, A. Profiling the antibody immune response against blood stage malaria vaccine candidates. Clin. Chem. 2007; 53:1244–1253.

Greenwood, B. Malaria vaccines. Evaluation and implementation. Acta Trop. 2005; 95:298–304.

Greenwood, B., Marsh, K. and Snow, R. Why do some African children develop severe malaria? Parasitol. Today (Personal ed.) 1991; 7:277–281.

Gregson, A.L., Oliveira, G., Othoro, C., Calvo-Calle, J.M., Thorton, G.B., Nardin, E. and Edelman, R. Phase I trial of an alhydrogel adjuvanted hepatitis B core virus-like particle containing epitopes of Plasmodium falciparum circumsporozoite protein. PLoS ONE 2008; 3:e1556.

Griffith, K.S., Lewis, L.S., Mali, S. and Parise, M.E. Treatment of malaria in the United States: a systematic review. JAMA 2007; 297:2264–2277.

Grimberg, B.T., Udomsangpetch, R., Xainli, J., McHenry, A., Panichakul, T., Sattabongkot, J., Cui, L., Bockarie, M., Chitnis, C., Adams, J., Zimmerman, P.A. and King, C.L. *Plasmodium vivax* invasion of human erythrocytes inhibited by antibodies directed against the Duffy binding protein. PLoS Med. 2007; 4:e337.

Group, B.C. 2005. www.malariavaccine.org/files/Market-Assessment-18 Jan 05-LB-BOS.pdf.

Gruner, A.C., Mauduit, M., Tewari, R., Romero, J.F., Depinay, N., Kayibanda, M., Lallemand, E., Chavatte, J.M., Crisanti, A., Sinnis, P., Mazier, D., Corradin, G., Snounou, G. and Renia, L. Sterile protection against malaria is independent of immune responses to the circumsporozoite protein. PLoS ONE 2007; 2:e1371.

Gruner, A.C., Snounou, G., Brahimi, K., Letourneur, F., Renia, L. and Druilhe, P. Pre-erythrocytic antigens of *Plasmodium falciparum*: from rags to riches? Trends Parasitol. 2003; 19:74–78.

Guerra, C.A., Gikandi, P.W., Tatem, A.J., Noor, A.M., Smith, D.L., Hay, S.I. and Snow, R.W. The limits and intensity of *Plasmodium falciparum* transmission: implications for malaria control and elimination worldwide. PLoS Med. 2008; 5:e38.

Guerra, C.A., Hay, S.I., Lucioparedes, L.S., Gikandi, P.W., Tatem, A.J., Noor, A.M. and Snow, R.W. Assembling a global database of malaria parasite prevalence for the Malaria Atlas Project. Malar. J. 2007; 6:17.

Guerra, C.A., Snow, R.W. and Hay, S.I. Mapping the global extent of malaria in 2005. Trends Parasitol. 2006; 22:353–358.

Guinovart, C. and Alonso, P.L. Methods for determining vaccine efficacy and effectiveness and the main barriers to developing a fully deployable malaria vaccine. Am. J. Trop. Med. Hyg. 2007; 77:276–281.

Gupta, S., Snow, R.W., Donnelly, C.A., Marsh, K. and Newbold, C. Immunity to non-cerebral severe malaria is acquired after one or two infections. Nat. Med. 1999; 5:340–343.

Guyatt, H.L. and Snow, R.W. Malaria in pregnancy as an indirect cause of infant mortality in sub-Saharan Africa. Trans. R. Soc. Trop. Med. Hyg. 2001; 95:569–576.

Guyatt, H.L. and Snow, R.W. Impact of malaria during pregnancy on low birth weight in sub-Saharan Africa. Clin. Microbiol. Rev. 2004; 17:760–769, Table of contents.

Hall, N., Karras, M., Raine, J.D., Carlton, J.M., Kooij, T.W., Berriman, M., Florens, L., Janssen, C.S., Pain, A., Christophides, G.K., James, K., Rutherford, K., Harris, B., Harris, D., Churcher, C., Quail, M.A., Ormond, D., Doggett, J., Trueman, H.E., Mendoza, J., Bidwell, S.L., Rajandream, M.A., Carucci, D.J., Yates, J.R., 3rd, Kafatos, F.C., Janse, C.J., Barrell, B., Turner, C.M., Waters, A.P. and Sinden, R.E. A comprehensive survey of the Plasmodium life cycle by genomic, transcriptomic, and proteomic analyses. Sciences 2005; 307:82–86.

Hans, D., Pattnaik, P., Bhattacharyya, A., Shakri, A.R., Yazdani, S.S., Sharma, M., Choe, H., Farzan, M. and Chitnis, C.E. Mapping binding residues in the *Plasmodium vivax* domain that binds Duffy antigen during red cell invasion. Mol. Microbiol. 2005; 55:1423–1434.

Hanscheid, T., Egan, T.J. and Grobusch, M.P. Haemozoin: from melatonin pigment to drug target, diagnostic tool, and immune modulator. Lancet Infect. Dis. 2007; 7:675–685.

Hansen, D.S., D'Ombrain, M.C. and Schofield, L. The role of leukocytes bearing natural killer complex receptors and killer immunoglobulin-like receptors in the immunology of malaria. Curr. Opin. Immunol. 2007; 19:416–423.

Hatz, C., Soto, J., Nothdurft, H.D., Zoller, T., Weitzel, T., Loutan, L., Bricaire, F., Gay, F., Burchard, G.D., Andriano, K., Lefevre, G., De Palacios, P.I. and Genton, B. Treatment of acute uncomplicated falciparum malaria with artemether-lumefantrine in nonimmune populations: a safety, efficacy, and pharmacokinetic study. Am. J. Trop. Med. Hyg. 2008; 78:241–247.

Hay, S.I., Guerra, C.A., Tatem, A.J., Noor, A.M. and Snow, R.W. The global distribution and population at risk of malaria: past, present, and future. Lancet Infect. Dis. 2004; 4:327–336.

Hay, S.I. and Snow, R.W. The malaria Atlas Project: developing global maps of malaria risk. PLoS Med. 2006; 3:e473.

Healer, J., Murphy, V., Hodder, A.N., Masciantonio, R., Gemmill, A.W., Anders, R.F., Cowman, A.F. and Batchelor, A. Allelic polymorphisms in apical membrane antigen-1 are responsible for evasion of antibody-mediated inhibition in *Plasmodium falciparum*. Mol. Microbiol. 2004; 52:159–168.

Hedstrom, R.C., Sedegah, M. and Hoffman, S.L. Prospects and strategies for development of DNA vaccines against malaria. Res. Immunol. 1994; 145:476–483.

Heppner, D.G., Cummings, J.F., Ockenhouse, C., Kester, K.E., Lyon, J.A. and Gordon, D.M. New World monkey efficacy trials for malaria vaccine development: critical path or detour? Trends Parasitol. 2001; 17:419–425.

Heppner, D.G., Gordon, D.M., Gross, M., Wellde, B., Leitner, W., Krzych, U., Schneider, I., Wirtz, R.A., Richards, R.L., Trofa, A., Hall, T., Sadoff, J.C., Boerger, P., Alving, C.R., Sylvester, D.R., Porter, T.G. and Ballou, W.R. Safety, immunogenicity, and efficacy of *Plasmodium falciparum* repeatless circumsporozoite protein vaccine encapsulated in liposomes. J. Infect. Dis. 1996; 174:361–366.

Heppner, D.G., Jr., Kester, K.E., Ockenhouse, C.F., Tornieporth, N., Ofori, O., Lyon, J.A., Stewart, V.A., Dubois, P., Lanar, D.E., Krzych, U., Moris, P., Angov, E., Cummings, J.F., Leach, A., Hall, B.T., Dutta, S., Schwenk, R., Hillier, C., Barbosa, A., Ware, L.A., Nair, L., Darko, C.A., Withers, M.R., Ogutu, B., Polhemus, M.E., Fukuda, M., Pichyangkul, S., Gettyacamin, M., Diggs, C., Soisson, L., Milman, J., Dubois, M.C., Garcon, N., Tucker, K., Wittes, J., Plowe, C.V., Thera, M.A., Duombo, O.K., Pau, M.G., Goudsmit, J., Ballou, W.R. and Cohen, J. Towards an RTS,S-based, multi-stage, multi-antigen vaccine against falciparum malaria: progress at the Walter Reed Army Institute of Research. Vaccine 2005; 23:2243–2250.

Hermsen, C.C., de Vlas, S.J., van Gemert, G.J., Telgt, D.S., Verhage, D.F. and Sauerwein, R.W. Testing vaccines in human experimental malaria: statistical analysis of parasitemia measured by a quantitative real-time polymerase chain reaction. Am. J. Trop. Med. Hyg. 2004; 71:196–201.

Hermsen, C.C., Verhage, D.F., Telgt, D.S., Teelen, K., Bousema, J.T., Roestenberg, M., Bolad, A., Berzins, K., Corradin, G., Leroy, O., Theisen, M. and Sauerwein, R.W. Glutamate-rich protein (GLURP) induces antibodies that inhibit in vitro growth of *Plasmodium falciparum* in a phase 1 malaria vaccine trial. Vaccine 2007; 25:2930–2940.

Hernandez-Valladares, M., Naessens, J. and Iraqi, F.A. Gene-knockout mice in malaria research: useful or misleading? Trends Parasitol. 2007; 23:522–526.

Herrera, S., Corradin, G. and Arevalo-Herrera, M. An update on the search for a *Plasmodium vivax* vaccine. Trends Parasitol. 2007; 23:122–128.

Herrera, S., Gomez, A., Vera, O., Vergara, J., Valderrama-Aguirre, A., Maestre, A., Mendez, F., Wang, R., Chitnis, C.E., Yazdani, S.S. and Arevalo-Herrera, M. Antibody response to *Plasmodium vivax* antigens in Fy-negative individuals from the Colombian Pacific coast. Am. J. Trop. Med. Hyg. 2005; 73:44–49.

Herrington, D.A., Clyde, D.F., Losonsky, G., Cortesia, M., Murphy, J.R., Davis, J., Baqar, S., Felix, A.M., Heimer, E.P., Gillessen, D. et al. Safety and immunogenicity in man of a synthetic peptide malaria vaccine against *Plasmodium falciparum* sporozoites. Nature 1987; 328:257–259.

Hill, A.V. Aspects of genetic susceptibility to human infectious diseases. Ann. Rev. Genet. 2006a; 40:469–486.

Hill, A.V. Pre-erythrocytic malaria vaccines: towards greater efficacy. Nat. Rev. Immunol. 2006b; 6:21–32.

Hill, A.V., Elvin, J., Willis, A.C., Aidoo, M., Allsopp, C.E., Gotch, F.M., Gao, X.M., Takiguchi, M., Greenwood, B.M., Townsend, A.R. et al. Molecular analysis of the association of HLA-B53 and resistance to severe malaria. Nature 1992; 360:434–439.

Hill, J. and Kazembe, P. Reaching the Abuja target for intermittent preventive treatment of malaria in pregnancy in African women: a review of progress and operational challenges. Trop. Med. Int. Health 2006; 11:409–418.

Hillier, C.J., Ware, L.A., Barbosa, A., Angov, E., Lyon, J.A., Heppner, D.G. and Lanar, D.E. Process development and analysis of liver-stage antigen 1, a preerythrocyte-stage protein-based vaccine for *Plasmodium falciparum*. Infect. Immun. 2005; 73:2109–2115.

Hisaeda, H., Saul, A., Reece, J.J., Kennedy, M.C., Long, C.A., Miller, L.H. and Stowers, A.W. Merozoite surface protein 3 and protection against malaria in *Aotus nancymai* monkeys. J. Infect. Dis. 2002; 185:657–664.

Hisaeda, H., Tetsutani, K., Imai, T., Moriya, C., Tu, L., Hamano, S., Duan, X., Chou, B., Ishida, H., Aramaki, A., Shen, J., Ishii, K.J., Coban, C., Akira, S., Takeda, K., Yasutomo, K., Torii, M. and Himeno, K. Malaria parasites require TLR9 signaling for immune evasion by activating regulatory T cells. J. Immunol. 2008; 180:2496–2503.

Hoffman, S.L. Experimental challenge of volunteers with malaria. Ann. Intern. Med. 1997; 127:233–235.

Hoffman, S.L., Rustama, D., Punjabi, N.H., Surampaet, B., Sanjaya, B., Dimpudus, A.J., McKee, K.T., Jr., Paleologo, F.P., Campbell, J.R., Marwoto, H. et al. High-dose dexamethasone in quinine-treated patients with cerebral malaria: a double-blind, placebo-controlled trial. J. Infect. Dis. 1988; Aug; 158(2):325–331.

Hoffman, S.L., Goh, L.M., Luke, T.C., Schneider, I., Le, T.P., Doolan, D.L., Sacci, J., de la Vega, P., Dowler, M., Paul, C., Gordon, D.M., Stoute, J.A., Church, L.W., Sedegah, M., Heppner, D.G., Ballou, W.R. and Richie, T.L. Protection of humans against malaria by immunization with radiation-attenuated *Plasmodium falciparum* sporozoites. J. Infect. Dis. 2002; 185:1155–1164.

Hollingdale, M.R., Ballou, W.R., Aley, S.B., Young, J.F., Pancake, S., Miller, L.H. and Hockmeyer, W.T. *Plasmodium falciparum*: elicitation by peptides and recombinant circumsporozoite proteins of circulating mouse antibodies inhibiting sporozoite invasion of hepatoma cells. Exp. Parasitol. 1987; 63:345–351.

Hu, J., Chen, Z., Gu, J., Wan, M., Shen, Q., Kieny, M.P., He, J., Li, Z., Zhang, Q., Reed, Z.H., Zhu, Y., Li, W., Cao, Y., Qu, L., Cao, Z., Wang, Q., Liu, H., Pan, X., Huang, X., Zhang, D., Xue, X. and Pan, W. Safety and immunogenicity of a malaria vaccine, Plasmodium falciparum AMA-1/MSP-1 chimeric protein formulated in montanide ISA 720 in healthy adults. PLoS ONE 2008; 3:e1952.

Huaman, M.C., Martin, L.B., Malkin, E., Narum, D.L., Miller, L.H., Mahanty, S. and Long, C.A. Ex vivo cytokine and memory T cell responses to the 42-kDa fragment of *Plasmodium falciparum* merozoite surface protein-1 in vaccinated volunteers. J. Immunol. 2008; 180:1451–1461.

Hurd, H., Al-Olayan, E. and Butcher, G.A. In vitro methods for culturing vertebrate and mosquito stages of *Plasmodium*. Microb. Infect./Institut Pasteur 2003; 5:321–327.

Hutchings, C.L., Birkett, A.J., Moore, A.C. and Hill, A.V. Combination of protein and viral vaccines induces potent cellular and humoral immune responses and enhanced protection from murine malaria challenge. Infect. Immun. 2007; 75:5819–5826.

Hviid, L. Naturally acquired immunity to *Plasmodium falciparum* malaria in Africa. Acta Trop. 2005; 95:270–275.

Hviid, L. Adhesion specificities of *Plasmodium falciparum*-infected erythrocytes involved in the pathogenesis of pregnancy-associated malaria. Am. J. Pathol. 2007; 170:1817–1819.

Ishino, T., Yano, K., Chinzei, Y. and Yuda, M. Cell-passage activity is required for the malarial parasite to cross the liver sinusoidal cell layer. PLoS Biol. 2004; 2:E4.

Ismail, M.R., Ordi, J., Menendez, C., Ventura, P.J., Aponte, J.J., Kahigwa, E., Hirt, R., Cardesa, A. and Alonso, P.L. Placental pathology in malaria: a histological, immunohistochemical, and quantitative study. Hum. Pathol. 2000; 31:85–93.

Jafari-Guemouri, S., Ndam, N.T., Bertin, G., Renart, E., Sow, S., Le Hesran, J.Y. and Deloron, P. Demonstration of a high level of parasite population homology by quantification of *Plasmodium falciparum* alleles in matched peripheral, placental, and umbilical cord blood samples. J. Clin. Microbiol. 2005; 43:2980–2983.

Jahiel, R.I., Vilcek, J. and Nussenzweig, R.S. Exogenous interferon protects mice against *Plasmodium berghei* malaria. Nature 1970; 227:1350–1351.

Janse, C.J., Franke-Fayard, B. and Waters, A.P. Selection by flow-sorting of genetically transformed, GFP-expressing blood stages of the rodent malaria parasite, *Plasmodium berghei*. Nat. Protoc. 2006; 1:614–623.

Jin, Y., Kebaier, C. and Vanderberg, J. Direct microscopic quantification of dynamics of *Plasmodium berghei* sporozoite transmission from mosquitoes to mice. Infect. Immun. 2007; 75:5532–5539.

Jobe, O., Lumsden, J., Mueller, A.K., Williams, J., Silva-Rivera, H., Kappe, S.H., Schwenk, R.J., Matuschewski, K. and Krzych, U. Genetically attenuated *Plasmodium berghei* liver stages induce sterile protracted protection that is mediated by major histocompatibility complex Class I-dependent interferon-gamma-producing CD8+ T cells. J. Infect. Dis. 2007; 196:599–607.

John, C.C., Moormann, A.M., Sumba, P.O., Ofulla, A.V., Pregibon, D.C. and Kazura, J.W. Gamma interferon responses to *Plasmodium falciparum* liver-stage antigen 1 and thrombospondin-related adhesive protein and their relationship to age, transmission intensity, and protection against malaria. Infect. Immun. 2004; 72: 5135–5142.

John, C.C., Tande, A.J., Moormann, A.M., Sumba, P.O., Lanar, D.E., Min, X.M. and Kazura, J.W. Antibodies to pre-erythrocytic *Plasmodium falciparum* antigens and risk of clinical malaria in Kenyan children. J. Infect. Dis. 2008; 197:519–526.

Jones, T.R., Narum, D.L., Gozalo, A.S., Aguiar, J., Fuhrmann, S.R., Liang, H., Haynes, J.D., Moch, J.K., Lucas, C., Luu, T., Magill, A.J., Hoffman, S.L. and Sim, B.K. Protection of *Aotus* monkeys by *Plasmodium falciparum* EBA-175 region II DNA prime-protein boost immunization regimen. J. Infect. Dis. 2001; 183:303–312.

Jones, T.R., Obaldia, N., 3rd, Gramzinski, R.A., Charoenvit, Y., Kolodny, N., Kitov, S., Davis, H.L., Krieg, A.M. and Hoffman, S.L. Synthetic oligodeoxynucleotides containing CpG motifs enhance immunogenicity of a peptide malaria vaccine in *Aotus* monkeys. Vaccine 1999; 17:3065–3071.

Kaiser, K., Matuschewski, K., Camargo, N., Ross, J. and Kappe, S.H. Differential transcriptome profiling identifies *Plasmodium* genes encoding pre-erythrocytic stage-specific proteins. Mol. Microbiol. 2004; 51:1221–1232.

Kappe, S.H., Buscaglia, C.A. and Nussenzweig, V. Plasmodium sporozoite molecular cell biology. Annu. Rev. Cell Dev. Biol. 2004; 20:29–59.

Kariu, T., Ishino, T., Yano, K., Chinzei, Y. and Yuda, M. CelTOS, a novel malarial protein that mediates transmission to mosquito and vertebrate hosts. Mol. Microbiol. 2006; 59:1369–1379.

Karunajeewa, H.A., Manning, L., Mueller, I., Ilett, K.F. and Davis, T.M. Rectal administration of artemisinin derivatives for the treatment of malaria. JAMA 2007; 297:2381–2390.

Kasehagen, L.J., Mueller, I., Kiniboro, B., Bockarie, M.J., Reeder, J.C., Kazura, J.W., Kastens, W., McNamara, D.T., King, C.H., Whalen, C.C. and Zimmerman, P.A. Reduced *Plasmodium vivax* erythrocyte infection in PNG Duffy-negative heterozygotes. PLoS ONE 2007; 2:e336.

Kaslow, D.C. Transmission-blocking vaccines. Chem. Immunol. 2002; 80:287–307.

Kaushal, D.C., Kaushal, N.A., Narula, A., Kumar, N., Puri, S.K., Dutta, S. and Lanar, D.E. Biochemical and immunological characterization of E. coli expressed 42 kDa fragment of *Plasmodium vivax* and *P. cynomolgi bastianelli* merozoite surface protein-1. Indian J. Biochem. Biophys. 2007; 44:429–436.

Kemp, D.J., Coppel, R.L., Cowman, A.F., Saint, R.B., Brown, G.V. and Anders, R.F. Expression of *Plasmodium falciparum* blood-stage antigens in *Escherichia coli*: detection with antibodies from immune humans. Proc. Nat. Acad. Sci. U.S.A. 1983; 80: 3787–3791.

Kester, K.E., McKinney, D.A., Tornieporth, N., Ockenhouse, C.F., Heppner, D.G., Hall, T., Krzych, U., Delchambre, M., Voss, G., Dowler, M.G., Palensky, J., Wittes, J., Cohen, J. and Ballou, W.R. Efficacy of recombinant circumsporozoite protein vaccine regimens against experimental *Plasmodium falciparum* malaria. J. Infect. Dis. 2001; 183:640–647.

Khusmith, S., Charoenvit, Y., Kumar, S., Sedegah, M., Beaudoin, R.L. and Hoffman, S.L. Protection against malaria by vaccination with sporozoite surface protein 2 plus CS protein. Science 1991; 252:715–718.

Khusmith, S., Sedegah, M. and Hoffman, S.L. Complete protection against *Plasmodium yoelii* by adoptive transfer of a CD8+ cytotoxic T-cell clone recognizing sporozoite surface protein 2. Infect. Immun. 1994; 62:2979–2983.

Kihara, M., Carter, J.A. and Newton, C.R. The effect of *Plasmodium falciparum* on cognition: a systematic review. Trop. Med. Int. Health 2006; 11:386–397.

Kilama, W.L., Chilengi, R. and Wanga, C.L. Towards an African-driven malaria vaccine development program: history and activities of the African Malaria Network Trust (AMANET). Am. J. Trop. Med. Hyg. 2007; 77:282–288.

Kochar, D.K., Saxena, V., Singh, N., Kochar, S.K., Kumar, S.V. and Das, A. *Plasmodium vivax* malaria. Emerg. Infect. Dis. 2005; 11:132–134.

Kocken, C.H., Withers-Martinez, C., Dubbeld, M.A., van der Wel, A., Hackett, F., Valderrama, A., Blackman, M.J. and Thomas, A.W. High-level expression of the malaria blood-stage vaccine candidate *Plasmodium falciparum* apical membrane antigen 1 and induction of antibodies that inhibit erythrocyte invasion. Infect. Immun. 2002; 70:4471–4476.

Kooij, T.W. and Matuschewski, K. Triggers and tricks of *Plasmodium* sexual development. Curr. Opin. Microbiol. 2007; 10:547–553.

Krzych, U., Lyon, J.A., Jareed, T., Schneider, I., Hollingdale, M.R., Gordon, D.M. and Ballou, W.R. T lymphocytes from volunteers immunized with irradiated *Plasmodium falciparum* sporozoites recognize liver and blood stage malaria antigens. J. Immunol. 1995; 155:4072–4077.

Kumar, K.A., Sano, G., Boscardin, S., Nussenzweig, R.S., Nussenzweig, M.C., Zavala, F. and Nussenzweig, V. The circumsporozoite protein is an immunodominant protective antigen in irradiated sporozoites. Nature 2006; 444:937–940.

Kumar, S., Epstein, J.E. and Richie, T.L. Vaccines against asexual stage malaria parasites. Chem. Immunol. 2002; 80:262–286.

Kumar, S., Melzer, M., Dodds, P., Watson, J. and Ord, R. *P. vivax* malaria complicated by shock and ARDS. Scand. J. Infect. Dis. 2007; 39:255–256.

Kumar, S. and Miller, L.H. Cellular mechanisms in immunity to blood stage infection. Immunol. Lett. 1990; 25:109–114.

Kwiatkowski, D.P. How malaria has affected the human genome and what human genetics can teach us about malaria. Am. J. Hum. Genet. 2005; 77:171–192.

Kyes, S.A., Kraemer, S.M. and Smith, J.D. Antigenic variation in *Plasmodium falciparum*: gene organization and regulation of the *var* multigene family. Eukaryotic Cell 2007; 6:1511–1520.

Labaied, M., Harupa, A., Dumpit, R.F., Coppens, I., Mikolajczak, S.A. and Kappe, S.H. *Plasmodium yoelii* sporozoites with simultaneous deletion of P52 and P36 are completely attenuated and confer sterile immunity against infection. Infect. Immun. 2007; 75:3758–3768.

LaCount, D.J., Vignali, M., Chettier, R., Phansalkar, A., Bell, R., Hesselberth, J.R., Schoenfeld, L.W., Ota, I., Sahasrabudhe, S., Kurschner, C., Fields, S. and Hughes, R.E. A protein interaction network of the malaria parasite *Plasmodium falciparum*. Nature 2005; 438:103–107.

Laddy, D.J. and Weiner, D.B. From plasmids to protection: a review of DNA vaccines against infectious diseases. Int. Rev. Immunol. 2006; 25:99–123.

Lanar, D.E., Tine, J.A., de Taisne, C., Seguin, M.C., Cox, W.I., Winslow, J.P., Ware, L.A., Kauffman, E.B., Gordon, D., Ballou, W.R., Paoletti, E. and Sadoff, J.C. Attenuated vaccinia virus-circumsporozoite protein recombinants confer protection against rodent malaria. Infect. Immun. 1996; 64:1666–1671.

Lasonder, E., Ishihama, Y., Andersen, J.S., Vermunt, A.M., Pain, A., Sauerwein, R.W., Eling, W.M., Hall, N., Waters, A.P., Stunnenberg, H.G. and Mann, M. Analysis of the *Plasmodium falciparum* proteome by high-accuracy mass spectrometry. Nature 2002; 419:537–542.

Laufer, M.K., Djimde, A.A. and Plowe, C.V. Monitoring and deterring drug-resistant malaria in the era of combination therapy. Am. J. Trop. Med. Hyg. 2007; 77:160–169.

LeBlanc, R., Vasquez, Y., Hannaman, D. and Kumar, N. Markedly enhanced immunogenicity of a Pfs25 DNA-based malaria transmission-blocking vaccine by in vivo electroporation. Vaccine 2008; 26:185–192.

Le Roch, K.G., Zhou, Y., Blair, P.L., Grainger, M., Moch, J.K., Haynes, J.D., De La Vega, P., Holder, A.A., Batalov, S., Carucci, D.J. and Winzeler, E.A. Discovery of gene function by expression profiling of the malaria parasite life cycle. Sciences 2003; 301:1503–1508.

Le, T.P., Coonan, K.M., Hedstrom, R.C., Charoenvit, Y., Sedegah, M., Epstein, J.E., Kumar, S., Wang, R., Doolan, D.L., Maguire, J.D., Parker, S.E., Hobart, P., Norman, J. and Hoffman, S.L. Safety, tolerability and humoral immune responses after intramuscular administration of a malaria DNA vaccine to healthy adult volunteers. Vaccine 2000; 18:1893–1901.

Lesko, C.R., Arguin, P.M. and Newman, R.D. Congenital malaria in the United States: a review of cases from 1966 to 2005. Arch. Pediatr. Adolesc. Med. 2007; 161:1062–1067.

Li, S., Locke, E., Bruder, J., Clarke, D., Doolan, D.L., Havenga, M.J., Hill, A.V., Liljestrom, P., Monath, T.P., Naim, H.Y., Ockenhouse, C., Tang, D.C., Van Kampen, K.R., Viret, J.F., Zavala, F. and Dubovsky, F. Viral vectors for malaria vaccine development. Vaccine 2007; 25:2567–2574.

Lindblade, K.A., Walker, E.D., Onapa, A.W., Katungu, J. and Wilson, M.L. Highland malaria in Uganda: prospective analysis of an epidemic associated with El Nino. Trans. R. Soc. Trop. Med. Hyg. 1999; 93:480–487.

Lomar, A.V., Vidal, J.E., Lomar, F.P., Barbas, C.V., Matos, G.J. and Boulos, M. Acute respiratory distress syndrome due to vivax malaria: case report and literature review. Braz. J. Infect. Dis. 2005; 9:425–430.

Long, C.A., Daly, T.M., Kima, P. and Srivastava, I. Immunity to erythrocytic stages of malarial parasites. Am. J. Trop. Med. Hyg. 1994; 50:27–32.

Lufesi, N.N., Andrew, M. and Aursnes, I. Deficient supplies of drugs for life threatening diseases in an African community. BMC Health Serv. Res. 2007; 7:86.

Luke, T.C. and Hoffman, S.L. Rationale and plans for developing a non-replicating, metabolically active, radiation-attenuated *Plasmodium falciparum* sporozoite vaccine. J. Exp. Biol. 2003; 206:3803–3808.

Lundquist, R., Nielsen, L.K., Jafarshad, A., Soesoe, D., Christensen, L.H., Druilhe, P. and Dziegiel, M.H. Human recombinant antibodies against *Plasmodium falciparum* merozoite surface protein 3 cloned from peripheral blood leukocytes of individuals with immunity to malaria demonstrate antiparasitic properties. Infect. Immun. 2006; 74:3222–3231.

Lusingu, J.P., Vestergaard, L.S., Alifrangis, M., Mmbando, B.P., Theisen, M., Kitua, A.Y., Lemnge, M.M. and Theander, T.G. Cytophilic antibodies to *Plasmodium falciparum* glutamate rich protein are associated with malaria protection in an area of holoendemic transmission. Malaria J. 2005; 4:48.

Lysenko, A.J. and Beljaev, A.E. An analysis of the geographical distribution of *Plasmodium* ovale. Bull. World Health Organ. 1969; 40:383–394.

Magill, A.J. Malaria: diagnosis and treatment of falciparum malaria in travelers during and after travel. Curr. Infect. Dis. Rep. 2006; 8:35–42.

Maguire, J.D., Sumawinata, I.W., Masbar, S., Laksana, B., Prodjodipuro, P., Susanti, I., Sismadi, P., Mahmud, N., Bangs, MJ. and Baird, J.K. Chloroquine-resistant *Plasmodium malariae* in south Sumatra, Indonesia. Lancet 2002; 360:58–60.

Mahajan, R.C., Farooq, U., Dubey, M.L. and Malla, N. Genetic polymorphism in *Plasmodium falciparum* vaccine candidate antigens. Indian J. Pathol. Microbiol. 2005; 48:429–438.

Mahanty, S., Saul, A. and Miller, L.H. Progress in the development of recombinant and synthetic blood-stage malaria vaccines. J. Exp. Biol. 2003; 206:3781–3788.

Malkin, E., Long, C.A., Stowers, A.W., Zou, L., Singh, S., MacDonald, N.J., Narum, D.L., Miles, A.P., Orcutt, A.C., Muratova, O., Moretz, S.E., Zhou, H., Diouf, A., Fay, M., Tierney, E., Leese, P., Mahanty, S., Miller, L.H., Saul, A. and Martin, L.B. Phase 1 study of two merozoite surface protein 1 (MSP1(42)) vaccines for *Plasmodium falciparum* malaria. PLoS Clin. Trials 2007; 2:e12.

Malkin, E.M., Diemert, D.J., McArthur, J.H., Perreault, J.R., Miles, A.P., Giersing, B.K., Mullen, G.E., Orcutt, A., Muratova, O., Awkal, M., Zhou, H., Wang, J., Stowers, A., Long, C.A., Mahanty, S., Miller, L.H., Saul, A. and Durbin, A.P. Phase 1 clinical trial of apical membrane antigen 1: an asexual blood-stage vaccine for *Plasmodium falciparum* malaria. Infect. Immun. 2005; 73:3677–3685.

Marussig, M., Renia, L., Motard, A., Miltgen, F., Petour, P., Chauhan, V., Corradin, G. and Mazier, D. Linear and multiple antigen peptides containing defined T and B epitopes of the *Plasmodium yoelii* circumsporozoite protein: antibody-mediated protection and boosting by sporozoite infection. Int. Immunol. 1997; 9:1817–1824.

Matuschewski, K. and Mueller, A.K. Vaccines against malaria: an update. FEBS J. 2007; 274:4680–4687.

Mbogo, C.N., Snow, R.W., Khamala, C.P., Kabiru, E.W., Ouma, J.H., Githure, J.I., Marsh, K. and Beier, J.C. Relationships between *Plasmodium falciparum* transmission by vector populations and the incidence of severe disease at nine sites on the Kenyan coast. Am. J. Trop. Med. Hyg. 1995; 52:201–206.

McCarthy, V.C. and Clyde, D.F. *Plasmodium vivax*: correlation of circumsporozoite precipitation (CSP) reaction with sporozoite-induced protective immunity in man. Exp. Parasitol. 1977; 41:167–171.

McConkey, S.J., Reece, W.H., Moorthy, V.S., Webster, D., Dunachie, S., Butcher, G., Vuola, J.M., Blanchard, T.J., Gothard, P.,

Watkins, K., Hannan, C.M., Everaere, S., Brown, K., Kester, K.E., Cummings, J., Williams, J., Heppner, D.G., Pathan, A., Flanagan, K., Arulanantham, N., Roberts, M.T., Roy, M., Smith, G.L., Schneider, J., Peto, T., Sinden, R.E., Gilbert, S.C. and Hill, A.V. Enhanced T-cell immunogenicity of plasmid DNA vaccines boosted by recombinant modified vaccinia virus Ankara in humans. Nat. Med. 2003; 9:729–735.

McCormick, M.C. The contribution of low birth weight to infant mortality and childhood morbidity. N. Engl. J. Med. 1985; 312:82–90.

McHenry, A.M. and Adams, J.H. The crystal structure of *P. knowlesi* DBPalpha DBL domain and its implications for immune evasion. Trends Biochem. Sci. 2006; 31:487–491.

McIntosh, R.S., Shi, J., Jennings, R.M., Chappel, J.C., de Koning-Ward, T.F., Smith, T., Green, J., van Egmond, M., Leusen, J.H., Lazarou, M., van de Winkel, J., Jones, T.S., Crabb, B.S., Holder, A.A. and Pleass, R.J. The importance of human FcgammaRI in mediating protection to malaria. PLoS Pathog. 2007; 3:e72.

McKelvie, N.D., Khan, S.A., Karavolos, M.H., Bulmer, D.M., Lee, J.J., Demarco, R., Maskell, D.J., Zavala, F., Hormaeche, C.E. and Khan, C.M. Genetic detoxification of an *aroA Salmonella enterica* serovar Typhimurium vaccine strain does not compromise protection against virulent *Salmonella* and enhances the immune responses towards a protective malarial antigen. FEMS Immunol. Med. Microbiol. 2008; 52:237–246.

Mendis, K., Sina, B.J., Marchesini, P. and Carter, R. The neglected burden of *Plasmodium vivax* malaria. Am. J. Trop. Med. Hyg. 2001; 64:97–106.

Menendez, C. Malaria during pregnancy: a priority area of malaria research and control. Parasitol. Today (Personal ed.) 1995; 11:178–183.

Menendez, C. Malaria during pregnancy. Curr. Mol. Med. 2006; 6:269–273.

Menendez, C., Ordi, J., Ismail, M.R., Ventura, P.J., Aponte, J.J., Kahigwa, E., Font, F. and Alonso, P.L. The impact of placental malaria on gestational age and birth weight. J. Infect. Dis. 2000; 181:1740–1745.

Meraldi, V., Nebie, I., Tiono, A.B., Diallo, D., Sanogo, E., Theisen, M., Druilhe, P., Corradin, G., Moret, R. and Sirima, B.S. Natural antibody response to *Plasmodium falciparum* Exp-1, MSP-3 and GLURP long synthetic peptides and association with protection. Parasite Immunol. 2004; 26:265–272.

Merckx, A., Nivez, M.P., Bouyer, G., Alano, P., Langsley, G., Deitsch, K., Thomas, S., Doerig, C. and Egee, S. *Plasmodium falciparum* regulatory subunit of cAMP-dependent PKA and anion channel conductance. PLoS Pathog. 2008; 4:e19.

Mettens, P., Dubois, P.M., Demoitie, M.A., Bayat, B., Donner, M.N., Bourguignon, P., Stewart, V.A., Heppner, D.G., Jr., Garcon, N. and Cohen, J. Improved T cell responses to *Plasmodium falciparum* circumsporozoite protein in mice and monkeys induced by a novel formulation of RTS,S vaccine antigen. Vaccine 2008; 26:1072–1082.

Migot-Nabias, F., Deloron, P., Ringwald, P., Dubois, B., Mayombo, J., Minh, T.N., Fievet, N., Millet, P. and Luty, A. Immune response to *Plasmodium falciparum* liver stage antigen-1: geographical variations within Central Africa and their relationship with protection from clinical malaria. Trans. R. Soc. Trop. Med. Hyg. 2000; 94:557–562.

Mikolajczak, S.A., Aly, A.S., Dumpit, R.F., Vaughan, A.M. and Kappe, S.H. An efficient strategy for gene targeting and phenotypic assessment in the *Plasmodium yoelii* rodent malaria model. Mol. Biochem. Parasitol. 2008; 158:213–216.

Miller, L.H., Mason, S.J., Clyde, D.F. and McGinniss, M.H. The resistance factor to *Plasmodium vivax* in blacks. The Duffy-blood-group genotype, FyFy. N. Engl. J. Med. 1976; 295:302–304.

Mitchell, G.H., Thomas, A.W., Margos, G., Dluzewski, A.R. and Bannister, L.H. Apical membrane antigen 1, a major malaria vaccine candidate, mediates the close attachment of invasive merozoites to host red blood cells. Infect. Immun. 2004; 72:154–158.

Miura, K., Keister, D.B., Muratova, O.V., Sattabongkot, J., Long, C.A. and Saul, A. Transmission-blocking activity induced by malaria vaccine candidates Pfs25/Pvs25 is a direct and predictable function of antibody titer. Malar. J. 2007a; 6:107.

Miura, K., Zhou, H., Muratova, O.V., Orcutt, A.C., Giersing, B., Miller, L.H. and Long, C.A. In immunization with *Plasmodium falciparum* apical membrane antigen 1, the specificity of antibodies depends on the species immunized. Infect. Immun. 2007b; 75:5827–5836.

Mlambo, G., Maciel, J. and Kumar, N. A murine model for assessment of *Plasmodium falciparum* transmission blocking vaccine using transgenic *Plasmodium berghei* parasites expressing the target antigen Pfs25. Infect. Immun. 2008; 76:2018–2024.

Mokashi, A.J., Shirahatti, R.G., Prabhu, S.K. and Vagholkar, K.R. Pathological rupture of malarial spleen. J. Postgrad. Med. 1992; 38:141–142.

Molineaux, L., Trauble, M., Collins, W.E., Jeffery, G.M. and Dietz, K. Malaria therapy reinoculation data suggest individual variation of an innate immune response and independent acquisition of antiparasitic and antitoxic immunities. Trans. R. Soc. Trop. Med. Hyg. 2002; 96:205–209.

Moore, A.C. and Hill, A.V. Progress in DNA-based heterologous prime-boost immunization strategies for malaria. Immunol. Rev. 2004; 199:126–143.

Moorthy, V.S., Imoukhuede, E.B., Milligan, P., Bojang, K., Keating, S., Kaye, P., Pinder, M., Gilbert, S.C., Walraven, G., Greenwood, B.M. and Hill, A.S. A randomised, double-blind, controlled vaccine efficacy trial of DNA/MVA ME-TRAP against malaria infection in Gambian adults. PLoS Med. 2004; 1:e33.

Moran, M.G.J., Ropars, A., Jorgensen, M., McDonald, A., Potter, S. and Haile-Selassie, H. The malaria product pipeline: planning for the future. London, UK: The George Institute for International Health, 2007.

Morosan, S., Hez-Deroubaix, S., Lunel, F., Renia, L., Giannini, C., Van Rooijen, N., Battaglia, S., Blanc, C., Eling, W., Sauerwein, R., Hannoun, L., Belghiti, J., Brechot, C., Kremsdorf, D. and Druilhe, P. Liver-stage development of *Plasmodium falciparum*, in a humanized mouse model. J. Infect. Dis. 2006; 193:996–1004.

Mota, M.M., Pradel, G., Vanderberg, J.P., Hafalla, J.C., Frevert, U., Nussenzweig, R.S., Nussenzweig, V. and Rodriguez, A. Migration of *Plasmodium* sporozoites through cells before infection. Sciences 2001; 291:141–144.

Mota, M.M. and Rodriguez, A. Migration through host cells: the first steps of *Plasmodium* sporozoites in the mammalian host. Cell. Microbiol. 2004; 6:1113–1118.

Mouchet, J., Manguin, S., Sircoulon, J., Laventure, S., Faye, O., Onapa, A.W., Carnevale, P., Julvez, J. and Fontenille, D. Evolution of malaria in Africa for the past 40 years: impact of climatic and human factors. J. Am. Mosq. Control Ass. 1998; 14:121–130.

Mueller, A.K., Camargo, N., Kaiser, K., Andorfer, C., Frevert, U., Matuschewski, K. and Kappe, S.H. *Plasmodium* liver stage developmental arrest by depletion of a protein at the parasite–host interface. Proc. Nat. Acad. Sci. U.S.A. 2005a; 102:3022–3027.

Mueller, A.K., Labaied, M., Kappe, S.H. and Matuschewski, K. Genetically modified *Plasmodium* parasites as a protective experimental malaria vaccine. Nature 2005b; 433:164–167.

Muller, H.M., Scarselli, E. and Crisanti, A. Thrombospondin related anonymous protein (TRAP) of *Plasmodium falciparum* in parasite–host cell interactions. Parassitologia 1993; 35(Suppl.):69–72.

Munga, S., Minakawa, N., Zhou, G., Mushinzimana, E., Barrack, O. O., Githeko, A.K. and Yan, G. Association between land cover and habitat productivity of malaria vectors in western Kenyan highlands. Am. J. Trop. Med. Hyg. 2006; 74:69–75.

Mwangi, T.W., Ross, A., Snow, R.W. and Marsh, K. Case definitions of clinical malaria under different transmission conditions in Kilifi District, Kenya. J. Infect. Dis. 2005; 191:1932–1939.

Nardin, E.H., Calvo-Calle, J.M., Oliveira, G.A., Nussenzweig, R. S., Schneider, M., Tiercy, J.M., Loutan, L., Hochstrasser, D. and Rose, K. A totally synthetic polyoxime malaria vaccine containing Plasmodium falciparum B cell and universal T cell epitopes elicits immune responses in volunteers of diverse HLA types. J. Immunol. 2001; 166:481–489.

Nardin, E.H., Oliveira, G.A., Calvo-Calle, J.M., Castro, Z.R., Nussenzweig, R.S., Schmeckpeper, B., Hall, B.F., Diggs, C., Bodison, S. and Edelman, R. Synthetic malaria peptide vaccine elicits high levels of antibodies in vaccines of defined HLA genotypes. J. Infect. Dis. 2000 Nov; 182(5):1486–1496.

Nardin, E.H., Oliveira, G.A., Calvo-Calle, J.M., Wetzel, K., Maier, C., Birkett, A.J., Sarpotdar, P., Corado, M.L., Thornton, G.B. and Schmidt, A. Phase I testing of a malaria vaccine composed of hepatitis B virus core particles expressing Plasmodium falciparum circumsporozoite epitopes. Infect. Immun. 2004; 72:6519–6527.

Narum, D.L., Kumar, S., Rogers, W.O., Fuhrmann, S.R., Liang, H., Oakley, M., Taye, A., Sim, B.K. and Hoffman, S.L. Codon optimization of gene fragments encoding *Plasmodium falciparum* merzoite proteins enhances DNA vaccine protein expression and immunogenicity in mice. Infect. Immun. 2001; 69:7250–7253.

Narum, D.L., Ogun, S.A., Thomas, A.W. and Holder, A.A. Immunization with parasite-derived apical membrane antigen 1 or passive immunization with a specific monoclonal antibody protects BALB/c mice against lethal *Plasmodium yoelii yoelii* YM blood-stage infection. Infect. Immun. 2000; 68:2899–2906.

Nebie, I., Diarra, A., Ouedraogo, A., Soulama, I., Bougouma, E.C., Tiono, A.B., Konate, A.T., Chilengi, R., Theisen, M., Dodoo, D., Remarque, E., Bosomprah, S., Milligan, P. and Sirima, S.B. Humoral responses to *Plasmodium falciparum* blood-stage antigens and association with incidence of clinical malaria in children living in an area of seasonal malaria transmission in Burkina Faso, West Africa. Infect. Immun. 2008; 76:759–766.

Newton, P.N., Fernandez, F.M., Plancon, A., Mildenhall, D.C., Green, M.D., Ziyong, L., Christophel, E.M., Phanouvong, S., Howells, S., McIntosh, E., Laurin, P., Blum, N., Hampton, C.Y., Faure, K., Nyadong, L., Soong, C.W., Santoso, B., Zhiguang, W., Newton, J. and Palmer, K. A collaborative epidemiological investigation into the criminal fake artesunate trade in South East Asia. PLoS Med. 2008; 5:e32.

Nixon, C.P., Friedman, J.F., Knopf, P.M., Duffy, P.E. and Kurtis, J.D. Protective human immunity as a vaccine discovery tool for falciparum malaria. Transfusion 2005; 45:81S–87S.

Ntoumi, F., Kwiatkowski, D.P., Diakite, M., Mutabingwa, T.K. and Duffy, P.E. New interventions for malaria: mining the human and parasite genomes. Am. J. Trop. Med. Hyg. 2007; 77:270–275.

Nussenzweig, R.S., Vanderberg, J., Most, H. and Orton, C. Protective immunity produced by the injection of x-irradiated sporozoites of *Plasmodium berghei*. Nature 1967; 216:160–162.

Oaks, S.C.M., V.S; Pearson, G.W. and Carpenter, C.C.J. Malaria: Obstacles and Opportunities. A Report of the Institute of Medicine. Washington, DC: National Academy Press, 1991.

Obonyo, C.O., Vulule, J., Akhwale, W.S. and Grobbee, D.E. In-hospital morbidity and mortality due to severe malarial anemia in western Kenya. Am. J. Trop. Med. Hyg. 2007; 77:23–28.

Ocana-Morgner, C., Mota, M.M. and Rodriguez, A. Malaria blood stage suppression of liver stage immunity by dendritic cells. J. Exp. Med. 2003; 197:143–151.

Ockenhouse, C.F., Angov, E., Kester, K.E., Diggs, C., Soisson, L., Cummings, J.F., Stewart, A.V., Palmer, D.R., Mahajan, B., Krzych, U., Tornieporth, N., Delchambre, M., Vanhandenhove, M., Ofori-Anyinam, O., Cohen, J., Lyon, J.A. and Heppner, D.G. Phase I safety and immunogenicity trial of FMP1/AS02A, a *Plasmodium falciparum* MSP-1 asexual blood stage vaccine. Vaccine 2006a; 24:3009–3017.

Ockenhouse, C.F., Hu, W.C., Kester, K.E., Cummings, J.F., Stewart, A., Heppner, D.G., Jedlicka, A.E., Scott, A.L., Wolfe, N.D., Vahey, M. and Burke, D.S. Common and divergent immune response signaling pathways discovered in peripheral blood mononuclear cell gene expression patterns in presymptomatic and clinically apparent malaria. Infect. Immun. 2006b; 74:5561–5573.

Ockenhouse, C.F., Sun, P.F., Lanar, D.E., Wellde, B.T., Hall, B.T., Kester, K., Stoute, J.A., Magill, A., Krzych, U., Farley, L., Wirtz, R.A., Sadoff, J.C., Kaslow, D.C., Kumar, S., Church, L.W., Crutcher, J.M., Wizel, B., Hoffman, S., Lalvani, A., Hill, A.V., Tine, J.A., Guito, K.P., de Taisne, C., Anders, R., Ballou, W.R. et al. Phase I/IIa safety, immunogenicity, and efficacy trial of NYVAC-Pf7, a pox-vectored, multiantigen, multistage vaccine candidate for *Plasmodium falciparum* malaria. J. Infect. Dis. 1998; 177:1664–1673.

Oeuvray, C., Bouharoun-Tayoun, H., Gras-Masse, H., Bottius, E., Kaidoh, T., Aikawa, M., Filgueira, M.C., Tartar, A. and Druilhe, P. Merozoite surface protein-3: a malaria protein inducing antibodies that promote *Plasmodium falciparum* killing by cooperation with blood monocytes. Blood 1994; 84:1594–1602.

Okitsu, S.L., Silvie, O., Westerfeld, N., Curcic, M., Kammer, A.R., Mueller, M.S., Sauerwein, R.W., Robinson, J.A., Genton, B., Mazier, D., Zurbriggen, R. and Pluschke, G. A virosomal malaria peptide vaccine elicits a long-lasting sporozoite-inhibitory antibody response in a phase 1a clinical trial. PLoS ONE 2007; 2:e1278.

Okoko, B.J., Enwere, G. and Ota, M.O. The epidemiology and consequences of maternal malaria: a review of immunological basis. Acta Trop. 2003; 87:193–205.

Okoko, B.J., Ota, M.O., Yamuah, L.K., Idiong, D., Mkpanam, S.N., Avieka, A., Banya, W.A. and Osinusi, K. Influence of placental malaria infection on foetal outcome in the Gambia: twenty years after Ian Mcgregor. J. Health Popul. Nutr. 2002; 20:4–11.

Oliveira, G.A., Wetzel, K., Calvo-Calle, J.M., Nussenzweig, R., Schmidt, A., Birkett, A., Dubovsky, F., Tierney, E., Gleiter, C.H., Boehmer, G., Luty, A.J., Ramharter, M., Thornton, G.B., Kremsner, P.G. and Nardin, E.H. Safety and enhanced immunogenicity of a hepatitis B core particle Plasmodium falciparum malaria vaccine formulated in adjuvant Montanide ISA 720 in a phase I trial. Infect. Immun. 2005; 73:3587–3597.

Ono, T., Tadakuma, T. and Rodriguez, A. *Plasmodium yoelii yoelii* 17XNL constitutively expressing GFP throughout the life cycle. Exp. Parasitol. 2007; 115:310–313.

Osier, F.H., Polley, S.D., Mwangi, T., Lowe, B., Conway, D.J. and Marsh, K. Naturally acquired antibodies to polymorphic and conserved epitopes of Plasmodium falciparum merozoite surface protein 3. Parasite Immunol. 2007; 29:387–394.

Osier, F.H., Fegan, G., Polley, S.D., Murungi, L., Verra, F., Tetteh, K. K., Lowe, B., Mwangi, T., Bull, P.C., Thomas, A.W., Cavanagh, D.R., McBride, J.S., Lanar, D.E., Mackinnon, M.J., Conway, D.J. and Marsh, K. Breadth and magnitude of antibody responses to multiple Plasmodium falciparum merozoite antigens are associated with protection from clinical malaria. Infect. Immun. 2008; 76:2240–2248.

Outchkourov, N.S., Roeffen, W., Kaan, A., Jansen, J., Luty, A., Schuiffel, D., van Gemert, G.J., van de Vegte-Bolmer, M., Sauerwein, R.W. and Stunnenberg, H.G. Correctly folded Pfs48/45 protein of *Plasmodium falciparum* elicits malaria transmission-blocking immunity in mice. Proc. Nat. Acad. Sci. U.S.A. 2008; 105:4301–4305.

Owusu-Agyei, S., Binka, F., Koram, K., Anto, F., Adjuik, M., Nkrumah, F. and Smith, T. Does radical cure of asymptomatic *Plasmodium falciparum* place adults in endemic areas at increased risk of recurrent symptomatic malaria? Trop. Med. Int. Health 2002; 7:599–603.

Ozsoy, M.F., Oncul, O., Pekkafali, Z., Pahsa, A. and Yenen, O.S. Splenic complications in malaria: report of two cases from Turkey. J. Med. Microbiol. 2004; 53:1255–1258.

Pan, W., Huang, D., Zhang, Q., Qu, L., Zhang, D., Zhang, X., Xue, X. and Qian, F. Fusion of two malaria vaccine candidate antigens enhances product yield, immunogenicity, and antibody-mediated inhibition of parasite growth in vitro. J. Immunol. 2004; 172:6167–6174.

Panichakul, T., Sattabongkot, J., Chotivanich, K., Sirichaisinthop, J., Cui, L. and Udomsangpetch, R. Production of erythropoietic cells in vitro for continuous culture of *Plasmodium vivax*. Int. J. Parasitol. 2007; 37:1551–1557.

Patz, J.A. and Olson, S.H. Malaria risk and temperature: influences from global climate change and local land use practices. Proc. Natl. Acad. Sci. U S A. 2006; 103:5635–5636.

Perlaza, B.L., Zapata, C., Valencia, A.Z., Hurtado, S., Quintero, G., Sauzet, J.P., Brahimi, K., Blanc, C., Arevalo-Herrera, M., Druilhe, P. and Herrera, S. Immunogenicity and protective efficacy of *Plasmodium falciparum* liver-stage Ag-3 in *Aotus lemurinus griseimembra* monkeys. Eur. J. Immunol. 2003; 33:1321–1327.

Persson, C., Oliveira, G.A., Sultan, A.A., Bhanot, P., Nussenzweig, V. and Nardin, E. Cutting edge: a new tool to evaluate human pre-erythrocytic malaria vaccines: rodent parasites bearing a hybrid *Plasmodium falciparum* circumsporozoite protein. J. Immunol. 2002; 169:6681–6685.

Peters, W., Chance, M.L., Lissner, R., Momen, H. and Warhurst, D.C. The chemotherapy of rodent malaria, XXX. The enigmas of the 'NS lines' of *P. berghei*. Ann. Trop. Med. Parasitol. 1978; 72:23–36.

Pichyangkul, S., Kum-Arb, U., Yongvanitchit, K., Limsalakpetch, A., Gettayacamin, M., Lanar, D.E., Ware, L.A., Stewart, V.A., Heppner, D.G., Mettens, P., Cohen, J.D., Ballou, W.R. and Fukuda, M.M. Preclinical evaluation of the safety and immunogenicity of a vaccine consisting of *Plasmodium falciparum* liver-stage antigen 1 with adjuvant AS01B administered alone or concurrently with the RTS,S/AS01B vaccine in rhesus primates. Infect. Immun. 2008; 76:229–238.

Pinzon-Charry, A. and Good, M.F. Malaria vaccines: the case for a whole-organism approach. Expert Opin. Biol. Ther. 2008; 8:441–448.

Plebanski, M., Lee, E.A. and Hill, A.V. Immune evasion in malaria: altered peptide ligands of the circumsporozoite protein. Parasitology 1997; 115(Suppl.):S55–S66.

Polhemus, M.E., Magill, A.J., Cummings, J.F., Kester, K.E., Ockenhouse, C.F., Lanar, D.E., Dutta, S., Barbosa, A., Soisson, L., Diggs, C.L., Robinson, S.A., Haynes, J.D., Stewart, V.A., Ware, L.A., Brando, C., Krzych, U., Bowden, R.A., Cohen, J.D., Dubois, M.C., Ofori-Anyinam, O., De-Kock, E., Ballou, W.R. and Heppner, D.G., Jr. Phase I dose escalation safety and immunogenicity trial of *Plasmodium falciparum* apical membrane protein (AMA-1) FMP2.1, adjuvanted with AS02A, in malaria-naive adults at the Walter Reed Army Institute of Research. Vaccine 2007; 25:4203–4212.

Polley, S.D., Mwangi, T., Kocken, C.H., Thomas, A.W., Dutta, S., Lanar, D.E., Remarque, E., Ross, A., Williams, T.N., Mwambingu, G., Lowe, B., Conway, D.J. and Marsh, K. Human antibodies to recombinant protein constructs of Plasmodium falciparum Apical Membrane Antigen 1 (AMA1) and their associations with protection from malaria. Vaccine 2004; 23:718–728.

Polley, S.D., Tetteh, K.K., Lloyd, J.M., Akpogheneta, O.J., Greenwood, B.M., Bojang, K.A. and Conway, D.J. Plasmodium falciparum merozoite surface protein 3 is a target of allele-specific immunity and alleles are maintained by natural selection. J. Infect. Dis. 2007; 195:279–287.

Pombo, D.J., Lawrence, G., Hirunpetcharat, C., Rzepczyk, C., Bryden, M., Cloonan, N., Anderson, K., Mahakunkijcharoen, Y., Martin, L.B., Wilson, D., Elliott, S., Elliott, S., Eisen, D.P., Weinberg, J.B., Saul, A. and Good, M.F. Immunity to malaria after administration of ultra-low doses of red cells infected with *Plasmodium falciparum*. Lancet 2002; 360:610–617.

Pradel, G. Proteins of the malaria parasite sexual stages: expression, function and potential for transmission blocking strategies. Parasitology 2007; 134:1911–1929.

Price, R.N., Dorsey, G., Ashley, E.A., Barnes, K.I., Baird, J. K., d'Alessandro, U., Guerin, P.J., Laufer, M.K., Naidoo, I., Nosten, F., Olliaro, P., Plowe, C.V., Ringwald, P., Sibley, C.H., Stepniewska, K. and White, N.J. World Antimalarial Resistance Network I: clinical efficacy of antimalarial drugs. Malar. J. 2007; 6:119.

Prieur, E., Gilbert, S.C., Schneider, J., Moore, A.C., Sheu, E.G., Goonetilleke, N., Robson, K.J. and Hill, A.V. A *Plasmodium falciparum* candidate vaccine based on a six-antigen polyprotein encoded by recombinant poxviruses. Proc. Nat. Acad. Sci. U.S.A. 2004; 101:290–295.

Purnomo, Solihin, A., Gomez-Saladin, E. and Bangs, M.J. Rare quadruple malaria infection in Irian Jaya Indonesia. J. Parasitol. 1999; 85:574–579.

Ramjanee, S., Robertson, J.S., Franke-Fayard, B., Sinha, R., Waters, A.P., Janse, C.J., Wu, Y., Blagborough, A.M., Saul, A. and Sinden, R.E. The use of transgenic *Plasmodium berghei* expressing the *Plasmodium vivax* antigen P25 to determine the transmission-blocking activity of sera from malaria vaccine trials. Vaccine 2007; 25:886–894.

Ranque, S., Poudiougou, B., Traore, A., Keita, M., Oumar, A.A., Safeukui, I., Marquet, S., Cabantous, S., Diakite, M., Mintha, D., Cisse, M.B., Keita, M.M., Dessein, A.J. and Doumbo, O.K. Life-threatening malaria in African children: a prospective study in a mesoendemic urban setting. Pediatr. Infect. Dis. J. 2008; 27:130–135.

Rapeah, S. and Norazmi, M.N. Immunogenicity of a recombinant *Mycobacterium bovis* bacille Calmette-Guerin expressing malarial and tuberculosis epitopes. Vaccine 2006; 24:3646–3653.

Remarque, E.J., Faber, B.W., Kocken, C.H. and Thomas, A.W. Apical membrane antigen 1: a malaria vaccine candidate in review. Trends Parasitol. 2008; 24:74–84.

Reyburn, H., Mbatia, R., Drakeley, C., Bruce, J., Carneiro, I., Olomi, R., Cox, J., Nkya, W.M., Lemnge, M., Greenwood, B.M. and Riley, E.M. Association of transmission intensity and age with clinical manifestations and case fatality of severe *Plasmodium falciparum* malaria. JAMA 2005; 293:1461–1470.

Reyes-Sandoval, A., Sridhar, S., Berthoud, T., Moore, A.C., Harty, J.T., Gilbert, S.C., Gao, G., Ertl, H.C., Wilson, J.C. and Hill, A.V. Single-dose immunogenicity and protective efficacy of simian adenoviral vectors against *Plasmodium berghei*. Eur. J. Immunol. 2008; 38:732–741.

Richie, T.L. Malaria vaccines for travelers. Travel Med. Infect. Dis. 2004; 2:193–210.

Richie, T.L. and Saul, A. Progress and challenges for malaria vaccines. Nature 2002; 415:694–701.

Rickman, L.S., Gordon, D.M., Wistar, R., Jr., Krzych, U., Gross, M., Hollingdale, M.R., Egan, J.E., Chulay, J.D. and Hoffman, S.L. Use of adjuvant containing mycobacterial cell-wall skeleton, monophosphoryl lipid, A, and squalane in malaria circumsporozoite protein vaccine. Lancet 1991; 337:998–1001.

Rieckmann, K.H. The chequered history of malaria control: are new and better tools the ultimate answer? Ann. Trop. Med. Parasitol. 2006; 100:647–662.

Riley, E.M., Allen, S.J., Wheeler, J.G., Blackman, M.J., Bennett, S., Takacs, B., Schonfeld, H.J., Holder, A.A. and Greenwood, B.M. Naturally acquired cellular and humoral immune responses to the major merozoite surface antigen (PfMSP1) of *Plasmodium falciparum* are associated with reduced malaria morbidity. Parasite Immunol. 1992; 14:321–337.

Roadmap, M.V.T. 2005. www.malariavaccineroadmap.net.

Rodrigues, M.M., Cordey, A.S., Arreaza, G., Corradin, G., Romero, P., Maryanski, J.L., Nussenzweig, R.S. and Zavala, F. CD8+ cytolytic T cell clones derived against the *Plasmodium yoelii* circumsporozoite protein protect against malaria. Int. Immunol. 1991; 3:579–585.

Rodriguez, A., Goudsmit, J., Companjen, A., Mintardjo, R., Gillissen, G., Tax, D., Sijtsma, J., Weverling, G.J., Holterman, L., Lanar, D.E., Havenga, M.J. and Radosevic, K. Impact of recombinant adenovirus serotype 35 priming versus boosting of a *Plasmodium falciparum* protein: characterization of T- and B-cell responses to liver-stage antigen 1. Infect. Immun. 2008; 76:1709–1718.

Rodriguez, R., Moreno, A., Guzman, F., Calvo, M. and Patarroyo, M.E. Studies in owl monkeys leading to the development of a synthetic vaccine against the asexual blood stages of *Plasmodium falciparum*. Am. J. Trop. Med. Hyg. 1990; 43:339–354.

Rogers, W.O., Atuguba, F., Oduro, A.R., Hodgson, A. and Koram, K.A. Clinical case definitions and malaria vaccine efficacy. J. Infect. Dis. 2006; 193:467–473.

Rogers, W.O., Weiss, W.R., Kumar, A., Aguiar, J.C., Tine, J.A., Gwadz, R., Harre, J.G., Gowda, K., Rathore, D., Kumar, S. and Hoffman, S.L. Protection of rhesus macaques against lethal *Plasmodium knowlesi* malaria by a heterologous DNA priming and poxvirus boosting immunization regimen. Infect. Immun. 2002; 70:4329–4335.

Rogerson, S.J., Chaiyaroj, S.C., Ng, K., Reeder, J.C. and Brown, G.V. Chondroitin sulfate A is a cell surface receptor for *Plasmodium falciparum*-infected erythrocytes. J. Exp. Med. 1995; 182:15–20.

Rogerson, S.J., Hviid, L., Duffy, P.E., Leke, R.F. and Taylor, D.W. Malaria in pregnancy: pathogenesis and immunity. Lancet Infect. Dis. 2007; 7:105–117.

Romero, J.F., Ibrahim, G.H., Renggli, J., Himmelrich, H., Graber, P. and Corradin, G. IL-12p40-independent induction of protective immunity upon multiple *Plasmodium berghei* irradiated sporozoite immunizations. Parasite Immunol. 2007; 29:541–548.

Roussilhon, C., Oeuvray, C., Muller-Graf, C., Tall, A., Rogier, C., Trape, J.F., Theisen, M., Balde, A., Perignon, J.L. and Druilhe, P. Long-term clinical protection from falciparum malaria is strongly associated with IgG3 antibodies to merozoite surface protein 3. PLoS Med. 2007; 4:e320.

Sabchareon, A., Burnouf, T., Ouattara, D., Attanath, P., Bouharoun-Tayoun, H., Chantavanich, P., Foucault, C., Chongsuphajaisiddhi, T. and Druilhe, P. Parasitologic and clinical human response to immunoglobulin administration in falciparum malaria. Am. J. Trop. Med. Hyg. 1991; 45:297–308.

Sachs, J. and Malaney, P. The economic and social burden of malaria. Nature 2002; 415:680–685.

Salazar, L.M., Alba, M.P., Torres, M.H., Pinto, M., Cortes, X., Torres, L. and Patarroyo, M.E. Protection against experimental malaria associated with AMA-1 peptide analogue structures. FEBS Lett. 2002; 527:95–100.

Saul, A. Efficacy model for mosquito stage transmission blocking vaccines for malaria. Parasitology 2008; 7:1–10.

Sauzet, J.P., Perlaza, B.L., Brahimi, K., Daubersies, P. and Druilhe, P. DNA immunization by *Plasmodium falciparum* liver-stage antigen 3 induces protection against *Plasmodium yoelii* sporozoite challenge. Infect. Immun. 2001; 69:1202–1206.

Schaecher, K., Kumar, S., Yadava, A., Vahey, M. and Ockenhouse, C.F. Genome-wide expression profiling in malaria infection reveals transcriptional changes associated with lethal and nonlethal outcomes. Infect. Immun. 2005; 73:6091–6100.

Schneider, J., Gilbert, S.C., Blanchard, T.J., Hanke, T., Robson, K.J., Hannan, C.M., Becker, M., Sinden, R., Smith, G.L. and Hill, A.V. Enhanced immunogenicity for CD8+ T cell induction and complete protective efficacy of malaria DNA vaccination by boosting with modified vaccinia virus Ankara. Nat. Med. 1998; 4:397–402.

Schofield, L. and Grau, G.E. Immunological processes in malaria pathogenesis. Nat. Rev. Immunol. 2005; 5:722–735.

Schofield, L., Hewitt, M.C., Evans, K., Siomos, M.A. and Seeberger, P.H. Synthetic GPI as a candidate anti-toxic vaccine in a model of malaria. Nature 2002; 418:785–789.

Schofield, L. and Mueller, I. Clinical immunity to malaria. Curr. Mol. Med. 2006; 6:205–221.

Schofield, L., Villaquiran, J., Ferreira, A., Schellekens, H., Nussenzweig, R. and Nussenzweig, V. Gamma interferon, CD8+ T cells and antibodies required for immunity to malaria sporozoites. Nature 1987; 330:664–666.

Sedegah, M., Charoenvit, Y., Minh, L., Belmonte, M., Majam, V.F., Abot, S., Ganeshan, H., Kumar, S., Bacon, D.J., Stowers, A., Narum, D.L., Carucci, D.J. and Rogers, W.O. Reduced immunogenicity of DNA vaccine plasmids in mixtures. Gene Ther. 2004; 11:448–456.

Sedegah, M., Hedstrom, R., Hobart, P. and Hoffman, S.L. Protection against malaria by immunization with plasmid DNA encoding circumsporozoite protein. Proc. Nat. Acad. Sci. U.S.A. 1994; 91:9866–9870.

Senn, N., D'Acremont, V., Landry, P. and Genton, B. Malaria chemoprophylaxis: what do the travelers choose, and how does pretravel consultation influence their final decision. Am. J. Trop. Med. Hyg. 2007; 77:1010–1014.

Shanks, D. and Karwacki, J.J. Malaria as a military factor in Southeast Asia. Military Med. 1991; 156:684–686.

Sharp, B.L. and le Sueur, D. Malaria in South Africa – the past, the present and selected implications for the future. S. Afr. Med. J. 1996; 86:83–89.

Sherman, I.W., Eda, S. and Winograd, E. Cytoadherence and sequestration in *Plasmodium falciparum*: defining the ties that bind. Microbes Infect./Institut Pasteur 2003; 5:897–909.

Sherwood, J.A., Copeland, R.S., Taylor, K.A., Abok, K., Oloo, A.J., Were, J.B., Strickland, G.T., Gordon, D.M., Ballou, W.R., Bales, J.D., Jr., Wirtz, R.A., Wittes, J., Gross, M., Que, J.U., Cryz, S.J., Oster, C.N., Roberts, C.R. and Sadoff, J.C. *Plasmodium falciparum* circumsporozoite vaccine immunogenicity and efficacy trial with natural challenge quantitation in an area of endemic human malaria of Kenya. Vaccine 1996; 14:817–827.

Sibley, L.D. Intracellular parasite invasion strategies. Sciences 2004; 304:248–253.

Siddiqui, W.A., Tam, L.Q., Kramer, K.J., Hui, G.S., Case, S.E., Yamaga, K.M., Chang, S.P., Chan, E.B. and Kan, S.C. Merozoite surface coat precursor protein completely protects *Aotus* monkeys against *Plasmodium falciparum* malaria. Proc. Nat. Acad. Sci. U.S.A. 1987; 84:3014–3018.

Silvie, O., Franetich, J.F., Renia, L. and Mazier, D. Malaria sporozoite: migrating for a living. Trends Mol. Med. 2004; 10:97–100, discussion 100-1.

Sinden, R.E. Mitosis and meiosis in malarial parasites. Acta Leiden. 1991; 60:19–27.

Singh, A.P., Puri, S.K. and Chitnis, C.E. Antibodies raised against receptor-binding domain of *Plasmodium knowlesi* Duffy binding protein inhibit erythrocyte invasion. Mol. Biochem. Parasitol. 2002; 121:21–31.

Singh, R.K. Hyperreactive malarial splenomegaly in expatriates. Travel Med. Infect. Dis. 2007; 5:24–29.

Singh, S., Soe, S., Mejia, J.P., Roussilhon, C., Theisen, M., Corradin, G. and Druilhe, P. Identification of a conserved region of *Plasmodium falciparum* MSP3 targeted by biologically active antibodies to improve vaccine design. J. Infect. Dis. 2004; 190:1010–1018.

Singh, S.K., Hora, R., Belrhali, H., Chitnis, C.E. and Sharma, A. Structural basis for Duffy recognition by the malaria parasite Duffy-binding-like domain. Nature 2006; 439:741–744.

Sinnis, P. and Coppi, A. A long and winding road: the *Plasmodium* sporozoite's journey in the mammalian host. Parasitol. Int. 2007; 56:171–178.

Sinnis, P. and Nardin, E. Sporozoite antigens: biology and immunology of the circumsporozoite protein and thrombospondin-related anonymous protein. Chem. Immunol. 2002; 80:70–96.

Sirima, S.B., Nebie, I., Ouedraogo, A., Tiono, A.B., Konate, A.T., Gansane, A., Derme, A.I., Diarra, A., Ouedraogo, A., Soulama, I., Cuzzin-Ouattara, N., Cousens, S. and Leroy, O. Safety and immunogenicity of the *Plasmodium falciparum* merozoite surface protein-3 long synthetic peptide (MSP3-LSP) malaria vaccine in healthy, semi-immune adult males in Burkina Faso, West Africa. Vaccine 2007; 25:2723–2732.

Snow, R.W., Guerra, C.A., Noor, A.M., Myint, H.Y. and Hay, S.I. The global distribution of clinical episodes of *Plasmodium falciparum* malaria. Nature 2005; 434:214–217.

Snow, R.W. and Marsh, K. The consequences of reducing transmission of *Plasmodium falciparum* in Africa. Adv. Parasitol. 2002; 52:235–264.

Soe, S., Theisen, M., Roussilhon, C., Aye, K.S. and Druilhe, P. Association between protection against clinical malaria and antibodies to merozoite surface antigens in an area of hyperendemicity in Myanmar: complementarity between responses to merozoite surface protein 3 and the 220-kilodalton glutamate-rich protein. Infect. Immun. 2004; 72:247–252.

Sokhna, C., Cisse, B., Ba el, H., Milligan, P., Hallett, R., Sutherland, C., Gaye, O., Boulanger, D., Simondon, K., Simondon, F., Targett, G., Lines, J., Greenwood, B. and Trape, J.F. A trial of the efficacy, safety and impact on drug resistance of four drug regimens for seasonal intermittent preventive treatment for malaria in senegalese children. PLoS ONE 2008; 3:e1471.

Song, H.H., O, S.O., Kim, S.H., Moon, S.H., Kim, J.B., Yoon, J.W., Koo, J.R., Hong, K.S., Lee, M.G., Kim, D.J., Shin, D.H., Kang, S.H., Choi, M.G. and Lee, K.H. Clinical features of *Plasmodium vivax* malaria. Korean J. Intern. Med. 2003; 18:220–224.

Sponaas, A.M., Cadman, E.T., Voisine, C., Harrison, V., Boonstra, A., O'Garra, A. and Langhorne, J. Malaria infection changes the ability of splenic dendritic cell populations to stimulate antigen-specific T cells. J. Exp. Med. 2006; 203:1427–1433.

Sridhar, S., Reyes-Sandoval, A., Draper, S.J., Moore, A.C., Gilbert, S.C., Gao, G.P., Wilson, J.M. and Hill, A.V. Single-dose protection against *Plasmodium berghei* by a simian adenovirus vector using a human cytomegalovirus promoter containing intron A. J. Virol. 2008; 82:3822–3833.

Steers, N., Schwenk, R., Bacon, D.J., Berenzon, D., Williams, J. and Krzych, U. The immune status of Kupffer cells profoundly influences their responses to infectious *Plasmodium berghei* sporozoites. Eur. J. Immunol. 2005; 35:2335–2346.

Steketee, R.W., Nahlen, B.L., Parise, M.E. and Menendez, C. The burden of malaria in pregnancy in malaria-endemic areas. Am. J. Trop. Med. Hyg. 2001; 64:28–35.

Stewart, V.A., McGrath, S.M., Dubois, P.M., Pau, M.G., Mettens, P., Shott, J., Cobb, M., Burge, J.R., Larson, D., Ware, L.A., Demoitie, M.A., Weverling, G.J., Bayat, B., Custers, J.H., Dubois, M.C., Cohen, J., Goudsmit, J. and Heppner, D.G., Jr. Priming with an adenovirus 35-circumsporozoite protein (CS) vaccine followed by RTS,S/AS01B boosting significantly improves immunogenicity to *Plasmodium falciparum* CS compared to that with either malaria vaccine alone. Infect. Immun. 2007; 75:2283–2290.

Stoeckert, C.J., Jr., Fischer, S., Kissinger, J.C., Heiges, M., Aurrecoechea, C., Gajria, B. and Roos, D.S. PlasmoDB v5: new looks, new genomes. Trends Parasitol. 2006; 22:543–546.

Stoute, J.A., Gombe, J., Withers, M.R., Siangla, J., McKinney, D., Onyango, M., Cummings, J.F., Milman, J., Tucker, K., Soisson, L., Stewart, V.A., Lyon, J.A., Angov, E., Leach, A., Cohen, J., Kester, K.E., Ockenhouse, C.F., Holland, C.A., Diggs, C.L., Wittes, J. and Heppner, D.G., Jr. Phase 1 randomized double-blind safety and immunogenicity trial of *Plasmodium falciparum* malaria merozoite surface protein FMP1 vaccine, adjuvanted with AS02A, in adults in western Kenya. Vaccine 2007; 25:176–184.

Stoute, J.A., Slaoui, M., Heppner, D.G., Momin, P., Kester, K.E., Desmons, P., Wellde, B.T., Garcon, N., Krzych, U. and Marchand, M. A preliminary evaluation of a recombinant circumsporozoite protein vaccine against *Plasmodium falciparum* malaria. RTS,S Malaria Vaccine Evaluation Group. N. Engl. J. Med. 1997; 336:86–91.

Stowers, A.W., Kennedy, M.C., Keegan, B.P., Saul, A., Long, C.A. and Miller, L.H. Vaccination of monkeys with recombinant *Plasmodium falciparum* apical membrane antigen 1 confers protection against blood-stage malaria. Infect. Immun. 2002; 70:6961–6967.

Stowers, A.W. and Miller, L.H. Are trials in New World monkeys on the critical path for blood-stage malaria vaccine development? Trends Parasitol. 2001; 17:415–419.

Stubbs, J., Simpson, K.M., Triglia, T., Plouffe, D., Tonkin, C.J., Duraisingh, M.T., Maier, A.G., Winzeler, E.A. and Cowman, A.F. Molecular mechanism for switching of *P. falciparum* invasion pathways into human erythrocytes. Sciences 2005; 309:1384–1387.

Sturm, A., Amino, R., van de Sand, C., Regen, T., Retzlaff, S., Rennenberg, A., Krueger, A., Pollok, J.M., Menard, R. and Heussler, V.T. Manipulation of host hepatocytes by the malaria parasite for delivery into liver sinusoids. Sciences 2006; 313:1287–1290.

Sundaresh, S., Doolan, D.L., Hirst, S., Mu, Y., Unal, B., Davies, D.H., Felgner, P.L. and Baldi, P. Identification of humoral immune responses in protein microarrays using DNA microarray data analysis techniques. Bioinformatics 2006; 22:1760–1766.

Suwanabun, N., Sattabongkot, J., Tsuboi, T., Torii, M., Maneechai, N., Rachapaew, N., Yim-amnuaychok, N., Punkitchar, V. and Coleman, R.E. Development of a method for the in vitro production of *Plasmodium vivax* ookinetes. J. Parasitol. 2001; 87:928–930.

Takala, S.L., Coulibaly, D., Thera, M.A., Dicko, A., Smith, D.L., Guindo, A.B., Kone, A.K., Traore, K., Ouattara, A., Djimde, A.A., Sehdev, P.S., Lyke, K.E., Diallo, D.A., Doumbo, O.K. and Plowe, C.V. Dynamics of polymorphism in a malaria vaccine antigen at a vaccine-testing site in Mali. PLoS Med. 2007; 4:e93.

Tao, D., Barba-Spaeth, G., Rai, U., Nussenzweig, V., Rice, C.M. and Nussenzweig, R.S. Yellow fever 17D as a vaccine vector for microbial CTL epitopes: protection in a rodent malaria model. J. Exp. Med. 2005; 201:201–209.

Tarun, A.S., Dumpit, R.F., Camargo, N., Labaied, M., Liu, P., Takagi, A., Wang, R. and Kappe, S.H. Protracted sterile protection with Plasmodium yoelii pre-erythrocytic genetically attenuated parasite malaria vaccines is independent of significant liver-stage persistence and is mediated by CD8+ T cells. J. Infect. Dis. 2007; 196:608–616.

Tarun, A.S., Peng, X., Dumpit, R.F., Ogata, Y., Silva-Rivera, H., Camargo, N., Daly, T.M., Bergman, L.W. and Kappe, S.H. A combined transcriptome and proteome survey of malaria parasite liver stages. Proc. Natl. Acad. Sci. U.S.A. 2008; 105:305–310.

Thera, M.A., Doumbo, O.K., Coulibaly, D., Diallo, D.A., Kone, A.K., Guindo, A.B., Traore, K., Dicko, A., Sagara, I., Sissoko, M.S., Baby, M., Sissoko, M., Diarra, I., Niangaly, A., Dolo, A., Daou, M., Diawara, S.I., Heppner, D.G., Stewart, V.A., Angov, E., Bergmann-Leitner, E.S., Lanar, D.E., Dutta, S., Soisson, L., Diggs, C.L., Leach, A., Owusu, A., Dubois, M.C., Cohen, J., Nixon, J.N., Gregson, A., Takala, S.L., Lyke, K.E. and Plowe, C.V. Safety and immunogenicity of an AMA-1 malaria vaccine in Malian adults: results of a phase 1 randomized controlled trial. PLoS ONE 2008; 3:e1465.

Thera, M.A., Doumbo, O.K., Coulibaly, D., Diallo, D.A., Sagara, I., Dicko, A., Diemert, D.J., Heppner, D.G., Jr., Stewart, V.A., Angov, E., Soisson, L., Leach, A., Tucker, K., Lyke, K.E. and Plowe, C.V. Safety and allele-specific immunogenicity of a malaria vaccine in Malian adults: results of a phase I randomized trial. PLoS Clin. Trials 2006; 1:e34.

Thompson, F.M., Porter, D.W., Okitsu, S.L., Westerfeld, N., Vogel, D., Todryk, S., Poulton, I., Correa, S., Hutchings, C., Berthoud, T., Dunachie, S., Andrews, L., Williams, J.L., Sinden, R., Gilbert, S.C., Pluschke, G., Zurbriggen, R. and Hill, A.V. Evidence of blood stage efficacy with a virosomal malaria vaccine in a phase iia clinical trial. PLoS ONE 2008a; 3:e1493.

Thompson, J., Millington, O.R., Garside, P. and Brewer, J.M. What can transgenic parasites tell us about the development of *Plasmodium*-specific immune responses? Parasite Immunol. 2008b; 30:223–233.

Timmann, C., Evans, J.A., Konig, I.R., Kleensang, A., Ruschendorf, F., Lenzen, J., Sievertsen, J., Becker, C., Enuameh, Y., Kwakye, K.O., Opoku, E., Browne, E.N., Ziegler, A., Nurnberg, P. and Horstmann, R.D. Genome-wide linkage analysis of malaria infection intensity and mild disease. PLoS Gen. 2007; 3:e48.

Tine, J.A., Lanar, D.E., Smith, D.M., Wellde, B.T., Schultheiss, P., Ware, L.A., Kauffman, E.B., Wirtz, R.A., De Taisne, C., Hui, G.S., Chang, S.P., Church, P., Hollingdale, M.R., Kaslow, D.C., Hoffman, S., Guito, K.P., Ballou, W.R., Sadoff, J.C. and Paoletti, E. NYVAC-Pf7: a poxvirus-vectored, multiantigen, multistage vaccine candidate for *Plasmodium falciparum* malaria. Infect. Immun. 1996; 64:3833–3844.

Trager, W. and Jensen, J.B. Human malaria parasites in continuous culture. Sciences 1976; 193:673–675.

Trager, W. and Most, H. A long-delayed primary attack of ovale malaria. Am. J. Trop. Med. Hyg. 1963; 12:837–839.

Trampuz, A., Jereb, M., Muzlovic, I. and Prabhu, R.M. Clinical review: severe malaria. Crit. Care 2003; 7:315–323.

Udomsangpetch, R., Somsri, S., Panichakul, T., Chotivanich, K., Sirichaisinthop, J., Yang, Z., Cui, L. and Sattabongkot, J. Short-term in vitro culture of field isolates of *Plasmodium vivax* using umbilical cord blood. Parasitol. Int. 2007; 56:65–69.

Uneke, C.J. Impact of placental *Plasmodium falciparum* malaria on pregnancy and perinatal outcome in sub-Saharan Africa: I: introduction to placental malaria. Yale J. Biol. Med. 2007; 80: 39–50.

Usynin, I., Klotz, C. and Frevert, U. Malaria circumsporozoite protein inhibits the respiratory burst in Kupffer cells. Cell. Microbiol. 2007; 9:2610–2628.

Uthaipibull, C., Aufiero, B., Syed, S.E., Hansen, B., Guevara Patino, J.A., Angov, E., Ling, I.T., Fegeding, K., Morgan, W.D., Ockenhouse, C., Birdsall, B., Feeney, J., Lyon, J.A. and Holder, A.A. Inhibitory and blocking monoclonal antibody epitopes on merozoite surface protein 1 of the malaria parasite *Plasmodium falciparum*. J. Mol. Biol. 2001; 307:1381–1394.

VanBuskirk, K.M., Sevova, E. and Adams, J.H. Conserved residues in the *Plasmodium vivax* Duffy-binding protein ligand domain are critical for erythrocyte receptor recognition. Proc. Natl. Acad. Sci. U.S.A. 2004; 101:15754–15759.

Vanderberg, J., Mueller, A.K., Heiss, K., Goetz, K., Matuschewski, K., Deckert, M. and Schluter, D. Assessment of antibody protection against malaria sporozoites must be done by mosquito injection of sporozoites. Am. J. Pathol. 2007; 171:1405–1406, author reply 1406.

van Dijk, M.R., Douradinha, B., Franke-Fayard, B., Heussler, V., van Dooren, M.W., van Schaijk, B., van Gemert, G.J., Sauerwein, R.W., Mota, M.M., Waters, A.P. and Janse, C.J. Genetically attenuated, P36p-deficient malarial sporozoites induce protective immunity and apoptosis of infected liver cells. Proc. Natl. Acad. Sci. U.S.A. 2005; 102:12194–12199.

Vanderberg, J., Nussenzweig, R. and Most, H. Protective immunity produced by the injection of x-irradiated sporozoites of *Plasmodium berghei*, V. In vitro effects of immune serum on sporozoites. Military Med. 1969; 134:1183–1190.

Vanderberg, J.P. and Frevert, U. Intravital microscopy demonstrating antibody-mediated immobilisation of *Plasmodium berghei* sporozoites injected into skin by mosquitoes. Int. J. Parasitol. 2004; 34:991–996.

van Hensbroek, M.B., Palmer, A., Onyiorah, E., Schneider, G., Jaffar, S., Dolan, G., Memming, H., Frenkel, J., Enwere, G., Bennett, S., Kwiatkowski, D. and Greenwood, B. The effect of a monoclonal antibody to tumor necrosis factor on survival from childhood cerebral malaria. J. Infect. Dis. 1996; 174:1091–1097.

Vedadi, M., Lew, J., Artz, J., Amani, M., Zhao, Y., Dong, A., Wasney, G.A., Gao, M., Hills, T., Brokx, S., Qiu, W., Sharma, S., Diassiti, A., Alam, Z., Melone, M., Mulichak, A., Wernimont, A., Bray, J., Loppnau, P., Plotnikova, O., Newberry, K., Sundararajan, E., Houston, S., Walker, J., Tempel, W., Bochkarev, A., Kozieradzki, I., Edwards, A., Arrowsmith, C., Roos, D., Kain, K. and Hui, R. Genome-scale protein expression and structural biology of *Plasmodium falciparum* and related Apicomplexan organisms. Mol. Biochem. Parasitol. 2007; 151:100–110.

Verhage, D.F., Telgt, D.S., Bousema, J.T., Hermsen, C.C., van Gemert, G.J., van der Meer, J.W. and Sauerwein, R.W. Clinical outcome of experimental human malaria induced by *Plasmodium falciparum*-infected mosquitoes. Neth. J. Med. 2005; 63:52–58.

Vestergaard, L.S. and Ringwald, P. Responding to the challenge of antimalarial drug resistance by routine monitoring to update national malaria treatment policies. Am. J. Trop. Med. Hyg. 2007; 77:153–159.

Villard, V., Agak, G.W., Frank, G., Jafarshad, A., Servis, C., Nebie, I., Sirima, S.B., Felger, I., Arevalo-Herrera, M., Herrera, S., Heitz, F., Backer, V., Druilhe, P., Kajava, A.V. and Corradin, G. Rapid identification of malaria vaccine candidates based on alpha-helical coiled coil protein motif. PLoS ONE 2007; 2:e645.

Vythilingam, I., Tan, C.H., Asmad, M., Chan, S.T., Lee, K.S. and Singh, B. Natural transmission of *Plasmodium knowlesi* to humans by *Anopheles* latens in Sarawak, Malaysia. Trans. R. Soc. Trop. Med. Hyg. 2006; 100:1087–1088.

Walters, M. and Sperandio, V. Autoinducer 3 and epinephrine signaling in the kinetics of locus of enterocyte effacement gene expression in enterohemorrhagic *Escherichia coli*. Infect. Immun. 2006; 74:5445–5455.

Walther, M., Thompson, F.M., Dunachie, S., Keating, S., Todryk, S., Berthoud, T., Andrews, L., Andersen, R.F., Moore, A., Gilbert, S. C., Poulton, I., Dubovsky, F., Tierney, E., Correa, S., Huntcooke, A., Butcher, G., Williams, J., Sinden, R.E. and Hill, A.V. Safety, immunogenicity, and efficacy of prime-boost immunization with recombinant poxvirus FP9 and modified vaccinia virus Ankara encoding the full-length Plasmodium falciparum circumsporozoite protein. Infect. Immun. 2006; 74:2706–2716.

Wang, R., Charoenvit, Y., Corradin, G., De La Vega, P., Franke, E.D. and Hoffman, S.L. Protection against malaria by *Plasmodium yoelii* sporozoite surface protein 2 linear peptide induction of CD4+ T cell- and IFN-gamma-dependent elimination of infected hepatocytes. J. Immunol. 1996; 157:4061–4067.

Wang, R., Charoenvit, Y., Corradin, G., Porrozzi, R., Hunter, R.L., Glenn, G., Alving, C.R., Church, P. and Hoffman, S.L. Induction of protective polyclonal antibodies by immunization with a *Plasmodium yoelii* circumsporozoite protein multiple antigen peptide vaccine. J. Immunol. 1995; 154:2784–2793.

Wang, R., Doolan, D.L., Le, T.P., Hedstrom, R.C., Coonan, K.M., Charoenvit, Y., Jones, T.R., Hobart, P., Margalith, M., Ng, J., Weiss, W.R., Sedegah, M., de Taisne, C., Norman, J.A. and Hoffman, S.L. Induction of antigen-specific cytotoxic T lymphocytes in humans by a malaria DNA vaccine. Science 1998; 282:476–480.

Wang, R., Epstein, J., Baraceros, F.M., Gorak, E.J., Charoenvit, Y., Carucci, D.J., Hedstrom, R.C., Rahardjo, N., Gay, T., Hobart, P., Stout, R., Jones, T.R., Richie, T.L., Parker, S.E., Doolan, D.L., Norman, J. and Hoffman, S.L. Induction of CD4(+) T cell-dependent CD8(+) type 1 responses in humans by a malaria DNA vaccine. Proc. Natl. Acad. Sci. U.S.A. 2001; 98:10817–10822.

Wang, R., Epstein, J., Charoenvit, Y., Baraceros, F.M., Rahardjo, N., Gay, T., Banania, J.G., Chattopadhyay, R., de la Vega, P., Richie, T.L., Tornieporth, N., Doolan, D.L., Kester, K.E., Heppner, D.G., Norman, J., Carucci, D.J., Cohen, J.D. and Hoffman, S.L. Induction in humans of CD8+ and CD4+ T cell and antibody responses by sequential immunization with malaria DNA and recombinant protein. J. Immunol. 2004; 172:5561–5569.

Wang, R., Richie, T.L., Baraceros, M.F., Rahardjo, N., Gay, T., Banania, J.G., Charoenvit, Y., Epstein, J.E., Luke, T., Freilich, D.A., Norman, J. and Hoffman, S.L. Boosting of DNA vaccine-elicited gamma interferon responses in humans by exposure to malaria parasites. Infect. Immun. 2005; 73:2863–2872.

Warrell, D.A., White, N.J. and Warrell, M.J. Dexamethasone deleterious in cerebral malaria. Br. Med. J. (Clin. Res. Ed.). 1982; 285:1652.

Watanabe, J., Suzuki, Y., Sasaki, M. and Sugano, S. Full-malaria 2004: an enlarged database for comparative studies of full-length cDNAs of malaria parasites, *Plasmodium* species. Nucleic Acids Res. 2004; 32:D334–D338.

Watanabe, J., Wakaguri, H., Sasaki, M., Suzuki, Y. and Sugano, S. Comparasite: a database for comparative study of transcriptomes of parasites defined by full-length cDNAs. Nucleic Acids Res. 2007; 35:D431–D438.

Weatherall, D.J., Miller, L.H., Baruch, D.I., Marsh, K., Doumbo, O.K., Casals-Pascual, C. and Roberts, D.J. Malaria and the Red Cell. Hematology, Education Book of the American Society of Hematology. American Society of Hematology, 2002, pp. 35–57.

Webster, D.P., Dunachie, S., Vuola, J.M., Berthoud, T., Keating, S., Laidlaw, S.M., McConkey, S.J., Poulton, I., Andrews, L., Andersen, R.F., Bejon, P., Butcher, G., Sinden, R., Skinner, M.A., Gilbert, S.C. and Hill, A.V. Enhanced T cell-mediated protection against malaria in human challenges by using the recombinant poxviruses FP9 and modified vaccinia virus Ankara. Proc. Natl. Acad. Sci. U.S.A. 2005; 102:4836–4841.

Weidanz, W.P., Melancon-Kaplan, J. and Cavacini, L.A. Cell-mediated immunity to the asexual blood stages of malarial parasites: animal models. Immunol. Lett. 1990; 25:87–95.

Weiss, W.R., Ishii, K.J., Hedstrom, R.C., Sedegah, M., Ichino, M., Barnhart, K., Klinman, D.M. and Hoffman, S.L. A plasmid encoding murine granulocyte-macrophage colony-stimulating factor increases protection conferred by a malaria DNA vaccine. J. Immunol. 1998; 161:2325–2332.

Whitty, C.J., Lalloo, D. and Ustianowski, A. Malaria: an update on treatment of adults in non-endemic countries. BMJ 2006; 333:241–245.

Wide, A., Capaldo, J., Zerpa, N., Pabon, R., Noda, A., Noya, B., Gonzalez, J. and Noya, O. Sharing of antigens between *Plasmodium falciparum* and *Anopheles albimanus*. Rev. Inst. Med. Trop. Sao Paulo 2006; 48:327–332.

Winkler, A.S., Salmhofer, G., Helbok, R., Kalokola, F., Matuja, W. and Schmutzhard, E. In-hospital risk estimation in children with malaria--early predictors of morbidity and mortality. J. Trop. Pediatr., 2008; 54:184–191.

Winzeler, E.A. Applied systems biology and malaria. Nat. Rev. 2006; 4:145–151.

Wisse, E., De Zanger, R.B., Charels, K., Van Der Smissen, P. and McCuskey, R.S. The liver sieve: considerations concerning the structure and function of endothelial fenestrae, the sinusoidal wall and the space of Disse. Hepatology 1985; 5:683–692.

Withers, M.R., McKinney, D., Ogutu, B.R., Waitumbi, J.N., Milman, J.B., Apollo, O.J., Allen, O.G., Tucker, K., Soisson, L.A., Diggs, C., Leach, A., Wittes, J., Dubovsky, F., Stewart, V.A., Remich, S.A., Cohen, J., Ballou, W.R., Holland, C.A., Lyon, J.A., Angov, E., Stoute, J.A., Martin, S.K. and Heppner, D.G., Jr. Safety and reactogenicity of an MSP-1 malaria vaccine candidate: a randomized phase Ib dose-escalation trial in Kenyan children. PLoS Clin. Trials 2006; 1:e32.

Wizel, B., Houghten, R., Church, P., Tine, J.A., Lanar, D.E., Gordon, D.M., Ballou, W.R., Sette, A. and Hoffman, S.L. HLA-A2-restricted cytotoxic T lymphocyte responses to multiple *Plasmodium falciparum* sporozoite surface protein 2 epitopes in sporozoite-immunized volunteers. J. Immunol. 1995a; 155: 766–775.

Wizel, B., Houghten, R.A., Parker, K.C., Coligan, J.E., Church, P., Gordon, D.M., Ballou, W.R. and Hoffman, S.L. Irradiated sporozoite vaccine induces HLA-B8-restricted cytotoxic T lymphocyte responses against two overlapping epitopes of the *Plasmodium falciparum* sporozoite surface protein 2. J. Exp. Med. 1995b; 182:1435–1445.

Wootton, J.C., Feng, X., Ferdig, M.T., Cooper, R.A., Mu, J., Baruch, D.I., Magill, A.J. and Su, X.Z. Genetic diversity and chloroquine selective sweeps in *Plasmodium falciparum*. Nature 2002; 418:320–323.

World Health Organization. World Malaria Report, 2005.

World Health Organization (2004)Note:World Health Organization Rainbow Table: http://www.who.int/vaccine_research/documents/en/Status_Table.pdf. The malaria vaccine section of this table, which lists malaria vaccines in clinical development, updates a more comprehensive table published in October, 2004 that additionally includes malaria vaccines in research and preclinical development, entitled "Portfolio of candidate malaria vaccines currently in development" (the rainbow table), http://www.who.int/vaccine_research/documents/malaria_table.pdf

Worrall, E., Morel, C., Yeung, S., Borghi, J., Webster, J., Hill, J., Wiseman, V. and Mills, A. The economics of malaria in pregnancy—a review of the evidence and research priorities. Lancet Infect. Dis. 2007; 7:156–168.

Yadava, A., Sattabongkot, J., Washington, M.A., Ware, L.A., Majam, V., Zheng, H., Kumar, S. and Ockenhouse, C.F. A novel chimeric *Plasmodium vivax* circumsporozoite protein induces biologically functional antibodies that recognize both VK210 and VK247 sporozoites. Infect. Immun. 2007; 75:1177–1185.

Yagmur, Y., Kara, I.H., Aldemir, M., Buyukbayram, H., Tacyildiz, I.H. and Keles, C. Spontaneous rupture of malarial spleen: two case reports and review of literature. Crit. Care 2000; 4:309–313.

Yoshida, N., Nussenzweig, R.S., Potocnjak, P., Nussenzweig, V. and Aikawa, M. Hybridoma produces protective antibodies directed against the sporozoite stage of malaria parasite. Sciences 1980; 207:71–73.

Zevering, Y., Khamboonruang, C. and Good, M.F. Human and murine T-cell responses to allelic forms of a malaria circumsporozoite protein epitope support a polyvalent vaccine strategy. Immunology 1998; 94:445–454.

Zheng, C., Xie, P. and Chen, Y. Recombinant *Mycobacterium bovis* BCG producing the circumsporozoite protein of *Plasmodium falciparum* FCC-1/HN strain induces strong immune responses in BALB/c mice. Parasitol. Int. 2002; 51:1–7.

Zhou, Y., Ramachandran, V., Kumar, K.A., Westenberger, S., Refour, P., Zhou, B., Li, F., Young, J.A., Chen, K., Plouffe, D., Henson, K., Nussenzweig, V., Carlton, J., Vinetz, J.M., Duraisingh, M.T. and Winzeler, E.A. Evidence-based annotation of the malaria parasite's genome using comparative expression profiling. PLoS ONE 2008; 3:e1570.

Zhu, J. and Hollingdale, M.R. Structure of *Plasmodium falciparum* liver stage antigen-1. Mol. Biochem. Parasitol. 1991; 48: 223–226.

Zingman, B.S. and Viner, B.L. Splenic complications in malaria: case report and review. Clin. Infect. Dis. 1993; 16:223–232.

CHAPTER

66

Hookworm Infection

Peter J. Hotez[1,2] *and David Diemert*[1,2]

[1]Department of Microbiology, Immunology and Tropical Medicine, The George Washington University, Washington, DC

[2]Albert B. Sabin Vaccine Institute, Washington, DC

OUTLINE

ABSTRACT

Hookworm infection is a chronic intestinal nematode infection that affects approximately 576 million people worldwide, predominantly those living in poor rural areas of the tropics. The two hookworms that infect humans—*Necator americanus* and *Ancylostoma duodenale*—are transmitted via skin contact with fecally contaminated soil. Hookworm infection causes significant morbidity due to chronic anemia and protein malnutrition. Current control efforts based on periodic mass anthelmintic administration are inadequate and unsustainable. Development of vaccines against hookworm infection is therefore a global health priority. Vaccine development has so far concentrated on the more prevalent *N. americanus*. Antigens expressed during the larval stage of this worm's life cycle, such as *Na*-ASP-2 and *Na*-SAA-2, have been identified and are being developed as candidate vaccine antigens. The first candidate vaccine—*Na*-ASP-2 formulated with Alhydrogel®—is currently in phase 1 clinical testing in an endemic area of Brazil. Antigens such as *Na*-GST-1 that are expressed during the adult stage of the parasite's life cycle and are essential to the worm's use of host hemoglobin as an energy source are also being developed as vaccine components. An eventual antihookworm vaccine will combine at least two hookworm antigens, one targeting the larval stage and another targeting the adult worm living in the gastrointestinal tract. Such a vaccine will be incorporated into existing deworming programs that target preschool and school-aged children living in areas of high transmission, to reduce the morbidity—primarily anemia and protein malnutrition—attributed to this parasite. Antihookworm vaccine development is being conducted in partnership with vaccine manufacturers in middle-income endemic countries such as Brazil, in order to ensure that the populations who need the vaccine the most will have timely access to it once efficacy is shown in phase 3 testing.

INTRODUCTION: HOOKWORM DISEASE BURDEN

Hookworm infection, together with ascariasis and trichuriasis, comprise a group of human parasitic diseases known as the soil-transmitted helminth infections (STHIs). The STHIs are among the most common infections of humans, with an estimated 807 million, 604 million, and 576 million people infected with either the roundworm, *Ascaris lumbricoides*, the whipworm, *Trichuris trichiura*, or the hookworms *Necator americanus* and *Ancylostoma duodenale*, respectively (Bethony et al., 2006). STHIs occur almost exclusively in areas of poverty where they inflict great human suffering by causing chronic infections in children that result in retardation of physical, intellectual, and cognitive development (WHO, 2005). These health effects result in the loss of up to 39 million disability-adjusted life years (DALYs) annually (Chan, 1997), a disease burden roughly equivalent to that of malaria. In addition, the STHIs exert a powerful negative influence on economic development and disrupt the education of children (Miguel and Kremer,

FIGURE 66.1 Goegraphic distribution of hookworm infection in the world.

2004; Bleakley, 2007). The STHIs belong to a unique category of *neglected tropical diseases* that have the following features: (1) they are ancient afflictions that have plagued humankind for centuries; (2) they occur predominantly in rural areas of low-income countries; (3) they cause long-term suffering and chronic disability; (4) not only do they occur in settings of poverty but, due to their chronic negative health and educational effects, the STHIs actually perpetuate poverty; (5) despite their considerable health, educational, and economic impact, they are frequently overlooked by the global medical and public health communities; and (6) because the STHIs afflict primarily the most impoverished, pharmaceutical companies have no significant commercial incentives to develop new drugs, vaccines, or diagnostics that target these infections (Hotez et al., 2006d; Hotez et al., 2007).

Among the STHIs, human hookworm infection is the most important in terms of disease burden and cause of death. The epidemiology and impact of hookworm disease has been reviewed (Brooker et al., 2004). Briefly, it has been estimated that up to 65,000 deaths and 22 million DALYs are lost annually worldwide due to hookworm (Chan, 1997; WHO, 2002), although other investigators have derived lower values (WHO, 2004), depending on the value of the disability weights assigned to the severity of anemia, malnutrition, and cognitive impairment that result from hookworm infection (Bethony et al., 2006; Hotez et al., 2006b). Currently, the greatest numbers of cases occur in sub-Saharan Africa (Bethony et al., 2006), where hookworm is frequently co-endemic with malaria (Brooker et al., 2006; Hotez et al., 2006c), and in Asia (Bethony et al., 2006). However, significant pockets of endemic hookworm transmission also occur in Central America, Brazil, and other tropical regions of the Americas (de Silva et al., 2003; Hotez, 2004; Fleming et al., 2006). As recently as the late 1980s and early 1990s, some of the highest prevalence and intensity of hookworm infection occurred in the People's Republic of China (Hotez et al., 1997; Brooker et al., 2004). However, as a result of aggressive economic reforms and the resultant reduction in poverty and improved water and sanitation, significant progress has been made in the control of this disease in China (Hotez, 2002, 2007). Of the two major hookworms, *N. americanus* and *A. duodenale*, the former accounts for more than 70% of infections worldwide (Hotez et al., 2003a) and is especially prevalent in Africa, Southeast Asia, and the Americas (Chan et al., 1997), making it the major target of interventions; all current hookworm vaccine development is focused on *N. americanus*. Other members of the genus *Ancylostoma,* including *Ancylostoma caninum, Ancylostoma ceylanicum,* and *Ancylostoma braziliense,* primarily infect animals but may also cause human disease, although none have a significant public health impact (Brooker et al., 2004).

Hookworms cause injury through intestinal blood loss that results from attachment of adult parasites to the mucosa and submucosa of the small intestine with resultant rupture of capillaries in the mucosa and arterioles in the submucosa (reviewed by Hotez et al., 2004a, 2004b). Adult hookworms inject anticoagulants, antiplatelet aggregating agents, and connective tissue hydrolases into the attachment site (Cappello et al., 1995; Ledizet et al., 2005), which results in a free flow of host blood into the intestinal lumen, some of which is ingested by the adult hookworms, which then lyse the red blood cells (Don et al., 2004) and degrade the ingested hemoglobin by an orchestrated cascade of hemoglobinases, to be used by the worms as an energy source (Williamson et al., 2004).

Hookworm-related anemia results when blood loss is sufficient to deplete the iron reserves of the host. This depends on a number of factors including the number of adult hookworms present (Lwambo et al., 1992), the species of infecting hookworm—each adult *A. duodenale* hookworm causes greater blood loss than *N. americanus* (Albonico et al., 1998)—and the preexisting iron reserves of the host. Children and women of reproductive age, especially pregnant women, typically have lower iron reserves and are therefore particularly susceptible to developing this condition (Bundy et al., 1995; Stoltzfus et al., 1997b; Brooker et al., 2004). The hallmarks of hookworm disease, therefore, are iron deficiency and iron deficiency anemia (Hotez et al., 2004a). In some regions of sub-Saharan Africa, hookworm disease in children and pregnant women is almost as important as malaria as a cause of anemia (Brooker et al., 1999), and is a major cause of anemia in Brazil (Brooker et al., 2007). In addition, heavy hookworm infection can lead to the loss of significant amounts of plasma protein, resulting in hypoalbuminemia, hypoproteinemia, and protein malnutrition. The basis by which chronic hookworm infection and disease results in growth and cognitive delays in children is not fully understood, but is conjectured to be related to the nutritional consequences of the chronic blood loss (Brooker et al., 2004). Hookworm-associated malnutrition may also be a contributing factor in the observed adverse impact on childhood education, including school performance and attendance (Miguel and Kremer, 2004), and economic development (Bleakley, 2007) that is associated with this disease.

IMPACT OF ANTHELMINTIC CHEMOTHERAPY ("DEWORMING")

New awareness of the important health, educational, and economic consequences of chronic hookworm infection and other STHIs have led the UN agencies to develop disease-control programs in endemic countries. In children, hookworm infection frequently occurs concomitantly with ascariasis, trichuriasis, and schistosomiasis, and there is evidence that hookworm may promote susceptibility to these other helminth infections (Fleming et al., 2006). Because school-aged children harbor the highest helminth burdens of any single group (Bundy, 1988; World Bank, 2003), the UN agencies have developed a strategy for targeting this population by administering a single dose of a benzimidazole agent (i.e., mebendazole or albendazole) and a single dose of praziquantel annually in order to "deworm" the children of their STH and schistosome infections, respectively (Savioli et al., 2002). The health and educational benefits of frequent and periodic deworming have recently been reviewed (World Bank, 2003; WHO, 2005; Hotez et al., 2004b, 2006b) and include improvements in physical growth, cognition, and host iron and hemoglobin status. In addition, deworming has the potential to interrupt transmission of *A. lumbricoides* and *T. trichiura* (de Silva et al., 2003). Accordingly, in 2001, the 54th World Health Assembly adopted resolution 54.19 urging member states to deworm at least 75% of all at-risk school-aged children annually by the year 2010 (*www.who.int/wormcontrol*).

Whether resolution 54.19 will be successful in controlling hookworm-related morbidity is not as obvious as it is for some of the other helminthiases (Hotez et al., 2004a, 2005, 2006a). There are multiple reasons for this concern:

1. As opposed to the other STHs, high-intensity hookworm infections occur in both adult and pediatric populations, so that school-based deworming programs are not expected to significantly reduce hookworm transmission (Chan, 1997) nor have an impact on other vulnerable populations such as pregnant women (Christian et al., 2004).
2. High rates of posttreatment hookworm reinfection occur in areas of intense transmission. In some cases, infection returns to pretreatment levels within 4–12 months (Albonico et al., 1995).
3. Single-dose mebendazole has a high failure rate in reducing hookworm burdens (Bennett and Guyatt, 2000; de Clercq et al., 1997).
4. The efficacy of mebendazole has been shown to diminish with frequent use (Albonico et al., 2003), and there are concerns that anthelmintic drug resistance could emerge because of large-scale and widespread use (Albonico et al., 2004). Resistance to benzimidazoles has been well documented for the nematode parasites of sheep and cattle (Albonico et al., 2004).

These concerns have brought into question the sustainability of hookworm control programs that rely on regular administration of currently available anthelmintic drugs (WHO, 2005), and have stimulated the search for alternative or complementary control tools such as new anthelmintic drugs and vaccines.

THE CASE FOR A VACCINE AGAINST HOOKWORM

The high rates of hookworm reinfection that occur following anthelmintic treatment with benzimidazoles suggest that previous infection does not confer protective immunity. The absence of naturally acquired immunity is also supported by epidemiological studies in endemic regions of China and Brazil, which show increasing community prevalence and intensity of infection with age (Liu et al., 1999; Gandhi et al., 2001; Bethony et al., 2002; Fleming et al., 2006).

The lack of naturally acquired protective immunity suggests that successful vaccine development could prove to be more challenging than it was for existing vaccines that employ live-attenuated viruses such as those for measles and polio. In this sense, hookworm vaccine development could be considered as complex as the development of vaccines for malaria, tuberculosis, or even HIV-AIDS. However, during the first-half of the twentieth century, William Cort's group at the John Hopkins School of Hygiene and Public Health successfully developed vaccines containing live hookworm larvae that protected laboratory animals against challenge infections with the canine hookworm *A. caninum* (reviewed by Hotez et al., 1996). This provided "proof-of-principle" that development of a hookworm vaccine was feasible. Both humans and dogs become infected with hookworm when third-stage infective larvae (L3) penetrate the skin or are ingested orally. Subsequently, the larvae enter the peripheral circulation and migrate to the lungs where they break into the alveolar space, ascend the bronchial tree, pass over the epiglottis, and then enter the gastrointestinal tract where they develop into adult hookworms. Studies by McCoy (1931) and Foster (1935) showed that dogs could be rendered resistant to larval challenge by first administering multiple doses of small numbers of live L3. Dogs "immunized" with live L3 tolerated challenges of several thousand larvae without developing anemia, even though the dogs did not develop complete or sterilizing immunity. Instead, vaccine-induced protection was evidenced by significant reductions in host hookworm burden (as measured by fecal egg counts), to the point where vaccinated dogs were protected against significant blood loss and canine hookworm disease.

Ultimately, the canine challenge model was standardized through the use of subcutaneously injected L3 (Otto and Kerr, 1939; Otto, 1941); using this model, vaccination with irradiation-attenuated L3 (irL3) reliably resulted in protection as measured by significant reductions in hookworm burden (Miller, 1971). These studies prompted the development of a commercial canine hookworm vaccine consisting of irradiation-attenuated live *A. caninum* L3, which was licensed and marketed initially in Florida and then in the Eastern United States during the 1970s (Miller, 1971, 1978). Although the irL3 vaccine was effective in preventing canine hookworm disease, it was ultimately removed from the market because of high production costs, the need for the L3 to remain alive during storage, and difficulties in explaining to pet owners and veterinarians how the vaccine could be effective without eliciting sterilizing immunity (Miller, 1978).

The immunological basis of the protection derived from live L3 and irL3 vaccines was subjected to intense study during the decades between 1930 and 1970 (reviewed by Hotez et al., 1987, 1996). These studies pointed to the importance of antibodies directed against antigens secreted by invading L3 (Cort and Otto, 1940). A partial breakthrough in the understanding of the humoral immunity elicited by hookworm vaccines was the discovery and description of the "Sarles phenomenon" by Merrit Sarles and William Taliaferro at the University of Chicago who demonstrated that serum recovered from rodents immunized with *Nippostrongylus brasiliensis* L3 formed immune precipitations around the oral openings of L3 when incubated with them in vitro (Sarles and Taliaferro, 1936; Sarles, 1938). This phenomenon, together with studies demonstrating protection of nonvaccinated dogs by the passive transfer of antibodies obtained from dogs immunized with either *A. caninum* L3 or irL3 vaccines (Otto, 1940; Miller, 1967), highlighted the important role of antibodies directed against L3 antigens. A recent reexamination of the immunity induced by the *A. caninum* irL3 vaccine confirmed that protection is associated with Th2-type responses against specific L3 secreted antigens and that anti-L3 antibody inhibits larval tissue penetration and migration in vitro (Fujiwara et al., 2006).

LARVAL ANTIGEN DISCOVERY AND SELECTION

The mid-20th century studies highlighting the importance of immune responses to L3 antigens in mediating vaccine protection identified them as potential human hookworm vaccine antigens (Hotez et al., 1996, 1999, 2003). With respect to secreted antigens, *A. caninum* L3 have been shown to release enzymes with protease and hyaluronidase activities, which likely facilitate larval migration through tissues (Hotez et al., 1990, 1992). However, the isolation and biochemical characterization of the major L3 secreted antigens were hampered by the minute quantities of protein produced in L3 secretions. This limitation was overcome by Hawdon and Hotez (1996), who were able to obtain abundant amounts of secreted antigens by incubating *A. caninum* L3 in vitro with a cocktail of glutathione (or its derivatives) and an ultrafiltrate of canine serum. It was presumed that these in vitro conditions mimicked the milieu encountered by larvae during invasion of human tissues.

Three of the major L3 macromolecules released after in vitro stimulation have been cloned and sequenced. Two of these proteins, known as *Ancylostoma* secreted proteins-1 (ASP-1) and -2 (ASP-2), are members of the pathogenesis-related protein (PRP) superfamily (Hotez et al., 1999). The PRPs are produced in response to infection with pathogens and other stresses, and consist of unique cysteine-rich PR domains folded in an alpha helix–beta sheet–alpha helix sandwich (Asojo et al., 2005a). ASP-1 is a 45-kDa protein containing two PR domains (Hawdon et al., 1996; Asojo et al., 2005b), whereas ASP-2 is a 20–22 kDa single PR domain protein (Hawdon et al., 1999; Asojo et al., 2005a). The function of the ASPs is unknown, although X-ray crystallographic studies suggest the possibility that they are proteases or immunomodulatory proteins (Asojo et al., 2005a). A third secreted antigen is the 62-kDa metalloprotease known as MTP-1 (Zhan et al., 2002a, 2002b). All three antigens, which were originally described in *A. caninum*, are also conserved in the human hookworms, *N. americanus* and *A. duodenale* (Zhan et al., 2000; Asojo et al., 2005a, 2005b; Goud et al., 2005).

In addition to the secreted L3 antigens, two proteins found on the surface of invading larvae, *Ac*-SAA-1 and *Ac*-16 (also referred to as *Ac*-SAA-2), have also shown potential as larval-stage vaccine antigens and are currently being developed as alternative larval vaccine antigens (Diemert et al., 2008; Fujiwara et al., 2007; Zhan et al., 2004). Although it is a component of both the larval and adult stages of the hookworm life cycle, *Ac*-16 has been shown to be an immunodominant larval surface protein that results in reduced blood loss and fecal egg counts when used as a vaccine in the canine parasite challenge model (Fujiwara et al., 2007). The *N. americanus* homolog of *Ac*-16 has already been identified and work is currently underway to identify an ideal expression system to produce it as a recombinant protein for clinical trials.

Studies have been undertaken to determine whether ASP-1, ASP-2, MTP-1, SAA-1, and SAA-2 might be suitable vaccine candidates for subsequent process development, manufacture, and clinical testing. Because inadequate quantities of natural products are available for immunological studies, it was necessary to produce these proteins as recombinant antigens in genetically engineered expression hosts. Expression of cysteine-rich ASPs in bacterial expression systems resulted in production of insoluble recombinant protein products that could not be resolubilized or refolded. Similarly, recombinant MTP-1 as expressed by *Escherichia coli* was an insoluble product. Instead, with the exception of the two surface antigens, successful expression of soluble and properly folded larval recombinant proteins required eukaryotic expression systems such as yeast (Asojo et al., 2005a, 2005b; Goud et al., 2004, 2005) or insect cells (Bethony et al., 2005).

Studies of the induced humoral immune responses in dogs vaccinated with irL3 *A. caninum* determined that among the secreted antigens, the resultant antibody responses are preferentially directed against ASP-2, suggesting that this is the immunodominant antigen associated with such vaccines (Fujiwara et al., 2006). This evidence advanced ASP-2 as a plausible lead candidate antigen. When recombinant *A. caninum* ASP-2 (*Ac*-ASP-2) was formulated with the GlaxoSmithKline Adjuvant System 03 (AS03) and used to vaccinate dogs (Bethony et al., 2005), or when recombinant *A. ceylanicum* ASP-2 (*Ay*-ASP-2) was formulated with Quil A and used to vaccinate hamsters (Goud et al., 2004; Mendez et al., 2005a, 2005b), high levels of antigen-specific antibody were elicited, which recognized the corresponding native antigen (Bethony et al., 2005). In both cases, the vaccines elicited high levels of protection in terms of reduction in adult worm burdens, fecal egg counts, and host blood loss, as compared to animals immunized only with adjuvant (Bethony et al., 2005; Goud et al., 2004; Mendez et al., 2005b). Additionally, both anti-ASP-2 and anti-L3 antisera from dogs were shown to inhibit larval invasion in vitro (Bethony et al., 2005; Fujiwara et al., 2005). The proposed mechanism of action of ASP-2 vaccines is that they elicit antibody responses that inhibit larval invasion and thereby reduce the number of L3 that develop into adult worms that inhabit the host's gastrointestinal tract, resulting in reduced host worm burdens, fecal egg counts, and blood loss (Bethony et al., 2005; Loukas et al., 2006).

The animal hookworm studies outlined above suggested that ASP-2 is a promising vaccine antigen. Additionally, studies of hookworm infection conducted in endemic areas of Brazil and China determined that individuals who naturally acquire anti-ASP-2 antibody responses following chronic exposure to hookworm larvae exhibit a significant reduction in the risk of acquiring heavy hookworm infections (Bethony et al., 2005). Therefore, ASP-2 was the first larval antigen selected for further development based on the promising evidence from the cumulative in vitro, animal challenge, and human immunoepidemiological studies (Hotez et al., 2005, 2006a). Based on its efficacy in the canine model and the ease with which it can be produced as a recombinant protein (Fujiwara et al., 2007), *Na*-SAA-2 (the *Necator* orthologue of *Ac*-16) has been selected as the second-line larval antigen and is currently undergoing early-stage process development and scale-up expression.

PROCESS DEVELOPMENT AND MANUFACTURE OF *Na*-ASP-2 AND *Na*-SAA-2

After selecting the *N. americanus* ortholog of ASP-2 (*Na*-ASP-2) and SAA-2 (*Na*-SAA-2) as the lead larval antigens for further development, an organization had to be created in order to oversee process development and cGMP manufacture of these two recombinant protein vaccines. As with other neglected tropical diseases, the development of a vaccine against hookworm infection is impeded by the absence of a viable commercial market (Hotez and Ferris, 2006). Although North American and European vaccine manufacturers may be sympathetic to developing vaccines for neglected tropical diseases, most are not in a financial position to invest in such development given their need to produce a profit and satisfy shareholders. Ultimately the development of *N. americanus* larval antigens into vaccines for humans required the creation of an entity now known as a product development public private partnership (PD-PPP) (Hotez et al., 2005; Bottazzi et al., 2006). PD-PPPs began to emerge during the late 1990s in response to a growing demand for nonprofit organizations with the capacity to produce health products for developing countries (Hotez et al., 2006d). Similar to traditional pharmaceutical manufacturers, the PD-PPPs use industry practices and maintain a portfolio of products, but with the goal of improving public health rather than generating profits or reporting to shareholders (Widdus and White, 2004). The PD-PPP known as the Human Hookworm Vaccine Initiative (HHVI) was established in 2000, and is based at the Sabin Vaccine Institute in Washington, DC. The HHVI's process development and quality control is conducted at the laboratories of the Department of Microbiology, Immunology, and Tropical Medicine at The George Washington University (GWU), whereas cGMP manufacture is performed at the Walter Reed Army Institute of Research (WRAIR) and Instituto Butantan, a public sector vaccine manufacturer located in São Paulo, Brazil. The Oswaldo Cruz Foundation (FIOCRUZ) in Brazil was selected to undertake clinical development of the vaccine, with additional support from the London School of Hygiene and Tropical Medicine. The Sabin Vaccine Institute provides sponsorship of the product and program management of the HHVI.

Na-ASP-2 was selected as the first larval antigen for process development and cGMP manufacture based on the reasons listed above. Early efforts to express *Na*-ASP-2 in *E. coli* resulted in the production of insoluble inclusion bodies that resisted attempts at either resolubilization or refolding. A similar experience has been reported for antigen 5, a PR protein from yellowjackets and paper wasps (Monsalve et al., 1999), which may result from the protein's high cysteine content and subsequent aberrant disulfide bond formation. Early success was achieved with insect cell expression of *Ac*-ASP-2 (Bethony et al., 2005) although the yields and expense of this expression host were considered inappropriate for further vaccine development. Both *Ay*-ASP-2 and *Na*-ASP-2 were successfully expressed in the methanol-utilizing yeast, *Pichia pastoris* (Goud et al., 2004, 2005).

The recombinant ASP-2 proteins were shown to be soluble and secreted by *P. pastoris* at high yields; in the case of *Na*-ASP-2, approximately 400 mg of purified protein were obtained per liter of fermentation broth (Goud et al., 2005). Recombinant *Na*-ASP-2 is a 21.3 kDa protein with an N-terminal acidic tag (EAEAEAF), the presence of which increases expression yield. The recombinant protein has a single O-linked monosaccharide, but no N-linked carbohydrates. Antibodies generated against the recombinant protein in rats both recognize native *Na*-ASP-2 on Western blots and inhibit larval invasion in vitro (Goud et al., 2005). Moreover, X-ray crystallographic analysis of the recombinant protein reveals that it contains a PR-1 domain with an alpha helix–beta sheet–alpha helix sandwich, which is characteristically found in other PR proteins (Asojo et al., 2005a). Together, these data suggest that recombinant *Na*-ASP-2 is correctly folded and displays conformational epitopes in a manner similar to the native molecule.

Because *Na*-ASP-2 is highly stable and is secreted in abundance by *P. pastoris*, a purification process was developed that is relatively straightforward and simple. The purification steps include ultrafiltration and microfiltration, two ion-exchange chromatography steps, and a final purification step using gel filtration chromatography (Goud et al., 2005).

Process development at the 10 L scale was followed by a 60 L fermentation and 1/12th-scale purification before the research seed stocks were transferred from GWU to WRAIR for production of the master and production cell banks, and cGMP manufacture at the 60 L scale. The final drug substance passed lot release criteria for identity, sterility, purity, immune recognition, and stability using qualified assays, as well as additional testing for residual DNA, monosaccharide analysis, residual methanol, N-terminal amino acid sequencing, host cell protein, and endotoxin content. Recombinant *Na*-ASP-2 was formulated with Alhydrogel® based on previous studies showing that *Na*-ASP-2 bound to this aluminum-containing substrate was immunogenic in rats and other laboratory animals (Goud et al., 2005; Fujiwara et al., 2005). The formulated drug product also passed lot release criteria for color and appearance, pH, identity, stability, and immunological potency in rats. An investigational new drug (IND) application was submitted to the Food and Drug Administration (FDA) in December 2004, and a phase 1 human trial of the vaccine was completed in 2006.

CLINICAL DEVELOPMENT OF THE Na-ASP-2 HOOKWORM VACCINE

The overall goal of the HHVI program is to develop a safe, efficacious, and low-cost human hookworm vaccine that will contribute to reducing the global burden of disease caused by this infection by reducing the likelihood of developing heavy infections in individual vaccine recipients. The pathway chosen by the HHVI for the clinical development of candidate hookworm vaccines is to first test them in uninfected, healthy adults in the United States, followed by age de-escalation studies in hookworm-endemic areas, culminating in "proof-of-principle" efficacy studies in children exposed to hookworm infection. Following demonstration of a meaningful biologic effect on hookworm infection, further product and clinical development will be transferred to partners in endemic countries with the capacity to develop and manufacture new vaccines, such as Brazil, India, or China (Bottazzi et al., 2006; Hotez et al., 2006a; Loukas et al., 2006). The HHVI has chosen Brazil as its first such partner, and an agreement is now in place that provides a blueprint for Instituto Butantan, one of Brazil's largest public sector vaccine manufacturers, to continue product development and for the Brazilian government to continue the clinical development of a human hookworm vaccine once proof-of-principle is established in a phase 2 clinical trial. The agreement also includes provisions for the eventual industrial-scale manufacture of the vaccine and sponsorship of pivotal phase 3 efficacy studies.

Following submission of the IND application for the recombinant *Na*-ASP-2 Hookworm Vaccine to the FDA in December 2004, and a clinical trial evaluating the safety, reactogenicity, and immunogenicity of this vaccine in healthy adults without evidence of hookworm infection and living in the United States was started in March 2005 and completed in June 2006. Thirty-six adults were enrolled in this randomized, double-blind, dose-escalation phase 1 study comparing three different concentrations of the *Na*-ASP-2 protein (10, 50, or 100 μg of recombinant protein) adjuvanted with Alhydrogel® to a saline placebo (Diemert et al., 2008). Individuals were administered according to a 0-, 2-, and 4-month schedule by intramuscular injection and participants were followed until 6 months after the final vaccination. The *Na*-ASP-2 Hookworm Vaccine was well tolerated and did not induce any significant vaccine-related adverse events. The most frequently observed adverse events were injection site reactions including mild-to-moderate pain, swelling, erythema, and pruritus. Additionally, significant antigen-specific antibody and cellular immune responses were observed.

Because this initial phase 1 study of the *Na*-ASP-2 Hookworm Vaccine demonstrated an acceptable safety profile and encouraging immunogenicity in a naïve population, a phase 1 trial of the *Na*-ASP-2 vaccine conducted in a hookworm-endemic area of Brazil in healthy adult volunteers with documented evidence of previous hookworm infection was initiated in June 2007 and is currently underway. In this study, the same dose concentrations and vaccination schedule will be used as in the initial phase 1 trial conducted in the United States. This first study of a hookworm vaccine in an endemic area will also be a randomized, double-blind study comparing the safety and immunogenicity of the *Na*-ASP-2 vaccine to the hepatitis B vaccine. A second phase 1 study in adults is necessary as neither the safety nor the immunogenicity of the vaccine in individuals chronically exposed to hookworm can be confidently extrapolated from studies performed in unexposed populations.

Assuming the phase 1 trial in Brazilian hookworm-exposed adults yields no significant safety concerns, clinical testing of the *Na*-ASP-2 vaccine will proceed into the pediatric population in which an eventual proof-of-principle study will be conducted; that is, in preschool and/or primary school-aged children. This target age group has been chosen for a number of reasons, but primarily because studies have shown that although high intensities of infection and presumably intestinal blood loss occur in both adults and children, the health impact of hookworm infection in terms of host blood loss resulting in anemia is greatest in children (Stoltzfus et al., 1997a, 1997b). In addition, future implementation of an effective human hookworm vaccine would likely be integrated into existing helminth control programs, which are currently based on the delivery of anthelmintics to preschool and school-aged children (Savioli et al., 1992). For these reasons, we believe that children are an appropriate study population for a phase 2 study of the *Na*-ASP-2 vaccine.

The study design of the proof-of-principle phase 2 trial will consist of assessing the rate and intensity of hookworm infection in vaccinated children, in comparison to those administered an active comparator such as the hepatitis B vaccine. An important component of the study design of the phase 2 trial will be pretreatment of infected individuals with either albendazole or mebendazole in order to eliminate the adult hookworms that are already present in the gastrointestinal tract of study participants, prior to vaccination (Diemert et al., 2008).

Estimation of vaccine efficacy will be assessed by measuring the impact on worm burden, which will be indirectly assessed by quantitative fecal egg counts. Because a larval hookworm vaccine is unlikely to provide completely sterilizing immunity, but may instead limit the number of invading L3 that reach the gastrointestinal tract and subsequently develop into adult worms, the worm burden (i.e., number of worms per individual) represents the most appropriate efficacy endpoint (rather than the presence or absence of infection). Due to the difficulties in obtaining adult worm expulsions, a commonly employed indirect measure of infection intensity is the quantitative fecal egg count. Measuring fecal egg counts is considered an appropriate indicator of the health impact of hookworm infection because of the demonstrated correlation between egg counts and host worm burden, as well as the relationship between egg counts and host blood loss as estimated by measuring the quantity of heme in fecal samples (Stoltzfus et al., 1997a). Fecal egg counts, host hematological indices such as hemoglobin concentration, and fecal heme measurements will accordingly be critical endpoints for assessing the potential efficacy of an experimental hookworm vaccine in a proof-of-principle study (Brooker et al., 2005; Diemert et al., 2008). Following demonstration of proof-of-principle, a phase 3 study will be required to demonstrate the impact of vaccination on clinical endpoints such as episodes of anemia.

In summary, the aim of the HHVI's clinical development of the *Na*-ASP-2 hookworm vaccine is focused upon showing, in a proof-of-principle study, that the vaccine is effective in reducing the intensity of reinfection with hookworm and the clinical outcomes of hookworm disease (e.g., intestinal blood loss and iron deficiency anemia) following anthelmintic treatment of hookworm-infected children from an endemic area of Brazil. The scope, scale, and design of the project will potentially provide sufficient evidence for a middle-income country such as Brazil to continue the clinical development of the vaccine. *Na*-SAA-2 is at an earlier stage of development and will be advanced depending on the results of the initial clinical trials of the *Na*-ASP-2 antigen in Brazil.

DEVELOPMENT OF ANTIGENS TARGETING ADULT HOOKWORMS

As outlined above, it is anticipated that a vaccine based on a larval antigen such as *Na*-ASP-2 will reduce the number of L3 entering the host gastrointestinal tract, thereby reducing host worm burdens, fecal egg counts, and blood loss (Goud et al., 2005). However, in the event that the vaccine is only partially effective in preventing infection, some L3 will still reach the gastrointestinal tract and develop into blood-feeding adult hookworms. In anticipation of this scenario, the HHVI has initiated a second antigen discovery program to identify macromolecules that are required by adult hookworms in order to utilize human blood as a food source (reviewed by Hotez et al., 2003). It is hypothesized that antibodies against these antigens, some of which are enzymes, would interfere with parasite feeding and thereby augment the reductions in blood loss and fecal egg counts resulting from vaccination with *Na*-ASP-2. This hypothesis is a modern-day version of the antienzyme antibody hypothesis first proposed by Professor Asa Chandler in the 1930s (Hotez and Cerami, 1983; Hotez et al., 1987). It is proposed that the second antigen would be added to a larval antigen in order to develop a bivalent hookworm vaccine.

From the HHVI antigen-discovery efforts, three candidate antigens have emerged. Two of these candidate antigens are proteases (hemoglobinases) that line the parasite alimentary canal and function to degrade host hemoglobin (Williamson et al., 2004). The third antigen is a glutathione S-transferase (GST-1) that detoxifies host heme (Zhan et al., 2005). The aspartic protease-hemoglobinase, APR-1, and the cysteine protease-hemoglobinase, CP-2, have both shown promise as protective antigens in laboratory dogs in terms of reduced host blood loss and fecal egg counts (Loukas et al., 2004, 2005). In addition, vaccination of dogs with APR-1 or GST-1 (with GlaxoSmithKline's AS03 adjuvant) or hamsters with these same antigens formulated with Alhydrogel® has also resulted in significant host worm burden reductions (Loukas et al., 2005; Xiao et al., 2007). Activities are now being concentrated on advancing at least one adult-stage antigen, such as either *Na*-APR-1 or *Na*-GST-1, through process development, cGMP manufacture, and phase 1 clinical testing. Currently, GST-1 has been manufactured according to cGMP at the 60 L scale, and should be ready for clinical testing in early 2009.

INDICATIONS AND USE OF THE HUMAN HOOKWORM VACCINE

The current aim of the HHVI is to develop a bivalent Human Hookworm Vaccine that contains an antigen targeting the larval stage of the life cycle of *N. americanus* and a second antigen targeting the adult worm. This two-component vaccine is being developed as an intramuscular injectable product for the prevention of disease—primarily anemia and protein malnutrition—related to infection with this helminth. The vaccine will most likely be used as part of an integrated hookworm control program, with areas of high transmission being targeted to obtain the greatest public health impact against this disease. The WHO defines high-transmission areas as those in which the prevalence equals or exceeds 70% or when the prevalence of moderate or heavy hookworm infections (defined as greater than or equal to 2000 hookworm eggs per gram of feces) is equal to or greater than 10% (WHO, 2002).

Administration of the Human Hookworm Vaccine to at-risk populations following deworming with a benzimidazole anthelmintic agent (e.g., albendazole or mebendazole) is being explored. The idea of prevaccination deworming with an anthelmintic is derived from observations that: (1) larval-stage vaccines will not target adult hookworms already present in the intestine (Zhan et al., 2003), and (2) in laboratory animals the presence of adult hookworms in the intestine reduces the immunogenicity of ASP-2 vaccines (Ghosh et al., 2006). Therefore, incorporation of the Human Hookworm Vaccine into existing deworming programs is under consideration (Hotez et al., 2005; Loukas et al., 2006; Hotez et al., 2006b). The concept of linking anthelmintic deworming with anthelmintic vaccination has also been proposed for the future control of schistosomiasis (Bergquist et al., 2005). In the major STHI-endemic countries, deworming is either conducted in schools or as a part of "child health days" in order to target school-aged and preschool children, respectively.

ENSURING GLOBAL ACCESS TO A HUMAN HOOKWORM VACCINE

Critical to the success of the HHVI is the ability to not only complete proof-of-principle testing in humans, but also develop plans for the licensure, industrial-scale manufacture, and distribution of hookworm vaccines. Given the large number of individuals at risk for hookworm infection and the observation that almost all of them live in rural areas without easy access to the cold chain, these goals will not be easily attainable. Equally challenging is the absence of a viable commercial market for hookworm vaccines, given that the vaccine recipients would be comprised almost exclusively of individuals who live on less than $2 per day (Hotez et al., 2005). Mahoney and Maynard (1999) point out that there was a lag time of more than two decades before hepatitis B vaccine was widely used from the time of its initial development, so it is highly desirable to identify mechanisms that could potentially shorten this time frame.

An important component of the global access strategy of the HHVI is to develop partnerships with vaccine manufacturers in middle-income countries with endemic hookworm infections (Hotez et al., 2005, 2006a; Bottazzi et al., 2006). Several of these nations are known as *innovative developing countries* (IDCs), which are characterized by evidence of innovation that exceeds expectations based on economic growth (Morel et al., 2005). Among the criteria for innovation are patents, peer-reviewed papers, and ability to manufacture drugs and vaccines. The HHVI has focused its initial efforts on tropical regions of the Americas, especially in Brazil where proof-of-principle testing of experimental hookworm vaccines will take place. Brazil is an IDC with two major vaccine manufacturers, FIOCRUZ Bio-Manguinhos and Instituto Butantan. Through signed memoranda of understanding, the HHVI is transferring hookworm vaccine development and manufacturing technology to Instituto Butantan, with clinical testing to be conducted by FIOCRUZ. In turn, the Brazilian government will commit to the purchase and distribution of the hookworm vaccine for its own country with an option to export the vaccine throughout the Americas. To date, Instituto Butantan has successfully produced GMP-grade *Na*-ASP-2.

Also of relevance to global access is the importance of improving health delivery systems (Loukas et al., 2006). As noted above, a Human Hookworm Vaccine could be integrated into existing deworming programs, even though it would mean that vaccination would be conducted outside of the infant-based Expanded Program on Immunization framework. Therefore, discussions with the WHO and other UN agencies are already underway to explore the compatibility of vaccination and deworming, thereby ensuring that an eventual Human Hookworm Vaccine is integrated into large-scale control programs, either as a part of school-based deworming programs or preschool based child health days, as soon as possible after its development. Such early-stage efforts are considered critical in order to avoid the hurdles that prevented the timely deployment of the hepatitis B vaccine in resource-limited settings. Despite the enormous challenges in developing, testing, and distributing a vaccine for a neglected tropical disease such as hookworm, success in this endeavor could be a significant public health breakthrough. Moreover, this would represent an important step in achieving several Millennium Development Goals (MDGs) for sustainable development including specific MDGs in the areas of infectious diseases, poverty reduction, and child and maternal health (Hotez and Ferris, 2006).

KEY ISSUES

- Over 500 million people worldwide are infected with the hookworms *Necator americanus* and *Ancylostoma duodenale*.
- The major health impact of hookworm infection is due to the resultant iron deficiency anemia and protein malnutrition.
- Current disease control efforts that focus on periodic mass administration of anthelmintic medications in endemic areas are not sustainable.
- High rates of anthelmintic drug failure and the potential emergence of drug resistance necessitate the development of new and alternative control tools.
- An efficacious vaccine would provide a powerful control tool when integrated into existing mass chemotherapy programs.

- Antigen discovery has concentrated on identifying targets found in both the larval and adult stages of the worm's life cycle.
- The goal of current hookworm vaccine development activities is to produce a vaccine that is a combination of a larval with an adult-stage antigen.
- Hookworm vaccine antigens are being produced as recombinant proteins expressed in either *Pichia pastoris* or *Escherichia coli*.
- The first hookworm vaccine formulation to enter clinical trials is a larval antigen adjuvanted with Alhydrogel, which is currently undergoing phase 1 testing in an endemic area of Brazil.

ACKNOWLEDGMENTS

Both Peter Hotez and David Diemert are supported by the HHVI, which is sponsored by the Sabin Vaccine Institute through financial support from the Bill and Melinda Gates Foundation.

References

Albonico, M., Bickle, Q., Ramsan, M., Montresor, A., Savioli, L. and Taylor, M. Efficacy of mebendazole and levamisole alone or in combination against intestinal nematode infections after repeated targeted mebendazole treatment in Zanzibar. Bull. World Health Organ. 2003; 81:343–352.

Albonico, M., Engels, D. and Savioli, L. Monitoring drug efficacy and early detection of drug resistance in human soil-transmitted nematodes: a pressing public health agenda for helminth control. Int. J. Parasitol. 2004; 34:1205–1210.

Albonico, M., Smith, M.P.G., Ercole, E., Hall, A., Chwaya, H.M., Alawi, K.S. and Savioli, L. Rate of reinfection with intestinal nematodes after treatment of children with mebendazole or albendazole in a highly endemic area. Trans. R. Soc. Trop. Med. Hyg. 1995; 89:538–541.

Albonico, M., Stolzfus, R.J., Savioli, L., Tielsch, J.M., Chwaya, H.M., Ercole, E. and Cancrini, G. Epidemiological evidence for a differential effect of hookworm species, *Ancylostoma duodenale* or *Necator americanus*, on iron status of children. Int. J. Epidemiol. 1998; 27:530–537.

Asojo, O.A., Goud, G., Dhar, K., Loukas, A., Zhan, B., Deumic, V., Liu, S., Borgstahl, G. and Hotez, P. X-ray structure of *Na*-ASP-2, a pathogenesis-related-1 protein from the nematode parasite *Necator americanus* and a vaccine antigen for human hookworm infection. J. Mol. Biol. 2005a; 346:801–814.

Asojo, O.A., Loukas, A., Inan, M., Barent, R., Huang, J., Plantz, B., Swanson, A., Gouthro, M., Megher, M.M. and Hotez, P.J. Crystallization and preliminary X-ray analysis of *Na*-ASP-1, a multi-domain pathogenesis-related-1 protein from the human hookworm parasite *Necator americanus*. Acta Crystallogr. Sect. F. Struct. Biol. Cryst. Commun. 2005b; 61:391–394.

Bennett, A. and Guyatt, H. Reducing intestinal nematode infection: efficacy of albendazole and mebendazole. Parasitol. Today 2000; 16:71–74.

Bergquist, N.R., Leonardo, L.R. and Mitchell, G.F. Vaccine-linked chemotherapy: can schistosomiasis control benefit from an integrated approach? Trends Parasitol. 2005; 21:112–117.

Bethony, J., Brooker, S., Albonico, M., Geiger, S.M., Loukas, A., Diemert, D. and Hotez, P.J. Soil-transmitted helminth infections: ascariasis, trichuriasis, and hookworm. Lancet 2006; 367:1521–1532.

Bethony, J., Chen, J.Z., Lin, S.X., Xiao, S.H., Zhan, B., Li, S.W., Xue, H.C., Xing, F.Y., Humphries, D., Chen, C., Foster, V., Wang, Y., Hawdon, J.M. and Hotez, P.J. Epidemiology of *Necator americanus* hookworm infection in Diaocong Village, Qiongshan County, Hainan Province, People's Republic of China. I: age, gender and household risk factors. Clin. Infect. Dis. 2002; 35:1336–1344.

Bethony, J.M., Loukas, A., Smout, M.J., Mendez, S., Wang, Y., Bottazzi, M.E., Zhan, B., Williamson, A.L., Lustigman, S., Correa-Oliveira, R., Xiao, S.H. and Hotez, P.J. Antibodies against a secreted protein from hookworm larvae reduce the intensity of infection in humans and vaccinated laboratory animals. FASEB J. 2005; 19:1743–1745.

Bleakley, H. Disease and development: evidence from hookworm eradication in the American South. Q. J. Econ. 2007; 122:73–117.

Bottazzi, M.E., Miles, A.P., Diemert, D.J. and Hotez, P.J. An ounce of prevention on a budget: a non-profit approach to developing vaccines against neglected diseases. Expert Rev. Vaccines 2006; 5:189–198.

Brooker, S., Bethony, J. and Hotez, P.J. Human hookworm infection in the 21st Century. Adv. Parasitol. 2004; 58:197–288.

Brooker, S., Bethony, J.M., Rodrigues, L., Alexander, N., Geiger, S. and Hotez, P.J. Epidemiological, immunological and practical considerations in developing and evaluating a human hookworm vaccine. Expert Rev. Vaccines 2005; 4:35–50.

Brooker, S., Clements, A.C., Hotez, P.J., Hay, S.I., Tatem, A., Bundy, D.A.P. and Snow, R.W. The co-distribution of *Plasmodium falciparum* and hookworm among African school children. Malar. J. 2006; 5:99.

Brooker, S., Jardim-Botelho, A., Quinnell, R.F., Geiger, S.M., Caldas, I.R., Fleming, F., Hotez, P.J., Correa-Oliveira, R., Rodrigues, L.C. and Bethony, J.M. Age-related changes in hookworm infection, anaemia and iron deficiency in an area of high *Necator americanus* hookworm transmission in south-eastern Brazil. Trans. R. Soc. Trop. Med. Hyg. 2007; 101:146–154.

Brooker, S., Peshu, N., Warn, P.A., Mosobo, M., Guyatt, H.L., Marsh, K. and Snow, R.W. The epidemiology of hookworm infection and its contribution to anaemia among pre-school children on the Kenyan coast. Trans. R. Soc. Trop. Med. Hyg. 1999; 93:240–246.

Bundy, D.A., Chan, M.S. and Savioli, L. Hookworm infection in pregnancy. Trans. R. Soc. Trop. Med. Hyg. 1995; 89:521–522.

Bundy, D.A.P. Population ecology of intestinal helminth infections in human communities. Phil. Trans. R. Soc. Lond. B. Biol. Sci. 1988; 321:405–420.

Cappello, M., Vlasuk, G.P., Bergum, P.W., Huang, S. and Hotez, P.J. *Ancylostoma caninum* anticoagulant peptide (AcAP): a novel hookworm derived inhibitor of human coagulation factor Xa. Proc. Natl. Acad. Sci. USA 1995; 92:6152–6156.

Chan, M.S. The global burden of intestinal nematode infections—fifty years on. Parasitol. Today 1997; 13:438–443.

Chan, M.S., Bradley, M. and Bundy, D.A.P. Transmission patterns and the epidemiology of hookworm. Int. J. Parasitol. 1997; 26:1392–1400.

Christian, P., Khatry, S.K. and West, J.P., Jr. Antenatal anthelminthic treatment, birthweight, and infant survival in rural Nepal. Lancet 2004; 364:981–983.

Cort, W. and Otto, G.F. Immunity in hookworm disease. Rev. Gastorenterol. 1940; 7:2–14.

De Clercq, D., Sacko, M., Behnke, J., Gilbert, F., Dorny, P. and Vercruysse, J. Failure of mebendazole in treatment of human hookworm infections in the southern region of Mali. Am. J. Trop. Med. Hyg. 1997; 57:25–30.

De Silva, N., Brooker, S., Hotez, P., Montresor, A., Engels, D. and Savioli, L. Soil-transmitted helminth infections: updating the global picture. Trends Parasitol. 2003; 12:547–551.

Diemert, D.J., Bethony, J.M. and Hotez, P.J. Hookworm vaccines. Clin. Infect. Dis. 2008; 46:282–288.

Don, T.A., Jones, M.K., Smyth, D., O-Donoghue, P.O., Hotez, P. and Loukas, A. A pore-forming hemolysin from the hookworm, *Ancylostoma caninum*. Int. J. Parasitol. 2004; 34:1029–1035.

Fleming, F., Brooker, S., Geiger, S., Caldas, I., Corea-Oliveira, R., Hotez, P. and Bethony, J. Synergistic associations between hookworm and other helminth species in a rural community in Brazil. Trop. Med. Int. Health 2006; 11:56–64.

Foster, A.O. The immunity of dogs to *Ancylostoma caninum*. Am. J. Hyg. 1935; 22:65–105.

Fujiwara, R., Zhan, B., Mendez, S., Loukas, A., Bueno, L.L., Wang, Y., Plieskatt, J., Oksov, Y., Lustigman, S., Bottazzi, A., Hotez, P. and Bethony, J.M. Reduction of worm fecundity and canine host blood loss mediates protection against hookworm infection elicited by vaccination with recombinant *Ac*-16. Clin. Vaccine Immunol. 2007; 14:281–287.

Fujiwara, R.T., Bethony, J., Bueno, L.L., Wang, Y., Ahn, S.Y., Samuel, A., Bottazzi, M.E., Hotez, P. and Mendez, S. Immunogenicity of the hookworm *Na*-ASP-2 vaccine candidate. Hum. Vaccines 2005; 1:123–128.

Fujiwara, R.T., Loukas, A., Mendez, S., Williamson, A.L., Bueno, L.L., Wang, Y., Samuel, A., Zhan, B., Bottazzi, M.E., Hotez, P.J. and Bethony, J.M. Vaccination with irradiated *Ancylostoma caninum* third stage larvae induces a Th-2-like response in dogs. Vaccine 2006; 24:501–509.

Gandhi, N.S., Chen, J.Z., Khoshnood, K. et al. Epidemiology of *Necator americanus* hookworm infections in Xiulongkan Village, Hainan Province, China: high prevalence and intensity among middle aged and elderly residents. J. Parasitol. 2001; 87: 739–743.

Ghosh, K., Wu, W.H., Antoine, A.D., Bottazzi, M.E., Valenzuela, J.G., Hotez, P.J. and Mendez, S. The impact of concurrent and treated *Ancylostoma ceylanicum* hookworm infection on the immunogenicity of a recombinant hookworm vaccine in hamsters. J. Infect. Dis. 2006; 193:155–162.

Goud, G.N., Bottazzi, M.E., Zhan, B., Mendez, S., Deumic, V., Pleiskatt, J., Liu, S., Wang, Y., Bueno, L., Fujiwara, R., Samuel, A., Ahn, S.Y., Solanki, M., Asojo, O., Wen, J., Saul, A., Bethony, J.M., Loukas, A., Roy, M. and Hotez, P.J. Expression of the *Necator americanus* hookworm larval antigen *Na*-ASP-2 in *Pichia pastoris* and purification of the recombinant protein for use in human clinical trials. Vaccine 2005; 23:4754–4764.

Goud, G.N., Zhan, B., Ghosh, K., Loukas, A., Hawdon, J., Dobardzic, A., Deumic, V., Liu, S., Dobardzic, R., Zook, R.C., Qun, J., Liu, Y.Y., Hoffman, L., Chung-Debose, D., Patel, R., Mendez, S. and Hotez, P.J. Cloning, yeast expression, isolation and vaccine testing of recombinant *Ancylostoma* secreted protein 1 (ASP-1) and ASP-2 from *Ancylostoma ceylanicum*. J. Infect. Dis. 2004; 189:919–929.

Hawdon, J.M. and Hotez, P.J. Hookworm: developmental biology of the infectious process. Curr. Opin. Genet. Dev. 1996; 6:618–623.

Hawdon, J.M., Jones, B.F., Hoffman, D. and Hotez, P.J. Cloning and expression of *Ancylostoma* secreted protein: a polypeptide associated with the transition to parasitism by infective hookworm larvae. J. Biol. Chem. 1996; 271:6672–6678.

Hawdon, J.M., Narasimhan, S. and Hotez, P.J. *Ancylostoma* secreted protein 2 (ASP-2): cloning and characterization of a second member of a novel family of nematode secreted proteins from *Ancylostoma caninum*. Mol. Biochem. Parasitol. 1999; 99:149–165.

Hotez, P., Brooker, S., Bethony, J., Bottazzi, M., Loukas, A. and Xiao, S. Current concepts: hookworm infection. N. Engl. J. Med. 2004a; 351:799–807.

Hotez, P., Ghosh, K., Hawdon, J.M. et al. Experimental approaches to the development of a recombinant hookworm vaccine. Immunol. Rev. 1999; 171:163–171.

Hotez, P.J. China's hookworms (research report). China Q. 2002; 172:1029–1041.

Hotez, P.J. The National Institutes of Health roadmap and the developing world. J. Investig. Med. 2004; 52:246–247.

Hotez, P.J. Hookworm and poverty. Ann. N. Y. Acad. Sci. 2007; 1136:38–44.

Hotez, P.J., Bethony, J., Bottazzi, M.E., Brooker, S. and Buss, P. Hookworm—"the great infection of mankind". PLoS Med. 2005; 2:e67, 177–181.

Hotez, P.J., Bethony, J., Bottazzi, M.E., Brooker, S., Diemert, D. and Loukas, A. New technologies for the control of human hookworm infection. Trends Parasitol. 2006a; 22:327–331.

Hotez, P.J., Bundy, D.A.P., Beegle, K., Brooker, S., DeSilva, N., Montresor, A., Engles, D., Drake, L., Chitsulo, L., Michaud, C., Bethony, J.M., Oliveira, R., Xiao, S.H., Fenwick, A. and Savioli, L. Helminth infections. In: Disease Control Priorities in Developing Countries, second edition, pp. 467–482. Washington, DC: The World Bank; and, New York: Oxford University Press (co-publication), 2006b.

Hotez, P.J. and Cerami, A. Secretion of a proteolytic anticoagulant by *Ancylostoma* hookworms. J. Exp. Med. 1983; 157:1594–1603.

Hotez, P.J., Feng, Z., Xu, L.Q. et al. Emerging and reemerging helminthiases and the public health of China. Emerg. Infect. Dis. 1997; 3:303–310.

Hotez, P.J. and Ferris, M. The anti-poverty vaccines. Vaccine 2006; 24:5787–5799.

Hotez, P.J., Haggerty, J., Hawdon, J. et al. Infective *Ancylostoma* hookworm larval metalloproteases and their possible functions in tissue invasion and ecdysis. Infect. Immun. 1990; 58:3883–3892.

Hotez, P.J., Hawdon, J.M., Cappello, M. et al. Molecular approaches to vaccinating against hookworm disease. Pediatr. Res. 1996; 40:515–521.

Hotez, P.J., Molyneux, D.H., Fenwick, A., Kumaresan, J., Ehrlich Sachs, S., Sachs, J.D. and Savioli, L. Control of neglected tropical diseases. N. Engl. J. Med. 2007; 357:1018–1027.

Hotez, P.J., Molyneux, D.H., Fenwick, A., Ottesen, E., Ehrlich Sachs, S. and Sachs, J.D. Incorporating a rapid-impact package for neglected tropical diseases with programs for HIV/AIDS, tuberculosis, and malaria. PLoS Med. 2006c; 3:e102.

Hotez, P.J., Narasimhan, S., Haggerty, J. et al. A hyaluronidase from *Ancylostoma* hookworm larvae and its function as a virulence factor in cutaneous larva migrans. Infect. Immun. 1992; 60:1018–1023.

Hotez, P.J., Ottesen, E., Fenwick, A. and Molyneux, D. The neglected tropical diseases: the ancient afflictions of stigma and poverty and the prospects for their control and elimination. Adv. Exp. Med. Biol. 2006d; 582:23–33.

Hotez, P.J., Remme, H., Buss, P., Alleyne, G., Morel, C. and Breman, J.G. Combating tropical communicable diseases: workshop report of the disease control priorities project (DCPP). Clin. Infect. Dis. 2004b; 38:871–878.

Hotez, P.J., Trang, N.L. and Cerami, A. Hookworm antigens: the potential for vaccination. Parasitol. Today 1987; 3:247–249.

Hotez, P.J., Zhan, B., Bethony, J.M., Loukas, A., Williamson, A., Goud, G.N., Hawdon, J.M., Dobardzic, A., Dobardzic, R., Ghosh, K., Bottazzi, M.E., Mendez, S., Zook, B., Wang, Y., Liu, S.,

Essiet-Gibson, I., Chung-Debose, S., Xiao, S.H., Knox, D., Megher, M., Inan, M., Correa-Oliveira, R., Vilk, P., Shepherd, H.R., Brandt, W. and Russell, P.K. Progress in the development of a recombinant vaccine for human hookworm disease: the human hookworm vaccine initiative. Int. J. Parasitol. 2003; 33:1245–1258.

Ledizet, M., Harrison, L.M., Koski, R.A. and Cappello, M. Discovery and pre-clinical development of antithrombotics from hematophagous invertebrates. Curr. Med. Chem. Cardiovasc. Hematol. Agents 2005; 3:1–10.

Liu, C.G., Zhang, X.R., Qiu, D.C., Xiao, S.H., Hotez, P.J. et al. Epidemiology of human hookworm infections among adult rural villagers in Heijiang and Santai Counties, Sichuan Province, China. Acta Trop. 1999; 73:255–265.

Loukas, A., Bethony, J., Brooker, S. and Hotez, P. Hookworm vaccines—past, present and future. Lancet Infect. Dis. 2006; 6:733–741.

Loukas, A., Bethony, J.M., Mendez, S., Fujiwara, R.T., Goud, G.N., Ranjit, N., Zhan, B., Jones, B., Bottazzi, M.E. and Hotez, P.J. Vaccination with recombinant aspartic hemoglobinase reduces parasite load and blood loss after hookworm infection. PLoS Med. 2005; 2:e295.

Loukas, A., Bethony, J.M., Williamson, A.L., Goud, G.N., Mendez, S., Zhan, B., Hawdon, J.M., Bottazzi, M.E., Brindley, P.J. and Hotez, P.J. Vaccination of dogs with recombinant cysteine protease from the intestine of canine hookworms diminishes fecundity and growth of worms. J. Infect. Dis. 2004; 189: 1952–1961.

Lwambo, N.J., Bundy, D.A. and Medley, G.F. A new approach to morbidity risk assessment in hookworm endemic communities. Epidemiol. Infect. 1992; 108:469–481.

Mahoney, R.T. and Maynard, J.E. The introdcution of new vaccines into developing countries. Vaccine 1999; 17:646–652.

McCoy, O.R. Immunity reactions of the dog against hookworm (*Ancylostoma caninum*) under conditions of repeated infection. Am. J. Hyg. 1931; 14:268–303.

Mendez, S., Valenzuela, J., Wu, W.H. and Hotez, P. Host cytokine production, lymphoproliferation, and antibody responses during the course of *Ancylostoma ceylancium* infection in the Golden Syrian hamster. Infect. Immun. 2005a; 73:3402–3407.

Mendez, S., Zhan, B., Goud, G., Ghosh, K., Dobardzic, A., Wu, W.H., Liu, S., Deumic, V., Dobardzic, R., Liu, Y.Y., Bethony, J. and Hotez, P.J. Effect of combining the larval antigens *Ancylostoma* secreted protein 2 (ASP-2) and metalloprotease 1 (MTP-1) in protecting hamsters against hookworm infection and disease caused by *Ancylostoma ceylanicum*. Vaccine 2005b; 23:3123–3130.

Miguel, E.A. and Kremer, M. Worms: identifying impacts on education and health in the presence of treatment externalities. Econometrica 2004; 72:159–217.

Miller, T.A. Transfer of immunity to *Ancylostoma caninum* infection in pups by serum and lymphoid cells. Immunology 1967; 12:231–241.

Miller, T.A. Vaccination against the canine hookworm disease. Adv. Parasitol. 1971; 9:153–183.

Miller, T.A. Industrial development and field use of the canine hookworm vaccine. Adv. Parasitol. 1978; 16:333–342.

Monsalve, R.I., Lu, G. and King, T.P. Expression of recombinant venom allergen, antigen 5 of yellowjacket (*Vespula vulgaris*) and paper wasp (*Polistes annularis*), in bacteria or yeast. Protein Exp. Purif. 1999; 16:410–416.

Morel, C.M., Acharya, T., Broun, D., Dangi, A., Elias, C., Ganguly, N.K., Gardner, C.A., Gupta, R.K., Haycock, J., Heher, A.D., Keusch, G.T., Hotez, P.J., Krattiger, A.F., Kerutz, F.T., Lall, S., Lee, K., Mahoney, R., Martinez-Palomo, A., Mashelkar, R.A., Matlin, S., Min, H.K., Mzimba, M., Oehler, J., Pick, W., Ridley, R.G., Senanayake, P., Singer, P. and Yun, M.Y. Health innovation: the neglected capacity of developing countries to address neglected diseases. Science 2005; 309:401–404.

Otto, G.F. A serum antibody in dogs actively immunized against the hookworm *Ancylostoma caninum*. Am. J. Hyg. 1940; 31(D): 23–27.

Otto, G.F. Further observations on the immunity induced in dogs by repeated infection with the hookworm *Ancylostoma caninum*. Am. J. Hyg. 1941; 33(D):39–57.

Otto, G.F. and Kerr, K.B. The immunization of dogs against hookworm, *Ancylostoma caninum*, by subcutaneous injection of graded doses of larvae. Am. J. Hyg. 1939; 29(D):25–45.

Sarles, M.P. The in vitro action of immune rat serum on the nematode, *Nippostrongylus muris*. J. Infect. Dis. 1938; 62:337–348.

Sarles, M.P. and Taliaferro, W.H. The local points of defense and the passive transfer of acquired immunity to *Nippostrongylus muris* in rats. J. Infect. Dis. 1936; 59:207–220.

Savioli, L., Bundy, D. and Tomkins, A. Intestinal parasitic infections: a soluble public health problem. Trans R. Soc. Trop. Med. Hyg. 1992; 86:353–354.

Savioli, L., Stansfield, S., Bundy, D.A., Mitchell, A., Bhatia, R., Engels, D., Montresor, A., Neira, M. and Shein, A.M. Schistosomiasis and soil-transmitted helminth infections: forging control efforts. Trans R. Soc. Trop. Med. Hyg. 2002; 96:577–579.

Stoltzfus, R.J., Chwaya, H.M., Tielsch, J.M., Schulze, K.J., Albonico, M. and Savioli, L. Epidemiology of iron deficiency anemia in Zanzibari schoolchildren: the importance of hookworms. Am. J. Clin. Nutr. 1997a; 65:153–159.

Stoltzfus, R.J., Dreyfuss, M.L., Chwaya, H.M. and Albonico, M. Hookworm control as a strategy to prevent iron deficiency. Nutr. Rev. 1997b; 55:223–232.

Widdus, R. and White, K. Combating Diseases Associated with Poverty. Geneva: Initiative on Public Private Partnerships for Health, 2004.

Williamson, A.L., Lecchi, P., Turk, B.E., Choe, Y., Hotez, P.J., McKerrow, J.H., Cantley, L.C., Sajid, M. and Loukas, A. A multi-enzyme cascade of hemoglobin proteolysis in the intestine of blood-feeding hookworms. J. Biol. Chem. 2004; 279:3560–3567.

World Bank. School Deworming at a Glance. Geneva: World Bank. http://go.worldbank.org/F1760HE7H0, 2003.

World Health Organization. Prevention and Control of Schistosomiasis and Soil-Transmitted Helminthiases. WHO Technical Report Series, number 912. Geneva: World Health Organization, 2002.

World Health Organization. Deworming for Health and Development. Report of the Third Global Meeting of the Partners for Parasite Control. Geneva, 29–30 November 2004, 2005.

World Health Organization. The World Health Report 2004: Changing History. Geneva: World Health Organization, 2004.

Xiao, S., Zhan, B., Xue, J., Goud, G.N., Loukas, A., Liu, Y., Williamson, A., Liu, S., Deumic, V. and Hotez, P. The evaluation of recombinant hookworm antigens as vaccines in hamsters (*Mesocricetus auratus*) challenged with human hookworm, *Necator americanus*. Exp. Parasitol. 2008; 118:32–40.

Zhan, B., Badamchian, M., Bo, M.H., Ashcom, J., Feng, J.J., Hawdon, J., Xiao, S.H. and Hotez, P.J. Molecular cloning and purification of *Ac*-TMP a developmentally regulated putative tissue inhibitor of metalloprotease released in relative abundance by adult *Ancylostoma* hookworms. Am. J. Trop. Med. Hyg. 2002a; 66:238–244.

Zhan, B., Hawdon, J., Shan, Q., Ren, H., Qiang, H., Xiao, S.H., Li, T.H., Feng, Z. and Hotez, P. Construction and analysis of cDNA library of *Necator americanus* third stage larvae. Zhongguo Ji Sheng Chong Xue Yu Ji Sheng Chong Bing Za Zhi 2000; 18:26–28.

Zhan, B., Hotez, P.J., Wang, Y. and Hawdon, J.M. A developmentally regulated metalloprotease secreted by host-stimulated *Ancylostoma caninum* third-stage infective larvae is a member of the astacin family of proteases. Mol. Biochem. Parasitol. 2002b; 120:291–296.

Zhan, B., Liu, S., Perally, S., Fujiwara, R., Brophy, P., Liu, Y.Y., Feng, J.J., Williamson, A., Wang, Y., Bueno, L.L., Mendez, S., Goud, G., Bethony, J.M., Hawdon, J.M., Loukas, A., Jones, K. and Hotez, P.J. Biochemical characterization and vaccine potential of a heme binding glutathione S transferase (GST) from the adult hookworm *Ancylostoma caninum*. Infect. Immun. 2005; 73:6903–6911.

Zhan, B., Liu, Y.Y., Badamchian, M., Williamson, A., Feng, J.J., Loukas, A., Hawdon, J.M. and Hotez, P.J. Molecular characterization of the *Ancylostoma* secreted protein (ASP) family from the adult stage of *Ancylostoma caninum*. Int. J. Parasitol. 2003; 33:897–907.

Zhan, B., Wang, Y., Liu, Y.Y., Williamson, A., Loukas, A., Hawdon, J.M., Xue, H.C., Xiao, S.H. and Hotez, P.J. *Ac*-SAA-1, an immunodominant 16 kDa surface-associated antigen of infective larvae and adults of *Ancylostoma caninum*. Int. J. Parasitol. 2004; 34:1037–1045.

CHAPTER

67

Onchocerciasis

Sara Lustigman[1] *and David Abraham*[2]

[1]Laboratory of Molecular Parasitology, Lindsley F. Kimball Research Institute, New York Blood Center, New York, NY, USA

[2]Department of Microbiology and Immunology, Thomas Jefferson University, Philadelphia, PA, USA

OUTLINE

ABSTRACT

Human onchocerciasis is a serious neglected tropical disease caused by *Onchocerca volvulus* and an important cause of blindness and chronic disability in the developing world. Through mass drug administration of ivermectin, onchocerciasis has been recognized by the World Health Organization as a potential candidate for global elimination. However, formidable technical and logistical obstacles must be overcome before the goal of elimination in Africa can be attained. In addition to difficulties of compliance in this region, evidence is building for the existence of *Onchocerca* resistance to the drug ivermectin, which is at present the only drug used for the mass treatment of this population. Therefore, additional tools are critically needed and include the need for a vaccine against onchocerciasis to "complement" the present control measures and thus potentially eliminate this infection from humans. An *Onchocerca* vaccine targeting the infective stage larvae could reduce adult worm burden and thus the number of microfilariae produced by the adult female worms, and the causative stage of pathology. Importantly, protective immunity against *Onchocerca* larvae has now been definitively demonstrated in humans, cattle, and mice, thereby providing the conceptual underpinnings that a vaccine can be produced against this infection. Moreover, several recombinant *Onchocerca* antigens have been identified that are capable of inducing a partial but significant reduction in the survival of a larval challenge using an experimental mouse model. Thus, the research community is in an ideal position to proceed with the preclinical research needed in order to accelerate the product development of an *Onchocerca* vaccine for human use.

INTRODUCTION

Human onchocerciasis (river blindness) is a serious neglected tropical disease caused by the filarial nematode parasite *Onchocerca volvulus* and an important cause of blindness and chronic disability in the developing world. Presently, it is estimated that 37 million people carry *O. volvulus*, with 90 million at risk in Africa. Morbidity at present is estimated at 987,000 disability-adjusted life-years and there are 46,000 new cases of blindness annually. Through mass drug administration of the donated drug ivermectin (Mectizan®), onchocerciasis has been recognized by the World Health Organization as a potential candidate for global elimination, with a World Health Assembly goal to establish community-based sustainable treatments in areas of meso- and hyperendemicity by 2010 (www.apoc.bf/en). However, formidable technical and logistical obstacles must be overcome before the goal of elimination can be attained, and additional tools are critically needed to support the control measures and thus potentially eliminate this infection as a public health problem. These new control tool requirements include a need for a vaccine against onchocerciasis to "complement" the present control measures and further the goals of the existing control programs (Cook et al., 2001; Dadzie et al., 2003; Hoerauf and Steel, 2004).

The rationale for why a vaccine is in critical need includes the observations that: (a) the present control measures are not going to succeed in the elimination of *O. volvulus* infection by themselves (Dadzie et al., 2003); (b) evidence for emerging resistance to ivermectin, the only drug used for mass treatment of onchocerciasis (Osei-Atweneboana et al., 2007); and (c) practical considerations in treating people for 14–35 years (Winnen et al., 2002; Boatin and Richards, 2006). Importantly, protective immunity against *Onchocerca* infective larval stages, also known as third-stage larvae or L3, has now been definitively demonstrated in humans, cattle, and mice, thereby proving the conceptual underpinnings that a vaccine can be produced against this infection (Abraham et al., 2002; Lustigman et al., 2003; Tchakoute et al., 2006).

The value of developing a vaccine against *O. volvulus* infection was recognized by the Edna McConnell Clark Foundation. Their 15 years of support (1985–1999) provided the research foundation for antigen discovery and development of animal models for testing the efficacy of vaccine candidates. This chapter will highlight the current status of vaccine development against onchocerciasis and the potential for being able to accelerate vaccine development to provide this crucial element required for the successful elimination of onchocerciasis. Reduction in worm burden after vaccination, even if not absolute, will reduce the number of microfilariae produced by the adult female worms, and thereby reduce both pathology and the rates of transmission within endemic regions.

DISEASE

Epidemiology

Onchocerciasis or "river blindness" is a human disease caused by the filarial worm *Onchocerca volvulus* (Nematoda: Secernentea; Spirurida; Filariidae; Onchocerca). Adult worms can live for over a decade in nodules under the skin and release millions of microfilariae that are responsible for the morbidity associated with infection. When the microfilariae die, they cause skin rashes, lesions, intense and debilitating itching, and changes in skin pigmentation. Microfilariae also migrate to the eye and can cause ocular lesions resulting in blindness (Richards et al., 2001).

The distribution of onchocerciasis has been determined using rapid epidemiological mapping of onchocerciasis in which the levels of endemicity are assessed by onchocercal nodule prevalence in the endemic adult population (Ngoumou et al., 1994; Noma et al., 2002). More than 22,000 villages in Africa (outside the Onchocerciasis Control Programme (OCP) in West Africa) had been surveyed, allowing the identification of many new foci (Basáñez et al., 2006). Presently, it is estimated that 37 million people carry *O. volvulus*, with 90 million at risk in Africa (APOC, 2005). The geographic distribution includes 28 countries in tropical Africa, where 99% of infected people live, isolated foci in 6 countries in Latin America and in Yemen (WHO, 2002) (Fig. 67.1). Morbidity at present is estimated at 987,000 disability-adjusted life-years (www.who.int/tdr/diseases/oncho/diseaseinfo.htm). Onchocerciasis causes 46,000 new cases of blindness annually (www.apoc.bf/en) resulting in 270,000 individuals being blinded and an additional 500,000 with visual impairment, making onchocerciasis the second leading cause of infectious blindness worldwide. Although the importance of *Onchocerca*-induced blindness has long been recognized, it was only in 1995 that research funded by The Special Programme for Research and Training in Tropical Diseases demonstrated that *Onchocerca* skin disease had an even greater impact on human health. Severe itching now accounts for 60% disability-adjusted life-years. Other skin manifestations are not included in the disability-adjusted life-years estimates, even though they are highly prevalent and have a major psychosocial and economic impact.

Based on the complexity of the life cycle and the chronic nature of the disease progression, it is judged that *O. volvulus* is not a potential biothreat agent.

Life Cycle

Onchocerciasis is associated with fast flowing rivers where *Simulium* blackfly vectors breed. When a parasitized female blackfly takes a blood-meal from a human host, the infective *Onchocerca* larvae, pass from the blackfly into the host's skin and enter the host's subcutaneous tissue, where they migrate, induce nodule formation, and mature into adult worms over a period of 6–12 months (Fig. 67.2). Female worms measure 33–50 cm in length and 270–400 μm in diameter, while males measure 19–42 mm by 130–210 μm. The lifespan of adult worms is up to 15 years. Developing female adult worms either induce new nodules or enter existing nodules and cluster together with the resident female worms, while the smaller male worms migrate between nodules to mate. Fertilized eggs develop into microfilariae that actively exit from the female worm. A female worm may produce up to 1000 microfilariae per day and the microfilarial lifespan is 1–2 years, which results in millions of microfilariae residing in the subcutaneous tissues. When the infected host is bitten by female blackfly, microfilariae are transferred from the host to the blackfly where they develop into

FIGURE 67.1 World distribution of onchocerciasis. Currently 35 countries around the world are affected by onchocerciasis; 28 in tropical Africa, isolated foci in 6 countries in Latin America and Yemen.

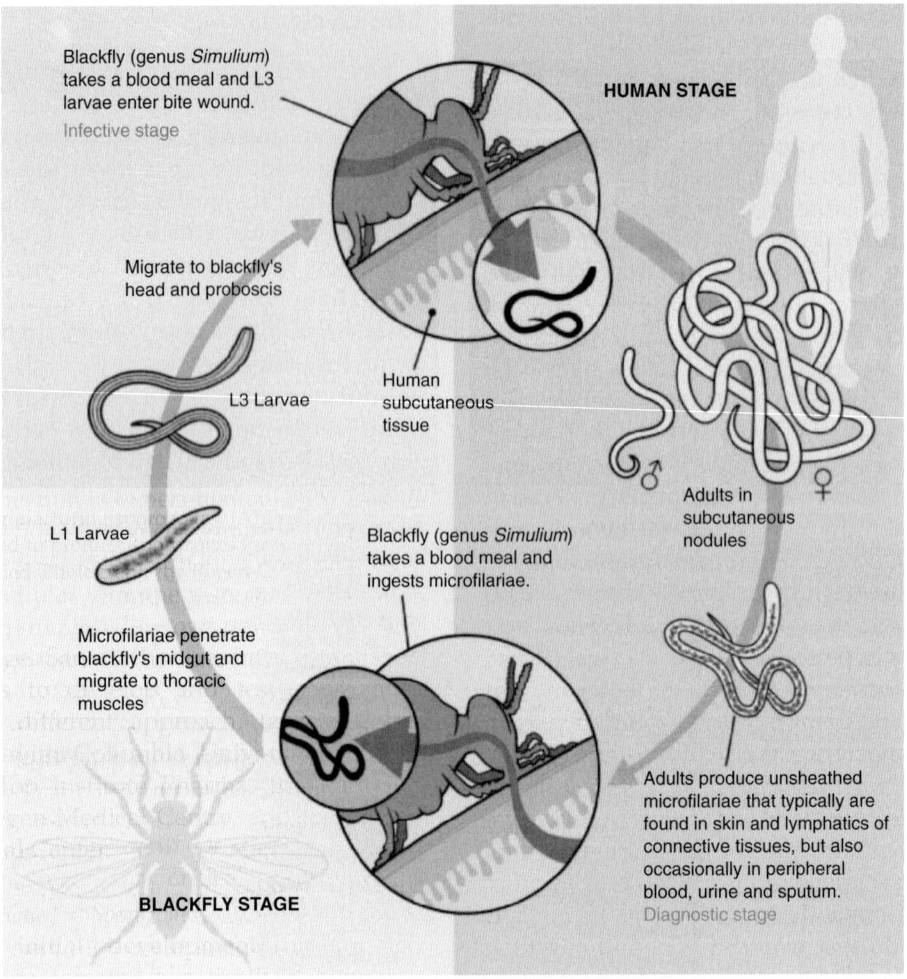

FIGURE 67.2 Life cycle of *O. volvulus*. The illustrator of this figure is Giovanni Maki (*Source*: Basáñez et al. (2006)). From: River Blindness: A Success Story under Threat? Basáñez, M.G., Pion, S.D.S., Churcher, T.S., Breitling, L.P., Little, M.P., et al. *PLos Medicine* Vol. 3, No. 9, e371 doi:10.1371/journal.pmed.0030371

infective larvae over a period of 6–12 days, which can then be transmitted to a new human host when the blackfly feeds again (www.dpd.cdc.gov/dpdx/html/Filariasis.asp?body = Frames/A-F/Filariasis/body_Filariasis_o_volvulus.htm).

Pathogenesis

Disease manifestations in the skin, lymphatics, and eye are all caused by immune responses to the parasite (Ottesen, 1995; Hoerauf and Brattig, 2002; Brattig, 2004; Pearlman and Gillette-Ferguson, 2007). Adult worms remain in subcutaneous nodules, which do not lead to morbidity or significant disfigurement in the human host. Nodule development occurs over a period of years and persists for decades. Nodules measure between 2mm and 6cm depending on the immune status of the host and the number of worms within the nodule. The principal cells of the granuloma surrounding the nodule are macrophages that mature into epithelioid cells and fuse to multinucleate foreign body giant cells. In addition, the granuloma is comprised of neutrophils, eosinophils, mast cells, lymphocytes, and NK cells (Buttner and Racz, 1983; Brattig, 2004).

The pathology of chronic onchocerciasis is a consequence of long-lasting host inflammatory responses directed at the microfilarial stage and the obligatory symbiont *Wolbachia* (Ottesen, 1995; Pearlman and Gillette-Ferguson, 2007). The skin is the principal site of infection, containing the mass of microfilariae. Live microfilariae appear to be unnoticed by host cells, whereas degenerating microfilariae are often surrounded by inflammatory cells that adhere and degranulate resulting in fragments of the worms being phagocytized by the adjacent cells (Buttner and Racz, 1983). The disease seen in the skin typically consists of intense itching, swelling, and inflammation. A grading system was developed to categorize the degree of skin involvement: acute papular dermatitis—scattered pruritic papules; chronic papular dermatitis—larger papules, resulting in hyperpigmentation; lichenified dermatitis—hyperpigmented papules and plaques, with edema, lymphadenopathy, pruritus, and common secondary bacterial infections; skin atrophy—loss of elasticity, skin resembles tissue paper, "lizard skin" appearance; depigmentation—"leopard skin" appearance, usually on anterior lower leg. Onchocerciasis represents a disease comprising a spectrum of manifestations and host responsiveness. The majority of *O. volvulus* infected persons present the common generalized form of onchocerciasis presenting mostly varying mild or moderate skin dermatitis. These patients are considered to be hyporeactive and patients with high microfilarial loads show no obvious capacity to effectively reduce the number of microfilariae and adult filariae, while patients with lower microfilarial densities intermittently present dermatitis with signs of antimicrofilarial defense reactivities. Patients may also have the rare hyperreactive form (sowda) which exhibits extensive reactivity and skin pathology. These patients show a severe chronic papular dermatitis and hyperpigmentation, lichenification, and lymphadenopathy. The number of parasites in these patients is very low and histological studies reveal massive perivascular infiltration in the large nodules, the regional lymph nodes and the affected skin sites (Buttner and Racz, 1983; Mackenzie et al., 1985; Brattig, 2004).

Most importantly, visual impairment and blindness develop as a result of microfilariae migrating into the conjunctiva, the cornea, and the posterior regions of the eye. Increasing numbers of degenerating microfilariae in the eye leads to inflammatory responses that result in bilateral blindness. In the area that is damaged, punctate keratitis occurs, which clears up as the inflammation subsides. However, if the infection is chronic, sclerosing keratitis can occur, making the affected area become opaque. Over time the entire cornea may become opaque, thus leading to blindness (Brattig, 2004).

All developmental stages and individual parasites of *O. volvulus* are infected with an obligatory symbiont rickettsia, *Wolbachia*. The adult worms harbor the *Wolbachia* in the hypodermis and in the female reproductive organs, where the bacteria are essential for the development of microfilariae within the female adult worm (Hoerauf et al., 2003; Taylor et al., 2005). The role of *Wolbachia* in the pathogenesis of filarial disease has been implicated by observations made after antifilarial therapy. Elevated *Wolbachia* DNA and even intact *Wolbachia* are detected in the blood and are associated with pro-inflammatory cytokines seen in patients with posttreatment side effects (Cross et al., 2001; Keiser et al., 2002). Furthermore, *Wolbachia* are required for recruitment of neutrophils to the *Onchocerca* nodules, as the number of neutrophils in nodules from individuals treated with doxycycline to eliminate *Wolbachia* is greatly reduced compared to untreated individuals (Brattig et al., 2001). Production of TNF-α and nitric oxide by mouse macrophages stimulated with filarial extracts is clearly associated with the presence of *Wolbachia* (Taylor et al., 2000). Using a mouse model of *O. volvulus* keratitis, it has been demonstrated that the endosymbiotic *Wolbachia* bacteria are essential for the pathogenesis of *O. volvulus* keratitis, as extracts from *O. volvulus* worms depleted of *Wolbachia* by antibiotic treatment do not induce corneal inflammation (Saint Andre et al., 2002).

Epidemiological, clinical, and genetic data all suggest that *O. volvulus* exists as two strains in Africa with the severe strain inducing ocular disease in a large proportion of the infected population, while the mild strain induces little ocular disease. It appears that the pathogenic differences seen in severe and mild strains *O. volvulus* are a function of the relative *Wolbachia* burden in the parasite with the median ratio of *Wolbachia* DNA to nuclear DNA significantly greater in the severe strain of parasites than in the mild strain (Higazi et al., 2005). Therefore, the pathogenesis of onchocerciasis is dependent on the host immune response to the microfilarial stage and to the endosymbiotic *Wolbachia* bacteria. The degree of pathology that develops is dependent on the number parasites and on the intensity and type of immune response that the patient develops to the infection. Finally, the pathology develops predominantly in chronic infections and is not associated with the infective stage larvae.

Diagnosis

The demonstration of microfilariae in skin biopsies (WHO, 1995) is the classic and definitive method for diagnosis of onchocerciasis as well as determining the intensity of infection. The use of techniques that are based on polymerase chain reaction (PCR) using species specific probes for the "Oncho-150" repeat have been shown to be effective for diagnosing *O. volvulus* infection by detecting parasite DNA in routine skin snips (Zimmerman et al., 1994). The technique is more sensitive in detecting low-level infections than standard skin-snip methods, but does not give any indication of the number of skin microfilariae. PCR methods are also used for surveillance of control measures, by identifying *O. volvulus* infections in the blackfly or the human host (Pischke et al., 2002; Rodriguez-Perez et al., 2006). Immunodiagnostic tests that are capable of detecting early infections have also developed (Harnett et al., 1998). These tests were of special interest for identifying new infections in areas under vector control (particularly the OCP area) and as a tool for onchocerciasis surveillance (Lipner et al., 2006). For larger epidemiological surveys such as "rapid epidemiological mapping of onchocerciasis," palpation for nodules, a less invasive diagnostic procedure, is being used (Ngoumou et al., 1994; Noma et al., 2002). The diethylcarbamazine patch test has been also shown to constitute a valuable tool to evaluate the levels of endemicity of onchocerciasis and to follow the intensity of transmission of *O. volvulus* (Ozoh et al., 2007).

Treatment and Control Measures

International programs supported by the World Health Organization and many other organizations have worked to control the impact of onchocerciasis using vector control with insecticides beginning in 1974 and mass drug administration with ivermectin (Mectizan) beginning in 1987 (Peters and Phillips, 2004). Ivermectin is a highly effective microfilaricide and inhibits female worm microfilarial production for several months. Annual mass drug administration reduces morbidity (Ejere et al., 2001; Tielsch and Beeche, 2004) and lowers transmission (Collins et al., 1992; Boussinesq et al., 1997). The Onchocerciasis Control Programme (OCP) in West Africa (1974–2002) greatly decreased *O. volvulus* transmission in the 11 OCP countries (Molyneux, 1995; Basáñez et al., 2006; Boatin and Richards, 2006). Ivermectin without vector control has been the principal tool for the Onchocerciasis Elimination Program in the Americas (OEPA, 1992–present) (Boatin and Richards, 2006) and the African Programme for Onchocerciasis Control (APOC, 1995–present). In the Americas, where *O. volvulus* is less common, OEPA has substantially reduced transmission and is on track to eliminate the disease (Boatin and Richards, 2006). APOC has extended treatment to 19 countries beyond OCP through sustainable community-directed ivermectin treatment (Remme, 1995; Richards et al., 2001). By the end of 2005, 400 million treatments had been supplied in Africa by Merck's Mectizan Donation Program (www.mectizan.org), with an estimated 40 million people treated by nearly 300,000 community distributors (www.apoc.bf).

Although vector control and periodic administration of ivermectin have resulted in significant reductions in disease rates in many regions of Africa (Abiose et al., 2000), the present programs, relying only on periodic administration of ivermectin, still have severe limitations. Ivermectin is only effective against microfilariae, which cause blindness and dermatitis, but has little activity against the long-lived adult worms. Drugs that eliminate *O. volvulus* adult worms by targeting the endosymbiotic *Wolbachia* bacteria have been developed, but they have limited populations in which they can be used (Taylor, 2000; Richards et al., 2001). Moreover, it is estimated that at least 25 years of annual treatment at 65% total population coverage would be necessary to eliminate infection in areas of medium to high levels of infection, and more dire model predictions suggest that durations exceeding 35 years would be required if there was heterogeneity in exposure to *S. damnosum* vector bites (Winnen et al., 2002; Boatin and Richards, 2006). The predictions that annual ivermectin treatments will need to be given for anything from 14 to 35 years or more, not only creates significant logistical obstacles, but also has lead to a concern that resistance to ivermectin in *O. volvulus* will eventually emerge. The potential emergence of drug-resistant strains of the *O. volvulus* parasite have been suggested initially by reports that showed incidence of patients with onchocerciasis who failed to respond to ivermectin treatment (Awadzi et al., 2004a, 2004b). Significantly, a recent report from Ghana has provided the first proof of ivermectin resistance in *O. volvulus* in humans (Osei-Atweneboana et al., 2007). Additionally, studies have associated ivermectin resistance with genetic markers (Huang and Prichard, 1999; Kohler, 2001; Ardelli and Prichard, 2004; Ardelli et al., 2005, 2006a, 2006b; Eng and Prichard, 2005; Bourguinat et al., 2006), particularly the β-tubulin gene in human *O. volvulus* and the livestock nematode parasite *Haemonchus contortus* (Eng and Prichard, 2005; Eng et al., 2006). A report by Bourguinat et al. (2007) showed that not only the *β-tubulin* genotype has changed after 4 or 13 ivermectin treatments but also female fertility was affected. The finding that ivermectin treatment selected for *β-tubulin* heterozygotes and that this selection was dependent on dose raises important concerns for the current river blindness control programs.

PROTECTIVE IMMUNITY AGAINST *O. volvulus* IN HUMANS AND ANIMAL MODELS

Human

Protective Immunity in Humans is Associated with Th1 and Th2 Responses against O. volvulus L3

Two distinctive expressions of anti-L3 protective immunity within the *O. volvulus* endemic population have been described: (a) which impedes the development of a patent infection (microfilaria positive) in the putatively immune (PI) individuals (Gallin et al.,

1988; Ward et al., 1988; Elson et al., 1994; Turaga et al., 2000); and (b) concomitant immunity to *O. volvulus* L3 that develops in the patently infected (INF) individuals with increasing age, and which is independent of the immune responses that are induced by the adult worms and microfilaria associated with patent infection. Concomitant immunity is believed to prevent most of the newly acquired L3 infections from developing, and results in a stable adult worm burden in the INF (Duke and Moore, 1968; Schulz-Key, 1990; MacDonald et al., 2002).

Studies of the mechanisms of acquired immunity against *O. volvulus* infection in humans have focused on identification of defined cellular and humoral responses in the PI individuals as compared to responses in the INF as well as in those associated with concomitant immunity (reviewed in McCarthy and Nutman (1996) and Hocrauf and Brattig (2002)). Some of the initial studies produced conflicting results (Turaga et al., 2000; Hoerauf and Brattig, 2002) in terms of Th1 or Th2 dominance of cellular responses in the PI to adult worm antigens. This may have been due to examining immune responses against adult worm antigens, which contain female, male, and microfilariae stage antigens. However, when antigens from L3, L3 in the process of molting into the fourth-stage larvae (mL3), isolated female worms (F-OvAg), and skin microfilariae (Smf) were used in studies with PI and INF, an interesting differential cytokine profile emerged. First, although an IL-5 response was present in both the PI and the INF toward L3 and mL3 antigens, the PI had significantly elevated IL-5 responses to both L3 antigens. Second, a subgroup of these individuals also had significantly elevated IFN-γ responses toward L3 (27%) and mL3 (54.5%) antigens, thus indicating an overall response of a mixed Th1/Th2 profile. This raises the possibility that some of the PI have more than one mechanism of anti-L3 protective immunity. In contrast, the F-OvAg and Smf antigens induced only a Th2 type response in the PI. However, the IL-5 response to F-OvAg in this group was significantly higher than that in the INF, a group that as a whole, exhibited only a Th2 phenotype in response to all antigens. Moreover, while the PI produced significantly higher levels of GM-CSF toward L3 and mL3 compared to the INF, there was no difference in the GM-CSF response between the two groups toward the other antigens.

These observations demonstrated that the two groups, PI and INF, although similar in their anti-F-OvAg and Smf responses, are distinct in their anti-L3 and mL3 immunity. It also appeared that the PI could be divided into two subgroups; those with elevated anti-L3/mL3 IL-5 (e.g., Th2) and GM-CSF responses, and those with elevated anti-L3/mL3 IL-5 and IFN-γ (e.g., mixed Th1/Th2), and GM-CSF responses. These studies confirmed the hypothesis that immunological studies of the PI could lead to the identification of immune mechanisms that play a role in anti-L3 protective immunity (Turaga et al., 2000).

Age-dependent resistance to super-infection (Day et al., 1991), and acquired concomitant humoral (Kurniawan-Atmadja et al., 1998) and cellular (Sartono et al., 1997) immunity in infected individuals has been described in lymphatic filariasis. The existence of anti-larval immunity that is separate and distinct from that directed against adult worms and microfilariae (concomitant immunity) has also been observed in individuals infected with *O. volvulus*. Studies have shown that the proliferative responses to L3 increased significantly with age, while there was a downward trend in response to F-OvAg. The responses in the INF group against L3, mL3, F-OvAg, and Smf antigens in the infected group were predominantly of the Th2-type (higher frequency of IL-5 producers than IFN-γ producers). However, when the cytokine responses were analyzed in relation to age of the PBMC donors (=years of exposure to *O. volvulus*), it appeared that a differential stage-specific IL-5 response in the infected group was observed. It appeared that a distinct IL-5 response was acquired with age against the L3 stages and this response was independent of that induced by F-OvAg and Smf antigens, which are associated with patent infection. The responses against F-OvAg and Smf antigens significantly decreased with age. Moreover, in INF individuals who had GM-CSF responses to L3, the amounts secreted were positively correlated with IL-5. Thus, older INF, like the PI subjects as a whole (Turaga et al., 2000), had predominantly IL-5 responses to L3 and mL3 antigens, which were accompanied by elevated GM-CSF. Similarly, a Th1 response was also present in a subgroup of the INF (MacDonald et al., 2002).

A Role for Antibodies in Protective Immunity Against O. volvulus Larvae in Humans

Data from several studies support the view that antibodies are part of the effector mechanism against *O. volvulus* larvae. These include the correlation between elevated antigen-specific IgG1 and/or IgG3 responses and the immune status of exposed individuals (Boyer et al., 1991; Soboslay et al., 1994, 1997; Ottesen, 1995; Stewart et al., 1995; Faulkner et al., 2001). When sera from INF and PI individuals were analyzed by Western blots, significant differences between their antibody responses to a broad range of crude *O. volvulus* larval and adult stage antigens were also observed (Boyer et al., 1991; Luder et al., 1996; Irvine et al., 1997; Soboslay et al., 1997). Whereas both groups recognized many common antigens, the PI sera reacted with some stage-specific antigens uniquely (Irvine et al., 1997). In addition, a positive correlation was seen between anti-L3-specific IgG3 and IgE cytophilic antibody responses and increasing age of the infected subjects as well as elevated IgG1 response regardless of age (MacDonald et al., 2002). L3 surface-specific antibodies were seen in 80% of putatively immune individuals and in 83% of those infected. Moreover, antibodies that bind specifically to the surface of L3 but not to the surface of microfilariae were found in sera from infected subjects, and the intensity of L3 surface recognition was positively correlated with their age, thus suggesting a role for antibodies also in concomitant immunity (MacDonald et al., 2002). Similar results were also obtained in individuals infected with lymphatic filariasis (Kurniawan-Atmadja et al., 1998) where concomitant immunity is more clearly defined (Day et al., 1991). In vitro antibody-dependent killing of *O. volvulus* L3 and inhibition of molting of L3 to fourth-stage larvae has been observed to be dependent on granulocytes and antibody from putatively immune and/or infected individuals (Greene et al., 1985; Leke et al., 1989; Johnson et al., 1994).

Animal

O. volvulus in Chimpanzees

The chimpanzee is the only species aside from humans that is susceptible to infection with *O. volvulus*, although without the development of significant pathology (Duke, 1980). Studies have been performed in chimpanzees to determine the immunological responses that develop following primary and challenge infections. Antibody responses were measured after primary infections with L3 against antigens solubilized from adult worms and against recombinant *O. volvulus* antigens (Weiss et al., 1986; Eberhard et al., 1991, 1995; Soboslay et al., 1992; Luder et al., 1993). When the infections developed, the number of individual antigens recognized increased as determined by Western blot analyses. In addition, antibody responses in animals with microfilariae in the skin could be distinguished from responses in nonpatent animals (Soboslay et al., 1992). Lymphocyte proliferation assays were performed with cells from infected chimpanzees and it was observed that, after a strong early response to parasite antigens, a reversible cellular hyporesponsiveness developed (Soboslay et al., 1991; Luder et al., 1993).

A single study was performed to determine whether or not chimpanzees could be immunized against infection with *O. volvulus*. Four chimpanzees were immunized with irradiated L3 followed by infection with L3. After 11–26 months, three out of the four immunized chimpanzees developed microfilariae in the skin, whereas all four of the control animals became patent. There were no significant differences in the number of microfilariae found in the skin of the immunized chimpanzees when compared with the infected controls (Prince et al., 1992). This experiment is difficult to interpret because adult worm burdens cannot be determined in live chimpanzees, hence protective effects could be obscured by microfilariae production from very few adult worms. The reason why the one vaccinated animal did not develop microfilariae in the skin could have been a result of a failure to get infected by the challenge L3 (not all chimpanzees develop patent infections after primary challenge (Eberhard et al., 1991; Soboslay et al., 1991)) or could have been the result of some immune protection. This animal developed an antibody response to 11–12 kDa L3 antigens of *A. viteae* (Luder et al., 1993). Serum from this immunized and protected chimpanzee was used for immunoscreening and resulted in the cloning and identification of *Ov*-CPI-2 (Lustigman et al., 1991, 1992a), one of the 15 protective recombinant *O. volvulus* antigens (r*Ov*Ags) (Lustigman et al., 2002).

Protective Immunity to *Onchocerca ochengi* in Cattle

O. ochengi infection of cattle is an excellent analog of human *Onchocerca* infection (Trees et al., 2000). Epidemiological evidence demonstrates crossreactivity between antigens from *O. volvulus* and *O. ochengi* that extends to cross-protection of humans against *O. volvulus*, by exposure to infected larvae of *O. ochengi* (Wahl et al., 1998). The superficial location of the adult worms of *O. ochengi* contained within nodules in the infected cows has enabled studies measuring adult parasite worm burdens. Treatment of cattle infected with *O. ochengi* with ivermectin (Njongmeta et al., 2004), melarsomine (Tchakoute et al., 2006), or oxytetracycline (Nfon et al., 2007) to control the infection did not effect the susceptibility of these cattle to re-infection. Field studies, however, have provided evidence that immunity develops in cattle to both microfilariae and larval stages. Whereas adult burdens increased with age in naturally infected cattle, microfilarial loads decreased in older cattle despite the continuing fecundity of excised, adult, female worms (Trees et al., 1992). Cattle have been identified in a number of studies that have naturally developed protective immunity to infection and are considered putatively immune. The cattle were maintained in endemic areas yet were not infected with *O. ochengi*, based on the absence of nodules with adult worms and microfilariae in the skin (Hoch et al., 1992; Trees et al., 1992; Wahl et al., 1994). Putatively immune cattle were significantly less susceptible to challenge infections as compared to naïve cattle based on acquisition of *O. ochengi* nodules. The density of microfilariae in the skin was substantially lower in the putatively immune cattle as compared to their naïve counterparts. Furthermore, cattle immunized with *O. ochengi* irradiated third-stage larvae were significantly protected against both experimental (100% reduction in median nodule load) and field challenge (64% reduction in median nodule load) infections (Tchakoute et al., 2006). Finally, calves immunized with live *O. volvulus* larvae and subsequently challenged with *O. ochengi* larvae had significantly fewer adult *O. ochengi* parasites as compared controls (Achukwi et al., 2007). In summary, these experiments demonstrate the feasibility of vaccinating a natural host against *Onchocerca* infection and support the feasibility of developing a vaccine against the infection in humans.

Immunity against Onchocerca spp. Microfilariae

Vaccination of cattle with microfilariae of *O. lienalis* was shown to enhance the clearance of microfilariae subsequently transplanted into the same animal (Townson and Bianco, 1982a). Moreover in natural infections of cattle with *O. ochengi*, skin microfilarial density falls with age in spite of increasing numbers of fecund female worms, which suggests that a level of stage-specific microfilarial immunity may develop (Trees et al., 1992). Mice have also been used as surrogate hosts to investigate immunity against *Onchocerca* spp. microfilariae. Mice can support the survival of inoculated microfilariae for 3–4 months. Microfilariae injected subcutaneously at the nape of the neck, concentrate in the ears, therefore recovery of microfilariae from the ears post-mortem can be used as a measure of parasite survival (Townson and Bianco, 1982b; Carlow et al., 1986). Using this model with *O. lienalis*, significant protective immunity was conferred by live, homologous primary infection, immunization with microfilariae extracts and with live infections that were abbreviated by drug treatment (Townson et al., 1984; Carlow et al., 1986). Some protection was also conferred by heterologous stages and species but at a lower level (Townson et al., 1985; Carlow and Bianco, 1987). Significantly, protection against *O. volvulus* microfilariae was also demonstrated in mice immunized with live homologous or heterologous (*O. lienalis*) microfilariae (Bianco et al., 1991). The mechanisms of protection have been extensively studied using *O. lienalis*. Eosinophils are crucial and IL-5 is dominant in the expression of microfilarial immunity (Folkard et al., 1996; Hogarth and Bianco, 1999b). Using gene knockout mice, clearance of microfilariae following primary and secondary infections was independent of IL-4 (Hogarth et al., 1995, 1998). Exogenous IL-12 was used to ablate parasite-induced IL-4 responses without eliminating the protective IL-5-dependent responses (Hogarth and Bianco, 1999a).

Protective Immunity to O. volvulus Larvae in Mice

A mouse model has been developed to study protective immunity against *O. volvulus* larvae. Although *O. volvulus* L3 do not develop into adult worms in mice, it is possible to study the effects of anti-L3 immunity by placing *O. volvulus* L3 in subcutaneous implantable diffusion chambers, which are porous to host antibody and cells. L3 develop equally well, for a short period of time, in resistant hosts (such as mice) as they do in the susceptible chimpanzee host when implanted within diffusion chambers (Abraham et al., 1993). To study protective immunity in this mouse model, mice were immunized by injection of normal, irradiated, or freeze-thaw-killed *Onchocerca* spp. larvae, and then received challenge infections of L3 contained in diffusion chambers. There was a significant reduction of ~50% in the survival of challenge parasites in all the experimentally immunized mice (Lange et al., 1993).

The mouse model was also used to establish the mechanism of immune-mediated killing of *O. volvulus* L3 in diffusion chambers in mice immunized with irradiated L3. IL-5, but not IFN-γ, was found in diffusion chambers of immunized mice coincident with the time of parasite killing. Elimination of either IL-5 or IL-4 by monoclonal antibody treatment significantly reduced the protective effects of vaccination against *O. volvulus* L3 (Lange et al., 1994). The dependency on Th2 cytokines was further corroborated in cytokine-deficient mice. Immunized IL-4-deficient mice did not develop protective immunity, whereas IFN-γ deficient animals developed immunity comparable to immunized wild-type mice (Johnson et al., 1998). The finding that immunity was dependent on IL-4 and IL-5 suggested a dependency on the antibody isotypes IgG1 and/or IgE. Two mouse strains, μMT mice that lack all mature B cells and Xid mice that have deficiencies in their B-1 cells, were used to determine if antibody was required for protective immunity in mice immunized with irradiated L3. Immunity did not develop in μMT mice but did develop in the Xid mice. Furthermore, mice treated to eliminate IgE from the blood did not develop protective immunity. The role of granulocytes in protective immunity was also determined in mice immunized with irradiated L3. Immunity did not develop in mice if granulocytes, including both neutrophils and eosinophils, were eliminated nor did it develop if only eosinophils were eliminated (Abraham et al., 2004).

Wolbachia from *O. volvulus* contains molecules with LPS-like activity that induce an inflammatory response from monocytes (Brattig et al., 2000). Various studies suggest that TLR4 is an important mediator in the response to the *Wolbachia* endosymbiont and that immunity to *Wolbachia* may be an integral component of the immune response to the L3 of *O. volvulus* (Taylor et al., 2000; Brattig et al., 2004). TLR4-mutant mice were immunized against *O. volvulus* with irradiated L3 and it was observed that antibody responses to *Wolbachia* were absent in both wild-type and TLR4-mutant mice. Protective immunity, however, did not develop in the TLR4-mutant mice even though the Th2 immune response was intact (Kerepesi et al., 2005). It can therefore be concluded that immunity induced by irradiated L3 in mice is dependent on: (a) a Th2 response through IL-4 and IL-5; (b) a B-2 cell response through production of parasite-specific IgE; (c) eosinophils as effector cells; and (d) signaling through TLR4. Protective immunity to the larvae of *O. volvulus* is not, however, dependent on an immune response to *Wolbachia.*

VACCINES

The development of an *Onchocerca* vaccine has been the focus of research supported by the Edna McConnell Clark Foundation who funded from 1985 to 1999 research projects aimed at the development of a vaccine against infection with *O. volvulus* L3 (Cook et al., 2001; Richards et al., 2001; Abraham et al., 2002; Hoerauf and Brattig, 2002; Lustigman et al., 2002; Nutman, 2002; Williams et al., 2002). This approach has also been endorsed in 2002 during a "Conference on the Eradicability of Onchocerciasis" at the Carter Center (Dadzie et al., 2003) and later at a conference convened in 2004 on the global program to eliminate lymphatic filariasis (Hoerauf and Steel, 2004). Vaccines are still the most economical, efficient, and effective tool to control infectious diseases and might be the only way to guarantee elimination of onchocerciasis as initially envisioned by the World Health Organization. The ultimate goal of such a vaccine would be to complement the existing control programs and thus to combine it with drug treatments in a program of vaccine-linked chemotherapy (Hotez and Ferris, 2006; Hotez, 2007). The objective is to induce protection that is not necessarily sterile against the incoming infective third-stage larvae but rather that reduces adult worm burden. This will result in a reduction in the number of microfilariae produced by the adult female worms, a reduction in the development of pathology in the skin and in the eye, and a decrease in the potential for disease transmission. It is anticipated that continual natural exposure of the immunized individuals to infective larvae from black flies would sustain the protective immunity induced by immunization with recombinant vaccine proteins, and that ultimately the combination of vaccination with long-term chemotherapy with ivermectin could accelerate the eradication of onchocerciasis from sub-Saharan Africa.

Importantly, protective immunity against *O. volvulus* L3 has now been definitively demonstrated in humans, cattle, and mice, thereby proving the conceptual underpinnings that a vaccine can be produced against this infection (Abraham et al., 2002; Lustigman et al., 2003; Tchakoute et al., 2006). Studies of other filarial worms have clearly shown that the targets of the protective host immune response were the L3 and the molting L3 (Eisenbeiss et al., 1994; Le Goff et al., 2000), and that excretory–secretory proteins from the developing parasites induce protection (Lucius et al., 1991). These studies also suggested that the protective immune responses could inhibit the development of L3 to L4 (Horauf and Fleischer, 1997; Le Goff et al., 2000; Abraham et al., 2001). Assuming that there are common protective mechanisms against filariae and that the target antigens are similar, it advocated pursuing *O. volvulus* L3, mL3, and excretory–secretory antigens, and to also include larval proteins that are homologous to the other filarial protective antigens.

Although irradiated larvae serve as successful immunogens for inducing protective immunity in mice, it is clear that these larvae could never be used in a vaccine for use in humans because of safety and practicality issues. A more reasonable approach for immunization would be to use recombinant antigens as the basis for the vaccine. The experimental *O. volvulus* mouse model was subsequently used to screen recombinant antigens and protection was defined by significant reduction in the survival of L3 following two immunizations with 25μg of respective purified recombinant protein in the presence of an adjuvant.

While vaccine-induced immunity may function by a different constellation of mechanisms than those seen in PI and the individuals with concomitant immunity, it is likely that some of the mechanisms developed naturally over time in humans will be also induced by the recombinant *O. volvulus* protective antigens. The objective of an *Onchocerca* vaccine will be to elicit in individuals living in endemic areas defined anti-larvae immune responses similar to those present in the PI and the individuals with acquired concomitant immunity that will enable them to control effectively incoming *O. volvulus* L3 infections.

Identification of Protective *O. volvulus* Larval Antigens

Two basic strategies were used to identify and clone *Onchocerca* larval target-antigens to be tested as possible vaccine candidates. The first strategy was based on the potential involvement of antibodies in protective immunity. Attempts were made to identify the target proteins by immunoscreening cDNA libraries using immune sera from human or animal hosts. Although this approach was used with success, it did not take into account carbohydrate and other nonprotein determinants shown to be important in other helminth systems (Ellis et al., 1994; McVay et al., 1998, 2000; van Der Kleij et al., 1999), and which could be crucial to the generation of protective immunity. The success of the immunoscreening effort relied mostly on the source and specificity of the antibodies used and, hence, was done mostly with serum samples from PI with the goal of identifying antigens that were preferentially recognized by the PI in comparison with the INF. In addition, sera from vaccinated or immune animals (chimpanzees, mice, or cows), polyclonal antibodies raised against *O. volvulus* L3 or monoclonal antibodies developed against specific parasite antigens were used to screen the cDNA libraries, initially using cDNA libraries constructed from adult worm stages of *O. volvulus* and later using cDNA libraries constructed from *O. volvulus* larval stages, L3, mL3, and L4 (Joseph et al., 1997, 1998; Lizotte-Waniewski et al., 2000).

Immunoscreening of the adult cDNA libraries resulted in the identification and characterization of >26 recombinant proteins. Because some of these proteins were also expressed in larval stages as determined by in situ hybridization analysis, immunolocalization, stage-specific Western blot or PCR, they were tested in the *O. volvulus* experimental mouse model for vaccine efficacy. Immunoscreening of the larval cDNA libraries resulted in the cloning of additional proteins that were successful in stimulating significant protection in mice (Abraham et al., 2001). Altogether, out of 26 recombinant antigens that were identified by immunoscreening and tested in the *O. volvulus* mouse model, 12 induced partial but significant protection (39–69%) in the presence of Block Copolymer, alum, or Freund's Complete Adjuvant (Table 67.1). Eight of them, *Ov*-CPI-2, *Ov*-RAL-2, *Ov*-RBP-1, *Ov*-103, *Ov*-B20, OI5/OI3, *Ov*-CAL-1, and *Ov*-TMY-1 were cloned from the adult library, and four of them *Ov*-ALT-1, *Ov*-FBA-1, *Ov*-ASP-1, and *Ov*-B8 were cloned from the L3 cDNA library. Only two of the protective antigens, *Ov*-CPI-2 and *Ov*-RAL-2, were cloned from both libraries. Five of the protective antigens (*Ov*-CPI-2, *Ov*-ALT-1, *Ov*-RAL-2, *Ov*-FBA-1, and *Ov*-B8) were protective in multiple experiments. The level of protection with each recombinant protein was in the same range as that induced by immunization with irradiated L3 (Lange et al., 1993; Taylor et al., 1994; Johnson et al., 1998). Moreover, it appeared that immunization with the recombinant antigens *Ov*-CPI-2, *Ov*-ALT-1, and *Ov*-B8 also induced significant reduction in the molting of L3 (Abraham et al., 2001).

The second strategy was to isolate molecules thought to be vital in the infection process. These would include proteins with vital metabolic functions or defense properties, which permit the parasite to survive in immunocompetent hosts. Targeting such molecules would block the establishment of the parasite in the host. In addition, antigens that are not normally seen by the host but that are nevertheless accessible to host immune-effector molecules and cells, the "hidden antigens," were also thought to be potentially useful as vaccine targets (Sher, 1988). Isolation of the gene encoding a protein of interest was achieved by either: (a) screening a cDNA library using a heterologous probe (Henkle-Duhrsen and Kampkotter, 2001); (b) PCR cloning using degenerate primers (Henkle-Duhrsen and Kampkotter, 2001); (c) purifying the protein followed by partial amino acid sequencing and molecular cloning (Wu et al., 1996); or (d) identifying the gene of interest by searching the *O. volvulus* expressed sequence tag (EST) database or the EST databases generated by the Filarial Genome Project (Lizotte-Waniewski et al., 2000).

Out of 18 recombinant antigens that have been cloned using the second strategy and tested in the *O. volvulus* mouse model, four (*Ov*-ALT-1, *Ov*-CHI-1, *Av*-ABC, and *Av*-UBI) induced partial but significant protection (48–63%). *Ov*-ALT-1, *Av*-ABC, and *Av*-UBI were protective in the presence of alum or Freund's Complete Adjuvant (Table 67.1). Chitinase, *Ov*-CHI-1, effectively induced protection using DNA immunization (Harrison et al., 1999). *Av*-ABC and *Av*-UBI that were used for immunization have been cloned from the rodent filarial parasite *Acanthocheilonema viteae* as the *Onchocerca* homologue of *Av*-ABC has not been cloned as yet and the *Av*-UBI of *A. viteae* is completely identical to *Ov*-UBI (Wu et al., unpublished).

TABLE 67.1 Portfolio of the O. *volvulus* protective larval proteins

Characteristics of the ***O. volvulus*** protective protein (Lustigman et al., 2002)						Protection in animal models		
Antigen (kDa)	**Identity (Function)**	**Localization[a]**	**Immunogenicity[b]**	**Number of ESTs[c] L3/mL3**	**In vitro killing assays[d]**	**Protection in *Ov* mouse model (%)[e] (adjuvant)**	**Protection in lymphatic filariae models**	**Protection in other helminth models**
Ov-CPI-2 (17)	Onchocystatin, (cysteine protease inhibitor)	Hypodermis; basal layer of cuticle; separation of L3/L4 cuticles; secretory vesicles; ES	Chimpanzee anti-*Ov*-xL3; PI sera; CI sera	59/9	Anti-*Ov*-CPI-2 and *Ov* L3 (Lustigman, unpublished)	43–49% (alum)	*Ls*-cystatin (Pfaff et al., 2002)	*Ac*-cystatin (Hotez, unpublished)
Ov-ASP-1 (25)	Novel, homolog of vespid venom allergen 5 and the PR-1 protein family	Granules of glandular esophagus; ES	PI sera CI sera mice anti-*Ov*-xL3	50/1	anti-*Bm*-ASP-1 and *Bm* L3 or Mf (Anand et al., 2006)	44% (alum) 42% (FCA)	*Bm*-ASP-1 (Anand et al., 2006)	*Ac*-ASP-2 (Goud et al., 2004)
Ov-RAL-2 (17)	Novel, nematode specific	Hypodermis	PI sera CI sera mice anti-*Ov*-xL3	6/4	Anti-r*As*16 and Ascaris L3 (Tsuji et al., 2004)	51–60% (FCA)	r*Wb*-SXP (Ramachandran et al., 2004) r*Bm*-SXP-1 (Wang et al., 1997)	r*As*16 (Tsuji et al., 2001, 2003, 2004) r*Ac*-16 (Hotez, unpublished)
Ov-ALT-1 (15)	Novel, filariae specific	Granules of glandular esophagus; cuticle; channels	PI sera CI sera	223/18	Anti-*Bm*-ALT-2 and *Bm* L3 or Mf	39–62% (alum)	*Bm*-ALT-1 (Gregory et al., 2000) *Bm*-ALT-2 (Anand et al., 2006)	ALT-1 is a filariae-specific protein
Ov-103 (15)	Novel, nematode specific	In L3: Basal layer of the cuticle; hypodermis; basal lamina; channels; multivesicular bodies. In Mf: surface	PI sera	5/0	Anti-*Ov*-103 and *Ov* Mf (Lustigman et al., 1992b)	30–69% (alum)	ND[f]	*Ac*-SAA-1 (Zhan et al., 2004)
Ov-B20 (52/65)	Novel, nematode specific	Cuticle; hypodermis; ES product	Cattle anti-*Ol*-xL3	3/2	ND	39% (alum)	*Ov*-B20 in *Av* model (Taylor et al., 1995; Jenkins et al., 1996)	ND
Ov-RBP-1 (20/22)	Novel, nematode specific; retinoid binding protein	Body wall; ES product	ND	1/2	ND	42% (BC)	*Ov*-RBP-1 in *Av* model (Taylor et al., 1995; Jenkins et al., 1996)	ND
Ov-CHI-1 (75)	Chitinase	Cuticle, granules of glandular esophagus	ND	0/0	ND	53% (DNA)	*Bm*-chitinase (Adam et al., 1996; Wang et al., 1997)	ND
Ov-FBA-1(41)	Fructose 1,6 biphosphate aldolase	Basal layer of cuticle; channels; basal lamina	PI sera mice anti-*Ov*-xL3	29/7	ND	51–54% (FCA)	ND	ND
Ov-TMY-1 (42)	Tropomyosin	Body wall; ES product	Cattle anti-*Ol*-xL3	15/3	ND	44% (alum)	ND	ND

OI5/OI3 (>200)	Novel	Hypodermis; cuticle	PI sera	0/0	ND	35% (BC)	ND	ND
Ov-B8 (66)	PHD-finger protein	Basal lamina; secretory vesicles	PI sera	0/0	ND	37–46% (alum)	ND	ND
Av-ABC (18)	ATP-binding-cassette transporter	Cuticle; glandular esophagus	ND	0/0	ND	63% (alum)	ND	ND
Av-UBI (30)	Ubiquitin	Glandular esophagus; body wall	ND	0/0	ND	48% (FCA)	ND	ND
Ov-CAL-1 (45)	Calponin	Longitudinal muscles	ND	0/2	ND	42% (BC)	ND	ND

Abbreviations: *Ov*, *O. volvulus*; *Ol*, *O. lienalis*; *Bm*, *B. malayi*; *Wb*, *W. branconfti*; *Ls*, *L. sigmodontis*; *Av*, *A. viteae*; *Ac*, *A. ceylanicum*; *As*, *Ascaris Suum*; BC, block copolymer; ES, excretory–secretory product.

[a] Localization based on the native protein in larval stages (L3 and mL3) as determined by IEM.

[b] Immunogenicity based on data obtained from protected humans, putatively immune individuals (PI), infected individuals who developed concomitant immunity (CI), and/or antibodies from xL3 animal models (xL3 mouse model, cows, or chimpanzees).

[c] The number of ESTs were determined by BLAST-searching the L3 and mL3 EST datasets (3510 and 5165 entries, respectively) using each individual gene sequence. A gene was considered upregulated if the ESTs occurred at least five times in a particular stage.

[d] Using in vitro cytotoxicity assays few antigens were shown to be a target for antibodies raised against the recombinant antigens of ***Ov*-CPI-2** (Lustigman, unpublished) and ***Ov*-103** (Lustigman et al., 1992b). Although antibodies against the other vaccine candidates were not tested in vitro for their ability to inhibit molting or kill larvae, it appeared that immunization with the recombinant antigens ***Ov*-CPI-2** and ***Ov*-ALT-1** also induced significant reduction in the molting of L3 (Abraham et al., 2001). Moreover, studies using antibodies from mice immunized with the *B. malayi* homologous recombinant proteins of ***Ov*-ALT-1 (*Bm*-ALT-2)** and ***Ov*-ASP-1** (***Bm*-ASP-1)** have shown that anti-***Bm*-ALT-2** antibodies elicited 71–72% cytotoxicity in vitro against both L3 and Mf, while anti-***Bm*-ASP-1** antibodies induced 61–62% cytotoxicity in vitro against both L3 and Mf (Anand et al., 2006). Interestingly, antibodies against the homologous proteins of ***Ov*-103** in hookworms (*Ac*-SAA-1) and ***Ov*-RAL-2** in Ascaris (rAs-16), when used in vitro inhibit invasion of L3 through dog skin or caused cytotoxicity in vitro against L3, respectively (Zhan et al., 2004; Hotez, unpublished).

[e] Protection was determined in mice after two immunization with 25μg of protein in the presence of an adjuvant or using a DNA vaccine , followed by challenge with 25 L3 within diffusion chambers, and is defined by a significant ($p < 0.05$) percentage of reduction of L3 survival in the immunized mice vs. control mice.

[f] ND, not determined.

Protection Induced by Recombinant O. *volvulus* Antigens in Mice is Associated with Th2 and/or Th1 Responses

The mechanism of adaptive protective immunity induced by immunization with r*Ov*Ags in mice was studied by observing the type of adjuvant that effectively induced protective immunity and by analyzing the isotypes of the antibody response (Abraham et al., 2001; Lustigman et al., 2002, 2003). Freund's Complete Adjuvant has been shown to preferentially induce a Th1 response characterized by IgG2a antibodies, and alum preferentially induces a Th2 response characterized by IgG1, IgE, and eosinophilia (Kenney et al., 1989; Forsthuber et al., 1996; Yip et al., 1999). Four antigens, *Ov*-CPI-2, *Ov*-ALT-1, *Ov*-B8, and *Ov*-103, induced protective immunity only in the presence of alum. Unlike mice immunized with irradiated L3, there was no increase in either eosinophil numbers or levels of eotaxin in the diffusion chambers. The antibody responses to each of these antigens were dominated by IgG1 and IgE antibodies, which suggested a Th2 response (Abraham et al., 2001). Two other antigens, *Ov*-FBA-1 (McCarthy et al., 2002) and *Ov*-RAL-2 (Lustigman et al., 2002), induced protective immunity of ~50%, only when injected in Freund's Complete Adjuvant, with an apparent Th1 response based on the adjuvant type and an elevated IgG2a antibody response. *Ov*-ASP-1 induced significant levels of protective immunity when administered with alum (42%) or Freund's Complete Adjuvant (44%) (MacDonald et al., 2004). The mechanisms of protective immunity induced by the two adjuvants and *Ov*-ASP-1 differed in that IgG1 dominated the response induced by alum and IgG2a dominated that induced by Freund's Complete Adjuvant. Hence, individual r*Ov*Ags can induce immunity through Th1, Th2, or both Th1 and Th2 responses.

Based on these results it can be concluded that the mechanisms of protective immunity induced with the protective r*Ov*Ags are different from those induced with irradiated L3; some of the r*Ov*Ags utilize a Th1 or a mixed Th1/Th2 response, and those that use a Th2 response do not appear to require the collaboration of eosinophils. The protective immunity elicited by irradiated L3 is Th2 and eosinophil-dependent. Although the discrepancies between responses to irradiated L3 and r*Ov*Ags may be due to differences in the immunization protocols, it was hypothesized that they may reflect intrinsic properties of these antigens. This is also supported by observing that some of the target L3 native proteins recognized by r*Ov*Ag-specific antibodies were not labeled by anti-irradiated L3 antibodies (Abraham et al., 2001). However, they are recognized by antibodies from PI or individuals who developed concomitant immunity (Lustigman et al., 2003), thus confirming that the r*Ov*Ags induce responses in mice that are more similar to those operating in humans.

The protection induced by DNA immunization corresponding to *Ov*-CHI-1 was associated with an IgG1-dominant response. Moreover, antibodies from the vaccinated mice reacted with the cuticle of postinfective L3 larvae, implying that this may be the site of immune attack following secretion of chitinase (Harrison et al., 1999).

Characteristics of the Protective r*Ov*Ags

The characteristics of the parasite proteins corresponding to 15 protective r*Ov*Ags have been reviewed previously (Abraham et al., 2001; Lustigman et al., 2002) and are also described in Table 67.1. Eight of the proteins, *Ov*-ALT-1, *Ov*-B8, *Ov*-RAL-2, *Ov*-B20, OI5/OI3, *Ov*-CHI-1, *Ov*-RBP-1, and *Ov*-103 are parasite-specific antigens, whereas *Ov*-ASP-1 is a member of the vespid venom allergen-like protein family (Tawe et al., 2000). Six of the protective proteins are homologs to recognized proteins of higher organisms. *Ov*-CPI-2 (onchocystatin), *Ov*-TMY-1 (tropomyosin), *Ov*-FBA-1 (aldolase), *Ov*-CAL-1 (calponin), *Av*-ABC (ATP-binding cassette protein transporter), and *Av*-UBI (ubiquitin) have 32%, 31%, 69%, 42%, 71%, and 98% amino acid identity, respectively, with human proteins. An important concern associated with vaccine antigens belonging to conserved gene families (e.g., enzymes, muscle proteins) is the risk of crossreactions with host or environmental antigens.

Whereas it was expected that many of the protecting antigens would be larval stage specific, only two were actually L3 specific (*Ov*-ALT-1 and *Ov*-CHI-1) (Wu et al., 1996; Joseph et al., 1998) and one (*Ov*-B20) was limited to L1–L4 stages (Taylor et al., 1995); all the other proteins were expressed in all stages of the parasite. Many of the protective antigens are potentially upregulated in L3 and/or mL3 as determined by the abundance of the transcripts of the protein in stage-specific *O. volvulus* cDNA libraries, following large-scale EST sequencing and clustering (Lizotte-Waniewski et al., 2000), which suggests their importance during the initial steps of infection in humans. Moreover, localization of their corresponding native proteins in L3 and mL3 has indicated that all the protective antigens are found in one or more of the regions in the larvae where antibodies from irradiated L3-immunized mice and/or PI bound. These include the different layers of the cuticle, channels connecting the esophagus to the cuticle, and the basal lamina surrounding the body cavity (Abraham et al., 2001). In addition, many of the proteins were found in secretory vesicles within the hypodermis and/or in the granules of the glandular esophagus (Table 67.1) which is significant because the glandular esophagus is thought to be important for the molting process (Joseph et al., 1998).

When the antibody responses to some of the recombinant proteins were analyzed with PI and INF serum, it appeared that both groups recognized all protecting antigens similarly. Recognition of antigens by PI and INF groups should not eliminate these antigens as vaccine components because it appears that in the INF concomitant immunity is acquired with age. Importantly, analysis of the immunoglobulin class and subclass responses to some of the protecting antigens revealed that these proteins induced mainly cytophilic antibodies, IgG1, IgG3, and/or IgE in PI and INF (Bradley et al., 1995; MacDonald et al., 2002). For example, elevated cytophilic antibodies (IgG1, IgG3, IgE) and complement-fixing (IgG3) responses were observed in infected subjects to the L3-specific recombinant antigen *Ov*-ALT-1, of which the IgG3 and IgE responses increased with age (MacDonald et al., 2002). In additional unpublished studies the IgG1, IgG3 against *Ov*-ASP-1, and IgG3 against *Ov*-CPI-2 and *Ov*-RAL-2 increased with age in the infected human population (Lustigman et al., unpublished results). Further support to the possibility that such antibodies may have a role in antibody-dependent cellular cytotoxicity effector mechanisms, came from observing that human neutrophils inhibited molting of L3 by 96–100% in the presence of purified human antibodies against *Ov*-CPI-2 (Lustigman et al., unpublished results). *Ov*-CPI-2 is known to be essential for molting of L3 (Lustigman et al., 1996), and the molting of the L3 challenge was inhibited in protected mice (Abraham et al., 2001).

Importantly, many of the protective r*Ov*Ags have homologs that have been shown to also induce protection in other filariae host–parasite systems: (a) the *Ov*-ALT-1 homolog from *D. immitis* is specifically recognized by antibodies from protected dogs (Frank et al., 1996); the *Bm* homolog (*Bm*-ALT-1) induced 76% reduction in parasite survival in the presence of Freund's Complete Adjuvant (Gregory et al., 2000); the closely related r*Bm*-ALT-2 induced in the presence of alum 74% reduction of L3 viability within a diffusion chamber implanted in jirds. When r*Bm*-ALT-2 was combined with r*Bm*-ASP-1 (the homolog of *Ov*-ASP-1) for immunization they induced in jirds a 76% reduction in worm burden that was associated with increased IL-4 production and the presence of cytotoxic antibodies (Ramachandran et al., 2004; Anand et al., 2006); (b) the *L. sigmodontis Ov*-CPI-2 homolog, *Ls*-cystatin, induced in mice in the presence of alum or Pam_3Cys reduction in the numbers of patent infections, i.e., less infections with microfilaremia were observed in vaccinated animals (Pfaff et al., 2002). Previous studies in *O. volvulus* have shown that cystatin is present in the eggshell surrounding the microfilariae and thus potentially functions during microfilarial development (Lustigman et al., 1991, 1992a), which may explain the results obtained in the *L. sigmodontis* model; (c) the *Bm* homolog of *Ov*-RAL-2, *Bm*-SXP-1, in the presence of Freund's Complete Adjuvant induced in mice 30% reduction in *B. malayi* L3 survival when the challenge infections were contained in diffusion chambers (Ramachandran et al., 2004). In another study, immunization of jirds with *Bm*-SXP-1 in the presence of Freund's Complete Adjuvant or alum reduced microfilaremia and, in some experiments, also adult worm burdens by 35% (Wang et al., 1997); (d) the *A. viteae* chitinase was identified as the main target for antibodies from protected jirds after vaccination with irradiated L3 (Adam et al., 1996). Moreover, immunization of jirds with the *Bm*-chitinase induced 48% reduction in worm burden and >90% in the number of circulating microfilariae (Wang et al., 1997); and (e) the *O. volvulus* recombinant proteins *Ov*-RBP-1 and *Ov*-B20 were shown to also induce protection against a challenge with *A. viteae* in jirds (Taylor et al., 1995; Jenkins et al., 1996).

Moreover, homologs of four of the r*Ov*Ags were tested for protection in other nematode host–parasite systems, such as hookworm infection in dogs and *Ascaris* in pigs, and were shown to elicit reduction in worm burden or other protective measures: (a) the *Ov*-ASP-1 homolog in hookworms, *Na*-ASP-2, is a highly protective excretory–secretory protein (Ghosh et al., 1996), and is in the process of testing for efficacy in humans as part of the Human Hookworm Vaccine Initiative (see Chapter 66) (Goud et al., 2005; Hotez et al., 2006); (b) the hookworm homolog of *Ov*-103, Ac-SAA-1, induced partial protection against hookworm infection (25%), 64% reduction in egg count and significant reduction in blood loss (Zhan et al., 2004); (c) *Ac*-cystatin, the hookworm homolog of *Ov*-CPI-2 elicited 22% reduction in worm burden in dogs (Hotez, unpublished); and (d) rAc-16, the hookworm homolog of *Ov*-RAL-2 induced in dogs 25% reduction in worm burden, 64% reduction in egg counts, and a significant reduction in anemia (Hotez, unpublished). Moreover, a homolog of *Ov*-RAL-2 from *A. suum*, rAs16, also induced in mice a 64% reduction of recovery of larvae when coupled with cholera toxin

B subunit for immunization (Tsuji et al., 2001, 2003). Pigs vaccinated with rAs16 coupled with cholera toxin were also protected from migration of *A. suum* larvae through the lungs, as indicated by a 58% reduction in the recovery of lung-stage L3, compared with that in nonvaccinated controls (Tsuji et al., 2004). Notably, anti-rAs16 antibodies inhibited survival of L3 and interrupted the molting of lung-stage L3 (Tsuji et al., 2004).

PROSPECTS FOR THE FUTURE

Although it was previously considered that *O. volvulus* infection was on the cusp of being controlled using ivermectin distribution, it is becoming increasingly clear that without additional modalities, such as drugs which kill or permanently sterilize the adult worms or a vaccine, elimination of onchocerciasis from sub-Saharan Africa will remain an unfulfilled goal. A vaccine aimed at preventing infection (anti-L3) or blocking transmission and/or pathology (anti-microfilariae) could be the essential supplement for the successful control or elimination of onchocerciasis. The process would be to link the vaccine with drug treatments in a program of vaccine-linked chemotherapy (Hotez and Ferris, 2006; Hotez, 2007). Although the ultimate goal of a vaccine against onchocerciasis has not yet been achieved, fortunately, through the commitment of the Edna McConnell Clark Foundation and researchers in the field of filariasis, appropriate scientific infrastructure is now in place to meet this challenge. Research on vaccine development has provided the foundation for antigen discovery and development of animal models for testing the efficacy of vaccine candidates, and consequently resulted with at least 15 vaccine candidates that were shown to induce partial but significant protection against an L3 challenge using an experimental mouse model. Moreover, protective immunity against *O. volvulus* L3 has now been definitively demonstrated in humans, cattle, and mice thereby proving the concept that a vaccine can be produced against this infection (Abraham et al., 2002; Lustigman et al., 2003).

Future steps for vaccine development would initially use the *O. volvulus* L3 mouse model to select a subset of highly immunogenic antigens from the 15 recombinant vaccine candidates for further development and assessment. This subset of r*Ov*Ags would be then moved to product development and scale up applicable for human use. The bovine *O. ochengi* model has proven the feasibility of immunoprophylaxis against *Onchocerca* using irradiated L3 (Tchakoute et al., 2006) and would therefore be a superb secondary screen to verify the potency of recombinant vaccines developed in the mouse model.

The availability of the *O. volvulus* EST database (Williams et al., 2002), the complete sequencing of the *Brugia* genome (Ghedin et al., 2004) and *Caenorhabditis elegans* (www.wormbase), as well as the rapidly expanding technologies in proteomics, functional genomics, and bioinformatics could provide the needed tools for integrated comparative functional genomics and result in the identification of additional *Onchocerca* larval proteins that could be used as potential vaccine candidates (Hashmi et al., 2001). For example, 47 essential *C. elegans* molting genes (>90% reduction in molting due to gene-specific RNA interference) belonging to the categories: novel, proteases, protease inhibitors, peroxidases, extracellular matrix, sterol-sensing domain, DNA binding, nucleic acid interacting, signaling, WD domains, and others have been identified (Frand et al., 2005). It appears that 46 of them are also potentially encoded by filarial transcripts (Lustigman, unpublished), and may prove to be excellent vaccine antigens.

An alternative approach that could be taken would be to develop a vaccine targeting the microfilaria stage. The microfilaria is responsible for the development of pathology and for transmission of the disease. Therefore, a vaccine against this stage would be a terrific asset for control of the infection. A caveat for this vaccine is the possibility that vaccination against microfilariae would exacerbate disease development caused by microfilariae. A possible way to circumvent this issue would be to immunize people with parasite antigens and not with *Wolbachia*-derived antigens.

It is anticipated that the vaccine against onchocerciasis will be administered to populations that may have been treated for many years with ivermectin. Therefore, longitudinal studies examining the immune responses following long-standing ivermectin treatment need to be performed. Attempts to define the nature of the targets of the posttreatment responses should be undertaken by surveying all possible available antigens. With the advent of antibiotic therapy against *Wolbachia*, studies following the changes in immune responses following this treatment should also be executed.

In conclusion, the human studies of onchocerciasis patients have suggested that protective immunity can develop in humans. The experimental and natural infections of animals have demonstrated that protective immunity does develop and that vaccines can protect animals from infection under natural conditions. The foundation studies on antigen screening have been accomplished and now the challenge is to optimize and formulate vaccines suitable for human usage. The development of drug resistance to the only treatment available for control of onchocerciasis has rejuvenated the impending critical need for a vaccine against onchocerciasis.

KEY ISSUES

- Human onchocerciasis is a serious neglected tropical disease caused by *O. volvulus* and an important cause of blindness and chronic disability in the developing world.
- Great progress has been made in reducing disease burden through mass drug administration of ivermectin.
- Technical and logistical obstacles must be overcome before the goal of elimination of this infection as a public health problem can be attained.
- Evidence for the existence of *Onchocerca* unresponsiveness or resistance to the drug ivermectin is building up. Therefore, additional tools are critically needed and include the need for a vaccine against onchocerciasis to "complement" the present control measures and thus further the goals of the existing control programs.
- An *Onchocerca* vaccine targeting the infective stage larvae could reduce adult worm burden and thus the number of microfilariae produced by the adult female worms.
- Protective immunity against *Onchocerca* larvae has now been definitively demonstrated in humans, cattle, and mice, thereby providing the conceptual underpinnings that a vaccine can be produced against this infection.
- Several recombinant *Onchocerca* antigens have been identified that are capable of inducing a significant but partial reduction in the survival of a larval challenge using an experimental mouse model.
- The foundation studies on antigen screening have been accomplished and now the challenge is to optimize and formulate vaccines suitable for human usage.
- It is anticipated that continual natural exposure of the immunized individuals to infective larvae from black flies would sustain the protective immunity induced by immunization with recombinant vaccine proteins, and that ultimately the combination of vaccination aimed at preventing infection (anti-L3) with long-term chemotherapy with ivermectin could accelerate the eradication of onchocerciasis from sub-Saharan Africa.
- An alternative approach will be to develop a vaccine against microfilariae, which is aimed at blocking transmission and/or pathology.
- It is anticipated that the vaccine against onchocerciasis will be administered to populations that may have been treated for many years with ivermectin. Therefore, longitudinal studies examining the immune responses following long-standing ivermectin treatment need to also be performed.

References

Abiose, A., Homeida, M., Liese, B., Molyneux, D. and Remme, H. Onchocerciasis control strategies. Lancet 2000; 356:1523–1524.

Abraham, D., Lange, A.M., Yutanawiboonchai, W., Trpis, M., Dickerson, J.W., Swenson, B. and Eberhard, M.L. Survival and development of larval *Onchocerca volvulus* in diffusion chambers implanted in primate and rodent hosts. J. Parasitol. 1993; 79:571–582.

Abraham, D., Leon, O., Leon, S. and Lustigman, S. Development of a recombinant antigen vaccine against infection with the filarial worm *Onchocerca volvulus*. Infect. Immun. 2001; 69:262–270.

Abraham, D., Leon, O., Schnyder-Candrian, S., Wang, C.C., Galioto, A.M., Kerepesi, L.A., Lee, J.J. and Lustigman, S. Immunoglobulin E and eosinophil-dependent protective immunity to larval *Onchocerca volvulus* in mice immunized with irradiated larvae. Infect. Immun. 2004; 72:810–817.

Abraham, D., Lucius, R. and Trees, A.J. Immunity to *Onchocerca* spp. in animal hosts. Trends Parasitol. 2002; 18:164–171.

Achukwi, M.D., Harnett, W., Enyong, P. and Renz, A. Successful vaccination against *Onchocerca ochengi* infestation in cattle using live *Onchocerca volvulus* infective larvae. Parasite Immunol. 2007; 29:113–116.

Adam, R., Kaltmann, B., Rudin, W., Friedrich, T., Marti, T. and Lucius, R. Identification of chitinase as the immunodominant filarial antigen recognized by sera of vaccinated rodents. J. Biol. Chem. 1996; 271:1441–1447.

Anand, S.B., Krithika, K.N., Murugan, V., Reddy, M.V. and Kaliraj, P. Comparison of immuno prophylactic efficacy of Bm rALT2 or Bm rVAH or rALT + VAH by single and multiple antigen vaccination mode. Am. Soc. Trop. Med. Hyg. 2006; 75:295.

Ardelli, B.F., Guerriero, S.B. and Prichard, R.K. Genomic organization and effects of ivermectin selection on *Onchocerca volvulus* P-glycoprotein. Mol. Biochem. Parasitol. 2005; 143:58–66.

Ardelli, B.F., Guerriero, S.B. and Prichard, R.K. Ivermectin imposes selection pressure on P-glycoprotein from *Onchocerca volvulus*: linkage disequilibrium and genotype diversity. Parasitology 2006a; 132:375–386.

Ardelli, B.F., Guerriero, S.B. and Prichard, R.K. Characterization of a half-size ATP-binding cassette transporter gene which may be a useful marker for ivermectin selection in *Onchocerca volvulus*. Mol. Biochem. Parasitol. 2006b; 145:94–100.

Ardelli, B.F. and Prichard, R.K. Identification of variant ABC-transporter genes among *Onchocerca volvulus* collected from ivermectin-treated and untreated patients in Ghana, West Africa. Ann. Trop. Med. Parasitol. 2004; 98:371–384.

Awadzi, K., Attah, S.K., Addy, E.T., Opoku, N.O., Quartey, B.T., Lazdins-Helds, J.K., Ahmed, K., Boatin, B.A., Boakye, D.A. and Edwards, G. Thirty-month follow-up of sub-optimal responders to multiple treatments with ivermectin, in two onchocerciasis-endemic foci in Ghana. Ann. Trop. Med. Parasitol. 2004a; 98:359–370.

Awadzi, K., Boakye, D.A., Edwards, G., Opoku, N.O., Attah, S.K., Osei-Atweneboana, M.Y., Lazdins-Helds, J.K., Ardrey, A.E., Addy, E.T., Quartey, B.T., Ahmed, K., Boatin, B.A. and Soumbey-Alley, E.W. An investigation of persistent microfilaridermias despite multiple treatments with ivermectin, in two onchocerciasis-endemic foci in Ghana. Ann. Trop. Med. Parasitol. 2004b; 98:231–249.

Basáñez, M.G., Pion, S.D., Churcher, T.S., Breitling, L.P., Little, M.P. and Boussinesq, M. River blindness: a success story under threat? PLoS Med. 2006; 3(9):e371, doi:10.1371/journal.pmed.0030371.

Bianco, A.E., Luty, A., Whitworth, J. and Taylor, D. Immunity to *Onchocerca volvulus* microfilariae in mice and the induction of cross-protection with *O. lienalis*. Trop. Med. Parasitol. 1991; 42:188–190.

Boatin, B.A. and Richards, F.O., Jr. Control of onchocerciasis. Adv. Parasitol. 2006; 61:349–394.

Bourguinat, C., Pion, S.D.S., Kamgno, J., Gardon, J., Duke, B.O., Boussinesq, M. and Prichard, R.K. Genetic selection of low fertile *Onchocerca volvulus* by ivermectin treatment. PLoS Negl. Trop. Dis., 2007; Aug 30; 1(1):e72

Bourguinat, C., Pion, S.D., Kamgno, J., Gardon, J., Gardon-Wendel, N., Duke, B.O., Prichard, R.K. and Boussinesq, M. Genetic polymorphism of the beta-tubulin gene of *Onchocerca volvulus* in ivermectin naive patients from Cameroon, and its relationship with fertility of the worms. Parasitology 2006; 132:1–8.

Boussinesq, M., Prod'hon, J. and Chippaux, J.P. *Onchocerca volvulus*: striking decrease in transmission in the Vina valley (Cameroon) after eight annual large scale ivermectin treatments. Trans. R. Soc. Trop. Med. Hyg. 1997; 91:82–86.

Boyer, A.E., Tsang, V.C., Eberhard, M.L., Zea-Flores, G., Hightower, A., Pilcher, J.B., Zea-Flores, R., Zhou, W. and Reimer, C.B. Guatemalan human onchocerciasis. II. Evidence for IgG3 involvement in acquired immunity to *Onchocerca volvulus* and identification of possible immune-associated antigens. J. Immunol. 1991; 146:4001–4010.

Bradley, J.E., Elson, L., Tree, T.I., Stewart, G., Guderian, R., Calvopina, M., Paredes, W., Araujo, E. and Nutman, T.B. Resistance to *Onchocerca volvulus*: differential cellular and humoral responses to a recombinant antigen, OvMBP20/11. J. Infect. Dis. 1995; 172:831–837.

Brattig, N.W. Pathogenesis and host responses in human onchocerciasis: impact of *Onchocerca* filariae and *Wolbachia* endobacteria. Microbes Infect. 2004; 6:113–128.

Brattig, N.W., Bazzocchi, C., Kirschning, C.J., Reiling, N., Buttner, D.W., Ceciliani, F., Geisinger, F., Hochrein, H., Ernst, M., Wagner, H., Bandi, C. and Hoerauf, A. The major surface protein of *Wolbachia* endosymbionts in filarial nematodes elicits immune responses through TLR2 and TLR4. J. Immunol. 2004; 173:437–445.

Brattig, N.W., Buttner, D.W. and Hoerauf, A. Neutrophil accumulation around *Onchocerca* worms and chemotaxis of neutrophils are dependent on *Wolbachia* endobacteria. Microbes Infect. 2001; 3:439–446.

Brattig, N.W., Rathjens, U., Ernst, M., Geisinger, F., Renz, A. and Tischendorf, F.W. Lipopolysaccharide-like molecules derived from *Wolbachia endobacteria* of the filaria *Onchocerca volvulus* are candidate mediators in the sequence of inflammatory and anti-inflammatory responses of human monocytes. Microbes Infect. 2000; 2:1147–1157.

Buttner, D.W. and Racz, P. Macro- and microfilariae in nodules from onchocerciasis patients in the Yemen Arab Republic. Tropenmed. Parasitol. 1983; 34:113–121.

Carlow, C.K. and Bianco, A.E. Resistance to *Onchocerca lienalis* microfilariae in mice conferred by egg antigens of homologous and heterologous *Onchocerca* species. Parasitology 1987; 94:485–496.

Carlow, C.K., Muller, R. and Bianco, A.E. Further studies on the resistance to *Onchocerca* microfilariae in CBA mice. Trop. Med. Parasitol. 1986; 37:276–281.

Collins, R.C., Gonzales-Peralta, C., Castro, J., Zea-Flores, G., Cupp, M.S., Richards, F.O., Jr. and Cupp, E.W. Ivermectin: reduction in prevalence and infection intensity of *Onchocerca volvulus* following biannual treatments in five Guatemalan communities. Am. J. Trop. Med. Hyg. 1992; 47:156–169.

Cook, J.A., Steel, C. and Ottesen, E.A. Towards a vaccine for onchocerciasis. Trends Parasitol. 2001; 17:555–558.

Cross, H.F., Haarbrink, M., Egerton, G., Yazdanbakhsh, M. and Taylor, M.J. Severe reactions to filarial chemotherapy and release of *Wolbachia* endosymbionts into blood. Lancet 2001; 358:1873–1875.

Dadzie, Y., Neira, M. and Hopkins, D. Final report of the Conference on the Eradicability of Onchocerciasis. Filaria J. 2003; 2:2.

Day, K.P., Gregory, W.F. and Maizels, R.M. Age-specific acquisition of immunity to infective larvae in a bancroftian filariasis endemic area of Papua New Guinea. Parasite Immunol. 1991; 13:277–290.

Duke, B.O. Observations on *Onchocerca volvulus* in experimentally infected chimpanzees. Tropenmed. Parasitol. 1980; 31:41–54.

Duke, B.O. and Moore, P.J. The contributions of different age groups to the transmission of Onchocerciasis in a Cameroon forest village. Trans. R. Soc. Trop. Med. Hyg. 1968; 62:22–28.

Eberhard, M.L., Dickerson, J.W., Boyer, A.E., Tsang, V.C., Zea-Flores, R., Walker, E.M., Richards, F.O., Zea-Flores, G. and Strobert, E. Experimental *Onchocerca volvulus* infections in mangabey monkeys (*Cercocebus atys*) compared to infections in humans and chimpanzees (*Pan troglodytes*). Am. J. Trop. Med. Hyg. 1991; 44:151–160.

Eberhard, M.L., Dickerson, J.W., Tsang, V.C., Walker, E.M., Ottesen, E.A., Chandrashekar, R., Weil, G.J., Trpis, M., Strobert, E., Constantinidis, I. and Swenson, R.B. *Onchocerca volvulus*: parasitologic and serologic responses in experimentally infected chimpanzees and mangabey monkeys. Exp. Parasitol. 1995; 80:454–462.

Eisenbeiss, W.F., Apfel, H. and Meyer, T.F. Protective immunity linked with a distinct developmental stage of a filarial parasite. J. Immunol. 1994; 152:735–742.

Ejere, H., Schwartz, E. and Wormald, R. Ivermectin for onchocercal eye disease (river blindness). Cochrane Database Syst. Rev. 2001; 2001:CD002219.

Ellis, L.A., Reason, A.J., Morris, H.R., Dell, A., Iglesias, R., Ubeira, F.M. and Appleton, J.A. Glycans as targets for monoclonal antibodies that protect rats against *Trichinella spiralis*. Glycobiology 1994; 4:585–592.

Elson, L.H., Guderian, R.H., Araujo, E., Bradley, J.E., Days, A. and Nutman, T.B. Immunity to onchocerciasis: identification of a putatively immune population in a hyperendemic area of Ecuador. J. Infect. Dis. 1994; 169:588–594.

Eng, J.K., Blackhall, W.J., Osei-Atweneboana, M.Y., Bourguinat, C., Galazzo, D., Beech, R.N., Unnasch, T.R., Awadzi, K., Lubega, G.W. and Prichard, R.K. Ivermectin selection on beta-tubulin: evidence in *Onchocerca volvulus* and *Haemonchus contortus*. Mol. Biochem. Parasitol. 2006; 150:229–235.

Eng, J.K. and Prichard, R.K. A comparison of genetic polymorphism in populations of *Onchocerca volvulus* from untreated- and ivermectin-treated patients. Mol. Biochem. Parasitol. 2005; 142:193–202.

Faulkner, H., Gardon, J., Kamgno, J., Enyong, P., Boussinesq, M. and Bradley, J.E. Antibody responses in onchocerciasis as a function of age and infection intensity. Parasite Immunol. 2001; 23:509–516.

Folkard, S.G., Hogarth, P.J., Taylor, M.J. and Bianco, A.E. Eosinophils are the major effector cells of immunity to microfilariae in a mouse model of onchocerciasis. Parasitology 1996; 112:323–329.

Forsthuber, T., Yip, H.C. and Lehmann, P.V. Induction of TH1 and TH2 immunity in neonatal mice. Science 1996; 271:1728–1730.

Frand, A.R., Russel, S. and Ruvkun, G. Functional genomic analysis of *C. elegans* molting. PLoS Biol. 2005; 3:e312.

Frank, G.R., Tripp, C.A. and Grieve, R.B. Molecular cloning of a developmentally regulated protein isolated from excretory-secretory products of larval *Dirofilaria immitis*. Mol. Biochem. Parasitol. 1996; 75:231–240.

Gallin, M., Edmonds, K., Ellner, J.J., Erttmann, K.D., White, A.T., Newland, H.S., Taylor, H.R. and Greene, B.M. Cell-mediated immune responses in human infection with *Onchocerca volvulus*. J. Immunol. 1988; 140:1999–2007.

Ghedin, E., Wang, S., Foster, J.M. and Slatko, B.E. First sequenced genome of a parasitic nematode. Trends Parasitol. 2004; 20:151–153.

Ghosh, K., Hawdon, J. and Hotez, P. Vaccination with alum-precipitated recombinant Ancylostoma-secreted protein 1 protects mice against challenge infections with infective hookworm (*Ancylostoma caninum*) larvae. J. Infect. Dis. 1996; 174:1380–1383.

Goud, G.N., Bottazzi, M.E., Zhan, B., Mendez, S., Deumic, V., Plieskatt, J., Liu, S., Wang, Y., Bueno, L., Fujiwara, R., Samuel, A., Ahn, S.Y., Solanki, M., Asojo, O.A., Wang, J., Bethony, J.M., Loukas, A., Roy, M. and Hotez, P.J. Expression of the *Necator americanus* hookworm larval antigen Na-ASP-2 in *Pichia pastoris* and purification of the recombinant protein for use in human clinical trials. Vaccine 2005; 23:4754–4764.

Goud, G.N., Zhan, B., Ghosh, K., Loukas, A., Hawdon, J., Dobardzic, A., Deumic, V., Liu, S., Dobardzic, R., Zook, B.C., Jin, Q., Liu, Y., Hoffman, L., Chung-Debose, S., Patel, R., Mendez, S. and Hotez, P.J. Cloning, yeast expression, isolation, and vaccine testing of recombinant Ancylostoma-secreted protein (ASP)-1 and ASP-2 from *Ancylostoma ceylanicum*. J. Infect. Dis. 2004; 189:919–929.

Greene, B.M., Gbakima, A.A., Albiez, E.J. and Taylor, H.R. Humoral and cellular immune responses to *Onchocerca volvulus* infection in humans. Rev. Infect. Dis. 1985; 7:789–795.

Gregory, W.F., Atmadja, A.K., Allen, J.E. and Maizels, R.M. The abundant larval transcript-1 and -2 genes of *Brugia malayi* encode stage-specific candidate vaccine antigens for filariasis. Infect. Immun. 2000; 68:4174–4179.

Harnett, W., Bradley, J.E. and Garate, T. Molecular and immunodiagnosis of human filarial nematode infections. Parasitology 1998; 117(Suppl):S59–S71.

Harrison, R.A., Wu, Y., Egerton, G. and Bianco, A.E. DNA immunisation with *Onchocerca volvulus* chitinase induces partial protection against challenge infection with L3 larvae in mice. Vaccine 1999; 18:647–655.

Hashmi, S., Tawe, W. and Lustigman, S. *Caenorhabditis elegans* and the study of gene function in parasites. Trends Parasitol. 2001; 17:387–393.

Henkle-Duhrsen, K. and Kampkotter, A. Antioxidant enzyme families in parasitic nematodes. Mol. Biochem. Parasitol. 2001; 114:129–142.

Higazi, T.B., Filiano, A., Katholi, C.R., Dadzie, Y., Remme, J.H. and Unnasch, T.R. *Wolbachia* endosymbiont levels in severe and mild strains of *Onchocerca volvulus*. Mol. Biochem. Parasitol. 2005; 141:109–112.

Hoch, B., Wahl, G., Enyong, P., Luder, C.G.K., Harnett, W. and Renz, A. Serological recognition of specific and crossreactive of *O. ochengi* and *O. volvulus* by infected cattle and humans. Trop. Med. Parasitol. 1992; 43:206–207.

Hoerauf, A. and Brattig, N. Resistance and susceptibility in human onchocerciasis—beyond Th1 vs. Th2. Trends Parasitol. 2002; 18:25–31.

Hoerauf, A., Mand, S., Volkmann, L., Buttner, M., Marfo-Debrekyei, Y., Taylor, M., Adjei, O. and Buttner, D.W. Doxycycline in the treatment of human onchocerciasis: kinetics of *Wolbachia* endobacteria reduction and of inhibition of embryogenesis in female *Onchocerca* worms. Microbes Infect. 2003; 5:261–273.

Hoerauf, A. and Steel, C. Towards a strategic plan for research to support the global program to eliminate lymphatic filariasis-3.2 protective immunity-vaccines. Am. J. Trop. Med. Hyg. 2004; 71(Suppl):34–36.

Hogarth, P.J. and Bianco, A.E. Interleukin-12 modulates T-cell responses to microfilariae but fails to abrogate interleukin-5-dependent immunity in a mouse model of onchocerciasis. Immunology 1999a; 98:406–412.

Hogarth, P.J. and Bianco, A.E. IL-5 dominates cytokine responses during expression of protective immunity to *Onchocerca lienalis* microfilariae in mice. Parasite Immunol. 1999b; 21:81–88.

Hogarth, P.J., Folkard, S.G., Taylor, M.J. and Bianco, A.E. Accelerated clearance of *Onchocerca* microfilariae and resistance to reinfection in interleukin-4 gene knockout mice. Parasite Immunol. 1995; 17:653–657.

Hogarth, P.J., Taylor, M.J. and Bianco, A.E. IL-5-dependent immunity to microfilariae is independent of IL-4 in a mouse model of onchocerciasis. J. Immunol. 1998; 160:5436–5440.

Horauf, A. and Fleischer, B. Immune responses to filarial infection in laboratory mice. Med. Microbiol. Immunol. (Berl.) 1997; 185:207–215.

Hotez, P.J. Control of onchocerciasis—the next generation. Lancet 2007; 369:1979–1980.

Hotez, P.J., Bethony, J., Bottazzi, M.E., Brooker, S., Diemert, D. and Loukas, A. New technologies for the control of human hookworm infection. Trends Parasitol. 2006; 22:327–331.

Hotez, P.J. and Ferris, M.T. The antipoverty vaccines. Vaccine 2006; 24:5787–5799.

Huang, Y.J. and Prichard, R.K. Identification and stage-specific expression of two putative P-glycoprotein coding genes in *Onchocerca volvulus*. Mol. Biochem. Parasitol. 1999; 102:273–281.

Irvine, M., Johnson, E.H. and Lustigman, S. Identification of larval-stage-specific antigens of *Onchocerca volvulus* uniquely recognized by putative immune sera from humans and vaccination sera from animal models. Ann. Trop. Med. Parasitol. 1997; 91:67–77.

Jenkins, R.E., Taylor, M.J., Gilvary, N. and Bianco, A.E. Characterization of a secreted antigen of *Onchocerca volvulus* with host-protective potential. Parasite Immunol. 1996; 18:29–42.

Johnson, E.H., Irvine, M., Kass, P.H., Browne, J., Abdullai, M., Prince, A.M. and Lustigman, S. *Onchocerca volvulus*: in vitro cytotoxic effects of human neutrophils and serum on third-stage larvae. Trop. Med. Parasitol. 1994; 45:331–335.

Johnson, E.H., Schynder-Candrian, S., Rajan, T.V., Nelson, F.K., Lustigman, S. and Abraham, D. Immune responses to third stage larvae of *Onchocerca volvulus* in interferon-gamma and interleukin-4 knockout mice. Parasite Immunol. 1998; 20:319–324.

Joseph, G.T., Huima, T. and Lustigman, S. Characterization of an *Onchocerca volvulus* L3-specific larval antigen, Ov-ALT-1. Mol. Biochem. Parasitol. 1998; 96:177–183.

Joseph, G.T., McCarthy, J.S., Huima, T., Mair, K.F., Kass, P.H., Boussinesq, M., Goodrick, L., Bradley, J.E. and Lustigman, S.

Onchocera volvulus: characterization of a highly immunogenic Gln-rich protein. Mol. Biochem. Parasitol. 1997; 90:55–68.

Keiser, P.B., Reynolds, S.M., Awadzi, K., Ottesen, E.A., Taylor, M.J. and Nutman, T.B. Bacterial endosymbionts of *Onchocerca volvulus* in the pathogenesis of posttreatment reactions. J. Infect. Dis. 2002; 185:805–811.

Kenney, J.S., Hughes, B.W., Masada, M.P. and Allison, A.C. Influence of adjuvants on the quantity, affinity, isotype and epitope specificity of murine antibodies. J. Immunol. Methods 1989; 121:157–166.

Kerepesi, L.A., Leon, O., Lustigman, S. and Abraham, D. Protective immunity to the larval stages of *Onchocerca volvulus* is dependent on toll-like receptor 4. Infect. Immun. 2005; 73:8291–8297.

Kohler, P. The biochemical basis of anthelmintic action and resistance. Int. J. Parasitol. 2001; 31:336–345.

Kurniawan-Atmadja, A., Sartono, E., Partono, F., Yazdanbakhsh, M. and Maizels, R.M. Antibody responses to filarial infective larvae are not dominated by the IgG4 isotype. Parasite Immunol. 1998; 20:9–17.

Lange, A.M., Yutanawiboonchai, W., Lok, J.B., Trpis, M. and Abraham, D. Induction of protective immunity against larval *Onchocerca volvulus* in a mouse model. Am. J. Trop. Med. Hyg. 1993; 49:783–788.

Lange, A.M., Yutanawiboonchai, W., Scott, P. and Abraham, D. IL-4- and IL-5-dependent protective immunity to *Onchocerca volvulus* infective larvae in BALB/cBYJ mice. J. Immunol. 1994; 153:205–211.

Le Goff, L., Martin, C., Oswald, I.P., Vuong, P.N., Petit, G., Ungeheuer, M.N. and Bain, O. Parasitology and immunology of mice vaccinated with irradiated *Litomosoides sigmodontis* larvae. Parasitology 2000; 120:271–280.

Leke, R.G., Boto, W.M., Lando, G. and Ngu, J.L. Immunity to *Onchocerca volvulus*. Serum mediated leucocyte adherence to infective larvae in vitro. Trop. Med. Parasitol. 1989; 40:39–41.

Lipner, E.M., Dembele, N., Souleymane, S., Alley, W.S., Prevots, D.R., Toe, L., Boatin, B., Weil, G.J. and Nutman, T.B. Field applicability of a rapid-format anti-Ov-16 antibody test for the assessment of onchocerciasis control measures in regions of endemicity. J. Infect. Dis. 2006; 194:216–221.

Lizotte-Waniewski, M., Tawe, W., Guiliano, D.B., Lu, W., Liu, J., Williams, S.A. and Lustigman, S. Identification of potential vaccine and drug target candidates by expressed sequence tag analysis and immunoscreening of *Onchocerca volvulus* larval cDNA libraries. Infect. Immun. 2000; 68:3491–3501.

Lucius, R., Textor, G., Kern, A. and Kirsten, C. *Acanthocheilonema viteae*: vaccination of jirds with irradiation-attenuated stage-3 larvae and with exported larval antigens. Exp. Parasitol. 1991; 73:184–196.

Luder, C.G., Schulz-Key, H., Banla, M., Pritze, S. and Soboslay, P.T. Immunoregulation in onchocerciasis: predominance of Th1-type responsiveness to low molecular weight antigens of *Onchocerca volvulus* in exposed individuals without microfilaridermia and clinical disease. Clin. Exp. Immunol. 1996; 105:245–253.

Luder, C.G., Soboslay, P.T., Prince, A.M., Greene, B.M., Lucius, R. and Schulz-Key, H. Experimental onchocerciasis in chimpanzees: cellular responses and antigen recognition after immunization and challenge with *Onchocerca volvulus* infective third-stage larvae. Parasitology 1993; 107:87–97.

Lustigman, S., Brotman, B., Huima, T. and Prince, A.M. Characterization of an *Onchocerca volvulus* cDNA clone encoding a genus specific antigen present in infective larvae and adult worms. Mol. Biochem. Parasitol. 1991; 45:65–75.

Lustigman, S., Brotman, B., Huima, T., Prince, A.M. and McKerrow, J.H. Molecular cloning and characterization of onchocystatin, a cysteine proteinase inhibitor of *Onchocerca volvulus*. J. Biol. Chem. 1992a; 267:17339–17346.

Lustigman, S., Brotman, B., Johnson, E.H., Smith, A.B., Huima, T. and Prince, A.M. Identification and characterization of an *Onchocerca volvulus* cDNA clone encoding a microfilarial surface-associated antigen. Mol. Biochem. Parasitol. 1992b; 50:79–93.

Lustigman, S., James, E.R., Tawe, W. and Abraham, D. Towards a recombinant antigen vaccine against *Onchocerca volvulus*. Trends Parasitol. 2002; 18:135–141.

Lustigman, S., MacDonald, A.J. and Abraham, D. CD4+ dependent immunity to *Onchocerca volvulus* third-stage larvae in humans and the mouse vaccination model: common ground and distinctions. J. Int. Parasitol. 2003; 33:1161–1171.

Lustigman, S., McKerrow, J.H., Shah, K., Lui, J., Huima, T., Hough, M. and Brotman, B. Cloning of a cysteine protease required for the molting of *Onchocerca volvulus* third stage larvae. J. Biol. Chem. 1996; 271:30181–30189.

MacDonald, A.J., Tawe, W., Leon, O., Cao, L., Liu, J., Oksov, Y., Abraham, D. and Lustigman, S. Ov-ASP-1, the *Onchocerca volvulus* homologue of the activation associated secreted protein family is immunostimulatory and can induce protective anti-larval immunity. Parasite Immunol. 2004; 26:53–62.

MacDonald, A.J., Turaga, P.S., Harmon-Brown, C., Tierney, T.J., Bennett, K.E., McCarthy, M.C., Simonek, S.C., Enyong, P.A., Moukatte, D.W. and Lustigman, S. Differential cytokine and antibody responses to adult and larval stages of *Onchocerca volvulus* consistent with the development of concomitant immunity. Infect. Immun. 2002; 70:2796–2804.

Mackenzie, C.D., Williams, J.F., Sisley, B.M., Steward, M.W. and O'Day, J. Variations in host responses and the pathogenesis of human onchocerciasis. Rev. Infect. Dis. 1985; 7:802–808.

McCarthy, J.S. and Nutman, T.B. Perspective: prospects for development of vaccines against human helminth infections. J. Infect. Dis. 1996; 174:1384–1390.

McCarthy, J.S., Wieseman, M., Tropea, J., Kaslow, D., Abraham, D., Lustigman, S., Tuan, R., Guderian, R.H. and Nutman, T.B. *Onchocerca volvulus* glycolytic enzyme fructose-1,6-bisphosphate aldolase as a target for a protective immune response in humans. Infect. Immun. 2002; 70:851–858.

McVay, C.S., Bracken, P., Gagliardo, L.F. and Appleton, J. Antibodies to tyvelose exhibit multiple modes of interference with the epithelial niche of *Trichinella spiralis*. Infect. Immun. 2000; 68:1912–1918.

McVay, C.S., Tsung, A. and Appleton, J. Participation of parasite surface glycoproteins in antibody-mediated protection of epithelial cells against *Trichinella spiralis*. Infect. Immun. 1998; 66:1941–1945.

Molyneux, D.H. Onchocerciasis control in West Africa: current status and future of the Onchocerciasis Control Programme. Parasitol. Today 1995; 11:399–402.

Nfon, C.K., Makepeace, B.L., Njongmeta, L.M., Tanya, V.N. and Trees, A.J. Lack of resistance after re-exposure of cattle cured of *Onchocerca ochengi* infection with oxytetracycline. Am. J. Trop. Med. Hyg. 2007; 76:67–72.

Ngoumou, P., Walsh, J.F. and Mace, J.M. A rapid mapping technique for the prevalence and distribution of onchocerciasis: a Cameroon case study. Ann. Trop. Med. Parasitol. 1994; 88: 463–474.

Njongmeta, L.M., Nfon, C.K., Gilbert, J., Makepeace, B.L., Tanya, V.N. and Trees, A.J. Cattle protected from onchocerciasis by ivermectin are highly susceptible to infection after drug withdrawal. Int. J. Parasitol. 2004; 34:1069–1074.

Noma, M., Nwoke, B.E., Nutall, I., Tambala, P.A., Enyong, P., Namsenmo, A., Remme, J., Amazigo, U.V., Kale, O.O. and

Seketeli, A. Rapid epidemiological mapping of onchocerciasis (REMO): its application by the African Programme for Onchocerciasis Control (APOC). Ann. Trop. Med. Parasitol. 2002; 96(Suppl 1):S29–S39.

Nutman, T.B. Future directions for vaccine-related onchocerciasis research. Trends Parasitol. 2002; 18:237–239.

Osei-Atweneboana, M.Y., Eng, J.K., Boakye, D.A., Gyapong, J.O. and Prichard, R.K. Prevalence and intensity of *Onchocerca volvulus* infection and efficacy of ivermectin in endemic communities in Ghana: a two-phase epidemiological study. Lancet 2007; 369:2021–2029.

Ottesen, E.A. Immune responsiveness and the pathogenesis of human onchocerciasis. J. Infect. Dis. 1995; 171:659–671.

Ozoh, G., Boussinesq, M., Bissek, A.C., Kobangue, L., Kombila, M., Mbina, J.R., Enyong, P., Noma, M., Seketeli, A. and Fobi, G. Evaluation of the diethylcarbamazine patch to evaluate onchocerciasis endemicity in Central Africa. Trop. Med. Int. Health 2007; 12:123–129.

Pearlman, E. and Gillette-Ferguson, I. *Onchocerca volvulus*, *Wolbachia* and river blindness. Chem. Immunol. Allergy 2007; 92:254–265.

Peters, D.H. and Phillips, T. Mectizan Donation Program: evaluation of a public-private partnership. Trop. Med. Int. Health 2004; 9:A4–A15.

Pfaff, A.W., Schulz-Key, H., Soboslay, P.T., Taylor, D.W., MacLennan, K. and Hoffmann, W.H. *Litomosoides sigmodontis* cystatin acts as an immunomodulator during experimental filariasis. Int. J. Parasitol. 2002; 32:171–178.

Pischke, S., Buttner, D.W., Liebau, E. and Fischer, P. An internal control for the detection of *Onchocerca volvulus* DNA by PCR-ELISA and rapid detection of specific PCR products by DNA detection test strips. Trop. Med. Int. Health 2002; 7:526–531.

Prince, A.M., Brotman, B., Johnson, E.H., Jr., Smith, A., Pascual, D. and Lustigman, S. *Onchocerca volvulus*: immunization of chimpanzees with X-irradiated third-stage (L3) larvae. Exp. Parasitol. 1992; 74:239–250.

Ramachandran, S., Kumar, M.P., Rami, R.M., Chinnaiah, H.B., Nutman, T., Kaliraj, P. and McCarthy, J. The larval specific lymphatic filarial ALT-2: induction of protection using protein or DNA vaccination. Microbiol. Immunol. 2004; 48:945–955.

Remme, J.H.F. The African Programme for Onchocerciasis Control: preparing to launch. Parasitol. Today 1995; 11:403–406.

Richards, F.O., Jr., Boatin, B., Sauerbrey, M. and Seketeli, A. Control of onchocerciasis today: status and challenges. Trends Parasitol. 2001; 17:558–563.

Rodriguez-Perez, M.A., Katholi, C.R., Hassan, H.K. and Unnasch, T.R. Large-scale entomologic assessment of *Onchocerca volvulus* transmission by poolscreen PCR in Mexico. Am. J. Trop. Med. Hyg. 2006; 74:1026–1033.

Saint Andre, A., Blackwell, N.M., Hall, L.R., Hoerauf, A., Brattig, N.W., Volkmann, L., Taylor, M.J., Ford, L., Hise, A.G., Lass, J.H., Diaconu, E. and Pearlman, E. The role of endosymbiotic *Wolbachia* bacteria in the pathogenesis of river blindness. Science 2002; 295:1892–1895.

Sartono, E., Kruize, Y.C., Kurniawan, A., Maizels, R.M. and Yazdanbakhsh, M. Depression of antigen-specific interleukin-5 and interferon-gamma responses in human lymphatic filariasis as a function of clinical status and age. J. Infect. Dis. 1997; 175: 1276–1280.

Schulz-Key, H. Observations on the reproductive biology of *Onchocerca volvulus*. Acta Leiden. 1990; 59:27–44.

Sher, A. Vaccination against parasites: special problems imposed by the adaptation of parasitic organisms to the host immune response. In: The Biology of Parasitism (P.T. Englund and A. Sher, Eds.), pp. 169–182. Alan R. Liss, New York, USA, 1988.

Soboslay, P.T., Dreweck, C.M., Taylor, H.R., Brotman, B., Wenk, P. and Greene, B.M. Experimental onchocerciasis in chimpanzees. Cell-mediated immune responses, and production and effects of IL-1 and IL-2 with *Onchocerca volvulus* infection. J. Immunol. 1991; 147:346–353.

Soboslay, P.T., Geiger, S.M., Weiss, N., Banla, M., Luder, C.G., Dreweck, C.M., Batchassi, E., Boatin, B.A., Stadler, A. and Schulz-Key, H. The diverse expression of immunity in humans at distinct states of *Onchocerca volvulus* infection. Immunology 1997; 90:592–599.

Soboslay, P.T., Luder, C.G., Hoffmann, W.H., Michaelis, I., Helling, G., Heuschkel, C., Dreweck, C.M., Blanke, C.H., Pritze, S., Banla, M. et al. Ivermectin-facilitated immunity in onchocerciasis; activation of parasite-specific Th1-type responses with subclinical *Onchocerca volvulus* infection. Clin. Exp. Immunol. 1994; 96:238–244.

Soboslay, P.T., Weiss, N., Dreweck, C.M., Taylor, H.R., Brotman, B., Schulz-Key, H. and Greene, B.M. Experimental onchocerciasis in chimpanzees. Antibody response and antigen recognition after primary infection with *Onchocerca volvulus*. Exp. Parasitol. 1992; 74:367–380.

Stewart, G.R., Elson, L., Araujo, E., Guderian, R., Nutman, T.B. and Bradley, J.E. Isotype-specific characterization of antibody responses to *Onchocerca volvulus* in putatively immune individuals. Parasite Immunol. 1995; 17:371–380.

Tawe, W., Pearlman, E., Unnasch, T.R. and Lustigman, S. Angiogenic activity of *Onchocerca volvulus* recombinant proteins similar to vespid venom antigen 5. Mol. Biochem. Parasitol. 2000; 109:91–99.

Taylor, M.J. *Wolbachia* bacteria of filarial nematodes in the pathogenesis of disease and as a target for control. Trans. R. Soc. Trop. Med. Hyg. 2000; 94:596–598.

Taylor, M.J., Abdel-Wahab, N., Wu, Y., Jenkins, R.E. and Bianco, A.E. *Onchocerca volvulus* larval antigen, OvB20, induces partial protection in a rodent model of onchocerciasis. Infect. Immun. 1995; 63:4417–4422.

Taylor, M.J., Cross, H.F. and Bilo, K. Inflammatory responses induced by the filarial nematode *Brugia malayi* are mediated by lipopolysaccharide-like activity from endosymbiotic *Wolbachia* bacteria. J. Exp. Med. 2000; 191:1429–1436.

Taylor, M.J., Makunde, W.H., McGarry, H.F., Turner, J.D., Mand, S. and Hoerauf, A. Macrofilaricidal activity after doxycycline treatment of *Wuchereria bancrofti*: a double-blind, randomised placebo-controlled trial. Lancet 2005; 365:2116–2121.

Taylor, M.J., van Es, R.P., Shay, K., Folkard, S.G., Townson, S. and Bianco, A.E. Protective immunity against *Onchocerca volvulus* and *O. lienalis* infective larvae in mice. Trop. Med. Parasitol. 1994; 45:17–23.

Tchakoute, V.L., Graham, S.P., Jensen, S.A., Makepeace, B.L., Nfon, C.K., Njongmeta, L.M., Lustigman, S., Enyong, P.A., Tanya, V.N., Bianco, A.E. and Trees, A.J. In a bovine model of onchocerciasis, protective immunity exists naturally, is absent in drug-cured hosts, and is induced by vaccination. Proc. Natl. Acad. Sci. U.S.A. 2006; 103:5971–5976.

The African Programme for Onchocerciasis Control (APOC). Final communiqué of the 11th session of the Joint Action Forum (JAF) of APOC, Paris, France, 6–9 December, 2005. Ouagadougou (Burkina Faso), 2005.

Tielsch, J.M. and Beeche, A. Impact of ivermectin on illness and disability associated with onchocerciasis. Trop. Med. Int. Health 2004; 9:A45–A56.

Townson, S. and Bianco, A.E. Immunization of calves against the microfilariae of *Onchocerca lienalis*. J. Helminthol. 1982a; 56:297–303.

Townson, S. and Bianco, A.E. Experimental infection of mice with the microfilariae of *Onchocerca lienalis*. Parasitology 1982b; 85:283–293.

Townson, S., Bianco, A.E., Doenhoff, M.J. and Muller, R. Immunity to *Onchocerca lienalis* microfilariae in mice. I. Resistance induced by the homologous parasite. Tropenmed. Parasitol. 1984; 35:202–208.

Townson, S., Nelson, G.S. and Bianco, A.E. Immunity to *Onchocerca lienalis* microfilariae in mice. II. Effects of sensitization with a range of heterologous species. J. Helminthol. 1985; 59:337–346.

Trees, A.J., Graham, S.P., Renz, A., Bianco, A.E. and Tanya, V. *Onchocerca ochengi* infections in cattle as a model for human onchocerciasis: recent developments. Parasitology 2000; 120:S133–S142.

Trees, A.J., Wahl, G., Klager, S. and Renz, A. Age-related differences in parasitosis may indicate acquired immunity against microfilariae in cattle naturally infected with *Onchocerca ochengi*. Parasitology 1992; 104:247–252.

Tsuji, N., Miyoshi, T., Islam, M.K., Isobe, T., Yoshihara, S., Arakawa, T., Matsumoto, Y. and Yokomizo, Y. Recombinant Ascaris 16-kilodalton protein-induced protection against *Ascaris suum* larval migration after intranasal vaccination in pigs. J. Infect. Dis. 2004; 190:1812–1820.

Tsuji, N., Suzuki, K., Kasuga-Aoki, H., Isobe, T., Arakawa, T. and Matsumoto, Y. Mice intranasally immunized with a recombinant 16-kilodalton antigen from roundworm Ascaris parasites are protected against larval migration of *Ascaris suum*. Infect. Immun. 2003; 71:5314–5323.

Tsuji, N., Suzuki, K., Kasuga-Aoki, H., Matsumoto, Y., Arakawa, T., Ishiwata, K. and Isobe, T. Intranasal immunization with recombinant *Ascaris suum* 14-kilodalton antigen coupled with cholera toxin B subunit induces protective immunity to *A. suum* infection in mice. Infect. Immun. 2001; 69:7285–7292.

Turaga, P.S., Tierney, T.J., Bennett, K.E., McCarthy, M.C., Simonek, S.C., Enyong, P.A., Moukatte, D.W. and Lustigman, S. Immunity to onchocerciasis: cells from putatively immune individuals produce enhanced levels of interleukin-5, gamma interferon, and granulocyte-macrophage colony-stimulating factor in response to *Onchocerca volvulus* larval and male worm antigens. Infect. Immun. 2000; 68:1905–1911.

van Der Kleij, D., Tielens, A.G. and Yazdanbakhsh, M. Recognition of schistosome glycolipids by immunoglobulin E: possible role in immunity. Infect. Immun. 1999; 67:5946–5950.

Wahl, G., Achu-Kwi, M.D., Mbah, D., Dawa, O. and Renz, A. Bovine onchocercosis in north Cameroon. Vet. Parasitol. 1994; 52:297–311.

Wahl, G., Enyong, P., Ngosso, A., Schibel, J.M., Moyou, R., Tubbesing, H., Ekale, D. and Renz, A. *Onchocerca ochengi*: epidemiological evidence of cross-protection against *Onchocerca volvulus* in man. Parasitology 1998; 116:349–362.

Wang, S.H., Zheng, H.J., Dissanayake, S., Cheng, W.F., Tao, Z.H., Lin, S.Z. and Piessens, W.F. Evaluation of recombinant chitinase and SXP1 antigens as antimicrofilarial vaccines. Am. J. Trop. Med. Hyg. 1997; 56:474–481.

Ward, D.J., Nutman, T.B., Zea-Flores, G., Portocarrero, C., Lujan, A. and Ottesen, E.A. Onchocerciasis and immunity in humans: enhanced T cell responsiveness to parasite antigen in putatively immune individuals. J. Infect. Dis. 1988; 157:536–543.

Weiss, N., van Den Ende, M.C., Albiez, E.J., Barbiero, V.K., Forsyth, K. and Prince, A.M. Detection of serum antibodies and circulating antigens in a chimpanzee experimentally infected with *Onchocerca volvulus*. Trans. R. Soc. Trop. Med. Hyg. 1986; 80:587–591.

WHO Onchocerciasis and its control. Report of a WHO Expert Committee on Onchocerciasis Control. World Health Organ. Tech. Rep. Ser. 1995; 852:1–104.

WHO. Onchocerciasis: strategic direction for research. Report of a WHO Expert Committee on Onchocerciasis Control. World Health Organ. Tech. Rep. Ser. 2002.

Williams, S.A., Laney, S.J., Lizotte-Waniewski, M., Bierwert, L.A. and Unnasch, T.R. The River Blindness Genome Project. Trends Parasitol. 2002; 18:86–90.

Winnen, M., Plaisier, A.P., Alley, E.S., Nagelkerke, N.J., van Oortmarssen, G., Boatin, B.A. and Habbema, J.D. Can ivermectin mass treatments eliminate onchocerciasis in Africa? Bull. World Health Organ. 2002; 80:384–391.

Wu, Y., Adam, R., Williams, S.A. and Bianco, A.E. Chitinase genes expressed by infective larvae of the filarial nematodes, *Acanthocheilonema viteae* and *Onchocerca volvulus*. Mol. Biochem. Parasitol. 1996; 75:207–219.

Yip, H.C., Karulin, A.Y., Tary-Lehmann, M., Hesse, M.D., Radeke, H., Heeger, P.S., Trezza, R.P., Heinzel, F.P., Forsthuber, T. and Lehmann, P.V. Adjuvant-guided type-1 and type-2 immunity: infectious/noninfectious dichotomy defines the class of response. J. Immunol. 1999; 162:3942–3949.

Zhan, B., Wang, Y., Liu, Y., Williamson, A., Loukas, A., Hawdon, J.M., Xue, H.C., Xiao, S.H. and Hotez, P.J. Ac-SAA-1, an immunodominant 16 kDa surface-associated antigen of infective larvae and adults of *Ancylostoma caninum*. Int. J. Parasitol. 2004; 34:1037–1045.

Zimmerman, P.A., Guderian, R.H., Aruajo, E., Elson, L., Phadke, P., Kubofcik, J. and Nutman, T.B. Polymerase chain reaction-based diagnosis of *Onchocerca volvulus* infection: improved detection of patients with onchocerciasis. J. Infect. Dis. 1994; 169:686–689.

CHAPTER

68

Schistosomiasis

Christopher L. King
Center for Global Health and Diseases, Case Western Reserve University School of Medicine, Cleveland, OH, USA

OUTLINE

ABSTRACT

Schistosomiasis is one of the most important tropical diseases chronically infecting over 200 million people worldwide, and an estimated 774 million people are at risk. It is the most important human helminth infection in terms of morbidity and mortality. Frequently, individuals are coinfected with schistosomiasis, HIV/AIDS, tuberculosis, and/or malaria, and increasing evidence indicates that schistosomiasis impacts the course of disease and transmission of these important infections. A safe and highly effective drug for the treatment of schistosomiasis is praziquantel. The mainstay of schistosomiasis control has been mass treatment programs that have reduced the burden of disease. However, individuals become rapidly reinfected, and drug cost and the infrastructure to sustain repeated treatment may be unsustainable.

Currently, no effective vaccine exists for schistosomiasis, and only one vaccine candidate is being tested in humans with no evidence for any efficacy yet. In spite of this slow progress, there has been about 40 years of research toward development of a schistosome vaccine. A strong rationale exists for development of a schistosome vaccine. There are good animal models for studying human schistosomiasis. High levels of protective immunity can be induced in animals with radiation-attenuated infective larval stage of the parasite. Humans develop partial immunity to schistosomiasis. Importantly, the parasite does not replicate in humans, and a vaccine with partial efficacy can reduce the intensity of infection and therefore the burden of disease.

Initially, vaccine candidates were targeted toward preventing infection. Increasingly, attention has focused on vaccine candidates that impair the fecundity of adult worms since the release of ova that become trapped in host tissues is the primary cause of disease. It is the granulomatous responses to these ova and ensuing fibrotic changes and chronic inflammation that cause organ failure, particularly the liver and kidneys, and the chronic anemia that are the hallmarks of schistosome disease. Reducing fecundity will also reduce transmission.

Many different approaches have been undertaken to identify vaccine candidates producing a long list of antigens delivered as recombinant antigens and DNA constructs with a variety of adjuvant formulations. These antigens present a range of proteins from glycolytic enzymes (e.g., glutathione S-transferase), myofibrillar proteins since schistosomes are highly motile organisms (e.g., paramyosin), proteins generated by the parasite that may modulate the host response to the parasite (such as superoxide dismutases, which may impair neutrophil O_2—killing of parasites), and integral membrane proteins expressed on the parasite surface, such as tetraspanins. The levels of protection have rarely exceeded 50% to any of these molecules or formulations and most have been less than 40%, suggesting that there are fundamental characteristics about the parasite biology to avoid host immune responses that need to be better understood in order to generate a more effective vaccine. In spite of these limitations, the recent sequencing of the schistosome genome, and characterization of the parasite transcriptome and proteome in different life-cycle stages, has identified many novel proteins. The analysis of these proteins firmly based on the physiological process of the parasite has begun to yield a new generation of exciting vaccine candidates.

INTRODUCTION

Schistosomiasis is likely an ancient disease of humans. Earliest records of human infection document the presence of schistosome ova in tissues of Egyptian mummies ~1500 BC, and earlier etchings from Egyptian temples show individuals urinating blood, which is characteristic of human urinary schistosomiasis. It was not until 1852 that Theordor Bilharz, working in Egypt, first described the disease and the initial elements of its transmission. It was termed bilharzia and later referred to as schistosomiasis.

Five species infect humans; *Schistosoma haematobium*, *Schistosoma japonicum*, *Schistosoma mansoni*, *Schistosoma mekongi*, and *Schistosoma intercalatum*. *S. mansoni* and *S. haematobium* are predominately human parasites while *S. japonicum* is zoonotic in a wide range of domestic animals as well as humans these species. Closely related *Schistosoma bovis*, *Schistosoma spindale*, *Schistosoma margrebowiei*, and *Schistosoma mattheei* infect domestic livestock. Several of animal-specific species can accidentally infect humans, notably *S. bovis* and *S. margrebowiei*. The infective stage of the parasite, referred to as cercariae, from other nonhuman schistosome parasites, particularly bird parasite (e.g., *Trichobilharzia* sp.) may penetrate human skin but fail to penetrate deeper and die. This can give rise to an allergic condition called swimmer's itch, or cercarial dermatitis, caused by release of antigens in the skin.

Biology and Life Cycle

The life cycle is shown in Fig. 68.1. Unlike other trematode infections, schistosomes have separate sexes (dioecious). Adult parasites generally range from 0.8 to 1.6 cm in length, featuring two terminal suckers, a complex tegument, a blind digestive tract, and reproductive organs. The body of the male worm forms a grove, or gynaecophoric channel, into which the longer and thinner female body is embedded (Gryseels et al., 2006; Ross et al., 2002). The male worm provides important nutrients to the female and helps to anchor the worm pair inside the vessel with prominent anterior and ventral suckers.

The mammalian host becomes infected when free-swimming larvae released from freshwater snail intermediate hosts directly penetrate the skin. The cercariae forces its way through the initial squamous cells of epidermis, and further penetration of skin is facilitated by release of an elastase and a series of proteolytic enzymes produced by specific glands in the anterior

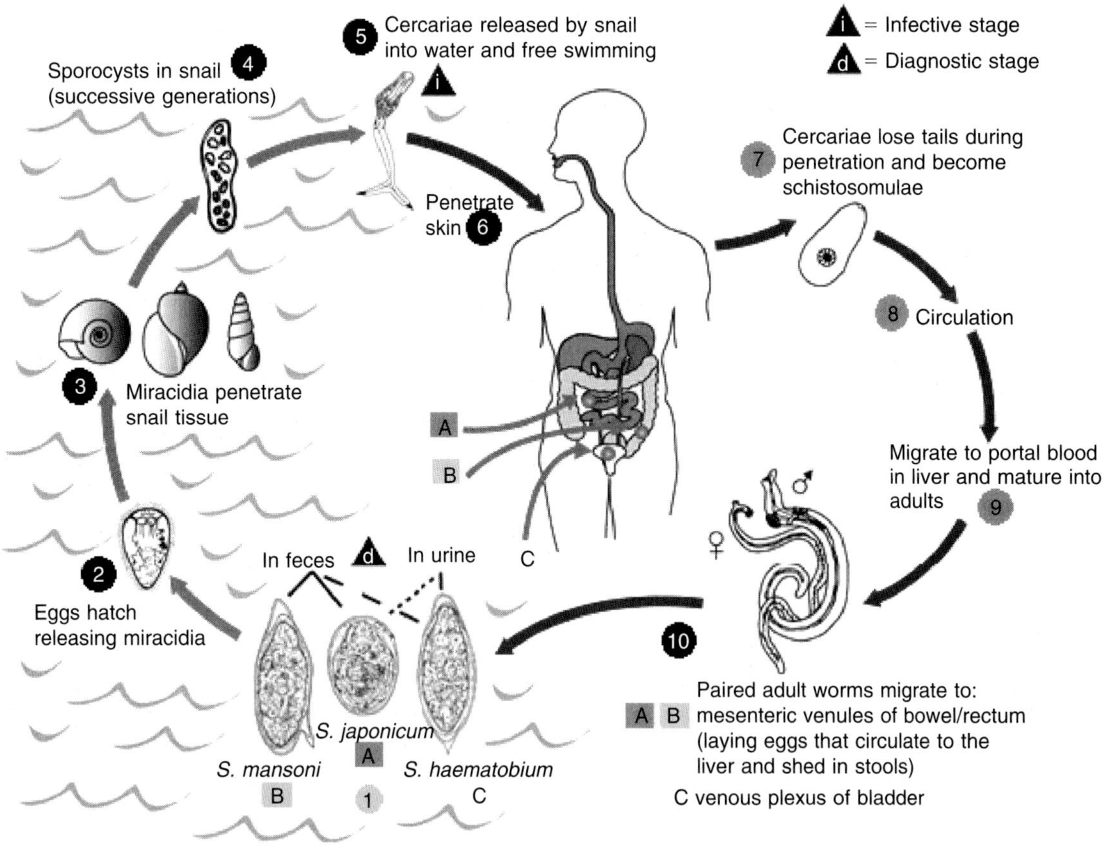

FIGURE 68.1 Life cycle of *S. masoni*, *S. japonicum*, and *S. haematobium* (Reprinted from DPDx, the Centers for Disease Control and Prevention website for parasite diseases, http://www.dpd.cdc.gov/dpdx/HTML/Schistosomiasis.htm).

portion of the parasite. After larvae successfully penetrate the skin, they transform into a tissue-migratory larval stage, or schistosomula, which are about 300 μm in size, whereupon the larvae undertake a complex intravascular migration in which they eventually accumulate in the vessels of the hepatic portal system. Initially, the parasite migrates in the skin for 1–4 days until they reach the venous circulation, are carried to the lungs where they shift to the arterial circulation and eventually concentrate in the liver, travel to the portal vein, and mature into adult worms. Migration of larvae from the skin to liver requires about 7–10 days, and maturation to adult worms requires 3–6 additional weeks. The worm pairs migrate distally into the mesenteric veins, that is, superior mesenteric veins for *S. mansoni* and inferior mesenteric and hemorrhoidal veins for *S. japonicum*. For *S. haematobium,* adult worms migrate past the hemorrhoidal veins into the venus plexus surrounding the bladder and adjacent organs. Females produce hundreds of ova each day (*S. mansoni* and *S. haematobium*) to thousands per day (for *S. japonicum*) for the life span of the worms that can be up to 30 years. These eggs release potent pro-inflammatory molecules that disrupt tissue barriers and enable the ova to transit from the endovascular space into the lumen of gut (*S. mansoni* and *S. japonicum*) or bladder (*S. haematobium*), where ova are excreted with stool or urine into the environment and then hatch and infect snails to complete the life cycle. Primarily the ova released into host tissues, especially those that fail to exit into the feces or urine, cause the pathology and disease associated with schistosomiasis. Infection intensity depends on exposure because the parasite does not replicate in the human host. The life cycle is completed when ova are passed in the urine or feces and come into contact with fresh water, whereupon they release miracidia, a ciliated highly motile form of the parasite, which infect freshwater snails. *S. haematobium* infects *Bulinus* species, *S. mansoni* infects *Biomphalaria* sp., and *S. japonicum* infects *Oncomelania* sp. The distribution of schistosome species depends on the ecology of their respective snail hosts. After infection of the snail, parasites undergo asexual replication to produce thousands of cercariae per snail that are shed in the water to complete the life cycle.

Schistosomiasis transmission is highly dependent upon water resource use and sanitation, usually associated with agricultural practices. Environmental changes linked to water resource development (e.g., large- and small-scale dams), and irrigation, migration, and population growth have led to an increase in the prevalence of schistosomiasis and spread into areas where it was not previously endemic (Gryseels et al., 2006; Ross et al., 2002; Li et al., 2007). The Diama dam on the Senegal River introduced *S. mansoni* into Senegal and Mauritania. The Aswan High Dam in Egypt spread *S. mansoni* into Upper Egypt, which was already infected with *S. haematobium* (Malek, 1975). The spread of small ponds for collection of seasonal rainfall throughout many areas of African have led to the spread of *S. haematobium* since its snail intermediate host is able to estivate in the mud, when the pond dries up, until the next rainfall. The soon-to-be-completed Three Gorges Dam on the Yangtze River in China is near to a *S. japonicum* endemic area that could lead to further spread of this parasite (Li et al., 2007).

Epidemiology and Significance

Schistosomiasis is estimated to have infected 207 million individuals in 74 countries worldwide, and ~1 billion individuals are at risk (Gryseels et al., 2006). It is likely these estimates significantly underestimate worldwide prevalence because of the poor sensitivity of existing diagnostic methods. Infections occur throughout much of topical and subtropical areas of the world (Fig. 68.2). The burden of infection occurs in rural communities; however, with increasing travel, residents of major cities in China and Brazil often become infected. North American and European travelers are often at risk and occasionally become infected. The heaviest burden of infection and disease is among adolescent children. Therefore, children under the age of 4 years, in whom schistosome infections are uncommon, would be an important target population for a schistosome vaccine.

Disease

Most diseases arise from inflammatory responses directed to ova trapped in tissues, especially the liver or gut (Wilson et al., 2007). These trapped eggs release 40 different proteins (Jang-Lee et al., 2007) that produce marked CD4+ T-cell-dependent inflammatory responses, resulting in the recruitment of primarily eosinophils, monocytes, and B cells to produce a granuloma around the egg. Importantly, the chronic granulomatous inflammation promotes collagen deposition with the consequence that ova trapped in the liver, bladder, or ureters can impair organ function, notably the liver, resulting in hepatic periportal

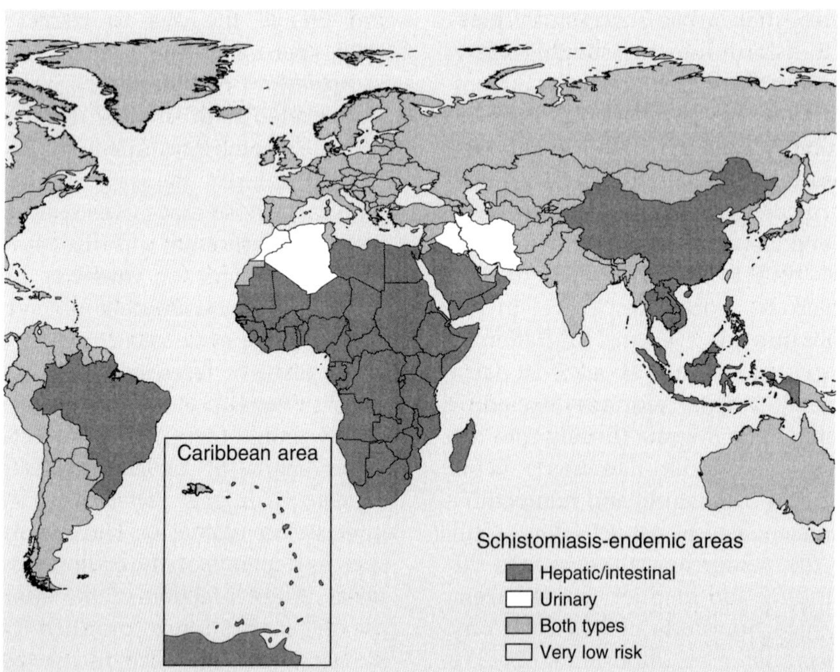

FIGURE 68.2 The global distribution of schistosomiasis. *S. mansoni*, *S. japonicum*, and *S. mekongi* represent hepatic/intestinal species and *S. haematobium* and *S. intercalatum* urinary schistosomiasis (Reprinted from DPDx, the Centers for Disease Control and Prevention website for parasite diseases, http://www.n.cdc.gov/travel/image.ashx?i=263).

(Symmer's) fibrosis and cirrhosis, or obstruction of the ureter, leading to obstructive hydronephrosis and renal failure (King et al., 1988). Chronic schistosomiasis has been associated with hepatocellular carcinoma and bladder cancer. An important aspect of disease is when adult worms develop or migrate to other organs. The residence of worms in veins surrounding the bladder, prostate, uterus, and vagina or near the central nervous system are notable. The release of ova from these aberrant sites produces inflammation in critical tissues with devastating effects. Transverse myelitis and seizures from the schistosomiasis, for example, can be difficult to diagnose and produce permanent neurological impairment (Pittella, 1997; Betting et al., 2005).

Recently, it has been appreciated that light infections can also produce significant morbidity. Symptoms such as diarrhea, dysentery, and dysuria, which cause chronic pain, can be disabling as well as the consequence of anema, undernutrition, exercise intolerance, stunting of growth, and learning and developmental delays (King et al., 2005). All these conditions can also lead to increased absences of schools, which compound learning and development. Schistosomiasis is a chronic infection, and coinfections are common with other important infections, notably HIV/AIDS, tuberculosis, and malaria. Since schistosomes can cause profound immune modulation in its host (Wilson et al., 2007; King et al., 1996), there is growing evidence that schistosomiasis can affect the natural course of these co-infections and potentially increase the transmission such as HIV/AIDS (Gallagher et al., 2005; Brown et al., 2005, 2006; Kallestrup et al., 2005, 2006; Booth et al., 2004; Lyke et al., 2005, 2006). Schistosomiasis also has an impact on the efficacy of vaccine of children born to infected women (Malhotra et al., 1999) and among older individuals infected with schistosomiasis (Elliott et al., 2007).

The number of deaths attributable to schistosomiasis in sub-Saharan Africa is estimated at 280,000 per year (van der Werf et al., 2003), with 1.7 million disability-adjusted life years (DALYs) from chronic infections (Mathers et al., 2007). Thus schistosomiasis has been recognized as the second most important human parasite disease, following malaria, as a cause of human suffering.

Treatment and Control and the Importance of a Vaccine

Schistosomiasis can be treated with a single dose of praziquantel (PZQ) to achieve over 90% cure rates as determined by individuals becoming egg negative. However, it is not known whether some worms survive treatment and transiently stop releasing ova. PZQ is safe with few side effects and has become increasingly less expensive. The widespread use of PZQ has led to reduction in severe disease; nonetheless, because individuals become rapidly infected, the overall prevalence of schistosomiasis has actually increased 7–12% in the past decade in spite of mass chemotherapy programs (Fenwick and Webster 2006). Although clinically significant PZQ resistance has not been reported in humans, resistance to drugs related to PZQ already exists in domestic animals (Bennett et al., 1997), and a low efficacy of PZQ treatment has been reported in Egypt and Senegal (Fallon et al., 1997; Ismail et al., 1999). No safe, broad-spectrum alternative drugs for treatment exist. Nearly all countries have an existing infrastructure for vaccine delivery but few for drug delivery. Therefore, a vaccine that produces even a partial reduction in worm burdens could considerably reduce pathology and limit parasite transmission. .

PROTECTIVE IMMUNITY

Animal Models

Most of what is known about protective immunity and vaccine testing has been obtained in the laboratory mice and baboons, which are the favored nonhuman primate models (Nyindo and Farah, 1999; Siddiqui et al., 2008). Human schistosome parasites can infect these animal models. Both of these animal models develop patent infections and pathology similar to that observed in humans. Studies of protective immunity and vaccine testing in rats and other nonhuman primates, notably the rhesus, have also been performed; however, both of these models spontaneously cure their infection (Moore and Meleney, 1952; Ritchie et al., 1963; Maddison et al., 1979). The mechanism by which these latter hosts can spontaneously eliminate infections may provide novel vaccine targets for parasite elimination.

Since schistosome parasites do not multiple in the mammalian host, protective immunity is primarily measured as the reduction in adult worm burden of vaccinated animals relative to controls. Typically, this is assessed by euthanasia of animals followed by perfusion of the portal vasculature to flush out the worms. Surrogates of worm burdens, such as assessing the fecal and urine egg numbers and circulating antigen concentrations, have been insensitive and only partially correlate with the actual worm burden (Wilson et al., 2006). This has been a major impediment for studying the levels of protective immunity in humans and will complicate assessment of vaccine efficacy in human trials. For a vaccine to be protective, it need not generate sterile immunity because a significant reduction in the number of worms would reduce the number of eggs produced, and thus morbidity and transmission. Indeed, even if vaccine had little effect on the adult worm population but markedly reduced worm fecundity, it would have a similar effect. This antifecundity effect can be measured by reduction in egg output in the feces or urine and tissue burden per adult worm pair. The effect on transmission would be assessed by reduced viability of released ova, using a miracidial hatching assay. Miracidia released from hatched ova in water infect the snail intermediate host to complete the transmission cycle.

A difficulty for measuring protective immunity in animal models is the long delay from the time an animal is challenged following immunization and measurement of protection by portal perfusion, which requires an interval of 5–6 weeks to allow maturation of parasites. An in vitro assay of immune-mediated protection would be of great value. One such assay measured antibody-dependent cellular cytotoxicity of schistosome larvae in vitro, using serum and cells from vaccinated animals, to predict vaccine potential (Butterworth et al., 1974, 1982). Although it has been used for decades, its ability to predict vaccine potential is limited, and the discordance of ultrastructural changes associated with larval killing in vivo and in vitro (McLaren et al., 1978; McLaren and James 1985) has led the majority of researchers in the field to consider this as a poor surrogate for protective immunity.

Radiation-Attenuated Parasites Induce Protective Immunity

The majority of studies that have evaluated mechanisms of protective immunity to schistosomiasis have focused on studies of radiation-attenuated (RA) cercariae that infect various animal models but fail to mature to sexually mature adults. Such studies have reliably demonstrated 60–70% reduction in worm burdens following challenge (Coulson, 1997). Certain mouse strains produce even higher levels of protection, and sterile immunity has been achieved when exposure to irradiated cercariae is combined with immunization with recombinant IL-12 (Wynn et al., 1995).

The mechanisms by which RA larvae confer protection in murine models have been evaluated in considerable detail (Coulson, 1997; Wilson et al., 1996). The optimal radiation dose permits parasites to infect the mouse skin but delays their migration to host vessels and, as a consequence, provokes more intensive inflammatory responses in the skin and draining lymph nodes compared to normal larvae. This has the effect of boosting cell-mediated immunity, particularly Th1-type immune responses. Upon challenge infection with normal larvae, the primary site of attrition is in the lung. Interestingly, this attrition does not appear to result from killing of the lung-stage larvae; instead, a primarily mononuclear infiltrate develops in the pulmonary capillaries upon challenge infection, which impairs onward migration of the larvae (Smythies et al., 1992). Some of the larvae enter pulmonary alveoli and thus ends their intravascular journey. The production of IFN-γ and TNF-α is essential for this effecter immune response (Wilson, 1998; Street, et al., 1999), and protection occurs in μMT mice, which cannot make an antibody response (Anderson et al., 1999), and emphasizes the cell-mediated basis for protection.

If mice are repeatedly immunized with RA cercariae, usually three times, similar levels of protective immunity are achieved compared to singly immunized mice; however, this tends to involve a substantial humoral component. Adoptive transfer of IgG antibodies from multiple immunized mice can confer protection to naive receipts, whereas singly vaccination mice cannot (Delgado and McLaren, 1990). The maximal levels of protection in nonhuman primates requires at least 5 exposure to RA cercariae producing up to 84% reduction in worm burden (Kariuki et al., 2004). The strong correlation of the IgG antibody level directed to larval stage antigens with levels of protection implies that humoral immunity is a critical component of protective immunity (Kariuki et al., 2004). Protection waned to an average of 72–53% if animals were challenged 3 versus 12 weeks following vaccination with irradiated cercariae. Although this was a single experiment, it highlights that the duration of immunity to a schistosome vaccine may be short, and approaches will be needed to sustain significant levels of protection.

The use of irradiated larvae as a human vaccine followed the demonstration of the effectiveness of similar irradiated sporozoite vaccines against malaria (Hoffman et al., 2002). The use of irradiated cercariae as a vaccine has two important drawbacks. First, if irradiation fails to attenuate all larvae and some develop into adults, this could result in patent infection, egg production, and ensuing disease as described above. This concern is significant because individuals would likely require exposure to thousands of larvae (Eberl et al., 2001), and even if only 1–5% were to mature to adults, this would represent a heavy infection with potentially adverse consequences. The second impediment has been the short life span of irradiated larvae and impracticality for their use in most endemic areas. Therefore, attempts have been made to cryopreserve RA larvae.

Cryopreserved larvae produce significantly lower and highly variable protection in nonhuman primates and mice (Bickle and James, 1978; Murrell et al., 1979; Damian et al., 1984; Lewis et al., 1984). The lack of reproducibility for cryopreservation eventually resulted in individuals loosing interest for this approach two decades ago. The recent success in the cryopreservation or lyophilization of RA *Theilera parva* sporozoites for subsequent successful vaccination suggests that reexamination of this approach for schistosomiasis is warranted (Marcotty et al., 2003; Mbao et al., 2006). The cryopreservation of unicellular *Theilera* parasites versus the more complex multicellular schistosome larvae may make this translation difficult.

Humans

Humans gradually acquire partial immunity to further infection following years of exposure such that adults generally have lower prevalence and intensity of infection compared to children and adolescents (Butterworth et al., 1985; Hagan et al., 1991; Roberts et al., 1987; Medhat et al., 1998; Karanja et al., 2002). Repeated treatment of individuals with PZQ accelerates naturally acquired protective immunity, suggesting that repeated priming of parasite antigens released upon worm killing can boost the host immune responses (Karanja et al., 2002). However, the level of protection acquired has been difficult to measure because of the surrogates to assess the presence and intensity of infections mentioned above, for example, fecal and urine egg counts, and serum markers of infection, are insensitive and imprecise. Another complication is the difficulty to measure human exposure. These same issues will also make vaccine testing in humans challenging and point to the necessity for developing additional tools to better assess exposure and biomarkers of infection.

In spite of these complications, human population studies have demonstrated increased levels of crude adult-driven IL-4, IL-5, IFN-γ, and other cytokines in human serum in recall responses by peripheral blood mononuclear cells with parasite protection to reinfection (Medhat et al., 1998; Roberts et al., 1993; Caldas et al., 2000). Other studies have shown that parasite-specific IgE directed to crude preparation of adult worms (Hagan et al., 1991; Rihet et al., 1991) and, in some cases, certain recombinant antigens, for example, Sm22.6 (Fitzsimmons et al., 2004) also correlated with protection. Despite the protective role of IgE, high levels of IgG_4 are also produced during infection and potentially blocking the protective effects of other immunoglobulins because IgG_4 lacks opsonizing and complement fixation capacity of other IgG subclasses (Hagan et al., 1991; Caldas et al., 2000; Rihet et al., 1992). The initial induction of IgE and IgG_4 require IL-4 and IL-13; however, the production of IgG_4 also requires IL-10. IL-10 plays a crucial role in the immunoregulation of host immune responses against schistosomiasis. Although the dissociation of IgE and IgG_4 production might be feasible by differential regulation of IL-4/IL-13 and IL-10, this might be difficult. Studies have also showed that eosinophilia, particularly activated eosinophils (assessed by increased expression of FcεRI-β chain), correlate with protection (Ganley-Leal et al., 2006a, 2006b). IgA antibodies may also have a protective role based on a series of correlation studies with protective immunity in different parts of the world (Caldas et al., 2000; Mutapi et al., 1997). Taken together, these studies support the development of a schistosome vaccine and indicate that such a vaccine will likely require both cellular and humoral effector mechanisms. The development of a vaccine that elicits IgE and eosinophilia is unlikely to be pursued because they risk developing atopic diseases. Because of obvious constrains for study of vaccines in humans, the bulk of work in schistosome vaccine development has centered on animals models.

SUBUNIT VACCINES

Given the difficulties for mass production, storage, and distribution of irradiated larvae and the complex heterogeneity of producing a vaccine based on crude parasite antigens, the focus has shifted to identification of individual parasite proteins as vaccine candidates.

Crude Parasite Extracts for Identification of Potential Protective Antigens

The immunization of mice with crude schistosomula larval extracts typically produced 51% protection against cercarial challenge when administered subcutaneously to C57BL/6 mice with BCG as an adjuvant (James, 1985). Immunity was primarily cell mediated and required potent adjuvants such as BCG to elicit protection (James, 1987). Similar levels of protection were observed with crude adult worm extracts.

Paramyosin

In order to identify which fractions were most important for eliciting protection, the crude antigen was fractionated by gel filtration to identify the antigens responsible for protection. This approach identified a fraction that contained a 97-kDa protein that was subsequently identified as paramyosin, an invertebrate muscle protein (Lanar et al., 1986). The native protein was subsequently purified and shown to induce about 30% protection with BCG as an adjuvant (Pearce et al., 1988) and similar levels of protection (up to 53%) without any adjuvant (Flanigan et al., 1989). We subsequently expressed the full-length Sm-97 using a vector provided by Dr. Charles Schoemaker but failed to achieve more than 30% protection using BCG or alum as an adjuvant (unpublished observations).

At about the same time, *S. japonicum* paramyosin obtained from a Philippines strain of the parasites demonstrated strong protection against cercarial challenge when administered intraperitoneally without adjuvants (62–86%) (Ramirez et al., 1996). Native and recombinant paramyosin proteins or DNA constructs from China confer protection against *S. japonicum* in mice, water buffaloes, and other mammalian hosts (Wu et al., 2005; Chen et al., 2006). Recent studies of human antibody (Nara et al., 2007) and Th2-type immune responses (Leenstra et al., 2006) to *S. japonicum* paramyosin correlate with protection providing support for the protein as a leading *S. japonicum* vaccine candidate. More recently, several studies have indicated that the exogenous form of paramyosin may act to bind Fc fragments of immunoglobulins (Loukas et al., 2001) and inhibit the terminal complement cascade (Deng et al., 2003, 2007), thus indicating an immunomodulatory role for the parasite. Therefore, paramyosin may further act as a potential vaccine candidate; unfortunately, the protein is poorly expressed in soluble form. To improve expression, various fragments have been cloned and the protein expressed, although protection has been less than the full-length molecule (Zhang et al., 2006).

GSTs

A different approach to identify antigens from crude extracts was used to identify glutathione S-transferases (GSTs) independently from two laboratories. In these experiments, crude adult worm extracts were used to immunize rats and then the immune sera to immunoprecipitate native proteins, one of which was the 28-kDa GST (Balloul et al., 1985). Antiserum raised to this protein was able to induce cytotoxic activity of schistosomula in vitro and subsequently shown to reduce worm burdens from 52% to 67% with cercarial challenge following immunization with Freund's complete adjuvant or aluminum hydroxide (Balloul et al., 1987a, 1987b). A great advantage for this protein is that the native protein can be readily prepared using immobilized glutathione to which it binds with high affinity. The 28-kDa GST was formulated with aluminum hydroxide and generated from 0% to 80% protection in baboons (mean 38%) (Boulanger et al., 1991). This study also noted for the first time an antifecundity effect of immunization with Sm28GST. There were a significantly reduced number of eggs per worm pair recovered from the tissues of the exposed animals. This antifecundity effect was confirmed in mice by immunization with recombinant protein as well as antigenic peptides (Charrier-Ferrara et al., 1992).

In a separate research study, a 26-kDa GST was identified from *S. japonicum*. It was cloned, expressed, and shown to elicit a small level of protection against adult worms (21%) but much greater antifecundity effect (59% reduction in number of eggs per adult worm pair) in pigs (Liu et al., 1995). The *S. japonicum* GSTs have subsequently shown inconsistent protection in mice (Scott and McManus, 2000) but obtained significant protection in sheep, water buffalo, and cattle with various formulations—including BCG, Freund's adjuvant, and/or keyhole limpet hemocyanin (KLH)—with levels of protection up to 70%, although most reduction in worm burdens were more in the range of 30–40% (Wu et al., 2005; Shi et al., 2001, 2002; Xu et al., 1995; Taylor et al., 1998).

The research focus for GST as a potential vaccine candidate shifted to an emphasis on its antifecundity effect, which would reduce both morbidity and transmission. It was also tested in the patas monkey, which is an animal model used for *S. haematobium* or urinary schistosome infections. Vaccination with the heterologous Sm28GST induced marked and persistent reduction in egg excretion in the urine and feces by 55% and 74%, respectively; however, it had no effect on worm burdens (Boulanger et al., 1995). When the *S. haematobium* homologue to GST was cloned and expressed, it also generated a similar reduction in egg excretion following vaccination of patas monkeys using BCG and Freund's complete adjuvant but no effect on worm burden (Boulanger et al., 1995). This antifecundity effect could also be observed following vaccination of cattle against *S. bovis* (Bushara et al., 1993), but not in goats or sheep, although it produced a significant reduction in worm burdens (Boulanger et al., 1994, 1999). The protective effect on different species, particularly the antifecundity effect of GST, has resulted in this antigen moving forward to human vaccine trials (described below).

Fatty Acid-Binding Proteins

Fasciola hepatica is a trematode parasite in the same family as schistosomes and shares cross-reactive antigens with schistosomes. A 12-kDa *F. hepatica* antigen was identified with cross-reactivity to *S. mansoni* and, when used to immunize mice, induced 52–77% protection against a heterologous challenge with *S. mansoni* (Hillyer et al., 1988). Subsequently, a fatty acid-binding protein (Sm14) was cloned using polyclonal sera from mice immunized with a saline extract of *S. mansoni* adult worms. This recombinant protein induced 37–67% reduction in worm burden in mice following cercarial challenge and 100% protection against *F. hepatica* challenge (Tendler et al., 1996). This protein was subsequently identified to have 44% sequence identity to the fasciola Sm12 antigen. This raised the possibilities that this vaccine could protect against both schistosomiasis and liver flukes (Hillyer, 1995). Although the *S. japonicum* homologue has 91% identity to Sm14, the recombinant antigen has consistently failed to elicit significant protection in mice against *S. japonicum* (Scott et al., 2000). A difficulty may be that the antigen is not surface expressed and appears to be localized in internal tissues.

Detection of Antigens Using Protective Monoclonal Antibodies

With the advent of hybridoma technology to develop monoclonal antibodies (MAbs), mice were injected with various preparations of schistosome extracts—eggs, larvae, adult worms—or vaccinated with irradiated cercariae. From these studies, MAbs were generated that could confer significant passive protection in vivo to cercarial challenge (Smithers et al., 1989). The antigens that these MAbs recognized were identified by immunoaffinity purification of parasite extracts. Three such antigen targets were subsequently evaluated in detail in protection experiments.

Sm23 and Tetraspanins

The MAbs that recognized Sm23 was made by cell fusion following immunization with schistosomula (Harn et al., 1985). The MAb reacted with newly transformed larvae or schistosomula, 5-day schistosomula, but not lung-stage larvae. The antigen was subsequently identified as a membrane tetraspanin with two highly immunogenic extracellular domains. Tetraspanins are four-transmembrane-domain proteins homologous to receptors on T and B cells. Mapping B and T epitopes (Reynolds et al., 1992) identified the immunogenic domains. The epitopes were synthesized as a multiple antigenic peptide (MAP), which was then shown to produce significant protection in mice (Harn et al., 1995). Subsequently, a recombinant form of the molecule administered with aluminum hydroxide failed to elicit protection in mice (Da'dara et al., 2003), although a DNA vaccine formulation injected alone or with plasmids encoding IL-4 or IL-12 elicited 21–44% protection (Da'dara et al., 2001). Administration of a *S. japonicum* homologue (Sj23) in various formulations either failed to elicit protection or induced variable protection levels of generally <40% (Waine et al., 1999; Zhu et al., 2003). However, a DNA formulation injected with plasmids encoding IL-12 was found to induce 51–59% protection in pigs to *S. japonicum* cercarial challenge (Zhu et al., 2004). Similar levels of protection were observed in water buffaloes, using a similar DNA vaccination formulation with IL-12. The focus for these studies was to develop a vaccine that reduced infection intensity in one of the major animal reservoirs for this infection in China (Williams et al., 2002). Recently, the extracellular domains of two additional tetraspanins, referred to as tsp1 and tsp2, were expressed and administered to mice with Freund's adjuvant, tsp2 induced 57–64% reduction and tsp1 34–52% reduction in worm burden (Tran et al., 2006). This antigen is expressed on the surface of adult worms but was not tested to determine if it was also surface expressed in the larval stages. Putatively resistant humans had significantly higher IgG antibodies directed to tsp2 compared to *S. mansoni* infected but not resistant individuals. Together these data support tetraspanins as potentially valuable vaccine candidates for future studies.

Triose Phosphate Isomerase (TPI)

This 28-kDa protein was identified as the target of the MAb that conferred partial protection in mice (41–49%) (Harn et al., 1992). This antigen is present on all stages of the parasite but only transiently expressed on schistosomula. Similar to Sm23, T and B cell epitopes were identified, and a MAP constructed (Reynolds et al., 1994) with 45–65% protection was reported (Harn et al., 1995). A DNA construct of *S. japonicum* homologue of TPI failed to induce protection in mice (Zhu et al., 2002). However, a DNA construct administered with or without a plasmid encoding IL-12 induced 46–48% reduction in worm burden following cercarial challenge, although this reduction in worm burden was greater for female worms (54–60%) (Zhu et al., 2006). TPI is a ubiquitous glycolytic enzyme, is intracellular, and is not surface expressed, however it undergoes exocrine release from the cercarial preacetabular glands, during skin penetration.

9B-Ag

The MAb that recognized this antigen was originally generated from mice chronically infected with

S. mansoni and boosted with a cercarial glycoprotein; the MAb was able confer ~42% protection upon passive transfer (Hazdai et al., 1985). Isolation of the native protein, which is multimeric and consists of 30- and 45-kDa subunits, conferred 40–50% protection in mice from cercarial challenge (Tarrab-Hazdai et al., 1998, 1999; Arnon et al., 2000). This antigen has not been cloned, but various formulations of peptides and mimotopes all gave similar levels of protection.

Identification of Vaccine Candidates from Expression Libraries

New technologies spawned additional efforts for vaccine candidate discovery, and when recombinant DNA technology became available in the early 1980s, a series of antigens were identified by screening schistosome cDNA expression libraries from different parasite stages and by immune sera obtained from mice vaccinated with RA sera or from putatively immune humans. Two notable antigens emerged from this approach.

IrV5

This antigen was identified using immune sera from mice vaccinated with irradiated cercariae and immunoaffinity-purified antigens (Dalton and Strand, 1987) that were then used to immunize rabbits (Soisson et al., 1992). One such antiserum recognized an antigen on the surface of newly transformed schistosomula and was then used to screen a larval expression library to identify a 62-kDa fragment of a larger 200-kDa protein. This protein was found to be a myosin heavy-chain protein, designated IrV5. The recombinant 62-kDa fragment administered with a meningococcal outer membrane protein as an adjuvant induced 75% protection in mice (Soisson et al., 1992). Vaccination of baboons with a similar formulation elicited a variable range of protection in animals from no protection to 54% (Soisson et al., 1993). The subsequent expression of the *S. japonicum* myosin protein fragment failed to induce significant protection in mice or pigs, thus holding out little promise for this molecule as a significant vaccine candidate (Zhang et al., 1998).

GAPDH

This antigen was initially identified by Western blot analysis of serum from putatively immune individuals (Dessein et al., 1988). The protein was subsequently purified, and antisera raised in rabbits were used to screen an adult expression library that identified a molecule as glyceraldehyde 3-phosphate dehydrogenase (GAPDH) (Goudot-Crozel et al., 1989). Interestingly, immune sera raised to this protein identified a motif on the larval surface, which is surprising considering that this is also an abundant glycolytic enzyme. It is possible that an isoform of this molecule may be expressed on the larval surface or possibly taken up from the host (Charrier-Ferrara et al., 1992). The dominant T and B cell epitopes for this molecule were identified and, when coupled to ovalbumin in Freund's adjuvant, produced modest protection (Argiro et al., 2000). Protection was boosted to 32–38% when the MAP was co-adsorbed to granulocyte-macrophage colony-stimulating factor (GM-CSF) and alum (Argiro et al., 1999).

Superoxide Dismutase (SOD)

Serum from putatively immune humans infected with schistosome was used to screen an adult cDNA library and found to recognize a gene encoding a schistosome superoxide dismutase with a signal peptide (SP-SOD) (Smyth et al., 2003). Prior studies have shown that granulocytes release oxygen radicals that have been shown to be toxic for schistosome ova. The addition of exogenous SOD can inhibit the granulocyte toxicity for schistosome ova as measured by their metabolic activity and hatching (Kazura et al., 1985). Therefore, it was reasoned that production of SOD by schistosomes might protect themselves from host granulocyte-mediated killing by oxygen radicals. Antibodies raised to cDNA encoding a cytosolic SOD (CT-SOD) were immunolocalized to the tegument and subtegumental tissues in adult worms (Mei and LoVerde, 1997). Proteomic studies of tegument of adult worms and larvae have shown that SOD is localized below the tegumental plasma membrane (van Balkom et al., 2005; Braschi et al., 2006). A DNA vaccine expressing the CT-SOD produced 44–60% reduction in worm burden following cercarial challenge (Shalaby et al., 2003; Cook et al., 2004). Although protection studies using a recombinant CT-SOD or the SP-SOD that is expressed on the tegument have not been performed, they show potential promise.

SECRETORY AND MEMBRANE PROTEINS AS VACCINE TARGETS

The availability of new technologies such as proteomics and microarray analysis along with the recent sequencing of the *S. mansoni* and *S. japonicum* genomes has opened new avenues for discovery of vaccine candidates (Ashton et al., 2001). This has led to a focus on identifying secreted antigens and tegumental antigens as possible vaccine candidates.

Secreted Proteins

Molecules that are secreted by the parasite are likely to be crucial enzymes for penetration of the epidermis and migration through host tissues. The acetabular glands at the anterior portion of the parasite release elastase, which has been cloned and expressed (McKerrow et al., 1985). More recently, proteomic analysis of cercarial secretions has also identified novel metalloproteinases, a dipeptidyl peptidase, a serine protease inhibitor, and several additional molecules that have homologies to potentially immunomodulatory molecules (Curwen et al., 2006). Although these proteins may be critical for invasion of the skin, they may not represent ideal vaccine candidates since they are primarily active in the epidermis before vascular entry, which is not immediately accessible to the host immune response. A potentially better target would be contents of the head glands, which appear to release their contents following skin penetration (Crabtree and Wilson, 1985) and transformation to schistosomula. Indeed, such proteins may facilitate migration within the dermis and penetration of the vessel wall (Curwen and Wilson, 2003). Although there is a lack of information on such secretions, examination of proteins released by larval stages of the parasite through proteomic analysis can potentially accomplish this task. Both the cercariae and schistosomula appear to release potentially immunomodulatory molecules, such as Sm16, which has been shown to inhibit neutrophil migration and lymphocyte proliferation in the dermis (Rao et al., 2002; Rao and Ramaswamy, 2000). Such molecules may be crucial for blocking host effector immune responses to larval stages of the parasites, and by inhibiting such molecules, host immunity may be much more effective in eliminating parasites. Immunomodulatory molecules released by adult worms, such as SOD as mentioned above, may also be important. Other potential immunomodulatory molecules, such as serine protease inhibitors that could block the host clotting mechanism on the adult surface (Blanton et al., 1994) or complement regulatory protein homologues that could inhibit host complement (Skelly, 2004).

Tegumental Antigens

Antigens expressed on the parasite tegument is an obvious source of putative protective antibodies since it is readily accessible to host immune responses and likely expresses molecules critical for the host survival within the host (Skelly and Wilson, 2006). Schistosome parasites are prodigious consumers of glucose, for example, and consume their dry weight in glucose in 4 h. Several schistosome glucose transporter molecules have been identified, cloned, and expressed on the larval surface (Skelly et al., 1994). These recombinant antigens could be useful vaccine candidates and have the further advantage that an in vitro assay could be developed to examine whether functional antibodies are generated following vaccination. Functional antibodies would be evaluated to determine if they block uptake of labeled glucose by parasites in culture.

The ability of tegumental antigens to induce protective immunity in mice has been approached by isolation of the tegument surface of parasites. The purity and degree of enrichment of these fractions is unclear. However, in one instance, the parasite surface was prepared by a freeze-thaw technique (Roberts et al., 1983), which in subsequent analyses was found to contain various membrane proteins but dominated by cytosolic and cytoskeletal proteins (Braschi et al., 2006). Overall, immunization of mice with these various fractions by different studies resulted in induction of highly variable levels of protection in mice ranging from 0% to ~50%, but most were less than 30% (Smithers et al., 1989; Maddison et al., 1978; Roberts et al., 1988; Hota-Mitchell et al., 1997). The generally unspectacular protection observed by this approach may relate to the high amount of variability in the antigen preparations used; however, a number of tegument-associated proteins have been identified (including the tetraspanins described earlier) by this approach.

Calpain (Smp80)

Calpain is a calcium-activated neutral cysteine protease. The large subunit of the molecule was first identified by immunoscreening of a cDNA expression library by sera from infected humans (Andresen et al., 1991). Calpain was immunolocalized to the tegument and underlying musculature of adult worms and shown to be involved in surface membrane turnover (Siddiqui et al., 1993) and inner tegumental membrane (Braschi and Wilson, 2006). This protein was first implicated in protective immunity as the target of a CD4+ T cell clone that could induce peritoneal macrophages to kill schistosomula in vitro (Jankovic et al., 1996). The same clone, when administered intraperitoneally, conferred 65% protection in irradiated mice challenged with cercariae and IL-2, making calpain the first vaccine antigen identified based on the T cell reactivity. The large subunit of calpain was expressed and shown to induce 29–39% protection in mice

(Hota-Mitchell et al., 1997). Subsequent efforts to improve this vaccine have focused on DNA vaccine constructs, with and without Th1-type cytokine cDNAs, which increased protection to 42–57% in mice and baboons (Siddiqui et al., 2003a, 2003b). The *S. japonicum* calpain has also been cloned and the recombinant protein shown to induce 41% protection in mice along with a demonstrable antifecundity effect (Zhang et al., 2001a, 2001b; Ohta et al., 2004). Calpain is expressed in all parasite stages, including the penetration glands and secretions of *S. japonicum* cercariae (Kumagai et al., 2005).

Additional Membrane Proteins

A group of tegumental proteins with sequence homology to cytoskeletal proteins have also been assessed for protective potential. Notable is a 22.6-kDa antigen that was recognized by IgE from patients infected with schistosomiasis (Santiago et al., 1998) and has homology to Sm22.6 (20.8-kDa protein) from *S. mansoni.* The gene, in the form of a DNA construct, elicited 29–30% protection in mice (Mohamed et al., 1998). These proteins, however, exist on the cytoplasmic side of the tegument and are not surface expressed. Aside from the tetraspanins previously described, another membrane-spanning protein that is expressed on the tegument surface has been recently identified using bioinformatics and in silico analysis (Cardoso et al., 2006a, 2006b), and preliminary results indicate the corresponding recombinant protein is efficacious in mice (Loukas et al., 2007). Additional surface-expressed proteins that warrant further study as potential vaccine candidates have been identified from the surface tegument of adult worms (Braschi and Wilson, 2006). These include structural membrane proteins with large extracellular regions, such as homologues to annexins and dysferlin, and other accessible proteins on the surface with no homologues of known function, such as Sm200.

Antigen Formulation and Vaccination with Multiple Antigens

In order to enhance vaccine efficacy, combinations of antigens as well as the use of more potent adjuvants to boost immunogenicity have been studied. Unfortunately, several attempts to combine antigens to boost protective efficacy have been generally unsuccessful. A DNA vaccine encoding paramyosin, Sm23, TPI, calpain, and Sm28GST as a single polyprotein containing dominant epitopes failed to induce significant protection in mice (Yang et al., 2000). Subsequently, a cocktail DNA vaccine encoding Sj28GST, Sj14-3-3, Sj23, and 62 kDa of myosin induced between 34% and 45% protection, irrespective of adjuvants used (Zhang et al., 2001). To date, a cocktail of recombinant antigens has not been used, but current data indicate that combining antigens does not improve vaccine efficacy. Perhaps these antigens induce similar protective mechanisms, the targets are not diverse enough, or are not consistently expressed on the parasite surface, or there are certain characteristics of the host immune response induced by chronic parasite infection that limits further protection (see below).

The success of a vaccine is also dependent on generating the optimum immune response. This can be accomplished either by varying the adjuvant or routine of immunization. This enhanced response is illustrated by administering the irradiated cercariae with recombinant IL-12 to generate sterile immunity in mice (Wynn et al., 1995) or soluble extracts of lung-stage larvae to give >90% protection (Mountford et al., 1996). With respect to a single vaccine candidate antigen, the greatest effort has been undertaken with Sm28GST. MAPs have been assembled and coupled to tetanous toxoid carriers (Auriault et al., 1988) or have been incorporated in live vectors, such as *Salmonella typhimurium* (Khan et al., 1994a, 1994b), *Bordetella pertussis* (Mielcarek et al., 1997), or BCG (Kremer, et al., 1996). The recombinant protein has also been formulated in biodegradable particles (Baras et al., 1999, 2000), linked to adjuvants to facilitate mucosal administrations, such as cholera toxin B subunit, and administered intranasally. A DNA vaccine has been formulated with genes encoding IL-12, IL-18, and GM-CSF, with the aim of enhancing the cellular immune response. Motifs have been included in DNA vaccines that contain CpG motifs to augment activation of antigen presentation cells or formulated with oligonucleotides corresponding to various CpG motifs. Overall, these various formulations have not markedly enhanced vaccine efficacy to Sm28GST (or to other schistosome antigens evaluated with other various formulations). These data suggest that beyond a certain point, additional protection may not be achieved, even with the most potent adjuvants to Sm28GST or perhaps to any one single antigen.

TRANSLATION FROM ANIMAL STUDIES TO HUMAN CLINICAL TRIALS

With the optimism of identifying a number of potential vaccine candidates, the World Health Organization sponsored independent testing of several of the most promising antigens, with the aim to move one or more to human trials. These included IrV5, Sm14, and Sm28GST as recombinant proteins, paramyosin as the native protein, and TPI and Sm23 as MAPs. The protein donors specified the adjuvants used. Unfortunately, none of the antigens achieved the target of 40% protection, and some lacked any significant protection in two independent laboratories (Anonymous, 1996). This study also highlighted the difficulty of producing protein of consistent quality in sufficient amounts for more extensive vaccine studies. In spite of these disappointing results, further studies of some of the candidate antigens were undertaken to correlate the cellular and humoral immune responses in humans with the rationale that murine studies might not predict immune responses in humans. Lymphocytes and serum were obtained from well-defined populations of putatively immune versus susceptible humans in Brazil and Egypt. Although there were distinct immune responses to these antigens, there were no clear markers for either susceptibility or resistance (Ribeiro de Jesus et al., 2000; Al-Sherbiny et al., 2003). Thus, the decision was made not to proceed with any of the candidates, putting a considerable damper on further development of schistosome vaccines.

In spite of the discouraging results, one vaccine candidate, Sh28GST or BILHVAX, progressed to clinical trials. This is the only vaccine currently studied in humans. The vaccine candidate was not tested in independent animal trials or human immune correlate studies. The rationale for moving this vaccine forward was the strong antifecundity effect it had in patas monkeys (Boulanger et al., 1995, 1999). In Phase I and II trials, the recombinant protein formulated with alum was safe and highly immunogenic (Capron et al., 2002). Phase III efficacy trials are reported to be underway (Capron et al., 2005); however, no published studies on its clinical efficacy are available at this time. It is unfortunate that more vaccine candidates have not progressed to human studies.

In order to facilitate the movement of more candidates to human studies, a better transition is required from vaccine efficacy studies in murine models to humans. This will require further testing in nonhuman primates, the best model being the baboon (Nyindo and Farah, 1999), followed by studies in chimpanzees (Eberl et al., 2001), although perfusion studies cannot be performed in the latter species. Studies of vaccine efficacy against *S. japonicum* in large animals, for example, water buffaloes, pigs, and cattle have demonstrated successful vaccine efficacy to several candidate antigens that demonstrated poor efficacy in mice. Several vaccine candidates have progressed to efficacy trials in nonhuman primates, for example, SOD and calpain, although results remain to be published.

WHY ARE HIGH LEVELS OF PROTECTION DIFFICULT TO ACHIEVE?

This question has been proposed in a recent review of schistosome vaccines by Wilson and Coulson (2006) and warrants careful consideration in approaches to a schistosome vaccine. Out of the many different vaccine candidates tested, a few have achieved greater than 40% protection under ideal conditions using potent adjuvants. It is rare for a candidate protein vaccine to break 50% protection, and such results are variable and require independent testing. Thus, for some reason, there seems to be a ceiling of about 40–50% protection induced by any single protein. There are several potential explanations. Many vaccine candidates represent cytosolic or subtegumental antigens, and it is hard to fathom how host effector immune responses recognize these parasite antigens. Protection may not actually involve direct parasite killing but rather impede parasite migration through the host, particularly through the pulmonary vasculature as has been observed in the case of RA cercariae described previously (Smythies et al., 1992). Any single antigen or group of antigens can elicit only so many T cell clones that are recruited to the lung and result in only partial blocking of their onward migration. This may also explain why individual antigens or cocktails of antigens fail to achieve the levels of protection observed with RA cercariae.

Among the antigens that are actually available on the surface of the parasite as vaccine targets, these may be only transiently expressed because the parasite goes through a series of transformations in the process of maturation to the adult stage. Once the parasite fully matures, it then rapidly turns over its tegument, shedding its surface and consistently renewing it (Skelly and Wilson, 2006). Therefore, antibodies or cells adhered to its surface may be shed before they could elicit a lethal blow to the parasite.

There are other mechanisms that the parasite might employ to mitigate the host effector immune responses. The parasite has been observed to acquire host proteins on its surface that might partially mask vaccine targets on its surface (Damian, 1965, 1987). The parasite also acquire host proteins on its surface that can mitigate the host immune responses. An example is the decay-accelerating factor (DAF), which is an inhibitor of the alternative complement pathway (Horta and Ramalho-Pinto, 1991; Pearce et al., 1990; Ramalho-Pinto, 1987; Ramalho-Pinto et al., 1992). Similarly, paramyosin has also been shown to bind human CD59, which inhibits activation of the terminal pathway of the complement (Deng et al., 2003). Interestingly, the parasite itself may also produce homologues of human complement regulatory proteins (Braschi et al., 2006).

The parasite produces other proteins that might impair recruitment or activation of host immunity. The cercarial preacetabular and, potentially, head gland contents contain several immunoregulatory proteins. As mentioned previously, Sm16 can suppress dermal immune responses (Rao and Ramaswamy, 2000), and it is likely that others exist. For example, penetrating parasites have been shown to promote eicosanoid synthesis in the skin (Fusco et al., 1985). The vaccine candidate Sm28GST has been shown to be responsible for producing prostaglandin D2 (Herve et al., 2003), which impairs migration of Langerhan's cells from the skin and consequently antigen presentation (Angeli et al., 2001).

Another possible mechanism why vaccine candidates may not be fully effective is because they produce the wrong type of antibody. In humans, the predominant isotype synthesized and directed against schistosome antigens is IgG_4, which recognizes similar epitopes to IgE (Hussain and Ottesen, 1988) and can block IgE-mediated effector mechanisms (Hagan et al., 1991). Several MAbs of different isotypes in both rats and mice were found to block cytotoxic activity of other isotypes (Grzych et al., 1984) and murine antibodies against the same targets (Yi et al., 1986) or human infected serum (Dunne et al., 1987) for schistosomula in vitro.

It is likely that vaccines will be initially administered to children or adults previously or already infected with schistosomiasis. A defining characteristic of human schistosomiasis is the capacity to undergo immune modulation. This is manifested by impaired lymphocyte recall responses as measured by reduced lymphocyte proliferation and cytokine production to a variety of schistosome antigens. This is most pronounced as the parasites mature into adult worms and begin release of ova (Wilson et al., 2007). This limitation may also serve to impair host immune responses to vaccination (Elliott et al., 2007).

Compared to many infections, schistosomiasis presents unique challenges to produce an effective vaccine because of the multiple mechanisms the parasite has evolved to survive within its host for many years. Therefore, development of a vaccine will likely have to accompany a better understanding of how they both limit the immune responses and provoke and circumvent the host effector immune responses.

PROSPECTS FOR THE FUTURE

Overall, prospects for developing a schistosome vaccine are encouraging. There is now a much broader international effort to develop a schistosome vaccine, particularly among scientists from China and Brazil, where the disease is endemic. These countries are providing substantial financial and personnel resources to move vaccine research forward. The research community is posed to identify a new generation of vaccine candidates now that the schistosomiasis genome, transcriptome, and proteome are being defined for all life stages of the parasite. The challenge is to integrate identification of the vaccine candidates in the context of physiological processes of the parasite. For example, vaccines that can generate antibodies capable of inhibiting the function of critical molecules on the parasite surface such as glucose transport proteins, or molecules that circumvent the host immune responses. Alternatively, a logical approach may be to generate antibodies to proteins in the parasite gut that are required for digestion. We emphasize that critical targets are likely to be proteins expressed on the surfaces of schistosomula and adult worms. Additional targets include secretory proteins, which may also serve in essential physiological processes for parasite migration in the host and avoidance of the host immune response. The most vulnerable stages of the parasites should also be identified, which are likely to be tissue-migratory larval stages of the parasite.

What can be reasonably achieved for a schistosome vaccine needs to be reconsidered. A high level of protective efficacy, as observed for most bacterial and viral vaccines currently in use, is unlikely to be achieved in the near future because of the complex and probably redundant mechanisms that the parasite has evolved to survive within its host. Yet, even partial protection conferred by a vaccine can be of great benefit because the parasite does not replicate within the host. It is also likely that greater effort will be put toward vaccines that limited parasite fecundity, for example, a vaccine that protects more against disease than infection, which might be a more achievable target. In this context, schistosome vaccines will be an important component for integrated control of schistosomiasis that complements existing strategies, including chemotherapy, health education, and vector control. It should be emphasized that PZQ is the only currently anti-schistosome drug in production, and with expanding mass treatment programs, it is likely that resistance will emerge. This makes development of schistosome vaccines ever more important.

KEY ISSUES

- Only a few vaccine candidates have achieved more than 50% efficacy, with most less than 40% under ideal conditions. Why there is this ceiling to protective efficacy needs to be better understood.
- A vaccine with an even partial efficacy will be of great benefit because the parasite does not replicate within its host.
- The points in the schistosome life cycle that are most susceptible to immune attack need to be better understood.
- A better understanding of what are the host effector mechanisms that lead to parasite elimination in large animal models following vaccination with irradiated cercariae.
- The genomes of *S. mansoni* and *S. japonicum* need to be fully assembled, analyzed, and annotated. The *S. haematobium* genome should be sequenced since it is now the most prevalent human schistosome species.
- A concerted effort should be made to complete the proteome and transcriptome for each stage of life cycle for the three major human species of schistosomiasis.
- Surface-expressed or -secreted antigens from this inventory of parasite proteins identified by these new technologies are likely to be preferred vaccine candidates.

References

Al-Sherbiny, M., Osman, A., Barakat, R., El Morshedy, H., Bergquist, R. and Olds, R. In vitro cellular and humoral responses to Schistosoma mansoni vaccine candidate antigens. Acta Trop. 2003; 88(2):117–130.

Anderson, S., Coulson, P.S., Ljubojevic, S., Mountford, A.P. and Wilson, R.A. The radiation-attenuated schistosome vaccine induces high levels of protective immunity in the absence of B cells. Immunology 1999; 96(1):22–28.

Andresen, K., Tom, T.D. and Strand, M. Characterization of cDNA clones encoding a novel calcium-activated neutral proteinase from Schistosoma mansoni. J. Biol. Chem. 1991; 266(23):15085–15090.

Angeli, V., Faveeuw, C., Roye, O., Fontaine, J., Teissier, E., Capron, A., Wolowczuk, I., Capron, M. and Trottein, F. Role of the parasite-derived prostaglandin D2 in the inhibition of epidermal Langerhans cell migration during schistosomiasis infection. J Exp. Med. 2001; 193(10):1135–1147.

Anonymous 1996. TDR News. The World Health Organization 50.

Argiro, L., Henri, S., Dessein, H., Dessein, A.J. and Bourgois, A. Induction of a protective immunity against Schistosoma mansoni with ovalbumin-coupled Sm37-5 coadsorbed with granulocyte-macrophage colony stimulating factor (GM-CSF) or IL-12 on alum. Vaccine 1999; 17(1):13–18.

Argiro, L.L., Kohlstadt, S.S., Henri, S.S., Dessein, H.H., Matabiau, V.V., Paris, P.P., Bourgois, A.A. and Dessein, A.J. Identification of a candidate vaccine peptide on the 37 kDa Schistosoma mansoni GAPDH. Vaccine 2000; 18(19):2039–2048.

Arnon, R., Tarrab-Hazdai, R. and Steward, M. A mimotope peptide-based vaccine against Schistosoma mansoni: synthesis and characterization. Immunology 2000; 101(4):555–562.

Ashton, P.D., Curwen, R.S. and Wilson, R.A. Linking proteome and genome: how to identify parasite proteins. Trends Parasitol. 2001; 17(4):198–202.

Auriault, C., Gras-Masse, H., Wolowczuk, I., Pierce, R.J., Balloul, J.M., Neyrinck, J.L., Drobecq, H., Tartar, A. and Capron, A. Analysis of T and B cell epitopes of the Schistosoma mansoni P28 antigen in the rat model by using synthetic peptides. J. Immunol. 1988; 141(5):1687–1694.

Balloul, J.M., Grzych, J.M., Pierce, R.J. and Capron, A. A purified 28,000 dalton protein from Schistosoma mansoni adult worms protects rats and mice against experimental schistosomiasis. J. Immunol. 1987; 138(10):3448–3453.

Balloul, J.M., Pierce, R.J., Grzych, J.M. and Capron, A. In vitro synthesis of a 28 kilodalton antigen present on the surface of the schistosomulum of Schistosoma mansoni. Mol. Biochem. Parasitol. 1985; 17(1):105–114.

Balloul, J.M., Sondermeyer, P., Dreyer, D., Capron, M., Grzych, J.M., Pierce, R.J., Carvallo, D., Lecocq, J.P. and Capron, A. Molecular cloning of a protective antigen of schistosomes. Nature 1987; 326(6109):149–153.

Baras, B., Benoit, M., Poulain-Godefroy, O., Schacht, A., Capron, A., Gillard, J. and Riveau, G. Vaccine properties of antigens entrapped in microparticles produced by spray-drying technique and using various polyester polymers. Vaccine 2000; 18(15):1495–1505.

Baras, B., Benoit, M.A., Dupre, L., Poulain-Godefroy, O., Schacht, A.M., Capron, A., Gillard, J. and Riveau, G. Single-dose mucosal immunization with biodegradable microparticles containing a Schistosoma mansoni antigen. Infect. Immun. 1999; 67(5):2643–2648.

Bennett, J.L., Day, T., Liang, F.T., Ismail, M. and Farghaly, A. The development of resistance to anthelmintics: a perspective with an emphasis on the antischistosomal drug praziquantel. Exp. Parasitol. 1997; 87(3):260–267.

Betting, L.E., Pirani, C., Jr., de Souza Queiroz, L., Damasceno, B. P. and Cendes, F. Seizures and cerebral schistosomiasis. Arch. Neurol. 2005; 62(6):1008–1010.

Bickle, Q.D. and James, E.R. Resistance against Schistosoma mansoni induced by immunization of mice with cryopreserved schistosomula. Trans. R. Soc. Trop. Med. Hyg. 1978; 72(6):677–678.

Blanton, R.E., Licate, L.S. and Aman, R.A. Characterization of a native and recombinant Schistosoma haematobium serine protease inhibitor gene product. Mol. Biochem. Parasitol. 1994; 63(1):1–11.

Booth, M., Vennervald, B.J., Butterworth, A.E., Kariuki, H.C., Amaganga, C., Kimani, G., Mwatha, J.K., Otedo, A., Ouma, J.H. and Dunne, D.W. Exposure to malaria affects the regression of hepatosplenomegaly after treatment for Schistosoma mansoni infection in Kenyan children. BMC Med. 2004; 2:36.

Boulanger, D., Reid, G.D., Sturrock, R.F., Wolowczuk, I., Balloul, J.M., Grezel, D., Pierce, R.J., Otieno, M.F., Guerret, S., Grimaud, J.A. et al. Immunization of mice and baboons with the recombinant Sm28GST affects both worm viability and fecundity after experimental infection with Schistosoma mansoni. Parasite Immunol. 1991; 13(5):473–490.

Boulanger, D., Trottein, F., Mauny, F., Bremond, P., Couret, D., Pierce, R.J., Kadri, S., Godin, C., Sellin, E., Lecocq, J.P. et al. Vaccination of goats against the trematode Schistosoma bovis with a recombinant homologous schistosome-derived glutathione S-transferase. Parasite Immunol. 1994; 16(8):399–406.

Boulanger, D., Warter, A., Sellin, B., Lindner, V., Pierce, R.J., Chippaux, J.P. and Capron, A. Vaccine potential of a recombinant glutathione S-transferase cloned from Schistosoma haematobium in primates experimentally infected with an homologous challenge. Vaccine 1999; 17(4):319–326.

Boulanger, D., Warter, A., Trottein, F., Mauny, F., Bremond, P., Audibert, F., Couret, D., Kadri, S., Godin, C., Sellin, E. et al. Vaccination of patas monkeys experimentally infected with Schistosoma haematobium using a recombinant glutathione S-transferase cloned from S. mansoni. Parasite Immunol. 1995; 17(7):361–369.

Braschi, S., Curwen, R.S., Ashton, P.D., Verjovski-Almeida, S. and Wilson, A. The tegument surface membranes of the human blood parasite Schistosoma mansoni: a proteomic analysis after differential extraction. Proteomics 2006; 6(5):1471–1482.

Braschi, S. and Wilson, R.A. Proteins exposed at the adult schistosome surface revealed by biotinylation. Mol Cell Proteomics 2006; 5(2):347–356.

Brown, M., Mawa, P.A., Joseph, S., Bukusuba, J., Watera, C., Whitworth, J.A., Dunne, D.W. and Elliott, A.M. Treatment of Schistosoma mansoni infection increases helminth-specific type 2 cytokine responses and HIV-1 loads in coinfected Ugandan adults. J. Infect. Dis. 2005; 191(10):1648–1657.

Brown, M., Miiro, G., Nkurunziza, P., Watera, C., Quigley, M.A., Dunne, D.W., Whitworth, J.A. and Elliott, A.M. Schistosoma mansoni, nematode infections, and progression to active tuberculosis among HIV-1-infected Ugandans. Am. J. Trop. Med. Hyg. 2006; 74(5):819–825.

Bushara, H.O., Bashir, M.E., Malik, K.H., Mukhtar, M.M., Trottein, F., Capron, A. and Taylor, M.G. Suppression of Schistosoma bovis egg production in cattle by vaccination with either glutathione S-transferase or keyhole limpet haemocyanin. Parasite Immunol. 1993; 15(7):383–390.

Butterworth, A.E., Capron, M., Cordingley, J.S., Dalton, P.R., Dunne, D.W., Kariuki, H.C., Kimani, G., Koech, D., Mugambi, M., Ouma, J.H. et al. Immunity after treatment of human schistosomiasis mansoni. II. Identification of resistant individuals, and analysis of their immune responses. Trans. R. Soc. Trop. Med. Hyg. 1985; 79(3):393–408.

Butterworth, A.E., Sturrock, R.F., Houba, V. and Rees, P.H. Antibody-dependent cell-mediated damage to schistosomula in vitro. Nature 1974; 252(5483):503–505.

Butterworth, A.E., Taylor, D.W., Veith, M.C., Vadas, M.A., Dessein, A., Sturrock, R.F. and Wells, E. Studies on the mechanisms of immunity in human schistosomiasis. Immunol. Rev. 1982; 61:5–39.

Caldas, I.R., Correa-Oliveira, R., Colosimo, E., Carvalho, O.S., Massara, C.L., Colley, D.G. and Gazzinelli, G. Susceptibility and resistance to Schistosoma mansoni reinfection: parallel cellular and isotypic immunologic assessment. Am. J. Trop. Med. Hyg. 2000; 62(1):57–64.

Capron, A., Capron, M. and Riveau, G. Vaccine development against schistosomiasis from concepts to clinical trials. Br. Med. Bull. 2002; 62:139–148.

Capron, A., Riveau, G., Capron, M. and Trottein, F. Schistosomes: the road from host-parasite interactions to vaccines in clinical trials. Trends Parasitol. 2005; 21(3):143–149.

Cardoso, F.C., Pacifico, R.N., Mortara, R.A. and Oliveira, S.C. Human antibody responses of patients living in endemic areas for schistosomiasis to the tegumental protein Sm29 identified through genomic studies. Clin. Exp. Immunol. 2006; 144(3):382–391.

Cardoso, F.C., Pinho, J.M., Azevedo, V. and Oliveira, S.C. Identification of a new Schistosoma mansoni membrane-bound protein through bioinformatic analysis. Genet. Mol. Res. 2006; 5(4):609–618.

Charrier-Ferrara, S., Caillol, D. and Goudot-Crozel, V. Complete sequence of the Schistosoma mansoni glyceraldehyde-3-phosphate dehydrogenase gene encoding a major surface antigen. Mol. Biochem. Parasitol. 1992; 56(2):339–343.

Chen, J.X., Liu, S.X., Cao, J.P., Song, G.C. and Xu, Y.X. [Anti-schistosomiasis effect induced by full length DNA vaccine coding paramyosin of Schistosoma japonicum in C57BL/6 mice]. Zhongguo Ji Sheng Chong Xue Yu Ji Sheng Chong Bing Za Zhi 2006; 24(2):81–85.

Cook, R.M., Carvalho-Queiroz, C., Wilding, G. and LoVerde, P. T. Nucleic acid vaccination with Schistosoma mansoni antioxidant enzyme cytosolic superoxide dismutase and the structural protein filamin confers protection against the adult worm stage. Infect. Immun. 2004; 72(10):6112–6124.

Coulson, P.S. The radiation-attenuated vaccine against schistosomes in animal models: paradigm for a human vaccine? Adv. Parasitol. 1997; 39:271–336.

Crabtree, J.E. and Wilson, R.A. Schistosoma mansoni: an ultrastructural examination of skin migration in the hamster cheek pouch. Parasitology 1985; 91(Pt 1):111–120.

Curwen, R.S., Ashton, P.D., Sundaralingam, S. and Wilson, R.A. Identification of novel proteases and immunomodulators in the secretions of schistosome cercariae that facilitate host entry. Mol. Cell. Proteomics 2006; 5(5):835–844.

Curwen, R.S. and Wilson, R.A. Invasion of skin by schistosome cercariae: some neglected facts. Trends Parasitol. 2003; 19(2):63–66, discussion 66–68.

Da'Dara, A.A., Skelly, P.J., Walker, C.M. and Harn, D.A. A DNA-prime/protein-boost vaccination regimen enhances Th2 immune responses but not protection following Schistosoma mansoni infection. Parasite Immunol. 2003; 25(8-9):429–437.

Da'dara, A.A., Skelly, P.J., Wang, M.M. and Harn, D.A. Immunization with plasmid DNA encoding the integral membrane protein, Sm23, elicits a protective immune response against schistosome infection in mice. Vaccine 2001; 20(3-4):359–369.

Dalton, J.P. and Strand, M. Schistosoma mansoni polypeptides immunogenic in mice vaccinated with radiation-attenuated cercariae. J. Immunol. 1987; 139(7):2474–2481.

Damian, R.T. Molecular Mimicry in Biological Adaptation. Science 1965; 147(3660):824.

Damian, R.T. Molecular mimicry revisited. Parasitol Today 1987; 3(9):263–266.

Damian, R.T., Roberts, M.L., Powell, M.R., Clark, J.D., Lewis, F.A. and Stirewalt, M.A. Schistosoma mansoni egg granuloma size reduction in challenged baboons after vaccination with irradiated cryopreserved schistosomula. Proc. Natl. Acad. Sci. U. S. A. 1984; 81(11):3552–3556.

Delgado, B. and McLaren, D.J. Evidence for enhancement of IgG1 subclass expression in mice polyvaccinated with radiation attenuated cercariae of *Schistosoma mansoni* and the role of this isotype in serum transferred immunity. Parasite Immunol. 1990; 12:15–32.

Deng, J., Gold, D., LoVerde, P.T. and Fishelson, Z. Inhibition of the complement membrane attack complex by Schistosoma mansoni paramyosin. Infect. Immun. 2003; 71(11):6402–6410.

Deng, J., Gold, D., LoVerde, P.T. and Fishelson, Z. Mapping of the complement C9 binding domain in paramyosin of the blood fluke Schistosoma mansoni. Int. J. Parasitol. 2007; 37(1):67–75.

Dessein, A.J., Begley, M., Demeure, C., Caillol, D., Fueri, J., dos Reis, M.G., Andrade, Z.A., Prata, A. and Bina, J.C. Human resistance to Schistosoma mansoni is associated with IgG reactivity to a 37-kDa larval surface antigen. J. Immunol. 1988; 140(8):2727–2736.

Dunne, D.W., Bickle, Q.D., Butterworth, A.E. and Richardson, B.A. The blocking of human antibody-dependent, eosinophil-mediated killing of Schistosoma mansoni schistosomula by monoclonal antibodies which cross-react with a polysaccharide-containing egg antigen. Parasitology 1987; 94(Pt 2):269–280.

Eberl, M., Langermans, J.A., Frost, P.A., Vervenne, R.A., van Dam, G.J., Deelder, A.M., Thomas, A.W., Coulson, P.S. and Wilson, R.A. Cellular and humoral immune responses and protection against schistosomes induced by a radiation-attenuated vaccine in chimpanzees. Infect. Immun. 2001; 69(9):5352–5362.

Elliott, A.M., Kizza, M., Quigley, M.A., Ndibazza, J., Nampijja, M., Muhangi, L., Morison, L., Namujju, P.B., Muwanga, M., Kabatereine, N. and Whitworth, J.A. The impact of helminths on the response to immunization and on the incidence of infection and disease in childhood in Uganda: design of a randomized, double-blind, placebo-controlled, factorial trial of deworming interventions delivered in pregnancy and early childhood [ISRCTN32849447]. Clin. Trials 2007; 4(1):42–57.

Fallon, P.G., Mubarak, J.S., Fookes, R.E., Niang, M., Butterworth, A.E., Sturrock, R.F. and Doenhoff, M.J. Schistosoma mansoni: maturation rate and drug susceptibility of different geographic isolates. Exp. Parasitol. 1997; 86(1):29–36.

Fenwick, A. and Webster, J.P. Schistosomiasis: challenges for control, treatment and drug resistance. Curr. Opin. Infect. Dis. 2006; 19(6):577–582.

Fitzsimmons, C.M., Stewart, T.J., Hoffmann, K.F., Grogan, J.L., Yazdanbakhsh, M. and Dunne, D.W. Human IgE response to the Schistosoma haematobium 22.6 kDa antigen. Parasite Immunol. 2004; 26(8-9):371–376.

Flanigan, T.P., King, C.H., Lett, R.R., Nanduri, J. and Mahmoud, A.A. Induction of resistance to Schistosoma mansoni infection in mice by purified parasite paramyosin. J. Clin. Invest. 1989; 83(3):1010–1014.

Fusco, A.C., Salafsky, B. and Kevin, M.B. Schistosoma mansoni: eicosanoid production by cercariae. Exp. Parasitol. 1985; 59(1):44–50.

Gallagher, M., Malhotra, I., Mungai, P.L., Wamachi, A.N., Kioko, J.M., Ouma, J.H., Muchiri, E. and King, C.L. The effects of maternal helminth and malaria infections on mother-to-child HIV transmission. AIDS 2005; 19(16):1849–1855.

Ganley-Leal, L.M., Mwinzi, P.N., Cetre-Sossah, C.B., Andove, J., Hightower, A.W., Karanja, D.M., Colley, D.G. and Secor, W.E. Correlation between eosinophils and protection against reinfection with Schistosoma mansoni and the effect of human immunodeficiency virus type 1 coinfection in humans. Infect. Immun. 2006; 74(4):2169–2176.

Ganley-Leal, L.M., Mwinzi, P.N., Cetre-Sossah, C.B., Andove, J., Hightower, A.W., Karanja, D.M., Colley, D.G. and Secor, W.E. Higher percentages of circulating mast cell precursors correlate with susceptibility to reinfection with Schistosoma mansoni. Am J. Trop. Med. Hyg. 2006; 75(6):1053–1057.

Goudot-Crozel, V., Caillol, D., Djabali, M. and Dessein, A.J. The major parasite surface antigen associated with human resistance to schistosomiasis is a 37-kD glyceraldehyde-3P-dehydrogenase. J. Exp. Med. 1989; 170(6):2065–2080.

Gryseels, B., Polman, K., Clerinx, J. and Kestens, L. Human schistosomiasis. Lancet 2006; 368(9541):1106–1118.

Grzych, J.M., Capron, M., Dissous, C. and Capron, A. Blocking activity of rat monoclonal antibodies in experimental schistosomiasis. J. Immunol. 1984; 133(2):998–1004.

Hagan, P., Blumenthal, U.J., Dunn, D., Simpson, A.J. and Wilkins, H.A. Human IgE, IgG4 and resistance to reinfection with Schistosoma haematobium. Nature 1991; 349(6306):243–245.

Harn, D.A., Gu, W., Oligino, L.D., Mitsuyama, M., Gebremichael, A. and Richter, D. A protective monoclonal antibody specifically recognizes and alters the catalytic activity of schistosome triosephosphate isomerase. J. Immunol. 1992; 148(2):562–567.

Harn, D.A., Mitsuyama, M., Huguenel, E.D. and David, J.R. Schistosoma mansoni: detection by monoclonal antibody of a 22,000-dalton surface membrane antigen which may be blocked by host molecules on lung stage parasites. J. Immunol. 1985; 135(3):2115–2120.

Harn, D.A., Reynolds, S.R., Chikunguwo, S., Furlong, S. and Dahl, C. Synthetic peptide vaccines for schistosomiasis. Pharm. Biotechnol. 1995; 6:891–905.

Hazdai, R.T., Levi-Schaffer, F., Brenner, V., Horowitz, S., Eshhar, Z. and Arnon, R. Protective monoclonal antibody against Schistosoma mansoni: antigen isolation, characterization, and suitability for active immunization. J. Immunol. 1985; 135(4):2772–2779.

Herve, M., Angeli, V., Pinzar, E., Wintjens, R., Faveeuw, C., Narumiya, S., Capron, A., Urade, Y., Capron, M., Riveau, G. and Trottein, F. Pivotal roles of the parasite PGD2 synthase and of the host D prostanoid receptor 1 in schistosome immune evasion. Eur. J. Immunol. 2003; 33(10):2764–2772.

Hillyer, G.V. Comparison of purified 12 kDa and recombinant 15 kDa Fasciola hepatica antigens related to a Schistosoma mansoni fatty acid binding protein. Mem. Inst. Oswaldo Cruz 1995; 90(2):249–253.

Hillyer, G.V., Garcia Rosa, M.I., Alicea, H. and Hernandez, A. Successful vaccination against murine Schistosoma mansoni infection with a purified 12 Kd Fasciola hepatica cross-reactive antigen. Am. J. Trop. Med. Hyg. 1988; 38(1):103–110.

Hoffman, S.L., Goh, L.M., Luke, T.C., Schneider, I., Le, T.P., Doolan, D.L., Sacci, J., de la Vega, P., Dowler, M., Paul, C., Gordon, D.M., Stoute, J.A., Church, L.W., Sedegah, M., Heppner, D.G., Ballou, W.R. and Richie, T.L. Protection of humans against malaria by immunization with radiation-attenuated Plasmodium falciparum sporozoites. J. Infect. Dis. 2002; 185(8):1155–1164.

Horta, M.F. and Ramalho-Pinto, F.J. Role of human decay-accelerating factor in the evasion of Schistosoma mansoni from the complement-mediated killing in vitro. J. Exp. Med. 1991; 174(6):1399–1406.

Hota-Mitchell, S., Siddiqui, A.A., Dekaban, G.A., Smith, J., Tognon, C. and Podesta, R.B. Protection against Schistosoma mansoni infection with a recombinant baculovirus-expressed subunit of calpain. Vaccine 1997; 15(15):1631–1640.

Hussain, R. and Ottesen, E.A. Parallel recognition of filarial antigens by IgE and IgG4 antibodies. N. Engl. Reg. Allergy. Proc. 1988; 9(1):51–56.

Ismail, M., Botros, S., Metwally, A., William, S., Farghally, A., Tao, L.F., Day, T.A. and Bennett, J.L. Resistance to praziquantel: direct evidence from Schistosoma mansoni isolated from Egyptian villagers. Am. J. Trop. Med. Hyg. 1999; 60(6):932–935.

James, S.L. Induction of protective immunity against Schistosoma mansoni by a nonliving vaccine is dependent on the method of antigen presentation. J. Immunol. 1985; 134(3):1956–1960.

James, S.L. Induction of protective immunity against Schistosoma mansoni by a non-living vaccine. V. Effects of varying the immunization and infection schedule and site. Parasite Immunol. 1987; 9(5):531–541.

Jang-Lee, J., Curwen, R.S., Ashton, P.D., Tissot, B., Mathieson, W., Panico, M., Dell, A., Wilson, R.A. and Haslam, S.M. Glycomics analysis of Schistosoma mansoni egg and cercarial secretions. Mol. Cell. Proteomics 2007; 6(9):1485–1499.

Jankovic, D., Aslund, L., Oswald, I.P., Caspar, P., Champion, C., Pearce, E., Coligan, J.E., Strand, M., Sher, A. and James, S.L. Calpain is the target antigen of a Th1 clone that transfers protective immunity against Schistosoma mansoni. J. Immunol. 1996; 157(2):806–814.

Kallestrup, P., Zinyama, R., Gomo, E., Butterworth, A.E., Mudenge, B., van Dam, G.J., Gerstoft, J., Erikstrup, C. and Ullum, H. Schistosomiasis and HIV-1 infection in rural Zimbabwe: effect of treatment of schistosomiasis on CD4 cell count and plasma HIV-1 RNA load. J. Infect. Dis. 2005; 192(11):1956–1961.

Kallestrup, P., Zinyama, R., Gomo, E., Butterworth, A.E., van Dam, G.J., Gerstoft, J., Erikstrup, C. and Ullum, H. Schistosomiasis and HIV in rural Zimbabwe: efficacy of treatment of schistosomiasis in individuals with HIV coinfection. Clin. Infect. Dis. 2006; 42(12):1781–1789.

Karanja, D.M., Hightower, A.W., Colley, D.G., Mwinzi, P.N., Galil, K., Andove, J. and Secor, W.E. Resistance to reinfection with Schistosoma mansoni in occupationally exposed adults and effect of HIV-1 co-infection on susceptibility to schistosomiasis: a longitudinal study. Lancet 2002; 360(9333):592–596.

Kariuki, T.M., Farah, I.O., Yole, D.S., Mwenda, J.M., Van Dam, G.J., Deelder, A.M., Wilson, R.A. and Coulson, P.S. Parameters of the attenuated schistosome vaccine evaluated in the olive baboon. Infect. Immun. 2004; 72(9):5526–5529.

Kazura, J.W., de Brito, P., Rabbege, J. and Aikawa, M. Role of granulocyte oxygen products in damage of Schistosoma mansoni eggs in vitro. J. Clin. Invest. 1985; 75(4):1297–1307.

Khan, C.M., Villarreal-Ramos, B., Pierce, R.J., Demarco de Hormaeche, R., McNeill, H., Ali, T., Chatfield, S., Capron, A., Dougan, G. and Hormaeche, C.E. Construction, expression, and immunogenicity of multiple tandem copies of the Schistosoma mansoni peptide 115-131 of the P28 glutathione S-transferase-expressed as C-terminal fusions to tetanus toxin fragment C in a live aro-attenuated vaccine strain of Salmonella. J. Immunol. 1994; 153(12):5634–5642.

Khan, C.M., Villarreal-Ramos, B., Pierce, R.J., Riveau, G., Demarco de Hormaeche, R., McNeill, H., Ali, T., Fairweather, N., Chatfield, S., Capron, A. et al. Construction, expression, and immunogenicity of the Schistosoma mansoni P28 glutathione S-transferase as a genetic fusion to tetanus toxin fragment C in a live Aro attenuated vaccine strain of Salmonella. Proc. Natl. Acad. Sci. U. S. A. 1994; 91(23):11261–11265.

King, C.H., Dickman, K. and Tisch, D.J. Reassessment of the cost of chronic helmintic infection: a meta-analysis of disability-related outcomes in endemic schistosomiasis. Lancet 2005; 365(9470):1561–1569.

King, C.H., Keating, C.E., Muruka, J.F., Ouma, J.H., Houser, H., Siongok, T.K. and Mahmoud, A.A. Urinary tract morbidity in schistosomiasis haematobia: associations with age and intensity of infection in an endemic area of Coast Province, Kenya. Am. J. Trop. Med. Hyg. 1988; 39(4):361–368.

King, C.L., Medhat, A., Malhotra, I., Nafeh, M., Helmy, A., Khaudary, J., Ibrahim, S., El-Sherbiny, M., Zaky, S., Stupi, R. J., Brustoski, K., Shehata, M. and Shata, M.T. Cytokine control of parasite-specific anergy in human urinary schistosomiasis. IL-10 modulates lymphocyte reactivity. J. Immunol. 1996; 156(12):4715–4721.

Kremer, L., Riveau, G., Baulard, A., Capron, A. and Locht, C. Neutralizing antibody responses elicited in mice immunized with recombinant bacillus Calmette-Guerin producing the Schistosoma mansoni glutathione S-transferase. J. Immunol. 1996; 156(11):4309–4317.

Kumagai, T., Maruyama, H., Hato, M., Ohmae, H., Osada, Y., Kanazawa, T. and Ohta, N. Schistosoma japonicum: localization of calpain in the penetration glands and secretions of cercariae. Exp. Parasitol. 2005; 109(1):53–57.

Lanar, D.E., Pearce, E.J., James, S.L. and Sher, A. Identification of paramyosin as schistosome antigen recognized by intradermally vaccinated mice. Science 1986; 234(4776):593–596.

Leenstra, T., Coutinho, H.M., Acosta, L.P., Langdon, G.C., Su, L., Olveda, R.M., McGarvey, S.T., Kurtis, J.D. and Friedman, J.F. Schistosoma japonicum reinfection after praziquantel treatment causes anemia associated with inflammation. Infect. Immun. 2006; 74(11):6398–6407.

Lewis, F.A., Stirewalt, M.A. and Leef, J.L. Schistosoma mansoni: vaccination of mice with 10-krad-irradiated, cryopreserved schistosomules. Exp. Parasitol. 1984; 57(3):261–268.

Li, Y.S., Raso, G., Zhao, Z.Y., He, Y.K., Ellis, M.K. and McManus, D.P. Large water management projects and schistosomiasis control, Dongting Lake region, China. Emerg. Infect. Dis. 2007; 13(7):973–979.

Liu, S.X., Song, G.C., Xu, Y.X., Yang, W. and McManus, D.P. Anti-fecundity immunity induced in pigs vaccinated with recombinant Schistosoma japonicum 26kDa glutathione-S-transferase. Parasite Immunol. 1995; 17(7):340–355.

Loukas, A., Jones, M.K., King, L.T., Brindley, P.J. and McManus, D. P. Receptor for Fc on the surfaces of schistosomes. Infect. Immun 2001; 69(6):3646–3651.

Loukas, A., Tran, M. and Pearson, M.S. Schistosome membrane proteins as vaccines. Int. J. Parasitol. 2007; 37(3-4):257–263.

Lyke, K.E., Dabo, A., Sangare, L., Arama, C., Daou, M., Diarra, I., Plowe, C.V., Doumbo, O.K. and Sztein, M.B. Effects of concomitant Schistosoma haematobium infection on the serum cytokine levels elicited by acute Plasmodium falciparum malaria infection in Malian children. Infect. Immun. 2006; 74(10):5718–5724.

Lyke, K.E., Dicko, A., Dabo, A., Sangare, L., Kone, A., Coulibaly, D., Guindo, A., Traore, K., Daou, M., Diarra, I., Sztein, M.B., Plowe, C.V. and Doumbo, O.K. Association of Schistosoma haematobium infection with protection against acute Plasmodium falciparum malaria in Malian children. Am. J. Trop. Med. Hyg. 2005; 73(6):1124–1130.

Maddison, S.E., Slemenda, S.B., Chandler, F.W. and Kagan, I.G. Studies on putative adult worm-derived vaccines and adjuvants for protection against Schistosoma mansoni infection in mice. J. Parasitol. 1978; 64(6):986–993.

Maddison, S.E., Slemenda, S.B., Hillyer, G.V., Chandler, F.W. and Kagan, I.G. Immune responses to Schistosoma mansoni in rhesus monkeys with multiple chronic and early primary infections. Infect. Immun. 1979; 25(1):249–254.

Malek, E.A. Effect of the Aswan High Dam on prevalence of schistosomiasis in Egypt. Trop. Geogr. Med. 1975; 27(4):359–364.

Malhotra, I., Mungai, P., Wamachi, A., Kioko, J., Ouma, J.H., Kazura, J.W. and King, C.L. Helminth- and Bacillus Calmette-Guerin-induced immunity in children sensitized in utero to filariasis and schistosomiasis. J. Immunol. 1999; 162(11): 6843–6848.

Marcotty, T., Berkvens, D., Besa, R.K., Losson, B., Dolan, T.T., Madder, M., Chaka, G., Van den Bossche, P. and Brandt, J. Lyophilisation and resuscitation of sporozoites of Theileria parva: preliminary experiments. Vaccine 2003; 22(2):213–216.

Mathers, C.D., Ezzati, M. and Lopez, A.D. Measuring the burden of neglected tropical diseases: the global burden of disease framework. PLoS Negl. Trop. Dis. 2007; 1(2):e114.

Mbao, V., Berkvens, D., Dolan, T., Speybroeck, N., Brandt, J., Dorny, P., Van den Bossche, P. and Marcotty, T. Infectivity of Theileria parva sporozoites following cryopreservation in four suspension media and multiple refreezing: evaluation by in vitro titration. Onderstepoort J. Vet. Res. 2006; 73(3):207–213.

McKerrow, J.H., Pino-Heiss, S., Lindquist, R. and Werb, Z. Purification and characterization of an elastinolytic proteinase secreted by cercariae of Schistosoma mansoni. J. Biol. Chem. 1985; 260(6):3703–3707.

McLaren, D.J. and James, S.L. Ultrastructural studies of the killing of schistosomula of Schistosoma mansoni by activated macrophages in vitro. Parasite Immunol. 1985; 7(3):315–331.

McLaren, D.J., Ramalho-Pinto, F.J. and Smithers, S.R. Ultrastructural evidence for complement and antibody-dependent damage to schistosomula of Schistosoma mansoni by rat eosinophils in vitro. Parasitology 1978; 77(3):313–324.

Medhat, A., Shehata, M., Bucci, K., Mohamed, S., Diab, A., Badary, S., Galal, H., Nafeh, M. and King, C.L. Increased interleukin-4 and interleukin-5 production in response to *Schistosoma haematobium* adult worm antigens correlates with lack of reinfection after treatment. J. Inf. Dis. 1998; 178:512–519.

Mei, H. and LoVerde, P.T. Schistosoma mansoni: the developmental regulation and immunolocalization of antioxidant enzymes. Exp. Parasitol. 1997; 86(1):69–78.

Mielcarek, N., Cornette, J., Schacht, A.M., Pierce, R.J., Locht, C., Capron, A. and Riveau, G. Intranasal priming with recombinant Bordetella pertussis for the induction of a systemic immune response against a heterologous antigen. Infect. Immun. 1997; 65(2):544–550.

Mohamed, M.M., Shalaby, K.A., LoVerde, P.T. and Karim, A. M. Characterization of Sm20.8, a member of a family of schistosome tegumental antigens. Mol. Biochem. Parasitol. 1998; 96(1-2):15–25.

Moore, D.V. and Meleney, H.E. Susceptibility of rice rats (Oryzomys Palustris) to Schistosoma mansoni. J. Parasitol. 1952; 38(1):76–79.

Mountford, A.P., Anderson, S. and Wilson, R.A. Induction of Th1 cell-mediated protective immunity to Schistosoma mansoni by co-administration of larval antigens and IL-12 as an adjuvant. J. Immunol. 1996; 156(12):4739–4745.

Murrell, K.D., Clark, S.S., Dean, D.A. and Vannier, W.E. Schistosoma mansoni: immunization of cynomolgus monkeys by injection of irradiated schistosomula. Exp. Parasitol. 1979; 48(3):415–420.

Mutapi, F., Ndhlovu, P.D., Hagan, P. and Woolhouse, M.E. A comparison of humoral responses to Schistosoma haematobium in areas with low and high levels of infection. Parasite Immunol. 1997; 19(6):255–263.

Nara, T., Iizumi, K., Ohmae, H., Sy, O.S., Tsubota, S., Inaba, Y., Tsubouchi, A., Tanabe, M., Kojima, S. and Aoki, T. Antibody isotype responses to paramyosin, a vaccine candidate for schistosomiasis, and their correlations with resistance and fibrosis in patients infected with Schistosoma japonicum in Leyte, The Philippines. Am. J. Trop. Med. Hyg. 2007; 76(2):384–391.

Nyindo, M. and Farah, I.O. The baboon as a non-human primate model of human schistosome infection. Parasitol. Today 1999; 15(12):478–482.

Ohta, N., Kumagai, T., Maruyama, H., Yoshida, A., He, Y. and Zhang, R. Research on calpain of Schistosoma japonicum as a vaccine candidate. Parasitol Int. 2004; 53(2):175–181.

Pearce, E.J., Hall, B.F. and Sher, A. Host-specific evasion of the alternative complement pathway by schistosomes correlates with the presence of a phospholipase C-sensitive surface molecule resembling human decay accelerating factor. J. Immunol. 1990; 144(7):2751–2756.

Pearce, E.J., James, S.L., Hieny, S., Lanar, D.E. and Sher, A. Induction of protective immunity against Schistosoma mansoni by vaccination with schistosome paramyosin (Sm97), a nonsurface parasite antigen. Proc. Natl. Acad. Sci. U. S. A. 1988; 85(15):5678–5682.

Pittella, J.E. Neuroschistosomiasis. Brain Pathol. 1997; 7(1):649–662.

Ramalho-Pinto, F.J. Decay accelerating factor (DAF) as the host antigen with protective activity to complement killing of schistosomula. Mem. Inst. Oswaldo Cruz 1987; 82(Suppl 4):213–216.

Ramalho-Pinto, F.J., Carvalho, E.M. and Horta, M.F. Mechanisms of evasion of Schistosoma mansoni schistosomula to the lethal activity of complement. Mem. Inst. Oswaldo Cruz 1992; 87(Suppl 4):111–116.

Ramirez, B.L., Kurtis, J.D., Wiest, P.M., Arias, P., Aligui, F., Acosta, L., Peters, P. and Olds, G.R. Paramyosin: a candidate vaccine antigen against Schistosoma japonicum. Parasite Immunol. 1996; 18(1):49–52.

Rao, K.V., He, Y.X. and Ramaswamy, K. Suppression of cutaneous inflammation by intradermal gene delivery. Gene. Ther. 2002; 9(1):38–45.

Rao, K.V. and Ramaswamy, K. Cloning and expression of a gene encoding Sm16, an anti-inflammatory protein from Schistosoma mansoni. Mol. Biochem. Parasitol. 2000; 108(1):101–108.

Reynolds, S.R., Dahl, C.E. and Harn, D.A. T and B epitope determination and analysis of multiple antigenic peptides for the Schistosoma mansoni experimental vaccine triose-phosphate isomerase. J. Immunol. 1994; 152(1):193–200.

Reynolds, S.R., Shoemaker, C.B. and Harn, D.A. T and B cell epitope mapping of SM23, an integral membrane protein of Schistosoma mansoni. J. Immunol. 1992; 149(12):3995–4001.

Ribeiro de Jesus, A., Araujo, I., Bacellar, O., Magalhaes, A., Pearce, E., Harn, D., Strand, M. and Carvalho, E.M. Human immune responses to Schistosoma mansoni vaccine candidate antigens. Infect. Immun. 2000; 68(5):2797–2803.

Rihet, P., Demeure, C.E., Bourgois, A., Prata, A. and Dessein, A. J. Evidence for an association between human resistance to *Schistosoma mansoni* and high anti-larval IgE levels. Eur. J. Immunol. 1991; 21:2679–2686.

Rihet, P., Demeure, C.E., Dessein, A.J. and Bourgois, A. Strong serum inhibition of specific IgE correlated to competing IgG4, revealed by a new methodology in subjects from a S. mansoni endemic area. Eur. J. Immnunol. 1992; 22:2063–2070.

Ritchie, L.S., Garson, S. and Knight, W.B. The Biology of Schistosoma Mansoni in Laboratory Rats. J. Parasitol. 1963; 49:571–577.

Roberts, M., Butterworth, A.E., Kimani, G., Kamau, T., Fulford, A.J., Dunne, D.W., Ouma, J.H. and Sturrock, R.F. Immunity after treatment of human schistosomiasis: association between cellular responses and resistance to reinfection. Infect. Immun. 1993; 61(12):4984–4993.

Roberts, S.M., Boot, C. and Wilson, R.A. Antibody responses of rodents to a tegument membrane preparation from adult Schistosoma mansoni. Parasitology 1988; 97(Pt 3):425–435.

Roberts, S.M., MacGregor, A.N., Vojvodic, M., Wells, E., Crabtree, J.E. and Wilson, R.A. Tegument surface membranes of adult Schistosoma mansoni: development of a method for their isolation. Mol. Biochem. Parasitol. 1983; 9(2):105–127.

Roberts, S.M., Wilson, R.A., Ouma, J.H., Kariuki, H.C., Koech, D., arap Siongok, T.K., Sturrock, R.F. and Butterworth, A.E. Immunity after treatment of human schistosomiasis mansoni quantitative and qualitative antibody responses to tegumental membrane antigens prepared from adult worms. Trans. R. Soc. Trop. Med. Hyg. 1987; 81(5):786–793.

Ross, A.G., Bartley, P.B., Sleigh, A.C., Olds, G.R., Li, Y., Williams, G. M. and McManus, D.P. Schistosomiasis. N. Engl. J. Med. 2002; 346(16):1212–1220.

Santiago, M.L., Hafalla, J.C., Kurtis, J.D., Aligui, G.L., Wiest, P. M., Olveda, R.M., Olds, G.R., Dunne, D.W. and Ramirez, B.L. Identification of the Schistosoma japonicum 22.6-kDa antigen as a major target of the human IgE response: similarity of IgE-binding epitopes to allergen peptides. Int. Arch. Allergy. Immunol. 1998; 117(2):94–104.

Scott, J.C., Kennedy, M.W. and McManus, D.P. Molecular and immunological characterisation of a polymorphic cytosolic fatty acid binding protein from the human blood fluke of humans, Schistosoma japonicum. Biochim. Biophys. Acta 2000; 1517(1):53–62.

Scott, J.C. and McManus, D.P. Molecular cloning and enzymatic expression of the 28-kDa glutathione S-transferase of Schistosoma japonicum: evidence for sequence variation but lack of consistent vaccine efficacy in the murine host. Parasitol. Int. 2000; 49(4):289–300.

Shalaby, K.A., Yin, L., Thakur, A., Christen, L., Niles, E.G. and LoVerde, P.T. Protection against Schistosoma mansoni utilizing DNA vaccination with genes encoding Cu/Zn cytosolic superoxide dismutase, signal peptide-containing superoxide dismutase and glutathione peroxidase enzymes. Vaccine 2003; 22(1):130–136.

Shi, F., Zhang, Y., Lin, J., Zuo, X., Shen, W., Cai, Y., Ye, P., Bickle, Q.D. and Taylor, M.G. Field testing of Schistosoma japonicum DNA vaccines in cattle in China. Vaccine 2002; 20(31-32):3629–3631.

Shi, F., Zhang, Y., Ye, P., Lin, J., Cai, Y., Shen, W., Bickle, Q.D. and Taylor, M.G. Laboratory and field evaluation of Schistosoma japonicum DNA vaccines in sheep and water buffalo in China. Vaccine 2001; 20(3-4):462–467.

Siddiqui, A.A., Ahmad, G., Damian, R.T. and Kennedy, R.C. Experimental vaccines in animal models for schistosomiasis. Parasitol. Res. 2008.

Siddiqui, A.A., Phillips, T., Charest, H., Podesta, R.B., Quinlin, M.L., Pinkston, J.R., Lloyd, J.D., Paz, M., Villalovos, R.M. and Pompa, J. Induction of protective immunity against Schistosoma mansoni via DNA priming and boosting with the large subunit ofcalpain (Sm-p80): adjuvant effects of granulocyte-macrophage colony-stimulating factor and interleukin-4. Infect. Immun. 2003; 71(7):3844–3851.

Siddiqui, A.A., Phillips, T., Charest, H., Podesta, R.B., Quinlin, M.L., Pinkston, J.R., Lloyd, J.D., Pompa, J., Villalovos, R.M. and Paz, M. Enhancement of Sm-p80 (large subunit of calpain) induced protective immunity against Schistosoma mansoni through co-delivery of interleukin-2 and interleukin-12 in a DNA vaccine formulation. Vaccine 2003; 21(21-22):2882–2889.

Siddiqui, A.A., Zhou, Y., Podesta, R.B., Karcz, S.R., Tognon, C.E., Strejan, G.H., Dekaban, G.A. and Clarke, M.W. Characterization of Ca(2+)-dependent neutral protease (calpain) from human blood flukes, Schistosoma mansoni. Biochim. Biophys. Acta 1993; 1181(1):37–44.

Skelly, P.J. Intravascular schistosomes and complement. Trends Parasitol. 2004; 20(8):370–374.

Skelly, P.J. and Alan Wilson, R. Making sense of the schistosome surface. Adv. Parasitol. 2006; 63:185–284.

Skelly, P.J., Kim, J.W., Cunningham, J. and Shoemaker, C.B. Cloning, characterization, and functional expression of cDNAs encoding glucose transporter proteins from the human parasite Schistosoma mansoni. J. Biol. Chem. 1994; 269(6):4247–4253.

Smithers, S.R., Hackett, F., Ali, P.O. and Simpson, A.J. Protective immunization of mice against Schistosoma mansoni with purified adult worm surface membranes. Parasite Immunol. 1989; 11(4):301–318.

Smyth, D., McManus, D.P., Smout, M.J., Laha, T., Zhang, W. and Loukas, A. Isolation of cDNAs encoding secreted and transmembrane proteins from Schistosoma mansoni by a signal sequence trap method. Infect. Immun. 2003; 71(5):2548–2554.

Smythies, L.E., Pemberton, R.M., Coulson, P.S., Mountford, A.P. and Wilson, R.A. T cell-derived cytokines associated with pulmonary immune mechanisms in mice vaccinated with irradiated cercariae of Schistosoma mansoni. J. Immunol. 1992; 148(5):1512–1518.

Soisson, L.A., Reid, G.D., Farah, I.O., Nyindo, M. and Strand, M. Protective immunity in baboons vaccinated with a recombinant antigen or radiation-attenuated cercariae of Schistosoma mansoni is antibody-dependent. J. Immunol. 1993; 151(9):4782–4789.

Soisson, L.M., Masterson, C.P., Tom, T.D., McNally, M.T., Lowell, G.H. and Strand, M. Induction of protective immunity in mice using a 62-kDa recombinant fragment of a Schistosoma mansoni surface antigen. J. Immunol. 1992; 149(11):3612–3620.

Street, M., Coulson, P.S., Sadler, C., Warnock, L.J., McLaughlin, D., Bluethmann, H. and Wilson, R.A. TNF is essential for the cell-mediated protective immunity induced by the radiation-attenuated schistosome vaccine. J. Immunol. 1999; 163(8):4489–4494.

Tarrab-Hazdai, R., Schechtman, D. and Arnon, R. Synthesis and characterization of a protective peptide-based vaccine against Schistosoma mansoni. Infect. Immun. 1998; 66(9):4526–4530.

Tarrab-Hazdai, R., Schechtman, D., Lowell, G., Pirak, E. and Arnon, R. Proteosome delivery of a protective 9B-antigen against Schistosoma mansoni. Int. J. Immunopharmacol. 1999; 21(3):205–218.

Taylor, M.G., Huggins, M.C., Shi, F., Lin, J., Tian, E., Ye, P., Shen, W., Qian, C.G., Lin, B.F. and Bickle, Q.D. Production and testing of Schistosoma japonicum candidate vaccine antigens in the natural ovine host. Vaccine 1998; 16(13):1290–1298.

Tendler, M., Brito, C.A., Vilar, M.M., Serra-Freire, N., Diogo, C.M., Almeida, M.S., Delbem, A.C., Da Silva, J.F., Savino, W., Garratt, R.C., Katz, N. and Simpson, A.S. A Schistosoma mansoni fatty acid-binding protein, Sm14, is the potential basis of a dual-purpose anti-helminth vaccine. Proc. Natl. Acad. Sci. U. S. A. 1996; 93(1):269–273.

Tran, M.H., Pearson, M.S., Bethony, J.M., Smyth, D.J., Jones, M.K., Duke, M., Don, T.A., McManus, D.P., Correa-Oliveira, R. and Loukas, A. Tetraspanins on the surface of Schistosoma mansoni are protective antigens against schistosomiasis. Nat. Med. 2006; 12(7):835–840.

van Balkom, B.W., van Gestel, R.A., Brouwers, J.F., Krijgsveld, J., Tielens, A.G., Heck, A.J. and van Hellemond, J.J. Mass spectrometric analysis of the Schistosoma mansoni tegumental sub-proteome. J. Proteome. Res. 2005; 4(3):958–966.

van der Werf, M.J., de Vlas, S.J., Brooker, S., Looman, C.W., Nagelkerke, N.J., Habbema, J.D. and Engels, D. Quantification of clinical morbidity associated with schistosome infection in sub-Saharan Africa. Acta Trop. 2003; 86(2-3):125–139.

Waine, G.J., Alarcon, J.B., Qiu, C. and McManus, D.P. Genetic immunization of mice with DNA encoding the 23 kDa transmembrane surface protein of Schistosoma japonicum (Sj23) induces antigen-specific immunoglobulin G antibodies. Parasite Immunol. 1999; 21(7):377–381.

Williams, G.M., Sleigh, A.C., Li, Y., Feng, Z., Davis, G.M., Chen, H., Ross, A.G., Bergquist, R. and McManus, D.P. Mathematical modelling of schistosomiasis japonica: comparison of control strategies in the People's Republic of China. Acta Trop. 2002; 82(2):253–262.

Wilson, M.S., Mentink-Kane, M.M., Pesce, J.T., Ramalingam, T.R., Thompson, R. and Wynn, T.A. Immunopathology of schistosomiasis. Immunol. Cell Biol. 2007; 85(2):148–154.

Wilson, R.A. Interferon gamma is a key cytokine in lung phase immunity to schistosomes but what is its precise role? Braz. J. Med. Biol. Res. 1998; 31(1):157–161.

Wilson, R.A. and Coulson, P.S. The Development of Schistosome Vaccine. Parasitic Flatworms: Molecular Biology, Bichemistry, Immunology and Physiology, pp. 303–317. Wallingsford, OX, UK: A. G. M. a. N. J. Marks, 2006, CAB International.

Wilson, R.A., Coulson, P.S., Betts, C., Dowling, M.A. and Smythies, L.E. Impaired immunity and altered pulmonary responses in mice with a disrupted interferon-gamma receptor gene exposed to the irradiated Schistosoma mansoni vaccine. Immunology 1996; 87(2):275–282.

Wilson, R.A., van Dam, G.J., Kariuki, T.M., Farah, I.O., Deelder, A.M. and Coulson, P.S. The detection limits for estimates of infection intensity in schistosomiasis mansoni established by a study in non-human primates. Int. J. Parasitol. 2006; 36(12):1241–1244.

Wu, Z.D., Lu, Z.Y. and Yu, X.B. Development of a vaccine against Schistosoma japonicum in China: a review. Acta Trop. 2005; 96(2-3):106–116.

Wynn, T.A., Jankovic, D., Hieny, S., Cheever, A.W. and Sher, A. IL-12 enhances vaccine-induced immunity to Schistosoma mansoni in mice and decreases T helper 2 cytokine expression, IgE production, and tissue eosinophilia. J. Immunol. 1995; 154(9):4701–4709.

Xu, S., Shi, F., Shen, W., Lin, J., Wang, Y., Ye, P., Tian, E., Qian, C., Lin, B., Shi, Y. et al. Vaccination of sheep against Schistosoma japonicum with either glutathione S-transferase, keyhole limpet haemocyanin or the freeze/thaw schistosomula/BCG vaccine. Vet. Parasitol. 1995; 58(4):301–312.

Yang, W., Jackson, D.C., Zeng, Q. and McManus, D.P. Multi-epitope schistosome vaccine candidates tested for protective immunogenicity in mice. Vaccine 2000; 19(1):103–113.

Yi, X.Y., Simpson, A.J., de Rossi, R. and Smithers, S.R. The presence of antibody in mice chronically infected with Schistosoma mansoni which blocks in vitro killing of schistosomula. J. Immunol. 1986; 137(12):3955–3958.

Zhang, D.M., Pan, W.Q., Qian, L., Duke, M., Shen, L.H. and McManus, D.P. Investigation of recombinant Schistosoma japonicum paramyosin fragments for immunogenicity and vaccine efficacy in mice. Parasite Immunol. 2006; 28(3):77–84.

Zhang, R., Yoshida, A., Kumagai, T., Kawaguchi, H., Maruyama, H., Suzuki, T., Itoh, M., El-Malky, M. and Ohta, N. Vaccination with calpain induces a Th1-biased protective immune response against Schistosoma japonicum. Infect. Immun. 2001; 69(1):386–391.

Zhang, Y., Taylor, M.G. and Bickle, Q.D. Schistosoma japonicum myosin: cloning, expression and vaccination studies with the homologue of the S. mansoni myosin fragment IrV-5. Parasite Immunol. 1998; 20(12):583–594.

Zhang, Y., Taylor, M.G., Johansen, M.V. and Bickle, Q.D. Vaccination of mice with a cocktail DNA vaccine induces a Th1-type immune response and partial protection against Schistosoma japonicum infection. Vaccine 2001; 20(5-6):724–730.

Zhu, Y., Ren, J., Da'dara, A., Harn, D., Xu, M., Si, J., Yu, C., Liang, Y., Ye, P., Yin, X., He, W., Xu, Y., Cao, G. and Hua, W. The protective effect of a Schistosoma japonicum Chinese strain 23 kDa plasmid DNA vaccine in pigs is enhanced with IL-12. Vaccine 2004; 23(1):78–83.

Zhu, Y., Ren, J., Harn, D.A., Si, J., Yu, C., Ming, X. and Liang, Y. Protective immunity induced with 23 kDa membrane protein dna vaccine of Schistosoma japonicum Chinese strain in infected C57BL/6 mice. Southeast Asian J. Trop. Med. Public Health 2003; 34(4):697–701.

Zhu, Y., Si, J., Ham, D.A., Yu, C., He, W., Hua, W., Yin, X., Liang, Y., Xu, M. and Xu, R. The protective immunity produced in infected C57BL/6 mice of a DNA vaccine encoding Schistosoma japonicum Chinese strain triose-phosphate isomerase. Southeast Asian J. Trop. Med. Public Health 2002; 33(2):207–213.

Zhu, Y., Si, J., Harn, D.A., Xu, M., Ren, J., Yu, C., Liang, Y., Yin, X., He, W. and Cao, G. Schistosoma japonicum triose-phosphate isomerase plasmid DNA vaccine protects pigs against challenge infection. Parasitology 2006; 132(Pt 1):67–71.

CHAPTER

69

American Trypanosomiasis

Vandanajay Bhatia[1], *Jian-Jun Wen*[1], *Michele A. Zacks*[2] *and Nisha Jain Garg*[3]

[1]Department of Microbiology and Immunology, University of Texas Medical Branch, Galveston, TX, USA

[2]Department of Pathology, University of Texas Medical Branch, Galveston, TX, USA

[3]Departments of Microbiology and Immunology and Pathology, Center for Biodefense and Emerging Infectious Diseases, Sealy Center for Vaccine Development, University of Texas Medical Branch, Galveston, TX, USA

OUTLINE

ABSTRACT

Trypanosoma cruzi is the etiologic agent of American trypanosomiasis or Chagas disease. It is normally transmitted by the reduviid insect vector, and also by blood transfusion. It is estimated that 16–18 million people are infected in the American continent, with >50,000 deaths reported every year. Acute infection by *T. cruzi* can be lethal, though the majority exhibit flu-like, nonspecific clinical symptoms. The disease usually evolves into a chronic stage, in which >30% of infected individuals exhibit severe debilitation of the heart. The pathology of Chagas disease presents a complicated and diverse picture in humans. The clinically relevant pathognomonic consequences of human infection by *T. cruzi* are dilation and hypertrophy of the left ventricle walls and thinning of the apex. The major complications and destructive evolutionary outcomes of chronic infection include ventricular fibrillation, thromboembolism, and congestive heart failure. American trypanosomiasis poses serious public health care and financial concerns. The currently available drugs, though effective against acute infection, are highly toxic and ineffective in arresting or attenuating clinical disease symptoms in chronic patients. No effective vaccines are available.

Studies in animal models and human patients have revealed several mechanisms, including parasite persistence, chronic inflammatory responses to self or parasite antigens, and sustained mitochondrial dysfunction and oxidative stress, that contribute to the pathogenic outcome of Chagasic cardiomyopathy. Protective immunity against *T. cruzi* infection requires the elicitation of Th1 cytokines, lytic antibodies and the concerted activities of macrophages (ϕs), T helper cells, and cytotoxic T lymphocytes (CTLs). Several antigens, antigen delivery vehicles, and adjuvants have been tested to elicit immune protection to *T. cruzi* in experimental animals; however, most of these attempts have met with limited success, and the development of an efficacious prophylactic vaccine faces many challenges. There is a critical lack of methods for prevention of infection or treatment of acute infection or chronic disease. This chapter will summarize what is known about the parasite and the current state of knowledge of pathogenesis and protective immunity. We will discuss the research efforts in vaccine development to date and the challenges faced in achieving an efficient prophylactic vaccine against human American trypanosomiasis, as well as the future perspectives.

HISTORICAL BACKGROUND OF CHAGAS DISEASE

A detailed history of American trypanosomiasis is presented in a review (Guerra, 1970). The chronicles of colonization of the New World contain many indirect references to Chagas disease, then referred to as *Mal de Bicho* in Brazil. The first description of the reduviid insect and its blood-sucking habits was made in 1590 by a missionary priest in Tucuman, Argentina. It was ~300 years later when reduviids were shown to be carriers of *Trypanosoma cruzi*. In 1908, a physician, Carlos Chagas, was commissioned by Oswaldo Cruz, then director of Manguinhos Institute, to the Lassance region of Amazon Basin in Brazil to control a malaria outbreak. During this visit, he noted a large number of reduviid insects, referred to as "barbeiros" or "kissing bug" infesting the poorly constructed rural dwellings. He considered that these insects might be important vector of human disease, and studied them and found flagellates, resembling *Crithidia*, in their hindgut. Armed with a vector and a possible agent, he allowed the reduviid insects to feed on laboratory animals, and subsequently found flagellated parasites in their blood. The parasite was named, in honor of Oswaldo Cruz, *Schyzotrypanum cruzi*, and later renamed *T. cruzi*. Chagas continued his studies and, in 1908, documented the first case of human *T. cruzi* infection in a 2-year-old girl. Within a period of less than 2 years, he described salient characteristics of the clinical disease, named Chagas disease after him (Lewinsohn, 1979). Subsequently, Vianna, a colleague of Chagas, made the first histopathological observation of amastigote nests in tissues, and Brunmpt delineated the life cycle of *T. cruzi*. Cases of *T. cruzi* infection were reported in Panama, Argentina, and Uruguay from 1931 to 1936. The first case of transfusion-mediated parasite transmission was reported in 1952 in Sao Paulo, Brazil. During the next 10 years, it became evident that the chronic infection is associated with pathologic lesions in the heart, extensively described by Mott and Hagstrom (1965). In 1974, the first observation of anti-heart antibodies in chronically infected patients was made (Cossio et al., 1974), and concurrently, others showed the destruction of heart cells by *T. cruzi*-sensitized lymphocytes in vitro (Santos-Buch and Teixeira, 1974). These observations initiated the hypothesis of Chagas disease as an autoimmune disease.

The earliest studies documenting the immunologic control of *T. cruzi* showed: the complement-dependent phagocytosis and cytotoxicity to *T. cruzi* (Kierszenbaum and Budzko, 1973); antibody-dependent, cell-mediated cytotoxicity to bloodform trypomastigotes (Abrahamsohn and Silva, 1977); macrophage activity against *T. cruzi* (Hoff, 1975); and genetic resistance-versus-susceptibility of different inbred mouse strains to *T. cruzi* (Trischmann et al., 1978). Much has been learned about the biology, pathogenesis, and immunity to *T. cruzi* since its first description in 1909, and the knowledge related to vaccine development against *T. cruzi* is summarized here.

VECTOR AND PARASITE

Vector Classification

The vectors of *T. cruzi* are insects of the order *Hemiptera*, family *Reduviidae*, and subfamily *Triatominae*. Of the 118 species of triatomines identified, only a few, e.g., *Triatoma infestans* and *Panstrongylus megistus* in South America; *Rhodnius prolixus* in Venezuela, Columbia, and Central America; and *T. barberi* and *T. dimidiata* in Mexico, are epidemiologically significant as vectors of *T. cruzi*. *T. sanguisuga* is the primary vector of *T. cruzi* in the southeastern US (Pung et al., 1998; Schofield and Dias, 1999; Guzman-Bracho, 2001) (Fig. 69.1).

Parasite Classification

T. cruzi is a hemoflagellate protozoan of the *Sarcomastigophora* phylum, *Mastigophora* subphylum, *Kinetoplastida* order, *Trypanosomatidae* family, and is characterized by the presence of one flagellum and a single mitochondrion, which contains the kinetoplast, a specialized DNA-containing organelle. *T. cruzi* isolates are grouped into two major phylogenetic groups, i.e., I and II, on the basis of zymodemes and several genetic markers. *T. cruzi* II consists of five related subgroups, IIa–IIe (Brisse et al., 2000). In South America, *T. cruzi* I is associated with a sylvatic transmission cycle and human disease north of the Amazon basin. *T. cruzi* II is predominantly distributed in the Southern Cone countries and also associated with a domestic transmission cycle and infection of placental mammals (Barnabe et al., 2001a). In Mexico and the US, the majority of human cases are linked to *T. cruzi* I (Barnabe et al., 2001b; Bosseno et al., 2002). No direct correlation between Chagas disease severity and parasite lineage has been established.

Life Cycle of *T. cruzi*

T. cruzi evolves during its life cycle into different forms that are morphologically distinct and can also be identified by the relative position of the kinetoplast in relation to the cell nucleus and flagellum emergence (Fig. 69.2) (Tyler et al., 2003). The infective, trypomastigote (15–20 × 2 μm) exists as a freely swimming, extracellular, nonreplicative form in the bloodstream of mammalian hosts. It may be ingested during a bloodmeal by the triatomine insect vector. In the midgut of the insect, trypomastigotes transform into

FIGURE 69.1 Geographical distribution of triatomines of clinical importance and *T. cruzi* infection in humans.

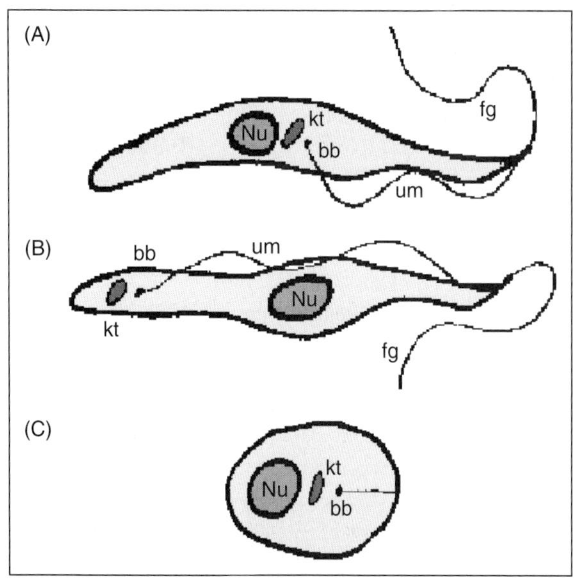

FIGURE 69.2 Different morphological forms of *T. cruzi*. (A) Epimastigote, (B) trypomastigote, (C) amastigote. Nu, nucleus; fg, flagellum; um, undulating membrane; kt, kinetoplast; bb, basal body.

epimastigotes (20 × 2 μm) that replicate by binary fission, and move to the hindgut and rectum, where they transform into infective metacyclic trypomastigotes. When an infected triatomine takes a bloodmeal and defecates, metacyclic trypomastigotes are released in the feces of the insect and may enter the body of the victim through the puncture wound caused by insect bite, or through mucosal membranes of the mouth, nose, or eyes. Trypomastigotes may also be transmitted by transplacental transmission, blood transfusion, or laboratory accident. There is local inflammation at the entry site where macrophages and/or local tissues are parasitized. Most cells can be parasitized, although there appears to be a tropism for striated muscle, smooth muscle, and neuroectodermal cells. Trypomastigotes escape from the parasitophorous vacuole into the host cell cytoplasm by an unusual mechanism involving transformation of trypomastigotes into spherical, intracellular amastigotes (2–4 μm diameter) that replicate by binary fission within the cytoplasm of the parasitized cells. Within 4–5 days, amastigotes of heavily parasitized cells convert back into trypomastigotes, which are released by cell lysis. The released trypomastigotes may infect other cells or be ingested by the vector to continue the life cycle (Fig. 69.3) (Tyler et al., 2003).

Diagnostic Measures

Parasitologic diagnosis of infection by *T. cruzi* depends upon detection of trypomastigotes by microscopic examination of blood samples. Trypomastigotes in patients may also be detected by indirect methods, e.g., xenodiagnosis and hemoculture (Chiari, 1992). The polymerase chain reaction (PCR) offers several advantages in the sense that it can be more sensitive, specific, and rapid than the above-mentioned methods. The majority of PCR assays used to detect *T. cruzi* target the amplification of kinetoplast DNA, ribosomal DNA, or mini-exon sequences (Gonzalez et al., 1994; Gomes et al., 1999; Pizarro et al., 2007).

Serological analysis is routinely conducted to determine exposure to *T. cruzi*. Generally, specific IgM to *T. cruzi* can be detected 1 week after infection, peaks by day 30, and decreases to undetectable levels in 3–4 months when IgG antibodies reach their maximum level. *T. cruzi*-specific IgG antibodies persist throughout the life of infected individuals. Anti-*T. cruzi* antibodies are detected by various methods, including complement fixation, enzyme-linked immunosorbent assay (ELISA), flow cytometry, direct agglutination or indirect hemagglutination, indirect immunoflourescence, and immunoblotting (Ferreira, 1992). *T. cruzi*-antigen-based serological analysis reagents are commercially available in the endemic countries and used for blood bank screening or seroepidemiological surveys. It is essential to perform multiple tests to confirm seropositivity. In the US, an ELISA assay that uses epimastigote lysate antigens for detection of antibodies to *T. cruzi* in serum and plasma (Ortho-Clinical Diagnostics, Raritan, NJ) has been licensed by the Food and Drug Administration, and used by American Red Cross and other blood banks for screening the donors for *T. cruzi* infection.

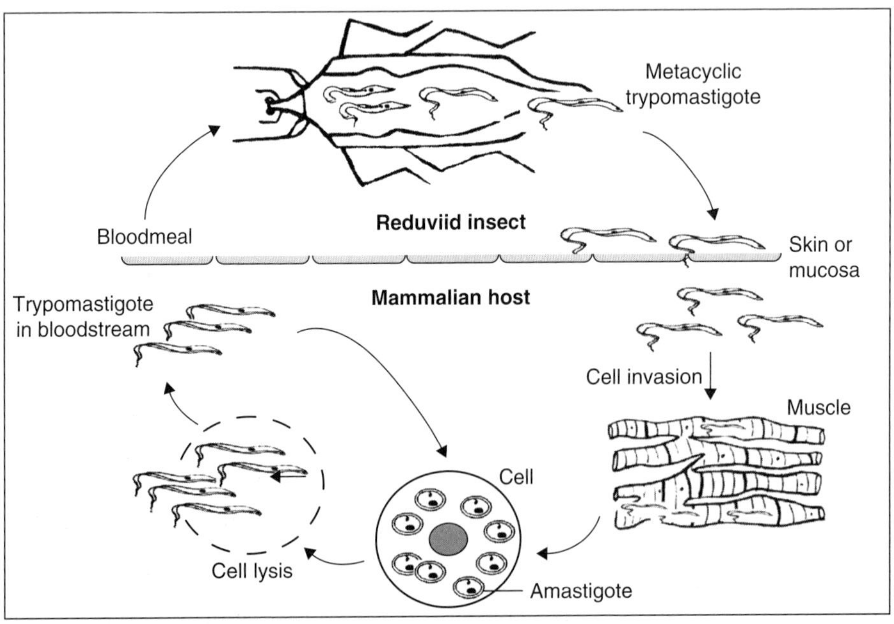

FIGURE 69.3 Life cycle of *T. cruzi*.

IMMUNE RESPONSE TO *T. CRUZI*

Protective Immune Responses

High parasitemia and wide distribution of *T. cruzi* in different organs and tissues after initial infection are a result of the parasites' ability to evade or suppress innate immune mechanisms. The acute phase ends when the host mounts a potent immune response that is largely effective in controlling the parasite. Sterilizing immunity does not exist in *T. cruzi* infection. The comparison in responses exhibited by susceptible and resistant experimental models has contributed to an understanding of protective immune responses to *T. cruzi*. Natural killer (NK) cells appear to be the first source of interferon-γ (IFN-γ) that, in turn, augments IL-12 synthesis by ϕs (Gazzinelli et al., 1992). Surface glycoproteins (mucins) and/or glycophospholipids (GIPLs) of *T. cruzi* are shown to stimulate the production of multiple cytokines, e.g., IFN-γ, TNF-α, IL-1, and IL-6; and chemokines, e.g., MCP-1, RANTES, and IP-10 by inflammatory ϕs (Almeida et al., 2000, 2001; Almeida and Gazzinelli, 2001). These cytokines induce ϕ activation and nitric oxide (NO) production that is important in the killing of *T. cruzi* (Martins et al., 1998). It is suggested that IFN-γ of NK origin and IL-12 of ϕ origin skew the differentiation of parasite-specific T helper cells toward a protective Th1 phenotype (Abrahamsohn and Coffman, 1996). Parasite-specific $CD4^+$ T cells may assist in the control of *T. cruzi* through secretion of Th1 cytokines (IFN-γ, IL-2), amplification of the phagocytic activity of macrophages, stimulation of B cell proliferation and antibody production, and differentiation and activation of $CD8^+$ T cells (Brener and Gazzinelli, 1997). *T. cruzi* antigen-specific $CD8^+$ T cells are frequently present in infected mice and humans (Wizel et al., 1997, 1998b); and may contribute to *T. cruzi* control, either by cytolysis of the infected cells or secretion of Th1 cytokines (IFN-γ) that induce trypanocidal activity (DosReis, 1997). A strong lytic antibody response enhances the opsonization, phagocytosis, and complement-dependent killing of the parasites (Krautz et al., 2000).

With progression to indeterminate and chronic disease, blood parasitemia is absent, and parasite nests are rarely detectable in cardiac tissue. The presence of parasite-specific antibodies is diagnostic of chronic *T. cruzi* infection. The mononuclear cells, ϕs, and IFN-γ-producing $CD8^+$ T cells represent the majority of the infiltrate in experimental models (Tarleton, 1996) and human cardiac biopsies (Reis et al., 1993). $CD4^+$ T cells are less prominent in the chronic myocardium of infected mice (DosReis, 1997) and human patients (Higuchi et al., 1997). It is likely that immune responses similar to those essential for control of acute parasitemia, i.e., substantial antibody response and activation of type 1 $CD4^+$ and $CD8^+$ T cells, are required to maintain control of *T. cruzi* during indeterminate and chronic phases of infection and disease. The evidence is provided from the experimental models and natural human infections demonstrating that absence or reduction in any of these immune responses through targeted depletion, immunosuppressive treatments, or infection-induced immunosuppression can result in an exacerbation of parasitemia (Sartori et al., 1995; D'Almeida et al., 1996; Tarleton et al., 1996).

Pathologic Immune Responses

Along with their immunoprotective role, $CD4^+$ and $CD8^+$ T cells are suggested to be involved in immunopathology. During acute infection in mice, $CD4^+$ T cells initiate unspecific, polyclonal activation of B cells from spleen and lymph nodes, leading to an increased number of nonspecific, IgG-secreting cells and hypergammaglobulinemia with disease progression (Minoprio et al., 1987, 1989). GIPLs, at least in part, are responsible for triggering of polyclonal B lymphocytes (Bento et al., 1996). $CD4^+$ T cells that persist in chronic hearts represent an autoreactive phenotype and are associated with myocyte death, tissue destruction, and increased animal mortality (Soares and Santos, 1999). In humans, most of the initial humoral response is dominated by antibodies to an epitope (Galα1-3Gal) expressed by glycosylphosphatidylinositol (GPI) anchored mucins of *T. cruzi* and is followed by polyclonal B cell activation and hypergammaglobulinemia with disease progression. $CD8^+$ T cells are the major T-cell subset found in the cardiac tissue of both experimental animals and human patients and are thought to be responsible for immunopathology during chronic Chagas disease (Higuchi, 1999; Higuchi et al., 2003).

Regulatory Immune Responses

Because *T. cruzi* infection elicits a strong activation of the immune system, it seems essential that immunoregulatory mechanisms be activated. IL-10 and TGF-β cytokines are important negative modulators of NO synthesis and trypanocidal activity of ϕs. A lack of IL-10 results in an overwhelming inflammatory response and animal deaths early in infection with *T. cruzi* (Hunter et al., 1997). Several mechanisms, e.g., regulation of IL-2 expression (Tarleton, 1988), ϕ activation and NO synthesis (Abrahamsohn and Coffman, 1995), and selective apoptosis of T helper cells (Lopes and Dosreis, 1996), may act independently, or in concert, to control the $CD4^+$ T cell population in infected mice. Parasite molecules (GIPLs) are shown to block $CD8^+$ T cell activation in vitro (DosReis et al., 2002); however, no regulatory mechanisms have been defined that may control $CD8^+$ T cell population in vivo. The occurrence of such regulatory mechanisms in human patients, though expected, has not been defined.

EPIDEMIOLOGY

Transmission and Geographic Distribution

The ancient sylvatic cycle of *T. cruzi* transmission involves interaction between wild vectors and hosts in different natural ecotypes of American continent. The domestic cycle results from human–vector contact. The reservoir hosts of *T. cruzi* include >150 mammal species, mainly small animals, that play an important role in maintaining both sylvatic and domestic cycles (WHO, 2002). Humans, followed by dogs, cats, and rodents are important domestic reservoirs. *T. cruzi* is widespread and endemic in almost all of South and Central America, and Mexico (Fig. 69.1) (Schofield et al., 2006; Cruz-Reyes and Pickering-Lopez, 2006). Several studies have shown the presence of the insect vector as well as infection of domestic dogs and wild animals in the US (Yabsley and Noblet, 2002; Beard et al., 2003; Hall et al., 2007; Hanford et al., 2007). Vector-transmitted human infections have been reported in the southern US (Herwaldt et al., 2000) (Fig. 69.1).

Profound economic and social changes have stimulated rural-to-urban migration within the endemic countries and towards the developed countries, e.g., the US and Canada. It is estimated that >300,000 infected individuals live in the city of Sao Paulo; >200,000 each in Rio de Janeiro and Buenos Aires (Dias, 1992b); and >100,000 infected immigrants from Latin America live in the US (Kirchhoff, 1993). These individuals may donate contaminated blood and thereby expose many others to the risk of *T. cruzi* infection (Dias, 1992a; Leiby et al., 1997). The infectivity risk, defined as the likelihood of infection from receiving an infected transfusion unit, is estimated to be ~20% for *T. cruzi* (WHO, 2002). Before serologic screening of blood donors for antibody to *T. cruzi* was implemented, infection rates through blood transfusion ranged from <0.1% to 4% in Argentina, Brazil, Chile, and Uruguay to >10% in Bolivia (Schmunis and Cruz, 2005). Transmission of *T. cruzi* infection by solid-organ transplantation has been evident in Latin America (Reis et al., 1995; Riarte et al., 1999).

In the US, five cases of *T. cruzi* transmission by organ transplantation from infected donors have been described (CDC, 2002a, 2006). Three blood banks, situated in California and Arizona states, have conducted a large-scale serosurvey, and documented that 1 in every 4665 blood donations was seropositive for *T. cruzi*-specific antibodies (CDC, 2002b). This study led to FDA approval of first blood donor screening test for Chagas disease in the US (CDC, 2007). Beginning January 2007, American Red Cross and Blood Systems Inc. have begun screening the blood donations for presence of *T. cruzi* antibodies and it is expected that all blood-collection establishments in the US will implement screening test for *T. cruzi* infection in near future. As of July 2007, 458 blood donations were repeat reactive by ELISA and 124 donations were confirmed positive by radioimmunoprecipitation test in the US (AABB Chagas' Biovigilance Network, http://www.aabb.org/Content/Programs_and_Services/Data_Center/Chagas/chagas.htm).

Disease Burden

According to World Health Organization estimates from 1998, >120 million people are exposed to risk of *T. cruzi* infection, ~18 million people are infected with *T. cruzi*, and ~50,000 children and adults die annually, because of the clinical complications of *T. cruzi*-induced heart disease and lack of effective treatments in endemic countries (WHO, 2002). Approximately 2–5% of fetuses carried by infected mothers are either aborted or born with congenital infection. Revenue loss, in terms of productivity declines due to sickness in prime years and medical costs, has an overwhelming effect on economic growth (WHO, 2002). It is currently estimated that the total annual cost of Chagas disease management in endemic countries is >8 billion US dollars. When considered from a global perspective, Chagas disease represents the third greatest tropical disease burden after malaria and schistosomiasis. American trypanosomiasis is recognized as a potentially emerging infectious disease in the US (Dodd and Leiby, 2004).

Potential as a Biothreat Agent

T. cruzi infection poses a minimal risk as a biothreat agent due to a latency period of incubation ranging ~10–30 years prior to development of clinical disease. Chagas disease is, however, an emerging infectious disease in the US and no longer considered a disease of the tropical, developing countries. Since January 2007, the American Red Cross and other blood banks across the United States have implemented the donor screening for *T. cruzi* infection, highlighting the urgent need for clinicians, laboratorians, and public health professionals to understand Chagas disease, its diagnosis, and treatment.

CLINICAL DISEASE AND CONTROL MEASURES

Clinical Manifestations of Chagas Disease

Infection by *T. cruzi* elicits nonspecific symptoms, e.g., fatigue, fever, enlarged liver or spleen, swollen lymph glands, and occasionally a characteristic swelling either at the site of entry in the skin, a *Chagoma*, or in the conjuctiva of eyelid, *Romana's* sign. Most acute patients develop heavy parasitism of muscles and some symptoms of cardiac involvement that is usually unrecognized. The clinical manifestations resemble myocarditis, with slight-to-moderate cardiac enlargement. In a small proportion of acute patients (<5%), more severe right- and left-sided heart failure may occur with pulmonary and systemic congestion, tachychardia, and other cardiac insufficiencies, occasionally resulting in fatal congestive heart failure. *T. cruzi* can also invade the gastrointestinal and central nervous systems, but, in general, damage to these systems is uncommon. The signs and symptoms of acute disease resolve spontaneously in 4–8 weeks postinfection, and the majority of patients enter an indeterminate phase of infection. This phase is characterized by very low or undetectable parasitemia, high parasite-specific antibody titers, and lack of clinical symptoms. Between 10 and 20 years after an initial infection, ~40% of infected individuals become symptomatic. The systemic changes during the chronic stage commonly develop in the heart, esophagus, and rectosigmoid colon, and less often in the central nervous system (Kirchhoff et al., 2004).

Heart problems in chronic Chagasic patients are evidenced by nonspecific symptoms such as palpitations, dizziness, and syncope. Typically, manifestations of chronic disease are thromboembolism in the brain, limbs, or lungs, congestive heart failure, and sudden death—mainly due to ventricular fibrillation. Once clinical symptoms appear, the prognosis is poor, and fatality occurs within a short period (Rassi et al., 2000). The principal defects in the conduction system present as a mixture of arrythmias, e.g., tachycardia, bradycardia, ventricular fibrillation, and electrical impulse blockages, e.g., right bundle branch and left anterior fasicle block. Myocardial structural defects are mainly cardiomegaly with hypertrophy and dilation of the chambers, apical aneurism in left ventricle, and clot formation with thromboembolization (Rossi et al., 2003). The histological examination of sections from both autopsy and biopsy specimens reveal that tissue fibrosis, inflammation, and hypertrophy of cardiac fibers are the chief pathological consequences of Chagas disease. Fibrosis is most pronounced in the left ventricle and apex, and may correspond to increased collagen fiber deposition around the muscle bundles. Inflammation varies in severity and its location may be interstitial and/or diffuse. Echocardiography, serial chest radiographs, and electrocardiography are the most accurate and commonly employed methods for detecting myocardial involvement in Chagasic patients (Rocha et al., 2003).

Clinical manifestations of Chagas disease may also be correlated with defective innervation and defective contraction within the myocardium. Pathogenic changes in the parasympathetic branch of the autonomic nervous system are sometimes observed at autopsy. Specifically, a reduction, or in some cases, a complete absence of cells in the neuronal ganglia has been described and may be indicative of a role for impaired heart innervation in cardiac dsyfunction associated with Chagas disease (Marin-Neto, 1998).

In some patients, clinical manifestations of Chagas disease include development of a megaesophagus and/or megacolon. Repeated episodes of aspiration pneumonitis are common in patients with severe esophageal dysfunction. Patients with a megacolon suffer from chronic constipation and abdominal pain. Ultrasonography, computed tomography, and magnetic resonance imaging are excellent imaging modalities used to determine the extent of tumor and enlargement, but are very expensive (Mattoso, 1981).

Control Measures

Vector Control

Three multinational, government-funded programs, initiated in 1991, are underway to control the endemicity of *T. cruzi* infection in Southern Cone nations of Argentina, Bolivia, Brazil, Chile, Paraguay, and Uruguay (Schofield, 1992; Schofield and Dias, 1999). The primary focus of these programs is the elimination of the domestic insect population through insecticide use, housing improvement, and education. These efforts have been very successful in reducing the domiciliary infestation by triatomines in vast regions of the endemic countries, and subsequently reducing the prevalence of acute infection in young children (Dias and Schofield, 1999; Schofield et al., 2006). The current goal is to eliminate vectorial *T. cruzi* transmission to humans throughout the Southern Cone by 2010. Venezuela and Colombia have also implemented parallel insecticide spray and housing improvement initiatives (WHO, 2002). However, the operational cost to sustain the insecticide spray programs, followed by constant surveillance, is one of the major obstacles in the success of the regional insecticide-based control programs. Insecticide use in the long-term may not be efficacious in blocking *T. cruzi* transmission, owing to the development of drug resistance by triatomines. Concern remains that reinfestation of homes by secondary sylvatic vectors, e.g., *Triatoma sordida*, in Brazil and other South American countries, will compromise the long-term efficacy of vector control measures (Moncayo, 2003). Successful elimination of disease as a public health problem through interruption of transmission would not eradicate the parasite due to the zoonotic nature of the disease.

Blood Supply Screening

The problem of transfusion-transmitted Chagas disease was apparent by the 1950s. However, it was only in the 1980s, with the emergence of AIDS that national and regional blood-screening programs were implemented in endemic countries. Among the 18 countries in Latin America for which data are available, the degree of blood bank screening for *T. cruzi* now reaches 100% of the donors in seven countries, ≥99% in three, 75–95% in four, and 25–34% in three (Schmunis and Cruz, 2005). In the US, an ELISA assay that uses epimastigote lysate antigens for detection of antibodies to *T. cruzi* in serum and plasma (Ortho-Clinical Diagnostics) has been licensed by the Food and Drug Administration in January 2007, and used by American Red Cross and other blood banks for screening the donors for *T. cruzi* infection.

Treatment

Two drugs, benznidazole and nifurtimox, are generally approved for the treatment of *T. cruzi* infection, and are effective in curing at least 50% of acute or recent cases of infection (Coura, 1996). Both of these drugs have serious and frequent side effects and are not available to many patients either because they are not registered in those countries or drug prices are high. Anti-parasitic drugs are not effective once the patients enter the chronic phase of infection and disease progression (Rodriques Coura and De Castro, 2002). Several other drugs, e.g., posaconazole (Liendo et al., 1998) and thioridazine (Lo Presti et al., 2004; Bustamante et al., 2007) have shown promise in curing *T. cruzi* infection in murine models, however, these drugs have not yet entered clinical trials possibly due to lack of financial support from the major drug companies for the development of new drugs against Chagas disease (Moncayo and Ortiz Yanine, 2006).

Treatment of chronic patients focuses on cardiac management (Rocha et al., 2003) that requires a specialized clinical infrastructure, and thus, is expensive and often beyond the reach of the patient. For example, in Brazil, considering that only 10% of infected individuals would develop chronic disease, the medical costs for the obligatory treatment could reach $250 million. Cardiac transplantation is an option for patients with severe cardiac disease; however, immunosuppression can lead to reactivation of infection with an intensity typical of acute Chagas disease, invasion of transplanted heart by parasites, and heart rejection.

No vaccines are currently available against *T. cruzi* infection.

PATHOGENIC MECHANISMS IN CHAGAS DISEASE DEVELOPMENT

Considerable debate exists regarding the pathomechanisms that are involved in Chagas disease development. No universally accepted model exists to explain the long latency period between infection and disease development, why a subset of individuals in the indeterminate phase progress to chronic disease, or what instigates or perpetuates the damage in the heart that ultimately results in Chagasic disease symptoms. We briefly discuss the pathomechanisms that are described in the literature.

Immune-Mediated Pathology: Autoimmunity or Parasite Persistence

A significant body of literature suggests that consistent inflammatory immune responses play a role in the development and/or propagation of pathological lesions in Chagasic hearts (Pathologic Immune Responses). In cardiac tissue, antibody- and cytotoxic T lymphocyte (CTL)-mediated responses, as well as the continuous production of cytokines and chemokines, may contribute to the selective destruction of cardiomyocytes, neurons, or endothelial cells, and the development of fibrosis in response to cytopathology (Brener and Gazzinelli, 1997). Two mechanisms, autoimmunity and parasite persistence, are proposed to sustain pathological inflammatory responses in Chagasic hearts (Soares et al., 2001).

The concept of autoimmunity is that polyclonal activation (Minoprio, 2001), molecular mimicry (Iwai et al., 2005), or the presence of cryptic epitopes shared by the host and parasites (Girones et al., 2001b) result in the recognition of self-antigens by the immune response elicited to control the parasite and, subsequently host tissue destruction. A growing number of studies have shown the pathogenic role of autoantibodies present in the serum of chronic mice and patients that react with various antigens in the heart, skeletal muscle, or nervous tissue (reviewed in Kierszenbaum (1999)). These include the β1-adrenoreceptor (Sterin-Borda and Borda, 2000), ribosomal P proteins (Kaplan et al., 1997), cruzipain (Giordanengo et al., 2000), and a novel Cha protein (Girones et al., 2001a). Common antigens between *T. cruzi* and human myocardial fibers (e.g., myosin) and auto-reactive T cells with specific cytotoxicity against myocardial fibers have been demonstrated in humans (Cunha-Neto et al., 1995; Cunha-Neto and Kalil, 2001). A cardiac myosin-specific autoimmune response and myocarditis can be induced by injecting several doses of subcellular antigens of *T. cruzi* in animals (Leon et al., 2004). The autoimmune theory, thus, suggests that the humoral and cellular immune reactions, elicited in response to *T. cruzi* infection, recognize self-antigens, are potentially harmful to the host, and contribute to the development and/or propagation of pathological lesions in Chagasic myocarditis (Leon and Engman, 2003). The lack of evidence that *T. cruzi*-induced self-reactive antibodies or T cells can induce disease upon transfer into a naïve host precludes us from defining chronic Chagasic cardiomyopathy as an autoimmune disease.

It is now recognized that human infection of many years' duration is associated with a nearly consistent low level of parasitosis (Higuchi et al., 2003). Recent studies using modern techniques, e.g., PCR, immunohistochemistry, and confocal microscopy, have detected parasite DNA or antigens in blood and heart tissue biopsies of experimental animals and chronic human patients (Anez et al., 1999; Mortara et al., 2000; Salomone et al., 2000; Caliari et al., 2002). It is proposed that parasite persistence provides sufficient, consistent antigens that work as a trigger for the hypersensitive response against myocardial fibers, leading to pathologic tissue injury, and, subsequently, to cardiac insufficiency.

Vascular Mediators in Endothelium

Yet, clinical-pathological severity is not universally correlated with the presence of parasites (Higuchi et al., 1993) and chronic inflammation (Palomino et al., 2000). This leads to the suggestion that other mechanisms must be involved. The hypothesis that vasculopathy contributes to Chagasic heart disease has developed from the observations indicating that both human and experimental infections cause a reduction in blood flow and an intense vasculitis. Bradykinin, a potent vasodilator, and endothelin 1 (ET-1), and thromboxane A2 (TXA2), two potent vasoconstrictors, have been identified in Chagasic hearts. Kinins convey the vasodilating response through their interaction with vascular receptors and lessen the detrimental effects of ischemia (Yang et al., 2001). Despite increased kinin expression, *T. cruzi*-mediated decrease in coronary flow was not normalized in infected rats, suggesting impaired kinin receptor function. Instead, activation of kinin receptors on cardiovascular cells was shown to contribute to an increased infectivity by *T. cruzi* (Scharfstein et al., 2000; Todorov et al., 2003). TXA2 and ET-1 may potentiate disease severity by promoting microvascular pathology, e.g., vasospasm, ischemia, and microthrombi (Petkova et al., 2001). Vascular mediators, in conjunction with cytokines induced by *T. cruzi* infection, may be significant to the development of microcirculatory abnormalities that are well documented in Chagasic hearts (reviewed in Mukherjee et al. (2003)).

Mitochondria Dysfunction and Oxidative Stress-Mediated Damage

Our studies suggest that Chagasic myocardium may be pre-disposed to sustained oxidative stress associated with mitochondrial dysfunction (reviewed in Garg (2005) and Zacks et al. (2005)). Cardiac tissue is extraordinarily dependent upon oxidative phosphorylation for energy to perform its functions. Accordingly, in cardiomyocytes, mitochondria are abundant. Mitochondria are the main producers of reactive oxygen species (ROS) in the heart and brain. Under normal conditions, NADH-ubiquinone oxidoreductase (CI) and ubiquinol-cytochrome c reductase (CIII) complexes of the respiratory chain release 2–4% of electrons to oxygen, resulting in superoxide (O_2^-) formation in mitochondria that is then converted to highly toxic hydroxyl ion (OH^-) and H_2O_2 by various reactions (Ide et al., 1999; Chen et al., 2003). An alteration in CI and CIII activities may result in increased ROS production in mitochondria (Ide et al., 1999; Chen et al., 2003), and is found in the myocardia of *T. cruzi*-infected mice (Vyatkina et al., 2004; Wen and Garg, 2004) and peripheral blood of seropositive Chagasic subjects (Wen et al., 2006b). Manganese superoxide dismutase (MnSOD) is the major O_2^- scavenger in mitochondria, and its decreased activity was noted during acute-to-chronic disease development in infected murine myocardium (Wen et al., 2004). It was, therefore, proposed that a functional decline in the respiratory chain, increased ROS generation, coupled with an inability to efficiently scavenge the mitochondrial free radicals, predisposes Chagasic hearts to sustained ROS during infection and disease development.

At low levels, ROS induces cytokine and chemokine production (Rahman and MacNee, 2000), thus providing a potential mechanistic link between mitochondrial ROS generation and acute and/or chronic inflammation in Chagasic cardiomyopathy. The cytotoxicity of sustained oxidative stress is related to ROS' ability to oxidize cell constituents, including proteins, lipids, and DNA, which lead to deterioration of cellular structure and function and, ultimately, to cell death (Tsutsui, 2001; Martindale and Holbrook, 2002). The sustained occurrence of oxidative damage, as evidenced by consistent presence of protein carbonylation and lipids peroxidation products, was shown in the myocardium of an experimental model of Chagas disease (Wen et al., 2004) and in peripheral blood of human Chagasic subjects (Wen et al., 2006b; de Oliveira et al., 2007). Oxidative damage during Chagas disease may not only be due to increased ROS formation but also exacerbated by inefficient glutathione antioxidant capacity (glutathione peroxidase-GSH-glutathione reductase), as is noted in infected mice (Wen et al., 2004) and human subjects (Perez-Fuentes et al., 2003). Subsequently, recent studies have suggested that combinatorial anti-parasite/antioxidant therapies would be useful in controlling Chagas disease pathology. Treatment of experimental animals with phenylbutylnitrone antioxidant (Wen et al., 2006a) and human Chagasic patients with vitamin C and vitamin E antioxidants (Macao et al., 2007) was shown to be effective in reducing the oxidative insult–associated pathology in Chagas disease.

In summary, mitochondria are targets of a variety of endogenous and exogenous insults, including inflammatory mediators, elicited by *T. cruzi* infection. Depending upon the extent and duration of mitochondrial dysfunction, oxidative stress may initiate or contribute to destruction of heart tissue and consequent dysfunction through sustenance of inflammatory responses and oxidative damage of the cardiac components.

Other Mechanisms

Less well explored is the potential for the *T. cruzi* strain or clonal variation to account for different clinical outcomes (Tibayrenc, 1998). Human genetic differences may also contribute to the dichotomy between indeterminate and chronic disease manifestations (Williams-Blangero et al., 2003), although no conclusive evidence for genetic differences is present to date.

Summary

We conclude that the contributions of parasite persistence, autoimmunity, vascular mediators, mitochondria dysfunction, oxidative stress, and potentially, parasite strain or clonal variation and host genetics are not mutually exclusive and could contribute in part and/or act at distinct time points during *T. cruzi* infection to initiate and sustain the observed, multifaceted cardiac pathology. The sum of these factors may also determine the degree of pathophysiology the infection may evoke and the severity of chronic disease. An important implication of these studies is that preventing infection or controlling the acute parasite load below a threshold level would be effective in decreasing the tissue damage imposed by multiple pathogenic mechanisms and lead to decreased disease severity. Obviously, these observations provided an impetus for vaccine development against *T. cruzi*, which is also favored by the fact that the effector mechanisms capable of controlling parasite burden were delineated.

VACCINE DEVELOPMENT AGAINST *T. CRUZI*

Given the current knowledge about the status of protective immunity from *T. cruzi* (see section Protective Immune Responses), a successful vaccine against the parasite is envisioned to elicit a long-term, lytic antibody response, type 1 cytokines, and CTLs. There are two stages of the parasite against which vaccines can be designed. Vaccines against trypomastigotes as they enter host cells following the bite of an infected triatomine, or the burst of an infected cell, will prevent the initiation or persistence of infection, and limit the parasitemia. Vaccines against intracellular replicative amastigotes would arrest the propagation of parasite in a host and prevent the parasite from entering the blood. Both types of vaccines would arrest or attenuate disease development in humans and the reservoir mammalian host. In addition, vaccines against either stage of the parasite would prevent triatomine infection and, thus, interrupt or reduce parasite transmission in both human and reservoir populations, as well as in insects.

Historical Perspective

Early efforts in developing prophylactic vaccines against *T. cruzi* tested killed parasite preparations as immunogens. Examples include *T. cruzi* killed by: (i) chemical treatment, e.g., thimerosal, formalin, glutaraldehyde; (ii) physical methods, e.g., pulverization, freezing and thawing, sonication, or pressure disruption; and (iii) irradiation. These antigenic preparations were tested for vaccine potential in various animal models, including mice, guinea pigs, dogs, and monkeys. Vaccination with chemically treated *T. cruzi* failed to provide any protection from lethal challenge infection. Other vaccination approaches generated a degree of resistance to *T. cruzi*; immunized animals controlled acute parasitemia and survived (70–100%) acute infection (Table 69.1). Similar efforts utilizing subcellular fractions, e.g., flagellar, soluble, or membranes, as vaccines also demonstrated an elicitation of partially protective immunity. For example, immunization with crude flagellar extracts enhanced resistance to *T. cruzi*, evidenced by a 90% survival rate of mice from challenge infection (Ruiz et al., 1986). Immunization of mice with *T. cruzi* soluble extract elicited IFN-γ and prevented death from an otherwise lethal infection (Garcia et al., 2000). Most of these approaches utilized the epimastigote, the insect stage of the parasite, that was later identified as expressing different antigens from those found in the infective and intracellular stages of *T. cruzi*. An absence of immunogenic proteins in epimastigotes, or a loss of protective epitopes during inactivation and fractionation, was believed to be the cause for the limited success met in these attempts in vaccine development.

The next series of efforts tested live vaccines, consisting of *T. cruzi* strains attenuated by treatment with drugs or pharmacological agents, or by serial passage in cultures, for their potency in experimental animals (mice and dogs). These vaccines were largely effective in controlling subsequent infections by virulent strains. Vaccinated animals exhibited decreased parasitemia and increased survival rates compared to unimmunized animals (Table 69.1). The danger of reversion of the attenuated strains to a virulent form and the likelihood of increased virulence of attenuated strains in immunocompromised individuals rendered these vaccines impractical. These studies, however, showed that a prophylactic vaccine capable of eliciting protective immunity with a minimal risk of biological reversion to virulent phenotype would be useful in controlling *T. cruzi* infection and disease, and provided a foundation for the identification of target antigens of the immune responses and the development of subunit vaccines.

TABLE 69.1 Traditional vaccination approaches against *T. cruzi*

Vaccination strategy	% Survival (dpi[a])	References
T. cruzi treated with		
Thimerosal	0	Kagan and Norman (1961)
Formalin	0	Hauschka (1949)
Glutaraldehyde	100 (60)	Basso et al. (2007)
Pulverization	100 (10–15)	Rego (1959)
Freezing and thawing	70 (120)	Basombrio (1990)
Sonication	0	Seneca and Peer (1966)
Pressure disruption	80–100 (120)	Gonzalez Cappa et al. (1974)
Irradiation	80–100	Okanla et al. (1982)
T. cruzi subcellular fractions		
Flagellar	90 (60)	Ruiz et al. (1986)
Soluble	85 (150)	Garcia et al. (2000)
Membranes	Not detected	Ruiz et al. (1985)
Live *T. cruzi* attenuated by		
Drug treatment		
Trypaflavine	100(30)	Collier (1931)
Actinomycin D	100 (13)	Fernandes et al. (1966)
Bayer 7602	100	Hauschka et al. (1950)
Primaquine	100	Pizzi and Prager (1952)
L-furaltodone	100 (11)	Brener and Chiari (1967)
Serial passage in culture	100 (13–77)	Menezes (1968, 1969)
	100 (30)	

[a]Experimental animals were observed for survival for *n* days postinfection.

Efforts Toward Identification of *T. cruzi* Antigenic Targets

The antigenic targets capable of eliciting the desired immune responses (see Protective Immune Response) at a level that confers protection from infection are the best choice for subunit vaccine development. Early studies employed random approaches, e.g., screening of expression libraries with immune sera from infected animals or human patients, screening with antigen-specific monoclonal antibodies, and amplification of the antigen-encoding genes (Ruiz et al., 1990; Malchiodi et al., 1993; Low and Tarleton, 1997) for the identification of putative vaccine candidates. These approaches, though successful in characterization of a variety of *T. cruzi* proteins, often led to the discovery of antigens capable of eliciting an antibody response only.

The diploid genome (106–110 Mb) of *T. cruzi*, distributed in 35–40 chromosomal bands, contains ~22,500 proteins encoded by genes (El-Sayed et al., 2005). The complexity of the *T. cruzi* genome has necessitated the development of logical strategies to determine which parasite proteins are the likely choice for immune activation. Garg et al. (Garg et al., 1997) employed a transfection system and expressed chicken ovalbumin (OVA), the classical immunogen, in different cellular compartments of *T. cruzi*. Antigen-presenting cells infected with *T. cruzi* expressing OVA as a GPI-anchored surface protein or as a secreted protein were recognized and lysed by the OVA-specific CTLs in vitro. Also, parasites expressing GPI-anchored or secreted OVA elicited the expansion of OVA-specific CTLs and antibodies in infected mice (Garg et al., 1997; Kumar and Tarleton, 2001). It was inferred that during the process of replication and/or differentiation, *T. cruzi* releases GPI-anchored and secretory proteins in host cell cytoplasm where they are degraded by proteosome enzymes. The resulting peptides that transported to endoplasmic reticulum are associated with MHC class I molecules, and displayed on the surface of infected cells, where they can be recognized by circulating $CD8^+$ T cells. On extracellular trypomastigotes and amastigotes, these same proteins expressed in a membrane-associated form, are engulfed and processed by phagocytic cells, displayed in association with MHC class II molecules, and can be recognized by $CD4^+$ T cells that provide help for the activation and proliferation of $CD8^+$ T and B cells. These studies concluded that GPI-anchored proteins, abundantly expressed in infective and intracellular stages of *T. cruzi*, and the secreted proteins are the most likely source of peptides for immune activation.

Accordingly, during the last 10 years, several surface proteins of *T. cruzi* have been identified, and their immunogenic potential examined. The extracellular and intracellular life cycle stages of *T. cruzi* adapt to sudden changes in the environment and survive in generally unfavorable conditions by changing the composition of surface glycoproteins. Most (but not all) of these surface proteins are attached to the plasma membrane by GPI anchor, their expression is developmentally regulated, and contain stage-specific modifications that ultimately reflect their functional importance in the life cycle and provide plausible targets for vaccine development.

We briefly describe the abundantly expressed and other surface proteins of *T. cruzi* and evidence that they are recognized as antigenic targets of immune responses elicited in infected experimental animals and humans.

Abundantly Expressed Surface Antigens

Many of the GPI-anchored surface proteins are encoded by genes belonging to large families, e.g., trans-sialidase (TS) super family (737 genes), mucins (662 genes), mucin-associated surface proteins (MASPs, 944 genes), and glycoprotein 63s (GP63s, 174 genes) (El-Sayed et al., 2005). These gene families are largely *T. cruzi*-specific and account for >18% of the total protein-encoding genome.

Trans-sialidases are a heterogenous group of GPI-anchored proteins, a majority of which are expressed in the trypomastigote stage. Most TS family members contain a conserved sialidase super family motif (VTVxNVxLYNR) but are not enzymatically active. The enzymatically active TS variants consist of 12-amino-acids long shed acute phase antigen (SAPA) repeats, are expressed in the trypomastigote stage, and catalyze the sialylation of parasite mucins through transfer of sialic acid from the host cell surface (Schenkman et al., 1994). This step is necessary for *T. cruzi* attachment and invasion of host cells, survival in and escape from the parasitophorous vacuole, and differentiation into a replicative amastigote form (Hall, 1993; Schenkman et al., 1994; Tan and Andrews, 2002). TC85, another member of the TS super family, binds to laminin and cytokeratin in vitro, and is hypothesized to facilitate movement of *T. cruzi* through the extracellular matrix, allowing it to traverse the tissues to invade other organs (Magdesian et al., 2001).

The substantial diversity of the TS gene family has likely developed in response to host immune pressure. Accordingly, several of the TS family members are identified to be targets of humoral and cell-mediated immune responses in Chagasic patients and experimental animals. A trypomastigote surface antigen, TSA1 (85 kDa), was the first *T. cruzi* protein shown to be the target of $CD8^+$ T lymphocytes in infected mice (Wizel et al., 1997) and humans (Wizel et al., 1998b). TSA1 peptide-specific $CD8^+$ T cell lines were cytotoxic against peptide-sensitized or infected cells, secreted IFN-γ and TNF-α, and were able, upon adoptive transfer, to confer substantial protection against challenge infection in mice (Wizel et al., 1997). Antibodies to enzymatically active TS (120–200 kDa) have been detected in the sera of Chagasic patients and experimental animals infected with *T. cruzi* (Pereira-Chioccola et al., 1994). The TS catalytic domain was identified to contain epitopes recognized by IFN-γ-producing type 1 cells and antibodies in Chagasic patients (Ribeirao et al., 2000).

Amastigote stage-specific proteins of TS family, i.e., ASP1 (Santos et al., 1997), ASP2 (Low and Tarleton, 1997), and ASP9 (Boscardin et al., 2003) have been identified, although their precise role is presently unknown. ASP1 (78 kDa) and ASP2 (83 kDa) were shown to be the targets of in vivo (mice)-generated CTLs that were parasite- and peptide-specific, MHC-restricted, and CD8-dependent (Low et al., 1998). Human circulating CTLs specific for ASP1- and ASP2-derived peptides were detected in HLA-A2$^+$ *T. cruzi*-infected patients (Wizel et al., 1998b).

Mucins contain a large proportion (85%) of *O*-glycosidic-linked carbohydrates. Mucins of 35–50 kDa are present in insect stage epimastigotes and metacyclic trypomastigotes, whereas larger (80–200 kDa) mucin proteins are expressed in bloodform trypomastigotes. The large heterogeneity of mucins is due to the presence of different core proteins along with the complexity and extent of oligosaccharide side chains (Schenkman et al., 1993). Mucins serve as acceptors of sialic acid in a trans-sialylation reaction catalyzed by TS and are essential for parasite protection and survival in a mammalian host (Schenkman et al., 1993; Almeida et al., 1999). *T. cruzi* mucins, such as GP35/50 and SSP3, expressed in metacyclic and trypomastigote stages, respectively, have been ascribed an analogous function in parasite ligand–host receptor binding (Schenkman et al., 1993). The GPI component of mucins is a powerful inducer of polyclonal B cells, inflammatory cytokines, and macrophages (Acosta-Serrano et al., 2001). The differences in the structure, composition, and abundance of the mucin proteins may contribute to infectivity, virulence, and tissue tropism of *T. cruzi* or constitute an immune-evasion mechanism, but this has not yet been experimentally proven.

MASP family members consist of chimeric domains, i.e., an N- or C-terminal conserved domain combined with the N- or C-terminal domain of mucin or the C-terminal domain from the TS super family (El-Sayed et al., 2005). The mechanism of generation of such chimeric MASPs, their functions in parasite life cycle or immunogenic potential are unknown.

GP63 family members were recently identified in *T. cruzi*. These are homologs of *Leishmania major* surface protease, GP63, and may serve analogous function as a surface protease (Cuevas et al., 2003). The immunologic potential of *T. cruzi* GP63s has not been examined to date.

Other Immunogenic Proteins

Complement regulatory protein (CRP), the 160 kDa surface antigen of trypomastigotes was named so because of its ability to protect *T. cruzi* from complement-mediated lysis (Norris et al., 1989). Antibodies to CRP have been detected in sera of human Chagasic patients (Norris et al., 1994).

Cruzipain, the major lysosomal cysteine proteinase (60 kDa) of *T. cruzi*, is encoded by multiple polymorphic tandemly organized genes located on different chromosomes (Campetella et al., 1992). Cruzipain consists of a catalytic moiety with high homology to cathepsins S and L, and a C-terminal domain, characteristic of type I cysteine proteinases of trypanosomatids (Cazzulo et al., 2001). The proteolytic function of cruzipain is essential for parasite differentiation and invasion of mammalian cells (Cazzulo et al., 2001). Irreversible inhibitors of cruzipain blocked parasite development, exhibited trypanocidal activity, and are currently examined for their therapeutic utility against *T. cruzi* (Santos et al., 2005). Cruzipain-derived epitopes were recognized by antibodies and IFN-γ-producing, CD8$^+$ immune cells in Chagasic patients, signifying its immunogenicity (Fonseca et al., 2005).

FCaBP, a 24 kDa flagellar calcium-binding protein, is expressed in all life cycle stages of *T. cruzi*. The N-terminal 24 amino acids, modified by palmitoylation and myristoylation, facilitate the flagellar localization of FCaBP (Godsel and Engman, 1999). Immune sera from *T. cruzi*-infected experimental animals and humans exhibit highly specific, sensitive reactivity with FCaBP (Godsel et al., 1995). FCaBP-specific CD8$^+$ T-cell are generated in mice infected with *T. cruzi* (Fralish and Tarleton, 2003).

GP90 (90 kDa) and GP82 (82 kDa), expressed on a plasma membrane of metacyclic trypomastigotes, interact with distinct host cell receptors. Attachment of the GP82-host cell receptor triggers a Ca^{2+} response in the target cell as well as in the parasite, an event that enables parasites to gain entry to the host cell (Yoshida et al., 2000). Binding of GP90 to its receptor failed to induce a Ca^{2+} response in parasites as well as host cells (Ruiz et al., 1998). Reduction in GP90 expression on the parasite surface resulted in increased infectivity, leading to a suggestion that GP90 may modulate parasite invasion by altering GP82-host receptor interactions (Malaga and Yoshida, 2001).

GP82-specific antibodies cross-reacted with heterologous antigens (GP90 and GP30/55), while antisera to GP90 reacted only with the homologous antigen (Yoshida et al., 1993). GP90 is specifically recognized by antisera from Chagasic, but not leishmaniasis, patients (Schechter et al., 1983), thus, suggesting its diagnostic importance. Immunization with GP82 or GP90 stimulated *T. cruzi*-specific CD4$^+$ T cell proliferation in lymph nodes of mice, signifying immunogenicity of the two glycoproteins (Yoshida et al., 1993).

A 11 kDa, cytoskeleton-associated, kinetoplastid membrane protein, KMP11, is highly conserved among trypanosomes (Thomas et al., 2000), though the biological role it plays remains unclear. The immunologic role of KMP11 was evidenced by the detection of KMP11-specific antibodies in sera of chronically infected patients (Trujillo et al., 1999). Spleen cells from KMP11-immunized C57BL/6$^{A2/Kb}$ transgenic mice, when incubated with an immunodominant A2-specific peptide from KMP11, lysed antigen, or peptide-pulsed Jurkat$^{A2/Kb}$ cells, suggesting the presence of a human CTL epitope in KMP11 (Maranon et al., 2001).

LYT1 (61 kDa), a secreted antigen, is expressed in all parasite life cycle stages. Genetic deletion studies showed that LYT1 is not required for viability of epimastigotes, but is essential for parasite development in mammalian cells. *LYT1*-deficient mutants exhibited accelerated in vitro development, reduced infectivity, and diminished hemolytic activity that was restored upon complementation with episomally expressed LYT1 (Manning-Cela et al., 2001). LYT1-derived H-2K^b peptides were presented in association with MHC I molecules. Antigen-specific CTLs generated in *T. cruzi*-infected mice showed high levels of lytic activity against peptide-pulsed target cells, thus signifying the immunogenicity of LYT1 (Fralish and Tarleton, 2003).

Paraflagellar rod proteins (PFR1-4, MW ~70 kDa; PFR5-6, 78–86 kDa) are highly conserved among different *T. cruzi* strains of clinical importance (Fouts et al., 1998; Clark et al., 2005). PFRs constitute a major structural component of flagellum and are critical for cell motility (Fouts et al., 1998; Clark et al., 2005). PFRs are highly immunogenic. PFR-specific CD8$^+$ T cells were elicited in infected mice and showed high levels of lytic activity against peptide-pulsed target cells and secreted IFN-γ in response to parasite-infected target cells (Wrightsman et al., 2002). PFRs are recognized by sera antibodies from indeterminate and chronic Chagasic patients (Michailowsky et al., 2003). Incubation of recombinant PFRs with PBMCs of Chagasic patients induced proliferation of antigen-specific, IFN-γ producing CD4$^+$ and CD8$^+$ lymphocytes (Michailowsky et al., 2003).

TC52, a 52 kDa trypanothione: glutathione disulfide thioltransferase enzyme, is essential for maintaining the intracellular thiol-disulphide redox balance in *T. cruzi* (Moutiez et al., 1997). *T. cruzi* heterozygous mutants deleted of one *TC52* allele exhibited low virulence and caused an attenuated form of Chagas disease in experimental animals. Deletion of both *TC52* alleles was lethal, indicating the importance of TC52 in parasite survival (Garzon et al., 2003). This enzyme is specific for the parasite (Moutiez et al., 1997) and targeted for therapeutics development. While TC52 was found to be immunologically silent in acutely infected animals, the immunomodulatory role of TC52 was suggested by the fact that recombinant TC52, in synergy with IFN-γ, stimulated gene expression for iNOS and IL-12, and NO production in ϕs (Fernandez-Gomez et al., 1998). A TC52-specific immune response appeared late in *T. cruzi* infection and may play a role in the modulation of its biological function(s) (Fernandez-Gomez et al., 1998).

The continuing progress in *T. cruzi* genomics, proteomics, and bioinformatics, followed by wet-lab testing of several genes in experimental models, is likely to further expand the pool of antigenic targets of *T. cruzi*.

Subunit Vaccines in Development (over the Last 10 Years)

A Brief Overview

Armed with data on protective immune mechanisms and their antigenic targets, researchers interested in vaccine development against *T. cruzi* were required to identify effective antigen delivery systems to elicit potent immune responses to the vaccine candidates. Initial studies utilized a protein immunization approach to test the vaccine potential of *T. cruzi* antigens. Most of these studies were designed such that the protective efficacy of putative vaccine candidates was determined based upon survival following a lethal challenge infection. Efforts were made to enhance the protective efficacy of protein vaccines by codelivery of adjuvants, use of alternative routes of antigen delivery, and by increasing the amount or the number of doses.

In subsequent studies, a DNA immunization approach was favored due to the ease of construction and production of the vectors, the stability of DNA, and the ability to enhance the immune response by the codelivery of genes encoding cytokines. Most importantly, this method of antigen delivery has proved to be efficient in eliciting antibodies, Th1 cytokines, and CD8$^+$ T cell immune responses to encoded antigens. Successful induction of humoral and/or cellular immune responses to the plasmid-encoded antigens using various routes of gene delivery has been shown to provide partial or complete protection against numerous infectious agents (reviewed in Donnelly et al. (2005)). *T. cruzi* proteins identified as targets of specific antibodies and CTLs in infected mice and humans (see section Efforts Toward Identification of *T. cruzi* Antigenic Targets) were tested as DNA vaccines in experimental models. Considering that complete genes were incorporated into the DNA vectors, epitopes capable of being presented by many MHC alleles were expected to be present. Accordingly, several of the genes were tested and afforded protective immunity in multiple mouse strains.

In parallel with the efforts toward identification of vaccine candidates, adjuvants were tested to enhance or skew the immune responses toward desirable Th1 type. Use of adjuvants to increase protective immunity against *T. cruzi* dates back to 1965 (Menezes, 1965) when saponin, a derivative from the bark of *Quillaja saponaria (Quil A)*, was injected to enhance the protective efficacy of immunogens. Co-inoculation of saponin with freeze-thaw-inactivated parasites stimulated a Th1 type immune response, and slightly increased protection from challenge infection (Johnson et al., 1963). Recent studies have examined the utility of IL-12, GM-CSF, CD40, HSP70, and CpG oligodeoxynucleotides in enhancing the Th1 responses to defined antigen vaccines. GM-CSF was chosen as a genetic adjuvant because it is a potent cytokine capable of enhancing the antigen presentation capability of antigen-presenting cells, such as dendritic cells. In addition, it facilitates B- and T-cell-mediated immunity (Warren and Weiner, 2000). IL-12 is a key cytokine involved in $CD8^+$ T cell activation and proliferation, and in directing the immune responses to type 1 (Pan et al., 1999). Similarly, other adjuvants were chosen for their ability to skew the immune response to a protective Th1 type through different mechanisms. We discuss the efforts to date focused on testing and enhancing the protective efficacy of subunit vaccines against *T. cruzi* (Table 69.2).

Subunit Protein Vaccines

Initial studies examined the vaccine potential of GPI-proteins of *T. cruzi*, selected for their abundant expression and immunogenicity (see section Efforts Toward Identification of *T. cruzi* Antigenic Targets).

GP90 Though recognized as an antigenic target in infected mice and humans, GP90 proved to be poorly immunogenic as a vaccine. Specific antibody and cell-mediated immunity were potentiated when GP90 was codelivered with saponin adjuvant, and these immune responses were effective against a range of *T. cruzi* strains. Further, the levels of acute tissue damage, as measured by production of auto antitissue immunoglobulins, were significantly reduced in immunized mice and marmosets (Scott et al., 1985). All animals, immunized and nonimmunized, remained xenodiagnosis-positive (tested at 60 weeks postinfection), which suggested sterile immunity was not achieved.

GP82 Mice immunized with a C-terminal peptide of GP82 (224–516 amino acids) plus alum elicited specific antibodies that were not protective; these antibodies lacked agglutinating or complement-dependent cytotoxicity and failed to neutralize parasite infectivity. The substantial resistance exhibited by immunized mice to acute infection was correlated with intense proliferation in the spleen of IFN-γ-producing $CD4^+$ T cells (Santori et al., 1996).

GP56 Immunization of mice with purified GP56, followed by challenge infection, provided increased survival by only 4 days compared to those not immunized before infection. Considering an epimastigote stage protein that is not expressed by infective trypomastigote and intracellular stages, limited protection afforded by GP56 was not surprising (Harth et al., 1994).

Cruzipain Cruzipain-immunized mice exhibited delayed death upon lethal challenge infection (Laderach et al., 1996). Efforts to enhance the protective efficacy of cruzipain utilized codelivery of IL-12 (plus neutralizing anti-IL-4 antibody) (Schnapp et al., 2002) or CpGODN (Frank et al., 2003), selected for their capacity to drive immunity toward a Th1 bias. Immunization of mice by either of these vaccines elicited high specific antibody titers. Spleen cells from immunized mice, when stimulated in vitro with antigen or infected ϕs, strongly proliferated and produced high levels of IL-2 and/or IFN-γ cytokines. All immunized mice exhibited substantial resistance to *T. cruzi*, evidenced by low parasitemia and 80% survival to acute infection. While cruzipain-specific Th1 responses were associated with protective immunity to *T. cruzi* in vitro and in vivo, adoptively transferred cruzipain-specific T cells failed to confer protection against challenge infection in mice, indicating that additional immune mechanisms were important to cruzipain-specific immunity. Recombinant attenuated *Salmonella typhimurium* serovar expressing cruzipain served to elicit immunity to mucosal *T. cruzi* infection in mice.

PFR The protective efficacy of PFR proteins has been tested in detail. First, PFRs purified from *T. cruzi* were shown to protect mice against lethal challenge infection that was markedly governed by an antigen delivery route. The subcutaneous route of antigen delivery elicited protective immunity, while intraperitoneal injections elicited a strong, nonprotective antibody response (Wrightsman et al., 1995). Similar observations were made with recombinant PFRs. The subcutaneous, but not intraperitoneal, administration of PFR1 and PFR2 (individually or in combination) with Freund or alum adjuvant provided significant protection in 100% and 83% mice, respectively, against

TABLE 69.2 Subunit vaccines against *T. cruzi*

Antigen	Adjuvant used	Experimental model mouse (*T. cruzi*) strain	% Survival (dpi[a])	References
Protein vaccine				
GP90	Saponin	CBA, Marmoset (Y)	60 (100)[c]	Scott et al. (1985)
GP82	Alum	Balb/c (CL)	Not detected[c]	Santori et al. (1996)
GP56	Freund adjuvant	Swiss-Webster (Y)	40 (12)	Harth et al. (1994)
Cruzipain[b]	IL-12, CpGODN	C3H/HeN (RA) Balb/c (Tulahuen)	67–80 (60–100)[c]	Laderach et al. (1996), Schnapp et al. (2002), Frank et al. (2003)
PFR1, PFR2[b]	Alum, Freund, IL-12	C57BL/6, Balb/c (Peru)	83–100 (30–60)[c]	Miller et al. (1996), Wrightsman and Manning (2000), Wrightsman et al. (2002), Luhrs et al. (2003)
TC52	Alum, *Bordtella pertusis*	Balb/c (Y)	62 (120)[c]	Ouaissi et al. (2002)
CRP	Freund adjuvant	Balb/c (Y)	10 (40)[e]	Sepulveda et al. (2000)
ASP2	Alum, CpGODN	A/Sn (Y)	53.3 (60)[d]	Araujo et al. (2005)
DNA vaccine				
CRP	None	Balb/c (Y)	100 (40)[c]	Sepulveda et al. (2000)
TSA1[b]	IL-12 + GM-CSF	Balb/c, C3H/HeSnJ, C57BL/6 (Brazil)	60 (140)[d]	Wizel et al. (1998a), Garg and Tarleton (2002)
ASP1[b]	IL-12 + GM-CSF	C3H/HeSnJ, C57BL/6 (Brazil)	<60 (140)[d]	Garg and Tarleton (2002)
ASP2[b]	IL-12 + GM-CSF	C3H/HeSnJ, C57BL/6 (Brazil)	80 (140)[c,f]	Garg and Tarleton (2002)
ASP2	None	A/Sn (Y)	63 (60)[d,f]	Vasconcelos et al. (2004)
ASP9	None	Balb/c (Y)	100 (60)[c]	Boscardin et al. (2003)
TS[b]	None	Balb/c (Y)	100 (50)[c,f]	Costa et al. (1998)
TSSA[b]	IL-12, RANK-L	Balb/c, C3H/HeJ, C57BL/6, B6 (Tulahuen)	80–100 (40)[c]	Katae et al. (2002), Miyahira et al. (2003)
KMP11	HSP-70	Balb/c (Y)	50 (70)[c]	Planelles et al. (2001)
LYT1	IL-12 + GM-CSF	C57BL/6 (Brazil)	80 (75)	Fralish and Tarleton (2003)
FCaBP	IL-12 + GM-CSF	C57BL/6 (Brazil)	0 (75)	Fralish and Tarleton (2003)
TCβ3	IL-12 + GM-CSF	C57BL/6 (Brazil)	0 (75)	Fralish and Tarleton (2003)
PFR2	HSP70	Balb/c (Y)	100 (35)[c,f]	Morell et al. (2006)
PFR3	HSP70	Balb/c (Y)	100 (35)[c]	Morell et al. (2006)
ASP1 + ASP2 + TSA1	IL-12 + GM-CSF	C3H/HeSnJ, C57BL/6 (Brazil)	83 (140)[c]	Garg and Tarleton (2002)
TS family members	None	C57BL/6 (Brazil)	75 (75)	Fralish and Tarleton (2003)
Mucin family members	None	C57BL/6 (Brazil)	25 (75)	Fralish and Tarleton (2003)
ASP2 + TSA1	None	A/Sn (Y)	86 (60)[c,f]	Vasconcelos et al. (2004)
Immunotherapeutic DNA vaccine				
TSA1[b]	None	Balb/c (H4), acute model	70 (45)[c,f]	Dumonteil et al. (2004)
TC24[b]	None	Balb/c (H4), acute model	100 (45)[c,f]	Dumonteil et al. (2004)
TSA1	None	CD1 (H1), chronic model	100 (140)[f]	Dumonteil et al. (2004)
TSA1	None	ICR (H1), acute model	Not detected[c,f]	Zapata-Estrella et al. (2006)
TSA1	None	ICR (H1), chronic model	Not detected[c,f]	Zapata-Estrella et al. (2006)
ASP9	None	ICR (H1), acute model	50 (50)[e]	Sanchez-Burgos et al. (2007)
TS	None	ICR (H1), acute model	50 (50)[e]	Sanchez-Burgos et al. (2007)
TSA1	None	ICR (H1), acute model	70 (50)[c,f]	Sanchez-Burgos et al. (2007)
Tc52	None	ICR (H1), acute model	75 (50)[c]	Sanchez-Burgos et al. (2007)
Tc24	None	ICR (H1), acute model	85 (50)[c,f]	Sanchez-Burgos et al. (2007)
Recombinant virus vaccine				
Ad-TSSA/MVA-TSSA	MVA-RANK-L	C57BL/6 (Tulahuen)	100 (50)[c]	Miyahira et al. (2005)
rAD-ASP2	None	Balb/c (Y)	80 (160)[c]	Machado et al. (2006)
rAD-TS	None	Balb/c (Y)	50 (160)[d]	Machado et al. (2006)
rAD-ASP2 + rAD-TS	None	Balb/c (Y)	100 (160)[c]	Machado et al. (2006)

[a] Experimental animals were observed for survival for *n* days postinfection.
[b] These antigens were shown to provide variable degree of protection in different mouse strains (data presented is from the animal model that exhibited best protection).
[c] Upon challenge infection, immunized animals exhibited very low (≤10%) parasitemia as detected in unimmunized/infected animals (data presented is from the animal model that exhibited best protection).
[d] Upon challenge infection, immunized animals exhibited moderate (~50%) parasitemia as detected in unimmunized/infected animals (data presented is from the animal model that exhibited best protection).
[e] Upon challenge infection, immunized animals exhibited similar parasitemia as detected in unimmunized/infected animals (data presented is from the animal model that exhibited best protection).
[f] Immunization with these antigens was effective in decreasing the severity of chronic disease, evaluated by histopathological analysis of cardiac tissue biopsies.

a lethal *T. cruzi* challenge. PFRs alone or delivered with other adjuvants (QS-21, Ribi-700, IL-12), irrespective of the route of antigen delivery, failed to control *T. cruzi* infection (Miller et al., 1996). Protection afforded by PFRs plus Freunds or alum adjuvants was associated with high levels of IFN-γ and IL-2 and low levels of IL-4 production in immunized mice. Using knockout mice depleted in specific immune functions, it was shown that PFR-induced protective immunity was MHC I- and $CD8^+$ T cell-dependent, and B cell responses were not necessary for resistance to *T. cruzi* infection (Miller et al., 1997). The function of $CD4^+$ T cells in PFR-immunized mice was associated with parasite clearance, as was also evidenced by the fact that these $CD4^+$ T cells released IFN-γ and stimulated NO production in infected ϕs, essential for killing of *T. cruzi* (Miller et al., 1997). Others showed that protection against *T. cruzi* in PFR-immunized mice was provided by induction of a highly polarized type 1 cytokine profile (Luhrs et al., 2003).

TC52 TC52 recombinant protein–elicited protection from infection in immunized mice was associated with TC52-mediated: (i) alleviation of immunosuppression otherwise presented by acute *T. cruzi* infection, and (ii) enhanced maturation of dendritic cells that play a central role in initiation of Th1 immunity (Ouaissi et al., 2002). The activation/maturation of dendritic cells from immunized mice was evidenced by increased CD83 and CD86 expression, inflammatory chemokines (IL-8, MCP-1, and MIP-1α) production, and presentation of potent co-stimulatory properties. Codelivery of alum or *Bordtella pertusis* adjuvants enhanced the protective efficacy of TC52 in immunized/infected mice.

CRP Immunization of mice with either the protein or DNA CRP vaccine elicited a Th1 type T cell response, comparable antibody titers, and similar immunoglobulin G isotype profiles. Only mice immunized with CRP-encoding plasmid produced antibodies that exhibited complement-dependent cytotoxicity to parasites and were protected against a lethal challenge infection with *T. cruzi*; thus suggesting the superiority of DNA immunization over protein immunization with recombinant CRP (Sepulveda et al., 2000).

ASP2 Protective efficacy of ASP2 was reported recently. Immunization of A/Sn mice with ASP2 recombinant protein alone or with *ASP2* DNA followed by ASP-2 recombinant protein resulted in 53% and 75 % survival, respectively (Araujo et al., 2005). Six ASP2-recombinant peptides, representing different segments of the ASP2 protein were tested for their efficacy as immunogen with alum and CPGODN 1826 adjuvants. ASP2 peptide consisting of the $CD8^+$ epitope provided 100% survival from challenge infection, indicating the importance of a $CD8^+$ T cell response in eliciting protection against *T. cruzi* infection (Araujo et al., 2005).

Subunit DNA Vaccines

Trans-Sialidase Genes The vaccine potential of several members of the TS gene family has been examined by the authors and others. *TSA1*, to our knowledge, was the first subunit DNA vaccine tested against *T. cruzi*. Intramuscular immunization of *TSA1* in C57BL/6 (H-2^b) and Balb/c (H-2^d) mice generated efficient antiparasite antibody responses and primed CTLs that lysed antigen-presenting cells in an antigen-specific, $CD8^+$ T cell-dependent manner. When challenged with *T. cruzi*, immunized mice (64% C57BL/6 and 89% Balb/c) survived lethal infection (Wizel et al., 1998a).

We have evaluated the protective efficacy of *ASP1* and *ASP2* DNA vaccines. C57BL/6 mice immunized with ASP1- or ASP2-encoding plasmids elicited protective levels of immune responses, evidenced by decreased parasitemia and 50% and 80% survival rates, respectively, from lethal infection (Garg and Tarleton, 2002). Codelivery of *IL-12*- and *GM-CSF*-encoding plasmids with *ASP1*- or *ASP2*-expression vectors resulted in splenomegaly, an indicator of the increase in lymphocyte activation and enhanced immune reactivity. In addition, an increase in the induction of immune responses, i.e., antigen-specific CTL activity, parasite-specific humoral immune responses, and secretion of type 1 cytokines (IFN-γ), correlated with improved resistance to challenge infection, was observed following co-administration of cytokines with *T. cruzi* antigen-encoding plasmids. Importantly, mice immunized with *ASP2* plus cytokine adjuvants exhibited better control of chronic inflammation and pathological lesions. This was the first observation to demonstrate that vaccination strategies capable of reducing the parasite burden below a threshold level will be successful in controlling the severity of chronic disease and provided a foundation for testing the vaccine potential of other *T. cruzi* antigenic targets as DNA vaccines.

Others have reported the protective efficacy of *ASP2* in A/Sn mice, which are highly susceptible to *T. cruzi*. Mice immunized with ASP2-encoding plasmid exhibited moderate parasitemia and minimal inflammatory foci in the heart and striated muscle. Subsequently, 63% of the immunized mice survived as compared to controls that succumbed to challenge infection. Protective immunity was associated with increased secretion of IFN-γ by spleen cells and the presence of ASP-2-specific $CD8^+$ T cells (Vasconcelos et al., 2004).

The immunogenic potential of a *TS* gene encoding the enzymatically active trans-sialidase was demonstrated by Costa et al. (1998). Mice immunized with *TS*-expressing plasmid were protected from challenge infection. The protective efficacy achieved by a DNA–DNA *TS* vaccine was not increased by DNA-recombinant protein vaccination approach (Vasconcelos et al., 2003). Along with the protection afforded by T-cell-mediated immune responses, a strong antibody response that was elicited in immunized mice contributed to protection from infection through inhibition of the enzymatic activity of the native enzyme. It was proposed that inhibition of TS activity prevented the sialylation of parasite surface mucins, a step essential for attachment, invasion, and survival of *T. cruzi* (details in section Abundantly Expressed Surface Antigens).

TSSA, another TS surface antigen, when delivered as a DNA vaccine in mice, elicited $CD4^+$ and $CD8^+$ T cell-dependent protective immunity against *T. cruzi*. Codelivery of IL-12 DNA improved the antigen-specific CD8+ T cell activity and vaccine efficacy of *TSSA*, as evidenced by the efficient control of parasitemia and tissue inflammation, and increased survival from infection in immunized mice (Katae et al., 2002). RANK-L, the ligand to receptor activator of NFκB, is implicated in CD40-CD40L-independent T cell priming by dendritic cells. Similar to IL-12, the co-administration of RANK-L-encoding gene enhanced the $CD8^+$ T cell-dependent vaccine efficacy of TSSA in mice (Miyahira et al., 2003).

KMP11, nonimmunogenic itself, was expressed in fusion with heat shock protein 70 (HSP70) that has adjuvant properties (Qazi et al., 2005). Murine immunization with *KMP11-HSP70* elicited antigen-specific, long-lasting IgG2a antibody response and $CD8^+$ CTL lytic activity, and subsequently, better control of parasitemia and 50% survival from lethal infection (Planelles et al., 2001).

A LYT1-encoding gene delivered with IL-12 and GM-CSF elicited antigen-specific $CD8^+$ CTLs and resulted in an 80% survival rate following lethal challenge infection (Fralish and Tarleton, 2003).

PFR2 and PFR3 were expressed in fusion with HSP70 for its adjuvant properties, and delivered to chronically infected mice. Immunization with PFR2-HSP70 and PFR3-HSP70 induced high levels of PFR-specific IgG2a antibodies. However, only PFR2-HSP70 elicited a statistically significant CTL response that was sufficient to provide protection against *T. cruzi* infection (Morell et al., 2006).

Not all of the antigenic targets were found to be useful as DNA vaccine candidates. For example, genes encoding FCaBP, TCβ3, mucins, and CCL4/MIP-1β failed to provide any protection from *T. cruzi* infection (Katae et al., 2002; Fralish and Tarleton, 2003; Roffe et al., 2006). It is interesting that FCaBP and TCβ3 were recognized by CTLs in infected mice and elicited a cell-mediated immune response in mice when delivered as a DNA vaccine, implying that mere elicitation of $CD8^+$ T cell responses by an antigen is not indicative of its vaccine potential. Similarly, all of the genes that provided protection in one mouse strain were not protective in other inbred strains of mice. Examples include *ASP1, ASP2, TSA1* that were tested individually or in combination (Garg and Tarleton, 1998), and *TSSA* (Katae et al., 2002). All of these genes failed to alter susceptibility of C3H mice to *T. cruzi* infection, suggesting host genetic restriction may also contribute to inefficacy of vaccine candidates.

Multicomponent Subunit Vaccines

Considering that an increase in the level and diversity of *T. cruzi*-specific immune responses could enhance the protective capacity of DNA vaccines, some investigators tested immunization with multiple genes encoding members of the TS or mucin families. The ability of the mixed genes to elicit protective immune responses depended upon: (i) the amount of a given plasmid sufficient to elicit protective responses, and (ii) the total amount of DNA that can be injected without toxicity. Various dilutions of TS family members (*ASP1, ASP2*, and *TSA1*) (0.001–33 μg/mouse) were tested and showed that codelivery of as little as 1 ng of each gene (plus IL-12 and GM-CSF adjuvants) resulted in the activation of a substantial, antigen-specific CTL response, while 10 ng of each DNA was needed to induce moderate levels of a parasite-specific antibody response (Garg and Tarleton, 1998). The level of resistance to *T. cruzi* infection correlated with the amount of DNA delivered, the maximal protection being obtained with 1 μg of each vaccine DNA plus cytokines. An inhibitory effect on the elicitation of antigen-specific immune responses was not observed when mice were immunized with the mixture of plasmids, indicating that multiple genes with or without cytokine adjuvants can be used in developing immunization strategies for control of *T. cruzi* infection. The level of protection from *T. cruzi* infection induced in mice immunized with mixture of TS family members (Garg and Tarleton, 2002; Fralish and Tarleton, 2003) was, however, not significantly better than that induced in mice immunized with individual family members (Table 69.2). Attempts to enhance the protective efficacy of TS family members by a DNA prime-protein boost approach were also not successful (Vasconcelos et al., 2003). No protection was observed in mice immunized with a pool of genes encoding mucin family members (Fralish and Tarleton, 2003). The findings of either no or little induction of protective immunity with a cocktail of antigen-encoding vectors was attributed to the fact that genes of large families may express shared epitopes that do not present any protective benefits in inbred mice. It was anticipated that the potential synergistic immunologic benefit of a combination of epitopes from multiple genes would induce a higher frequency of immune effectors in heterogeneous host populations and provide effective immunity against diverse parasite strains, both of which would likely be verified in future studies.

Genome-Based Vaccines

By using the sequence database of *T. cruzi* (El-Sayed et al., 2005), there is the potential to conduct a large-scale, unbiased screening of the *T. cruzi* genome for the identification of genes of interest. Sophisticated bioinformatics programs are designed to evaluate gene functions on the basis of homologies to genes characterized in other organisms and the presence of motifs predictive of targeting, cellular localization, surface expression, and functional characteristics of the gene product. Such programs have the ability to circumvent the time-consuming, laborious experimental techniques and allow us to directly proceed from sequence information to antigenic target identification and vaccine design. Web-based bioinformatics tools coupled with an experimental strategy have been employed for the identification of putative genes encoding GPI-anchored or secreted proteins in a *T. cruzi*-expressed sequence tag (EST) database. Molecular and biochemical characterization of eight of the sequences selected by this approach demonstrated that the encoded proteins were conserved in the genome of *T. cruzi* strains of clinical importance and expressed as surface proteins during different developmental stages of the parasite. When delivered as a DNA vaccine in mice, the selected antigens elicited a trypanolytic antibody response that was in agreement with the intensity of the surface expression of these proteins in infective and intracellular stages of the parasite (Bhatia et al., 2004). This study validated the hypothesis that a *T. cruzi* sequence database committed to appropriate screening strategies would be an efficient resource for the identification of potential vaccine candidates.

Therapeutic Vaccines

No approved postexposure immunoprophylactic or therapeutic treatment exists for *T. cruzi*. Several studies have shown improved survival and preservation of cardiac structure in acutely or chronically infected mice treated with TSA1-, Tc52-, and TC24-encoding genes (Dumonteil et al., 2004; Sanchez-Burgos et al., 2007) that was associated with vaccine-induced rapid increase in the number of $CD4^+$ and $CD8^+$ T cells (Zapata-Estrella et al., 2006). Conversely, treatment of acutely infected mice with ASP2-, and TS-encoding plasmids was not effective in controlling parasetimia or prolonging survival (Sanchez-Burgos et al., 2007; Dumonteil, 2007).

Recombinant Virus Vaccines

Replication-deficient recombinant human viruses have an unprecedented ability to induce strong Th1 type immune response (Rocha et al., 2004). Several studies have been performed to evaluate the ability of the recombinant viruses encoding *T. cruzi* genes in inducing a long-lasting and protective immunity against *T. cruzi* infection in experimental models. Priming of immune response in mice with adenovirus encoding a single $CD8^+$-T-cell epitope derived from *TSSA* antigen, followed by boosting with vaccinia virus encoding the same $CD8^+$-T-cell epitope along with vaccinia virus encoding RANK-L as adjuvant provided significant protection against lethal *T. cruzi* infection (Miyahira et al., 2005) (Table 69.2). Further, recombinant adenoviruses encoding *ASP2* and *TS* antigens, alone or in combination, elicited strong antibody and T-cell responses and provided high level of protection against lethal *T. cruzi* challenge in mice (Machado et al., 2006). These studies have opened another arena for developing and testing the potential of *T. cruzi* vaccines.

Summary

Taken together, the above studies established the immunogenic potential of several of the parasite surface antigens and provided evidence for the usefulness of a DNA immunization approach in eliciting protective immune responses against *T. cruzi*. Unfortunately, all of the tested antigen(s) that conferred protection for a short term (2–4 weeks after vaccination) failed to confer long-term immunity. Further, all of the defined antigens tested so far, individually or in combination, with or without adjuvants, failed to produce sterile immunity and/or prevent death from infection in 100% of the vaccinated animals. It is surmised that the immune responses elicited by the tested vaccine candidates were either below the threshold required for generation of sterile immunity or were not rapid enough in their kinetics to control *T. cruzi* efficiently during the acute phase of infection. The ability of parasites to utilize multiple immune-evasion mechanisms may also contribute to *T. cruzi* survival in the immunized host.

PROSPECTS FOR THE FUTURE

Studies in experimental models have delineated the effector mechanisms that are essential to provide resistance to *T. cruzi* infection. Several defined vaccine candidates are known to elicit partially protective immunity against *T. cruzi*. A systemic approach to vaccine development against *T. cruzi* in future studies would require further characterization of protective immune responses, identification of new antigenic targets of these protective responses, development of efficient antigen delivery systems, and use of adjuvants and vaccination regimens to enhance the protective responses to known vaccine candidates. Given the complexity of the *T. cruzi* genome, multiple life cycle stages in the mammalian host, and strain variations, it is essential that substantial efforts are employed in selecting the appropriate vaccine candidates. Microarray analysis has emerged as a powerful technology to assess differential gene expression in various stages or strains of pathogens, and identification of stage-specific virulence genes that could be potent drug or vaccine targets (Rathod et al., 2002; Grifantini et al., 2002). The development and free availability of complete genome arrays will facilitate the efforts of the research community in identifying stage-specific novel genes that may also be conserved across various lineages of *T. cruzi* and serve as potential vaccine candidates. Development of computational strategies for efficient screening of the *T. cruzi* EST database has proven to be useful in identifying novel vaccine candidates (Bhatia et al., 2004) and we anticipate that complete sequencing and annotation of *T. cruzi* genome (El-Sayed et al., 2005) would facilitate the unbiased identification and testing of additional vaccine candidates in the next 5 years. Once the selected antigens are shown to be effective in eliciting protective immunity in mice, testing and optimization of the gene-mixes consisting of protective candidates that synergize in their antigenic activity would facilitate formulation of the multivalent vaccines capable of providing maximally protective immunity against *T. cruzi*. Further development of small (mice) and large animal models (dogs) would be essential to conduct the field studies and adequately assess the protective efficacy of vaccines in providing short- and long-term immunity before a clinical trial in humans can be envisioned.

Given the parasites' ability to evade immune detection and survive long-term in an immuno-competent host, it is unlikely that anti-*T. cruzi* vaccines would be effective in preventing infection or in providing sterile immunity. It is, however, likely that vaccines capable of eliciting immune responses that are sufficient to keep the parasite burden below a threshold level would be effective in eliminating the parasite-mediated pathology and tissue injury and thus arrest the progression of Chagas disease severity.

KEY ISSUES

- Several studies have shown that control of acute parasitemia and tissue parasite burden is an effective approach in arresting the development and progression of disease.
- Some success with the first-generation vaccines, including fractionated or inactivated parasites, led the identification of effector immune mechanisms against *T. cruzi*.
- The use of new technologies and experimental systems has allowed a better understanding of the mechanisms underlying the protective immune response against *T. cruzi*.
- Secreted and GPI-anchored proteins of *T. cruzi* are recognized as the best choice for the development of subunit vaccines.
- The DNA vaccination approach has proved to be promising in eliciting both humoral and cellular immune responses and testing and identification of vaccine candidates against *T. cruzi*.
- Completion of *T. cruzi* genome sequencing and development of a variety of bioinformatics and computational tools would allow the global screening of the genome for the identification of novel vaccine candidates.
- The complexity of the parasite life cycle, expression of developmentally regulated proteins in different morphological stages, and multifaceted manifestations of Chagas disease caused by different *T. cruzi* strains necessitates the identification of multiple antigenic targets.
- The complexity of the protective immune responses constituted by lytic antibodies, CTLs, and type 1 cytokines complicates the selection of protective vaccine candidates.

- The expression of immune-suppressive membrane antigens (SAPA (Gorelik et al., 1998), AGC10 (Kierszenbaum et al., 1999)) and the parasite's ability to utilize multiple immune-evasion mechanisms (Sztein and Kierszenbaum, 1993; Ouaissi et al., 1995, 2001) may limit the immunoprotective capacity of anti-*T. cruzi* vaccines.
- The lack of interest of large pharmaceutical companies in developing vaccines against tropical diseases requires a major commitment from the health care agencies of developed and endemic countries.
- Concerted efforts are needed in testing the vaccine potential of known and novel antigenic targets, followed by the design of a multicomponent vaccine cocktail that is effective in eliciting protective immunity in multiple hosts against different parasite strains.
- Evaluation of route and mode of delivery, immunization dose and regimens, and the role of adjuvants in eliciting maximal protective immunity in experimental hosts would provide an impetus for future developments.

ACKNOWLEDGMENTS

The work in N. J. Garg's laboratory has been supported in part by grants from the American Heart Association, John Sealy Memorial Endowment Fund for Biomedical Research, American Health Assistance Foundation, and National Institutes of Health. V. Bhatia was an awardee of a Postdoctoral Fellowship from the Sealy Center of Vaccine Development. M. A. Zacks was awarded the James W. McLaughlin Predoctoral Fellowship and a Postdoctoral Fellowship in the Emerging Infectious Disease Program at the US Centers for Disease Control and Prevention. Our thanks are due to Ms. Mardelle Susman for proofreading and editing of the manuscript.

References

Abrahamsohn, I.A. and Coffman, R.L. Cytokine and nitric oxide regulation of the immunosuppression in *Trypanosoma cruzi* infection. J. Immunol. 1995; 155:3955–3963.

Abrahamsohn, I.A. and Coffman, R.L. *Trypanosoma cruzi*: IL-10, TNF, IFN-gamma, and IL-12 regulate innate and acquired immunity to infection. Exp. Parasitol. 1996; 84:231–244.

Abrahamsohn, I.A. and Silva, W.D. Antibody dependent cell-mediated cytotoxicity against *Trypanosoma cruzi*. Parasitology 1977; 75:317–323.

Acosta-Serrano, A., Almeida, I.C., Freitas-Junior, L.H., Yoshida, N. and Schenkman, S. The mucin-like glycoprotein super-family of *Trypanosoma cruzi*: structure and biological roles. Mol. Biochem. Parasitol. 2001; 114:143–150.

Aliberti, J.C., Souto, J.T., Marino, A.P., Lannes-Vieira, J., Teixeira, M.M., Farber, J., Gazzinelli, R.T. and Silva, J.S. Modulation of chemokine production and inflammatory responses in interferon-gamma- and tumor necrosis factor-R1-deficient mice during *Trypanosoma cruzi* infection. Am. J. Pathol. 2001; 158:1433–1440.

Almeida, I.C., Camargo, M.M., Procopio, D.O., Silva, L.S., Mehlert, A., Travassos, L.R., Gazzinelli, R.T. and Ferguson, M.A. Highly purified glycosylphosphatidylinositols from *Trypanosoma cruzi* are potent proinflammatory agents. EMBO J. 2000; 19:1476–1485.

Almeida, I.C. and Gazzinelli, R.T. Proinflammatory activity of glycosylphosphatidylinositol anchors derived from *Trypanosoma cruzi*: structural and functional analyses. J. Leukoc. Biol. 2001; 70:467–477.

Almeida, I.C., Gazzinelli, R., Ferguson, M.A. and Travassos, L.R. *Trypanosoma cruzi* mucins: potential functions of a complex structure. Mem. Inst. Oswaldo Cruz 1999; 94(Suppl. 1):173–176.

Anez, N., Carrasco, H., Parada, H., Crisante, G., Rojas, A., Fuenmayor, C., Gonzalez, N., Percoco, G., Borges, R., Guevara, P. and Ramirez, J.L. Myocardial parasite persistence in chronic chagasic patients. Am. J. Trop. Med. Hyg. 1999; 60:726–732.

Araujo, A.F., De Alencar, B.C., Vasconcelos, J.R., Hiyane, M.I., Marinho, C.R., Penido, M.L., Boscardin, S.B., Hoft, D.F., Gazzinelli, R.T. and Rodrigues, M.M. CD8 + -T-cell-dependent control of *Trypanosoma cruzi* infection in a highly susceptible mouse strain after immunization with recombinant proteins based on amastigote surface protein 2. Infect. Immun. 2005; 73:6017–6025.

Barnabe, C., Neubauer, K., Solari, A. and Tibayrenc, M. *Trypanosoma cruzi*: presence of the two major phylogenetic lineages and of several lesser discrete typing units (DTUs) in Chile and Paraguay. Acta Trop. 2001a; 78:127–137.

Barnabe, C., Yaeger, R., Pung, O. and Tibayrenc, M. *Trypanosoma cruzi*: a considerable phylogenetic divergence indicates that the agent of Chagas disease is indigenous to the native fauna of the United States. Exp. Parasitol. 2001b; 99:73–79.

Basombrio, M.A. *Trypanosoma cruzi*: partial prevention of the natural infection of guinea pigs with a killed parasite vaccine. Exp. Parasitol. 1990; 71:1–8.

Basso, B., Castro, I., Introini, V., Gil, P., Truyens, C. and Moretti, E. Vaccination with *Trypanosoma rangeli* reduces the infectiousness of dogs experimentally infected with *Trypanosoma cruzi*. Vaccine 2007; 25:3855–3858.

Beard, C.B., Pye, G., Steurer, F.J., Rodriguez, R., Campman, R., Peterson, A.T., Ramsey, J., Wirtz, R.A. and Robinson, L.E. Chagas disease in a domestic transmission cycle, southern Texas, USA. Emerging Infect. Dis. 2003; 9:103–105.

Bento, C.A., Melo, M.B., Previato, J.O., Mendonca-Previato, L. and Pecanha, L.M. Glycoinositolphospholipids purified from *Trypanosoma cruzi* stimulate Ig production in vitro. J. Immunol. 1996; 157:4996–5001.

Bhatia, V., Sinha, M., Luxon, B. and Garg, N. Utility of *Trypanosoma cruzi* sequence database for the identification of potential vaccine candidates: in silico and in vitro screening. Infect. Immun. 2004; 72:6245–6254.

Boscardin, S.B., Kinoshita, S.S., Fujimura, A.E. and Rodrigues, M.M. Immunization with cDNA expressed by amastigotes of *Trypanosoma cruzi* elicits protective immune response against experimental infection. Infect. Immun. 2003; 71:2744–2757.

Bosseno, M.F., Barnabe, C., Magallon Gastelum, E., Lozano Kasten, F., Ramsey, J., Espinoza, B. and Breniere, S.F. Predominance of *Trypanosoma cruzi* lineage I in Mexico. J. Clin. Microbiol. 2002; 40:627–632.

Brener, Z. and Chiari, E. [Susceptibility of different strains of *Trypanosoma cruzi* to various chemotherapeutic agents]. Rev. Inst. Med. Trop. Sao Paulo 1967; 9:197–207.

Brener, Z. and Gazzinelli, R.T. Immunological control of *Trypanosoma cruzi* infection and pathogenesis of Chagas' disease. Int. Arch. Allergy Immunol. 1997; 114:103–110.

Brisse, S., Dujardin, J.C. and Tibayrenc, M. Identification of six *Trypanosoma cruzi* lineages by sequence-characterised amplified region markers. Mol. Biochem. Parasitol. 2000; 111:95–105.

Bustamante, J.M., Presti, M.S., Rivarola, H.W., Fernandez, A.R., Enders, J.E., Fretes, R.E. and Paglini-Oliva, P. Treatment with benznidazole or thioridazine in the chronic phase of experimental Chagas disease improves cardiopathy. Int. J. Antimicrob. Agents 2007; 29:733–737.

Caliari, M.V., De Lana, M., Caja, R.A., Carneiro, C.M., Bahia, M.T., Santos, C.A., Magalhaes, G.A., Sampaio, I.B. and Tafuri, W.L. Immunohistochemical studies in acute and chronic canine chagasic cardiomyopathy. Virchows Arch. 2002; 441:69–76.

Campetella, O., Henriksson, J., Aslund, L., Frasch, A.C., Pettersson, U. and Cazzulo, J.J. The major cysteine proteinase (cruzipain) from *Trypanosoma cruzi* is encoded by multiple polymorphic tandemly organized genes located on different chromosomes. Mol. Biochem. Parasitol. 1992; 50:225–234.

Cazzulo, J., Stoka, V. and Turk, V. The major cysteine proteinase of *Trypanosoma cruzi*: a valid target for chemotherapy of Chagas disease. Curr. Pharm. Des. 2001; 7:1143–1156.

CDC (Centers for Disease Control and Prevention) Chagas disease after organ transplantation—United States, 2001. MMWR Morb. Mortal. Wkly. Rep. 2002a; 51:210–212.

CDC Chagas disease after organ transplantation—United States, 2001. JAMA 2002b; 287:1795–1796.

CDC Chagas disease after organ transplantation—Los Angeles, California, 2006. MMWR Morb. Mortal. Wkly. Rep. 2006; 55:798–800.

CDC Blood donor screening for Chagas disease—United States, 2006–2007. MMWR Morb. Mortal. Wkly. Rep. 2007; 56:141–143.

Chen, Q., Vazquez, E.J., Moghaddas, S., Hoppel, C.L. and Lesnefsky, E.J. Production of reactive oxygen species by mitochondria: central role of complex III. J. Biol. Chem. 2003; 278(38):36027–36031.

Chiari, E. Diagnostic tests for Chagas disease. In: Chagas Disease (American Trypanosomiasis): Its Impact on Transfusion and Clinical Medicine (S. Wendel, Z. Brener, M.E. Camargo and A. Rassi, Eds.). Sao Paulo, Brazil: ISBT, 1992.

Clark, A.K., Kovtunovych, G., Kandlikar, S., Lal, S. and Stryker, G.A. Cloning and expression analysis of two novel paraflagellar rod domain genes found in *Trypanosoma cruzi*. Parasitol Res. 2005; 96:312–320.

Collier, W.A. Uber immunitat bei der Chagas krankheit der weissen mau. Z. Hyg. Infektionskr. 1931; 112:88–92.

Cossio, P.M., Diez, C., Szarfman, A., Kreutzer, E., Candiolo, B. and Arana, R.M. Chagasic cardiopathy. Demonstration of a serum gamma globulin factor which reacts with endocardium and vascular structures. Circulation 1974; 49:13–21.

Costa, F., Franchin, G., Pereira-Chioccola, V.L., Ribeirao, M., Schenkman, S. and Rodrigues, M.M. Immunization with a plasmid DNA containing the gene of trans-sialidase reduces *Trypanosoma cruzi* infection in mice. Vaccine 1998; 16:768–774.

Coura, J.R. Current prospects of specific treatment of Chagas' disease. Bol. Chil. Parasitol. 1996; 51:69–75.

Cruz-Reyes, A. and Pickering-Lopez, J.M. Chagas disease in Mexico: an analysis of geographical distribution during the past 76 years—a review. Mem. Inst. Oswaldo Cruz 2006; 101:345–354.

Cuevas, I.C., Cazzulo, J.J. and Sanchez, D.O. gp63 homologues in *Trypanosoma cruzi*: surface antigens with metalloprotease activity and a possible role in host cell infection. Infect. Immun. 2003; 71:5739–5749.

Cunha-Neto, E., Duranti, M., Gruber, A., Zingales, B., De Messias, I., Stolf, N., Bellotti, G., Patarroyo, M.E., Pilleggi, F. and Kalil, J. Autoimmunity in Chagas disease cardiopathy: biological relevance of a cardiac myosin-specific epitope crossreactive to an immunodominant *Trypanosoma cruzi* antigen. Proc. Natl. Acad. Sci. U.S.A. 1995; 92:3541–3545.

Cunha-Neto, E. and Kalil, J. Heart-infiltrating and peripheral T cells in the pathogenesis of human Chagas' disease cardiomyopathy. Autoimmunity 2001; 34:187–192.

D'almeida, P., Keitel, E., Bittar, A., Goldani, J., Santos, A., Neumann, J. and Garcia, V. Long-term evaluation of kidney donors. Transplant Proc. 1996; 28:93–94.

De Oliveira, T.B., Pedrosa, R.C. and Filho, D.W. Oxidative stress in chronic cardiopathy associated with Chagas disease. Int. J. Cardiol. 2007; 116:357–363.

Dias, J. and Schofield, C. The evolution of Chagas disease (American trypanosomiasis) control after 90 years since Carlos Chagas discovery. Mem. Inst. Oswaldo Cruz 1999; 94(Suppl. 1):103–121.

Dias, J.A.P. Chagas disease and blood transfusion in endemic areas. In: Chagas Disease (American Trypanosomiasis): Its Impact on Transfusion and Clinical Medicine (S. Wendel, Z. Brener, M.E. Camargo and A. Rassi, Eds.). Sao Paulo, Brazil: ISBT, 1992a.

Dias, J.A.P. Epidemiology of Chagas disease. In: Chagas Disease (American Trypanosomiasis): Its Impact on Transfusion and Clinical Medicine (S. Wendel, Z. Brener, M.E. Camargo and A. Rassi, Eds.), Sao Paulo, Brazil: ISBT, 1992b.

Dodd, R.Y. and Leiby, D.A. Emerging infectious threats to the blood supply. Ann. Rev. Med. 2004; 55:191–207.

Donnelly, J.J., Wahren, B. and Liu, M.A. DNA vaccines: progress and challenges. J. Immunol. 2005; 175:633–639.

Dosreis, G.A. Cell-mediated immunity in experimental *Trypanosoma cruzi* infection. Parasitol. Today 1997; 13:335–342.

Dosreis, G.A., Pecanha, L.M., Bellio, M., Previato, J.O. and Mendonca-Previato, L. Glycoinositol phospholipids from *Trypanosoma cruzi* transmit signals to the cells of the host immune system through both ceramide and glycan chains. Microbes Infect. 2002; 4:1007–1013.

Dumonteil, E. DNA vaccines against protozoan parasites: advances and challenges. J. Biomed. Biotechnol. 2007; 90520.

Dumonteil, E., Escobedo-Ortegon, J., Reyes-Rodriguez, N., Arjona-Torres, A. and Ramirez-Sierra, M.J. Immunotherapy of *Trypanosoma cruzi* infection with DNA vaccines in mice. Infect. Immun. 2004; 72:46–53.

El-Sayed, N.M., Myler, P.J., Bartholomeu, D.C., Nilsson, D., Aggarwal, G., Tran, A.N., Ghedin, E., Worthey, E.A., Delcher, A.L., Blandin, G., Westenberger, S.J., Caler, E., Cerqueira, G.C., Branche, C., Haas, B., Anupama, A., Arner, E., Aslund, L., Attipoe, P., Bontempi, E., Bringaud, F., Burton, P., Cadag, E., Campbell, D.A., Carrington, M., Crabtree, J., Darban, H., Da Silveira, J.F., De Jong, P., Edwards, K., Englund, P.T., Fazelina, G., Feldblyum, T., Ferella, M., Frasch, A.C., Gull, K., Horn, D., Hou, L., Huang, Y., Kindlund, E., Klingbeil, M., Kluge, S., Koo, H., Lacerda, D., Levin, M.J., Lorenzi, H., Louie, T., Machado, C.R., Mcculloch, R., Mckenna, A., Mizuno, Y., Mottram, J.C., Nelson, S., Ochaya, S., Osoegawa, K., Pai, G., Parsons, M., Pentony, M., Pettersson, U., Pop, M., Ramirez, J.L., Rinta, J., Robertson, L., Salzberg, S.L., Sanchez, D.O., Seyler, A., Sharma, R., Shetty, J., Simpson, A.J., Sisk, E., Tammi, M.T., Tarleton, R., Teixeira, S., Van Aken, S., Vogt, C., Ward, P.N., Wickstead, B., Wortman, J., White, O., Fraser, C.M., Stuart, K.D. and Andersson, B. The genome sequence of *Trypanosoma cruzi*, etiologic agent of Chagas disease. Science 2005; 309:409–415.

Fernandes, J.F., Castellani, O. and Okumura, M. Histopathology of the heart and muscles in mice immunized against *Trypanosoma cruzi*. Rev. Inst. Med. Trop. Sao Paulo 1966; 8:151–156.

Fernandez-Gomez, R., Esteban, S., Gomez-Corvera, R., Zoulika, K. and Ouaissi, A. *Trypanosoma cruzi*: Tc52 released protein-induced increased expression of nitric oxide synthase and nitric oxide production by macrophages. J. Immunol. 1998; 160: 3471–3479.

Ferreira, A.W. Tests for Chagas disease serolodiagnosis: a review. In: Chagas Disease (American Trypanosomiasis): Its Impact on Transfusion and Clinical Medicine (S. Wendel, Z. Brener, M.E. Camargo and A. Rassi, Eds.). Sao Paulo, Brazil: ISBT, 1992.

Fonseca, S.G., Moins-Teisserenc, H., Clave, E., Ianni, B., Nunes, V.L., Mady, C., Iwai, L.K., Sette, A., Sidney, J., Marin, M.L., Goldberg, A.C., Guilherme, L., Charron, D., Toubert, A., Kalil, J. and Cunha-Neto, E. Identification of multiple HLA-A*0201-restricted cruzipain and FL-160 CD8+ epitopes recognized by T cells from chronically *Trypanosoma cruzi*-infected patients. Microbes Infect. 2005; 7:688–697.

Fouts, D.L., Stryker, G.A., Gorski, K.S., Miller, M.J., Nguyen, T.V., Wrightsman, R.A. and Manning, J.E. Evidence for four distinct major protein components in the paraflagellar rod of *Trypanosoma cruzi*. J. Biol. Chem. 1998; 273:21846–21855.

Fralish, B.H. and Tarleton, R.L. Genetic immunization with LYT1 or a pool of trans-sialidase genes protects mice from lethal *Trypanosoma cruzi* infection. Vaccine 2003; 21:3070–3080.

Frank, F.M., Petray, P.B., Cazorla, S.I., Munoz, M.C., Corral, R.S. and Malchiodi, E.L. Use of a purified *Trypanosoma cruzi* antigen and CpG oligodeoxynucleotides for immunoprotection against a lethal challenge with trypomastigotes. Vaccine 2003; 22:77–86.

Garcia, C.A., Oliveira, E.C., Sakurada, J.K. and Santos, L.M. Protective immunity induced by a *Trypanosoma cruzi* soluble extract antigen in experimental Chagas' disease. Role of interferon gamma. Immunol. Invest. 2000; 29:1–12.

Garg, N. Mitochondrial disorders in chagasic cardiomyopathy. Front. Biosci. 2005; 10:1341–1354.

Garg, N., Nunes, M.P. and Tarleton, R.L. Delivery by *Trypanosoma cruzi* of proteins into the MHC class I antigen processing and presentation pathway. J. Immunol. 1997; 158:3293–3302.

Garg, N. and Tarleton, R.L. Elicitation of protective cellular and humoral immune responses to *Trypanosoma cruzi* infection using DNA vaccines can be augmented with cytokines. Proceedings 10th International Congress of Immunology, New Delhi, India, 1998.

Garg, N. and Tarleton, R.L. Genetic immunization elicits antigen-specific protective immune responses and decreases disease severity in *Trypanosoma cruzi* infection. Infect. Immun. 2002; 70:5547–5555.

Garzon, E., Borges, M.C., Cordeiro-da-Silva, A., Nacife, V., Meirelles Mde, N., Guilvard, E., Bosseno, M.F., Guevara, A.G., Breniere, S.F. and Ouaissi, A. *Trypanosoma cruzi* carrying a targeted deletion of a Tc52 protein-encoding allele elicits attenuated Chagas' disease in mice. Immunol. Lett. 2003; 89:67–80.

Gazzinelli, R.T., Oswald, I.P., Hieny, S., James, S.L. and Sher, A. The microbicidal activity of interferon-gamma-treated macrophages against *Trypanosoma cruzi* involves an L-arginine-dependent, nitrogen oxide-mediated mechanism inhibitable by interleukin-10 and transforming growth factor-beta. Eur. J. Immunol. 1992; 22:2501–2506.

Giordanengo, L., Fretes, R., Diaz, H., Cano, R., Bacile, A., Vottero-Cima, E. and Gea, S. Cruzipain induces autoimmune response against skeletal muscle and tissue damage in mice. Muscle Nerve 2000; 23:1407–1413.

Girones, N., Rodriguez, C.I., Basso, B., Bellon, J.M., Resino, S., Munoz-Fernandez, M.A., Gea, S., Moretti, E. and Fresno, M. Antibodies to an epitope from the Cha human autoantigen are markers of Chagas' disease. Clin. Diagn. Lab. Immunol. 2001a; 8:1039–1043.

Girones, N., Rodriguez, C.I., Carrasco-Marin, E., Hernaez, R.F., De Rego, J.L. and Fresno, M. Dominant T- and B-cell epitopes in an autoantigen linked to Chagas' disease. J. Clin. Invest. 2001b; 107:985–993.

Godsel, L.M. and Engman, D.M. Flagellar protein localization mediated by a calcium-myristoyl/palmitoyl switch mechanism. EMBO J. 1999; 18:2057–2065.

Godsel, L.M., Tibbetts, R.S., Olson, C.L., Chaudoir, B.M. and Engman, D.M. Utility of recombinant flagellar calcium-binding protein for serodiagnosis of *Trypanosoma cruzi* infection. J. Clin. Microbiol. 1995; 33:2082–2085.

Gomes, M.L., Galvao, L.M., Macedo, A.M., Pena, S.D. and Chiari, E. Chagas' disease diagnosis: comparative analysis of parasitologic, molecular, and serologic methods. Am. J. Trop. Med. Hyg. 1999; 60:205–210.

Gonzalez, N., Galindo, I., Guevara, P., Novak, E., Scorza, J.V., Anez, N., Da Silveira, J.F. and Ramirez, J.L. Identification and detection of *Trypanosoma cruzi* by using a DNA amplification fingerprint obtained from the ribosomal intergenic spacer. J. Clin. Microbiol. 1994; 32:153–158.

Gonzalez Cappa, S.M., Pesce, U.J., Cantarella, A.I. and Schmunis, G.A. *Trypanosoma cruzi*: protection of mice with epimastigote antigens from immunologically different parasite strains. Exp. Parasitol. 1974; 35:179–186.

Gorelik, G., Cremaschi, G., Borda, E. and Sterin-Borda, L. *Trypanosoma cruzi* antigens down-regulate T lymphocyte proliferation by muscarinic cholinergic receptor-dependent release of PGE2. Acta Physiol. Pharmacol. Ther. Latinoam. 1998; 48:115–123.

Grifantini, R., Bartolini, E., Muzzi, A., Draghi, M., Frigimelica, E., Berger, J., Randazzo, F. and Grandi, G. Gene expression profile in *Neisseria meningitidis* and *Neisseria lactamica* upon host-cell contact: from basic research to vaccine development. Ann. N. Y. Acad. Sci. 2002; 975:202–216.

Guerra, F. American trypanosomiasis. An historical and a human lesson. J. Trop. Med. Hyg. 1970; 73:105–118.

Guzman-Bracho, C. Epidemiology of Chagas disease in Mexico: an update. Trends Parasitol. 2001; 17:372–376.

Hall, B.F. *Trypanosoma cruzi*: mechanisms for entry into host cells. Semin. Cell Biol. 1993; 4:323–333.

Hall, C.A., Polizzi, C., Yabsley, M.J. and Norton, T.M. *Trypanosoma cruzi* prevalence and epidemiologic trends in lemurs on St. Catherines Islands, Georgia. J. Parasitol. 2007; 93:93–96.

Hanford, E.J., Zhan, F.B., Lu, Y. and Giordano, A. Chagas disease in Texas: recognizing the significance and implications of evidence in the literature. Soc. Sci. Med. 2007; 65:60–79.

Harth, G., Mills, A.A., Lin, T. and Araujo, F.G. *Trypanosoma cruzi* glycoprotein of M(r) 56,000 characterization and assessment of its potential to protect against fatal parasite infections. Mol. Microbiol. 1994; 11:261–271.

Hauschka, T.S. Persistence of strain-specific behavior in two strains of *Trypanosoma cruzi* after prolonged transfer through inbred mice. J. Parasitol. 1949; 35:593–599.

Hauschka, T.S., Goodwin, M.B., Palmquist, J. and Brown, E. Immunological relationship between seven strains of *Trypanosoma cruzi* and its application in the diagnosis of Chagas' disease. Am. J. Trop. Med. Hyg. 1950; 30:1–16.

Herwaldt, B.L., Grijalva, M.J., Newsome, A.L., Mcghee, C.R., Powell, M.R., Nemec, D.G., Steurer, F.J. and Eberhard, M.L. Use of polymerase chain reaction to diagnose the fifth reported US case of autochthonous transmission of *Trypanosoma cruzi*, in Tennessee, 1998. J. Infect. Dis. 2000; 181:395–399.

Higuchi, M. Human chronic chagasic cardiopathy: participationof parasite antigens, subsets of lymphocytes, cytokines and microvascular abnormalities. Mem. Inst. Oswaldo Cruz 1999; 94:263–267.

Higuchi, M.D., Benvenuti, L.A., Martins Reis, M. and Metzger, M. Pathophysiology of the heart in Chagas' disease: current status and new developments. Cardiovasc. Res. 2003; 60:96–107.

Higuchi, M.D., Ries, M.M., Aiello, V.D., Benvenuti, L.A., Gutierrez, P.S., Bellotti, G. and Pileggi, F. Association of an increase in CD8+ T cells with the presence of *Trypanosoma cruzi* antigens in chronic, human, chagasic myocarditis. Am. J. Trop. Med. Hyg. 1997; 56:485–489.

Higuchi, M.D.L., Brito, T.D., Reis, M.M., Barbosa, A., Bellotti, G., Pereira-Barreto, A.C. and Pileggi, F. Correlation between *Trypanosoma cruzi* parasitism and mayocardial inflammatory infiltrate in human chronic chagasic myocarditis: light microscopy and immunohistochemical findings. Cardiovasc. Pathol. 1993; 2:101–106.

Hoff, R. Killing in vitro of *Trypanosoma cruzi* by macrophages from mice immunized with *T. cruzi* or BCG, and absence of cross-immunity on challege in vivo. J. Exp. Med. 1975; 142:299–311.

Hunter, C.A., Ellis-Neyes, L.A., Slifer, T., Kanaly, S., Grunig, G., Fort, M., Rennick, D. and Araujo, F.G. IL-10 is required to prevent immune hyperactivity during infection with *Trypanosoma cruzi*. J. Immunol. 1997; 158:3311–3316.

Ide, T., Tsutsui, H., Kinugawa, S., Utsumi, H., Kang, D., Hattori, N., Uchida, K., Arimura, K., Egashira, K. and Takeshita, A. Mitochondrial electron transport complex I is a potential source of oxygen free radicals in the failing myocardium. Circ. Res. 1999; 85:357–363.

Iwai, L.K., Juliano, M.A., Juliano, L., Kalil, J. and Cunha-Neto, E. T-cell molecular mimicry in Chagas disease: identification and partial structural analysis of multiple cross-reactive epitopes between *Trypanosoma cruzi* B13 and cardiac myosin heavy chain. J. Autoimmun. 2005; 24:111–117.

Johnson, P., Neal, R.A. and Gall, D. Protective effect of killed trypanosome vaccines with incorporated adjuvants. Nature 1963; 200:83.

Kagan, I.G. and Norman, L. Immunologic studies on *Trypanosoma cruzi*. III. Duration of acquired immunity in mice initially infected with a North American strain of *T. cruzi*. J. Infect. Dis. 1961; 108:213–217.

Kaplan, D., Ferrari, I., Bergami, P.L., Mahler, E., Levitus, G., Chiale, P., Hoebeke, J., Van Regenmortel, M.H. and Levin, M.J. Antibodies to ribosomal P proteins of *Trypanosoma cruzi* in Chagas disease possess functional autoreactivity with heart tissue and differ from anti-P autoantibodies in lupus. Proc. Natl. Acad. Sci. U.S.A. 1997; 94:10301–10306.

Katae, M., Miyahira, Y., Takeda, K., Matsuda, H., Yagita, H., Okumura, K., Takeuchi, T., Kamiyama, T., Ohwada, A., Fukuchi, Y. and Aoki, T. Coadministration of an interleukin-12 gene and a *Trypanosoma cruzi* gene improves vaccine efficacy. Infect. Immun. 2002; 70:4833–4840.

Kierszenbaum, F. Chagas' disease and the autoimmunity hypothesis. Clin. Microbiol. Rev. 1999; 12:210–223.

Kierszenbaum, F. and Budzko, D.B. Immunological properties of a trypanosomal lipopolysaccharide and its effects on the infection with *Trypanosoma (Schizotrypanum) cruzi*. Immunol. Commun. 1973; 2:265–276.

Kierszenbaum, F., De Diego, J.L., Fresno, M. and Sztein, M.B. Inhibitory effects of the *Trypanosoma cruzi* membrane glycoprotein AGC10 on the expression of IL-2 receptor chains and secretion of cytokines by subpopulations of activated human T lymphocytes. Eur. J. Immunol. 1999; 29:1684–1691.

Kirchhoff, L.V. American trypanosomiasis (Chagas' disease)—a tropical disease now in the United States. N. Engl. J. Med. 1993; 329:639–644.

Kirchhoff, L.V., Weiss, L.M., Wittner, M. and Tanowitz, H.B. Parasitic diseases of the heart. Front. Biosci. 2004; 9:706–723.

Krautz, G.M., Kissinger, J.C. and Krettli, A.U. The targets of the lytic antibody response against *Trypanosoma cruzi*. Parasitol. Today 2000; 16:31–34.

Kumar, S. and Tarleton, R.L. Antigen-specific Th1 but not Th2 cells provide protection from lethal *Trypanosoma cruzi* infection in mice. J. Immunol. 2001; 166:4596–4603.

Laderach, D., Cerban, F., Motran, C., Vottero de Cima, E. and Gea, S. *Trypanosoma cruzi*: the major cysteinyl proteinase (cruzipain) is a relevant immunogen of parasite acidic antigens (FIII). Int. J. Parasitol. 1996; 26:1249–1254.

Leiby, D.A., Read, E.J., Lenes, B.A., Yund, A.J., Stumpf, R.J., Kirchhoff, L.V. and Dodd, R.Y. Seroepidemiology of *Trypanosoma cruzi*, etiologic agent of Chagas' disease, in US blood donors. J. Infect. Dis. 1997; 176:1047–1052.

Leon, J.S., Daniels, M.D., Toriello, K.M., Wang, K. and Engman, D.M. A cardiac myosin-specific autoimmune response is induced by immunization with *Trypanosoma cruzi* proteins. Infect. Immun. 2004; 72:3410–3417.

Leon, J.S. and Engman, D.M. The significance of autoimmunity in the pathogenesis of Chagas heart disease. Front. Biosci. 2003; 8:e315–e322.

Lewinsohn, R. Carlos Chagas (1879–1934): the discovery of *Trypanosoma cruzi* and of American trypanosomiasis (foot-notes to the history of Chagas's disease). Trans. R. Soc. Trop. Med. Hyg. 1979; 73:513–523.

Liendo, A., Lazardi, K. and Urbina, J.A. In-vitro antiproliferative effects and mechanism of action of the bis-triazole D0870 and its S(-) enantiomer against *Trypanosoma cruzi*. J. Antimicrob. Chemother. 1998; 41:197–205.

Lopes, M.F. and Dosreis, G.A. *Trypanosoma cruzi*-induced immunosuppression: selective triggering of CD4+ T-cell death by the T-cell receptor-CD3 pathway and not by the CD69 or Ly-6 activation pathway. Infect. Immun. 1996; 64:1559–1564.

Lo Presti, M.S., Rivarola, H.W., Bustamante, J.M., Fernandez, A.R., Enders, J.E., Fretes, R., Gea, S. and Paglini-Oliva, P.A. Thioridazine treatment prevents cardiopathy in *Trypanosoma cruzi* infected mice. Int. J. Antimicrob. Agents 2004; 23:634–636.

Low, H.P., Santos, M.A., Wizel, B. and Tarleton, R.L. Amastigote surface proteins of *Trypanosoma cruzi* are targets for CD8+ CTL. J. Immunol. 1998; 160:1817–1823.

Low, H.P. and Tarleton, R.L. Molecular cloning of the gene encoding the 83 kDa amastigote surface protein and its identification as a member of the *Trypanosoma cruzi* sialidase superfamily. Mol. Biochem. Parasitol. 1997; 88:137–149.

Luhrs, K.A., Fouts, D.L. and Manning, J.E. Immunization with recombinant paraflagellar rod protein induces protective immunity against *Trypanosoma cruzi* infection. Vaccine 2003; 21:3058–3069.

Macao, L.B., Filho, D.W., Pedrosa, R.C., Pereira, A., Backes, P., Torres, M.A. and Frode, T.S. Antioxidant therapy attenuates oxidative stress in chronic cardiopathy associated with Chagas' disease. Int. J. Cardiol. 2007; 123(1):43–49.

Machado, A.V., Cardoso, J.E., Claser, C., Rodrigues, M.M., Gazzinelli, R.T. and Bruna-Romero, O. Long-term protective immunity induced against *Trypanosoma cruzi* infection after vaccination with recombinant adenoviruses encoding amastigote surface protein-2 and trans-sialidase. Hum. Gene Ther. 2006; 17:898–908.

Magdesian, M.H., Giordano, R., Ulrich, H., Juliano, M.A., Juliano, L., Schumacher, R.I., Colli, W. and Alves, M.J. Infection by

Trypanosoma cruzi. Identification of a parasite ligand and its host cell receptor. J. Biol. Chem. 2001; 276:19382–19389.

Malaga, S. and Yoshida, N. Targeted reduction in expression of *Trypanosoma cruzi* surface glycoprotein gp90 increases parasite infectivity. Infect. Immun. 2001; 69:353–359.

Malchiodi, E.L., Chiaramonte, M.G., Martinez, J.A., Zwirner, N.W., Margni, R.A. and Cazzulo, J.J. Identity of the major cysteine proteinase (cruzipain) from *Trypanosoma cruzi* and an antigen (Ag163B6) isolated with a monoclonal antibody. Immunol. Lett. 1993; 35:59–62.

Manning-Cela, R., Cortes, A., Gonzalez-Rey, E., Van Voorhis, W.C., Swindle, J. and Gonzalez, A. LYT1 protein is required for efficient in vitro infection by *Trypanosoma cruzi*. Infect. Immun. 2001; 69:3916–3923.

Maranon, C., Thomas, M.C., Planelles, L. and Lopez, M.C. The immunization of A2/K(b) transgenic mice with the KMP11-HSP70 fusion protein induces CTL response against human cells expressing the *T. cruzi* KMP11 antigen: identification of A2-restricted epitopes. Mol. Immunol. 2001; 38:279–287.

Marin-Neto, J.A. Cardiac dysautonomia and pathogenesis of Chagas' heart disease. Int. J. Cardiol. 1998; 66:129–131.

Martindale, J.L. and Holbrook, N.J. Cellular response to oxidative stress: signaling for suicide and survival. J. Cell. Physiol. 2002; 192:1–15.

Martins, G.A., Cardoso, M.A., Aliberti, J.C. and Silva, J.S. Nitric oxide-induced apoptotic cell death in the acute phase of *Trypanosoma cruzi* infection in mice. Immunol. Lett. 1998; 63:113–120.

Mattoso, L.F. (Ed.), Chagas Disease (American Trypanosomiasis). New York: Springer-Verlag, 1981.

Menezes, H. The use of adjuvants in the vaccination of mice with lyophilized "*Trypanosoma cruzi*". Hospital (Rio. J.) 1965; 68:1341–1346.

Menezes, H. Protective effect of an avirulent (cultivated) strain of *Trypanosoma cruzi* against experimental infection in mice. Rev. Inst. Med. Trop. Sao Paulo 1968; 10:1–4.

Menezes, H. Active immunization of mice with the avirulent Y strain of *Trypanosoma cruzi* against heterologous virulent strains of the same parasite. Rev. Inst. Med. Trop. Sao Paulo 1969; 11:335–342.

Michailowsky, V., Luhrs, K., Rocha, M.O., Fouts, D., Gazzinelli, R.T. and Manning, J.E. Humoral and cellular immune responses to *Trypanosoma cruzi*-derived paraflagellar rod proteins in patients with Chagas' disease. Infect. Immun. 2003; 71:3165–3171.

Miller, M.J., Wrightsman, R.A. and Manning, J.E. *Trypanosoma cruzi*: protective immunity in mice immunized with paraflagellar rod proteins is associated with a T-helper type 1 response. Exp. Parasitol. 1996; 84:156–167.

Miller, M.J., Wrightsman, R.A., Stryker, G.A. and Manning, J.E. Protection of mice against *Trypanosoma cruzi* by immunization with paraflagellar rod proteins requires T cell, but not B cell, function. J. Immunol. 1997; 158:5330–5337.

Minoprio, P. Parasite polyclonal activators: new targets for vaccination approaches? Int. J. Parasitol. 2001; 31:588–591.

Minoprio, P., Eisen, H., Joskowicz, M., Pereira, P. and Coutinho, A. Suppression of polyclonal antibody production in *Trypanosoma cruzi*-infected mice by treatment with anti-L3T4 antibodies. J. Immunol. 1987; 139:545–550.

Minoprio, P., Itohara, S., Heusser, C., Tonegawa, S. and Coutinho, A. Immunobiology of murine *T. cruzi* infection: the predominance of parasite-nonspecific responses and the activation of TCRI T cells. Immunol. Rev. 1989; 112:183–207.

Miyahira, Y., Akiba, H., Katae, M., Kubota, K., Kobayashi, S., Takeuchi, T., Garcia-Sastre, A., Fukuchi, Y., Okumura, K., Yagita, H. and Aoki, T. Cutting edge: a potent adjuvant effect of ligand to receptor activator of NF-kappa B gene for inducing antigen-specific CD8+ T cell response by DNA and viral vector vaccination. J. Immunol. 2003; 171:6344–6348.

Miyahira, Y., Takashima, Y., Kobayashi, S., Matsumoto, Y., Takeuchi, T., Ohyanagi-Hara, M., Yoshida, A., Ohwada, A., Akiba, H., Yagita, H., Okumura, K. and Ogawa, H. Immune responses against a single CD8 + -T-cell epitope induced by virus vector vaccination can successfully control *Trypanosoma cruzi* infection. Infect. Immun. 2005; 73:7356–7365.

Moncayo, A. Chagas disease: current epidemiological trends after the interruption of vectorial and transfusional transmission in the Southern Cone countries. Mem. Inst. Oswaldo Cruz 2003; 98:577–591.

Moncayo, A. and Ortiz Yanine, M.I. An update on Chagas disease (human American trypanosomiasis). Ann. Trop. Med. Parasitol. 2006; 100:663–677.

Morell, M., Thomas, M.C., Caballero, T., Alonso, C. and Lopez, M.C. The genetic immunization with paraflagellar rod protein-2 fused to the HSP70 confers protection against late *Trypanosoma cruzi* infection. Vaccine 2006; 24:7046–7055.

Mortara, R.A., Da Silva, S. and Taniwaki, N.N. Confocal fluorescence microscopy: a powerful tool in the study of Chagas' disease. Rev. Soc. Bras. Med. Trop. 2000; 33:79–82.

Mott, K.E. and Hagstrom, J.W. The pathologic lesions of the cardiac autonomic nervous system in chronic Chagas' myocarditis. Circulation 1965; 31:273–286.

Moutiez, M., Quemeneur, E., Sergheraert, C., Lucas, V., Tartar, A. and Davioud-Charvet, E. Glutathione-dependent activities of *Trypanosoma cruzi* p52 makes it a new member of the thiol: disulphide oxidoreductase family. Biochem. J. 1997; 322(Pt. 1): 43–48.

Mukherjee, S., Huang, H., Weiss, L.M., Costa, S., Scharfstein, J. and Tanowitz, H.B. Role of vasoactive mediators in the pathogenesis of Chagas' disease. Front. Biosci. 2003; 8:e410–e419.

Norris, K.A., Galvao, L.M., Schrimpf, J.E., Cancado, J.R. and Krettli, A.U. Humoral immune response to the *Trypanosoma cruzi* complement regulatory protein as an indicator of parasitologic clearance in human Chagas' disease. Infect. Immun. 1994; 62:4072–4074.

Norris, K.A., Harth, G. and So, M. Purification of a *Trypanosoma cruzi* membrane glycoprotein which elicits lytic antibodies. Infect. Immun. 1989; 57:2372–2377.

Okanla, E.O., Stumpf, J.L. and Dusanic, D.G. Resistance of mice immunized with irradiated and lyophilized stages of *Trypanosoma cruzi* to infections with metacyclics. Int. J. Parasitol. 1982; 12:251–256.

Ouaissi, A., Da Silva, A.C., Guevara, A.G., Borges, M. and Guilvard, E. *Trypanosoma cruzi*-induced host immune system dysfunction: a rationale for parasite immunosuppressive factor(s) encoding gene targeting. J. Biomed. Biotechnol. 2001; 1:11–17.

Ouaissi, A., Guevara-Espinoza, A., Chabe, F., Gomez-Corvera, R. and Taibi, A. A novel and basic mechanism of immunosuppression in Chagas' disease: *Trypanosoma cruzi* releases in vitro and in vivo a protein which induces T cell unresponsiveness through specific interaction with cysteine and glutathione. Immunol. Lett. 1995; 48:221–224.

Ouaissi, A., Guilvard, E., Delneste, Y., Caron, G., Magistrelli, G., Herbault, N., Thieblemont, N. and Jeannin, P. The *Trypanosoma cruzi* Tc52-released protein induces human dendritic cell maturation, signals via toll-like receptor 2, and confers protection against lethal infection. J. Immunol. 2002; 168:6366–6374.

Palomino, S.A., Aiello, V.D. and Higuchi, M.L. Systematic mapping of hearts from chronic chagasic patients: the association between the occurrence of histopathological lesions and *Trypanosoma cruzi* antigens. Ann. Trop. Med. Parasitol. 2000; 94:571–579.

Pan, C.H., Chen, H.W. and Tao, M.H. Modulation of immune responses to DNA vaccines by codelivery of cytokine genes. J. Formos. Med. Assoc. 1999; 98:722–729.

Pereira-Chioccola, V.L., Schenkman, S. and Kloetzel, J.K. Sera from chronic Chagasic patients and rodents infected with *Trypanosoma cruzi* inhibit trans-sialidase by recognizing its amino-terminal and catalytic domain. Infect. Immun. 1994; 62:2973–2978.

Perez-Fuentes, R., Guegan, J.F., Barnabe, C., Lopez-Colombo, A., Salgado-Rosas, H., Torres-Rasgado, E., Briones, B., Romero-Diaz, M., Ramos-Jimenez, J. and Sanchez-Guillen Mdel, C. Severity of chronic Chagas disease is associated with cytokine/antioxidant imbalance in chronically infected individuals. Int. J. Parasitol. 2003; 33:293–299.

Petkova, S.B., Huang, H., Factor, S.M., Pestell, R.G., Bouzahzah, B., Jelicks, L.A., Weiss, L.M., Douglas, S.A., Wittner, M. and Tanowitz, H.B. The role of endothelin in the pathogenesis of Chagas' disease. Int. J. Parasitol. 2001; 31:499–511.

Pizarro, J.C., Lucero, D.E. and Stevens, L. PCR reveals significantly higher rates of *Trypanosoma cruzi* infection than microscopy in the Chagas vector, *Triatoma infestans*: high rates found in Chuquisaca, Bolivia. BMC Infect. Dis. 2007; 7:66.

Pizzi, T. and Prager, R. Immunity to infection induced by culture of *Trypanosoma cruzi* of attenuated virulence; preliminary communication. Bol. Inf. Parasit. Chil. 1952; 7:20–21.

Planelles, L., Thomas, M.C., Alonso, C. and Lopez, M.C. DNA immunization with *Trypanosoma cruzi* HSP70 fused to the KMP11 protein elicits a cytotoxic and humoral immune response against the antigen and leads to protection. Infect. Immun. 2001; 69:6558–6563.

Pung, O.J., Spratt, J., Clark, C.G., Norton, T.M. and Carter, J. *Trypanosoma cruzi* infection of free-ranging lion-tailed macaques (*Macaca silenus*) and ring-tailed lemurs (*Lemur catta*) on St. Catherine's Island, Georgia, USA. J. Zoo Wildl. Med. 1998; 29:25–30.

Qazi, K.R., Wikman, M., Vasconcelos, N.M., Berzins, K., Stahl, S. and Fernandez, C. Enhancement of DNA vaccine potency by linkage of *Plasmodium falciparum* malarial antigen gene fused with a fragment of HSP70 gene. Vaccine 2005; 23:1114–1125.

Rahman, I. and MacNee, W. Regulation of redox glutathione levels and gene transcription in lung inflammation: therapeutic approaches. Free Radic. Biol. Med. 2000; 28:1405–1420.

Rassi, A., Jr., Rassi, A. and Little, W.C. Chagas' heart disease. Clin. Cardiol. 2000; 23:883–889.

Rathod, P.K., Ganesan, K., Hayward, R.E., Bozdech, Z. and Derisi, J.L. DNA microarrays for malaria. Trends Parasitol. 2002; 18:39–45.

Rego, S.F.M. Esatado das lesoes provacadas pelo *Trypanosoma cruzi* Chagas, 1909, no baco e no figado do camundongo branco ("*mus musculis*") com diversos graus de resistencia. J. Bras. Med. 1959; 1:599–674.

Reis, D.D., Jones, E.M., Tostes, S., Jr., Lopes, E.R., Gazzinelli, G., Colley, D.G. and Mccurley, T.L. Characterization of inflammatory infiltrates in chronic chagasic myocardial lesions: presence of tumor necrosis factor-alpha+ cells and dominance of granzyme A + , CD8+ lymphocytes. Am. J. Trop. Med. Hyg. 1993; 48:637–644.

Reis, M.A., Costa, R.S. and Ferraz, A.S. Causes of death in renal transplant recipients: a study of 102 autopsies from 1968 to 1991. J. R. Soc. Med. 1995; 88:24–27.

Riarte, A., Luna, C., Sabatiello, R., Sinagra, A., Schiavelli, R., De Rissio, A., Maiolo, E., Garcia, M.M., Jacob, N., Pattin, M., Lauricella, M., Segura, E.L. and Vazquez, M. Chagas' disease in patients with kidney transplants: 7 years of experience 1989–1996. Clin. Infect. Dis. 1999; 29:561–567.

Ribeirao, M., Pereira-Chioccola, V.L., Renia, L., Augusto Fragata Filho, A., Schenkman, S. and Rodrigues, M.M. Chagasic patients develop a type 1 immune response to *Trypanosoma cruzi* trans-sialidase. Parasite Immunol. 2000; 22:49–53.

Rocha, C.D., Caetano, B.C., Machado, A.V. and Bruna-Romero, O. Recombinant viruses as tools to induce protective cellular immunity against infectious diseases. Int. Microbiol. 2004; 7:83–94.

Rocha, M.O., Ribeiro, A.L. and Teixeira, M.M. Clinical management of chronic Chagas cardiomyopathy. Front. Biosci. 2003; 8:e44–e54.

Rodriques Coura, J. and de Castro, S.L. A critical review on Chagas disease chemotherapy. Mem. Inst. Oswaldo Cruz 2002; 97:3–24.

Roffe, E., Souza, A.L., Caetano, B.C., Machado, P.P., Barcelos, L.S., Russo, R.C., Santiago, H.C., Souza, D.G., Pinho, V., Tanowitz, H.B., Camargos, E.R., Bruna-Romero, O. and Teixeira, M.M. A DNA vaccine encoding CCL4/MIP-1beta enhances myocarditis in experimental *Trypanosoma cruzi* infection in rats. Microbes Infect. 2006; 8:2745–2755.

Rossi, M.A., Ramos, S.G. and Bestetti, R.B. Chagas' heart disease: clinical-pathological correlation. Front. Biosci. 2003; 8:e94–e109.

Ruiz, A.M., Esteva, M., Cabeza Meckert, P., Laguens, R.P. and Segura, E.L. Protective immunity and pathology induced by inoculation of mice with different subcellular fractions of *Trypanosoma cruzi*. Acta Trop. 1985; 42:299–309.

Ruiz, A.M., Esteva, M., Riarte, A., Subias, E. and Segura, E.L. Immunoprotection of mice against *Trypanosoma cruzi* with a lyophilized flagellar fraction of the parasite plus adjuvant. Immunol. Lett. 1986; 12:1–4.

Ruiz, A.M., Esteva, M., Subias, E., Moreno, M., Rosenstein de Campanini, A., Velazquez, E. and Segura, E.L. Monoclonal antibodies against the flagellar fraction of epimastigotes of *Trypanosoma cruzi*: immunoprotection against metacyclic trypomastigotes obtained by immunization of mice with an affinity-purified antigen. Mol. Biochem. Parasitol. 1990; 39:117–125.

Ruiz, R.C., Favoreto, S., Jr., Dorta, M.L., Oshiro, M.E., Ferreira, A.T., Manque, P.M. and Yoshida, N. Infectivity of *Trypanosoma cruzi* strains is associated with differential expression of surface glycoproteins with differential Ca^{2+} signalling activity. Biochem. J. 1998; 330(Pt 1):505–511.

Salomone, O.A., Juri, D., Omelianiuk, M.O., Sembaj, A., Aguerri, A.M., Carriazo, C., Barral, J.M. and Madoery, R. Prevalence of circulating *Trypanosoma cruzi* detected by polymerase chain reaction in patients with Chagas' cardiomyopathy. Am. J. Cardiol. 2000; 85:1274–1276.

Sanchez-Burgos, G., Mezquita-Vega, R.G., Escobedo-Ortegon, J., Ramirez-Sierra, M.J., Arjona-Torres, A., Ouaissi, A., Rodrigues, M.M. and Dumonteil, E. Comparative evaluation of therapeutic DNA vaccines against *Trypanosoma cruzi* in mice. FEMS Immunol. Med. Microbiol. 2007; 50:333–341.

Santori, F.R., Paranhos-Bacalla, G.S., Franco, D.A.S.J., Yamauchi, L.M., Araya, J.E. and Yoshida, N. A recombinant protein based on the *Trypanosoma cruzi* metacyclic trypomastigote 82-kilodalton antigen that induces and effective immune response to acute infection. Infect. Immun. 1996; 64:1093–1099.

Santos-Buch, C.A. and Teixeira, A.R. The immunology of experimental Chagas' disease. 3. Rejection of allogeneic heart cells in vitro. J. Exp. Med. 1974; 140:38–53.

Santos, C.C., Sant'anna, C., Terres, A., Cunha-E-Silva, N.L., Scharfstein, J. and De, A.L.A.P. Chagasin, the endogenous cysteine-protease inhibitor of *Trypanosoma cruzi*, modulates parasite differentiation and invasion of mammalian cells. J. Cell. Sci. 2005; 118:901–915.

Santos, M.A., Garg, N. and Tarleton, R.L. The identification and molecular characterization of *Trypanosoma cruzi* amastigote surface protein-1, a member of the trans-sialidase gene superfamily. Mol. Biochem. Parasitol. 1997; 86:1–11.

Sartori, A.M., Lopes, M.H., Caramelli, B., Duarte, M.I., Pinto, P.L., Neto, V. and Amato Shikanai-Yasuda, M. Simultaneous occurrence of acute myocarditis and reactivated Chagas' disease in a patient with AIDS. Clin. Infect. Dis. 1995; 21:1297–1299.

Scharfstein, J., Schmitz, V., Morandi, V., Capella, M.M., Lima, A.P., Morrot, A., Juliano, L. and Muller-Esterl, W. Host cell invasion by *Trypanosoma cruzi* is potentiated by activation of bradykinin B(2) receptors. J. Exp. Med. 2000; 192:1289–1300.

Schechter, M., Flint, J.E., Voller, A., Guhl, F., Marinkelle, C.J. and Miles, M.A. Purified *Trypanosoma cruzi* specific glycoprotein for discriminative serological diagnosis of South American trypanosomiasis (Chagas' disease). Lancet 1983; 2:939–941.

Schenkman, S., Eichinger, D., Pereira, M.E. and Nussenzweig, V. Structural and functional properties of Trypanosoma transsialidase. Annu. Rev. Microbiol. 1994; 48:499–523.

Schenkman, S., Ferguson, M.A., Heise, N., De Almeida, M.L., Mortara, R.A. and Yoshida, N. Mucin-like glycoproteins linked to the membrane by glycosylphosphatidylinositol anchor are the major acceptors of sialic acid in a reaction catalyzed by transsialidase in metacyclic forms of *Trypanosoma cruzi*. Mol. Biochem. Parasitol. 1993; 59:293–303.

Schmunis, G.A. and Cruz, J.R. Safety of the blood supply in Latin America. Clin. Microbiol. Rev. 2005; 18:12–29.

Schnapp, A.R., Eickhoff, C.S., Sizemore, D., Curtiss, R., III and Hoft, D. F. Cruzipain induces both mucosal and systemic protection against *Trypanosoma cruzi* in mice. Infect. Immun. 2002; 70:5065–5074.

Schofield, C.J. Eradication of *Triatoma infestans*: a new regional programme for southern Latin America. Ann. Soc. Belg. Med. Trop. 1992; 72(Suppl. 1):69–70.

Schofield, C.J. and Dias, J.C. The Southern Cone initiative against Chagas disease. Adv. Parasitol. 1999; 42:1–27.

Schofield, C.J., Jannin, J. and Salvatella, R. The future of Chagas disease control. Trends Parasitol. 2006; 22:583–588.

Scott, M.T., Neal, R.A. and Woods, N.C. Immunization of marmosets with *Trypanosoma cruzi* cell surface glycoprotein (GP90). Trans R. Soc. Trop. Med. Hyg. 1985; 79:451–454.

Seneca, H. and Peer, P.M. Immuno-biological properties of chagatoxin (lipopolysaccharide). Trans. R. Soc. Trop. Med. Hyg. 1966; 60:610–620.

Sepulveda, P., Hontebeyrie, M., Liegeard, P., Mascilli, A. and Norris, K.A. DNA-Based immunization with *Trypanosoma cruzi* complement regulatory protein elicits complement lytic antibodies and confers protection against *Trypanosoma cruzi* infection [in process citation]. Infect. Immun. 2000; 68:4986–4991.

Soares, M.B., Pontes-de-Carvalho, L. and Ribeiro-dos-Santos, R. The pathogenesis of Chagas' disease: when autoimmune and parasite-specific immune responses meet. An. Acad. Bras. Cienc. 2001; 73:547–559.

Soares, M.B. and Santos, R.R. Immunopathology of cardiomyopathy in the experimental Chagas disease. Mem. Inst. Oswaldo Cruz 1999; 94:257–262.

Sterin-Borda, L. and Borda, E. Role of neurotransmitter autoantibodies in the pathogenesis of chagasic peripheral dysautonomia. Ann. N. Y. Acad. Sci. 2000; 917:273–280.

Sztein, M.B. and Kierszenbaum, F. Mechanisms of development of immunosuppression during Trypanosoma infections. Parasitol. Today 1993; 9:424–428.

Tan, H. and Andrews, N.W. Don't bother to knock—the cell invasion strategy of *Trypanosoma cruzi*. Trends Parasitol. 2002; 18:427–428.

Tarleton, R.L. *Trypanosoma cruzi*-induced suppression of IL-2 production. II. Evidence for a role for suppressor cells. J. Immunol. 1988; 140:2769–2773.

Tarleton, R.L. Immunity to *Trypanosoma cruzi*. In: Host Response to Intracellular Pathogens (S. Kaufmann, Ed.). Austin, TX: R.G. Landes Company, 1996.

Tarleton, R.L., Grusby, M.J., Postan, M. and Glimcher, L.H. *Trypanosoma cruzi* infection in MHC-deficient mice: further evidence for the role of both class I- and class II-restricted T cells in immune resistance and disease. Int. Immunol. 1996; 8:13–22.

Thomas, M.C., Garcia-Perez, J.L., Alonso, C. and Lopez, M.C. Molecular characterization of KMP11 from *Trypanosoma cruzi*: a cytoskeleton-associated protein regulated at the translational level. DNA Cell Biol. 2000; 19:47–57.

Tibayrenc, M. Integrated genetic epidemiology of infectious diseases: the Chagas model. Mem. Inst. Oswaldo Cruz 1998; 93:577–580.

Todorov, A.G., Andrade, D., Pesquero, J.B., Araujo Rde, C., Bader, M., Stewart, J., Gera, L., Muller-Esterl, W., Morandi, V., Goldenberg, R.C., Neto, H.C. and Scharfstein, J. *Trypanosoma cruzi* induces edematogenic responses in mice and invades cardiomyocytes and endothelial cells in vitro by activating distinct kinin receptor (B1/B2) subtypes. FASEB J. 2003; 17:73–75.

Trischmann, T., Tanowitz, H., Wittner, M. and Bloom, B. *Trypanosoma cruzi*: role of the immune response in the natural resistance of inbred strains of mice. Exp. Parasitol. 1978; 45:160–168.

Trujillo, C., Ramirez, R., Velez, I.D. and Berberich, C. The humoral immune response to the kinetoplastid membrane protein-11 in patients with American leishmaniasis and Chagas disease: prevalence of IgG subclasses and mapping of epitopes. Immunol. Lett. 1999; 70:203–209.

Tsutsui, H. Oxidative stress in heart failure: the role of mitochondria. Intern. Med. 2001; 40:1177–1182.

Tyler, K.M., Olson, C.L. and Engman, D.M. The life cycle of *Trypanosoma cruzi*. In: American Trypanosomiasis (K.M. Tyler and M.A. Miles, Eds.). Dordecht, Netherlands: Kluwer Academic Publishers, 2003.

Vasconcelos, J.R., Boscardin, S.B., Hiyane, M.I., Kinoshita, S.S., Fujimura, A.E. and Rodrigues, M.M. A DNA-priming protein-boosting regimen significantly improves type 1 immune response but not protective immunity to *Trypanosoma cruzi* infection in a highly susceptible mouse strain. Immunol. Cell Biol. 2003; 81:121–129.

Vasconcelos, J.R., Hiyane, M.I., Marinho, C.R., Claser, C., Machado, A.M., Gazzinelli, R.T., Bruna-Romero, O., Alvarez, J.M., Boscardin, S.B. and Rodrigues, M.M. Protective immunity against *Trypanosoma cruzi* infection in a highly susceptible mouse strain after vaccination with genes encoding the amastigote surface protein-2 and trans-sialidase. Hum. Gene Ther. 2004; 15:878–886.

Vyatkina, G., Bhatia, V., Gerstner, A., Papaconstantinou, J. and Garg, N. Impaired mitochondrial respiratory chain and bioenergetics during chagasic cardiomyopathy development. Biochim. Biophys. Acta 2004; 1689:162–173.

Warren, T.L. and Weiner, G.J. Uses of granulocyte-macrophage colony-stimulating factor in vaccine development. Curr. Opin. Hematol. 2000; 7:168–173.

Wen, J.-J., Bhatia, V., Popov, V.L. and Garg, N.J. Phenyl-alpha-tert-butyl nitrone reverses mitochondrial decay in acute Chagas disease. Am. J. Pathol. 2006a; 169:1953–1964.

Wen, J.-J. and Garg, N. Oxidative modifications of mitochondrial respiratory complexes in response to the stress of *Trypanosoma cruzi* infection. Free Radic. Biol. Med. 2004; 37:2072–2081.

Wen, J.-J., Vyatkina, G. and Garg, N. Oxidative damage during chagasic cardiomyopathy development: role of mitochondrial oxidant release and inefficient antioxidant defense. Free Radic. Biol. Med. 2004; 37:1821–1833.

Wen, J.-J., Yachelini, P.C., Sembaj, A., Manzur, R.E. and Garg, N. Increased oxidative stress is correlated with mitochondrial dysfunction in chagasic patients. Free Radic. Biol. Med. 2006b; 41:270–276.

WHO. Control of Chagas disease: second report of the World Health Organization Expert Committee. *WHO Technical Report Series 905.* Geneva, Switzerland: UNDP/World Bank/WHO, 2002.

Williams-Blangero, S., Vandeberg, J.L., Blangero, J. and Correa-Oliveira, R. Genetic epidemiology of *Trypanosoma cruzi* infection and Chagas' disease. Front. Biosci. 2003; 8:e337–e345.

Wizel, B., Garg, N. and Tarleton, R.L. Vaccination with trypomastigote surface antigen 1-encoding plasmid DNA confers protection against lethal *Trypanosoma cruzi* infection. Infect. Immun. 1998a; 66:5073–5081.

Wizel, B., Nunes, M. and Tarleton, R.L. Identification of *Trypanosoma cruzi* trans-sialidase family members as targets of protective CD8+ TC1 responses. J. Immunol. 1997; 159:6120–6130.

Wizel, B., Palmieri, M., Mendoza, C., Arana, B., Sidney, J., Sette, A. and Tarleton, R. Human infection with *Trypanosoma cruzi* induces parasite antigen-specific cytotoxic T lymphocyte responses. J. Clin. Invest. 1998b; 102:1062–1071.

Wrightsman, R.A., Luhrs, K.A., Fouts, D. and Manning, J.E. Paraflagellar rod protein-specific CD8+ cytotoxic T lymphocytes target *Trypanosoma cruzi*-infected host cells. Parasite Immunol. 2002; 24:401–412.

Wrightsman, R.A. and Manning, J.E. Paraflagellar rod proteins administered with alum and IL-12 or recombinant adenovirus expressing IL-12 generates antigen-specific responses and protective immunity in mice against *Trypanosoma cruzi.* Vaccine 2000; 18:1419–1427.

Wrightsman, R.A., Miller, M.J., Saborio, J.L. and Manning, J.E. Pure paraflagellar rod protein protects mice against *Trypanosoma cruzi* infection. Infect. Immun. 1995; 63:122–125.

Yabsley, M.J. and Noblet, G.P. Seroprevalence of *Trypanosoma cruzi* in raccoons from South Carolina and Georgia. J. Wildl. Dis. 2002; 38:75–83.

Yang, X.P., Liu, Y.H., Mehta, D., Cavasin, M.A., Shesely, E., Xu, J., Liu, F. and Carretero, O.A. Diminished cardioprotective response to inhibition of angiotensin-converting enzyme and angiotensin II type 1 receptor in B(2) kinin receptor gene knockout mice. Circ. Res. 2001; 88:1072–1079.

Yoshida, N., Araya, J.E., Da Silveira, J.F. and Giorgio, S. *Trypanosoma cruzi*: antibody production and T cell response induced by stage-specific surface glycoproteins purified from metacyclic trypomastigotes. Exp. Parasitol. 1993; 77:405–413.

Yoshida, N., Favoreto, S., Ferreira, A.T. and Manque, P.M. Signal transduction induced in *Trypanosoma cruzi* metacyclic trypomastigotes during the invasion of mammalian cells. Braz. J. Med. Biol. Res. 2000; 33:269–278.

Zacks, M.A., Wen, J.-J., Vyatkina, G., Bhatia, V. and Garg, N. An overview of chagasic cardiomyopathy: pathogenic importance of oxidative stress. An. Acad. Bras. Cienc. 2005; 77:695–715.

Zapata-Estrella, H., Hummel-Newell, C., Sanchez-Burgos, G., Escobedo-Ortegon, J., Ramirez-Sierra, M.J., Arjona-Torres, A. and Dumonteil, E. Control of *Trypanosoma cruzi* infection and changes in T-cell populations induced by a therapeutic DNA vaccine in mice. Immunol. Lett. 2006; 103:186–191.

Index

Note: Page numbers in *italics* refer to figures or tables.

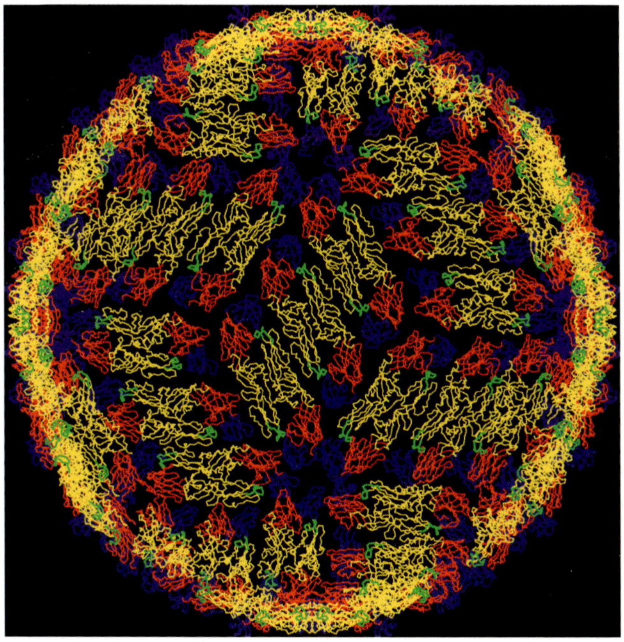

FIGURE 19.1 Arrangement of the E proteins on the surface of the mature virus (Kuhn et al., 2002). The virus surface is covered with 30 protein "rafts," each containing three anti-parallel dimers of E protein. Each monomer is colored as follows: red, domain I (central region); yellow, domain II (dimerization and fusion region); blue, domain III (receptor-binding region); green, fusion peptide.

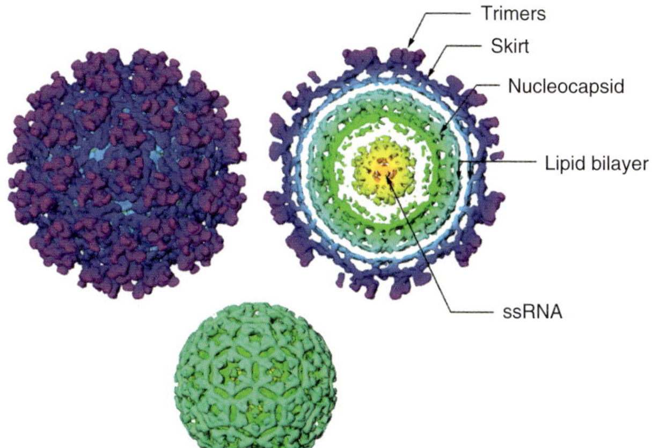

FIGURE 21.1 Cryo-electron microscopic reconstruction of the VEE virion at 14Å resolution. Upper left, external view of the virion showing the envelope glycoproteins on the surface; upper right, cross section of the virion showing the RNA genome at the center surrounded by 240 copies of the capsid protein, the bilipid envelope derived from the host cell plasma membrane, the E1 envelope glycoprotein lying parallel to the envelope, and the E2 envelope glycoprotein forming spikes on the surface; lower, nucleocapsid showing the capsid proteins in T = 4 icosahedral symmetry. Courtesy of Angel Paredes and Wah Chiu, Baylor College of Medicine, and Stanley Watowich and Scott Weaver, University of Texas Medical Branch.

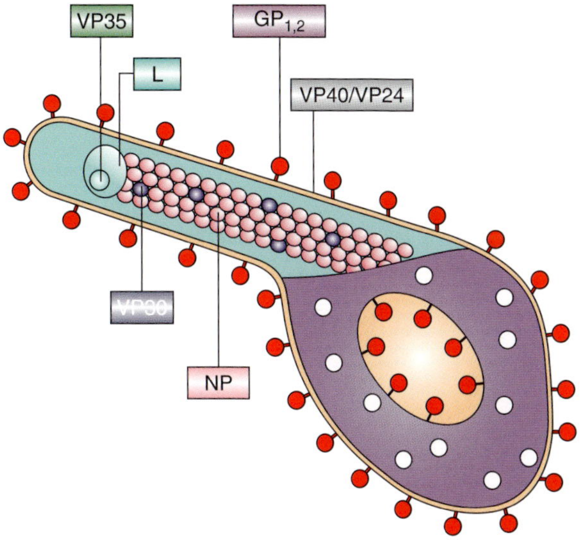

FIGURE 20.1 Schematic illustration of a filovirus particle. Four proteins form the RNP: polymerase or large (L) protein, NP, virion structural protein 30 (VP30), and virion structural protein 35 (VP35). The glycoprotein (GP) is anchored with the carboxy-terminal part in the virion membrane. Homotrimers of GP form the spikes on the surface of the virion. VP40 is the matrix protein and VP24 another membrane-associated protein. Adapted from Feldmann et al. (2003).

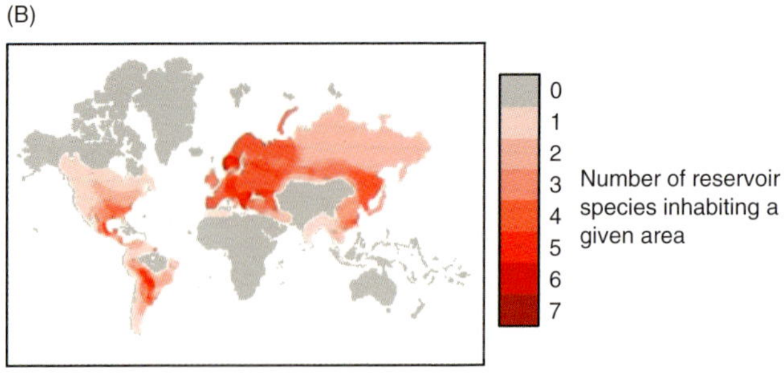

FIGURE 23.3 Distribution of hantavirus rodent reservoirs. **(B)** Estimate of the number of different rodent species in a given area that are reservoirs for hantaviruses known to cause human disease.

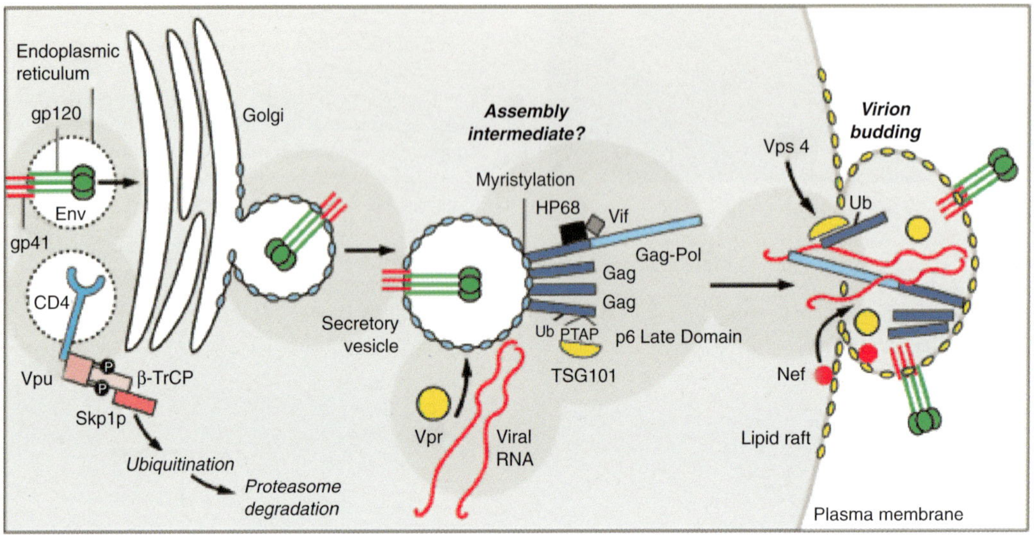

FIGURE 25.4 Viral assembly and budding. The figure depicts assembly of new virus particles, and budding from a cholesterol-rich raft in the plasma membrane. Adapted with permission from HIV InSite.

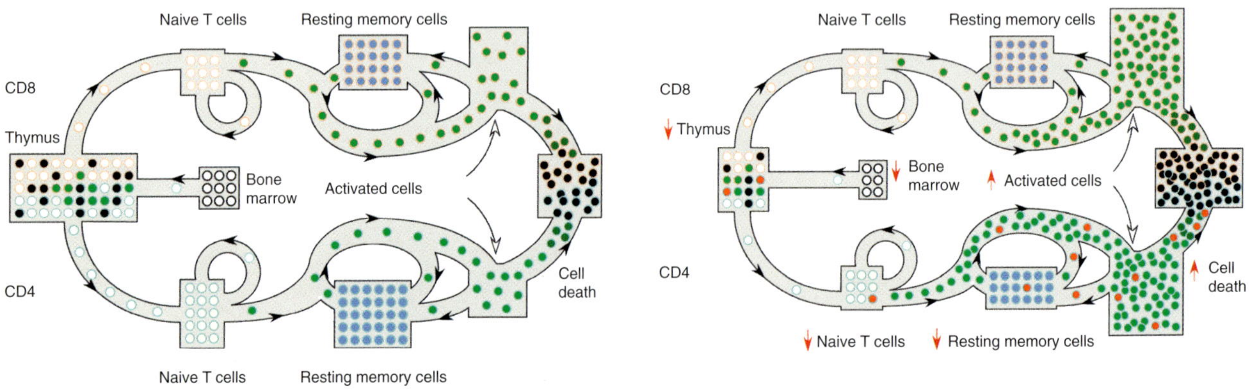

FIGURE 25.5 Disruption of normal T-cell homeostasis due to HIV infection. (A) Under normal conditions, losses from the resting memory CD4 and CD8 T lymphocyte pools through activation, and apoptosis are balanced by the input of new naïve T cells from the marrow and thymus. (B) During chronic HIV infection, increased activation of both CD4 and CD8 T cells results in increased cell turnover. CD4 T-cell homeostasis is relatively impaired because of reduced output from marrow and thymus, resulting in a smaller pool of na ï ve T cells and resting memory CD4 T cells. CD8 cells, orange outline; CD4 cells, blue outline; na ï ve cells, clear center; resting memory cells, blue center; activated cells, green center; cells commit-ted to apoptosis, dark green center; dying cells, black center; HIV-infected cells, red center. Processes/cell populations affected by HIV infection are highlighted by a red arrow. Adapted with permission from Douek (2003) .

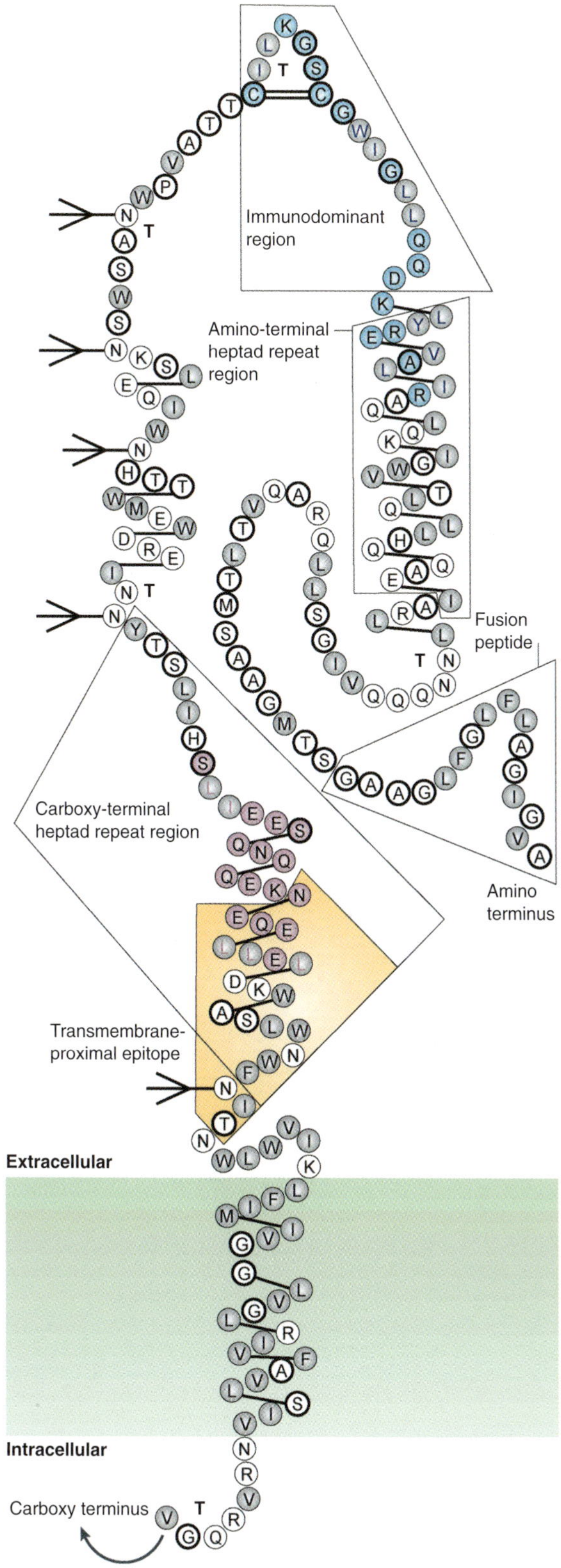

FIGURE 25.6 HIV gp41. Schematic diagram of HIV transmembrane protein gp41, depicting functional regions important for virus entry into the host cell, and sites important for immune responses. Hydrophobic amino acids shown in gray, charged amino acids shown as unfilled circles, neutral amino acids outlined in bold, potential N-linked glycosylation sites shown as branched stick figures. Important antibody epitopes are presented in turquoise, pink, and yellow. Adapted from Zolla-Pazner (2004)

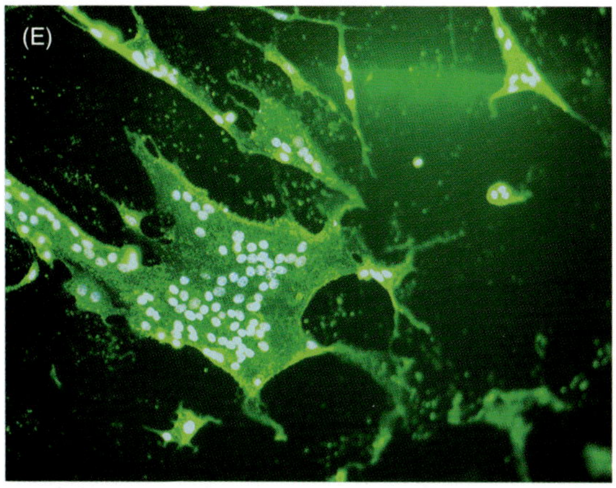

FIGURE 31.2 Microscopy of Nipah virus. (E) Syncytia of BHK-21 cells infected with Nipah virus (m.o.i. 0.025, 24h postinfection) and immunostained with anti-Nipah antibodies and counterstained with Hoechst (original magnification: 20×).

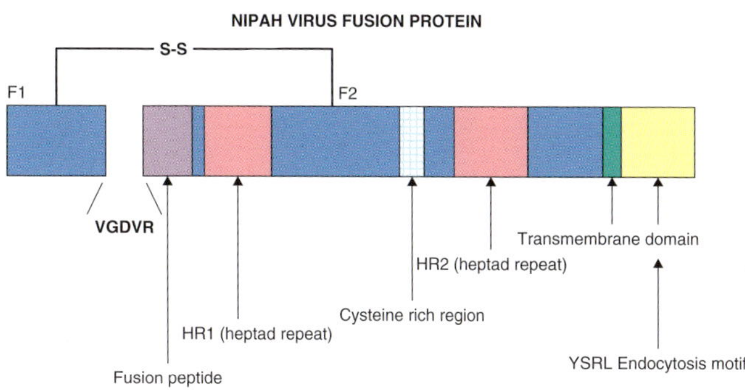

FIGURE 31.5 Henipavirus fusion F protein.

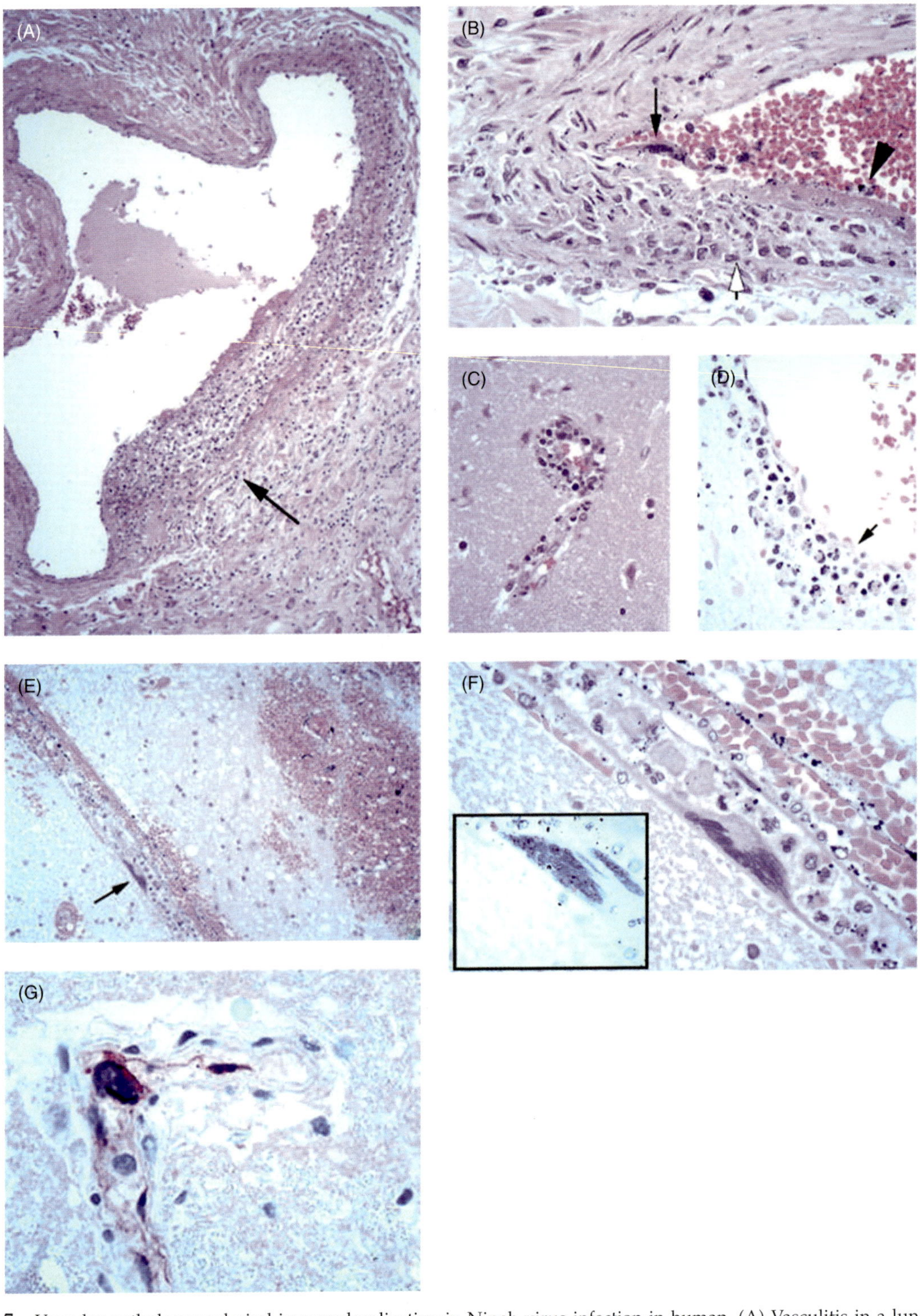

FIGURE 31.7 Vascular pathology and viral immunolocalization in Nipah virus infection in human. (A) Vasculitis in a lung artery. Focal and transmural mixed inflammatory infiltrate is visible (arrow). (B) High-power magnification of another pulmonary vessel showing endothelial syncytium (arrow) and ulceration (arrowhead). Note focal transmural inflammation (open arrow). (C) Vasculitis in a small cerebral vessel. (D) A cerebral venule showing endothelial ulceration (arrow) associated with inflammatory cellular debris. (E) Cerebral hemorrhage adjacent to a vessel with a multinucleated giant cell (arrow). (F) Higher-power magnification showing endothelial origin of syncytium seen in (E). (Inset) Intranuclear immunostaining of viral antigens in same cell. (G) Positive immunostaining is also seen in the cytoplasm of an endothelial syncytial cell protruding in lumen of a cerebral vessel. H&E stain: (A–F and inset in F) immunoalkaline phosphatase with napthol fast red substrate and hematoxylin counterstain (G). Original magnifications: 25× (A); 100× (B–D); 50× (E); 250× (F, inset); 158× (F, G) (from Wong et al., 2002).

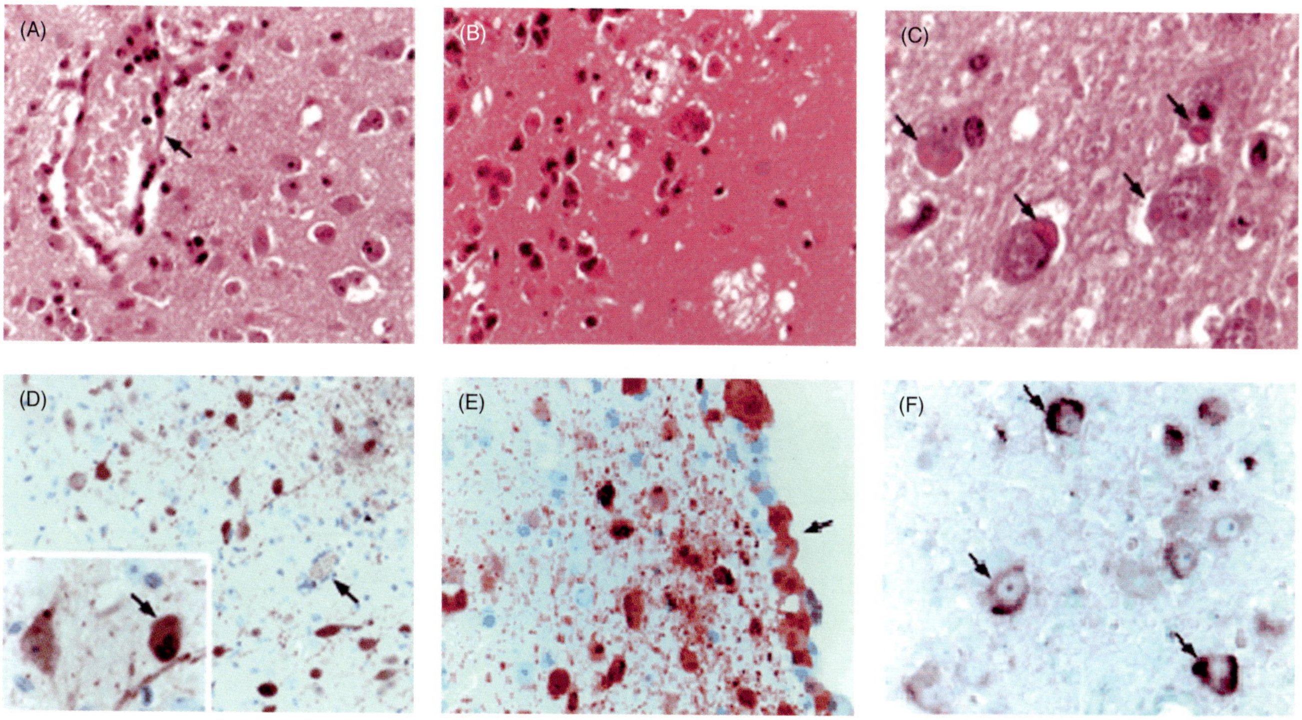

FIGURE 31.8 Cerebral pathology in acute Nipah infection in a hamster model. (A) Small vessel vasculitis (arrow) characterized by mild inflammation in the vicinity of infected neurons. (B) Focal areas of parenchymal ischemia, infarction, and edema. (C) Neurons with eosinophilic viral inclusions (arrows). (D) Immunolocalization of viral antigens to neurons in the nucleus (inset, arrow), cytoplasm, and processes near a vasculitic vessel (arrow). (E) Viral antigens localized to ependymal lining (arrow) and neurons. (F) Neurons demonstrating viral RNA in the cytoplasm (arrows). (A–C) Hemalin-phloxine-safranin stain; (D, E) IHC, hematoxylin counterstain; (F) in situ hybridization, hematoxylin counterstain. Original magnifications: 40× (A, B, E, D (inset); 20× (D); 100× (C, F) (from Wong et al., 2003).

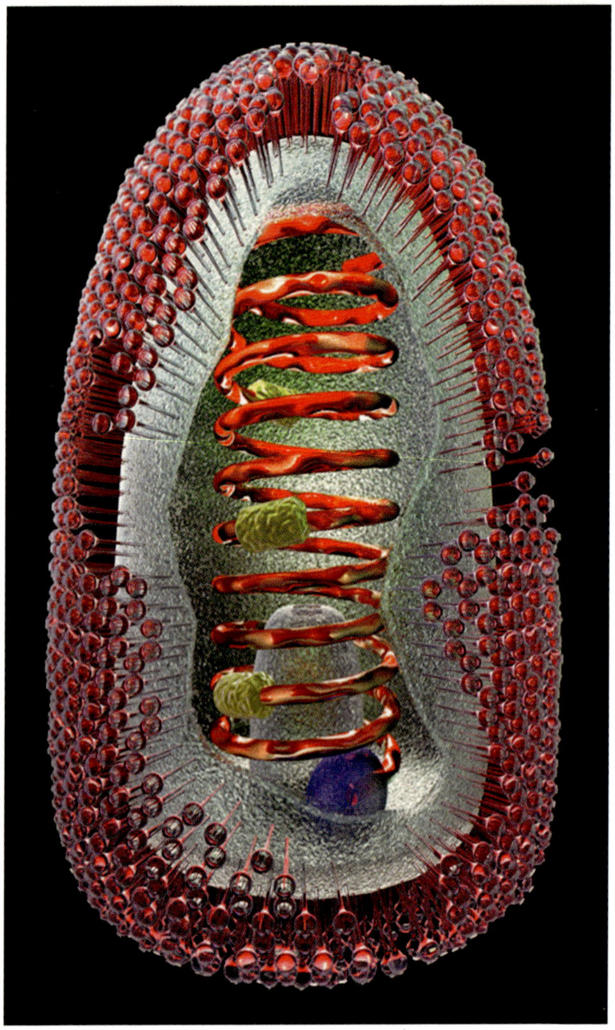

FIGURE 33.1 Schematic diagram of rabies virus.

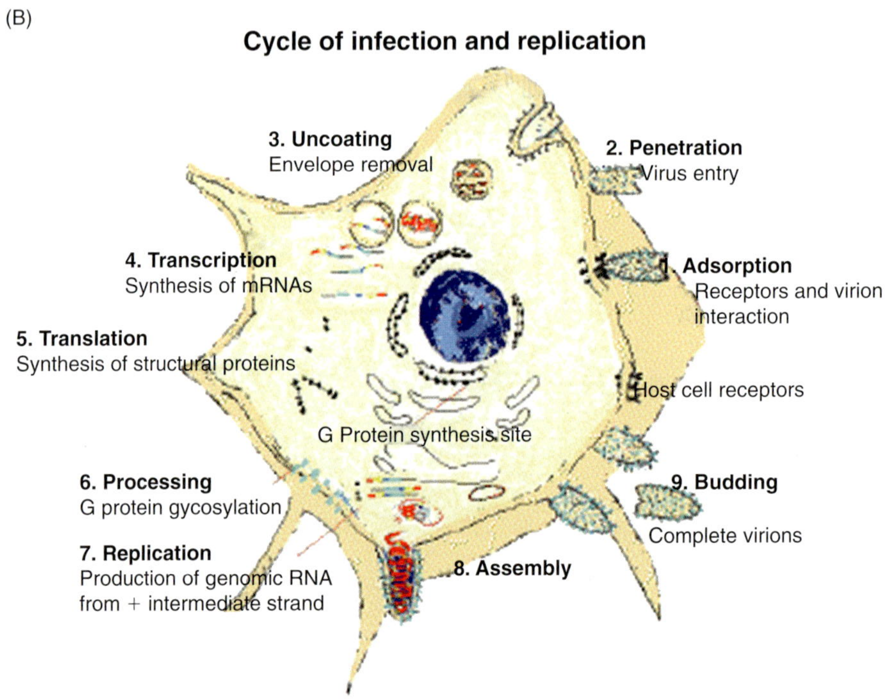

FIGURE 33.4 The infectious path of rabies virus. (B) Intraneuronal: from reception to assembly. .

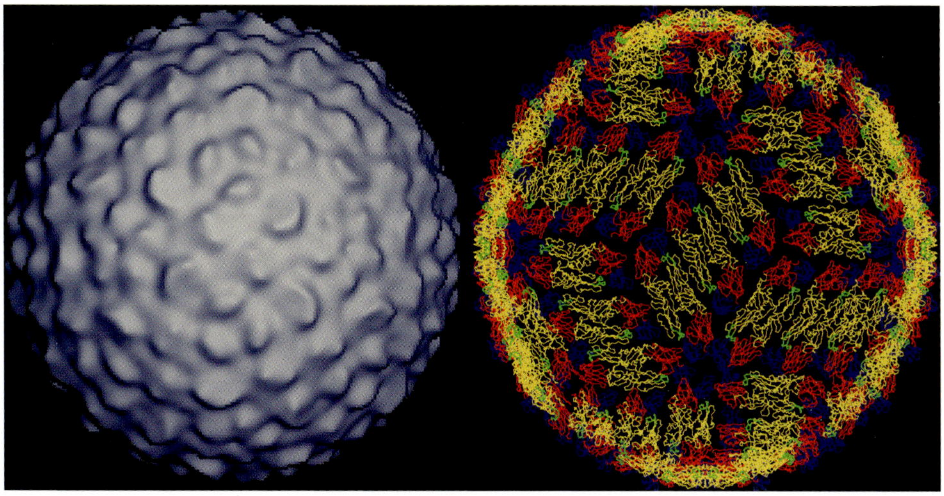

FIGURE 38.2 Cryo-electron microscopy surface representation of dengue-2 virus, a mosquito-borne flavivirus (left). The incorporation of the viral envelope protein into electron densities is observed by cryo-electron microscopy. Envelope protein domains I (red), II (yellow) and III (blue) are shown along with putative carbohydrate moieties. Adapted from Kuhn and Strauss (2003), with permission from Elsevier.

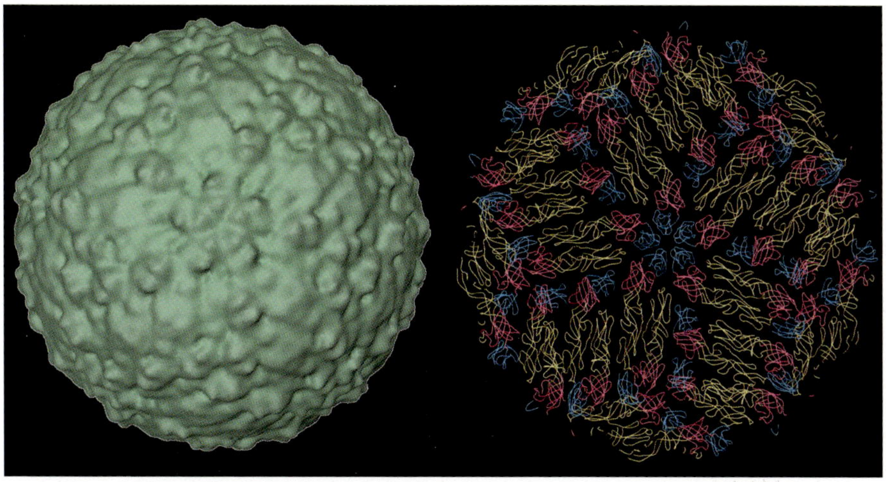

FIGURE 39.4 Cryoelectron microscopic structure of WNV. Panel A, surface shaded view of the WN virion looking down the fivefold axis. Panel B, C-alpha trace showing the fitted E-protein heterodimers looking down the fivefold axis. Structrural domains of monomers identified by color: red, Domain I; yellow, Domain II; blue, Domain III. Courtesy of Baerbel Kaufmann and Richard Kuhn, Purdue University.

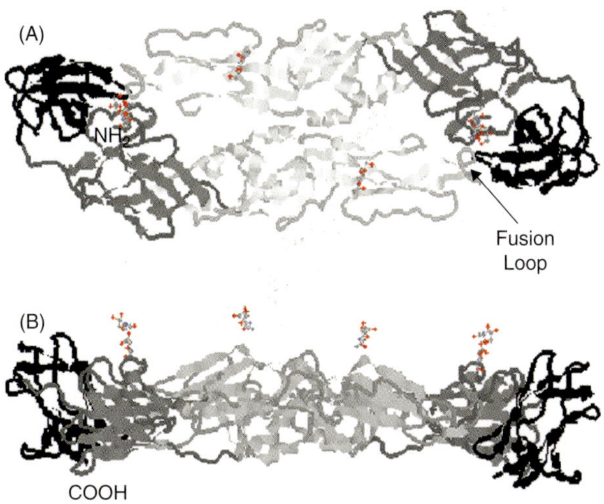

FIGURE 39.5 Crystallographic molecular structure of the DEN2 E-glycoprotein homodimer. Panel A, homodimer top view (aa 1–394). Panel B, homodimer lateral view. Structural domains of monomers identified by color: gray, Domain I; white, Domain II; black, domain III. Structural domains color-keyed as defined in Figure 4. N-linked glycosylation shown in ball-and-stick notation green. Fusion loop (aa 98–110) marked shown in orange in DII. Structural coordinates from Modis et al. (2003).

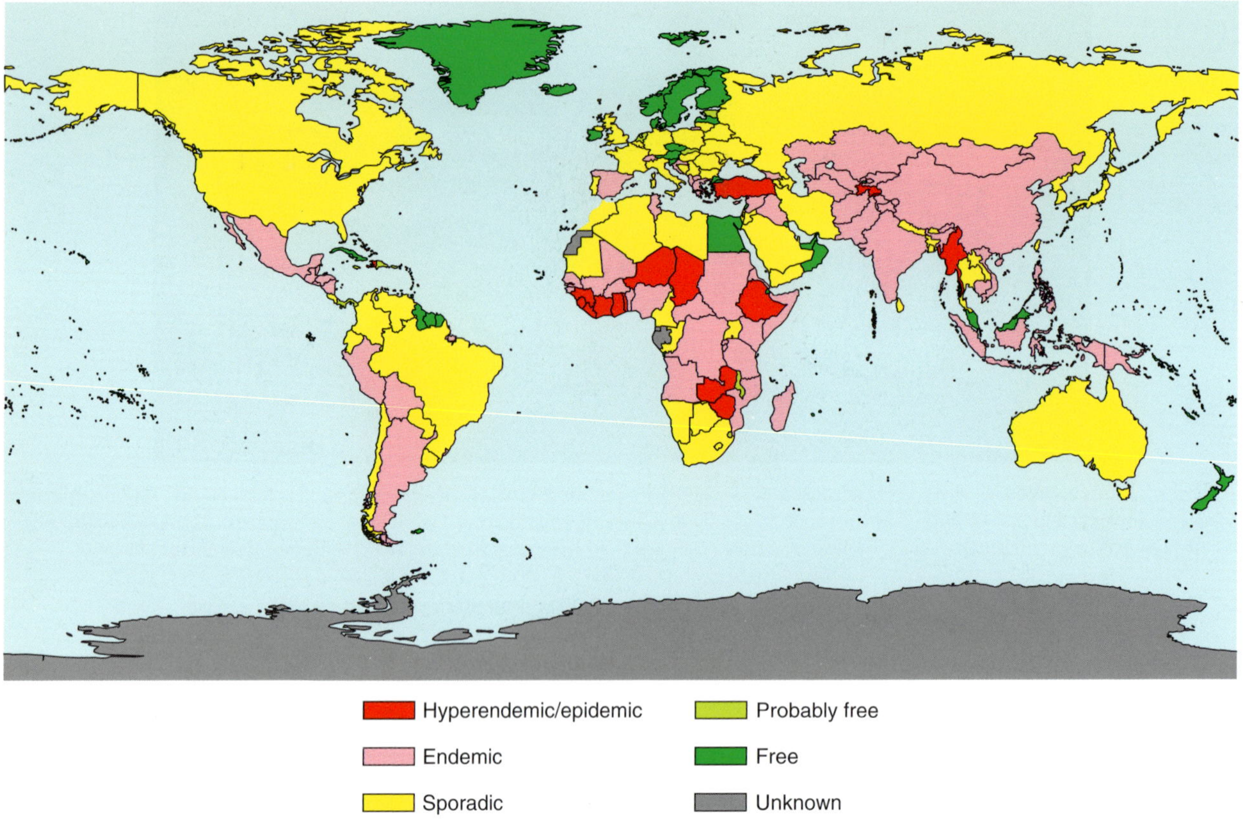

FIGURE 41.2 Worldwide occurrence of anthrax as of 2003. Map courtesy of Martin E. Hugh-Jones, Ph.D., WHO Collaborating Center for Remote Sensing and Geographic Information Systems for Public Health, Department of Epidemiology and Community Health, School of Veterinary Medicine, Louisiana State University, Baton Rouge, LA. (www.vetmed.lsu.edu/whocc/mp_world.htm.)

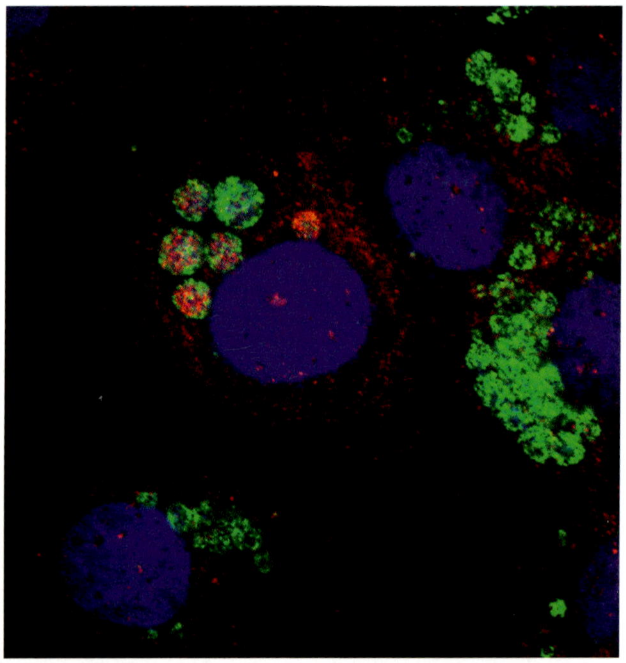

FIGURE 47.3 *E. chaffeensis* infected macrophages (DH82 cells) exhibiting differential expression of gp47 as visualized by three-color scanning laser confocal fluorescent microscopy. *E. chaffeensis* infected cells were dually stained with rabbit anti-*Ehrlichia* disulfide bond formation (Dsb) (green: Alexa Fluor 488), and mouse anti-*E. chaffeensis* gp47 (red: Alexa Fluor 568). Host cell nuclei were counterstained with 4′, 6′-diamidino-2-phenylindole, dihydrochloride (DAPI) and images merged. Infected macrophage (center) exhibits single (green: Dsb or red: gp47) and dual-labeled (yellow) ehrlichiae in individual morulae (reproduced with permission of J.W. McBride, University of Texas Medical Branch).

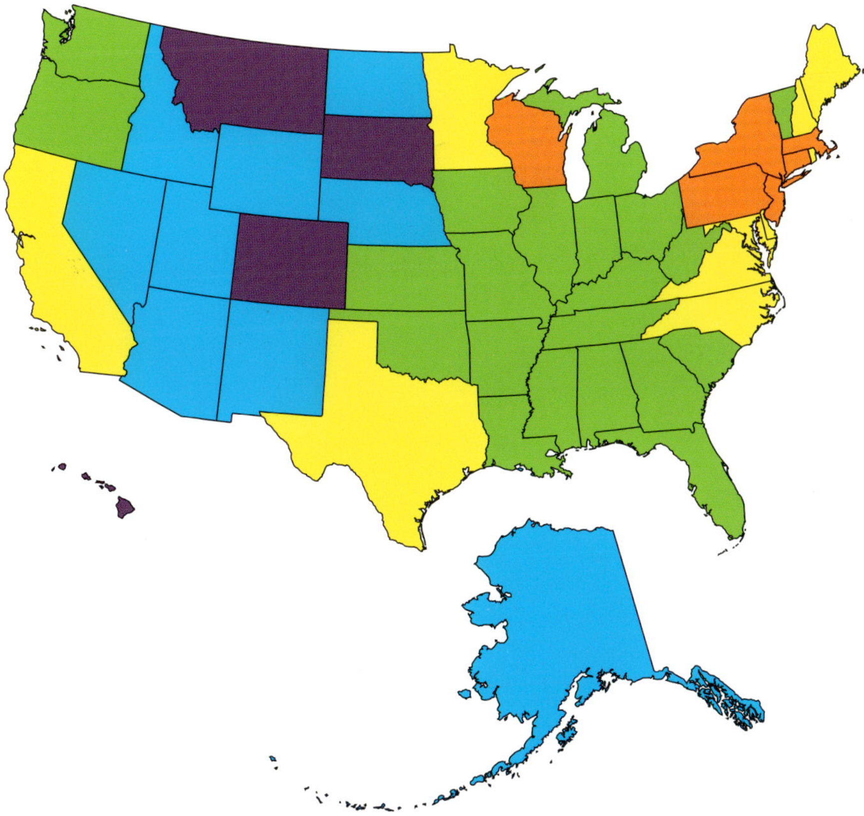

FIGURE 52.4 Reported cases of Lyme disease by state. The data represent the aggregate number of cases reported to the U.S. Centers for Disease Control and Prevention between 1992 and 2006. Each change in color represents a logarithmic increase in the total number of cases, shown as purple (0–10 cases), blue (10–99), green (100–999), yellow (1000–9999), and orange (>10,000). The greatest majority of Lyme disease cases occur in the northeast states and in Wisconsin, likely due to environmental conditions suitable for the tick vector.

FIGURE 52.5 Crystal structure of OspA. The ribbon diagram is derived from the OspA crystal structure 2G8C (Makabe et al., 2006). The conformationally determined epitope recognized by a protective mAb (LA-2) is noted in blue (Golde et al., 1997; Ding et al., 2000). The peptide that may be associated with autoimmune-inducing cross-reactivity with the LFA-1 protein is highlighted in red (see text). The N-terminus is at the upper right of the figure.

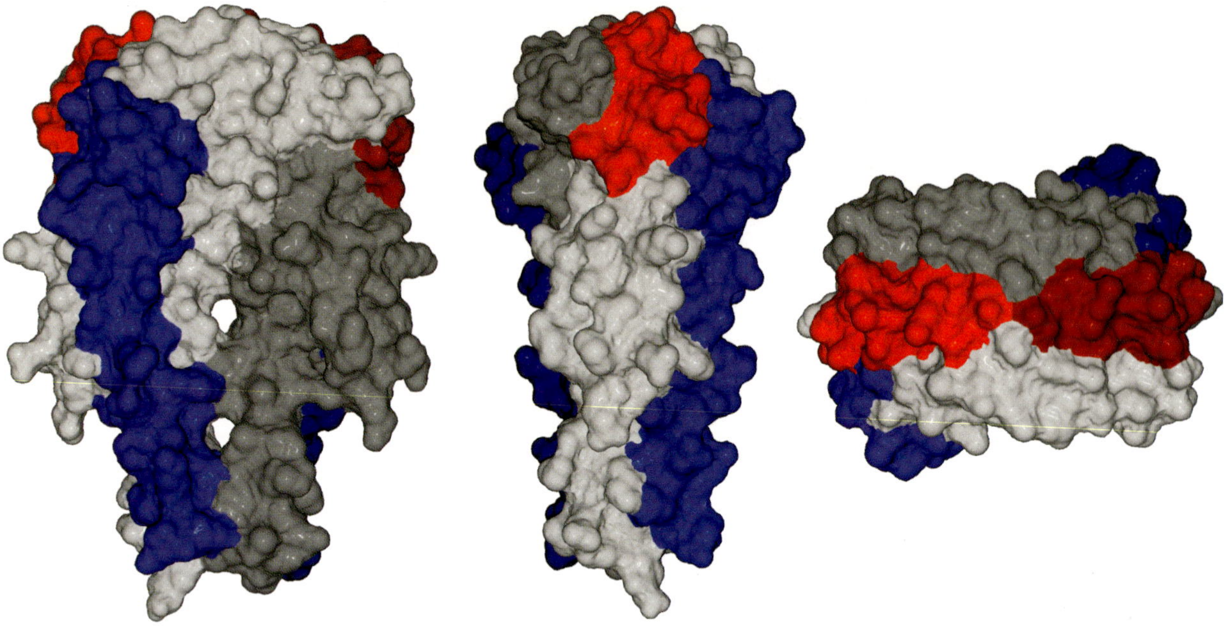

FIGURE 52.6 Crystal structure of outer surface protein C. The diagram is derived from the type A OspC crystal structure 1GGQ (Kumaran et al., 2001). A surface representation of the OspC molecule is shown in three rotations. The locations of the two regions in which epitopes have been mapped in multiple OspC types are denoted in red (loop 5 region) and blue (helix 5 region). The molecule is a homodimer; the component monomers are differentiated using darker and lighter color shades. A portion of both the N- and C-termini is omitted from the structure.

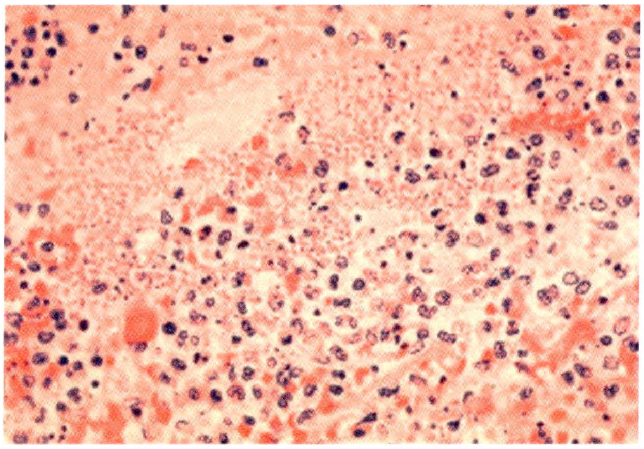

FIGURE 54.2 Pneumonic plague. This Giemsa stain reveals pneumonia and many *Yersinia pestis* bacteria in the lung. Image (PHIL ID NO. 741) courtesy of Dr. Marshall Fox/CDC; available at http://phil.cdc.gov/phil/.

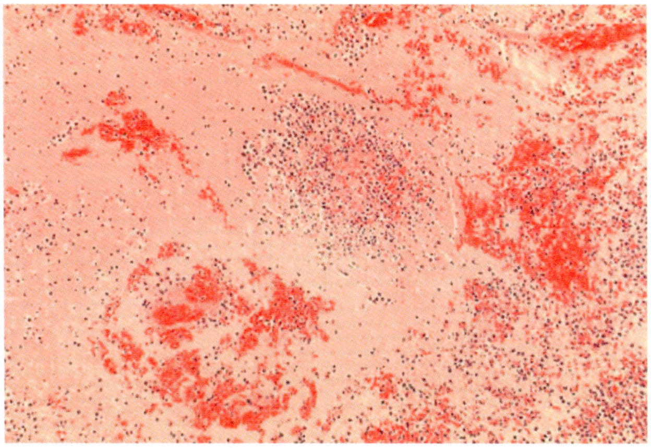

FIGURE 54.3 Evidence of plague in the lymph nodes. In this image, medullary necrosis with fluid and *Yersinia pestis* bacteria are evident. Image (PHIL ID NO. 731) courtesy of Dr. Marshall Fox/CDC; available at http://phil.cdc.gov/phil/.

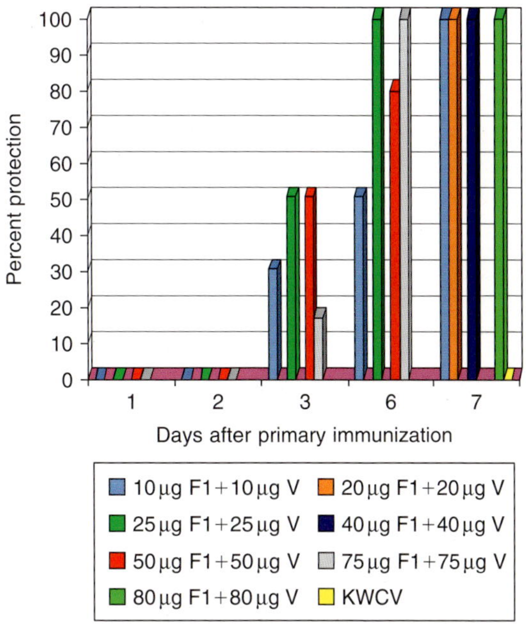

FIGURE 54.4 Time to immunity against pneumonic plague in the Balb/c mouse model; groups of five mice received a single immunizing dose of the rF1 + rV vaccine on day zero at escalating dose-levels of each subunit, followed by inhalational challenge with >300 MLD (3×10^5 cfu) *Y. pestis* strain GB on days 1–7. The figure shows percent protection at 14 days post-challenge.

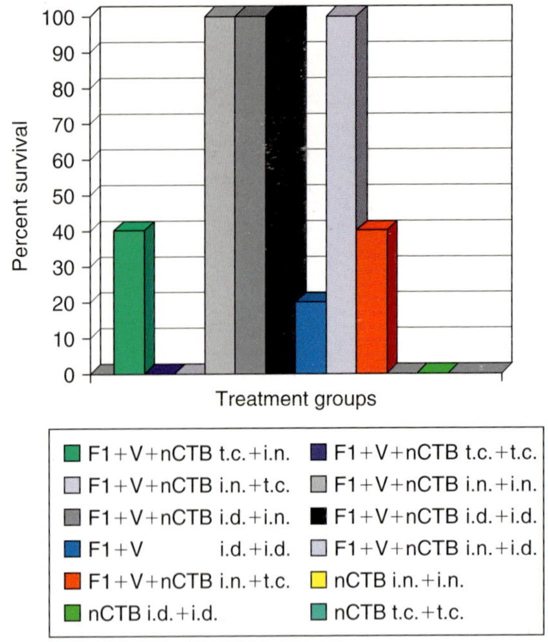

FIGURE 54.5 The protective efficacy of rF1 + rV coadministered with CTB by the transcutaneous, intradermal, or intranasal routes, or combinations of these. The figure shows the percentage survival for groups of 5 Balb/c mice immunized by the routes indicated (1 bar per group) on two occasions (days 1 and 21) with 10μg rF1 + 40μg rV, with or without 10μg CTB (Sigma, Poole, United Kingdom) in 50μl phosphate-buffered saline per mouse. All groups were challenged subcutaneously at day 90 with 5 × 106 MLD *Y. pestis* (GB strain) (Somavarapu et al., 2001).

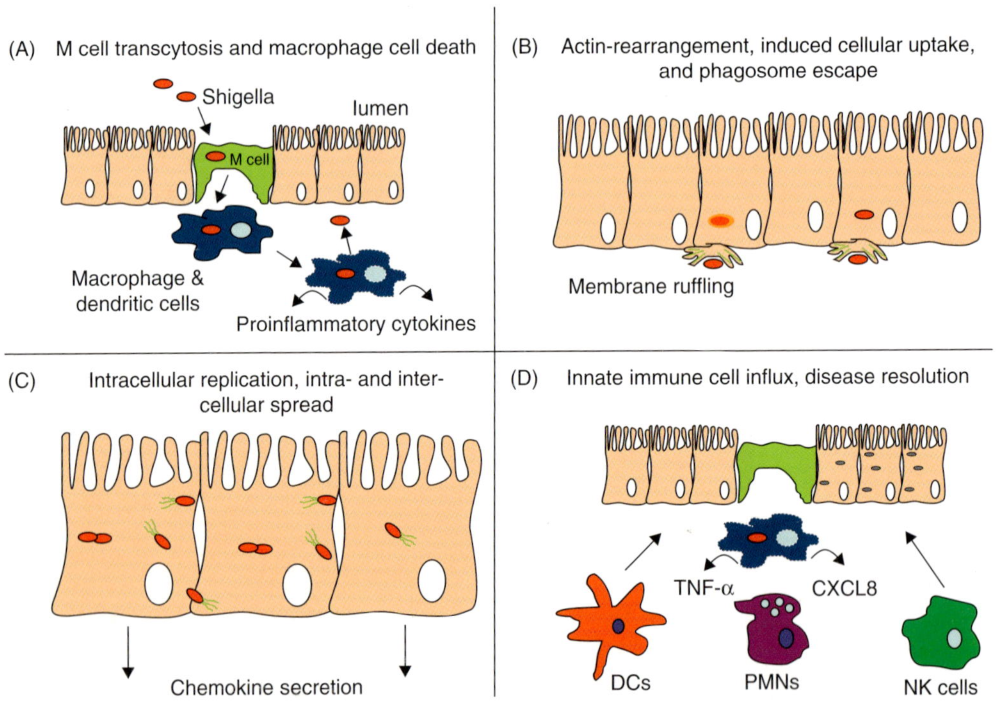

FIGURE 58.2 Major steps of *Shigella* pathogenesis.